Paul's Fundamental Immunology

EIGHTH EDITION

Paul's Fundamental Immunology

EIGHTH EDITION

Martin F. Flajnik, PhD
Professor
Department of Microbiology and Immunology
University of Maryland at Baltimore
Baltimore, Maryland

Nevil J. Singh, PhD
Assistant Professor
Department of Microbiology and Immunology
University of Maryland at Baltimore
Baltimore, Maryland

Steven M. Holland, MD
Director, Division of Intramural Research
National Institute of Allergy and Infectious Diseases
National Institutes of Health
Bethesda, Maryland

Wolters Kluwer Health | Lippincott Williams & Wilkins
Philadelphia • Baltimore • New York • London
Buenos Aires • Hong Kong • Sydney • Tokyo

Acquisitions Editor: Nicole Dernoski
Senior Development Editor: Ariel S. Winter
Editorial Coordinator: Christopher Rodgers
Marketing Manager: Kirsten Watrud
Production Project Manager: David Saltzberg
Manager, Graphic Arts & Design: Stephen Druding
Manufacturing Coordinator: Beth Welsh
Prepress Vendor: TNQ Technologies

8th Edition

9 8 7 6 5 4 3 2 1

Printed in Mexico.

Library of Congress Cataloging-in-Publication Data

ISBN-13: 978-1-975142-51-3

Cataloging in Publication data available on request from publisher

---

shop.lww.com

QUADM0822

William Paul at the microscope, drawing by Louis Du Pasquier

CONTRIBUTORS

Colin C. Anderson, PhD
Professor
Department of Surgery and Medical Microbiology and Immunology
University of Alberta
Edmonton, Alberta

Wael Awad, PhD
Research Fellow
Department of Biochemistry and Molecular Biology
Institution Monash University
Melbourne, Australia

Troy A. Baldwin, PhD
Professor
Department of Medical Microbiology and immunology
University of Alberta
Edmonton, Alberta

Nicole Baumgarth, DVM, PhD
Professor
Centre for Immunology and Infectious Diseases
University of California
Davis, California

Yasmine Belkaid, PhD
Director NIAID Microbiome Program
Chief Laboratory of Host Immunity and Microbiome
National Institute of Allergy and Infectious Diseases, NIH
Bethesda, Maryland

Avinash Bhandoola, MB, BS, PhD
Senior Investigator, National Institutes of Health
Laboratory of Genome Integrity
Bethesda, Maryland

Remy Bosselut, MD, PhD
Senior Investigator, Laboratory of Immune Cell Biology
Center for Cancer Research
National Cancer Institute
Bethesda, Maryland

Peter A. Bretscher, PhD
Professor
Department of Biochemistry Microbiology, and Immunology
University of Saskatchewan
Saskatchewan, Canada

Nicolas Buchon, PhD
Associate Professor
Cornell University
Ithaca, Italy

Francis R. Carbone, PhD
Professor
Department of Microbiology and Immunology
The University of Melbourne
Parkville, Victoria, Australia

Jean-Laurent Casanova, MD, PhD
Levy Family Professor, The Rockefeller University
Investigator, Howard Hughes Medical Institute
Head, St. Giles Laboratory of Human Genetics of Infectious Diseases
Senior Attending Physician, The Rockefeller University Hospital
New York, New York
Professor
University of Paris
Paris, France

Sara Cherry, PhD
Professor of Pathology and Laboratory Medicine and Microbiology
Scientific Director, High-throughput Screening Core
Director, Program for Chemogenomic Discovery
University of Pennsylvania
Philadelphia, Pennsylvania

Yueh-Hsiu Chien, PhD
Professor
Department of Microbiology and Immunology
Stanford University
Stanford, California

Danielle A. Chisolm, PhD
Independent Research Scholar
Molecular Immunology and Inflammation Branch
Bethesda, Maryland

Anita S. Chong, PhD
Professor
Department of Surgery
The University of Chicago
Chicago, Illinois

Gustaf Christoffersson, PhD
Associate Senior Lecturer
Science for Life Laboratory
Department of Medical Cell Biology
Uppsala University
Uppsala, Sweden

Marco Colonna, MD
Professor
Department of Pathology and Immunology
Washington University School of Medicine in St. Louis
St. Louis, Missouri

Alexandra J. Corbett, BSc (Hons), PhD
Associate Professor and Laboratory Head
Department of Microbiology and Immunology
Peter Doherty Institute for Infection and Immunity
The University of Melbourne
Melbourne, Victoria, Australia

Ken T. Coppieters, PhD
Research Project Vice President
Project and Alliance Management
Novo Nordisk A/S
Copenhagen, Denmark

Mark M. Davis, PhD
Professor
Department of Microbiology and Immunology
Stanford University
Stanford, California

Sidonia B. G. Eckle, Bsc, MSc, PhD
Australian National Health and Medical Research Council (NHMRC)
Emerging Leadership Fellow and Group Leader, MAIT Cell Program-Professor James McCluskey
Department of Microbiology and Immunology
Peter Doherty Institute for Infection and Immunity at The University of Melbourne
Melbourne, Victoria, Australia

Laurence C. Eisenlohr, VMD, PhD
Professor
Department of Pathology and Laboratory Medicine
Children's Hospital of Philadelphia and University of Pennsylvania
Philadelphia, Pennsylvania

Adrian Erlebacher, MD, PhD
Professor
Department of Laboratory Medicine
Institution University of California, San Francisco
San Francisco, California

Hildegund C. J. Ertl, MD
Professor
The Wistar Institute
Philadelphia, Pennsylvania

Martin F. Flajnik, PhD
Professor
Department of Microbiology and Immunology
University of Maryland at Baltimore
Baltimore, Maryland

Mary F. Fontana, PhD
Acting Assistant Professor
Department of Immunology
University of Washington
Seattle, Washington

Sebastian D. Fugmann, PhD
Professor
Department of Biomedical Sciences
Chang Gung University
Taoyuan City, Taiwan

Gaurav D. Gaiha, MD, DPhil
Assistant Professor
Department of Medicine
Harvard Medical School
Boston, Massachusetts

Thomas F. Gajewski, MD, PhD
Abbvie Foundation Professor of Cancer Immunotherapy
Institution University of Chicago
Chicago, Illinois

Stephen J. Galli, MD
Mary Hewitt Loveless, MD Professor of Pathology and of Microbiology and Immunology
Department of Pathology
Stanford University School of Medicine
Centre for Clinical Sciences Research
Stanford, California

Thomas Gebhardt, MD, PhD
Professor
Department of Microbiology and Immunology
The University of Melbourne
Melbourne, Victoria, Australia

Douglas R. Green, PhD
Member and Chair
Department of Immunology
St. Jude Children's Research Hospital
Memphis, Tennessee

Neil S. Greenspan, MD, PhD
Professor
Department of Pathology
Case Western Reserve University
Cleveland, Ohio

Pierre Guermonprez, PhD
CNRS DR2, CNRS ERL8252 INSERM 1149
University of Paris
Paris, France

Aaron Gupta, BS, PhD
Biomedical Fellow
Medical Scientist Training Program (MD-PhD)
Tri-Institutional MD-PhD Program
New York, New York

Adrian C. Hayday, PhD
Kay Glendinning Professor of Immunology
King's College London and Francis Crick Institute
London, United Kingdom

Matthias von Herrath, MD
Professor and Vice President (Novonordisk)
Type 1 Diabetes
Novonordisk
Seattle, Washington
La Jolla Institute
San Diego, California

Andres Hidalgo, PhD
Professor
Area of Cell and Developmental Biology
Centro Nacional De Investigaciones Cardiovasculares Carlos III
Madrid, Spain

Steven M. Holland, MD
Director, Division of Intramural Research
National Institute of Allergy and Infectious Diseases
National Institutes of Health
Bethesda, Maryland

Dirk Homann, MD, MA
Professor
Department of Medicine
Icahn School of Medicine at Mount Sinai
New York, New York

Morgan Huse, PhD
Member
Immunology Program
Memorial Sloan Kettering Cancer Centre
New York, New York

Roger W. Innes, PhD
Distinguished Professor Biology
Indiana University
Bloomington, Indiana

Elizabeth A. Jacobsen, PhD
Associate Consultant
Division of Allergy, Asthma and Clinical Immunology
Assistant Professor of Immunology
Department of Immunology
Mayo Clinic Scottsdale, Arizona

Marc K. Jenkins, PhD
Regents Professor
Center for Immunology
University of Minnesota Medical School
Minneapolis, Minnesota

Thirumala-Devi Kanneganti, PhD
Lab-Director
Kanneganti Lab
Department of Immunology
St. Jude Children's Research Hospital
Memphis, Tennessee

Mariana J. Kaplan, MD
Senior Investigator Chief, Systemic Autoimmunity Branch
Deputy Scientific Director
National Institute of Arthritis and Musculoskeletal and Skin Diseases
National Institutes of Health
Bethesda, Maryland

Rajendra Karki, PhD
Lab-Director
Kanneganti Lab
Department of Immunology
St. Jude Children's Research Hospital
Memphis, Tennessee

Achsah D. Keegan, PhD
Professor
Department of Microbiology and Immunology
University of Maryland School of Medicine
Baltimore, Maryland

Brian L. Kelsall, MD, PhD
Senior Investigator
Laboratory of Molecular Immunology
National Institute of Allergy and Infectious Diseases
National Institutes of Health
Bethesda, Maryland

Claudia Kemper, PhD
Senior Investigator, Immunology Centre
Chief, Complement and Inflammation Research Section
CIRS, DIR, NHLBI, NIH
Bethesda, Maryland

Mohamed Khass, MD, PhD
Assistant Professor
Medicine Clinical Immunology and Rheumatology
University of Alabama at Birmingham
Birmingham, Alabama

Jonathan Kipnis, PhD
Professor of Pathology and Immunology
Professor of Neurology, Neuroscience and Neurosurgery
Director, Center for Brain Immunology and Glia (BIG)
Washington University School of Medicine in St. Louis
St. Louis, Missouri

Hirohito Kita, MD
Professor of Medicine and Immunology
Walter and Leonore Annenberg Professor of Pulmonary Medicine
Division of Allergy, Asthma and Clinical Immunology
Mayo Clinic Arizona
Scottsdale, Arizona

Justin Kline, MD
Associate Professor
Department of Medicine
Section of Hematology/Oncology University of Chicago Medicine
Chicago, Illinois

Douglas S. Kwon, MD, PhD
Associate Professor of Medicine
Harvard Medical School
Ragon Institute of MGH, MIT and Harvard
Cambridge, Massachusetts

Stéphanie Lejeune, MD, PhD
Sean N. Parker Center for Allergy and Asthma Research
Stanford University
Stanford, California
Lille University, INSERM Unit 1019, CNRS UMR 9017, CHU Lille
Institute Pasteur de Lille
Centre for Infection and Immunity of Lille
Paediatric Pulmonology and Allergy Department
Hospital Jeanne de Flandre
Lille, France

Warren J. Leonard, MD
National Heart, Lung, and Blood Institute, NIH
Bethesda, Maryland

P'ng Loke, PhD
Senior Investigator
National Institute of Allergy and Infectious Diseases, NIH
Bethesda, Maryland

Markus Y. Mapara, MD, PhD
Professor of Medicine
Department of Medicine
Columbia University Irving Medical Center
Vagelos College of Physicians and Surgeons
New York, New York

David H. Margulies, MD, PhD
Chief, Molecular Biology Section
Laboratory of Immune System Biology
National Institute of Allergy and Infectious Diseases
National Institutes of Health
Bethesda, Maryland

James McCluskey, FAA
Redmond Barry Distinguished Professor
Department of Microbiology and Immunology
Peter Doherty Institute for Infection and Immunity
The University of Melbourne
Melbourne, Victoria, Australia

Isabelle Meyts, MD, PhD
Professor
Department of Microbiology, Immunology, Transplantation -Department
KU Leuven-University Hospitals Leuven
Leuven, Belgium

Joshua D. Milner, MD
Professor
Department of Pediatrics
Columbia University Medical Center
New York, New York

S. Celeste Morley, MD, PhD
Associate Professor, Pediatric Infectious Diseases
Associate Professor, Pathology and Immunology
Co-Director, Pediatric Infectious Diseases Fellowship
Washington University School of Medicine in St. Louis
St. Louis, Missouri

Edward J. Moticka, PhD (*In memoriam*)
Professor
Illinois University School of Medicine
Springfield, Illinois
AT Still University School of Osteopathic Medicine
Mesa, Arizona

Philip M. Murphy, MD
Chief, Molecular Signalling Section, Laboratory of Molecular Immunology
Chief, Laboratory of Molecular Immunology
National Institute of Allergy and Infectious Diseases
National Institutes of Health
Bethesda, Maryland

Kari C. Nadeau, MD, PhD
Nadissy Foundation Professor of Medicine and Pediatrics
Director of the Sean Parker Center for Allergy and Asthma Research at Stanford University
School of Medicine
Stanford University
Stanford, California

Moon H. Nahm, MD
Professor
Department of Medicine
University of Alabama at Birmingham
Birmingham, Alabama

Kannan Natarajan, PhD
Staff Scientist
Molecular Biology Section
Laboratory of Immune System Biology
National Institute of Allergy and Infectious Diseases
National Institutes of Health
Bethesda, Maryland

Falk Nimmerjahn, PhD
Professor of Genetics and Immunology
Friedrich-Alexander-University Erlangen-Nuremberg
Erlangen, Germany

Luigi D. Notarangelo, MD, PhD
Chief, Laboratory of Clinical Immunology and Microbiology
National Institute of Allergy and Infectious Diseases
National Institutes of Health
Bethesda, Maryland

John J. O'Shea, MD
NIAMS Scientific Director
Senior Investigator
Chief, Molecular Immunology and Inflammation Branch
Bethesda, Maryland

Chirag H. Patel, PhD
Post-Doctoral Fellow
Department of Oncology
Johns Hopkins University-School of Medicine
Baltimore, Maryland

Vincent Peng, MD, PhD
Candidate, Colonna Lab
Washington University School of Medicine
St. Louis, Missouri

Marion Pepper, PhD
Associate Professor and Chair
Department of Immunology
University of Washington
Seattle, Washington

Jonathan D. Powell, MD, PhD
Head of Oncology
Calico Life Sciences
San Francisco, California

Bali Pulendran, PhD
Professor
Department of Pathology, Microbiology, and Immunology
Stanford University
Stanford, California

Gwendalyn J. Randolph, PhD
Emil R. Unanue Distinguished Professor
Department of Pathology and Immunology
Washington University in St. Louis
St. Louis, Missouri

Jeffrey V. Ravetch, MD, PhD
Theresa and Eugene M. Lang Professor
The Rockefeller University
New York, New York

Claude-Agnès Reynaud, PhD
Group leader
Institut Necker-Enfants Malades - INSERM U1151/CNRS UMR8253
Université de Paris
Paris, France

Gabrielle Rizzuto, MD, PhD
Assistant Adjunct Professor
Department of Pathology
University of San Francisco
San Francisco, California

Paul Roche, PhD
Senior Investigator
Experimental Immunology Branch
National Cancer Institute
Bethesda, Maryland

Jamie Rossjohn, FRS
Lab Head and NHMRC Investigator Fellow
Department of Biochemistry and Molecular Biology
Monash University
Melbourne, Victoria, Australia

Nancy H. Ruddle, PhD
Professor Emerita
Epidemiology of Microbial Diseases
Yale School of Public Health
New Haven, Connecticut

Sebastian Ruehl, PhD
Postdoctoral Researcher
Department of Immunology
St. Jude Children's Research Hospital
Memphis, Tennessee

David Sacks, PhD
Senior Investigator
Laboratory of Parasitic Diseases
National Institute of Allergy and Infectious Diseases
National Institutes of Health
Bethesda, Maryland

Andrea Francesca Salvador, MSc, PhD
Master of Science
Washington University in St. Louis
St. Louis, Missouri
University of Virginia
Charlottesville, Virginia

Harry W. Schroeder, Jr, MD, PhD
John R. Irby Research Scholar
Professor of Medicine, Microbiology, and Genetics
Division of Clinical Immunology and Rheumatology
Director, UAB Program in Immunology
University of Alabama at Birmingham
Birmingham, Alabama

David W. Scott, PhD
Professor
Department of Medicine
Uniformed Services University of Health Sciences
Bethesda, Maryland

Pankaj Sharma, PhD
Post-Doctoral Fellow
Program in Cellular and Molecular Medicine
Boston Children's Hospital
Boston, Massachusetts

Ethan M. Shevach, MD
Chief Cellular Immunology Section
Laboratory of Immune System Biology
National Institute of Allergy and Infectious Diseases
National Institute of Health
Bethesda, Maryland

Nevil J. Singh, PhD
Assistant Professor
Department of Microbiology and Immunology
University of Maryland at Baltimore
Baltimore, Maryland

Megan Sykes, MD
Michael J. Friedlander Professor of Medicine
Professor of Microbiology & Immunology and Surgical Sciences (in Surgery)
Columbia University
Director, Columbia Center for Translational Immunology
Columbia University College of Physicians and Surgeons
New York, New York

Warren Strober, MD
Chief, Mucosal Immunity Section
Laboratory of Clinical Immunology and Microbiology
National Institute of Allergy and Infectious Diseases, NIH
Bethesda Maryland

Im-Hong Sun, PhD
Post-Doctoral Fellow
Department of Oncology
Johns Hopkins University-School of Medicine
Baltimore, Maryland

Yousuke Takahama, PhD
Senior Investigator, Thymus Biology Section
Experimental Immunology Branch
National Cancer Institute, National Institutes of Health
Bethesda, Maryland

Mindy Tsai, DMSc
Senior Research Scientist
Department of Pathology
Stanford University School of Medicine
Stanford, California

Roxane Tussiwand, PhD
Principal Investigator
NIDCR, NIH
Bethesda, Maryland

David Wald, MD, PhD
Associate Professor
Department of Pathology
Case Western Reserve University
Cleveland, Ohio

Carl F. Ware, PhD
Professor and Director
Infectious and Inflammatory Disease Centre
Laboratory of Molecular Immunology
Sanford Burnham Prebys Medical Discovery Institute
San Diego, California

Jean-Claude Weill, PhD
Professor
Institut Necker-Enfants Malades - INSERM U1151/CNRS UMR8253
Université de Paris
Paris, France

Erin E. West, PhD
Complement and Inflammation Research Section
NHLBI/NIH
Bethesda, Maryland

Florian Winau, MD
Program in Cellular and Molecular Medicine
Boston Children's Hospital
Department of Pediatrics
Harvard Medical School
Boston, Massachusetts

Gideon Wolf, MD, PhD
Department of Microbiology and Molecular Immunology
University of Maryland
Baltimore, Maryland

Jonathan W. Yewdell, MD, PhD
Chief
Cellular Biology Section
Laboratory of Viral Diseases
DIR, NIAID, NIH

Simon Yona, PhD
Lecturer
Faculty of Dental Medicine
Hebrew University
Jerusalem, Israel

FOREWORD

WILLIAN E. PAUL: AN IMMUNOLOGIST'S IMMUNOLOGIST

William Erwin Paul was born in 1936 in Brooklyn, the son of an immigrant father and a mother who was the daughter of immigrants, as part of the Jewish diaspora that came in a great wave to the United States in the early part of the last century and produced many of our greatest scientists. While his father Jack ran a car repair shop, his mother Sylvia came from a family that included a number of prominent scientists, including one older cousin on his mother's side of the family, who was a Professor of Neurology at Harvard. As a bright kid, Bill skipped two grades in high school and entered Brooklyn College at 16, where he majored in chemistry and premed studies. He notes that it was free and of high quality. He then attended the State University of New York's Downstate Medical School, also in Brooklyn, where he had his first taste of research working in the lab of George Talbot, studying growth hormone activity in rats. He also had the good fortune to marry Marilyn Heller during that time, an important source of support and advice for the rest of his life. He then left Brooklyn for an internship and residency in Boston at Boston University's Massachusetts Memorial Hospital. There he also turned to research in the lab of Alan Cohen, working on the biology of fibrils and their contribution to Alzheimer and other diseases. A pivotal moment came when he applied for a very competitive NIH clinical fellowship, since the alternative was likely being drafted into the army. Fortunately, he was awarded the fellowship and so in 1960 he and Marilyn moved to Bethesda, the main NIH campus, where they had their first child, Jonathan.

Here he worked with Bill Odell and Jack Wilbur on the first antibody-based assay for thyroid-stimulating hormone, which was the most sensitive assay available for detecting abnormalities of the thyroid. The subsequent paper received over 500 citations, quite an accomplishment for a newcomer who also had patient-care responsibilities. Bill also describes this time at NIH as quite exciting, and many of the brightest MDs at that time were gravitating toward a research stint at the NIH as an alternative to the doctor draft. This trend would accelerate dramatically as the Vietnam War ramped up, so much so that for the whole era through 1972, the "yellow berets" as they were called, included many famous scientists of later years, contributing to a veritable golden age of science at NIH. Yet, after 2 years at NIH, Bill moved on to NYU to work with Baruj Benacerraf, then a rising star in immunology; he'd hoped to work with Henry Kunkel at Rockefeller University but that laboratory had no open positions. Bill had been fascinated with the mysteries of immunology since college, when he happened on the work of Michael Heidelberger, who strove to introduce precise quantitation to the study

William (Bill) Paul in 1982, around the time that Mark Davis worked with him at NIH. Bill is shown with one of his mentors, Baruj Benacerraf (mentioned in the Foreword), and James Wyngaarden, the director of the NIH from 1982 to 1989. Bill was presenting the G. Burroughs Mider Lecture, inaugurated in 1968 in honor of the first NIH director of laboratories and clinics. The title of Bill's lecture was "Living with Lymphocytes. B Lymphocytes and How They Grow."

of antibody responses and specificity. By coincidence, Heidelberger, who lived to the ripe old age of 104, was doing a second postretirement stint next door to Bill's lab and they frequently interacted. In the Benacerraf group, Bill became immersed in the mysteries of the burgeoning field of cellular immunology, far removed from immunoglobulin chemistry, the pursuit of which would guide the rest of his career.

In 1968, Bill moved back to the NIH when Benacerraf was recruited to NIH to head the Laboratory of Immunology. Soon after they had arrived, however, Benacerraf was offered a Chaired Professorship at Harvard, and decided to go there. He asked Bill to go with him but Bill was ready for independence at this point, and decided to stay at NIH, probably remembering his exciting days there earlier. When a prominent immunologist declined to replace Benacerraf, the job of lab chief, essentially the department chair in other academic settings, was open; with Benacerraf's strong support, Bill was selected for this position, and would remain as lab chief for the rest of his life. While some NIH lab chiefs chose to devote all the resources of a laboratory to their own interests, Bill was much more egalitarian, more in the line of a modern academic department, where faculty-level scientists were independent, and had their own small groups of postdocs and technical staff (PhD students would only come later in a program with Oxford University developed by Michael Lenardo). The 1970s was a time of turmoil but also great insights in the field of cellular immunology. The importance of the MHC loci was being defined, first by Benacerraf and McDevitt, but also by Rosenthal/Shevach and Zinkernagel/Doherty who showed that the true function of the MHC proteins is to provide a context for T cell antigen recognition. Thymic selection was also being defined by Bevan and Zinkernagel and others. Then in the mid-1970s, recombinant DNA became mainstream and utterly transformed almost all of biology, especially immunology. Susumu Tonegawa, then at the Basel Institute for Immunology, showed the first evidence of DNA rearrangement in the formation of immunoglobulin genes, followed quickly by Phillip Leder at NIH, and Leroy Hood at Caltech, and others, in a flurry of remarkable papers. These studies resolved the most age-old questions about antibody diversity, class switching, membrane-bound and secreted forms of antibodies, and affinity maturation in rapid succession in just 4 years. As a graduate student at Caltech, I was in the thick of isolating and characterizing mouse Ig heavy chain genes in the Hood lab at the time, working with fellow PhD student Phillip Early. We were helped enormously by the arrival of Tom Maniatis and his group, who brought a wealth of recombinant DNA expertise from Cold Spring Harbor and Harvard, and developed many more key methods at Caltech.

Norman Davidson and Tom started what was essentially a recombinant DNA club, with regular meetings to disseminate the emerging technology. Tom later wrote the recombinant DNA "bible" of that era (*Molecular Cloning*, CSH press 1982) and started a summer course at Cold Spring Harbor on recombinant DNA methodologies that I was later honored to be part of. Phil and I were lucky to be included in the "club" and benefited enormously from what we learned, so much so that we ended up scooping Tonegawa's lab on several fundamental aspects of heavy chain genes. As I neared the end of my PhD in 1979, a crisis loomed as my then girlfriend and now spouse, Yueh-hsiu Chien, had finished up work in Norman Davidson's lab and taken a job at NIH, using her newly minted recombinant DNA skills to clone endocrine system genes. I asked Mitch Kronenberg, a fellow student in the Hood lab and the only one of us broadly knowledgeable about immunology beyond antibodies, if he could recommend a postdoctoral mentor at NIH. Mitch immediately said, "Bill Paul, he's really smart," or words to that effect. At that time immunology was clearly divided between the antibody people—biochemists and molecular biologists who studied the relatively straightforward world of antibodies—and cellular immunologists who did much more mysterious things like putting the thymus of one mouse into another mouse, and who discovered the phenomenon of MHC restriction despite having no idea of how it worked molecularly. Very weird but strangely wonderful stuff, to someone like myself who had only worked on, and thought about, molecules (mostly DNA and RNA). I then visited Bill and his department (AKA "Laboratory of Immunology") and was struck and fascinated that his group almost exclusively worked in this exotic area of cellular immunology. Originally I thought that I learning about this world would expand my immunological horizons, but encouraged by Bill, I also applied for postdoctoral fellowships. This was great advice and something that I encourage all of my postdoctoral applicants to do, since it forces one to think critically about what to study in a new environment. In my case this discipline worked exceptionally well, because I realized that the antibody gene work was reaching an end—the major questions had been answered or nearly so, and thus it was important to think about what might come next. One obvious route that many took was to study transcriptional regulation of whatever gene you had in hand, but that didn't excite me. Another possibility was to identify the recombinases that mediate V-D-J joining, but this brought two problems to my mind: (1) they would be very hard to clone because they would be very rare transcripts, since they would need to be tightly controlled to avoid deleterious rearrangements, and (2) they were probably related to topoisomerases which was a popular field at the time, but I never could understand the lectures. But even in ruling out that option, I asked myself, if I did want to clone these recombinases, how could it be done? Here I drew on my knowledge of nucleic acid reassociation kinetics and RNA handling that I had learned about earlier in the lab of Eric Davidson. If T and B cells shared most of their gene expression, I could subtract the T and B transcriptomes from each other to perhaps identify the rare T- or B-specific gene products. This strategy could result in the cloning of all sorts of genes that were required for B and T cell function. So, I wrote up these wild speculations into my grant proposal and submitted them to the proper authorities, finished up my work in the Hood lab, and made it out to NIH in the fall of 1980. I told Bill at some point about this idea, and he asked a few pertinent questions, as always, but also suggested that I might work on some projects that he was interested in.

When I told him that I preferred to work on my idea, Bill was totally supportive and introduced me to Jin Kim and Richard Asofsky who had various mouse B and T cell tumor lines that they could grow for me. I gratefully accepted, since at that time I had never grown any mammalian cell lines. From these cells I was able to purify cytoplasmic RNA from several B and T cell lines and by the end of 1980 I did my first subtractions of labeled B or T cell cDNAs with RNA from the other cell type. I found that very reproducibly, 98% of the cDNA of T cells would hybridize with RNA from B cells, and vice versa. Thus, purifying B or T cell-specific probes would result in a ~50-fold enrichment for these sequences, enabling even rare genes (most of them) to be detectable. This was even better than I could have hoped for and excited all of us. I think it had immediate appeal to Bill and other immunologists because it was standard practice at the time to absorb complex antisera with unwanted antigens, to obtain antibodies for specific target antigens. In contrast, molecular biologists tended to be dismissive, because they never appreciated (or even understood in many cases) DNA or RNA reassociation kinetics, considering it to be a relic of molecular biology in the precloning age, now past. Fashion is a surprisingly major influence, even in science, although it's clear that progress often comes from resisting this impulse. In any event the subtraction project progressed quite well, and was clearly working, when we found that the just recently cloned Thy-1 gene (by a former Hood lab colleague Jack Silver) was present in our T cell-specific collection, and we also helped Lee Hood and his group identify the location of class II E^k in their MHC cosmid clones. As this was going on, Bill told me something astonishing, which was that he didn't understand enough about the technical aspects of my work to be competent to supervise me, and that I should consider myself independent and he wouldn't put his name on my papers. In addition, he wanted me to stay in the LI and join the faculty.

He assigned to me a fantastic technician, Ellen Nielsen (now Bernstein), and encouraged two postdocs, David Cohen and then Steve Hedrick, to work with me, as they were eager to learn recombinant DNA technologies. It was also very helpful that Tom Sargent, a graduate school buddy who had also gone to NIH, shared with me a neat trick for making subtracted cDNA libraries. David Cohen used this procedure to make our B-subtracted T-specific library, later used by Steve and I to extract the first T cell receptor (TCR) gene. However, Bill's support wasn't entirely on faith, as much later Tasuku Honjo told me that Bill had asked him if what I was doing made any sense. Luckily, Honjo had done some DNA reassociation analysis himself earlier to define deletion as part of immunoglobulin heavy chain switching, and so he wasn't as prejudiced about the value of this methodology as most other molecular biologists. Bill was true to his word, and when my little group actually produced the first TCR clone (TCR beta) in July 1983 among our T cell-specific genes, he was incredibly supportive, even though by then I had accepted an offer from Hugh McDevitt at Stanford to join the newly reconstituted Department of Medical Microbiology (now Microbiology and Immunology). Bill was enormously generous to not insist on even a co-authorship on what would be the hottest papers in immunology at the time, but it shows the lengths he went to for the advancement of others' careers. As a famous and experienced immunologist in 1983, Bill knew that if his name were on the papers that he would reap most of the credit for this seminal discovery. He also declined an NIH-sponsored trip to China that year, suggesting me as a substitute, which I accepted, even though I had just arrived at Stanford and was setting up the lab with Chien, who volunteered to help (since she knew I would need all I could get!). It was a curious trip, since China had only recently opened to the West and the people were very poor, and the immunology in the country was largely focused on monoclonal antibodies.

The greatest value for me was that this trip was timed to end just as the International Congress of Immunology was set to begin in Kyoto Japan. I never would have gone to that meeting if it were not a convenient hop from Shanghai to Kyoto. Unfortunately, I didn't know that one needed a visa to enter at that time, and so I spent hours in the airport chatting with the immigration authorities about why I had arrived without a visa. I volunteered Dr. Honjo as a character witness, and while I don't know if they ever contacted him, the authorities eventually let me officially enter the country. At the meeting, I received a fresh infusion of data from my NIH lab. Steve and Ellen had been working away sequencing our new TCR genes, including some we isolated from a thymocyte library kindly provided by Christoph Benoist. And so I approached Harvey Cantor, who was chairing the session on T cell recognition, and asked if I could present our data. Somehow he said yes, and later I found myself in front of over 5000 immunologists telling them we had solved a puzzle that had frustrated the field for years. Definitive earlier work on the TCR heterodimer by Jim Allison, Ellis Reinherz, and Kappler and Marrack, presaged our work, but nothing was as tangible, or empowering, as a gene clone, in this case TCRb from mice. Our work was published in the following year, along with the unexpected human TCRb clone from Tak Mak, and later a paper from Susumu Tonegawa, to whom I had naively given my subtraction protocol. Later in 1984 both my group and Tonegewa's published on the TCRα gene, essentially launching the modern era of T cell biology.

Meanwhile Bill was working away to define a novel cytokine activity that Maureen Howard in his group had identified, initially identified as a novel B cell growth factor (Howard and Paul JEM 1982). This turned out to be IL-4, which was later shown to be crucial for driving the differentiation of the Th2 subset and for the switch to IgE by B cells (Paul and R Seder 1994). The identification of this important factor is perhaps Bill's greatest scientific achievement, and was the subject of many subsequent studies.

Bill also increasingly became a statesman of Immunology. In the early 1980s, he was elected to the governing council of the American Association of Immunologists, the major immunology society, and then its President from 1986 to 1987. During this time he also was the founding editor of the *Annual Reviews of Immunology* in 1983, where he served for 31 years.

Somehow he also found the time to edit an advanced text with up-to-date chapters written by experts in the various areas of immunology (*Fundamental Immunology* 1984). This book has been extremely valuable for graduate students and researchers in the field and went through seven editions in Bill's lifetime. This eighth edition, ably helmed by Martin Flajnik, is a small token of our esteem.

Also very important, in the AIDS crisis that began in late 1980s, there was increasing anger in the gay community at what was perceived as a lack of urgency on the part of NIH and the government. This created a crisis of trust that was corroding support for basic research, and someone of stature was needed to calm the waters. Bill was the natural choice, not because he had ever worked on HIV, but because of his ability to listen and render thoughtful responses to frustrated and angry people. Thus he agreed to become the first head of the NIH Office of AIDS research in 1994, to coordinate funding decisions and to liaise with the public. In this role he was very successful, and in this capacity helped usher in the life-saving protease inhibitors, which revoked the death sentence for HIV-infected individuals. Yet, in this capacity, Bill saw that an effective vaccine was also required, and that the basic science of HIV infection was needed to move the understanding of vaccines forward. Thus, he thought that what was needed was to establish a dedicated vaccine research center at NIH. After vetting this idea with colleagues he went to the White House with Anthony Fauci and others in 1997 to pitch the proposal directly to President Bill Clinton. Clinton was persuaded and, with subsequent congressional approval, this became a reality in just a short time. Thus the Dale and Betty Bumpers Vaccine Research Center open its doors in 2000, with a mission to both explore the science of vaccination and to manufacture vaccines and conduct early clinical trials. In these missions, the VRC has been very successful, developing effective vaccines for Ebola and RSV, and most recently led the first trials of the SARS-CoV-2 vaccine from Moderna. With these major accomplishments, Bill stepped down from his role at the Office of AIDS Research in 1997 and was able to devote more time to his own research and other pursuits.

In summarizing a remarkable life like Bill's, it seems clear that while his scientific achievements were remarkable, what really stands out was his concern for the field of immunology and humanity in general, as well as his unparalleled selflessness. While almost all of the attention in science deals with the achievements of individuals, talented people curious about the rules governing the natural world—in earlier times Newton, Darwin, and Einstein—all worked essentially alone to achieve their stunning insights. Modern science, on the other hand, is a team sport with single-author papers a rarity from physics to biology; multiple to dozens of authors are individuals, and are needed to achieve a recognizable advance in that field. Still, scientists are typically individualistic loners and thus are attracted to these icons of the past. We are typically not "joiners" or "followers" but take pride in thinking for ourselves and taking the path "less traveled." Yet, while singular visions and pioneering efforts are as important as ever, it is also critical to recognize that individuals can flourish scientifically with the support of a strong team and community, so that the vision of an individual or group can actually be realized. William Erwin Paul (1936-2015) recognized this reality long before most of us, and while pursuing his own scientific interests, he very successfully devoted more and more of his efforts to creating a supportive scientific community, both in his own department (the Laboratory of Immunology at NIH) but also more broadly at the NIH and in the entire country. This enabled many young immunologists he mentored, including this one, to achieve much more than they could have normally. Most scientists take on community-building activities only under duress, especially B.T. (before tenure). Their mission is to build up an important body of work as quickly as possible, and indeed, this is what they are rewarded for. I have an interesting thought experiment, which asks, "what would your colleagues think if you were second author on the 10 greatest papers in biology? The correct answer is "loser!". This common attitude is destructive to collaborations and the supporting scientific community—we need each other, but many factors work against this and to some extent we can entrap ourselves into how science was done long ago. Here again, we have Bill Paul as an example of science in the modern age, encouraging young scientists to do their best in his role as chief of the LI, always ready with sage advice, and not concerned with credit for himself. It's a wonderful example to us all, and a remarkable legacy.

Mark M. Davis

ACKNOWLEDGMENT

William Paul's wife Marilyn is acknowledged for providing the Paul family history and editing of the Foreword.

PREFACE

This is the eighth edition of *Fundamental Immunology* (now called *Paul's Fundamental Immunology*), a decade since the previous edition was compiled and the first edition after the untimely death of Bill Paul in 2015. As Dr. Paul wrote in the preface of the last edition: "*Fundamental Immunology has the goal of aiding in the education of a new generation of immunologists who can both probe more deeply into the organizing principles of the immune system and can translate this new information into effective treatments and preventatives that will extend and enlarge on the record of immunologic science in bettering the lot of humankind.*" We agree wholeheartedly with this sentiment but would add that seminal features in immunology such as the generation of antigen receptor diversity, germinal center reactions, antigen processing, lymphocyte development and migration, among many others, are fascinating in their own right, regardless of the therapeutic potential.

In this eighth edition, we go into some detail in the Introduction (Chapter 1) to familiarize readers with the topics covered in each of the following 49 chapters. The classical areas in immunology included in previous editions were all updated with the latest captivating discoveries in basic and translational immunology. Furthermore, in this edition there are new topics included that have come to the fore in the last decade including innate lymphocytes, maternal-fetal immunity, programmed cell death, interactions with microbiota, evolution of immunity, and neuroimmunology.

In the preface to the last edition Dr. Paul also wrote: "*The continued need for progress in immunology is clear. The epidemic of human immunodeficiency virus roars on. Glimmers of hope from vaccine trials have led to a redoubling of effort, and the struggle to design effective vaccines for the great infectious scourges goes on, with encouraging results but no breakthroughs yet. Highly effective therapeutic vaccines for cancers still elude us, but some immunologic therapies for cancer have met with encouraging results.*" Encouraging results indeed! The amazing progress we have witnessed in the last decade in cancer immunotherapy with checkpoint inhibitors and CAR-T cells has been revolutionary, to state the obvious. Permutations of these therapies, in combination with new strategies on how to manipulate the tumor microenvironment to render tumors susceptible to T cell/NK cell destruction, will be the goals for the next 10 years. Regarding vaccines, we were privileged to observe the taming of a viral pandemic with the first use of mRNA vaccines, in combination with approaches to activate the innate immune system and manipulate the structure of the immunogen. We predict that in the next decade, in addition to continued progress in cancer immunotherapy and vaccine development, our asymptotic knowledge on basic immunology will result in breakthroughs in treatment of autoimmunity and transplantation.

In a final nod to Dr. Paul, he wrote, "*I repeat a word of caution that has been in the Preface to each edition. Immunology is moving very fast. Each of the chapters is written by an expert in the field, but in some areas there may be differences of opinion expressed by equally accomplished authors. I ask the reader to take note of the differences and to follow developments in the field.*" Such differences of opinion are to be expected in any rapidly moving field, yet disagreements and frank discussions are *actually required* for a field to continue to progress. It is our hope that readers of this edition will not only come away with an appreciation of immunology, but also become privy to the controversies and areas that still befuddle us.

We would like to thank each of the authors for their contributions to this eighth volume of *Paul's Fundamental Immunology*. We appreciate the efforts to assemble their chapters amid a global pandemic, while the authors were trying to sustain their lab structure, and many were actually examining various aspects of immunity to COVID-19. It was a pleasure for us to edit each chapter, as we learned a great deal about every topic. We especially are grateful to those authors who stepped up late in the game (you know who you are!) when we recruited you to contribute to chapters that were essential for the volume. Special thanks to Mark Davis, who in addition to writing the chapter on T cell receptors with Yueh-Hsieh Chien wrote the foreword as a dedication to William Paul, one of his mentors.

Finally, we thank Wolters Kluwer for taking on the responsibility for publishing this new edition of *Fundamental Immunology*, with special recognition of Ariel S. Winter, Christopher Rodgers, and Ramkumar Soundararajan, who worked with us tirelessly to plan this volume, interact with all of the authors, and shape each chapter into its final form. This edition would not have been possible without their support.

CONTENTS

Section VI. Integrative Immunology

Section VII. Immune Responses to Pathogens

Section VIII. Immunological Disease States and Immunodeficiency

SECTION I

Introduction to Immunology

CHAPTER 1

Overview of the Immune System

Martin F. Flajnik, PhD • Steven M. Holland, MD • Nevil J. Singh, PhD

As we edit this book during a global pandemic, we are acutely aware of the remarkable prowess of the immune system. It is critical to keep us safe, in the face of constant threats from the myriad potentially pathogenic microorganisms that inhabit the world. A half-billion years of arms race between such microbes and the mechanisms used by the immune system to neutralize the threats they pose, has resulted in the evolution of a complex and multilayered system of defense (Chapters 3-5). It begins with layers that offer a nonspecific barrier to pathogens and becomes increasingly specific, crowned by the adaptive immune system that can discriminate narrowly between even minor variants of a particular microbe (Fig. 1.1). The tragic examples of acquired immunodeficiency syndrome (AIDS) or inherited immunodeficiencies illustrate the consequences of impaired adaptive immunity. Patients with AIDS (Chapter 45) and children with severe combined immunodeficiency (SCID) fall victim to infections that are of little or no consequence to those with functioning immune systems. The immune system also monitors threats that arise from within our bodies. It recognizes and rejects spontaneous tumors and is involved in healing wounds. Elegant recognition mechanisms allow the immune system to make very specific receptors against almost any target, a process exploited commercially to develop antibodies for many applications. Yet these extensive capabilities also force the immune system to constantly walk a tightrope, to avoid damaging healthy cells in the body—via autoimmunity—while still attacking diverse threats to its well-being. Rigorous control mechanisms have emerged to keep autoimmunity at bay, which can break down and result in autoimmune diseases (including insulin-dependent diabetes mellitus, multiple sclerosis, rheumatoid arthritis, etc). In addition, excessive inflammatory pathology from damaging healthy cells can occur while attacking an infection even in the absence of autoimmunity; thus, other layers of regulation have emerged to prevent or dampen such autoinflammation.

Fundamental Immunology was conceived in 1984 by Dr. William E. Paul to provide an authoritative presentation of the basic elements of the immune system; of the means by which the mechanisms of immunity act in a wide range of clinical conditions, including recovery from infectious diseases, rejection of tumors, transplantation of tissue and organs, autoimmunity and other immunopathologic conditions, and allergy; and how the mechanisms of immunity can be marshaled by vaccination or biologic therapies to confront tumors and pathogens. This edition perpetuates the long tradition championed by Dr. Paul, to recruit experts in each field to provide an in-depth contemporary summary of our understanding of these topics in the following 49 chapters.

The purpose of this opening chapter is to provide readers with a general introduction to the immune system. It should be of particular importance for those with a limited background in immunology, providing them with the preparation required for subsequent chapters of the book and the capacity to then follow the basic and translational immunological literature. Rather than providing extensive references in this chapter, each of the subjects will designate the chapters that deal in detail with the topic under discussion. Those chapters will not only provide an extended treatment of the topic but will also furnish the reader with a comprehensive reference list.

THE IMMUNE RESPONSE IN A NUTSHELL

Microbes gain access to our bodies through tissue surfaces that are exposed to the environment, such as the skin, lung epithelium, and inner linings of the intestines. The first layers of defense against pathogen entry are the **barriers** presented by these surfaces themselves (Fig. 1.1). These structural barriers represent the most effective protections we possess, and microbes which colonize the outside of the barrier surface can often enter a mutualistic relationship, becoming commensal flora. As long as a more distal layer of immunity is

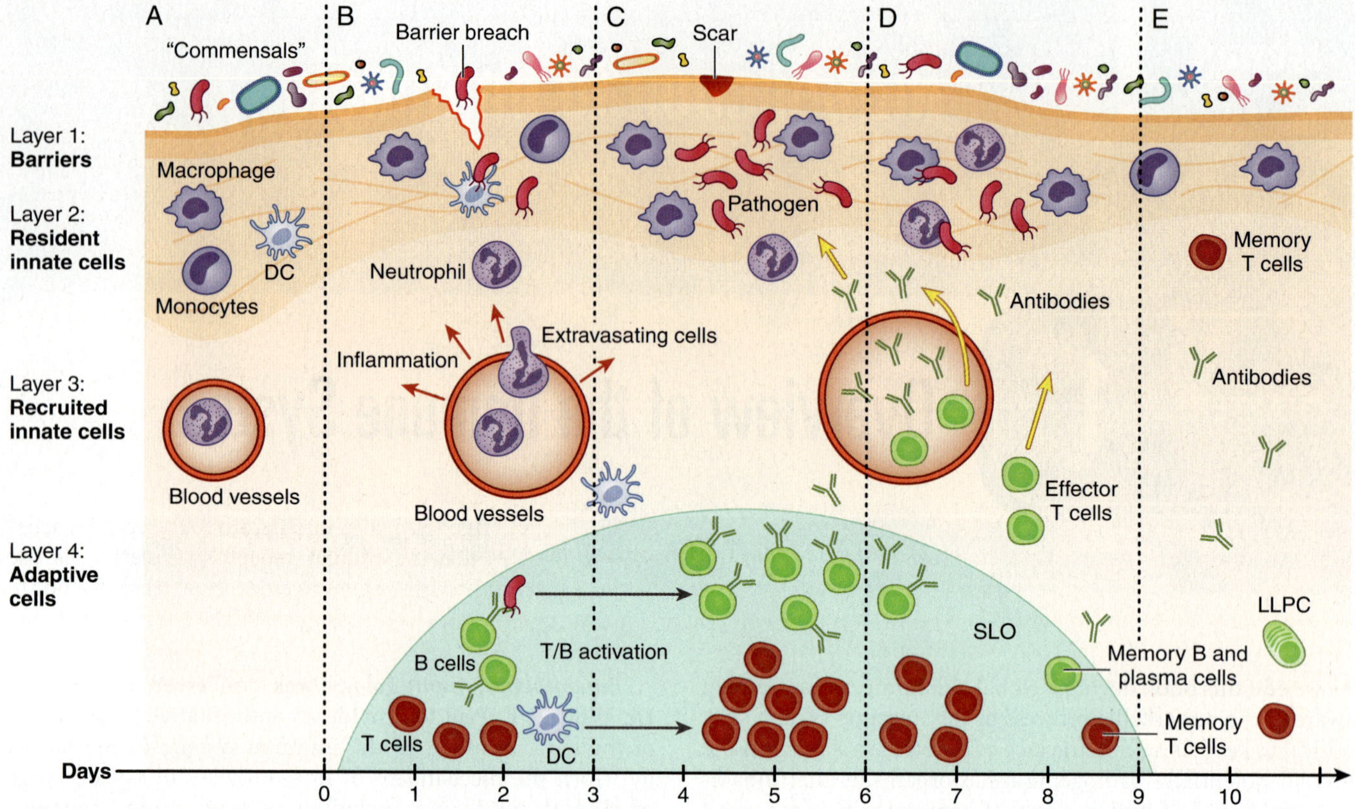

FIG. 1.1. Stages of an immune response. A: Prior to the infection, commensals exist in homeostasis with the host outside the barrier, such as skin. Macrophages, dendritic cells (DCs), and monocytes are resident innate cells within the tissue. **B:** When the barrier is breached by a microbe, resident cells are activated by pattern recognition receptors, sending signals such as cytokines and chemokines to recruit innate cells like neutrophils from the blood via extravasation (also called transendothelial migration). DCs begin to move through the lymph to the draining lymph node (SLO [secondary lymphoid organ]). **C:** Invading innate cells destroy the microbes, while the adaptive response is initiated with the activation of T cells (red) and B cells (green) in the SLO, followed by clonal proliferation and the first production of antibodies. Scars may be formed via homeostatic mechanisms at the breach site. **D:** Innate cells become armed with antibodies for microbe destruction and the breach is healed. Effector T cells and natural killer cells may also come to the site. **E:** Memory B and T cells are generated that may stay in the tissue for some time and disseminate to SLO all over the body. Plasma cells as long-lived plasma cells (LLPC) can persist for months or years.

successful in repelling pathogens, no more central layer in the immune system is activated. When a microbe breaches the barrier, **preformed effector molecules** such as the Alternative Complement (C′) Pathway or constitutively produced lectins might target the pathogen for destruction. The next layer of defense that is activated is **cells of the innate immune system.** Unlike barriers and preformed effectors, which are always ready to resist invasion (ie, "always on" or constitutively protective), this next layer consists of cells that are typically quiescent and positioned just below the barrier tissues. Once they detect pathogens, tumors, or the tissue-damage caused by either, these innate immune cells are activated and mount an effector response that includes phagocytosis, secretion of noxious compounds, and production of proteins that promote downstream immunity. Although these cells broadly sense invasion or damage (using pattern recognition receptors [PRRs]), they are relatively nonspecific in their activation. These PRRs are found in all organisms (Chapters 3-5, 12), not just vertebrates, and this early response is therefore the evolutionarily more ancient form of defense. During this initial response, the site of infection also starts generating **inflammation.** In addition to making the infection site inhospitable to some microbes, inflammation also promotes blood flow to the area, disrupts tight junctions of blood vessel endothelia, and **recruits another layer of circulating innate immune cells** to the venue. This early innate response (barriers, preformed effectors, locally activated cells, and recruited effector cells) is the major defense mechanism for the first 2 to 3 days after an infection. If the pathogen (or tumor or irritant) continues to pose a threat, **cells of the adaptive immune system** are then enlisted for defense. The reason why these cells (T cells and B cells) are delayed in the response is because they are initially activated away from the site of infection, in specialized secondary lymphoid organs (SLOs) (Chapter 8), and the frequency of cells specific for the pathogen/tumor is very low (so-called precursor frequency). After almost a week of activation, cellular proliferation, and maturation, these cells can contribute cellular effectors (T cells) or antibodies (made by B cells), which leave the SLO and circulate back to the tissue site where the pathogen resides (Fig. 1.1). A considerable upgrade that comes with the adaptive immune response is that, unlike the PRR used by the innate system to detect pathogen or damage, their receptors (T cell receptors [TCRs] and antibodies or immunoglobulins [Ig]) are extremely specific for their target. If these sequential (and cooperating) layers are eventually effective and the pathogen is cleared, some of the T cells and B cells that were activated in this process persist for a long time in the body as **memory cells**. Immunological memory allows the immune system to mount a much more rapid response when the same pathogen is again encountered (ie, it does not require the extra week of activation and maturation to mount the response). Obviously, this sequence of events requires an extensive choreography to ensure that all the "right cells" develop in the body, arrive at appropriate tissue sites at the required time, use specific receptors to be activated, deliver critical sequential effector responses, and incorporate sufficient control mechanisms to suppress the response at the appropriate time. We will now summarize these cell types, processes, and pathways, and discuss the overarching principles evolved in the immune system to operate or deploy all of these armaments appropriately.

KEY PRINCIPLES UNDERLYING AN IMMUNE RESPONSE

While we have historically viewed the immune system as a protective force charged with attacking and repelling pathogens and tumors, it is increasingly clear that it plays a more complex role in keeping the body healthy. In some cases, the latter mandate even requires the system to "ignore" some microbes (for example, commensals or even chronic infections [Chapter 38]) especially if the costs of continued inflammatory responses are high. It participates in healing wounds, facilitating organ development, reinforcing neurological learning, etc. Considering these broad functions, the central principles governing the operation of the mammalian immune system are summarized below. Developing an understanding of these general concepts can be helpful for students, as they try to digest the significance of each detailed mechanistic process in the chapters that follow.

1. *Barriers—The immune system relies on a strong system of passive barriers to avoid mounting an offense.* Resisting infection or the damage due to infection by offering structural barriers around the body is an evolutionarily conserved defense strategy. Some sensitive organs (eg, the brain) have additional layers of selective barriers. Most of the environmental threats the body faces are kept out by these barricades. In turn, the immune system is primarily organized around these barrier functions, to catch threats that manage to get past.
2. *Tolerance—Multiple mechanisms and cross-checks operate to avoid attacking healthy parts of the body.* If pathogens breach immunological barriers, subsequent protection requires immune cells to be activated to perform their defensive roles. Critically, such activation requires sensing the threat using cellular receptors. Innate cells use receptors to conserved molecules (patterns) found in many pathogens or unhealthy cells. The adaptive immune system uses many more specific receptors that surgically identify minute features on each threat. In both instances, it is important that immune cells are not activated in response to normal body cells or targets in healthy tissues. Such discrimination is not trivial since pathogens can also mutate the targets of immune attack to evade it. The innate PRRs are typically quite conserved over evolution (Fig. 1.2) and have presumably been optimized over eons to identify microbes or damage but spare healthy cells. T and B cell receptors (AKA antigen receptors) are

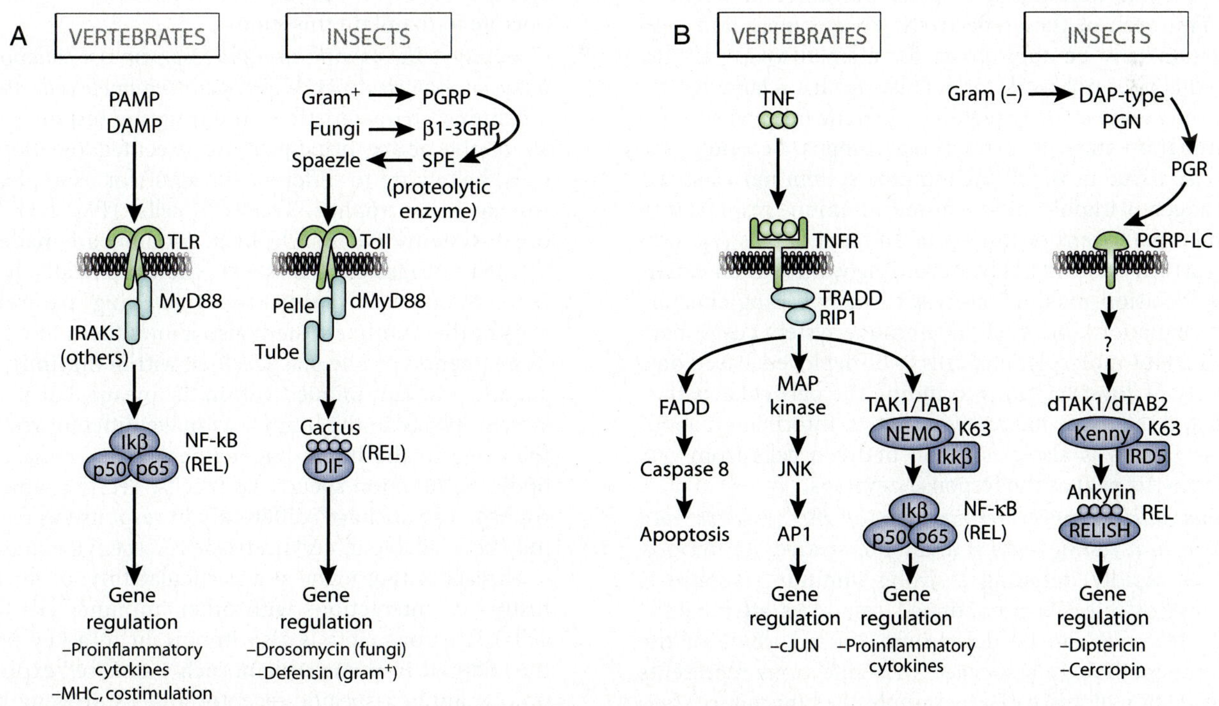

FIG. 1.2. Comparison of innate immune response induction, intracellular pathways, and immune outcome in insects (eg, *Drosophila*) and vertebrates (eg, human). Note that the initiation of the innate response and the outcome(s) in insects and vertebrates are quite different for both the toll/toll-like receptor pathway (*left*, **A**) and the immune deficiency/tumor necrosis factor pathways (*right*, **B**), but the intracellular signaling pathways are well conserved evolutionarily in all animals.

educated and regulated by a variety of processes to avoid strong reactivity against healthy self-tissues. While the term "immunological tolerance" (or self-tolerance) usually applies to the latter, the general principle of tolerating healthy tissues operates in recognition steps for most cells in the immune system.

3. *Layering—The immune system sequentially engages and coordinates the functioning of increasingly threat-specific tiers of defense.* As discussed above, if the pathogen evades one layer of defense, then the immune system has another one called to action (Fig. 1.1). Understanding this tiered approach to immune functions involves dissecting how each layer is sequentially activated. In general, cells in the earlier layers activate the subsequent ones. This process, however, also involves a "bridging" of the layers, such that all the early-activated cells continue to cooperate to neutralize the ongoing threat. For instance, adaptive immune cells (T cells) require critical cells in the preceding layer (that is, activated innate cells such as dendritic cells [DCs] [Chapter 34]) to be activated. But once fully activated, effector T cells look to the earlier layers by further recruiting and arming innate cells to respond more robustly. This cross-connectivity and coordination between different cell types requires sophisticated cell-to-cell communication mechanisms. Typically, this involves messenger proteins called cytokines and chemokines (Chapters 9-12) secreted and sensed by cells in different layers.
4. *Customization—Each immune response is tailored to both the nature of the threat and the specific tissue location affected.* There are a range of insults that can threaten our body including toxins, viruses, bacteria, parasites, tumors, etc, and the immune system deploys different weapons to fight each of these effectively. Phagocytosis and degradation may be appropriate for a small toxic particle, but this does not work for defense against a large worm. An intracellular virus requires different mechanisms for elimination than an extracellular fungus. Secondly, the precise tissue in which the immune system must battle a pathogen is important to consider: immune responses to similar pathogens or tumors in different organs (eg, skin vs gut) vary considerably. Accordingly, there is an extensive "decision-making" process (usually during effector-differentiation), in which the immune system customizes the various molecules and cells to be deployed, based on a variety of considerations including the particular pathogen and the tissue microenvironment. Information about these factors is also passed on between cells from one layer to the next as the response develops.
5. *Homeostasis—Timely termination of the response and return to a resting state is a key prerogative.* As important as rapidly initiating a strong immune response is to control infections; continued responses after pathogen clearance can be deleterious. Indeed, some of the immunopathology associated with infections represents dysfunction of this arm, for example, the lung failure associated with exacerbated responses to influenza or SARS-2. Accordingly, a variety of molecules and specialized cells have evolved to control this phase. The principles of tolerance and response-termination can often be difficult to separate out. One hallmark is that the former operates to prevent both the initiation and maintenance of a response; but termination mechanisms come to the fore after the effector phase has fulfilled its mandate by eliminating or controlling the threat. The impetus to preserve homeostasis can be thought of as a teleological objective of the immune system. Its participation in wound healing, avoiding metabolic imbalances, etc are likely to be driven by this goal. Homeostasis does not mean that the immune system (and the body) always returns to its "preinfection" or naïve state; instead after each infection or threat management, the system reaches a new resting state by processes which are still being defined.
6. *Memory—The immune system retains a memory of its previous responses*: The first time a pathogen is repulsed (called the "primary response") the adaptive system takes weeks to do so. This might give the pathogen longer to establish, grow, and damage the body. In contrast, the next time the same pathogen attacks, the "secondary" immune response is much more efficient (Chapter 30). Vaccinations are the clinical strategy to mimic this "first exposure," albeit usually using nonpathogenic formulations, so that when the infection does come about, the system uses this memory to mount a secondary response (Chapter 40). While memory is largely described in the context of the adaptive immune system, a process of "training" is also thought to condition cells of the innate immune system. As discussed in Principle #5, the formation of additional memory (or innate training) after each threat, is part of the new homeostatic state of the immune system. Multiple cellular and molecular mechanisms operate to maintain this state.
7. *Clonality—The overall immune response is a collaboration between many independently operating single cells.* Perhaps a defining characteristic of the immune response is that all of the above principles are executed by individual cells, belonging to different lineages (for example, macrophages, neutrophils, T cells, B cells) (Fig. 1.3). While organ systems like muscle, liver, and lung are made up of distinct lineages of cells as well, these are relatively hardwired to operate as a unit. The "free-living" nature of the cells in the immune system also allows them to be much more phenotypically plastic, albeit within the limits of the genetic program imbued within its lineage. For instance, a macrophage in the brain may make different responses than one in the heart, but neither can ever make antibodies, a function specific of B cells. Yet, the same macrophage can "behave" differently in response to cues that may be available in certain tissue microenvironments, or to threats in that niche at a particular time or during its history of interactions with other immune cells (like T cells). This principle is also important to a key tenet of the adaptive immune system: each T or B cell expresses a unique antigen-specific receptor and its progeny inherit that receptor. This leads to concepts based on clonality, such as clonal expansion, clonal deletion, clonal contraction discussed below.

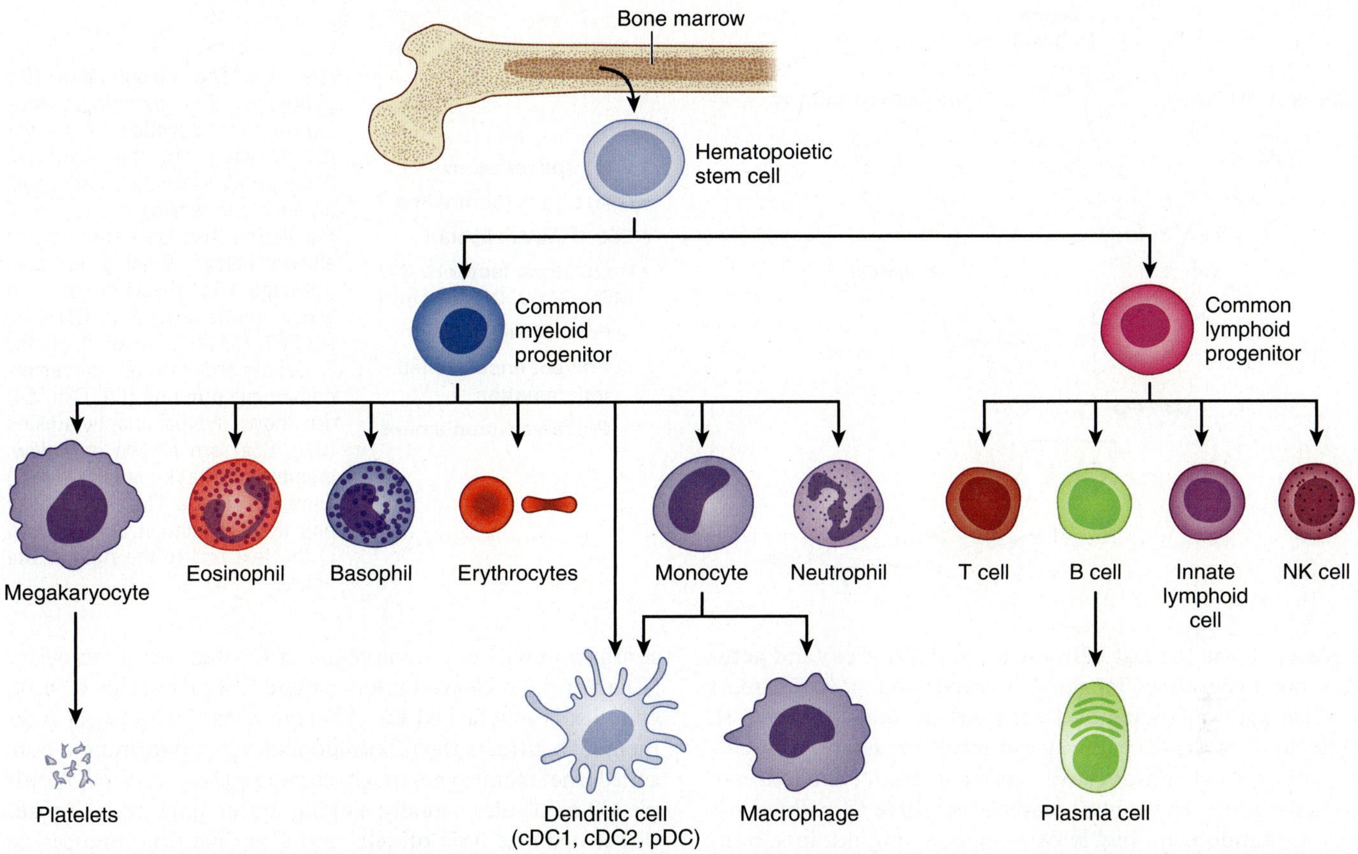

FIG. 1.3. Blood cell types. Hematopoietic stem cells (HSC, light blue) reside in the adult bone marrow and give rise to all of the blood cell types. The common myeloid progenitor (CMP, dark blue) gives rise to the cells of the myeloid lineage displayed as they appear with a histological staining procedure (except dendritic cells, DC, also light blue). DC can be derived from monocytes or from a dedicated precursor; they are divided into three types: cDC1, cDC2, and plasmacytoid DC (pDC). The common lymphoid progenitor (CLP, fuchsia) gives rise to cells of the lymphoid lineage, including lymphocytes with rearranging antigen receptors (T cell, red; B cells, green) and those that do not (natural killer [NK] cells, maroon; innate lymphoid cells [ILCs], purple). When B cells are stimulated by antigen, they become secretory plasma cells (also green). Note that the cells in the lymphoid lineage are artificially colored.

KEY STAGES OF THE IMMUNE RESPONSE

In this section, we discuss how the tiered approach to defense operates during a typical immune response (Fig. 1.1). For simplicity, we focus mostly on responses to a microbial pathogen, but a similar flow of events is triggered by toxins, tumors, and other threats as well.

Barrier Defenses

The initial obstacles to pathogen entry are provided by specializations in the linings of these entry sites that provide physical, chemical, and mechanical barriers (Fig. 1.1). Physical obstacles include dead cells on the epidermis of the skin and tight junctions between epithelial cells. In addition, specialized chemical secretions such as mucus (a carbohydrate-rich secretion that lines the gut, lung, and reproductive tracts) and antimicrobial peptides secreted by the barrier cells provide a chemical barrier. Finally, mechanical movements by cilia and tissues themselves help move larger pathogens out of the body. The features of the mucosal barrier and cells involved in mucosal immunity are discussed in Chapter 38. Barriers are a major mechanism of defense in plants (Chapter 5) and invertebrates (Chapter 3), which possess innate mechanisms of defense but no (or not much) adaptive immunity. In addition to blocking pathogen entry, the barriers also form a physical limit for a relatively continuous system that joins all the cellular structures in the body. In addition to blood, an additional system of fluid recirculation known as the lymphatic system (Chapter 8) allows the tissue fluid that bathes all cells to circulate. This lymphatic fluid drains via lymphatic vessels into specialized SLO called lymph nodes. This also carries to the subbarrier tissue the second layer of protection known as the complement system.

The Complement System (Chapter 13)

Complement (C′) is a complex system of at least 30 proteins, including binding proteins, proteolytic enzymes, regulatory and inflammatory molecules, and cell surface receptors capable of destroying pathogens (Fig. 1.4). C′ can be activated by immune complexes of IgM or IgG bound to pathogens in the Classical C′ Pathway, or by lectins such as mannose-binding lectin (MBL) or ficolin (or others) in the Lectin C′ Pathway. A third pathway, the Alternative C′ Pathway, is so-named

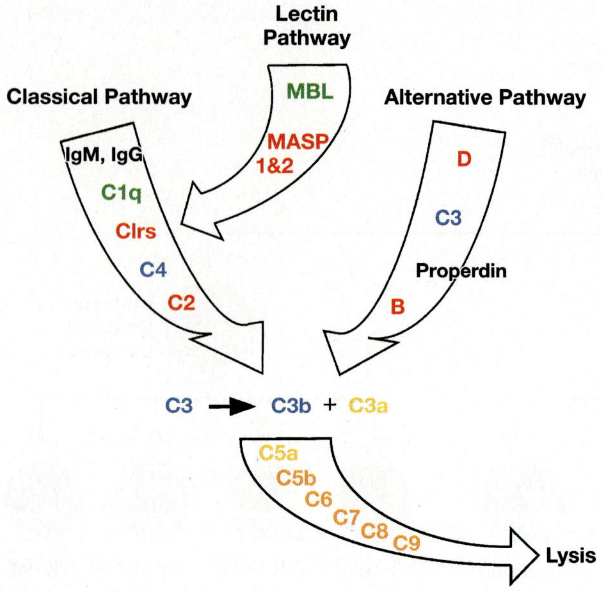

- **C3b: Opsonization**
- **C3a, C5a: Inflammation**
- **C5b-9 (MAC): Lysis**
- **Regulators: factor H, DAF, MCP, CD59, CR1, factor I**
  - **Protect Self**
  - **Prevent uncontrolled inflammation**
  - **Preserve Complement**

FIG. 1.4. **The complement (C′) pathways.** The pathways leading to the activation of C3 via its cleavage are the Classical, Lectin, and Alternative pathways, noted in the arrows at the top of the figure. The Lysis pathway is shown below. Binding proteins (IgM, IgG, C1q, MBL) are shown in green; serine proteases (C1r, C1s, MASP1, MASP2, Factor D, Factor B, C2) are shown in red; thioester-containing proteins (C3, C3b, C4) are shown in blue; anaphylatoxins (C3a, C5a) are shown in yellow; membrane-attack complex (MAC) shown in orange. The C′ cascades and their regulation is described in the text and to the right of the figure.

because it was the last pathway to be discovered (and actually not believed in for several years!), yet it is the most ancient pathway dating back (at least) to corals (Chapter 4). Activation of the C′ pathway can result in enhanced uptake of pathogens via phagocytosis, cell lysis, and/or induction of inflammation. The central player of all three C′ pathways is C3, a multidomain, highly expressed (>1 mg/mL in serum) serum protein that is the *only* extracellular defense molecule capable of forming a covalent attachment to cells via a thioester bond.

The ancient Alternative Pathway requires no other binding protein for its activation besides C3. The thioester bond is normally sequestered in the C3 molecule and when exposed it is hydrolyzed by water molecules and then degraded by C′ serum inhibitory proteins (the binder Factor H and the serine protease Factor I). However, if there is a biological surface such as that displayed by a bacterium devoid of regulatory proteins to block C′ activation, the thioester can make a covalent attachment to hydroxyl or amide groups on the pathogen. The bound C3 is protected from degradation by Factor H/I in the serum and by properdin, which is attracted to C3 on surfaces devoid of sialic acid, like those found on bacteria. The membrane-bound C3 molecule now attracts the serine protease Factor B, which itself then becomes a substrate for the serine protease factor D. When Factor B is cleaved into Bb (and another small fragment), its enzymatic activity is activated. The heterodimer C3b/Bb is now the active protease, AKA C3 convertase, of the Alternative Pathway, which can amplify the response by cleaving many C3 molecules into C3a and C3b to expose the thioester bond and densely coat the pathogen surface with covalently bound C3b. C3b receptors on neutrophils and macrophages bind to the molecules displayed on the bacterium and **phagocytosis is enhanced, a process called opsonization.** The smaller, soluble fragment of C3 released by cleavage, **C3a, induces an inflammatory response as described above.** If two C3b molecules are associated with the Bb, the protease specificity is modified and the serum protein C5, a homologue of C3 that lost its thioester (Chapter 4), is cleaved into C5a and C5b; thus, this $C3b_2Bb$ complex is now called the "C5 convertase." The larger fragment C5b attracts the C′ components C6-C9 forming a pore called the membrane-attack complex (MAC), with multiple C9 molecules actually making up the pore. **MAC causes direct osmotic lysis of cells** and C5a, like C3a, induces an inflammatory response. C3a and C5a are historically called anaphylatoxins, since at very high levels (especially C5a) they can produce symptoms of anaphylaxis.

The Lectin and Classical C′ Pathways follow the same pattern as described above, except these pathways are initiated by molecules that bind to the pathogen, carbohydrate-binding proteins (lectins) acting as PRR for the Lectin Pathway and IgM or IgG antibodies for the Classical Pathway. Enzymes associated with the lectins (MBL-associated serine protease 1, 2 or MASP 1, 2) cleave another C3 homologue, C4, into C4a and C4b exposing the thioester bond of C4 as well as a site for binding to the serine protease C2, a homologue of Factor B. C2 is then also cleaved by MASP and the generated C4bC2a heterodimer is the C3 convertase, homologous to C3bBb, which cleaves C3 to initiate the same downstream cascade events detailed above. A molecule similar in structure to MBL, C1q, binds to antibodies on the pathogen surface and initiates the Classical C′ Pathway. Enzymes homologous to MASPs, C1r and C1s, cleave C4 and C2 to form the C3 convertase C4bC2a (Fig. 1.4).

As mentioned, the process of activation of the C′ cascade is highly regulated. Because activated C3 can bind to any OH or NH_2 group, self-molecules can also be targeted. Besides Factor H and Factor I, other regulatory proteins either in the serum or at the cell surface (eg, C1 esterase inhibitor, decay accelerator factor, membrane cofactor protein, CD59) prevent uncontrolled C′ activation. In addition, the regulatory proteins control the C′ cascade once it is initiated to prevent overactivation leading to excessive inflammation, as mentioned above. Abnormalities in these regulatory proteins are associated with clinical disorders such as hereditary angioedema, paroxysmal nocturnal hemoglobinuria, and macular degeneration, to name just a few.

C′ activation also has profound effects on adaptive immunity. Antigens decorated with C3 fragments can be transported into SLO via several different pathways, for loading onto FDC and subsequent antigen-specific B cell activation (see below). In addition, C′-decorated antigen can bind to both BCR and C′ receptors on antigen-specific B cells, which lowers the threshold for B cell activation, that is, the adaptive system is alerted that the antigen has been declared "foreign" by the innate system. Finally, recent work has shown that C′ is important for regulating synapse formation in the nervous system, and in certain phases of T cell activation. Considering the complexity of the C′ system, it should come as no surprise that it can be recruited for other functions.

The C′ system is an elegant demonstration of the evolution of a biochemical process that seems to be "irreducibly complex"; yet we now understand the precise events in the vertebrate lineage that resulted in this seemingly intact unit. As discussed in Chapter 4, gene duplication and neofunctionalization of the loci encoding the binding proteins, serine proteases, pore-formers, and regulatory proteins sequentially welded together rigid biochemical modules. For example, the founding member C3 can be traced back to corals, while the vertebrate paralogues C4 and C5 have assumed new functions in the Classical Pathway and cell lysis, respectively.

Innate Immunity by Tissue-Resident Cells (Chapters 12, 15, 18, 34-38)

The cells that can detect the pathogen soon after barrier breach are those that are already present at or beneath the barrier tissues (Figs. 1.1 and 1.3). During development of the organism, a few lineages of immune cells (see below) are programmed to migrate to and then remain at these sites prior to pathogen attack. These cells include innate cells including subsets of macrophages and innate lymphoid cells (ILCs, Chapter 18) but also certain semiadaptive (Chapter 35) cells like γδ T cells. While innate cells are capable of rapid responses, they must be activated by sensing the intruders. This means that such cells require dedicated receptors to sense and trigger them. In the innate system, these sensors are families of PRR defined above (Fig. 1.2) (Chapter 12). Based on the tolerance principle (see Principle #2 in the above section) of recognizing threats while avoiding attacking healthy tissue, **PRRs sense two categories of molecules**. The first are some that tend to be highly conserved in groups of pathogenic microbes and hence referred to as **pathogen-associated molecular patterns (PAMPs)**. The second are patterns that become evident only when a cell is unhealthy (stressed, damaged, infected, or otherwise in "danger") and called **danger-associated molecular patterns (DAMPs)**. PRRs include several families of molecules, of which the most intensively studied are the toll-like receptors (TLRs) and the nucleotide oligomerization domain–like receptors (NLR). Each TLR recognizes a distinct set of substances: TLR4 recognizes lipopolysaccharides (LPS), TLR3 double-stranded ribonucleic acid (RNA), and TLR9 unmethylated CpG-containing DNA. The key is that substances like LPS are found in most gram-negative bacteria but not vertebrate hosts. In the case of RNA and DNA, the sensors are placed at locations where they would not be detected if a cell is healthy. Additionally, the recognized substances are generally indispensable to the infectious agent. So, the microbial sensors provide a highly efficient means to recognize potential pathogens. Vertebrates have inherited, *en masse*, the basic families of PRR found in the invertebrates (Chapters 3, 4). Plants also use PRR with similar domain structures as those found in vertebrates, but with extensive modifications to the point that even when the molecules appear to be derived from a common ancestor, the PRR were more likely to have been derived by convergence (eg, NLR, Chapter 5). Plant PRR can detect PAMPs but also metabolic changes within the cell that has been infected, which can also occur in vertebrates. The innate system per se can provide an effective means to control or eliminate pathogens. Indeed, invertebrates and plants rely almost entirely on the innate immune system to cope with microbial infection.

The interaction of a TLR with its ligand induces a series of intracellular signaling events, of which activation of the transcription factor NF-κB is pivotal (Fig. 1.2). Many genes that are part of the innate cell's response to infection or damage are controlled by promoters with NF-κB-response elements. Macrophage activation for instance involves the production of the cytokine tumor necrosis factor alpha (TNFα), which can kill infected cells but also activate local epithelia and endothelia to become inhospitable to microbes. TNFα also promotes the vascular changes critical for the next step of inflammation and enhances the macrophage's phagocytic activity and the induction of antimicrobial systems. In addition, the activated macrophage and irritated epithelia secrete chemokines which initiate a process of cellular recruitment from the circulation into the infected site. As mentioned above, induction of an inflammatory response after activation of the innate immune system recruits other cell types, including neutrophils, to the site.

Inflammation and the Recruitment of Circulating Innate Cells (Chapter 12)

The circulatory system is the major highway by which immune cells and messenger proteins (cytokines, chemokines described below) can be transported from one part of the body to another. Endothelial cells that line the blood vessels typically form a tight barrier, keeping immune cells in the circulation. Within minutes of activation of stromal cells and tissue-resident innate cells at the site of a barrier breach, mediators (cytokines and chemokines) made by these cells engage receptors on vascular endothelial cells. This triggers changes in the shape and rigidity of the blood vessel (via remodeling of the actin cytoskeleton) and alters blood flow in the vessels closest to the site of infection (Fig. 1.1C). Note that C3a/C5a described in the C′ section can also activate mast cells in the tissue to release mediators such as histamine that also causes vascular leakage. Cytokines like IL-1 and TNF produced by activated macrophages then further induce the expression of cell adhesion molecules in the inner lumen of the blood vessel and further dilate the vessel. The combination of the vascular leakage (which contributes to the localized swelling or edema in inflammation) and release of inflammatory mediators like prostaglandins drastically alters the microenvironment. Circulating neutrophils (Chapter 14), macrophages (Chapter 15), and natural killer (NK) cells (Chapter 18) are now attracted to this vascular niche by chemokines (for example, IL8 [CXCL8] acting on the receptor CXCR1), which further activates cell adhesion molecules on their membranes. The combination of adhesion receptors induces these cells to bind strongly to the endothelia at the site of inflammation and then crawl out into the extravascular space and eventually reach the site of infection, a process called diapedesis or extravasation. The newly recruited cells bring an additional arsenal to the fight. Among these, neutrophils are particularly effective against bacterial pathogens and can even extrude their genomic DNA as sticky NETs (neutrophil extracellular traps) which trap antibacterial peptides and then glom onto and kill microbes. Overall, the action of innate cells at this phase has three defining characteristics. First, although they are triggered broadly by threats (PRR ligands), they are not very specific to the pathogen. This also leads to the second characteristic: that is, they tend to be broad-acting in their effector mechanisms. Cytokines like IL-1 and TNFα as well as mechanisms like NETs can damage some healthy cells in the milieu as they pursue a scorched-earth policy to rid the system of threats. Maintaining such responses for prolonged periods can be deleterious. Third, this response is a one-and-done effort; it does not (and classically cannot) generate specific memory of the pathogen to respond better against the same threat. While there is emerging evidence for epigenetic programming of epithelia and some innate cells that modify their responses to subsequent infections, AKA trained immunity, it is not clear if this constitutes a true memory or rather a maturation process of the immune system.

Innate Myeloid-Derived Cells

Cells of the myeloid lineage that orchestrate innate immunity and alert adaptive immunity include macrophages, DCs, and the granulocytes comprising neutrophils, eosinophils, basophils, and mast cells (Fig. 1.3). One of the goals of cellular immunity is to energize macrophages to eliminate organisms that have established intracellular infections (Chapter 15). Nonactivated macrophages are inefficient in destroying intracellular microbes and are important for homeostasis by removing cellular debris such as apoptotic cells. However, when macrophages are stimulated by interferon (IFN)-γ and other mediators produced by T cells, their antimicrobial capacity is enhanced. These mechanisms include reactive oxygen species (ROS), nitric oxide (NO), proteolytic enzymes, and proinflammatory cytokines. Macrophages can be stimulated by other cytokines, such as IL-4, to adopt a different phenotype, one more programmed to stimulate a healing process to counter inflammation. Macrophages can also serve as APCs and enlist the "help" of activated, cytokine-producing CD4+ T cells (see below). Although macrophages function as APCs for stimulation of activated T cells, they are not particularly effective in the activation of naïve CD4 T cells. In instances in which they are the site of infection or have phagocytosed infectious agents or their proteins, antigens may be transferred to DCs, the major myeloid cell involved in antigen presentation (see below).

Neutrophils play critical roles in a wide range of inflammatory situations, as described below. In their absence, it is exceedingly difficult to clear infections with extracellular bacteria (Chapters 14, 43), and young children with a delay in neutrophil production suffer from recurrent bacterial infections in the blood. Both innate and adaptive responses play important roles in orchestrating the growth, differentiation, and mobilization of these crucial cells. For example, Th17 cells are particularly important because of their role in recruiting neutrophils to sites of immune responses to extracellular microbes.

Mast cells and basophils (Chapters 16, 47) play major roles in the induction of allergic inflammatory responses. They express cell surface receptors for the Fc portions of IgE (FcR) and for IgG subclasses (FcR). These receptors enable them to avidly bind antibody on their surfaces, and when antigens capable of reacting with those antibodies are encountered, the resultant cross-linking of FcεRI results in the immediate release of a series of potent mediators such as histamine, serotonin, and a variety of enzymes that initiate allergic and anaphylactic-type responses. Mast cells can also be activated via their anaphylatoxin receptors C3ar/C5aR. Such stimulation also induces these cells to produce cytokines including IL-3, IL-4, IL-13, IL-5, IL-6, granulocyte-macrophage colony-stimulating factor (GMCSF), and TNFα, which have important consequences in inflammation, and in overstimulation in allergic responses.

Eosinophils (Chapters 17, 42, 47) differentiate under the influence of the cytokine IL-5. They migrate to tissue sites in response to the chemokine eotaxin. Because Th2 cells can produce IL-5 and stimulate the production of eotaxin, eosinophil accumulation is often associated with Th2-mediated inflammation. Eosinophils store proteins in their secondary granules including major basic protein, eosinophil cationic protein, and eosinophil peroxidase. When released, these proteins are responsible for much of the damage that eosinophils mediate both to helminthic parasites—their presumed physiological function—and infected tissues—chronic allergic inflammation in conditions such as asthma.

Innate Cells Initiate Adaptive Immune Responses (Chapters 8, 34)

Following barriers, tissue-resident cells, and recruited innate cells, the fourth layer of immune protection is the adaptive system. Prior to their activation, T and B cells, which make up this system, are not found in peripheral tissue sites but migrate in and out of SLOs. This geographical separation between the initial site and these fourth layer cells necessitates a bridging mechanism. DCs, which also act as sentinel cells patrolling subbarrier tissue sites, perform this role for T cells (Figs. 1.1B and C). DCs use pinocytosis to continually sample their environment for antigens. These antigens are processed proteolytically and then pieces (peptides) displayed back on the surface of DC on specialized major histocompatibility complex (MHC) molecules (see below). MHC molecules serve as adaptors for peptide display on the cell surface and the TCR can recognize the amalgam formed by the peptide and an MHC protein (pMHC). If a DC were to encounter a PRR ligand (indicating that the antigen it just acquired may be derived from a threat), then the DC also gets activated by this signal as described above for macrophages. The activated DCs upregulate chemokine receptors such as CCR7 on their surface and home to the nearest AKA "draining" lymph nodes following the gradient of CCL19/21, chemokines made by lymph node stroma. The arrival of DC in the draining lymph nodes allows the SLO T cells to sample antigens that were in the original at-risk tissue site. In addition, the activated DCs also make cytokines that dictate T cell differentiation from a naïve precursor cell (Chapter 28). Depending on specific signals delivered to DCs educated by the specific pathogen and the original tissue niche, these DCs will produce cytokines that tailor T cell responses, as prescribed in the fourth principle above. Unlike T cells, B cells are activated by intact antigen, which can enter the SLO by fluid flow through the lymphatics or be carried and/or transported to SLO B cell zones by specialized macrophages.

The Primary Immune Response (Chapters 26, 28)

After the innate phase, cells of the adaptive immune system lymphocytes (T and B cells) are now recruited to the response. T cells, B cells, NK cells, and ILC are derived from a lymphoid precursor cell, and their development is described below. The first time a pathogen triggers adaptive immunity, the T and B cells are in a "naïve" state, that is, they have completed maturation in their developmental organs, but have never encountered their cognate antigen. This first response is therefore called a primary immune response. The term is relevant to adaptive cells since (as we discuss further in the next section) any subsequent reexposure to the pathogen usually elicits a significantly different response than the first one. Innate cells do not typically make that distinction.

Clonality of Lymphocytes

Adaptive immunity is mediated by lymphocytes that recognize antigen with narrow specificity. Each T cell or B cell bears a receptor that is very specific to small regions of an antigen (an epitope or antigenic determinant). The high degree of specificity allows their response to be very surgical as opposed to the brute force methods used by the innate system. It also means that the body must produce millions of different receptors for T and B cells, such that (ideally) every possible antigen can be detected. The fascinating genetic mechanism that makes this possible is unique to the immune system and involves the rearrangement of different gene segments in the developing lymphocyte (see below and Chapters 4, 22, 25, 27). By the time a mature T or B cell arrives in the peripheral SLO, each cell expresses only one of many possible receptors that was generated by VDJ rearrangement. Importantly, although the system has millions of receptors, each T or B cell just expresses only one of them, with ~10,000 copies at the cell surface. When such a mature T or B cell divides, all the daughter cells inherit this unique receptor, and this is the basis of the principle of clonal expansion in the immune system (Fig. 1.1C). When activated T or B cells divide, the number of cells expressing the specific receptor for a particular epitope (from a pathogen or tumor) also increases. Clonal selection also offers a mechanism for tolerance: if one of these receptors happens to be self-reactive, then the cell is induced to undergo apoptosis during lymphocyte development or even as a mature cell, and the self-reactivity is also purged from the system.

Burnet's Clonal Selection Hypothesis

These principles were first laid out as four bedrock tenets of the adaptive immune response by McFarlane Burnet in 1954. While the molecular details and mechanisms have evolved with new knowledge, the basic rules have stayed surprisingly consistent. Updated for current language, the four tenets of this theory are:

1. Each naïve lymphocyte expresses a single receptor with a unique specificity by the time it completes its development.
2. When the mature lymphocyte's receptor engages an antigen (under appropriate conditions), the lymphocyte will be activated, that is, clonal selection.
3. Once activated, the lymphocyte undergoes cell division extensively, such that its daughter cells all express the same receptor (ie, are clones of itself), that is, clonal expansion.
4. Any lymphocyte that can be activated to self-molecules should typically be subjected to tolerance. In Burnet's hypothesis, it was assumed that such cells would undergo cell death during development (a process called negative selection or clonal deletion); but we now know that negative selection is only one of many tolerance processes.

Antigen Recognition by T Cells and B Cells

As mentioned, T cells differ from B cells in their mechanisms of antigen recognition. Ig, the B cell's receptor (BCR), binds to individual antigenic epitopes on the surface of native molecules, be they on cell surfaces or in solution. In that sense the BCR is like any other conventional receptor binding a ligand based on structural complementarity. Antibody and BCRs have evolved in part to bind to and to protect against microorganisms in extracellular fluids.

By contrast, T cells recognize cell-associated molecules and mediate their functions by interacting with *antigen-presenting cells* (APCs) including DCs, macrophages, and B cells. As mentioned above, the TCR does not recognize antigenic determinants on intact, undenatured molecules. It recognizes peptides (p) from the microbial antigen that have been proteolyzed (known as antigen processing) within APCs. This is further described below and in Chapter 19. These peptides are then loaded onto specialized peptide presentation MHC molecules to form pMHC complexes. There are two kinds of T cells bearing TCR named helper (Th) and cytotoxic (CTL), based on their functional and phenotypic differences. CTLs are identified by surface expression of the protein CD8 and are concerned with killing any cells infected with pathogens or that have turned tumorous. Accordingly, the MHC allelic product that recognizes the proteins that infected/tumor cells express is adept at carrying cytosolic peptides to the surface to display for CD8 T cell recognition. These MHC genes encode for MHC-I (or class I) proteins that are expressed by all nucleated cells in the body. In contrast, Th cells, which express cell-surface CD4, regulate multiple immune cells. So, they are interested in proteins taken up by immune cells from their environment (eg, pathogen particles, extracellular microbes, tumor debris). These proteins are processed by lysosomal proteases and displayed on another set of MHC proteins (MHC-II or class-II) expressed predominantly by specialized APCs (see below).

The Significance of Two-Signal Models

The activation of T and B cells represents a crucial tipping point in the immune response, since that sets in motion a massive amplification of the cellular attack mechanisms. Once activated, these initially rare and quiescent (naïve) antigen-specific T cells can divide once in every 3 to 4 hours, leading to an ~10,000 fold expansion in cell numbers over the next 4 days. Each of these cells can secrete noxious cytokines, kill multiple target cells, or secrete huge amounts of specific antibodies, as the case may be. This is not a force to be unleashed flippantly. Accordingly, the immune system employs multiple checks and balances to make sure that the response is triggered only if necessary and against a threat rather than a healthy self-tissue. Although a bit simplistic, immunologists classify some of these control mechanisms as "two-signal" operations (Chapters 24, 32). This means that the TCR or BCR engaging a specific target antigen only provides the lymphocyte with the first half of the complete signaling required to drive a productive response (Fig. 1.5). Therefore, antigen-derived signals are termed Signal 1. The second signal, which is necessary to complete the process driving full activation,

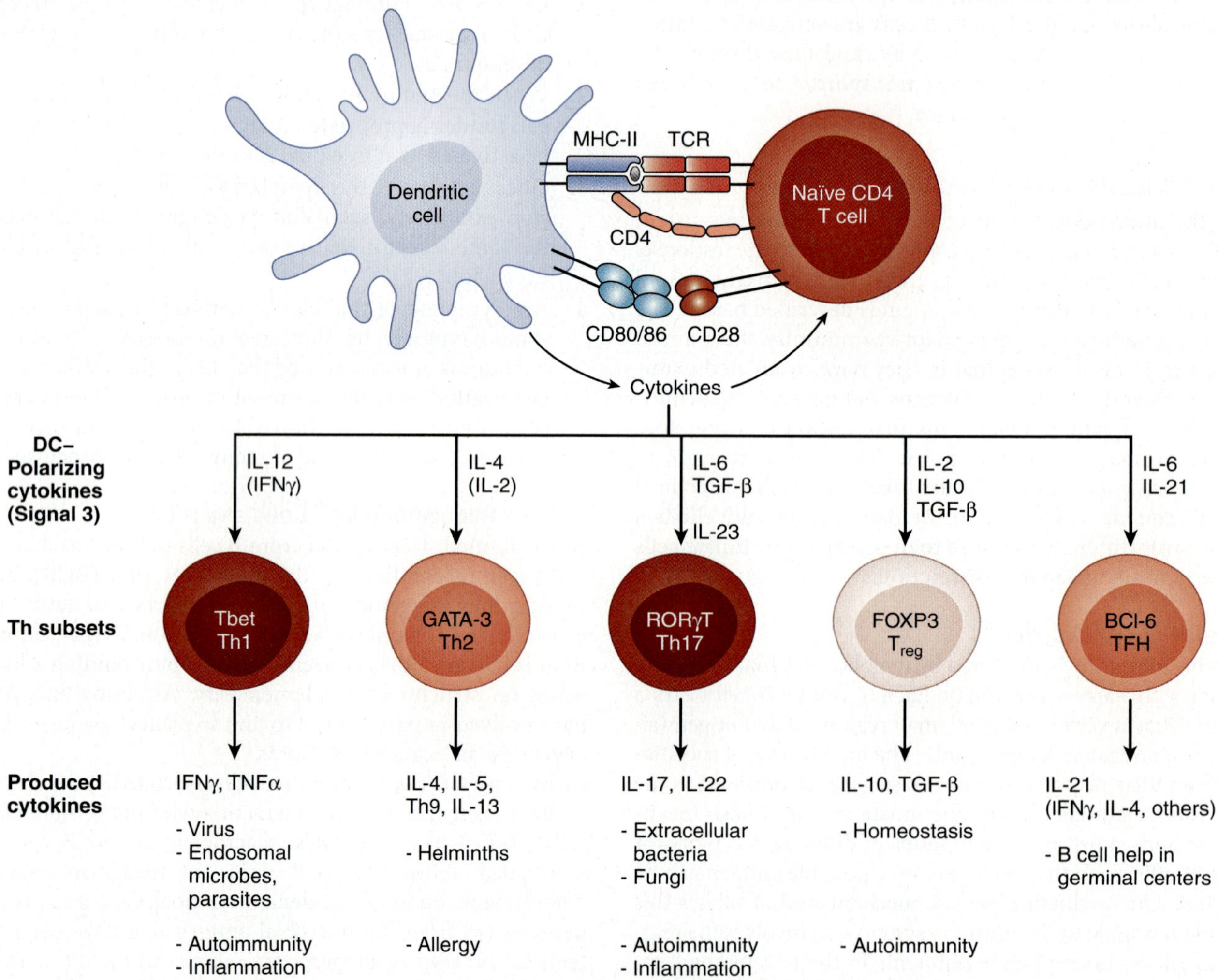

FIG. 1.5. T helper cell phenotypes and division of labor in the adaptive immune response. A DC (*light blue*) stimulates a naïve CD4 T cell (*red*) through its Signal 1 TCR, Signal 2 CD28, and DC-polarizing cytokines. Molecules produced by the DC are in light blue and by the T cell in red. The classic transcription factor for each Th cell is shown within the nucleus of the cell, and the classic cytokines produced, the physiological functions, and potential dysfunctions are shown below the cells. The Th types are colored by their extent of inflammation: Th1 and Th17 cells dark red; Th2 and TFH light red; T_{reg} whitish red.

differentiation, and survival of the T or B cells, differs for each cell type. In the case of the T cell, activation of the initial naïve, quiescent antigen-specific T cell requires DCs that have migrated to the SLO displaying pMHC on their surface, as described. However, in addition to the pMHC (which provides Signal 1 for T cells by engaging their TCR), any DC that was in the presence of a threat in the tissue also conveys the second signal. This second signal is provided by specific surface molecules upregulated on DCs that have been exposed to PAMPs or DAMPs in the tissue. These surface molecules (eg, B7 or CD80) are called costimulatory ligands and engage costimulatory receptors (eg, CD28) on T cells (Fig. 1.5). T cells which receive both Signal 1 and Signal 2 from DC become fully activated, while those that receive only Signal 1 are programmed to become inactivated (a process known as anergy) or die apoptotically. The teleological rationale here is that if the DCs were displaying pMHC without a costimulatory ligand, it likely acquired the antigen from a healthy uninfected tissue (that did not have PAMPs or DAMPs) and hence Signal 1 is a self-antigen against which an immune response would be autoimmune. So, Principle #2 (tolerance) kicks in.

In the case of B cells, the two-signal model operates slightly differently. Again, Signal 1 is still the BCR recognition of the specific target antigen. However, Signal 2 in this case must come from a helper T cell that is specific for the same pathogen or threat (Fig. 1.6). The fully activated T cell (following Signals 1 plus 2 from a DC) must interact with a Signal-1-activated B cell and provide a second signal in the form of surface molecules (eg, CD40L on the T cell and CD40 on the B cell) and cytokines (eg, IL-21) to fully induce B cell differentiation toward antibody secretion.

Activation of Helper T Cells

Migrating DCs from the site of infection settle into the T cell zone of the SLOs where circulating naïve T cells have an opportunity to sample pMHC with their TCR. CD4+ Th cells engaging pMHC-II and receiving costimulation then undergo a period of activation for about 32 hours before they enter the first cell division. Subsequently, the rapid cycle of cell divisions results in the generation of a clonal brigade of T cells. Helper T cells play a central role in driving and coordinating the subsequent immune response. Therefore, as discussed in Principle #4, these cells must choose one of many possible "weaponization" trajectories during this early period. This would match their eventual effector mechanisms with the nature of the original threat and the tissue affected. The information necessary for T cells to understand both threat and tissue comes from DCs as well (Fig. 1.5). While this communication of contextual information involves many different mechanisms (Chapters 12, 28) including some poorly understood, a major component is the category of cytokines made by DCs. For instance, faced with a bacterial infection (eg, *Mycobacterium leprae* which causes leprosy, Chapter 43), DCs secrete the cytokine IL-12, which in turn promotes the differentiation of Th cells to produce IFNγ. The IFNγ produced can activate macrophages to kill the intracellular bacterium. In contrast, if the DCs instruct T cells to differentiate into IL-4-producing effector cells, the bacterium will persist and form long-lasting lesions. Based on classic studies, it was suggested that early innate signals instruct Th cells to differentiate into one of many "classes" of effector cells which were named as Th1 (those producing IFNγ, TNF, etc), Th2 (producing IL-4, IL-5, IL-13, etc), Th17, and so on (Fig. 1.5). The

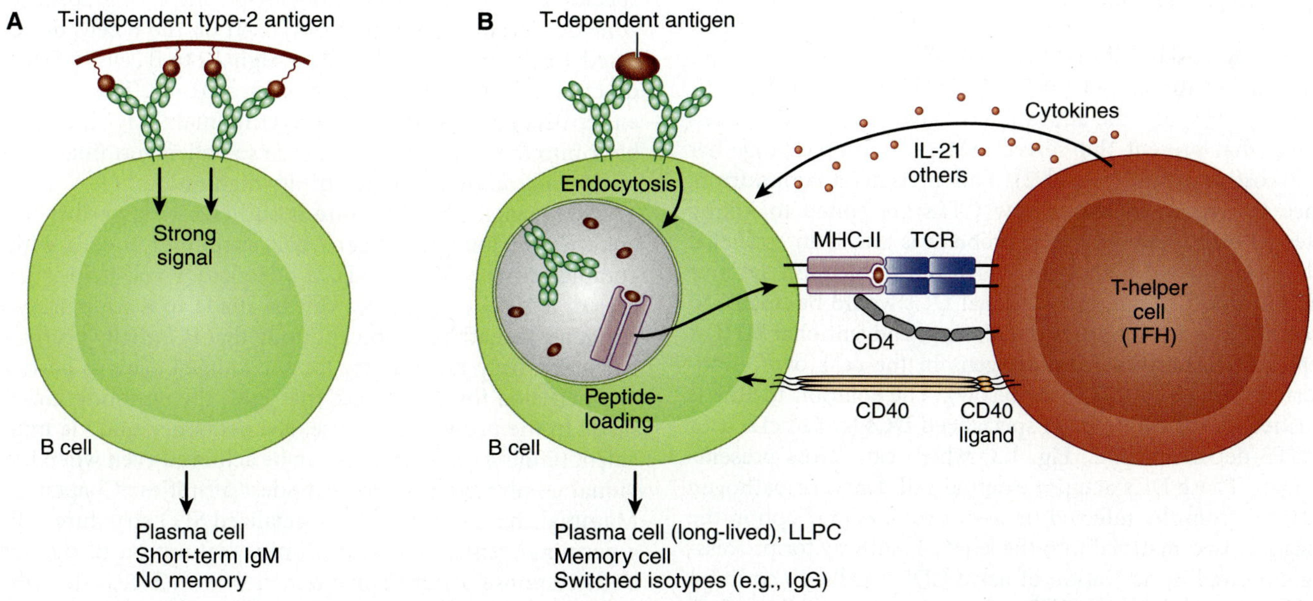

FIG. 1.6. Two forms of B-cell activation, T-independent and T-dependent. A: When a naïve B cell (*green*) is stimulated through its IgM BCR by an antigen (*dark red*) with repeating identical epitopes, the strong BCR signal can result in differentiation into short-term IgM-producing plasma cells. **B:** When a B cell (*green*) receives Signal 1 through its BCR with minimal cross-linking, it takes the antigen (*red*) into the cell where it is degraded. Peptides derived from the antigen bind to MHC class II in endosomes, which is then transported to the cell surface to activate primed TFH cells (red) that provide costimulatory signals in the form of CD40-Ligand (CD40L) and cytokines. These B cells undergo affinity maturation and class-switch recombination and go on to become long-lived plasma cells (LLPC) or memory cells.

innate-derived cytokines made by DC (eg, IL-12) that promote particular T cell classes were categorized as "Signal 3" or "innate differentiation cytokines." While these terminologies are still quite widely used, it is increasingly clear that Th classes are not as rigidly segregated in vivo as previous in vitro experiments had suggested. For instance, there are cases in which Th cells can differentiate to effectors secreting both IFNγ and IL-10, which seems to be a hybrid of Th1 and Th2. Further new classifications of effector Th states are still being described (Chapter 24) (Fig. 1.5). While a new consensus nomenclature to replace Th-class designations is yet to emerge, increasing availability for single cell measurements in vivo is likely to drive the field to adopt a new naming system.

In recent years, two major differentiation pathways of CD4 Th cell differentiation have gained special significance (Fig. 1.5). One is a subset of cells called T-follicular helper (TFH) cells, which play a critical role in providing Signal 2 for B cells as well as driving their further differentiation. A second is a subset known as regulatory T cells (T_{regs}) which seems to be primarily concerned with inhibiting or balancing aggressive immune responses. T_{regs} are identified by the expression of a transcription factor FoxP3, which is sufficient to drive their lineage differentiation (Chapter 29). Unlike all other Th cell classes, a significant fraction of them arises during development as predetermined T_{regs} (AKA "natural" T_{regs}) and are discussed further in the context of lymphocyte development below. But these can also arise in the periphery if naïve CD4+ T cells are stimulated by antigen in the presence of the cytokine transforming growth factor-beta (TGF-β) as Signal 3. The amelioratory role of FoxP3-expressing cells is underscored by patients defective in FOXP3, a congenital disease called immunodysregulation polyendocrinopathy enteropathy X-linked (IPEX).

Activation of CTL (Chapters 34, 44, 49)

Antigen-specific naïve CD8 T cells in the SLOs are also activated by DCs, like the activation of naïve Th cells discussed above, that present the correct pMHC-I complex together with costimulation (Signal 2). This presents a conundrum, since we discussed above how CTLs are suited to killing infected cells or tumors. The problem is that if an epithelial cell were infected by a virus which does not infect DCs or had become a tumor, conventional DCs would be unable to present those antigens on MHC-I, since conventional MHC-I peptides are derived from antigens in the cells own endoplasmic reticulum (ER) (see below). The solution to this is satisfied by the existence of specialized DCs (called cDC1 or BATF3-dependent DC, Fig. 1.3) which can "cross-present" antigen. These DCs acquire external cell debris or pathogen particles from the infected tissues or tumors and siphon the phagocytosed material into the MHC-I pathway for processing, followed by activation of naïve CD8 T cells. In addition to Signals 1 and 2 from PRR-activated cross-presenting DC, CD8 T cells can also be further activated by interactions with activated Th cells. This form of interaction is also called T cell help (for CD8 T cells) and can be delivered by cytokines like IL-2 and/or cell surface molecules, like the CD40L-CD40T cell activation of B cells mentioned above.

Activation of B Cells

Like naïve T cells, resting B cells must undergo activation, proliferation, and differentiation before they secrete their clonally selected antibodies. Antigen-specific IgG antibodies of high affinity are the result of such clonal selection; this response requires "help" from T cells (Fig. 1.6). Antigen enters the B cell zone or B cell follicle in the lymph nodes via the lymphatics and enters conduits where it can provide Signal 1 to antigen-specific B cells or be transported to specialized cells in the follicle called follicular dendritic cells (FDCs). Antigens that are "tagged" with C′ C3 fragments as described above can be transferred from the lymph to the SLO B cell zone via specialized macrophages called subcapsular sinus macrophages. From there, noncognate (non-antigen-specific) B cells can transport the antigen through their C′ receptors to FDC. In the spleen, specialized macrophages in the highly vascularized marginal zone (MZ) transport antigens to the FDC in splenic B cell follicles. Antigen-specific B cells in any SLO detect native antigens displayed on the FDC, which provides Signal 1 via the BCR. Simultaneously to BCR recognition, C′ fragments on the antigen can bind to C′ receptors on the antigen-specific B cells and lower the threshold for activation by orders of magnitude. Signal 1 induces B cells to upregulate T cell-specific chemokine receptors that attract them to the borders of the T and B cell zones, where B cells can then interact with activated T cells, which themselves have upregulated B cell-specific chemokine receptors. B cells use their BCR to take up antigen, process it in lysosomes, and display peptides from the antigen on MHC-II molecules (Fig. 1.6). Importantly, all other noncognate B cells, unlike DC, are incapable of taking up antigen except through the BCR; this focused uptake mechanism ensures that only the antigen-specific B cells will engage antigen-specific T cells at the T-B zone border. Recognition of pMHC-II on the B cells by activated T cells provides cell surface Signal 2 to B cells (CD40, as described above) and cytokines such as IL-21. The B cell can either become a plasma cell, secreting mainly IgM, or move back into the follicle to initiate the so-called germinal center (GC) reaction that leads to affinity maturation.

This primary Ig repertoire is sufficiently large that most epitopes on foreign antigens will encounter B cells with a complementary BCR. Thus, if adequate T cell help is generated, antibody responses can be made to a wide array of antigens. Nonetheless, the antibody that is initially produced has a relatively low affinity for the antigen. This is partially compensated for by the fact that IgM, the antibody initially made in the primary response, is a pentamer, and via multivalent binding high avidities can be achieved even when individual combining sites have modest affinities (Chapter 21). Mammals have developed a specialized SLO structure called a germinal center (GC) for affinity maturation of the antibody response. After their interaction with T cells described above, some B cells move back into the follicle to initiate the GC reaction. The B cells rapidly divide and upregulate the enzyme activation-induced cytidine deaminase (AID), which is required for somatic hypermutation (SHM) of the antibody genes and for class-switch recombination (CSR, described below) (Chapter 22). During the SHM process, mutational rates of 1/1000 base pairs/generation can be achieved. Thus, with each cell division approximately one mutation will occur in either the Ig heavy (H)- or light (L)-chain V region of an individual cell. Such a high rate of SHM creates an enormous increase in antibody diversity. Although most mutations will neither modify nor enhance affinity, some will increase it. Thus, some B cells emerge that can bind antigen more avidly than the initial responding cells. Because there is an active process of apoptosis in the GC from which B cells can be rescued by the binding of antigen on FDC with their BCR, cells with the most avid receptors have an advantage over other antigen-specific B cells and dominate the population of responding cells. Therefore, upon rechallenge, the affinity of the antibody produced will be greater than that in the initial response. As time after immunization elapses, the affinity of antibody produced increases. This results in immunized individuals with high-affinity antibodies that are more effective in protecting against microbial agents than the antibody initially produced. This process is also associated with CSR to other Ig isotypes, which is described below in the Nuts and Bolts of the Adaptive Immune System section. Following the GC reaction, plasma cells and memory cells are generated that provide substantial protection to reinfection.

B cells can also be stimulated in the absence of T cell help to produce antibodies, the so-called T-independent response (Fig. 1.6). When an antigen-specific B cell encounters an antigen with repeating identical epitopes, as are found on bacterial carbohydrates, the BCR can be extensively cross-linked and the strong Signal 1 alone can induce differentiation into plasma cells. Such stimulation results in the production of short-term protective IgM but does not induce a memory response.

The Effector Arms of the Primary Response

The initial activation, proliferation, and differentiation of T and B cells in the SLOs take 3 to 7 days (Fig. 1.1D). At this point, differentiated Th and CTL cells have altered the expression of chemokine receptors that now guide them from the SLO to the infection sites. As the cells are migrating out of SLO, they are fully armed to deliver effector functions (whether it be in secreting inflammatory cytokines like IFNγ or being able to kill other cells) but are not continuously making these effector products. The migration itself is also not well focused to just the initially threatened tissue sites, meaning that differentiated (but not actively secreting) effector T cells can be found in the circulation as well as in the healthy tissues. Two principles ensure that these cells deliver effector function maximally only at the relevant sites. First, although "casting a wide net" in terms of migrating out, the inflammatory changes in the infected tissues are quantitatively better at both recruiting more effector cells and at retaining them there. Second, and most importantly, effector T cells must reengage their target pMHC to deliver effector function, but at this point costimulatory molecules are not required for activation. For CD8 T cells, this means that any cells infected with the pathogen (or having mutant tumor proteins) would be the ones to display pMHC-I with the correct peptides. In the case of CD4 effector T cells, these could also be some infected cells which present pMHC-II in response to inflammatory stimuli. However, more often, the CD4s interact with macrophages or other innate cells, which present antigens from threats, as mentioned above. In this case, Th cells instruct the innate cells to either change their own activation status to fight the threat or alter the course of their ongoing response. Depending on the kind of pathogen involved, these instructions can be quite diverse (Fig. 1.5). For intracellular pathogens, effector Th-derived instructions can enhance the generation of ROS in macrophages, while for large parasites appropriately differentiated CD4s can instruct goblet cells to increase mucus production. T cells exploit and augment the abilities of the previous layers to effectively combat the threat.

There are two major mechanisms of CTL-mediated cytotoxicity (Chapter 31). One involves the production of perforin, a molecule that can insert into the membrane of target cells and promote their lysis. Perforin-mediated lysis depends on a series of enzymes produced by activated CTLs referred to as granzymes. Many active CTLs also express large amounts of the TNF superfamily member fas ligand on their surface. The binding of fas ligand with its receptor, fas, on the surface of the target cell initiates apoptosis in the target cell. CTL-mediated lysis is a major mechanism for the destruction of virus-infected cells (Chapter 44). If activated when a virus is in its eclipse phase, CTLs may help eliminate the virus with relatively limited cell destruction, as normally occurs as the killing mechanisms induce target cells to undergo noninflammatory programmed cell death (Chapter 7). However, vigorous CTL activity while a virus is widely disseminated may lead to substantial tissue injury. Thus, clinical disease may be caused by the destruction of tissue by CTLs rather than by the virus itself. One example is hepatitis B, in which much of the liver damage represents the attack of hepatitis B virus–specific CTLs on infected liver cells. CTLs are also the major mechanism involved in the destruction of tumor cells, as described below.

After activation, Ig-secreting plasma cells change their chemokine receptors and also move out of the SLO into bone marrow (BM) and other body sites (Fig. 1.1D). The effector arm of the B cell response is secreted antibodies which circulate in the blood and access most of the niches in the body including the site of infection. Antibodies can neutralize the pathogen by directly binding and blocking critical pathogen functions (such as invasion receptors, vital nutrient transporters). In addition, antibodies can also help macrophages and NK cells find the pathogen by using Fc receptors (FcR, Chapter 33).

The Secondary Immune Response and Immunological Memory (Chapters 26, 30)

The primary adaptive immune response takes about a week to fully engage and be maximally effective against a pathogen. While this is acceptable in many cases, if the invader causes a lot of pathology in this time or replicates too fast to overcome the slowly developing adaptive response, it can lead to severe morbidity or mortality. These limitations are circumvented if the pathogen invades a second time, due to the phenomenon of immunological memory. Memory refers to the ability of the system to respond faster and more robustly to the same pathogen during a reexposure (Fig. 1.1E). There are five main processes which form the biological basis of memory. First, after the initial response, a small fraction of the T cells that expanded clonally to fight the pathogen are retained in the body. This is significant because in the primary infection, pathogen-specific naïve T cells are quite rare (a so-called **"precursor frequency"** as low as 1 in 100,000). Even if a few of these T cells are retained as memory T cells (between 10 and 100 per 100,000 total T cells), this represents a significant increase in the frequency of pathogen-specific T cells available to respond to the infection. Secondly, naïve T cells need extensive checks and balances for full activation, resulting in a 32-hour lag time before their first proliferation. Memory T cells can be recruited more rapidly and with lower doses of stimulating antigen. Third, a subset of initial effector T cells are retained in tissue sites as **"resident memory" T cells**, so that they can respond at the same time as the innate cells beneath the barrier get activated. This is a massive advantage for the secondary response, since it takes the primary response about a week to dispatch effector cells to the tissues. Fourth, plasma cells called long-lived plasma cells (LLPC) remain after the primary response stationed in the BM to constantly secrete high levels of the antibody that they made in the primary response. These LLPC-produced antibodies are the source of antipathogen antibodies detected in the blood months and sometimes even years after an initial infection. These offer a definitive advantage to the host, since they can rapidly neutralize or opsonize the pathogen attempting to reinfect the host at any tissue site. Finally, a subset of B cells that mirror memory T cells are called memory B cells (MBC). These also await pathogen reappearance and can be rapidly activated to provide protective immunity. The difference between LLPC and MBC is that the former continually secrete antibodies regardless of pathogen presence, while the MBC must be reactivated during reinfection before they can start secreting antibodies. Taken together, the primed adaptive system now can repel the repeat invader much more efficiently than the first time. In many cases, this means that the second attack only leads to subclinical infections where the pathogen can often be cleared without the patient even becoming symptomatic. In other cases, this response still clears the microbe faster, leading to less morbidity or mortality than occurs in a primary infection.

This is the basis of most of the licensed vaccines in use today (Chapter 40). Exposing the immune system to components (proteins) of the pathogen generates a combination of all five of the above. In this case, a vaccinated individual encountering the cognate pathogen for the first time can already mount a secondary immune response, rather than going through the slower and less-effective primary response. The great power of vaccines is illustrated by the elimination of smallpox from the world and by the complete control of polio in the western hemisphere.

NUTS AND BOLTS OF THE ADAPTIVE IMMUNE SYSTEM

Immunoglobulins

Igs are antigen-specific membrane receptors (B cell receptors [BCRs]) and secreted products of B cells. They are members of a large family of proteins designated the Ig supergene family (IgSF). Members of this family have sequence homology, a common gene organization (generally one exon encoding each domain), and similarities in three-dimensional structure of the IgSF domain, consisting of seven-to-nine β-pleated sheets organized into two opposing sheets in an open-face sandwich (Fig. 1.7) (Chapters 4, 21). Many of the cell surface proteins that participate in immunologic recognition processes, including the TCR and BCR, the signaling molecules associated with the antigen receptors, and many more are members IgSF family members. A lymphocyte may bear as many as 30 such members on its surface.

The Igs themselves consist of two H and two L chains (Fig. 1.8). The H and L chains are composed of a series of IgSF domains, each consisting of approximately 100 amino acids. There are two L chains in mammals, κ and λ, which have two IgSF domains; in other vertebrates there are as many as five subclasses of L chains (Chapter 4). The carboxy-terminal domain is essentially identical among L chains of a given type and is referred to as the constant (C) region. The amino-terminal domain varies among L chains and contributes to the binding site of antibody: because of its variability it is called the V region. V domain diversity is largely concentrated in three segments, designated the hypervariable (HV) or complementarity-determining regions (CDRs) (Chapter 21). The three CDRs contain the amino acids that are the L chain's contribution to the antibody-combining site (Fig. 1.7) and are interspersed among four regions with much lower variability, designated framework regions (FR).

IgH chains are of several classes determined by their constant regions (μ, δ, γ [of which there are several subclasses], α, and ε) (Fig. 1.9). An assembled Ig molecule, consisting of one or more units of two identical H and L chains, derives its name from the C regions. Thus, there are IgM, IgD, IgG, IgA, and IgE antibodies. The H chains each consist of a single amino-terminal V region and three or four C regions. In many H chains, a hinge region separates the first and second C regions and conveys flexibility to the molecule, allowing the two identical combining sites of a single unit to move in relation to one another from nearly 0 to 180° (Fig. 1.8). Such divalent binding to a single antigenic structure and inherent flexibility results in a great gain in energy of interaction (Chapter 21). The H chain V region, like that of the L chain, contains three CDRs lining the combining site of the antibody and four FR. Thus, the antibody-combining sites are an amalgam of six CDRs, three each for VH and VL.

The C region of each H chain class conveys the functional attributes to the antibodies. Among the distinct biologic functions of each class of antibody are the following:

- IgM are potent neutralizers of microbes and activators of the C′ system (Chapters 4, 13).
- IgA are secreted into a variety of bodily fluids and are principally responsible for immunity and homeostasis at mucosal surfaces (Chapter 38).
- IgE are bound by specific receptors (FcεRI) on basophils and mast cells. When cross-linked by antigen, these IgE/FcεRI complexes cause the cells to release a set of mediators

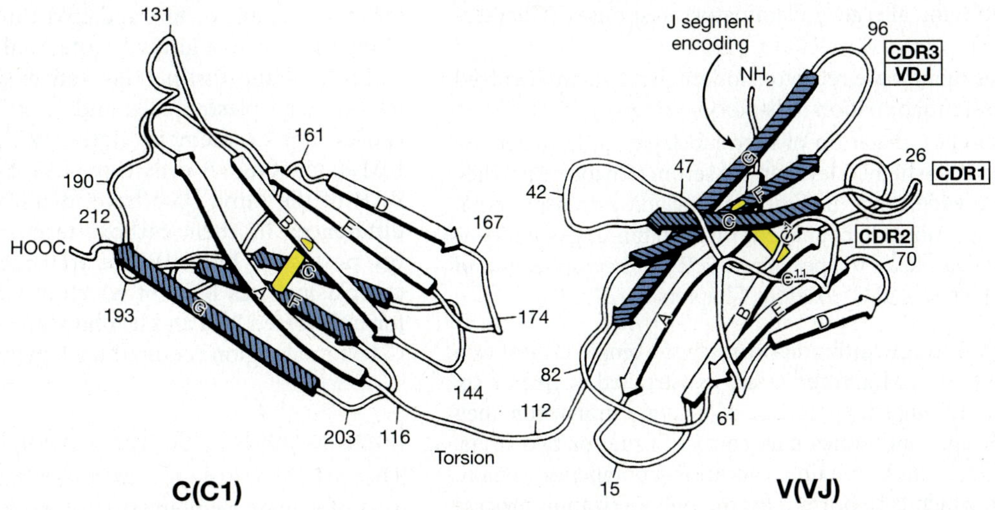

FIG. 1.7. IgSF V and C domains in an Ig L chain or TCR chain. V domains are made up of nine beta strands (A, B, C, C′, C″, D, E, F, G) forming two sheets (A, B, D, E and C, C′, C″, F, G). CDRs are found on the loops between strands B and C (CDR1), strands C′ and C″ (CDR2), and between strands F and G (CDR3), which is generated by the RAG-mediated rearrangement events. This V domain is called a VJ domain because its G strand has a conserved Gly-X-Gly motif involved in V domain dimerization in Ig and TCR. Dimerization occurs via the C, F, G sheet interface for V domains. The C domain is compact, with the "C1 domain" found in Ig, TCR, and MHC class I and class II being most compact. The strands and sheets are the same as in the V domain except C′ and C″ strands are not present in C domains. There is torsion in the loop connecting the V and C domains such that the C-C dimer interface is formed between the A, B, D, E sheet interface. Intradomian disulfide bonds shown in yellow. Note that the 'D' in CDR3 is only found in IgH and TCRB/TCRD.

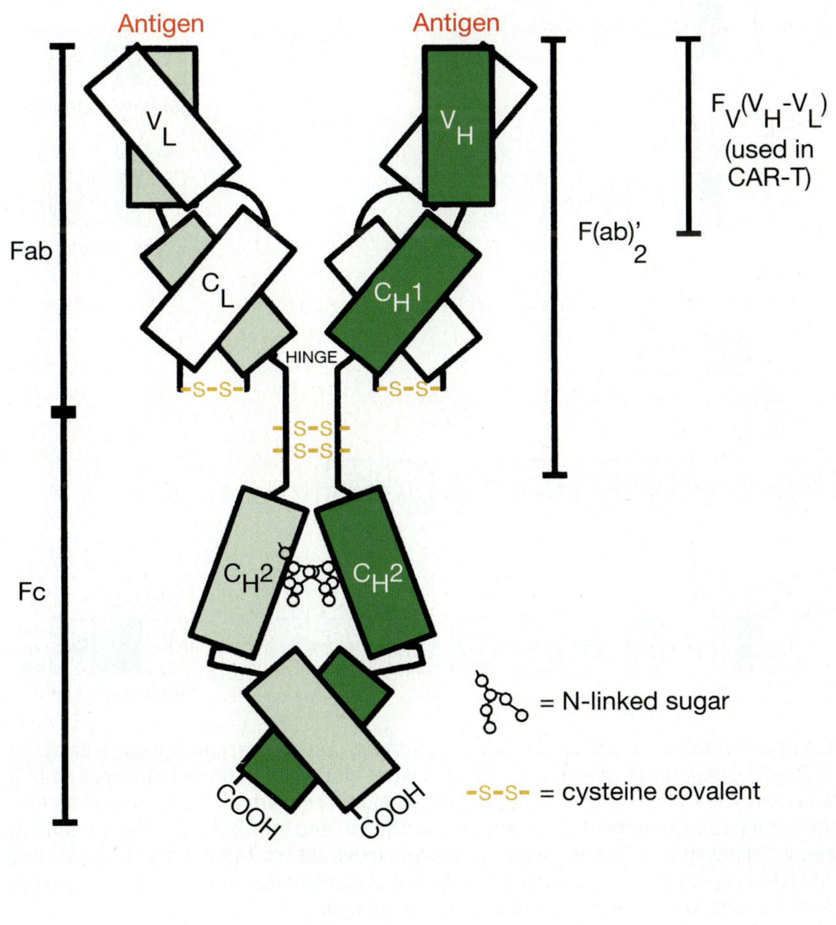

FIG. 1.8. IgG antibody molecule. An IgG molecule is made up of a "dimer of dimers," with a heavy chain (*H, green*) and light chain (*L, white*) making up each dimer. IgG H chains have an N-terminal V domain and three C domains, while L chains have and N-terminal V domain and one C domain. The four chains are covalently associated with interchain disulfide bonds as indicated (*yellow*). Fc, fragment crystallizable for effector functions; Fab, fragment-antigen binding made up of the N-terminal two domains of the H chain and the L chain; F(ab)′$_2$, fragment-antigen-binding-2, made up of two disulfide-linked Fabs that retain the bivalent combining sites; FV, fragment-variable that only contains the N-terminal V domains of VH and VL, often used in therapeutic antibodies or chimeric antigen receptors on T cells (CAR-T). The Asn-linked sugar on the CH2 domain is vital for Fc effector functions (Chapter 33). The hinge provides flexibility for antigen binding by the two Fabs.

responsible for allergic inflammatory responses (Chapters 16, 33, 47).

- IgD act as membrane receptors for antigen and are involved in mucosal inflammation (Chapters 4, 21, 26).
- IgG antibodies, made up of four subclasses in humans, are expressed at the highest levels in the serum (Chapter 21). They mediate a wide range of functions including transplacental passage, opsonization of antigens through binding of antigen/antibody complexes to specialized FcR on macrophages and other cell types, and C′ fixation (Chapters 13, 21, 33).

IgD, IgG, and IgE antibodies consist of a single unit of two H and L chains. IgM antibodies are constructed of five or six such units, although they consist of a single unit when they act as BCR. IgA antibodies may consist of one or two units. IgM and dimeric IgA contain an additional polypeptide chain, the J chain, which is important for the polymerization process during Ig biosynthesis in plasma cells. In addition, secreted IgA contains a secretory piece, that is derived from the receptor for polymeric IgA and IgM (J chain-dependent), which is required for the transport of IgA and IgM through the epithelial cells lining the lumen of lung and intestinal mucosae.

Each of the distinct Igs can exist as secreted antibodies made by plasma cells and as BCR on naïve (IgM/IgD-expressing) or memory (IgG-, IgA-, IgE-, and sometimes IgM-expressing) B cells (Fig. 1.9). Secreted antibodies and BCR of the same class have identical structures except for differences in their carboxy-terminal regions. Membrane Igs possess a hydrophobic transmembrane region and a cytoplasmic tail, both of which are lacking in the secretory form. Secreted IgM and IgA have an evolutionarily conserved C-terminal region required for J chain association.

Immunoglobulin Genetics (Chapters 4, 22)

The IgH chain gene of a mature lymphocyte is derived from a set of genetic elements that are separated from one another in the germline (Fig. 1.9). The V gene is composed of three types of genetic elements: *V*, *D*, and *J*. More than 100 V_H elements exist in humans, 9 *D* elements, and 6 J_H elements.

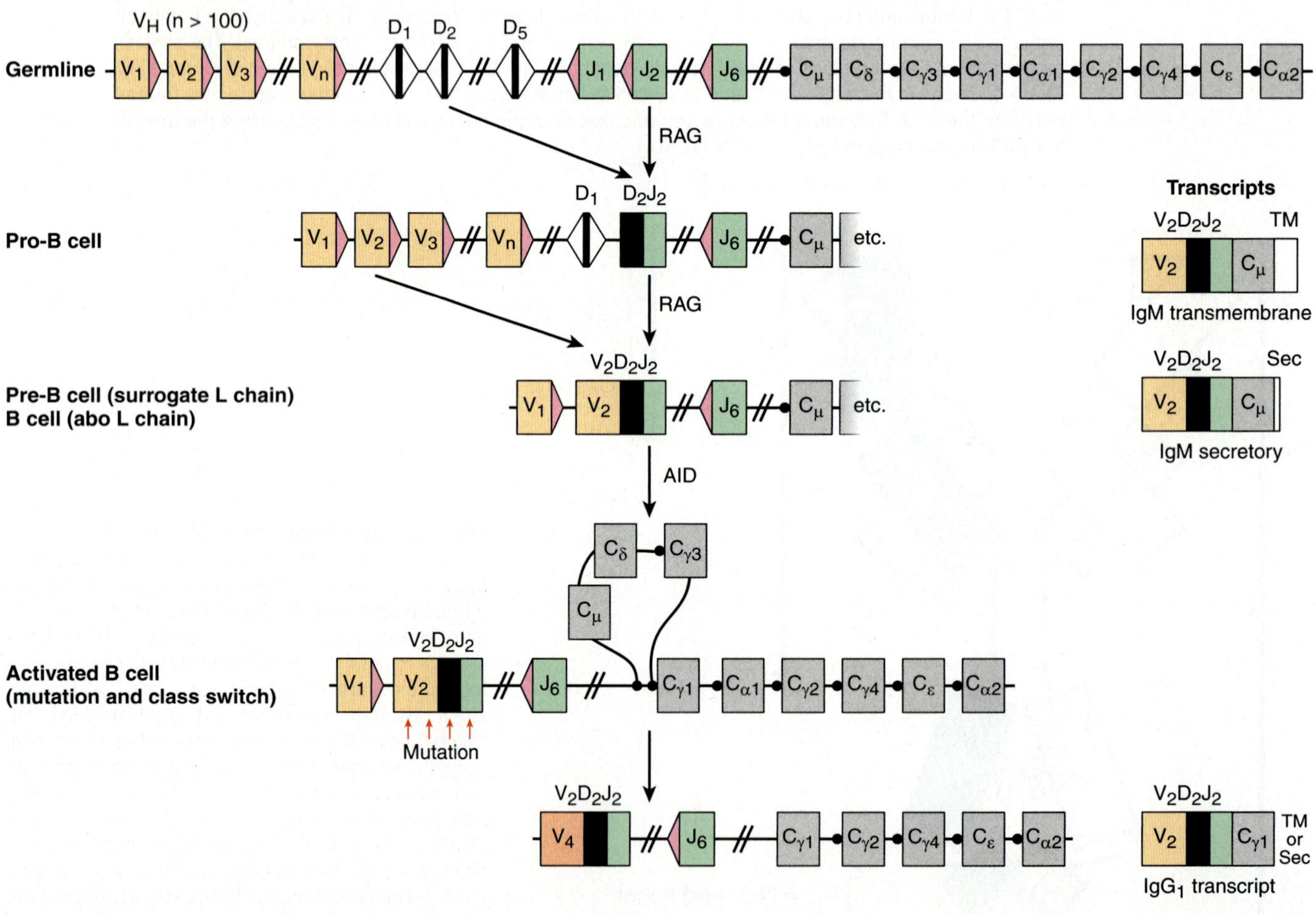

FIG. 1.9. The IgH locus, rearrangement, and class-switch recombination. V (*orange*), D (*black*), J (*green*), and C (*gray*) gene segments at the human IgH locus. RSS are shown as triangles flanking V, D, and J segments: pink triangle 23 bp spacer and white triangle 12 bp spacer. In a developing B cell in the bone marrow, first D joins to J and then V to that DJ in a RAG-dependent fashion. This permits production of an IgM molecule that associates with a surrogate light chain. Light chain rearrangement commences (not shown) and IgM BCRs are then produced by transitional and then mature B cells in the follicles of secondary lymphoid tissues. When B cells are activated in a T-dependent fashion, they upregulate AID which is required for somatic hypermutation (arrows in the V gene) and class-switch recombination, here, showing a switch from IgM to IgG1. IgH transcripts can either encode transmembrane or secretory IgH via alternative splicing.

An H-chain V_HDJ_H gene is created by the joining of one of the *D* elements on a given chromosome to one of the J_H elements on that chromosome, usually with the excision of the intervening DNA. The recombination-activating genes (RAG) are required for this process, recognizing recombination signal sequences (RSS) at the 3′ end of V segments, the 5′ end of J segments, and both sides of the D elements. This is followed by a second rearrangement event in which one of the V_H elements is joined with the assembled DJ_H element to create the V region (V_HDJ_H) gene (Fig. 1.9). Although the choice of the V_H, *D*, and J_H elements that are assembled is not entirely random, the combinatorial process allows the creation of a very large number of distinct H chain V-region genes. Additional diversity is created by the imprecision of the joining events and by the deletion of nucleotides, and especially by the addition of new, untemplated nucleotides between *D* and J_H and between V_H and *D*, forming "N regions" in these areas. This greatly increases the diversity of distinct IgH chains that can be generated from the relatively modest amount of genetic information present in the germline. The assembly of L-chain genes follows similar rules, but L chains are assembled only from V_L and J_L elements. Although there is junctional diversity, N regions are found at lower levels in L chains. Additional diversity is provided by the existence of two classes of L chains, κ and λ. An Ig molecule is assembled by the pairing of an IgH chain with an IgL chain. Although this process is not completely random, it allows the formation of an exceedingly large number of distinct Ig molecules having distinct specificities. The regulation of V(D)J rearrangement is discussed below in the B Cell Development section.

The RAG genes were acquired by vertebrates 500 million years ago from a microbial transposon that entered the genome (Chapter 4). The enzymes as well as the RSS directing rearrangement invaded a germline IgSF exon between the parts encoding the F and G stands (Fig. 1.7). The gene became thus inactivated and could only be stitched back together by the action of the RAG enzymes. This transposition event is believed to have been an innovative force in the emergence of adaptive immunity as almost all features of adaptive immunity described in this volume are found in all extant jawed vertebrates (starting with sharks), that is, the RAG transposon initiated a "Big Bang" of adaptive immunity. The active enzymatic fragment of RAG1, called transib, which is required to cleave DNA at the RSS/rearranging segment to initiate rearrangement, is found in transposons inserted into genomes throughout the animal kingdom.

Jawless vertebrates (lamprey and hagfish) have antigen receptors called variable lymphocyte receptors (VLRs) based on an entirely different gene family, the leucine-rich repeats (LRR). These receptors are also somatically generated by enzymes related to the AID and are clonally expressed. Proof of the existence of two sets of rearranging receptors in the vertebrates was one of the most exciting findings in basic immunology in the past 15 years (Chapter 4).

Class Switching (Chapters 22, 26)

An individual B cell continues to express the same IgH-chain V region as it matures, but it can switch its IgH-chain C region (Fig. 1.9). Thus, a cell that expresses receptors of the IgM and IgD classes may differentiate into a cell that expresses IgG, IgA, or IgE BCR and then into a cell secreting antibody of the same class that it expressed on the cell surface. This process allows the production of antibodies capable of mediating the distinct biologic functions described above yet retaining the same antigen-combining specificity. When linked with the process of affinity maturation of antibodies, Ig CSR provides antibodies of extremely high efficacy in preventing reinfection with microbes or in rapidly eliminating pathogens. The coupled phenomena of CSR and affinity maturation account for the high degree of effectiveness of antibodies produced in secondary immune responses.

The CSR process involves a genetic recombination event between specialized switch (S) regions, containing repetitive sequences and AID target sites, located upstream of each C region gene (except the δ C region). Thus, the S region upstream of the μ C_H region gene (Sμ) recombines with an S region upstream of a more 3′ isotype, such as Sγ1, to create a chimeric Sμ/Sγ1 region and in the deletion of the intervening DNA (Fig. 1.9). The genes encoding the C regions of all other Ig isotype genes are located 3′ of the Cμ and Cδ genes.

The induction of the switching process is dependent on the action of a specialized set of B-cell stimulants. Of these, the most widely studied are CD40L (CD154), expressed on the surface of activated T cells (Fig. 1.6), and TLR ligands, such as bacterial LPS. The targeting of the C region that will be expressed after CSR is determined by cytokines. For example, IL-4 determines that switch events in the human and mouse will be to the ε C region and to the γ4 (human) C regions. IFN-γ determines switching to γ1/γ3, and TGF-β determines switching to α.

As already noted, both CSR and SHM depend upon AID. Humans lacking AID fail to undergo both processes, and in humans this causes an immunodeficiency called hyper-IgM syndrome.

T Cell Receptors

The TCR is a disulfide-linked heterodimer (Fig. 1.10) whose constituent chains (α and β, or γ and δ) are IgSF members. The TCR is associated with a set of transmembrane proteins, collectively designated the CD3 complex, which plays an indispensable role in signal transduction. The CD3 complex consists of γ, δ (note that the CD3 γ and δ chains and the TCR γ and δ chains are distinct polypeptides that, unfortunately, have similar designations), and ε chains and is associated with a homodimer of two ζ chains. CD3 γ, δ, and ε consist of extracellular domains that are IgSF members. The cytosolic domains of CD3 γ, δ, and ε and of ζ and η contain one or more copies of the immunoreceptor tyrosine–based activation motif (ITAM) (D/ExxYxxLxxxxxxxYxxL/I) that is found in a variety of chains associated with immune recognition receptors, most notably the BCR, NK cell receptors, and FcR. This motif is important in the signal-transduction process and provides a site through which protein tyrosine kinases can interact with these chains to propagate signaling events (Chapter 24).

The TCR chains are organized much like Ig chains, with N-terminal V domains and C-terminal C domains.

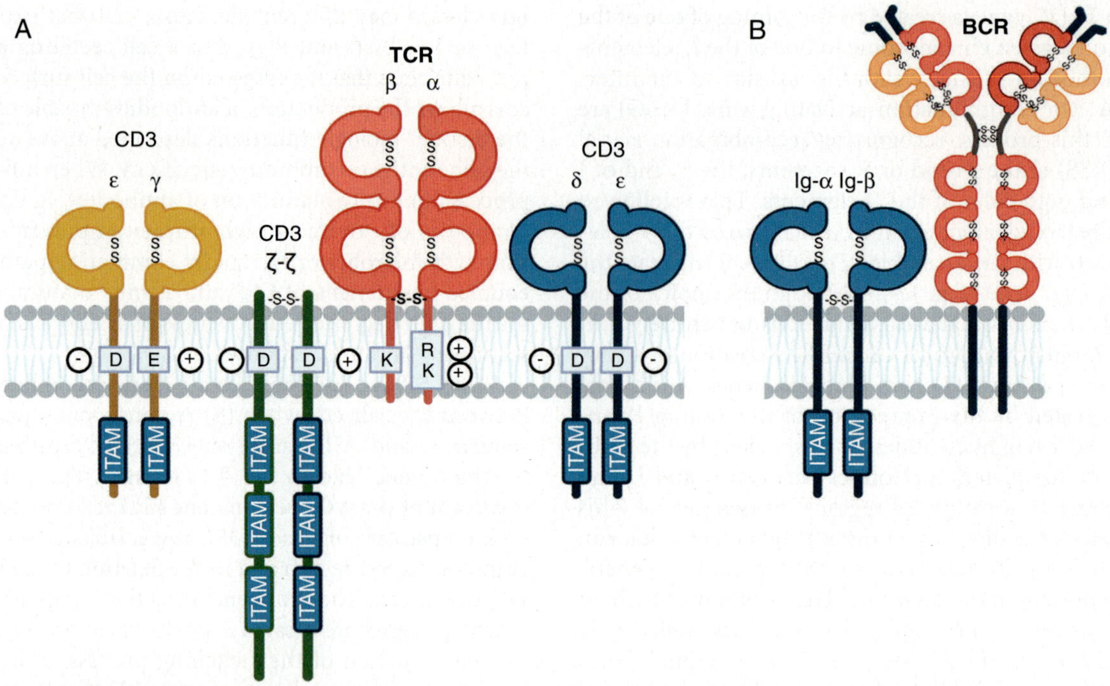

FIG. 1.10. The T cell receptor-CD3 complex. A: The TCR is made up two chains (α and β, red polypeptides) which have characteristic extracellular Ig folds that are stabilized by disulfide bonds (Chapter 23). The TCR does not have a cytosolic signaling domain. In order to couple extracellular ligand (peptide-MHC [pMHC] complexes) recognition to intracellular signaling, it constitutively associates with six polypeptides of the CD3 family—which exist as disulfide-linked dimers of CD3ε-CD3γ, CD3ε-CD3δ, CD3ζ-CD3ζ. The TCR-CD3 complex is stabilized by electrostatic interactions in the membrane-spanning region with the positive charges on Lysine (K) and Arginine (R) in the TCR α and β chains associating with the aspartic acid (D) on the CD3 molecules. In addition, glutamic acid residue (E) on the CD3γ chain stabilizes the extended lattice like interactions to form the holocomplex. The immunoreceptor tyrosine–based activation motifs (ITAMs) on the CD3 chains are marked as blue boxes. A total of 10 ITAMs are therefore associated with the TCR complex, the largest number for any receptor unit. The current consensus model for how pMHC can trigger the TCR are discussed in Chapters 23, 24. **B:** Signaling through the BCR uses the same strategy as through the TCR but with different associated molecules (Chapter 24).

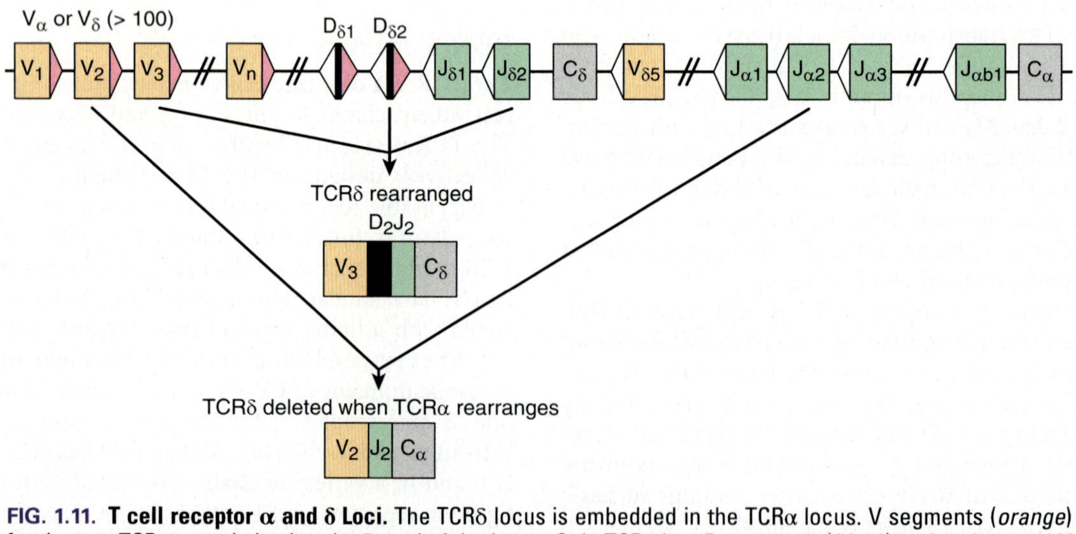

FIG. 1.11. T cell receptor α and δ Loci. The TCRδ locus is embedded in the TCRα locus. V segments (*orange*) for the two TCRs are admixed at the 5′ end of the locus. Only TCRδ has D segments (*black*) and an inverted V segment (Vδ5). J segments for both loci are in green, and the C gene segments in gray. RSS are shown as triangles flanking V, D, and J segments, pink triangle 23 bp spacer (*pink*) and white triangle 12 bp spacer. A functional TCRδ gene is produced upon VDJ rearrangement as shown. When a Vα segment joins to a Jα, then the entire TCRδ locus is deleted. Note the large number of Jα segments that can be used for receptor editing during thymus differentiation.

Furthermore, the same basic RAG-dependent rearrangement mechanisms are used to assemble the TCR V genes. Like IgH, the V region of the TCR β chain is encoded by a gene constructed from V_β, *D*, and J_β segments separated in the germline (Fig. 1.11 for TCR α and δ). Although the relative numbers of TCR rearranging gene segments differ from that for the IgH V region elements, the strategies for creation of a very large number of distinct genes by combinatorial assembly are the same. N-region addition further diversifies the genes and their encoded products. TCRβ has fewer *V* genes than IgH but much more diversity centered on the *D/J* region, which encodes CDR3. The α chain follows similar principles, except that like IgL chains it does not have *D* gene segments (Fig. 1.11).

The genes for TCR γ and δ chains are assembled in a similar manner, with only TCRδ having D segments (Fig. 1.11). The TCRδ D and J segments are located between the Vα/Vδ segments and the Jα/Cα segments, so that when TCR Vα genes are assembled during thymic differentiation the entire TCRδ locus is deleted.

The MHC and Antigen Presentation

As described above, the MHC governs T cell recognition of antigen-derived peptides bound to specialized grooves in MHC-I and MHC-II proteins, and they were first recognized because of the dominant role that MHC proteins play in transplantation immunity (Chapters 19, 48). The genetic basis of transplantation rejection between mice of distinct inbred strains is mediated by multiple genetic regions, but one region was shown to play the dominant role. Differences at this region alone would cause prompt (or acute) graft rejection, whereas any other individual difference usually resulted in a slow (or chronic) rejection. For this reason, the genetic region responsible for acute graft rejection was termed the *major histocompatibility complex*. Unfortunately, for 40 years the polymorphic MHC's role in transplantation obscured its true function: focusing the T cell recognition of microbial infection.

An MHC exists in all jawed vertebrates (Chapter 4). Its defining features are the transplantation antigens that it encodes, MHC-I and MHC-II molecules. MHC-I/II genes show a remarkable degree of polymorphism. The MHC also includes other genes, such as those for the C′ components C2, BF, and C4 as well as the cytokines TNFα and lymphotoxin in the so-called MHC class III region. In addition, genes encoding the proteasome subunits (LMP [low molecular mass polypeptide]) and ER transporters (TAP [transporter associated with antigen processing]), which produce and transport peptides for binding to class I, are also encoded in the MHC.

Class I MHC Molecules

Class I MHC molecules (MHC-1) are transmembrane glycoproteins expressed on most cells. They consist of an α chain of approximately 45,000 daltons noncovalently associated with β2-microglobulin, a 12,000-dalton molecule (Fig. 1.12A). The gene for the α chain is MHC-encoded, whereas that for β2-microglobulin is not. The membrane-proximal domain of the α chain and β2-microglobulin are IgSF members. The two membrane-distal domains form a groove called the

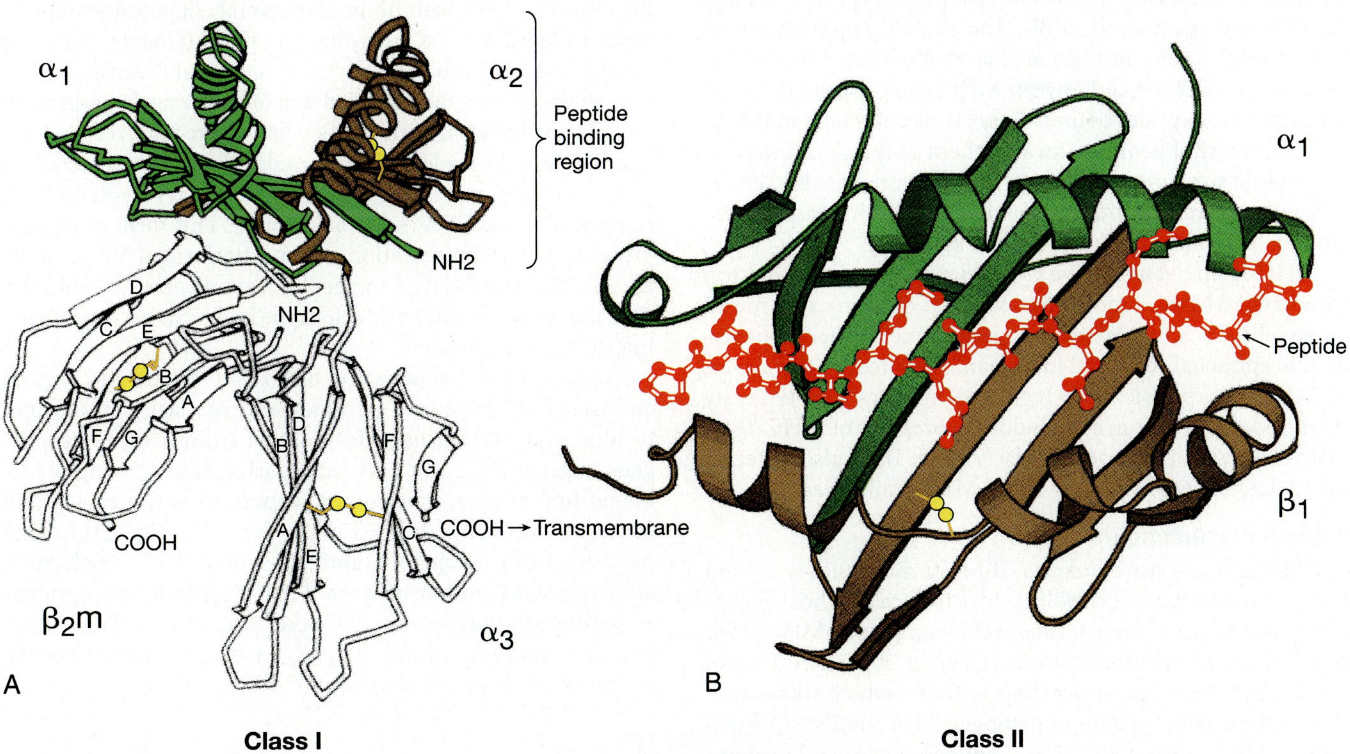

FIG. 1.12. MHC proteins. A: A class I molecule (MHC-I) in all vertebrates is composed of two N-terminal domains that make up the peptide-binding region (*green and brown*) and two C-terminal C1 domains that are members of the IgSF superfamily (Fig. 1.7). Intradomain disulfide bonds in yellow. **B:** Bird's-eye view of an MHC-II PBR with bound peptide. Each domain is made up of four contiguous beta strands topped by an alpha helix (α1 domain in green and α2 domain in light brown). The peptide (*red*) is bound as a beta strand and protrudes from both ends of the groove; for MHC-I the peptide is locked in at both ends of the groove (not shown). Intradomain disulfide bond in yellow.

peptide-binding region (PBR) made up of a β sheet floor topped with two α helices. The α chain is highly polymorphic with the diversity found in the PBR, specifically those amino acids that bind the antigen-derived peptides and contact sites for the TCR.

In the human, three loci encode classical class I molecules (or class Ia) and are designated human leukocyte antigen (HLA)-A, HLA-B, and HLA-C. HLA-A and HLA-B are highly polymorphic, and this high level of polymorphism is found in most vertebrate classical class I genes (Chapter 4). In addition, there is a second set of nonpolymorphic genes that encode class I–like molecules called nonclassical or class Ib molecules. One of the earliest of these to be discovered (M3) has antigen-presenting activity for formylated peptides, specialized to present prokaryotic antigens to CTL. The class Ib molecule CD1 has antigen-presenting function for glycolipids, providing a mechanism for generating T cells with special properties specific for such molecules (Chapter 36, see below). Another class Ib molecule, MR1, presents microbial metabolites to another subset of so-called "innate T cells" called mucosal-associated innate T cells (MAITs) (Chapter 37, see below).

Class II MHC Molecules

Class II MHC molecules (MHC-II) are also heterodimeric membrane glycoproteins whose chains are designated α and β, both encoded in the MHC (Fig. 1.12B). In the human MHC, there are three major sets of class II molecules, encoded in the DR, DQ, and DP regions of the HLA complex. The membrane-proximal domains (α2 and β2) of each chain are IgSF members, and the α1 and β1 domains form the PBR like that described for class I above. Class II genes are also polymorphic with most the diversity again in the PBR. The overall conformation of class II MHC molecules is quite like that of class I molecules. One distinct difference between MHC-I and MHC-II lies in the size of the peptides bound. Class I molecules bind to the N- and C-termini of peptides, locking them in the groove so that the peptide size ranges only from 8 to 10 amino acids. Class II molecules permit peptides to overhang on both ends, the size ranging from 11 to 22 amino acids (Fig. 1.12B).

MHC-II molecules have a restricted tissue distribution compared to MHC-I molecules. MHC-II molecules are found on B cells, DCs, epidermal Langerhans cells, macrophages, thymic epithelial cells, and in human, activated T cells. Levels of MHC-II expression are regulated in many cell types by IFNs. Indeed, IFNs can even induce expression of MHC-II on many cell types that normally lack them. IFNs also potently upregulate the expression of class I MHC molecules.

Antigen Presentation

The function of class I and class II MHC molecules is to bind and present antigen-derived peptides to T cells, whose receptors can recognize the peptide/MHC complex (pMHC) that is generated, as mentioned above (Chapter 20). There are two major types of antigen-processing pathways which are specialized to the distinct classes of pathogens that the T cell system must confront (Fig. 1.13). Extracellular bacteria and proteins may enter APCs by endocytosis or phagocytosis, enhanced by opsonization. Antigens within endosomes or lysosomes are fragmented in these organelles, and peptides derived from the antigen are loaded onto MHC-II molecules as these proteins traverse the vesicular compartments in which the peptides were formed. The loading of peptide stabilizes the MHC-II molecule, and the acidic pH of the compartments in which loading occurs facilitates the loading process. Once the peptide-loaded MHC-II molecules attain neutral pH, such as at the cell surface, the pMHC complex is stable. Peptide dissociation from such class II molecules is very slow, with a half-life measured in hours. The pMHC-II is recognized by CD4 T cells having complementary TCR. The tropism of CD4 T cells to the recognition of p/MHC-II complexes is partly due to the affinity of the CD4 molecule for monomorphic determinants on an MHC-II IgSF domain, focusing CD4 cells on those cells that normally express MHC-II molecules, namely DCs, B cells, and macrophages.

T cells also can recognize proteins that are produced within the cell (Fig. 1.13). The major pathogens recognized by this means are viruses and other obligate intracellular (nonendosomal/nonlysosomal) microbes that have infected cells (Chapter 44). In addition, proteins that are unique to tumors, such as mutant oncogenes, are overexpressed in tumors also can be recognized by T cells (Chapter 49). Endogenously produced proteins are fragmented in the cytosol by specialized proteases in the proteasome specific to adaptive immunity (immunoproteasomes). The resultant peptides are transported into the ER by the TAP where they are available for loading into MHC-I molecules. In contrast to the loading of MHC-II molecules, which is facilitated by the acid pH of the loading environment, the loading of class I molecules is controlled by interaction of the class I α chain with β2-m and several chaperones and catalysts called the "class I-loading complex" (Chapter 20). The peptide-loaded MHC-I molecule is stable and transported to the cell surface via the so-called default pathway. In contrast to peptide-loaded class II molecules that are recognized by CD4 T cells, peptide-loaded MHC-I molecules are recognized by CD8 T cells, in which the CD8 molecule plays a role in TCR recognition, like CD4 mentioned above. This form of antigen processing and presentation can be performed by virtually all cells because MHC-I molecules are universally expressed. This makes sense since any cell in the body can become virally infected or transformed.

Although the specialization of MHC-I to bind and present endogenously produced peptides and of class II molecules to bind and present peptides derived from exogenous antigens is generally correct, there are exceptions having physiologic importance. Particularly important is the capacity of some DCs (cDC1 or BATF3-dependent DC, Fig. 1.3) to load peptides from exogenous antigens into MHC-I molecules, allowing sensitization of naïve CD8 T cells to the antigens of pathogens that infect cells other than DCs. This "cross-priming" and "cross-presentation" was described above in the context of an immune response (Chapter 34).

Cytokines

Many of the functions of immune cells are mediated through cytokines (Chapters 9, 10, 12). These proteins can be divided

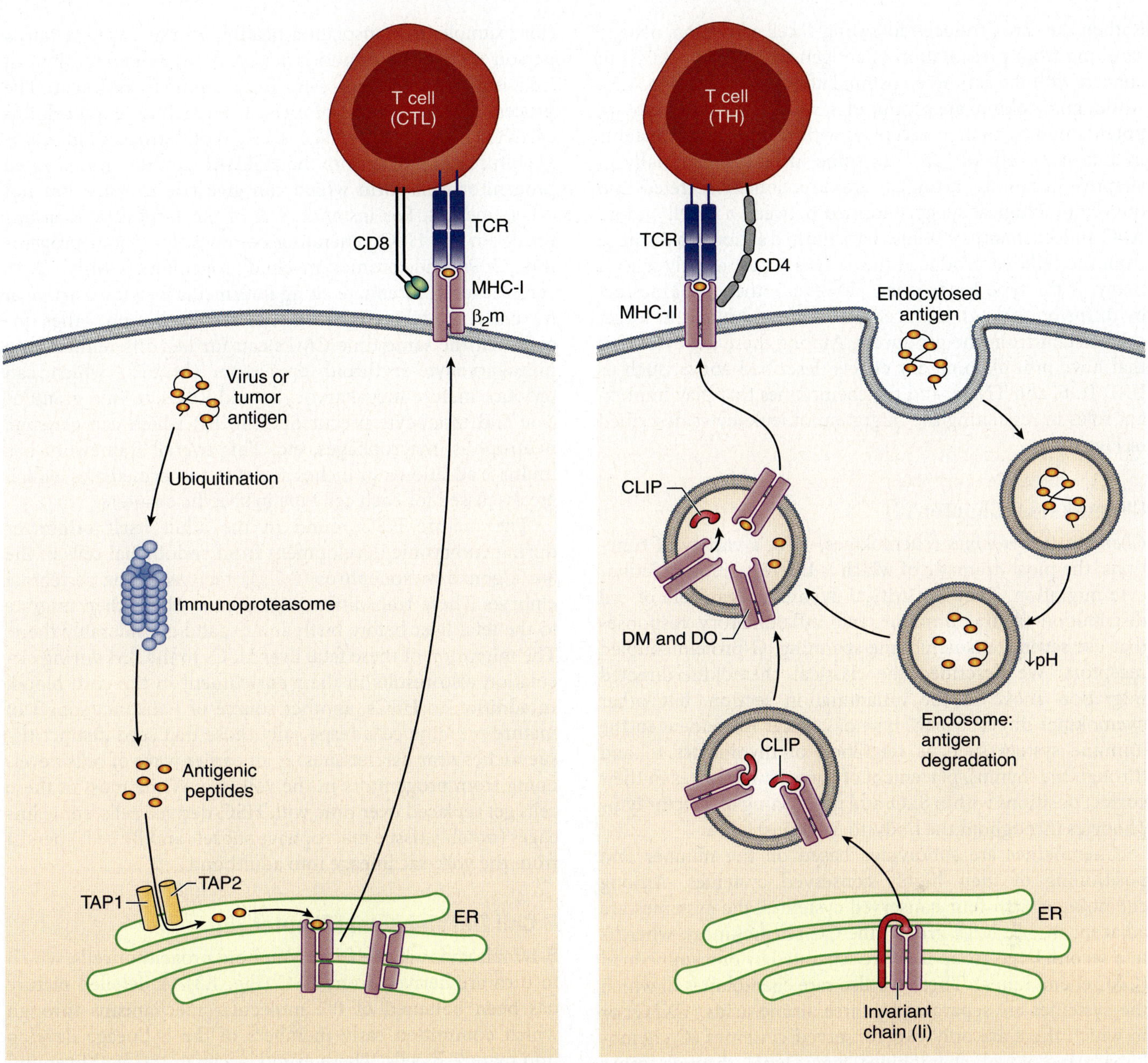

FIG. 1.13. Antigen presentation by MHC-I and MHC-II. A: Antigen (*orange*) in the cytosol misfolds and is conveyed to the immunoproteasome (*light blue*) after ubiquitination and then transported into the endoplasmic reticulum (ER) by the TAP (transporter associated with antigen processing) dimer (*light orange*). Peptides bind to class I molecules (*purple*) in the ER lumen. The class I molecules so-stabilized are transported to the cell surface where they are recognized by CD8 T cells (*red*). **B:** Antigens (*orange*) taken up into the endosome in antigen-presenting cells are degraded by cathepsins and other lysosomal proteases. MHC class II (*purple*) is deflected from the default pathway by the invariant chain (Ii) into endosomes containing the peptides (*orange*). CLIP, the last piece of Ii still attached to MHC class II, is removed from the MHC-II PBR which then binds peptides and moves to the cell surface where it is recognized by CD4 T cells (*red*). DM and DO class II molecules are involved in the removal of CLIP and selection of optimal peptides to bind class II.

into several families. Type I cytokines or hematopoeitins encompass many of the interleukins (ie, IL-2, IL-3, IL-4, IL-5, IL-6, IL-7, IL-9, IL-11, IL-12, IL-13, IL-15, IL-21, IL-23, and IL-27), as well as several hematopoietic growth factors. Type II cytokines include the IFNs and IL-10. The TNF superfamily molecules make up a very large family of regulatory molecules, the most recognizable being TNFα, lymphotoxin, CD40, and Fas ligand. The IgSF-based IL-1 family includes IL-1, IL-18, IL-33, IL-36, and IL-37. Chemokines (Chapter 11) are a large family of molecules that play critical roles in a wide variety of immune and inflammatory functions. IL-17 and its congeners (IL-17A-F), constitute a structurally unique set of cytokines.

Many of the cytokines are T cell products (Fig. 1.5), but most cytokines are not constitutive products of the T cell.

Rather, they are produced following T cell activation, usually resulting from presentation of antigen to T cells by APCs in concert with the action of costimulatory molecules (Fig. 1.5). Although cytokines are produced in small quantities, they are potent, binding to their receptors with equilibrium constants of approximately 10^{10} M^{-1}. In some instances, especially in adaptive immunity, cytokines are directionally secreted into the immunological synapse formed between a T cell and an APC; indeed, many cytokines have limited action at a distance from the cell that produced them. This is particularly true of many of the type I cytokines. However, other cytokines act by diffusion through extracellular fluids and blood to target cells distant from the producers. Among these are cytokines that have proinflammatory effects described above, such as IL-1, IL-6, and TNFα, and the chemokines that play important roles in regulating the migration of leukocytes described in Fig. 1.1.

Chemokines (Chapter 11)

Chemotactic cytokines (chemokines) have a variety of functions, the most dramatic of which is the regulation of leukocyte migration. They are critical dynamic organizers of cell distribution in the immune and inflammatory responses that use seven transmembrane-spanning, G-protein coupled receptors. We described the classical chemokine-directed migration above in the inflammation section, but other chemokines direct "every" type of cellular movement in the immune system such as migration of thymocytes to and through the thymus, placement of naïve lymphocytes in their correct positions within SLO, and positioning of effector lymphocytes throughout the body, to name just a few.

Chemokines are subdivided based on the number and positioning of their highly conserved cysteines. Among chemokines with four conserved cysteines, the cysteines are adjacent in one large group (the CC chemokines), whereas in a second large group they are separated by one amino acid (CXC chemokines). There are also rare chemokines in which the cysteines are separated by three amino acids (CX3C) or in which there are only two conserved cysteines (C chemokines). Individual chemokines may signal through more than one chemokine receptor, and individual receptors may interact with more than one chemokine, producing a complex set of chemokine/chemokine receptor pairs and providing opportunities for exceedingly fine regulation of cellular functions.

HEMATOPOIETIC CELL DEVELOPMENT AND INNATE LYMPHOCYTES

A striking feature of the immune system, with profound consequences to clinical immunology, is that the entire cellular catalog of innate, adaptive, and red blood cells can be generated from a single stem cell, known as the hematopoietic pluripotent stem cells (HSCs) (Fig. 1.3). These cells reside in the BM of adults and from there continuously generate all the numerous circulating lineages in healthy people. Indeed, this allows us to clinically transfer an entire immune system (for example, to transplant a healthy immune system into a person with an immunodeficiency) between persons by just isolating and injecting HSCs from donor to recipient. The process of development of each of the mature blood cell lineages (eg, RBC, macrophage, T cell, B cell) proceeds in multiple bifurcating steps from the HSC. Along the way dedicated progenitor cells form which can give rise to some but not other lineages. For instance, one of the earliest bifurcations results in the HSC generating common lymphoid progenitors (CLPs) and common myeloid progenitors (CMPs). CLPs can further differentiate along intermediate steps to arrive at mature ILCs, NK cells, T cells, or B cells, but not other lineages. At the same time CMPs can further differentiate into megakaryocyte-erythroid precursors (MkEPs) which can produce mature megakaryocytes and RBCs, or into granulocyte and monocyte precursors (GMPs) which can generate neutrophils, macrophages, etc. This overall framework has multiple additional branches and stable intermediates, which are discussed for each cell type in specific chapters.

The mature HSC found in the adult itself originates during embryonic development from endothelial cells in the aorta/gonad/mesonephros (AGM) regions of the vertebrate embryo. These transdifferentiated early HSCs then migrate to the fetal liver before birth and expand considerably there. The migration of these fetal liver HSCs to the BM during circulation also results in their enrichment in the cord blood. In addition to HSCs, another source of a distinct subset of mature immune cells (especially those that seed distinct tissue niches near barrier sites) is an earlier wave of cells developing from progenitors in the yolk sac. While most of these cells get replaced over time with HSC-derived cells, some lineages (notably tissue macrophage subsets and B1 cells) persist from the yolk-sac lineage into adulthood.

B Cell Development (Chapter 25)

B-lymphocytes derive from lymphoid progenitor cells, which in turn are derived from HSC (Fig. 1.3). A detailed picture has been obtained of the molecular mechanisms through which committed early members of the B lineage develop into mature B cells. These events occur in the fetal liver and adult BM. Interaction with specialized stromal cells and their products, including cytokines such as interleukin (IL)-7 and BAFF, is critical to the normal regulation of this process.

B cell development requires commitment to the B lineage and repression of differentiation to cells of other lineages, especially T cells. In pro-B cells and pre-B cells, the genetic segments that encode the BCRs are assembled. As described above, the genetic elements encoding the variable portions of Ig H and L chains are not contiguous in germline DNA nor in the DNA of nonlymphoid cells (Chapter 22) (Fig. 1.9). In pro- and pre-B cells, these genetic elements are rearranged to construct expressible V-region genes in a RAG-dependent fashion. The differentiation process is controlled at several steps by checks that determine whether prior steps have been successfully completed. These checks depend on the surface expression of appropriately constructed Ig or precursor-Ig molecules. The IgH V, D, and J elements rearrange first, and the H-chain V region is expressed along with the μ C-region gene to make the IgH chain of IgM associated with a surrogate light chain, VpreB/λ5. The V region of the surrogate chain is a complete germline V domain, not generated by V-J rearrangement. The surrogate chain "checks" whether the newly generated H chain can interact with typical L chains in a mature B cell. The successful completion of the process of IgH gene rearrangement and the expression of the resultant IgM on the cell surface marks the transition between the pre-B and B cell states, and the cell undergoes a proliferation stage. After proliferation, the developing B cell reexpresses RAGs and commencement of L chain gene rearrangement occurs, eventually resulting in an IgM BCR at the cell surface and the cell moves out of the BM as a "transitional B cell." B cells complete their maturation process by expressing on its surface a second class of Ig composed of the same L chain and the same V (VDJ) region but of a different H chain C region, the aforementioned IgD. Thus, through an alternative splicing mechanism, mature naïve B cells express both IgM and IgD surface molecules that share the same V region.

There are several checkpoints during B cell development to ensure a functioning, self-tolerant B cell population. First, if rearrangement of the H chain gene fails to produce a functioning H chain, either because the gene is out-of-frame, or the protein cannot associate with the surrogate L chain, the cell will die. Second, the L chain V-J rearrangement may be out-of-frame, or the L chain cannot associate with the H chain because of its sequence; in this case the L chain V segments upstream and J segments downstream of the rearranged V-J can undergo "receptor editing," generating a new V-J and deleting the previous one. Furthermore, if the nascent IgM is autoreactive, receptor editing can also occur in an attempt to produce a non-self-reactive BCR. These checks on tolerance are important because up to two-thirds of newly generated BCR are self-reactive. After the B cell satisfies these BM checkpoints and moves to the periphery, the transitional B cells are also tolerance-susceptible and will die when coming in contact with self-antigens of the body. Finally, mature IgM/IgD-positive B cells move into the B cell follicles of spleen, lymph node, or mucosal SLO and can respond to antigen as described above (Fig. 1.6).

B1 and Marginal Zone B Cells, AKA Innate B Cells (Chapter 25)

B cells consist of at least three distinct populations: conventional or follicular B cells (or B2 cells), B1 cells, and MZ B cells. B1 cells were initially recognized because a large subpopulation uniquely expresses the cell-surface protein CD5. In the adult mouse, B1 B cells are found in relatively high frequency in the peritoneal cavity and some mucosal surfaces cavity but at low levels in SLO. B1 cells undergo self-renewal throughout life, and their development is under different rules than conventional B cells, as they are derived from an early precursor derived from the yolk sac, not the AGM. B1 cells are not IL-7-dependent and do not use the surrogate L chain during their development. Their differentiation is positively selected by the BCR, which has a repertoire skewed toward recognition of conserved epitopes on pathogens, like the PAMP paradigm. Interestingly, expression of a B1 BCR in conventional B cells induces a change to a B1 phenotype, for example, capable of responding rapidly to antigen in a T-independent fashion.

MZ B cells are localized in the highly vascularized and antigen-filtering splenic MZ that surrounds the white pulp. Like B1 B cells, MZ B cells express a repertoire biased toward bacterial cell wall constituents and senescent self components. MZ and B1 cells respond very rapidly to antigenic challenge, often independently of T cells. Uniquely, among all populations of B cells, MZ B cells are dependent on Notch-2 signaling for their development.

B1 cells and MZ B cells are responsible for the secretion of the majority of serum IgM that exists in nonimmunized mice, often referred to as natural IgM. Among the antibodies found in such natural IgM are molecules that can combine with phosphatidylcholine (a component of pneumococcal cell walls) and with LPS and influenza virus. B1 B cells also produce autoantibodies, although they are generally of low affinity and not pathogenic. There is evidence that B1 B cells are important in resistance to several pathogens and may have a significant role in mucosal immunity.

T Cell Development (Chapter 27)

Upon entry into the thymus, T cell precursors do not express TCR, the CD3 complex, or the CD4 or CD8 molecules (Fig. 1.14A). Because these cells lack both CD4 and CD8, they are called double-negative (DN) T cells, which simultaneously rearrange TCR β, γ, and δ genes; if the cells have productive rearrangement of γ and δ and a functional receptor is generated, this provides a strong signal for the cell to develop into a γδ T cell that expresses neither CD4 nor CD8. This is a rare event after birth in many mammalian species, where most T cells are of the conventional αβ type. However, γ/δ cells are numerous early in fetal life and express TCR with specific V regions and no N-region diversity (Chapters 23, 35).

DN thymocytes that have rearranged their TCRβ genes express the protein with a surrogate receptor called Pre-Tα, which performs the same function as the surrogate chain in B cells. The cells undergo a proliferative phase and then express both CD4+ and CD8+ (double-positive [DP] cells) on their surface. The DP cells then commence rearrangement of the TCRα chain genes. Once TCR chains are expressed, these cells undergo two selection processes within the thymus, positive and negative selection. These events result in the deletion of potentially autoreactive (capable of high-affinity binding of self-peptide/self-MHC complexes) cells and the enrichment of T cells biased toward the recognition of peptides in association with self-MHC molecules. **Positive selection** occurs when T cells with receptors with "intermediate affinity" for self-peptides bound to self-MHC molecules expressed by the cortical epithelium are selected, thus forming the bias of the T cell repertoire for foreign peptides associated with self-MHC molecules (Figs. 1.14A and B). The capacity of the TCR and CD4 (or CD8) to bind to an MHC-II (or an MHC-I) molecule on the thymic epithelium leads either to the differentiation of DP thymocytes into CD4+ (or CD8+) single-positive (SP)

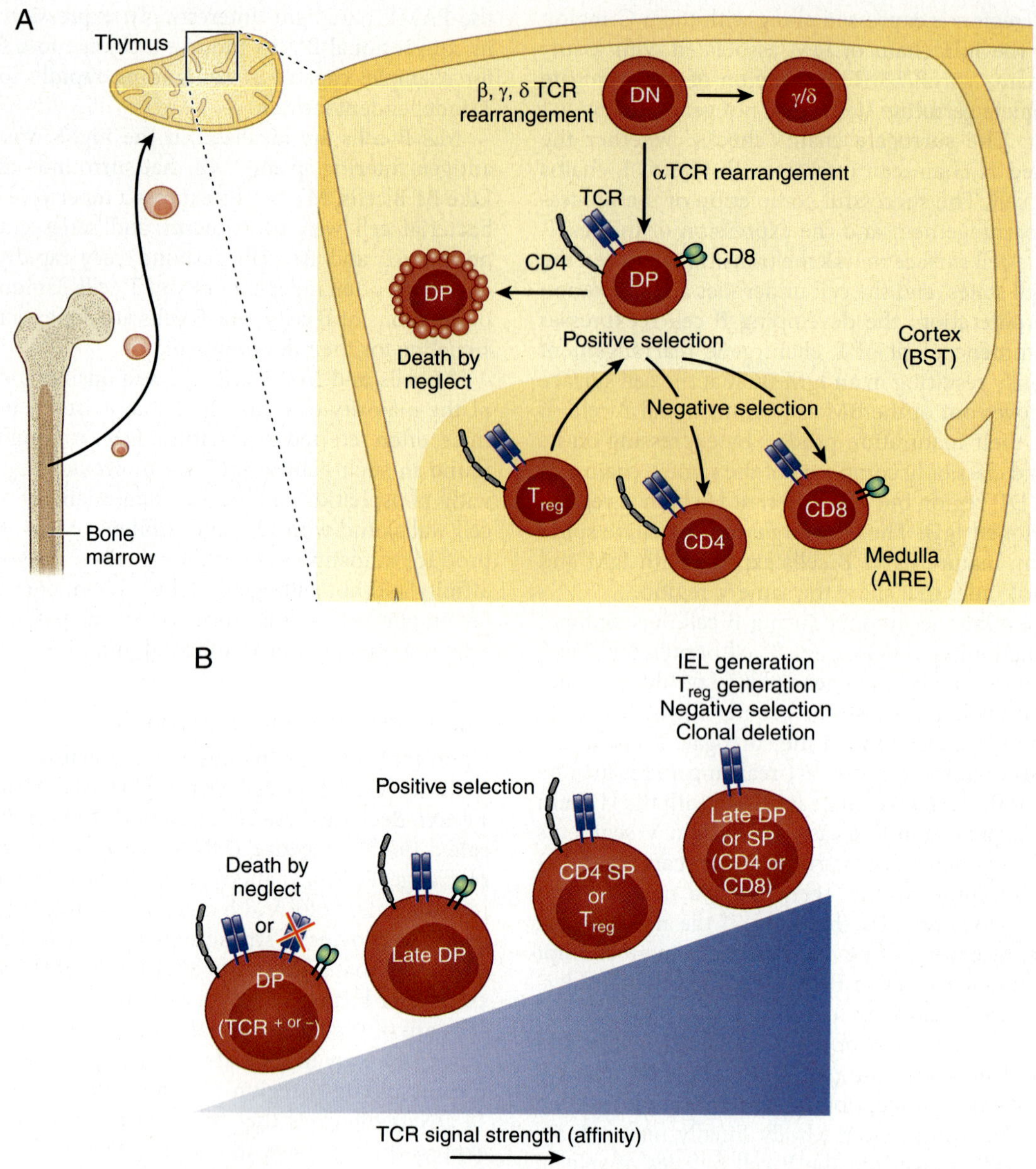

FIG. 1.14. T cell development. A: Differentiation of cells in the thymus progresses from the thymic cortex to medulla. Precursor cells (light red) move from the bone marrow to the thymus, where they are first found in the subcapsular region as double-negative cells (no CD4 or CD8, red). DN cells simultaneously rearrange β, γ, and δ genes. If the γ and δ rearrangements are productive, γδ T cells are generated. If they are nonproductive and TCRβ rearrangements are productive, then the cells undergo a proliferative phase followed by expression of CD4 and CD8 (DP), and TCRα rearrangement commences. If the TCR can be positively selected on the cortical epithelium, then the cells become single-positive (SP) CD4 or CD8; if not, then the cells are programmed to die. If the positively selected cells bind to MHC/peptide complexes too avidly at the corticomedullary junction or the medulla the cells undergo apoptosis, referred to as negative selection, or may be recruited as T_{reg} cells or intraepithelial T cells. **B:** The affinity of TCR binding to the thymic epithelium results in different T cell fates: Death by neglect if there is no interaction or weak interaction; positive selection of cells with intermediate affinity; and negative selection (clonal deletion) or selection of T_{regs} or intraepithelial lymphocytes (IEL) with high avidity interactions.

T cells or to the selection of cells that have "stochastically" differentiated down the CD4 (or CD8) pathway. T cells that receive no signal from the TCR are eliminated in the thymic cortex by apoptosis, called "death by neglect," much like death of B cells in the GC failing to recognize antigen on FDC.

Negative selection is the deletion of cells with high-affinity TCR for complexes of self-peptides associated with self-MHC molecules on APC at the thymic cortical-medullary junction or in the medulla on the epithelium (Fig. 1.14A). This is a major mechanism by which the T cell compartment develops immunologic unresponsiveness to self-antigens (Chapters 27, 32). T cells that are negatively selected are deleted through apoptosis. Medullary epithelial cells express a molecule called AIRE (Autoimmune Regulatory Element) that induces expression of self-molecules expressed throughout the body, AKA tissue-restricted antigens (TRAs) such as insulin (pancreas-specific) or C-reactive protein (liver-specific). Mutations of AIRE result in an autoimmune disease called autoimmune polyendocrinopathy candidiasis ectodermal dystrophy (APECED).

Regulatory T Cells (Chapter 29)

T cells can diminish as well as help immune responses, and the cells that mediate such inhibitory effects are called regulatory T cells (T_{regs}). T_{regs} typically express Foxp3, CD25, and the IL-2Rα chain. These cells inhibit both CD4 and CD8 T cell responses to their cognate antigens. How their inhibitory functions are mediated remains controversial. In some instances, cell/cell contact is required for suppression, whereas T_{reg} cytokines like IL-10 and TGF-β may inhibit others.

T_{regs} are critical in autoimmune diseases. In the absence of T_{regs}, conventional T cells cause several autoimmune conditions, including autoimmune gastritis, diabetes mellitus, adrenal insufficiency, and inflammatory bowel disease. T_{regs} normally recognize autoantigens and their responses result in the suppression of conventional T cell responses. Whether the receptor repertoire of T_{regs} and the conventional T cells are the same has not been fully determined, although T_{regs} may derive from a thymic CD4 T-cell population with relatively high affinity for self-antigen (Fig. 1.14B). As mentioned above, iT_{regs} can be derived in the periphery from naïve CD4 T cell populations when naïve cells are stimulated on DC by their cognate ligands in the presence of TGF-β and IL-2 (Fig. 1.5).

Innate Lymphoid Cells and NK Cells (Chapter 18)

NK cells play an important role in the immune system. Indeed, in mice that lack mature T and B cells the NK system still provides these animals a substantial measure of protection against infection. NK cells lack TCR (or Ig) but are closely related to T cells in their functions. They have a set of activating receptors (associated with ITAM-containing receptors) that allow them to recognize features associated with virally infected cells or tumor cells, such as "SOS molecules" that are only expressed at the cell surface in times of stress. They also express receptors with inhibitory cytosolic domains, such as ITIM (immunoreceptor tyrosine–based inhibitory motif) for MHC molecules that prevent NK lytic activity. Thus, virally infected cells or tumor cells that escape the surveillance of CTL by downregulating expression of MHC-I molecules become targets for efficient killing by NK cells because the cytotoxic activity of the latter cells is no longer inhibited by the recognition of particular MHC- I alleles, a phenomenon known as "missing self."

NK cells also express FcγRIII (CD16). Antibody-coated cells can be recognized by NK cells, and such cells can then be lysed (Chapter 33). This process is referred to as antibody-dependent cellular cytotoxicity (ADCC). NK cells are efficient producers of IFN-γ in response to recognition of virally infected cells and tumor cells, cross-linkage of FcγRIII, and stimulation by the cytokines IL-12 and IL-18.

Since the previous *FI* edition, a new subclass of lymphocytes came into prominence called innate lymphoid cells (ILCs) (Fig. 1.3) (Chapter 18). They were discovered over 20 years ago as lymphoid tissue-inducer cells (LTI) that organize SLO generation (for all tissues except the spleen). ILC types (ILC1-4) mimic the different Th cell types described above in the types of cytokines secreted. However, like NK cells (a type of ILC), ILCs do not have receptors generated by RAG-mediated rearrangement; instead, ILCs are activated by cytokines in the tissue in which they reside. ILCs have been best studied in mucosal tissues such as the lung and intestine in responses to parasites, but this is just the tip of the iceberg. ILCs' roles in infection and homeostasis will continue to be a fertile area of research.

Gamma Delta T Cells (Chapters 23, 35)

γ/δ T cells were discovered in the early 1980s as a complete surprise, with no underlying biology. A breakthrough in the 1990s was the realization that γ/δ T cells were not MHC-restricted and likely recognized antigen more like antibodies, as detailed above. For many years these cells were known for their innate phenotypes, including a major subset in human blood that recognizes isopentenyl pyrophosphate (IPP, and other such intermediates from microbes), and for waves of γ/δ cells in mice that are generated during fetal life and peripheralized to the skin and mucosal surfaces. Recently, members of the butyrophilin family have been shown to participate in the recognition of IPP and of other unknown antigens. Some γ/δ crystal structures show that the δV CDR3 recognizes the "true" antigen, suggesting that γ/δ T cells can also have adaptive characteristics. It has been suggested that a large proportion of T cells could have both innate (rapid fire) and adaptive (large TCR repertoire) characteristics, so-called "adaptate" T cells.

Natural Killer-T Cells (Chapter 36) and Mucosal-Associated Innate T Cells (Chapter 37)

The first "innate αβ T cells" to be discovered, natural killer-T (NKT) cells, recognize the class Ib molecule CD1, which does not present peptides like classical MHC-I but rather lipid or glycolipid antigens. These T cells are defined by their restricted TCRα usage and positive selection on CD1 expressed by thymocytes, not the thymic epithelium. NKT cells respond rapidly when confronted by pathogens, and different NKT subsets produce cytokines that can fight pathogens and skew adaptive responses. The first NKT cells discovered were shown to make IL-4, but subsets of cells secreting different cytokines are apparent, like conventional CD4 T cell subsets (Fig. 1.5), but these phenotypes are imprinted during thymus differentiation rather than in response to foreign antigens.

The class Ib molecule MR1 was discovered 20 years ago with no known functions for nearly a decade. The innate-like MAIT T cells were found to use MR1 as a restricting element, predominantly in mucosal tissues. The basic TCR usage, positive selection, and rapid activation are like NKT cells. Recently MR1 was shown to present microbial metabolites to MAITs, a substantial breakthrough in immunology.

Currently, we could think of the class I system as a trifecta: (1) the polymorphic class I molecules presenting peptides to conventional CD8 cells; (2) the CD1 molecules presenting lipid antigens to innate T cells; and (3) MR1 molecules presenting bacterial "PAMPs" to MAITs.

CLINICAL IMMUNOLOGY

Immunological Tolerance and Autoimmunity (Chapters 32, 46)

One of the central problems facing the immune system is how to mount highly effective immune responses to threats posed by potentially pathogenic agents while ignoring the host's own tissues. This is further complicated by the need to respond to some self (albeit with mutations) cells as in tumors. The mechanisms ensuring the lack of response to healthy self-antigens are complex and involve a series of strategies. Chief among them is elimination or inactivation of cells capable of self-reactivity. The encounter of immature, naïve B cells with antigens with repetitive epitopes capable of cross-linking membrane Ig can lead to elimination of the B cells, particularly if no T-cell help is provided simultaneously. This elimination of potentially self-reactive cells is often referred to as clonal elimination. As described above, many self-reactive cells, rather than dying upon encounter with self-antigens, undergo a further round of Ig gene rearrangement called receptor editing described above, which allows a self-reactive cell to substitute a new receptor and therefore avoid elimination.

There are many self-antigens that are not encountered by the developing B-cell population or that do not have the capacity to cross-link BCR sufficiently to elicit the receptor editing/clonal elimination process. Such cells, even when mature, may nonetheless be inactivated through a process that involves cross-linking of receptors without the administration of critical costimulatory signals. These inactivated cells may be retained in the body but are unresponsive to antigen and are referred to as anergic. When removed from the presence of the anergy-inducing stimulus, anergic cells may regain responsiveness.

The immune system can discriminate between antigenic determinants expressed on foreign substances, such as pathogenic microbes, and potential antigenic determinants expressed by the tissues of the host. The ability of the system to forego full-blown immune responses to self-antigens is referred to as *immunologic tolerance*, a complex concatenation of processes. Perhaps the most important element is the active destruction or inactivation of self-reactive cells. The encounter of antigens (such as self-antigens) in the absence of cues from the innate immune system may fail to engender a response, may lead to a minimal response, or may lead to immune cellular inactivation through a process referred to as *anergy*. Finally, the T_{regs} described above actively *suppress* responses against self-antigens. Indeed, mutations in the key transcription factor *FOXP3*, which determines T regulatory cells, lead to severe multiorgan autoimmunity (IPEX syndrome). The critical necessity to control self-reactivity is clearly shown by this multilayered system that involves elimination, inactivation, and suppression.

Failure to establish immunologic tolerance or unusual presentations of self-antigens can give rise to tissue-damaging immune responses against host molecules, resulting in autoimmune diseases (Chapter 46). Autoimmune diseases or those with major autoimmune components include systemic lupus erythematosus (SLE), rheumatoid arthritis, insulin-dependent diabetes mellitus, multiple sclerosis, myasthenia gravis, and inflammatory bowel disease. Successful therapeutic modulation of the (auto)immune response is a reality using inhibitors of antigen-receptor signaling (Chapter 24), cytokines, or transfer of special cell types.

Primary Immunodeficiency (Chapter 50)

Mutations in genes that control the development or function of different immune cells can cause different primary (congenital or hereditary) immunodeficiencies (PID). As discussed above, all mature immune cells develop in adults from HSCs. Different regulatory genes are critical for each step in HSC development and deleterious mutations in any of these can result in the absence of mature cells derived from these precursors. For instance, both B cells and T cells require VDJ recombination to generate their receptors. Without functional receptors, the precursors to B and T cells cannot achieve their next developmental step. As a result, when critical enzymes (eg, the RAGs) are mutated in people, they fail to make both B and T cells, resulting in SCID. Similarly, mutations in receptors for key cytokines that instruct development of a joint precursor for T and NK cells can lead to SCID through absence of these critical cells. In each case, patients are susceptible to a range of infections that are normally kept at bay by these functional cells. However, the fact that these cells can continuously develop from HSCs offers a clear therapeutic approach—the transfer of healthy BM cells (derived from matched donors) intravenously to patients allows these to take over and repopulate the mature immune cell compartments.

Tumor Immunity (Chapter 49)

In healthy individuals, constant exposure to mutagens (as simple as sunlight), viral infections, errors in DNA replication, etc generate somatic mutations every day. While the propensity of such mutations to transform into tumors varies with genetic background, in most cases, the immune system can police these aberrations and effectively eliminate them. Recognition of tumors presents a particular challenge to the immune receptors which (see Principle #2) must maintain tolerance of self-tissues. Especially in the case of nonviral tumors (where there is no infectious etiology), identification is challenging. The solutions appear to be manifold. In the case of innate immune cells, dedicated sensors for DNA (such as STING) released from necrotic tumor cells or receptors for stress-associated molecules (such as Rae1) allow a variety of effector cells to be turned on. Typically, T cells (primarily CD8s, but also CD4s) activated by tumor-associated antigens (TAA) or tumor-specific (neo)antigens (TSA) form the effector response. While these processes operate quite well in acute antitumor responses, immunological control of established tumors is challenging. For the most part, the immune system then begins to treat the slow-growing tumor as another self-tissue. Both immune termination signals (such as negative-regulatory checkpoints on T cells including PD1) and tolerance associated players (illustrated by the expansion of FOXP3+ regulatory T cells for instance) are now recruited to curtail robust antitumor effector responses. Over the past decade, several years of previous work defining the basic immunobiology of these processes has begun to bear clinical dividends in the form of promising immunotherapies.

First, it has become possible to sequence expressed genes from tumors and identify the mutations to detect those self-peptides that can bind to self-MHC allelic products and thus serve as targets for CD8 cells. Antigen-specific CD8 cells can be transferred into the patient or specific TCRs can be transduced into CTL before adoptive transfer. For virally induced tumors, T cells for common HLA alleles (eg, HLA-A02) can be taken "off the shelf" for adoptive transfer. Second, for hematological tumors (especially of B cells), chimeric antigen receptor (CAR)-T cells have been a major breakthrough. By combining the V regions of antibodies specific for the B cell markers CD19 or CD20 with TCR signaling machinery (Chapter 24) (hence the chimeric nature of the construct), transduced T cells can kill the tumor cells after adoptive transfer (Fig. 1.15). Of course, normal B cells are also killed off with this protocol, but patients can be treated with passive transfer of Igs to protect them, even for a lifetime. There is no end in sight in the mixing and matching of BCR/TCR/signaling components to perfect the CAR-T technique, making it applicable to different tumor types. Third, with the identification of the plethora of costimulatory molecules of either the activating or inhibitory type (Chapter 24), reagents that block certain inhibitory molecules, like CTLA4 and PD1, have been shown to unleash T cells specific for tumor-specific antigens, especially efficacious for highly mutable tumors like melanoma and renal cancer. Fourth, the tumor microenvironment has been scrutinized for suppressive cells (myeloid and lymphoid) and molecular mediators that inhibit T cell responses. Tumors can be made "hot" by blocking these suppressive mechanisms. There are other advances in this field (discussed in Chapter 49), but we just highlight those that have been successful in the clinic. It should be emphasized that *none of these breakthroughs* was in common usage in 2012, when the previous edition of *FI* was published.

Transplantation (Chapter 48)

The field of transplantation has a special historical significance to immunology, since the pivotal efforts in the area by Gorer, Medawar, Brent, Billingham, Hasek, and Owen in the 1940s and 1950s identified many of the key phenomena which then took years of discovery to explain. This phase of discovery laid the foundations of much of cellular immunology as we know today. The identification of MHC alleles, the appreciation of their polymorphisms, and their significance in graft rejection have allowed for much of the success in

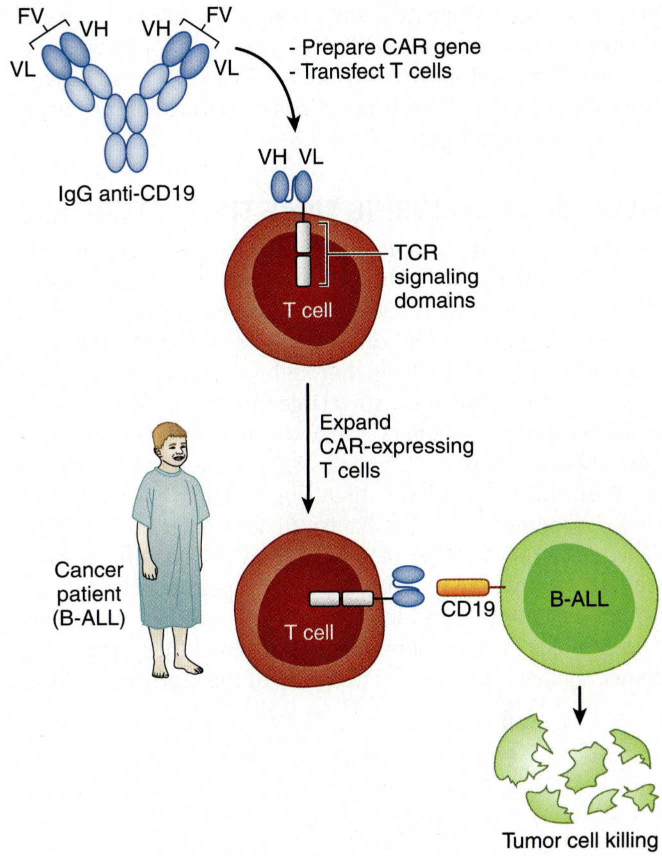

FIG. 1.15. Generation and use of CAR-T cells in the clinic. Chimeric antigen receptors (CAR) are generated with FV (VH/VL in *blue*, Fig. 1.7) from a monoclonal antibody with TCR signaling motifs from CD3 and costimulatory molecules (*white rectangles*, and see Fig. 1.10). The chimeric gene is transfected into the patient's T cells (*red*), which are then transferred to the patient. The CAR-T cells recognize the surface antigen on the cancer cells (CD19 in orange on B cell-acute lymphocytic leukemia [ALL] in this case, *green*) and kill them.

modern clinical organ transplantation. In the 1970s, development of the immunosuppressive drug cyclosporin by Borel, which targets antigen-receptor signaling in T cells, allowed transplants to be performed even when the donor and recipient were not fully matched (the next-generation cyclosporin-related drug is Tacrolimus, Chapter 24). With increasing appreciation of other cell types and processes involved in tolerating transplants, newer approaches are now in laboratory evaluation. Chapter 48 discusses BM transplants from the same solid organ donors to generate tolerance via blood cell chimerism, as well as ways to block costimulatory pathways or the priming of donor-specific T cells in SLO. While not surmounting the rejection puzzle, the problem of a short supply of organs such as hearts for transplantation may be circumvented by xenotransplantation. Over the past 20 years, the miniature swine genome has been modified to overcome the acute and chronic rejection episodes that occur when these organs are transplanted into humans. The promise of using embryonic stem cells for differentiation into solid organs is also an area of intense study. Finally, the possibility of inhibiting innate pathways in transplanted tissue to block activation of adaptive immunity is an active area of research. Before the next edition of FI in 10 years or so, like what has occurred in the tumor field, there will likely be significant breakthroughs in generating antigen-specific tolerance in the area of transplantation.

NEW TOPICS MATURING SINCE THE LAST EDITION

Thirty-seven of the chapters in FI2022 update contemporary advances of the same fields as in 2012, but 21 of those chapters have new authors, such as the chapter on the history of immunology (Chapter 2). The other chapters written by the same authors as in 2012 have been thoroughly updated. Twelve chapters cover new subfields, with a large focus on evolution of the immune system (Chapters 3-5) and innate lymphocytes, now covered in six chapters (Chapters 18, 35-37, and in Chapters 23, 25 in subsections). Other entirely new chapters cover the fields of Immunometabolism (Chapter 6) and Neuroimmunology (Chapter 39). The latter field has begun to make inroads into understanding how the immune system patrols neural tissues which were once considered privileged, but also how these two complex organ systems (the nervous and the immune systems) regulate each other. The ability of the mother to tolerate a fetus (which can bear many antigens encoded by the paternal genome which are "foreign" for the mother's immune system) remains one of the most intriguing puzzles of Immunology. The field of Maternal-Fetal Immunology has experienced major breakthroughs in the past decade and is presented in a new chapter (Chapter 41).

CONCLUSION

This introductory chapter has attempted to provide the reader with an appreciation of the overall organization of the immune system with its key cellular and molecular components. The immune system is highly complex, capable of a wide range of effector functions, and it has potent but only partially understood regulatory processes. The immune system is the key to the development of prevention and treatment of a broad range of diseases. Indeed, in addition to providing basic and conceptual knowledge of immunity, this edition also examines immunity to infectious agents and the immunologic mechanisms in disease. The introductory material provided here should orient the uninitiated reader in understanding the immunologic mechanisms brought into play in a wide range of clinical conditions.

CHAPTER 2

History of Immunology

Edward J. Moticka, PhD

OVERVIEW

Immunology is a young science—most historians date its beginnings to the late 18th century. During the approximately 225 years since Edward Jenner developed the smallpox vaccine in 1796, the knowledge obtained by immunologists has appreciably altered our views of other basic science disciplines as well as of health and medical practice.

Some immunologists consider Edward Jenner the father of immunology. Others honor Louis Pasteur or Robert Koch who validated the germ theory of disease in the 1860s. All three scientists relied on earlier observations that patients who survive systemic infectious diseases seldom contract that disease again.

Initially immunologists focused on mechanisms employed by the immune system to protect us from infectious diseases. As early as the 1880s, Paul Ehrlich and Ilya Metchnikoff disagreed about these mechanisms—Ehrlich championed the role of antibodies while Metchnikoff hypothesized that cells provided the body's defense. This disagreement influenced the development of immunology until the middle of the 20th century.

In 1890, Emil von Behring and Shibasaburo Kitasato developed serotherapy to help patients recover from diphtheria; this was the initial indication that immunology was clinically useful. A similar strategy is still being used against infectious microorganisms including Ebola and the coronavirus (SARS-CoV-2).

In 1902, Paul Portier and Charles Richet observed that the immune system induced pathology even in the absence of an infection. Soon investigators uncovered other forms of immune-mediated hypersensitivities. These highlighted a relationship between immunology and clinical medicine resulting in additional therapeutic interventions for allergic, transplant, and cancer patients.

The 20th century is replete with immunologic advances. In 1953, Rupert Billingham, Leslie Brent, and Peter Medawar induced immunologic tolerance in neonatal mice. Macfarlane Burnet proposed the clonal selection theory in 1957, a major conceptual advance in understanding the interaction between foreign antigens and lymphocytes. In 1959, Rodney Porter and Gerald Edelman revealed the structure of the antibody molecule while in 1975 Georges Köhler and César Milstein introduced the technology to produce monoclonal antibodies. Rolf Zinkernagel and Peter Doherty in 1974 demonstrated that cellular interactions in the immune system require the cells share major histocompatibility genes. Other experiments by a host of international scientists refined an understanding of the immune system.

This chapter describes these studies following the development of immunology as a discipline and underlining eight immunologic concepts.

1. Individuals who survive an infectious disease develop immunity to that disease.
2. Microorganisms (bacteria, viruses, fungi, parasites) cause disease.
3. Antibodies provide protection against extracellular pathogens: T lymphocytes provide protection against intracellular pathogens. Subsets of T lymphocytes respond to different classes of pathogens.
4. Cells of the immune system individually respond to a single antigen but collectively respond to an almost infinite number of antigens.
5. Progenitors of immune competent lymphocytes acquire specificity and diversity due to a unique set of somatic genetic mechanisms.
6. Antigen "selects" those cells that can respond by interacting with preexisting cell surface receptors. Lymphocytes potentially reactive to self are silenced during development (immunologic tolerance).
7. Immune responses may result in pathology.
8. Cells of the immune system interact both positively and negatively through interaction of cell surface–expressed and secreted molecules. A subset of lymphocytes exists that are uniquely positioned to suppress other lymphocytes.

PRE-GERM THEORY DEFENSE MECHANISMS

All living organisms possess defense mechanisms that protect them from infectious diseases. Many potential pathogens are destroyed by innate host defense mechanisms including physical barriers (skin, mucous membranes), cells (macrophages, granulocytes), and molecules (inflammatory cytokines, complement components). If pathogens circumvent these defenses, they trigger adaptive immune responses that are specific for that pathogen. The evidence that these mechanisms exist predates the identification of microorganisms and their role in causing disease. Thus, Thucydides, a Greek scholar, reported almost 2500 years ago that survivors of a plague epidemic could nurse others with the disease without fear of recurring illness.[1] Other commentators made similar conclusions about smallpox and other diseases.

Development of Vaccines

More than 3000 years ago, Chinese physicians employed these observations to develop a method to protect their patients from smallpox. They introduced powdered smallpox scabs through a scratch or by inhalation. They reasoned that induction of a mild case of smallpox would protect the individual from the disease during future epidemics.

Physicians around the world, including India, Africa, and the Middle East, employed a similar stratagem. Lady Mary Worley Montague, the wife of the British ambassador to Istanbul, observed this procedure in 1717 and recognized its potential.[2] The Turks called this method of providing protection "engraftment." Lady Montague had one of her children engrafted by Turkish physicians and subsequently introduced the technique to England where the procedure was termed variolation.

Variolation became fashionable in England during the 18th century. African slaves introduced a similar technique to North America. Onesimus, a slave owned by Cotton Mather, a preacher in Boston, demonstrated the technique to Zabdiel Boylston, a Boston physician, who performed the first inoculation during a smallpox epidemic in 1721. Variolation succeeded in decreasing smallpox mortality, although approximately 2% to 3% of variolated patients died of the procedure. This contrasts with a mortality rate of approximately 30% when the disease was contracted naturally.[3] In 1778 George Washington mandated that troops under his command be variolated thus decreasing smallpox morbidity and mortality. This mandate is credited as helping the American colonists defeat the British in the War of Independence.

In 1765, John Fewster presented a paper to the London Medical Society describing his observations that individuals naturally exposed to cowpox, a disease of cattle related to smallpox, developed protection to even a mild case of smallpox when variolated. Fewster, a surgeon from Gloucestershire, failed to investigate this observation further.[4] More than 30 years later, Edward Jenner explored similar anecdotal observations to develop a method for inducing protection against smallpox termed vaccination,[5] the inoculation of patients with material from animals with cowpox. Subsequent experimentation demonstrated that cowpox is caused by a virus cross-reactive with smallpox virus. Vaccination induces a mild disease that protects individuals from the more serious smallpox. Vaccination, with a mortality rate of between 1 and 2 per million, proved approximately 1000 times safer than variolation.

Anatomy of the Immune System

Beginning in the 17th century, scientists studied the anatomy of the immune system describing the lymphatic vascular system and many of the lymphatic organs. They concluded that the lymphatic system is separate from the blood vascular system but that the two are connected. Organs comprising the lymphatic system include organs where immunologically active cells mature (thymus, bone marrow/bursa of Fabricius) or reside (spleen, lymph nodes).

Hippocrates, a Greek physician born approximately 2500 years ago, palpated "glands" (lymph nodes) beneath the skin. During the next 2000 years, other scientists localized these lymph nodes, identified the thymus and spleen, and demonstrated the presence of lymphatic vessels including lacteals and the thoracic duct. By the middle of the 17th century, European anatomists defined the basic structure of the entire lymphatic system.

Between 1650 and 1653, Jean Pecquet, Thomas Bartholin, and Olof Rudbeck published information on the organization of the lymphatic system.[6,7] In 1651, Pecquet, from France, dissected dog hearts and noted a milky fluid in the superior vena cava. He traced this fluid back to the thoracic duct and discovered a structure (the cisterna chyli) that drained the intestinal lacteals. Although Galen originally claimed that the lacteals emptied into the liver, Pecquet corrected this misconception and demonstrated that the lacteals drain into the venous system.

Bartholin, while a student of anatomy at the University of Padua in Italy, confirmed that the lymphatics draining nonintestinal organs of the peritoneal cavity terminated in the thoracic duct. This conclusion challenged the long-held view that these systemic lymphatics drained into the liver. Bartholin also described lymphatic vessels throughout the body that he designated the *vasa lymphatica*.

Rudbeck reported his observations to the faculty at the University of Uppsala, Sweden, on the structure of the lymphatic system. He confirmed the findings of Bartholin and Pecquet that lymphatic vessels drain virtually all the organs of the body and that the lacteals and the systemic lymphatic vessels drain into the thoracic duct which subsequently empties into the left subclavian vein. He published these results in *Nova exercitatio anatomica exhibens Ductus Hepaticus Aquosus et Vasa Glandularum Serosa*.

In the 1660s, Anton van Leeuwenhoek in the Netherlands improved the existing microscopes and described many objects too small to be observed by the naked eye including "animalcules" (bacteria). In 1675, he reported red blood cells in humans and several other species that formed sediments upon sitting and lysed when exposed to water.

In the 1840s, Gabriel Andral in France and William Addison in England observed white cells in the peripheral blood (leukocytes) that increased during various diseases including sexually transmitted infections and tuberculosis.[8] Rudolf Virchow in Germany and John Hughes Bennett in Scotland described an increase in the number of cells (lymphocytes) in the peripheral blood of patients with leukemia.[9] The function of these cells became a major focus of research during the second half of the 19th century. These studies, performed following the elucidation of the germ theory of disease, inform the mechanisms responsible for elimination of disease-causing infectious agents.

THE GERM THEORY OF DISEASE

Between 1832 and the 1850s, cholera epidemics broke out in various locales in England. Physicians disagreed about the source of the contagion with some concluding that the condition was spread by "bad air" from rotting organic material (the miasma theory) while others concluded that transmission required contact between healthy and diseased individuals. In 1849, John Snow published a short treatise *On the Mode of Communication of Cholera*[10] outlining the arguments for transmission of the disease through the water supply.

In 1854, a cholera outbreak infected many of the residents of the Soho District of London. Snow investigated and demonstrated that the disease was spread by sewage dumped in rivers that were the source of public water supplies provided by wells. Snow's epidemiological study concluded that many of the cases of cholera in Soho clustered around a public well on Broad Street. These data convinced the local officials to remove the pump handle from this well. Within weeks, the cholera outbreak ended. Subsequent investigation demonstrated that the well was contaminated by garbage in a cesspit, including soiled diapers from a cholera-stricken baby.[11]

Snow presented his germ theory of disease to the Medical Society of London. Unfortunately, the theory was disputed by the medical community for the next 10 years. During the next 25 years, two scientists, Louis Pasteur in France and Robert Koch in Germany, provided experimental evidence that microorganisms exist, they are alive, and some of them cause disease.

Pasteur studied wine fermentation and demonstrated that the process was caused by a living organism. He showed that the microorganisms responsible for fermentation derived from other similar microorganisms rather than spontaneously from the liquid itself and that they could be killed by heating the liquid to between 50 °C and 60 °C (pasteurization).[12,13]

Koch demonstrated unequivocally that anthrax in cattle was caused by a microorganism (*Bacillus anthracis*) living in the soil.[14] He injected groups of mice with blood from cattle that had died from anthrax or, as a control, with blood from healthy cattle. The mice injected with blood from anthrax-infected cows developed the disease, while the control group failed to show evidence of disease. Based on these results, Koch developed a set of postulates that are required to demonstrate that a specific disease is caused by a distinct microorganism. These postulates are:

- The bacteria must be present in every case of the disease.
- The bacteria must be isolated from the host with the disease and grown in pure culture.
- The specific disease must be reproduced when a pure culture of the bacteria is inoculated into a healthy susceptible host.
- The bacteria must be recoverable from the experimentally infected host.

These postulates have been modified to include other forms of microorganisms. Robert Koch received the Nobel Prize in 1905 in Physiology or Medicine for this work.

IMMUNOLOGY POST-GERM THEORY

Jenner developed the smallpox vaccine more than 50 years prior to the elaboration of the germ theory by Snow in the 1850s.[5] Scientists in the second half of the 19th century pursued development of vaccines to protect the population against other infectious diseases. Louis Pasteur manipulated infectious microorganisms so that they retained their immunogenicity (ability to induce an immune response) while decreasing their pathogenicity. Using these methods, he developed a successful vaccine against the rabies virus in 1885.[15] Subsequently, scientists created vaccines to more than 25 infectious diseases (Table 2.1).[16]

MECHANISMS OF HOST DEFENSES AGAINST INFECTIOUS MICROORGANISMS

Innate Host Defenses

The response to most potential pathogens involves two separate but interrelated mechanisms, a nonspecific innate defense system and a specific adaptive immune response. The innate immune system, which eliminates most infectious microorganisms, comprises anatomical barriers such as the skin and mucus membranes, cells including macrophages, dendritic cells (DCs), and granulocytes, and non–antigen-specific cytokines and chemokines. The adaptive immune response includes the activity of antibodies and specifically sensitized B and T lymphocytes. Although investigators studied these defense mechanisms independently, immunologists now realize that the two systems are inextricably linked working together to eliminate infectious microorganisms.

The historical origin of the study of the innate host defense mechanisms occurred in the late 19th century. Over 150 years ago, Paul Langerhans described a cell type in human skin with a stellate morphology.[17] Originally thought to be part of the nervous system, these cells, now classified as DCs, play a critical role in initiating both innate and adaptive immune responses. DCs are present in the skin as well as in organs throughout the body. They originate from bone marrow precursors and recognize foreign antigens through specialized molecules known as pattern recognition receptors (PRRs). This interaction activates genes coding for cytokines and other mediators of the innate system. DCs simultaneously serve as antigen-presenting cells to activate T lymphocytes of the adaptive immune system.

Ilya Metchnikoff studied the response of starfish (sea stars) to rose thorns in the late 1800s.[18] He observed that the foreign material became surrounded by phagocytic cells (macrophages) that digested the thorn. Metchnikoff developed his theory that immune defenses depend on the action of cells from this observation. Subsequent investigation of DCs and macrophages provided crucial information concerning the initiation of innate and adaptive defense mechanisms.

Much of our contemporary knowledge about DCs derives from a collaboration between Ralph Steinman and Zanvil Cohn. This work, which spanned more than 20 years, was aimed at answering how the immune system is stimulated when antigen enters the body.[19]

In 1973, Steinman and Cohn reported the presence of a novel cell type in cultures of single cell suspensions from mouse lymphoid organs.[20] These large cells comprise between 0.1% and 1.6% of all nucleated cells in various organs and possess dendrite-like processes that move about as if sampling the environment. The investigators subsequently spent several years characterizing these DCs both morphologically and functionally.

Steinman and Cohn demonstrated that DCs fail to endocytose several foreign antigens even when the antigens are pretreated with specific antibodies, thus differentiating them from macrophages. DCs fail to express significant levels of either T cell receptors (TCRs) or immunoglobulin, although they do express both class I and class II major histocompatibility complex (MHC) molecules. Macrophages and DCs also differ in several characteristics including their sensitivity to steroids and radiation, their kinetics and turnover, and their response to antigenic stimulation.[21,22]

Functionally, DCs stimulate allogeneic lymphocytes *in vitro*[23] and in graft rejection.[24] Cultures lacking DCs fail to stimulate a cytotoxic T cell response[25] or a primary antibody response.[26] These results led to the conclusion that DCs are a potent accessory cell in the initiation of many different immune responses.

In 1989, Charles Janeway speculated that cells capable of initiating innate immune responses, including DCs and

TABLE 2.1 History of Vaccines for Preventable Infections

1796	Vaccinia virus	Smallpox, monkeypox	Available to special populations
1881	*Bacillus anthracis*	Anthrax	Available to exposed individuals
1885	Rabies virus	Rabies	Available to exposed individuals
1896	*Salmonella typhi*	Typhoid fever	Recommended for travelers to endemic areas
1921	*Bacillus calmette-guerin*	Tuberculosis	Not used in the United States: available for infants in other countries
1923	*Corynebacterium diphtheriae*	Diphtheria	Recommended pediatric vaccine
1926	*Bordetella pertussis*	Whooping cough	Recommended pediatric vaccine
1927	*Clostridium tetani*	Tetanus	Recommended pediatric vaccine
1935	Yellow fever virus	Yellow fever	Recommended for travelers to endemic areas
1945	Influenza viruses type A and B	Seasonal flu	Recommended yearly for all ages
1954	Measles virus	Measles	Recommended pediatric vaccine
1955	Polio virus	Poliomyelitis	Recommended pediatric vaccine
1967	Mumps virus	Mumps	Recommended pediatric vaccine
1970	Rubella virus	Rubella	Recommended pediatric vaccine
1981	Hepatitis B virus	Chronic hepatitis, cirrhosis, cancer	Recommended pediatric vaccine
1985	*Haemophilus influenza* type B	Meningitis, pneumonia	Recommended pediatric vaccine
1995	*Varicella zoster*	Chicken pox	Recommended pediatric vaccine
1996	Hepatitis A virus	Acute hepatitis	Recommended pediatric vaccine
2000	*Streptococcus pneumonia* (PPSV23)	Pneumonia, bacteremia, meningitis	Recommended adult vaccine
2005	*Neisseria meningitides*	Meningitis, septicemia	Recommended adolescent vaccine
2006	Rotavirus	Gastroenteritis	Recommended pediatric vaccine
2006	*Herpes zoster*	Shingles	Recommended senior vaccine
2006	Human papillomavirus	Genital warts, cervical cancer	Recommended adolescent vaccine
2009	Japanese encephalitis virus	Encephalitis	Recommended for travelers to endemic areas
2010	*Streptococcus pneumonia* (PCV-13)	Pneumonia, bacteremia, meningitis	Recommended senior vaccine
2021	SARS-CoV2	COVID	Recommended for adults and adolescents

macrophages, express cell surface molecules he called PRR. He further proposed that these receptors bind pathogen-associated molecular patterns (PAMPs) on foreign materials that might pose danger to the body.[27] Janeway based this hypothesis on the realization that the effector mechanisms (phagocytosis, inflammation) for removal of potential pathogens are similar in both innate and adaptive immune systems. During the next 10 years, several lines of evidence coalesced to indicate that this speculation was correct.

A major conceptual advance occurred in 1994 with the publication by Polly Matzinger of the danger model of immunologic recognition.[28] Before the development of the danger model, immunologists hypothesized that the immune system recognized self from non-self and responded only to the latter.[29] This new model argued that the immune system is stimulated by danger signals (DAMPs—danger-associated molecular patterns) released by pathogens or by dying or injured cells. These signals help the system discriminate between stimuli that may be dangerous from those that are safe.

PAMPs and DAMPs bind cell surface or intracellular molecules that fit the definition of PRRs as proposed by Janeway. Three groups of molecules are recognized including toll-like receptors (TLRs), NOD-like receptors (NLRs), and RIG-1–like receptors (RLRs). TLRs are components of the plasma membrane, while RLRs and NLRs are intracellular PRRs.

Investigations leading to the description of TLRs derived from studies on the mechanism by which endotoxin causes pathology. As early as the 1890s, physicians and scientists knew that many bacteria caused disease through the action of toxins.[30,31] Toxins are classified as either exotoxins or endotoxins depending on if they are secreted or are an integral component of the bacterial cell wall. In 1892, Pfeiffer described a toxin released upon death of the bacteria that kills guinea pigs.[32] This toxin is present in the cell walls of several different species of bacteria and is now termed an endotoxin. For nearly 100 years, the mechanism by which endotoxins produce damage was unknown.

In 1985, Christine Nüsslein-Volhard and co-workers studied fruit flies with gene mutations that are involved in the embryological development of the flies. One of the mutations disrupted the ventral-dorsal axis of the developing fly and was named Toll.[33] Ten years later, Jules Hoffmann and co-workers demonstrated that fruit flies required the wild-type Toll gene to control fungal infections.[34] Flies with mutated Toll genes synthesize decreased amounts of antifungal peptides. Further study of the molecular pathways leading to the synthesis of these peptides revealed the expression of a cell surface receptor, controlled by this gene and known as the toll receptor.[35] Stimulation of the toll receptor activates a signaling cascade resulting in the activation of NFκB.

Shortly after Hoffmann's group reported these findings, Ruslan Medzhitov and colleagues[36] cloned the mammalian homolog of the Toll gene, transfected human cells with this gene, and reported that these cells expressed a TLR. Stimulation of the TLR in transfected cells activated the NFκB pathway leading to the synthesis and secretion of proinflammatory cytokines including IL-1, IL-6, and IL-8 as well as the expression of the costimulatory molecule, B7.

Medzhitov and co-workers postulated that stimulation of the TLR plays a role in initiation of mammalian host defense mechanisms. However, the natural ligand for the TLR remained unknown until investigators identified the receptor for lipopolysaccharide (LPS). Injecting several mammals (mice, rabbits) with LPS leads to an often-fatal syndrome characterized by fever, secretion of proinflammatory cytokines leading to generalized inflammation, activation of complement, and initiation of coagulation. In 1986, Tobias and colleagues isolated and purified a serum protein from rabbits that bound LPS.[37] Subsequent studies identified a cell surface molecule on macrophages, CD14, which bound the LPS-binding protein. Interaction of CD14 with the LPS-binding protein activates macrophages. While investigators considered CD14 to be the receptor for LPS on cells, further studies ruled this out.

In parallel studies, Bruce Beutler and colleagues identified tumor necrosis factor (TNF) as a primary mediator of LPS toxicity.[38] Concentrating on the role of TNF in the response to LPS, Beutler's group mapped a gene locus, Lps, in mouse strains resistant to the action of LPS. They reported a mutation in this locus that coded for a protein that was identical to the receptor (TLR-4) identified by Medzhitov and his co-workers.[39] Mice possessing a mutation in TLR-4 fail to respond to LPS and are, therefore, susceptible to infection with gram-negative bacteria. Subsequent studies revealed that mammalian cells express several different receptors that bind ligands on potential pathogens. These receptors are the PRRs hypothesized by Janeway. The ligands they bind are PAMPs.

Binding of PAMPs by PRRs induces phagocytosis and intracellular destruction of the pathogen. As the pathogen is degraded, peptides are produced that bind to MHC molecules and relocate to the cell surface where they can interact with T lymphocytes to initiate the adaptive immune response. Thus, PRRs on DCs and macrophages detect foreign and potentially dangerous material in the system and activate both the innate and adaptive immune responses. For these insights, Jules Hoffmann and Bruce Beutler shared the 2011 Nobel Prize in Physiology or Medicine with Ralph Steinman.

Adaptive Immune Responses

In the late 1800s, information about the mechanism(s) by which the immune system eliminates infectious microorganisms was a mystery. Investigators in France and Germany studied this question and reached seemingly opposing conclusions. Paul Ehrlich[40] in Germany proposed that antibodies are responsible for providing protection while Ilya Metchnikoff[18] in France hypothesized that all defense mechanisms could be explained by the action of cells (phagocytes). Experiments to unravel this controversy continued well into the 20th century before resolution.

Antibody-Mediated Immunity

In the late 1880s, George Nuttall, Jozsef Fodor, Karl Flügge, and Hans von Buchner independently demonstrated that serum derived from animals injected with bacteria (*B. anthracis* or *Vibrio cholera*) contained a substance that killed the injected bacteria.[41,42] The substance is now termed antibody. This observation led to speculation that antibodies might be used to protect individuals against pathologies induced by bacterial infections.

Based on this prediction, Emil von Behring and Shibasaburo Kitasato in 1890 passively transferred antibodies to treat individuals infected with *Corynebacterium diphtheriae or Clostridium tetani*.[43] These investigators obtained horse antibodies specific for the toxins produced by *C. diphtheriae* or *C. tetani* and injected them into patients diagnosed with either diphtheria or tetanus. These antibodies effectively treated the diseases and established serotherapy as an important therapeutic intervention. This work resulted in the awarding of the initial Nobel Prize in Physiology or Medicine to von Behring in 1901 "for his work on serum therapy, especially its application against diphtheria by which he has opened a new road in the domain of medical science and thereby placed in the hands of the physician a victorious weapon against illness and deaths."[30] Passive transfer of antibodies continues to serve as therapy for several infectious diseases including rabies, Ebola, and COVID-19 induced by the severe acute respiratory syndrome coronavirus 2 (SARS-CoV-2).

Cell-Mediated Immunity

Ilya Metchnikoff, a Russian, worked at the Pasteur Institute in Paris starting in 1888. Prior to this appointment, he performed his most famous experiment involving the insertion of a rose thorn into the larvae of a starfish (sea star). The thorn stimulated the accumulation of phagocytic cells around the foreign body. These cells ingested and broke down the thorn. Metchnikoff hypothesized from his observations that phagocytosis was a mechanism by which animals could defend themselves against foreign materials including microorganisms. He postulated that this phagocytic theory[18] explained all aspects of immunology known at the time, and he and his students, known as cellularists, vehemently disagreed with Ehrlich and the humoralists.

The dispute between the humoralists and the cellularists continued for over 50 years. Ehrlich and his collaborators described the activity of antibodies in several clinical and laboratory situations.[40,44] He studied induced hemolysins as well as the therapeutic implications of passive transfer of antibodies. Subsequent investigations by Ehrlich and his group focused on characterization of antibodies including their specificity and quantitation, the method by which immunity is transferred from mother to offspring via breast milk, and the ability of the immune system to synthesize antibodies to self.

This dispute is seen as a continuation of the disagreements between Robert Koch and Louis Pasteur that started 20 years earlier. Both Koch and Pasteur studied *B. anthracis*, the causative agent of anthrax. Koch developed his set of postulates based on this work and felt that the chemical and biological characteristics of a microorganism were permanent. Pasteur employed a low-virulent strain of *B. anthracis* to develop a successful vaccine and felt that the virulence of a microorganism changed over time and could explain the appearance and disappearance of infectious diseases.[45]

Ehrlich and Metchnikoff shared the Nobel Prize in Physiology or Medicine in 1908 "in recognition of their work on immunity." The resolution of the controversy between humoralists and cellularists required additional understanding of the immune system including identification of cell-mediated hypersensitivities, division of lymphocytes into separate T and B populations, and observations of patients with immunodeficiencies.

ANTIBODIES

The identification of serum antibodies and the success of serotherapy stimulated numerous investigations in the first half of the 20th century including the mechanism by which antibodies are synthesized, the specificity and functional capabilities of antibodies, and the utility of using antibodies in treating a variety of clinical diseases.

Paul Ehrlich was the first scientist to use the term antibody (antikorper) in 1891. Currently antibodies are defined as glycoproteins induced by antigenic challenge. The detection of antibody in the late 19th century depended on the ability of serum from immunized animals to neutralize bacteria. Eventually, scientists devised several different systems for antibody detection including precipitation, agglutination, complement activation, and phagocytosis enhancement. In the early 1900s, immunologists assigned names to antibodies based on the *in vitro* assay by which they were detected. Thus, the literature of the time reports on precipitins, agglutinins, amboceptors (antibodies that activate the complement system and cause cell lysis), and opsonins (antibodies that enhance phagocytosis). Initially, investigators thought that the molecules responsible for each of these phenomena were distinct. In 1913, Hans Zinsser[46] demonstrated that antibodies capable of precipitating an antigen could also activate the complement system. Zinsser proposed the unitarian hypothesis of antibodies which states that antigen induces production of antibody that can perform a variety of the *in vitro* functions attributed previously to different molecules.

When Zinsser proposed the unitarian hypothesis, little was known about the structure of antibodies. In 1915, Oswald Avery demonstrated that antibody activity is present in the globulin fraction of serum.[47] In 1937, Michael Heidelberger and Kai Pedersen concluded that antibodies are modified serum proteins.[48] Elvin Kabat and Arne Tiselius demonstrated that antibody activity is present in two molecular forms as detected by serum electrophoresis; this was the initial description of different antibody molecules (isotypes) IgG and IgM.[49] Subsequent investigations by several immunochemists identified three additional isotypes—IgA by Joseph Heremans and collaborators in 1959,[50] IgD by David Rowe and John Fahey in 1965,[51,52] and IgE by Kimishige Ishizaka and co-workers in 1966[53] and independently by Gunnar Johansson and Hans Bennick in 1967.[54]

Rodney Porter[55] and Gerald Edelman[56] independently reported the chemical structure of antibody in 1959; these results led to the current four-chain model of the molecule originally proposed by Julian Fleischman and collaborators in 1963.[57] This model allowed immunologists to identify the location of the antigen-binding site as well as to determine the molecular areas responsible for differences between isotypes. Edelman and Porter received the Nobel Prize in Physiology or Medicine for their work in 1972.

Functional Capabilities of Antibodies

Antibody eliminates foreign antigens including extracellular bacteria by one of four mechanisms: neutralization, activation of the complement system, opsonization, and participation in antibody-dependent cell-mediated cytotoxicity (ADCC). In 1888, several investigators reported that serum taken from rabbits injected with *B. anthracis* killed or neutralized the bacteria *in vitro*.[42] These observations stimulated other investigators to employ antibodies to decrease the pathology caused by several different microorganisms. von Behring and Kitasato[29,43] applied this approach successfully to reverse the toxicity of *C. diphtheriae*. Other examples of antibody-mediated neutralization include vaccines directed to infectious microorganisms (ie, oral polio) and several of the biologics (ie, anti-TNF) that are used to decrease chronic inflammation.

Complement activation was one of the first mechanisms described by which antibody could protect against infectious microorganisms. George Nuttall in his original description of a substance in the serum derived from immunized animals showed that heating the serum destroyed the antibacterial activity. Similarly, Hans Buchner diluted serum from immunized animals with water and obtained two fractions: one that precipitated and the other that remained soluble.[42] Neither fraction alone contained the antibacterial activity of the serum; however, if the two fractions were combined, the antibacterial activity was restored.

In 1894, Richard Pfeiffer and co-workers presented additional evidence for the presence of a nonantibody component required for efficient killing of bacteria.[58,59] These investigators injected guinea pigs with cholera bacteria to induce specific antibodies. Serum from these animals inhibited the movement of the bacteria but failed to kill them. When antibody coated bacteria were injected into the peritoneal cavity of nonimmune guinea pigs, the bacteria disintegrated. This Pfeiffer phenomenon provided evidence that antibody required a complementary system to effectively eliminate bacteria.

Jules Bordet investigated guinea pigs infected with *V. cholera*. He demonstrated that serum from guinea pigs injected with *V. cholera* killed the bacteria. This activity was destroyed by heating the serum. Bordet further showed that the heated serum agglutinated the bacteria without killing them. Bordet concluded that the serum contained two substances that worked in concert to kill the microorganism—one, stable to heating, he called sensibilatrices (antibody) and the other, sensitive to heating, he called alexine (complement).[60,61]

In 1898, Bordet confirmed this conclusion when he treated erythrocytes with immune serum heated to destroy complement. The addition of nonheated control serum to these erythrocytes resulted in the lysis of the red cells and the release of hemoglobin.[62] This hemolysin test provided a method for the quantitation of complement in serum. Using this approach, Bordet showed that alexine is not antigen-specific and that erythrocytes from different species express different antigens on their surfaces.

Complement consists of at least 30 serum proteins. A group of 11 of these proteins act sequentially to trigger cell lysis and induce inflammation. The remaining proteins in the system act to regulate this activation. As early as 1907, A. Ferrata working in Paul Ehrlich's laboratory separated serum into two fractions by dialyzing it against water. Some of the proteins precipitated (euglobulins) while others remained soluble (pseudoglobulins). Neither of these fractions by themselves caused antibody-coated erythrocytes to lyse, although a combination of the two restored the lytic activity.[63] Further evaluation of these two fractions by E. Brand[64] demonstrated that the euglobulin interacted with the antibody on the red cell while the pseudoglobulin interacted with the antibody-euglobulin complex to affect lysis.

Several investigators described a third component of the system. L. Omorokow in 1911[65] and H. Ritz in 1912[66] both showed that treatment of serum with cobra venom removed a component that is now termed C3. In 1914, Arthur Coca reported[67] that treatment of serum with yeast removed this same complement component. A fourth component of the system, C4, was discovered by John Gordon and colleagues in 1926. These investigators added guinea pig serum to antibody-coated red blood cells, resulting in the lysis of the cells. Treatment of the serum with ammonia abrogated this activity. When ammonia-treated serum was mixed with serum that had been heated to 56 °C, the lytic activity was restored. These results suggest that ammonia inactivates a heat-stable component of the system.[68]

By the mid-1920s, immunochemists had identified four components of complement (C1-C4) but were not sure how these components interacted. While some advances were made in the interim, Manfred Mayer finally unraveled the order of the reaction in 1954.[69] Mayer added isolated complement proteins in various sequences to antibody-coated erythrocytes along with the addition of Ca^{++} and Mg^{++} as enzyme catalysts. These studies demonstrated that C1 and C4 interact with antibody bound to red blood cells in the presence of Ca^{++} followed by the activation of C2 in the presence of Mg^{++}. This complex then reacts with C3, leading to hemolysis.

Identification of the other complement components occurred during the 1950s and 1960s as a result of the introduction of several new techniques including ion exchange chromatography, gel filtration based on molecular size, electrophoresis, ultracentrifugation, radioisotope labeling, immunodiffusion, and availability of antibodies specific for isolated proteins. In 1963, Irwin Lepow and colleagues separated the first component of complement (C1) using DEAE (diethylaminomethyl) cellulose into three proteins (C1q, C1r, and C1s).[70] C1q interacts with the Fc portion of antibody and initiates the activation sequence. This is followed by activation of C1r and C1s.

In 1960, Hans Muller-Eberhard and colleagues isolated a β_2-globulin that lysed erythrocytes pretreated with antibody and the C1, C4, and C2 components of complement.[71] At first, this protein was called C3. Further study demonstrated that this globulin comprised six individual proteins. During the next decade, scientists isolated each of these proteins and demonstrated that they interacted with the early complement components. Accordingly, they were designated C3, C5, C6, C7, C8, and C9.

By the end of the 1960s, investigators had identified the proteins involved in antibody-mediated complement activation (the classical path). Between 1954 and 1972, several investigators noted that complement could be activated in the absence of antibody. Subsequent studies revealed two additional mechanisms of complement activation, the alternate (properdin) and the lectin pathways. These latter two mechanisms of complement activation are evolutionarily older than the classical pathway and are essential in protecting the host against infectious microorganisms.

Louis Pillemer and his co-workers are primarily responsible for our understanding of the alternate pathway. These investigators treated human serum with zymosan, a polysaccharide that had previously been shown to inactivate C3. Their attempt to elute C3 from the zymosan failed. Further investigation showed that zymosan inactivated C3 most likely by enzymatic cleavage. Mixing serum with zymosan at 37 °C followed by centrifugation resulted in a decreased amount of C3 in the supernatant. When the same experiment was performed at 17 °C, the amount of C3 was not decreased. Moreover, increasing the temperature of the supernatant to 37 °C and adding more zymosan failed to decrease the C3 activity.[72,73]

Pillemer and his colleagues interpreted these results by proposing the existence of a serum protein, properdin, that binds zymosan.[74] The complex of zymosan and properdin inactivates C3 at body temperature but not at 17 °C. Pillemer isolated properdin by treating this complex with a high ionic strength solution and demonstrated that it is a protein but is not an antibody or a component of the classical complement system. He postulated that properdin binds pathogens and activates the C3 component of complement, which then completes activation of the terminal components (C5-C9) to cause lysis and initiate an inflammatory response.

The initial description of the properdin (alternate) pathway of complement activation led to controversy. In 1958, Robert Nelson presented data that contradicted Pillemer conclusion.[75] Nelson proposed that normal serum contained a naturally occurring antibody to zymosan and that interaction of zymosan with this antibody inactivated the classical pathway of complement. This interpretation was accepted by much of the immunological community at the time despite Nelson being incorrect about the presence of the antibody. The scientific controversy may have contributed to Pillemer's death in 1957 at age 49 due to an overdose of barbiturates.

Subsequent investigators showed that an alternate pathway of complement activation exists and that Pillemer was essentially correct. While properdin does not bind to pathogens, it plays a role in this path by stabilizing spontaneously formed C3 convertase on the surface of a pathogen. This enzyme then activates C5 and the other terminal components of the complement system.

In the early 1970s, several investigators reported that serum binds and activates C1 in the absence of antibody.[76-80] Complement activation requires interaction of pathogens with naturally occurring lectins (carbohydrate-binding proteins) rather than with antibody. Two of these lectins, mannose-binding lectin and ficolin, recognize pathogen-associated carbohydrates and induce lectin-mediated complement activation.

Antibodies initiate two additional effector mechanisms—opsonization and ADCC. Almroth Wright studied phagocytosis of *Staphylococcus* by human white blood cells in 1903.[81,82] He and his co-workers observed minimal phagocytosis by these cells in the absence of serum while the addition of serum enhanced the uptake of the bacteria. They termed the serum substances responsible for this enhancement opsonins. Further study revealed two opsonins, one found in serum from unimmunized individuals that is heat-labile (destroyed by heating at 60 °C-65 °C for 10-15 minutes) and the other a heat-stable opsonin present in the serum of individuals exposed to *Salmonella*. Heat-labile opsonins are complement components while heat-stable opsonins are antibodies.

In 1965, Erna Möller studied lymphocyte-mediated cytotoxicity and described a previously unknown effector mechanism, ADCC.[83] This phenomenon occurs when cells (ie, bacteria, virally infected somatic cells, tumor cells) are coated by specific antibodies and exposed to a unique population of cytotoxic lymphocytes (natural killer lymphocytes) that exist constitutively in healthy animals and are not increased in number by immunization or infection.

Möller cultured lymphocytes activated against an MHC antigen with sarcoma cells expressing the same MHC molecules. The level of lysis of the sarcoma cells was increased by the addition of immune serum but not nonimmune serum to these cultures. Möller incorrectly hypothesized that the antibodies in these cultures increased contact between effector T lymphocytes and the sarcoma cells. The correct interpretation of these data came in 1968 when Ian MacLennan and G. Loewi studied immunologic nonspecific cytotoxicity of a human cell line (Chang cells) *in vitro*.[84] Like Möller they demonstrated that nonspecific cytotoxicity is increased when antibody directed to the target cells is included in the cultures.

DISCOVERY OF IMMUNE-MEDIATED HYPERSENSITIVITIES

Antibodies and T lymphocytes are crucial in protecting vertebrates against many pathogens. Unfortunately, antibodies and T lymphocytes also initiate some pathologies including hypersensitivity reactions and autoimmune diseases.

In the late 1800s and early 1900s, clinicians observed pathologies (hypersensitivities) related to abnormal immune responses. By the early 1960s, a variety of such hypersensitivities had been described. In 1963, Robin Coombs and Phillip Gell devised a classification system based on the mechanisms responsible for the pathology.[85] Based on this classification system, immunologists identify four types of hypersensitivity, three of which involve the activity of antibodies while the fourth involves T lymphocytes.

Type I Hypersensitivities

In 1894, Emil von Behring reported that some animals repeatedly injected with antibody specific for diphtheria toxin developed cardiovascular or pulmonary distress.[86] He called this reaction hypersensitivity. Other investigators reported similar observations, including death, in animals injected repeatedly with small doses of other foreign antigens. Originally, they attributed these reactions, erroneously, to an increase in the sensitivity of the animals to the toxicity of the antigens.

In 1902, two French scientists Paul Portier and Charles Richet described the mechanism of this reaction.[87] Prince Albert I of Monaco, an amateur scientist, outfitted a series of yachts with scientific equipment and invited scientists to investigate topics of interest to the Prince. In 1901, the Prince requested Portier and Richet to sail on one of these yachts. Prince Albert knew of von Behring's success using antibody to treat patients infected with diphtheria. He asked Portier and Richet to determine if an antibody to the toxin from the Portuguese man-of-war could counteract the response, including breathing difficulties and cardiac arrest, experienced by individuals who were exposed to the toxin.

Portier and Richet prepared an extract of the tentacles of the Portuguese man-of-war and injected it into various animals. They observed respiratory distress and other reactions that duplicated the effect of the antigen itself. At summer's end, the two scientists returned to their laboratories in Paris where they continued this work. Since they were unable to obtain Portuguese man-of-war, they switched to a similar toxin from the sea anemone, *Anemonia sulcata*. Portier and Richet injected dogs with this toxin to induce what they thought would be a protective antibody. None of the dogs showed any adverse reactions. After several weeks, the investigators reinjected the dogs with a second dose of the toxin to assess the level of protection. Many of the dogs now developed sudden onset diarrhea and vomiting and several of them succumbed to this second injection.

Portier and Richet coined the term aphylaxis to differentiate this reaction from phylaxis, a term used to describe antibody's ability to protect against infection; aphylaxis was subsequently changed to anaphylaxis. Charles Richet received the Nobel Prize in Physiology or Medicine in 1913 "in recognition of his work on anaphylaxis."

The mechanism responsible for anaphylaxis was unknown in the early 1900s and investigators proposed two disparate hypotheses—a humoral theory that postulated the release of a cleavage product from the toxin more rapidly on second injection[88] and a cellular theory that proposed the synthesis of an antibody taken up by cells in the target organs prior to the second challenge.[89] Attempts to detect specific antibody in the serum of injected animals met with failure, although the transfer of the reaction from sensitized to control animals was sporadically reported.[90]

In vitro models of anaphylaxis produced more convincing results. In 1910, Werner Schultz developed an *in vitro* model that eventually led to understanding the mechanism of tissue damage induced by anaphylaxis. Schultz isolated muscle strips from the ileum of guinea pigs injected with horse serum and placed them in isotonic saline. The addition of horse serum (but not serum from other species) to these preparations induced muscle contraction.[91] In 1911, Henry Hallett Dale and Patrick Laidlaw used a similar model to demonstrate the role of histamine in the reaction. These investigators used uterine muscle from nonsensitized guinea pigs to which they added histamine. These muscles contracted identically to muscles from sensitized animals to which they added the sensitizing antigen.[92]

Despite never collaborating, the phenomenon that these investigators studied became known as the Schultz-Dale reaction. Investigators used this *in vitro* model for almost 50 years to study antibody-mediated type I hypersensitivity.

In 1921, Otto Prausnitz and Heinz Küstner reported the serum-mediated transfer of type 1 hypersensitivity from an allergic to a healthy individual. They collected serum from Küstner, who was allergic to fish, and transferred it subcutaneously to Prausnitz. Twenty-four hours later, a fish extract applied to Prausnitz' skin elicited a wheal and flare like that seen in allergic individuals.[93] This P-K reaction proved useful for the study of the pathophysiology of the reaction in humans and, in fact, was used 40 years later to identify the role of IgE antibodies in ragweed pollen allergy.[53,54]

Two groups of investigators are credited with identifying the isotype of antibody responsible for the allergic reaction. Kimishige Ishizaka, working with his wife Teruka and colleagues, isolated antigens from ragweed pollen and identified the most potent of the antigens which they called antigen E. They subsequently sought antibodies that would bind this antigen from serum of highly allergic individuals. They isolated this antibody and characterized it demonstrating that it was different from the other know isotypes—accordingly they called it IgE.[53]

In 1967, Gunnar Johansson and Hans Bennich described a myeloma protein that failed to react with antibodies to the heavy chains of IgG, IgM, IgA, or IgD. The protein contained light chains identical to those found in these other isotypes. The heavy chain differed in that it had a molecular weight of approximately 200,000 Da and a sedimentation rate of 8S. They called this protein ND (from the patient's initials).[54]

In 1968, an international meeting to compare protein ND and IgE concluded that the two are identical and the participants agreed that this new immunoglobulin isotype should be called IgE.[94]

Type II Hypersensitivities

Type II hypersensitivities involve the action of antibodies directed against one's own cells or tissues. Many of these antibodies are important mediators of autoimmune diseases. In 1900, Karl Landsteiner demonstrated the clinical utility of antibodies specific for cells when he described the ABO blood groups.[95] He mixed blood from different individuals and noted that in some of the mixtures, red cells agglutinated. Further study revealed that the agglutination resulted from the presence of naturally occurring antibodies in one of the individuals that bound to antigens expressed by the second individual's erythrocytes. By testing many blood samples, Landsteiner identified three groups of individuals based on their blood types—A, B, and O. Blood type AB was discovered a year later. Landsteiner also showed that transfusion of blood between two individuals succeeded if they shared blood types. This resulted in the clinical use of blood transfusions, a major improvement in survival in surgery patients.

Investigators described additional blood group antigens in the 1920s and 1930s. None of these induce naturally occurring antibodies and thus are not clinically important unless a patient receives multiple transfusions. In 1939, Phillip Levine and Rufus E. Stetson described a 25-year-old woman who delivered a stillborn infant during the 33rd week of gestation. Following delivery of the baby, the woman required a transfusion; since she and her husband were both blood group O, he served as the blood donor. Almost immediately the woman suffered a transfusion reaction. The next day she received a compatible blood transfusion from an unrelated donor who was blood group O without any adverse reaction. Levine and Stetson subsequently demonstrated that the woman's serum contained antibodies that agglutinated her husband's red blood cells (as well as the red cells of many other blood group O donors).[96]

Levine and Stetson hypothesized that the antibody responsible for both the death of the infant and the transfusion reaction was induced by antigens present in the fetus that had been inherited from the father. This antibody appeared to be the same as one produced by rabbits injected with red cells from rhesus monkeys.[97] Accordingly, the antigen was named rhesus factor (Rh)—subsequent studies revealed that the two antibodies differed however the nomenclature remained.

Hemolytic disease of the newborn (HDN), induced by an incompatibility at the Rh locus, leads to a type II hypersensitivity. Antibodies formed by an Rh-negative woman pregnant with an Rh-positive fetus cross the placenta, bind to the fetal red cells, and cause their destruction. The release of hemoglobin results in a buildup of bilirubin, resulting in a variety of symptoms ranging from jaundice to death. Levine studied numerous cases of HDN and reported in 1943 that severe disease occurred more commonly when mother and father were ABO compatible than when they were incompatible.[98] In 1956, Israel Davidsohn and colleagues injected Rh-negative male volunteers with Rh-positive erythrocytes and demonstrated that more antibody to the Rh antigen was produced if the donor and recipient were ABO compatible than if they were incompatible.[99] The observation that ABO incompatibility provided some protection against formation of antibodies to the rhesus antigen led directly to the postulate that HDN might be prevented by treating the mother with antibodies specific for the Rh antigen, thus removing the antigenic stimulus prior to induction of antibody synthesis.

In 1961, Ronald Finn and colleagues demonstrated that injecting Rh-negative volunteers with Rh-positive cells followed by antibodies to the Rh antigen produced a significant decrease in the number of Rh-positive cells circulating as soon as 2 days after the injections.[100] Vincent Freda and colleagues[101] showed that the injection of antibodies to the Rh antigen also decreased subsequent production of Rh antibodies by Rh-negative individuals injected with Rh-positive cells. These results led to clinical trials in which Rh-negative women who delivered their first Rh-positive baby were injected with antibody to the Rh antigen. Initial results reported by Freda and collaborators in 1967 demonstrated that none of the injected women synthesized antibodies to the Rh antigen while almost 15% of the controls became sensitized. When these women became pregnant for a second time, none of the babies born of injected mothers developed HDN while 11 of 27 babies born of control mothers developed the disease.[102]

Autoimmune diseases comprise other examples of type II hypersensitivities. In 1904, Julius Donath and Karl Landsteiner identified antibodies that bound a patient's red blood cells in the extremities where the temperature is below 37 °C. The antibodies activate complement and destroy the red cells when the blood returns to the body core.[103] Other examples of antibody-mediated autoimmune disease include myasthenia gravis, Grave disease, and multiple sclerosis.

Type III Hypersensitivities

Investigators described type III hypersensitivities as an adverse reaction to serotherapy. In 1905, Clemens von Pirquet and Bela Schick reported that some patients treated with horse antibodies specific for diphtheria toxin developed a (usually) self-limiting disease involving fever, skin rashes, joint pain, and (occasionally) kidney damage.[104] These symptoms occurred 10 to 14 days following injection with the foreign antibody. Von Pirquet and Schick observed similar reactions in patients who received a second injection of the antitoxin and termed these reactions serum sickness.

In 1911, von Pirquet realized that these symptoms occurred simultaneously with the appearance of antibodies against the foreign serum.[105] He reasoned that the antibodies combined with the foreign serum and formed a "toxic compound" that caused the disease. The identification of this "toxic compound" required 50 years of further study and the development of radioactive tracers. In 1951, David Talmage and colleagues[106] injected rabbits with radiolabeled bovine gamma globulin (BGG) and followed the elimination of the radioactive antigen over time. Simultaneously, they measured the appearance of antibodies to BGG in the serum of the injected rabbits and correlated the rate of antigen elimination with the appearance of antibody. The results indicate that antibodies appear during the rapid phase of antigen elimination. The antibodies formed might explain the occurrence of serum sickness seen in patients injected with horse antibodies and led to the concept of immune complex–mediated disease.

Immune complexes cause pathology in several other diseases. These include some autoimmune diseases (ie, systemic lupus erythematosus), drug hypersensitivities (ie, allopurinol), and some allergies. In addition, the antibodies formed to certain infectious microorganisms may lead to type III hypersensitivity. Study of sequelae of group A streptococcal infection provided compelling information about the mechanism of this pathology.

In the first decade of the 20th century, Schick and von Pirquet independently suggested that the kidney pathology seen in patients with scarlet fever following streptococcal infections mimics the pathology observed in patients suffering from serum sickness induced by injection with horse antibodies to diphtheria toxin.[107,108] In 1961, Frank Dixon and colleagues developed a model to study immune complex–mediated disease when they injected rabbits daily with one of several foreign proteins.[109] Some of these rabbits developed clinical signs and morphological changes like those seen in patients with glomerulonephritis. The severity of the pathology correlated with the concentration of antibodies in the rabbit's serum. Immunohistochemical staining of kidney biopsies revealed that complexes of foreign protein along with specific antibodies were deposited along the glomerular basement membranes.

Type IV Hypersensitivities

In 1921, Hans Zinsser described hypersensitivity reactions now classified as type IV.[110] Zinsser compared the skin reactions of guinea pigs injected intraperitoneally with either horse serum or an extract of *Mycobacterium tuberculosis*. Animals primed with horse serum reacted to an intradermal injection with a rapidly appearing skin reaction that was like the reaction seen in other antibody-mediated responses. By contrast, the skin reaction induced by an intradermal injection of *M. tuberculosis* developed more slowly (48-72 hours). Zinsser demonstrated that this tuberculin reaction is not due to the presence of serum antibodies.

For 20 years following the initial description of the tuberculin reaction, investigators studied the mechanism of the skin reaction. Many of these studies involved transfer of serum or skin-bound antibodies from sensitized animals to naïve animals. The results were inconclusive. Finally, in 1942, Merrill Chase and Karl Landsteiner demonstrated that the reaction could be transferred by cells isolated from a sensitized animal but not by serum.[111]

Chase and Landsteiner employed a model of delayed hypersensitivity induced by synthetic chemicals such as picryl chloride, 2,3-dinitrochlorobenzene, and *o*-chlorobenzoyl chloride. They attempted several protocols to transfer this reactivity to naïve animals using antibodies without success. Chase postulated that the antibodies might be cell-bound and attempted to isolate such antibodies from cells derived from the peritoneal cavities of the sensitized animals. Once again, the experiments failed. However, in one experiment, Chase accidently used a preparation that was contaminated with some of the cells from the peritoneal cavity. Injection of this cell containing preparation into nonsensitized animals resulted in positive skin reactions.[112]

Publication of these studies came at a time when the identity of the antigen-reactive cell was unknown. The prevailing hypothesis in the 1940s was that antibodies were formed by cells of the reticuloendothelial system (phagocytic cells including macrophages, monocytes, specialized endothelial cells, and reticular cells). Conclusive evidence that the antigen-reactive cell is the lymphocyte only derived from the seminal investigations of James Gowans in the 1950s.

LYMPHOCYTES AS ANTIGEN-REACTIVE CELLS

As a young investigator, Gowans joined the laboratory of Howard Florey who suggested that Gowans investigate the "disappearing lymphocyte phenomenon." This phenomenon referred to the observation that chronic drainage of the thoracic duct depletes an animal of most of his lymphocytes.[113] Gowans performed several experiments during the late 1950s and early 1960s that demonstrated that lymphocytes migrate between the peripheral blood and the thoracic duct[114,115] and that these same cells populate lymph nodes and other lymphatic tissues.[116]

Further studies in the 1960s revealed the functions of lymphocytes in the immune response. Jacque Miller reported that removal of the thymus of mice during the neonatal period depressed the ability of the thymectomized animals to respond to antigenic challenge.[117] Gowans theorized that a similar immunodeficient state would be achieved by chronic thoracic duct drainage. Adult rats depleted of lymphocytes by chronic thoracic duct drainage had decreased numbers of peripheral blood lymphocytes, a reduction of lymph node weight, and a lymphopenia in their spleen and lymph nodes.[118] Subsequent studies showed that these lymphopenic rats responded to antigens (sheep erythrocytes or tetanus toxoid) with decreased antibody production. The antibody response could be restored by injecting the depleted rats with lymphocytes derived from thoracic duct lymphocytes from intact, syngeneic rats.

Parallel investigations demonstrated that thoracic duct lymphocytes participate in allograft rejection and in the initiation of graft-versus-host reactions.[119,120] Although the division of lymphocytes into separate T and B cells had not yet been described, these studies from Gowans' laboratory provided knowledge that lymphocytes are immunocompetent. The next question to be addressed by immunologists was the mechanism by which lymphocytes recognize antigen. This question intrigued several early investigators including Paul Ehrlich who proposed a solution with the publication of his side-chain theory in 1900.[121]

The side-chain theory postulated that the cells responsible for antibody formation expressed surface receptors that mimicked the antibodies those cells could produce. Pathogens bound these receptors causing the cell to release the side chains. This in turn induced the cell to produce more molecules with the identical configuration that were then secreted into the blood to act as antibodies.

When proposed, most investigators criticized this idea in part because the structure of antibodies was unknown. In addition, the heterogeneity of antibodies that an individual animal could synthesize was not appreciated. Most antibodies were cataloged based on the functional assays used to detect them—precipitation, hemagglutination, hemolysis, etc. While these initial descriptions included only antibodies specific for potential pathogens, Karl Landsteiner and others demonstrated subsequently that antibodies could be produced against an enormous variety of foreign substances including synthetic molecules.[122] These observations led many investigators to question the validity of a model such as that proposed by Ehrlich that required the immune system to synthesize large numbers of potential antibodies prior to experiencing antigen that would bind it.

As a result, many immunologists in the first half of the 20th century believed that antigen provided a template that instructed the cells of the immune system to produce specific antibody. These instruction models hypothesized the presence of intracellular antibody precursor molecules that were altered by the antigen to allow the antibody to bind antigen. These changes were then somehow communicated to the antibody-forming cell to allow the synthesis of additional, specific antibodies.[123,124]

Several characteristics of the antibody response argued against instruction theories including the synthesize of increased antibody during a memory response, the increase in average avidity of the antibody as the response matures, the rate of antibody formation, the production of antibodies in the absence of antigenic stimulation (so-called natural antibodies), the amount of antibody produced, the occurrence of a lag phase prior to the appearance of antibodies in the serum, and the lack of antibodies directed at self.[125,126] Similarly, the understanding that the antibody-binding capacity of the molecule is dependent on the amino acid sequence of the binding site and the realization that amino acid sequence is determined by immunoglobulin genes resulted in an abandoning of instruction theories by most, but not all, investigators.[127]

CLONAL SELECTION HYPOTHESIS

Immunologists developed three new models to explain antibody formation during the 1940s and 1950s. The modified enzyme concept proposed by Macfarlane Burnet and Frank Fenner in 1949[125] is based on the Nobel Prize winning studies performed by Jacques Monod, Francois Jacob, and Andre Lwoff on inducible enzymes in bacteria and yeast. Burnet and Fenner hypothesized that phagocytes which remove antigen contain enzymes involved in degrading dead cells and cellular debris. When these phagocytes detect and engulf antigen, new enzymes are induced that are specific for the antigen and help degrade it. These modified enzymes are secreted after which they are modified further to act as antibodies. This hypothesis explained the continued production of antibody in the absence of antigen and the memory response. To explain tolerance to self, Burnet and Fenner proposed that phagocytes contained a self-marker enzyme that could not be modified into antibody.

Niels Jerne proposed a natural selection hypothesis of antibody formation in 1955.[126] This is the first model to propose a selection function rather than an instruction function for antigen. Jerne's hypothesis postulated that the serum of an animal contained antibodies of every specificity that it could make. These antibodies were spontaneously synthesized during the production of serum globulins. Exposure of the animal to a pathogen led to binding of antigens associated with the pathogen with one or more of these natural antibodies. Phagocytosis of the antigen-antibody complexes resulted in a signal to the cell to synthesize additional globulin molecules possessing the identical binding capability. Although this model answered several of the criticisms of the instruction theories, it was rejected by many immunologists due to the requirement for the synthesis of many different antibodies in the absence of antigenic challenge.

The models developed by Burnet and Fenner and by Jerne both depended on the antibody-forming cell being a phagocytic cell. This is a natural extension of the earlier instruction theories; however, studies performed between 1930 and 1960 on the identity of the antibody-producing cells by Gowans and others focused on the lymphocytes as the important mediator of immune responses. Stephen McMaster and Phillip Hudack demonstrated in 1935 that antibody appeared to be produced in the lymph nodes,[128] a structure composed primarily of lymphocytes. Seven years later, William Ehrich and T. N. Harris showed that injection of antigen into rabbit footpads induced morphological changes, including an increase in lymphocyte numbers in the draining lymph node that correlated with the appearance of specific antibodies in the cells as well as in the draining lymph.[129] These same scientists demonstrated that macrophages contain insignificant amounts of antibodies during the height of the response to antigen.[130]

Building on these results, Astrid Fagraeus investigated the cellular source of antibodies and demonstrated that plasma cells produced most antibody.[131] Wilhelm von Hartz Waldeyer introduced the term plasma cell in 1875 based on the histological appearance of the cell.[132] These cells have an eccentrically placed nucleus with a cart-wheel pattern of chromatin along with a clear area near the nucleus that accommodates an expanded Golgi apparatus in activated cells. While Fagraeus originally concluded that plasma cells derived from the reticuloendothelial system, passive transfer studies performed by Harris and colleagues demonstrated that lymphocytes derived from the lymph nodes of antigen-sensitized animals and injected intravenously into naive animals resulted in the production of specific antibodies in the recipients.[133]

Introduction of bacterial immobilization and immunofluorescence provided confirmation that many antibodies are synthesized by plasma cells.[134,135] Finally, Gowans and his group demonstrated that small lymphocytes differentiate into plasma cells as they are stimulated to synthesize antibodies.[118,119]

As these studies proceeded, Burnet developed the clonal selection theory. He initially published an outline of this theory in 1957[136] and expanded on it in a series of lectures given at Vanderbilt University in 1959.[137] Burnet postulated that antigen selects the antibody to be produced by binding to a population of preexisting antibody molecules expressed on the surface of lymphocytes circulating in the peripheral blood or in tissues. The binding of antigen to the cell surface molecules induces the lymphocytes to synthesize additional molecules of an identical specificity that are then secreted to eliminate the pathogen.

Burnet envisioned that the antigen-binding receptors expressed on lymphocytes are produced prior to antigenic stimulation and that a given lymphocyte expresses antibodies of a single specificity. To explain immunologic tolerance, Burnet proposed that lymphocytes go through a development stage during which they are exquisitely sensitive to elimination if their antibody receptors bind self-antigens. This hypothesis helped explain the experiments published in 1953 by Billingham and co-workers who demonstrated that mice could be manipulated to accept an allograft if they were injected with foreign lymphocytes during fetal life or early in the neonatal period.[138]

Simultaneously with the publication of Burnet's clonal selection theory, David Talmage published a similar theory.[139] He modified Ehrlich's side-chain hypothesis and proposed that antigen selected cells for activation based on their ability to bind to these side chains. This proposal introduced the concept that antigen selection occurs at the level of the cell. While it has been argued that Talmage should receive more credit for developing the clonal selection theory (Burnet was awarded the Nobel Prize in 1960), it is of note that Talmage did not include the idea that an individual cell produced antibody of a single specificity.

In 1958, Gus Nossal and Joshua Lederberg proposed an experiment designed to disprove the clonal selection theory.[140] They reasoned that if they could demonstrate that antibody-forming cells produced antibodies of more than one specificity, the clonal selection theory would be refuted. These investigators immunized animals repeatedly with two species of *Salmonella*, isolated lymph node cells from these animals and incubated individual cells in microdrops. They assayed these microdrops for antibody by introducing one of the *Salmonella* species. If antibody was detected, they added the second strain to the microdrop to determine if individual cells produced antibody to both strains. In numerous experiments, no cells producing antibodies against both species of bacteria could be detected. The authors concluded that antibody-forming cells were committed to the production of a single antibody specificity.

ANTIGEN RECOGNITION BY ANTIBODY-FORMING (B) LYMPHOCYTES

In 1961, Goran Möller treated mouse lymphocytes with fluorescein-conjugated antibodies to rabbit anti-mouse immunoglobulin.[141] While this was a negative control for other experiments being performed, Möller noted a small percentage of the lymphocytes in his preparations bound to antibody to immunoglobulin. Despite efforts to eliminate this binding, Möller finally concluded that some lymphocytes express immunoglobulin on their surface. During the next decade, several other immunologists confirmed these results and demonstrated that the immunoglobulin molecules expressed on some lymphocytes bind antigen.[142]

Martin Raff and colleagues[143] established that the immunoglobulin receptors on an individual lymphocyte all possess the identical specificity. They treated lymphocytes with fluoresceinated antigen at 37 °C, causing the cell surface antibody molecules to migrate (cap) to one pole of the cell. When they exposed these same cells to an antibody specific for immunoglobulin at 4 °C, no additional cell surface immunoglobulins were detected.

In 1976, Gus Nossal and Beverly Pike isolated B lymphocytes from animals injected with a hapten-carrier complex. They demonstrated that lymphocytes purified by binding to the hapten secreted antibodies specific for the hapten but failed to secrete antibodies to the carrier. These results confirmed the hypothesis that the immunoglobulin receptors expressed on antibody-forming lymphocytes share specificity with the antibody synthesized by that cell.[144]

GENERATION OF ANTIBODY DIVERSITY

The B lymphocyte repertoire as identified by the diversity of cell surface immunoglobulin receptors exceeds one million. During the 1960s and 1970s, a persistent debate occurred among immunologists about the mechanism by which this diversity developed. Leroy Hood and David Talmage (among others) argued that the germline DNA contained a gene for every possible polypeptide chain the individual could synthesize.[145] Alternatively, Niels Jerne[146] argued for the existence of a limited number of genes in the germline that diversified somatically as the B lymphocytes matured. As reviewed by Susumu Tonegawa, both hypotheses had difficulties.[147] The germline theory of Hood and Talmage implied that almost the entire human genome coded for antibodies. In addition, this hypothesis failed to explain the inheritance of immunoglobulin allotypes. The somatic diversification hypothesis of Jerne required a unique genetic mechanism for the generation of diversity. Resolution of this debate depended on studies on the amino acid sequence of immunoglobulin chains as well as on the organization of immunoglobulin genes in the germline and in mature lymphocytes.

By the mid-1960s, investigators sequenced several immunoglobulin light chains from patients diagnosed with multiple myeloma, a malignancy of B lymphocytes. These patients excrete Bence Jones (BJ) proteins in their urine. BJ proteins are light chains identical to those included in the myeloma proteins these patients synthesize and secrete. Comparison of the amino acid sequences of BJ proteins revealed that light chains contain a portion that varies between proteins (the variable (V) portion located in the amino half of the chain) and a constant (C) portion in the carboxy terminus.[148-150] Other investigators reported a similar division of heavy chains into C and V regions.[151,152]

In the 1970s, Susumu Tonegawa and collaborators approached the problem of antibody diversity by hybridizing mRNA specific for a light chain with DNA isolated from liver cells. Tonegawa reasoned that the DNA from liver cells represented the genes in their germline configuration. If the germline hypothesis of Hood and Talmage were correct, the RNA would bind to many genes all containing the C region gene. Alternatively, if Jerene's somatic diversification postulate were correct, the mRNA would bind to one or a few C region genes. Based on the results of this study, the authors concluded that the number of germline genes was too small to account for the variability in V regions.[153] In additional studies, Tonegawa and his group revealed that the germline DNA contained a limited number of light chain genes and that variability in antibody genes resulted from a somatic diversification mechanism in both heavy and light chains.[154,155]

The conclusion that one or a few immunoglobulin genes are inherited that then undergo somatic diversification raised a question about how variability could be introduced into nucleotides coding for part of a protein (the V region) while sparing another portion of the molecule (the C region). One possible solution came from the speculation made by William Dreyer and J. Claude Bennett in 1965.[156] These immunologists hypothesized that a single immunoglobulin polypeptide chain may be coded by two separated genes.

At the time this was a highly unorthodox hypothesis and investigators produced a variety of alternative hypotheses to explain these findings. Among these was the suggestion by Sydney Brenner and Cesar Milstein of a mechanism that inserts nongermline mutations into the V regions of antibody genes.[157]

In the mid-1970s, Tonegawa and collaborators analyzed DNA coding for immunoglobulin chains with restriction enzymes to test the Dreyer-Bennett hypothesis. They compared DNA fragments derived from murine myeloma cell lines with similar fragments obtained from embryonic tissue. When hybridized with mRNA probes specific for either the entire light chain or the C region of the light chain, two different patterns emerged. The embryonic DNA had two components selected by the probes—one that hybridized with the C region sequences and the other that hybridized with V gene sequences. DNA from myeloma cells had a single component that hybridized with both V and C probes.[158] These results validated the hypothesis that V and C regions are coded by separate genes that are joined during differentiation of B lymphocytes to code for a single polypeptide chain. Based on these studies, Susumu Tonegawa was awarded the Nobel Prize in Physiology or Medicine in 1987.

Additional study in the 1970s revealed that V and C regions of both heavy and light chains are coded for by two separate genes. A third gene segment, a joining (J) gene, exists that codes for 10 to 15 amino acids between the V and C regions. Finally, in the genes coding for the heavy immunoglobulin chains, a separate diversity (D) gene exists that codes for a short segment of amino acids between the V and J regions.[159-161]

IDENTIFICATION OF TWO POPULATIONS OF LYMPHOCYTES

A convergence of clinical observations and basic science investigations during the 1950s and 1960s lead to our current division of lymphocytes into two main populations, B (responsible for synthesis of antibody) and T (responsible for cell-mediated immunity and regulation of the immune system).

In 1952, Ogden Bruton described an 8-year-old boy with a long history of recurrent infections.[162] Testing the boy's serum with the newly introduced technique of electrophoresis resulted in the finding that he lacked detectable gamma globulin. Since antibodies were known to be gamma globulins,[49] Bruton postulated that the boy lacked antibodies and antibody-forming cells. Confirmation of this conclusion came when Bruton passively transferred immunoglobulins from normal donors that alleviated the patient's immunodeficiency.

Four years later, Bruce Glick removed a hindgut, lymphoepithelial organ from newly hatched chickens to ascertain its function.[163] This organ, the bursa of Fabricius, exists in many avian species but not in mammals. When challenged with *Salmonella* type O antigen, these bursectomized chickens failed to produce antibodies. These results, published in the journal *Poultry Science* after initially being rejected by *Science*, were unknown to most immunologists until other scientists published confirming studies almost 5 years later.

In the early 1960s, Jacques Miller demonstrated that a second lymphoepithelial organ, the thymus, plays an important role in the maturation of the immune response.[117] Miller originally studied the role of the thymus in development of virally induced leukemia in mice. Mice injected with virus at birth and thymectomized at weaning failed to develop leukemia. When Miller transplanted thymus cells into these mice several months later, leukemia developed.[164]

Continuing these studies, Miller thymectomized additional mice at various ages prior to virus injection. While not relevant to the studies on leukemia, a key finding of this work was that approximately 50% of the mice thymectomized at birth died within 2 to 4 months whether they were injected with virus or not. Evaluation of mice thymectomized during the first day of life revealed decreased numbers of peripheral blood lymphocytes as well as diminished histologic development of lymph nodes. In addition, many of the mice thymectomized in the neonatal period accepted foreign skin grafts indefinitely.[117]

Observations of immunoincompetent patients made in the early 1960s paralleled these results. Christian Nezelof reported a case of a patient who presented with a congenital absence of the thymus in 1964.[165] Thymic dysplasia (Nezelof syndrome) is characterized by severe recurrent infections primarily with opportunistic microorganisms including viruses and fungi. These patients lack peripheral T lymphocytes and any thymic tissue present lacks Hassall corpuscles.

Angelo DiGeorge and colleagues reported on another patient lacking a functional thymus in 1967.[166] Patients with DiGeorge syndrome (thymic aplasia) present at birth with life-threatening tetany due to congenital absence of the parathyroid glands. While the tetany can be controlled by exogenous calcium, these patients suffer from recurrent infections. All four of the original patients described succumbed most likely due to microbial infections. Autopsies revealed a total lack of thymic tissue. These patients lack T lymphocytes, fail to mount a normal delayed-type hypersensitivity reaction, and fail to synthesize antibodies following injection with various antigens.

In 1964, Noel Warner and Alex Szenberg first proposed the existence of two functionally different populations of lymphocytes.[167] They postulated that those lymphocytes dependent on the bursa of Fabricius produced antibodies while those dependent on the thymus participated in cell-mediated immune responses including graft rejection. Max Cooper investigated this hypothesis and integrated data from clinical immunodeficiency syndromes with experimental observations.

Cooper and his colleagues studied the immune system of chickens following bursectomy, thymectomy, or both bursectomy and thymectomy.[168] Initially, they confirmed that chickens bursectomized shortly after hatching failed to produce antibodies. However, they were unable to demonstrate a deficit in graft rejection following thymectomy. Addition of sublethal irradiation to the surgical manipulations provided evidence that two distinct immune modalities existed, each mediated by discrete lymphocytes dependent on different organs for development.[169] Comparing these experimental results with clinical observations, Cooper and his coinvestigators divided various immune responses into those dependent on lymphocytes maturing in the bursa (antibody production) and those dependent on the thymus (graft rejection, delayed-type hypersensitivity).[170]

By the late 1960s, several other investigators described two functionally distinct populations of lymphocytes with discrete developmental histories. In 1969, Ivan Roitt and his colleagues reviewed the cellular basis of immunological responses and introduced the terms T lymphocytes (referring to those lymphocytes maturing in the thymus) and B lymphocytes (referring to lymphocytes dependent on the bursa of Fabricius for development).[171]

In addition to the unique responses performed by T and B lymphocytes, evidence accumulated in the late 1960s that some functions of the immune system, particularly optimal antibody synthesis, require interaction between T and B lymphocytes.

LYMPHOCYTE INTERACTIONS IN ANTIBODY RESPONSES

In 1961, Jacques Miller demonstrated that thymectomy of neonatal mice resulted in a deficient immune response.[117] While this affected primarily responses later attributed to activity of T lymphocytes, these mice also had diminished antibody responses to several antigens. Other investigators made similar observations.[172] The explanation of this phenomenon depended on the studies of three independent groups of immunologists.

In 1966, Henry Claman and his colleagues studied sublethally irradiated mice that were reconstituted with lymphocytes from various anatomical sites (thymus, bone marrow, spleen).[173] These investigators immunized the mice with sheep erythrocytes (SRBC) and measured antibody responses. Irradiated mice injected with thymus cells synthesized more antibodies than did irradiated control mice. Addition of adult bone marrow cells to the thymocytes enhanced this antibody response to that seen in unmanipulated animals. Claman concluded that these results supported the hypothesis that bone marrow and thymus cells interact to generate optimal antibody responses to SRBC.

Anthony Davies and colleagues reached a similar conclusion when they transferred lymphocytes to irradiated mice.[174] Optimal antibody response required a mixture of bone marrow cells and thymocytes. This group further demonstrated that the bone marrow–derived cells but not the thymus-derived cells synthesized the antibody.

A third group of immunologists led by Jacques Miller confirmed and extended these observations.[175,176] These investigators reconstituted the ability of neonatally thymectomized mice to produce antibodies to SRBC with an injection of thymus cells or cells from the thoracic duct. Bone marrow cells by themselves did not reconstitute the response. Thymic or thoracic duct cells from syngeneic or semi-allogeneic (F_1 hybrids) reconstituted the response to the same extent. Treatment of spleen cells from the reconstituted mice with antibodies against the histocompatibility antigens of either the host or the donor permitted Miller's group to identify the source of the cells producing the antibody to SRBC. These studies revealed that the antibody-forming lymphocytes derived from the thymectomized mice and not from the cells used for reconstitution. The authors concluded that cells from the thymus or thoracic duct could interact with antigen and induce differentiation of antibody-forming cells derived from the thymectomized host. Similar studies performed by injecting thymectomized, irradiated mice with syngeneic bone marrow identified the bone marrow as the source of the antibody-forming cells.

Avrion Mitchison and co-workers used the response to hapten-carrier complexes to confirm the requirement for T-B lymphocyte collaboration in antibody synthesis. In addition, they demonstrated that both cell types needed to bind the antigen.[177-179] Haptens are simple chemicals that by themselves fail to induce antibodies. When conjugated to protein carriers, experimental animals produce antibodies to the hapten. Mitchison and co-workers transferred lymphocytes from mice primed with the hapten NIP conjugated to ovalbumin (OVA) as a carrier to irradiated, syngeneic recipients. Mice reconstituted with such lymphocytes produced a memory (IgG) antibody response when subsequently injected with NIP-OVA. Similar mice developed a primary (IgM) response when immunized with NIP conjugated to different protein carriers (chicken gamma globulin [CGG] or bovine serum albumin [BSA]). Further studies revealed that recipients reconstituted with lymphocytes from donors primed with NIP-OVA plus either CGG or BSA produced primarily IgM to an injection of NIP-CGG or NIP-BSA.

Additional investigations identified the cells that bound the carrier protein as T lymphocytes while B lymphocytes bound hapten. The investigators hypothesized that T and B lymphocytes involved in initiation of optimal antibody responses needed to be physically linked by the hapten-carrier conjugate.

RECOGNITION OF ANTIGEN BY T LYMPHOCYTES

T lymphocytes interact with antigen with the same degree of specificity as do B lymphocytes; however, the identity of the TCR remained unclear until 1982. Prior to the identification of the TCR, several immunologists described the presence of immunoglobulin-like molecules on the surface of thymocytes and T lymphocytes.[180-182] Detection of the TCR involved two complementary approaches—isolation of cell surface glycoproteins from T lymphocyte clones and lymphomas and identification of genes coding for these proteins.

In the early 1980s, three groups of investigators used similar methodology to isolate putative receptors expressed on T lymphocyte surfaces.[183-185] One group led by James Allison[183] immunized BALB/c mice with lymphomas derived from C57Bl mice. They subsequently used spleen cells from the immunized BALB/c mice to develop hybridomas secreting monoclonal antibodies specific for the lymphoma cells. Reasoning that such a monoclonal antibody would identify the antigen-specific receptor expressed by the lymphoma cells, Allison and his co-workers isolated a glycoprotein from these cells consisting of two polypeptide chains of 39 and 41 kD.

Kathryn Haskins and colleagues developed monoclonal antibodies specific for T lymphocyte clones that interfered with the binding of antigen to the clones[184] while Stefan Meuer and coworkers developed a monoclonal antibody that inhibited the effector function of a cytotoxic clone.[185] Meuer's group used the monoclonal antibody to isolate a cell surface glycoprotein from the surface of the cytotoxic clone. This glycoprotein was composed of two polypeptide chains like the polypeptides reported by Allison's group. Based on these structural studies, the chains of the TCR were termed α and β.

In 1984, two groups of investigators reported the isolation of genes coding for one of the chains of the TCR.[186,187] Both groups cloned DNA complementary to lymphocyte RNA, isolated those clones that were specific for T lymphocytes, and deduced the proteins that could be coded. The gene detected by this procedure coded for a protein hybridized to a region of the genome that was known to be rearranged in a T lymphocyte lymphoma as well as in several T lymphocyte clones. Based on the deduced amino acid sequence coded by this gene, the authors concluded that it coded for the β-chain of the TCR.

Susumu Tonegawa led a group of investigators who isolated the genes coding for two chains of the TCR.[188] Tonegawa's group deduced the amino acid sequence of two polypeptide chains specific for T lymphocyte clones and demonstrated that both chains are expressed and rearranged in T lymphocytes but not in other cell types. Based on these data, this group proposed that the structure of the TCR consists of two chains that are disulfide linked and that are integral membrane proteins. Comparison of the amino acid sequences of TCR molecules from several T lymphocyte clones led to the conclusion that the chains of the TCR contain both variable and constant regions much like the H and L chains of the immunoglobulin molecules that serve as B cell receptors. To further the analogy, the variable regions of these two chains contain the antigen-binding regions of the TCR.

The structure of the α-chain reported by Tonegawa's group lacked sites for N-glycosylation, one of the features of the protein noted by Allison and his colleagues in their original report.[183] Two groups of investigators cloned additional T lymphocyte–specific genes and discovered a gene that coded the α-chain.[189,190]

The gene originally thought to code for the α-chain of the TCR codes for a different T lymphocyte–specific molecule, expressed on approximately 10% of T lymphocytes. This chain designated the γ-chain is linked with a fourth polypeptide chain, a δ-chain to form a second TCR. The identity of the δ-chain was uncovered in 1986 by a group of investigators led by Michael Brenner,[191] while the gene for this chain was discovered by Yueh-hsiu Chien and collaborators.[192]

Gene transfer studies provided conclusive evidence that the α and β chains comprise the antigen-specific receptor for T lymphocytes.[193] This group isolated genes coding for the two polypeptides from a cytotoxic T lymphocyte clone specific for the hapten fluorescein. They transferred these genes to another T lymphocyte clone specific for a different hapten. Analysis of the transfected clone revealed that the cells expressed the transferred genes and were specific for fluorescein and that there was no mixing of specificities in the transfected cell.

THE MAJOR HISTOCOMPATIBILTY COMPLEX

In the first decades of the 20th century, investigators studied whether spontaneously arising tumors could be transferred between members of the same species. In 1902, Jensen reported transferring a tumor through 19 generations of outbred mice. The tumors grew in approximately 50% of the inoculated animals.[194] In the following decade, Clarence Little and Ernest Tizzer performed similar studies using inbred strains of mice as well as F_1 and F_2 hybrids. Based on their data, they concluded that the rejection of tumors is an immunological phenomenon[195] that is regulated by genes differing between mouse strains.[196,197]

In the 1930s, Peter Gorer, a student of J.B.S. Haldane, demonstrated antigenic differences between the erythrocytes of three inbred strains of mice. Haldane is important to the field of genetics due to his development of the principles of population genetics. Gorer identified three antigens (I, II, and III) that are uniquely distributed in the mouse strains.[198] Tumors transferred between two mice differing at antigen II are rapidly rejected and the recipient mice produce serum antibodies to antigen II.[199] Gorer collaborated with George Snell who defined histocompatibility genes based on his work on tumor rejection in inbred strains of mice. Together, these investigators showed that antigen II is coded for by one of these histocompatibility genes and they renamed the gene and the protein it codes for H-2.[200]

Snell continued his studies on tumor transplantation in inbred strains of mice and eventually published a series of observations that led to the laws of transplantation.[201] In recognition of this accomplishment, Snell was awarded the Nobel Prize in Physiology or Medicine in 1980.

In 1958, Jean Dausset identified the gene complex in humans analogous to the histocompatibility complex described in mice.[202] He demonstrated that the peripheral blood from patients who had received multiple blood transfusions contained antibodies that agglutinated leukocytes from their donors. Two other groups of investigators reported similar findings in the late 1950s. Jon von Rood and co-workers in the Netherlands and Rose Payne and Mary Rolfs in California both described the presence of serum antibodies in multiparous women that agglutinate leukocytes from their children and their spouses.[203,204] In many of these women, there was no history of blood transfusions suggesting that the fetus was the source of antigenic stimulation.

By the late 1950s, the rudimentary genetics of the histocompatibility genes and the proteins for which they code were established in mice, humans, and several other species. Many early studies on the role of histocompatibility antigens on graft rejection involved studies of tumor transplantation. Beginning in the 1940s, Peter Medawar and others turned their attention to transplantation of nonmalignant tissues. In 1944, Medawar reported that the rejection of grafts between two unrelated individuals shares characteristics (specificity, memory) with other adaptive immune responses.[205] In 1958, Medawar and his collaborators demonstrated that injection of lymphocytes, but not antibodies, from an animal that had rejected a skin graft into the skin of the original donor elicited a local inflammatory response reminiscent of a tuberculin reaction.[206] Additional studies confirmed that the rejection of nonmalignant tissue grafts is identical to the destruction of transplanted tumors and that the target of both reactions is the proteins coded for by the histocompatibility (MHC) genes of the donor.

Role of the MHC in T Lymphocyte Activation

In 1963, Bernard Levine and co-workers initially described the genetic control of immune responses in inbred and outbred guinea pigs injected with the hapten-carrier conjugate, dinitrophenyl-poly-L-lysine (DNP-PLL).[207] Approximately 40% of outbred guinea pigs produced antibody to the hapten (DNP). When investigators injected offspring from two responder animals with the antigen, 82% responded. None of the offspring of two nonresponders produced antibodies. The investigators obtained similar results when they injected outbred guinea pigs with a second hapten, benzylpenicilloyl, conjugated to PLL.

Inbred guinea pigs injected with either hapten conjugated to PLL revealed that all strain 2 animals, but no strain 13 guinea pigs, produced antibodies to the haptens. Levine and colleagues postulated that the immune response to PLL is under control of a gene they termed *Ir*. Over the next decade, investigators demonstrated that several other immune phenomena also are under genetic regulation. Examples include delayed hypersensitivity and *in vitro* lymphocyte proliferation. Identification of separate T and B lymphocytes in the late 1960s revealed that many of the phenomena under *Ir* gene control are due to the activity of T lymphocytes.

Several observations in the 1960s and 1970s clarified the mechanism responsible for genetic control of immune responses. In 1965, Elizabeth Leuchars and her colleagues reported that allogeneic thymic grafts failed to restore immune functions in thymectomized mice.[208] Alan Aisenberg confirmed this observation in 1970.[209] While these results could be explained by rejection of the allogeneic thymus grafts by the recipient, experiments performed in 1972 by Bernice Kindred and Donald Schreffler[210] argue against this interpretation. Athymic (nude) mice reconstituted with syngeneic thymocytes synthesized antibodies when injected with sheep erythrocytes or the bacteriophage T4. However, nude mice reconstituted with allogeneic thymus cells failed to produce antibodies to these antigens.

When Kindred and Schreffler reported these results, immunologists knew that optimal antibody synthesis requires cooperation between T and B lymphocytes.[173-176] They also understood the role of histocompatibility antigens in graft and tumor rejection.[200,206] Thus, it was reasonable for Kindred and Schreffler to conclude that the H-2 complex plays a role in the collaboration between B and T lymphocytes. However, studies performed by Donald Mosier in 1967 demonstrated that optimal antibody formation requires interaction between cells adherent to plastic (ie, macrophages) and nonadherent cells (lymphocytes).[211] A year later, Martin Cline and Virginia Swett reported that the interaction of T lymphocytes with macrophages during the initiation of cell proliferation requires compatibility at the MHC.[212]

In 1972, Baruj Benacceraf and Hugh McDevitt[213] postulated that most, if not all, genes that regulate adaptive immune responses are linked to the MHC. Investigations of the mechanism by which these genes function revealed that several of them function by controlling T lymphocyte activation. Additional studies using inbred guinea pigs showed that antibodies specific for histocompatibility gene products interfered with T lymphocyte activation and proliferation,[214] leading to the conclusion that immune response genes within the MHC code for cell surface molecules that bind to TCRs and interfere with antigen recognition.

Ethan Shevach and collaborators incubated purified lymphocytes from immunized inbred guinea pigs with macrophages preincubated with the immunizing antigen.[215,216] Results showed that presentation of antigen by macrophages required that the two cell types share MHC molecules. These investigators also demonstrated that the requirements for activation of T and B lymphocytes are different.

Rolf Zinkernagel and Peter Doherty expanded our understanding of the role of MHC-coded molecules when they demonstrated that a T lymphocyte function (killing of virally infected cells) also requires histocompatibility between the cytotoxic cell and its target.[217] These investigators radiolabeled fibroblasts infected with lymphocytic choriomeningitis virus and added spleen lymphocytes from mice immunized with the same virus. Following overnight incubation, Zinkernagel and Doherty measured the amount of radioactivity released as an indication that the lymphocytes killed the virally infected cells. Optimal release of the radioisotope required that the spleen cells and the fibroblasts share at least one set of H2 proteins. Based on these findings, Zinkernagel and Doherty received the Nobel Prize in Physiology or Medicine in 1996.

Based on studies in experimental animals, immunologists speculated that some immune-mediated human diseases might also be associated with the expression of histocompatibility antigens. In 1973, two groups of investigators provided evidence of such an association.[218,219] Both groups studied patients with the autoimmune disease, ankylosing spondylitis (AS). They demonstrated that more than 90% of AS patients expressed the class I MHC molecule HLA-B27 on their cells. In the general population, between 4% and 8% of individuals express this protein. Numerous other associations between HLA genes and the development of pathology have been reported although none as strong as that between AS and HLA-B27.

T LYMPHOCYTE SUBPOPULATIONS

Differentiation of functional subpopulations of T and B lymphocytes occurred in the 1960s. Activated B lymphocytes synthesize and secrete antibodies. Activated T lymphocytes induce several different functions including stimulation of inflammation and other lymphocytes, rejection of transplants, destruction of malignantly transformed cells, elimination of infected cells, and regulation of ongoing immune responses.

Investigators employed two techniques to identify different subpopulations of T lymphocytes, detection of unique cell surface molecules using antibodies and characterization of secreted cytokines by purified cell populations. Mouse lymphocytes express cell surface molecules that differ between strains. Injecting one inbred strain with lymphocytes from a second inbred mouse strain induces antibodies to these molecules. In 1968, Edward Boyse and his colleagues characterized antigens on mouse lymphoid leukemia lines using this approach.[220] These antibodies bound normal lymphocytes from the immunizing strain as well as the leukemia cells. Based on the reaction of lymphocytes with various antibodies, Boyse and co-workers identified three systems of antigens (Ly), each of which has two alleles that are coded for by different genetic loci.[221,222]

Harvey Cantor and Boyse treated mouse lymphocytes with these antibodies plus complement to eliminate cells expressing the relevant antigen. They then tested the remaining cells in functional assays to determine which populations might help B lymphocytes synthesize antibodies and which might be cytotoxic. Results revealed that peripheral lymphocytes enriched for Ly-2/3–expressing cells failed to provide help to B lymphocytes but did generate cytotoxic lymphocytes *in vitro*. Conversely, peripheral lymphocytes enriched for Ly-1–expressing cells helped B lymphocytes but failed to generate cytotoxic cells.[223] Additional studies revealed that Ly-2/3 lymphocytes function as T suppressor cells while Ly-1 lymphocytes aid other T lymphocytes.[224] In 1975, Georges Kohler and Cèsar Milstein described a method to produce monoclonal antibodies.[225] In the following years, several groups used this technique to obtain monoclonal antibodies specific for antigens on T lymphocytes. Patrick Kung and colleagues produced a series of antibodies (OKT1, OKT3, and OKT4) specific for human T lymphocytes.[226] These investigators determined that human lymphocytes expressing OKT4 include those cells that assist B lymphocytes to produce antibodies, while populations devoid of OKT4-bearing cells become cytotoxic when stimulated with allogeneic lymphocytes.[227,228] In 1980, Ellis Reinherz and his colleagues performed additional studies using newly available monoclonal antibodies and demonstrated that cytotoxic T lymphocytes express a molecule defined by OKT8 antibodies.[229]

Over the next several years, various laboratories developed their own monoclonal antibodies to these T lymphocyte markers. This led to confusion since each laboratory named their antibodies and antigens differently. To reduce this confusion, the International Union of Immunology Societies and the World Health Organization convened a workshop on Human Leukocyte Differentiation Antigens in 1982.[230] Participants in this workshop defined 15 cluster of differentiation (CD) antigens that identify markers on different leukocytes including T lymphocytes. Populations of T lymphocytes reacting with the OKT4 antibody express a molecule called CD4, while those that react with the OKT8 antibody express

CD8 molecules. Most peripheral T lymphocytes express either CD4 or CD8 molecules.

In the 1970s, immunologists observed that interactions between T lymphocytes and other cells of the adaptive immune system require the cells to be from animals that share histocompatibility genes.[210,217] These data led to the concept that T lymphocytes are activated only by antigens presented to them by molecules coded for by the histocompatibility genes of that organism. During the early 1970s, a debate about two possible mechanisms responsible for MHC restriction ensued. Some immunologists postulated that the TCR binds antigen while the MHC molecules expressed on the T lymphocyte and other cells (ie, B lymphocytes, antigen-presenting cells, virally infected target cells) interacted independently. Alternatively, other investigators hypothesized that antigen and the MHC molecules interacted and that the TCR recognized this altered self. The resolution of this debate occurred in 1985 when Bruce Babbitt and co-workers demonstrated that antigenic peptides bind to class II histocompatibility molecules and that the complex is recognized by the TCR.[231]

Differentiation of T lymphocytes into CD4+ and CD8+ populations led to investigations of whether both cell types are activated similarly. Studies using human T lymphocytes demonstrated that CD4+ lymphocytes are activated by antigens presented by MHC class II molecules, while CD8+ lymphocytes are stimulated by antigens presented by MHC class I molecules.[232]

Interaction of the TCR with antigen presented by either class I or class II MHC molecules is insufficient to activate the T lymphocyte and, in fact, might cause inactivation.[233] In 1975, Kevin Lafferty and A.J. Cunningham hypothesized that activation of T lymphocytes requires two signals based on their observation that generation of cytotoxic T lymphocytes requires the stimulating cell to be viable.[234] In 1982, Lawerence Lum and colleagues identified a molecule expressed on all CD4+ T lymphocytes and on some CD8+ T lymphocytes.[235] These investigators cloned this molecule which subsequently received designation as CD28. Several experiments provided evidence that CD28 serves as a costimulatory molecule.

Although Lum and co-workers described CD28 in 1982, it was not until 1990 that the natural ligand for this molecule was identified. Peter Linsley and co-workers transfected Chinese hamster ovary cells with the gene for CD28 and used this construct to detect tumor lines and normal cells to which CD28 bound.[236] Several lymphoblastoid and leukemic cell lines as well as activated B lymphocytes bound the transfected cells. The investigators identified the ligand for CD28 by blocking this binding with monoclonal antibodies. One antibody labeled B7/BB-1 was most efficient in this blocking and the molecule was tentatively called B7. Subsequently, the gene for B7 has been isolated and cloned, and the cell surface ligand for CD28 is now termed CD80. Additional studies revealed that T lymphocytes interact with CD80 on B lymphocytes or other antigen-presenting cells, leading to cell activation and transcription of the gene coding for a cytokine termed interleukin-2 (IL-2).[237]

IL-2 was initially described in supernatants from mixed leukocyte cultures.[238,239] Subsequent studies revealed that IL-2 is secreted by T lymphocytes and functions by activating other T lymphocytes.[240,241]

The study of cytokines such as IL-2 led to the division of functionally diverse T lymphocyte populations. By the early 1980s, investigators separated CD4+ T lymphocytes into populations that helped B lymphocytes synthesize antibodies and others that enhanced inflammation and activated macrophages. In 1986, Robert Coffman, Timothy Mossman, and colleagues identified cytokines secreted by long-term clones of helper T lymphocytes specific for a variety of antigens.[242] Clones designated Th1 secreted IL-2, interferon-γ, granulocyte-macrophage colony-stimulating factor, and IL-3, while those designated Th2 secreted IL-3, IL-4, IL-6, and a mast cell growth factor. Continued study during the next 2 decades led to a consensus that Th1 lymphocytes secrete cytokines that control cell-mediated immunity, while Th2 lymphocytes secrete cytokines responsible for activating B lymphocytes.[243]

Ongoing studies have divided CD4+ T lymphocytes into at least three more subpopulations based on function and cytokines secreted. In 2000, two groups of investigators reported that some helper lymphocytes preferentially migrated to lymphoid follicles where they help activated B lymphocytes switch from the production of IgM to other isotypes.[244,245] These T_{FH} lymphocytes secrete unique cytokines (IL-21, IL-6) and express cell surface molecules (CD154) important in this function.

Several autoimmune and autoinflammatory diseases are characterized by the presence of CD4+ T lymphocytes secreting IL-17 and IL-22. In 2005, Heon Park and colleagues[246] and Harrington and co-workers[247] characterized these lymphocytes and described a new subpopulation of CD4+ lymphocytes. These Th17 lymphocytes secrete proinflammatory cytokines and are involved in the activation of neutrophils and the production of antimicrobial peptides.

The fifth subpopulation of CD4+ lymphocytes identified is the T_{reg} cells. In 1969, Yasuaki Nishizuka and Teruyo Sakakura[248] reported that thymectomy of 3-day-old female mice resulted in ovarian dysgenesis once the animals matured. Similar results are not observed in mice thymectomized at birth or on the seventh day of life. Subsequent investigations of this model revealed that the phenomenon is not restricted to mice, and other organs than the ovaries may be affected. The evolving consensus was that a cell type exists that is necessary to regulate tolerance to self and to inhibit the development of potential autoimmune and autoinflammatory reactions. Characterization of these cells demonstrated that they express both CD4 and CD25 but not CD8.[249] As such they differ from T suppressor lymphocytes Richard Gershon and collaborators identified in the 1970s.[250,251] Extensive study of these CD8+ T suppressor T lymphocytes eventually led to a general realization that such cells do not exist.[252]

Both clinical and experimental findings support the role of CD4+ T_{reg} lymphocytes in regulating the immune response. Male mice possessing the scurfy mutation are reported to have one or more of several autoimmune/inflammatory diseases including dermatological problems, anemia, cachexia, and death in the first month of life. Evaluation of these mice revealed hematological and immunological defects as well. Many of the organs of these mice are infiltrated with activated

CD4+ T lymphocytes that are responsible for the autoimmune process. In 2001, Mary Brunkow and colleagues mapped the gene responsible for the pathology in these mice.[253] The gene codes for a protein (Foxp3) that serves as a transcriptional regulator. A frame shift mutation results in loss of function of the protein and is responsible for the disorder in scurfin mice.

A similar dysregulation of self-tolerance is observed in the clinical syndrome IPEX.[254,255] In 2003, Jason Fontenot and coworkers studied the expression of Foxp3 in CD4+ lymphocytes that were either CD25+ or CD25−.[256] Foxp3 is expressed in significantly greater amounts in CD4+ CD25+ T lymphocytes than in CD4+ CD25− cells. Additional investigations showed that T_{reg} cells are Fox3P-positive and that the absence of these cells in patients with IPEX as well as in scurfin mice explains the unregulated lymphocyte proliferation and the abrogation of self-tolerance leading to the unrestrained autoimmune phenomena.

CONTROL OF IMMUNE RESPONSES

In addition to T_{reg} lymphocytes, investigators have discovered two additional mechanisms by which the immune response is naturally regulated—immunological tolerance and the activity of immune checkpoints. These physiological mechanisms are important in maintaining homeostasis.

Tolerance

One of the hallmarks of the adaptive immune response is the ability to discriminate self from non-self. In the late 1800s, most investigators studied the ability of the immune system to produce antibodies to potentially pathogenic microorganisms and considered the ability of the immune system to produce antibodies to self-tissues counterintuitive. In 1901, Paul Ehrlich and Julius Morgenroth reported experiments during which they injected goats with their own red blood cells, with red blood cells from other goats, or with red blood cells from other species.[257] Goats failed to produce antibodies against their own erythrocytes, although they produced antibodies to red cells from other goats as well as to red cells from other species. Based on these results, the investigators developed the concept of "horror autotoxicus"—literally the fear of self-toxicity.

Shortly after Ehrlich and Morgenroth introduced this concept, Julius Donath and Karl Landsteiner demonstrated that autoantibodies are responsible for the pathology observed in the human disease, paroxysmal cold hemoglobinuria (PCH).[103] PCH patients present with hemolytic anemia and the presence of large amounts of hemoglobin in their urine. These patients possess antibodies that bind their red blood cells in the periphery and cause lysis when the cells return to the body core.

Between 1900 and 1945, several other investigators reported anecdotal evidence of pathology due to autoantibodies including sympathetic ophthalmia, renal pathology in rabbits injected with homologous kidney, and neurological pathology following injection with the rabies vaccine. In the 1950s, Noel Rose and Ernest Witebsky demonstrated that specific autoantibodies could result in pathology to the thyroid.[258] Initially, these investigators injected rabbits with purified thyroglobulin—the rabbits synthesized antibodies to foreign thyroglobulin but not to rabbit thyroglobulin. In subsequent experiments, Rose, known as the father of autoimmunity, and his colleagues injected thyroidectomized rabbits with their own thyroglobulin incorporated in complete Freund's adjuvant and demonstrated that these animals produced antibodies specific for the thyroid. Finally, they removed just one lobe of the thyroid before the injection and showed that the rabbits produced antibodies and that this antibody produced pathology in the remaining thyroid tissue.

Histological observations of the thyroid glands from these rabbits revealed a picture reminiscent of that seen in clinical cases of Hashimoto thyroiditis. Further investigation showed that Hashimoto patients possessed antibodies specific for thyroglobulin in their serum.[259] This initial description of an autoimmune disease was followed by reports of other autoimmune states and suggested that the immune system possessed a mechanism to inhibit the synthesis of autoantibodies.

A hint of the mechanism responsible for this phenomenon derived from observations made by Ray Owens on freemartin dizygotic twin cattle.[260] Freemartin cattle share circulatory systems during gestation. Owen studied the blood types of these cattle and demonstrated that many of the twins have identical blood types. Since the incidence of identical twins in cattle is rare, Owen argued that the twins exchanged hematopoietic stem cells as embryos and that these stem cells provided blood cells throughout the animal's life.

Peter Medawar and his group extended these observations by exchanging skin grafts between freemartin twins.[261,262] Most of the grafts were accepted even when the twins were known to be dizygotic and sired by different bulls. These authors concluded that the exposure of the developing immune response to foreign antigens altered the reactivity of the immune competent cells.

In 1949, Macfarlane Burnet and Frank Fenner reported that animals failed to respond to an immunologic stimulus until late in embryogenesis or shortly after birth[125] and predicted that the immune response would not respond to antigens introduced during the fetal period. An extension of this postulate was that any immunogenic material the cells of the developing immune system interacted with during embryogenesis would be seen as self. The idea that the immune system could "learn" the difference between self and non-self was a major shift in the contemporary understanding of the system. In the late 1940s, most immunologists still accepted instruction models of antibody formation. In these models, antigen determined the structure of the antibody molecules that would be synthesized. It took an additional 8 years before Burnet proposed his clonal selection theory that included the idea that lymphocytes proceed through a stage in their development during which they could be inactivated by contact with antigen.[136,137]

Medawar and his colleagues provided conclusive evidence that lymphocytes can be inactivated by contact with antigen during their development.[138] These investigators injected a mixture of cells from the testis, kidney and spleen of one inbred strain of mice into embryonic mice of a second strain. Eight weeks after birth, the mice that had been treated as embryos were transplanted with skin from the cell donor strain. Normally such skin grafts survive an average of 11 days. In this experiment, 3 of 5 grafts to treated mice survived until the study was terminated at 50, 77, and 101 days, respectively. One mouse received a second skin graft from the original donor strain—this graft survived as well. The investigators treated the other two mice bearing surviving grafts with lymph node cells from mice that had rejected a graft. In these instances, the tolerated grafts became inflamed and necrotic within 2 to 3 days.

Chicken embryos treated in ovo with blood from a second strain confirmed these studies. Five of seven treated chickens accepted skin grafts for a prolonged period. Contemporaneously with this report, Milan Hašek published his studies[263] on chickens parabiosed during embryonic development. When hatched, these chickens failed to produce antibodies against their partner's red blood cells.

Burnet's prediction that antigen-reactive lymphocytes go through a developmental stage during which they are sensitive to being inactivated by exposure to foreign antigen was proven correct by the experimental work of Medawar and colleagues and of Hašek. In 1960, the Nobel Prize in Physiology or Medicine was awarded to Medawar and Burnet "for discovery of acquired immunological tolerance."

Immune Checkpoints

In 1987, Joan Brunet and colleagues isolated DNA derived from cytotoxic T lymphocyte clones.[264] The sequence they studied coded for a membrane associated protein they called cytotoxic T lymphocyte antigen 4 (CTLA-4). CTLA-4 is expressed on cytotoxic T lymphocytes and activated T lymphocytes. CTLA-4 is now known as CD152.

Three years later, Jane Gross and her collaborators isolated a similar gene from a murine lymphoma library.[265] The isolated clone had 61% homology with human CD28 and the protein was expressed on cytotoxic T lymphocytes.

Peter Linsley and co-workers demonstrated that human cytotoxic T lymphocytes also express CD152.[266] In functional studies, these investigators showed that CD152 bound specifically to CD80 (on antigen-presenting cells) and inhibited T lymphocyte activation *in vivo*. Jeffrey Bluestone and his colleagues studied transplant rejection in mice and observed that CD152 inhibited the rejection of xenogeneic pancreatic islet cells.[267]

In 1995, two groups of investigators described massive lymphoproliferation leading to early death in mice in which the CTLA-4 gene had been knocked out.[268,269] The conclusion of these studies was that CD152 is a negative regulator of T lymphocyte activation. This result was confirmed by Matthew Krummell and James Allison who revealed that CD28 and CD152 compete for binding to CD80 and have opposing effects on the eventual fate of the T lymphocyte.[270]

Similar studies by Tasuku Honjo and his collaborators revealed another inhibitory molecule expressed on T lymphocytes and other cells called programmed cell death 1 (PD-1). Honjo's group isolated the gene for PD-1 from lymphocytes undergoing apoptosis.[271,272] PD-1 binds to PD-L1 (CD274) and other cell surface receptors included in the PD-L family. Inactivation of the PD-1:PD-L1 binding results in lupus-like autoimmune disease.

Both CD152 and CD274 are expressed by some tumor cells where they appear to downregulate tumor-specific T lymphocytes. Several novel cancer immunotherapies have been developed using this information. In 2018, James Allison and Tasuku Honjo were awarded the Nobel Prize in Medicine or Physiology "for their discovery of cancer therapy by inhibition of negative immune regulation."

CONCLUSION

Immunology represents a major discipline in the biological and biomedical sciences. Some historians identify Edward Jenner (1749-1823) as the father of immunology based on his development of the smallpox vaccine in 1796. Others give this honor to Louis Pasteur (1822-1895) or Robert Koch (1843-1910) for their studies validating the germ theory of disease. However, anecdotal evidence for the existence of host defenses against disease predates all three of these individuals.

The study of immunology began as a subdiscipline of microbiology due to an early focus on understanding the mechanisms responsible for protecting vertebrates from potentially pathogenic microorganisms. Many of the studies performed in the late 1800s and the first half of the 1900s depended on models in which microbes served as antigens. The development of the clonal selection theory by Macfarlane Burnet in 1957 initiated an explosion of investigations leading to a better understanding of the cells and molecules of the immune response. These studies fostered the development of immunology as an independent discipline.

Beginning in the late 1960s, several prominent immunologists reported that all major breakthroughs in immunology have been made and that future generations of scientists would merely complete a picture that was 95% finished.[27,273] These predictions seem premature as shown by this brief historical review. To paraphrase Mark Twain, reports of the death of immunology are greatly exaggerated.

TIMELINE

1796: Edward Jenner develops a vaccine against smallpox.

1849: John Snow postulates a germ theory of disease based on his study of a cholera epidemic.

1864: Louis Pasteur provides experimental evidence in support of the germ theory of disease.

1883: Ilya Metchnikov develops the phagocytic theory of immunity.

1890: Robert Koch publishes his postulates concerning the relationship between a microbe and an infectious disease.

1890: Emil von Behring and Shibasaburo Kitasato develop serotherapy for diphtheria and tetanus.

1900: Karl Landsteiner discovers the ABO blood groups.

1901: Paul Ehrlich publishes his side-chain theory postulating that antibody-forming cells possess receptors by which they identify antigens.

1902: Paul Portier and Charles Richet describe anaphylaxis and conclude that it is a pathological counterpart of a protective antibody response.

1905: Clemens von Pirquet and Bela Schick describe serum sickness.
1936: Peter Gorer reports antigenic difference on cells of different mouse strains.
1953: Rupert Billingham, Leslie Brent, and Peter Medawar demonstrate the induction of immunological tolerance in neonatal mice.
1956: Bruce Glick reports that removal of the bursa of Fabricius in newly hatched chickens severely depresses subsequent antibody formation.
1957: F. Macfarlane Burnet proposes the clonal selection theory of antibody formation.
1959: Gerald Edelman and Rodney Porter present data leading to deciphering the structure of antibody.
1961: Jacque Miller describes the role of the thymus in development of immune competence.
1963: Baruj Benacceraf and colleagues provide evidence that immune responses in guinea pigs are under genetic control.
1966: Max Cooper and colleagues differentiate thymus-dependent and bursa-dependent lymphocytes.
1966: Henry Claman shows that a combination of thymus and bone marrow lymphocytes reconstitutes a primary antibody response in irradiated mice better than either cell alone.
1974: Ralph Steinman and co-workers present evidence that dendritic cells function by presenting antigens to T lymphocytes.
1975: Georges Köhler and Cesar Milstein develop hybridoma technology to produce monoclonal antibodies.
1976: Susumu Tonegawa publishes hybridization data supporting the presence of a limited number of germline light chain genes that undergo somatic diversification.
1980: The World Health Organization declares smallpox eradicated from the world.
1995: James Allison and co-workers describe roles of CD28 and CTLA-4 on T lymphocyte activation.
1996: Jules Hoffman and colleagues report on the role of Toll genes in controlling antifungal defense mechanisms in insects.
1998: Bruce Beutler and collaborators identify the receptor for endotoxin produced by gram-negative bacteria and demonstrate that it is coded for by a gene analogous to Toll.
1999: Tasuku Honjo and collaborators describe inhibitory effect of PD-1 on T lymphocyte activation.

REFERENCES

1. Thucydides. History of the Peloponnesian War. In: Crawley R, trans-ed. *Fourth and Fifth Years of the War*. (Chapter 9). 1904. Accessed May 11, 2021. https://www.gutenberg.org/ebooks/7142
2. Montagu LMW. *Letter to Sarah Criswell*. 1717. Accessed May 11, 2021. http://pyramid.spd.louisville.edu/~eri/fos/lady_mary_montagu.html
3. Farmer L. The smallpox inoculation controversy and the Boston press 1721-2. *Bull N Y Acad Med*. 1958;34:599-601, 608.
4. Jesty R, Williams G, Who invented vaccination? *Malta Med J*. 2011;23:29-32.
5. Jenner E. *An Inquiry into the Causes and Effects of the Variolae Vaccine: A Disease Discovered in Some of the Western Counties of England, Particularly Gloucestershire, and Known by the Name Cow Pox*. Sampson Low, London, 1798. Referenced in Jesty and Williams. 2011.
6. Ambrose CT, Immunology's first priority dispute – an account of the 17th century Rudbeck-Bartholin feud. *Cell Immunol*. 2006;242:1-8.
7. Sklavem JH. The scientific life of Thomas Bartholin. *Ann Med Hist*. 1921;3:67-81.
8. Hajdu SI. A note from history: the discovery of blood cells. *Clin Lab Sci*. 2003;33:237-238.
9. Degos L. John Hughes Bennett, Rudolf Virchow…and Alfred Donné: the first description of leukemia. *Hematol J*. 2001;2:1.
10. Snow J. *On the Mode of Communication of Cholera*. John Churchill; 1849. Accessed May 11, 2021. https://collections.nlm.nih.gov/ext/cholera/PDF/0050707.pdf
11. Snow J. *On the Mode of Communication of Cholera*. 2nd ed. John Churchill; 1855. https://archive.org/stream/b28985266#page/n3/mode/2up
12. Pasteur L. Mémoire sur la fermentation appelée lactique. *C R Chimie*. 1857;45:913-916.
13. Pasteur L. Mémoire sur la fermentation alcoolique. *C R Chimie*. 1857;45:1032-1036.
14. Koch R. Untersuchungen über Bakterien: V. Die Ätiologie der Milzbrand-Krankheit, begründet auf die Entwicklungsgeschichte des Bacillus anthracis. *Cohns Beitrage zur Biologie der Pflanzen*. 1876;2:277-310.
15. Pasteur L. Méthode pour preventir la rage après morsure. *C R Acad Sci*. 1885;101:765-844.
16. Moticka EJ. *A Historical Perspective on Evidence-Based Immunology*. Elsevier; 2016.
17. Langerhans P. Über die nerven der menschlichen haut. *Virchows Arch (B)*. 1868;44:325-337.
18. Metchnikov I. *On the present state of the question of immunity in infectious diseases*. In: *Nobel Lectures, Physiology or Medicine 1901-1921*. Elsevier; 1967. Accessed December 11, 1908. http://nobelprize.org/nobel_prizes/medicine/laureates/1908/mechnikov-lecture.html
19. Steinman RM. Decisions about dendritic cells: past, present and future. *Annu Rev Immunol*. 2012;30:1-22.
20. Steinman RM, Cohn ZA. Identification of a novel cell type in peripheral lymphoid organs of mice. I. Morphology, quantitation, tissue distribution. *J Exp Med*. 1973;137:1142-1162.
21. Steinman RM, Cohn AA. Identification of a novel cell type in peripheral lymphoid organs of mice. II. Functional properties in vitro. *J Exp Med*. 1974;139:380-397.
22. Steinman RM, Lustig DS, Cohn ZA. Identification of a novel cell type in peripheral lymphoid organs of mice. III. Functional properties in vivo. *J Exp Med*. 1974;139:1431-1445.
23. Steinman RM, Witmer MD. Lymphoid dendritic cells are potent stimulators in the primary mixed leukocyte reaction in mice. *Proc Natl Acad Sci U S A*. 1978;75:5132-5136.
24. Lechler RI, Batchelor JR. Restoration of immunogenicity to passenger cell-depleted kidney allografts by the addition of donor strain dendritic cells. *J Exp Med*. 1982;155:31-41.
25. Nussenzweig MC, Steinman RM, Gutchinov B, et al. Dendritic cells are accessory cells for the development of anti-trinitrophenyl cytotoxic T lymphocytes. *J Exp Med*. 1980;152:1070-1084.
26. Inaba K, Steinman RM, Van Voorhis WC, et al. Dendritic cells are critical accessory cells for thymus-dependent antibody responses in mouse and in man. *Proc Natl Acad Sci U S A*. 1983;80:6041-6045.
27. Janeway CA. Approaching the asymptote? Evolution and revolution in immunology. *Cold Spring Harb Symp Quant Biol*. 1989;54:1-13.
28. Matzinger P. Tolerance, danger and the extended family. *Annu Rev Immunol*. 1994;12:991-1045.
29. Matzinger P. The danger model: a renewed sense of self. *Science*. 2002;296:301-305.
30. Behring EA, Kitasato S. Über das Zustandekommen der Diphtheria-immunität and der Tetanus-immunität bei Thieren. *Deutsch Med Wochenschr*. 1890;49:1113-1114.
31. https://www.nobelprize.org/prizes/medicine/1901/behring/facts/
32. Pfeiffer R. Untersuchungen über des Choleragift. *Z Hyg*. 1892;11:393-411.
33. Anderson KV, Jürgens G, Nüsslein-Volhard C. Establishment of dorsal-ventral polarity in the drosophila embryo: genetic studies on the role of the Toll gene product. *Cell*. 1985;42:779-789.
34. Lemaitre B, Nicolas E, Michaut L, et al. The dorsoventral regulatory gene cassette spätzle/Toll/cactus controls the potent antifungal response in Drosophila adults. *Cell*. 1996;86:973-983.
35. Hoffman JA. Immune responsiveness in vector insects. *Proc Natl Acad Sci U S A*. 1997;94:11152-11153.
36. Medzhitov R, Preston-Hurlburt P, Janeway CA. A human homologue of the Drosophila Toll protein signals activation of adaptive immunity. *Nature*. 1997;388:394-397.
37. Tobias PS, Soldau K, Ulevitch RJ. Isolation of a lipopolysaccharide-binding acute phase reactant from rabbit serum. *J Exp Med*. 1986;164:777-793.
38. Beutler B, Greenswald D, Hulmes JD, et al. Identity of tumor necrosis factor and the macrophage-secreted factor cachectin. *Nature*. 1985;316:552-554.
39. Poltorak A, He X, Smirnova I, et al. Defective LPS signaling in C3H/HeJ and C57BL/10ScCr mice: mutations in Tlr4 gene. *Science*. 1998;282:2085-2088.
40. Ehrlich P. *Partial cell functions*. In: *Nobel Lectures, Physiology or Medicine, 1901-1921*. Elsevier; 1967. Accessed December 11, 1908. http://nobelprize.org/nobel_prizes/medicine/laureates/1908/ehrlich-lecture.html
41. VonFodor J. Die Fähigkeit des Bluts Bakterien zu vernichten. *Deutsch Med Wochenschr*. 1887;13:745-746.
42. Schmalstieg FC, Goldman AS, Jules Bordet (1870-1961): a bridge between early and modern immunology. *J Med Biogr*. 2009;17:217-224.
43. Behring EA. Untersuchungen über das Zustandekommen der Diphtherie-immunität bei Thieren. *Deutsch Med Wochenschr*. 1891;50:1145-1148.
44. Bolduan CF *Studies in Immunity by Paul Ehrlich and His Collaborators*. 2nd ed. John Wiley and Sons; 1910. Accessed May 11, 2021. https://babel.hathitrust.org/cgi/pt.?id=ucl.b4210298&view=1up&seq=11
45. Gachelin G. The designing of anti-diphtheria serotherapy at the Institut Pasteur (1881-1900): the role of a supranational network of microbiologists. *Dynamis*. 2007;27:45-62.
46. Zinsser H. Further studies on the identity of precipitins and protein sensitizers (albuminolysins). *J Exp Med*. 1913;18:219-227.
47. Avery O. The distribution of the immune bodies occurring in antipneumococcus serum. *J Exp Med*. 1915;21:133-145.
48. Heidelberger M, Pedersen KO. The molecular weight of antibodies. *J Exp Med*. 1937;65:393-414.
49. Tiselius A, Kabat EA. An electrophoretic study of immune serum and purified antibody preparations. *J Exp Med*. 1939;69:119-131.
50. Heremans JF, Heremans M-T, Schultze HE. Isolation and description of a few properties of the β2A–globulin of human serum. *Clin Chim Acta*. 1959;4:96-102.
51. Rowe DS, Fahey JL. A new class of human immunoglobulin. I. A unique myeloma protein. *J Exp Med*. 1965;121:171-184.
52. Rowe DS, Fahey JL. A new class of human immunoglobulin. II. Normal serum IgD. *J Exp Med*. 1965;121:185-199.
53. Ishizaka K, Ishizaka T, Hornbrook M. Physiochemical properties of reaginic antibody. IV. Presence of a unique immunoglobulin as a carrier of reaginic activity. *J Immunol*. 1966;97:75-85.
54. Johansson SGO, Bennich H. Immunological studies of an atypical (myeloma) immunoglobulin. *Immunology*. 1967;13:381-394.
55. Porter RR. The hydrolysis of rabbit γ-globulin and antibodies with crystalline papain. *Biochem J*. 1959;73:119-126.
56. Edelman GM. Dissociation of γ-globulin. *J Am Chem Soc*. 1959;81:3155-3156.
57. Fleischman JB, Porter RR, Press EM. The arrangement of the peptide chains in γ-globulin. *Biochem J*. 1963;88:220-228.
58. Pfeiffer R. Weitere Untersuchungen über das Wesen der Choleraimmunität and über specifische bactericide Processe. *Z Hyg Infektionskr*. 1894;18:1-16.
59. Pfeiffer R, Issaeff R. Über die specifische Bedeutung der Choleraimmunitat. *Z Hyg Infectionskr*. 1894;17:355-400.
60. Bordet J. Les leucocytes et les proprieties actives du serum chez les vaccines. *Ann Inst Pasteur*. 1895;9:462-506.
61. Bordet J. Sur le mode d'action des serums preventives. *Ann Inst Pasteur*. 1896;10:193-219.
62. Bordet J. Sur l'agglutination et la dissolution des globules rouges par le serum d'animaux injecties de sang defibriné. *Ann Inst Pasteur*. 1898;12:688-695.
63. Ferrata A. Die Unwirksamkeit der complexen Hämolysine in salzfreien Lösungen and ihre Unsache. *Berlin Klin Wochsch*. 1907;44:366.

64. Brand E. Über das Verhalten der Komplemente bei der Dialyse. *Berlin Klin Wochsch.* 1907;44:1075.
65. Omorokow L. Über die Wirkung des Cobragiftes auf der Komplemente. *Z Immunitätsforsch.* 1911;10:285.
66. Ritz H. Über due Wirkung des Cobragiftes auf die Komplemente. *Z Immunitätsforsch.* 1912;13:62-63.
67. Coca AF. A study of the anticomplementary action of yeast, of certain bacteria and of cobra venom. *Z Immunitätsforsch.* 1914;21:604-610.
68. Gordon J, Whitehead HR, Wormall A. The action of ammonia on complement. The fourth component. *Biochem J.* 1926;20:1028-1035.
69. Mayer MM, Levine L, Rapp HJ, et al. Kinetic studies on immune hemolysis VII. Decay of EAC'1, 4, 2, fixation of C'3 and other factors influencing the hemolytic action of complement. *J Immunol.* 1954;73:443-454.
70. Lepow IH, Naff GB, Todd EW, et al. Chromatographic resolution of the first component of human complement into three activities. *J Exp Med.* 1963;117:983-1008.
71. Müller-Eberhard HJ, Nilsson U. Relation of a β1 glycoprotein of human serum to the complement system. *J Exp Med.* 1960;111:217-234.
72. Pillemer L, Blum L, Pensky J, et al. The requirement for magnesium ions in the inactivation of the third component of human complement (C'3) by insoluble residues of yeast cells (zymosan). *J Immunol.* 1953;71:331-338.
73. Pillemer L, Lepow IH, Blum L. The requirement for a hydrazine-sensitive serum factor and heat-labile serum factors in the inactivation of human C'3 by zymosan. *J Immunol.* 1953;71:339-345.
74. Pillemer L, Blum L, Lepow IH, et al. The properdin system and immunity. I. Demonstration and isolation of a new serum protein, properdin, and its role in immune phenomenon. *Science.* 1954;120:279-285.
75. Nelson RA. An alternative mechanism for the properdin system. *J Exp Med.* 1958;108:515-535.
76. Natali PG, Tan EM. Precipitin reactions between polyribonucleotides and heat labile serum factors. *J Immunol.* 1972;108:318-324.
77. Pinkard RN, Olson MS, Kelley RE, et al. Antibody-independent activation of human C1 after interaction with heart subcellular membranes. *J Immunol.* 1973;110:1376-1382.
78. Cooper NR, Jensen FC, Welsh RM, et al. Lysis of RNA tumor viruses by human serum: direct antibody-independent triggering of the classical complement pathway. *J Exp Med.* 1976;144:970-984.
79. Cooper NR, Morrison DC. Binding and activation of the first component of human complement by the lipid A region of lipopolysaccharide. *J Immunol.* 1978;120:1862-1868.
80. Bartholomew RM, Esser AF. Mechanism of antibody-independent activation of the first component of complement (C1q) on retrovirus membranes. *Biochemistry.* 1980;19:2847-2853.
81. Wright AE, Douglas SR. An experimental investigation of the role of the blood fluids in connection with phagocytosis. *Proc R Soc Lond.* 1903;72:357-370.
82. Wright AE, Douglas SR. Further observations on the role of the blood fluids in connection with phagocytosis. *Proc R Soc Lond.* 1904;73:128-142.
83. Möller E. Contact-induced cytotoxicity by lymphoid cells containing foreign isoantigens. *Science.* 1965;147:873-874, 879.
84. MacLennan ICM, Loewi G. Effect of specific antibody to target cells on their specific and non-specific interactions with lymphocytes. *Nature.* 1968;219:1069-1070.
85. Coombs RRA. *Gell PGH.* In: *Clinical Aspects of Immunology.* F. A. Davis; 1963:317.
86. Behring E. *Infektion and Desinfektion.* Thomas; 1894.
87. Portier P, Richet C. De l'action anaphylactique de certains venins. *Compt Rend Soc de Biol.* 1902;54:170-172.
88. Vaughan VC, Wheeler SM. The effects of egg white and its split products on animals: a study of susceptibility and immunity. *J Infect Dis.* 1907;4:476-508.
89. Besredka A. Du mecanisme de l'anaphylaxie vis-à-vis du derum de cheval. *Comp Rend Soc de Biol.* 1907;59:294-296.
90. Anderson JF, Frost WH. Studies upon anaphylaxis with special reference to the antibodies concerned. *J Med Res.* 1910;23:31-69.
91. Schultz WH. Physiological studies in anaphylaxis: I. The reaction of smooth muscle of the guinea pig sensitized with horse serum. *J Pharmacol Exp Ther.* 1910;1:549-567.
92. Dale HH, Laidlaw PP. Further observations on the physiological action of β-iminapolylethylamine. *J Physiol.* 1911;43:182-195.
93. Praustnitz C, Küstner H. Studien über Uberempfindlicht. *Centralb Bakterial I Abt Orig.* 1921;86:160-169.
94. Bennich HH, Ishizaka K, Johansson SGO. et al. Immunoglobulin E. A new class of human immunoglobulin. *Immunochem.* 1968;5:327-328.
95. Landsteiner K. Zur Kenntnis der antifermentativen, lytischen und agglutinierenden Wirkungen des Blutserums und der Lymphe. *Centralblatt Bakteriol Parasiten u Infektions.* 1900;27:357-362.
96. Levine P, Stetson RE. An unusual case of intra-group agglutination. *J Am Med Assoc.* 1939;113:126-127.
97. Landsteiner K, Wiener AS. An agglutinable factor in human blood recognized by immune sera for rhesus blood. *Proc Soc Exp Biol.* 1940;43:223.
98. Levine P. Serological factors as possible causes of spontaneous abortion. *J Hered.* 1943;34:71-80.
99. Davidsohn I, Masaitis L, Stern K. Experimental studies on Rh immunization. *Am J Clin Pathol.* 1956;26:833-843.
100. Finn R, Clarke CA, Donohoe WTA. et al. Experimental studies on the prevention of Rh haemolytic disease. *Br Med J.* 1961;1:1486-1490.
101. Freda VJ, Gorman JG, Pollack W. Successful prevention of experimental Rh sensitization in man with an anti-Rh gamma2globulin antibody preparation: a preliminary report. *Transfus.* 1964;4:26-32.
102. Freda VJ, Gorman JG, Pollack W. et al. Prevention of Rh isoimmunization. Progress report of the clinical trial in mothers. *J Am Med Assoc.* 1967;199:383-389.
103. Donath J, Landsteiner K. Uber paroxysmale Hamoglobinurie. *Munch Med Wochenschr.* 1904;51:1590-1593.
104. von Pirquet CE, Schick B. *Die Serumkrankheit.* Franz Deuticke; 1905.
105. von Pirquet CE. Allergy. *Arch Intern Med.* 1911;7:259-288.
106. Talmage DW, Dixon FJ, Bukantz SC. et al. Antigen elimination from blood as an early manifestation of the immune response. *J Immunol.* 1951;67:243-255.
107. Michael AF, Drummond KN, Good RA. et al. Acute poststreptococcal glomerulonephritis: immune deposit disease. *J Clin Invest.* 1966;45:237-248.
108. Huber B. 100 Jahre Allergie: Clemens von Pirquet-sein Allergiebegriff und das ihn zugrunde liegende Krankheitsverständnis. *Wien Klin Wochen.* 2006;118:573-579.
109. Dixon FJ, Feldman JD, Vazquez JJ. Experimental glomerulonephritis. The pathogenesis of a laboratory model resembling the spectrum of human glomerulonephritis. *J Exp Med.* 1961;113:899-920.
110. Zinsser H. Studies on the tuberculin reaction and on specific hypersensitivities in bacterial infection. *J Exp Med.* 1921;34:495-524.
111. Landsteiner K, Chase MW. Experiments on transfer of cutaneous sensitivity to simple compounds. *Proc Soc Exp Biol Med.* 1942;49:688-690.
112. Chase MW. The cellular transfer of cutaneous hypersensitivity to tuberculin. *Proc Soc Exp Biol Med.* 1945;59:134-135.
113. Bollman JL, Cain JC, Grindlay JH. Techniques for the collection of lymph from the liver, small intestine or thoracic duct of the rat. *J Lab Clin Med.* 1948;33:1349-1352.
114. Gowans JL. The effect of the continuous re-infusion of lymph and lymphocytes on the output of lymphocytes from the thoracic duct of unanaesthetized rats. *Br J Exp Pathol.* 1957;38:67-78.
115. Gowans JL. The recirculation of lymphocytes from blood to lymph in the rat. *J Physiol.* 1959;146:54-69.
116. Gowans JL, Knight EJ. The route of re-circulation of lymphocytes in the rat. *Pro R Soc Lond B Biol Sci.* 1964;159:257-282.
117. Miller JFAP. Immunological function of the thymus. *Lancet.* 1961;278:748-749.
118. McGregor DD, Gowans JL. The antibody response of rats depleted of lymphocytes by chronic drainage from the thoracic duct. *J Exp Med.* 1963;117:303-320.
119. Gowans JL. The fate of parental strain lymphocytes in F1 hybrid rats. *Ann N Y Acad Sci.* 1962;99:432-455.
120. Gowans JL. The lymphocyte – a disgraceful gap in medical knowledge. *Immunol Today.* 1996;17:288-291.
121. Ehrlich P. On immunity with special reference to cell life. *Proc R Soc Lond.* 1900;66:424-428.
122. Landsteiner K. *The Specificity of Serological Reactions.* Harvard University Press; 1945.
123. Haurowitz F. The immunological response. *Ann Rev Micro.* 1953;7:389-414.
124. Pauling L. A theory of the structure and process of formation of antibodies. *J Am Chem Soc.* 1940;62:2643-2657.
125. Burnet FM, Fenner F. *The Production of Antibodies.* 2nd ed. Macmillan; 1949.
126. Jerne NK. The natural selection theory of antibody formation. *Proc Natl Acad Sci U S A.* 1955;41:849-857.
127. Haurowitz F. Mechanism of antigen-induced antibody biosynthesis from antibody precursors, the heavy and light immunoglobulin chains. *Proc Natl Acad Sci U S A.* 1978;75:2434-2438.
128. McMaster PD, Hudack SS. The formation of agglutinins within lymph nodes. *J Exp Med.* 1935;61:783-805.
129. Ehrich WE, Harris TN. The formation of antibodies in the popliteal lymph node in rabbits. *J Exp Med.* 1942;76:335-347.
130. Ehrich WE, Harris TN, Mertens E. The absence of antibody in the macrophages during maximum antibody formation. *J Exp Med.* 1946;83:373-381.
131. Fagraeus A. The plasma cellular reaction and its relation to the formation of antibodies in vitro. *J Immunol.* 1948;58:1-13.
132. Ribatti D. The discovery of plasma cell: an historical note. *Immunol Lett.* 2017;188:64-67.
133. Harris S, Harris TN, Farber MB. Studies on the transfer of lymph node cells. I. Appearance of antibody in recipients of cells from donor rabbits injected with antigen. *J Immunol.* 1954;72:148-160.
134. Reiss E, Marten E, Ehrich WE. Agglutiniation of bacteria by lymphoid cells in vitro. *Proc Soc Exp Biol Med.* 1950;74:732-735.
135. Coons AH, Leduc EH, Connolly JM. Studies on antibody production. I. A method for the histochemical demonstration of specific antibody and its application to a study of the hyperimmune rabbit. *J Exp Med.* 1955;102:49-60.
136. Burnet FM. A modification of Jerne's theory of antibody production using the concept of clonal selection. *Aust J Sci.* 1957;20:67-68.
137. Burnet FM. *The Clonal Selection Theory of Acquired Immunity*, Vanderbilt University Press; 1959.
138. Billingham RE, Brent L, Medawar PB. 'Actively acquired tolerance' of foreign cells. *Nature.* 1953;172:603-606.
139. Talmage DW. Allergy and immunology. *Annu Rev Med.* 1957;8:239-256.
140. Nossal GJV, Lederberg J. Antibody production by single cells. *Nature.* 1958;181:1419-1420.
141. Möller G. Demonstration of mouse isoantigens at the cellular level by the fluorescent antibody technique. *J Exp Med.* 1961;114:415-434.
142. Davie JM, Paul WE. Receptors on immunocompetent cells II. Specificity and nature of receptors of dinitrophenylated Guinea pig albumin 125I-binding lymphocytes of normal Guinea pigs. *J Exp Med.* 1971;134:495-516.
143. Raff MC, Feldman M, de Petris S. Monospecificity of bone marrow-derived lymphocytes. *J Exp Med.* 1973;137:1024-1030.
144. Nossal GJV, Pike BL. Single cell studies on the antibody-forming potential of fractionated, hapten-specific B lymphocytes. *Immunology.* 1976;30:189-202.
145. Hood L, Talmage DW. Mechanism of antibody diversity: germ line basis for variability. *Science.* 1970;168:325-334.
146. Jerne NK. The somatic generation of immune recognition. *Eur J Immunol.* 1971;1:1-9.
147. Tonegawa S. Somatic generation of immune diversity. *Biosci Rep.* 1988;8:3-26.
148. Hilschmann N, Craig LC. Amino acid sequence studies with Bence-Jones proteins. *Proc Natl Acad Sci U S A.* 1965;53:1403-1409.
149. Titani K, Whitley E Jr, Avogardo L. et al. Immunoglobulin structure: partial amino acid sequence of a Bence Jones protein. *Science.* 1965;149:1090-1092.
150. Putnam FW. Immunoglobulin structure: variability and homology. *Science.* 1969;163:633-644.
151. Press EM, Hogg NM. Comparative study of two immunoglobulin G Fd-fragments. *Nature.* 1969;223:807-810.
152. Edelman GM, Cunningham BA, Gall WE. et al. The covalent structure of an entire γG immunoglobulin molecule. *Proc Natl Acad Sci U S A.* 1969;63:78-85.
153. Tonegawa S, Steinberg C, Dube S. et al. Evidence for somatic generation of antibody diversity. *Proc Natl Acad Sci U S A.* 1974;71:4027-4031.
154. Tonegawa S. Reiteration frequency of immunoglobulin light chain genes: further evidence for somatic generation of antibody diversity. *Proc Natl Acad Sci U S A.* 1976;73:203-207.
155. Tonegawa S, Brack C, Hozumi N. et al. Cloning of an immunoglobulin variable region gene from mouse embryo. *Proc Natl Acad Sci U S A.* 1977;74:3518-3522.
156. Dreyer WJ, Bennett JC. The molecular basis of antibody formation: a paradox. *Proc Natl Acad Sci U S A.* 1965;54:864-869.
157. Brenner S, Milstein C. Origin of antibody variation. *Nature.* 1966;211:242-243.
158. Hozumi N, Tonegawa S. Evidence for somatic rearrangement of immunoglobulin genes coding for variable and constant regions. *Proc Natl Acad Sci U S A.* 1976;73:3628-3632.
159. Weigert M, Gatmaitan L, Loh E, et al. Rearrangement of genetic information may produce immunoglobulin diversity. *Nature.* 1978;276:785-790.
160. Seidman JG, Leder P. The arrangement and rearrangement of antibody genes. *Nature.* 1978;276:790-795.
161. Sakano H, Maki R, Kurosawa Y. et al. Two types of somatic recombination are necessary for the generation of complete immunoglobulin heavy-chain genes. *Nature.* 1980;286:676-683.
162. Bruton OC. Agammaglobulinemia. *Pediatrics.* 1952;9:722-728.
163. Glick B, Chang TS, Jaap RG. The bursa of Fabricius and antibody production in the domestic fowl. *Poult Sci.* 1956;35:224-225.
164. Miller JFAP. Role of the thymus in murine leukemia. *Nature.* 1959;183:1069.
165. Nezelof C, Jammet ML, Lortholary P. et al. Hereditary thymic aplasia: its place and responsibility in a case of lymphocytic, normoplasmocytic and normoglobulinemia aplasia in an infant. *Arch Fr Pediatr.* 1964;21:897-920.
166. Lischner HW, Punnett HH, DiGeorge AM. Lymphocytes in the absence of the thymus. *Nature.* 1967;214:580-582.
167. Warner NL, Szenberg A. The immunological function of the bursa of Fabricius in chickens. *Annu Rev Microbiol.* 1964;18:253-266.
168. Cooper MD, Peterson RDA, Good RA. Delineation of the thymic and bursal lymphoid systems in the chicken. *Nature.* 1965;205:143-146.
169. Cooper MD, Peterson RDA, South MA, et al. The functions of the thymus system and the bursa system in the chicken. *J Exp Med.* 1966;123:75-102.
170. Cooper MD, Gabrielsen AE, Good RA. Role of the thymus and other central lymphoid tissues in immunological disease. *Annu Rev Med.* 1967;18:113-138.
171. Roitt IM, Greaves MF, Torrigiani G, et al. The cellular basis of immunological responses. A synthesis of some current views. *Lancet.* 1969;294:367-371.
172. Miller JFAP, Osoba D. Current concepts of the immunological functions of the thymus. *Physiol Rev.* 1967;47:437-520.
173. Claman H, Chaperon EA, Triplett RF. Immunocompetence of transferred thymus-marrow cell combinations. *J Immunol.* 1966;97:828-832.
174. Davies AJS, Leuchars E, Wallis V, et al. The failure of thymus-derived cells to produce antibody. *Transplantation.* 1967;5:222-231.
175. Miller JFAP, Mitchell GF. Cell to cell interaction in the immune response I. Hemolysin-forming cells in neonatally thymectomized mice reconstituted with thymus or thoracic duct lymphocytes. *J Exp Med.* 1968;128:801-820.
176. Mitchell GF, Miller JFAP. Cell to cell interaction in the immune response II. The source of hemolysin-forming cells

in irradiated mice given bone marrow and thymus or thoracic duct lymphocytes. *J Exp Med.* 1968;128:821-837.
177. Mitchison NA. The carrier effect in the secondary response to hapten-protein conjugates. I. Measurement of the effect with transferred cells and objections to the local environment hypothesis. *Eur J Immunol.* 1971;1:10-17.
178. Mitchison NA. The carrier effect in the secondary response to hapten-protein conjugates. II. Cellular cooperation. *Eur J Immunol.* 1971;1:18-27.
179. Boak JL, Mitchison NA, Pattisson PH. The carrier effect in the secondary response to hapten-protein conjugates. 3. The anatomical distribution of helper cells and antibody-forming-cell-precursors. *Eur J Immunol.* 1971;1:63-65.
180. Bankhurst AD, Warner NL, Sprent J. Surface immunoglobulins on thymus and thymus-derived lymphoid cells. *J Exp Med.* 1971;134:1005-1015.
181. Lesley JF, Kettman JR, Dutton RW. Immunoglobulins on the surface of thymus-derived cells engaged in the initiation of a humoral immune response. *J Exp Med.* 1971;134:618-629.
182. Marchalonis JJ, Cone RE, Atwell JL. Isolation and partial characterization of lymphocyte surface immunoglobulins. *J Exp Med.* 1972;135:956-971.
183. Allison JP, McIntyre BW, Bloch D. Tumor specific antigens of murine T-lymphoma defined with monoclonal antibody. *J Immunol.* 1982;129:2293-2300.
184. Haskins K, Kubo R, White J, et al. The major histocompatibility complex restricted antigen receptor on T cells. I. Isolation with a monoclonal antibody. *J Exp Med.* 1983;157:1149-1169.
185. Meuer SC, Fitzgerald KA, Hussey RE, et al. Clonotypic structures involved in antigen-specific human T cell function: relationship to the T3 molecular complex. *J Exp Med.* 1983;157:705-719.
186. Yanagi Y, Yoshikai Y, Leggett K, et al. A human T cell-specific cDNA clone encodes a protein having extensive homology to immunoglobulin chains. *Nature.* 1984;308:145-149.
187. Hedrick SM, Cohen DI, Nielsen EA, et al. Isolation of cDNA clones encoding T cell-specific membrane-associated proteins. *Nature.* 1984;308:149-153.
188. Saito H, Kranz DM, Takagaki Y, et al. Complete primary structure of a heterodimeric T-cell receptor deduced from cDNA sequences. *Nature.* 1984;309:757-762.
189. Chien Y-H, Becker DM, Lindsten T, et al. A third type of murine T-cell receptor gene. *Nature.* 1984;312:31-35.
190. Saito H, Kranz DM, Takagaki Y, et al. A third rearranged and expressed gene in a clone of cytotoxic T lymphocytes. *Nature.* 1984;312:36-40.
191. Brenner MB, McLean J, Dialynes DP, et al. Identification of a putative second T-cell receptor. *Nature.* 1986;322:145-149.
192. Chien YH, Iwashima M, Kaplan KB, et al. A new T cell receptor gene located within the alpha locus and expressed early in T cell differentiation. *Nature.* 1987;327:677-682.
193. Dembic Z, Haas W, Weiss S, et al. Transfer of specificity by murine α and β T-cell receptor genes. *Nature.* 1986;320:232-238.
194. Billingham RE. Transplantation: past, present and future. *J Invest Dermatol.* 1963;41:165-180.
195. Little CC. A possible Mendelian explanation for a type of inheritance apparently non-Mendalian in nature. *Science.* 1914;40:904-906.
196. Little CC, Tyzzer EE. Further experimental studies on the inferitance of susceptibility to a transplantable tumor, carcinoma (J.W.A.) of the Japanese waltzing mouse. *J Med Res.* 1916;33:393-453.
197. Tyzzer EE, Little CC. Studies on the inheritance of susceptibility to a transplantable sarcoma (J.W.B.) of Japanese waltzing mice. *J Cancer Res.* 1916;1:387-389.
198. Gorer P. The detection of antigenic differences in mouse erythrocytes by the employment of immune sera. *Br J Exp Pathol.* 1936;17:42-50.
199. Gorer P. The genetic and antigenic basis of tumor transplantation. *J Pathol Bacteriol.* 1937;44:691-697.
200. Gorer P, Lyman S, Snell GD. Studies on the genetic and antigenic basis of tumor transplantation. Linkage between a histocompatibility gene and "fused" in mice. *Proc R Soc Lond B Biol Sci.* 1948;135:499-505.
201. Snell GD. Methods for the study of histocompatibility genes. *J Genet.* 1948;49:87-103.
202. Dausset J. Iso-leuco-anticorps. *Acta Haematol.* 1958;20:156-166.
203. van Rood JJ, Eernisse JG, van Leeuwen A. Leucocyte antibodies in sera from pregnant women. *Nature.* 1958;181:1735-1736.
204. Payne R, Rolfs MR. Fetomaternal leukocyte incompatibility. *J Clin Invest.* 1958;37:1756-1762.
205. Medawar PB. The behavior and fate of skin autografts and skin homografts in rabbits. A report to the War Wounds Committee of the Medical Research Council. *J Anat.* 1944;78:176-199.
206. Brent L, Brown J, Medawar PB. Skin transplantation immunity in relation to hypersensitivity. *Lancet.* 1958;272:561-564.
207. Levine BB, Ojeda A, Benacceraf B. Studies on artificial antigens. III. The genetic control of the immune response to hapten poly-L-lysine conjugates in Guinea pigs. *J Exp Med.* 1963;118:953-957.
208. Leuchars E, Cross AM, Dukor P. The restoration of immunological function by thymus grafting in thymectomized irradiated mice. *Transplantation.* 1965;3:28-35.
209. Aisenberg AC. Allogeneic thymus grafts and the restoration of immune function in irradiated thymectomized mice. *J Exp Med.* 1970;131:275-286.
210. Kindred B, Schreffler DC. H-2 dependence of co-operation between T and B cells in vivo. *J Immunol.* 1972;109:940-943.
211. Mosier DE. A requirement for two cell types for antibody formation in vitro. *Science.* 1967;158:1573-1575.
212. Cline MJ, Swett VC. The interaction of human monocytes and lymphocytes. *J Exp Med.* 1968;128:1309-1325.
213. Benacceraf B, McDevitt HO. Histocompatibility immune response genes. *Science.* 1972;175:273-279.
214. Shevach EM, Paul WE, Green I. Histocompatibility-linked immune response gene function in Guinea pigs. Specific inhibition of antigen-induced lymphocyte proliferation by alloantiserum. *J Exp Med.* 1972;136:1207-1221.
215. Rosenthal AS, Shevach EM. Function of macrophages in antigen recognition by Guinea pig T lymphocytes. I. Requirement for histocompatible macrophages and lymphocytes. *J Exp Med.* 1973;138:1194-1212.
216. Shevach EM, Rosenthal AS. Function of macrophages in antigen recognition by Guinea pig T lymphocytes. II. Role of the macrophage in the regulation of genetic control of the immune response. *J Exp Med.* 1973;138:1213-1229.
217. Zinkernagel RM, Doherty PC. Restriction of in vitro T cell-mediated cytotoxicity in lymphocytic choriomeningitis within a syngeneic or semiallogeneic system. *Nature.* 1974;248:701-702.
218. Caffrey MFP, James DCO. Human lymphocyte antigen association in ankylosing spondylitis. *Nature.* 1973;242:121.
219. Scholsstein L, Teresaki PI, Bluestone R, et al. High association of an HL-A antigen, W27, with ankylosing spondylitis. *N Eng J Med.* 1973;288:704-706.
220. Boyse EA, Miyazawa M, Aoki T, et al. Ly-A and Ly-B: two systems of lymphocyte isoantigens in the mouse. *Proc R Soc Lond B Biol Sci.* 1968;170:175-193.
221. Boyse EA, Itakura K, Stockert E, et al. Ly-C: a third locus specifying alloantigen expressed only on thymocytes and lymphocytes. *Transplantation.* 1971;11:351-352.
222. Shiku H, Kisielow P, Bean MA, et al. Expression of T cell differentiation antigens on effector cells in cell-mediated cytotoxicity in vitro. Evidence for functional heterogeneity related to the surface phenotype of T cells. *J Exp Med.* 1975;141:227-241.
223. Cantor H, Boyse EA. Functional subclasses of T lymphocytes bearing different Ly antigens. I. The generation of functionally distinct T-cell subclasses is a differentiative process independent of antigen. *J Exp Med.* 1975;141:1376-1389.
224. Cantor H, Boyse EA. Functional subclasses of T lymphocytes bearing different Ly antigens. II. Cooperation between different subclasses of Ly+ cells in the generation of killer activity. *J Exp Med.* 1975;141:1390-1399.
225. Kohler G, Milstein C. Continuous cultures of fused cells secreting antibody of predefined specificity. *Nature.* 1975;256:495-497.
226. Kung PC, Goldstein G, Reinherz EL, et al. Monoclonal antibodies defining distinctive human T cell surface antigens. *Science.* 1979;206:347-349.
227. Reinherz EL, Kung PC, Goldstein G, et al. Further characterization of the human inducer T cell subset defined by monoclonal antibody. *J Immunol.* 1979;123:2894-2896.
228. Reinherz EL, Kung PC, Goldstein G, et al. Separation of functional subsets of human T cells by a monoclonal antibody. *Proc Natl Acad Sci U S A.* 1979;76:4061-4065.
229. Reinherz EL, Kung PC, Goldstein G, et al. A monoclonal antibody reactive with the human cytotoxic/suppressor T cell subset previously defined by a heteroantiserum termed TH2. *J Immunol.* 1980;124:1301-1307.
230. IUIS-WHO Nomenclature Subcommittee. Nomenclature for clusters of differentiation (CD) of antigens defined on human leukocyte populations. *Bull World Health Organ.* 1984;62:809-811.
231. Babbitt BP, Matsueda G, Haber E, et al. Antigenic competition at the level of peptide-Ia binding. *Proc Natl Acad Sci U S A.* 1986;83:4509-4513.
232. Engleman EG, Benike CJ, Grumet FC, et al. Activation of human T lymphocyte subsets: helper and suppressor/cytotoxic T cells recognize and respond to distinct histocompatibility antigens. *J Immunol.* 1981;127:2124-2129.
233. Mueller DJ, Jenkins MK, Schwartz RH. Clonal expansion versus clonal inactivation: a co-stimulatory signaling pathway determines the outcome of T cell antigen receptor occupancy. *Annu Rev Immunol.* 1989;7:445-480.
234. Lafferty KJ, Cunningham AJ. A new analysis of allogeneic interactions. *Aust J Exp Biol Med Sci.* 1975;53:27-42.
235. Lum LG, Orcutt-Thordarson N, Seigneuret MC, et al. In vitro regulation of immunoglobulin synthesis by T-cell subpopulations defined by a new human T-cell antigen (9.3). *Cell Immunol.* 1982;72:122-129.
236. Linsley PS, Clark EA, Ledbetter JA. T-cell antigen CD28 mediates adhesion with B cells by interacting with activation antigen B7/BB-1. *Proc Natl Acad Sci U S A.* 1990;87:5031-5035.
237. Linsley PS, Brady W, Grosmarie L, et al. Binding of the B cell activation B7 to CD28 costimulates T cell proliferation and interleukin 2 mRNA accumulation. *J Exp Med.* 1991;173:721-730.
238. Kasakura S, Lowenstein L. A factor stimulating DNA synthesis derived from the medium of leukocyte cultures. *Nature.* 1965;208:794-795.
239. Gordon J, MacLean D. A lymphocyte-stimulating factor produced in vitro. *Nature.* 1965;208:795-796.
240. Morgan DA, Ruscetti FW, Gallo R. Selective in vitro growth of T lymphocytes from normal human bone marrow. *Science.* 1976;193:1007-1008.
241. Ruscetti FW, Morgan DA, Gallo RC. Functional and morphological characterization of human T cells continuously grown in vitro. *J Immunol.* 1977;119:131-138.
242. Mossman TR, Cherwinski H, Bond MW, et al. Two types of murine helper T cell clone. I. Definition according to profiles of lymphokine activities and secreted proteins. *J Immunol.* 1986;136:2348-2357.
243. McGhee JR. The world of Th1/Th2 subsets: first proof. *J Immunol.* 2005;175:3-4.
244. Breitfeld D, Ohl L, Kremmer E, et al. Follicular B helper T cells express CXC chemokine receptor 5, localize to B cell follicles, and support immunoglobulin production. *J Exp Med.* 2000;192:1545-1551.
245. Schaerli P, Willimann K, Lang AB, et al. CXC chemokine receptor 5 expression defines follicular homing T cells with B cell helper function. *J Exp Med.* 2000;192:1553-1562.
246. Park H, Li Z, Yang XO, et al. A distinct lineage of CD4 T cells regulated tissue inflammation by producing interleukin 17. *Nat Immunol.* 2005;6:1133-1141.
247. Harrington LE, Hatton RD, Mangan PR, et al. Interleukin 17-producing CD4+ effector T cells develop via a lineage distinct from the T helper type 1 and 2 lineages. *Nat Immunol.* 2005;6:1123-1132.
248. Nishizuka Y, Sakakura T. Thymus and reproduction: sex-linked dysgenesis of the gonad after neonatal thymectomy in mice. *Science.* 1969;166:753-755.
249. Sakaguchi S, Sakaguchi N, Asano M, et al. Immunologic self-tolerance maintained by activated T cells expressing IL-2 receptor α-chains (CD25). Breakdown of a single mechanism of self-tolerance causes various autoimmune diseases. *J Immunol.* 1995;155:1151-1164.
250. Gershon RK, Kondo K. Cell interactions in the induction of tolerance: the role of thymic lymphocytes. *Immunology.* 1970;18:723-737.
251. Gershon RK, Cohen P, Hencin R, et al. Suppressor T cells. *J Immunol.* 1972;108:586-590.
252. Möller G. Do suppressor T cells exist? *Scand J Immunol.* 1988;27:247-250.
253. Brunkow ME, Jeffery EW, Hjerrild KA, et al. Disruption of a new forkhead/winged-helix protein, scurfin, results in the fatal lymphoproliferative disorder of the scurfy mouse. *Nat Genet.* 2001;27:68-73.
254. Powell BR, Buist NRM, Stenzel P. An X-linked syndrome of diarrhea, polyendocrinopathy, and fatal infection in infancy. *J Pediatr.* 1982;100:731-737.
255. Wildin RS, Smyk-Pearson S, Filipovich AH. Clinical and molecular features of the immunodysregulation, polyendocrinopathy, enteropathy, X-linked (IPEX) syndrome. *J Med Genet.* 2001;39:537-545.
256. Fontenot JD, Gavin MA, Rudensky AY. Foxp3 programs the development and function of CD4+CD25+ regulatory T cells. *Nat Immunol.* 2003;4:330-336.
257. Ehrlich P, Morgenroth J. Uber Hämolysine. *Berl Klin Wochenschr.* 1901;38:251-257.
258. Witebsky E, Rose NR, Terplan K, et al. Chronic thyroiditis and autoimmunization. *J Am Med Assoc.* 1957;164:1439-1447.
259. Rose NR. The discovery of thyroid autoimmunity. *Immunol Today.* 1991;12:167-168.
260. Owen RD. Immunogenetic consequences of vascular anastomoses between bovine twins. *Science.* 1945;102:400-401.
261. Anderson D, Billingham RE, Lampkin GH, et al. The use of skin grafting to distinguish between monozygotic and dizygotic twins in cattle. *Heredity.* 1951;5:379-397.
262. Billingham RE, Lampkin GH, Medawar PB, et al. Tolerance to homografts, twin diagnosis, and the freemartin condition in cattle. *Heredity.* 1952;6:201-212.
263. Hašek M. Vegetative hybridization of animals by joint blood circulation during embryonal development. 1953. Translated from Czech by Ivanyi J. *Transplantation.* 2003;76:1412-1421.
264. Brunet J-F, Denizot F, Luciani M-F, et al. A new member of the immunoglobulin superfamily – CTLA-4. *Nature.* 1987;238:267-270.
265. Gross JA, St John T, Allison JP. The murine homologue of the T lymphocyte antigen CD28. Molecular cloning and cell surface expression. *J Immunol.* 1990;144:3201-3210.
266. Linsley PS, Brady W, Urnes M, et al. CTLA-4 is a second receptor for the B cell activation antigen B7. *J Exp Med.* 1991;174:561-569.
267. Lenschow DJ, Zeng Y, Thistlethwaite JR, et al. Long-term survival of xenogeneic pancreatic islet grafts induced by CTLA4Ig. *Science.* 1992;257:789-792.
268. Waterhouse P, Penninger JM, Timms E, et al. Lymphoproliferative disorders with early lethality in mice deficient in Ctla-4. *Science.* 1995;270:985-988.
269. Tivol EA, Borriello F, Schweitzer AN, et al. Loss of CTLA-4 leads to massive lymphoproliferation and fatal multiorgan tissue destruction revealing a critical negative regulatory role of CTLA-4. *Immunity.* 1995;3:541-547.
270. Krummel MF, Allison JP. CD28 and CTLA-4 have opposing effects on the response of T cells to stimulation. *J Exp Med.* 1995;182:459-465.
271. Ishida Y, Agata Y, Shibihara K, et al. Induced expression of PD-1, a novel member of the immunoglobulin gene superfamily, upon programmed cell death. *EMBO J.* 1992;2:3887-3895.
272. Nishimura H, Nose M, Hlai H, et al. Development of lupus-like autoimmune diseases by disruption of the PD-1 gene encoding an ITIM motif-carrying immunoreceptor. *Immunity.* 1999;11:141-151.
273. Jerne NK. Summary: waiting for the end. *Cold Spring Harb Symp Quant Biol.* 1967;32:591-603.

SECTION II

Evolution of the Immune System

CHAPTER 3

The Insect Immune System

Sara Cherry, PhD • Nicolas Buchon, PhD

OVERVIEW OF THE INSECT IMMUNE SYSTEM

Insects, like all organisms on earth, live in a microbe-rich environment and are therefore constantly exposed to microbes, both pathogenic and benign.[1] While they lack an acquired immune system, insects have evolved an innate immune system that can readily deal with such challenges.[2,3] While many have suggested that "lower organisms" have short lifespans and thus do not require as sophisticated an immune system as vertebrates, this is not the case. Some invertebrates live decades while many vertebrates have short lifespans. This chapter will discuss the mechanisms used by insects and other invertebrates to protect themselves from infection.

Pathogens can colonize different organs of the insect, including the surface of the external cuticle and the lumen of the digestive tract, and colonization can lead to invasion and subsequent infection of internal tissues. Insect pathogens include all classes of microbes as well as eukaryotic parasites. Bacteria, fungi, viruses, and nematodes can all be pathogenic to insects. Some pathogens, called entomopathogens, can infect a broad range of insects.[4] For instance, the fungi *Beauveria bassiana* and *Metarhizium anisopliae* are both able to infect multiple insect species by breaching the insect cuticle. Other pathogens are specialized to a small subset of insect species. This is particularly true for viruses. The natural virus of *Drosophila melanogaster*, Drosophila C virus, cannot infect even closely related drosophilids. Vector-borne viruses such as Zika virus are transmitted by some species of mosquitoes (eg, *Aedes aegypti*) but not others (eg, *Anopheles gambiae*).[5,6] In addition to pathogens, insects are associated with diverse nonpathogenic microbes that compose their microbiota. The association of insects with microbiota ranges from very specific interactions to a loose association with environmental microbes.[7] Finally, most insects harbor endosymbiotic microbes that are transmitted vertically in the germline. For instance, Wolbachia is found in more than 40% of all insect species, making it one of the most successful endosymbionts on earth, and has recently been shown to provide protection from viral infections leading to the use of this bacterium to control human arthropod-borne viral infections.[8,9]

To control infection by pathogens, insects have multiple layers of defense, including physical and chemical barriers, as well as innate immune responses, including both the humoral and cellular immune systems[10] (Fig. 3.1). A first physical barrier that any pathogen has to cross is the insect cuticle.[11] The exoskeleton of insects covers all outside surfaces, including the digestive tract and respiratory system (tracheae). Some entomopathogenic microbes, including the fungus *B. bassiana*, can penetrate this barrier by secreting enzymes capable of degrading chitin, the core constituent of the cuticle. Other pathogens can penetrate the insect body cavity following physical injury. Many pathogens can be transmitted orally.[1] However, in the digestive tract, chemical barriers (pH) and digestive enzymes add a layer of protection to the presence of physical barriers, such as the pharyngeal armatures, or additional chitinous barriers, such as the peritrophic matrix.[12] Several entomopathogens have evolved potent mechanisms to pass through the gut barrier. Mechanisms include resistance to digestive proteases, ability to degrade the secreted matrix (peritrophic matrix), and possibly the secretion of toxins able to perturb the gut epithelium, thus disrupting gut barrier integrity. A classic example of such toxins is the Cry toxins of *Bacillus thuringiensis* that are now used as pesticides in transgenic plants.[13]

In addition to physical and chemical barriers, the insect has epithelial barriers lining the digestive, respiratory, and reproductive tracts. The epithelium is directly in contact with microbes and can mount local antimicrobial responses as well as inducing systemic responses. In addition, infection triggers conserved innate cellular and humoral immune responses. The cellular response is carried out by dedicated immune cells

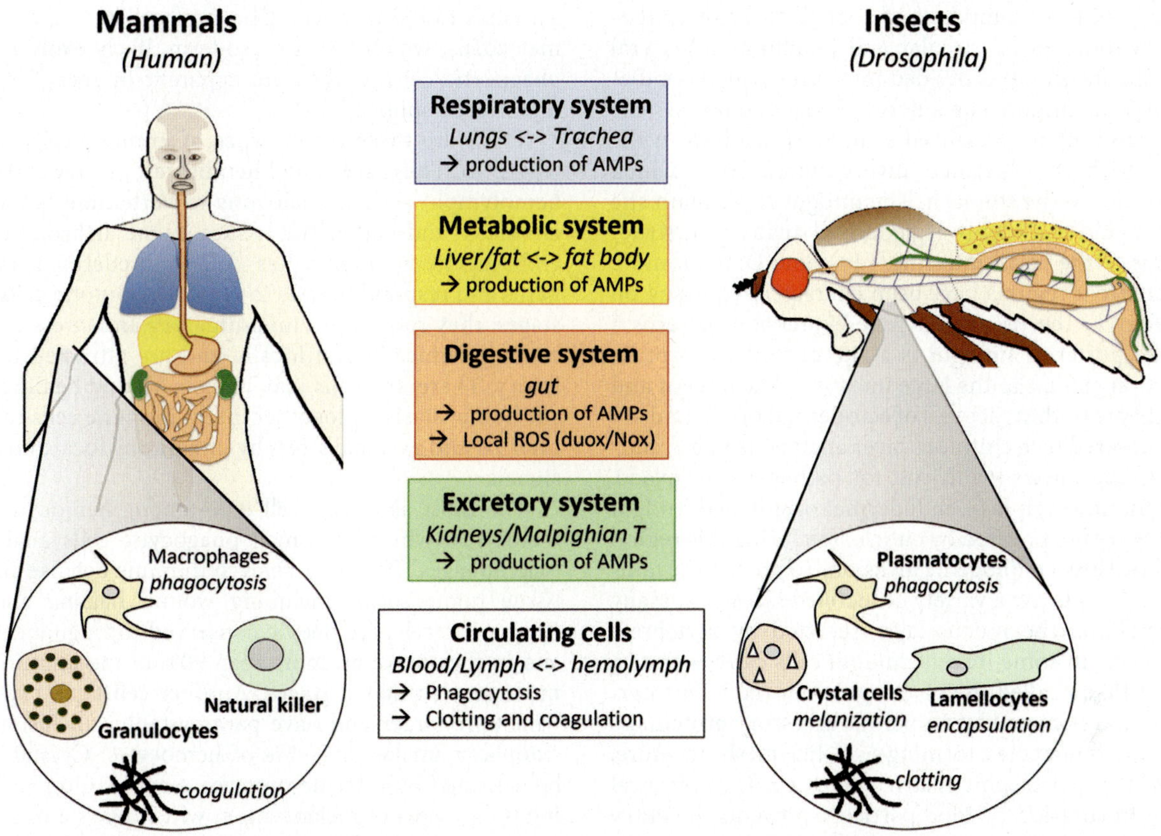

FIG. 3.1. Comparative anatomical structure of mammals and insects. We highlight parallel systems and structures as well as immune cells. Several organ systems are well conserved between insects and mammals. Insects possess an open circulatory system that consists of tracheal tubes that bring oxygen directly to tissues. This interface with the outside world can produce antimicrobial peptides (AMPs). Nutrients are acquired in the gut and stored in the fat body, which is a functional analog of the liver and fat tissues. The excretory system is composed of malpighian tubules that assume a similar function than the mammalian kidney. Most insect tissues have been shown to mount immune responses. The malpighian tubules and fat body secrete AMPs in circulation, especially in response to septicemia (called systemic infection). Barrier epithelia such as the gut and the trachea also secrete AMPs. The midgut epithelium also generates reactive oxygen species (ROS) to target gut microbes. As in mammals, circulating cells called hemocytes circulate in insect hemolymph and act as professional immune cells, responsible for phagocytosing pathogens, coagulation, and melanization. (Redrawn by permission from Springer Nature: Buchon N, Silverman N, Cherry S. Immunity in Drosophila melanogaster-from microbial recognition to whole-organism physiology. *Nat Rev Immunol.* 2014;14(12):796-810. doi:10.1038/nri3763.)

that are capable of phagocytosis of small microbes or encapsulation of bigger pathogens, as well as alerting the organism to the invaders.[14,15] A second arm is the induction of secreted immune effector molecules, termed humoral defenses.[3] A specialized organ, the fat body, produces many of these mediators. Epithelial cells can also mount a response and produce many secreted factors that impact systemic responses.[1] Major immune effectors include reactive oxygen species (ROS), antimicrobial peptides (AMPs), and a melanization reaction.

BARRIERS: INSECT CUTICLE AND EPITHELIAL BARRIERS

All organisms have external protective barriers (Fig. 3.1). The exoskeleton, or cuticle, of insects represents the first barrier to injury and infection by pathogens.[16] The exoskeleton is composed of multiple layers of distinct compositions, often comprising polymers of chitin associated with accessory proteins and lipids. The outer layer of the cuticle, also called the epicuticle, is often formed by a mixture of free fatty acids and wax esters that build a layer of wax. A specific cell type, the oenocytes, synthesizes these cuticular hydrocarbons responsible for repulsion of hydrophilic molecules.[17] The development and maintenance of the cuticle parallels that of the mammalian dermis, as cuticle is synthesized by a basal layer of epidermal cells that sit underneath the endocuticle. This is a formidable and impermeable barrier for pathogens as well as providing protection from injury.[18]

To breathe, insects use a branched epithelial tubular system called the tracheal system[19] (Fig. 3.1). Oxygen is conveyed directly to tissues through openings in the cuticle called spiracles and a network of primary, secondary, and tertiary tracheae. The cells that compose the tracheal epithelium have the ability to respond to invading pathogens such as fungal spores or airborne bacteria.

Another surface that faces the outside world is the digestive tract[1,20] (Fig. 3.1). The digestive tract of insects has many parallels with the vertebrate gut[21] (Figs. 3.1 and 3.9A).

It is a complex tube composed of a single layer of epithelial cells surrounded by circular and longitudinal visceral muscles. The insect gut is divided into three regions of distinct embryonic origin. The anterior foregut comprises the esophagus and often specialized structures involved in the storage of food (for instance, diverticulum, crop) which has similarities to the stomach. The midgut is the main site of digestion and is a highly regionalized anatomical structure similar to the small intestine.[22] In some Diptera, more than 10 different regions have been characterized based on the histology of the tissue and gene expression patterns.[22] Finally, the posterior hindgut is a site of both absorption and excretion similar to the large intestine. The foregut and the hindgut are both structures of ectodermal origin and are therefore covered by a chitinous layer analogous to the cuticle. This lining makes it difficult for pathogens to invade. The midgut, however, is of endodermal origin, and midgut enterocytes are not covered by cuticle, rendering this region vulnerable to direct pathogenic invasion. To protect the midgut surface, insects use a variety of secreted factors, in many cases resembling the mucus layer secreted by vertebrate intestinal cells. In some insects, midgut cells can generate a network of fibers called the peritrophic matrix.[23] This peritrophic matrix is coated by numerous accessory proteins as well as digestive enzymes, forming a gel-like mesh structure. Altogether this peritrophic matrix serves a role of physical protection to abrasion by food particles, prevents the entry of pathogens, and filters toxins and viruses.[23] Some insects do not produce a peritrophic matrix but instead secrete a layer of lipoproteins called the perimicrovillar membrane that serves a similar purpose. In addition, all insects encode mucin-like proteins that are expressed in the gut and are likely secreted into the lumen for increased protection from pathogenic challenge.[24]

Many insects subsist on a diet of rotting material that is high in fungal and bacterial organisms. Therefore, oral ingestion of microbes is a very common route of infection. In addition, all metazoans, including insects, harbor a microbiota which largely resides in the intestinal tract.[25] For most insects, the midgut contains a complex microbiota that plays roles in development, immunity, and homeostasis. The midgut epithelium thus integrates diverse signals including those from commensals and pathobionts, requiring both robust immune responsiveness and immune tolerance to symbiotic microbiota. Maintaining this fine balance between the positive and negative regulation of the production of antimicrobial effectors and ROS in the gut is key to an insect's immune defense and homeostasis.[26,27]

ACTORS OF THE INVERTEBRATE IMMUNE SYSTEM

Cellular Immunity and Hematopoiesis

The development of specialized immune cells allows organisms to use surveillance mechanisms to defend from pathogenic challenge and damage. Insects have developed a plethora of immune cells that play fundamental roles in development, wound healing, and immune defense. The primary cell types are functionally similar to myeloid lineages and play important roles in phagocytosis and other specialized cell biological activities to kill invaders. Choanoflagellate-like ancestors of metazoans, which lack a mesoderm, likely evolved the first phagocytic cells as they are apparent in several species of diploblastic sponges.

There are three major types of immune cells in insects (Fig. 3.2). These are called hemocytes, as they reside in the hemolymph, a fluid analogous to vertebrate blood, which circulates and bathes the tissues in the arthropod. During embryogenesis, hemocytes follow predefined migratory paths and respond to cues to remove apoptotic cells. In later stages, they respond to inflammatory and stress conditions both systemically and locally and are attracted to sites of injury. There, the cells play roles in wound healing and tissue homeostasis. Altogether insect immune cells respond to internal and external stress by integrating local and systemic signals.

The most abundant cell type among hemocytes is plasmatocytes, which are small phagocytic cells analogous to macrophages. They play roles in immune defense along with tissue homeostasis, including wound healing and repair. Phagocytic cells are deeply conserved throughout evolution and in flies make up more than 90% of the immune cells at most developmental stages. Another cell type, crystal cells, which are larger and have paracrystalline inclusions in the cytoplasm, makes up ~5% of hemocytes. Crystal cells can be activated by a serine protease cascade, ultimately inducing the process of melanization, which plays a major role in clotting and wound healing (Fig. 3.3). A third hemocyte type, the lamellocyte, is a large, flat, and adhesive cell that are not present under homeostatic conditions, but rather are induced by particular challenges.[28] They can be stress- or microbe-induced and their role is to encapsulate large objects including insect eggs that are deposited in the body cavity by other insects.[29,30] Thus, the role of lamellocytes is to capture foreign particles larger than those that can be phagocytosized by plasmatocytes. The products of this encapsulation reaction resemble granulomas in vertebrates.[31,32] In addition, there is crosstalk between hemocytes. For example, lamellocytes induce crystal cells to produce melanin and together they orchestrate the killing of parasitic invaders.

There are two major waves of hemocyte development.[33,34] During embryogenesis, the hemangioblast, which is a common precursor of hematopoietic and vascular cells[35-37] from flies to vertebrates, produces a population of immune cells largely comprised of plasmatocytes. These develop in the head mesoderm and migrate from their point of origin to disperse along a preprogrammed pathway across the embryo. The first plasmatocytes produced during embryogenesis are long-lived, persisting for the lifespan of the animal. This parallels some tissue-resident myeloid cells produced during embryogenesis in vertebrates. Crystal cells, making up only 5% of the differentiated cells, also originate in the embryo, differentiating from the same precursor population, and remain tissue resident.

A second wave of hematopoiesis occurs later in embryogenesis, when the paired lobes of the lymph gland develop in the dorsal thoracic mesoderm. The lymph gland cells arise from two populations of cells, one of which is stem cell–like and the other that can differentiate either into a cardiac cell or

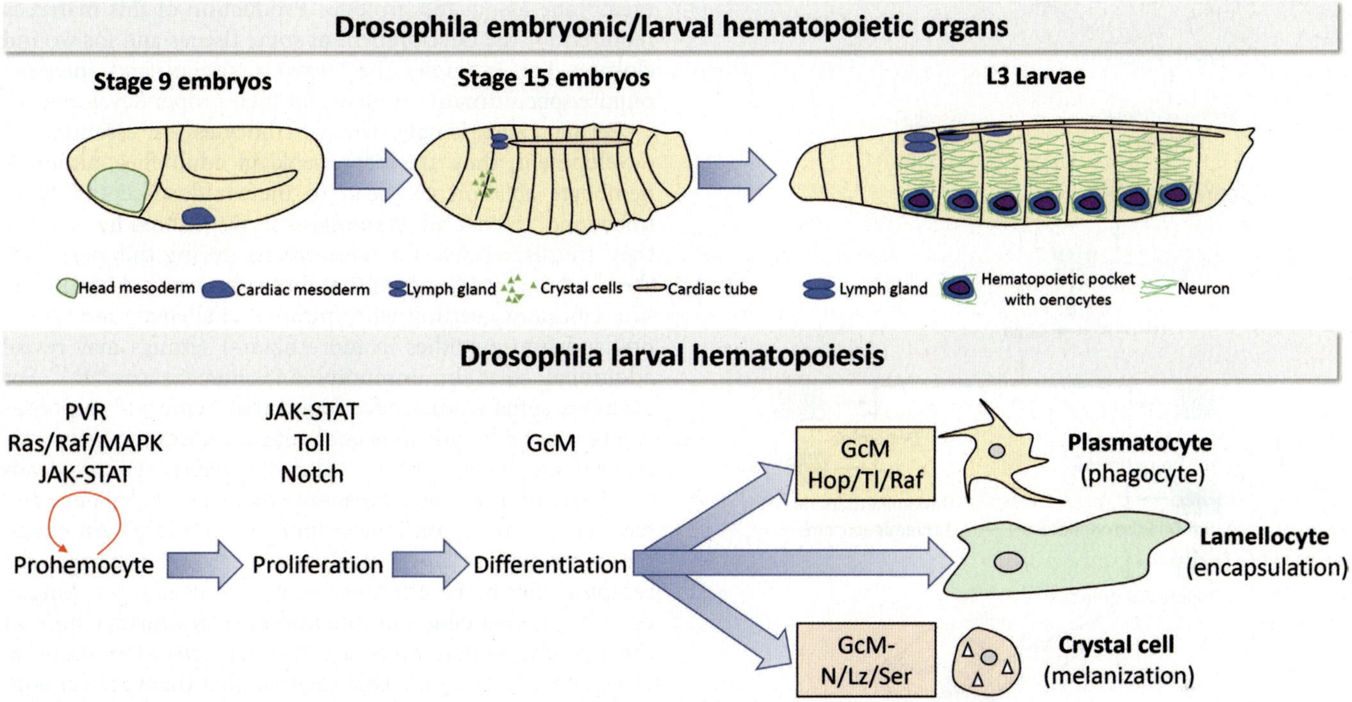

FIG. 3.2. Insect hematopoiesis and hematopoietic organs. Top panel: During embryogenesis and larval development, waves of immune cells develop. Prohemocytes derive from the head mesoderm, while lymph gland cells derive from the cardiogenic mesoderm. In the embryo, crystal cells form a cluster in the anterior segments of the embryo. The lymph gland is located in proximity of the heart in both embryo and larvae. In larvae, hemocytes circulate in hemolymph and accumulate in proximity of peripheral sensory neurons where they form hematopoietic pockets. Bottom panel: Conserved signaling pathways drive immune cell development. The PVR, JAK–STAT, and MAPK pathways are involved in the maintenance of prohemocytes while the JAK–STAT, Notch, and Toll pathways promote their proliferation. These pathways, in conjunction with key transcription factors such as GcM, also determine differentiation of prohemocytes in different hemocyte classes. (Redrawn from Morin-Poulard I, Tian Y, Vanzo N, Crozatier M. Drosophila as a model to study cellular communication between the hematopoietic niche and blood progenitors under homeostatic conditions and in response to an immune stress. *Front Immunol.* 2021;12:3203. doi:10.3389/FIMMU.2021.719349/BIBTEX.)

a lymph gland cell regulated by canonical cell fate pathways, including Notch signaling. Once the lymph gland forms, it continues to proliferate during larval development and is the main site of hematopoiesis during this developmental phase. There are three structurally distinct zones within the lymph gland similar to secondary hematopoietic organs in higher organisms. There is an outer cortical zone, an inner medullary zone that houses the immature cells, and a posterior signaling center. As the immature cells differentiate, they move into the outer zone. The transcription factors that govern these lineage specifications in flies are conserved in mammals. More recent single-cell sequencing studies of larval hemocytes have revealed additional complexity and developmental fates with parallels to mammals.[38]

The lymph gland is a reservoir of plasmatocytes, while the circulating plasmatocytes and those that line the body cavity in the larvae are largely embryonic. However, upon parasitization by wasps that lay eggs in the body cavity of flies, large numbers of hemocytes are released to attack the foreign body. In addition, during metamorphosis, the lymph glands release hemocytes for tissue remodeling as the organs are rebuilt in the adult. Interestingly, there are no hematopoietic organs in adult flies. Thus, all hemocytes in the adult arose in earlier stages in development. Embryonic hemocytes make up most immune cells upon eclosion of the adult from the pupa.[39,40] There is no evidence of de novo immune cell development in the adult, even upon immune challenge.[39] Consistent with this observation, there is a decline in hemocyte number as insects age as well as a decline in the ability to phagocytose foreign bodies.[41,42] These immune cell lineages not only play roles in immunity but also tissue-specific roles in development and homeostasis.[29,43] Moreover, many activities of innate immune cell lineages are defined by their local microenvironments.[44] In vertebrates, there are known functional distinctions in myeloid cells that can be driven by developmental stage,[45,46] injury, and immune challenge.[38,47-49] Hemocytes also display this specificity and play similar roles across development in insects.

Across organisms, innate immune cells are migratory and use similar mechanisms for motility including polarized actin-rich protrusions such as filopodia and lamellipodia.[34] During development, hemocytes migrate by sensing chemotactic signals leading to directional movement in stereotypic waves to clear apoptotic debris. Plasmatocytes also respond to signals from changes in the extracellular matrix and signals produced by stressed, infected, or wounded cells. Hemocytes migrate toward and engulf many diverse targets, but it remains unclear how many of these complex decisions are made. Efforts to define the receptors have led to the discovery of many factors involved. For example,

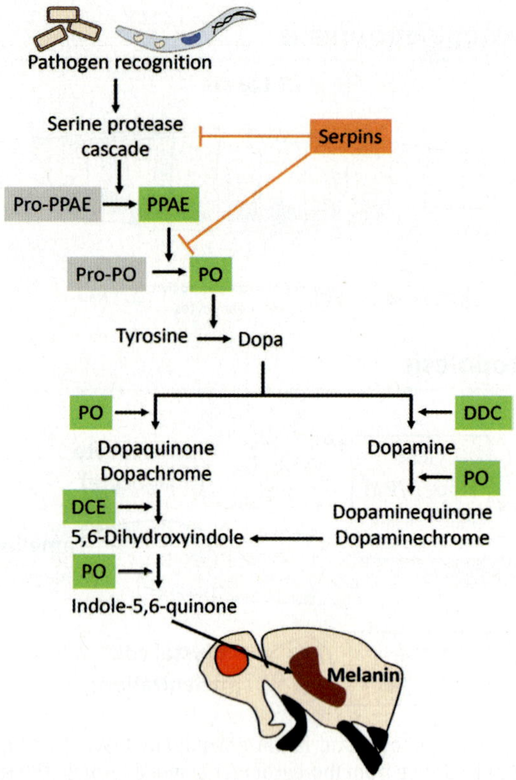

FIG. 3.3. The insect melanization cascade. Redrawn from Ref. 92. Pathogen recognition leads to the induction of an enzymatic (proteolytic) cascade that leads to the cleavage and maturation of prophenoloxidase in phenol oxidase. This enzyme converts tyrosine in DOPA, which will result in the generation of pigments (melanin) that accumulate at the site of infection or injury. Melanization is involved in the control of infection and wound healing.

the chemotactic signals during development are known to involve canonical PDGF/VEGF signals. Some signals from invading microbes are dependent on scavenger receptors, yet signals from many stimuli including from the wound site remain unclear.

At the wound site, hemocytes induce wound healing and coagulation.[50] Coagulation occurs in three phases. First, flies encode transglutaminase which is a homolog of the vertebrate clotting factor XIIIa. Transglutaminase uses the protein Fondue as its substrate to form a clot matrix. This is the first step in wound closure. Next, there is a scab and syncytium formation. Here the initial clot is further hardened by crosslinking and the action of melanization. This produces a substrate to allow the epidermal cells surrounding the wound to migrate across the tissue and fuse to form a large syncytium. As additional epidermal cells are attracted to the initial clot, a larger syncytium forms that is the final form of the wound site, which remains pigmented due to the melanin deposits.

A highly conserved role of immune cells is to produce extracellular matrix proteins and this is the case in insects. Hemocytes produce a diverse set of extracellular matrix proteins, including collagens, proteoglycans, and other basement membrane-associated proteins. Production of this matrix is required for the development of some tissues and for wound closure. For example, the nervous system and intestine require specific matrix proteins for their proper development.

Perhaps surprisingly, while hemocytes are essential for development, they are dispensable in adult flies, although hemocyte ablation can lead to increased susceptibility to infection.[51-54] Since adult animals survive without hemocytes, they are dispensable for homeostasis during this period. It should, however, be noted that these studies were conducted in a laboratory setting with minimal challenges and stressors, and future studies in more natural settings may reveal additional roles in immunity and homeostasis.[39,51-54] For example, some studies have found that hemocyte responses can be altered by previous encounters, including phagocytosis or tissue damage.[55-59] This altered behavior typically leads to a faster or more robust response to a second challenge and has been termed "immune priming or training." An example is the expression of an altered repertoire of phagocytic receptors for more effective secondary challenges. Single-cell RNA sequencing and functional studies further support the hypothesis that there are distinct activation states in Drosophila hemocytes. This suggests that there are subpopulations with distinct roles in immune defense, metabolism, and tissue homeostasis.[38,45-49] It is also possible that hemocytes play more important or complex roles in longer-lived insects.

The Fat Body and Systemic Immunity

The fat body is an insect-specific organ, functionally analogous to a combination of mammalian adipose tissue and liver, that has a central role in the systemic immune response, as well as energy storage and expenditure[60](Fig. 3.1). The fat body also has a key role during insect reproduction in females, as it sustains metabolically demanding oogenesis.[61] As a secretory organ, the fat body is the main site of the production and secretion of AMPs, which are a key player in defense against the systemic invasion of bacterial and fungal pathogens.[2,3] The fat body lines the body cavity between the hemolymph and the outside cuticle all along the dorsal side of the animal.[61] Fat body cells are bathed in circulating hemolymph, allowing the fat body to intimately interact with systemic circulation. While multiple cell types have been described in the fat body of different insects, adipocytes and AMP production are conserved in all insect fat bodies, suggesting that the basal functions of this organ are metabolic homeostasis and immune control. However, the fat body is not a homogenous tissue, and functional specialization of different regions of the fat body has been described in some insects.[62]

Crosstalk Between Hemocytes and the Fat Body

There is crosstalk between the hemocytes, which are largely responsible for cellular immunity, and the fat body, which largely controls humoral immunity. For example, a hemocyte-produced cytokine, Upd-3, can activate downstream signaling in the fat body to promote immune defense and wound healing.[63] Hemocytes also produce the cytokine Spätzle, which is processed by a serine protease cascade in the hemolymph and required for the Toll signaling pathway that induces AMP synthesis in the fat body.[64] However, while hemocyte ablation impacts humoral responses to bacterial infection in larvae, it does not have this effect in adults.[52,53,64] Thus, Spätzle is produced by other cell types in the adult. This difference likely reflects the distinct lifestyle of the larva, which is submersed in rotting food and is thus continually challenged by a greater quantity of microbes than the adult. Another classical instance of crosstalk involves ROS. Intestinal stress induces ROS in the gut, which signals to hemocytes via a nitric oxide (NO)-dependent pathway, leading to increased AMP production by the fat body.[65] Hemocytes can also amplify antiviral pathways in some cases; however, ablation of hemocytes has only modest or no impact on many viral infections in Drosophila.

IMMUNE RECOGNITION AND SIGNAL TRANSDUCTION

Innate immune responses are initiated by the recognition of microbes, which depends for the most part on the detection of conserved microbial molecules or patterns called PAMPs[66] (pathogen-associated molecular patterns) (Fig. 3.4). For bacteria and fungi, major PAMPs are conserved molecules that compose the outer wall of these organisms (eg, bacterial peptidoglycan [PGN] and fungal beta glucan).[67] Since viruses present fewer targets on their outer surface, viral nucleic acids are major PAMPS discussed below. To recognize these PAMPs, organisms encode pattern recognition receptors (PRRs) that can bind PAMPs and trigger immune signaling to activate the immune system.[67] In addition to PAMP detection, cells can detect virulence factors. Classic examples are pathogen-encoded proteases that cleave and activate immune sensors such as serine proteases.[68]

Bacterial Recognition: Peptidoglycan and PGRPs

Bacterial cell walls are chiefly composed of PGN, which is a polymeric molecule comprising a backbone of glycan chains formed by β-(1,4)-*N*-acetylglucosamine (NAG) and *N*-acetylmuramic acid (NAM), linked together by linker molecules.[69] The linker molecules vary for different types of bacteria, but in most cases the linker includes either a lysine for Gram-positive bacteria (Lys-type PGN) or a diaminopimelic acid for Gram-negative bacteria (DAP-type PGN).[70,71] The recognition of PGN by PGRPs involves the binding of the PGN linker, which results in selective binding of either DAP-type containing bacteria or Lys-type containing bacteria. In Drosophila, DAP-type PGN leads to the activation of the Imd pathway while Lys-type PGN triggers the activation of the Toll pathway[72] (Fig. 3.5). Engagement of these different signaling pathways leads to the activation of overlapping and distinct groups of effector molecules.[3] The recognition of different PGN types exists in most insects; however, their downstream responses are less well characterized.

PGN recognition proteins (PGRPs) are a diverse family with homology to lysozymes.[69] These proteins bind PGN in distinct compartments: some are secreted, others are membrane-bound and the last group is cytoplasmic. PGRPs can be classified with respect to their enzymatic activity. Some PGRPs have retained their amidase activity and the ability to cleave PGN, while other non-enzymatic PGRPs can bind to PGN without degrading it, enabling them to act as PRRs.[73,74]

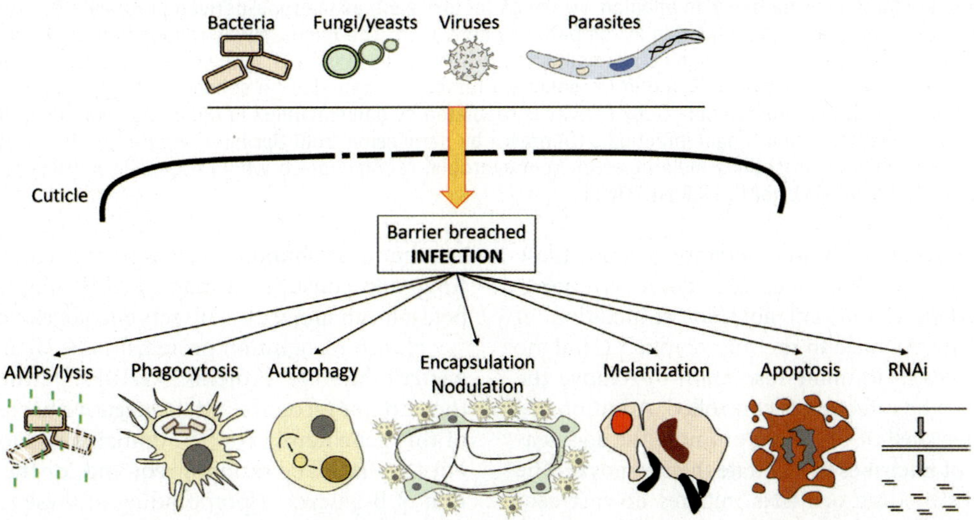

FIG. 3.4. Overview of insect immune effector mechanisms and antimicrobial defenses. Pathogens of insects include bacteria, fungi and yeasts, viruses, and parasites. Infection proceeds when these pathogens can cross the cuticular barriers of insects. The main antimicrobial effector mechanisms include the secretion of antimicrobial peptides (AMPs) leading to the lysis of pathogens, the phagocytosis of pathogens by circulating heocytes, autophagy, the encapsulation and nodulation of bigger invaders such as eggs of parasitic wasps by hemocytes, the melanization reaction, apoptosis, and RNA interference. (Redrawn from Hillyer JF. Insect immunology and hematopoiesis. *Dev Comp Immunol.* 2016;58:102-118. doi:10.1016/J.DCI.2015.12.006, with permission from Elsevier.)

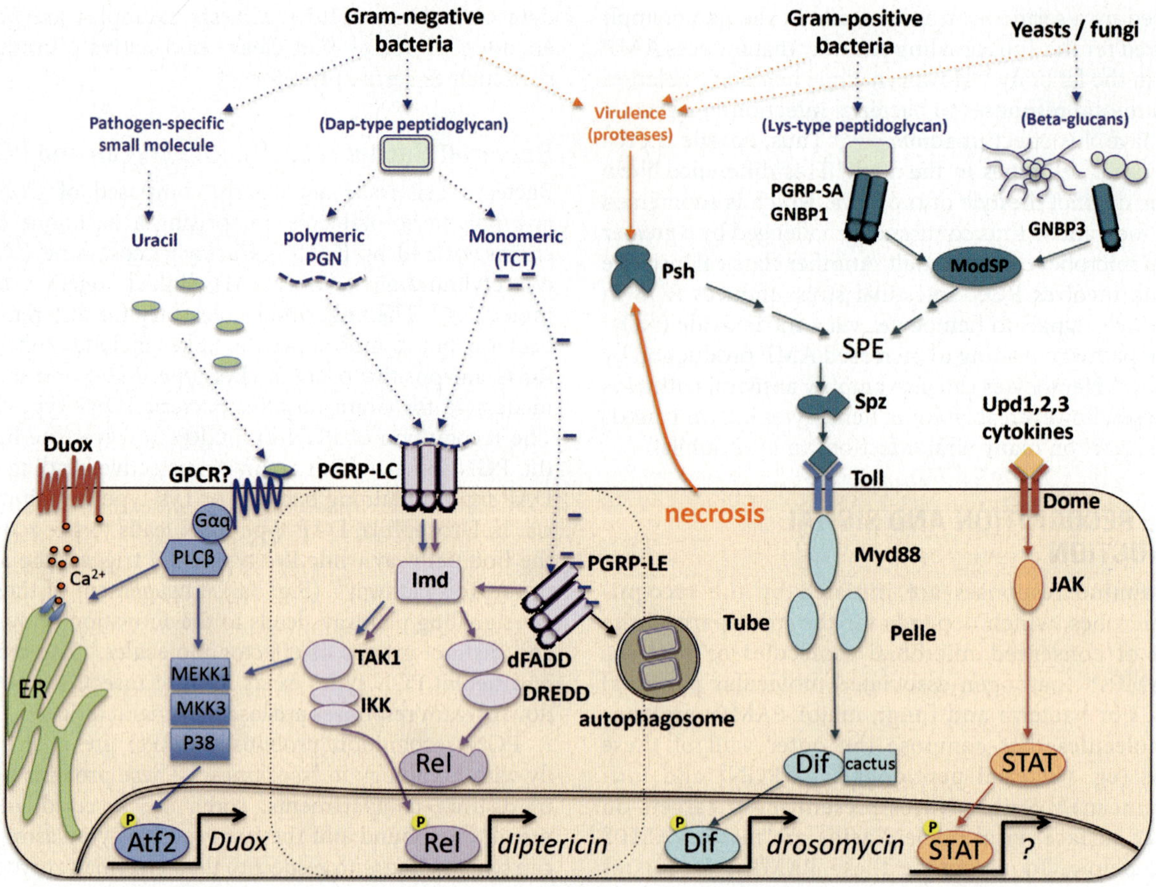

FIG. 3.5. Regulation of insect immune responses by the Toll and Imd pathways in Drosophila. PAMPs are sensed in the hemolymph by PRRs that can either induce downstream cascades converging on Toll activation or can directly induce the Imd pathway. In Drosophila, the recognition of Gram-positive bacteria and yeasts/fungi leads to the activation of the serine protease modSP that activates a cascade of serine proteases resulting in the maturation of the cytokine Spätzle and the induction of the Toll pathway. In addition, virulence factors such as microbe-derived proteases can process the Persephone serine protease, also resulting in Toll pathway activation. The Imd pathway responds to detection of DAP-type peptidoglycan by dedicated recognition proteins, the PGRPs. The JAK–STAT pathway is activated in response to the presence of cytokines such as the Upd in Drosophila. Metabolites can also be sensed leading to the activation of MAPK cascades and Duox transcription. Duox activity is regulated by calcium influx in the cytoplasm. Altogether, these pathways control bacterial and fungal infections. (Redrawn by permission from Springer Nature; Buchon N, Silverman N, Cherry S. Immunity in Drosophila melanogaster-from microbial recognition to whole-organism physiology. *Nat Rev Immunol.* 2014;14(12):796-810. doi:10.1038/NRI3763.)

Catalytic PGRPs hydrolyze PGN and therefore, degrade these immunostimulatory molecules.[75] Such degradation is important at the gut surface to prevent aberrant stimulation of immune pathways in response to the microbiota.[76] Catalytic PGRPs also play roles in immune resolution to remove the innate stimulus after the infection is controlled. Additionally, removal of the outer shell of PGN can promote the aggregation/agglutination of microbes to facilitate their removal. The noncatalytic PGRPs binding of PGNs initiates downstream antimicrobial signaling pathways in diverse insects.

Fungal Recognition: Beta-Glucan, Lectins, and Opsonins

The cell wall of fungi is composed of glucans of various structures. While alpha glucans provide a mesh structure to the fungal wall, beta-glucans can be linear or branched and constitute microfibrillar structures in the cell wall.[77] They are the most abundant cell wall molecules of most types of fungi and constitute a major PAMP (Fig. 3.5). To recognize beta-glucan molecules, insects encode two classes of PRRs, β-1,3-glucan recognition proteins (βGRPs) and Gram-negative bacteria-binding proteins (GNBPs, termed improperly as they do not recognize Gram-negative bacteria).[78,79] Members of this gene family contain domains including β-glucanases, but they lack key residues required for the enzymatic digestion of β-glucans. Upon binding of these receptors to fungal walls or components, serine proteases are activated. This can lead to two types of classical responses. First, the activation of prophenoloxidase leading to a melanization reaction[80] (Fig. 3.3). Second, some serine proteases process the Toll "cytokine" ultimately leading to Toll pathway activation.[81] Vertebrates also encode β-glucan recognition proteins (the dectins) to orchestrate antifungal defenses.[82]

C-type lectins (CTLs) are a large, conserved family, characterized by C-type lectin-like domains (CTLD) capable of binding carbohydrates.[83] CTLs play essential roles in anti-fungal responses across organisms. Insect CTLs have the ability to bind more diverse carbohydrates than mammalian CTLs.[84] Moreover, calcium is required for mammalian CTLs to bind pathogens, while this is not always the case for insects, possibly because calcium promotes protein oligomerization of insect CTLs.[85,86] CTLs act as opsonins, binding pathogens to promote immune activation, leading to either melanization or phagocytosis.[83] Surprisingly, some CTLs have been co-opted by pathogens to promote their own survival. For example, in the case of the infection of *A. gambiae* by Plasmodium parasites, two lectins (CTLMA2 and CTL4) have been shown to protect Plasmodium parasites from melanization.[87,88] CTLs can also be encoded by insect viruses such as those contained in endoparasitoid wasps. The virus-derived CTLs promote parasitization by the wasp, decreasing immune activation by masking surface recognition.[89]

Fibrinogen-related proteins (FREPs) are a large and conserved family of proteins related to mammalian fibrinogen.[90] In vertebrates, fibrinogen is converted into fibrin to provide the framework for clotting. Invertebrates encode a large family of these proteins, many of which play essential roles in defense. FREPs are highly polymorphic and can be diversified by multiple mechanisms including exon loss, alternative splicing, and random somatic mutations, including gene conversion and point mutations.[91] Thus, FREPs are the only known invertebrate molecules that undergo somatic changes. This diversified family possesses lectin-like qualities and can agglutinate bacteria.

Innate Immune Signaling Pathways

Many of the signaling pathways that control innate immune responses in mammals also play key roles in insects. Most insects encode at least two NF-kB transcription factors that mediate the activation of two parallel NF-kB pathways essential for immunity: the Toll and Imd pathways[2,3,92] (Fig. 3.5). Mechanistic studies from Drosophila have shown that these two pathways recognize different classes of microbes and lead to the transcriptional activation of hundreds of effector molecules. Specifically, PGRP receptors recognize DAP-type PGN of Gram-negative bacteria to activate the Imd pathway, while recognition of β-1,3-glucans by GNBPs and Lys-type PGN by PGRPs leads to the activation of the Toll pathway. However, this very stereotypical example may be quite specific to Drosophila, while the two pathways are conserved in most insects, the degree of differentiation of the immune response to classes of pathogens is unclear. In addition, the level of crosstalk and interactions between these pathways in different insect species remains unexplored. In Drosophila, it was demonstrated that gene targets of both the Toll and Imd pathways are numerous, suggesting that coregulation of immune genes can occur.[93] The Toll and Imd signaling pathways are conserved in mammals, and pathway components share parallels with the interleukin-1 TLR/NLR and TNF pathways, respectively (see below). The JAK–STAT pathway is another innate immune pathway well conserved from mammals to insects.[94] Additional pathways are involved in the response to infection including stress-activated protein kinases such as the JNK and MAPK pathways.[95]

The Toll Pathway

The Toll pathway is one of the two main NF-kB pathways of insects and has roles in immunity, development, and neurobiology[2,96-100] (Fig. 3.5). The Toll pathway is activated by the binding of Toll receptors to the circulating cytokine Spätzle. For activation, Spätzle is processed by proteolytic cleavage by the CLIP domain Spätzle-processing enzyme (SPE). Upon binding of Spätzle, the Toll receptor triggers the activation of intracellular signaling that involves conserved molecules such as Myd88, Pelle, and Tube.[97-100] As in mammals, the degradation of the insect homolog of IkB Cactus releases the NF-kB transcription factors to translocate to the nucleus and transactivate the expression of immune effectors. The Toll pathway results in expression of AMP-encoding genes, as well as other immune effectors and Cactus itself, creating negative feedback loop to suppress pathway activation.[3] The Toll pathway can be activated in response to bacteria, fungi, viruses, and parasites such as Plasmodium in mosquitoes. In addition to the expression of immune effectors, the Toll pathway controls the expression of genes involved in secretion and metabolism, allowing for physiological adaptations alongside AMP expression. While insect genomes encode multiple Toll receptors and multiple Spätzle cytokines, the functions of these other receptors and cytokines in insect immunity are incompletely understood.

While many PRRs are membrane bound, there are also circulating PRRs that have key roles in immunity and a subset of these activate upstream serine protease cascades important for Toll pathway induction. There are two major upstream pathways leading to Toll pathway activation. The autoactivation of a modular serine protease (modSP) leads to the activation of a cascade of CLIP-SPs that leads to SPE and Spätzle maturation.[101,102] In parallel, the serine protease Persephone can also initiate a cascade of proteolytic events leading to Toll pathway activation; however, Persephone's maturation is not dependent on PAMP recognition. Rather, proteases secreted by invading pathogens or dying cells can cleave Persephone. Thus, Persephone acts as a sensor for virulence or the damage associated with infection (danger-associated molecular patterns [DAMPs]).[103] These cascades lead to amplification of a signal and the rapid relay of this information that converges on the induction of diverse antimicrobial effectors. Surprisingly, the induction of these Toll-dependent immune responses in insects is sexually dimorphic, differing in males and females (example in Ref. 104).

The Imd Pathway

The Imd pathway is a second signaling pathway that leads to the activation of NF-kB transcription factors and can be activated by both bacterial and viral infections (Fig. 3.5). This signaling cascade was identified by a mutation in the immune deficiency gene (Imd), which is homologous to the mammalian RIP protein of the TNF pathway. Imd is activated upon recognition by PGRP receptors, which leads to receptor aggregation and the initiation of intracellular signaling.[105,106] Activated Imd leads to cleavage of the caspases FADD and DREDD which subsequently induces the cleavage of the ankyrin domain of the NF-kB transcription factor Relish, allowing for its translocation to the nucleus.[3,107] In parallel, the pathway leads to the activation of the Kinase TAK1, the activation of the IKKbeta/IKKgamma complex, and the phosphorylation of Relish. Both phosphorylation and cleavage of the cytoplasmic anchor are required for Relish to translocate to the nucleus and transactivate target genes. When activated, the Imd pathway leads to the transcriptional activation of a set of antimicrobial effectors including AMPs.[108] Targets of the Imd pathway also include amidase PGRPs, alongside a suite of negative regulators of this pathway.[109] Negative regulators such as amidase PGRPs have two functions: (1) they establish a negative feedback loop on Imd pathway activation essential for the resolution of immune induction; and (2) they set a threshold for initial Imd pathway activation. This threshold is higher in the intestine and lower in the hemolymph; this would allow for tolerance to the microbiota in the gut, while allowing for sensitive detection in the blood.[1] In addition to negative regulators of the Imd pathway, hemocytes and the excretory system prevent accumulation of PGN in hemolymph and thus aberrant immune stimulation.[42,110] While the Imd pathway is well conserved, it is not perfectly conserved in all insect species where some species have lost components. However, it is likely that they have evolved other proteins to function in the pathway. The kissing bug, for example, lacks orthologs for major pathway components, but the expression program remains functional.[111] However, it is also possible that some insects have lost the pathway due to evolutionary selection. For example, the pea aphid, which feeds mostly on sterile sap, may no longer require this innate sensing pathway due to its microbe-poor lifestyle.[112]

The JAK–STAT Pathway

In parallel to the NF-kB pathways, the JAK–STAT pathway is another key regulator of immune response in insects. This pathway has roles in development and immunity, as well as tissue repair mechanisms that can be central to survival during host–microbe interactions.[94] The JAK–STAT pathway is activated by the binding of the receptor Domeless (JAK) to cytokines called unpaired genes (Upd-1, Upd-2, Upd-3) in Drosophila.[113] It is likely that other ligands exist as JAK and STAT are conserved in many insects, but the unpaired genes are not. Upon cytokine binding, receptor dimerization induces phosphorylation of the Janus Kinase, JAK, that is associated with the receptor, which in turn phosphorylates the transcription factor STAT. Phosphorylation of STAT leads to dimerization and translocation into the nucleus where STAT induces the expression of numerous genes, including antimicrobial effectors such as nitric oxide synthase (NOS),[114] but also genes involved in stem cell activation and proliferation.[26,115] Moreover, STAT also induces negative regulators of the Toll and Imd pathway, such as the PIAS and SOCS families of inhibitors, promoting a negative feedback loop. This pathway is important in response to most classes of pathogens, including viruses, parasites (eg, Plasmodium[114]), bacteria, and fungi. It has been shown that the induction is downstream of the damage associated with these infections to promote homeostasis.

ANTIMICROBIAL IMMUNE EFFECTOR MECHANISMS

One goal of innate immunity is to limit the growth of microbes. Diverse mechanisms are used including cellular immunity, humoral immunity, and homeostatic regulation of tissue integrity (Fig. 3.4). Classical cellular responses including the capture and removal of microbes are deeply conserved. This results in the elimination of microbes, whether small (eg, bacteria and viruses) through phagocytosis or bigger parasites through a phenomenon of encapsulation. Humoral responses include the secretion of immune effector molecules in the hemolymph or at the surface of epithelial cells. The additional conserved effector responses include (1) killing by AMP; (2) killing by free radicals produced during melanization or via oxidases; (3) autophagy; and (4) host cell death. These pathways are often regulated by common signaling pathways and multiple mechanisms are involved in the response to different pathogens.

Phagocytosis, Nodulation, Encapsulation

Phagocytosis: Phagocytosis is a conserved arm of the cellular response of insects, quite similar to phagocytosis in mammals. Most insects have circulating hemocytes that have a strong phagocytic capability (eg, plasmatocytes), and phagocytes represent the major type of hemocytes in insects.[116,117] However, each insect species can have specialized immune cells. Phagocytes can be broadly separated into two groups: plasmatocytes that are specific to Diptera and granulocytes that are found in mosquitoes, Hemiptera and Lepidoptera.[92,117] Phagocytosis is stimulated by the recognition of microbes or foreign objects at the surface of phagocytes by a diverse set of phagocytic receptors.[92] There are many parallels between the mechanisms of phagocytosis in hemocytes and macrophages. Phagocytic receptors are well conserved from humans to insects and include PRRs that bind to the PAMPs of microbes. Scavenger receptors are required for recognition and cytoskeletal rearrangements for proper engulfment.[42] These receptors include some families such PGRPs, Nimrod proteins, thioester-containing proteins (TEPs), and β-integrins.[118-121] Drosophila hemocytes express several families that recognize PAMPs: Eater, Nimrods, Draper, Croquemort, and other receptors coordinate the downstream responses.[122] Dscam (Down syndrome cell adhesion molecule) is a member of the Ig superfamily with an essential function in neuron interconnection. Dscam can also bind to the surfaces of some microorganisms to facilitate phagocytosis. It has been suggested that this receptor can form 18,000 isoforms through alternative splicing in hemocytes and fat body.[123] The existence of a potential repertoire of thousands of Ig-domain–containing proteins suggests diversity in the recognition of pathogens in Drosophila, opening the possibility of "adaptive" immunity across animal phyla previously considered to only have innate responses.[124,125] These receptors have specific interactions with the outer layer or envelope of microbes and multiple PRRs can be involved in the recognition of each microbe.

Pathogens can by opsonized by proteins binding to the surface of these microbes and, upon recognition of these surface receptors, the microbe can be phagocytosed by the hemocyte.[126] The cargo is internalized into a structure called the phagosome. The phagosome eventually matures, acidifies, generates ROS, and fuses with the lysosome. The resulting phagolysosome contains multiple hydrolytic enzymes that, in synergy with AMPs and ROS, are responsible for the killing of the phagocytosed microbes. Thus, insect hemocytes are professional microbial killers having the capacity to repeatedly internalize hundreds of microbes and eliminate them.

Nodulation is another cellular response that requires the coordinated action of hemocytes.[127] Nodulation often targets larger particles than single microbes and captures assemblies of microbes or bigger parasites. The main actors of nodulation are granulocytes that detect the infection and in response to detection, adhere together and induce degranulation of cytoplasmic contents. The targeted particle is therefore isolated from the host environment and the nodule, formed of pathogens and granulocytes, is often subsequently melanized. In mosquitoes, nodulation has been shown to be activated by the eicosanoid pathway, suggesting that this process could require additional cellular actors.[128] Encapsulation is a very similar pathway and in Drosophila, specialized hemocytes called lamellocytes encapsulate large parasites, such as eggs of parasitic wasps.[129] Importantly, the introduction of large lipid bubbles in the body cavity of Drosophila, via oil injection, was found to be sufficient to promote encapsulation, suggesting that this process is regulated by physical disruption rather than by recognition of microbial PAMPs. Lamellocytes are not present basally, but upon larval infection, a complex sequence of events results in the activation of the JAK–STAT pathway in the lymph gland and an alteration of hematopoiesis and cell differentiation to generate new lamellocytes.[130]

Melanization and the Role of Phenoloxidases

Melanin is a free radical oxidative by-product that can directly kill microorganisms and facilitate wound healing.[131-133] It involves formation of black pigmentation resulting from de novo synthesis and deposition of melanin. Melanization is an insect-specific reaction involved in the immune response, as well as cuticular hardening and response to wounding.[131-133] Melanization manifests as the buildup of a dark diffuse or punctate matrix at the site of injury or surrounding pathogens. The dark matrix originates from an enzymatically controlled multistep conversion of tyrosine in melanin, directly at the site of injury. The final step in the melanization reaction is tightly controlled by a cascade of serine endopeptidases[134] (Fig. 3.3). Upon recognition of pathogens by PRRs such as PGRPs, βGRPs or CTLs, serine proteases are activated, leading to a chain reaction in which each serine protease activates the next serine protease in the cascade.[135,136] Such serine protease cascades allow amplification of the signal and leads to the maturation of the final enzyme in the melanization reaction, the prophenoloxidase in its active form, phenoloxidase. While all components required for melanization are in circulation, activation of melanization remains a local response, due to the action of inhibitors called serpins.[88,134] Serpins act as suicide substrates for serine proteases, blocking their activity to prevent aberrant activation of the cascade, and allowing termination upon pathogen clearance. Altogether, the enzymatic reactions required to produce melanin generate a local oxidative stress that damages the pathogen. In addition, oxidative stress allows cross-linking of proteins involved in clotting, creating a protein layer that insulates pathogens from the insect hemolymph and leading to their death by starvation. Melanization often acts in concert with hemocytes to coordinate encapsulation and nodulation of parasites.

Reactive Oxygen Species and Nitric Oxide

Across kingdoms, organisms have harnessed ROS as a means of warfare. These include short-lived molecules that are highly reactive (eg, hydrogen peroxide and hypochlorous acid, ie, bleach) and reactive nitrogen species include derivatives of nitric oxide. These molecules have high reactivity with proteins, nucleic acids, and other biomolecules, making them highly toxic for microbes. Indeed, they also present toxicity to their hosts and thus these molecules need to be confined either in cellular compartments or outside the body. The best characterized localized function of ROS in immunity is the oxidative burst that takes place in phagocytes in the phagosome and results in the destruction of the phagocytosed organisms.[137] Another conserved location outside the body is the gut: ROS are produced by epithelial cells in the midgut to control pathogens that engage at the surface of the intestine. This induces lytic activity in the extracellular environment keeping the microbes far from the surface. In the midgut of *D. melanogaster*, two enzymes are able to generate ROS at the surface of enterocytes; dual oxidase (Duox) generates HOCl while Nox generates hydrogen peroxide.[137,138] A well-described example of the use of NO takes place in the midgut of *A. gambiae*, the mosquito vector of Plasmodium.[139,140] When Plasmodium parasites cross the midgut epithelial barrier, enterocytes produce NO from the oxidation of l-Arginine to l-Citrulline, and the nitrification of Plasmodium is an important signal for its recognition by hemocytes and effector molecules in circulation. NO is also part of the systemic response and is produced by hemocytes where it acts both as an antibacterial effector and a mediator of immune signaling.[141] These activities are conserved in vertebrates.

AMPs and Microbial Lysis

AMPs are small proteins with immune activity against pathogens and are encoded in genomes from insects to hum ans[108,142,143] (Fig. 3.6). They mainly act as direct effectors often promoting the lysis of targeted pathogens by creating pores in the outer membrane. AMPs have also been shown to participate in additional mechanisms of action including the neutralization of microbial-derived toxins and the regulation of other immune responses. While AMPs are one of the most

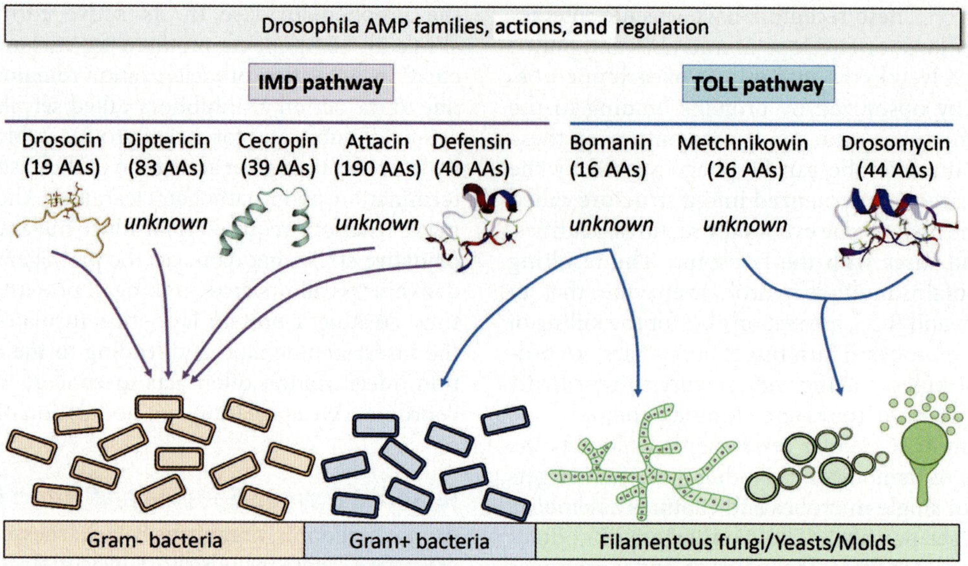

FIG. 3.6. Insect antimicrobial peptides (AMPs) and their modes of action. AMPs play fundamental roles in innate immunity to diverse pathogens and fall into a number of distinct classes. These are transcriptionally induced by conserved NF-kB–dependent signaling pathways (Toll and Imd pathways) and are largely secreted into the hemolymph by the fat body. AMPs are short peptides ranging from a few amino acids to a few hundreds. The specificity of AMPs varies from broad spectrum action to specific action on subtypes of pathogens. In Drosophila, it has been demonstrated that their specificity matches overall the specificity of the immune pathways that control them. (Redrawn from Hillyer JF. Insect immunology and hematopoiesis. *Dev Comp Immunol.* 2016;58:102-118. doi:10.1016/J.DCI.2015.12.006, with permission from Elsevier.)

important line of defense against invading pathogens, the number of AMPs encoded in the genome of insects varies dramatically, ranging from insects apparently entirely devoid of AMPs (eg, the pea aphid) to species that encode more than 50 AMPs.[142,143] The high diversity of lifestyles and ecosystems encountered by insects is likely the determinant of such variation. The increasing number of insect genomes sequenced in the past few years has dramatically increased the repertoire of known AMPs, and overall insects encode many more AMPs than any other taxonomic group.[144] The identification of insect AMPs remains a difficult task as AMPs diversify at a fast evolutionary rate. Several gene families exist with multiple paralogues in each species, allowing for greater plasticity. Such plasticity likely facilitates rapid adaptation to new pathogenic threats encountered in the wild.

AMPs can be grouped into different categories depending on their structure (Fig. 3.6). Three main categories cover most AMPs in insects: (1) α-helical AMPs that lack cysteine residues; (2) globular AMPs with a β-sheet and numerous and essential disulfide bridges; and (3) small peptides enriched with glycine or proline.[143] The mode of action of AMPs is still poorly characterized in regard to the diversity of peptides encoded. Thus, most AMPs are classified based on the spectrum of pathogens they recognize. Some AMPs have broad antimicrobial activity on most classes of pathogens (often bacteria and fungi), while some have specificity to classes of microbes such as Gram-negative bacteria, Gram-positive bacteria or yeasts/fungi.[145] Recent work in Drosophila suggests that some AMPs could even be specific to a limited set of bacterial species.[108]

An archetypal AMP of the α-helical peptide class is Cecropin.[146] Cecropin was the first peptide isolated from insects, as it accumulates in the hemolymph of the giant silk moth in response to infection. Multiple Cecropins have been identified in other insects (called Cecropin-like peptides) and they show activity against a broad range of bacterial pathogens.[147] Archetypes of the AMPs with a β-sheet are encoded by the defensins.[148] Defensins are thought to be ancient AMPs as they are present in most Orders of insects.[149] Their specificity is mostly against Gram-positive bacteria but a few defensins have some level of moderate activity against Gram-negative bacteria and filamentous fungi.[149] Two subcategories of AMPs enriched in specific amino acids have been identified: AMPs rich in either glycine or proline.[149] Glycine-rich AMPs encode both long and short peptides that differ in their specificity. Short peptides of less than 20 amino acids show activity against Gram-negative bacteria while long peptides are more active against Gram-positive bacteria. These peptides include the diptericins and attacins of Diptera, holotricins and tenecins, or the hemiptericins of Hemiptera. Most of these AMPs have activity against Gram-negative bacteria, and sometimes specific species of Gram-negative bacteria. Proline-rich AMPs include Hemiptera's metalnikowin and pyrrhocoricin, Drosophila's drosocin and metchnikowin, Lepidoptera's lebocin, and Hymenoptera's abaecins, apidaecins, and formaecins.

Basally AMPs are not expressed, but upon bacterial or fungal detection are transcriptionally activated and secreted. Upon systemic infection, AMPs are secreted in the hemolymph largely by the Drosophila fat body reaching very high concentrations.[3] AMPs can also be produced by local epithelia of the intestine, trachea, and reproductive tract, at lower levels in response to the commensal microbiota, and at high levels upon pathogenic infection.[21,26,115] AMPs are controlled by two main immune pathways: the Toll and the Imd pathway (see above). There is heterogeneity in the response since insects exposed to similar infections do not all survive infections. Stochastic variation in the early AMP response to infection results in binary outcomes upon systemic infections.[150]

Most insect AMPs exert their antimicrobial effect by perforating the microbial cell wall, leading to growth inhibition or cell death. Their high positive charge allows interaction with microbial surfaces that are negatively charged by electrostatic interactions.[149] The cell membrane of most microbes is broadly negatively charged, in contrast to the cell membrane of most eukaryotic cells that are more neutral allowing AMPs to bind microbial cells with high affinity and to interact with eukaryotic cells only minimally.[149] Classically, once bound to the surface of microbes, hydrophobic residues promote the insertion of AMPs into the microbial membrane, leading to pore formation and eventually cell lysis.[151,152] However, this simple mechanism of action has rarely been directly demonstrated, for instance, through dye leakage assays of microbes treated with AMPs. Thus, a number of questions remain unanswered about AMPs' mechanisms of action. For instance, how do AMPs reach the bacterial membrane and pass the barrier of the cell wall? It is likely that AMP action requires complementary activity of multiple enzymes and opsonins that together promote bacterial killing in vivo. It has been recently proposed that AMPs can synergize for bacterial killing and show combinatorial activity against parasites.[153] In addition, not all AMP families generate pores. For instance, proline-rich AMPs target intracellular microbial proteins (eg, the translation machinery, or enzymes involved in cell wall physiology).[153]

In addition to their role in the control of microbial infection or shaping the intestinal microbiota, insect AMPs have additional functions.[108] First, AMPs can be involved in antitumoral immunity in Drosophila.[154,155] Recently, AMPs have been implicated in neurobiology, both in physiological and pathological situations.[108] For example, the Drosophila AMP Diptericin promotes long-term neurological memory.[156] In addition, the AMP demuri has sequence similarity to both neuropeptides, and the mammalian AMP cathelicidin acts as a regulator of sleep upon infection.[156] It has also been suggested that amyloid-beta, the peptide involved in Alzheimer's disease, acts as an AMP.[156] In all these examples, however, the clear mechanism of action is still lacking, requiring further studies.

Additional molecules can promote microbial lysis[92] including enzymes that digest the cell wall of microbes. Chitinases are upregulated by infection and have activity against the cell wall of fungi and yeasts. The ability to digest PGN is an important component of the digestive system of insects that eat microbial-rich foods.[159] Lysozymes are another family of enzymes with lytic activity[92] and insects have multiple lysozymes in their genome that are often expressed in the digestive tract.[157,158] Lysozymes are also transcriptionally upregulated upon infection and have a clear lytic function against microbes. Surprisingly, they are also involved in parasitic immunity and in the melanization response, two immune functions in which lysozyme's enzymatic role is not understood.[159]

Autophagy

When pathogens invade a cell, the cell recognizes the threat and induces antimicrobial effector mechanisms. Autophagy is a highly evolutionarily conserved process required for homeostasis and responses to nutrient stress. Autophagy and the factors that regulate this pathway orchestrate a degradative process that proceeds through a series of defined stages to ultimately capture cargo for decay. First the pathway is activated, in many cases by nutrient depravation, followed by initiation and elongation of a membranous compartment called an autophagosome that, upon closure, captures cytoplasmic components for ultimate degradation upon fusion with the lysosome. Autophagy plays a housekeeping role by removing aging or damaged organelles, as well as aggregated or misfolded proteins too large for the proteasome. The conserved mechanism by which cytoplasmic cargo is delivered to the endolysosomal system coordinates a powerful host response that can also capture diverse intracellular pathogens, including viruses.

The intracellular Gram-positive bacterium L. monocytogenes can invade the cytoplasm of cells and is protected from the action of secreted effectors, including AMPs. From vertebrates to insects, autophagy restricts cytoplasmic L. monocytogenes replication.[160-162] In Drosophila, Listeria infects hemocytes which also encode an intracellular PRR, PGRP-LE, which recognizes the PAMP tracheal cytotoxin to induce antimicrobial autophagy.[105,163] The canonical PGRP, PGRP-LC, can also respond to Listeria infection; however, this receptor faces the extracellular space and only recognizes Listeria when outside of the cell. Thus, these receptors have nonredundant functions: PGRP-LE restricts intracellular bacterial replication while PGRP-LC signaling controls extracellular bacteria. Clear parallels can be drawn between PGRP-LE in flies and intracellular NLRs Nod1 and Nod2 in mammals that reside in the cytoplasm and recognize degradation products of PGN, similar to PGRP-LE, to induce autophagy. Recognition of bacteria by additional PRRs, including TLRs, can lead to the induction of antimicrobial autophagy.[164]

In addition to Listeria, other bacteria including Wolbachia, a common intracellular bacterium in arthropods, are controlled by autophagy. Wolbachia levels are both negatively and positively regulated by autophagy: attenuated by autophagy in the soma but promoted by autophagy in the germline.[165] Autophagy can also impact the growth of the intracellular bacteria *Mycobacterium marinum* and filarial nematodes.[166] Interestingly, mutants in ird1, a component of the autophagy-initiating complex, have enhanced susceptibility to bacterial infections.[167]

Autophagy plays important roles in the homeostasis of the intestinal epithelium from vertebrates to insects. Autophagy dysfunction in the gut during aging can lead to increased DNA damage, cell cycle arrest, and apoptosis.[168] Defects in autophagy genes in the presence of particular intestinal microbes cause a disruption of homeostasis, associated with inflammatory bowel disease and Crohn's disease. Enterocytes secrete ROS for two functions: to control commensal bacteria in the intestine and to suppress regeneration. Thus, dysregulation and chronic activation of autophagy leads to increased barrier dysfunction, systemic inflammation, and ultimately a shorter lifespan. Across organisms, aging and turnover of enterocytes are dependent on autophagy.[169-172]

Autophagy is antiviral in many systems and, in flies, controls systemic infection of two negative-sense RNA viruses, the rhabdovirus vesicular stomatitis virus (VSV) and the bunyavirus Rift Valley fever virus (RVFV).[173-175] Loss of function of autophagy genes leads to increased infection and mortality of VSV downstream of nutrient signaling. As was observed for Listeria infection, viral PAMPs induce antiviral autophagy—the viral glycoprotein VSV-G triggers autophagy. And the PRR is Toll-7 (a Toll receptor) that recognizes VSV to elicit antiviral autophagy.[174] Identification of Toll-7 as a PRR triggering antimicrobial autophagy defined a role for a second Toll receptor in flies and was the first fly Toll to directly interface with a PAMP. Further study found that RVFV is also controlled by the Toll-7 autophagy pathway in insects, and that this antiviral autophagy pathway is conserved in mammals. Altogether there is clear conservation between TLRs and antiviral autophagy in all animals.[176]

Autophagy is tightly controlled by nutrient signaling, which also regulates other aspects of immunity. Likewise, immune activation can impact nutrient signaling. Redistribution of the energy is crucial to initiate an immune response.[177-179] Across evolution, excessive immune responses result in dysregulation of metabolic stores.[180] Conversely, it has been observed that the loss of metabolic homeostasis results in weakening of immune responses, presumably due to the lack of energy stores. Studies in flies have converged on the understanding that the Drosophila fat body integrates these signals and alters its energy storage capabilities in response to infection.[2,60] In addition to alterations in storage, the fat body can switch transcriptional programs from anabolic to more immune stimulatory.[181] In addition to systemic control, local change in nutrient availability in the intestinal tract changes dramatically with feeding and influences virus infections.[182] When mosquitoes take a blood meal, this is a huge nutrient influx and has an influence on arboviral infections. Vertebrate blood contains growth factors including insulin, which can bind to insect insulin receptors to impact enteric biology and infection.[183] Indeed, insects fed vertebrate insulin at the levels present in a blood meal protected them from enteric viral infection through antiviral signaling in enterocytes.[182,184] The role of additional nutrients in enteric immunity is poorly understood, but clearly important.

Cell Death and Immune Defense

The immune system also protects its host from invading microbes by inducing cell death programs. The relationship between cell death and immunogenicity can be complicated due to the complexity of mechanisms by which cells can die.[185,186] Indeed, there are many different pathways that can converge on cell death. Some are silent while others are very lytic and immunogenic.[187-189] For example, apoptotic mechanisms are thought to be less immunogenic than autophagy-dependent death pathways. Moreover, depending on the cell type and timing, there can be distinct immunological outcomes.[190] In some cases, the same signaling components can lead to distinct death pathways, depending on the stimulus and cell type.[191] In Drosophila, the molecular mechanisms of cell death pathways have been uncovered in diverse genetic screens, and the signaling pathways are highly conserved in mammals.[2,192,193] Apoptotic death pathways are developmentally regulated, and many of the genes required are known. As part of this program, phagocytosis is required to scavenge dead cells. Other death pathways also impact immunity and infections. Upon viral infection, cell damage caused by viral infection can release (DAMPs) which induce the JAK/STAT pathway. JAK/STAT mutants challenged with these viruses show increased mortality in the absence of altered viral loads demonstrating that the pathway is involved in tolerance.[194,195] Some characterized DAMPs include components of the cortical cytoskeleton and can be released upon tissue injury inducing the JAK/STAT pathway.[196,197] The NF-kB Imd pathway also regulates apoptosis.[198] There is also some evidence that programmed cell death and apoptosis can regulate viral infections in Drosophila, including studies that suggest that phagocytosis of dying cells reduces the spread of viral infections.[199,200]

Viruses have complex interactions with cell death pathways, since cell death is thought to be a major response to viral infection limiting replication and spread. Viruses encode proteins that perturb various steps in cell death programs. Some viruses block cell death by expressing proteins that interfere with cell death transduction pathways. Baculoviruses are a large class of dsDNA insect viruses that encode proteins that block cell death. P35s and additional inhibitors of apoptosis (IAPs) block cell death during viral infection and during other developmental programs through the inhibition of caspases, the executioners of cell death programs.[201,202]

Antiviral Immunity and Nucleic Acid Recognition

All organisms can be infected by viruses. And given that insects are major crop pests, there have been significant efforts to identify viral pathogens and, potentially, use this as an insect control strategy. For other insects which are important to ecosystems, such as bees, efforts have made to identify viral pathogens to determine if they are playing a role in the collapse of bee populations worldwide. The third area of viral infection and insects relates to the group of insects that directly impact human health, as the vectors of human disease. A wide variety of insects can transmit human diseases, with arthropod-borne viruses representing a large class. These arboviruses are transmitted between blood-feeding insects and vertebrates. There are hundreds of millions of these infections worldwide each year. In addition, Drosophila harbors its own cadre of viruses, including RNA and DNA viruses.[5] Next-generation sequencing has expanded our understanding of the abundance and diversity of insect-associated viruses. Importantly, Drosophila can be used to model infection of both vertebrates and invertebrates; many facets of antiviral immunity active in vectors and vertebrates are conserved in flies. Flies can be experimentally infected with many human arboviruses. While early studies challenged flies systemically with diverse viruses, more recent studies have explored infection at barrier sites including the gut. As obligate intracellular pathogens that use host machinery to replicate, viruses present with few non–self-targets. Across the tree of life, viral nucleic acids are a major target of antiviral innate immunity as these have features that are unique and thus can be distinguished from self.

RNA viruses are a major group of pathogens in insects and include all of medically important human arboviruses. These viruses encode RNA genomes as well as additional RNAs, including dsRNA intermediates. Virus-derived dsRNAs can be recognized by diverse RNA-binding proteins across the tree of life. An important discovery was that the RNA silencing or RNA interference (RNAi) pathway is potently antiviral in plants and was subsequently shown to be involved in Drosophila immunity and in other insects including vectors.[203] RNAi is one type of RNA silencing pathway. There are three highly conserved small RNA silencing pathways: (1) small interfering RNA (siRNA); (2) microRNA (miRNA); and (3) PIWI-interacting RNA (piRNA) pathways.[204] At the core of all three pathways are Argonaute (Ago) family proteins that silence target RNA.[205] The Agos themselves are also directed by small RNAs (sRNAs) in a sequence-specific manner. Each pathway has distinct protein components, and these pathways can be differentiated by the length and origin of the guiding sRNAs and functional outcome of the RNA target.[205] More recent studies have shown that another class of nucleic acid–binding proteins, the cGAS orthologs, can also bind viral nucleic acids and induce antiviral responses across evolution.[206]

Antiviral siRNA Pathway

The siRNA pathway is an essential component of the systemic immune response of insects, with demonstrated control of a broad array of RNA viruses in diverse species[207] (Fig. 3.7). This pathway begins with the recognition of viral dsRNA or viral RNAs in the cytoplasm by the RNaseIII enzyme Dicer-2 (Dcr-2), which then cleaves the target RNA into approximately 21 bp duplexes.[208] Subsequently one strand is loaded into the RNA-induced silencing complex (RISC). Argonaute-2 (Ago2) is the catalytic core of the RISC complex, which binds one strand of the siRNA duplex and uses complementarity for target recognition. If the siRNA is derived from the viral RNA, then the cognate target is viral RNA as well. This leads to the cleavage of the viral RNA, which then can be further degraded by exonucleases in the cell. Thus, this pathway has two levels of antiviral effector functions, dicing by Dcr-2 and slicing by Ago2.

The antiviral RNAi machinery in Drosophila was discovered using a variety of genetic approaches.[209] Since orthologs are highly conserved, core members of the pathway that are required in plants for this activity were also found to be essential in insects. In addition, large-scale genetic screens also identified antiviral factors required for RNA silencing. This includes many accessory proteins that regulate the function of the pathway (eg, R2D2, Ars2). Dicer2, a dsRNA-binding protein, canonically binds virus-derived dsRNA, which is produced as a replication intermediate for all RNA viruses and does so in a complex with other factors.[210] dsRNA recognition is a widespread innate immune strategy used from yeast to humans. Dcr-2 cleavage of the viral RNA leads to the production of 21 to 23 bp siRNAs which are then loaded into the RISC complex with Ago-2 as the core enzyme, which slices the viral RNAs.[211] Deep sequencing of virus-derived small RNAs has revealed virus-specific complexity to the targeting of viral RNA by Dcr-2.[212] Structured regions of single-stranded viral RNA can also be targeted by Dcr-2. In addition to RNA viruses, many DNA viruses produce RNAs that are recognized: these can be produced from convergent promoters leading to overlapping transcripts that form dsRNA regions and DNA viruses produce RNAs with structured elements.[213] Therefore, DNA viruses can also be controlled by antiviral RNA silencing.[214] Perhaps not surprisingly, many

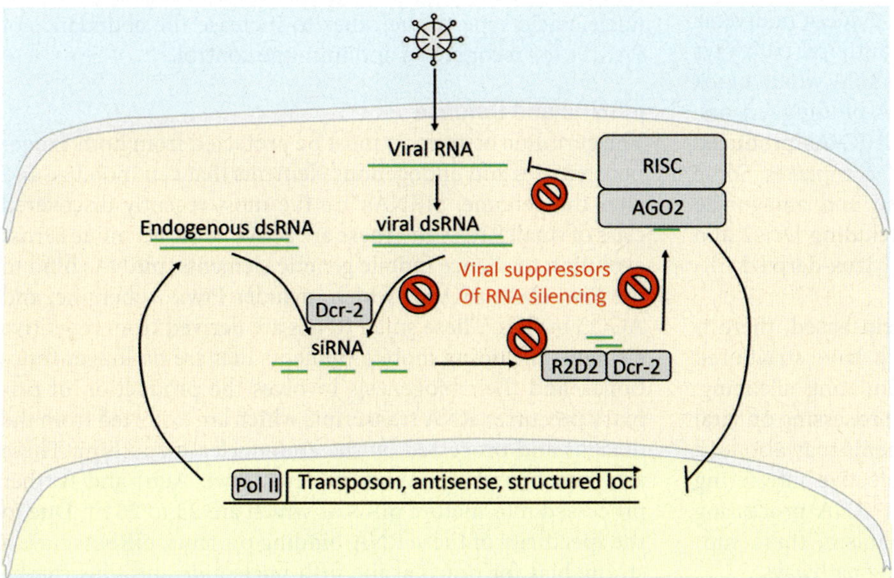

FIG. 3.7. Insect antiviral immunity and RNAi. Viral dsRNA intermediates or other endogenous RNAs with double-stranded character are processed into siRNAs by Dcr-2. These siRNAs are incorporated into an AGO2-containing RISC complex that then targets cognate viral RNAs for decay. Viruses encode suppressors of RNA silencing that can block every step of the silencing pathway. (Redrawn from Morin-Poulard I, Tian Y, Vanzo N, Crozatier M. Drosophila as a model to study cellular communication between the hematopoietic niche and blood progenitors under homeostatic conditions and in response to an immune stress. *Front Immunol.* 2021;12:3203. doi:10.3389/FIMMU.2021.719349/BIBTEX)

endogenous transposons also produce structured RNAs, and these are targeted by the RNA silencing pathway. Thus, genes that are antiviral against RNA viruses also impact the levels of transposons and retroelements[215] (Fig. 3.7).

While RNA silencing is clearly antiviral against systemic viral infections, recent studies have suggested that this pathway not be as active in barrier tissues. This is particularly important during oral infections. Following a viremic blood meal, arboviruses infect enterocytes in the mosquito gastrointestinal tract and are subsequently released into the body cavity, leading to systemic infection and, ultimately, infection of the salivary glands, which is required for transmission to vertebrates during the next blood meal. While RNAi is antiviral during systemic infection, recent studies have suggested the RNAi pathway is dispensable for viral clearance during oral infection in the fly intestine.[216] Moreover, in mosquitoes, RNAi does not appear to be universally antiviral in the intestine as loss of the RNAi pathway had no effect on o'nyong'nyong virus infection in *A. gambiae*. Moreover, RNAi mutants also do not show altered infection by diverse viruses in another barrier epithelial organ, the salivary gland. Together, these data suggest that antiviral RNAi is key during systemic infections, and that there are distinct immune defenses active in the barrier tissues.

Suppressors of RNA Silencing

Reflecting the key role for this pathway in antiviral defense, many viruses have evolved mechanisms to evade or antagonize antiviral RNAi[217] (Fig. 3.7). These virally encoded proteins are termed viral suppressors of RNA silencing. Viruses have independently evolved unique strategies to counter this RNA silencing response. Both RNA and DNA viruses carry suppressors, but they lack sequence or structural similarity, suggesting convergent evolution. Viral suppressors can act through a variety of mechanisms, many of which have parallels in vertebrates. One major mechanism involves binding of virus-derived dsRNA by a virally encoded dsRNA-binding protein that competes for binding with the dsRNA-binding protein Dcr-2, thereby limiting the target for recognition and dicing. This strategy is evident in diverse viruses, likely because there are multiple antiviral pathways that recognize dsRNA, and thus binding dsRNA would block recognition by Dcr-2 as well as other dsRNA-binding sensors. Additional suppressors can bind the small RNAs produced by Dcr-2 to compete for loading into RISC complexes. Some viruses encode proteins that directly bind and antagonize the enzymes involved in RNA silencing including Dcr-2 and Ago2. A third strategy involves degrading virus-derived siRNAs to limit their activity.

While the previous examples are protein-based, there is also evidence of RNA-based decoys which have structured elements that recruit Dcr-2, thereby attenuating silencing. Additionally, there are hotspots for siRNA processing on viral RNAs, and the accumulation of these elements may also lead to competition. RNA viruses produce defective interfering particles which may act as decoys for the RNA processing machinery.[218,219] Depending on the mechanism, these suppressors can also impact additional silencing pathways.

Immune Memory Through Nucleic Acids

Innate immune memory has been observed in a variety of organisms including invertebrates. A primary infection can confer protection to a secondary challenge and has been termed "immune priming."[220-224] Recent studies in Drosophila have shown that RNA virus infection can result in the detection of virus-derived DNA. Virus-derived DNA accumulates during RNA viral infections in flies, and these likely serve as sources of siRNAs remaining detectable even after clearance of the infection. RNA viruses do not carry the machinery to produce DNA from RNA and studies have implicated reverse transcriptases from endogenous retrotransposons in the synthesis of the DNA forms. It is suggested that these DNA sequences are then copied by unknown mechanisms into dsRNAs that are themselves subject to the antiviral RNA silencing machinery. The production of DNA forms from RNA viruses has additional implications; these DNA forms can potentially be integrated into the genome. This process may explain how RNA virus sequences can be found in the genomes of diverse organisms. How and where these DNA forms are generated, and how they are transcribed back into dsRNA remains unclear. Interestingly, many RNA viruses produce defective viral genomes (DVGs) during replication, and these seemingly give rise to many of the viral DNAs observed.[225] Thus, treatments that alter the levels of DVGs may also impact the antiviral response, since conditions presenting with high levels of DVGs are associated with increased vDNA accumulation and antiviral activity.[225] Whether vDNAs play a role in shaping downstream responses to previously encountered viruses remains unclear.[216,226,227] Future investigations will clarify the role of vDNAs in immune priming. In addition, it is unclear whether the route of infection impacts the formation of these species and how tissue tropism impacts the activity of viral DNAs during secondary infections. The recognition of RNA viruses through DNA intermediates has parallels in mammalian systems where DNA viruses are transcribed by host RNA polymerase III into RNA PAMPs which activate RNA-based innate immune sensors. In both cases, the innate immune system converts the nucleic acid signal from one nucleic acid type to the other to increase the abundance of PAMPs for recognition and immune control.

piRNAs and Defense

The germline of animals must be protected from both exogenous viruses and endogenous elements that can mobilize and alter the genome. piRNAs are the most recently discovered class of small RNAs and these are active in the germline across evolution to silence mobile genetic elements. piRNAs bind to the Piwi clade of Agos which includes Piwi, Aubergine, and AGO3 in flies. These small RNAs are derived from repetitive elements including mobile elements that are no longer functional, and their biogenesis involves the production of primary precursor RNA transcripts which are exported from the nucleus and processed by the endonuclease zucchini. These are then bound by a piwi protein (Piwi, Aub) and further processed into mature piRNAs which are 23 to 26 nt. Due to the specificity of these RNA-binding proteins, piRNAs have a strong bias for uracil at the fifth nucleotide and thus can be distinguished from siRNA and miRNAs by length, sequence, and post-translational modifications, as siRNAs and piRNAs are 2′Ome at 3′ end. The primary pathway production of piRNAs is amplified by the secondary piRNA biogenesis pathway called the ping-pong pathway. This involves recognition of complementary RNA by Aub-bound primary piRNAs and catalyzes the cleavage of target RNA that is loaded onto an argonaute (Ago3). The initial piRNA directs the cleavage event of the secondary piRNA, and thus structurally, these are antisense to the primary piRNA.[228] Ago3-bound secondary piRNAs direct cleavage of complementary RNA targets. These products are subsequently processed into primary piRNAs.[228] Thus, the ping-pong cycle is a series of reciprocal cleavage events that ultimately performs post-transcriptional silencing and amplifies the initial piRNA-based response initiated on active repetitive elements. This ultimately protects the germline from mobile elements and viruses as they are silenced.[228,229] By protecting the germline from transposon activation, these RNA-binding proteins maintain the fidelity of the genome. As transposon and virus-derived DNA sequences are integrated into host genomes, inherited, and retained, there must be mechanisms to induce piRNA production from new threats. The mechanism by which new elements are silenced is poorly understood. It would be beneficial to harness this biology for creating new antivirals by reprogramming the genome with these sequences for pest control.

miRNA and Antiviral Activities

The third small RNA silencing pathway is the highly conserved miRNA pathway. miRNAs are produced from endogenous genes by RNA pol II, and these primary RNA (pri-miRNA) transcripts encode structured hairpins that are bound by the RNA cleavage enzyme Drosha in the nucleus and are processed into ~60-nt bp precursor miRNAs (pre-miRNAs).[228,229] These pre-miRNAs are exported to the cytoplasm where they are bound and cleaved by Dicer-1 to form the mature 20-24 bp miRNA duplex.[230] Drosha, Dicer-1, and Dicer-2 are all RNASE III enzymes that recognize differing RNA targets based on structure. From the miRNA duplex, one strand is loaded onto Argonaute-1 (Ago1) within the miRNA-induced silencing complex.[231] This silencing complex represses translation and can increase the decay of the RNA target through deadenylation and decapping.[232] While the core siRNA and miRNA pathway components are distinct in insects, some accessory factors, such as Ars2, impact siRNA and miRNA pathways and are antiviral.[233] In addition, miRNAs can impact infection indirectly by altering the transcriptome of the cell, and viruses can also impact the transcriptome including pri-miRNAs, both impacting viral replication.[234] Finally, some viruses encode miRNAs that target host or viral mRNA to promote infection.

cGAS/STING and Nucleic Acids

There are additional innate antimicrobial pathways that have evolved to recognize microbial nucleic acids. An important example involves the recognition of DNA, and defense against invading DNA occurs across the tree of life. To defend from foreign DNA, bacteria utilize diverse nucleic acid–based immune pathways which include restriction-modification systems and CRISPRs (clustered regularly interspaced short palindromic repeats).[235] Another strategy involves cGAS, which is a nucleic acid–binding protein that is deeply conserved. In bacteria, these enzymes synthesize a wide range of cyclic oligonucleotides that act as second messengers in response to viral infections. These second messengers are sensed and activate downstream effectors to block bacteriophage infection. Some bacteria encode homologs of STING, thus the cGAS–STING pathway has been conserved across evolution (Fig. 3.8). Indeed, organisms from sea anemones to

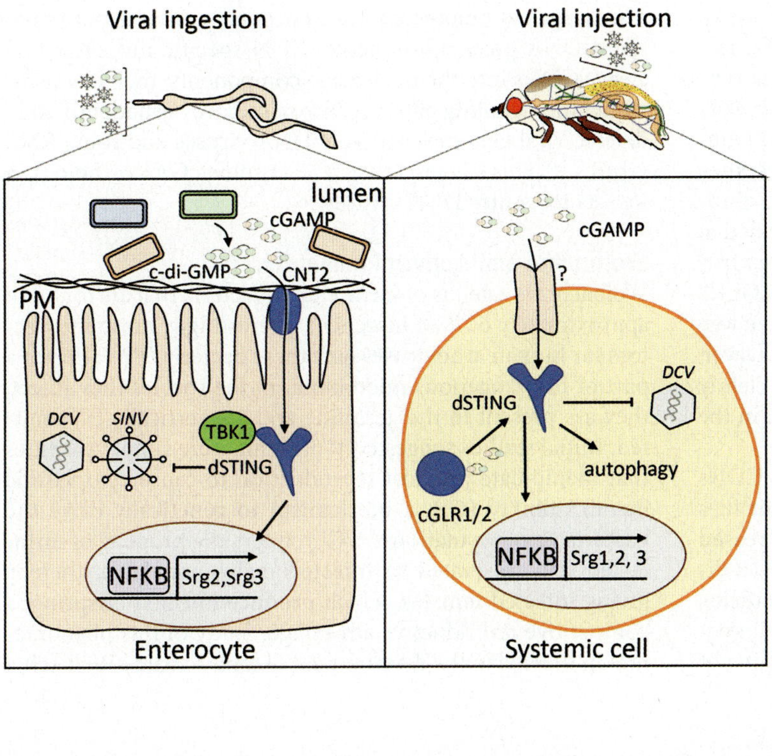

FIG. 3.8. Drosophila Sting in antiviral immunity. Orally acquired CDNs can induce a STING-TBK1–dependent antiviral response active against diverse enteric viruses. Injection of cGAMP can also induce a STING-dependent antiviral response with restricted activity. cGAS orthologs cGLR1 and cGLR2 can activate STING and impact infection systemically. STING can induce NF-kB and autophagy to control infections.

Drosophila to humans encode cGAS-STING. In mammals, cGAS is a major sensor of dsDNA; dsDNA binding induces the enzymatic production of a second messenger cyclic dinucleotide (CDN). There are diverse arrays of CDNs produced by endogenous enzymes as well as bacteria. Stimulator of interferon genes (STING) binds CDNs either produced endogenously by cGAS or exogenously by bacterial pathogens. Binding of CDNs activates STING, which subsequently induces diverse downstream antimicrobial effector pathways including interferon, NF-kB, and autophagy in mammals.

cGAS is a Mab21 domain containing protein which consists of a deeply conserved nucleotidyl transferase fold.[236] The CDN synthesized by mammalian cGAS is 2′3′ cGAMP.[237] Drosophila encodes several Mab21-containing cGAS-like enzymes, and two were shown to produce CDNs: cGLR1 (CG12970) and cGLR2 (CG30424).[238,239] Interestingly, another Mab21 containing protein CG7194, with the highest sequence identity to human cGAS did not produce CDNs in vitro, despite conservation of key catalytic residues.[240,241] In vitro assays found that cGLR2 (CG30424) produces both 2′3′ cGAMP and 3′2′cGAMP and that cGLR1 (CG12970) produces 3′2′ cGAMP and 2′3′ cdiAMP (Fig. 3.8). In contrast to mammalian cGAS, cGLR1 is activated by dsRNA while the ligand for cGLR2 is unknown and not likely DNA. This opens the door to a new understanding of the evolution of these sensors and how they may sense antiviral immunity.

Genetic studies found that ectopic expression of cGLR1 or cGLR2 decreases VSV replication.[238,239] Ectopic expression of cGLR2 reduced DCV infection and mortality, but cGLR1 reduced mortality but not replication. Perhaps surprisingly, only cGLR1/2 double-knockout flies show differences in DCV replication and mortality.[239] Despite the lack of a role of cGLR1 in virus infection, it is required for DCV-induced expression of a STING-dependent gene, srg1. cGLR2 only influences virus-induced gene expression changes and DCV replication or lethality when overexpressed, not when deleted.[239] This suggests that cGLR2 can be antiviral when ectopically expressed, but that it is not in a context of infection. In contrast to ectopic studies, there was no phenotype for viral load or survival when cGLR knockout animals were challenged with VSV or a DNA virus IIV6. While these studies identify endogenous sources of cyclic dinucleotides, they do not reveal how cGLRs produce endogenous cyclic dinucleotides in the context of infection as cGLR1 is expressed at low levels, and endogenous cGLR2 does not influence virus infection in vivo. The lack of antiviral activity during cGLR2 knockdown may indicate that viruses have evolved some way to block cGLR-dependent activation of STING. However, while cGLR2 is not antiviral against DCV, STING clearly is.[242-244] A critical question of how dSTING is activated in the context of virus infection remains unanswered.

STING is conserved in Drosophila as it binds CDNs and is antimicrobial.[243-245] Moreover, dSTING-deficient flies are more susceptible to Listeria infection due to decreased expression of AMPs, which are induced by the CDN (c-di-GMP).[245] While dSTING is antiviral, there is some specificity, as STING mutants are more susceptible to systemic infection by two related RNA viruses DCV and CrPV but not two other RNA viruses FHV and SINV or the dsDNA virus IIV6. Additionally, there are differences in CDN antiviral activity during systemic infection. For example, 2′3′cGAMP, 3′2′cGAMP, and c-di-AMP protect flies against systemic virus infection, but c-di-GMP does not.[238,242] Moreover, infection of different tissues can be regulated by dSTING, although perhaps by distinct mechanisms as dSTING is antiviral against oral DCV and SINV infection of the intestine but dSTING does not control systemic infection of SINV (Fig. 3.8). But there appears to be differences in the regulation of the dSTING pathway in these tissues. In addition, since CDNs are not cell permeable, there are efforts to identify the transporters involved. Oral c-di-GMP feeding protects from SINV infection and this is dependent on CNT2, an apically expressed nucleoside transporter.

CDN binding to STING activates downstream signaling pathways, which includes NF-kB, IRF3, and autophagy in mammals.[246] Drosophila STING has been shown to induce both NF-kB and autophagy. STING regulation of autophagy may be its ancestral antiviral function.[5] STING-dependent gene induction is IKKbeta and Relish dependent but not dependent on other upstream Imd components suggesting noncanonical activation.[243] Indeed, ectopic expression of STING, CDN feeding, or injection of flies with CDNs can induce NF-kB–dependent gene expression.[242] And there appears to be differences in the breadth of genes induced by endogenous CDNs and exogenous bacterial CDNs systemically and within the gut. Survival analysis shows that injection of flies with CDNs can protect animals from DCV-induced mortality and that this was independent of ATG7, a canonical autophagy component, and Ago2, a component of RNAi.[242] The gene expression program induced by CDN injection includes a number of genes with known roles in antiviral defense.

These data demonstrate a deeply conserved role for STING in antiviral immunity. As viruses evolve evasion mechanisms against important host-encoded antiviral pathways, the dsDNA poxviruses encode CDN-specific nucleases that cleave CDNs into the nucleoside components, thereby inactivating the signaling pathway.[247] Poxins have been identified in diverse viral genomes of insect DNA viruses and some RNA viruses.[248] This suggests that at least some cGAS orthologs in insects recognize DNA viruses.

Wolbachia and Antiviral Immunity

Wolbachia is a genus of intracellular bacteria that are found in approximately 60% of insects.[249] Interestingly, mosquito vectors for human arboviruses lack these bacteria.[249] While not a part of the exogenous microbiota, in the species they infect, they are present in the germline and are vertically transmitted. Initial studies suggested Wolbachia were simply parasites that manipulate the host reproduction to cause cytoplasmic incompatibility (CI), a mechanism to genetically drive the bacteria into populations.[250] CI renders the progeny of uninfected females mated to infected males nonviable, thereby giving infected females, which produce infected offspring, a competitive reproductive advantage. Study of this phenomenon in flies led to the development of tools to study Wolbachia and landmark studies that showed that Wolbachia can protect Drosophila against RNA viruses.[251,252] Further studies in mosquitoes, using Wolbachia derived from Drosophila, showed that introduction of Wolbachia into the germline can block dengue virus in mosquitoes.[253,254] This immediately led to the development of Wolbachia as a transmission-blocking agent to protect vectors from diverse arboviruses, including dengue virus.[253] While this protection is not universal, as some Wolbachia do not block infection with some viruses, trials have shown that this strategy is a useful approach to block the spread of dengue virus in endemic regions. To date, the mechanisms underlying the decrease in vectoring capacity by Wolbachia remain poorly characterized, but it has been proposed that Wolbachia could alter the host's immunity, modify cholesterol metabolism, or alter the insect epigenome.

GUT IMMUNITY

Across metazoans, the gut is a highly regionalized structure which is exposed to the outside world, and therefore in constant contact with microbes[1] (Fig. 3.9). The gut of insects is no exception and, while it faces exposure to pathogens, it is also associated with gut microbes that have beneficial functions for insects.[20] This imposes a specific pressure on midgut immunity as the immune response needs to achieve control of pathogens while still tolerating commensal microbes. The structure of the insect gut is very similar to the digestive tract of mammals. In most species, the midgut is analogous to the small intestine and is a highly regionalized organ, with up to 14 regions identified in the midgut of *D. melanogaster*.[22,255] Multiple cell types compose the midgut epithelium, including progenitor and stem cells that serve a homeostatic function by constantly producing new mature cells, including enterocytes and enteroendocrine cells.[21,253,256] The size and structure of the midgut adjust to food availability and nutritional conditions.[257]

Midgut Physical and Chemical Barriers to Infection

In the digestive tract of mammals, the gut lining is separated from the lumen by a layer of mucus. In insects, this function is assumed by the peritrophic matrix (see section on cuticle and PM) in conjunction with proteins associated with the PM, including multiple mucin-like proteins.[12] In addition, the midgut of insects is a naturally hostile environment to microbes (Fig. 3.9). Digestive enzymes such as lipases and

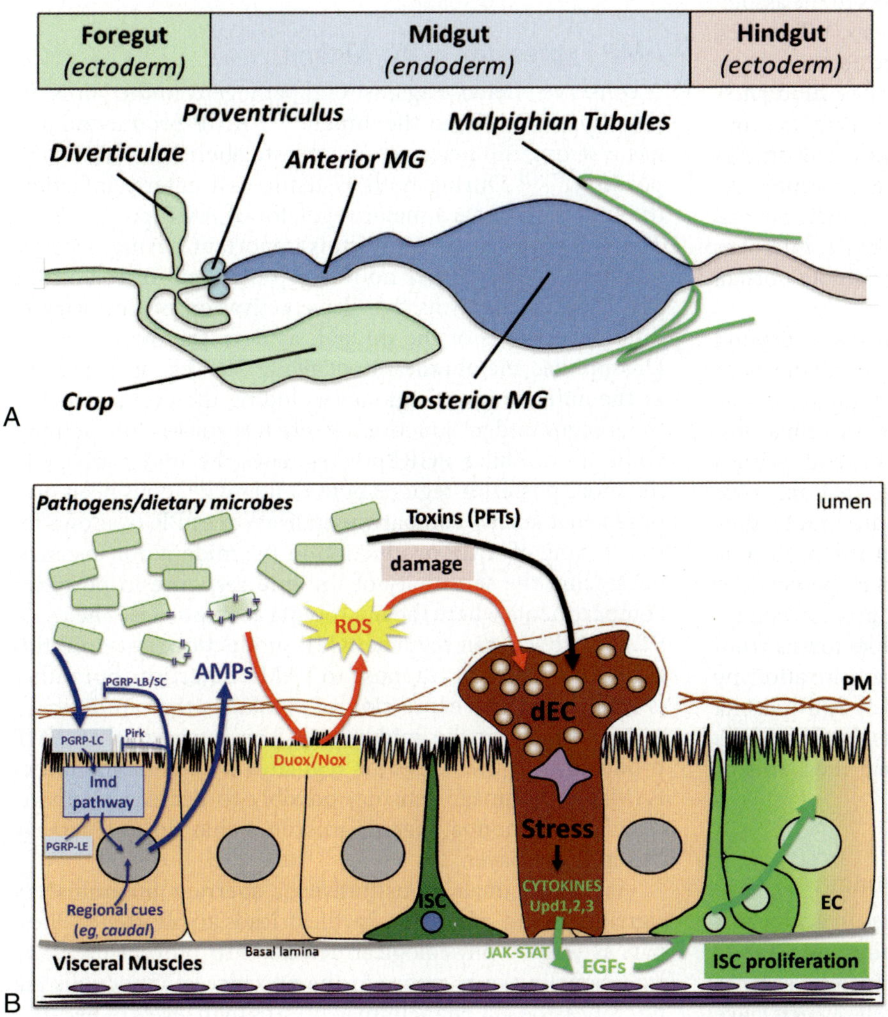

FIG. 3.9. The digestive tract of insects, gut immunity, and tissue repair. A. Schematic of the organization of the insect gut. The anterior foregut and posterior hindgut are of ectodermal origin. In multiple insects, the crop serves as a storage for food. Food is further passed onto the midgut, the main site of nutrient digestion and absorption. **B.** Homeostatic and tissue repair pathways are regulated to maintain the barrier and replenish damaged cells. In addition, while tolerance to commensals is maintained, pathogens induce strong immune pathways to protect from infection and breaches of the barrier. (Redrawn by permissionfrom Springer Nature; Buchon N, Silverman N, Cherry S. Immunity in Drosophila melanogaster-from microbial recognition to whole-organism physiology. *Nat Rev Immunol.* 2014;14(12):796-810. doi:10.1038/NRI3763.)

proteases can attack microbes and act as a natural defense. The pH of the midgut of multiple insects varies greatly, from species with a largely alkaline lumen to species hosting very acidic regions which also limit pathogens. Some regions of the insect midgut are also low in oxygen and close to anoxic, for instance, the larval midgut of scarabaeid beetles.

The Insect (Gut) Microbiota

The association of animals with gut microbes is an evolutionary conserved feature, as most animals contain communities of microbes associated with their digestive tract.[25] Insects inhabit multiple terrestrial environments, including the most extreme environments (from deserts to permafrost), their lifestyles are highly variable.[25] A consequence of such extreme lifestyles is that some insects face imbalanced diets and extreme nutrition. Such insect species rely for their survival on association with gut microbes, including bacteria and yeast/fungi.[20] For instance, insects such as whiteflies and aphids that feed on sap obtain mostly carbohydrates from their plant host and rely on nutrient exchange with endosymbiotic bacteria for other macronutrients.[258,259] The pest *Drosophila suzukii* relies on its gut microbes to develop in the sugar-rich, but nitrogen-deprived fruit.[260] Herbivorous insects, such as leaf-cutting ants, survive thanks to the association with fungi that they grow in fungal gardens.[261] Termites, which eat mostly wood particles, thrive due to their partnership with microorganisms they house in their hindgut.[262] More generally, most insects harbor some diverse communities of microbes in their digestive tract, and recent studies indicate that these microbes, when they are not essential for the survival of the insect, help at least to shape their normal physiology.[263] For instance, Drosophila harbors gut microbes that are not essential on rich diets but help promote normal development on deficient diets.[264]

The role of the insect microbiota extends well beyond nutrition. One important function of the microbiota is to improve resistance when facing invading pathogens. The microbiota occupies a specific niche in the gut, thus preventing the colonization of pathogens. The endogenous microbiota also secretes bioactive molecules that are toxic to invading pathogens and stimulates host immunity thus limiting pathogen colonization. Examples of the protective actions of microbiota include the secretion of polyketides such as pederin from Pseudomonas that protect beetles from predation[265] or the secretion of shiga-like toxins from Hamiltonella to protect pea aphids from parasitization by wasps.[266] Endosymbionts such as Wolbachia or Spiroplasma have been demonstrated to protect insects against virus infection or parasitic attack[266] through unclear mechanisms (see section on Wolbachia).

Reactive Oxygen Species and Gut Immunity

Some of the most important effector molecules to act against pathogens at the surface of the midgut lining are ROS.[266] As mentioned earlier, the enzymes Duox and Nox synthesize hypochlorous acid and H_2O_2, respectively, which have direct antimicrobial function.[27] In addition to their role as effectors, ROS can act as secondary messengers during infection, initiating processes ranging from tissue repair to hemocyte migration or hematopoiesis. The regulation of the activity of NADPH oxidases has been mostly studied in Drosophila, where Nox responds to microbiota while Duox responds robustly to pathogens and only minimally to the microbiota.[267-269] ROS are essential effector molecules as flies deficient in Duox are highly sensitive to oral infection with bacterial pathogens. Importantly, both the lack of ROS detoxification enzymes and excessive ROS production are highly detrimental for insects, shortening their lifespan, and demonstrating that a tight regulation is needed to control ROS production in the midgut.[270] The calcium-dependent activity of Duox is regulated by G protein–coupled receptors that trigger phospholipase C beta pathways, leading to the release of Ca2+ from the endoplasmic reticulum.[271] The metabolite Uracil has been proposed to be the PAMP that activates Duox activity.[267] Interestingly, uracil is abundantly secreted by oral pathogens of Drosophila such as *Erwinia carotovora*, but not released from the indigenous microbiota, and thus can be sensed to induce antimicrobial defenses in response. ROS and NADPH oxidases are conserved and play roles in immune defense across evolution.

AMP Expression in the Midgut

A conserved defense against oral pathogens in the gut is the secretion of AMPs in the lumen.[26,272] AMP production also has a strong influence on microbiota abundance and composition.[273,274] During both systemic and enteric infection, the Imd pathway is a major regulator of AMP production.[26] The Toll pathway, by contrast, is important during systemic infection but has either no role or only a minor role in the gut.[1] The mechanisms that drive activation of Imd vary in different regions of the midgut, as does the output.[22,275] In Drosophila, membrane-associated PGRP-LC is expressed in the anterior-most regions (including the proventriculus, the foregut midgut junction) where it regulates Imd activity, while intracellular PGRP-LE regulates the Imd pathway in the more posterior regions of the midgut.[276] It has been proposed that such a compartmentalization of PRRs allows for fine-tuning of AMP production in the midgut, for instance, decreasing the sensitivity of the Imd response in posterior compartments where the microbiota accumulates. The JAK–STAT pathway can regulate AMP production; however, this regulation is not in response to PAMP detection, but rather is induced by the expression of cytokines that are released upon damage of the midgut.[26,115] This suggests that AMP production in the midgut is regulated by a combination of PAMP and damage sensing, possibly to decrease immune reactivity to nonpathogenic microbes that do not damage the epithelial layer.

As in mammals, constitutive or aberrant inflammatory signaling (Imd activation in flies) leads to shortened lifespan as well as physiological defects.[76] To prevent excessive Imd signaling in response to the presence of innocuous gut microbes, the gut epithelium expresses high levels of negative regulators of the Imd pathway, including amidase PGRPs that degrade PGN.[75,76,275,277] In both insects and mammals, PGRP molecules are required for the proper management of the gut microbiota and to prevent dysbiosis.[1,276] However, the exact sequence of events linking PGRPs, immune regulation, and dysbiosis remains unclear.

Cell Sloughing and Tissue Repair in the Gut

In response to the ingestion of oral pathogens, toxin production by microbes combined with PAMP detection leads to the secretion of AMPs and the production of ROS.[26,27] These signals trigger the activation of a massive stress response in midgut enterocytes leading to a high proportion of enterocytes extruded from the epithelium and sloughed off into the lumen.[21,26,115,278,279] An essential aspect of insects' infection survival is the repair of this cell loss.[115,280,281] Tissue repair is mediated by the activation and proliferation of intestinal stem cells that regenerate the midgut.[26,282] Importantly, this process of stem cell–mediated repair is controlled by the JAK–STAT pathway and involves the coordination of multiple cell types.[113,115,278,280,282] Repair is initiated by the stressed enterocyte which synthesizes Upd cytokines that trigger the JAK–STAT pathway in visceral muscles and enteroblast progenitor cells. In response to JAK–STAT activation, these cells release growth factors (EGF) that promote proliferation in the stem cells.[278] In addition to this epithelial network, hemocytes participate in the regulation of ISCs by releasing TGF-β molecules.[257] In larvae, the trade-off between growth and repair upon bacterial infection leads to limited survivability.[281] Bacteria that can overwhelm tissue repair mechanisms lead to midgut disruption and insect death.[283] Basally, the commensal microbiota stimulates epithelial turnover at low levels, consistent with the low levels of damage it inflicts on the epithelium.[115,273] This regulated turnover plays important roles in homeostatic control and limiting PAMP escape from the gut into systemic circulation.[284]

Antiviral Defense in the Gut

All metazoans can be infected orally by enteric pathogens, including viruses. Arthropod-borne viruses are acquired by insects through an infected blood meal, and these viruses are thus orally acquired. Within the intestinal tract, arboviruses infect enterocytes from the lumen of the gut and are released basolaterally for systemic infection that ultimately is transmitted to the next vertebrate host.[285] The intestine presents a high barrier to infection. For example, feeding mosquitoes viruses that are not normally transmitted is typically nonproductive; however, bypassing the intestine through experimental systemic inoculation results in robust infection that can frequently be transmitted.[286,287] This phenomenon has parallels in mammals, as human enteric viruses can infect mice systemically but typically not orally. Thus, successful enteric viruses co-evolve with their hosts to evade intestinal immune defenses. Drosophila also harbor natural enteric viral pathogens. As arboviruses can infect mosquitoes and humans, it is perhaps not surprising that Drosophila can also be systemically infected by human arboviruses.[182,288] However, the fly gut also presents a high barrier to infection by these arboviruses, as they readily replicate in flies upon systemic infection, but innate immune pathways must be disabled to observe infection in the gut.[182,288] Therefore, this system has been used to discover of innate pathways and mechanisms that control enteric immunity in insects.

The intestinal epithelium is highly responsive to enteric infection and innate signaling pathways are active against orally acquired viruses. The pathways involved are seemingly distinct; while antiviral RNAi plays important roles in controlling systemic infection, RNAi does not control oral infection, as flies can clear DCV infection in the absence of this pathway.[216] Rather, other inducible innate pathways are essential. In mosquitoes, the JAK–STAT and NF-kB pathways play roles in enteric defense against viruses by activating specific antiviral gene expression programs upon infection.[289-292] Moreover, there is a clear interplay with the microbiota. The microbiota influence NF-kB signaling, and studies found that, depending on the composition, the microbiota can promote or restrict viral infection, likely through regulation of NF-kB.[289,291,293] In some cases, antibiotic treatment led to increased dengue infection, and this was through microbiota-dependent activation of Toll–NF-kB signaling pathways.[289] In contrast, *Serratia marcescens* promotes dengue virus replication in the gut.[294] In flies, the microbiota promotes inflammatory NF-kB signaling in enterocytes that induced an antiviral cytokine to control diverse RNA viruses.[288] In mammals, the microbiota can also have distinct roles in enteric viral infection depending on the microbiota and virus.

IMMUNE PRIMING, MEMORY, AND SPECIFICITY

Immune Priming

Insects rely on innate immune responses; however, this does not mean that the secondary response to a pathogen parallels the primary response to the same pathogen.[257] This "immune priming," in which the insect is exposed to a nonlethal microbe or PAMP and later challenged with a pathogenic infection, suggests that the response to a pathogen in a primed state is different from naïve individuals. This priming or "education" demonstrates that the primary challenge can provide protection against a secondary infection. While this phenomenon is readily observed, the mechanisms underlying such priming and the specificity of the protection conferred by priming remain unclear.[257] Two main mechanisms have been proposed to explain the protective effect of priming. First, the stability of AMPs could lead to their persistence in the host. Accordingly, it was found that some of the most stable AMPs can still be detected in hemolymph a week after immune stimulation. Another proposed mechanism involves an increase in the number of circulating hemocytes upon priming.[295,296] As in most insects, hematopoiesis is restricted to larvae. Therefore, the increase in hemocyte numbers at the site of infection is not due to proliferation, but rather to the release of sessile or tissue-bound hemocytes and their recruitment to new sites.

Another potential mechanism involves altered gene expression. Exposure to either *Candida albicans* or *Saccharomyces cerevisiae* increases the performance of Galleria against a secondary infection with *C. albicans*.[297] In the primed state, the expression of AMPs gallerimycin and galiomicin were elevated, suggesting that the stimulation of the NF-kB response led to upregulation of immune effectors broadly. Priming does not only occur in response to live microbes. Injection of β-glucan and laminarin, two fungal PAMPs, led to a similar priming effect and protection against *C. albicans* infection, reinforcing the idea that a first stimulation of the NF-kB pathway led to the protective state.[297] However, the priming does not always require an increased transcription of effector molecules concomitant to the secondary infection. For instance, *Bombus terrestris* was protected against secondary infection up to 22 days post priming, at a time when effectors were no longer upregulated.[298] In Drosophila, priming with a nonlethal Streptococcus pneumoniae protected against a lethal injection of this pathogen. Injection of beads to inhibit phagocytosis decreased the priming effect, suggesting that cellular immunity mediates this phenotype.[299] And the microbiota can prime immunity by producing low levels of PAMPs, constitutively leading to basal activation of antimicrobial gene expression programs. Priming is not only protective, but also costly for the host, explaining why investment in the constant production of immune effectors is avoided.[300,301] Indeed, priming was shown to be deleterious to insects when on nutrient-poor diets, suggesting that immune priming is energetically costly.

Transgenerational Immunity

Transgenerational immunity occurs when the parent is infected and confers increased resistance to infection on their progeny.[220] This may be due to enhanced immunocompetence of the offspring that can potentially be transmitted over multiple generations.[302] Increased immunity and survival upon challenge of the offspring of the mother has been demonstrated in diverse taxa including coleopteran, crustacean, hymenopteran, orthopteran, and mollusk species (for a comprehensive review, see Ref. 220). In *Caenorhabditis elegans*, there are diverse transgenerational mechanisms. For example, worms' learned avoidance of the pathogenic bacteria *Pseudomonas aeruginosa* can be transmitted by small RNAs across four generations.[303] And this pathway is dependent on virus-like particles from an endogenous retrotransposon.[304] In another example, transgenic expression of an RNA virus flock house virus results in the production of virus-derived small RNAs that are transgenerationally transmitted in the germline to silence the viral genome in subsequent generations.[305] The small RNAs that were transmitted were associated with epigenetic changes of a set of host genes, leading to transgenerational histone changes.[306] A recent study in flies found evidence for transgenerational immunity that was RNA-dependent but RNAi-independent.[304] The progeny of primed flies displays distinct gene expression profiles, and future studies will define the prevalence and mechanisms involved.

CONCLUSION

Insects are an important group of organisms that are not only abundant species but also have important impacts on human health. They can negatively impact us by transmitting disease and destroying crops. However, they also provide important services, including pollinating crops, cleaning up waste, and keeping some pests in check. Thus, a better understanding of insect physiology and immunity can benefit us in a myriad of ways. Importantly, many facets of innate immunity are conserved, and thus insects can teach us about our own defenses. We have discovered much about immunity through comparative approaches and insects have contributed to these understandings as the first Toll receptors were discovered in flies and will continue to inform new facets of immunity to infection.

TABLE OF TERMS

Adult Sexually mature stage of insect development

Antimicrobial peptides (AMPs) Peptides that function as part of the immune system and are broad spectrum antibiotics

Apoptosis Conserved programmed cell death pathway

Argonaute protein (Ago) Family of RNA-binding proteins that play a central role in RNA silencing processes, as essential components of the small RNA silencing pathways. Small RNAs guide Argonaute proteins to their specific targets through sequence complementarity (base pairing), which then leads to mRNA cleavage or translation inhibition

Arthropod Invertebrates with an exoskeleton, a segmented body, and paired jointed appendages

Autophagy Conserved cell biological process to degrade intracellular components

cGAS Cyclic GMP-AMP synthase (*cGAS*, cGAMP synthase) belonging to the nucleotidyltransferase family of nucleic acid sensors that produces a cyclic dinucleotide second messenger that activates STING and thus downstream antimicrobial pathways

Chitin A long-chain polymer of *N*-acetylglucosamine, an amide derivative of glucose

Commensals Constituents of the microbiota that benefit the microbe and host

Crop Enlarged part of the insect gastrointestinal tract that can store food

Crystal cell Insect hemocyte that has crystalline inclusions for melanization

Cuticle Any of a variety of tough but flexible, nonmineral outer coverings of an organism, or parts of an organism that provide protection

Cyclic dinucleotide (CDN) Second messengers synthesized directly from two nucleosides with diverse chemical linkages

Cytoplasmic incompatibility (CI) A phenomenon that results in sperm and eggs being unable to form viable offspring. The effect arises from changes in the gamete cells caused by intracellular parasites like *Wolbachia*, which infect a wide range of insect species

Damage-associated molecular pattern (DAMP) Molecules that are present in cells but not normally exposed to the extracellular milieu, released from damaged or dying cells

Defective viral genomes (DVGs) Replication-defective viral products generated during viral replication that are typically deletion mutants and have been identified across many RNA viral families

Down syndrome cell adhesion molecule (Dscam) A conserved PRR that is differentially spliced in insects leading to thousands of distinct isoforms

dsRNA Double-stranded RNA is RNA with two complementary strands

Eclosion The emergence of an adult insect from a pupal case

Embryo Unhatched but developing offspring

Entomopathogens Pathogens that infect insects

Fibrinogen-related protein (FREP) Secreted proteins in invertebrates that are diversified by mutation in snails

Gram-negative bacteria-binding protein (GNBP) Pattern recognition receptor for lipopolysaccharide and β-1,3-glucan

Hemocytes Specialized immune cell in insects

Hemolymph The fluid in the body cavity of insects akin to vertebrate blood and lymph

IMD pathway Broadly conserved NF-κB immune signaling pathway of insects and some arthropods (and other invertebrates) that regulates a potent antimicrobial defense response that can be induced by diverse PRRs

Immune priming A memory-like phenomenon that demonstrates a more robust secondary response to a pathogen

Insect Pancrustacean hexapod invertebrates of the class Insecta, the largest group within the arthropod phylum

Invertebrate Animals lacking a backbone

JAK–STAT pathway The Janus kinase (*JAK*)-signal transducer and activator of transcription (*STAT*) *pathway Janus* kinase *(JAK)* is conserved signaling pathway that transduces cytokine-mediated signals

Lamellocyte Insect hemocyte that encapsulates large bodies

Larvae The active immature form of an insect that typically looks like a caterpillar between egg and pupae

Lectin Carbohydrate-binding proteins that are highly specific for sugar groups that are usually part of other molecules

Melanin Dark pigment produced in insects for immunity and wound healing

Melanization Enzymatic reaction in insects that culminates in the synthesis of melanin required for various immune responses

Microbiota The range of microorganisms that are found in or on multicellular organisms. Microbiota include bacteria, archaea, protists, fungi, and viruses. These may be commensal or pathogenic

Midgut The gastrointestinal tract is made up of the foregut, midgut, and hindgut. The midgut is the region of intestinal absorption and most pathogen invasion

miRNA MicroRNAs are a class of small noncoding RNAs approximately 21 to 25 nucleotides in length functions in post-transcriptional regulation of gene expression

Nitric oxide (NO) A free radical with the formula NO that is a gaseous signaling molecule

Nuclear factor kappa-light-chain-enhancer of activated B cells (NF-kB) Highly conserved protein complex that controls transcription of DNA

Oogenesis The production of an egg

Opsonins Extracellular proteins that, when bound to substances or cells, induce phagocytes to phagocytose the substances or cells with the opsonin bound

Parasitization Infestation by a parasite

Pathobionts Microorganisms that only harm organisms under certain circumstances

Pathogen-associated molecular pattern (PAMP) Molecules with evolutionarily conserved motifs that are present on groups of pathogens

Pattern recognition receptor (PRR) Germline-encoded sensors that detect molecules frequently found in pathogens

Peptidoglycan (PGN) A substance forming the cell walls of many bacteria, consisting of glycosaminoglycan chains interlinked with short peptides

Peptidoglycan recognition protein (PGRP) A group of highly conserved pattern *recognition* receptors with at least one *peptidoglycan-recognition* domain

Peritrophic matrix Semipermeable, noncellular structure that surrounds the food bolus in the midgut

Phenyloxidase Terminal enzyme in the melanization reaction

Piwi-interacting RNAs (piRNA) A large class of small noncoding RNAs 26 to 31 nucleotide molecules expressed in animal cells and suppress expression of mobile genetic elements in the germline

Plasmatocyte Insect hemocytes that phagocytose

Pupae Insect in an inactive form between larvae and adult

Reactive oxygen species (ROS) Highly reactive chemicals formed from oxygen and including peroxides, superoxide, hydroxyl radical, singlet oxygen, and alpha-oxygen

RISC RNA-induced silencing complex is a ribonucleoprotein complex containing an Argonaute family protein that functions in gene silencing using siRNAs or miRNAs to post-transcriptionally silence cognate RNAs

RNA interference (RNAi) A biological process in which RNA molecules are involved in sequence-specific suppression of gene expression

siRNA Short interfering RNAs are a class of small noncoding RNAs typically 20 to 24 nucleotides in length, operating within the RNA interference (RNAi) pathway

Spiracle Opening in the exoskeletons of some arthropods for air exchange

Stimulator of interferon genes (STING) An ER-associated protein that binds CDNs to activate downstream antimicrobial pathways

Toll pathway A broadly conserved NF-κB immune signaling pathway of insects and some arthropods (and other invertebrates) that regulates a potent antimicrobial defense response downstream of the Toll receptor

Trachea Part of the respiratory tract of insects that are tubes

Transgenerational immunity A memory-like immune response transmitted from parents to offspring in many invertebrates

Wolbachia Genus of intracellular bacteria that infects mainly arthropod species and is one of the most common parasitic microbes and possibly the most common reproductive parasite in the biosphere

β-Glucan A group of β-D-glucose polysaccharides naturally occurring in the cell walls of plants, bacteria, and fungi

REFERENCES

1. Buchon N, Broderick NA, Lemaitre B. Gut homeostasis in a microbial world: insights from Drosophila melanogaster. *Nat Rev Microbiol.* 2013;11(9):615-626. doi:10.1038/nrmicro3074
2. Buchon N, Silverman N, Cherry S. Immunity in Drosophila melanogaster-from microbial recognition to whole-organism physiology. *Nat Rev Immunol.* 2014;14(12):796-810. doi:10.1038/nri3763
3. Lemaitre B, Hoffmann J. The host defense of *Drosophila melanogaster*. *Annu Rev Immunol.* 2007;25(1):697-743. doi:10.1146/annurev.immunol.25.022106.141615
4. Ebani VV, Mancianti F. Entomopathogenic fungi and bacteria in a veterinary perspective. *Biology (Basel).* 2021;10(6):479. doi:10.3390/BIOLOGY10060479
5. Schneider J, Imler JL. Sensing and signalling viral infection in drosophila. *Dev Comp Immunol.* 2021;117:103985. doi:10.1016/J.DCI.2020.103985
6. Schneider WM, Hoffmann H-H. Flavivirus-host interactions: an expanding network of proviral and antiviral factors. *Curr Opin Virol.* 2021;52:71-77. doi:10.1016/J.COVIRO.2021.11.007
7. Douglas AE. Multiorganismal insects: diversity and function of resident microorganisms. *Annu Rev Entomol.* 2015;60:17-34. doi:10.1146/ANNUREV-ENTO-010814-020822
8. Ferreira AG, Fairlie S, Moreira LA. Insect vectors endosymbionts as solutions against diseases. *Curr Opin insect Sci.* 2020;40:56-61. doi:10.1016/J.COIS.2020.05.014
9. Landmann F. The Wolbachia endosymbionts. *Microbiol Spectr.* 2019;7(2). doi:10.1128/MICROBIOLSPEC.BAI-0018-2019
10. Kanost MR, Jiang H, Yu XQ. Innate immune responses of a lepidopteran insect, Manduca sexta. *Immunol Rev.* 2004;198:97-105. doi:10.1111/J.0105-2896.2004.0121.X
11. Ashida M, Brey PT. Role of the integument in insect defense: pro-phenol oxidase cascade in the cuticular matrix. *Proc Natl Acad Sci U S A.* 1995;92(23):10698. doi:10.1073/PNAS.92.23.10698
12. Hegedus D, Erlandson M, Gillott C, Toprak U. New insights into peritrophic matrix synthesis, architecture, and function. *Annu Rev Entomol.* 2009;54(1):285-302. doi:10.1146/annurev.ento.54.110807.090559
13. Liu L, Li Z, Luo X, et al. Which is stronger? A continuing battle between Cry toxins and insects. *Front Microbiol.* 2021;12:665101. doi:10.3389/FMICB.2021.665101
14. Browne N, Heelan M, Kavanagh K. An analysis of the structural and functional similarities of insect hemocytes and mammalian phagocytes. *Virulence.* 2013;4(7):597-603. doi:10.4161/VIRU.25906
15. Meister M. Blood cells of Drosophila: cell lineages and role in host defence. *Curr Opin Immunol.* 2004;16(1):10-15. doi:10.1016/J.COI.2003.11.002
16. Balabanidou V, Grigoraki L, Vontas J. Insect cuticle: a critical determinant of insecticide resistance. *Curr Opin insect Sci.* 2018;27:68-74. doi:10.1016/J.COIS.2018.03.001
17. Chung H, Carroll SB. Wax, sex and the origin of species: dual roles of insect cuticular hydrocarbons in adaptation and mating. *Bioessays.* 2015;37(7):822-830. doi:10.1002/BIES.201500014
18. Vincent JFV, Wegst UGK. Design and mechanical properties of insect cuticle. *Arthropod Struct Dev.* 2004;33(3):187-199. doi:10.1016/J.ASD.2004.05.006
19. Pass G. Beyond aerodynamics: the critical roles of the circulatory and tracheal systems in maintaining insect wing functionality. *Arthropod Struct Dev.* 2018;47(4):391-407. doi:10.1016/J.ASD.2018.05.004
20. Caccia S, Casartelli M, Tettamanti G. The amazing complexity of insect midgut cells: types, peculiarities, and functions. *Cell Tissue Res.* 2019;377(3):505-525. doi:10.1007/S00441-019-03076-W
21. Bonfini A, Liu X, Buchon N. From pathogens to microbiota: how Drosophila intestinal stem cells react to gut microbes. *Dev Comp Immunol.* 2016;64:22-38. doi:10.1016/j.dci.2016.02.008
22. Buchon N, Osman D, David FPA, et al. Morphological and molecular characterization of adult midgut compartmentalization in Drosophila. *Cell Rep.* 2013;3(5):1725-1738. doi:10.1016/j.celrep.2013.04.001
23. Erlandson MA, Toprak U, Hegedus DD. Role of the peritrophic matrix in insect-pathogen interactions. *J Insect Physiol.* 2019;117. doi:10.1016/J.JINSPHYS.2019.103894
24. Gutiérrez-Cabrera AE, Córdoba-Aguilar A, Zenteno E, Lowenberger C, Espinoza B. Origin, evolution and function of the hemipteran perimicrovillar membrane with emphasis on Reduviidae that transmit Chagas disease. *Bull Entomol Res.* 2016;106(3):279-291. doi:10.1017/S0007485315000929
25. Douglas AE. Lessons from studying insect symbioses. *Cell Host Microbe.* 2011;10(4):359-367. doi:10.1016/j.chom.2011.09.001
26. Buchon N, Broderick NA, Poidevin M, Pradervand S, Lemaitre B. Drosophila intestinal response to bacterial infection: activation of host defense and stem cell proliferation. *Cell Host Microbe.* 2009;5(2):200-211. doi:10.1016/j.chom.2009.01.003
27. Bae YS, Choi MK, Lee WJ. Dual oxidase in mucosal immunity and host-microbe homeostasis. *Trends Immunol.* 2010;31(7):278-287. doi:10.1016/j.it.2010.05.003
28. Lanot R, Zachary D, Holder F, Meister M. Postembryonic hematopoiesis in Drosophila. *Dev Biol.* 2001;230(2):243-257. doi:10.1006/DBIO.2000.0123
29. Gold KS, Brückner K. Macrophages and cellular immunity in *Drosophila melanogaster*. *Semin Immunol.* 2015;27(6):357-368. doi:10.1016/j.smim.2016.03.010
30. Banerjee U, Girard JR, Goins LM, Spratford CM. Drosophila as a genetic model for hematopoiesis. *Genetics.* 2019;211(2):367-417. doi:10.1534/GENETICS.118.300223
31. Sorrentino RP, Carton Y, Govind S. Cellular immune response to parasite infection in the Drosophila lymph gland is developmentally regulated. *Dev Biol.* 2002;243(1):65-80. doi:10.1006/DBIO.2001.0542
32. Lee MJ, Kalamarz ME, Paddibhatla I, Small C, Rajwani R, Govind S. Virulence factors and strategies of Leptopilina spp.: selective responses in Drosophila hosts. *Adv Parasitol.* 2009;70:123-145. doi:10.1016/S0065-308X(09)70005-3
33. Wang L, Kounatidis I, Ligoxygakis P. Drosophila as a model to study the role of blood cells in inflammation, innate immunity and cancer. *Front Cell Infect Microbiol.* 2013;3(JAN). doi:10.3389/FCIMB.2013.00113
34. Wood W, Jacinto A. Drosophila melanogaster embryonic haemocytes: masters of multitasking. *Nat Rev Mol Cell Biol.* 2007;8(7):542-551. doi:10.1038/nrm2202
35. Mandal L, Banerjee U, Hartenstein V. Evidence for a fruit fly hemangioblast and similarities between lymph-gland hematopoiesis in fruit fly and mammal aorta-gonadal-mesonephros mesoderm. *Nat Genet.* 2004;36(9):1019-1023. doi:10.1038/NG1404
36. Ema M, Rossant J. Cell fate decisions in early blood vessel formation. *Trends Cardiovasc Med.* 2003;13(6):254-259. doi:10.1016/S1050-1738(03)00105-1
37. Medvinsky A, Dzierzak E. Definitive hematopoiesis is autonomously initiated by the AGM region. *Cell.* 1996;86(6):897-906. doi:10.1016/S0092-8674(00)80165-8
38. Tattikota SG, Cho B, Liu Y, et al. A single-cell survey of Drosophila blood. *Elife.* 2020;9:1-35. doi:10.7554/ELIFE.54818
39. Sanchez Bosch P, Makhijani K, Herboso L, et al. Adult Drosophila lack hematopoiesis but rely on a blood cell reservoir at the respiratory epithelia to relay infection signals to surrounding tissues. *Dev Cell.* 2019;51(6):787-803.e5. doi:10.1016/J.DEVCEL.2019.10.017
40. Holz A, Bossinger B, Strasser T, Janning W, Klapper R. The two origins of hemocytes in Drosophila. *Development.* 2003;130(20):4955-4962. doi:10.1242/DEV.00702
41. Horn L, Leips J, Starz-Gaiano M. Phagocytic ability declines with age in adult Drosophila hemocytes. *Aging Cell.* 2014;13(4):719-728. doi:10.1111/ACEL.12227
42. Guillou A, Troha K, Wang H, Franc NC, Buchon N. The Drosophila CD36 homologue croquemort is required to maintain immune and gut homeostasis during development and aging. *PLoS Pathog.* 2016;12(10):e1005961. doi:10.1371/JOURNAL.PPAT.1005961
43. Okabe Y, Medzhitov R. Tissue biology perspective on macrophages. *Nat Immunol.* 2016;17(1):9-17. doi:10.1038/NI.3320
44. Lavin Y, Mortha A, Rahman A, Merad M. Regulation of macrophage development and function in peripheral tissues. *Nat Rev Immunol.* 2015;15(12):731-744. doi:10.1038/NRI3920
45. Cho B, Yoon SH, Lee D, et al. Single-cell transcriptome maps of myeloid blood cell lineages in Drosophila. *Nat Commun.* 2020;11(1). doi:10.1038/S41467-020-18135-Y
46. Cattenoz PB, Sakr R, Pavlidaki A, et al. Temporal specificity and heterogeneity of Drosophila immune cells. *EMBO J.* 2020;39(12). doi:10.15252/EMBJ.2020104486
47. Ramond E, Dudzic JP, Lemaitre B. Comparative RNA-Seq analyses of Drosophila plasmatocytes reveal gene specific signatures in response to clean injury and septic injury. *PLoS One.* 2020;15(6). doi:10.1371/JOURNAL.PONE.0235294
48. Fu Y, Huang X, Zhang P, van de Leemput J, Han Z. Single-cell RNA sequencing identifies novel cell types in Drosophila blood. *J Genet Genomics.* 2020;47(4):175-186. doi:10.1016/J.JGG.2020.02.004
49. Coates JA, Brittle A, Armitage EL, Zeidler MP, Evans IR. Identification of functionally-distinct macrophage subpopulations regulated by efferocytosis in Drosophila. Published online April 18, 2020. bioRxiv. 2020.04.17.047472. doi:10.1101/2020.04.17.047472
50. Haine ER, Rolff J, Siva-Jothy MT. Functional consequences of blood clotting in insects. *Dev Comp Immunol.* 2007;31(5):456-464.doi:10.1016/j.dci.2006.08.004
51. Arefin B, Kucerova L, Krautz R, Kranenburg H, Parvin F, Theopold U. Apoptosis in hemocytes induces a shift in effector mechanisms in the Drosophila immune system and leads to a pro-inflammatory state. *PLoS One.* 2015;10(8):e0136593. doi:10.1371/JOURNAL.PONE.0136593
52. Defaye A, Evans I, Crozatier M, Wood W, Lemaitre B, Leulier F. Genetic ablation of Drosophila phagocytes reveals their contribution to both development and resistance to bacterial infection. *J Innate Immun.* 2009;1(4):322-334. doi:10.1159/000210264
53. Charroux B, Royet J. Elimination of plasmatocytes by targeted apoptosis reveals their role in multiple aspects of the Drosophila immune response. *Proc Natl Acad Sci U S A.* 2009;106(24):9797-9802. doi:10.1073/PNAS.0903971106
54. Braun A, Hoffmann JA, Meister M. Analysis of the Drosophila host defense in domino mutant larvae, which are devoid of hemocytes. *Proc Natl Acad Sci U S A.* 1998;95(24):14337-14342. doi:10.1073/PNAS.95.24.14337
55. Neves J, Zhu J, Sousa-Victor P, et al. Immune modulation by MANF promotes tissue repair and regenerative success in the retina. *Science.* 2016;353(6294):aaf3646. doi:10.1126/SCIENCE.AAF3646
56. Chakrabarti S, Visweswariah SS. Intramacrophage ROS primes the innate immune system via JAK/STAT and toll activation. *Cell Rep.* 2020;33(6). doi:10.1016/J.CELREP.2020.108368
57. Roddie HG, Armitage EL, Coates JA, Johnston SA, Evans IR. Simu-dependent clearance of dying cells regulates macrophage function and inflammation resolution. *PLoS Biol.* 2019;17(5):e2006741 doi:10.1371/JOURNAL.PBIO.2006741
58. Nonaka S, Ando Y, Kanetani T, et al. Signaling pathway for phagocyte priming upon encounter with apoptotic cells. *J Biol Chem.* 2017;292(19):8059-8072. doi:10.1074/JBC.M116.769745
59. Weavers H, Evans IR, Martin P, Wood W. Corpse engulfment generates a molecular memory that primes the macrophage inflammatory response. *Cell.* 2016;165(7):1658-1671. doi:10.1016/J.CELL.2016.04.049
60. Arrese EL, Soulages JL. Insect fat body: energy, metabolism, and regulation. *Annu Rev Entomol.* 2010;55:207-225. doi:10.1146/ANNUREV-ENTO-112408-085356
61. Kerkut GA, Gilbert LI. Comprehensive insect physiology, biochemistry and pharmacology. Vol. 3. Integument, respiration and circulation. *Compr insect Physiol Biochem Pharmacol Vol 3 Integument, Respir Circ.* 1985;3:289-353. Accessed December 18, 2021. https://books.google.ca/books?hl=en&lr=&id=RnXYBAAAQBAJ&oi=fnd&pg=PA355&dq=ephemeroptera+hemolymph&ots=r-f8rwoMt2&sig=465-Oxiy4R3l8bc3CGcQxZle8jU#v=onepage&q=ephemeroptera hemolymph&f=false
62. Haunerland NH, Shirk PD. Regional and functional differentiation in the insect fact body. 2003;40(1):121-145. doi:10.1146/ANNUREV.EN.40.010195.001005
63. Agaisse H, Petersen UM, Boutros M, Mathey-Prevot B, Perrimon N. Signaling role of hemocytes in Drosophila JAK/STAT-dependent response to septic injury. *Dev Cell.* 2003;5(3):441-450. doi:10.1016/S1534-5807(03)00244-2
64. Shia AKH, Glittenberg M, Thompson G, Weber AN, Reichhart JM, Ligoxygakis P. Toll-dependent antimicrobial responses in Drosophila larval fat body require Spätzle secreted by haemocytes. *J Cell Sci.* 2009;122(pt 24):4505-4515. doi:10.1242/JCS.049155
65. Wu SC, Liao CW, Pan RL, Juang JL. Infection-induced intestinal oxidative stress triggers organ-to-organ immunological communication in Drosophila. *Cell Host Microbe.* 2012;11(4):410-417. doi:10.1016/J.CHOM.2012.03.004

66. Newton K, Dixit VM. Signaling in innate immunity and inflammation. *Cold Spring Harbor Perspect Biol.* 2012;4(3):a006049. doi:10.1101/CSHPERSPECT.A006049
67. Stokes BA, Yadav S, Shokal U, Smith LC, Eleftherianos I. Bacterial and fungal pattern recognition receptors in homologous innate signaling pathways of insects and mammals. *Front Microbiol.* 2015;6(JAN):19. doi:10.3389/FMICB.2015.00019/BIBTEX
68. Issa N, Guillaumot N, Lauret E, et al. The circulating protease persephone is an immune sensor for microbial proteolytic activities upstream of the Drosophila toll pathway. *Mol Cell.* 2018;69(4):539. doi:10.1016/J.MOLCEL.2018.01.029
69. Wang Q, Ren M, Liu X, Xia H, Chen K. Peptidoglycan recognition proteins in insect immunity. *Mol Immunol.* 2019;106:69-76. doi:10.1016/J.MOLIMM.2018.12.021
70. Schleifer KH, Kandler O. Peptidoglycan types of bacterial cell walls and their taxonomic implications. *Bacteriol Rev.* 1972;36(4):407-477. doi:10.1128/BR.36.4.407-477.1972/ASSET/1742D96F-CC8B-4E5C-801A-EBC6DBC899B3/ASSETS/BR.36.4.407-477.1972.FP.PNG
71. Kurata S, Ariki S, Kawabata S. Recognition of pathogens and activation of immune responses in Drosophila and horseshoe crab innate immunity. *Immunobiology.* 2006;211(4):237-249. doi:10.1016/J.IMBIO.2005.10.016
72. Valanne S, Wang J-H, Rämet M. The Drosophila toll signaling pathway. *J Immunol.* 2011;186(2):649-656. doi:10.4049/jimmunol.1002302
73. Reiser JB, Teyton L, Wilson IA. Crystal structure of the Drosophila peptidoglycan recognition protein (PGRP)-SA at 1.56 Å resolution. *J Mol Biol.* 2004;340(4):909-917. doi:10.1016/J.JMB.2004.04.077
74. Royet J, Gupta D, Dziarski R. Peptidoglycan recognition proteins: modulators of the microbiome and inflammation. *Nat Rev Immunol.* 2011 1112. 2011;11(12):837-851. doi:10.1038/nri3089
75. Zaidman-Rémy A, Hervé M, Poidevin M, et al. The Drosophila amidase PGRP-LB modulates the immune response to bacterial infection. *Immunity.* 2006;24(4):463-473. doi:10.1016/j.immuni.2006.02.012
76. Paredes JC, Welchman DP, Poidevin M, Lemaitre B. Negative regulation by Amidase PGRPs shapes the drosophila antibacterial response and protects the Fly from innocuous infection. *Immunity.* 2011;35(5):770-779. doi:10.1016/j.immuni.2011.09.018
77. Ruiz-Herrera J, Ortiz-Castellanos L. Cell wall glucans of fungi. A review. *Cell Surf.* 2019;5:100022. doi:10.1016/J.TCSW.2019.100022
78. Rao XJ, Zhan MY, Pan YM, et al. Immune functions of insect βGRPs and their potential application. *Dev Comp Immunol.* 2018;83:80-88. doi:10.1016/J.DCI.2017.12.007
79. Warr E, Das S, Dong Y, Dimopoulos G. The Gram-negative bacteria-binding protein gene family: its role in the innate immune system of anopheles gambiae and in anti-Plasmodium defence. *Insect Mol Biol.* 2008;17(1):39-51. doi:10.1111/J.1365-2583.2008.00778.X
80. Wang Y, Jiang H. Interaction of beta-1,3-glucan with its recognition protein activates hemolymph proteinase 14, an initiation enzyme of the prophenoloxidase activation system in Manduca sexta. *J Biol Chem.* 2006;281(14):9271-9278. doi:10.1074/JBC.M513797200
81. Roh KB, Kim CH, Lee H, et al. Proteolytic cascade for the activation of the insect toll pathway induced by the fungal cell wall component. *J Biol Chem.* 2009;284(29):19474-19481. doi:10.1074/JBC.M109.007419
82. Brown GD, Herre J, Williams DL, Willment JA, Marshall ASJ, Gordon S. Dectin-1 mediates the biological effects of beta-glucans. *J Exp Med.* 2003;197(9):1119-1124. doi:10.1084/JEM.20021890
83. Xia X, You M, Rao XJ, Yu XQ. Insect C-type lectins in innate immunity. *Dev Comp Immunol.* 2018;83:70-79. doi:10.1016/J.DCI.2017.11.020
84. Dodd RB, Drickamer K. Lectin-like proteins in model organisms: implications for evolution of carbohydrate-binding activity. *Glycobiology.* 2001;11(5):71R-79R. doi:10.1093/GLYCOB/11.5.71R
85. Yu XQ, Ma Y. Calcium is not required for Immulectin-2 binding, but protects the protein from proteinase digestion. *Insect Biochem Mol Biol.* 2006;36(6):505-516. doi:10.1016/J.IBMB.2006.03.010
86. Yu XQ, Kanost MR. Immulectin-2, a lipopolysaccharide-specific lectin from an insect, manduca sexta, is induced in response to gram-negative bacteria. *J Biol Chem.* 2000;275(48):37373-37381. doi:10.1074/JBC.M003021200
87. Warr E, Lambrechts L, Koella JC, Bourgouin C, Dimopoulos G. Anopheles gambiae immune responses to Sephadex beads: involvement of anti-Plasmodium factors in regulating melanization. *Insect Biochem Mol Biol.* 2006;36(10):769-778. doi:10.1016/J.IBMB.2006.07.006
88. Osta MA, Christophides GK, Kafatos FC. Effects of mosquito genes on Plasmodium development. *Science.* 2004;303(5666):2030-2032. doi:10.1126/SCIENCE.1091789
89. Nalini M, Choi JY, Je YH, Hwang I, Kim Y. Immunoevasive property of a polydnaviral product, CpBV-lectin, protects the parasitoid egg from hemocytic encapsulation of Plutella xylostella (Lepidoptera: Yponomeutidae). *J Insect Physiol.* 2008;54(7):1125-1131. doi:10.1016/J.JINSPHYS.2008.04.023
90. Hanington PC, Zhang SM. The primary role of fibrinogen-related proteins in invertebrates is defense, not coagulation. *J Innate Immun.* 2011;3(1):17-27. doi:10.1159/000321882
91. Adema CM. Fibrinogen-related proteins (FREPs) in mollusks. *Results Probl Cell Differ.* 2015;57:111-129. doi:10.1007/978-3-319-20819-0_5
92. Hillyer JF. Insect immunology and hematopoiesis. *Dev Comp Immunol.* 2016;58:102-118. doi:10.1016/J.DCI.2015.12.006
93. Busse MS, Arnold CP, Towb P, Katrivesis J, Wasserman SA. A κB sequence code for pathway-specific innate immune responses. *EMBO J.* 2007;26(16):3826. doi:10.1038/SJ.EMBOJ.7601798
94. Zeidler MP, Bausek N. The Drosophila JAK-STAT pathway. *JAK-STAT.* 2013;2(3):e25353. doi:10.4161/JKST.25353
95. Li Z, Wu C, Ding X, Li W, Xue L. Toll signaling promotes JNK-dependent apoptosis in Drosophila. *Cell Div.* 2020;15(1):7. doi:10.1186/S13008-020-00062-5
96. Sim S, Jupatanakul N, Dimopoulos G. Mosquito immunity against arboviruses. *Viruses.* 2014;6(11):4479-4504. doi:10.3390/V6114479
97. Lindsay SA, Wasserman SA. Conventional and non-conventional Drosophila Toll signaling. *Dev Comp Immunol.* 2014;42(1):16-24. doi:10.1016/J.DCI.2013.04.011
98. Evans JD, Aronstein K, Chen YP, et al. Immune pathways and defence mechanisms in honey bees *Apis mellifera*. *Insect Mol Biol.* 2006;15(5):645-656. doi:10.1111/J.1365-2583.2006.00682.X
99. Clayton AM, Dong Y, Dimopoulos G. The Anopheles innate immune system in the defense against malaria infection. *J Innate Immun.* 2014;6(2):169-181. doi:10.1159/000353602
100. Cao X, He Y, Hu Y, et al. The immune signaling pathways of Manduca sexta. *Insect Biochem Mol Biol.* 2015;62:64-74. doi:10.1016/J.IBMB.2015.03.006
101. Buchon N, Poidevin M, Kwon HM, et al. A single modular serine protease integrates signals from pattern-recognition receptors upstream of the Drosophila Toll pathway. *Proc Natl Acad Sci U S A.* 2009;106(30):12442-12447. doi:10.1073/pnas.0901924106
102. Jang IH, Chosa N, Kim SH, et al. A Spätzle-processing enzyme required for toll signaling activation in Drosophila innate immunity. *Dev Cell.* 2006;10(1):45-55. doi:10.1016/J.DEVCEL.2005.11.013
103. Chamy LEl, Leclerc V, Caldelari I, Reichhart JM. Sensing of "danger signals" and pathogen-associated molecular patterns defines binary signaling pathways "upstream" of Toll. *Nat Immunol.* 2008;9(10):1165-1170. doi:10.1038/ni.1643
104. Duneau DF, Kondolf HC, Im JH, et al. The Toll pathway underlies host sexual dimorphism in resistance to both Gram-negative and Gram-positive bacteria in mated Drosophila. *BMC Biol.* 2017;15(1):124. doi:10.1186/s12915-017-0466-3
105. Kaneko T, Yano T, Aggarwal K, et al. PGRP-LC and PGRP-LE have essential yet distinct functions in the drosophila immune response to monomeric DAP-type peptidoglycan. *Nat Immunol.* 2006;7(7):715-723. doi:10.1038/ni1356
106. Chang CI, Chelliah Y, Borek D, Mengin-Lecreulx D, Deisenhofer J. Structure of tracheal cytotoxin in complex with a heterodimeric pattern-recognition receptor. *Science.* 2006;311(5768):1761-1764. doi:10.1126/science.1123056
107. Zhai Z, Huang X, Yin Y. Beyond immunity: the Imd pathway as a coordinator of host defense, organismal physiology and behavior. *Dev Comp Immunol.* 2018;83:51-59. doi:10.1016/J.DCI.2017.11.008
108. Hanson MA, Lemaitre B. New insights on Drosophila antimicrobial peptide function in host defense and beyond. *Curr Opin Immunol.* 2020;62:22-30. doi:10.1016/J.COI.2019.11.008
109. Charroux B, Capo F, Kurz CL, et al. Cytosolic and secreted peptidoglycan-degrading enzymes in Drosophila respectively control local and systemic immune responses to microbiota. *Cell Host Microbe.* 2018;23(2):215-228.e4. doi:10.1016/j.chom.2017.12.007
110. Troha K, Nagy P, Pivovar A, Lazzaro BP, Hartley PS, Buchon N. Nephrocytes remove microbiota-derived peptidoglycan from systemic circulation to maintain immune homeostasis. *Immunity.* 2019;51(4):625-637.e3. doi:10.1016/j.immuni.2019.08.020
111. Salcedo-Porras N, Guarneri A, Oliveira PL, Lowenberger C. Rhodnius prolixus: identification of missing components of the IMD immune signaling pathway and functional characterization of its role in eliminating bacteria. *PLoS One.* 2019;14(4):e0214794. doi:10.1371/JOURNAL.PONE.0214794
112. Gerardo NM, Altincicek B, Anselme C, et al. Immunity and other defenses in pea aphids, *Acyrthosiphon pisum*. *Genome Biol.* 2010;11(2):R21. doi:10.1186/GB-2010-11-2-R21
113. Osman D, Buchon N, Chakrabarti S, et al. Autocrine and paracrine unpaired signaling regulate intestinal stem cell maintenance and division. *J Cell Sci.* 2012;125(24):5944-5949. doi:10.1242/jcs.113100
114. Bahia AC, Kubota MS, Tempone AJ, et al. The JAK-STAT pathway controls Plasmodium vivax load in early stages of Anopheles aquasalis infection. *PLoS Negl Trop Dis.* 2011;5(11):e1317. doi:10.1371/JOURNAL.PNTD.0001317
115. Buchon N, Broderick NA, Chakrabarti S, Lemaitre B. Invasive and indigenous microbiota impact intestinal stem cell activity through multiple pathways in Drosophila. *Genes Dev.* 2009;23(19):2333-2344. doi:10.1101/gad.1827009
116. Honti V, Csordás G, Kurucz É, Márkus R, Andó I. The cell-mediated immunity of Drosophila melanogaster: hemocyte lineages, immune compartments, microanatomy and regulation. *Dev Comp Immunol.* 2014;42(1):47-56. doi:10.1016/J.DCI.2013.06.005
117. Hillyer JF, Strand MR. Mosquito hemocyte-mediated immune responses. *Curr Opin Insect Sci.* 2014;3:14-21. doi:10.1016/J.COIS.2014.07.002
118. Kurucz É, Márkus R, Zsámboki J, et al. Nimrod, a putative phagocytosis receptor with EGF repeats in Drosophila plasmatocytes. *Curr Biol.* 2007;17(7):649-654. doi:10.1016/j.cub.2007.02.041
119. Levashina EA, Moita LF, Blandin S, Vriend G, Lagueux M, Kafatos FC. Conserved role of a complement-like protein in phagocytosis revealed by dsRNA knockout in cultured cells of the mosquito, *Anopheles gambiae*. *Cell.* 2001;104(5):709-718. doi:10.1016/S0092-8674(01)00267-7
120. Mamali I, Lamprou I, Karagiannis F, Karakantza M, Lampropoulou M, Marmaras VJ. A β integrin subunit regulates bacterial phagocytosis in medfly haemocytes. *Dev Comp Immunol.* 2009;33(7):858-866. doi:10.1016/J.DCI.2009.02.004
121. Rämet M, Manfruelli P, Pearson A, Mathey-Prevot B, Ezekowitz RAB. Functional genomic analysis of phagocytosis and identification of a Drosophila receptor for *E. Coli*. *Nature.* 2002;416(6881):644-648. doi:10.1038/NATURE735
122. Ferrandon D, Imler JL, Hetru C, Hoffmann JA. The Drosophila systemic immune response: sensing and signalling during bacterial and fungal infections. *Nat Rev Immunol.* 2007;7(11):862-874. doi:10.1038/NRI2194
123. Dong Y, Taylor HE, Dimopoulos G. AgDscam, a hypervariable immunoglobulin domain-containing receptor of the *Anopheles gambiae* innate immune system. *PLoS Biol.* 2006;4(7):1137-1146. doi:10.1371/journal.pbio.0040229
124. Schmucker D, Chen B. Dscam and DSCAM: complex genes in simple animals, complex animals yet simple genes. *Genes Dev.* 2009;23(2):147-156. doi:10.1101/GAD.1752909
125. Watson FL, Püttmann-Holgado R, Thomas F, et al. Extensive diversity of Ig-superfamily proteins in the immune system of insects. *Science.* 2005;309(5742):1874-1878. doi:10.1126/SCIENCE.1116887
126. Moita LF, Wang-Sattler R, Michel K, et al. In vivo identification of novel regulators and conserved pathways of phagocytosis in *A. gambiae*. *Immunity.* 2005;23(1):65-73. doi:10.1016/j.immuni.2005.05.006
127. Satyavathi VV, Minz A, Nagaraju J. Nodulation: an unexplored cellular defense mechanism in insects. *Cell Signal.* 2014;26(8):1753-1763. doi:10.1016/J.CELLSIG.2014.02.024
128. Carton Y, Frey F, Stanley DW, Vass E, Nappi AJ. Dexamethasone inhibition of the cellular immune response of *Drosophila melanogaster* against a parasitoid. *J Parasitol.* 2002;88(2):405-407. doi:10.1645/0022-3395(2002)088[0405:diotci]2.0.co;2
129. Russo J, Dupas S, Frey F, Carton Y, Parasitology MB-, 1996 undefined. *Insect Immunity: Early Events in the Encapsulation Process of Parasitoid (Leptopilina Boulardi) Eggs in Resistant and Susceptible Strains of Drosophila*. cambridge.org. Accessed December 18, 2021. https://www.cambridge.org/core/journals/parasitology/article/insect-immunity-early-events-in-the-encapsulation-process-of-parasitoid-leptopilina-boulardi-eggs-in-resistant-and-susceptible-strains-of-drosophila/E5F9CDD43F53CD3E560F510FF9716F6F
130. Morin-Poulard I, Tian Y, Vanzo N, Crozatier M. Drosophila as a model to study cellular communication between the hematopoietic niche and blood progenitors under homeostatic conditions and in response to an immune stress. *Front Immunol.* 2021;12:3203. doi:10.3389/FIMMU.2021.719349/BIBTEX
131. Nappi AJ, Christensen BM. Melanogenesis and associated cytotoxic reactions: applications to insect innate immunity. *Insect Biochem Mol Biol.* 2005;35(5):443-459. doi:10.1016/J.IBMB.2005.01.014
132. Christensen BM, Li J, Chen CC, Nappi AJ. Melanization immune responses in mosquito vectors. *Trends Parasitol.* 2005;21(4):192-199. doi:10.1016/J.PT.2005.02.007
133. Cerenius L, Lee BL, Söderhäll K. The proPO-system: pros and cons for its role in invertebrate immunity. *Trends Immunol.* 2008;29(6):263-271. doi:10.1016/J.IT.2008.02.009
134. An C, Budd A, Kanost MR, Michel K. Characterization of a regulatory unit that controls melanization and affects longevity of mosquitoes. *Cell Mol Life Sci.* 2010;68(11):1929-1939. doi:10.1007/S00018-010-0543-Z
135. Wang JL, Zhang Q, Tang L, Chen L, Liu XS, Wang YF. Involvement of a pattern recognition receptor C-type lectin 7 in enhancing cellular encapsulation and melanization due to its carboxyl-terminal CRD domain in the cotton bollworm, Helicoverpa armigera. *Dev Comp Immunol.* 2014;44(1):21-29. doi:10.1016/J.DCI.2013.11.002
136. Matskevich AA, Quintin J, Ferrandon D. The Drosophila PRR GNBP3 assembles effector complexes involved in antifungal defenses independently of its Toll-pathway activation function. *Eur J Immunol.* 2010;40(5):1244-1254. doi:10.1002/EJI.200940164
137. Slauch JM. How does the oxidative burst of macrophages kill bacteria? Still an open question. *Mol Microbiol.* 2011;80(3):580-583. doi:10.1111/J.1365-2958.2011.07612.X
138. Ha EM, Oh CT, Bae YS, Lee WJ. A direct role for dual oxidase in Drosophila gut immunity. *Science.* 2005;310(5749):847-850. doi:10.1126/science.1117311

139. Luckhart S, Vodovotz Y, Vodovotz I, et al. 1998 undefined. The mosquito *Anopheles stephensi* limits malaria parasite development with inducible synthesis of nitric oxide. *Natl Acad Sci.* 1998;95:5700-5705. Accessed December 19, 2021. https://www.pnas.org/content/95/10/5700.short
140. Gupta L, Molina-Cruz A, Kumar S, et al. The STAT pathway mediates late-phase immunity against Plasmodium in the mosquito *Anopheles gambiae*. *Cell Host Microbe.* 2009;5(5):498-507. doi:10.1016/J.CHOM.2009.04.003
141. Hillyer JF, Estévez-Lao TY. Nitric oxide is an essential component of the hemocyte-mediated mosquito immune response against bacteria. *Dev Comp Immunol.* 2010;34(2):141-149. doi:10.1016/J.DCI.2009.08.014
142. Hancock REW, Sahl HG. Antimicrobial and host-defense peptides as new anti-infective therapeutic strategies. *Nat Biotechnol.* 2006;24(12):1551-1557. doi:10.1038/nbt1267
143. Bulet P, Stöcklin R, Menin L. Anti-microbial peptides: from invertebrates to vertebrates. *Immunol Rev.* 2004;198(1):169-184. doi:10.1111/J.0105-2896.2004.0124.X
144. Vilcinskas A. Evolutionary plasticity of insect immunity. *J Insect Physiol.* 2013;59(2):123-129. doi:10.1016/J.JINSPHYS.2012.08.018
145. Pretzel J, Mohring F, Rahlfs S, et al. Antiparasitic peptides. *Adv Biochem Eng Biotechnol.* 2013;135:157-192. doi:10.1007/10_2013_191
146. Steiner H, Hultmark D, Engström Å, Bennich H, Boman HG. Sequence and specificity of two antibacterial proteins involved in insect immunity. *Nature.* 1981;292(5820):246-248. doi:10.1038/292246a0
147. Pöppel AK, Vogel H, Wiesner J, Vilcinskas A. Antimicrobial peptides expressed in medicinal maggots of the blow fly *Lucilia sericata* show combinatorial activity against bacteria. *Antimicrob Agents Chemother.* 2015;59(5):2508-2514. doi:10.1128/AAC.05180-14/SUPPL_FILE/ZAC005153891SO1.PDF
148. Zhao BC, Lin HC, Yang D, Ye X, Li ZG. Disulfide bridges in Defensins. *Curr Top Med Chem.* 2015;16(2):206-219. doi:10.2174/1568026615666150701115911
149. Mylonakis E, Podsiadlowski L, Muhammed M, Vilcinskas A. Diversity, evolution and medical applications of insect antimicrobial peptides. *Philos Trans R Soc B Biol Sci.* 2016;371(1695). doi:10.1098/RSTB.2015.0290
150. Duneau D, Ferdy JB, Revah J, et al. Stochastic variation in the initial phase of bacterial infection predicts the probability of survival in *D. melanogaster*. *Elife.* 2017;6:e28298. doi:10.7554/eLife.28298
151. Wiesner J, Vilcinskas A. Antimicrobial peptides: the ancient arm of the human immune system. 2010;1(5):440-464. doi:10.4161/VIRU.1.5.12983
152. Brogden KA. Antimicrobial peptides: pore formers or metabolic inhibitors in bacteria? *Nat Rev Microbiol.* 2005;3(3):238-250. doi:10.1038/nrmicro1098
153. Rahnamaeian M, Cytryńska M, Zdybicka-Barabas A, et al. Insect antimicrobial peptides show potentiating functional interactions against Gram-negative bacteria. *Proc R Soc B Biol Sci.* 2015;282(1806). doi:10.1098/RSPB.2015.0293
154. Parvy JP, Yu Y, Dostalova A, et al. The antimicrobial peptide defensin cooperates with tumour necrosis factor to drive tumour cell death in Drosophila. *Elife.* 2019;8. doi:10.7554/ELIFE.45061
155. Araki M, Kurihara M, Kinoshita S, et al. Anti-tumour effects of antimicrobial peptides, components of the innate immune system, against haematopoietic tumours in Drosophila mxc mutants. *DMM Dis Model Mech.* 2019;12(6):dmm037721. doi:10.1242/DMM.037721
156. Barajas-Azpeleta R, Wu J, Gill J, et al. Antimicrobial peptides modulate long-term memory. *PLoS Genet.* 2018;14(10):e1007440. doi:10.1371/JOURNAL.PGEN.1007440
157. Kwon H, Bang K, Lee M, Cho S. Molecular cloning and characterization of a lysozyme cDNA from the mole cricket Gryllotalpa orientalis (Orthoptera: Gryllotalpidae). *Mol Biol Rep.* 2014;41(9):5745-5754. doi:10.1007/S11033-014-3446-5
158. Elmogy M, Bassal T, et al, 2015 undefined. *Isolation, Characterization, Kinetics, and Enzymatic and Nonenzymatic Microbicidal Activities of a Novel C-type Lysozyme from Plasma of Schistocerca gregaria*. academic.oup.com. Accessed December 19, 2021. https://academic.oup.com/jinsectscience/article-abstract/15/1/57/2583151
159. Ito Y, Nakamura M, Hotani T, Imoto T. Insect lysozyme from house fly (Musca domestica) larvae: possible digestive function based on sequence and enzymatic properties. *J Biochem.* 1995;118(3):546-551. doi:10.1093/OXFORDJOURNALS.JBCHEM.A124943
160. Dortet L, Mostowy S, Louaka AS, et al. Recruitment of the major vault protein by InlK: a Listeria monocytogenes strategy to avoid autophagy. *PLoS Pathog.* 2011;7(8):e1002168. doi:10.1371/JOURNAL.PPAT.1002168
161. Yoshikawa Y, Ogawa M, Hain T, et al. Listeria monocytogenes ActA-mediated escape from autophagic recognition. *Nat Cell Biol.* 2009;11(10):1233-1240. doi:10.1038/NCB1967
162. Yano T, Mita S, Ohmori H, et al. Autophagic control of listeria through intracellular innate immune recognition in drosophila. *Nat Immunol.* 2008;9(8):908-916. doi:10.1038/NI.1634
163. Goto A, Yano T, Terashima J, Iwashita S, Oshima Y, Kurata S. Cooperative regulation of the induction of the novel antibacterial Listericin by peptidoglycan recognition protein LE and the JAK-STAT pathway. *J Biol Chem.* 2010;285(21):15731-15738. doi:10.1074/JBC.M109.082115
164. Anand PK, Tait SWG, Lamkanfi M, et al. TLR2 and RIP2 pathways mediate autophagy of Listeria monocytogenes via extracellular signal-regulated kinase (ERK) activation. *J Biol Chem.* 2011;286(50):42981-42991. doi:10.1074/JBC.M111.310599
165. Voronin D, Cook DAN, Steven A, Taylor MJ. Autophagy regulates Wolbachia populations across diverse symbiotic associations. *Proc Natl Acad Sci U S A.* 2012;109(25):E1638-E1646. doi:10.1073/PNAS.1203519109
166. Kim JJ, Lee HM, Shin DM, et al. Host cell autophagy activated by antibiotics is required for their effective antimycobacterial drug action. *Cell Host Microbe.* 2012;11(5):457-468. doi:10.1016/J.CHOM.2012.03.008
167. Wu J, Randle KE, Wu LP. ird1 is a Vps15 homologue important for antibacterial immune responses in Drosophila. *Cell Microbiol.* 2007;9(4):1073-1085. doi:10.1111/J.1462-5822.2006.00853.X
168. Nagy P, Sándor GO, Juhász G. Autophagy maintains stem cells and intestinal homeostasis in Drosophila. *Sci Rep.* 2018;8(1):1-9. doi:10.1038/s41598-018-23065-3
169. Guo T, Nan Z, Miao C, et al. The autophagy-related gene Atg101 in Drosophila regulates both neuron and midgut homeostasis. *J Biol Chem.* 2019;294(14):5666-5676. doi:10.1074/JBC.RA118.006069
170. Wen JK, Wang YT, Chan CC, et al. Atg9 antagonizes TOR signaling to regulate intestinal cell growth and epithelial homeostasis in Drosophila. *Elife.* 2017;6. doi:10.7554/ELIFE.29338
171. Nagy P, Szatmári Z, Sándor GO, Lippai M, Hegedűs K, Juhász G. Drosophila atg16 promotes enteroendocrine cell differentiation via regulation of intestinal Slit/Robo signaling. *Dev.* 2017;144(21):3990-4001. doi:10.1242/DEV.147033/264189/AM/DROSOPHILA-ATG16-PROMOTES-ENTEROENDOCRINE-CELL
172. Tusco R, Jacomin AC, Jain A, et al. Kenny mediates selective autophagic degradation of the IKK complex to control innate immune responses. *Nat Commun.* 2017;8(1):1-15. doi:10.1038/s41467-017-01287-9.
173. Moy RH, Gold B, Molleston JM, et al. Antiviral autophagy restricts Rift Valley fever virus infection and is conserved from flies to mammals. *Immunity.* 2014;40(1):51. doi:10.1016/J.IMMUNI.2013.10.020
174. Nakamoto M, Moy RH, Xu J, et al. Virus recognition by Toll-7 activates antiviral autophagy in Drosophila. *Immunity.* 2012;36(4):658-667. doi:10.1016/J.IMMUNI.2012.03.003
175. Shelly S, Lukinova N, Bambina S, Berman A, Cherry S. Autophagy is an essential component of Drosophila immunity against vesicular stomatitis virus. *Immunity.* 2009;30(4):588-598. doi:10.1016/J.IMMUNI.2009.02.009
176. Delgado MA, Elmaoued RA, Davis AS, Kyei G, Deretic V. Toll-like receptors control autophagy. *EMBO J.* 2008;27(7):1110-1121. doi:10.1038/EMBOJ.2008.31
177. Lee KA, Lee WJ. Immune-metabolic interactions during systemic and enteric infection in Drosophila. *Curr Opin insect Sci.* 2018;29:21-26. doi:10.1016/J.COIS.2018.05.014
178. Alwarawrah Y, Kiernan K, MacIver NJ. Changes in nutritional status impact immune cell metabolism and function. *Front Immunol.* 2018;9(MAY). doi:10.3389/FIMMU.2018.01055
179. Odegaard JI, Chawla A. The immune system as a sensor of the metabolic state. *Immunity.* 2013;38(4):644-654. doi:10.1016/J.IMMUNI.2013.04.001
180. DiAngelo JR, Bland ML, Bambina S, Cherry S, Birnbaum MJ. The immune response attenuates growth and nutrient storage in Drosophila by reducing insulin signaling. *Proc Natl Acad Sci.* 2009;106(49):20853-20858. doi:10.1073/PNAS.0906749106
181. Clark RI, Tan SWS, Péan CB, et al. MEF2 is an in vivo immune-metabolic switch. *Cell.* 2013;155(2):435. doi:10.1016/J.CELL.2013.09.007
182. Xu J, Hopkins K, Sabin L, et al. ERK signaling couples nutrient status to antiviral defense in the insect gut. *Proc Natl Acad Sci.* 2013;110(37):15025-15030. doi:10.1073/pnas.1303193110
183. Kang MA, Mott TM, Tapley EC, Lewis EE, Luckhart S. Insulin regulates aging and oxidative stress in *Anopheles stephensi*. *J Exp Biol.* 2008;211(Pt 5):741-748. doi:10.1242/JEB.012955
184. Ahlers LRH, Trammell CE, Carrell GF, et al. Insulin potentiates JAK/STAT signaling to broadly inhibit flavivirus replication in insect vectors. *Cell Rep.* 2019;29(7):1946-1960.e5. doi:10.1016/J.CELREP.2019.10.029
185. Fuchs Y, Steller H. Live to die another way: modes of programmed cell death and the signals emanating from dying cells. *Nat Rev Mol Cell.* 2015;16(6):329-344. doi:10.1038/nrm3999
186. Galluzzi L, Vitale I, Aaronson SA, et al. Molecular mechanisms of cell death: recommendations of the nomenclature Committee on cell death 2018. *Cell Death Differ.* 2018;25(3):486-541. doi:10.1038/s41418-017-0012-4
187. Krysko DV, Garg AD, Kaczmarek A, Krysko O, Agostinis P, Vandenabeele P. Immunogenic cell death and DAMPs in cancer therapy. *Nat Rev Cancer.* 2012;12(12):860-875. doi:10.1038/NRC3380
188. Zitvogel L, Kepp O, Kroemer G. Decoding cell death signals in inflammation and immunity. *Cell.* 2010;140(6):798-804. doi:10.1016/J.CELL.2010.02.015
189. Green DR, Ferguson T, Zitvogel L, Kroemer G. Immunogenic and tolerogenic cell death. *Nat Rev Immunol.* 2009;9(5):353-363. doi:10.1038/NRI2545
190. Medzhitov R, Schneider DS, Soares MP. Disease tolerance as a defense strategy. *Science.* 2012;335(6071):936-941. doi:10.1126/SCIENCE.1214935
191. Nikoletopoulou V, Markaki M, Palikaras K, Tavernarakis N. Crosstalk between apoptosis, necrosis and autophagy. *Biochim Biophys Acta.* 2013;1833(12):3448-3459. doi:10.1016/J.BBAMCR.2013.06.001
192. Bhattacharjee A, Szabó Á, Csizmadia T, Laczkó-Dobos H, Juhász G. Understanding the importance of autophagy in human diseases using Drosophila. *J Genet Genomics.* 2019;46(4):157-169. doi:10.1016/J.JGG.2019.03.007
193. Xu D, Woodfield SE, Lee TV, Fan Y, Antonio C, Bergmann A. Genetic control of programmed cell death (apoptosis) in Drosophila. *Fly.* 2009;3(1):78. doi:10.4161/FLY.3.1.7800
194. West C, Rus F, Chen Y, et al. IIV-6 inhibits NF-κB responses in Drosophila. *Viruses.* 2019;11(5):409. doi:10.3390/V11050409
195. Merkling SH, Bronkhorst AW, Kramer JM, Overheul GJ, Schenck A, Van Rij RP. The epigenetic regulator G9a mediates tolerance to RNA virus infection in Drosophila. *PLoS Pathog.* 2015;11(4):e1004692. doi:10.1371/JOURNAL.PPAT.1004692
196. Gordon O, Henry CM, Srinivasan N, et al. α-actinin accounts for the bioactivity of actin preparations in inducing STAT target genes in *Drosophila melanogaster*. *Elife.* 2018;7:e38636 doi:10.7554/ELIFE.38636
197. Srinivasan N, Gordon O, Ahrens S, et al. Actin is an evolutionarily-conserved damage-associated molecular pattern that signals tissue injury in *Drosophila melanogaster*. *Elife.* 2016;5(NOVEMBER2016):e19662. doi:10.7554/ELIFE.19662
198. Georgel P, Naitza S, Kappler C, et al. Drosophila immune deficiency (IMD) is a death domain protein that activates antibacterial defense and can promote apoptosis. *Dev Cell.* 2001;1(4):503-514. doi:10.1016/S1534-5807(01)00059-4
199. Nainu F, Tanaka Y, Shiratsuchi A, Nakanishi Y. Protection of insects against viral infection by apoptosis-dependent phagocytosis. *J Immunol.* 2015;195(12):5696-5706. doi:10.4049/JIMMUNOL.1500613
200. Lamiable O, Arnold J, de Faria IJ da S, et al. Analysis of the contribution of hemocytes and autophagy to Drosophila antiviral immunity. *J Virol.* 2016;90(11):5415-5426. doi:10.1128/JVI.00238-16/SUPPL_FILE/ZJV999181672SO3.PDF
201. Nainu F, Shiratsuchi A, Nakanishi Y. Induction of apoptosis and subsequent phagocytosis of virus-infected cells as an antiviral mechanism. *Front Immunol.* 2017;8(SEP):1220. doi:10.3389/FIMMU.2017.01220/BIBTEX
202. Hay BA, Wassarman DA, Rubin GM. Drosophila homologs of baculovirus inhibitor of apoptosis proteins function to block cell death. *Cell.* 1995;83(7):1253-1262. doi:10.1016/0092-8674(95)90150-7
203. Guo Z, Li Y, Ding SW. Small RNA-based antimicrobial immunity. *Nat Rev Immunol.* 2018;19(1):31-44. doi:10.1038/s41577-018-0071-x
204. Ghildiyal M, Zamore PD. Small silencing RNAs: an expanding universe. *Nat Rev Genet.* 2009;10(2):94-108. doi:10.1038/nrg2504
205. Siomi H, Siomi MC. On the road to reading the RNA-interference code. *Nature.* 2009;457(7228):396-404. doi:10.1038/NATURE07754
206. Cai H, Imler JL. cGAS-STING: insight on the evolution of a primordial antiviral signaling cassette. *Fac Rev.* 2021;10:54. doi:10.12703/R/10-54
207. Bonning BC, Saleh MC. The interplay between viruses and RNAi pathways in insects. 2021;66:61-79. doi:10.1146/ANNUREV-ENTO-033020-090410.
208. Bernstein E, Caudy AA, Hammond SM, Hannon GJ. Role for a bidentate ribonuclease in the initiation step of RNA interference. *Nature.* 2001;409(6818):363-366. doi:10.1038/35053110
209. Mussabekova A, Daeffler L, Imler JL. Innate and intrinsic antiviral immunity in Drosophila. *Cell Mol Life Sci.* 2017;74(11):2039-2054. doi:10.1007/S00018-017-2453-9
210. Tomari Y, Du T, Zamore PD. Sorting of Drosophila small silencing RNAs. *Cell.* 2007;130(2):299-308. doi:10.1016/J.CELL.2007.05.057
211. Martinez J, Patkaniowska A, Urlaub H, Lührmann R, Tuschl T. Single-stranded antisense siRNAs guide target RNA cleavage in RNAi. *Cell.* 2002;110(5):563-574. doi:10.1016/S0092-8674(02)00908-X
212. Sabin LR, Zheng Q, Thekkat P, Yang J, Hannon GJ. Dicer-2 processes diverse viral RNA species. *PLoS One.* 2013;8(2):55458. doi:10.1371/journal.pone.0055458
213. Schuster S, Miesen P, van Rij RP. Antiviral RNAi in insects and mammals: parallels and differences. *Viruses.* 2019;11(5):448. doi:10.3390/V11050448
214. Bronkhorst AW, Van Cleefa KWR, Vodovar N, et al. The DNA virus invertebrate iridescent virus 6 is a target of the Drosophila RNAi machinery. *Proc Natl Acad Sci U S A.* 2012;109(51):E3604-E3613. doi:10.1073/PNAS.1207213109/-/DCSUPPLEMENTAL
215. Buchon N, Vaury C. RNAi: a defensive RNA-silencing against viruses and transposable elements. *Heredity (Edinb).* 2006;96(2):195-202. doi:10.1038/sj.hdy.6800789
216. Mondotte JA, Gausson V, Frangeul L, Blanc H, Lambrechts L, Saleh MC. Immune priming and clearance of orally acquired RNA viruses in Drosophila. *Nat Microbiol.* 2018;3(12):1394-1403. doi:10.1038/S41564-018-0265-9
217. Bronkhorst AW, Van Rij RP. The long and short of antiviral defense: small RNA-based immunity in insects. *Curr Opin Virol.* 2014;7(1):19-28. doi:10.1016/J.COVIRO.2014.03.010

218. Levi LI, Rezelj VV, Henrion-Lacritick A, et al. Defective viral genomes from chikungunya virus are broad-spectrum antivirals and prevent virus dissemination in mosquitoes. *PLoS Pathog.* 2021;17(2):e1009110. doi:10.1371/JOURNAL.PPAT.1009110
219. Rezelj VV, Carrau L, Merwaiss F, et al. Defective viral genomes as therapeutic interfering particles against flavivirus infection in mammalian and mosquito hosts. *Nat Commun.* 2021;12(1). doi:10.1038/S41467-021-22341-7
220. Tetreau G, Dhinaut J, Gourbal B, Moret Y. Trans-generational immune priming in invertebrates: Current knowledge and future prospects. *Front Immunol.* 2019;10(AUG). doi:10.3389/FIMMU.2019.01938
221. Sadd BM, Kleinlogel Y, Schmid-Hempel R, Schmid-Hempel P. Trans-generational immune priming in a social insect. *Biol Lett.* 2005;1(4):386-388. doi:10.1098/RSBL.2005.0369
222. Milutinović B, Kurtz J. Immune memory in invertebrates. *Semin Immunol.* 2016;28(4):328-342. doi:10.1016/J.SMIM.2016.05.004
223. Little TJ, Kraaijeveld AR. Ecological and evolutionary implications of immunological priming in invertebrates. *Trends Ecol Evol.* 2004;19(2):58-60. doi:10.1016/J.TREE.2003.11.011
224. Contreras-Garduño J, Lanz-Mendoza H, Franco B, Nava A, Pedraza-Reyes M, Canales-Lazcano J. Insect immune priming: ecology and experimental evidences. *Ecol Entomol.* 2016;41(4):351-366. doi:10.1111/EEN.12300
225. Poirier EZ, Goic B, Tomé-Poderti L, et al. Dicer-2-Dependent generation of viral DNA from defective genomes of RNA viruses modulates antiviral immunity in insects. *Cell Host Microbe.* 2018;23(3):353-365.e8. doi:10.1016/J.CHOM.2018.02.001
226. Mondotte JA, Saleh MC. Antiviral immune response and the route of infection in *Drosophila melanogaster*. *Adv Virus Res.* 2018;100:247-278. doi:10.1016/BS.AIVIR.2017.10.006
227. Goic B, Stapleford KA, Frangeul L, et al. Virus-derived DNA drives mosquito vector tolerance to arboviral infection. *Nat Commun.* 2016;7. doi:10.1038/NCOMMS12410
228. Brennecke J, Aravin AA, Stark A, et al. Discrete small RNA-generating loci as master regulators of transposon activity in Drosophila. *Cell.* 2007;128(6):1089-1103. doi:10.1016/J.CELL.2007.01.043
229. Czech B, Hannon GJ. One loop to rule them all: the ping-pong cycle and piRNA-guided silencing. *Trends Biochem Sci.* 2016;41(4):324-337. doi:10.1016/J.TIBS.2015.12.008
230. Lund E, Güttinger S, Calado A, Dahlberg JE, Kutay U. Nuclear export of microRNA precursors. *Science.* 2004;303(5654):95-98. doi:10.1126/SCIENCE.1090599
231. Förstemann K, Horwich MD, Wee LM, Tomari Y, Zamore PD. Drosophila microRNAs are sorted into functionally distinct argonaute complexes after production by dicer-1. *Cell.* 2007;130(2):287-297. doi:10.1016/J.CELL.2007.05.056
232. Iwakawa HO, Tomari Y. The functions of MicroRNAs: mRNA decay and translational repression. *Trends Cell Biol.* 2015;25(11):651-665. doi:10.1016/J.TCB.2015.07.011
233. Gruber JJ, Zatechka DS, Sabin LR, et al. Ars2 links the nuclear Cap-binding complex to RNA interference and cell proliferation. *Cell.* 2009;138(2):328-339. doi:10.1016/j.cell.2009.04.046
234. Monsanto-Hearne V, Johnson KN. miRNAs in insects infected by animal and plant viruses. *Viruses.* 2018;10(7):354. doi:10.3390/V10070354.
235. Gasiunas G, Sinkunas T, Siksnys V. Molecular mechanisms of CRISPR-mediated microbial immunity. *Cell Mol Life Sci.* 2014;71(3):449. doi:10.1007/S00018-013-1438-6
236. Sun L, Wu J, Du F, Chen X, Chen ZJ. Cyclic GMP-AMP synthase is a cytosolic DNA sensor that activates the type I interferon pathway. *Science.* 2013;339(6121):786-791. doi:10.1126/SCIENCE.1232458
237. Zhang X, Shi H, Wu J, et al. Cyclic GMP-AMP containing mixed phosphodiester linkages is an endogenous high-affinity ligand for STING. *Mol Cell.* 2013;51(2):226-235. doi:10.1016/J.MOLCEL.2013.05.022
238. Slavik KM, Morehouse BR, Ragucci AE, et al. cGAS-like receptors sense RNA and control 3′2′-cGAMP signalling in Drosophila. *Nature.* 2021;597(7874):109-113. doi:10.1038/S41586-021-03743-5
239. Holleufer A, Winther KG, Gad HH, et al. Two cGAS-like receptors induce antiviral immunity in Drosophila. *Nature.* 2021;597(7874):114-118. doi:10.1038/S41586-021-03800-Z
240. Wu X, Wu FH, Wang X, et al. Molecular evolutionary and structural analysis of the cytosolic DNA sensor cGAS and STING. *Nucleic Acids Res.* 2014;42(13):8243-8257. doi:10.1093/NAR/GKU569
241. Kranzusch PJ, Wilson SC, Lee ASY, Berger JM, Doudna JA, Vance RE. Ancient origin of cGAS-STING reveals mechanism of universal 2′,3′ cGAMP signaling. *Mol Cell.* 2015;59(6):891-903. doi:10.1016/J.MOLCEL.2015.07.022
242. Cai H, Holleufer A, Simonsen B, et al. 2′3′-cGAMP triggers a STING- and NF-κB-dependent broad antiviral response in Drosophila. *Sci Signal.* 2020;13(660). doi:10.1126/SCISIGNAL.ABC4537
243. Goto A, Okado K, Martins N, et al. The kinase IKKβ regulates a STING- and NF-κB-Dependent antiviral response pathway in Drosophila. *Immunity.* 2018;49(2):225-234.e4. doi:10.1016/J.IMMUNI.2018.07.013
244. Liu Y, Gordesky-Gold B, Leney-Greene M, Weinbren NL, Tudor M, Cherry S. Inflammation-induced, STING-dependent autophagy restricts Zika virus infection in the Drosophila brain. *Cell Host Microbe.* 2018;24(1):57-68.e3. doi:10.1016/J.CHOM.2018.05.022
245. Martin M, Hiroyasu A, Guzman RM, Roberts SA, Goodman AG. Analysis of Drosophila STING reveals an evolutionarily conserved antimicrobial function. *Cell Rep.* 2018;23(12):3537-3550.e6. doi:10.1016/J.CELREP.2018.05.029
246. Ahn J, Barber GN. STING signaling and host defense against microbial infection. *Exp Mol Med.* 2019;51(12):1-10. doi:10.1038/s12276-019-0333-0
247. Motesi CMDe. Poxvirus cGAMP nucleases: clues and mysteries from a stolen gene. *PLoS Pathog.* 2021;17(3):e1009372. doi:10.1371/JOURNAL.PPAT.1009372
248. Eaglesham JB, McCarty KL, Kranzusch PJ. Structures of diverse poxin cgamp nucleases reveal a widespread role for cgas-sting evasion in host–pathogen conflict. *Elife.* 2020;9:1-22. doi:10.7554/ELIFE.59753
249. Pimentel AC, Cesar CS, Martins M, Cogni R. The antiviral effects of the symbiont bacteria Wolbachia in insects. *Front Immunol.* 2021;11:3690. doi:10.3389/FIMMU.2020.626329/BIBTEX
250. Shropshire JD, Leigh B, Bordenstein SR. Symbiont-mediated cytoplasmic incompatibility: what have we learned in 50 years? *Elife.* 2020;9:1-36. doi:10.7554/ELIFE.61989
251. Hedges LM, Brownlie JC, O'Neill SL, Johnson KN. Wolbachia and virus protection in insects. *Science.* 2008;322(5902):702. doi:10.1126/SCIENCE.1162418
252. Teixeira L, Ferreira Á, Ashburner M. The bacterial symbiont Wolbachia induces resistance to RNA viral infections in *Drosophila melanogaster*. *PLoS Biol.* 2008;6(12):2753-2763. doi:10.1371/JOURNAL.PBIO.1000002
253. Moreira LA, Iturbe-Ormaetxe I, Jeffery JA, et al. A Wolbachia symbiont in *Aedes aegypti* limits infection with dengue, Chikungunya, and Plasmodium. *Cell.* 2009;139(7):1268-1278. doi:10.1016/J.CELL.2009.11.042
254. Bian G, Xu Y, Lu P, Xie Y, Xi Z. The endosymbiotic bacterium Wolbachia induces resistance to dengue virus in *Aedes aegypti*. *PLoS Pathog.* 2010;6(4):1-10. doi:10.1371/JOURNAL.PPAT.1000833
255. Buchon N, Osman D. All for one and one for all: regionalization of the Drosophila intestine. *Insect Biochem Mol Biol.* 2015;67:2-8. doi:10.1016/j.ibmb.2015.05.015
256. Liu X, Hodgson JJ, Buchon N. Drosophila as a model for homeostatic, antibacterial, and antiviral mechanisms in the gut. Kline KA, ed. *PLoS Pathog.* 2017;13(5):e1006277. doi:10.1371/journal.ppat.1006277
257. Ayyaz A, Li H, Jasper H. Haemocytes control stem cell activity in the Drosophila intestine. *Nat Cell Biol.* 2015;17(6):736-748. doi:10.1038/NCB3174
258. Sloan DB, Moran NA. Endosymbiotic bacteria as a source of carotenoids in whiteflies. *Biol Lett.* 2012;8(6):986-989. doi:10.1098/RSBL.2012.0664
259. Macdonald SJ, Lin GG, Russell CW, Thomas GH, Douglas AE. The central role of the host cell in symbiotic nitrogen metabolism. *Proc R Soc B Biol Sci.* 2012;279(1740):2965-2973. doi:10.1098/RSPB.2012.0414
260. Bing XL, Gerlach J, Loeb G, Buchon N. Nutrient-dependent impact of microbes on Drosophila suzukii development. McFall-Ngai MJ, ed. *MBio.* 2018;9(2). doi:10.1128/mBio.02199-17
261. Currie CR. A community of ants, fungi, and bacteria: a multilateral approach to studying symbiosis. 2003;55:357-380. doi:10.1146/ANNUREV.MICRO.55.1.357
262. Brune A. Symbiotic digestion of lignocellulose in termite guts. *Nat Rev Microbiol.* 2014;12(3):168-180. doi:10.1038/nrmicro3182
263. Douglas AE. The molecular basis of bacterial–insect symbiosis. *J Mol Biol.* 2014;426(23):3830-3837. doi:10.1016/J.JMB.2014.04.005
264. Grenier T, Leulier F. How commensal microbes shape the physiology of *Drosophila melanogaster*. *Curr Opin insect Sci.* 2020;41:92-99. doi:10.1016/J.COIS.2020.08.002
265. Piel J. A polyketide synthase-peptide synthetase gene cluster from an uncultured bacterial symbiont of Paederus beetles. *Proc Natl Acad Sci U S A.* 2002;99(22):14002-14007. doi:10.1073/PNAS.222481399
266. Degnan PH, Moran NA. Diverse phage-encoded toxins in a protective insect endosymbiont. *Appl Environ Microbiol.* 2008;74(21):6782-6791. doi:10.1128/AEM.01285-08
267. Lee KA, Kim SH, Kim EK, et al. Bacterial-derived uracil as a modulator of mucosal immunity and gut-microbe homeostasis in drosophila. *Cell.* 2013;153(4):797-811. doi:10.1016/j.cell.2013.04.009
268. Ha EM, Lee KA, Seo YY, et al. Coordination of multiple dual oxidase-regulatory pathways in responses to commensal and infectious microbes in drosophila gut. *Nat Immunol.* 2009;10(9):949-957. doi:10.1038/ni.1765
269. Jones RM, Luo L, Ardita CS, et al. Symbiotic lactobacilli stimulate gut epithelial proliferation via Nox-mediated generation of reactive oxygen species. *EMBO J.* 2013;32(23):3017-3028. doi:10.1038/emboj.2013.224
270. Biteau B, Hochmuth CE, Jasper H. JNK activity in somatic stem cells causes loss of tissue homeostasis in the aging Drosophila gut. *Cell Stem Cell.* 2008;3(4):442-455. doi:10.1016/j.stem.2008.07.024
271. Ha EM, Lee KA, Park SH, et al. Regulation of DUOX by the gαq-phospholipase Cβ-Ca 2+ pathway in Drosophila gut immunity. *Dev Cell.* 2009;16(3):386-397. doi:10.1016/j.devcel.2008.12.015
272. Royet J. Epithelial homeostasis and the underlying molecular mechanisms in the gut of the insect model *Drosophila melanogaster*. *Cell Mol Life Sci.* 2011;68(22):3651-3660. doi:10.1007/s00018-011-0828-x
273. Broderick NA, Buchon N, Lemaitre B. Microbiota-induced changes in *Drosophila melanogaster* host gene expression and gut morphology. McFall-Ngai MJ, ed. *MBio.* 2014;5(3). doi:10.1128/mBio.01117-14
274. Marra A, Hanson MA, Kondo S, Erkosar B, Lemaitre B. Drosophila antimicrobial peptides and lysozymes regulate gut microbiota composition and abundance. *MBio.* 2021;12(4):e0082421. doi:10.1128/MBIO.00824-21
275. Ryu JH, Kim SH, Lee HY, et al. Innate immune homeostasis by the homeobox gene Caudal and commensal-gut mutualism in Drosophila. *Science.* 2008;319(5864):777-782. doi:10.1126/science.1149357
276. Capo F, Charroux B, Royet J. Bacteria sensing mechanisms in Drosophila gut: local and systemic consequences. *Dev Comp Immunol.* 2016;64:11-21. doi:10.1016/J.DCI.2016.01.001
277. Lhocine N, Ribeiro PS, Buchon N, et al. PIMS modulates immune tolerance by negatively regulating Drosophila innate immune signaling. *Cell Host Microbe.* 2008;4(2):147-158. doi:10.1016/j.chom.2008.07.004
278. Buchon N, Broderick NA, Kuraishi T, Lemaitre B. Drosophila EGFR pathway coordinates stem cell proliferation and gut remodeling following infection. *BMC Biol.* 2010;8(1):152. doi:10.1186/1741-7007-8-152
279. Amcheslavsky A, Jiang J, Ip YT. Tissue damage-induced intestinal stem cell division in Drosophila. *Cell Stem Cell.* 2009;4(1):49-61. doi:10.1016/j.stem.2008.10.016
280. Houtz P, Bonfini A, Liu X, et al. Hippo, TGF-β, and Src-MAPK pathways regulate transcription of the upd3 cytokine in Drosophila enterocytes upon bacterial infection. Lee WJ, ed. *PLoS Genet.* 2017;13(11):e1007091. doi:10.1371/journal.pgen.1007091
281. Houtz P, Bonfini A, Bing X, Buchon N. Recruitment of adult precursor cells underlies limited repair of the infected larval midgut in Drosophila. *Cell Host Microbe.* 2019;26(3):412-425.e5. doi:10.1016/j.chom.2019.08.006
282. Jiang H, Patel PH, Kohlmaier A, Grenley MO, McEwen DG, Edgar BA. Cytokine/jak/stat signaling mediates regeneration and homeostasis in the Drosophila midgut. *Cell.* 2009;137(7):1343-1355. doi:10.1016/j.cell.2009.05.014
283. Chakrabarti S, Liehl P, Buchon N, Lemaitre B. Infection-induced host translational blockage inhibits immune responses and epithelial renewal in the Drosophila gut. *Cell Host Microbe.* 2012;12(1):60-70. doi:10.1016/j.chom.2012.06.001
284. Rera M, Clark RI, Walker DW. Intestinal barrier dysfunction links metabolic and inflammatory markers of aging to death in Drosophila. *Proc Natl Acad Sci U S A.* 2012;109(52):21528-21533. doi:10.1073/pnas.1215849110
285. Weaver SC, Barrett ADT. Transmission cycles, host range, evolution and emergence of arboviral disease. *Nat Rev Microbiol.* 2004;2(10):789-801. doi:10.1038/NRMICRO1006
286. Salazar MI, Richardson JH, Sánchez-Vargas I, Olson KE, Beaty BJ. Dengue virus type 2: replication and tropisms in orally infected Aedes aegypti mosquitoes. *BMC Microbiol.* 2007;7. doi:10.1186/1471-2180-7-9
287. Bennett KE, Olson KE, Muñoz ML, et al. Variation in vector competence for dengue 2 virus among 24 collections of *Aedes aegypti* from Mexico and the United States. *Am J Trop Med Hyg.* 2002;67(1):85-92. doi:10.4269/AJTMH.2002.67.85
288. Sansone CL, Cohen J, Yasunaga A, et al. Microbiota-dependent priming of antiviral intestinal immunity in Drosophila. *Cell Host Microbe.* 2015;18(5):571-581. doi:10.1016/j.chom.2015.10.010
289. Xi Z, Ramirez JL, Dimopoulos G. The *Aedes aegypti* toll pathway controls dengue virus infection. *PLoS Pathog.* 2008;4(7). doi:10.1371/JOURNAL.PPAT.1000098
290. Souza-Neto JA, Sim S, Dimopoulos G. An evolutionary conserved function of the JAK-STAT pathway in anti-dengue defense. *Proc Natl Acad Sci U S A.* 2009;106(42):17841-17846. doi:10.1073/PNAS.0905006106
291. Carissimo G, Pondeville E, McFarlane M, et al. Antiviral immunity of *Anopheles gambiae* is highly compartmentalized, with distinct roles for RNA interference and gut microbiota. *Proc Natl Acad Sci U S A.* 2015;112(2):E176-E185. doi:10.1073/PNAS.1412984112
292. Angleró-Rodríguez YI, MacLeod HJ, Kang S, Carlson JS, Jupatanakul N, Dimopoulos G. Aedes aegypti molecular responses to Zika virus: modulation of infection by the toll and jak/stat immune pathways and virus host factors. *Front Microbiol.* 2017;8(OCT):2050. doi:10.3389/FMICB.2017.02050

293. Barletta ABF, Nascimento-Silva MCL, Talyuli OAC, et al. Microbiota activates IMD pathway and limits Sindbis infection in *Aedes aegypti*. *Parasit Vectors*. 2017;10(1). doi:10.1186/S13071-017-2040-9
294. Wu P, Sun P, Nie K, et al. A gut commensal bacterium promotes mosquito permissiveness to arboviruses. *Cell Host Microbe*. 2019;25(1):101-112.e5. doi:10.1016/J.CHOM.2018.11.004
295. Matha V, Áček Z. Changes in haemocyte Counts in Galleria mellonella (L.) (Lepidoptera – galleriidae) larvae infected with Steinernema sp. (nematoda: Steinernematidae). *Nematologica*. 1984;30(1):86-89. doi:10.1163/187529284X00482
296. Fallon JP, Troy N, Kavanagh K. Pre-exposure of Galleria mellonella larvae to different doses of Aspergillus fumigatus conidia causes differential activation of cellular and humoral immune responses. *Virulence*. 2011;2(5):413-421. doi:10.4161/VIRU.2.5.17811
297. Bergin D, Murphy L, Keenan J, Clynes M, Kavanagh K. Pre-exposure to yeast protects larvae of Galleria mellonella from a subsequent lethal infection by Candida albicans and is mediated by the increased expression of antimicrobial peptides. *Microb Infect*. 2006;8(8):2105-2112. doi:10.1016/J.MICINF.2006.03.005
298. Sadd BM, Schmid-Hempel P. Insect immunity shows specificity in protection upon secondary pathogen exposure. *Curr Biol*. 2006;16(12):1206-1210. doi:10.1016/J.CUB.2006.04.047
299. Pham LN, Dionne MS, Shirasu-Hiza M, Schneider DS. A specific primed immune response in Drosophila is dependent on phagocytes. *PLoS Pathog*. 2007;3(3):e26. doi:10.1371/JOURNAL.PPAT.0030026
300. Ardia DR, Gantz JE, Schneider BC, Strebel S. Costs of immunity in insects: an induced immune response increases metabolic rate and decreases antimicrobial activity. *Funct Ecol*. 2012;26(3):732-739. doi:10.1111/J.1365-2435.2012.01989.X
301. Moret Y, Schmid-Hempel P. Survival for immunity: the price of immune system activation for bumblebee workers. *Science*. 2000;290(5494):1166-1168. doi:10.1126/SCIENCE.290.5494.1166
302. Moret Y. "Trans-generational immune priming": specific enhancement of the antimicrobial immune response in the mealworm beetle, Tenebrio molitor. *Proceedings Biol Sci*. 2006;273(1592):1399-1405. doi:10.1098/RSPB.2006.3465
303. Kaletsky R, Moore RS, Vrla GD, Parsons LR, Gitai Z, Murphy CT. *C. elegans* interprets bacterial non-coding RNAs to learn pathogenic avoidance. *Nature*. 2020;586(7829):445-451. doi:10.1038/S41586-020-2699-5
304. Moore RS, Kaletsky R, Lesnik C, et al. The role of the Cer1 transposon in horizontal transfer of transgenerational memory. *Cell*. 2021;184(18):4697-4712.e18. doi:10.1016/J.CELL.2021.07.022
305. Rechavi O, Minevich G, Hobert O. Transgenerational inheritance of an acquired small RNA-based antiviral response in *C. elegans*. *Cell*. 2011;147(6):1248-1256. doi:10.1016/J.CELL.2011.10.042
306. Lev I, Seroussi U, Gingold H, Bril R, Anava S, Rechavi O. MET-2-Dependent H3K9 methylation suppresses transgenerational small RNA inheritance. *Curr Biol*. 2017;27(8):1138-1147. doi:10.1016/J.CUB.2017.03.008

CHAPTER 4

Evolution of the Vertebrate and Lower Deuterostome Immune System

Martin F. Flajnik, PhD

INTRODUCTION: GENERAL PRINCIPLES OF IMMUNE SYSTEM EVOLUTION

Most studies defining the field of immunology have been done in mammals, especially mice and humans, and most of our paradigms are derived from these models. Yet, one of our new laws developed by Janeway[1] and Matzinger,[2] that adaptive immunity is called to action not by foreignness but rather by external or internal danger via pattern recognition receptors (PRRs), has its origins in the study of the invertebrate *Drosophila* Toll receptors.[3] Additionally, over 50 years ago, the divergence of two major subsets of lymphocytes (B and T cells) was partially revealed in studies of the bursa of Fabricius in birds.[4] Breakthrough discoveries by studying nonmammalian animals have shown a growing awareness of immune origins and alerted us to new possibilities.[5]

A convenient way to appreciate the evolution of immunity is to compartmentalize the immune system into features that are conserved over time versus those that change rapidly, as described below. Throughout this chapter, I will rely on this unifying concept.

Common Ancestor Hypothesis

Figure 4.1 displays the animal phyla ranging from the single-celled protozoa to the metazoan protostome and deuterostome lineages, the latter is the focus of this chapter (the former in Chapter 3, and plants in Chapter 5). It is often assumed by those unfamiliar with thinking in evolutionary terms that molecules or mechanisms found in living protostomes, like the well-studied arthropod *Drosophila*, are ancestral to similar molecules/mechanisms in deuterostome mouse/human. While this is true in some cases, one must appreciate that *Drosophila* and humans have taken just as long (>900 million years) to evolve from a common ancestral triploblastic coelomate (three germ layers and a mesoderm-lined body cavity, features shared by protostomes and deuterostomes) that was nothing like a fly or a human, and thus *Drosophila* is not our ancestor, ie, the manner by which flies and humans use certain families of defense molecules may be different and both may be unlike the common ancestor. Thus, deciphering how model animals perform certain immune tasks is an important first step, one can only deduce what is primordial or derived when we have examined similar immune mechanisms/molecules in species from a wide range of taxa (animal groups), employing the principle of parsimony. I will touch upon several defense molecule families and emphasize those which have been conserved and those that have evolved rapidly.[6,7]

Rapid Evolution of Defense Mechanisms and "Trade-Offs"

Immune systems are often compared to the Red Queen in *Alice in Wonderland* (Red Queen Hypothesis: one must continually keep moving just to avoid falling behind[8]). Because of the perpetual conflict with pathogens and the capacity to "get along with" commensals, the immune system is in perpetual evolutionary flux. This is exemplified by great differences in the immune systems of animals that are even within the same phylogenetic group (eg, bony fish species, or human and mouse, both mammals). In fact, defense mechanisms are extremely diverse throughout invertebrate phyla and Kepler et al. have aptly and succinctly described the situation in a review: "not homogeneous, not simple, not well understood."[9] In the jawed vertebrate (or *gnathostomes*), the most rapidly evolving system is an innate system, natural killer (NK) cell recognition, governed by different classes (superfamilies) of receptors in mice and humans, and also extremely plastic even within the same species, illustrated by the large number of killer immunoglobulin superfamily (KIR) haplotypes found in humans.[10,11] Studies of Ig superfamily (IgSF) genes expressed in the nervous system and immune system showed definitively that immune system molecules evolve at a faster rate.[12] Finally, rapid evolution of immune system molecules and mechanisms is the general rule, but molecules functioning at different levels of immune defense (recognition, signaling, or effector) can evolve at widely varying rates.

Conservation of Defense Mechanisms

While the immune system is the most rapidly evolving physiologic system, nevertheless there is also deep conservation of defense families and mechanisms. Klein has compared this dichotomy to the two-headed god Janus, the major idea being that certain basic mechanisms/functions are obligatory for immune systems to function, yet they still must evolve rapidly to avoid pathogen subterfuge.[7] For example, MHC class I molecules have similar structure/function/features in all gnathostomes, but even within groups of primates class I genes are not orthologous, ie, they can be derived from different ancestral class I genes. So, the "intent" is to preserve vital immune functions but rapidly modify the gene or pathway to evade pathogens. Additionally, certain features are conserved (eg, general development and function of conventional αβ T cells), but a second, similar system can be exploited in very different ways in closely related species (eg, the function[s] of γδ T cells). The "Janus paradigm," therefore, can be quite useful

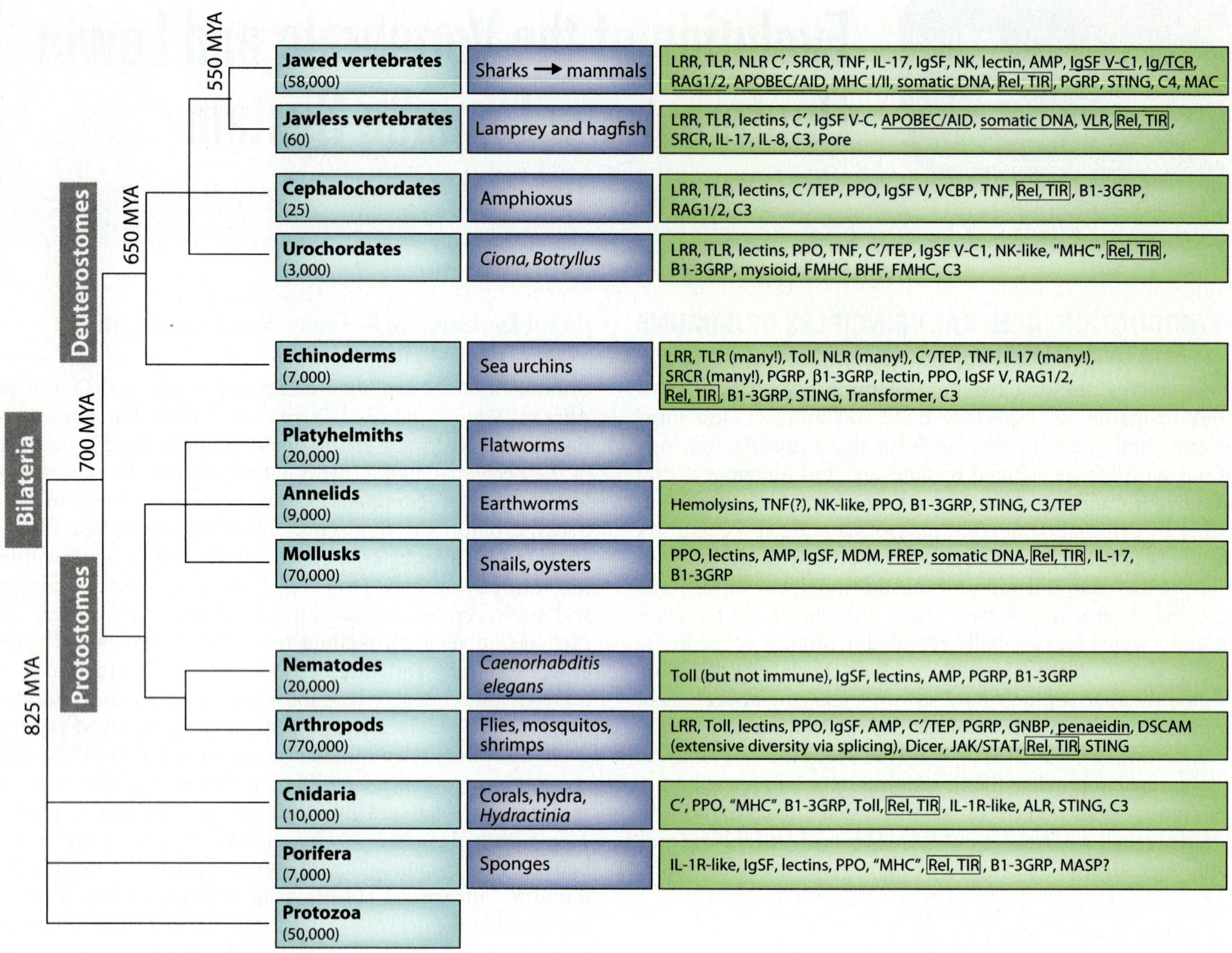

FIG. 4.1. **Major animal groups and immune mechanisms/molecules described to date in each group**. The first box on the left in each row describes the animal taxon and the approximate number of species in that group. The next box shows specific examples of species or subgroups. The third box lists molecules/mechanisms found in each group: *underlined terms* indicate somatic changes to antigen receptors or secreted molecules. Timings are in millions of years. (Modified from Hibino T, Loza-Coll M, Messier C, et al. The immune gene repertoire encoded in the purple sea urchin genome. *Dev Biol.* 2006;300(1):349-65. doi:10.1016/j.ydbio.2006.08.065 and Flajnik MF, Du Pasquier L. Evolution of innate and adaptive immunity: can we draw a line? *Trends Immunol.* 2004;25(12):640-644. doi:10.1016/j.it.2004.10.001, with permission from Elsevier.)

when examining any pathway in the immune system. A recent survey of defense molecules in mammals focusing on innate T cells showed that by analyzing replacement/silent mutations in a large set of immune genes one could determine which pathways were essential and others that could be redundant.[6]

The study of the evolution of immunity has resulted in a fundamental appreciation of the heart of immunity, both innate and adaptive. Especially now, with studies in many plants and animals, we see which features have been conserved and when they arose. Interestingly, major pathways of defense known to all immunologists, such as the TLR and JAK-signal transducer and activator of transcription (STAT) pathways, clearly arose in an early animal ancestor and were perpetuated in derived ways in all major taxa. Almost all the innate pathways arising in protostomes or earlier were adopted and maintained by vertebrates. We shall see how pathways are manipulated in different animals, always drawing upon the best-known mammalian models as a foundation.

Convergent Evolution

Early in the comparative study of immunity, it was assumed that the same features found in different taxa proved that they were also present in the common ancestor (ie, divergent evolution). While this standard holds true and establishes one of the dogmas of comparative immunology, later we discovered, because of the aforementioned rapid evolution of immune systems, *convergence* of similar functions occurs in evolution: the same function or even molecular conformation has arisen independently in different organisms, sometimes in closely related species. While we will discuss several cases of convergent evolution throughout the chapter, for frame of reference, the NK cells, which use different receptor families in primates and rodents to achieve the same ends of recognizing polymorphic MHC class I molecules, are a striking example of convergence[10] (Chapter 18). Additionally, the emergence of a lymphocyte-based somatic generation with two entirely different receptor families for the same functions in jawless and jawed vertebrates is another remarkable illustration of convergent evolution.[13] There is fascinating convergence of mechanisms to protect the vulnerable single-cell layer of mucosal epithelium throughout the evolution of deuterostomes.[14] Single-domain antibodies arose independently in camels and sharks.[15] In innate immunity, the cytosolic nucleotide-binding domain leucine-rich repeat (NLR) proteins, despite their striking similarity in deuterostomes and plants, arose (at least) twice in evolution (Chapter 5). This feature in the study of evolution provides insight into the pressures exerted on a system to attain a similar end via different means.

Multigene Families

Genes involved in immunity are often found in clusters, with extensive contraction and expansion via so-called "birth and death" processes.[16] It is well known that such gene clusters can change rapidly over evolutionary time due to unequal recombination crossovers and gene conversion (and not only in the immune system, but also in any *cis*-duplicating gene family). Often, families of related immune genes—especially those involved in recognition events—are found near the telomeres of chromosomes, likely because this further promotes gene-shuffling events by recombination. Nonclassical MHC class I loci, NK receptor (NKR), and NLRs are good examples of this phenomenon, again believed to be a consequence of the race against pathogens. I will discuss several examples of interesting multigene families in different species.

Gene duplication, either in the clusters mentioned earlier or as a consequence of *en bloc* duplications, certainly has been a major feature of immune system diversity and plasticity. The two types of duplications are not equivalent, the former being more taxon-specific and the latter having a lasting impact on the entire system. It is now universally accepted that two genome-wide duplications (the so-called 2R hypothesis) occurred early in vertebrate history, tracking very well with the emergence of the Ig/TCR/MHC-based adaptive immune system.[17-19] This theory forms the basis for much that will be discussed concerning the evolution of the vertebrate adaptive immunity as one can track emergence of new immune mechanisms and fine-tuning of old ones by examining paralogous synteny. These genome-wide duplications were as crucial as the "RAG transposon" in the development of the Ig/TCR/MHC-based adaptive immunity. Recent data also suggest how genome-wide duplications may explain why the gnathostome immune system is much more advanced over that of agnathans.[20,21] In addition, the common ancestor of the major group of bony fish (teleosts) underwent a third round of genome-wide duplication (Fig. 4.2), which many believe played the major role in their outlier status among vertebrates regarding primordial immune system genetics and physiology.[22]

Somatic Generation of Diversity

Somatic modifications can take place at multiple levels to generate immune system diversity. Long believed to be the sole domain of gnathostomes, modifications at the deoxyribonucleic acid (DNA) level via somatic hypermutation (SHM), gene conversion, and gene rearrangement (primary and secondary [eg, receptor editing]) irreversibly modify genes within an individual. The well-known V(D)J joining, class switch recombination (CSR), and SHM are examples of these processes in the IgSF receptors of jawed vertebrates, but modifications to genomic DNA can also occur in the jawless fish via gene shuffling and mutation and also in some invertebrates[23] (Fig. 4.1).

Major Gene Families Involved in Immunity

Defense molecules can be composed of a very large number of protein folds, some of which are clearly used to a large extent. Some of the most common families (Fig. 4.3) are IgSF, leucine-rich repeats (LRRs), C-type lectins, and the TNF superfamily, and certain other domains in immune recognition (eg, scavenger receptor cysteine-rich [SRCR]). A nifty review from Luke O'Neil proposes that immune system recognition and signaling can largely be explained by mixing and matching nine early-emerging domains.[24] All the domains discussed in this chapter are found in Table 4.1 and a few are displayed in Fig. 4.3. As a means of introduction, two of these families, which constitute the "top two" quantitatively, will be described in the following.

Leucine-Rich Repeats

LRRs consist of 2 to 45 motifs of 20 to 30 amino acids in length (XLXXLXLXXNXHXXHXXXXFXXLX) that fold into an arc shape (Fig. 4.3). Both the concave and convex parts of the domain have been shown to interact with ligands. Molecular modeling suggests that the conserved pattern LxxLxL is sufficient to impart the characteristic horseshoe curvature (AKA "the horseshoe thingee") to proteins with 20- to 30-residue repeats.[25,26] LRRs are often flanked by cysteine-rich domains. LRRs occur in proteins ranging from viruses to eukaryotes and are found most famously in the toll/TLRs, as well as tyrosine kinase receptors, cell adhesion molecules, resistance (R) factors in plants found at the cell surface and in the cytosol, extracellular matrix (ECM)–binding glycoproteins (eg, peroxidasin), and are involved in a variety of protein–protein interactions: signal transduction, cell adhesion, DNA repair, recombination, transcription, RNA processing, disease resistance, apoptosis, and the immune response. LRR-containing proteins can be associated with a variety of other domains, whether they are extracellular (LRR associated with IgSF or fibronectin [FN] type III) or intracellular (LRR associated with a variety of effector domains). In these chimeric molecules, the LRR moiety is involved in recognition, most likely due to its malleable structure. There are at least six families of LRR proteins, characterized by different lengths and consensus sequences of the repeats.

LRR-containing proteins are involved in immunity in plants and animals. Functions in immunity range from control of motility of hematopoietic cells to specific recognition of antigens via a novel system of gene rearrangement (the variable lymphocyte receptors [VLR] described below; Fig. 4.8). LRRs can occur in soluble forms, the ECM, in the cytosol, or as TM forms, either integral membrane proteins or glycophosphatidylinositol (GPI)-anchored. Because of its basic structure and

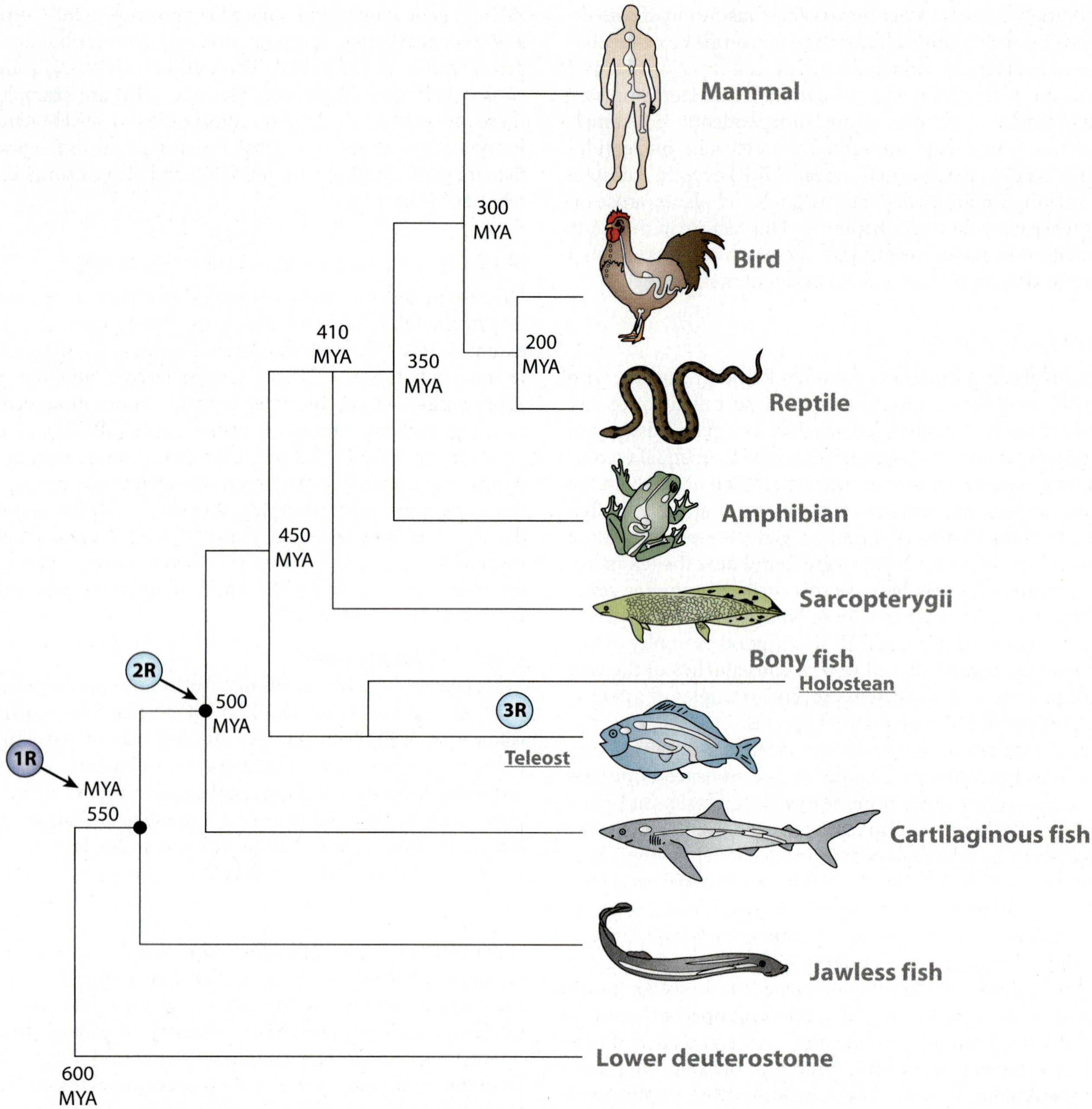

FIG. 4.2. Major classes of vertebrates described throughout the text. Timings are in millions of years. "1R" is the first genome-wide duplication that occurred before the emergence of all vertebrates and "2R" occurred before the emergence of all jawed vertebrates or gnathostomes. 3R indicates a third genome-wide duplication that occurred at the origins of the Euteleostei, but not the Holosteans (gar, bowfin). Note there are three major groups of bony fish, the Euteleostei that make up the majority of this group, the gar/bowfin group that is an older branch, and the sarcopterygians (lungfish and coelacanth) that are most closely related to amphibians.

plasticity, the LRR module was locked in early in evolution as an ideal motif for recognition of essentially any ligand.

Immunoglobulin Superfamily

IgSF domains are encountered in a very large number of molecules in the animal kingdom[27] (Fig. 4.3). They are found intracellularly (eg, connectin) or as cell adhesion molecules, many of which are in the nervous system (eg, the neural cell adhesion molecule, NCAM), coreceptors and costimulatory molecules of the immune system (eg, cluster of differentiation [CD]79, CD80), molecules involved in antigen presentation to lymphocytes (eg, class I molecules), certain classes cytokine receptors (eg, interleukin [IL]-1R), and of course Ig (and TCR), where they were first characterized and bestowed with their name (Ig). They can be associated with other domains such as fibrinogen (FN) (eg, titin) and LRRs, or they can be the sole constitutive elements of the polypeptide chain often associated to a transmembrane segment and a cytoplasmic

TABLE 4.1 Molecules and Abbreviations Found Throughout the Text

Acronym/Defense Molecule	Full Name	Description
AID (AICDA)	Activation-induced cytidine deaminase	Required for SHM/gene conversion/CSR
APOBEC	Apolipoprotein B mRNA editing enzyme catalytic polypeptide	Innate immunity (antiviral); includes AID, CDA
AGM	Aorta/gonad/mesonephros	Intraembryonic origin of hematopoietic cells
AgR	Antigen receptor	Ig, TCR, VLR
ALR	Allorecognition	Histocompatibility complex in *Hydractinia* (cnidarian)
AMP	Antimicrobial peptide	Innate immunity molecule, eg, defensins
APAR	Agnathan paired antigen receptor	Similarities to Ig/TCR and NKR in lamprey
AVR	Avirulence protein	Pathogen effector recognized by plant NLR
Bf	Factor B	Enzyme of C′ cascade (alt)
B1-3GNP	Beta 1-3 glucan-recognizing protein	Binds to gram-negative bacteria
BCR	B cell receptor	Ig on surface of B cell with transmembrane region
BM	Bone marrow	Tissue where precursor cells reside; different tissue in various vertebrates
C′	Complement	Innate/adaptive immunity
CARD	Caspase-recruitment domain	Domain in intracellular defense molecules
CATERPILLER or CLR	CARD, transcription enhancer, R(purine)-binding, pyrin, lots of leucine repeats	Apoptosis/immunity/inflammation innate immune molecule
CDA	Cytidine deaminase	Lamprey enzyme for GOD; CDA1 T cell, CDA2 B cell
CDR	Complementarity-determining region	Portion of Ig/TCR that binds to antigen, HVR synonym
CRP	C-reactive protein	Pentraxin acute phase protein
CSR	Class-switch recombination	Adaptive humoral immunity modification of IgH class
DD	Death domain	Cytosolic interacting domain
DSCAM	Down syndrome cell-adhesion molecule	Insect immune (adaptive?) defense and neuron specification
ECM	Extracellular matrix	
ETI	Effector-triggered immunity	Immunity in plants triggered by NLR
FBA	F box–associated domain	Intracellular domain
FcRN	Fc receptor neonatal	MHC class I–like FcR
Ficolin		Lectin that fixes C′
FN3	Fibronectin type III repeat	Domain found in many innate molecules
FREP	Fibrinogen-related protein	Mollusk (adaptive?) defense
FuHC	Fusion histocompatibility	Histocompatibility locus in tunicates
GALT	Gut-associated lymphoid tissue	Organized only in endotherms; present in all vertebrates
GNBP	Gram-negative binding protein	Binds bacteria
GPI	Glycophosphatidylinositol	Lipid linkage to cell membrane (eg, VLR)
hemolysin		Cell lysis
HVR	Hypervariable region	Synonymous with CDR, parts of AgR that bind antigen
ICE	Interleukin-converting enzyme	IL-1β processing enzyme
Ig	Immunoglobulin	Adaptive immunity AgR produced by B cells
IgSF	Immunoglobulin superfamily	Innate/adaptive immunity domain in hundreds of molecules (90–100 aa)
IFN	Interferon	Innate (type I)/adaptive (type II) immunity cytokine

(Continued)

TABLE 4.1 Molecules and Abbreviations Found Throughout the Text (Continued)

Acronym/Defense Molecule	Full Name	Description
IMD	Immune deficiency	Insect innate defense
IRF	Interferon regulatory factor	Innate (transcription factor)
IRG	Immunity-related GTPases	innate immunity
ITAM	Immunoreceptor tyrosine-based activation motif	Signaling motif for NKR and AgR
ITIM	Immunoreceptor tyrosine-based inhibitory motif	signaling motif for NKR and AgR
JAK	Janus kinase	Signaling molecule associated with cytokine receptors
KIR	Killer IgSF receptor	NK cell receptor primates
Lectins	Galectin, C-type, S-type	Many (eg, NKR, selectins, collectins, etc.); classically bind sugar but some do not
LITR	Leukocyte immune-type receptors	Fish NK-like receptors of the IgSF
LMP	low molecular weight protein	Proteasome subunit (PSMB)
LRC	Leukocyte receptor complex	Gene complex encoding KIR and many IgSF molecules (huchr 19q)
LRR	Leucine-rich repeat	Innate/adaptive immunity module found in hundreds of proteins
MAC	Membrane-attack complex	C′, pore-forming
MACPF	MAC-perforin domain	Potential pore former (perforin-1, perforin-2, MAC C5b-9)
MASP	MBP-associated serine protease	Enzyme C′ lectin pathway
MBP (or MBL)	Mannose-binding protein (lectin)	Binder C′ lectin pathway
MDM	Mollusk defense molecule	IgSF defense molecule
MHC	Major histocompatibility complex	T cell recognition; innate immunity; 4Mb in human
MIF	Macrophage inhibitory factor	innate immunity; inflammation
MyD88	Myeloid differentiation primary response gene 88	TLR and IL-1R adaptor
NITR	Novel immune-type receptors	Teleost fish NK-like receptors of the IgSF (VJ)
NK cell	Natural killer cell	Vertebrate innate cellular immunity
NKC	Natural killer cell complex	Gene complex with many C-type lectin genes (esp NK cells)
NKR	Natural killer cell receptor	Receptor on NK cells
NALP	NACHT leucine-rich repeat and PYD-containing protein	Intracellular PRR
NBD-LRR	Nucleotide-binding domain LRR	Motif of intracellular defense molecules
NF-κB	Nuclear factor-κ B (Rel homology domain)	Evolutionarily conserved transcription factor
NLR	NACHT leucine-rich repeat protein	Intracellular PRR
NOD	Nucleotide oligomerization domain protein	Intracellular PRR
NOS	Nitric oxide synthase	Intracellular killing innate defense molecule
PAMP	Pathogen-associated molecular pattern	Evolutionarily conserved target epitopes on pathogens
PCD	Programmed cell death	Many pathways
Penaedins		Defense molecule in shrimp
PGRP	Peptidoglycan-recognition protein	Gram-positive bacteria defense family; receptor and effector
PRR	Pattern-recognition receptor	Recognize PAMP, innate/adaptive immunity

TABLE 4.1 Molecules and Abbreviations Found Throughout the Text (Continued)

Acronym/Defense Molecule	Full Name	Description
PMSB	Proteasome subunit beta subunit	Proteolytic member of 20S proteasome (see LMP)
Polμ	DNA polymerase μ	Error-prone polymerase (related to TdT)
PPO	Prophenoloxidase	Invertebrate defense molecule
PYD	Pyrin domain	Domain in intracellular defense molecules
RAG	Recombination-activating gene	Required for Ig/TCR gene rearrangement
RFP-Y	Restriction fragment polymorphism-Y	Chicken nonclassical MHC gene cluster
RFX	Regulatory factor X	Transcription factor, class I regulation
RIG	Retinoic acid–inducible gene	Intracellular double-stranded RNA recognition
RSS	Recombination signal sequence	DNA element next to Ig/TCR gene segments necessary for RAG-mediated rearrangement
RXR	Retinoid X receptor	Transcription factor encoded in MHC
SHM	Somatic hypermutation	Adaptive humoral immunity
SPE	Spaetzle-processing enzyme	Insect defense molecule in Toll cascade
SRCR	Scavenger receptor cysteine-rich	Innate immunity recognition molecules
STING	Stimulator of interferon genes	Ancient PRR for nucleotide recognition
TAK	TGF-beta activated kinase	Ubiquitin-dependent kinase of innate pathways
TAP (and TAP-L)	Transporter associated with antigen processing	Transports peptides from cytosol to ER lumen; TAPL endosomal
TAPBP	TAP-binding protein	Tethers TAP to class I
TCR	T cell receptor	Adaptive defense AgR T cells
TdT	Terminal deoxynucleotidyl transferase	Involved in Ig/TCR rearrangement trmplstr-independent nuc addition
TEP	Thioester-containing protein	Opsonization (like C3)
TGF	Transforming growth factor	Immunosuppressive cytokine
TLR	Toll-like receptor	Innate receptor on the cell surface or in endosomes
TM	Transmembrane	
TNF	Tumor necrosis factor	Proinflammatory cytokine (and family)
TRIM	Tripartite motif-containing (TRIM) proteins	Large family of cytosolic innate defense molecules
UPD	Unpaired domain	Protostome cytokine induced by viral infection (JAK-STAT)
V, C1, C2, I	Variable, constant 1 and 2, intermediate IgSF domain	IgSF domain types
VJ-	Variable J segment domain	IgSF domain in Ig/TCR rearranging; in many other proteins non-rearranging
VAV	Guanine exchange factor (GEF), The "onc F" proto-oncogene	Encoded in MHC, involved in adaptive signaling pathways
VCBP	Variable domain chitin binding	Amphioxus defense molecule
VLR	Variable lymphocyte receptor	Agnathan adaptive defense molecule
WKRY		Plant transcription factor used to upregulate defense genes (analog of NFκB)
XMIV	*Xenopus* MHC-linked IgSF V region	*Xenopus* MHC-linked NKR-like genes
XNC	*Xenopus* nonclassical	*Xenopus* class Ib cluster
185/333 Transformer	Sea urchin defense molecule	(Adaptive?) defense in echinoderms

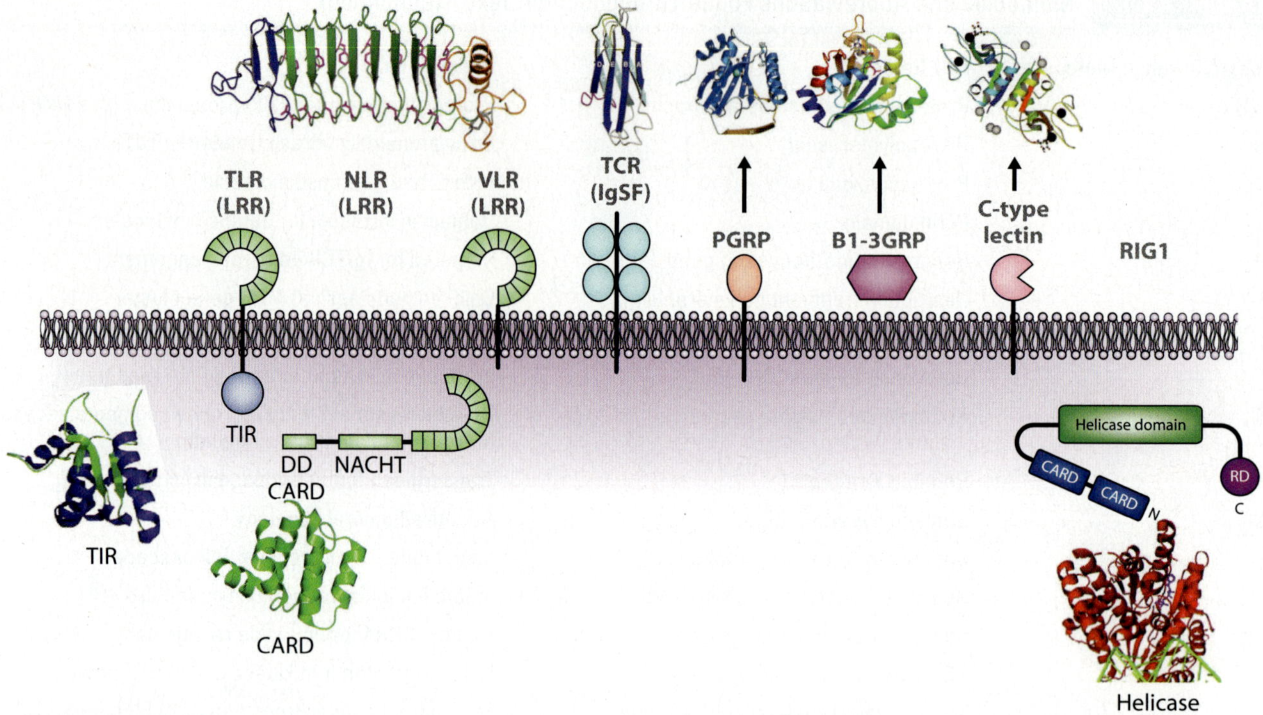

FIG. 4.3. Major molecular families described in the text and representatives of each family. Leucine-rich repeats (LRRs), immunoglobulin superfamily (IgSF), peptidoglycan recognition protein (PGRP), β1-3 glucan recognition protein (β1-3GRP), C-type lectins, TIR domains, CARD domains, and helicase domains. Representative structures are shown above for LRR (toll-like receptor, nucleotide-binding domain leucine-rich repeat [NLR], variable lymphocyte receptor [VLR]), IgSF (TCR), PGRP, β1-3GRP, C-type lectin, and RIG1. Other acronyms are defined in Table 4.1. For the NLR model, the echinoderms have N-terminal death domains, whereas all other animals have caspase-recruitment domains. (Modified from Buckley KM, Rast JP. Diversity of animal immune receptors and the origins of recognition complexity in the deuterostomes. *Dev Comp Immunol.* 2015;49(1):179-89. doi:10.1016/j.dci.2014.10.013 [published Online First: Epub Date and Hibino T, Loza-Coll M, Messier C, et al. The immune gene repertoire encoded in the purple sea urchin genome. *Dev Biol.* 2006;300(1):349-365. doi:10.1016/j.ydbio.2006.08.065. with permission from Elsevier.)

tail (or GPI-linked). The β barrel IgSF structure was adopted independently in other families such as cadherins, calycins, lipocalin, etc., and the superfamily has hundreds of members and has been co-opted for many functions. The vertebrate lymphocyte surface can express 30 different IgSF members.

IgSF domains are commonly classified according to domain constitution of their β strands and loops. All have a stable β barrel consisting of two interfacing β sheets, usually linked by a disulfide bridge forming an "open-faced sandwich."[27] There are three types of domains: variable (V) and two types of constant (C1 and C2); the so-called I set domain is intermediate between the C1 and C2.[28] The V domain is most complex with more strands (C′ and C″), which make up complementarity determining region (CDR) 2 in conventional Igs and TCRs. C1 domains lack these strands entirely, and C2/I domains have varying sizes in the C′/C″ region. V domains, either alone (eg, the new antigen receptor [NAR]) or in association with another V domain (eg, Ig H/L), recognize the antigenic epitope. Domains with the typical V fold, whether belonging to the V-set or the I-set, have been found from sponges to insects (eg, amalgam) and even in bacteria. The mollusk FN-related proteins (FREPs) have one or two V-like domains at their distal end, associated with a FN-like domain. For V domains, the interface between dimers is the beta strand bearing the C, C′, C″, F, and G strands, so that in Ig/TCR CDR3 are in the center of the binding site; for C1 domains, the A, B, D, and E beta strands form the interface.

The binding by V domains in molecules besides Ig/TCR can reside in different areas of the molecule (interstrand loops, A-A′ strand, F strand), while in Ig/TCR, CD8, and certain NKR, these regions are targets for variation in shape and charge. The binding capacities can be modulated whether one domain acts as a single receptor unit (eg, IgNAR[29]) or whether it is associated with a contiguous domain (eg, KIR, variable domain chitin-binding protein [VCBP][30]) or with another polypeptide chain (eg, TCR, Ig). In the case of a dimer, the binding capacity can again be modulated by the presence or the absence in the G strand (J segment of Ig/TCR) of a diglycine bulge, which can modify the space between the faces of the Ig domain.[31] In several cases, the sites responsible for binding are known (KIR, Ig, TCR); in many other cases, they are not known but inferred from crystal structures (leukocyte immune-type receptor [LITR]).

Evolution of the Deuterostomes

The animal kingdom (Fig. 4.1) is comprised of the Protozoa (single-celled) and Metazoa (many-celled). Metazoa consists of

Porifera (sponges), Cnidaria (two germ layers), Platyhelminthes (three germ layers), and protostomes and deuterostomes (three germ layers with a mesoderm-lined coelom). The latter two groups are defined by whether the embryonic blastopore becomes the mouth (protostomes) or anus (deuterostomes). Protostomes are all invertebrates and their immune system is examined in Chapter 3. Deuterostomes, the subjects of this chapter, are split into three major groups, the echinoderms, with pentradiate symmetry as adults but bilateral symmetry as larvae; the invertebrate chordates (with a notochord); and the vertebrate chordates also with a notochord but also a dorsal hollow nerve cord, gills (some only during embryonic life), a vertebral column, and a postanal tail (again, some only during embryonic life). Vertebrates are further split into the earlier-arising jawless vertebrates (agnathans) and the jawed vertebrates or gnathostomes (Fig. 4.2). The non-vertebrate deuterostomes were not subject to the genome-wide duplications early in vertebrate evolution and serve as touchstones for the preduplicated deuterostome genome.[32] Neither do they have described adaptive immunity, although their innate immunity can be complex.

There are only two extant groups of jawless vertebrates, lampreys and hagfish, which have been separated by hundreds of millions of years and are quite different in body type and lifestyle compared to fossil agnathans.[33] Yet, the lamprey and hagfish immune systems are similar as far as they have been studied and clearly they have a robust adaptive immune system. The agnathans emerged after the first genome-wide duplication found in all vertebrates and then underwent three agnathan-specific large-scale duplications, which have confounded studies of synteny in this taxon.[20,21] The gnathostomes emerged after the second genome-wide duplication ~500 million years ago and are comprised of the extinct Placodermi, Chondrichthyes (cartilaginous fish), Actinopterygii (ray-finned fish, including the teleosts and holosteans), lobe-finned fish or Sarcopterygii (coelacanth and lungfish), Amphibia, Reptilia, Avia, and Mammalia, the latter four taxa referred to as tetrapods (Fig. 4.2). Gnathostomes form a natural group having a thymus, spleen, Ig/TCR, MHC, RAG, AID, as well as complex chemokine and cytokine networks. Prior to the discovery of the agnathan VLR, Jack Marchalonis referred to the emergence of full-blown adaptive immunity in gnathostomes as the "Immunological Big Bang."[34] Recently, I have modified this idea and suggested that there were "two Big Bangs of adaptive immunity," one giving rise to the VLR-based adaptive immunity and a second superseding the first, resulting in the Ig/TCR/MHC-based adaptive system.[35] This theme will be expanded later in the chapter.

A Note on This Chapter

In the past 10 years, there has been an explosion of data from transcriptomes, genome sequencing, etc., that has (nearly) filled our bingo card for the genes and gene families found in different animal taxa. In contrast, functional work, as expected, has lagged on how all of these immune genes are called to action. Due to limited space, the referencing in the chapter will focus on recent reviews where the reader can harvest the latest *detailed* information on each gene family in particular deuterostome groups.

INNATE IMMUNE RESPONSES

Immune responses can be partitioned into their recognition, signaling, and effector phases, which are subjected to different pressures defined by whether othology is preserved and gene divergence rates involved in the various phases. Recognition molecules are derived from evolutionarily conserved families, but their genes are subjected to rapid duplication/deletion so that orthology may not be observed.[36] By contrast, signaling pathways are often conserved (eg, Fig. 4.4), although the genes can be divergent. Effector molecules can either be extremely conserved (eg, reactive oxygen species, ROS) or divergent to the point of being species-specific (eg, antimicrobial peptides, AMP[37]) (Chapter 3). Here, I break down the immune response into these three phases, beginning with the recognition phase. Innate immune mechanisms will be discussed first but the same can be said for adaptive immunity, eg, the ITAM/ITIM paradigm for antigen receptors[38] or the JAK/STAT pathway for cytokine receptors[39] (Chapter 9).

Initiation of an immune reaction can involve the recognition of nonself, altered self, or the absence of self. Nonself recognition is initiated with receptors (PRRs) that detect PAMPs, which were defined by Janeway as evolutionarily conserved epitopes displayed by pathogens but not host cells (Chapter 12). Danger-associated molecular patterns (DAMPs), championed by Matzinger, are self-molecules, usually found inside cells but when released during inflammation or necrotic cell death also interact with PRR to initiate immunity. The second mode, altered self, is typified by molecules that are induced in self-cells during infections and recognized by conserved defense molecules, like SOS systems described in the MHC section or by peptide presentation by MHC molecules. A third mechanism, "am I still myself," depends on recognition of self-tags and their changes in expression (eg, NK recognition of self-MHC molecules through NKR).

Whether the invader is related to its host (cells from individuals of the same species or a parasitoid) or is distant from the host (fungi, bacteria), there are different principles of recognition. PAMPs have been identified on very different organisms—sugars like β1-3 glucan of fungi, lipopolysaccharide (LPS) and bacterial peptidoglycans, phosphoglycans of parasites, and especially nucleic acids of bacteria and viruses (we find new ones all the time!)—and they can trigger similar cascades of events (Table 4.1). Note that there are innate cytokines, but they are covered in the cytokines of the adaptive section.

Cell Types and Phagocytosis

Cell Types in Deuterostomes

Echinoderm coelomocytes express a diversity of effector functions, but few studies of lineages have been performed. In echinoderms, the number of different coelomocytes may vary according to the particular family (see more below in the Transformer section). The sea urchin is endowed with at least four hematopoietic cell types, only one of which is phagocytic and corresponds to a filiform form. Another type is described as the round vibrating cell involved in clotting. Pigment cells (red spherule cells) have been detected ingesting bacteria;

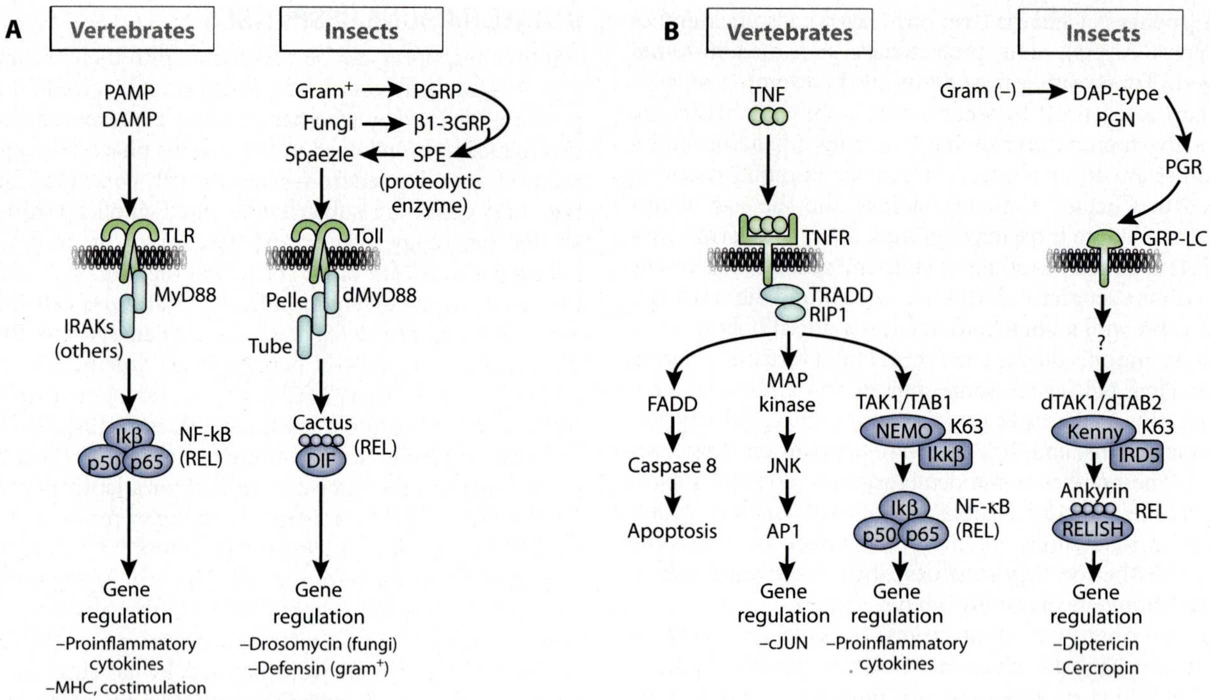

FIG. 4.4. Comparison of immune response induction, intracellular pathways, and immune outcome in insect (eg, *Drosophila*) and vertebrates (eg, human). Note that the initiation of the response and the outcome(s) in insects and vertebrates are quite different for both the toll/toll-like receptor pathway (*left*, **A**) and the immune deficiency/tumor necrosis factor pathways (*right*, **B**), but the intracellular signaling pathways are well conserved evolutionarily in insect and vertebrate. (Modified with permission of American Association for the Advancement of Science from Ronald PC, Beutler B. Plant and animal sensors of conserved microbial signatures. Science. 2010;330(6007):1061-1064; permission conveyed through Copyright Clearance Center, Inc.)

the morphology of phagocytic cells can vary enormously, precluding any easy classification. Recent work by FACS has permitted separation of these cells via light scatter and cell sorting with specific antisera.[40]

In tunicates, ameboid cells circulate in the blood and are involved in many processes, such as clotting, excretion, nutrition budding, and immunity. Large numbers of blood cells are present (up to 10^7/mL) in the blood of ascidians such as *Ciona*. Blood cells in ascidians proliferate in the connective tissue next to the atrium.[41] The pharyngeal hematopoietic nodule of this animal contains many hyaline and granular cells called "leukocytes" with supposed intermediary forms of differentiation between blast and granular mature types. Recent work in the colonial tunicate *Botryllus* has demonstrated different cell types of the myeloid (bur not lymphoid) lineage, some implicated in the histocompatibility reaction described below.[42]

In *amphioxus*, cells with phagocytic capacity have been identified in the coelom. Both free cells and the lining of the perivisceral coelom can phagocytose bacteria. Cells with the morphologic appearance of lymphocytes and expression of lymphocyte-specific genes were detected in this species, the earliest identification of such cells in phylogeny.[43]

Vertebrate Cell Types

The vertebrate innate immune system is defined by cells of the myeloid lineage and NK cells/innate lymphoid cells (ILCs) of the lymphoid lineage. Evolutionarily, there is much to learn about the diversity and function of innate cells, but the basic classes of mammalian myeloid cells, granulocytes (neutrophils, eosinophils, and basophils), macrophages, and thrombocytes are present in all nonmammalian vertebrates. Each of these cell types is covered in a dedicated FI chapter and here I briefly focus on features in nonmammalian vertebrates.[44]

Granulocytes: Neutrophils have been identified by their histological staining and eccentric nuclei and are found in all vertebrates. These cells are generally present in the blood and perform the same basic functions as in mammals. In cartilaginous fish, neutrophils are present but cells with eosinophilic coarse granules are most efficient at phagocytosis.[45]

Zebrafish models have provided the touchstone of hematological origins of myeloid cells outside of mammals. The ease of transgenesis and crispr technology and transparency of the animals have permitted lineage tracing from embryonic and definitive hematopoietic sites, as well as migration and functions of cells after infection or other inflammatory conditions such as injury. As expected, neutrophils play a protective role in infection models with the production of NO and NETS, while during injury neutrophils generally perform a clean-up function without inducing inflammation.[46]

Eosinophils are instrumental in parasite immunity and allergy in mammals and are present in other vertebrates but have been poorly studied. Shark "eosinophils" come in at least two types with either fine or coarse granules, the latter being highly phagocytic, as mentioned more like neutrophils in other vertebrates.[45]

Basophils are present in ectotherms but poorly studied (see below).[47] It would be of interest to define their functions in animals devoid of IgE, present only in mammals, but IgD is also an important regulator of these cells and their functions.

Macrophages: Macrophages have been the best studied myeloid cells in nonmammalian vertebrates.[48] In all cold-blooded vertebrates (besides sharks) they are found in the blood as monocytes and in the peritoneal cavity and spleen as distinctive esterase-positive cells, as well as throughout the body. At this time, the mammalian literature is showing a high diversity of macrophage subsets[49] and how this relates to macrophages in nonmammals is not well understood. In zebrafish and amphibians, the M1/M2 paradigm of inflammatory and tissue remodeling cells seems to hold up.[50] In *Xenopus*, stimulation of cells with either CSF or IL-34 separates the cells into functional subsets that have distinct functions. The chemokine receptor CSF1R is a reliable marker for macrophages in fish and amphibians.[51]

Thrombocytes: Rather than platelets, all nonmammalian vertebrates have thrombocytes, found mainly in the blood. They are generally "annoying" to cellular immunologists because they look like lymphocytes at first glance and thus confound one's lymphocyte counts. The cells are involved in blood clotting and also have been suggested to have immune properties not well understood.[52]

Phagocytosis

To obtain phagocytosis at the site of microorganism invasion implies recruitment of cells via chemoattraction. In vertebrates, this can be done by several categories of molecules such as proinflammatory chemokines/cytokines or the complement fragments C3a and C5a (in the following section, C3a fragments like in mammals may be found in tunicates but not in other non-vertebrates; yet, C3 may be cleaved in different ways in the invertebrates). C3b, mannose-binding lectin (MBL), and many other lectins can function as opsonins. Ingestion follows phagocytosis, and then killing occurs by an oxidative mechanism with the production of ROS and nitric oxide (NO). These mechanisms are conserved in phylogeny.[44] Unique to all gnathostomes, activation of phagocytes also leads to upregulation of antigen processing machinery, costimulatory molecules, and proinflammatory cytokines to enhance adaptive immunity. Phagocytosis can also be performed by B cells, as described below.

Integral Membrane (and Sometimes Secreted) Proteins

C-Type Lectins

Lectins were originally defined by their ability to bind carbohydrates in a calcium-dependent manner (how C-type lectins got their name and some have been described previously) (Fig. 4.3). They are found in many phyla in both the deuterostome and protostome lineages in both membrane and secreted forms (eg, MBL in the following section). Some C-type lectins are encoded in the natural killer complex (NKC), including the Ly49 (in mice) and NKG2 families, as well as CD94 and several other members of the family that are central to NK cell function in endotherms. Another large gene family that is implicated in the response of the sea urchin to immune challenge includes 100 small C-type lectins, consistent with the enormous expansion of innate immune defense families in this animal.[53] We describe functions/genetics of C-type lectins in the NK cell sections.

Other lectin families, particularly the galectins, have been studied in bony fish.[54] Recent studies have shown how galectins can either serve as defense molecules or be subverted to aid in viral infection. In mammals, galectins have been shown to serve many different roles and future studies on their evolution are warranted.[55]

Scavenger Receptors

The SRCR superfamily is an ancient and conserved group of cell surface and/or secreted proteins, some of which are involved in the development of the immune system as well as the regulation of innate and adaptive immune responses; they are well known for their function in macrophages.[36,56] Group B SRCR domains contain eight regularly spaced cysteines that allow the formation of a well-defined intradomain disulfide bond pattern. Scavenger receptors are best known for their housekeeping role of taking up lipids modified by oxidation or acetylation, but they have many other functions, such as uptake of apoptotic bodies (eg, croquemort in *Drosophila* of the CD36 subfamily).

SRCRs have been studied mainly in the coelomocytes of echinoderms.[57] Within a few hours after bacterial challenge, coelomocytes upregulate a variety of genes, including diverse SRCRs. A very large number of SRCR domains are present (approximately 1200), but each individual may express different groups of SRCR genes at different levels (and even with differential splicing). To assume that they are all involved in defense is premature, as SRCR genes are both upregulated and downregulated after bacterial infection. This elevated level of innate gene duplication is a general rule in echinoderms.

In vertebrates, PGRPs are all secreted and have direct microbicidal activity.[36] Best studied in zebrafish, PGRPs are expressed in many tissues such as the gills, skin, and intestine, providing immune defense. They are expressed before development of adaptive immunity and likely provide a protective role. All detected splice-variant isoforms bind to bacteria and peptidoglycan. Like the fish molecules, mammalian PGRPs are also positioned at epithelial surfaces and promote intestinal homeostasis by discriminating somewhat between commensal and pathogenic bacteria.[58] Knockout mice have increased pathogenic bacteria on mucosal surfaces that induce colitis after injury in the dextran sulfate sodium autoimmune assay.

Toll and Toll-Like Receptors

The toll receptors were originally described in *Drosophila* as genes involved in dorsoventral patterning during early development (Chapter 3). They were later shown to be essential sensors of infection, initiating antimicrobial responses.[3] In humans, many are on human chromosome 4p and q (TLR 1, 2, 3, 6, 10) but the others are distributed on chromosomes 9q33 (TLR 4), 1q41 (TLR 5), 3p21 (TLR 9), and Xp22 (TLR 7,8). TLR 11, 12, and 13 are related and lost in humans, but are present in many ectothermic vertebrates. Ectotherm-specific TLR are related to the 11/12/13 and 1/2/6/10 lineages, while TLR3, 4, 5, and 7/8/9 are retained as single-copy genes in most vertebrates (see Ref. 59 for the latest review). TLR ectodomains are comprised of 19 to 25 tandem LRR domains made of 20 to 29 aa capped by characteristic N- and C-terminal sequences. TLR all share the TIR domain, which is the intracellular segment shared with the IL-1/-18/-33 receptors of vertebrates. TIR domains associate with Myd88 to initiate signaling cascades culminating in the activation of NF-κB/Rel (Fig. 4.4).

The arsenal of TLRs in vertebrates is endowed with specific and diverse capacities (Chapter 12). Each vertebrate TLR has a range of specificities and, in addition, combinations of different TLR can create different binding specificities, for example, the association of TLR2 with TLR6 or TLR1 and TLR2 or even TLR2 homodimers in regulatory T cells. Toll and related proteins from insects and mammals cluster separately in the analysis, indicating independent generation of the major families in protostomes and vertebrates.[60] Consistent with the expansion of SRCR and NLR genes in sea urchins, hundreds of TLRs also were found in this species. TLRs of the protostome type are in small numbers (three members) while the vertebrate type has been enormously amplified, all within a single family (>200 members) and most without introns.[36,61] TLRs have also been expanded in amphioxus.[43] Cartilaginous fish (and likely some amphibians) have lost TLR4 and its associated molecules MD2 and CD14 and pseudogenes can still be detected in the genome.[62] Sharks have been shown to have commensal bacteria within their internal organs and the peritoneal cavity is exposed to seawater.[63] Such special physiology may have necessitated TLR4 loss, as well as potentially the loss of the alternative C′ pathway.

Intracellular Recognition

Nucleotide-Binding Domain Leucine-Rich Repeat (NLR)

A major group of intracellular sensors in animals and plants is the NLR family[64,65] (Fig. 4.3). Each of the family members has a central NB/NACHT (nucleotide-binding domain) and C-terminal LRR used for recognition and a unique N-terminal domain. The subfamilies are defined by their N-terminal domains, baculoviral inhibitory repeat, caspase-recruitment domain (CARD), pyrin domain (PYD), and activation domain in animals (Chapter 12).

Best described in mammals, the nucleotide oligomerization domain (NOD)/NLR recognizes PAMPs such as peptidoglycan and induces an autophagy-mediated destruction of intracellular pathogens as well as production and secretion of proinflammatory cytokines; additionally, indirect recognition of cellular metabolites can stimulate NLR function, like the situation in plants (Chapter 5). The NLRP and NAIP NLRs are activated by various PAMPs or DAMPs and form inflammasomes, best known for the activation of caspase 1 and the processing of pro–IL-1beta (or mature IL-18) for release from cells. The founding member of the family, CIITA, has long been known to upregulate class II genes (and associated genes, such as cathepsins and invariant chain), and its function is outside the norm. Another member, NLRC5, has been shown to upregulate MHC class I expression, but the mechanism is poorly understood.

NLRs are expressed by echinoderm coelomocytes, again representing a highly diversified family (>200 members, like the TLRs and SRCRs).[36,53] In the teleosts like zebrafish, a large number of NLR sequences have been found, like the situation in plants.[64] In humans, most NLR genes are encoded in clusters on chromosomes 11p15, 16p12, and 19q13, where six sequences are found in a single telomeric region.

Rig-I-Like Receptors (RLR)

The retinoic acid–inducible gene (RIG)-I is an intracellular defense molecule that is unrelated to the NOD proteins, with N-terminal CARD and C-terminal helicase domains (Fig. 4.3). With the helicase domain, RIG-I binds to an uncapped 5′ phosphate group, which is diagnostic of viral RNAs. RIG-I also recognizes short double-stranded RNAs, while a second member of this family MDA5 recognizes long double-stranded RNAs (Chapter 12). These molecules contain two CARD domains at the N-terminus, a DEXDc domain, a helicase domain, and a regulatory domain. Ligands bind to the regulatory domain, inducing a conformational change leading to interaction with the adaptor protein MAVS (or IPS-1) and ultimately to the induction of type I interferons (IFNs). A third member of the RLR family is LPG2, which lacks the CARD domain.

This family is found in all vertebrates and in lower deuterostomes, such as amphioxus and echinoderms.[66] Somewhat surprisingly, the RLR family is only mildly expanded in sea urchins (12 members).[36]

Cytosolic Deoxyribonucleic Acid Sensors

There are four mechanisms of cytosolic DNA sensing, three of them, the DNA-dependent activator of IFN-regulatory factor, IFI16, and RNA polymerase III (which converts viral DNA into RNA recognized by RIG-I), induce type I IFN production through the intermediate STING, a protein associated with the endoplasmic reticulum (ER) (Chapter 12). In addition to being an intermediate in IFN upregulation (through IFN regulatory factor-3), STING is also a PRR in its own right, responding to the PAMP cyclic dinucleotides produced by intracellular bacteria like *Listeria*; this suggests that STING was originally a PRR and then was co-opted by several other PRR sensors to induce effector functions (discussed in great detail in Chapter 3). This hypothesis has been born out with the discovery of STING-like molecules in bacteria.[67] IFI16 is part of the AIM2-like receptor family; the founding member, AIM2, like the inflammasome, activates caspase-1 to process pro–IL-1beta.[68]

Tripartite Motifs

Tripartite motif (TRIM) proteins belong to a family induced by type I and II IFNs, with 68 members in the human genome (Chapter 12). TRIMs are involved in resistance against pathogens in mammals, especially lentivirus (eg, human Trim 5α is a retroviral restriction factor with activity against HIV). TRIMs contain an N-terminal moiety composed of three modules: RING (with an E3 ubiquitinase activity)–Bbox-coiled coil motif followed by different C-terminal domains. TRIMs fit into two major categories by the function of their C-terminal domain: Category 1 with a PHD, MATH, ARF, FNIII, exoII, or NHL domains, and Category 2 with a B30.2 domain shared with butyrophilins and other proteins and essential for ligand binding.

The family has been greatly diversified in vertebrates and in a taxon-specific manner, as observed for many multigenic immune families.[69] The zebrafish genome harbors a striking diversity of a subset of Category 2 TRIMs not encountered in mammals, called finTRIM, with 84 genes distributed in clusters on different chromosomes. This subset, specific of teleosts, is overexpressed after virus infection in the trout.

Finally, *trim* genes encoding Category 2 proteins are preferentially located in the vicinity of MHC or MHC gene paralogs in fish, birds, and human, suggesting that they were part of the ancestral MHC.[70,71] The B30.2 domains most closely related to finTRIM are found among NLRs, indicating that the evolution of TRIMs and NLRs was intertwined by exon shuffling. Exon shuffling was likely responsible for the presence of the B30.2 domain in butyrophilin and TRIM genes where it was perhaps favored by the proximity of the genes in the MHC. It has been argued that during evolution the combination of SPRY and PRY motifs that build up the B30.2 domain was selected and maintained for immune defense.

The Complement System

The best-studied immune proteolytic cascade that is surprisingly well conserved in the animal kingdom is complement (Fig. 4.5) (Chapter 13). In contrast to the other defense molecules that we have discussed, orthologous complement genes

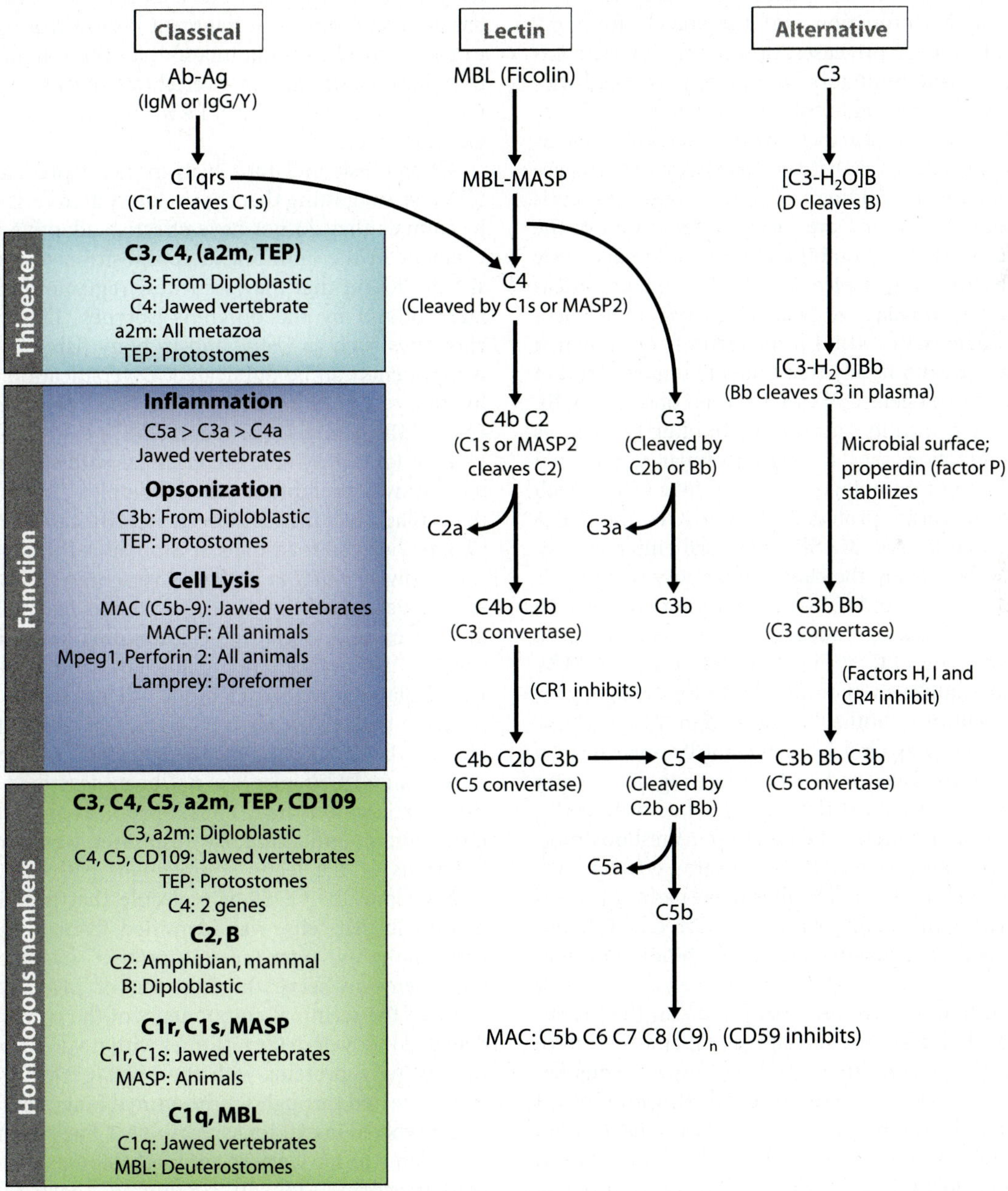

FIG. 4.5. Evolution of the complement system. The general pathways and appearance of the various components in the phylogenetic tree are emphasized in a self-explanatory way.

can be detected in all deuterostomes generally without a great deal of expansion/contractions of the gene family.[72] The three major functions of complement in jawed vertebrates (see Chapter 13) are (1) coating of pathogens to promote uptake by phagocytes (opsonization); (2) initiation of inflammatory responses by stimulating smooth muscle contraction, vasodilatation, and chemoattraction of leukocytes; and (3) lysis of pathogens via membrane disruption. Additionally, in the vertebrates, C′ is vital for the removal of immune complexes from the body and the elicitation of adaptive humoral immunity. The focal point of C′ is C3, which lies at the intersection of the alternative, classical, and lectin pathways of C′ activation.[73] It is the only known secreted immune recognition molecule (besides its homolog C4) that makes a covalent bond with biologic surfaces via a thioester linkage. C3 has a nonspecific recognition function, and it interacts with many other proteins, including proteases, opsonic receptors, complement activators, and inhibitors. In the alternative pathway, C3 exposes its thioester bond in solution, and in the presence of cell surfaces lacking regulatory proteins that block C3 activation (by cleaving it into iC3b), it associates with the protease factor B (B or Bf). After binding to C3, B becomes susceptible to cleavage by the factor D, resulting in formation of the active protease Bb that in combination with the covalently attached C3 cleaves many molecules of C3 in an amplification step. Another nonadaptive recognition system, the lectin pathway, initiates with MBL (or the lectin ficolin), which is a PRR of the collectin family that binds mannose residues on the surface of pathogens and can act as an opsonin. MBL is analogous to C1q with its high-avidity binding to surfaces by multiple interaction sites through globular C-terminal domains, but it is not homologous to C1q. Like C1q, which associates with the serine proteases C1r and C1s, the MBL-associated serine proteases (MASPs) physically interact with MBL and not only activate the classical pathway of complement by splitting of C4 and C2 (the same function as C1s; MASP2 is active protease). Indeed, MASP-1 and MASP-2 are homologs of C1r and C1s (Fig. 4.5). Both C1q and MBL can promote the uptake of apoptotic bodies by phagocytes, via collectin receptors. Another lectin, ficolin, can also initiate the MASP pathway, and it would not be surprising if other activators were discovered in the future (eg, the ancient molecule C-reactive protein is also capable of activating C′). Finally, the classical pathway, dependent upon antibody molecules bound to a surface, results in the same effector outcomes described previously for the alternative pathway. Novel molecules initiating this pathway are C1q, C1r, C1s, C4, and C2, as well as specific regulators such as C4-binding protein (Chapter 13).

C3 and MBL (and ficolin) are vital players in the immediate innate immune response in vertebrates, and both have been described in non-vertebrate deuterostomes.[72] Thus far, the best-studied invertebrate systems for investigation of C3 evolution are the sea urchin and the ascidians *Halocynthia* and *Ciona*, in which C3 and B molecules and genes have been analyzed in some detail. In contrast to the very high levels of C3 found in the plasma of gnathostomes, sea urchin C3 is not expressed at high levels but is induced in response to infection in coelomocytes.[74] The C3 opsonic function has been identified, but so far initiation of inflammatory or lytic responses has not been detected. Receptors involved in opsonization in echinoderms have not been identified, but in the ascidian gene fragments related to the C3 integrin receptor CR3 were identified, and antisera raised to one of the receptors inhibited C3-dependent opsonization.

Hagfish and lamprey C3-like genes were thought to be ancestral C3/C4 genes because the sequence predicts two processing sites (leading to a three-chain molecule), like C4, but a C3-like properdin-binding site is clearly present.[75] However, like C3 in other animals, the hagfish protein is composed of only two chains of 115 and 72 kDa, and sea urchin and ascidian C3 sequences predict only two chains (one proteolytic processing site). The lamprey, but not sea urchin C3, has a recognizable C3a fragment known from gnathostomes to be involved in inflammation, so the role of complement in inflammation may be a vertebrate invention. Interestingly, recent data suggest that VLRB can fix C′ similar to Igs in gnathostomes.[76]

Certain fish and amphibian species express more than one C3 gene, suggesting that the innate system might compensate in animals that do not have effective adaptive immunity.[72,77] Changes in the amino acid composition of the C3-binding site are found that may somehow regulate the types of surfaces bound by the different isotypes. Likewise, in lower chordates such as *Ciona* and in bony fish, C3 and other C′ components can be duplicated. Diversification of the carbohydrate recognition domains has been observed also in the *Ciona* MBP family (nine members).[72]

Like Ig/TCR/MHC, the classical pathway and the terminal pathway membrane-attack complex (MAC) appear first in cartilaginous fish. However, as MBL can activate the classical pathway in mammals, it is possible that some portion of this pathway exists in prejawed vertebrates. Nevertheless, C4 and C2 genes have not been detected to date in jawless fish or invertebrates. A bona fide C2 homolog has only been identified to the level of amphibians, although duplicate B genes were isolated from cartilaginous fish and teleost fish that may function both in the classical and alternative pathways.[72] The lytic or MAC pathway, which is initiated by the cleavage of C5 into C5a and C5b, also has not been described in taxa older than cartilaginous fish. Thus, opsonization and perhaps the induction of inflammatory responses were the primordial functions of the lectin/complement pathways. However, a cDNA clone for CD59, a molecule that inhibits MAC formation in self-cells, was identified from a hagfish library, and a lamprey "pore-forming protein" has been uncovered that seems to serve the function of MAC.[76] In addition, some of the terminal components of the pathway have been detected in lower deuterostomes with no described functions. Interestingly, proteins with the MAC/perforin domain have been detected throughout the animal kingdom, and some are even involved in cytotoxic reactions.[78] The perforin gene itself also seems to have arisen in gnathostomes, from an ancient MAC/perforin domain–containing gene, macrophage-expressed gene 1 protein (MPEG1, also called perforin-2), which dates back to sponges; thus, cellular cytotoxic reactions in the invertebrates described in the following must use novel cytotoxic effector molecules.

C3, C4, C5, and α2m (and TEP) are members of the same small family.[79] A cell surface–expressed (GPI-linked) member of this family, CD109, has been shown to associate with the transforming growth factor (TGF)-β receptor and modulate its expression. The protease inhibitor, α2m, clearly present in invertebrates (protostomes and deuterostomes) and vertebrates is thought to be the oldest along with C3, but obviously this must be viewed with caution considering the data in cnidarians. Along with its ability to bind to and inactivate proteases of all known specificities through a "bait region," it also has been shown to be opsonic in some situations. α2m, C3, C4, and CD109 (as well as the TEPs) have internal thioester sites, so this feature is primordial; C5 subsequently lost the site. The first divergence probably occurred between α2m and C3, with C5 and then C4 emerging later in the jawed vertebrates,[72] although other scenarios are possible, with C5 emerging first in gnathostomes and C4 and C3 after gene duplication.[80] Consistent with Ohno vertebrate polyploidization scheme is the fact that C3, C4, and C5 genes are located on three of the four previously described paralogous clusters in mammals, and this also fits with the absence of classical (no antibody) and lytic (no true MAC) pathways in phyla older than cartilaginous fish. α2m is encoded at the border of the NKC in mice and human, and there are similarities between these regions and the other MHC paralogs (see below). There are two C4 genes in many vertebrates with different binding specificities, dating back to the cartilaginous fish.[81] The C3a and C5a receptors that promote the inflammatory responses upon complement activation have been identified in several vertebrates; like the chemokine receptors they are G protein–coupled receptors whose genes may also be found on the ohnologs (C3aR, chr 12p13; C5aR, chr 19q13). If indeed such receptors are found in the prejawed vertebrates as suggested by studies in *Ciona* and *Styela*, it will be interesting to determine whether they are involved in some type of inflammation, thought to be the domain of the vertebrates.[82]

Lower Deuterostome–Specific Molecules and Mechanisms

Transformer (185/333)

There are additional expanded gene families in the sea urchin genome that encode proteins with immune-related functions. The *Transformer* (formerly *185/333*) genes were first noted because they are highly upregulated in coelomocytes after exposure to LPS, constituting up to 7% of the mRNA in such cells. Subsequently, transcripts were shown to be upregulated by different types of PAMPs.[83] The encoded proteins have no detectable similarity to any other gene family, but they are highly diversified and are produced (at least) by a subset of coelomocytes. This seems to be a gene family that bridges innate and adaptive immunity.

There are estimated to be at least 50 185/333 Trf genes in the sea urchin genome. The gene is composed of two exons, one encoding the leader and the other encoding the mature protein. The second exon is made up of so-called elements and repetitive sequences that are quite different from gene to gene, which can to a large extent explain the diversity of expressed loci.[84] However, there are hints of RNA editing, a unique form of alternative splicing, somatic mutation (perhaps targeting cytosine residues), and other novel mechanisms of diversity to explain the incredible number of different isoforms; additionally, like many other defense molecules, there is evidence of protein multimerization. A recombinant Trf protein was shown to bind to b1-3 glucan, LPS, and flagellin, and LPS binding inhibits binding to other PAMPS.[83] Trf undergoes modifications of structure from disordered to alpha helical after PAMP binding, hence the name Transformer.[85] Trf is expressed on the surface of phagocytes despite not having a TM or gpi-linked motif. Recent work suggests that only one Trf is expressed/cell, which is surprising considering the number of variants potentially produced.[86,87] The presence of this unique multigene immune family in sea urchins is consistent with the aforementioned great expansion of other immune gene families in this group, including TLRs, SRCRs, and NLRs.[36]

Variable Domain Chitin-Binding Proteins

VCBP, first discovered not only in *amphioxus* but also present in *Ciona*, consists of two Ig domains of the V type but with a different folding motif when compared to Ig or TCR V domains followed by a chitin-binding domain.[30] The chitin-binding domain resembles chitinases found throughout the animal kingdom, and like dedicated chitinases, VCBP is usually expressed in the gut. Apparently, there are no cell surface–expressed forms and thus all VCBPs are likely to be secreted, effector molecules. In *amphioxus*, their diversity is enormous, apparently entirely because of polymorphism and polygeny.[88] Each individual can carry up to five genes per haplotype, and in limited studies (11 individuals) no identical haplotype has been encountered. The general structure of the V domain is like that of the vertebrate rearranging antigen receptors, but with unusual properties including packing in a "head-to-tail" dimeric fashion, totally unlike Ig and TCR.[30] VCBP diversity does not reside in the Ig/TCR CDR resides, but rather in the A, A′, and B strands, like in DSCAM (Chapter 3). VCBP binds to chitin in the gut mucous and seems to be important in forming biofilms and protecting mucosae, like vertebrate IgA. If true, this would provide a link to the regulation of commensals in early deuterostomes.[89,90] Soluble VCBPs bind to bacteria and induce opsonization in coelomocytes.

Natural Killing Activity

The word "cytotoxicity" encompasses vastly different protocols of cell killing by different cell types. It can be an effector function of cells of the adaptive (cytotoxic T lymphocyte [CTL] or NKT/MAIT cells) or of the innate arm (bona fide NK) of the jawed vertebrate immune system (Chapter 18). Likewise, the term "NK cells" covers different cell types and functions. Vertebrate NK cells can "recognize" missing self-MHC class I, but also ligands induced on stressed cells following virus infection, transformation, or stress. They can also have an immunoregulatory role by interactions with antigen-presenting cells (APCs). Interestingly, all the human homologs of those lectin genes are present either in the NK complex on chr 12p13 (CD94, CD69) or on the lymphocyte receptor complex (LRC) on chr 19q (CD209). Taken together with studies of the chicken MHC which encodes some C-type lectins (see later), the data suggest that a conserved MHC-linked region contains several lectin genes.[91]

In addition, genes encoding membrane proteins with extracellular C-type lectin or immunoreceptor tyrosine-based inhibitory motifs (ITIMs) and immunoreceptor tyrosine-based activation motifs (ITAMs) (plus their associated signal transduction molecules) were identified in *Ciona*, which suggests that activating and inhibitory receptors have an MHC class I– and class II–independent function and an early evolutionary origin.[32]

Natural Killer and Natural Killer–Like Cells

NK cells express both activating and inhibitory cell surface receptors; in fact, the paradigm for positive and negative signaling via such receptors began with these cells; however, activating and inhibitory receptors (often paired) are conserved throughout vertebrates and invertebrate deuterostomes and are expressed in hematopoietic cells of all types. In NK cells, stimulation of the activating receptors, which associate with proteins having an intracellular ITAM (CD3ζ, DAP12, or DAP10 conserved in deuterostomes), results in killing of target cells. Inhibitory signaling receptors all possess cytoplasmic ITIMs, which recruit phosphatases and generally are dominant over the activating receptors. These receptors fall into two categories: IgSF and C-type lectin group V(II). In general, NKRs recognize MHC class I molecules of either the classical or nonclassical type, the latter sometimes encoded by viruses.

General Evolution of Natural Killer Receptor Families

As mentioned, NK cells in mammals can use different families of receptors, IgSF (KIR) in primates and C-type lectins (Ly49) in rodents.[11] A few receptors are conserved but others are highly polymorphic. Very few families show conservation of domains throughout the jawed vertebrates. The question remains as to whether NK cells, or NK-like cells, preceded the emergence of T lymphocytes and B lymphocytes, or are even present in agnathans.[92]

NKRs are the most rapidly evolving molecular feature of the gnathostome immune system. Most (but not all) ligands for these diverse NKRs are MHC class I molecules or molecules of host or pathogen origin related to MHC class I. The KIR families are divergent, as very few genes are conserved even between chimpanzees and humans, and there are different numbers of genes in KIR haplotypes within a species.[93] Humans have two KIR haplotypes A and B, one encoding many activating receptors and the other very few. It is speculated that the haplotypes are under balancing selection within the population both for defense against virus and for maternal/fetal interactions. By contrast, CD94 and NKG2 receptors are conserved in mammals and birds. So, while receptors for polymorphic class I molecules are divergent, those for nonpolymorphic or stress-induced class I molecules seem relatively conserved. NK cell recognition of virus-infected cells engages the activating KIR and Ly49 receptors and NKG2D. Thus, viruses are hypothesized to supply the evolutionary pressure on diversification of NKRs.[10] Generally speaking, inhibitory receptors are older and more conserved, whereas activating receptors evolve more rapidly and can be derived from inhibitory receptors via mutations that result in loss of the ITIM.[10]

Comparative Studies of Natural Killer Function

NK cell function has been detected in species of all vertebrate classes, but it has been difficult to identify their receptors (note, this was also the case in mammals in the 1990s).[11] NKRs evolve at a very fast rate, which has been clear for many years. This rapid evolutionary rate has made it difficult to detect NKRs in ectotherms. Large gene families of NKR-like IgSF proteins have been detected in amphibians[94] and fish,[95] but for the most part their role in NK recognition per se has not been determined, and many certainly have other functions. Conversely, clearly there are lymphocytes in fish and amphibians from thymectomized animals, or not bearing TCRs of any type, that are capable of directing cytolysis.[96]

As mentioned, the MHC can harbor NKRs of both the IgSF and C-type lectin families, depending on the taxon studied. Birds have two C-type lectin NKRs related (even orthologous) to mammalian NKRs in the MHC.[70] NKp30, a specialized IgSF member which maps to the MHC in humans, is the most conserved NKR in gnathostomes and is also found in cartilaginous fish.[97] Amphibians have a direct homolog of NKp30 in the class III region of the MHC, called XMIV, while NKp30 has translocated outside the MHC and expanded on the telomere of another chromosome.[98] The ligand for NKp30, B7H6, is also found in the shark MHC; interestingly, in species in which NKp30 has been deleted, B7H6 also has been deleted; conversely, when NKp30 genes are expanded, B7H6 genes are expanded as well.[99]

Xenopus NK cells were detected by in vitro ^{51}Cr-release assays.[100] Splenocytes from thymectomized frogs spontaneously lyse allogeneic thymus tumor cell lines that lack MHC antigen expression.[96] This activity is increased after the injection of tumor cells or after treating the splenocytes in vitro with mitogens, suggesting cytokine activation of the killers. Splenocytes isolated with an anti-NK monoclonal antibody (mAb) revealed large lymphoid cells with distinct pseudopodia. Immunohistology indicated that each anti-NK mAb routinely labeled cells within the gut epithelium but not in the spleen.[101]

NK cells have also been described in a number of teleost fish with the most in-depth studies in catfish, in which there are clonal likes of cytotoxic cells, some that clearly lack TCR expression.[102] A subset of the fish NK cell bears a high-affinity FcR that can be utilized for antibody-dependent cellular cytotoxicity.

Phylogeny of Immunoglobulin Superfamily Natural Killer Receptors

There are other bony fish–specific IgSF NKR families. One family, the novel immune-type receptors (NITRs), has one or two Ig domains with a charged residue in the TM and could therefore be associated to an ITAM DAP12 equivalent and indeed was shown to interact with mammalian

DAP12.[103] Like NKp30, the NITR N-terminal V domain is also of the VJ type. NITRs can be expressed by cells of the hematopoietic lineage. NITRs have been found in all bony fish, with rapid contraction and expansion of the gene family, with most proteins predicted to have inhibitory ITIMs in their cytoplasmic tails. One catfish NITR was shown to interact with MHC class II molecules.[104] Thus, the sequence analysis, signaling properties, involvement in cytotoxicity, and recognition of MHC molecules identify the NITRs as candidates for teleost NKRs.

Genes encoding the classical FcRs on the long arm of human chromosome 1 (1q21-23) are linked to other FcR-like genes. A large multigene family, which includes genes encoding the FcγR and the NK cell Ig-like receptors, is located in the LRC (human chromosome 19q13). This region was originally associated with the MHC in a region called MHC-trans.[105] Their phylogenetic conservation in birds, amphibians, and bony fish suggests a biologic importance even though the size of the families, their expression pattern, and the specific nature of the receptors vary greatly among species.[95]

Innate-Like Lymphocytes

An exciting current area of immunology research is the study of innate-like lymphocytes, such as NKT cells, mucosal-associated invariant T (MAIT) cells, B1 cells, marginal zone B cells, and ILCs, including NK cells. It is fitting that studies of their evolution should be a new priority, considering that it is tempting to propose that ILCs (especially) may have predated lymphocytes bearing antigen receptors.[92]

Most immunologists believed that innate T cells (NKT cells and MAIT cells) were late evolutionary additions in mammals, similarly to germinal centers (GC). This notion was proven wrong when Robert and colleagues detected NKT-like cells in the amphibian *Xenopus* that were specific for one of the nonclassical class I molecules in the XNC.[106] Like the mammalian NKT cells, the frog NKT cells bear an invariant TCR α chain and have an effector cell phenotype. Such cells have been shown to be effective against tumors, virus, and mycobacteria.[107] Larval *Xenopus* express low levels of classical MHC class I molecules[108] and preliminary evidence suggests that the larval TCR repertoire is predominantly made up of clones bearing these invariant TCR α chains.[109] This work in amphibians, and the identification of CD1 molecules in reptiles and birds,[110-112] has changed the paradigm for NKT emergence. Not only did NKT cells arise early in evolution, but one could propose that animals with relatively few lymphocytes predominantly use T cells that can fire rapidly, and large TCR repertoires that favor clonal selection are not used in all cases. Robert's studies suggest that NKT cells will be found throughout the vertebrate phylum, since as described above lineages of nonclassical class I genes are present in all vertebrate classes.[113]

ILCs have not been examined in any detail yet in species other than humans and rodents. Lamprey lymphocytes have been detected that bear no antigen receptors and these are candidates for NK cells and ILCs, especially ILC2 via transcriptomics.[92]

Other Immunoglobulin Superfamily Families to Explore Further

In the sea urchin genome, many IgSF await study and will certainly contribute to a better understanding of the evolution and origin of Ig/TCR. Families of LRR-IgSF in amphioxus (besides VCBP) could perhaps represent interesting intermediaries in the genesis of either VLR in agnathans or Ig/TCR in gnathostomes (Fig. 4.1). In hagfish, the discovery of leukocyte expressed receptors agnathan-paired receptors (APARs) revealed what might have been a precursor of Ig or TCR.[114,115] APARs resemble Ag receptors and are expressed in leukocytes and predicted to encode membrane glycoproteins with organizations characteristic of paired Ig-like receptors. Based on their transmembrane regions, APAR-A molecules may associate with an adaptor molecule with an ITAM and function as activating receptors and APAR-B molecules with an ITIM may function as inhibitory receptors. APAR V domains have a J region and are more closely related to those of TCR/B cell receptor (BCR) than any other V-type domain identified to date outside of jawed vertebrates.

In recent years, there have been a large number of genes in gnathostomes with the so-called VJ domain besides the teleost NITR and NKp30 described above.[116] These molecules are examined in more detail in the Conclusions.

VERTEBRATE ADAPTIVE IMMUNITY

For many years it was believed that only jawed vertebrates had an adaptive immune system. From our discussion of lower deuterostome immune responses, clearly mechanisms exist to generate levels of immune diversity, one of the hallmarks of an adaptive response. Many agree that the boundary between innate and adaptive immunity is artificial (eg, Transformer in sea urchins), and it may not be a useful dichotomy when studying immune responses in diverse organisms.[117] Nevertheless, all vertebrates have clear adaptive immunity—unless it was lost—based on somatic generation of the antigen receptors, although jawless fish and gnathostomes use antigen receptors in different families.

As an introduction to this section,[118] while the jawless fish have cells like T cells and B cells as well as a thymus equivalent, their MHC is elusive; like the agnathan antigen receptors, the VLRs, an "MHC" may have arisen in this group via convergent evolution. Mucosal adaptive immunity is present and unique for each group.[14] Cartilaginous fish, bony fish, amphibians, and birds not only display features that distinguish the entire vertebrate phylum but also (in some cases) have distinctive features that are found within taxon. Prominent in bony fish, and consistent with the general rapid evolution of this vertebrate class, the immune system has unique features such as the loss of an MHC class II system and certain Ig isotypes. Cartilaginous fish have a distinctive Ig organization, the cluster-type, that has permitted the emergence of antigen receptor genes with novel functions. Ancient lineages of class I, immunoproteasome, and TAP are found in the MHC of all nonmammalian vertebrates so far studied, but in mammals this prototype has been superseded by a system-less rigid for the most part in MHC class I peptide specificity.[119] A large leap forward in evolution is clear in amphibians, which show canonical antibody class-switching and an IgG class (IgY) involved in typical memory immune responses. The last major advance in vertebrate evolution was the advent of lymph nodes and GC in endotherms; the ectothermic vertebrates allow for the study of immunity that preceded the emergence of follicular dendritic cells (FDCs), a major player in mammalian affinity maturation.[118] This section will delve into these characteristics of immunity, focusing on basic questions of interest to all immunologists.

Hematopoiesis and Transcription Factors in the Vertebrates

Transcription factors of the family PAX 2/5/8; GATA 1, 2, 3; ets/erg; and runt domain–containing factors have been cloned in several invertebrates. One plausible model to explain the genesis of true lymphocytes in vertebrates is that closely related members of transcription factor families are the result of a relatively late divergence in lineage pathways followed by specialization of duplicated genes.[120] These duplications occurred during the history of chordates (see MHC and "Origins" section). Within deuterostomes, the generation of true GATA 2 and 3 probably occurred after echinoderms diverged from the chordate branch, and the GATA, ets, early B cell factor (BCF), and Pax5-dependent pathways of T cell/B cell differentiation are thus specific to vertebrates.[121] Expression has been detected in the gut of sea urchins, which may be related to the fundamental nature of "gut-associated lymphoid tissue (GALT)" as a lymphoid cell–producing organ.[122]

In vertebrates, the generation of T-, B-, and NK lymphocyte lineages from pluripotent hematopoietic stem cells depends on the early and tissue-specific expression of *Ikaros* and related loci, which by means of alternative splicing produces a variety of zinc-finger DNA-binding transcription factors. The orthologs of *Ikaros*, *Aiolos*, *Helios*, and *Eos* have been identified in the skate, where two of the four *Ikaros* family members are expressed in their specialized hematopoietic tissues (epigonal and Leydig organs; see subsequent discussion) like in mammals. In lower deuterostomes, single genes that seem to be related to the ancestor of the Ikaros and Ets family of transcription factors exist, further suggesting that the division of labor between the family members in the jawed vertebrates was a result of en bloc duplications.[121] A recent work describes all of the known transcription factors involved in the different hematopoietic lineages and their expression in different vertebrates.[123]

The bone marrow as a site of hematopoietic stem cell differentiation first emerged in the amphibians[124] (Fig. 4.11). Cartilaginous fish use either Leydig gland, associated with the esophagus, or the epigonal organ, associated with the gonad (some species have both). In teleost fish, the head kidney serves as the bone marrow–equivalent. An old theory suggested that bone marrow became the site of HSC differentiation to protect these precious cells from the effects of ultraviolet (UV) light.[124] This idea was put to the test in zebrafish and was shown to hold water. In aquatic organisms, a "melanocyte shield" protects the HSC from UV light, as demonstrated in cellular ablation studies.[125]

Antigen Receptors (AgR): VLR, Ig, TCR

Gnathostome Igs and TCRs are members of the IgSF and are derived from an unknown precursor, although models have been proposed, discussed below.[116] By contrast, the jawless fish VLRs belong to the ancient LRR family, possibly originating from (most related to) a cell surface receptor expressed on platelets.[126] Unlimited diversity can be generated in both types of antigen receptor during lymphocyte development, but the gene rearrangement mechanisms used to generate diversity are entirely different.

Variable Lymphocyte Receptors

In the 1960s, jawless vertebrates were reported to mount humoral responses to foreign antigens.[127] However, for anti-streptococcal antigens, the hagfish "antibodies" were actually the C′ component C3. Lamprey and hagfish were long known, however, to possess cells resembling lymphocytes and plasma cells with expression of lymphocyte- or at least leukocyte-specific genes,[128] but the quest for RAG or Ig/TCR/MHC genes was a complete failure. Reports of specific memory in allograft rejection and other specific humoral responses were difficult to reconcile with absence of the RAG-based rearranging machinery and the possibility of generating specific lymphocyte clones. Thus, despite the earlier functional work, there was no adaptive immune system in agnathans. This assumption was debunked by Pancer and Cooper, who discovered many LRR-containing proteins with varying numbers of internal repeats from immunized lamprey.[13] These LRR-containing receptors were clonally expressed and generated by rearrangement during lymphocyte ontogeny and were christened VLRs. The first gene studied was VLRB, which is expressed in a subset of lamprey lymphocytes. The VLRB protein is GPI-anchored on the surface of naive lymphocytes in lampreys and then secreted (through an unknown mechanism) as a pentamer of dimers following antigenic stimulation (Figs. 4.8 and 4.9). At first it was thought that stimulation of the lamprey lymphocytes was like a T-independent immune response in mammals, that is, crosslinking of antigen receptors on lymphocytes, likely in combination with another signal through PRRs, might activate B cells to induce VLRB secretion.[129] However, it was later found that a second rearranging VLR locus, VLRA, was expressed by another subset of lymphocytes, with a transcriptome like gnathostome T cells.[130] Further analysis revealed a third antigen receptor VLRC[131] and thus agnathan T cells may be split into two subsets like gnathostome $\gamma\delta$ and $\alpha\beta$ cells (Fig. 4.9) and more below.

VLR homologs were found in two additional lamprey species and in hagfish, the only other order of living agnathans (Fig. 4.2). As in the sea lamprey, the incomplete hagfish germline VLR generate somatically highly diverse repertoires. Interestingly, the *amphioxus* genome harbors many intronless VLR-like sequences that could represent an alternative germline VLR diversity[132] akin to the echinoderm gene families (TLR, SRCR) described previously; one could speculate that they are related to ancestral VLR before invasion of its analogous "RAG transposon" (see below).

Immunoglobulins

A typical Ig molecule is composed of four polypeptide chains (two heavy [H] and two light [L]) joined into a macromolecular complex via several disulfide bonds (Fig. 4.6). Each chain is composed of a linear combination of IgSF domains, and almost all molecules can be expressed in secreted or transmembrane forms (Chapter 21).

Immunoglobulin heavy (H) chain isotypes: Like all other building blocks of the adaptive immune system, Ig is present in all jawed vertebrates. Consistent with studies of most molecules of the immune system, the sequences of IgH chain C region genes are not well conserved in evolution and insertions and deletions in loop segments occur more often in C than in V domains. Consequently, relationships among non-μ isotypes (and even μ isotypes among divergent taxa) are difficult to establish. Despite these obstacles, a working evolutionary tree now has been developed among all isotypes.[133]

IgM antibodies: Two antibody isotypes are found in (almost) all gnathostomes. IgM has been known for decades to be the primordial antibody class, although there are unique features of IgM in some vertebrate classes. In mammals, amphibians, and retiles, IgM is secreted as a pentamer (or hexamer) associated with the J chain. IgM in cartilaginous fish is found in two forms: a multimeric form, as is present in all other vertebrates, and a monomeric form, which is most common in shark antigen-specific responses and is likely produced in a T-dependent manner.[134,135] Bony fish IgM is a tetramer, and the J chain gene has been lost from IgM in this group (like several other bedrock immune genes), as it is present in the older cartilaginous fish.[136] The extent of disulfide bonding in secreted teleost IgM is modified over the course of an antigen-specific response in a type of affinity maturation that is so far unique to this antibody class.[137] In addition, the transmembrane form of bony fish IgM is alternatively spliced so that it only has three constant domains (unlike the usual four).[138] In summary, IgM is the major serum antibody found in bony fish and produced by these animals in responses to conventional antigens; cartilaginous fish use IgM as an innate antibody class as well as in adaptive responses, but also use other antibody isotypes in adaptive immunity; and amphibian and bird IgM functions much like its mammalian ortholog.

Some modifications apparently related to the particular environment are found in IgM of the Antarctic fish *Trematomus bernacchii*. There are two remarkable insertions, one at the V_H-C_H1 boundary and another at the C_H2-C_H3 boundary; the latter insertion results in a very long CH2-CH3 hinge region. Rates of replacement substitutions were high in the modified regions, suggesting strong selection for these modifications. These unusual features (also unique glycosylation sites) may permit flexibility of this IgM at very low temperatures.[139]

Immunoglobulin M_{1gj}: Nurse shark, *Ginglymostoma cirratum*, expresses an IgM subclass in neonates[140] (Fig. 4.6). The V_H gene underwent V-D-J rearrangement in germ cells ("germline-joined" or "gj," see below). Expression of H_{1gj} is detected in primary and secondary lymphoid tissues early in life, but in adults only in the primary lymphoid tissue, the epigonal organ (Fig. 4.11). H_{1gj} associates covalently with L chains and is most similar in sequence to IgM H chains, but like mammalian IgG, it has three rather than the typical four IgM C domains, with deletion of the ancestral IgM second domain defining both IgG and IgM_{1gj}. Because sharks are in the oldest vertebrate class known to have antibodies, specialized antibodies expressed early in ontogeny in sharks and other vertebrates were likely present at the inception of the adaptive immune system. This isotype may interact with a common determinant on pathogens and/or a self-waste product.

IgD and IgW antibodies: IgD, long believed to be a recent addition to the vertebrates, is actually very old, also going back to gnathostome origins as an isotype found in sharks called IgW[141,142] (Fig. 4.6). An Ig-like IgD was found first in bony fish, which was very surprising considering to that point IgD had been found only in some mammals and not in birds.[143] Subsequently, IgD was uncovered in the genomic database of the amphibian *Xenopus* and shown to be expressed, like in mammals, via alternative splicing of IgM and IgD mRNA in naïve B cells. The cartilaginous fish homolog of IgD, IgW, was also found in lungfish, and in coelacanths it seems to be the major isotype, this species having lost IgM.[144] Note that the connection of IgD and IgW is primarily through the CH1 domain, the region that interacts with L chains; recent work suggests that there is also an affiliation of IgW with bony fish IgT in other C domains.[145] Data in bony fish and especially humans suggest that IgD is involved in mucosal responses, binding to a receptor (CD44/galectin 9 in humans) on basophils.[146] Compared to IgM, IgD is *extremely* plastic in both structure and function; IgD is highly variable for the numbers of constant (C) domains (and even presence of IgD itself and potential connections to other isotypes), and likely in functions, for the transmembrane (largely unknown) and secretory (arming of innate cells) forms.[142]

IgG and IgY antibodies: Mammalian IgG emerged as an isotype called IgY in amphibians that arose concurrently with the first canonical class switch recombination (CSR) (Fig. 4.6). Like IgM, the IgY H chain has four C domains. Mammalian IgG and IgE are both related to an IgY common ancestor, with the IgE H chain maintaining four C domains and the IgG H chain losing CH2.[147] In *Xenopus*, the switch to IgY, and hence its expression, is entirely T cell-dependent (lacking in thymectomized animals) and typical B cell memory generation as defined in mammals appeared in this group. The cytoplasmic tail of both transmembrane IgY and IgG has a novel signaling motif that contributes to the proliferative burst of B cells in a memory response.[148] Besides its T-dependence, very little is known about the conditions for CSR (see below) outside of mammals. An isotype related to IgY, IgF, but lacking the two C-terminal domains was found in the *Xenopus* genome[149]; its function is unknown, and turtle IgY has some unique properties such as asymmetry.[150] IgY was first uncovered in birds,[151] where it got its name, and is present in amphibians, reptiles, and birds.

Immunoglobulin new antigen receptor (IgNAR): A dimer found in the serum of nurse sharks and so far restricted to elasmobranchs, IgNAR, is composed of two H chains each containing a V domain generated by rearrangement and five constant C1 domains.[152] IgNAR was originally found in sera, but TM forms exist as cDNA, and cell surface staining is detected with specific mAbs. The single V resembles a fraction of camel/llama (camelid) IgG that binds to antigen in a monovalent fashion with a single V region, clearly derived by convergent evolution.[15] In phylogenetic trees, NAR V domains cluster with TCR and L chain V domains rather with V_H. A molecule with similar characteristics has also been reported in ratfish and elephant shark, although it was independently derived from an ancestral IgM,[62,153] like the camelid molecule emerged from IgG. IgNAR V region genes accumulate a high frequency of somatic mutations (see below). Another single-domain Ig was recently discovered in a chimera, with unusual properties such as a highly positively charged binding site.[154]

The crystal structure of a type I IgNAR V regions showed that, in contrast to typical V regions, they lacked CDR2 and had a connection between the two IgSF sheets much like an IgSF C1 domain.[155] The domain wraps around its antigen with the CDR3 penetrating the active site. The structure of a type II V region has a disulfide bond between CDR1 and CDR3 that forces the most

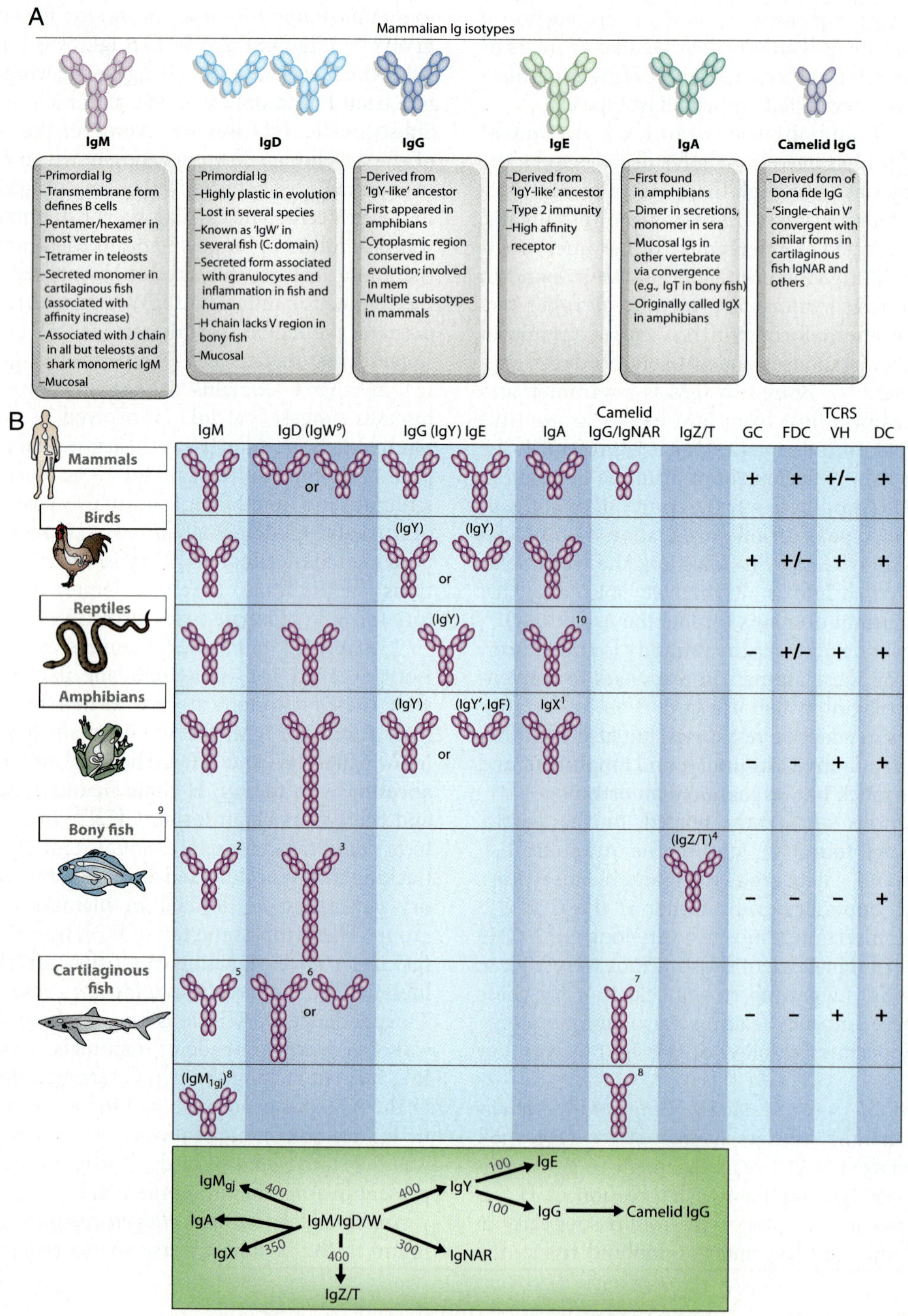

diverse regions of the molecule to form raised loop, like camelid V domains.[156] Thus, differential placement of disulfide bonds forces major changes in the orientations of CDR1 and CDR3 and provides two major conformations for antigen binding. Several other crystal structures have been obtained, with a plasticity in how the IgNAR V can interact with antigens.[157]

Other IgH isotypes: There are other "dead end" heavy chain isotypes that arose in various vertebrate taxa, especially in lungfish,[158] which have been recently reviewed in great detail, whose functions are unexplored.[133] Mucosal isotypes, like the classical IgA in mammals, including amphibian IgX/A and teleost IgT, are described below.

Immunoglobulin Light Chains

Unlike what was thought for many years, the split of L chains into κ and λ goes back to gnathostome origins.[159,160] In most ectotherms, there is a third primordial L chain, σ, which was lost in endotherms.[161] There are L chain subgroups in some taxa, also "dead end" like the H chains, especially in fish.[59] In all of the cartilaginous fish, the λ genes are "germline-joined," which allowed for a definitive analysis of SHM of all CDRs in sharks.[162] Almost all L chains in bony fish can be categorized as either κ or σ, despite the substantial number of gene expansions and contractions in this group. As expected, λ exists in some bony fish species,[163] consistent with the work in sharks. Genes in different species can be found either in the cluster type organization, translocon, or some intermediate type, and this organization is especially well suited to allow for receptor editing,[164,165] either to generate the repertoire or as a consequence of binding to self-ligand.[166]

In summary, all vertebrate groups except birds have two to four L chain isotypes, all of which can be categorized as κ, λ, or σ (or related forms).[160] However, we still only speculate on the significance of possessing multiple isotypes as there is scant evidence that L chains have any effector functions.[160,167,168] In all vertebrates, L chain preferences for certain H chain isotypes have been noted, one potential purpose for the multiple L chains. The different L chain isotypes might permit very different conformations of CDRs when paired with similar H chains. There are differential CDR sizes for σ as compared to κ/λ, which suggest that the conformation model is plausible. Finally, multiple L chains allow for receptor editing in cells with a functioning H chain, but when the L chain association provides a self-reactive receptor or the first L chain gene rearrangement is nonproductive.

V_H Evolution

Diversity of the Ig repertoire depends on germline V segment variety and upon the further diversification by rearrangement (CDR3 only) and SHM (all CDR). Early in life, the repertoire depends chiefly on inheritable genes as one finds little N-region diversity and SHM (eg, 169). Perhaps environmental antigens played a role in shaping the germline repertoire and selected some V_H/V_L germline sequences used by neonates. V_H families arose in a bony fish lineage and have been conserved for hundreds of millions of years. Conserved regions defining families are found on solvent-exposed faces of the V_H, at some distance from the antibody-combining site (Chapter 21). Phylogenetic analyses show clustering of V_H into groups A, B, C, D, and E. All cartilaginous fish V_H belong to group E; and bony fish V_H genes to cluster into all groups (D unique only to them). By contrast, group C includes bony fish sequences and V_H from all other tetrapod classes. Another phylogenetic analysis classifies mammalian V_H genes in three "clans" (I, II, and III), coexisting in the genome for >400 million years.[170] Only in cartilaginous fish have V_H gene families been subjected to concerted evolution that homogenized members (except IgM_{1gj}).[171] It has been debated whether IgV genes could be under positive selection or not because these genes hypermutate somatically. However, several features (eg, codon bias) and discovery of high replacement/silent (R/S) ratios in *germline* CDR codons argue for positive selection.[172]

In summary, much of the V_H germline repertoire has been conserved over extremely long periods of vertebrate evolution. Birds and some mammalian species that rely on GALT to generate Ig diversity (below) are exceptions with a reduced germline repertoire, but gene conversion and SHM compensate for this situation in formation of the primary repertoire (see below). Even in the cartilaginous fish where there is a single V_H family, there is nonetheless heterogeneity in CDR1 and CDR2 sequences (and SHM) that boost diversity in the expressed repertoire.

The J Chain

The joining (J) chain is a small polypeptide expressed by plasma cells that regulates polymer formation of IgA and IgM. J chain incorporation into dimeric IgA and pentameric IgM endows IgM/A with the ability to be transported across epithelial cell barriers.[136] J chain facilitates creation of the binding site for pIgR (secretory component in Ig polymers), not only by regulating the polymeric structure but also by

FIG. 4.6. Immunoglobulin (Ig) isotypes in the jawed vertebrates. A: Mammalian isotypes and their relationships to Igs in vertebrates from other classes are detailed in the boxes. Each oval represents an Ig superfamily V or C1 domain. IgD is shown in two forms, mouse (*left*) and human (*right*). **B:** Major Ig isotypes in all vertebrates. The *bottom panel* displays the approximate divergence times of all isotypes. IgM/D/W was found at the inception of adaptive immunity. 1, IgX is in the IgA column because it is preferentially expressed in the intestine, and IgA seems to have been derived from an IgX ancestor; IgX itself seems to have been derived from both IgM and IgY ancestors; 2, secreted IgM in teleost fish is a tetramer (and the J chain has been lost in that taxon), and the transmembrane (TM) form only has three C domains; 3, most teleost fish IgD H chains incorporate the μC1 domain via alternative splicing in the TM form, and the secretory form does not have a V region and does not associate with L chains; 4, the bony fish isotype, IgZ/T, may not be found in all fish species and is the mucosal Ig of teleosts; 5, the secreted form of cartilaginous fish IgM is present as a pentamer and monomer at approximately equal levels; 6, the TM form of IgW has four C domains, like IgM; 7, a major TM form of IgNAR has three C domains, and IgNAR is related to camelid IgG and other cartilaginous fish single-domain Igs by convergent evolution; 8, no TM form has been found (to date) for IgM_{1gj}; 9, there is at least one other single domain class in cartilaginous fish, so far only found in holocephalans; 10, the relationship of IgW to IgD is mainly in the C1 domains and other domains have affiliations with teleost IgT; 11, there are other "dead end" IgH classes that are not shown, especially in the lungfish. The *bottom panel* displays the approximate divergence times of all isotypes. IgM/D/W was found at the inception of adaptive immunity. On the right are other features of humoral immunity: germinal centers (GC), follicular dendritic cells (FDCs), the usage of IgV with TCRδ chains (TCRδVH), and dendritic cells (DC).

interacting directly with the receptor.[173,174] Therefore, both the J chain and the pIgR/secretory component are key proteins in secretory immunity. J chain is found in all jawed vertebrates except teleost fish, which have lost it.[62] *Xenopus* IgX is a pentamer, but the H chain has a stop codon before the Cys of C_{H4} domain and thus cannot make a covalent attachment to J chain.[175] Elasmobranch J chain is highly similar to the N-terminal half of other vertebrate J chains but is divergent or absent in other regions.[176] This suggests that shark J chain function may be solely for IgM polymerization, and the transport function arose later[177]; consistent with this idea, *Xenopus*, but not shark, J chain is capable if interacting with human IgA and pIgR.[178] J chain is expressed only in a subpopulation of shark plasma cells, those expressing a relatively lower amount of IgM than the BLIMP1-hi plasma cells and presumably secreting the multimeric form.[179] The loss of J chain in bony fish does not preclude the interaction of IgT with "pIgR" and transport across mucosal and skin epithelia, implying an intense pressure to maintain a mucosal-specific Ig.[180]

T Cell Receptors

Ig, TCRs, and MHC class I and class II are all composed of IgSF domains. The membrane-proximal domains of each Ig/TCR/MHC chain are IgSF C1-set domains, while the N-terminal domains of Ig and TCR proteins are V-set domains encoded by genes generated via rearrangement of two or three gene segments during ontogeny (Chapter 23). In all vertebrates studied to date, TCRs are membrane-bound and never secreted, while almost all Ig proteins have TM and secreted forms. Generally, αβTCRs are well conserved across evolution.[181] In cases where it is technically feasible, early thymectomy has shown that effective adaptive immunity is dependent on T cells, including relatively high affinity, class-switched antibodies, and cellular immunity to allografts and viral infection.[182] TCR diversity is high in all animals studied, except in cells resembling natural killer T (NKT) cells. NGS in certain bony fish has shown a large diversity of αβ TCRs involved in their responses to antigens, such as viruses.[183]

α/β Constant domains: Genes encoding the two types of TCR, α/β (accounting for all known MHC-restricted functions) and γ/δ (to the best of our knowledge recognizes antigens in an Ig-like manner and can play immunoregulatory or homeostatic roles during certain infections), existed in the earliest jawed vertebrates. The α/β TCR is considered rather "boring" evolutionarily, as all gnathostomes from the basal cartilaginous fish appear to have this MHC-restricted form, but new work in sharks has changed this view somewhat (below).

While TCR genes have been cloned from representatives of all vertebrate classes, few biochemical data are available, except in birds where α/β and γ/δ TCR have been identified with mAbs,[184] and amphibians, in which the *Xenopus* α/β TCR was coimmunoprecipitated with cross-reactive antibodies raised against human CD3ε chains[185] (note: this same antibody identifies CD3 epsilon from all gnathostomes save sharks!). Alpha chains from diverse vertebrates are poorly conserved and the structure of the Cα IgSF domain itself is unusual: only strands A, B, C, E, and F are identified, although strands E and F are shorter than those of mammals and strand D is absent; this modification has an important role in TCR dimerization and subsequent signaling.[186] The lack of conservation in this extracellular domain as well as deletions found in bird and teleost fish TCR (especially the connecting peptide) suggest that the coreceptor may be structurally distinct from mammalian CD3 complex components.

Pre-Tα, which associates with TCR β chains during thymocyte development, has been identified only in endotherms (and perhaps reptiles).[187] The pre-Tα protein has no V domain, and its interaction with the TCRβ chain is unique. Unlike other TCR chains, pre-Tα has a long cytoplasmic tail, which seems to be important for T cell differentiation. The pre-Tα gene is linked to the MHC in mammals, and more phylogenetic studies should be performed to determine whether this linkage group is ancient (see below).

The TM region and cytoplasmic tail of Cα are the most conserved parts of the molecule. Cα and Cβ TM segments in all species have the so-called conserved antigen receptor transmembrane motif (CART) motif,[188] in which conserved amino acids form an interacting surface with CD3. Besides CART, the opposite TM face with conserved residues Ile-Lys-Leu interacts with other CD3 components. There are also remarkable differences: the solvent-exposed segment 98 to 120 in mammals, which is important in negative selection, is absent in all other vertebrates.

The number of Cβ genes varies in varied species. Like Cα, Cβ sequences are not well conserved in evolution. Indeed, damselfish Cβ was shown to be encoded by two polymorphic genes, and this feature extends to other teleosts.[189] As the polymorphic sites likely interact with associated CD3 signaling molecules, signals might be transduced to T cells in different ways depending on the expressed TCR allele.

α/β T cell receptor variable domains: Because T cell recognition is MHC-restricted, TCR V regions have been evolutionarily selected for different properties compared to Ig; indeed, TCR V regions are much less similar to each other than are Ig V regions. Furthermore, TCR Vs, unlike V_H, have conserved CDR3 lengths, likely a restriction on size for recognition of MHC-peptide complexes.[190] Alpha loci in all vertebrates examined have many J segments, and consistent with the mammalian paradigm, the absence of D and the large numbers of J segments favors potential receptor editing during thymic positive selection.[191] Some Vβ gene families are another evolutionarily conserved feature. There are 12 families in nurse sharks and 19 in *Xenopus*. There is evidence that the TCR has coevolved with MHC, displayed by the canonical types of interactions seen in crystal structures. Vβ sequences from many different vertebrates share residues in CDR2 that interact with class II and CD1 MHC molecules (Y-46 and Y-48, especially).[192]

γδTCRs: During the TCR discovery years, a new rearranging antigen receptor was discovered for which there was no known function that was later named TCRγ.[193] After the discovery of its partner TCRδ a few years later, γδ T cells became the least understood features of adaptive immunity, and only now are we beginning to grasp their recognition of antigen at the molecular level, at least for γδ T cells involved in innate immunity (Chapter 35). In all ectotherms except bony fish (and placental mammals), a large percentage of TCRδ genes bear IgVH (or a relative) as the recognition element.[35] The IgH and TCRδ loci are linked in several vertebrates, suggesting

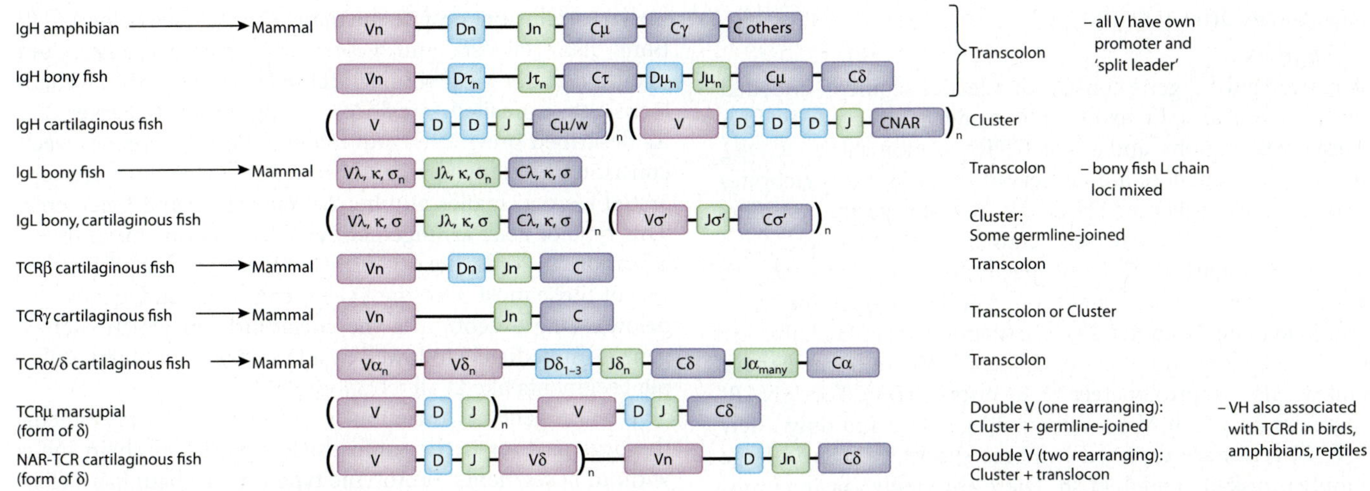

FIG. 4.7. Organization of immunoglobulin (Ig) heavy and light (L) chain and T cell receptor (TCR) genes in all jawed vertebrates. All cold-blooded gnathostomes except bony fish have three L chain isotypes, κ, λ, and σ; mammals have κ and λ; and birds have only λ. RSS are found at the 3′ end of V segments, 5′ end of J segments, and on both sides of D segments. Marsupial μ TCR is related to Ig in its V regions and TCRδ in its C region (and is not related to the μ of IgM).

that the loci were derived from a *cis* duplication, and it has been long recognized that TCRVδ and IgVH share attributes, especially a large size range of CDR3 sequences.[190,194] The relationship between IgH and TCRδ is both ancient, likely via the *cis* duplication, and ongoing, via the transfer of VH into the TCRδ locus.[195] Additionally, transrearrangements of shark IgM and IgW V segments to TCRδ DJ segments from nearby Ig clusters are found at high levels.[181] These data suggest that many γδ T cells act in an adaptive fashion, which has been borne out in several studies in mammals.[196] Future work should be done to analyze γ/δ adaptive responses in ectotherms, which is at its inception.[197] This is especially of interest in cartilaginous fish, where the TCRγ V genes have been definitively shown to hypermutate, similar to what has been detected in shark Ig genes.[198]

While analyzing the TCRVδ repertoire in nurse sharks, a new form of this chain, which encodes three domains, V-V-C, was detected.[199] The C is encoded by the single-copy Cδ gene, and the membrane-proximal V is encoded by a Vδ gene that rearranges to the DJδ elements. The membrane distal V domain is encoded by a gene in the NAR family, found in a rearranging VDJ cluster typical of all cartilaginous fish Ig clusters (Fig. 4.7). The NAR-TCR V genes, unlike IgNAR V genes which have three D segments, only have a single D region in each cluster. The particular Vδ loci linked to each NAR-TCR gene—called NAR-TCR–supporting Vδ—encode a cysteine in CDR1 that likely makes a disulfide bridge with the NAR-TCR V domain. The J segment of the rearranged NAR-TCR V gene splices at the RNA level directly to the supporting Vδ segment, which has lost its leader exon. This organization likely arose from an IgNAR V cluster that translocated to the TCRδ locus upstream of a Vδ gene segment, although the opposite is possible since holocephalans (older than elasmobranchs) have NAR-TCR but not IgNAR. After modifications of the supporting Vδ genes, this entire V-V gene set duplicated and diverged several times in different shark species. About 25% of the expressed nurse shark TCR δ repertoire is composed of this TCR, and likely the typical γ/δ TCR acts as a scaffold upon which sits the single-chain NAR V, like the marsupial molecule described below.

Our interpretation is that, consistent with the proposal that γ/δ TCRs interact with free antigen, NAR V provides a binding site that can interact with antigen in a different way than conventional heterodimeric Vs. Thus, this is the first case in which a V region family has been shown to be associated with a BCR and TCR; in the case of the BCR, the function resides within the Fc portion of IgNAR and for the TCR the function (cytokine secretion, killing) lies within the T cell itself. Interestingly, a second TCRδ chain locus has also been described in marsupials and monotremes with properties like NAR-TCR. In this case, there are also two V domains, but in marsupials one (proximal to the membrane) is germline-joined, and only the membrane-distal domain undergoes rearrangement.[200] In monotremes, both V regions undergo rearrangement, like for the NAR-TCR. These TCRδ loci in monotremes and marsupials are preferentially expressed early in development and have recently been shown in structural studies to have the predicted domain interactions described above for NARTCR.[201]

In ruminants and chickens (so-called GALT species), the γ/δ repertoires are quite diverse, and there seems to be ligand-mediated selection of γ/δ cells during ontogeny.[184] Scavenger receptors serve as coreceptors with γ/δ TCR that permit responses to different microbes.[202,203] The degree of γ/δ expression is correlated with the evolution of the TCR V families in warm-blooded vertebrates. Indeed, mammals can be classified into "γ/δ low" (humans and mice, in which γ/δ T cells constitute limited portion of the T cell population) or "γ/δ high" (chicken, sheep, cattle, and rabbits, in which such γ/δ cells comprise up to 60% of T cells).

AgR Gene Organization

V_H Regions

A rearranged V_H gene consists of a leader segment, encoded by a canonical split exon (split leader), followed by four framework regions and three CDRs. Canonical V_H CDR1 nucleotide sequences are conserved in all jawed vertebrates and serve as targets for SHM.[172,204] A major germline difference is the lack of conserved octamers and TATA box in the 5′ region of shark Vs.[205] In all species, functional V genes are assembled by rearrangement and joining of germline V, D, and J elements (Chapter 22). Cartilaginous fish H chains are encoded by large numbers of clusters (>100 in horned shark and dogfish; approximately 15 in nurse shark). For IgNAR, there are only four V genes/haploid genome and only a few IgW V genes are detected in nurse sharks,[206,207] but a large number in skates and dogfish. There are widely varying numbers of V genes in different species; importantly, V_H complexity does not seem to limit diversity of the antibody repertoire in any ectothermic vertebrate studied to date.

Chondrichthyan Germline-Joined (gj) Genes

In all vertebrates, functional Ig genes are assembled by rearranging DNA segments scattered on the chromosome. However, in cartilaginous fish, some V genes are the products of V(D) J rearrangement in eggs/sperm, as mentioned. Type I L chain (σ) genes are all gj in skates but split in horned sharks, and the piecemeal gj (eg, VD, VDD, VDDJ) found in many horned shark H chain gene clusters and in nurse shark L chain type I (σ) clusters strongly suggest that the gj is a derived feature.[205,208] Definitive proof came from a study of a gj nurse shark type III (κ) L chain gene, shown by phylogenetic analysis to have been joined within the last 10 million years[208]; this was followed by the identification of the nurse shark gj IgM_{1gj} described earlier.[140] When there is a *mixture* of gj and conventional genes, split genes are expressed in adults, while gj genes are expressed at significant levels only early in ontogeny. When *all* genes in a particular family are gj (eg, skate type I L chain genes and type II [λ] L chains in all elasmobranchs), they continue to be expressed into adult life.[162,209] It is possible that the surrogate L chain gene VpreB is the product of a gj event in the line leading to mammals (Chapter 25). Additionally, as described previously, there is a marsupial gj V that serves as the membrane-proximal domain of the unusual γ/δ TCR.

Organization of Rearranging Genes

Shark IgH-chain genes are structured into tens to hundreds of clusters, each consisting of V, D, J, and C elements (Fig. 4.7); all evidence from studies in horned shark, nurse shark, skate, and sandbar shark (and holocephalan ratfish) suggest that V, D, and J genes only rearrange within one cluster (but switch may occur between clusters[210]). While there is extensive N-region diversity and sometimes usage of two D segments (three Ds in IgNAR), and there are V_H subfamilies having substantial CDR1/2 heterogeneity, diversity of the primary repertoire is lower than in other vertebrates as there is no combinatorial diversity. The special constitution of the shark H and L chain loci suggests an exclusion mechanism like that of mouse TCR γ loci, also found in clusters. Only one V_H transcript is expressed in each shark lymphocyte, consistent with isotypic exclusion, despite the many clusters (see below).[211,212] Bony fish (teleosts and holosteans), frogs, reptiles, and mammals have very similar architectures of their H chain locus—the so-called translocon configuration (Chapter 22). As described previously, multiple families of V_H genes, each consisting of many apparently functional elements, are separated from a smaller number of genomic D and J elements. This type of gene arrangement enables combinatorial diversification. In birds, the organization is similar but all V genes except those most 3′ to the D elements are pseudogenes (see below). One exception is the coelacanth, in which each V gene is immediately followed by a D segment and then multiple J segments like in all tetrapods.[144]

L chain gene organization is variable. In elasmobranchs, the organization is the same clusters as the H-chain locus without D segments. Prototypic type I (σ) L chain has a cluster organization in which V, J, and C segments are closely linked. As mentioned, teleost L-chain genes have the shark cluster-type organization, but some species have multiple V genes in some clusters, demonstrating a rapid evolution of not only sequence but also gene organization. Shark L chain genes are not allelically excluded as single-cell PCR demonstrated expression of multiple loci in mature B cells.[213] Only one of the expressed genes/cell seems to be used as the others are either out of frame or have a high S/R mutation ratio suggesting no selection on antigen.

D Segments

D segments are always present in one of the two loci encoding an Ig/TCR heterodimer (IgH and TCR β,δ). Sharks have two D genes/H chain cluster,[205] and there are only minor variations among the clusters; one recombination signal sequence (RSS) follows the 12-12 paradigm seen in tetrapods and the other is like the TCR 12-23. In teleosts, amphibians, and reptiles where the organization of the H chain locus is like humans, the number of Ds deduced from cDNAs ranges from 10 to 16. In birds, there are 15 very similar D_H.

There are several reasons why D segments have been preserved in evolution (Figs. 4.7). Incorporation of D segments obviously augments CDR3 diversity and size, directly influencing the combining site.[214] Three different Ds contribute to IgNAR CDR3, and besides generating great diversity, CDR3 length and amino acid composition fulfills tertiary structure requirements: D-encoded cysteine residues bond with cysteine(s) in the body of the V domain, thereby stabilizing a loop involved in the antigen binding of the unusual monomeric receptor.[155,215] Finally, rearrangement of one (IgH, TCRβ) locus "locks it in" during development and allows the second locus to undergo receptor editing, for negative selection of the B cell repertoire[166] and positive selection of the T cell repertoire.[216] Finally, one of the mouse TCRδ D segments encodes a section that interacts with the ligand, the nonclassical class I molecule TL.[217]

Agnathan VLR Diversity

As mentioned earlier, our view of the jawless fish immune system was radically transformed when Pancer et al. prepared cDNA libraries of naïve lymphocyte RNA subtracted from lymphocyte RNA derived from lamprey larvae (Ammocoetes) that had been immunized to a bacterial/PAMP mixture and found a highly diverse set of LRR sequences enriched in the immunized animals.[218] The clones were not only diverse in sequence, but also in the number of LRR "cassettes" found between invariant 5′ (LRR NT) and 3′ (LRR CT) cassettes (Fig. 4.8). Unlike what has been discussed earlier for immune genes in many lower deuterostome species, there were few germline NT and CT cassettes, suggesting that a somatic recombination process convergent with Ig/TCR rearrangement occurred in developing lamprey lymphocytes. Germline VLR have 5′ and 3′ LRR cassettes separated by an intron. Upstream and downstream of the invariant exons are many LRRs, which become inserted between the 5′ and 3′ cassettes during lymphocyte differentiation. Individual lymphocytes apparently express a uniquely rearranged VLR gene in a monoallelic fashion. The potential VLR repertoire can be as great as 10^{14}, vastly outnumbering the lymphocytes within an individual.[126]

Fragments of homology between the LRR cassettes allow for joining and then priming of synthesis of a copy of the transferred cassette.[219] Because these small regions of homology are found throughout the cassettes, "hybrid" LRR can also be formed. The insertion of LRRs can occur at either end of the recombinant, but it is always 5′-to-3′ on the growing strand. This type of genomic modification resembles the initial stage of gene conversion, but since there is not a complete transfer of genetic material between homologous gene segments but an addition of sequence, the assembly of the mature VLR is most likely a recombinational mechanism described in yeast called "copy choice." The enzymology of "copy choice" has not been examined; because a gene conversion-like process occurs, it is likely that an APOBEC family member is involved. Indeed, APOBEC family members were detected in the lamprey genome project and are expressed in a lymphocyte-specific fashion; the two members that have been discovered, CDA1 and CDA2, are differentially expressed in agnathan lymphocyte subsets[126] (see below, Fig. 4.9). A knockout of CDA2 resulted in loss of VLRB-positive cells, but not the VLRA/C-positive cells.[220] Because APOBEC family members are involved in repertoire building in all vertebrates, mutation/gene conversion may have predated RAG-mediated repertoire building of the repertoire.

The independent development of two different strategies for receptor somatic diversification at the dawn of vertebrate evolution approximately 500 million years ago reveals the magnitude of the selective pressure applied on to the immune system and the emergence of individualized adaptive responses (Fig. 4.9). Are VLR responsible for the graft rejection results seen in the old experiments?[127] If so, this is most likely a "convergent MHC" coopted for the VLR system. Skin grafting experiments (or other tests of alloreactivity) should be repeated to study potential VLR involvement. The VLR system is discussed below as well in the lymphoid tissues section and the Conclusion.

RAG and AID in AgR Diversity Generation

The Ig/TCR Rearranging Machinery

Most models propose that the generation of somatically rearranging receptors occurred abruptly in evolution via the generation of the RAG machinery made of two lymphocyte-specific proteins, RAG1 and RAG2.[221,222] *RAG* genes have so far been isolated in all classes of jawed vertebrates and have been quite conserved. In every case examined, *RAG1* and *RAG2* genes are closely linked and in opposite transcriptional orientation.[223] Some regions of RAG1 and RAG2 are like bacterial recombinases or molecules involved in DNA repair or the regulation

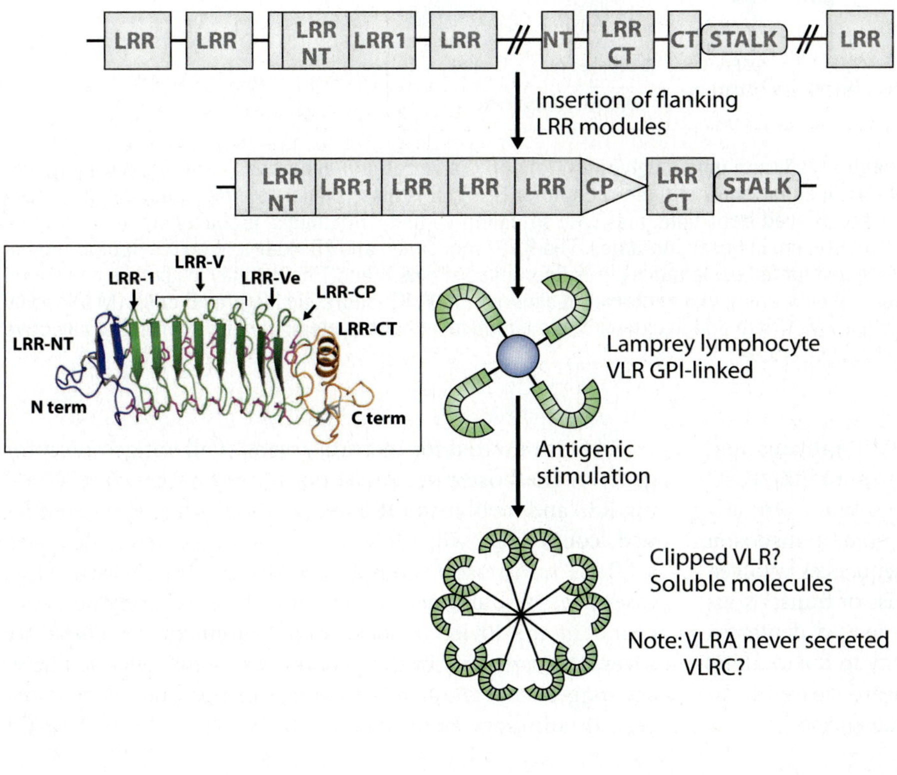

FIG. 4.8. Genetics, generation of diversity, and speculative cell biology of the agnathan variable lymphocyte receptor (VLR) system. The *top line* shows an incomplete VLR gene (NT and CT, N-terminal and C-terminal cassettes, respectively). During lymphocyte ontogeny, upstream and downstream LRR cassettes are inserted between the NT and CT gene segments, resulting in an intronless, mature VLR gene (*second line*). VLRB proteins are attached to the lymphocyte surface via a glycophosphatidylinositol linkage and are released into the blood upon antigenic stimulation as a pentamer (also see Fig. 4.9). *Boxed* is a hagfish VLR protein, the first structure to be elucidated by Kasahara and colleagues. Note the loop at the C-terminal leucine-rich repeat, which is mentioned in the text as a prominent region inserting into antigens, like single-domain antibodies.

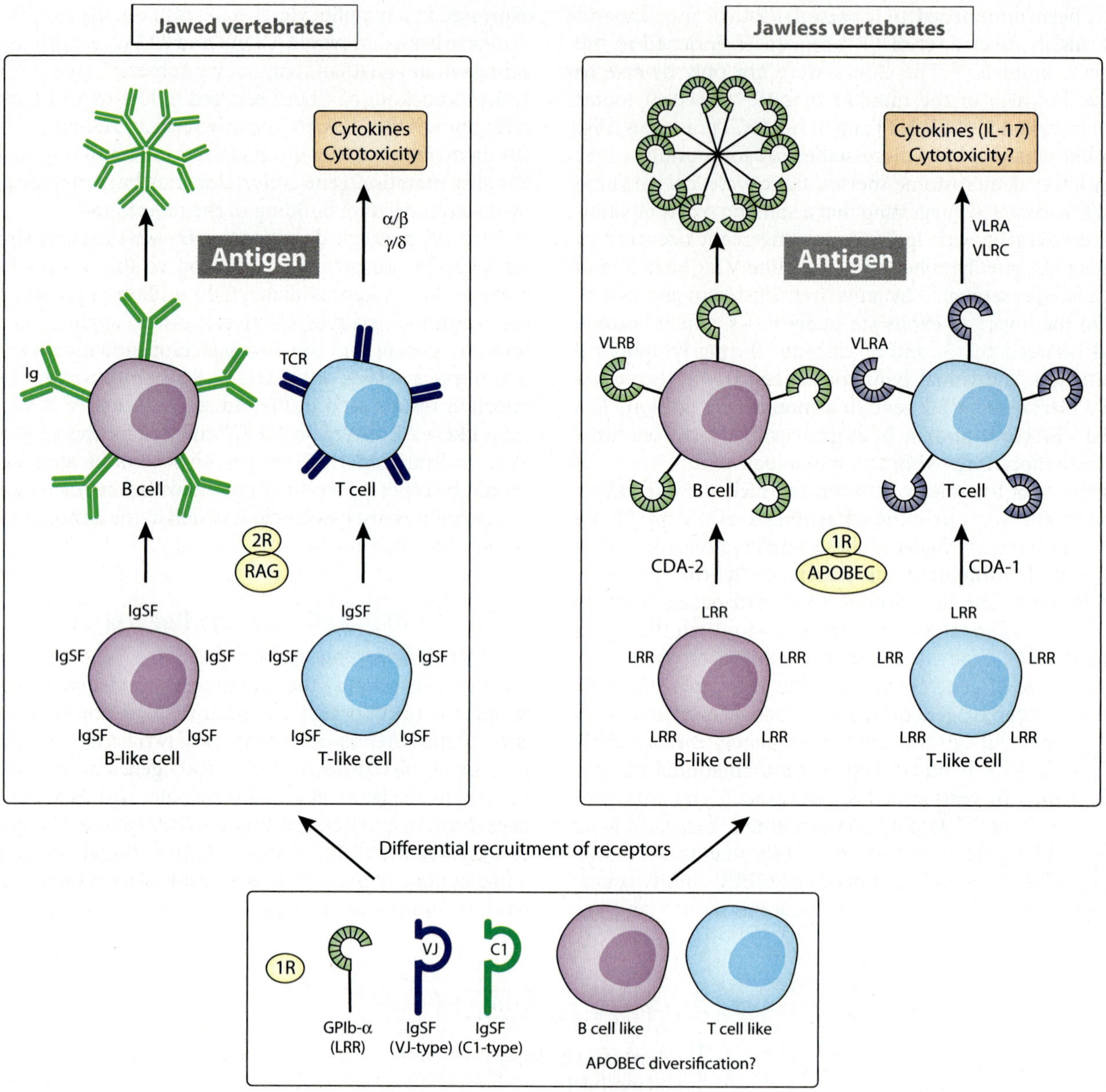

FIG. 4.9. Emergence of variable lymphocyte receptor (VLR) and immunoglobulin (Ig) early in vertebrate evolution. T and B cells likely preceded the divergence of the leucine-rich repeat and Ig superfamily antigen receptors. The VLR is most similar to a molecule called GPIb-α present in *amphioxus*, and IgSF antigen receptors are derived from molecules with so-called VJ and C1 domains, probably in the proto-MHC (see text for candidates). Both precursor genes were present in basal chordates. The RAG transposon and 2R likely ignited the appearance of the Ig superfamily receptors (Ig/T cell receptor) (see text for full explanation). This figure also shows B and T cells today in jawless and jawed vertebrates, with secretion of IgM from gnathostome B cells and a VLR pentamer of dimers from VLRB-secreting lamprey B cells. (Modified by permission from Springer Nature: Flajnik MF, Kasahara M. Origin and evolution of the adaptive immune system: genetic events and selective pressures. *Nat Rev Genet.* 2010;11(1):47-59 doi:10.1038/nrg2703.)

of gene expression. Similarities to prokaryotic proteins and the gene structure suggest that vertebrates acquired the RAG machinery by horizontal transfer and transposition from bacteria. Indeed, *RAG* genetic organization has some transposon characteristics as the RSS are reminiscent of sequences involved in targeting excision of transposons.[221] A class of transposons called transib detected in several protostome and deuterostome invertebrate species that shows similarity to the catalytic domain of RAG1.[224] It is believed that RSS were derived from the terminal inverted repeats (TIR) of this transposon.

RAG is required for rearrangement of all antigen receptor genes in gnathostomes. RAG1 does "most of the work" cleaving RSS and involved in RSS recognition, while RAG2 guides and cooperates with RAG1 in rearrangement. Recently, a "RAG transposon" complete with RSS (or at least TIR), insertion sites, and genes encoding the RAG enzymes were found in the lower deuterostome amphioxus.[225] There are several transposons in the genome, and phylogenetic analyses suggest that they are still active in this and related species. In addition, as mentioned, the catalytic core of RAG1 (transib)[224] is found encoded in several invertebrate species, showing that this transposon is quite old. Questions remain as to whether RAG2 was part of the original transposon or if it was an existing gene in the genome recruited to assist RAG1 in gene rearrangement; recent data suggest that RAG2 can be found in the transposon of several divergent invertebrates[226] (Fig. 4.10). Note that the RAGs *are not* used for VLR gene rearrangement (see below), yet as mentioned RAG1 and RAG2-like genes have been found in the genomes of lower deuterostomes whose ancestors predated the vertebrates, like sea urchin and amphioxus, where their functions (if any) are unknown.

Another source of somatic antigen receptor diversity shared by all gnathostomes is a unique DNA polymerase, TdT, which diversifies CDR3 during Ig and TCR gene rearrangement through the addition of nucleotides in a template-independent fashion (Chapter 22). Furthermore, its expression serves as an unambiguous developmental marker for sites of lymphopoiesis.[227] TdT has been highly conserved in both sequence (<70% amino acid similarity, <50 amino acid identity) and overall structure during vertebrate evolution. An amino acid alignment of all TdT sequences reveals that some structural motifs believed to be critical for TdT activity are well conserved. TdT protein alignments, and the crystal structure for rat β-polymerase, support the hypothesis that both evolved from an ancestral DNA repair gene. Four protein kinase C phosphorylation sites are conserved and may be involved in TdT regulation. Homologs related to the ancestor of polymerase β and TdT have been found in sea urchin and other lower deuterostomes. Thus, unlike RAG, TdT has evolved by gradual evolution from a polymerase family and was recruited for immune system function.[228]

The APOBEC Family

Activation-induced cytidine deaminase (AID) was discovered in mouse and human in 2000 as the enzyme required for both SHM and CSR in mammals.[229,230] Not long afterward, AID was detected in all gnathostomes and shown in several species to be capable of mutation and expression in secondary lymphoid tissues. The AID N- and C-termini are required for SHM and CSR, respectively; thus it was surprising that AID from bony fish, a phylogenetic group clearly lacking CSR, was capable of inducing CSR in in vitro systems.[229] One possibility is that the bony fish lost the CSR capacity since cartilaginous fish, formerly believed incapable of CSR because of their Ig gene cluster-type organization, indeed can undergo CSR between different IgM clusters as well as between IgW and IgM clusters.[206,210] No typical switch boxes with either the repetitive elements or the targeting RGYW motifs are found in the shark noncoding regions, so the switch mechanism is unknown as well as the relationship to typical CSR.

AID is a member of the APOBEC family of enzymes, which were first discovered as a modifier of mRNA splicing. Later, other APOBEC members were shown to be involved in viral defense and protection of the genome from retroviral invasion.[231] As mentioned, two members of the APOBEC family are expressed in lamprey lymphocytes, one in developing T cells (CDA1) and the other in B cells (CDA2), and these enzymes are likely required for the generation of VLR diversity[126] (Fig. 4.9). As mentioned, crispr of CDA2 results in a B cell deficiency in lamprey.[220] Apparently, the emergence of two enzymes, one specific for rearrangement in T cells and the other in B cells, obviated the requirement for complex regulation required for the one-enzyme complex of RAG responsible for rearrangement of all gnathostome

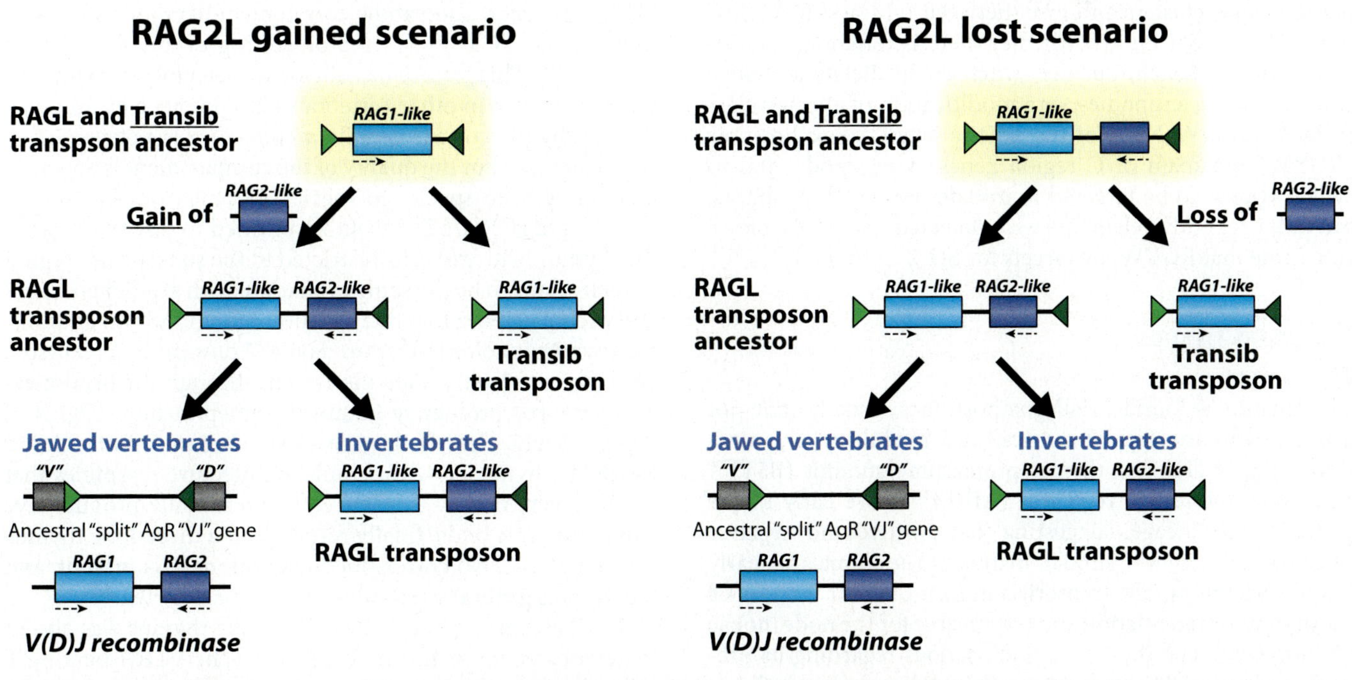

FIG. 4.10. Origins of RAG. RAG1 is related to a transposable element called transib, found in many invertebrates. RAG2-like genes previously were only found in deuterostomes, but now have been found in protostomes. The two scenarios show how the RAG transposon may have emerged over evolutionary time (see text for details). (Republished with permission of Annual Reviews, Inc, from Neely HR, Flajnik MF. Emergence and evolution of secondary lymphoid organs. *Annu Rev Cell Dev Biol.* 2016;32:693-711; permission conveyed through Copyright Clearance Center, Inc.)

Ig/TCR genes. CDA has been shown to be a mutator in in vitro systems, and CDA1 has been fused to a CRISPR cassette for in vivo mutagenesis and knockout experiments.[232] The discovery of AID and these related molecules has heralded an exciting foray into adaptive immunity, innate immunity, and general cellular homeostasis. For example, the APOBEC family is also involved in scanning for retroelements for general preservation of the genome as well as antiviral defense. This field is in its infancy, and a comparative approach to APOBEC function in non(typical) immune function holds much promise.[233,234]

CSR in tetrapods results from fusing switch regions upstream of the μ gene and another 3′ isotype gene, accompanied by the deletion of the intervening sequences. Mammalian and bird class switch regions are GC-rich and contain tandem repeats in which motifs such as TGGGG, GGGGT, AGCT, and GGCT are abundant.[235] Because of the GC richness of these regions, transcription generates stable R loops, which provide single-strand substrates for AID activity. The first comparative studies on CSR were done in the amphibian *Xenopus* where the switch regions are not GC-rich but adenine-thymine (AT)–rich and cannot form R loops.[236] Replacement in mouse of its switch region with a *Xenopus* switch region showed that it mediated efficient CSR and that the junctions were associated with the short palindromic AGCT motifs already recognized as the main component in *Xenopus* switch as predicted from the absence of R loops. The breakpoints were located in the AGCT-rich region of the switch box. Other motifs have been identified in the other switch regions, such as in IgX and chicken isotypes; all of these correspond to the RGYW hotspot consensus.[141] AID-mediated deamination in the context of these motifs may be the conserved major event in the initiation of CSR.

As mentioned, an isotype switch can occur in sharks, between IgM clusters and even between IgM and IgW clusters.[206,210] It is not known whether there is functional significance to the switch, but the switch could dictate a change from 19S to 7S antibodies, or a modification of effector class in IgM and IgW. None of the AID-targeted classic hotspots (RGYW) upstream of C region genes in tetrapods vital to switch appear to be targeted to initiate the switch in sharks, but other repetitive elements were detected that might play a role (note that RGYW *are* targets for SHM in sharks).[237]

Lymphoid Tissues

The Thymus

The thymus is found in all gnathostomes, usually (but not always) with the typical cortical and medullary organization[238](Fig. 4.11). The thymic proteasome subunit (B5T)[239] and the autoimmune regulator (AIRE)[240] arose early in the gnathostome lineage, suggesting that positive and negative selection occurs in a similar manner to mammals in early jawed vertebrates. The transcription factor FoxN1 is required for thymus differentiation and responsible for the nude mouse phenotype.[241] The thymus is promiscuous regarding its precise developmental origin, arising from the endoderm of pharyngeal pouches two to six in sharks, from the second pouch in frogs, from the second and third in reptiles, and from the third and/or fourth in bony fish, birds, and mammals. The final number of thymi can also be variable, ranging from five or more pairs of organs in sharks, to four in caecilian amphibians, to three in urodeles, and to one in other vertebrates. The thymus of reptiles varies in term of location and number of lobes, reflecting variation in embryonic origin. The crocodilian thymus is an elongated chain-like structure, not unlike that of birds. The discovery of a second (cervical) functional thymus in mice has raised interesting ontogenetic and immunologic questions.[242] Regardless of the underlying mechanism leading to the development of this cervical thymus, the result is suggestive of the secondary "thymus spreading" in birds, but it differs regarding an uneven final size distribution.

As a rule, the thymus is the first organ to become lymphoid during development (see below). Another emerging rule is that, excepting sharks, development of the thymus-dependent MHC-restricted T cell repertoire is similar in all species, and this is reflected in the evolution of TCR gene organization described previously as well as a core set of molecules that are required for thymocyte migration and differentiation; in contrast, B cell repertoire generation differs dramatically among different species, at times even within the same class of vertebrates (see below).

Like all other major adaptive immune system components, cartilaginous fish are the first in evolution to possess a prototypic thymus[181] (Fig. 4.11). As in mammals, it has a distinct cortex/medulla structure, and RAG/TdT expression was detected in thymocytes by in situ hybridization, where it is found throughout the cortex. Unlike most other vertebrates, age is not an indicator as to the size of the cartilaginous fish thymus; it can be small or large at any stage of development.[243] Interestingly, a recent study showed that in addition to V(D)J rearrangement in the thymus, the TCR alpha gene undergoes SHM. AID expression at the corticomedullary junction likely mutates the TCR after receptor editing events for positive selection.[181] This feature has obviously been lost in mammals but its presence in other vertebrates has not been examined.

The thymic cortex/medulla architecture is not precise in all vertebrates, but the duality of the compartment is apparent and varies from species to species. Recent work has shown that the g/dTCR/a/bTCR split is confined to discrete areas of the thymus, with g/d cells restricted to the subcapsular region (which seems to be the same in sharks), with gradients of IL-7 and cxcl12 helping to guide the differentiation.[244] In anurans, the thymus is colonized from days 6/7 onward by precursors derived from lateral plate mesoderm through the head mesenchyme and proliferate in situ as the epithelium begins to express MHC class II but not classical class I molecules.[108] By day 8, thymic cortex/medulla architecture resembles that of other vertebrates. Urodele embryos initially produce five pairs of thymic buds, finally resulting in a three-lobe thymus in *Ambystoma*. No cortex–medulla boundary is present, and the thymus generally resembles a canonical cortex.[245]

Until recently, it was believed that the thymus was absent in agnathans, but with the discovery of the VLRA-bearing T cells, this question was reexamined. In situ probes for delta ligand (for the notch receptor) and FoxN1 defined a structure lining the pharynx of larval lampreys.[246,247] Associated with

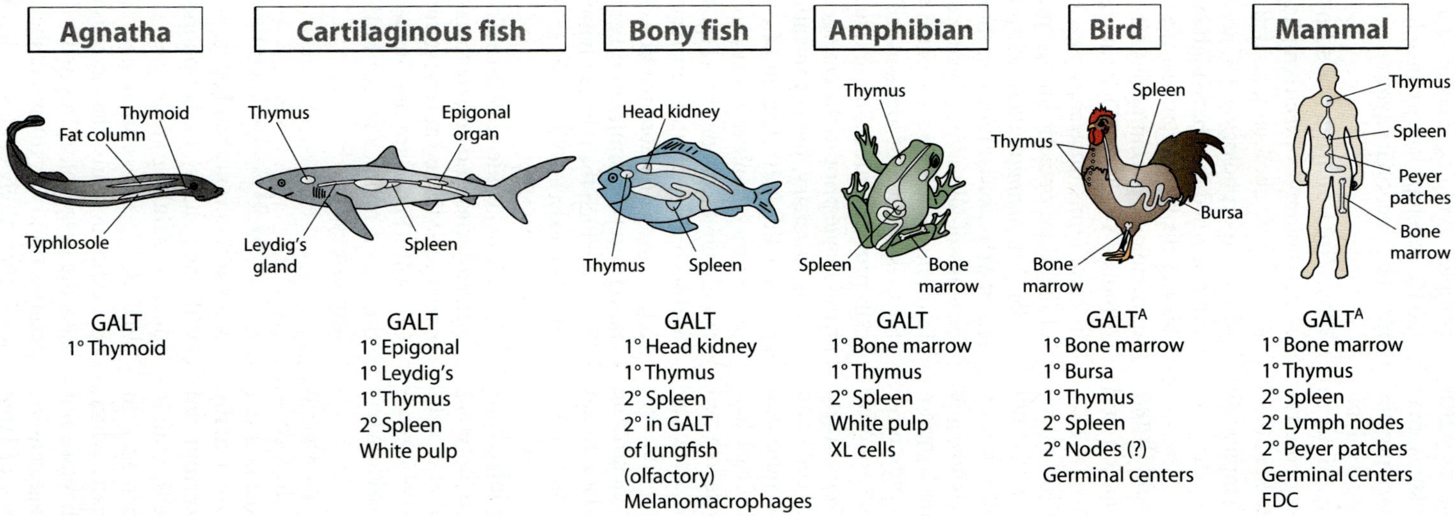

FIG. 4.11. **Evolution of lymphoid tissues in the vertebrates**. All jawed vertebrates have a thymus and a spleen with demarcated T cell and B cell zones. Fish have different bone marrow equivalents (epigonal, Leydig gland, head kidney), and amphibians are the oldest group with lymphopoietic bone marrow (also the first to have a typical immunoglobulin [Ig]H chain class switch, although sharks seem to have switch as well, despite the IgH cluster organization). Germinal centers are found only in warm-blooded vertebrates. A thymus equivalent (thymoid) has been discovered in agnathans. Reptiles are not shown, but they may have T cell zones (PALS) like mammals.[249] (Modified from Schatz et al. [Martin EC, Vicari C, Tsakou-Ngouafo L, Pontarotti P, Petrescu AJ, Schatz DG. Identification of RAG-like transposons in protostomes suggests their ancient bilaterian origin. *Mob DNA*. 2020;11:17 doi:10.1186/s13100-020-00214-y.)

these epithelia were lymphocytes expressing VLRA (and in later studies VLRC) and the APOBEC enzyme CDA1, mentioned above as the molecule important in VLRA and VLRC rearrangement. This structure was named the thymoid and it is likely the agnathan thymus equivalent. Thus far, no B5T (no specialized proteasome components, actually) or AIRE genes have been found in agnathans, so further study may provide the original rationale for physically separating the development of T and B cells into different primary lymphoid tissues, especially if there is no MHC restriction per se in agnathans. Structures in the gill region of lower deuterostomes like amphioxus also express FoxN1 and delta ligand, so it will be of interest to study their role (if any) in lymphocyte differentiation.[248]

The Spleen as the Primordial Secondary Lymphoid Organ

With the advent of clonal selection, the accumulation and segregation of T and B cells in specialized organs for antigen presentation became necessary and indeed the spleen as such an organ is found in all jawed vertebrates, but not in agnathans or invertebrates[238] (Fig. 4.11). Only warm-blooded vertebrates have lymph nodes, Peyer patches, and GC, all of which are dependent on the cytokine lymphotoxin for their development. Almost all gnathostomes, however, do have a spleen (unless it was lost) in which adaptive immune responses are generated, concentrating the antigen for interactions between antigen-specific T cells, B cells, and APCs.[249] The spleen can be partitioned into red pulp and white pulp in representative species to the level of cartilaginous fish, and B cells are also partitioned into segregated areas in many vertebrates (note, this segregated structure has been lost in most bony fish and some amphibians). During development, B cells are attracted to the chemokine CXCL13, which is expressed by the splenic vasculature, and this forms the nascent white pulp.[250] In mammals (and probably reptiles), the B cells are displaced into follicles while the T cells now surround the vessel in the periarteriolar lymphocytic sheath. In amphibians and fish, the B cells retain this embryonic characteristic. It is interesting that B cells form segregated structures in evolution prior to the formation of T cell zones, discussed extensively in a classic review article.[251]

In bony fish, the spleen contains the basic elements seen in other vertebrates—blood vessels, red pulp, and white pulp—but white pulp is poorly developed or undetectable and not associated with vasculature.[252] Ellipsoids, which are actually terminal capillaries, have a thin endothelial layer surrounded by fibrous reticulum and an accumulation of cells, mainly macrophages. Lymphocyte accumulations are often seen in their vicinity, especially during immune responses, which have been suggested to be primitive GCs, but they are not homologous. Red pulp is rich in melanomacrophage centers, groups of pigment-containing cells at bifurcations of large blood vessels, which may regulate immune responses in the bony fish. Antigen collects at these centers and B cells found there express AID and likely undergo SHM. This type of interaction among cells is reminiscent of tertiary lymphoid tissues (TLT) in mammals.[253]

The spleen in reptiles has well-defined white and red pulp regions. In *Chrysemys scripta*, white pulp is composed of two lymphoid compartments: lymphoid tissue surrounds central arterioles and thick layers of reticular tissue (ellipsoids). In *Python reticulatus*, dendritic cells (DCs) involved in immune complex trapping may be related to mammalian FDCs.[254] GALT develops later than spleen during development, and it may be a secondary lymphoid tissue.

When Did Dendritic Cells Appear?

The dichotomy between lymphocytes and myeloid cells appears in the lower chordates, predating adaptive immunity.[255] Cells with the morphology of DC are found in all gnathostomes, and where studied they have high expression of MHC class II.[256-258] These cells have been best described in bony fish, at least by morphology, peanut agglutinin (PNA) staining, and ability to stimulate T cells.[259] FDCs, in contrast, have not been reported in ectotherms. Instead, it seems that typical DCs (or APCs) of the hematopoietic lineage present antigen to both T and B cells.[118] In *Xenopus*, these cells, called XL cells,[260] have high levels of MHC class II for presentation of antigens to T cells and bear Igs of all three isotypes on their surface, presumably in immune complexes and coated with complement components for presentation of antigen to B cells. It is proposed that XL cells provide a model for antigen presentation to both types of lymphocyte prior to the emergence of FDCs.[258]

In summary, the organization of the lymphoid tissues is perhaps the only element of the immune system that shows increasing complexity that can be superimposed on the vertebrate phylogenetic tree (Fig. 4.11). The absence of primary and secondary lymphoid tissues (thymus and spleen) is correlated with the absence of a rearranging and/or hypermutating receptor family in other animals, except the agnathan VLR in which the relevant lymphoid tissues are being defined (if SLO exist). While all gnathostomes have a true SLO (spleen), ectotherms lack lymph nodes and organized GALT. In addition, while ectotherms clearly have B cell zones resembling follicles, and despite the ability for ectothermic B cells to undergo SMH and at least some degree of affinity maturation, GCs with FDCs are not formed after immunization; clearly, this was a major advance in the evolution of the vertebrate immune system.[258]

Ontogeny of Adaptive Immunity

The cartilaginous fish Leydig and epigonal organs (associated with the gonads) are lymphopoietic and erythropoietic, producing mainly granulocytes and lymphocytes, and there is high RAG expression in these tissues (see the following).[238,257] In addition, many plasma cells are found peppered throughout the epigonal, fitting with the mammalian bone marrow connection.[179] In the nurse shark, neonatal splenic white pulp consists entirely of class II–negative B cells and no detectable marginal zone separates red pulp from white pulp. Class II–positive dendritic-like cells are found throughout the white pulp in adults.[257]

In the skate, Ig and TCR expression is sharply upregulated relatively late in development (8 weeks). IgM expression is first detected in the spleen of young skates, but IgW is expressed first in the gonad, liver, Leydig organ, and thymus. In adults, Leydig organ and spleen are sites of the highest IgM and IgW expression.[261] In nurse sharks, IgM1gj and 19S IgM appear in the serum before 7S IgM and IgNAR, and this profile is reflected in the lack of IgNAR+ cells in the spleen until 2 months after birth.[140] RAG and TdT expression in the thymus and epigonal organ of the nurse shark suggests that lymphopoiesis is ongoing in adult life. In contrast to most other vertebrates, N-region diversity is detected, albeit reduced by approximately 50%, in skate and nurse shark IgM and IgNAR CDR3 from the earliest stages analyzed (neonates), suggesting that a diverse repertoire is important for young elasmobranchs.[169,262] As mentioned previously, a subset of Ig genes is prearranged in the germline of chondrichthyans, and many of those germline-joined genes are transcribed in embryos and hatchlings, but not in the adult. This pattern fits with the expression of the nurse shark IgM_{1gj} with its germline-joined V region and suggests that some germline-joined genes "take advantage" of their early transcriptional edge and thus some clusters can be selected for specialized tasks in early development. With many gene clusters, it is not known how "clusteric exclusion" is achieved at the molecular level (and why the germline-joined gene expression is extinguished in adult life), but as mentioned, studies in two cartilaginous fish species have shown that only one H chain cluster is expressed in each lymphocyte.[211,212] It was suggested, based on the paucity of rearranged gene segments at the nonexpressed IgH loci, that there are limiting factors regulating accessibility to RAG proteins and a short time window for rearrangement of accessible loci in B cells.

The architecture of shark Ig loci allows greatest diversity only in CDR3, because the CDR1 and CDR2 are always encoded in the germline and V segments do not combine with (D)J segments from other clusters. Yet the diversity of possible CDR3 is essentially limitless, and the number of germline clusters is also high (at least 15 genes in all species and as many as 100 genes; and usually three rearrangement events take place because of two D segments in each cluster)[263] (Fig. 4.7).

As described previously, studies of mature B cell activation and homing have been done in trout. Immunization of animals results in the production of short-term Ig-secreting cells in the blood and spleen and long-lived plasma cells in the head kidney (pronephros).[264] Further analysis with B cell–specific transcription factors like PAX5 and BLIMP1 reinforced the functional studies and showed that the blood contains primarily "resting" B cells and the head kidney both plasma cells and B cell precursors.[265] These findings appear to be true of sharks as well, with the epigonal organ as the primary lymphoid tissue-equivalent. Interestingly, conclusive studies have shown trout B cells to be efficient at phagocytosis,[266] raising questions about myeloid/lymphoid lineage commitment in the vertebrates. It is predicted that B cells from all ectothermic vertebrates are capable of phagocytosis, and the work has been extended to mouse B1 cells.[267] It certainly will be of interest to study these innate characteristics of B cell subsets in the future, considering the early appearance of B1 cells and macrophages derived from the yolk sac of mouse embryos.[268]

In *Xenopus*, rearrangement starts on day 5 after fertilization for the V_H locus, and within 9 days, all V_H families are used. For VL, the κ locus is the first to rearrange on day 7 (2 days after V_H), a situation similar to mice. During this early phase, B cells are present in the liver, where their number increases to approximately 500 cells.[269] Later in larval life, rearrangement resumes at metamorphosis, as suggested by the low incidence of pre–B cells and by the reexpression of RAG during the second histogenesis of the lymphoid system. T cells show a similar type of RAG expression/cell renewal during ontogeny as the B cells, and the larval and adult Vβ T cell repertoires differ significantly.[106] This is consistent with a degeneration of the thymus at metamorphosis followed by rapid thymocyte repopulation.[270]

In *Xenopus*, no lymphoid organ apart from the thymus is detectable until day 12 when the spleen appears and with it the ability to respond to antigen. RAG expression together with the detection of DNA rearrangement circles in the adult bone marrow suggests that rearrangement is ongoing throughout life and is not restricted to an early period, like in birds and rabbits.[269,271] Tadpole rearrangements are characterized by a lack of N-region diversity, like in mammals, and thus very short CDR3.[272] During ontogeny, TdT appears in significant amounts in the thymus of tadpoles at metamorphic climax, but little expression is detected in tadpoles, which correlates well with the paucity of N-region addition in larval IgH chain sequences.[227] Studies of the ontogeny of the *Xenopus* immune system have revealed a less efficient tadpole immune response (skin graft rejection and Ig heterogeneity and affinity); the absence of TdT expression during tadpole life fits well with the findings of lower larval Ig (and perhaps TCR) diversity. The restricted repertoire in larvae is consistent with having a "useful" repertoire in animals with very few lymphocytes.[273]

The bursa of Fabricius is a primary lymphoid organ unique to birds in which B cells are produced. It arises at day 5 of development and involutes 4 weeks later. T-B heterogeneity is obviously well defined in birds (indeed, the "B" in B cell stands for bursa.) The effects of thymectomy—T cell and B cell collaboration and generation of MHC-restricted helper and killer cells—are very similar to mammals,[4] the other class of warm-blooded vertebrates. During the embryonic period, chicken stem cells found in yolk sac and blood rearrange their IgH and LV genes simultaneously over a very restricted period of time, and very few cells colonize each bursal follicle (~10^4 follicles).[274] Three weeks after hatching, these cells have differentiated in the bursa and then seed the secondary lymphoid tissues, after which time B cells are no longer generated from multipotent stem cells; thus, only approximately 2×10^4 productive Ig rearrangements occur in the life of the chicken. When an antiserum specific for chicken IgM was administered *in ovo* to block early bursal immigration, there were no stem cells arising later in development colonizing the bursa, and these chickens lacked B cells for their entire lives.[275] Although the general Ig locus architecture is like that of frogs/mammals, only one rearrangement is possible as there is only one functional V_L or V_H on each allele.

Diversity is created during bursal ontogeny by a hyperconversion mechanism in which a pool of pseudogenes (25 ψL and approximately 80 ψH) act as donors and the unique rearranged gene acts as an acceptor during a proliferative phase in bursal follicles. For H chains, the situation is more complex as there are multiple D elements. During ontogeny, selection of productive rearrangements parallels the selection of a single D reading frame, suggesting that the many D segments favor D-D joins to provide junctions that are diversified by gene conversion. The gene conversion process requires AID, which is also required for SHM and CSR.[276]

Because diversification by gene conversion occurs after Ig rearrangement and cellular entry into bursal follicles, and there is only a single germline V_H and V_L expressed on all developing B cells, it was tempting to implicate a bursal ligand binding to cell surface IgM to initiate and sustain cellular proliferation and gene conversion. However, surface expression of IgM devoid of V regions permitted the typical B cell developmental progression, demonstrating that such receptor/ligand interactions are not required.[277] Thus, currently we know little of how cells enter the bursa, which signals induce them to proliferate/convert, and how cells arrest their development and seed the periphery, but the microbiota may be involved.[278]

Mechanisms leading to the generation of diversity vary among mammalian species. Categories can be made depending upon the mode of B cell development: rabbits, cattle, swine, and chickens, unlike fish, amphibians, reptiles, and primates/rodents, use a single V_H family, of which only a few members (sometimes only one) are functional. To diversify their antibody repertoire, this group uses gene conversion or hypermutation in hindgut follicles of GALT early in life, rather than bone marrow or its equivalent throughout life, and is therefore known as the "GALT group."[279] At the rabbit H locus, as in the chicken, a single V_H is expressed in most peripheral B cells. During development, B cells that have rearranged this particular V in the bone marrow (and other sites) migrate to the appendix where this rearranged gene is diversified by gene conversion using upstream donor V segments.[280] This development of B cells in rabbits is dependent on the intestinal microflora, and efforts have been made to define the potential bacterial superantigens involved, as well as binding sites on IgM necessary for the differentiation.[281] In ruminants, the ileal Peyer patches are the bursa-like primary B cell–generating tissues. Although bursa, appendix, and sheep ileal Peyer patches show morphologic similarities, the mechanisms generating diversity are different: conversion in the chicken and hypermutation in sheep, and both in the rabbit. As described previously, most of the "GALT group" also lacks IgD; thus, IgD might serve some purpose in repertoire development in some groups of mammals and not others.[142]

Adaptive Immune Responses in Vertebrates

T and B Cells in Agnathans

It was exciting when it was revealed that VLRA and VLRB were differentially expressed on lymphocytes, with transcriptomes consistent with T cells and B cells, respectively (Fig. 4.9).[130] VLRA could not be found in plasma and was expressed exclusively as[131,282] a cell surface receptor, while, as described, VLRB was present as both a lymphocyte receptor and a secreted molecule (Fig. 4.9). The lamprey adaptive scheme, therefore, parallels the situation in Gnathostomes where Ig is found both as a cell surface receptor on B cells and a secreted serum effector molecule, whereas the TCR is present only as a cell surface receptor on T cells; effector functions in T cells, like cytokine secretion or production of cytotoxic mediators, are properties of the T cells themselves. Currently, it is not clear how VLRC fits into this paradigm, but their transcriptome and tissue distribution suggest a g/d T cell phenotype.[131,282,283]

As expected for members of this family, the framework, or backbone, of the LRR is very similar between the cassettes. Diversity between cassettes is concentrated in the concave surface, often the region contacting antigen. Crystal structures of VLRB molecules have been determined, for example, one specific for HEL[284] and another for H-trisaccharide.[285] In both cases, the concave surface makes contacts as well as a loop in the LRR-CT, which in the case of HEL inserts into the active site, like what has been found for single-domain camelid and shark Vs. One VLR-A crystal structure was determined, also specific for HEL. This VLRA was derived from a HEL-hyperimmunized adult lamprey, which apparently underwent affinity maturation with induced molecules having affinities in the picomolar range.[286] The antigen is bound in the region of the molecule that is most diverse among all VLRAs; surprisingly, the LLRNT among all VLRA are not diverse, suggesting that this region might bind to a self-"restricting element" and the rest of the binding site to the true antigen.[287,288]

Comparisons to date suggest that VLRA and VLRB recognize antigen in different ways. Like gnathostome BCR, VLRB is expressed at the cell surface and is secreted as a pentamer of dimers, much like IgM (Fig. 4.8). The affinities detected from immunized animals are in the micromolar range, like low-affinity pentameric IgM.[129,287] By contrast, VLRA has never been detected in the sera or secreted in culture, fitting with its similarity to TCR. The affinity of VLRA for soluble antigen was high (in the nanomolar range) for the HEL-immunized animal, which appeared to undergo affinity maturation. Combining the information with the conserved and diverse regions of the VLRA biding site, perhaps VLRA recognizes antigen (budding virus?) associated with a self-protein (MHC analog?) at the cell surface.[35,288] It remains to be seen whether VLRA clones will be found from other immunized animals, and perhaps the HEL experiment was an artifact of the immunization schedule.

Adaptive Immune Responses in Gnathostomes

The quality of T cell and B cell responses depends on the heterogeneity and the diversity of the antigen receptor repertoires and the ability to select cells in SLO.[249] Because of the indefinite number of potential Ig/TCR V region sequences in most taxa, potential diversity exceeds the number of available lymphocytes; furthermore, all jawed vertebrates are capable of AID-dependent SHM or gene conversion after antigen stimulation. Yet, while potential repertoires are diverse in all vertebrate classes, and polymorphic MHC class I and II and TCR genes have been isolated from all classes, antibody diversity in nonmammalian vertebrates is relatively low.[118,273] The expressed repertoire has been studied via structural studies, affinity measurements during the maturation of the immune response, enumeration of antigen-binding Igs by isoelectric focusing (IEF), and idiotypic analysis. Sequences of Ig and TCR genes expressed in the course of a response estimate diversity at another level, allowing study of V genes diversified by gene conversion and/or somatic mutation.

Cartilaginous fish: Natural antibodies binding many antigens have been detected at surprisingly high levels in chondrichthyans and other vertebrates.[289] In older experiments, after immunization, the horned shark mounted a low-affinity 19S (pentameric) IgM antibody response, which varied little among individuals and did not increase in affinity after prolonged immunization.[290] The relative homogeneity and large number of V genes hindered SHM studies until a single unique reference horned shark IgM V_H gene was found. Mutations in this gene were slightly more frequent than those in *Xenopus* (see below). This first study proved that SHM preceded diversity obtained by combinatorial association of gene segments in evolution.[171] In contrast to mutations in the horned shark IgM V_H genes, unusual patterns of somatic mutation were detected in nurse shark IgNAR and type II (lambda) germline-joined L chains.[152,162] Half of the mutations (338/631) occur in tandem without the GC bias seen in *Xenopus*. Tandem and point mutations that take place simultaneously were not generated by gene conversion as there are no repeated patterns or potential donor genes. The germline-joined L chain genes can only diversify through SHM, perhaps like the hypothetical prototypic V region gene prior to RAG-mediated rearrangement (ie, SHM may have preceded gene rearrangement as the primordial somatic diversification mechanism). Lastly, a reappraisal of mutation at H chain and other L chain loci in the nurse shark showed that the mutations were not so different from H chains.[291] Mutations in randomly isolated IgNAR clones appeared to be under positive selection in IgNAR secretory but not TM clones, strongly suggesting that mutations do not generate the primary repertoire like in sheep but arise after antigenic stimulation.[292] In total, the shark mutations seem quite mammalian-like but with unusual characteristics.

Affinity maturation and memory generation can be detected in sharks.[135,293] Soon after immunization with HEL, an IgM response can be detected, primarily of the pentameric class. Over time, 7S IgM and IgNAR responses develop, and the 7S antibodies have a higher binding strength than those of the 19S class. When titers were permitted to drop to baseline, a memory response was induced by immunization of antigen without adjuvant. However, unlike responses in mice, titers do not increase over those in the primary response, suggesting a unique type of regulation of antigen-specific 7S IgM and IgNAR. Nevertheless, these data strongly suggest that the hallmarks of an adaptive response occur in sharks, and the memory can last for years.[294] This type of response is reminiscent of IgA responses of intestinal lymphocytes in mice.[295] In another study, a family of HEL-specific IgNAR clones was followed over time after immunization, and a 10-fold increase in the affinity of an already high-affinity germline clone (10^{-9} M) was observed. These results suggest that affinity maturation, memory, and "switch" to the monomeric IgM isotype occur, but it takes much longer to attain these adaptive hallmarks compared to mammals, perhaps a paradigm for ectotherms.

The role of T cells in shark immune responses has not been studied in detail. No thymectomy experiments have been performed, and T cells have not been monitored during an immune response. Shark mixed lymphocyte responses (MLRs) and graft rejection have been attempted, MLR with little success (probably for technical reasons) and grafts with the demonstration of a chronic type of rejection for which the genetics has not been analyzed (unpublished). However, from the MHC and TCR studies, all molecular components are available for proper antigen presentation in sharks and skates, and studies of splenic architecture suggestive class II + DC in the white pulps[257] argue for a prominent T cell regulatory role in adaptive immunity. Furthermore, an increase in binding strength and memory response, as well as mutation and now switch, also strongly suggests a T cell involvement in humoral immunity.[118] In summary, it appears that all the components of adaptive immunity in mammals occur in the cartilaginous fish, although with different features.

Bony fish: There are high levels of low-affinity natural antibody to nitrophenyl acetate in some bony fish. Natural antibodies in catfish have been correlated with resistance to virus infection or furunculosis.[296] As a rule, and similar to cartilaginous fish, little affinity maturation has been detected in fish, although some changes in fine specificities were noticed in the trout with a sensitive ELISA-based test.[137] The mild increase in trout antibody affinity (similar to that found in *Xenopus* and shark IgNAR) is attributed to selection of either minor preexisting B cell populations or somatic mutants. A large literature deals with vaccination attempts in teleost fish, due to their economic importance.[297] The availability of catfish B cell, macrophage, and T cell lines have been instrumental in analyses of antibody production. There are puzzling differences in responses from different teleost groups, much like differences between urodeles and anurans (amphibians). Cod, for example, do not respond well to specific antigen and have very high levels of "natural antibodies"[298]; recent studies showing that cod have lost class II genes provide an explanation for their poor humoral responses (see below).

Like the sharks, isolation of TCR genes and the existence of a polymorphic class I and class II molecules suggest that antigen presentation is operative teleosts, but unlike sharks, functional experiments examining mammalian-like T-APC interactions have been performed.[299] TCR messenger RNAs are selectively expressed, and specific TCR rearrangements have been detected in catfish clonal cell lines, which produce factor(s) with leukocyte growth–promoting activity (reviewed in Ref. 102). Modifications of the trout T cell repertoire during an acute viral infection (rhabdovirus) have also been followed. After an acute infection with viral hemorrhagic septicemia virus, CDR3 size profiles were skewed for several Vβ/Jβ combinations, corresponding to T cell clonal expansions.

Both "public" and "private" T cell expansions were detected in the infected genetically identical individuals. The "public" response resulted in expansion of Vβ4/Jβ1-positive T cells that appeared first in the primary response and were boosted during the secondary response. Further work examined fine specificity of the viral T cell response, which is a model for studies in cold-blooded animals.[300]

Recent results suggest that, despite the fact that IgM is the major Ig expressed in an immune response, high- and low-affinity IgM appears to be a function of the degree of polymerization of the tetramer.[137] It is suggested that high-affinity interactions of TM IgM on B cells modify the enzymes involved in disulfide bond formation, resulting in modifications of secreted IgM. In addition, expression of transcription factors in different subpopulations of trout B cells suggests that the typical populations of naïve, memory, and long- and short-lived plasma cells exist in bony fish; these populations have been followed in the blood, spleen, and head kidney by measuring proliferation and B cell–specific transcription factors, as mentioned earlier.[265]

Studies of SHM in teleost fish have been hampered by the lack of good reference genes, but it has been clear that AID-targeting motifs are found in Ig V genes.[301,302] Furthermore, AID is expressed in the spleen during immune responses, again consistent with its role in SHM. A recent study examined mutations in a reference zebrafish L chain gene and detected the typical mutational pattern described previously (without the tandems seen in sharks). As mentioned, AID-expressing cells are associated with melanomacrophage centers, where immune responses are believed to be initiated, perhaps like TLT in mammals.[252,253] Also as mentioned earlier, AID from several bony fish species is capable of inducing CSR in mammalian B cells, despite the fact that teleosts do not undergo CSR.[303]

As touched on, species living in extreme cold develop adaptive structural differences in their Igs.[139] At the level of global immune response, temperature exerts a strong influence in ectothermic vertebrates in general, low temperature generally being immunosuppressive. Lowering the water temperature from 23°C to 11°C over a 24-hour period suppresses both B cell and T cell functions of catfish for 3 to 5 weeks as assessed by in vitro responses. Virgin T cells are most sensitive to this cold-induced suppression, a property shared with mammals when tested appropriately. Fish have developed ways to adapt to the lack of fluidity of their B cell membranes by altering the composition of fatty acid by using more oleic acid at low temperatures.[304]

In the NK section, we briefly discussed fish cytotoxic cells.[102,305] In vitro studies have now shown that leukocytes from immunized fish specifically kill a variety of target cells (allogeneic erythrocytes and lymphocytes, hapten-coupled autologous cells); fish CTL of the αβ (and perhaps γδ) lineages as well as NK cells were found (see above). Naïve catfish leukocytes spontaneously kill allogeneic cells and virally infected autologous cells without sensitization, and allogeneic cytotoxic responses were greatly enhanced by in vitro alloantigen stimulation. Cloned cytotoxic cells contain granules and likely induce apoptosis in sensitive targets via a perforin/granzyme or Fas/FasL-like interactions. All catfish cytotoxic cell lines express a signal transduction molecule with homology to the Fcγ chain of mammals; this chain with an ITAM is an accessory molecule for several activating receptors on mammalian NK cells.

Amphibians: Differences in immune system features between urodele (axolotl) and anuran (*Xenopus*) amphibians are also seen in immune responses. Rarely is such divergence seen within one vertebrate class (although the two groups diverged over 250 million years ago!). Urodeles express a restricted antibody repertoire in response to specific antigen that peaks at 40 days postimmunization and is entirely of the IgM class, even though the serum also contains IgY.[306] They do not respond well to thymus-dependent antigens, which may be due to lack of T cell help, yet their expressed TCR diversity looks normal.

Anuran larvae can respond specifically (with only one million lymphocytes) to many antigens, with a modest affinity maturation of the IgM anti-dinitrophenol (DNP) response.[273,307] In adults, the number of different anti-DNP antibodies does not exceed 40, versus 500 in mammals. In secondary responses, the peak of the response is about 10-fold higher and is reached in 2 weeks; there are no major changes in affinity over this initial rise. Isogeneic *Xenopus* produce homogenous antibodies to DNP or other antigens with identical or similar IEF spectrotypes and idiotypes, while outbred individuals differ. Both IEF spectrotypes and idiotypes are heritable, suggesting that diversity is a reflection of the germline repertoire without a major contribution from somatic mutations.[308] Thus, somatic mutations were followed during the course of an antigen-specific immune response at the peak of the modest affinity maturation. Of the 32 members of the V_H1 family involved in the anti-DNP response, expression of only 5[309] was detected, indicating that immunization was being monitored. Few mutations were detected, and there was not a strong preference for mutations in CDR1 and CDR2 and virtually none in CDR3.[310] Like in the horn shark IgM study noted previously (but not IgNAR or type II L chains), the mutations were targeted to GC bases, and such a pattern has been suggested to be the first phase of the SHM phase in mouse/human; perhaps *Xenopus* has lost the second phase of the process that results in an evening of mutation frequency for all bases. There is no shortage of variants, and the low heterogeneity and poor affinity maturation is likely due to less than optimal selection of the mutants, discussed below.

Essential T cell functions in anurans have been demonstrated with in vitro assays for T-B collaboration and MHC restriction, demonstrating the similarity of the role of MHC in *Xenopus* and mammals.[311] Regulatory T cells have been shown indirectly in hematopoietic/thymic chimaeras for control of CTL generation and in antibody responses.[312] Thymectomy early in life totally precludes CSR from IgM to IgY, but not IgX synthesis; thus, T cells are absolutely required for the switch to the IgG-like isotype and for high-affinity IgM responses, but switch to the mucosal Ig can be T-independent.[313] The switch is also temperature-dependent, and as described previously for channel catfish, ectotherm T cells are quite temperature-sensitive. AID is expressed in lymphocytes in the spleen as well as in secretory cells. However, consistent with many previous studies, there is no evidence for a typical GC response to date in amphibians or any other ectothermic species. The continued expression of AID in plasma cells warrants further study.[314]

Similar to studies in mammals, the chaperone gp96 has been shown to shuttle peptides into cells making them targets for MHC-restricted CTL lysis. Immunization of frogs with gp96 from a thymic tumor results in the elicitation of CTL that display antitumor activity.[315] As mentioned previously, NK cells have been characterized in *Xenopus* with mAbs that recognize non-B/T cells. Those cells kill MHC class I–negative target tumor cells but not class I–positive lymphocytes, and after thymectomy these cells are enriched in the spleen. CD8+ cells expressing TCR were isolated with the same mAb, suggesting the existence of amphibian NKT cells. Robert and colleagues have developed one of the best-defined model of innate and adaptive immunity to viral infection in cold-blooded vertebrates, the ranavirus FV3 in *Xenopus*. Involvement of CD8 cytotoxic cells and humoral responses have been studied over the course of primary and secondary infections with this virus.

Reptiles: Lack of an increase in affinity and homogeneity of IEF spectrotypes suggest low-antibody heterogeneity in reptiles.[316] In the turtle *Pseudemys scripta*, a number of genomic VH sequences were isolated, as was a genomic Cμ, all shown to be encoded at a single locus.[317] For T cells, there are no data on T effector function, but there are studies on the behavior of T cell population changes due to seasonal and hormonal variations. Thymocytes from the turtle *Mauremys caspica* proliferate in response to phytohemagglutinin (PHA) and ConA and can kill tumor target cells by both antibody-dependent cellular cytotoxicity–mediated and NK-mediated cytotoxicity.[318]

Birds and mammals: The poor increase in affinity of chicken anti-DNP antibodies again indicates lower heterogeneity in chickens.[319] Few changes occur after immunization, even if one waits 1 year after several injections. Hyperconversion and SHM in Ig genes have been found in splenic GC B cells after immunization.[320] The relatively poor affinity maturation of the chicken response may be due to a balance between gene conversion and somatic mutation. Indeed, modification of V genes with segments of DNA is not an optimal strategy for fine-tuning antibody responses. In the rabbit, there is also conversion/mutation by B cells in GCs after immunization.[276] Within mammals, large variations are found from marsupials with no obvious secondary response,[321] to mouse with 1000-fold increases in affinity, but the basis for the relatively poor responses has not been established.

In summary, studies over the past 50 years have revealed that, by and large, specific antibody responses in cold-blooded vertebrates were of low affinity that did not mature to the high levels seen in mammals over time. Before the molecular era, this was attributed perhaps to a lack of SHM in nonmammals.[322] However, mutations were discovered in cartilaginous fish and amphibian Ig genes (and now bony fish), in the 1990s as described earlier. Despite this, affinity increases do not exceed 100-fold in any immunization study of ectotherms to date, nevertheless showing at least some level of selection. As described in the text, at least to the taxonomic level of amphibians, FDC are not present and recent work suggests that conventional, hematopoietically derived APCs stimulate both T and B cells.[258] Thus, emergence of FDCs is believed to have been provided for the emergence of GC formation and the selecting environment for generation of high levels of affinity maturation.[310] As described in the previous review on the evolution of Ig genes and function, the discovery of AID has provided insight into the evolution of lymphoid tissues, but now should be used more extensively to study adaptive responses in lymphoid tissues over time.[252,314] In addition, the jawless fish apparently have no SLO, so the locations in which cells contact antigen and interact with each other are unknown; the APOBEC family members implicated in generation of diversity and likely mutation should be useful for such analyses.

Mucosal Immunity

GALT is present in all vertebrates, but lymphoid tissue in the small intestine does not have typical secondary lymphoid tissue structure seen in mammals with Peyer patches and isolated lymphoid follicles[238] (Fig. 4.11). Instead, lymphoid cells are found throughout the lamina propria and the gut epithelium. However, defined lymphoid tissue does exist in the olfactory epithelium and this is an active area of research in lower vertebrates.[323] It appears first in the lungfish and is also found in all amphibians (except the model species *Xenopus*, where it was apparently lost). Despite this lack of defined organization in the intestine of all lower vertebrates, dedicated mucosal immunity has been shown for all vertebrates examined.[14]

Mucosal Antibodies Specific for Each Vertebrate Group

Amphibian IgX: Dimeric IgA is the mammalian/avian mucosal immunoglobulin and it has well-defined roles in mucosal immunity, such as in the coating of commensal organisms to prevent their intimate association with mucosal epithelia[295] (Fig. 4.6). In amphibians there is a third antibody isotype called IgX which forms pentamers (or hexamers) like IgM but is preferentially expressed in mucosae.[175,313] The first bioinformatics analysis showed IgX to be most similar to IgM in sequence and thus it appeared to be one of the "dead end" isotypes mentioned earlier. As more Igs were cloned, however, it became clear that IgX is orthologous to IgA, reestimating the emergence of this isotype to the origins of tetrapods.[309] IgX levels increase after immunization with mucosal antigens.[324] In addition, unlike IgY but similar to mammalian IgA, switching to IgX can occur in the absence of T cells. Studies have not been carried out to determine whether IgX, IgM, or both (like in mammals) can coat commensal organisms in the gut lumen and elsewhere.

Immunoglobulin Z/T: Bony fish have a unique mucosal antibody isotype, originally called IgZ/IgT and now settled on IgT (IgTeleost) (Fig. 4.6). Like the mammalian TCR αδ locus, the D, J, and C segments for this isotype are embedded in the IgM locus, so that the IgT gene is deleted upon

IgM rearrangement.[325,326] Excellent work has shown that IgT is expressed primarily at mucosal or epithelial sites, and it can be induced specifically in response to mucosal pathogens, and both IgT and IgM coat commensal pathogens.[327] A molecule related to the pIgR transports both isotypes,[328] despite the fact that the J chain has been lost in bony fish.[62] Recent studies showed that depletion of IgT-expressing B cells leads to disruption of the mucosal barrier and a poor response to mucosal pathogens.[329] In sum, IgT, while displaying unique evolutionary features, has attained almost all of the characteristics of IgA via convergent evolution.

As mentioned earlier, cartilaginous fish have an isotype called IgW, most related to IgD (Fig. 4.6). It has several isoforms expressed in different tissues, but its high expression in the pancreas suggests that it may be a mucosal isotype of cartilaginous fish.[207]

The Major Histocompatibility Complex

T cells distinguish self from nonself through the presentation of small peptides bound to MHC class I and class II molecules (ie, MHC restriction) (Chapter 19). The genetic restriction of T cell–APC collaboration, processing of antigen by professional APCs, and T cell education in the thymus described in mice hold true (or is assumed) for most jawed vertebrate classes. For technical reasons, no MHC-regulated T cell responses have been documented in cartilaginous fish, but the identification of polymorphic class I and II and rearranging TCRα/β genes and segregated B cell zones in spleens strongly suggest that functional analyses will reveal MHC restriction of adaptive responses. By contrast, urodele amphibians and teleost cod are notorious for their poor immune responses, but biochemical and molecular evidence suggests that class II polymorphism is low in the axolotl and cod have lost the entire class II–based immune system. Recent results in agnathans and other vertebrates have opened our eyes to new functions in MHC. The MHC, complete with class I, class II, and class III regions, is also first found in the cartilaginous fish. The levels of polymorphism of classical MHC class I and MHC class II molecules in the species examined are very high. Furthermore, nonclassical MHC class I genes are also found in all gnathostomes tested (Fig. 4.12).

Class I/II Structure Through Evolution

The three-dimensional structures of class I and class II are remarkably similar: the two membrane-distal domains of both molecules form a PBR composed of two antiparallel α helices resting on a floor of eight β strands, and the two membrane-proximal domains are IgSF C1. Although sequence identity among class I and class II genes in vertebrates is low (like most immune genes involved in recognition), the four extracellular domain organization and other conserved features are likely to be found in the ancestral class I/II gene.[330] An intrachain disulfide bridge exists within the class I PBR α2 and class IIβ1 domains, but not the class I/II PBR α1 domains, and phylogenetic trees show that these respective domains are most similar. Bony fish class II α1 domains, like class II DMα molecules, have a disulfide bridge. The exon/intron structure of class I and class II extracellular domains is also well conserved. Other conserved features of class I genes include a glycosylation site on the loop between the α1 and α2 domains important in biosynthesis (shared with class II β chains), a Tyr and one to three Ser in the cytoplasmic regions that can be phosphorylated in mammals, as well as several stabilizing ionic bonds. Class II with its two TM regions differs from class I with only one; conserved residues in the class II α and β TM/cytoplasmic regions facilitate dimerization. In summary, because sequence similarity is very low among MHC genes in different taxa, these conserved features are important for function, biosynthesis, and maintenance of structure.[330,331]

MHC Class II Molecules

MHC class II alpha and beta chain genes are found in almost all gnathostomes. Generally, two or three isotypes are present, with high polymorphism. Class II molecules have nearly invariant and evolutionarily conserved residues that bind to main-chain atoms of peptides in the center of the groove. Thus, tight binding to main-chain peptide atoms occurs in the center of the class II PBR, and peptides can protrude from both ends. Consistent with a slower evolution than class I, mammalian MHC class II isotypes can be detected in ectotherms.[332] The DO class II molecules that regulate the binding of peptides to MHC class II molecules through DM molecules are not found in any ectotherm.[333,334] The DM genes appear first in amphibians (Fig. 4.12). It will be of interest to study how the lack of these catalysts impacts the association of peptides with MHC class II molecules; much work has been done on the biochemistry of MHC class II molecules in amphibians,[335] but very little in bony or cartilaginous fish. However, the invariant chain, which is essential for the stable assembly and transport of MHC class II molecules, is found in all ectotherms.[336]

Several teleost species have lost the MHC class II system, as first shown in the cod and later in other species; in fact, it was lost several times via convergent evolution.[337] It was known for a long time that specific antibody responses could not be generated in cod after immunization, essentially producing the same repertoire of IgM antibodies to each antigen.[298] How lifestyle has impacted this lack of class II has been speculated on—these animals live in a cold-water environment and a lack of pathogen pressure may have contributed to a "use it or lose it" scenario.

MHC Class I Molecules

The classical and nonclassical MHC class I paradigm is clearly found in all cold-blooded gnathostomes[338,339] (Fig. 4.12). Classical class I molecules are recognized by their high levels of polymorphism, ubiquitous tissue expression and their nine defined peptide-binding residues that lock in the N- and C-termini of the bound peptides. Thus, tight binding of peptides, a likely source of conformational changes in class I allowing transport through the endoplasmic reticulum and cell surface expression, is an evolutionarily conserved trait.[330] In nonmammalian vertebrates, one of these residues at the C-terminus is lysine rather than tyrosine, which permits the C-terminal end of the peptide to protrude from the groove, like for class II molecules.[340]

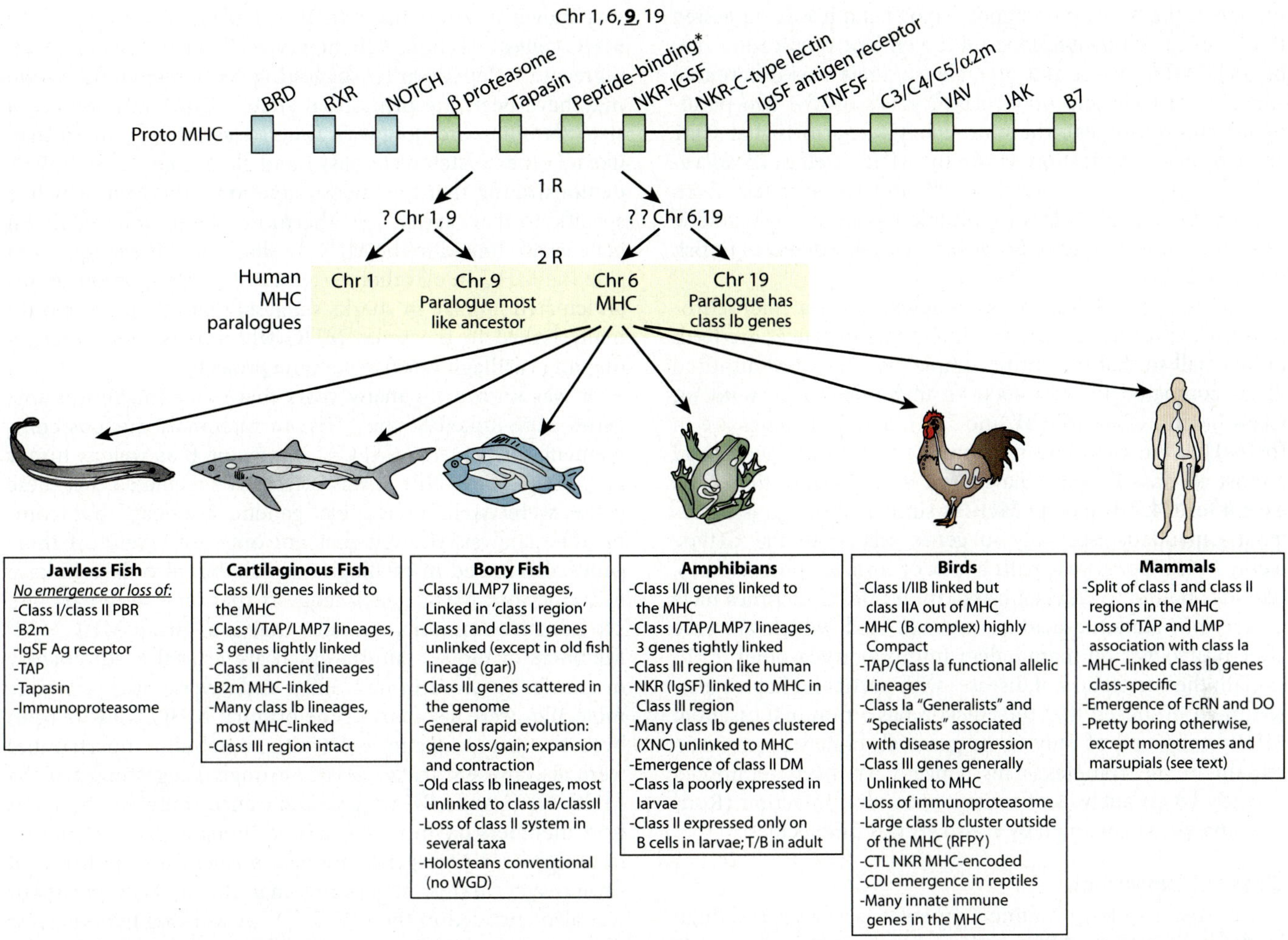

FIG. 4.12. Genome-wide duplication (2R) model of major histocompatibility complex (MHC) evolution and specialties of MHC in the different vertebrate classes. Displayed are the MHC paralogous regions and their human chromosomal designations, first identified by Kasahara and Pontarotti and modified over the years. Note that such paralogous regions have also been studied for chemokine receptors, tumor necrosis factor and tumor necrosis factor receptors, immunoglobulin superfamily molecules involved in cell–cell interactions, and others—see text for details. Note that the natural killer cell complex and leukocyte receptor complex, respectively, are likely to have been in the Ur MHC based upon linkages seen today in various taxa.

Unlike what is found in mammals, the antigen processing genes TAP and immunoproteasome (especially LMP7) genes are closely linked to class Ia genes, generally in lineages.[341,342] No functional studies have been carried out in ectotherms, but in birds TAP and MHC, class Ia alleles have been shown definitively to have co-evolved in peptide transport and binding studies.[343] Biochemical work was done long ago in amphibians, in which class I proteins were isolated using cross-reactive xenoantisera, alloantisera, and monoclonal antibodies.[344] Now, it is time to step up with studies examining association with peptides and the significance of linkage.

Nonclassical MHC or class Ib genes are also present in all gnathostomes (Fig. 4.12). Amphibians have a large cluster of nonclassical genes downstream of the bona fide MHC (called *Xenopus* nonclassical or XNC), at the telomere.[345] Among *Xenopus* species the number of genes found in the XNC varies greatly, as is the case for many large multigene families. Early studies showed interesting tissue distributions of the XNC isotypes in the lung, intestine, and spleen, with expression of all isotypes in the thymus, consistent with a positive selection of class Ib–reactive T cells.[346] Recent work focusing on one of the isotypes, xnc10, has shown that NKT cells, bearing an invariant TCR alpha chain, use this class Ib molecule as a restriction element, aiding in responses to tumor and viral antigens[107] (see below for clarification, mammalian NKT cells use the nonclassical class I molecules CD1 and MR1 as restriction elements). Ancient nonclassical class I lineages in bony fish and cartilaginous fish are candidates for recognition by NKT-like cells, or for other nonimmune functions as shown in mammals. For example, despite general rapid evolution of the bony fish, there are deep lineages of class I molecules, especially the so-called Z lineage in which the PBR is extremely conserved in all species.[339] A new lineage in cartilaginous fish, UDA, has[105] similar properties.[347]

Surprisingly, CD1 genes are closely linked to the reptile and chicken MHC.[110-112] CD1 genes in mammals seem to be on one of the MHC paralogous regions and it was suggested that it arose as a consequence of the en bloc duplication (see below). While this is still possible (ie, there was differential silencing of CD1 and other class I genes on the two paralogous chromosomes), the more likely scenario is that CD1 arose by gene duplication within the MHC itself in an ancestor of warm-blooded vertebrates[105] and translocated. Even though it is an "old" class I molecule based on phylogenetic analysis,[331] no bona fide CD1 genes have been detected in fish or amphibians.

The chicken MHC, the B complex, is on a microchromosome, and intron sizes and intergenic distances are both quite small so that the entire complex is only a few hundred kb as compared to over 4000 kb in humans and *Xenopus*. Class Ia (BF), class IIβ (BL and DM), and TAP genes are in the MHC, but there are no immunoproteasome genes and almost all class III genes have been moved out of the MHC except for C4.[70] The quail MHC is similar, although is somewhat expanded, especially in genes related to the C-type lectin NKR. Consistent with a lack of immunoproteasomes, peptides bound to chicken class I molecules sometimes have C-terminal glutamic acid or aspartic acid, which are rare after proteolysis by mammalian immunoproteasomes.[348] To explain the correlation of diseases with particular haplotypes, Kaufman proposed that the chicken has a minimal essential MHC composed of only those genes absolutely required to remain in the complex. This concept has been reinforced recently by an analysis of resistance to viral infection (Rous sarcoma virus) governed by classical class I molecules.[119]

Class I/II Expression

In *Xenopus* species, immunocompetent larvae express high levels of class II on APCs such as B cells, but express only very low levels of class Ia molecules on all hematopoietic cells until metamorphosis.[108] Expression of the immunoproteasome element *lmp7* and all identified class Ib isotypes is also low. Larval skin and gut, organs with epithelia in contact with the environment, appear to coexpress class I (transcripts) and class II.[349] Such expression may provide immune protection during larval life; perhaps expression of class Ia is limited to organs that undergo massive destruction and remodeling at metamorphosis. Class II molecules also change their distribution after metamorphosis and are highly expressed by unstimulated T cells. Axolotl class II molecules are also regulated differentially during ontogeny, expressed in young animals on B cells, and then expanded to all hematopoietic cells, including erythrocytes, later in life.[350] Changes in MHC expression are not correlated with cryptic metamorphosis in axolotls, but class II expression by erythrocytes is correlated with the switch from larval to adult globins.[350]

Significance of MHC Linkage

In addition to the lineages of antigen processing and presenting genes found in ectotherms and the conservation of linkage groups, comparative analysis of the MHC has also permitted an understanding of the MHC prior to the emergence of adaptive immunity. Linkage of MHC class I and class II genes (graft rejection and mixed leukocyte reaction cosegregation) was shown in amphibians long ago, prior to cloning of the MHC genes.[351] In bony fish, however, class I and class II genes segregate independently, suggesting that this configuration may have been the primordial state.[352] However, studies of shark families and now with the first genomes showed that, like all other vertebrates, class I and class II genes are linked, demonstrating that the teleost situation is derived, which is specific to that taxon.[353] Furthermore, the β2-microglobulin gene is also linked to the MHC in sharks and is encoded outside the MHC in all other vertebrates.[354] Preservation of this anticipated linkage in sharks suggests that other genes originally linked to the proto-MHC will also be uncovered in ongoing cartilaginous fish genome projects.

It was known for many years that other innate immune genes were linked to the MHC in mammals, such as complement components C4, C2, and factor B as well as tumor necrosis factor (TNF). It was originally speculated that these genes were MHC-linked by "genetic accident," but comparative analysis showed that not only are several of these genes also linked in cartilaginous fish, but also their linkage to framework MHC genes suggests that they were part of a pre-adaptive immune complex, that is a "proto-MHC."[355,356] Members of the B7 family, which are crucial costimulatory molecules involved in immunity, and several other immune molecules were also part of the original MHC.[98] NKRs from both the C-type lectin and immunoglobulin superfamilies were also present in the MHC, strongly suggesting that the MHC, NKC, and LRC were syntenic ancestrally. Additionally, IgSF members having a specialized domain (IgSF VJ) found in antigen receptor genes are also encoded in the MHC of several ectotherms, suggesting that the Ig/TCR precursor was also encoded in the MHC.[105,357] A working hypothesis is that the proto-MHC was a gene cluster already dedicated to immunity prior to the advent of adaptive immunity, and the major components of adaptive immunity, namely Igs, TCRs, and MHC class I/II, "piggybacked" onto a region already programmed to respond with enhanced transcription upon infection.[31,91,332]

An MHC in Jawless Fish?

When it was revealed that lampreys had both B and T cells, the obvious question became: what are the restricting elements for VLRA recognition? Extensive database searches, as well as various molecular approaches yielded no evidence for MHC class I, MHC class II, TAP, or an immunoproteasome, showing that either the genes are present but unrecognizable at the protein level or they indeed are lost from the lamprey genome.[128] If truly lost in agnathans, there are several possibilities. There may be a convergent system to MHC class I/II that presents peptides to agnathan T cells; a polymorphic cell surface antigen called NICIR3/ALA generates a strong alloantibody response in hagfish and may play such a convergent MHC role.[114,115] Alternatively, VLRA might recognize whole antigen associated with a cell surface molecule such as an Fc or complement receptor. Mentioned above, in one set of experiments, recombinant VLRA from expression libraries was capable of binding to an immunizing antigen, hen egg lysozyme (HEL)[284]; while TAP is not present in lampreys, a related molecule TAPL that is involved in cross-presentation in mammals is found in the lamprey genome and also in invertebrates.[358] It is possible that TAPL may have been co-opted for antigen presentation in agnathans.

Cytokines and Chemokines

Many cytokines/chemokines and their receptors, like most molecules of the immune system, evolve rapidly. However, consistent with the "Big Bang" theory, it is an emerging picture that the majority of cytokines and chemokines found in mouse and human are also found in the genome and EST projects of nonmammalian jawed vertebrates, best studied in chickens and bony fish models, but now extending to the cartilaginous fish as well. This suggests that whatever the initiating pressures in the evolution of the Ig/TCR adaptive immune system, the network of cytokines and chemokines emerged (practically) full blown early in evolution after 2R.[39,359] A picture starts to emerge; this is, in part, because the situation in mouse human is clouded by the plasticity of T helper cell lineages, as well as studies of ILCs that produce cytokines previously thought to be within the domain of adaptive lymphocytes.[92]

The Proinflammatory Cytokines: Interleukin-1 Family Members, Interleukin-6, Interleukin-8, and Tumor Necrosis Factor α

IL-1 and related family members IL-18 and IL-33 (and others), IL-6, TNF-α, and IL-8 (CXCL8) are the prototypic cytokines associated with inflammatory responses, which are defined by induction of vasodilatation and vascular permeability, and upregulation of innate immune system–specific molecules that have direct functions or that costimulate/attract T and B cells (Chapter 12). Classically, many of these activities can be assayed in supernatants from PAMP (eg, LPS)-stimulated phagocytes by determining whether thymocytes are induced to proliferate when one also adds suboptimal concentrations of T cell mitogens. It was reasonable to hypothesize that such cytokines, which act both at a distance and in a cognate fashion, might be found in the invertebrates. Indeed, IL-1–like *activities* have been described for several invertebrates, but unfortunately no molecular data revealing the structures of the active invertebrate cytokine/cytokine receptor have been reported. In fact, no ortholog has been detected in the genome projects from Protostomic invertebrates (see Fig. 4.1), and only IL-17 and TNF homologs have been detected in non-vertebrate deuterostomes and protostomes[360] and IL-8 in agnathans[130]; thus, we may consider these as primordial cytokines related to the vertebrate versions.

IL-1 activity as measured by costimulation assays or as a consequence of PAMP stimulation has been detected in all nonmammalian vertebrates. IL-1β upregulation has been detected after treatment of macrophages with LPS, consistent with its inflammatory function in mammals. In addition, injection of gram-negative bacteria into trout induced IL-1β expression in many tissues. Identity with the mammalian IL-1β gene in all other species ranges from 28% to 40% (identity between mammalian IL-1α and IL-1β is about 25%). In nonmammalian species, IL-1β lacks the so-called ICE cleavage site, important for function in mouse/human. IL-18, an IL-1–related cytokine, in contrast to IL-1 seems more focused in its function of potentiating TH1 and NK cell responses. IL-18 has been detected in birds and fish, but the tissue distribution in fish seems to be expanded as compared to mammals. Additionally, IL-18 in nonmammalian vertebrates contains the ICE cleavage site, unlike IL-1β. Other IL-1–related cytokines, including IL-33 and the IL-1F series, have also been found in several jawed vertebrates, but with little study of functional activity.[39] This will be of great interest to uncover, considering the IL-1 family members' roles in regulating T-helper differentiation.[361]

Both chicken IL-1β and the IL-1R were identified and have been expressed as recombinant proteins. The IL-1R homology to mammalian orthologs is quite high (61% identity), but the highest similarity is found in the cytoplasmic domains.[362] In addition, there are four blocks of high similarity to the cytoplasmic tail of toll/TLR proteins, and IL-1R and TLR use similar signal transduction cascades (see previous discussion).

As mentioned, IL-8 (actually, the CXC chemokine CXCL8) has been identified in the lamprey and seems to be expressed specifically in B cells, whereas the IL-8 receptor is found in T cells; this has been one of the mechanisms proposed to permit the cells to interact in a cognate fashion.[130] IL-8 has also been found in various gnathostomes such as trout, flounder, and perhaps chicken; a chicken CXC chemokine called K60 clusters with IL-8 in phylogenetic trees and is upregulated in macrophages stimulated with LPS, IL-1β, and IFN. Interestingly, Marek disease virus expresses an IL-8 homolog (v-IL-8), which may be involved in inducing immune deviation.[363]

The "immediate" TNF family in mammals includes the canonical TNF-α, lymphotoxin (LT)α, and LTβ, all encoded at the distal end of the MHC class III region, as well as a large number of other members with diverse immune functions (Chapter 10). TNF-α is the best studied of these cytokines, and it is one of the key regulators of innate and adaptive immunity. The other two cytokines have a more limited tissue distribution and function and are noted in their roles in lymphoid tissue development, especially in the formation of splenic white pulp and segregation of T and B cells in lymph nodes.[362,363] In contrast to IL-1, TNF homologs have been detected in cnidarians, protostomes (see previous discussion), sea urchin, *Ciona*, and *amphioxus*, consistent with the idea that such a multifunctional cytokine would predate the jawed vertebrate adaptive immune system.[356,364] Homologs have also been cloned from several teleost species, and TNF-α expression in leukocytes is upregulated within 4 hours after treatment with LPS, IL-1β, and PMA. While there is good phylogenetic support for orthology of fish TNF-α to that of mammals, the other TNF genes are teleost-specific duplicates rather than the LT genes. Conversely, in *Xenopus*, the three TNF family members described previously in mammals are closely linked in the class III region of the MHC.[39] This is a bit of a surprising result as all of these family members seem to be lacking in chickens, consistent with a lack of lymph nodes in these animals; however, the unusual nature of the MHC in birds (see above) rather is consistent with a loss of TNF genes in these animals, which is indeed surprising, especially for TNF-α.[365]

An excellent review detailing the evolution of all TNF and TNF receptor family members demonstrated that all of the genes are found in paralogs on four chromosomes, consistent with the 2R hypothesis.[364] This work demonstrated that the earliest member was linked to the proto MHC, and thus it is not surprising that TNF is an ancient gene family.

IL-6 and its related members IL-11 and IL-31 are also found in most gnathostomes examined, with few functional data.

Interleukin-2 and the γC Family of Cytokines

The family of cytokines that signals through the γC receptor includes IL-2, IL-4, IL-7, IL-9, IL-15, and IL-21. Like what was described for most molecules involved in adaptive immunity, all of these cytokines (except perhaps IL-9) are present in all gnathostomes studied. Costimulation assays of thymocytes, as described previously for IL-1, and perpetuation of T cell lines with stimulated T cell supernatants are performed to detect IL-2 or "T cell growth factor" activities. Unlike IL-1, IL-2–like factors generally stimulate cells only from the same species,[366] and it is a "cognate" cytokine, meant for release only between closely opposed cells, or as an autocrine factor. From teleost fish to mammals, stimulated T cell supernatants costimulate thymocyte proliferation or can maintain the growth of T cell blasts. The chicken IL-2 protein is only 24% identical to human IL-2 and only 70% identical to a near cousin, the turkey. IL-15, a relative of IL-2, has also been cloned in the chicken.[367] A candidate IL-2R in chicken was identified by a mAb recognizing a 50-kDa molecule only on stimulated T cells (thus an IL-2Rα homolog). This mAb blocks costimulation by IL-2–like molecules in chicken T cell supernatants. IL-2 has been studied in the bony fish *Fugu*. In both chicken and the deduced *Fugu* IL-2 protein, there is a second set of cysteine residues, which are found in IL-15 and thus is a primordial feature.

As mentioned, γC is the signaling subunit of the IL-2, IL-4, IL-7, IL-9, IL-15, and IL-21 receptors; absence of this chain in mammals leads to major defects in lymphocyte development ("boy in the bubble"). In fish, a γC homolog was cloned in rainbow trout with unusually high identity (44% to 46%) to mouse/human genes. Since then, many other orthologs have been uncovered in fish, often duplicated in various species. IL-21, which is encoded adjacent to IL-2 in mammals, has been found in ectothermic vertebrates. Because one of its major functions is the differentiation of T-follicular cells and the GC response, it will be of interest to study its role in those animals. IL-7 is involved in lymphocyte differentiation as well as homeostasis of mature lymphocytes. Recently, mutations in Il-7R and downstream signaling molecules in the zebrafish have shown a block in T cell development. These mutants will permit future study of lymphocyte production in this crucial developmental model.[368]

TH2 Cytokines: Interleukin-4, Interleukin-5, Interleukin-9, and Interleukin-13

Responses to extracellular pathogens, especially filarial worms, are largely regulated by TH2 cytokines in mammals. IL-4 is the most pleiotropic cytokine in this regard, stimulating the production of neutralizing antibodies, stimulation of eosinophils and mast cells, and an antagonism of TH1 responses. IL-4, IL-5, and IL-13 (and granulocyte macrophage colony-stimulating factor) are encoded in the so-called TH2 cytokine complex in mammals, and all of the genes are coordinately upregulated after stimulation of a nearby locus-control region.[369] This same "TH2 complex" is found in chickens but with a pseudogene for IL-5, but in *Xenopus* only the IL-4 gene has been identified to date. Fish have genes in which the ancestor of IL-4 and IL-13 seems to be in a preduplicated state, syntenic with genes in the TH2 complex. Again, duplicates of IL-4/-13 and their receptor have been found in various fish.[39] Cartilaginous fish have TH2 homologs in the TH2 complex as well.

Interferons

IFNs, classically known for their antiviral properties, are divided into three groups: type I, type II, and type III. Perhaps surprisingly, this entire group has only been uncovered in jawed vertebrates to date, put into context in a recent review.[370] However, viral immunity in insects is partially regulated via a JAK/STAT response. Types I (α and β) and III (λ) IFN are widely expressed in cells of many types and induces inhibition of viral replication in neighboring cells, as well as molecules of the innate immune system such as inducible nitric oxide synthase (iNOS) and IFN regulatory factor-1. In contrast, type II IFN (IFNγ or immune IFN) is synthesized by activated T cells, activates macrophages, and upregulates class I, class II, immune proteasome subunits, and TAP, and many other genes (see Chapter 9).

Antiviral activity is detected in supernatants from virally infected fish fibroblasts, epithelial cell lines, and leukocytes. All of the biochemical properties of mammalian type I IFN (eg, acid-stable, temperature-resistant) are present in these fish supernatants, and the putative IFN reduces viral cytopathic effects in homologous cell lines infected with virus. In vivo, passive transfer of serum from virally infected fish protects naïve fish from acute viral pathogenesis. There appears to be two lineages of type I IFNs in fish that are specific to this group. In chickens there are up to 10 closely related, intronless type I IFN genes. Sequence identity to human type I IFN ranges from 25% to 80%, with the apparent functional gene having highest similarity.[39] Interestingly, bats have greatly expanded their type I and III IFN genes and receptors (as well as many other genes associated with viral immunity). These animals are highly infected with (and susceptible to) virus, and some IFN forms are constitutively expressed at low levels.[371]

While type I/III IFNs can be highly duplicated in some vertebrates, this is not true of IFN receptors, and in some cases many cytokines will use two to four receptors within a species. It appears that even single amino acid changes in the cytokine can induce differential conformational changes in the receptor that can modify downstream signaling. This paradigm will be exciting to follow in a phylogenetic context.[39,372]

Type II or IFNγ has been cloned in most jawed vertebrates. The chicken gene is 35% identical to human type II IFN and only 15% identical to chicken type I IFN. Recombinant chicken IFN stimulates nitric oxide production and class II expression by macrophages. The gene has been cloned in many fish and has been studied mostly in the trout where it has been shown to upregulate CXCL10 and activate protein kinase C. Thus, at least in chickens and teleost fish, type II IFN seems to have the same function as in mammals, suggesting that the TH1-type responses emerged early in vertebrate evolution.

Heterodimeric Cytokines

The IL-12 family are composed of heterodimers that share chains (p19, p40, p35, and EB13) and include IL-12, IL-23, IL-27, and IL-35. IL-12 is generally considered to be a proinflammatory cytokine produced by APCs that promote a TH1 response after exposure to intracellular pathogens (Chapter 28). Consistent with the ancient derivation of TH1 responses, the two subunits of IL-12 p70 (p35 and p40) have been found in chickens and several teleost species.[373] The p40 subunit of IL-12 can also associate with p19 to form the cytokine IL23, which is involved in the perpetuation of TH17 cells. IL-27 generally inhibits all TH subsets but expands regulatory T cells and is composed of p28 and EB13; IL-35 is a relatively new regulatory T cell–produced cytokine composed of p35 and EB13. All of these chains have been detected in gnathostomes and several studies have been done to examine their expression at the RNA level; however, to date no reagents have been prepared to the heterodimers for functional analysis.[39]

Transforming Growth Factor-β and Interleukin-10

TGF forms a large family with pleiotropic effects in many developmental systems. For the immune system, TGF-β isoforms are best known for their capacity to suppress adaptive immune responses (even across species barriers), although they can also stimulate lymphocytes under certain conditions, especially in mucosal surfaces (see Chapter 38). TGF-β inhibits macrophage activation in trout and growth of T cell lines in *Xenopus* species. Recombinant *Xenopus* TGF-β, like the mammalian form, also can inhibit IL-2–like dependent growth of splenic lymphoblasts. Four TGF-β isoforms were isolated from chickens, as opposed to three major forms in mammals. The three major forms of the cytokine have been isolated in several teleost species.[39]

IL-10 is often considered along with TGF-β because it is mostly an immunosuppressive cytokine with multiple effects; both cytokines are expressed in subsets of "regulatory T" cells in mammals (Chapter 9). This cytokine was originally discovered by its ability to suppress TH1 responses, but now is known to have a much-expanded role in immunity. IL-10 has been found in chickens and many teleosts. As usual, functional experiments have lagged the molecular work, but recombinant chicken IL-10 can block IFNγ production by splenocytes. IL-10 is a class II cytokine, related to IFNγ, IL-19, IL-20, IL-24, and IL-26.[374] IL-22, a cytokine released by TH17 cells and ILCs in mammals, is upregulated in fish vaccinated with pathogens and is correlated with immune protection.[375]

Note that there have been very few "regulatory T" experiments reported in the comparative literature, but many older experiments suggest that such cells exist. Over 25 years ago, experiments in *Xenopus* showed that graft rejection could be delayed when lymphocytes from metamorphosing animals were adoptively transferred into adult frogs. This result suggested that a "wave" of suppressor cells emerged at metamorphosis that protect animals from autoimmunity when adult-specific molecules were expressed.[376] It is time to reexamine such experiments with modern tools. In a recent experiment in zebrafish, immunization of animals with central nervous system (CNS) antigens induced an autoimmune reaction. Overproduction of foxp3 inhibited the production of IL-17 and IFNγ in these animals, ameliorating the disease.[377]

Interleukin-17

TH17 cells in mammals are important for inflammatory responses to extracellular bacteria via indirect stimulation of neutrophils (Chapter 28). Despite its late appearance in the immunologic literature, IL-17 appears to be the most ancient cytokine, with homologs in protostomes and lower deuterostomes, and extensively amplified in sea urchins.[36] Even different isoforms seem to be conserved among the jawed vertebrates, which should initiate many new avenues of study. The family is composed of IL-17A/F, with the A and F forms being the best studied to date. IL-17A/F genes have been found in all mammals studied, as well as IL-17C/E. The most ancient IL-17 form is IL-17D, present in both oysters and agnathans. As mentioned, this form is expressed in lamprey T cells and has been proposed as a helper factor for cognate interactions.[378] Finally, like many other immune genes, as many as 20 IL-17 genes have been found in sea urchins, again suggesting expanded innate immunity in this species, especially along mucosal surfaces.[379]

Despite our "big picture" knowledge of the emergence of cytokines, obviously functional experiments have lagged. Some genes may have been lost, as described previously for the chicken—again, further studies of amphibians and cartilaginous fish should help in our understanding. In mammals, the LT TNF family members are important in the development of lymph nodes and splenic white pulp, and yet the three genes are present in lymph node–less *Xenopus*—what are their functions in ectotherms? If we can uncover these other roles in nonmammalian vertebrates, it could be quite informative to reinvestigate such functions in mammals. Finally, chickens and fish have evolved their own paralogs of some of the well-known cytokine genes; their study should reveal selection pressures on particular species that could also be quite informative in our understanding of the gestalt of the cytokine network.

TH Cell Subsets

Classically, mammalian T-helper cell phenotypes over the past 10 years have expanded from the classic Th1/Th2 paradigm to that founding pair as well as Th17, T-follicular helper (Tfh), Treg, and several others (Chapters 28 and 29). While much work must be done to study T cell function in ectotherms, it is an even bet that such phenotypes will be found in all gnathostome classes. A previous suggestion that cartilaginous fish lacked most of these phenotypes as well as a functional CD4 molecule was proven incorrect[62]; several cytokines that were not detected in the genome project were uncovered with structurally based search programs[380]; in addition, at least the expression levels of the shark CD4 molecule also appear to be conventional (Ohta and Flajnik, unpublished). To date, jawless fish only have been shown to have IL-17 (in T cells) and IL-17R (in B cells), proposed to be involved in T-B collaboration, but these are early days and other cytokines will likely be uncovered.[92]

Evolution of the Treg lineage is of special interest. Studies in bony fish suggest that foxp3-positive cells act as Tregs, perhaps tTregs (thymic-derived or natural Tregs). As mentioned, studies carried out long ago in amphibians suggested that a population of "suppressor cells" was peripheralized at metamorphosis, likely involved in regulating adaptive responses to newly arising adult-specific self-antigens. Rudensky and colleagues have suggested that iTregs (adaptive or inducible Tregs) arose in evolution with the advent of placentation, consistent with a foxp3 genetic regulatory element (CNS1) only found in placental mammals and no other vertebrates.[381] They propose that after the emergence of iTregs for this purpose, such cells were co-opted to suppress adaptive immunity in a variety of situations. It will, of course, be of interest to determine whether this theory holds water, both in maternal–fetal interactions and generally in ectothermic vertebrates (and birds).

General Evolution of Chemokines and Their Receptors

Chemokines and chemokine receptors are essential for many aspects of the immune system, including inflammation, the differentiation of lymphoid tissues, trafficking of hematopoietic cells during ontogeny and immune responses, and even stimulation of cells under various conditions (Chapter 11). With the advent of the genome and EST projects, a seminal paper showed that, consistent with the Big Bang theory of adaptive immune system evolution, all of the major classes of chemokines and their receptors were present in the bony fish lineage and probably will be found in cartilaginous fish as well when the genomes are completed.[359] Similar to most of the immune-related genes discussed throughout the chapter, bony fish have more chemokines/receptors than any other vertebrate, including amphibians and mammals.[373]

Interestingly, no CC (so-called homeostatic) or CXC (so-called inflammatory) chemokines/receptors were detected in lower deuterostomes like *Ciona* and sea urchin. The chemokine receptors are member so the G protein–coupled receptor family, and of course such molecules are found in all animals. However, no members of the specialized family (G protein–coupled receptor γ) to which chemokine receptors belong were found in lower deuterostomes.[382] The chemokine receptor genes are found on four chromosomes in mammals, four of which contain the HOX genes, one of the gene families that provided evidence for the 2R hypothesis early in the reawakening of this theory. The distribution of chemokine receptor genes on these chromosomes correlates very well with 2R and helps to account for the large-scale gene expansion of this family in the jawed vertebrates.[383] Thus far, very few chemokines/receptors (eg, IL-8 [CXCL8] and CXCR4) have been detected in agnathans, but as mentioned previously, agnathan T cells express IL-8R and B cells IL-8, believed to be a factor in promoting the cognate interactions between the cells.[291] CXCL12, a chemokine vital for thymic differentiation, is also present in lamprey, consistent with new work on the agnathan thymus discussed previously. In summary, it appears that a few chemokines/receptors arose at the dawn of vertebrate adaptive immunity, and they were greatly expanded by both en bloc and *cis* duplications and became "full-blown" shortly after the 2R Big Bang. Currently, this is our most complete story of the evolution of a family of immune genes in deuterostomes.

Summary of Adaptive Immunity

Several major questions must be addressed in the comparative immunology adaptive immunity field, touched on here. The discovery of the adaptive system in jawless fish has provided a goldmine of interesting questions including the primordial role of the thymus (selection or isolation?), origins of lymphocytes (ILC or antigen receptor–bearing?), emergence of IgSF antigen receptors (present in which form?), and the (convergent?) "MHC" for presentation to VLRA (T)-bearing lymphocytes. Studies of fish and amphibians, with new reagents and powerful sequencing and knockout procedures, will provide more insight into the study of antigen presentation prior to the appearance of FDC and GC, the multifaceted nature of γδ T cells, NKRs and their functions, and the origins of mucosal immunity; especially for the latter, study of this system in animals lacking mucosal secondary lymphoid tissues may provide a nice framework for understanding the highly complex mammalian GALT. We have just scratched the surface in our understanding of the evolution of the APOBEC family. There is no doubt that future studies of immunity in nonmammalian animals will surprise us and more importantly continue to alter our views of the adaptive immune system.

EVOLUTION OF ALLORECOGNITION

Histocompatibility Reactions in Protostomes

The study of MHC in mammals began with alloreactivity (Chapters 19 and 48). Graft rejection was an early indication of immunity in several animal models, but such studies revealed *physiological* histocompatibility rather than the artifactual histocompatibility found in vertebrates. Scrutinizing specific memory in non-vertebrate systems was a major tool to establish the universality of adaptive immunity, as championed by Burnet in a famous early paper, and it fueled speculations on the origins of the vertebrate MHC.[384] However, genetic regions and mechanisms responsible and molecular underpinnings of some alloreactions are now coming to fruition, with no (or little) resemblance to the vertebrate MHC, although regions of the extended vertebrate MHC have been identified in some organisms but without functional correlation.

Colonies of *Porifera*, *Cnidaria*, *Bryozoa*, and *Tunicata* (Fig. 4.1) often compete for space and can develop histocompatibility reactions in the zone of contact. In addition, cell lineage parasitism, in which the somatic (including HSC) and/or germ cell lineage of one partner replaces that of the other, may ensue if colonies fuse into a chimera.[385] Thus, effector functions following allorecognition can also protect the genetic integrity of the individual. In some species, fusion is apparently restricted to tissues of the same individual (complete matching); in other species, such as bryozoans, fusion also occurs between genetically distinct individuals if they share kinship (partial matching). Two divergent invertebrate phyla have been studied in detail: *Cnidaria* and *Tunicata*. Note that Cnidaria is in the Protostomic lineage but is still considered in this chapter as one of the better understood alloreactions in animals.

The existence of highly polymorphic histocompatibility loci was demonstrated long ago in various corals where apoptosis induction was responsible for the death of partners in incompatible combinations, whereas in sea anemone nematocyte discharge was induced between incompatible individuals. The colonial cnidarian *Hydractinia* was the only species providing a model to analyze genetic control of such reactions. In 1950, Hauenschild noticed that allorecognition was under the control of a single genetic region with multiple alleles.[386] Colonies of *Hydractinia* encrust the shells of hermit crabs where they grow by elongation and branching of stolons. Embryos and larvae fuse indiscriminately. However, when two or more larvae are recruited to the same substratum, colonial forms may come into contact through their stolons. If the two colonies are histocompatible, stolon tips adhere and fuse, establishing gastrovascular connections and a permanent genetic chimera. If tips of incompatible colonies fail to adhere, they swell (hyperplastic stolons) with the migration of nematocysts, which discharge and damage the tissues. In addition, transitory fusions can also occur.[385] Similar to the genetics of other non-vertebrates, colonies fuse if they share one or two haplotypes, reject if they share no haplotypes, and display transitory fusion if they share only one allele at one haplotype and no alleles at the other; this type of reaction is similar to what is found for NK cell activation described earlier. Examination of the polymorphic locus governing this reaction (*alr*) revealed the involvement of two closely linked polymorphic loci 1.7 cm apart, *alr1* and *alr2*. *alr1* encodes a three IgSF domain surface molecule, not unlike other metazoan nectin-like molecules (Fig. 4.13), with a cytoplasmic tail equipped with multiple signaling motifs. In the first domain, the molecule shows a high level of variability at particular residues, suggesting positive selection. *alr1* is embedded in a genetic region consisting of multiple IgSF members.[387] *alr2* encodes another highly polymorphic gene with three IgSF domains with the distal V-like domain being the most variable. A high level of allelic diversity was found at *alr2*. Alr1 and Alr2 alleles undergo a homophilic interaction, suggesting that such recognition is involved in the productive fusion reaction and the lack of this interaction (missing self) somehow results in rejection.[388] The interaction occurs across one face of the IgSF domain where the polymorphic residues are found. Besides these two genes, there are many other IgSF genes related to *alr* that are likely playing roles in the reaction.

The Famous *Botryllus* System

Compared to *Cnidaria*, tunicates are deuterostomes (Fig. 4.1) and thus might be expected to have a histocompatibility system more related to MHC. Allorecognition has been studied in both colonial and solitary ascidians. In *Botryllus*, a colonial ascidian, extra attention was afforded to this system because the locus controlling histocompatibility was linked to (or was the same locus as) the locus controlling fertilization by preventing self-fertilization, discovered by Oka and Watanabe.[389]

Metamorphosis in *Botryllus* is followed by budding that eventually gives rise to a large colony of asexually derived genetically identical individuals (zooids), united through a vascular network.[390] At the periphery of the colony, the vasculature ends in ampullae, which are the sites of interaction when two colonies meet during their expansion. The interaction results in either fusion of the two ampullae to form a single chimeric colony sharing a common blood supply or a rejection reaction during which the interacting ampullae are destroyed, thus preventing vascular fusion. Hemocytes (morula cells, mentioned earlier) are involved in the reaction.[391] Fusion or rejection is governed by a single highly polymorphic (tens to hundreds of alleles) locus called the FuHC (for fusion/histocompatibility). Like what was described earlier, when two colonies share one or both FuHC alleles, they will fuse; rejection occurs if no alleles are in common. However, like in *Hydractinia*, intermediary pathways have been reported in the past and have not been fully elucidated.

The C-terminal region of the FuHC molecule consists of a nectin-like segment with three Ig domains, showing most similarity to one of the *Ciona* nectins and perhaps CD155 and other nectins (Fig. 4.13).[392] Those members are found in a tetrad of paralogs in vertebrates but not linked directly to the tetrad of MHC paralogs. The N-terminal region contains an epidermal growth factor domain and some other unrecognizable regions not conserved among ascidians. Any two FuHC alleles differ by an average of 4% at the nucleotide level. Unlike MHC class I and class II, polymorphic residues in FuHC alleles are not concentrated in particular regions.[393]

Some 200 kb away from FuHC is a second polymorphic (and polygenic) locus, *fester*, which is inherited in distinct haplotypes.[394] Diversified through extensive alternative splicing, with each individual expressing a unique repertoire of splice forms (each individual expressing up to three splice variants in addition to the regular full-length product), it potentially exists in both membrane-bound and secreted forms, all expressed in tissues intimately associated with histocompatibility. After *fester* knockdown, the histocompatibility reaction is blocked at the stage of initiation as if the colonies ignored each other. By contrast, when fester was blocked with specific mAbs, fusion reactions were unaffected, but rejection reactions became fusions in an allele-dependent manner. These data combined with the genetic location suggest that fester encodes the FuHC binding partner. Fester contains a short consensus repeat (or sushi domain) found in vertebrate complement receptors. Besides these two polymorphic products, another *Botryllus* cell surface–expressed nonpolymorphic molecule related to fester, uncle fester, is required for incompatibility reactions, while fester seems to

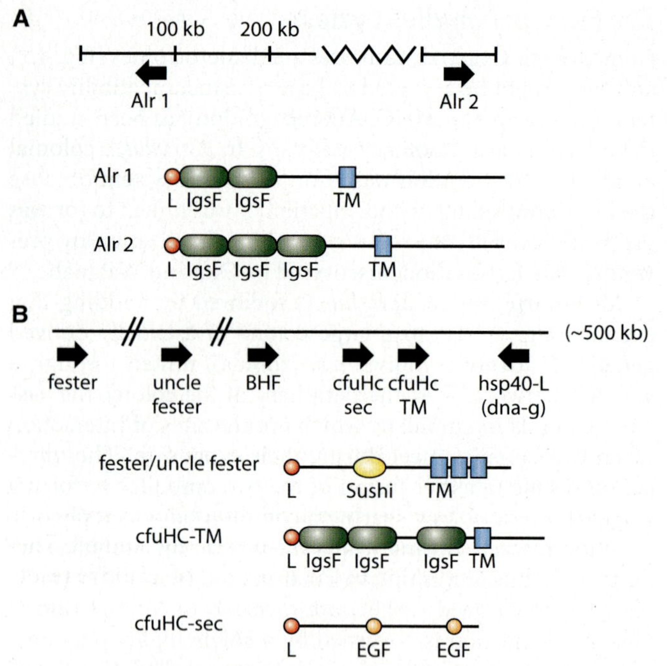

FIG. 4.13. Histocompatibility loci in the cnidarian *Hydractinia* and the urochordate *Botryllus*. A: *Alr* loci in *Hydractinia* encode IgSF molecules (green domains). The loci are very polymorphic and interact in a like-like fashion. Arrows show the transcription orientation. L (leader, red); TM (transmembrane, blue). **B:** FuHC in the colonial tunicate *Botryllus* showing its "MHC," in an ~500 kb region. The loci shown have been demonstrated to be involved in the histocompatibility reaction, except for hsp40-L, which has not been examined functionally but is polymorphic and expressed at the site of rejection. Sushi domain (yellow); epidermal growth factor domain (EGF, orange). (Reprinter from Nicotra ML. Invertebrate allorecognition. *Curr Biol.* 2019;29(11):R463-R467. doi:10.1016/j.cub.2019.03.039, with permission from Elsevier.)

be required for allele discrimination and inhibition of killing. So, several independent pathways seem to control the final outcome of the interaction between individuals. Additionally, another polymorphic locus at the FuHC, BHF, which has no similarity to any described gene family, was found to be the best predictor of fusion/rejection reactions.[395] BHF has no transmembrane domain and transfection studies suggest it is not found at the cell surface but rather in the cytoplasm and nucleus.[396] Another polymorphic gene in the hsp40 (DNAj) family is expressed in ampullae and also could play a role in histocompatibility[397] (Fig. 4.13).

The morula cells involved in the cytotoxic reaction have been studied in some detail with a panel of monoclonal antibodies. They were shown to be cytotoxic with a transcriptome strongly suggestive of the myeloid lineage.[42] This is a big step forward in analysis of the system and should allow a better study of how the different molecules encoded at FuHC interact to regulate histocompatibility.

Solitary Ascidians

In *Halocynthia roretzi*, a "polymorphism of color" has been observed and histocompatibility reset by a mixed hemocyte technique in vitro resulting in a melanization reaction likely to involve the PPO cascade. Depending on the strains, the percentage of positive reaction varied from approximately 55% to 70%, indicative of polymorphism. Grafting experiments had already shown the existence of allorecognition in solitary ascidians, and investigation at the cellular level had demonstrated the occurrence of cytotoxic cells in such organisms.[398] To shed light on allorecognition in urochordates and on the molecules involved in preventing self-fertilization, gonadal cDNAs of three genetically unrelated *Ciona* individuals were compared by suppression subtractive hybridization. This led to the discovery of the highly polymorphic variable C′ receptor–like 1 gene coding for a transmembrane protein with several short consensus repeat domains (short consensus repeat/complement control protein [CCP]), a motif shared with the variable fester receptor of *Botryllus* described previously.[399] Genes encoding variable C′ receptor–like are in the same linkage group as a set of IgSF domains with homology to nectin (CD155/poliovirus receptor [PVR], cortical thymocyte protein [CTX]/junctional adhesion molecule [JAM] family members, etc.) and other adhesion molecules of which related members can be found also on one genetic region in vertebrates, the 19q34 human chromosome segment with the extended LRC.[32]

The Meaning of Histocompatibility Reactions in the Invertebrates

The association between allorecognition in *Botryllus* and fertilization led to the proposal that histocompatibility systems were selected during evolution to avoid inbreeding. The hypothesis made sense in the case of sessile colonial invertebrates that might have difficulty dispersing their gametes and therefore are susceptible to inbreeding depression. Indeed, the partial matching mentioned previously is a general characteristic of fusion compatibility in colonial invertebrates, perhaps driven by selection operating on an error-prone genetic system for self-recognition that is perhaps constrained by derivation from a gametic function selected to reduce inbreeding.[41] Furthermore, even in mammals there is a large literature suggesting a selection both at the mate choice and pregnancy levels for preserving heterozygosity at the MHC. However, the possibility of a common genetic system, or linked systems, governing fusion and gametic compatibility awaits confirmation. Moreover, inbreeding avoidance can only explain the selection of histocompatibility alleles if the loci are genetically linked to a large fraction of its genome. This is inconsistent with the tight linkage of histocompatibility genes to a single major locus, especially in invertebrates.

Another hypothesis is that alloimmunity was selected because it avoided intraspecific parasitism and/or competition for attachment sites. Indeed, after fusion of compatible colonies, bloodborne germline or totipotent stem cells are transferred between colonies and can expand and differentiate in the newly arising, asexually derived individuals of the vascular partner.[400] This can result in a situation where only one genotype is represented in the gametic output of the fused individuals. The FuHC polymorphism in *Botryllus* could function to restrict to compatible individuals the vascular fusion and the germline parasitism. The high allelic polymorphism characteristic of all invertebrate recognition systems may have evolved in response to selection for fusion with self rather than kin. Fusion with self will allow the development of a colony that benefits from fusion while eliminating the possible cost of somatic cell parasitism and thus would be the raison d'etre of the allorecognition mechanisms.

Both previous hypotheses are essentially based on observation made in colonial tunicates, but colonial living has evolved several times independently. Urochordates are most related to the vertebrates, but the data argue against any of the genes involved in their histocompatibility reactions being ancestral to MHC class I and class II. It is clear in all these studies that a strong pressure for polymorphism is the rule, but the evidence from all of these fascinating reactions suggests that such "pressures" select for similar systems via convergence, at least at the level of the receptor. The selecting molecular environment can perhaps show more conservation (see below). On the other hand, as described previously, genetic regions with homology to the vertebrate MHC have been detected in the invertebrates, yet are (most likely) unrelated to the reactions detailed here (see Conclusions). As described, in all of these allorecognition systems, convergent mechanisms have been encountered to reach an analogous end, and thus it is unlikely that a unique "cause" arose for selecting them in diverse organisms.

Recently, innate histocompatibility has also been uncovered in mammals. The activating IgSF receptor PIRA was shown to be responsible for a "memory"-type rejection of kidney grafts.[401] This innate memory may be the holdover searched for in connecting the invertebrate histocompatibility responses with responses in the vertebrates, but we are in VERY early stages. These new studies also are consistent with epigenetic programming described for "trained immunity."[402]

CONCLUSIONS

Summary of Comparative Immunology

- For the last 15 years, the sea urchin has provided an essential molecular and functional model to understand immunity in pre-vertebrate deuterostomes.[36] Generally, they have large classical innate immune gene families and the ancient cytokine IL-17, the latter shown to be important for barrier immunity in these echinoderms. In addition, a unique immune family, Transformer, provides surprise after surprise regarding protein structure, PAMP specificity, and "generation of diversity," perhaps offering a connection between innate and adaptive immunity.
- The cephalochordate amphioxus has provided the best genetic and functional model to understand the basic "0R" genome of the vertebrates. A functional RAG transposon was uncovered in this species, changing our view of RAG origins.[403] Transcriptome analysis suggests that the origin of lymphocytes might be revealed in this species. Functional studies have lagged the molecular work (true of most species over the last 10 years), but study of one special IgSF family VCBP suggests its important role in mucosal homeostasis.[89] Additionally, another VJ-type IgSF member has been implicated in systemic and mucosal immunity.[404]
- Generally, the vertebrates have inherited, *en masse*, the innate immune repertoire from invertebrate ancestors. Besides type I interferons, there are very few molecules involved in innate immunity not found in the protostomes. This is also true of signaling mechanisms such as those used for TLR, TNF, JAK-STAT (Chapter 3), SYK,[405] and many others. What has occurred, no doubt, is the expansion of the genes involved in the pathways and in regulation. This has occurred via the genome-wide duplications early in vertebrate history, but also by cis duplications of genes within clusters, for example, the TNF superfamily.[364] If you take anything away from this chapter, take this!
- Agnathans have an adaptive immune system centered on a different gene family than the gnathostomes.[406] VLR are composed of LRR (not IgSF) domains, assembled by APOBEC family members during the development of lymphocytes. There are three VLR loci, VLRA, VLRB, and VLRC, which are expressed by cells that seem to be expressed by the gnathostome equivalents of αβ T cells, B cells, and γδ T cells, respectively.
- Agnathans also have VJ-type IgSF molecules expressed in lymphocytes that have not been extensively studied. They could offer a connection to the two types of adaptive receptors in vertebrates.[114,115] Such VJ domains have also been detected in the lower deuterostomes like amphioxus and *Botryllus.*
- Cartilaginous fish (and other vertebrates) have lost the TLR4 gene, along with MD2 and CD14, suggesting that there is a different type of interaction between the microbiota and the immune system in these animals.[62] Bacteria have been cultured from shark tissues, which may fit with this proposal.[63] Additionally, the alternative C′ pathway may also have been lost in sharks, also consistent with a different type of interaction with commensals.
- Cartilaginous fish Ig H and L chain genes are in the so-called "cluster organization," with V, (D), J, and C elements found in each cluster.[407] Despite this, each B cell shows antigen receptor exclusion that is the expression of only one H chain (but not L chain) per cell.[263] This gene organization has allowed rapid evolution of distinct types of Igs, including a single-domain V region receptor called IgNAR, intimate associations of Ig genes with TCR genes (eg, NAR-TCR), as well as clusters which have been "germline-joined" and believed to serve unique functions early in ontogeny and into adult life. A recent study showed that mutation of the TCRαV gene in thymocytes contributes to diversity of T cells, suggesting that this feature was lost in more advanced vertebrates.[191] Sharks are da bomb.
- Teleost fish have been shown again and again to have rapidly evolving immune systems, often expanding gene families in unprecedented ways (eg, NLR, TRIM), generating specific paralogs and losing others (eg, cytokines, chemokines, etc.), losing canonical syntenies found in all other vertebrates (eg, MHC class I and class II nonlinkage), and having unique forms of ancient molecules (eg, tetrameric IgM with no J chain). The third genome-wide duplication (3R) specific to teleosts has been a major force in this instability and rapid evolution.[22,59] Certain bony fish taxonomic lineages have lost class II genes as well as the invariant chain and CD4 molecules, that is apparently the entire T-helper system. In cod, the animals might compensate for the lack of class II by expressing an inordinate number of nonclassical class I genes, perhaps

selecting for a system with many NKT-like cells. Indeed, there are several families of innate-type receptors in teleosts that are likely involved in extensive innate immune reactions.[95] Additionally, fish with different lifestyles, including anglerfish species in which the male attaches to and makes vascular connection with the female, have lost cardinal molecules of adaptive immunity such as RAG, AID, class II, and others.[408] This group provides an opportunity to study the plasticity of immune systems in related organisms over a short period of evolutionary time. Note that studies in the older bony fish lineage, the holosteans, which did not undergo the 3R genome-wide duplication, suggest that they are much more like other vertebrates in their genomic organization[59]; in this recent paper, a comparison of the MHCs of the holostean bowfin, the teleost zebrafish, and human demonstrates this dichotomy of teleosts with other vertebrates.

- Antarctic fish Igs show clear evidence of selection of their Igs and other immune molecules, especially in their hinge regions, to allow for preservation of molecular flexibility in extremely cold temperatures.[139]
- An immune hypothesis has been proposed for the migration of salmon for spawning back to their original hatching grounds. During the stressful terminal migration, naïve lymphocytes are depleted but plasma cells are spared, likely producing antibodies specific for pathogens where the fish first were exposed.[409]
- The genomic sequence of the famous sarcopterygian coelacanth revealed a loss of IgM, the only vertebrate species to date with this feature.[144] IgW loci may have taken over the function of IgM, yet their sequence is highly unusual with many C domains. Transcriptome analyses of these famous animals would be quite useful since they, in addition to lungfish, serve as crucial intermediates in our understanding of adaptive immunity in the transition from sea to land (Fig. 4.2). The emergence of dedicated mucosal SLO is found in sarcopterygians, the lungfish.[323]
- Another sarcopterygian, the lungfish, live for long periods of time in a cocoon to prevent desiccation. This tissue recruits granulocytes that protect the hibernating fish.[410] Such recruitment of the immune system to meet certain challenges in other vertebrates will certainly be found in the future.
- Many fish and amphibians are polyploid, and adaptive immune genes are under pressure to become diploidized over evolutionary time (especially MHC class I/class II).[411,412] This has been an active area of research for decades and has been reinvigorated in the era of genome sequencing. This is especially interesting considering the polyploidization events at the origins of the agnathans and gnathostomes.[20,21]
- While sharks are the oldest living vertebrates with Ig/TCR/MHC/SHM/spleen/thymus, another major advance in immune evolution occurred in amphibians—the first true tetrapods—with the emergence of CSR and classic memory with an IgG (IgY) class,[413] as well as the bone marrow niche for HSC. Recent work has provided strong evidence for an old theory that bone marrow is required for protection against harmful radiation of HSC in land animals.[125]
- Amphibians undergo metamorphosis, with extensive changes in adaptive immunity occurring at this transition. Terminal deoxynucleotidyl transferase (TdT) is not expressed in larvae, so antigen receptor junctions in Igs and TCR lack N-regions and thus diversity is quite low.[227,272] In addition, classical class I expression is also low and class II expression is distinct in tadpoles (B cells and APCs) and adults (all lymphocytes). After metamorphosis, a second wave of lymphocytes develops rapidly, now with a large amount of CDR3 diversity. A working hypothesis is that in larvae, humoral immunity is regulated by CD4 cells, yet with low diversity of the B cell Ig repertoire[182]; cellular immunity might be the domain of NKT-like cells[106]; note that this hypothesis also holds for recent studies of very small fish with very few lymphocytes.[414] Suppressor cells may emerge at amphibian metamorphosis to prevent autoimmune reactions to newly arising adult-specific self-antigens.[376]
- Amphibians are in decline, and studies have found a correlation with innate immune mechanisms and their suppression by pathogens.[415] Several studies, however, have also implicated adaptive immune mechanisms, specifically MHC polymorphisms that correlate with susceptibility or resistance to the chytrid fungus.
- Most nonmammalian vertebrates have lineages of classical class I, TAP, and immunoproteasome (birds lost the immunoproteasome).[332] These genes, again unlike mammals, are tightly linked with class II genes. In birds, the functional link between class I and TAP alleles has been proven. LMP7 (β5i) but not LMP2 or MECL1 are included in these lineages, consistent with a critical function for LMP7 in the organization of immunoproteasomes. Note that both LMP7 and beta5t (thymic proteasome) are paralogs of the constitutive LMPX (β5), again suggesting a crucial role for this member of the β-proteasome ring.[416]
- TCRδ in cartilaginous fish, amphibians, coelacanth, archaic mammals, and birds utilize V regions from IgH, and the genes are found at the TCR αδ locus.[194] This phenomenon was first discovered for the cartilaginous fish IgNAR and soon after in marsupials and other species as a single V domain, or VH domain, associated with TCR-Cδ. Consistent with old theories of γδ TCR function, these findings suggest that γδ cells in many vertebrates have adaptive functions.[35]
- B cells, in all ectotherms tested, are capable of phagocytosis of particles and microorganisms.[266] This feature also applies to mammalian B1 cells, suggesting an ancient connection between ectotherm B cells and mammalian B1 cells, as well as between B cells and myeloid cells.[267] The functional significance of this phenomenon must be explored.
- Birds, like the bony fish, offer another example of rapid evolution, perhaps because of attack by viruses.[417] For the MHC, as mentioned, birds have lost immunoproteasomes and the thymic-specific proteasome and have modified their class I groove to be more promiscuous in peptide binding for one set of alleles, called "generalists," the other set being "specialists" and fastidious in peptide binding.[119] Chickens (but not ducks) have lost the RIG-1 gene, apparently making them susceptible to influenza.[365] Clearly, birds have fundamental and unique features of their immune systems to teach us about how pathogens (most likely viruses in this case) can compel a remodeling of innate and adaptive immunity.

- Mammals have dedicated SLO in their mucosa, and thus it is often difficult to tease apart the innate from adaptive responses in these tissues, except in contrived knockout or transgenic models. By contrast, ectothermic vertebrates have no Peyer patches or mesenteric lymph nodes and thus are excellent models to examine responses in the lamina propria, which may be predominantly T-independent. In addition, studies in all deuterostomes attest to a strong selection for mucosal immunity and rapid evolution so that similar mechanisms for homeostasis and defense have arisen via convergent evolution of the effector molecules.[14] Bony fish have led the way in this regard, with studies of the IgT isotype.
- Cytokines such as IFNγ, IL-2, and IL-4 and other adaptive cytokines (and type I/III IFNs) and their receptors are not present outside of jawed vertebrates (so far), nor is the vast array of chemokines seen in agnathans. However, now it appears that (almost) all of the cytokines/chemokines discovered in mammals are also present in basal jawed vertebrates.[39] IL-17, a cytokine garnering great attention in basic immunology, and TNF, seems to be the first cytokine to have emerged in evolution based on studies of lower deuterostomes and protostomes. If, as suggested from studies in mammals, IL-17 is important for responses to extracellular bacteria, non-vertebrate deuterostomes may hold the key to understanding the physiology of the entire system (ie, in the absence of the competing TH1, TH2, and regulatory T cells), and one may be able to study the dynamics of IL-17 development and function.[379]
- FDC and GC, with robust affinity maturation, arose in warm-blooded vertebrates.[238,249,258] The spleen is the oldest SLO and the only dedicated tissue for generation of adaptive immunity in ectotherms. B cell follicles surrounding a central arteriole make up splenic white pulp and they arose early in evolution, in the cartilaginous fish, and are also seen in amphibians and reptiles. In the absence of FDC, myeloid-derived APCs present antigen to B cells (and likely T cells), which has been shown in amphibians[258] and predicted to function similarly in sharks and reptiles. Bony fish have lost white pulp (like many other things), but recent evidence suggests that splenic melanomacrophages serve as a reservoir for antigen and generate immune responses by recruitment of T and B cells, perhaps like the development of TLT in mammals.[252,253,418] This field has advanced in the last decade and a comprehensive picture should emerge soon.
- A working hypothesis proposes that the divergence of lymphocytes into T and B cells occurred early in vertebrate evolution (Fig. 4.9). LRR- and IgSF-based antigen receptors existed simultaneously as well at this time (in the same organism or organisms in the same epoch), with VLR likely predominating in early vertebrates.[118] The RAG transposon insertion into an IgSF receptor gene (perhaps one already diversifying by mutation or conversion), along with a genome-wide duplication, resulted in the Ig/TCR/MHC-based system of gnathostomes, with all of its accompanying sophistication in all aspects of adaptive immunity, including development of the thymus and spleen, complexity of cytokine and chemokine networks, and intimate interactions of cells with primary and secondary lymphoid tissues.[35]
- The long-awaited molecular mechanisms of histocompatibility in non-vertebrates have been at least partially uncovered in *Botryllus* and *Hydractinia*.[385,390] The common themes thus far are that the genes encoding the polymorphic receptors/ligands are closely linked and that the genes all *appear to be* unrelated to the vertebrate MHC. However, there are provocative links to immune gene clusters found throughout the animal kingdom that suggest there may be an underlying fundamental similarity in recognition, perhaps related to the surface receptors and the signaling cascades.
- Translationally, the VLR and shark single-domain antibodies are novel platforms for diagnostic/therapeutic antibodies. The great phylogenetic distance between agnathans/cartilaginous fish and humans allows for immune responses to evolutionarily conserved epitopes on human targets. How useful will such reagents prove in our pharmaceutical arsenal?

Musings

Rearrangement or Somatic Hypermutation First?

Did IgSF- and LRR-containing antigen receptors co-exist in a common ancestor? What drove the gnathostome system to emerge? Was the original "generation of diversity" system for Ig/TCR actually APOBEC-based (ie, mutation-based) and then superseded by the RAG-based (ie, rearranging) system? And how do other APOBEC family members function throughout the vertebrate lineages and even in protostomes?

Because all antigen receptor genes use somatic rearrangement of V genes to generate diversity in CDR3 as well as to promote combinatorial diversity, there is little doubt that this mechanism is at the heart of adaptive immunity. Indeed, as described, most investigators believe that the introduction of the transposable element into a V gene was *the* driving force in the abrupt appearance of vertebrate adaptive immunity[221,403]; the finding of the RAG genes in sea urchins and amphioxus challenges this notion at some level. However, it cannot be overemphasized that SHM is also at the origins of the immune system; furthermore, all evidence to date suggests a gradual evolution of the SHM machinery (the AID/APOBEC family and associated polymerases/mismatch repair proteins) rather than the "hopeful monster" generated by the famous RAG transposon.[233,234] Thus, diversity generated via SHM or gene conversion may have existed in an adaptive immune system prior to rearrangement, and V gene rearrangement was *superimposed* onto this already existing system. Indeed, the presence of APOBEC family members in the jawless fish (CDA1 and CDA2)[126] and in invertebrates suggest that diversity generated via mutation/conversion preceded rearrangement. The RAG-induced rearrangement break and subsequent repair, however, provided something new in gnathostomes, not only diversity in sequence but also heterogeneity in size; this was a remarkable innovation and likely heralded the sophistication of jawed vertebrate adaptive immunity.[255]

Which Antigen Receptor First?

Phylogenetic analyses have suggested that among IgSF receptors, γ/δ TCR-like ancestor may have predated α/β TCR and Ig H/L.[419] This would suggest that direct antigen recognition, perhaps by a cell surface receptor, arose first in evolution followed by a secreted molecule and then an MHC-restricted one. Hood and colleagues argue that phylogenetic analyses over such large evolutionary distances obscure true relationships among the antigen receptor genes (eg, the relationships of the molecules in the phylogenetic trees have to impose multiple loss/gain of D segments in the different antigen receptor families) and suggest a model based upon genomic organization.[420] They propose an alternative phylogeny in which an ancestral chromosomal region with linked genes encodes both chains of an ancestral antigen receptor heterodimer, one having D segments and the other not (eg, IgH/IgL). The α and δ TCR loci are still closely linked in all gnathostomes (on human chromosome 14), and a pericentric inversion is suggested to have separated the TCR β and γ loci (linked on human chromosome 7). This model predicts that D segments only emerged once and also explain the existence of inverted V elements in both the TCR β and δ loci. This model does not predict which antigen receptor is oldest, but does provide a view of receptor evolution, consistent with *en bloc* duplications. Recent data have added to this scenario. First, as mentioned previously, VH elements have been found at the TCR-δ locus in many vertebrates, including sharks, amphibians, birds, and marsupial mammals.[194] In part, this is due to the close proximity of the IgH/L (lambda) and TCR alpha/delta loci in many living vertebrates (especially *Xenopus*, consistent with the Hood hypothesis of ancient *en bloc* duplications).[195] Additionally, the idea that most TCR-δ receptors do not recognize MHC-restricted antigen shows that they can continue to "borrow" elements from Ig loci over recent evolutionary time.

Thus, an alternative scenario is that MHC-restricted antigen recognition arose first, perhaps derived from a NKR with a "VJ" IgSF domain[105] that recognized SOS proteins, like the class I–like ULBPs or MIC today. The rearrangement break induced by RAG occurs in CDR3, which is in the center of the antigen receptor–combining site, in perfect position to interact with peptide bound to MHC,[421] which could have been another innovation to detect foreign antigens within cells. Thus, such cells would be most like extant αβ T cells with an MHC-restricted antigen receptor encoded by linked genes. Once rearrangement was introduced, the combining site may have been "relaxed" so that it was no longer required to be MHC-restricted. Duplications then would provide two types of TCR, one MHC-restricted and the other not, like γδ TCR today. One locus, according to the Hood model, underwent a cis duplication to give rise to the IgH/L loci. Ig, then, developed a new differentiation pathway with an alternative splicing mechanism for the transmembrane and secreted forms of BCRs. Recent evidence suggests that all the gnathostome antigen receptors arose after 2R, from an MHC paralogous region.[105,357]

V(J) and C1 Domains

C1 domains are found in the antigen receptors, MHC class I and class II, and very few other molecules.[422] This IgSF domain is so far most prevalent in gnathostomes, as if C1 domains arose concurrently with the adaptive immune system and coevolved with it. What was the value of the C1 domain, and why is it found almost exclusively in adaptive immune system–related molecules? All of these molecules interact with coreceptors such as CD3 (TCR), Igα and β chains (Ig on B cells), CD4, and CD8 (with MHC on opposing cells), and it is conceivable that in sections of IgSF domain in which C1 differs from the C2 there is a specific region favoring interaction with other molecules.

The G strand of Ig/TCR V domains is encoded by the J gene segment, separated from the V region–encoded A-F strands, and rearrangement is necessary to assemble a complete V gene. The primary structure of each Ig/TCR chain bears hallmarks of the dimeric nature of the receptor in which they participate. A diglycine bulge (Gly-X-Gly), present in all of these V domains, is thought either to be a beneficial adaptation or to promote dimer formation by inducing a twist in the G strand that results in V domain pairing to orient the CDR. In V genes that do not somatically rearrange, the G stand is an integral part of the V exon. There are several nonrearranging genes in deuterostome genomes that have this basic structure that are called "VJ domains"; such genes, or related genes, are candidates for invasion by the RAG transposon.[105]

Interestingly, many VJ and C1 genes are present in the MHC class III region (human chromosome 6p21) or its paralogs (see Fig. 4.12). Two MHC-linked gene segments stand out: a single VJ, NKP30, and a gene containing a VJC1 core, tapasin (TAP-binding protein), involved in antigen processing. NKP30, made of a single Ig domain of the VJ type, is an NK cell–activating receptor, and it may offer a link to cell types in invertebrates. It could be a relative of an ancient receptor whose history is linked to the emergence of MHC class II and class I; unlike most NKRs, NKp30 is evolutionarily conserved, present from sharks to mammals.[98] In order to resemble an ancestor, the NKP30 V domain need to be only associated with a C1 domain. In fact, a C1 single-domain gene, pre–TCR-α, is also encoded in the MHC. Besides Ig and TCR, tapasin is one of the rare cases of a gene segment with a VJC1 structure existing on several paralogous linkage groups (6, 9q33, 19q13).[116] While this gene is related to the rearranging receptor structure, it is undoubtedly very old and probably predated the ancient block (genome-wide) duplications. It could have acted as a donor of C1 to a VJ domain–containing gene in the MHC class III region (like the XMIV), which then could have been the first substrate of the rearrangement.

Other members resemble "primitive" antigen receptors and several of them map to at least two tetrads of paralogous genes, the MHC on chromosomes 1, 6, 9 (12), 19 and the LRC on chromosomes 1, 3, 11, 19 (21). Many form dimers and are expressed in lymphocytes, where they form a family of adhesion molecules. Perhaps such nonrearranging VJC1 genes that regulated cytotoxicity/phagocytosis were predecessors of the antigen receptors. Some molecules with VJ/C1 cores involved in cell–cell interactions cab serve as ports of entry for viruses.[31,422] In a move from "property to function," an arms race consequence could result in a virus receptor developing into an immune receptor. The best examples of such molecules are in the CTX/JAM family, in which receptor interactions with viruses may trigger an apoptosis, a primitive form of antiviral immunity. Additionally, there are VJ exons associated with the MHC paralogous regions of unknown function, which are found in amphibians and reptiles but lost in mammals.[105,357] These genes have been proposed by two groups related to the IgSF gene invaded by the RAG transposon. Was an IgSF family member encoded in the proto MHC closely related to the precursor gene that was invaded by the RAG transposon to become a primordial antigen receptor?

Emergence of T and B Cells

The transcriptional network for lymphocyte development emerged prior to advent of adaptive immunity. Was this system ILC-like, or did classical lymphocytes emerge first and then lose their antigen receptors? Note that the "adaptive cytokines," such as IL-2, IL-4, γIFN, and many others, to this point, have been found only in gnathostomes.

As described previously, agnathans have now been shown to have cells in the T and B lineages, like what is found in gnathostomes. Because the jawless and jawed vertebrates have antigen receptors derived from different gene families, at first glance it seemed that the T-B split occurred twice in evolution. However, the gene expression profiles for the lamprey T and B cells are too like the patterns in gnathostome lymphocyte populations to be derived by convergent evolution; furthermore, the development of lamprey T cells in a thymus-like structure also suggests evolution from a common ancestor. Thus, it seems most likely that both the LRR and IgSF receptor families were present in the common ancestor, with the VLR system predominating early in vertebrate evolution (Fig. 4.9). Such a scenario exists today with NKRs in mammals, with different gene families (C-type lectin or IgSF) predominating in different species.

Because APOBEC family members are most likely involved in VLR assembly and perhaps in mutation, there may have been a primordial IgSF receptor that was also modified somatically by an AID-like molecule, which functioned side-by-side with VLRs. The rearrangement break caused by the RAG transposon in a primordial IgSF gene offered a great advantage,[255] as was detailed previously, and the IgSF receptors then could have supplanted VLRs in gnathostomes, while maintaining the AID-dependent machinery to diversify it further. It is difficult to imagine how a rearrangement break in an antigen receptor gene generated by LRR insertion could be advantageous; conversely, as mentioned previously, a break in the CDR3 loop in the center of the binding site is in perfect position to interact with peptide bound to MHC.[421] While lampreys have a thymoid for development of their "T cells," as described previously, there was a great leap in complexity of thymus structure in the gnathostomes, consistent with a new TCR/MHC system in which positive and negative selection became a requirement for a functioning adaptive immune system. In addition, the general transcriptional and cytokine/chemokine networks for T cell (and B cell) development would have also been in place in the jawless vertebrates; in gnathostomes, this network became much more complex in the Big Bang (eg, see the chemokine section in which the leap in sophistication can be examined with precision[383]).

Why did the thymus evolve? If the agnathan thymoid is a model, then perhaps the thymus emerged not for positive or negative selection in the gnathostome thymus, but rather to sequester developing T cells away from the microenvironment where B cells develop (bone marrow equivalent). It (potentially) follows that, once a sequestered environment for T cell development was acquired, sophisticated and complex mechanisms for positive and negative selection emerged, perhaps in conjunction with T cell recognition of peptide/MHC complexes.[244]

Furthermore, from comparative studies, Boehm and colleagues have proposed a simple model for the most fundamental elements required for thymocyte differentiation, which has been superimposed onto the lamprey thymus-like structures.[368] Further study of this minimal network of transcription factors, chemokines, and cytokines could uncover the origins of thymus differentiation in lower deuterostomes that clearly have no T/B cells.

Potpourri

Studies of how the microbiota interact with the immune system have filled journals to the gills over the past 5 years, pun intended. McFall-Ngai suggested that the driving force for adaptive immunity was not so much to fight off invaders, but rather to tolerate commensals.[423] This framework of interactions between leukocytes, and between leukocytes and commensals/pathogens, may be the key to understanding the genesis of adaptive immunity. Even if this hypothesis will not hold water, clearly the study of mucosal immunity throughout the animal kingdom is crucial for our basic understanding of homeostasis and defense.

How do lamprey T cells recognize their antigen? Is there a convergent peptidic recognition of foreign antigen, or an entirely different mechanism? The VLR system continues to amaze us. All these discoveries force us to acknowledge that we should not be surprised by future discoveries in the evolution of immunity, such as a "convergent MHC," or an antigen presentation system not based on peptide presentation, or even a primordial type of adaptive immune system in lower deuterostomes or protostomes.

With the genome projects and advances in molecular biology, we have made strides in understanding old problems in vertebrate adaptive immunity, such as (1) when the L chains emerged and the significance of more than one isotype[160]; (2) which antigen receptor (if any of the extant ones) came first, and all of them came to evolve to their present state; (3) when IgD emerged and what its function is in different vertebrate phyla, especially the transmembrane form; and (4) how γ/δ T cells recognize antigen. We propose that are two arms of the γ/δ T cell lineage, one innate and the other adaptive, like B cells and α/β T cells.[35]

Did adaptive Tregs (iTregs) truly emerge as a subpopulation of cells that regulate paternal-specific immunity in placental mammals,[381] ie, do ectothermic vertebrates (and birds) only have thymic-derived Tregs (tTregs)?

ACKNOWLEDGMENTS

Thanks to Louis Du Pasquier for working with me on all previous editions of the chapters. Thanks as well to Hanover Matz for critical reading.

REFERENCES

1. Janeway CA Jr. Approaching the asymptote? Evolution and revolution in immunology. *Cold Spring Harb Symp Quant Biol.* 1989;54(Pt 1):1-13. doi:10.1101/sqb.1989.054.01.003
2. Matzinger P. Tolerance, danger, and the extended family. *Annu Rev Immunol.* 1994;12:991-1045. doi:10.1146/annurev.iy.12.040194.005015
3. Lemaitre B, Nicolas E, Michaut L, Reichhart JM, Hoffmann JA. The dorsoventral regulatory gene cassette spätzle/Toll/cactus controls the potent antifungal response in Drosophila adults. *Cell.* 1996;86(6):973-983. doi:10.1016/s0092-8674(00)80172-5
4. Cooper MD, Raymond DA, Peterson RD, South MA, Good RA. The functions of the thymus system and the bursa system in the chicken. *J Exp Med.* 1966;123(1):75-102. doi:10.1084/jem.123.1.75
5. Grčević D. Robert Jack and Louis Du Pasquier: Evolutionary concepts in immunology. *Croat Med J.* 2019;60(5):483-484. doi:10.3325/cmj.2019.60.483
6. Mayassi T, Barreiro LB, Rossjohn J, Jabri B. A multilayered immune system through the lens of unconventional T cells. *Nature.* 2021;595(7868):501-510. doi:10.1038/s41586-021-03578-0
7. Klein J, Nikolaidis N. The descent of the antibody-based immune system by gradual evolution. *Proc Natl Acad Sci U S A.* 2005;102(1):169-174. doi:10.1073/pnas.0408480102
8. Daugherty MD, Malik HS. Rules of engagement: molecular insights from host-virus arms races. *Annu Rev Genet.* 2012;46:677-700. doi:10.1146/annurev-genet-110711-155522
9. Loker ES, Adema CM, Zhang SM, Kepler TB. Invertebrate immune systems—not homogeneous, not simple, not well understood. *Immunol Rev.* 2004;198:10-24. doi:10.1111/j.0105-2896.2004.0117.x.
10. Abi-Rached L, Parham P. Natural selection drives recurrent formation of activating killer cell immunoglobulin-like receptor and Ly49 from inhibitory homologues. *J Exp Med.* 2005;201(8):1319-1332. doi:10.1084/jem.20042558
11. Yoder JA, Litman GW. The phylogenetic origins of natural killer receptors and recognition: relationships, possibilities, and realities. *Immunogenetics.* 2011;63(3):123-141. doi:10.1007/s00251-010-0506-4
12. Hughes AL. Rapid evolution of immunoglobulin superfamily C2 domains expressed in immune system cells. *Mol Biol Evol.* 1997;14(1):1-5. doi:10.1093/oxfordjournals.molbev.a025694
13. Pancer Z, Amemiya CT, Ehrhardt GR, Ceitlin J, Gartland GL, Cooper MD. Somatic diversification of variable lymphocyte receptors in the agnathan sea lamprey. *Nature.* 2004;430(6996):174-180. doi:10.1038/nature02740
14. Flajnik MF. All GOD's creatures got dedicated mucosal immunity. *Nat Immunol.* 2010;11(9):777-779. doi:10.1038/ni0910-777
15. Flajnik MF, Deschacht N, Muyldermans S. A case of convergence: why did a simple alternative to canonical antibodies arise in sharks and camels? *PLoS Biol.* 2011;9(8):e1001120. doi:10.1371/journal.pbio.1001120
16. Nei M, Gu X, Sitnikova T. Evolution by the birth-and-death process in multigene families of the vertebrate immune system. *Proc Natl Acad Sci U S A.* 1997;94(15):7799-7806. doi:10.1073/pnas.94.15.7799
17. Ohno S. Gene duplication and the uniqueness of vertebrate genomes circa 1970-1999. *Semin Cell Dev Biol.* 1999;10(5):517-522. doi:10.1006/scdb.1999.0332
18. Kasahara M, Hayashi M, Tanaka K, et al. Chromosomal localization of the proteasome Z subunit gene reveals an ancient chromosomal duplication involving the major histocompatibility complex. *Proc Natl Acad Sci U S A.* 1996;93(17):9096-9101. doi:10.1073/pnas.93.17.9096
19. Vienne A, Shiina T, Abi-Rached L, et al. Evolution of the proto-MHC ancestral region: more evidence for the plesiomorphic organisation of human chromosome 9q34 region. *Immunogenetics.* 2003;55(7):429-436. doi:10.1007/s00251-003-0601-x
20. Simakov O, Marlétaz F, Yue JX, et al. Deeply conserved synteny resolves early events in vertebrate evolution. *Nat Ecol Evol.* 2020;4(6):820-830. doi:10.1038/s41559-020-1156-z.
21. Nakatani Y, Shingate P, Ravi V, et al. Reconstruction of proto-vertebrate, proto-cyclostome and proto-gnathostome genomes provides new insights into early vertebrate evolution. *Nat Commun.* 2021;12(1):4489. doi:10.1038/s41467-021-24573-z
22. Postlethwait JH. The zebrafish genome in context: ohnologs gone missing. *J Exp Zool B Mol Dev Evol.* 2007;308(5):563-577. doi:10.1002/jez.b.21137
23. Criscitiello MF, de Figueiredo P. Fifty shades of immune defense. *PLoS Pathog.* 2013;9(2):e1003110. doi:10.1371/journal.ppat.1003110
24. Pålsson-McDermott EM, O'Neill LA. Building an immune system from nine domains. *Biochem Soc Trans.* 2007;35(Pt 6):1437-1444. doi:10.1042/bst0351437
25. Enkhbayar P, Kamiya M, Osaki M, Matsumoto T, Matsushima N. Structural principles of leucine-rich repeat (LRR) proteins. *Proteins.* 2004;54(3):394-403. doi:10.1002/prot.10605
26. Batkhishig D, Enkhbayar P, Kretsinger RH, Matsushima N. A crucial residue in the hydrophobic core of the solenoid structure of leucine rich repeats. *Biochim Biophys Acta Proteins Proteom.* 2021;1869(6):140631. doi:10.1016/j.bbapap.2021.140631
27. Williams AF, Barclay AN. The immunoglobulin superfamily—domains for cell surface recognition. *Annu Rev Immunol.* 1988;6:381-405. doi:10.1146/annurev.iy.06.040188.002121
28. Harpaz Y, Chothia C. Many of the immunoglobulin superfamily domains in cell adhesion molecules and surface receptors belong to a new structural set which is close to that containing variable domains. *J Mol Biol.* 1994;238(4):528-539. doi:10.1006/jmbi.1994.1312
29. Greenberg AS, Hughes AL, Guo J, Avila D, McKinney EC, Flajnik MF. A novel "chimeric" antibody class in cartilaginous fish: IgM may not be the primordial immunoglobulin. *Eur J Immunol.* 1996;26(5):1123-1129. doi:10.1002/eji.1830260525
30. Hernández Prada JA, Haire RN, Allaire M, et al. Ancient evolutionary origin of diversified variable regions demonstrated by crystal structures of an immune-type receptor in amphioxus. *Nat Immunol.* 2006;7(8):875-882. doi:10.1038/ni1359
31. Du Pasquier L, Zucchetti I, De Santis R. Immunoglobulin superfamily receptors in protochordates: before RAG time. *Immunol Rev.* 2004;198:233-248. doi:10.1111/j.0105-2896.2004.00122.x
32. Zucchetti I, De Santis R, Grusea S, Pontarotti P, Du Pasquier L. Origin and evolution of the vertebrate leukocyte receptors: the lesson from tunicates. *Immunogenetics.* 2009;61(6):463-481. doi:10.1007/s00251-009-0373-z
33. Telford MJ, Budd GE, Philippe H. Phylogenomic insights into animal evolution. *Curr Biol.* 2015;25(19):R876-R887. doi:10.1016/j.cub.2015.07.060
34. Bernstein RM, Schluter SF, Bernstein H, Marchalonis JJ. Primordial emergence of the recombination activating gene 1 (RAG1): sequence of the complete shark gene indicates homology to microbial integrases. *Proc Natl Acad Sci U S A.* 1996;93(18):9454-9459. doi:10.1073/pnas.93.18.9454
35. Flajnik MF. Re-evaluation of the immunological Big Bang. *Curr Biol.* 2014;24(21):R1060-R1065. doi:10.1016/j.cub.2014.09.070
36. Buckley KM, Rast JP. Diversity of animal immune receptors and the origins of recognition complexity in the deuterostomes. *Dev Comp Immunol.* 2015;49(1):179-189. doi:10.1016/j.dci.2014.10.013
37. Zasloff M. Antimicrobial peptides of multicellular organisms: my perspective. *Adv Exp Med Biol.* 2019;1117:3-6. doi:10.1007/978-981-13-3588-4_1
38. Wcisel DJ, Ota T, Litman GW, Yoder JA. Spotted gar and the evolution of innate immune receptors. *J Exp Zool B Mol Dev Evol.* 2017;328(7):666-684. doi:10.1002/jez.b.22738
39. Secombes CJ, Wang T, Bird S. Chapter 5—vertebrate cytokines and their evolution. In: Malagoli D, ed. *The Evolution of the Immune System.* Academic Press; 2016:87-150.
40. Smith LC, Hawley TS, Henson JH, Majeske AJ, Oren M, Rosental B. Methods for collection, handling, and analysis of sea urchin coelomocytes. *Methods Cell Biol.* 2019;150:357-389. doi:10.1016/bs.mcb.2018.11.009
41. Wright RK. Protochordate immunity. I. Primary immune response of the tunicate *Ciona intestinalis* to vertebrate erythrocytes. *J Invertebr Pathol.* 1974;24(1):29-36. doi:10.1016/0022-2011(74)90160-8
42. Rosental B, Raveh T, Voskoboynik A, Weissman IL. Evolutionary perspective on the hematopoietic system through a colonial chordate: allogeneic immunity and hematopoiesis. *Curr Opin Immunol.* 2020;62:91-98. doi:10.1016/j.coi.2019.12.006
43. Wang W, Wang C, Chen W, Ding S. Advances in immunological research of amphioxus. *Dev Comp Immunol.* 2021;118:103992. doi:10.1016/j.dci.2020.103992
44. Barreda DR, Neely HR, Flajnik MF. Evolution of myeloid cells. *Microbiol Spectr.* 2016;4(3). doi:10.1128/microbiolspec.MCHD-0007-2015
45. Arnold JE. Hematology of the sandbar shark, Carcharhinus plumbeus: standardization of complete blood count techniques for elasmobranchs. *Vet Clin Pathol.* 2005;34(2):115-123. doi:10.1111/j.1939-165x.2005.tb00023.x
46. Wattrus SJ, Zon LI. Blood in the water: recent uses of zebrafish to study myeloid biology. *Curr Opin Hematol.* 2021;28(1):43-49. doi:10.1097/moh.0000000000000627
47. Chen K, Xu W, Wilson M, et al. Immunoglobulin D enhances immune surveillance by activating antimicrobial, proinflammatory and B cell-stimulating programs in basophils. *Nat Immunol.* 2009;10(8):889-898. doi:10.1038/ni.1748
48. Grayfer L, Kerimoglu B, Yaparla A, Hodgkinson JW, Xie J, Belosevic M. Mechanisms of fish macrophage antimicrobial immunity. *Front Immunol.* 2018;9:1105. doi:10.3389/fimmu.2018.01105
49. Jakubzick CV, Randolph GJ, Henson PM. Monocyte differentiation and antigen-presenting functions. *Nat Rev Immunol.* 2017;17(6):349-362. doi:10.1038/nri.2017.28
50. Yaparla A, Koubourli DV, Popovic M, Grayfer L. Exploring the relationships between amphibian (*Xenopus laevis*) myeloid cell subsets. *Dev Comp Immunol.* 2020;113:103798. doi:10.1016/j.dci.2020.103798
51. Yaparla A, Docter-Loeb H, Melnyk MLS, Batheja A, Grayfer L. The amphibian (*Xenopus laevis*) colony-stimulating factor-1 and interleukin-34-derived macrophages possess disparate pathogen recognition capacities. *Dev Comp Immunol.* 2019;98:89-97. doi:10.1016/j.dci.2019.04.011
52. Tanizaki Y, Ichisugi M, Obuchi-Shimoji M, et al. Thrombopoietin induces production of nucleated thrombocytes from liver cells in *Xenopus laevis. Sci Rep.* 2015;5:18519. doi:10.1038/srep18519
53. Hibino T, Loza-Coll M, Messier C, et al. The immune gene repertoire encoded in the purple sea urchin genome. *Dev Biol.* 2006;300(1):349-365. doi:10.1016/j.ydbio.2006.08.065
54. Vasta GR. Galectins in host-pathogen interactions: structural, functional and evolutionary aspects. *Adv Exp Med Biol.* 2020;1204:169-196. doi:10.1007/978-981-15-1580-4_7
55. Blidner AG, Méndez-Huergo SP, Cagnoni AJ, Rabinovich GA. Re-wiring regulatory cell networks in immunity by galectin-glycan interactions. *FEBS Lett.* 2015;589(22):3407-3418. doi:10.1016/j.febslet.2015.08.037
56. Alquraini A, El Khoury J. Scavenger receptors. *Curr Biol.* 2020;30(14):R790-r95. doi:10.1016/j.cub.2020.05.051
57. Pancer Z, Rast JP, Davidson EH. Origins of immunity: transcription factors and homologues of effector genes of the vertebrate immune system expressed in sea urchin coelomocytes. *Immunogenetics.* 1999;49(9):773-786. doi:10.1007/s002510050551
58. Wolf AJ, Underhill DM. Peptidoglycan recognition by the innate immune system. *Nat Rev Immunol.* 2018;18(4):243-254. doi:10.1038/nri.2017.136
59. Thompson AW, Hawkins MB, Parey E, et al. The bowfin genome illuminates the developmental evolution of ray-finned fishes. *Nat Genet.* 2021;53(9):1373-1384. doi:10.1038/s41588-021-00914-y
60. Nie L, Cai SY, Shao JZ, Chen J. Toll-like receptors, associated biological roles, and signaling networks in non-mammals. *Front Immunol.* 2018;9:1523. doi:10.3389/fimmu.2018.01523
61. Brennan JJ, Gilmore TD. Evolutionary origins of toll-like receptor signaling. *Mol Biol Evol.* 2018;35(7):1576-1587. doi:10.1093/molbev/msy050
62. Venkatesh B, Lee AP, Ravi V, et al. Elephant shark genome provides unique insights into gnathostome evolution. *Nature.* 2014;505(7482):174-179. doi:10.1038/nature12826
63. Grimes DJ, Stemmler J, Hada H, et al. Vibrio species associated with mortality of sharks held in captivity. *Microb Ecol.* 1984;10(3):271-282. doi:10.1007/bf02010940
64. Meunier E, Broz P. Evolutionary convergence and divergence in NLR function and structure. *Trends Immunol.* 2017;38(10):744-757. doi:10.1016/j.it.2017.04.005
65. Jones JD, Vance RE, Dangl JL. Intracellular innate immune surveillance devices in plants and animals. *Science.* 2016;354:6316. doi:10.1126/science.aaf6395
66. Brisse M, Ly H. Comparative structure and function analysis of the RIG-I-like receptors: RIG-I and MDA5. *Front Immunol.* 2019;10:1586. doi:10.3389/fimmu.2019.01586
67. Cai H, Imler JL. cGAS-STING: insight on the evolution of a primordial antiviral signaling cassette. *Fac Rev.* 2021;10:54. doi:10.12703/r/10-54
68. Kumar V. The trinity of cGAS, TLR9, and ALRs guardians of the cellular galaxy against host-derived self-DNA. *Front Immunol.* 2020;11:624597. doi:10.3389/fimmu.2020.624597
69. Langevin C, Levraud JP, Boudinot P. Fish antiviral tripartite motif (TRIM) proteins. *Fish Shellfish Immunol.* 2019;86:724-733. doi:10.1016/j.fsi.2018.12.008
70. Kaufman J. Innate immune genes of the chicken MHC and related regions. *Immunogenetics.* 2021. doi:10.1007/s00251-021-01229-2
71. Miller MM, Taylor RL Jr. Brief review of the chicken Major Histocompatibility Complex: the genes, their distribution on chromosome 16, and their contributions to disease resistance. *Poult Sci.* 2016;95(2):375-392. doi:10.3382/ps/pev379
72. Nonaka M. Evolution of the complement system. *Subcell Biochem.* 2014;80:31-43. doi:10.1007/978-94-017-8881-6_3
73. Ricklin D, Reis ES, Mastellos DC, Gros P, Lambris JD. Complement component C3—the "Swiss Army Knife" of innate immunity and host defense. *Immunol Rev.* 2016;274(1):33-58. doi:10.1111/imr.12500
74. Smith LC. The complement system in sea urchins. *Adv Exp Med Biol.* 2001;484:363-372. doi:10.1007/978-1-4615-1291-2_35
75. Matsushita M. The complement system of agnathans. *Front Immunol.* 2018;9:1405. doi:10.3389/fimmu.2018.01405
76. Wu F, Feng B, Ren Y, et al. A pore-forming protein implements VLR-activated complement cytotoxicity in lamprey. *Cell Discov.* 2017;3:17033. doi:10.1038/celldisc.2017.33
77. Najafpour B, Cardoso JCR, Canário AVM, Power DM. Specific evolution and gene family expansion of complement 3 and regulatory factor H in fish. *Front Immunol.* 2020;11:568631. doi:10.3389/fimmu.2020.568631
78. Bayly-Jones C, Pang SS, Spicer BA, Whisstock JC, Dunstone MA. Ancient but not forgotten: new insights into MPEG1, a macrophage perforin-like immune effector. *Front Immunol.* 2020;11:581906. doi:10.3389/fimmu.2020.581906
79. Smith LC. Thioester function is conserved in SpC3, the sea urchin homologue of the complement component C3. *Dev Comp Immunol.* 2002;26(7):603-614. doi:10.1016/s0145-305x(02)00017-4
80. Hughes AL. Phylogeny of the C3/C4/C5 complement-component gene family indicates that C5 diverged first. *Mol Biol Evol.* 1994;11(3):417-425. doi:10.1093/oxfordjournals.molbev.a040123

81. Nonaka MI, Terado T, Kimura H, Nonaka M. Evolutionary analysis of two complement C4 genes: ancient duplication and conservation during jawed vertebrate evolution. *Dev Comp Immunol.* 2017;68:1-11. doi:10.1016/j.dci.2016.11.009
82. Yano K, Gale D, Massberg S, et al. Phenotypic heterogeneity is an evolutionarily conserved feature of the endothelium. *Blood.* 2007;109(2):613-615. doi:10.1182/blood-2006-05-026401
83. Smith LC, Lun CM. The SpTransformer gene family (formerly Sp185/333) in the purple sea urchin and the functional diversity of the anti-pathogen rSpTransformer-E1 protein. *Front Immunol.* 2017;8:725. doi:10.3389/fimmu.2017.00725
84. Oren M, Barela Hudgell MA, D'Allura B, et al. Short tandem repeats, segmental duplications, gene deletion, and genomic instability in a rapidly diversified immune gene family. *BMC Genom.* 2016;17(1):900. doi:10.1186/s12864-016-3241-x
85. Lun CM, Bishop BM, Smith LC. Multitasking immune Sp185/333 protein, rSpTransformer-E1, and its recombinant fragments undergo secondary structural transformation upon binding targets. *J Immunol.* 2017;198(7):2957-2966. doi:10.4049/jimmunol.1601795
86. Majeske AJ, Oren M, Sacchi S, Smith LC. Single sea urchin phagocytes express messages of a single sequence from the diverse Sp185/333 gene family in response to bacterial challenge. *J Immunol.* 2014;193(11):5678-5688. doi:10.4049/jimmunol.1401681
87. Oren M, Rosental B, Hawley TS, et al. Individual sea urchin coelomocytes undergo somatic immune gene diversification. *Front Immunol.* 2019;10:1298. doi:10.3389/fimmu.2019.01298
88. Ghosh J, Lun CM, Majeske AJ, Sacchi S, Schrankel CS, Smith LC. Invertebrate immune diversity. *Dev Comp Immunol.* 2011;35(9):959-974. doi:10.1016/j.dci.2010.12.009
89. Dishaw LJ, Leigh B, Cannon JP, et al. Gut immunity in a protochordate involves a secreted immunoglobulin-type mediator binding host chitin and bacteria. *Nat Commun.* 2016;7:10617. doi:10.1038/ncomms10617
90. Liberti A, Cannon JP, Litman GW, Dishaw LJ. A soluble immune effector binds both fungi and bacteria via separate functional domains. *Front Immunol.* 2019;10:369. doi:10.3389/fimmu.2019.00369
91. Flajnik MF, Kasahara M. Origin and evolution of the adaptive immune system: genetic events and selective pressures. *Nat Rev Genet.* 2010;11(1):47-59. doi:10.1038/nrg2703
92. Vivier E, van de Pavert SA, Cooper MD, Belz GT. The evolution of innate lymphoid cells. *Nat Immunol.* 2016;17(7):790-794. doi:10.1038/ni.3459
93. Parham P, Moffett A. Variable NK cell receptors and their MHC class I ligands in immunity, reproduction and human evolution. *Nat Rev Immunol.* 2013;13(2):133-144. doi:10.1038/nri3370
94. Guselnikov SV, Ramanayake T, Erilova AY, et al. The Xenopus FcR family demonstrates continually high diversification of paired receptors in vertebrate evolution. *BMC Evol Biol.* 2008;8:148. doi:10.1186/1471-2148-8-148
95. Wang J, Belosevic M, Stafford JL. Identification of distinct LRC- and Fc receptor complex-like chromosomal regions in fish supports that teleost leukocyte immune-type receptors are distant relatives of mammalian Fc receptor-like molecules. *Immunogenetics.* 2021;73(1):93-109. doi:10.1007/s00251-020-01193-3
96. Horton TL, Stewart R, Cohen N, et al. Ontogeny of Xenopus NK cells in the absence of MHC class I antigens. *Dev Comp Immunol.* 2003;27(8):715-726. doi:10.1016/s0145-305x(03)00040-5
97. Kinlein A, Janes ME, Kincer J, et al. Analysis of shark NCR3 family genes reveals primordial features of vertebrate NKp30. *Immunogenetics.* 2021;73(4):333-348. doi:10.1007/s00251-021-01209-6
98. Flajnik MF, Tlapakova T, Criscitiello MF, Krylov V, Ohta Y. Evolution of the B7 family: co-evolution of B7H6 and NKp30, identification of a new B7 family member, B7H7, and of B7's historical relationship with the MHC. *Immunogenetics.* 2012;64(8):571-590. doi:10.1007/s00251-012-0616-2
99. Ohta Y, Flajnik MF. Coevolution of MHC genes (LMP/TAP/class Ia, NKT-class Ib, NKp30-B7H6): lessons from cold-blooded vertebrates. *Immunol Rev.* 2015;267(1):6-15. doi:10.1111/imr.12324
100. Watkins D, Harding F, Cohen N. In vitro proliferative and cytotoxic responses against Xenopus minor histocompatibility antigens. *Transplantation.* 1988;45(2):499-501. doi:10.1097/00007890-198802000-00054
101. Horton TL, Minter R, Stewart R, Ritchie P, Watson MD, Horton JD. Xenopus NK cells identified by novel monoclonal antibodies. *Eur J Immunol.* 2000;30(2):604-613 doi:10.1002/1521-4141(200002)30:2<604::Aid-immu604>3.0.Co;2-x
102. Shen L, Stuge TB, Zhou H, et al. Channel catfish cytotoxic cells: a mini-review. *Dev Comp Immunol.* 2002;26(2):141-149. doi:10.1016/s0145-305x(01)00056-8
103. Yoder JA, Turner PM, Wright PD, et al. Developmental and tissue-specific expression of NITRs. *Immunogenetics.* 2010;62(2):117-122. doi:10.1007/s00251-009-0416-5
104. Cannon JP, Haire RN, Magis AT, et al. A bony fish immunological receptor of the NITR multigene family mediates allogeneic recognition. *Immunity.* 2008;29(2):228-237. doi:10.1016/j.immuni.2008.05.018
105. Ohta Y, Kasahara M, O'Connor TD, Flajnik MF. Inferring the "primordial immune complex": origins of MHC class I and antigen receptors revealed by comparative genomics. *J Immunol.* 2019;203(7):1882-1896. doi:10.4049/jimmunol.1900597
106. Edholm ES, Albertorio Saez LM, Gill AL, et al. Nonclassical MHC class I-dependent invariant T cells are evolutionarily conserved and prominent from early development in amphibians. *Proc Natl Acad Sci U S A.* 2013;110(35):14342-14347. doi:10.1073/pnas.1309840110
107. Edholm ES, Banach M, Hyoe Rhoo K, Pavelka MS Jr, Robert J. Distinct MHC class I-like interacting invariant T cell lineage at the forefront of mycobacterial immunity uncovered in *Xenopus. Proc Natl Acad Sci U S A.* 2018;115(17):E4023-e31. doi:10.1073/pnas.1722129115
108. Flajnik MF, Kaufman JF, Hsu E, Manes M, Parisot R, Du Pasquier L. Major histocompatibility complex-encoded class I molecules are absent in immunologically competent *Xenopus* before metamorphosis. *J Immunol.* 1986;137(12):3891-3899.
109. Edholm ES, Banach M, Robert J. Evolution of innate-like T cells and their selection by MHC class I-like molecules. *Immunogenetics.* 2016;68(8):525-536. doi:10.1007/s00251-016-0929-7
110. Yang Z, Wang C, Wang T, et al. Analysis of the reptile CD1 genes: evolutionary implications. *Immunogenetics.* 2015;67(5-6):337-346. doi:10.1007/s00251-015-0837-2
111. Salomonsen J, Sørensen MR, Marston DA, et al. Two CD1 genes map to the chicken MHC, indicating that CD1 genes are ancient and likely to have been present in the primordial MHC. *Proc Natl Acad Sci U S A.* 2005;102(24):8668-8673. doi:10.1073/pnas.0409213102
112. Miller MM, Wang C, Parisini E, et al. Characterization of two avian MHC-like genes reveals an ancient origin of the CD1 family. *Proc Natl Acad Sci U S A.* 2005;102(24):8674-8679. doi:10.1073/pnas.0500105102
113. D'Souza MP, Adams E, Altman JD, et al. Casting a wider net: Immunosurveillance by nonclassical MHC molecules. *PLoS Pathog.* 2019;15(2):e1007567. doi:10.1371/journal.ppat.1007567
114. Suzuki T, Shin IT, Fujiyama A, Kohara Y, Kasahara M. Hagfish leukocytes express a paired receptor family with a variable domain resembling those of antigen receptors. *J Immunol.* 2005;174(5):2885-2891. doi:10.4049/jimmunol.174.5.2885
115. Takaba H, Imai T, Miki S, et al. A major allogenic leukocyte antigen in the agnathan hagfish. *Sci Rep.* 2013;3:1716. doi:10.1038/srep01716
116. Dornburg A, Wcisel DJ, Zapfe K, et al. Holosteans contextualize the role of the teleost genome duplication in promoting the rise of evolutionary novelties in the ray-finned fish innate immune system. *Immunogenetics.* 2021. doi:10.1007/s00251-021-01225-6
117. Flajnik MF, Du Pasquier L. Evolution of innate and adaptive immunity: can we draw a line? *Trends Immunol.* 2004;25(12):640-644. doi:10.1016/j.it.2004.10.001
118. Flajnik MF. A cold-blooded view of adaptive immunity. *Nat Rev Immunol.* 2018;18(7):438-453. doi:10.1038/s41577-018-0003-9
119. Kaufman J. Generalists and specialists: a new view of how MHC class I molecules fight infectious pathogens. *Trends Immunol.* 2018;39(5):367-379. doi:10.1016/j.it.2018.01.001
120. Anderson MK, Sun X, Miracle AL, Litman GW, Rothenberg EV. Evolution of hematopoiesis: three members of the PU.1 transcription factor family in a cartilaginous fish, Raja eglanteria. *Proc Natl Acad Sci U S A.* 2001;98(2):553-558. doi:10.1073/pnas.021478998
121. Anderson MK, Pant R, Miracle AL, et al. Evolutionary origins of lymphocytes: ensembles of T cell and B cell transcriptional regulators in a cartilaginous fish. *J Immunol.* 2004;172(10):5851-5860. doi:10.4049/jimmunol.172.10.5851
122. Rast JP, Smith LC, Loza-Coll M, Hibino T, Litman GW. Genomic insights into the immune system of the sea urchin. *Science.* 2006;314(5801):952-956. doi:10.1126/science.1134301
123. Redmond AK, Macqueen DJ, Dooley H. Phylotranscriptomics suggests the jawed vertebrate ancestor could generate diverse helper and regulatory T cell subsets. *BMC Evol Biol.* 2018;18(1):169. doi:10.1186/s12862-018-1290-2
124. Cooper EL. Comparative immunology. *Integr Zool.* 2006;1(1):32-43. doi:10.1111/j.1749-4877.2006.00010.x
125. Kapp FG, Perlin JR, Hagedorn EJ, et al. Protection from UV light is an evolutionarily conserved feature of the haematopoietic niche. *Nature.* 2018;558(7710):445-448. doi:10.1038/s41586-018-0213-0
126. Rogozin IB, Iyer LM, Liang L, et al. Evolution and diversification of lamprey antigen receptors: evidence for involvement of an AID-APOBEC family cytosine deaminase. *Nat Immunol.* 2007;8(6):647-656. doi:10.1038/ni1463
127. Finstad J, Good RA. The evolution of the immune response. 3. Immunologic responses in the lamprey. *J Exp Med.* 1964;120(6):1151-1168. doi:10.1084/jem.120.6.1151
128. Uinuk-Ool T, Mayer WE, Sato A, Dongak R, Cooper MD, Klein J. Lamprey lymphocyte-like cells express homologs of genes involved in immunologically relevant activities of mammalian lymphocytes. *Proc Natl Acad Sci U S A.* 2002;99(22):14356-14361. doi:10.1073/pnas.212527699
129. Alder MN, Herrin BR, Sadlonova A, et al. Antibody responses of variable lymphocyte receptors in the lamprey. *Nat Immunol.* 2008;9(3):319-327. doi:10.1038/ni1562
130. Guo P, Hirano M, Herrin BR, et al. Dual nature of the adaptive immune system in lampreys. *Nature.* 2009;459(7248):796-801. doi:10.1038/nature08068
131. Li J, Das S, Herrin BR, Hirano M, Cooper MD. Definition of a third VLR gene in hagfish. *Proc Natl Acad Sci U S A.* 2013;110(37):15013-15018. doi:10.1073/pnas.1314540110
132. Cao DD, Liao X, Cheng W, et al. Structure of a variable lymphocyte receptor-like protein from the amphioxus *Branchiostoma floridae. Sci Rep.* 2016;6:19951. doi:10.1038/srep19951
133. Sun Y, Huang T, Hammarström L, Zhao Y. The immunoglobulins: new insights, implications, and applications. *Annu Rev Anim Biosci.* 2020;8:145-169. doi:10.1146/annurev-animal-021419-083720
134. Marchalonis J, Edelman GM. Polypeptide chains of immunoglobulins from the smooth dogfish (*Mustelus canis*). *Science.* 1966;154(3756):1567-1568. doi:10.1126/science.154.3756.1567
135. Dooley H, Flajnik MF. Shark immunity bites back: affinity maturation and memory response in the nurse shark, Ginglymostoma cirratum. *Eur J Immunol.* 2005;35(3):936-945. doi:10.1002/eji.200425760
136. Castro CD, Flajnik MF. Putting J chain back on the map: how might its expression define plasma cell development? *J Immunol.* 2014;193(7):3248-3255. doi:10.4049/jimmunol.1400531
137. Ye J, Bromage ES, Kaattari SL. The strength of B cell interaction with antigen determines the degree of IgM polymerization. *J Immunol.* 2010;184(2):844-850. doi:10.4049/jimmunol.0902364
138. Wilson MR, Marcuz A, van Ginkel F, et al. The immunoglobulin M heavy chain constant region gene of the channel catfish, Ictalurus punctatus: an unusual mRNA splice pattern produces the membrane form of the molecule. *Nucleic Acids Res.* 1990;18(17):5227-5233. doi:10.1093/nar/18.17.5227
139. Rosaria Coscia M, Oreste U. Exploring Antarctic teleost immunoglobulin genes. *Mar Genomics.* 2009;2(1):19-25. doi:10.1016/j.margen.2009.03.002
140. Rumfelt LL, Avila D, Diaz M, Bartl S, McKinney EC, Flajnik MF. A shark antibody heavy chain encoded by a nonsomatically rearranged VDJ is preferentially expressed in early development and is convergent with mammalian IgG. *Proc Natl Acad Sci U S A.* 2001;98(4):1775-1780. doi:10.1073/pnas.98.4.1775
141. Ohta Y, Flajnik M. IgD, like IgM, is a primordial immunoglobulin class perpetuated in most jawed vertebrates. *Proc Natl Acad Sci U S A.* 2006;103(28):10723-10728. doi:10.1073/pnas.0601407103
142. Wan Z, Zhao Y, Sun Y. Immunoglobulin D and its encoding genes: an updated review. *Dev Comp Immunol.* 2021;124:104198. doi:10.1016/j.dci.2021.104198
143. Wilson M, Bengtén E, Miller NW, Clem LW, Du Pasquier L, Warr GW. A novel chimeric Ig heavy chain from a teleost fish shares similarities to IgD. *Proc Natl Acad Sci U S A.* 1997;94(9):4593-4597. doi:10.1073/pnas.94.9.4593
144. Amemiya CT, Alföldi J, Lee AP, et al. The African coelacanth genome provides insights into tetrapod evolution. *Nature.* 2013;496(7445):311-316. doi:10.1038/nature12027
145. Mirete-Bachiller S, Olivieri DN, Gambón-Deza F. Immunoglobulin T genes in Actinopterygii. *Fish Shellfish Immunol.* 2021;108:86-93. doi:10.1016/j.fsi.2020.11.027
146. Shan M, Carrillo J, Yeste A, et al. Secreted IgD amplifies humoral T helper 2 cell responses by binding basophils via galectin-9 and CD44. *Immunity.* 2018;49(4):709-724.e8. doi:10.1016/j.immuni.2018.08.013
147. Zhang X, Calvert RA, Sutton BJ, Doré KA. IgY: a key isotype in antibody evolution. *Biol Rev Camb Philos Soc.* 2017;92(4):2144-2156. doi:10.1111/brv.12325
148. Mussmann R, Wilson M, Marcuz A, Courtet M, Du Pasquier L. Membrane exon sequences of the three *Xenopus* Ig classes explain the evolutionary origin of mammalian isotypes. *Eur J Immunol.* 1996;26(2):409-414. doi:10.1002/eji.1830260221
149. Zhao Y, Pan-Hammarström Q, Yu S, et al. Identification of IgF, a hinge-region-containing Ig class, and IgD in *Xenopus tropicalis. Proc Natl Acad Sci U S A.* 2006;103(32):12087-12092. doi:10.1073/pnas.0600291103
150. Work TM, Dagenais J, Breeden R, et al. Green turtles (*Chelonia mydas*) have novel asymmetrical antibodies. *J Immunol.* 2015;195(11):5452-5460. doi:10.4049/jimmunol.1501332
151. Leslie GA, Clem LW. Phylogen of immunoglobulin structure and function. 3. Immunoglobulins of the chicken. *J Exp Med.* 1969;130(6):1337-1352. doi:10.1084/jem.130.6.1337
152. Greenberg AS, Avila D, Hughes M, Hughes A, McKinney EC, Flajnik MF. A new antigen receptor gene family that undergoes rearrangement and extensive somatic diversification in sharks. *Nature.* 1995;374(6518):168-173. doi:10.1038/374168a0
153. Rast JP, Amemiya CT, Litman RT, Strong SJ, Litman GW. Distinct patterns of IgH structure and organization in a divergent lineage of chrondrichthyan fishes. *Immunogenetics.* 1998;47(3):234-245. doi:10.1007/s002510050353

154. Anumukonda K, Francis M, Currie P, Tulenko F, Hsu E. Heavy chain-only antibody genes in fish evolved to generate unique CDR3 repertoire. *Eur J Immunol.* 2021. doi:10.1002/eji.202149588
155. Stanfield RL, Dooley H, Flajnik MF, Wilson IA. Crystal structure of a shark single-domain antibody V region in complex with lysozyme. *Science.* 2004;305(5691):1770-1773. doi:10.1126/science.1101148
156. Stanfield RL, Dooley H, Verdino P, Flajnik MF, Wilson IA. Maturation of shark single-domain (IgNAR) antibodies: evidence for induced-fit binding. *J Mol Biol.* 2007;367(2):358-372. doi:10.1016/j.jmb.2006.12.045
157. Kovalenko OV, Olland A, Piché-Nicholas N, et al. Atypical antigen recognition mode of a shark immunoglobulin new antigen receptor (IgNAR) variable domain characterized by humanization and structural analysis. *J Biol Chem.* 2013;288(24):17408-17419. doi:10.1074/jbc.M112.435289
158. Zhang T, Tacchi L, Wei Z, Zhao Y, Salinas I. Intraclass diversification of immunoglobulin heavy chain genes in the African lungfish. *Immunogenetics.* 2014;66(5):335-351. doi:10.1007/s00251-014-0769-2
159. Greenberg AS, Steiner L, Kasahara M, Flajnik MF. Isolation of a shark immunoglobulin light chain cDNA clone encoding a protein resembling mammalian kappa light chains: implications for the evolution of light chains. *Proc Natl Acad Sci U S A.* 1993;90(22):10603-10607. doi:10.1073/pnas.90.22.10603
160. Criscitiello MF, Flajnik MF. Four primordial immunoglobulin light chain isotypes, including lambda and kappa, identified in the most primitive living jawed vertebrates. *Eur J Immunol.* 2007;37(10):2683-2694. doi:10.1002/eji.200737263
161. Schwager J, Bürckert N, Schwager M, Wilson M. Evolution of immunoglobulin light chain genes: analysis of Xenopus IgL isotypes and their contribution to antibody diversity. *EMBO J.* 1991;10(3):505-511.
162. Lee SS, Tranchina D, Ohta Y, Flajnik MF, Hsu E. Hypermutation in shark immunoglobulin light chain genes results in contiguous substitutions. *Immunity.* 2002;16(4):571-582. doi:10.1016/s1074-7613(02)00300-x
163. Edholm ES, Wilson M, Sahoo M, et al. Identification of Igsigma and Iglambda in channel catfish, *Ictalurus punctatus*, and Iglambda in Atlantic cod, *Gadus morhua*. *Immunogenetics.* 2009;61(5):353-370. doi:10.1007/s00251-009-0365-z
164. Hsu E, Criscitiello MF. Diverse immunoglobulin light chain organizations in fish retain potential to revise B cell receptor specificities. *J Immunol.* 2006;177(4):2452-2462. doi:10.4049/jimmunol.177.4.2452
165. Lange MD, Waldbieser GC, Lobb CJ. Patterns of receptor revision in the immunoglobulin heavy chains of a teleost fish. *J Immunol.* 2009;182(9):5605-5622. doi:10.4049/jimmunol.0801013
166. Nemazee D. Mechanisms of central tolerance for B cells. *Nat Rev Immunol.* 2017;17(5):281-294. doi:10.1038/nri.2017.19
167. Rast JP, Anderson MK, Ota T, et al. Immunoglobulin light chain class multiplicity and alternative organizational forms in early vertebrate phylogeny. *Immunogenetics.* 1994;40(2):83-99. doi:10.1007/bf00188170
168. Zhang N, Zhang XJ, Song YL, et al. Preferential combination between the light and heavy chain isotypes of fish immunoglobulins. *Dev Comp Immunol.* 2016;61:169-179. doi:10.1016/j.dci.2016.04.001
169. Diaz M, Stanfield RL, Greenberg AS, Flajnik MF. Structural analysis, selection, and ontogeny of the shark new antigen receptor (IgNAR): identification of a new locus preferentially expressed in early development. *Immunogenetics.* 2002;54(7):501-512. doi:10.1007/s00251-002-0479-z
170. Kirkham PM, Mortari F, Newton JA, Schroeder HW Jr. Immunoglobulin VH clan and family identity predicts variable domain structure and may influence antigen binding. *EMBO J.* 1992;11(2):603-609.
171. Hinds-Frey KR, Nishikata H, Litman RT, Litman GW. Somatic variation precedes extensive diversification of germline sequences and combinatorial joining in the evolution of immunoglobulin heavy chain diversity. *J Exp Med.* 1993;178(3):815-824. doi:10.1084/jem.178.3.815
172. Chang B, Casali P. The CDR1 sequences of a major proportion of human germline Ig VH genes are inherently susceptible to amino acid replacement. *Immunol Today.* 1994;15(8):367-373. doi:10.1016/0167-5699(94)90175-9
173. Kumar N, Arthur CP, Ciferri C, Matsumoto ML. Structure of the secretory immunoglobulin A core. *Science.* 2020;367(6481):1008-1014. doi:10.1126/science.aaz5807
174. Kumar N, Arthur CP, Ciferri C, Matsumoto ML. Structure of the human secretory immunoglobulin M core. *Structure.* 2021;29(6):564-571.e3. doi:10.1016/j.str.2021.01.002
175. Hsu E, Flajnik MF, Du Pasquier L. A third immunoglobulin class in amphibians. *J Immunol.* 1985;135(3):1998-2004.
176. Hohman VS, Stewart SE, Rumfelt LL, et al. J chain in the nurse shark: implications for function in a lower vertebrate. *J Immunol.* 2003;170(12):6016-6023. doi:10.4049/jimmunol.170.12.6016
177. Flowers EM, Neely HR, Guo J, et al. Identification of the Fc-alpha/mu receptor in *Xenopus* provides insight into the emergence of the poly-Ig receptor (pIgR) and mucosal Ig transport. *Eur J Immunol.* 2021. doi:10.1002/eji.202149383
178. Braathen R, Hohman VS, Brandtzaeg P, Johansen FE. Secretory antibody formation: conserved binding interactions between J chain and polymeric Ig receptor from humans and amphibians. *J Immunol.* 2007;178(3):1589-1597. doi:10.4049/jimmunol.178.3.1589
179. Castro CD, Ohta Y, Dooley H, Flajnik MF. Noncoordinate expression of J-chain and Blimp-1 define nurse shark plasma cell populations during ontogeny. *Eur J Immunol.* 2013;43(11):3061-3075. doi:10.1002/eji.201343416
180. Zhang YA, Salinas I, Li J, et al. IgT, a primitive immunoglobulin class specialized in mucosal immunity. *Nat Immunol.* 2010;11(9):827-835. doi:10.1038/ni.1913
181. Criscitiello MF, Ohta Y, Saltis M, McKinney EC, Flajnik MF. Evolutionarily conserved TCR binding sites, identification of T cells in primary lymphoid tissues, and surprising trans-rearrangements in nurse shark. *J Immunol.* 2010;184(12):6950-6960. doi:10.4049/jimmunol.0902774
182. Du Pasquier L, Schwager J, Flajnik MF. The immune system of *Xenopus*. *Annu Rev Immunol.* 1989;7:251-275. doi:10.1146/annurev.iy.07.040189.001343
183. Castro R, Navelsaker S, Krasnov A, Du Pasquier L, Boudinot P. Describing the diversity of Ag specific receptors in vertebrates: contribution of repertoire deep sequencing. *Dev Comp Immunol.* 2017;75:28-37. doi:10.1016/j.dci.2017.02.018
184. Cooper MD, Chen CL, Bucy RP, Thompson CB. Avian T cell ontogeny. *Adv Immunol.* 1991;50:87-117. doi:10.1016/s0065-2776(08)60823-8
185. Göbel TW, Meier EL, Du Pasquier L. Biochemical analysis of the *Xenopus laevis* TCR/CD3 complex supports the "stepwise evolution" model. *Eur J Immunol.* 2000;30(10):2775-2781. doi:10.1002/1521-4141(200010)30:10<2775::Aid-immu2775>3.0.Co;2-u
186. Sibener LV, Fernandes RA, Kolawole EM, et al. Isolation of a structural mechanism for uncoupling T cell receptor signaling from peptide-MHC binding. *Cell.* 2018;174(3):672-687.e27. doi:10.1016/j.cell.2018.06.017
187. Smelty P, Marchal C, Renard R, et al. Identification of the pre-T-cell receptor alpha chain in nonmammalian vertebrates challenges the structure-function of the molecule. *Proc Natl Acad Sci U S A.* 2010;107(46):19991-19996. doi:10.1073/pnas.1010166107
188. Campbell KS, Bäckström BT, Tiefenthaler G, Palmer E. CART: a conserved antigen receptor transmembrane motif. *Semin Immunol.* 1994;6(6):393-410. doi:10.1006/smim.1994.1049
189. Criscitiello MF, Wermenstam NE, Pilström L, McKinney EC. Allelic polymorphism of T-cell receptor constant domains is widespread in fishes. *Immunogenetics.* 2004;55(12):818-824. doi:10.1007/s00251-004-0652-7
190. Rock EP, Sibbald PR, Davis MM, Chien YH. CDR3 length in antigen-specific immune receptors. *J Exp Med.* 1994;179(1):323-328. doi:10.1084/jem.179.1.323
191. Ott JA, Castro CD, Deiss TC, Ohta Y, Flajnik MF, Criscitiello MF. Somatic hypermutation of T cell receptor α chain contributes to selection in nurse shark thymus. *Elife.* 2018;7. doi:10.7554/eLife.28477
192. Garcia KC. Reconciling views on T cell receptor germline bias for MHC. *Trends Immunol.* 2012;33(9):429-436. doi:10.1016/j.it.2012.05.005
193. Saito H, Kranz DM, Takagaki Y, Hayday AC, Eisen HN, Tonegawa S. A third rearranged and expressed gene in a clone of cytotoxic T lymphocytes. *Nature.* 1984;312(5989):36-40. doi:10.1038/312036a0
194. Ott JA, Ohta Y, Flajnik MF, Criscitiello MF. Lost structural and functional inter-relationships between Ig and TCR loci in mammals revealed in sharks. *Immunogenetics.* 2021;73(1):17-33. doi:10.1007/s00251-020-01183-5
195. Parra ZE, Ohta Y, Criscitiello MF, Flajnik MF, Miller RD. The dynamic TCRδ: TCRδ chains in the amphibian Xenopus tropicalis utilize antibody-like V genes. *Eur J Immunol.* 2010;40(8):2319-2329. doi:10.1002/eji.201040515
196. Luoma AM, Castro CD, Mayassi T, et al. Crystal structure of Vδ1 T cell receptor in complex with CD1d-sulfatide shows MHC-like recognition of a self-lipid by human γδ T cells. *Immunity.* 2013;39(6):1032-1042. doi:10.1016/j.immuni.2013.11.001
197. Hayday AC. Γδ T cell update: adaptate orchestrators of immune surveillance. *J Immunol.* 2019;203(2):311-320. doi:10.4049/jimmunol.1800934
198. Chen H, Kshirsagar S, Jensen I, et al. Characterization of arrangement and expression of the T cell receptor gamma locus in the sandbar shark. *Proc Natl Acad Sci U S A.* 2009;106(21):8591-8596. doi:10.1073/pnas.0811283106
199. Criscitiello MF, Saltis M, Flajnik MF. An evolutionarily mobile antigen receptor variable region gene: doubly rearranging NAR-TcR genes in sharks. *Proc Natl Acad Sci U S A.* 2006;103(13):5036-5041. doi:10.1073/pnas.0507074103
200. Wang X, Parra ZE, Miller RD. Platypus TCRμ provides insight into the origins and evolution of a uniquely mammalian TCR locus. *J Immunol.* 2011;187(10):5246-5254. doi:10.4049/jimmunol.1101113
201. Morrissey KA, Wegrecki M, Praveena T, et al. The molecular assembly of the marsupial γμ T cell receptor defines a third T cell lineage. *Science.* 2021;371(6536):1383-1388. doi:10.1126/science.abe7070
202. Hsu H, Chen C, Nenninger A, Holz L, Baldwin CL, Telfer JC. WC1 is a hybrid γδ TCR coreceptor and pattern recognition receptor for pathogenic bacteria. *J Immunol.* 2015;194(5):2280-2288. doi:10.4049/jimmunol.1402021
203. Baldwin CL, Damani-Yokota P, Yirsaw A, Loonie K, Teixeira AF, Gillespie A. Special features of γδ T cells in ruminants. *Mol Immunol.* 2021;134:161-169. doi:10.1016/j.molimm.2021.02.028
204. Hsu E, Schwager J, Alt FW. Evolution of immunoglobulin genes: VH families in the amphibian Xenopus. *Proc Natl Acad Sci U S A.* 1989;86(20):8010-8014. doi:10.1073/pnas.86.20.8010
205. Kokubu F, Hinds K, Litman R, Shamblott MJ, Litman GW. Complete structure and organization of immunoglobulin heavy chain constant region genes in a phylogenetically primitive vertebrate. *EMBO J.* 1988;7(7):1979-1988.
206. Zhang C, Du Pasquier L, Hsu E. Shark IgW C region diversification through RNA processing and isotype switching. *J Immunol.* 2013;191(6):3410-3418. doi:10.4049/jimmunol.1301257
207. Rumfelt LL, Diaz M, Lohr RL, Mochon E, Flajnik MF. Unprecedented multiplicity of Ig transmembrane and secretory mRNA forms in the cartilaginous fish. *J Immunol.* 2004;173(2):1129-1139. doi:10.4049/jimmunol.173.2.1129
208. Lee SS, Fitch D, Flajnik MF, Hsu E. Rearrangement of immunoglobulin genes in shark germ cells. *J Exp Med.* 2000;191(10):1637-1648. doi:10.1084/jem.191.10.1637
209. Anderson MK, Shamblott MJ, Litman RT, Litman GW. Generation of immunoglobulin light chain gene diversity in Raja erinacea is not associated with somatic rearrangement, an exception to a central paradigm of B cell immunity. *J Exp Med.* 1995;182(1):109-119. doi:10.1084/jem.182.1.109
210. Zhu C, Lee V, Finn A, et al. Origin of immunoglobulin isotype switching. *Curr Biol.* 2012;22(10):872-880. doi:10.1016/j.cub.2012.03.060
211. Eason DD, Litman RT, Luer CA, Kerr W, Litman GW. Expression of individual immunoglobulin genes occurs in an unusual system consisting of multiple independent loci. *Eur J Immunol.* 2004;34(9):2551-2558. doi:10.1002/eji.200425224
212. Malecek K, Lee V, Feng W, et al. Immunoglobulin heavy chain exclusion in the shark. *PLoS Biol.* 2008;6(6):e157. doi:10.1371/journal.pbio.0060157
213. Iacoangeli A, Lui A, Haines A, Ohta Y, Flajnik M, Hsu E. Evidence for Ig light chain isotype exclusion in shark B lymphocytes suggests ordered mechanisms. *J Immunol.* 2017;199(5):1875-1885. doi:10.4049/jimmunol.1700762
214. Schroeder HW Jr. The evolution and development of the antibody repertoire. *Front Immunol.* 2015;6:33. doi:10.3389/fimmu.2015.00033
215. Roux KH, Greenberg AS, Greene L, et al. Structural analysis of the nurse shark (new) antigen receptor (NAR): molecular convergence of NAR and unusual mammalian immunoglobulins. *Proc Natl Acad Sci U S A.* 1998;95(20):11804-11809. doi:10.1073/pnas.95.20.11804
216. Petrie HT, Livak F, Schatz DG, Strasser A, Crispe IN, Shortman K. Multiple rearrangements in T cell receptor alpha chain genes maximize the production of useful thymocytes. *J Exp Med.* 1993;178(2):615-622. doi:10.1084/jem.178.2.615
217. Adams EJ, Chien YH, Garcia KC. Structure of a gammadelta T cell receptor in complex with the nonclassical MHC T22. *Science.* 2005;308(5719):227-231. doi:10.1126/science.1106885
218. Alder MN, Rogozin IB, Iyer LM, Glazko GV, Cooper MD, Pancer Z. Diversity and function of adaptive immune receptors in a jawless vertebrate. *Science.* 2005;310(5756):1970-1973. doi:10.1126/science.1119420
219. Nagawa F, Kishishita N, Shimizu K, et al. Antigen-receptor genes of the agnathan lamprey are assembled by a process involving copy choice. *Nat Immunol.* 2007;8(2):206-213. doi:10.1038/ni1419
220. Morimoto R, O'Meara CP, Holland SJ, et al. Cytidine deaminase 2 is required for VLRB antibody gene assembly in lampreys. *Sci Immunol.* 2020;5(45). doi:10.1126/sciimmunol.aba0925
221. Sakano H, Hüppi K, Heinrich G, Tonegawa S. Sequences at the somatic recombination sites of immunoglobulin light-chain genes. *Nature.* 1979;280(5720):288-294. doi:10.1038/280288a0
222. Agrawal A, Eastman QM, Schatz DG. Transposition mediated by RAG1 and RAG2 and its implications for the evolution of the immune system. *Nature.* 1998;394(6695):744-751. doi:10.1038/29457
223. Oettinger MA, Schatz DG, Gorka C, Baltimore D. RAG-1 and RAG-2, adjacent genes that synergistically activate V(D)J recombination. *Science.* 1990;248(4962):1517-1523. doi:10.1126/science.2360047
224. Kapitonov VV, Jurka J. RAG1 core and V(D)J recombination signal sequences were derived from Transib transposons. *PLoS Biol.* 2005;3(6):e181. doi:10.1371/journal.pbio.0030181
225. Huang S, Tao X, Yuan S, et al. Discovery of an active RAG transposon illuminates the origins of V(D)J recombination. *Cell.* 2016;166(1):102-114. doi:10.1016/j.cell.2016.05.032
226. Martin EC, Vicari C, Tsakou-Ngouafo L, Pontarotti P, Petrescu AJ, Schatz DG. Identification of RAG-like transposons in protostomes suggests their ancient bilaterian origin. *Mob DNA.* 2020;11:17. doi:10.1186/s13100-020-00214-y.

227. Lee A, Hsu E. Isolation and characterization of the *Xenopus* terminal deoxynucleotidyl transferase. *J Immunol.* 1994;152(9):4500-4507.
228. Bartl S, Miracle AL, Rumfelt LL, et al. Terminal deoxynucleotidyl transferases from elasmobranchs reveal structural conservation within vertebrates. *Immunogenetics.* 2003;55(9):594-604. doi:10.1007/s00251-003-0608-3
229. Muramatsu M, Kinoshita K, Fagarasan S, Yamada S, Shinkai Y, Honjo T. Class switch recombination and hypermutation require activation-induced cytidine deaminase (AID), a potential RNA editing enzyme. *Cell.* 2000;102(5):553-563. doi:10.1016/s0092-8674(00)00078-7
230. Revy P, Muto T, Levy Y, et al. Activation-induced cytidine deaminase (AID) deficiency causes the autosomal recessive form of the Hyper-IgM syndrome (HIGM2). *Cell.* 2000;102(5):565-575. doi:10.1016/s0092-8674(00)00079-9
231. Marin M, Rose KM, Kozak SL, Kabat D. HIV-1 Vif protein binds the editing enzyme APOBEC3G and induces its degradation. *Nat Med.* 2003;9(11):1398-1403. doi:10.1038/nm946
232. Bae SJ, Park BG, Kim BG, Hahn JS. Multiplex gene disruption by targeted base editing of *Yarrowia lipolytica* genome using cytidine deaminase combined with the CRISPR/Cas9 system. *Biotechnol J.* 2020;15(1):e1900238. doi:10.1002/biot.201900238
233. Liu MC, Liao WY, Buckley KM, Yang SY, Rast JP, Fugmann SD. AID/APOBEC-like cytidine deaminases are ancient innate immune mediators in invertebrates. *Nat Commun.* 2018;9(1):1948. doi:10.1038/s41467-018-04273-x
234. Holland SJ, Berghuis LM, King JJ, et al. Expansions, diversification, and interindividual copy number variations of AID/APOBEC family cytidine deaminase genes in lampreys. *Proc Natl Acad Sci U S A.* 2018;115(14):E3211-E20. doi:10.1073/pnas.1720871115
235. Huber MD, Duquette ML, Shiels JC, Maizels N. A conserved G4 DNA binding domain in RecQ family helicases. *J Mol Biol.* 2006;358(4):1071-1080. doi:10.1016/j.jmb.2006.01.077
236. Zarrin AA, Alt FW, Chaudhuri J, et al. An evolutionarily conserved target motif for immunoglobulin class-switch recombination. *Nat Immunol.* 2004;5(12):1275-1281. doi:10.1038/ni1137
237. Diaz M, Velez J, Singh M, Cerny J, Flajnik MF. Mutational pattern of the nurse shark antigen receptor gene (NAR) is similar to that of mammalian Ig genes and to spontaneous mutations in evolution: the translesion synthesis model of somatic hypermutation. *Int Immunol.* 1999;11(5):825-833. doi:10.1093/intimm/11.5.825
238. Zapata A, Amemiya CT. Phylogeny of lower vertebrates and their immunological structures. *Curr Top Microbiol Immunol.* 2000;248:67-107. doi:10.1007/978-3-642-59674-2_5
239. Sutoh Y, Kondo M, Ohta Y, et al. Comparative genomic analysis of the proteasome β5t subunit gene: implications for the origin and evolution of thymoproteasomes. *Immunogenetics.* 2012;64(1):49-58. doi:10.1007/s00251-011-0558-0
240. Saltis M, Criscitiello MF, Ohta Y, et al. Evolutionarily conserved and divergent regions of the autoimmune regulator (Aire) gene: a comparative analysis. *Immunogenetics.* 2008;60(2):105-114. doi:10.1007/s00251-007-0268-9
241. Nehls M, Pfeifer D, Schorpp M, Hedrich H, Boehm T. New member of the winged-helix protein family disrupted in mouse and rat nude mutations. *Nature.* 1994;372(6501):103-107. doi:10.1038/372103a0
242. Terszowski G, Müller SM, Bleul CC, et al. Evidence for a functional second thymus in mice. *Science.* 2006;312(5771):284-287. doi:10.1126/science.1123497
243. Wyffels JT, Walsh CJ, Luer CA, Bodine AB. In vivo exposure of clearnose skates, Raja eglanteria, to ionizing X-radiation: acute effects on the thymus. *Dev Comp Immunol.* 2005;29(4):315-331. doi:10.1016/j.dci.2004.08.005
244. Aghaallaei N, Dick AM, Tsingos E, et al. αβ/γδ T cell lineage outcome is regulated by intrathymic cell localization and environmental signals. *Sci Adv.* 2021;7(29). doi:10.1126/sciadv.abg3613
245. Henry M, Charlemagne J. Development of amphibian thymus. I. Morphological differentiation, multiplication, migration and lysis of thymocytes in the urodele Pleurodeles waltlii. *J Embryol Exp Morphol.* 1980;57:219-232.
246. Bajoghli B, Guo P, Aghaallaei N, et al. A thymus candidate in lampreys. *Nature.* 2011;470(7332):90-94. doi:10.1038/nature09655
247. Flajnik M. Immunology: the origin of sweetbreads in lampreys? *Curr Biol.* 2011;21(6):R218-R220. doi:10.1016/j.cub.2011.01.066
248. Bajoghli B, Aghaallaei N, Hess I, et al. Evolution of genetic networks underlying the emergence of thymopoiesis in vertebrates. *Cell.* 2009;138(1):186-197. doi:10.1016/j.cell.2009.04.017
249. Neely HR, Flajnik MF. Emergence and evolution of secondary lymphoid organs. *Annu Rev Cell Dev Biol.* 2016;32:693-711. doi:10.1146/annurev-cellbio-111315-125306
250. Neely HR, Flajnik MF. CXCL13 responsiveness but not CXCR5 expression by late transitional B cells initiates splenic white pulp formation. *J Immunol.* 2015;194(6):2616-2623. doi:10.4049/jimmunol.1401905
251. Hofmann J, Greter M, Du Pasquier L, Becher B. B-cells need a proper house, whereas T-cells are happy in a cave: the dependence of lymphocytes on secondary lymphoid tissues during evolution. *Trends Immunol.* 2010;31(4):144-153. doi:10.1016/j.it.2010.01.003
252. Muthupandian A, Waly D, Magor BG. Do ectothermic vertebrates have a home in which to affinity mature their antibody responses? *Dev Comp Immunol.* 2021;119:104021. doi:10.1016/j.dci.2021.104021
253. Sunyer JO, Shibasali Y, Takizawa F, Yang D, Boudinot P, Krasnov A. Identification of primordial organized lymphoid structure in the spleen of teleost FISH. *J Immunol.* 2020;204(1 Supplement):92.40-92.40.
254. Kroese FG, Van Rooijen N. Antigen trapping in the spleen of the turtle, Chrysemys scripta elegans. *Immunology.* 1983;49(1):61-68.
255. Hsu E. The invention of lymphocytes. *Curr Opin Immunol.* 2011;23(2):156-162. doi:10.1016/j.coi.2010.12.007
256. Lewis KL, Del Cid N, Traver D. Perspectives on antigen presenting cells in zebrafish. *Dev Comp Immunol.* 2014;46(1):63-73. doi:10.1016/j.dci.2014.03.010
257. Rumfelt LL, McKinney EC, Taylor E, Flajnik MF. The development of primary and secondary lymphoid tissues in the nurse shark Ginglymostoma cirratum: B-cell zones precede dendritic cell immigration and T-cell zone formation during ontogeny of the spleen. *Scand J Immunol.* 2002;56(2):130-148. doi:10.1046/j.1365-3083.2002.01116.x
258. Neely HR, Guo J, Flowers EM, Criscitiello MF, Flajnik MF. "Double-duty" conventional dendritic cells in the amphibian *Xenopus* as the prototype for antigen presentation to B cells. *Eur J Immunol.* 2018;48(3):430-440. doi:10.1002/eji.201747260
259. Lugo-Villarino G, Balla KM, Stachura DL, Bañuelos K, Werneck MB, Traver D. Identification of dendritic antigen-presenting cells in the zebrafish. *Proc Natl Acad Sci U S A.* 2010;107(36):15850-15855. doi:10.1073/pnas.1000494107
260. Baldwin WM3rd, Cohen N. A giant cell with dendritic cell properties in spleens of the anuran amphibian *Xenopus laevis*. *Dev Comp Immunol.* 1981;5(3):461-473. doi:10.1016/s0145-305x(81)80058-4
261. Miracle AL, Anderson MK, Litman RT, et al. Complex expression patterns of lymphocyte-specific genes during the development of cartilaginous fish implicate unique lymphoid tissues in generating an immune repertoire. *Int Immunol.* 2001;13(4):567-580. doi:10.1093/intimm/13.4.567
262. Malecek K, Brandman J, Brodsky JE, Ohta Y, Flajnik MF, Hsu E. Somatic hypermutation and junctional diversification at Ig heavy chain loci in the nurse shark. *J Immunol.* 2005;175(12):8105-8115. doi:10.4049/jimmunol.175.12.8105
263. Hsu E. Assembly and expression of shark Ig genes. *J Immunol.* 2016;196(9):3517-3523. doi:10.4049/jimmunol.1600164
264. Ma C, Ye J, Kaattari SL. Differential compartmentalization of memory B cells versus plasma cells in salmonid fish. *Eur J Immunol.* 2013;43(2):360-370. doi:10.1002/eji.201242570
265. Zwollo P. Dissecting teleost B cell differentiation using transcription factors. *Dev Comp Immunol.* 2011;35(9):898-905. doi:10.1016/j.dci.2011.01.009
266. Li J, Barreda DR, Zhang YA, et al. B lymphocytes from early vertebrates have potent phagocytic and microbicidal abilities. *Nat Immunol.* 2006;7(10):1116-1124. doi:10.1038/ni1389
267. Parra D, Rieger AM, Li J, et al. Pivotal advance: peritoneal cavity B-1 B cells have phagocytic and microbicidal capacities and present phagocytosed antigen to CD4+ T cells. *J Leukoc Biol.* 2012;91(4):525-536. doi:10.1189/jlb.0711372
268. Montecino-Rodriguez E, Fice M, Casero D, Berent-Maoz B, Barber CL, Dorshkind K. Distinct genetic networks orchestrate the emergence of specific waves of fetal and adult B-1 and B-2 development. *Immunity.* 2016;45(3):527-539. doi:10.1016/j.immuni.2016.07.012
269. Mussmann R, Courtet M, Du Pasquier L. Development of the early B cell population in *Xenopus*. *Eur J Immunol.* 1998;28(9):2947-2959. doi:10.1002/(sici)1521-4141(199809)28:09<2947::Aid-immu2947>3.0.Co;2-a
270. Flajnik MF, Hsu E, Kaufman JF, Pasquier LD. Changes in the immune system during metamorphosis of Xenopus. *Immunol Today.* 1987;8(2):58-64. doi:10.1016/0167-5699(87)90240-4
271. Greenhalgh P, Olesen CE, Steiner LA. Characterization and expression of recombination activating genes (RAG-1 and RAG-2) in *Xenopus laevis*. *J Immunol.* 1993;151(6):3100-3110.
272. Schwager J, Bürckert N, Courtet M, Du Pasquier L. The ontogeny of diversification at the immunoglobulin heavy chain locus in *Xenopus*. *EMBO J.* 1991;10(9):2461-2470.
273. Du Pasquier L. Antibody diversity in lower vertebrates—why is it so restricted? *Nature.* 1982;296(5855):311-313. doi:10.1038/296311a0
274. Reynaud CA, Anquez V, Grimal H, Weill JC. A hyperconversion mechanism generates the chicken light chain preimmune repertoire. *Cell.* 1987;48(3):379-388. doi:10.1016/0092-8674(87)90189-9
275. Ratcliffe MJ, Ivanyi J. Allotype suppression in the chicken. III. Analysis of the recovery form suppression by neonatally injected or maternal antibodies. *Eur J Immunol.* 1981;11(4):301-306. doi:10.1002/eji.1830110407
276. Schiaffella E, Sehgal D, Anderson AO, Mage RG. Gene conversion and hypermutation during diversification of VH sequences in developing splenic germinal centers of immunized rabbits. *J Immunol.* 1999;162(7):3984-3995.
277. Pike KA, Ratcliffe MJ. Ligand-independent signaling during early avian B cell development. *Immunol Res.* 2006;35(1–2):103-116. doi:10.1385/ir:35:1:103
278. Davani D, Pancer Z, Ratcliffe MJ. Ligation of surface Ig by gut-derived antigen positively selects chicken bursal and peripheral B cells. *J Immunol.* 2014;192(7):3218-3227. doi:10.4049/jimmunol.1302395
279. Reynaud CA, Mackay CR, Müller RG, Weill JC. Somatic generation of diversity in a mammalian primary lymphoid organ: the sheep ileal Peyer's patches. *Cell.* 1991;64(5):995-1005. doi:10.1016/0092-8674(91)90323-q
280. Becker RS, Knight KL. Somatic diversification of immunoglobulin heavy chain VDJ genes: evidence for somatic gene conversion in rabbits. *Cell.* 1990;63(5):987-997. doi:10.1016/0092-8674(90)90502-6
281. Severson KM, Mallozzi M, Driks A, Knight KL. B cell development in GALT: role of bacterial superantigen-like molecules. *J Immunol.* 2010;184(12):6782-6789. doi:10.4049/jimmunol.1000155
282. Kasahara M. Déjà vu: the identity of a third lineage of lymphocytes in lampreys. *Immunol Cell Biol.* 2013;91(10):599-600. doi:10.1038/icb.2013.52
283. Das S, Hirano M, Aghaallaei N, Bajoghli B, Boehm T, Cooper MD. Organization of lamprey variable lymphocyte receptor C locus and repertoire development. *Proc Natl Acad Sci U S A.* 2013;110(15):6043-6048. doi:10.1073/pnas.1302500110
284. Velikovsky CA, Deng L, Tasumi S, et al. Structure of a lamprey variable lymphocyte receptor in complex with a protein antigen. *Nat Struct Mol Biol.* 2009;16(7):725-730. doi:10.1038/nsmb.1619
285. Han BW, Herrin BR, Cooper MD, Wilson IA. Antigen recognition by variable lymphocyte receptors. *Science.* 2008;321(5897):1834-1837. doi:10.1126/science.1162484
286. Deng L, Velikovsky CA, Xu G, et al. A structural basis for antigen recognition by the T cell-like lymphocytes of sea lamprey. *Proc Natl Acad Sci U S A.* 2010;107(30):13408-13413. doi:10.1073/pnas.1005475107
287. Deng L, Luo M, Velikovsky A, Mariuzza RA. Structural insights into the evolution of the adaptive immune system. *Annu Rev Biophys.* 2013;42:191-215. doi:10.1146/annurev-biophys-083012-130422
288. Sutoh Y, Kasahara M. Copy number and sequence variation of leucine-rich repeat modules suggests distinct functional constraints operating on variable lymphocyte receptors expressed by agnathan T cell-like and B cell-like lymphocytes. *Immunogenetics.* 2014;66(6):403-409. doi:10.1007/s00251-014-0773-6
289. Adelman MK, Schluter SF, Marchalonis JJ. The natural antibody repertoire of sharks and humans recognizes the potential universe of antigens. *Protein J.* 2004;23(2):103-118. doi:10.1023/b:jopc.0000020077.73751.76
290. Mäkelä O, Litman GW. Lack of heterogeneity in antihapten antibodies of a phylogenetically primitive shark. *Nature.* 1980;287(5783):639-640. doi:10.1038/287639a0
291. Iacoangeli A, Lui A, Naik U, Ohta Y, Flajnik M, Hsu E. Biased immunoglobulin light chain gene usage in the shark. *J Immunol.* 2015;195(8):3992-4000. doi:10.4049/jimmunol.1501426
292. Diaz M, Greenberg AS, Flajnik MF. Somatic hypermutation of the new antigen receptor gene (NAR) in the nurse shark does not generate the repertoire: possible role in antigen-driven reactions in the absence of germinal centers. *Proc Natl Acad Sci U S A.* 1998;95(24):14343-14348. doi:10.1073/pnas.95.24.14343
293. Dooley H, Stanfield RL, Brady RA, Flajnik MF. First molecular and biochemical analysis of in vivo affinity maturation in an ectothermic vertebrate. *Proc Natl Acad Sci U S A.* 2006;103(6):1846-1851. doi:10.1073/pnas.0508341103
294. Eve O, Matz H, Dooley H. Proof of long-term immunological memory in cartilaginous fishes. *Dev Comp Immunol.* 2020;108:103674. doi:10.1016/j.dci.2020.103674
295. Macpherson AJ, Geuking MB, Slack E, Hapfelmeier S, McCoy KD. The habitat, double life, citizenship, and forgetfulness of IgA. *Immunol Rev.* 2012;245(1):132-146. doi:10.1111/j.1600-065X.2011.01072.x
296. Baldo BA, Uhlenbruck G. Cross-reactive human blood group H-active polysaccharide from Helix pomatia. I. Detection with catfish anti-H and eel sera. *Immunology.* 1973;25(4):649-661.
297. Salgado-Miranda C, Loza-Rubio E, Rojas-Anaya E, García-Espinosa G. Viral vaccines for bony fish: past, present and future. *Expert Rev Vaccines.* 2013;12(5):567-578. doi:10.1586/erv.13.38
298. Magnadóttir B, Jónsdóttir H, Helgason S, Björnsson B, Solem ST, Pilström L. Immune parameters of immunised cod (*Gadus morhua* L.). *Fish Shellfish Immunol.* 2001;11(1):75-89. doi:10.1006/fsim.2000.0296
299. Vallejo AN, Miller NW, Jørgensen T, Clem LW. Phylogeny of immune recognition: antigen processing/presentation in channel catfish immune responses to hemocyanins. *Cell Immunol.* 1990;130(2):364-377. doi:10.1016/0008-8749(90)90279-z

300. Castro R, Bernard D, Lefranc MP, Six A, Benmansour A, Boudinot P. T cell diversity and TcR repertoires in teleost fish. *Fish Shellfish Immunol.* 2011;31(5):644-654. doi:10.1016/j.fsi.2010.08.016
301. Barreto VM, Magor BG. Activation-induced cytidine deaminase structure and functions: a species comparative view. *Dev Comp Immunol.* 2011;35(9):991-1007. doi:10.1016/j.dci.2011.02.005
302. Yang F, Waldbieser GC, Lobb CJ. The nucleotide targets of somatic mutation and the role of selection in immunoglobulin heavy chains of a teleost fish. *J Immunol.* 2006;176(3):1655-1667. doi:10.4049/jimmunol.176.3.1655
303. Wakae K, Magor BG, Saunders H, et al. Evolution of class switch recombination function in fish activation-induced cytidine deaminase, AID. *Int Immunol.* 2006;18(1):41-47. doi:10.1093/intimm/dxh347
304. Bly JE, Quiniou SM, Clem LW. Environmental effects on fish immune mechanisms. *Dev Biol Stand.* 1997;90:33-43.
305. Spencer DA, Quiniou SMA, Crider J, Musungu B, Bengten E, Wilson M. Insights into the dynamics of memory, effector and apoptotic cytotoxic T lymphocytes in channel catfish, *Ictalurus punctatus. Dev Comp Immunol.* 2019;92:116-128. doi:10.1016/j.dci.2018.11.001
306. Charlemagne J. Antibody diversity in amphibians. Noninbred axolotls used the same unique heavy chain and a limited number of light chains for their anti-2,4-dinitrophenyl antibody responses. *Eur J Immunol.* 1987;17(3):421-424. doi:10.1002/eji.1830170319
307. Du Pasquier L, Blomberg B, Bernard CC. Ontogeny of immunity in amphibians: changes in antibody repertoires and appearance of adult major histocompatibility antigens in Xenopus. *Eur J Immunol.* 1979;9(11):900-906. doi:10.1002/eji.1830091112
308. Wabl MR, Du Pasquier L. Antibody patterns in genetically identical frogs. *Nature.* 1976;264(5587):642-644. doi:10.1038/264642a0
309. Mashoof S, Goodroe A, Du CC, et al. Ancient T-independence of mucosal IgX/A: gut microbiota unaffected by larval thymectomy in Xenopus laevis. *Mucosal Immunol.* 2013;6(2):358-368. doi:10.1038/mi.2012.78
310. Wilson M, Hsu E, Marcuz A, Courtet M, Du Pasquier L, Steinberg C. What limits affinity maturation of antibodies in *Xenopus*–the rate of somatic mutation or the ability to select mutants? *EMBO J.* 1992;11(12):4337-4347.
311. Blomberg B, Bernard CC, Du Pasquier L. In vitro evidence for T-B lymphocyte collaboration in the clawed toad, *Xenopus. Eur J Immunol.* 1980;10(11):869-876. doi:10.1002/eji.1830101112
312. Flajnik MF, Du Pasquier L, Cohen N. Immune responses of thymus/lymphocyte embryonic chimeras: studies on tolerance and major histocompatibility complex restriction in Xenopus. *Eur J Immunol.* 1985;15(6):540-547. doi:10.1002/eji.1830150603
313. Mussmann R, Du Pasquier L, Hsu E. Is *Xenopus* IgX an analog of IgA? *Eur J Immunol.* 1996;26(12):2823-2830. doi:10.1002/eji.1830261205
314. Marr S, Morales H, Bottaro A, Cooper M, Flajnik M, Robert J. Localization and differential expression of activation-induced cytidine deaminase in the amphibian *Xenopus* upon antigen stimulation and during early development. *J Immunol.* 2007;179(10):6783-6789. doi:10.4049/jimmunol.179.10.6783
315. Robert J, Ménoret A, Basu S, Cohen N, Srivastava PR. Phylogenetic conservation of the molecular and immunological properties of the chaperones gp96 and hsp70. *Eur J Immunol* 2001;31(1):186-195 doi:10.1002/1521-4141(200101)31:1<186::Aid-immu186>3.0.Co;2-d
316. Herbst LH, Klein PA. Monoclonal antibodies for the measurement of class-specific antibody responses in the green turtle, *Chelonia mydas. Vet Immunol Immunopathol.* 1995;46(3-4):317-335. doi:10.1016/0165-2427(94)05360-5
317. Turchin A, Hsu E. The generation of antibody diversity in the turtle. *J Immunol.* 1996;156(10):3797-3805.
318. Keller JM, McClellan-Green PD, Lee AM, et al. Mitogen-induced lymphocyte proliferation in loggerhead sea turtles: comparison of methods and effects of gender, plasma testosterone concentration, and body condition on immunity. *Vet Immunol Immunopathol.* 2005;103(3-4):269-281. doi:10.1016/j.vetimm.2004.09.029
319. Takeo K, Tanaka T, Nakamura K, Suzuno R. Studies on the heterogeneity of anti-hapten antibodies by means of two-dimensional affinity electrophoresis. *Electrophoresis.* 1989;10(12):818-824. doi:10.1002/elps.1150101204
320. Mehr R, Edelman H, Sehgal D, Mage R. Analysis of mutational lineage trees from sites of primary and secondary Ig gene diversification in rabbits and chickens. *J Immunol.* 2004;172(8):4790-4796. doi:10.4049/jimmunol.172.8.4790
321. Waugh C, Gillett A, Polkinghorne A, Timms P. Serum antibody response to koala retrovirus antigens varies in free-ranging koalas (Phascolarctos cinereus) in Australia: implications for vaccine design. *J Wildl Dis.* 2016;52(2):422-425. doi:10.7589/2015-09-257
322. Matsunaga T, Rahman A. What brought the adaptive immune system to vertebrates?—The jaw hypothesis and the seahorse. *Immunol Rev.* 1998;166:177-186. doi:10.1111/j.1600-065x.1998.tb01262.x
323. Sepahi A, Salinas I. The evolution of nasal immune systems in vertebrates. *Mol Immunol.* 2016;69:131-138. doi:10.1016/j.molimm.2015.09.008
324. Du CC, Mashoof SM, Criscitiello MF. Oral immunization of the African clawed frog (*Xenopus laevis*) upregulates the mucosal immunoglobulin IgX. *Vet Immunol Immunopathol.* 2012;145(1-2):493-498. doi:10.1016/j.vetimm.2011.10.019
325. Hansen JD, Landis ED, Phillips RB. Discovery of a unique Ig heavy-chain isotype (IgT) in rainbow trout: implications for a distinctive B cell developmental pathway in teleost fish. *Proc Natl Acad Sci U S A.* 2005;102(19):6919-6924. doi:10.1073/pnas.0500027102
326. Danilova N, Bussmann J, Jekosch K, Steiner LA. The immunoglobulin heavy-chain locus in zebrafish: identification and expression of a previously unknown isotype, immunoglobulin Z. *Nat Immunol.* 2005;6(3):295-302. doi:10.1038/ni1166
327. Parra D, Korytář T, Takizawa F, Sunyer JO. B cells and their role in the teleost gut. *Dev Comp Immunol.* 2016;64:150-166. doi:10.1016/j.dci.2016.03.013
328. Xu Z, Parra D, Gómez D, et al. Teleost skin, an ancient mucosal surface that elicits gut-like immune responses. *Proc Natl Acad Sci U S A.* 2013;110(32):13097-13102. doi:10.1073/pnas.1304319110
329. Xu Z, Takizawa F, Casadei E, et al. Specialization of mucosal immunoglobulins in pathogen control and microbiota homeostasis occurred early in vertebrate evolution. *Sci Immunol.* 2020;5(44). doi:10.1126/sciimmunol.aay3254
330. Kaufman J, Salomonsen J, Flajnik M. Evolutionary conservation of MHC class I and class II molecules – different yet the same. *Semin Immunol.* 1994;6(6):411-424. doi:10.1006/smim.1994.1050
331. Dijkstra JM, Yamaguchi T, Grimholt U. Conservation of sequence motifs suggests that the nonclassical MHC class I lineages CD1/PROCR and UT were established before the emergence of tetrapod species. *Immunogenetics.* 2018;70(7):459-476. doi:10.1007/s00251-017-1050-2
332. Kaufman J. Unfinished business: evolution of the MHC and the adaptive immune system of jawed vertebrates. *Annu Rev Immunol.* 2018;36:383-409. doi:10.1146/annurev-immunol-051116-052450
333. Dijkstra JM, Grimholt U, Leong J, Koop BF, Hashimoto K. Comprehensive analysis of MHC class II genes in teleost fish genomes reveals dispensability of the peptide-loading DM system in a large part of vertebrates. *BMC Evol Biol.* 2013;13:260. doi:10.1186/1471-2148-13-260
334. Almeida T, Gaigher A, Munoz-Merida A, et al. Cartilaginous fish class II genes reveal unprecedented old allelic lineages and confirm the late evolutionary emergence of DM. *Mol Immunol.* 2020;128:125-138. doi:10.1016/j.molimm.2020.10.003
335. Kaufman JF, Flajnik MF, Du Pasquier L, Riegert P. *Xenopus* MHC class II molecules. I. Identification and structural characterization. *J Immunol.* 1985;134(5):3248-3257.
336. Criscitiello MF, Ohta Y, Graham MD, Eubanks JO, Chen PL, Flajnik MF. Shark class II invariant chain reveals ancient conserved relationships with cathepsins and MHC class II. *Dev Comp Immunol.* 2012;36(3):521-533. doi:10.1016/j.dci.2011.09.008
337. Malmstrøm M, Matschiner M, Tørresen OK, et al. Evolution of the immune system influences speciation rates in teleost fishes. *Nat Genet.* 2016;48(10):1204-1210. doi:10.1038/ng.3645
338. Almeida T, Ohta Y, Gaigher A, et al. A highly complex, MHC-linked, 350 million-year-old shark nonclassical class I lineage. *J Immunol.* 2021;207(3):824-836. doi:10.4049/jimmunol.2000851
339. Grimholt U, Tsukamoto K, Azuma T, Leong J, Koop BF, Dijkstra JM. A comprehensive analysis of teleost MHC class I sequences. *BMC Evol Biol.* 2015;15:32. doi:10.1186/s12862-015-0309-1
340. Xiao J, Xiang W, Zhang Y, et al. An invariant arginine in common with MHC class II allows extension at the C-terminal end of peptides bound to chicken MHC class I. *J Immunol.* 2018;201(10):3084-3095. doi:10.4049/jimmunol.1800611
341. Nonaka M, Namikawa C, Kato Y, Sasaki M, Salter-Cid L, Flajnik MF. Major histocompatibility complex gene mapping in the amphibian *Xenopus* implies a primordial organization. *Proc Natl Acad Sci U S A.* 1997;94(11):5789-5791. doi:10.1073/pnas.94.11.5789
342. Tsukamoto K, Miura F, Fujito NT, Yoshizaki G, Nonaka M. Long-lived dichotomous lineages of the proteasome subunit beta type 8 (PSMB8) gene surviving more than 500 million years as alleles or paralogs. *Mol Biol Evol.* 2012;29(10):3071-3079. doi:10.1093/molbev/mss113
343. Walker BA, Hunt LG, Sowa AK, et al. The dominantly expressed class I molecule of the chicken MHC is explained by coevolution with the polymorphic peptide transporter (TAP) genes. *Proc Natl Acad Sci U S A.* 2011;108(20):8396-8401. doi:10.1073/pnas.1019496108
344. Flajnik MF, Kaufman JF, Riegert P, Du Pasquier L. Identification of class I major histocompatibility complex encoded molecules in the amphibian *Xenopus. Immunogenetics.* 1984;20(4):433-442. doi:10.1007/bf00345617
345. Courtet M, Flajnik M, Du Pasquier L. Major histocompatibility complex and immunoglobulin loci visualized by in situ hybridization on *Xenopus* chromosomes. *Dev Comp Immunol.* 2001;25(2):149-157. doi:10.1016/s0145-305x(00)00045-8
346. Salter-Cid L, Nonaka M, Flajnik MF. Expression of MHC class Ia and class Ib during ontogeny: high expression in epithelia and coregulation of class Ia and lmp7 genes. *J Immunol.* 1998;160(6):2853-2861.
347. Almeida T, Esteves PJ, Flajnik MF, Ohta Y, Verissimo A. An ancient, MHC-linked, nonclassical class I lineage in cartilaginous fish. *J Immunol.* 2020;204(4):892-902. doi:10.4049/jimmunol.1901025
348. Chappell P, Meziane el K, Harrison M, et al. Expression levels of MHC class I molecules are inversely correlated with promiscuity of peptide binding. *Elife.* 2015;4:e05345. doi:10.7554/eLife.05345
349. Flajnik MF, Ferrone S, Cohen N, Du Pasquier L. Evolution of the MHC: antigenicity and unusual tissue distribution of Xenopus (frog) class II molecules. *Mol Immunol.* 1990;27(5):451-462. doi:10.1016/0161-5890(90)90170-5
350. Völk H, Charlemagne J, Tournefier A, et al. Wide tissue distribution of axolotl class II molecules occurs independently of thyroxin. *Immunogenetics.* 1998;47(5):339-349. doi:10.1007/s002510050368
351. Du Pasquier L, Chardonnens X, Miggiano VC. A major histocompatibility complex in the toad *Xenopus laevis* (Daudin). *Immunogenetics.* 1974;1(1):482-494. doi:10.1007/BF01564086
352. Sato A, Figueroa F, Murray BW, et al. Nonlinkage of major histocompatibility complex class I and class II loci in bony fishes. *Immunogenetics.* 2000;51(2):108-116. doi:10.1007/s002510050019
353. Ohta Y, Okamura K, McKinney EC, Bartl S, Hashimoto K, Flajnik MF. Primitive synteny of vertebrate major histocompatibility complex class I and class II genes. *Proc Natl Acad Sci U S A.* 2000;97(9):4712-4717. doi:10.1073/pnas.97.9.4712
354. Ohta Y, Shiina T, Lohr RL, et al. Primordial linkage of β2-microglobulin to the MHC. *J Immunol.* 2011;186(6):3563-3571. doi:10.4049/jimmunol.1003933
355. Goshima M, Sekiguchi R, Matsushita M, Nonaka M. The complement system of elasmobranches revealed by liver transcriptome analysis of a hammerhead shark, *Sphyrna zygaena. Dev Comp Immunol.* 2016;61:13-24. doi:10.1016/j.dci.2016.03.009
356. Ohta Y, Goetz W, Hossain MZ, Nonaka M, Flajnik MF. Ancestral organization of the MHC revealed in the amphibian *Xenopus. J Immunol.* 2006;176(6):3674-3685. doi:10.4049/jimmunol.176.6.3674
357. Fu Y, Yang Z, Huang J, et al. Identification of two nonrearranging IgSF genes in chicken reveals a novel family of putative remnants of an antigen receptor precursor. *J Immunol.* 2019;202(7):1992-2004. doi:10.4049/jimmunol.1801305
358. Uinuk-ool TS, Mayer WE, Sato A, et al. Identification and characterization of a TAP-family gene in the lamprey. *Immunogenetics.* 2003;55(1):38-48. doi:10.1007/s00251-003-0548-y
359. Nomiyama H, Yoshie O. Functional roles of evolutionary conserved motifs and residues in vertebrate chemokine receptors. *J Leukoc Biol.* 2015;97(1):39-47. doi:10.1189/jlb.2RU0614-290R
360. Sodergren E, Weinstock GM, Davidson EH, et al. The genome of the sea urchin *Strongylocentrotus purpuratus. Science.* 2006;314(5801):941-952. doi:10.1126/science.1133609
361. Guo L, Wei G, Zhu J, et al. IL-1 family members and STAT activators induce cytokine production by Th2, Th17, and Th1 cells. *Proc Natl Acad Sci U S A.* 2009;106(32):13463-13468. doi:10.1073/pnas.0906988106
362. Gibson MS, Kaiser P, Fife M. The chicken IL-1 family: evolution in the context of the studied vertebrate lineage. *Immunogenetics.* 2014;66(7-8):427-438. doi:10.1007/s00251-014-0780-7
363. Haertle S, Alzuheir I, Busalt F, Waters V, Kaiser P, Kaufer BB. Identification of the receptor and cellular ortholog of the Marek's disease virus (MDV) CXC chemokine. *Front Microbiol.* 2017;8:2543. doi:10.3389/fmicb.2017.02543
364. Collette Y, Gilles A, Pontarotti P, Olive D. A co-evolution perspective of the TNFSF and TNFRSF families in the immune system. *Trends Immunol.* 2003;24(7):387-394. doi:10.1016/s1471-4906(03)00166-2
365. Magor KE, Miranzo Navarro D, Barber MR, et al. Defense genes missing from the flight division. *Dev Comp Immunol.* 2013;41(3):377-388. doi:10.1016/j.dci.2013.04.010
366. Dijkstra JM. A method for making alignments of related protein sequences that share very little similarity; shark interleukin 2 as an example. *Immunogenetics.* 2021;73(1):35-51. doi:10.1007/s00251-020-01191-5
367. Kaiser P, Rothwell L, Avery S, Balu S. Evolution of the interleukins. *Dev Comp Immunol.* 2004;28(5):375-394. doi:10.1016/j.dci.2003.09.004
368. Lawir DF, Hess I, Sikora K, et al. Evolutionary transition from degenerate to nonredundant cytokine signaling networks supporting intrathymic T cell development. *Proc Natl Acad Sci U S A.* 2019;116(52):26759-26767. doi:10.1073/pnas.1915223116
369. Lee GR, Fields PE, Griffin TJ, Flavell RA. Regulation of the Th2 cytokine locus by a locus control region. *Immunity.* 2003;19(1):145-153. doi:10.1016/s1074-7613(03)00179-1

370. Redmond AK, Zou J, Secombes CJ, Macqueen DJ, Dooley H. Discovery of all three types in cartilaginous fishes enables phylogenetic resolution of the origins and evolution of interferons. *Front Immunol.* 2019;10:1558. doi:10.3389/fimmu.2019.01558
371. Zhou P, Tachedjian M, Wynne JW, et al. Contraction of the type I IFN locus and unusual constitutive expression of IFN-α in bats. *Proc Natl Acad Sci U S A.* 2016;113(10):2696-2701. doi:10.1073/pnas.1518240113
372. Mendoza JL, Escalante NK, Jude KM, et al. Structure of the IFNγ receptor complex guides design of biased agonists. *Nature.* 2019;567(7746):56-60. doi:10.1038/s41586-019-0988-7
373. Bird S, Tafalla C. Teleost chemokines and their receptors. *Biology.* 2015;4(4):756-784. doi:10.3390/biology4040756
374. Wu Z, Hu T, Rothwell L, et al. Analysis of the function of IL-10 in chickens using specific neutralising antibodies and a sensitive capture ELISA. *Dev Comp Immunol.* 2016;63:206-212. doi:10.1016/j.dci.2016.04.016
375. Huo HJ, Chen SN, Laghari ZA, et al. Specific bioactivity of IL-22 in intestinal cells as revealed by the expression of IL-22RA1 in Mandarin fish, *Siniperca chuatsi*. *Dev Comp Immunol.* 2021;121:104107. doi:10.1016/j.dci.2021.104107
376. Du Pasquier L, Bernard CC. Active suppression of the allogeneic histocompatibility reactions during the metamorphosis of the clawed toad *Xenopus*. *Differentiation.* 1980;16(1):1-7. doi:10.1111/j.1432-0436.1980.tb01052.x
377. Kasheta M, Painter CA, Moore FE, et al. Identification and characterization of T reg-like cells in zebrafish. *J Exp Med.* 2017;214(12):3519-3530. doi:10.1084/jem.20162084
378. Han Q, Das S, Hirano M, et al. Characterization of lamprey IL-17 family members and their receptors. *J Immunol.* 2015;195(11):5440-5451. doi:10.4049/jimmunol.1500892
379. Buckley KM, Ho ECH, Hibino T, et al. IL17 factors are early regulators in the gut epithelium during inflammatory response to Vibrio in the sea urchin larva. *Elife.* 2017;6. doi:10.7554/eLife.23481
380. Dijkstra JM. TH2 and Treg candidate genes in elephant shark. *Nature.* 2014;511(7508):E7-E9. doi:10.1038/nature13446
381. Samstein RM, Josefowicz SZ, Arvey A, Treuting PM, Rudensky AY. Extrathymic generation of regulatory T cells in placental mammals mitigates maternal-fetal conflict. *Cell.* 2012;150(1):29-38. doi:10.1016/j.cell.2012.05.031
382. Nomiyama H, Osada N, Yoshie O. A family tree of vertebrate chemokine receptors for a unified nomenclature. *Dev Comp Immunol.* 2011;35(7):705-715. doi:10.1016/j.dci.2011.01.019
383. DeVries ME, Kelvin AA, Xu L, Ran L, Robinson J, Kelvin DJ. Defining the origins and evolution of the chemokine/chemokine receptor system. *J Immunol.* 2006;176(1):401-415. doi:10.4049/jimmunol.176.1.401
384. Burnet FM. "Self-recognition" in colonial marine forms and flowering plants in relation to the evolution of immunity. *Nature.* 1971;232(5308):230-235. doi:10.1038/232230a0
385. Nicotra ML. Invertebrate allorecognition. *Curr Biol.* 2019;29(11):R463-r67. doi:10.1016/j.cub.2019.03.039
386. Hauenschild C. [Not available]. *Wilhelm Roux Arch Entwickl Mech Org.* 1954;147(1):1-41. doi:10.1007/bf00576821
387. Rosa SF, Powell AE, Rosengarten RD, et al. Hydractinia allodeterminant alr1 resides in an immunoglobulin superfamily-like gene complex. *Curr Biol.* 2010;20(12):1122-1127. doi:10.1016/j.cub.2010.04.050
388. Karadge UB, Gosto M, Nicotra ML. Allorecognition proteins in an invertebrate exhibit homophilic interactions. *Curr Biol.* 2015;25(21):2845-2850. doi:10.1016/j.cub.2015.09.030
389. Oka H, Watanabe H. Vascular budding in botrylloides. *Biol. Bull.* 1959;117(2):340-346. doi:10.2307/1538913.
390. Mueller WA, Rinkevich B. Cell communication-mediated nonself-recognition and -intolerance in representative species of the animal kingdom. *J Mol Evol.* 2020;88(6):482-500. doi:10.1007/s00239-020-09955-z
391. Mandujano-Tinoco EA, Sultan E, Ottolenghi A, Gershoni-Yahalom O, Rosental B. Evolution of cellular immunity effector cells; perspective on cytotoxic and phagocytic cellular lineages. *Cells.* 2021;10(8). doi:10.3390/cells10081853
392. Taketa DA, De Tomaso AW. Botryllus schlosseri allorecognition: tackling the enigma. *Dev Comp Immunol.* 2015;48(1):254-265. doi:10.1016/j.dci.2014.03.014
393. De Tomaso AW, Nyholm SV, Palmeri KJ, et al. Isolation and characterization of a protochordate histocompatibility locus. *Nature.* 2005;438(7067):454-459. doi:10.1038/nature04150
394. Nyholm SV, Passegue E, Ludington WB, et al. *fester*, A candidate allorecognition receptor from a primitive chordate. *Immunity.* 2006;25(1):163-173. doi:10.1016/j.immuni.2006.04.011
395. Voskoboynik A, Newman AM, Corey DM, et al. Identification of a colonial chordate histocompatibility gene. *Science.* 2013;341(6144):384-387. doi:10.1126/science.1238036
396. Taketa DA, Nydam ML, Langenbacher AD, Rodriguez D, Sanders E, De Tomaso AW. Molecular evolution and in vitro characterization of *Botryllus* histocompatibility factor. *Immunogenetics.* 2015;67(10):605-623. doi:10.1007/s00251-015-0870-1
397. Nydam ML, Hoang TA, Shanley KM, De Tomaso AW. Molecular evolution of a polymorphic HSP40-like protein encoded in the histocompatibility locus of an invertebrate chordate. *Dev Comp Immunol.* 2013;41(2):128-136. doi:10.1016/j.dci.2013.03.004
398. Fuke MT, Numakunai T. Allogeneic cellular reactions between intra-specific types of a solitary ascidian, Halocynthia roretzi. *Dev Comp Immunol.* 1982;6(2):253-261. doi:10.1016/s0145-305x(82)80008-6
399. Kürn U, Sommer F, Hemmrich G, Bosch TC, Khalturin K. Allorecognition in urochordates: identification of a highly variable complement receptor-like protein expressed in follicle cells of Ciona. *Dev Comp Immunol.* 2007;31(4):360-371. doi:10.1016/j.dci.2006.06.008
400. Laird DJ, De Tomaso AW, Weissman IL. Stem cells are units of natural selection in a colonial ascidian. *Cell.* 2005;123(7):1351-1360. doi:10.1016/j.cell.2005.10.026
401. Dai H, Lan P, Zhao D, et al. PIRs mediate innate myeloid cell memory to nonself MHC molecules. *Science.* 2020;368(6495):1122-1127. doi:10.1126/science.aax4040
402. Netea MG, Domínguez-Andrés J, Barreiro LB, et al. Defining trained immunity and its role in health and disease. *Nat Rev Immunol.* 2020;20(6):375-388. doi:10.1038/s41577-020-0285-6
403. Liu C, Zhang Y, Liu CC, Schatz DG. Structural insights into the evolution of the RAG recombinase. *Nat Rev Immunol.* 2021. doi:10.1038/s41577-021-00628-6
404. Chen R, Zhang L, Qi J, et al. Discovery and analysis of invertebrate IgV(J)-C2 structure from amphioxus provides insight into the evolution of the Ig superfamily. *J Immunol.* 2018;200(8):2869-2881. doi:10.4049/jimmunol.1700906
405. Srinivasan N, Gordon O, Ahrens S, et al. Actin is an evolutionarily-conserved damage-associated molecular pattern that signals tissue injury in *Drosophila melanogaster*. *Elife.* 2016;5. doi:10.7554/eLife.19662
406. Boehm T, Hirano M, Holland SJ, Das S, Schorpp M, Cooper MD. Evolution of alternative adaptive immune systems in vertebrates. *Annu Rev Immunol.* 2018;36:19-42. doi:10.1146/annurev-immunol-042617-053028
407. Hinds KR, Litman GW. Major reorganization of immunoglobulin VH segmental elements during vertebrate evolution. *Nature.* 1986;320(6062):546-549. doi:10.1038/320546a0
408. Swann JB, Holland SJ, Petersen M, Pietsch TW, Boehm T. The immunogenetics of sexual parasitism. *Science.* 2020;369(6511):1608-1615. doi:10.1126/science.aaz9445
409. Zwollo P. Why spawning salmon return to their natal stream: the immunological imprinting hypothesis. *Dev Comp Immunol.* 2012;38(1):27-29. doi:10.1016/j.dci.2012.03.011
410. Heimroth RD, Casadei E, Benedicenti O, Amemiya CT, Muñoz P, Salinas I. The lungfish cocoon is a living tissue with antimicrobial functions. *Sci Adv.* 2021;7(47):eabj0829. doi:10.1126/sciadv.abj0829
411. Grimholt U, Lukacs M. Fate of MHCII in salmonids following 4WGD. *Immunogenetics.* 2021;73(1):79-91. doi:10.1007/s00251-020-01190-6
412. Session AM, Uno Y, Kwon T, et al. Genome evolution in the allotetraploid frog *Xenopus laevis*. *Nature.* 2016;538(7625):336-343. doi:10.1038/nature19840
413. Good RA, Finstad J, Gewurz H, Cooper MD, Pollara B. The development of immunological capacity in phylogenetic perspective. *Am J Dis Child.* 1967;114(5):477-497. doi:10.1001/archpedi.1967.02090260065004
414. Giorgetti OB, Shingate P, O'Meara CP, et al. Antigen receptor repertoires of one of the smallest known vertebrates. *Sci Adv.* 2021;7(1). doi:10.1126/sciadv.abd8180
415. Rollins-Smith LA. Amphibian immunity-stress, disease, and climate change. *Dev Comp Immunol.* 2017;66:111-119. doi:10.1016/j.dci.2016.07.002
416. Kasahara M, Flajnik MF. Origin and evolution of the specialized forms of proteasomes involved in antigen presentation. *Immunogenetics.* 2019;71(3):251-261. doi:10.1007/s00251-019-01105-0
417. Davison TF. The immunologists' debt to the chicken. *Br Poult Sci.* 2003;44(1):6-21. doi:10.1080/0007166031000085364
418. Steinel NC, Bolnick DI. Melanomacrophage centers as a histological indicator of immune function in fish and other poikilotherms. *Front Immunol.* 2017;8:827. doi:10.3389/fimmu.2017.00827
419. Richards MH, Nelson JL. The evolution of vertebrate antigen receptors: a phylogenetic approach. *Mol Biol Evol.* 2000;17(1):146-155. doi:10.1093/oxfordjournals.molbev.a026227
420. Glusman G, Rowen L, Lee I, et al. Comparative genomics of the human and mouse T cell receptor loci. *Immunity.* 2001;15(3):337-349. doi:10.1016/s1074-7613(01)00200-x
421. Davis MM, Bjorkman PJ. T-cell antigen receptor genes and T-cell recognition. *Nature.* 1988;334(6181):395-402. doi:10.1038/334395a0
422. Du Pasquier L. Speculations on the origin of the vertebrate immune system. *Immunol Lett.* 2004;92(1-2):3-9. doi:10.1016/j.imlet.2003.10.012
423. McFall-Ngai M. Adaptive immunity: care for the community. *Nature.* 2007;445(7124):153. doi:10.1038/445153a
424. Ronald PC, Beutler B. Plant and animal sensors of conserved microbial signatures. *Science.* 2010;330(6007):1061-1064. doi:10.1126/science.1189468

CHAPTER 5

The Plant Immune System

Roger W. Innes, PhD

INTRODUCTION

Like animals, plants possess a robust immune system that features both cell surface and intracellular innate immune receptors. Unlike vertebrates, however, they lack an adaptive immune system and motile cells. In fact, plant cells are cemented in place by a cellulosic cell wall. Despite lacking motile cells, plants are still able to send systemic alarm signals from cell to cell through their vascular systems and, in at least some cases, from one plant to another via volatile chemicals. Because plants lack both motile cells and cells with specialized immune function, all cells within a plant must possess an ability to detect pathogens and activate immune responses. This, in turn, means that individual cells must be able to detect a huge diversity of potential pathogens, ranging from viruses to multicellular parasites, and this must be done with a relatively limited number of immune receptors and no ability to increase immune receptor diversity through somatic recombination. Understanding how plant cells overcome this incredible challenge is a major goal of the field of molecular plant–microbe interactions (MPMI).

Given that plants generally grow with their roots in the ground, where they are surrounded by a very diverse collection of microbes, many of which are potentially pathogenic, a major challenge faced by plants is how to distinguish between friend and foe. Indeed, given their role in absorbing nutrients, plant roots can be considered analogous to the human intestine in which the immune system must be tightly regulated so that immune responses are activated only when the microbial balance is upset. This chapter presents our current understanding of how plants detect pathogens, how detection is translated into robust immune responses, and how these responses are regulated in a manner that enables beneficial microbes to flourish, while limiting the damage from pathogenic ones.

AN OVERVIEW OF THE PLANT IMMUNE SYSTEM

Although the last common ancestor of plants and humans was almost certainly a single-celled organism,[1,2] the innate immune systems of plants and humans bear striking similarities in overall design. Chief among these are the use of both cell surface and intracellular receptors to detect the presence of pathogens (Fig. 5.1). In the plant field, the cell surface receptors are referred to as pattern recognition receptors (PRRs), whereas the intracellular receptors are referred to as nucleotide-binding leucine-rich repeat receptors (NLRs). Note that this terminology differs from that used in humans, in which NLRs are considered a subclass of PRRs (see Chapter 20). This terminology difference is more than semantic, because unlike NLRs in vertebrates, NLRs in plants do not participate in detecting conserved molecules such as bacterial peptidoglycan or double-stranded RNA ("patterns"). Instead, they detect effector proteins produced by pathogens that are either translocated into host cells or, in the case of viruses, that are produced inside host cells. Disease resistance induced by NLR proteins is thus commonly referred to as effector-triggered immunity (ETI), whereas resistance induced by PRRs is referred to as PRR-triggered immunity (PTI).

Activation of PTI in plants results in induction of multiple signaling pathways via a cascade of phosphorylation events (Fig. 5.1). Most PRRs contain cytoplasmic kinase domains that, upon activation, phosphorylate "receptor-like cytoplasmic kinases" (RLCKs[3]), which are phylogenetically related to the kinase domains of PRRs, but lack a transmembrane domain.[4] Once activated by PRRs, RLCKs phosphorylate and activate multiple target proteins, including plasma membrane–localized NADPH/respiratory burst oxidase homologs (Rboh) similar to those activated in mammalian cells,[5] anion channels,[6] mitogen-associated protein kinase kinase kinases (MAPKKKs),[7] and plasma membrane–localized calcium channels.[8-10] Opening of the latter leads to an influx of Ca^{2+} ions that then activate calcium-dependent protein kinases (CPKs[11,12]). CPKs constitute a plant-specific family of protein kinases with 34 members in *Arabidopsis thaliana*.[13] These various signaling pathways ultimately lead to dramatic changes in gene expression, with approximately 30% of the transcriptome responding to a single pathogen molecule such as flagellin,[14] which consequently induces a major shift in cellular resources away from growth and toward defense.

To be successful, a plant pathogen must suppress the PRR signaling pathway. This is usually accomplished by translocation of effector proteins into host cells (for extracellular pathogens) or by production of effector proteins inside host cells (for intracellular pathogens such as viruses). Most plant pathogen effector proteins function to suppress immune signaling triggered by PRRs. The detection of effectors by NLRs is thus usually described as a second layer of pathogen recognition that backs up the PRR pathways in case PTI is suppressed. As detailed below, however, recent research has revealed that there is an intimate interplay between the PRR and NLR signaling pathways in plants, with each being dependent on the other for proper immune function.[15,16]

A notable difference between plant and vertebrate immune systems is the lack of any system analogous to the major histocompatibility complex (MHC), which is considered part of the adaptive immune system in vertebrates (see Chapter 20). MHC-encoded proteins are highly polymorphic between individuals and are responsible, in part, for rejection of transplanted tissue or organs when donor and recipient do not closely match. Consistent with the lack of an adaptive immune system, and particularly the lack of an MHC, grafting parts of one individual plant (eg, the root system) to parts of another individual plant (eg, the shoot system or scion) are routinely done, especially in production of fruit trees and grapes, but also increasingly common with vegetables such as tomato. As long the two individuals are from the same species, such grafts almost always work.[17] Indeed, it is often possible to graft different closely related species together (eg, one can generate a plant with potato roots and tomato shoots, thus one can harvest both potatoes and tomatoes from a single plant!). However, attempts at grafting species that are more distantly related (eg, in different genera) typically fail. The reasons for such failures remain unclear, but do not appear to be related to immunity. Rather, graft failure appears to be caused by an inability of vascular systems from two distantly related plant species to fuse, likely due to structural and physiological differences between species.[17] Without vascular fusion, nutrients, hormones, RNAs, and other metabolites cannot move properly.

CELL SURFACE RECEPTORS AND THEIR LIGANDS: MAMPS AND DAMPS

All characterized PRRs in plants reside on the cell surface, where they detect ligands in the extracellular space, which, in plants, is referred to as the apoplast (Fig. 5.1). Most plant PRRs are receptor-like kinases (RLKs) that contain a variable extracellular (ecto) domain, a single-pass transmembrane domain, an intracellular juxtamembrane domain, and a cytoplasmic kinase domain. RLKs in plants belong to the same group of protein kinases as the Pelle family of kinases in animals, whose members include INTERLEUKIN RECEPTOR-ASSOCIATED KINASEs (IRAKs) (see Chapter 20). However, the RLK family has greatly expanded in plants compared to animals, with the Arabidopsis genome containing over 400 RLK genes,[18] which represents nearly all transmembrane

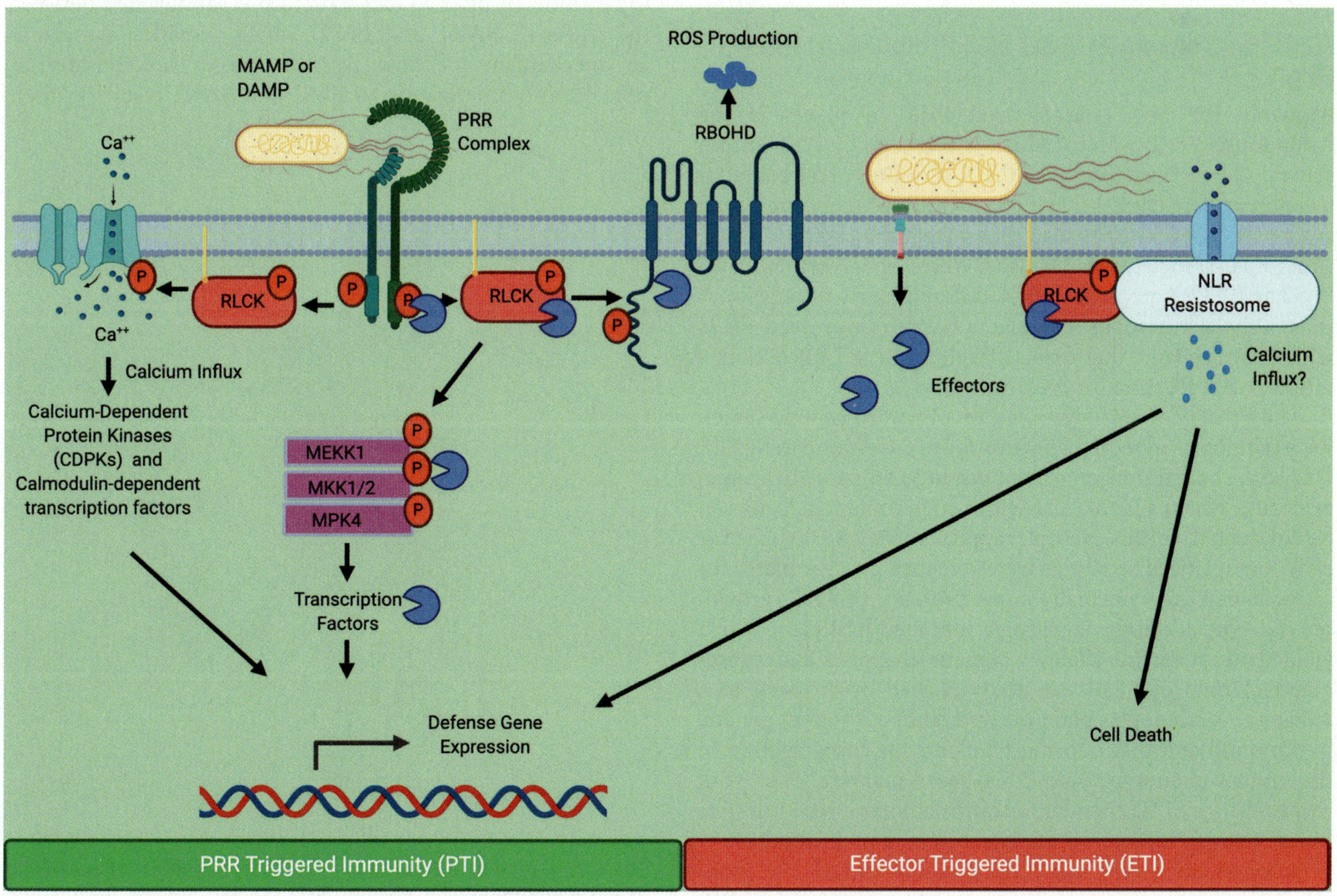

FIG. 5.1. Overview of plant immune system. Cell surface receptors (pattern recognition receptors) detect MAMPs and DAMPs, which activates intracellular kinase domains leading to phosphorylation (P) of multiple proteins, including other kinases (RLCKs, CPKs, and MAPKs). Pathogen effector proteins are translocated into host cells, where they target multiple steps in this signaling pathway. Effectors can be detected by host NLR proteins, which trigger ETI. CPKs, calcium-dependent protein kinases; DAMPs, damage-associated molecular patterns; ETI, effector-triggered immunity; MAMPs, microbial-associated molecular patterns; MAPKs, mitogen-associated protein kinases; RLCKs, receptor-like cytoplasmic kinases. (Figure created with BioRender.com.)

protein kinases in Arabidopsis. Some RLKs play important roles in plant growth and development, but the majority of RLK family expansion appears to be associated with roles in defense.[19]

The ectodomains of RLKs are diverse, but the most common by far is the leucine-rich repeat (LRR) domain, with over 230 LRR-containing RLKs in Arabidopsis.[18] Each unit of the LRR domain consists of approximately 24 amino acids with a consensus sequence of LXXL/CXXLXXLXϕXXNXL/FTGXIPXX where ϕ indicates an aliphatic amino acid residue and X represents any amino acid. Each LRR repeat forms a section of β-sheet and a section of α-helix, with the β-sheet face predicted to be solvent exposed.[20] The prevalence of LRR domains in immune receptors is likely a consequence of the versatility of LRRs in generating ligand-binding surfaces. It is noteworthy that extracellular LRR domains are also used extensively by animal innate immune receptors for ligand binding, as seen in Toll-like receptors (TLRs; see Chapter 13), and mutations in LRR domains are often associated with human diseases.[21] Thus, plants and animals appear to have co-opted many of the same building blocks when designing PRRs, even if the receptors evolved independently.[2]

Flagellin-Sensing 2 (FLS2) as a Prototypical Plant PRR

Arguably the best characterized PRR in plants is the Arabidopsis FLS2 protein, which is activated by a 22–amino acid peptide derived from near the N-terminus of the bacterial flagellin protein (flg22). The flg22 peptide is highly conserved among bacterial species and thus meets the definition of a "microbial-associated molecular pattern" or MAMP. Notably, FLS2 and TLR5 are activated by different parts of flagellin;[22] thus, although humans and plants have both evolved PRRs with LRR domains that detect flagellin, these LRR domains interact with flagellin differently.[23,24]

The *FLS2* gene was identified in a forward genetic screen for Arabidopsis mutant plants that failed to respond to flg22.[25] One of the benefits of using plant systems compared to human or mouse systems is the ease with which mutant screens can be conducted. In this example, approximately 80,000 chemically mutagenized seedlings were screened on agar plates for mutants that grew normally in the presence of flg22 (growth of wild-type seedlings is strongly inhibited by flg22), which yielded multiple *fls2* alleles.[25] Genetic mapping and subsequent cloning of *FLS2* was accomplished by crossing to a wild-type Arabidopsis plant derived from a different genetic background and then scoring F2 plants for cosegregation of the mutant phenotype with DNA-based markers. This work revealed that *FLS2* encoded a transmembrane receptor–like kinase with an LRR ectodomain containing 28 LRR motifs.[25]

After the discovery that *FLS2* encoded an LRR receptor-like kinase, a reverse genetic screen targeting other LRR-RLKs that were induced at a transcriptional level by flg22 was pursued, with the goal of identifying additional receptors that might contribute to flg22 perception.[26] This screen revealed that a previously characterized LRR-RLK named BRI1-ASSOCIATED RECEPTOR KINASE 1 (BAK1) was required for the full response to flg22. Although BAK1 is an LRR-RLK like FLS2, it only has five LRR repeats, and thus its ectodomain is much smaller than FLS2's. BAK1 was originally identified based on its association with a different LRR-RLK, BRASSINOSTEROID INSENSITIVE 1(BRI1), which mediates perception of the plant hormone brassinolide (a steroid-like molecule[27]), rather than perception of a microbial-derived molecule. BAK1 is required for a full response to brassinolide;[27] thus BAK1 contributes to RLK signaling from both MAMP receptors (PRRs) and endogenous plant hormone receptors.

The application of flg22 to Arabidopsis seedlings strongly promotes the association of FLS2 with BAK1 within seconds, with both proteins associating with flg22.[26,28] These observations suggested that BAK1 and FLS2 form a co-receptor complex in response to binding flg22. This was subsequently confirmed by X-ray crystallography, which revealed that flg22 binds to FLS2 LRR repeats 3 through 16, while repeats 18 to 20 and 23 to 26 make direct contacts with the LRR domain of BAK1[24] (Fig. 5.2). The LRR repeats of FLS2 form an open solenoid corkscrew-like structure (Fig. 5.2), which provides a large solvent-exposed surface for ligand binding. Notably, the C-terminus of flg22 in this structure is sandwiched between the LRRs of BAK1 and FLS2, which provides a molecular mechanism for how flg22 promotes heterodimerization. Heterodimerization of FLS2 and BAK1 results in their

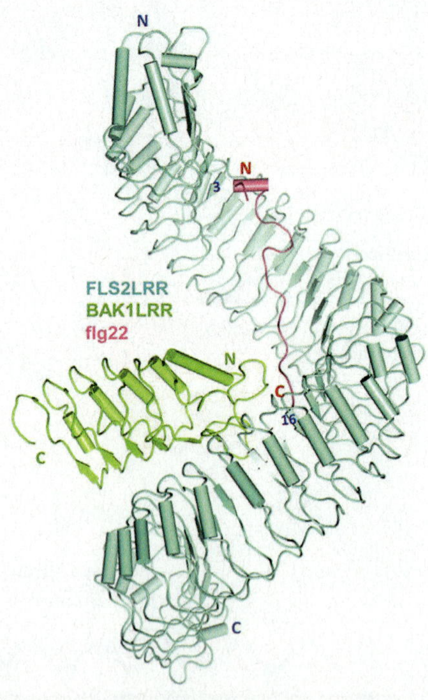

FIG. 5.2. Structure of the ectodomains of the flagellin receptor bound to the flagellin-derived peptide flg22. Shown are the leucine-rich repeat (LRR) domains of FLS2 (*blue*) and BAK1 (*green*) in association with flg22 (*red*). The positions of LRR3 and LRR16 are indicated by blue numbers. "N" and "C" represent the N and C terminus, respectively. (From Sun Y, Li L, Macho AP, et al. Structural basis for flg22-induced activation of the Arabidopsis FLS2-BAK1 immune complex. Science. 2013;342:624-628. Reprinted with permission from AAAS.)

transphosphorylation, which then leads to BAK1-mediated phosphorylation of RLCK proteins such as Botrytis-induced kinase 1 (BIK1[29]) (Fig. 5.1), which then release from the complex and phosphorylate multiple downstream signaling proteins.

BAK1 is now known to function as a co-receptor with many different LRR-RLKs (at least nine different RLKs in Arabidopsis alone) that mediate perception of a diverse collection of ligands, including peptides of both microbial and plant origin and steroid-like molecules of plant origin (Table 5.1).[46] BAK1 itself is a member of a small subfamily of LRR-RLKs called SERKs (somatic embryogenesis receptor kinases; five members in Arabidopsis), some of which have partially overlapping functions with BAK1.[30]

Plant PRRs Detect Diverse MAMPs

Although FLS2 and flg22 are the best-studied PRR:microbial ligand pair, many other plant PRRs and their corresponding ligands have been described (Table 5.1). A second bacterial MAMP recognized by Arabidopsis is elf18, which is an 18-amino acid peptide derived from the N-terminus of bacterial translation elongation factor Tu (EF-Tu).[47] EF-Tu is a highly conserved protein among bacteria and is extremely abundant in growing bacteria (5%-9% of total protein), thus constituting an ideal MAMP from a plant's perspective. Note, however, that EF-Tu, or its breakdown products, must be released from bacterial cells to access plant cell surface receptors, which presumably happens at some frequency due to bacterial cell lysis.

The receptor for elf18 was identified using a reverse genetics approach in which Arabidopsis plants with knockout mutations in paralogs of FLS2 were screened for their response to elf18 using the same sort of seedling growth assay as that used to identify *fls2* mutants.[31] One of these paralogs, which was named *EFR* for EF-tu receptor, was found to be required for elf18 perception. Significantly, mutations in *FLS2* have no effect on elf18 perception, and mutations in *EFR* have no effect on *flg22* perception; thus, these two receptors function independently, despite activating largely the same responses.[31] The EFR protein contains an ectodomain with 21 tandem LRR repeats of 24 amino acids each, a single

TABLE 5.1 Arabidopsis Pattern Recognition Receptors (PRRs) and Their Ligands

PRR	Ectodomain Type	Co-receptors	Ligand Name	Ligand Type	References
Microbial-derived ligands					
FLS2	Leucine-rich repeat (LRR)	BAK1 or SERK4	flg22 (from bacterial flagellin)	Peptide	24,30
EFR	LRR	BAK1 or SERK4	elf18 (from bacterial EF-Tu protein)	Peptide	30,31
LORE	Lectin	Unknown	Medium-chain 3-OH-fatty acids	Hydroxylated fatty acid	32,33
LYM1-LYM3[a]	Lysin motif	CERK1	Peptidoglycan (PGN)	Polysaccharide	34
LYK4/LYK5	Lysin motif	CERK1	Chitin	Polysaccharide	35,36
RLP23 plus SOBIR1[b]	LRR	BAK1	nlp20 (a member of the microbial necrosis and ethylene-inducing peptide 1–like protein family)	Peptide	37
Host-derived ligands (DAMPs)					
DORN1 (LecRK-I.9)	Lectin	Unknown	Extracellular ATP	Metabolite	38,39
LecRK-I.8	Lectin	BAK1	Extracellular NAD	Metabolite	40,41
LecRK-VI.2	Lectin	BAK1	Extracellular NADP	Metabolite	40,42
PEPR1/2	LRR	BAK1	AtPEP1 and related peptides (derived from endogenous PROPEP proteins)	Peptide	28,43
RLK7	LRR	BAK1	AtPIP1 and related peptides (derived from endogenous PrePIP proteins)	Peptide	44
WAK1	EGF-like repeats	Unknown	Oligogalacturonides derived from cell wall	Polysaccharide	45

[a]LYM1 and LYM3 contain LysM ectodomains and a transmembrane domain, but lack a cytoplasmic kinase domain. CERK1 contains all three domains and is required for both PGN recognition and chitin recognition, whereas LYM1 and LYM3 are specific to PGN.
[b]RLP23 contains an LRR ectodomain and transmembrane domain, but lacks a cytoplasmic kinase domain. Kinase activity is providing by pairing with SOBIR1, which is an RLK with a short LRR ectodomain.

transmembrane domain, and an intracellular kinase domain and thus is very similar in structure to FLS2. Unlike *FLS2*, however, the *EFR* gene is not widely distributed among flowering plants, appearing to be restricted to members of the mustard family (Brassicaceae). Significantly, expression of *EFR* in the tobacco relative *Nicotiana benthamiana* confers the ability to recognize elf18; thus, the kinase domain of EFR can engage downstream signaling components that must be conserved between the mustard family and the tobacco family (Solanaceae), including BAK1. This observation suggested that it may be possible to enhance disease resistance properties of crop plants by addition of PRRs from other plant species that confer new pathogen recognition abilities.

A third type of MAMP from bacteria that induces robust defense responses in Arabidopsis is medium-chain 3-hydroxy fatty acids (3-OH-FAs), typically 8 to 12 carbons long.[32] Notably, 3-OH-FAs are components of bacterial lipopolysaccharide (LPS) from many bacterial species, and free 3-OH-FAs are contaminants of many LPS preparations. Initial studies assumed that LPS was the active MAMP (as it is in mammalian systems), rather than free fatty acids. Because of this, the receptor for 3-OH-FAs was named LIPOOLIGOSACCHARIDE-SPECIFIC REDUCED ELICITATION (LORE). Similar to FLS2 and EFR, LORE was identified using a genetic screen, but in this case, the phenotype assayed was an increase in cytosolic calcium following MAMP addition, rather than root growth inhibition.[33] Also, similar to FLS2 and EFR, the transcription of LORE was found to be inducible by various MAMPs. Unlike FLS2 and EFR, however, the ectodomain of LORE does not contain LRRs, but instead contains an amino-terminal B-type lectin domain, followed by an epidermal growth factor (EGF)–like domain and a plasminogen apple nematode domain. There are 32 RLKs in Arabidopsis with this type of ectodomain, but their ligands and their function remain mostly unknown. Similar to EFR, LORE function appears to be absent from Solanaceae species, but transfer of LORE to *N. benthamiana* confers the ability to recognize LPS preparations (presumably 3-OH-FA).[33]

A fourth MAMP from bacteria that is recognized by Arabidopsis (and humans) is peptidoglycan, which is a major component of the cell walls of both Gram-positive and Gram-negative bacteria. It is composed of alternating β1,4-linked *N*-acetylglucosamine and *N*-acetylmuramic acid residues crosslinked by short oligopeptides to form a lattice. Notably, the ectodomains of PRRs in plants that detect peptidoglycan contain lysin motifs, rather than LRRs. The lysin motif (LysM; ~40 amino acids) is an ancient protein domain originally identified in bacterial autolysin and is generally found in bacterial enzymes involved in bacterial cell wall degradation.[48] The combination of a LysM ectodomain fused to a cytosolic kinase domain, however, appears to be unique to plants.[49] The Arabidopsis genome encodes five LysM-containing receptor-like kinases (CERK1 [CHITIN ELICITOR RECEPTOR KINASE 1] and LYK2 through LYK5), but in addition, contains three LsyM-containing receptor-like proteins (RLPs) that lack a cytosolic kinase domain (LYM1-3).[34] Both LYM1 and LYM3 have been shown to bind directly to peptidoglycan, and significantly, knockout of either *LYM1* or *LYM3* blocks recognition of peptidoglycan.[34] These results indicate that LYM1 and LYM3 function together, likely as a heterodimer, to detect peptidoglycan.[34] Significantly, single and double *lym1/lym3* mutants have enhanced susceptibility to bacterial pathogens, revealing the importance of peptidoglycan recognition for plant immunity to bacterial pathogens.[34]

Because both LYM1 and LYM3 lack kinase domains, it was expected that these proteins might interact with a third protein that could supply this function. To test this hypothesis, Willmann et al[34] tested knockout mutants of the *CERK1* gene, which encodes a LysM-containing RLK that had previously been shown to be required for chitin perception (see next paragraph). As hoped, the *cerk1* mutation blocked perception of peptidoglycan, indicating the LYM1, LYM3, and CERK1 may form a heterotrimeric PRR, with the kinase domain of CERK1 providing the intracellular signaling capacity.[34]

CERK1 is also required for perception of chitin,[50,51] which shares structural similarities to peptidoglycan in that it is a polymer of β-(1,4)-linked *N*-acetylglucosamine. Chitin is a constituent of fungi, nematodes, and insects, all of which represent important pathogens of plants, and thus it makes good sense that plants have evolved a PRR to detect this molecule, especially given that plants, themselves, do not produce chitin. Notably, there is a narrow window of chitin oligomer size that activates immune responses in plants, with octamers having the highest activity, and oligomers shorter than six subunits having no activity.[35,52]

Similar to CERK1's role in peptidoglycan perception, it appears to function as a co-receptor rather than the primary receptor for chitin. In Arabidopsis, the primary receptor is LysM-RECEPTOR LIKE KINASE5 (LYK5). Application of chitin octamers to Arabidopsis seedlings induces association of LYK5 and CERK1, similar to how flg22 induces association of FLS2 and BAK1.[35] Significantly, the affinity of the ectodomain of LYK5 for chitin octamers is ~200-fold higher than that of CERK1, confirming that LYK5 is the primary receptor. Knockout of *LYK5* reduces chitin-induced defense responses about 90% in Arabidopsis, with the remaining response eliminated by knockout of the *LYK5* paralog *LYK4*.[35] CERK1 is also required for chitin perception in rice, but interestingly, the function of LYK5 has been replaced by a LysM-containing receptor like protein that lacks a kinase domain, CEBiP (chitin elicitor–binding protein).[53] Chitin octamers induce dimerization of CEBiP via a sandwich-like mechanism in which chitin acts as the glue holding two molecules of CEBiP together.[53]

The last class of MAMP we will discuss in this chapter is necrosis and ethylene-inducing peptide 1–like proteins (NLPs), which are ~24 kD proteins produced by bacteria, fungi, and oomycetes (a very important class of filamentous pathogens of plants). As their name implies, many NLPs induce cell death when injected into plant leaves, but this response is restricted to eudicot species (eg, tomatoes and soybeans) and is not observed in monocots (eg, grasses such as corn and rice).[54] The basis for this difference between eudicots and monocots appears to be differences in the sugar moieties of specific sphingolipids (glycosyl inositol phosphoryl ceramides) in the outer leaflet of their respective plasma membranes, with only the eudicot sphingolipids being efficiently bound by NLPs.[55] Cytotoxicity of NLPs correlates with their ability to disrupt membranes and thus directly lysing cells.[56] Although inducing cell death may be advantageous to pathogens that live on dead tissue (necrotrophs), many pathogens require living host tissue for survival, and such pathogens typically produce NLPs that are noncytotoxic.[54] The biological function of NLPs in this context remains enigmatic, but their widespread distribution in pathogenic oomycetes, fungi, and bacteria indicates they play an important role in virulence.

Given both their common occurrence in diverse pathogens and their contribution to virulence, it is not surprising that plants have evolved signaling systems to detect NLPs. Current data indicate that plants can detect NLPs both directly through PRRs and indirectly by sensing cell lysis and subsequent release of damage-associated molecular patterns (DAMPs; see next section). The best characterized example of direct sensing of NLPs by PRRs is the detection of noncytotoxic NLPs from the oomycete *Hyaloperonospora arabidopsidis,* a biotrophic pathogen of Arabidopsis.[37,57] These studies revealed that a 20-amino acid motif (nlp20) from the conserved central region of HaNLP3 functions as a potent elicitor of defense responses in Arabidopsis. Notably, the same 20-amino acid motif from both bacterial (*Bacillus subtilis*) and fungal (*Botrytis cinerea*) NLPs also activate defense responses in Arabidopsis; thus, this MAMP is conserved across three kingdoms of life.

To identify the corresponding PRR(s) involved in detecting NLP-derived MAMPs, Albert and colleagues screened a collection of Arabidopsis lines with mutations in RLKs and RLPs for loss of response to nlp20 and identified two independent lines with knockout mutations in *RLP23.* This gene encodes an RLP with an LRR ectodomain containing 27 LRR repeats and a transmembrane domain, but no cytoplasmic kinase domain.[37] These rlp23 mutations also blocked perception of nlp20 MAMPs from *B. subtilis* and *B. cinerea,* confirming that a single PRR can detect multiple nlp20 peptides from very diverse pathogens.

The lack of a kinase domain on RLP23 suggested that it must partner with one or more co-receptors to induce downstream responses. To identify these, Albert and colleagues used immunoprecipitation of RLP23 from plants treated, or untreated, with nlp20, followed by mass spectrometry to identify the coprecipitating proteins.[37] These analyses revealed that RLP23 associates with the RLK protein SOBIR1 (suppressor of BAK1-interacting receptor kinase 1) in both the presence and absence of nlp20, indicating that these two proteins form a pre-activation complex. In addition, however, these analyses revealed that BAK1 associates with this complex only in the presence of nlp20; thus, similar to other PRRs, the addition of ligand promotes association of the LRR domains of BAK1 and the primary ligand-binding protein (in this case RLP23). Mutations in either SOBIR1 or BAK1 abrogate nlp20-induced defense responses, confirming their role in the PRR that detects nlp20.[37] Significantly, transgenic expression of Arabidopsis RLP23 in potato, which lacks an RLP23 ortholog, conferred the ability to detect nlp20 and, most importantly, enhanced the resistance of potato to the economically important pathogens *Phytophthora infestans* (an oomycete and causal agent of the Irish potato famine) and *Sclerotinia sclerotiorum* (a fungal pathogen). These findings raise the possibility that it may be possible to significantly enhance disease resistance traits of crop plants by stacking multiple PRRs from various unrelated plant species, including noncrop species such as Arabidopsis.

Plant PRRs Also Detect Diverse DAMPs

In addition to microbial-derived ligands, plants express multiple PRRs that are activated by host-derived molecules that are released into the extracellular space only when cells are damaged or when their expression is induced by perception of MAMPs. These host-derived signals are usually referred to as DAMPs[58-60] and induce many of the same responses as MAMPs, including upregulation of PRR gene expression. Examples of DAMPs include extracellular ATP (eATP), extracellular NAD(P), oligogalacturonides (OGs) released from cell walls, nonsecreted peptides that are released by cell breakage (PEPs), and secreted PAMP-induced peptides (PIPs) (Table 5.1).

Exogenous application of ATP to plant leaves induces many of the same responses as observed with MAMPs such as an increase in cytosolic calcium levels and production of reactive oxygen species (ROS).[38,61] Notably, infection with bacteria causes an increase in eATP levels, as does application of flg22; thus, plant cells have a mechanism for exporting ATP in response to MAMPs, with this eATP then amplifying defense responses.[61] The PRR that mediates eATP perception was identified in a screen of 50,000 chemically mutagenized seedlings.[62] To score for loss of response to eATP, the investigators used transgenic Arabidopsis that expressed the calcium-sensing protein aequorin, identifying seedlings that failed to show elevated [Ca^{2+}]cyt following ATP addition. The mutants were named Does not Respond to Nucleotides 1 (*dorn1*). The *DORN1* gene was identified by genetic mapping and whole genome sequencing and was shown to encode a lectin receptor kinase (LecRK-I.9), but from a different subfamily than that of the LORE protein described above. It is worth noting that mammals express two distinct classes of eATP receptors, but neither is similar to DORN1; thus, eATP perception has evolved independently in plants and mammals.[39]

To confirm that DORN1 functions as the receptor for eATP, the DORN1 ectodomain was purified and then shown to bind to ATP with nanomolar affinity.[39] Notably, *dorn1* mutations also blocked ROS production, while increasing susceptibility to infection by the bacterial pathogen *Pseudomonas syringae.*[39,61] Consistent with these observations, DORN1 was shown to directly phosphorylate RESPIRATORY BURST OXIDASE HOMOLOG D (RBOHD) upon binding to ATP and thus induce production of ROS.[61] Based on this finding, it appears that DORN1 does not require a co-receptor to activate downstream signaling pathways, which distinguishes it from the MAMP-induced PRRs described above.

A second DAMP that is similar to eATP in function and in its mechanism of perception is extracellular nicotinamide adenine dinucleotide (eNAD+). Exogenous application of eNAD+ to the leaf apoplast induces defense gene expression and enhances resistance to *P. syringae*.[40] Similar to eATP, extracellular NAD+ levels increase in response to pathogen infection and mechanical wounding.[40] Among the defense genes induced by eNAD+ application are several lectin receptor kinases (LecRK). A reverse genetic screen of knockout mutations in 14 of these genes revealed that knockout of *LecRK-1.8* compromised (but did not eliminate) NAD+-induced resistance to *P. syringae*.[41] Consistent with LecRK-1.8 functioning as one PRR that mediates NAD perception, the ectodomain of LecRK-1.8 was found to bind NAD+ with dissociation constant (Kd) of 437 nM. Notably, knockout of *LecRK-1.8* does not block perception of the closely related DAMP, NADP+, or eATP, indicating that *LecRK-1.8* is highly specific for NAD+,[41] and that there was likely to exist at least one other PRR in Arabidopsis that mediated NADP+ perception.

A second receptor was identified in 2019, when the same set of *LecRK* mutants were tested for loss of response to NADP+,[42] which revealed LecRK-VI.2 as the likely receptor. *In vitro* assays showed that the ectodomain of this PRR binds to both NAD+ and NADP+ with a small preference for NADP+.[42] Of note, loss of LecRK-VI.2 function resulted in loss of NAD(P)+-induced systemic resistance to *P. syringae*.

Similar to many MAMP receptors, LecRK-VI.2 was found to form a complex with BAK1 and to require BAK1 for eNAD(P)+-induced signaling.[42] The requirement for BAK1 as a co-receptor appears to differ from DORN1, which was shown to directly phosphorylate the downstream target RBOHD.[61] However, a role for BAK1 in DORN1-mediated signaling cannot be ruled out, as *bak1* mutants have not been tested for their response to eATP.

eNAD(P)+ and eATP represent DAMPs that are normally found inside of plant cells and are either specifically pumped out in response to MAMP perception or leak out following cellular damage. A second class of DAMPs that plays an important role in plant immunity is plant cell wall degradation products such as OGs, which are derived from the breakdown of pectin. Many plant pathogens secrete pectinases as part of their infection process, and thus release of OGs is an indicator of pathogen presence. Although OGs were among the earliest plant-derived DAMPs to be identified,[63] identification of their corresponding PRRs was complicated by the twin banes of geneticists, functionally redundant genes, and essential genes, which precluded their identification through standard forward and reverse genetic screens. Ultimately, the PRRs WALL-ASSOCIATED KINASE 1 and WALL-ASSOCIATED KINASE 2 (WAK1 and WAK2) were shown to be the functional receptors for OGs using a chimeric receptor strategy. WAK1 was a particularly attractive candidate for testing by this approach because its ectodomain had previously been shown to bind OGs,[64] and its transcription was specifically induced by exogenous application of OGs to Arabidopsis leaves.[65] To establish WAK1's function in OG perception, the ectodomains of WAK1 (contains EGF repeats) and EFR1 (contains LRRs) were swapped, transiently expressed in tobacco plants (which lacks an EFR receptor), and then tested for responsiveness to either OGs or the ligand for EFR (elf18).[45] Remarkably, addition of OGs to the receptor with the WAK1 ectodomain plus EFR transmembrane and cytosolic domain induced classic elf18 responses (upregulation of ethylene production), whereas addition of elf18 to the reciprocal chimera induced OG-like responses (ROS burst without ethylene induction). Thus, despite the different classes of ectodomain, ligand binding was still able to activate signaling by the intracellular kinase domains, and these domains retained their specificity for downstream activation events.[45] Notably, WAK1 and WAK2 also contribute to normal development in Arabidopsis, as inactivation of both genes is lethal,[45] which explains why forward genetic screens had failed to identify OG receptors.

The remaining plant DAMPs we will cover in this chapter are all peptides and thus bear similarity to mammalian cytokines in that these function in intercellular communication and are released in response to pathogen infection and related cellular damage. The best characterized class of peptide DAMPs are referred to as plant elicitor peptides (Peps). These are produced inside plant cells in response to danger signals (eg, wounding or MAMP perception) and are released to the extracellular space upon loss of plasma membrane integrity (ie, they lack signal peptides and are not secreted through the classical secretory system). Peps are generated from the C-termini of precursor proteins (PROPEPs) by proteolysis.[66] In Arabidopsis, PROPEPS are encoded by eight homologous genes named *PROPEP1* to *PROPEP8*.[67] Transcription of *PROPEP* genes is induced by diverse danger signals, including other DAMPs (eg, OGs), MAMPs, herbivore chewing, and wounding.[68,69] Most significantly, PROPEP proteins are tethered to the membrane surrounding the central vacuole of plant cells, and release of Peps requires the action of a cytosolic protease, METACASPASE4 (MC4).[70] Metacaspases are class of cysteine proteases that are found in plants, fungi, protists, and bacteria that display a certain degree of structural homology to mammalian caspases. However, they differ in activity and substrate specificity. Importantly, most metacaspases require millimolar amounts of [Ca^{2+}] for *in vitro* activity, thus remain inactive at normal cytosolic levels of calcium. Recent work in Arabidopsis has shown that wounding, and/or MAMP perception, causes an increase in $[Ca^{2+}]_{cyt}$ levels that is sufficient to activate MC4, which then releases Peps from their precursor proteins (Fig. 5.3).[70] Provided plasma membrane integrity is compromised, these Peps can then access cell surface receptors on adjacent cells and even enter the plant circulatory system to induce responses systemically.

Once released from damaged cells, Peps induce immune responses by binding to PEPR1 and PEPR2, which are LRR-type receptor kinases (Table 5.1).[71,72] Unlike most PRRs in Arabidopsis, PEPR1 was identified using a biochemical approach rather than a genetic approach. Specifically, PEPR1 was tagged with radioactive iodine ($^{[125]}$I) by incubating Arabidopsis suspension-cultured cells with $^{[125]}$I-labeled azido-Cys-AtPep1 peptide. The labeled protein was then purified to homogeneity and identified by mass spectrometry. Confirmation that PEPR1 was a functional receptor for AtPep1 was obtained by expressing PEPR1 in tobacco cells and demonstrating gain of the ability to respond to AtPep1.[72]

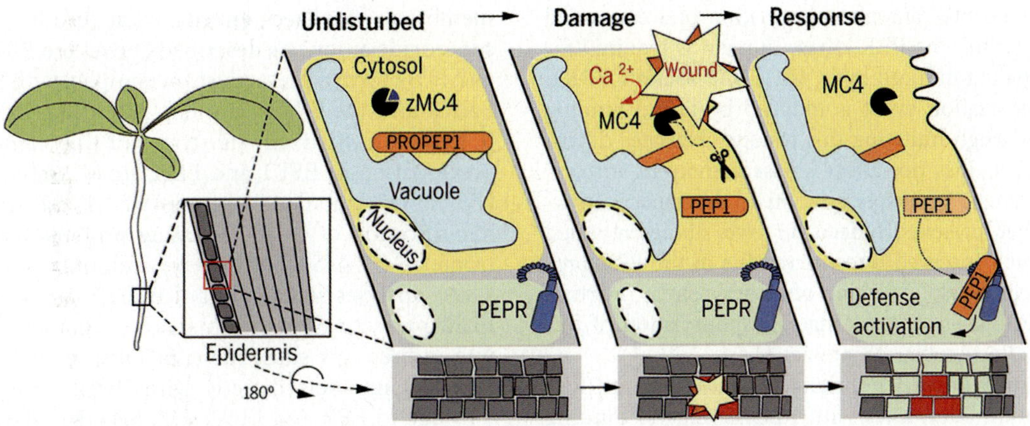

FIG. 5.3. Sustained increases in cytosolic calcium induce maturation of DAMPs. Elevated cytosolic calcium activates the metacaspase MC4, which then cleaves PROPEP1, releasing the mature DAMP protein PEP1. If the plasma membrane is disrupted, PEP1 is released into the extracellular space, where it can activate the cell surface receptor PEPR on adjacent cells. DAMPs, damage-associated molecular patterns. (Image reproduced from Hander T, Fernandez-Fernandez AD, Kumpf RP, et al. Damage on plants activates Ca(2+)-dependent metacaspases for release of immunomodulatory peptides. *Science*. 2019;363:eaar7486.)

PEPR1 was not identified in initial genetic screens because Arabidopsis expresses a functionally redundant homolog named PEPR2 that has 76% amino acid similarity to PEPR1.[71] Significantly, knockout of both PEPR1 and PEPR2 completely eliminates responses to AtPep1, while each single mutant only partially reduces responses.[71] Photoaffinity labeling with other Arabidopsis Peps revealed that PEPR1 binds Pep1-6, whereas PEPR2 binds only Pep1 and Pep2, and thus there has been some divergence in ligand binding specificity between these two receptors. Similar to other LRR-RLKs, signaling by PEPR1 and PEPR2 requires BAK1 as a co-receptor,[30] and ligand binding induces the association of PEPR1 and PEPR2 with BAK1.[28,43] There thus appear to be no fundamental mechanistic differences between PRRs that detect MAMPs versus those that detect DAMPs.

In addition to PEPs, plants also produce PIPs, which are peptide DAMPs that are derived from precursor proteins (prePIPs) that are secreted into the extracellular space via the standard secretory system.[44] These prePIPs thus contain classic signal peptides. Release of active PIPs (~13 amino acids) from prePIPs requires the activity of extracellular proteases. *prePIP* gene expression is induced by MAMP perception whereas exogenous application of PIPs upregulates expression of MAMP receptors (PRRs); thus, the primary role of PIPS appears to be to form a positive feedback loop that enhances MAMP-induced defense responses.[44] Arabidopsis has at least 11 prePIP genes and orthologs are found across flowering plant species, including soybean, maize, grape, and rice.

The PRR that mediates PIP perception was identified using a reverse genetics approach.[44] Because PIPs are similar in structure to PEPs, knockout mutations in LRR-RLKs related to PEPR1 and PEPR2 that are MAMP-inducible were screened for loss of PIP1 response. This screen revealed that the LRR-RLK protein RLK7 is required for PIP1 perception.[44] RLK7 was shown to bind to PIP1 by pulling down RLK7 from plant extracts using biotin-tagged PIP1, as well as by $^{[125]}$I-labeled PIP1 photoaffinity labeling. Similar to other LRR-RLKs, signaling induced by RLK7 is dependent on BAK1.

In summary, the initial recognition of diverse pathogens by plants is mediated by a large set of cell-surface localized PRRs that have evolved to detect conserved microbial-derived molecules of diverse structures including oligosaccharides, fatty acids and oligopeptides (Fig. 5.1). This collection of MAMP receptors is integrated with a collection of DAMP receptors that recognize host-produced molecules that are released during pathogen infection, either as a result of cellular damage, or in response to initial MAMP perception. Ligand binding to these PRRs most commonly induces heterodimerization with a co-receptor that contains a functional intracellular kinase domain, inducing transphosphorylation events. Collectively, these plant PRRs serve the same role as Toll-like receptors and other cell surface PRRs in mammals, but plants contain many more such receptors, and perceive a greater number and diversity of signaling molecules than mammals. This large number of receptors increases the robustness of the plant immune system by enabling detection of multiple different molecules per potential pathogen, making it more challenging for a potential pathogen to evade detection. The smaller number of such receptors in mammals is, perhaps, compensated for by the presence of the adaptive immune system in mammals (Chapters 21-34). In the next section, we will discuss the details of how activation of these diverse PRRs lead to activation of immune responses.

RECEPTOR-LIKE CYTOPLASMIC KINASES LINK PRR ACTIVATION TO MULTIPLE PARALLEL DOWNSTREAM RESPONSES

Unlike animal PRRs, which often require adaptor proteins to associate with cytoplasmic kinases,[1] plant PRRs transduce signals by directly phosphorylating their downstream substrates (Fig. 5.1).[73] As described above, most plant PRRs function as heterodimers, with at least one member of the dimer containing a cytoplasmic kinase domain.[74] Of the PRR kinase domains that have been characterized, most appear to phosphorylate serine, threonine, and tyrosine residues (ie, they are dual specificity protein kinases).[74-77] Significantly, autophosphorylation of a conserved tyrosine residue within the PRR kinase domain is required for downstream signaling induced by LORE1, BAK1, and CERK1, and this phosphorylation event is induced by ligand binding (Fig. 5.4).[75-78] Although mutating this residue abrogates downstream signaling, it does not affect kinase activity *in vitro*,[75,78] indicating that tyrosine phosphorylation likely impacts interactions with other proteins. Indeed, *in vivo* studies revealed that mutating this conserved tyrosine residue in the PRR protein LORE1 blocks phosphorylation of its downstream partner RLCK protein PBL34, indicating that phosphorylation of this tyrosine residue affects substrate interactions.[77]

As indicated in Figures 5.1 and 5.4, all known plant PRRs form pre-activation complexes with RLCKs. Ligand binding to the PRR triggers transphosphorylation of the RLCK within its activation loop,[79] which triggers further autophosphorylation of the RLCK. This autophosphorylation appears to be required for subsequent release of the RLCK from the PRR complex, as kinase inactive mutants of RLCKs fail to release from the PRR complex even though they are phosphorylated by the PRR.[77] Release of RLCKs from the PRR pre-activation complex appears to be a critical step in propagation of signaling further downstream, potentially because this is required for RLCKs to engage their substrates.

RLCKs Phosphorylate RBOHD to Induce ROS Production

Multiple targets of RLCKs have been identified, but the best characterized of these is RBOHD,[5,80] which is a plasma membrane–localized enzyme that produces extracellular superoxide anion. As described above, the PRR proteins EFR and FLS2 form a pre-activation complex with the LRR-kinase BAK1 and the RLCK protein BIK1 (Fig. 5.4). To identify potential downstream substrates of this complex, immunoprecipitation of EFR1 and FLS2 from Arabidopsis seedlings (EFR1) or Arabidopsis protoplasts (FLS2) was followed by identification of co-precipitating proteins using mass spectrometry (IP-MS), both before and after addition of ligand. These analyses revealed that RBOHD is part of the EFR and FLS2 pre-activation complexes.[5,80] Notably, this interaction was indirect, as subsequent *in vitro* analyses revealed that RBOHD interacts directly with BIK1, with BIK1 forming a bridge to EFR and FLS2 (Fig. 5.4). RBOHD was found to preferentially associate with the autophosphorylated form of BIK1, whereas EFR and FLS2 preferentially associate with the nonphosphorylated form, providing a molecular explanation for how ligand binding induces release of BIK1 and promotes phosphorylation of downstream targets by BIK1.[5] *In vitro* and *in vivo* kinase assays showed that BIK1 directly phosphorylates RBOHD, whereas the kinase domains of EFR and BAK1 have very little activity on RBOHD, establishing that BIK1 plays a central role in PRR signal transduction.[5,80] Analysis of phosphosites by mass spectrometry revealed that BIK1 phosphorylates RBOHD on residues S39, S339, and S343 in response to ligand binding.[5,80] Mutation of these sites to alanine blocks elf18- and flg22-induced production of ROS, indicating that phosphorylation of these residues by BIK1 plays a critical role in PRR-induced ROS burst.

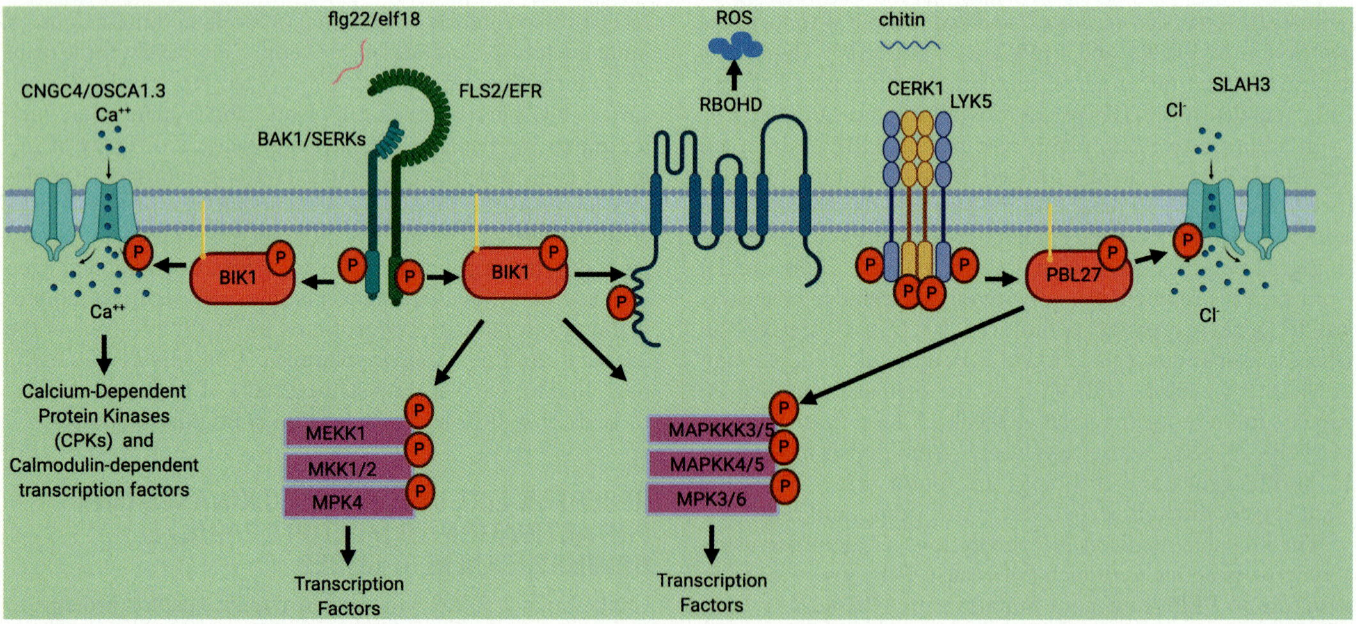

FIG. 5.4. Receptor-like cytoplasmic kinases (RLCKs) activate multiple defense signaling pathways in parallel. Phosphorylation (P) of RLCK proteins such as BIK1 and PBL27 by pattern recognition receptors activates the kinase activity of RLCKs. Activated RLCKs then phosphorylate multiple additional target proteins to simultaneously activate multiple independent signaling pathways. These RLCK target proteins include RBOHD (to activate ROS production), CNGC4 and OSCA1.3 (to induce calcium influx from the extracellular space and activation of CPKs), MEKK1 and MAPKKK5 (to activate MAPK pathways leading to activation of transcription factors such as WRKY33), and SLAH3 to open anion channels. BIK1 and PBL27 represent just 2 of 46 RLCK family VII proteins in Arabidopsis, many of which have overlapping functions. (Figure created with BioRender.com.)

RLCKs Phosphorylate Calcium Channels to Increase Cytoplasmic Ca^{2+} Levels

In addition to RBOHD, BIK1 also phosphorylates two different classes of calcium channels, the cyclic nucleotide-gated channel protein CNGC4[9] and the OSCA/TMEM63 family channels OSCA1.3 and OSCA1.7 (Fig. 5.4).[10] These phosphorylation events induce opening of these channels,[9,10] and thus BIK1 provides a direct link between MAMP/DAMP perception and opening of plasma membrane–localized calcium channels (Fig. 5.4). The resulting increase in $[Ca^{2+}]_{cyt}$ then activates multiple responses, including activation of several CPKs.

CPKs play an essential role in regulating plant immune responses, including activation of RBOHD to produce ROS. As aforementioned, RBOHD induction requires direct phosphorylation by RLCK proteins,[5] which occurs independent of calcium influx. Although RLCK-mediated phosphorylation of RBOHD is required for ROS production, it is not sufficient, and ROS production can be blocked by addition of an extracellular calcium chelator, EGTA,[5] indicating that RBOHD is also regulated by $[Ca^{2+}]_{cyt}$. This regulation is mediated, at least in part, by phosphorylation by several different CPKs.[5] Notably, the RBOHD amino acids phosphorylated by CPKs differ from those phosphorylated by RLCKs, with RBOHD being activated only when both sets are phosphorylated. This requirement for dual modification by separate signaling pathways likely prevents inappropriate activation of RBOHD, thus assuring that ROS production only occurs when needed.

One of the CPKs that phosphorylates RBOHD in Arabidopsis is CPK6. Notably, this CPK also phosphorylates SLAC1, which functions as a slow anion channel in guard cells to regulate closure of stomata, the microscopic pores in plant leaves that regulate efflux and influx of gasses.[81] Phosphorylation of SLAC1 by CPKs thus links PRR activation to closure of stomata via increases in $[Ca^{2+}]_{cyt}$. Closure of stomata is an important defense response, as these natural openings are often exploited by pathogens to gain entry into the intercellular spaces of leaves.

RLCKs Phosphorylate Anion Channels to Close Stomata

In addition to regulation by CPKs, slow anion channels can also be directly regulated by RLCKs. In Arabidopsis, the slow anion channel SLAH3 is phosphorylated by the RLCK protein PBL27 (Fig. 5.4).[6] Opening of the SLAH3 channel reduces membrane potential, which leads to a loss of turgor in guard cells and closing of stomata. In the work of Liu and colleagues,[6] they showed using single mutants that stomatal closure in response to chitin perception requires the chitin receptor LYK5, the co-receptor CERK1, the RLCK protein PBL27, and SLAH3. They were able to show that PBL27 directly phosphorylates SLAH3 at three specific positions. Significantly, mutating one of these positions to alanine (S127) blocks PBL27-induced channel opening and chitin-induced stomatal closure.[6] Notably, *slah3* mutants are compromised in their resistance to the fungal pathogen *B. cinerea*, indicating that anion channel function plays a key role in immunity.[6]

RLCKs Phosphorylate MAP Kinase Kinase Kinases to Activate Classic MAP Kinase Cascades, Leading to Activation of Transcription Factors

A fourth important target of RLCKs is MAPKKKs, which function at the top of classic mitogen activated protein kinase (MAPK) cascades found in all eukaryotes. In Arabidopsis, MEKK1, MAPKKK3, and MAPKKK5 interact with and are phosphorylated by multiple different RLCKs belonging to the same subfamily as BIK1, subfamily VII, which contains 46 members (Fig. 5.4).[82] MEKK1 appears to function in a MAPK signaling pathway that is parallel to that controlled by MAPKKK3 and MAPKK5, with MEKK1 activation leading to activation of the MAP kinase kinases MKK1 and MKK2, which then activate the MAP kinase MPK4,[83-85] while MAPKKK3/5 activation leads to activation of MKK4 and MKK5, which then activate the MAP kinases MPK3 and MPK6.[82,86] Significantly, mutation of both *MAPKKK3* and *MAPKKK5* nearly eliminates activation of MPK3 and MPK6 by multiple different MAMPs and DAMPs, including flg22 (FLS2), elf18 (EFR), chitin (LYK5), and Pep2 (PEPR1/2), indicating that these two MAPKKK proteins are the key targets of PRR signaling with regards to activating MPK3 and MPK6.[82] Similarly, mutation of *MEKK1* eliminates activation of MPK4, but has no effect on activation of MPK3/6.[84] Notably, a systematic analysis of higher order knockout mutations in RLCK family members revealed that six different RLCK proteins contribute additively to chitin-triggered MPK activation,[82,87] but are not required for flg22-triggered activation, establishing that different PRRs likely employ distinct, but overlapping, sets of RLCK proteins to activate downstream responses.

Although MPK3, MPK4, and MPK6 are all activated in parallel by PRRs, they are also activated by other inputs independent of PRRs.[88] Similar to MAP kinase signaling in animals, these MPKs play central roles in many aspects of growth and development beyond immune responses.[88] Specificity of signal outputs must therefore be achieved via precise spatiotemporal regulation of protein complexes that include both distinct activators at the top of MAP kinase cascades and distinct targets at the bottom of these cascades. In the context of PRR signaling, MPK3 and MPK6 appear to play the most central role in immunity. Simultaneous inactivation of MPK3 and MPK6 in adult Arabidopsis leaves using a chemical genetic approach renders plants highly susceptible to the bacterial pathogen *P. syringae*.[86] This enhanced susceptibility is correlated with an inability to close the stomata on the leaf surface. The MAMP flg22 normally induces closure of stomata, which blocks entry of bacterial pathogens into the leaf interior, but this response is lost when MPK3/6 activity is blocked.

MPK3 and MPK6 have diverse targets,[88] consistent with their roles in diverse aspects of plant development and defense. In the context of immunity, two key targets are the ethylene biosynthesis enzyme 1-aminocyclopropane-1-carboxylic acid synthase (ACS) and the transcription factor WRKY33 (Fig. 5.4).[89-91] Ethylene functions as a hormone in plants and induces various immune-related responses, along with ripening of fruit and senescence of leaves. Phosphorylation of

ACS by MPK3/MPK6 stabilizes the protein, thus increasing its cellular concentration and raising ethylene production.[92] WRKY33 is a member of a large family of plant-specific transcription factors, the majority of which regulate stress response genes. Phosphorylation of WRKY33 increases its transcriptional activation activity, including transcription of ACS-encoding genes, thus further elevating ethylene levels.[91] Consistent with this, mutation of *WRKY33* mostly eliminates MPK-induced ethylene production.[91]

Summary of PRR-Triggered Immune Responses

The above studies illustrate the diverse ligands sensed by PRRs in plants and the diverse responses triggered by PRRs. This diversity enables plants to detect most any potential pathogen via multiple signals (eg, via both MAMPs and DAMPs) and assures that responses are not easily suppressed due to redundancies built into the system (eg, two independent pathways for shutting stomata). The PRR-triggered immune system is thus quite robust, which enables any individual plant to resist the great majority of potential pathogens.

SUPPRESSION OF PTI BY PATHOGEN EFFECTOR PROTEINS

Most potential pathogens fail to colonize a given host plant due to induction of a diverse set of defense responses by PRRs, including production of ROS, closing of stomata, and induction of a broad array of defense-related genes that lead to production and secretion of antimicrobial compounds. To be successful on a given host, a pathogen must suppress these PRR-induced defenses. This is also true for most human pathogens, including many viruses that actively suppress immune responses (see Chapter 47). For the vast majority of pathogens, and perhaps all pathogens, suppression of PRR-induced signaling is accomplished primarily by translocating proteins from the pathogen into host plant cells, where they target specific components of PRR signaling pathways. Such pathogen proteins are commonly referred to as "effector proteins." The identification and characterization of pathogen effector proteins constitutes a major ongoing effort in the MPMI field. In addition to providing critical insights into how pathogens cause disease, the identification of effector targets in the host has provided key insights into how the plant immune system works.[93]

All characterized extracellular pathogens, including bacteria, fungi, oomycetes, and nematodes, translocate a cocktail of effectors into their hosts. Many hundreds of putative effectors have been identified through genome sequencing, and dozens have been carefully characterized in terms of their biochemical activities and targets.[93] For brevity, I will provide just a few representative examples of well-characterized effectors.

Bacterial Effector Proteins Typically Are Enzymes That Modify Their Targets

The best characterized effectors are from the bacterial plant pathogens *P. syringae* and *Xanthomonas* species. These are gram-negative pathogens that colonize the extracellular spaces within plant leaves. To translocate effectors into host cells, both bacterial species employ a type III secretion system (TTSS) that is evolutionarily conserved across many gram-negative pathogens of animals and plants, including the animal pathogens Salmonella, Shigella, and *Pseudomonas aeruginosa*. A TTSS effector of particular relevance to this chapter is the *P. syringae* effector AvrPphB. Consistent with a role in suppressing PTI, AvrPphB functions as a cysteine protease that targets multiple RLCKs in the BIK1 subfamily, including BIK1,[79] cleaving them within their activation loops. Cleavage of BIK1 and related RLCKs suppresses signaling by multiple PRRs, including FLS2 (flg22), EFR (elf18), and LYK5 (chitin).[79] Thus, by targeting a family of protein kinases that are collectively required for signaling by multiple PRRs, this one effector significantly inhibits PTI.

RLCKs may represent an "Achilles heel" of the PTI system, as they are targeted by unrelated effectors from multiple different pathogens. For example, the TTSS effector AvrAC from *Xanthomonas campestris* also targets BIK1 and related kinases, but instead of cleaving them, it modifies their activation loops with the addition of uridine monophosphate, concealing phosphorylation sites required for their activation.[94] Similarly, the NIS1 effector protein, which is conserved across diverse fungal species, binds to BIK1 and inhibits its kinase activity and its ability to associate with RBOHD.[95] Deletion of NIS1 from *Magnaporthe oryzae* (the most important fungal pathogen of rice) strongly attenuates its virulence, demonstrating the central role that this effector plays in suppressing PTI.[95]

A second common effector target class is represented by the kinases BAK1 and CERK1, which function as co-receptors for many different PRRs (Table 5.1) and thus represent another point of vulnerability in the PTI signaling system. Examples of effectors that target BAK1 and/or CERK1 include the *P. syringae* effectors AvrPtoB, HopF2, and HopB1 and the *X. oryzae* effector Xoo2875.[96-99] Notably, these are unrelated effectors and each modifies BAK1/CERK1 by a different mechanism. AvrPtoB functions as an E3 ubiquitin ligase, triggering proteasome-mediated degradation of target proteins.[100,101] HopB1 functions as protease, cleaving its targets,[99] and HopF2 functions as an ADP-ribosyltransferase, with this modification inhibiting the kinase activity of BAK1.[102] This diversity of mechanisms for inhibiting BAK1/CERK1 function illustrates the selective pressure placed on pathogens to attenuate PTI signaling and confirms the central roles that BAK1 and CERK1 play in PTI.

A third common effector target class is MAP kinase pathway components, with Arabidopsis MKK4, MKK5, MPK3, MPK4, and MPK6 all targeted by various effectors.[102-104] For example, the *P. syringae* effector HopAI1 targets MPK3 and MPK6, functioning as a phosphothreonine lyase to remove the phosphate that activates MPKs and thus inactivating them.[103] Notably, a homolog of HopAI1 is present in the human pathogen *Shigella flexneri*, and it targets the human MAP kinases ERK1/2, JNK, and P38;[105] hence, bacterial pathogens of plants and animals share common approaches to inhibiting innate immune responses (see Chapter 46 for a discussion of bacterial pathogens of animals).

Oomycete Effectors Appear to Suppress Target Function by Direct Binding, Rather Than Enzymatic Modification

As can be deduced from the above examples, most bacterial effectors function as enzymes that modify host targets in a manner that blocks their signaling activities and often results in their degradation. For reasons that are not yet clear, this is not the case for most effectors produced by fungal and oomycete pathogens. Instead, these effectors are typically small proteins that lack apparent enzymatic activity and thus must inhibit the function of their targets by direct binding.

The best characterized class of oomycete effectors are known as RXLR effectors, which refers to an N-terminal motif found in all members of this family that is required for translocation into host cells. Precisely how the RXLR motif mediates translocation is controversial, but recent work suggests that it is removed prior to secretion from the pathogen and thus is not directly involved in the translocation step.[106] Most plant pathogenic oomycetes express hundreds of proteins that contain RXLR motifs during infection,[107] but the functions of most of these remain unknown.[108]

Beyond the RXLR motif, RXLR effectors are very diverse in terms of primary amino acid sequence, but many (~40% of all RXLR effectors) contain "WY domains," which fold into a four helix bundle, with individual effector proteins often containing tandem WY domains.[107,109] These domains appear to function primarily in mediating protein:protein interactions,[110,111] with binding of these effectors to target proteins and then altering the subcellular location of target proteins, or their ability to interact with other proteins. Expression of 52 different fluorescent protein-tagged RXLR effectors in plant leaves revealed that they locate to diverse subcellular organelles and structures, including many that localize to the nucleus.[112] Functional studies have shown that RXLR effectors target a wide range of defense pathways, including MAP kinase signaling,[113] small RNA production (to inhibit host-induced gene silencing [HIGS]),[114,115] histone acetylation (promoting changes in gene expression favoring the pathogen),[116] protein secretion (blocking export of antimicrobial enzymes),[117] and various transcription factors (blocking nuclear import, thereby blocking defense gene expression). The large number of effectors secreted by oomycetes and their diverse targets suggests that there has been an arms race between plants and oomycetes, which has led to a proliferation of weapons on both sides of the interaction.

Fungi Secrete a Diverse Array of Effector Proteins

Similar to oomycetes, fungal pathogens of plants also secrete a large number of small protein effectors (<250 amino acids).[118] However, these are more challenging to identify because they do not carry well-defined translocation signals equivalent to the RXLR motif of many oomycete effectors and appear to be more diverse in structure and function than oomycete effectors. Indeed, most fungal effectors have been identified via their recognition by NLR proteins (ie, by induction of ETI), rather than by bioinformatics or their contribution to virulence.[118] Because of this history, many fungal effectors carry the name "Avr" for "avirulence," as their presence renders a fungal strain avirulent on a host plant expressing an NLR protein capable of recognizing it (see next section). One of the first fungal effectors to be characterized was AVR-Pita from the rice pathogen *Magnaporthe grisea*, which causes major losses to the rice crop each year.[119] *AVR-Pita* was cloned based on its ability to activate the rice disease resistance protein Pi-ta.[119,120] *AVR-Pita* encodes a putative metalloprotease that physically binds to the LRR domain of Pi-ta. However, host targets of AVR-Pita have not been described, nor has protease activity been confirmed.

Although AVR-Pita is predicted to have protease function, most other fungal effectors lack any recognizable catalytic sites. Among the best characterized fungal effectors is AvrL567 from the flax rust fungus *Melampsora lini*.[121] Avr567 is just 127 amino acids (after removal of the signal peptide) and binds directly to and activates the flax NLR protein L.[121,122] Structural analysis of AvrL567 revealed that it forms a novel fold that is relatively compact.[123] Although cloned in 2004, the first putative virulence target of Avr567 was not described until 2019.[124] In that work, the authors used yeast two hybrid analyses and bifluorescence complementation analyses to show that Avr567 binds a cytokinin oxidase, which is an enzyme that regulates the level of cytokinins, an important family of plant hormones that regulate growth and cell division. However, transgenic expression of Avr567 did not consistently alter *in vivo* cytokinin levels and thus the function of Avr567 remains obscure.

When considering the virulence functions of fungal effectors that function inside the host cell, the best understood effector is AvrPiz-t from *M. oryzae*. This effector is just 108 amino acids, but appears to target at least five different proteins in rice, including two different E3 ubiquitin ligases, a potassium channel, a nucleoporin, and a protease inhibitor.[125-129] In each case, silencing or knockout of the target protein increases susceptibility to *M. oryzae*, suggesting these are functionally relevant targets. Although it is not yet clear how these targets regulate defense responses, the observation that a single effector can target multiple unrelated proteins in the host cell, and that fungi express a large number of small secreted proteins during infection,[130] indicates that fungi, similar to oomycetes, suppress PTI by targeting many different signaling steps in parallel.

EFFECTOR-TRIGGERED IMMUNITY IS A CRITICAL SECOND LAYER OF THE PLANT IMMUNE SYSTEM

Given that plant pathogens have evolved a large number of effector proteins that collectively suppress PTI, it is not surprising that plants have evolved a second layer of defense that functions to detect the presence of pathogen effectors and thus back up PTI. As noted in the introduction to this chapter, ETI is mediated by NLR proteins, which are intracellular receptors rather than cell surface receptors. Although plant NLR proteins share structural similarities to animal NLR proteins in that both contain a characteristic nucleotide binding domain (NBD) and have C-terminal LRRs, phylogenetic analyses have shown that plant and animal NLRs evolved independently.[1] Perhaps reflecting this independent origin, plant NLRs all appear to be activated by effector proteins, rather than MAMPs such as double-stranded RNA or peptidoglycan, which are known to activate vertebrate NLRs (Chapter 13).

Plant NLR proteins can be divided into three broad classes based on their N-terminal domains: coiled coil domain (CC), Toll, interleukin receptor, and R gene domain (TIR), and RPW8 domain.[131] Phylogenetic trees based on just the NBDs of each of these classes clearly separate them; thus, these classes appear to have evolved independently during land plant evolution and were present in the common ancestor of all flowering plants (angiosperms).[131] Notably, the entire TIR class is missing from plants belonging to the grass family, which includes corn, rice, and wheat, and was apparently lost at an early stage in the evolution of this group. In contrast, TIR-NLR proteins predominate in terms of number and diversity in Arabidopsis and are common in most eudicot species such as soybean and mustard.[131] RPW8-NLRs appear to function downstream of TIR-NLRs and typically number fewer than 20 family members in individual plant genomes. Arabidopsis variety Columbia-0 contains 106 TIR-NLRs, 52 CC-NLRs, and 7 RPW8-NLRs.[131] Thus, NLRs appear to play a much larger role in the immune system of plants than in humans, which have only 22 NLR genes.[132]

Structures of Plant NLR "Resistosomes"

Structural analyses have revealed that activated CC-NLRs and TIR-NLRs form oligomeric structures (resistosomes) that differ from each other and from that formed by vertebrate NLR proteins. While vertebrate NLR proteins form a wheel-like structure containing 10 or 11 individual NLR subunits,[133,134] plant CC-NLRs form a star-shaped structure with 5 NLR subunits,[135] and TIR-NLRs form a cloverleaf structure with 4 NLR subunits[136] (Fig. 5.5). In all activated NLR structures, however, the LRR domains point out toward the periphery, with the N-terminal domains located in the center, but sticking out from the overall plane of the structure (Fig. 5.5). Another major difference between plant NLRs and vertebrate NLRs is their N-terminal domains, which bear no similarity. Despite their lack of similarity, it is the N-terminal domains that function in downstream signaling in all NLR proteins.

Signaling by the N-terminal domains of plant NLR proteins is activated by an induced proximity mechanism in which multiple copies of the N-terminal domain must come together to form an active complex. Induced proximity is mediated by oligomerization of the nucleotide binding-APAF1, resistance gene, and CED4 (NB-ARC) domain, which is equivalent to the nucleotide oligomerization domain (NOD) of mammalian NLR proteins. For both plant and animal NLR proteins, oligomerization is triggered by exchange of ADP for ATP, with only the ATP-bound form able to form higher order oligomers. Thus, in both classes of NLR proteins, the key step in activating signaling is induction of nucleotide exchange.

Nucleotide exchange is triggered by a large conformational change within the NB-ARC domain.[137] The NB-ARC domain is made up of three discrete subdomains, each connected by a short flexible hinge. These are referred to as the NBD, the helix domain 1 (HD1), and the winged helix domain (WHD), with the former connected to the N-terminal domain and the latter connected to the LRR domain.[137] In the inactive monomer state, ADP binds to a pocket formed by all three domains, with the first three LRR motifs interacting with the WHD domain. Together, the WHD and LRR domains form a lid over the ADP-binding pocket, effectively preventing nucleotide exchange. Based on a comparison of inactive ZAR1 monomer and activated ZAR1 pentamer structures, activation of ZAR1 involves a 180° rotation at the hinge connecting the HD1 domain to the WHD

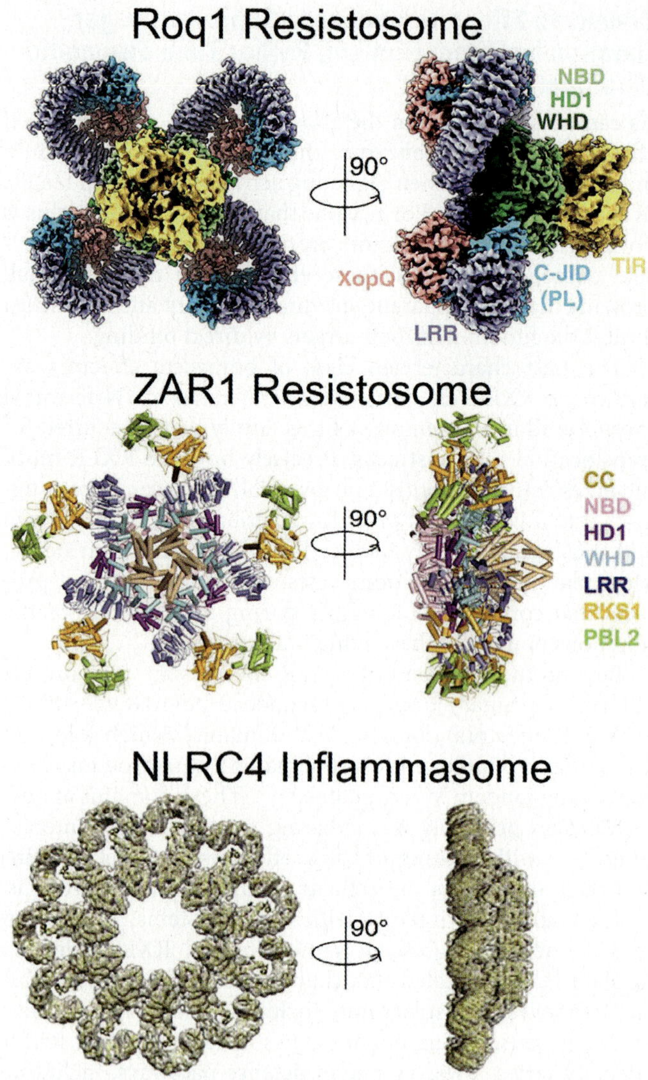

FIG. 5.5. Structures of resistosomes ZAR1 from Arabidopsis[133] and Roq1 from *Nicotiana benthamiana* (a tobacco relative)[134] and the NLRC4 inflammasome from mouse.[131,132] In addition to the ZAR1 NLR protein, the ZAR1 resistosome includes the RKS1 and PBL2 proteins, which are RLCK protein kinases that associate with and activate ZAR1 upon modification of PBL2 by the effector protein AvrAC, which uridylylates PBL2. This is an example of indirect activation of an NLR protein. The Roq1 structure shows the Roq1 NLR protein bound to the effector protein XopQ (in pink). This is an example of direct activation. Note that all structures are disk-shaped, but the number of NLR protomers in each structure differs, with Roq1 containing 4, ZAR1 containing 5, and NLRC4 containing 11.

domain, which opens the nucleotide binding pocket, enabling release of ADP and binding of ATP.[135] The γ-phosphate of the bound ATP stabilizes the active form via formation of a hydrogen bond to a conserved arginine residue in the WHD domain. This conformational change then enables the NBD-HD1 surface of one protomer to intercalate with the NBD-WHD surface of its neighbor.[135] A similar structure has been described for the Roq1 resistosome;[136] thus, assembly of CC-NLR and TIR-NLR complexes occurs by a common mechanism.

Mutations in Plant NLR Genes Are Often Associated With Autoimmune Phenotypes

Similar to humans, misregulation of immune signaling in plants can cause autoimmune phenotypes. Autoimmune phenotypes in plants typically include constitutive expression of defense-related genes, reduced growth (dwarfism), and spontaneous necrotic lesions, especially in leaves. Although genetic screens have uncovered mutations in many different genes that can give rise to such autoimmune phenotypes, of most relevance to human immunology is the frequent identification of mutations in genes encoding NLR proteins. As described in Chapter 20, many human autoimmune diseases have been associated with mutations in NLR-encoding genes (eg, mutations in *NOD2* are associated with Crohn disease), which presumably lead to inappropriate formation of inflammasomes. In plants, the best-studied autoimmune-associated NLR gene is the Arabidopsis *SNC1* gene, which encodes a TIR-NLR. A single amino acid change from a negatively charged glutamate to a positively charged lysine in the region between the WHD domain and the LRR domain causes constitutive activation of defense genes and severe dwarfing,[138] suggesting that this residue may play a key role in positioning the LRR domain relative to the ATP/dATP binding pocket, which is needed to maintain NLR proteins in the off state. In support of this model, deletion of the LRR domain often leads to constitutive activation of plant NLR proteins.[139,140] This has also been observed in human NLR proteins, as deletion of the LRR domain from the human NOD2 protein constitutively activates NOD2 signaling,[141] pointing to the similarity in activation mechanisms between animal and plant NLRs.

Misexpression of Specific NLR Alleles Can Cause Autoimmunity

The majority of autoimmune studies in plants has been performed on mutants that were induced in the lab. However, there is a very interesting phenomenon in plants, referred to as hybrid incompatibility, that may share an underlying mechanism with some cases of human autoimmune disease. Hybrid incompatibility is well known among plant breeders and most often occurs when two distantly related plant varieties, or even two different species, are crossed. F1 progeny in such crosses constitutively expresses defense genes, are dwarfed, and, in severe cases, develop spontaneous necrotic lesions and may even die as young seedlings.

The cell death phenotype and constitutive expression of defense genes indicates that hybrid incompatibility is likely caused by autoimmunity triggered by combining incompatible genotypes. To uncover the molecular nature of this incompatibility, researchers have used classic genetic mapping to identify the specific genes that mediate hybrid incompatibility. In most cases, two loci, one from each parent, are required. Notably, in nearly every case, at least one of these loci encodes an NLR protein, or cluster of NLR proteins,[142-147] indicating that specific NLR alleles can trigger autoimmunity when placed in a mismatched genetic background. In several cases of hybrid necrosis, F1 plants display elevated accumulation of multiple NLR proteins compared to their accumulation in their parents,[142,146,148] suggesting that autoimmunity can be caused by overaccumulation of specific NLRs and that hybrid incompatibility results when NLR expression becomes misregulated in specific genetic backgrounds.

Signaling by NLR Resistosomes

Assembly of the resistosome complex brings the N-termini of each NLR protomer in close proximity, enabling formation of functional complexes. However, CC domains and TIR domains form completely different structures and have unrelated functions. For ZAR1, assembly of the CC domains results in the N-terminal α-helix rotating away from the body of the resistosome to form a five-sided funnel (Fig. 5.5). This α-helix is amphipathic, and it has been proposed that this funnel inserts into the plasma membrane, possibly forming a pore that can function as a calcium channel.[135] If this model is correct, it would provide a direct link between activation of a plant NLR protein and influx of extracellular calcium. Consistent with this model, blocking of plasma membrane–localized calcium channels using lanthanum chloride ($LaCl_3$) is known to block defense signaling by a CC-NLR protein.[149]

In contrast to CC-NLR proteins such as ZAR1, most TIR-NLR proteins do not associate with the plasma membrane, and TIR domains do not form a channel-like structure. Instead, assembly of TIR domains creates a binding pocket for nicotinamide adenine dinucleotide (NAD),[136] and activated TIR-NLR complexes display NADase activity.[136,150] Furthermore, mutation of the putative catalytic residue, which blocks NADase activity, blocks activation of resistance, indicating that enzymatic activity is required for signaling, presumably via a second messenger created by digestion of NAD, possibly cyclic adenosine diphosphate ribose (cADPR).[136,150]

Although it is not yet clear how cADPR (or other NAD breakdown product) activates resistance, one clue is the observation that all TIR-NLR proteins appear to require an RPW8-NLR protein to induce resistance responses.[151-153] This observation suggests that cADPR somehow activates an RPW8-NLR, either directly or indirectly, and that it is the N-terminal domain of the RPW8-NLR that activates resistance. No structure is yet available for an RPW8-NLR resistosome, but RPW8 domains are similar in sequence to fungal HeLo/HELL domains, which act as membrane pore-forming toxins;[154] thus, RPW8-NLRs may employ a mechanism similar to ZAR1 to trigger defense signaling. Notably, however, RPW8-NLR proteins appear to localize to the endoplasmic reticulum and general endomembrane system, rather than the plasma membrane[151] and thus are unlikely to mediate extracellular calcium influx.

A second clue for how cADPR may trigger resistance is that all TIR-NLR proteins, but not CC-NLR proteins, require two related lipase-like proteins to activate resistance, EDS1 and PAD4.[155,156] These two proteins form a heterodimer.[157] Significantly, auto-activating mutations in TIR-NLR proteins are fully suppressed by loss-of-function mutations in EDS1, while auto-activating mutations in RPW8-NLR proteins are not.[151] This observation suggests that cADPR functions upstream of EDS1, while RPW8-NLRs function downstream of EDS1. It is thus unlikely that cADPR interacts directly with RPW8-NLRs and also unlikely that there is a physical association between TIR-NLRs and RPW8-NLRs.

Perception of Pathogen Signals by Plant NLR Proteins

The primary function of NLR proteins is to detect the presence of pathogens. The molecular mechanisms by which plant NLR proteins detect pathogens, however, is surprisingly diverse. These mechanisms can be grouped into two broad categories, direct and indirect. Direct recognition involves a physical interaction between the NLR protein and a pathogen effector protein and does not require the effector protein to have an enzymatic activity.[122] In contrast, indirect recognition involves a second host protein that is modified by the effector (ie, the effector target), with modification of the effector target causing activation of the NLR protein.[139] In such indirect recognition, the NLR protein never physically associates with the effector.[158]

Whether a pathogen effector is recognized directly or indirectly by an NLR protein is expected to have a large impact on the co-evolution of NLRs and the effectors they recognize.[159] In the case of direct recognition, there will be selection on the pathogen for mutations that alter the surface of the effector to prevent binding to the NLR protein. Such mutations would not necessarily affect the function of the effector in terms of binding to and modifying its target and thus would be expected to rapidly evolve. To "keep up" with effectors in this arms race, the matching NLRs would need to rapidly co-evolve and thus are expected to display high allelic diversity and bear hallmarks of positive selection. Based on genomic analyses of 62 Arabidopsis ecotypes, this exactly is what is found.[160] All Arabidopsis NLR proteins that are known to bind effector proteins directly were found to be rapidly evolving. This is true regardless of whether the NLR is from the CC class or the TIR class.[160]

In contrast to direct-detecting NLRs, those that detect effectors indirectly all appear to be evolving slowly.[160] This is expected because indirect recognition involves detection of effector activity, rather than effector appearance. Indirect detection is analogous to a burglar alarm that detects a broken window or door lock, while direct detection is analogous to a facial recognition system. One can evade the latter by putting on a mask, but to evade the former system, one has to devise an alternative entry mechanism (eg, a different enzymatic activity), which is much more difficult to evolve. Thus, it should be more challenging for a pathogen to evolve around an NLR protein that employs indirect recognition, and thus such NLR proteins should be more stable in terms of sequence change over time.

Activation Mechanisms of Plant NLR Proteins

The structure of the Roq1 TIR-NLR protein bound to the effector protein XopQ provides the first structural insights into how direct-detecting NLR proteins are activated (Fig. 5.5). In that structure, the effector protein binds to the LRR domain and a domain immediately C-terminal to the LRR domain that displays structural similarities to the antigen-binding domain of immunoglobulins.[136] This post-LRR domain forms a β-sandwich structure, with loops connecting the β-strands that interact with multiple surfaces and pockets of the effector. Binding of the effector causes the LRR and WHD domains of the NLR protein to rotate at the HD1-WHD hinge, enabling the release of ADP as discussed above. The C-terminal domains of NLR proteins are quite variable, which likely reflects their selection to bind diverse effectors.

The structure of the ZAR1 CC-NLR protein in complex with the effector target protein PBL2 and the kinase RKS1 provides insight into how NLR proteins can be activated indirectly by effectors (Fig. 5.6). In the absence of effector, ZAR1 forms a pre-activation complex with RKS1 and does not oligomerize.[137] In this heterodimeric form, the WHD and LRR domains form a lid over the ADP/ATP-binding pocket, effectively trapping ADP and keeping ZAR in an off state.[137] The ZAR1-RKS1 complex is activated by association with a PBL2 protein, but this association is dependent on uridylation of PBL2 by the bacterial effector protein AvrAC.[135] Uridylated PBL2 binds to the activation loop of RKS1 and, in the process, induces a 60° rotation of the NBD, which enables release of ADP and binding of ATP. The ATP-bound form can then oligomerize, forming the pentameric structure shown in Figure 5.6. As described above, this oligomerization event causes the N-terminal α-helix of the CC domain of ZAR1 to flip outward, forming a five-sided funnel-shaped structure.

Downstream Signaling Mechanisms Employed by CC-NLR Proteins Are Likely to Be Diverse

Figure 5.6 suggests that the ZAR1 CC-NLR protein may activate downstream signaling by forming an ion channel in the plasma membrane. However, this mechanism is likely to be used by only a subset of CC-NLR proteins. This is because many CC-NLR proteins do not localize to the plasma membrane in either their active or inactive forms. For example, the Mla10 NLR protein from barley, which is activated by an effector from a fungal pathogen that causes powdery mildew disease, appears to shuttle between the cytoplasm and nucleus.[161] Similarly, the CC-NLR protein Rx1 from potato, which is activated by the coat protein of potato virus X, also localizes to both the cytoplasm and nucleus.[162] As a third

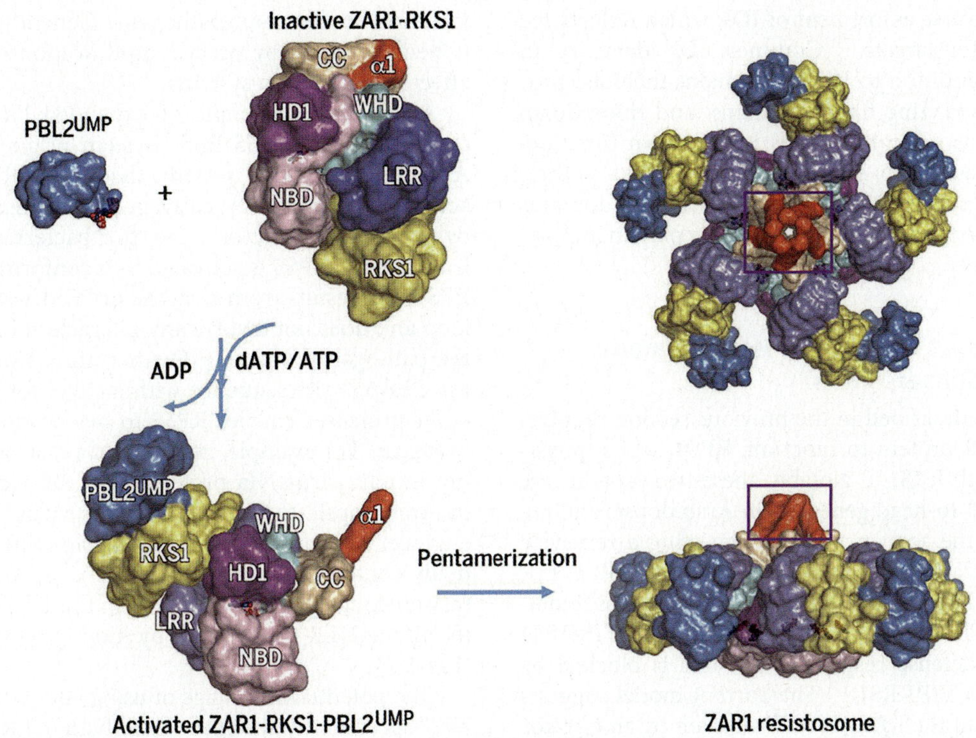

FIG. 5.6. PBL2^{UMP}-induced assembly of the ZAR1 resistosome. Interaction of PBL2^{UMP} (*blue*) with the preformed ZAR1-RKS1 complex (inactive ZAR1-RKS1) triggers conformational changes in ZAR1 nucleotide-binding domain and ADP release, allowing the complex to bind dATP or ATP. dATP or ATP binding induces structural remodeling of ZAR1, resulting in full activation of ZAR1 (activated ZAR1-RKS1-PBL2^{UMP}) and formation of the pentameric ZAR1 resistosome (shown in two orientations). The very N-terminal alpha helix (α1) (*red*) of ZAR1 buried in the inactive ZAR1-RKS1 complex becomes solvent-exposed in the activated ZAR1-RKS1-PBL2^{UMP} complex and forms a funnel-shaped structure (highlighted within the *purple frame*) in the ZAR1 resistosome that is required for ZAR1 PM association, cell death triggering, and disease resistance. (From Wang J, Hu M, Wang J, et al. Reconstitution and structure of a plant NLR resistosome conferring immunity. Science. 2019;364. Reprinted with permission from AAAS.)

example, the R3a protein from potato, which is activated by an effector from the oomycete pathogen *P. infestans*, localizes to both the cytoplasm and late endosomes.[163] This diversity of locations likely reflects the diversity of locations to which pathogen effectors are targeted. This in turn may impact the molecular mechanisms by which NLR proteins activate resistance.

Even among CC-NLR proteins that localize to the plasma membrane, it is unlikely that all employ a signaling mechanism similar to that of ZAR, as many CC-NLR proteins do not have an N-terminal amphipathic α-helix, which would be required to form a transmembrane channel.[164] A major unanswered question in the plant immunology field is thus how such CC-NLR proteins activate signaling.

A Subset of Plant NLR Proteins Contain Integrated Effector Target Domains

As discussed above, recognition of pathogen effectors by NLR proteins is broadly divided into direct and indirect recognition mechanisms. However, many NLR proteins have evolved an intermediate mechanism in which the NLR protein appears to have fused with a new domain that mimics an effector target.[165] Modification of this "integrated domain" (ID) by an effector then activates the NLR protein. Formally, this amounts to direct recognition since it involves direct binding of an effector to an NLR protein, but mechanistically it is more similar to indirect recognition, since the effector is modifying a target motif and this modification then activates the NLR protein. This recognition mechanism should provide the evolutionary stability associated with indirect recognition, while assuring co-segregation and co-evolution of NLR and effector target genes, since they are the same gene. An example of a well-characterized NLR with an ID is the RRS1 protein of Arabidopsis, which contains a C-terminal domain that mimics a WRKY family transcription factor, and is targeted by effectors from at least two different bacterial species.[166,167] WRKY family transcription factors are plant-specific transcription factors that regulate various stress response genes, including numerous genes associated with immunity (see below). It thus appears that RRS1 evolved by capturing a WRKY-encoding gene, which enables its C-terminus to function as a decoy target for effectors from diverse pathogens. Notably, both CC-NLR and TIR-NLR proteins can have IDs,[167] and thus this strategy of capturing effector target domains is shared by both classes of plant NLRs.

Consistent with the model that IDs function as decoy effector targets, genomic analyses across flowering plants have revealed a diverse assortment of IDs, which reflects the diversity of effector targets.[165] Common IDs identified in these analyses in addition to WRKY domains included protein kinase domains, zinc finger domains, and thioredoxin domains, as well as several domains of unknown function. The latter will be especially interesting to investigate as their presence in NLRs suggests that proteins with these domains that are unlinked to NLR-encoding genes may play an important role in immunity.

Some NLR Proteins Function as Heterodimeric "Helper" and "Sensor" Pairs

The RRS1 protein described in the previous section requires a second TIR-NLR protein to function, RPS4, which physically associates with RRS1.[168] Notably, these two proteins are encoded by a head-to-head gene pair in Arabidopsis and are closely related at the sequence level, suggesting a relatively recent gene duplication. The association of the RRS1 TIR domain with the RPS4 TIR domain appears to suppress signaling by the RPS4 TIR domain, as overexpression of TIRRPS4 by itself induces defense responses, but this is blocked by co-expression with TIRRRS1.[168] The current model suggests that modification of the RRS1 WRKY domain by an effector promotes its interaction with the RRS1 TIR domain, which then releases the TIRRPS4 domain from suppression.[169] Presumably activation also requires oligomerization of RPS4, but this has not yet been established.

A second example of a sensor–helper NLR pair is provided by the CC-NLR proteins RGA4 and RGA5 from rice, which mediate recognition of effectors from the rice blast fungus *M. oryzae*.[170] Similar to *RPS4* and *RRS1*, *RGA4* and *RGA5* are arranged as a head-to-head gene pair. Also similar to RPS4 and RRS1, only one of member of the pair (RGA5) contains an ID, and this member suppresses signaling by the other. Binding of fungal effectors to the ID of RGA5 then releases RGA4 from inhibition, enabling signaling by the CC domain of RGA4 by an unknown mechanism.[170] It thus appears that sensor–helper pairs may commonly function by a mechanism whereby the sensor NLR acts as a suppressor of the helper NLR, with suppression being relieved upon modification by an effector.

Engineering Novel Immune Traits in Plants

The emerging knowledge of NLR activation mechanisms is now enabling researchers to modify the recognition specificity of plant NLRs, which should enable plant breeders to improve the disease resistance traits of crop plants in the not-too-distant future. This is an important goal in agriculture because plant diseases currently cause large yield losses, unless large amounts of pesticides are used. The heavy dependence of commercial agriculture on pesticides has many negative consequences, including large losses in biodiversity, pollution of waterways and ground water, increased expenses for farmers, and exposing humans, especially farm workers, to toxic and carcinogenic compounds. It is hoped that the need for such pesticides can be dramatically reduced by developing crop varieties that are genetically resistant to infection to common pathogens. Generation of such crops is being enabled by precise modification of NLR-mediated effector recognition systems.

A promising example of expanded NLR-mediated recognition is the RPS5-PBS1 system originally described in Arabidopsis. In that system, the CC-NLR protein RPS5 is activated indirectly by cleavage of the RLCK protein PBS1 by a protease effector from the bacterium *P. syringae*.[171] Importantly, RPS5 is activated by a conformational change in PBS1 that results from cleavage of PBS1 within its activation loop and does not involve any interaction between RPS5 and the pathogen protease.[172] Furthermore, swapping the protease cleavage site sequence within PBS1 for cleavage sites for other proteases enables RPS5 to be activated by these other proteases. For example, insertion of a cleavage site for the turnip mosaic virus NIa protease into PBS1 confers resistance in Arabidopsis to infection by this virus.[172] Thus, one can engineer resistance to a virus using an NLR protein that normally confers resistance to a bacterium. Also notable is that resistance specificity was altered in this example, not by modifying the NLR protein, but by modifying the effector target (Fig. 5.7).

One potential advantage of using effector targets to modify NLR specificity, rather than modifying NLR proteins directly, is that effector targets tend to be more highly conserved across species than are NLR proteins. Such conservation makes it more straightforward to translate work done in a model species such as Arabidopsis into crop plants. It also makes it more likely that such engineered resistance traits will be more stable in terms of how long it takes for pathogens to evolve ways around these novel resistance systems. Here, again, the RPS5-PBS1 system is informative. The RPS5 protein is not conserved outside the mustard family to which Arabidopsis belongs. However, PBS1, which regulates MAMP-triggered immune responses, is very highly conserved across flowering plants,[174] with clear orthologs found in important crop species such as soybean, barley, and wheat.[175,176] Significantly, cleavage of PBS1 in soybean activates a robust defense response; thus, despite lacking an RPS5 ortholog, soybean has evolved the ability to "guard" PBS1.[175] Consistent with this conclusion, engineering of a soybean PBS1 ortholog so that it can be cleaved by the NIa protease of soybean mosaic virus confers complete resistance in soybean to infection by this virus.[173] This modification required changing just five amino acids in the soybean genome, which can now be accomplished using CRISPR-based genome editing. This is important to note because, in the United States, crops produced in this manner are treated the same as traditionally bred crops by regulatory agencies, rather than as "genetically modified organisms" (GMOs). Avoiding the GMO label will greatly accelerate commercialization and will reduce costs to farmers and consumers.

PRR-TRIGGERED IMMUNITY AND EFFECTOR-TRIGGERED IMMUNITY ARE TIGHTLY INTEGRATED

In the above sections, PTI and ETI were discussed as distinct branches of the plant immune system. Recent work, however, has revealed that these immune signaling pathways are highly integrated, with both branches potentiating each other and

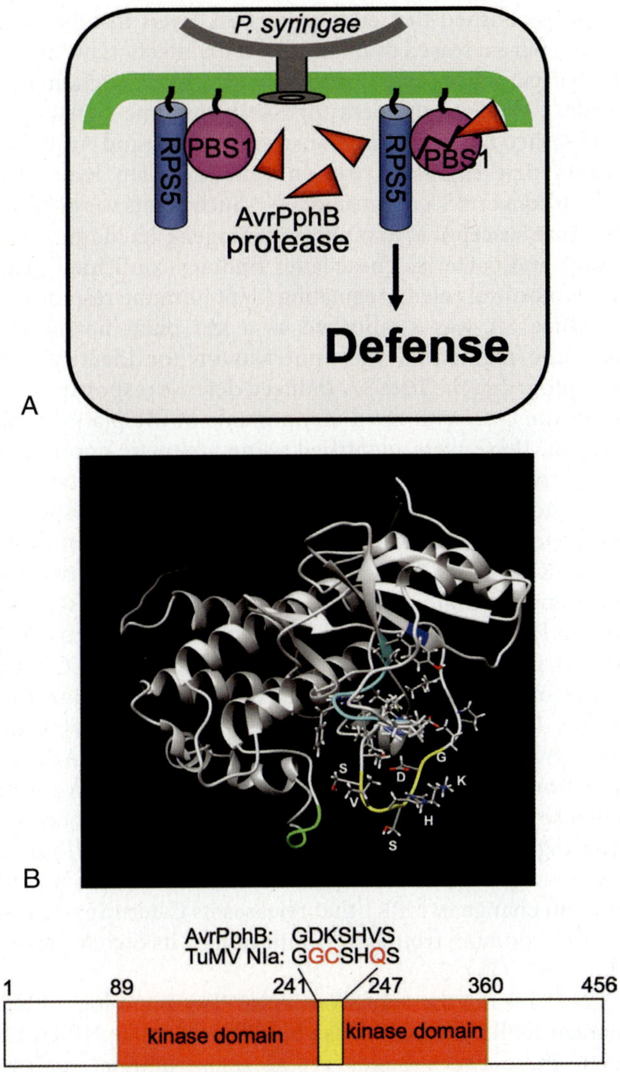

FIG. 5.7. Changing nucleotide-binding leucine-rich repeat receptor (NLR) specificity by modifying effector targets. A: The NLR protein RPS5 from Arabidopsis is normally activated by cleavage of the RLCK protein PBS1 by the AvrPphB protease from *Pseudomonas syringae*.[171] **B:** The predicted structure of the PBS1 protein, with residues flanking the site cleaved by AvrPphB (GDKSHVS) shown in stick format. Note its location on an exposed loop of PBS1. *Green* highlight indicates a loop required for recognition of PBS1 by RPS5. *Light blue* highlight indicates the P-loop, which binds the alpha phosphate of ATP. **C:** Schematic showing the replacement of the AvrPphB cleavage motif within PBS1 with a motif cleavable by the NIa protease of turnip mosaic virus (TuMV). This modification requires changing just three amino acids (highlighted in *red letters*) and enables RPS5 to be activated by a viral protease instead of a bacterial protease,[172] resulting in complete resistance to TuMV.[173]

both being required for successful defense.[15,16] For example, induction of ROS production by the MAMP flg22 is greatly increased by simultaneous activation of the TIR-NLR protein RPS4, even though activation of RPS4 by itself does not induce ROS. Similarly, neither RPS4 activation by itself nor flg22 treatment induces programmed cell death, but simultaneous activation does,[16] illustrating the synergistic effect of these two branches of the immune system. The potentiation of PTI by ETI results from transcriptional upregulation of PTI signaling components, including RLCKs, MAPKs, and RBOHD.[15,16]

The integration of PTI and ETI signaling pathways is further illustrated by the impact of mutations in PTI signaling components on ETI-triggered resistance. For example, simultaneous mutation of the MAMP receptor genes *FLS2*, *EFR*, and *CERK1*, which should block perception of flagellin, EFTu, and peptidoglycan, strongly abrogates resistance normally induced by NLR receptors that recognize bacterial effectors.[15] The requirement of MAMP perception for NLR-mediated resistance is mediated, in part, by co-activation of RBOHD, with both pathways independently contributing to RBOHD activation and full ROS production achieved only when both are functioning.[15]

TRANSCRIPTIONAL REGULATION OF PLANT IMMUNE RESPONSES

The immune responses triggered by both PTI and ETI include dramatic changes in gene expression, with activation of a single PRR causing approximately 30% of all transcripts to change in abundance.[14] These wholesale changes in gene expression require the activity of numerous transcription factors.[177,178] Two transcription factors that play central roles in mediating early responses following pathogen perception in Arabidopsis are SARD1 (SYSTEMIC-ACQUIRED RESISTANCE DEFICIENT 1) and CBP60g (CALMODULIN-BINDING PROTEIN 60g), which both belong to a plant-specific family of transcription factors.[179-181] CBP60g, but not SARD1, binds to CALMODULIN, and this binding is required for its ability to induce transcription;[179,180] thus, CBP60g potentially provides a direct link between PTI and ETI-induced increases in cytosolic calcium and activation of defense genes (Fig. 5.8). Consistent with this model, transcriptional responses induced by PTI and ETI overlap extensively, suggesting that independent signaling events converge on the same sets of transcriptions factors.[178]

SARD1 and CBP60g are partially redundant in function, as simultaneous mutation of both genes has a much stronger impact on defense responses than mutation of either gene alone.[180,181] Additionally, the double mutant is approximately 100-fold more susceptible to infection by the bacterium *P. syringae* than wild-type plants. Notably, MAMP perception strongly induces transcription of SARD1 and CBP60g, indicating that there exists a positive feedback loop for the activity of these two transcription factors, likely through binding to their own promoters.[181] SARD1 and CBP60g bind to a motif in the promoters of defense genes with a consensus sequence of GAAATTT, and this sequence is highly enriched in known MAMP-inducible genes.[180,181]

SARD1 AND CBP60g REGULATE BIOSYNTHESIS OF DEFENSE HORMONES

SARD1 and CBP60g appear to exert their dramatic effect on plant immune responses largely by regulating biosynthesis of the plant hormones salicylic acid (SA; a naturally occurring nonsteroidal anti-inflammatory drug that is closely related to

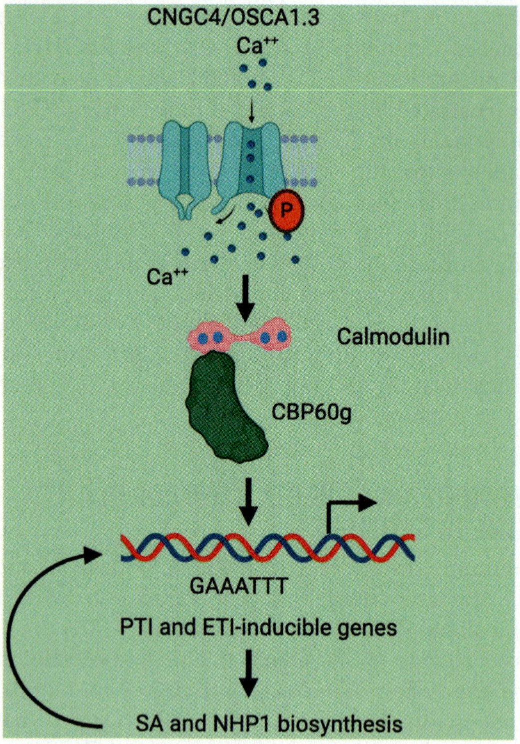

FIG. 5.8. PTI and ETI-triggered cytosolic calcium increases induce CALMODULIN-dependent transcription factor CBP60g. Ca^{++} ions (*blue dots*) bind to CALMODULIN, which then binds to the N-terminus of CBP60g. The latter then binds to numerous PTI and ETI-inducible genes via binding to the consensus DNA sequence GAAATTT. These genes include SA and NHP1 biosynthetic genes, setting up a feedback loop that increases levels of CBP60g and SARD1 that further induces defense genes. ETI, effector-triggered immunity; PTI, PRR-triggered immunity. (Created with BioRender.com.)

aspirin, which is acetylsalicylic acid) and N-hydroxy-pipecolic acid (NHP; an amino acid derived from L-lysine).[180-182] Both of these hormones are dramatically upregulated during PTI and ETI, and exogenous application of these hormones impacts the expression of thousands of plant genes (see next section).

A key step in the biosynthesis of SA is catalyzed by the enzyme isochorismate synthase (ICS1), expression of which is directly regulated by CBP60g and SARD1.[180,181] Similarly, *FMO1* (*FLAVIN MONOOXYGENASE 1*) and *ALD1* (*AGD2-LIKE DEFENSE RESPONSE PROTEIN 1*), which encode enzymes required for synthesis of NHP, are also directly regulated by SARD1 and CBP60g.[183] The resulting increase in these hormones then leads to activation of numerous additional transcription factors and a complex network of both positive and negative feedback loops that regulate the duration and intensity of the defense response.

Salicylic Acid Is Perceived by the SA Receptors NPR1, NPR3, and NPR4

SA was identified as an inducer of plant defense response in the 1979, when it was observed that injection of aspirin into tobacco plants conferred protection against infection by tobacco mosaic virus (TMV).[184] A pair of landmark papers in 1990 established that endogenous SA levels in tobacco and cucumber increased dramatically upon infection by virus and fungal pathogens, implicating SA as a likely signaling molecule.[185,186] A requirement for SA in immune signaling was established by generating transgenic tobacco and Arabidopsis plants that expressed the enzyme salicylate hydroxylase, which converts SA to catechol.[187,188] Such plants were shown to be more susceptible to multiple pathogens, including viruses, fungi, and bacteria. These latter findings confirmed that SA plays a central role in regulating plant immune responses.

Once SA was established as a key plant hormone for immune regulation, the hunt was on for identifying the receptors that mediate SA-induced defense responses. As has been the case with most components of the plant immune system, these were identified using a genetic approach. In brief, researchers screened for Arabidopsis mutants that failed to induce expression of defense genes in response to exogenous application of SA. This led to the identification of NPR1 (NONEXPRESSER OF PR GENES) by two independent laboratories.[189,190] Subsequent work established that NPR1 binds to SA,[191,192] and that NPR1 functions as a transcriptional co-activator in association with TGA transcription factors, which belong to the basic leucine zipper family (bZIP).[193,194] Specifically, NPR1 interacts with three TGA transcription factors, TGA2, TGA5, and TGA6, which function redundantly in positive regulation of SA-induced defense gene expression and pathogen resistance.[195] SA binding is not required for the interaction of NPR1 with TGA transcription factors.[191] Rather, SA induces a confirmation change in NPR1 that releases its C-terminal transactivation domain from auto-inhibition by its own N-terminal domain[192] (Fig. 5.9).

NPR1 is not the only SA receptor. Two homologs of NPR1 named NPR3 and NPR4 also bind SA.[191] Unlike NPR1, however, these two proteins act as transcriptional repressors rather than activators and are partially redundant. In the absence of SA, they associate with TGA transcription factors and keep them from initiating transcription via a C-terminal transcriptional repression domain. In the presence of SA, this repressor domain is neutralized, enabling TGA transcription factors to transcribe (Fig. 5.9).[191] Thus, most SA-inducible genes are under both negative and positive regulation by different NPR family members.

One of the genes that is strongly induced by NPR-TGA co-receptors is *SARD1*.[191] This is noteworthy because SARD1 is a transcription factor that strongly upregulates transcription of *ICS1*, a key step in the upregulation of SA production. Hence, there appears to be a strong positive feedback loop in this system with SA increasing transcription of *SARD1* and increased SARD1 activity leading to increasing levels of SA.

Full activation of immune responses requires the combined activation of both SA and NHP-signaling pathways, with both compounds regulating the synthesis of the other. For example, knocking out SA perception by combining an *npr1* loss-of-function mutation with a dominant negative form of NPR4 nearly eliminates NHP production due to lack of expression of CBP60g and SARD1,[196] which are required for expression of NHP biosynthetic enzymes.[182] On the other

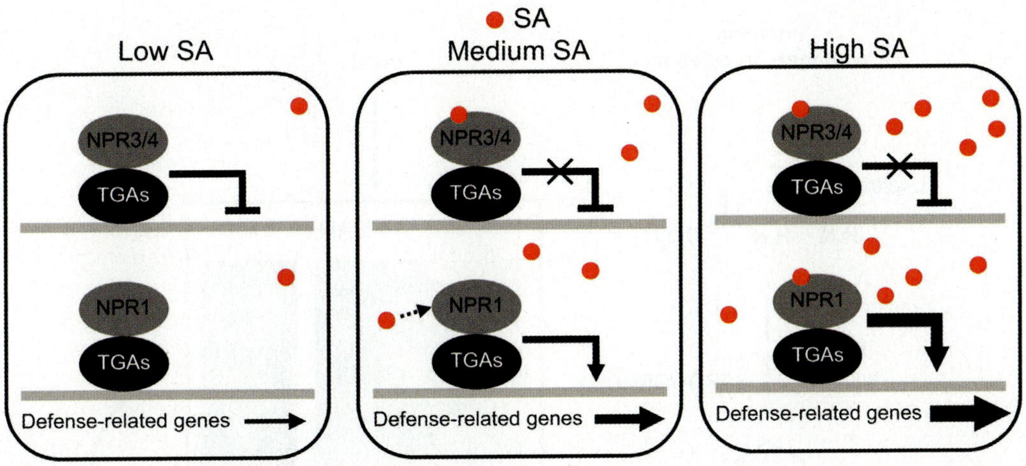

FIG. 5.9. A working model of NPR1/NPR3/NPR4 in SA-induced defense activation.[196] When SA level is low under an uninfected state, NPR3/NPR4 interacts with TGA2/TGA5/TGA6 to inhibit the expression of defense-related gene expression. As the SA level increases during pathogen infection, SA binds to NPR3/NPR4 to release the transcriptional repression of defense genes. Meanwhile, binding of SA to NPR1 promotes activation of the transcription of defense genes. (Liu Y, Sun T, Sun Y, et al. Diverse roles of salicylic acid receptors NPR1 and NPR3/NPR4 in plant immunity. *Plant Cell.* 2020 by permission of Oxford University Press.)

hand, NHP induces synthesis of SA,[197] demonstrating the circular nature of the SA and NHP signaling circuits.

Simultaneous elimination of both SA and NHP-induced signaling greatly compromises both PTI and ETI. This is especially evident in the *npr1 npr4* dominant negative mutant mentioned above that eliminates both SA perception and NHP production.[196] In that line, induction of defense genes by PTI-inducing agents is nearly eliminated, while disease resistance triggered by the CC-NLR protein RPS2 and TIR-NLR protein RPS4 is nearly abolished. Thus, immunity triggered by PRRs and NLRs requires amplification of transcriptional responses controlled by the defense hormones SA and NHP.

SYSTEMIC IMMUNE SIGNALING IS MEDIATED BY SA AND NHP

Infection of one leaf on a plant will induce immune responses in uninfected leaves on the same plant, and thus plants have evolved mechanisms for systemic signaling. Such signaling is often referred to as systemic-acquired resistance (SAR), because the systemic leaves become immune to subsequent infection due to the heightened expression of some defense genes, along with "priming" of many other genes. In this context, priming refers to the ability of genes to be more rapidly and strongly expressed in response to an infection. Identification of the molecules that mediate systemic immune signaling has been an important goal of the MPMI field over the last 30 years, as such molecules could potentially be very useful for controlling diseases on crop plants.

The first strong candidate for mediating SAR was SA, which has been introduced above. Because it is a small molecule, it should rapidly travel through the plant vasculature. Consistent with SA playing a role in SAR, mutations that block pathogen-induced SA biosynthesis block SAR,[198] as does artificial depletion of endogenous SA levels via expression of an SA hydroxylase enzyme.[199]

Although SA was initially the favored molecule for functioning as the primary systemic signaling molecule for immune regulation, subsequent work revealed that although SA is required in both source and target tissues for SAR to be effective, SA is unlikely to be the translocated signal. Instead, the primary translocated signal appears to be NHP or its glucosylated derivative N-OGlc-Pip[197,200] (Fig. 5.10). Exogenously applied NHP is rapidly converted to N-OGlc-Pip, which then rapidly accumulates in distal leaves. Importantly, this distal accumulation occurs even in plants that lack FMO1, which is required for the synthesis of NHP.[200] This observation demonstrates that the exogenously added NHP is transported. Similar to SA, NHP application causes rapid global changes in defense gene expression and metabolic pathways that lead to enhanced resistance to pathogens, including induction of *ICS1* and elevation of SA. Consistent with NHP/N-OGlc-Pip functioning as the primary systemic signal rather than SA, it accumulates in systemic leaves approximately 24 hours prior to SA accumulation.[197] Furthermore, mutation of *FMO1* blocks SAR (Fig. 5.10). Notably, NHP also appears to play a role in ETI-mediated defense signaling, as mutations in *FMO1* also suppress autoimmunity in Arabidopsis that is associated with activation of the NLR proteins DSC1 and DSC2.[201,202]

RNA INTERFERENCE PLAYS A CENTRAL ROLE IN ANTIVIRAL IMMUNE RESPONSES IN PLANTS

Plant Researchers Discovered RNA Interference

The phenomenon of RNA-mediated gene silencing was first observed in 1990 in plants, several years before it was reported in the nematode worm *Caenorhabditis elegans.* This discovery was accidental and occurred when researchers at a small start-up plant biotechnology company were attempting to modify the pigments in petunia flowers by overexpressing the *chalcone synthase* (*CHS*) gene, which encodes an enzyme involved in isoflavanoid production.[203] Instead of obtaining flowers that were deep purple (the original goal), the researchers obtained flowers with patches of unpigmented white cells. The lack of pigment correlated with a strong reduction (~50-fold) in *CHS* mRNA in these cells. Thus, in their attempt to

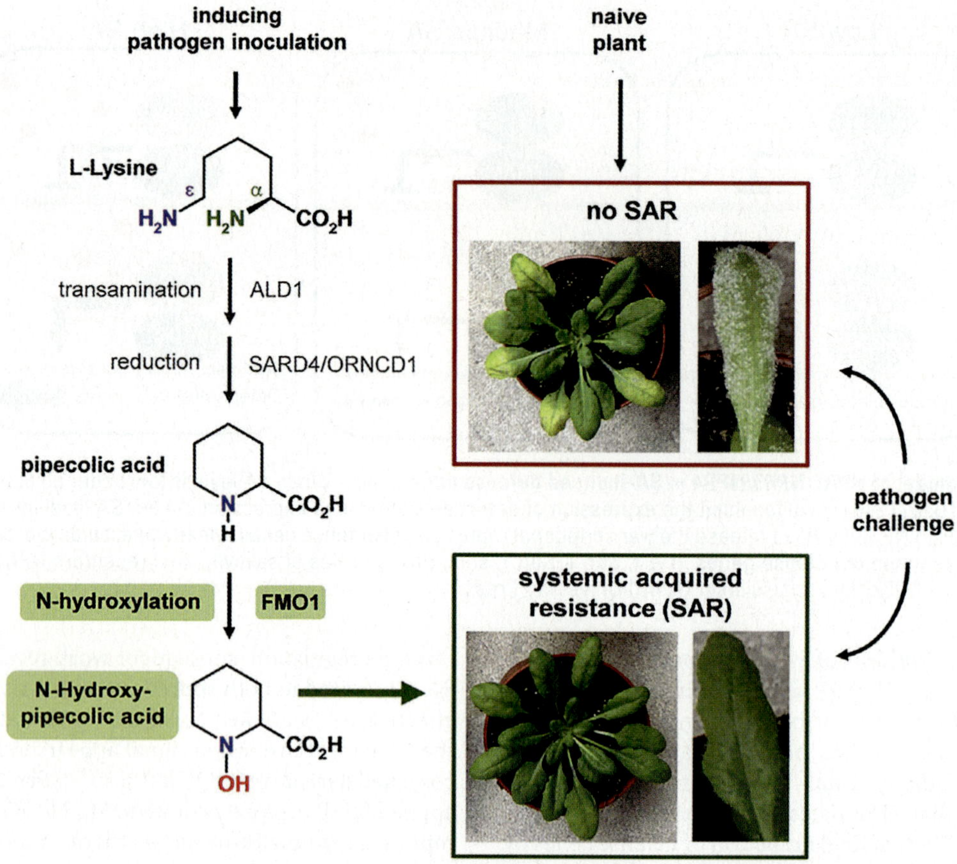

FIG. 5.10. N-hydroxy pipecolic acid is a systemic signal for inducing immunity in noninfected leaves. Shown is the biosynthetic pathway from generation of *N*-hydroxy pipecolic acid from the amino acid lysine, which requires the enzymes ALD1, SARD4, and FMO1. On the right are shown two Arabidopsis plants inoculated with oomycete pathogen *Hyaloperonospora arabidopsidis*. (Figure reproduced from Hartmann M, Zeier T, Bernsdorff F, et al. Flavin monooxygenase-generated N-hydroxypipecolic acid is a critical element of plant systemic immunity. *Cell.* 2018;173:456-469 e416.)

overexpress the *CHS* gene by introducing a second copy under a strong promoter, they had somehow induced the elimination of the full-length mRNA produced by the endogenous copy. This phenomenon was quickly replicated in multiple plant species, and ultimately in animals, including in human cells, and was subsequently named "RNA interference" (RNAi).

A key breakthrough in our understanding of the molecular mechanisms of RNAi came in 1998 when it was shown in tobacco and in *C. elegans* that double-stranded RNA was more effective in silencing homologous genes than either sense strand or antisense strand alone.[204,205] Shortly thereafter, Hamilton and Baulcombe showed that dsRNAs induced the accumulation of small RNA molecules (~25 nt) homologous to both the sense and antisense strands of the gene being silenced.[206] This discovery was, again, made in plants. Importantly, this latter work also showed that the RNAi mechanism likely involved an RNA-dependent RNA polymerase (RdRP) to amplify dsRNA species.

Infection of Plants by RNA Viruses Induces RNAi

The paper by Hamilton and Baulcombe[206] also established that infection of plants with a positive strand RNA virus (potato virus X) induced accumulation of small RNAs homologous to the infecting virus, which are now known as small interfering RNAs (siRNAs). Baulcombe's group proposed that this process was associated with an RNA-based immune response that occurs even in nontransgenic plants.[207] This was based on the observation that plants frequently recover from a virus infection, whereby newly emerging leaves on a virus-infected plant display greatly enhanced resistance to re-infection by the same virus and viruses with closely related sequences, but not to more distantly related viruses.[207] This sequence specificity, combined with the observation that siRNA levels correlated strongly with virus resistance and gene silencing, led to the conclusion that transgene silencing in plants was a consequence of an antiviral immune system that was somehow mistaking transgene mRNAs for viral RNAs. In support of this, it was found that insertion of an endogenous plant gene antisense cDNA sequence into an RNA virus can induce silencing of that gene upon viral infection.[208] Such "virus-induced gene silencing" has become a useful tool in plant biotechnology, as it provides a rapid method for systemic silencing of genes in plants.

Molecular Components of the RNAi Pathway in Plants

The discovery of RNAi in the late 1990s led to intensive genetic and biochemical studies in both plant and animal systems to understand the molecular mechanisms underlying this process. These studies revealed that the molecular mechanisms that mediate RNAi are mostly conserved between plants and animals, and thus plant studies have been highly informative for understanding RNAi in animals (including humans), and vice versa. Figure 5.11 provides an overview of the RNAi pathway (for a recent review of the plant RNAi literature, see reference[209]). The key components of this pathway are the Dicer-like (DCL) enzymes that cleave dsRNAs into 21 to 24 nt long dsRNA fragments and Argonaute proteins (AGO), which employ one strand of these small RNA duplexes to target complementary mRNAs for cleavage (slicing). AGO proteins are part of a larger multiprotein complex called the RNA-induced silencing complex (RISC). The mRNA cleavage products produced by RISC activity can be copied by RdRPs, initiating an amplification cycle that leads to elevated siRNA levels and robust reduction of target RNAs.

In plants, the *DCLs*, *AGOs*, and *RdRP* genes have undergone multiple rounds of duplication, leading to specialization for different aspects of small RNA–mediated gene regulation.[210] Although not covered in this chapter, small RNAs are now known to play a major role in epigenetic gene regulation in both animals and plants, completely independent of their roles in antiviral defenses.[211] Nevertheless, evolutionary studies indicate that the original role of RNAi was as a defense system against viral infection.[210]

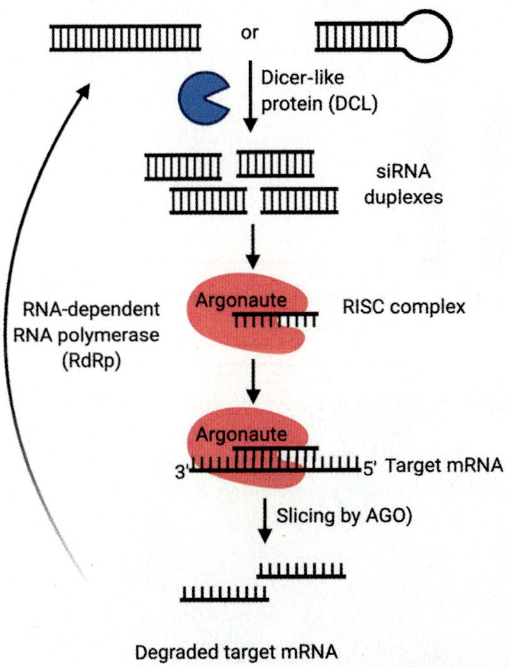

FIG. 5.11. Schematic of RNA interference–mediated gene silencing in eukaryotes. Double-stranded RNAs or hairpin RNAs generate small siRNA duplexes by the action of Dicer. These short dsRNAs are bound by Argonaute (Ago) and other proteins to form an RNA-induced silencing complex (RISC). One strand of the dsRNA is sliced by the Ago protein and is destabilized, leaving a single-stranded siRNA that functions as a guide. The siRNA/RISC complex then binds to the complementary sequence of the target mRNA, resulting in the degradation of the target transcript or inhibition of translation. The components of siRNA/mRNA complex can be recycled to the RISC complex or generate siRNA duplexes by the action of RNA-dependent RNA polymerase (RdRP).

RNAi-Mediated Defense Systems Are Suppressed by Pathogen Effectors

To overcome RNAi-mediated defenses, most viruses have evolved proteins that can suppress various steps in the RNAi process.[212] For example, the P19 protein of TMV and the 2b protein of cucumber mosaic virus bind double-stranded siRNAs, preventing their loading into AGO proteins.[213,214] As another example, the HC-Pro protein of tobacco etch virus inhibits 3′-methylation of siRNAs, which leads to their degradation.[215] In addition to targeting siRNAs, many viral RNAi suppressors target AGO proteins, either directly or indirectly. For example, the P0 protein of *Poleroviruses* destabilizes AGO proteins by inducing ubiquitination of AGOs by a host E3 ubiquitin ligase.[216] Lastly, RdRP activity is also frequently targeted, as typified by the V2 protein of tomato yellow leaf curl virus (TYLCV), which directly interacts with SGS3, a cofactor of RDR6, to block amplification of viral siRNAs.[217]

Viruses are not the only pathogens that suppress host cell RNAi as part of their virulence strategy. Oomycete and bacterial pathogens have also been shown to employ effectors that target the plant RNAi machinery, which indicates that RNAi plays a more general role in regulating immune responses beyond simply targeting viral RNAs.[218] For example, the HopT1-1 effector from *P. syringae* has been shown to suppress the activity of AGO1, especially its ability to inhibit translation of target mRNAs.[219] Similarly, the PSR1 effector from *Phytophthora sojae* inhibits production of both miRNAs and siRNAs in plants by targeting a conserver nuclear protein named PINP1, which is required for proper localization of DCL1 complexes.[220] A second *P. sojae* effector (PSR2) specifically suppresses production of so-called secondary siRNAs in host plant cells.[115] Secondary siRNAs are produced from host cell mRNAs that are targeted by specific host cell miRNAs. The roles of these secondary siRNAs are incompletely understood, but at least some of them appear to be translocated from host plant cells into Phytophthora, where they silence genes and reduce virulence; thus, PSR2 likely evolved to suppress siRNA production to prevent this form of cross-kingdom RNAi.[115]

Naturally occurring cross-kingdom RNAi has also been shown for plant–fungus interactions, in both directions, from fungus to plant host[221] and from plant host to fungus.[222] In both of these examples, siRNAs were shown to be transferred from one organism to the other. In the case of fungus to plant transfer, siRNAs produced in the fungus (*B. cinerea*) could be immunoprecipitated using antibodies against a host AGO protein; thus, the fungal siRNAs exploit the host cell RNAi machinery to target host genes. Conversely, host cell siRNAs can be identified inside *B. cinerea* cells during infection, and these siRNAs target specific fungal genes that are required for full virulence. Thus, there appears to be an ongoing arms race between plants and pathogens at the level of siRNA exchange.

Exploiting Cross-Kingdom RNAi to Engineer Disease Resistant Crop Plants

The possibility that small RNAs might be capable of moving from host cells into pathogen cells first came to light in 2010, when it was shown that expression of a long hairpin RNA construct in a host plant cell with homology to a pathogen virulence gene could reduce virulence of the pathogen.[223] This phenomenon was given the name of HIGS. In this first example, the host plant was barley and the pathogen was the powdery mildew fungus *Blumeria graminis*.

Although multiple HIGS experiments in plants established that expression of dsRNA targeting fungal genes could reduce virulence of the fungus, strong evidence of target gene silencing was not shown until 2013, when Koch and colleagues showed silencing of cytochrome P450 genes in the fungus *Fusarium graminearum* (causal agent of head blight) by expression of a long hairpin RNA in Arabidopsis.[224] As shown in Figure 5.12, these transgenic plants were highly resistant to infection by this fungus. Significantly, resistance correlated with a strong reduction in transcript levels of the three target genes in the fungus, *CYP51A*, *CYP51B*, and *CYP51C*, which encode cytochrome P450s involved in sterol synthesis that are necessary for fungal membrane integrity. This work also showed that expression of the long hairpin RNA induced production of siRNAs homologous to the *CYP51* genes, suggesting that siRNAs were being translocated from the host plant cell into the fungal cells.

The Mechanism of siRNA Translocation Between Host and Pathogen Remains Enigmatic

The above examples of siRNA movement between filamentous plant pathogens (oomycetes and fungi) and host plant cell raise the question of how siRNAs are transported. Both classes of pathogens employ specialized feeding structures called haustoria. Importantly, in haustoria, both the host cell plasma membrane, which is invaginated, and the fungal plasma membrane are intact. Thus, siRNA translocation from host to pathogen requires that siRNAs cross two plasma membranes to enter pathogen cells (Fig. 5.13). Since siRNAs carry a strong negative charge due to their phosphate backbones, this translocation is unlikely to occur by passive diffusion.

Currently, the preferred model for siRNA translocation invokes packaging of siRNAs into extracellular vesicles

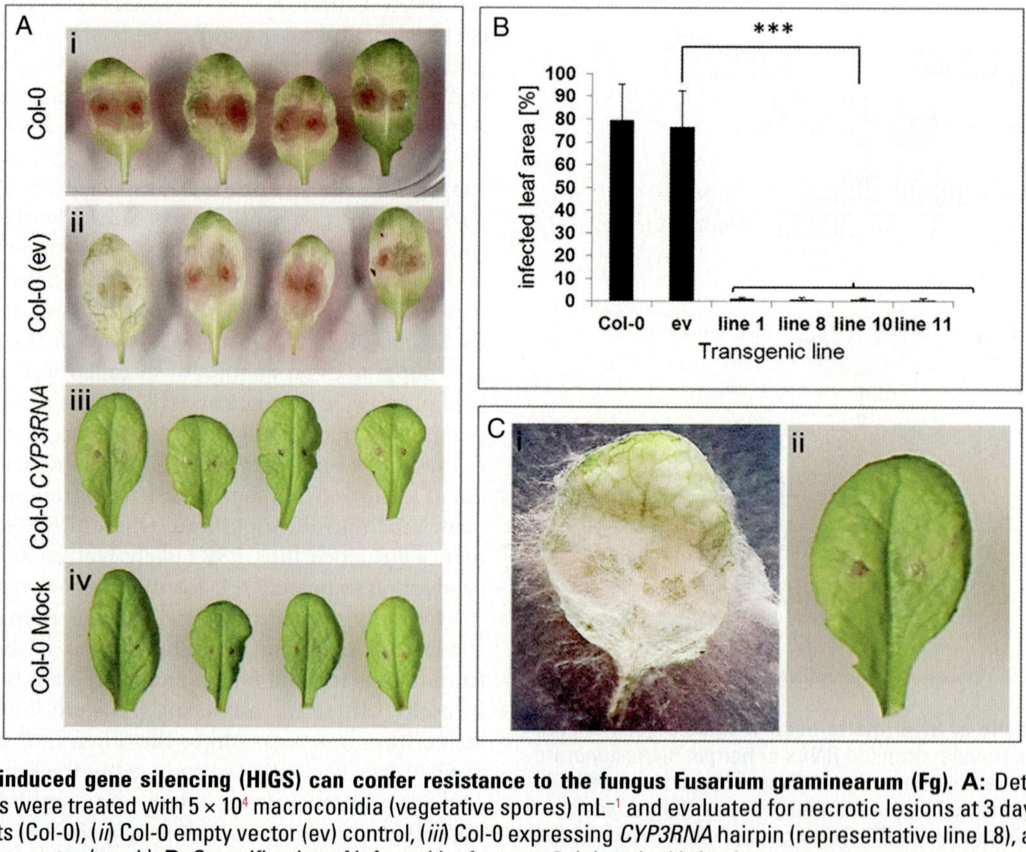

FIG. 5.12. Host-induced gene silencing (HIGS) can confer resistance to the fungus Fusarium graminearum (Fg). A: Detached leaves of 5-week-old plants were treated with 5×10^4 macroconidia (vegetative spores) mL^{-1} and evaluated for necrotic lesions at 3 days post infection. (*i*) wild-type plants (Col-0), (*ii*) Col-0 empty vector (ev) control, (*iii*) Col-0 expressing *CYP3RNA* hairpin (representative line L8), and (*iv*) wild-type treated with Tween water (mock). **B:** Quantification of infected leaf area at 3 dpi; typical infection symptoms are recorded as a percent of the total leaf area. Bars represent mean values ±SDs of three independent experiments, each using 20 leaves collected from 15 different plants of each transgenic lines, as well as wild-type and Col-0 ev plants. The reduction in infection symptoms on *CYP3RNA*-expressing leaves compared with wild-type and Col-0 ev control was statistically significant (***$P < 0.0001$; Student's *t* test). **C:** *Fg*-inoculated *Arabidopsis* leaves at 5 dpi. (*i*) The Col-0 ev leaf is heavily infected with *Fg*; (*ii*) Col-0 expressing *CYP3RNA* does not show infection symptoms. (From Koch A, Kumar N, Weber L, et al. Host-induced gene silencing of cytochrome P450 lanosterol C14a-demethylase-encoding genes confers strong resistance to Fusarium species. Proc Natl Acad Sci U S A. 2013;110:19324-19329.)

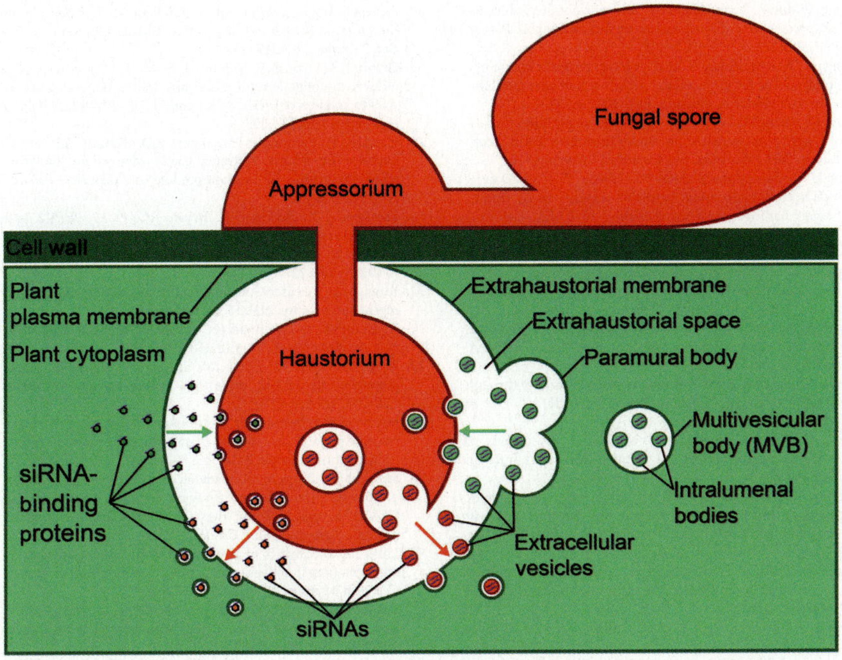

FIG. 5.13. Cross-kingdom RNA interference (RNAi) requires transport of siRNAs across two cellular membranes. Diagram depicts a fungal haustorium (feeding structure) pressed into a plant epidermal cell. The mechanisms of siRNA transport between cells is not yet understood, but may involve extracellular vesicles, which are frequently observed around haustoria. Endocytosis of naked siRNAs or siRNA-protein particles has not been ruled out, however. (Courtesy of Brian Rutter.)

(EVs), which would then be taken up by recipient cells through endocytosis and eventually release from endosomes in the recipient cell.[222] In support of this model, plants have been shown to produce EVs,[225] and these EVs have been shown to carry diverse small RNAs.[226] However, more recent studies on mammalian EVs has shown that sRNAs can associate with the outside of EVs, and that a combination of protease and RNase treatment is required to ascertain whether EV-associated RNA is located on the inside or outside of EVs.[227] Such analyses have not yet been reported for plant EVs. It will be important to do so, however, in order to establish the cellular pathways by which siRNAs are released from host cells.

OUTSTANDING QUESTIONS IN PLANT IMMMUNOLOGY

Our understanding of the molecular and cellular mechanisms underlying the plant immune system has increased dramatically over the last 20 years. As detailed above, we now have clear insights into the basic mechanisms by which pathogens are detected by plants and the primary signaling pathways by which immune responses are induced. However, there are still multiple foundational questions remaining to be answered. Below I have listed my personal top six unanswered questions in plant immunology. The first three of these also apply to human immunology:

1. How do plants, which are surrounded by microbes, including inside roots and leaves, distinguish between friend and foe, so that friends do not induce an adverse immune response?
2. How do microbe–microbe interactions affect the plant immune system?
3. How are siRNAs translocated between plants and pathogens, and vice versa?
4. How do TIR-NLR proteins mechanistically link to downstream RPW8-NLR proteins that are required for their function?
5. How do NLR proteins induce cell death?
6. How is the plant hormone N-hydroxy-pipecolic acid (NHP) perceived by plants?

ACKNOWLEDGMENTS

Research in the Innes laboratory is currently supported by grants from the United States National Science Foundation Plant Biotic Interactions Program and Plant Genome Research Program and the United States Department of Energy Office of Science.

REFERENCES

1. Ausubel FM. Are innate immune signaling pathways in plants and animals conserved? *Nat Immunol*. 2005;6:973-979.
2. Urbach JM, Ausubel FM. The NBS-LRR architectures of plant R-proteins and metazoan NLRs evolved in independent events. *Proc Natl Acad Sci U S A*. 2017;114:1063-1068.
3. Liang X, Zhou JM. Receptor-like cytoplasmic kinases: central players in plant receptor kinase-mediated signaling. *Annu Rev Plant Biol*. 2018;69:267-299.
4. Shiu SH, Bleecker AB. Receptor-like kinases from Arabidopsis form a monophyletic gene family related to animal receptor kinases. *Proc Natl Acad Sci U S A*. 2001;98:10763-10768.
5. Kadota Y, Sklenar J, Derbyshire P, et al. Direct regulation of the NADPH oxidase RBOHD by the PRR-associated kinase BIK1 during plant immunity. *Mol Cell*. 2014;54:43-55.
6. Liu Y, Maierhofer T, Rybak K, et al. Anion channel SLAH3 is a regulatory target of chitin receptor-associated kinase PBL27 in microbial stomatal closure. *Elife*. 2019;8:44474.
7. Bi G, Zhou JM. MAP kinase signaling pathways: a hub of plant-microbe interactions. *Cell Host Microbe*. 2017;21:270-273.
8. Ranf S, Eschen-Lippold L, Frohlich K, et al. Microbe-associated molecular pattern-induced calcium signaling requires the receptor-like cytoplasmic kinases, PBL1 and BIK1. *BMC Plant Biol*. 2014;14:374.
9. Tian W, Hou C, Ren Z, et al. A calmodulin-gated calcium channel links pathogen patterns to plant immunity. *Nature*. 2019;572:131-135.

10. Thor K, Jiang S, Michard E, et al. The calcium-permeable channel OSCA1.3 regulates plant stomatal immunity. *Nature.* 2020;585:569-573. doi:10.1038/s41586-020-2702-1
11. Boudsocq M, Willmann MR, Mccormack M, et al. Differential innate immune signalling via Ca(2+) sensor protein kinases. *Nature.* 2010;464:418-422.
12. Boudsocq M, Sheen J. CDPKs in immune and stress signaling. *Trends Plant Sci.* 2013;18:30-40.
13. Cheng SH, Willmann MR, Chen HC, et al. Calcium signaling through protein kinases. The Arabidopsis calcium-dependent protein kinase gene family. *Plant Physiol.* 2002;129:469-485.
14. Hillmer RA, Tsuda K, Rallapalli G, et al. The highly buffered Arabidopsis immune signaling network conceals the functions of its components. *PLoS Genet.* 2017;13:e1006639.
15. Yuan M, Jiang Z, Bi G, et al. Pattern-recognition receptors are required for NLR-mediated plant immunity. *Nature.* 2021; 592:105-109.
16. Ngou BPM, Ahn H-K, Ding P, et al. Mutual potentiation of plant immunity by cell-surface and intracellular receptors. *Nature.* 2021;592:110-115.
17. Goldschmidt EE. Plant grafting: new mechanisms, evolutionary implications. *Front Plant Sci.* 2014;5:727.
18. Shiu SH, Bleecker AB. Expansion of the receptor-like kinase/Pelle gene family and receptor-like proteins in Arabidopsis. *Plant Physiol.* 2003;132:530-543.
19. Shiu SH, Karlowski WM, Pan R, et al. Comparative analysis of the receptor-like kinase family in Arabidopsis and rice. *Plant Cell.* 2004;16:1220-1234.
20. Kajava AV, Vassart G, Wodak SJ. Modeling of the three-dimensional structure of proteins with the typical leucine-rich repeats. *Structure.* 1995;3:867-877.
21. Matsushima N, Takatsuka S, Miyashita H, et al. Leucine rich repeat proteins: sequences, mutations, structures and diseases. *Protein Pept Lett.* 2019;26:108-131.
22. Smith KD, Andersen-Nissen E, Hayashi F, et al. Toll-like receptor 5 recognizes a conserved site on flagellin required for protofilament formation and bacterial motility. *Nat Immunol.* 2003;4:1247-1253.
23. Yoon SI, Kurnasov O, Natarajan V, et al. Structural basis of TLR5-flagellin recognition and signaling. *Science.* 2012;335: 859-864.
24. Sun Y, Li L, Macho AP, et al. Structural basis for flg22-induced activation of the Arabidopsis FLS2-BAK1 immune complex. *Science.* 2013;342:624-628.
25. Gomez-Gomez L, Boller T. FLS2: an LRR receptor-like kinase involved in the perception of the bacterial elicitor flagellin in Arabidopsis. *Mol Cell.* 2000;5:1003-1011.
26. Chinchilla D, Zipfel C, Robatzek S, et al. A flagellin-induced complex of the receptor FLS2 and BAK1 initiates plant defence. *Nature.* 2007;448:497-500.
27. Li J, Wen J, Lease KA, et al. BAK1, an Arabidopsis LRR receptor-like protein kinase, interacts with BRI1 and modulates brassinosteroid signaling. *Cell.* 2002;110:213-222.
28. Schulze B, Mentzel T, Jehle AK, et al. Rapid heteromerization and phosphorylation of ligand-activated plant transmembrane receptors and their associated kinase BAK1. *J Biol Chem.* 2010;285:9444-9451.
29. Lu D, Wu S, Gao X, et al. A receptor-like cytoplasmic kinase, BIK1, associates with a flagellin receptor complex to initiate plant innate immunity. *Proc Natl Acad Sci U S A.* 2010;107:496-501.
30. Roux M, Schwessinger B, Albrecht C, et al. The Arabidopsis leucine-rich repeat receptor-like kinases BAK1/SERK3 and BKK1/SERK4 are required for innate immunity to hemibiotrophic and biotrophic pathogens. *Plant Cell.* 2011;23:2440-2455.
31. Zipfel C, Kunze G, Chinchilla D, et al. Perception of the bacterial PAMP EF-Tu by the receptor EFR restricts agrobacterium-mediated transformation. *Cell.* 2006;125:749-760.
32. Kutschera A, Dawid C, Gisch N, et al. Bacterial medium-chain 3-hydroxy fatty acid metabolites trigger immunity in Arabidopsis plants. *Science.* 2019;364:178-181.
33. Ranf S, Gisch N, Schaffer M, et al. A lectin S-domain receptor kinase mediates lipopolysaccharide sensing in *Arabidopsis thaliana. Nat Immunol.* 2015;16:426-433.
34. Willmann R, Lajunen HM, Erbs G, et al. Arabidopsis lysin-motif proteins LYM1 LYM3 CERK1 mediate bacterial peptidoglycan sensing and immunity to bacterial infection. *Proc Natl Acad Sci U S A.* 2011;108:19824-19829.
35. Cao Y, Liang Y, Tanaka K, et al. The kinase LYK5 is a major chitin receptor in Arabidopsis and forms a chitin-induced complex with related kinase CERK1. *Elife.* 2014;3:03766.
36. Liu T, Liu Z, Song C, et al. Chitin-induced dimerization activates a plant immune receptor. *Science.* 2012;336: 1160-1164.
37. Albert I, Bohm H, Albert M, et al. An RLP23-SOBIR1-BAK1 complex mediates NLP-triggered immunity. *Nat Plants.* 2015;1:15140.
38. Kim SY, Sivaguru M, Stacey G. Extracellular ATP in plants. Visualization, localization, and analysis of physiological significance in growth and signaling. *Plant Physiol.* 2006;142:984-992.
39. Choi J, Tanaka K, Cao Y, et al. Identification of a plant receptor for extracellular ATP. *Science.* 2014;343:290-294.
40. Zhang X, Mou Z. Extracellular pyridine nucleotides induce PR gene expression and disease resistance in Arabidopsis. *Plant J.* 2009;57:302-312.
41. Wang C, Zhou M, Zhang X, et al. A lectin receptor kinase as a potential sensor for extracellular nicotinamide adenine dinucleotide in *Arabidopsis thaliana. Elife.* 2017;6:25474.
42. Wang C, Huang X, Li Q, et al. Extracellular pyridine nucleotides trigger plant systemic immunity through a lectin receptor kinase/BAK1 complex. *Nat Commun.* 2019;10:4810.
43. Yamada K, Yamashita-Yamada M, Hirase T, et al. Danger peptide receptor signaling in plants ensures basal immunity upon pathogen-induced depletion of BAK1. *EMBO J.* 2016;35:46-61.
44. Hou S, Wang X, Chen D, et al. The secreted peptide PIP1 amplifies immunity through receptor-like kinase 7. *PLoS Pathog.* 2014;10:e1004331.
45. Brutus A, Sicilia F, Macone A, et al. A domain swap approach reveals a role of the plant wall-associated kinase 1 (WAK1) as a receptor of oligogalacturonides. *Proc Natl Acad Sci U S A.* 2010;107:9452-9457.
46. Yasuda S, Okada K, Saijo Y. A look at plant immunity through the window of the multitasking coreceptor BAK1. *Curr Opin Plant Biol.* 2017;38:10-18.
47. Kunze G, Zipfel C, Robatzek S, et al. The N terminus of bacterial elongation factor Tu elicits innate immunity in Arabidopsis plants. *Plant Cell.* 2004;16:3496-3507.
48. Ponting CP, Aravind L, Schultz J, et al. Eukaryotic signalling domain homologues in archaea and bacteria. Ancient ancestry and horizontal gene transfer. *J Mol Biol.* 1999;289: 729-745.
49. Zhang XC, Wu X, Findley S, et al. Molecular evolution of lysin motif-type receptor-like kinases in plants. *Plant Physiol.* 2007;144:623-636.
50. Miya A, Albert P, Shinya T, et al. CERK1, a LysM receptor kinase, is essential for chitin elicitor signaling in Arabidopsis. *Proc Natl Acad Sci U S A.* 2007;104:19613-19618.
51. Kouzai Y, Mochizuki S, Nakajima K, et al. Targeted gene disruption of OsCERK1 reveals its indispensable role in chitin perception and involvement in the peptidoglycan response and immunity in rice. *Mol Plant Microbe Interact.* 2014;27:975-982.
52. Zhang B, Ramonell K, Somerville S, et al. Characterization of early, chitin-induced gene expression in Arabidopsis. *Mol Plant Microbe Interact.* 2002;15:963-970.
53. Hayafune M, Berisio R, Marchetti R, et al. Chitin-induced activation of immune signaling by the rice receptor CEBiP relies on a unique sandwich-type dimerization. *Proc Natl Acad Sci U S A.* 2014;111:E404-E413.
54. Seidl MF, Van Den Ackerveken G. Activity and phylogenetics of the broadly occurring family of microbial Nep1-like proteins. *Annu Rev Phytopathol.* 2019;57:367-386.
55. Lenarcic T, Albert I, Bohm H, et al. Eudicot plant-specific sphingolipids determine host selectivity of microbial NLP cytolysins. *Science.* 2017;358:1431-1434.
56. Ottmann C, Luberacki B, Kufner I, et al. A common toxin fold mediates microbial attack and plant defense. *Proc Natl Acad Sci U S A.* 2009;106:10359-10364.
57. Oome S, Raaymakers TM, Cabral A, et al. Nep1-like proteins from three kingdoms of life act as a microbe-associated molecular pattern in Arabidopsis. *Proc Natl Acad Sci U S A.* 2014;111:16955-16960.
58. Li Q, Wang C, Mou Z. Perception of damaged self in plants. *Plant Physiol.* 2020;182:1545-1565.
59. Land W. Allograft injury mediated by reactive oxygen species. From conserved proteins of Drosophila to acute and chronic rejection of human transplants. Part III: interaction of (oxidative) stress-induced heat shock proteins with toll-like receptor-bearing cells of innate immunity and its consequences for the development of acute and chronic allograft rejection. *Transplant Rev.* 2003;17:67-86.
60. Matzinger P. The danger model: a renewed sense of self. *Science.* 2002;296:301-305.
61. Chen D, Cao Y, Li H, et al. Extracellular ATP elicits DORN1-mediated RBOHD phosphorylation to regulate stomatal aperture. *Nat Commun.* 2017;8:2265.
62. Choi J, Tanaka K, Liang Y, et al. Extracellular ATP, a danger signal, is recognized by DORN1 in Arabidopsis. *Biochem J.* 2014;463:429-437.
63. Cote F, Hahn MG. Oligosaccharins: structures and signal transduction. *Plant Mol Biol.* 1994;26:1379-1411.
64. Decreux A, Thomas A, Spies B, et al. In vitro characterization of the homogalacturonan-binding domain of the wall-associated kinase WAK1 using site-directed mutagenesis. *Phytochemistry.* 2006;67:1068-1079.
65. Denoux C, Galletti R, Mammarella N, et al. Activation of defense response pathways by OGs and flg22 elicitors in Arabidopsis seedlings. *Mol Plant.* 2008;1:423-445.
66. Huffaker A, Pearce G, Ryan CA. An endogenous peptide signal in Arabidopsis activates components of the innate immune response. *Proc Natl Acad Sci U S A.* 2006;103:10098-10103.
67. Bartels S, Lori M, Mbengue M, et al. The family of Peps and their precursors in Arabidopsis: differential expression and localization but similar induction of pattern-triggered immune responses. *J Exp Bot.* 2013;64:5309-5321.
68. Bartels S, Boller T. Quo vadis, Pep? Plant elicitor peptides at the crossroads of immunity, stress, and development. *J Exp Bot.* 2015;66:5183-5193.
69. Gravino M, Locci F, Tundo S, et al. Immune responses induced by oligogalacturonides are differentially affected by AvrPto and loss of BAK1/BKK1 and PEPR1/PEPR2. *Mol Plant Pathol.* 2017;18:582-595.
70. Hander T, Fernandez-Fernandez AD, Kumpf RP, et al. Damage on plants activates Ca(2+)-dependent metacaspases for release of immunomodulatory peptides. *Science.* 2019;363:eaar7486.
71. Yamaguchi Y, Huffaker A, Bryan AC, et al. PEPR2 is a second receptor for the Pep1 and Pep2 peptides and contributes to defense responses in Arabidopsis. *Plant Cell.* 2010;22:508-522.
72. Yamaguchi Y, Pearce G, Ryan CA. The cell surface leucine-rich repeat receptor for AtPep1, an endogenous peptide elicitor in Arabidopsis, is functional in transgenic tobacco cells. *Proc Natl Acad Sci U S A.* 2006;103:10104-10109.
73. Tang D, Wang G, Zhou JM. Receptor kinases in plant-pathogen interactions: more than pattern recognition. *Plant Cell.* 2017;29:618-637.
74. Macho AP, Schwessinger B, Ntoukakis V, et al. A bacterial tyrosine phosphatase inhibits plant pattern recognition receptor activation. *Science.* 2014;343:1509-1512.
75. Suzuki M, Shibuya M, Shimada H, et al. Autophosphorylation of specific threonine and tyrosine residues in Arabidopsis CERK1 is essential for the activation of chitin-induced immune signaling. *Plant Cell Physiol.* 2016;57:2312-2322.
76. Liu J, Liu B, Chen S, et al. A tyrosine phosphorylation cycle regulates fungal activation of a plant receptor ser/thr kinase. *Cell Host Microbe.* 2018;23:241-253.e6.
77. Luo X, Wu W, Liang Y, et al. Tyrosine phosphorylation of the lectin receptor-like kinase LORE regulates plant immunity. *EMBO J.* 2020;39:e102856.
78. Perraki A, Defalco TA, Derbyshire P, et al. Phosphocode-dependent functional dichotomy of a common co-receptor in plant signalling. *Nature.* 2018;561:248-252.
79. Zhang J, Li W, Xiang T, et al. Receptor-like cytoplasmic kinases integrate signaling from multiple plant immune receptors and are targeted by a *Pseudomonas syringae* effector. *Cell Host Microbe.* 2010;7:290-301.
80. Li L, Li M, Yu L, et al. The FLS2-associated kinase BIK1 directly phosphorylates the NADPH oxidase RbohD to control plant immunity. *Cell Host Microbe.* 2014;15:329-338.
81. Brandt B, Munemasa S, Wang C, et al. Calcium specificity signaling mechanisms in abscisic acid signal transduction in Arabidopsis guard cells. *Elife.* 2015;4:03599.
82. Bi G, Zhou Z, Wang W, et al. Receptor-like cytoplasmic kinases directly link diverse pattern recognition receptors to the activation of mitogen-activated protein kinase cascades in Arabidopsis. *Plant Cell.* 2018;30:1543-1561.
83. Gao M, Liu J, Bi D, et al. MEKK1, MKK1/MKK2 and MPK4 function together in a mitogen-activated protein kinase cascade to regulate innate immunity in plants. *Cell Res.* 2008;18:1190-1198.
84. Suarez-Rodriguez MC, Adams-Phillips L, Liu Y, et al. MEKK1 is required for flg22-induced MPK4 activation in Arabidopsis plants. *Plant Physiol.* 2007;143:661-669.
85. Qiu JL, Zhou L, Yun BW, et al. Arabidopsis mitogen-activated protein kinase kinases MKK1 and MKK2 have overlapping functions in defense signaling mediated by MEKK1, MPK4, and MKS1. *Plant Physiol.* 2008;148:212-222.
86. Su J, Zhang M, Zhang L, et al. Regulation of stomatal immunity by interdependent functions of a pathogen-responsive MPK3/MPK6 cascade and abscisic acid. *Plant Cell.* 2017;29: 526-542.
87. Rao S, Zhou Z, Miao P, et al. Roles of receptor-like cytoplasmic kinase VII members in pattern-triggered immune signaling. *Plant Physiol.* 2018;177:1679-1690.
88. Zhang M, Su J, Zhang Y, et al. Conveying endogenous and exogenous signals: MAPK cascades in plant growth and defense. *Curr Opin Plant Biol.* 2018;45:1-10.
89. Liu Y, Zhang S. Phosphorylation of 1-aminocyclopropane-1-carboxylic acid synthase by MPK6, a stress-responsive mitogen-activated protein kinase, induces ethylene biosynthesis in Arabidopsis. *Plant Cell.* 2004;16:3386-3399.
90. Mao G, Meng X, Liu Y, et al. Phosphorylation of a WRKY transcription factor by two pathogen-responsive MAPKs drives phytoalexin biosynthesis in Arabidopsis. *Plant Cell.* 2011;23:1639-1653.
91. Li G, Meng X, Wang R, et al. Dual-level regulation of ACC synthase activity by MPK3/MPK6 cascade and its downstream WRKY transcription factor during ethylene induction in Arabidopsis. *PLoS Genet.* 2012;8:e1002767.
92. Joo S, Liu Y, Lueth A, et al. MAPK phosphorylation-induced stabilization of ACS6 protein is mediated by the non-catalytic C-terminal domain, which also contains the cis-determinant for rapid degradation by the 26S proteasome pathway. *Plant J.* 2008;54:129-140.
93. Toruno TY, Stergiopoulos I, Coaker G. Plant-pathogen effectors: cellular probes interfering with plant defenses in spatial and temporal manners. *Annu Rev Phytopathol.* 2016;54: 419-441.

94. Feng F, Yang F, Rong W, et al. A Xanthomonas uridine 5'-monophosphate transferase inhibits plant immune kinases. *Nature.* 2012;485:114-118.
95. Irieda H, Inoue Y, Mori M, et al. Conserved fungal effector suppresses PAMP-triggered immunity by targeting plant immune kinases. *Proc Natl Acad Sci U S A.* 2019;116:496-505.
96. Shan L, He P, Li J, et al. Bacterial effectors target the common signaling partner BAK1 to disrupt multiple MAMP receptor-signaling complexes and impede plant immunity. *Cell Host Microbe.* 2008;4:17-27.
97. Yamaguchi K, Nakamura Y, Ishikawa K, et al. Suppression of rice immunity by *Xanthomonas oryzae* type III effector Xoo2875. *Biosci Biotechnol Biochem.* 2013;77:796-801.
98. Zhou J, Wu S, Chen X, et al. The *Pseudomonas syringae* effector HopF2 suppresses Arabidopsis immunity by targeting BAK1. *Plant J.* 2014;77:235-245.
99. Li L, Kim P, Yu L, et al. Activation-dependent destruction of a co-receptor by a *Pseudomonas syringae* effector dampens plant immunity. *Cell Host Microbe.* 2016;20:504-514.
100. Janjusevic R, Abramovitch RB, Martin GB, et al. A bacterial inhibitor of host programmed cell death defenses is an E3 ubiquitin ligase. *Science.* 2006;311:222-226.
101. Gimenez-Ibanez S, Hann DR, Ntoukakis V, et al. AvrPtoB targets the LysM receptor kinase CERK1 to promote bacterial virulence on plants. *Curr Biol.* 2009;19:423-429.
102. Wang Y, Li J, Hou S, et al. A *Pseudomonas syringae* ADP-ribosyltransferase inhibits Arabidopsis mitogen-activated protein kinase kinases. *Plant Cell.* 2010;22:2033-2044.
103. Zhang J, Shao F, Li Y, et al. A *Pseudomonas syringae* effector inactivates MAPKs to suppress PAMP-induced immunity in plants. *Cell Host Microbe.* 2007;1:175-185.
104. Cui H, Wang Y, Xue L, et al. *Pseudomonas syringae* effector protein AvrB perturbs Arabidopsis hormone signaling by activating MAP kinase 4. *Cell Host Microbe.* 2010;7:164-175.
105. Li H, Xu H, Zhou Y, et al. The phosphothreonine lyase activity of a bacterial type III effector family. *Science.* 2007;315:1000-1003.
106. Wawra S, Trusch F, Matena A, et al. The RxLR motif of the host targeting effector AVR3a of *Phytophthora infestans* is cleaved before secretion. *Plant Cell.* 2017;29:1184-1195.
107. Mcgowan J, Fitzpatrick DA. Genomic, network, and phylogenetic analysis of the oomycete effector arsenal. *mSphere.* 2017;2:e00408. doi:10.1128/mSphere.00408-00417
108. Boevink PC, Birch PRJ, Turnbull D, et al. Devastating intimacy: the cell biology of plant-Phytophthora interactions. *New Phytol.* 2020;228:445-458. doi:10.1111/nph.16650
109. Win J, Krasileva KV, Kamoun S, et al. Sequence divergent RXLR effectors share a structural fold conserved across plant pathogenic oomycete species. *PLoS Pathog.* 2012;8:e1002400.
110. Du Y, Mpina MH, Birch PR, et al. *Phytophthora infestans* RXLR effector AVR1 interacts with exocyst component Sec5 to manipulate plant immunity. *Plant Physiol.* 2015;169:1975-1990.
111. King SR, Mclellan H, Boevink PC, et al. *Phytophthora infestans* RXLR effector PexRD2 interacts with host MAPKKK epsilon to suppress plant immune signaling. *Plant Cell.* 2014;26:1345-1359.
112. Wang S, Mclellan H, Bukharova T, et al. Phytophthora infestans RXLR effectors act in concert at diverse subcellular locations to enhance host colonization. *J Exp Bot.* 2019;70:343-356.
113. Ren Y, Armstrong M, Qi Y, et al. *Phytophthora infestans* RXLR effectors target parallel steps in an immune signal transduction pathway. *Plant Physiol.* 2019;180:2227-2239.
114. Zhang P, Jia Y, Shi J, et al. The WY domain in the Phytophthora effector PSR1 is required for infection and RNA silencing suppression activity. *New Phytol.* 2019;223:839-852.
115. Hou Y, Zhai Y, Feng L, et al. A Phytophthora effector suppresses trans-kingdom RNAi to promote disease susceptibility. *Cell Host Microbe.* 2019;25:153-165.e5.
116. Li H, Wang H, Jing M, et al. A Phytophthora effector recruits a host cytoplasmic transacetylase into nuclear speckles to enhance plant susceptibility. *Elife.* 2018;7:40039.
117. Bozkurt TO, Schornack S, Win J, et al. Phytophthora infestans effector AVRblb2 prevents secretion of a plant immune protease at the haustorial interface. *Proc Natl Acad Sci U S A.* 2011;108:20832-20837.
118. Selin C, De Kievit TR, Belmonte MF, et al. Elucidating the role of effectors in plant-fungal interactions: progress and challenges. *Front Microbiol.* 2016;7:600.
119. Orbach MJ, Farrall L, Sweigard JA, et al. A telomeric avirulence gene determines efficacy for the rice blast resistance gene Pi-ta. *Plant Cell.* 2000;12:2019-2032.
120. Jia Y, Mcadams SA, Bryan GT, et al. Direct interaction of resistance gene and avirulence gene products confers rice blast resistance. *EMBO J.* 2000;19:4004-4014.
121. Dodds PN, Lawrence GJ, Catanzariti AM, et al. The Melampsora lini AvrL567 avirulence genes are expressed in haustoria and their products are recognized inside plant cells. *Plant Cell.* 2004;16:755-768.
122. Dodds PN, Lawrence GJ, Catanzariti AM, et al. Direct protein interaction underlies gene-for-gene specificity and coevolution of the flax resistance genes and flax rust avirulence genes. *Proc Natl Acad Sci U S A.* 2006;103:8888-8893.
123. Wang CI, Guncar G, Forwood JK, et al. Crystal structures of flax rust avirulence proteins AvrL567-A and-D reveal details of the structural basis for flax disease resistance specificity. *Plant Cell.* 2007;19:2898-2912.
124. Wan L, Koeck M, Williams SJ, et al. Structural and functional insights into the modulation of the activity of a flax cytokinin oxidase by flax rust effector AvrL567-A. *Mol Plant Pathol.* 2019;20:211-222.
125. Zhang C, Fang H, Shi X, et al. A fungal effector and a rice NLR protein have antagonistic effects on a Bowman-Birk trypsin inhibitor. *Plant Biotechnol J.* 2020;18:2354-2363. doi:10.1111/pbi.13400
126. Shi X, Long Y, He F, et al. The fungal pathogen *Magnaporthe oryzae* suppresses innate immunity by modulating a host potassium channel. *PLoS Pathog.* 2018;14:e1006878.
127. Tang M, Ning Y, Shu X, et al. The Nup98 homolog APIP12 targeted by the effector AvrPiz-t is involved in rice basal resistance against Magnaporthe oryzae. *Rice.* 2017;10:5.
128. Park CH, Shirsekar G, Bellizzi M, et al. The E3 ligase APIP10 connects the effector AvrPiz-t to the NLR receptor Piz-t in rice. *PLoS Pathog.* 2016;12:e1005529.
129. Park CH, Chen S, Shirsekar G, et al. The Magnaporthe oryzae effector AvrPiz-t targets the RING E3 ubiquitin ligase APIP6 to suppress pathogen-associated molecular pattern-triggered immunity in rice. *Plant Cell.* 2012;24:4748-4762.
130. Zhao S, Shang X, Bi W, et al. Genome-wide identification of effector candidates with conserved motifs from the wheat leaf rust fungus *Puccinia triticina*. *Front Microbiol.* 2020;11:1188.
131. Shao ZQ, Xue JY, Wu P, et al. Large-scale analyses of angiosperm nucleotide-binding site-leucine-rich repeat genes reveal three anciently diverged classes with distinct evolutionary patterns. *Plant Physiol.* 2016;170:2095-2109.
132. Ting JP, Lovering RC, Alnemri ES, et al. The NLR gene family: a standard nomenclature. *Immunity.* 2008;28:285-287.
133. Zhang L, Chen S, Ruan J, et al. Cryo-EM structure of the activated NAIP2-NLRC4 inflammasome reveals nucleated polymerization. *Science.* 2015;350:404-409.
134. Hu Z, Zhou Q, Zhang C, et al. Structural and biochemical basis for induced self-propagation of NLRC4. *Science.* 2015;350:399-404.
135. Wang J, Hu M, Wang J, et al. Reconstitution and structure of a plant NLR resistosome conferring immunity. *Science.* 2019;364:eaav5870.
136. Martin R, Qi T, Zhang H, et al. Structure of the activated ROQ1 resistosome directly recognizing the pathogen effector XopQ. *Science.* 2020;370. doi:10.1126/science.abd9993
137. Wang J, Wang J, Hu M, et al. Ligand-triggered allosteric ADP release primes a plant NLR complex. *Science.* 2019;364:eaav5868.
138. Zhang Y, Goritschnig S, Dong X, et al. A gain-of-function mutation in a plant disease resistance gene leads to constitutive activation of downstream signal transduction pathways in suppressor of npr1-1, constitutive 1. *Plant Cell.* 2003;15:2636-2646.
139. Ade J, Deyoung BJ, Golstein C, et al. Indirect activation of a plant nucleotide binding site-leucine-rich repeat protein by a bacterial protease. *Proc Natl Acad Sci U S A.* 2007;104:2531-2536.
140. Bendahmane A, Farnham G, Moffett P, et al. Constitutive gain-of-function mutants in a nucleotide binding site-leucine rich repeat protein encoded at the Rx locus of potato. *Plant J.* 2002;32:195-204.
141. Tanabe T, Chamaillard M, Ogura Y, et al. Regulatory regions and critical residues of NOD2 involved in muramyl dipeptide recognition. *EMBO J.* 2004;23:1587-1597.
142. Barragan AC, Collenberg M, Wang J, et al. A truncated singleton NLR causes hybrid necrosis in Arabidopsis thaliana. *Mol Biol Evol.* 2021;38:557-574. doi:10.1093/molbev/msaa245
143. Chae E, Bomblies K, Kim ST, et al. Species-wide genetic incompatibility analysis identifies immune genes as hot spots of deleterious epistasis. *Cell.* 2014;159:1341-1351.
144. Alcazar R, Von Reth M, Bautor J, et al. Analysis of a plant complex resistance gene locus underlying immune-related hybrid incompatibility and its occurrence in nature. *PLoS Genet.* 2014;10:e1004848.
145. Smith LM, Bomblies K, Weigel D. Complex evolutionary events at a tandem cluster of Arabidopsis thaliana genes resulting in a single-locus genetic incompatibility. *PLoS Genet.* 2011;7:e1002164.
146. Bomblies K, Lempe J, Epple P, et al. Autoimmune response as a mechanism for a Dobzhansky-Muller-type incompatibility syndrome in plants. *PLoS Biol.* 2007;5:e236.
147. Deng J, Fang L, Zhu X, et al. A CC-NBS-LRR gene induces hybrid lethality in cotton. *J Exp Bot.* 2019;70:5145-5156.
148. Atanasov KE, Liu C, Erban A, et al. NLR mutations suppressing immune hybrid incompatibility and their effects on disease resistance. *Plant Physiol.* 2018;177:1152-1169.
149. Grant M, Brown I, Adams S, et al. The RPM1 plant disease resistance gene facilitates a rapid and sustained increase in cytosolic calcium that is necessary for the oxidative burst and hypersensitive cell death. *Plant J.* 2000;23:441-450.
150. Wan L, Essuman K, Anderson RG, et al. TIR domains of plant immune receptors are NAD(+)-cleaving enzymes that promote cell death. *Science.* 2019;365:799-803.
151. Wu Z, Li M, Dong OX, et al. Differential regulation of TNL-mediated immune signaling by redundant helper CNLs. *New Phytol.* 2019;222:938-953.
152. Castel B, Ngou PM, Cevik V, et al. Diverse NLR immune receptors activate defence via the RPW8-NLR NRG1. *New Phytol.* 2019;222:966-980.
153. Saile SC, Jacob P, Castel B, et al. Two unequally redundant "helper" immune receptor families mediate *Arabidopsis thaliana* intracellular "sensor" immune receptor functions. *PLoS Biol.* 2020;18:e3000783.
154. Barragan CA, Wu R, Kim ST, et al. RPW8/HR repeats control NLR activation in *Arabidopsis thaliana*. *PLoS Genet.* 2019;15:e1008313.
155. Falk A, Feys BJ, Frost LN, et al. EDS1, an essential component of R gene-mediated disease resistance in Arabidopsis has homology to eukaryotic lipases. *Proc Natl Acad Sci U S A.* 1999;96:3292-3297.
156. Jirage D, Tootle TL, Reuber TL, et al. *Arabidopsis thaliana PAD4* encodes a lipase-like gene that is important for salicylic acid signaling. *Proc Natl Acad Sci U S A.* 1999;96:13583-13588.
157. Feys BJ, Moisan LJ, Newman MA, et al. Direct interaction between the Arabidopsis disease resistance signaling proteins, EDS1 and PAD4. *EMBO J.* 2001;20:5400-5411.
158. Deyoung BJ, Qi D, Kim SH, et al. Activation of a plant nucleotide binding-leucine rich repeat disease resistance protein by a modified self protein. *Cell Microbiol.* 2012;14:1071-1084.
159. Van Der Biezen EA, Jones JD. Plant disease-resistance proteins and the gene-for-gene concept. *Trends Biochem Sci.* 1998;23:454-456.
160. Prigozhin DM, Krasileva KV. Analysis of intraspecies diversity reveals a subset of highly variable plant immune receptors and predicts their binding sites. *Plant Cell.* 2021. doi:10.1093/plcell/koab013
161. Shen QH, Saijo Y, Mauch S, et al. Nuclear activity of MLA immune receptors links isolate-specific and basal disease-resistance responses. *Science.* 2007;315:1098-1103.
162. Slootweg E, Roosien J, Spiridon LN, et al. Nucleocytoplasmic distribution is required for activation of resistance by the potato NB-LRR receptor Rx1 and is balanced by its functional domains. *Plant Cell.* 2010;22:4195-4215.
163. Engelhardt S, Boevink P, Armstrong M, et al. Relocalization of late blight resistance protein R3a to endosomal compartments is associated with effector recognition and required for the immune response. *Plant Cell.* 2012;24:5142-5158.
164. Pottinger SE, Innes RW. RPS5-mediated disease resistance: fundamental insights and translational applications. *Annu Rev Phytopathol.* 2020;58:139-160. doi:10.1146/annurev-phyto-010820-012733
165. Sarris PF, Cevik V, Dagdas G, et al. Comparative analysis of plant immune receptor architectures uncovers host proteins likely targeted by pathogens. *BMC Biol.* 2016;14:8.
166. Le Roux C, Huet G, Jauneau A, et al. A receptor pair with an integrated decoy converts pathogen disabling of transcription factors to immunity. *Cell.* 2015;161:1074-1088.
167. Sarris PF, Duxbury Z, Huh SU, et al. A plant immune receptor detects pathogen effectors that target WRKY transcription factors. *Cell.* 2015;161:1089-1100.
168. Williams SJ, Sohn KH, Wan L, et al. Structural basis for assembly and function of a heterodimeric plant immune receptor. *Science.* 2014;344:299-303.
169. Guo H, Ahn HK, Sklenar J, et al. Phosphorylation-regulated activation of the Arabidopsis RRS1-R/RPS4 immune receptor complex reveals two distinct effector recognition mechanisms. *Cell Host Microbe.* 2020;27:769-781.e6.
170. Cesari S, Kanzaki H, Fujiwara T, et al. The NB-LRR proteins RGA4 and RGA5 interact functionally and physically to confer disease resistance. *EMBO J.* 2014;33:1941-1959.
171. Shao F, Golstein C, Ade J, et al. Cleavage of Arabidopsis PBS1 by a bacterial type III effector. *Science.* 2003;301:1230-1233.
172. Kim SH, Qi D, Ashfield T, et al. Using decoys to expand the recognition specificity of a plant disease resistance protein. *Science.* 2016;351:684-687.
173. Pottinger SE, Bak A, Margets A, et al. Optimizing the PBS1 decoy system to confer resistance to potyvirus Infection in Arabidopsis and soybean. *Mol Plant Microbe Interact.* 2020;33:932-944.
174. Caldwell KS, Michelmore RW. *Arabidopsis thaliana* genes encoding defense signaling and recognition proteins exhibit contrasting evolutionary dynamics. *Genetics.* 2009;181:671-684.
175. Helm M, Qi M, Sarkar S, et al. Engineering a decoy substrate in soybean to enable recognition of the soybean mosaic virus NIa protease. *Mol Plant Microbe Interact.* 2019;32:760-769.
176. Carter ME, Helm M, Chapman AVE, et al. Convergent evolution of effector protease recognition by Arabidopsis and barley. *Mol Plant Microbe Interact.* 2019;32:550-565.
177. Ng DW, Abeysinghe JK, Kamali M. Regulating the regulators: the control of transcription factors in plant defense signaling. *Int J Mol Sci.* 2018;19.
178. Jacob F, Kracher B, Mine A, et al. A dominant-interfering *camta3* mutation compromises primary transcriptional outputs mediated by both cell surface and intracellular immune receptors in *Arabidopsis thaliana*. *New Phytol.* 2018;217:1667-1680.
179. Wang L, Tsuda K, Sato M, et al. Arabidopsis CaM binding protein CBP60g contributes to MAMP-induced SA accumulation and is involved in disease resistance against *Pseudomonas syringae*. *PLoS Pathog.* 2009;5:e1000301.

180. Zhang Y, Xu S, Ding P, et al. Control of salicylic acid synthesis and systemic acquired resistance by two members of a plant-specific family of transcription factors. *Proc Natl Acad Sci U S A*. 2010;107:18220-18225.
181. Wang L, Tsuda K, Truman W, et al. CBP60g and SARD1 play partially redundant critical roles in salicylic acid signaling. *Plant J*. 2011;67:1029-1041.
182. Sun T, Busta L, Zhang Q, et al. TGACG-BINDING FACTOR 1 (TGA1) and TGA4 regulate salicylic acid and pipecolic acid biosynthesis by modulating the expression of SYSTEMIC ACQUIRED RESISTANCE DEFICIENT 1 (SARD1) and CALMODULIN-BINDING PROTEIN 60g (CBP60g). *New Phytol*. 2018;217:344-354.
183. Sun T, Zhang Y, Li Y, et al. ChIP-seq reveals broad roles of SARD1 and CBP60g in regulating plant immunity. *Nat Commun*. 2015;6:10159.
184. White RF. Acetylsalicylic acid (aspirin) induces resistance to tobacco mosaic virus in tobacco. *Virology*. 1979;99:410-412.
185. Malamy J, Carr JP, Klessig DF, et al. Salicylic acid: a likely endogenous signal in the resistance response of tobacco to viral infection. *Science*. 1990;250:1002-1004.
186. Metraux JP, Signer H, Ryals J, et al. Increase in salicylic acid at the onset of systemic acquired resistance in cucumber. *Science*. 1990;250:1004-1006.
187. Gaffney T, Friedrich L, Vernooij B, et al. Requirement of salicylic acid for the induction of systemic acquired resistance. *Science*. 1993;261:754-756.
188. Delaney TP, Uknes S, Vernooij B, et al. A central role of salicylic acid in plant disease resistance. *Science*. 1994;266:1247-1250.
189. Cao H, Bowling SA, Gordon AS, et al. Characterization of an Arabidopsis mutant that is nonresponsive to inducers of systemic acquired resistance. *Plant Cell*. 1994;6:1583-1592.
190. Delaney TP, Friedrich L, Ryals JA. Arabidopsis signal transduction mutant defective in chemically and biologically induced disease resistance. *Proc Natl Acad Sci U S A*. 1995;92:6602-6606.
191. Ding Y, Sun T, Ao K, et al. Opposite roles of salicylic acid receptors NPR1 and NPR3/NPR4 in transcriptional regulation of plant immunity. *Cell*. 2018;173:1454-1467 e1415.
192. Wu Y, Zhang D, Chu JY, et al. The Arabidopsis NPR1 protein is a receptor for the plant defense hormone salicylic acid. *Cell Rep*. 2012;1:639-647.
193. Zhang Y, Fan W, Kinkema M, et al. Interaction of NPR1 with basic leucine zipper protein transcription factors that bind sequences required for salicylic acid induction of the PR-1 gene. *Proc Natl Acad Sci U S A*. 1999;96:6523-6528.
194. Despres C, Delong C, Glaze S, et al. The Arabidopsis NPR1/NIM1 protein enhances the DNA binding activity of a subgroup of the TGA family of bZIP transcription factors. *Plant Cell*. 2000;12:279-290.
195. Zhang Y, Tessaro MJ, Lassner M, et al. Knockout analysis of Arabidopsis transcription factors TGA2, TGA5, and TGA6 reveals their redundant and essential roles in systemic acquired resistance. *Plant Cell*. 2003;15:2647-2653.
196. Liu Y, Sun T, Sun Y, et al. Diverse roles of salicylic acid receptors NPR1 and NPR3/NPR4 in plant immunity. *Plant Cell*. 2020;32:4002-4016.
197. Hartmann M, Zeier T, Bernsdorff F, et al. Flavin monooxygenase-generated N-hydroxypipecolic acid is a critical element of plant systemic immunity. *Cell*. 2018;173:456-469 e416.
198. Wildermuth MC, Dewdney J, Wu G, et al. Isochorismate synthase is required to synthesize salicylic acid for plant defence. *Nature*. 2001;414:562-565.
199. Lawton K, Weymann K, Friedrich L, et al. Systemic acquired resistance in Arabidopsis requires salicylic acid but not ethylene. *Mol Plant Microbe Interact*. 1995;8:863-870.
200. Chen YC, Holmes EC, Rajniak J, et al. N-hydroxy-pipecolic acid is a mobile metabolite that induces systemic disease resistance in Arabidopsis. *Proc Natl Acad Sci U S A*. 2018;115:E4920-E4929.
201. Lolle S, Greeff C, Petersen K, et al. Matching NLR immune receptors to autoimmunity in *camta3* mutants using antimorphic NLR alleles. *Cell Host Microbe*. 2017;21:518-529.e4.
202. Sun T, Huang J, Xu Y, et al. Redundant CAMTA transcription factors negatively regulate the biosynthesis of salicylic acid and N-hydroxypipecolic acid by modulating the expression of SARD1 and CBP60g. *Mol Plant*. 2020;13:144-156.
203. Napoli C, Lemieux C, Jorgensen R. Introduction of a chimeric chalcone synthase gene into petunia results in reversible co-suppression of homologous genes in trans. *Plant Cell*. 1990;2:279-289.
204. Waterhouse PM, Graham MW, Wang MB. Virus resistance and gene silencing in plants can be induced by simultaneous expression of sense and antisense RNA. *Proc Natl Acad Sci U S A*. 1998;95:13959-13964.
205. Fire A, Xu S, Montgomery MK, et al. Potent and specific genetic interference by double-stranded RNA in Caenorhabditis elegans. *Nature*. 1998;391:806-811.
206. Hamilton AJ, Baulcombe DC. A species of small antisense RNA in posttranscriptional gene silencing in plants. *Science*. 1999;286:950-952.
207. Ratcliff F, Harrison BD, Baulcombe DC. A similarity between viral defense and gene silencing in plants. *Science*. 1997;276:1558-1560.
208. Kumagai MH, Donson J, Della-Cioppa G, et al. Cytoplasmic inhibition of carotenoid biosynthesis with virus-derived RNA. *Proc Natl Acad Sci U S A*. 1995;92:1679-1683.
209. Rosa C, Kuo YW, Wuriyanghan H, et al. RNA interference mechanisms and applications in plant pathology. *Annu Rev Phytopathol*. 2018;56:581-610.
210. Borges F, Martienssen RA. The expanding world of small RNAs in plants. *Nat Rev Mol Cell Biol*. 2015;16:727-741.
211. D'ario M, Griffiths-Jones S, Kim M. Small RNAs: big impact on plant development. *Trends Plant Sci*. 2017;22:1056-1068.
212. Csorba T, Kontra L, Burgyan J. Viral silencing suppressors: tools forged to fine-tune host-pathogen coexistence. *Virology*. 2015;479-480:85-103.
213. Silhavy D, Molnar A, Lucioli A, et al. A viral protein suppresses RNA silencing and binds silencing-generated, 21-to 25-nucleotide double-stranded RNAs. *EMBO J*. 2002;21:3070-3080.
214. Duan CG, Fang YY, Zhou BJ, et al. Suppression of Arabidopsis ARGONAUTE1-mediated slicing, transgene-induced RNA silencing, and DNA methylation by distinct domains of the Cucumber mosaic virus 2b protein. *Plant Cell*. 2012;24:259-274.
215. Lozsa R, Csorba T, Lakatos L, et al. Inhibition of 3' modification of small RNAs in virus-infected plants require spatial and temporal co-expression of small RNAs and viral silencing-suppressor proteins. *Nucleic Acids Res*. 2008;36:4099-4107.
216. Csorba T, Lozsa R, Hutvagner G, et al. Polerovirus protein P0 prevents the assembly of small RNA-containing RISC complexes and leads to degradation of ARGONAUTE1. *Plant J*. 2010;62:463-472.
217. Glick E, Zrachya A, Levy Y, et al. Interaction with host SGS3 is required for suppression of RNA silencing by tomato yellow leaf curl virus V2 protein. *Proc Natl Acad Sci U S A*. 2008;105:157-161.
218. Zhao JH, Hua CL, Fang YY, et al. The dual edge of RNA silencing suppressors in the virus-host interactions. *Curr Opin Virol*. 2016;17:39-44.
219. Navarro L, Jay F, Nomura K, et al. Suppression of the microRNA pathway by bacterial effector proteins. *Science*. 2008;321:964-967.
220. Qiao Y, Shi J, Zhai Y, et al. Phytophthora effector targets a novel component of small RNA pathway in plants to promote infection. *Proc Natl Acad Sci U S A*. 2015;112:5850-5855.
221. Weiberg A, Wang M, Lin FM, et al. Fungal small RNAs suppress plant immunity by hijacking host RNA interference pathways. *Science*. 2013;342:118-123.
222. Cai Q, Qiao L, Wang M, et al. Plants send small RNAs in extracellular vesicles to fungal pathogen to silence virulence genes. *Science*. 2018;360:1126-1129.
223. Nowara D, Gay A, Lacomme C, et al. HIGS: host-induced gene silencing in the obligate biotrophic fungal pathogen Blumeria graminis. *Plant Cell*. 2010;22:3130-3141.
224. Koch A, Kumar N, Weber L, et al. Host-induced gene silencing of cytochrome P450 lanosterol C14a-demethylase-encoding genes confers strong resistance to *Fusarium* species. *Proc Natl Acad Sci U S A*. 2013;110:19324-19329.
225. Rutter BD, Innes RW. Extracellular vesicles isolated from the leaf apoplast carry stress-response proteins. *Plant Physiol*. 2017;173:728-741.
226. Baldrich P, Rutter BD, Karimi HZ, et al. Plant extracellular vesicles contain diverse small RNA species and are enriched in 10- to 17-Nucleotide "Tiny" RNAs. *Plant Cell*. 2019;31:315-324.
227. Rutter BD, Innes RW. Growing pains: addressing the pitfalls of plant extracellular vesicle research. *New Phytol*. 2020;228:1505-1510. doi:10.1111/nph.16725

SECTION III

Cellular Pathways and Processes in Immunity

CHAPTER 6

Immunometabolism

Jonathan D. Powell, MD, PhD • Chirag H. Patel, PhD • Im-Hong Sun, PhD

INTRODUCTION

Over the last 20 years, immunologists have recognized the extraordinary proliferative capacity and biosynthetic function of the immune response, which necessitates specialized metabolic reprogramming to facilitate such demanding tasks. Appreciation and strong interest of these early critical observations established the growing field of "immunometabolism" as it has become clear that this metabolic reprogramming is an integral component of immune cell activation, differentiation, and function. Indeed, immune cells do not become activated *and then* reprogram their metabolism to support these highly active processes. Rather, metabolic reprogramming is indistinct from the activation process.[1] Upon stimulation, immune cells activate critical nutrient-sensing pathways, such as mammalian/mechanistic target of rapamycin (mTOR), that simultaneously regulate differentiation and the metabolic pathways necessary to carry out their function. Likewise, metabolic transcription factors such as HIF-1 (hypoxia inducible factor) regulate the genes required for the proper metabolic reprogramming to support expansion and function but also directly mediate transcription of genes essential for immune function such as cytokines.[2-4] It is now evident that metabolites themselves, the products generated from elevated metabolic activity, not only play a central role in fueling anabolic processes but also directly regulate cellular differentiation and function at the molecular level.[5,6] This intimate relationship between metabolic reprogramming and immune cell differentiation and function has revealed that phenotypically diverse subsets of immune cells can be distinguished by their differential metabolic properties.[7] Consequently, it has been observed that targeting select metabolic pathways can differentially regulate specific immune cells and thus represents a potentially powerful means of developing novel immune-regulating therapeutics.[8] This inaugural chapter will provide an overview of our current understanding of essential metabolic processes that influence immune cell differentiation and function (Fig. 6.1).

THE WARBURG EFFECT

The vast majority of nonproliferating somatic cells employ glycolysis to fuel oxidative phosphorylation (OXPHOS) to generate energy in the form of adenosine triphosphate (ATP) in the presence of oxygen. That is, very diverse cell types will engage canonical metabolic pathways that are classically described in biochemical texts.[9] In 1924, Otto Warburg observed that cancer cells did not appear to follow these basic rules seen in normal cells.[10] Rather, Warburg discovered that cancer cells even in the presence of sufficient oxygen levels primarily utilize glycolysis as their main source of energy generation. At the time, it was thought that this unusual type of metabolism reflected dysfunctional mitochondria in cancer cells. Subsequently, it was proposed that this observation, increased lactate-forming glycolysis even in the presence of oxygen known as the Warburg effect or lactate-forming aerobic glycolysis (often referred to as "aerobic glycolysis"), facilitated extensive cellular growth and proliferation required for cancer progression.[11] As seen in Figure 6.1, robust aerobic glycolysis, while perhaps an inefficient means of generating ATP, relies on the conversion of pyruvate to lactate by lactate dehydrogenase (LDH) to regenerate sufficient NAD^+ levels required for glyceraldehyde 3-phosphate dehydrogenase (GAPDH) activity and ensure glycolysis remains uninterrupted. Concomitantly, rather than employing glucose to primarily fuel the more energy efficient tricarboxylic acid (TCA) cycle and oxidative phosphorylation (OXPHOS) to generate more ATP, the substrates from the glycolytic pathway are siphoned off in order to provide critical biosynthetic substrates for the prodigious anabolic demands of cancer cell growth. Over the years, the Warburg effect has not only provided a valuable foundation for the study of cancer metabolism

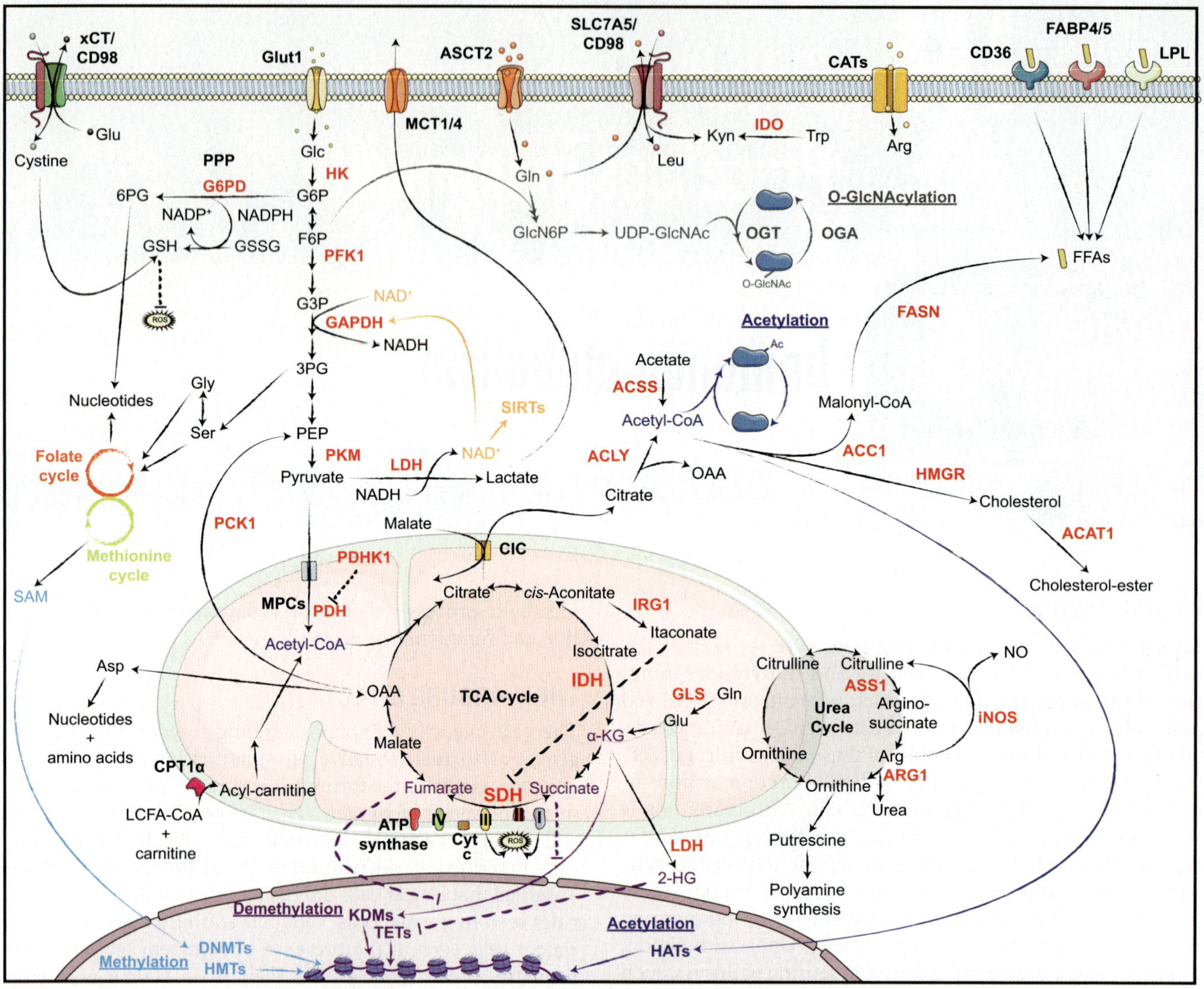

FIG. 6.1. Metabolic programs regulate immune activation, differentiation, and function. Major metabolic pathways described in this chapter are presented here. In brief, **glycolysis** converts glucose into pyruvate, which can then either be converted into lactate or enter into the mitochondria to fuel the tricarboxylic acid (**TCA**) cycle for oxidative phosphorylation **(OXPHOS)**. Glycolytic intermediates can also be diverted into the pentose phosphate pathway (**PPP**) for nucleotide synthesis and to maintain redox balance; the **hexosamine** pathway to generate substrates for protein glycosylation; the **one-carbon metabolism** pathway to synthesize nucleotides. An active TCA cycle generates ATP through the electron transport chain (ETC) and important metabolic substrates that can be utilized for biosynthetic processes or facilitating epigenetic modifications. Additional fuels can sustain the TCA cycle and OXPHOS. **Fatty acid oxidation (FAO)** results in breakdown of fatty acids to generate acetyl-CoA, while **glutaminolysis** converts glutamine to glutamate and then α-ketoglutarate. The TCA cycle intermediate citrate can be exported into the cytoplasm to generate sufficient levels of acetyl-CoA for **fatty acid synthesis (FAS)** or provide substrates for acetylation. The expression of several cell surface or mitochondrial transporters and metabolic enzymes play a critical role in regulating the flux of metabolites to sustain metabolic activity. 2-HG, 2-hydroxyglutarate; 3PG, glycerate 3-phosphate; 6PG, 6-P-gluconate; α-KG, α-ketoglutarate; Ac, Acetylation; ACAT1, acetyl-CoA acetyltransferase; ACC1, acetyl-CoA carboxylase 1; acetyl-CoA, acetyl coenzyme A; ACSS, acetyl-CoA synthetase; Arg, arginine; Arg1, arginase 1; ASCT2, sodium dependent neutral amino acid transporter type 2 (SLC1A5); Asp, aspartate; ASS1, argininosuccinate synthase 1; CATs, cationic amino acid transporters; CIC, mitochondrial citrate carrier or SLC25A1; CPT1α, carnitine palmitoyltransferase 1A; Cyt C, cytochrome C; DNMT, DNA methyltransferase; F6P, fructose 6-phosphate; FABP, fatty acid binding protein; FASN, fatty acid synthase; FFAs, free fatty acids; G3P, glyceraldehyde 3-phosphate; G6P, glucose 6-phosphate; G6PD, glucose-6-phosphate dehydrogenase; GAPDH, glyceraldehyde 3-phosphate dehydrogenase; Glc, glucose; GlcN6P, glucosamine-6-phosphate; Gln, glutamine; GLS, glutaminase; Glu, glutamate; Glut1, glucose transporter 1 or SLC2A1; Gly, glycine; GSH, glutathione; GSSG, glutathione disulfide; HATs, histone acetyltransferases; HK, hexokinase; HMGR, HMG-CoA reductase; HMT, histone methyltransferase; IDH, isocitrate dehydrogenase; IDO, indoleamine 2,3-dioxygenase; iNOS, nitric oxide synthase; IRG1, immune responsive gene 1 or ACOD1; KDM, lysine demethylases; Kyn, kynurenine; LCFA-CoA, long-chain fatty acid-CoA; LDH, lactate dehydrogenase; Leu, leucine; LPL, lipoprotein lipase; MCT1/4, monocarboxylate transporter 1/4; MPCs, mitochondrial pyruvate carriers; NAD, nicotinamide adenine dinucleotide; NADH, reduced nicotinamide adenine dinucleotide; NO, nitric oxide; OAA, oxaloacetate; OGA, protein O-GlcNAcase; OGT, UDP-*N*-acetylglucosamine–peptide *N*-acetylglucosaminyltransferase; PCK1, phosphoenolpyruvate carboxykinase 1; PDH, pyruvate dehydrogenase; PDHK1, pyruvate dehydrogenase kinase isoform 1; PEP, phosphoenolpyruvate; PFK1, phosphofructokinase 1; PKM, pyruvate kinase; PPP, Pentose Phosphate Pathway; ROS, reactive oxygen species; SAM, S-adenosylmethionine; SDH, succinate dehydrogenase; Ser, serine; SIRT, sirtuin; TETs, ten–eleven translocation family members; Trp, tryptophan; UDP-GlcNAc, uridine diphosphate *N*-acetylglucosamine; xCT, cystine/glutamate transporter.

but also to the field of immunometabolism, though emerging theories suggest the Warburg effect might not be as straightforward.[12,13] However, while an exaggerated reliance on glycolysis even in the presence of oxygen is certainly one distinct property of cancer cells, it is now clear that: (1) this does not preclude the importance of OXPHOS in cancer cell growth and (2) within tumors, there exists (regional) heterogeneity in terms of the metabolic programming of cancer cells.[14] That said, this concept of exaggerated lactate-forming glycolysis (the Warburg effect) has been applied to not just cancer cells but highly active and proliferating immune cells.[15,16] As will be delineated throughout this chapter, this increased glycolytic state seen in numerous immune cells upon stimulation plays an essential role not only in providing sufficient energy and supporting anabolic processes for biosynthetic materials but also facilitating the immune activation and differentiation process itself.

SIGNAL 1 + 2: CONNECTING THE IMMUNE RESPONSE TO METABOLISM

In as much as the adaptive immune response is a consequence of the stochastic generation of post–germline-encoded receptors, a regulatory mechanism must be in place to dictate the outcome of immune recognition with regard to T cell activation versus tolerance. The two-signal model provides a paradigm for distinguishing between "dangerous" antigens for which recognition should result in T cell activation and innocuous (self-antigens) for which recognition should result in tolerance.[17,18] In such a model, Signal 1 refers to TCR-antigen recognition on MHC molecules, while Signal 2 refers to costimulation provided by antigen-presenting cells (Chapter 29).[19] Antigen recognition in the absence of costimulation leads to a state of tolerance in the form of activation-induced cell death or anergy, while recognition in the presence of costimulation leads to full T cell activation. As such, the presence of costimulation is associated with rapid proliferation and the sufficient production of cytokines (ie, IL-2). In 2002, Thompson and colleagues made the insightful observation that metabolic reprogramming was tightly linked to this two-signal paradigm.[1] They observed that the significant surface upregulation of the glucose transporter, Glut1 (SLC2A1), and subsequently elevated glucose uptake critical to fuel glycolysis were not induced by TCR engagement alone (Signal 1) but rather required the addition of costimulation (Signal 1 + 2) (Fig. 6.2). Mechanistically, this critical step in metabolic reprogramming was attributed to the ability of costimulation to not only activate but also prolong phosphoinositide 3-kinase (PI3K)-Akt signaling given its role in the synthesis

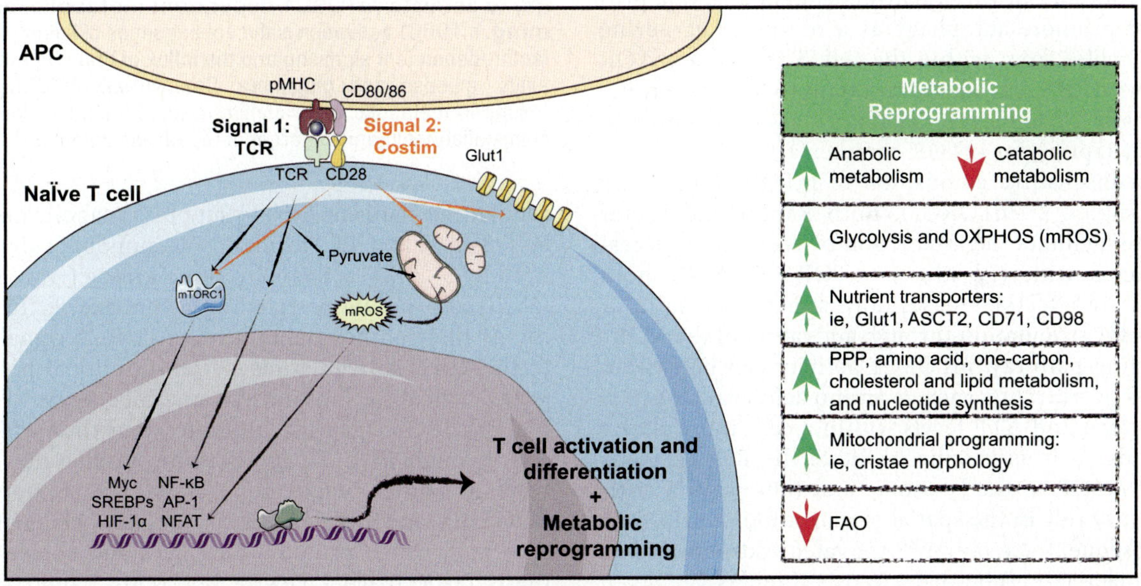

FIG. 6.2. The "two-signal" hypothesis and metabolic reprogramming. The "two-signal" model provides the basic paradigm for understanding how the adaptive immune response initiates tolerogenic versus active immune responses. For T cells, canonical signaling downstream of TCR-antigen recognition (Signal 1) and costimulation (Signal 2) results in full activation as exemplified by cytokine production and effector differentiation. However, a critical component of this response is sustained activation of mTORC1 signaling resulting in expression of essential metabolic transcription factors (ie, Myc, SREBPs) needed for metabolic reprogramming. Signals derived from CD28-PI3K-Akt directly controls Glut1 expression to enhance glycolysis and thus facilitate increased anabolic metabolism while shutting down catabolic metabolism. AP-1, activator protein-1; ASCT2, sodium-dependent neutral amino acid transporter type 2 (SLC1A5); FAO, fatty acid oxidation; Glut1, glucose transporter 1 or SLC2A1; HIF-1α, hypoxic inducible factor 1; mROS, mitochondrial reactive oxygen species; mTORC1, mammalian/mechanistic target of rapamycin complex-1; NFAT, nuclear factor of activated T cells; NF-κB, nuclear factor kappa B; pMHC, peptide-major histocompatibility complex; PPP, Pentose Phosphate Pathway; SREBPs, sterol regulatory element-binding proteins; TCR, T cell receptor.

and surface trafficking of glucose transporters.[20] The ability of costimulation to promote full T cell activation and prevent anergy has been tightly linked to sustained mTOR signaling (Chapter 25). Indeed, costimulation in the absence of mTOR stimulation leads to anergy and anergic T cells fail to upregulate the metabolic machinery necessary for full T cell activation.[21,22] More recently, it has been demonstrated that costimulation promotes early mitochondrial priming that instills T cells with "fitter" mitochondria cristae organization as T cells undergo differentiation.[23] Overall, these studies mechanistically connect the two-signal paradigm with early metabolic reprogramming. These studies also suggest that one mechanism by which T cell tolerance is enforced is through the inability of T cells to metabolically reprogram.

mTOR: INTEGRATING ENVIRONMENTAL CUES TO COORDINATE GROWTH AND METABOLIC REPROGRAMMING

As a brief overview, the mTOR is a conserved serine/threonine kinase that senses both intra- and extracellular nutrient availability to regulate anabolic processes. Evolutionarily, from yeast to mammalian cells, mTOR activation promotes growth and proliferation by upregulating expression of proteins involved in anabolic metabolism.[24] Alternatively, nutrient- and energy-depleted environments shut down mTOR-dependent anabolic processes and promote autophagy as a means of garnering macromolecules from within the cell. The kinase mTOR interacts with different adaptor proteins to form two functionally distinct complexes: Regulatory-associated protein of mTOR (Raptor) for mTOR complex 1 (mTORC1) and rapamycin-insensitive companion of mTOR (Rictor) for mTOR complex 2 (mTORC2). Both Raptor and Rictor function to facilitate the interaction between other accessory proteins with the kinase mTOR.[24-27] While both complexes are effector kinases downstream of insulin or growth factor stimulation through activation of the PI3K-Akt signaling pathway, mTORC1 activity can be further modulated by nutrients such as amino acid availability as well as oxygen and ATP levels within a cell. Intracellular amino acids such as leucine facilitate the Rag GTPases to recruit mTOR to the lysosome and thus plays a critical regulatory role in the spatial organization of mTORC1 for its subsequent activation.[28] Activated Akt phosphorylates and inactivates tuberous sclerosis complex 2 (TSC2), which is part of the tuberous sclerosis complex formed with TSC1. Together, they act as a GTPase-activating protein (GAP) that negatively regulates the GTPase Ras homolog enriched in brain (Rheb). Inactivation of the TSC complex allows Rheb to promote mTORC1 activity at the lysosomal surface.

The intricate signaling cascade that leads to full activation of mTORC1 pathway has been discussed extensively in several reviews.[24-27,29] Active mTORC1 signaling regulates cellular growth and proliferation by promoting anabolic processes while simultaneously inhibiting catabolic processes. However, in a stressful low-energy state,

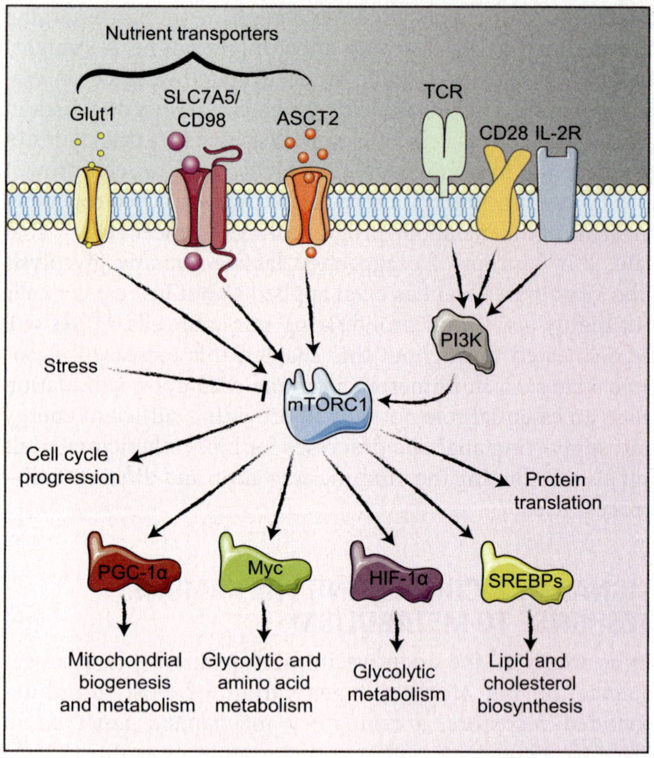

FIG. 6.3. Integrating immunologic and metabolic signaling to activate mTORC1-dependent metabolic reprogramming. mTORC1 activation is the consequence of ligand- and growth factor–dependent signaling and the influx of branched-chain amino acids, glucose and glutamine. Downstream mTORC1 signaling results in metabolic programming critical for activation. mTORC1, mammalian/mechanistic target of rapamycin complex-1.

cellular mechanisms exist to suppress anabolic metabolism by sensing a low ATP to adenosine monophosphate (AMP) ratio by 5′ AMP-activated protein kinase (AMPK), which inactivates mTORC1 signaling by enhancing TSC2 activity. At the transcriptional level, mTORC1 signaling leads to the direct upregulation of transcriptional programs of genes critical for biosynthesis and bioenergy.[30] For example, mTORC1 signaling promotes mitochondrial biogenesis through PGC-1α (peroxisome proliferator activated receptor gamma coactivator 1-alpha) and the upregulation of SREBP1 and SREBP2 (sterol regulatory element-binding proteins), which promote fatty acid and cholesterol synthesis, respectively. Likewise, Myc (cMyc) and HIF-1α, two transcription factors that are critical for metabolic reprogramming, are regulated by mTOR in a translational manner[24] (Fig. 6.3).

Furthermore, a critical mechanism by which mTORC1 regulates protein expression is by promoting translation through its direct regulation of eIF4E-binding proteins (4EBPs). 4EBPs are translation repressor proteins that normally bind to eukaryotic initiation factor 4E (eIF4E), a major component of the eIF4F complex. mTORC1-mediated hyperphosphorylation of 4EBPs triggers dissociation from eIF4E and allows eIF4E to interact with eIF4G1 and, together with eIF4A1 (an RNA helicase), form the heterotrimetric translation initiation complex eIF4F that regulates a specialized subclass of mRNAs for protein translation, notably the 5′ terminal oligopyrimidine motif (5′ TOP) at the 5′ untranslated region of mRNA (5′ UTR).[31] Many of the proteins that result from translation of these mRNA classes are important in promoting ribosomal and mitochondrial biogenesis, metabolic programs such as glycolysis, apoptosis, cell cycle control, and proliferation. S6K is a serine/threonine kinase that is part of the AGC kinase family and plays a critical role in regulating protein synthesis. Thus, active mTORC1 phosphorylates S6K and promotes its kinase activity. Activated S6K further phosphorylates three different proteins that ultimately allow for translation: eIF4B, which activates eIF4F; ribosomal protein S6 (S6), which is a component of 40S ribosomal subunit; and programmed cell death 4 (PDCD4), which is normally an inhibitor of translation initiation that binds to eIF4A1 and inhibits its RNA binding.

The upstream signaling that leads to activation of mTORC2 has not been as precisely defined; however, activation of PI3K through insulin receptor signaling activates mTORC2. Activated mTORC2 phosphorylates Akt, serum and glucocorticoid-regulated kinase 1 (SGK1), and protein kinase C (PKC).[32,33] Both Akt and SGK1 phosphorylate and prevent the nuclear translocation of forkhead box transcription factors of the class 1 and 3a (Foxo1/3a), thus preventing their transcriptional activity. Foxo family proteins have been shown to be important in regulating cell cycle control, apoptosis, and metabolism, and especially important in stem cell and T cell homeostasis.[34,35] As another downstream substrate of mTORC2, PKC family kinases regulate multiple processes related to cytoskeletal remodeling and cell migration.[24,27]

mTOR REGULATES ACTIVATION, DIFFERENTIATION, AND FUNCTION

Rapamycin, a product of *Streptomyces hygroscopicus*, originally isolated from soil samples taken from the Island Rapa Nui (Easter Island), was initially developed as an antibiotic and eventually as an anticancer compound through its strong ability to inhibit cell growth.[36,37] The mechanism of action of rapamycin was attributed to its ability to bind to FKBP12 and inhibit an evolutionarily conserved serine threonine kinase, which was aptly named the "target of rapamycin (TOR)". Based on its potent ability to arrest cell cycle and to inhibit cellular proliferation, rapamycin was also developed as an immunosuppressive agent to prevent strong effector T cell responses.[38-40] Along these lines, it was found that rapamycin could promote CD4$^+$ T cell anergy even in the presence of costimulation.[41,42] Mechanistically, it was initially thought that anergy was induced as a consequence of TCR engagement and rapamycin's ability to halt proliferation.[43] However, subsequent studies revealed that mTOR activity independent of proliferation was responsible for activating T cells and preventing anergy.[21] Anergic CD4$^+$ T cells are characterized by a marked decrease in IL-2 production and proliferation but also fail to upregulate the metabolic machinery required to support full T cell activation. This observation demonstrated that one mechanism of maintaining T cell anergy is by inhibiting proper metabolic reprogramming.[22]

When mTOR is genetically deleted from naïve CD4$^+$ T cells, they fail to become Th1, Th2, and Th17 cells when activated under appropriate polarizing conditions.[44] Rather, antigen recognition in the absence of mTOR leads to the enhanced generation of regulatory T cells (T_{reg}). These findings are consistent with the ability of rapamycin to promote the generation of T_{regs}.[45,46] However, while mTOR inhibition promotes the generation of T_{regs}, it has become clear that effector T_{reg} (eT_{reg}) cells require mTOR activity for full suppressive activity in vivo.[47,48] In an analogous fashion, it has been shown that knocking down mTOR or Raptor prevents Th1 differentiation but promotes T follicular helper cell (Tfh) differentiation.[49] Furthermore, the nutrient-sensing kinase, AMPK, which regulates mTOR activity, is critical for CD4$^+$ T cells to adapt under nutrient-limiting conditions. As such, AMPK plays an essential role in allowing CD4$^+$ T cells to become metabolically flexible and in promoting Th1 and Th17 effector responses especially during acute immune responses when nutrients may be limiting.[50]

mTOR signaling also plays a critical role in regulating CD8$^+$ T cell differentiation and function. While rapamycin was originally developed as an immunosuppressive agent, it has become clear that even though rapamycin can blunt clonal expansion, it can also promote the generation of more long-lived memory CD8$^+$ T cells.[51] Genetic studies have clarified the role of mTOR in CD8$^+$ T cells. While mTORC1 activity is critical in promoting effector T cell generation and function, mTORC2 activity prevents the formation of memory CD8$^+$ T cells.[52,53] CD8$^+$ T cells lacking Rheb (and hence mTORC1 activity) fail to become effector CD8$^+$ T cells, while CD8$^+$ T cells lacking either TSC1 or TSC2 (thus having constitutively enhanced mTORC1 activity) demonstrate potent effector function but fail to become memory CD8$^+$ T cells.[52,54] Alternatively, Rictor-deficient CD8$^+$ T cells (lacking mTORC2 activity) readily differentiate into memory CD8$^+$ T cells. Importantly, the enhanced effector CD8$^+$ T cell differentiation observed in TSC2-deficient CD8$^+$ T cells and the enhanced memory generation observed in the Rictor-deficient CD8$^+$ T cells are tightly associated with metabolic reprogramming.[52,53] To this end, it has been further demonstrated that the asymmetric division of naïve CD8$^+$ T cells, which results in one daughter cell destined to become an effector CD8$^+$ T cell and one daughter cell destined to become a memory CD8$^+$ T cell, is a consequence of the asymmetric partitioning of mTORC1 activity.[55,56] That is, the daughter CD8$^+$ T cell destined to become the effector CD8$^+$ T cell has increased mTORC1 activity when compared to the daughter CD8$^+$ T cell destined to become the memory CD8$^+$ T cell with both daughter cells displaying individualized metabolic programs.

mTOR has also been shown to play an important role in integrating cues from the immune microenvironment for innate immune cells. mTORC1 has been implicated in promoting activation, differentiation, and function of natural killer (NK) cells through its ability to promote metabolic reprogramming.[57] In terms of myeloid cells, the role of mTOR signaling is as diverse as the nuanced differentiation of these cells in vivo. Early studies demonstrated that somewhat surprisingly, activation of mTORC1 in monocytes and dendritic cells led to a decrease in proinflammatory cytokines such as IL-12 with a concomitant increase in more suppressive cytokines such as IL-10.[58-60] However, treatment of monocytes and dendritic cells with rapamycin leads to an increase in proinflammatory cytokines. That said, mTORC1 activity was shown to be necessary for IRF7-induced type 1 interferon (IFN) production by plasmacytoid dendritic cells.[61] The diverse role of mTOR in regulating myeloid cell differentiation and function is further demonstrated by observations that mTORC2 signaling promotes M2 macrophage function both in vitro and in vivo, while selective inhibition of mTORC2 signaling promotes the generation of GATA6$^+$ peritoneal-resident macrophages.[62-64]

THE METABOLISM OF T LYMPHOCYTES

T Cell Development and Metabolism

T cell development in the thymus involves a series of quiescent (DN1-DN3, DP) and highly anabolic (DN4, ISP) metabolic states before ultimately leading to the selection of mature T cells (Chapter 28).[65] To support the highly active growth and proliferation seen in immature T cells, proper metabolic reprogramming must take place through the different developmental stages. As such, survival of pre-T cells during beta-selection in the DN3 stage is mediated by the Notch-PI3K-Akt signaling axis, which drives T cells out of quiescence and promotes expression of nutrient transporters such as CD71 (iron uptake) and CD98 (amino acid uptake) and also Glut1 to increase glucose uptake and glycolysis.[66,67] Likewise, signaling pathways such as mTOR and the metabolic checkpoint kinase, LKB1 (liver kinase B1), contribute to thymocyte development in part by regulating metabolic programs.[68-70] In particular, DN4 thymocytes engage in rapid proliferation to transition into DP thymocytes. Not surprisingly, glucose transport by increased Glut1 expression and glycolytic activity are markedly elevated at this stage of thymocyte development.[65,71] In the absence of Myc or Glut1, mice have reduced overall number of thymocytes at the DP stage, leading to a reduction in SP CD4$^+$ T cells thus demonstrating the essential need for elevated metabolism during development.[65,72] Interestingly, once SP mature T cells exit the thymus into the periphery, both cellular and metabolic activity are markedly dampened as quiescent naïve T cells. Also, mitochondrial metabolism during T cell development is critical since deletion of mitochondrial pyruvate carrier 1 (MPC1), which reduces pyruvate transport into the mitochondria, results in reduced peripheral naïve T cells.[73] All of these studies suggest that metabolism is critical for proper T cell development and maturation into SP T cells.

Basics of T Cell Activation

Naïve or resting T cells display minimal cellular and metabolic activity as they reside in a quiescent G0 state.[74-76] These T cells rely on catabolic metabolism as they breakdown amino acids, fatty acids through fatty acid oxidation (FAO), and glucose to generate intermediates that fuel the TCA cycle and subsequently OXPHOS through the electron transport chain (ETC) to generate a stable pool of ATP.[76] Homeostatic prosurvival growth factors such as IL-7 or tonic TCR signaling are essential to maintain this low-level catabolic state by regulating glucose import and overall metabolic activity.[77-79] Therefore, proper regulation of mTORC1 activity is needed to maintain T cell homeostasis and metabolic quiescence as hyperactive mTORC1 signaling via TSC deletion leads to a heightened baseline activation state leading to more terminally differentiated effector T cells and a subsequently diminished naïve T cell pool.[52,54]

Antigen recognition and costimulation of naïve T cells leads to calcineurin-mediated nuclear factor of activated T cells (NFAT) activation, MAPK-mediated AP-1 activation, and PKC-mediated NF-κB activation resulting in exit from T cell quiescence (Chapter 25). In addition, the PI3K and mTOR pathways become activated–both of which play important roles in coordinating cell cycle and the metabolic demands necessary to support differentiation, proliferation, and function.[1,38,80-82] To this end, activation of naïve T cells is metabolically defined by the rapid redirection of carbons from one pathway (ie, TCA cycle) into glycolysis-generated lactate (aerobic glycolysis)[76] (Fig. 6.2). Green and colleagues were the first to demonstrate the detailed mechanisms involved in metabolic reprogramming by studying metabolic gene expression in parallel with metabolic radiolabeling assays.[76] Their elegant study revealed that T cell activation induced metabolic reprogramming from FAO-fueled OXPHOS to anabolic metabolism by upregulating glycolysis, glutaminolysis, and the synthesis of biomolecules such as nucleotides, cholesterol, and fatty acids.[76] The majority of these early anabolic processes are highly dependent on the increased expression of Myc that is readily observed within hours of T cell activation to mediate transcription of essential genes involved in both glucose and glutamine metabolism. While HIF-1α signaling is ultimately critical in metabolic reprogramming, stable HIF-1 expression does not appear until much later and hence is not required for activation-induced metabolic reprogramming. To undergo such rapid reprogramming upon T cell activation, naïve T cells already reside in a poised translational state by possessing sufficient mRNAs associated with glycolysis (ie, Glut1) and fatty acid synthesis (ie, ACC1) for rapid protein translation mediated by mTORC1-eIF4E.[83] Therefore, within a few hours of naïve T cell stimulation (8-12 hours), visual assessment of T cells under a microscope or analysis with flow cytometry reveals the consequence of this reprogramming in the form of large blasting T cells (Fig. 6.4). This increase in cell size is indicative of previously quiescent naïve or resting T cells having exited the G0 phase of cell cycle and entering into the G1 phase, accruing essential biomass required for subsequent clonal expansion. For example, activated T cells treated with rapamycin or T cells lacking proteins involved in the mTOR signaling (ie, mTOR) or metabolic associated proteins (ie, Glut1, Myc) display a significant defect in initial cell size thereby immediately foreshadowing a "metabolically" mitigated T cell response (Fig. 6.4).

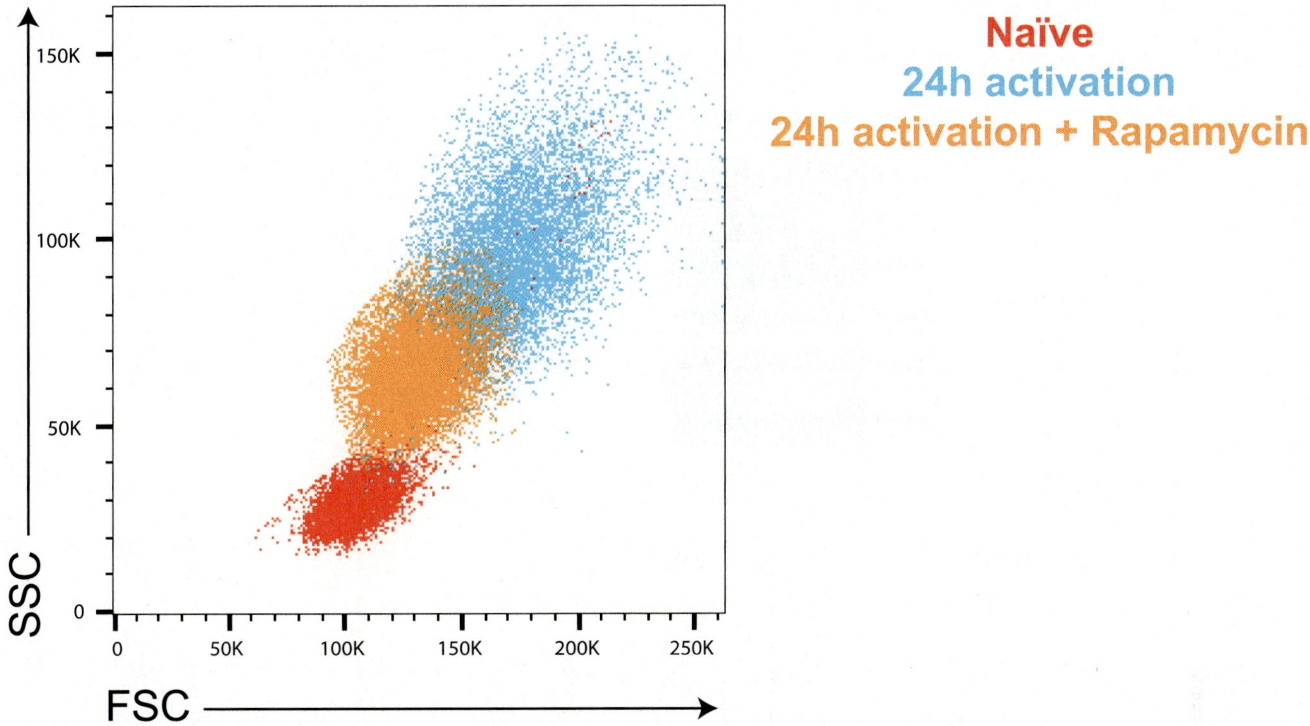

FIG. 6.4. Naïve T cells undergo mTORC1-dependent growth upon activation. Naïve CD8+ T cells activated with anti-CD3 and anti-CD28 for 24 hours plus or minus Rapamycin (100 nM) are then analyzed by flow cytometry. Assessing forward scatter and side scatter reveals the role of mTORC1 in promoting cell growth (blasting). mTORC1, mammalian/mechanistic target of rapamycin complex-1.

Within the early minutes after T cell activation, the strength of TCR stimulation (Signal 1) is tightly linked to the initial rapid burst of glycolysis observed as lactate production[84] (Fig. 6.5 and see Tools and Trades section on Seahorse assay). Activation of downstream TCR signaling molecules ZAP70 and Lck results in direct interaction with pyruvate dehydrogenase kinase 1 (PDHK1) to inhibit pyruvate dehydrogenase (PDH) function. The PDHK1-PDH axis plays a critical regulatory node in glucose-pyruvate metabolism by dictating the directional flow of pyruvate. Mechanistically, PDHK1 phosphorylates PDH, which inhibits its ability to catalyze pyruvate to acetyl-coenzyme A(CoA), which would normally enter the TCA cycle (Fig. 6.1). As such, active PDHK1 reduces the flow of pyruvate entering the TCA cycle but allows for increased conversion of pyruvate into lactate. While Signal 1 is sufficient for early glycolysis, Signal 2 from costimulatory (ie, CD28) and growth factor signaling are necessary to quickly mobilize Glut1 to the surface and to increase protein expression to sustain elevated glycolytic programming.[1,85,86] Signal 2 also promotes strong mitochondrial priming to form tight cristae, which is critical to support efficient ETC activity for OXPHOS.[23,87]

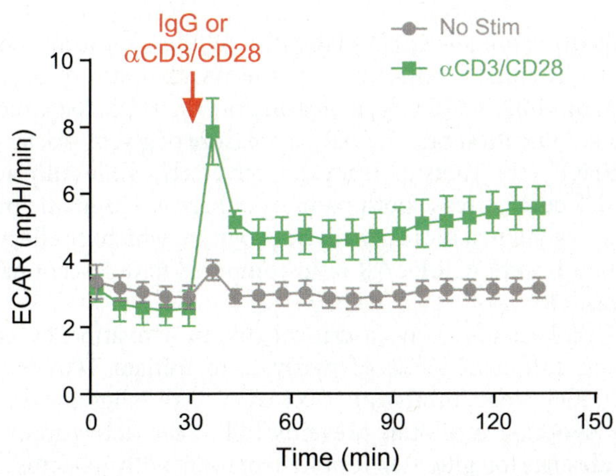

FIG. 6.5. Naïve T cells engage in glycolysis immediately post–TCR activation. Analysis of extracellular acidification rate (ECAR) upon T cell stimulation of naive CD8+ T cells stimulated with anti-CD3 and anti-CD28 reveals an early switch to increased glycolysis. TCR, T cell receptor.

In order to generate ample ATP, glycolysis is inefficient compared to OXPHOS. However, it is believed that the increased metabolism of glucose facilitates the highly anabolic aspects of T cell activation and differentiation by feeding several pathways involved in the biosynthesis of macromolecules. For example, diverting the glycolytic intermediate (ie, glucose-6-phosphate) into the pentose phosphate pathway (PPP) promotes the production of nucleotides and helps maintain redox status.[76] Likewise, glucose and glutamine derived metabolites are required as important intermediates for the hexosamine pathway that generates uridine diphosphate *N*-acetylglucosamine (UDP-GlcNAc) for O-GlcNAcylation posttranslational modifications of proteins such as Myc to prevent degradation by glycogen synthase kinase 3 (GSK3β)[88] (Fig. 6.1).

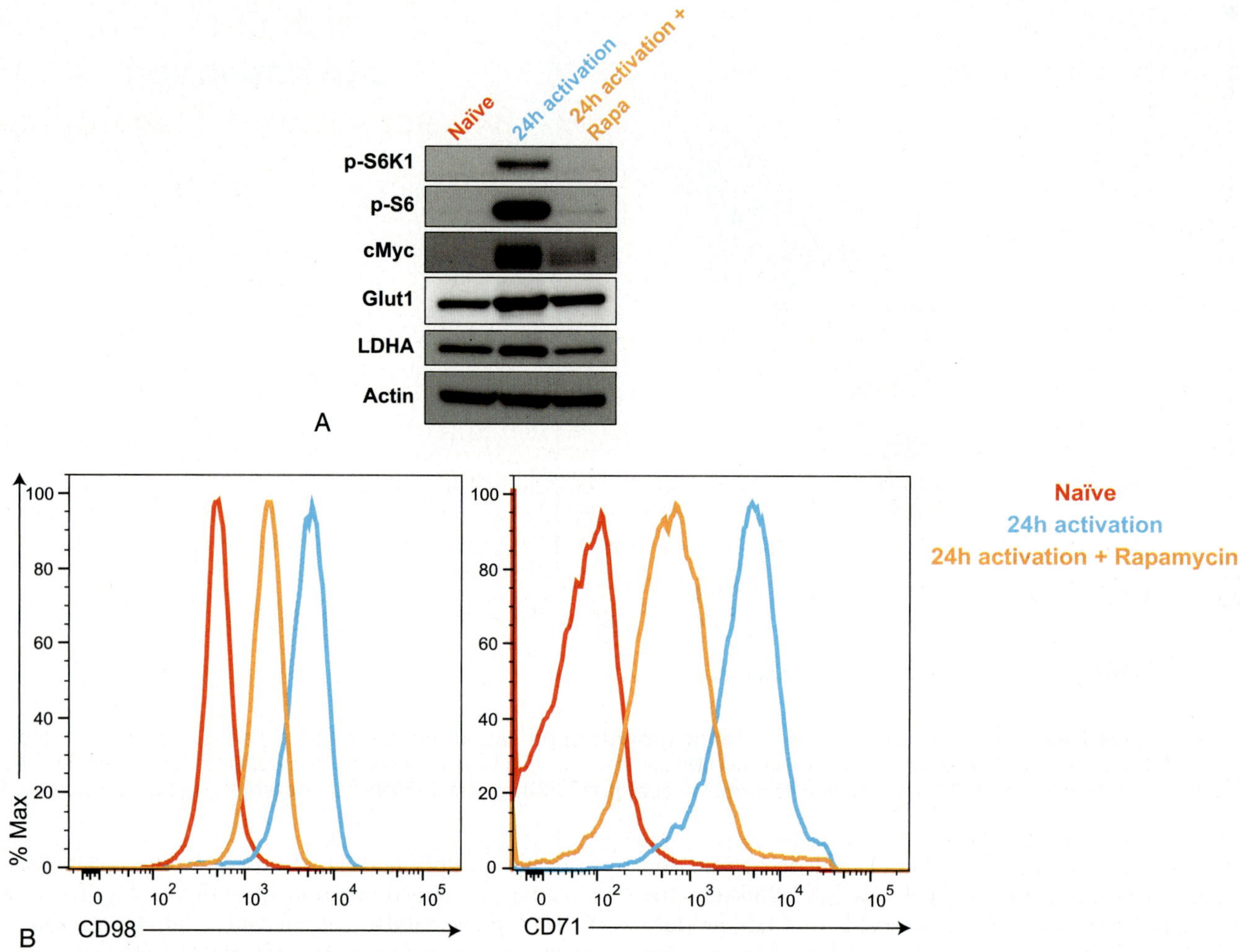

FIG. 6.6. Naïve T cells undergo mTORC1-dependent metabolic reprogramming upon activation. Naïve CD8+ T cells activated with anti-CD3/28 for 24 hours plus or minus Rapamycin (100 nM). **A:** Immunoblots showing changes in mTORC1 signaling (pS6 and pS6K) and the expression of proteins involved in glycolytic metabolism that are Myc-dependent (Glut1 and LDHA). **B:** Flow cytometry showing changes in metabolic solute carrier component (CD98) and the transferrin receptor (CD71). mTORC1, mammalian/mechanistic target of rapamycin complex-1.

Early glycolytic reprogramming is tightly regulated by mTORC1-dependent upregulation of Myc expression and Myc-dependent expression of metabolic enzymes (ie, HK II, PKM2, LDHA) involved in glycolysis along with solute carriers (ie, CD98, CD71)[76] (Fig. 6.6). While, HIF-1α–mediated metabolic reprogramming seems to be dispensable early on during activation, effector CD8+ T cells responding to IL-2 induce high levels of glucose uptake and maintain glycolytic metabolism dependent on the mTORC1-HIF-1 signaling axis.[80] As such, the initial strength of TCR engagement determines the levels of Myc expression, which controls the early metabolic reprogramming of activated T cells.[89,90] Lactate dehydrogenase (LDH) protein expression is significantly increased in activated T cells to keep up with the high demand for NAD+ required for the GAPDH enzymatic reaction.[5] Furthermore, lactate secretion through highly expressed solute carriers (MCT1 and MCT4) is necessary to prevent intracellular lactate buildup.[90,91] It is important to note that, in addition to glycolysis, activated T cells also markedly upregulate OXPHOS.[92,93] While quiescent naïve T cells minimally utilize both pathways, the ratio of oxygen consumption rate (OCR), a measure of OXPHOS, to extracellular acidification rate (ECAR), a measure of glycolysis, favors OXPHOS (see Tools of Trades). Conversely, while fully activated T cells increase both pathways, there is a dramatic metabolic switch to a more glycolytic program, which is reflective by their lower OCR:ECAR ratio compared to quiescent naïve T cells.[92,93]

Glycolysis also plays a critical role in activation by generating sufficient levels of pyruvate to initiate TCA-fueled OXPHOS since inhibiting OXPHOS with oligomycin, an ATP synthase inhibitor, prevents full T cell activation.[92] In the absence of glucose, supplementation with galactose or pyruvate is sufficient to fuel the TCA cycle and subsequently OXPHOS for full T cell activation.[92,94] Signal 1– and 2–dependent Ca^{2+} influx into the mitochondria induces the production of mitochondrial reactive oxygen species (mROS) by complex III of the ETC to activate NFAT-dependent transcription (ie,

IL-2) necessary for full T cell activation[94] (Fig. 6.2). Therefore, mitochondria play a critical role in facilitating proper full T cell activation. After initial T cell activation, T cells can use either glycolysis or OXPHOS to fuel proliferation and survival.[92] TCR-induced mitochondria localization to the immunological synapse is also important to provide an additional source of Ca^{2+} and ATP to facilitate the energetic demands required to sustain the T cell signaling cascade.[95-97] Also as part of their metabolic reprogramming, newly activated T cells undergo mitochondrial biogenesis and proteome remodeling to support enhanced OXPHOS and folate-mediated one-carbon metabolism needed for *de novo* purine and thymidine synthesis.[98,99] After T cell activation, rapid elevation of metabolic activity generates additional levels of intracellular ROS. Increased ROS levels activate the antioxidative glutathione (GSH) response pathway as a means to scavenge additional ROS and to prevent cell damage, which can lead to cell death.[100] Compared to naïve T cells, activated T cells express higher levels of the cystine transporter xCT (SLC7A11) to actively import cystine (oxidized dimer of cysteine) for glutathione synthesis and redox balance.[101,102] Furthermore, TCA-derived metabolites such as acetyl-CoA, α-ketoglutarate (α-KG), 2-hydroxglutarate (2-HG), and S-adenosylmethionine (SAM) regulate various epigenetic modifying enzymes (ie, histone acetyltransferases [HATs] or demethylases) to alter the transcriptional program during T cell differentiation and function.[103]

While highly activated and proliferating T cells continue to accumulate biomass, it should be noted that glucose and glutamine contribute only about 20% of all carbons in T cells while the remaining carbons are mostly derived from other amino acids.[104] This is somewhat surprising since glutamine (the most abundant free amino acid in blood) can adequately supply the necessary building blocks to synthesize proteins and to supply the nitrogen pool for *de novo* nucleotide synthesis.[105] As a consequence of glutamine's low usage, other individual amino acids (ie, alanine), through their synthesis or import into cells, also play important and distinct roles during T cell activation, differentiation, and function.[106] Naïve or resting T cells express minimal levels of amino acid transporters with many sharing subunits such as CD98 (SLC3A2) and SLC7A5 to form the heterodimeric transporter for large amino acid uptake (ie, leucine).[76,81,90,107] Within hours of T cell activation, there is significant upregulation of these various amino acid and solute transporters supported by the mTORC1-Myc axis[76] (Fig. 6.6). As part of a feed-forward loop, sufficient amino acid levels like leucine and glutamine maintained by constant nutrient uptake play a critical role in sustaining mTORC1 activity and Myc protein expression.[107] In addition, compared to other amino acids, intracellular arginine is the most rapidly metabolized and plays an essential role in promoting T cell proliferation.[108,109] Furthermore, amino acids involved in one-carbon metabolism are also critical for T cells. One-carbon metabolism through the folate and the methionine cycle refers to the utilization of single carbon units, primarily derived from serine and glycine, in order to synthesize nucleotides, to promote DNA methylation, and to generate critical reducing cofactors. Upon T cell activation, rapid induction of one-carbon metabolism takes place and requires exogenous serine from the environment or through *de novo* synthesis from glucose by upregulating critical enzymes such as serine hydroxymethyltransferase 2 (SHMT2).[99,110,111] Both mechanisms provide a steady supply of serine for one-carbon metabolism to support T cell proliferation rather than directly modulating T cell activation or cytokine production.

An additional node involved in the metabolic reprogramming of activated T cells is *de novo* lipid and sterol synthesis as part of lipid metabolism.[76,112] Prior to T cell activation, a steady pool of cholesterol is important for maintaining membrane lipid composition. Membrane integrity plays a vital role in controlling the degree of TCR clustering thus enhancing TCR avidity for antigen.[113,114] Upon T cell activation, FAO is reduced in favor of promoting lipid biosynthesis, which is necessary for the generation of cellular membrane synthesis that accompanies rapid clonal expansion.[76,112] Therefore, the generation of plasma membranes depends on acetyl-CoA carboxylase I (ACC1) activity to initiate long-chain fatty acid synthesis (FAS), which is crucial for antigen-driven clonal expansion but also homeostatic proliferation.[115] Overall, newly activated T cells undergo metabolic reprogramming to upregulate both glycolysis and OXPHOS to support growth and proliferation. However, as activated $CD4^+$ and $CD8^+$ T cells differentiate into different subsets their overall metabolic programs adapt to be more tailored to their specialized immune function which will be discussed in the following section (Chapters 29-32)[7,116] (Fig. 6.7).

Th1 Metabolism

One of the most striking differences in metabolic profiles of $CD4^+$ T cells is readily observed when comparing Th1 and T_{reg} cells.[7] Th1 differentiation leads to enhanced glycolysis with increased upregulation of the glucose transporter Glut1. Indeed, Glut1 knockout $CD4^+$ T cells fail to differentiate into Th1 cells.[72] This link between Th1 differentiation and glycolytic programming is further demonstrated by the fact that Tbet, the key transcription factor driving Th1 differentiation, also promotes the transcription of genes involved in glycolysis.[117] By slowing down glycolysis using galactose as a fuel, the ability of Th1 cells to produce IFN-γ significantly decreases. This was found to be regulated at the posttranscriptional level.[92] GAPDH, the glycolytic enzyme involved in converting glyceraldehyde 3-phosphate (G3P) to glycerate 1,3-biphosphate, is critical in regulating the translation of *Ifng* mRNA.[92,118] When not actively engaged in glycolysis, GAPDH "moonlights" in a secondary role and binds onto the 3′ UTR of *Ifng* mRNA thereby inhibiting the translation of *Ifng*. Also, highly active aerobic glycolysis facilitated by LDHA has also been shown to promote IFN-γ production but at the transcriptional level through epigenetic modifications of histones.[5] Genetic deletion of LDHA in Th1 cells shifts cells to adopt an OXPHOS-dependent program leading to a noticeable reduction in the cytosolic pool of acetyl-CoA. This in turn results in less acetylation of the transcriptionally active H3K9 mark at the *Ifng* loci thus preventing transcription. Other aspects of

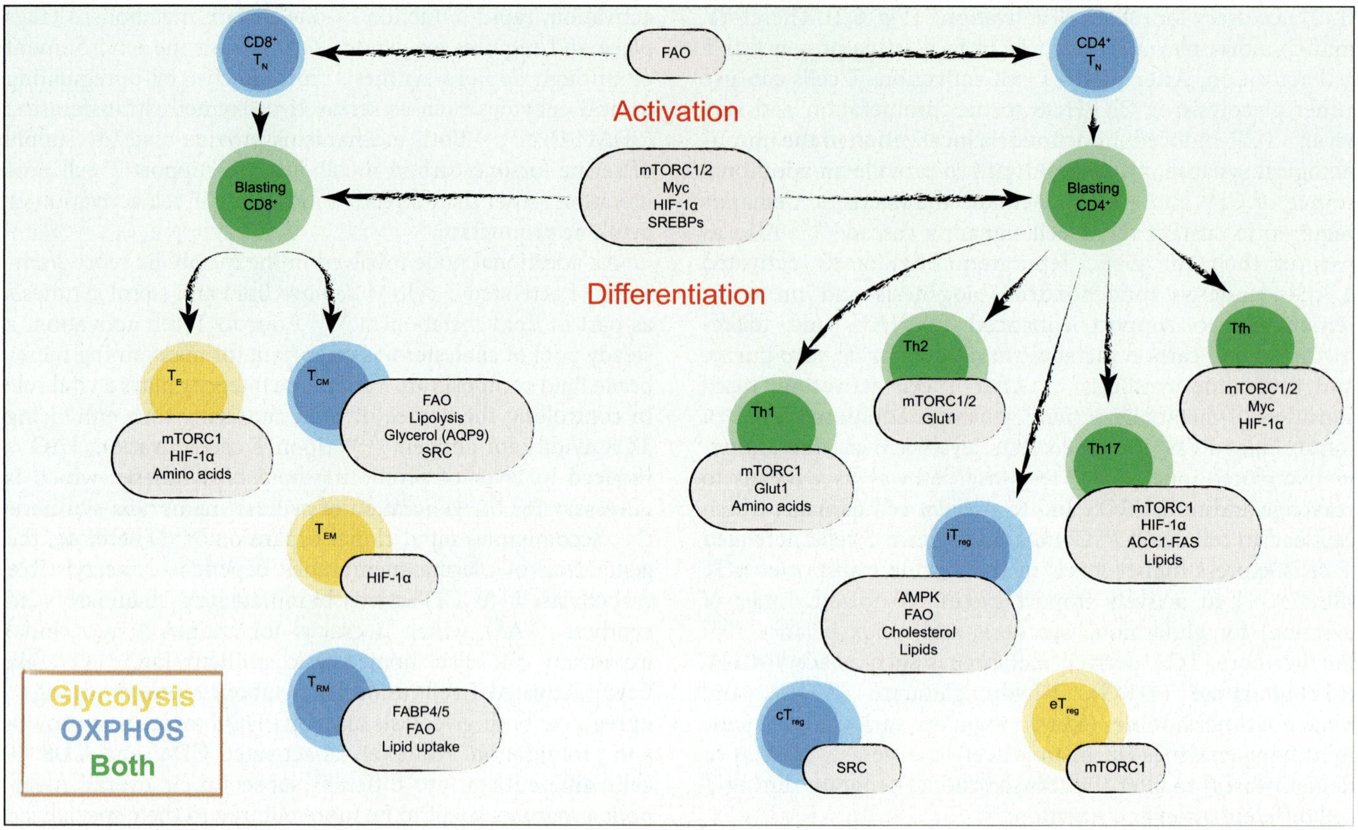

FIG. 6.7. Metabolic phenotypes of specific CD4+ and CD8+ T cells subsets. Individual T cell subsets display specialized metabolic profiles that influence differentiation and function. ACC1, acetyl-CoA carboxylase 1; AQP9, aquaporin 9; cT_{reg} cell, central regulatory T cell; eT_{reg} cell, effector regulatory T cell; FABP4, fatty acid–binding protein 4; FAS, fatty acid synthesis; Glut1, glucose transporter type 1; HIF-1α, hypoxia inducible factor-1α; iT_{reg} cell, induced regulatory T cell; mTORC1, mammalian/mechanistic target of rapamycin complex 1; mTORC2, mammalian/mechanistic target of rapamycin complex 2; ROS, reactive oxygen species; SRC, spare respiratory capacity; SREBP, sterol regulatory element–binding protein; T_{CM}, central memory T cell; T_E, effector T cell; T_{EM}, effector memory T cell; Tfh, T follicular helper cell; Th1 cell, type 1 T helper cell; Th17 cell, IL-17-producing T helper cell; Th2 cell, type 2 T helper cell; TRM, tissue resident memory T cell.

glycolytic metabolism have also been found to impact IFN-γ production. The glycolytic metabolite, phosphoenolpyruvate (PEP), blocks the resorption of cytosolic Ca^{2+} into the endoplasmic reticulum (ER) through sarcoendoplasmic reticulum calcium transport ATPase (SERCA), thus sustaining TCR induced NFAT signaling and CD4+ IFN-γ production.[119]

The coordination of both glycolysis and Th1 differentiation is facilitated in part by mTORC1 activation, which is necessary for Th1 differentiation.[120] Similarly, the amino acid leucine, which plays a critical role in the localization and activation of mTORC1, is necessary for Th1 differentiation without perturbing T_{reg} differentiation.[82,107] In the absence of SLC7A5, which facilitates leucine import, T cell development and peripheral T cell subpopulations are unaffected at steady state.[107] However, upon activation, SLC7A5-deficient T cells fail to proliferate and differentiate into Th1 cells due to an inability to sustain Myc expression necessary for proper metabolic reprogramming. Activated CD4+ T cells significantly increase glutamine uptake to support proliferation.[121,122] Like glucose uptake, costimulation further enhances glutamine uptake compared to Signal 1 alone.[121] Mechanistically, increased glutamine import through the amino acid transporter ASCT2 (SLC1A5) supports TCR induced mTORC1 activity required for Th1 differentiation and function.[122] Indeed, culturing CD4+ T cells under glutamine-deprived conditions reduces the generation of Th1 cells.[122,123] However, these observations may be a consequence of glutamine's role in supporting biosynthesis required for proliferation and the critical role of proliferation in promoting Th1 differentiation.[124] Through a process known as glutaminolysis, glutamine is catabolized into intermediates that can enter the TCA cycle and replenish critical intermediates, which is critical for both continued biosynthesis as well as regulation of Th1 differentiation. The activity of glutaminase (GLS), the enzyme that catalyzes the conversion of glutamine to glutamate (glutaminolysis), does not play a role in maintaining T cell homeostasis but can hinder Th1 differentiation.[125]

Th2 Metabolism

Unlike Th1 cells, the metabolic factors and detailed metabolic pathways that regulate Th2 differentiation and function have yet to be fully elucidated. Like Th1 cells, Th2 cells are more glycolytic than T_{reg} cells.[7] In fact, the canonical Th2 transcription factor GATA3 has been shown to upregulate Myc.[126] To this end, Th2 cells isolated from asthmatic patients have elevated Myc transcripts compared to Th2 cells from healthy controls.[127] Blocking glycolysis at the first step with the hexokinase (HK) inhibitor and glucose analogue, 2-DG (2-deoxy-D-glucose), reduces both GATA3 and IL-4 receptor expression as well as IL-4 production.[82,128] Likewise, inhibition of glucose uptake by deleting Glut1 reduces Th2 differentiation while, conversely, overexpression of Glut1 elevates IL-4 production.[7,72] These studies emphasize the importance of glucose metabolism in promoting Th2 differentiation and function. In addition, peroxisome proliferator–activated receptor family (PPAR) proteins play an important role in promoting Th2 differentiation and function by increasing FAO-fueled OXPHOS.[129,130] Furthermore, consistent with the role of PPARγ in promoting mitochondrial biogenesis, Th2 cells derived from asthmatic patients show an increase in CPT1α (carnitine palmitoyltransferase 1A) expression, the rate-limiting enzyme in long-chain FAO, compared to Th2 cells from healthy controls, suggesting that FAO is also important in fueling the pathogenicity of Th2 cells.

T_{reg} Metabolism

Initial studies demonstrated that inhibition of mTOR either pharmacologically with rapamycin or through genetic deletion led to the enhanced generation of T_{reg} upon in vitro activation.[44-46] These observations led to a paradigm whereby TCR engagement in the absence of mTOR signaling readily promoted the differentiation of Foxp3+ inducible T_{reg} (iT_{reg}). From a metabolic perspective, these data were consistent with observations that unlike Th1, Th2, and Th17 CD4+ helper T cells, T_{reg} cells rely less on lactate-forming glycolysis but rather use alternative fuels such as lipid or pyruvate oxidation in the mitochondria for OXPHOS.[7,131] These observations were further corroborated by experiments demonstrating that pharmacologically inhibiting glycolysis with 2-DG or inhibiting HIF-1–dependent glycolytic programming promoted T_{reg} generation.[3] In general, stable HIF-1α protein expression is regulated by oxygen-sensing mechanisms mediated by prolyl hydroxylation via proly-4-hydroxylase (PHD) proteins.[132] When oxygen is sufficient (normoxia), PHD proteins hydroxylate HIF-1α, leading to ubiquitination by the von Hippel-Lindau (VHL) E3 ligase complex for rapid proteasomal degradation. Therefore, sustained HIF-1α expression through VHL deficiency consequently results in increased glycolysis and mitigated T_{reg} differentiation leading to more IFN-γ production.[133] Interestingly, HIF-1 expression has also been reported to antagonize iT_{reg} differentiation by promoting proteasomal degradation of FoxP3.[2] Similarly, hyperactive PI3K signaling through deletion of the negative regulator phosphatase and tensin homolog (PTEN) leads to an upregulation of glycolytic metabolism and mTORC2 activity and also promotes increased methylation of the T_{reg} lineage-defining transcription factor *Foxp3* locus, leading to its instability and thus a lack of maintaining homeostasis.[134,135]

In addition to its role in promoting T_{reg} identity and function, FoxP3 influences its own metabolic programming. FoxP3 itself has been shown to suppress genes related to glucose metabolism (ie, Glut1, HK II) and instead promote the expression of genes that regulate mitochondrial OXPHOS (ie, CPT1α, cytochrome C).[136,137] To this end, disrupting mitochondrial function with rotenone, a drug that inhibits complex I of the mitochondrial ETC, selectively reduces proliferation, differentiation, survival, and function of iT_{reg} without affecting other helper CD4+ T cells.[131,136,138] Furthermore, T_{reg} rendered deficient of RISP, a component of complex III in the ETC, also demonstrate decreased suppressive abilities further emphasizing the role of intact OXPHOS in supporting T_{reg} function.[139] The addition of exogenous fatty acids such as palmitate to support lipid oxidation enhances the differentiation of iT_{reg} but suppresses the generation of other helper CD4+ T cells.[7] Inhibiting FAO with the CPT1α inhibitor, etomoxir, prevents the generation and function of iT_{reg} while increasing FAO through AMPK activation with metformin increases T_{reg} generation.[7,140] However, given the potential off-target effects of etomoxir observed in CPT1α-deficient T_{reg},[141] long-chain fatty acids are likely not the sole source of lipids for FAO to support OXPHOS. It is quite possible that short-chain or medium-chain fatty acids that can independently transport into the mitochondria might be oxidized for mitochondrial fuel. Overall, iT_{reg} cells require some levels of lipid and pyruvate oxidation to sustain increased levels of mitochondrial OXPHOS activity compared to effector helper CD4+ T cells.[131]

While the *induction* of T_{regs} seems to be enhanced by mTOR inhibition and the subsequent decrease in glycolytic programming, mTORC1 signaling is still required to promote cholesterol and lipid metabolism needed for T_{reg} suppressive function.[3,44,142] Thus, functionally speaking, similar to their CD4+ T helper counterparts, effector T_{reg} (eT_{reg}) cells display increased mTORC1 activity and glycolytic metabolism to support their suppressive ability.[47,48] To this end, deletion of mTOR or the mTORC1 signaling component, Raptor, leads to deficient effector T_{reg} function in vivo. However, resting central T_{reg} (cT_{reg}) cells, similar to memory CD8+ T cells (see below), rely on an OXPHOS-dependent metabolic program with reduced mTORC1 activity to promote a long-lived phenotype. Likewise, increased mTOR activity and glycolysis have been observed in human T_{reg} cells compared to conventional CD4+ T cells.[143] An increase in glycolysis allows binding of the glycolytic enzyme, enolase-1, to *Foxp3* regulatory regions to promote the expression of the FOXP3 variant FOXP3-E2, which is important to regulate suppressive ability in human T_{regs}.[144]

Th17 Metabolism

Like Th1 and Th2 cells, Th17 cells take up more glucose than T_{regs} as they are highly dependent on a glycolytic program for differentiation and function.[7,131] Early findings show that by inhibiting glycolysis with 2-DG, CD4+ T cells fail to differentiate into Th17 cells both in vitro and in vivo.[3,131] Deletion of enzymes involved in glycolytic metabolism such as pyruvate kinase isozyme M2 (PKM2) or PDHK in CD4+ T cells also mitigate Th17 cell differentiation.[131,145] As a result, silencing of PDHK1 or its inhibition with dichloroacetate (DCA) disrupt the balance between Th17 and T_{reg} development and thus promotes T_{reg} generation.[131] PKM2 is dually involved in the glycolytic reaction of converting PEP to pyruvate but also moonlights with a transcriptional role to promote gene transcription and ultimately HIF-1α stabilization. Preventing PKM2 nuclear translocation through forced tetramerization by the allosteric activator TEPP-46 reduces the glycolytic program and mTORC1 activity that are essential for Th17 cell differentiation.[145] Furthermore, HIF-1α is highly expressed in in vitro generated Th17 cells to promote a strong glycolytic program, which consequently inhibits the generation of iT_{reg}.[3] $CD4^+$ T cells that are deficient in HIF-1α fail to differentiate into Th17 cells but default into generating T_{reg}.[2,3] One mechanism by which HIF-1α promotes Th17 differentiation is through its interaction with RORγt and p300 to promote a Th17-based transcriptional program.[2] In addition, besides regulating Th17 cell differentiation, mTORC1 also promotes the transdifferentiation of Th17 cells into Th1-like cells that drive autoimmune disease by regulating Myc expression[146] (Fig. 6.3).

However, other studies suggest that glycolytic metabolism alone is not sufficient to drive Th17 function. Experiments using siRNA to knockdown expression of HIF-1α and PDHK1 within in vitro differentiated Th17 cells show no defect in IL-17A production upon restimulation.[147] Alternatively, inhibiting OXPHOS with oligomycin, which inhibits ATP synthase in the ETC, completely abrogates IL-17A production in Th17 cells generated in vivo but not in vitro generated Th17 cells. Such observations suggest that the maintenance and function of Th17 cells in vivo relies on ATP generation from the mitochondria rather than a strong glycolytic program. This study highlights the potential differences in the metabolism of cells generated using in vitro based culture conditions versus physiologic in vivo conditions. It is thought that perhaps in vitro culture conditions with ample or even excessive nutrient and oxygen availability at a neutral pH does not reflect the physiologic state observed in vivo.[111] The in vivo heterogeneous "regional" immune microenvironment may have scarce nutritional resources thus influencing activation and differentiation of T cells. These divergent works in terms of the metabolic programming of Th17 cells serve to highlight the potential limitations of interpreting solely in vitro metabolic observations especially given the recent findings into the heterogenous pool of pathogenic versus nonpathogenic Th17 cells observed in vivo compared to the more stable Th1 and Th2 cells.

Glucose-dependent *de novo* fatty acid synthesis is also critically involved in the differentiation of Th17 cells.[148] Activation of naïve $CD4^+$ T cells with soraphen A, an inhibitor of ACC1 and ACC2, prevents the differentiation of Th17 cells by reducing proliferation and IL-17A production, while reciprocally promoting T_{reg} cell development. Both isoforms of ACCs catalyze the ATP-dependent carboxylation of acetyl-CoA to malonyl-CoA, a critical rate-limiting step for *de novo* fatty acid synthesis. T cells that are deficient in ACC1 recapitulate the same phenotype as soraphen A-treated $CD4^+$ T cells demonstrating that an additional driver of Th17 cell differentiation is *de novo* fatty acid synthesis to sustain the glycolytic-lipogenic axis.

Furthermore, the regulation of cholesterol biosynthesis is important in promoting Th17 cell differentiation. The activity of the master transcription factor Rorgt that drives Th17 differentiation requires binding to metabolic intermediates derived from the cholesterol biosynthetic pathway.[149-152] This is mediated in part through the scavenger receptor CD5L that mediates lipid homeostasis. As such, CD5L regulation of fatty acid metabolism during Th17 differentiation influences the development of pathogenic versus nonpathogenic Th17 cells.[153] CD5L expression on Th17 cells regulates lipid composition by favoring polyunsaturated fatty acids over saturated fatty acids in turn limiting RORγt ligand availability. In response to IL-23, Th17 cells downregulate CD5L to alter the balance in favor of more intracellular saturated fatty acids, leading to increased cholesterol pathway intermediates and thus promoting the development of pathogenic Th17 cells that drive diseases. Other extrinsic factors such as excess salt concentrations induce SGK1 expression to induce expression of the IL-23 receptor through Foxo1 to promote the development of pathogenic Th17 cells.[154,155]

Tfh Metabolism

During germinal center (GC) reactions, $CD4^+$ follicular helper T cells (Tfh) provide essential help to B cells with affinity maturation, class switching, and memory B cell generation. The study of Tfh metabolism is complex due to the unique tissue microenvironment of the GC composed of hypoxic regions and likely competition for nutrients with other cells. During acute LCMV infection, Tfh cells are less proliferative compared to Th1 cells and do not rely on mTORC1-Myc driven anabolic metabolism that is associated with Th1 cells.[49] Therefore, Tfh cells take up less glucose and are less glycolytic compared to Th1 cells. This finding supports the prior observations—IL-2 through STAT5 signaling is necessary for Th1 differentiation but negatively regulates Tfh differentiation[156-158]; and Bcl-6, the major transcription factor for Tfh, represses the transcriptional program for glycolytic metabolism.[117] Likewise, IL-2 signaling during $CD4^+$ differentiation inhibits Bcl-6 as well as TCF-1, another critical transcription factor required for Tfh formation.[159] TCF-1 is critical to promote mitochondrial function and to induce a transcriptional program associated with the PPP, fatty acid metabolism, and the TCA cycle.

In contrast, other studies suggest that GC Tfh cells are dependent on glycolytic programming. In addition to Myc, mTOR-dependent HIF-1α expression also plays an important role in promoting glycolytic programming especially under hypoxic conditions typically seen within the light zone in GCs. Upregulation and stabilization of HIF-1α expression in VHL-deficient Tfh cells promote glycolysis as it is required for the differentiation and function of Tfh cells.[160,161] In the absence of HIF-1α, Tfh cells cannot induce GCs and antigen specific B cell antibody production. Indeed, overexpression of the glucose transporter Glut1, which enables more glucose uptake, in $CD4^+$ T cells promotes Tfh differentiation.[162] Furthermore, GC Tfh cells require both mTORC1 and mTORC2 signaling to promote glycolysis for differentiation and direct localization of activated Tfh cells to the GCs.[162,163] Stimulation through ICOS signaling on Tfh cells promotes glucose-dependent lipogenesis through the mTORC1-Myc axis. The differences in Tfh metabolism described here might be due to their anatomical location within the B cell follicle as Tfh cells reside either outside at the T:B cell zone or inside the germinal center, which may lead to nutrient competition with other highly proliferating B and T cells. These contrasting studies suggest further work needs to be done to fully elucidate the role of metabolism in Tfh cells with an emphasis on the microenvironment in guiding development.

Effector CD8+ T Cell Metabolism

Similar to the generation of effector CD4+ T cells, naïve CD8+ T cells upon activation and differentiation into effector CD8+ T cells metabolically reprogram to become highly dependent on glycolysis over OXPHOS[84,116] (Fig. 6.8). More so, mitochondria in effector CD8+ T cells dynamically undergo fission, which facilitates glycolysis, but fission also yields looser mitochondrial cristae and less efficient ETC for OXPHOS.[87] The initial activation of naïve CD8+ T cells is dominated by Myc-mediated metabolic reprogramming.[76,90] During differentiation into effector CD8+ T cells, mTORC1-mediated expression of HIF-1α dominates in promoting and sustaining an elevated glycolytic metabolic program necessary for proper effector differentiation and cytotoxic function.[80] To this end, deletion of HIF-1β mitigates effector function, while sustained HIF-1α and HIF-2α expression (secondary to VHL deletion) markedly enhances effector CD8+ T cell responses.[80,164] Consistent with the role of mTORC1 in promoting HIF-1α expression and glycolytic metabolism, pharmacologic or genetic deletion of mTORC1 signaling inhibits effector CD8+ T cell generation and function.[52] Alternatively, TSC2-deficient CD8+ T cells with constitutive hyperactive mTORC1 activity display enhanced CD8+ T cell effector function.[52] Pharmacologic inhibition of glycolysis with the hexokinase inhibitor 2-DG inhibits the generation and function of effector CD8+ T cells.[165] However, even though blocking glucose uptake by deleting the major glucose transporter Glut1 inhibits clonal expansion of effector CD8+ T cells, these CD8+ T cells still preserve effector function due to increased compensatory expression of other glucose transporters (ie, Glut3 and Glut6).[72] Under glucose-restricted conditions, the short-chain fatty acid, acetate, can promote histone acetylation in CD8+ T cells to promote IFN-γ effector function by upregulating acetyl-CoA synthetase (ACSS), which converts cytosolic acetate to acetyl-CoA.[166]

As clonal expansion is a major hallmark of potent effector CD8+ T cell responses, it is not surprising that in addition to the upregulation of glycolysis, cholesterol and fatty acid biosynthesis are also critical. SREBP1 and SREBP2 interact with the sterol-regulated chaperon protein SCAP and, upon

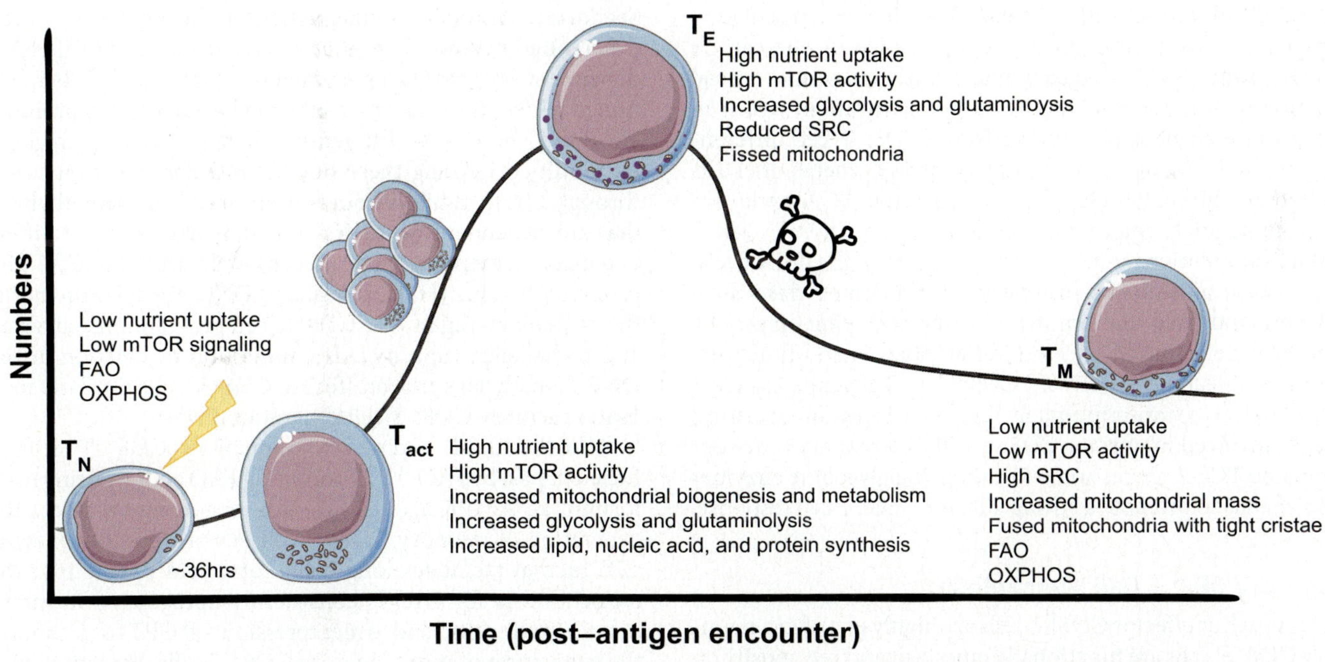

FIG. 6.8. The dynamic changes of CD8+ T cell metabolism during activation and differentiation. Naïve CD8+ T cells (T_N) reside in a quiescent state displaying minimal metabolic activity reflected by their dependency on fatty acid oxidation (FAO) to fuel oxidative phosphorylation (OXPHOS) as a main source of energy. Upon antigen recognition, activated CD8+ T cells (T_{act}) rapidly undergo metabolic reprogramming and switch to a highly glycolytic state. Increased nutrient uptake facilitates the generation and buildup of essential biosynthetic material for clonal expansion and the generation of potent effector CD8+ T cells (T_E). After antigen clearance, effector CD8+ T cells undergo contraction mediated by activation induced cell death (denoted by skull and cross bones) as CD8+ T cells transition into the memory phase and generate long-lived memory CD8+ T cells (T_M). Like naïve CD8+ T cells, memory CD8+ T cells remain in a quiescent state and rely on FAO and OXPHOS for survival. Mitochondrial reprogramming early on instills memory CD8+ T cells with important metabolic properties like increased spare respiratory capacity (SRC) and robust mitochondrial metabolism for future antigen rechallenge.

cleavage and activation, are translocated to the nucleus to promote gene expression for lipogenesis.[112] CD8+ T cells lacking SCAP demonstrate a marked reduction in effector CD8+ T cell differentiation secondary to an inability to produce sufficient amounts of lipids, especially cholesterol, needed for plasma membrane synthesis to sustain clonal proliferation.[112] Similarly, regulation of *de novo* fatty acid synthesis is critical for survival of effector CD8+ T cells. In response to an acute infection, the absence of ACC1 results in a defect in the expansion of effector CD8+ T cells due to increased cell death of proliferating CD8+ T cells.[115] In contrast, inhibiting cholesterol esterification with the pharmacological inhibitors, CP113,818 or K604, to acyl-CoA:cholesterol acyltransferases 1 and 2 (ACAT1/2), enhances CD8+ T cell effector function.[114] Mechanistically, inhibition of cholesterol esterification redirects cholesterol away from lipoprotein generation, favoring plasma membrane cholesterol incorporation, which promotes better TCR clustering and signaling with a more efficient formation of the immunological synapse.

Amino acid metabolism also critically regulates effector CD8+ T cell differentiation. In this regard, similar to Th1 cells, effector CD8+ T cells require SLC7A5 to support proliferation and differentiation by facilitating sufficient import of amino acids such as leucine.[107] In addition, an exogenous supply of serine is critical for CD8+ T cells as *de novo* synthesis of serine is not sufficient to support one-carbon metabolism for nucleotide biosynthesis required during massive clonal expansion.[110,111] Furthermore, the metabolites of one-carbon metabolism can directly regulate function by modulating epigenetic programming. As a consequence of one-carbon metabolism, SAM is generated, which provides critical methyl donors for DNA and histone methylation, which in turn can epigenetically guide effector CD8+ T cell differentiation and function.[110,167] Similarly, various metabolites (ie, fumarate, succinate, etc) derived from the TCA cycle can directly regulate epigenetic programming in CD8+ T cells.[103] Glutamine-derived α-KG, an intermediate of the TCA cycle, is a critical substrate for histone and DNA demethylases such as Tet2 and Jmjc that are involved in modifying epigenetic marks for effector CD8+ T cell differentiation and function.[6] For example, highly activated CD8+ T cells generate high levels of α-KG to transcriptionally regulate expression of critical genes involved in effector CD8+ T cell differentiation by suppressing TCF-1 expression and enhancing glycolytic enzymes as a means to promote a strong effector CD8+ T cell response.

Memory CD8+ T Cell Metabolism

In as much as effector CD8+ T cells are highly glycolytic, memory CD8+ T cells are functionally quiescent and rely mostly on OXPHOS and FAO to maintain long-term persistence.[116,168] To this end, while mTORC1 signaling promotes glycolytic programming to support effector CD8+ T cell differentiation and function, inhibition of mTORC1 either pharmacologically or through genetic deletion enhances the number and quality of memory CD8+ T cells.[51,52] Indeed, it has been shown that the asymmetric division of CD8+ T cells, which promotes the early generation of both memory and effector CD8+ T cells is a consequence in part of the asymmetric partitioning of mTORC1 signaling and Myc.[55,56] Furthermore, inhibition of glycolysis with 2-DG is able to promote memory CD8+ T cells at the expense of effector CD8+ T cell generation.[165] Alternatively, CD8+ T cells with constitutively increased mTORC1 activity (through TSC2 deletion) or increased HIF-1 (through VHL deletion) glycolytic metabolism readily differentiate into potent effector CD8+ T cells at the expense of generating long-lived central memory CD8+ T cells (T_{CM}), which was the first memory CD8+ T cell subset to be metabolically characterized.[52,116,169] However, an enhanced glycolytic phenotype seen in VHL-deficient CD8+ T cells is in fact sufficient to fuel long-term effector memory CD8+ T cells (T_{EM}) suggesting that different memory CD8+ T cell subsets can adapt and survive using different types of metabolism.[169]

While the metabolic programming of short-lived effector CD8+ T cells is uniquely suited for their explosive energy and anabolic demands, so too is the programming of memory CD8+ T cells tailored for their function (Fig. 6.7). The metabolic programming of memory CD8+ T cells enables them to be long-lived in a state of ready preparedness for rapid antigen reencounter. Along these lines, central memory CD8+ T cells have increased spare respiratory capacity (SRC) endowing them with the extra ability to quickly produce energy compared to effector CD8+ T cells.[116] This is a measurement of the maximal capacity of mitochondrial function and can be calculated by determining the difference between the baseline OCR and the maximal OCR (see Tools of Trades under Seahorse). Arginine supplementation during CD8+ T cell differentiation strongly pushes T cells to adopt an OXPHOS-dependent program favoring memory formation.[109] As mentioned before, the TCA cycle intermediate α-KG can promote effector CD8+ T cell differentiation by regulating epigenetic modifiers.[6] Along these lines, 2-HG derived from α-KG through LDHA inhibits both histone and DNA demethylases that are dependent on α-KG and thus promotes the differentiation of memory CD8+ T cells.[170] In this model, 2-HG promotes histone (H3K27me3) and DNA methylation marks thereby enhancing in vivo CD8+ T cell proliferation, survival, and rechallenge capacity. Also, inhibition or deletion of the DNA demethylase enzyme Tet2 in CD8+ T cells similarly promotes memory CD8+ T cell generation.[170,171]

Mitochondrial respiration in memory CD8+ T cells is fueled in part by FAO.[116,168] Promoting FAO activity using metformin, an AMPK agonist, or with rapamycin enhances the generation of memory CD8+ T cells.[51,168] Hence, the enzyme CPT1α that facilitates long-chain fatty acid uptake into the mitochondria for FAO is consistently upregulated in memory CD8+ T cells, and overexpression of CPT1α promotes the generation of more memory CD8+ T cells.[116] Surprisingly, genetic deletion of CPT1α in CD8+ T cells does not impact long-term memory formation since these deficient CD8+ T cells may readily utilize short- or medium-chain fatty acids to fuel FAO.[141] Interestingly, it has been noted that rather than taking up extracellular long-chain fatty acids to fuel FAO, memory CD8+ T cells use extracellular glucose to generate fatty acids to fuel FAO through lipolysis.[172] Memory CD8+ T cells in response to IL-7 exposure upregulate expression of aquaporin 9 (AQP9) to facilitate glycerol uptake from the environment for triglyceride synthesis and storage.[173] In addition, autophagy has been critically linked to proper memory $CD8^+$ T cell generation. During an acute response, inhibiting autophagy by deleting ATG5/7 in effector $CD8^+$ T cells leads to dysregulation of critical metabolic pathways that involve fatty acid transport and oxidation thus leading to reduced formation and survival of memory $CD8^+$ T cells.[174]

In light of the critical role of OXPHOS and FAO, it is not surprising that mitochondrial health and function also play an important role in maintaining and promoting memory $CD8^+$ T cells. An increase in mitochondrial mass in memory $CD8^+$ T cells compared with naïve $CD8^+$ T cells is important for survival and provides them with a bioenergetic advantage to respond during secondary recall.[116,175] The integrity of the mitochondria thus plays an important role in supporting the longevity of memory $CD8^+$ T cells since accumulation of ROS can result in damaged and aged mitochondria. Consequently, mitochondrial clearance by mitophagy is critical to support the self-renewal ability of memory T cells.[176] In addition, mitochondrial morphology governs memory $CD8^+$ T cell formation.[87] In memory $CD8^+$ T cells, a fused mitochondrial network driven by Opa1 drives tight cristae structure to enhance the formation of supramolecular structures of the ETC complexes thereby allowing efficient electron transfer and ATP generation. As mentioned earlier, naïve T cell activation via CD28 signaling promotes tight remodeling of its cristae to help sustain efficient oxidative metabolism by regulating CPT1α expression to develop SRC.[23] In the absence of CD28 signaling, $CD8^+$ T cells are unable to remodel their mitochondria to generate the necessary metabolic capacity for future recall responses.

In this context, mechanisms involved in maintaining proper redox balance are also critical in maintaining the longevity of memory $CD8^+$ T cells. Memory $CD8^+$ T cells upregulate the enzyme phosphoenolpyruvate carboxykinase 1 (PCK1) to catalyze the conversion of oxaloacetate (OAA) generated from both glucose and glutamine to PEP, which leads to a buildup of a glycogen reservoir.[177] Memory $CD8^+$ T cells mitigate excessive ROS buildup by breaking down glycogen to generate G6P to fuel the PPP pathway thereby increasing reduced forms of nicotinamide adenine dinucleotide phosphate (NADPH) levels and more reduced glutathione to balance oxidative stress. Compared with naïve $CD8^+$ T cells, resting memory $CD8^+$ T cells possess a higher glycolytic capacity to support their ability to rapidly carry out effector function during rechallenge.[178] It has recently been shown that acetate metabolism plays an essential role in this ability. During bacterial infections, extracellular acetate levels increase.[179] Uptake of extracellular acetate in memory $CD8^+$ T cells generates an increased pool of intracellular acetyl-CoA, which facilitates the acetylation of GAPDH. This important posttranslational modification enhances the enzymatic activity of the GAPDH, which supports the increased glycolytic switch to boost a rapid memory $CD8^+$ T cell recall response.[178,179]

A subset of memory $CD8^+$ T cells that reside in peripheral tissues, called tissue-resident memory (T_{RM}) $CD8^+$ T cells, provide immediate protective tissue responses in response to antigen. These $CD8^+$ T cells have a unique metabolic phenotype compared to more conventional central or effector memory $CD8^+$ T cells.[180] Interestingly, skin T_{RM} $CD8^+$ T cells express high levels of the fatty acid–binding protein 4 and 5 (FABP4/5) and the lipid scavenger receptor CD36 to facilitate the uptake of exogenous fatty acids to fuel FAO for long-term survival in the tissue. Notably, central memory $CD8^+$ T cells lacking FABP4 and FABP5 show no significant survival disadvantage. Additionally, these data suggest that T_{RM} from different tissues express varying FABP isoforms, highlighting the tissue-specific programming of T_{RM} that enable selective uptake of unique fatty acids depending on the immune and tissue microenvironment.[181]

Metabolism of Exhausted $CD8^+$ T Cells

Persistent antigen stimulation during chronic infection or during antitumor responses has been linked to the generation of exhausted $CD8^+$ T cells.[182] Exhausted $CD8^+$ T cells express a marked increase in checkpoint molecules such as PD-1 and a reduction in polyfunctionality in terms of cytokine production. Likewise, exhausted $CD8^+$ T cells possess dysregulated metabolism that contributes to their dysfunctional state. During chronic LCMV infection (Clone 13), exhausted $CD8^+$ T cells sustain constant activation of mTORC1 activity, but this increase in mTOR signaling is not accompanied by productive glycolytic function and anabolic processes.[183] Suppressing mTOR signaling with rapamycin is able to partially reverse this exhaustive state. In tumors, it has been noted that a subset of exhausted $CD8^+$ T cells possesses mitochondrial dysfunction due to increased Akt signaling, which in turn leads to the inhibition of PGC-1α.[184] This transcription factor is critical for mitochondrial biogenesis and maintaining overall mitochondrial health. Overexpression of PGC-1α in antitumor $CD8^+$ T cells can partially reverse the exhausted phenotype and improve effector function again emphasizing the interconnection between metabolic programming and T cell function. This observation highlights how correcting metabolic dysfunction in exhausted cells can lead to enhanced immunologic function.

B CELL METABOLISM

Development

Akin to naïve T cells, antigen stimulation of naïve B lymphocytes results in activation and clonal expansion.[185] Thus, metabolic reprogramming of B cells closely resembles what is observed in T cells due to the high demand for energy these processes require. Bone marrow–derived hemopoietic cells through a series of critical lineage guiding steps give rise to immature B cells before they exit the bone marrow and into the periphery for further maturation (Chapter 26).[186] The various microenvironmental niches and heterogeneous pool of growth factor secreting cells within the bone marrow play a critical role in guiding B cell development. Metabolically speaking, proper nutrient availability is essential as competition with stromal cells may play a critical role in B cell

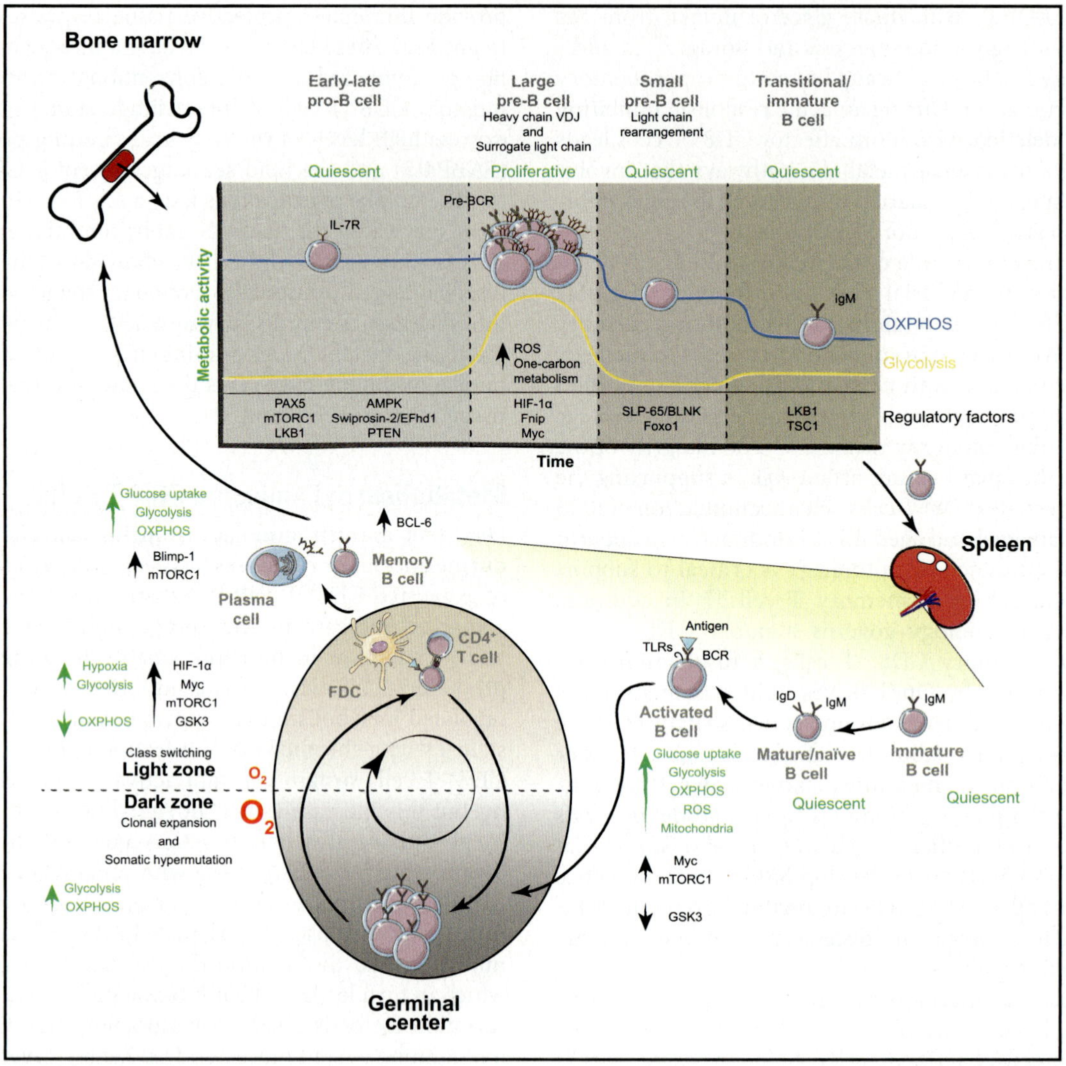

FIG. 6.9. Dynamic changes in metabolic activity during B cell development, maturation, and differentiation. The different stages of B cell differentiation in the bone marrow are characterized by metabolic reprogramming with differential anabolic and catabolic needs. Likewise, B cells undergo metabolic reprogramming during antigen activation and subsequent differentiation to memory cells and antibody-producing plasma cells.

development (see Fig. 6.9). During lineage differentiation, the essential developmental transcription factors PAX5 and IKAROS (IKZF1) act as metabolic gatekeepers by controlling glucose metabolism through the transcriptional repression of genes involved in glucose transport (ie, Glut1, Glut3) and glycolytic enzymes (ie, HK II) leading to the restriction of metabolites fueling the TCA cycle and OXPHOS.[187] The expression of these transcription factors to restrict metabolism possess important tumor suppressor function to prevent the malignant transformation of pre-B cells.

Early stages of development (pro-B cell to large pre-B cell) are associated with various phases of early active cell growth and proliferation but also later stages of quiescence (small pre-B cell) as a functional B cell receptor (BCR) is being formed and selectively tested. Therefore, dynamic expression of the IL-7 receptor and signaling through the pre-BCR play a critical role in regulating cell growth, differentiation, and survival.[188] PTEN-mediated suppression of PI3K signaling induces higher IL-7 receptor expression on pro-B cells to respond to copious amounts of IL-7 produced by bone marrow stromal cells to activate mTORC1 signaling and Myc expression, which are both necessary to drive anabolic programming.[188,189] The pro-B and pre-B cell stages display the most active mTORC1 signaling while activity subsides as B cells acquire expression of a functional BCR at later immature and mature B cell stages.[189] The absence of mTORC1 signaling or Myc expression, but not mTORC2 signaling, halts the pro-B to pre-B cell transition especially since high Myc expression corresponds with the highly proliferative developmental stages for the generation of mature B cells.[188,190] During the pre-B cell transition (large to small), diminishing expression of the IL-7 receptor and pre-BCR signaling induces expression of SLP-65/BLNK to suppress PI3K-Akt signaling and allowing light-chain rearrangement in a Foxo1 dependent manner.[191] This lower signaling state correlates with the low metabolic and quiescent state observed in small pre-B cells.[192]

During development, the transition of pro-B cells to large pre-B cells results in major metabolic programming.[188,192] Pre-BCR signaling in pro-B cells downregulates expression of Swiprosin-2/EFhd1, a mitochondrial Ca^{2+}-binding protein, as a means to increase glycolysis but reduce oxygen consumption in preparation for the large pre-B cell expansion for light-chain rearrangement.[192] Like clonally expanding T cells, to support their energetic and biosynthetic demands for cell growth and massive proliferation, large pre-B cells increase glucose uptake, oxygen consumption, and also generate more ROS compared to quiescent small pre-B cells.[188,192] Glucose metabolism as a whole is important in development since B cell precursors are more susceptible to glycolysis inhibition with 2-DG, and Glut1 deficiency hinders the generation of mature B cells.[93,193] Low metabolic activity corresponds well with small pre-B cells since they may be devoting resources to actively rearranging their light chain and undergoing selection for self-antigen. Furthermore, as B cells migrate in the bone marrow in response to various soluble cues, the differential low oxygen tension levels within certain regions of the bone marrow may influence mitochondrial respiration and stabilize HIF-1α expression.[193,194] During development, stabilization of HIF-1α is important to maintain high levels of glycolysis for the transition of pro-B cells to pre-B cells since deficiency in HIF-1α leads to the generation of abnormal peripheral B-1-like lymphocytes.[195] HIF-1α directly controls glycolysis in precursor B cells by increasing gene expression of key glycolytic enzymes (ie, Pfkfb3) and glucose transporters (Glut1 and Glut3).[193] Interestingly, HIF-1α–deficient progenitor B cells are metabolically flexible as they are able to adapt to the microenvironment by efficiently using pyruvate and by inducing expression of genes related to the TCA cycle and the ETC. Not surprisingly, glycolysis is dispensable for the transition of pre-B cells to IgM^+ immature B cells as they are less metabolically active due to their nonproliferative quiescent state.[193]

Finally, developing progenitor B cells utilize different energy and nutrient-sensing mechanisms as metabolic checkpoints to restrain excessive growth in the absence of sufficient energy and nutrient availability by controlling mTORC1 signaling. LKB1 activation of AMPK in response to energy deprivation is necessary for normal B cell development at the pre-B cell stage.[187] The AMPK interacting protein, folliculin-interacting protein 1 (Fnip1) acts as critical checkpoint to facilitate pre-B cell differentiation by regulating metabolic homeostasis.[196] Fnip1 helps confirm that large pre-B cells meet the minimal metabolic requirements to support division and development. In the absence of these critical regulatory nodes, B cell development is compromised. Immature B cells with a rearranged BCR will then exit the bone marrow into the periphery and advance through additional transitional stages before becoming a mature follicular B cell. Transitional B cells increase mTORC1 activity compared to mature follicular B cells as they are more metabolically active with increased mitochondrial content, ROS generation, and OXPHOS along with glycolytic activity.[197] Patients with gain of function mutations in PI3Kδ or the lack of TSC1 in murine B cells with elevated mTORC1 signaling are unable to induce metabolic quiescence as a checkpoint for final B cell maturation.[198,199] Consequently, an inability to repress mTORC1 signaling and metabolic activity results in an absence of mature B cells.

Activation, Differentiation, and Function

Naïve or resting B cells are metabolically quiescent and utilize catabolic metabolism to sustain energy homeostasis through both glycolysis and TCA-fueled mitochondrial respiration.[93,200] B cell activation begins with various cell surface receptor-ligand interactions such as CD40, TLRs, and the BCR to drive proliferation, class switch recombination (CSR), and differentiation into antibody secreting B cells and memory B cells (Chapter 27). Signaling events downstream of these various nodes of activation simultaneously induce cell cycle progression, survival, and metabolic reprogramming.[93,201] The primary role of B cells to rapidly produce antibodies necessitates a robust energetic metabolic program requiring a rich source of nutrients to generate sufficient biosynthetic substrates for organelle biogenesis such as the endoplasmic reticulum for antibody generation and membranes and nucleotides for subsequent clonal expansion.[202,203] Therefore, nutrient availability and metabolic programming strongly influence B cell activation, differentiation, and function. The soluble prosurvival and growth factor, B cell–activating factor (BAFF), helps to maintain mature naïve B cells in part by controlling metabolic fitness in preparation for future antigen stimulation. BAFF exposure initiates PKCβ-Akt signaling to induce a glycolytic transcriptional program.[204] Inhibition of mTORC1 signaling with rapamycin ablates BAFF-dependent signaling to maintain cell growth.[205] Likewise, IL-4 critically promotes naïve B cell survival by mediating glucose metabolism in a PI3K-independent but STAT6-dependent fashion.[206]

Lipopolysaccharide (LPS) or BCR signaling and subsequent activation of the PI3K-Akt pathway leads to glucose uptake and an increase in Glut1 expression, promoting glucose uptake and entry into the PPP to produce NADPH for redox stabilization[93,201] (Fig. 6.9). Initial B cell metabolic reprogramming is HIF-1α–independent but requires Myc to upregulate Glut1 for glycolysis and to support glutamine oxidation.[93] Altering glucose and pyruvate metabolism with 2-DG or DCA but also mTORC1 signaling with rapamycin suppresses LPS induced B cell proliferation and antibody generation.[93,207] Activated B cells over time significantly upregulate enzymes involved in glycolysis (ie, GAPDH, α-enolase) and amino acid metabolism (ie, OAT, PSAT) in preparation for rapid antibody secretion.[202] Increased glucose metabolism and ATP-citrate lyase (ACLY) expression are critical to support *de novo* lipogenesis of cholesterol, free fatty acids, and phospholipids to support the endomembrane expansion of key organelles such as the endoplasmic reticulum for antibody production.[203] Interestingly, naïve B cell and T cell metabolic programming upon activation slightly differ.[93] BCR (anti-IgM) or LPS-stimulated B cells maintain a constant OCR:ECAR ratio relative to naïve B cells meaning activation similarly increases both glycolytic and mitochondrial OXPHOS activity to the same degree. However, recall, activated T cells compared with naïve T cells have a lower OCR:ECAR ratio as they heavily favor glycolysis relative to OXPHOS. To support metabolic reprogramming, LPS-activated B cells within six hours rapidly increase their mitochondrial mass, which is consistent with the increase in OXPHOS.[93]

Upon BCR stimulation, early downstream PKCβ signaling is essential to mediate early mTORC1 signaling and thus expression of solute transport carriers and their subunits (ie, Glut1 and CD98).[208] Importantly, PKCβ signaling promotes mTORC1-dependent mitochondrial remodeling to generate sufficient levels of heme to inhibit BACH2 repression of Blimp-1 as a means to promote germinal center (GC) responses and plasma cell differentiation. Furthermore, adequate expression of CD98 is absolutely necessary for generating plasma cell antibody responses and clonal expansion.[209] Similar to TCR activation, BCR stimulation induces a rapid burst of ROS generation that occurs within minutes through NADPH oxidase (NOX) and is maintained at later stages by increased mitochondrial respiration.[210] Prolonged ROS sustains BCR-induced PI3K signaling, which is needed for optimal B cell activation and proliferation.

To maintain metabolic and B cell homeostasis, diverse mechanisms exist as metabolic checkpoints to prevent B cell hyperactivation or malignant transformation. Expression of the let-7adf microRNA cluster and GSK3 activity are negative regulators of anabolic metabolism.[197,211] The let-7adf microRNA cluster suppresses T cell–independent IgM antibody production by directly targeting *Hk2* transcripts and also suppressing Myc expression and consequently glutamine uptake.[211] GSK3 activity in naive and light zone GC B cells is an essential metabolic checkpoint under nutrient-restricted states to promote survival and to regulate proliferation.[197] GSK3 also regulates metabolic activity in response to CD40 costimulation and IL-4 signaling during GC T cell–dependent help. During the GC reaction, B cells undergoing somatic hypermutation will randomly edit their immunoglobulin genes with the potential to develop autoreactive clones. B cell activation following antigen engagement via the BCR induces initial metabolic programming but requires a secondary signal (ie, TLR9) to subsequently sustain metabolic activity over time.[212] A secondary signal within hours will inhibit the buildup of excess intracellular Ca^{2+} and ROS levels to maintain mitochondrial integrity for OXPHOS and ultimately prevent activation-induced cell death. Thus, like T cells, this two-signal model assures B cells do not activate in response to self-antigen unless an immune response elicits proper T cell help or a release of TLR agonists and other danger signals.

Within the GC, B cells undergoing somatic hypermutation are massively proliferating in the dark zone area while, in the light zone area, undergoing CSR and selection based on BCR affinity. These two critical events allow B cells the ability to produce high-affinity antibodies of different isotype classes. The outcome of the GC reaction is the generation of both long-lived plasma cells and memory B cells. mTORC1 activity is essential to form GC B cells by inducing Bcl-6 expression.[213] The absence of mTORC1 signaling reduces the number of GCs, antigen-specific memory, and plasma cells and critically reduces the induction of AID and somatic hypermutation thereby hindering the primary humoral response.[214] Likewise, during influenza vaccine immunizations, inhibiting mTORC1 activity with rapamycin prevents the formation of GCs and inhibits class switching thereby generating a pool of B cells with a unique repertoire of IgM antibodies that can mediate broader protection.[215] Similarly, mTOR inhibition with the clinical grade rapamycin, RAD001, in elderly people with declining immune function enhances antibody titers to heterologous influenza strains.[216]

Highly proliferative GC B cells expressing HIF-1α are more metabolically active than naïve B cells defined by larger cell size with increased protein content, glucose uptake, and mitochondrial mass.[197] Repeatedly transitioning between low and high anabolic growth states is directly dependent on costimulatory signals (ie, via CD40) delivered by helper T cells in the light zone. In the light zone, dynamic mTORC1 signaling and GSK3 inhibition play influential roles by integrating various selection signals to induce expression of Myc to promote cell growth and to build a strong anabolic program in GC B cells needed to provide enough fuel in preparation for massive expansion upon reentry into the dark zone.[197,217-219] In this way, Myc is dynamically expressed due to its regulatory role in cell cycling, and thus Myc expression is absolutely required for germinal center formation. Upon leaving the dark zone, stable HIF-1α expression in B cells due to hypoxia in the light zone decreases mTORC1 activity, proliferation, cell survival, class switching to IgG2c, and the generation of high-affinity antibodies.[220] Microenvironmental factors like limited oxygen (hypoxia) and nutrient availability create this additional metabolic checkpoint to select for the fittest germinal B cells to ensure survival and to continue the GC reaction cycle.[220]

The generation and maintenance of high-affinity antibodies produced by plasma cells is a vital component of humoral immunity. Terminal differentiation of B cells into either mature short- or long-lived plasma cells both require profound metabolic changes to support their biosynthetic demands. Since plasma cells cease to divide and their translational capacities are largely devoted toward producing immunoglobulins, B cells primarily use amino acids and glucose to facilitate the robust generation of antibodies rather than diverting biomass for other cellular processes.[221] Blimp-1 expression in plasma cells transcriptionally promotes expression of the CD98 amino acid transporter to mediate mTORC1-dependent protein synthesis.[222] Short-lived plasma cells import significantly less glucose than long-lived plasma cells.[223] In long-lived plasma cells, large amounts of consumed glucose (~90%) enter into the hexosamine pathway for antibody glycosylation; however, under low-ATP-generating conditions, glucose can be diverted towards glycolysis in order to fuel OXPHOS for ATP.[223] Similar to memory $CD8^+$ T cells, long-lived plasma B cells have greater mitochondrial SRC fueled by mitochondrial pyruvate. Mitochondrial function in activated B cells also plays a critical role in promoting plasma cell differentiation. Reduced mitochondrial mass in activated B cells lowers mROS levels and thus increases heme synthesis to block BACH2 and to increase Blimp-1 to promote plasma cell differentiation.[224] Additionally, as antibody production imposes tremendous cellular stress, proper B cell memory formation and maintenance are dependent on autophagy to remove damaged organelles or toxic byproducts.[225,226]

INNATE CELLS

Neutrophils

As a first-line defense to pathogens, short-lived neutrophils sense chemoattractants at the site of inflammation to quickly contain the infection from spreading (Chapter 15).[227] Functionally, neutrophils phagocytize and eliminate pathogens primarily through ROS, degranulation, and NETosis by forming neutrophil extracellular traps (NETs). In fact, the NETosis process is highly dependent on ROS.[228] During homeostasis, granulopoiesis leads to the remarkable production of approximately 1×10^{11} neutrophils per day from the bone marrow into the periphery; therefore, it is an energy-demanding process. As this burst of proliferation drastically escalates energy demands, autophagy is critical for early neutrophil differentiation from myeloblasts to mature neutrophils by limiting glycolysis and promoting lipolysis to generate free fatty acids for FAO-fueled OXPHOS.[229] In response to stimuli, mature neutrophils adopt a more glycolytic phenotype as their primary source of ATP.[230] In contrast, OXPHOS activity is dispensable for activation and some function as a mitochondrial ETC uncoupler such as FCCP (2-[2-[4-(trifluoromethoxy) phenyl]hydrazinylidene]-propanedinitrile) fail to inhibit phagocytosis and the respiratory burst.[231,232] Rather, a strong glycolytic phenotype is necessary to carry out both phagocytic function and the oxidative burst. In case of the latter, glycolysis is critical to fuel the oxidative phase of the PPP to produce NADPH from NADPH oxidase 2 (NOX2).[233] As part of the oxidative burst, neutrophils require oxygen not to fuel OXPHOS but rather sustain NOX2 activity and produce superoxides and other types of ROS to combat phagocytosed pathogens.[234] In fact, patients with impaired neutrophil function due to mutations in NADPH are prone to severe bacterial and fungal infections called chronic granulomatous disease.[235] NETosis—the next major neutrophil effector mechanism—is divided into two metabolic phases with differing requirements for exogenous glucose with little dependency on glutamine and mitochondrial function.[236] Notably, mTOR-dependent HIF-1α expression is critical for NETosis and the production of granule proteases and antimicrobial peptides, as well as for survival under hypoxia, a common feature within sites of inflammation.[237-239] Neutrophils are classically known to possess few mitochondria. As such, mature neutrophils have considerably less ETC activity stemming from a lack of mitochondrial supercomplex organization. That said, any few functional mitochondria and the TCA enzymes contained within are essential for chemotaxis and cell morphology.[231,240] Additionally, these mitochondria aid in the maintenance of membrane potential and contain an important supply of proapoptotic proteins (Smac/DIABLO and HtrA2/Omi) that are poised to promote apoptosis if released into the cytosol.[240] While NOX2 production of ROS plays a significant role in NET formation, mROS can also contribute to NETosis when NOX2 is unable to generate sufficient ROS.[228,241]

Macrophages: M1/M2 Paradigm

As part of the rapid innate response, macrophages require the ability to rapidly escalate their metabolic activity to permit prompt functional responses. While imperfect, the in vitro M1 and M2 model of macrophage activation and polarization remains a useful tool to categorize macrophages based on function (Chapter 16). This in vitro categorization likely reflects the extrema of the tissue-induced differentiation that naturally occurs along a continuum in vivo.[242] Additionally, as with T cells, macrophage differentiation is also intricately linked to metabolic reprogramming. Classically activated macrophages, or M1 macrophages, are generated in response to IFN-γ or TLR agonists such as LPS, whereas alternatively activated macrophages, or M2 macrophages, are generated in response to type II cytokines such as IL-4 or IL-13. Interestingly, canonical definition of these two subsets has usually included expression of enzymes involved in arginine metabolism. That is, inducible nitric oxide synthase (iNOS) expression in defining M1 macrophages and arginase 1 (Arg1) expression as an important marker for M2 macrophage identification and function (Fig. 6.10A).

M1 Macrophages

Activation of M1 macrophages is associated with a concomitant increase in glycolysis. A prominent feature of M1 activation is the increased expression of the glucose transporter Glut1.[243] Glut1-deficient macrophages not only demonstrate decreased glycolysis, but they also decrease iNOS expression.[244] Interestingly, IL-6 and tumor necrosis factor (TNF) production remain unaffected. Alternatively, Glut1 overexpression augments glycolytic capacity with enhanced M1 gene expression and function.[245] Inhibiting GAPDH function in macrophages with heptiledic acid reduces expression of IL-1β and iNOS expression.[246] Blocking glycolysis with 2-DG also abrogates LPS-induced upregulation of IL-1β by inhibiting the upregulation of HIF-1α, which binds to the IL-1β promoter to regulate gene expression.[4,247] HIF-1α plays an important role in both upregulating the metabolic programs necessary to support M1 function as well as a direct role in the upregulation of various effector molecules.[248,249] Consistent with this finding, HIF-1α–deficient macrophages have diminished levels of PGK1 and Glut1, which are associated with decreased bactericidal killing, cell aggregation, and motility.[249] Further macrophage regulation by HIF-1α of macrophage function is facilitated by the glycolytic enzyme PKM2, which converts PEP to pyruvate. This multifunctional metabolic enzyme with additional "moonlighting" roles can form dimers and interact with stabilized HIF-1α in the nucleus leading to enhanced expression of glycolytic genes and IL-1β production.[250] Thus, M1 macrophages mainly use glycolysis for ATP production while relying on the oxidative PPP to produce sufficient NADPH levels for ROS generation.[251,252] Disrupting the oxidative PPP through deletion of enzymes involved in the PPP leads to a reduced proinflammatory phenotype.[253] Likewise,

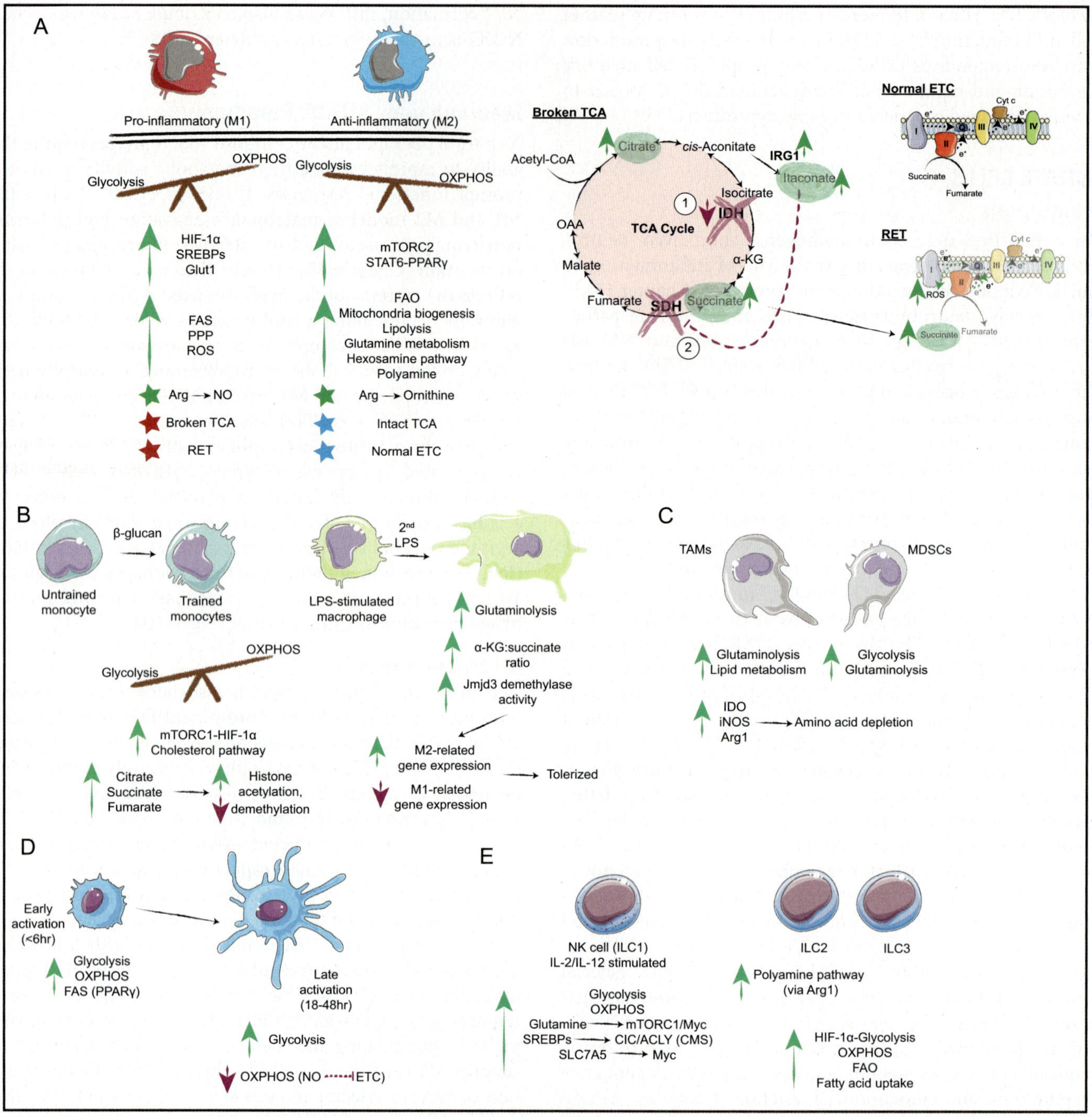

FIG. 6.10. Metabolic phenotypes of innate immune cell subsets. Innate immune cells display specialized metabolic profiles and activity that influence their activation and function. **A:** Overview of metabolic differences between in vitro-generated M1 (proinflammatory) and M2 (anti-inflammatory) bone marrow–derived macrophages (BMDMs) on the left. Graphical depiction of the broken tricarboxylic acid (TCA) cycle and reverse electron transport (RET) observed in M1 macrophages (*right*). **B:** Overview of metabolic differences between trained and tolerized immune cells. **C:** Within the tumor microenvironment, tumor-associated macrophages (TAMs) and myeloid-derived suppressor cells (MDSCs) adopt a unique metabolic phenotype. **D:** Metabolic differences between early and late activation in dendritic cells. **E:** Metabolic differences between different subsets of innate lymphoid cells (ILCs).

enhancing the reductive PPP by overexpressing carbohydrate kinase–like protein (CARKL), a sedoheptulose kinase that catalyzes the formation of sedoheptulose-7-phosphophate, also reduces M1 function.[254] CARKL-deficient macrophages increase proinflammatory cytokine production, which is consistent with the role of CARKL in negatively regulating M1 differentiation since macrophages downregulate CARKL upon LPS stimulation.

A classic metabolic hallmark of M1 macrophages is their low OXPHOS state, which can be attributed to the role of increased NO build up inhibiting the components in the ETC.[251,255,256] However, the mitochondria act as a central hub in response to TLR signaling to support their function. TLR activation induces mROS production and plays a critical role in mediating bactericidal activity.[257] However, while glycolysis takes on the primary role of generating ATP generation, mitochondrial TCA cycle intermediates, citrate and succinate, play a critical role in directly regulating macrophage function.[4,252] LPS stimulation upregulates expression of the mitochondrial citrate transporter (CIC or SLC25A1), which allows citrate to move into the cytosol in exchange for malate.[258] Inhibition of citrate transport reduces the production of NO and ROS as well as the production of prostaglandin E2 (PGE2) and pro-IL-1β.[258-260] LPS-stimulated macrophages convert cytosolic citrate to generate acetyl-CoA through ACLY for histone modifications to upregulate the proinflammatory gene program.[261,262] Besides modifying histones, acetyl-CoA can be converted into malonyl-CoA by ACC1, which is an essential step for fatty acid synthesis. In addition to being a substrate for fatty acid synthesis, malonyl-CoA is also involved in the posttranslational modification of GAPDH at lysine residues.[263] Similar to the "moonlighting" role in T cells, GAPDH binds to the 3′-UTR of *Tnfa* mRNA, thereby inhibiting its translation.[264] Upon undergoing malonylation, GAPDH dissociates from *Tnfa* mRNA and allows enhanced translation of TNF-α.[263] Furthermore, acetyl-CoA conversion to acetoacetyl-CoA by fatty acid synthase (FASN) is critical to increase cholesterol levels to enable TLR4 recruitment to lipid rafts in response to LPS.[265]

Succinate accumulation—primarily derived from glutamine anaplerosis (the process of forming metabolic intermediates to fuel the TCA cycle)—promotes M1 macrophage function by driving HIF-1α–dependent IL-1β production in response to LPS stimulation.[4,247] Mechanistically, succinate leads to HIF-1α stabilization by directly inhibiting the PHD proteins, a class of proteins that degrades HIF-1α in an oxygen-dependent manner, thereby promoting IL-1β production.[4] Furthermore, oxidation of succinate to fumarate through succinate dehydrogenase (SDH), also referred to as complex II of the ETC, generates electrons that are directed retrograde back to complex I, a process that is known as reverse electron transport (RET), thereby generating mROS to sustain the proinflammatory M1 phenotype.[266] As such, accumulation of mROS also inhibits PHD activity and further stabilizes HIF-1α expression. Along these lines, inhibition of mROS generated from both mitochondrial complex I and II limits IL-1β production from M1 macrophages. Besides its intracellular role, extracellular succinate levels sustain inflammation by activating the G-protein–coupled receptor GPR91.[267] Mechanistically, LPS-induced upregulation of GPR91 acts as an autocrine and paracrine succinate sensor to further promote IL-1β production through HIF-1α.

In response to LPS, proinflammatory macrophages significantly rewire their TCA cycle by introducing two critical breaks in the cycle as a late-stage anti-inflammatory measure to prevent exuberant inflammation (Fig. 6.10A). During LPS stimulation, the first break occurs due to reduced expression and function of the TCA cycle enzyme isocitrate dehydrogenase (IDH), an enzyme that converts isocitrate to α-KG.[252] As a result, citrate buildup ultimately leads to a significant increase in itaconate production by immunoresponsive gene 1 (IRG1). IRG1 is significantly upregulated in M1 macrophages and is responsible for the production of itaconate from *cis*-aconitate.[268] Genetic deletion of IRG1 in macrophages leads to a marked diminution of itaconate accumulation and subsequently an elevated proinflammatory response.[269] The generation of itaconate in the first break facilitates the second break of the TCA cycle resulting in succinate accumulation.[252] Mechanistically, itaconate competitively inhibits SDH function to prevent succinate oxidation and reduces the reverse-electron-transport production of mROS, which is critical for HIF-1α stabilization and IL-1β expression.[269] Itaconate also activates the transcription factor NRF2 to initiate a potent antioxidant measure to combat ROS buildup and inhibit IL-1β production.[270] To this end, itaconate is emerging as a potent modulator of M1 macrophage function that helps prevent an overly inflammatory response.

In addition to the TCA cycle, amino acids are able to regulate M1 macrophage function. Arginine metabolism plays a critical role in facilitating both M1 and M2 macrophages function. M1 macrophages break down arginine to form nitric oxide (NO) and citrulline through increased iNOS expression.[271] Consistent with this finding, arginine uptake through cationic amino acid transporters (CATs) promotes M1 macrophage function.[272] CAT2-deficient macrophages fail to upregulate NO in response to LPS and IFN-γ.[272] Furthermore, macrophages can also produce arginine through the citrulline-nitric oxide cycle by two enzymes: argininosuccinate synthase (ASS1) and argininosuccinate lyase (ASL).[273,274] Mice deficient in ASS1 are unable to control bacterial infections, highlighting the importance of arginine metabolism for NO production in M1 macrophages.[273] Serine is also important in sustaining IL-1β production in M1 macrophages by providing a steady glycine supply to generate glutathione and thus maintain cellular redox balance.[275]

Fatty acid metabolism in proinflammatory macrophages is tightly regulated, partly due to a variety of byproducts that can be produced by lipid and cholesterol metabolism. In M1 macrophages, SREBP1-dependent fatty acid synthesis facilitates membrane remodeling required for phagocytosis and aids in the production of inflammatory mediators.[276,277] Upon LPS stimulation, macrophages upregulate SREPB1 signaling through mTORC1 activation, leading to an increase in expression of FASN.[276] Macrophages deficient in SREBP1a, an isoform that promotes the transcription of genes involved in fatty acid synthesis and the Nlrp1a inflammasome subunit, have reduced proinflammatory cytokine production and lipid biosynthesis.[277] Blocking fatty acid synthesis either by pharmacological inhibition or siRNA silencing compromises proinflammatory cytokine production followed by LPS stimulation.[265] SREBP1-deficient macrophages also fail to upregulate anti-inflammatory fatty acids after LPS stimulation, leading to a sustained proinflammatory response when macrophages should be in the resolution phase (12-24 hours).[278] During antiviral responses, type I IFN signaling inhibits cholesterol biosynthesis but rather increases import of cholesterol and lipids.[279] SCAP-deficient macrophages, which lack both SREBP1 and SREBP2, or silencing SREBP2 promote type I IFN signaling through the repressing the cholesterol mevalonate pathway and enhancing the activation of the TBK1-STING pathway. However, in response to type I IFN signaling, cholesterol intermediates such as oxysterols (ie, 25-hydroxycholesterol [25-HC]) serve to play a critical feedback regulatory role in inhibiting IL-1β inflammation by limiting inflammasome activation.[280,281]

The metabolic pathways mentioned above are partly regulated by mTORC1 signaling. However, the role of mTORC1 in M1 macrophage function is somewhat paradoxical. In T cells, mTORC1 signaling plays an important role in effector differentiation by promoting glycolytic metabolic programming. Interestingly, in spite of the fact that glycolysis is important for M1 function, inhibition of mTOR, either pharmacologically or genetically, leads to enhanced production of proinflammatory cytokines. Mechanistically, this has been attributed to the regulation of NF-κB signaling; however, the precise mechanisms for this finding are not still clear. mTOR signaling positively regulates STAT3 signaling, which is responsible for promoting IL-10 expression while negatively regulating NF-κB signaling needed to promote proinflammatory cytokine production such as TNF-α and IL-12.[58,282] In fact, IL-10, as an anti-inflammatory measure, alters metabolic reprogramming of LPS-activated macrophages by inhibiting glucose uptake and glycolysis while promoting mitophagy to remove dysfunctional mitochondria.[283] Thus, inhibition of mTOR signaling with rapamycin augments NF-κB activity and inhibits STAT3 signaling, which results in increased production of proinflammatory cytokines while suppressing the expression of IL-10. However, hyperactivation of mTORC1, through the deletion of TSC1, also promotes proinflammatory cytokine production.[284,285] One study suggests that the hyperinflammation from TSC1-deficient macrophages is due to increased ERK signaling.[284] These studies illustrate the complexity of mTORC1 signaling in promoting M1 macrophage function and highlight that this area remains under investigation.

M2 Macrophages

Although M2 macrophages demonstrate increased glycolytic function relative to unpolarized M0 macrophages, the glucose consumption and lactate production of M2 macrophages appear to be much lower than M1 macrophages[62] (Fig. 6.10A). Inhibition of glycolysis by 2-DG abrogates upregulation of M2 markers such as PD-L2 and resistin-like molecule α (RELMα) under M2-polarizing condition.[62,286,287] Likewise, inhibition of pyruvate transport to the mitochondria by UK5099, which inhibits MPCs, also inhibits M2 polarization and oxygen consumption. Glucose-derived pyruvate is critical in promoting M2 gene expression during early polarization.[62] As such, knocking down PDHK1 during early M2 polarization, thus promoting pyruvate entry into the TCA cycle, enhances M2 gene expression such as YM1, FIZZ1, Arg1, and CD206/MRC1.[286] Glucose metabolism mediated by IL-4 activation of mTORC2 signaling is critical for M2 polarization and function both in vitro and in vivo.[62,63] For instance, mice with selective Rictor deletion in macrophages demonstrate compromised immune responses to parasites in vivo as well as an inability to maintain core body temperature in hypothermia-inducing conditions.

In addition to glucose, glutamine is also a pivotal nutrient for M2 macrophage polarization and function. In the absence of glutamine, M2 macrophages fail to upregulate CD206/MRC1, CD301, and RELMa expression.[252] These observations reflect the important role of glutamine (along with glucose) to fuel the hexosamine pathway and generate UDP-GlcNAC to produce M2 cell surface proteins. In addition, histone 3 lysine 27 trimethylation (H3K27me3) is a repressive epigenetic mark that inhibits the transcription of M2 markers such as *Arg1*, *Ym1*, *Retnla*, and *Mrc1*, and IL-4 stimulation promotes Jmjd3-dependent demethylation to promote the expression of M2 genes.[288] One of the glutamine-derived metabolites, α-KG, is known to promote Jmjd3 demethylase activity while succinate inhibits it. As such, glutaminolysis supports a high α-KG to succinate ratio to induce Jmjd3-dependent demethylation of H3K27. Consequently, polarization of macrophages with IL-4 decreases H3K27me3 at promoter regions of these genes in a glutamine-dependent manner.[289] In contrast, glutamine-derived α-KG suppresses M1 activation. In addition, α-KG promotes PHD-mediated posttranslational modification of IKKβ thereby inhibiting M1 function.[289] Thus, the ratio of α-KG to succinate regulates the balance of M1 and M2 polarization. Also, a steady pool of acetyl-CoA generated by ACLY plays a critical role for histone acetylation and promoting M2 gene expression such as *Retnla* and *Arg1* in murine macrophages.[290] However, another study suggests ACLY is dispensable during human M2 macrophage differentiation and histone modifications.[291] Not surprising, there are many sources to supply an intracellular pool of acetyl-CoA required for epigenetic modifications that can be generated independent of ACLY such as from fatty acids.[290]

As mentioned above, long-chain FAO is an integral part of M2 metabolic reprogramming and function by providing acetyl-CoA for histone modifications. IL-4 signaling through STAT6 increases expression of PGC-1β, which in turn promotes mitochondrial biogenesis.[292] STAT6 also activates the nuclear receptor PPARγ to facilitate fatty acid uptake by inducing the expression of the transporter, CD36.[293] Uptake of lipids by CD36 and subsequently lipolysis by lysosomal acid lipase (LAL) provides additional substrate for OXPHOS that is critical for proper M2 macrophage polarization.[294] Interrupting this pathway, either by knocking down CD36, or inhibiting or knocking down LAL, all reduce M2 polarization.

Arginine metabolism also plays an important role in M2 function. Arg1, which is a defining marker of M2 macrophages (in mouse), facilitates hydrolysis of arginine to ornithine and urea. Arg1 is critical in maintaining M2 macrophage function by diverting arginine away from NO production. Ornithine itself is an important substrate for collagen synthesis and thus is essential for wound healing and tissue remodeling.[295] Furthermore, ornithine is converted to putrescine by ornithine decarboxylase (ODC) and feeds into the polyamine pathway. Macrophages that are deficient in ODC possess epigenetic changes that lead to increased expression of M1 proinflammatory genes.[296] Supplementation of putrescine to ODC-deficient macrophages reverses this M1 proinflammatory phenotype, demonstrating that the polyamine pathway helps to maintain the M1 and M2 balance. To this end, putrescine is involved in the polyamine-dependent posttranslational modification (hypusination) of hypusine, an uncommon amino acid, on eIF5A, which regulates protein translation of OXPHOS-related genes in M2 macrophages.[297]

Trained Immunity Versus Tolerant Macrophages

Trained immunity was first observed in mice injected with β-glucan providing protection against a subsequent lethal injection of *Staphylococcus aureus*.[298] Trained immune cells demonstrate enhanced cytokine production upon secondary challenge, as opposed to a tolerized phenotype, which has been observed in LPS prestimulated immune cells. Mechanistically, trained immunity has been attributed to epigenetic changes induced upon initial stimulation, with responsive genes acquiring marks at histone 3 lysine 27 acetylation (H3K27Ac) and histone 3 lysine 4 trimethylation (H3K4me3), leading to enhanced gene expression associated with a proinflammatory phenotype such as increased production of TNF-α.[299-301] Recent studies have suggested that these epigenetic changes are in fact metabolically regulated (Fig. 6.10B). Trained human monocytes display an enhanced glycolytic program that is dependent on the mTORC1-HIF-1α pathway.[300] Deletion of HIF-1α in murine myeloid cells abrogates β-glucan-induced trained immunity and cannot protect mice against a secondary lethal dose of infection. β-glucan-trained monocytes also have reduced oxygen consumption compared to untrained monocytes; however, there is a significant increase in TCA cycle intermediates such as citrate, succinate, and fumarate—a phenomenon that is also associated with proinflammatory macrophages. Both succinate and fumarate inhibit demethylases such as Tets and KDMs.[300] The intermediate, mevalonate, within the cholesterol biosynthesis pathway, plays a crucial role in promoting trained immunity by increasing IGF1-R and mTOR activity for the necessary epigenetic modifications.[302] Statins, which inhibit mevalonate production, hinder trained immunity. However, the precise mechanism by which IGF1-R promotes trained immunity has not been fully elucidated. Furthermore, mevalonate-trained macrophages and β-glucan-trained macrophages display completely different epigenetic landscapes, further suggesting the different spectrum of macrophage responding to environmental signals.

On the opposite end of the spectrum of trained immunity is tolerance, a state in which LPS-primed macrophages are refractory to secondary stimuli and unable to produce proinflammatory cytokines. Glutamine-fueled α-KG production during the initial LPS priming phase plays a critical role inducing endotoxin tolerance of proinflammatory genes by impairing the NF-κB pathway[289] (Fig. 6.10B). Consequently, macrophages primed in the absence of glutamine or by blocking glutaminolysis, but then challenged again with LPS, show higher expression of proinflammatory genes compared to glutamine replete conditions. Similarly, β-glucan stimulation can also reverse the tolerized state induced by LPS priming, potentially rewiring the epigenetic landscape that leads to an unresponsive state.[301]

Tissue-Resident Macrophages

While the M1 and M2 paradigm is instructive in dissecting macrophage polarization and function, it is clear that in vivo, there is a great spectrum of differentiation based on the specific tissue microenvironment. It has become clear that some subsets of tissue-resident macrophages are established during early embryonic and neonatal development from yolk sac or fetal liver even before the bone marrow is developed.[303] Their primary role is to maintain tissue homeostasis, while replenishing themselves through self-renewal depending on the tissue. Tissue-resident macrophages have unique transcriptional and metabolic programs that support their function within distinct tissues.[304]

Peritoneal tissue-resident macrophages are defined by the transcription factor GATA6.[305,306] Functionally, these cells are responsible for maintaining tissue homeostasis in the peritoneal cavity by controlling infection and also supporting the resolution phase after an infection by engulfing apoptotic cells. Interestingly, although peritoneal tissue-resident macrophages express many markers found on M2 macrophages, metabolically they are quite distinct.[64,307] Indeed, while mTORC2 signaling is necessary for M2 macrophage differentiation, inhibition of mTORC2 actually promotes the generation of peritoneal tissue-resident macrophages.[64] Also, homeostasis of skin-resident Langerhans cells depends on mTORC1 signaling to maintain normal turnover rates and to prevent apoptosis.[308] TSC2-deficient macrophages increase expression of cyclin-dependent kinase (CDK4), leading to enhanced proliferation and glycolytic program driving granuloma formation in the lung.[309] Peritoneal macrophages possess higher basal OXPHOS compared to monocytic macrophages.[64,307] In addition, the oxidative burst in peritoneal macrophages is primarily fueled by glutaminolysis.[307] Along these lines, lipid metabolism has been shown to be important in some tissue-resident macrophages. One function of macrophages is to take up lipoproteins and free fatty acids. These lipids are then digested in lysosomes to generate intracellular free cholesterol and fatty acids. FAO is regulated by genetic programs through the PPAR family, LXR family, C/EBP family, and SREBP transcription factors.[310] Genetic disruption of these programs leads to reduced splenic marginal zone macrophages indicating the importance of lipid metabolism in regulating the homeostasis of tissue-specific macrophages. Furthermore, deletion of PPARγ in macrophages disrupts lung homeostasis and induces mild inflammation under homeostatic conditions.[311] PPARδ-deficient Kupffer cells (liver-resident macrophage) and macrophages residing in adipose tissues show a defect in M2-like features, further emphasizing the importance of PPAR-dependent lipid metabolism in regulating tissue-resident macrophages.[312] Besides upregulating CD36 to facilitate the uptake of fatty acids, PPARγ also induces upregulation of genes essential for cholesterol efflux such as ABCA1.[313] Thus, PPARγ plays an important role particularly in adipose tissue-resident macrophages for the uptake, processing, and export of cholesterol metabolism to regulate fatty acid and cholesterol homeostasis.

Tumor-Associated Macrophages and Myeloid-Derived Suppressor Cells

Tumor-associated macrophages (TAMs) reside in the TME and promote tumor growth and immune evasion.[314] They are generally thought to adapt a M2-like phenotype by suppressing the antitumor inflammatory response. For many heterogeneous tumors, the vast majority of the solid tumor mass itself consists of macrophages. The TME is greatly influenced by the high anabolic metabolic demand of the cancer cells. The TME is thus hypoxic, acidic, and often depleted of essential nutrients.[315] Consequently, the metabolism of TAMs must be well suited to promote tumor growth while maintaining their own survival within the TME. In fact, tumor produced lactic acid promotes M2-like polarization to increase Arg1 expression mediated by HIF-1α.[316]

TAMs express high levels of the immunosuppressive enzyme, indoleamine 2,3-dioxygenase (IDO), which converts extracellular tryptophan to kynurenine.[317] Kynurenine can act as a ligand and bind to AhR to suppress effector T cell function and promote T_{reg} generation.[318] In addition, TAMs express Arg1 and iNOS, both of which utilize arginine as substrates and thus deplete arginine levels from the TME depriving T cells of this critical nutrient.[319,320] TAMs also produce high levels of itaconate that also plays a role in promoting tumor growth.[321] Peritoneal macrophages isolated from mice bearing peritoneal tumors express higher levels of IRG1 and consequently increases levels of itaconate, while genetic deletion of *Irg1* from peritoneal macrophages reduces tumor growth.[321] Metabolically speaking, similar to M2 macrophages, glutamine metabolism is essential to form TAMs. Inhibiting glutamine metabolism in TAMs promotes a proinflammatory phenotype with enhanced TNF-α production[322] (Fig. 6.10C).

Similar to TAMs, myeloid-derived suppressor cells (MDSCs) promote tumor growth by suppressing potent antitumor immune responses. Metabolic programs of MDSCs not only serve to promote their own survival within the TME but also facilitate the inhibition of the metabolism and function of tumor-specific T cells. In addition to expressing Arg1, IDO, and iNOS, MDSCs also express xCT, a transporter that imports cystine, which depletes extracellular pools of cysteine that support T cell growth and function.[323,324] Both glycolysis and glutaminolysis are important to support the recruitment and function of MDSCs[322,325] (Fig. 6.10C). Indeed, targeting glycolysis with 2-DG or glutamine metabolism with the antagonist DON can inhibit the recruitment, generation, and function of MDSCs within the TME and thus enhance cancer immunotherapy.

Dendritic Cells

Mature dendritic cells (DCs) play a central role in T cell priming and activation by bridging the innate and adaptive arms of the immune response. The DC field is complex when discussing murine and human DC nomenclature; however, DCs can be classified into major subsets and even subtypes as described more in depth in Chapter 25. However, most of our current understanding of DC metabolism is through in vitro studies using TLR agonist (ie, LPS) stimulation of murine bone marrow–derived DCs (BMDCs) using either granulocyte-macrophage colony-stimulating factor (GM-CSF) or FLT3L. While these methods of in vitro differentiation are much needed, it is important to keep in mind that BMDC cultures are quite heterogeneous. Therefore, these culture conditions can only provide us with general insight and not specific details into DC metabolism of specific subsets. Immature DC activation in response to cytokines, PAMPs, and receptor-ligand interactions results in the production of cytokines (ie, IL-12) and chemokines for trafficking as well as increased antigen processing activity and costimulatory molecule expression. Although activation does not result in robust proliferation, resting DCs dramatically alter their metabolism to support the bioenergetic and biosynthetic requirements to support maturation and function.[326]

Regulation of PI3K-Akt-mTORC1 signaling is a critical metabolic checkpoint for the development, homeostasis, and differentiation of both murine and human DCs. PTEN mediated inhibition of PI3K-mTOR signaling downstream of FLT3L is critical for the normal development of $CD8\alpha^+$ cDC1s (conventional DCs) and has been shown to be required for human pDC (plasmacytoid DC) differentiation.[327,328] Inhibiting mTORC1 signaling in vitro with rapamycin impairs FLT3-mediated DC development whereas in vivo PTEN deletion expands $CD8\alpha^+$ and $CD103^+$ cDCs. Furthermore, unrestrained mTORC1 activity via TSC1 deletion and consequent enhanced metabolic programming via Myc leads to a marked defect in DC development in vivo due to spontaneous maturation and impaired survival and proliferation.[329] L-Myc, a Myc paralog, is also abundantly expressed in common DC precursor cells, pre-cDCs, and mature splenic cDC1s ($CD8\alpha^+$).[330] L-Myc is required to transcriptionally support cDC-T cell priming responses. Moreover, mTOR is critical in programming homeostasis and function of tissue-resident DCs in various tissues like the lung and the liver by promoting SREBP-dependent lipid synthesis.[331] Rapamycin exposure during human moDC (monocyte-derived DC) differentiation reduces the number of DCs and expression of costimulatory molecules. DCs generated under such conditions are actually tolerogenic and hinder T cell activation.[332] Interestingly however, differentiation of DCs with rapamycin promotes IL-12 production by inhibiting the autocrine secretion of IL-10, which typically acts a negative feedback mechanism to IL-12.[59,60] These findings seem inconsistent since suppressing mTORC1 activity is generally associated with reduced glycolytic programming and effector function in other cell types especially T cells.

Resting BMDCs rely on FAO to fuel the TCA cycle, but upon stimulation, DCs undergo two phases of metabolic programming: early (<6 hours) and late (18-48 hours)[326,333] (Fig. 6.10D). Pattern recognition receptor (ie, TLR4) stimulation within minutes quickly elevates both mitochondrial activity and aerobic glycolysis of resting DCs in various in vitro and in vivo/ex vivo murine culture settings. During acute LPS stimulation, GM-CSF–derived BMDCs breakdown intracellular stored glycogen (via glycogen phosphorylase [PYG]) to generate a steady source of carbons to fuel the TCA cycle and glycolytic programming to support early DC maturation.[334] In the absence of glycogen metabolism by inhibiting its synthesis or breakdown, there is significant impairment in DC maturation and survival.

Increased glucose uptake and expression of glycolytic enzymes are essential for the maturation and function of GM-CSF–derived BMDCs and cDCs to support the expression of costimulatory molecules (CD86), antigen presentation molecules (MHC I and II), and cytokines (IL-6, IL-12p70, and TNF-α) production.[335] LPS stimulates the TBK1-IKKε-Akt signaling axis to induce early glycolytic metabolism by phosphorylating HK II to promote its association with the mitochondria which subsequently elevates glycolytic activity. In fact, activation of DCs in hypoxia stabilizes HIF-1α expression and enhances glycolytic metabolism thus promoting the expression of costimulatory molecules and cytokine function.[336] Conversely, pDC stimulation via TLR9 by CpG does not immediately increase glycolysis but rather relies on PPARα induced FAO and OXPHOS through autocrine type I IFN signaling, which is essential for full activation.[337] Also, as activated DCs begin to mature, HIF-1α and glycolysis help to induce cytoskeletal changes needed for reshaping their morphology to support DC motility and CCR7 oligomerization to facilitate DC migration into lymph nodes.[338,339] Under resting conditions, analysis of splenic cDC1s ($CD8\alpha^+$) reveal higher glycolytic (ECAR) and mitochondrial metabolism (OCR and mitochondrial mass) compared with cDC2s ($CD8\alpha^-$).[340] Unlike cDC2s, cDC1s require this enhanced metabolic state to perform their functional role in priming T cells.

However, after prolonged LPS stimulation, DCs dramatically reduce their mitochondrial metabolism as they commit to glycolytic metabolism[333] (Fig. 6.10D). This commitment to glycolysis is critical as iNOS expressing DCs generate excess levels of nitric oxide (NO), thereby inhibiting the ETC and OXPHOS metabolism. Interestingly, rapamycin inhibition of mTORC1 prevents induction of iNOS expression and NO production that promotes survival of LPS activated GM-CSF–derived BMDCs by preserving mitochondrial function and activity.[341] Rapamycin treatment of murine DCs during activation enhances their ability to activate naïve $CD8^+$ T cells.[342] Such a strategy was shown to significantly improve therapeutic efficacy of an autologous vaccine to a B16 melanoma model.[342] During human moDC differentiation, early PGC-1α and later both TFAM and NRF1 expression induces mitochondrial biogenesis and provides DCs with potent antioxidant protection to manage ROS levels due to higher OXPHOS.[343] Furthermore, increased expression of the mitochondrial fusion proteins, MFN2 and Opa1, facilitates dynamic mitochondrial changes during DC differentiation that support their ability to migrate.[344]

In addition to glucose metabolism, fatty acid and lipid metabolism play an essential role during DC development, activation, and function. PPARγ promotes early moDC differentiation by inducing transcription of lipid metabolism– and transport-related genes before genes related to immune function are even increased.[345] Fatty acid synthesis is necessary for in vivo murine dendropoiesis as treating mice with the FASN inhibitor C75 suppresses the number of cDCs in the liver and other organs without impacting other immune cell counts.[346] Upon LPS stimulation, the rapid glycolytic burst generates pyruvate not only for lactate formation, but it also feeds into the mitochondrial TCA cycle to generate citrate.[335] Citrate is then transported into the cytosol through the mitochondrial citrate transporter CIC and is converted into acetyl-CoA by ACLY to provide additional carbons for *de novo* synthesize of fatty acids. Increased fatty acids facilitate the expansion of the ER and the Golgi network to support the buildup of organelles necessary for antigen processing and presentation but also production of cell surface molecules (ie, CD86, MHC II) and cytokines (ie, IL-12, TNF-α). Interestingly, blocking FAO or glutaminolysis does not affect DC activation demonstrating glucose-derived citrate is the primary fuel for the TCA cycle.

Compared with mature human moDCs, tolerogenic DCs possess a unique metabolic phenotype associated with high OXPHOS and more SRC but also increased glycolytic capacity and reserve without any changes in basal glycolytic activity.[347] This metabolic phenotype depends on enhanced FAO-based metabolism since inhibiting FAO activity in tolerogenic DCs partially rescues T cell responses.[347] Tumor-associated DCs reside in a glucose-deprived environment thereby potentially impairing DC activation and function and rendering them in a tolerogenic state preventing potent T cell responses. Furthermore, accumulation of cytosolic lipids by uptake through the Msr1 scavenger receptor negatively alters antigen processing and presentation in both murine and human tumor-residing DCs, thus hindering T cell activation.[348] Elevated ROS in tumor-associated DCs increases byproducts derived from intracellular lipid oxidation such as 4-hydroxynonenal (4-HNE). The byproducts then activate the ER stress response mediated by XBP1 and potentiate the accumulation of fatty acids and triglycerides, which subsequently suppress protective antitumor immunity.[349]

Innate Lymphoid Cell Metabolism

The discovery of a new subset of lymphocytes (non-T and non-B) led to the classification of what is now called innate lymphoid cells (ILCs) (Chapter 19).[350] ILCs are composed of a heterogeneous family of lymphocytes (ILC1, ILC2, and ILC3) residing in various lymphoid and nonlymphoid tissues primarily enriched at the epithelial interface such as the intestine or skin. Tissue-residing ILCs play critical developmental and biological roles in initiating immune responses as a first-line defense but also regulating tissue homeostasis and systemic metabolism of the host.

ILC1s

Natural killer (NK) cells, classified within the ILC1 subset, compose a significant arm of the innate system as highly plastic responders against viruses and cancer. NK cell development in the bone marrow requires highly anabolic processes as cells undergo many rounds of proliferation. For rapidly proliferating pre-NK cells ($CD11b^{lo}CD27^{hi}$),[351] mTOR signaling is absolutely required to coordinate proper metabolic reprogramming and to promote optimal surface expression of CD122 and CD132 as these subunits make up the IL-15 receptor to drive IL-15-driven proliferation and survival in the bone marrow and periphery. Increased mTOR activity in pre-NK cells results in higher levels of CD71 and CD98 expression and also enhanced glucose uptake compared to mature NK ($CD11b^{hi}CD27^{lo}$) cells. Metabolic gene expression linked to cellular growth is upregulated in developing NK cells while gene expression related to cellular quiescence such as autophagy or fatty acid catabolism is found in resting NK cells. In the absence of mTOR, there is a defect in differentiation of $CD27^{hi}$ NK cells into more mature $CD11b^{hi}$ NK cells found in the periphery. Hyperactivation (though TSC1 deletion) of mTORC1 signaling in developing NK cells results in elevated proliferation and exhaustion in response to IL-15; therefore, tight regulation of mTOR activity is necessary for proper NK cell development.[352] However, intact mTORC2 signaling during development is also critical to limit hyper mTORC1 activity by restraining STAT5-mediated regulation of SLC7A5 expression.[353]

As NK cells integrate various inflammatory signals, sufficient nutrient availability and metabolic activity are critical in dictating NK cell effector responses which are characterized by rapid proliferation, a marked increase in cytokine production and enhanced cytotoxic function. Resting murine NK cells have a low basal metabolic profile by maintaining minimal rates of glycolysis and OXPHOS.[351,354] Acute activation (<4-6 hours) by receptor-ligand (NK1.1 or Ly49D) interaction or cytokine stimulation (IL-12/15, IL-12/18) results in no immediate changes in metabolic programming based on ECAR or OXPHOS even though these cells can still produce copious amounts of IFN-γ.[354] This contrasts to T cells in which upregulation of glycolysis is necessary to generate IFN-γ.[92] However, newly activated NK cells still rely on glucose to fuel OXPHOS for ATP. Inhibiting OXPHOS with oligomycin or glycolysis with 2-DG inhibits IFN-γ and granzyme B expression.[57,354] Indeed, 2-DG treatment impairs the NK cell response to MCMV infection in vivo.[355] Also, NK cells can be flexible in the type of fuel used to support function depending on the activating stimuli.[354] For example, IFN-γ production by receptor-mediated stimulation (NK1.1) is more sensitive to glycolysis and OXPHOS blockade compared to cytokine-mediated stimulation (IL-12/18).

In particular, glycolysis and OXPHOS are both critical to support NK cell function during prolonged infections.[57,351,356] Resting NK cells have minimal mTORC1 activity, but cytokine-mediated stimulation with IL-2, IL-15, IL-12, or IL-18 and certain combinations of cytokines results in an increase in mTORC1 activity.[57,351,356] Activated NK cells undergo dramatic mTOR-dependent metabolic reprogramming as they increase expression of enzymes (HK II, LDHA, etc) along with nutrient transporters (CD71, CD98, Glut1) to support nutrient uptake but also increase mitochondrial mass to support elevated OXPHOS activity and respiratory capacity.[57] As such, impairment in mTOR signaling prevents the upregulation of the glycolytic machinery, glycolysis, and solute transporters thereby inhibiting NK cell function. Myc is absolutely essential to control NK cell metabolism and function whereas HIF-1α is dispensable.[357] During MCMV infection, sufficient mTORC1 signaling is required to sustain NK cell proliferation and effector function for viral clearance.[351,355] With prolonged cytokine stimulation (ie, IL-15 or IL-2/12), NK cells further elevate their basal metabolism (ECAR and OCR) but with a decreased OCR:ECAR ratio, demonstrating greater dependence on glycolysis similar to activated T cells.[57,351]

NK cells stimulated with IL-2/12 utilize glucose to generate lactate but also divert portions of pyruvate into the mitochondria.[358] However, this pyruvate is not metabolized in the TCA cycle to fuel OXPHOS but rather shuttled out of the mitochondria as citrate through CIC as part of the citrate-malate shuttle (CMS) (Fig. 6.10E). This shuttle system helps maintain sufficient levels of cytosolic NAD^+ to facilitate GAPDH activity for glycolysis and importantly allows ACLY to generate cytosolic acetyl-CoA to be used in acetylation modifications or lipid synthesis. NK cells heavily depend on SREBPs to promote expression of the citrate transporter CIC and ACLY, in contrast to the more traditional role SREBPs play in T cells and macrophages regulating synthesis of fatty acids and cholesterol.[112,276,277] Interestingly, since NK cells do not use the TCA cycle to fuel OXPHOS, glutaminolysis of glutamine is not necessary for NK cells. Rather glutamine levels are essential to sustain high rates of Myc translation by facilitating amino acid transport through SLC7A5 to maintain Myc protein levels by balancing translation and degradation mediated by GSK3.[357]

Human NK cells can be classified by their maturation status based on the surface expression of CD56 ($CD56^{bright}$ or $CD56^{dim}$).[356] Mature $CD56^{dim}$ NK cells can readily induce cytotoxic killing of target cells whereas immature $CD56^{bright}$ NK cells readily produce IFN-γ but can later upregulate their cytotoxic machinery to kill targets. Like murine NK cells, human NK cells upregulate anabolic metabolism upon cytokine stimulation increasing both glycolysis and OXPHOS.[356] In response to cytokine exposure (IL-2 or IL-12/15), $CD56^{bright}$ NK cells are more metabolically active as they strongly upregulate mTORC1 activity and nutrient transporters (CD71, CD98, ASCT2, Glut1) compared to $CD56^{dim}$ NK cells.[356,359] Interestingly, only IL-2-mediated metabolic reprogramming is sensitive to rapamycin, which suggests IL-12/15 signaling drives metabolic activity independently of mTOR signaling.[356] Hence, metabolic restriction greatly impacts function of $CD56^{bright}$ NK cells in contrast to $CD56^{dim}$ NK cells. As such, increased glycolysis is absolutely essential to meet the metabolic demands required to rapidly produce maximal amounts of IFN-γ by $CD56^{bright}$ NK cells while sufficient amounts of ATP derived from OXPHOS is critical in both subsets of NK cells to support cytotoxicity and IFN-γ production.[356] Furthermore, human NK cells responding to NKG2D stimulation require IL-2 priming to induce mTORC1 activity and expression of ASCT2 (SLC1A5) and CD98 for degranulation and IFN-γ production.[359]

Like memory T cells, NK cells undergo metabolic reprogramming to maintain a self-renewing pool of long-lived memory or adaptive NK cells. During MCMV infection, $Ly49H^+$ NK cells respond to the viral ligand, m157 leading to massive clonal expansion during the acute stage.[360] The early clonal burst (day 7) leads to mitochondrial depolarization and an increase in mROS. As NK cells transition into memory, compromised or dysfunctional mitochondria are cleared through mitophagy mediated by BCL2 family members BNIP3 and BNIP3L. Induction of autophagy by inhibiting mTOR (via rapamycin) or activating AMPK (via metformin) during contraction enhances NK cell survival and overall numbers similar to CD8+ T cell memory.[51,168,360] HCMV-specific adaptive human NK cells demonstrate enhanced OXPHOS and mitochondrial mass and activity including higher SRC as a means to support long-term survival.[361] This long-lived phenotype is supported by the chromatin-modifying protein ARID5B to regulate expression genes involved in the ETC such as UQCRB and ATP synthase.

Impaired NK cell functional responses observed in chronic diseases such as cancer and obesity can be attributed in part to dysfunctional metabolic programs. The dependency on glucose as the primary fuel places NK cells at a disadvantage since microenvironments deprived of glucose (ie, tumors) will significantly hinder NK cell function.[57,119] Furthermore, a lactic acid rich environment such as that found in tumors leads to mitochondrial dysfunction, apoptosis, and impaired NFAT expression thereby inhibiting NK cell activation and IFN-γ production.[362] In obesity, PPAR family protein expression drives strong lipid metabolic programming, which consequently inhibits mTOR activity and impairs both glycolysis and OXPHOS leading to NK cell loss of cytotoxic function thereby hindering antitumor immunity.[363] Furthermore, tumors themselves can promote dysfunctional NK cell responses by inducing the expression of the enzyme fructose 1,6-bisphophatase (FBP1), which ultimately leads to impaired glycolysis and cell viability in NK cells.[364] Similarly, TGFβ, which can be highly enriched in the TME, can directly inhibit mTOR-dependent metabolic reprogramming in response to IL-15 stimulation.[365,366]

ILC2s and ILC3s

Like Th2 and Th17 cells, ILC2 and ILC3 cells play critical roles in type 2 and type 3 immunity, respectively.[350] ILC2s canonically express type 2 cytokines IL-4, IL-5, and IL-13. They can be found residing in the lungs where they might respond to allergic inflammation or the intestines during parasitic infections. ILC3s reside at mucosal barriers and respond to extracellular and commensal bacteria. Not surprisingly, the metabolic profiles of these ILCs are largely influenced by specific stimuli and tissue localization.

Based on transcriptional profiling of naïve intestinal ILCs, ILC2s demonstrate increased expression of genes related to steroid biosynthesis along with sphingolipid and amino acid (alanine and glutamate) metabolism.[367,368] Conversely, transcriptional analysis of intestinal ILC3s reveals the upregulation of genes involved in glycolysis along with fructose, mannose, arginine, and pyruvate metabolism.[368] More so, maturation of lung ILC2s requires proper VHL-mediated HIF-1α regulation to balance glycolysis and OXPHOS to epigenetically maintain expression of the IL-33 receptor ST2 and other ILC2-related genes.[369] In response to IL-33 cytokine exposure in the lung, activated ILC2s readily utilize both glycolysis and OXPHOS[370] (Fig. 6.10E). Moreover, constant Arg1 expression is readily observed in both precursor and mature ILC2s and plays a critical role in ILC2 responses to lung inflammation. Arg1 expression in activated ILC2s facilitates the generation of polyamines from arginine and to support maximal glycolytic capacity, which are both necessary for ILC2 proliferation and proinflammatory function.[370] The contrasting metabolic profiling of intestinal ILC2s versus lung ILC2s emphasizes the role of the tissue microenvironment in tailoring metabolic reprogramming of tissue ILCs.

Interestingly, both ILC2s and ILC3s acquire long-chain fatty acids from their extracellular environment especially within adipose tissue[371] (Fig. 6.10E). Helminth infections cause ILC2s to increase uptake of fatty acids to fuel FAO to expand in numbers and to support cytokine function.[371] In ILC3s, the essential dietary micronutrient vitamin A activates the retinoic acid receptors to bind to the *Il22* and *Rorc* promoters along with other genes necessary for development and function.[372] Deficiency in vitamin A leads to reduced numbers of intestinal ILC3s and enhances susceptibility to bacterial infections. However, vitamin A deficiency enhances type II immunity mediated by ILC2s providing even better protection from helminth infections. As mentioned above, ILC3s are glycolytic and as a consequence activated ILC3s require mTORC1 activity and HIF-1α–mediated metabolic reprogramming to promote proliferation and IL-17A and IL-22 cytokine production.[351,373] Furthermore, an increase in mitochondrial metabolism and the generation of mROS helps to stabilize HIF-1α glycolytic reprogramming to further support ILC3 proliferation, RORγt expression, and cytokine production.[373]

Invariant Natural Killer T Cells

Invariant natural killer T (iNKT) cells are a unique subset of T cells that possess a T cell receptor to recognize self and foreign lipids presented by nonpolymorphic MHC CD1d.[374] Like CD4+ T helper cell subsets, iNKT cells can be also classified based on the secretion of diverse cytokines and expression of transcription factors that can influence protective or pathogenic immune responses. iNKT cells are similar to NK cells as they can rapidly respond to antigen and can also express memory markers. In the thymus, mTORC1 signaling is critical during early iNKT cell development by localizing the essential transcription factor promyelocytic leukemia zinc finger (PLZF) transcription factor to its nuclear compartment.[375] Furthermore, intact mTORC1 signaling is necessary for proper iNKT cell activation and differentiation into iNKT-1 cells in response to stimulation.[376] In addition, tight regulation of mTORC1 activity by TSC1 is critical for terminal iNKT cell differentiation of iNKT-1 cells over iNKT-17 cells by promoting T-bet expression.[377] mTORC2 activity is also necessary for proper iNKT development by promoting proliferation of cells found in stage 1 of development and also subsequent stages.[378] Like Th2 cells, mTORC2 activity is necessary to promote iNKT-2 differentiation by controlling GATA3 expression and also IL-4 cytokine production. Furthermore, the AMPK-associated adaptor protein, folliculin-interacting protein (Fnip1), plays a critical role in thymic iNKT cell development at stage 2.[379] Fnip1 helps restrain mTORC1 activity and promotes mitochondrial biogenesis to sufficiently supply enough ATP and to prevent apoptosis during development.

Interestingly, neither resting nor activated iNKT cells are able to efficiently take up significant quantities of glucose compared to conventional $CD4^+$ T cells.[380] Furthermore, the carbons derived from glucose are not readily converted to lactate but rather enter the PPP and TCA cycle. As such, glucose restriction does not compromise iNKT cell survival and proliferation unlike $CD4^+$ T cells but does inhibit optimal cytokine expression. Rather than relying on a dominant glycolytic program, iNKT heavily depend on OXPHOS to maintain survival, proliferation, and effector function.[380] iNKT cell stimulation increases PPARγ expression to synergize with PLZF to enhance SREBP1 transcription for lipid synthesis.[381] Like T cells, lipids especially cholesterol enhance TCR signaling at the immunological synapse and thus supports optimal activation and IFN-γ production. In the absence of antigen, peripheral iNKT cells with higher ROS levels are more sensitive to oxidative stress and undergo apoptosis to maintain homeostasis and eliminate unwarranted inflammation. However, upon activation, iNKT cells decrease ROS levels to prevent cell death and then initiate an inflammatory response.[382]

TARGETING METABOLISM AS A MEANS OF REGULATING IMMUNE RESPONSES

The intimate link between metabolic programming and immune function provides the opportunity to regulate immune responses by targeting metabolism. By exploiting the differential metabolic programs of various immune cells, a tailored targeting approach can avoid indiscriminate suppression or activation of immune responses. Rather, even an inhibitor of a ubiquitous metabolic process can selectively regulate immune responses based on the differential metabolic demands of individual immune cells. For example, a generic inhibitor of glycolysis might mitigate T effector cells while simultaneously enhancing T_{reg} cells. That is, the inhibitor demonstrates cellular selectivity not based on its biochemical target but rather on the metabolic demand of the responding immune cells.

TARGETING METABOLISM TO ENHANCE CANCER IMMUNOTHERAPY

Over the last several years, cancer immunotherapy in the form of checkpoint blockade and adoptive cellular therapy (ACT) has revolutionized our approach to treating cancer (Chapter 51).[383] Cancer cells are metabolically programmed to support their prodigious growth.[11,384] As a consequence of their robust metabolism, cancer cells create an inhospitable tumor microenvironment (TME) that promotes the evasion of potent antitumor immune responses. The TME is highly hypoxic, acidic, and nutrient depleted. This environment thwarts the metabolism and hence the function of effector T cells while simultaneously promoting the recruitment and generation of suppressive macrophages and T_{regs}[385,386] (Fig. 6.11). In this

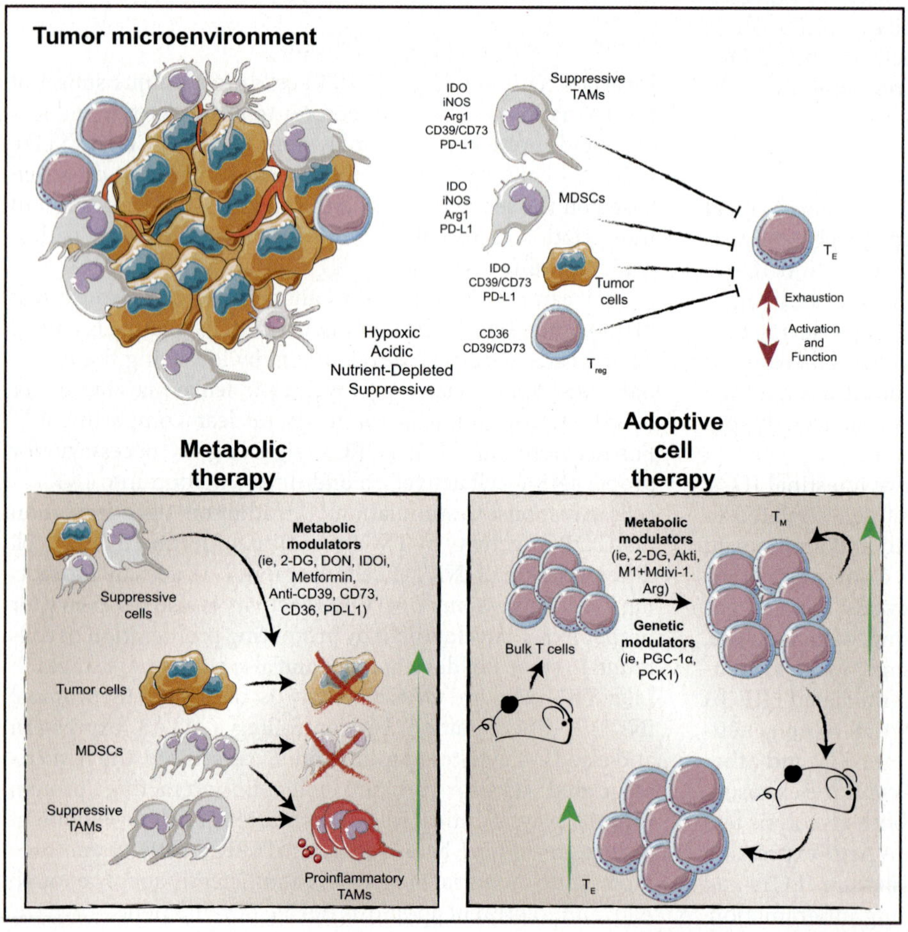

FIG. 6.11. Strategies to improve immunotherapy by altering the metabolism of immune cells. Targeting metabolism has the potential to enhance immunotherapy for cancer. Metabolic therapy can simultaneously inhibit tumor growth while also enhancing the robustness of T cell memory and function. Likewise, metabolic therapy can alter the tumor microenvironment (TME) by inhibiting IDO and the ectonucleotidases CD39 and CD73. Finally, targeting metabolism can enhance the persistence and efficacy of adoptively transferred T cells. 2-DG, 2-deoxy-d-glucose; Akti, Akt inhibitor; Arg, arginine; DON, 6-diazo-5-oxo-L-norleucine; IDO, indoleamine 2,3-dioxygenase; IDOi, IDO inhibitor.

context, targeting metabolism has the potential to simultaneously inhibit tumor growth, favorably alter the TME, and inhibit the metabolism and hence the function of suppressive immune cells, while also enhancing the metabolism and function of antitumor effector T cells.

Checkpoint blockade in the form of anti-PD-1, anti-PD-L1, and anti-CTLA-4 has been approved for the treatment of a variety of tumors. This therapeutic approach relies on revitalizing and expanding endogenous antitumor immune responses. Thus, enhancing these responses by both inhibiting tumor metabolism and enhancing the metabolism of effector T cells is desirable. Interestingly, anti-PD-L1 in addition to blocking an inhibitory signal to T cells also reduces the tumor's demand for glucose in the TME.[387] Consequently, there is an increase in glucose available for T cells to utilize to support glycolysis and effector function. Both cancer and T cells rely on glutamine metabolism to support their rapid growth.[76,105,122,123] Interestingly, broadly blocking glutamine metabolism with a glutamine analogue (DON) leads to the marked inhibition of tumor growth, while concomitantly reprogramming $CD8^+$ T cells to adopt the desirable robust memory cell phenotype for long-term tumor control.[388] Inhibiting glutamine metabolism disables tumor anabolic processes, but $CD8^+$ T cells are able to adapt (reprogram) by upregulating acyl-CoA synthetase short-chain family member 1 and 2 (ACSS1 and ACSS2) and increasing their metabolism of acetate as their primary fuel. Similarly, inhibiting glutamine metabolism in tumors prevents the formation of suppressive myeloid cells but rather promotes the generation of proinflammatory macrophages to enhance antitumor immunity.[322] Likewise, inhibiting cholesterol esterification by blocking ACAT1 with avasimibe potentiates $CD8^+$ T cell TCR clustering, signaling, and effector function resulting in better antitumor immunity especially combined with anti-PD-1 therapy.[114]

In addition to directly enhancing antitumor effector function, metabolic therapy can enhance immunotherapy by altering the TME. For example, tumor metabolism and poor vascularization can create hypoxic regions, which can inhibit T cell function.[389] It was found that the diabetes drug metformin could enhance immunotherapy in combination with anti-PD-1 by inhibiting mitochondrial complex I of the ETC in tumor cells and thus decreasing OXPHOS and overall TME hypoxia[389] (Fig. 6.11). Likewise, expression of IDO in immune and cancer cells suppresses antitumor immunity by depleting the TME of tryptophan and promoting the buildup of kynurenine, an immunosuppressive metabolite itself.[390-392] Inhibiting IDO was shown to enhance immunotherapy in preclinical cancer models but thus far have not shown great promise in clinical trials.[393,394] In a similar fashion, the ectonucleotidases CD39 and CD73 expressed tumor cells and suppressive immune cells have been shown to metabolize ATP (a proinflammatory molecule) into adenosine, which inhibits antitumor immune responses by signaling through the adenosine A2a receptor (A2aR) expressed on activated antitumor T cells.[395] Currently, a number of clinical agents are being developed to enhance immunotherapy by either blocking the production of adenosine by inhibiting CD39 or CD73 or by antagonizing A2aR. Finally, directly targeting T_{reg} metabolism within the TME has also proved to be a potentially promising approach especially since T_{reg} cells are able to thrive in low-glucose and high-lactate environments.[136] T_{reg} cells thrive in the TME by upregulating the lipid scavenger receptor CD36 to support mitochondrial fitness and lipid metabolism.[396] Selective targeting of T_{reg} cells with an antibody to CD36 inhibits intratumoral T_{reg} cells, thus enhancing antitumor immunity and providing a synergistic benefit with anti-PD-1 therapy.

Adoptive cell therapy (ACT) in the form of chimeric antigen receptor (CAR) T cells is currently approved as treatment for a number of cancers.[397] While such therapy has shown great promise, one of the important hurdles to this approach has been a lack of persistence of the adoptively transferred T cells. To this end, the ex vivo generation and expansion of CAR T cells lends this approach to pharmacologic metabolic intervention. In a murine model of ACT against melanoma, inhibiting glycolysis with the glucose analogue 2-DG, in transgenic PMEL $CD8^+$ T cells promoted a switch to an OXPHOS-based metabolic program[165] (Fig. 6.11). This promoted $CD8^+$ T cells to adopt a transcriptional profile associated with long-lived memory $CD8^+$ T cells, leading to better T cell persistence and hence antitumor activity in vivo. Similarly, inhibiting Akt signaling with an allosteric inhibitor (Akti) during ex vivo activation and expansion of tumor infiltrating lymphocytes promotes a transcriptional and metabolic profile with enhanced FAO associated with memory $CD8^+$ T cells to improve antitumor efficacy in mouse models.[398] Furthermore, the design and the use of certain CAR costimulatory signaling domains mediate metabolic reprogramming thus playing a critical role in the efficacy of CAR-T cells. The 4-1BB signaling domain in CAR T cells promotes the massive expansion of central memory T (T_{CM}) cells with increased mitochondrial biogenesis and oxidative metabolism compared to the more glycolytic CD28 signaling domain CAR T cells.[399]

In addition, exogenous arginine supplementation during T cell activation and differentiation shifts $CD8^+$ T cells into OXPHOS over glycolysis and generates memory-like T cells with better in vivo survival capacity and antitumor activity in a model of ACT[109] (Fig. 6.11). Ex vivo expansion also affords the opportunity for genetic manipulation prior to transfer. Overexpression of PCK1, which converts OAA into PEP, enhances $CD4^+$ adoptive T cell therapy in a model of melanoma.[119] Furthermore, hyperpolarized mitochondria, excessive ROS, and loss of mitochondria in $CD8^+$ tumor-infiltrating lymphocytes contribute to the dysfunctional state of exhausted T cells.[184,400] Forced overexpression of PGC-1α in donor $CD8^+$ T cells to promote mitochondrial biogenesis improves overall mitochondrial fitness and hence better functionality and tumor burden control with ACT.[184] Finally, pharmacologic interventions to promote mitochondrial fusion with the small molecule, M1, and to inhibit mitochondrial fission with Mdivi-1 during $CD8^+$ T cell differentiation impose a memory $CD8^+$ T cell phenotype with increased OXPHOS and SRC.[87] As such, altering donor $CD8^+$ T cell mitochondrial dynamics with M1+Mdivi-1 leads to superior tumor control in the setting of ACT.

TARGETING METABOLISM TO SUPPRESS IMMUNE-MEDIATED DISEASE

Treatments against cancer require metabolically reprogramming immune cells to boost antitumor immunity as described above. Alternatively, it is possible to control different autoimmune diseases or prolong transplant organ lifetime by selectively manipulating the metabolism of the immune cells responsible for exacerbating the disease (Chapters 48-50) (Fig. 6.12).

Systemic Lupus Erythematosus

Systemic lupus erythematosus (SLE) is a multifaceted autoimmune disease due to immune dysregulation of various adaptive and innate immune cells. The transgenic B6.Sle1.Sle2.Sle3 (TC) lupus-prone murine model is commonly used to study SLE as it closely resembles the human disease including anti–double-stranded DNA (anti-dsDNA) IgG titers and immune complex–mediated fatal glomerulonephritis.[401] Morel and colleagues found that $CD4^+$ T cells from TC lupus mice display enhanced glycolytic metabolism and OXPHOS associated with effector $CD4^+$ T cells. By treating mice with 2-DG to block glycolysis and metformin to inhibit OXPHOS through complex I, they were able to normalize $CD4^+$ T cell metabolism and suppress disease pathology in terms of decreased splenomegaly and anti-dsDNA IgG without any long-term side effects.[401] Interestingly, blocking glycolysis with 2-DG or glutamine metabolism with DON alone is able to reduce autoreactive Tfh cells in various lupus-prone murine models.[402] Similarly, treating lupus-prone mice with glutaminase 1 inhibitor, bis-2-(5-phenylacetamido-1,3,4-thiadiazol-2-yl) ethyl sulfide (BPTES), suppresses Th17 differentiation and hence autoimmune pathology.[403] Furthermore, in the (NZB × NZW) F1 murine model of lupus, direct inhibition of mitochondrial F_1F_0-ATPase with 1,4-benzodiazepine (Bz-423) induces mitochondrial superoxide generation and initiates apoptosis of B cells thereby improving overall survival by controlling B cell germinal hyperplasia and reducing glomerulonephritis.[404] Elevated ROS and concomitantly increased mTORC1 activity play a critical role in mediating lupus disease pathology in mouse models but also in SLE patients.[405,406] Treatment with the ROS scavenger, *N*-acetyl-cysteine (NAC), reduces ROS levels and mTOR activity in murine models and in a double-blind placebo controlled clinical trial leading to decreased disease pathology.[406] Similarly, direct inhibition of mTORC1 signaling with rapamycin in both a murine model and clinical trials diminishes SLE disease.[407-409]

Rheumatoid Arthritis

In rheumatoid arthritis (RA), the disease pathogenesis is mediated in part by Th1 and Th17 effector $CD4^+$ cells and can be mitigated by T_{reg} cells.[410] Since $CD4^+$ effector T cells and T_{reg} cells prefer and utilize different metabolic programs, an effective strategy to inhibit Th1 and Th17 differentiation and function but simultaneously promote T_{reg} cells represents a potentially potent course of treatment.[7,131,148] In a collagen-induced arthritis model, it has been shown that metformin can inhibit Th1 and Th17 differentiation and function, while simultaneously promoting T_{reg} cells.[411] In a similar fashion, inhibiting glycolysis with either 3-bromopyruvate (BrPA) or 2-DG has also been shown to mitigate disease.[412,413] Targeting downstream glycolysis with dichloroacetate (DCA), which inhibits PDHK1 and alters pyruvate metabolism, prevents cartilage joint destruction, delays disease onset, and slows disease progression by suppressing Th17 cells and promoting T_{reg} cells.[414]

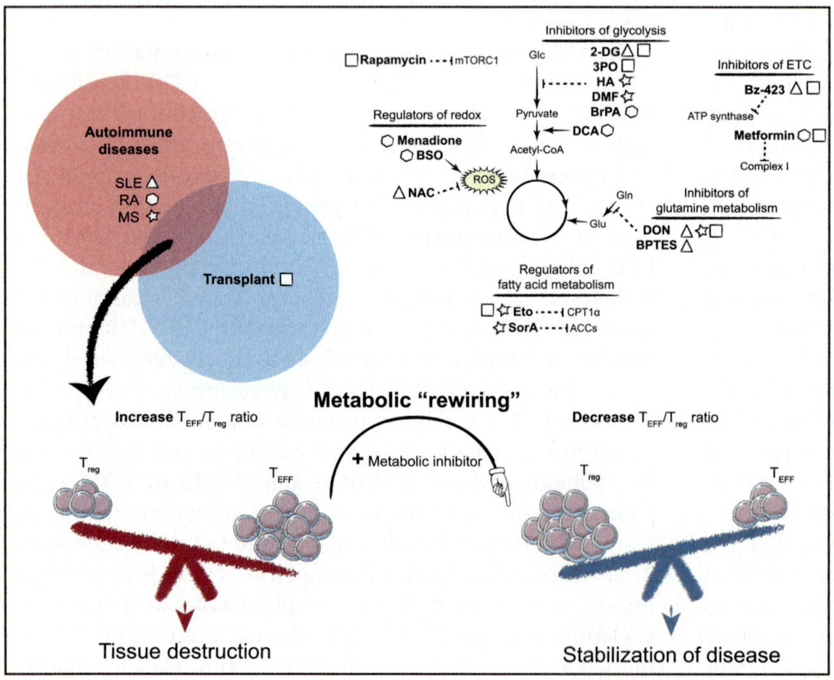

FIG. 6.12. Strategies to target metabolism to mitigate immune mediated pathology. Inhibition of specific metabolic pathways can regulate destructive immune responses by both blocking effector function and enhancing regulatory cell generation and function. The upper right corner shows inhibitors that can target select enzymes or metabolic pathways. The shapes denote the inhibitors that have been used in specific murine disease models. Triangle: systemic lupus erythematosus (SLE); hexagon: rheumatoid arthritis (RA); star: multiple sclerosis (MS); square: transplantation.

Interestingly, RA-inducing $CD4^+$ T cells upregulate the expression of glucose-6-phosphate dehydrogenase (G6PD), which is the rate-limiting enzyme involved in the PPP.[415] By shunting glucose into the PPP, RA-causing $CD4^+$ T cells generate additional NADPH and glutathione to scavenge ROS, which enhances the proliferation and differentiation of autoreactive Th1 and Th17 cells. In a humanized mouse model of RA, menadione (vitamin K3) or buthionine sulfoximine (BSO) treatment were able to restore intracellular ROS levels to normal.[415] Redox rewiring in RA in $CD4^+$ T cells reduces overall inflammation by generating less IFN-γ, IL-17, TNF-α, IL-6, and other proinflammatory cytokines.

Multiple Sclerosis

Multiple sclerosis (MS) is an autoimmune disease of the central nervous system (CNS) with severe neurologic impairment caused by the destruction of the myelin sheath by pathogenic Th17 cells. Experimental autoimmune encephalomyelitis (EAE) is a murine model of MS. As mentioned before, Th17 cells require the synthesis of *de novo* lipids through ACC1.[148] Specific inhibition of ACC1 using soraphen A (SorA) in an in vivo model of EAE is able to significantly delay disease onset and diminish the severity of EAE by repressing Th1 and Th17 cells but promoting T_{reg}. Likewise, regulating pyruvate metabolism with DCA treatment in EAE is able to prevent disease.[131] Indeed, the clinical metabolic and immunomodulatory drug, dimethyl fumarate (DMF), is used to treat psoriasis and MS patients.[416] Interestingly, DMF promotes the succination of cysteine residues on GAPDH, thereby inhibiting its enzymatic activity and consequently glycolysis required for Th1 and Th17 cell differentiation and function.[246] During EAE, in vivo GAPDH inhibition with heptelidic acid, a similar inhibitor to DMF, attenuates disease. Fatty acid metabolism is also critical in the development of EAE disease pathology. Treatment with the CPT1α inhibitor etomoxir to alter FAO reduces demyelination of the CNS and hence disease progression with less immune infiltration and proinflammatory IFN-γ production in EAE.[417] Broad targeting of glutamine metabolism with the glutamine analogue DON decreases MS disease severity in a prevention model and also in a model with established disease.[418]

Transplantation

Traditionally, powerful immunosuppressant agents such as calcineurin inhibitors, steroids, and T cell-depleting antibodies have been employed as both prophylaxis and treatment of graft rejection. Such agents truly "suppress" immune responses in that they inhibit both effector and regulatory T cells. The differential metabolic programming of different immune cells offers the opportunity to "regulate" immune responses in this setting. For solid organ transplantation, studies have demonstrated the ability of metabolic inhibitors to prevent solid organ transplant rejection.[419,420] In one study, a combination therapy of a glutamine antagonist DON, metformin, and 2-DG was able to abrogate rejection in a completely mismatched heart transplantation in mice with the absence of any conventional immunosuppressive agents. Furthermore, since "metabolic" therapy does not inhibit regulatory mechanisms (but in fact supports them), it has subsequently been shown that this combination along with CTLA-4-Ig can promote long-term tolerance in the absence of long-term immunosuppression in such models.[420]

In murine models of graft-versus-host disease (GVHD) during bone marrow transplantation (BMT), alloreactive donor T cells undergo mTORC1-dependent glycolytic programming to fuel their potent effector response.[421] Thus, alloreactive donor T cells display greater glycolytic activity due to increased gene expression of key enzymes such as HK II and LDHA along with Glut1 surface expression. In a murine GVHD model, in vivo blockade of glycolysis with 3-(3-pyridinyl)-1-(4-pyridinyl)-2-propen-1-one (3-PO), a specific inhibitor of PFKFB3 to reduce pyruvate fructose kinase 1 (PFK1) activity, prolongs long-term survival with less disease pathology.[421] Also, metformin treatment suppresses GVHD symptoms and mTORC1 activity, leading to a decrease in Th1 and Th17 cells and conversely enhancing T_{reg}.[422] Studies utilizing other models of GVHD emphasize the role of glutamine-fueled OXPHOS and fatty acid metabolism in alloreactive donor T cells.[423-425] As such, alloreactive T cells not only generate more acylcarnitines but also take up fatty acids to fuel FAO for more mitochondrial activity.[424,425] Thus, in vivo treatment of mice with the ATP synthase inhibitor Bz-423 or the FAO inhibitor etomoxir induces apoptosis of alloreactive donor T cells and improves overall survival in the setting of BMT without any adverse side effects even after stopping treatment, which signifies a long-lasting effect.

TOOLS OF THE TRADE

The intimate connection between metabolic programs and immune cell activation, differentiation, and function has led to the integration of several biochemical techniques into the immunologist's repertoire. Likewise, routine yet powerful immunologic techniques and strategies have been adapted to interrogate metabolic programs.

Seahorse XF Analyzer

While attending talks or reading the literature on immunometabolism, an immunologist is bound to come across data generated with the Seahorse (Agilent), an extracellular metabolic flux analyzer.[426] This machine, in either 24-well or 96-well format, facilitates measuring the extracellular acidification rate (ECAR) and oxygen consumption rate (OCR). The 96-well format even allows accurate measurements with limited cell numbers, thus facilitating the analysis of rarer immune cells such as $CD8^+$ T cells from tumors. ECAR, which measures released protons in the media, is often referred to as a surrogate readout for active aerobic glycolysis (the measure of lactic acid production), while OCR measures oxygen consumption as cells undergo

OXPHOS. The adoption of the Seahorse by immunologists has greatly facilitated the analysis of basic energetic profiles of immune cells without the need for complex technology like mass spectrometry. Overall, the Seahorse enables immunologists to measure glycolysis and OXPHOS for different cell types previously cultured under different in vitro culture conditions or ex vivo cells. By selectively employing several well-defined inhibitors through individual port injections, it is possible to gain great insight into glycolytic and mitochondrial capacity and function of various immune cells (Fig. 6.13A). For example, the glycolytic stress test assay measures the cell's ability to utilize glycolysis in response to acute glucose exposure while also determining their glycolytic capacity and reserve. This involves culturing cells in nonbuffered, glucose- and serum-free media. The nonbuffered media is essential to accurately assess ECAR as it is determined by measuring protons released over time. After determining baseline measurements, glucose injection (Port A) leads to an increase in ECAR, thus demonstrating elevated glycolysis as cells quickly metabolize glucose. The injection of oligomycin (Port B), which inhibits Complex V (ATPase) of the ETC, maximally forces all available glucose to be pushed through glycolysis to measure glycolytic reserve. Alternatively, the injection (Port C) of the hexokinase inhibitor, 2-DG, completely inhibits glycolysis thus permitting the measurement of total glycolytic capacity (Fig. 6.13B).

The mitochondrial stress test is performed by culturing cells in nonbuffered, serum-free culture media in the presence of glucose, glutamine, and pyruvate. First, both basal ECAR and OCR are measured over time. By injecting oligomycin (Port A), the disruption in the ETC inhibits oxygen consumption, and thus, overall OCR drops. This reflects the oxygen consumption that is linked to ATP synthesis derived from the basal rate of OXPHOS. By injecting FCCP (Port B), this ionophore disrupts the proton gradient across the inner mitochondrial membrane, thereby maximizing oxygen consumption (OCR). This reflects the maximal OCR of the cells as they attempt to restore the ETC. The difference between the maximum OCR post-FCCP injection and the basal OCR measures the spare respiratory capacity (SRC). The injection of rotenone and antimycin A (Port C), which inhibit complex I and III of the ETC, respectively, completely inhibits mitochondrial OXPHOS. The remaining OCR observed post rotenone and antimycin A injection reflects nonmitochondrial respiration (Fig. 6.13C).

The palmitate oxidation stress test can be employed to measure the contribution of fatty acids to fuel OXPHOS by FAO. Initially cells are cultured in fatty acid–free media with minimal nutrients. Next, a bovine serum albumin (BSA)-conjugated fatty acid such as palmitate is injected into the assay. Conjugating the fatty acid to BSA allows for the passive uptake of fatty acids independent of specific transporters. The difference between basal OCR and the maximum OCR after palmitate injection provides information of the cell's ability to use palmitate as a fuel for OXPHOS.

In addition, by culturing cells in full Seahorse media conditions supplemented with or without glutamine, glucose, and pyruvate and by sequentially injecting different inhibitors, it is possible to test the essential nutrients that fuel the TCA cycle and hence sustain OXPHOS activity. For example, by injecting UK5099, an inhibitor of MPCs, etomoxir, an inhibitor of FAO, or BPTES, a glutaminase inhibitor, it is possible to identify the substrates that fuel OXPHOS by determining the changes in OCR during the course of the assay.

Flow Cytometry–Based Methods and Microscopy

Flow cytometry is proving to be a powerful means of interrogating metabolic function on a single cell level requiring less material. For example, rather than using radiolabeled glucose, glucose uptake can now be measured by the uptake of the fluorescence glucose analogue 2-[*N*-(7-nitrobenz-2-oxa-1,3-diazol-4-yl) amino]-2-deoxy-D-glucose, or 2-NBDG, which can be excited by a blue laser and detected in the fluorescein-5-isothiocyanate (FITC) channel.[427] 2-NBDG is not cell permeable and is commonly used to detect glucose uptake through the Glut family of glucose transporters. Once in the cell, 2-NBDG is phosphorylated by hexokinases and cannot be metabolized further. Thus, the accumulation of intracellular 2-NBDG is a useful tool to measure how much glucose a cell can uptake within a given timeframe typically 20 to 30 minutes. Figure 6.14A shows the differences in glucose uptake between naïve and activated T cells as demonstrated by 2-NBDG labeling. However, developing studies suggest that 2-NBDG uptake in lymphocytes might not truly reflect GLUT-dependent import of glucose.

Mitochondrial mass and function can also be measured by employing flow cytometry. Fluorescent dyes such as tetramethylrhodamine ethyl ester (TMRE) or tetramethylrhodamine methyl ester (TMRM) are commonly used to measure mitochondrial activity.[428] Both dyes are cell permeable and can be sequestered within the mitochondria to measure mitochondrial membrane potential due to electron transfer. Alternatively, mitochondrial mass can be determined using MitoTracker dyes that are cell permeable and accumulate within the mitochondria.[429] Depending on the specific MitoTracker dye, the accumulation of the dye can be dependent on the membrane potential of the mitochondria. Figure 6.14B shows the differences in mitochondrial potential (as denoted by TMRM) and mitochondrial mass (as denoted by MitoTracker) between naïve and activated T cells. In addition to flow cytometry, most of the mitochondrial dyes can also be used to image mitochondrial morphology with microscopy. However, electron microscopy still remains the gold standard technique to visualize individual mitochondria and its intricate cristae organization. In addition to staining the mitochondria, other organelle dyes are available to label lysosomes, the ER, and peroxisomes, which can all influence the metabolic state of immune cells.

The relative quantification of ROS within the cell can also be measured by flow cytometry through commercially

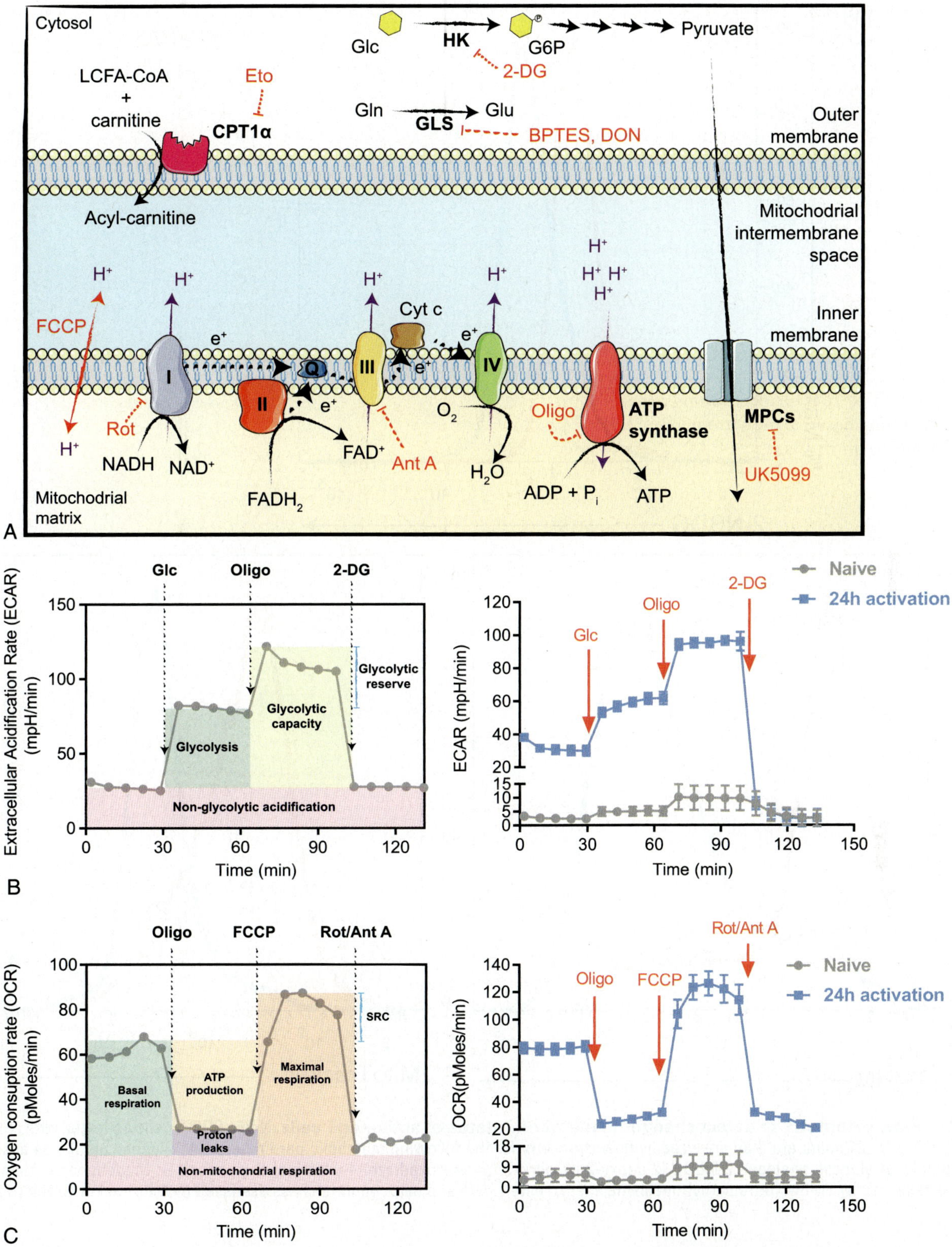

FIG. 6.13. Assessing metabolism using Seahorse Analyzer. A: A graphical overview summarizing the effects of the various inhibitors to target major metabolic pathways and their fuel. 2-DG, 2-deoxy-d-glucose; Ant A, antimycin A; BPTES, bis-2-(5-phenylacetamido-1,3,4-thiadiazol-2-yl) ethyl sulfide; DON, 6-diazo-5-oxo-L-norleucine; Eto, etomoxir; FCCP, 2-[2-[4-(trifluoromethoxy)phenyl]hydrazinylidene]-propanedinitrile; Glu, glucose; Oligo, oligomycin; Rot, rotenone. **B:** A schematic displaying an overview of the glycolytic stress test (*left*) and extracellular acidification rate (ECAR, *right*) analysis of CD8+ T cells undergoing glycolytic stress test. **C:** A schematic displaying an overview of the mitochondrial stress test (*left*) and oxygen consumption rate (OCR, *right*) of CD8+ T cells undergoing mitochondrial stress test.

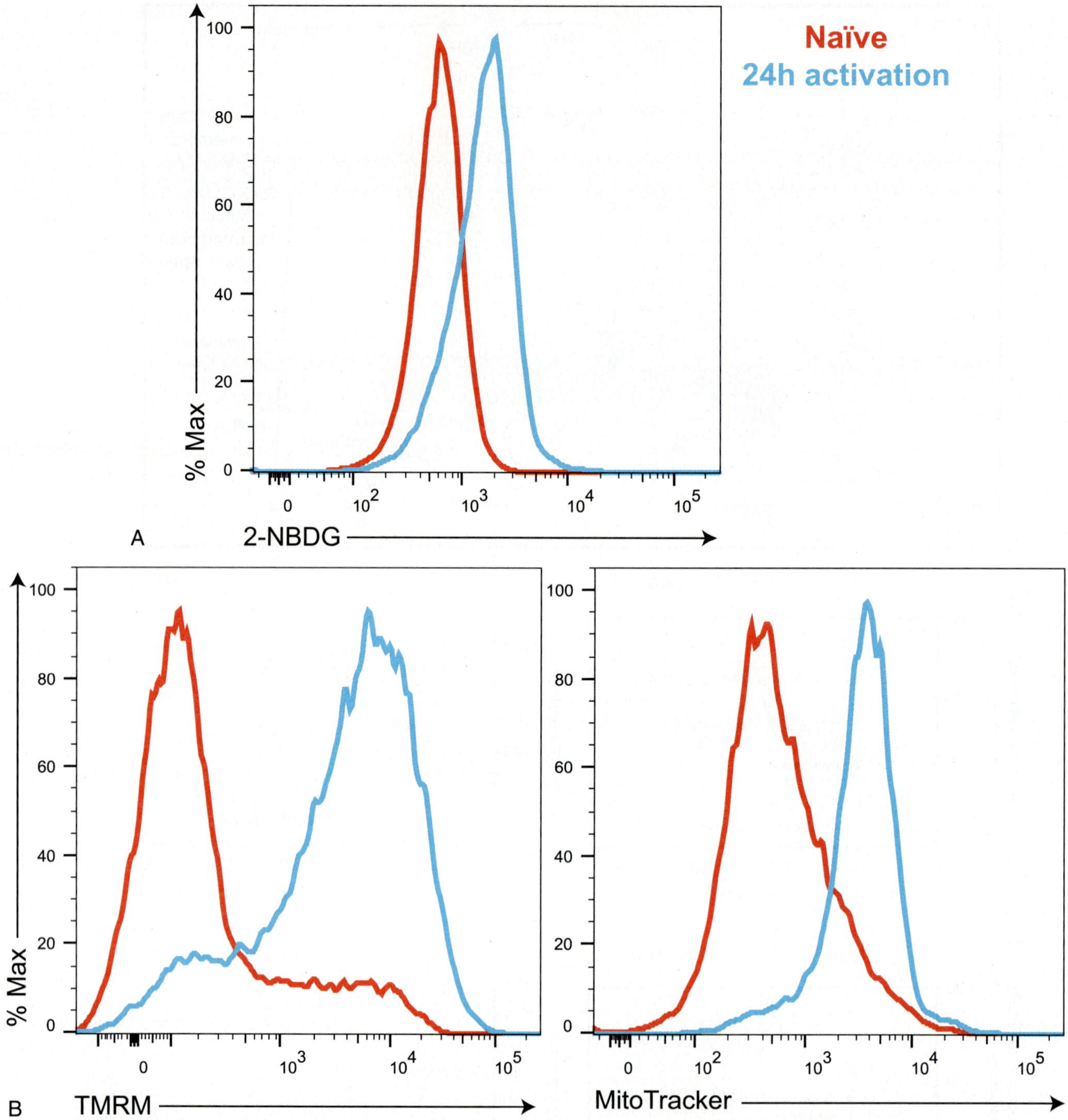

FIG. 6.14. Flow cytometry to detect changes in metabolic status of activated T cells. Naïve CD8+ T cells activated with anti-CD3 and anti-CD28 for 24 hours are then analyzed by flow cytometry for the following metabolic parameters: **(A)** Glucose uptake as assessed by the uptake of glucose analogue 2-NBDG (2-deoxy-2-[(7-nitro-2,1,3-benzoxadiazol-4-yl)amino]-d-glucose). **B:** Mitochondrial membrane potential assessed by TMRM (tetramethylrhodamine, methyl ester) (*left*) and mitochondrial mass assessed by Mitotracker Green (*right*).

available probes. 2′,7′-dichlorofluorescin diacetate (DCFDA) is a widely used agent to detect cytosolic ROS. DCFDA diffuses into the cell and is then deacetylated by cellular esterases to a nonfluorescent compound, which can later be oxidized by ROS into the fluorescent compound 2′,7′-dichlorofluorescein (DCF). Various fluorescent forms of CellROX can also be used to detect subcellular and/or nuclear superoxides. For example, MitoSOX is a dye that can be used to detect ROS specifically generated within the mitochondria.

Flow cytometry can also be employed to detect cellular lipid content. BODIPY is a dye that stains intracellular lipids. Both the uptake of and total lipid content can be determined by staining the cells with specific BODIPY probes. LipidTOX can be used to specifically measure neutral lipid content of cells. In addition, fillipin III binds to nonesterified cholesterol to measure cholesterol content within a cell.

Finally, one way to query the energetic status of cells is by measuring the NADH/NAD^+ ratio or ATP levels. There are a number of commercially available colorimetric kits that readily facilitate such measurements. Interestingly, intracellular NADH levels contribute to the autofluorescence signal in cells. Thus, a simple strategy to measure NADH is using a flow cytometer with a UV laser and measuring autofluorescence.[430]

Metabolomics

Metabolomics refers to the high-dimensional analysis of metabolites at a cellular or organ level.[431] This technique offers a snapshot of the metabolome at a specific point in time.

The metabolome contains thousands of metabolites with a wide range of physical and chemical properties. From an analytical perspective, the two most commonly employed detection techniques are nuclear magnetic resonance (NMR) and mass spectrometry (MS). The advantages of NMR include its simple quantitation capability, nondestructive analysis, and ability to identify stereoisomers (ie, R or S enantiomers).[170] The major limitation of this technique is its relatively low sensitivity, which means that this technique is not suitable for the detection of metabolites that are present in low concentrations. Compared to NMR, MS offers higher sensitivity, broader selectivity, and a wider dynamic range for metabolite analysis. Because of these advantages, MS is more suitable for studying metabolites on the "omics" level and it has emerged as a very powerful analytical tool in the metabolomics field. Often, however, in order to achieve the comprehensive coverage of the metabolites, a combination of these two analytical technologies is applied.

In general, to address global metabolite differences within cells treated under different conditions, the metabolites are extracted and then separated by chromatography. These separated metabolites are then sequentially ionized for mass spectrometry. Based on their mass to charge ratio and fragmentation pattern, the ions can be analyzed and detected by the mass analyzer and detector. Due to the unique physical and chemical properties of each metabolite, the retention time, mass to charge ratio, and the fragmentation pattern are the key parameters for metabolomics analysis. There are two major approaches used in metabolomic studies: untargeted (global) and targeted. The untargeted approach aims to find the differentiated metabolites based on their relative quantitation and annotation of thousands of chromatographic or spectroscopic peaks. This approach usually generates a large amount of data and requires efficient data reduction by using appropriate bioinformatics tools. The untargeted approach is mainly used for hypothesis generation and biomarker discovery. Targeted metabolomics specifically examines metabolites from selected pathways using a curated data set to identify specific metabolites. Since this approach is hypothesis driven, the metabolic pathways of interest must be known prior to analysis. Targeted metabolomics is particularly powerful when interrogating specific pathways. By using metabolomics, principal component analysis (PCA) plots depict metabolic differences among different subsets of $CD4^+$ T cells and M1 versus M2 polarized macrophages as shown in PCA component plots (Fig. 6.15A). Note the differences in metabolite concentrations between in vitro–generated M1 and M2 macrophages that supports the function of the two individual subsets (Fig. 6.15B).

Tracing

Metabolomics can only provide a snapshot of metabolite content.[432] For example, if a particular metabolite is elevated, metabolomics alone cannot tell you if it is increased due to a robust forward reaction or an inhibited distal reaction. Tracing can provide a more dynamic picture of cellular metabolism within a given timeframe. This method relies on heavy (nonradioactive) isotopes typically of carbon, nitrogen, or hydrogen. The mass difference of the heavy isotopes facilitates tracking their contribution to the generation of metabolites within a cell. For example, by culturing naïve or activated T cells in ^{13}C-glucose, tracing enables the interrogation of glycolysis and TCA cycle activity by tracking the incorporation of the six heavy ^{13}C derived from glucose into each of the metabolites of these pathways. Figure 6.16 demonstrates metabolomics tracing of ^{13}C-glucose uniformly labelled with carbon mass of 13 (U-^{13}C-glucose) between naïve T cells and T cells activated with anti-CD3 and anti-CD28 for 24 hours demonstrating both upregulation of aerobic glycolysis (ie, lactate) and the TCA cycle (ie, citrate) for OXPHOS.

ACKNOWLEDGMENTS

First, we would like to sincerely apologize to all of the investigators whose work we were not able to cite in this chapter. The field of immunometabolism continues to rapidly develop, and this inaugural chapter was our attempt to bring it all together. As this young discipline continues to mature, future editions will be able to hone a clearer picture of this field. In addition, we would like to thank Min-hee Oh, Robert Leone, Wei Xu, Samuel Warner, Liang Zhao, Debebe Theodros, Hong Yu, Elizabeth Thompson, Vinicius de Andrade Oliveira, Im-Meng Sun, and Richard Blosser for their helpful advice and critical reading of this chapter. Scientific research and training in the laboratory are supported by the National Institutes for Health and The Bloomberg~Kimmel Institute for Cancer Immunotherapy.

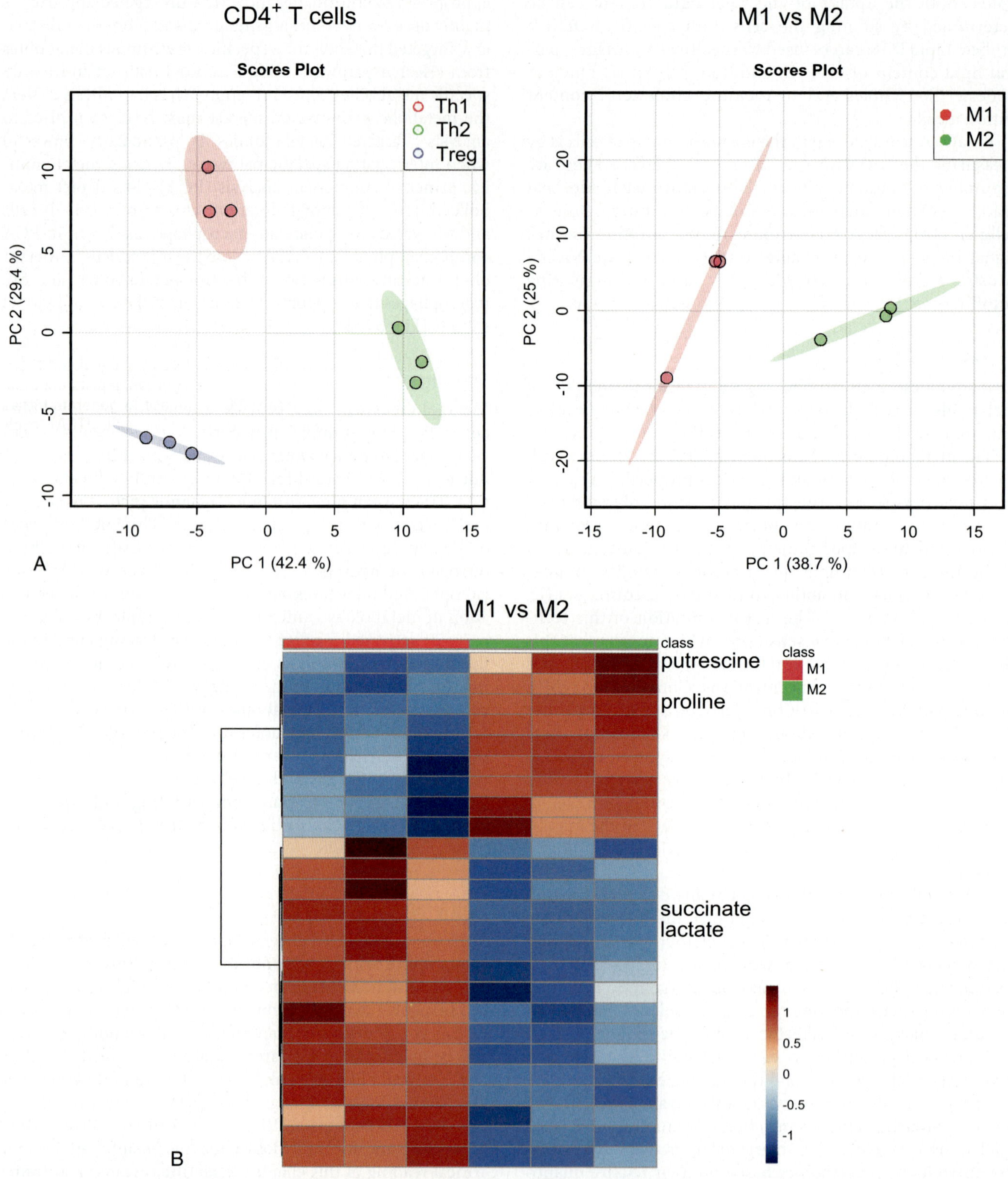

FIG. 6.15. T cells and macrophages can be identified by their distinct metabolic profile. Targeted metabolite analysis of in vitro differentiated CD4+ T cells and polarized macrophages performed with liquid chromatography (LC)-tandem mass spectrometry (MS). **A:** Principal component analysis (PCA) comparing in vitro differentiated murine CD4+ Th1 (*red*), Th2 (*green*), and induced T_{regs} (*blue*) (*left*) and in vitro polarized bone marrow–derived macrophages, M1 ([LPS+IFN-γ] [*red*]) or M2 ([IL-4] [*green*]) (*right*). **B:** Heat map visualization of key metabolites between M1 and M2 macrophages. For simplicity and focus, four major metabolites are highlighted in the heat map.

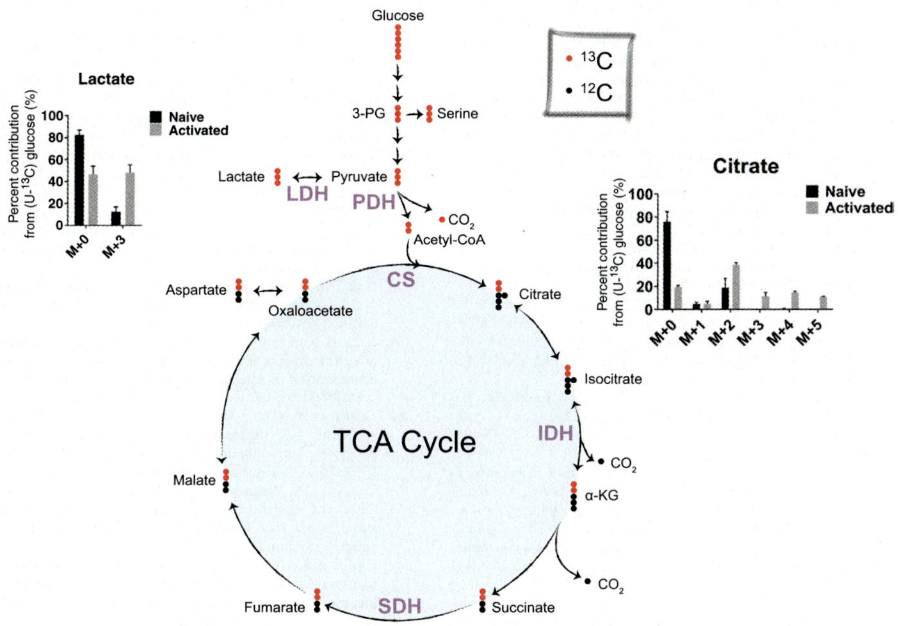

FIG. 6.16. Metabolic tracing of activated T cells. Graphic showing stable isotope distribution by using ^{13}C-U-glucose (red dots) as tracer compared to unlabeled (*black dots*). Naïve murine CD8+ T cells rested in IL-7 or activated with anti-CD3 and anti-CD28 for 24 hours then pulsed with ^{13}C-U-glucose for 6 hours. Labeled metabolites were extracted and assessed by mass spectrometry. Top left graph depicts contribution of ^{13}C-U-glucose to generate lactate (M+3) between naïve and activated CD8+ T cells. Top right graph depicts contribution of ^{13}C-U-glucose to generate citrate of the tricarboxylic acid (TCA) cycle (M+2) between naïve and activated CD8+ T cells.

REFERENCES

1. Frauwirth KA, Riley JL, Harris MH, et al. The CD28 signaling pathway regulates glucose metabolism. *Immunity*. 2002;16:769-777.
2. Dang EV, Barbi J, Yang H-Y, et al. Control of T(H)17/T(reg) balance by hypoxia-inducible factor 1. *Cell*. 2011;146:772-784. doi:10.1016/j.cell.2011.07.033
3. Shi LZ, Wang R, Huang G, et al. HIF1α–dependent glycolytic pathway orchestrates a metabolic checkpoint for the differentiation of TH17 and Treg cells. *J Exp Med*. 2011;208:1367-1376. doi:10.1084/jem.20110278
4. Tannahill GM, Curtis AM, Adamik J, et al. Succinate is an inflammatory signal that induces IL-1beta through HIF-1alpha. *Nature*. 2013;496:238-242. doi:10.1038/nature11986
5. Peng M, Yin N, Chhangawala S, Xu K, Leslie CS, Li MO. Aerobic glycolysis promotes T helper 1 cell differentiation through an epigenetic mechanism. *Science*. 2016;354:481-484. doi:10.1126/science.aaf6284
6. Chisolm DA, Savic D, Moore AJ, et al. CCCTC-binding factor translates interleukin 2-and α-ketoglutarate-sensitive metabolic changes in T cells into context-dependent gene programs. *Immunity*. 2017;47:251-267.e7. doi:10.1016/j.immuni.2017.07.015
7. Michalek RD, Gerriets VA, Jacobs SR, et al. Cutting edge: distinct glycolytic and lipid oxidative metabolic programs are essential for effector and regulatory CD4+ T cell subsets. *J Immunol*. 2011;186:3299-3303. doi:10.4049/jimmunol.1003613
8. Patel CH, Leone RD, Horton MR, Powell JD. Targeting metabolism to regulate immune responses in autoimmunity and cancer. *Nat Rev Drug Discov*. 2019;18:669-688. doi:10.1038/s41573-019-0032-5
9. Nelson DL, Cox MM, Lehninger AL. *Lehninger Principles of Biochemistry*. 7th ed. W.H. Freeman and Company; Macmillan Higher Education; 2017.
10. Warburg O, Posener K, Negelein E. Ueber den stoffwechsel der tumoren. *Biochem Z*. 1924;152:319-344.
11. Vander Heiden MG, Cantley LC, Thompson CB. Understanding the Warburg effect: the metabolic requirements of cell proliferation. *Science*. 2009;324:1029-1033. doi:10.1126/science.1160809
12. DeBerardinis RJ, Chandel NS. We need to talk about the Warburg effect. *Nat Metab*. 2020;2:127-129. doi:10.1038/s42255-020-0172-2
13. Vander Heiden MG, DeBerardinis RJ. Understanding the intersections between metabolism and cancer biology. *Cell*. 2017;68:657-669. doi:10.1016/j.cell.2016.12.039
14. Kim J, DeBerardinis RJ. Mechanisms and implications of metabolic heterogeneity in cancer. *Cell Metab*. 2019;30:434-446. doi:10.1016/j.cmet.2019.08.013
15. Buck MD, Sowell RT, Kaech SM, Pearce EL. Metabolic instruction of immunity. *Cell*. 2017;169:570-586. doi:10.1016/j.cell.2017.04.004
16. Fox CJ, Hammerman PS, Thompson CB. Fuel feeds function: energy metabolism and the T-cell response. *Nat Rev Immunol*. 2005;5:844-852. doi:10.1038/nri1710
17. Lafferty KJ, Warren HS, Woolnough JA, Talmage DW. Immunological induction of T lymphocytes: role of antigen and the lymphocyte costimulator. *Blood Cells*. 1978;4:395-406.
18. Matzinger P. Tolerance, danger, and the extended family. *Annu Rev Immunol*. 1994;12:991-1045. doi:10.1146/annurev.iy.12.040194.005015
19. Schwartz RH. Costimulation of T lymphocytes: the role of CD28, CTLA-4, and B7/BB1 in interleukin-2 production and immunotherapy. *Cell*. 1992;71:1065-1068. doi:10.1016/s0092-8674(05)80055-8
20. Mueckler M, Thorens B. The SLC2 (GLUT) family of membrane transporters. *Mol Aspects Med*. 2013;34:121-138. doi:10.1016/j.mam.2012.07.001
21. Zheng Y, Collins SL, Lutz MA, et al. A role for mammalian target of rapamycin in regulating T cell activation versus anergy. *J Immunol*. 2007;178:2163-2170.
22. Zheng Y, Delgoffe GM, Meyer CF, Chan W, Powell JD. Anergic T cells are metabolically anergic. *J Immunol*. 2009;183:6095-6101. doi:10.4049/jimmunol.0803510
23. Klein Geltink RI, O'Sullivan D, Corrado M, et al. Mitochondrial priming by CD28. *Cell*. 2017;171:385-397.e11. doi:10.1016/j.cell.2017.08.018
24. Saxton RA, Sabatini DM. mTOR signaling in growth, metabolism, and disease. *Cell*. 2017;168:960-976. doi:10.1016/j.cell.2017.02.004
25. Condon KJ, Sabatini DM. Nutrient regulation of mTORC1 at a glance. *J Cell Sci*. 2019;132:jcs222570. doi:10.1242/jcs.222570
26. Liu GY, Sabatini DM. mTOR at the nexus of nutrition, growth, ageing and disease. *Nat Rev Mol Cell Biol*. 2020;21:183-203. doi:10.1038/s41580-019-0199-y. PMID: 31937935.
27. Shimobayashi M, Hall MN. Making new contacts: the mTOR network in metabolism and signalling crosstalk. *Nat Rev Mol Cell Biol*. 2014;15:155-162. doi:10.1038/nrm3757
28. Wolfson RL, Sabatini DM. The dawn of the age of amino acid sensors for the mTORC1 pathway. *Cell Metab*. 2017;26:301-309. doi:10.1016/j.cmet.2017.07.001
29. Valvezan AJ, Manning BD. Molecular logic of mTORC1 signalling as a metabolic rheostat. *Nat Metab*. 2019;1:321-333. doi:10.1038/s42255-019-0038-7
30. Düvel K, Yecies JL, Menon S, et al. Activation of a metabolic gene regulatory network downstream of mTOR complex 1. *Mol Cell*. 2010;39:171-183. doi:10.1016/j.molcel.2010.06.022
31. Thoreen CC. The molecular basis of mTORC1-regulated translation. *Biochem Soc Trans*. 2017;45:213-221. doi:10.1042/BST20160072
32. Dai H, Thomson AW. The "other" mTOR complex: new insights into mTORC2 immunobiology and their implications. *Am J Transplant*. 2019;19:1614-1621. doi:10.1111/ajt.15320
33. Knudsen JR, Fritzen AM, James DE, Jensen TE, Kleinert M, Richter EA. Growth factor-dependent and-independent activation of mTORC2. *Trends Endocrinol Metab*. 2020;31:13-24. doi:10.1016/j.tem.2019.09.005
34. Eijkelenboom A, Burgering BM. FOXOs: signalling integrators for homeostasis maintenance. *Nat Rev Mol Cell Biol*. 2013;14:83-97. doi:10.1038/nrm3507
35. Kerdiles YM, Beisner DR, Tinoco R, et al. Foxo1 links homing and survival of naive T cells by regulating L-selectin, CCR7 and interleukin 7 receptor. *Nat Immunol*. 2009;10:176-184. doi:10.1038/ni.1689
36. Vezina C, Kudelski A, Sehgal SN. Rapamycin (AY-22,989), a new antifungal antibiotic. I. Taxonomy of the producing streptomycete and isolation of the active principle. *J Antibiot (Tokyo)*. 1975;28:721-726. doi:10.7164/antibiotics.28.721
37. Abraham RT, Wiederrecht GJ. Immunopharmacology of rapamycin. *Annu Rev Immunol*. 1996;14:483-510. doi:10.1146/annurev.immunol.14.1.483
38. Terada N, Lucas JJ, Szepesi A, Franklin RA, Domenico J, Gelfand EW. Rapamycin blocks cell cycle progression of activated T cells prior to events characteristic of the middle to late G1 phase of the cycle. *J Cell Physiol*. 1993;154:7-15. doi:10.1002/jcp.1041540103
39. Terada N, Franklin RA, Lucas JJ, Blenis J, Gelfand EW. Failure of rapamycin to block proliferation once resting cells have entered the cell cycle despite inactivation of p70 S6 kinase. *J Biol Chem*. 1993;268:12062-12068.
40. Martel RR, Klicius J, Galet S. Inhibition of the immune response by rapamycin, a new antifungal antibiotic. *Can J Physiol Pharmacol*. 1977;55:48-51. doi:10.1139/y77-007
41. Jenkins MK. The role of cell division in the induction of clonal anergy. *Immunol Today*. 1992;13:69-73. doi:10.1016/0167-5699(92)90137-V
42. Powell JD. The induction and maintenance of T cell anergy. *Clin Immunol*. 2006;120:239-246. doi:10.1016/j.clim.2006.02.004
43. Powell JD, Lerner CG, Schwartz RH. Inhibition of cell cycle progression by rapamycin induces T cell clonal anergy even in the presence of costimulation. *J Immunol*. 1999;162:2775-2784.
44. Delgoffe GM, Kole TP, Zheng Y, et al. The mTOR kinase differentially regulates effector and regulatory T cell lineage commitment. *Immunity*. 2009;30, 832-844. doi:10.1016/j.immuni.2009.04.014
45. Valmori D, Tosello V, Souleimanian NE, et al. Rapamycin-mediated enrichment of T cells with regulatory activity in stimulated CD4+ T cell cultures is not due to the selective expansion of naturally occurring regulatory T cells but to the induction of regulatory functions in conventional CD4+ T cells. *J Immunol*. 2006;177:944-949. doi:10.4049/jimmunol.177.2.944
46. Battaglia M, Stabilini A, Migliavacca B, et al. Rapamycin promotes expansion of functional CD4+CD25+FOXP3+ regulatory T cells of both healthy subjects and type 1 diabetic patients. *J Immunol*. 2006;177:8338-8347. doi:10.4049/jimmunol.177.12.8338
47. Sun IH, Oh M-H, Zhao L, et al. mTOR complex 1 signaling regulates the generation and function of central and effector Foxp3(+) regulatory T cells. *J Immunol*. 2018;201:481-492. doi:10.4049/jimmunol.1701477
48. Chapman NM, Zeng H, Nguyen TM, et al. mTOR coordinates transcriptional programs and mitochondrial metabolism of activated Treg subsets to protect tissue homeostasis. *Nat Commun*. 2018;9:2095. doi:10.1038/s41467-018-04392-5

49. Ray JP, Staron MM, Shyer JA, et al. The interleukin-2-mTORc1 kinase axis defines the signaling, differentiation, and metabolism of T helper 1 and follicular B helper T cells. *Immunity.* 2015;43:690-702. doi:10.1016/j.immuni.2015.08.017
50. Blagih J, Coulombe F, Vincent EE, et al. The energy sensor AMPK regulates T cell metabolic adaptation and effector responses in vivo. *Immunity.* 2015;42:41-54. doi:10.1016/j.immuni.2014.12.030
51. Araki K, Turner AP, Shaffer VO, et al. mTOR regulates memory CD8 T-cell differentiation. *Nature.* 2009;460, 108-112. doi:10.1038/nature08155
52. Pollizzi KN, Patel CH, Sun I-H, et al. mTORC1 and mTORC2 selectively regulate CD8+ T cell differentiation. *J Clin Invest.* 2015;125:2090-2108. doi:10.1172/JCI77746
53. Zhang L, Tschumi BO, Lopez-Mejia IC, et al. Mammalian target of rapamycin complex 2 controls CD8 T cell memory differentiation in a Foxo1-dependent manner. *Cell Rep.* 2016;14:1206-1217. doi:10.1016/j.celrep.2015.12.095
54. Shrestha S, Yang K, Wei J, Karmaus PWF, Neale G, Chi H. Tsc1 promotes the differentiation of memory CD8+ T cells via orchestrating the transcriptional and metabolic programs. *Proc Natl Acad Sci U S A.* 2014;111:14858-14863. doi:10.1073/pnas.1404264111
55. Pollizzi KN, Sun I-H, Patel CH, et al. Asymmetric inheritance of mTORC1 kinase activity during division dictates CD8(+) T cell differentiation. *Nat Immunol.* 2016;17:704-711. doi:10.1038/ni.3438
56. Verbist KC, Guy CS, Milasta S, et al. Metabolic maintenance of cell asymmetry following division in activated T lymphocytes. *Nature.* 2016;532:389-393. doi:10.1038/nature17442
57. Donnelly RP, Loftus RM, Keating SE, et al. mTORC1-dependent metabolic reprogramming is a prerequisite for NK cell effector function. *J Immunol.* 2014;193, 4477-4484. doi:10.4049/jimmunol.1401558
58. Weichhart T, Costantino G, Poglitsch M, et al. The TSC-mTOR signaling pathway regulates the innate inflammatory response. *Immunity.* 2008;29:565-577. doi:10.1016/j.immuni.2008.08.012
59. Ohtani M, Nagai S, Kondo S, et al. Mammalian target of rapamycin and glycogen synthase kinase 3 differentially regulate lipopolysaccharide-induced interleukin-12 production in dendritic cells. *Blood.* 2008;112:635-643. doi:10.1182/blood-2008-02-137430
60. Turnquist HR, Cardinal J, Macedo C, et al. mTOR and GSK-3 shape the CD4+ T-cell stimulatory and differentiation capacity of myeloid DCs after exposure to LPS. *Blood.* 2010;115:4758-4769. doi:10.1182/blood-2009-10-251488
61. Cao W, Manicassamy S, Tang H, et al. Toll-like receptor-mediated induction of type I interferon in plasmacytoid dendritic cells requires the rapamycin-sensitive PI(3)K-mTOR-p70S6K pathway. *Nat Immunol.* 2008;9:1157-1164. doi:10.1038/ni.1645
62. Huang SC-C, Smith AM, Everts B, et al. Metabolic reprogramming mediated by the mTORC2-IRF4 signaling axis is essential for macrophage alternative activation. *Immunity.* 2016;45, 817-830. doi:10.1016/j.immuni.2016.09.016
63. Hallowell RW, Collins SL, Craig JM, et al. mTORC2 signalling regulates M2 macrophage differentiation in response to helminth infection and adaptive thermogenesis. *Nat Commun.* 2017;8:14208. doi:10.1038/ncomms14208
64. Oh M-H, Collins SL, Sun I-H, et al. mTORC2 signaling selectively regulates the generation and function of tissue-resident peritoneal macrophages. *Cell Rep.* 2017;20:2439-2454. doi:10.1016/j.celrep.2017.08.046
65. Yang K, Blanco DB, Chen X, et al. Metabolic signaling directs the reciprocal lineage decisions of alphabeta and gammadelta T cells. *Sci Immunol.* 2018;3:eaas9818. doi:10.1126/sciimmunol.aas9818.
66. Kelly AP, Finlay DK, Hinton HJ, et al. Notch-induced T cell development requires phosphoinositide-dependent kinase 1. *EMBO J.* 2007;26:3441-3450. doi:10.1038/sj.emboj.7601761
67. Ciofani M, Zuniga-Pflucker JC. Notch promotes survival of pre-T cells at the beta-selection checkpoint by regulating cellular metabolism. *Nat Immunol.* 2005;6:881-888. doi:10.1038/ni1234
68. Tamás P, Macintyre A, Finlay D, et al. LKB1 is essential for the proliferation of T-cell progenitors and mature peripheral T cells. *Eur J Immunol.* 2010;40:242-253. doi:10.1002/eji.200939677
69. MacIver NJ, Blagih J, Saucillo DC, et al. The liver kinase B1 is a central regulator of T cell development, activation, and metabolism. *J Immunol.* 2011;187:4187-4198. doi:10.4049/jimmunol.1100367
70. Hoshii T, Kasada A, Hatakeyama T, et al. Loss of mTOR complex 1 induces developmental blockage in early T-lymphopoiesis and eradicates T-cell acute lymphoblastic leukemia cells. *Proc Natl Acad Sci U S A.* 2014;111:3805-3810. doi:10.1073/pnas.1320265111
71. Swainson L, Kinet S, Manel N, Battini J-L, Sitbon M, Taylor N. Glucose transporter 1 expression identifies a population of cycling CD4+ CD8+ human thymocytes with high CXCR4-induced chemotaxis. *Proc Natl Acad Sci U S A.* 2005;102:12867-12872. doi:10.1073/pnas.0503603102
72. Macintyre AN, Gerriets VA, Nichols AG, et al. The glucose transporter Glut1 is selectively essential for CD4 T cell activation and effector function. *Cell Metab.* 2014;20:61-72. doi:10.1016/j.cmet.2014.05.004
73. Ramstead AG, Wallace JA, Lee S-H, et al. Mitochondrial pyruvate carrier 1 promotes peripheral T cell homeostasis through metabolic regulation of thymic development. *Cell Rep.* 2020;30:2889-2899.e6. doi:10.1016/j.celrep.2020.02.042
74. Sprent J, Surh CD. Normal T cell homeostasis: the conversion of naive cells into memory-phenotype cells. *Nat Immunol.* 2011;12:478-484. doi:10.1038/ni.2018
75. Surh CD, Sprent J. Homeostasis of naive and memory T cells. *Immunity.* 2008;29:848-862. doi:10.1016/j.immuni.2008.11.002
76. Wang R, Dillon CP, Shi LZ, et al. The transcription factor Myc controls metabolic reprogramming upon T lymphocyte activation. *Immunity.* 2011;35:871-882.
77. Rathmell JC, Vander Heiden MG, Harris MH, Frauwirth KA, Thompson CB. In the absence of extrinsic signals, nutrient utilization by lymphocytes is insufficient to maintain either cell size or viability. *Mol Cell.* 2000;6:683-692. doi:10.1016/s1097-2765(00)00066-6
78. Rathmell JC, Farkash EA, Gao W, Thompson CB. IL-7 enhances the survival and maintains the size of naive T cells. *J Immunol.* 2001;167:6869-6876. doi:10.4049/jimmunol.167.12.6869
79. Jacobs SR, Michalek RD, Rathmell JC. IL-7 is essential for homeostatic control of T cell metabolism in vivo. *J Immunol.* 2010;184:3461-3469. doi:10.4049/jimmunol.0902593
80. Finlay DK, Rosenzweig E, Sinclair LV, et al. PDK1 regulation of mTOR and hypoxia-inducible factor 1 integrate metabolism and migration of CD8+ T cells. *J Exp Med.* 2012;209:2441-2453. doi:10.1084/jem.20112607
81. Howden AJM, Hukelmann JL, Brenes A, et al. Quantitative analysis of T cell proteomes and environmental sensors during T cell differentiation. *Nat Immunol.* 2019;20:1542-1554. doi:10.1038/s41590-019-0495-x
82. Yang K, Shrestha S, Zeng H, et al. T cell exit from quiescence and differentiation into Th2 cells depend on Raptor-mTORC1-mediated metabolic reprogramming. *Immunity.* 2013;39:1043-1056. doi:10.1016/j.immuni.2013.09.015
83. Ricciardi S, Manfrini N, Alfieri R, et al. The translational machinery of human CD4(+) T cells is poised for activation and controls the switch from quiescence to metabolic remodeling. *Cell Metab.* 2018;28:895-906.e5. doi:10.1016/j.cmet.2018.08.009
84. Menk AV, Scharping NE, Moreci RS, et al. Early TCR signaling induces rapid aerobic glycolysis enabling distinct acute T cell effector functions. *Cell Rep.* 2018;22:1509-1521. doi:10.1016/j.celrep.2018.01.040
85. Jacobs SR, Herman CE, MacIver NJ, et al. Glucose uptake is limiting in T cell activation and requires CD28-mediated Akt-dependent and independent pathways. *J Immunol.* 2008;180:4476-4486. doi:10.4049/jimmunol.180.7.4476
86. Wieman HL, Wofford JA, Rathmell JC. Cytokine stimulation promotes glucose uptake via phosphatidylinositol-3 kinase/Akt regulation of Glut1 activity and trafficking. *Mol Biol Cell.* 2007;18:1437-1446. doi:10.1091/mbc.e06-07-0593
87. Buck MD, O'Sullivan D, Klein Geltink RI, et al. Mitochondrial dynamics controls T cell fate through metabolic programming. *Cell.* 2016;166:63-76. doi:10.1016/j.cell.2016.05.035
88. Swamy M, Pathak S, Grzes KM, et al. Glucose and glutamine fuel protein O-GlcNAcylation to control T cell self-renewal and malignancy. *Nature Immunol.* 2016;17:712-720. doi:10.1038/ni.3439
89. Preston GC, Sinclair LV, Kaskar A, et al. Single cell tuning of Myc expression by antigen receptor signal strength and interleukin-2 in T lymphocytes. *EMBO J.* 2015;34:2008-2024. doi:10.15252/embj.201490252
90. Marchingo JM, Sinclair LV, Howden AJ, Cantrell DA. Quantitative analysis of how Myc controls T cell proteomes and metabolic pathways during T cell activation. *Elife.* 2020;9:e53725. doi:10.7554/eLife.53725
91. Murray CM, Hutchinson R, Bantick JR, et al. Monocarboxylate transporter MCT1 is a target for immunosuppression. *Nat Chem Biol.* 2005;1:371-376. doi:10.1038/nchembio744
92. Chang C-H, Curtis JD, Maggi LB, et al. Posttranscriptional control of T cell effector function by aerobic glycolysis. *Cell.* 2013;153:1239-1251. doi:10.1016/j.cell.2013.05.016
93. Caro-Maldonado A, Wang R, Nichols AG, et al. Metabolic reprogramming is required for antibody production that is suppressed in anergic but exaggerated in chronically BAFF-exposed B cells. *J Immunol.* 2014;192:3626-3636. doi:10.4049/jimmunol.1302062
94. Sena LA, Li S, Jairaman A, et al. Mitochondria are required for antigen-specific T cell activation through reactive oxygen species signaling. *Immunity.* 2013;38:225-236. doi:10.1016/j.immuni.2012.10.020
95. Quintana A, Schwindling C, Wenning AS, et al. T cell activation requires mitochondrial translocation to the immunological synapse. *Proc Natl Acad Sci U S A.* 2007;104:14418-14423. doi:10.1073/pnas.0703126104
96. Gómez-Cabañas L, López-Cotarelo P, Criado-García O, Murphy MP, Boya P, Rodríguez-Fernández JL. Immunological synapse formation induces mitochondrial clustering and mitophagy in dendritic cells. *J Immunol.* 2019;202:1715-1723. doi:10.4049/jimmunol.1800575
97. Baixauli F, Martín-Cófreces NB, Morlino G, et al. The mitochondrial fission factor dynamin-related protein 1 modulates T-cell receptor signalling at the immune synapse. *EMBO J.* 2011;30:1238-1250. doi:10.1038/emboj.2011.25
98. Tan H, Yang K, Li Y, et al. Integrative proteomics and phosphoproteomics profiling reveals dynamic signaling networks and bioenergetics pathways underlying T cell activation. *Immunity.* 2017;46:488-503. doi:10.1016/j.immuni.2017.02.010
99. Ron-Harel N, Santos D, Ghergurovich JM, et al. Mitochondrial biogenesis and proteome remodeling promote one-carbon metabolism for T cell activation. *Cell Metab.* 2016;24:104-117. doi:10.1016/j.cmet.2016.06.007
100. Mak TW, Grusdat M, Duncan GS, et al. Glutathione primes T cell metabolism for inflammation. *Immunity.* 2017;46:675-689. doi:10.1016/j.immuni.2017.03.019
101. Siska PJ, Kim B, Ji X, et al. Fluorescence-based measurement of cystine uptake through xCT shows requirement for ROS detoxification in activated lymphocytes. *J Immunol Methods.* 2016;438:51-58. doi:10.1016/j.jim.2016.08.013
102. Arensman MD, Yang XS, Leahy DM, et al. Cystine-glutamate antiporter xCT deficiency suppresses tumor growth while preserving antitumor immunity. *Proc Natl Acad Sci U S A.* 2019;116:9533-9542. doi:10.1073/pnas.1814932116
103. Chisolm DA, Weinmann AS. Connections between metabolism and epigenetics in programming cellular differentiation. *Annu Rev Immunol.* 2018;36:221-246. doi:10.1146/annurev-immunol-042617-053127
104. Hosios AM, Hecht VC, Danai LV, et al. Amino acids rather than glucose account for the majority of cell mass in proliferating mammalian cells. *Dev Cell.* 2016;36:540-549. doi:10.1016/j.devcel.2016.02.012
105. Altman BJ, Stine ZE, Dang CV. From Krebs to clinic: glutamine metabolism to cancer therapy. *Nat Rev Cancer.* 2016;16:619-634. doi:10.1038/nrc.2016.71
106. Ron-Harel N, Ghergurovich JM, Notarangelo G, et al. T cell activation depends on extracellular alanine. *Cell Rep.* 2019;28:3011-3021.e4. doi:10.1016/j.celrep.2019.08.034
107. Sinclair LV, Rolf J, Emslie E, Shi Y-B, Taylor PM, Cantrell DA. Control of amino-acid transport by antigen receptors coordinates the metabolic reprogramming essential for T cell differentiation. *Nat Immunol.* 2013;14:500-508. doi:10.1038/ni.2556
108. Rodriguez PC, Quiceno DG, Ochoa AC. L-arginine availability regulates T-lymphocyte cell-cycle progression. *Blood.* 2007;109:1568-1573. doi:10.1182/blood-2006-06-031856
109. Geiger R, Rieckmann JC, Wolf T, et al. L-arginine modulates T cell metabolism and enhances survival and anti-tumor activity. *Cell.* 2016;167:829-842.e13. doi:10.1016/j.cell.2016.09.031
110. Ma EH, Bantug G, Griss T, et al. Serine is an essential metabolite for effector T cell expansion. *Cell Metab.* 2017;25:345-357. doi:10.1016/j.cmet.2016.12.011
111. Ma EH, Verway MJ, Johnson RM, et al. Metabolic profiling using stable isotope tracing reveals distinct patterns of glucose utilization by physiologically activated CD8(+) T cells. *Immunity.* 2019;51:856-870.e5. doi:10.1016/j.immuni.2019.09.003
112. Kidani Y, Elsaesser H, Hock MB, et al. Sterol regulatory element-binding proteins are essential for the metabolic programming of effector T cells and adaptive immunity. *Nat Immunol.* 2013;14:489-499. doi:10.1038/ni.2570
113. Molnár E, Swamy M, Holzer M, et al. Cholesterol and sphingomyelin drive ligand-independent T-cell antigen receptor nanoclustering. *J Biol Chem.* 2012;287:42664-42674. doi:10.1074/jbc.M112.386045
114. Yang W, Bai Y, Xiong Y, et al. Potentiating the antitumour response of CD8(+) T cells by modulating cholesterol metabolism. *Nature.* 2016;531:651-655. doi:10.1038/nature17412
115. Lee J, Walsh MC, Hoehn KL, James DE, Wherry EJ, Choi Y. Regulator of fatty acid metabolism, acetyl coenzyme a carboxylase 1, controls T cell immunity. *J Immunol.* 2014;192:3190-3199. doi:10.4049/jimmunol.1302985
116. van der Windt G, Everts B, Chang CH, et al. Mitochondrial respiratory capacity is a critical regulator of CD8+ T cell memory development. *Immunity.* 2012;36:68-78. doi:10.1016/j.immuni.2011.12.007
117. Oestreich KJ, Read KA, Gilbertson SE, et al. Bcl-6 directly represses the gene program of the glycolysis pathway. *Nat Immunol.* 2014;15:957-964. doi:10.1038/ni.2985
118. Villarino AV, Katzman SD, Gallo E, et al. Posttranscriptional silencing of effector cytokine mRNA underlies the anergic phenotype of self-reactive T cells. *Immunity.* 2011;34:50-60. doi:10.1016/j.immuni.2010.12.014
119. Ho P-C, Bihuniak JD, Macintyre AN, et al. Phosphoenolpyruvate is a metabolic checkpoint of anti-tumor T cell responses. *Cell.* 2015;162:1217-1228. doi:10.1016/j.cell.2015.08.012
120. Delgoffe GM, Pollizzi KN, Waickman AT, et al. The kinase mTOR regulates the differentiation of helper T cells through the selective activation of signaling by mTORC1 and mTORC2. *Nat Immunol.* 2011;12:295-303. doi:10.1038/ni.2005

121. Carr EL, Kelman A, Wu GS, et al. Glutamine uptake and metabolism are coordinately regulated by ERK/MAPK during T lymphocyte activation. *J Immunol.* 2010;185:1037-1044. doi:10.4049/jimmunol.0903586
122. Nakaya M, Xiao Y, Zhou X, et al. Inflammatory T cell responses rely on amino acid transporter ASCT2 facilitation of glutamine uptake and mTORC1 kinase activation. *Immunity.* 2014;40:692-705. doi:10.1016/j.immuni.2014.04.007
123. Klysz D, Tai X, Robert PA, et al. Glutamine-dependent alpha-ketoglutarate production regulates the balance between T helper 1 cell and regulatory T cell generation. *Sci Signal.* 2015;8:ra97. doi:10.1126/scisignal.aab2610
124. Bird JJ, Brown DR, Mullen AC, et al. Helper T cell differentiation is controlled by the cell cycle. *Immunity.* 1998;9:229-237. doi:10.1016/s1074-7613(00)80605-6
125. Johnson MO, Wolf MM, Madden MZ, et al. Distinct regulation of Th17 and Th1 cell differentiation by glutaminase-dependent metabolism. *Cell.* 2018;175:1780-1795.e19. doi:10.1016/j.cell.2018.10.001
126. Wang Y, Misumi I, Gu A-D, et al. GATA-3 controls the maintenance and proliferation of T cells downstream of TCR and cytokine signaling. *Nat Immunol.* 2013;14:714-722. doi:10.1038/ni.2623
127. Seumois G, Zapardiel-Gonzalo J, White B, et al. Transcriptional profiling of Th2 cells identifies pathogenic features associated with asthma. *J Immunol.* 2016;197:655-664. doi:10.4049/jimmunol.1600397
128. Yang J-Q, Kalim KW, Li Y, et al. RhoA orchestrates glycolysis for TH2 cell differentiation and allergic airway inflammation. *J Allergy Clin Immunol.* 2016;137:231-245.e4. doi:10.1016/j.jaci.2015.05.004
129. Angela M, Endo Y, Asou HK, et al. Fatty acid metabolic reprogramming via mTOR-mediated inductions of PPARgamma directs early activation of T cells. *Nat Commun.* 2016;7:13683. doi:10.1038/ncomms13683
130. Chen T, Tibbitt CA, Feng X, et al. PPAR-gamma promotes type 2 immune responses in allergy and nematode infection. *Sci Immunol.* 2017;2:eaal5196. doi:10.1126/sciimmunol.aal5196
131. Gerriets VA, Kishton RJ, Nichols AG, et al. Metabolic programming and PDHK1 control CD4+ T cell subsets and inflammation. *J Clin Invest.* 2015;125:194-207. doi:10.1172/JCI76012
132. Semenza GL. Targeting HIF-1 for cancer therapy. *Nat Rev Cancer.* 2003;3:721-732. doi:10.1038/nrc1187
133. Lee JH, Elly C, Park Y, Liu YC. E3 ubiquitin ligase VHL regulates hypoxia-inducible factor-1alpha to maintain regulatory T cell stability and suppressive capacity. *Immunity.* 2015;42:1062-1074. doi:10.1016/j.immuni.2015.05.016
134. Huynh A, DuPage M, Priyadharshini B, et al. Control of PI(3) kinase in Treg cells maintains homeostasis and lineage stability. *Nat Immunol.* 2015;16:188-196. doi:10.1038/ni.3077
135. Shrestha S, Yang K, Guy C, Vogel P, Neale G, Chi H. Treg cells require the phosphatase PTEN to restrain TH1 and TFH cell responses. *Nat Immunol.* 2015;16:178-187. doi:10.1038/ni.3076
136. Angelin A, Gil-de-Gómez L, Dahiya S, et al. Foxp3 reprograms T Cell metabolism to function in low-glucose, high-lactate environments. *Cell Metab.* 2017;25:1282-1293.e7. doi:10.1016/j.cmet.2016.12.018
137. Gerriets VA, Kishton RJ, Johnson MO, et al. Foxp3 and Toll-like receptor signaling balance Treg cell anabolic metabolism for suppression. *Nat Immunol.* 2016;17:1459-1466. doi:10.1038/ni.3577
138. Wheaton WW, Weinberg SE, Hamanaka RB, et al. Metformin inhibits mitochondrial complex I of cancer cells to reduce tumorigenesis. *Elife.* 2014;3:e02242. doi:10.7554/eLife.02242
139. Weinberg SE, Singer BD, Steinert EM, et al. Mitochondrial complex III is essential for suppressive function of regulatory T cells. *Nature.* 2019;565:495-499. doi:10.1038/s41586-018-0846-z
140. Beier UH, Angelin A, Akimova T, et al. Essential role of mitochondrial energy metabolism in Foxp3(+) T-regulatory cell function and allograft survival. *FASEB J.* 2015;29:2315-2326. doi:10.1096/fj.14-268409
141. Raud B, Roy DG, Divakaruni AS, et al. Etomoxir actions on regulatory and memory T cells are independent of Cpt1a-mediated fatty acid oxidation. *Cell Metab.* 2018;28:504-515.e7. doi:10.1016/j.cmet.2018.06.002
142. Zeng H, Yang K, Cloer C, Neale G, Vogel P, Chi H. mTORC1 couples immune signals and metabolic programming to establish T(reg)-cell function. *Nature.* 2013;499:485-490. doi:10.1038/nature12297
143. Procaccini C, De Rosa V, Galgani M, et al. An oscillatory switch in mTOR kinase activity sets regulatory T cell responsiveness. *Immunity.* 2010;33:929-941. doi:10.1016/j.immuni.2010.11.024
144. De Rosa V, Galgani M, Porcellini A, et al. Glycolysis controls the induction of human regulatory T cells by modulating the expression of FOXP3 exon 2 splicing variants. *Nat Immunol.* 2015;16:1174-1184. doi:10.1038/ni.3269
145. Angiari S, Runtsch MC, Sutton CE, et al. Pharmacological activation of pyruvate kinase M2 inhibits CD4(+) T cell pathogenicity and suppresses autoimmunity. *Cell Metab.* 2020;31:391-405.e8. doi:10.1016/j.cmet.2019.10.015
146. Karmaus PWF, Chen X, Lim SA, et al. Metabolic heterogeneity underlies reciprocal fates of TH17 cell stemness and plasticity. *Nature.* 2019;565:101-105. doi:10.1038/s41586-018-0806-7
147. Franchi L, Monteleone I, Hao L-Y, et al. Inhibiting oxidative phosphorylation in vivo restrains Th17 effector responses and ameliorates murine colitis. *J Immunol.* 2017;198:2735-2746. doi:10.4049/jimmunol.1600810
148. Berod L, Friedrich C, Nandan A, et al. De novo fatty acid synthesis controls the fate between regulatory T and T helper 17 cells. *Nat Med.* 2014;20:1327-1333. doi:10.1038/nm.3704
149. Santori FR, Huang P, van de Pavert SA, et al. Identification of natural RORgamma ligands that regulate the development of lymphoid cells. *Cell Metab.* 2015;21:286-298. doi:10.1016/j.cmet.2015.01.004
150. Jin L, Martynowski D, Zheng S, Wada T, Xie W, Li Y. Structural basis for hydroxycholesterols as natural ligands of orphan nuclear receptor RORgamma. *Mol Endocrinol.* 2010;24:923-929. doi:10.1210/me.2009-0507
151. Wang Y, Kumar N, Crumbley C, Griffin PR, Burris TP. A second class of nuclear receptors for oxysterols: regulation of RORalpha and RORgamma activity by 24S-hydroxycholesterol (cerebrosterol). *Biochim Biophys Acta.* 2010;1801:917-923. doi:10.1016/j.bbalip.2010.02.012
152. Soroosh P, Wu J, Xue X, et al. Oxysterols are agonist ligands of RORgammat and drive Th17 cell differentiation. *Proc Natl Acad Sci U S A.* 2014;111:12163-12168. doi:10.1073/pnas.1322807111
153. Wang C, Yosef N, Gaublomme J, et al. CD5L/AIM regulates lipid biosynthesis and restrains Th17 cell pathogenicity. *Cell.* 2015;163. 1413-1427. doi:10.1016/j.cell.2015.10.068
154. Wu C, Yosef N, Thalhamer T, et al. Induction of pathogenic TH17 cells by inducible salt-sensing kinase SGK1. *Nature.* 2013;496:513-517. doi:10.1038/nature11984
155. Kleinewietfeld M, Manzel A, Titze J, et al. Sodium chloride drives autoimmune disease by the induction of pathogenic TH17 cells. *Nature.* 2013;496:518-522. doi:10.1038/nature11868
156. Johnston RJ, Choi YS, Diamond JA, Yang JA, Crotty S. STAT5 is a potent negative regulator of TFH cell differentiation. *J Exp Med.* 2012;209:243-250. doi:10.1084/jem.20111174
157. Ballesteros-Tato A, León B, Graf BA, et al. Interleukin-2 inhibits germinal center formation by limiting T follicular helper cell differentiation. *Immunity.* 2012;36:847-856. doi:10.1016/j.immuni.2012.02.012
158. Liao W, Lin JX, Wang L, Li P, Leonard WJ. Modulation of cytokine receptors by IL-2 broadly regulates differentiation into helper T cell lineages. *Nat Immunol.* 2011;12, 551-559. doi:10.1038/ni.2030
159. Wu T, Shin HM, Moseman EA, et al. TCF1 is required for the T follicular helper cell response to viral infection. *Cell Rep.* 2015;12:2099-2110. doi:10.1016/j.celrep.2015.08.049
160. Zhu Y, Zhao Y, Zou L, et al. The E3 ligase VHL promotes follicular helper T cell differentiation via glycolytic-epigenetic control. *J Exp Med.* 2019;216:1664-1681. doi:10.1084/jem.20190337
161. Cho SH, Raybuck AL, Blagih J, et al. Hypoxia-inducible factors in CD4(+) T cells promote metabolism, switch cytokine secretion, and T cell help in humoral immunity. *Proc Natl Acad Sci U S A.* 2019;116:8975-8984. doi:10.1073/pnas.1811702116
162. Zeng H, Cohen S, Guy C, et al. mTORC1 and mTORC2 kinase signaling and glucose metabolism drive follicular helper T cell differentiation. *Immunity.* 2016;45:540-554. doi:10.1016/j.immuni.2016.08.017
163. Yang J, Lin X, Pan Y, et al. Critical roles of mTOR Complex 1 and 2 for T follicular helper cell differentiation and germinal center responses. *Elife.* 2016;5:e17936. doi:10.7554/eLife.17936
164. Doedens AL, Phan AT, Stradner MH, et al. Hypoxia-inducible factors enhance the effector responses of CD8(+) T cells to persistent antigen. *Nat Immunol.* 2013;14:1173-1182. doi:10.1038/ni.2714
165. Sukumar M, Liu J, Ji Y, et al. Inhibiting glycolytic metabolism enhances CD8+ T cell memory and antitumor function. *J Clin Invest.* 2013;123:4479-4488. doi:10.1172/JCI69589
166. Qiu J, Villa M, Sanin DE, et al. Acetate promotes T cell effector function during glucose restriction. *Cell Rep.* 2019;27:2063-2074.e5. doi:10.1016/j.celrep.2019.04.022
167. Sinclair LV, Howden AJ, Brenes A, et al. Antigen receptor control of methionine metabolism in T cells. *Elife.* 2019;8:e44210. doi:10.7554/eLife.44210
168. Pearce EL, Walsh MC, Cejas PJ, et al. Enhancing CD8 T-cell memory by modulating fatty acid metabolism. *Nature.* 2009;460:103-107. doi:10.1038/nature08097
169. Phan AT, Doedens AL, Palazon A, et al. Constitutive glycolytic metabolism supports CD8(+) T cell effector memory differentiation during viral infection. *Immunity.* 2016;45:1024-1037. doi:10.1016/j.immuni.2016.10.017
170. Tyrakis PA, Palazon A, Macias D, et al. S-2-hydroxyglutarate regulates CD8(+) T-lymphocyte fate. *Nature.* 2016;540:236-241. doi:10.1038/nature20165
171. Carty SA, Gohil M, Banks LB, et al. The loss of TET2 promotes CD8(+) T cell memory differentiation. *J Immunol.* 2018;200:82-91. doi:10.4049/jimmunol.1700559
172. O'Sullivan D, van der Windt GJ, Huang SC, et al. Memory CD8+ T cells use cell-intrinsic lipolysis to support the metabolic programming necessary for development. *Immunity.* 2014;41:75-88. doi:10.1016/j.immuni.2014.06.005
173. Cui G, Staron MM, Gray SM, et al. IL-7-induced glycerol transport and TAG synthesis promotes memory CD8+ T cell longevity. *Cell.* 2015;161:750-761. doi:10.1016/j.cell.2015.03.021
174. Xu X, Araki K, Li S, et al. Autophagy is essential for effector CD8(+) T cell survival and memory formation. *Nat Immunol.* 2014;15:1152-1161. doi:10.1038/ni.3025
175. vanderWindt GJW, O'Sullivan D, Everts B, et al. CD8 memory T cells have a bioenergetic advantage that underlies their rapid recall ability. *Proc Natl Acad Sci U S A.* 2013;110:14336-14341. doi:10.1073/pnas.1221740110
176. Adams WC, Chen Y-H, Kratchmarov R, et al. anabolism-associated mitochondrial stasis driving lymphocyte differentiation over self-renewal. *Cell Rep.* 2016;17:3142-3152. doi:10.1016/j.celrep.2016.11.065
177. Ma R, Ji T, Zhang H, et al. A Pck1-directed glycogen metabolic program regulates formation and maintenance of memory CD8+ T cells. *Nat Cell Biol.* 2018;20:21-27. doi:10.1038/s41556-017-0002-2
178. Gubser PM, Bantug GR, Razik L, et al. Rapid effector function of memory CD8+ T cells requires an immediate-early glycolytic switch. *Nature Immunol.* 2013;14:1064-1072. doi:10.1038/ni.2687
179. Balmer ML, Ma EH, Bantug GR, et al. Memory CD8(+) T cells require increased concentrations of acetate induced by stress for optimal function. *Immunity.* 2016;44:1312-1324. doi:10.1016/j.immuni.2016.03.016
180. Pan Y, Tian T, Park CO, et al. Survival of tissue-resident memory T cells requires exogenous lipid uptake and metabolism. *Nature.* 2017;543:252-256. doi:10.1038/nature21379
181. Frizzell H, Fonseca R, Christo SN, et al. Organ-specific isoform selection of fatty acid-binding proteins in tissue-resident lymphocytes. *Sci Immunol.* 2020;5:eaay9283. doi:10.1126/sciimmunol.aay9283
182. McLane LM, Abdel-Hakeem MS, Wherry EJ. CD8 T cell exhaustion during chronic viral infection and cancer. *Annu Rev Immunol.* 2019;37:457-495. doi:10.1146/annurev-immunol-041015-055318
183. Bengsch B, Johnson AL, Kurachi M, et al. Bioenergetic insufficiencies due to metabolic alterations regulated by the inhibitory receptor PD-1 are an early driver of CD8(+) T cell exhaustion. *Immunity.* 2016;45:358-373. doi:10.1016/j.immuni.2016.07.008
184. Scharping NE, Menk AV, Moreci RS, et al. The tumor microenvironment represses T cell mitochondrial biogenesis to drive intratumoral T cell metabolic insufficiency and dysfunction. *Immunity.* 2016;45:374-388. doi:10.1016/j.immuni.2016.07.009
185. Cyster JG, Allen CDC. B cell responses: cell interaction dynamics and decisions. *Cell.* 2019;177:524-540. doi:10.1016/j.cell.2019.03.016. PMID: 31002794.
186. Nagasawa T. Microenvironmental niches in the bone marrow required for B-cell development. *Nat Rev Immunol.* 2006;6:107-116. doi:10.1038/nri1780
187. Chan LN, Chen Z, Braas D, et al. Metabolic gatekeeper function of B-lymphoid transcription factors. *Nature.* 2017;542:479-483. doi:10.1038/nature21076 PMID: 28192788.
188. Zeng H, Yu M, Tan H, et al. Discrete roles and bifurcation of PTEN signaling and mTORC1-mediated anabolic metabolism underlie IL-7–driven B lymphopoiesis. *Sci Adv.* 2018;4:eaar5701. doi:10.1126/sciadv.aar5701. PMID: 29399633.
189. Iwata TN, Ramírez JA, Tsang M, et al. Conditional disruption of raptor reveals an essential role for mTORC1 in B cell development, survival, and metabolism. *J Immunol.* 2016;197:2250-2260. doi:10.4049/jimmunol.1600492. PMID: 27521345.
190. Huang C-Y, Bredemeyer AL, Walker LM, Bassing CH, Sleckman BP. Dynamic regulation ofc-Myc proto-oncogene expression during lymphocyte development revealed by aGFP-c-Myc knock-in mouse. *Eur J Immunol.* 2008;38:342-349. doi:10.1002/eji.200737972. PMID: 18196519.
191. Herzog S, Hug E, Meixlsperger S, et al. SLP-65 regulates immunoglobulin light chain gene recombination through the PI(3)K-PKB-Foxo pathway. *Nat Immunol.* 2008;9:623-631. doi:10.1038/ni.1616. PMID: 18488031.
192. Stein M, Dütting S, Mougiakakos D, et al. A defined metabolic state in pre B cells governs B-cell development and is counterbalanced by Swiprosin-2/EFhd1. *Cell Death Differ.* 2017;24:1239-1252. doi:10.1038/cdd.2017.52. PMID: 28524857.
193. Kojima H, Kobayashi A, Sakurai D, et al. Differentiation stage-specific requirement in hypoxia-inducible factor-1α–regulated glycolytic pathway during murine B cell development in bone marrow. *J Immunol.* 2010;184:154-163. doi:10.4049/jimmunol.0800167. PMID: 19949104.

194. Spencer JA, Ferraro F, Roussakis E, et al. Direct measurement of local oxygen concentration in the bone marrow of live animals. *Nature*. 2014;508:269-273. doi:10.1038/nature13034
195. Kojima H, Gu H, Nomura S, et al. Abnormal B lymphocyte development and autoimmunity in hypoxia-inducible factor 1alpha-deficient chimeric mice. *Proc Natl Acad Sci U S A*. 2002;99:2170-2174. doi:10.1073/pnas.052706699. PMID: 11854513.
196. Park H, Staehling K, Tsang M, et al. Disruption of Fnip1 reveals a metabolic checkpoint controlling B lymphocyte development. *Immunity*. 2012;36:769-781. doi:10.1016/j.immuni.2012.02.019. PMID: 22608497.
197. Jellusova J, Cato MH, Apgar JR, et al. Gsk3 is a metabolic checkpoint regulator in B cells. *Nat Immunol*. 2017;18:303-312. doi:10.1038/ni.3664
198. Farmer JR, Allard-Chamard H, Sun N, et al. Induction of metabolic quiescence defines the transitional to follicular B cell switch. *Sci Signal*. 2019;12:eaaw5573. doi:10.1126/scisignal.aaw5573
199. Ci X, Kuraoka M, Wang H, et al. TSC1 promotes B cell maturation but is dispensable for germinal center formation. *PLoS One*. 2015;10:e0127527. doi:10.1371/journal.pone.0127527. PMID: 26000908.
200. Waters LR, Ahsan FM, Wolf DM, Shirihai O, Teitell MA. Initial B cell activation induces metabolic reprogramming and mitochondrial remodeling. *iScience*. 2018;5:99-109. doi:10.1016/j.isci.2018.07.005. PMID: 30240649.
201. Doughty CA, Bleiman BF, Wagner DJ, et al. Antigen receptor-mediated changes in glucose metabolism in B lymphocytes: role of phosphatidylinositol 3-kinase signaling in the glycolytic control of growth. *Blood*. 2006:107:4458-4465. doi:10.1182/blood-2005-12-4788
202. Anken EV, Romijn EP, Maggioni C, et al. Sequential waves of functionally related proteins are expressed when B cells prepare for antibody secretion. *Immunity*. 2003;18:243-253. doi:10.1016/s1074-7613(03)00024-4. PMID: 12594951.
203. Dufort FJ, Gumina MR, Ta NL, et al. Glucose-dependent de novo lipogenesis in B lymphocytes: a requirement for atp-citrate lyase in lipopolysaccharide-induced differentiation. *J Biol Chem*. 2014;289:7011-7024. doi:10.1074/jbc.m114.551051. PMID: 24469453.
204. Patke A, Mecklenbräuker I, Erdjument-Bromage H, Tempst P, Tarakhovsky A. BAFF controls B cell metabolic fitness through a PKCβ-and Akt-dependent mechanism. *J Exp Med*. 2006;203:2551-2562. doi:10.1084/jem.20060990. PMID: 17060474.
205. Woodland RT, Fox CJ, Schmidt MR, et al. Multiple signaling pathways promote B lymphocyte stimulator–dependent B-cell growth and survival. *Blood*. 2008;111:750-760. doi:10.1182/blood-2007-03-077222. PMID: 17942753.
206. Dufort FJ, Bleiman BF, Gumina MR, et al. Cutting edge: IL-4-mediated protection of primary B lymphocytes from apoptosis via Stat6-dependent regulation of glycolytic metabolism. *J Immunol*. 2007;179:4953-4957. doi:10.4049/jimmunol.179.8.4953. PMID: 17911579.
207. Limon JJ, So L, Jellbauer S, et al. mTOR kinase inhibitors promote antibody class switching via mTORC2 inhibition. *Proc Natl Acad Sci*. 2014;111:E5076-E5085. doi:10.1073/pnas.1407104111. PMID: 25385646.
208. Tsui C, Martinez-Martin N, Gaya M, et al. Protein kinase C-beta dictates B cell fate by regulating mitochondrial remodeling, metabolic reprogramming, and heme biosynthesis. *Immunity*. 2018;48:1144-1159.e5. doi:10.1016/j.immuni.2018.04.031
209. Cantor J, Browne CD, Ruppert R, et al. CD98hc facilitates B cell proliferation and adaptive humoral immunity. *Nat Immunol*. 2009;10:412-419. doi:10.1038/ni.1712
210. Wheeler ML, DeFranco AL. Prolonged production of reactive oxygen species in response to b cell receptor stimulation promotes B cell activation and proliferation. *J Immunol*. 2012;189:4405-4416. doi:10.4049/jimmunol.1201433. PMID: 23024271.
211. Jiang S, Yan W, Wang SE, Baltimore D. Let-7 suppresses B cell activation through restricting the availability of necessary nutrients. *Cell Metab*. 2018;27:393-403.e4. doi:10.1016/j.cmet.2017.12.007. PMID: 29337138.
212. Akkaya M, Traba J, Roesler AS, et al. Second signals rescue B cells from activation-induced mitochondrial dysfunction and death. *Nat Immunol*. 2018;19:871-884. doi:10.1038/s41590-018-0156-5. PMID: 29988090.
213. Raybuck AL, Cho SH, Li J, et al. B cell-intrinsic mTORC1 promotes germinal center-defining transcription factor gene expression, somatic hypermutation, and memory B cell generation in humoral immunity. *J Immunol*. 2018;200:2627-2639. doi:10.4049/jimmunol.1701321. PMID: 29531165.
214. Chiu H, Jackson LV, Oh KI, et al. The mTORC1/4E-BP/eIF4E axis promotes antibody class switching in B lymphocytes. *J Immunol*. 2019;202:579-590. doi:10.4049/jimmunol.1800602
215. Keating R, Hertz T, Wehenkel M, et al. The kinase mTOR modulates the antibody response to provide cross-protective immunity to lethal infection with influenza virus. *Nat Immunol*. 2013;14:1266-1276. doi:10.1038/ni.2741
216. Mannick JB, Del Giudice G, Lattanzi M, et al. mTOR inhibition improves immune function in the elderly. *Sci Transl Med*. 2014;6:268ra179. doi:10.1126/scitranslmed.3009892
217. Ersching J, Efeyan A, Mesin L, et al. Germinal center selection and affinity maturation require dynamic regulation of mTORC1 kinase. *Immunity*. 2017;46:1045-1058.e6. doi:10.1016/j.immuni.2017.06.005. PMID: 28636954.
218. Calado DP, Sasaki Y, Godinho SA, et al. The cell-cycle regulator c-Myc is essential for the formation and maintenance of germinal centers. *Nat Immunol*. 2012;13:1092-1100. doi:10.1038/ni.2418. PMID: 23001146.
219. Dominguez-Sola D, Victora GD, Ying CY, et al. The proto-oncogene MYC is required for selection in the germinal center and cyclic reentry. *Nat Immunol*. 2012;13:1083-1091. doi:10.1038/ni.2428. PMID: 23001145.
220. Cho SH, Raybuck AL, Stengel K, et al. Germinal centre hypoxia and regulation of antibody qualities by a hypoxia response system. *Nature*. 2016;537:234.
221. Garcia-Manteiga JM, Mari S, Godejohann M, et al. Metabolomics of B to plasma cell differentiation. *J Proteome Res*. 2011;10:4165-4176. doi:10.1021/pr200328f
222. Tellier J, Shi W, Minnich M, et al. Blimp-1 controls plasma cell function through the regulation of immunoglobulin secretion and the unfolded protein response. *Nat Immunol*. 2016;17:323-330. doi:10.1038/ni.3348
223. Lam WY, Becker AM, Kennerly KM, et al. Mitochondrial pyruvate import promotes long-term survival of antibody-secreting plasma cells. *Immunity*. 2016;45:60-73. doi:10.1016/j.immuni.2016.06.011. PMID: 27396958.
224. Jang KJ, Mano H, Aoki K, et al. Mitochondrial function provides instructive signals for activation-induced B-cell fates. *Nat Commun*. 2015;6:6750. doi:10.1038/ncomms7750
225. Pengo N, Scolari M, Oliva L, et al. Plasma cells require autophagy for sustainable immunoglobulin production. *Nat Immunol*. 2013;14:298-305. doi:10.1038/ni.2524. PMID: 23354484.
226. Chen M, Hong MJ, Sun H, et al. Essential role for autophagy in the maintenance of immunological memory against influenza infection. *Nat Med*. 2014;20:503-510. doi:10.1038/nm.3521. PMID: 24747745.
227. Ng LG, Ostuni R, Hidalgo A. Heterogeneity of neutrophils. *Nat Rev Immunol*. 2019;19:255-265. doi:10.1038/s41577-019-0141-8
228. Fuchs TA, Abed U, Goosmann C, et al. Novel cell death program leads to neutrophil extracellular traps. *J Cell Biol*. 2007;176:231-241. doi:10.1083/jcb.200606027
229. Riffelmacher T, Clarke A, Richter FC, et al. Autophagy-dependent generation of free fatty acids is critical for normal neutrophil differentiation. *Immunity*. 2017;47:466-480.e5. doi:10.1016/j.immuni.2017.08.005. PMID: 28916263.
230. Borregaard N, Herlin T. Energy metabolism of human neutrophils during phagocytosis. *J Clin Invest*. 1982;70:550-557. doi:10.1172/jci110647. PMID: 7107894.
231. Fossati G, Moulding DA, Spiller DG, Moots RJ, White MRH, Edwards SW. The mitochondrial network of human neutrophils: role in chemotaxis, phagocytosis, respiratory burst activation, and commitment to apoptosis. *J Immunol*. 2003;170:1964-1972. doi:10.4049/jimmunol.170.4.1964. PMID: 12574365.
232. Reiss M, Roos D. Differences in oxygen metabolism of phagocytosing monocytes and neutrophils. *J Clin Invest*. 1978;61:480-488. doi:10.1172/jci108959. PMID: 202614.
233. Azevedo EP, Rochael NC, Guimarães-Costa AB, et al. A metabolic shift toward pentose phosphate pathway is necessary for amyloid fibril-and phorbol 12-myristate 13-acetate-induced neutrophil extracellular trap (NET) formation. *J Biol Chem*. 2015;290:22174-22183. doi:10.1074/jbc.m115.640094. PMID: 26198639.
234. Segal AW, Abo A. The biochemical basis of the NADPH oxidase of phagocytes. *Trends Biochem Sci*. 1993;18:43-47. doi:10.1016/0968-0004(93)90051-n
235. Segal BH, Veys P, Malech H, Cowan MJ. Chronic granulomatous disease: lessons from a rare disorder. *Biol Blood Marrow Transplant*. 2011;17:S123-S131. doi:10.1016/j.bbmt.2010.09.008
236. Rodríguez-Espinosa O, Rojas-Espinosa O, Moreno-Altamirano MMB, López-Villegas EO, Sánchez-García FJ. Metabolic requirements for neutrophil extracellular traps formation. *Immunol*. 2015;145:213-224. doi:10.1111/imm.12437. PMID: 25545227.
237. Walmsley SR, Print C, Farahi N, et al. Hypoxia-induced neutrophil survival is mediated by HIF-1α–dependent NF-κB activity. *J Exp Med*. 2005;201:105-115. doi:10.1084/jem.20040624. PMID: 15630139.
238. Peyssonnaux C, Datta V, Cramer T, et al. HIF-1α expression regulates the bactericidal capacity of phagocytes. *J Clin Invest*. 2005;115:1806-1815. doi:10.1172/jci23865. PMID: 16007254.
239. McInturff AM, Cody MJ, Elliott EA, et al. Mammalian target of rapamycin regulates neutrophil extracellular trap formation via induction of hypoxia-inducible factor 1 alpha. *Blood*. 2012;120:3118-3125. doi:10.1182/blood-2012-01-405993
240. Maianski NA, Geissler J, Srinivasula SM, et al. Functional characterization of mitochondria in neutrophils: a role restricted to apoptosis. *Cell Death Differ*. 2004;11:143-153. doi:10.1038/sj.cdd.4401320. PMID: 14576767.
241. Lood C, Blanco LP, Purmalek MM, et al. Neutrophil extracellular traps enriched in oxidized mitochondrial DNA are interferogenic and contribute to lupus-like disease. *Nat Med*. 2016;22:146-153. doi:10.1038/nm.4027. PMID: 26779811.
242. Xue J, Schmidt SV, Sander J, et al. Transcriptome-based network analysis reveals a spectrum model of human macrophage activation. *Immunity*. 2014;40:274-288. doi:10.1016/j.immuni.2014.01.006
243. Fukuzumi M, Shinomiya H, Shimizu Y, Ohishi K, Utsumi S. Endotoxin-induced enhancement of glucose influx into murine peritoneal macrophages via GLUT1. *Infect Immun*. 1996;64:108-112. doi:10.1128/IAI.64.1.108-112.1996
244. Freemerman AJ, Zhao L, Pingili AK, et al. Myeloid Slc2a1-deficient murine model revealed macrophage activation and metabolic phenotype are fueled by GLUT1. *J Immunol*. 2019;202:1265-1286. doi:10.4049/jimmunol.1800002
245. Freemerman AJ, Johnson AR, Sacks GN, et al. Metabolic reprogramming of macrophages: glucose transporter 1 (GLUT1)-mediated glucose metabolism drives a proinflammatory phenotype. *J Biol Chem*. 2014;289:7884-7896. doi:10.1074/jbc.M113.522037
246. Kornberg MD, Bhargava P, Kim PM, et al. Dimethyl fumarate targets GAPDH and aerobic glycolysis to modulate immunity. *Science*. 2018;360:449-453. doi:10.1126/science.aan4665
247. Lachmandas E, Boutens L, Ratter JM, et al. Microbial stimulation of different toll-like receptor signalling pathways induces diverse metabolic programmes in human monocytes. *Nat Microbiol*. 2016;2:16246. doi:10.1038/nmicrobiol.2016.246
248. Fang H-Y, Hughes R, Murdoch C, et al. Hypoxia-inducible factors 1 and 2 are important transcriptional effectors in primary macrophages experiencing hypoxia. *Blood*. 2009;114:844-859. doi:10.1182/blood-2008-12-195941
249. Cramer T, Yamanishi Y, Clausen BE, et al. HIF-1alpha is essential for myeloid cell-mediated inflammation. *Cell*. 2003;112:645-657. doi:10.1016/s0092-8674(03)00154-5
250. Palsson-McDermott EM, Curtis AM, Goel G, et al. Pyruvate kinase M2 regulates HIF-1alpha activity and IL-1beta induction and is a critical determinant of the warburg effect in LPS-activated macrophages. *Cell Metab*. 2015;21:65-80. doi:10.1016/j.cmet.2014.12.005
251. Xu Q, Choksi S, Qu J, et al. NADPH oxidases are essential for macrophage differentiation. *J Biol Chem*. 2016;291:20030-20041. doi:10.1074/jbc.M116.731216
252. Jha AK, Huang SC-C, Sergushichev A, et al. Network integration of parallel metabolic and transcriptional data reveals metabolic modules that regulate macrophage polarization. *Immunity*. 2015;42:419-430. doi:10.1016/j.immuni.2015.02.005
253. Baardman J, Verberk SGS, Prange KHM, et al. A defective pentose phosphate pathway reduces inflammatory macrophage responses during hypercholesterolemia. *Cell Rep*. 2018;25:2044-2052.e5. doi:10.1016/j.celrep.2018.10.092
254. Haschemi A, Kosma P, Gille L, et al. The sedoheptulose kinase CARKL directs macrophage polarization through control of glucose metabolism. *Cell Metab*. 2012;15:813-826. doi:10.1016/j.cmet.2012.04.023
255. Granger DL, Lehninger AL. Sites of inhibition of mitochondrial electron transport in macrophage-injured neoplastic cells. *J Cell Biol*. 1982;95:527-535. doi:10.1083/jcb.95.2.527
256. Hard GC. Some biochemical aspects of the immune macrophage. *Br J Exp Pathol*. 1970;51:97-105.
257. West AP, Brodsky IE, Rahner C, et al. TLR signalling augments macrophage bactericidal activity through mitochondrial ROS. *Nature*. 2011;472:476-480. doi:10.1038/nature09973
258. Infantino V, Iacobazzi V, Menga A, Avantaggiati ML, Palmieri F. A key role of the mitochondrial citrate carrier. SLC25A1) in TNFalpha-and IFNgamma-triggered inflammation. *Biochim Biophys Acta*. 2014;1839:1217-1225. doi:10.1016/j.bbagrm.2014.07.013
259. Infantino V, Convertini P, Cucci L, et al. The mitochondrial citrate carrier: a new player in inflammation. *Biochem J*. 2011;438:433-436. doi:10.1042/BJ20111275
260. Zaslona Z, Palsson-McDermott EM, Menon D, et al. The Induction of Pro-IL-1beta by lipopolysaccharide requires endogenous prostaglandin E2 production. *J Immunol*. 2017;198:3558-3564. doi:10.4049/jimmunol.1602072
261. Lauterbach MA, Hanke JE, Serefidou M, et al. Toll-like receptor signaling rewires macrophage metabolism and promotes histone acetylation via ATP-citrate lyase. *Immunity*. 2019;51:997-1011.e7. doi:10.1016/j.immuni.2019.11.009
262. Langston PK, Nambu A, Jung J, et al. Glycerol phosphate shuttle enzyme GPD2 regulates macrophage inflammatory responses. *Nat Immunol*. 2019;20:1186-1195. doi:10.1038/s41590-019-0453-7
263. Galvan-Pena S, Carroll RG, Newman C, et al. Malonylation of GAPDH is an inflammatory signal in macrophages. *Nat Commun*. 2019;10:338. doi:10.1038/s41467-018-08187-6
264. Millet P, Vachharajani V, McPhail L, Yoza B, McCall CE. GAPDH binding to TNF-alpha mRNA contributes to posttranscriptional repression in monocytes: a novel mechanism of communication between inflammation and

metabolism. *J Immunol.* 2016;196:2541-2551. doi:10.4049/jimmunol.1501345
265. Carroll RG, Zasłona Z, Galván-Peña S, et al. An unexpected link between fatty acid synthase and cholesterol synthesis in proinflammatory macrophage activation. *J Biol Chem.* 2018;293:5509-5521. doi:10.1074/jbc.RA118.001921
266. Mills EL, Kelly B, Logan A, et al. Succinate dehydrogenase supports metabolic repurposing of mitochondria to drive inflammatory macrophages. *Cell.* 2016;167:457-470.e13. doi:10.1016/j.cell.2016.08.064
267. Littlewood-Evans A, Sarret S, Apfel V, et al. GPR91 senses extracellular succinate released from inflammatory macrophages and exacerbates rheumatoid arthritis. *J Exp Med.* 2016;213:1655-1662. doi:10.1084/jem.20160061
268. Michelucci A, Cordes T, Ghelfi J, et al. Immune-responsive gene 1 protein links metabolism to immunity by catalyzing itaconic acid production. *Proc Natl Acad Sci U S A.* 2013;110:7820-7825. doi:10.1073/pnas.1218599110
269. Lampropoulou V, Sergushichev A, Bambouskova M, et al. Itaconate links inhibition of succinate dehydrogenase with macrophage metabolic remodeling and regulation of inflammation. *Cell Metab.* 2016;24:158-166. doi:10.1016/j.cmet.2016.06.004
270. Mills EL, Ryan DG, Prag HA, et al. Itaconate is an anti-inflammatory metabolite that activates Nrf2 via alkylation of KEAP1. *Nature.* 2018;556:113-117. doi:10.1038/nature25986
271. Wu GY, Brosnan JT. Macrophages can convert citrulline into arginine. *Biochem J.* 1992;281(pt 1):45-48. doi:10.1042/bj2810045
272. Yeramian A, Martin L, Serrat N, et al. Arginine transport via cationic amino acid transporter 2 plays a critical regulatory role in classical or alternative activation of macrophages. *J Immunol.* 2006;176:5918-5924. doi:10.4049/jimmunol.176.10.5918
273. Qualls JE, Subramanian C, Rafi W, et al. Sustained generation of nitric oxide and control of mycobacterial infection requires argininosuccinate synthase 1. *Cell Host Microbe.* 2012;12:313-323. doi:10.1016/j.chom.2012.07.012
274. Nussler AK, Billiar TR, Liu ZZ, Morris SM Jr. Coinduction of nitric oxide synthase and argininosuccinate synthetase in a murine macrophage cell line. Implications for regulation of nitric oxide production. *J Biol Chem.* 1994;269:1257-1261.
275. Rodriguez AE, Ducker GS, Billingham LK, et al. Serine metabolism supports macrophage IL-1 beta production. *Cell Metab.* 2019;29:1003-1011.e4. doi:10.1016/j.cmet.2019.01.014
276. Lee J-H, Phelan P, Shin M, et al. SREBP-1a-stimulated lipid synthesis is required for macrophage phagocytosis downstream of TLR4-directed mTORC1. *Proc Natl Acad Sci U S A.* 2018;115:E12228-E12234. doi:10.1073/pnas.1813458115
277. Im S-S, Yousef L, Blaschitz C, et al. Linking lipid metabolism to the innate immune response in macrophages through sterol regulatory element binding protein-1a. *Cell Metab.* 2011;13:540-549. doi:10.1016/j.cmet.2011.04.001
278. Oishi Y, Spann NJ, Link VM, et al. SREBP1 contributes to resolution of pro-inflammatory TLR4 signaling by reprogramming fatty acid metabolism. *Cell Metab.* 2017;25:412-427. doi:10.1016/j.cmet.2016.11.009
279. York AG, Williams KJ, Argus JP, et al. Limiting cholesterol biosynthetic flux spontaneously engages type I IFN signaling. *Cell.* 2015;163:1716-1729. doi:10.1016/j.cell.2015.11.045
280. Dang EV, McDonald JG, Russell DW, Cyster JG. Oxysterol restraint of cholesterol synthesis prevents AIM2 inflammasome activation. *Cell.* 2017;171:1057-1071.e11. doi:10.1016/j.cell.2017.09.029
281. Reboldi A, Dang EV, McDonald JG, Liang G, Russell DW, Cyster JG. Inflammation. 25-Hydroxycholesterol suppresses interleukin-1-driven inflammation downstream of type I interferon. *Science.* 2014;345:679-684. doi:10.1126/science.1254790
282. Ivanov SS, Roy CR. Pathogen signatures activate a ubiquitination pathway that modulates the function of the metabolic checkpoint kinase mTOR. *Nat Immunol.* 2013;14:1219-1228. doi:10.1038/ni.2740
283. Ip WKE, Hoshi N, Shouval DS, Snapper S, Medzhitov R. Anti-inflammatory effect of IL-10 mediated by metabolic reprogramming of macrophages. *Science.* 2017;356:513-519. doi:10.1126/science.aal3535
284. Zhu L, Yang T, Li L, et al. TSC1 controls macrophage polarization to prevent inflammatory disease. *Nat Commun.* 2014;5:4696. doi:10.1038/ncomms5696
285. Byles V, Covarrubias AJ, Ben-Sahra I, et al. The TSC-mTOR pathway regulates macrophage polarization. *Nat Commun.* 2013;4:2834. doi:10.1038/ncomms3834
286. Tan Z, Xie N, Cui H, et al. Pyruvate dehydrogenase kinase 1 participates in macrophage polarization via regulating glucose metabolism. *J Immunol.* 2015;194:6082-6089. doi:10.4049/jimmunol.1402469
287. Zhao Q, Chu Z, Zhu L, et al. 2-Deoxy-D-glucose treatment decreases anti-inflammatory M2 macrophage polarization in mice with tumor and allergic airway inflammation. *Front Immunol.* 2017;8:637. doi:10.3389/fimmu.2017.00637
288. Ishii M, Wen H, Corsa CAS, et al. Epigenetic regulation of the alternatively activated macrophage phenotype. *Blood.* 2009;114:3244-3254. doi:10.1182/blood-2009-04-217620
289. Liu P-S, Wang H, Li X, et al. α-ketoglutarate orchestrates macrophage activation through metabolic and epigenetic reprogramming. *Nat Immunol.* 2017;18:985-994. doi:10.1038/ni.3796
290. Covarrubias AJ, Aksoylar HI, Yu J, et al. Akt-mTORC1 signaling regulates acly to integrate metabolic input to control of macrophage activation. *Elife.* 2016;5:e11612. doi:10.7554/eLife.11612
291. Namgaladze D, Zukunft S, Schnütgen F, et al. Polarization of human macrophages by interleukin-4 does not require ATP-citrate lyase. *Front Immunol.* 2018;9:2858. doi:10.3389/fimmu.2018.02858
292. Vats D, Mukundan L, Odegaard JI, et al. Oxidative metabolism and PGC-1beta attenuate macrophage-mediated inflammation. *Cell Metab.* 2006;4:13-24. doi:10.1016/j.cmet.2006.05.011
293. Szanto A, Balint BL, Nagy ZS, et al. STAT6 transcription factor is a facilitator of the nuclear receptor PPARgamma-regulated gene expression in macrophages and dendritic cells. *Immunity.* 2010;33:699-712. doi:10.1016/j.immuni.2010.11.009
294. Huang SC-C, Everts B, Ivanova Y, et al. Cell-intrinsic lysosomal lipolysis is essential for alternative activation of macrophages. *Nat Immunol.* 2014;15:846-855. doi:10.1038/ni.2956
295. Albina JE, Abate JA, Mastrofrancesco B. Role of ornithine as a proline precursor in healing wounds. *J Surg Res.* 1993;55:97-102. doi:10.1006/jsre.1993.1114
296. Hardbower DM, Asim M, Luis PB, et al. Ornithine decarboxylase regulates M1 macrophage activation and mucosal inflammation via histone modifications. *Proc Natl Acad Sci U S A.* 2017;114:E751-E760. doi:10.1073/pnas.1614958114
297. Puleston DJ, Buck MD, Klein Geltink RI, et al. Polyamines and eIF5A hypusination modulate mitochondrial respiration and macrophage activation. *Cell Metab.* 2019;30:352-363 e8. doi:10.1016/j.cmet.2019.05.003
298. Netea MG, Domínguez-Andrés J, Barreiro LB, et al. Defining trained immunity and its role in health and disease. *Nat Rev Immunol.* 2020;20, 375-388. doi:10.1038/s41577-020-0285-6
299. Quintin J, Saeed S, Martens JHA, et al. Candida albicans infection affords protection against reinfection via functional reprogramming of monocytes. *Cell Host Microbe.* 2012;12:223-232. doi:10.1016/j.chom.2012.06.006
300. Saeed S, Quintin J, Kerstens HHD, et al. Epigenetic programming of monocyte-to-macrophage differentiation and trained innate immunity. *Science.* 2014;345:1251086. doi:10.1126/science.1251086
301. Novakovic B, Habibi E, Wang S-Y, et al. β-glucan reverses the epigenetic state of LPS-induced immunological tolerance. *Cell.* 2016;167:1354-1368.e14. doi:10.1016/j.cell.2016.09.034
302. Bekkering S, Arts RJW, Novakovic B, et al. Metabolic induction of trained immunity through the mevalonate pathway. *Cell.* 2018;172:135-146.e9. doi:10.1016/j.cell.2017.11.025
303. Epelman S, Lavine KJ, Randolph GJ. Origin and functions of tissue macrophages. *Immunity.* 2014;41:21-35. doi:10.1016/j.immuni.2014.06.013
304. Lavin Y, Winter D, Blecher-Gonen R, et al. Tissue-resident macrophage enhancer landscapes are shaped by the local microenvironment. *Cell.* 2014;159:1312-1326. doi:10.1016/j.cell.2014.11.018
305. Okabe Y, Medzhitov R. Tissue-specific signals control reversible program of localization and functional polarization of macrophages. *Cell.* 2014;157:832-844. doi:10.1016/j.cell.2014.04.016
306. Rosas M, Davies LC, Giles PJ, et al. The transcription factor Gata6 links tissue macrophage phenotype and proliferative renewal. *Science.* 2014;344:645-648. doi:10.1126/science.1251414
307. Davies LC, Rice CM, Palmieri EM, et al. Peritoneal tissue-resident macrophages are metabolically poised to engage microbes using tissue-niche fuels. *Nat Commun.* 2017;8:2074. doi:10.1038/s41467-017-02092-0
308. Kellersch B, Brocker T. Langerhans cell homeostasis in mice is dependent on mTORC1 but not mTORC2 function. *Blood.* 2013;121:298-307. doi:10.1182/blood-2012-06-439786. PMID: 23212520.
309. Linke M, Pham HTT, Katholnig K, et al. Chronic signaling via the metabolic checkpoint kinase mTORC1 induces macrophage granuloma formation and marks sarcoidosis progression. *Nat Immunol.* 2017;18:293-302. doi:10.1038/ni.3655
310. Remmerie A, Scott CL. Macrophages and lipid metabolism. *Cell Immunol.* 2018;330:27-42. doi:10.1016/j.cellimm.2018.01.020
311. Gautier EL, Chow A, Spanbroek R, et al. Systemic analysis of PPARgamma in mouse macrophage populations reveals marked diversity in expression with critical roles in resolution of inflammation and airway immunity. *J Immunol.* 2012;189:2614-2624. doi:10.4049/jimmunol.1200495
312. Odegaard JI, Ricardo-Gonzalez RR, Red Eagle A, et al. Alternative M2 activation of Kupffer cells by PPARdelta ameliorates obesity-induced insulin resistance. *Cell Metab.* 2008;7:496-507. doi:10.1016/j.cmet.2008.04.003
313. Chawla A, Boisvert WA, Lee C-H, et al. A PPAR gamma-LXR-ABCA1 pathway in macrophages is involved in cholesterol efflux and atherogenesis. *Mol Cell.* 2001;7:161-171. doi:10.1016/s1097-2765(01)00164-2
314. Pathria P, Louis TL, Varner JA. Targeting tumor-associated macrophages in cancer. *Trends Immunol.* 2019;40:310-327. doi:10.1016/j.it.2019.02.003
315. Scharping NE, Delgoffe GM. Tumor microenvironment metabolism: a new checkpoint for anti-tumor immunity. *Vaccines (Basel).* 2016;4:46. doi:10.3390/vaccines4040046
316. Colegio OR, Chu N-Q, Szabo AL, et al. Functional polarization of tumour-associated macrophages by tumour-derived lactic acid. *Nature.* 2014;513:559-563. doi:10.1038/nature13490
317. Munn DH, Mellor AL. IDO in the tumor microenvironment: inflammation, counter-regulation, and tolerance. *Trends Immunol.* 2016;37:193-207. doi:10.1016/j.it.2016.01.002
318. Mezrich JD, Fechner JH, Zhang X, et al. An interaction between kynurenine and the aryl hydrocarbon receptor can generate regulatory T cells. *J Immunol.* 2010;185:3190-3198. doi:10.4049/jimmunol.0903670
319. Rodriguez PC, Quiceno DG, Zabaleta J, et al. Arginase I production in the tumor microenvironment by mature myeloid cells inhibits T-cell receptor expression and antigen-specific T-cell responses. *Cancer Res.* 2004;64:5839-5849. doi:10.1158/0008-5472.CAN-04-0465
320. Rodríguez D, Silvera R, Carrio R, et al. Tumor microenvironment profoundly modifies functional status of macrophages: peritoneal and tumor-associated macrophages are two very different subpopulations. *Cell Immunol.* 2013;283:51-60. doi:10.1016/j.cellimm.2013.06.008
321. Weiss JM, Davies LC, Karwan M, et al. Itaconic acid mediates crosstalk between macrophage metabolism and peritoneal tumors. *J Clin Invest.* 2018;128:3794-3805. doi:10.1172/JCI99169
322. Oh M-H, Sun I-H, Zhao L, et al. Targeting glutamine metabolism enhances tumor-specific immunity by modulating suppressive myeloid cells. *J Clin Invest.* 2020;130:3865-3884. doi:10.1172/JCI131859
323. Ostrand-Rosenberg S, Sinha P. Myeloid-derived suppressor cells: linking inflammation and cancer. *J Immunol.* 2009;182:4499-4506. doi:10.4049/jimmunol.0802740
324. Srivastava MK, Sinha P, Clements VK, Rodriguez P, Ostrand-Rosenberg S. Myeloid-derived suppressor cells inhibit T-cell activation by depleting cystine and cysteine. *Cancer Res.* 2010;70:68-77. doi:10.1158/0008-5472.CAN-09-2587
325. Li W, Tanikawa T, Kryczek I, et al. Aerobic glycolysis controls myeloid-derived suppressor cells and tumor immunity via a specific CEBPB isoform in triple-negative breast cancer. *Cell Metab.* 2018;28:87-103.e6. doi:10.1016/j.cmet.2018.04.022
326. Krawczyk CM, Holowka T, Sun J, et al. Toll-like receptor–induced changes in glycolytic metabolism regulate dendritic cell activation. *Blood.* 2010;115:4742-4749. doi:10.1182/blood-2009-10-249540. PMID: 20351312.
327. Sathaliyawala T, O'Gorman WE, Greter M, et al. Mammalian target of rapamycin controls dendritic cell development downstream of Flt3 ligand signaling. *Immunity.* 2010;33:597-606. doi:10.1016/j.immuni.2010.09.012. PMID: 20933441.
328. van de Laar L, van den Bosch A, Boonstra A, et al. PI3K-PKB hyperactivation augments human plasmacytoid dendritic cell development and function. *Blood.* 2012;120:4982-4991. doi:10.1182/blood-2012-02-413229. PMID: 23091295.
329. Wang Y, Huang G, Zeng H, et al. Tuberous sclerosis 1 (Tsc1)-dependent metabolic checkpoint controls development of dendritic cells. *Proc Natl Acad Sci.* 2013;110:E4894-E4903. doi:10.1073/pnas.1308905110. PMID: 24282297.
330. Kc W, Satpathy AT, Rapaport AS, et al. L-Myc expression by dendritic cells is required for optimal T-cell priming. *Nature.* 2014;507:243-247. doi:10.1038/nature12967. PMID: 24509714.
331. Sinclair C, Bommakanti G, Gardinassi L, et al. mTOR regulates metabolic adaptation of APCs in the lung and controls the outcome of allergic inflammation. *Science.* 2017;357:1014-1021. doi:10.1126/science.aaj2155. PMID: 28798047.
332. Haidinger M, Poglitsch M, Geyeregger R, et al. A versatile role of mammalian target of rapamycin in human dendritic cell function and differentiation. *J Immunol.* 2010;185:3919-3931. doi:10.4049/jimmunol.1000296. PMID: 20805416.
333. Everts B, Amiel E, van der Windt GJW, et al. Commitment to glycolysis sustains survival of NO-producing inflammatory dendritic cells. *Blood.* 2012;120:1422-1431. doi:10.1182/blood-2012-03-419747. PMID: 22786879.
334. Thwe PM, Pelgrom LR, Cooper R, et al. Cell-intrinsic glycogen metabolism supports early glycolytic reprogramming required for dendritic cell immune responses. *Cell Metab.* 2017;26:558-567.e5. doi:10.1016/j.cmet.2017.08.012
335. Everts B, Amiel E, Huang SC-C, et al. TLR-driven early glycolytic reprogramming via the kinases TBK1-IKKvarepsilon supports the anabolic demands of dendritic cell activation. *Nat Immunol.* 2014;15:323-332. doi:10.1038/ni.2833
336. Jantsch J, Chakravortty D, Turza N, et al. Hypoxia and hypoxia-inducible factor-1 alpha modulate lipopolysaccharide-induced dendritic cell activation and function. *J Immunol.* 2008;180:4697-4705. doi:10.4049/jimmunol.180.7.4697. PMID: 18354193.
337. Wu D, Sanin DE, Everts B, et al. Type 1 interferons induce changes in core metabolism that are critical for immune function. *Immunity.* 2016;44:1325-1336. doi:10.1016/j.immuni.2016.06.006. PMID: 27332732.

338. Guak H, Al Habyan S, Ma EH, et al. Glycolytic metabolism is essential for CCR7 oligomerization and dendritic cell migration. *Nat Commun.* 2018;9:2463. doi:10.1038/s41467-018-04804-6. PMID: 29941886.
339. Köhler T, Reizis B, Johnson RS, Weighardt H, Förster I. Influence of hypoxia-inducible factor 1α on dendritic cell differentiation and migration. *Eur J Immunol.* 2012;42:1226-1236. doi:10.1002/eji.201142053. PMID: 22539295
340. Du X, Wen J, Wang Y, et al. Hippo/Mst signalling couples metabolic state and immune function of CD8α+ dendritic cells. *Nature.* 2018;558:141-145. doi:10.1038/s41586-018-0177-0. PMID: 29849151.
341. Amiel E, Everts B, Fritz D, et al. Mechanistic target of rapamycin inhibition extends cellular lifespan in dendritic cells by preserving mitochondrial function. *J Immunol.* 2014;193:2821-2830. doi:10.4049/jimmunol.1302498. PMID: 25108022.
342. Amiel E, Everts B, Freitas TC, et al. Inhibition of mechanistic target of rapamycin promotes dendritic cell activation and enhances therapeutic autologous vaccination in mice. *J Immunol.* 2012;189:2151-2158. doi:10.4049/jimmunol.1103741
343. Zaccagnino P, Saltarella M, Maiorano S, et al. An active mitochondrial biogenesis occurs during dendritic cell differentiation. *Int J Biochem Cell Biol.* 2012;44:1962-1969. doi:10.1016/j.biocel.2012.07.024
344. Ryu S-W, Han EC, Yoon J, Choi C. The mitochondrial fusion-related proteins Mfn2 and OPA1 are transcriptionally induced during differentiation of bone marrow progenitors to immature dendritic cells. *Mol Cell.* 2014;38:89-94. doi:10.14348/molcells.2015.2285. PMID: 25387754.
345. Szatmari I, Töröcsik D, Agostini M, et al. PPARγ regulates the function of human dendritic cells primarily by altering lipid metabolism. *Blood.* 2007;110:3271-3280. doi:10.1182/blood-2007-06-096222. PMID: 17664351.
346. Rehman A, Hemmert KC, Ochi A, et al. Role of fatty-acid synthesis in dendritic cell generation and function. *J Immunol.* 2013;190:4640-4649. doi:10.4049/jimmunol.1202312. PMID: 23536633.
347. Malinarich F, Duan K, Hamid RA, et al. High mitochondrial respiration and glycolytic capacity represent a metabolic phenotype of human tolerogenic dendritic cells. *J Immunol.* 2015;194:5174-5186. doi:10.4049/jimmunol.1303316. PMID: 25917094.
348. Herber DL, Cao W, Nefedova Y, et al. Lipid accumulation and dendritic cell dysfunction in cancer. *Nat Med.* 2010;16:880-886. doi:10.1038/nm.2172. PMID: 20622859.
349. Cubillos-Ruiz JR, Silberman PC, Rutkowski MR, et al. ER stress sensor XBP1 controls anti-tumor immunity by disrupting dendritic cell homeostasis. *Cell.* 2015;161:1527-1538. doi:10.1016/j.cell.2015.05.025. PMID: 26073941.
350. Vivier E, Artis D, Colonna M, et al. Innate lymphoid cells: 10 years on. *Cell.* 2018;174:1054-1066. doi:10.1016/j.cell.2018.07.017. PMID: 30142344.
351. Marçais A, Cherfils-Vicini J, Viant C, et al. The metabolic checkpoint kinase mTOR is essential for IL-15 signaling during the development and activation of NK cells. *Nat Immunol.* 2014;15:749-757. doi:10.1038/ni.2936. PMID: 24973821.
352. Yang M, Chen S, Du J, et al. NK cell development requires Tsc1-dependent negative regulation of IL-15-triggered mTORC1 activation. *Nat Commun.* 2016;7:12730. doi:10.1038/ncomms12730. PMID: 27601261.
353. Wang F, Meng M, Mo B, et al. Crosstalks between mTORC1 and mTORC2 variagate cytokine signaling to control NK maturation and effector function. *Nat Commun.* 2018;9:4874. doi:10.1038/s41467-018-07277-9. PMID: 30451838.
354. Keppel MP, Saucier N, Mah AY, Vogel TP, Cooper MA. Activation-specific metabolic requirements for NK cell IFN-γ production. *J Immunol.* 2015;194:1954-1962. doi:10.4049/jimmunol.1402099. PMID: 25595780.
355. Mah AY, Rashidi A, Keppel MP, et al. Glycolytic requirement for NK cell cytotoxicity and cytomegalovirus control. *JCI Insight.* 2017;2:e95128. doi:10.1172/jci.insight.95128. PMID: 29212951.
356. Keating SE, Zaiatz-Bittencourt V, Loftus RM, et al. Metabolic reprogramming supports IFN-γ production by CD56bright NK cells. *J Immunol.* 2016;196:2552-2560. doi:10.4049/jimmunol.1501783. PMID: 26873994.
357. Loftus RM, Assmann N, Kedia-Mehta N, et al. Amino acid-dependent cMyc expression is essential for NK cell metabolic and functional responses in mice. *Nat Commun.* 2018;9:2341. doi:10.1038/s41467-018-04719-2
358. Assmann N, O'Brien KL, Donnelly RP, et al. Srebp-controlled glucose metabolism is essential for NK cell functional responses. *Nat Immunol.* 2017;18:1197-1206. doi:10.1038/ni.3838
359. Jensen H, Potempa M, Gotthardt D, Lanier LL. Cutting edge: IL-2–induced expression of the amino acid transporters SLC1A5 and CD98 is a prerequisite for NKG2D-mediated activation of human NK cells. *Nat Immunol.* 2017;199:1967-1972. doi:10.4049/jimmunol.1700497. PMID: 28784848.
360. O'Sullivan TE, Johnson LR, Kang HH, Sun JC. BNIP3-and BNIP3L-mediated mitophagy promotes the generation of natural killer cell memory. *Immunity.* 2015;43:331-342. doi:10.1016/j.immuni.2015.07.012. PMID: 26253785.
361. Cichocki F, Wu C-Y, Zhang B, et al. ARID5B regulates metabolic programming in human adaptive NK cells. *J Exp Med.* 2018;215:2379-2395. doi:10.1084/jem.20172168
362. Brand A, Singer K, Koehl GE, et al. LDHA-associated lactic acid production blunts tumor immunosurveillance by T and NK cells. *Cell Metab.* 2016;24:657-671. doi:10.1016/j.cmet.2016.08.011
363. Michelet X, Dyck L, Hogan A, et al. Metabolic reprogramming of natural killer cells in obesity limits antitumor responses. *Nat Immunol.* 2018;19:1330-1340. doi:10.1038/s41590-018-0251-7. PMID: 30420624.
364. Cong J, Wang X, Zheng X, et al. Dysfunction of natural killer cells by FBP1-induced inhibition of glycolysis during lung cancer progression. *Cell Metab.* 2018;28:243-255.e5. doi:10.1016/j.cmet.2018.06.021. PMID: 30033198.
365. Viel S, Marçais A, Guimaraes FS-F, et al. TGF-β inhibits the activation and functions of NK cells by repressing the mTOR pathway. *Sci Signal.* 2016;9:ra19. doi:10.1126/scisignal.aad1884. PMID: 26884601.
366. Zaiatz-Bittencourt V, Finlay DK, Gardiner CM. Canonical TGF-β signaling pathway represses human NK cell metabolism. *J Immunol.* 2018;200:3934-3941. doi:10.4049/jimmunol.1701461. PMID: 29720425.
367. Robinette ML, Fuchs A, Cortez VS, et al. Transcriptional programs define molecular characteristics of innate lymphoid cell classes and subsets. *Nat Immunol.* 2015;16:306-317. doi:10.1038/ni.3094. PMID: 25621825.
368. Gury-BenAri M, Thaiss CA, Serafini N, et al. The spectrum and regulatory landscape of intestinal innate lymphoid cells are shaped by the microbiome. *Cell.* 2016;166:1231-1246.e13. doi:10.1016/j.cell.2016.07.043. PMID: 27545347.
369. Li Q, Li D, Zhang X, et al. E3 ligase VHL promotes group 2 innate lymphoid cell maturation and function via glycolysis inhibition and induction of interleukin-33 receptor. *Immunity.* 2018;48:258-270.e5. doi:10.1016/j.immuni.2017.12.013. PMID: 29452935.
370. Monticelli LA, Buck MD, Flamar A-L, et al. Arginase 1 is an innate lymphoid-cell-intrinsic metabolic checkpoint controlling type 2 inflammation. *Nat Immunol.* 2016;17:656-665. doi:10.1038/ni.3421. PMID: 27043409.
371. Wilhelm C, Harrison OJ, Schmitt V, et al. Critical role of fatty acid metabolism in ILC2-mediated barrier protection during malnutrition and helminth infection. *J Exp Med.* 2016;213:1409-1418. doi:10.1084/jem.20151448. PMID: 27432938.
372. Spencer SP, Wilhelm C, Yang Q, et al. Adaptation of innate lymphoid cells to a micronutrient deficiency promotes type 2 barrier immunity. *Science.* 2014;343:432-437. doi:10.1126/science.1247606. PMID: 24458645.
373. Luccia BD, Gilfillan S, Cella M, Colonna M, Huang SC-C. ILC3s integrate glycolysis and mitochondrial production of reactive oxygen species to fulfill activation demands. *J Exp Med.* 2019;216:2231-2241. doi:10.1084/jem.20180549. PMID: 31865388.
374. Crosby CM, Kronenberg M. Tissue-specific functions of invariant natural killer T cells. *Nat Rev Immunol.* 2018;18:559-574. doi:10.1038/s41577-018-0034-2. PMID: 29967365.
375. Shin J, Wang S, Deng W, et al. Mechanistic target of rapamycin complex 1 is critical for invariant natural killer T-cell development and effector function. *Proc Natl Acad Sci.* 2014;111:E776-E783. doi:10.1073/pnas.1315435111. PMID: 24516149.
376. Zhang L, Tschumi BO, Corgnac S, et al. Mammalian target of rapamycin complex 1 orchestrates invariant NKT cell differentiation and effector function. *J Immunol.* 2014;193:1759-1765. doi:10.4049/jimmunol.1400769. PMID: 25015820.
377. Wu J, Yang J, Yang K, et al. iNKT cells require TSC1 for terminal maturation and effector lineage fate decisions. *J Clin Invest.* 2014;124:1685-1698. doi:10.1172/jci69780. PMID: 24614103.
378. Prevot N, Pyaram K, Bischoff E, Sen JM, Powell JD, Chang C-H. Mammalian target of rapamycin complex 2 regulates invariant NKT cell development and function independent of promyelocytic leukemia zinc-finger. *J Immunol.* 2014;194:223-230. doi:10.4049/jimmunol.1401985. PMID: 25404366.
379. Park H, Tsang M, Iritani BM, Bevan MJ. Metabolic regulator Fnip1 is crucial for iNKT lymphocyte development. *Proc Natl Acad Sci.* 2014;111:7066-7071. doi:10.1073/pnas.1406473111. PMID: 24785297.
380. Kumar A, Pyaram K, Yarosz EL, et al. Enhanced oxidative phosphorylation in NKT cells is essential for their survival and function. *Proc Natl Acad Sci.* 2019;116:7439-7448. doi:10.1073/pnas.1901376116. PMID: 30910955.
381. Fu S, He K, Tian C, et al. Impaired lipid biosynthesis hinders anti-tumor efficacy of intratumoral iNKT cells. *Nat Commun.* 2020;11:438. doi:10.1038/s41467-020-14332-x. PMID: 31974378.
382. Kim Y-H, Kumar A, Chang C-H, Pyaram K. Reactive oxygen species regulate the inflammatory function of NKT cells through promyelocytic leukemia zinc finger. *J Immunol.* 2017;199:3478-3487. doi:10.4049/jimmunol.1700567. PMID: 29021374.
383. Waldman AD, Fritz JM, Lenardo MJ. A guide to cancer immunotherapy: from T cell basic science to clinical practice. *Nat Rev Immunol.* 2020;20:651-668. doi:10.1038/s41577-020-0306-5
384. Pavlova NN, Thompson CB. The emerging hallmarks of cancer metabolism. *Cell Metab.* 2016;23:27-47. doi:10.1016/j.cmet.2015.12.006
385. Leone RD, Powell JD. Metabolism of immune cells in cancer. *Nat Rev Cancer.* 2020;20:516-531. doi:10.1038/s41568-020-0273-y
386. Li X, Wenes M, Romero P, et al. Navigating metabolic pathways to enhance antitumour immunity and immunotherapy. *Nat Rev Clin Oncol.* 2019;16:425-441. doi:10.1038/s41571-019-0203-7
387. Chang C-H, Qiu J, O'Sullivan D, et al. Metabolic competition in the tumor microenvironment is a driver of cancer progression. *Cell.* 2015;162:1229-1241. doi:10.1016/j.cell.2015.08.016
388. Leone RD, Zhao L, Englert JM, et al. Glutamine blockade induces divergent metabolic programs to overcome tumor immune evasion. *Science.* 2019;366:1013-1021. doi:10.1126/science.aav2588
389. Scharping NE, Menk AV, Whetstone RD, Zeng X, Delgoffe GM. Efficacy of PD-1 blockade is potentiated by metformin-induced reduction of tumor hypoxia. *Cancer Immunol Res.* 2017;5:9-16. doi:10.1158/2326-6066.CIR-16-0103
390. Friberg M, Jennings R, Alsarraj M, et al. Indoleamine 2,3-dioxygenase contributes to tumor cell evasion of T cell-mediated rejection. *Int J Cancer.* 2002;101:151-155. doi:10.1002/ijc.10645
391. Sharma MD, Baban B, Chandler P, et al. Plasmacytoid dendritic cells from mouse tumor-draining lymph nodes directly activate mature Tregs via indoleamine 2,3-dioxygenase. *J Clin Invest.* 2007;117:2570-2582. doi:10.1172/JCI31911
392. Holmgaard RB, Zamarin D, Munn DH, Wolchok JD, Allison JP. Indoleamine 2,3-dioxygenase is a critical resistance mechanism in antitumor T cell immunotherapy targeting CTLA-4. *J Exp Med.* 2013;210:1389-1402. doi:10.1084/jem.20130066
393. Gu T, Rowswell-Turner RB, Kilinc MO, Egilmez NK. Central role of IFNgamma-indoleamine 2,3-dioxygenase axis in regulation of interleukin-12-mediated antitumor immunity. *Cancer Res.* 2010;70:129-138. doi:10.1158/0008-5472.CAN-09-3170
394. Labadie BW, Bao R, Luke JJ. Reimagining IDO pathway inhibition in cancer immunotherapy via downstream focus on the tryptophan-kynurenine-aryl hydrocarbon axis. *Clin Cancer Res.* 2019;25:1462-1471. doi:10.1158/1078-0432.CCR-18-2882
395. Bono MR, Fernandez D, Flores-Santibanez F, Rosemblatt M, Sauma D. CD73 and CD39 ectonucleotidases in T cell differentiation: beyond immunosuppression. *FEBS Lett.* 2015;589:3454-3460. doi:10.1016/j.febslet.2015.07.027
396. Wang H, Franco F, Tsui Y-C, et al. CD36-mediated metabolic adaptation supports regulatory T cell survival and function in tumors. *Nat Immunol.* 2020;21:298-308. doi:10.1038/s41590-019-0589-5
397. Guedan S, Ruella M, June CH. Emerging cellular therapies for cancer. *Annu Rev Immunol.* 2019;37:145-171. doi:10.1146/annurev-immunol-042718-041407
398. Crompton JG, Sukumar M, Roychoudhuri R, et al. Akt inhibition enhances expansion of potent tumor-specific lymphocytes with memory cell characteristics. *Cancer Res.* 2015;75:296-305. doi:10.1158/0008-5472.CAN-14-2277
399. Kawalekar OU, O'Connor RS, Fraietta JA, et al. Distinct signaling of coreceptors regulates specific metabolism pathways and impacts memory development in CAR T cells. *Immunity.* 2016;44:380-390. doi:10.1016/j.immuni.2016.01.021
400. Siska PJ, Beckermann KE, Mason FM, et al. Mitochondrial dysregulation and glycolytic insufficiency functionally impair CD8 T cells infiltrating human renal cell carcinoma. *JCI Insight.* 2017;2:e93411. doi:10.1172/jci.insight.93411
401. Yin Y, Choi SC, Xu Z, et al. Normalization of CD4+ T cell metabolism reverses lupus. *Sci Transl Med.* 2015;7:274ra18. doi:10.1126/scitranslmed.aaa0835
402. Choi SC, Titov AA, Abboud G, et al. Inhibition of glucose metabolism selectively targets autoreactive follicular helper T cells. *Nat Commun.* 2018;9:4369. doi:10.1038/s41467-018-06686-0
403. Kono M, Yoshida N, Maeda K, et al. Glutaminase 1 inhibition reduces glycolysis and ameliorates lupus-like disease in MRL/lpr mice and experimental autoimmune encephalomyelitis. *Arthritis Rheum.* 2019;71:1869-1878. doi:10.1002/art.41019
404. Blatt NB, Bednarski JJ, Warner RE, et al. Benzodiazepine-induced superoxide signals B cell apoptosis: mechanistic insight and potential therapeutic utility. *J Clin Invest.* 2002;110:1123-1132. doi:10.1172/JCI16029
405. Suwannaroj S, Lagoo A, Keisler D, Lupus M-RW. Antioxidants suppress mortality in the female NZB× NZW F1 mouse model of systemic lupus erythematosus (SLE). *Lupus.* 2001;10:258-265.
406. Lai Z-W, Hanczko R, Bonilla E, et al. N-acetylcysteine reduces disease activity by blocking mammalian target of rapamycin in T cells from systemic lupus erythematosus patients: a randomized, double-blind, placebo-controlled trial. *Arthritis Rheum.* 2012;64:2937-2946. doi:10.1002/art.34502

407. Lai Z-W, Kelly R, Winans T, et al. Sirolimus in patients with clinically active systemic lupus erythematosus resistant to, or intolerant of, conventional medications: a single-arm, open-label, phase 1/2 trial. *Lancet.* 2018;391:1186-1196. doi:10.1016/S0140-6736(18)30485-9
408. Warner LM, Adams LM, Sehgal SN. Rapamycin prolongs survival and arrests pathophysiologic changes in murine systemic lupus erythematosus. *Arthritis Rheum.* 1994;37:289-297. doi:10.1002/art.1780370219
409. Fernandez D, Bonilla E, Mirza N, Niland B, Perl A. Rapamycin reduces disease activity and normalizes T cell activation-induced calcium fluxing in patients with systemic lupus erythematosus. *Arthritis Rheum.* 2006;54:2983-2988. doi:10.1002/art.22085
410. Imboden JB. The immunopathogenesis of rheumatoid arthritis. *Annu Rev Pathol.* 2009;4:417-434. doi:10.1146/annurev.pathol.4.110807.092254
411. Son H-JJ, Lee J, Lee SY, et al. Metformin attenuates experimental autoimmune arthritis through reciprocal regulation of Th17/Treg balance and osteoclastogenesis. *Mediators Inflamm.* 2014;2014:973986. doi:10.1155/2014/973986
412. Okano T, Saegusa J, Nishimura K, et al. 3-bromopyruvate ameliorate autoimmune arthritis by modulating Th17/Treg cell differentiation and suppressing dendritic cell activation. *Sci Rep.* 2017;7:42412. doi:10.1038/srep42412
413. Abboud G, Choi SC, Kanda N, et al. Inhibition of glycolysis reduces disease severity in an autoimmune model of rheumatoid arthritis. *Front Immunol.* 2018;9:1973. doi:10.3389/fimmu.2018.01973
414. Bian L, Josefsson E, Jonsson IM, et al. Dichloroacetate alleviates development of collagen II-induced arthritis in female DBA/1 mice. *Arthritis Res Ther.* 2009;11:R132. doi:10.1186/ar2799v
415. Yang Z, Shen Y, Oishi H, et al. Restoring oxidant signaling suppresses proarthritogenic T cell effector functions in rheumatoid arthritis. *Sci Transl Med.* 2016;8:331ra38. doi:10.1126/scitranslmed.aad7151
416. Linker RA, Haghikia A. Dimethyl fumarate in multiple sclerosis: latest developments, evidence and place in therapy. *Ther Adv Chronic Dis.* 2016;7:198-207. doi:10.1177/2040622316653307
417. Shriver LP, Manchester M. Inhibition of fatty acid metabolism ameliorates disease activity in an animal model of multiple sclerosis. *Sci Rep.* 2011;1:79. doi:10.1038/srep00079
418. Hollinger KR, Smith MD, Prchalova E, Kirby LA. Glutamine antagonism attenuates physical and cognitive deficits in a model of MS. *Neuro Inflamm.* 2019;6:e609. doi:10.1212/NXI.0000000000000609
419. Lee C-F, Lo Y-C, Cheng C-H, et al. Preventing allograft rejection by targeting immune metabolism. *Cell Rep.* 2015;13:760-770. doi:10.1016/j.celrep.2015.09.036
420. Cheng CH, Lee CF, Oh BC, et al. Targeting metabolism as a platform for inducing allograft tolerance in the absence of long-term immunosuppression. *Front Immunol.* 2020;11:572. doi:10.3389/fimmu.2020.00572
421. Nguyen HD, Chatterjee S, Haarberg KMK, et al. Metabolic reprogramming of alloantigen-activated T cells after hematopoietic cell transplantation. *J Clin Invest.* 2016;126:1337-1352. doi:10.1172/JCI82587
422. Park M-J, Lee S-Y, Moon S-J, et al. Metformin attenuates graft-versus-host disease via restricting mammalian target of rapamycin/signal transducer and activator of transcription 3 and promoting adenosine monophosphate-activated protein kinase-autophagy for the balance between T helper 17 and Tregs. *Transl Res.* 2016;173:115-130. doi:10.1016/j.trsl.2016.03.006
423. Glick GD, Rossignol R, Lyssiotis CA, et al. Anaplerotic metabolism of alloreactive T cells provides a metabolic approach to treat graft-versus-host disease. *J Pharmacol Exp Ther.* 2014;351:298-307. doi:10.1124/jpet.114.218099
424. Gatza E, Wahl DR, Opipari AW, et al. Manipulating the bioenergetics of alloreactive T cells causes their selective apoptosis and arrests graft-versus-host disease. *Sci Transl Med.* 2011;3:67ra8. doi:10.1126/scitranslmed.3001975
425. Byersdorfer CA, Tkachev V, Opipari AW, et al. Effector T cells require fatty acid metabolism during murine graft-versus-host disease. *Blood.* 2013;122:3230-3237. doi:10.1182/blood-2013-04-495515
426. van der Windt GJW, Chang CH, Pearce EL. Measuring bioenergetics in T cells using a seahorse extracellular flux analyzer. *Curr Protoc Immunol.* 2016;113:3 16B 11-13 16B 14. doi:10.1002/0471142735.im0316bs113
427. Yamamoto N, Ueda-Wakagi M, Sato T, et al. Measurement of glucose uptake in cultured cells. *Curr Protoc Pharmacol.* 2015;71:12 14 11-12 14 26. doi:10.1002/0471141755.ph1214s71
428. Chazotte B. Labeling mitochondria with TMRM or TMRE. *Cold Spring Harb Protoc.* 2011;2011:895-897. doi:10.1101/pdb.prot5641
429. Chazotte B. Labeling mitochondria with MitoTracker dyes. *Cold Spring Harb Protoc.* 2011;2011:990-992. doi:10.1101/pdb.prot5648
430. Mayevsky A, Rogatsky GG. Mitochondrial function in vivo evaluated by NADH fluorescence: from animal models to human studies. *Am J Physiol Cell Physiol.* 2007;292:C615-C640. doi:10.1152/ajpcell.00249.2006
431. Jang C, Chen L, Rabinowitz JD. Metabolomics and isotope tracing. *Cell.* 2018;173:822-837. doi:10.1016/j.cell.2018.03.055
432. Long CP, Antoniewicz MR. High-resolution (13)C metabolic flux analysis. *Nat Protoc.* 2019;14:2856-2877. doi:10.1038/s41596-019-0204-0

CHAPTER 7

Regulated Cell Death in the Immune System

Douglas R. Green, PhD • Sebastian Ruehl, PhD

INTRODUCTION: SUICIDE AND SABOTAGE IN REGULATED CELL DEATH

Cells in our bodies die, as does any living thing. While cell death may be a passive event that occurs when the "slings and arrows of outrageous fortune" simply overwhelm cellular integrity, most cells die in a manner that is regulated by molecular processes. Such regulated cell death comes in two general forms. Cells can undergo a process of suicide, in which molecular mechanisms are engaged that likely evolved to mediate cell death. Alternatively, molecular processes can be perturbed in such a way that the cell dies, a process we refer to as sabotage. Cell death by sabotage is an active process, in the same sense that a speeding train but not a standing one derails if the track is faulty. There are relatively few bona fide mechanisms of regulated cell suicide, while many forms of regulated cell sabotage are likely to exist.

Another distinction is made between forms of regulated cell death that disrupt the plasma membrane, which are called necrosis, and cell death that occurs while the plasma membrane remains intact. The latter is best exemplified by apoptosis. Here we consider apoptosis and four forms of regulated necrosis: necroptosis, pyroptosis, ferroptosis, and netosis. Each has distinct roles and consequences in the immune system.

CHARACTERISTICS OF APOPTOSIS

Apoptotic cell death is probably the best studied form of regulated cell death. It is characterized by distinct morphological changes like cell shrinkage, nuclear condensation (pyknosis), DNA fragmentation, scrambling of plasma membrane lipids, such as phosphatidylserine (PS), and membrane blebbing.[1] Through this blebbing, the dying cell is packaged into membranous vesicles, so-called apoptotic bodies, which contain intracellular content, and hence early apoptosis is usually not accompanied by permeabilization of the plasma membrane. Apoptotic cell death is generally considered an "immunologically silent cell death," that is, not inducing a pro-inflammatory immune response, and both maintenance of plasma membrane integrity and externalization of PS are critical features which contribute to efficient clearance of apoptotic cells and the subsequent anti-inflammatory effects.[2]

ACTIVE CASPASES CAUSE APOPTOTIC MORPHOLOGY

Apoptosis is both initiated and executed by a set of cysteine proteases with specificity for aspartate residues (caspases), which cleave cellular substrates, leading to the morphological changes and characteristics described above. Caspases can be divided into apoptotic (caspase-3, -6, -7, -8, -9, -10) and inflammatory caspases (caspase-1 and caspase-11 in mice and caspase-1, -4, -5 in humans).[3] The role of inflammatory caspases in innate immunity and cell death is discussed in greater detail below as well as in Chapter 12. Apoptotic caspases can be further subdivided into initiator (or apical) caspases (8, 9, 10) and executioner caspases (3, 6, 7).[4]

One important substrate of the executioner caspases-3 and -7 is the inhibitor of a caspase-activated deoxyribonuclease (iCAD). In resting cells, iCAD acts as a chaperone to properly fold caspase-activated deoxyribonuclease (CAD), which it then inhibits through direct binding. Upon cleavage by caspase-3, iCAD liberates CAD, which now is free to degrade nuclear DNA.[5,6] This produces the characteristic DNA fragmentation observed during apoptotic cell death.

Executioner caspases also directly regulate the exposure of PS on the plasma membrane. In living cells, a PS flippase, ATP11c, maintains PS at the inner leaflet of the plasma membrane. During apoptosis, executioner caspases cleave and inactivate ATP11c.[7] In addition, a phospholipid scramblase, XK-related protein 8 (Xkr8), which has low basal activity in living cells, is cleaved by caspase-3, and this activates the scramblase to randomize phospholipids in the plasma membrane.[8] Both enhanced externalization of PS by activated Xkr8 and reduced internalization through inactivated ATP11c lead to robust accumulation of PS on the outer leaflet of the plasma membrane. This serves as the key "eat-me" signal for professional phagocytes (macrophages and dendritic cells [DCs]), which clear apoptotic cells effectively via a process called efferocytosis.[2] Mutation or dysfunction in this pathway can lead to several autoimmune diseases, such as systemic lupus erythematosus (SLE), as we discuss in greater detail below.

CASPASES ARE ACTIVATED BY PROTEOLYSIS

All caspases are synthesized as pro-caspases or zymogens, exhibiting little protease activity, and contain a large (p20) and a small (p10) subunit connected by a linker with a caspase cleavage site. Caspases are active only when homodimeric, that is, when two large and two small subunits assemble to create two active sites. There are important distinctions between the types of caspases: the zymogen form of the executioner caspases exists as an inactive dimer, while the zymogens of the initiator and inflammatory caspases are monomeric and contain protein–protein interaction motifs in the large pro-domains that precede the large subunit in the zymogen. Cleavage of the linker between p10 and p20 is essential to allow for structural rearrangement of these domains of executioner caspases into fully active homodimers. During apoptosis, this cleavage is mediated by initiator caspases.[9]

Activation of the initiator caspases, caspase-8 and caspase-9, occurs via induced proximity, that is, their dimerization (and often oligomerization) on adapter proteins that form caspase activation platforms. In their N-terminal prodomains, caspase-8 contains two death effector domains (DEDs) and caspase-9 contains a caspase activation and recruitment domain (CARD). These motifs bind to DED and CARD domains of the respective caspase activation platforms. In the case of caspase-8, autocleavage in the linker region further stabilizes the active homodimers.[10] In contrast, linker region cleavage of caspase-9 reduces the affinity of the mature caspase for its caspase activation platform, so that it is replaced by additional caspase-9 monomers that are then activated in a "tick-over" process.[11,12] An additional cleavage also occurs in both caspase-8 and caspase-9 to dissociate the pro-domain from the p20.

These principles have a practical value. Cleavage of the linker region of an executioner caspase is required for and indicative of its activation. In contrast, while cleavage of the linker region occurs upon initiator caspase activation, this is not a demonstration that the caspase was, indeed, activated. Cleavage of an initiator caspase does not activate it and in some cases can prevent its activation on the appropriate caspase activation platform.

SOME CASPASES ARE REPRESSED BY INHIBITORS OF APOPTOSIS PROTEINS

Activation of an initiator caspase does not always result in cleavage and activation of executioner caspases. Important regulators of caspase activation are cellular inhibitors of apoptosis (c-IAPs), which are characterized by the presence of one or more baculovirus IAP repeat (BIR). First characterized in insect baculoviruses,[13,14] BIRs ensure replication of viruses through inhibition of caspase activation and, hence, activation of host apoptotic programs. Of the vertebrate BIR-containing proteins, X-lined inhibitor of apoptosis (XIAP), c-IAP1, c-IAP2, NLR family apoptosis inhibitory protein (NAIP), and survivin, only XIAP is able to inhibit active caspases through direct binding.[15] XIAP contains three BIR domains: BIR3 inhibits caspase-9, while BIR2 inhibits the executioner caspases-3 and -7. XIAP, c-IAP1, and c-IAP2 can also promote ubiquitination of caspases and, by this mechanism, also inhibit caspase activation through protein degradation.[16,17] IAPs are not known to inhibit caspase-8, although, as outlined below, XIAP prevents the activation of executioner caspases by caspase-8.

ACTIVATION OF CASPASE-9 OCCURS ON THE APOPTOSOME

During apoptosis, caspase-9 is recruited to a large cytosolic protein complex, called the apoptosome. The apoptosome consists of pro–caspase-9, apoptotic protease activating factor 1 (APAF-1), and heme-containing cytochrome C (holocytochrome C).[18] APAF-1 binds cytochrome C, released from the mitochondrial intermembrane space (IMS), at a 1:1 stoichiometry via its WD40 repeats, and both assemble into a heptamer with a cartwheel-like structure. Deoxy-ATP (dATP) binding to the central nucleotide-binding domain of APAF-1 is required for this assembly.[19] The C-terminal CARD domains of APAF-1 are perpendicular to the plane of the WD40 cartwheel and are the recruitment site of pro–caspase-9.[20] Binding of pro–caspase-9 to the APAF-1 apoptosome, through homotypic CARD–CARD interactions of both partners, results in proximity-induced activation of caspase-9 and subsequent processing and activation of caspase-3 and -7.

MITOCHONDRIAL INTERMEMBRANE SPACE PROTEINS REGULATE CASPASE ACTIVATION

Mature holocytochrome C is not found in the cytosol of healthy cells and hence the apoptosome is in its auto-inhibited state. As cytochrome C is an essential component of the electron transport chain and mitochondrial respiration, it is found in the mitochondrial IMS. Other proteins such as SMAC (second mitochondrial activator of caspases) and OMI/Htr2A are also IMS proteins. SMAC and OMI, once released into the cytosol, antagonize the action of XIAP and its inhibitory effect on caspase activation.[21,22] They also inhibit cIAP1 and cIAP2 and promote their degradation (Fig. 7.1).

Bcl-2 FAMILY PROTEINS AS THE KEY REGULATORS OF MITOCHONDRIAL APOPTOSIS

The IMS proteins are released through a tightly regulated process termed mitochondrial outer membrane permeabilization (MOMP). This process marks, in most cases, the point of no return for a cell and its commitment to apoptosis execution. The regulation and execution of MOMP is a function of the Bcl-2 family of proteins.[23] This protein family, named after its founding member Bcl-2, is at the core of the intrinsic or mitochondrial pathway of apoptosis. Bcl-2 proteins are characterized by the presence of one or multiple Bcl-2 homology (BH) domains and can be divided into three subgroups. The proapoptotic effector proteins BAX, BAK, and BOK are able to permeabilize the mitochondrial outer membrane through oligomerization and pore formation. In healthy cells, the active effectors are inhibited by the antiapoptotic Bcl-2 family proteins (Bcl-2, Mcl-1, Bcl-xL, Bcl-W, and Bcl2A1) (reviewed in Ref. 24). All effectors and antiapoptotic proteins have a C-terminal transmembrane domain,

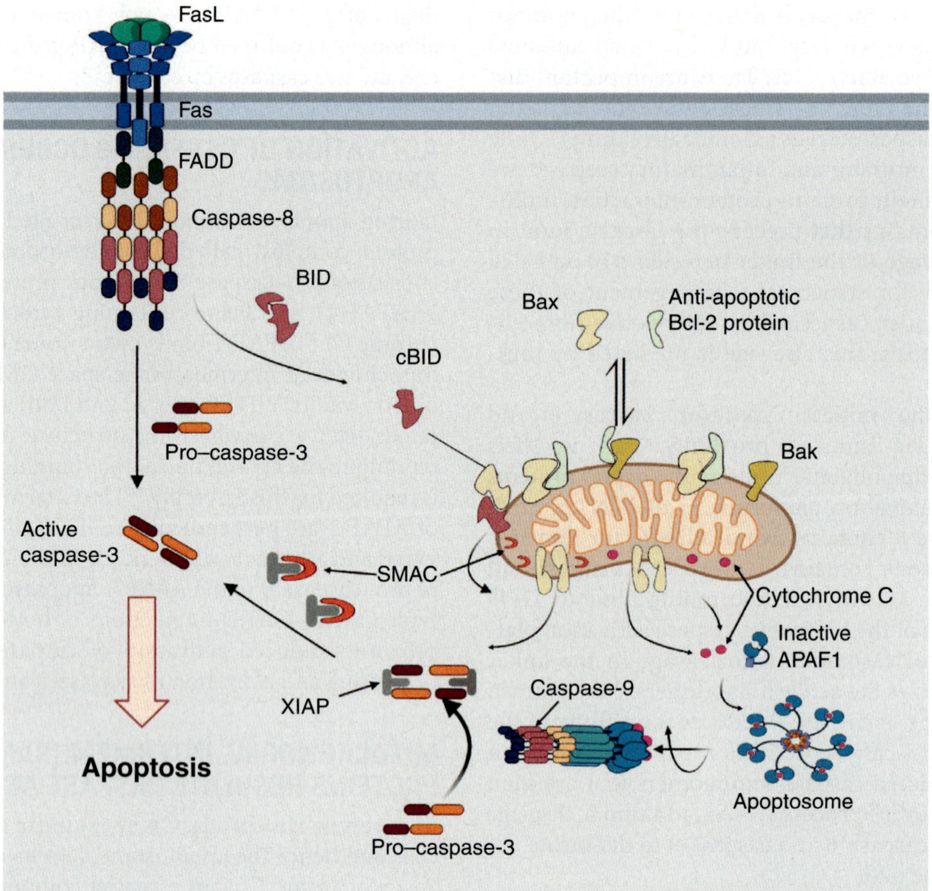

FIG. 7.1. Extrinsic and intrinsic apoptosis. Death receptor stimulation (eg, Fas) leads to the recruitment of FADD (Fas-associated death domain) and caspase-8, which can cleave and activate caspase-3. In the absence of XIAPs (X-lined inhibitors of apoptosis), caspase-3 precipitates apoptosis (type I cells, see text). Caspase-8 also cleaves BID to generate cBID, a potent mediator of mitochondrial apoptosis. Proapoptotic effectors (Bax and Bak) are inhibited by anti-apoptotic Bcl-2 proteins through direct binding and retrotranslocation from the mitochondria to the cytosol. cBID disrupts Bcl-2/Bax complexes and promotes oligomerization of Bax, which causes mitochondrial outer membrane permeabilization (MOMP). Release of Cytochrome C into the cytosol oligomerizes APAF-1; this activates caspase-9, which subsequently cleaves and activates caspase-3. SMAC, also released from the intermembrane space during MOMP, relieves inhibition of caspase-3 by XIAPs, allowing caspase-3 to be fully active. Hence, in type II cells, death receptor signaling requires MOMP to induce apoptosis. (Created with BioRender.com.)

which targets them to the mitochondria, either constitutively or upon activation. While the antiapoptotic proteins contain four BH domains (BH1–BH4), the effector protein contains three of them (BH1–BH3). The BH3 domain is a crucial domain through which members of this family interact with each other. This domain primarily determines the specificity and affinity of the Bcl-2 protein for other family members.[25]

To understand how the effector proteins are activated upon induction of apoptosis, we need to understand the action of the third class of Bcl-2 family proteins, the proapoptotic BH3-only proteins. As the name suggests, these proteins have a BH3 domain (but not other BH domains), and some also contain a C-terminal transmembrane domain. BH3-only proteins can be classified biochemically into sensitizer (BAD, BID, BMF, NOXA) and direct activator (BIM, BID, PUMA) proteins. Both types of BH3-only proteins are capable of neutralizing the antiapoptotic proteins, with some specificity constraints dictated by the BH3 sequence. Direct activator proteins, in turn, are also capable of activating BAX and BAK directly, through transient interaction. All of these interactions between BH3-only proteins and the other classes of Bcl-2 proteins are understood at the structural level.

Models of Bcl-2 Family Protein Interactions

There are currently at least two models of how BAX and BAK are activated, the "neutralization model" and a "direct activator/sensitizer model." The neutralization model postulates that, in unstressed cells, Bax and Bak are inhibited by the antiapoptotic Bcl-2 proteins. As levels of BH3-only proteins increase during cell stress (we discuss how this happens below), the BH3-only proteins displace the antiapoptotic proteins from Bax and Bak, which are now free to induce MOMP.

In this model, activation of the effectors occurs spontaneously on the mitochondrial outer membrane. The direct activator/sensitizer model suggests that activation of the effectors Bax and Bak by the direct activator BH3-only proteins is a crucial step toward the decision to engage MOMP. Antiapoptotic proteins neutralize direct activator proteins (eg, BIM) in resting cells, and increasing concentrations of sensitizing BH3-only proteins (eg, BAD release of direct activator proteins from the antiapoptotic proteins. The released direct activator BH3-only proteins activate Bax and/or Bak, causing MOMP. There is evidence in support of each of these models.[26-28]

Unified models integrate the concepts of both theories, stating that both processes occur at the same time in a given cell. In one of these unified models, neutralization of direct activator BH3-only proteins by antiapoptotic proteins is referred to as Mode 1 inhibition. Upon derepression, the released direct activator proteins activate the effector proteins, which can then be sequestered by antiapoptotic proteins to prevent cell death (referred to as Mode 2 inhibition). Only if the binding capacity of antiapoptotic proteins is overwhelmed by the levels of BH3-only proteins and activated effectors, MOMP and apoptosis will ensue. A functional consequence of these interactions is that cells skewed toward Mode 1 are poised to rapidly undergo MOMP, whereas cells in Mode 2 undergo MOMP more slowly, which could be due to a stronger binding of antiapoptotic proteins to activated effectors in this scenario.[29]

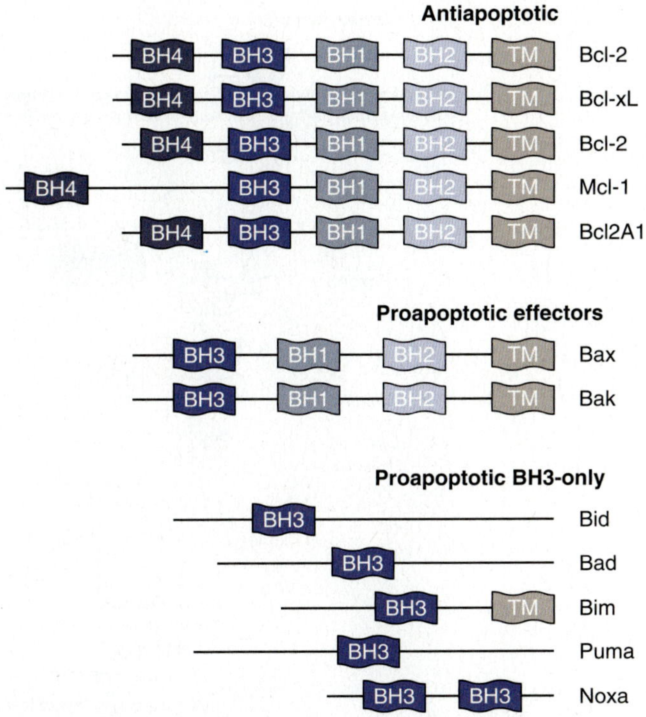

FIG. 7.2. Schematic representation of Bcl-2 family members. This list is not exhaustive as only proteins mentioned in the text are displayed.

Integration of Cellular Signals by the Bcl-2 Family

PUMA and NOXA Are Expressed in Response to Stress

The active forms of BH3-only proteins are usually not constitutively present in cells but rather are induced by cell stress. In many cases, this is a transcriptional event of stress response pathways of transcription factor activation. For example, the expression of the BH3-only proteins, PUMA and NOXA, is engaged by p53 in response to DNA damage[30,31] and also by ATF4, the transcription factor induced in the integrated stress response pathway.[32,33] In the absence of these BH3-only proteins (in particular, PUMA), the induction of apoptosis by a variety of stressors is curtailed.

The induction of BH3-only proteins is not only limited to transcriptional upregulation but also includes posttranslational modifications such as cleavage and phosphorylation. We discuss examples of these below.

BID Is a Cytosolic Protease Sensor

Full-length BID is constitutively expressed in cells, but inactive. Unlike other BH3-only proteins, full-length BID is structurally similar to other Bcl-2 family members. It acts as a cytosolic protease sensor as its flexible loop can be cleaved by lysosomal proteases (cathepsins), calcium-activated proteases (calpains[34]), granzyme B,[35-38] and several of the caspases.[39] Upon cleavage, the C-terminal fragment of BID (truncated BID or cleaved BID [cBID]) can bind to mitochondrial membranes, where it activates Bax and Bak and/or neutralizes antiapoptotic proteins[40] (Fig. 7.2).

Growth Factor Signaling Regulates Apoptosis Through Akt/PKB Activation

BIM and BAD are, among other proteins, sensors for growth factor deprivation, but the mechanisms of their regulation differ. The presence of growth factors promotes Akt (also called protein kinase B) activation, which phosphorylates (among other substrates) FOXO3a, a transcription factor. Phosphorylated FOXO3a is sequestered by a 14-3-3 protein in the cytosol. Upon growth factor withdrawal, Akt activity and hence FOXO3a phosphorylation are reduced, the 14-3-3 protein dissociates, and FOXO3a is free to translocate into the nucleus, where one of its target genes is BIM.[41] BIM is regulated both transcriptionally and posttranslationally. Phosphorylation by ERK (at Ser69) targets BIM for degradation,[42] whereas phosphorylation by JNK (primarily at Tyr58) stabilizes it.[43]

Growth factor signaling also regulates BAD activity. As we have seen, growth factor withdrawal reduces Akt activity, and BAD is a direct Akt target. While Akt phosphorylation of BAD promotes complex formation of BAD with a 14-3-3 protein, unphosphorylated BAD does not form this complex and is free to inhibit antiapoptotic proteins at the mitochondria, promoting apoptosis.[44-46]

Growth factor signaling also regulates the antiapoptotic proteins, especially those with a short half-life (Mcl-1 and BCL2A1). A target of Akt is glycogen synthase kinase 3 (GSK3), which is inactive in the presence of growth factors and high Akt activity. If GSK3 is not inhibited by Akt, it phosphorylates Mcl-1, which targets this antiapoptotic protein

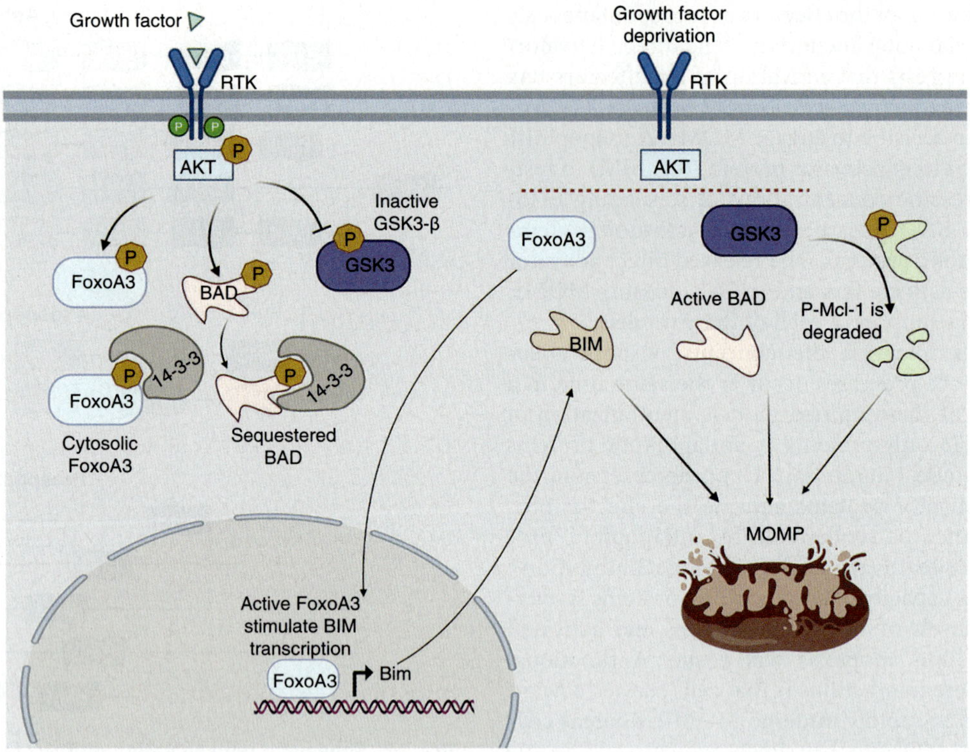

FIG. 7.3. **Regulation mitochondrial apoptosis by growth factor signaling.** In the presence of extracellular growth factors, the respective receptor tyrosine kinases (RTKs) inhibit apoptosis through various mechanisms: (a) cytosolic sequestration of FoxoA3, (b) inactivation of BAD by phosphorylation, and (c) inactivation of GSK3-β. Upon growth factor withdrawl, FoxoA3 activates BIM transcription, dephosphorylated (active) BAD is available, and GSK3-β activation phosphorylates Mcl-1 to promote its degradation. All these events prime cells for the initiation of mitochondrial apoptosis. MOMP, mitochondrial outer membrane permeabilization. (Created with BioRender.com.)

for degradation by the ubiquitin E3 ligase FBW7, and hence primes the cell for apoptosis.[47]

These are only examples, and a more comprehensive summary of how Bcl-2 family proteins are regulated has been reviewed elsewhere.[48-51] Nevertheless, we can see that there are multiple ways in which the decision to execute apoptosis via the mitochondrial pathway can be regulated (Fig. 7.3).

DEATH RECEPTOR SIGNALING

Besides the mitochondrial or "intrinsic" pathway, there is also an "extrinsic" or death receptor pathway eliciting apoptosis. Both pathways are crucial for maintaining proper immune function and immune homeostasis in different cell types and different stages of an immune response. The death receptor pathway is triggered by, as the name suggests, members of the death receptor family expressed on the surfaces of many mammalian cells (and those of many other animals). These receptors are activated by their cognate extracellular ligands, which are all closely related members of the tumor necrosis factor (TNF) (see Chapter 10 for a detailed discussion of these receptors) family and the receptors are part of the larger family of tumor necrosis factor receptors (TNFRs). The death receptor members of this family are TNFR1, CD95 (also called FAS or APO-1), the TRAIL (TNF-related apoptosis-inducing ligand) receptors (TRAIL R1, also called DR4, and TRAIL R2, also called DR5 in humans; rodents only have one TRAIL R, resembling DR5), DR3, and DR6. Their ligands include TNF, CD95 ligand (CD95L, also called FAS ligand), and TRAIL (also called APO2L). The ligand for DR6 has been identified as cell-bound amyloid precursor protein (APP),[52] which is not a member of the TNF family, and the roles of APP and DR6 in cell death are somewhat controversial.

CD95 SIGNALING AS AN EXAMPLE FOR DEATH RECEPTOR–INDUCED APOPTOSIS

CD95 is present as a trimer in the plasma membrane. While the extracellular portion of CD95 is important for binding to its ligand, CD95L, the intracellular region mediates signal transduction. The latter contains one death domain (DD), which adopts a prototypic "death-fold," also seen in related protein domains important for cell death: the CARD, DED, and pyrin domain (PYD). These protein domains, while not sharing sequence homology, adopt a similar structure and are efficient in promoting homotypic protein–protein interactions. Upon binding of trimeric CD95L or antibodies which mimic this stimulation, the intracellular DD of CD95 undergoes structural rearrangement and recruits the small cytosolic protein FADD (Fas-associated death domain), which contains both a DD and a DED domain. Upon recruitment by CD95, FADD exposes its DED domain and functions as an adaptor molecule to recruit pro–caspase-8 via one of the latter's two DED domains, DED1. When the DED1 of caspase-8 is bound by FADD, the DED2 is exposed and able to bind to a DED1 of another pro–caspase-8 molecule.[53] This sequence of events can be propagated and leads to formation of caspase-8 filaments, unless "something else" stops filament formation. We discuss this "something else" below. The bound caspase-8 is now active and is able to cleave caspase-3 and caspase-7 to induce apoptosis. The complex of CD95, FADD, and caspase-8 is referred to as the death-inducing signaling complex (DISC).[54] DISC formation occurs within seconds after receptor ligation. Subsequent caspase activation and progression of apoptosis can often be detected after tens of minutes, limited by the abundance of activated executioner caspases.

cFLIP MODULATES CASPASE-8 ACTIVITY IN A CONCENTRATION-DEPENDENT MANNER

Activation of pro–caspase-8 does not necessarily lead to immediate activation of caspase-3 and caspase-7. FLICE-like inhibitory protein (cFLIP), the product of the *cflar* gene, is similar to caspase-8 but lacks the catalytic cysteine required for protease activity. The cFLIP mRNA can be spliced into a short $cFLIP_S$ or a long $cFLIP_L$ isoform.[54,55] The short isoform of cFLIP ($cFLIP_S$) is truncated and contains two DEDs and a C-terminal tail. $cFLIP_S$ inhibits caspase-8 activation through competitive binding to the DED1 of caspase-8. The long isoform of cFLIP similarly inhibits caspase-8 filament formation by the incorporation of $cFLIP_L$ subunits, preventing full caspase-8 activation.[53,54,56] Therefore, cFLIP is the "something else," which can prevent robust caspase-8 activation. Of note, caspase-8 bound to $cFLIP_L$ is catalytically active, and while not promoting apoptosis, it efficiently blocks a necrotic form of cell death, necroptosis, discussed in greater detail below.

TRAIL RECEPTOR SIGNALING AND APOPTOSIS

Signaling of DR4 and DR5 upon stimulation by soluble TRAIL largely follows the principles we have illustrated for CD95-induced apoptosis. Like CD95, not all cells expressing TRAIL receptors are sensitive to TRAIL treatment. The resistance mechanisms are not well understood but seem to involve FLIP or receptor-specific modifications.[57,58] DR5 can also induce apoptosis independently of TRAIL stimulation. Upon endoplasmic reticulum (ER) stress, DR5 spontaneously dimerizes and activates caspase-8 to induce apoptosis.[59]

TNFR1 SIGNALING IS COMPLEX

TNFR1 has two closely related ligands, TNF and lymphotoxin. Binding of either trimeric ligand to TNFR1 induces exposure of its intracellular DD domain. However, in contrast to CD95 or TRAIL receptors, this does not bind FADD, but instead another adapter protein TNF receptor-associated factor 2 (TRAF2) is recruited. The DD of TRAF2 recruits TNF receptor B–associated death domain (TRADD),[60] which stabilizes the binding of TRAF2 and recruits other proteins, including cIAP1, cIAP2, and receptor-interacting protein kinase 1 (RIPK1) to the complex.

RIPK1 is a kinase, but the kinase function is dispensable for the processes we are describing in this section.[61] We explore RIPK1 kinase function in greater detail in our discussion of necroptosis below.

cIAP1 and cIAP2 function in this setting is to ubiquitinate RIPK1 via lysine (K)63–linked ubiquitin chains, which have a function different from that of degradative K48-linked ubiquitin chains. Ubiquitinated RIPK1 then recruits a protein complex called LUBAC, which conjugates linear ubiquitin chains to RIPK1. This leads to recruitment of NEMO (also known as I-kB kinase γ, IKK-γ), which in turn activates the IKK (IκB kinase) complex and promotes NF-κB activation.[62] This results in expression of pro-inflammatory cytokines as well as genes important for cell survival (eg, cFLIP). Stimulation of TNFR1 can lead to resistance against apoptosis stimulated by other death receptor through upregulation of cFLIP. Does this mean that TNFR1 ligation does not induce apoptosis? As the title of this section states, it is complex.

If NF-κB activity or the expression of its target genes is inhibited, a subsequent signaling cascade can lead to apoptosis. CYLD (a deubiquitinase, or DUB) removes K63-linked ubiquitin chains from RIPK1.[63] The TRADD, RIPK1 complex (commonly referred to TNFR1 signaling complex I) dissociates and the TRADD DD, which was previously bound to TNFR1, is exposed and can bind FADD, which leads to caspase-8 recruitment, activation, and apoptosis. We call the TRADD, RIPK1, FADD, pro–caspase-8 complex TNFR1 signaling complex IIa.

This signaling complex IIa can also form in response to other stimuli, including DNA damage and ligation of Toll-like receptors (TLRs) 3 and 4 to induce apoptosis. In such settings, complex IIa is sometimes referred to as the RIPoptosome.[64-66]

Animals deficient in NF-κB signaling die during embryonic development due to excessive death in the liver. This lethality can be prevented by co-deletion of TNFR1, indicating that the apoptosis and prenatal lethality is mediated by TNFR signaling and not by signaling of other death receptors.

ACTIVE CASPASE-8 ENGAGES THE MITOCHONDRIAL PATHWAY OF APOPTOSIS

Caspase-8 is very efficient in cleaving BID within its flexible loop, which generates the C-terminal fragment, cBID, a potent proapoptotic BH3-only protein.[39] Activation of death receptors therefore induces MOMP via cleavage of BID and subsequent activation of BAX and BAK. In some cells, termed type I cells, inhibition of MOMP (through genetic deletion of BID or overexpression of antiapoptotic Bcl-2 proteins) does not affect apoptosis induced by death receptor stimulation. Lymphocytes and lymphoid tumors are examples of type I cells. In type II cells, such as hepatocytes, inhibition of MOMP profoundly impacts the ability of death receptors to induce apoptosis.

The crucial difference between these two cell types is the expression of XIAP.[67] As we have seen, XIAP can inhibit the executioner caspases-3 and -7 but not caspase-8. This means that although caspase-8 is successfully activated

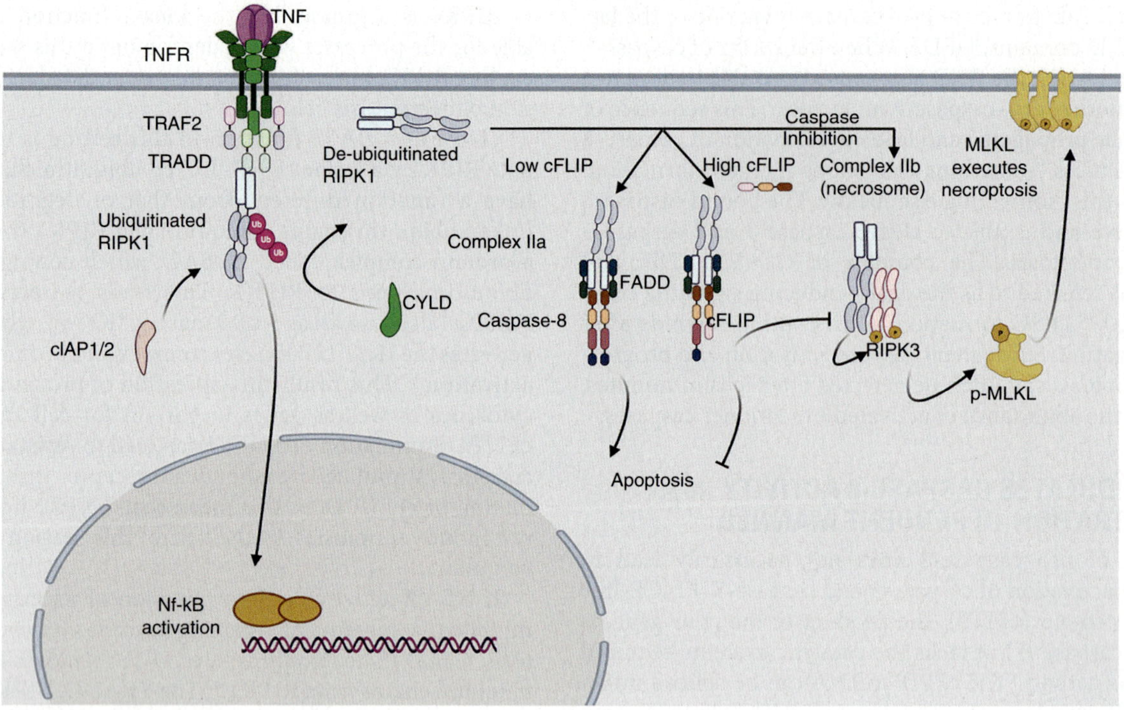

FIG. 7.4. TNF receptor signaling. TNF receptors (TNFRs) are trimeric plasma membrane receptors, which, upon binding of extracellular TNF-α, recruit the cytosolic adaptor proteins TRAF2 and TRADD as well as the signaling molecule RIPK1. This complex, referred to as TNFR signaling complex I, allows for NF-kB activation. The ubiquitination status of RIPK1 (controlled by cIAP1/2 and CYLD, see text for details) determines if RIPK1 dissociates from TRADD into the cytosol, where it can recruit FADD (Fas-associated death domain) and caspase-8 to form complex IIa. If cFLIP levels are low, caspase-8 is fully activated and induces apoptosis through subsequent cleavage of caspase-3 (not shown). If cFLIP levels are high, cFLIP inhibits caspase-8 filament formation and apoptosis does not proceed. The caspase-8/cFLIP heterodimer is however sufficiently active to cleave RIPK1 and RIPK3, which prevents execution of necroptosis. In conditions of caspase-8 inhibition or deficiency, RIPK1 phosphorylates RIPK3, which phosphorylates MLKL. Phosphorylated MLKL migrates to the plasma membrane to execute necroptosis. (Created with BioRender.com.)

upon death receptor ligation in type II cells, high XIAP levels inhibit apoptosis execution by inhibition of caspase-3 and caspase-7. Cleavage of BID into cBID to trigger MOMP and the release of the XIAP antagonists, SMAC and OMI, are required for neutralization of XIAP activity and progression of apoptosis. Due to low XIAP levels, MOMP induction via cBID is dispensable for death receptor–induced apoptosis in type I cells.

Injection of CD95L or antagonistic anti-CD95 antibody into mice causes massive apoptosis in the liver and death of the animal, which is not observed in mice lacking BID.[68] Mice deficient in both BID and XIAP again succumb to injection of CD95L, showing that neutralization of XIAP expression is a requisite step in CD95-mediated lethality[67] (Fig. 7.4).

REGULATED NECROSIS

The modes of cell death we discuss in this section all lead to permeabilization of the plasma membrane, but compared to apoptosis, this permeabilization occurs early after cell death induction and can be considered a "defining" feature of these modes of cell death. We discuss the impact of lytic modes of cell death on immune function in a separate section.

Necroptosis

The FADD-Caspase-8-cFLIP Complex Inhibits Necroptosis

The realization that another form of programmed cell death (PCD), besides apoptosis, exists downstream of TNFR1 stimulation was fueled by early findings on caspase-8, which were not easily reconciled with its known function in the apoptotic pathway. Mice deficient for either caspase-8[69] or FADD[70] die during embryonic development, and this could not be ascribed to any failure to elicit apoptosis. Embryonic lethality is prevented by co-ablation of either RIPK1 (discussed above)[71-73] or a related kinase, RIPK3.[74,75] This can be recapitulated *in vitro*, as cells treated with TNF die if caspase-8 is absent or pharmacologically inhibited. This TNF-induced death in the absence of caspase-8 activity, however, is not apoptosis, but instead a form of regulated necrosis called necroptosis. This demonstrates that caspase-8 negatively regulates this necrotic form of regulated cell death, which in turn is dependent on RIPK1 and RIPK3 upon TNFR1 ligation. It turns out that FADD-dependent caspase-8 activation formation downstream of death receptor signaling allows caspase-8 to cleave either RIPK1 or RIPK3. Caspase-8 bound to $cFLIP_L$, which as we have seen only does not induce apoptosis, is capable of cleaving RIPK1[76] and RIPK3[77] and prevents progression of necroptotic cell death. This identifies $cFLIP_L$ as a central regulator of cell survival downstream of death receptor signaling as the FADD-caspase-8-$cFLIP_L$ complex successfully inhibits necroptosis without engaging apoptosis.[75] The DUB CYLD which destabilizes TNFRI complex I through removal of ubiquitin from RIPK1 is also a substrate for caspase-8-$cFLIP_L$, and the cleavage of CYLD also negatively regulates necroptosis through favoring complex I formation.[78]

Regulation of Necroptosis Downstream of TNFRI

As we have discussed, complex IIa formation downstream of TNFR1 engagement can lead to caspase-8–dependent apoptosis. This happens, for example, if NF-κB signaling is compromised and TRADD, TRAF2, and RIPK1 dissociate from the receptor to recruit FADD and caspase-8 in the cytosol. If $cFLIP_L$ levels are high, neither apoptosis nor necroptosis occurs, as $cFLIP_L$ inhibits caspase-8 (see above).[79] Low $cFLIP_L$ levels allow for caspase-8 activation, hence apoptosis proceeds, but as caspase-8 still cleaves RIPK1 and RIPK3 necroptosis is inhibited. Under conditions when caspase-8 activity is blocked, RIPK1 autophosphorylates, which switches it to an active confirmation. This active confirmation of RIPK1 now is able to bind RIPK3 via RIP homotypic interaction motif (RHIM) domain homotypic interactions between the RHIMs of both proteins.[80-82] This leads to oligomerization and activation of RIPK3, which phosphorylates both itself and the mixed lineage kinase domain-like (MLKL) pseudokinase, the effector of necroptosis. It should be noted that for this pathway, both the kinase activity of RIPK1 (in contrast to NF-κB and apoptotic signaling) and RIPK3 as well as their RHIM domains are required for cell death.[73-75]

MLKL Is the Central Effector of Necroptosis

Cells deficient in MLKL successfully engage all signaling steps up to phosphorylation of RIPK3 but then fail to undergo cell death.[83,84] Upon phosphorylation by RIPK3, MLKL undergoes a conformational change, which allows interaction of its N-terminal "bundle" region with the plasma membrane. Forced oligomerization and plasma membrane translocation of the N-terminal region of MLKL is sufficient to induce necroptosis in cells.[85,86] Animals lacking caspase-8 and MLKL develop normally,[87] emphasizing the importance of MLKL for necroptosis.

As in apoptosis, the activation of MLKL causes exposure of PS on the outer leaflet of the plasma membrane before membrane integrity is lost. However, this externalization is independent of the known PS scramblases. PS externalization could be a direct effect of MLKL functionally acting as a phospholipid scramblase, which leads to disruption of the membrane and cell death.[88,89] It is not entirely clear if MLKL is the final step of necroptosis, as some studies have suggested it could open ion channels and promote cell death through this mechanism.

Other RHIM-Containing Proteins Can Engage RIPK3-Dependent Necroptosis

While RIPK1 is essential for necroptosis engagement downstream of TNFR1 receptor ligation, RIPK3 can be activated by other RHIM domain–containing proteins through other stimuli. The TIR domain–containing adaptor-inducing interferon (TRIF) functions as an adaptor molecule downstream of TLR-3 and TLR-4.[90,91] This protein contains a RHIM domain, which is able to oligomerize RIPK3 upon TLR stimulation, leading to phosphorylation and activation of MLKL and necroptosis. The cytosolic protein Z-DNA binding protein 1 (ZBP1 also called DAI), which can bind viral nucleic acids, is also able to oligomerize RIPK3. It is again the RHIM domain of ZBP1 which interacts with RIPK3 to promote its activation. Either ZBP1 or TRIF can only induce RIPK3-dependent necroptosis if caspase-8 function is compromised, for example, in the presence of a pan-caspase inhibitor.[92-95]

RIPK1 Is a Signaling Hub at the Interface of Apoptosis, Necroptosis, and Cell Survival

As illustrated above, RIPK1 scaffolding function is important to elicit NF-κB activity upon TNFR1 engagement, and hence inhibits TNFR-induced apoptosis.[62] RIPK1 can also promote apoptosis via complex IIa (as discussed above). At the same time, in scenarios where caspase-8 function is compromised (eg, by addition of caspase inhibitors), RIPK1 promotes activation and oligomerization of RIPK3 (complex IIb or "necrosome") and activates necroptosis. The activation of RIPK3 by RIPK1 is dependent on the kinase activity of both proteins. Inhibition of the kinase activity of RIPK1 (eg, with pharmacologic inhibitors) prevents TNFR-induced necroptosis.[96-98]

In the absence of RIPK1, TRIF- or ZBP1-induced RIPK3-mediated necroptosis proceeds independently of the activity of caspase-8.[90,91,99] In the presence of a functional RIPK1, RIPK1 is able to recruit the FADD-caspase-8-$cFLIP_L$ complex, which inhibits necroptosis, perhaps via cleavage of RIPK3, TRIF, ZBP1, or another cofactor.

Mice that lack the caspase cleavage site in RIPK1 die during gestation, highlighting the importance of caspase-8–mediated cleavage of RIPK1. This lethality is prevented by co-ablation of caspase-8 and either RIPK3 or MLKL.[100] Humans who are heterozygous for the corresponding mutation in human RIPK1 develop severe autoimmune disease, likely through loss of FADD-caspase-8-cFLIP cleavage-mediated control of the mutant RIPK1.[101]

Pyroptosis

Pyroptosis originates from the Greek words for fever ("pyretós") and to fall ("ptósi") indicating the "hot" features of this type of cell death. Pyroptosis is morphologically distinct from apoptosis and was first described in macrophages infected with *Salmonella typhimurium*.[102] Characteristic features of pyroptosis include swelling of the cell, presumably due to water influx, condensation of the nucleus, rupture of the cell membrane, and subsequent release of cytosolic content. A significant body of work has outlined multiple signaling pathways which detect cellular insults and activate pyroptosis. These pathways are elicited by pattern recognition receptors (PRRs) and converge on the activation of the inflammatory caspases-1 and -11 in mice and caspases-1, -4, and -5 in humans.[103,104] The exact PRRs, their ligands, and mechanisms of activation are discussed in detail in Chapter 12. We focus our discussion on events in the pathway upon inflammatory caspase activation.

Inflammatory Caspases Are Activated by Induced Proximity

The architecture of inflammatory caspases resembles that of caspase-9, as these caspases have an N-terminal CARD followed by a large (p20) and small subunit (p10).[4] Pro-caspase-1 is recruited to activated inflammasomes through its CARD domain, where it is processed by another pro-caspase-1 molecule between the p20 and p10 subunits. This form of the caspase (CARD-p20, also called p33), which transiently associates with the p10 subunit, represents the active form of the caspase. Subsequent cleavage between the CARD and the p20 subunit leads to dissociation of the p20/p10 heterodimer from the inflammasome and leads to loss of proteolytic activity. Hence, detection of the fully processed (p10 and p20 fragments) caspase-1 is only an indication that the caspase may have been activated, but not necessarily active[105] (Fig. 7.5; Table 7.1).

Caspase-11, in turn, does not require an adaptor or receptor protein as its CARD domain is capable of directly binding its ligand lipopolysaccharide (LPS) from gram-negative bacteria.[106] Activation of caspase-11 requires dimerization by LPS, which promotes activation and is sufficient for autoprocessing. Autoprocessing occurs in the linker between the large and the small subunits leading to full activation of caspase-11, which, like fully active caspase-1, executes pyroptosis. The human homologue of caspase-11 is caspase-4, which is also activated by binding of LPS to its CARD domain. Processing of caspase-4 (at least *in vitro*) seems to enhance activation similar to caspase-11 proteolysis.[107,108] Humans also have caspase-5, which is thought to be similar to caspase-4. Purified caspase-5 is activated by LPS *in vitro* but seems not to be activated by cytosolic LPS in human THP-1 cells. It however induces pyroptosis in response to infection with the gram-negative bacterium *S. typhimurium* in THP-1 cells.[109]

Inflammatory Caspases Cleave Gasdermin D to Induce Pyroptosis

Gasdermin D (GSDMD) is a substrate common to caspases-1, -11, -4, and -5 and promotes rapid permeabilization of the plasma membrane and pyroptosis.[110,111] Humans have six gasdermin genes: GSDMA, GSDMB, GSDMC, GSDMD, GSDME or DFNA5, and DFNB59 or PJVK. Mice do not have GSDMB but posses multiple copies of GSDMA(1-3) and GSDMC(1-4). GSDMD, GSDME, and DFNB59 are conserved in mice. Their name derives from their expression in the gastrointestinal tract and the dermis, but we now know that some of these, such as GSDMD, are more ubiquitously expressed (reviewed in Ref. 112). GSDMD is a cytosolic protein, comprised of an N-terminal and a C-terminal domain connected by a flexible linker, which contains the caspase cleavage site around Asp_{274} (Asp_{275} in humans). The C-terminus ($GSDMD^{Cter}$) acts as a solubilizer or inhibitor when bound to the N-terminus ($GSDMD^{Nter}$). Upon cleavage by inflammatory caspases, the C-terminal domain dissociates and the N-terminal fragment targets cellular membranes, where it forms pores and permeabilizes these membranes.[113-116] Permeabilization of the plasma membrane by the $GSDMD^{Nter}$ leads to collapse of the

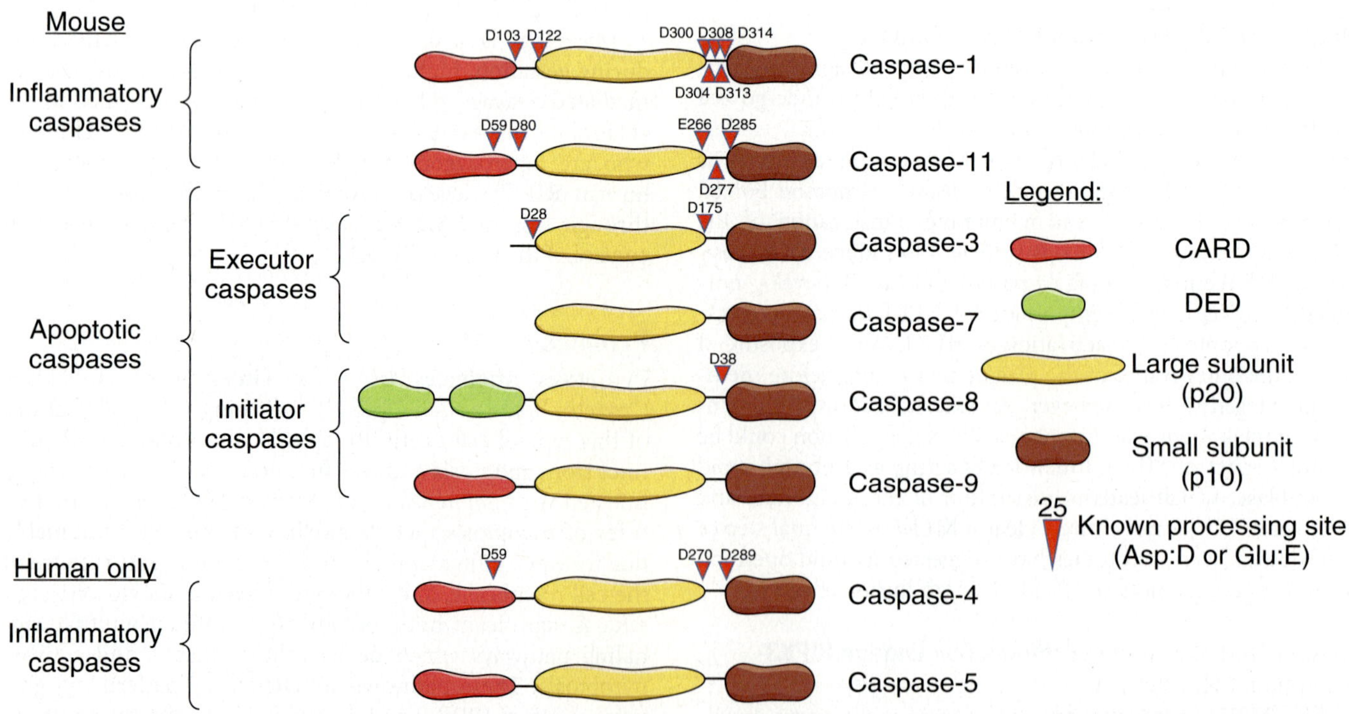

FIG. 7.5. Schematic representation of different caspases. CARD, caspase activation and recruitment domain; DED, death effector domain.

TABLE 7.1 Summary of Caspases, Their Substrates, and Activation Mechanisms

Caspase	Stimulation	Relevant Substrates	Activation by
1	Canonical inflammasomes	Gasdermin D, IL-1 cytokines	Induced proximity
2	Intrinsic/extrinsic apoptosis	iCAD, ATP11c, Xkr8	Cleavage by caspase-8, caspase-9, granzyme b
3	Cytosolic lipopolysaccharide (LPS)	Gasdermin D	Induced proximity
4	Cytosolic LPS, *Salmonella typhimurium*	Gasdermin D	Induced proximity
7	Intrinsic/extrinsic apoptosis	iCAD, ATP11c, Xkr8	Cleavage by caspase-8, caspase-9, granzyme b
8	Extrinsic apoptosis	BID, caspase-3, caspase-8, RIPK3, RIPK1, gasdermin D	Induced proximity
9	Intrinsic apoptosis	Caspase-3	Induced proximity
11	Cytosolic LPS	Gasdermin D, caspase-11	Induced proximity

electrochemical gradient, water influx, swelling, and eventually rupture of the cell. Cleavage of the cytoskeletal protein vimentin by calpain, induced as calcium levels increase in the dying cell, sensitizes pyroptotic cells to extremely low levels of shear stress (similar to the forces in tissues or the blood stream), which ruptures cells and releases their cytosolic content.[117] The N-terminal fragments of different GSDM family members possess pore-forming ability;[113] however, their mechanisms of activation (except GSDMD and DFNA5/GSDME, discussed below) remain elusive.

Unlike the cleavage by inflammatory caspases, executioner caspases-3 and -7 cleave GSDMD at another site, and this cleavage inactivates the pore-forming function. Caspase-1 can cleave and thereby activate these executioner caspases, and therefore there appears to be a competition between GSDMD activation and inhibition in cells in which caspase-1 is activated.[118]

Gasdermin D Forms Pores in Cellular Membranes Upon Cleavage by Inflammatory Caspases

Recently, the mechanism of recognition of GSDMD by inflammatory caspases has been described. Surprisingly, the sequence of the tetrapeptide motif ($_{272}$FLTD$_{275}$ in hGSDMD, $_{273}$LLSD$_{276}$ in mGSDMD) is not essential for cleavage of the intradomain linker, unless the aspartic acid is mutated. Both caspase-1 and caspase-11 bind to the C-terminus of GSDMD at a hydrophobic groove in the protein. Mutants within the binding interface of either the caspase or GSDMD disrupt interaction, cleavage, and pyroptosis induction. The N-terminal portion of GSDMD is not essential for recognition and cleavage by inflammatory caspases as replacing the GSDMD^{Nter} with either the GSDME^{Nter} or glutathione S-transferase (GST) maintained cleavage by caspase-11. Structurally, the two p20/p10 dimers of caspase-4 associate with two GSDMD molecules to process them.[108]

Structural studies of active mGSDM-A3^{Nter} suggest that gasdermins form 27-fold oligomeric membrane pores, with an inner diameter of 180 Å and an outer diameter of 280 Å. The pore is composed of 108 beta-strands, which, with a height of 50 Å can span a complete lipid bilayer. One gasdermin A3 protomer is similar to a left hand where the fingers extend as beta sheets into the membrane. The globular region, or the "palm" of the hand, shapes the rim of the pore and contains both α- and β-elements.[119]

Gasdermin E Induces Secondary Necrosis During Apoptotic Caspase Activation

GSDME (the gene product of DFNA5) has been shown to induce lytic cell death in response to cleavage by activated caspase-3, hence rapidly converting apoptotic cell death into necrotic cell death. Cleavage occurs at Asp270, and caspase-3, but not caspase-7, can cleave gasdermin E. Similar to GSDMD, the GSDME^{Cter} dissociates from the N-terminal fragment, and the pore-forming activity of GSDME^{Nter} is released upon cleavage.[120] Primary cells or tumor cells expressing GSDME rapidly succumb to this lytic form of cell death upon engagement of apoptosis, in a process referred to as secondary necrosis.[121] Necrotic cell death induced by gasdermins stimulates the priming of cytotoxic lymphocytes (CTLs)[122] and synergizes with PD1 blockade in cancer therapy.[123]

Ferroptosis

Ferroptotic cell death was discovered as a nonapoptotic form of cell death in response to erastin treatment, an inhibitor of the glutamate/cystine antiporter system X_C. The name arises from its dependency on iron. Ferroptosis is a regulated form of cell death as it is under the control of cellular processes.

Elucidation of the molecular mechanisms showed that ferroptosis is a form of cell death driven by the oxidation of polyunsaturated fatty acids (PUFAs) in the plasma membrane. It is regulated by the redox state of the cell, amino acids, iron metabolism, and fatty acid synthesis pathways. The lipid peroxidase glutathione peroxidase 4 (GPX4) is a key inhibitor of ferroptosis.[124] This selenoenzyme detoxifies oxidized lipid species into nontoxic alcohols, a mechanism in mammalian cells to prevent PUFA peroxidation in the plasma membrane. This is a chain reaction occurring in the presence of oxygen and the absence of GPX4 activity. This chain reaction destabilizes the plasma membrane to such a degree that the electrochemical gradient cannot be maintained, and cells eventually burst upon influx of water. GPX4 oxidizes two molecules of glutathione (GSH) to GSSG to detoxify PUFA peroxides into alcohols.[125] Pathways which either reduce GPX4 activity or PUFA availability can induce or repress ferroptosis, respectively. The small molecule, erastin, for example, through inhibition of system X_c^-, depletes intracellular cysteine, which is a precursor of GSH synthesis, and therefore the substrate pool of active GPX4 is depleted.[126] Extracellular glutamate prevents the exchange, by system X_c^-, of intracellular glutamate for extracellular cystine, therefore reducing intracellular cysteine levels. This could be one reason why high extracellular glutamate levels induce ferroptosis in neurons.[127] Direct inhibition of GPX4 by RSL3 or ML210 induces ferroptosis in some, but not all cells. Several antioxidations such as ferrostatins, liproxstatins, coenzyme Q10 (CoQ10), and vitamin E are able to inhibit ferroptosis through the sequestration of PUFA peroxides, thereby inhibiting the initiation of membrane lipid peroxidation. Depleting cells of PUFAs through deletion of one of the enzymes in the PUFA synthesis pathway, acyl-CoA synthetase long-chain family member 4 (ACSL4), also protect cells from ferroptosis.[128-130] Recently, another lipid peroxidase was identified and possibly explains why some cells are resistant to GPX4 inhibition. The gene product of *AIFM2*, renamed ferroptosis suppressor protein 1 (FSP1), is a GSH-independent lipid peroxidase, which uses ubiquinone (CoQ10) and produces its reduced form ubiquinol in order to prevent lipid peroxidation at the plasma membrane. This pathway parallels the GSH-dependent regulation of ferroptosis by GPX4.[131,132]

Ferroptosis seems to occur in a variety of scenarios, such as stroke, heart attack, and renal ischemia reperfusion injury (reviewed in Ref. 133). Cancer cells, which "persist" following treatment with chemotherapy drugs or targeted therapy, become highly dependent on GPX4 activity, even when this is not the case for the parental, chemotherapy-sensitive population. This identifies ferroptosis as a potential Achilles heel of chemotherapy-resistant cancer cells.[134,135]

Ferroptosis can be considered a regulated form of cell death. Many cellular signaling and metabolic pathways control the initiation of ferroptosis, and its induction and progression can be genetically and pharmacologically modified. However, unlike the other forms of regulated cell death we have discussed, ferroptosis is not a process which is actively executed by the cell but happens rather due to the absence of an essential cellular pathway. It is death due to "the lack of something necessary" rather than death due to "the activation of something lethal." At least for now, we can define ferroptosis as a form of cell sabotage, rather than an active cell suicide mechanism. However, this may be reconsidered as we learn more about the mechanisms and physiological importance of this cell death process (Fig. 7.6).

NETosis

Neutrophils (their biology is discussed in Chapter 14) represent a crucial component of our innate immune system due to their sheer mass (up to 70% of white blood cells), their exquisite phagocytic capabilities, and their ability to efficiently release antimicrobial substances by degranulation. Another antibacterial function of neutrophils is NETosis. NETosis is a process of regulated cell suicide of neutrophils, which leads to the expulsion of neutrophil extracellular traps (NETs). NET formation is induced by production of reactive oxygen species (ROS) by the NADPH oxidase, which leads to

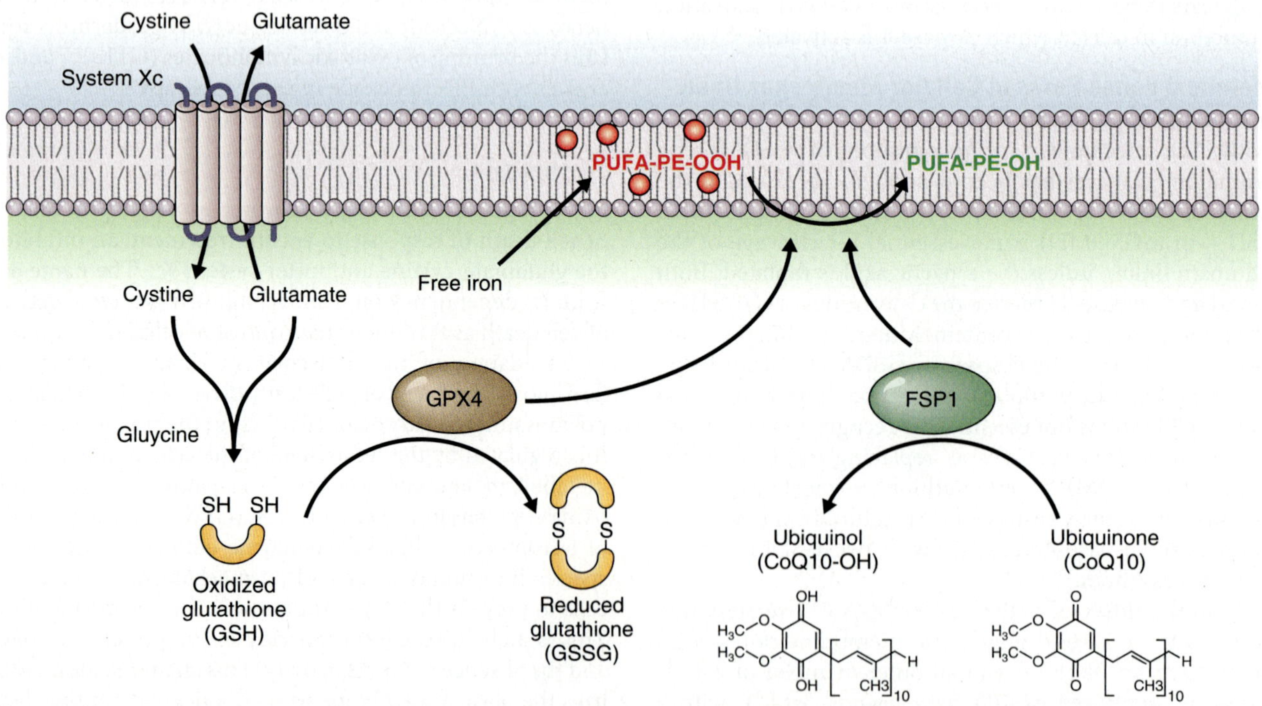

FIG. 7.6. Ferroptosis. Free iron promotes oxidation of polyunsaturated fatty acids (PUFAs) on the plasma membrane, a process spontaneously occurring in the presence of oxygen. GPX4 and FSP1 are able to detoxify oxidized PUFAs to alcohols using either glutathione or ubiquinone as electron acceptors (respectively). The cellular glutathione pool needs to be replenished through import of cystine by system Xc.

activation of several key enzymes allowing initiation of NET formation.[136,137] In a first step, protein arginine deiminase 4 (PAD4) is activated, which converts arginine to citrulline on histones, causing chromatin decondensation. At the same time, myeloid peroxidase (MPO) and neutrophil elastase (NE) are released from granules into the cytosol and translocated to the nucleus, where they further promote chromatin unfolding. As a consequence, the nuclear membrane breaks down and the unfolded chromatin structures are exposed to the cytosol, where they are decorated further with granular proteins, such as MPO and NE.[138,139] NE also cleaves and activates GSDMD, and the resulting cell lysis is required for NETosis.[140] Neutrophils lyse and expel the NETs, which due to the presence of histones, MPO, and NE are strongly antibacterial, trapping and killing bacteria.

NET formation does not always result in cell death, as NETs can be released by nuclear envelope blebbing and vesicular transport. This occurs independently of ROS production upon stimulation with bacteria or bacterial products, TLR4-activated platelets, or complement proteins in tandem with TLR2 ligands. "Vital NETosis" allows neutrophils to perform other functions after deposition of NETs.[141-144] Suicidal NETosis appears to be an essential antibacterial function of our immune system but can also contribute to disease, as defective clearance of NETs can promote autoimmunity (reviewed in Ref. 145) (Table 7.2).

THE IDENTIFICATION, ENGULFMENT, AND CLEARANCE OF DEAD CELLS

Dead and dying cells are rapidly removed from tissues by phagocytosis, which is called efferocytosis ("cell burial").[146] There are several redundant steps in this process and here we consider these in the terms of the clearance of apoptotic cells, which is best understood. While it is likely that the same or similar events relate to other modes of cell death, it remains possible that differences exist, as the clearance of dead, nonapoptotic cells is understudied.

The steps in efferocytosis can be summarized as (1) the elaboration of "find me" signals, (2) downregulation or blockade of "don't eat me" signals, (3) expression and recognition of "eat me" signals, (4) engulfment and digestion, and (5) effects on phagocytes and the surrounding tissue (Fig. 7.7). These are considered in turn, as are the pathological consequences of defective efferocytosis.

While dying cells can be engulfed and cleared by many cell types, including epithelia, endothelia, and other cells, macrophages and DCs are the major cells responsible for efferocytosis. The extent to which the mechanisms of efferocytosis may differ in cells other than macrophages and DCs remains unclear. Therefore, our focus is on efferocytosis by macrophages and predominantly the engulfment of apoptotic cells.

"Find Me" Signals

Cells that undergo apoptosis recruit macrophages to engulf them, the so-called "find me" signals. These may differ depending on the type of cell that dies,[147] and there is controversy as to which are most important. These include the modified membrane lipids lysophosphatidylcholine (LPC) and sphingosine-1-phosphate (S-1-P),[148,149] the nucleotides ATP (adenosine triphosphate) and UTP,[150] the chemokine, CX3CL1,[151] and the eicosinoid, leukotriene-β4 (LTB4).[152]

Caspase-dependent events mediate the production and/or release of these signals, and in some cases, the mechanisms are well understood. For example, the cleavage of calcium-independent phospholipase A2 by executioner caspases activates it to produce LPC from phosphatidylcholine.[148] Similarly, caspase cleavage of the pannexin-1 channel protein opens this channel to release ATP and UTP.[153]

Nonapoptotic cell death causes the release of DAMPs (danger-associated molecular patterns), as discussed later, and some of these may act as "find me" signals as well. In addition, cells undergoing necroptosis produce the chemoattractant chemokines CXCL1 and CXCL10.[154] This appears to be a direct response to plasma membrane damage, and therefore these may also be produced in other forms of cell death, such as pyroptosis, although this has not been shown.

"Eat Me" and "Don't Eat Me" Signals

As discussed above, the exposure of PS on the surface of dying and dead cells serves as the major "eat me" signal. However, PS can also be exposed in response to other events, such as calcium influx.[155] Living cells express several "don't eat me" signals that prevent their engulfment despite PS exposure. Two of these are CD24 and CD47, which bind to the inhibitory receptors SIRPα and Siglec-10, respectively, inhibit engulfment.[156,157] There is also evidence that homotypic interaction of CD31 can prevent engulfment,[158] as can the binding of class I MHC to the inhibitory receptor LIRB1.[159]

It is likely that the situation with "don't eat me" signals is more complex than this. First, there is no evidence that such signals are indeed lost as cells undergo apoptosis. Secondly, blockade of such a signal, for example, with a neutralizing CD47 antibody, results in engulfment of living tumor cells but not normal cells, and this engulfment appears to depend not on PS recognition, but rather the recognition of cell surface calreticulin (perhaps as a consequence of ER stress) by CD91 on macrophages.[160] Finally, constitutive exposure of PS on living B lymphocytes by ablation of ATP11c results in their clearance *in vivo*,[161] despite the presumed presence of the "don't eat me" signals.

Therefore, while the "don't eat me" signals appear to prevent engulfment of living cells bearing cell surface calreticulin, it is likely that it is not only PS exposure but also disruption of ATP11 function that is important for the recognition of dying cells (either by caspase cleavage in apoptotic cells, or in necrotic cells by loss of ATP, needed for its function). In addition to "flipping" PS from the outer to the inner leaflet of the plasma membrane, ATP11a and ATP11c function to restrict the lateral movement of PS in the membrane.[162] It is possible that freeing this movement is necessary for the uptake of the dying cell by efferocytosis.

TABLE 7.2 Comparison of Features of Different Cell Death Pathways

Cell Death Pathway	Plasma Membrane Rupture	Cell Types	Triggers	Receptors (Sensors)	Signaling Pathway	Immunogenicity	Physiological Examples	Pathological Role
Extrinsic apoptosis	Not always or late	All cells, death receptor expression required	Death receptor ligands	Death receptors	FADD, caspase-8, caspase-3, caspase-7	No, immunologically silent	AICD, cytotoxic T cells	Failure to induce extrinsic apoptosis (especially by Fas/CD95) results in ALPS
Intrinsic apoptosis	Yes, late	All cells	Various: eg, DNA damage, ER stress, growth factor withdrawal	Bcl-2 family	Cytochrome c, APAF1, caspase-9, caspase-3, caspase-7	Mostly immunologically silent, immunogenic apoptosis can occur under special circumstances	AICD, T cell and B cell development, development of the nervous system	Failure in intrinsic apoptosis can contribute to development of cancer
Necroptosis	Yes, early	All cells, death receptor or TLR expression required	Death receptor ligands, TLR ligands, Z-nucleic acids	Death receptors TLRs (via TRIF), ZBP1	RIPK1, RIPK3, MLKL (inhibited by caspase-8)	Yes	Organ development, clearance of virus infected cells	Possible role in organ damage after ischemia reperfusion injury. Failure to engage the necroptosis pathway can sensitize for lethal virus infection
Pyroptosis	Yes, early	Predominantly myeloid, endothelial, and epithelial cells	Various cytosolic PAMPs and DAMPs, eg, dsDNA, uric acid crystals, bacterial toxins, flagellin	NLRs and ALRs	Caspases-1, caspase-11 in mice; caspase-1, caspase-4, and caspase-5 in humans; and gasdermin D	Yes	Clearance of virus or bacteria infected cells, attraction of neutrophils and other immune cells	Excess pyroptosis leads to organ failure during sepsis. Pyroptosis-associated IL-1 cytokines promote joint inflammation in rheumatoid arthritis
NETosis	Not always	Neutrophils	Extracellular PAMPs or DAMPs	Unknown	Gasdermin D	Unknown	Killing of extracellular bacteria	Failure to clear NETs can induce autoimmunity
Ferroptosis	Yes, early	Variable, although many transformed cells are sensitive	Cysteine deprivation, inhibition of system X_c^-, inhibition of GPX4	None	Fenton reaction	No	Response to pharmacologic inhibitors of system X_c^-, GPX4, possibly killing of tumor cells by cytotoxic lymphocytes	Possible role in ischemia reperfusion injury

AICD, activation-induced cell death; ALPS, autoimmune lymphoproliferative disease; ALR, AIM-like receptor; APAF1, apoptosis protease activating factor 1; DAMP, danger-associated molecular pattern; FADD, Fas-associated protein with death domain; GPX4, glutathione peroxidase 4; IL-1, interleukin-1; MLKL, mixed lineage kinase domain-like pseudokinase; NET, neutrophil extracellular trap; NLR, NOD-like receptor; PAMP, pathogen-associated molecular pattern; RIPK, receptor-interacting serine/threonine protein kinase 1; TLR, Toll-like receptor; TRIF, TIR domain–containing adapter-inducing interferon-beta; ZBP1, Z-DNA binding protein 1.

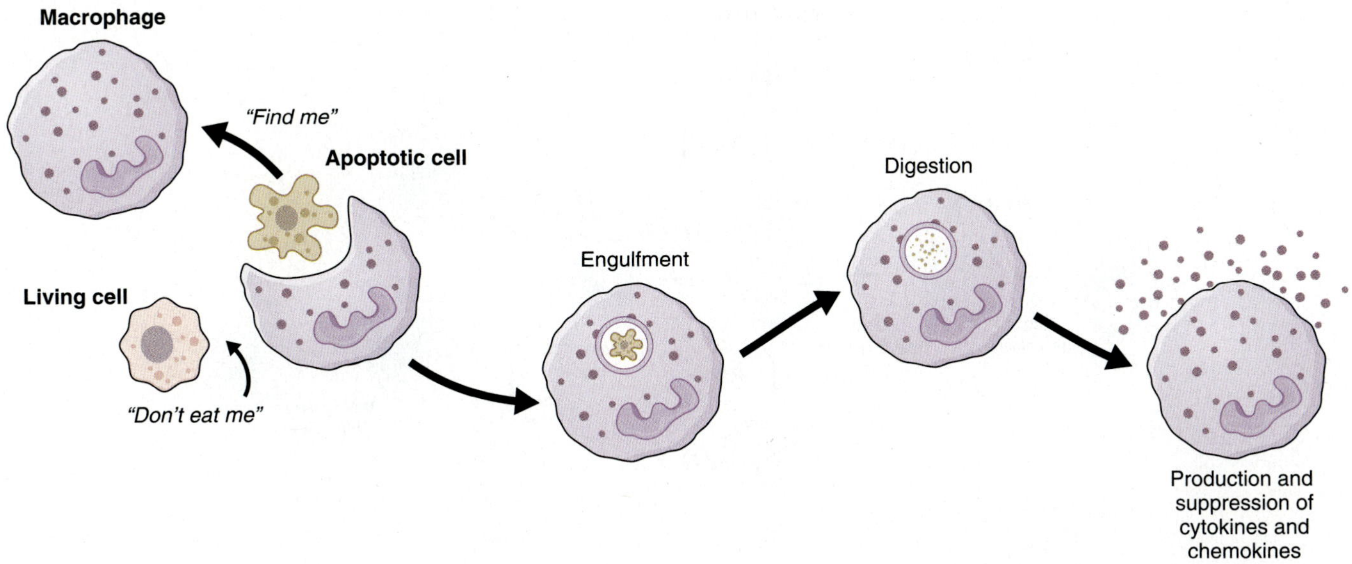

FIG. 7.7. Efferocytosis. Living cells display "do not eat me" (eg, CD24 and CD47), which prevents them from engulfment by professional phagocytic cells (eg, macrophages). Apoptotic cells instead loose these signals, release "find me" (eg, lysophosphatidylcholine and sphingosine-1-phosphate), and display "eat me" signals (phosphatidylserine), which facilitates engulfment of dead cells by macrophages. Successful engulfment and digestion of dead cells by macrophages leads to reprogramming of the macrophage and production of cytokines.

Phosphatidylserine Is Recognized in Multiple Ways

Several receptors on macrophages and other cells directly recognize exposed PS on dead and dying cells. These include TIM1, TIM4,[163] BAI1,[164] stabilin-2,[165] and CD300f.[166] In addition, there are several soluble "bridge" molecules that bind to PS and to receptors on phagocytes. MFG-E8 (also called lactadherin) binds to PS and to the αvβ3 and αvβ5 integrins on phagocytes.[167] Similarly, protein S and Gas6 bind to both PS and to the TAM receptors, Mertk, Axl, and Tyrol.[168] Which mechanism predominantly functions depends on the phagocyte; while macrophages of the germinal center predominantly employ MFG-E8 to engulf dying B cells,[167] peritoneal macrophages predominantly employ TIM4.[169]

Remarkably, ablation of any of several of these mechanisms results in pathology, most notably an autoimmune SLE-like syndrome in mice, among others. This is discussed in more detail below (Fig. 7.8).

Engulfment and Digestion of Dead Cells

Once PS receptors are engaged, the phagocyte engulfs the dead or dying cell by the process of macropinocytosis. This process, unlike opsonization, takes up both the cell and surrounding fluid. Regardless of the PS recognition mechanism employed, reorganization of the actin cytoskeleton is mediated via activation of the Rac1 GTPase,[170] resulting in protrusions that "reach around" the dead cell, fuse, and draw the corpse into the membrane-bound phagosome (Fig. 7.7).

Fusion of the phagosome with lysosomes then allows digestion of the corpse. This fusion is facilitated by the lipidation of the autophagy ATG8 family of small proteins, including LC3 proteins, on the phagosome membrane by the process of LC3-associated phagocytosis.[171] This noncanonical function of autophagy proteins involves some, but not all of the autophagy machinery, as well as other proteins that are not involved in autophagy.[172]

The digestion of the corpse represents a metabolic challenge to the macrophage, which in taking up a dead cell has nearly doubled its contents, including cholesterol. The activation of liver X receptors (LXRs) by cholesterol-derived oxysterols in the macrophage increases the expression of the cholesterol transporter, ABC1, permitting cholesterol efflux from the cell.[173] In addition, processed fatty acids stimulate the peroxisome proliferator-activated receptors PPARδ and PPARγ[174,175] and an increase in mitochondrial free fatty acid oxidation.[176] Meanwhile, the resultant oxidative stress is offset by the expression and activation of the uncoupling protein, UCP2.[177]

Immune Consequences of Efferocytosis

Upon engulfment of apoptotic cells, macrophages produce several cytokines, including TGFβ, IL-10, and platelet-activating factor, as well as prostaglandin E2, which function in wound healing and anti-inflammation, and suppress the production of IL-12, IL-6, TNF, and other inflammatory cytokines. These effects appear to be due to the activity of PPARδ,[174] although other studies have implicated additional transcription factors involved in the expression of IL-10[178] and the inhibition of IL-12.[179]

Defects in efferocytosis can result in inflammatory disease. Macrophages from patients with SLE often display defects in efferocytosis.[177,178] Conversely, SLE-like disease is observed in mice lacking any of several of the PS receptors or bridge molecules, including Mertk,[180,181] Tim4,[169] or MFG-E8,[182]

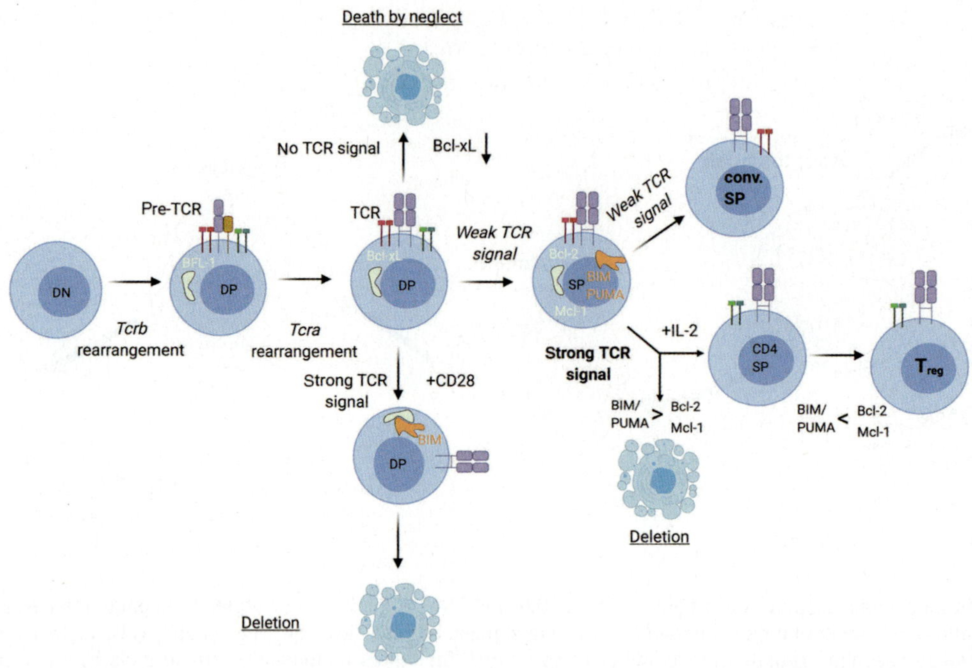

FIG. 7.8. **Cell death in T cell development.** Transition from the double-negative (CD4⁻, CD8⁻) stage to the double-positive (CD4⁺, CD8⁺) stage of T cell development is accompanied by the upregulation of the antiapoptotic protein BFL-1, which promotes survival of these cells. *Tcra* rearrangement allows for expression of a T cell receptor (TCR) and expression of Bcl-xL. If a strong TCR is received at this stage (through high affinity recognition of self-MHC proteins), these potentially autoreactive T cells are eliminated by upregulation of BIM. If the TCR does not recognize self-MHC, Bcl-xL half-life determines survival of these cells. Weak to intermediate TCR signaling upregulates Mcl-1 and Bcl-2 as well as BIM and PUMA and developing T cell transition to the single-positive (CD4⁺ or CD8⁺) stage. Another round of weak to intermediate TCR stimulation promotes egress from the thymus as conventional T cells. If CD4⁺ cells receive a strong TCR signal, they can be deleted, as seen for double-positive cells. This happens if the proapoptotic activity of BIM and PUMA overwhelms Bcl-2 and Mcl-1 levels. In the presence of IL-2 and sufficient antiapoptotic proteins, these cells become regulatory T cells. (Adapted from Daley SR, Teh C, Hu DY, Strasser A, Gray DHD. Cell death and thymic tolerance. Immunol Rev. 2017;277:9-20.

mice lacking the apoptotic PS-scramblase Xkr8,[183] mice deficient in myeloid LC3-associated phagocytosis,[184] and in mice deficient in PPARδ or PPARγ.[174,175] While most studies focus on macrophage-mediated efferocytosis, mice in which Rac1 is deficient in pulmonary epithelial cells display inflammatory lung injury upon exposure to apoptotic cells.[185] Defects in efferocytosis have also been implicated in a wide range of other inflammatory diseases.[186]

REGULATED CELL DEATH IN IMMUNE HOMEOSTASIS

Regulated Cell Death in Lymphocyte Development

Development of T cells and B cells is a complex multistep process, which involves DNA recombination to produce the variety of antigen receptors. In a successful immune response, the proliferation and function of antigen-specific lymphocytes eliminates the pathogen, following which homeostasis must be restored. During such "contraction," many activated lymphocytes die. Apoptosis is the main mechanism by which lymphocyte quality and numbers are controlled. The following section discusses cell death pathways regulating lymphocyte development, selection, expansion, and contraction mechanisms.

THE INTRINSIC PATHWAY OF APOPTOSIS REGULATES T CELL DEVELOPMENT

As discussed in greater detail in Chapter 27, the development of T cells is a spatially- and timely regulated process occurring in the thymus, which involves several checkpoints that the developing T cells must pass in order to mature and egress to the periphery. Here, we focus our discussion on the cell death mechanisms governing these checkpoints.

The first checkpoint thymocytes pass is beta-selection.[187] This refers to the evaluation, in the developing T cell, of the functional rearrangement and expression of the *Tcrb* gene to produce a functional T cell receptor (TCR) beta chain. In conjunction with the pre–T alpha chain, this forms the pre-TCR. This rearrangement occurs at the CD4⁻, CD8⁻ (DN, double-negative) stage of T cell development. Engagement of the pre-TCR leads to NF-κB activation, which upregulates Bcl2-A1 (also called BFL-1), a member of the antiapoptotic Bcl-2 family proteins.[188] This inhibits mitochondrial apoptosis and promotes survival of cells beyond this stage of T cell development.[187,189] Deletion of both apoptotic effectors Bax and Bak results in a failure of productive beta-selection, that is cells with a nonproductive *Tcrb* rearrangement progress to the CD4⁺, CD8⁺ (DP, for double-positive) stage, corroborating the importance of intrinsic apoptosis in this process.[190]

Other apoptotic mechanisms also appear to be in place, which regulate beta-selection. Overexpression of a dominant-negative form of FADD leads to dysfunctional beta-selection, suggesting a potential involvement of death receptor–mediated apoptosis.[191] To date, beta-selection is the least well-understood step of T cell development and we do not know which signals activate apoptosis at this stage.

Upon successful *Tcrb* rearrangement and pre-TCR expression, T cells become CD4⁺, CD8⁺ DP cells, where they rearrange and express the *Tcra* gene. Upon completion of this process, DP cells express a TCR which is capable of recognizing MHC-I or MCH-II complexes presenting self-peptides.[192] If the TCR expressed does not recognize these complexes, cells undergo "death by neglect." The antiapoptotic protein Bcl-xL is upregulated during transition from the DN to the DP stage. Bcl-xL, with its half-life of 20 hours, is thought to act as a molecular timer.[193,194] In the absence of TCR stimulation, Bcl-xL half-life limits the survival of cells without functional TCR expression, which engage intrinsic apoptosis, as the half-life of the Bcl-2 effectors is longer than that of Bcl-xL. This fits with the observation that most DP thymocytes survive for about 3 days.[195] Enforced expression of Bcl-2 rescues DP thymocytes from this "death by neglect," suggesting that the intrinsic apoptotic pathway regulates this mechanism of thymocyte selection.[196,197]

DP thymocytes, which express a TCR with a high affinity for MHC self-peptide complexes, are potentially dangerous and can cause autoimmune responses. Stimulation of the TCR induces upregulation of Bcl-2, Mcl-1, and BIM. If cells receive CD28 co-stimulation in combination with a strong TCR signal, BIM expression leads to apoptosis induction and a first wave of deletion of potentially autoreactive thymocytes.[198-200]

Cells which receive a weak to intermediate TCR signal transition to the single-positive stage (CD4 or CD8 SP). It is thought that these cells migrate from the thymic cortex to the medulla, where they might undergo another round of negative deletion, similar to the one at the DP stage. Again, cells receiving a weak TCR stimulus by self-peptide MHC complexes are positively selected and become naïve, peripheral T cells. Apoptosis upon strong stimulation of the TCR by self-MHC complexes is mediated by BIM and PUMA at the SP stage, as mice deficient for both BIM and PUMA have more mature $CD24^{low}$ CD4⁺ SP thymocytes than mice only deficient for BIM.[201]

SP CD4⁺ CCR7⁺ T cells with a high affinity for self-antigens can survive intermediate to strong TCR stimulation in the presence of strong NF-κB activity and exogenous IL-2 (presumably from thymic DCs), which promotes their differentiation into regulatory T cells (T_{regs}).[201-207] The exact mechanisms whereby cells commit to becoming T_{regs} or fall victim to deletion are currently not fully understood.

These two waves of deletion are commonly grouped under the term negative selection. The process of negative selection occurs at both the DP and SP stages. Mice lacking BIM and PUMA display some autoimmune syndromes, but it is not clear that this is due to the maturation of autoreactive T cells[201] (Fig. 7.8).

DEATH RECEPTOR SIGNALING CONTROLS T CELL CONTRACTION

After mounting an immune response against a foreign antigen, expansion of both CD4 and CD8 T cells, and successful elimination of the immunological insult, most T cells die in a controlled and regulated fashion and only some memory T cells survive. This contraction of the T cell population is mediated by both the death receptor and mitochondrial pathways of apoptosis.

Early studies showed that T cell hybridomas stimulated through their TCR die by apoptosis, a phenomenon later named activation-induced cell death (AICD).[208,209] This is a feature of activated T cells, that is, naïve T cells do not die by AICD. Restimulation of activated T cells *in vitro* in the absence of CD28 co-stimulation reproduces these early findings.[210] Stimulation through the TCR leads to upregulation of CD95L in these cells, and CD95-mediated apoptosis has been identified as the primary mechanism of AICD. Membrane-bound CD95L can kill cells via suicide (ie, the same cell) or fratricide (ie, kill neighboring cells through cell–cell contact).[211] In CD8⁺ T cells activated in the absence of CD4⁺ T cells, so-called helpless CD8⁺ cells, restimulation similarly induces apoptosis. In this case, TRAIL receptor signaling is an important mediator of AICD.[212] It should be noted that expression of CD95L on T cells can mediate AICD, but *in vivo* AICD can be dependent on expression of CD95L in the nonhemopoietic compartment.[213,214]

Besides AICD, activated cell autonomous cell death (ACAD) can also play a role in peripheral deletion. This death is mediated by the intrinsic pathway of apoptosis and is induced by cytokine deprivation.[199,215,216] As we have discussed above, growth factor deprivation (in this case cytokine deprivation) can induce BIM transcription, stabilization, and Bax/Bak-dependent apoptosis. Animals lacking BIM do not display contraction after T cell expansion[216] and this effect is more pronounced in animals lacking both BIM and PUMA.

Interestingly, some mature T cell lineages are protected from AICD. T_{regs} have survived an intermediate to strong agonistic selection in the thymus, which in combination with IL-2 signaling leads to upregulation of the central lineage determining transcription factor Foxp3. Foxp3 prevents upregulation of CD95L upon (re)stimulation of the TCR in T_{regs}, preventing AICD in these cells, despite expressing sufficient levels of CD95 on the surface, making them resistant to AICD.[217]

Defects in CD95 or CD95L induce a disease called autoimmune lymphoproliferative syndrome (ALPS), characterized by massive expansion of an unusual population of CD3⁺B220⁺ T cells and autoantibody production. Mice with defects in CD95 or CD95 have a similar phenotype.[218-220] As noted above, ablation of caspase-8 or FADD is embryonically lethal in mice, and this lethality is prevented by co-deletion of either RIPK3 or MLKL. Such animals not only mature normally but also manifest the ALPS-like syndrome.[73-75,87,92] This suggests that CD95-induced signaling, dependent on caspase-8 and FADD, and therefore probably apoptosis, controls expansion of CD3⁺B220⁺ T cells and autoimmunity.

IMMUNE MURDER BY CYTOTOXIC LYMPHOCYTES

As described in detail in Chapter 31, the main function of $CD8^+$ T cells is to kill infected cells as well as tumor cells. Here, we discuss the molecular mechanisms whereby cell death can be induced by CTLs. It should be noted that the cell death effector mechanisms described are similar in $CD8^+$ T cells and natural killer cells, while the inhibitory and stimulatory signals for either cell type are different.

Upon recognition of target cells by activated CTLs, the latter secrete cytotoxic granules at the immunological synapse which leads to target cell apoptosis via two distinct mechanisms: perforin and granzyme-induced apoptosis and death receptor ligation-induced apoptosis.

These secretory granules are coated with trimeric membrane-inserted CD95L.[221,222] Death receptor signaling through CD95 on target cells represents one mechanism whereby CTLs induce death of infected cells and tumor cells. Cell death through CD95 ligation follows the mechanism we have outlined above.

These granules are also loaded with perforin and granzymes, which collaborate to induce apoptosis in target cells. Perforin is a protein, similar to the 9th component of the complement system,[223] which inserts into the plasma membrane of target cells.[224,225] Pore formation licenses delivery of granzymes, a family of lysosomal proteases, into the cytosol of target cells. Humans express 5 granzymes (mice express 10 granzymes), but granzyme B (and partly A) are primarily used by CTLs.[224,226,227] Perforin-mediated granzyme B delivery into target cells can activate the mitochondrial pathway of apoptosis through cleavage of BID into cBID. Granzyme B also cleaves and activates the apoptotic executioner caspases.[35-37] Interestingly, it is not cytochrome C release which is required for apoptosis to progress but the release of SMAC to relieve XIAP inhibition from caspase-3.

Intriguingly, CTLs may also be capable of inducing ferroptosis in T cells.[228] In this case this appears to be via the action of interferon-γ to inhibit the expression of system X_c^- in some tumors. Depletion of cystine and cysteine in the tumor microenvironment increases the efficacy of checkpoint blockade therapy mouse models, and in human tumors the expression of system X_c^- negatively correlates with $CD8^+$ T cell infiltration and patient outcome.

THE EXPOSURE OF DANGER-ASSOCIATED MOLECULAR PATTERNS DURING CELL DEATH MODULATES THE IMMUNE RESPONSE

PRRs, as discussed in Chapter 12, have evolved to allow our innate immune system to distinguish "self" from "non-self" that is to recognize conserved molecules present in viruses, bacteria, and parasites called pathogen-associated molecular patterns (PAMPs). The PAMP theory, however, does not explain the strong immune response observed in sterile conditions such as ischemic injuries, trauma, tumors, tissue transplants, and autoimmune diseases. Analogous to PAMPs, the DAMP theory was proposed, stating that injured cells release intracellular molecules upon death, which elicit an immune response.[229] This remained a theory until uric acid crystals and high mobility group box 1 (HMGB1) were identified as the first DAMPs.[230,231] DAMPs can be released by cells undergoing lytic cell death (eg, necroptosis or pyroptosis) but can also be secreted through active mechanisms prior to cell death. DAMPs usually have "day jobs" inside the cell but upon release signal to alert surrounding cells of the immune system. Here we discuss the release mechanisms and functions of two well defined DAMPs, HMGB1 and ATP.

HMGB1

HMGB1 is expressed by almost all cells and functions in the nucleus, where it facilitates nucleosome formation, acts as a DNA chaperone, and modulates binding of transcription factors.[232] The release of HMGB1 is not fully understood, but thought to involve two steps: acetylation, which leads to translocation of HMGB1 from the nucleus to the cytoplasm, and subsequent release or secretion. Actively secreted HMGB1 is hyperacetylated, whereas HMGB1 passively released by dead cells is not.[233,234] HMGB1 can be released during necroptosis and pyroptosis upon cell lysis and through an unknown mechanism during immunogenic cell death (ICD, discussed below). HMGB1 signals primarily through the receptor for advanced glycation end products (RAGE), which leads to NF-κB, JNK, and p38 activation as well as MAP kinase signaling.[235,236] Signaling through RAGE promotes to cell migration, expression of adhesion molecules, and expression of CXCL12, which is able to form a heterocomplex with HMGB1.[237,238] HMGB1 can also complex PAMPs and signal through several TLRs.[239,240] Signaling of TLR4 by stimulation with HMGB1 has been demonstrated in several experimental models.[239-244]

HMGB1 contains three cysteine residues (C23, C45, and C106). C23 and C45 form a disulfide bond, while C106 is unpaired.[245] The disulfide form of HMGB1 acts as a pro-inflammatory cytokine, signaling through TLR4,[246] whereas the "all-thiol" form forms a heterocomplex with CXCL12 and functions as a chemoattractant.[237] C106 is required for binding of disulfide HMGB1 to MD2 (myeloid differentiation factor 2) and subsequently TLR4.[247] Priming of macrophages by cell surface TLRs, but not endosomal TLRs, leads to oxidation of HMGB1 into the disulfide form, and subsequent pyroptosis releases this pro-inflammatory form of HMGB1, whereas pyroptosis of unprimed cells releases the reduced form of HMGB1.[248] We can conclude that it is more important to identify "how" HMGB1 is released (ie, its redox state prior to release) than only to measure its release *in vitro* or *in vivo*.[249]

HMGB1 can not only function as a pro-inflammatory signal but also can promote tissue regeneration and repair. Release of HMGB1 promotes the migration of mesenchymal stem cells and monocytes.[250-253] HMGB1 can also promote vascularization through upregulation of integrins and sprouting of endothelial cells in a RAGE-dependent manner.[254,255] It has been demonstrated that apical application of reduced HMGB1 to wounds in diabetic mice promotes wound healing.[256] Additional beneficial effects of HMGB1 include the activation of cardiac stem cells in models of heart failure[257-259] and stimulatory effects on muscle regeneration.[260-264] It should be noted that for most of the studies on the regenerative effects of HMGB1, we do not have information on the redox state of the protein, which can strongly modulate its activity. Nevertheless, it seems remarkable that one molecule can elicit two different responses, both necessary to resolve a pathogenic or sterile insult.

Adenosine Triphosphate

ATP serves as a universal supply of energy for many cellular processes, but once released to the extracellular environment, it alerts neighboring cells. Different, context-dependent mechanisms of ATP release have been described, including secretion via the classical ER–Golgi route by dying cancer cells,[265] lysosome exocytosis by astrocytes, pannexin- or connexin-dependent release by neutrophils, and passive release through lytic cell death (reviewed in Ref. 266). ATP in the extracellular space is intrinsically short-lived. ATP is degraded by the ecto-nucleoside triphosphate diphosphohydrolase CD39 (from ATP/ADP to AMP) and the ecto-5′-nucleosidase CD73 (from AMP to adenosine).[267] ATP signals through the P2 purinergic receptors, more specifically through the metabotropic P2YRs, which are G protein–coupled receptors, and ionotropic P2XR2, which is a nucleotide gated ion channel.[268]

Signaling through P2Y2R can act on macrophages as a "find-me" signal (discussed above) to stimulate phagocytic clearance of dead cells by macrophages.[269] Neutrophil-dependent ATP release can also act as a chemoattractant in a paracrine manner.[270] However, ATP signaling through P2Y2R has also been implicated in the induction of chronic inflammation (reviewed in Ref. 267) as it induces pro-allergic mediators such as IL-33 and IL-8 during allergic airway disease.[271,272]

Signaling through P2XRs, predominantly P2X7R which is expressed on immune cells, has pro-inflammatory properties. Activation of P2X7R increases cytosolic Ca^{2+} concentrations to activate p38/MAPK signaling.[271,273] Concomitant potassium efflux leads to activation of the NLRP3 inflammasome, one well established pro-inflammatory function of extracellular ATP.[274,275] P2X7R stimulation also has other effects, which include stimulation of killing of *Mycobacterium tuberculosis* by macrophages[276] and the stimulation of DCs, to prime interferon-γ–positive CD8 T cells.[277] ATP signaling via P2X7R on DCs also has negative effects such as an involvement in the sensitization phase of allergic contact hypersensitivity[278] and it contributes to transplant rejection by eliciting a Th1 response.[279]

While signaling through P2X7R is considered to be pro-inflammatory, nucleotide signaling through P2Y2R promotes resolution of inflammation and wound healing (except for some settings such as asthma, mentioned above). Release of ATP and UTP from apoptotic cells acts as "find-me" signals, which recruit macrophages through P2Y2R signal, which promotes resolution of inflammation and wound healing.

P2YR signaling in general has mitogenic and proliferation stimulating effects on multiple cell types, including brain capillary endothelial cells,[280] cardiac endothelial cells,[281] fibroblasts,[282] and HUVEC cells.[283] P2Y2R-deficient mice have strongly impaired wound healing[282] as extracellular nucleotides have been suggested to facilitate the migration of endothelial cells after wounding.[284,285] Although ATP can be considered a bone fide DAMP, with well-described immune stimulatory and pro-inflammatory functions, the same molecule is repurposed to promote recovery and wound healing.

IMMUNOGENIC CELL DEATH

Apoptosis is usually considered an immunologically silent cell death in the absence of gasdermin E–mediated secondary necrosis.[122] There are however apoptotic stimuli which induce a form of PCD termed immunogenic cell death (ICD).[286] Tumor cells killed via this cell death pathway are able to elicit an immune response in syngeneic mice. Mice vaccinated with these dead cells are protected from subsequent growth of tumors while cells which were killed via other apoptotic stimuli do not elicit an immune response and protection of mice upon rechallenge with the same tumor cells.[287] Inducers of ICD have been shown to include some specific chemotherapeutics (mitoxantrone,[288] anthracyclines,[289] oxaliplatin,[290,291] cyclophosphamide[292]), the proteasome inhibitor bortezomib,[293] gamma- and UV irradiation,[287] and others (reviewed in Ref. 294). Interestingly, cisplatin, which is very similar to oxaliplatin, does not induce ICD by itself, but only in combination with tunicamycin or thapsigargin.[295] These experiments show that one critical step which seems to distinguish ICD from "nonimmunogenic apoptosis" is the induction of ER stress, which has also been demonstrated for other inducers of ICD.[290] However, we do not fully understand the molecular pathways which decide if a cell undergoes ICD or not.

ICD is elicited by the intrinsic pathway of apoptosis[290,296] but is characterized by additional exposure of ER proteins on the cell surface, such as calreticulin and ERp57, prior to apoptosis.[288] Moreover, HMGB1 and ATP can be released during this form of cell death.[289,297] As aforementioned, both molecules are bona fide DAMPs, capable of stimulating innate immune cells to elicit an antitumor response. It has been shown that the exposure of calreticulin on the cell surface and the release of HMGB1 prior to apoptosis are partially required for successful vaccination of mice with doxorubicin-treated cancer cells.[296]

The anti-inflammatory functions of apoptosis usually dominate the impact of this form of cell death on the immune system. Therefore, ICD probably involves additional processes, which to date is not fully understood.

NECROPTOSIS AND PYROPTOSIS STIMULATE ANTITUMOR IMMUNITY

Lytic forms of cell death, necroptosis and pyroptosis, have also been shown to potentiate the immune response to tumors and synergize with immune checkpoint blockade. Necroptosis was shown to stimulate cross presentation of antigens associated with dying cells by Batf3-expressing DCs.[298] This property of necroptotic cells could be exploited to stimulate antitumor immunity by co-injection of necroptotic fibroblasts with murine B16-F10 melanoma cells. The effect was potentiated if the necroptotic fibroblasts expressed a nonsecretable form of ovalbumin (OVA) and B16-F10 OVA-expressing cells.[299] Induction of RIPK3-dependent necroptosis (in the studies above elicited through chemical dimerization of RIPK3) also stimulated RIPK1-dependent NF-κB activation, leading to expression of pro-inflammatory cytokines.[99] These publications showed that NF-κB activation is required[298] and is even sufficient[299] for priming of DCs, cross-presentation, and induction of tumor-specific CD8$^+$ T cells. Two other publications failed to either reproduce these findings or did not find a correlation between induction of antitumor immunity by vaccination with necroptotic cells and NF-κB activation in these cells.[300,301] To date, we do not know if NF-κB activation is necessarily required for necroptosis-induced cross-presentation in DCs.

Pyroptosis has been shown to also stimulate an antitumor immune response. Tumor cells, which expressed gasdermin E and therefore underwent lytic cell death upon CTL attack, showed defective growth in mice compared to gasdermin E–deficient tumors, because of CTL infiltration and killing induced release of DAMPs, which presumably stimulates infiltration of more lymphocytes.[122] Similar results were obtained when the N-terminal fragment of gasdermin A3 (which induces cell lysis) was delivered into cells conjugated to nanoparticles.[123] Stimulation of antitumor immunity through either necroptosis or pyroptosis synergizes with anti–PD-1 blockade, highlighting the therapeutic potential of these finding.[123,299]

While ferroptosis appears to be inflammatory,[302] to date it is not clear if ferroptosis can also promote cross-priming and CD8$^+$ T cell immunity.

REGULATION OF CELL DEATH BY MEMBRANE REPAIR MECHANISMS

Mammalian cells have the capability to seal small (<10-20 nm) to medium (<100 nm) sized holes in their plasma membrane through various mechanisms (reviewed in Ref. 303). Two pathways are understood at the mechanistic level: acid sphingomyelinase (ASM)–dependent endocytosis and ESCRT (endosomal sorting complex required for protein transport)–mediated membrane repair.

Pores formed by bacterial toxins have been shown to trigger the exocytosis of secretory vesicles,[304-307] which is stimulated by Ca^{2+} microdomains formed at the site of pore formation.[308] These secretory lysosomes contain ASM, which is secreted upon fusion with the plasma membrane. ASM hydrolyses sphingomyelin to ceramide on the outer leaflet of the plasma membrane, which promotes local formation of lipid rafts.[309,310] These lipid rafts favor membrane invagination and subsequent endocytosis.[311-314] This mechanism allows for enhanced endocytosis around the site of plasma membrane damage, therefore removing stable transmembrane pores such as those induced by streptolysin O. Once internalized, these pores travel along the endo-lysosomal pathway to be degraded in multivesicular bodies.[315] As mentioned, cells primarily use this mechanism to defend themselves against pore-forming toxins from bacteria, but it has also been observed that exocytosis of LAMP1-positive vesicles occurs upon CTL attack and addition of purified perforin. The calcium flux elicited by perforin and subsequent endocytosis was required for internalization of granzyme B and execution of apoptosis.[316]

The ESCRT complex was discovered in yeast, where it regulates the trafficking of proteins along the secretory pathway, and the formation of intraluminal vesicles in multivesicular bodies. The ESCRT machinery is a multiprotein complex composed of several subcomplexes (ESCRT 0, I, II, III) and the only machinery known in mammalian cells to deform membranes away from the cytosol, leading to vesicle generation at the distal side of the membrane, from where the ESCRT complex is active.[317] Other functions of the ESCRT machinery, including cytokinesis,[318,319] and a function for restoration of nuclear envelope restoration have been demonstrated.[320,321] Some enveloped viruses hijack the ESCRT machinery to promote budding of mature virions.[322] A more recently discovered function of the ESCRT machinery is the repair of small wounds of the plasma membrane.[323,324] It was first described that ESCRT components were enriched in membrane fractions of cells treated with the ionophore ionomycin.[324] ESCRT components are also recruited to the plasma membrane upon treatment with mild detergents, bacterial pore-forming toxins, and wounding with a laser. Components of the ESCRT 0, I, and II subcomplexes are dispensable (except ALIX and ALG-2, which are probably required to sense Ca^{2+} influx) for membrane repair, and ESCRT-III primarily mediates plasma membrane repair.[323,324] It was subsequently shown that both plasma membrane damage caused by MLKL, during necroptosis, and GSDMD during pyroptosis can be repaired by the ESCRT system. Necroptotic cells, which already had exposed PS on the outer leaflet of the plasma membrane, could be resuscitated in the presence of ESCRT-III–mediated membrane repair, but failed to survive if the ESCRT machinery was inactivated.[154] Plasma membrane repair during necroptosis allowed for proper NF-κB activation, which was essential for cross-priming of CD8$^+$ T cells. GSDMD pores can also be repaired by the ESCRT system dependent on calcium influx in response to caspase-11 or caspase-1 activation. There is little evidence that cells, once they engaged caspase-1 or caspase-11, could be resuscitated; however, titration experiments with low levels of cytosolic LPS delivery showed that inhibiting calcium signaling (required for ESCRT-mediated repair) converted otherwise sublethal doses of LPS transfection into lethal doses in primary murine bone marrow–derived macrophages. During pyroptosis, ESCRT-III–mediated membrane repair negatively regulates secretion of pro-inflammatory cytokines.[325]

To date we do not entirely understand which signals, besides calcium influx, regulate initiation of membrane repair mechanisms and whether ASM-dependent membrane repair and ESCRT-mediated shedding of damaged plasma membrane are mutually exclusive pathways or whether they can be engaged at the same time in the same cell.

CONCLUSIONS

Cell death is a fundamental biological process, and therefore it should come as no surprise that cells of the immune system undergo multiple forms of cell death. Clearly, however, the coevolution of cell death mechanisms and immunity has created many ways in which different types of cell death influence the immune response. While apoptosis is not inflammatory, necrosis (whether regulated or not) is generally pro-inflammatory, and this has important implications for normal homeostasis and disease pathology. In mammals (and probably other vertebrates), the mode of cell death can also influence the extent to which antigens associated with or, in the vicinity of the dying cell, stimulate adaptive immune responses, with important consequences for anticancer immunity and autoimmunity. Similarly, the clearance of dead and dying cells influences immune (and autoimmune) responses and plays important roles in tissue repair.

As in any vibrant field, there are many areas of agreement as well as controversy. Many DAMPs have been identified, but the roles and extent to which specific DAMPs influence the immune response remain unclear. A compelling argument has been made that the most important molecules released from dying cells are members of the interleukin-1 family.[326] Similarly, studies have suggested that metabolites, rather than only proteins, released from dying cells have important roles in immune regulation and wound repair.[327] Many questions remain in this field,[328] and the coming decade is certain to reveal aspects of the biology of cell death and the ways dying cells influence the immune system and its effects.

REFERENCES

1. Kerr JFR. History of the events leading to the formulation of the apoptosis concept. *Toxicology*. 2002;181-182:471-474. doi:10.1016/S0300-483X(02)00457-2
2. Green DR, Oguin TH, Martinez J. The clearance of dying cells: table for two. *Cell Death Differ*. 23;2016:915-926. doi:10.1038/cdd.2015.172
3. Alnemri ES. Mammalian cell death proteases: a family of highly conserved aspartate specific cysteine proteases. *J Cell Biochem*. 64;1997:33-42. doi:10.1002/(sici)1097-4644(199701)64:1<33::aid-jcb6>3.0.co;2-0
4. Ramirez MLG, Salvesen GS. A primer on caspase mechanisms. *Semin Cell Dev Biol*. 82;2018:79-85. doi:10.1016/j.semcdb.2018.01.002
5. Sakahira H, Enari M, Nagata S. Cleavage of CAD inhibitor in CAD activation and DNA degradation during apoptosis. *Nature*. 2003;426:6967. 1998;391:96-99. doi:10.1038/34214
6. Enari M, Sakahira H, Yokoyama H, Okawa K, Iwamatsu A, Nagata S. A caspase-activated DNase that degrades DNA during apoptosis, and its inhibitor ICAD. *Nature*. 2003;426:6967. 1998;391:43-50. doi:10.1038/34112
7. Segawa K, Kurata S, Yanagihashi Y, Brummelkamp TR, Matsuda F, Nagata S. Caspase-mediated cleavage of phospholipid flippase for apoptotic phosphatidylserine exposure. *Science*. 2014;344:1164-1168. doi:10.1126/science.1252809
8. Suzuki J, Denning DP, Imanishi E, Horvitz HR, Nagata S. Xk-related protein 8 and CED-8 promote phosphatidylserine exposure in apoptotic cells. *Science*. 2013;341:403-406. doi:10.1126/science.1236758
9. Stennicke HR, Jürgensmeier JM, Shin H, et al. Pro-caspase-3 is a major physiologic target of caspase-8. *J Biol Chem*. 1998;273:27084-27090. doi:10.1074/jbc.273.42.27084
10. Boatright KM, Renatus M, Scott FL, et al. A unified model for apical caspase activation. *Mol Cell*. 2003;11:529-541. doi:10.1016/S1097-2765(03)00051-0
11. Malladi S, Malladi MC, Fearnhead HO, Bratton SB. The Apaf-1*procaspase-9 apoptosome complex functions as a proteolytic-based molecular timer. *EMBO J*. 2009;28:1916-1925. doi:10.1038/emboj.2009.152
12. Pop C, Timmer J, Sperandio S, Salvesen GS. The apoptosome activates caspase-9 by dimerization. *Mol Cell*. 2006;22:269-275. doi:10.1016/j.molcel.2006.03.009
13. Clem RJ, Fechheimer M, Miller LK. Prevention of apoptosis by a baculovirus gene during infection of insect cells. *Science*. 1991;254:1388-1390. doi:10.1126/science.1962198
14. Hawkins CJ, Ekert PG, Uren AG, Holmgreen SP, Vaux DL. Anti-apoptotic potential of insect cellular and viral IAPs in mammalian cells. *Cell Death Differ*. 1998;5:569-576. doi:10.1038/sj.cdd.4400389
15. Deveraux QL, Takahashi R, Salvesen GS, Reed JC. X-linked IAP is a direct inhibitor of cell-death proteases. *Nature*. 2003;426:6967. 1997;388:300-304. doi:10.1038/40901
16. Duckett CS, Nava VE, Gedrich RW, et al. A conserved family of cellular genes related to the baculovirus iap gene and encoding apoptosis inhibitors. *EMBO J*. 1996;15:2685-2694. doi:10.1002/j.1460-2075.1996.tb00629.x
17. Vaux DL, Silke J. IAPs—the ubiquitin connection. *Cell Death Differ*. 2005;12:1205-1207. doi:10.1038/sj.cdd.4401696
18. Zou H, Li Y, Liu X, Wang X. An APAF-1.cytochrome c multimeric complex is a functional apoptosome that activates procaspase-9. *J Biol Chem*. 1999;274:11549-11556. doi:10.1074/jbc.274.17.11549
19. Li P, Nijhawan D, Budihardjo I, et al. Cytochrome c and dATP-dependent formation of Apaf-1/caspase-9 complex initiates an apoptotic protease cascade. *Cell*. 1997;91:479-489. doi:10.1016/S0092-8674(00)80434-1
20. Qin H, Srinivasula SM, Wu G, Fernandes-Alnemri T, Alnemri ES, Shi Y. Structural basis of procaspase-9 recruitment by the apoptotic protease-activating factor 1. *Nature*. 2003;426:6967. 1999;399:549-557. doi:10.1038/21124
21. Du C, Fang M, Li Y, Li L, Wang X. Smac, a mitochondrial protein that promotes cytochrome c–dependent caspase activation by eliminating IAP inhibition. *Cell*. 2000;102:33-42. doi:10.1016/S0092-8674(00)00008-8
22. Suzuki Y, Imai Y, Nakayama H, Takahashi K, Takio K, Takahashi R. A serine protease, HtrA2, is released from the mitochondria and interacts with XIAP, inducing cell death. *Mol Cell*. 2001;8:613-621. doi:10.1016/S1097-2765(01)00341-0
23. Kalkavan H, Green DR. MOMP, cell suicide as a BCL-2 family business. *Cell Death Differ*. 2018;25:46-55. doi:10.1038/cdd.2017.179
24. Tait SWG, Green DR. Mitochondrial regulation of cell death. *Cold Spring Harb Perspect Biol*. 2013;5:a008706. doi:10.1101/cshperspect.a008706
25. Kale J, Osterlund EJ, Andrews DW. BCL-2 family proteins: changing partners in the dance towards death. *Cell Death Differ*. 2018;25:65-80. doi:10.1038/cdd.2017.186
26. Chen H-C, Kanai M, Inoue-Yamauchi A, et al. An interconnected hierarchical model of cell death regulation by the BCL-2 family. *Nat Cell Biol*. 2012;14:6. 2015;17:1270-1281. doi:10.1038/ncb3236
27. Moldoveanu T, Grace CR, Llambi F, et al. BID-induced structural changes in BAK promote apoptosis. *Nat Struct Mol Biol*. 2009;16:11. 2013;20:589-597. doi:10.1038/nsmb.2563
28. Huang K, O'Neill KL, Li J, et al. BH3-only proteins target BCL-xL/MCL-1, not BAX/BAK, to initiate apoptosis. *Cell Res*. 2019;29:942-952. doi:10.1038/s41422-019-0231-y
29. Llambi F, Moldoveanu T, Tait SWG, et al. A unified model of mammalian BCL-2 protein family interactions at the mitochondria. *Mol Cell*. 2011;44:517-531. doi:10.1016/j.molcel.2011.10.001.
30. Villunger A, Michalak EM, Coultas L, et al. p53-and drug-induced apoptotic responses mediated by BH3-only proteins Puma and Noxa. *Science*. 2003;302:1036-1038. doi:10.1126/science.1090072
31. Flinterman M, Guelen L, Ezzati-Nik S, et al. E1A activates transcription of p73 and Noxa to induce apoptosis. *J Biol Chem*. 2005;280:5945-5959. doi:10.1074/jbc.M406661200
32. Wang Q, Mora-Jensen H, Weniger MA, et al. ERAD inhibitors integrate ER stress with an epigenetic mechanism to activate BH3-only protein NOXA in cancer cells. *Proc Natl Acad Sci U S A*. 2009;106:2200-2205. doi:10.1073/pnas.0807611106
33. Galehdar Z, Swan P, Fuerth B, Callaghan SM, Park DS, Cregan SP. Neuronal apoptosis induced by endoplasmic reticulum stress is regulated by ATF4–CHOP-mediated induction of the bcl-2 homology 3-only member PUMA. *J Neurosci*. 2010;30:16938-16948. doi:10.1523/JNEUROSCI.1598-10.2010
34. Chen M, He H, Zhan S, Krajewski S, Reed JC, Gottlieb RA. Bid is cleaved by calpain to an active fragment in vitro and during myocardial ischemia/reperfusion. *J Biol Chem*. 2001;276:30724-30728. doi:10.1074/jbc.M103701200
35. Heibein JA, Goping IS, Barry M, et al. Granzyme B-mediated cytochrome C release is regulated by the bcl-2 family members Bid and bax. *J Exp Med*. 2000;192:1391-1402. doi:10.1084/jem.192.10.1391
36. Pinkoski MJ, Waterhouse NJ, Heibein JA, et al. Granzyme B-mediated apoptosis proceeds predominantly through a Bcl-2-inhibitable mitochondrial pathway. *J Biol Chem*. 2001;276:12060-12067. doi:10.1074/jbc.M009038200
37. Heibein JA, Barry M, Motyka B, Bleackley RC. Granzyme B-induced loss of mitochondrial inner membrane potential (Delta Psi m) and cytochrome c release are caspase independent. *J Immunol*. 1999;163:4683-4693.
38. Barry M, Heibein JA, Pinkoski MJ, et al. Granzyme B short-circuits the need for caspase 8 activity during granule-mediated cytotoxic T-lymphocyte killing by directly cleaving Bid. *Mol Cell Biol*. 2000;20:3781-3794. doi:10.1128/MCB.20.11.3781-3794.2000
39. Li H, Zhu H, Xu C-J, Yuan J. Cleavage of BID by caspase 8 mediates the mitochondrial damage in the Fas pathway of apoptosis. *Cell*. 1998;94:491-501. doi:10.1016/S0092-8674(00)81590-1
40. Eskes R, Desagher S, Antonsson B, Martinou JC. Bid induces the oligomerization and insertion of Bax into the outer mitochondrial membrane. *Mol Cell Biol*. 2000;20:929-935. doi:10.1128/mcb.20.3.929-935.2000
41. Biswas SC, Shi Y, Sproul A, Greene LA. Pro-apoptotic Bim induction in response to nerve growth factor deprivation requires simultaneous activation of three different death signaling pathways. *J Biol Chem*. 2007;282:29368-29374. doi:10.1074/jbc.M702634200
42. Luciano F, Jacquel A, Colosetti P, et al. Phosphorylation of Bim-EL by Erk1/2 on serine 69 promotes its degradation via the proteasome pathway and regulates its proapoptotic function. *Oncogene*. 2006;26:16. 2003;22:6785-6793. doi:10.1038/sj.onc.1206792
43. Lei K, Davis RJ. JNK phosphorylation of Bim-related members of the Bcl2 family induces Bax-dependent apoptosis. *Proc Natl Acad Sci U S A*. 2003;100:2432-2437. doi:10.1073/pnas.0438011100
44. Yang E, Zha J, Jockel J, Boise LH, Thompson CB, Korsmeyer SJ. Bad, a heterodimeric partner for Bcl-xL and Bcl-2, displaces bax and promotes cell death. *Cell*. 1995;80:285-291. doi:10.1016/0092-8674(95)90411-5
45. Datta SR, Katsov A, Hu L, et al. 14-3-3 proteins and survival kinases cooperate to inactivate BAD by BH3 domain phosphorylation. *Mol Cell*. 2000;6:41-51. doi:10.1016/S1097-2765(05)00012-2
46. Hockings C, Anwari K, Ninnis RL, et al. Bid chimeras indicate that most BH3-only proteins can directly activate Bak and Bax, and show no preference for Bak versus Bax. *Cell Death Dis*. 2018;9:2. 2015;6:e1735-e1735. doi:10.1038/cddis.2015.105
47. Inuzuka H, Shaik S, Onoyama I, et al. SCF FBW7 regulates cellular apoptosis by targeting MCL1 for ubiquitylation and destruction. *Nature*. 2003;426:6967. 2011;471:104-109. doi:10.1038/nature09732
48. Shamas-Din A, Kale J, Leber B, Andrews DW. Mechanisms of action of bcl-2 family proteins. *Cold Spring Harb Perspect Biol*. 2013;5:a008714. doi:10.1101/cshperspect.a008714
49. Shamas-Din A, Brahmbhatt H, Leber B, Andrews DW. BH3-only proteins: orchestrators of apoptosis. *Biochim Biophys Acta*. 2011;1813:508-520. doi:10.1016/j.bbamcr.2010.11.024
50. Lomonosova E, Chinnadurai G. BH3-only proteins in apoptosis and beyond: an overview. *Oncogene*. 2006;26:16. 2008;27:S2-S19. doi:10.1038/onc.2009.39
51. Chipuk JE, Moldoveanu T, Llambi F, Parsons MJ, Green DR. The BCL-2 family reunion. *Mol Cell*. 2010;37:299-310. doi:10.1016/j.molcel.2010.01.025

52. Nikolaev A, McLaughlin T, O'Leary DDM, Tessier-Lavigne M. APP binds DR6 to trigger axon pruning and neuron death via distinct caspases. *Nature*. 2003;426:6967. 2009;457:981-989. doi:10.1038/nature07767
53. Fu T-M, Li Y, Lu A, et al. Cryo-EM structure of caspase-8 tandem DED filament reveals assembly and regulation mechanisms of the death-inducing signaling complex. *Mol Cell*. 2016;64:236-250. doi:10.1016/j.molcel.2016.09.009
54. Hughes MA, Powley IR, Jukes-Jones R, et al. Co-operative and hierarchical binding of c-FLIP and caspase-8: a unified model defines how c-FLIP isoforms differentially control cell fate. *Mol Cell*. 2016;61:834-849. doi:10.1016/j.molcel.2016.02.023
55. Scaffidi C, Schmitz I, Krammer PH, Peter ME. The role of c-FLIP in modulation of CD95-induced apoptosis, *J Biol Chem*. 1999;274:1541-1548. doi:10.1074/jbc.274.3.1541
56. Schleich K, Buchbinder JH, Pietkiewicz S, et al. Molecular architecture of the DED chains at the DISC: regulation of procaspase-8 activation by short DED proteins c-FLIP and procaspase-8 prodomain. *Cell Death Differ*. 2016;23:681-694. doi:10.1038/cdd.2015.137
57. Lee H, Oh Y, Jeon Y-J, et al. DR4-Ser424 O-GlcNAcylation promotes sensitization of TRAIL-tolerant persisters and TRAIL-resistant cancer cells to death. *Cancer Res*. 2019;79:2839-2852. doi:10.1158/0008-5472.CAN-18-1991
58. Kavuri SM, Geserick P, Berg D, et al. Cellular FLICE-inhibitory protein (cFLIP) isoforms block CD95-and TRAIL death receptor-induced gene induction irrespective of processing of caspase-8 or cFLIP in the death-inducing signaling complex. *J Biol Chem*. 2011;286:16631-16646. doi:10.1074/jbc.M110.148585
59. Lu M, Lawrence DA, Marsters S, et al. Opposing unfolded-protein-response signals converge on death receptor 5 to control apoptosis. *Science*. 2014;345:98-101. doi:10.1126/science.1254312
60. Hsu H, Xiong J, Goeddel DV. The TNF receptor 1-associated protein TRADD signals cell death and NF-kappa B activation. *Cell*. 1995;81:495-504. doi:10.1016/0092-8674(95)90070-5
61. Lee TH, Shank J, Cusson N, Kelliher MA. The kinase activity of Rip1 is not required for tumor necrosis factor-alpha-induced IkappaB kinase or p38 MAP kinase activation or for the ubiquitination of Rip1 by Traf2. *J Biol Chem*. 2004;279:33185-33191. doi:10.1074/jbc.M404206200
62. Ea C-K, Deng L, Xia Z-P, Pineda G, Chen ZJ. Activation of IKK by TNFα requires site-specific ubiquitination of RIP1 and polyubiquitin binding by NEMO. *Mol Cell*. 2006;22:245-257. doi:10.1016/j.molcel.2006.03.026
63. Moquin DM, McQuade T, Chan FK-M. CYLD deubiquitinates RIP1 in the TNFα-induced necrosome to facilitate kinase activation and programmed necrosis. *PLoS One*. 2013;8:e76841. doi:10.1371/journal.pone.0076841
64. Geserick P, Wang J, Schilling R, et al. Absence of RIPK3 predicts necroptosis resistance in malignant melanoma. *Cell Death Dis*. 2018;9:2. 2015;6:e1884-e1884. doi:10.1038/cddis.2015.240
65. Koo G-B, Morgan MJ, Lee D-G, et al. Methylation-dependent loss of RIP3 expression in cancer represses programmed necrosis in response to chemotherapeutics. *Cell Res*. 2015;25:707-725. doi:10.1038/cr.2015.56
66. Tenev T, Bianchi K, Darding M, et al. The ripoptosome, a signaling platform that assembles in response to genotoxic stress and loss of IAPs. *Mol Cell*. 2011;43:432-448. doi:10.1016/j.molcel.2011.06.006
67. Jost PJ, Grabow S, Gray D, et al. XIAP discriminates between type I and type II FAS-induced apoptosis. *Nature*. 2003;426:6967. 2009;460:1035-1039. doi:10.1038/nature08229
68. Yin X-M, Wang K, Gross A, et al. Bid-deficient mice are resistant to Fas-induced hepatocellular apoptosis. *Nature*. 2003;426:6967. 1999;400:886-891. doi:10.1038/23730
69. Yeh W-C, de la Pompa JL, McCurrach ME, et al. FADD: essential for embryo development and signaling from some, but not all, inducers of apoptosis. *Science*. 1998;279:1954-1958. doi:10.1126/science.279.5358.1954
70. Varfolomeev EE, Schuchmann M, Luria V, et al. Targeted disruption of the mouse caspase 8 gene ablates cell death induction by the TNF receptors, Fas/Apo1, and DR3 and is lethal prenatally. *Immunity*. 1998;9:267-276. doi:10.1016/S1074-7613(00)80609-3
71. Kaiser WJ, Daley-Bauer LP, Thapa RJ, et al. RIP1 suppresses innate immune necrotic as well as apoptotic cell death during mammalian parturition. *Proc Natl Acad Sci U S A*. 2014;111:7753-7758. doi:10.1073/pnas.1401857111
72. Rickard JA, O'Donnell JA, Evans JM, et al. RIPK1 regulates RIPK3-MLKL-driven systemic inflammation and emergency hematopoiesis. *Cell*. 2014;157:1175-1188. doi:10.1016/j.cell.2014.04.019
73. Dillon CP, Weinlich R, Rodriguez DA, et al. RIPK1 blocks early postnatal lethality mediated by caspase-8 and RIPK3. *Cell*. 2014;157:1189-1202. doi:10.1016/j.cell.2014.04.018
74. Kaiser WJ, Upton JW, Long AB, et al. RIP3 mediates the embryonic lethality of caspase-8-deficient mice. *Nature*. 2003;426:6967. 2011;471:368-372. doi:10.1038/nature09857
75. Oberst A, Dillon CP, Weinlich R, et al. Catalytic activity of the caspase-8–FLIP L complex inhibits RIPK3-dependent necrosis. *Nature*. 2003;426:6967. 2011;471:363-367. doi:10.1038/nature09852
76. Lin Y, Devin A, Rodriguez Y, Liu ZG. Cleavage of the death domain kinase RIP by caspase-8 prompts TNF-induced apoptosis. *Genes Dev*. 1999;13:2514-2526. doi:10.1101/gad.13.19.2514
77. Feng S, Yang Y, Mei Y, et al. Cleavage of RIP3 inactivates its caspase-independent apoptosis pathway by removal of kinase domain. *Cell Signal*. 2007;19:2056-2067. doi:10.1016/j.cellsig.2007.05.016
78. O'Donnell MA, Perez-Jimenez E, Oberst A, et al. Caspase 8 inhibits programmed necrosis by processing CYLD. *Nat Cell Biol*. 2012;14:6. 2011;13:1437-1442. doi:10.1038/ncb2362
79. Linkermann A, Green DR. Necroptosis. *N Engl J Med*. 2014;370:455-465. doi:10.1056/NEJMra1310050
80. Cho Y, Challa S, Moquin D, et al. Phosphorylation-driven assembly of the RIP1-RIP3 complex regulates programmed necrosis and virus-induced inflammation. *Cell*. 2009;137:1112-1123. doi:10.1016/j.cell.2009.05.037
81. He S, Wang L, Miao L, et al. Receptor interacting protein kinase-3 determines cellular necrotic response to TNF-α. *Cell*. 2009;137:1100-1111. doi:10.1016/j.cell.2009.05.021
82. Zhang D-W, Shao J, Lin J, et al. RIP3, an energy metabolism regulator that switches TNF-induced cell death from apoptosis to necrosis. *Science*. 2009;325:332-336. doi:10.1126/science.1172308
83. Murphy JM, Czabotar PE, Hildebrand JM, et al. The pseudokinase MLKL mediates necroptosis via a molecular switch mechanism. *Immunity*. 2013;39:443-453. doi:10.1016/j.immuni.2013.06.018
84. Sun L, Wang H, Wang Z, et al. Mixed lineage kinase domain-like protein mediates necrosis signaling downstream of RIP3 kinase. *Cell*. 2012;148:213-227. doi:10.1016/j.cell.2011.11.031
85. Chen X, Li W, Ren J, et al. Translocation of mixed lineage kinase domain-like protein to plasma membrane leads to necrotic cell death. *Cell Res*. 2014;24:105-121. doi:10.1038/cr.2013.171
86. Cai Z, Jitkaew S, Zhao J, et al. Plasma membrane translocation of trimerized MLKL protein is required for TNF-induced necroptosis. *Nat Cell Biol*. 2012;14:6. 2014;16:55-65. doi:10.1038/ncb2883
87. Alvarez-Diaz S, Dillon CP, Lalaoui N, et al. The pseudokinase MLKL and the kinase RIPK3 have distinct roles in autoimmune disease caused by loss of death-receptor-induced apoptosis. *Immunity*. 2016;45:513-526. doi:10.1016/j.immuni.2016.07.016
88. Dondelinger Y, Declercq W, Montessuit S, et al. MLKL compromises plasma membrane integrity by binding to phosphatidylinositol phosphates, *Cell Rep*. 2014;7:971-981. doi:10.1016/j.celrep.2014.04.026
89. Wang H, Sun L, Su L, et al. Mixed lineage kinase domain-like protein MLKL causes necrotic membrane disruption upon phosphorylation by RIP3. *Mol Cell*. 2014;54:133-146. doi:10.1016/j.molcel.2014.03.003
90. Kaiser WJ, Sridharan H, Huang C, et al. Toll-like receptor 3-mediated necrosis via TRIF, RIP3, and MLKL. *J Biol Chem*. 2013;288:31268-31279. doi:10.1074/jbc.M113.462341
91. He S, Liang Y, Shao F, Wang X. Toll-like receptors activate programmed necrosis in macrophages through a receptor-interacting kinase-3–mediated pathway. *Proc Natl Acad Sci U S A*. 2011;108:20054-20059. doi:10.1073/pnas.1116302108
92. Newton K, Wickliffe KE, Maltzman A, et al. RIPK1 inhibits ZBP1-driven necroptosis during development. *Nature*. 2003;426:6967. 2016;540:129-133. doi:10.1038/nature20559
93. Rebsamen M, Heinz LX, Meylan E, et al. DAI/ZBP1 recruits RIP1 and RIP3 through RIP homotypic interaction motifs to activate NF-κB. *EMBO Rep*. 2009;10:916-922. doi:10.1038/embor.2009.109
94. Upton JW, Kaiser WJ, Mocarski ES. DAI/ZBP1/DLM-1 complexes with RIP3 to mediate virus-induced programmed necrosis that is targeted by murine cytomegalovirus vIRA. *Cell Host Microbe*. 2012;11:290-297. doi:10.1016/j.chom.2012.01.016
95. Jiao H, Wachsmuth L, Kumari S, et al. Z-nucleic-acid sensing triggers ZBP1-dependent necroptosis and inflammation. *Nature*. 2003;426:6967. 2020;580:391-395. doi:10.1038/s41586-020-2129-8
96. Polykratis A, Hermance N, Zelic M, et al. Cutting edge: RIPK1 kinase inactive mice are viable and protected from TNF-induced necroptosis in vivo. *J Immunol*. 2014;193:1539-1543. doi:10.4049/jimmunol.1400590
97. Newton K, Dugger DL, Wickliffe KE, et al. Activity of protein kinase RIPK3 determines whether cells die by necroptosis or apoptosis. *Science*. 2014;343:1357-1360. doi:10.1126/science.1249361
98. Berger SB, Kasparcova V, Hoffman S, et al. Cutting edge: RIP1 kinase activity is dispensable for normal development but is a key regulator of inflammation in SHARPIN-deficient mice. *J Immunol*. 2014;192:5476-5480. doi:10.4049/jimmunol.1400499
99. Cook WD, Moujalled DM, Ralph TJ, et al. RIPK1-and RIPK3-induced cell death mode is determined by target availability. *Cell Death Differ*. 2014;21:1600-1612. doi:10.1038/cdd.2014.70
100. Newton K, Wickliffe KE, Dugger DL, et al. Cleavage of RIPK1 by caspase-8 is crucial for limiting apoptosis and necroptosis. *Nature*. 2003;426:6967. 2019;574:428-431. doi:10.1038/s41586-019-1548-x
101. Lalaoui N, Boyden SE, Oda H, et al. Mutations that prevent caspase cleavage of RIPK1 cause autoinflammatory disease. *Nature*. 2003;426:6967. 2020;577:103-108. doi:10.1038/s41586-019-1828-5
102. Brennan MA, Cookson BT. Salmonella induces macrophage death by caspase-1-dependent necrosis. *Mol Microbiol*. 2000;38:31-40. doi:10.1046/j.1365-2958.2000.02103.x
103. Broz P, Dixit VM. Inflammasomes: mechanism of assembly, regulation and signalling. *Nat Rev Immunol*. 2016;16:407-420. doi:10.1038/nri.2016.58
104. Kesavardhana S, Kanneganti T-D. Mechanisms governing inflammasome activation, assembly and pyroptosis induction. *Int Immunol*. 2017;29:201-210. doi:10.1093/intimm/dxx018
105. Boucher D, Monteleone M, Coll RC, et al. Caspase-1 self-cleavage is an intrinsic mechanism to terminate inflammasome activity. *J Exp Med*. 2018;215:827-840. doi:10.1084/jem.20172222
106. Shi J, Zhao Y, Wang Y, et al. Inflammatory caspases are innate immune receptors for intracellular LPS. *Nature*. 2003;426:6967. 2014;514:187-192. doi:10.1038/nature13683
107. Ross C, Chan AH, Von Pein J, Boucher D, Schroder K. Dimerization and auto-processing induce caspase-11 protease activation within the non-canonical inflammasome. *Life Sci Alliance*. 2018;1:e201800237. doi:10.26508/lsa.201800237
108. Wang K, Sun Q, Zhong X, et al. Structural mechanism for GSDMD targeting by autoprocessed caspases in pyroptosis. *Cell*. 2020;180:941-955.e20. doi:10.1016/j.cell.2020.02.002
109. Baker PJ, Boucher D, Bierschenk D, et al. NLRP3 inflammasome activation downstream of cytoplasmic LPS recognition by both caspase-4 and caspase-5. *Eur J Immunol*. 2015;45:2918-2926. doi:10.1002/eji.201545655
110. Kayagaki N, Stowe IB, Lee BL, et al. Caspase-11 cleaves gasdermin D for non-canonical inflammasome signalling. *Nature*. 2003;426:6967. 2015;526:666-671. doi:10.1038/nature15541
111. Shi J, Zhao Y, Wang K, et al. Cleavage of GSDMD by inflammatory caspases determines pyroptotic cell death. *Nature*. 2003;426:6967. 2015;526:660-665. doi:10.1038/nature15514
112. Kovacs SB, Miao EA. Gasdermins: effectors of pyroptosis. *Trends Cell Biol*. 2017;27:673-684. doi:10.1016/j.tcb.2017.05.005
113. Ding J, Wang K, Liu W, et al. Pore-forming activity and structural autoinhibition of the gasdermin family. *Nature*. 2003;426:6967. 2016;535:111-116. doi:10.1038/nature18590
114. Sborgi L, Rühl S, Mulvihill E, et al. GSDMD membrane pore formation constitutes the mechanism of pyroptotic cell death. *EMBO J*. 2016;35:1766-1778. doi:10.15252/embj.201694696
115. Liu X, Zhang Z, Ruan J, et al. Inflammasome-activated gasdermin D causes pyroptosis by forming membrane pores. *Nature*. 2003;426:6967. 2016;535:153-158. doi:10.1038/nature18629
116. Aglietti RA, Estevez A, Gupta A, et al. GsdmD p30 elicited by caspase-11 during pyroptosis forms pores in membranes. *Proc Natl Acad Sci U S A*. 2016;113:7858-7863. doi:10.1073/pnas.1607769113
117. Davis MA, Fairgrieve MR, Den Hartigh A, et al. Calpain drives pyroptotic vimentin cleavage, intermediate filament loss, and cell rupture that mediates immunostimulation. *Proc Natl Acad Sci U S A*. 2019;116:5061-5070. doi:10.1073/pnas.1818598116
118. Chen KW, Demarco B, Heilig R, et al. Extrinsic and intrinsic apoptosis activate pannexin-1 to drive NLRP3 inflammasome assembly. *EMBO J*. 2019;38:e101638. doi:10.15252/embj.2019101638
119. Ruan J, Xia S, Liu X, Lieberman J, Wu H. Cryo-EM structure of the gasdermin A3 membrane pore. *Nature*. 2003;426:6967. 2018;557:62-67. doi:10.1038/s41586-018-0058-6
120. Rogers C, Fernandes-Alnemri T, Mayes L, Alnemri D, Cingolani G, Alnemri ES. Cleavage of DFNA5 by caspase-3 during apoptosis mediates progression to secondary necrotic/pyroptotic cell death. *Nat Commun*. 2017;8:1-14. doi:10.1038/ncomms14128
121. Wang Y, Gao W, Shi X, et al. Chemotherapy drugs induce pyroptosis through caspase-3 cleavage of a gasdermin. *Nature*. 2003;426:6967. 2017;547:99-103. doi:10.1038/nature22393
122. Zhang Z, Zhang Y, Xia S, et al. Gasdermin E suppresses tumour growth by activating anti-tumour immunity. *Nature*. 2003;426:6967. 2020;579:415-420. doi:10.1038/s41586-020-2071-9
123. Wang Q, Wang Y, Ding J, et al. A bioorthogonal system reveals anti-tumour immune function of pyroptosis. *Nature*. 2003;426:6967. 2020;579:421-426. doi:10.1038/s41586-020-2079-1
124. Yang WS, SriRamaratnam R, Welsch ME, et al. Regulation of ferroptotic cancer cell death by GPX4. *Cell*. 2014;156:317-331. doi:10.1016/j.cell.2013.12.010
125. Ursini F, Maiorino M, Valente M, Ferri L, Gregolin C. Purification from pig liver of a protein which protects liposomes and biomembranes from peroxidative degradation and exhibits glutathione peroxidase activity on phosphatidylcholine hydroperoxides. *Biochim Biophys Acta*. 1982;710:197-211. doi:10.1016/0005-2760(82)90150-3
126. Dixon SJ, Patel DN, Welsch M, et al. Pharmacological inhibition of cystine-glutamate exchange induces endoplasmic reticulum stress and ferroptosis. *Elife*. 2014;3:175. doi:10.7554/eLife.02523
127. Gao M, Monian P, Quadri N, Ramasamy R, Jiang X. Glutaminolysis and transferrin regulate ferroptosis. *Mol Cell*. 2015;59:298-308. doi:10.1016/j.molcel.2015.06.011

128. Yuan H, Li X, Zhang X, Kang R, Tang D. Identification of ACSL4 as a biomarker and contributor of ferroptosis. *Biochem Biophys Res Commun.* 2016;478:1338-1343. doi:10.1016/j.bbrc.2016.08.124
129. Kagan VE, Mao G, Qu F, et al. Oxidized arachidonic and adrenic PEs navigate cells to ferroptosis. *Nat Chem Biol.* 2017;13:81-90. doi:10.1038/nchembio.2238
130. Doll S, Proneth B, Tyurina YY, et al. ACSL4 dictates ferroptosis sensitivity by shaping cellular lipid composition. *Nat Chem Biol.* 2017;13:91-98. doi:10.1038/nchembio.2239
131. Bersuker K, Hendricks JM, Li Z, et al. The CoQ oxidoreductase FSP1 acts parallel to GPX4 to inhibit ferroptosis. *Nature.* 2003;426:6967. 2019;575:688-692. doi:10.1038/s41586-019-1705-2
132. Doll S, Freitas FP, Shah R, et al. FSP1 is a glutathione-independent ferroptosis suppressor. *Nature.* 2003;426:6967. 2019;575:693-698. doi:10.1038/s41586-019-1707-0
133. Stockwell BR, Friedmann Angeli JP, Bayir H, et al. Ferroptosis: a regulated cell death nexus linking metabolism, redox biology, and disease. *Cell.* 2017;171:273-285. doi:10.1016/j.cell.2017.09.021
134. Viswanathan VS, Ryan MJ, Dhruv HD, et al. Dependency of a therapy-resistant state of cancer cells on a lipid peroxidase pathway. *Nature.* 2003;426:6967. 2017;547:453-457. doi:10.1038/nature23007
135. Hangauer MJ, Viswanathan VS, Ryan MJ, et al. Drug-tolerant persister cancer cells are vulnerable to GPX4 inhibition. *Nature.* 2003;426:6967. 2017;551:247. doi:10.1038/nature24297
136. Takei H, Araki A, Watanabe H, Ichinose A, Sendo F. Rapid killing of human neutrophils by the potent activator phorbol 12-myristate 13-acetate (PMA) accompanied by changes different from typical apoptosis or necrosis. *J Leukoc Biol.* 1996;59:229-240. doi:10.1002/jlb.59.2.229
137. Brinkmann V, Reichard U, Goosmann C, et al. Neutrophil extracellular traps kill bacteria. *Science.* 2004;303:1532-1535. doi:10.1126/science.1092385
138. Papayannopoulos V, Metzler KD, Hakkim A, Zychlinsky A. Neutrophil elastase and myeloperoxidase regulate the formation of neutrophil extracellular traps. *J Cell Biol.* 2010;191:677-691. doi:10.1083/jcb.201006052
139. Metzler KD, Goosmann C, Lubojemska A, Zychlinsky A, Papayannopoulos V. A myeloperoxidase-containing complex regulates neutrophil elastase release and actin dynamics during NETosis. *Cell Rep.* 2014;8:883-896. doi:10.1016/j.celrep.2014.06.044
140. Chen KW, Monteleone M, Boucher D, et al. Noncanonical inflammasome signaling elicits gasdermin D–dependent neutrophil extracellular traps. *Sci Immunol.* 2018;3:eaar6676. doi:10.1126/sciimmunol.aar6676
141. Rochael NC, Guimarães-Costa AB, Nascimento MTC, et al. Classical ROS-dependent and early/rapid ROS-independent release of Neutrophil Extracellular Traps triggered by Leishmania parasites. *Sci Rep.* 2015;5:1-11. doi:10.1038/srep18302
142. Yipp BG, Petri B, Salina D, et al. Infection-induced NETosis is a dynamic process involving neutrophil multitasking in vivo. *Nat Med.* 2012;18:1386-1393. doi:10.1038/nm.2847
143. Pilsczek FH, Salina D, Poon KKH, et al. A novel mechanism of rapid nuclear neutrophil extracellular trap formation in response to *Staphylococcus aureus*. *J Immunol.* 2010;185:7413-7425. doi:10.4049/jimmunol.1000675
144. Clark SR, Ma AC, Tavener SA, et al. Platelet TLR4 activates neutrophil extracellular traps to ensnare bacteria in septic blood. *Nat Med.* 2007;13:463-469. doi:10.1038/nm1565
145. Jorch SK, Kubes P. An emerging role for neutrophil extracellular traps in noninfectious disease. *Nat Med.* 2017;23:279-287. doi:10.1038/nm.4294
146. Henson PM, Bratton DL, Fadok VA. Apoptotic cell removal. *Curr Biol.* 2001;11:R795-R805. doi:10.1016/s0960-9822(01)00474-2
147. Collins RF, Schreiber AD, Grinstein S, Trimble WS. Syntaxins 13 and 7 function at distinct steps during phagocytosis. *J Immunol.* 2002;169:3250-3256. doi:10.4049/jimmunol.169.6.3250
148. Lauber K, Bohn E, Kröber SM, et al. Apoptotic cells induce migration of phagocytes via caspase-3-mediated release of a lipid attraction signal. *Cell.* 2003;113:717-730. doi:10.1016/s0092-8674(03)00422-7
149. Gude DR, Alvarez SE, Paugh SW, et al. Apoptosis induces expression of sphingosine kinase 1 to release sphingosine-1-phosphate as a "come-and-get-me" signal. *FASEB J.* 2008;22:2629-2638. doi:10.1096/fj.08-107169
150. Elliott MR, Chekeni FB, Trampont PC, et al. Nucleotides released by apoptotic cells act as a find-me signal to promote phagocytic clearance. *Nature.* 2009;461:282-286.
151. Truman LA, Ford CA, Pasikowska M, et al. CX3CL1/fractalkine is released from apoptotic lymphocytes to stimulate macrophage chemotaxis. *Blood.* 2008;112:5026-5036. doi:10.1182/blood-2008-06-162404
152. Lämmermann T, Afonso PV, Angermann BR, et al. Neutrophil swarms require LTB4 and integrins at sites of cell death in vivo. *Nature.* 2003;426:6967. 2013;498:371-375. doi:10.1038/nature12175
153. Chekeni FB, Elliott MR, Sandilos JK, et al. Pannexin 1 channels mediate "find-me" signal release and membrane permeability during apoptosis. *Nature.* 2003;426:6967. 2010;467:863-867. doi:10.1038/nature09413
154. Gong Y-N, Guy C, Olauson H, et al. ESCRT-III acts downstream of MLKL to regulate necroptotic cell death and its consequences. *Cell.* 2017;169:286-300.e16. doi:10.1016/j.cell.2017.03.020
155. Suzuki J, Umeda M, Sims PJ, Nagata S. Calcium-dependent phospholipid scrambling by TMEM16F. *Nature.* 2003;426:6967. 2010;468:834-838. doi:10.1038/nature09583
156. Oldenborg PA, Zheleznyak A, Fang YF, Lagenaur CF, Gresham HD, Lindberg FP. Role of CD47 as a marker of self on red blood cells. *Science.* 2000;288:2051-2054. doi:10.1126/science.288.5473.2051
157. Barkal AA, Brewer RE, Markovic M, et al. CD24 signalling through macrophage Siglec-10 is a target for cancer immunotherapy. *Nature.* 2003;426:6967. 2019;572:392-396. doi:10.1038/s41586-019-1456-0
158. Brown S, Heinisch I, Ross E, Shaw K, Buckley CD, Savill J. Apoptosis disables CD31-mediated cell detachment from phagocytes promoting binding and engulfment. *Nature.* 2003;426:6967. 2002;418:200-203. doi:10.1038/nature00811
159. Barkal AA, Weiskopf K, Kao KS, et al. Engagement of MHC class I by the inhibitory receptor LILRB1 suppresses macrophages and is a target of cancer immunotherapy. *Nat Immunol.* 2018;19:76-84. doi:10.1038/s41590-017-0004-z
160. Chao MP, Jaiswal S, Weissman-Tsukamoto R, et al. Calreticulin is the dominant pro-phagocytic signal on multiple human cancers and is counterbalanced by CD47. *Sci Transl Med.* 2010;2:63ra94. doi:10.1126/scitranslmed.3001375
161. Segawa K, Yanagihashi Y, Yamada K, Suzuki C, Uchiyama Y, Nagata S. Phospholipid flippases enable precursor B cells to flee engulfment by macrophages. *Proc Natl Acad Sci U S A.* 2018;115:12212-12217. doi:10.1073/pnas.1814323115
162. Takatsu H, Tanaka G, Segawa K, et al. Phospholipid flippase activities and substrate specificities of human type IV P-type ATPases localized to the plasma membrane. *J Biol Chem.* 2014;289:33543-33556. doi:10.1074/jbc.M114.593012
163. Kobayashi N, Karisola P, Peña-Cruz V, et al. TIM-1 and TIM-4 glycoproteins bind phosphatidylserine and mediate uptake of apoptotic cells. *Immunity.* 2007;27:927-940. doi:10.1016/j.immuni.2007.11.011
164. Park D, Tosello-Trampont A-C, Elliott MR, et al. Bai1 is an engulfment receptor for apoptotic cells upstream of the ELMO/Dock180/Rac module. *Nature.* 2003;426:6967. 2007;450:430-434. doi:10.1038/nature06329
165. Park S-Y, Jung M-Y, Kim H-J, et al. Rapid cell corpse clearance by stabilin-2, a membrane phosphatidylserine receptor. *Cell Death Differ.* 2008;15:192-201. doi:10.1038/sj.cdd.4402242
166. Tian L, Choi S-C, Murakami Y, et al. p85α recruitment by the CD300f phosphatidylserine receptor mediates apoptotic cell clearance required for autoimmunity suppression. *Nat Commun.* 2014;5:3146-3215. doi:10.1038/ncomms4146
167. Hanayama R, Tanaka M, Miwa K, Shinohara A, Iwamatsu A, Nagata S. Identification of a factor that links apoptotic cells to phagocytes. *Nature.* 2003;426:6967. 2002;417:182-187. doi:10.1038/417182a
168. Geng K, Kumar S, Kimani SG, et al. Requirement of gamma-carboxyglutamic acid modification and phosphatidylserine binding for the activation of Tyro3, Axl, and Mertk receptors by growth arrest-specific 6. *Front Immunol.* 2017;8:1521. doi:10.3389/fimmu.2017.01521
169. Rodriguez-Manzanet R, Sanjuan MA, Wu HY, et al. T and B cell hyperactivity and autoimmunity associated with niche-specific defects in apoptotic body clearance in TIM-4-deficient mice. *Proc Natl Acad Sci U S A.* 2010;107:8706-8711. doi:10.1073/pnas.0910359107
170. Cox D, Chang P, Zhang Q, Reddy PG, Bokoch GM, Greenberg S. Requirements for both Rac1 and Cdc42 in membrane ruffling and phagocytosis in leukocytes. *J Exp Med.* 1997;186:1487-1494. doi:10.1084/jem.186.9.1487
171. Martinez J, Almendinger J, Oberst A, et al. Microtubule-associated protein 1 light chain 3 alpha (LC3)-associated phagocytosis is required for the efficient clearance of dead cells. *Proc Natl Acad Sci U S A.* 2011;108:17396-17401. doi:10.1073/pnas.1113421108
172. Heckmann BL, Boada-Romero E, Cunha LD, Magne J, Green DR. LC3-associated phagocytosis and inflammation. *J Mol Biol.* 2017;429:3561-3576. doi:10.1016/j.jmb.2017.08.012
173. Kiss RS, Elliott MR, Ma Z, Marcel YL, Ravichandran KS. Apoptotic cells induce a phosphatidylserine-dependent homeostatic response from phagocytes. *Curr Biol.* 2006;16:2252-2258. doi:10.1016/j.cub.2006.09.043
174. Mukundan L, Odegaard JI, Morel CR, et al. PPAR-delta senses and orchestrates clearance of apoptotic cells to promote tolerance. *Nat Med.* 2009;15:1266-1272. doi:10.1038/nm.2048
175. Roszer T, Menéndez-Gutiérrez MP, Lefterova MI, et al. Autoimmune kidney disease and impaired engulfment of apoptotic cells in mice with macrophage peroxisome proliferator-activated receptor gamma or retinoid X receptor alpha deficiency. *J Immunol.* 2011;186:621-631. doi:10.4049/jimmunol.1002230
176. Zhang S, Weinberg S, DeBerge M, et al. Efferocytosis fuels requirements of fatty acid oxidation and the electron transport chain to polarize macrophages for tissue repair. *Cell Metab.* 2019;29:443-456.e5. doi:10.1016/j.cmet.2018.12.004
177. Park D, Han CZ, Elliott MR, et al. Continued clearance of apoptotic cells critically depends on the phagocyte Ucp2 protein. *Nature.* 2003;426:6967. 2011;477:220-224. doi:10.1038/nature10340
178. Chung EY, Liu J, Homma Y, et al. Interleukin-10 expression in macrophages during phagocytosis of apoptotic cells is mediated by homeodomain proteins Pbx1 and Prep-1. *Immunity.* 2007;27:952-964. doi:10.1016/j.immuni.2007.11.014
179. Kim S, Elkon KB, Ma X. Transcriptional suppression of interleukin-12 gene expression following phagocytosis of apoptotic cells. *Immunity.* 2004;21:643-653. doi:10.1016/j.immuni.2004.09.009
180. Cohen PL, Caricchio R, Abraham V, et al. Delayed apoptotic cell clearance and lupus-like autoimmunity in mice lacking the c-mer membrane tyrosine kinase. *J Exp Med.* 2002;196:135-140. doi:10.1084/jem.20012094
181. Scott RS, McMahon EJ, Pop SM, et al. Phagocytosis and clearance of apoptotic cells is mediated by MER. *Nature.* 2003;426:6967. 2001;411:207-211. doi:10.1038/35075603
182. Hanayama R, Tanaka M, Miyasaka K, et al. Autoimmune disease and impaired uptake of apoptotic cells in MFG-E8-deficient mice. *Science.* 2004;304:1147-1150. doi:10.1126/science.1094359
183. Kawano M, Nagata S. Lupus-like autoimmune disease caused by a lack of Xkr8, a caspase-dependent phospholipid scramblase. *Proc Natl Acad Sci U S A.* 2018;115:2132-2137. doi:10.1073/pnas.1720732115
184. Martinez J, Cunha LD, Park S, et al. Noncanonical autophagy inhibits the autoinflammatory, lupus-like response to dying cells. *Nature.* 2003;426:6967. 2016;533:115-119. doi:10.1038/nature17950
185. Juncadella IJ, Kadl A, Sharma AK, et al. Apoptotic cell clearance by bronchial epithelial cells critically influences airway inflammation. *Nature.* 2003;426:6967. 2013;493:547-551. doi:10.1038/nature11714
186. Boada-Romero E, Martinez J, Heckmann BL, Green DR. The clearance of dead cells by efferocytosis. *Nat Rev Mol Cell Biol.* 2020;21:398-414. doi:10.1038/s41580-020-0232-1
187. Mandal M, Borowski C, Palomero T, et al. The BCL2A1 gene as a pre-T cell receptor-induced regulator of thymocyte survival. *J Exp Med.* 2005;201:603-614. doi:10.1084/jem.20041924
188. Voll RE, Jimi E, Phillips RJ, et al. NF-κB activation by the pre-T cell receptor serves as a selective survival signal in T lymphocyte development. *Immunity.* 2000;13:677-689. doi:10.1016/S1074-7613(00)00067-4
189. Ottina E, Grespi F, Tischner D, et al. Targeting antiapoptotic A1/Bfl-1 by in vivo RNAi reveals multiple roles in leukocyte development in mice. *Blood.* 2012;119:6032-6042. doi:10.1182/blood-2011-12-399089
190. Rathmell JC, Lindsten T, Zong W-X, Cinalli RM, Thompson CB. Deficiency in Bak and Bax perturbs thymic selection and lymphoid homeostasis. *Nat Immunol.* 2002;3:932-939. doi:10.1038/ni834
191. Newton K, Harris AW, Bath ML, Smith KGC, Strasser A. A dominant interfering mutant of FADD/MORT1 enhances deletion of autoreactive thymocytes and inhibits proliferation of mature T lymphocytes. *EMBO J.* 1998;17:706-718. doi:10.1093/emboj/17.3.706
192. Petrie HT, Livak F, Schatz DG, Strasser A, Crispe IN, Shortman K. Multiple rearrangements in T cell receptor alpha chain genes maximize the production of useful thymocytes. *J Exp Med.* 1993;178:615-622. doi:10.1084/jem.178.2.615
193. Ma A, Pena JC, Chang B, et al. Bclx regulates the survival of double-positive thymocytes. *Proc Natl Acad Sci U S A.* 1995;92:4763-4767. doi:10.1073/pnas.92.11.4763
194. Mason KD, Carpinelli MR, Fletcher JI, et al. Programmed anuclear cell death delimits platelet life span. *Cell.* 2007;128:1173-1186. doi:10.1016/j.cell.2007.01.037
195. Egerton M, Scollay R, Shortman K. Kinetics of mature T-cell development in the thymus. *Proc Natl Acad Sci U S A.* 1990;87:2579-2582. doi:10.1073/pnas.87.7.2579
196. Strasser A, Harris AW, Corcoran LM, Cory S. Bcl-2 expression promotes B-but not T-lymphoid development in scid mice. *Nature.* 2003;426:6967. 1994;368:457-460. doi:10.1038/368457a0
197. Strasser A, Harris AW, von Boehmer H, Cory S. Positive and negative selection of T cells in T-cell receptor transgenic mice expressing a bcl-2 transgene. *Proc Natl Acad Sci U S A.* 1994;91:1376-1380. doi:10.1073/pnas.91.4.1376
198. Hu Q, Sader A, Parkman JC, Baldwin TA. Bim-mediated apoptosis is not necessary for thymic negative selection to ubiquitous self-antigens. *J Immunol.* 2009;183:7761-7767. doi:10.4049/jimmunol.0902181
199. Bouillet P, Metcalf D, Huang DCS, et al. Proapoptotic bcl-2 relative Bim required for certain apoptotic responses, leukocyte homeostasis, and to preclude autoimmunity. *Science.* 1999;286:1735-1738. doi:10.1126/science.286.5445.1735
200. Bouillet P, Purton JF, Godfrey DI, et al. BH3-only Bcl-2 family member Bim is required for apoptosis of autoreactive thymocytes. *Nature.* 2003;426:6967. 2002;415:922-926. doi:10.1038/415922a
201. Gray DHD, Kupresanin F, Berzins SP, et al. The BH3-only proteins Bim and Puma cooperate to impose deletional tolerance of organ-specific antigens. *Immunity.* 2012;37:451-462. doi:10.1016/j.immuni.2012.05.030

202. Tai X, Erman B, Alag A, et al. Foxp3 transcription factor is proapoptotic and lethal to developing regulatory T cells unless counterbalanced by cytokine survival signals. *Immunity.* 2013;38:1116-1128. doi:10.1016/j.immuni.2013.02.022
203. Burchill MA, Yang J, Vang KB, et al. Linked T cell receptor and cytokine signaling govern the development of the regulatory T cell repertoire. *Immunity.* 2008;28:112-121. doi:10.1016/j.immuni.2007.11.022
204. Fontenot JD, Rasmussen JP, Gavin MA, Rudensky AY. A function for interleukin 2 in Foxp3-expressing regulatory T cells. *Nat Immunol.* 2005;6:1142-1151. doi:10.1038/ni1263
205. Lio C-WJ, Hsieh C-S. A two-step process for thymic regulatory T cell development, *Immunity.* 2008;28:100-111. doi:10.1016/j.immuni.2007.11.021
206. Cowan JE, Parnell SM, Nakamura K, et al. The thymic medulla is required for Foxp3+ regulatory but not conventional CD4+ thymocyte development. *J Exp Med.* 2013;210:675-681. doi:10.1084/jem.20122070
207. Hu DY, Yap JY, Wirasinha RC, Howard DR, Goodnow CC, Daley SR. A timeline demarcating two waves of clonal deletion and Foxp3 upregulation during thymocyte development. *Immunol Cell Biol.* 2016;94:357-366. doi:10.1038/icb.2015.95
208. Shi YF, Sahai BM, Green DR. Cyclosporin A inhibits activation-induced cell death in T-cell hybridomas and thymocytes. *Nature.* 2003;426:6967. 1989;339:625-626. doi:10.1038/339625a0
209. Shi YF, Szalay MG, Paskar L, et al. Activation-induced cell death in T cell hybridomas is due to apoptosis. Morphologic aspects and DNA fragmentation. *J Immunol.* 1990;144:3326-3333.
210. Mercep M, Weissman AM, Frank SJ, Klausner RD, Ashwell JD. Activation-driven programmed cell death and T cell receptor zeta eta expression. *Science.* 1989;246:1162-1165. doi:10.1126/science.2531464
211. Dhein J, Walczak H, Bäumler C, Debatin K-M, Krammer PH. Autocrine T-cell suicide mediated by APO-1/(Fas/CD95). *Nature.* 2003;426:6967. 1995;373;438-441. doi:10.1038/373438a0
212. Janssen EM, Droin NM, Lemmens EE, et al. CD4+ T-cell help controls CD8+ T-cell memory via TRAIL-mediated activation-induced cell death. *Nature.* 2003;426:6967. 2005;434:88-93. doi:10.1038/nature03337
213. Bonfoco E, Stuart PM, Brunner T, et al. Inducible nonlymphoid expression of Fas ligand is responsible for superantigen-induced peripheral deletion of T cells. *Immunity.* 1998;9:711-720. doi:10.1016/S1074-7613(00)80668-8
214. Pinkoski MJ, Droin NM, Lin T, Genestier L, Ferguson TA, Green DR. Nonlymphoid Fas ligand in peptide-induced peripheral lymphocyte deletion. *Proc Natl Acad Sci U S A.* 2002;99:16174-16179. doi:10.1073/pnas.262660999
215. Davey GM, Kurts C, Miller JFAP, et al. Peripheral deletion of autoreactive CD8 T cells by cross presentation of self-antigen occurs by a Bcl-2-inhibitable pathway mediated by Bim. *J Exp Med.* 2002;196:947-955. doi:10.1084/jem.20020827
216. Hildeman DA, Zhu Y, Mitchell TC, et al. Activated T cell death in vivo mediated by proapoptotic bcl-2 family member Bim. *Immunity.* 2002;16:759-767. doi:10.1016/S1074-7613(02)00322-9
217. Weiss E-M, Schmidt A, Vobis D, et al. Foxp3-Mediated suppression of CD95L expression confers resistance to activation-induced cell death in regulatory T cells. *J Immunol.* 2011;187:1684-1691. doi:10.4049/jimmunol.1002321
218. Morgan GM, McKenzie IF. Cell membrane alloantigenic determinants of several unusual mouse strains. *J Immunogenet.* 1980;7:393-399. doi:10.1111/j.1744-313x.1980.tb00733.x
219. Roths JB, Murphy ED, Eicher EM. A new mutation, gld, that produces lymphoproliferation and autoimmunity in C3H/HeJ mice. *J Exp Med.* 1984;159:1-20. doi:10.1084/jem.159.1.1
220. Watanabe-Fukunaga R, Brannan CI, Copeland NG, Jenkins NA, Nagata S. Lymphoproliferation disorder in mice explained by defects in Fas antigen that mediates apoptosis. *Nature.* 2003;426:6967. 1992;356:314-317. doi:10.1038/356314a0
221. Lee J, Dieckmann NMG, Edgar JR, Griffiths GM, Siegel RM. Fas ligand localizes to intraluminal vesicles within NK cell cytolytic granules and is enriched at the immune synapse. *Immun Inflamm Dis.* 2018;6:312-321. doi:10.1002/iid3.219
222. Martínez-Lorenzo MJ, Anel A, Gamen S, et al. Activated human T cells release bioactive Fas ligand and APO2 ligand in microvesicles. *J Immunol.* 1999;163:1274-1281.
223. Shinkai Y, Takio K, Okumura K. Homology of perforin to the ninth component of complement (C9). *Nature.* 2003;426:6967. 1988;334:525-527. doi:10.1038/334525a0
224. Podack ER, Young JD, Cohn ZA. Isolation and biochemical and functional characterization of perforin 1 from cytolytic T-cell granules. *Proc Natl Acad Sci U S A.* 1985;82:8629-8633. doi:10.1073/pnas.82.24.8629
225. Masson D, Tschopp J. Isolation of a lytic, pore-forming protein (perforin) from cytolytic T-lymphocytes. *J Biol Chem.* 1985;260:9069-9072.
226. Masson D, Nabholz M, Estrade C, Tschopp J. Granules of cytolytic T-lymphocytes contain two serine esterases. *EMBO J.* 1986;5:1595-1600. doi:10.1002/j.1460-2075.1986.tb04401.x
227. Masson D, Tschopp J. A family of serine esterases in lytic granules of cytolytic T lymphocytes. *Cell.* 1987;49:679-685. doi:10.1016/0092-8674(87)90544-7
228. Wang W, Green M, Choi JE, et al. CD8+ T cells regulate tumour ferroptosis during cancer immunotherapy. *Nature.* 2003;426:6967. 2019;569:270-274. doi:10.1038/s41586-019-1170-y
229. Matzinger P. Tolerance, danger, and the extended family. *Annu Rev Immunol.* 1994;12:991-1045. doi:10.1146/annurev.iy.12.040194.005015
230. Shi Y, Evans JE, Rock KL. Molecular identification of a danger signal that alerts the immune system to dying cells. *Nature.* 2003;426:6967. 2003;425:516-521. doi:10.1038/nature01991
231. Scaffidi P, Misteli T, Bianchi ME. Release of chromatin protein HMGB1 by necrotic cells triggers inflammation. *Nature.* 2003;426:6967. 2002;418:191-195. doi:10.1038/nature00858
232. Bianchi ME, Agresti A. HMG proteins: dynamic players in gene regulation and differentiation. *Curr Opin Genet Dev.* 2005;15:496-506. doi:10.1016/j.gde.2005.08.007
233. Antoine DJ, Jenkins RE, Dear JW, et al. Molecular forms of HMGB1 and keratin-18 as mechanistic biomarkers for mode of cell death and prognosis during clinical acetaminophen hepatotoxicity. *J Hepatol.* 2012;56:1070-1079. doi:10.1016/j.jhep.2011.12.019
234. Bonaldi T, Talamo F, Scaffidi P, et al. Monocytic cells hyperacetylate chromatin protein HMGB1 to redirect it towards secretion. *EMBO J.* 2003;22:5551-5560. doi:10.1093/emboj/cdg516
235. Kokkola R, Andersson Å, Mullins G, et al. RAGE is the major receptor for the proinflammatory activity of HMGB1 in rodent macrophages. *Scand J Immunol.* 2005;61:1-9. doi:10.1111/j.0300-9475.2005.01534.x
236. Parkkinen J, Raulo E, Merenmies J, et al. Amphoterin, the 30-kDa protein in a family of HMG1-type polypeptides. Enhanced expression in transformed cells, leading edge localization, and interactions with plasminogen activation. *J Biol Chem.* 1993;268:19726-19738.
237. Schiraldi M, Raucci A, Muñoz LM, et al. HMGB1 promotes recruitment of inflammatory cells to damaged tissues by forming a complex with CXCL12 and signaling via CXCR4. *J Exp Med.* 2012;209:551-563. doi:10.1084/jem.20111739
238. Fiuza C, Bustin M, Talwar S, et al. Inflammation-promoting activity of HMGB1 on human microvascular endothelial cells. *Blood.* 2003;101:2652-2660. doi:10.1182/blood-2002-05-1300
239. Urbonaviciute V, Fürnrohr BG, Meister S, et al. Induction of inflammatory and immune responses by HMGB1–nucleosome complexes: implications for the pathogenesis of SLE. *J Exp Med.* 2008;205:3007-3018. doi:10.1084/jem.20081165
240. Tian J, Avalos AM, Mao S-Y, et al. Toll-like receptor 9–dependent activation by DNA-containing immune complexes is mediated by HMGB1 and RAGE. *Nat Immunol.* 2007;8:487-496. doi:10.1038/ni1457
241. Maroso M, Balosso S, Ravizza T, et al. Toll-like receptor 4 and high-mobility group box-1 are involved in ictogenesis and can be targeted to reduce seizures. *Nat Med.* 2010;16:413-419. doi:10.1038/nm.2127
242. Weng H, Deng Y, Xie Y, Liu H, Gong F. Expression and significance of HMGB1, TLR4 and NF-κB p65 in human epidermal tumors. *BMC Cancer.* 2013;13:1-10. doi:10.1186/1471-2407-13-311
243. Yang Z, Deng Y, Su D, et al. TLR4 as receptor for HMGB1-mediated acute lung injury after liver ischemia/reperfusion injury. *Lab Invest.* 2013;93:792-800. doi:10.1038/labinvest.2013.66
244. Wang X, Sun R, Wei H, Tian Z. High-mobility group box 1 (HMGB1)-toll-like receptor (TLR)4-interleukin (IL)-23-IL-17A axis in drug-induced damage-associated lethal hepatitis: interaction of γδ T cells with macrophages. *Hepatology.* 2013;57:373-384. doi:10.1002/hep.25982
245. Antoine DJ, Harris HE, Andersson U, Tracey KJ, Bianchi ME. A systematic nomenclature for the redox states of high mobility group box (HMGB) proteins. *Mol Med.* 2014;20:135-137. doi:10.2119/molmed.2014.00022
246. Yang H, Wang H, Ju Z, et al. MD-2 is required for disulfide HMGB1–dependent TLR4 signaling. *J Exp Med.* 2015;212:5-14. doi:10.1084/jem.20141318
247. Vénéreau E, Casalgrandi M, Schiraldi M, et al. Mutually exclusive redox forms of HMGB1 promote cell recruitment or proinflammatory cytokine release. *J Exp Med.* 2012;209:1519-1528. doi:10.1084/jem.20120189
248. Nyström S, Antoine DJ, Lundbäck P, et al. TLR activation regulates damage-associated molecular pattern isoforms released during pyroptosis. *EMBO J.* 2013;32:86-99. doi:10.1038/emboj.2012.328
249. Vénéreau E, Ceriotti C, Bianchi ME. DAMPs from cell death to new life. *Front Immunol.* 2015;6:1. doi:10.3389/fimmu.2015.00422
250. Tamai K, Yamazaki T, Chino T, et al. PDGFRα-positive cells in bone marrow are mobilized by high mobility group box 1 (HMGB1) to regenerate injured epithelia. *Proc Natl Acad Sci U S A.* 2011;108:6609-6614. doi:10.1073/pnas.1016753108
251. Xie HL, Zhang Y, Huang YZ, et al. Regulation of high mobility group box 1 and hypoxia in the migration of mesenchymal stem cells. *Cell Biol Int.* 2014;38:892-897. doi:10.1002/cbin.10279
252. Lotfi R, Eisenbacher J, Solgi G, et al. Human mesenchymal stem cells respond to native but not oxidized damage associated molecular pattern molecules from necrotic (tumor) material. *Eur J Immunol.* 2011;41:2021-2028. doi:10.1002/eji.201041324
253. Meng E, Guo Z, Wang H, et al. High mobility group box 1 protein inhibits the proliferation of human mesenchymal stem cells and promotes their migration and differentiation along osteoblastic pathway. *Stem Cells Dev.* 2008;17:805-814. doi:10.1089/scd.2007.0276. https://Home.Liebertpub.com/Scd
254. Mitola S, Belleri M, Urbinati C, et al. Cutting edge: extracellular high mobility group box-1 protein is a proangiogenic cytokine. *J Immunol.* 2006;176:12-15. doi:10.4049/jimmunol.176.1.12
255. Chavakis E, Hain A, Vinci M, et al. High-mobility group box 1 activates integrin-dependent homing of endothelial progenitor cells. *Circ Res.* 2007;100:204-212. doi:10.1161/01.RES.0000257774.55970.f4
256. Straino S, Di Carlo A, Mangoni A, et al. High-mobility group box 1 protein in human and murine skin: involvement in wound healing. *J Invest Dermatol.* 2008;128:545-1553. doi:10.1038/sj.jid.5701212
257. Volz HC, Seidel C, Laohachewin D, et al. HMGB1: the missing link between diabetes mellitus and heart failure. *Basic Res Cardiol.* 2010;105:805-820. doi:10.1007/s00395-010-0114-3
258. Limana F, Esposito G, Fasanaro P, et al. Transcriptional profiling of Hmgb1-induced myocardial repair identifies a key role for notch signaling. *Mol Ther.* 2013;21:1841-1851. doi:10.1038/mt.2013.137
259. Takahashi K, Fukushima S, Yamahara K, et al. Modulated inflammation by injection of high-mobility group box 1 recovers post-infarction chronically failing heart. *Circulation.* 2008;118:S106-S114. doi:10.1161/CIRCULATIONAHA.107.757443
260. Palumbo R, Sampaolesi M, De Marchis F, et al. Extracellular HMGB1, a signal of tissue damage, induces mesoangioblast migration and proliferation. *J Cell Biol.* 2004;164:441-449. doi:10.1083/jcb.200304135
261. Sorci G, Riuzzi F, Arcuri C, Giambanco I, Donato R. Amphoterin stimulates myogenesis and counteracts the antimyogenic factors basic fibroblast growth factor and S100B via RAGE binding. *Mol Cell Biol.* 2004;24:4880-4894. doi:10.1128/MCB.24.11.4880-4894.2004
262. Degryse B, Bonaldi T, Scaffidi P, et al. The high mobility group (Hmg) boxes of the nuclear protein Hmg1 induce chemotaxis and cytoskeleton reorganization in rat smooth muscle cells. *J Cell Biol.* 2001;152:1197-1206. doi:10.1083/jcb.152.6.1197
263. De Mori R, Straino S, Di Carlo A, et al. Multiple effects of high mobility group box protein 1 in skeletal muscle regeneration. *Arterioscler Thromb Vasc Biol.* 2007;27:2377-2383. doi:10.1161/ATVBAHA.107.153429
264. Campana L, Santarella F, Esposito A, et al. Leukocyte HMGB1 is required for vessel remodeling in regenerating muscles. *J Immunol.* 2014;192:5257-5264. doi:10.4049/jimmunol.1300938
265. Garg AD, Krysko DV, Verfaillie T, et al. A novel pathway combining calreticulin exposure and ATP secretion in immunogenic cancer cell death. *EMBO J.* 2012;31:1062-1079. doi:10.1038/emboj.2011.497
266. Junger WG. Immune cell regulation by autocrine purinergic signalling. *Nat Rev Immunol.* 2011;11:201-212. doi:10.1038/nri2938
267. Idzko M, Ferrari D, Eltzschig HK. Nucleotide signalling during inflammation. *Nature.* 2003;426:6967. 2014;509:310-317. doi:10.1038/nature13085
268. Eltzschig HK, Sitkovsky MV, Robson SC. Purinergic signaling during inflammation. *N Engl J Med.* 2012;367:2322-2333. doi:10.1056/NEJMra1205750
269. Elliott MR, Chekeni FB, Trampont PC, et al. Nucleotides released by apoptotic cells act as a find-me signal to promote phagocytic clearance. *Nature.* 2003;426:6967. 2009;461:282-286. doi:10.1038/nature08296
270. Chen Y, Corriden R, Inoue Y, et al. ATP release guides neutrophil chemotaxis via P2Y2 and A3 receptors. *Science.* 2006;314:1792-1795. doi:10.1126/science.1132559
271. Idzko M, Hammad H, van Nimwegen M, et al. Extracellular ATP triggers and maintains asthmatic airway inflammation by activating dendritic cells. *Nat Med.* 2007;13:913-919. doi:10.1038/nm1617
272. Kouzaki H, Iijima K, Kobayashi T, O'Grady SM, Kita H. The danger signal, extracellular ATP, is a sensor for an airborne allergen and triggers IL-33 release and innate Th2-type responses. *J Immunol.* 2011;186:4375-4387. doi:10.4049/jimmunol.1003020
273. Surprenant A, Rassendren F, Kawashima E, North RA, Buell G. The cytolytic P2Z receptor for extracellular ATP identified as a P2X receptor (P2X7). *Science.* 1996;272:735-738. doi:10.1126/science.272.5262.735
274. Muñoz-Planillo R, Kuffa P, Martínez-Colón G, Smith BL, Rajendiran TM, Núñez G. K+ efflux is the common trigger of NLRP3 inflammasome activation by bacterial toxins and particulate matter. *Immunity.* 2013;38:1142-1153. doi:10.1016/j.immuni.2013.05.016
275. Mariathasan S, Weiss DS, Newton K, et al. Cryopyrin activates the inflammasome in response to toxins and ATP. *Nature.* 2003;426:6967. 2006;440:228-232. doi:10.1038/nature04515

276. Placido R, Auricchio G, Falzoni S, et al. P2X7 purinergic receptors and extracellular ATP mediate apoptosis of human monocytes/macrophages infected with *Mycobacterium tuberculosis* reducing the intracellular bacterial viability. *Cell Immunol.* 2006;244:10-18. doi:10.1016/j.cellimm.2007.02.001
277. Ghiringhelli F, Apetoh L, Tesniere A, et al. Activation of the NLRP3 inflammasome in dendritic cells induces IL-1beta-dependent adaptive immunity against tumors. *Nat Med.* 2009;15:1170-1178. doi:10.1038/nm.2028
278. Weber FC, Esser PR, Müller T, et al. Lack of the purinergic receptor P2X7 results in resistance to contact hypersensitivity. *J Exp Med.* 2010;207 (2609-2619. doi:10.1084/jem.20092489
279. Wilhelm K, Ganesan J, Müller T, et al. Graft-versus-host disease is enhanced by extracellular ATP activating P2X 7 R. *Nat Med.* 2010;16:1434-1438. doi:10.1038/nm.2242
280. Albert JL, Boyle JP, Roberts JA, Challiss RAJ, Gubby SE, Boarder MR. Regulation of brain capillary endothelial cells by P2Y receptors coupled to Ca2+, phospholipase C and mitogen-activated protein kinase. *Br J Pharmacol.* 1997;122:935-941. doi:10.1038/sj.bjp.0701453
281. Satterwhite CM, Farrelly AM, Bradley ME. Chemotactic, mitogenic, and angiogenic actions of UTP on vascular endothelial cells. *Am J Physiol Heart Circ Physiol.* 1999;276:H1091-H1097. doi:10.1152/ajpheart.1999.276.3.H1091
282. Jin H, Seo J, Eun SY, et al. P2Y2R activation by nucleotides promotes skin wound-healing process. *Exp Dermatol.* 2014;23:480-485. doi:10.1111/exd.12440
283. Erlinge D. Extracellular ATP: a growth factor for vascular smooth muscle cells. *Gen Pharmacol-Vasc S.* 1998;31:1-8. doi:10.1016/S0306-3623(97)00420-5
284. Schattauer GmbH; Kaczmarek E, Erb L, Koziak K, et al. Modulation of endothelial cell migration by extracellular nucleotides. *Thromb Haemost.* 2017;93:735-742. doi:10.1160/TH04-09-0576
285. Taboubi S, Milanini J, Delamarre E, et al. Gα(q/11)-coupled P2Y2 nucleotide receptor inhibits human keratinocyte spreading and migration. *FASEB J.* 2007;21:4047-4058. doi:10.1096/fj.06-7476com
286. Inoue H, Tani K. Multimodal immunogenic cancer cell death as a consequence of anticancer cytotoxic treatments. *Cell Death Differ.* 2014;21:39-49. doi:10.1038/cdd.2013.84
287. Obeid M, Panaretakis T, Joza N, et al. Calreticulin exposure is required for the immunogenicity of γ-irradiation and UVC light-induced apoptosis. *Cell Death Differ.* 2007;14:1848-1850. doi:10.1038/sj.cdd.4402201
288. Panaretakis T, Joza N, Modjtahedi N, et al. The co-translocation of ERp57 and calreticulin determines the immunogenicity of cell death. *Cell Death Differ.* 2008;15:1499-1509. doi:10.1038/cdd.2008.67
289. Apetoh L, Ghiringhelli F, Tesniere A, et al. Toll-like receptor 4–dependent contribution of the immune system to anticancer chemotherapy and radiotherapy. *Nat Med.* 2007;13:1050-1059. doi:10.1038/nm1622
290. Panaretakis T, Kepp O, Brockmeier U, et al. Mechanisms of pre-apoptotic calreticulin exposure in immunogenic cell death. *EMBO J.* 2009;28:578-590. doi:10.1038/emboj.2009.1
291. Tesniere A, Schlemmer F, Boige V, et al. Immunogenic death of colon cancer cells treated with oxaliplatin. *Oncogene.* 2006;26:16. 2010;29:482-491. doi:10.1038/onc.2009.356
292. Schiavoni G, Sistigu A, Valentini M, et al. Cyclophosphamide synergizes with type I interferons through systemic dendritic cell reactivation and induction of immunogenic tumor apoptosis. *Cancer Res.* 2011;71;768-778. doi:10.1158/0008-5472.CAN-10-2788
293. Spisek R, Charalambous A, Mazumder A, Vesole DH, Jagannath S, Dhodapkar MV. Bortezomib enhances dendritic cell (DC)–mediated induction of immunity to human myeloma via exposure of cell surface heat shock protein 90 on dying tumor cells: therapeutic implications. *Blood.* 2007;109:4839-4845. doi:10.1182/blood-2006-10-054221
294. Vandenabeele P, Vandecasteele K, Bachert C, Krysko O, Krysko DV. Immunogenic apoptotic cell death and anticancer immunity. *Adv Exp Med Biol.* 2016;930:133-149. doi:10.1007/978-3-319-39406-0_6
295. Martins I, Kepp O, Schlemmer F, et al. Restoration of the immunogenicity of cisplatin-induced cancer cell death by endoplasmic reticulum stress. *Oncogene.* 2006;26:16. 2011;30:1147-1158. doi:10.1038/onc.2010.500
296. Casares N, Pequignot MO, Tesniere A, et al. Caspase-dependent immunogenicity of doxorubicin-induced tumor cell death. *J Exp Med.* 2005;202:1691-1701. doi:10.1084/jem.20050915
297. Martins I, Wang Y, Michaud M, et al. Molecular mechanisms of ATP secretion during immunogenic cell death. *Cell Death Differ.* 2014;21:79-91. doi:10.1038/cdd.2013.75
298. Yatim N, Jusforgues-Saklani H, Orozco S, et al. RIPK1 and NF-κB signaling in dying cells determines cross-priming of CD8+ T cells. *Science.* 2015;350:328-334. doi:10.1126/science.aad0395
299. Snyder AG, Hubbard NW, Messmer MN, et al. Intratumoral activation of the necroptotic pathway components RIPK1 and RIPK3 potentiates antitumor immunity. *Sci Immunol.* 2019;4:eaaw2004. doi:10.1126/sciimmunol.aaw2004
300. Ren J, Jia X, Zhao Y, et al. The RIP3-RIP1-NF-κB signaling axis is dispensable for necroptotic cells to elicit cross-priming of CD8 + T cells. *Cell Mol Immunol.* 2017;14:639-642. doi:10.1038/cmi.2017.31
301. Aaes TL, Kaczmarek A, Delvaeye T, et al. Vaccination with necroptotic cancer cells induces efficient anti-tumor immunity. *Cell Rep.* 2016;15:274-287. doi:10.1016/j.celrep.2016.03.037
302. Li W, Feng G, Gauthier JM, et al. Ferroptotic cell death and TLR4/Trif signaling initiate neutrophil recruitment after heart transplantation. *J Clin Invest.* 2019;129:2293-2304. doi:10.1172/JCI126428
303. Andrews NW, Corrotte M. Plasma membrane repair. *Curr Biol.* 2018;28:R392-R397. doi:10.1016/j.cub.2017.12.034
304. Bi GQ, Alderton JM, Steinhardt RA. Calcium-regulated exocytosis is required for cell membrane resealing. *J Cell Biol.* 1995;131:1747-1758. doi:10.1083/jcb.131.6.1747
305. Miyake K, McNeil PL. Vesicle accumulation and exocytosis at sites of plasma membrane disruption. *J Cell Biol.* 1995;131:1737-1745. doi:10.1083/jcb.131.6.1737
306. Roy D, Liston DR, Idone VJ, et al. A process for controlling intracellular bacterial infections induced by membrane injury. *Science.* 2004;304:1515-1518. doi:10.1126/science.1098371
307. Forestier C-L, Machu C, Loussert C, Pescher P, Späth GF. Imaging host cell-Leishmania interaction dynamics implicates parasite motility, lysosome recruitment, and host cell wounding in the infection process. *Cell Host Microbe.* 2011;9:319-330. doi:10.1016/j.chom.2011.03.011
308. Walev I, Bhakdi SC, Hofmann F, et al. Delivery of proteins into living cells by reversible membrane permeabilization with streptolysin-O. *Proc Natl Acad Sci U S A.* 2001;98:3185-3190. doi:10.1073/pnas.051429498
309. Tam C, Idone V, Devlin C, et al. Exocytosis of acid sphingomyelinase by wounded cells promotes endocytosis and plasma membrane repair. *J Cell Biol.* 2010;189:1027-1038. doi:10.1083/jcb.201003053
310. Babiychuk EB, Monastyrskaya K, Draeger A. Fluorescent annexin A1 reveals dynamics of ceramide platforms in living cells. *Traffic.* 2008;9:1757-1775. doi:10.1111/j.1600-0854.2008.00800.x
311. Grassmé H, Jendrossek V, Bock J, Riehle A, Gulbins E. Ceramide-rich membrane rafts mediate CD40 clustering. *J Immunol.* 2002;168:298-307. doi:10.4049/jimmunol.168.1.298
312. Schissel SL, Jiang X, Tweedie-Hardman J, et al. Secretory sphingomyelinase, a product of the acid sphingomyelinase gene, can hydrolyze atherogenic lipoproteins at neutral pH. Implications for atherosclerotic lesion development. *J Biol Chem.* 1998;273:2738-2746. doi:10.1074/jbc.273.5.2738
313. van Blitterswijk WJ, van der Luit AH, Veldman RJ, Verheij M, Borst J. Ceramide: second messenger or modulator of membrane structure and dynamics? *Biochem J.* 2003;369:199-211. doi:10.1042/BJ20021528
314. Gulbins E. Regulation of death receptor signaling and apoptosis by ceramide. *Pharmacol Res.* 2003;47:393-399. doi:10.1016/s1043-6618(03)00052-5
315. Corrotte M, Fernandes MC, Tam C, Andrews NW. Toxin pores endocytosed during plasma membrane repair traffic into the lumen of MVBs for degradation. *Traffic.* 2012;13:483-494. doi:10.1111/j.1600-0854.2011.01323.x
316. Keefe D, Shi L, Feske S, et al. Perforin triggers a plasma membrane-repair response that facilitates CTL induction of apoptosis. *Immunity.* 2005;23:249-262. doi:10.1016/j.immuni.2005.08.001
317. Katzmann DJ, Odorizzi G, Emr SD. Receptor downregulation and multivesicular-body sorting. *Nat Rev Mol Cell Biol.* 2002;3:893-905. doi:10.1038/nrm973
318. Morita E, Sandrin V, Chung H-Y, et al. Human ESCRT and ALIX proteins interact with proteins of the midbody and function in cytokinesis. *EMBO J.* 2007;26:4215-4227. doi:10.1038/sj.emboj.7601850
319. Carlton JG, Martin-Serrano J. Parallels between cytokinesis and retroviral budding: a role for the ESCRT machinery. *Science.* 2007;316:1908-1912. doi:10.1126/science.1143422
320. Denais CM, Gilbert RM, Isermann P, et al. Nuclear envelope rupture and repair during cancer cell migration. *Science.* 2016;352:353-358. doi:10.1126/science.aad7297
321. Raab M, Gentili M, de Belly H, et al. ESCRT III repairs nuclear envelope ruptures during cell migration to limit DNA damage and cell death. *Science.* 2016;352:359-362. doi:10.1126/science.aad7611
322. Morita E, Sandrin V, McCullough J, Katsuyama A, Baci Hamilton I, Sundquist WI. ESCRT-III protein requirements for HIV-1 budding. *Cell Host Microbe.* 2011;9:235-242. doi:10.1016/j.chom.2011.02.004
323. Jimenez AJ, Maiuri P, Lafaurie-Janvore J, Divoux S, Piel M, Perez F. ESCRT machinery is required for plasma membrane repair. *Science.* 2014;343:1247136. doi:10.1126/science.1247136
324. Scheffer LL, Sreetama SC, Sharma N, et al. Mechanism of Ca 2+-triggered ESCRT assembly and regulation of cell membrane repair. *Nat Commun.* 2014:5:1-12. doi:10.1038/ncomms6646
325. Rühl S, Shkarina K, Demarco B, Heilig R, Santos JC, Broz P. ESCRT-dependent membrane repair negatively regulates pyroptosis downstream of GSDMD activation. *Science.* 2018;362:956-960. doi:10.1126/science.aar7607
326. Martin SJ. Cell death and inflammation: the case for IL-1 family cytokines as the canonical DAMPs of the immune system. *FEBS J.* 2016;283:2599-2615. doi:10.1111/febs.13775
327. Medina CB, Mehrotra P, Arandjelovic S, et al. Metabolites released from apoptotic cells act as tissue messengers. *Nature.* 2003;426:6967. 2020;580:130-135. doi:10.1038/s41586-020-2121-3
328. Green DR. The coming decade of cell death research: five riddles. *Cell.* 2019;177:1094-1107. doi:10.1016/j.cell.2019.04.024
329. Daley SR, Teh C, Hu DY, Strasser A, Gray DHD. Cell death and thymic tolerance. *Immunol Rev.* 2017;277:9-20.

CHAPTER 8

Lymphoid Tissues and Organs

Martin F. Flajnik, PhD • Nancy H. Ruddle, PhD

INTRODUCTION

The vertebrate immune system defends against invading pathogens, by both innate and adaptive mechanisms (Chapter 1). Although cells that can respond to pathogens are found in tissues throughout the body, the optimal structures for the adaptive response to pathogens are organized, compartmentalized, cellular aggregates that facilitate antigen concentration and presentation to antigen-specific lymphocytes having an unlimited receptor repertoire. The primary lymphoid organs, the yolk sac (YS), fetal liver, thymus, and bone marrow (BM), are the sites where diverse populations of naïve lymphocytes mature to disperse throughout the body to await foreign invaders (Fig. 8.1). These remarkable differentiation processes occur in a foreign antigen-independent fashion. The secondary lymphoid organs (SLOs), including the lymph nodes (LNs), spleen, Peyer's patches, and other mucosal-associated lymphoid tissues (MALTs), are discrete sites in which naïve, antigen-specific T and B lymphocytes encounter invaders to generate an adaptive response. LNs and spleen are responsive to environmental influences and undergo remarkable changes during antigenic challenge. Precise programs control the development of the spleen, LN, Peyer's patches, tonsils and adenoids, and (in the mouse and rat) the nasal-associated lymphoid tissue (NALT). Somewhat less anatomically restricted tissues that are even more sensitive to the environment facilitate and include accumulations of lymphoid cells that are organized but less discretely defined: the bronchus-associated lymphoid tissues (BALTs) and inducible lymphoid follicles (ILFs). Tertiary lymphoid organs (TLOs), or more accurately, tertiary lymphoid tissues (TLTs), are accumulations of lymphoid cells that arise ectopically, in sites that are not anatomically restricted, and are not regulated by typical SLO developmental programs. TLTs respond to environmental stimuli and arise during chronic inflammation subsequent to microbial infection, graft rejection, autoimmunity, or cancer by the process of lymphoid neogenesis.[1,2] In this chapter, the structure, function, trafficking patterns, and developmental signals that regulate the hierarchy of lymphoid organs will be described.

PRIMARY LYMPHOID ORGANS

The primary lymphoid organs are the sites where pre-B and pre-T lymphocytes mature into naïve B and T cells in the absence of foreign antigen (Chapters 25 and 27). Each T cell or B cell expresses a unique antigen receptor that can recognize and respond to exogenous antigen and, in most cases, discriminate between self and foreign antigens. Naive cells leave the primary lymphoid organs having received and responded to developmental cues that result in the rearrangement of their antigen receptor gene segments to generate a repertoire capable of recognizing and responding to a wide variety of foreign antigens. During maturation in the primary lymphoid organs, the naïve lymphocytes express various chemokine receptors and adhesion molecules that direct them to discrete areas in the SLO.

Yolk Sac

For decades, the YS has been known to provide the first wave of fetal erythropoiesis beginning at D10.5. Fetal hemoglobin is expressed in this first wave of cells, which is largely replaced at birth.[3] Over the past decade, leukocyte lineages were shown to originate in the YS, in most cases cells that self renew for the life of the individual. In the macrophage lineage, Langerhans cells, microglia, Kupffer cells, among others are YS-derived (Fig. 8.1).[4] These macrophages are tissue restricted and continually self renew without replenishment from new precursors. B1 cells, which have differentiative properties unlike B2 cells (eg, no dependence on IL-7) are also derived from YS precursors.[5] The first thymocyte precursors that give rise to the early waves of gamma-delta T cells that seed the skin, vagina, and intestine are also YS derived.[6] These cells have defined variable (V) regions and differentiate prior to the expression of terminal deoxynucleotidyl transferase and thus have no diversity at the rearrangement junctions (Chapter 25). These early waves of leukocyte development have provided a new paradigm for the development of self-renewing white blood cell lineages.[7]

Fetal Liver

The earliest lymphoid cell precursors derive from self-renewing hematopoietic precursors called hematopoietic stem cells (HSCs). During ontogeny, these cells occupy several niches. In the fetal mouse, the first wave of hematopoiesis occurs in the YS and aorta-gonad-mesonephros (AGM) region at E10.5[8] (Fig. 8.1). The placenta also contains HSC activity[9,10] (reviewed in 11). HSCs leave the YS and placenta and migrate to the fetal liver, and then the BM and thymus and spleen under the influence of chemokines and adhesion molecules (Fig. 8.1). The cells in the fetal liver respond to CXCL12 (stromal cell derived factor or SDF-1) and, in contrast to those in the BM, also respond to Steel factor.[12] The fetal liver is also the source of CD4+CD3− lymphoid tissue

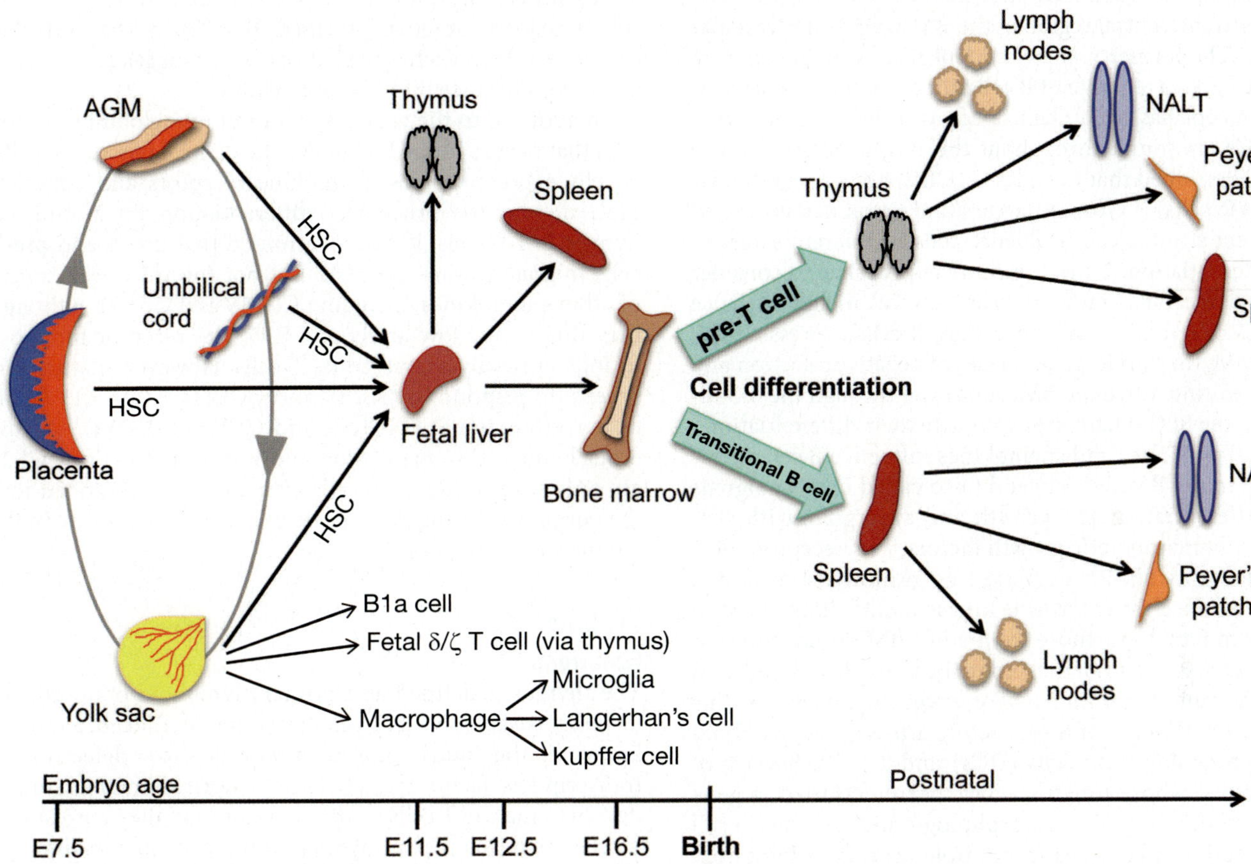

FIG. 8.1. Overview of embryonic and postnatal development of hematopoietic cells in mouse primary and secondary lymphoid organs. AGM, aorta-gonad-mesonephros; HSC, hematopoietic stem cell; NALT, nasal-associated lymphoid tissue.

inducer (LTi) cells that express lymphotoxin α (LTα) (also called tumor necrosis factor [TNF] β) and LTβ. LTi cells and NK cells were the first-known lineages of innate-lymphoid cells (ILCs). The requirement of LTi for the development of SLOs is described below. As mentioned, differentiation of B cells first occurs in the YS, predominately B1a cells (CD5-positive in mice). The ligands for P-selectin, E-selectin, and VCAM-1 are required for HSC to leave the fetal liver and home to the BM.[13,14]

Bone Marrow

Functions

The BM is the source of self-renewing populations of stem cells. These cells include hematopoietic precursor cells, HSCs, and endothelial progenitor cells, which may derive from a single precursor.[15] The adult BM contains stem cells that can differentiate into adipocytes, chondrocytes, osteocytes, and myoblasts.[16] Collectively these cells are called HSPCs or hematopoietic/stem progenitor cells. In the adult, cells leave the BM and seed the thymus. Additional factors enable the differentiation of immature B cells from HSC.

In addition to serving as a primary lymphoid organ where B cell differentiation and development occur, the BM is also a niche for antibody-secreting cells.[17] After naïve B cells interact with antigen in the SLOs, such as the LNs, spleen, and Peyer's patches, and differentiate into plasma cells, they enter the bloodstream and travel to the BM. Thus, this organ serves not only as a primary lymphoid organ but also as a reservoir for fully differentiated plasma cells, known as long-lived plasma cells (LLPCs).

Architecture: Cellular and Functional Niches

The microenvironment of the BM, contained in the central cavity of bone, is a complex 3-dimensional structure, with cellular niches that influence B cells during their development and later, as plasma cells, upon their return to the BM (Chapter 25). BM has a rich blood supply with a nutrient artery that branches into ascending and descending arteries further dividing into cortical capillaries, periosteal capillaries, and endosteal capillaries, finally merging into a sinus.[18] The previous understanding of B cell differentiation in the BM was that primitive HSCs were in close contact with the endosteum near osteoblasts (the "endosteal niche"). During the course of differentiation into mature B cells, these cells moved into the central region of the BM cavity (the "vascular niche").[19] The former niche was identified as the location of HSC; the latter, as the site of B cell differentiation.[20] This anatomic concept has been challenged, as it has been reported that HSCs are found throughout the BM; a reticular niche has been described that includes CXCL12-abundant reticular cells (CAR cells).[21] Growth factors and cytokines produced by different stromal cells influence cells at different stages in their differentiation. Thus, it is more appropriate to consider functional or cellular, rather than anatomical, niches.[18,22] Once the HSCs differentiate into immature B cells expressing cell surface IgM, they undergo processes of negative selection and receptor editing, leave the BM, and travel through the bloodstream to the SLO where they complete their differentiation.

Several cytokines and chemokines influence B cell differentiation in the BM. Flt-3 ligand (also called Flk-2L) signals B cell differentiation and growth and synergizes with several other hematopoietic growth factors.[23] Its receptor, Flt-3 (also known as Flk-2), expressed by primitive HSCs, is a member of the class II tyrosine kinase family. In contrast to HSCs from fetal liver, those from adult BM do not respond to Steel factor. Chemokines contribute to BM B cell and define the functional niches. For excellent reviews of this topic, see.[18,21,24] Many of these factors affect other lymphoid cells, such as dendritic cells (DCs) and T cells. Several of these factors, whose functions have been identified in gene deletion studies in mice, in morphologic analysis, and in cell culture studies, play roles in multiple aspects of lymphoid organ development; their activities, although important in the BM, are not limited to that organ. CXCL12, also known as stromal cell-derived factor-1 (SDF1), is a chemokine that is crucial for recruitment of HSC to the BM. It is widely expressed by osteoblasts, reticular cells,[22] and endothelial cells.[25] In fact, the interaction of HSC expressing CXCR4, the receptor for CXCL12, with that chemokine on the endothelial surface is the first step in the HSC's exit from the circulation into the marrow (summarized in 26). CXCL12 is also essential for the earliest stage of B cell development (pre-pro-B cells). CXCR4 is expressed by early B cells and is downregulated in pre-B cells. CXCR4 remains at low levels in immature B cells and mature B cells SLO but is upregulated after B cells interact with antigen and differentiate into LLPC.[17] Once a pre-pro-B cell has interacted with CXCL12, it moves on to a different stromal cell that expresses IL-7. In B cell development, IL-7 acts later than CXCL12 in a narrow window between pro-B cells and immature B cells in the scheme proposed by Nagasawa.[18]

Traffic In and Out: Chemokines and Adhesion Molecules

The extensive vascularization of the BM allows entrance of hematopoietic precursors and plasma cells and egress of mature cells. Hematopoietic progenitor recruitment to the mouse BM is dependent on the interaction of a variety of chemokines, integrins, and selectins, and their receptors, counter receptors, and vascular cell adhesion molecules. These include α_4 integrin (VLA4 or $\alpha_4\beta_1$), VCAM-1,[27] P-selectin glycoprotein ligand-1 (PSGL-1), E-selectin,[28] $\alpha 4\beta 7$, and mucosal addressin cell adhesion molecule (MAdCAM-1).[29] A small subpopulation of newly formed B cells in the BM that expresses L-selectin has been described,[30] suggesting a mechanism for entrance into LN (see below).

In addition to the acquisition of immunoglobulin expression that occurs in the BM under the control of stromal cells, B cells express various chemokine receptors and adhesion molecules in the process of differentiation. In a study of human BM B cells, it was determined that pro-B and pre-B cells migrate toward CXCL12 but not toward a wide range of other chemokines, including CCL19 and CCL21, although they do express low levels of CCR7, the receptor for those ligands normally expressed by T cells. However, mature BM B cells do respond to CCL19 and CCL 21,[31] CXCL13 (indicating a functional CXCR5), and CCL20 (MIP3α). On the other hand, CCR6 expression and responsiveness to CCL20 are only seen in mouse B cells after they have migrated into the periphery during the process of maturation and are in the circulating B cell pool.[32]

Thymus

Functions

The thymus is defined as a primary lymphoid organ due to its indispensable role in T cell development. Indeed, a mutant known as the "nude" (*nu/nu*) mouse due to a defect in the transcription factor FoxN1, lacks a normal thymus and is devoid of mature T cells.[33] Although some studies suggest that T cells can develop extrathymically in organs such as the "cervical thymus"[34] and gut epithelium,[35] the thymus remains the primary site for T cell maturation, education, and selection. Developing T cells or thymocytes represent more than 95% of total cells in the thymus and give rise to mature T cells that migrate to the SLO.

The T cell antigen receptor repertoire is shaped during development in the thymus by the processes of positive and negative selection. These selection processes are so stringent that nearly 95% of all T cell precursors die within the thymus.[36] Negative selection ensures that T cells, which can recognize the organism's self-antigens presented on major histocompatibility complex (MHC) class I or II molecules with high affinity, are eliminated prior to their export to the periphery. In contrast, positive selection generates T cells recognizing self-antigen with low to medium affinity to eventually protect the organism against invading pathogens (Chapters 27 and 32).

While negative selection is highly efficient in eliminating most self-reactive T cells, some of these cells nevertheless exit the thymus. These autoreactive cells impose a threat to various organs as their activation may result in the development of autoimmune diseases. The immune system has evolved several ways to prevent the activation of autoreactive T cells in the periphery. One mechanism that protects against autoimmunity is the generation of a T cell population capable of suppressing activation of self-reactive T cells (Chapter 29). These thymus-derived protective T cells, known as regulatory cells (T_{reg}), recognize self-antigens with relatively high affinity.

Architecture

The thymus in the mouse consists of two symmetric lobes located above the heart, while in humans the thymus is multilobed. In other vertebrates there can be multiple lobes that run down the length of the neck (Chapter 4). The thymus comprises lobules divided into three distinct regions, capsule, cortex, and medulla, that harbor thymocytes at various maturation stages (Fig. 8.2A). Although maturing T cells constitute the majority of cells in the thymus, other cell types such as macrophages, DCs, B cells, and epithelial cells are also present[37] (Fig. 8.2B). Histological analysis of the thymus reveals a clear distinction between the thymic cortex and medulla, which are separated by a corticomedullary junction. The thymic cortex appears darker because it is more densely populated with T cell precursors, whereas the medulla appears considerably lighter and contains smaller numbers of T cells relative to other cell types (Fig. 8.2A). Blood vessels and small blood capillaries are found throughout the thymus. The fact that T cell progenitors are found in the more highly vascularized corticomedullary border suggests that blood vessels in this region facilitate the entry of T cell progenitors into the thymic parenchyma.[38] In the thymic medulla, the close association between medullary thymic epithelial cells (mTECs) and thymic blood vessels[39] suggests that these vessels may act as organizers of the medullary thymic compartment. Lymphatic vessel distribution coincides with that of blood vessels and capillaries.[40] The majority of lymphatic vessels are located in the thymic medulla, although some can also be found in the cortex. The role of lymphatic vessels in thymic function is unclear, although it has been proposed that these vessels deliver extrathymic antigens into the thymus and/or export mature T cells from the thymus into the circulation.

Cellular Composition and Functions

T cell precursors at various stages of differentiation represent the majority of cells in the thymus. Antigen-presenting cells (APCs) of either hematopoietic or stromal origin mediate the differentiation and selection of T cells in the thymus. The different stages in T cell selection and maturation that take place in distinct regions of the thymus (Fig. 8.2C) are discussed at length elsewhere in this volume (Chapter 27). Briefly, lymphoid progenitors enter the thymus at the corticomedullary junction and are identified as early thymic progenitors still capable of differentiating into myeloid cells or B cells.[38] The next stage of differentiation involves the double-negative (DN) T cells that lack expression of the cell surface molecules CD4 and CD8. DN cells undergo four maturation steps termed DN1-DN4 distinguished by the expression of the cell surface molecules CD25 and CD44.[41,42] DN3 cells migrate to the subcapsular zone while rearranging their TCRβ chain genes and expressing the β chain protein in combination with a surrogate α chain, AKA pre-Tα. The cells expressing this receptor undergo a proliferative phase and then differentiate into cells expressing CD4 and CD8 (double-positive [DP] cells) and begin to rearrange the TCRα chain genes. In the cortex, DP cells undergo positive selection under the influence of the cortical thymic epithelial cells (cTECs). If the T cell receptor (TCR) is never formed due to failure to express a functional TCR, or if the cell fails to recognize self MHC for selection, the cell dies via "death by neglect." This is the major source of cell wastage in the thymus. Positively selected DP cells further differentiate into single-positive (SP) cells expressing either CD4 or CD8 depending on whether the TCR is specific for class II or class I molecules, respectively. SP cells migrate to the medulla where they interact with cDCs and mTECs and can be negatively selected if they recognize self antigen with high avidity. mTECs express two transcription factors, AIRE and FEZF2, that induce the transcription of genes encoding tissue-restricted antigens (TRAs) that can engender tolerance (see below[43]). Self-tolerant SP cells are then exported out of the thymus and migrate into the T cell zones of SLO.[44]

DN precursors can also give rise to T cells expressing the γδTCR, described above as the earliest cells that differentiate in the thymus. γδ T cells arising in the adult are distinct from TCRαβ T cells in their tissue distribution and recognition of antigens. TCRαβ DP cells control the development of TCRγδ cells via the production of LTαβ.[45]

The thymic parenchyma consists of a complex three-dimensional structure comprising thymic epithelial cells (TECs) (Fig. 8.2B). cTEC and mTEC are phenotypically and functionally distinct and support different stages of T cell maturation, with cTEC involved in positive selection and mTEC in negative selection. Several cell surface markers are used to distinguish MTECs from CTECs. Among these markers are the cytokeratins K5 and K8, the adhesion molecule Ep-CAM, and the glycoprotein Ly-51. K8 and Ly-51 are expressed by cTECs, while K5 and Ep-CAM are expressed by mTECs.[46,47] TECs are unique cells in the body in that they express MHC II constitutively, like professional APCs. cTECs express a unique catalytic subunit of the proteasome called β5-thymus (β5t), which produces a set of peptides specific to the thymic cortex. These peptides permit positive selection on peptides that are not found in any other tissue in the body and optimize the selection of CD8 T cells in ways that are not understood.[48] As mentioned, the role of mTECs in T cell selection was first elucidated by their expression of the transcription factor autoimmune regulator (AIRE) and later FEZF2. These transcription factors play important roles in T cell selection and prevention of autoimmunity as illustrated by the fact that humans with a mutated form of the *AIRE* gene exhibit polyendocrine autoimmunity due to inadequate T cell selection.[49,50] AIRE controls the expression of certain TRAs in the thymus, such as insulin, a protein that is unique to the β cells of the islets of Langerhans in the pancreas.[47,51] The expression of TRA in the thymus facilitates the negative selection of maturing T cells, which would otherwise result in the peripheralization of autoreactive T cell clones. The thymic medulla includes distinct structures also known as Hassall's corpuscles. Originally described by Arthur Hill Hassall in 1849, this structure consists of concentric stratified keratinizing epithelium. Hassall's corpuscles are implicated in several processes of the thymus[52] including the expression of TRA, such as insulin, immunoglobulins, and filaggrin (key antigens in rheumatoid arthritis)[53,54] and serving as a prominent site for T cell apoptosis.[55] Specialized cTECs, called thymic nurse cells (TNCs) are found at the corticomedullary junction.

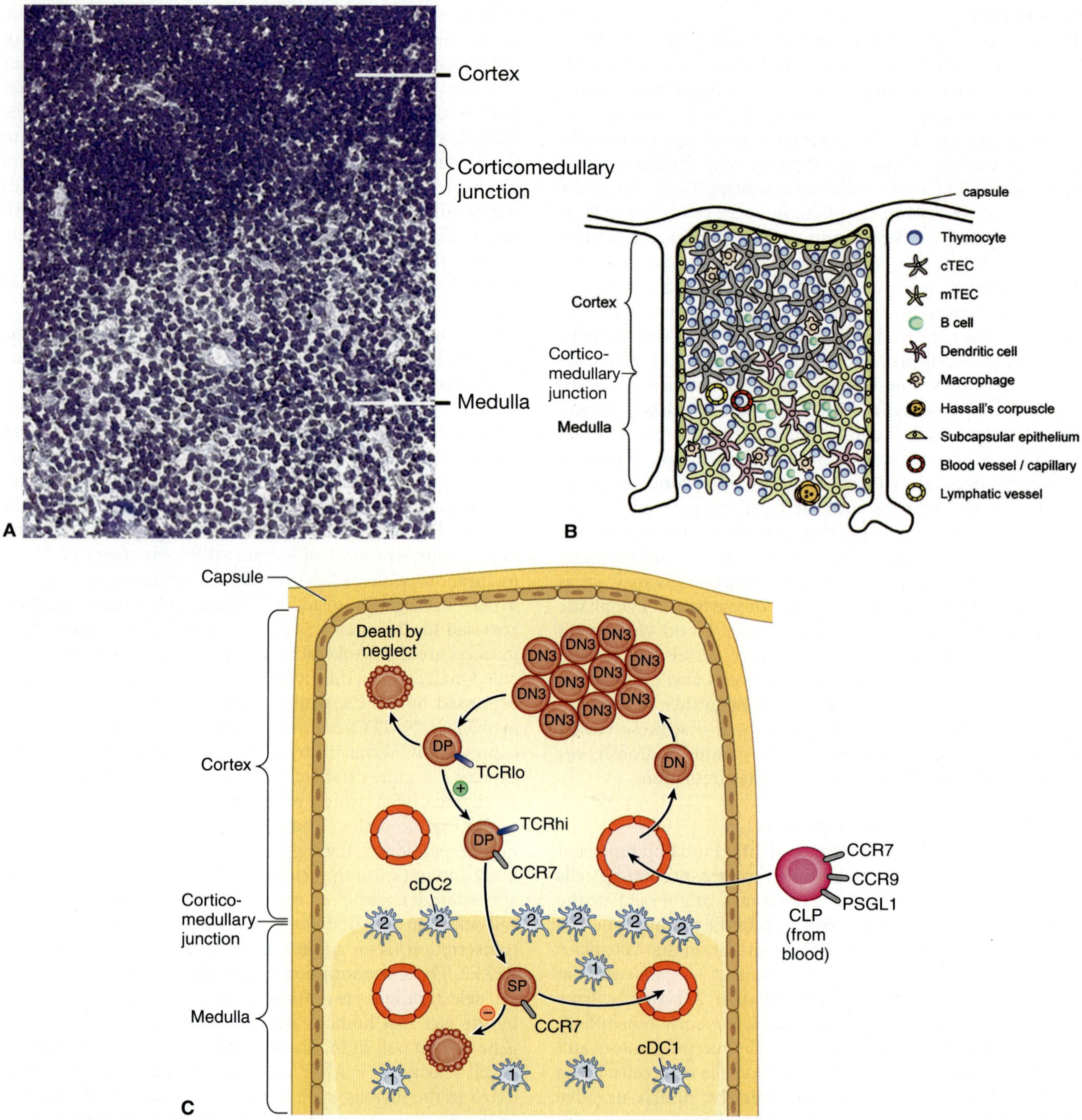

FIG. 8.2. **Thymic structure and function. A:** Low-power magnification of a hematoxylin-stained frozen section of a mouse thymus. The cortex is tightly packed with lymphocytes; the medulla is less tightly packed and thus stains less intensely. **B:** Diagram of mouse thymic cellular populations. cTEC, cortical thymic epithelial cell; mTEC, medullary thymic epithelial cell. **C:** Precursor cells (CLP, common lymphocyte progenitor, blue) enter through blood vessels at the corticomedullary junction. They progress to the subcapsular region as double-negative T cell precursors (DN, CD4/CD8-negative). DN3 cells rearrange the T cell receptor (TCR)β locus and undergo proliferation and progress to double-positive (DP) cells expressing both CD4 and CD8 and rearrange the TCRα chain. Thymic stromal cells provide growth factors, and TCR-expressing DP cells undergo positive selection (+) under the influence of cortical thymic epithelial cells (cTEC). Those cells that fail positive selection undergo "death by neglect." Single-positive (SP) CD4 or CD8 cells migrate into the medulla where they undergo negative selection (clonal deletion), mainly through AIRE-expressing mTECs and dendritic cells (DCs). cDC2 cells are generally found at the corticomedullary junction and cDC1 in the medulla. Single-positive (SP) thymocytes, having undergone differentiation, exit through blood vessels to the periphery.

TNCs surround up to 50 thymocytes and serve as a dedicated area for positive selection, specifically where DP thymocytes undergo high levels of receptor editing of the TCRα locus.[56]

DCs are also found in the thymus, either originating from a thymic precursor[57,58] or partially mature peripheral DCs that continuously enter the thymus from the circulation.[59] DCs are not involved in positive selection but have been known for decades to contribute to negative selection.[60-63] The three major classes of DC—cDC1, cDC2, and plasmacytoid DC (pDC)—are found in discrete locations within the thymus. cDC2 are found at the corticomedullary junction, and cDC1 and pDC are predominantly localized to the medulla.[64] cDC1 can take up TRA from mTEC and cross-present peptides on class I molecules to delete autoreactive CD8 SP cells.[65] DCs are also involved in the selection of regulatory T cells in humans,[66] and in the mouse, peripheral DCs can migrate to the thymus and serve as efficient inducers of T_{reg}s.[67] Activation of thymic DCs can be mediated in part by the IL-7-like cytokine, thymic stromal lymphopoietin (TSLP). In the human thymus, Hassall's corpuscles produce TSLP, and TSLP-activated DCs can alter the fate of self-reactive T cell from deletion to positive selection of regulatory T cells.[68] pDCs can induce the generation of particularly potent T_{reg}s from CD4+CD8-CD25- thymocytes.[69] These findings highlight the heterogeneity of DCs and emphasize their ability to fulfill different roles during T cell development in the thymus.

Macrophages and B cells[70] also can act as APCs in the thymus. In contrast to DCs, thymic macrophages are located throughout that organ and do not play a significant role in T cell selection[71] but are important for homeostatic uptake of apoptotic thymocytes. B cells are detected in human and mouse thymus at relatively low numbers.[72,73] They are capable of producing antibodies of several different isotypes.[74] It has been suggested that thymic B cells induce negative selection in developing thymocytes. Mice whose B cells are experimentally manipulated to express a myelin oligodendrocyte glycoprotein (MOG) peptide show deletion of MOG-specific T effector cells.[75] One mechanism for B cells' effects on T cell selection is suggested by the observation that B cells play a role in the control of TRA. Studies done in mice deficient in B cells showed fewer TECs with reduced expression of insulin and MOG, native antigens of the pancreas and brain, respectively.[76]

Traffic In and Out: Adhesion Molecules and Chemokines

Adhesion molecules mediate the extravasation of leukocytes from blood and lymphatic vessels into the thymus. These molecules also play an important role in facilitating lymphocyte homing into various regions of the thymus. Indeed, thymic blood vessels express the adhesion molecules ICAM-1, VCAM-1, CD34, peripheral node addressin (PNAd), and vascular adhesion protein-1 (VAP-1).[77] The expression of high levels of ICAM-1, VCAM-1, and VAP-1 on venules near the corticomedullary border suggests that these molecules may play a role in the recruitment of thymocyte progenitors. More specifically, VAP-1, which is restricted to the venules surrounding the sites of progenitor homing, can mediate the extravasation of leukocytes. The regional distribution of these adhesion molecules further illustrates the importance of a distinct anatomical separation between cortex and medulla and represents their individual functions.

Chemokine-mediated T cell migration and traffic to and within the thymus is crucial for normal T cell selection. The compartmentalization of the thymus is orchestrated by a milieu of chemokines. As mentioned, the thymic cortex is involved in the maturation of DN cells to DP cells and the positive selection of DP T cells capable of recognizing self MHC:peptide complexes and elimination of cells that fail to express a TCR that can interact with self MHC. The thymic medulla is the site of negative selection of SP T cells and positive selection of T_{reg}. Various chemokines that are produced by the thymic cortex and medulla allow T cells expressing different chemokine receptors to home to specific regions of the thymus. This differential expression of chemokines is complemented by the fact that T cells at different maturation states express different chemokine receptors. During development, the entry of lymphoid progenitors into the thymus is highly dependent on CCL21 and CCL25, which bind the chemokine receptors CCR7 and CCR9, respectively. Mice lacking CCR7 or CCL21[78,79] show a transient delay in thymus colonization by lymphocytes (day 14.5), and this delay is further extended (day 17.5) in mice lacking CCR9.[80] Lymphoid progenitors enter the thymus at the corticomedullary border and commence their migration outward toward the subcapsular region of the cortex as DN3 cells. The expression of the chemokine receptors CXCR4 and CCR7 by DN cells is important in directing cell migration.[81,82] In the subcapsular region, DN thymocytes that have successfully rearranged their TCRαβ chains progress to the DP cell stage. Positively selected DP cells move inward toward the thymic medulla for further differentiation into SP cells. The ligands for CCR7 are crucial in mediating the migration of positively selected DP cells into the medulla as illustrated by the fact that a deficiency in CCR7 or its ligands, CCL19 or CCL21, prevents DP cell relocation from the cortex to the medulla resulting in abnormal central tolerance.[83,84] Interesting, antigen itself can control T cell migration speed from the cortex to the medulla. In the presence of a negative selection ligand, T cells slow down considerably and are limited to a confined zone 30μ in diameter allowing for a more prolonged selection and induction of developmental arrest.[85]

The export of positively selected SP T cells out of the thymus is also dependent on chemokines. Chemokines involved in T cell emigration are CXCL12 and its receptor CXCR4, which repel SP cells out of thymus,[86] and CCL19, which promotes T cell emigration from thymus of newborn mice and migration into SLO T cell zones.[87] The chemoattractant sphingosine 1-phosphate (S1P) is an additional mediator of T cell emigration. SP T cells express the S1P receptor ($S1P_1$) and are attracted to the high levels of S1P present in the serum promoting their egress.[88,89]

Although the role of different thymic compartments and chemokines in the maturation of naïve SP T cells has been extensively studied, the thymic regions and chemokines that control the selection of regulatory T cells remain largely unknown.

Development

The initial development of the thymus at mid-gestation in the mouse is independent of vascularization or BM-derived cells. The mouse thymus rudiment first evident on day 11 of gestation is derived from the endoderm of the third pharyngeal pouch.[90] This gives rise to the thymic lobes as well as to the parathyroid gland; development of both tissues is defective in the human congenital deficiency DiGeorge syndrome, which impacts the differentiation of the pharyngeal arches. On day 12.5 of gestation, a separation of the primordium is observed and by day 13.5 a distinct thymus is apparent. Evidence has suggested that not only the pharyngeal pouch endoderm but also the ectoderm is involved in thymic development.[91,92] cTEC and mTEC are derived solely from the pharyngeal pouch endoderm[93]; although the endoderm and ectoderm are found in close proximity between gestational day 10.5 and 11, only the endoderm actively contributes directly to thymic development, with ectoderm playing a supportive role.[94] These findings support the so-called single origin model of thymus development. Both cTEC and mTEC are derived from a cTEC precursor expressing β5t, with the divergence of the two populations occurring late in mouse fetal life.[95]

The contribution of lymphocytes to the normal development of thymic cortex and medulla is well recognized. In models of T cell deficiency, cTEC and mTEC development is blocked at different stages depending on the stage of T cell arrest. If T cell development is arrested at the DN stage, as in the RAG-deficient mouse, the thymic medulla is greatly reduced while the thymic cortex remains unaffected.[96,97] A more severe phenotype is observed in transgenic mice that overexpress the human CD3 signaling molecule. In these mice, T cell arrest occurs earlier than in RAG knockout mice, leading to a loss of both cortex and medulla and to a shift from a three-dimensional to a two-dimensional structure of the thymic epithelia.[97]

LTα and LTβ have been identified as master regulators of mTEC development and expression of TRAs. LTα and LTβ are members of the TNF super family and mediate the processes of SLO development and inflammation.[98] In the absence of LT, TRA and AIRE expression are reduced and certain mTEC subpopulations fail to develop.[99,100] T cells, owing to their vast numbers appear to be the main source for LT production in the thymus[45]; however, additional resident cells also serve as a source of LT. Resident thymic B cells are the highest source of LTα and LTβ on a per cell basis[76] revealing yet another aspect of thymic B cells in thymus development and function. The TNF family member RANK produced by thymocytes is also involved in the development of the thymic medulla.[101]

The development of the thymus in humans closely follows the model of thymic development in the mouse. As in the mouse, human thymic colonization by HSCs occurs relatively early, at week 8.2 of gestation. During this stage, the thymic medulla and cortex are organized, suggesting that thymocytes are required for normal thymic development. Between gestation weeks 9.5 and 10, the first signs of thymocyte negative selection are evident, and by gestation week 10 to 12.75 the gradual onset of positive selection is detected.[102]

SECONDARY LYMPHOID ORGANS AND TISSUES

Naïve T and B cells express their receptors for specific antigen, leave primary lymphoid organs, circulate through the blood stream, migrate into the tissues, and lodge in SLOs. The frequency of naïve cells specific for an individual antigen is quite low (estimates range from 1 in 10^5 to 1 in 10^6). Thus, the chance that an individual T or B cell will encounter its specific antigen in the circulation is rather low. SLOs are strategically located in anatomically distinct sites where foreign antigen and APCs efficiently concentrate and activate rare antigen-specific lymphocytes, thus leading to the initiation of adaptive immune responses and generation of long-lived protective immunity. These organs include highly organized, compartmentalized, and mostly encapsulated tissues such as LNs, spleen, appendix, tonsils, murine NALTs, and Peyer's patches. Naïve cells are also primed in less discrete tissues throughout the body, including the BALTs, cryptopatches, and isolated lymphoid follicles (ILFs).

Lymph Nodes

LNs are bean-shaped structures dispersed along lymphatic vessels (Fig. 8.3A and B). Lymphatic vessels play important roles in tissue fluid balance, fat transport, and the immune response. In contradistinction to blood vessels, which form a closed recirculating system, lymphatic vessels comprise a blind-end, unidirectional transportation system. The absorbing lymphatic vessels, or lymphatic capillaries, remove interstitial fluid, macromolecules, and cells from extracellular spaces and transport the collected lymph through the primary collector. The collected lymph and its cellular contents are transported into the thoracic duct and returned to the blood circulation. In humans, lymph collected from the entire lower body region, the left head, and arm region accumulates in the thoracic duct and returns to blood circulation via the left subclavian vein; lymph collected from right head and arm region returns to blood via the right subclavian vein.

LNs, usually embedded in fat, are located at vascular junctions and are served by lymphatic vessels that bring in antigen and connect them to other LNs. Although most LNs are classified as peripheral LNs, a few (cervical, mesenteric, and sacral), termed mucosal nodes, express a slightly different complement of endothelial adhesion molecules, cooperate with the mucosal system, and are regulated somewhat differently in development (see below). Although all LNs are vascularized, and thus can receive antigens from the bloodstream, they are also served by a rich lymphatic vessel system and are thus particularly effective in mounting responses to antigens that are present in tissues. These antigens may be derived from foreign invaders that are transported by APCs or can be derived from self-antigens. Thus, LNs extend the role of the primary lymphoid organs and discriminate between dangerous foreign antigens and benign self-antigens. This capacity relies on the APCs and their state of activation in the LN, and the recognition capacity of the naïve T and B cells.

LNs can also function as niches for generating peripheral tolerance, an additional mechanism to minimize the effects of those self-reactive T cells that escape central tolerance in the thymus.[103] DCs constitutively sample self-antigens and migrate to draining LNs even in the steady state.[104-106] Because most self-antigen-bearing DCs in LNs are immature[107] and have low levels of costimulatory molecules, they are not effective at activating naïve cells. They regulate self-reactive T cells by inducing anergy, clonal deletion, and/or expanding regulatory T cells.[106-110] In addition, populations of cells in SLO that can present TRAs to CD8 cells[111-115] include two different populations of AIRE-expressing stromal cells, DCs, and AIRE-negative lymphatic endothelial cells. These cells may play a role in self-tolerance, or perhaps they can prime for autoimmunity in inflammatory conditions.

Structure and Organization

The LN, surrounded by a collagen capsule, is divided into discrete areas, the cortex, the paracortex, and medulla, which contains cells specialized for innate and adaptive functions, supported by a network of stromal cells and blood and

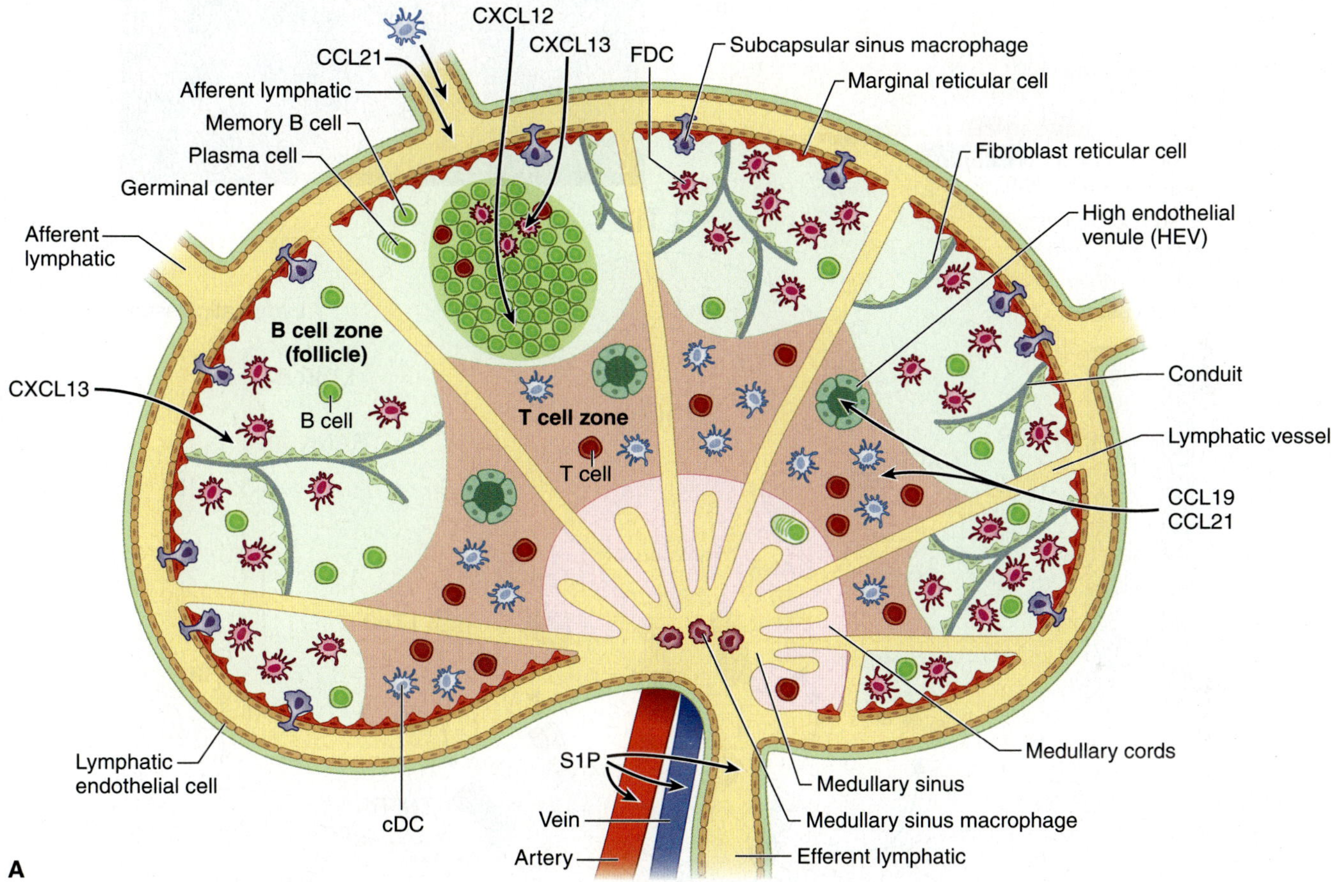

A

FIG. 8.3. A: Lymph node structure, functional regions, and development. A: The lymph node is divided into an outer cortex and inner medulla surrounded by a capsule and lymphatic sinus (*yellow*). The cortex includes B cells (*blue*) and follicular dendritic cells (FDC, *orange*) in B cell follicles (*blue shade*). The paracortical region (T cell zone, *green shade*) includes T cells (*green*) and conventional dendritic cells (cDC, *yellow*). Macrophages are found in the subcapsular sinus (subcapsular sinus macrophages [SSM], *pink*; marginal reticular cells [MRC], *red*) and medullary cords (medullary sinus macrophages [MSM], *purple*). Lymphocytes enter the lymph node through an artery at the hilus region and into the parenchyma through high endothelial venules (HEVs, *green circles* in the T cells zones) expressing PNAd and CCL19 and CCL21. Activated cDCs enter the LN via the afferent lymphatics. Stromal cells also produce these chemokines. T cells and cDCs are directed to the paracortical region by CCL19 and CCL21. B cells are directed to the cortex by CXCL13 expressed by FDC and other stromal cells. After interaction with antigen and T cells, germinal centers (GC) develop (circle in one follicle, filled with activated B cells); FDCs move to one pole of the GC and plasma cells and memory B cells are generated. Antigen and DCs drain into the lymph node from the tissues through afferent lymphatic vessels. Antigen continues to percolate through the node via conduits (*light blue*) that are lined with fibroblast reticular cells (FRC), *light green*. Activated lymphocytes leave through efferent lymphatic vessels. **B:** Flow of blood and *lymph* into and through the lymph node. Note the arteriovenous communications (AVC), the venous sphincters (VS), and cells leaving the HEVs into the parenchyma. **C:** Immunofluorescence of a lymph node section. T cell zone stained in red and B cell zone in *green*. **D:** LN development. Schematic of LN development. (From *Inducible Lymphoid Organs*. Chapter Basics of Inducible Lymphoid Organs Fig. 2A K. Kabashima and G. Egawa editors. Springer Nature Switzerland, 2020 with permission.)

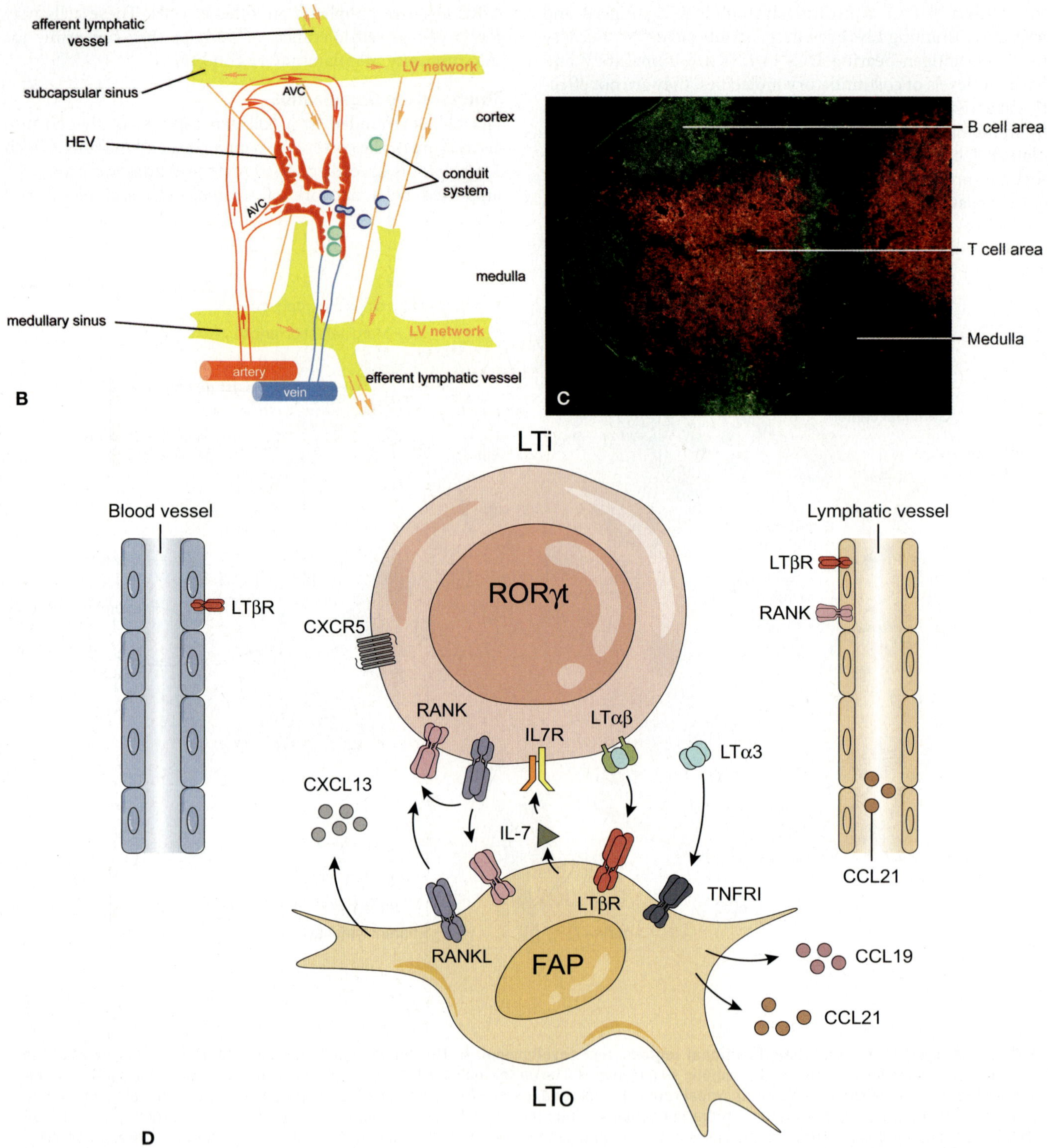

FIG. 8.3. Cont'd

lymphatic vessels (Fig. 8.3). The LN is highly compartmentalized with B cells with their APCs including subcapsular sinus macrophages (SSMs) and follicular dendritic cells (FDCs) in the cortex; T cells and DCs and their APCs predominate in the paracortical region (Fig. 8.3A and C). Efferent lymphatic vessels enter the LN carrying lymphocytes and antigen-containing DCs. Below the capsule is the subcapsular sinus, whose floor consists of lymphatic endothelial cells. Just below the sinus is a population of CD169+ cells termed the SSMs. These cells can produce a variety of cytokines including type I interferon (IFN), which is crucial in activating both innate and adaptive responses. In addition, the

SSMs transport antigens into the cortical region that includes discrete clusters of cells called primary follicles consisting of densely packed naïve B cells and FDCs. They appear to be particularly adept at presenting antigen-antibody complexes to B cells.[116] The SSMs pass their antigens to FDCs, which in turn present the soluble antigen to B cells. After B cells encounter their cognate antigen, they are activated and proliferate and secondary follicles and germinal centers (GCs) develop. T cells and DCs distribute in the paracortex. FDCs, a population of mesenchymal origin, support B cell follicles or GCs under stimulation.[1] Plasma cells are also concentrated in the medulla as they prepare to leave the LN and circulate to the BM. A network composed of reticular fibers, fibrous extracellular matrix bundles, and another mesenchymal cell population, the fibroblastic reticular cells (FRCs), supports the entire LN.[117]

After naïve T and B cells encounter antigen, they undergo extensive changes in expression of chemokine receptors and adhesion molecules that result in their movement to different areas of the LN or leaving it all together.[118,119] Sphingosine-1-phosphate receptor 1 ($S1P_1$) facilitates lymphocyte egress from LNs as they move toward the ligand, S1P in the lymph,[120,121] as discussed in more detail below.

Lymph Node Vasculature: Blood Vessels and Lymphatic Vessels

Soluble antigen and APCs enter LNs via afferent lymphatic vessels. After surveying antigens in the LNs, lymphocytes leave those organs via efferent lymphatic vessels that can connect to the next LN in the chain and finally return to blood circulation.[122-124] In this manner, HEVs and lymphatic vessels maintain lymphocyte homeostasis during the steady state.

Blood endothelial cells play a crucial role in lymphocyte trafficking in the LN. One or two arteries enter the LN at the hilus (Fig. 8.3A). These arteries branch and pass through the medulla area, enter the cortex, and sometimes continue in the subcapsular area. Beneath the subcapsular sinus, the branching capillaries form loops and some of them become arteriovenous communications (AVCs). AVCs become high endothelial venules (HEVs) in the cortex area and occasionally extend from the subcapsular sinus to the medulla. HEVs, so called because of their cuboidal endothelial cells, constitute a specialized postcapillary network in the LN, playing a critical role in lymphocyte recirculation. Each main HEV trunk receives three to five branches lined with high endothelial cells and two or three branches lined with flat endothelial cells. The luminal diameters of HEVs progressively increase from cortex to medulla. Finally, HEVs merge into segmental veins in the medulla area and join larger veins in the hilus[122] (Fig. 8.3A). Intravital microscopy has revealed that the entire venular tree consists of five branching orders with the higher orders in the paracortex and the lower orders located in the medulla and hilus areas. Only the higher-order venules, located in the T cell area, are specialized into HEVs and are recognized by the monoclonal antibody MECA-79.[123,124] Recirculating naive lymphocytes leave the bloodstream via HEVs, specialized vessels with a high cuboidal endothelium, and migrate into the LN parenchyma. Peripheral node addressin (PNAd), defined by the MECA-79 antigen, is an L-selectin ligand and is a characteristic HEV adhesion molecule. PNAd is composed of a variety of core glycoproteins, including GlyCAM-1, CD34, Sgp200, and podocalyxin; these proteins must be sialylated, sulfated, and fucosylated to become functional L-selectin ligands. The several enzymes that mediate these posttranslational modification events include FucT-IV, FucT-VII, and GlcNAc6ST2 (also called HEC-6ST, LSST, GST-3, HEC-GlcNAc6ST, gene name *Chst4*),[125-129] which with the exception of a population of cells in the intestine[130,131] is uniquely expressed in high endothelial cells. Together with another sulfotransferase, GlcNAC6ST1, that is expressed at sites in addition to HEVs, it sulfates glycoproteins in the Golgi apparatus[132] to generate the MECA 79 epitope, PNAd.

Lymphatic vessels also play critical roles in the immune response. Human and murine lymphatic vessels express several characteristic markers: lymphatic vessel endothelial hyaluronan receptor (LYVE-1), Prox-1, podoplanin, CCL21 the vascular endothelial growth factor receptor-3 (VEGFR-3), and neuropilin-2.[133] The collected lymph and cell contents enter the LN via several afferent lymphatic vessels and filter through the node where they again are concentrated in the medullary sinus. Lymphatic vessels are concentrated in the subcapsular sinus and medullary area.[122] Factors from afferent lymph can be either transported deep into the LN cortex or move via the subcapsular sinus and leave the LN through efferent lymphatic vessels.[117] In this manner, soluble antigen and APCs from peripheral tissues are efficiently concentrated in the draining LN and initiate an adaptive immune response. Lymphocytes can also enter LNs through afferent lymphatic vessels and leave via efferent vessels and move to the next LN in the chain. This is accomplished in part through a gradient of sphingosine-1-phosphate, which is at a high concentration in the lymph and blood and low concentration in the LN. Lymphocytes in the LN downregulate this receptor ($S1P_1$) and then upregulate it as they prepare to leave and migrate toward the higher concentration in the efferent lymph.[120,121]

Conduit System

A conduit system in the LNs physically connects the lymphatic sinus with the walls of blood vessels and enables the incoming factor(s) from lymph to move rapidly deep into the paracortical area[117,122] (Fig. 8.3A). The conduit system consists of four layers: (1) a core of type I and type III collagen bundles; (2) a microfibrillar zone composed largely of fibrillins; (3) a basement membrane abundant with laminins 8 and 10, perlecan, and type IV collagen that provides a supportive structure; and (4) FRC that embraces the entire conduit system.[134,135] This conduit system enables incoming lymph to penetrate deep into the T cell area. A special subset of immature DCs, called conduit-associated DCs, can take up and process antigens moving along the conduit.[135] In this manner, the conduit system probably provides a physical support for rapid initiation of adaptive immune responses after immunization.

Cell Trafficking Into and Within the LN

Naïve lymphocytes enter the LN through interactions with HEVs through a series of steps termed rolling, tight binding, and diapedesis (or transendothelial migration). PNAd, expressed on the endothelial surface, slows down (tethers) naïve L-selectin^{hi} lymphocytes in their progress through the blood vessels. After this initial interaction, CCL19 and CCL21 on the HEVs are instrumental in activating the lymphocyte integrin LFA-1. This results in tight binding of LFA-1 to ICAM-1 on the HEV, facilitating diapedesis of lymphocytes that migrate between or through the endothelial cells toward the chemokines located in the paracortical region (T cells, DCs) or cortex (B cells). Data suggest that the HEV-lymphocyte interaction is not random in different lymphoid tissues in the mouse.[136,137] T lymphocytes preferentially adhere to peripheral LN HEVs, B cells preferentially adhere to HEVs in Peyer's patches, and T and B cells exhibit an intermediate pattern of adhesion to mesenteric LN HEVs.[36,136,138] The selective adhesion of naïve lymphocytes to HEVs is at least partially controlled by the differentially expressed adhesion molecules in different lymphoid organs. PNAd rapidly replaces MAdCAM-1 after birth in mouse peripheral LNs[139] but is expressed in mucosal LNs together with MAdCAM-1, the ligand for the integrin $\alpha_4\beta_7$.

Compartmentalization of cells in the LN is orchestrated by lymphoid chemokines CCL19, CCL21, and CXCL13. Stromal cells in the paracortical region produce CCL19, and the protein is transported to the surface of HEVs.[140] CCL21 is encoded by several genes[141]; CCL21-leu is expressed by lymphatic vessels outside the LN[142]; CCL21-ser is made by stromal cells and HEVs in the LN. CCL19 and CCL21 recruit CCR7-expressing cells across the HEVs to the paracortical region. CXCL13, produced by stromal cells in the B cell follicles, attracts CXCR5-expressing B cells.[143] Movement through the LN is facilitated by P-selectin glycoprotein ligand 1 (PSGL-1), a molecule on the surface of many cells of the immune system that reacts with a variety of receptors, originally defined as P-selectin, E-selectin, and L-selectin. PSGL-1's originally described role was to allow entrance of activated cells into inflamed tissues through its binding to those selectins. It is now known that PSGL-1 has a wide variety of binding partners[144] including SLO chemokines. A seminal paper described an important role of PSGL-1 on naïve T cells, namely, facilitation of their entrance into LNs through its interaction with the canonical lymphoid organ chemokines, CCL19 and CCL21.[145] This enhanced chemotactic response is lost on activated cells, although it is present on memory cells. PSGL-1 expression is low on circulating B cells,[146] indicating that B cells do not generally use this pathway to enter LNs. Observations with PSGL-1^{null} mice indicates that PSGL-1 plays additional roles in homeostasis and supports egress of T cells from LNs in that PSGL-1^{null} T cells remain in LNs longer than WT cells.[145]

After naïve T and B cells encounter antigen, they undergo extensive changes in expression of chemokine receptors and adhesion molecules that result in their movement to different areas of the LN or leaving it all together.[118,119] After naïve T and B cells encounter antigen, they undergo extensive changes in expression of chemokine receptors and adhesion molecules that result in their movement to different areas of the LN or leaving it all together.[118,119] Sphingosine-1-phosphate receptor 1 ($S1P_1$) facilitates lymphocyte egress from LNs as the cells move toward the ligand S1P in the lymph.[120,121]

The Intimate Relationship Between Lymph and HEVs

Incoming lymph is necessary for the maintenance of HEV phenotype and function. After afferent lymphatic vessels are severed, dramatic changes occur in HEVs. These include flattening of the endothelium, a decrease in the uptake of ^{35}S-sulfate[147,148] (a functional marker of GlcNAC6ST-2), a reduction of lymphocyte adherence to the vessels,[149-151] and decreased expression of PNAd and the HEV genes, *Glycam-1* and *Fuc-TVII*.[150-152] An increase in MAdCAM-1 expression[150] suggests that these events are not simply due to a general downregulation of blood vessel gene expression and indicates that continual accumulation of afferent lymph factor(s) in LNs is necessary for HEV maintenance.

Topographic relations between HEVs and the lymphatic sinus have been described in the rat mesenteric LNs. Some HEVs are located in the medulla area and positioned closely in relation to the lymphatic sinus.[122] Occasionally, HEVs are separated from an adjacent lymphatic sinus by only a thin layer of collagen bundles. The closely apposed HEV and lymphatic sinus in the medulla provide the physical support for the intimate relationship between lymph and HEV. However, most HEVs are in the paracortical area and are separated from the lymphatic sinus by lymphocytes. The conduit system physically connects the subcapsular sinus and HEVs and allows incoming lymph factor(s) to migrate rapidly to the wall of HEVs.[122,153,154] Low-molecular-weight fluorescent tracers (below 70 kD) move rapidly via the conduit system and lead directly to the wall of HEVs. In this manner, low-molecular-weight tracers can migrate with lymph within minutes to HEVs and enter the HEV lumen.[117] Lymph-borne chemokines likely adopt this route to regulate HEV function. IL-8 administration via afferent lymph increases lymphocyte HEV transmigration within minutes.[122,155] In addition, lymph-borne chemokines, such as MIP-1α also enter the conduit system and move rapidly to HEVs.[117] These data suggest that lymph factor(s) can quickly access and regulate HEVs.

Development

Despite their distinctions in the adult regarding morphology and expressed genes, a close association of blood vessels and lymphatic vessels is seen during embryogenesis. The generation of embryonic lymphatic vessels from preexisting veins in pig embryos was first described in the early 1900s by Sabin and has recently been molecularly defined (reviewed in 133). A variety of transcription factors have been identified that contribute to the lymphatic specification and maintenance of the lymphatic vessels phenotype (reviewed in 156). As early as mouse E9, expression of Sox18 and CoupTFII is apparent in the cardinal vein. These induce Prox1 in the dorsolateral side of the vein resulting in polarization and lymphatic-biased endothelial cells. Prox-1 generates a feedback signal for the further maintenance of budding and migration of endothelial cells. Targets of Prox1 include Prox1 itself, podoplanin, VEGFR3, integrin-α, and Nrp-2. Prox1 expression in blood endothelial cells also represses expression of markers of those cells. NfatC1, Foxc2, and Tbx1 are additional transcription factors that are expressed later in lymphatic vessel development and contribute to patterning, pericyte covering of lymphatic vessels, lymphatic vessel valves, and lymphatic vessel maturation. At E11.5 to 12.0, CCL21 is expressed in these lymphatic-biased endothelial cells, as is VEGFR-3, which is reduced in blood endothelial cells. The endothelial cells expressing LYVE-1, PROX-1, VEGFR-3, and CCL21 become irreversibly committed toward a lymphatic pathway.[157,158] The separation of lymphatic endothelial cells from venous endothelium requires a Syk/SLP-76 signal[159] that is provided by platelets.[160,161] Thus, during early lymphangiogenesis, some endothelial cells express both blood vessel and lymphatic vessel markers, indicating the close association of these two vascular systems. Mesenchymal lymphangioblasts may also contribute to early lymphangiogenesis.[162] The lymphatic venous junction remains in adults in only limited regions but plays an essential role in connecting the function of the two vascular systems.

Studies in the mouse have taken advantage of transgenic, knockout, and imaging studies to provide a mechanistic understanding of the process of LN development. Although lymph sacs per se do not appear to be essential for the initiation of LNs, a lymphatic vessel system is necessary for the later development and cellular population of LNs.[163] Furthermore, an interaction between mesenchymal cells and primitive endothelial cells is apparent in the early LN anlage.[164]

Cytokines, Chemokines, and Transcription Factors in Lymphoid Organogenesis

The TNF family members and several chemokines play key roles in SLO development as has been summarized in several reviews (Fig. 8.3D). These include LTα3, signaling through TNFRI and TNFRII,[165-167] and membrane-bound $LT\alpha_1\beta_2$, signaling through the LTβ receptor (LTβR)[164-166] and the extended TNF family members RANKL and RANK.[168,169] The relative importance of the different cytokines and chemokines in the development of individual lymphoid organs has been summarized in several reviews.[170,171] The cytokines include members of the immediate LT/TNF family and the extended family TRANCE or TRANCER, also called RANKL and RANK.[172,173] Studies of mice deficient in these cytokines and chemokines have provided insight into this process. The precise signals for the development of individual types of SLOs (peripheral LNs, mesenteric LNs, Peyer's patches, spleen) vary somewhat, but all depend on hematopoietic cells that express members of the LT/TNF family that interact with stromal cells expressing receptors for that family.[170]

LNs develop at defined locations throughout the body. The precise signals of initiation are unclear, but the earliest stages appear to be dependent on retinoic acid produced by nerves in the vicinity of the developing node,[172] which activates mesenchymal cells to produce CXCL13 in a location that is determined by LTβR on endothelial cells.[173] Further development depends on interactions between mesenchymal stromal cells, called lymphoid tissue organizer (LTo) cells, and hematopoietic-derived lymphoid tissue inducer cells (LTi), a type of innate lymphoid cell (ILC3). Additional blood and lymphatic endothelial cells also play organizing roles through their expression of cytokines, cytokine receptors, chemokines, and adhesion molecules and their transport of lymphoid cells (Fig 8.3D).

In the mouse, LTo cells, derived from FAP+ precursors,[174] express IL-7, LTβR, TNFR, VCAM, and MAdCAM-1. Pre-LTi cells (CD45+CD3-CD4+Rorγt-) derived from fetal liver progenitors[175-179] arrive via a vein between E14.5 and 15.0 at the inguinal LN anlagen that is already enmeshed in lymphatic vessels whose temporal interaction with the developing LN has been elucidated.[180] At later times (E15.0-18.5), additional LTi cells (CD45+CD4+CD3-a4b7+RORγt+ IL7Rα+LTα+LTβ+ CXCR5+RANK+RANKL+) are drawn into the region via expression of CXCL13, CCL19, and CCL21 and IL7 by the LTo cells and enter the LN anlagen via lymphatic vessels. Lymphatic vessels contribute in crucial ways to LN development, as *initiators* and/or as *facilitators* of the process.[181-183]

LTi cells express RANK and RANKL resulting in the autocrine activation of LTα and further activation of LTo cells and lymphatic endothelial cells.[184] CXCL13 produced by the LTo cells induces additional LTαβ from LTi cells. LTo cells expressing TNFRI and LTβR respond to LTα and LTαβ produced by the LTi cells and produce additional chemokines, including CCL19 and CCL21. Eventually the lymphatic vessels contribute to formation of the LN capsule and subcapsular sinus. Even before birth there exists a small population of blood vessels that has some characteristics of HEVs and allows further accumulation of LTi cells and presumably B and T cells.[185] Later as the structure of the LN emerges, LTα3 and LTα1β2 induce lymphoid chemokines from stromal cells and lymphatic vessels and cause partitioning of T cells and DCs in the paracortical regions and B cells and FDCs into the cortex, whereas PNAd expression is regulated through the LTβR in part through LTαβ induction of Chst4,[125] a crucial glucosyl transferase that modifies the scaffold proteins of PNAd. The prolonged interaction between lymphoid tissue–inducer cells and stromal organizer cells promotes the development of HEVs, which support the entry of naïve lymphocytes.[186]

It is unclear how HEVs differentiate from the flat blood vessels during early lymphoid organogenesis. At birth, HEVs of all LNs express MAdCAM-1, which is replaced in the first few days in peripheral LNs by PNAd.[187] Both MAdCAM-1 and PNAd are expressed in mucosal LNs. LTα alone can induce MAdCAM-1, but PNAd requires LTαβ.[2,188,189] In the remaining mesenteric LNs of *Ltβ−/−* mice, PNAd expression is impaired,[189] indicating that optimal LNHEV PNAd expression requires $LT\alpha_1\beta_2$ signaling through the LTβR and the alternative NF-κB pathway (see below)[190] in part through LTαβ induction of Chst4.[125] Continual signaling through the LTβR is necessary for maintenance of HEV gene expression.[191,192] Data suggest that the actual physical presence of lymphocytes in the close vicinity of HEVs contributes in important ways to the physical cobble-stone appearance of these vessels.[193]

The NF-κB signaling pathways, downstream of the TNF family receptors, play important roles in lymphoid organ development.[194] The alternative pathway, characterized by NF-κB-inducing kinase (NIK) and IKKα is particularly important. *aly/aly* mice, which have a point mutation in *Nik*, lack all LNs and Peyer's patches.[195,196] In these mice, LTβR, mediated signaling between NIK and members the TRAF family appears to be disrupted.[196-198] LTβR signaling induces gene expression via both the classical and alternative NF-κB pathways in mouse embryo fibroblasts. The classical pathway mediated by p50:p65 heterodimers induces expression of proinflammatory genes (*Vcam-1, Mip1b, Mip2*). Intraperitoneal injection of an agonistic LTβR antibody induces splenic chemokines (CCL19, CCL21 and CXCL13) and requires NIK activity and subsequent p100 processing.[98] IKKα is a critical component in alternative NF-κB pathway. Mice with a mutated form of the *ikka* gene have reduced HEV expression of HEC-6ST (GlcNAc6ST2) and GlyCAM-1, further confirming that the LTβR signal regulates HEVs through the alternative NF-κB pathway.[190] Several other signaling pathways that contribute to lymphoid organogenesis include the helix-loop-helix transcription factor inhibitor (Id2) and retinoid acid–related orphan receptors (RORs) RORγ and RORγt.[177,199,200]

Changes in Lymph Nodes After Immunization

LNs undergo dramatic changes and remodeling after immunization. Early after a variety of immunogenic exposures, such as skin painting with oxazolone, injection of ovalbumin or sheep red blood cells in adjuvant, or bacterial or viral infection, remodeling occurs. This remodeling is apparent as a complex kinetics of changes in lymph flow, lymph cell content, blood flow, HEV gene expression, and lymphatic vessels.[192] Afferent lymph flow and lymph cell content increase soon after initial inflammation and eventually return to preimmunization levels.[201-205] LN lymphangiogenesis occurs, which eventually resolves.[192,206] Blood flow and lymphocyte migration into LNs peak at 72 to 96 hours,[204,207-209] accompanied by an increase in HEV number and dilation,[210,211] accounting for the significant LN enlargement apparent at 72 to 96 hours.[207] Efferent lymph flow also increases soon after immunization, but lymph cell content in the efferent lymph drops during the first several hours, indicating the first wave of accumulation of lymphocytes in the draining LN. The cell content of the efferent lymph later increases and peaks at 72 to 96 hours.[212] These events, taken together, contribute to the significant enlargement of the draining LN at day 4 after immunization. During the early times after immunization, despite the increase in the number of HEVs, the expression of genes that contribute to L-selectin ligand, including *Fuct-vii, Glycam-1, Sgp200*, and *Chst4*, is initially downregulated followed by a recovery.[152,192,213] However, despite the downregulation of chemokines CCL21 and CXCL12, some genes such as those encoding CXCL9, CCL3, and E-selectin are upregulated,[214] as is MAdCAM-1,[192] suggesting a reversion to an immature phenotype before the eventual recovery of the mature phenotype. Transcriptomic single cell analysis of HEVs at homeostasis, after treatment with agonistic LTβR-Ig or during inflammation, again reveals remarkable phenotypic diversity, heterogeneity, and plasticity of HEVs.[215]

In rodents, HEV maturation is coincident with the continuing development of the LN and population of lymphocytes after birth, indicating that the mature HEV phenotype relies on the LN microenvironment.[139,216] Plasticity of the mature LN is also seen after injection of LTβR-Ig. LTβR-Ig treatment reduces LN cellularity, reverts the HEV phenotype to the immature state, inhibits FDC function, and disrupts immune responses to foreign antigens.[191,192,217]

Spleen

Function

The spleen is a large reddish organ located beneath the diaphragm near the stomach and pancreas (Fig. 8.4). It is the main filter of the blood and integrates innate and adaptive immune responses. Its structure in well-defined different compartments, red pulp (RP) and white pulp (WP), determines a variety of functions. The RP makes up ~75% of the spleen and is a source of hematopoiesis in the embryo that can continue in adult life under stress. Clearance of blood-borne damaged platelets, aged erythrocytes, and dead cells occurs in that region. An important function, consequent to its destruction of effete erythrocytes (which also occurs in the liver), is its role in iron recycling. Aged erythrocytes are phagocytosed by RP macrophages and free ferrous iron is scavenged from the hemoglobin. In some species, such as horses or dogs, the spleen stores erythrocytes that can be released after stress. The WP and marginal zone (MZ) constitute the highly organized lymphoid compartment of the spleen (Fig. 8.4). Owing to these organized lymphoid structures and the special vasculature and circulation, the spleen is a crucial site for blood-borne antigen clearance and presentation to T and B cells. The spleen is crucial in defense against blood-borne pathogens and contributes most significantly to defense against bacterial and fungal infections.[218]

Architecture and Cellular Composition

A fibrous capsule and trabeculae of fibrous connective tissue maintain the structure of the spleen. The general principles of splenic organization are similar across all jawed vertebrates (sharks through mammals), although specifics differ[219-223] (Chapter 4). As noted above, the spleen is divided into two areas: red pulp (RP) and white pulp (WP). A highly vascularized transit area, the marginal zone (MZ), surrounds the WP (Fig. 8.4). These compartments are anatomically and functionally distinct. As mentioned, the RP, in its activities as a hematogenous organ, removes damaged cells and acts as a site for iron storage and turnover. The WP is an organized SLO that shares basic features with the LN regarding T cell zones and B cell follicles. The complex structure of the spleen is directly related to the complexity of the vasculature of this organ. The afferent splenic artery branches into central arterioles, surrounded by WP areas, and ends in cords in the RP. Blood then collects in venous sinuses that determine the engulfment of erythrocytes by RP macrophages. Finally, the sinuses empty into the efferent splenic vein. Some

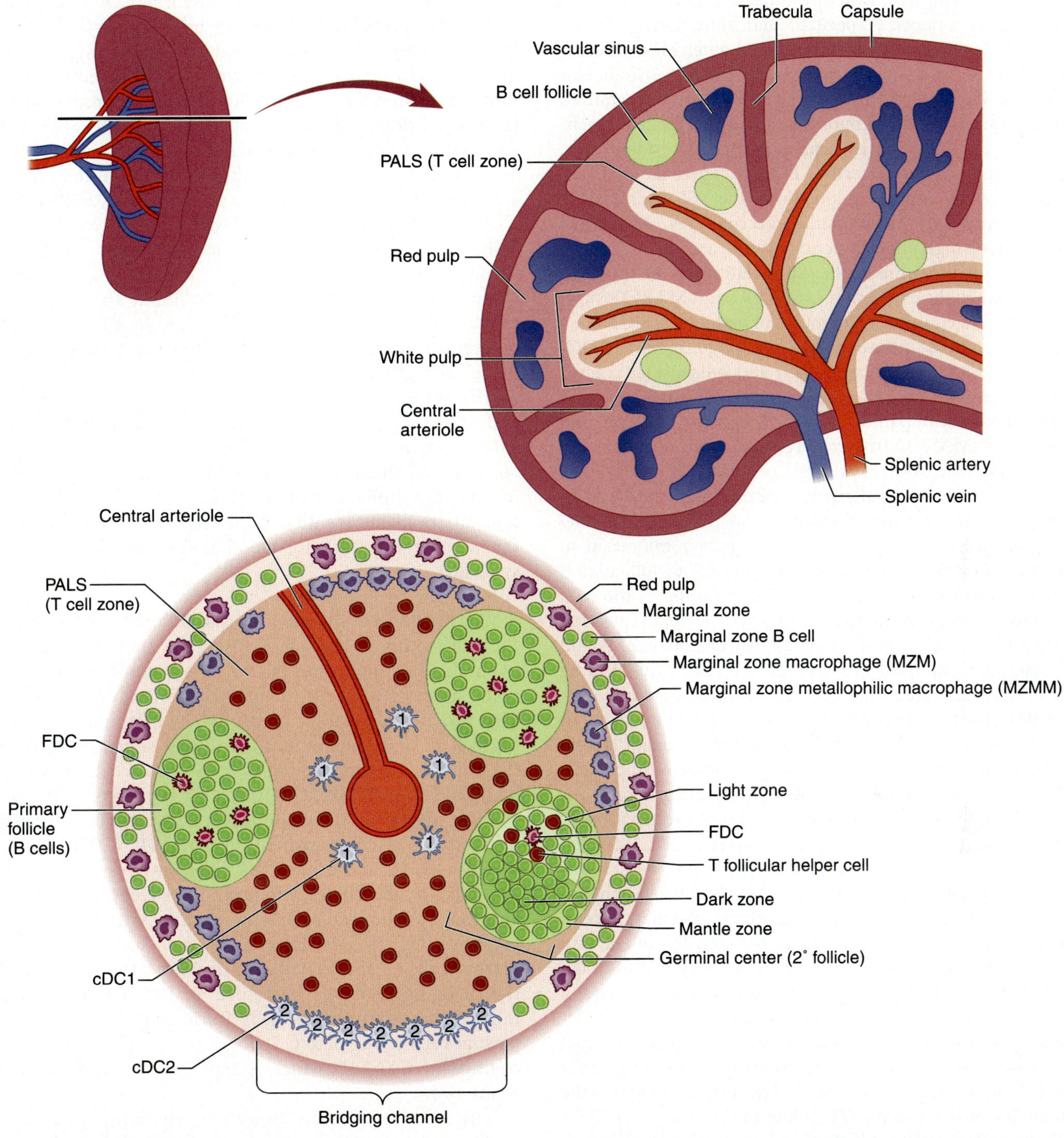

FIG. 8.4. **Structure of mouse spleen.** Left, the spleen as an organ with high vascularization (*red*, arteries; *blue*, veins). Top, section from the spleen showing general organization of red pulp (*light red*) and white pulp containing the periarteriolar lymphocytic sheath containing T cells (*green*) encompassing an arteriole and B cell follicles in blue. Also shown are the vascular sinuses (*light blue*) and splenic capsule/trabeculae (*brown*). Bottom, cross section of a white pulp. The outer, vascularized marginal zone (MZ) contains MZ macrophages (MZM, *purple*) and MZ B cells (*blue*). MZ metallophilic macrophages (MZMM, *pink*) are found at the inner border of the MZ. Three B cell zones (*shaded light blue*) with B cells (*blue*) and follicular dendritic cells (FDC), *orange*. One B cell zone shows a germinal center (GC) where active immune responses occur. B cells proliferate in the dark zone of the GC and are selected in the light zone on antigen displayed on FDC. T-follicular helper cells (Tfh), green, are also found in the light zone. Unstimulated B cells in the follicle are pushed into a mantle zone in the GC. The central arteriole penetrates the T cell zone (*shaded green*) containing T cells (*green*) and conventional dendritic cells-1 (cDC1), yellow. Bridging channels from the red pulp are lined with cDC2.

small arterial branches end in the MZ, which demarcates the RP and WP. The MZ structure differs somewhat between humans and rodents. A perifollicular zone surrounds the human MZ, which consists of inner and outer MZs, whereas the rodent MZ is a single structure and has no perifollicular zone. The MZ is a transition between the innate and acquired immune systems and an important transit area for cells reaching the WP. It also contains many particular resident cells, which have been best described in mice. There are two specialized MZ macrophage populations, the MZ macrophages (MZMs) and MZ metallophilic macrophages (MZMMs). MZMs have a high phagocytic activity and are phenotypically different from other macrophage populations in the spleen as exemplified by the expression of SIGNR1 and MARCO, molecules implicated in the recognition of pathogens. SIGNR1, a C-type lectin in the mouse that is a homologue of human DC-SIGN,[224] is also found in medullary and subcapsular macrophages (SSM) in the LN. MZMs are crucial for the capture of a wide variety of pathogens, including yeast, bacteria, and viruses.[225,226] MARCO (macrophage-associated receptor with collagenous structure), expressed constitutively on MZM,[227] is a class A scavenger pattern recognition receptor. In addition to the recognizing blood-borne pathogens, MARCO interplays with MZ B cells, modulating their migration to the WP with transferred antigen to pass off to follicular B cells.[228] MZMMs are located at the inner border of the MZ, in contact with marginal sinus lining cells, stromal cells that express MadCAM-1.[229] MZMMs differ from the MZMs and RP macrophages in that they express the sialic acid–binding Ig-like lectin sialoadhesin (Siglec-1), which binds to oligosaccharide ligands present on many cells.[230] Little is known about the specific roles of this population of macrophages during the immune response; however, the phenotype of mice deficient in sialoadhesin implies a role of MZMMs in T cell activation.[231,232] MZMMs are essential for cross-presentation of blood-borne adenovirus antigens to splenic cDC1 (which may also reside in the MZ), activating cytotoxic T cells.[233] MZ B cells are a specialized population of "innate B cell," phenotypically and functionally different from follicular B cells; they can be considered as a bridge between the innate and adaptive immune systems (Chapter 25). They express high levels of IgM, low levels of IgD, and the nonclassical class I molecule CD1d. MZB cells bind antigen through complement- and Fc-Receptors in the MZ directly and/or through interactions with MZM[234] and migrate to the WP, where they present blood-borne antigens.[235] The organization of the remainder of the splenic WP is like that of the other SLOs with compartmentalized B and T cell areas. The WP consists of a central arteriole that is surrounded by T cells, also known as the periarteriolar lymphoid sheath (PALS) that is surrounded by B cell follicles (Fig. 8.4). T cells interact with DCs and B cells sequentially after immunization. B cells migrate within the follicles where they interact with FDCs. At the T:B cell border, T cells, especially T follicular helper cells,[236] interact with B cells. GCs are the site of somatic hypermutation and Ig class switching, as described in the LN section. Plasma cells generated in the GC are found mainly in the RP and then can migrate to the BM as LLPC. cDCs are found in the WP in discrete locations.[237] A few cross-presenting cDC1 are found in the MZ, probably transiting as migratory DC from the blood. Resident cDC1 are present predominantly scattered in the T cell zone, as they are strongly influenced by CCR7 responding to CCL19/21 produced by the stroma. cDC2 localize to bridging channels that connect the RP to the T cell zone, one egress site for naïve and effector lymphocytes back into the blood. cDC2 positioning in this area is influenced by a combination of GPCR such as CCR7, EBI-2, S1PR1, and CD97.[238,239] Interactions of CD97 with CD55 on erythrocytes from the RP helps to set up a "shear effect" important for the homeostatic localization of cDC2 in the bridging channel.[240] Upon immunization cDC2 can reposition to the T cell zone to present antigen, and this is influenced by a lack of CD47 on erythrocytes (don't eat me signal) as a missing self signal for cell activation.

The spleen does not have an afferent lymphatic system, and initial antigen transport must occur through the blood vasculature. A conduit system has been described that allows antigens and chemokines to be transported through the WP in a manner similar to that described for the LN.[241] This conduit differs in the identity of the transported molecules by the fact that it contacts the blood, rather than the lymphatic system.

Traffic In and Out: Chemokines and Adhesion Molecules

Cell trafficking in the spleen resembles that which occurs in the LN in some respects and differs in others. The MZ is the main transit area for blood cells that enter the WP. This process seems to be controlled by marginal sinus lining cells[242] and chemokines, involving signaling through G-coupled protein receptors.[243] Stromal cell–produced lymphoid chemokines, CXCL13, CCL19, and CCL21, control the migration of lymphocytes and their organization in the various compartments in the WP. CXCL13, produced by FDC and other FRC, is essential to positioning of B cells in the follicles, whereas T cells respond to CCL19 and CCL21 produced by splenic stroma. After the immune response, GC B cells differentiate into memory cells or plasma cells in the WP, then migrate to the RP through the bridging channels or return to the bloodstream to migrate to peripheral tissues.[244] In the migration to the RP, plasma cells upregulate CXCR4, a receptor that binds the chemokine CXCL12, which is expressed in the RP as well as the BM.

The role of adhesion molecules regarding lymphocyte traffic in the spleen is not fully understood. Lymphocytes enter the WP through the marginal sinus, and cells lining that structure express MAdCAM-1 and ICAM-1.[229,245] MAdCAM-1 expression in these cells seems to be involved in splenic structure development and therefore migration of B and T cells.[242] However, treatment with anti-MAdCAM-1 or anti-$\alpha 1\beta 7$ antibodies does not totally inhibit homing to the spleen, suggesting that this ligand-receptor pair is not the only receptor required for lymphocyte entrance into the WP. Early studies indicating that treatment with an antibody that blocks the αL-integrin chain of LFA-1 inhibits homing by only 20%, and that lymphocytes deficient in LFA-1 enter the WP,[245] suggested that LFA-1 is not essential for entry of all cells into the WP. However, both $\alpha_4\beta_1$ and LFA-1 integrins are necessary for B cell retention in the MZ,[246] indicating a role for VCAM-1 and ICAM-1 in cell trafficking in the spleen. Once lymphocytes have encountered antigen, they undergo changes in chemokine receptor and adhesion molecule expression similar to those noted in the LNs, leave the WP, and enter the RP (eg, plasma cells or effector T cells) or the circulation. The mechanisms by which lymphocytes leave the WP are not totally understood, although there is movement out of the WP via the bridging channels.[244] In the LNs, the process of egress of activated lymphocytes through the efferent lymphatic vessels is mediated by upregulation of the expression of $S1P_1$. In the spleen, $S1P_1$ is required for B cell localization in the MZ, and the interplay with CXCR5 regulates the constant trafficking of these cells between the marginal zone and the WP,[235,247] as MZ B cells from $S1P_1$-deficient mice are not found in the MZ but instead in the follicles.

Although it is still unclear which molecules are involved in the egress of lymphocytes from the spleen, and splenic egress cannot be compared with egress in the LNs, two studies attribute an important role to the transcription factor Nkx2.3 in the regulation of spleen vasculature and homing of lymphocytes.[248,249]

Development

Because the spleen has characteristics of both a hematopoietic organ during early life and an SLO, the genes that regulate its development include, in addition to LT, others that are concerned with patterning and hematopoiesis. Several genes that contribute to the development of the spleen are also crucial for normal development of other nonlymphoid organs. The first sign of splenic development in the mouse occurs at E10.5-11. Segregation of RP and WP starts to be regulated during embryogenesis and continues after birth. Some structures, such as the MZ, develop during the first 3 weeks after birth. At E10.5-11, progenitor cells begin to condense within the dorsal mesogastrium, adjacent to the stomach and dorsal pancreas. The spleen and pancreas are so intimately associated that it is difficult to distinguish between them at these early stages. In fact, many genes that affect splenic development also contribute to pancreatic development. Several of these are homeobox genes and transcription factors that are expressed in the splenopancreatic mesenchyme at E10.5 (summarized in 218). The effects of just one of these genes, *Hox11* (now called *Tlx1*[218]), are on the spleen. The others affect multiple organs. *Tlx1* was originally described as an oncogene in T cell childhood acute leukemias involving the (10;14) translocation breakpoint.[250] *Tlx1*-deficient mice are asplenic,[251] and the product of this gene is a cell survival factor.[252] Several additional genes that are important for development of LNs, NALT, and Peyer's patches also contribute to splenic development. Mice deficient in members of the LT/TNF ligand receptor family and their downstream signaling molecules in the classical and alternative pathways exhibit defects in splenic development. However, none of these molecules is required for development of the early splenic anlagen, as mice deficient in any of the chemokines or cytokines retain a spleen. Studies have addressed a role for Glce, an enzyme that modifies heparan sulfate, in early lymphoid tissue morphogenesis, as fetal spleens in Glce-deficient mice exhibit a reduced size.[253] The changes in lymphoid tissue organization in mice deficient in LT/TNF ligand receptor family members are due in part to a reduction or near absence of lymphoid chemokines (CXCL13, CCL19, CCL21).[254] Lymphoid chemokine mRNA expression is reduced in the spleens of mice deficient in TNFR1, TNF, LTα, or LTβ, although CXCL12 mRNA levels are normal. However, treatment with an agonistic LTβR antibody induces expression of the lymphoid chemokines and CXCL12, suggesting that signaling through both the classical and alternative NF-κB pathways is responsible for organization and maintenance of splenic WP architecture. Mice deficient in LTα exhibit a disorganized WP with loss of T and B cell compartmentalization and loss of MZM, MZMM, MZ B cells, MAdCAM-1 sinus lining cells, and GC. *Ltb−/−* mice exhibit similar characteristics except that the disorganization is somewhat less pronounced. However, these lymphoid cytokines are dispensable during late fetal spleen development (see below), when segregation between WP and RP begins, but essential for maintenance of this structure during postnatal WP development.[255] *Tnf−/−* and *Tnfr1−/−* mice show defects similar to LT-deficient mice, with the exception of MZ B cells. T cells, B cells, and CD4+CD3− cells produce the cytokines necessary for maintenance of the splenic architecture.[256,257] In mice, CXCL13 is expressed in the splenic vasculature at embryonic day 17 (E17).[258] One day later a large number of transitional (T) B cells are found in circulation and within the spleen, all RP at this stage. At birth, B cells coalesce around the splenic vasculature, forming the nascent WP.[255] T1 and T2 immature B cells both express CXCR5, but only T2 cells are attracted by CXCL13, in vitro and presumably in vivo, and this dichotomy of the transitional B cell response to the chemokine persists throughout life. Thus, B cells form the WP at birth, forming the early PALS, and are displaced by T cells at about 7 days after birth, forming the definitive PALS and adult WP. Cold-blooded vertebrates follow the same developmental scheme of WP formation, but B cells continue to form the PALS throughout life with T cells being rather scattered in the RP.[259,260] After immunization, T cells migrate around and enter the B cell follicle, forming an inducible "T cell zone" for several weeks.

Plasticity After Virus Infection

Although much is known regarding changes in the LN after immunization, the spleen has not been studied as extensively in this regard. However, after infection with cytomegalovirus, WP T:B compartmentalization is disrupted.[261] The spleens of *Ltα−/−*-deficient mice exhibit a marked reduction in expression of CCL21-ser. This is even further reduced in cytomegalovirus infection, indicating that, in the adult, LT-independent pathways can contribute to maintenance of expression of lymphoid chemokines.

Mucosal-Associated Lymphoid Tissues

General Features

MALT covers all mucosal surfaces in the gut, the oropharyngeal and lacrimal mucosae, the nasal and bronchial airways, and the genitourinary tracts. MALT protects a huge surface area with approximately half of the lymphocytes of the entire immune system.[262] MALT is considered to be the body's gatekeeper because it is in intimate contact with the flora at the mucosal surfaces.

Although the location of some mucosal lymphoid tissues such as the palatine tonsils and appendix are fixed, all MALT are somewhat plastic because of their constant exposure to environmental antigens that induces them to change and remodel. Tonsils, adenoids, and their equivalent in rodents, the NALT, have a fixed location; Peyer's patches are found on the surface of the small intestine, but their number and size vary according to antigen exposure (Fig. 8.5). BALT in the lung and ILFs in the colon are the most plastic MALT, and their number and location are subject to environmental influences.

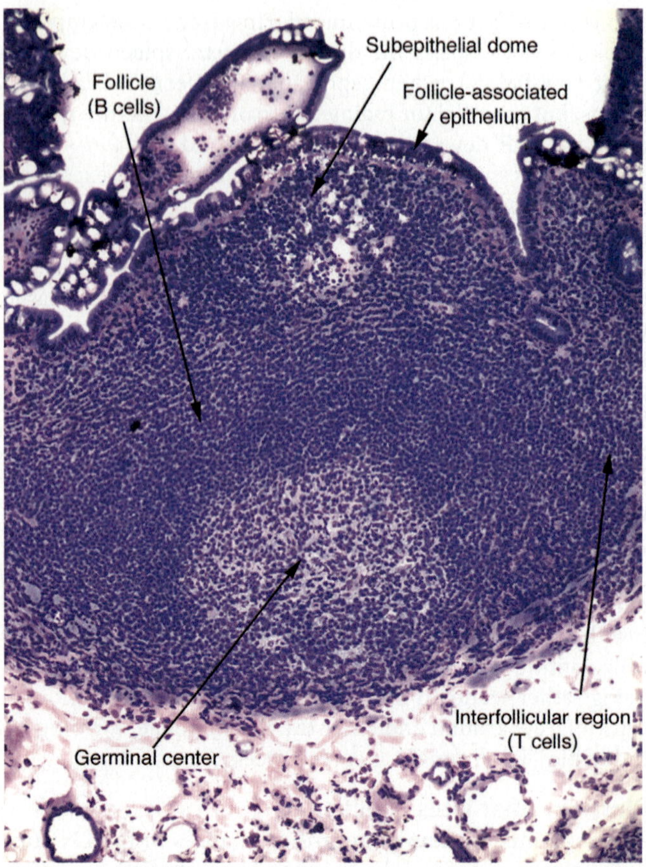

FIG. 8.5. Organization of the Peyer's patch. Photomicrograph of Peyer's patch. The Peyer's patches are located in the intestine near intestinal villi. The follicle-associated epithelium (FAE) is in contact with the gut lumen. M cells (not shown) in the FAE transport antigen into the subepithelial dome, populated by DCs and T cells. The interfollicular T cell–rich region surrounds the B cell follicle and germinal center. (Courtesy of A. Iwasaki Yale University School of Medicine, New Haven, CT.)

The mucosal epithelium that surrounds MALT is populated by a dense network of DCs, plasma cells, and intraepithelial lymphocytes that helps to maintain the epithelial barrier. These intraepithelial leukocytes provide retinoic acid and cytokines like IL-10 and TGF-β that condition the gut to become tolerant of antigens produced by the harmless commensal bacteria.[263,264] They are also responsible for inducing and maintaining tolerance to food antigens.[265]

The structurally defined MALT organs like the tonsils, NALT, and Peyer's patches share many features with LNs. Like those organs, MALT contains HEVs for the entry of naïve lymphocytes and stromal cells that secrete chemokines that direct lymphocyte traffic in the MALT (CCL19, CCL21 and its receptor CCR7,[266] CCL25 and CCL28 and their receptors CCR9 and CCR10,[267] and CXCL13 and its receptor CXCR5).[268,269] These organs also contain separate T cell and B cell compartments with B cell follicles, FDCs, GCs, and interfollicular T cells and DCs.[270]

MALT differs importantly from other SLOs with regard to its capsules and afferent lymphatic vessels. Unlike the spleen and LNs, which are surrounded by a dense fibrous capsule, MALT often has no capsule, with the exception of the tonsil that has a partial capsule that separates it from the pharyngeal constrictor muscles. Because MALT can sample antigens directly at the epithelial surface it has no afferent lymphatics. MALT epithelium, especially that of the Peyer's patch, is replaced by specialized lymphoepithelial cells called microfold (M) cells, which because of their high transcytotic capacity, transport antigens to the underlying lymphoid tissue. Tonsils and adenoids are covered by a squamous epithelium and have few M cells. They can sample surface antigens using epithelial DCs that push dendrites through the epithelial surface.[271,272] M cells have not been characterized in human BALT.[273]

The proximity of MALT to the epithelium is an important point, because it helps differentiate MALT from TLOs (see below). MALT can be defined as organized lymphoid structures in the mucosa that are in direct contact with the epithelium. Using these criteria, a lymphoid aggregate in the submucosa that is not directly in contact with the epithelium (ie, below the lamina propria) is more correctly defined as a TLO.[274]

The MALT Immune Response: Inductive and Effector Sites

MALT can simultaneously be a site for the induction of an immune response and an effector organ. The MALT surface containing M cells and DCs efficiently samples and transports antigens across the epithelium. Immediately below this single-cell epithelium is a dense network of DCs that can extend dendrites through the epithelium to grasp antigen to present to lymphocytes. Two distinct types of DCs have been identified in MALT, CD103+ DC and CX3CR1+ DC.[275] Both types of DC produce tolerogenic responses to commensal bacteria by enhancing the differentiation of Foxp3+ regulatory T cells and inhibiting that of inflammatory Th17 cells. CD103+ DCs express αE integrin, make retinoic acid, induce T regulatory cells, and increase T cell expression of the two gut homing receptors CCR9 and α4β7. CD103⁺ DCs exit MALT via the lymphatics and present antigen in the draining LNs. CX3CR1⁺ DCs that express the chemokine receptor for fractalkine do not migrate to the LN. CX3CR1+ DCs are longer-lived mucosal resident DCs that produce less retinoic acid than CD103+ DCs.[275]

MALT has three effector roles: local antibody secretion, systemic antibody secretion, and effector lymphocyte dispersal. MALT plays an important role in defense against pathogens by generating cells that migrate to other sites. MALT can dispatch lymphocytes to LNs, and the spleen and plasma cells to the BM. MALT also sends lymphocytes to other MALT effector sites including the salivary and lacrimal glands, the lactating breast,[276] and the vagina. There are some differences between the effector cells made at different MALT sites. NALT-derived B cells express CCR7 and do not home back to the gut. Instead, cells induced in the NALT home to the salivary glands and the vaginal mucosa LNs,[277] whereas lymphocytes leaving the Peyer's patch express $\alpha_4\beta_7$ and migrate to the gut mucosa. These "preferred pathways" for effector cells coming out of the different MALT has led to the suggestion that MALT is compartmentalized and cells circling in the BALT are in a separate compartment from the GALT. It is more likely that the nature of antigen influences the homing of effector cells. There needs to be flexibility and there is much overlap and redundancy in the system. For example, after the removal of the tonsils and adenoids in humans, or the NALT in the mouse, the cervical LN can act as an inductive site.[278] Likewise, the mouse BALT can mount an immune response in the absence of SLOs.[279,280]

Tonsils, Adenoids, and Nasal-Associated Lymphoid Tissue

The Waldeyer's ring is a group of lymphoid tissues encircling the wall of the throat that are the first defense against pathogens entering through the mouth or nose.[281] Humans have several tonsils with indistinct borders: one pharyngeal tonsil (adenoid), two tubal, two palatine, and one lingual. The tonsils contain, in addition to T cells, a large complement of B cells, many of which are positive for IgA.[281] The tonsil epithelium produces the polymeric IgA receptor, or secretory component,[282] crucial for transport of IgA dimers across the epithelium. sIgA provides an early form of defense against pathogens and toxins. Overall, the adenoids produce more secreted IgA (sIgA) than the tonsils.

The common cold virus uses ICAM-1 as a receptor to invade the nasal mucosa. ICAM-I is also expressed by HEVs in tonsils.[281] As noted above, HEVs are specialized vessels that express adhesion molecules that allow naïve lymphocytes enter the tonsil. Naïve L-selectin^{hi} lymphocytes adhere to peripheral node addressin (PNAd) on HEVs.[268,269] L-selectin^{hi} cells also bind to MAdCAM-1 predominantly found in the gut HEVs. Considering the "mucosal name" MAdCAM-1, it is surprising that tonsil HEVs express MAdCAM-1 only weakly or not at all.[283] In mice, the HEVs also manufacture the chemokine CCL21 that can recruit naïve T cells and mature DCs that express CCR7. Human tonsil HEVs do not manufacture CCL21 but are able to display chemokines that have been made and secreted by fibroblasts.[284] During tonsillitis, inflammatory chemokines, cytokines, and adhesion molecules such as CCL19,[266] VCAM-1, and E- and P-selectin are upregulated.[285]

The NALT consists of a pair of lymphoid organs above the soft palate in mice and rats that are considered analogous to the human Waldeyer's ring.[286] Despite being anatomically separate from the genitourinary tract, NALT has an important "effector" role concerning the generation of immune responses in that region.[287] After nasal immunization with HPV16 or ovalbumin, HPV16-specific or ovalbumin-specific IgA is detected in vaginal washes[167,288,289] and cytotoxic T cells are found in vaginal draining LNs.[277] However, the NALT itself is clearly an inductive site for both humoral and cellular immune responses[290] and supports class switching to IgA.[291] The stromal cells in murine NALT produce both CCL19 and CXCL13, whereas, in contrast to LNs, only HEVs transcribe CCL21 mRNA.[167] NALT HEVs express high levels of luminal and abluminal PNAd and HEC-6ST.[167] Similar to the tonsils, the major homing receptor-ligand pair in NALT is L-selectin and PNAd, rather than MAdCAM-1, as might have been expected of this mucosal-associated tissue.[292]

Human adenoids and tonsils appear early in fetal development and are populated with cells and have primary follicles. After birth, secondary follicles and GCs appear in response to bacterial antigens.[276] In contrast, the NALT in the mouse and rat is hypocellular at birth and undergoes dramatic changes after weaning, strongly suggesting that bacterial colonization is important for NALT development in these species. Id2 is required for initiation of the rodent NALT,[293] although RORγT, LTα, and LTβ are not required for this step.[166] The expansion of rodent NALT at weaning includes changes in the expression of LTα and LTβ and lymphoid chemokines, leading to T and B cell compartmentalization and HEV maturation.[166,167] The alternative NF-κB pathway is required for the expression of chemokines and the HEV genes *glycam-1* and *chst4* in NALT.[190] LTα, LTβ, IL7R, and the NIK signaling pathways are required for NALT organization and function, and mice that are deficient in these cytokines have a hypocellular NALT.[167,190,294,295]

Bronchus-Associated Lymphoid Tissue

BALT is less organized but more responsive to environmental antigens than the NALT. The number of lymphoid aggregates varies depending on the level of microbial exposure, and germ-free pigs have no BALT.[262,296] BALT is found commonly in rabbits and rats, is less frequent in guinea pigs and pigs, and is absent in cats.[274] BALT is not a prominent structure in the laboratory mouse, and its presence varies by strain and age.[297]

An inducible form of the BALT (iBALT) has been described in *Ltα−/−* mice after infection with influenza.[298] The term iBALT is misleading because these aggregates have no contact with the bronchial lumen and cannot sample inspired antigens. Therefore, inducible BALT should be considered as a TLO.[299] Splenectomized, lethally irradiated LTα−/− mice, reconstituted with normal BM, have no lymphoid organs. However, these mice can generate immunological memory in the iBALT.[279,280]

BALT is rare in normal adult human lungs, but TLOs are common in chronic pulmonary diseases.[294] BALT is frequent in children and is found in fetuses after infections in utero.[300,301] Like NALT, the HEVs in human BALT express PNAd and not MAdCAM-1.[302]

Gut-Associated Lymphoid Tissue

Peyer's Patches

Peyer's patches are the major organized tissue of gut-associated lymphoid tissue (GALT), which also includes smaller ILFs and cryptopatches. In the large intestine there is the appendix, cecal patches, and lymphoglandular complexes.[303] Peyer's patches are lymphoid aggregates aligned on the antimesenteric border of the small intestine (Fig. 8.5A). They are present in most species, although their number and location vary. They are dome shaped and are covered by a specialized epithelium that lacks surface microvilli and goblet cells and have numerous M cells. Because the Peyer's patches lack afferent lymphatic vessels, these M cells are critical transporters of antigen. Pathogens, including HIV and *Salmonella*, take advantage of the M cells to facilitate their own invasion.[304,305] Below the epithelium is a diffuse area, the subepithelial dome, divided into three to six parts. The Peyer's patch organization is similar to that of the LN (Fig. 8.5). Peyer's patches contain 6 to 12 basally located GCs, but the B cell areas are larger with a T:B cell ratio of 0.2 and CD4+ cells predominate over CD8+ cells.[306] Although M cells transport antigen across the epithelial barrier, they are not believed to have a crucial role in processing or presenting antigen. Peyer's patches are populated by several subsets of DCs.[307,308]

Cells enter the Peyer's patches through HEVs. In contrast to peripheral LNs and the NALT, the HEVs of Peyer's patches express MAdCAM-1[309] and homing depends on MAdCAM-1's interaction with the integrin $\alpha_4\beta_7$ on lymphocytes. Luminal PNAd is rarely found in Peyer's patch HEVs; only occasional abluminal expression is detected. This pattern is identical to that seen in LN HEVs in *chst4−/−* mice.[125] However, lymphocytes from mice that lack both L-selectin ligand and $\alpha_4\beta_7$ home less efficiently to Peyer's patches than do lymphocytes from mice that lack only one or the other of the ligands,[310,311] suggesting that an L-selectin ligand does contribute to homing to Peyer's patches. Even though P-selectin is expressed only weakly on HEVs in Peyer's patches, cells from mice deficient in P-selectin show reduced rolling and adhesion in vivo. This suggests that P-selectin, as well as MAdCAM-1, is important for homing to Peyer's patches.[310] Single cell transcriptomic comparison of peripheral LN HEVs with those of Peyer's patch HEVs reveal a wide array of gene differences.[312] These studies also reveal a mechanism for B cell homing to PPs through CD22 recognition of gycan ligands that are uniquely decorated in Peyer's patch HEVs.

Lymphoid chemokines CCL19, CCL21, and CXCL13 are found in the T and B cell areas[313] (Fig. 8.5). Stromal cells make CXCL12 mRNA, but the protein is also found on HEVs, presumably transported in a manner similar to CCL19.[140] The CXCL16/CXCR6 chemokine/receptor pair is important in Peyer's patches but not in LNs. Even under germfree conditions, CXCL16 is constitutively produced by the dome epithelium.[306] CD4+ and CD8+ cells that express CXCR6 are found in the Peyer's patches and require CXCL16 to home to the subepithelial dome. CCL20 and its receptor CCR6 are also important in Peyer's patches. CCL20 is made by the intestinal epithelium and plays a crucial role in DC trafficking to Peyer's patches.[314] Peyer's patch HEVs express CCL25, the ligand for CCR9, which has been defined as a mucosal homing chemokine receptor for intraepithelial lymphocytes and plasma cells.[80]

Several cytokines and chemokines have been shown to be crucial for Peyer's patch development, and their regulation is similar to that of LNs in studies with various knockout mice. *Ltα−/−* and *Ltβ−/−*mice completely lack Peyer's patches[165,315] as do *Cxcr5−/−* mice.[313,316,317] Since mice deficient in the LTβR also lack Peyer's patches,[318] these effects are mediated in large part through $LT\alpha_1\beta_2$. However, some *Tnfr1−/−* mice lack organized Peyer's patches, suggesting that either LTα3 or TNFα also influence generation or later stages in maintenance of Peyer's patches.[319] Mice deficient in IL7, or RORγt, Id2, NIK, and factors in the classical and alternative NF-κB pathways,[194,318,320] lack Peyer's patches, although RANKL is not required.[168]

A model for the embryonic development of Peyer's patches[321-323] provided the framework for studies in that organ. The first sign of Peyer's patch development at E15.5 in the mouse is the appearance of regions on the small intestine that stain positively with antibodies to ICAM-1, VCAM-1, and LTβR. Lymphoid tissue organizer cells (Lti), like those involved in LN development, express CXCL13, CCL19, CCL21, and IL-7. At mouse E17.5, clusters of IL-7R+CD4+CD3+ inducer cells are found. These express LTα and LTβ, CXCR5 and CCR7, Id2 and RORγ. They also express α4β1 integrin activated by CXCR5,[324] which allows interaction with the VCAM-1+ organizer cells. At E18.5, mature T and B cells enter through HEVs, CCL20 is produced by the FAE, and DCs expressing CCR7 and CCR6 are found. By day 4 after birth the typical microarchitecture is apparent, with M cells and T and B cell compartmentalization.[325]

Lymphocytes influence the maintenance of Peyer's patches. Mice that lack mature T and B cells have either undetectable or small Peyer's patches that lack follicles and GCs.[326,327] B cell–deficient mice retain some M cells, suggesting that T cells may regulate M cell maintenance. When B cell–deficient mice are reconstituted by a membrane IgM transgene, M cells are found at levels comparable with those of normal mice,[326] indicating that cells in addition to CD4+CD3− LTi are crucial for the maintenance of mature, functioning Peyer's patches.

Cryptopatches and Isolated Lymphoid Follicles

The small intestine of a mouse contains 100 to 200 ILFs and more numerous cryptopatches. The cryptopatches in the lamina propria containing LTi and DCs can be precursors of ILFs. The cryptopatches are composed of lin−c-kit+ cells, DCs, and VCAM-1+ stromal cells with few or no mature T and B cells. The cryptopatch cells express RORγt, IL-7R, and CCR6,[328] and their development is dependent on IL-7, CCR6, and its receptor CCL20. Cryptopatches are quite plastic, and although the LT family is necessary for their development, they can be restored by administration of wild-type BM to adult *Lta−/−* mice.

After mice are exposed to microbes or during some forms of autoimmunity, cryptopatches give rise to ILF,[329,330] which resemble small Peyer's patches and usually only have a single dome. ILFs require B cells that express LTa, but not T cells, for their formation.[331] In contrast to LN, they also require the TNFR1.[332] The location of ILFs is antigen driven, and they resolve completely after mice are treated with antibiotics or cytokine inhibitors.[118,177,329,331,332]

Appendix

Both rabbits and humans have an appendix,[333] a vestigial organ distal to the ileocecal junction that divides the small intestine from the colon. The appendix has epithelial M cells and dense lymphoid accumulations and GCs that are similar to Peyer's patches. Similar to the tonsils and adenoids the appendix involutes and regresses after puberty.

Colon Lymphoglandular Complexes

The colon has the highest bacterial load of any of the mucosa MALT, and the distribution of the MALT is more random than in the small intestine. Lymphoglandular complexes are smaller than Peyer's patches. As their name suggests, the lymphoid cells of the lymphoglandular complexes surround the mucous glands.[303]

TERTIARY LYMPHOID TISSUES

TLTs, also termed tertiary lymphoid organs (TLOs) or tertiary lymphoid structures (TLS), are ectopic accumulations of organized lymphoid cells that resemble LNs and arise in nonlymphoid organs during chronic inflammation in autoimmunity, microbial infection, graft rejection, and cancer, through a process termed "lymphoid neogenesis" (also "lymphoid neoorganogenesis").[2] When one uses this definition, iBALT or any organized lymphoid cell accumulation could be considered as a TLT. This semantic issue epitomizes the plasticity of all lymphoid organs and the fact that lymphoid organ regulation represents a continuum from ontogeny through chronic inflammation. A notable difference between SLOs and TLTs is the fact that the latter can arise in almost any organ in the adult. Nonetheless, TLTs exhibit remarkable morphologic, cellular, chemokine, and vasculature similarities to SLOs. They exhibit T and B cell compartmentalization; naïve T and B cells; DCs; FDCs; Tfh; GCs; plasma cells; lymphoid chemokines CCL19, CCL21, and CXCL13[1]; HEVs[132,189,334]; and conduits[335] and other stromal cells. Lymphatic vessels have also been noted in TLT in chronic graft rejection[336,337] and mouse models of Sjögren's syndrome[134] (Fig. 8.6) and type 1 diabetes.[338] The fact that the TLTs respond to S1P inhibitors suggest that their lymphatic vessels function similarly to those in LNs.[338]

Although, as noted above, there are striking similarities between TLTs and SLOs, there are differences. TLTs rarely have a capsule, although such a structure has been noted in at least one study of glioblastomas.[339] The lack of a capsule suggests the absence of SSMs, so important in LNs in transfer of antigen to B cells. HEVs are present, but in at least one study of tumor HEVs, they expressed a different complement of adhesion molecules.[340] Perhaps the most important difference between TLTs and LNs is the degree of plasticity. Both undergo remodeling in the presence of antigen, but the actual TLT can be resolved if the antigen is removed, whereas the lymphoid organ remains even if depleted of cells.

Lymphoid neogenesis has been noted in humans in autoimmunity, microbial infection, chronic allograft rejection, and cancer. A few examples are shown in Table 8.1 and described in more detail in 1. These accumulations also occasionally occur in atherosclerotic plaques with FDCs, organized B cell follicles, HEVs, and CCL19 and CCL21 in addition to those chemokines more often associated with acute inflammation.[344] TLTs have also been noted in non–small-cell lung cancer[345] and many other solid tumors.[342] The tendency for TLTs to develop into lymphomas has been noted in many studies (summarized in 346).

TLTs are apparent in several animal models. The pancreatic infiltrates in early stages of diabetes in the nonobese diabetic (NOD) mouse express lymphoid chemokines[347] and HEVs expressing MAdCAM-1,[348,349] PNAd, and HEC-6ST.[132] The brain in experimental autoimmune encephalomyelitis, a model of multiple sclerosis, has HEVs, CCL19, CCL21, CXCL13, and FDCs.[350-352] The thyroid in the BB rat has T cell and B cell compartmentalization and DCs,[353] and the gut in autoimmune gastritis in the mouse has HEVs and CXCL13.[354] Atherosclerotic plaques of apoprotein-E-deficient mice exhibit a marked increase in T and B cells, expression of CCL19 and CCL21, and PNAd+HEVs.[355-357] Lymphoid neogenesis also occurs in chronic mouse heart allograft rejection.[358,359] TLTs have been induced in several transgenic mouse models with the use of tissue-specific promoters driving the expression of inflammatory cytokines or lymphoid chemokines.[1] These mouse models, in addition to serving as examples of human disease, have provided valuable insight into the regulation of SLO development.

From Chronic Inflammation to Organized Lymphoid Microenvironments

Data generated from analyzing the cellular and molecular requirements for SLO development have provided a paradigm for understanding the development of TLTs in chronic inflammation. This paradigm proposes that the processes and molecules governing SLOs are also the basis of TLT development[2] and informs understanding of both secondary and tertiary lymphoid tissues. The precise physiologic event(s) that precipitate lymphoid neogenesis vary with individual circumstances. Data obtained from experiments in knockout and transgenic mice and clinical observations indicate that cooperative activities of TNF/LT family members and the lymphoid chemokines play central roles in this process.

Inflammation is a localized response to tissue injury, irritation, or infection often marked by tissue damage. Acute inflammation is an early innate immune response that is generally short-lived and self-limiting. However, in some situations, acute inflammation transitions to a

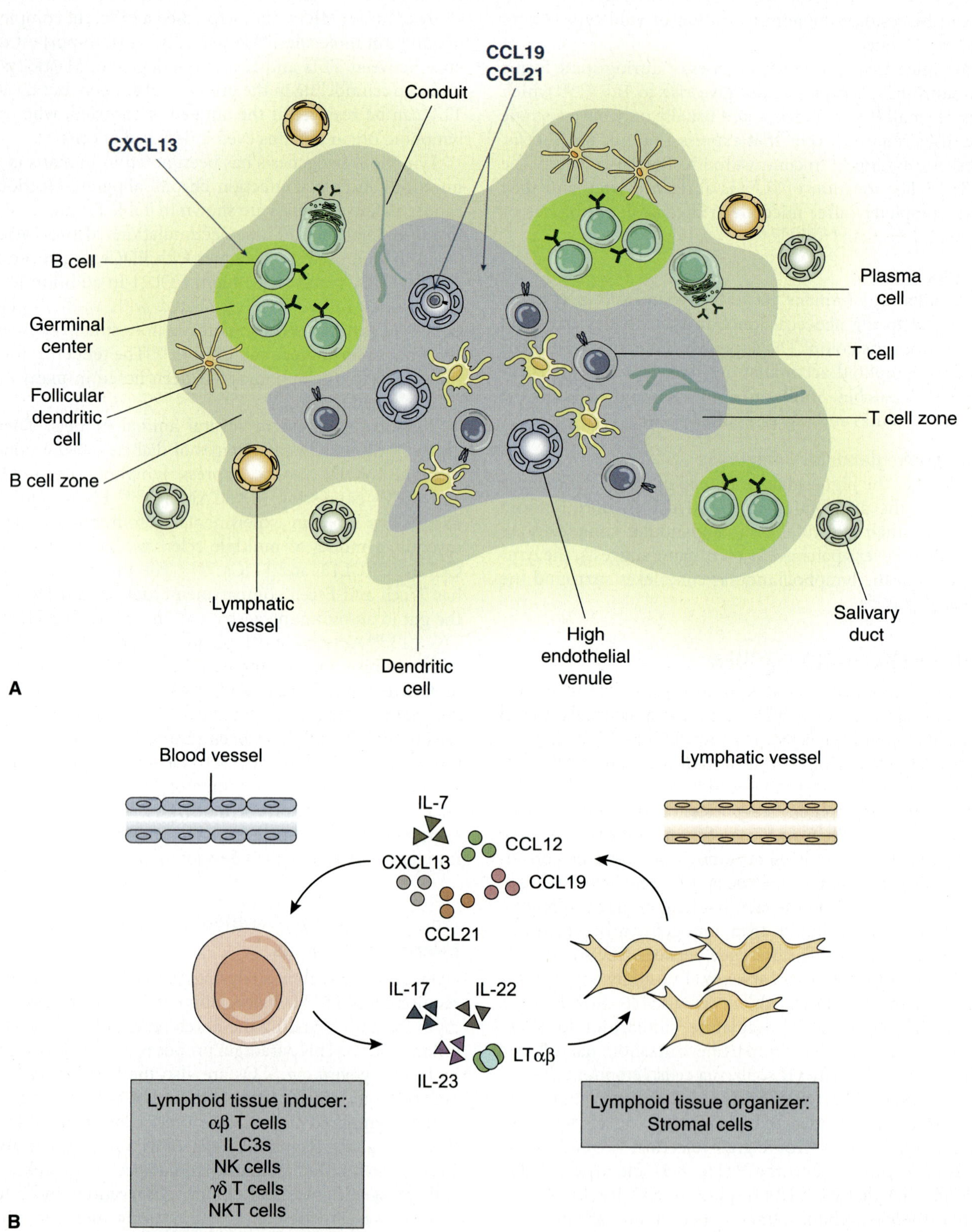

FIG. 8.6. Tertiary lymphoid tissue (TLT). A: A diagrammatic rendering of an organized lymphoid accumulation in the salivary gland of a Sjögren syndrome patient. **B:** Development of TLT. (A: From *Immunoepidemiology.* Chapter 2. Figure 2.10 P. J. Krause, P. B. Kavathas, and N. H. Ruddle, Editors. Springer Nature Switzerland 2019 © Ruddle 2020; B: From *Inducible Lymphoid Organs.* Chapter Basics of Inducible Lymphoid Organs Fig. 2B K. Kabashima and G. Egawa editors. Springer Nature Switzerland, 2020 with permission.)

TABLE 8.1 Representative Examples of Lymphoid Neogenesis in Human Autoimmunity, Infectious Diseases, Graft Rejection, and Cancer

Disease	Affected Tissue	Characteristics
Autoimmunity		
Rheumatoid arthritis	Synovial membrane	T cells and B cells, plasma cells, GCs, FDCs, CXCL13, CCL21, HEVs (PNAd, HEC-6ST)
Sjögren syndrome	Salivary glands	T cells and B cells, plasma cells, GCs, FDCs, CCL21, CXCL12, CXCL13 HEVs (PNAd)
Myasthenia gravis	Thymus	T cells and B cells, GCs, FDCs
Hashimoto thyroiditis	Thyroid	T cells and B cells, GCs, FDCs CCL21, CXCL13, CXCL12, plasma cells, HEVs (PNAd)
Grave disease	Thymus	T cells and B cells, GCs, FDCs, CCL21 CXCL13, CXCL12, HEVs (PNAd)
Multiple sclerosis	Brain	Lymphatic capillaries, B cell follicles and centroblasts, GCs, CCL19, CCL21, CXCL12, CXCL13
Ulcerative colitis	Colon	CXCL13
Inflammatory bowel disease (Crohn disease)	Bowel	T cell-B compartments, LVs, HECA-452^{+} HEV
Psoriatic arthritis	Joint	T cells and B cells, CXCL13, CCL19, CCL21, HEVs (PNAd)
Infectious diseases		
Borrelia burgdorferi/Lyme disease	Joints	T and B cells, FDCs, HEVs
Borrelia burgdorferi/neuroborreliosis	Central nervous system/ cerebrospinal fluid	CXCL13
Hepatitis C virus	Liver	T cell-B cell compartments, MAdCAM-1
Bartonella henselae/cat scratch disease	Granuloma	CXCL13
Graft Rejection		
Organ		
Heart		Germinal centers
Kidney		Germinal centers, lymphatic vessels
Cancer		
Ovarian cancer	Ovary	HEVs (Martinet et al. 2011)
Melanoma	Skin	CXCR5, CXCL13, B follicles[341]
Breast cancer	Breast	HEVs, B, T cells[342]
Non–small cell lung cancer	Lung	Germinal centers[343]

Original references up to Cancer are in Drayton et al[1]

chronic inflammatory response that is long-lived and self-perpetuating. TLTs arise under conditions of constitutive cytokine and/or chemokine expression, and multiple signals initiate their development. By integrating studies of lymphoid neogenesis in human pathologies and in animal models, it is becoming clear that at least three critical events promote TLT formation: inflammatory (eg, TNF/LT) cytokine expression, lymphoid chemokine production by stromal cells, and HEV development. Thus, there are similarities to the events in SLO development. However, many other cells than the canonical LTi cells can likely substitute for those effectors, even though LTi cells have been noted in the mouse models of lymphoid neogenesis.[360] The many different cell types that have been demonstrated to function in place of LTi cells are noted in Figure 8.6 and summarized in a recent review.[361]

Functions of TLTs

Ectopic accumulation of lymphoid cells has been considered the hallmark of destructive inflammation. Indeed, some TLTs are accompanied by tissue damage. However, it is likely that TLTs in chronic inflammation have roles in addition to tissue destruction. In the case of microbial infection, it is likely that lymphoid neogenesis occurs to sequester pathogens and to prevent their access to the other parts of the body. This may represent a primitive form of immunity. Local antigen presentation within the TLT itself likely prevents bacteremia or viremia. Enhanced long-term survival has been noted for those individuals whose lung or breast tumors included TLTs.[342,345] The propensity for TLTs to develop into lymphomas, as noted above; the ability of lymphatic vessels to serve as conduits for tumor metastasis; and the ability of TLTs to serve as sites of prion accumulation are obvious manifestations of detrimental functions.[362] Furthermore, the development of TLTs in autoimmunity may perpetuate clinical disease through epitope spreading.

Data from human and mouse studies provide compelling evidence that TLTs are permissive microenvironments for the induction of antigen-specific immune responses. Extensive immunohistochemical analyses of TLTs in autoimmunity and other chronic inflammatory states have established the presence of GCs and FDC networks in these tissues; several groups have demonstrated that TLT GCs can support B cell differentiation. Microdissection of discrete lymphocytic foci and subsequent DNA sequence analysis of GC B cells from the inflamed synovial tissue of patients with rheumatoid arthritis revealed a restricted number of Vκ gene rearrangements, a result consistent with oligoclonal B cell expansion in the synovial tissue.[363] Somatic hypermutation is apparent in synovial GC B cells.[364] Furthermore, synovial B cells exhibit a limited number of heavy and light chain gene rearrangements consistent with local clonal expansion of these cells. The molecular analysis of TLT GCs from the salivary glands of patients with primary Sjögren syndrome or the thymus of patients with myasthenia gravis demonstrates oligoclonal B cell proliferation in these tissues in addition to somatic hypermutation of Ig variable genes.[365-367] Together, these studies indicate that TLT GCs in several autoimmune pathologies can support antigen-driven clonal expansion and extensive diversification.

Another important hallmark of antigen-driven B cell responses in TLTs is the terminal differentiation of activated B cells into Ig-secreting plasma cells. Plasma cells have been detected in TLTs associated with GCs in rheumatoid arthritis[368] and in Sjogren syndrome. Mice expressing LTα under the control of the insulin promoter (RIPLTα) exhibit TLT at the sites of transgene expression (pancreas, kidney, and skin). After immunization with sheep red blood cells, evidence of isotype switching is apparent in these cellular infiltrates.[2] Although the presence of plasma cells in TLTs is consistent with local antigen presentation, it is occasionally unclear whether these cells develop in the TLTs themselves or migrate from canonical secondary lymphoid tissues. Nonetheless, taken together, these studies indicate that TLTs in several human pathologies and animal models support antigen-driven B cell differentiation marked by somatic hypermutation of immunoglobulin variable genes, affinity maturation, isotype switching, and terminal differentiation into antibody-secreting plasma cells.

T cell priming occurs in TLTs as suggested by the presence of isotype-switched plasma cells in the RIPLTα TLTs,[2] accelerated graft rejection,[369] and T cell epitope spreading in the central nervous system during experimental allergic encephalomyelitis.[370] The restricted TCR repertoire in a melanoma-associated TLT[371] further supports the concept that TLTs can act as priming sites. The demonstration of naïve T cell proliferation in the islets of NOD mice after surgical removal of pancreatic LNs[372] suggests, together with evidence noted above, that TLTs present antigen to naïve cells at the local site and generate an immune response. Determinant or epitope spreading, a phenomenon that arises in several autoimmune diseases, occurs when epitopes other than the inducing antigen become major targets of an ongoing immune response. It is considered to occur subsequent to the tissue damage induced by the initiating autoreactive T cells and therefore is the result of the presentation of new antigens.[373] Data generated in murine models of central nervous system inflammation support the possibility that intermolecular and intramolecular epitope spreading occur in TLTs in the central nervous system.[370] More recently it has become apparent that T_{reg}s can populate TLTs,[357] again suggesting that manipulating the immune response at the local site is an avenue of control.

Plasticity and Adaptability of TLTs

TLTs are the most plastic and adaptable of the lymphoid tissues. First, lymphoid neogenesis can be induced by a variety of stimuli. Their nimbleness in this regard suggests that they might represent the most primitive tissues in the immune system. TLTs can be "turned off" (ie, resolve) upon removal of the initial stimulus or after therapeutic intervention. The destruction of the islets of Langerhans' β cells in type I diabetes mellitus is an example of a situation in which removal of the antigen stimulus is accompanied by TLT resolution. Antibiotic treatment results in the resolution of TLTs and even MALT lymphomas.[374] Treatment has been shown to resolve some established TLTs, reversing insulitis and protecting against diabetes in NOD mice.[375] Such treatment can also "turn off" established TLTs in a mouse model of collagen-induced arthritis.[376] These studies are similar but not identical in outcome to those described above regarding the plasticity of LNs after mice are immunized or treated with LTβR-Ig,[191,192] again emphasizing the commonality of these tissues.

The Role of TLTs in Cancer

As noted above, there is a tendency for TLTs to develop into lymphomas as has been noted in Sjögren's syndrome[377] and *Helicobacter pylori* infection.[378] Thus, in that context, they may be considered as detrimental. On the other hand, several observations point to their beneficial role in that disease. In the case of several cancers, the presence of TLTs in a tumor biopsy as evaluated by a variety of markers such as extent of HEVs or FDCs can be a positive predictor of disease-free survival.[377,379] Several studies have indicated that the presence of TLTs with B cell follicles is associated with a positive response to check point inhibitor blockade.[341,380,381] A recent paper extends this point and indicates that HEV content in melanomas at initial biopsy predicts response to combination check point inhibitor blockade. In an experimental mouse model triple treatment (anti CTLA-4, anti-PD-1, agonistic LTbR) resulted in more mature tumor TLTs HEVs and enhanced survival.[340]

EVOLUTION OF LYMPHOID TISSUES (CHAPTER 4 FOR MORE DETAIL)

Bone Marrow

BM is found in the first land animals, the amphibians (Fig. 8.7). The BM equivalent in bony fish is the head kidney (pronephros), and LLPC can also be found there.[382] Sharks have primary lymphoid structures associated with the gonad (epigonal) and/or esophagus (Leydig gland) that contain HSC and LLPC.[383,384] An old idea that BM emerged in evolution to shield HSC from harmful UV radiation gained support in a recent study of zebrafish hematopoiesis.[385] The YS and AGM contribute to hematopoiesis as described above for mammals wherever it has been examined, especially in *Xenopus* and zebrafish.[386]

Thymus

The thymus is found in all jawed vertebrates (gnathostomes), usually with the typical cortical/medullary organization (Figs. 8.2 and 8.7). It can range from one lobule (eg, *Xenopus*, bilaterally) to a multilobed and/or discontinuous structure, depending on the species (Chapter 4). In addition, the thymic proteasome and AIRE arose early in the gnathostome lineage (first seen in sharks, the oldest vertebrates with mammalian-like adaptive immunity), suggesting that positive and negative selection occurs in a manner as described above.[387,388] For many years, it was believed that the thymus was absent in jawless fish (hagfish and lamprey), but with the discovery of the lamprey T cells, this was proven to be incorrect. *In situ* probes for a notch ligand (delta) and FoxN1, the transcription factor essential in thymus development, a defined structure lining the pharynx of larval lampreys dubbed the "thymoid" was identified.[389,390] Associated with these epithelia were lymphocytes expressing VLRA and VLRC, the analogues of jawed vertebrate TCRs. In addition, the APOBEC enzyme CDA1, which is required for VLRA/C rearrangement, is also expressed by the thymoid. Structures in the gill region of lower chordates like Amphioxus also express FonN1, so it will be of interest to study their role (if any) in lymphocyte differentiation.[391]

Spleen as the Primordial SLO

Only warm-blooded vertebrates have LNs, Peyer's patches, and GCs (Fig. 8.7), all dependent on LT (although some birds have lost the LT genes[392]). All jawed vertebrates, however, do have a spleen in which immune responses are generated (Fig. 8.7). The spleen is partitioned into RP and WP in representative species to the level of sharks, and B cells are also partitioned into segregated areas in many vertebrates, from sharks to mammals. It is of interest that the formation of organized SLO first appeared only for B cells, with T cell zones emerging in reptiles.[393] As mentioned above, during development, mammalian B cells are attracted to the chemokine cxcl13 expressed by the splenic vasculature and form the nascent WP. In mammals (and probably reptiles) the B cells are displaced into follicles while the T cells now surround the vessel in the PALS.[259] In amphibians and cartilaginous fish, the B cells retain this embryonic characteristic, and T cells are drawn to the WP only during immune responses, like in the development of TLT. Many bony fish have lost the splenic WP entirely, and immune responses are organized around a cell

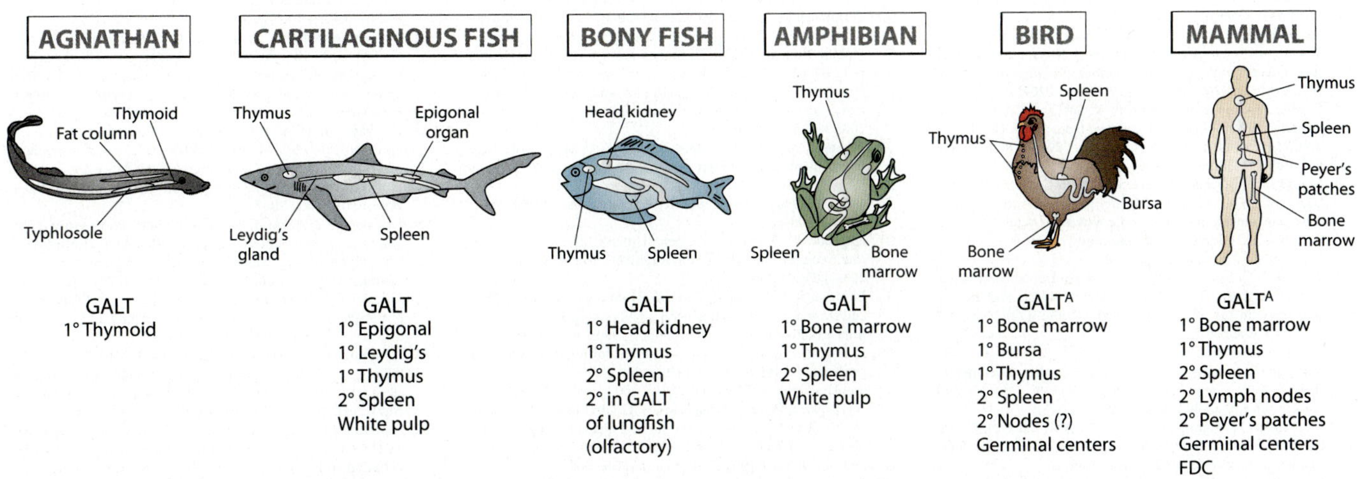

FIG. 8.7. Evolution of lymphoid tissues in the vertebrates. All jawed vertebrates have a thymus and a spleen with demarcated T- and B-cell zones, except where this feature has been lost in some taxa like bony fish and some amphibians. Fish have different bone marrow equivalents than tetrapods (epigonal, Leydig gland, head kidney), and amphibians are the oldest group with lymphopoietic bone marrow (also the first to have a typical immunoglobulin [Ig]H chain class switch, although sharks seem to have switch as well, despite the IgH cluster organization). Germinal centers are found only in warm-blooded vertebrates. A thymus equivalent (thymoid) has recently been discovered in agnathans, but no true SLO has been uncovered in jawless fish.

called a melanomacrophage.[394] After immunization, both T cells and B cells coalesce around the melanomacrophages that bear native antigen on their surface, entirely reminiscent of the formation of TLT in mammals.[395]

The dichotomy between lymphocytes and myeloid cells appears in the lower chordates (eg, Amphioxus), predating the emergence of adaptive immunity. Cells with the morphology of DC are found in all jawed vertebrates, and they have high expression of MHC class II. These cells have been best described in bony fish, at least by morphology, PNA staining, and ability to stimulate T cells.[396] FDC, in contrast, have not been reported in ectotherms. Instead, it seems that typical DC (or APC) of the hematopoietic lineage present antigen to both T and B cells. These cells, called XL cells in amphibians, have high levels of class II and bear immunoglobulins of all three isotypes on their surface, presumably in immune complexes.[260,397,398]

Early studies revealed that, by and large, specific antibody responses in cold-blooded vertebrates were of low affinity that did not mature over time.[399] Before the molecular era this was attributed perhaps to a lack of SHM in nonmammals. However, mutations were discovered in cartilaginous fish and amphibian Ig genes, sometimes at levels greater than in mammals.[400-402] Despite this, affinity increases do not exceed 100-fold in any study to date, nevertheless showing at least some level of selection. As mentioned, at least to the level of amphibians, FDCs are not present, and recent work suggests that conventional, hematopoietically derived APCs stimulate both T and B cells. Thus, it is believed that the emergence of FDC and complex GCs in warm-blooded vertebrates was a great leap forward in the generation of matured antibody responses (reviewed in 403).

CONCLUSIONS

The immune system depends on a remarkable organization of tissues and cells. The organs have defined functions that include generation of an immune repertoire (primary lymphoid organs) and responding to antigen (SLOs and TLTs). Development of primary and secondary lymphoid organs depends on precisely regulated expression of cytokines, chemokines, and adhesion molecules. Similar signals regulate the transition from inflammation to TLTs. Chemokines and adhesion molecules regulate trafficking in and out of lymphoid organs. SLOs are remarkably plastic in their response to antigenic assault and adapt with changes in expression of chemokines and adhesion molecules to maximize encounter of antigen with antigen-specific cells. TLTs, characteristic of many pathologic states, may represent the most primitive form of lymphoid tissues in their even greater plasticity and ability to develop directly at the site of antigen exposure.

ACKNOWLEDGMENTS

The authors acknowledge the contributions of Eitan Akirav, Noelia Alonso Gonzalez, and Lucy Truman to an earlier version of this chapter and thank these individuals.

REFERENCES

1. Drayton DL, Liao S, Mounzer RH, Ruddle NH. Lymphoid organ development: from ontogeny to neogenesis. *Nat Immunol.* 2006;7(4):344-353.
2. Kratz A, Campos-Neto A, Hanson MS, Ruddle NH. Chronic inflammation caused by lymphotoxin is lymphoid neogenesis. *J Exp Med.* 1996;183(4):1461-1472.
3. Soares-da-Silva F, Freyer L, Elsaid R, et al. Yolk sac, but not hematopoietic stem cell-derived progenitors, sustain erythropoiesis throughout murine embryonic life. *J Exp Med.* 2021;218(4):e20201729. doi:10.1084/jem.20201729
4. Gomez Perdiguero E, Klapproth K, Schulz C, et al. Tissue-resident macrophages originate from yolk-sac-derived erythro-myeloid progenitors. *Nature.* 2015;518(7540):547-551. doi:10.1038/nature13989
5. Montecino-Rodriguez E, Fice M, Casero D, Berent-Maoz B, Barber CL, Dorshkind K. Distinct genetic networks orchestrate the emergence of specific waves of fetal and adult B-1 and B-2 development. *Immunity.* 2016;45(3):527-539. doi:10.1016/j.immuni.2016.07.012
6. Gentek R, Ghigo C, Hoeffel G, et al. Epidermal gammadelta T cells originate from yolk sac hematopoiesis and clonally self-renew in the adult. *J Exp Med.* 2018;215(12):2994-3005. doi:10.1084/jem.20181206
7. Cumano A, Berthault C, Ramond C, et al. New molecular insights into immune cell development. *Annu Rev Immunol.* 2019;37:497-519. doi:10.1146/annurev-immunol-042718-041319
8. Medvinsky AL, Samoylina NL, Muller AM, Dzierzak EA. An early pre-liver intraembryonic source of CFU-S in the developing mouse. *Nature.* 1993;364(6432):64-67.
9. Gekas C, Dieterlen-Lievre F, Orkin SH, Mikkola HK. The placenta is a niche for hematopoietic stem cells. *Dev Cell.* 2005;8(3):365-375.
10. Ottersbach K, Dzierzak E. The murine placenta contains hematopoietic stem cells within the vascular labyrinth region. *Dev Cell.* 2005;8(3):377-387.
11. Medvinsky A, Rybtsov S, Taoudi S. Embryonic origin of the adult hematopoietic system: advances and questions. *Development* 2011;138(6):1017-1031. doi:138/6/1017 [pii] 10.1242/dev.040998
12. Christensen JL, Wright DE, Wagers AJ, Weissman IL. Circulation and chemotaxis of fetal hematopoietic stem cells. *PLoS Biol.* 2004;2(3):E75.
13. Frenette PS, Subbarao S, Mazo IB, von Andrian UH, Wagner DD. Endothelial selectins and vascular cell adhesion molecule-1 promote hematopoietic progenitor homing to bone marrow. *Proc Natl Acad Sci U S A.* 1998;95(24):14423-14428.
14. Vermeulen M, Le Pesteur F, Gagnerault MC, Mary JY, Sainteny F, Lepault F. Role of adhesion molecules in the homing and mobilization of murine hematopoietic stem and progenitor cells. *Blood.* 1998;92(3):894-900.
15. Orkin SH, Zon LI. Hematopoiesis and stem cells: plasticity versus developmental heterogeneity. *Nat Immunol.* 2002;3(4):323-328.
16. Pittenger MF, Mackay AM, Beck SC, et al. Multilineage potential of adult human mesenchymal stem cells. *Science.* 1999;284(5411):143-147.
17. Cyster JG. Homing of antibody secreting cells. *Immunol Rev.* 2003;194:48-60.
18. Nagasawa T. Microenvironmental niches in the bone marrow required for B-cell development. *Nat Rev Immunol.* 2006;6(2):107-116.
19. Jacobsen K, Osmond DG. Microenvironmental organization and stromal cell associations of B lymphocyte precursor cells in mouse bone marrow. *Eur J Immunol.* 1990;20(11):2395-2404.
20. Heissig B, Hattori K, Dias S, et al. Recruitment of stem and progenitor cells from the bone marrow niche requires MMP-9 mediated release of kit-ligand. *Cell.* 2002;109(5):625-637.
21. Nagasawa T, Omatsu Y, Sugiyama T. Control of hematopoietic stem cells by the bone marrow stromal niche: the role of reticular cells. *Trends Immunol.* 2011;32(7):315-320. doi:S1471-4906(11)00054-8 [pii] 10.1016/j.it.2011.03.009
22. Tokoyoda K, Egawa T, Sugiyama T, Choi BI, Nagasawa T. Cellular niches controlling B lymphocyte behavior within bone marrow during development. *Immunity.* 2004;20(6):707-718.
23. Banu N, Deng B, Lyman SD, Avraham H. Modulation of haematopoietic progenitor development by FLT-3 ligand. *Cytokine.* 1999;11(9):679-688.
24. Mazo IB, Massberg S, von Andrian UH. Hematopoietic stem and progenitor cell trafficking. *Trends Immunol* 2011;32(10):493-503. doi:S1471-4906(11)00117-7 [pii] 10.1016/j.it.2011.06.011
25. Heissig B, Ohki Y, Sato Y, Rafii S, Werb Z, Hattori K. A role for niches in hematopoietic cell development. *Hematology.* 2005;10(3):247-253.
26. Lataillade JJ, Domenech J, Le Bousse-Kerdiles MC. Stromal cell-derived factor-1 (SDF-1)\CXCR4 couple plays multiple roles on haematopoietic progenitors at the border between the old cytokine and new chemokine worlds: survival, cell cycling and trafficking. *Eur Cytokine Netw.* 2004;15(3):177-188.
27. Papayannopoulou T, Craddock C, Nakamoto B, Priestley GV, Wolf NS. The VLA4/VCAM-1 adhesion pathway defines contrasting mechanisms of lodgement of transplanted murine hemopoietic progenitors between bone marrow and spleen. *Proc Natl Acad Sci U S A.* 1995;92(21):9647-9651.
28. Katayama Y, Hidalgo A, Furie BC, Vestweber D, Furie B, Frenette PS. PSGL-1 participates in E-selectin-mediated progenitor homing to bone marrow: evidence for cooperation between E-selectin ligands and alpha4 integrin. *Blood.* 2003;102(6):2060-2067.
29. Katayama Y, Hidalgo A, Peired A, Frenette PS. Integrin alpha-4beta7 and its counterreceptor MAdCAM-1 contribute to hematopoietic progenitor recruitment into bone marrow following transplantation. *Blood.* 2004;104(7):2020-2026.
30. Lindsley RC, Thomas M, Srivastava B, Allman D. Generation of peripheral B cells occurs via two spatially and temporally distinct pathways. *Blood.* 2006;109(6):2521-2528.
31. Honczarenko M, Glodek AM, Swierkowski M, Na IK, Silberstein LE. Developmental stage-specific shift in responsiveness to chemokines during human B-cell development. *Exp Hematol.* 2006;34(8):1093-1100.
32. Bowman EP, Campbell JJ, Soler D, et al. Developmental switches in chemokine response profiles during B cell differentiation and maturation. *J Exp Med.* 2000;191(8):1303-1318.
33. Pantelouris EM. Absence of thymus in a mouse mutant. *Nature.* 1968;217:370-371.
34. Terszowski G, Muller SM, Bleul CC, et al. Evidence for a functional second thymus in mice. *Science.* 2006;312(5771):284-287.
35. Bubanovic IV. Crossroads of extrathymic lymphocytes maturation pathways. *Med Hypotheses.* 2003;61(2):235-239.
36. Scollay RG, Butcher EC, Weissman IL. Thymus cell migration. Quantitative aspects of cellular traffic from the thymus to the periphery in mice. *Eur J Immunol.* 1980;10(3):210-218.

37. Blackburn CC, Manley NR. Developing a new paradigm for thymus organogenesis. *Nat Rev Immunol.* 2004;4:278-289.
38. Lind EF, Prockop SE, Porritt HE, Petrie HT. Mapping precursor movement through the postnatal thymus reveals specific microenvironments supporting defined stages of early lymphoid development. *J Exp Med.* 2001;194(2):127-134.
39. Anderson M, Anderson SK, Farr AG. Thymic vasculature: organizer of the medullary epithelial compartment? *Int Immunol.* 2000;12(7):1105-1110.
40. Odaka C, Morisada T, Oike Y, Suda T. Distribution of lymphatic vessels in mouse thymus: immunofluorescence analysis. *Cell Tissue Res.* 2006;325(1):13-22.
41. Pearse M, Wu L, Egerton M, Wilson A, Shortman K, Scollay R. A murine early thymocyte developmental sequence is marked by transient expression of the interleukin 2 receptor. *Proc Natl Acad Sci USA.* 1989;86(5):1614-1618.
42. Shinkai Y, Rathbun G, Lam KP, et al. RAG-2-deficient mice lack mature lymphocytes owing to inability to initiate V(D)J rearrangement. *Cell.* 1992;68(5):855-867.
43. Klein L. Aire gets company for immune tolerance. *Cell.* 2015;163(4):794-795. doi:10.1016/j.cell.2015.10.057
44. Egerton M, Scollay R, Shortman K. Kinetics of mature T-cell development in the thymus. *Proc Natl Acad Sci USA.* 1990;87(7):2579-2582.
45. Silva-Santos B, Pennington DJ, Hayday AC. Lymphotoxin-mediated regulation of gammadelta cell differentiation by alphabeta T cell progenitors. *Science.* 2005;307(5711):925-928.
46. Klug DB, Carter C, Crouch E, Roop D, Conti CJ, Richie ER. Interdependence of cortical thymic epithelial cell differentiation and T-lineage commitment. *Proc Natl Acad Sci USA.* 1998;95(20):11822-11827.
47. Derbinski J, Gabler J, Brors B, et al. Promiscuous gene expression in thymic epithelial cells is regulated at multiple levels. *J Exp Med.* 2005;202(1):33-45.
48. Kondo K, Takada K, Takahama Y. Antigen processing and presentation in the thymus: implications for T cell repertoire selection. *Curr Opin Immunol.* 2017;46:53-57. doi:10.1016/j.coi.2017.03.014
49. Aaltonen J, Bjorses P, Sandkuijl L, Perheentupa J, Peltonen L. An autosomal locus causing autoimmune disease: autoimmune polyglandular disease type I assigned to chromosome 21. *Nat Genet.* 1994;8(1):83-87.
50. Akirav EM, Ruddle NH, Herold KC. The role of AIRE in human autoimmune disease. *Nat Rev Endocrinol* 2011;7(1):25-33. doi:nrendo.2010.200[pii] 10.1038/nrendo.2010.200
51. Anderson MS, Venanzi ES, Klein L, et al. Projection of an immunological self shadow within the thymus by the aire protein. *Science.* 2002;298(5597):1395-1401.
52. Patel DD, Whichard LP, Radcliff G, Denning SM, Haynes BF. Characterization of human thymic epithelial cell surface antigens: phenotypic similarity of thymic epithelial cells to epidermal keratinocytes. *J Clin Immunol.* 1995;15(2):80-92.
53. Girbal-Neuhauser E, Durieux JJ, Arnaud M, et al. The epitopes targeted by the rheumatoid arthritis-associated antifilaggrin autoantibodies are posttranslationally generated on various sites of (pro)filaggrin by deimination of arginine residues. *J Immunol.* 1999;162(1):585-594.
54. Favre A. Identification of filaggrin in Hassall's corpuscle by histochemical and immunohistochemical methods. *Acta Anat (Basel).* 1989;135(1):71-76.
55. Douek DC, Altmann DM. T-cell apoptosis and differential human leucocyte antigen class II expression in human thymus. *Immunology.* 2000;99(2):249-256.
56. Nakagawa Y, Ohigashi I, Nitta T, et al. Thymic nurse cells provide microenvironment for secondary T cell receptor alpha rearrangement in cortical thymocytes. *Proc Natl Acad Sci U S A.* 2012;109(50):20572-20577. doi:10.1073/pnas.1213069109
57. Ardavin C, Wu L, Li CL, Shortman K. Thymic dendritic cells and T cells develop simultaneously in the thymus from a common precursor population. *Nature.* 1993;362(6422):761-763.
58. Foss DL, Donskoy E, Goldschneider I. The importation of hematogenous precursors by the thymus is a gated phenomenon in normal adult mice. *J Exp Med.* 2001;193(3):365-374.
59. Donskoy E, Goldschneider I. Two developmentally distinct populations of dendritic cells inhabit the adult mouse thymus: demonstration by differential importation of hematogenous precursors under steady state conditions. *J Immunol.* 2003;170(7):3514-3521.
60. Anderson G, Owen JJ, Moore NC, Jenkinson EJ. Thymic epithelial cells provide unique signals for positive selection of CD4+CD8+ thymocytes in vitro. *J Exp Med.* 1994;179(6):2027-2031.
61. Anderson G, Partington KM, Jenkinson EJ. Differential effects of peptide diversity and stromal cell type in positive and negative selection in the thymus. *J Immunol.* 1998;161(12):6599-6603.
62. Brocker T, Riedinger M, Karjalainen K. Targeted expression of major histocompatibility complex (MHC) class II molecules demonstrates that dendritic cells can induce negative but not positive selection of thymocytes in vivo. *J Exp Med.* 1997;185(3):541-550.
63. Heino M, Peterson P, Sillanpaa N, et al. RNA and protein expression of the murine autoimmune regulator gene (Aire) in normal, RelB-deficient and in NOD mouse. *Eur J Immunol.* 2000;30(7):1884-1893.
64. Klein L, Kyewski B, Allen PM, Hogquist KA. Positive and negative selection of the T cell repertoire: what thymocytes see (and don't see). *Nat Rev Immunol.* 2014;14(6):377-391. doi:10.1038/nri3667
65. Gallegos AM, Bevan MJ. Central tolerance to tissue-specific antigens mediated by direct and indirect antigen presentation. *J Exp Med.* 2004;200(8):1039-1049. doi:10.1084/jem.20041457
66. Watanabe N, Wang YH, Lee HK, et al. Hassall's corpuscles instruct dendritic cells to induce CD4+CD25+ regulatory T cells in human thymus. *Nature.* 2005;436(7054):1181-1185.
67. Proietto AI, van Dommelen S, Zhou P, et al. Dendritic cells in the thymus contribute to T-regulatory cell induction. *Proc Natl Acad Sci U S A.* 2008;105(50):19869-19874. doi:0810268105 [pii] 10.1073/pnas.0810268105
68. Liu YJ, Soumelis V, Watanabe N, et al. TSLP: an epithelial cell cytokine that regulates T cell differentiation by conditioning dendritic cell maturation. *Annu Rev Immunol.*2007;25:193-219.
69. Hanabuchi S, Ito T, Park WR, et al. Thymic stromal lymphopoietin-activated plasmacytoid dendritic cells induce the generation of FOXP3+ regulatory T cells in human thymus. *J Immunol.* 2010;184(6):2999-3007. doi:jimmunol.0804106 [pii]10.4049/jimmunol.0804106
70. Yamano T, Nedjic J, Hinterberger M, et al. Thymic B cells are licensed to present self antigens for central T cell tolerance induction. *Immunity* 2015;42(6):1048-1061. doi:10.1016/j.immuni.2015.05.013
71. Miyazaki T, Suzuki G, Yamamura K. The role of macrophages in antigen presentation and T cell tolerance. *Int Immunol.* 1993;5(9):1023-1033.
72. Ferrero I, Anjuere F, Martin P, et al. Functional and phenotypic analysis of thymic B cells: role in the induction of T cell negative selection. *Eur J Immunol.* 1999;29(5):1598-1609.
73. Flores KG, Li J, Hale LP. B cells in epithelial and perivascular compartments of human adult thymus. *Hum Pathol.* 2001;32(9):926-934.
74. Fukuba Y, Inaba M, Taketani S, et al. Functional analysis of thymic B cells. *Immunobiology.* 1994;190(1-2):150-163.
75. Frommer F, Waisman A. B cells participate in thymic negative selection of murine auto-reactive CD4+ T cells. *PLoS One.* 2010;5(10):e15372. doi:10.1371/journal.pone.0015372
76. Akirav EM, Xu Y, Ruddle NH. Resident B cells regulate thymic expression of myelin oligodendrocyte glycoprotein. *J Neuroimmunol.* 2011;235(1-2):33-39. doi:S0165-5728(11)00097-X [pii] 10.1016/j.jneuroim.2011.03.013
77. Lepique AP, Palencia S, Irjala H, Petrie HT. Characterization of vascular adhesion molecules that may facilitate progenitor homing in the post-natal mouse thymus. *Clin Dev Immunol.* 2003;10(1):27-33.
78. Bleul CC, Boehm T. Chemokines define distinct microenvironments in the developing thymus. *Eur J Immunol.* 2000;30(12):3371-3379.
79. Liu C, Ueno T, Kuse S, et al. The role of CCL21 in recruitment of T-precursor cells to fetal thymi. *Blood.* 2005;105(1):31-39.
80. Wurbel MA, Malissen M, Guy-Grand D, et al. Mice lacking the CCR9 CC-chemokine receptor show a mild impairment of early T- and B-cell development and a reduction in T-cell receptor gammadelta(+) gut intraepithelial lymphocytes. *Blood.* 2001;98(9):2626-2632.
81. Misslitz A, Pabst O, Hintzen G, et al. Thymic T cell development and progenitor localization depend on CCR7. *J Exp Med.* 2004;200(4):481-491.
82. Plotkin J, Prockop SE, Lepique A, Petrie HT. Critical role for CXCR4 signaling in progenitor localization and T cell differentiation in the postnatal thymus. *J Immunol.* 2003;171(9):4521-4527.
83. Kurobe H, Liu C, Ueno T, et al. CCR7-dependent cortex-to-medulla migration of positively selected thymocytes is essential for establishing central tolerance. *Immunity.* 2006;24(2):165-177.
84. Ueno T, Saito F, Gray DH, et al. CCR7 signals are essential for cortex-medulla migration of developing thymocytes. *J Exp Med.* 2004;200(4):493-505.
85. Le Borgne M, Ladi E, Dzhagalov I, et al. The impact of negative selection on thymocyte migration in the medulla. *Nat Immunol.* 2009;10(8):823-830. doi:ni.1761 [pii] 10.1038/ni.1761
86. Poznansky MC, Olszak IT, Evans RH, et al. Thymocyte emigration is mediated by active movement away from stroma-derived factors. *J Clin Invest.* 2002;109(8):1101-1110.
87. Ueno T, Hara K, Willis MS, et al. Role for CCR7 ligands in the emigration of newly generated T lymphocytes from the neonatal thymus. *Immunity.* 2002;16(2):205-218.
88. Edsall LC, Spiegel S. Enzymatic measurement of sphingosine 1-phosphate. *Anal Biochem.* 1999;272(1):80-86.
89. Slifka MK, Matloubian M, Ahmed R. Bone marrow is a major site of long-term antibody production after acute viral infection. *J Virol.* 1995;69(3):1895-1902.
90. Manley NR. Thymus organogenesis and molecular mechanisms of thymic epithelial cell differentiation. *Semin Immunol.* 2000;12(5):421-428.
91. Gordon J, Bennett AR, Blackburn CC, Manley NR. Gcm2 and Foxn1 mark early parathyroid- and thymus-specific domains in the developing third pharyngeal pouch. *Mech Dev.* 2001;103(1-2):141-143.
92. Manley NR, Blackburn CC. A developmental look at thymus organogenesis: where do the non-hematopoietic cells in the thymus come from? *Curr Opin Immunol.* 2003;15(2):225-232.
93. Le Douarin NM, Jotereau FV. Tracing of cells of the avian thymus through embryonic life in interspecific chimeras. *J Exp Med.* 1975;142(1):17-40.
94. Gordon J, Wilson VA, Blair NF, et al. Functional evidence for a single endodermal origin for the thymic epithelium. *Nat Immunol.* 2004;5(5):546-553.
95. Ohigashi I, Kozai M, Takahama Y. Development and developmental potential of cortical thymic epithelial cells. *Immunol Rev.* 2016;271(1):10-22. doi:10.1111/imr.12404
96. Klug DB, Carter C, Gimenez-Conti IB, Richie ER. Cutting edge: thymocyte-independent and thymocyte-dependent phases of epithelial patterning in the fetal thymus. *J Immunol.* 2002;169(6):2842-2845.
97. van Ewijk W, Hollander G, Terhorst C, Wang B. Stepwise development of thymic microenvironments in vivo is regulated by thymocyte subsets. *Development.* 2000;127(8):1583-1591.
98. Dejardin E, Droin NM, Delhase M, et al. The lymphotoxin-beta receptor induces different patterns of gene expression via two NF-kappaB pathways. *Immunity.* 2002;17(4):525-535.
99. Boehm T, Scheu S, Pfeffer K, Bleul CC. Thymic medullary epithelial cell differentiation, thymocyte emigration, and the control of autoimmunity require lympho-epithelial cross talk via LTbetaR. *J Exp Med.* 2003;198(5):757-769.
100. Chin RK, Lo JC, Kim O, et al. Lymphotoxin pathway directs thymic Aire expression. *Nat Immunol.* 2003;4(11):1121-1127.
101. Roberts NA, White AJ, Jenkinson WE, et al. Rank signaling links the development of invariant gammadelta T cell progenitors and Aire(+) medullary epithelium. *Immunity* 2012;36(3):427-437. doi: 10.1016/j.immuni.2012.01.016
102. Haynes BF, Heinly CS. Early human T cell development: analysis of the human thymus at the time of initial entry of hematopoietic stem cells into the fetal thymic microenvironment. *J Exp Med.* 1995;181(4):1445-1458.
103. Bouneaud C, Kourilsky P, Bousso P. Impact of negative selection on the T cell repertoire reactive to a self-peptide: a large fraction of T cell clones escapes clonal deletion. *Immunity.* 2000;13(6):829-840.
104. Steinman RM. The dendritic cell system and its role in immunogenicity. *Annu Rev Immunol.* 1991;9:271-296.
105. von Andrian UH, Mempel TR. Homing and cellular traffic in lymph nodes. *Nat Rev Immunol.* 2003;3(11):867-878.
106. Wilson NS, El-Sukkari D, Belz GT, et al. Most lymphoid organ dendritic cell types are phenotypically and functionally immature. *Blood.* 2003;102(6):2187-2194.
107. Wilson NS, El-Sukkari D, Villadangos JA. Dendritic cells constitutively present self antigens in their immature state in vivo and regulate antigen presentation by controlling the rates of MHC class II synthesis and endocytosis. *Blood.* 2004;103(6):2187-2195.
108. Cavanagh LL, Von Andrian UH. Travellers in many guises: the origins and destinations of dendritic cells. *Immunol Cell Biol.* 2002;80(5):448-462.
109. Steinman RM, Hawiger D, Liu K, et al. Dendritic cell function in vivo during the steady state: a role in peripheral tolerance. *Ann N Y Acad Sci.* 2003;987:15-25.
110. Stoitzner P, Tripp CH, Douillard P, Saeland S, Romani N. Migratory Langerhans cells in mouse lymph nodes in steady state and inflammation. *J Invest Dermatol.* 2005;125(1):116-125.
111. Cohen JN, Guidi CJ, Tewalt EF, et al. Lymph node-resident lymphatic endothelial cells mediate peripheral tolerance via Aire-independent direct antigen presentation. *J Exp Med* 2010;207(4):681-688. doi:10.1084/jem.20092465
112. Fletcher AL, Lukacs-Kornek V, Reynoso ED, et al. Lymph node fibroblastic reticular cells directly present peripheral tissue antigen under steady-state and inflammatory conditions. *J Exp Med.* 2010;207(4):689-697. doi:jem.20092642 [pii] 10.1084/jem.20092642
113. Gardner JM, Devoss JJ, Friedman RS, et al. Deletional tolerance mediated by extrathymic Aire-expressing cells. *Science.* 2008;321(5890):843-847. doi:321/5890/843 [pii] 10.1126/science.1159407
114. Lee JW, Epardaud M, Sun J, et al. Peripheral antigen display by lymph node stroma promotes T cell tolerance to intestinal self. *Nat Immunol.* 2007;8(2):181-190. doi:ni1427 [pii] 10.1038/ni1427
115. Poliani PL, Kisand K, Marrella V, et al. Human peripheral lymphoid tissues contain autoimmune regulator-expressing dendritic cells. *Am J Pathol.* 2010;176(3):1104-1112. doi:ajpath.2010.090956 [pii] 10.2353/ajpath.2010.090956
116. Phan TG, Green JA, Gray EE, Xu Y, Cyster JG. Immune complex relay by subcapsular sinus macrophages and noncognate B cells drives antibody affinity maturation. *Nat Immunol.* 2009;10(7):786-793. doi:10.1038/ni.1745
117. Gretz JE, Norbury CC, Anderson AO, Proudfoot AE, Shaw S. Lymph-borne chemokines and other low molecular weight molecules reach high endothelial venules via specialized conduits while a functional barrier limits access to the lymphocyte microenvironments in lymph node cortex. *J Exp Med.* 2000;192(10):1425-1440.
118. Ebert LM, Schaerli P, Moser B. Chemokine-mediated control of T cell traffic in lymphoid and peripheral tissues. *Mol Immunol.* 2005;42(7):799-809.

119. Rodrigo Mora J, Von Andrian UH. Specificity and plasticity of memory lymphocyte migration. *Curr Top Microbiol Immunol.* 2006;308:83-116.
120. Cyster JG. Chemokines, sphingosine-1-phosphate, and cell migration in secondary lymphoid organs. *Annu Rev Immunol.* 2005;23:127-159.
121. Lo CG, Xu Y, Proia RL, Cyster JG. Cyclical modulation of sphingosine-1-phosphate receptor 1 surface expression during lymphocyte recirculation and relationship to lymphoid organ transit. *J Exp Med.* 2005;201(2):291-301.
122. Anderson AO, Anderson ND. Studies on the structure and permeability of the microvasculature in normal rat lymph nodes. *Am J Pathol.* 1975;80(3):387-418.
123. M'Rini C, Cheng G, Schweitzer C, et al. A novel endothelial L-selectin ligand activity in lymph node medulla that is regulated by alpha(1,3)-fucosyltransferase-IV. *J Exp Med.* 2003;198(9):1301-1312.
124. von Andrian UH, M'Rini C. In situ analysis of lymphocyte migration to lymph nodes. *Cell Adhes Commun.* 1998;6(2-3):85-96.
125. Hemmerich S, Bistrup A, Singer MS, et al. Sulfation of L-selectin ligands by an HEV-restricted sulfotransferase regulates lymphocyte homing to lymph nodes. *Immunity.* 2001;15(2):237-247.doi:10.1016/s1074-7613(01)00188-1
126. Hemmerich S, Butcher EC, Rosen SD. Sulfation-dependent recognition of HEV-ligands by L-selectin and MECA-79, an adhesion-blocking mAb. *J Exp Med.* 1994;180:2219-2226.
127. Hiraoka N, Kawashima H, Petryniak B, et al. Core 2 branching beta1,6-N-acetylglucosaminyltransferase and high endothelial venule-restricted sulfotransferase collaboratively control lymphocyte homing. *J Biol Chem.* 2004;279(4):3058-3067.
128. Hiraoka N, Petryniak B, Nakayama J, et al. A novel, high endothelial venule-specific sulfotransferase expresses 6-sulfo sialyl Lewis(x), an L-selectin ligand displayed by CD34. *Immunity.* 1999;11(1):79-89.
129. Homeister JW, Thall AD, Petryniak B, et al. The alpha(1,3) fucosyltransferases FucT-IV and FucT-VII exert collaborative control over selectin-dependent leukocyte recruitment and lymphocyte homing. *Immunity.* 2001;15(1):115-126.
130. Kawashima H, Hirakawa J, Tobisawa Y, Fukuda M, Saga Y. Conditional gene targeting in mouse high endothelial venules. *J Immunol.* 2009;182(9):5461-5468. doi:10.4049/jimmunol.0802327
131. Liao S, Bentley K, Lebrun M, Lesslauer W, Ruddle FH, Ruddle NH. Transgenic LacZ under control of Hec-6st regulatory sequences recapitulates endogenous gene expression on high endothelial venules. *Proc Natl Acad Sci U S A.* 2007;104(11):4577-4582. doi:0700334104 [pii]10.1073/pnas.0700334104
132. Bistrup A, Tsay D, Shenoy P, et al. Detection of a sulfotransferase (HEC-GlcNAc6ST) in high endothelial venules of lymph nodes and in high endothelial venule-like vessels within ectopic lymphoid aggregates: relationship to the MECA-79 epitope. *Am J Pathol.* 2004;164(5):1635-1644.
133. Oliver G. Lymphatic vasculature development. *Nat Rev Immunol.* 2004;4(1):35-45.
134. Anderson AO, Shaw S. Conduit for privileged communications in the lymph node. *Immunity.* 2005;22(1):3-5.
135. Sixt M, Kanazawa N, Selg M, et al. The conduit system transports soluble antigens from the afferent lymph to resident dendritic cells in the T cell area of the lymph node. *Immunity.* 2005;22(1):19-29.
136. Kraal G, Mebius RE. High endothelial venules: lymphocyte traffic control and controlled traffic. *Adv Immunol.* 1997;65:347-395.
137. Miyasaka M, Tanaka T. Lymphocyte trafficking across high endothelial venules: dogmas and enigmas. *Nat Rev Immunol.* 2004;4(5):360-370.
138. Butcher EC, Scollay RG, Weissman IL. Lymphocyte adherence to high endothelial venules: characterization of a modified in vitro assay, and examination of the binding of syngeneic and allogeneic lymphocyte populations. *J Immunol.* 1979;123(5):1996-2003.
139. Mebius RE, Streeter PR, Michie S, Butcher EC, Weissman IL. A developmental switch in lymphocyte homing receptor and endothelial vascular addressin expression regulates lymphocyte homing and permits CD4+ CD3- cells to colonize lymph nodes. *Proc Natl Acad Sci USA.* 1996;93(20):11019-11024.
140. Baekkevold ES, Yamanaka T, Palframan RT, et al. The CCR7 ligand elc (CCL19) is transcytosed in high endothelial venules and mediates T cell recruitment. *J Exp Med.* 2001;193:1105-1112.
141. Nakano H, Gunn MD. Gene duplications at the chemokine locus on mouse chromosome 4: multiple strain-specific haplotypes and the deletion of secondary lymphoid-organ chemokine and EBI-1 ligand chemokine genes in the plt mutation. *J Immunol.* 2001;166(1):361-369.
142. Vassileva G, Soto H, Zlotnik A, et al. The reduced expression of 6Ckine in the plt mouse results from the deletion of one of two 6Ckine genes. *J Exp Med.* 1999;190(8):1183-1188.
143. Cyster JG. Chemokines and cell migration in secondary lymphoid organs. *Science.* 1999;286:2098-2102.
144. DeRogatis JM, Viramontes KM, Neubert EN, Tinoco R. PSGL-1 immune checkpoint inhibition for CD4(+) T cell cancer immunotherapy. *Front Immunol* 2021;12:636238. doi:10.3389/fimmu.2021.636238
145. Veerman KM, Williams MJ, Uchimura K, et al. Interaction of the selectin ligand PSGL-1 with chemokines CCL21 and CCL19 facilitates efficient homing of T cells to secondary lymphoid organs. *Nat Immunol.* 2007;8(5):532-539. doi:10.1038/ni1456
146. Silvan J, Gonzalez-Tajuelo R, Vicente-Rabaneda E, et al. Deregulated PSGL-1 expression in B cells and dendritic cells may Be implicated in human systemic sclerosis development. *J Invest Dermatol.* 2018;138(10):2123-2132. doi:10.1016/j.jid.2018.04.003
147. Drayson MT, Ford WL. Afferent lymph and lymph borne cells: their influence on lymph node function. *Immunobiology.* 1984;168(3-5):362-379.
148. Hendriks HR, Eestermans IL. Disappearance and reappearance of high endothelial venules and immigrating lymphocytes in lymph nodes deprived of afferent lymphatic vessels: a possible regulatory role of macrophages in lymphocyte migration. *Eur J Immunol.* 1983;13:663-669.
149. Hendriks HR, Duijvestijn AM, Kraal G. Rapid decrease in lymphocyte adherence to high endothelial venules in lymph nodes deprived of afferent lymphatic vessels. *Eur J Immunol.* 1987;17:1691-1695.
150. Mebius RE, Bauer J, Twisk AJ, Breve J, Kraal G. The functional activity of high endothelial venules: a role for the subcapsular sinus macrophages in the lymph node. *Immunobiology.* 1991;182(3-4):277-291.
151. Mebius RE, Breve J, Kraal G, Streeter PR. Developmental regulation of vascular addressin expression: a possible role for site-associated environments. *Int Immunol.* 1993;5(5):443-449.
152. Swarte VV, Joziasse DH, Van den Eijnden DH, et al. Regulation of fucosyltransferase-VII expression in peripheral lymph node high endothelial venules. *Eur J Immunol.* 1998;28(10):3040-3047.
153. Gretz JE, Anderson AO, Shaw S. Cords, channels, corridors and conduits: critical architectural elements facilitating cell interactions in the lymph node cortex. *Immunol Rev.* 1997;156:11-24.
154. Gretz JE, Kaldjian EP, Anderson AO, Shaw S. Sophisticated strategies for information encounter in the lymph node: the reticular network as a conduit of soluble information and a highway for cell traffic. *J Immunol.* 1996;157(2):495-499.
155. Larsen CG, Anderson AO, Appella E, Oppenheim JJ, Matsushima K. The neutrophil-activating protein (NAP-1) is also chemotactic for T lymphocytes. *Science.* 1989;243(4897):1464-1466.
156. Francois M, Harvey NL, Hogan BM. The transcriptional control of lymphatic vascular development. *Physiology (Bethesda)* 2011;26(3):146-155. doi:26/3/146 [pii] 10.1152/physiol.00053.2010
157. Wigle JT, Harvey N, Detmar M, et al. An essential role for Prox1 in the induction of the lymphatic endothelial cell phenotype. *EMBO J.* 2002;21(7):1505-1513.
158. Wigle JT, Oliver G. Prox1 function is required for the development of the murine lymphatic system. *Cell.* 1999;98(6):769-778.
159. Abtahian F, Guerriero A, Sebzda E, et al. Regulation of blood and lymphatic vascular separation by signaling proteins SLP-76 and Syk. *Science.* 2003;299(5604):247-251.
160. Bertozzi CC, Schmaier AA, Mericko P, et al. Platelets regulate lymphatic vascular development through CLEC-2-SLP-76 signaling. *Blood.* 2010;116(4):661-670. doi:blood-2010-02-270876 [pii] 10.1182/blood-2010-02-270876
161. Uhrin P, Zaujec J, Breuss JM, et al. Novel function for blood platelets and podoplanin in developmental separation of blood and lymphatic circulation. *Blood.* 2010;115(19):3997-4005. doi:blood-2009-04-216069 [pii] 10.1182/blood-2009-04-216069
162. Scavelli C, Weber E, Agliano M, et al. Lymphatics at the crossroads of angiogenesis and lymphangiogenesis. *J Anat.* 2004;204(6):433-449.
163. Vondenhoff MF, van de Pavert SA, Dillard ME, et al. Lymph sacs are not required for the initiation of lymph node formation. *Development.* 2009;136(1):29-34. doi:136/1/29 [pii] 10.1242/dev.028456
164. Benezech C, White A, Mader E, et al. Ontogeny of stromal organizer cells during lymph node development. *J Immunol.* 2010;184(8):4521-4530. doi:jimmunol.0903113 [pii] 10.4049/jimmunol.0903113
165. De Togni P, Goellner J, Ruddle NH, et al. Abnormal development of peripheral lymphoid organs in mice deficient in lymphotoxin. *Science.* 1994;264(5159):703-707.
166. Harmsen A, Kusser K, Hartson L, et al. Cutting edge: organogenesis of nasal-associated lymphoid tissue (NALT) occurs independently of lymphotoxin-alpha (LT alpha) and retinoic acid receptor-related orphan receptor-gamma, but the organization of NALT is LT alpha dependent. *J Immunol.* 2002;168(3):986-990.
167. Ying X, Chan K, Shenoy P, Hill M, Ruddle NH. Lymphotoxin plays a crucial role in the development and function of nasal-associated lymphoid tissue through regulation of chemokines and peripheral node addressin. *Am J Pathol.* 2005;166(1):135-146.
168. Dougall WC, Glaccum M, Charrier K, et al. RANK is essential for osteoclast and lymph node development. *Genes Dev.* 1999;13(18):2412-2424.
169. Kim D, Mebius RE, MacMicking JD, et al. Regulation of peripheral lymph node genesis by the tumor necrosis factor family member TRANCE. *J Exp Med.* 2000;192(10):1467-1478.
170. Ruddle NH, Akirav EM. Secondary lymphoid organs: responding to genetic and environmental cues in ontogeny and the immune response. *J Immunol.* 2009;183(4):2205-2212. doi:10.4049/jimmunol.0804324
171. Jeucken CM, Konig JJ, Mebius RE, Tas SW. The role of endothelial cells and TNF-receptor superfamily membes in lymphoid organogenesis and function during health and inflammation. *Front Immunol.*2019;10:1-12. doi:10.3389/fimmun.2019.02700
172. van de Pavert SA, Olivier BJ, Goverse G, et al. Chemokine CXCL13 is essential for lymph node initiation and is induced by retinoic acid and neuronal stimulation. *Nat Immunol.* 2009;10(11):1193-1199. doi:10.1038/ni.1789
173. Onder L, Morbe U, Pikor N, et al. Lymphatic endothelial cells control initiation of lymph node organogenesis. *Immunity.* 2017;47(1):80-92 e4. doi:10.1016/j.immuni.2017.05.008
174. Denton AE, Carr EJ, Magiera LP, Watts AJB, Fearon DT. Embryonic FAP(+) lymphoid tissue organizer cells generate the reticular network of adult lymph nodes. *J Exp Med.* 2019;216(10):2242-2252.doi:10.1084/jem.20181705
175. Mebius RE. Organogenesis of lymphoid tissues. *Nat Rev Immunol.* 2003;3(4):292-303.
176. Mebius RE, Rennert P, Weissman IL. Developing lymph nodes collect CD4+CD3- LTbeta+ cells that can differentiate to APC, NK cells, and follicular cells but not T or B cells. *Immunity.* 1997;7(4):493-504.
177. Eberl G, Marmon S, Sunshine MJ, Rennert PD, Choi Y, Littman DR. An essential function for the nuclear receptor RORgamma(t) in the generation of fetal lymphoid tissue inducer cells. *Nat Immunol.* 2004;5(1):64-73.
178. Yoshida H, Honda K, Shinkura R, et al. IL-7 receptor alpha+ CD3(-) cells in the embryonic intestine induces the organizing center of Peyer's patches. *Int Immunol.* 1999;11(5):643-655.
179. Yoshida H, Kawamoto H, Santee SM, et al. Expression of alpha(4)beta(7) integrin defines a distinct pathway of lymphoid progenitors committed to T cells, fetal intestinal lymphotoxin producer, NK, and dendritic cells. *J Immunol.* 2001;167(5):2511-2521.
180. Bovay E, Sabine A, Prat-Luri B, et al. Multiple roles of lymphatic vessels in peripheral lymph node development. *J Exp Med.* 2018;215(11):2760-2777. doi:10.1084/jem.20180217
181. Onder L, Ludewig B. A fresh view on lymph node organogenesis. *Trends Immunol.* 2018;39(10):775-787. doi:10.1016/j.it.2018.08.003
182. Koning JJ, Mebius RE. Complexity of lymphoid tissue organizers: a response to onder and ludewig. *Trends Immunol.* 2018;39(12):951-952. doi:10.1016/j.it.2018.10.006
183. Onder L, Ludewig B. Redefining the nature of lymphoid tissue organizer cells: response to 'complexity of lymphoid tissue organizers' by Koning and Mebius. *Trends Immunol.* 2018;39(12):952-953. doi:10.1016/j.it.2018.10.007
184. Camara A, Cordeiro OG, Alloush F, et al. Lymph node mesenchymal and endothelial stromal cells cooperate via the RANK-RANKL cytokine Axis to shape the sinusoidal macrophage niche. *Immunity.* 2019;50(6):1467-1481 e6. doi:10.1016/j.immuni.2019.05.008
185. Wang Z, Chai Q, Zhu M. Differential roles of LTbetaRin endothelial cell subsets for lymph node organogenesis and maturation. *J Immunol.* 2018;201(1):69-76. doi:10.4049/jimmunol.1701080
186. Girard JP, Springer TA. High endothelial venules (HEVs): specialized endothelium for lymphocyte migration. *Immunol Today.* 1995;16(9):449-457.
187. Mebius RE, Schadee-Eestermans IL, Weissman IL. MAdCAM-1 dependent colonization of developing lymph nodes involves a unique subset of CD4+CD3- hematolymphoid cells. *Cell Adhes Commun.* 1998;6(2-3):97-103.
188. Cuff CA, Schwartz J, Bergman CM, Russell KS, Bender JR, Ruddle NH. Lymphotoxin alpha3 induces chemokines and adhesion molecules: insight into the role of LT alpha in inflammation and lymphoid organ development. *J Immunol.* 1998;161(12):6853-6860.
189. Drayton DL, Ying X, Lee J, Lesslauer W, Ruddle NH. Ectopic LT alpha beta directs lymphoid organ neogenesis with concomitant expression of peripheral node addressin and a HEV-restricted sulfotransferase. *J Exp Med.* 2003;197(9):1153-1163.
190. Drayton DL, Bonizzi G, Ying X, Liao S, Karin M, Ruddle NH. I kappa B kinase complex alpha kinase activity controls chemokine and high endothelial venule gene expression in lymph nodes and nasal-associated lymphoid tissue. *J Immunol.* 2004;173(10):6161-6168.
191. Browning JL, Allaire N, Ngam-Ek A, et al. Lymphotoxin-beta receptor signaling is required for the homeostatic control of HEV differentiation and function. *Immunity.* 2005;23(5):539-550. doi:10.1016/j.immuni.2005.10.002
192. Liao S, Ruddle NH. Synchrony of high endothelial venules and lymphatic vessels revealed by immunization. *J Immunol.* 2006;177(5):3369-3379.
193. Mionnet C, Sanos SL, Mondor I, et al. High endothelial venules as traffic control points maintaining lymphocyte population homeostasis in lymph nodes. *Blood.* 2011;118(23):6115-6122. doi:10.1182/blood-2011-07-367409

194. Weih F, Caamano J. Regulation of secondary lymphoid organ development by the nuclear factor-kappaB signal transduction pathway. *Immunol Rev*. 2003;195:91-105.
195. Miyawaki S, Nakamura Y, Suzuka H, et al. A new mutation, aly, that induces a generalized lack of lymph nodes accompanied by immunodeficiency in mice. *Eur J Immunol*. 1994;24(2):429-434.
196. Shinkura R, Kitada K, Matsuda F, et al. Alymphoplasia is caused by a point mutation in the mouse gene encoding Nf-kappa b-inducing kinase. *Nat Genet*. 1999;22(1):74-77.
197. Senftleben U, Cao Y, Xiao G, et al. Activation by IKKalpha of a second, evolutionary conserved, NF-kappa B signaling pathway. *Science*. 2001;293(5534):1495-1499.
198. Yilmaz ZB, Weih DS, Sivakumar V, Weih F. RelB is required for Peyer's patch development: differential regulation of p52-RelB by lymphotoxin and TNF. *EMBO J*. 2003;22(1):121-130.
199. Eberl G, Littman DR. The role of the nuclear hormone receptor RORgammat in the development of lymph nodes and Peyer's patches. *Immunol Rev*. 2003;195:81-90.
200. Yokota Y, Mansouri A, Mori S, et al. Development of peripheral lymphoid organs and natural killer cells depends on the helix-loop-helix inhibitor Id2. *Nature*. 1999;397(6721):702-706.
201. Hall JG, Hopkins J, Reynolds J. Studies of efferent lymph cells from nodes stimulated with oxazolone. *Immunology*. 1980;39(2):141-149.
202. Hall JG, Smith ME. Studies on the afferent and efferent lymph of lymph nodes draining the site of application of fluorodinitrobenzene (FDNB). *Immunology*. 1971;21(1):69-79.
203. Hay JB, Cahill RN, Trnka Z. The kinetics of antigen-reactive cells during lymphocyte recruitment. *Cell Immunol*. 1974;10(1):145-153.
204. He C, Young AJ, West CA, Su M, Konerding MA, Mentzer SJ. Stimulation of regional lymphatic and blood flow by epicutaneous oxazolone. *J Appl Physiol*. 2002;93(3):966-973.
205. West CA, He C, Su M, et al. Stochastic regulation of cell migration from the efferent lymph to oxazolone-stimulated skin. *J Immunol*. 2001;166(3):1517-1523.
206. Angeli V, Ginhoux F, Llodra J, et al. B cell-driven lymphangiogenesis in inflamed lymph nodes enhances dendritic cell mobilization. *Immunity*. 2006;24(2):203-215.
207. Hay JB, Johnston MG, Vadas P, Chin W, Issekutz T, Movat HZ. Relationships between changes in blood flow and lymphocyte migration induced by antigen. *Monogr Allergy*. 1980;16:112-125.
208. Ottaway CA, Parrott DM. Regional blood flow and its relationship to lymphocyte and lymphoblast traffic during a primary immune reaction. *J Exp Med*. 1979;150(2):218-230.
209. Soderberg KA, Linehan MM, Ruddle NH, Iwasaki A. MAdCAM-1 expressing sacral lymph node in the lymphotoxin beta-deficient mouse provides a site for immune generation following vaginal herpes simplex virus-2 infection. *J Immunol*. 2004;173(3):1908-1913.
210. Mebius RE, Breve J, Duijvestijn AM, Kraal G. The function of high endothelial venules in mouse lymph nodes stimulated by oxazolone. *Immunology*. 1990;71(3):423-427.
211. Myking AO. Morphological changes in paracortical high endothelial venules to single and repeated application of oxazolone to mouse skin. *Virchows Arch B Cell Pathol Incl Mol Pathol*. 1980;35(1):63-71.
212. Cahill RN, Frost H, Trnka Z. The effects of antigen on the migration of recirculating lymphocytes through single lymph nodes. *J Exp Med*. 1976;143(4):870-888.
213. Hoke D, Mebius RE, Dybdal N, et al. Selective modulation of the expression of L-selectin ligands by an immune response. *Curr Biol*. 1995;5(6):670-678.
214. Yoneyama H, Matsuno K, Zhang Y, et al. Evidence for recruitment of plasmacytoid dendritic cell precursors to inflamed lymph nodes through high endothelial venules. *Int Immunol*. 2004;16(7):915-928.
215. Veerman K, Tardiveau C, Martins F, Coudert J, Girard JP. Single-cell analysis reveals heterogeneity of high endothelial venules and different regulation of genes controlling lymphocyte entry to lymph nodes. *Cell Rep*. 2019;26(11):3116-3131 e5. doi:10.1016/j.celrep.2019.02.042
216. Eikelenboom P, Nassy JJ, Post J, Versteeg JC, Langevoort HL. The histogenesis of lymph nodes in rat and rabbit. *Anat Rec*. 1978;190(2):201-215.
217. Mackay F, Browning JL. Turning off follicular dendritic cells. *Nature*. 1998;395(6697):26-27.
218. Brendolan A, Rosado MM, Carsetti R, Selleri L, Dear TN. Development and function of the mammalian spleen. *Bioessays*. 2007;29(2):166-177.
219. Blue J, Weiss L. Electron microscopy of the red pulp of the dog spleen including vascular arrangements, periarterial macrophage sheaths (ellipsoids), and the contractile, innervated reticular meshwork. *Am J Anat*. 1981;161(2):189-218.
220. Blue J, Weiss L. Periarterial macrophage sheaths (ellipsoids) in cat spleen—an electron microscope study. *Am J Anat*. 1981;161(2):115-134.
221. Snook T. A comparative study of the vascular arrangements in mammalian spleens. *Am J Anat*. 1950;87(1):31-77.
222. Waksman BH. *Atlas of Experimental Immunobiology and Immunopathology*. Yale University Press; 1970.
223. Mebius RE, Kraal G. Structure and function of the spleen. *Nat Rev Immunol*. 2005;5(8):606-616.
224. Geijtenbeek TB, Groot PC, Nolte MA, et al. Marginal zone macrophages express a murine homologue of DC-SIGN that captures blood-borne antigens in vivo. *Blood*. 2002;100(8):2908-2916.
225. Koppel EA, Wieland CW, van den Berg VC, et al. Specific ICAM-3 grabbing nonintegrin-related 1 (SIGNR1) expressed by marginal zone macrophages is essential for defense against pulmonary Streptococcus pneumoniae infection. *Eur J Immunol*. 2005;35(10):2962-2969.
226. Kraal G, Mebius R. New insights into the cell biology of the marginal zone of the spleen. *Int Rev Cytol*. 2006;250:175-215.
227. Ito S, Naito M, Kobayashi Y, et al. Roles of a macrophage receptor with collagenous structure (MARCO) in host defense and heterogeneity of splenic marginal zone macrophages. *Arch Histol Cytol*. 1999;62(1):83-95.
228. Karlsson MC, Guinamard R, Bolland S, Sankala M, Steinman RM, Ravetch JV. Macrophages control the retention and trafficking of B lymphocytes in the splenic marginal zone. *J Exp Med*. 2003;198(2):333-340. doi:10.1084/jem.20030684
229. Kraal G, Schornagel K, Streeter PR, Holzmann B, Butcher EC. Expression of the mucosal vascular addressin, MAdCAM-1, on sinus-lining cells in the spleen. *Am J Pathol*. 1995;147(3):763-771.
230. Crocker PR, Gordon S. Mouse macrophage hemagglutinin (sheep erythrocyte receptor) with specificity for sialylated glycoconjugates characterized by a monoclonal antibody. *J Exp Med*. 1989;169(4):1333-1346.
231. Jiang HR, Hwenda L, Makinen K, Oetke C, Crocker PR, Forrester JV. Sialoadhesin promotes the inflammatory response in experimental autoimmune uveoretinitis. *J Immunol*. 2006;177(4):2258-2264.
232. Oetke C, Vinson MC, Jones C, Crocker PR. Sialoadhesin-deficient mice exhibit subtle changes in B- and T-cell populations and reduced immunoglobulin M levels. *Mol Cell Biol*. 2006;26(4):1549-1557. doi:10.1128/MCB.26.4.1549-1557.2006
233. Backer R, Schwandt T, Greuter M, et al. Effective collaboration between marginal metallophilic macrophages and CD8+ dendritic cells in the generation of cytotoxic T cells. *Proc Natl Acad Sci U S A*. 2010;107(1):216-221. doi:10.1073/pnas.0909541107
234. You Y, Myers RC, Freeberg L, et al. Marginal zone B cells regulate antigen capture by marginal zone macrophages. *J Immunol*. 2011;186(4):2172-2181. doi:10.4049/jimmunol.1002106
235. Cinamon G, Zachariah MA, Lam OM, Foss FW,Jr, Cyster JG. Follicular shuttling of marginal zone B cells facilitates antigen transport. *Nat Immunol*. 2008;9(1):54-62. doi:10.1038/ni1542
236. Poholek AC, Craft J. Competing for help: new insights into the function of follicular helper T cells. *Immunol Cell Biol*.2009;87(6):438-439. doi:10.1038/icb.2009.26
237. Eisenbarth SC. Dendritic cell subsets in T cell programming: location dictates function. *Nat Rev Immunol*. 2019;19(2):89-103. doi:10.1038/s41577-018-0088-1
238. Yi T, Li J, Chen H, et al. Splenic dendritic cells survey red blood cells for missing self-CD47 to trigger adaptive immune responses. *Immunity*. 2015;43(4):764-775. doi:10.1016/j.immuni.2015.08.021
239. Liu D, Duan L, Cyster JG. Chemo- and mechanosensing by dendritic cells facilitate antigen surveillance in the spleen. *Immunol Rev*. 2022;306(1):25-42. doi:10.1111/imr.13055
240. Liu D, Duan L, Rodda LB, et al. CD97 promotes spleen dendritic cell homeostasis through the mechanosensing of red blood cells. *Science* 2022;375(6581):eabi5965. doi:10.1126/science.abi5965
241. Nolte MA, Belien JA, Schadee-Eestermans I, et al. A conduit system distributes chemokines and small blood-borne molecules through the splenic white pulp. *J Exp Med*. 2003;198(3):505-512.
242. Pabst O, Forster R, Lipp M, Engel H, Arnold HH. NKX2.3 is required for MAdCAM-1 expression and homing of lymphocytes in spleen and mucosa-associated lymphoid tissue. *EMBO J*. 2000;19(9):2015-2023. doi:10.1093/emboj/19.9.2015
243. Cyster JG, Goodnow CC. Pertussis toxin inhibits migration of B and T lymphocytes into splenic white pulp cords. *J Exp Med*. 1995;182(2):581-586.
244. Lewis SM, Williams A, Eisenbarth SC. Structure and function of the immune system in the spleen. *Sci Immunol*. 2019;4(33) doi:10.1126/sciimmunol.aau6085
245. Nolte MA, Hamann A, Kraal G, Mebius RE. The strict regulation of lymphocyte migration to splenic white pulp does not involve common homing receptors. *Immunology*. 2002;106(3):299-307.
246. Lu TT, Cyster JG. Integrin-mediated long-term B cell retention in the splenic marginal zone. *Science*. 2002;297(5580):409-412.
247. Cinamon G, Matloubian M, Lesneski MJ, et al. Sphingosine 1-phosphate receptor 1 promotes B cell localization in the splenic marginal zone. *Nat Immunol*. 2004;5(7):713-720.
248. Czompoly T, Labadi A, Kellermayer Z, Olasz K, Arnold HH, Balogh P. Transcription factor Nkx2-3 controls the vascular identity and lymphocyte homing in the spleen. *J Immunol*. 2011;186(12):6981-6989. doi:10.4049/jimmunol.1003770
249. Kellermayer Z, Labadi A, Czompoly T, Arnold HH, Balogh P. Absence of Nkx2-3 homeodomain transcription factor induces the formation of LYVE-1-positive endothelial cysts without lymphatic commitment in the spleen. *J Histochem Cytochem*. 2011;59(7):690-700. doi:10.1369/0022155411410061
250. Hatano M, Roberts CW, Minden M, Crist WM, Korsmeyer SJ. Deregulation of a homeobox gene, HOX11, by the t(10;14) in T cell leukemia. *Science*. 1991;253(5015):79-82.
251. Roberts CW, Shutter JR, Korsmeyer SJ. Hox11 controls the genesis of the spleen. *Nature*. 1994;368(6473):747-749.
252. Dear TN, Colledge WH, Carlton MB, et al. The Hox11 gene is essential for cell survival during spleen development. *Development*. 1995;121(9):2909-2915.
253. Reijmers RM, Vondenhoff MF, Roozendaal R, et al. Impaired lymphoid organ development in mice lacking the heparan sulfate modifying enzyme glucuronyl C5-epimerase. *J Immunol*. 2010;184(7):3656-3664. doi:10.4049/jimmunol.0902200
254. Ngo VN, Korner H, Gunn MD, et al. Lymphotoxin alpha/beta and tumor necrosis factor are required for stromal cell expression of homing chemokines in B and T cell areas of the spleen. *J Exp Med*. 1999;189(2):403-412. doi:10.1084/jem.189.2.403
255. Vondenhoff MF, Desanti GE, Cupedo T, et al. Separation of splenic red and white pulp occurs before birth in a LTalphabeta-independent manner. *J Leukoc Biol*. 2008;84(1):152-161. doi:10.1189/jlb.0907659
256. Ngo VN, Cornall RJ, Cyster JG. Splenic T zone development is B cell dependent. *J Exp Med*. 2001;194(11):1649-1660. doi:10.1084/jem.194.11.1649
257. Kim MY, McConnell FM, Gaspal FM, et al. Function of CD4+CD3- cells in relation to B and T zone stroma in spleen. *Blood*. 2006;109:1602-1610.
258. Neely HR, Flajnik MF. CXCL13 responsiveness but not CXCR5 expression by late transitional B cells initiates splenic white pulp formation. *J Immunol*. 2015;194(6):2616-2623. doi:10.4049/jimmunol.1401905
259. Neely HR, Flajnik MF. Emergence and evolution of secondary lymphoid organs. *Annu Rev Cell Dev Biol*. 2016;32:693-711. doi:10.1146/annurev-cellbio-111315-125306
260. Neely HR, Guo J, Flowers EM, Criscitiello MF, Flajnik MF. "Double-duty" conventional dendritic cells in the amphibian Xenopus as the prototype for antigen presentation to B cells. *Eur J Immunol*. 2018;48(3):430-440. doi:10.1002/eji.201747260
261. Benedict CA, De Trez C, Schneider K, Ha S, Patterson G, Ware CF. Specific remodeling of splenic architecture by cytomegalovirus. *PLoS Pathog*. 2006;2(3):e16.
262. Croitoru K, Bienenstock J. Characteristics and functions of mucosa-associated lymphoid tissue. In: Ogra PL, Mestecky J, Lamm ME, Strober W, McGheee JR, Bienenstock J, eds. *Handbook of Mucosal Immunology*. Academic Press; 1994:141-151.
263. Iwata M, Yokota A. Retinoic acid production by intestinal dendritic cells. *Vitam Horm*. 2011;86:127-152. doi:10.1016/B978-0-12-386960-9.00006-X
264. Weiner HL, da Cunha AP, Quintana F, Wu H. Oral tolerance. *Immunol Rev*.2011;241(1):241-259. doi:10.1111/j.1600-065X.2011.01017.x
265. Mowat AM. Anatomical basis of tolerance and immunity to intestinal antigens. *Nat Rev Immunol*. 2003;3(4):331-341.
266. Dieu MC, Vanbervliet B, Vicari A, et al. Selective recruitment of immature and mature dendritic cells by distinct chemokines expressed in different anatomic sites. *J Exp Med*. 1998;188(2):373-386.
267. Meurens F, Whale J, Brownlie R, Dybvig T, Thompson DR, Gerdts V. Expression of mucosal chemokines TECK/CCL25 and MEC/CCL28 during fetal development of the ovine mucosal immune system. *Immunology*. 2007;120(4):544-555.
268. Puri KD, Finger EB, Gaudernack G, Springer TA. Sialomucin CD34 is the major L-selectin ligand in human tonsil high endothelial venules. *J Cell Biol*. 1995;131(1):261-270.
269. Berg EL, Mullowney AT, Andrew DP, Goldberg JE, Butcher EC. Complexity and differential expression of carbohydrate epitopes associated with L-selectin recognition of high endothelial venules. *Am J Pathol*. 1998;152(2):469-477.
270. van Kempen MJ, Rijkers GT, Van Cauwenberge PB. The immune response in adenoids and tonsils. *Int Arch Allergy Immunol*. 2000;122(1):8-19.
271. Karchev T, Kabakchiev P. Electron-microscope observations on the tonsillar epithelium in children with recurrent tonsillitis. *Int J Pediatr Otorhinolaryngol*. 1982;4(2):149-156.
272. Winther B, Innes DJ. The human adenoid. A morphologic study. *Arch Otolaryngol Head Neck Surg*.1994;120(2):144-149.
273. Gebert A, Pabst R. M cells at locations outside the gut. *Semin Immunol*.1999;11(3):165-170. doi:10.1006/smim.1999.0172
274. Pabst R, Gehrke I. Is the bronchus-associated lymphoid tissue (BALT) an integral structure of the lung in normal mammals, including humans? *Am J Respir Cell Mol Biol*. 1990;3(2):131-135.
275. Schulz O, Jaensson E, Persson EK, et al. Intestinal CD103+, but not CX3CR1+, antigen sampling cells migrate in lymph and serve classical dendritic cell functions. *J Exp Med*. 2009;206(13):3101-3114. doi:10.1084/jem.20091925
276. Brandtzaeg P. Potential of nasopharynx-associated lymphoid tissue for vaccine responses in the airways. *Am J Respir Crit Care Med*. 2011;183(12):1595-1604. doi:10.1164/rccm.201011-1783OC
277. Dupuy C, Buzoni-Gatel D, Touze A, Bout D, Coursaget P. Nasal immunization of mice with human papillomavirus type 16 (HPV-16) virus-like particles or with the HPV-16 L1 gene elicits specific cytotoxic T lymphocytes in vaginal draining lymph nodes. *J Virol*. 1999;73(11):9063-9071.

278. Wiley JA, Tighe MP, Harmsen AG. Upper respiratory tract resistance to influenza infection is not prevented by the absence of either nasal-associated lymphoid tissue or cervical lymph nodes. *J Immunol.* 2005;175(5):3186-3196.
279. Moyron-Quiroz JE, Rangel-Moreno J, Hartson L, et al. Persistence and responsiveness of immunologic memory in the absence of secondary lymphoid organs. *Immunity.* 2006;25(4):643-654.
280. Constant SL, Brogdon JL, Piggott DA, et al. Resident lung antigen-presenting cells have the capacity to promote Th2 T cell differentiation in situ. *J Clin Invest.* 2002;110(10):1441-1448.
281. Perry M, Whyte A. Immunology of the tonsils. *Immunol Today.* 1998;19(9):414-421.
282. Brandtzaeg P. Regionalized immune function of tonsils and adenoids. *Immunol Today.* 1999;20(8):383-384.
283. Rebelatto MC, Mead C, HogenEsch H. Lymphocyte populations and adhesion molecule expression in bovine tonsils. *Vet Immunol Immunopathol.* 2000;73(1):15-29.
284. Carlsen HS, Haraldsen G, Brandtzaeg P, Baekkevold ES. Disparate lymphoid chemokine expression in mice and men: no evidence of CCL21 synthesis by human high endothelial venules. *Blood.* 2005;106(2):444-446. doi:10.1182/blood-2004-11-4353
285. Andoh N, Ohtani H, Kusakari C, et al. Expression of E- and P-selectins by vascular endothelial cells in human tonsils. *Acta Otolaryngol Suppl.* 1996;523:52-54.
286. Goeringer GC, Vidic B. The embryogenesis and anatomy of Waldeyer's ring. *Otolaryngol Clin.* 1987;20(2):207-217.
287. Wu HY, Russell MW. Induction of mucosal immunity by intranasal application of a streptococcal surface protein antigen with the cholera toxin B subunit. *Infect Immun.* 1993;61(1):314-322.
288. Balmelli C, Roden R, Potts A, Schiller J, De Grandi P, Nardelli-Haefliger D. Nasal immunization of mice with human papillomavirus type 16 virus-like particles elicits neutralizing antibodies in mucosal secretions. *J Virol.* 1998;72(10):8220-8229.
289. Liu XS, Abdul-Jabbar I, Qi YM, Frazer IH, Zhou J. Mucosal immunisation with papillomavirus virus-like particles elicits systemic and mucosal immunity in mice. *Virology.* 1998;252(1):39-45.
290. Zuercher AW, Coffin SE, Thurnheer MC, Fundova P, Cebra JJ. Nasal-associated lymphoid tissue is a mucosal inductive site for virus-specific humoral and cellular immune responses. *J Immunol.* 2002;168(4):1796-1803.
291. Shikina T, Hiroi T, Iwatani K, et al. IgA class switch occurs in the organized nasopharynx- and gut-associated lymphoid tissue, but not in the diffuse lamina propria of airways and gut. *J Immunol.* 2004;172(10):6259-6264.
292. Csencsits KL, Jutila MA, Pascual DW. Mucosal addressin expression and binding-interactions with naive lymphocytes vary among the cranial, oral, and nasal-associated lymphoid tissues. *Eur J Immunol.* 2002;32(11):3029-3039.
293. Fukuyama S, Nagatake T, Kim DY, et al. Cutting edge: uniqueness of lymphoid chemokine requirement for the initiation and maturation of nasopharynx-associated lymphoid tissue organogenesis. *J Immunol.* 2006;177(7):4276-4280.
294. Rangel-Moreno J, Hartson L, Navarro C, Gaxiola M, Selman M, Randall TD. Inducible bronchus-associated lymphoid tissue (iBALT) in patients with pulmonary complications of rheumatoid arthritis. *J Clin Invest.* 2006;116(12):3183-3194.
295. Fukuyama S, Hiroi T, Yokota Y, et al. Initiation of NALT organogenesis is independent of the IL-7R, LTbetaR, and NIK signaling pathways but requires the Id2 gene and CD3(-)CD4(+)CD45(+) cells. *Immunity.* 2002;17(1):31-40.
296. Delventhal S, Hensel A, Petzoldt K, Pabst R. Effects of microbial stimulation on the number, size and activity of bronchus-associated lymphoid tissue (BALT) structures in the pig. *Int J Exp Pathol.* 1992;73(3):351-357.
297. Kolopp-Sarda MN, Bene MC, Massin N, Moulin JJ, Faure GC. Immunohistological analysis of macrophages, B-cells, and T-cells in the mouse lung. *Anat Rec.* 1994;239(2):150-157.
298. Moyron-Quiroz JE, Rangel-Moreno J, Kusser K, et al. Role of inducible bronchus associated lymphoid tissue (iBALT) in respiratory immunity. *Nat Med.* 2004;10(9):927-934.
299. Pabst R, Tschernig T. Bronchus-associated lymphoid tissue: an entry site for antigens for successful mucosal vaccinations? *Am J Respir Cell Mol Biol.* 2010;43(2):137-141. doi:10.1165/rcmb.2010-0152RT
300. Ersch J, Tschernig T, Stallmach T. Frequency and potential cause of bronchus-associated lymphoid tissue in fetal lungs. *Pediatr Allergy Immunol.* 2005;16(4):295-298. doi:10.1111/j.1399-3038.2005.00269.x
301. Tschernig T, Pabst R. Bronchus-associated lymphoid tissue (BALT) is not present in the normal adult lung but in different diseases. *Pathobiol.* 2000;68(1):1-8.
302. Csencsits KL, Jutila MA, Pascual DW. Nasal-associated lymphoid tissue: phenotypic and functional evidence for the primary role of peripheral node addressin in naive lymphocyte adhesion to high endothelial venules in a mucosal site. *J Immunol.* 1999;163(3):1382-1389.
303. Cesta MF. Normal structure, function, and histology of mucosa-associated lymphoid tissue. *Toxicol Pathol* 2006;34(5):599-608. doi:10.1080/01926230600865531
304. Fotopoulos G, Harari A, Michetti P, Trono D, Pantaleo G, Kraehenbuhl JP. Transepithelial transport of HIV-1 by M cells is receptor-mediated. *Proc Natl Acad Sci U S A.* 2002;99(14):9410-9414.
305. Jones BD, Ghori N, Falkow S. *Salmonella typhimurium* initiates murine infection by penetrating and destroying the specialized epithelial M cells of the Peyer's patches. *J Exp Med.* 1994;180(1):15-23.
306. Hase K, Murakami T, Takatsu H, et al. The membrane-bound chemokine CXCL16 expressed on follicle-associated epithelium and M cells mediates lympho-epithelial interaction in GALT. *J Immunol.* 2006;176(1):43-51.
307. Iwasaki A, Kelsall BL. Localization of distinct Peyer's patch dendritic cell subsets and their recruitment by chemokines macrophage inflammatory protein (MIP)-3alpha, MIP-3beta, and secondary lymphoid organ chemokine. *J Exp Med.* 2000;191(8):1381-1394.
308. Iwasaki A, Kelsall BL. Unique functions of CD11b+, CD8 alpha+, and double-negative Peyer's patch dendritic cells. *J Immunol.* 2001;166(8):4884-4890.
309. Williams MB, Butcher EC. Homing of naive and memory T lymphocyte subsets to Peyer's patches, lymph nodes, and spleen. *J Immunol.* 1997;159(4):1746-1752.
310. Kunkel EJ, Ramos CL, Steeber DA, et al. The roles of L-selectin, beta 7 integrins, and P-selectin in leukocyte rolling and adhesion in high endothelial venules of Peyer's patches. *J Immunol.* 1998;161(5):2449-2456.
311. Steeber DA, Tang ML, Zhang XQ, Muller W, Wagner N, Tedder TF. Efficient lymphocyte migration across high endothelial venules of mouse Peyer's patches requires overlapping expression of L-selectin and beta7 integrin. *J Immunol.* 1998;161(12):6638-6647.
312. Lee M, Kiefel H, LaJevic MD, et al. Transcriptional programs of lymphoid tissue capillary and high endothelium reveal control mechanisms for lymphocyte homing. *Nat Immunol.* 2014;15(10):982-995. doi:10.1038/ni.2983
313. Okada T, Ngo VN, Ekland EH, et al. Chemokine requirements for B cell entry to lymph nodes and Peyer's patches. *J Exp Med.* 2002;196(1):65-75.
314. Cook DN, Prosser DM, Forster R, et al. CCR6 mediates dendritic cell localization, lymphocyte homeostasis, and immune responses in mucosal tissue. *Immunity.* 2000;12(5):495-503.
315. Koni PA, Sacca R, Lawton P, Browning JL, Ruddle NH, Flavell RA. Distinct roles in lymphoid organogenesis for lymphotoxins alpha and beta in lymphotoxi-beta deficient mice. *Immunity.* 1997;6:491-500.
316. Ansel KM, Ngo VN, Hyman PL, et al. A chemokine-driven positive feedback loop organizes lymphoid follicles. *Nature.* 2000;406(6793):309-314.
317. Forster R, Mattis AE, Kremmer E, Wolf E, Brem G, Lipp M. A putative chemokine receptor, BLR1, directs B cell migration to defined lymphoid organs and specific anatomic compartments of the spleen. *Cell.* 1996;87(6):1037-1047.
318. Kiyono H, Fukuyama S. NALT- versus Peyer's-patch-mediated mucosal immunity. *Nat Rev Immunol.* 2004;4(9):699-710.
319. Neumann B, Luz A, Pfeffer K, Holzmann B. Defective Peyer's patch organogenesis in mice lacking the 55-kD receptor for tumor necrosis factor. *J Exp Med.* 1996;184(1):259-264.
320. Adachi S, Yoshida H, Honda K, et al. Essential role of IL-7 receptor alpha in the formation of Peyer's patch anlage. *Int Immunol.* 1998;10(1):1-6.
321. Nishikawa S, Nishikawa S, Honda K, Hashi H, Yoshida H. Peyer's patch organogenesis as a programmed inflammation: a hypothetical model. *Cytokine Growth Factor Rev.* 1998;9(3-4):213-220.
322. Nishikawa SI, Hashi H, Honda K, Fraser S, Yoshida H. Inflammation, a prototype for organogenesis of the lymphopoietic/hematopoietic system. *Curr Opin Immunol.* 2000;12(3):342-345.
323. Nishikawa S, Honda K, Vieira P, Yoshida H. Organogenesis of peripheral lymphoid organs. *Immunol Rev.* 2003;195:72-80.
324. Finke D, Acha-Orbea H, Mattis A, Lipp M, Kraehenbuhl J. CD4+CD3- cells induce Peyer's patch development: role of alpha4beta1 integrin activation by CXCR5. *Immunity.* 2002;17(3):363-373.
325. Finke D, Kraehenbuhl JP. Formation of Peyer's patches. *Curr Opin Genet Dev.* 2001;11(5):561-567.
326. Golovkina TV, Shlomchik M, Hannum L, Chervonsky A. Organogenic role of B lymphocytes in mucosal immunity. *Science.* 1999;286(5446):1965-1968.
327. Debard N, Sierro F, Browning J, Kraehenbuhl JP. Effect of mature lymphocytes and lymphotoxin on the development of the follicle-associated epithelium and M cells in mouse Peyer's patches. *Gastroenterology.* 2001;120(5):1173-1182.
328. Lugering A, Kucharzik T, Soler D, Picarella D, Hudson JTIIIrd, Williams IR. Lymphoid precursors in intestinal cryptopatches express CCR6 and undergo dysregulated development in the absence of CCR6. *J Immunol.* 2003;171(5):2208-2215.
329. Lorenz RG, Newberry RD. Isolated lymphoid follicles can function as sites for induction of mucosal immune responses. *Ann N Y Acad Sci.* 2004;1029:44-57.
330. Eberl G. Inducible lymphoid tissues in the adult gut: recapitulation of a fetal developmental pathway? *Nat Rev Immunol.* 2005;5(5):413-420.
331. Newberry RD, Lorenz RG. Organizing a mucosal defense. *Immunol Rev.* 2005;206:6-21.
332. Lorenz RG, Chaplin DD, McDonald KG, McDonough JS, Newberry RD. Isolated lymphoid follicle formation is inducible and dependent upon lymphotoxin-sufficient B lymphocytes, lymphotoxin beta receptor, and TNF receptor I function. *J Immunol.* 2003;170(11):5475-5482.
333. Haley PJ. Species differences in the structure and function of the immune system. *Toxicology.* 2003;188(1):49-71.
334. Pablos JL, Santiago B, Tsay D, et al. A HEV-restricted sulfotransferase is expressed in rheumatoid arthritis synovium and is induced by lymphotoxin-alpha/beta and TNF-alpha in cultured endothelial cells. *BMC Immunol.* 2005;6(1):6.
335. Link A, Hardie DL, Favre S, et al. Association of T-zone reticular networks and conduits with ectopic lymphoid tissues in mice and humans. *Am J Pathol.* 2011;178(4):1662-1675. doi:S0002-9440(11)00039-3 [pii] 10.1016/j.ajpath.2010.12.039
336. Kerjaschki D. Lymphatic neoangiogenesis in human neoplasia and transplantation as experiments of nature. *Kidney Int.* 2005;68(5):1967-1968.
337. Kerjaschki D, Regele HM, Moosberger I, et al. Lymphatic neoangiogenesis in human kidney transplants is associated with immunologically active lymphocytic infiltrates. *J Am Soc Nephrol.* 2004;15(3):603-612.
338. Penaranda C, Tang Q, Ruddle NH, Bluestone JA. Prevention of diabetes by FTY720-mediated stabilization of peri-islet tertiary lymphoid organs. *Diabetes.* 2010;59(6):1461-1468. doi:10.2337/db09-1129
339. van de Walle T, Vaccaro A, Ramachandran M, Pietila I, Essand M, Dimberg A. Tertiary lymphoid structures in the central nervous system: implications for glioblastoma. *Front Immunol.* 2021;12:724739. doi:10.3389/fimmu.2021.724739
340. Asrir A, Tardiveau C, Coudert J, et al. Tumor-associated high endothelial venules mediate lymphocyte entry into tumors and predict response to PD-1 plus CTLA-4 combination immunotherapy. *Cancer Cell.* 2022;40(3):318-334 e9. doi:10.1016/j.ccell.2022.01.002
341. Cabrita R, Lauss M, Sanna A, et al. Tertiary lymphoid structures improve immunotherapy and survival in melanoma. *Nature.* 2020;577(7791):561-565. doi:0.1038/s41586-019-1914-8
342. Martinet L, Garrido I, Filleron T, et al. Human solid tumors contain high endothelial venules: association with T- and B-lymphocyte infiltration and favorable prognosis in breast cancer. *Cancer Res.* 2011;71(17):5678-5687. doi:10.1158/0008-5472.CAN-11-0431
343. Germain C, Gnjatic S, Tamzalit F, et al. Presence of B cells in tertiary lymphoid structures is associated with a protective immunity in patients with lung cancer. *Am J Respir Crit Care Med.* 2014;189(7):832-844. doi:10.1164/rccm.201309-1611OC
344. Reape TJ, Rayner K, Manning CD, et al. Expression and cellular localization of the CC chemokines PARC and ELC in human atherosclerotic plaques. *Am J Pathol.* 1999;154(2):365-374.
345. Dieu-Nosjean MC, Antoine M, Danel C, et al. Long-term survival for patients with non-small-cell lung cancer with intratumoral lymphoid structures. *J Clin Oncol.* 2008;26(27):4410-4417. doi:10.1200/JCO.2007.15.0284
346. Hjelmstrom P. Lymphoid neogenesis: de novo formation of lymphoid tissue in chronic inflammation through expression of homing chemokines. *J Leukoc Biol.* 2001;69(3):331-339.
347. Hjelmstrom P, Fjell J, Nakagawa T, Sacca R, Cuff CA, Ruddle NH. Lymphoid tissue homing chemokines are expressed in chronic inflammation. *Am J Pathol.* 2000;156(4):1133-1138.
348. Hanninen A, Jaakkola I, Jalkanen S. Mucosal addressin is required for the development of diabetes in nonobese diabetic mice. *J Immunol.* 1998;160(12):6018-6025.
349. Yang XD, Sytwu HK, McDevitt HO, Michie SA. Involvement of beta 7 integrin and mucosal addressin cell adhesion molecule-1 (MAdCAM-1) in the development of diabetes in obese diabetic mice. *Diabetes.* 1997;46(10):1542-1547.
350. Cannella B, Cross AH, Raine CS. Upregulation and coexpression of adhesion molecules correlate with relapsing autoimmune demyelination in the central nervous system. *J Exp Med.* 1990;172(5):1521-1524.
351. Columba-Cabezas S, Serafini B, Ambrosini E, Aloisi F. Lymphoid chemokines CCL19 and CCL21 are expressed in the central nervous system during experimental autoimmune encephalomyelitis: implications for the maintenance of chronic neuroinflammation. *Brain Pathol.* 2003;13(1):38-51.
352. Magliozzi R, Columba-Cabezas S, Serafini B, Aloisi F. Intracerebral expression of CXCL13 and BAFF is accompanied by formation of lymphoid follicle-like structures in the meninges of mice with relapsing experimental autoimmune encephalomyelitis. *J Neuroimmunol.* 2004;148(1-2):11-23.
353. Mooij P, de Wit HJ, Drexhage HA. An excess of dietary iodine accelerates the development of a thyroid-associated lymphoid tissue in autoimmune prone BB rats. *Clin Immunol Immunopathol.* 1993;69:189-198.
354. Katakai T, Hara T, Sugai M, Gonda H, Shimizu A. Th1-biased tertiary lymphoid tissue supported by CXC chemokine ligand 13-producing stromal network in chronic lesions of autoimmune gastritis. *J Immunol.* 2003;171(8):4359-4368.
355. Galkina E, Kadl A, Sanders J, Varughese D, Sarembock IJ, Ley K. Lymphocyte recruitment into the aortic wall before and during development of atherosclerosis is partially L-selectin dependent. *J Exp Med.* 2006;203(5):1273-1282.

356. Damas JK, Smith C, Oie E, et al. Enhanced expression of the homeostatic chemokines CCL19 and CCL21 in clinical and experimental atherosclerosis. Possible pathogenic role in plaque destabilization. *Arterioscler Thromb Vasc Biol.* 2006;27(3):614-620.
357. Grabner R, Lotzer K, Dopping S, et al. Lymphotoxin beta receptor signaling promotes tertiary lymphoid organogenesis in the aorta adventitia of aged ApoE-/- mice. *J Exp Med.* 2009;206(1):233-248. doi:10.1084/jem.20080752
358. Baddoura FK, Nasr IW, Wrobel B, Li Q, Ruddle NH, Lakkis FG. Lymphoid neogenesis in murine cardiac allografts undergoing chronic rejection. *Am J Transplant.* 2005;5(3):510-516.
359. Motallebzadeh R, Rehakova S, Conlon TM, et al. Blocking lymphotoxin signaling abrogates the development of ectopic lymphoid tissue within cardiac allografts and inhibits effector antibody responses. *Faseb J.* 2011;26(1):51-62. doi:10.1096/fj.11-186973
360. Luther SA, Ansel KM, Cyster JG. Overlapping roles of CXCL13, interleukin 7 receptor alpha, and CCR7 ligands in lymph node development. *J Exp Med.* 2003;197(9):1191-1198.
361. Ruddle NH. Basics of inducible lymphoid organs. *Curr Top Microbiol Immunol.* 2020;426:1-19. doi:10.1007/82_2020_218
362. Heikenwalder M, Zeller N, Seeger H, et al. Chronic lymphocytic inflammation specifies the organ tropism of prions. *Science.* 2005;307(5712):1107-1110.
363. Gause A, Gundlach K, Zdichavsky M, et al. The B lymphocyte in rheumatoid arthritis: analysis of rearranged V kappa genes from B cells infiltrating the synovial membrane. *Eur J Immunol.* 1995;25(10):2775-2782.
364. Schroder AE, Greiner A, Seyfert C, Berek C. Differentiation of B cells in the nonlymphoid tissue of the synovial membrane of patients with rheumatoid arthritis. *Proc Natl Acad Sci USA.* 1996;93:221-225.
365. Dorner T, Hansen A, Jacobi A, Lipsky PE. Immunglobulin repertoire analysis provides new insights into the immunopathogenesis of Sjogren's syndrome. *Autoimmun Rev.* 2002;1(3):119-124.
366. Sims GP, Shiono H, Willcox N, Stott DI. Somatic hypermutation and selection of B cells in thymic germinal centers responding to acetylcholine receptor in myasthenia gravis. *J Immunol.* 2001;167(4):1935-1944.
367. Stott DI, Hiepe F, Hummel M, Steinhauser G, Berek C. Antigen-driven clonal proliferation of B cells within the target tissue of an autoimmune disease. The salivary glands of patients with Sjogren's syndrome. *J Clin Invest.* 1998;102(5):938-946.
368. Kim HJ, Krenn V, Steinhauser G, Berek C. Plasma cell development in synovial germinal centers in patients with rheumatoid and reactive arthritis. *J Immunol.* 1999;162(5):3053-3062.
369. Nasr IW, Reel M, Oberbarnscheidt MH, et al. Tertiary lymphoid tissues generate effector and memory T cells that lead to allograft rejection. *Am J Transplant.* 2007;7(5):1071-1079. doi:AJT1756 [pii] 10.1111/j.1600-6143.2007.01756.x
370. McMahon EJ, Bailey SL, Castenada CV, Waldner H, Miller SD. Epitope spreading initiates in the CNS in two mouse models of multiple sclerosis. *Nat Med.* 2005;11(3):335-339.
371. Schrama D, thor Straten P, Fischer WH, et al. Targeting of lymphotoxin-alpha to the tumor elicits an efficient immune response associated with induction of peripheral lymphoid-like tissue. *Immunity.* 2001;14(2):111-121.
372. Lee Y, Chin RK, Christiansen P, et al. Recruitment and activation of naive T cells in the islets by lymphotoxin beta receptor-dependent tertiary lymphoid structure. *Immunity.* 2006;25(3):499-509.
373. Kaufman DL, Clare-Salzler M, Tian J, et al. Spontaneous loss of T-cell tolerance to glutamic acid decarboxylase in murine insulin-dependent diabetes. *Nature.* 1993;366(6450):69-72.
374. Wotherspoon AC, Doglioni C, Diss TC, et al. Regression of primary low-grade B-cell gastric lymphoma of mucosa-associated lymphoid tissue type after eradication of *Helicobacter pylori. Lancet.* 1993;342(8871):575-577.
375. Wu Q, Salomon B, Chen M, et al. Reversal of spontaneous autoimmune insulitis in nonobese diabetic mice by soluble lymphotoxin receptor. *J Exp Med.* 2001;193(11):1327-1332.
376. Fava RA, Notidis E, Hunt J, et al. A role for the lymphotoxin/LIGHT axis in the pathogenesis of murine collagen-induced arthritis. *J Immunol.* 2003;171(1):115-126.
377. Zintzaras E, Voulgarelis M, Moutsopoulos HM. The risk of lymphoma development in autoimmune diseases: a meta-analysis. *Arch Intern Med.* 2005;165(20):2337-2344. doi:10.1001/archinte.165.20.2337
378. Mazzucchelli L, Blaser A, Kappeler A, et al. BCA-1 is highly expressed in Helicobacter pylori-induced mucosa-associated lymphoid tissue and gastric lymphoma. *J Clin Invest.* 1999;104(10):R49-R54. doi:10.1172/JCI7830
379. Sautes-Fridman C, Petitprez F, Calderaro J, Fridman WH. Tertiary lymphoid structures in the era of cancer immunotherapy. *Nat Rev Cancer.* 2019;19(6):307-325. doi: 10.1038/s41568-019-0144-6
380. Helmink BA, Reddy SM, Gao J, et al. B cells and tertiary lymphoid structures promote immunotherapy response. *Nature.* 2020;577(7791):549-555. doi:10.1038/s41586-019-1922-8
381. Petitprez F, de Reynies A, Keung EZ, et al. B cells are associated with survival and immunotherapy response in sarcoma. *Nature.* 2020;577(7791):556-560. doi:10.1038/s41586-019-1906-8
382. Bromage ES, Kaattari IM, Zwollo P, Kaattari SL. Plasmablast and plasma cell production and distribution in trout immune tissues. *J Immunol.* 2004;173(12):7317-7323. doi:10.4049/jimmunol.173.12.7317
383. Rumfelt LL, McKinney EC, Taylor E, Flajnik MF. The development of primary and secondary lymphoid tissues in the nurse shark *Ginglymostoma cirratum*: B-cell zones precede dendritic cell immigration and T-cell zone formation during ontogeny of the spleen. *Scand J Immunol.* 2002;56(2):130-148. doi:10.1046/j.1365-3083.2002.01116.x
384. Castro CD, Ohta Y, Dooley H, Flajnik MF. Noncoordinate expression of J-chain and Blimp-1 define nurse shark plasma cell populations during ontogeny. *Eur J Immunol.* 2013;43(11):3061-3075. doi:10.1002/eji.201343416
385. Kapp FG, Perlin JR, Hagedorn EJ, et al. Protection from UV light is an evolutionarily conserved feature of the haematopoietic niche. *Nature.* 2018;558(7710):445-448. doi:10.1038/s41586-018-0213-0
386. Wattrus SJ, Zon LI. Blood in the water: recent uses of zebrafish to study myeloid biology. *Curr Opin Hematol.* 2021;28(1):43-49. doi:10.1097/moh.0000000000000627
387. Saltis M, Criscitiello MF, Ohta Y, et al. Evolutionarily conserved and divergent regions of the autoimmune regulator (Aire) gene: a comparative analysis. *Immunogenetics.* 2008;60(2):105-114. doi:10.1007/s00251-007-0268-9
388. Sutoh Y, Kondo M, Ohta Y, et al. Comparative genomic analysis of the proteasome β5t subunit gene: implications for the origin and evolution of thymoproteasomes. *Immunogenetics.* 2012;64(1):49-58. doi:10.1007/s00251-011-0558-0
389. Bajoghli B, Guo P, Aghaallaei N, et al. A thymus candidate in lampreys. *Nature.* 2011;470(7332):90-94. doi:10.1038/nature09655
390. Flajnik M. Immunology: the origin of sweetbreads in lampreys? *Curr Biol.* 2011;21(6):R218-R220. doi:10.1016/j.cub.2011.01.066
391. Bajoghli B, Aghaallaei N, Hess I, et al. Evolution of genetic networks underlying the emergence of thymopoiesis in vertebrates. *Cell.* 2009;138(1):186-197. doi:10.1016/j.cell.2009.04.017
392. Magor KE, Miranzo Navarro D, Barber MR, et al. Defense genes missing from the flight division. *Dev Comp Immunol.* 2013;41(3):377-388. doi:10.1016/j.dci.2013.04.010
393. Hofmann J, Greter M, Du Pasquier L, Becher B. B-cells need a proper house, whereas T-cells are happy in a cave: the dependence of lymphocytes on secondary lymphoid tissues during evolution. *Trends Immunol.* 2010;31(4):144-153. doi:10.1016/j.it.2010.01.003
394. Muthupandian A, Waly D, Magor BG. Do ectothermic vertebrates have a home in which to affinity mature their antibody responses? *Dev Comp Immunol.* 2021;119:104021. doi:10.1016/j.dci.2021.104021
395. Sunyer JO, Shibasali Y, Takizawa F, Yang D, Boudinot P, Krasnov A. Identification of primordial organized lymphoid structure in the spleen of teleost fish. *J Immunol.*2020;204(1 suppl):92.40-92.40.
396. Lewis KL, Del Cid N, Traver D. Perspectives on antigen presenting cells in zebrafish. *Dev Comp Immunol.* 2014;46(1):63-73. doi:10.1016/j.dci.2014.03.010
397. Horton JD, Manning MJ. Effect of early thymectomy on the cellular changes occurring in the spleen of the clawed toad following administration of soluble antigen. *Immunology.* 1974;26(4):797-807.
398. Baldwin WM,IIIrd, Cohen N. A giant cell with dendritic cell properties in spleens of the anuran amphibian *Xenopus laevis. Dev Comp Immunol.* 1981;5(3):461-473. doi:10.1016/s0145-305x(81)80058-4
399. Du Pasquier L Antibody diversity in lower vertebrates—why is it so restricted? *Nature.* 1982;296(5855):311-313. doi:10.1038/296311a0
400. Hinds-Frey KR, Nishikata H, Litman RT, Litman GW. Somatic variation precedes extensive diversification of germline sequences and combinatorial joining in the evolution of immunoglobulin heavy chain diversity. *J Exp Med.* 1993;178(3):815-824. doi:10.1084/jem.178.3.815
401. Wilson M, Hsu E, Marcuz A, Courtet M, Du Pasquier L, Steinberg C. What limits affinity maturation of antibodies in Xenopus--the rate of somatic mutation or the ability to select mutants? *EMBO J.* 1992;11(12):4337-4347.
402. Greenberg AS, Avila D, Hughes M, Hughes A, McKinney EC, Flajnik MF. A new antigen receptor gene family that undergoes rearrangement and extensive somatic diversification in sharks. *Nature.* 1995;374(6518):168-173. doi:10.1038/374168a0
403. Flajnik MF. A cold-blooded view of adaptive immunity. *Nat Rev Immunol.* 2018;18(7):438-453. doi:10.1038/s41577-018-0003-9

CHAPTER 9

Type I Cytokines, Interferons, and Their Receptors

Achsah D. Keegan, PhD • Warren J. Leonard, MD

OVERVIEW

Introduction

Cytokines are proteins that are secreted by cells and transduce signals via specific cell surface receptors on either the cytokine-producing cell (autocrine actions) or on other target cells (paracrine actions).[1] From this operational type of description, it is clear that the distinction between cytokines, growth factors, and hormones may be imprecise. In general, cytokines and growth factors are similar, except that the term "cytokine" most often refers to molecules involved in host defense that have actions on leukocytes, whereas the term "growth factor" more often refers to molecules acting on other somatic cell types.

Cytokine secretion is generally brief and self-limiting. It typically requires *de novo* gene transcription and protein synthesis, although in some cases cytokines are produced and stored in granules and subsequently released upon cellular degranulation. Some cytokine mRNAs are often unstable, with short half-lives, resulting in limited protein synthesis. Cytokines can elicit very potent biological responses by binding to specific cell surface receptor complexes and inducing cellular responses, sometimes (depending on the cytokine) at very low concentrations and with low receptor occupancy (Fig. 9.1). The binding of a cytokine to its receptor stimulates the formation of a receptor complex that can activate signal transduction pathways responsible for the ultimate biological response; these are major sites for therapeutic intervention (cytokine binding, receptor complex formation, signaling pathway activation).

Cytokines most frequently act through autocrine and paracrine mechanisms in highly localized environments. For example, in the interaction between a T cell and an antigen-presenting cell, cytokines are produced and usually exert potent actions locally, with rather limited biologic half-lives in the circulation. In contrast, hormones are released and then disseminate via the bloodstream throughout the body, with actions on distal target organs. Nevertheless, this distinction between cytokines and hormones is not absolute, with certain cytokines acting at longer distances as well.

Cytokines exhibit the feature of *pleiotropy*. This means that a single cytokine may act on many different cell types and mediate diverse biological effects. Historically, this feature limited the use of many cytokines therapeutically because of unwanted side effects. Cytokines also exhibit the feature of *redundancy*. This means that multiple cytokines can elicit the same or similar functional effects. Because of this feature, strategies to antagonize a single cytokine may not eliminate pathological responses. Cytokines can at times *synergize* with each other to amplify a response, or they can *antagonize* each other's actions. During a complex immune response to a pathogen, many cytokines can be produced simultaneously, and thus these four key features of cytokine biology must be considered in any targeted strategy.

Issues of Nomenclature

In the immune system, terms such as *interferons* (*IFNs*), *monokines*, and *lymphokines* were originally devised to identify the activity or cellular source for a protein found in cell-free supernatants.[2-4] IFN was the first member of the cytokine family to be identified in 1957 by Isaacs and Lindenmann as a secreted factor that *interfered* with viral infection of cells.[4] This turned out to be type I IFN (IFNα/β). Type II IFN (IFNγ) was discovered in 1965 as an antiviral activity produced by leukocytes stimulated with phytohemagglutinin.[5] Monokines included molecules such as interleukin (IL)-1, which was first recognized to be produced by monocytes, while lymphokines included molecules such as IL-2, which was first described as a T cell growth factor produced by T lymphocytes. The monokine/lymphokine nomenclature can be problematic, however, when a cytokine has more than one function or is synthesized by more than one type of cell. This resulted in the adoption of the term "cytokine," as proposed by Stanley Cohen in 1974.[3] The term cytokine was coined to indicate a protein made by a cell ("cyto") that acts ("kine") on target cells.

Many cytokines are referred to as *interleukins* to indicate molecules that are produced by one leukocyte and act on another.[6] However, this term is also problematic as some interleukins (eg, IL-1 and IL-6) are additionally produced by cells other than leukocytes and/or exert actions on cell types beyond the immune system. For example, IL-7 is produced by stromal and epithelial cells rather than by leukocytes. Furthermore, nomenclature can be inconsistent. For example, IL-7 is highly related to another cytokine that is denoted as thymic stromal lymphopoietin (TSLP) rather than as an interleukin, even though both can be produced by stromal and epithelial cells and share a receptor component as well as select actions on lymphocytes.

Among the many classes of cytokines, the "type I" cytokines (Table 9.1) share a similar four α-helical bundle structure (Fig. 9.2) as detailed below,[7-10] distinguishing them from other groups of cytokines such as IL-1 or IL-17 family

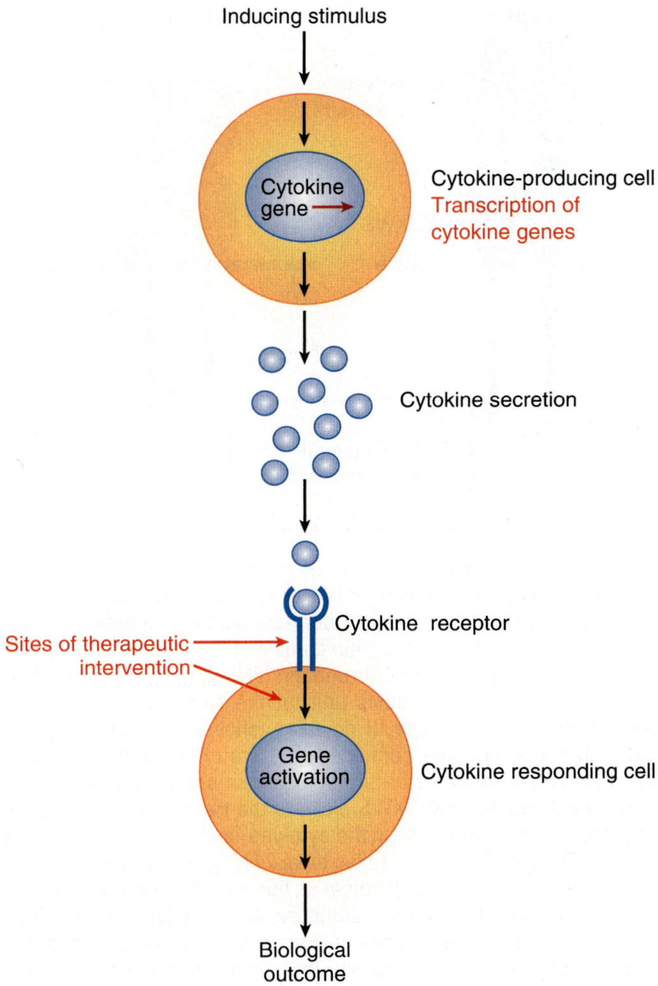

FIG. 9.1. Basic cytokine/cytokine receptor paradigm. A cytokine-producing cell responds to a stimulus with new gene transcription of cytokines and secretion of mature protein into the extracellular space. The cytokine then binds to specific receptor complexes found on the surface of a cytokine-responsive cell. This stimulates signal transduction event(s) that lead to changes in gene transcription patterns and elicitation of the biological outcome. The sites of cytokine/cytokine receptor interaction and signal transduction pathway are the focus for therapeutic interventions.

members. Correspondingly, type I cytokine receptors share characteristic features that have led to their description as the cytokine receptor superfamily or the hematopoietin receptor superfamily (Fig. 9.3, left).[8,11-13] Although many interleukins are type I cytokines, some are not. For example, two of the major "proinflammatory cytokines," IL-1 and IL-6, are interleukins, but IL-6 is a type I cytokine, whereas IL-1 is not. One interleukin, IL-8, is a CXC family chemokine, an entirely different type of molecule involved in chemotaxis. Over time, it was recognized that type I cytokines and IFNs share a number of common features, including signaling pathways. Moreover, IL-10, IL-19, IL-20, IL-22, IL-24, IL-26, IL-28A, IL-28B, and IL-29 are more similar to IFNs[14] and are denoted as type II cytokines.

TABLE 9.1 Four α-Helical Bundle Cytokines, also Known as Type I Cytokines

Short-Chain Cytokines	Long-Chain Cytokines
IL-2	IL-6
IL-4	IL-11
IL-7	Oncostatin M
IL-9	Leukemia inhibitory factor
IL-13	CNTF
IL-15	Cardiotrophin-1
IL-21	NNT-1/BSF-3
TSLP	
IL-3	Growth hormone
IL-5[a]	Prolactin
GM-CSF	Erythropoietin
	Thrombopoietin
M-CSF[a,b]	Leptin
SCF[b]	G-CSF

BSF-3, B cell–stimulating factor 3; CNTF, ciliary neurotrophic factor; G-CSF, granulocyte colony-stimulating factor; GM-CSF, granulocyte-macrophage colony-stimulating factor; IL, interleukin; M-CSF, macrophage colony-stimulating factor; NNT-1, novel neurotrophin-1; SCF, stem cell factor; TSLP, thymic stromal lymphopoietin.

[a]Dimers.

[b]Different from the other four helical bundle cytokines in that the M-CSF and SCF receptors (CSF-1R and c-kit, respectively) have intrinsic tyrosine kinase activity and are not type I cytokine receptors.

The term "interleukin" indicates a relationship to leukocytes, whereas the identification of a cytokine as a type I cytokine indicates general properties of its three-dimensional structure. Knowing that a molecule is a type I cytokine is instructive as it indicates a likely general structure for the cytokine receptor as well as the mechanism of signal transduction. In contrast, the identification of a molecule as an interleukin provides little information other than that it often, but not always, is a type I or type II cytokine of immunologic interest.

In addition to molecules of primarily immunologic interest, other important proteins, such as growth hormone, prolactin, erythropoietin (Epo), thrombopoietin, and leptin, are type I cytokines and their receptors are type I cytokine receptors. Despite having their major actions outside the immune system, these cytokines nevertheless share important signal transduction pathways with type I cytokines of immunologic interest. By presenting the broad topic of type I cytokines and IFNs and their receptors, this chapter necessarily focuses on cytokines that are evolutionarily related and share common signaling pathways, instead of focusing on common functions per se. For example, although IL-6 has overlapping actions with IL-1 and tumor necrosis factor (TNF) α, these latter proinflammatory cytokines are not discussed in this chapter because they are not type I cytokines, and the signaling pathways they use are distinct from those used by IL-6.

An unfortunate nomenclature issue exists that should be noted. There are specialized populations of T cells that include

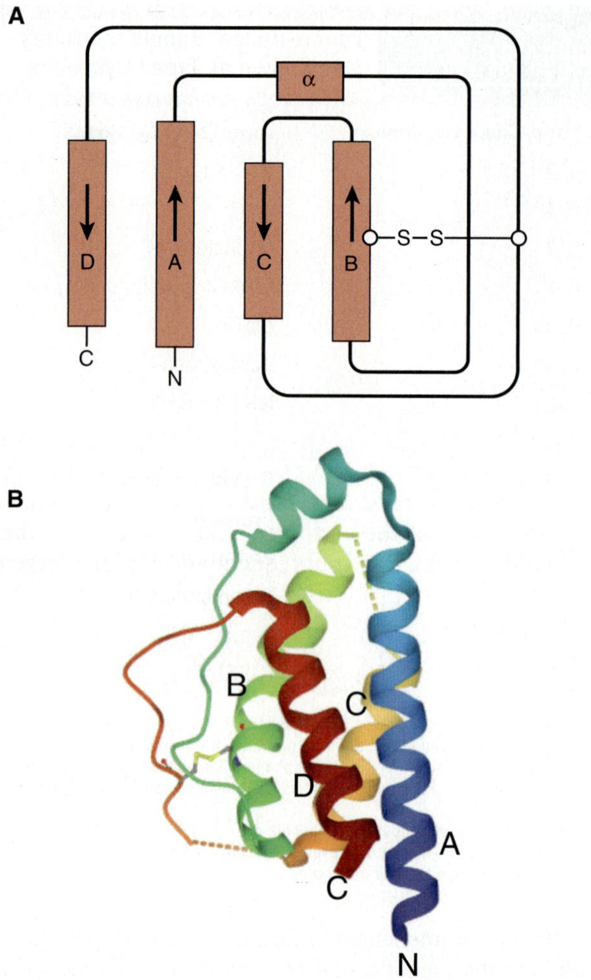

FIG. 9.2. Structure of the short chain four α-helical bundle cytokine IL-2. A: Schematic drawing showing the typical short-chain four helical bundle cytokine human IL-2. All cytokines in both the short chain and long chain groups exhibit an "up-up-down-down" topology to their four α-helices labeled A, B, C, and D. A short stretch of helix is labeled "α." **B:** Ribbon diagram of human IL-2 crystal structure (PDB ID 1M4C) published in Ref. 15 labeled as in **A.**

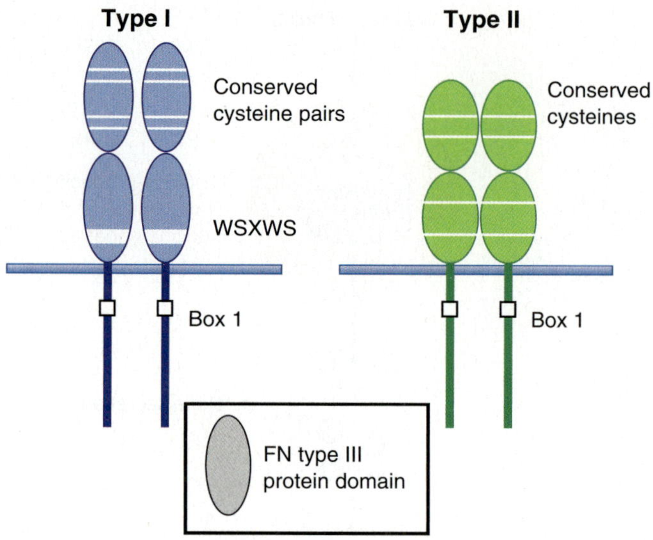

FIG. 9.3. Basic cytokine receptor structure. The basic cytokine receptor complex consists of at least two transmembrane proteins, which may be identical to each other (homodimers) or different (heterodimers). The elemental protein domain of the extracellular region of cytokine receptors is the fibronectin (FN) type III domain protein fold shown in *gray*. The basic unit for the type I cytokine receptors consists of 2 FN type III domains (shown in *blue*). The most amino-terminal of these contains two pairs of conserved cysteine residues while the carboxy-terminal domain contains a consensus amino acid sequence WSXWS close to the transmembrane region. Together, these domains form the cytokine-binding homology region or CHR. The type II receptors also have 2 FN type III domains in their extracellular region (*green*); however, the conserved cysteines have different spacing than type I receptors, and they lack the WSXWS motif. The cytoplasmic regions contain consensus sequence motifs termed Box 1 (shown in *white box*) that act as docking sites for tyrosine kinases of the Janus family. In addition, the cytoplasmic regions contain tyrosine residues that become phosphorylated after cytokine binding and receptor subunit dimerization/multimerization.

T helper 1 (Th1) and T helper 2 (Th2) cells (see Chapter 29 and following discussion), which produce specialized sets of cytokines, such as IFNγ by Th1 and IL-4 by Th2 cells. These are sometimes called type I and type II cytokines (for Th1 and Th2 cytokines); unfortunately, this jargon can result in confusion as IL-4 is a type I (ie, four α-helical bundle) cytokine that is functionally a type II cytokine (in that it is produced by Th2 cells), where IFNγ is a type II structure cytokine produced by Th1 cells.

CYTOKINES—STRUCTURAL CONSIDERATIONS

Type I cytokines typically share only limited amino acid sequence identity, but strikingly, all type I cytokines whose structures have been solved by nuclear magnetic resonance and/or X-ray crystallographic methods achieve similar three-dimensional structures.[7-10] Type I cytokines contain four α-helices and thus are designated as four α-helical bundle cytokines. As illustrated by human IL-2 (Fig. 9.2),[15] the first two and last two α-helices are each connected by long overhand loops, resulting in an "up-up-down-down" topological structure, as the first two helices (A and B) can be oriented in an "up" orientation and the last two helices (C and D) can be oriented in a "down" orientation, as viewed from the N- to C-terminal direction. As shown in Figure 9.2, the N- and C-termini of the cytokines are positioned on the same part of the molecule.

Type I cytokines are either "short-chain" or "long-chain" four α-helical bundle cytokines based on the lengths of the α-helices.[9,11] Some of the short-chain cytokines include IL-2, IL-3, IL-4, IL-5, granulocyte-macrophage colony-stimulating factor (GM-CSF), IL-7, IL-9, IL-13, IL-15, IL-21, macrophage colony-stimulating factor (M-CSF), stem cell factor (SCF), and TSLP, whereas long-chain cytokines include growth hormone, prolactin, Epo, thrombopoietin, leptin, IL-6, IL-11, leukemia inhibitory factor (LIF), oncostatin M (OSM), ciliary neurotrophic factor (CNTF), cardiotrophin-1 (CT-1), novel neurotrophin-1 (NNT-1)/B cell–stimulating factor 3 (BSF-3)/ cardiotrophin-like cytokine factor 1 (CLCF1), and granulocyte colony-stimulating factor (G-CSF) (Table 9.1).[10,11,16] The α-helices are approximately 15 amino acids long in short-chain helical cytokines and 25 amino acids long in long-chain cytokines. Additional differences include differences in the angles between the pairs of helices, and the AB loop is "under" the CD loop in the short cytokines but "over" the CD loop in the long cytokines. Moreover, short-chain cytokines have β structures in the AB and CD loops, whereas long-chain cytokines do not.[9,11]

The division of type I cytokines into short-chain and long-chain cytokines has evolutionary considerations and correlates with grouping of their receptor chains for these two subfamilies of type I cytokines. An analysis of short-chain cytokines has revealed a common structural framework, including most of the 31 residues that contribute to the buried inner core and 30 exposed residues.[10] The similarities and differences in the structures of IL-2, IL-4, and GM-CSF have been analyzed.[7] Among these cytokines, there is considerable variation in the intrachain disulfide bonds that stabilize the structures. For example, IL-4 has three intrachain disulfide bonds, GM-CSF has two, and IL-2 has only one. In IL-4, the first disulfide bond (between Cys 24 and Cys 65) connects loop AB to BC, the second disulfide bond (between Cys 46 and Cys 99) connects helix B and loop CD, and the third disulfide bond (between Cys 3 and Cys 127) connects the residue preceding helix B with helix D. In GM-CSF, the N-terminus of helix B and the N-terminus of β strand CD are connected by one disulfide bond, whereas the other disulfide bond connects the C-terminus of helix C and a strand following helix D. In IL-2, a single essential disulfide bond between Cys 58 and Cys 105 connects helix B to strand CD. Thus, each cytokine has evolved distinctive disulfide bonds to stabilize its structure, although it is typical that helix B is connected to the loop between helices C and D. The structures formed by helices A and D are more conserved than those formed by helices B and C, primarily due to the interhelical angles; helix D and the connecting region are the most highly conserved elements among the three cytokines.[7] This is of particular interest, as the regions of type I cytokines that are most important for cytokine receptor interactions include helices A and D and residues in the AB and CD loops, whereas helices B and C do not form direct contacts.

Variations on these typical four α-helical bundle structures can occur. For example, IL-5 is unusual in that it is a dimer, wherein the ends containing the N- and C-termini are juxtaposed, and helix D is "exchanged" between the two covalently attached monomers so that helix D of each molecule actually forms part of the four helix bundle of the other.[17] M-CSF is also a dimer, but no exchange of helix D occurs.[16]

The IFNs form related albeit distinctive structures from type I cytokines and are designated as type II cytokines.[11] IFNβ has an extra helix that is positioned in place of the CD strand while IFNγ is a dimer, each of which consists of six α-helices. Two of these helices are interchanged, including one from each four α-helical bundle. IL-10, which is closely related to IFNγ, has a similar structure[18] as do the other IL-10 family cytokines, including IL-19, IL-20, IL-22, IL-24, IL-26, as well as IL-28A, IL-28B, and IL-29.[14] Interestingly, although not universally the case, the genes encoding most four α-helical bundle cytokines have four exons, with helix A in exon 1, helices B and C in exon 3, and helix D in exon 4.[10] A related organization is found in the gene encoding IFNγ, as well as for the long-chain helical cytokines, such as growth hormone, and G-CSF. However, there are exceptions, for example, the IL-15 gene is divided into nine exons, whereas the IFNαs and IFNβ genes are encoded by single exons.

TYPE I AND TYPE II (IFN) CYTOKINE RECEPTORS

Basic Paradigm

The first report suggesting that type I cytokines interacted with receptors with similar features identified similarities in the sequences of the erythropoietin receptor and the IL-2 receptor β chain,[19] and subsequent analysis of a large number of type I cytokine receptors established these receptors as members of a superfamily.[8] The structure of receptors for type II cytokines, which include the IFNs and IL-10 family members, was found to be related (Fig. 9.3). The basic cytokine receptor complex consists of at least two transmembrane glycoproteins with the N-terminus extracellular and the C-terminus intracellular. The elemental protein domain of the extracellular region of cytokine receptors is the fibronectin type III (FNIII) domain protein fold. The basic unit for the type I cytokine receptors consists of 2 FNIII domains. The most amino-terminal of these contains two pairs of conserved cysteine residues while the carboxy-terminal domain contains a consensus amino acid sequence WSXWS (Trp-Ser-X-Trp-Ser) close to the transmembrane region. Together, these domains form the cytokine-binding homology region or CHR. There are some exceptions; for example, in the growth hormone receptor, the WSXWS motif is a substantially different YGEFS (Tyr-Gly-Glu-Phe-Ser) sequence, and in the IL-23R, it is WQPWS (Trp-Gln-Pro-Trp-Ser) (Table 9.2). The two pairs of conserved cysteine residues are typically encoded in two adjacent exons, and the exon containing the WSXWS motif is typically just 5′ to the exon encoding the transmembrane domain. Although serines can be encoded by six different codons (ie, sixfold degeneracy in codon usage), only two of these (AGC and AGT) dominate as the codons used for the serines in WSXWS. All of these features indicate a common ancestral type I cytokine receptor.

The type II receptors typically also have two FNIII domains in their extracellular region. However, the conserved cysteines have different spacing than type I receptors, and they lack the WSXWS motif. The cytoplasmic regions of both type I and II receptors contain consensus sequence motifs termed Box 1 that act as docking sites for tyrosine kinases of the Janus family.[20] In addition, the cytoplasmic regions contain tyrosine residues that become phosphorylated after cytokine binding and receptor subunit dimerization/multimerization.

TABLE 9.2 Features Common to Type I Cytokine Receptors

Extracellular domain
1. Four conserved cysteine residues, involved in intrachain disulfide bonds
2. WSXWS motif
3. Fibronectin type III modules

Cytoplasmic domain
1. Box 1/Box 2 regions—The Box 1 region is a proline-rich region that is involved in mediating the interaction of Janus family tyrosine kinases.

Depending on the cytokine, functional receptor complexes are often homodimers, heterodimers, or heterotetramers,[12] Not all receptor chains are transmembrane; for example, the CNTF receptor α chain lacks a cytoplasmic domain and instead has a glycosylphosphatidylinositol (GPI) anchor. In some cases, such as the common cytokine receptor β chain, βc, shared by the IL-3, IL-5, and GM-CSF receptors, the extracellular domain is extended, with a duplication of the CHR domain and addition of immunoglobulin (Ig)-like domains.

TYPE I CYTOKINE RECEPTORS

Structures

Overall, analogous to limited sequence identity between type I cytokines, there is only limited sequence identity among type I receptor molecules. Nevertheless, they appear to form similar structures. The first cytokine receptor structure solved was that for growth hormone (Fig. 9.4A).[21] Prior to X-ray crystallographic analysis, it was believed that growth hormone bound to its receptor with a stoichiometry of 1:1, but the X-ray crystal solution structure revealed that a single growth hormone molecule interacts with a dimer of the growth hormone receptor, in which each receptor monomer contributes a total of seven β strands. Perhaps the most striking finding is that totally different parts of growth hormone interact with the same general region of each receptor monomer. Solving the structure also clarified the basis for the assembly of the growth hormone receptor complex. Growth hormone first interacts with one receptor monomer via a relatively large and high-affinity interaction surface (site I), spanning ~1230 Å². A second receptor monomer then interacts with the growth hormone/growth hormone receptor complex via two contact points: one on growth hormone (spanning ~900 Å² (site II) and the other on the first receptor monomer spanning ~500 Å² (site III), located much more proximal to the cell membrane. Thus, three extracellular interactions are responsible for the formation and stabilization of the growth hormone/growth hormone receptor complex. Mutations in critical residues in site I should prevent growth hormone binding to its receptor, whereas inactivating mutations in site II can be predicted to prevent dimerization and signal transduction, suggesting a basis by which different classes of antagonists might be identified. Over time, analysis of additional crystal structures of type I cytokines bound to their receptors revealed that this three-site strategy is a common feature (Fig. 9.4).[13,22]

The growth hormone/growth hormone receptor structure revealed that the growth hormone receptor extracellular domain is composed of two FNIII modules, each of which is approximately 100 amino acids long and contains seven β strands, resulting in the formation of an Ig-like structure. The contact surface between ligand and receptor occurs in the hinge region that separates these two FNIII modules. Analysis of a growth hormone-prolactin receptor complex revealed a similar structure for the prolactin receptor.[23]

The growth hormone/growth hormone receptor structure served as a paradigm for the structures of other type I cytokine receptors. As a receptor homodimer, it immediately served as a model for other homodimers, such as the Epo receptor, whose structure was solved[24] using a small protein mimetic (20 amino acid long peptide) of Epo. The Epo receptor structure is similar to that of the growth hormone receptor, although the "site III" stem region interaction surface in the Epo receptor is much smaller than that in the growth hormone receptor, comprising only 75 Å².

In addition to the structural similarities for the growth hormone and Epo receptors and perhaps other homodimeric type I cytokine receptors, a similar structure was achieved by the heterodimeric growth hormone/growth hormone receptor/prolactin receptor structure,[23] in which one of the growth hormone receptor monomers is replaced by a prolactin receptor molecule. Thus, the two surfaces of growth hormone interact either with two identical monomers of growth hormone receptor or with two nonidentical monomers in the case of the growth hormone receptor/prolactin receptor heterodimer.

It seems reasonable that cytokine receptor systems with a homodimeric receptor are evolutionarily older than those with heterodimeric receptors, and that the coordination of two different receptor chains in heterodimeric receptors would have evolved in order to allow higher levels of specialization. In this regard, it is interesting that growth hormone and Epo, whose actions are vital for growth and erythropoiesis, bind to receptor homodimers, whereas the heterodimeric structures that typify the immune system are perhaps more "specialized" functions that arose later in evolution.

Heterodimeric receptors are involved when site II on a cytokine has evolved to a point where it interacts with a different receptor molecule than site I does. This latter situation is the case for many cytokines, including all short-chain type I cytokines except for SCF and M-CSF. An example of such a complex is the IL-4 receptor complex (Fig. 9.4B) consisting of the IL-4Rα chain and the common γ chain (γc) termed the type I IL-4 receptor (discussed below).[25] This overall structure is similar in topology to the growth hormone receptor complex; IL-4 makes contact with loops emanating from the

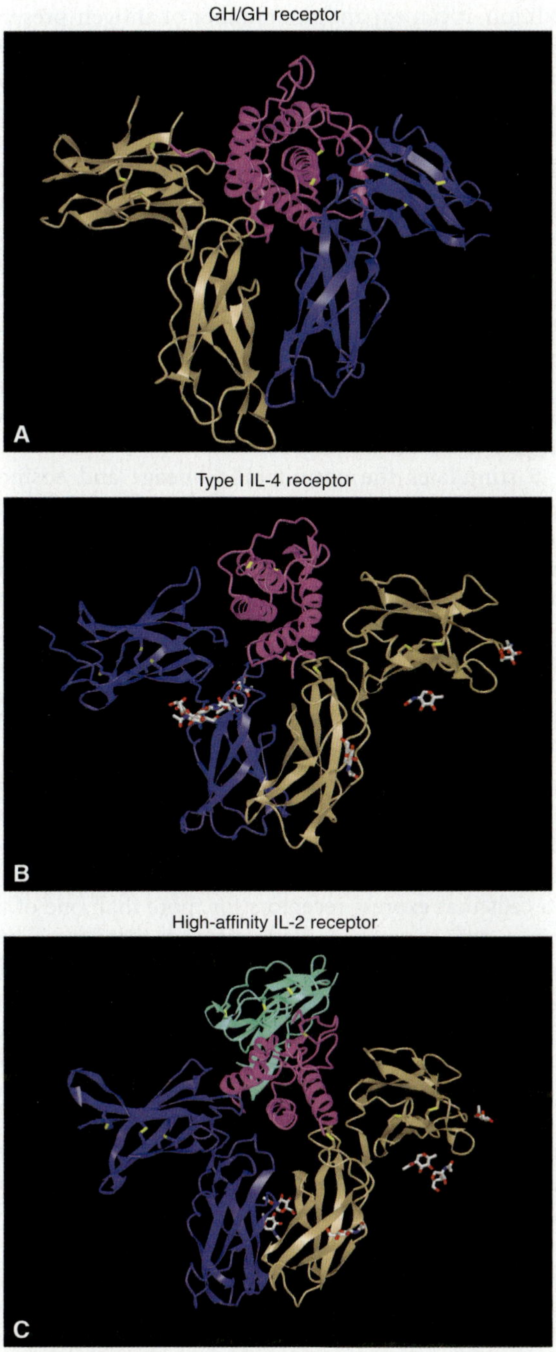

FIG. 9.4. Structures of cytokine/cytokine receptor complexes. A: Shown are ribbon diagrams of the structure of the human growth hormone receptor, the first reported structure of a type I cytokine receptor bound to its ligand. Growth hormone is shown in pink, while the growth hormone receptor homodimer chains are shown in blue and tan. See text for discussion of the structures. The growth hormone/growth hormone receptor structure is from PDB ID 3HHR and reported in de Vos et al.[21] **B:** Shown are ribbon diagrams of the type I IL-4 receptor ternary complex consisting of IL-4 (*pink*), the IL-4Rα (*blue*), and the common gamma chain (γc, tan). This structure is from PDB ID 3BPL and reported in Ref. 25. **C:** Shown are ribbon diagrams of the high-affinity IL-2 receptor complex consisting of IL-2 (*pink*), IL-2Rβ (*blue*), the γc (*tan*), and the sushi-domain containing IL-2Rα chain (*teal*). This structure is from PDB ID 2B51 and reported in Wang et al.[26] Of note is the consistent arrangement of the two main signaling chains and the cytokine.

hinge regions of the two FNIII domains of IL-4Rα through charge complementarity centered on residues Glu9 and Arg88 on the IL-4 A and C helices, respectively, which form polar contacts with Tyr13, Tyr183, Ser70, and Asp72 on IL-4Rα while it interacts with γc via residues in A and D helices. The membrane proximal FNIII domains of IL-4Rα and γc interact with each other.

Other receptor complexes show additional features including a third receptor chain. An example of this is the high affinity IL-2 receptor (Fig. 9.4C).[26] This complex includes IL-2Rα (also known as CD25), IL-2Rβ (CD122), and γc (CD132). IL-2Rα, which binds IL-2 with low affinity, is not a type I cytokine receptor protein but instead is a distinctive sushi domain–containing protein. Both the IL-2Rβ and γc are type I cytokine receptor proteins. In the structure of IL-2 bound to its high-affinity receptor (Fig. 9.4C), there is a long peptide connecting the IL-2Rα globular head and transmembrane segment, allowing the binding site on this protein to extend relatively far from the cell surface in order to bind the dorsal surface of IL-2. Both the IL-2/IL-2Rα and IL-2/IL-2Rβ contacts are independent, and IL-2Rα does not appear to contact either IL-2Rβ or γc; however, IL-2/IL-2Rβ forms a composite surface with γc, somewhat analogous to the composite surface formed by growth hormone and one growth hormone receptor monomer for binding to a second receptor monomer. As anticipated, the surface interaction between IL-2 and γc is smaller than that between IL-2 and either of the other chains. In addition to the structures shown here, numerous other type I cytokine receptor structures have been solved.[13,22]

TYPE I CYTOKINE RECEPTORS SHARE SUBUNITS: FAMILIES AND THEIR RELATIONS

The Common Beta Chain (βc) Family of Cytokines and Cytokine Receptors

The common beta (βc) family of cytokine receptors all utilize the same protein chain in their receptor complexes, the βc chain (Fig. 9.5). The human βc chain consists of two CHR domains in the extracellular region. This chain is an important signaling chain for GM-CSF, IL-3, and IL-5.[12,27-30] There are unique GM-CSF, IL-3, and IL-5 receptor alpha chains that are the principal ligand binding proteins. These chains consist of one CHR and an amino-terminal Ig domain in their extracellular domains. In the mouse, IL-3 signals through IL-3Rα plus either βc (AIC2A) or an alternative unique IL-3Rβ that shares 91% amino acid identity with βc (AIC2B) and appears to be a completely functionally redundant protein for IL-3 signaling. This receptor, which is not present in humans, contains sequence substitutions that enhance IL-3Rα chain–independent binding to IL-3.[31] However, murine IL-3Rβ cannot substitute for βc in the context of IL-5 or GM-CSF signaling. Analysis of the crystal structure of the GM-CSF receptor showed that the active form of the GM-CSF receptor is a large complex comprised of two βc chains surrounded by two GM-CSFRα chains with two molecules of GM-CSF bound between them.[30,32] This hexameric complex is thought to dimerize, forming a larger complex responsible

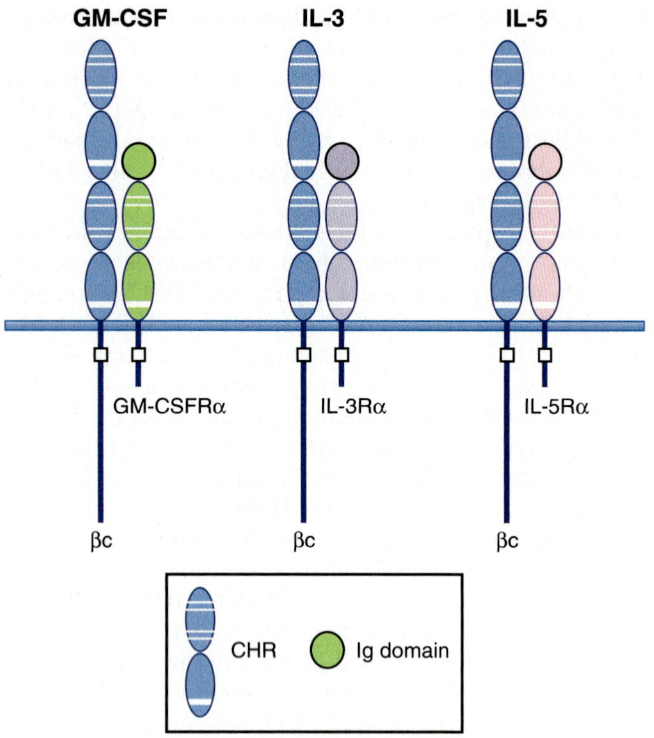

FIG. 9.5. Schematic of the common β chain family of cytokine receptors. The cytokines that all utilize the common beta chain (βc, *blue*) as one of the critical signaling chains in their receptor complexes include GM-CSF, IL-3, and IL-5. The human common beta chain consists of two CHR domains in the extracellular region. This chain is an important signaling chain for the GM-CSF, IL-3, and IL-5 receptors. There are unique GM-CSF, IL-3, and IL-5 receptor alpha chains that bear affinity for its ligand. These chains consist of one CHR and an amino-terminal Ig domain (*circle*) in their extracellular domains.

for signaling. The IL-3 receptor complex is thought to behave similarly. Because IL-5 functions as a dimer,[17] the IL-5 receptor complex was proposed to be an octomer with four IL-5 molecules, two IL-5Rα chains, and two βc chains.[29]

These hematopoietic cytokines, IL-3, IL-5, and GM-CSF (Table 9.3) are all synthesized by T cells and exert effects on cells of hematopoietic lineage.[29,33,34] These cytokines are vital for proliferation as well as differentiation of myeloid precursor cells. Of these three cytokines, IL-3 is the most pluripotent and historically was also called multi-CSF, reflecting the large number of lineages on which it can act. It can act to promote proliferation, survival, and development of multipotent hematopoietic progenitor cells and of cells that have become dedicated to a range of different lineages, including granulocyte, macrophage, eosinophil, mast cell, basophil, megakaryocyte, and erythroid lineages. It induces a range of effects, including, for example, inducing the production off IL-4 by basophils.[35] IL-3 also can exert end-function effects, such as enhancing phagocytosis and cytotoxicity.

GM-CSF is mainly restricted to the granulocyte and monocyte/macrophage lineages, but its actions are nevertheless still quite broad.[29,33] It is both a growth and survival factor. In addition, it can expand the number of antigen-presenting cells, such as dendritic cells (DCs), and thereby may greatly expand the ability of the host to respond to antigen. IL-23 and the transcription factor RORγt can augment GM-CSF production in Th cells, whereas IL-12, IFNγ, and IL-27 negatively regulate its expression, and GM-CSF is now known to be required for the initiation of autoimmune neuroinflammation in experimental autoimmune encephalitis. Collectively, these results suggest that GM-CSF is important for encephalitogenic actions of both Th1 and Th17 cells.[36,37] GM-CSF mediates autoimmune effects at least in part by enhancing IL-6–dependent Th17 cell development and survival.[38] While IL-3 and GM-CSF can act on eosinophils, they act at much earlier stages than IL-5, presumably expanding the number of eosinophil-committed precursor cells.

IL-5 stimulates the eosinophilic lineage and eosinophil release from the bone marrow and is essential for expanding eosinophils after helminth infections and can mediate the killing of *Schistosoma mansoni*.[28,29,39,40] IL-5 can also induce Ig production in B cells activated by contact with activated Th cells in mouse systems, and IL-5 and IL-5Rα knockout mice have diminished CD5+ B1 cells and decreased thymocytes until approximately 6 weeks of age.[39,40] Interestingly, the production of IL-5 by memory Th2 cells is driven by GATA3 but downregulated by expression of transcription factor eomesodermin (Eomes), which interacts with GATA3. Most memory Th2 cells produce Eomes, but the $CD62^{loCXCR3lo}$ population has decreased Eomes and is the population of cells that produces IL-5.[41]

On cells that express receptors for more than one of these cytokines, such as eosinophilic progenitors that express receptors for IL-3, IL-5, and GM-CSF, or on mouse pre–B cells, which express receptors for IL-3 and IL-5, the signals induced are indistinguishable.[29,33,34] Thus, sharing of receptor subunits is one explanation for cytokine redundancy. The differential lineage specificities of these cytokines are determined by the cellular distribution of their receptors rather than by fundamental differences in the signals that are induced by each cytokine.

As aforementioned, each of these three cytokines has its own unique 60 to 80 kDa α chain (ie, IL-3Rα, IL-5Rα, and GM-CSFRα) (Fig. 9.5) in addition to the 120 to 130 kDa βc. The α chains are the principal binding proteins for the cytokines, whereas the shared βc subunit augments binding affinity but does not exhibit binding activity in the absence of the proper α chain.[28,29] The α chains have relatively short cytoplasmic domains (approximately 55 amino acids long for IL-3Rα, IL-5Rα, and GM-CSFRα), whereas βc, with its cytoplasmic domain of 432 amino acids, contains the majority of signaling information. As a result, there is a relative compartmentalization of binding and signaling function for these cytokines, although the cytoplasmic domains of the α chains appear to contribute to the growth signals in transfected cells.[29]

An interesting feature of the βc family of cytokines is that there appears to be considerable redundancy of function so that knockout mice that lack the ability to respond to all three cytokines (GM-CSF, IL-3, and IL-5) due to deletion of βc as well as the mouse IL-3–specific βc-like protein nevertheless

TABLE 9.3 Features of Cytokines Whose Receptors Share the Common β Chain, βc

Cytokine	Major Source	Size	Cellular Targets	Chromosome Location (h/m)	Exons
IL-3	T cells	h152/19aa m166/26 aa 22-34 kDa	Multiple lineages	5q31.1/11	5
IL-5	T cells	h139/22aa m133/21 45 kDa dimer	Eosinophils B cells	5q31.1/11	4
GM-CSF	T cells	h144/17aa m141/17aa 23 kDa	Granulocytes Macrophages	5q31.1/11	4

GM-CSF, granulocyte-macrophage colony-stimulating factor; IL, interleukin.

exhibit relatively normal hematopoiesis.[34,42] These observations do not minimize the potency of these particular cytokines but instead underscore a substantial redundancy for a particularly important set of functions. It is also noteworthy that βc-deficient mice exhibit defective host responses to infectious challenge, suggesting that these hematopoietic cytokines play a vital role in promoting immune function.

The Common γ Chain (γc) Family of Cytokines and Cytokine Receptors

The receptors for six different immunologically important cytokines, IL-2, IL-4, IL-7, IL-9, IL-15, and IL-21, share the common cytokine receptor γ chain, γc (CD132) (Fig. 9.6).[13,43-45] These cytokines are all short-chain four α-helical bundle cytokines; basic features of these cytokines are summarized in Table 9.4. The properties of these cytokines and their distinctive receptor chains are summarized in the following text, followed by a discussion of the discovery that they share a common receptor component and the implications thereof.

IL-2 is important not only for its function but also historically, as it was the first type I cytokine that was cloned,[46] the first type I cytokine for which a receptor component was cloned,[47,48] and the first short-chain type I cytokine whose receptor structure was solved.[26] Many general principles have derived from studies of this cytokine, including it being the first cytokine demonstrated to act in a growth factor–like fashion through specific high-affinity receptors, analogous to the growth factors being studied by endocrinologists and biochemists.[43]

Mature IL-2 is a 133 amino acid long peptide that can act as a major T cell growth factor, in keeping with its original discovery in 1976 as a T cell growth factor.[49] Although IL-2 is not produced by resting T cells, it is rapidly and potently induced following antigen encounter with resting CD4+ T cells, and transcription and synthesis of IL-2 are often used as indicators of T cell receptor (TCR)–mediated cellular activation. Although the antigen determines the specificity of the T cell immune response, the interaction of IL-2 with high-affinity IL-2 receptors regulates the magnitude and duration of the subsequent response, based on the amount of IL-2 produced, the levels of high-affinity receptors expressed, and the duration of IL-2 production and receptor expression. IL-2 can act in either an autocrine or paracrine fashion, depending on whether the producing cell is also the responding cell or whether the responding cell is a nonproducing cell. The gene encoding IL-2 is located on chromosome 4, and like many other helical cytokines, its gene consists of four exons.[10] IL-2 binds to three different classes of receptors (Table 9.5). These are formed by different combinations of three different chains, IL-2Rα, IL-2Rβ,[50] and a protein initially called IL-2Rγ[51] but now known as the common cytokine receptor γ chain, γc.[44] These different classes of IL-2 receptors are discussed in the following text.

In addition to acting as a T cell growth factor, IL-2 has other important actions as well (see Table 9.4). For example, it can increase Ig synthesis and J chain transcription in B cells, potently augment the cytolytic activity of natural killer (NK) cells, induce the cytolytic activity of lymphokine-activated killer cells, promote the elimination of autoreactive cells in a process known as antigen-induced (or activation-induced) cell death, and promote the differentiation of regulatory T (T_{reg}) cells, cells that suppress inappropriate responses and are important for immunologic tolerance.[43,50,52-55] This latter activity is an essential function of IL-2 (see Chapter 30). Low-dose IL-2 is sufficient to promote T_{reg} cell survival and, thereby, for example, can protect mice from developing autoimmune diabetes.[56] Interestingly, IL-2 can also prime CD8+ T cells during a primary response to undergo enhanced proliferation *in vivo* during a secondary response and autocrine IL-2 is believed to be required for secondary expansion of CD8+ memory T cells.[57,58] There appears to be a complex interplay between IL-2 and inflammatory signals to regulate effector and memory cytolytic T lymphocytes generation in lymphocytic choriomeningitis virus infection, with persistent

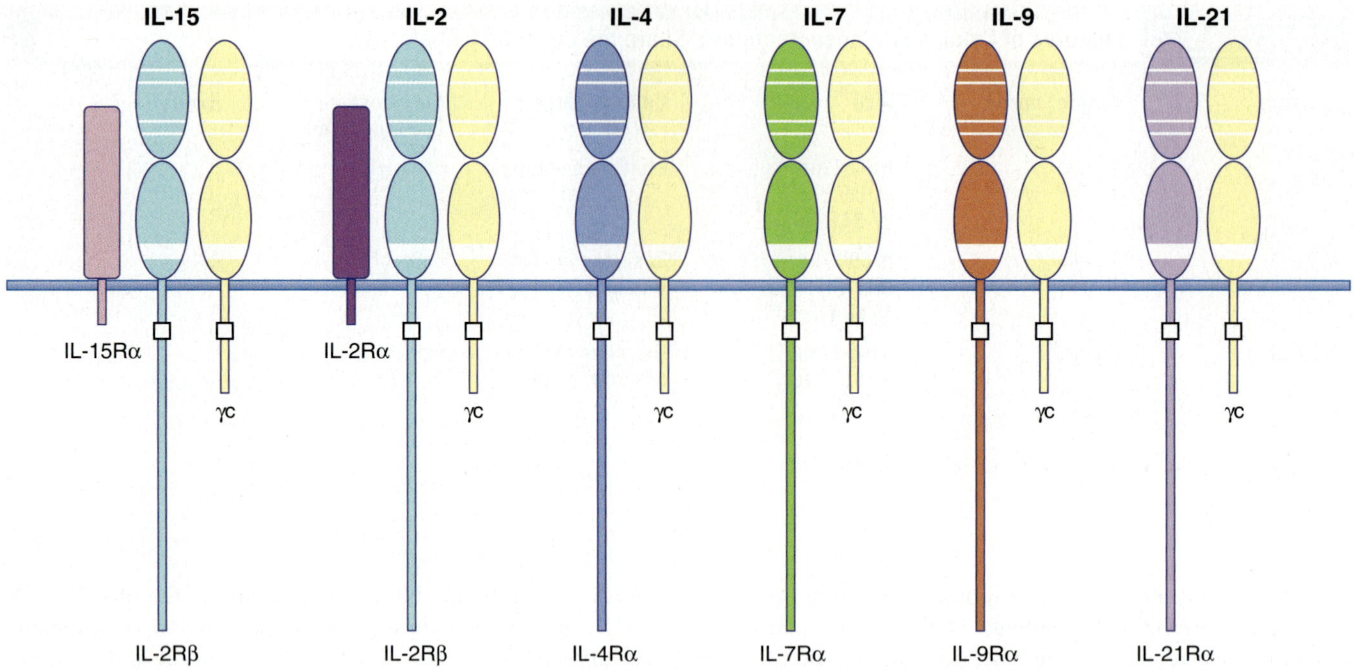

FIG. 9.6. **Schematic of the common γ chain family of cytokine receptors.** The cytokines that all utilize the common γ chain (γc, *yellow*) as one of the critical signaling chains in their receptor complexes include IL-2, IL-4, IL-7, IL-9, IL-15, and IL-21. Mutations in γc cause X-linked severe combined immunodeficiency or more moderate forms of X-linked immunodeficiency. The Box 1 region in the cytoplasmic domain (see white boxes) of γc is specific for the Janus kinase JAK3. Mutations in JAK3 cause an autosomal recessive form of severe combined immunodeficiency (see text). This schematic also shows sharing of the IL-2Rβ chain (*teal*) in the receptor complexes for IL-2 and IL-15. For these receptors, there are also unique cytokine binding chains denoted as IL-2Rα and IL-15Rα.

IL-2 promoting effector rather than memory cytotoxic T lymphocyte development.[59,60] In *Listeria monocytogenes* infection, Th1 effector memory cells highly express the transcription factor T-bet and IL-2Rα, and IL-2Rα appears to be critical for the development of these cells.[61] Because of its critical role in T_{reg} development, the absence of IL-2, IL-2Rα, or IL-2Rβ leads to autoimmunity.[55] Interestingly, the production of IL-2 by mast cells has been reported to contribute to the suppression of chronic allergic dermatitis by increasing the relative number of T_{reg} cells at the site of inflammation.[62]

IL-2 is also important in inducing or inhibiting Th differentiation. It is required for efficient Th1 and Th2 differentiation,[63,64] populations of T helper cells first described based on patterns of cytokine production by mouse T cells.[65] This theme was then extended to human cells as well.[66] IFNγ is the cytokine most reliably produced by Th1 cells; IL-2 is also produced by these cells, although without as rigorous an association. In contrast, IL-4, IL-5, IL-6, IL-9, IL-10, and IL-13 are produced by Th2 cells. In both species, certain type I cytokines (eg, IL-3 and GM-CSF) are produced by both Th1 and Th2 cells. IL-12 and the transcription factors signal transducer and activator of transcription (STAT) 4 and T-bet are the major transcription factors promoting Th1 differentiation, whereas IL-4 and the transcription factor STAT6 drive Th2 differentiation. For Th1 cells, IL-2 acts by STAT5-dependent induction of IL-12Rβ1/IL-12Rβ2 and T-bet, whereas for Th2 differentiation, IL-2 via STAT5 promotes IL-4 and IL-4Rα expression.[67] In addition, IL-2 also affects chromatin remodeling at the *Ifng* locus during Th1 differentiation.[68] IL-2 also inhibits Th17 differentiation in part via downregulation of IL-6Rα/gp130 expression and by inducing T-bet which prevents Runx1-mediated activation of RORγt, a transcription factor that drives Th17 differentiation[67-69] (see Chapter 29).

Like IL-2, IL-4 is produced primarily by activated CD4+ T cells.[70,71] IL-4 can also be produced by CD4+NK1.1+ "natural" T cells, denoted as natural killer T (NKT) cells, mast cells, basophils, eosinophils, macrophages, and under some circumstances by type 2 innate lymphoid cells (ILC2s).[71] IL-4 was initially described as a major B cell growth factor, and a promoter of Ig class switch, enhancing the production and secretion of mouse IgG1 (human IgG4).[72,73] Subsequent work established that IL-4 is essential for the production of IgE, the physiologic response to parasites, including helminths, and for allergen sensitization.[71,74,75] IL-4 induces expression of class II major histocompatibility complex molecules and increases cell surface expression of the CD23 (the low-affinity IgE receptor) on B cells. In addition to its actions on B cells, IL-4 can also act as a T cell growth factor, inducing proliferation in both human and mouse T cells, and is critical for normal differentiation of Th2 cells.[76] Moreover, when IL-4 and transforming growth factor (TGF) β are combined, cells that

TABLE 9.4 Features of Cytokines Whose Receptors Share γc

Cytokine	Major Source	Size	Actions	Chromosome Location (h/m)	Genomic Org
IL-2	Activated T cells	h153 aa/20aa m169aa/20aa 15.5 kDa[a]	T cell growth factor B cell growth, Ig production, J chain expression Induces LAK activity Induces tumor-infiltrating lymphocyte activity Augments NK activity Critical roles in antigen-induced cell death Stimulates macrophage/monocyte Antitumor effects Promotes T_{reg} development Promotes Th1 and Th2 differentiation Inhibits Th17 differentiation	4q26-27/3	Four exons
IL-4	Activated T cells (Th2 cells) CD4+NK1.1+ natural T cells	h153/24 aa m140 aa/20 aa 18 kDa	B cell proliferation Ig class switch: IgG1, IgE production Augment MHC II, Fcε receptors, IL-4Rα, and IL-2Rβ expression Th2 cell differentiation Antitumor effects	5q31.1/11	Four exons
IL-7	Stromal cells	h177 aa/25aa m154aa/25aa 17-25 kDa	Thymocyte growth T cell growth Pre–B cell growth in mice but not human Survival and growth of peripheral T cells $CD4^+$ and $CD8^+$ T cell homeostasis	8q12-13/3	Six exons
IL-9	Activated Th cells (Th9)	h144aa/18aa m144aa/18aa 14 kDa	Th helper clones Erythroid progenitors B cells Mast cells/allergic responses Fetal thymocytes Antitumor activity	5q31-35/13	Five exons
IL-15	Monocytes and many cells outside the immune system[b]	h162aa/48aa m162aa/48aa 14-15 kDa	Mast cell growth NK cell development and activity T cell proliferation Memory $CD8^+$ T cell homeostasis	4q31/8	Nine exons
IL-21	Activated CD4+ T cells	h162aa/31 aa m146aa/24 aa	Comitogen for T cell proliferation Inhibits B-cell proliferation to anti-IgM + IL-4 Augments B cell proliferation to anti-CD40 Can activate and expand NK cells Cooperates with IL-7 and IL-15 to expand CD8 cells Antitumor effects Drives terminal B cell differentiation to the plasma cells Proapoptotic for B and NK cells Promotes T_{FH} differentiation Promotes Th17 differentiation	4q67-27/3	Five exons

CD, cluster of differentiation; Ig, immunoglobulin; IL, interleukin; LAK, lymphokine-activated killer; MHC, major histocompatibility complex; NK, natural killer; T_{FH}, T follicular helper; Th, T helper; T_{reg}, regulatory T cell.

[a]h and m refer to human and mouse, respectively. The number of amino acids refers to the length of the open reading frame/length of signal peptide. The number of amino acids in the mature protein is therefore the difference between these numbers. Note that for IL-15, residues 1 to 29 have been identified as a signal peptide and 30 to 48 as a propeptide.

[b]More IL-15 mRNA is produced in the skeletal muscle, kidney, placenta, and lung than in the thymus or spleen. It is important to note, however, that IL-15 mRNA is widely expressed without concomitant production of IL-15 protein so that the source of biologically meaningful IL-15 may be more limited.

TABLE 9.5 Classes of Interleukin-2 Receptors

Affinity	K_d	Where Expressed	Composition	Functional
Low	10^{-8} M	Activated lymphocytes; T_{regs}	IL-2Rα	No
Intermediate	10^{-9} M	Resting cells, NK, CD8+ T cells	IL-2Rβ and γc	Yes
High	10^{-11}M	Activated lymphocytes, T_{regs}	IL-2Rα, IL-2Rβ, and γc	Yes

IL, interleukin.

produce IL-9 (Th9) cells are induced (see below). When combined with phorbol 2-myristate 3-acetate, IL-4 is also a potent comitogen for thymocytes. Moreover, IL-4 can exert actions on macrophages, hematopoietic precursor cells, stromal cells, and fibroblasts[70,71] and acts as a potent antiapoptotic factor for many cell types.[77,78] In recent years, additional functions of IL-4 have been described including contributions to T_{reg} development, regulation of metabolism and thermogenesis, and formation of memory.[79-83]

The gene encoding IL-4 is located on human chromosome 5 (5q23.3-31.2) and mouse chromosome 11 in the same region as IL-3, IL-5, IL-13, and GM-CSF.[71] The type I IL-4 receptor is expressed on T cells and other hematopoietic cells consists of the 140 kDa IL-4Rα protein and γc.[84,85] Expression of IL-4Rα tends to be quite low, and cells that potently respond to IL-4 often express only a few hundred receptors per cell. In addition to the type I IL-4 receptor, an alternate form of the receptor (the type II IL-4 receptor), containing IL-4Rα and IL-13Rα1 (discussed below), although not expressed on mature T cells, is expressed on many other cell types and can transduce IL-4 signals into these cells.[85] For example, IL-13Rα1 is expressed on neonatal Th1 cells and the type II IL-4 receptor has been implicated as mediating apoptosis of these cells.[86]

IL-7 is not produced by lymphocytes but instead is a 152 amino acid long cytokine that is produced by stromal cells and certain other cells.[87-89] The major role of IL-7 is to enhance thymocyte survival, growth, and differentiation and is essential for T cell development in humans. In addition, IL-7 promotes homeostatic proliferation of naive and memory CD8+ T cells.[90,91] IL-7 can also regulate the homeostasis of CD4+ memory T cells and stimulate the growth of mature T cells.[91] In the mouse, IL-7 is vital for the growth of mouse pre–B cells.[87,88] In contrast to its requirement for B cell development in the mouse, there is normal B cell development in patients with defective IL-7 signaling, as is found in patients with X-linked severe combined immunodeficiency disease (SCID), JAK3-deficient SCID, and IL-7Rα–deficient SCID (see following discussion). Thus, human B cells can develop normally in the absence of IL-7 responsiveness, demonstrating that in humans, IL-7 is not vital for the growth of human pre–B cells. Interestingly, IL-7 also acts on DCs, and this appears to limit the homeostatic proliferation of T cells under lymphopenic conditions.[89] In recent years, it has become clear that IL-7 is required for the development and maintenance of ILCs and thus makes a critical contribution to barrier defense.[92]

Clinically, IL-7 has been of interest in terms of its ability to induce T cell proliferation and to thereby increase T cell numbers, with a greater effect on CD8+ than on CD4+ T cells, as well as to augment the diversity of the TCR repertoire; IL-7 also can increase antigen-specific responses following vaccination[89] and can promote antiviral immunity. The gene encoding IL-7 is located on human chromosome 8q12 to 8q13 and mouse chromosome 3. The functional IL-7 receptor contains the 75 kDa IL-7Rα and γc.[44] Interestingly, signaling via the TCR, IL-2 or IL-7 can downregulate IL-7Rα expression with phosphatidylinositol 3-kinase (PI3K)/AKT and GFI1B being implicated in its downregulation in T cells.[93,94] The downregulation of IL-7Rα not only can decrease responsiveness of cells but also can increase the availability of the cytokine for other cells that are poised to respond.

IL-9 is efficiently produced by CD4+ helper T cells termed Th9 cells.[95-97] IL-9 can also be produced by ILC2s, mast cells, CD8+ T cells, and NKT cells under specific activating conditions. IL-9 was originally described as a mouse T cell growth factor produced by activated T cells to support the growth of T helper clones but not of cytolytic clones.[97,98] In the mouse, IL-9 can exert proliferative effects on erythroid progenitors, B cells, B-1 cells, mast cells, and fetal thymocytes. It also enhances class switching to IgE in B cells. IL-9 is identical to mast cell growth–enhancing activity, a factor present in conditioned medium from splenocytes and synergizes with IL-3 for maximal mast cell proliferation. IL-9 is highly expressed in the lungs of patients with asthma and overexpression leads to airway inflammation and Th2 cytokine production. Consistent with the action of IL-9 on thymocytes *in vitro*, IL-9 transgenic mice develop thymic lymphomas, and IL-9 is a major antiapoptotic factor for such tumors.[99] Nevertheless, IL-9 knockout mice have normal T cell development[100]; instead, they exhibit a defect in pulmonary goblet cell hyperplasia and mastocytosis following challenge with *S. mansoni* eggs, a synchronous pulmonary granuloma formation model. However, there was no defect in eosinophilia or granuloma formation. Mice expressing an IL-9 transgene in the lung exhibit airway inflammation and bronchial hyperresponsiveness; nevertheless, IL-$9^{-/-}$ mice exhibit normal eosinophilia and airway hyperreactivity in an ovalbumin-induced inflammatory model.[101] Thus, although IL-9 can contribute to allergic/pulmonary responses, there are compensatory cytokines that substitute for IL-9 in at least certain settings. IL-9 can be produced by the Th9 populations of cells, which can be induced by IL-4 plus TGF-β in a fashion that is dependent on PU.1 and interferon regulatory factor (IRF) 4.[96] Interestingly, the addition of IL-25 further enhances the production of IL-9.[97] IL-9–producing cells are typically IL-9+IL-10+Foxp3− effector T cells and can be derived from Th2 cells.[95-97] Overall, IL-9 is believed to play a key role in allergic responses; additionally, given the ability of both Th17 and T_{reg} cells to produce IL-9 during autoimmunity and transplantation, it will be interesting to further clarify the role of IL-9 in autoimmunity.[96] IL-9 also appears to contribute to the pathogenesis of inflammatory bowel disease as well.[102,103]

Human IL-9 is located on chromosome 5q31 to 5q35, which is also the location for the genes encoding IL-3, IL-4, IL-5, IL-13, and GM-CSF. In contrast, mouse IL-9 is isolated on chromosome 13, while IL-3, IL-4, IL-5, IL-13, and GM-CSF are clustered on chromosome 11. IL-9 binds to the 64 kDa IL-9Rα binding protein, similar in size to γc, and the functional IL-9 receptor consists of IL-9Rα plus γc.[43-45]

IL-15 was identified as a T cell growth factor that also unexpectedly was expressed in the supernatant of a human T lymphotropic virus type I (HTLV-I)–transformed T cell line.[104-106] Although IL-15 RNA is produced by a range of nonlymphocytic cell types, it is difficult to detect physiological levels of IL-15 protein. It is produced mainly by DCs and monocytes, and unlike IL-2, IL-15 is not produced by activated T cells.[107] IL-15 receptors are widely expressed; although IL-15 is perhaps most important for the development of NK cells and CD8+ memory T cells, it also has both paracrine and autocrine actions on DCs. The receptor for IL-15 on T cells contains IL-2Rβ, γc, and an IL-15–specific protein, IL-15Rα (Fig. 9.7).[43-45,108] IL-15Rα shares a number of structural similarities with IL-2Rα, including that it is a sushi domain–containing protein (IL-2Rα has two sushi domains, whereas IL-15Rα has one) and the *IL2RA* and *IL15RA* genes are closely positioned on human chromosome 10p14. A distinctive feature of IL-15 signaling is that it signals substantially by a process called *transpresentation*, wherein IL-15Rα on the surface of DCs, monocytes, or epithelial cells will transpresent IL-15 to responding cells such as CD8+ T cells or NK cells that express IL-2Rβ + γc.[107,109] Interestingly, transpresentation can occur with IL-2 as well.[110] In contrast to IL-2, which is a growth factor as well as a mediator of antigen-induced (or activation-induced) cell death and promoter of T_{reg} differentiation, the role of IL-15 appears to be more focused on the growth of CD8+ T cells and maintaining long-lasting, high-avidity T cell responses to foreign pathogens (ie, CD8+ T cell memory).[110] Overall, the fact that IL-15 promotes the proliferation and differentiation of B, T, and NK cells, the cytolytic activity of CD8 T cells, and the maturation of DCs suggest potential as an anticancer therapeutic agent.[111]

IL-21 was the final member of the IL-2 family of cytokines to be identified.[112] It was discovered as the ligand for an orphan type I cytokine receptor with similarity to IL-2Rβ before its biologic activity was defined.[113,114] IL-21 is produced by T cells including Th17, NKT, and T follicular helper (T_{FH}) cells.[112] It drives the expansion of CD8+ T cells in cooperation with IL-7 or IL-15, enhances CD8 memory,

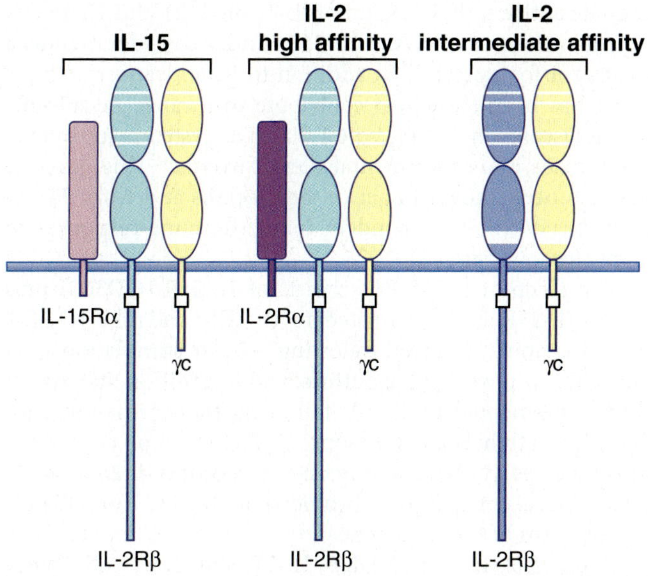

FIG. 9.7. Sharing of the IL-2Rβ chain. The IL-2Rβ chain is a critical component of three receptor complexes including the IL-15 receptor, the high-affinity IL-2 receptor, and the intermediate-affinity IL-2 receptor. In addition to the IL-2Rβ and γc chain, the IL-15 receptor includes a unique IL-15Rα chain, while the high-affinity IL-2 receptor includes the IL-2Rα chain.

and augments the antitumor activity of CD8+ T cells.[115,116] As compared to IL-2, it exerts different metabolism effects, with striking effects on B16 melanoma in adoptive transfer-mediated immunotherapy.[117,118] IL-21 promotes the differentiation of T_{FH} cells by inducing the transcription factor Bcl-6 and Th17 cells as part of an IL-6 to IL-21 to IL-23 signaling cascade.[119-121] IL-21 also exerts activities on B cells. It augments B cell proliferation when combined with anti-CD40 or lipopolysaccharide (LPS) but inhibits proliferation in response to anti-IgM + IL-4; this inhibition is reversed if anti-CD40 is additionally provided. It induces apoptosis of incompletely activated B cells, perhaps serving a role analogous to that of IL-2 in activation-induced cell death of T cells to eliminate incompletely activated cells.[112] IL-21 promotes the production of IgG1 while suppressing IgE.[122] It also drives the terminal differentiation of B cells to plasma cells.[123] In IL-21R knockout mice, following immunization, IgG1 is diminished related to a role for IL-21 in class switching to IgG1 and IgG3, whereas IgE is elevated, related to the ability of IL-21 to inhibit Cε transcription and/or to its ability to augment the apoptosis of IgE-producing B cells.[112] The IL-21R/IL-4 double knockout mice showed a more pronounced effect than the IL-21R single knockout revealing that IL-21 cooperates with IL-4 to globally regulate Ig production; these mice exhibit a pan-hypogammaglobulinemia, mimicking the B cell phenotype in humans with X-linked SCID.[122]

Interestingly, IL-21 exerts potent antitumor effects, including against large established solid tumors.[111,112] In the pMEL-1 CD8 TCR transgenic system, the effector cells recognize a melanoma antigen. When cells are adoptively transferred into tumor-bearing mice and animals are vaccinated with the cognate antigen, the addition of IL-21 or IL-21 plus IL-15 has potent antitumor effects.[115] IL-21 is under consideration as a treatment for cancer.[111] In addition to its anticancer activity, IL-21 has been shown to contribute to a range of autoimmune diseases including type I diabetes, systemic lupus erythematosus, and experimental allergic uveitis.[44] This cytokine has also been shown to act on neutrophils and with IFN to regulate neutrophil-dependent innate immune responses to *Staphylococcus aureus*.[124]

The receptor for IL-21 consists of IL-21R (CD360) plus γc.[43-45] IL-21R is most related to IL-2Rβ, and like IL-2Rβ, its expression is induced following cellular stimulation with anti-CD3 or phytohemagglutinin, and in addition, its expression is augmented in T cells following transformation with HTLV-I. Both human and mouse IL-21 can act on cells of the other species. IL-21 is on human chromosome 4q26 to 4q27, while its receptor is on chromosome 16p11, immediately downstream of the *IL4R* gene.

Thus, IL-2, IL-4, IL-7, IL-9, IL-15, and IL-21 collectively exhibit partially overlapping roles related to T cells, NK cells, B cells, and mast cells and together play vital roles for normal development and/or function of these cellular lineages. As discussed below, IL-2, IL-7, IL-9, and IL-15 utilize primarily STAT5A and STAT5B; IL-4 activates primarily STAT6; and IL-21 activates STAT1, STAT3, STAT5A, and STAT5B, with STAT3 being the dominant STAT for this cytokine.[43-45] The fact that these six cytokines share γc is of particular interest, especially as the gene encoding γc is mutated in patients with the most common form of severe combined immunodeficiency in humans.[125]

The situation for the γc family of receptors is distinct from the βc family. In the γc family, the chains with the largest cytoplasmic domains (IL-2Rβ, IL-4Rα, IL-7Rα, IL-9Rα, and IL-21R) not only contribute most to signaling specificity but also are the proteins principally involved in ligand binding (note that for IL-2 and IL-15, IL-2Rα and IL-15Rα, respectively, serve important roles as well). The shared chain, γc, serves a vital accessory function, but other than recruitment of JAK3 (see below), it contributes less to cytokine binding and does not in and of itself provide an obvious basis for signaling specificity. The opposite is true in the βc family, where the specific ligand binding chains have small cytoplasmic regions with the majority of signaling information encoded in the large cytoplasmic region of the shared βc (Fig. 9.5).

γc Subfamily; Receptors That Share IL-2Rβ

The IL-2Rβ chain is shared by three receptor complexes (Fig. 9.7), including the IL-15 receptor and the high- and intermediate-affinity IL-2 receptor complexes. There is a third type of IL-2 receptor that binds IL-2 with low affinity and does not contain IL-2Rβ (Table 9.5). In addition to the high-affinity receptor (IL-2Rα + IL-2Rβ + γc, $K_d \approx 10^{-11}$ M), there are both low-affinity (IL-2Rα alone, $K_d \approx 10^{-8}$ M) and intermediate-affinity (IL-2Rβ + γc, $K_d \approx 10^{-9}$ M) receptors.[22,43-45] Low-affinity and high-affinity receptors are expressed on activated lymphocytes and T_{regs}, whereas intermediate-affinity receptors are found on resting lymphocytes, particularly on NK cells. Both intermediate-affinity and high-affinity receptors can signal, and indeed both IL-2Rβ and γc are necessary for signaling, in keeping with the theme of dimerization indicated above.

Given that the intermediate-affinity form is functional, what then is the rationale for having a high-affinity IL-2 receptor that also contains IL-2Rα? This is particularly interesting given that IL-2Rα has an extremely short cytoplasmic domain that does not appear to play a role in signaling. The importance of IL-2Rα is clearly demonstrated by the severely abnormal phenotype of IL-2Rα–deficient mice, which exhibit autoimmunity, inflammatory bowel disease, and premature death[126] and by the observation that IL-2Rα mutations can cause an autoimmune syndrome in humans as well.[127] A clue to the importance of IL-2Rα comes from the kinetics of association of IL-2 with each chain. Although IL-2Rα appears to lack a direct signaling function, it has a very fast on-rate for IL-2 binding. Thus, the combination of this rapid on-rate with the slow off-rate from IL-2Rβ/γc dimers results in high-affinity binding that is vital for responding to the very low concentrations of IL-2 that are physiologically present *in vivo*.[22] Moreover, as activated T cells express approximately 10 times as many low-affinity than high-affinity receptors, IL-2Rα may serve as an efficient means of recruitment and concentration of IL-2 on the cell surface, allowing more efficient formation of IL-2/IL-2Rβ/γc signaling complexes. Indeed, binding of IL-2 to IL-2Rα results in a conformational change in the structure of IL-2, promoting its interaction with IL-2Rβ. These detailed binding parameters are critically important in the design of IL-2/IL-2 receptor targeting strategies discussed below.

As noted above, IL-2Rα is not a type I cytokine receptor and like IL-15Rα contains "sushi" domains, which contribute to ligand binding. The fact that IL-2 and IL-15 both have related α chains is consistent with the close relationship between IL-2 and IL-15, and that the receptors for both IL-2 and IL-15 contain both IL-2Rβ and γc.

As IL-2Rα cannot transduce a signal by itself, the detection of IL-2Rα on the cell surface does not necessarily reflect IL-2 responsiveness. Because IL-2Rα was discovered before IL-2Rβ and γc, many early papers in the literature evaluated IL-2 receptor expression based on IL-2Rα expression alone, and this may not reflect IL-2 responsiveness. IL-2Rα is also expressed on a subpopulation of double-negative thymocytes, but it remains unclear if IL-2 physiologically acts on these cells. Each of the components of the IL-2 receptor is located on a different chromosome: human IL-2Rα is located on chromosome 10p14 to 10p15, IL-2Rβ is at 22q, and γc is at Xq13.1 while the mouse homologs are located at chromosomes 2, 15, and X.

γc Subfamily; Receptors That Share IL-4Rα

As noted above, on T cells, IL-4 acts through a receptor comprising IL-4Rα and γc, termed the type I IL-4 receptor complex (Fig. 9.8A)[84]; however, IL-4 can also signal through type II IL-4 receptors comprising IL-4Rα and IL-13Rα1 but not γc.[85,128,129] IL-13 is another cytokine that shares some actions with IL-4 and was originally described as a T cell–derived cytokine capable of inhibiting inflammation,[130] although it is now clear that other cells such as NK cells, ILC2 cells, and mast cells can also produce IL-13. IL-13 can induce identical signals to IL-4 on non–T cells that respond to IL-4 but lacks effects on mature T cells, as these cells do not typically express IL-13Rα1 and cannot bind IL-13.[85] The shared actions of IL-4 and IL-13 include the ability to (1) decrease expression of inflammatory cytokines, (2) induce major histocompatibility complex class II expression, (3) induce CD23 expression and IgE production by B cells in humans, (4) inhibit IL-2–induced proliferation of chronic lymphocytic leukemia cells of B cell origin, (5) costimulate with anti-CD40 antibodies, and (6) drive the alternative activation of macrophages. This is explained by the observation that the type II IL-4 receptor consists of IL-4Rα plus IL-13Rα1 with both IL-4 and IL-13 inducing similar signals on cells expressing this receptor. However, there are differences in signaling kinetics and strength thay may lead to differential responses in cells.[25] Interestingly, IL-4 binds primarily to IL-4Rα followed by engagement of the IL-13Rα1 while IL-13 binds primarily to IL-13Rα1 followed by engagement of the IL-4Rα. An additional IL-13 binding protein, IL-13Rα2, has much higher binding affinity for IL-13 than does IL-13Rα1.[131,132] IL-13Rα2 does not signal through the JAK/STAT pathway and in many cases instead acts as a "decoy" receptor for IL-13.

Although IL-13 was originally believed to be substantially redundant with IL-4, it has important distinctive actions as well. IL-13 differentially modulates human DC differentiation when compared to IL-4.[133] IL-13 is an essential effector cytokine in asthma, as blocking IL-13 can inhibit pathophysiologic changes of asthma.[75,134,135] It also is clear that IL-13 regulates eosinophilic infiltration, airway hyperresponsiveness, and mucus secretion, and that it is a potent mediator of fibrosis.[136] Moreover, IL-13 that is produced by ILC2s in the lung contributes to asthma.[75] The phenotype of IL-13 knockout mice suggested a role for IL-13 in the ability to expel helminthes[137] and it is clear that IL-13 specifically can modulate resistance to intracellular organisms, including *Leishmania major* and *L. monocytogenes*.[136] IL-13 has also been identified as a factor that is secreted by and can stimulate the growth of Hodgkin lymphoma Reed-Sternberg cells.[138] Although IL-4 and IL-13 are both critical for host defense to helminths, the use of reporter mice revealed differential expression of these two cytokines.[139] T_{FH} cells in lymph nodes as well as basophils produce IL-4, whereas ILC2s secrete IL-13. Thus, although both are produced during helminth infection, there are distinctive sites of production, correlating with distinctive roles.[139]

γc Subfamily; Receptors That Share IL-7Rα

In addition to IL-7, a second stromal factor called TSLP was identified that shares at least some actions with IL-7.[44,140] The TSLP receptor is a heterodimer of TSLPR and IL-7Rα (Fig. 9.8B).[141-143] Interestingly, TSLPR is 24% identical to γc, making it the cytokine receptor most like γc.[141] Human and mouse TSLP share only 43% amino acid identity, and human and mouse TSLP receptors share only 39% identity. This is very low for human and mouse orthologues of cytokine receptors; for example, human and mouse γc are 70% identical. In addition to their wide sequence divergence, mouse

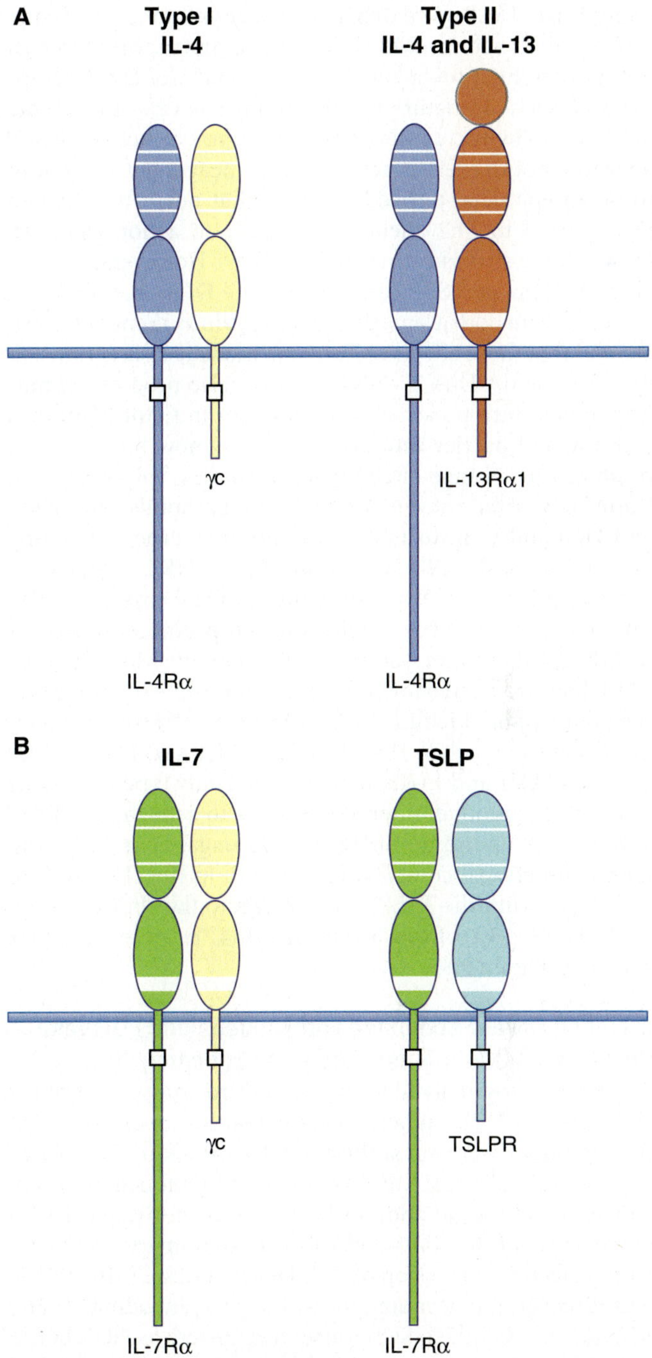

FIG. 9.8. Schematic of receptors that contain IL-4Rα or IL-7Rα. A: The receptor comprised of IL-4Rα (*blue*) and γc (*yellow*) is termed the type I IL-4 receptor and uniquely responds to IL-4. The IL-4Rα chain is also part of the type II IL-4 receptor comprised of the IL-4Rα and the IL-13Rα1 (*orange*). This receptor complex responds to either IL-4 or IL-13, although the initial ligand binding chain is different. IL-4 initially binds to the IL-4Rα with high affinity and stimulates heterodimerization with IL-13Rα1. IL-13 initially binds to IL-13Rα1 with intermediate affinity and stimulates dimerization with IL-4Rα. **B:** The IL-7Rα chain (*green*) is a critical component of the IL-7 and TSLP receptors. IL-7 initially binds to the IL-7Rα (*green*) chain and stimulates dimerization with γc (*yellow*). TSLP initially binds to the TSLPR (*sky blue*) and stimulates dimerization with the IL-7Rα (*green*). While this pattern of sharing is similar to the IL-4Rα–containing complexes shown in **(A)**, there is an important difference. IL-7 binding to IL-7Rα does not support dimerization with TSLPR.

and human TSLP were originally suggested to substantially differ in their actions. Mouse TSLP was first reported to be a B cell differentiation factor that is important for the development of IgM+ immature B cells from pre–B cells and to also be a weak thymic comitogen.[140] In contrast, human TSLP appeared not to exert effects on these lineages but was found to be an epithelial-derived cytokine that potently activated DCs related to Th2 allergic responses, an action not then known to be shared by mouse TSLP.[144,145] However, it is now clear that mouse TSLP can also activate DCs, that it plays a critical role in both atopic disease and asthma in mouse models, and that these effects of TSLP are mediated via actions on both DCs and CD4+ T cells. TSLP can also promote immunity to helminth parasites and is a major initiator of allergic priming and barrier integrity.[143] TSLP is now known to be produced by a range of cell types, including epithelial cells, fibroblasts, keratinocytes, stromal cells, basophils, mast cells, and DCs and to in turn act on multiple lineages, including DCs, CD4+ and CD8+ T cells, mast cells, NKT cells, eosinophils, and B cells.[143-145] Interestingly, TSLP promotes Th2 differentiation and has been shown to promote metastasis in breast cancer and pancreatic cancer.[146,147] More recenty, TSLP has also been shown to act on neutrophils, promoting the killing of methicillin resistant *S. aureus*.[148] From a signaling perspective, while IL-7 activates JAK1 and JAK3, TSLP activates JAK1 and JAK2, making it the only type I cytokine to use this combination of JAK kinases to activate STAT5.[149] While TSLP bound to the TSLPR can engage the IL-7Rα to form a functional complex, IL-7 bound to the IL-7Rα does not engage with the TSLPR, in contrast to the ability of both IL-4 and IL-13 to form a type II "IL-4" receptor complex shown in Figure 9.8A.

X-Linked Severe Combined Immunodeficiency Disease Results From Mutations in the Gene Encoding γc

IL-2Rγ was originally identified as a third component of the IL-2 receptor[51] after it became clear that IL-2 receptor α and β chains alone were not sufficient to transduce an IL-2 signal. The hypothesis that IL-2Rγ was a shared component of receptors for cytokines in addition to IL-2 was motivated from a comparison of the clinical phenotypes in humans that result from defective expression of IL-2 versus IL-2Rγ.[150] In 1993, it was discovered that mutations in the gene encoding IL-2Rγ resulted in X-linked SCID also designated as SCIDX1.[125] X-linked SCID is characterized by profoundly diminished numbers of T cells and NK cells[150,151] (Table 9.6). Although the B cells are normal in number, they are nonfunctional, apparently due to a lack of T cell help as well as an intrinsic B cell defect likely due to loss of IL-4 and IL-21 responsiveness. In contrast to the profoundly decreased number of T cells in patients with X-linked SCID, IL-2–deficient patients and mice have normal numbers of T cells.[152,153] This observation indicated that defective IL-2 signaling was unlikely to be responsible for X-linked SCID, making the finding that the gene encoding the γ chain was mutated in X-linked SCID all the more unexpected. Thus, the conundrum was why a defect in a component of a receptor would cause a more severe disease than a defect in the corresponding cytokine.

TABLE 9.6 Features of X-Linked Severe Combined Immunodeficiency

1. Absent or profoundly diminished numbers of T cells and mitogen responses.
2. Absence of NK cells.
3. Normal numbers of B cells, but defective B cell responses.
4. IgM can be normal but greatly diminished Igs of other classes.
5. XSCID carrier females exhibit nonrandom X-inactivation patterns in their T cells and NK cells; the X-inactivation pattern is random in surface IgM-positive B cells but nonrandom in more terminally differentiated B cells.

Ig, immunoglobulin; NK, natural killer; XSCID, X-linked severe combined immunodeficiency.

This led to the hypothesis that the γ chain was critical for other cytokine receptors as well.[125] In this model, defective IL-2 signaling either did not contribute to the defects in X-linked SCID or these defects were explained by the simultaneous inactivation of multiple signaling pathways. Indeed, when it was found that the IL-2Rγ chain was also an essential component of both the IL-4 and IL-7 receptors on T cells, it was renamed as the common cytokine receptor γ chain, γc.[150] IL-9, IL-15, and IL-21 were subsequently also shown to share γc[43] (Fig. 9.6).

The sharing of γc by six different cytokine receptors revealed that X-linked SCID is a disease of defective cytokine signaling. The major deficiencies in X-linked SCID can be attributed to defects related to different cytokines. Most if not all of the defect in T cell development in patients with X-linked SCID can be attributed to defective IL-7 signaling; there was dramatically diminished T cell development in IL-7–deficient or IL-7Rα–deficient mice and in *IL-7R*–deficient humans with T−B+NK+ SCID yet normal T cell development in mice deficient in IL-2, IL-4, IL-9, IL-15, IL-15Rα, and IL-21R.[43,150] In addition to profoundly diminished numbers of T cells, humans with X-linked SCID lack NK cells. NK cell development is defective in IL-15– and IL-15Rα–deficient mice, indicating that it is defective IL-15 signaling that is responsible for the defective NK cell development in X-linked SCID.[106,150]

In contrast to the greatly diminished number of T cells and absent NK cells in patients with X-linked SCID, B cell numbers are normal. This is in contrast to the greatly diminished numbers of B cells in γc-deficient mice[154] as well as mice deficient in either IL-7 or IL-7Rα, indicating that IL-7 is not required for pre–B cell development in humans and underscoring a major difference for IL-7 in human versus mouse biology. Indeed, patients with *IL-7R*–deficient SCID have a T−B+NK+ form of SCID.[155] Although B cells develop in patients with X-linked SCID, they are nonfunctional. This is due in part to a lack of T cell help (given the near absence of T cells in X-linked SCID), but a variety of data indicate an intrinsic B cell defect as well.[150] As discussed previously, analysis of IL-4/IL-21R double knockout mice indicates that defective signaling by IL-4 and IL-21 appears to explain the intrinsic B cell defect in X-linked SCID,[122] and other data indicate the particularly key role played by IL-21.[156,157] As a result of T cell, NK cell, and B cell defects, XSCID patients are highly susceptible to life-threatening infections beginning during the first year of life. With the development of the bone marrow and later stem cell transplantation strategies, XSCID has been successfully treated by gene therapy with γc.[158]

An analysis of mice deficient in IL-2, IL-2Rα, IL-2Rβ, and γc[126,153,154,159] provides the interesting observation that although the mice lacking γc have defective signaling in six different cytokine pathways, mice deficient in IL-2, IL-2Rα, and IL-2Rβ appear to be less healthy than the γc-knockout mice with more markedly activated T cells and autoimmunity that is not evident in γc-deficient mice. This is at least in part explained by the fact that IL-2–deficient mice lack T_{reg} cells while effector T cells are still present and thus develop autoimmune disease[55] whereas γc-deficient mice lack all T cells due to defective IL-7 signaling and thus lack T_{reg} and T effectors and ILCs. These and other data indicate that γc plays a major role in regulating many aspects of lymphoid homeostasis (see Chapter 30).

The gp130 Family of Cytokines and Cytokine Receptors

There are now ten cytokines that are known to utilize glycoprotein 130 (gp130) (CD130) as a signal transducing molecule, some of which are shown in Figure 9.9.[160,161] Some of the properties of these cytokines are summarized in Table 9.7. This family is often referred to as the IL-6 family of cytokines and includes IL-6, IL-11, OSM, LIF, CNTF, CT-1, CLCF1, IL-27, IL-35, and IL-39. This group of cytokines comprises molecules with a diverse range of actions, ranging beyond the hematopoietic and immune systems to the central nervous and cardiovascular systems.

IL-6 was originally identified and then cloned as a B cell differentiation factor that stimulated terminal differentiation/maturation of B cells into antibody-producing plasma cells.[160-163] However, IL-6 also can exert effects for T cell growth and differentiation (and thus is a thymocyte "comitogen"), induce myeloid differentiation into macrophages, induce acute-phase protein synthesis of hepatocytes, and exert actions on keratinocytes, mesangial cells, hematopoietic stem cells, the development of osteoclasts, and neural differentiation of PC12 cells. IL-6 is also a mediator of amplified lymphocyte trafficking during febrile inflammatory responses, which is dependent on IL-6–mediated activation of L-selectin.[164] IL-6 binds to an 80 kDa IL-6 binding protein, denoted IL-6Rα, which has a comparatively short 82 amino acid long cytoplasmic domain.[165] This IL-6-IL-6Rα complex then interacts with and recruits the 130 kDa signal transducing molecule, gp130, which together with IL-6Rα can form a functional IL-6 receptor (Fig. 9.9).[160,161] From a structural perspective, gp130 contains a total of five FNIII modules and an amino terminal Ig domain. The CHR comprises the first and second of the FNIII modules, starting from the N-terminus. As such, these regions are topologically positioned at greater distance external to the cell membrane than is the case for the other type I cytokine receptors discussed previously.

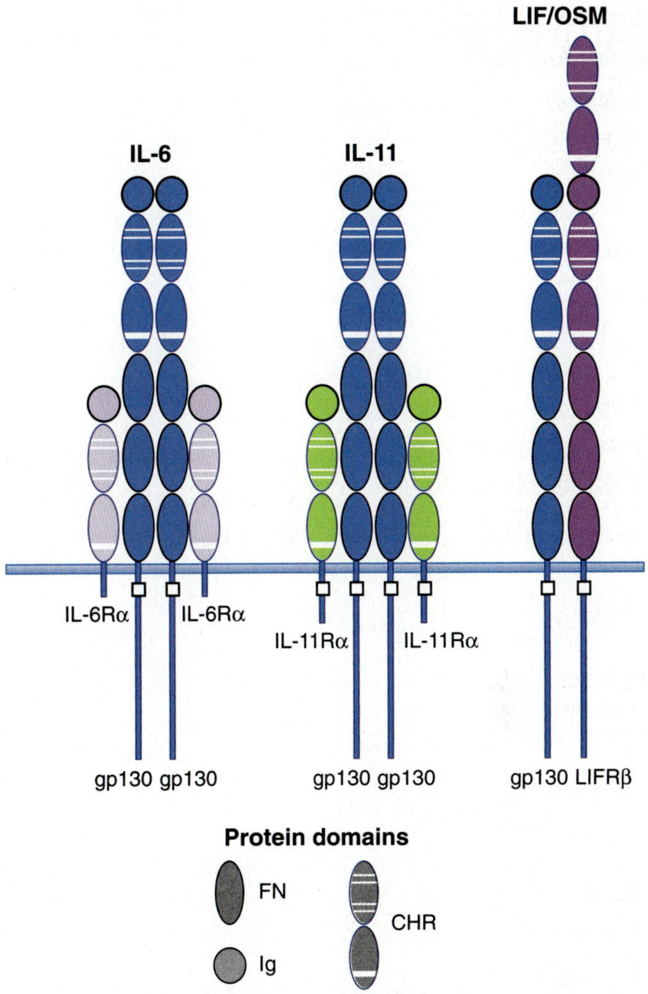

FIG. 9.9. Schematic of several members of the gp130 family of receptors. Schematic showing receptors for IL-6, IL-11, and LIF/OSM. In all cases, there is at least one chain of gp130 (*dark blue*) and a unique ligand-binding chain. The gp130 chain contains multiple FNIII domains that move the cytokine receptor homology region farther from the transmembrane region than in other type I cytokine receptors. The overall functional IL-6 receptor complex is a hexamer consisting of two molecules each of IL-6, IL-6Rα, and gp130.[167-169] The OSM receptor depicted here is referred to as the type I OSM receptor complex. See Table 9.7.

The IL-6 system illustrates a novel twist related to the properties of their principal binding proteins: The cytoplasmic domain of IL-6Rα is superfluous for signaling; a soluble form of the IL-6Rα extracellular domain is sufficient for ligand binding and coordination with gp130, a process termed *IL-6 transsignaling*.[166] Thus, in the presence of soluble IL-6Rα and IL-6, many cell types that express gp130 but not IL-6Rα are capable of signaling in response to IL-6. It was observed that IL-6 signaling requires the dimerization of gp130 and that the overall complex containing two molecules each of IL-6, IL-6Rα, and gp130 (a dimer of a trimer or a hexamer)[167-169] provides a paradigm for the stoichiometry of subunits for other members of the IL-6 family of cytokines.

TABLE 9.7 Cytokines Whose Receptors Share gp130

Cytokine	Chromosome Location (h/m)
IL-6	7p21/5
IL-11	19q13.3-13.4/7
LIF	22q12.1-12.2/11
OSM	22q12.1-12.2/11
CNTF	11q12.2/19
CT-1	16p11.1-11.2/7
CLCF1/NNT-1/BSF-3/CLC	11q13/19
IL-27 (IL27A:EBI3 dimer)	16p12.1-11.2:19p13.3/17
IL-35 (IL12A:EBI3 dimer)	3q25.33/3:19p13.3/17
IL-39[a] (IL-23A:EBI3 dimer)	[b]/10:[b]/17

Overlapping Actions of Several gp130 Cytokines

	IL-6	IL-11	LIF	OSM	CNTF	CT-1
Growth of myeloma cells	+	−	+	+	+	?
Maintenance of embryonic stem cell pluripotency	−	−	+	+	+	+
Induction of hepatic acute-phase proteins	+	+	+	+	+	+
Induction of cardiac hypertrophy	−	+	+	+	±	+
Induction of osteoclast formation	−	+	+	+	−	+
Enhanced neuronal survival/ differentiaion	+	+	+	+	+	+
Inhibit adipogenesis	+	+	+	+	−	−

BSF-3, B cell–stimulating factor 3; CLCF1, cardiotrophin-like cytokine factor; CNTF, ciliary neurotrophic factor; CT-1, cardiotrophin-1; IL, interleukin; LIF, leukemia inhibitory factor; NNT-1, novel neurotrophin-1; OSM, oncostatin M ; ?, uncertain.
[a]Not yet identified in humans.
[b]Human IL-39 has not yet been identified. Therefore only mouse chromosome locations for IL-23A and EB13 are indicated here.

IL-11 was identified as a factor produced by a stromal cell line in response to stimulation with IL-1.[170] It has a number of effects on hematopoiesis, particularly in combination with IL-3 and SCF. Because IL-11 exhibits "IL-6–like activities," a complementary deoxyribonucleic acid (cDNA) was isolated based on the presence of IL-6–like activity in the presence of antibodies to IL-6.[171] Other actions of IL-11 include the ability to stimulate the proliferation of lymphoid and hematopoietic progenitor cells, stimulate megakaryocytic progenitors and megakaryocyte maturation, regulate adipogenesis, and stimulate erythroid progenitors. Overall, it acts on T cells, B cells, hematopoietic cells, epithelial cells, endothelial cells, and osteoclasts.[170,171] Like IL-6, IL-11 induces acute-phase proteins and augments antigen-specific B cell responses, but it does not stimulate human myeloma cells. IL-11 is also produced by lung eosinophils and various structural cells in the lung and is expressed in patients with modest-to-severe asthma.[172] Similar to IL-6, IL-11 signals via a receptor complex containing a specific binding chain, IL-11Rα, and homodimers of gp130 (Fig. 9.9).[161] Interestingly, IL-11Rα messenger RNA can be alternatively spliced to yield a form lacking the cytoplasmic domain, and like IL-6Rα, a soluble form of IL-11Rα can coordinate with IL-11 to transsignal in cells expressing gp130.[173]

LIF is another multifunctional cytokine originally cloned based on the activity associated with its name.[174] LIF can suppress the differentiation of pluripotent embryonic stem cells, inhibit adipogenesis (like IL-11), and induce monocyte

TABLE 9.8 Composition of Receptors for the Interleukin-6 Family of Cytokines

Cytokines whose receptors do not contain LIFRβ	
IL-6	IL-6Rα + gp130
IL-11	IL-11Rα + gp130
OSM	OSMRβ + gp130
Cytokines whose receptors contain LIFRβ	
LIF	LIFRβ + gp130
OSM	LIFRβ + gp130
CNTF	CNTFRα + LIFRβ + gp130
CT-1	LIFRβ + gp130 + CT1Rα
CLCF1/CLCF1 + CLF-1/CLCF1	CNTFRα + LIFRβ + gp130
Cytokines whose receptors contain OSMRβ	
OSM	OSMRβ + gp130
IL-31	IL-31R + OSMRβ

BSF-3, B cell–stimulating factor-3; CLCF1, cardiotrophin-like cytokine factor; CLF-1, cytokine-like factor 1; CNTF, ciliary neurotrophic factor; CNTFR, CNTF receptor; CT-1, cardiotrophin-1; IL, interleukin; LIF, leukemia inhibitory factor; NNT-1, novel neurotrophin-1; OSM, oncostatin M.
Note that the sharing of CNTFRα by CNTF and the dimeric CLCF1–CLF-1 ligand helps to explain why the phenotype of $CNTF^{-/-}$ mice is less severe than $CNTFR\alpha^{-/-}$ mice.

differentiation of the M1 murine leukemia cell line, thus mimicking a number of the actions of IL-6.[175] In addition, it exerts a number of actions in the central nervous system; LIF was shown to be identical to cholinergic neural differentiation factor and can induce acetylcholine synthesis while simultaneously suppressing catecholamine production, thereby inducing cholinergic function while suppressing noradrenergic function. LIF has been shown to be essential for embryo implantation. LIF binds to a receptor (LIFRβ, CD118) that is structurally related to gp130,[176] but the functional LIF receptor (LIFR) receptor requires the heterodimerization of LIFRβ and gp130 as well.[177] Interestingly, while IL-6 promotes Th17 differentiation, this is inhibited by LIF, apparently based on LIF's ability to activate ERKs and to augment the expression of SOCS-3, which can inhibit STAT3 activation.[178] LIF can also promote the differentiation of T_{reg} cells and may exhibit therapeutic potential in multiple sclerosis.[179] LIF maintains mouse embryonic stem cells, but LIF is not capable of maintaining human embryonic stem cells or mouse epiblast stem cells in the pluripotent state.[180]

CNTF was discovered based on its ability to promote neuronal survival.[181] CNTF signals through a receptor comprising LIFRβ and gp130 but requires a specific binding protein[182] now denoted CNTFRα. Interestingly, CNTFRα lacks transmembrane and cytoplasmic domains and instead is a GPI-linked receptor molecule. CNTFRα appears to provide a receptor-cytokine surface with which gp130 and LIFRβ can interact. Thus, CNTF is like IL-6 in that each requires initial binding to a receptor component (CNTFRα or IL-6Rα) that does not require its own cytoplasmic domain for signaling. While IL-6 signaling involves homodimerization of gp130, CNTF signaling involves the heterodimerization of LIFRβ and gp130. In fact, the functional CNTFR appears to be a hexameric structure containing two molecules of CNTF, two of CNTFRα, and one each of gp130 and LIFRβ.[183] The receptor is expressed largely within the nervous system and in skeletal muscle, accounting for largely restricted actions of CNTF.[161]

OSM is a growth regulator that was originally identified based on its ability to inhibit the growth of A375 human melanoma cells.[184] OSM is a potent growth factor for Kaposi sarcoma in patients with acquired immunodeficiency syndrome.[185,186] Not only can OSM bind directly to gp130 and signal through a receptor combination of gp130 and LIFRβ, but it also can use an alternative receptor comprising a specific OSM receptor subunit (OSMRβ) and gp130.[177,187] These are now known as the type I and type II OSM receptors (OSMRs), respectively. OSM can enhance the development of both endothelial cells and hematopoietic cells, possibly by increasing hemangioblasts, a common precursor for endothelial and hematopoietic cells.[188]

CT-1 was initially isolated based on its actions on cardiac muscle cells.[189] However, it is now clear that it is a multifunctional cytokine with hematopoietic, neuronal, and developmental effects, in addition to its effects on cardiac development and hypertrophy.[190] Like OSM and LIF, CT-1 can also signal through a heterodimer of LIFRβ and gp130.[161]

CLCF1, also known as NNT-1, BSF-3, and CLC, can also support the survival of chicken embryonic sympathetic and motor neurons.[161,191] Interestingly, in mice, CLCF1 can augment the effects of IL-1 and IL-6 and is a B cell–stimulating factor (hence the original name BSF-3). CLCF1 forms homodimers and interacts with the receptor containing LIFRβ and gp130.[161] CLCF1 can also form a heterodimer with a soluble receptor protein known as cytokine-like factor (CLF). Together, this complex is a second ligand for

CNTFR.[192] CLCF1 can also interact with soluble CNTFR to form a related cytokine.[191]

There are additional cytokines included in the IL-6 family that are also members of the IL-12 family of cytokines.[161] Members of this cytokine family are heterodimers, with one of the dimeric proteins showing the CHR structure[193] (Chapter 12). IL-27 represents a dimer of the p28 protein (also known as IL-30) and EBi3 (Epstein-Barr virus–induced gene 3) (also known as IL-27B), which can enhance proliferation of naive CD4+ T cells.[194] p28 can exist by itself independently of EBI3 and is capable of antagonizing gp130-mediated signaling by IL-6, and indeed, overexpression of p28 in transgenic mice prevents normal development of germinal centers and antibody production.[195,196] IL-35 is a dimer of EBi3 and p35, a component of IL-12.[197,198] Together with IL-12, IL-23, CLCF1/CLF, and CLCF1/sCNTFR, these cytokines represent dimers including a type I cytokine and a soluble receptor-like protein. IL-27 signals via gp130 and the IL-27R (WSX-1) while IL-35 signals via gp130 homodimers, gp130 and IL-12Rβ2 heterodimers, or IL-12Rβ2 homodimers.[198]

IL-27 has both proinflammatory and anti-inflammatory effects, while IL-35 is anti-inflammatory.[198] Although it was suggested that IL-27 promotes Th1 responses, it is clear that it can also antagonize such responses, and IL-27 can promote effector responses of both CD4+ and CD8+ T cells and of NK cells, and it additionally stimulates mast cells. The anti-inflammatory effects of IL-27 are indicated by the absence of WSX-1, resulting in increased mast cell and macrophage responses.[196,199] IL-35 is produced by wide range of regulatory lymphocytes and plays a role in immune suppression.[197] In addition, IL-35 can block the development of Th1 and Th17 cells by limiting early T cell proliferation.[198]

IL-39 is a recently described murine cytokine consisting of IL-23p19 and EBi3 that has been implicated in development of lupus-like symptoms in murine models.[200] IL-39 signals through an IL-23R, gp130 receptor complex expressed on B cells. The existence of a human IL-39 homolog has not been established and remains theoretical for the time being. However, while not all human cytokines exist in mouse, all other confirmed members of the IL-12 family have been shown to exist in both mice and man.

Thus, ten cytokines (IL-6, IL-11, LIF, CNTF, OSM, CT-1, CLCF1, IL-27, IL-35, and IL-39) all have receptors that contain gp130.[161,201] These can be divided into two sets of cytokines: those known to not require LIFRβ (IL-6, IL-11, IL-27, IL-35, and IL-39) and those that use both gp130 and LIFRβ (LIF, CT-1, OSM, CNTF, and CLCF1), with OSM having two forms of receptors, each of which contains gp130 but only one of which contains LIFRβ. When two cytokines share essentially the same receptor, the cytokines might exert similar actions on cells that can respond to both cytokines; however, subtle differences in complex formation may alter nuances of the biological outcome.[201] It is clear that the presence of IL-6Rα, IL-11Rα, and CNTFRα (either on the cell surface or as a soluble receptor form, discussed subsequently) determines whether a cell can respond to IL-6, IL-11, and CNTF.

Other Receptors With Similarities to gp130

As noted previously, LIFRβ and OSMRβ bear some similarities to gp130.[187] In addition, the G-CSF receptor, the leptin receptor (also denoted OB-R, for obesity receptor), the IL-31R, and IL-12/IL-23/IL-35/IL-39 receptor components all resemble gp130. The amino acid identity among these different receptors, compared pairwise, ranges from 18% to 32%, with LIFRβ and OSMRβ being the most similar.

Leptin is the product of the obesity (LEP or OB) gene, an adipose tissue–derived cytokine that plays a role in body weight homeostasis.[202] Additionally, leptin affects thymic homeostasis, increasing thymocyte number and having anti-apoptotic effects in the thymus and periphery. It is proinflammatory, promoting Th1 differentiation and augmenting the production of TNF-α, IL-1, and IL-6 by monocytes/macrophages. The leptin receptor, LEP-R or OB-R, is most closely related to gp130, G-CSF receptor, and LIFRβ.[203] Interestingly, this receptor is encoded by the "diabetes gene," which is mutated in db/db mice.[204]

IL-31 is a four α-helical bundle cytokine that fits into the greater IL-6 extended family in that its receptor consists of IL-31RA and the OSMRβ.[205] IL-31R is a gp130-like type I receptor, with four splice variants.[206] IL-31 is produced primarily by activated T cells and when overexpressed results in pruritis, alopecia, and skin lesions, indicating a potential role in dermatitis.[205]

As noted previously, the IL-12 family of cytokines and receptors overlaps with the IL-6/gp130 family.[193-200] Their critical roles in immune responses are well-documented[193,198] and are presented in detail in Chapter 12. Structurally, IL-12 is a covalently linked dimer of 35 (IL-12A) and 40 (IL-12B) kDa peptides; thus, successful production of IL-12 requires that a cell transcribes both the p35 and p40 genes. Interestingly, while p35 bears sequence similarity to IL-6 and G-CSF, p40 is homologous to the extracellular domains of IL-6Rα, CNTFRα, and G-CSF receptor and bears some of the features typical of type I receptors, including four conserved cysteines, a conserved tryptophan, and a WSEWAS motif, which has some similarity to the typical WSXWS motif. Moreover, as both IL-12 receptor (IL-12Rβ1 and IL-12Rβ2) chains bear some similarity to gp130, one can think of p40 as a functional homolog of the soluble 80 kDa IL-6Rα chain. Thus, for this cytokine, part of the "receptor" has become part of the cytokine. Interestingly, all cells that produce IL-12 synthesize much more p40 than p35, suggesting that the careful control of signaling is at the level of the "primordial" p35 cytokine part of IL-12. p40 is on human chromosome 5q31 to 5q33 while p35 is on 3p12 to 3p13.2.

IL-23 is similar to IL-12 in that it also contains p40.[198] However, rather than also containing p35, IL-23 is a dimer of p40 and p19 (IL-23A). IL-23 signals via a receptor that contains IL-12Rβ1 but not IL-12Rβ2. Instead, another receptor chain, denoted IL-23R, is the second component of the IL-23 receptor; both IL-12Rβ2 and IL-23R are located on chromosome 1 within 150 kb of each other. IL-27 is a dimer of p28 (IL-27A) and EBi3 (IL-27B) that signals via a gp130 and IL-27R complex. IL-35 is produced by T_{reg} cells and plays a role in immunosuppression.[197] It is a heterodimer of

TABLE 9.9 Type II Cytokines

		Chromosomal Location (h/m)	Receptor Usage R1:R2
Type I IFNs			
IFNα	Many genes	9p22/4	IFNAR-1:IFNAR-2
IFNβ	Single gene	9p21/4	IFNAR-1:IFNAR-2
IFNω	Single gene	9p21/4	IFNAR-1:IFNAR-2
IFNτ	Many genes		IFNAR-1:IFNAR-2
Type II IFN			
IFNγ	Single gene	12q14/10	IFNGR-1:IFNGR-2
IL-10 family cytokines			
IL-10	Single gene	1q31-32/1	IL-10R1:IL-10R2
IL-19	Single gene	1q32/1	IL-20R1:IL-20R2
IL-20	Single gene	1q32/1	IL-20R1:IL-20R2
IL-22	Single gene	12q14-15/10	IL-22R1:IL-10R2
IL-24	Single gene	1q32/1	IL-20R1:IL-20R2
IL-26	Single gene	12q14-15/	IL-20R1:IL-10R2
Type III IFN/IL-10 family cytokines			
IL-28A IFNλ2	Single gene	19q13.2/7	IL-28R1:IL-10R2
IL-28B IFNλ3	Single gene	19q13.2/7	IL-28R1:IL-10R2
IL-29 IFNλ1	Single gene	19q13	IL-28R1:IL-10R2

IFN, interferon; IL, interleukin. IL-28A, IL-28B, and IL-29 are type III IFNs also termed IFNλ. IL-28R1 is also known as the IFNλR. Note that IL-20 and IL-24 can signal through an IL-22R1/IL-20R2 complex as well as the IL20R1:IL-20R2 complex indicated above.

IL-12A (p35) and IL-27B (EBi3). The receptors for IL-35 include gp130-IL12Rβ2, gp130-gp130, and IL-12Rβ2-IL-12Rβ2 complexes.[161,198] IL-39 consists of IL-23p19 and EBi3 and signals via the IL-23R-gp130 complex.

Significance of the Sharing of Receptor Chains

Interestingly, γc, βc, and gp130 all contribute to signaling, but none of these shared cytokine receptor proteins has primary binding activity for any known cytokine. Instead, they each increase binding affinity in the context of the primary binding protein for each cytokine. Consequently, the capacity of a cell to respond to a given cytokine is determined by the relative amounts of their unique binding chain(s) and dimerization partners, but signaling pathways can be shared. This concept is critical to the design of rational strategies to target cytokine/cytokine receptors in disease.

INTERFERONS (TYPE II CYTOKINES) AND THEIR RECEPTORS

IFNs represent an evolutionarily conserved family (Table 9.9) of cytokines (IFNs I-III) that are related to the IL-10 family of type II cytokines.[14] IFNs were discovered in 1957 on the basis of their antiviral activity and were the first cytokines that were discovered.[4] IFNs are known as type I, type II, or type III IFNs where type I IFNs include IFNα (originally known as leukocyte IFN) and IFNβ (originally known as fibroblast IFN), and IFNε, IFNκ, IFNω, IFNδ, IFNτ, and IFNζ. IFNω is closely related to the IFNαs and was formerly designated as an IFNα. There are multiple (at least 12) IFNαs in mice and humans. There are multiple pseudogenes most closely related to IFNα and IFNω. The type I IFNs are clustered on human chromosome 9 and mouse chromosome 4. Type II IFN is IFNγ[5,207] and is encoded by a single gene on human chromosome 12 and mouse chromosome 10. IFNα/β is produced after viral infection in many cell types, although plasmacytoid DCs appear to be the largest producers of IFNα/β. These type I IFNs inhibit viral replication and induce the apoptosis of virally infected cells. In addition, IFNα/β mediate the activation of macrophages and NK cells, and they additionally affect the proliferation and survival of CD8+ T cells.

The grouping of the IFNαs and IFNβ together as type I IFNs is logical not only because of the similar amino acid sequences and structures of these IFNs but also based on the fact that they share the same receptor comprised of IFNAR-1 and IFNAR-2 to induce signaling.[208-210] Although gene expression array analysis revealed some differences in the genes induced by IFNα and IFNβ, the basis for these differences is unclear.[211] Recent structural studies suggest that differential complex stability influences the response outcomes of the various type I IFNs.[212-214] These signals include not only antiproliferative and antiviral activities but also the ability to stimulate cytolytic activity in lymphoyctes, NK cells, and macrophages. In contrast, IFNγ has a distinct receptor comprised of IFNGR-1 and IFNGR-2[215] (Table 9.9). Type I and type II IFN receptors share a sufficient degree of similarity to each other so as to form a family.[216] The structure of the multimeric IFNγ receptor[217] is shown in Figure 9.10. IFN receptors are type II cytokine receptors, and there are substantial differences between these receptors and the type I cytokine receptors.[11] Because both type I and type II IFNs bind to type II cytokine receptors, IFNs are occasionally referred to as type II cytokines but more generally are referred to as IFNs. Based on the similarity of the IL-10 receptor to the IFNγ receptors,[11] as noted above, IL-10 was designated as a type II cytokine,

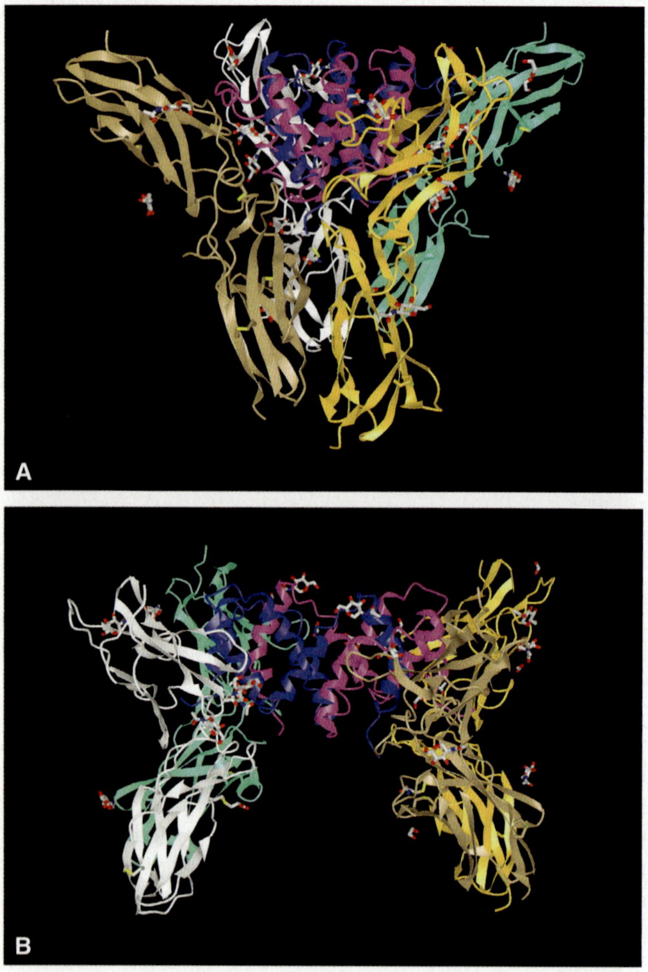

FIG. 9.10. Structure of the IFNγ receptor complex. A: Shown are ribbon diagrams of the structures of the IFNγ/IFNγR complex showing a heterotetrameric receptor complex interacting with dimeric IFNγ. This structure consists of two IFNGR-1 chains (*tan and white*) and two IFNGR-2 chains (*yellow and teal*) complexed to the IFNγ dimer (*blue and pink*). Of note is the similar arrangement of the FN domains in IFNGR-1 and IFNGR-2 and IFNγ to the cytokine receptor structures shown in Figure 9.4. **B:** Rotation of the complex by 90°. These structures are from PDB ID 6E3K reported by Mendoza et al.[217] (Reprinted by permission from Springer Nature: Mendoza JL, Escalante NK, Jude KM, et al. Structure of the IFNγ receptor complex guides design of biased agonists. *Nature.* 2019;567(7746):56-60.)

and indeed when its X-ray crystal was determined, IL-10 was found to be topologically related to IFNγ.[18]

As noted previously, a series of IL-10–related cytokines has been identified, including IL-19, IL-20, IL-22, IL-24, IL-26, IL-28A, IL-28B, and IL-29, and these are also designated as type II cytokines, with IL-28A, IL-28B, and IL-29 also known as IFNλs or type III IFNs.[14] Among type II cytokines, IFNγ has helices similar to those of the type I short-chain helical cytokines, but its short helices that occupy the AB and CD loops positions exhibit the long-chain cytokine-like AB over CD topology. IL-10 and IFNαβ have long-chain structures.[11] Thus, the theme of short-chain and long-chain type I cytokines also extends to the IFNs and the IL-10 family of type II cytokines.

Type I IFNs signal through a receptor known as the type I IFN receptor.[14,209-211] The receptor consists of two chains. In contrast to the α and β chain nomenclature typical for type I cytokine receptors, IFN receptor chains are officially denoted as IFNAR-1 (previously also denoted IFN-αR1, IFNAR-1, and IFN-Rα) and IFNAR-2 (previously also known as IFNα/β receptor, IFNαR2, IFNAR-2, and IFNRβ).[218] IFNAR-2 has both short and long forms as well as a soluble form.[219] The long form has a much larger cytoplasmic domain and serves a more important role in signal transduction. While IFNAR-1 cannot bind IFNα, IFNAR-2 binds with low affinity, and the combination of both chains results in high-affinity binding and function. In addition to these cellular receptors, it is interesting that vaccinia virus and other orthopoxviruses encode a soluble form of type I IFN receptor that is related to the IL-1 receptors and that is capable of binding IFNα, IFNβ, and IFNω.[220,221] This form of IFN receptor is therefore not a member of the type II cytokine family but instead is an Ig superfamily protein. IFNαs are typically species specific, although IFNα8 is a human type I IFN that can bind to the mouse receptor. IFNAR-1 confers species specificity of binding. As noted previously, type I IFNs are induced by viral infection. Infection with viruses results in the generation of nonself RNAs in the cell, which are recognized by retinoic acid inducible gene-I (RIG-I)–like receptors, which include RIG-I, MDA5, and LGP2, each of which contains a helicase domain.[222] RIG-I–like receptors are essential for the production of type I IFNs as well as IL-6, and cells lacking both RIG-I and MDA5 are completely defective in the production of type I IFNs (see Chapter 13).

IFNγ was first recognized in 1965 as immune IFN, named as IFNγ in 1980, cloned in 1982, and then shown to be the same as primary macrophage activating factor.[5,215,223] In addition to its antiviral activity, IFNγ also has a range of actions on proinflammatory and anti-inflammatory host responses.[224] For example, IFNγ can prime macrophages to manifest antimicrobial and antitumor effects. Moreover, following IL-12–mediated differentiation of Th1 cells, IFNγ is a major secreted product. This production is critical for host response to a range of infectious pathogens, including, for example, *Toxoplasmosis gondii.* IFNγ inhibits the generation of Th2 cells as well as Th17 cells. IFNγ is produced by a range of cells, including NK cells, and can mediate NK-dependent lysis of tumor cells, augment IgG2a isotype switching of B cells, and activate DCs. Moreover, IFNγ can either contribute to disease progression or can be protective, depending on the context.[215]

IFNγ is encoded by four exons on chromosome 12. IFNγ forms a functional homodimer with an apparent molecular weight of 34 kDa, whereas little of the monomeric form can be detected and it is not biologically active. Each IFNγ monomer has six α-helices, four of which resemble the short-chain helical cytokines, and there is no β sheet structure. The subunits interact in an antiparallel fashion. In contrast to the ability of many different cells to produce IFNα, IFNγ is more restricted, with it being produced primarily by NK cells, CD8+ T cells, and the Th1 subclass of CD4+ T cells.[207] Many signals, including antigen stimulation, IL-12, and IL-18, can induce the production of IFNγ. IFNγ exerts its effects through specific receptors that are expressed on all types of cells except erythrocytes. Interestingly, even platelets express IFNγ receptors, raising the possibility that they can serve a function in transporting IFNγ in the circulation.[207] The functional receptor consists of two chains[224]: IFNGR-1, formerly also denoted as IFNγR1 or IFNγRα, a 90 kDa protein whose gene is located on human chromosome 6q16 to 6q22 and mouse chromosome 10, and IFNGR-2, also denoted as IFNγRβ, located on human chromosome 21q22.1 and mouse chromosome 16.[207] IFNGR-1 is required for ligand binding, whereas IFNGR-2 plays a role in signaling. The fact that IFNγ is a homodimer explains how its binding induces the homodimerization of IFNGR-1, which then allows the recruitment of IFNGR-2. Crystal structure analysis reveals that the functional IFNγ receptor consists of two molecules each of IFNGR-1 and IFNGR-2[217] (Fig. 9.10). Normal IFNγ production is dependent on IL-12, and defective IFNγ signaling is associated with failure to appropriately clear mycobacterial and other infections. Both type I and type II IFNs are important for cancer immunoediting, a process in which neoplastic growth is suppressed by the immune system and tumor immunogenicity is shaped.[224] IFNγ can suppress tumor development, but type I IFNs are required to initiate the antitumor response, at least in part via actions on DCs.

The IL-10 family of cytokines includes five type II cytokines and three type III IFNs (IFNλ).[14] IL-10 was originally identified as cytokine synthesis inhibitory factor produced by Th2 cells.[225-229] It was subsequently observed to be produced by T_{reg} cells as well, and it is now clear that IL-10 is broadly expressed, including by Th1, Th17 cells, B cells, macrophages, NKT, NK, T_{reg}, monocytes, and myeloid DCs, keratinocytes, as well as mast cells and eosinophils.[228] IL-10 production by T_{reg} cells helps to explain their suppressive effects, and IL-10 production by T_{reg} cells is required for their suppression of Th17-mediated inflammation.[230] The IL-10 gene contains five exons and is located on chromosome 1 in both mice and humans. Human IL-10 receptor maps to 11q23.3. IL-10 can inhibit the production of a number of cytokines, including IL-2, IL-3, IFNγ, GM-CSF, and TNF-α. IL-10 production is induced by pathogen-derived products with TLR2 signaling being critical for its induction by antigen-presenting cells and macrophages, although TLR-independent induction of IL-10 has also been observed.[228] Interestingly, IL-21 has been shown to be a potent inducer of IL-10 and to mediate IL-27 and part of the IL-6–mediated induction of IL-10 as well.[231] IL-10 inhibits monocyte/macrophage/DC-dependent T cell proliferation, in part, by markedly decreasing synthesis of a variety of cytokines, expression of costimulatory molecules, and chemokines, to name just some of its inhibitory effects.[225] In addition to these indirect effects on T cells, IL-10 can exert direct stimulatory effects on thymocytes and T cells *in vitro* and promote the development of B1 cells and activity of NK cells. IL-10–deficient mice develop a form of inflammatory bowel disease that is similar to Crohn's disease.[232] Interestingly, the BCRF1 protein that is encoded by Epstein-Barr virus is very similar to IL-10 and shares many of its biologic properties as a macrophage "deactivating" factor and as a costimulator of proliferation of B cells.[233] The Epstein-Barr virus IL-10 homolog is a selective agonist, although its binding to the IL-10 receptor is somewhat impaired. IL-10 is an inhibitor of Th1 functions and plays a major role in limiting and terminating inflammatory responses in general.[227]

The IL-10 receptor is most closely related to IFN receptors, making it a type II cytokine receptor and corresponding to the close structural relationship of IL-10 to IFNγ.[234,235]The receptor for IL-10 consists of an IL-10R1 chain and IL-10R2[227]; this latter protein is the same as CRF2-4, which was first identified as an "orphan" IFN receptor family member that is located on chromosome 21 within 35 kb of IFNGR-2.[236,237]

A series of IL-10–related cytokines have been identified, including IL-19, IL-20, IL-22, IL-24, IL-26, IL-28A, IL-28B, and IL-29.[14,225,226] IL-19 was discovered as a gene that was induced in LPS-stimulated monocytes.[238] IL-19 has been suggested to potentially be involved in the pathogenesis of chronic inflammatory diseases, such as psoriasis. Like IL-19, IL-20 is produced by myeloid cells; it was also found to be produced by epidermal cells, with overexpression resulting in aberrant epidermal differentiation.[239,240] IL-20 appears to have a role in psoriasis, and in psoriatic skin, expression of IL-20 is primarily detected in DCs. IL-20 also appears to contribute to rheumatoid arthritis, atherosclerosis, and stroke; recently, IL-20 has been implicated in bone loss, with elevated IL-20 in patients with osteopenia and osteoporosis in ovariectomized mice. Moreover, IL-20R1–deficient mice have increased bone density as did treatment with anti–IL-20.[241] Both IL-19 and IL-20 signal through the IL-20R1/IL-20R2 chains.

IL-22 was discovered as an IL-9–induced cytokine that can in turn induce acute-phase reactant production by hepatocytes and accordingly that it might be involved in inflammatory responses.[242-244] It was first observed to be produced by activated Th1 cells but was subsequently noted to be more broadly produced by CD4+ and CD8+ T cells, NK cells, lymphoid tissue–inducer cells, and mucosal RORγt+NKp46+ cells in humans and mice and to be produced by Th17 cells. IL-22 is often coexpressed with IL-10, and STAT3, the Notch pathway, and the aryl hydrocarbon receptor promoted production of these cytokines; in contrast, TGF-β, ICOS, and IL-27 augment IL-10 production yet inhibit IL-22.[229] IL-22 mediates cross talk between leukocytes and tissue epithelia, with target cells primarily in the digestive and respiratory systems, but it also can act on keratinocytes, can promote innate immunity, and can cooperate with IL-17A to effect epithelial and healing responses. In particular, IL-22 can contribute to epithelial resistance to injury after microbial infection of the gut or lungs and thereby may play a role in host defense and tissue repair. IL-22 signals through the IL-22R1 and IL-10R2 chains.[14]

IL-24 was originally discovered as melanoma differentiation-associated antigen 7.[14,225,226] IL-24 is induced in peripheral blood mononuclear cells by mitogen stimulation and is produced not only by Th2 cells but also by myeloid cells. Overexpression of IL-24 can induce apoptosis in a wide range of tumor cells, including, for example, melanoma and malignant gliomas, and infection with adenoviral-driven melanoma differentiation-associated antigen 7 can sensitize tumors to ionizing radiation. Like IL-19, IL-20, and IL-22, IL-24 is expressed in psoriatic but not normal skin, and like these other cytokines, functional receptors are highly expressed on keratinocytes. IL-10, IL-19, IL-20, IL-22, and IL-24 are all increased in synovial fluid cells in rheumatoid arthritis and thus are potential mediators of this autoimmune disease. IL-20 and IL-22 in particular may be associated with lupus. IL-26 was originally discovered as AK155, a cationic amphipathic protein that kills extracellular bacteria via membrane pore formation.[14,225,226] IL-26, produced by Th17 cells, acts as an innate sensor by forming complexes with bacterial DNA and self-DNA released by dying bacteria and host cells. Both IL-24 and IL-26 signal through the IL-20R1 chain, but differ in their use of IL-20R2 or IL-10R2 as the second chain, respectively.

IL-28A and IL-28B (IFNλ2 and IFNλ3) and IL-29 (IFNλ1) are also called type III IFNs and are induced in plasmacytoid DCs by viral infection and exhibit antiviral activity, analogous to type I IFNs[14]; nevertheless, they bind to a receptor consisting of IL-10R2 and IL-28R.[245] IL-28A, IL-28B, and IL-29 induce relatively weak antiviral responses, and the IL-28R chain is relatively restricted in its expression, as compared to IFNα and the IFNα receptor, respectively. IL-28A appears to enhance antiviral activity induced by TLR3 and TLR9 and to be important for host defense to influenza A.[225]

Interestingly, the genes encoding IL-10, IL-19, IL-20, and IL-24 colocalize at human chromosome 1q32, whereas those encoding IL-22 and IL-26 are at 12q14 to 12q15 near the *IFNG* gene, and IL-28 and IL-29 are at 19q13. Thus, these type II cytokines can be subdivided into three different groups based at least in part on chromosome localization. Overall, IL-10 family cytokines are diverse and can be subdivided into the IL-19, IL-20, IL-22, IL-24, and IL-26 subfamily that acts on tissue epithelial cells and in the case of IL-20 in wound healing, whereas IL-28A, IL-28A, and IL-29 exhibit antiviral activity and are more IFN-like in their biology.[14]

OTHER FEATURES OF CYTOKINES AND IFNS

Species Specificity

There are no general rules for the species specificity of human and mouse cytokines and how the cytokines and their receptor chains have coevolved. To provide illustrative examples of each situation, human IL-2 can stimulate both human and mouse cells, whereas mouse IL-2 exhibits little action on human cells.[50] Conversely, human IL-12 does not work on mouse cells, whereas mouse IL-12 is biologically active on both mouse and human cells.[246] This selective property of IL-12 is dependent on the species origin of p35. Finally, IL-4 exhibits rather strict specificity so that human and mouse IL-4 only induce responses on human and mouse cells, respectively.[70] As noted previously, IFNαs are generally species specific, although IFNα8 is not. In addition to these examples, varying degrees of relative species specificity have been demonstrated, depending on the cytokine. Thus, a cytokine may or may not be restricted in its specificity to its own species, and if it does occur, cross-species activity may have attenuated potency.

Cytokine Receptor Sharing: Pleiotropy and Redundancy

As aforementioned, many cytokines exhibit the phenomena of cytokine "pleiotropy" and "redundancy."[247] Cytokine pleiotropy refers to the ability of a cytokine to exert many different types of responses, often on different cell types, whereas cytokine redundancy refers to the fact that many different cytokines can induce similar actions. One set of cytokines that exhibit cytokine pleiotropy is the γc family of cytokines. For example, IL-2 can induce T cell growth, augment B cell Ig synthesis, increase the cytolytic activity of lymphokine-activated killer and NK cells, and play an essential role in mediating activation-induced cell death; IL-4 can induce B cell growth, Ig class switch, Th2 differentiation, and alternative macrophage activation; and IL-7 not only plays a major role in thymocyte development but also can stimulate mature T cells and, at least in the mouse, can act as a pre–B cell growth factor. The gp130 family cytokines also exhibit broad actions. For example, IL-6 exerts effects ranging from that of a comitogen for thymocyte activation to that of a mediator of the acute-phase response in liver. Regarding cytokine redundancy, it has already been highlighted that IL-2, IL-4, IL-7, and IL-15 receptors containing γc can all act as T cell growth factors and that IL-3 has actions that overlap with IL-5 and GM-CSF. These characteristics are in part explained by receptor expression on many cell types and by subunit sharing.[247]

Cytokine receptor pleiotropy can be explained by the ability of a single cytokine receptor subunit to function in more than one receptor. Thus, examples include the sharing of γc, βc, gp130, IL-4Rα, LIFRβ, and OSMRβ, as summarized previously, as well as the sharing of IL-2Rβ by IL-2 and IL-15 receptors, the sharing of IL-4Rα and IL-13Rα in type II IL-4 receptors and IL-13 receptors, and the sharing of IL-7Rα by the IL-7 and TSLP receptors. Another way of viewing receptor pleiotropy is that certain receptor chains are useful "modules" that function in more than one context.

Likewise, cytokine receptor redundancy can be explained by the existence of alternative receptor chains. For example, in the mouse there is an alternative IL-3Rβ that shares 91%

amino acid identity with βc and appears to be a redundant protein for IL-3 signaling.[31] Other potential examples exist. For example, in type I and type II IL-4 receptors, IL-4Rα coordinates with either γc or IL-13Rα1, respectively (Fig. 9.8A), and there are two types of OSMRs, both of which contain gp130, but one of which contains a specific OSMR while the other contains LIFRβ. What remains unknown, however, is whether the signals mediated by these different types of receptors are truly identical so that there is redundancy, or whether there are distinctive features to the signals that IL-4 and OSM induce via the different receptors. Indeed, mice lacking IL-13Rα1 have exacerbated Th2 responses, while IL-13Rα1 is essential for allergen-induced airway hyperreactivity, suggesting that type II IL-4 receptor signaling (either via IL-4 or IL-13) is not identical to that mediated by the type I IL-4 receptor (via IL-4).[85,248] Additional studies of signaling pathways activated by IL-4 and IL-13 indicate there are indeed subtle differences in activation that impact cellular responses such as those elicited in DC for example.[133]

In addition to these examples related to type I cytokines, the IL-10 subfamily of type II cytokines is interesting in that IL-10Rβ (IL-10R1), IL-20Rα (IL-10R1), IL-20Rβ (IL-20R2), and IL-22Rα (IL-22R) are each shared receptor components, collectively affecting signaling in response to IL-10, IL-19, IL-20, IL-22, and IL-24 (see Table 9.9).[14,225] Specifically, IL-10 signals via a receptor containing IL-10R1 and IL-10R2, IL-19 signals via a receptor containing IL-20R1 and IL-20R2, IL-20 signals via receptors containing IL-20R2 and either IL-20R1 or IL-22R, IL-22 signals via receptors containing IL-22R and IL-10R2, and IL-24 signals through receptors containing IL-20R2 plus either IL-20R1 or IL-22R. Two other IL-10 family members, IL-28 and IL-29, both signal via IL-28R + IL-10R2.[14] Interestingly, IL-10 is believed to be a dimer; IL-26 may also be a dimer, but the other family members are monomers.[225,226]

Soluble Receptors

Soluble forms of many cytokine receptors have been identified, including those for IL-1, IL-2, IL-4, IL-5, IL-6, IL-7, IL-13Rα2, GM-CSF, type I and type II IFNs, IL-22, and TNF.[249-253] As is clear from this list of cytokines, soluble receptors are not restricted to receptors for type I cytokines, and in the case of IL-2, the principal soluble receptor protein is IL-2Rα, which is not a type I cytokine receptor. Soluble receptors can be created by alternative splicing that truncates the protein N-terminal to the transmembrane domain, resulting in a secreted protein rather than a membrane-anchored membrane in the case of IL-4Rα, IL-5Rα, IL-6Rα, IL-7Rα, IFNAR-2, and GM-CSFRα. Alternatively, they can be created by proteolytic cleavage of the membrane receptor as is found for IL-2Rα (Table 9.10).[249] The major questions to consider related to these soluble receptors are as follows: (1) Do they have physiologic or pathophysiologic functions? (2) How do their affinities compare to the corresponding cell surface receptor? (3) Do they have diagnostic, prognostic, and therapeutic applications?

Unfortunately, there is little information available on the *in vivo* role of soluble receptors. In general, when analyzed using *in vitro* studies, soluble receptors can compete with their corresponding cell surface receptors, thereby serving negative regulatory roles. For example, this appears to be the case for the IL-22 soluble receptor. However, soluble IL-6Rα exerts an agonistic role because, as summarized previously, IL-6 signaling occurs equally well via gp130 when the soluble rather than transmembrane form of IL-6Rα interacts with IL-6. Nevertheless, a mutated form of IL-6Rα that cannot interact with gp130 but still binds IL-6 can effectively inhibit the actions of IL-6.[252] In addition, soluble gp130 is found in human serum and functions as an inhibitor of IL-6/sIL-6R complexes.[253] In the case of IL-2Rα, there is no reported physiologic function for soluble IL-2Rα, as the affinity of the released receptor is, as expected, similar to that of the low-affinity receptor ($K_d \approx 10{-}^{8}$ M), making it unlikely to effectively compete with the high-affinity cell surface receptor ($K_d \approx 10^{-11}$ M). However, this and other soluble receptors could serve as cytokine carrier proteins and potentially could increase stability of a cytokine by protecting it from proteolysis.[250] This theoretically could be similar to the role of IL-2/anti–IL-2 conjugates in exhibiting higher activity.[254] Moreover, there are potential diagnostic and prognostic uses for measuring the level of shed receptors (Table 9.10).

SIGNALING THROUGH INTERFERON AND CYTOKINE RECEPTORS

Signaling Overview

Our understanding of signaling through IFN and cytokine receptors has tremendously increased in the past 30 years. Multiple signaling pathways/molecules have been observed for various cytokines. Collectively, these include the JAK-STAT pathway, IRF family proteins, Ras/mitogen-activated protein (MAP) kinase pathway, Src and Zap70 and related proteins, PI3K, IRS1 and IRS2, and phosphatases. Each of these pathways will be discussed in turn. The JAK-STAT pathways are important for both type I and type II cytokines, while the IRF proteins play central functions more so for the IFNs than type I cytokines.

JAKs and STATs

The JAK-STAT signaling pathway is particularly exciting in that it serves as a rapid mechanism by which signals can be transduced from the membrane to the nucleus.[255-260] JAK kinases are known as Janus family tyrosine kinases, and the acronym "STAT" denotes signal transducer and activator of transcription. STAT proteins are substrates for JAK kinases. A tremendous amount of information is now available on JAKs and STATs that demonstrate their importance in transmitting signals initiated by type I cytokines and IFNs. This signaling pathway relates to development, differentiation, proliferation, cellular transformation, and tumorigenesis.

TABLE 9.10 Soluble Interleukin-2 Receptors in Human Disease

Malignancies

Hematologic

- Adult T cell leukemia
- Hairy cell leukemia
- Acute lymphocytic leukemia
- Chronic lymphocytic leukemia (B cell)
- Acute myelogenous leukemia
- Chronic myelogenous leukemia (especially in blast crisis)
- Malignant lymphomas
- Hodgkin disease
- Non-Hodgkin lymphomas

Nonhematologic

- Adenocarcinoma of the lung, breast, and pancreas
- Small cell bronchogenic carcinoma
- Ovarian, cervical, and endometrial cancers
- Nasopharyngeal carcinoma
- Melanoma

Infections

- Human immunodeficiency virus
- Tuberculosis
- Rubeola
- Infectious mononucleosis

Other diseases

- End-stage renal disease
- Rheumatoid arthritis
- Systemic lupus erythematosus
- Scleroderma
- Sarcoidosis

After transplantation

After interleukin-2 administration

In adults, the mean levels of sIL-2Rα are 280 ± 161 U/mL (levels tend to be higher in pediatric populations). The situations where the levels exceed 5000 U/mL are adult T cell leukemia, hairy cell leukemia, chronic myelogenous leukemia, and after interleukin-2 administration. The situations where levels are between 1000 and 5000 U/mL include acute myelogenous leukemia, chronic lymphocytic leukemia, non-Hodgkin lymphomas, acquired immunodeficiency syndrome associated with Kaposi sarcoma, tuberculosis, rubeola, and end-stage renal disease.

Data are from Kurman C, Rubin L, Nelson D. *Article: Soluble Products of Immune Activation. Soluble Interleukin-2 Receptor (sIL-2R, Tac Protein).* American Society for Microbiology; 1992. and Fernandez-Botran R, Chilton PM, Ma Y. Soluble cytokine receptors: their roles in immunoregulation, disease, and therapy. *Adv Immunol.* 1996;63:269-336.

JAKs

The JAK kinases are 116 to 140 kDa and comprise approximately 1150 amino acids.[258-260] The seven regions of conserved amino acid sequences in JAK kinases, denoted as Janus homology (JH) 1 to JH7,[261] are depicted in Figure 9.11A. One of the hallmark features of these kinases is that in addition to the presence of a catalytic tyrosine kinase domain (JH1), there is also a pseudokinase region (JH2). The name Janus kinase reflects the two faces of the mythological Roman god, with one face representing the true kinase and the other the pseudokinase. Modeling has indicated four important structural regions, including an N-terminal/FERM (4.1, ezrin, radixin, moesin) domain (JH4-JH7), an SH2-like domain (JH3), a pseudokinase domain (JH2), and the tyrosine kinase domain (JH1). JAKs contain conserved tyrosine residues in the JH1 domain that are necessary for maximal JAK activation. It is thought that phosphorylation of these tyrosines leads to the conformational changes in the JAK protein that facilitate its function. The pseudokinase, JH2 domain has no catalytic activity; however, it regulates the activity of the JH1 kinase. The FERM-SH2 domains (JH3-7) are responsible for association of JAKs with the cytoplasmic regions of receptor subunits in part through association with the proline-rich Box 1 motif.[20,262,263]

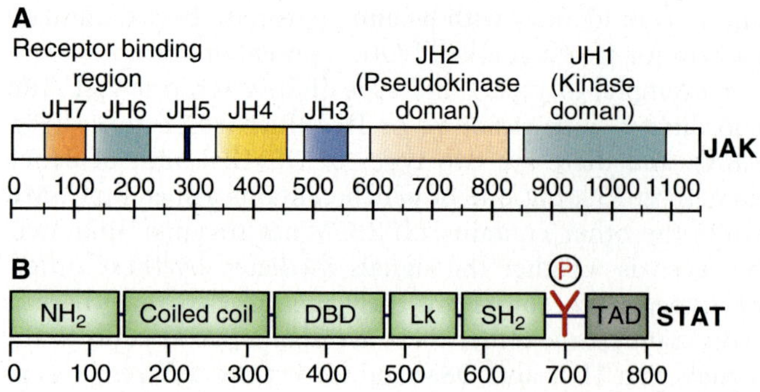

FIG. 9.11. Schematic of JAK family kinases and STAT family proteins. A: Shown are the locations of the seven Janus homology (JH) domains found in JAKs. JH1 is the catalytic domain. JH2 is the pseudokinase domain, the presence of which prompted the naming of this family as Janus family tyrosine kinases, with the JH1 and JH2 domains representing the two faces of the Roman mythological god. As noted in the text, the JH nomenclature has limitations and in fact masks the presence of an SH2 domain that spans parts of the JH4 and JH3 domains. **B:** Shown are the critical domains of STAT family members including the transactivating domain (TAD), the SH2 domain that participates in receptor binding and STAT dimer formation, the DNA-binding domain (DBD), the coiled coil domain, and the amino-terminal domain. There is also a small linker domain (Lk). There is a critical tyrosine residue (Y) in the C-terminal region of STATs (Y) that is phosphorylated by JAKs and enables reciprocal STAT–STAT interaction via SH2 domain interactions. The amino-terminal, coiled coil, and TAD domains are sites for other posttranslational modifications such as serine phosphorylation that controls STAT–protein interactions and STAT functionality. Note that this structure is typical of that for STAT1, STAT3, STAT4, STAT5A, and STAT5B. The main features are conserved in STAT2 and STAT6, but these are approximately 50 to 100 amino acids longer.

There are four mammalian JAK kinases, JAK1, JAK2, JAK3, and TYK2, each of which was identified as part of a study intended to identify new kinases.[258] At least one JAK kinase is activated by every IFN and cytokine, and some cytokines activate two or three JAK kinases.[255-260] Table 9.11 lists a number of features of each JAK kinase, whereas Table 9.12 summarizes the JAK kinases that are activated by a variety of

TABLE 9.11 Features of Janus Family Kinases

Kinase	Inducible vs Constitutive	Size	Chromosomal Location (H/M)
JAK1	Constitutive	135 kDa	1p31.3/4
JAK2	Constitutive	130 kDa	9p23-24/19
JAK3	Inducible	116 kDa	19p13/8
TYK2	Constitutive	140 kDa	19p13.2/9

cytokines.[258] As each functional cytokine receptor is a dimer or higher order oligomer, at least two JAK molecules will be activated since one JAK is associated with each receptor chain. In accordance with their ubiquitous expression, JAK1, JAK2, and TYK2 are activated by a variety of different cytokines (Table 9.12). For example, JAK1 is activated not only by IFNs but also by the γc family of cytokines, whereas JAK2 is activated not only by IFNγ but also by growth hormone, Epo, prolactin, and the hematopoietic cytokines IL-3, IL-5, and GM-CSF. TYK2 is somehat more restricted in that it is activated by IFNα/β and IL-12/IL-23. Interestingly, JAK1, JAK2, and TYK2 are recruited by each of the gp130 receptor. In the case of IL-6–induced responses, JAK1 is critical.[264]

The JAKs become activated once the cytokine has bound to its binding subunit and induced receptor multimerization. Bringing the receptor subunits together brings the JAKs bound to the cytoplasmic tails of the receptor subunits together leading to the stimulation of their kinase activity by an as-yet-undefined mechanism.[265] It is thought that transphosphorylation of JAK kinases is a mechanism for the amplification of catalytic activity. Unless other kinases are involved, however, implicit to this idea is that the JAK kinases themselves must exhibit some basal activity that is amplified to a higher level by autophosphorylation or transphosphorylation. It is reasonable to speculate that both models may be operative, depending on the system. Consistent with JAK kinases being activated by phosphorylation, mutagenesis of a critical tyrosine in TYK2[266] or JAK2[267] in the activation loop of the kinase domain inhibits activity. It is also conceivable that phosphorylation of other

TABLE 9.12 Cytokines and the JAKs They Activate

Cytokine	JAK Kinase(s) Activated
IFNα/β	JAK1, TYK2
IFNγ	JAK1, JAK2
Growth hormone	JAK2
Prolactin	JAK2
Erythopoietin	JAK2
Thrombopoietin	JAK2
IL-10	JAK1, TYK2
IL-12, IL-23	JAK2, TYK2
G-CSF	JAK1, JAK2
TSLP	JAK1, JAK2
IL-13	JAK1, JAK2
γc family	
IL-2, IL-4, IL-7, IL-9, IL-15, IL-21	JAK1, JAK3
βc family	
IL-3, IL-5, GM-CSF	JAK2, JAK1
gp130 family	
IL-6, IL-11, CNTF, LIF, OSM, CT-1	JAK1, JAK2, TYK2

CNTF, ciliary neurotrophic factor; CT-1, cardiotrophin-1; G-CSF, granulocyte colony-stimulating factor; GM-CSF, granulocyte-macrophage colony-stimulating factor; IFN, interferon; IL, interleukin; LIF, leukemia inhibitory factor; OSM, oncostatin M; TSLP, thymic stromal lymphopoietin.

[a]Note that IL-4 activates JAK1 and JAK3 when it acts through the type I IL-4 receptor (IL-4Rα + γc, found for example on T cells). However, JAK3 is not activated when IL-4 signals through the type II IL-4 receptors (IL-4Rα + IL-13Rα1, a form of receptor that is expressed on a number of non–T cells, including fibroblasts).

tyrosines on the JAKs may create appropriate motifs for the recruitment of additional signaling molecules. In receptors with only two chains, the direct transphosphorylation of one JAK by the other seems likely to occur. In more complex receptors, such as the IFNγ system, because the receptor is a heterotetramer with two IFNGR-1 chains that each bind JAK1 and two IFNGR-2 chains that each bind JAK2, it is not clear if JAK1 and JAK2 transactivate each other or if one of the JAK kinases plays a dominant role. Indeed, one study suggests that JAK2 may phosphorylate both itself and JAK1, thereby increasing the catalytic activities of both kinases.[268] JAK1 in turn then phosphorylates IFNGR-1, allowing the recruitment of STAT1. In this model, it is additionally suggested that JAK2 phosphorylates STAT1.

The pseudokinase domain is thought to regulate activation of the kinase domain. The JH2 lacks the third glycine in the critical Gly-X-Gly-X-X-Gly motif, is missing an aspartic acid that serves as the proton acceptor that is typically conserved in the catalytic loop of both tyrosine and serine kinases, and is missing the conserved phenylalanine in the Asp-Phe-Glu motif that binds adenosine triphosphate together, explaining the lack of catalytic function of the JH2 pseudokinase domain.[258] Despite the lack of catalytic activity, there are data in support of vital functions for this region. Mutation in the JAK kinase JH2 domain can hyperactivate the *Drosophila* (*hop*$^{Tum\text{-}l}$/DSTAT) JAK-STAT pathway, and the corresponding Glu695 to Lys mutation in mouse JAK2 also results in increased autophosphorylation of JAK2 and phosphorylation of STAT5.[269]

Given that JAK kinases are associated with a wide range of IFN and cytokine receptors, it was predicted that JAKs would mediate signals involved in multiple processes. These include important roles in development, signaling in response to cytokines that are mitogenic growth factors, and in the antiviral response. Studies in zebrafish revealed not only a role for JAK1 in early vertebrate development but also revealed that during early development JAK1 kinase was exclusively of maternal origin.[270] These developmental roles for JAK1 in zebrafish are consistent with the importance of a JAK kinase in early *Drosophila* development as well.[271]

JAK3 is different from the other three JAK kinases in that it is much more inducible. Moreover, JAK3 is only activated by cytokines whose receptors contain γc.[150,151] Interestingly, each cytokine whose receptor contains γc activates not only JAK3 but also JAK1. The basis for the activation of both JAK1 and JAK3 by IL-2, IL-4, IL-7, IL-9, IL-15, and IL-21 is that JAK1 associates with each of the distinctive signaling chains (IL-2Rβ, IL-4Rα, IL-7Rα, IL-9Rα, and IL-21R), whereas JAK3 associates with γc. Although this could be a coincidental result of coevolution of these cytokine systems, these observations raise the possibility that JAK1 is the JAK kinase that most efficiently functionally cooperates with JAK3. Given the wide range of cytokines that activate any particular JAK kinase and that in some cases multiple cytokines can activate the same set of JAKs, it is clear that the JAK kinases by themselves do not determine signaling specificity. For example, IL-2, IL-4, IL-7, IL-9, IL-15, and IL-21 all activate JAK1 and JAK3 but induce a range of actions. Moreover, JAK2 is the only JAK kinase that is activated by growth hormone and erythropoietin, cytokines with different target cells and biologic functions. Interestingly, JAK2 is the JAK that interacts with all cytokine receptors that form homodimers. Given that homodimeric receptors are likely the oldest cytokine receptors in evolution, this suggests that JAK2 might be the "primordial" JAK kinase from which others evolved.

Importance of JAK Kinases in Signaling

In addition to the activation of JAK kinases by multiple cytokines and IFNs, a variety of other data indicate their importance for signaling. One of the vital series of experiments that led to the establishment of the critical role of JAK kinases in IFN signaling involved a group of mutant cell lines that were defective for IFN signaling, but wherein signaling could be rescued by genetic complementation.[260,272] Defective signaling in response to IFNα and IFNβ was found in a mutant fibroblast cell line (U1 cells) lacking TYK2; defective signaling in response to IFNα, IFNβ, and IFNγ was found in a mutant cell line lacking JAK1 (U4 cells), and defective IFNγ signaling was found in cells lacking JAK2. A variety of other data have indicated the importance of JAK kinases in cytokine signaling pathways. First, dominant negative JAK2 inhibits signaling by erythropoietin and growth hormone[273,274] while a dominant negative JAK3 inhibits signaling in response to IL-2[275]; as noted previously, JAK1 is vital for IL-6 signaling. Second, humans and mice deficient in JAK3 exhibit developmental and signaling defects.[276,277] Third, humans with an activating mutation in JAK2 develop polycythemia vera,[278] and a JAK2 inhibitor inhibited the growth of acute lymphoblastic leukemia cells *in vitro*.[279] Fourth, a patient with mutation in TYK2 developed a form of immunodeficiency.[280] Fifth, in *Drosophila*, the *hopscotch* gene encodes a JAK kinase, wherein loss-of-function alleles result in lethality and decreased proliferation, whereas a gain-of-function allele, *hopscotch Tumorous-lethal*, results in melanotic tumors and hypertrophy of the hematopoietic organs.[281,282] Sixth, in zebrafish, JAK1 is vital for normal cell migration and anterior specification.[270] Seventh, as discussed in the following, JAK kinases are constitutively activated in many cell lines infected with a number of viruses, including HTLV-I and v-Abl.[283,284] These data together underscore the vital roles of JAKs in cytokine signaling. Depending on the function of the particular cytokine (eg, development, differentiation, or proliferation), the particular JAK kinases potentially may be involved in a variety of processes, and when dysregulated, in at least certain settings, they appear to contribute to cellular transformation.

JAK3 Mutations Result in an Autosomal Recessive Form of Severe Combined Immunodeficiency

A very large number of different mutations in γc have been observed in X-linked SCID, with more than 200 mutations now having been identified (for example see the Human Gene Mutation Database; http://www.hgmd.cf.ac.uk/ac/gene.php?gene=IL2RG).[150,151] As might be expected, in cases where it has been examined, amino acid substitutions in the extracellular domain result in defective cytokine binding. In contrast, mutations or truncations in the γc cytoplasmic domain result in defective signaling. Analysis of a family in which a number of males exhibit a moderate form of X-linked combined immunodeficiency revealed that this disease also resulted from a mutation in γc (Leu 271 → Gln) that resulted in a decrease, but not total loss of JAK3 association,[285] in contrast to the loss of JAK3 interaction seen with mutations in the γ_c cytoplasmic domain that cause X-linked SCID. Thus, the severity of the immunodeficiency inversely correlated with the degree of JAK3 activation. Moreover, it was predicted that JAK3 was required for T cell and NK cell development and that mutations in JAK3 might result in a clinical phenotype indistinguishable from that in X-linked SCID.[285] Indeed, this is the case in humans.[276] As one would hypothesize, many mutations in JAK3 have been identified.[258] Presumably, any mutation that interferes with its ability to interact with γc, with its catalytic activity, or with recruitment of substrates could result in clinical disease, ranging from moderate-to-severe immunodeficiency.

Analogous to the similarity of X-linked SCID and JAK3-deficient SCID, mice deficient in either γc or JAK3 also have indistinguishable phenotypes.[154,277] These *in vivo* data underscore the vital role of JAK3 in mediating γc-dependent functions and suggested that JAK3 is essential for most, if not all, γc functions. Because JAK3 deficiency is not clinically or phenotypically more severe than γc deficiency in humans and mice, it seems likely that γc is the major protein with which JAK3 associates. Although one study reported that JAK3 associates with CD40, a functional role for JAK3 in CD40 signaling is controversial.[286] The phenotypes of mice lacking each of the four JAK kinases are summarized in Table 9.13.

JAK2 Mutations and Translocations

The JAK2V617F mutations have been identified in most patients with polycythemia vera,[287] some of whom have homozygous and some have heterozygous mutations. The homozygous patients had duplications of the mutated allele. This mutation results in a constitutively activated kinase and confers factor-independent growth in transfected cell lines. Moreover, patients with polycythemia exhibit erythropoietin-independent BFU-E. The *JAK2* locus is involved in a chromosomal translocation to create the TEL-JAK2 protein that is causally related to a human leukemia,[288] further underscoring the relationship of JAK2 to the growth of hematopoietic cells.

TYK2 Mutations and Human Immunodeficiency

A number of TYK2 variants have been associated with primary immunodeficiency, inflammation, and autoimmune diseases.[280,289] In some cases, the clinical defects were more severe than anticipated from TYK2-deficient mice, with increased susceptibility to viral infections and atypical mycobacterial infections and defective signaling in response to IL-23, IL-6, and IL-10. Additionally, hyper-IgE production was observed, which perhaps corresponds to enhanced Th2 cell–mediated allergic inflammation in *Tyk2*$^{-/-}$ mice.[290]

STATs

STAT proteins are classically considered to be latent transcription factors that initially exist in the cytosol but then must be activated by phosphorylation, with subsequent translocation to the nucleus.[255-260] A schematic of STAT proteins is shown in Figure 9.11B. Conserved structural features common to all STATs include an SH2 domain, a conserved tyrosine residue, a DNA-binding domain, a C-terminal transactivation domain, and an N-terminal STAT tetramerization region (Fig. 9.11B). The classic, canonical model is that in order for STATs to be "activated" and to be able to regulate transcription, a number of cellular events must occur. They first bind to phosphorylated tyrosine-based motifs on cytokine receptors, are tyrosine phosphorylated, dissociate, dimerize via SH2-SH2 domain interaction, translocate from the cytosol to the nucleus, bind to target DNA sequences via the DNA-binding domain, and regulate gene expression via the transactivation domain. In addition, there is now evidence for noncanonical function of STATs that are not phosphorylated, as discussed subsequently.

There are seven mammalian STAT proteins including STAT1, STAT2, STAT3, STAT4, STAT5A, STAT5B, and STAT6.[255-260] The STATs can be divided into two basic groups: those that are longer (STAT2 and STAT6, approximately 850 amino acids) and those that are shorter (STAT1, STAT3, STAT4, STAT5A, and STAT5B, between 750 and 800 amino acids). The STAT genes cluster in three locations. Mouse STAT2 and STAT6 are both located on chromosome 10; STAT1 and STAT4 are located on chromosome 1; and STAT3, STAT5A, and STAT5B are located on chromosome 11. Correspondingly, human STAT3, STAT5A, and STAT5B are closely positioned on chromosome 17q.

Some STATs (eg, STAT3) are known to have more than one isoform with distinct functions.[291] Although the STATs conserve a reasonable level of homology, STAT5A and STAT5B are unusually closely related, being 91% identical at the amino acid level. It is interesting that mouse and human STAT5A are more related than STAT5A and STAT5B within a single species. The same is true for mouse and human STAT5B, suggesting that there has been evolutionary pressure to maintain the difference between STAT5A and STAT5B and that these two proteins might have certain distinctive functions. In this regard, STAT5A and STAT5B knockout mice exhibit both similarities and some differences in their phenotypes. STAT5A is required for prolactin-induced mammary gland development, while STATB is essential for growth hormone–induced responses.[292,293] Both STAT5A and STAT5B participate in T cell responses to a number of cytokines including IL-2 and IL-7.[294]

Most STAT proteins when activated form homo- or hetero-dimers.[295] The ability of some STATs to form heterodimers or heterotetramers further increases the number of different complexes that can form.[258,260,295,296] Formation of tetramers is mediated by the N-terminal domain interaction of two STAT dimers allowing cooperative binding to tandem DNA motifs.[297-299] N-terminal domain interactions are not dependent on tyrosine phosphorylation leading to unphosphorylated noncanonical STAT dimers.[300,301] The specific role of

TABLE 9.13 Phenotypes of Mice Deficient in Type I and Type II Cytokines, Their Receptors, JAKs and STATs

Type I cytokines and their receptors	
γc family	
IL-2[43,153]	Normal thymic and peripheral T cell development. Decreased polyclonal T cell responses *in vitro*, but more normal *in vivo* responses to pathogenic challenges. Autoimmunity with marked changes in levels of serum immunoglobulin isotypes. Ulcerative colitis–like inflammatory bowel disease.
IL-4[380]	Defective Th2 cytokine responses and class switch; defective IgG1 and IgE production.
IL-2/IL-4[381]	Some features of both IL-2 and IL-4 knockout mice. No gross abnormalities of T cell development.
IL-7[382]	Greatly diminished thymic and peripheral T cell development and B lymphopoiesis, resulting in profound lymphopenia.
IL-9[100]	No T cell defect. Defect in pulmonary goblet cell hyperplasia and mastocytosis following challenge with *Schistosoma mansoni* eggs. No defect in eosinophilia or granuloma formation.
IL-15[383]	Defective NK cell development. Defect in CD8 memory T cell homeostasis.
IL-2Rα[126]	Normal initial lymphoid development, but massive enlargement of peripheral lymphoid organs, polyclonal T and B cell expansions, and activated T cells, with impaired activation-induced cell death. Autoimmunity with increasing age, including hemolytic anemia and inflammatory bowel disease.
IL-2Rβ[159]	Severe autoimmunity including autoimmune hemolytic anemia. Death within approximately 3 mo. Deregulated T cell activation. Dysregulated B cell differentiation and altered Ig profile.
γc[154]	Greatly diminished thymic development but double negative, double positives, and single positives all represented. Age-dependent accumulation of peripheral CD4+ T cells with an activated memory phenotype. Greatly diminished numbers of conventional B cells, although B1 cells are present. No NK cells or γδ cells. Absent gut-associated lymphoid tissue, including Peyer patches.
IL-4Rα[384]	Like *Il4*$^{-/-}$ mice, they have defective Th2 cytokine responses and class switch; defective IgG1 and IgE production. In addition, they cannot expel *Nippostrongylus brasiliensis*, presumably because of defective IL-13 signaling.
IL-7Rα[385]	Greatly diminished thymic and peripheral T cell development and B lymphopoiesis, resulting in profound lymphopenia.
IL-15Rα[386]	Defective NK cell development. Defect in CD8 memory T cell homeostasis.
IL-13[137]	Defective Th2 cell development and the ability to expel helminths. Impaired diarrhea to OVA.
IL-4/IL-13 DKO[387]	Produce almost no IgE, highly resistant to OVA-induced diarrhea.
IL-13Rα1[248]	Exacerbated Th2 responses, with diminished mortality after infection with *S. mansoni* and greater susceptibility to *N. brasiliensis*. However, partially impaired allergic diarrhea to OVA.
IL-21R[122]	Decreased IgG1, elevated IgE.
IL-21R/IL-4 DKO[122]	Pan-hypogammaglobulinemia (mimics the B cell phenotype in humans with XSCID).
βc family	
IL-5[39]	Decreased basal level of eosinophils and defective induction of eosinophils following infectious challenge. Developmental defect in CD5+ B1 cells. Normal antibody and cytotoxic T cell responses.
GM-CSF[388]	Normal basal hematopoiesis. Unexpected abnormalities of the lung; abnormal pulmonary homeostasis.
IL-3Rβ[389]	No defects (due to redundant function of βc).
βc[42,389]	Defective responses to IL-5 and GM-CSF but normal responses to IL-3 (due to redundant function of IL-3Rβ). Diminished eosinophils: both basal levels and in responses to infectious challenge. Unexpected abnormalities of the lung characterized by pulmonary proteinosis and reduced phagocytosis by alveolar macrophages. In other words, the defects are a combination of those found in the IL-5– and GM-CSF–deficient mice.
IL-3Rβ + βc DKO[389]	Same phenotype as βc-deficient mice, except that they cannot respond to IL-3.
gp130 family	
IL-6[390]	Impaired acute-phase responses following infection or tissue damage. Decreased numbers of hematopoietic progenitor cells.
gp130[161,391]	Embryonic lethal. Extreme hypoplastic development of the myocardium; although the ventricular wall was very thin, trabeculation within the ventricle chamber was normal. Hematologic abnormalities characterized by greatly reduced CFU-S and somewhat reduced CFU-Gm and BFU-E. Markedly diminished size of thymus and number of thymocytes. Reduced primordial germ cells in embryonic gonads. Diminished size of the placenta.

TABLE 9.13 Phenotypes of Mice Deficient in Type I and Type II Cytokines, Their Receptors, JAKs and STATs (Continued)

LIF[161,392,393]	Decreased hematopoietic progenitor cells. Normal sympathetic neurons but deficient neurotransmitter switch *in vitro*. Defective blastocyst implantation.
LIFRβ[393]	Postnatal lethality. Normal hematopoietic and germ cell compartments but multiple neurologic, skeletal, placental, and metabolic defects. The greater severity than found in LIF-deficient mice reflects that LIFRβ is shared by several cytokines, including CNTF, LIF, OSM, and CT-1.
CNTF[394]	Progressive atrophy and loss of motor neurons.
CNTFRα[395]	Severe motor neuron deficient resulting in perinatal mortality. The more severe phenotype than in CNTF-deficient mice was unexpected and suggests another cytokine may utilize CNTFRα.
IL-11Rα[396]	Blastocysts can implant but decidualization cannot occur, associated with failure of pregnancy. Fetal lethal phenotype.
IL-12 p40[397]	Impaired but not completely lacking in their ability to produce IFNγ and to mount a Th1 response *in vivo*. Elevated secretion of IL-4; normal production of IL-2 and IL-10. Substantially decreased CTL responses. Resistant to infection with intracellular pathogens.
IL-12Rβ1[398]	Defective IL-12 signaling. IL-2 but not IL-12 could augment NK activity. Defective IFNγ production in response to ConA or anti-CD3. Severe defect in Th1 differentiation.
IL-12 p35[399]	Selective loss of IL-12 function without loss of IL-23. Less resistant to infection than p40 and IL-12Rβ1 KO.
IL-12Rβ2[400]	Selective loss of IL-12 function without loss of IL-23. Less resistant to infection than p40 and IL-12Rβ1 KO.
IL-23 p19[401]	Complete resistance to EAE, defective DTH response.
EBI3[402]	Increased pathologic features of colitis and shorter survival than in p28 KO mice, indicating that IL-35 rather IL-27 is protective.
IL-27 p28[402]	Decreased pathologic features of colitis and longer survival than in EBI3 KO mice, indicating that IL-35 rather than IL-27 is protective.
Epo, Tpo, G-CSF, and M-CSF	
Epo[403]	Embryonic lethal. Complete block of fetal liver erythropoiesis, resulting in severe anemia, yet normal development of BFU-E and CFU-E progenitor cells.
EpoR[404]	Same as Epo-deficient mice.
TpoR[405]	Decreased megakaryocytes and platelets, but other hematopoietic cells are present in normal numbers.
G-CSF[406]	Neutropenia and impaired neutrophil mobility. Diminished granulocytes and macrophage precursors.
M-CSF[407]	Osteopetrosis, absence of teeth. Females are infertile, suggesting an unexpected role for M-CSF.
M-CSF + GM-CSF[407]	A combination of defects of both M-CSF and GM-CSF, with osteopetrosis and alveolar proteinosis. Early death of pneumonia.
Type II cytokines and their receptors	
IFNAR-1[408]	Normal development. Defective immune defense against most viral infections tested, including lymphocytic choriomeningitis virus, Semliki Forest virus, Theiler virus, vesicular stomatitis virus. Normal resistance to *Listeria monocytogenes*, *Leishmania major*, *Mycobacterium bovis*, and *Mycobacterium avium*.
IFNγ[409]	Normal lymphoid development. Impaired resistance to *L. monocytogenes*, *L. major*, *M. bovis*, and *M. avium*. Can mount curative responses to a number of viruses. CD4+ effector cells default to the Th2 pathway after infection with *Leishmania*. Succumb to infection with *Toxoplasmosis gondii*.
IFNGR-1[410]	Normal lymphoid development. Impaired resistance to *L. monocytogenes*, *L. major*, *M. bovis*, and *M. avium*. Can mount curative responses to a number of viruses.
IL-10[232]	Normal lymphocyte development and antibody responses. Chronic enterocolitis, anemia, and growth retardation. Augmented inflammatory responses.
IL-19[411]	Increased susceptibility to dextran sulfate sodium–induced colitis, increased accumulation of macrophages, and production of proinflammatory cytokines and IFNγ.
IL-20R1[240,241]	Increased bone density with defective osteoclast differentiation.
IL-20R2[412]	Anti-TCR stimulation results in elevated IL-2 and IFNγ production, with increased antigen-specific CD4 and CD8 cells that produce IFNγ, but diminished IL-10.
IL-22[413]	Increased susceptibility to infection with *Citrobacter rodentium*, with augmented bacterial burden, intestinal epithelial damage, and mortality.
IL-28A[414]	Normal clearance of a range of viruses but diminished antiviral activity in response to TLR3 and TLR9 agonists. Expression of IL-28α on nonhematopoietic cells appears to be the most critical.

(continued)

TABLE 9.13 Phenotypes of Mice Deficient in Type I and Type II Cytokines, Their Receptors, JAKs and STATs (Continued)

STATs and JAKs	
STAT1[345,346]	Defective responses to both type I and type II IFNs. Defective response to certain viruses and bacterial antigens.
STAT2[347]	Defective signaling in response to type I IFNs; interestingly, the defect in not as severe in STAT2-deficient macrophages as it is in STAT2-deficient fibroblasts.
STAT3[348-351]	Embryonic lethal. Embryos implant but cannot grow. The fact that this phenotype is even more severe than that seen with gp130 suggests a role for STAT3 via a gp130-independent cytokine. By Cre-lox P methodology, *Stat3* was also selectively targeted in T cells, which exhibit defective IL-2–induced proliferation that correlates with a defect in IL-2–induced IL-2Rα expression; as expected, T cells also exhibit defective signaling in response to IL-6. *Stat3*-deficient neutrophils and macrophages show defective IL-10 signaling. STAT3 is also essential for normal involution of the mammary epithelium, for wound healing, and for normal hair cycle processes. *STAT3*-deficient DCs exhibit defective Flt3L-dependent expansion of DCs.
STAT4[336,337]	Defective Th1 development. Essentially, the same phenotype as in IL-12–deficient mice, including defective IL-12–mediated boosting of NK cell cytolytic activity.
STAT5A[292,340]	Defective lobuloalveolar development in the mammary gland, a syndrome resulting from defective prolactin signaling. Defective IL-2–induced IL-2Rα expression and associated defects in IL-2–induced T cell proliferation. Defective superantigen-induced expansion of Vβ8 T cells. Defective antigen-induced recruitment of eosinophils into the lung as well as antigen-induced IgG1 production.
STAT5B[293,341]	Defective growth analogous to Laron dwarfism, a disease of defective growth hormone signaling. Defective IL-2–induced IL-2Rα expression. More severe defects in IL-2–induced T cell proliferation and NK cell proliferation. Defective antigen-induced recruitment of eosinophils into the lung as well as antigen-induced IgG1 production. Defective NK cytolytic activity.
Partial STAT5A/ STAT5B[294]	Defective signaling in response to prolactin and growth hormone. Absent NK cell development. Major defect in T cell proliferation and TCR signaling. Anemia.
Complete STAT5A/ STAT5B[342]	Fetal lethality; absent T cell development.
STAT6[338,339]	Defective Th2 development. Essentially, the same phenotype as IL-4–deficient mice. Defective B cell proliferation.
JAK1[415]	Perinatal lethality, with defective signaling by gp130-dependent cytokines (IL-6, IL-11, CNTF, OSM, LIF, CT-1, and NNT-1/BSF-3). Defective signaling by γc-dependent cytokines (IL-2, IL-4, IL-7, IL-9, IL-15, IL-21; of these, IL-2 and IL-7 were formally evaluated).
JAK2[416]	Fetal lethality with profound anemia due to defective signaling in response to Epo.
JAK3[277]	Very similar and possibly identical to γc-deficient mice. Defective signaling in response to γc-dependent cytokines (IL-2, IL-4, IL-7, IL-9, IL-15, IL-21). Greatly diminished T cells in thymus and spleen, but then age-dependent peripheral expansion of CD4+ T cells. Unlike humans with *Jak3* mutations, *Jak3*-deficient mice have greatly diminished B cell numbers as well.
TYK2[417,418]	Mice lacking TYK2 exhibit diminished signaling in response to IFNα/β and IL-12, but it is not abrogated. Primarily STAT3 activation is diminished, even though STAT1/STAT2 and STAT4, respectively, are the STAT proteins that are primarily activated by IFNα/β and IL-12. Additionally, there are diminished responses to IFNγ and IL-18.

BFU, burst forming unit; BSF-3, B cell–stimulating factor 3; CD, cluster of differentiation; CFU, colony forming unit; CNTF, ciliary neurotrophic factor; CT-1, cardiotrophin-1; CTL, cytotoxic T lymphocyte; DC, dendritic cell; DKO, double knockout; DTH, delayed type hypersensitivity; EAE, experimental autoimmune encephalomyelitis; Epo, erythropoietin; G-CSF, granulocyte colony-stimulating factor; GM-CSF, granulocyte-macrophage colony-stimulating factor; IFN, interferon; Ig, immunoglobulin; IL, interleukin; KO, knockout; LIF, leukemia inhibitory factor; M-CSF, macrophage colony-stimulating factor; NK, natural killer; NNT-1, novel neurotrophin-1; OSM, oncostatin M; STAT, signal transducer and activator of transcription; TCR, T cell receptor; Th, T helper; Tpo, thrombopoietin; XSCID, X-linked severe combined immunodeficiency disease.

tetramer formation has been examined for STAT1 and STAT5 by generating mutations in the N-terminal domains which allows normal tyrosine phosphorylation and STAT dimer formation.[302-304] STAT1-tetramer–deficient mice showed defective IFNγ responses, but normal type I IFN responses; STAT5-tetramer-deficient mice had fewer CD8+ T cells, NK cells, and T_{reg} expansion. Mice lacking these STATs had more profound phenotypes indicating that tetramers have distinct roles in cellular responses.[45] Further complexity can be generated by the ability of at least some of the STATs to exist in alternatively spliced forms, some of which are inactive.[305]

STATs Are Evolutionarily Old

Just as *Drosophila* has a JAK kinase, there is a *Drosophila* STAT, denoted as either DSTAT or STAT92E.[306,307] The existence of JAK kinases and STAT proteins in lower organisms suggests that the system is evolutionarily old. A STAT has been identified in *Dictyostelium* that recognizes the sequence TTGA has highest sequence similarity to STAT5B and can bind mammalian IFN-stimulated response elements.[308] Interestingly, *Saccharomyces cerevisiae* do not appear to have STATs, as no SH2 domains have been identified in the entire *S. cerevisiae* genome.

Docking of STATs

Each STAT protein has an SH2 domain that plays two important roles: (1) receptor docking, for example, STAT1 docking on IFNGR-1, STAT2 docking on IFNAR-1, STAT3 docking on gp130, STAT5A and STAT5B docking on IL-2Rβ and IL-7Rα, and STAT6 docking on IL-4Rα; and (2) STAT dimerization that is mediated by the SH2 of one STAT interacting with the conserved phosphorylated tyrosine of another STAT protein.[255-260,309] It is also possible that STATs can dock on phosphorylated JAKs, given the ability to directly coprecipitate JAKs and STATs.[310] After docking has occurred, a conserved tyrosine (tyrosine 701 in STAT1, tyrosine 694 in STAT5A, etc.) can be phosphorylated. This phosphorylation is required for SH2 domain–mediated STAT dimerization, and the phosphorylation likely occurs while the STAT is docked on the receptor in physical proximity to receptor-associated JAK kinases. Following STAT phosphorylation, the STAT protein dissociates from the receptor, and its dimerization with itself or another STAT is then favored over its reassociation with the cytokine receptor chain given that STAT dimerization involves two reciprocal phosphotyrosine–SH2 interactions, whereas docking on a receptor involves only one interaction. Thus, efficient activation of STATs requires the presence in STATs of a conserved SH2 domain and a critical tyrosine.

While some receptor proteins such as IFNGR-1 and IL-7Rα have a single STAT docking site (for STAT1 and STAT5, respectively), a number of receptor molecules, including IL-2Rβ, IL-4Rα, gp130, Epo receptor, and IL-10R1, have more than one docking site for their respective STATs.[258] The presence of more than one site not only provides functional redundancy but also potentially could allow the simultaneous activation of two STAT molecules, providing a high local concentration of phosphorylated STATs to facilitate their dimerization.

STAT Nuclear Translocation and DNA Binding

Following dimerization, the STATs translocate into the nucleus where they can bind to DNA as transcription factors. The mechanism for nuclear translocation was originally mysterious given the absence of an obvious nuclear localization signal.[311] However, it was shown that tyrosine-phosphorylated STAT1 dimers can directly interact with importin-α5, allowing internalization. This suggested that there indeed was a nuclear localization signal that is normally masked, and mutation of Leu407 does not interfere with tyrosine phosphorylation, dimerization, or DNA binding but prevents nuclear localization.[311] Following its dephosphorylation, nuclear STAT1 is exported to the cytosol by a process that is dependent on the chromosome region maintenance 1 (CRM1) export reporter. Thus, both import and export of STAT1 are regulated processes.

Most STAT dimers directly bind to DNA via their DNA-binding domains.[312,313] In the case of IFN-α/β, STAT1-STAT2 heterodimers are formed, and these bind DNA in conjunction with a 48 kDa DNA-binding protein known as IRF9; the STAT1-STAT2-IRF9 complex is known as IFN-stimulated gene factor 3 (ISGF3) (see Fig. 9.12).[14,255] In the case of other STAT dimers, accessory proteins are not required for DNA binding. The motif recognized by ISGF3 complexes is AGTTTNCNTTTCC (known as an ISRE, for IFN-stimulated response element), whereas the other STAT complexes bind more semipalindromic TTCNmGAA motifs that are generally denoted as gamma-activated site (GAS) motifs for IFNγ-activated sequences, reflecting their original discovery in the context of IFNγ.[255-260]

A series of chimeric STAT proteins were used to delineate a DNA-binding domain of approximately 180 amino acids, with two conserved subdomains. Although multiple STATs bind to the same motifs, their relative efficiencies vary, indicating fine specificities conferred by the different DNA-binding domains.

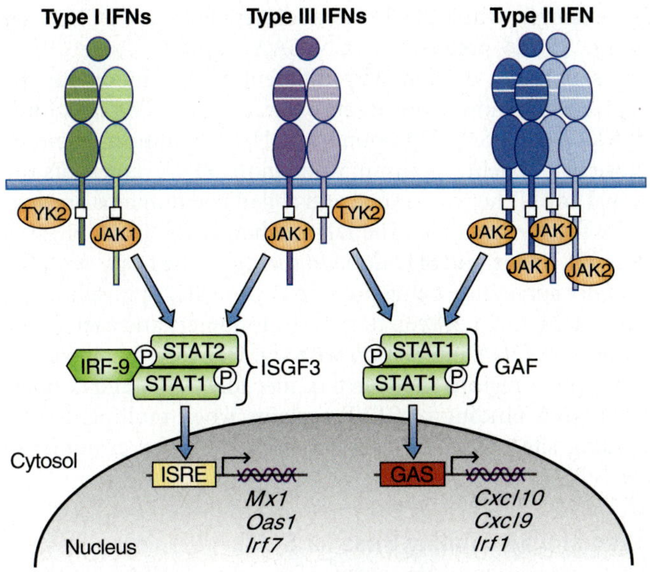

FIG. 9.12. Schematic of signaling by type I, II, and III interferons. Type I interferons all utilize IFNAR-1 and IFNAR-2 dimers as their functional receptors (*green*). IFNγ, the sole type II IFN member, functions as a dimer and recruits two IFNGR-1 and two IFNGR-2 chains into the receptor complex (*blue*). The type III interferons (also known as IL-28/IL-29 or IFNλ) utilize the IL-28R and the IL-10R2 as the functional receptor (*lavender*). The IL-10R2 chain is also shared by other receptors in the IL-10 family. All three of these different IFN receptor complexes induce phosphorylation of STAT1; however, there are a number of differences in their signaling pathways. All type I IFNs engage IFNAR-1 and IFNAR-2 resulting in the activation of TYK2 and JAK2 bound to the cytoplasmic tails of these chains, respectively. Subsequent phosphorylation of tyrosine residues in the tails produces docking sites for the SH2 domain of STAT1. STAT1 then becomes tyrosine phosphorylated and interacts with STAT2 and IRF9 to form a complex called the interferon stimulated gene factor (ISGF) 3. This complex translocates to the nucleus where it binds to a DNA element termed the interferon stimulated response element (ISRE) located in genes that mediate an antiviral state including mx1 and oas. The type II IFN member, IFNγ dimers, engages IFNGR-1 and IFNGR-2 resulting in the activation of JAK1 and JAK2 bound to their receptor tails resulting in the tyrosine phosphorylation of STAT1. Phosoph-STAT1 forms homodimers initially termed gamma-activated factor (GAF). This dimer translocates to the nucleus where it binds the canonical STAT palindrome initially termed gamma-activated site (GAS) for IFNγ activated site. This leads to induction of proinflammatory genes such as Cxcl10 and Cxcl9. The type III IFNs are able to activate both the ISGF-3 and GAF pathways by engaging the IL-28R and IL-10R2 and activating JAK1 and TYK2.

For example, while STAT1 homodimers favor a $TTCN_3GAA$ motif, STAT6 prefers a $TTCN_4GAA$ motif.[314] These differences partially explain why different STATs modulate the expression of nonidentical sets of target genes. The structures of STAT1 and STAT3β bound to DNA[312,313] almost resemble those of a vertebral column, wherein the DNA represents the spinal cord. The N-terminal and coiled coil domains are spatially the furthest from the DNA, whereas the DNA-binding domain, linker, and SH2 domain surround the DNA, with the stability apparently being provided by the SH2–phosphotyrosine interaction between the STAT monomers and each STAT monomer–DNA interaction with the DNA. As noted above, N-terminal regions mediate tetramer formation and cooperative DNA binding of STAT proteins when multiple STAT-binding sites occur in close proximity,[297-299] for example, in the IFNγ gene.

Optimal DNA-Binding Sites for STATs

The optimal binding motifs for several STAT proteins have been determined. For STAT1, STAT3, and STAT4, a TTCCSGGAA motif was defined[255,297] while STAT5A and STAT5B optimally bind a TTCYNRGAA motif,[299] and STAT6 binds a TTCNTNGGAA motif (where Y is C or T and R is G or A).[299,314] While dimeric STAT protein binding strongly preferred canonical motifs, the range of sequence motifs recognized by STAT5 tetramers is broader, with binding often occurring to two imperfect motifs or with one canonical motif with an associated "TTC" or "GAA" half GAS motif. The optimal inter-GAS motif spacing is 11 to 12 bp, with relative secondary peaks of 6 to 7 bp and 16 bp as well.[302] The presence of suboptimal GAS motifs spaced at appropriate distances to allow tetrameric binding may allow greater specificity via cooperative binding of STAT oligomers. A subset of genes bind STAT5 tetramers and STAT5 tetramers are critical for cytokine responses and normal immune function.[299]

Transcriptional Regulation by STATs

The classical function of STAT dimers is to regulate gene transcription.[260] Typically, tyrosine phosphorylation of STATs is needed for efficient SH2-SH2-domain–mediated dimerization. However, in addition to tyrosine phosphorylation, some STATs can be phosphorylated on serine, and STAT1 and STAT3 phosphorylation at serine 727 in the C-terminal transactivation domain is required for full transcriptional activity.[315-317] The serine 727 region resembles a MAP kinase recognition site. Numerous studies have shown that serine kinase activity is required for STAT-induced gene expression, especially for STAT1.[255,316-319] Serine 727 mediates the interaction of STAT1 with MCM5, a member of the minichromosome maintenance family of proteins; this interaction is thought to be important for maximal transcriptional activation.[319] In contrast, STAT2 is not serine phosphorylated. STAT5A and STAT5B are serine phosphorylated, and this phosphorylation may be important in cellular transformation.[320,321] A number of other posttranslational modifications of STATs have been described including methylation, acetylation, sumoylation, and addition of O-linked sugars that can variably modify their ability to engage in protein–protein interactions and to regulate gene expression.[260]

In addition to regulated modification of the STAT proteins, STATs can also interact with other factors to cooperatively regulate transcription.[260] For example, as noted previously, STAT1-STAT2 heterodimers bind DNA only after interacting with IRF9 to form ISGF3.[14,255] STAT1 interacts with and synergizes with Sp1 for transcriptional activation in the ICAM-1 gene[322] and with TRADD to influence IFNγ signaling.[323] An alternatively spliced shorter form of STAT3, denoted as STAT3β, associates with c-Jun to enhance transcriptional activity, and full-length STAT3 can also associate with c-Jun.[324,325] Moreover, certain STATs can interact with the potent transcriptional coactivators CBP/p300.[255,326] In the case of STAT1, this is mediated by interactions involving both the N- and C-terminal regions of STAT1 and the CREB and E1A binding regions of CBP, respectively. STAT5A has been shown to associate with the glucocorticoid receptor in the β-casein promoter.[327] Additionally, the IL-2 response element in the IL-2Rα gene requires not only STAT5 binding but also the binding of Elf-1, an Ets family transcription factor, to a nearby site.[328] Thus, active STAT complexes appear to involve the coordination of STAT proteins with other factors. STATs not only stimulate gene transcription; in some scenarios, STATs can mediate repression of transcription by recruiting transcriptional silencing proteins.[260] For example, the corepressor silencing mediator for retinoic acid receptor and thyroid hormone receptor was identified as a STAT5-binding partner that binds to both STAT5A and STAT5B and potentially plays a negative regulatory role.[329] In addition, IFNγ-activated STAT1 functions as a transcriptional repressor of the c-myc gene, an activity dependent upon its serine phosphorylation.[330]

Certain cytokine-regulated genes contain regions known as superenhancers, regions containing multiple STAT binding elements as well as other binding sites for other transcription factors.[331,332] When such superenhancers were ranked on STAT-binding intensity, STAT5-bound superenhancers were identified as being highly IL-2 responsive in the IL-2Rα gene (*Il2ra*) in CD4+ and CD8+ T cells.[333] The mouse *Il2ra* superenhancer has 13 STAT5-binding elements as well as p300 binding elements. Deletion of three of the STAT5 elements separately substantially reduced *IL2ra* gene expression demonstrating nonredundant function of certain elements. IL-2–responsive STAT5 superenhancers were identified in a number of other genes including *Lta*, *Lif*, *Cdk6*, and *Cish*.

Functions Mediated by STATs

STAT proteins were initially discovered by analyzing the signaling mechanism by which IFN induced the expression of genes controlling the development of the antiviral state.[260] In addition to roles in cellular differentiation, STAT proteins can contribute, directly or indirectly, to survival and/or mitogenic/proliferative responses that typify hematopoietic and immunologic cytokines, such as IL-3, IL-5, GM-CSF, IL-2, and IL-4.[258-260] A number of *in vitro* systems demonstrated that viruses or viral oncogenes are associated with activated

JAK-STAT pathways, indicating a role for STATs in cellular transformation. Indeed, STAT3 and STAT5 act as oncogenes, with compelling data for transformation capability of STAT3 in cell lines, persistently tyrosine-phosphorylated STATs in several types of leukemia and lymphoma, and the development of T lymphoblastic lymphomas in STAT5 transgenic mice.[334,335] The role of STAT3 in the tumor microenvironment makes it an attractive target for cancer immunotherapy and indeed there are a number of STAT3 inhibitors being tested in preclinical and clinical trials.[335] In addition, there is diminished proliferation of cells obtained from a number of the STAT knockout mice. For example, STAT4-deficient cells exhibit diminished cellular proliferation to IL-12[336,337]; STAT6-deficient cells exhibit diminished proliferation to IL-4[338,339]; and STAT5-deficient mice have diminished T cell proliferation to IL-2.[340,341] Some of these effects are indirect, based on modulation of receptor expression. For example, STAT6-deficient mice exhibit decreased IL-4Rα expression. Similarly, the absence of STAT5 results in decreased IL-2–induced IL-2Rα expression as well as decreased proliferation.[333,334] Thus, in some cases, decreased proliferation results at least in part from decreased expression of receptor components. STAT5A and STAT5B regulate BCL-XL induction, indicating their ability to affect cell survival.[342] On the other hand, STAT1 has been linked to cell growth arrest and induction of the cdk inhibitor p21WAF^{1}/CIP^{1},[343] and activation of STAT1 occurs in thanatophoric dysplasia type II dwarfism as the result of a mutant fibroblast growth factor receptor.[344] In this chondrodysplasia, the mutant fibroblast growth factor receptor induces nuclear translocation of STAT1, expression of p21WAF^{1}/CIP^{1}, and growth arrest, suggesting a relationship to the disease. Thus, different STATs can mediate either growth expansion or growth arrest. Moreover, STATs play many other types of roles, as well.

The phenotypes of mice lacking expression of each of the seven STAT proteins are known, and these knockout models have helped to clarify which cytokines critically depend on a given STAT protein (Table 9.13). STAT1 knockout mice exhibit defects that are very selective for the actions of type I and type II IFNs,[345,346] suggesting that STAT1 is most vital for the actions of IFNs, even though a variety of other cytokines have been reported to activate STAT1. STAT2-deficient mice exhibit defects consistent with selective inactivation of IFNα/β signaling.[347] STAT3-deficient mice exhibit fetal lethality; the embryos implant, but they exhibit defective development and growth.[348] Deletion of STAT3 within specific lineages has revealed that mice lacking STAT3 in T cells have normal lymphoid development but exhibit a defect in IL-2–induced IL-2Rα expression, somewhat analogous to what is seen in STAT5A- and STAT5B-deficient mice.[349] They also exhibit defective Th17 differentiation due to defective signaling by IL-6 and IL-21.[121] Neutrophils and macrophages lacking STAT3 exhibit defective signaling to IL-10, and it is known that STAT3 is important for the normal involution of the mammary epithelium, for wound healing, and for normal hair cycle.[349] DCs lacking STAT3 exhibit a defect in Flt3L-dependent differentiation, and antigen-presenting cells that are deficient in STAT3 exhibit disruption of priming of antigen-specific CD4+ T cells.[350,351] STAT4-deficient mice exhibit a phenotype similar to that of IL-12–deficient mice (ie, defective Th1 development), consistent with the observation that STAT4 is activated by IL-12.[336,337] Analogously, STAT6-deficient mice exhibit a phenotype similar to that of IL-4–deficient mice (ie, defective Th2 development)[338,339] in keeping with the observation that STAT6 is only activated by IL-4 and the closely related cytokine, IL-13. Interestingly, mice lacking STAT5A exhibit a defect in prolactin-mediated effects, including defective lobuloalveloar proliferation,[292] whereas mice lacking STAT5B have defective growth similar to that found in Laron dwarfism.[293] Thus, although STAT5A and STAT5B appear to always be coordinately induced, each of these STATs shows unique function *in vivo*. In addition to these defects, both STAT5A-deficient and STAT5B-deficient mice have defects in T cell development and signaling.[340,341] STAT5A-deficient mice have diminished numbers of splenocytes and exhibit a defect in IL-2–induced IL-2Rα expression.[340] STAT5B-deficient mice not only have similar defects but also have diminished thymocytes.[341] Most dramatically, these mice have a major defect in the proliferation of freshly isolated splenocytes and defective NK cell development. Mice lacking both STAT5A and STAT5B in lymphoid cells exhibit a dramatic defect in T cell development,[342] and even hypomorphic expression of both STAT5A and STAT5B is sufficient to result in a lack of NK cell development.[294,341] The lack of T cell development is attributed to defective IL-7 signaling, whereas the lack of NK cell development relates to defective IL-15–dependent STAT5 activation. STAT5A/STAT5B double-knockout mice exhibit lethality that is associated with severe anemia that develops in these mice.[352]

A more complete understanding of STAT-mediated cytokine responses has been facilitated by compilation of gene sets that are regulated by each STAT. Progress in this area has been made for a number of cytokines/STATs initially by microarray approaches and more recently by sophisticated analyses of chromatin landscape changes, gene network interactions, and single-cell RNA-seq.[211,353-355] It is important to recognize that not all cytokine signals are dependent on STATs, as cytokines activate a number of other signaling pathways in addition to STATs that influence the networks of transcription factors activated and thus the state of chromatin and gene expression patterns in a cell.

Noncanonical Functions of JAKs and STATs

While the canonical JAK-STAT pathway is well known to be critical for transmitting cytokine signals (Table 9.13), other activities outside of this main function have been described.[260,354] JAKs are potent cytosolic tyrosine kinases and do more than just phosphorylate tyrosine residues on receptors where STAT proteins dock or the STATs themselves. *In vitro* data indicate that JAK kinases also can phosphorylate receptor tyrosines other than the docking sites for STATs. For example, in the case of IL-2Rβ, JAK1 can phosphorylate not only tyrosines 392 and 510, which are STAT docking sites, but also tyrosine 338, which is a docking site for SHC,[356] all of which are required for maximal proliferation. JAK kinases are known to autophosphorylate themselves and/or transphosphorylate other JAKs. Other molecules can be phosphorylated by JAK kinases, including the STAM adapter molecule[357] and the p85 subunit of PI3K.[358]

Nonphosphorylated STATs Can Be Nuclear

In addition to the classical tyrosine phosphorylation–mediated dimerization and nuclear translocation, it is now clear that STAT proteins can exist in the nucleus even without being tyrosine phosphorylated and in that context modulate gene expression.[354,359-362] Moreover, nonphosphorylated STATs (eg, STAT4) can form N-domain–mediated dimers.[300] Unphosphorylated (U-) STAT1, STAT2, STAT3, and STAT5A have been localized to the nuclei of mammalian cells and are thought to regulate gene expression in unique ways.[363-365] For example, U-STAT1 mediates constitutive gene expression (ie, LMP2) and prolongs expression of IFN-induced regulatory genes,[359,363] while U-STAT5 interacts with CTCF to modulate a transcriptional program suppressing hematopoietic differentiation.[365] Studies in *Drosophila* demonstrated that U-STAT interacts with heterochromatin protein 1 (HP1) to stabilize heterochromatin leading to genome stability, longevity, and tumor suppression,[366] and at least in one study U-STAT5A was shown to stabilize heterochromatin and suppress tumor formation in a xenogenic mouse model.[367]

STATs in Mitochondria

Several STATs, including STAT1, STAT2, STAT3, STAT5, and STAT6, have been detected in mitochondria where they appear to mediate unique functions.[368] The function of these STATs may vary; STAT1 has been repored to limit mitochondrial biogenesis while both STAT1 and STAT2 suppress mitochondrial RNA expression. STAT3 has been reported to participate in numerous activities in mitochondria including cellular respiration, mitochondrial membrane potential, reactive oxygen species production, and ATP production.

Mitochondrial STAT3 is localized to the inner membrane and matrix and is required for optimal function of the electron transport chain (ETC).[369] This localization does not depend on STAT3 tyrosine phosphorylation. Rather, serine phosphorylation of STAT3 on S727 is required for STAT3 to bind to a mitochondrial protein termed genes associated with retinoid-IFN-induced mortality (GRIM)-19, a component of the ETC.[370-372] Interaction of STAT3 with GRIM-19 in the cytosol facilitates their cotranslocation to mitochondria. Mitochondrial STAT3 participates in Ras-mediated oncogenic transformation by maintaining glycolysis and oxidative phosphorylation needed to support tumor cell growth.[373] Much less is known about the function of mitochondrial STAT5 and STAT6.

CYTOKINE SIGNALING SPECIFICITY

Given that there are only four JAK kinases and seven STATs but scores of cytokines with distinct biological outcomes, it is clear that JAKs and STATs by themselves cannot fully determine cytokine specificity. Indeed, different cytokines with different actions activate the same JAKs and STATs. While clearly the JAK-STAT pathway is essential for the vast majority of cytokine-mediated effects (Table 9.13), other modifying pathways come in to play in different cell types to direct the specific response.

Contribution of JAKs to Specificity

The four JAKs are recruited to the cytoplasmic tails of different cytokine receptor subunits (Table 9.12) through the interaction of the JAK NH2-FERM domains with Box 1 sequences in the receptor tails.[20,262] As noted above, JAK3 uniquely interacts with γc, thus encoding a small level of specificity.[44] However, all members of the γc family utilize the JAK1 and JAK3 pair of kinases and yet mediate distinct functions. JAK2 interacts with cytokine receptors that mediate a wide array of functions. To directly address the contribution of JAKs to signaling specificity, chimeric JAKs were developed that replaced the NH2-FERM domain of one JAK for another to alter receptor-binding preference.[374,375] Such studies revealed that the identity of the JAK contributed little specificity to the signaling outcome.

Another small contribution to specificity comes from the fact that the same JAKs will have access to different substrates, depending on the receptor. The best characterized substrates for JAKs are tyrosine residues in the receptor tails, thus recruiting a diverse array of potential substrates, including the STAT proteins.[255-260]

Contribution of STATs to Specificity

There are only seven different STATs and the same STATs are activated by multiple cytokines (Table 9.14). Some of the STATs have limited receptor associations such as STAT1, STAT2, STAT4, and STAT6, while others are shared by many different receptors including STAT3, STAT5A, and STAT5B. Thus, similar to JAKs, STATs cannot completely explain specificity of signals. Nevertheless, as noted previously, their major, unique contributions can be revealed by phenotypes of gene-targeted mice (Table 9.13).

In addition, molecular studies demonstrate that some specificity is encoded in the SH2-binding preference of a STAT molecule. Once an active receptor complex is formed, the JAKs phosphorylate tyrosine residues in the cytoplasmic tails of the receptor subunits. Some of these phosphotyrosines reside in consensus sites for STAT–SH2 domain interaction. For example, STAT1–SH2 domain interacts with a phosphorylated ^{P}YDXXH consensus motif, while STAT3–SH2 domain and STAT6–SH2 domain prefer a ^{P}YXXQ motif and ^{P}YXXF motif, respectively.[309,376,377] These three amino acid sequence motifs are found in IFNγ, gp130, and IL-4Rα receptor subunits, respectively. SH2 domain exchange experiments, similar to those described above for JAKs, demonstrated that the specific SH2 domain is required for appropriate receptor docking and signal transduction.[378] Interestingly, grafting of regions of the IL-4Rα containing the STAT6-docking consensus sites onto IL-2Rβ resulted in an IL-4–like activation of STAT6 by IL-2[379] while grafting of the STAT3-docking site onto the EpoR lead to the activation of STAT3 by Epo.[309]

TABLE 9.14 Cytokines and the STATs They Activate

Cytokine	STATs Activated
IFNα/β	STAT1, STAT2, STAT4
IFNγ	STAT1
Growth hormone	STAT5A, STAT5B
Prolactin	STAT5A, STAT5B
Erythopoietin	STAT5A, STAT5B
Thrombopoietin	STAT5A, STAT5B
IL-10	STAT3
IL-12	STAT4, STAT3
G-CSF	STAT3
γc family	
IL-2, IL-7, IL-9, IL-15, IL-21	STAT5A, STAT5B, STAT3, STAT1
IL-4	STAT6, STAT5A, STAT5B
IL-13	STAT6
TSLP	STAT5A, STAT5B
βc family	
IL-3, IL-5, GM-CSF	STAT5A, STAT5B
gp130 family	
IL-6, IL-11, CNTF, LIF, OSM, CT-1	STAT3
Leptin	STAT3

CNTF, ciliary neurotrophic factor; CT-1, cardiotrophin-1; G-CSF, granulocyte colony-stimulating factor; GM-CSF, granulocyte-macrophage colony-stimulating factor; IFN, interferon; IL, interleukin; LIF, leukemia inhibitory factor; STAT, signal transducer and activator of transcription; TSLP, thymic stromal lymphopoietin.

Note that for IL-2, IL-7, IL-9, and IL-15, STAT5A and STAT5B appear to be the major STATs that are activated, although STAT3 in particular and STAT1 to a lesser degree can also be activated. For IL-21, STAT3 is the dominant STAT protein activated, followed by STAT1 and STAT5A/STAT5B.

Thus, a significant level of specificity is encoded by the presense of consensus docking sites in the cytoplasmic tails of receptor subunits and the binding preference of STAT–SH2 domains. Further, STATs have preferences for alterations to the standard DNA-binding motifs found in promoters of cytokine-inducible genes that help to specify the set of genes induced; for example, STAT6 uniquely binds the N-4-GAS DNA-binding motif.[314]

However, STATs still cannot fully explain cytokine specificity. There are many examples of cytokines activating the same set of STAT dimers but eliciting quite diverse biologic effects.[296] For example, IL-6, IL-10, and IL-21 all induce the tyrosine phosphorylation of STAT3 but overall these three cytokines have quite different biologic effects.[43-45,160,161]

Other Signaling Molecules Important for Cytokines

It is important to underscore that cytokine receptors stimulate a range of signaling pathways in addition to the JAK-STAT pathway including the activation of other tyrosine kinases besides JAKS. In addition to their activation of JAK kinases, a number of cytokines have been reported to activate Src family kinases. For example, IL-2 was reported to activate p56lck in T cells and p59fyn and p53/p56lyn in B cell lines.[43-45] Another tyrosine kinase, Syk, was reported to associate with IL-2Rβ. The G-CSF receptor also can form a complex with Lyn and Syk, βc has also been reported to interact with Src family kinases, and gp130 has been reported to associate with a number of other kinases, including Btk, Tec, and Fes[161,258]; and IL-4Rα has been shown to interact with Fes.[84] However, the significance of tyrosine kinases besides JAKs in cytokine signaling is less well understood. Additionally, PI3K, p38, ERK, and JNK can each become activated by various cytokines, leading to activation of many downstream transcription factors to mediate a program of gene expression.

Interferon Regulatory Factor (IRF) Family Proteins

IRFs form a family of nine proteins, IRF1–IRF9, that are regulated by type I IFNs.[419,420] They each have a well-conserved N-terminal DNA-binding domain that forms a helix-loop-helix structure as well as a C-terminal interaction domain. IRFs are critical for a number of actions, including type I IFN–dependent gene transcription. Select IRFs play critical roles in TLR-mediated IFN induction. IRF family members play important and diverse roles in immunity and oncogenesis. IRFs can have negative as well as positive effects. For example, IFNγ-mediated repression of IL-4 is mediated by IRF protein(s).[421] The expression of IRF4 is limited to the immune system and its abundance is increased by T cell activation.[419,420] Using genome-wide chromatin immunoprecipitation coupled to DNA sequencing (ChIP-Seq), it was shown that IRF4 cooperates with STAT3 to control IL-21–induced *Prdm1* expression in T cells and that these factors together globally regulate IL-21–mediated gene expression.[422] IRF4 expression was required for normal STAT3 binding *in vivo* and for development of an additional T cell effector population, T_{FH} cells.[422] In addition, it was shown that IRF4 could cooperate with activator protein 1 (AP-1) complexes to bind to AP-1/IRF4 composite (TGAnTCA/GAAA) motifs denoted as AP-1/IRF composite elements (AICEs).[423] Moreover, AP-1 complexes cooperate with IRF4 in binding to AICEs in CD4+ T cells stimulated with IL-21 and in Th17 cells. AP-1 and IRF complexes also cooperatively promoted transcription of the *Il10* gene, which is expressed in Th17 cells and potently regulated by IL-21. Similar interactions were shown to regulate DC and Th17 differentiation as well.[424-426] Thus, the IRF pathway coordinates with the STAT pathway and the AP-1 pathway.

Insulin Receptor Substrate (IRS) Proteins

IRS1, a large cytoplasmic adapter protein, was discovered as a tyrosine phosphorylated substrate of the insulin receptor; subsequently, other members of this family were identified and characterized.[78,427] IRS family proteins have a large number of phosphotyrosine docking sites, particularly for the p85 subunit of PI3K, and they serve to recruit important accessory molecules and stimulate the Akt pathway as well as other serine/threonine kinases. Interestingly, both insulin and IL-4 could induce tyrosine phosphorylation of an IRS1-like molecule in hematopoietic cells; this protein was initially termed IL-4–induced phosphorylated substrate (4PS) and was later identified as IRS2.[78] The 32D myeloid progenitor cells lack IRS1 and could only signal in response to insulin or IL-4 when they were transfected with IRS1.[428] Both the insulin receptor and IL-4Rα proteins contain NPXY-containing sequences that are important for IRS1 or IRS2 binding; in IL-4Rα, this is contained within a sequence denoted as the insulin-IL-4 receptor (I4R) motif.[429]

The IRS1 and IRS2 adapters are important for response to insulin and IGF-1.[427] A role for the IRS2/PI3K pathway in IL-4–induced cell survival, proliferation, and gene expression was delineated in cell lines and primary cell populations *in vitro* including B cells and macrophages.[78,85,430] However, studies in IRS2-deficient mice indicate that IL-4–activated IRS2 mediates negative regulatory feedback for the PI3K pathway in macrophages by targeting the phosphorylation of IRS1.[431] In addition to IL-4 and IL-13, other cytokines have been shown to activate IRS1 and/or IRS2 *in vitro*. For example, growth hormone can induce phosphorylation of IRS1 and IRS2, IFNγ and LIF induce phosphorylation of IRS2, and the γc-dependent cytokines IL-2, IL-7, and IL-15 can induce tyrosine phosphorylation of IRS1 and IRS2 in T cells.[432-434] The significance of these *in vitro* findings remains unclear, although the smaller size of IRS1-deficient mice supports a role for IRS1 in GH-induced responses *in vivo*.[78,427]

Phosphatidylinositol 3-Kinase

PI3K is a lipid kinase that consists of an 85 kDa regulatory subunit and a 110 kDa catalytic subunit.[435] PI3K phosphorylation and activation can be induced by a number of cytokines, and the use of inhibitors, such as wortmannin or LY294002, has demonstrated its importance in signaling for certain cytokines. In some cases, the cytokine receptor tails contain the canonical p85 docking site YXXM and recruit p85 directly after tyrosine phosphorylation of this site.[436,437] However, some cytokine receptors do not have this docking site. IRS1 has multiple docking sites for PI3K (YXXM motifs), and thus for some cytokines, such as IL-4, the association of IRS1 with the IL-4Rα chain leads to PI3K recruitment and activation.[78,85,427] As another expample, STAT3 has been reported to serve as an adapter to couple PI3K to the IFNAR-1 component of type I IFN receptors.[438]

The Ras/Mitogen-Activated Protein Kinase Pathway

Another major signaling pathway for a number of cytokines is the Ras/MAP kinase pathway.[439] This pathway is used by cytokines whose receptors recruit the SHC adapter molecule, which in turn mediates the recruitment of Grb2 and Sos, eventually leading to the activation of Ras. In turn, Ras couples to the MAP kinase pathway through a well-defined signaling cascade. Certain cytokines, such as IL-2 (Fig. 9.13) and IL-3, appear to robustly use this pathway, whereas others, such as IL-4, typically do not, indicating that this pathway is differentially important depending on the cytokine that is being used.

Modulation of STATs by Other Signaling Pathways

The concomitant activation of the JAK-STAT pathway and other signaling pathways described above leads to modification of STATs themselves through posttranslational modifications, changes in protein binding partners, and interactions with other transcription factors in the context of a complex promoter element. As one example, during IFNγ signaling, STAT1 becomes tyrosine phosphorylated. In addition, the ERK and p38 pathways are also activated, leading to serine phosphorylation of the transcription factor C/EBPβ and serine phosphorylation of STAT1, respectively.[215] STAT1 and cEBPβ can then act together to efficiently activate certain target genes that have binding elements for both factors in their promoters. Examples of signal transduction by type I and type II cytokines are shown in Figures 9.13 and 9.12, respectively.

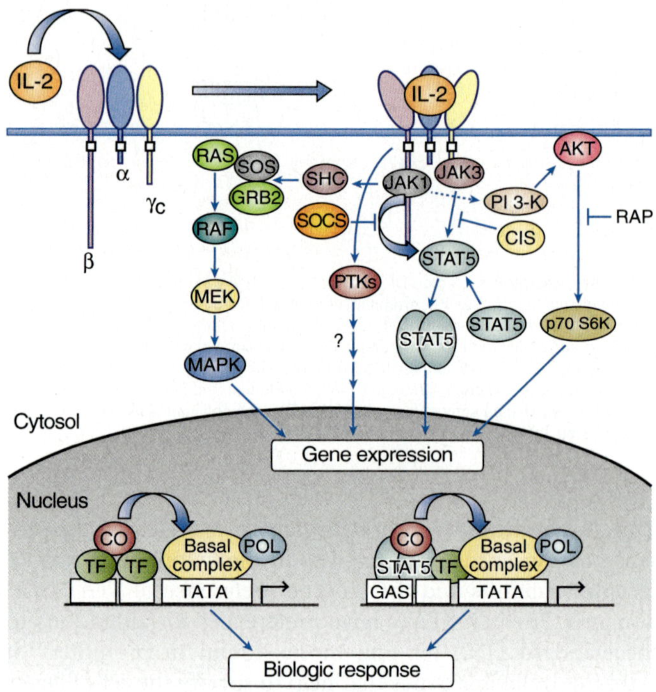

FIG. 9.13. Schematic of cytokine signaling showing multiple signaling pathways activated by IL-2. Shown is the association of JAK1 and JAK3 with different chains of the IL-2 receptor. Activation of JAK kinases results in tyrosine phosphorylation of IL-2Rβ. This allows the docking of STAT5 proteins via their SH2 domain. The STATs themselves are tyrosine phosphorylated, dimerize, and translocate to the nucleus where they modulate expression of target genes. The schematic also indicates that another phosphotyrosine mediates recruitment of SHC, which then can couple to the Ras/Raf/MEK/mitogen-activated protein kinase pathway. Also shown is the important phosphatidylinositol 3-kinase pathway. These and other pathways are activated by many type I cytokines and cooperate with the STAT pathway by mediating posttranslational modifications of STATs themselves and/or activating other transcription factors that cooperate with STATs to regulate gene transcription.

Pathway Integration Leads to Specificity

There are multiple parameters that ultimately lead to cytokine specificity. As discussed above, the regulation of STAT recruitment to receptor subunits by specific recognition of amino acid residues flanking phosphotyrosines in receptor tails by the SH2 domains in STATs is a well-recognized mechanism. Furthermore, STATs can be posttranslationally modified, impacting their interaction with other STATs and/or accessory factors to modulate DNA binding and transcriptional activation. Other signaling pathways modify STATs and activate other transcription factors that confer context-specific promoter recruitment and gene activation in response to the cytokine (see Fig. 9.13). Of course, the cell type and differentiation state of the responding cell will also influence the specific outcome of cytokine signaling. Thus, the specific response of a given cell type is mediated by integration of multiple signals.

Other factors that have been shown to control the magnitude and specificity of cytokine-induced responses include receptor complex stability and dimer orientation. As discussed above, the contribution of cytokine/receptor complex stability in controlling the ultimate cytokine-induced response has been best shown for the type I IFNs.[212-214,440] All of the IFN subtypes bind to the same heterodimeric receptor complex consisting of the IFNAR-1 and IFNAR-2 chains and can activate STATs. While the different IFN subtypes activate STATs to a similar extent, they lead to distinct biological outcomes relating to antiproliferative and antiviral functions. The impact of dimer orientation in controlling specificity of outcome is best studied in the Epo/EpoR system. Synthetic anti-EpoR antibodies termed diabodies (DA) were used as surrogate ligands to dimerize the EpoR in non-native geometries.[441] DA-induced signaling strength varied from full to minimal agonism and activated a differential set of signaling molecules and ultimately gene expression profiles compared to the natural ligand EPO. Analysis of the structures of the DA/EpoR complexes demonstrated EpoR dimer orientation and proximity differed from the Epo/EpoR complexes. Thus, the physical geometry of the cytokine/receptor complex and the longevity of the receptor complex control to some extent the cytokine-induced response.

DOWNMODULATION OF CYTOKINE SIGNALS

Much of the previous discussion has centered on the mechanisms by which cytokines induce signals. However, the mechanisms by which cytokine signals can be terminated are also extremely important. There are multiple levels at which negative regulation can occur. These include (1) regulating a balance between the production (transcriptional and translational control) of the cytokine, its receptor, and/or downstream signaling molecules and the degradation of these same molecules and (2) regulation of the activation state of the receptor and downstream signaling molecules. For example, these can be mediated by transcriptional repressors or molecules that either inhibit cytokine signaling (eg, SOCS proteins) or reverse activated states (eg, phosphatases).

Transcriptional control of cytokine production is a widely used mechanism. Many T cell–derived cytokines such as IL-2, IL-3, and IL-4 are only produced by activated T cells, and their production is lost with the loss of activation. IL-15 provides an example where translation of the protein is carefully regulated, in part by the existence of multiple upstream antithymocyte globulins.[106] Most cytokine receptor chains are constitutively expressed, but some like IL-21R and IL-2Rβ are regulated in part by signals that act through the TCR. The most regulated receptor chain may be IL-2Rα, whose expression is absent on resting lymphocytes but strongly induced following stimulation with antigens, mitogens, and certain cytokines. While the transcriptional/translational control of most cytokine receptors is known to occur, the mechansims of regulation are still not completely characterized.

Tyrosine Phosphatases

Because phosphorylation events are vital for the creation of phosphotyrosine docking sites, dephosphorylation is an obvious mechanism of control. Indeed, two tyrosine phosphatases, Shp-1 (formerly also known as SHP, HCP, SH-PTP1, and PTP1C) and Shp-2 (formerly also known as Syp and PTP1D), have been shown to play roles related to cytokine signaling.[442] Shp-1 mutations cause the motheaten (*me*) and viable motheaten (*mev*) phenotypes in mice.[443,444] The viable motheaten mouse had a less severe phenotype that is associated with increased numbers of erythroid progenitor cells and hyperresponsiveness to Epo, suggesting that Shp-1 might normally diminish responsiveness to Epo.[445] Indeed, it was demonstrated that Shp-1 binds directly to the EpoR when Y429 is phosphorylated.[446] This tyrosine is located in a "negative" regulatory region of the EpoR, and when mutated, Epo-responsive cells can grow in lower concentrations of Epo. Following Shp-1 binding to Y429, dephosphorylation and inactivation of JAK2 is facilitated. Thus, the negative regulation of Epo signaling appears to be at the level of a receptor-dependent inactivation of a JAK kinase. Shp-1 has also been shown to interact with tyrosine phosphorylated βc and IL-4Rα to downregulate signaling[447,448] and to be able to associate with both TYK2 and JAK2.[449,450]

Shp-2 has generally been considered primarily an "activating" phosphatase, by removing phosphates from negative regulatory tyrosines; it is therefore interesting that it can also interact with JAK1, JAK2, and TYK2.[451] In addition to the dephosphorylation of JAK kinases by phosphatases, STAT proteins also appear to be regulated at the level of tyrosine dephosphorylation[452]; interestingly, dephosphorylation of phosphotyrosine on STAT1 dimers requires extensive spatial reorientation of the two monomers within the dimer, something that is facilitated by the N-terminal domain that is involved in tetramerization.[453] Finally, another type of phosphatase, namely the lipid phosphatases, known as SHIP and SHIP2, can act as negative regulators of cytokine signals.[454]

In addition to Shp-1 and Shp-2, CD45, PTP1B, and T cell PTP (TCPTP) have been reported to regulate JAK kinases. CD45 appears to play roles related to Epo and IFN signaling. PTP1B has been reported to dephosphorylate JAK2 and TYK2, and TCPTP has been reported to dephosphorylate JAK1 and JAK3.[455] In the cytosol, Shp-2 and PTP1B have also been reported to dephosphorylate cytosolic STAT5 and nuclear STAT1 and STAT3, whereas TCPTP has been reported to dephosphorylate STAT1 and STAT3 in both cytosol and nucleus, but additional work is needed to determine whether these events are occurring physiologically *in vivo*.[455]

Another mode of negative regulation is by protein degradation. In addition to the degradation of receptor molecules, STAT1 itself is a target of the ubiquitin-proteasome pathway.[456] Finally, it is possible that regulation also can occur at the level of alternative splicing. In this regard, alternatively spliced versions of some of the STATs[257,291,305] have been reported.

The CIS/SOCS/JAB/SSI Family of Inhibitory Adapter Proteins

In 1995, the protoype molecule for an interesting class of proteins was discovered. The prototype molecule was named CIS, for cytokine inducible, SH2-containing protein, and it was shown to negatively regulate the actions of a set of cytokines.[457,458] CIS (also called CISH) is rapidly induced by a variety of cytokines, including IL-2, IL-3, GM-CSF, and Epo, to physically associate with both the βc and Epo receptors. Subsequently, a related protein, variably denoted as SOCS-1, JAB (JAK-binding protein), and SSI-1 (STAT-induced STAT inhibitor 1), was identified that could negatively regulate the activity of other cytokines, including IL-6 and IL-2, among others.[459] Interestingly, this protein could associate with JAK family kinases, whereas this function has not been reported for CIS. A total of eight CIS/SOCS/JAB/SSI family members have been identified that collectively regulate signals in response to multiple cytokines.[459] These proteins are now generally known as SOCS proteins and share a central SH2 domain and a C-terminal region known as a SOCS box. The SH2 domains of SOCS protiens can interact with phosphorylated JAKs and in some cases with phospotyrosines in the receptor tails. The SOCS box domain recruits the multisubunit E3 ubiquitin ligase, leading to ubiquitination and proteolytic degradation of targeted proteins. In addition to mediating degradation, binding of the SOCS proteins to JAKs can block substrate access and thus downregulate signaling.

Additional SOCS box–containing proteins lack SH2 domains and include proteins known as ASBs (ankyrin repeat–containing proteins with a SOCS box), SSBs (SPRY domain–containing proteins with a SOCS box), and WSBs (WD40 repeat–containing proteins with a SOCS box). SOCS proteins tend to be expressed at very low levels in nonactivated cells and to be induced by cytokines and pathogens. Following their induction, they negatively influence cytokine signaling, serving as negative feedback regulators.

Knockout mice for multiple SOCS proteins have been prepared, alone and in combination.[459] For example, CIS knockout mice are relatively normal, but CIS transgenic mice exhibit low body weight, failure of lactation, and diminished numbers of NK and NKT cells as well as altered TCR-mediated responses, presumably related to its ability to inhibit STAT5-dependent responses including those related to growth hormone, prolactin, IL-15, and IL-2, as well as others. *Socs1* knockout mice exhibit multiorgan inflammation and neonatal lethality as the result of augmented responsiveness to IFNγ, and this lethality can be prevented if the mice are crossed to IFNγ-deficient mice. Interestingly, SOCS-1$^{-/-}$ mice also indicate an essential role for SOCS-1 in thymocyte differentiation, and there are diminished numbers of maturing B cells. Deletion of *Socs1* within thymocytes/T cells/NKT cells results in multiple lymphoid abnormalites with increased CD8+ T cells due to increased sensitivity to γc-dependent cytokines, including IL-7.[460] *Socs2* knockout mice exhibit gigantism, ostensible due to dysregulated growth hormone signaling. *Socs2* knockout T cells exhibit enhanced Th2 differentiation, and knockout mice exhibit elevated Th2 responses after helminth infection.[461] *Socs3* knockout mice exhibit embryonic lethality due to placental insufficiency and dysregulated responses to LIF and IL-6 as well as hematopoietic defects. Conditional *Socs3* knockout mice have revealed a physiologic role for SOCS-3 in regulating G-CSF signaling in myeloid cells and for "emergency" granulopoiesis as well as for IL-6/gp130 signaling.[462,463] *Socs6* knockout have mild growth retardation.[459]

Protein Inhibitors of Activated Signal Transducer and Activator of Transcription Proteins

Protein inhibitors of activated STAT (PIAS) proteins are other negative regulators of cytokine actions[455] with PIAS-1 being an inhibitor of STAT1-binding activity and PIAS-3 being a similar inhibitor of STAT3-binding activity. These PIAS proteins block the DNA-binding activity of STAT dimers. Interaction of PIAS-1 occurs with STAT1 molecules that lack amino-terminal methylation.[464] Enhanced methylation of STAT1, which occurs downstream of IFNγ signaling, diminishes interaction with PIAS-1. Two other members, PIASx (which has both α and β splice variants) and PIASy, have also been described, with PIASx inhibiting STAT4 and PIASY inhibiting STAT1. They appear to act at least in part as transcriptional corepressors by recruiting other proteins such as histone deacetylase. In addition, PIASx and PIASy have added actions not restricted to the context of STAT inhibition. For example, PIASy is a nuclear matrix-associated SUMO E3 ligase (a ubiquitin-related protein) that can repress the activity of the Wnt-responsive transcription factor, LEF1.[465] PIASy coexpression results in the covalent modification of LEF1 by SUMO. Thus, as a class, PIAS proteins have more than one type of action.[455]

DISEASES OF CYTOKINE RECEPTORS AND RELATED MOLECULES

A Range of Cytokine-Related Causes of Severe Combined Immunodeficiency

As detailed previously, mutations in the γc cause X-linked SCID, and mutations in JAK3 cause a similar T−B+NK− SCID. A number of γc-dependent cytokines activate primarily STAT5A and STAT5B as signaling molecules downstream of the JAKs, and interestingly *STAT5B* mutations have been found in human disease, resulting in growth failure, immunodeficiency, and pulmonary disease.[466] It can be predicted that human immunodeficiencies might also result from mutations in some of the cytokines whose receptors contain γc or from mutations in other components of the receptors for these cytokines. Indeed, as noted previously, patients with IL-2 deficiency exhibit a SCID-like syndrome, due to inadequate function of their T cells, and an unusual immunodeficiency has been found to result from a mutation in IL-2Rα.[127] One patient with defective IL-2Rβ expression also had an immunodeficiency syndrome characterized by autoimmunity somewhat analogous to IL-2Rβ–deficient mice.[467] Other patients with single amino acid changes in IL-2Rβ had defects in immunity and peripheral tolerance. Patients with mutations in IL-21 and IL-21R have also been identified and these individuals have severe disease. The IL-21R-deficient individuals had cryptosporidial infections and chronic cholangitis with liver disease, recurrent repiratory infecitons, reduced T cell effector function, and impaired immunologulin class switch; the individual with IL21 deficiency also had impaired class switch and inflammatory bowel disease.[468,469] Given that mutations in IL-7Rα in humans cause T−B+NK+ SCID, mutations in IL-7 might be predicted to cause a similar syndrome.[155] Interestingly, however, in humans IL7 deficiency has been associated with generalized verrucosis associated with CD4+ T cells deficiency.[470,471] Defective expression or function of IL-2, IL-2Rα, IL-2Rβ, IL-7Rα, γc, and JAK3 has been found to cause SCID, with TYK2 being implicated in another form of immunodeficiency. It is likely that additional mutations in other cytokines, cytokine receptors, JAKs, or STATs will also be linked to SCID.

Defects in the Ability to Clear Mycobacterial Infections and Chronic Mucocutaneous Candidiasis

A number of immunodeficiencies have been characterized where affected individuals cannot properly clear mycobacterial infections. These have also turned out to be diseases of defective cytokine signaling. Mutations have been found in the components of either IL-12 itself or in the IFN or IL-12/IL-23 receptors, with mutations having been found in either the gene encoding the p40 subunit of IL-12 (which as noted previously is also a component of IL-23), in IL-12Rβ1 (a component of both the IL-12 and IL-23 receptors), or in the IFNGR-1 or IFNGR-2 components of IFNγ receptors.[472-474] The critical role of IL-12 for Th1 cell–mediated differentiation and production of IFNγ provides the explanation for finding similar clinical syndromes in humans lacking the p40, IL-12Rβ1, IFNGR-1, or IFNGR-2. Moreover, a patient with a mutation in the *STAT1* gene was also identified with a similar clinical syndrome indicating that as anticipated, STAT1 is a critical mediator of IFNγ signaling.[475] Interestingly, this patient had a mutation on only one *STAT1* allele, but the mutation was a dominant negative mutation that selectively inhibits the formation of STAT1 dimers (hence abrogating IFNγ signaling) but yet had at most only a modest effect on the ability to form ISGF3, hence leaving signaling in response to IFNα/β relatively intact. Chronic mucocutaneous candidiasis can be caused by autosomal-dominant IL-17F or autosomal-recessive IL-17RA deficiency or by gain-of-function heterozygous STAT1 mutant alleles that impair IL-17 immunity.[476] Somatic mutations that constitutively activate STAT3 occur in benign liver adenomas, as well as other cancers.[477] On the other hand, autosomal-dominant *STAT3* deficiency results in the susceptibility to bacterial and fungal diseases with very high serum IgE levels making up a subset of patients with hyper-IgE syndrome.[478]

Other Diseases Associated With Cytokine Receptors

There are many examples of other diseases that are linked to cytokine receptors and their downstream signaling pathways. For example, mutations in the growth hormone receptor have been found in a form of dwarfism (Laron dwarfism) in which target cells cannot respond to growth hormone.[479] Interestingly, some aspects of STAT5B deficiency are related to this syndrome. A kindred of patients with familial erythrocytosis has truncation in the erythropoietin receptor, resulting in hypersensitivity to Epo.[480] In addition, gain-of-function mutations in *IL7R* and genomic aberrations of *TSLPR* (formally denoted as *CRLF2*) have been found in childhood acute lymphoblastic leukemias.[481,482]

THERAPEUTIC STRATEGIES BASED ON ADAPTIVE CYTOKINE BIOLOGY

The accumulation of knowledge on the biologic effects of cytokines and the composition of their receptor complexes and their mechanisms of signal transduction has led to an explosion of clinical trials testing a variety of therapeutic strategies to treat disease. Such strategies include administration of recombinant cytokines, JAK inhibitors, monoclonal antibodies against cytokines or cytokine receptor subunits that block their function, soluble receptor modules, mutant cytokines (called muteins and neokines) that have altered functions, and strategic formation of cytokine/anticytokine complexes that modify activity (Table 9.15; Fig. 9.14). The fact that cytokines exhibit pleiotropy and redundancy and the sharing of receptor subunits by some cytokines can make specific targeting strategies especially tricky.

Cytokines in the Clinic

Epo was the first cytokine that was biochemically purified and is vital for erythropoiesis, whereas thrombopoietin is critical for thrombopoiesis. Epo has been used clinically since the 1990s for the treatment of anemia associated with renal insufficiency and chemotherapy.[483] Although it is dominantly associated with erythrocyte production, Epo receptors have been found more broadly on immune cells. Epo has been shown to have effects on a number of cell types, blocking NF-κB p65 and resulting in inhibition of the induction of TNF-α and inducible nitric oxide synthase in activated macrophages, with diminished control of *Salmonella* infection; in a model of chemical-induced colitis, such inhibition of NF-κB–dependent immune modulators limited tissue damage and disease severity.[484] Thus, administration of Epo for anemia may elicit immunologic side effects.

IL-2 was one of the first cytokines to be used clinically to treat cancer.[485] Other cytokines including IL-9, IL-15, and IL-21 have also been reported to have anticancer activity. The initial goal for IL-2 therapy was to expand T cells capable of recognizing and killing tumor cells. Some clinical benefit was observed in a subset of patients with malignant melanoma and renal cell carcinoma resulting in approval of IL-2 therapy for clinical use. Very high doses of IL-2 were needed to see an effect, and side

TABLE 9.15 Examples of Therapeutic Strategies

Therapy	Target(s)/effects
Exogeneous cytokines	Boost cytokine action
JAK inhibitors	JAK1, JAK2, JAK3, TYK2
Blocking monoclonal antibodies	Cytokines, cytokine receptors
Mutant cytokines (muteins)	Alters subunit dimerization or binding preference
Neokines	Engineered cytokines to enhance stability and affinity
Anticytokine antibodies as carriers	Alters receptor subunit binding preference
Engineered cytokine receptor complexes	Alters receptor subunit binding preference

effects could be severe (vascular leak syndrome). In some cases, increases in T_{reg} cell numbers were observed, which would not be beneficial in an anticancer setting.[486]

IL-2 is also used to expand the numbers of tumor-infiltrating lymphocytes (TILs) *in vitro* prior to their adoptive transfer back into patients with or without genetic modification.[485] However, T cells derived from tumors may show a dysfunctional phenotype and may not be the most effective source for enhancing antitumor activity. The use of IL-7, IL-15, or IL-21 during the expansion phase is being tested to preserve optimal T cell function and persistence after adoptive transfer.[487]

Because T_{reg} cells constitutively express the high affinity IL-2 receptor complex and IL-2 is critical for T_{reg} homeostasis, low-dose IL-2 therapy is being tested for the treatment of autoimmune diseases.[486] Low-dose IL-2 therapy is now in clinical trials to specifically increase T_{reg} numbers and suppress pathologic immune activation observed in hepatitis C–related vasculitis, graft-versus-host disease, and others to increase the number of T_{regs}.[486]

Type II cytokines can also be used therapeutically. IFNα is approved for the treatment of melanoma, hairy cell leukemia, and chronic hepatitis B and C infections to name a few.[488,489] This antiviral cytokine also showed some efficacy against the severe acute respiratory syndrome (SARS) outbreak in Toronto[490] and is being tested as a possible agent against SARS-coronavirus (CoV) 2, the virus responsible for the coronavirus disease 2019 (COVID-19) pandemic.[491] IFNβ is approved for use in multiple sclerosis; its action may be due to a shift in balance in IL-10 and IL-12 production as well as effects on stabilizing the blood–brain barrier disrupted by the disease.[492] Furthermore, IFNγ is used to treat chronic granulomatous disease.[493]

JAK Inhibitors

All type I and type II cytokines use the JAK/STAT pathway for signal transduction. They represent a major target for therapeutic intervention with small molecule inhibitors since they are in the tyrosine kinase class. There are a number of FDA-approved JAK inhibitors, with many more in the pipeline.[494] One of the first JAK inhibitors characterized, tofacitinib, most potently inhibits JAK3, although it can also inhibit JAK1 and JAK2.[495] It is approved for the treatment of rheumatoid arthritis (RA), psoriatic arthritis, and ulcerative colitis. Baricitinib is a JAK1 and JAK2 inhibitor approved to treat RA. Ruxolitinib is also a JAK1/JAK2 inhibitor and is approved for the treatment of myeloproliferative disorders such as polycythemia vera.

Each inhibitor was initially designed to be selective for a particular JAK; however, like most tyrosine kinase inhibitors, the specificity was not absolute leading to adverse events including cytopenia and venous thromboembolisms. Thus, a number of more selective inhibitors are being developed. Several JAK3 selective inhibitors including PF-06651600 are being evaluated as second-generation therapeutics.[496,497] More uses for these approved drugs and additional drugs are in clinical trials and preclinical development.

Blocking Monoclonal Antibodies

Certain diseases are associated with increased levels of cytokines or other situations where treatment with an anticytokine receptor antibody that prevents cytokine binding and/or receptor subunit dimerization is a rationale therapy. One example is Castleman disease, which is associated with overproduction of IL-6 by lymph node cells, leading to the successful treatment of this disease with IL-6 receptor blockade.[160,161,498] An anti-IL-6R based therapeutic, tocilizumab, is used worldwide to treat a variety of chronic and acute inflammatory diseases including RA and giant cell arteritis.[498] More recently, this anti–IL-6R was approved by the FDA to treat the cytokine storm syndrome observed after chimeric antigen receptor (CAR) T cell therapy[499] and is currently being tested in clinical trials to determine safety and efficacy as a treatment for severe forms of COVID-19 caused by SARS-Cov2 infection also shown to induce a potent cytokine storm.[500]

The IL-2 receptor blocking monoclonal antibody humanized anti-Tac (IL-2Rα), marketed under the name daclizumab, was approved for the treatment of T cell–mediated responses including renal allograft rejection and the relapsing form of multiple sclerosis.[501] This antibody was eventually withdrawn from use due to adverse events such as encephalitis. Similar to the effects of IL-2 treatment, blocking of the high-affinity IL-2 receptor with anti–IL-2Rα impacts both T effector cells and T_{reg} responses leading to mixed clinical outcomes.

Blocking the IL-4 and IL-13 receptors has been used in the treatment for a variety of allergic diseases.[502] Dupilumab is a

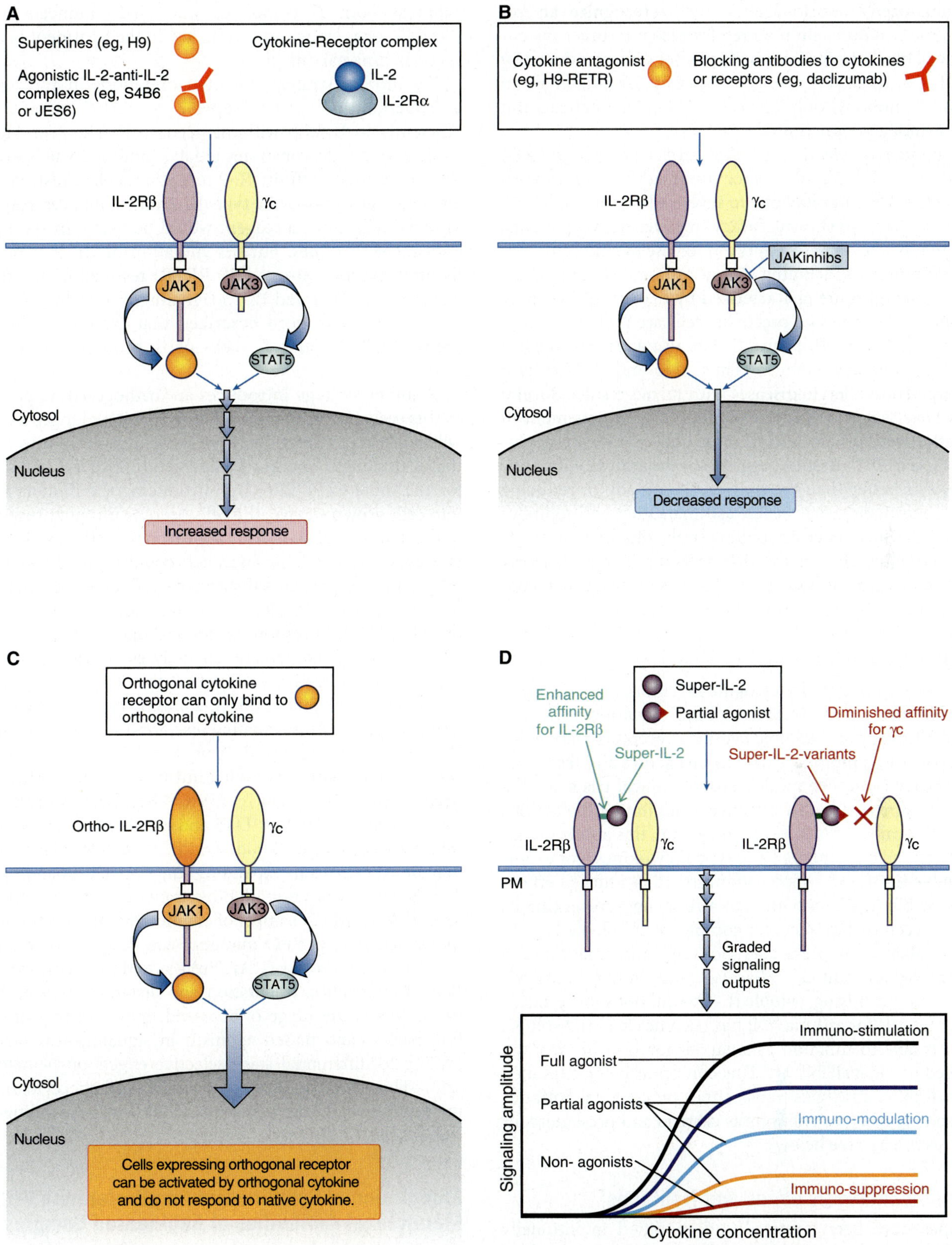

FIG. 9.14. Schematic of IL-2/IL-2 receptor targeting strategies. IL-2 as well as other cytokine action can be manipulated by a variety of targeting strategies. **A:** Illustration of the effects of changing several amino acids of IL-2 to enhance binding (H-9, superagonist) and fine-tuning of IL-2 action by the use of anti–IL-2 antibody complexed with IL-2 (S4B6-IL-2 or JES6-IL-2) or by IL-2-IL-2Rα complexes to enhance responses. **B:** IL-2 action can be inhibited by an engineered IL-2 molecule that acts as an antagonist (H9-RETR), by blocking antibody against the receptor (daclizumab) or by JAK inhibitors. **C:** Orthogonal IL-2Rβ only responds to orthogonal IL-2, thus potentially limiting *in vivo* side effects. **D:** Mutants of IL-2 have been generated with enhanced binding to IL-2Rβ but with various levels of attenuated γc interaction, ranging from full agonism to partial agonism, to potent antagonism (inhibitors), representing a spectrum of signaling amplitudes and different biological actions.

fully humanized monoclonal antibody that recognizes an epitope on the IL-4Rα chain required for receptor dimer formation. It will block the formation of receptor dimers induced by either IL-4 or IL-13 (Fig. 9.8A). In constrast, lebrikizumab, an anti–IL-13 antibody, only blocks IL-13–induced effects but not IL-4–induced signaling.[502]

Dupilumab is effective in adult atopic dermatitis (AD) patients and is FDA-approved for use in adults with moderate to severe AD and moderate to severe allergic asthma.[503,504] Dupilumab has also shown efficacy in pediatric AD patients, although it is not yet FDA-approved for use in children.[505]

Allergic diseases including allergic asthma, AD, and eosinophilic esophagitis are characterized by tissue infiltration by eosinophils. Eosinophilic precursor cells are mobilized from the bone marrow by IL-5, and IL-5 is also a critical survival factor for tissue eosinophils.[29] Thus, IL-5 and IL-5Rα have been targeted in allergic diseases with mixed results. Several studies have demonstrated airway function improvement in asthmatics treated with anti–IL-5 or anti–IL-5Rα monoclonal antibodies.[506,507] Results suggest that the anti–IL-5Rα antibody works better than the anti–IL-5 antibody in reducing oral glucocorticoid use in severe asthma. Other eosinophilic-rich diseases such as eosinophilic esophagitis may be treatable by targeting IL-5 or the IL-5 receptor. There are many more cases where antibodies to cytokines or cytokine receptor subunits are being developed as therapeutics.

Soluble Receptors

As discussed above, many cytokine receptors exist naturally as soluble forms and are able to bind cytokines with reasonable affinity. Thus, soluble cytokine receptors were engineered for use as inhibitors of cytokine action and tested for their efficacy in *in vivo* models and in clinical trials. Such a strategy has proved highly effective at inhibiting the actions of TNF clinically.[488,489] However, in general, this approach has not yet panned out for type I and type II cytokines. For example, soluble IL-4Rα was shown to inhibit IL-4–induced effects *in vitro* by binding IL-4 with high affinity and preventing its binding to cell surface receptor complexes.[250] Soluble IL-4Rα was also able to suppress allerigic lung inflammation in a mouse model for asthma.[508] However, while initial clinical trials appeared promising, soluble IL-4Rα did not show efficacy in clinical trials for asthma.[509] It has become clear that soluble IL-4Rα is able to function as a carrier for IL-4 *in vivo* and enhanced its bioavailabilty.[250] This concept is being exploited purposefully in strategies to use either monoclonal antibodies or soluble receptor subunits as carriers and presenters for cytokines *in vivo* (see below).

Mutant Cytokines (Muteins and Neokines)

Cytokines have been mutated and modified intentionally to alter their receptor subunit binding preferences to adjust their biological action. For example, pitrakinra is a mutant of human IL-4 (called a mutein) with strategic mutations in amino acid residues in helix D that are important for dimer formation with γc or IL-13Rα1 chains (see Fig. 9.8A). Thus, this mutant IL-4 can bind to the IL-4Rα chain but will not support dimerization and will not induce signaling.[85] It will suppress both IL-4–induced and IL-13–induced signals. Pitrakinra has shown some efficacy in the treatment of a subset of asthma patients.[510]

As another strategy, targeted mutations in IL-2 created a "superkine."[22,511] This "super-2" or "H9" mutant binds efficiently to IL-2Rβ without needing IL-2Rα (see Fig. 9.7; Table 9.5). H9 demonstrates greater binding to cells expressing intermediate-affinity IL-2 receptors such as resting T cells and NK cells than wild type IL-2 and enhances responses of these cells. Such a reagent would be useful in settings for anticancer therapies. Further mutation of H9 that prevents its dimerization with γc (H9-RETR) resulted in an efficient antagonist of IL-2 and IL-15 signaling.[512] Recently, IL-2 partial agonists have been described that exhibit the ability to preferentially expand T_{reg} cells or to promote CD8+ T cell stemness.[513,514]

Another strategy introduces an "orthogonal" approach.[515] In this case, IL-2Rβ was mutated so that it no longer bound to wild-type IL-2. Mutations were then introduced into IL-2 so that it no longer bound to wild-type IL-2Rβ but had affinity for the mutated IL-2Rβ. By introducing the mutant IL-2Rβ into cells grown ex vivo, the cells become uniquely responsive to the orthogonal IL-2. This approach was effective in mouse models of cancer. This approach could be used therapeutically with the goal of substantially reducing side effects, as only the mutant IL-2Rβ–transduced and adoptively transferred cells would respond to the orthogonal IL-2.

Sophisticated computational analyses of cytokine/cytokine receptor structures have supported the design of novel cytokines (neokines). Since both IL-2 and IL-15 utilize the IL-2Rβ-γc pair as part of their signaling receptors (see Fig. 9.7), a hybrid IL-2/IL-15 cytokine called neoleukin-2/15 was designed and shown to bind with very high affinity to IL-2Rβ-γc without the need for either IL-2Rα or IL-15Rα.[516] Further, this neokine was more stable than IL-2 or IL-15 and exhibited improved antitumor activity in mouse models with less toxicity. This neokine may be efficacious in expanding functional TILs and/or CAR T cells prior to or after adoptive transfer for the treatment of cancer. As another example of engineering novel cytokine-like molecules, a set of EPO mimetics was designed based on the ankyrin repeat protein (DARPin) scaffold to be able to control EpoR dimerization orientation and distance between monomers. Across the range of achieved EpoR dimer geometries, full, partial, and biased agonism in signaling was observed leading to differential stage-selective effects on hematopoiesis.[517] Thus, the capacity to generate synthetic cytokine-like molecules to fine-tune desired responses while avoiding other responses is within reach.

Modulation of Cytokine Binding Preference With Monoclonal Antibodies or Engineered Complexes

Strategies to use anticytokine antibodies to modulate binding preferences to receptor subunits are also being developed. The most advanced example is the targeting of IL-2, although similar approaches are in development for other cytokines. It was first noted that the half-life of IL-2 could be prolonged by two mouse monoclonal antibodies called S4B6 and JES6.[518] S4B6 was shown to block interaction of IL-2 with IL-2Rα but allowed interaction with IL-2Rβ, thus favoring the expansion of cells with low levels of IL-2Rα such as CD8+ T cells and NK cells. In contrast, JES6 preferentially inhibits the binding of IL-2 to IL2Rβ while allowing interaction with IL-2Rα. This binding profile results in the preferential stimulation of IL-2Rα high cells which include T_{regs}. Treatment of mice with JES6-IL-2 complexes results in the increase in FoxP3+ T_{regs} by at least twofold. Studies using such anti–IL-2/IL-2 complexes to specifically target the high-affinity IL-2 receptor expressed at high levels on T_{reg} cells in states of autoimmunity or to specifically target the intermediate affinity IL-2 receptor expressed on NK cells and CD8+ T cells during cancer immunotherapy have shown promise in preclinical testing.[519]

Recently, a human anti–IL-2 antibody with reactivity similar to JES6 (F5111.2) has been identified.[520] F5111.2-human IL-2 complexes have shown efficacy in treating type I diabetes in the NOD mouse model, reducing disease severity in the experimental autoimmune encephalomyelitis model of multiple sclerosis, and protecting against graft-versus-host disease. In these cases, the therapeutic effect is most likely due to the preferential expansion of T_{regs} *in vivo*.

Engineered cytokine receptor complexes have also been designed to modify cytokine binding preference. A single-chain fusion protein combining IL-15 and IL-15Rα was generated that acts as a superagonist to preferentially enhance proliferation of memory CD8+ T cells, thus augmenting antitumor immunity.[521] A single-chain fusion protein combining IL-2 and IL-2Rα has also been developed.[522] This fusion protein is long-lived *in vivo* and preferentially presents IL-2 to IL-2Rα on cells to selectively stimulate expansion of T_{regs}. This fusion protein reduced the incidence of spontaneous diabetes in female NOD mice. Such studies support the design of other cytokine-cytokine receptor subunit fusion proteins to modulate biological outcomes through receptor complexes with shared subunits.

CONCLUSION

Type I cytokines and IFNs are involved in the regulation of an enormous number of immunologic and nonimmunologic processes. There has been a progressive transition from viewing these as discrete molecules with special actions to sets of molecules that can be grouped according to shared receptor components and common signaling pathways. Signaling is one area where our understanding has greatly expanded; the pathways that are activated are similar for many cytokines, even when the biologic functions they induce are dramatically different. Although some of the differences can be explained by "compartmentalization" according to which cells produce the cytokine and which cells express receptors that allow them to respond to the cytokine, a tremendous amount still needs to be learned about how distinctive signals are triggered as well as more regarding the sets of genes that are induced by each cytokine. These will provide vital information important to the quest to completely understand the mechanisms by which type I cytokines and IFNs can affect their actions. At the same time, the generation of knockout mice for most cytokines and their receptors, as well as many signaling molecules, has provided *in vivo* clues as to vital functions served by these cytokines. Caution is clearly needed, however, in generalizing from these findings to human biology, given some apparently major differences in roles served, such as the essential role played by IL-7 in both humans and mice for T cell development, whereas IL-7 is also essential for B cell development in mice but not in humans. The identification of so many human disorders associated with cytokines and cytokine receptors has tremendously helped to teach us more about normal human biology as well. The accumulation of knowledge on the biologic effects of cytokines and the composition of their receptor complexes and their mechanisms of signal transduction has led to an explosion of clinical trials testing targeting strategies. Many of these have shown robust clinical efficacy and are now approved for clinical use.

REFERENCES

1. Paul WE. Pleiotropy and redundancy: T cell-derived lymphokines in the immune response. *Cell.* 1989;57:521-524.
2. Dumonde DC, Wolstencroft RA, Panayi GS, et al. "Lymphokines": non-antibody mediators of cellular immunity generated by lymphocyte activation. *Nature.* 1969;224(214):38-42.
3. Cohen S, Bigazzi PE, Yoshida T. Commentary. Similarities of T cell function in cell- mediated immunity and antibody production. *Cell Immunol.* 1974;12(1):150-159.
4. Isaacs A, Lindemann J. Virus interference. I. The interferon. *Proc R Soc Lond B Biol Sci.* 1957;147(927):258-267.
5. Biliau A, Matthys P. Interferon-gamma: a historical perspective. *Cytokine Growth Factor Rev.* 2009;20:97-113.
6. Oppenheim JJ, Gery I. From lymphodrek to interleukin 1 (IL-1). *Immunol Today.* 1993;14(5):232-234.
7. Wlodawer A, Pavlovsky A, Gustchina A. Hematopoietic cytokines: similarities and differences in the structures, with implications for receptor binding. *Protein Sci.* 1993;2(9):1373-1382.
8. Bazan JF. Structural design and molecular evolution of a cytokine receptor superfamily. *Proc Natl Acad Sci U S A.* 1990;87(18):6934-6938.
9. Bazan JF. Neuropoietic cytokines in the hematopoietic fold. *Neuron.* 1991;7(2):197-208.
10. Rozwarski DA, Gronenborn AM, Clore GM, et al. Structural comparisons among the short-chain helical cytokines. *Structure.* 1994;2(3):159-173.
11. Sprang SR, Bazan JF. Cytokine structural taxonomy and mechanisms of receptor engagement. *Curr Opin Struct Biol.* 1993;3:815.
12. Boulay JL, O'Shea JJ, Paul WE. Molecular phylogeny within type I cytokines and their cognate receptors. *Immunity.* 2003;19:159-163.
13. Wang X, Lupardus P, Laporte SL, Garcia KC. Structural biology of shared cytokine receptors. *Annu Rev Immunol.* 2009;27:29-60.
14. Donnelly RP, Kotenko SV. Interferon-lamda: a new addition to an old family. *J Interferon Cytokine Res.* 2010;30:555-564.
15. Arkin MR, Randal M, Delano WL, et al. Binding of small molecules to an adaptive protein-protein interface. *Proc Natl Acad Sci U S A.* 2003;100:1603-1608.
16. Davies DR, Wlodawer A. Cytokines and their receptor complexes. *FASEB J.* 1995;9(1):50-56.
17. Milburn MV, Hassell AM, Lambert MH, et al. A novel dimer configuration revealed by the crystal structure at 2.4 A resolution of human interleukin-5. *Nature.* 1993;363(6425):172-176.
18. Zdanov A, Schalk-Hihi C, Gustchina A, Tsang M, Weatherbee J, Wlodawer A. Crystal structure of interleukin-10 reveals the functional dimer with an unexpected topological similarity to interferon gamma. *Structure.* 1995;3(6):591-601.
19. D'Andrea AD, Fasman GD, Lodish HF. Erythropoietin receptor and interleukin-2 receptor beta chain: a new receptor family. *Cell.* 1989;58(6):1023-1024.
20. Murakami M, Narazaki M, Hibi M, et al. Critical cytoplasmic region of the interleukin 6 signal transducer gp130 is conserved in the cytokine receptor family. *Proc Natl Acad Sci U S A.* 1991;88(24):11349-11353.
21. de Vos AM, Ultsch M, Kossiakoff AA. Human growth hormone and extracellular domain of its receptor: crystal structure of the complex. *Science.* 1992;255(5042):306-312.
22. Spangler JB, Morago I, Mendoza JL, et al. Insights into cytokine-receptor interactions from cytokine engineering. *Annu Rev Immunol.* 2015;33:139-167.
23. Somers W, Ultsch M, De Vos AM, Kossiakoff AA. The X-ray structure of a growth hormone-prolactin receptor complex. *Nature.* 1994;372(6505):478-481.
24. Livnah O, Stura EA, Johnson DL, et al. Functional mimicry of a protein hormone by a peptide agonist: the EPO receptor complex at 2.8 A. *Science.* 1996;273(5274):464-471.
25. LaPorte SL, Juo SZ, Vaclavikova J, et al. Molecular and structural basis of cytokine receptor pleiotropy in the interleukin-4/13 system. *Cell.* 2008;132(2):259-272.
26. Wang X, Rickert M, Garcia KC. Structure of the quaternary complex of interleukin-2 with its alpha, beta, and gammac receptors. *Science.* 2005;310(5751):1159-1163.
27. Miyajima A, Kitamura T, Harada N, Yokota T, Arai K. Cytokine receptors and signal transduction. *Annu Rev Immunol.* 1992;10:295-331.
28. D'Andrea RJ, Gonda TJ. A model for assembly and activation of the GM-CSF, IL-3, and IL-5 receptors: insights from activated mutants of the common beta subunit. *Exp Hematol.* 2000;28:231-243.
29. Dougan M, Dranoff G, Dougan SK. GM-CSF, IL-3, and IL-5 family of cytokines: regulators of inflammation. *Immunity.* 2019;50:796-811.

30. Hansen G, Hercus TR, McClure BJ, et al. The structure of the GM-CSF receptor complex reveals a distinct mode of cytokine receptor activation. *Cell.* 2008;134:496-507.
31. Hara T, Miyajima A. Two distinct functional high affinity receptors for mouse interleukin-3 (IL-3). *EMBO J.* 1992;11(5):1875-1884.
32. Broughton SE, Hercus TR, Nero TL, et al. Conformational changes in the GM-CSF receptor suggest a molecular mechanism for affinity conversion and receptor signaling. *Structure.* 2016;24:1271-1281.
33. Metcalf D. *The Hematopoietic Colony Stimulating Factors.* Elsevier; 1984.
34. Hara T, Miyajima A. Function and signal transduction mediated by the interleukin 3 receptor system in hematopoiesis. *Stem Cells.* 1996;14(6):605-618.
35. Hida S, Yamasaki S, Sakamoto Y, et al. Fc receptor gamma-chain, a constitutive component of the IL-3 receptor, is required for IL-3-induced IL-4 production in basophils. *Nat Immunol.* 2009;10(2):214-222.
36. Codarri L, Gyulveszi G, Tosevski V, et al. RORγt drives production of the cytokine GM-CSF in helper T cells, which is essential for the effector phase of autoimmune neuroinflammation. *Nat Immunol.* 2011;12(6):560-567.
37. El-Behi M, Ciric B, Dai H, et al. The encephalitogenicity of T(H)17 cells is dependent on IL-1-and IL-23-induced production of the cytokine GM-CSF. *Nat Immunol.* 2011;12(6): 568-575.
38. Sonderegger I, Iezzi G, Maier R, Schmitz N, Kurrer M, Kopf M. GM-CSF mediates autoimmunity by enhancing IL-6-dependent Th17 cell development and survival. *J Exp Med.* 2008;205(10):2281-2294.
39. Kopf M, Brombacker F, Hodgkin PD, et al. IL-5-deficient mice have a developmental defect in CD5+ B-1 cells and lack eosinophilia but have normal antibody and cytotoxic T cell responses. *Immunity.* 1996;4(1):15-24.
40. Yoshida T, Ikuta K, Sugaya H, et al. Defective B-1 cell development and impaired immunity against Angiostrongylus cantonensis in IL-5R alpha-deficient mice. *Immunity.* 1996;4(5): 483-494.
41. Endo Y, Iwamura C, Kuwahara M, et al. Eomesodermin controls interleukin-5 production in memory T helper 2 cells through inhibition of activity of the transcription factor GATA3. *Immunity.* 2011;35(5):733-745.
42. Nishinakamura R, Miyajima A, Mee PJ, et al. Hematopoiesis in mice lacking the entire granulocyte-macrophage colony-stimulating factor/interleukin-3/interleukin-5 functions. *Blood.* 1996;88(7):2458-2464.
43. Lin JX, Leonard WJ. The common cytokine receptor γ chain family of cytokines. *Cold Spring Harb Perspect Biol.* 2018;10(9):a028449.
44. Leonard WJ, Lin JX, O'Shea JJ. The γc family of cytokines: basic biology to therapeutic ramifications. *Immunity.* 2019;50:832-850.
45. Lin JX, Leonard WJ. Fine-tuning cytokine signals. *Annu Rev Immunol.* 2019;37:295-324.
46. Taniguchi T, Matsui H, Fujita T, et al. Structure and expression of a cloned cDNA for human interleukin-2. *Nature.* 1983;302(5906):305-310.
47. Leonard WJ, Depper JM, Crabtree GR, et al. Molecular cloning and expression of cDNAs for the human interleukin-2 receptor. *Nature.* 1984;311(5987):626-631.
48. Nikaido T, Shimizu A, Ishida N, et al. Molecular cloning of cDNA encoding human interleukin-2 receptor. *Nature.* 1984;311(5987):631-635.
49. Morgan DA, Ruscetti FW, Gallo R. Selective in vitro growth of T lymphocytes from normal human bone marrows. *Science.* 1976;193(4257):1007-1008.
50. Waldmann TA. The multi-subunit interleukin-2 receptor. *Annu Rev Biochem.* 1989;58:875-911.
51. Takeshita T, Asao H, Ohtani K, et al. Cloning of the gamma chain of the human IL-2 receptor. *Science.* 1992;257(5068):379-382.
52. Janssen RA, Mulder NH, The TH, de Leij L. The immunobiological effects of interleukin-2 in vivo. *Cancer Immunol Immunother.* 1994;39(4):207-216.
53. Lenardo M, Cham KM, Hornung F, et al. Mature T lymphocyte apoptosis—immune regulation in a dynamic and unpredictable antigenic environment. *Ann Rev Immunol.* 1999;17:221-253.
54. Malek TR, Yu A, Vincek V, Scibelli P, Kong L. CD4 regulatory T cells prevent lethal autoimmunity in IL-2Rbeta-deficient mice. Implications for the nonredundant function of IL-2. *Immunity.* 2002;17(2):167-178.
55. Malek TR, Castro I. Interleukin-2 receptor signaling: at the interface between tolerance and immunity. *Immunity.* 2010;33(2):153-165.
56. Tang Q, Adams JY, Penaranda C, et al. Central role of defective interleukin-2 production in the triggering of islet autoimmune destruction. *Immunity.* 2008;28(5):687-697.
57. Williams MA, Tyznik AJ, Bevan MJ. Interleukin-2 signals during priming are required for secondary expansion of CD8+ memory T cells. *Nature.* 2006;441(7095): 890-893.
58. Feau S, Arens R, Togher S, Schoenberger SP. Autocrine IL-2 is required for secondary population expansion of CD8(+) memory T cells. *Nat Immunol.* 2011;12(9):908-913.
59. Pipkin ME, Sacks JA, Cruz-Guilloty F, et al. Interleukin-2 and inflammation induce distinct transcriptional programs that promote the differentiation of effector cytolytic T cells. *Immunity.* 2010;32(1):79-90.
60. Kalia V, Sarkar S, Subramaniam S, Haining WN, Smith KA, Ahmed R. Prolonged interleukin-2Ralpha expression on virus-specific CD8+ T cells favors terminal-effector differentiation in vivo. *Immunity.* 2010;32(1): 91-103.
61. Pepper M, Pagan AJ, Igyarto BZ, Taylor JJ, Jenkins MK. Opposing signals from the Bcl6 transcription factor and the interleukin-2 receptor generate T helper 1 central and effector memory cells. *Immunity.* 2011;35(4):583-595.
62. Hershko AY, Suzuki R, Charles N, et al. Mast cell interleukin-2 production contributes to suppression of chronic allergic dermatitis. *Immunity.* 2011;35(4):562-571.
63. Cote-Sierra J, Foucras G, Guo L, et al. Interleukin 2 plays a central role in Th2 differentiation. *Proc Natl Acad Sci U S A.* 2004;101(11):3880-3885.
64. Liao W, Lin JX, Leonard WJ. IL-2 family cytokines: new insights into the complex roles of IL-2 as a broad regulator of T helper cell differentiation. *Curr Opin Immunol.* 2011;23(5):598-604.
65. Mosmann TR, Cherwinski H, Bond MW, et al. Two types of murine helper T cell clone. I. Definition according to profiles of lymphokine activities and secreted proteins. *J Immunol.* 1986;136(7):2348-2357.
66. Romagnani S. TH1/TH2 interleukins. In: Oppenheim JJ, Feldman M. *Cytokine Reference.* Academic Press; 2001:99-112.
67. Liao W, Lin JX, Wang L, Li P, Leonard WJ. Modulation of cytokine receptors by IL-2 broadly regulates differentiation into helper T cell lineages. *Nat Immunol.* 2011;12(6): 551-559.
68. Shi M, Lin TH, Appell KC, Berg LJ. Janus-kinase-3-dependent signals induce chromatin remodeling at the Ifng locus during T helper 1 cell differentiation. *Immunity.* 2008;28(6):763-773.
69. Lazarevic V, Chen X, Shim JH, et al. T-bet represses T(H)17 differentiation by preventing Runx1-mediated activation of the gene encoding RORγt. *Nat Immunol.* 2011;12(1):96-104.
70. Paul WE. Interleukin-4: a prototypic immunoregulatory lymphokine. *Blood.* 1991;77(9):1859-1870.
71. Paul WE. History of IL-4. *Cytokine.* 2015;75:3-7.
72. Howard M, Farrar J, Hilfiker M, et al. Identification of a Tcell-derived B cell growth factor distinct from IL-2. *J Exp Med.* 1982;155:914-923.
73. Isakson PC, Pure E, Vitetta ES, et al. T cell-derived B cell differentiation factor(s). Effect on the isotype switch of murine B cells. *J Exp Med.* 1982;155:734-748.
74. Finkelman FD, Shea-Donohue T, Morris SC, et al. Interleukin-4-and interleukin-13-mediated host protection against intestinal nematode parasites. *Immunol Rev.* 2004;201:139-155.
75. Gour N, Wills-Karp M. IL-4 and IL-13 signaling in allergic airway disease. *Cytokine.* 2015;75(1):68-78.
76. Zhu J, Yamane H, Paul WE. Differentiation of effector CD4 T cell populations. *Annu Rev Immunol.* 2010;28:445-489.
77. Zamorano J, Wang HY, Wang LM, et al. IL-4 protects cells from apoptosis via the Insulin Receptor Substrate pathway and a second independent signaling pathway. *J Immunol.* 1996;157(11):4926-4934.
78. Keegan AD, Zamorano J, Keselman A, Heller NM. IL-4 and IL-13 receptor signaling from 4PS to insulin receptor substrate 2: there and back again, a historical view. *Front Immunol.* 2018;9:1037.
79. Keegan AD. IL-4 and IL-13: from "supe" to nuts. *Cytokine.* 2015;75:1-3.
80. Owen DL, Mahmud SA, Sjaastad LE, et al. Thymic regulatory T cells arise via two distinct developmental programs. *Nat Immunol.* 2019;20:195-205.
81. Ricardo-Gonzalez RR, Red Eagle A, Odegaard JI, et al. IL-4/STAT6 immune axis regulates peripheral nutrient metabolism and insulin sensitivity. *Proc Natl Acad Sci U S A.* 2010;107(52):22617-22622.
82. Nguyen KD, Qiu Y, Cui X, et al. Alternatively activated macrophages produce catecholamines to sustain adaptive thermogenesis. *Nature.* 2011;480(7375):104-108.
83. Gadani SP, Cronk JC, Norris GT, Kipnis J. IL-4 in the brain: a cytokine to remember. *J Immunol.* 2012;189(9): 4213-4219.
84. Nelms K, Keegan AD, Zamorano J, Ryan JJ, Paul WE. The IL-4 receptor: signaling mechanisms and biologic functions. *Annu Rev Immunol.* 1999;17:701-738.
85. Heller NM, Dasgupta P Dorsey NJ, et al. Chapter 3: the Type I and Type II receptor complexes for IL-4 and IL-13 differentially regulate allergic lung inflammation. In: Pereira C, ed. *Allergic Diseases. Highlights in the Clinic, Mechanisms and Treatments.* InTechOpen; 2012:43-82. NIHMS 405572.
86. Li L, Lee HH, Bell JJ, et al. IL-4 utilizes an alternative receptor to drive apoptosis of Th1 cells and skews neonatal immunity toward Th2. *Immunity.* 2004;20(4):429-440.
87. Namen AE, Lupton S, Hjerrild K, et al. Stimulation of B-cell progenitors by cloned murine interleukin-7. *Nature.* 1988;333(6173):571-573.
88. Goodwin RG, Lupton S, Schmierer A, et al. Human interleukin 7: molecular cloning and growth factor activity on human and murine B-lineage cells. *Proc Natl Acad Sci U S A.* 1989;86(1):302-306.
89. Mackall CL, Fry TJ, Gress RE. Harnessing the biology of IL-7 for therapeutic application. *Nat Rev Immunol.* 2011;11(5): 330-342.
90. Schluns KS, Kieper WC, Jameson SC, Lefrancois L. Interleukin-7 mediates the homeostasis of naive and memory CD8 T cells in vivo. *Nat Immunol.* 2000;1(5):426-432.
91. Schluns KS, Lefrancois L. Cytokine control of memory T-cell development and survival. *Nat Rev Immunol.* 2003;3(4):269-279.
92. Barata JT, Durum SK, Seddon B. Flip the coin: IL-7 and IL-7R in health and disease. *Nat Immunol.* 2019;20:1584-1593.
93. Park JH, Yu Q, Erman B, et al. Suppression of IL7Ralpha transcription by IL-7 and other prosurvival cytokines: a novel mechanism for maximizing IL-7-dependent T cell survival. *Immunity.* 2004;21(2):289-302.
94. Chandele A, Joshi NS, Zhu J, et al. Formation of IL-7Ralphahigh and IL-7Ralphalow CD8 T cells during infection is regulated by the opposing functions of GABPalpha and Gfi-1. *J Immunol.* 2008;180(8):5309-5319.
95. Goswami R, Kaplan MH. A brief history of IL-9. *J Immunol.* 2011;186:3283-3288.
96. Kaplan MH, Hufford MM, Olson MR. The development and in vivo function of TH9 cells. *Nat Rev Immunol.* 2015;15:295-307.
97. Noelle RJ, Nowak EC. Cellular sources and immune functions of interleukin-9. *Nat Rev Immunol.* 2010;10(10):683-687.
98. Van Snick J, Goethals A, Renauld JC, et al. Cloning and characterization of a cDNA for a new mouse T cell growth factor (P40). *J Exp Med.* 1989;169(1):363-368.
99. Renauld JC, Vink A, Louahed J, Van Snick J. Interleukin-9 is a major anti-apoptotic factor for thymic lymphomas. *Blood.* 1995;85(5):1300-1305.
100. Townsend JM, Fallon GP, Matthews JD, et al. IL-9-deficient mice establish fundamental roles for IL-9 in pulmonary mastocytosis and goblet cell hyperplasia but not T cell development. *Immunity.* 2000;13(4):573-583.
101. Renauld JC. New insights into the role of cytokines in asthma. *J Clin Pathol.* 2001;54(8):577-589.
102. Gerlach K, Hwang Y, Nikolaev A, et al. TH9 cells that express the transcription factor PU.1 drive T-cell mediated colitis via IL-9 receptor signaling in intestinal epithelial cells. *Nat Imunol.* 2014;15:676-686.
103. Nalleweg N, Chiriac MT, Podstawa E, et al. IL-9 and its receptor are predominantly involved in the pathogenesis of UC. *Gut.* 2015;64:743-755.
104. Grabstein KH, Eisenman J, Shaneback K, et al. Cloning of a T cell growth factor that interacts with the beta chain of the interleukin-2 receptor. *Science.* 1994;264(5161):965-968.
105. Burton JD, Bamford RN, Peters C, et al. A lymphokine, provisionally designated interleukin T and produced by a human adult T-cell leukemia line, stimulates T-cell proliferation and the induction of lymphokine-activated killer cells. *Proc Natl Acad Sci U S A.* 1994;91(11):4935-4939.
106. Waldmann TA, Tagaya Y. The multifaceted regulation of interleukin-15 expression and the role of this cytokine in NK cell differentiation and host response to intracellular pathogens. *Annu Rev Immunol.* 1999;17:19-49.
107. Waldmann TA. The biology of interleukin-2 and interleukin-15: implications for cancer therapy and vaccine design. *Nat Rev Immunol.* 2006;6(8):595-601.
108. Bamford RN, Grant AJ, Burton JD, et al. The interleukin (IL) 2 receptor beta chain is shared by IL-2 and a cytokine, provisionally designated IL-T, that stimulates T-cell proliferation and the induction of lymphokine-activated killer cells. *Proc Natl Acad Sci U S A.* 1994;91(11):4940-4944.
109. Dubois S, Mariner J, Waldmann TA, Tagaya Y. IL-15Ralpha recycles and presents IL-15 in trans to neighboring cells. *Immunity.* 2002;17(5):537-547.
110. Waldmann TA, Dubois S, Tagaya Y. Contrasting roles of IL-2 and IL-15 in the life and death of lymphocytes: implications for immunotherapy. *Immunity.* 2001;14(2):105-110.
111. Conlon KC, Miljkovic MD, Waldmann TA. Cytokines in the treatment of cancer. *J Interferon Cytokine Res.* 2019;39(1):6-21.
112. Spolski R, Leonard WJ. Interleukin-21: basic biology and implications for cancer and autoimmunity. *Annu Rev Immunol.* 2008;26:57-79.
113. Parrish-Novak J, Dillon SR, Nelson A, et al. Interleukin 21 and its receptor are involved in NK cell expansion and regulation of lymphocyte function. *Nature.* 2000;408(6808):57-63.
114. Ozaki K, Kikly K, Michalovich D, et al. Cloning of a type I cytokine receptor most related to the IL-2 receptor beta chain. *Proc Natl Acad Sci U S A.* 2000;97(21):11439-11444.
115. Zeng R, Spolski R, Finkelstein SE, et al. Synergy of IL-21 and IL-15 in regulating CD8+ T cell expansion and function. *J Exp Med.* 2005;201(1):139-148.

116. Novy P, Huang X, Leonard WJ, et al. Intrinsic IL-21 signaling is critical for CD8 T cell survival and memory formation in response to vaccinia viral infection. *J Immunol.* 2011;186:2729-2738.
117. Hinrichs CS, Spolski R, Paulos CM, et al. IL-2 and IL-21 confer opposing differentiation programs to CD8+ T cells for adoptive immunotherapy. *Blood.* 2008;111(11):5326-5333.
118. Hermans D, Gautam S, García-Cañaveras JC, et al. Lactate dehydrogenase inhibition synergizes with IL-21 to promote CD8+ T cell stemness and antitumor immunity. *Proc Natl Acad Sci U S A.* 2020;117(11):6047-6055.
119. Nurieva RI, Chung Y, Hwang D, et al. Generation of T follicular helper cells is mediated by interleukin 21 but independent of T helper 1, 2, or 17 cell lineages lineages. *Immunity.* 2008;29:138-149.
120. Nurieva R, Yang XP, Martinez G, et al. Essential autocrine regulation by IL-21 in the generation of inflammatory T cells. *Nature.* 2007;448:480-483.
121. Zhou L, Ivanov IL, Spolski R, et al. IL-6 programs T(H)-17 cell differentiation by promoting sequential engagement of the IL-21 and IL-23 pathways. *Nat Immunol.* 2007;8(9): 967-974.
122. Ozaki K, Spolski R, Feng CG, et al. A critical role for IL-21 in regulating immunoglobulin production. *Science.* 2002;298(5598):1630-1634.
123. Ozaki K, Spolski R, Ettinger R, et al. Regulation of B cell differentiation and plasma cell generation by IL-21, a novel inducer of Blimp-1 and Bcl-6. *J Immunol.* 2004;173(9):5361-5371.
124. Spolski R, West EE, Li P, et al. IL-21/type I interferon interplay regulates neutrophil-dependent innate immune responses to *Staphylococcus aureus*. *Elife.* 2019;8:e45501.
125. Noguchi M, Yi H, Rosenblatt HM, et al. Interleukin-2 receptor gamma chain mutation results in X-linked severe combined immunodeficiency in humans. *Cell.* 1993;73(1):147-157.
126. Willerford DM, Chen J, Ferry JA, et al. Interleukin-2 receptor alpha chain regulates the size and content of the peripheral lymphoid compartment. *Immunity.* 1995;3(4):521-530.
127. Sharfe N, Dadi HK, Shahar M, Roifman CM. Human immune disorder arising from mutation of the alpha chain of the interleukin-2 receptor. *Proc Natl Acad Sci U S A.* 1997;94(7):3168-3171.
128. Hilton DJ, Zhang JG, Metcalf D, et al. Cloning and characterization of a binding subunit of the interleukin 13 receptor that is also a component of the interleukin 4 receptor. *Proc Natl Acad Sci U S A.* 1996;93(1):497-501.
129. Kelly-Welch AE, Hanson EM, Boothby MR, et al. Interleukin-4 and interleukin-13 signaling connections maps. *Science.* 2003;300(5625):1527-1528.
130. Zurawski G, de Vries JE. Interleukin 13, an interleukin 4-like cytokine that acts on monocytes and B cells, but not on T cells. *Immunol Today.* 1994;15(1):19-26.
131. Caput D, Laurent P, Kaghad M, et al. Cloning and characterization of a specific interleukin (IL)-13 binding protein structurally related to the IL-5 receptor alpha chain. *J Biol Chem.* 1996;271(28):16921-16926.
132. Donaldson DD, Whitters MJ, Fitz LJ, et al. The murine IL-13 receptor alpha 2: molecular cloning, characterization, and comparison with murine IL-13 receptor alpha 1. *J Immunol.* 1998;161(5):2317-2324.
133. Lutz MB, Schnare M, Menges M, et al. Differential functions of IL-4 receptor types I and II for dendritic cell maturation and IL-12 production and their dependency on GM-CSF. *J Immunol.* 2002;169:3574-3580.
134. Grunig G, Warnock M, Wakil AE, et al. Requirement for IL-13 independently of IL-4 in experimental asthma. *Science.* 1998;282(5397):2261-2263.
135. Wills-Karp M, Luyimvazi J, Xu X, et al. Interleukin-13: central mediator of allergic asthma. *Science.* 1998;282(5397):2258-2261.
136. Wynn TA. IL-13 effector functions. *Annu Rev Immunol.* 2003;21:425-456.
137. McKenzie GJ, Emson C, Bell SE, et al. Impaired development of Th2 cells in IL-13-deficient mice. *Immunity.* 1998;9(3):423-432.
138. Suzuki A, Leland P, Joshi BH, et al. Targeting of IL-4 and IL-13 receptors for cancer therapy. *Cytokine.* 2015;75(1):79-88.
139. Bao K, Reinhardt RL. The differential expression of IL-4 and IL-13 and its impact on type-2 immunity. *Cytokine.* 2015;75:25-37.
140. Friend SL, Hosier S, Nelson A, et al. A thymic stromal cell line supports in vitro development of surface IgM+ B cells and produces a novel growth factor affecting B and T lineage cells. *Exp Hematol.* 1994;22(3):321-328.
141. Pandey A, Ozaki K, Baumann H, et al. Cloning of a receptor subunit required for signaling by thymic stromal lymphopoietin. *Nat Immunol.* 2000;1(1):59-64.
142. Park LS, Martin U, Garka K, et al. Cloning of the murine thymic stromal lymphopoietin (TSLP) receptor: formation of a functional heteromeric complex requires interleukin 7 receptor. *J Exp Med.* 2000;192(5):659-670.
143. Ziegler SF, Artis D. Sensing the outside world: TSLP regulates barrier immunity. *Nat Immunol.* 2010;11(4):289-293.
144. Leonard WJ. TSLP: finally in the limelight. *Nat Immunol.* 2002;3(7):605-607.
145. Liu YJ, Soumelis V, Watanabe N, et al. TSLP: an epithelial cell cytokine that regulates T cell differentiation by conditioning dendritic cell maturation. *Annu Rev Immunol.* 2007;25:195-219.
146. Pedroza-Gonzalez A, Xu K, Wu TC, et al. Thymic stromal lymphopoietin fosters human breast tumor growth by promoting type 2 inflammation. *J Exp Med.* 2011;208(3):479-490.
147. De Monte L, Reni M, Tassi E, et al. Intratumor T helper type 2 cell infiltrate correlates with cancer-associated fibroblast thymic stromal lymphopoietin production and reduced survival in pancreatic cancer. *J Exp Med.* 2011;208(3):469-478.
148. West EE, Spolski R, Kazemian M, et al. A TSLP-complement axis mediates neutrophil killing of methicillin-resistant *Staphylococcus aureus*. *Sci Immunol.* 2016;1(5):eaaf8471.
149. Rochman Y, Kashyap M, Robinson GW, et al. Thymic stromal lymphopoietin-mediated STAT5 phosphorylation via kinases JAK1 and JAK2 reveals a key difference from IL-7-induced signaling. *Proc Natl Acad Sci U S A.* 2010;107(45):19455-19460.
150. Leonard WJ. Cytokines and immunodeficiency diseases. *Nat Rev Immunol.* 2001;1(3):200-208.
151. Leonard WJ. The molecular basis of X-linked severe combined immunodeficiency: defective cytokine receptor signaling. *Annu Rev Med.* 1996;47:229-239.
152. Weinberg K, Parkman R. Severe combined immunodeficiency due to a specific defect in the production of interleukin-2. *N Engl J Med.* 1990;322(24):1718-1723.
153. Schorle H, Holtschke T, Hunig T, Schimpl A, Horak I. Development and function of T cells in mice rendered interleukin-2 deficient by gene targeting. *Nature.* 1991;352(6336):621-624.
154. Cao X, Shores EW, Hu-Li J, et al. Defective lymphoid development in mice lacking expression of the common cytokine receptor gamma chain. *Immunity.* 1995;2(3):223-238.
155. Puel A, Leonard WJ. Mutations in the gene for the IL-7 receptor result in T(−)B(+)NK(+) severe combined immunodeficiency disease. *Curr Opin Immunol.* 2000;12(4):468-473.
156. Recher M, Berglund LJ, Avery DT, et al. IL-21 is the primary common γ chain-binding cytokine required for human B-cell differentiation in vivo. *Blood.* 2011;118(26):6824-6835.
157. Miggelbrink AM, Logan BR, Buckley RH, et al. B-cell differentiation and IL-21 response in IL2RG/JAK3 SCID patients after hematopoietic stem cell transplantation. *Blood.* 2018;131(26):2967-2977.
158. Cavazzana M, Six E, Lagresle-Peyrou C, et al. Gene therapy for X-linked severe combined immunodeficiency: where do we stand? *Hum Gene Ther.* 2016;27(2):108-116.
159. Suzuki H, Kindug TM, Surlonger C, et al. Deregulated T cell activation and autoimmunity in mice lacking interleukin-2 receptor beta. *Science.* 1995;268(5216):1472-1476.
160. Rose-John S. Interleukin-6 family cytokines. *Cold Spring Harb Perspect Biol.* 2018;10:a028415.
161. Kang S, Narazaki M, Metwally H, et al. Historical overview of the interleukin-6 family cytokine. *J Exp Med.* 2020;217:e20190347.
162. Kishimoto T, Ishizaka K. Regulation of antibody response in vitro. X. Biphasic effect of cyclic AMP on the secondary anti-hapten antibody response to anti-immunoglobulin and enhancing soluble factor. *J Immunol.* 1976;116:534-541.
163. Hirano T, Yaukawa K, Harada H, et al. Complementary DNA for a novel human interleukin (BSF-2) that induces B lymphocytes to produce immunoglobulin. *Nature.* 1986;324(6092):73-76.
164. Chen Q, Wang WC, Bruce R, et al. Central role of IL-6 receptor signal-transducing chain gp130 in activation of L-selectin adhesion by fever-range thermal stress. *Immunity.* 2004;20(1):59-70.
165. Yamasaki K, Taga T, Hirata Y, et al. Cloning and expression of the human interleukin-6 (BSF-2/IFN beta 2) receptor. *Science.* 1988;241(4867):825-828.
166. Rose-John S, Neurath MF. IL-6 trans-signaling: the heat is on. *Immunity.* 2004;20(1):2-4.
167. Ward LD, Howlett GJ, Discolo G, et al. High affinity interleukin-6 receptor is a hexameric complex consisting of two molecules each of interleukin-6, interleukin-6 receptor, and gp-130. *J Biol Chem.* 1994;269(37):23286-23289.
168. Boulanger MJ, Chow DC, Brevnova EE, et al. Hexameric structure and assembly of the interleukin-6/IL-6 alpha-receptor/gp130 complex. *Science.* 2003;300:2101-2104.
169. Skiniotis G, Boulanger MJ, Garcia KC, Walz T. Signaling conformations of the tall cytokine receptor gp130 when in complex with IL-6 and IL-6 receptor. *Nat Struct Mol Biol.* 2005;12(6):545-551.
170. Du XX, Williams DA. Interleukin-11: a multifunctional growth factor derived from the hematopoietic microenvironment. *Blood.* 1994;83(8):2023-2030.
171. Putoczki T, Ernst M. More than a sidekick: the IL-6 family cytokine IL-11 links inflammation to cancer. *J Leukoc Biol.* 2010;88(6):1109-1117.
172. Zheng T, Zhu Z, Wang J, Homer RJ, Elias JA. IL-11: insights in asthma from overexpression transgenic modeling. *J Allergy Clin Immunol.* 2001;108(4):489-496.
173. Baumann H, Wang Y, Morella KK, et al. Complex of the soluble IL-11 receptor and IL-11 acts as IL-6-type cytokine in hepatic and nonhepatic cells. *J Immunol.* 1996;157(1):284-290.
174. Gearing DP, Gough NM, King JA, et al. Molecular cloning and expression of cDNA encoding a murine myeloid leukaemia inhibitory factor (LIF). *EMBO J.* 1987;6(13):3995-4002.
175. Nicola NA, Babon JJ. Leukemia inhibitory factor (LIF). *Cytokine Growth Factor Rev.* 2015;26(5):533-544.
176. Gearing DP, Thut CJ, VandeBos T, et al. Leukemia inhibitory factor receptor is structurally related to the IL-6 signal transducer, gp130. *EMBO J.* 1991;10(10):2839-2848.
177. Gearing DP, Corneau MR, Friend DJ, et al. The IL-6 signal transducer, gp130: an oncostatin M receptor and affinity converter for the LIF receptor. *Science.* 1992;255(5050):1434-1437.
178. Cao W, Yang Y, Wang Z, et al. Leukemia inhibitory factor inhibits T helper 17 cell differentiation and confers treatment effects of neural progenitor cell therapy in autoimmune disease. *Immunity.* 2011;35(2):273-284.
179. Metcalfe SM. LIF in the regulation of T-cell fate and as a potential therapeutic. *Gene Immun.* 2011;12(3):157-168.
180. Pera MF, Tam PP. Extrinsic regulation of pluripotent stem cells. *Nature.* 2010;465(7299):713-720.
181. Lin LF, Mismer D, Lile JD, et al. Purification, cloning, and expression of ciliary neurotrophic factor (CNTF). *Science.* 1989;246(4933):1023-1025.
182. Davis S, Aldrich TH, Stahl N, et al. LIFR beta and gp130 as heterodimerizing signal transducers of the tripartite CNTF receptor. *Science.* 1993;260(5115):1805-1808.
183. De Serio A, Graziani R, Laufer R, Ciliberto G, Paonessa G. In vitro binding of ciliary neurotrophic factor to its receptors: evidence for the formation of an IL-6-type hexameric complex. *J Mol Biol.* 1995;254(5):795-800.
184. Malik N, Kallestad JC, Gunderson NL, et al. Molecular cloning, sequence analysis, and functional expression of a novel growth regulator, oncostatin M. *Mol Cell Biol.* 1989;9(7):2847-2853.
185. Nair BC, DeVico AL, Nakamura S, et al. Identification of a major growth factor for AIDS-Kaposi's sarcoma cells as oncostatin M. *Science.* 1992;255(5050):1430-1432.
186. Miles SA, Martinez-Maza O, Rezai A, et al. Oncostatin M as a potent mitogen for AIDS-Kaposi's sarcoma-derived cells. *Science.* 1992;255(5050):1432-1434.
187. Mosley B, De Imus C, Friend D, et al. Dual oncostatin M (OSM) receptors. Cloning and characterization of an alternative signaling subunit conferring OSM-specific receptor activation. *J Biol Chem.* 1996;271(51):32635-32643.
188. Miyajima A, Kinshita T, Tanaka M, et al. Role of oncostatin M in hematopoiesis and liver development. *Cytokine Growth Factor Rev.* 2000;11(3):177-183.
189. Pennica D, King KL, Shaw KJ, et al. Expression cloning of cardiotrophin 1, a cytokine that induces cardiac myocyte hypertrophy. *Proc Natl Acad Sci U S A.* 1995;92(4):1142-1146.
190. Latchman DS. Cardiotrophin-1: a novel cytokine and its effects in the heart and other tissues. *Pharmacol Ther.* 2000;85(1):29-37.
191. Vlotides G, Zitzmann K, Stalla GK, Auernhammer CJ. Novel neurotrophin-1/B cell-stimulating factor-3 (NNT-1/BSF-3)/cardiotrophin-like cytokine (CLC) – a novel gp130 cytokine with pleiotropic functions. *Cytokine Growth Factor Rev.* 2004;15(5):325-336.
192. Elson GC, Lelievre E, Guillet C, et al. CLF associates with CLC to form a functional heteromeric ligand for the CNTF receptor complex. *Nat Neurosci.* 2000;3(9):867-872.
193. Trinchieri G, Pflanz S, Kastelein RA. The IL-12 family of heterodimeric cytokines: new players in the regulation of T cell responses. *Immunity.* 2003;19(5):641-644.
194. Pflanz S, Timans JC, Cheung J, et al. IL-27, a heterodimeric cytokine composed of EBI3 and p28 protein, induces proliferation of naive CD4(+) T cells. *Immunity.* 2002;16(6):779-790.
195. Stumhofer JS, Tait ED, Quinn WJ III, et al. A role for IL-27p28 as an antagonist of gp130-mediated signaling. *Nat Immunol.* 2010;11(12):1119-1126.
196. Hunter CA. New IL-12-family members: IL-23 and IL-27, cytokines with divergent functions. *Nat Rev Immunol.* 2005;5(7):521-531.
197. Collison LW, Delgoffe GM, Guy CS, et al. The composition and signaling of the IL-35 receptor are unconventional. *Nat Immunol.* 2012;13:290-299.
198. Vignali DAA, Kuchroo VK. IL-12 family cytokines: immunological playmakers. *Nat Immunol.* 2012;13(8):722-728.
199. Kastelein RA, Hunter CA, Cua DJ. Discovery and biology of IL-23 and IL-27: related but functionally distinct regulators of inflammation. *Annu Rev Immunol.* 2007;25:221-242.
200. Wang X, Wei Y, Xiao H, et al. A novel IL-23p19/Ebi3 (IL-39) cytokine mediates inflammation in Lupus-like mice. *Eur J Immunol.* 2016;46:1343-1350.
201. Boulanger MJ, Garcia KC. Shared cytokine signaling receptors: structural insights from the gp130 system. *Adv Protein Chem.* 2004;68:107-146.
202. La Cava A, Matarese G. The weight of leptin in immunity. *Nat Rev Immunol.* 2004;4(5):371-379.
203. Tartaglia LA, Dembski M, Weng X, et al. Identification and expression cloning of a leptin receptor, OB-R. *Cell.* 1995;83(7):1263-1271.
204. Chen H, Charlat O, Tartaglia LA, et al. Evidence that the diabetes gene encodes the leptin receptor: identification of a mutation in the leptin receptor gene in db/db mice. *Cell.* 1996;84(3):491-495.

205. Dillon SR, Sprecher C, Hammond A, et al. Interleukin 31, a cytokine produced by activated T cells, induces dermatitis in mice. *Nat Immunol.* 2004;5(7):752-760.
206. Zhang Q, Putheti P, Zhou Q, Liu Q, Gao W. Structures and biological functions of IL-31 and IL-31 receptors. *Cytokine Growth Factor Rev.* 2008;19(5-6):347-356.
207. Farrar MA, Schreiber RD. The molecular cell biology of interferon-gamma and its receptor. *Annu Rev Immunol.* 1993;11:571-611.
208. Taniguchi T, Takaoka A. The interferon-alpha/beta system in antiviral responses: a multimodal machinery of gene regulation by the IRF family of transcription factors. *Curr Opin Immunol.* 2002;14(1):111-116.
209. Vilcek J. Fifty years of interferon research: aiming at a moving target. *Immunity.* 2006;25(3):343-348.
210. van Boxel-Dezaire AH, Rani MR, Stark GR. Complex modulation of cell type-specific signaling in response to type I interferons. *Immunity.* 2006;25(3):361-372.
211. de Veer MJ, Holko M, Frevel M, et al. Functional classification of interferon-stimulated genes identified using microarrays. *J Leukoc Biol.* 2001;69(6):912-920.
212. Kalie E, Jaitin DA, Podoplelova Y, et al. The stability of the ternary interferon-receptor complex rather than the affinity to the individual subunits dictates differential biological activities. *J Biol Chem.* 2002;283:32925-32936.
213. Thomas C, Moraga I, Levin D, et al. Structural linkage between ligand discrimination and receptor activation by type I interferons. *Cell.* 2011;146:621-632.
214. Piehler J, Thomas C, Garcia JC, et al. Structural and dynamic determinants of type I interferon receptor assembly and their functional interpretation. *Immunol Rev.* 2012;250:317-334.
215. Bach EA, Aguet M, Schreiber RD. The IFN gamma receptor: a paradigm for cytokine receptor signaling. *Annu Rev Immunol.* 1997;15:563-591.
216. Bazan JF. Shared architecture of hormone binding domains in type I and II interferon receptors. *Cell.* 1990;61(5):753-754.
217. Mendoza JL, Escalante NK, Jude KM, et al. Structure of the IFNγ receptor complex guides design of biased agonists. *Nature.* 2019;567(7746):56-60.
218. ISICR-Newsletter. Nomenclature for interferon receptors and interferon-delta. *J Interferon Cytokine Res.* 1997;4:1.
219. Domanski P, Colamonici OR. The type-I interferon receptor. The long and short of it. *Cytokine Growth Factor Rev.* 1996;7(2):143-151.
220. Symons JA, Alcami A, Smith GL. Vaccinia virus encodes a soluble type I interferon receptor of novel structure and broad species specificity. *Cell.* 1995;81(4):551-560.
221. Spriggs MK. Poxvirus-encoded soluble cytokine receptors. *Virus Res.* 1994;33(1):1-10.
222. Kato H, Takahasi K, Fujita T. RIG-I-like receptors: cytoplasmic sensors for non-self RNA. *Immunol Rev.* 2011;243(1):91-98.
223. Gray PW, Leung DW, Pennica D, et al. Expression of human immune interferon cDNA in *E. coli* and monkey cells. *Nature.* 1982;295(5849):503-508.
224. Dunn GP, Koebel CM, Schreiber RD. Interferons, immunity and cancer immunoediting. *Nat Rev Immunol.* 2006;6(11):836-848.
225. Ouyang W, Rutz S, Crellin NK, Valdez PA, Hymowitz SG. Regulation and functions of the IL-10 family of cytokines in inflammation and disease. *Annu Rev Immunol.* 2011;29:71-109.
226. Pestka S, Krause CD, Sarkar D, et al. Interleukin-10 and related cytokines and receptors. *Annu Rev Immunol.* 2004;22:929-979.
227. Moore KW, de Waal Malefyt R, Coffman RL, O'Garra A. Interleukin-10 and the interleukin-10 receptor. *Annu Rev Immunol.* 2001;19:683-765.
228. Saraiva M, O'Garra A. The regulation of IL-10 production by immune cells. *Nat Rev Immunol.* 2010;10(3):170-181.
229. Rutz S, Ouyang W. Regulation of interleukin-10 and interleukin-22 expression in T helper cells. *Curr Opin Immunol.* 2011;23(5):605-612.
230. Chaudhry A, Samstein RM, Treuting P, et al. Interleukin-10 signaling in regulatory T cells is required for suppression of Th17 cell-mediated inflammation. *Immunity.* 2011;34(4):566-578.
231. Spolski R, Leonard WJ. IL-21 is an immune activator that also mediates suppression via IL-10. *Crit Rev Immunol.* 2009;30(6):559-570.
232. Kuhn R, Lohler J, Rennick D, Rajewsky K, Muller W. Interleukin-10-deficient mice develop chronic enterocolitis. *Cell.* 1993;75(2):263-274.
233. Liu Y, de Waal Malefyt R, Briere F, et al. The EBV IL-10 homologue is a selective agonist with impaired binding to the IL-10 receptor. *J Immunol.* 1997;158(2):604-613.
234. Ho AS, Liu Y, Khan TA, et al. A receptor for interleukin 10 is related to interferon receptors. *Proc Natl Acad Sci U S A.* 1993;90(23):11267-11271.
235. Liu Y, Wei SH, Ho AS, de Waal Malefyt R, Moore KW. Expression cloning and characterization of a human IL-10 receptor. *J Immunol.* 1994;152(4):1821-1829.
236. Lutfalla G, Gardiner K, Uze G. A new member of the cytokine receptor gene family maps on chromosome 21 at less than 35 kb from IFNAR. *Genomics.* 1993;16(2):366-373.
237. Gibbs VC, Pennica D. CRF2-4: isolation of cDNA clones encoding the human and mouse proteins. *Gene.* 1997;186(1):97-101.
238. Gallagher G, Dickensheets H, Eskdale J, et al. Cloning, expression and initial characterization of interleukin-19 (IL-19), a novel homologue of human interleukin-10 (IL-10). *Genes Immun.* 2000;1(7):442-450.
239. Blumberg H, Conklin D, Xu WF, et al. Interleukin 20: discovery, receptor identification, and role in epidermal function. *Cell.* 2001;104(1):9-19.
240. Wegenka UM. IL-20: biological functions mediated through two types of receptor complexes. *Cytokine Growth Factor Rev.* 2010;21(5):353-363.
241. Hsu YH, Chen WY, Chan CH, et al. Anti-IL-20 monoclonal antibody inhibits the differentiation of osteoclasts and protects against osteoporotic bone loss. *J Exp Med.* 2011;208(9):1849-1861.
242. Colonna M. Interleukin-22-producing natural killer cells and lymphoid tissue inducer-like cells in mucosal immunity. *Immunity.* 2009;31(1):15-23.
243. Ouyang W. Distinct roles of IL-22 in human psoriasis and inflammatory bowel disease. *Cytokine Growth Factor Rev.* 2010;21(6):435-441.
244. Vivier E, Spits H, Cupedo T. Interleukin-22-producing innate immune cells: new players in mucosal immunity and tissue repair? *Nat Rev Immunol.* 2009;9(4):229-234.
245. Sheppard P, Kindsvogel W, Xu W, et al. IL-28, IL-29 and their class II cytokine receptor IL-28R. *Nat Immunol.* 2003;4(1):63-68.
246. Trinchieri G. Interleukin-12: a proinflammatory cytokine with immunoregulatory functions that bridge innate resistance and antigen-specific adaptive immunity. *Annu Rev Immunol.* 1995;13:251-276.
247. Ozaki K, Leonard WJ. Cytokine and cytokine receptor pleiotropy and redundancy. *J Biol Chem.* 2002;277(33):29355-29358.
248. Ramalingam TR, Pesce JT, Sheikh F, et al. Unique functions of the type II interleukin 4 receptor identified in mice lacking the interleukin 13 receptor alpha1 chain. *Nat Immunol.* 2008;9(1):25-33.
249. Kurman C, Rubin L, Nelson D. *Article: Soluble Products of Immune Activation. Soluble Interleukin-2 Receptor (sIL-2R, Tac Protein).* American Society for Microbiology; 1992.
250. Fernandez-Botran R, Chilton PM, Ma Y. Soluble cytokine receptors: their roles in immunoregulation, disease, and therapy. *Adv Immunol.* 1996;63:269-336.
251. Kotenko SV, Izotova LS, Mirochnitchenko OV, et al. Identification, cloning, and characterization of a novel soluble receptor that binds IL-22 and neutralizes its activity. *J Immunol.* 2001;166(12):7096-7103.
252. Salvati AL, Lahm A, Paonessa G, et al. Interleukin-6 (IL-6) antagonism by soluble IL-6 receptor alpha mutated in the predicted gp130-binding interface. *J Biol Chem.* 1995;270(20):12242-12249.
253. Narazaki M, Yasukawa K, Saito T, et al. Soluble forms of the interleukin-6 signal transducing receptor component gp130 in human serum possessing a potential to inhibit signals through membrane anchored gp130. *Blood.* 1993;82:1120-1126.
254. Webster KE, Walters S, Kohler RE, et al. In vivo expansion of T reg cells with IL-2-mAb complexes: induction of resistance to EAE and long-term acceptance of islet allografts without immunosuppression. *J Exp Med.* 2009;206(4):751-760.
255. Darnell JE Jr, Kerr IM, Stark GR. Jak-STAT pathways and transcriptional activation in response to IFNs and other extracellular signaling proteins. *Science.* 1994;264(5164):1415-1421.
256. Lin JX, Migone TS, Tsang M, et al. The role of shared receptor motifs and common Stat proteins in the generation of cytokine pleiotropy and redundancy by IL-2, IL-4, IL-7, IL-13, and IL-15. *Immunity.* 1995;2(4):331-339.
257. Horvath CM, Darnell JE. The state of the STATs: recent developments in the study of signal transduction to the nucleus. *Curr Opin Cell Biol.* 1997;9(2):233-239.
258. Leonard WJ, O'Shea JJ. Jaks and STATs: biological implications. *Annu Rev Immunol.* 1998;16:293-322.
259. Schindler C, Plumlee C. Interferons pen the JAK-STAT pathway. *Semin Cell Dev Biol.* 2008;19(4):311-318.
260. Stark GR, Darnell JE Jr. The JAK-STAT pathway at twenty. *Immunity.* 2012;36:503-514.
261. Giordanetto F, Kroemer RT. Prediction of the structure of human Janus kinase 2 (JAK2) comprising JAK homology domains 1 through 7. *Protein Eng.* 2002;15(9):727-737.
262. Tanner JW, Chen W, Young RL, et al. The conserved box 1 motif of cytokine receptors is required for association with JAK kinases. *J Biol Chem.* 1995;270(12):6523-6530.
263. Zhao Y, Wagner F, Frank SJ, Kraft AS. The amino-terminal portion of the JAK2 protein kinase is necessary for binding and phosphorylation of the granulocyte-macrophage colony-stimulating factor receptor beta c chain. *J Biol Chem.* 1995;270(23):13814-13818.
264. Guschin D, Rogers N, Briscoe J, et al. A major role for the protein tyrosine kinase JAK1 in the JAK/STAT signal transduction pathway in response to interleukin-6. *EMBO J.* 1995;14(7):1421-1429.
265. Lupardus PJ, Skiniotis G, Rica AJ, et al. Structural snapshots of full-length Jak1, a transmembrane gp130/IL-6/IL-6Ralpha cytokine receptor complex, and the receptor-Jak1 holocomplex. *Structure.* 2011;19(1):45-55.
266. Gauzzi MC, Valezquez L, McKendry R, Mogensen KE, Fellous M, Pellegrini S. Interferon-alpha-dependent activation of Tyk2 requires phosphorylation of positive regulatory tyrosines by another kinase. *J Biol Chem.* 1996;271(34):20494-20500.
267. Feng J, Witthuhn BA, Matsuda T, Kohlhuber F, Kerr IM, Ihle JN. Activation of Jak2 catalytic activity requires phosphorylation of Y1007 in the kinase activation loop. *Mol Cell Biol.* 1997;17(5):2497-2501.
268. Briscoe J, Rogers NC, Witthuhn BA, et al. Kinase-negative mutants of JAK1 can sustain interferon-gamma-inducible gene expression but not an antiviral state. *EMBO J.* 1996;15(4):799-809.
269. Luo H, Rose P, Barber D, et al. Mutation in the Jak kinase JH2 domain hyperactivates *Drosophila* and mammalian Jak-Stat pathways. *Mol Cell Biol.* 1997;17(3):1562-1571.
270. Conway G, Margoliath A, Wong-Madden S, et al. Jak1 kinase is required for cell migrations and anterior specification in zebrafish embryos. *Proc Natl Acad Sci U S A.* 1997;94(7):3082-3087.
271. Binari R, Perrimon N. Stripe-specific regulation of pair-rule genes by hopscotch, a putative Jak family tyrosine kinase in *Drosophila. Genes Dev.* 1994;8(3):300-312.
272. Velazquez L, Fellous M, Stark GR, Pellegrini S. A protein tyrosine kinase in the interferon alpha/beta signaling pathway. *Cell.* 1992;70(2):313-322.
273. Frank SJ, Yi W, Zhao Y, et al. Regions of the JAK2 tyrosine kinase required for coupling to the growth hormone receptor. *J Biol Chem.* 1995;270(24):14776-14785.
274. Zhuang H, Patel SV, He TC, et al. Inhibition of erythropoietin-induced mitogenesis by a kinase-deficient form of Jak2. *J Biol Chem.* 1994;269(34):21411-21414.
275. Kawahara A, Minami Y, Miyazaki T, Ihle JN, Taniguchi T. Critical role of the interleukin 2 (IL-2) receptor gamma-chain-associated Jak3 in the IL-2-induced c-fos and c-myc, but not bcl-2, gene induction. *Proc Natl Acad Sci U S A.* 1995;92(19):8724-8728.
276. Russell SM, Tavebi N, Nakajima H, et al. Mutation of Jak3 in a patient with SCID: essential role of Jak3 in lymphoid development. *Science.* 1995;270(5237):797-800.
277. Thomis DC, Gurniak CB, Tivol E, Sharpe AH, Berg LJ. Defects in B lymphocyte maturation and T lymphocyte activation in mice lacking Jak3. *Science.* 1995;270(5237):794-797.
278. Levine RL, Wadleigh M, Cools J, et al. Activating mutation in the tyrosine kinase JAK2 in polycythemia vera, essential thrombocythemia, and myeloid metaplasia with myelofibrosis. *Cancer Cell.* 2005;7(4):387-397.
279. Meydan N, Grunberger T, Dadi H, et al. Inhibition of acute lymphoblastic leukaemia by a Jak-2 inhibitor. *Nature.* 1996;379(6566):645-648.
280. Minegishi Y, Saito M, Morio T, et al. Human tyrosine kinase 2 deficiency reveals its requisite roles in multiple cytokine signals involved in innate and acquired immunity. *Immunity.* 2006;25(5):745-755.
281. Harrison DA, Binari R, Nahreini TS, Gilman M, Perrimon N. Activation of a *Drosophila* Janus kinase (JAK) causes hematopoietic neoplasia and developmental defects. *EMBO J.* 1995;14(12):2857-2865.
282. Luo H, Hanratty WP, Dearolf CR. An amino acid substitution in the *Drosophila* hopTum-l Jak kinase causes leukemia-like hematopoietic defects. *EMBO J.* 1995;14(7):1412-1420.
283. Danial NN, Pernis A, Rothman PB. Jak-STAT signaling induced by the v-abl oncogene. *Science.* 1995;269(5232):1875-1877.
284. Migone TS, Lin JX, Cereseto A, et al. Constitutively activated Jak-STAT pathway in T cells transformed with HTLV-I. *Science.* 1995;269(5220):79-81.
285. Russell SM, Johnston JA, Noguchi M, et al. Interaction of IL-2R beta and gamma c chains with Jak1 and Jak3: implications for XSCID and XCID. *Science.* 1994;266(5187):1042-1045.
286. Jabara HH, Buckley RH, Roberts JL, et al. Role of JAK3 in CD40-mediated signaling. *Blood.* 1998;92(7):2435-2440.
287. Levine RL, Gilliland DG. JAK-2 mutations and their relevance to myeloproliferative disease. *Curr Opin Hematol.* 2007;14(1):43-47.
288. Lacronique V, Boureux A, Valle VD, et al. A TEL-JAK2 fusion protein with constitutive kinase activity in human leukemia. *Science.* 1997;278(5341):1309-1312.
289. Li Z, Rotival M, Patin E, et al. Two common disease-associated TYK2 variants impact exon splicing and TYK2 dosage. *PLoS One.* 2020;5(1):e0225289.
290. Seto Y, Nakajima H, Suto A, et al. Enhanced Th2 cell-mediated allergic inflammation in Tyk2-deficient mice. *J Immunol.* 2003;170(2):1077-1083.
291. Maritano D, Sugrue ML, Tininini S, et al. The STAT3 isoforms alpha and beta have unique and specific functions. *Nat Immunol.* 2004;5(4):401-409.
292. Liu X, Robinson GW, Wagner KU, et al. Stat5a is mandatory for adult mammary gland development and lactogenesis. *Genes Dev.* 1997;11(2):179-186.
293. Udy GB, Towers RP, Snell RG, et al. Requirement of STAT5b for sexual dimorphism of body growth rates and liver gene expression. *Proc Natl Acad Sci U S A.* 1997;94(14):7239-7244.
294. Teglund S, McKay C, Schuetz E, et al. Stat5a and Stat5b proteins have essential and nonessential, or redundant, roles in cytokine responses. *Cell.* 1998;93(5):841-850.
295. Levy DE, Darnell JE Jr. Stats: transcriptional control and biological impact. *Nat Rev Mol Cell Biol.* 2002;3:651-662.
296. Delgoffe GM, Vignali DA. STAT heterodimers in immunity: a mixed message or a unique signal? *JAKSTAT.* 2013;2(1):e23060.

297. Xu X, Sun YL, Hoey T. Cooperative DNA binding and sequence-selective recognition conferred by the STAT amino-termial domain. *Science*. 1996;273:794-797.
298. Vinkemeier U, Cohen SL, Moarefi I, et al. DNA binding of an in vitro activated STAT1α, STAT1β and truncated Stat1: interaction between NH2-terminal domains stabilizes binding of two dimers to tandem DNA sites. *EMBO J*. 1996;15:5616-5626.
299. Soldaini E, John S, Moro S, et al. DNA binding site selection of dimeric and tetrameric Stat5 proteins reveals a large repertoire of divergent tetrameric Stat5a binding sites. *Mol Cell Biol*. 2000;20:389-401.
300. Ota N, Brett TJ, Murphy TL, et al. N-domain-dependent nonphosphorylated STAT4 dimers required for cytokine-driven activation. *Nat Immunol*. 2004;5:208-215.
301. Mao X, Ren Z, Parker GN, et al. Structural bases of unphosphorylated STAT1 association and receptor binding. *Mol Cell*. 2005;17:761-771.
302. Begitt A, Droescher M, Meyer T, et al. STAT1-cooperative DNA-binding distinguishes type 1 from type 2 interferon signaling. *Nat Immunol*. 2014;15:168-176.
303. Lin JX, Li P, Liu D, et al. Critical role of STAT5 transcription factor tetramerization for cytoine responses and normal immune function. *Immunity*. 2012;36:586-599.
304. Lin JX, Du N, Li P, et al. Critical functions for STAT5 tetramers in the maturation and survival of natural killer cells. *Nat Commun*. 2017;8(1):1320.
305. Wang D, Stravopodis D, Teglund S, et al. Naturally occurring dominant negative variants of Stat5. *Mol Cell Biol*. 1996;16(11):6141-6148.
306. Yan R, Small S, Desplan C, Dearolf CR, Darnell JE Jr. Identification of a Stat gene that functions in *Drosophila* development. *Cell*. 1996;84(3):421-430.
307. Hou XS, Melnick MB, Perrimon N. Marelle acts downstream of the *Drosophila* HOP/JAK kinase and encodes a protein similar to the mammalian STATs. *Cell*. 1996;84(3):411-419.
308. Kawata T, Shevchenko A, Fukuzawa M, et al. SH2 signaling in a lower eukaryote: a STAT protein that regulates stalk cell differentiation in dictyostelium. *Cell*. 1997;89(6):909-916.
309. Stahl N, Farruggella TK, Boulton TG, et al. Choice of STATs and other substrates specified by modular tyrosine-based motifs in cytokine receptors. *Science*. 1995;267(5202):1349-1353.
310. Fujitani Y, Hibi M, Fukada T, et al. An alternative pathway for STAT activation that is mediated by the direct interaction between JAK and STAT. *Oncogene*. 1997;14(7):751-761.
311. Reich NC. STAT dynamics. *Cytokine Growth Factor Rev*. 2007;18(5-6):511-518.
312. Chen X, Veinkemeier U, Zhao Y, et al. Crystal structure of a tyrosine phosphorylated STAT-1 dimer bound to DNA. *Cell*. 1998;93(5):827-839.
313. Becker S, Groner B, Muller CW. Three-dimensional structure of the Stat3beta homodimer bound to DNA. *Nature*. 1998;394(6689):145-151.
314. Schindler U, Wu P, Rothe M, et al. Components of a Stat recognition code: evidence for two layers of molecular selectivity. *Immunity*. 1995;2(6):689-697.
315. Wen Z, Zhong Z, Darnell JE Jr. Maximal activation of transcription by Stat1 and Stat3 requires both tyrosine and serine phosphorylation. *Cell*. 1995;82(2):241-250.
316. Zhang X, Blenis J, Li HC, et al. Requirement of serine phosphorylation for formation of STAT-promoter complexes. *Science*. 1995;267(5206):1990-1994.
317. Horvath CM, Darnell JE Jr. The antiviral state induced by alpha interferon and gamma interferon requires transcriptionally active Stat1 protein. *J Virol*. 1996;70(1):647-650.
318. David M, Petricoin E III, Benjamin C, et al. Requirement for MAP kinase (ERK2) activity in interferon alpha-and interferon beta-stimulated gene expression through STAT proteins. *Science*. 1995;269(5231):1721-1723.
319. Zhang JJ, Zhao Y, Chait BT, et al. Ser727-dependent recruitment of MCM5 by Stat1alpha in IFN-gamma-induced transcriptional activation. *EMBO J*. 1998;17(23):6963-6971.
320. Xue HH, Fink DW Jr, Zhang X, et al. Serine phosphorylation of Stat5 proteins in lymphocytes stimulated with IL-2. *Int Immunol*. 2002;14(11):1263-1271.
321. Friedbichler K, Hoelbl A, Li G, et al. Serine phosphorylation of the Stat5a C-terminus is a driving force for transformation. *Front Biosci*. 2012;17:3043-3056.
322. Look DC, Pelletier MR, Tidwell RM, et al. Stat1 depends on transcriptional synergy with Sp1. *J Biol Chem*. 1995;270(51):30264-30267.
323. Wesemann DR, Qin H, Kokorina N, Benveniste EN. TRADD interacts with STAT1-alpha and influences interferon-gamma signaling. *Nat Immunol*. 2004;5(2):199-207.
324. Schaefer TS, Sanders LK, Nathans D. Cooperative transcriptional activity of Jun and Stat3 beta, a short form of Stat3. *Proc Natl Acad Sci U S A*. 1995;92(20):9097-9101.
325. Zhang X, Wrzeszczynska MH, Horvath CM, Darnell JE Jr. Interacting regions in Stat3 and c-Jun that participate in cooperative transcriptional activation. *Mol Cell Biol*. 1999;19(10):7138-7146.
326. Zhang JJ, Vinkemeier U, Gu W, et al. Two contact regions between Stat1 and CBP/p300 in interferon gamma signaling. *Proc Natl Acad Sci U S A*. 1996;93(26):15092-15096.
327. Stocklin E, Wissler M, Gouilleux F, Groner B. Functional interactions between Stat5 and the glucocorticoid receptor. *Nature*. 1996;383(6602):726-728.
328. John S, Robbins CM, Leonard WJ. An IL-2 response element in the human IL-2 receptor alpha chain promoter is a composite element that binds Stat5, Elf-1, HMG-I(Y) and a GATA family protein. *EMBO J*. 1996;15(20):5627-5635.
329. Nakajima H, Brindle PK, Handa M, Ihle JN. Functional interaction of STAT5 and nuclear receptor co-repressor SMRT: implications in negative regulation of STAT5-dependent transcription. *EMBO J*. 2001;20(23):6836-6844.
330. Ramana CV, Grammatikakis N, Chernov M, et al. Regulation of c-myc expression by IFN-gamma through Stat1-dependent and independent pathways. *EMBO J*. 2000;19(2):263-272.
331. Hnisz D, Abraham BJ, Lee TI, et al. Super-enhancers in the control of cell identity and disease. *Cell*. 2013;155:934-947.
332. Vahedi G, Kanno Y, Furumoto Y, et al. Super-enhancers delineate disease-associated regulatory nodes in T cells. *Nature*. 2015;520:558-562.
333. Li P, Mitra S, Spolski R, et al. STAT5-mediated chromatin interactions in superenhancers activate IL-2 highly inducible genes: functional dissection of the Il2ra gene locus. *Proc Natl Acad Sci U S A*. 2017;114:12111-12119.
334. O'Shea JJ, Holland SM, Staudt LM. JAKs and STATs in immunity, immunodeficiency, and cancer. *N Engl J Med*. 2013;368(2):161-170.
335. Bharadwaj U, Kasembeli MM, Robinson P, Tweardy DJ. Targeting Janus kinases and signal transducer and activator of transcription 3 to treat inflammation, fibrosis, and cancer: rationale, progress, and caution. *Pharmacol Rev*. 2020;72(2):486-526.
336. Kaplan MH, Sun YL, Hoey T, Grusby MJ. Impaired IL-12 responses and enhanced development of Th2 cells in Stat4-deficient mice. *Nature*. 1996;382(6587):174-177.
337. Thierfelder WE, van Deursen JM, Yamamoto K, et al. Requirement for Stat4 in interleukin-12-mediated responses of natural killer and T cells. *Nature*. 1996;382(6587):171-174.
338. Takeda K, Tanaka T, Shi W, et al. Essential role of Stat6 in IL-4 signalling. *Nature*. 1996;380(6575):627-630.
339. Kaplan MH, Schindler U, Smiley ST, Grusby MJ. Stat6 is required for mediating responses to IL-4 and for development of Th2 cells. *Immunity*. 1996;4(3):313-319.
340. Nakajima H, Liu XW, Wynshaw-Boris A, et al. An indirect effect of Stat5a in IL-2-induced proliferation: a critical role for Stat5a in IL-2-mediated IL-2 receptor alpha chain induction. *Immunity*. 1997;7(5):691-701.
341. Imada K, Bloom ET, Nakajima H, et al. Stat5b is essential for natural killer cell-mediated proliferation and cytolytic activity. *J Exp Med*. 1998;188(11):2067-2074.
342. Yao Z, Cui Y, Watford WT, et al. Stat5a/b are essential for normal lymphoid development and differentiation. *Proc Natl Acad Sci U S A*. 2006;103(4):1000-1005.
343. Chin YE, Kitagawa M, Su WC, et al. Cell growth arrest and induction of cyclin-dependent kinase inhibitor p21 WAF1/CIP1 mediated by STAT1. *Science*. 1996;272(5262):719-722.
344. Su WC, Kitagawa M, Xue N, et al. Activation of Stat1 by mutant fibroblast growth-factor receptor in thanatophoric dysplasia type II dwarfism. *Nature*. 1997;386(6622):288-292.
345. Durbin JE, Hackenmiller R, Simon MC, Levy DE. Targeted disruption of the mouse Stat1 gene results in compromised innate immunity to viral disease. *Cell*. 1996;84(3):443-450.
346. Meraz MA, White JM, Sheehan KC, et al. Targeted disruption of the Stat1 gene in mice reveals unexpected physiologic specificity in the JAK-STAT signaling pathway. *Cell*. 1996;84(3):431-442.
347. Park C, Li S, Cha E, Schindler C. Immune response in Stat2 knockout mice. *Immunity*. 2000;13(6):795-804.
348. Takeda K, Noguchi K, Shi W, et al. Targeted disruption of the mouse Stat3 gene leads to early embryonic lethality. *Proc Natl Acad Sci U S A*. 1997;94(8):3801-3804.
349. Akira S. Roles of STAT3 defined by tissue-specific gene targeting. *Oncogene*. 2000;19(21):2607-2611.
350. Laouar Y, Welte T, Fu XY, Flavell RA. STAT3 is required for Flt3L-dependent dendritic cell differentiation. *Immunity*. 2003;19(6):903-912.
351. Cheng F, Wang HW, Cuenca A, et al. A critical role for Stat3 signaling in immune tolerance. *Immunity*. 2003;19(3):425-436.
352. Socolovsky M, Fallon AE, Wang S, et al. Fetal anemia and apoptosis of red cell progenitors in Stat5a$^{-/-}$5b$^{-/-}$ mice: a direct role for Stat5 in Bcl-X(L) induction. *Cell*. 1999;98(2):181-191.
353. Golnaz V, Takahashi H, Nakayamada S, et al. STATs shape the active enhancer landscape of T cell populations. *Cell*. 2012;151(5):981-993.
354. Platanitis E, Decker T. Regulatory networks involving STATs, IRFs, and NFκB in inflammation. *Front Immunol*. 2018;9:2542.
355. Cano-Gamez E, Soskic B, Roumeliotis TI, et al. Single-cell transcriptomics identifies an effectorness gradient shaping the response of CD4+ T cells to cytokines. *Nat Commun*. 2020;11:1801.
356. Friedmann MC, Migone TS, Russell SM, Leonard WJ. Different interleukin 2 receptor beta-chain tyrosines couple to at least two signaling pathways and synergistically mediate interleukin 2-induced proliferation. *Proc Natl Acad Sci U S A*. 1996;93(5):2077-2082.
357. Takeshita T, Arita T, Higuchi M, et al. STAM, signal transducing adaptor molecule, is associated with Janus kinases and involved in signaling for cell growth and c-myc induction. *Immunity*. 1997;6(4):449-457.
358. Migone TS, Rodig S, Cacalano NA, et al. Functional cooperation of the interleukin-2 receptor beta chain and Jak1 in phosphatidylinositol 3-kinase recruitment and phosphorylation. *Mol Cell Biol*. 1998;18(11):6416-6422.
359. Kumar A, Commane M, Flickinger TW, et al. Defective TNF-alpha-induced apoptosis in STAT1-null cells due to low constitutive levels of caspases. *Science*. 1997;278:1630-1632.
360. Chatterjee-Kishore M, Wright KL, Ting JP, Stark GR. How Stat1 mediates constitutive gene expression: a complex of unphosphorylated Stat1 and IRF1 supports transcription of the LMP2 gene. *EMBO J*. 2000;19(15):4111-4122.
361. Meyer T, Begitt A, Lodige I, et al. Constitutive and IFN-gamma-induced nuclear import of STAT1 proceed through independent pathways. *EMBO J*. 2002;21(3):344-354.
362. Yang J, Chatterjee-Kishore M, Staugaitis SM, et al. Novel roles of unphosphorylated STAT3 in oncogenesis and transcriptional regulation. *Cancer Res*. 2005;65(3):939-947.
363. Cheon H, Stark GR. Unphosphorylated STAT1 prolongs the expression of interferon-induced immune regulatory genes. *Proc Natl Acad Sci U S A*. 2009;106:9373-9378.
364. Cheon H, Holvey-Bates EG, Schoggins JW, et al. IFNb-dependent increases in STAT1, STAT2, and IRF9 mediate resistance to viruses and DNA damage. *EMBO J*. 2013;32:2751-2763.
365. Park HJ, Hannah R, Biddie S, et al. Cytokine-induced megakaryocytic differentiation is regulated by genome-wide loss of an U-STAT transcriptional program. *EMBO J*. 2016;35:580-594.
366. Silver-Morse L, Li WX. JAK-STAT in heterochromatin and genome stability. *Landes Biosci*. 2013;2(3):e26090.
367. Hu X, Dutta P, Tsurumi A, et al. Unphosphorylated STAT5A stabilizes heterochromatin and suppresses tumor growth. *Proc Natl Acad Sci U S A*. 2013;110:10213-10218.
368. Meier JA, Larner AC. Toward a new STATe: the role of STATs in mitochondrial function. *Semin Immunol*. 2014;26:20-28.
369. Wegrzyn J, Potla R, Chwae YJ, et al. Function of mitochondrial Stat3 in cellular respiration. *Science*. 2009;323:793-797.
370. Lufei C, Ma J, Huang G, et al. GRIM-19, a death-regulatory gene product, suppresses Stat3 activity via functional interaction. *EMBO J*. 2003;22:1325-1335.
371. Zhang J, Yang J, Roy SK, et al. The cell death regulator GRIM-19 is an inhibitor of signal transducer and activator of transcription 3. *Proc Natl Acad Sci U S A*. 2003;100:9342-9347.
372. Tammineni P, Anugula C, Mohammed F, et al. The import of the transcription factor STAT3 into mitochondria depends on GRIM-19, a component of the electron transport chain. *J Biol Chem*. 2013;288:4723-4732.
373. Gough DJ, Corlett A, Schlessinger K, et al. Mitochondrial STAT3 supports Ras-dependent oncogenic transformation. *Science*. 2009;324:1713-1716.
374. Chen M, Cheng A, Chen YQ, et al. The amino terminus of JAK3 is necessary and sufficient for binding to the common gamma chain and confers the ability to transmit interleukin 2-mediated signals. *Proc Natl Acad Sci U S A*. 1997;94(13):6910-6915.
375. Cacalano NA, Migone TS, Bazan F, et al. Autosomal SCID caused by a point mutation in the N-terminus of Jak3: mapping of the Jak3-receptor interaction domain. *EMBO J*. 1999;18(6):1549-1558.
376. Greenlund AC, Morales MO, Viviano BL, et al. Stat recruitment by tyrosine-phosphorylated cytokine receptors: an ordered reversible affinity-driven process. *Immunity*. 1995;2(6):689-697. *Immunity*. 1995;2(6):677-687.
377. Ryan JJ, McReynolds LJ, Keegan A, et al. Growth and gene expression are predominantly controlled by distinct regions of the human IL-4 receptor. *Immunity*. 1996;4(2):123-132.
378. Hemmann U, Gerhartz C, Heesel B, et al. Differential activation of acute phase response factor/Stat3 and Stat1 via the cytoplasmic domain of the interleukin 6 signal transducer gp130. II. Src homology SH2 domains define the specificity of Stat factor Activation. *J Biol Chem*. 1996;271(22):12999-13007.
379. Wang HY, Paul WE, Keegan AD. IL-4 function can Be transferred to the IL-2 receptor by tyrosine containing sequences found in the IL-4 receptor alpha chain. *Immunity*. 1996;4(2):113-121.
380. Kuhn R, Rajewsky K, Muller W. Generation and analysis of interleukin-4 deficient mice. *Science*. 1991;254(5032):707-710.
381. Sadlack B, Kuhn R, Schorle H, et al. Development and proliferation of lymphocytes in mice deficient for both interleukins-2 and 4. *Eur J Immunol*. 1994;24(1):281-284.
382. von Freeden-Jeffry U, Vieiea P, Lucian LA, et al. Lymphopenia in interleukin (IL)-7 gene-deleted mice identifies IL-7 as a nonredundant cytokine. *J Exp Med*. 1995;181(4):1519-1526.
383. Kennedy MK, Glaccum M, Brown SN, et al. Reversible defects in natural killer and memory CD8 T cell lineages in interleukin 15-deficient mice. *J Exp Med*. 2000;191(5):771-780.

384. Noben-Trauth N, Shultz LD, Brombacher F, et al. An interleukin 4 (IL-4)-independent pathway for CD4+ T cell IL-4 production is revealed in IL-4 receptor-deficient mice. *Proc Natl Acad Sci U S A*. 1997;94(20):10838-10843.
385. Peschon JJ, Morrissey PJ, Grabstein KH, et al. Early lymphocyte expansion is severely impaired in interleukin 7 receptor-deficient mice. *J Exp Med*. 1994;180(5):1955-1960.
386. Lodolce JP, Boone DL, Chai S, et al. IL-15 receptor maintains lymphoid homeostasis by supporting lymphocyte homing and proliferation. *Immunity*. 1998;9(5):669-676.
387. Brandt EB, Munitz A, Orekov T, et al. Targeting IL-4/IL-13 signaling to alleviate oral allergen-induced diarrhea. *J Allergy Clin Immunol*. 2009;123(1):53-58.
388. Dranoff G, Crawford AD, Sadelain M, et al. Involvement of granulocyte-macrophage colony-stimulating factor in pulmonary homeostasis. *Science*. 1994;264(5159):713-716.
389. Nishinakamura R, Nakayama N, Hirabayashi Y, et al. Mice deficient for the IL-3/GM-CSF/IL-5 beta c receptor exhibit lung pathology and impaired immune response, while beta IL3 receptor-deficient mice are normal. *Immunity*. 1995;2(3):211-222.
390. Kopf M, Baumann H, Freer G, et al. Impaired immune and acute-phase responses in interleukin-6-deficient mice. *Nature*. 1994;368(6469):339-342.
391. Yoshida K, Taga T, Saito M, et al. Targeted disruption of gp130, a common signal transducer for the interleukin 6 family of cytokines, leads to myocardial and hematological disorders. *Proc Natl Acad Sci U S A*. 1996;93(1):407-411.
392. Stewart CL, Kaspar P, Brunet LJ, et al. Blastocyst implantation depends on maternal expression of leukaemia inhibitory factor. *Nature*. 1992;359(6390):76-79.
393. Rao MS, Sun Y, Escary JL, et al. Leukemia inhibitory factor mediates an injury response but not a target-directed developmental transmitter switch in sympathetic neurons. *Neuron*. 1993;11(6):1175-1185.
394. Masu Y, Wolf E, Holtmann B, et al. Disruption of the CNTF gene results in motor neuron degeneration. *Nature*. 1993;365(6441):27-32.
395. DeChiara TM, Vejsada R, Poueymirou WT, et al. Mice lacking the CNTF receptor, unlike mice lacking CNTF, exhibit profound motor neuron deficits at birth. *Cell*. 1995;83(2):313-322.
396. Robb L, Li R, Hartley L, Nandurkar HH, Koentgen F, Begley CG. Infertility in female mice lacking the receptor for interleukin 11 is due to a defective uterine response to implantation. *Nat Med*. 1998;4(3):303-308.
397. Magram J, Connaughton SE, Warrier RR, et al. IL-12-deficient mice are defective in IFN gamma production and type 1 cytokine responses. *Immunity*. 1996;4(5):471-481.
398. Wu C, Ferrante J, Gately MK, Magram J. Characterization of IL-12 receptor beta1 chain (IL-12Rbeta1)-deficient mice: IL-12Rbeta1 is an essential component of the functional mouse IL-12 receptor. *J Immunol*. 1997;159(4):1658-1665.
399. Gran B, Zhang GX, Yu S, et al. IL-12p35-deficient mice are susceptible to experimental autoimmune encephalomyelitis: evidence for redundancy in the IL-12 system in the induction of central nervous system autoimmune demyelination. *J Immunol*. 2002;169(12):7104-7110.
400. Zhang GX, Gran B, Yu S, et al. Induction of experimental autoimmune encephalomyelitis in IL-12 receptor-beta 2-deficient mice: IL-12 responsiveness is not required in the pathogenesis of inflammatory demyelination in the central nervous system. *J Immunol*. 2003;170(4):2153-2160.
401. Ghilardi N, Kljavin N, Chen Q, et al. Compromised humoral and delayed-type hypersensitivity responses in IL-23-deficient mice. *J Immunol*. 2004;172(5):2827-2833.
402. Wirtz S, Billmeier U, McHedlidze T, Blumberg RS, Neurath MF. Interleukin-35 mediates mucosal immune responses that protect against T-cell-dependent colitis. *Gastroenterology*. 2011;141(5):1875-1886.
403. Wu H, Liu X, Jaenisch R, Lodish HF. Generation of committed erythroid BFU-E and CFU-E progenitors does not require erythropoietin or the erythropoietin receptor. *Cell*. 1995;83(1):59-67.
404. Lin CS, Lim SK, D'Agati V, Costantini F. Differential effects of an erythropoietin receptor gene disruption on primitive and definitive erythropoiesis. *Genes Dev*. 1996;10(2):154-164.
405. Gurney AL, Carver-Moore K, de Sauvage FJ, Moore MW. Thrombocytopenia in c-mpl-deficient mice. *Science*. 1994;265(5177):1445-1447.
406. Lieschke GJ, Grail D, Hodgson G, et al. Mice lacking granulocyte colony-stimulating factor have chronic neutropenia, granulocyte and macrophage progenitor cell deficiency, and impaired neutrophil mobilization. *Blood*. 1994;84(6):1737-1746.
407. Metcalf D. The granulocyte-macrophage regulators: reappraisal by gene inactivation. *Exp Hematol*. 1995;23(7):569-572.
408. van den Broek MF, Muller U, Huang S, Zinkernagel RM, Aguet M. Immune defence in mice lacking type I and/or type II interferon receptors. *Immunol Rev*. 1995;148:5-18.
409. Dalton DK, Pitts-Meek S, Keshav S, Figari IS, Bradley A, Stewart TA. Multiple defects of immune cell function in mice with disrupted interferon-gamma genes. *Science*. 1993;259(5102):1739-1742.
410. Huang S, Hendriks W, Althage A, et al. Immune response in mice that lack the interferon-gamma receptor. *Science*. 1993;259(5102):1742-1745.
411. Azuma YT, Matsuo Y, Kuwamura M, et al. Interleukin-19 protects mice from innate-mediated colonic inflammation. *Inflamm Bowel Dis*. 2010;16(6):1017-1028.
412. Wahl C, Muller W, Leithauser F, et al. IL-20 receptor 2 signaling down-regulates antigen-specific T cell responses. *J Immunol*. 2009;182(2):802-810.
413. Zheng Y, Valdez PA, Danilenko DM, et al. Interleukin-22 mediates early host defense against attaching and effacing bacterial pathogens. *Nat Med*. 2008;14(3):282-289.
414. Ank N, Iversen MB, Bartholdy C, et al. An important role for type III interferon (IFN-lambda/IL-28) in TLR-induced antiviral activity. *J Immunol*. 2008;180(4):2474-2485.
415. Rodig SJ, Meraz MA, White JM, et al. Disruption of the Jak1 gene demonstrates obligatory and nonredundant roles of the Jaks in cytokine-induced biologic responses. *Cell*. 1998;93(3):373-383.
416. Parganas E, Wang D, Stravopodis D, et al. Jak2 is essential for signaling through a variety of cytokine receptors. *Cell*. 1998;93(3):385-395.
417. Karaghiosoff M, Meubauer H, Lassnig C, et al. Partial impairment of cytokine responses in Tyk2-deficient mice. *Immunity*. 2000;13(4):5 49-560.
418. Shimoda K, Kato K, Aoki K, et al. Tyk2 plays a restricted role in IFN alpha signaling, although it is required for IL-12-mediated T cell function. *Immunity*. 2000;13(4):561-571.
419. Savitsky D, Tamura T, Yanai H, Taniguchi T. Regulation of immunity and oncogenesis by the IRF transcription factor family. *Cancer Immunol Immunother*. 2010;59(4):489-510.
420. Negishi H, Taniguchi T, Yanai H. The interferon (IFN) class of cytokines and the IFN regulatory factor (IRF) transcription factor family. *Cold Spring Harb Perspect Biol*. 2018;10(11):a028423.
421. Elser B, Lohoff M, Kock S, et al. IFN-gamma represses IL-4 expression via IRF-1 and IRF-2. *Immunity*. 2002;17(6):703-712.
422. Kwon H, Thierry-Mieg D, Thierry-Mieg J, et al. Analysis of interleukin-21-induced Prdm1 gene regulation reveals functional cooperation of STAT3 and IRF4 transcription factors. *Immunity*. 2009;31:941-952.
423. Li P, Spolski R, Liao W, et al. BATF-JUN is critical for IRF4-mediated transcription in T cells. *Nature*. 2012;490(7421):543-546.
424. Tussiwand R, Lee WL, Murphy TL, et al. Compensatory dendritic cell development mediated by BATF-IRF interactions. *Nature*. 2012;490(7421):502-507.
425. Glasmacher E, Agrawal S, Chang AB, et al. A genomic regulatory element that directs assembly and function of immune-specific AP-1-IRF complexes. *Science*. 2012;338(6109):975-980.
426. Ciofani M, Madar A, Galan C, et al. A validated regulatory network for Th17 cell specification. *Cell*. 2012;151(2):289-303.
427. Copps KD, White MF. Regulation of insulin sensitivity by serine/threonine phosphorylation of insulin receptor substrate proteins IRS1 and IRS2. *Diabetologia*. 2012;55(10):2565-2582.
428. Wang LM, Myers MG Jr, Sun XJ, et al. IRS-1: essential for insulin-and IL-4-stimulated mitogenesis in hematopoietic cells. *Science*. 1993;261(5128):1591-1594.
429. Keegan AD, Nelms K, White M, et al. An IL-4 receptor region containing an insulin receptor motif is important for IL-4-mediated IRS-1 phosphorylation and cell growth. *Cell*. 1994;76(5):811-820.
430. Heller NM, Qi X, Junttila IS, et al. Type I IL-4Rs selectively activate IRS-2 to induce target gene expression in macrophages. *Sci Signal*. 2008;1(51):ra17.
431. Dasgupta P, Dorsey NJ, Li J, et al. The adaptor protein insulin receptor substrate 2 inhibits alternative macrophage activation and allergic lung inflammation. *Sci Signal*. 2016;9(433):ra63.
432. Ridderstrale M, Degerman E, Tornqvist H. Growth hormone stimulates the tyrosine phosphorylation of the insulin receptor substrate-1 and its association with phosphatidylinositol 3-kinase in primary adipocytes. *J Biol Chem*. 1995;270(8):3471-3474.
433. Argetsinger LS, Norstedt G, Billestrup N, et al. Growth hormone, interferon-gamma, and leukemia inhibitory factor utilize insulin receptor substrate-2 in intracellular signaling. *J Biol Chem*. 1996;271(46):29415-29421.
434. Johnston JA, Wang LM, Hanson EP, et al. Interleukins 2, 4, 7, and 15 stimulate tyrosine phosphorylation of insulin receptor substrates 1 and 2 in T cells. Potential role of JAK kinases. *J Biol Chem*. 1995;270(48):28527-28530.
435. Fruman DA, Meyers RE, Cantley LC. Phosphoinositide kinases. *Annu Rev Biochem*. 1998;67:481-507.
436. Truitt KE, Mills GB, Turck CW, Imboden JB. SH2-dependent association of phosphatidylinositol 3'-kinase 85-kDa regulatory subunit with the interleukin-2 receptor beta chain. *J Biol Chem*. 1994;269(8):5937-5943.
437. Damen JE, Cutler RL, Jiao H, Yi T, Krystal G. Phosphorylation of tyrosine 503 in the erythropoietin receptor (EpR) is essential for binding the P85 subunit of phosphatidylinositol (PI) 3-kinase and for EpR-associated PI 3-kinase activity. *J Biol Chem*. 1995;270(40):23402-23408.
438. Pfeffer LM, Mullersman JE, Pfeffer SR, et al. STAT3 as an adapter to couple phosphatidylinositol 3-kinase to the IFNAR1 chain of the type I interferon receptor. *Science*. 1997;276(5317):1418-1420.
439. Dong C, Davis RJ, Flavell RA. MAP kinases in the immune response. *Annu Rev Immunol*. 2002;20:55-72.
440. Moraga I, Spangler J, Mendoza JL, Garcia KC. Multifarious determinants of cytokine receptor signaling specificity. *Adv Immunol*. 2014;121:1-39.
441. Moraga I, Wernig G, Wilmes S, et al. Tuning cytokine receptor signaling by re-orienting dimer geometry with surrogate ligands. *Cell*. 2015;160(6):1196-1208.
442. Tonks NK. Protein tyrosine phosphatases: from genes, to function, to disease. *Nat Rev Mol Cell Biol*. 2006;7(11):833-846.
443. Tsui HW, Siminovitch KA, de Souza L, Tsui FW. Motheaten and viable motheaten mice have mutations in the haematopoietic cell phosphatase gene. *Nat Genet*. 1993;4(2):124-129.
444. Shultz LD, Schweitzer PA, Pajan TV, et al. Mutations at the murine motheaten locus are within the hematopoietic cell protein-tyrosine phosphatase (Hcph) gene. *Cell*. 1993;73(7):1445-1454.
445. Van Zant G, Shultz L. Hematologic abnormalities of the immunodeficient mouse mutant, viable motheaten (mev). *Exp Hematol*. 1989;17(2):81-87.
446. Klingmuller U, Lorenz U, Cantley LC, Neel BG, Lodish HF. Specific recruitment of SH-PTP1 to the erythropoietin receptor causes inactivation of JAK2 and termination of proliferative signals. *Cell*. 1995;80(5):729-738.
447. Yi T, Mui AL, Krystal G, Ihle JN. Hematopoietic cell phosphatase associates with the interleukin-3 (IL-3) receptor beta chain and down-regulates IL-3-induced tyrosine phosphorylation and mitogenesis. *Mol Cell Biol*. 1993;13(12):7577-7586.
448. Kashiwada M, Giallourakis CC, Pan PY, Rothman PB. Immunoreceptor tyrosine-based inhibitory motif of the IL-4 receptor associates with SH2-containing phosphatases and regulates IL-4-induced proliferation. *J Immunol*. 2001;167(11):6382-6387.
449. Yetter A, Uddin S, Krolewski JJ, et al. Association of the interferon-dependent tyrosine kinase Tyk-2 with the hematopoietic cell phosphatase. *J Biol Chem*. 1995;270(31):18179-18182.
450. Jiao H, Berrada K, Yang W, Tabrizi M, Platanias LC, Yi T. Direct association with and dephosphorylation of Jak2 kinase by the SH2-domain-containing protein tyrosine phosphatase SHP-1. *Mol Cell Biol*. 1996;16(12):6985-6992.
451. Yin T, Shen R, Feng GS, Yang YC. Molecular characterization of specific interactions between SHP-2 phosphatase and JAK tyrosine kinases. *J Biol Chem*. 1997;272(2):1032-1037.
452. Haspel RL, Salditt-Georgieff M, Darnell JE Jr. The rapid inactivation of nuclear tyrosine phosphorylated Stat1 depends upon a protein tyrosine phosphatase. *EMBO J*. 1996;15(22):6262-6268.
453. Mertens C, Zhong M, Krishnaraj R, et al. Dephosphorylation of phosphotyrosine on STAT1 dimers requires extensive spatial reorientation of the monomers facilitated by the N-terminal domain. *Genes Dev*. 2006;20(24):3372-3381.
454. Krystal G. Lipid phosphatases in the immune system. *Semin Immunol*. 2000;12(4):397-403.
455. Shuai K, Liu B. Regulation of JAK-STAT signalling in the immune system. *Nat Rev Immunol*. 2003;3(11):900-911.
456. Kim TK, Maniatis T. Regulation of interferon-gamma-activated STAT1 by the ubiquitin-proteasome pathway. *Science*. 1996;273(5282):1717-1719.
457. Yoshimura A, Ohkubo T, Kigucki T, et al. A novel cytokine-inducible gene CIS encodes an SH2-containing protein that binds to tyrosine-phosphorylated interleukin 3 and erythropoietin receptors. *EMBO J*. 1995;14(12):2816-2826.
458. Matsumoto A, Masuhara M, Mitsui K, et al. CIS, a cytokine inducible SH2 protein, is a target of the JAK-STAT5 pathway and modulates STAT5 activation. *Blood*. 1997;89(9):3148-3154.
459. Linossi EM, Babon JJ, Hilton DJ, Nicholson SE. Suppression of cytokine signaling: the SOCS perspective. *Cytokine Growth Factor Rev*. 2013;24(3):241-248.
460. Chong MM, Cornish AL, Darwiche R, et al. Suppressor of cytokine signaling-1 is a critical regulator of interleukin-7-dependent CD8+ T cell differentiation. *Immunity*. 2003;18(4):475-487.
461. Knosp CA, Carroll HP, Elliott J, et al. SOCS2 regulates T helper type 2 differentiation and the generation of type 2 allergic responses. *J Exp Med*. 2011;208(7):1523-1531.
462. Croker BA, JMetcalf D, Robb L, et al. SOCS3 is a critical physiological negative regulator of G-CSF signaling and emergency granulopoiesis. *Immunity*. 2004;20(2):153-165.
463. Croker BA, Krebs DL, Zhang JG, et al. SOCS3 negatively regulates IL-6 signaling in vivo. *Nat Immunol*. 2003;4(6):540-545.
464. Mowen KA, Tang J, Zhu W, et al. Arginine methylation of STAT1 modulates IFNα/β-induced transcription. *Cell*. 2001;104:731-739.
465. Palvimo JJ. PIAS proteins as regulators of small ubiquitin-related modifier (SUMO) modifications and transcription. *Biochem Soc Trans*. 2007;35(pt 6):1405-1408.
466. Nadeau K, Hwa V, Rosenfeld RG. STAT5b deficiency: an unsuspected cause of growth failure, immunodeficiency, and severe pulmonary disease. *J Pediatr*. 2011;158(5):701-708.
467. Gilmour KC, Fujii H, Cranston T, et al. Defective expression of the interleukin-2/interleukin-15 receptor beta subunit leads to a natural killer cell-deficient form of severe combined immunodeficiency. *Blood*. 2001;98(3):877-879.

468. Zhang Z, Gothe F, Pennamen P, et al. Human interleukin-2 receptor β mutations associated with defects in immunity and peripheral tolerance. *J Exp Med.* 2019;216(6):1311-1327.
469. Kotlarz D, Zietara N, Milner JD, and Klein C. Human IL-21 and IL-21R deficiencies: two novel entities of primary immunodeficiency. *Curr Opin Pediatr.* 2014;26(6):704-712.
470. Horev L, Unger S, Molho-Pessach, et al. Generalized verrucosis and HPV-3 susceptibility associated with CD4 T-cell lymphopenia caused by inherited human interleukin-7 deficiency. *J Am Acad Dermatol.* 2015;72(6):1082-1084.
471. Kosumi H, Natsuga K, Takashima S, et al. Two cases of interleukin-7-deficient generalized verrucosis. *Cli Infect Dis.* 2020;71(6):1561-1563.
472. Altare F, Lammas D, Revy P, et al. Inherited interleukin 12 deficiency in a child with bacille Calmette-Guerin and Salmonella enteritidis disseminated infection. *J Clin Invest.* 1998;102(12):2035-2040.
473. Altare F, Durandy A, Lammas D, et al. Impairment of mycobacterial immunity in human interleukin-12 receptor deficiency. *Science.* 1998;280(5368):1432-1435.
474. Jouanguy E, Lamhamedi-Cherradi S, Altare F, et al. Partial interferon-gamma receptor 1 deficiency in a child with tuberculoid bacillus Calmette-Guerin infection and a sibling with clinical tuberculosis. *J Clin Invest.* 1997;100(11):2658-2664.
475. Dupuis S, Dargemont C, Fieschi C, et al. Impairment of mycobacterial but not viral immunity by a germline human STAT1 mutation. *Science.* 2001;293(5528):300-303.
476. Liu L, Okada S, Kong XF, et al. Gain-of-function human STAT1 mutations impair IL-17 immunity and underlie chronic mucocutaneous candidiasis. *J Exp Med.* 2011;208(8):1635-1648.
477. Pilati C, Amessou M, Bihl MP, et al. Somatic mutations activating STAT3 in human inflammatory hepatocellular adenomas. *J Exp Med.* 2011;208(7):1359-1366.
478. Zhang Q, Boisson B, Béziat V, Puel A, Casanova JL. Human hyper-IgE syndrome: singular or plural? *Mamm Genome.* 2018;29(7-8):603-617.
479. Amselem S, Duquesnoy P, Attree O, et al. Laron dwarfism and mutations of the growth hormone-receptor gene. *N Engl J Med.* 1989;321(15):989-995.
480. de la Chapelle A, Traskelin AL, Juvonen E. Truncated erythropoietin receptor causes dominantly inherited benign human erythrocytosis. *Proc Natl Acad Sci U S A.* 1993;90(10):4495-4499.
481. Shochat C, Tal N, Bandapalli OR, et al. Gain-of-function mutations in interleukin-7 receptor-alpha (IL7R) in childhood acute lymphoblastic leukemias. *J Exp Med.* 2011;208(5):901-908.
482. Mullighan CG, Collins-Underwood JR, Phillips LA, et al. Rearrangement of CRLF2 in B-progenitor-and Down syndrome-associated acute lymphoblastic leukemia. *Nat Genet.* 2009;41(11):1243-1246.
483. Nekoui A, Blaise G. Erythropoietin and nonhematopoietic effects. *Am J Med Sci.* 2017;353(1):76-81.
484. Nairz M, Schroll A, Moschen AR, et al. Erythropoietin contrastingly affects bacterial infection and experimental colitis by inhibiting nuclear factor-kappaB-inducible immune pathways. *Immunity.* 2011;34(1):61-74.
485. Rosenberg SA. IL-2: the first effective immunotherapy for human cancer. *J Immunol.* 2014;192:5451-5458.
486. Choudhry H, Helmi N, Abdulaal WH, et al. Prospects of IL-2 in cancer immunotherapy. *Biomed Res Int.* 2018;2018: 9056173.
487. Waldmann TA. Cytokines in cancer immunotherapy. *Cold Spring Harb Perspect Biol.* 2018;10:a028472.
488. Kinch MS. An overview of FDA-approved biologics medicines. *Drug Discov Today.* 2015;20:393-398.
489. Rider P, Carmi Y, Cohen I. Biologics for targeting inflammatory cytokines, clinical uses, and limitations. *Int J Cell Biol.* 2016;2016:9259646.
490. Loutfy MR, Blatt LM, Siminovitch KA, et al. Interferon alphacon-1plus corticosteroids in severe acute respiratory syndrome: a preliminary study. *J Am Med Assoc.* 2003;290(24):3222-3228.
491. Wang BX, Fish EN. Global virus outbreaks: interferons as 1st responders. *Semin Immunol.* 2019;43:101300.
492. Kieseier BC. The mechanism of action of interferon-β in relapsing multiple sclerosis. *CNS Drugs.* 2011;25(6): 491-502.
493. Todd PA, Goa KL. Interferon gamma-1b. A review of its pharmacology and therapeutic potential in chronic granulomatous disease. *Drugs.* 2012;43(1):111-122.
494. O'Shea JJ, Gadina M. Selective Janus kinase inhibitors come of age. *Nat Rev Rheumatol.* 2019;15:74-75.
495. Changelian PS, Flanagan ME, Ball DJ, et al. Prevention of organ allograft rejection by a specific Janus kinase inhibitor. *Science.* 2003;302:875-878.
496. Thorarensen A, Dowty ME, Banker ME, et al. Design of a Janus Kinase 3 (JAK3) specific inhibitor PF-06651600 allowing for the interrogation of JAK3 signaling in humans. *J Med Chem.* 2017;60:1971-1993.
497. Gadina M, Le MT, Schwartz DM, et al. Janus kinases to jakinibs: from basic insights to clinical practice. *Rheumatology.* 2019;58(suppl 1):i4-i16.
498. Kang S, Tanaka T, Narazaki M, et al. Targeting interleukin-6 signaling in clinic. *Immunity.* 2019;50:1007-1023.
499. Grupp SA, Kalos M, Barrett D, et al. Chimeric antigen receptor-modified T cells for acute lymphoid leukemia. *N Engl J Med.* 2013;368:1509.
500. Moore JB, June CH. Cytokine release syndrome in severe COVID-19. *Science.* 2020;268:473-474.
501. Cohan SL, Lucassen EB, Romba MC, Linch SN. Daclizumab: mechanisms of action, therapeutic efficacy, adverse events and its uncovering the potential role of innate immune system recruitment as a treatment strategy for relapsing multiple sclerosis. *Biomedicines.* 2019;7(1):18.
502. May RD, Fung M. Strategies targeting the IL-4/IL-13 axes in disease. *Cytokine.* 2015;75(1):89-116.
503. Beck L, Thaçi D, Hamilton JD, et al. Dupilumab treatment in adults with moderate-to-severe-atopic dermatitis. *N Engl J Med.* 2014;371:130-139.
504. Castro M, Corren J, Pavord ID, et al. Dupliumab efficacy and safety in moderate to severe uncontrolled asthma. *N Engl J Med.* 2018;378:2486-2496.
505. Johnson BB, Beck LA, Mustafa SS. Case report: remarkable response to dupilumab in a 5-year old patient with severe, recalcitrant atopic dermatitis. *JAAD Case Rep.* 2019;5(7):605-608.
506. Bel EH, Wenzel SE, Thompson PJ, et al. Oral glucocorticoid-sparing effect of mepolizumab in eosinophilic asthma. *N Engl J Med.* 2014;371(13):1189-1197.
507. Nair P, Wenzel S, Rabe KF, et al. Oral glucocorticoid-sparing effect of benralizumab in severe asthma. *N Engl J Med.* 2017;376:2448-2458.
508. Sato TA, Widmer MB, Finkelman FD, et al. Recombinant soluble murine IL-4 receptor can inhibit or enhance IgE responses in vivo. *J Immunol.* 1993;150(7):2717-2723.
509. Steinke JW, Negri J, Enelow R, et al. Proinflammatory effects of IL-4 antagonism. *J Allergy Clin Immunol.* 2006;118(3):756-758.
510. Slager RE, Otulana BA, Hawkins GA, et al. IL-4 receptor polymorphisms predict reduction in asthma exacerbations during response to an anti-IL-4 receptor α antagonist. *J Allergy Clin Immunol.* 2012;130(2):516-522.e4.
511. Levin AM, Bates DL, Ring AM, et al. Exploiting a natural conformational switch to engineer an interleukin-2 "superkine." *Nature.* 2012;484:529-533.
512. Mitra S, Ring AM, Amarnath S, et al. Interleukin-2 activity can be fine tuned with engineered receptor signaling clamps. *Immunity.* 2015;42:826-838.
513. Glassman CR, Su L, Majri-Morrison SS, et al. Calibration of cell-intrinsic interleukin-2 response thresholds guides design of a regulatory T cell biased agonist. *Elife.* 2021;10:e65777.
514. Mo F, Yu Z, Li P, et al. An engineered IL-2 partial agonist promotes CD8+ T cell stemness. *Nature.* 2021;597:544-548.
515. Sockoloskyv JT, Trotta E, Parisi G, et al. Selective targeting of engineered T cells using orthogonal IL-2 cytokine-receptor complexes. *Science.* 2018;359:1037-1042.
516. Silva DA, Yu S, Ulge UY, et al. De novo design of potent and selective mimics of IL-2 and IL-15. *Nature.* 2019;565:186-191.
517. Mohan K, Ueda G, Kim AR, et al. Topological control of cytokine receptor signaling induces differential effects in hematopoiesis. *Science.* 2019;364(6442):eaav7532.
518. Boyman O, Kovar M, Rubinstein MP, et al. Selective stimulation of T cell subsets with antibody-cytokine immune complexes. *Science.* 2006;311:1924-1927.
519. Arenas-Ramirez N, Zhou C, Popp S, et al. Improved cancer immunotherapy by a CD25-mimobody conferring selectivity to human IL-2. *Sci Transl Med.* 2016;8:367ra166.
520. Trotta E, Bessette PH, Silveria SL, et al. A human anti-IL-2 antibody that potentiates regulatory T cells by a structure-based mechanism. *Nat Med.* 2018;24(7):1005-1014.
521. Desbois M, LeVue P, Coutzac C, et al. IL-15 trans-signaling with the superagonist RLI promotes effector/memory CD8+ T cell responses and enhances anti-tumor activity of PD-1 antagonists. *J Immunol.* 2016;197:168-178.
522. Ward NC, Yu A, Moro A, et al. IL-2/CD25: a long-acting fusion protein that promotes immune tolerance by selectively targeting the IL-2 receptor on regulatory T cells. *J Immunol.* 2018;201:2579-2592.

CHAPTER 10

Tumor Necrosis Factor Superfamily of Cytokines and Receptors

Carl F. Ware, PhD • Nancy H. Ruddle, PhD

INTRODUCTION

Lymphocytes are highly mobile cells that transit the vascular and lymphatic circulatory systems, making limited stopovers in organized lymphoid tissues (lymph nodes, spleen, and Peyer patches) in which they commune with other cell types to initiate immune responses. Cytokines, both secreted and membrane-anchored proteins, serve as the communication media of the immune system. Cytokines in the tumor necrosis factor (TNF) superfamily help orchestrate the development, homeostasis, and effector actions of cells in the immune system and in the environments they inhabit. The diversity of the TNF superfamily is evidenced by additional roles of family members in regulating cells of neuronal, skeletal, and ectodermal origin. This chapter's focus is on members of the TNF superfamily (TNFSF) that are primarily involved in regulating the immune system.

Ligands belonging to the TNFSF initiate intercellular communication by binding to specific receptors on the surface of the receiving target cell. Many of the TNF-related ligands are membrane-bound and require cell-cell contact to initiate signaling. Other TNF-related ligands are secreted or released in soluble form, influencing cellular responses at locations distant from the source of production. The TNF receptors (TNFR) form a corresponding superfamily of cognate membrane proteins that activate intracellular signaling pathways to influence cellular growth, differentiation, survival, and death. Each ligand-receptor pair forms a "system." More than 40 distinct ligand-receptor systems are currently recognized. Many of the ligands and receptors engage more than one cognate, thereby forming "circuits" of signaling systems. These circuits often function together as coordinated or integrated networks with other cytokines to regulate a specific cellular process. The expression patterns of TNFSF ligand and receptor circuits form "cellular networks" that mediate communication among various cell types. TNFR-activated signals are delivered to the cytosol to activate transcription factors such as NFκB and AP1, which in turn initiate the expression of hundreds of genes that alter cellular differentiation. Additionally, some TNFR signaling activates cell death pathways that mediate apoptosis and necroptosis, terminating cellular life. Genetic mutations have revealed the roles of individual TNFSF members in regulating human immune responses and organ development. The TNFSF ligands and receptors are known to be critical factors in mediating host-defense responses to pathogens. Conversely, they also contribute to pathologic processes in cancers and autoimmune diseases. Therapeutics directed at modulating TNF action have proven clinical benefits, best exemplified by the successful use of TNF antagonists in patients with certain autoimmune diseases. This clinical impact brings the need to understand the complexities of the TNFSF into sharp focus.

The nomenclature of the TNFSF is a bit of a morass for students new to the field, although introduction of a numerical system for standardizing gene names (http://www.genenames.org) helps in searching the literature and genomic databases. Genes encoding the ligands are designated *TNFSF*, while receptor genes are denoted as *TNFRSF*, in each case followed by a numerical identifier. Exceptions to these standardizations are represented by several ligand and receptor genes with historical acronyms, such as TNF, lymphotoxin-α, LTβ, CD40L, FAS, CD27, nerve growth factor receptor p75, and ectodysplasin (EDA), among others (see Tables 10.1 and 10.2). These more common, often colorful, acronyms for proteins continue to be widely used. The name and genomic identification for each ligand and receptor are tabulated along with additional pertinent information for human and mouse genes (Tables 10.1 and 10.2). Detailed genetic and structural information is available at the National Center for Biotechnology Information (NCBI).

TNF LIGAND SUPER FAMILY

Most of the TNF-related cytokines are type II transmembrane proteins (intracellular N-terminus) with a short cytoplasmic tail and a larger extracellular region (~150 amino acids) containing the signature TNF homology domain in which the receptor binding sites are located[1] (Fig. 10.1). The TNF homology domain assembles into trimers, representing the functional, receptor binding unit of the ligand. An exception is the dimeric mouse GITR ligand.[3] Sequence and structural analyses of several members of the TNFSF ligand family have revealed three structural groups: (1) the conventional (TNF, LTα, LTβ, Apo2L/TRAIL, TL1A, LIGHT, FASL, RANKL, and CD40L) with a pyramidal shape of the trimer; (2) the globular-shaped E-F disulfide bond variants (APRIL, BAFF, TWEAK, and EDA); and (3) the divergent (CD27L, CD30L, GITRL, 4-1BBL, and OX40L) ligands that exhibit little conservation in sequence and tertiary structure.[4-11] The conserved residues defining the TNF homology domain are primarily located within the internal β-strands that form the molecular scaffold of the extracellular domain and promote its assembly into trimers. Residues in specific loops between

TABLE 10.1 TNF Ligand Superfamily

Protein Name	Chromosome Position Human	Chromosome Position Mouse	Gene Symbol	TNFSF #
TNF	6p21.3	17 (19.06)	*TNF*	TNFSF1A
LTa	6p21.3	17 (19.06)	*LTa*	TNFSF1B
LTβ	6p21.3	17 (19.06)	*LTβ*	TNFSF3
OX40-L	1q25	1 (84.90)	*TNFSF4*	TNFSF4
CD40-L, CD154	Xq26	X (18.0)	*CD40LG*	TNFSF5
FASL	1q23	1 (85.0)	*FASLG*	TNFSF6
CD27-L, CD70	19p13	17 (20.0)	*CD70*	TNFSF7
CD30-L, CD153	9q33	4 (32.20)	*TNFSF8*	TNFSF8
4-1BB-L	19p13	17 (20.0)	*TNFSF9*	TNFSF9
TRAIL	3q26	3 (10.82)	*TNFSF10*	TNFSF10
RANK-L	13q14	14 (45.0)	*TNFSF11*	TNFSF11
TWEAK	17p13	11 (42.86)	*TNFSF12*	TNFSF12
APRIL	17P13.1	13 (42.86)	*TNFSF13*	TNFSF13
BAFF	13q32-q34	8 (4.55)	*TNFSF13B*	TNFSF13B
LIGHT	19p13.3	17 (29.71)	*TNFSF14*	TNFSF14
TL1A	9q33	4 (31.80)	*TNFSF15*	TNFSF15
GITRL	1q23	1 (69.75)	*TNFSF18*	TNFSF18
EDA1	Xq12-q13.1	X (37.0)	*EDA1*	EDA1
EDA2	Xq12-q13.1	X (37.0)	*EDA2*	EDA2

Protein abbreviations: 4-1BB, a cDNA clone; April, A proliferation-inducing ligand; BAFF, B-cell activating factor; CD40, Cluster of Differentiation(CD)-40; EDA, ectodysplasin-1 and 2; FASL, First apoptosis signaling ligand; GITR-L, glucocorticoid-induced TNF receptor protein-ligand; LIGHT, homologous to *L*ymphotoxin, exhibits inducible expression, and competes with HSV *g*lycoprotein-D for binding to *h*erpesvirus entry mediator, a receptor expressed on *T* lymphocytes; LT, lymphotoxin; OX40, Oxford-40; RANK-L, Receptor activator of nuclear factor κ B-Ligand; TL1A, TNF-like protein 1A; TNF, tumor necrosis factor; TRAIL, TNF-related apoptosis-inducing ligand; TWEAK, TNF-related weak inducer of apoptosis.

the external β-strands make contact with the receptor. Although most of the TNF ligands self-assemble into homotrimers, heterotrimers can also form between LTα and LTβ[12] and between APRIL and BAFF.[13] The stoichiometry of the LT heterotrimer of 1:2 (LTα1β2) imparts receptor specificity for the LTβR which is distinct from the binding of the LTα homotrimer to TNFR1, TNFR2, and herpesvirus entry mediator (HVEM). Alternate splicing can also generate distinct ligands, such as the splice form joining TWEAK and APRIL (TWE-PRI)[14]; the alternate ligands for ectodysplasin receptor (EDAR)[15]; and a cytosolic form of LIGHT.[16] Several of the ligands are soluble, either lacking a transmembrane sequence (LTα; APRIL) or else shed by proteolytic cleavage from the cell surface (eg, FASL, TNF, LIGHT, BAFF, TL1A, and EDA). These functional forms are capable of activating target cells in a systemic fashion.

The genetic organization of TNF ligands is highly conserved. They are typically encoded in three or four exons, with the fourth exon encoding most of the extracellular TNF homology domain (Table 10.1). The TNF, LTα, and LTβ genes reside adjacent to each other in the gene-dense major histocompatibility complex (MHC), class III region (human chromosome 6p21.3). Here, they are sandwiched between the antigenic-peptide presenting MHC-I and II and other immune regulatory genes including members of the complement system and NFκB transcription factors.[12,17] Three other genetic clusters of TNF-related cytokines are found within the corresponding MHC paralogous regions, located on human chromosomes 1, 9, and 19. These genetic clusters share a remarkably conserved structure and transcriptional orientation that also encode proteins regulating other cellular immune responses.[16] The selection pressures responsible for retaining this genetic configuration of the MHC in general is encompassed in the paralogy theory of genome evolution.[18] Conservation is less pronounced in the TNF ligand genes residing outside these MHC paralogs.

TNF RECEPTOR SUPERFAMILY

Members of the TNF receptor superfamily include proteins of vertebrate and viral origin that are primarily defined by their ligand specificity but can also be distinguished by their cytosolic signaling domains (Table 10.2). Most of the signaling receptors in the TNFRSF are type I transmembrane

TABLE 10.2 TNF Receptor Super Family

	Chromosome Position			
Protein Name	**Human**	**Mouse**	**Gene Symbol**	**TNFRSF #**
TNFR-1	12p13.2	6 (60.55)	*TNFR1A*	TNFRSF1A
TNFR2	1p36.3-36.2	4 (75.5)	*TNFR1B*	TNFRSF1B
LTβR	12p13	6 (60.4)	*LTβR*	TNFRSF3
OX40	1p36	4 (79.4)	*TNFRSF4*	TNFRSF4
CD40	20q12-q13.2	2 (97.0)	*CD40*	TNFRSF5
FAS	10q24.1	19 (23.0)	*FAS*	TNFRSF6
DcR3	20q13	not present	*TNFRSF6B*	TNFRSF6B
CD27	12p13	6 (60.35)	*CD27*	TNFRSF7
CD30	1p36	4 (75.5)	*TNFRSF8*	TNFRSF8
4-1BB	1p36	4 (75.5)	*TNFRSF9*	TNFRSF9
TRAILR-1, DR4	8p21	not present	*TNFRSF10A*	TNFRSF10A
TRAIL-R2, DR5	8p22-p21	14 (D1)	*TNFRSF10B*	TNFRSF10B
TRAILR3, DcR1	8p22-p21	7 (69.6)	*TNFRSF10C*	TNFRSF10C
TRAILR4, DcR2	8p21	7 (69.6)	*TNFRSF10D*	TNFRSF10D
RANK	18q22.1	1	*TNFRSF11A*	TNFRSF11A
OPG	8q24	15	*TNFRSF11B*	TNFRSF11B
FN14	16p13.3	17	*TNFRSF12A*	TNFRSF12A
DR3	1p36.3	4 (E1)	*TNFRSF25*	TNFRSF25
BCMA	16.1	16.82	*TNFRSF17*	TNFRSF17
TACI	17p11.2	11	*TNFRSF13B*	TNFRSF13B
BAFFR	22q13.1-q13.31	15	*TNFRSF13C*	TNFRSF13C
HVEM	1p36.3-p36.2	4	*TNFRSF14*	TNFRSF14
p75NTR, NGFR	17q12-q22	11 (55.6)	*NGFR*	TNFRSF16
BCMA	16p13.1	16 (B3)	*TNFRSF17*	TNFRSF17
GITR	1p36.3	4 (E)	*TNFRSF18*	TNFRSF18
RELT	11q13.2	7	*TNFRSF19L*	TNFRSF19L
TROY	13q12.11-q12.3	14	*TNFRSF19*	TNFRSF19
EDAR	2q11-q13	10	*EDAR1*	EDAR1
EDA2R	Xq11.1	X	*EDAR2*	EDA2R
DR6	6P12.2-21.1	17	*TNFRSF21*	TNFRSF21
IGFLR1	19q13.12	7	*IGFLR1*	IGFLR1

Protein Abbreviations: BCMA, B-cell maturation antigen; DcR, decoy receptor; DR, death receptor; FN14, TWEAK receptor; HVEM, herpesvirus entry mediator; IGFLR1, insulin growth factor-like receptor-1; OPG, osteoprotegerin; p75NTR, 75 kDa neurotrophin receptor; RELT, receptor expressed in lymphoid tissues; TACI, transmembrane activator and CAML interactor; TROY, TNFRSF expressed on the mouse embryo; (see Table 10.1 for corresponding names not abbreviated here).

glycoproteins (N-terminus exterior to the cell). However, several TNFRSF members (DcR3; TRAILR3) lack conventional transmembrane or cytosolic signaling domains acting as a decoy to sequester the ligand from binding its signaling receptor.

A cysteine-rich domain (CRD) in the extracellular, ligand-binding region defines membership in the TNF receptor superfamily (Fig. 10.1). Each CRD typically contains six cysteine residues that form three disulfide bonds. Among TNFRSF members, this motif may be repeated from one to six times. Based on crystal structures solved for several TNFRSF members, the CRD confers an elongated shape and sidedness to the receptor ectodomain.[19,20] The crystal structure of the complex between TNFR1 and one of its ligands, the LTα homotrimer, reveals that residues in both CRD2 and 3 are in contact with the ligand.[19] Variations in ligand-binding CDRs range from a single domain in BAFFR[5,21] to three different CRDs in HVEM with distinct binding sites for multiple ligands.

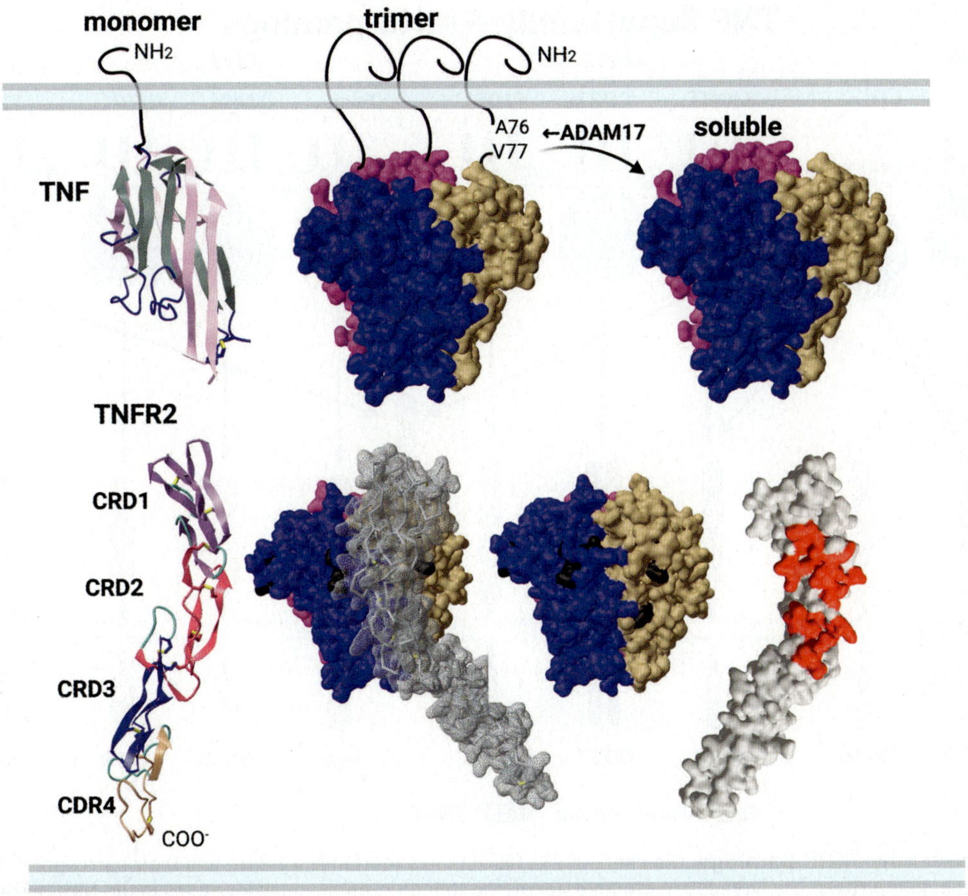

FIG. 10.1. Structure of tumor necrosis factor (TNF),TNF receptor, and the ligand-receptor complex (*top panel*), The β-sandwich of the TNF monomer shown as a ribbon diagram contains two-stacked β-pleated sheets each formed by five antiparallel β strands (wide ribbon; alternating from N-terminus *silver* to *pink*) that fold into a Greek key or "jelly-roll" topology. The inner β-sheet forms contacts between adjacent subunits, which promotes assembly into a trimer. The trimer is formed such that one edge of each subunit is packed against the inner sheet of its neighbor. The surface-exposed outer β sheet loops (*blue*) and strands are involved in receptor-binding. The trimeric structure is characteristic of the TNF-related ligands. The type II configuration of TNF (N-terminus inside the cell) and three transmembrane domains anchor TNF in the membrane. The TNF trimer is ~60 Å in height with relatively flat base residing close to the membrane creating a pyramidal shape (shown as a molecular surface representation with different colors for each subunit of the trimer). TNF is cleaved by TNFα-converting enzyme, ADAM17 (aka TACE), a member of the ADAM family of metalloproteinases involved in processing of many cell surface proteins. ADAM17 is a type 1 transmembrane protein that cleaves TNF between residues Val77 and Ala76, releasing TNF into soluble phase when all three sites in the trimer are cleaved (*bottom panel*), TNFR2 and the ligand complex. The ectodomain of TNFR2 forms an elongated molecule with cysteine-rich domain (CRD1) proximal to the N-terminus (ribbon diagram; CRDs color-coded; disulfide bonds in *yellow*). The molecular surface of TNFR2 (wire frame in *silver*, same orientation as the ribbon diagram) in complex with one receptor binding site, depicted as two subunits of TNF. Contact residues in TNF (*black*) are present in two adjacent subunits forming a composite receptor-binding site. Correspondingly, TNFR2 is rotated 180° to expose residues contacting TNF (in *red*) in CRD2 and CRD3. In the ligand-receptor complex, CRD2 and 3 of TNFR2 contact the cleft formed between adjacent subunits of TNF with three TNFR2 bind to a single-TNF trimer. The TNFR2 N-terminus points upward, closest to the base of the ligand in the transorientation as occurring at junctions formed at sites of cell-to-cell contact. Structures derived from 3ALQ pdb[2](MMDB ID:86,206) visualized iCN3D (NCBI, ncbi.nlm.nih.gov/Structure). (Created with BioRender.com.)

LIGAND-RECEPTOR COMPLEX

Formation of the ligand-receptor complex is the first step in initiating TNFRSF signaling pathways. Several features of TNF ligands determine their ability to initiate receptor signaling. These include anchorage of the ligand in the membrane, secretion of the ligand as a soluble protein, avidity of the ligand-receptor interaction, the presence of decoy receptors, and the identity of the cell types expressing the ligand and receptor. In the TNF-TNFR2 complex, the surface-exposed loops β strands contain many of the amino acid residues that make contact with the receptor, with the receptor-binding site formed as a composite of adjoining subunits in the trimeric ligand[2] (see Fig. 10.1). Although the binding specificities for most members of the TNF ligand and receptor superfamilies result in monotypic interactions, several members interact with multiple partners[22] to create complex networks of signaling pathways (Figs. 10.2 and 10.3). Binding interactions of the soluble ligands typically exhibit high avidity, with equilibrium binding constants measured in the high pM to low nM range. The soluble forms of some TNFSF members (eg, OX40L) have

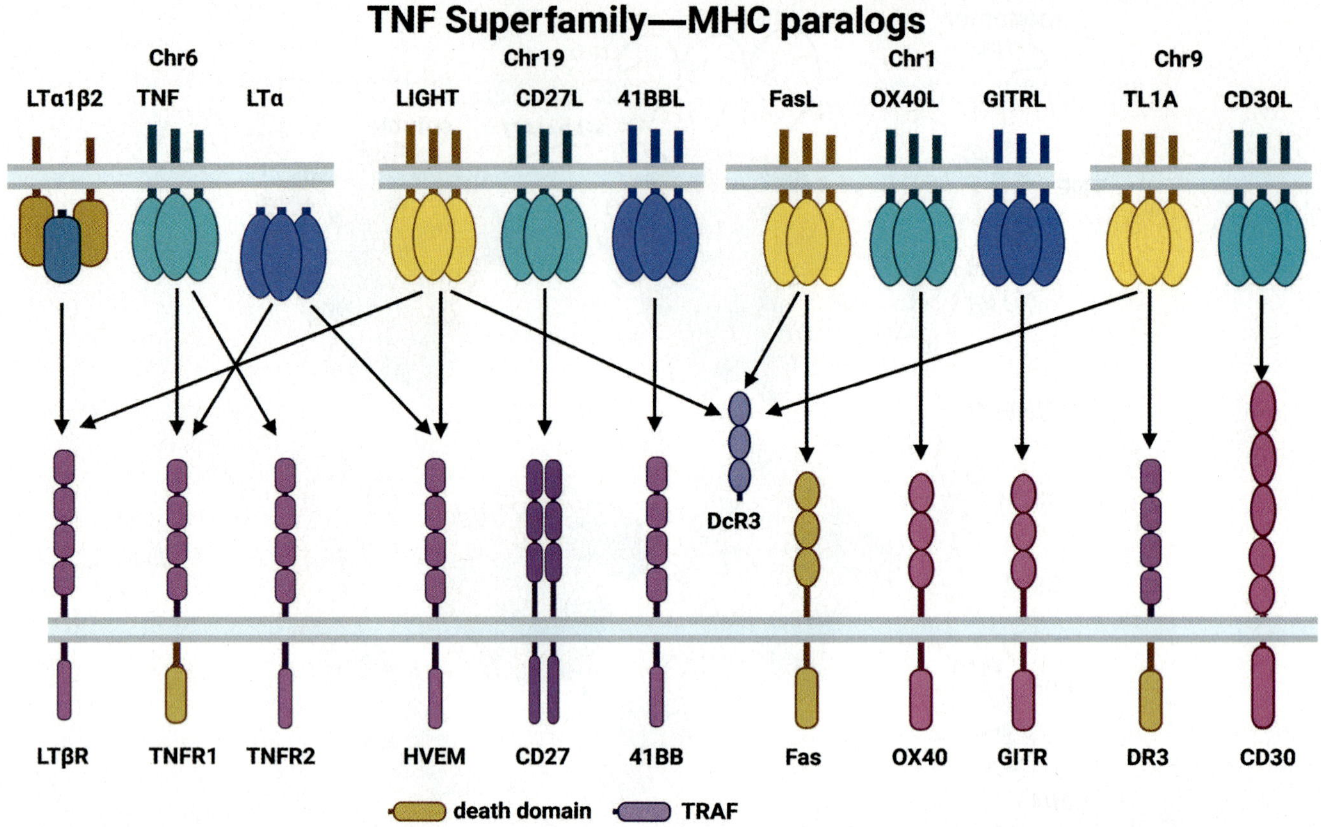

FIG. 10.2. **TNF superfamily–MHC paralogs.** Members of the TNF ligand superfamily (*above*) and their corresponding receptors (*below*) are identified by connecting *arrows*. The ligands are grouped according to their chromosomal locations in the MHC paralogous regions. The number of cysteine-rich domains are depicted for each TNF receptor. (Figure generated using BioRender [BioRender.com].)

only modest affinity (50-200 nM) for their respective receptors, which would seem to require supraphysiologic levels of ligand to activate signaling. This suggests that the membrane-bound form of the ligand may represent the predominant functional species. For membrane-anchored ligands, the ligands and receptors must be *in trans* configuration concentrated in a signaling complex at the junction between interacting cells. For example, CD40 ligand expressed by T helper cells is recruited and concentrated in the immune synapse formed with the CD40 receptor expressed by B cells.[23-25] Receptor activation by the membrane-bound ligands limits their action to the local microenvironment, whereas soluble ligands can diffuse to activate receptor signaling in systemic fashion.

Structural studies reveal both conserved and divergent features of the TNFSF ligand-receptor complex. The trimeric architecture of the TNF ligands, containing three equivalent receptor-binding sites, provides the basis for initiating signaling via clustering of receptors, creating a platform for assembly of signaling complexes on the cytosolic domains. This concept is supported by the finding that receptor-specific bivalent antibodies can act as agonists that cluster receptors, thus replacing the signaling activity of the natural ligand.[26] Indeed, agonist TNFR antibodies can simultaneously act as an antagonist by blocking the ability of the natural ligand to bind the receptor.[27] TNFRI can exist in monomeric and dimeric receptor units. Upon binding TNF, the receptors cluster into trimers that build into higher order oligomers increasing the avidity of the receptor-ligand complex.[28,29] The TNF ligands, FASL, TRAIL, and BAFF also form higher-ordered oligomers of the basic trimer. These higher-ordered ligand oligomers promote supraclustering of receptors, enhancing or sustaining signaling pathways in the receptor-bearing cell.[30,31] Overexpression of TNFR in cells can also lead to ligand-independent signaling, a phenomenon due in part to the propensity of the cytosolic tails to self-associate.[32] Under physiological conditions, the expression level and compartmentalization of these receptors are tightly controlled. A subregion in the first CRD of TNFR1, known as the preligand assembly domain (PLAD), may restrict the orientation of the unligated receptor to prevent spontaneous activation.[33] It is not known if this mechanism applies to all TNFRSF members.

Diversification of TNFSF Ligands and Receptors

Among members of the TNFSF and TNFRSF, there is significant divergence in both the structural features of the ligands and receptors and the mechanisms of ligand-receptor binding. Evolutionary diversification of TNFSF is exemplified by the complement component C1q, which contains a TNF homology domain that folds in strikingly familiar trimeric fashion despite having low-sequence homology. The C1q/

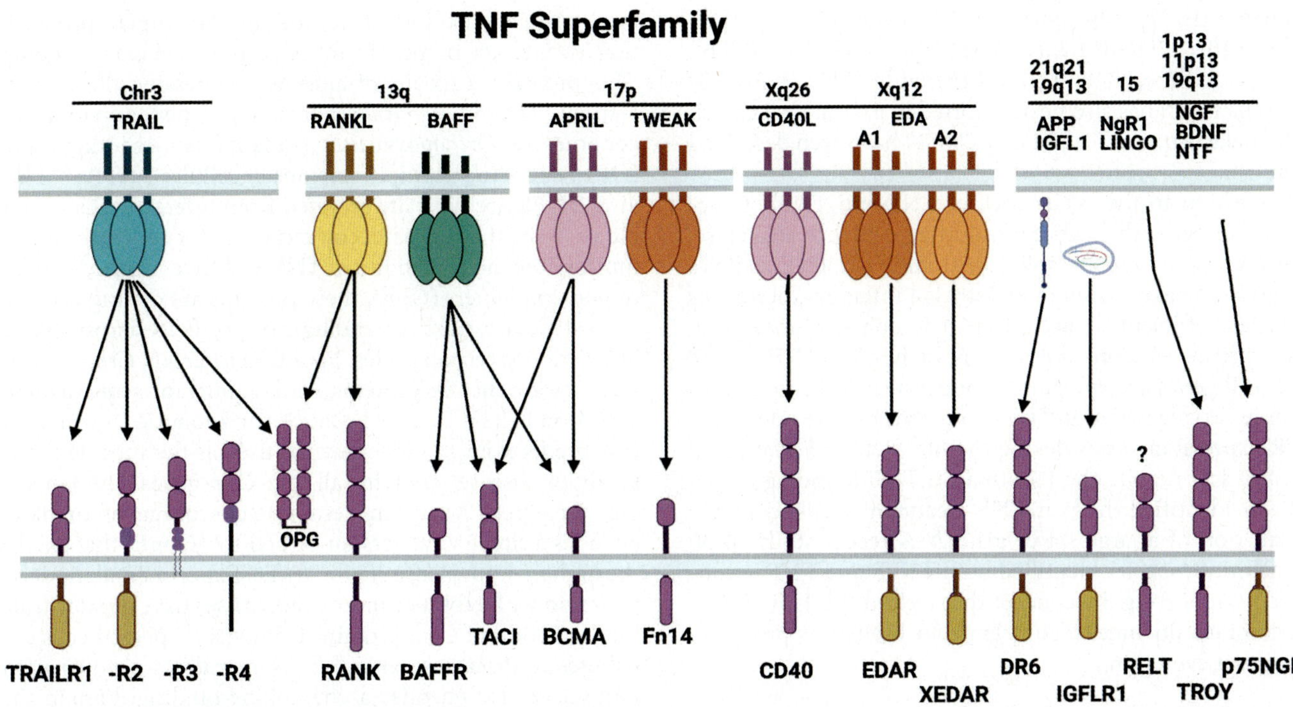

FIG. 10.3. The tumor necrosis factor (TNF) superfamily—part 2. Members of the TNF ligand superfamily (*above*) and their corresponding receptors (*below*) are identified by connecting *arrows*. The ligands are grouped according to their chromosomal locations. The number of cysteine-rich domains are depicted for each TNF receptor, and the cytoplasmic tail containing a DD are identified by a *gold* cylinder and the TNF receptor-associated factors (TRAF) domain in *purple*. The non-TNFSF ligands are APP, amyloid precursor protein; BDNF, brain-derived neurotrophic factor; IGFL1, insulin growth factor-like 1; LINGO, leucine-rich repeat and Ig-containing Nogo receptor-interacting protein-1; NGF, nerve growth factor; NgR1, Nogo-66 receptor 1; NTF, neurotrophic factors (neurotrophins). (Figure generated using BioRender [BioRender.com].)

TNF-related proteins (CTRP) form a distinct, 15-member family of proteins that bind non-TNFRSF members. This family is primarily involved in regulation of sugar and fat metabolism.[34,35]

The nerve growth factor receptor, p75 neurotrophin receptor (p75NTR),[36] binds nerve growth factor and three other neurotrophins (BDNF, NT3, and NT4), all of which are structurally unrelated to TNFSF ligands. Molecular contacts between NGF and p75NTR occur through two spatially separate binding regions located at the first and second CRD and at the junction between the CRD3 and CRD4.[37] The p75NTR signaling complex involves several other components including sortilin and SorCS2. Ligands and coreceptors interacting with p75NTR contribute to both stimulatory and inhibitory regulation of axonal regeneration, as well as allergic and inflammatory processes in nonneuronal sites.[38]

Similar to p75NTR, other TNFRSF members, DR6, ILGF1R, RELT, and TROY do not bind any of the known TNFSF ligands but nevertheless have immune functions as well as roles in neural and other organs[39,40](see Fig. 10.3). IGFL1R exhibits conserved positioning of cysteines delineating a CRD1 yet contains atypical CRD2/3 regions. TNFRSF engagement of ligands distinct from TNFSF members indicates a higher level of integration with other signaling pathways.

TNF in Evolution

Studies of the evolution of the TNF and TNF receptor superfamilies yield fascinating insights into the importance of these cytokines and receptors in natural selection.[41] TNF is present in all vertebrates, and TNF and TNFR genes have also been described in many invertebrates[42] including fruit flies, shrimps, abalones, sea urchins, sea squirts, mollusks, sea anemones, corals, hydra, and schistosomes. For example, the *Eiger* system in *Drosophila melanogaster* is an evolutionarily conserved TNFSF pathway.[43,44] While the predicted proteomes of Cnidarians, Choanoflagellates, sponges, Ctenophores, bacteria, and plants contain an extracellular CRD, similar to the TNFR family, they do not contain a death domain (DD) and thus would not be expected to be involved in apoptosis.[45] Interestingly, both DD and CRD are present in certain species of coral; one such Cnidaria has 44 proteins with a CRD domain and 13 of these contain a DD.[45]

Homologues of the various TNFRs have been described in early evolution as recently summarized.[46] TNFR homologues present in invertebrates include EDA2R (TNFSF27) in the Zhikong scallop, NGR (TNFRSF16; p75), HVEM (TNFRSF14), LTBR (TNFRSF3), TROY (TNFRSF19), and EDA2R (TNFRSF27) in the purple sea urchin, HVEM (TNFRSF14) in amphioxus, NGR (TNFRSF16; p75), in the sea anemone, and European sea squirt. TNFR homologues have been reported in several teleosts. These include TNFR1 (TNFRSF1A), TRAILR1 (TNFRSF10A), TRAILR2 (TNFRSF10B), FAS (TNFRSF6), DR6 (TNFRSF21), NGR (TNFRSF16), and CD40 (TNFRSF5) in zebrafish. FAS has also been reported in the channel catfish and Medaka fish, and TNFR1 and TNFR2 have been described in goldfish.

In addition to the occurrence of TNF and TNFR structural homologues throughout evolution, evidence also exists for at least some comparable functions. Fish TNF induces changes in the expression of endothelial inflammatory genes but is less efficient at activating macrophages, indicating some differences from the actions of human TNF.[47] TNF and TNFR genes are upregulated under heat stress in corals[48] and in hydra regeneration[49] processes that involve apoptosis. TNFR expression varies during the life cycle of *Schistosoma mansoni*, and despite the fact that no TNF homologue had yet been identified, human TNF treatment resulted in the upregulation of a large set of genes.[50] A recent study of 30 platyhelminth species identified both TNFR and TNF homologues,[51] supporting the concept that indeed the TNF/TNFR superfamilies do indeed contribute to biological processes throughout evolution.

Viral TNFSF Orthologs

As part of their evolving strategies for immune evasion, multiple viral pathogens have captured genes coding for various TNFSF members (Table 10.3).[52] Both receptor and ligand-like viral proteins exhibit structural or functional properties that counteract or mimic those of their cellular orthologs. This provides a means of altering/suppressing the immune response to promote viral replication and dissemination. The genomes of *Orthopoxviruses* contain open-reading frames (ORF) encoding multiple versions of cellular TNFRs.[53] These viral TNFR are typically secreted from infected cells as soluble proteins that act as decoy receptors for sequestering and limiting the bioavailability of TNF and LTα. Termed *cytokine response modifiers* (*Crm*), these proteins are distinguished by their structure, number, and ligand specificity across the various *Orthopoxvirus* species. Four ORFs (*CrmB, CrmC, CrmD, and CrmE*) encode proteins with significant sequence conservation to TNFR2. For instance, the zoonotic Cowpoxvirus (the predecessor to vaccinia virus used as the vaccine against smallpox disease) contains all four *Crm* genes. Human variola (smallpox) virus, the eradicated scourge of mankind, encodes a single viral version of TNFR2 (*CrmB*) that exhibits the highest-known affinity for TNF.[54] Experimental models of infection with rabbit or mouse poxviruses have demonstrated that secretion of Crm proteins from virus-infected cells contributes to disease severity.[55-57] As a strategic ploy, *CrmD* in mousepox also encodes a chemokine-binding domain that, together with the TNF decoy domain, forms a major determinant of virulence.[58] The variation and widespread presence of the viral TNFRs reflects the evolutionary adaptation that Orthopoxviruses have utilized to evade the immune responses of their host species.

TABLE 10.3 Viral orthologs and Modulators of the TNF Superfamily

Virus	Name/ORF	Ortholog/Target	Mechanism
Poxvirus:			
Myxoma	T2	Soluble TNFR2	TNF and LTα Decoy
Variola	crm B (G2R)	Soluble TNFR2	TNF and LTα Decoy
Cowpox	vCD30	Soluble CD30	CD30 ligand inhibitor
Herpesvirus:			
HSV1&2	Envelope gD	BTLA	HVEM blockade
HCMV	UL144	HVEM	BTLA activation
EBV	LMP1	CD40 intracellular	TRAF activation
HHV6b	Envelope gQ	OX40	entry
KSHV	vFLIP K13	FLIP/NEMO	Caspase 8 blockade
Adenovirus	E3-10.4,14.5,6.7	FAS; TRAILR	Downmodulation
Influenza	Hemagglutinin	OX40L	Entry
Retrovirus:			
EIAV	Envelope gp90	HVEM	Entry factor
FIV	Envelope gp95	Ox40	Entry factor
ASLV	Envelope gp85	TRAILR	Entry factor
Rabies virus	Envelope RVG	NTRp75	Entry factor

ASLV, Avian sarcosis and leukosis virus; EBV, Epstein-Barr virus; EIAV, equine infectious anemia virus; FIV, feline immunodeficiency virus; HCMV, human cytomegalovirus; HHV6b, human herpesvirus-6; HSV1&2, herpes simplex virus-1 and 2; KSHV, Kaposi sarcoma herpesvirus. Relationship determined by sequence or structural homology to the indicated TNF superfamily member; cellular target of viral protein.

Herpesviridae directly modulate members of the TNFSF. The virion glycoprotein D (gD) of herpes simplex virus type 1 and 2 (in the α-herpesvirus subgroup) binds CRD1 of HVEM in a manner that mimics the Ig superfamily member BTLA.[59-61] In mice infected with HSV1, gD modulates HVEM signaling to inhibit antibody class switching to IgG2a/c.[62] The *UL144* ORF present in clinical isolates of human and primate cytomegalovirus (β-herpesvirus) encodes a structural ortholog of HVEM containing CRD1 and 2 with additional mutations restricting binding exclusively with BTLA.[61,63,64] UL144 expressed during viral infection induces secretion of selected chemokines that promote responses in type 2 helper and regulatory T cells, thus avoiding the effects of antiviral T and NK effector cells.[65] The inhibitory effects of BTLA on several effector T cell subsets implicate UL144 as a player in key immune evasion strategies used by human cytomegalovirus.

Herpesviruses also utilize pseudoligands to modulate cellular TNFR. Human (*UL141*) and mouse (*m166*) CMV encode modulators of the TRAIL death receptors. These viral ligands bind the ectodomain of TRAILR in the endoplasmic reticulum-blocking receptor transport to the cell surface of infected cells. *M166* is required to suppress the cytolytic activity of innate lymphoid cell-type1 (ILC1), which are important effector cells in the salivary glands and the site of virus dissemination.[66] Another example of a pseudoligand is the virion protein gQ of human β-herpesvirus HHV6b that contains a structurally distinct mimic of Ox40 ligand. gQ uses TNFRSF Ox40 as an entry factor, exhibiting a cellular tropism for T cells.[67,68]

Retroviruses, such as the Avian leukosis-sarcoma viruses, require the TNFRSF member *CAR1* for viral fusion and entry through the viral surface subunit (SU). CAR1 is a homolog of the TRAIL receptor DR5 (TNFRSF10b) that contains a cytoplasmic DD associated with cytopathic activity. Interestingly, human syncytin1 and 2 are homologs of the SU proteins derived from endogenous retroviruses. The syncytins are involved in the formation of trophoblast and macrophage syncytia.[69] The SU gp90 of the lentivirus equine infectious anemia virus targets the HVEM homolog (encoded by *ELR1*) in horses.[70,71] While several single nucleotide polymorphic variants in the ELR1 ectodomain are targeted by SU gp90, the functional residues that are critical for cellular responses are retained, providing a selective advantage to the specific function of this TNFRSF member.[72]

Signaling pathways activated by TNFRSF are impacted by a plethora of virally encoded proteins that act to resist or skew immune responses initiated by death receptor or CD40 signaling. Examples include the latent membrane protein LMP1, the main oncoprotein of Epstein-Barr virus (HHV4).[73,74] LMP1 behaves as a constitutively active TNFR that mimics signaling by CD40 and other TNFRSF members. LMP1 induces signaling through the protein kinases controlled by the TNF receptor-associated factors (TRAF) family of E3 ubiquitin ligases.[75] These kinases in turn activate the NFκB transcriptome to enhance cell survival and promote lymphomagenesis. The β-coronaviruses, including SARS-CoV2, also target members of the TRAF family. This may account for some pathologic aspects of immune dysregulation that lead to release of excess cytokines such as LIGHT in severe COVID patients.[76,77]

Investigations into fundamental mechanisms of viral modulation of the TNFSF provide heuristic insights into the design of drugs, such as etanercept, a soluble dimeric fusion protein of TNFR2[78] used to treat patients with rheumatoid arthritis.[79] Inspired by the actions of viral immune modulators, additional approaches currently under development are directed at modulating immune system responses in cancer, autoimmune diseases, and vaccines.[62,80,81]

SIGNALING PATHWAYS AND CELLULAR RESPONSES

Regulation

The cellular response activated by a TNF-related cytokine depends on several factors, including the temporal patterns of expression of the ligands and receptors in interacting cells and the cellular context (the state of differentiation of the responding cell). A "cellular response network" is defined by expression patterns of the ligands and receptors in various cell types. Signaling via membrane-anchored ligands and receptors serves to restrict responses to the local environment of the interacting cells, whereas secreted ligands or receptors will more broadly affect both the local and regional tissue microenvironments. For example, contact between marginal zone B cells, expressing membrane-anchored LTαβ, and stromal cells in lymphoid organs, displaying LTβR, creates a localized cellular network. In contrast, both local and systemic response networks are generated by membrane and secreted TNF. Organ and cell-type–specific expression patterns of TNFSF members are available at human protein Atlas.org and for mouse, Immgen.org.

Transcriptional, translational, and posttranslational regulatory mechanisms control the expression of the TNFSF ligands and receptors. Expression patterns in certain cell lineages may also be impacted by nonheritable epigenetic modifications of TNFSF and TNFRSF genes that are observed during embryonic and postnatal development.[82] For some ligands, transcriptional activation is a critical feature controlling the duration of mRNA expression. TNF, for example, exhibits inducible, transient expression of mRNA following signals from antigen receptors or innate receptor recognition systems. The half-life of mRNA for several TNF family members is short, controlled by an AU-rich element in the 3′ region.[83] Deletion of this AU-rich mRNA element in mice leads to sustained TNF production that causes profound inflammatory disease.[83,84] TNF mRNA is inducible in macrophages by multiple pathways, particularly innate activation pathways such as toll-like receptor signaling. T and B lymphocytes also require activation prior to expression of TNF. Antigen receptor signaling utilizing NFAT transcription factors is needed for the induction of TNF transcription. Other ligands like 41BBL, OX40L, and CD40L are constitutively expressed in differentiated antigen-presenting dendritic cells. Inducible or constitutive patterns of expression are also observed with some TNFR family members.

Posttranslational regulation of signaling is achieved via proteolytic cleavage of the ligand or receptor from the cell surface. This releases the protein as a soluble fragment whose half-life may be dramatically reduced. TNF and FAS ligand for example are shed from the cell surface by membrane proteases. ADAM17 (also known as TNFα converting enzyme, TACE), the enzyme that releases TNF in soluble form, is involved in the shedding of multiple cell surface proteins including TGFα, L-selectin, and TNFR1 and 2. Production of soluble TNFR1 and 2 may be important in regulating TNF bioavailability.

Many of the TNFSF ligands are expressed in dendritic cells, activated lymphocytes, and myeloid cells, particularly macrophages, and some can also be produced by nonlymphoid cells. TNF itself is an example of a ligand that can be expressed by many cell types, depending on the type of stimulus. Expression of TNFR1 is also widespread, whereas TNFR2 expression is limited to cells of hematopoietic origin, endothelial cells,[85] and activated B or T cells. Macrophages are a primary source of TNF in response to TLR signaling, and T cells produce TNF when activated by antigenic stimuli. Fibroblasts also produce TNF in response to virus infection, and nonlymphoid tumor cells may ectopically express TNF. FASL is another example of a ligand with a varied cellular expression pattern. FASL is expressed by effector T cells and NK cells as a key element of their cell lytic activity. Yet FASL mRNA is also detected in reproductive organs and epithelium of the eye, sites at which this TNFSF member may be used to kill organ-infiltrating inflammatory cells as a means of damping inflammation.

Signal Transduction Pathways

TNFRs initiate signaling pathways that alter gene expression patterns to change the differentiation status of a cell. They also activate apoptotic pathways that terminate cellular life.[86] The propagation of signals from receptor to enzymatically active proteins is mediated by two distinct types of signaling motifs in the cytosolic domain of the TNFR: death domains and TRAF-binding motifs. TNFR family members that contain a death domain (DD) include FAS, TNFR1, DR3, and TRAILR1 and 2, and others (see Fig. 10.3). Other TNFR have TRAF recruitment motifs or lack signaling domains. Three basic schemes are used by TNFRs to activate enzymatically driven signaling pathways. Adaptor molecules are required to establish connections between the receptor and downstream signaling enzymes. The DD connects TNFR to cytosolic proteins containing a death effector domain (DED), which in turn provide a link to caspases (cysteine-based, aspartic acid–specific proteinases). Alternatively, TRAF proteins connect to the cytosolic domain of the TNFR, altering ubiquitin-dependent pathways that regulate key protein kinases needed to activate NFκB and AP1 transcription factors. The third scheme involves an indirect link between the DD and TRAFs via the adaptor TNFR-associated death domain (TRADD).

Cell death and survival pathways activated by the TNFR family help regulate cellular homeostasis by controlling cell death and survival. In the immune system, apoptosis is essential for homeostasis and for eliminating pathogen-bearing cells from the host. Most TNFRs can induce activation of survival and/or death pathways depending in part on the differentiation status of that cell. In all nucleated cells, apoptosis is the default pathway; that is, all of the constituents of the pathway are expressed and ready to be activated. In contrast, cellular survival requires transcriptional control of genes encoding regulatory proteins that suppress the progression of the apoptotic pathway.

Cell Death Signaling

Apoptosis is the process of programmed cell death, well known to occur during embryonic development. Apoptosis results in membrane blebbing, cellular shrinkage, DNA degradation, chromatin condensation, and eventually phagocytosis to clear cell debris.[87] Apoptosis can also be induced by several members of the TNF family whose receptors contain DDs, including TNFR1, FAS, and TRAILR1and 2, among others (Fig. 10.4). Ligand binding of the DD domain containing TNFRs promotes cytosolic assembly of complex I. Apoptotic signaling by FAS recruits the adaptor FAS-associated death domain (FADD), which directly links to caspase 8. TNFR1 signaling uses TNFR-associated death domain (TRADD) as the adaptor. These adaptors promote dimerization of procaspase 8 to form an active enzyme complex,[88] called complex II or cell death-inducing signaling complex (DISC). Receptor-interacting protein kinase 1 (RIPK1) is also included in complex II. It was originally thought that RIPK1 was not involved in apoptosis, but recent data suggest that it does play a role in apoptosis in embryogenesis[89] and in complex IIb. Activated caspase 8 acts directly to cleave procaspases 3 and 7 that are known as the executioner caspases as they directly cleave critical cellular substrates leading to apoptotic death. Caspase 8 also cleaves BID, a crucial connector to the mitochondria-associated death mechanism, which greatly amplifies the apoptotic process.[90] A cell must be capable of actively transcribing and translating genes to resist apoptosis signaling. A variety of genes can inhibit apoptosis and cell death pathways. For example, the cellular inhibitor of apoptosis (XIAP) is a direct caspase inhibitor that is regulated at the level of gene expression by transcription factor NFκB. Another regulator, FLIP (FLICE inhibitory protein), is also an NFκB-regulated gene that contains a DED and a pseudocaspase domain that attenuates the apoptotic pathway by blocking conversion of proCaspase 8 to the active form.[91] Many viral pathogens parasitize the transcriptional capabilities of the cell and prevent the cell from making new survival proteins, allowing apoptosis to proceed. As expected, viruses have evolved many different strategies to alter proapoptotic pathways; for example, viral orthologs of XIAP and FLIP prevent premature death of the cell.

Necroptosis is another form of programmed cell death that is mediated by several members of the TNFR family. These include TNFR1, FAS, and TRAILR in addition to other activating receptors such as TLR3, TLR4, IFNγR, and the T cell receptor (see Fig. 10.4). Necroptotic cells exhibit swelling and rupture of cell membranes and thus are morphologically distinct from apoptotic cells. Necroptosis occurs predominately

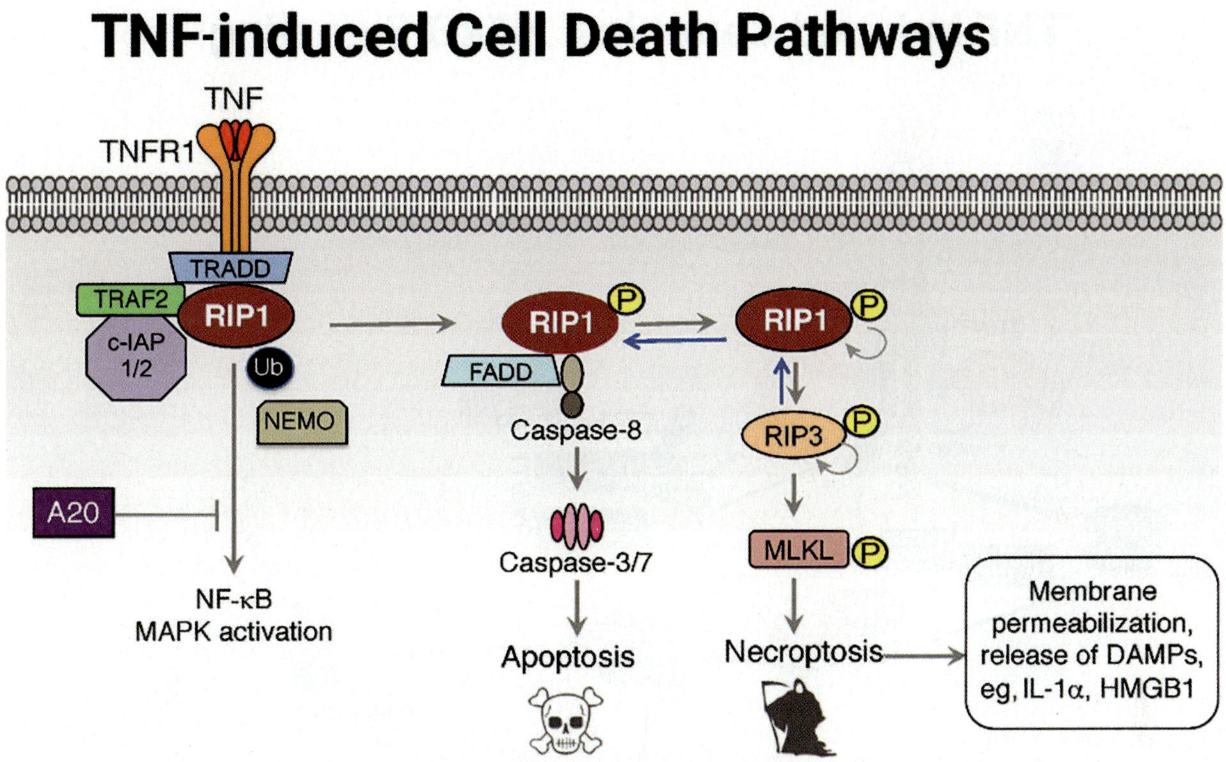

FIG. 10.4. **Tumor necrosis factor (TNF)-induced cell death pathways.** Inhibition of NF-κB and MAPK signaling can divert TNF-mediated signaling to the formation of an intracellular complex II centered on FAS-associated death domain (FADD) and caspase-8 in a RIP1-dependent apoptotic cell death. This cell death pathway can be augmented by A20 or by the absence of the E3 ligases c-IAP1/2 or LUBAC, thereby eliminating the ubiquitination of complex I components and promoting the switch to complex II. Activation of RIP1-dependent cell death under caspase-8 inhibited or deficient conditions can lead to a necroptotic form of cell death that is mediated by kinase activity of RIP1 and RIP3 and results in the activation of mixed lineage kinase domain-like protein (MLKL) and membrane permeabilization. (Figure courtesy of Joshua D. Webster and Domagoj Vucic and reproduced with permission from Webster JD, Vucic D. The balance of TNF mediated pathways regulates inflammatory cell death signaling in healthy and diseased tissues. *Front Cell Dev Biol.* 2020;8:365.)

under conditions in which the apoptotic pathway fails due to the absence of caspase 8 activity. The process is mediated by the "necrosome" consisting of RIPK1 and RIPK3. The phosphorylation of RIPK1 and RIPK3 results in activation of mixed lineage kinase domain-like protein (MLKL). Once activated, MLKL trimerizes and is inserted into the cell membrane.[92] The mechanism of the ensuing cell death is not completely settled. The pore may release cellular contents and/or act as an ion channel. Although apoptosis appears to be the default pathway, a role for necroptosis has been suggested in cancer defense, embryonic development, resistance to some pathogens, and some pathologies.[93]

Cell Survival Signaling

TRAF regulators. Death is the default pathway for cells that respond to TNFR1 signaling and signaling by other DD-containing TNFR that are unable to activate cell survival pathways. Cellular survival signaling by TNFRSF in innate and adaptive immune responses is initiated by the TRAF family of ubiquitin E3 ligases (Fig. 10.5).[75,94,95] Control of survival signaling is accomplished through a cascade of ubiquitin (Ub)-conjugating enzymes and protein kinases. Protein conjugation with ubiquitin (ubiquitination) is a dynamic process that plays an essential role in regulating signal transduction by TNFRSF members[96,97] and by various pathogen recognition receptors and cytokines as part of innate host defenses.[98] There are seven members of the TRAF family (given numerical designations) that play key roles in regulating TNFR signaling and activation of host defenses. Each TRAF protein appears to play a different role in modulating signaling either to promote or limit signaling pathways that activate transcription factors. TRAF proteins contain a TRAF homology domain and a RING and/or zinc finger domain characteristic of E3 ubiquitin ligases.[99,100] The TRAF homology domain directly engages the cytosolic tail of ligand-bound receptor.[101] The receptor-binding groove in TRAF3 also serves to recruit protein substrates to the E3 Ub ligase complex. The ubiquitination status of the various protein substrates determines whether the protein undergoes degradation or stabilization. E3 ubiquitin ligase activity within the RING domain targets specific substrates for proteasomal degradation using K48-polyubiquitin linkages or stabilization with K63-polyubiquitin linkages for participation in specific signaling pathways.

Ligand-driven TNFR clustering promotes the recruitment of TRAFs into a cytosolic complex with the receptor through two different mechanisms. TNFR1 uses TRADD to couple with TRAF2 and TRAF5. In non–DD-containing TNFR,

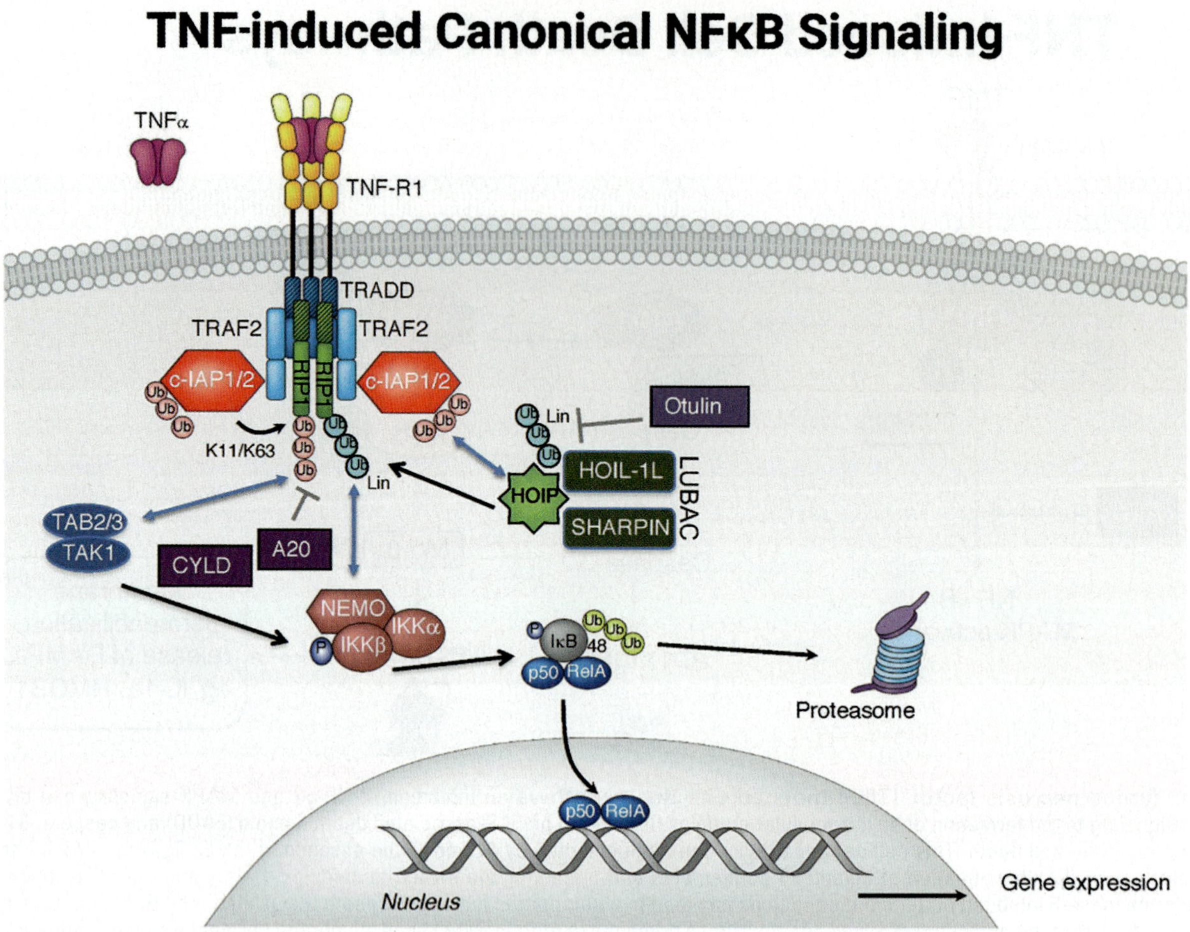

FIG. 10.5. Tumor necrosis factor (TNF)-induced canonical NFkB pathway. TNF clustering of TNFR1 triggers the recruitment of TRADD, TRAF2, RIP1, and c-IAP1/2 to cytosolic domain. E3 ligases c-IAP1/2 polyubiquitinate themselves and RIP1 with K11 and K63 ubiquitin linkages, creating a platform for further recruitment of LUBAC. LUBAC mediates linear polyubiquitin, resulting in gene expression via the IKK complex. Several deubiquitinases (DUBs) have been implicated in the regulation of TNFR1-associated complex I by removing linear (CYLD and OTULIN) or K63-linked polyubiquitin chains (A20 and CYLD). (Figure courtesy of Joshua D. Webster and Domagoj Vucic and reproduced with permission from Webster JD, Vucic D. The balance of TNF mediated pathways regulates inflammatory cell death signaling in healthy and diseased tissues. *Front Cell Dev Biol.* 2020;8:365.)

TRAF2 and TRAF3 bind directly to the receptor's cytosolic domain via recognition of a consensus PXQS/T motif. TRAFs are preassociated with other E3 Ub-ligases in complexes involved in ubiquitin conjugation and deubiquitination. For example, TRAF2 and/or TRAF3 form complexes with cellular inhibitor of apoptosis protein-1 and 2 (cIAP1/2) that target protein kinases NIK and RIP1 for proteosome degradation. The constitutive degradation of these kinases prevents activation of transcription factors in the absence of receptor activation.[102] Several other Ub E3 ligases serve as negative regulators of the TNFR1 pathway. Genetic deficiencies in the negative regulators such as *A20* and *HOIL* lead to autoinflammatory syndromes.[103]

NFκB is major control point of transcriptional regulation. Most TNFRSF members activate the nuclear factor-κB (NFκB) family of transcription factors that regulate expression of genes critical for cell survival, inflammation, immune responses, and lymphoid organ development.[104] In mammalian cells, the NFκB family consists of five members: RelA, RelB, c-Rel, p50/NFκB B1, and p52/NFκB B2 (proteolytic processing of p105 and p100 yields the active forms p50 and p52, respectively).[105] The activation processes triggered by TNFRSF members release NFκB family members from inhibitors that restrict their nuclear translocation.[106] Homo and heterodimers of NFκB family members are held inactive in the cytosol by the inhibitors of κB (IκB), such as IκBα, which mask nuclear localization motifs on the NFκB dimers. An IKK complex consisting of the kinase catalytic subunits IKKα and IKKβ and the regulatory/scaffold subunit IKKγ form the IKK complex that mediates phosphorylation and ubiquitination of IκB leading to its degradation and subsequent release of the active transcription factor for translocation into the nucleus. IκB is the common target of a variety of signals that control the activation of the RelA NFκB transcription factor (often referred to as the canonical pathway), which in turn regulates expression of many proinflammatory genes.[107] A distinct pathway (the noncanonical pathway) regulates the activation of RelB NFκB through the NFκB-inducing kinase

(NIK) and IKKα kinase. In unstimulated cells, NIK levels are maintained at vanishingly low levels by ubiquitination and proteosome degradation via an E3 ligase complex comprised of TRAF3, TRAF2, and cIAP1/2.[99] Receptor ligation and recruitment of TRAF3 to the receptor competitively displaces NIK from the shared binding groove in TRAF3, preventing further ubiquitination and degradation of NIK. Additionally, TRAF3 also undergoes ubiquitin-mediated degradation by cIAP and TRAF2 allowing accumulation of NIK.[108] NIK in turn phosphorylates IKKα, which induces the proteosome-dependent processing of p100 to p52. Degradation of the inhibitory domain of p100 allows nuclear translocation of the RelB/p52 complex for transcriptional regulation.

As examples, TNFR1 activates the canonical RelA/p50 pathway, whereas LTβR signaling primarily activates the noncanonical RelB/p52 pathway.[109] Each form of NFκB promotes activation of a large number of genes with distinct roles in physiology. TNFR1 signaling rapidly mobilizes (within minutes) the nuclear translocation of RelA/p50 complex that drives inflammatory responses for host defense. In contrast, the processing of p100 and associated accumulation of nuclear RelB/p52 occurs over hours after the initial stimulus. RelB-dependent genes are often involved in lymphoid tissue organogenesis and homeostasis. NIK is also required for NFκB RelB and RelA activation by CD27, CD40, and BAFF-R, but not by TNFRI, which is restricted to activating the RelA/p50 complex. Components of the NFκB pathway including TRAF3, cIAP, NIK, and A20 are frequently mutated in cancer, leading to constitutive expression of survival factors such as the BCL2 family (Fig. 10.5).[110]

TNF-LYMPHOTOXIN-LIGHT NETWORK

TNF; formerly TNFα and lymphotoxin-α (*LTα*; formerly TNFβ) were originally discovered and characterized as inducers of tumor cell death, holding promise as antitumor therapeutics.[111-114] However, the potent inflammatory action of TNF, particularly in the cardiovasculature where it causes a shock-like syndrome, was quickly realized in cancer patients injected with recombinant TNF protein. TNF and LTα are now recognized as two components of an interconnected network of "signaling circuits" that include LTαβ, LIGHT (TNFSF14), and their specific receptors and regulatory proteins. Each individual pathway has unique as well as cooperative signaling activities with other members of this network. This signaling controls immunological processes ranging from the development and homeostasis of lymphoid organs, the mobilization of innate defense systems, and cosignaling activity that promotes and limits innate and adaptive immune responses.[115]

The trimeric forms of membrane-bound TNF and soluble LTα both bind and activate two distinct receptors, TNFR1 (p55-60, *TNFR1A*) and TNFR2 (p75-80, *TNFR1B*). Soluble TNF binds both receptors but does not stimulate TNFR2.[116] The heterotrimeric LTα1β2 complex exclusively binds and activates the LTβR. LIGHT bridges signaling via the LTβR and the HVEM, TNFRSF14, serving as a hub between inflammatory and inhibitory pathways.

TNF, Initiator of Inflammation

TNF mediates a diverse range of cellular and physiological responses linked to acute and chronic inflammatory processes. This diversity of responses is due in part to the broad expression pattern of TNFR1 and the release of TNF as a soluble mediator that can act in a systemic fashion. TNF production is triggered by many of the innate recognition systems, such as the toll-like receptors, and by T and B cells. In chronic inflammation, TNF production can lead to tissue damage and organ failure. Genes expressed in response to TNF signaling coordinate the physiological responses that occur during inflammation. Specific cellular responses reflect the characteristic of the inflamed organ, the local or systemic source of TNF, and the duration of TNF signaling. Acute inflammatory responses involve rapid changes in hemodynamics (plasma leakage and edema) and leukocyte adherence, extravasation, and organ infiltration. TNF production during chronic inflammation may contribute to systemic metabolic derangements and wasting (cachexia) or loss of organ structure and function (bone erosion in joints of patients with rheumatoid arthritis). In some models, TNF can promote an inflammatory environment that may enhance tumor formation[117] (see below).

Elucidation of the functions associated with this TNF/LT signaling network has been aided by studies with genetically modified mice engineered for null or transgenic expression of the cytokine or receptor (Table 10.4). Deficiencies in TNF or TNFR1, but not TNFR2, have similar phenotypes with alterations in host defense against intracellular bacterial pathogens like *Listeria monocytogenes* and *Mycobacterium tuberculosis* but surprisingly modest effects on susceptibility to some viral pathogens. These results demonstrate a role for TNF in acute phase response of the host-defense system. In contrast, LTα, LTβ, and LTβR gene-deficient mice exhibit failure in the development and organization of lymphoid organs, a phenotype not observed in mice deficient in either TNFR1 or 2 or TNF. This phenotype demonstrates the LTαβ complex signaling through the LTβR which is a key developmental pathway for lymphoid organogenesis.

LTαβ-LTβR: A Mammalian Organ Developmental Pathway

Mice that exhibit profound defects in lymphoid organs have revealed the framework of a signaling pathway involved in mammalian development. LTα, LTβ, and LTβR-deficient mice lack all or most lymph nodes and Peyer patches and exhibit defects in nasal-associated lymphoid tissue development, splenic organization, and germinal center formation.[126,127] Several other knockout mice including those lacking the transcriptional regulators Ikaros, ID2, and RORγt also lack lymph nodes.[128-132] This phenotype is shared with mice deficient in components of the NFκB activation pathway, including TRAF6, NIK, IKKα, and RelB and their gene targets. Mice with deficiencies in target genes including CXCR5, the receptor for CXCL13, exhibit defective lymphoid organ structure, as do CCR7-/- mice (receptor for CCL19/CCL21). The developmental program of secondary lymphoid organ

TABLE 10.4 Developmental Phenotypes of Mice Deficient in the TNF-LT-LIGHT Network

	Gene Deletion		Phenotypes:			
	LN	**PP**	**Architecture**	**NK**	**NKT**	**DC**
LTα	–	–	Disrupted	impaired	impaired	CD8-DC
LTβ	–[a]	–	Disrupted	impaired	impaired	CD8-DC
LIGHT	+	+	+	+	+	+
LTβ-B	+	+	Disrupted	nr	nr	Nr
LTβ-T	+	+	+	nr	nr	Nr
TNF	+	+	+	+	+	Maturation
LTβR	–	–	Disrupted	–	–	CD8-DC
TNFR1	+	–	disrupted MZ	+	+	Maturation
TNFR2	+	+	+	+	+	+

LN, lymph nodes[118]; PP, peyer patches[119,120]; architecture of splenic white pulp includes T- and B-zone segregation; marginal zone (MZ), germinal center and follicular dendritic cell network[119,120]; Natural killer (NK)[121]; NK-T cells Vα14 subset[122]; Dendritic Cell[123]; [a]LTb–/– retains mesenteric and sacral LN; LTβ conditionally deleted in B cells or T cells[124]; absent, –; normal, +; nr, not reported.

Modified with permission from Ware CF. Network communications: Lymphotoxins, LIGHT, and Tnf. *Annu Rev Immunol.* 2005;23:787-81.

formation in mice initiates at 9 days post coitus, progressing in organized fashion from posterior to anterior, with intestinal Peyer patches and NALT forming during the first postnatal week.[133] This phenotype is irreversible in LTα-deficient mice in that transferring LTαβ-sufficient bone marrow into an adult LT-mutant mouse fails to induce lymph node formation. However, the defect in lymph node formation can be corrected using replacement therapy *in utero* with an agonist mAb to LTβR.[134] The formation of lymphoid organs involves at least two distinct cell types: (1) an embryonic LTβR expressing mesenchymal stromal cell that responds to LTαβ expressed in a hematopoietically derived lineage, termed the lymphoid tissue inducer (LTi) cell and (2) a mesenchyme-derived stromal organizer cell (LTo) (Fig. 10.6). Additional stromal cells have been implicated both in normal lymphoid development and in chronic inflammation.[135,136] The LTαβ-expressing LTi belongs to the innate lymphoid cell family (ILC3). This CD4+CD3–IL7R+ID2+ RORγt+

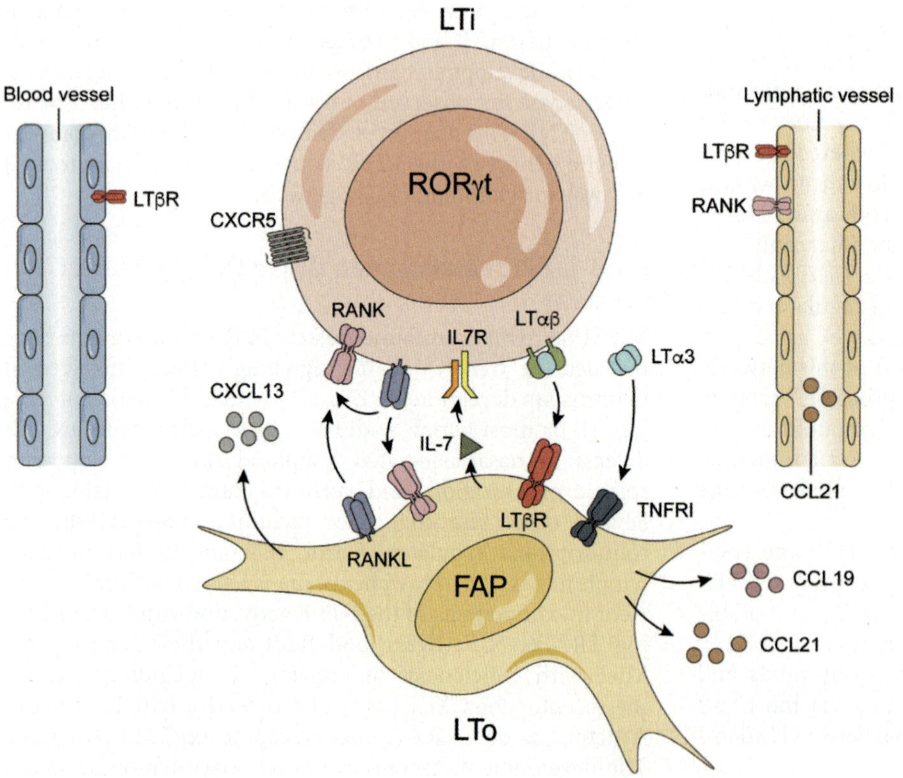

FIG. 10.6. Lymph node development. Note the interactions between the lymphoid tissue inducing cell (LTi) and the lymphoid tissue organizing cell (LTo) and contributions of lymphatic vessels and high endothelial venules. (Modified from Ruddle NH. Basics of inducible lymphoid organs. *Curr Top Microbiol Immunol.* 2020;426:1-19.)

cell participates in a pathway dependent on several cytokines including autocrine production of RANKL-RANK and IL7, produced by the LTo. These cytokines induce surface LTαβ in LTi cells that engenders formation of lymph nodes and Peyer patches.[137] LTi lineage cells are maintained in the adult tissues at low levels (0.5%), aiding in the homeostasis of lymphoid organs.[138,139]

The microarchitecture of the white pulp in the spleen but not its early development[140] (Fig. 10.7) is also disrupted in LT- and TNF-deficient mice.[141] Multiple additional features are observed in LT- and TNF-deficient mice, including the absence of differentiated macrophages in the marginal sinus, the loss of positional segregation of T cells and B cells into discrete zones, and failure of germinal center formation during adaptive immune responses. The segregation of T and B cells into discrete compartments depends on expression of the tissue organizing chemokines CXCL13, a chemotactic factor for B cells, and CCL19 and CCL21, which attract T cells and dendritic cells. CCL19 and CCL21 act through the chemokine receptor CCR7 expressed on T cells and dendritic cells, while CXCL13 binds CXCR5 on B cells to promote localization in the follicles. An LT-chemokine circuit is formed by migration of B cells to LTβR+ stromal cells expressing CXCL13, which in turn induces B cell expression of LTαβ.[142-144] Circulating B cells lack surface LTαβ, but expression is regained upon reentry into the CXCL13 rich microenvironment.[144] Formation of the splenic microarchitecture depends on B cell expression of LTαβ, which induces differentiation of specialized stromal cells (eg, secretion of CCL20 and CXCL13 chemokines) in the spleen during postnatal maturation. Remodeling of the microarchitecture in the secondary lymphoid organs occurs during immune responses, requiring signaling via both TNF and LT pathways in fibroblastic reticular cells.[145]

The LTβR and TNFR pathways facilitate lymphocyte entry into lymphoid tissues in part by modulating expression of lymphoid chemokines and adhesion molecules, such as peripheral node and mucosal addressins on high endothelial venules.[118] LTβR signaling is necessary for maintaining networks of follicular dendritic cells involved in capturing antigen and immune complexes that aid in activating B cells. These cellular interactions are further enhanced by LTβR signals that enhance growth of some conventional myeloid DC ($CD8^-$ subsets) within the lymphoid organ. In contrast, TNF plays a role in differentiation of DC progenitors in bone marrow.[123]

T cells and B cells can communicate with stromal and myeloid cells via the LTαβ-LTβR pathway and in doing so are able to modify their immediate microenvironment during the course of an immune response. Nonlymphoid tissues impacted by chronic inflammation associated with autoimmune disease, graft rejection, microbial infection, or cancer often contain organized accumulations of lymphocytes with many characteristics of secondary lymphoid organs.[135] These cellular accumulations are called tertiary lymphoid organs (TLOs), tertiary lymphoid tissues, tertiary lymphoid structures, ectopic lymphoid tissues, or inducible lymphoid tissues (Fig. 10.8). Activated T and B cells provide the source of LTαβ that helps drive the formation of these structures. These structures lack the permanence of lymph nodes since they resolve as antigen is cleared and immune responsiveness subsides. Another difference from lymph nodes is the absence of a capsule. TLOs exhibit a gradation of features, including the presence of dendritic cells, follicular dendritic cells, chemokines, high endothelial venules, lymphatic vessels, segregated regions of T and B cells, and germinal centers. TNF also contributes to formation of granuloma tissue that assists in walling off bacteria but can contribute to organ pathology.[146]

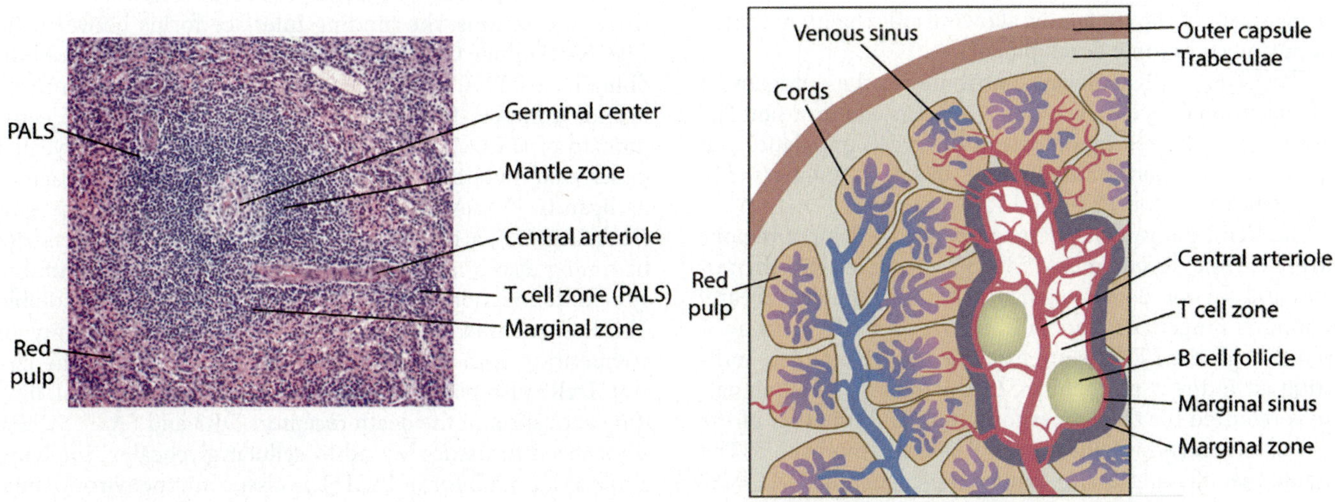

FIG. 10.7. Microarchitecture of the spleen (*left panel*), Staining with hematoxylin (*blue*, nuclei) and eosin (*red*, cytoplasm) of a normal human spleen. The staining shows the demarcation of the *red* and *white* pulp regions of the spleen. The *red* pulp is primarily RBC which lack nuclei and stain with eosin, whereas in the *white* pulp, the nuclei of lymphocytes stain *blue*. (*right panel*), Diagrammatic rendering of the *white* pulp of the spleen demonstrating the site of cell entry through the central arteriole. T cells are located in the periarteriolar lymphoid sheath (PALS), B cells in follicles. (Left panel, Photograph courtesy of David Hudnall [reproduced with permission from Krause PJ, Kavathas PB, Ruddle NH. *Immunoepidemiology.* 2020;35. © Ruddle 2020.] and Right panel, reproduced with permission from Krause PJ, Kavathas PB, Ruddle NH. *Immunoepidemiology.* 2020;35. © Ruddle 2020.)

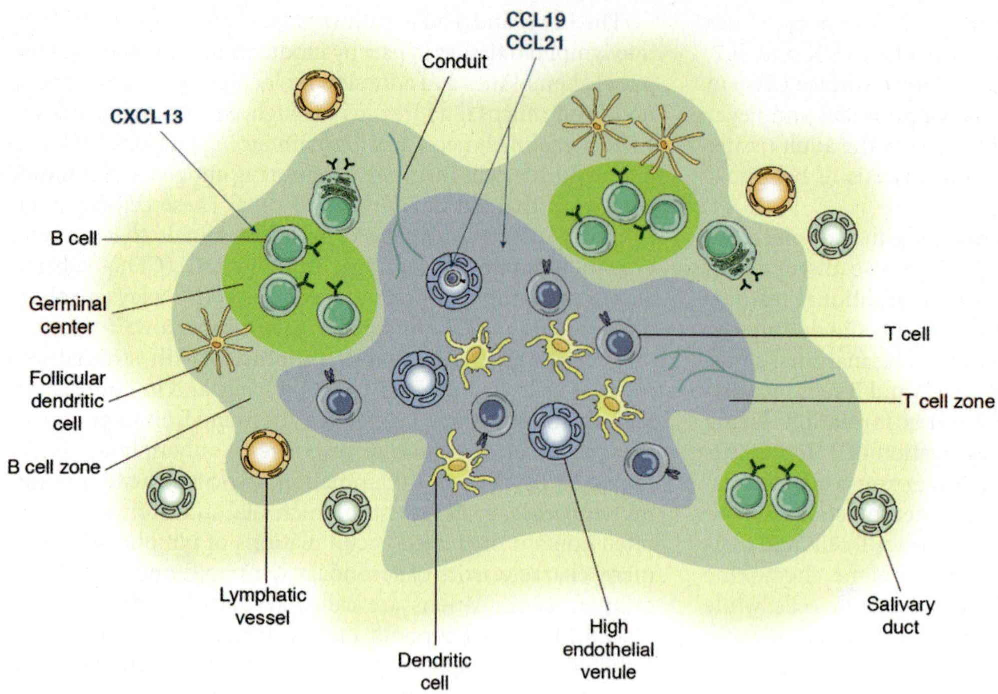

FIG. 10.8. Tertiary lymphoid tissue. A diagram of an organized accumulation of lymphoid cells in the salivary gland of an individual affected with Sjögren syndrome, an autoimmune condition. There are extensive similarities in cellular content and organization to that of a lymph node. (Reproduced with permission from Krause PJ, Kavathas PB, Ruddle NH. *Immunoepidemiology.* 2020;35. © Ruddle 2020.)

These lymphoid accumulations are capable of presenting and responding to antigen and can be either detrimental in the case of autoimmunity or beneficial in the case of cancer. Thus, the LTαβ pathways that operate during embryonic life also play critical roles in the formation of TLOs in adults. Note that in the adult, additional cytokines, such as TNF and LIGHT, and downstream chemokine signals can substitute for members of the LT family, allowing a flexibility not present during embryonic development.

The LTαβ-LTβR pathway contributes to the ontogeny of unconventional T cells, including γδ T cells and invariant NK T cells, as evidenced by deficits in these cell populations in knockout mice. Conventional T cell subsets are unaffected in mice deficient in the TNF and LTβR pathways.

The LTβR pathway appears to operate in distinct regions during thymic development.[147] Double positive (DP) thymocytes and thymic B cells regulate the differentiation of early thymocyte progenitors and γδ T cells via the LTβR pathway,[148] even though the LTβR is not expressed in thymocytes, indicating an indirect mechanism. In this regard, LTβR signaling is required for the proper formation and function of the thymic stroma, which influences T cell development.[149] The thymic medulla appears to control the export of iNKT cells from the thymus.[150,151] In addition, LTβR signaling in thymic stroma affects central tolerance to peripherally restricted antigens, which may be either dependent or independent upon the autoimmune regulator *Aire*.[152,153] Thymic differentiation depends in part on the LTβR pathway for mediating cellular communication between lymphoid and stromal compartments.

LIGHT-HVEM-BTLA Circuit: Signaling for Effector Cell Homeostasis

HVEM engages two TNFSF members, LIGHT and LTα in CRD2 and 3. HVEM also binds three members of the immunoglobulin superfamily, B and T lymphocyte attenuator (BTLA), CD160 and its namesake, virion glycoprotein D (gD) of herpes simplex virus 1 and 2 (HSV1/2) (Fig. 10.9).[115,154] In these interactions, the binding interface forms between the first N-terminal CRD in HVEM and the immunoglobulin domain in BTLA,[60] CD160,[155] and gD.[156] Even though these ligands compete for the same general surface in the ectodomain of HVEM, the contact residues are distinct revealed in selected mutations that distinguish the binding of each of its ligands.[80,157] SALM5 (encoded by *leucine-rich fibronectin III-5(LRFN5)*), a neuron-specific protein linked to dendrite trimming also binds HVEM indicating a neuronal-immune system connection.[158] Decoy receptor-3 (DcR3) is a soluble TNFRSF member that binds LIGHT with high affinity in competition with HVEM and LTβR.[10,159] Extending the circuit, DcR3 also binds TNFSF members TL1A and FASL limiting activation of the death receptors DR3 and FAS.[9,11] DcR3 is localized in tissues bound to cellular glycocalyx, implying a role as an inhibitor of LIGHT in tissue microenvironments. DcR3 is associated with pathological processes, including cancer and autoimmune diseases.[160]

HVEM serves as a regulatory hub between the inflammatory action of LIGHT and inhibitory signaling through BTLA. The broad expression patterns of HVEM and BTLA in the hematopoietic system indicate wide-ranging regulatory roles in immune responses. BTLA contains a

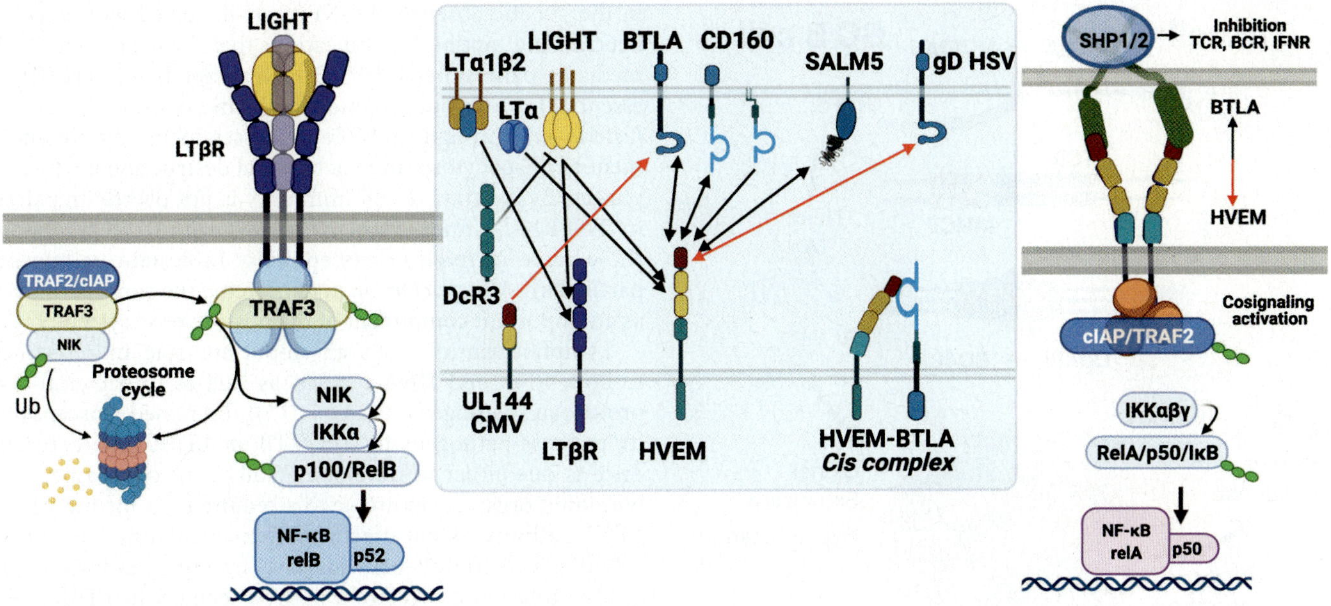

FIG. 10.9. LIGHT-HVEM-BTLA: Signaling for effector cell homeostasis. (*central panel*), Arrows indicate the direction and specificity of receptor-ligand interactions, bidirectional signaling indicated with dual arrowheads. (*left panel*), LIGHT-LTβR signaling via TRAF2, TRAF3, and cIAP activate the formation of noncanonical NFkB relB/p52 transcription complex. NIK is held at low levels by a Ubiquitin-proteosome–dependent cycle that is redirected by LTβR activation. LTβR induces TRAF2 and TRAF3 degradation allowing accumulation of NIK and in turn activates the IKK kinase-driven NFκB-forming cascade. (*right panel*), HVEM engages BTLA-activating SHP1/2 phosphatases which in turn attenuates signaling by T and B cell antigen receptors and IFN-I signaling. Bidirectional signaling by LIGHT, LTα, BTLA, or CD160 can activate HVEM inducing the canonical relA/p50 NFκB transcription factor. Herpesviruses proteins gD (HSV) and UL144 (HCMV) mimic BTLA and HVEM, respectively (*red arrows*); binding domains of HVEM denoted for TNFSF (*gold*) and Ig ligands (*red*). (Figure generated using BioRender [BioRender.com].)

cytosolic immunoreceptor tyrosine inhibitory motif (ITIM) that recruits SHP1/2 tyrosine phosphatases to mediate signaling through the IFNαβ receptor, IL7-receptor, and T and B cell antigen receptors.[115,161]

HVEM is also coexpressed in T and B lymphocytes forming stable complexes *in cis* with either BTLA or CD160. The HVEM-BTLA complex formed *in cis* limits HVEM-mediated activation of the TRAF2-NFκB inflammatory pathway, which is thought to maintain lymphocytes a naïve state within an inflammatory environments.[162] The high avidity of membrane LIGHT for HVEM disassociates BTLA from the cis complex, terminating the inhibitory signaling of BTLA. In turn, LIGHT initiates HVEM signaling to enhance effector cell responses.

During cell to cell contact, HVEM and BTLA mediate bidirectional signaling that initiates NFκB activation in HVEM+ cells and inhibitory signaling in BTLA+ cells.[163] For example, signaling *in trans* occurs between T and B cells during the germinal center reaction that generates class-switching and high-affinity antibodies (Fig. 10.10).[25] Cell type-specific gene-deficient mice have revealed that HVEM limits the T follicular helper signals delivered to B cells to influence germinal center response. The inhibitory pathway is formed by B cell HVEM engagement of BTLA in T cells. BTLA in T cells utilizes SHP1 phosphatase to negatively regulate the TCR signaling. Limiting TCR signaling restricts the quantity of preformed CD40L that is mobilized to the T cell surface to activate CD40 in B cells, a requirement for the germinal center antibody response.

CD160 participates in the HVEM-BTLA circuit to regulate homeostasis of effector cells.[165] CD160 is coexpressed with BTLA in T effector memory and γd T cells, in NK and NK T cells, and in other inflammatory effector cell populations. CD160 lacks a signaling domain but competes with BTLA for binding to HVEM, limiting BTLA's inhibitory signaling to allow enhanced responsiveness of innate effector populations. Additionally, human CD160 contains a cytosolic domain that activates HVEM in NK cells, whereas HVEM activates BTLA to limit NK-induced inflammatory activities.

The homeostatic features of BTLA are observed in γdT cells, which act as innate effector cells designed to mobilize rapidly in response to pathogens at tissue barrier sites. Pathologic inflammation caused by γd T cells is limited by the coordinated expression of BTLA via the action of RORγt and IL-7. BTLA transcription in γδ T cells is limited by the transcription factor RORγt, whereas IL-7 increases BTLA protein expression. BTLA signaling limits γδ·T·cell numbers and sustains normal γδ·T cell·subset abundance by restricting IL-7 signaling and expansion of the RORγt(+) population. BTLA plays a key role in balancing the suppressive and activating stimuli that regulate effector T cell homeostasis and inflammatory responses. The HVEM-BTLA pathway is required to

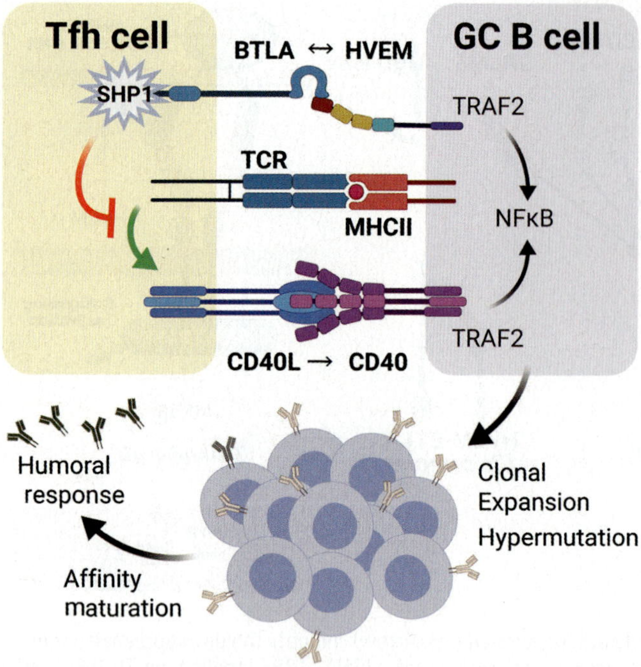

FIG. 10.10. HVEM and BTLA regulate T and B cell collaboration in the germinal center. Visual description of antigen-activated T follicular helper cells (Tfh) engaging antigen-presenting B cells in the germinal center in lymph organs. Antigen recognition between the T cell receptor (TCR) and B cell MHC-II forms an immune synapse promoting TCR tyrosine kinase signaling. TCR signals mobilize CD40 ligand to the cell surface activating CD40 in B cells. CD40 signaling is essential for activated B cells to proliferate and undergo Ig class-switching and somatic mutation of IgV regions for selection of high-affinity antibodies. HVEM expression in B cells activates SHP1 tyrosine phosphatase limiting TCR signals and decreasing the mobilization of CD40L to the cell surface. Bidirectional signaling from BTLA to HVEM promotes survival and proliferative responses in B cells via activation of the NFκB transcriptome. Together, the HVEM-BTLA serves as a homeostatic mechanism that limits CD40L-CD40 signals to restrict the degree of B cell clonal expansion and hypermutation. Mutations in HVEM, often present in follicular lymphoma,[164] can cause a loss of BTLA inhibitory signaling that may contribute to lymphomagenesis.[25]

promote survival of memory and effector T cells following clonal expansion. This pathway appears particularly effective in controlling CD4 T cell activation[166] but is also important in regulating CD8 T cell memory.[167-169]

DISEASE ASSOCIATIONS OF THE TNF-LT-LIGHT NETWORK

Infectious Diseases

TNF is a major inflammatory cytokine required for orchestrating the acute phase response to pathogens, especially bacteria (Table 10.5). For instance, lipopolysaccharide (LPS) in gram-negative bacteria is a potent and rapid inducer of TNF secretion by macrophages via the TLR4 innate recognition system. In mice, LPS induces a systemic shock syndrome that is rapidly lethal owing to profound changes in the cardiovascular system. However, mice survive LPS treatment in the genetic absence of TNFR1 or if treated with a TNF-neutralizing antibody, indicating that host-derived TNF mediates pathogenesis.[188,189] On the other hand, TNFR1 is essential for resistance to infection by live organisms, such as *Listeria m.*, through multiple processes including enhancement of phagocytosis and bactericidal destruction by macrophages. By contrast, T cell immunity is not overtly impaired in TNFR1−/− mice. Humans treated with TNF inhibitors show some increase in susceptibility to certain pathogens, particularly *Mycobacterium t.*, reinforcing the concept of TNF as an important component of the host-defense system.[190]

Lymphotoxin-αβ plays an important role in resistance to both RNA and DNA viruses, as well as to bacterial and protozoan pathogens (Table 10.5). Increased susceptibility to these pathogens in LTα, LTβ, or LTβR gene-deficient mice is due either to developmentally controlled aspects of lymphoid organ structure or to a requirement for the LTαβ-LTβR pathway as an effector system regulating the innate IFNαβ production[191] and adaptive immune responses (eg, IgA).[192] Interferon regulation in the spleen by the LTβR pathway occurs via the differentiation of macrophages and stromal cells in the marginal zone.[193] The innate-like marginal zone B cells in the spleen that constitutively express LTαβ are required for the differentiation of LTβR+ macrophages (CD169+) and chemokine-secreting stromal cells. This is a feature of the organized architecture of the white pulp that provides a highly efficient mechanism for capturing and responding to blood-borne pathogen.[194] Upon viral infection, the differentiated LTβR responsive cells rapidly secrete IFNβ to initiate the major antiviral defense system.[191,195] Adaptive immunity in barrier organs such as the lungs and intestine requires immunoglobulin class-switching to IgA via cellular interactions between innate lymphoid cells expressing LTα1β2 and LTβR+ subepithelial dendritic cells that in turn contact antigen-responsive B cells.[196]

The LIGHT-HVEM pathway is intimately involved in host mechanisms needed for innate and adaptive defenses underlying the immune evasion strategies of viral pathogens. Mice deficient in HVEM show increased susceptibility to viral pathogens.[197,198] For example, HSV envelope glycoprotein D effectively disrupts HVEM signaling inhibiting the production of Fcγ receptor-activating IgG2, the antibody subclass required for virus clearing antibody-dependent cytotoxicity and phagocytosis.[62] Moreover, LIGHT-HVEM signaling is necessary for the FcγR activation involved in killing HSV-infected cells. Elimination of the HVEM immune-modulating function in HSV, a strategy in which gD is deleted from the HSV-1 genome, has been used to generate a single cycle candidate herpes simplex vaccine. This vaccine generates high levels of antibodies and T cell responses with complete protection from infection and latency against clinical strains of HSV1 and 2.[199]

In a subset of patients infected with the SARS coronavirus 2 (CoV-2), that is responsible for coronavirus infectious disease (COVID-19), progress to severe respiratory disease and mortality associated with cytokine dysfunction, including elevated serum levels of LIGHT among other cytokines.[76,77,200] While the significance of increased serum LIGHT is unknown

TABLE 10.5 Lymphotoxins in Host Defense: Mouse Models

Pathogen	Mouse Model	Susceptibility	Mechanism	Reference
Herpesvirus:				
HSV-1	LTα−/−	Increased	Decreased effector CD8+ T	170
MCMV	LTα−/−; B-LTβ	Increased	IFN-I defect	171
MCMV	LTβR-Fc Tg	Increased	Poor innate defenses	172
LCMV	LTβ−/−; LTα−/−	Increased	Defective architecture	173-175
LCMV	LTβR-Fc	Decreased	Decreased CD8+/IFNγ	176
Theiler virus	LTα−/−LTβR-Fc	Increased	Defective architecture	177
Influenza	LTα−/−	Minimal	Delayed response	178
VSV	LTβ−/−	Increased	Defective architecture	175
Bacteria and parasites:				
M. tuberculosis	LTβR−/−	Increased	NO_2 synthase decreased	179
M. tuberculosis	LTα−/−	Increased	No T cells in granuloma	180
M. bovis	LTβR-Fc	Increased	Poor granuloma formation	181
Listeria m.	LTβR−/−	Increased	Deficient IFN-I response	182
Leishmania m.	LTβ−/−	Increased	Defective architecture	183
Leishmania d.	LIGHT−/−	Increased	Defective T cell	184
Toxoplasma g.	LTα−/−;LTβR−/−	Increased	NO_2 synthase decreased	185,186
Plasmodium b.	LTα−/−	Decreased	Decreased inflammation	187

Pathogens: Viral, MHV68, mouse γ-herpesvirus-68; HSV, herpes simplex virus; LCMV, lymphocytic choriomeningitis virus; MCMV, mouse cytomegalovirus; VSV, vesicular stomatitis virus (VSV). Bacteria and parasites, *Mycobacterium; Listeria monocytogenes.* Parasite: *Leishmania major; Leishmania donovani; Toxoplasma gondii; Plasmodium berghei.* Studies conducted in gene-deficient mice (−/−); LTβR-Fc Tg, mice expressing LTβR-Fc as a transgene; LTβR-Fc, mice injected with protein.

Modified with permission from Ware CF. Network communications: Lymphotoxins, LIGHT, and Tnf. *Annu Rev Immunol.* 2005;23:787-81.

at this time, these high levels may override the control by DcR3 levels causing dysregulation of inflammatory processes.

Viewed from an evolutionary perspective, immune evasion mechanisms used by various pathogens provide a recognizable validation of the TNF-LT-LIGHT network as a host-defense network.[161]

Autoimmunity

Dysregulated expression of several members of the TNFSF leads to autoimmune-like diseases in humans and animal models. For example, forced expression of TNF or LIGHT, which overrides the normal transient expression, causes severe inflammatory processes in mice.[201-203] Transgenic expression of LTα or LTαβ in the pancreas leads to insulitis and the formation of tertiary lymphoid structures.[204] Genetic analysis of MHC-linked TNFSF genes revealed *Ltα, Ltβ,* and LTβR genes that are associated with rheumatoid arthritis.[205,206] In a mouse model of persistent lung inflammation, LIGHT is an important driver of fibrosis and leukocyte similar to asthma.[207] The LTβR pathway is linked to smoking-associated chronic obstructive pulmonary disease (COPD).[208] Following exposure to smoke, LTαβ and LIGHT expressed in innate and CD4 T cells induce LTβR target gene expression in lung epithelial cells with formation of inducible bronchus-associated lymphoid tissue. Blockade of LTαβ/LIGHT inhibited TGFβ airway fibrosis, systemic muscle wasting, and promoted regeneration of lung tissue. Polymorphic and mutational variants in the MHC class III region, encoding genes for TNF, LTα, and LTβ are associated with autoimmune diseases.[209] Genetic variants of *TNFSF14*[159] and *DcR3*[210] may also contribute to autoimmune diseases[211] while somatic mutations in *TNFRSF14* are frequent in certain cancers.[164] These types of results have identified these members of the TNFSF as immune regulators and support the notion that TNF, LTαβ, and LIGHT pathways contribute to inflammation and disease processes in humans.

Cancer

Cancer is a disease caused by somatic mutations that impact cell growth, resistance to cell death, and susceptibility to immune surveillance. Signaling via LTαβ and LIGHT pathways appears to play important roles in controlling cancer. The presence of tertiary lymphoid structures, which most generally are associated with activation by TNF family members, especially LTβR, is associated with more favorable outcomes in multiple tumor types[212,213] and, remarkably, in the context of drugs that target checkpoint inhibitors.[214,215]

LTαβ-LTβR signaling in the liver increases and becomes constitutive during chronic viral infection that promotes hepatocyte proliferation and progression of hepatocellular carcinoma.[216,217] Intriguingly, LTβR signaling encourages formation of a unique microenvironment for hepatocyte proliferation. This mechanism is required for liver regeneration following partial hepatectomy[218] and may also be responsible for promoting hepatocellular carcinoma.

TNFRSF14 is one of the most frequently mutated genes in human follicular lymphomas and diffuse large B cell lymphomas that arise in the germinal center. Complete deletions and nonsynonymous mutations in this gene selectively impact ligand binding of HVEM and are associated with poor prognosis.[219,220] In mice, mutations in HVEM block inhibition of T helper cells by BTLA allowing enhanced B cell survival and proliferation. Thus, HVEM-BTLA pathway may act as an extrinsic suppressor of B cell lymphomagenesis.[221]

In several experimental cancer models, LIGHT signaling via LTβR and HVEM-mediated activation of T effector cells can induce tumor eradication through formation of tertiary lymphoid structures within the tumor microenvironment.[222] Utilizing LIGHT-activated pathways may be an effective strategy for cancer immunotherapy.[164,223]

COSIGNALING TNFSF IN ADAPTIVE IMMUNITY

Antigen recognition together with activation of cooperative "cosignaling" systems determines the quality of adaptive immune responses. Lymphocyte responses to antigens are dynamic processes that start with the activation of naïve cells and transition through effector and memory phases. Cosignaling systems assist these phase transitions by promoting more efficient responses to TCR signaling. Cosignaling TNFRs enhance initial cell activation and cell division (clonal expansion), augment cell survival (clonal contraction or memory cell differentiation), and induce effector functions such as cytokine secretion or cytotoxic killing. Negative signals (inhibitory cosignaling) may also be delivered to T cells depending upon the particular system. This may prevent initial cellular activation or eliminate excess activated cells to damp inflammation. Cosignaling can be either quantitative, modifying thresholds of common signaling intermediates, or qualitative, involving signals distinct from other cosignaling systems or the TCR. Cosignaling receptors and ligands can be up- or down-regulated at both the transcriptional and translational levels depending on the stage of the T cell response and the inflammatory milieu. In the absence of cosignaling, T cells may become unresponsive (anergic) or die. The TNF receptor superfamily is one of two major families of cosignaling regulators that modulate T cells. The other cosignaling systems belong to the immunoglobulin superfamily, exemplified by CD28 B7 family.[224]

T Cells: OX40, 41BB, DR3, CD27, CD30, and GITR Systems

TNF receptor superfamily members involved in T cell cosignaling include OX40, 41BB, DR3, CD27, CD30, and HVEM,[225-229] whereas CD40 and BAFFR are more involved in cosignaling in B lymphocytes.[230,231] However, considerable crossover of activities in both lymphocyte populations can be demonstrated. Death receptors such as FAS and TNFR1 are thought to be involved in clonal contraction through apoptosis of activated effector cells, although TNF acting through TNFR2 also exhibits costimulatory action in naïve T cells. In tissue culture models, several of the TNF-related signaling pathways exhibit costimulatory activities in T cells. This probably reflects the common induction of NFκB-dependent survival genes, a common trait among TNFRSF members. However, more physiological analyses using genetically deficient mice reveal distinct roles for these molecules in the life cycle of T cells. Cosignaling systems are emerging as important targets for enhancing immune responses to tumors or for attenuating immune responses in autoimmune diseases.[222,232-236]

TNFRSF cosignaling molecules are important during primary and memory T cell responses. Antigen-presenting dendritic cells and monocytes exhibit complex patterns of TNFSF ligand expression that are required for the antiviral activity of CD4 T cells.[237] The TNFR encoded on Chr 1p36 share a common but not exclusive function as cosignaling systems for T cells.[227,229] This region (human 1p36.33-1p36.21; syntenic with mouse 4E2-D3) contains the genes encoding GITR, OX40, HVEM, DR3, 41BB, CD30, and TNFR2 and is paralogous to the region on Chr 12p13 where genes for TNFR1, LTβR, and CD27 reside. The TNF-related ligands for these receptors are linked in the MHC paralogs on Chr 1, 6, 9, and 19, reflecting a common functional link as costimulatory molecules that modulate T cell differentiation.

Expression of TNFRSF members by T cells follows several distinct patterns during T cell activation. CD27 and HVEM are constitutively expressed on naïve T cells in peripheral lymphoid tissues, whereas OX40, 4-1BB, TNFR2, and CD30 are expressed by T cells after initial activation.[227] TNFSF ligands display a complex pattern of expression on antigen-presenting dendritic cells and monocytes that are necessary for CD4 T cell antiviral activity.[237] For example, chronic LCMV infection downmodulates GITRL on APC, resulting in impaired CD8 T cell activation. However, an agonist mAb against GITR can supplement signaling to restore antiviral immunity.[238] In another example, induction of OX40L expression in bronchiolar progenitors during influenza infection enhances viral attachment through viral hemagglutinin and α-2,6 sialic acid modification of OX40L.[239] Combinations of cosignaling TNFR are useful in monitoring activated T cell differentiation, for example, Ox40 and CD40L in CD4 T memory cells.[240] CD27, which is expressed at very early stages of T cell activation, mediates initial cytokine production, cell division, and/or survival,[241,242] whereas OX40 and 4-1BB, which are expressed after cellular activation, promote T cell survival and proliferation during the later stages of the primary response.[243-245] CD30-CD30L interactions promote primary CD8+ T cell expansion and with OX40 act to promote CD4+ T cell memory suggesting functional parallels with other TNFRSF molecules.[246] Cowpox virus encodes a specific CD30 mimic to block the action of CD30L as part of its immune evasion strategy (Table 10.6).

TABLE 10.6 TNFSF Cosignaling Pathways in Infectious Disease Models

	Co-Signaling Pathway				
Pathogen	**CD27-CD27L**	**Ox40-Ox40L**	**4-1BB-41BBL**	**CD30-CD30L**	**GITR-GITRL**
Influenza	CD4/8[242,247]	CD4[248]	CD8[249]	No role[250]	CD4/8 Teff[251]
LCMV	CD8[252]	CD4[253]	No role[249]	No role[250]	APC[254]
MCMV	NK/CD8[255]	CD4+[256,257]	CD8[258]	CD8[259]	nd
Leishmania	Th1[260]	Th2[261]	Th2[262]	No role[262]	nd
Helminths	Nd	Th2[263]	Th2[264]	nd	Th2[265]
Cryptococcus	Nd	Th2[266]	nd	nd	nd
M. avium	No role[267]	No role[267]	No role[267]	CD4/8[267]	nd
Listeria m.	CD4/8[268]	Th1-DC[269]	CD8+[270]	CD8+[271]	nd

Unless indicated the functions of T cells were impaired. LCMV, lymphocytic choriomeningitis virus; *Leishmania major*, *Cryptococcus neoformans*; MCMV, mouse cytomegalovirus; *Mycobacterium avium*; *Listeria monocytogenes*. nd, not done.

Death receptor-3 (DR3) engages TL1A, promoting IFNγ expression by proinflammatory Th1 cells in mouse[272] and humans,[273] especially in intestinal tissues.[274] Polymorphisms in TL1A are associated with inflammatory bowel disease.[275,276] Similar cosignaling activities are reported for GITR,[277] with the potential of these systems to affect different subsets of T cells, such as GITR expression on T regulatory subsets. The role of these TNFRSF in regulating different T cell subsets is an active area of research.

The TNFSF costimulatory pathways contribute to the pathology seen in autoimmune disease models. Blockade of TNF and LTα[278] or OX40-Ox40L[279,280] reduces pathology in experimental autoimmune encephalomyelitis (EAE). Inhibition of the CD27-CD70 pathway also ameliorates EAE through the suppression of TNF induction, but not T cell priming.[281] Additionally, in several experimental models blockade of OX40, 4-1BB, CD27, or CD30 costimulation all serve to reduce transplant rejection and/or graft-versus-host disease.[282-285]

Cosignaling through several TNFRSF members promotes the differentiation of memory T cells. Mice deficient in OX40-OX40L, CD27-CD70, CD30-CD30L, or HVEM pathways all display defective memory responses to antigen.[242,248,249,286,287] The OX40 system is illustrative of the costimulatory signals provided to T cells by these TNFR. Pharmacological stimulation of OX40 with an agonist antibody or ligand during a primary antigen-specific response in mice enhances the generation of memory CD4 T cells.[288] The enhanced response of T cells includes increased accumulation of antigen-specific effector T cells producing cytokines such as IFNγ at the site of inflammation. Moreover, blockade of OX40-OX40L interactions during recall responses ameliorates inflammation in models of both collagen-induced arthritis[289] and experimental allergic lung inflammation.[290] In OX40-deficient mice, CD8 T cells expand normally, but their accumulation and survival at later times in the primary response is significantly impaired. T cells from OX40-deficient mice also fail to fully differentiate, as evidenced by loss of expression of effector cell surface markers and decreased synthesis of cytokines and cytotoxic activity. OX40 signaling induces transcription of survival genes such as BCL2 and surviving in T cells, which may contribute to memory cell differentiation.[291]

The expression of TNFRSF molecules such as GITR or OX40 on T regulatory cells suggests that costimulatory pathways may limit the suppressive function of Treg cells during immune responses. CD30 signals can, in some circumstances, suppress cytotoxic T lymphocyte activity.[292] 4-1BB–deficient CD4+ T cells in some models are hyper-responsive,[293] whereas stimulation of 4-1BB can limit immune responses in several models of autoimmune disease.[294,295]

Stimulation of these pathways may also be appropriate for the treatment of diseases such as cancer and infectious diseases where antigen-specific T cell responses are ineffective. Stimulation of OX40, 4-1BB, CD30, and CD27 with agonist antibodies augments T cell–mediated killing of a variety of tumors. Similar to the immediate TNF family, the role of a specific cosignaling pathway in mediating protective T cell responses during infection depends on the pathogen (Table 10.6).

B Cells: CD40L-CD40, BAFF-APRIL Systems

CD40L-CD40 System

CD40 ligand (CD40L) and BAFF play major roles in coordinating a range of costimulatory signals important for B-cell function. These include clonal expansion and differentiation, antibody isotype switching, affinity maturation, and immunoglobulin secretion.[230,296-298] In common with other TNFSF members, the CD40 system is a key communication mechanism in host defense, autoimmunity, and cancer. These pathogenic processes are often centered on the function of B cells. CD40 ligand is transcriptionally regulated by NFAT in T helper cells to promote humoral immunity. CD40L is also expressed in a variety of other cells including activated B cells, endothelial cells, basophils, mast cells, and platelets. CD40 is expressed prominently on B cells and utilizes both TRAF3 and 6 to activate NFκB and AP1 transcription factors, among other signaling pathways.[299,300]

Rare inherited disorders in CD40 signaling can lead to primary immunodeficiency characterized by immunoglobulin class-switch recombination defects associated with hyper-IgM (HIGM) syndromes.[301] The mutated genes that underlie these syndromes include CD40 ligand and CD40,[302,303] the RelA NFκB activating kinase subunit, IKKγ (NEMO), and activation-induced cytidine deaminase (AID), among others.[304] Patients with these mutations have elevated levels of IgM, often with no serum IgG, IgE, or IgA, and display increased susceptibility to bacterial and opportunistic infections. Treatment of these patients with passive transfer of human immune IgG rescues the response to bacterial infection, but not to opportunistic fungal pathogens such as *Pneumocystis carinii*. This latter result reflects the importance of the CD40 system in T cell–macrophage activation, which controls these opportunistic pathogens. In some patients, significant amounts of IgA and IgE are present in serum, even in the complete absence of CD40 ligand, indicating that the BAFF/APRIL system can induce Ig class-switch independent of the CD40 system.[305]

BAFF-APRIL Circuit

Since the discovery of B-cell activating factor of the TNFSF (BAFF, TNFSF13B; BLyS) and a ProlifeRation-inducing ligand (APRIL, TNFSF13), much has been elucidated with regard to CD40L-independent Ig isotype-switching and B cell development and survival. In addition, BAFF-APRIL dysregulation has been implicated in autoimmune disease and lymphoma development.[231,306] Although mapping to different chromosomes (BAFF Chr 13q34; APRIL to 17p13.1), these proteins are strikingly conserved (~50% homology). The gene for TWEAK (TNF-related ligand with weak apoptosis activity) maps adjacent to that of APRIL and the genetically similar members ectodermal dysplasin (EDA1/EDA2) system. BAFF and APRIL form homo and heterotrimers that interact with multiple receptors to form a complex signaling network. Membrane-anchored BAFF is shed via a furin protease at the cell surface, while APRIL is processed in the golgi and secreted only in soluble form.[307] Interestingly, the receptor-binding domain of APRIL can be membrane-anchored via an unusual splice variant with TWEAK cytosolic and transmembrane domain (TWE-PRIL).[308] The significance of membrane-restricted local BAFF expression versus systemic availability of soluble BAFF and APRIL remains to be established. It is notable that BAFF-APRIL heterotrimers are detected in serum from patients with systemic rheumatic diseases.[13] Lymphoid tissue stromal cell expression of BAFF maintains the major mature B cell pool, whereas inducible BAFF expression in myeloid cells aids in local B cell survival at sites of inflammation. BAFF and APRIL expression is induced in monocytes, macrophages, dendritic cells, and T cells in the spleen and lymph nodes in different cell types in response to certain cytokines such as type I and II IFN, IL-10, and other TNF family members like CD40L and LTαβ. BAFF can be expressed in some B cell–derived chronic lymphocytic leukemia cells. In addition, some epithelial-derived cancer cell lines and primary tumor tissues express APRIL.

Three cognate receptors mediate the biological actions of BAFF. BAFFR (TNFRSF13C), TACI (transmembrane activator and calcium signal modulating cyclophilin ligand interactor, TNFRSF13B), and BCMA (B cell maturation antigen, TNFRSF17) are expressed on B cells and on other cell types. APRIL also engages matrix proteoglycans with relatively weak binding, which may serve to enhance access to TACI and BCMA or may function as a distinct receptor-ligand interaction.[309] Signaling through BAFFR and BCMA are required for B cell survival and differentiation at different stages of the B cell–mediated humoral response. Shedding of BAFF by furin proteases controls the formation of higher-ordered oligomers which influences BAFFR signaling.[310] BAFF initiates shedding of BAFFR by ADAM10 and in B cells coexpressing TACI, BAFFR shedding requires ADAM17. BAFFR processing by ADAM10 cleavage decreases survival of primary human B cells, whereas ADAM17 decreases BAFFR affecting germinal center B cells.

BAFFR−/− mice display an almost complete loss of mature and marginal zone B cells beginning from the late transitional phase of B cell maturation, whereas BCMA deficiency impairs the survival of long-lived bone marrow plasma cells.[311,312] By contrast, APRIL−/− mice develop normally, but exhibit altered class-switching to IgA,[313,314] although APRIL expression in gut eosinophils is insufficient for IgA secretion, rather IL1β dominates the IgA response.[315] TACI signaling may antagonize BAFFR and BCMA, as TACI−/− mice have elevated numbers of mature B cells.[316] Both classical and alternative NFκB pathways have been implicated in BAFFR and BCMA survival signals, with transgenic BCL2 expression rescuing the B cell defect in BAFFR−/− mice. BAFFR engagement has been shown to provide the necessary signals for CD40L-independent class-switch recombination to IgG and IgE, while TACI engagement provides CD40L-independent signals for switching to IgG, IgE, and IgA. There are indications that IgA production may specifically depend upon APRIL-TACI interaction.[317] BAFFR is also connected to promoting formation of B cell coreceptor complex CD19/CD21/CD81, which binds to C3b opsonized antigens to enhance rapid T-independent antibody response. Overall impairment of T-independent antibody response appears to be greatest in TACI−/− mice and is dependent upon APRIL-TACI interactions in B1 B cells. BAFFR, but not TACI, has been proven to be important in T cell costimulation in *in vitro* and *in vivo* mouse models of allograft rejection. BAFFR-/- and BAFF−/− but not TACI−/− or BCMA−/− mouse recipients prolong graft survival due to weakened alloproliferative responses.[318] TACI is mutated in patients with common variable immune deficiency, an antibody deficiency syndrome characterized by loss of memory B cells and plasma cells.[319]

Elevated levels of BAFF and APRIL have been detected in the serum, and especially in the synovial fluid, of patients suffering from inflammatory autoimmune disorders such as systemic lupus erythematosus, rheumatoid arthritis, Sjögren syndrome, and multiple sclerosis.[320] These levels tend to correlate with increasing amounts of autoreactive antibodies in serum; in mouse models of lupus and collagen-induced arthritis, treatment with soluble TACI-Ig can lessen the severity of disease symptoms and progression.[321] It is hypothesized that these excess quantities of BAFF and APRIL may contribute to the formation of autoreactive B lymphocytes and loss of tolerance during development, as well as to perpetuating their survival as plasmablasts. Emerging evidence connects the BAFF and APRIL system to the pathology of B cell–derived cancers like B-chronic lymphocytic leukemia, Non-Hodgkin lymphoma, and multiple myeloma. Malignant myeloma cells express high levels of BAFF and APRIL and receive ligand-induced survival signals, with upregulation of antiapoptotic Bcl-2 and Mcl-1 molecules.[322] It is possible that BAFF and APRIL may bolster B cell tumor survival via autocrine and paracrine signaling mechanisms.

DEATH RECEPTORS AND LIGANDS

FAS-FAS Ligand System

The first apoptosis signal receptor (FAS, CD95, *TNFRSF6*)-FAS Ligand (*TNFSF6*) system is a direct signaling pathway leading to apoptotic cell death but can also activate non-death signals.[323,324] The importance of the FAS-FASL system in immune regulation was revealed by the recognition that mice and humans with autoimmune-like disorders, *lymphoproliferative* (*lpr*) and *generalized lymphoproliferative disorder* (*gld*) exhibit mutations in FAS and FASL.[325-327] Mice with *lpr* or *gld* mutations display autoimmune phenotypes characterized by high accumulations of activated T cells in the periphery and large numbers of CD4−CD8− T cells in the lymph nodes (lymphadenopathy) and high levels of autoantibody production. In humans, heterozygous mutations in *TNFRSF6* cause autoimmune lymphoproliferative syndrome (ALPS), an inherited nonmalignant lymphoproliferative disorder.[328,329] This disease in humans and mice reflects the apoptosis-inducing activity of the FAS-FASL system to limit lymphocyte clonal expansion and survival as a component of immune homeostasis.

The FAS-FASL system is involved in proinflammatory responses, immune tolerance, tumor survival, and nonimmune tissue homeostasis including osteoclastogenesis and angiogenesis.[330,331] Activated lymphocytes are particularly sensitive to FAS-induced apoptosis, suggesting a role for FAS in the clonal contraction phase of the immune response. FAS-FASL is one of several proapoptotic systems used as a killing mechanism by cytotoxic T cells and NK cells.[332] Epithelial cells in the eye express FASL, which participates in limiting inflammation in organs sensitive to immune damage by inducing apoptosis of activated effector cells.[333,334]

FASL can undergo cleavage by matrix metalloproteases to produce soluble FASL, which may induce markedly different events, such as loss of cytotoxicity, than membrane-bound FASL upon engagement of FAS.[335,336] Other findings suggest that soluble FASL may act as an immunosuppressive agent to counteract the proinflammatory properties of membrane-associated FASL in the eye and tumor environments.[333,337] Similar to LIGHT and TL1A, FASL binds DcR3 as a mechanism that limits the ability of FASL to activate cell death.[9,160]

Essential biological roles of FAS-FASL include regulating lymphocyte homeostasis (via contraction of clonally expanded effector T and NK cells), directing CTL-mediated apoptosis and lysis (immune surveillance), and establishing immune-privileged organs and sites. FAS is expressed in a wide variety of tissues such as the thymus, spleen, heart, and liver, while FASL is expressed in activated T cells, NK cells, tumor cells, and immune-privileged sites including lung and other tissues.[330,338,339] In mature-activated T and NK cells, FAS-FASL signaling plays a critical role in inducing apoptosis during the latter stages of inflammation as a means of shifting lymphocyte population numbers back to baseline.[340] In certain organs like the liver, lung, and small intestines, FASL expressed in epithelial cells can be upregulated to control the overabundance of activated T cells after an inflammatory response, ultimately reducing tissue damage. Alternatively, dysregulation during hepatitis C infections can lead to death of hepatocytes and acute liver damage.[341] The FAS system has also been implicated in curtailing autoantibody production by damping T helper–induced B cell activation and by direct elimination of activated B cells, situations in which CD40 stimulation upregulates FAS expression.[330]

The FAS-FASL system, along with perforin and granzymes, is a mechanism vital to CTL-mediated destruction and clearance of tumor cells and virally infected cells. Viruses have evolved multiple mechanisms for suppressing FAS signaling,[342] including downmodulation of FAS[343] or blockade of caspase 8 activation.[344] Tumor genome instability may lead to mutations that aid in escaping or resisting death pathways, alterations that are conducive to developing the metastatic phenotype.[345] Some uterine and breast cancer patients have elevated levels of soluble FASL and FAS, which may aid in neutralizing CTL- and NK cell–induced lysis or even killing FAS-expressing effector cells. In cell culture models, HSV-1 infection of human neonatal neutrophils, critical effectors in innate and humoral immunity upregulates FAS and FASL surface expression leading to apoptosis. The modulation of FAS-FASL exemplifies a theme concerning pathogenic modulation of apoptotic signaling events as a means of evading or attenuating the immune response.[346] FASL is constitutively expressed in the eye, CNS, testis, fetal trophoblasts, and placenta, creating immune-privileged sites that protect against lymphocyte infiltration that could lead to unrestricted, irreversible bystander damage to these tissues.[347] The implications of immune privilege have been applied to tissue transplantation rejection therapies; however, conflicting data as to whether FASL expression reduces the likelihood of allograft rejection[348,349] or potentates effector response,[350,351] emphasizes the fact that induction of immune privilege by the FASL system is complex. FAS-FASL interactions are at the core of graft-versus-host disease (GVHD), in which contaminating donor effector T cells in an allograft expand to damage the recipient tissues. In a mouse model of acute GVHD, p53-dependent upregulation of FAS on host stem cells leads to subsequent bone marrow depletion through FASL-mediated apoptosis.[352] With regard to tissue homeostasis, FASL has been reported to enhance RANKL-mediated osteoclastogenesis and differentiation in mouse bone marrow–derived macrophages.[353] FAS is also implicated in pulmonary inflammatory diseases resulting in fibrosis after acute lung injury.[338]

TRAIL and TRAIL Receptor Network

TRAIL (TNF-related apoptosis-inducing ligand, TNFSF10) is closely related to FASL and TNF based on sequence homology. Interest in TRAIL was spurred by its ability to selectively induce apoptosis in tumor cells.[354] TRAIL forms a complex network through five different receptors: TRAILR1 (DR4, TNFRSF10A), TRAILR2 (DR5, TNFRSF10B), TRAILR3 (DcR1, TNFRSF10C), TRAILR4 (DcR2, TNFRSF10D), and osteoprotegerin (OPG, OCIF, TNFRSF11B). TRAILR1, 2, and 4 are type I transmembrane proteins with signaling domains, while the more distantly related OPG is a dimeric, soluble decoy receptor, and TRAILR3 lacks a cytoplasmic domain and is anchored to the membrane via glycophosphatidyl inositol linkage.[355,356] TRAILR1 and 2 contain highly homologous cytoplasmic DD that form the death-inducing signaling complex responsible for apoptosis,[357,358] whereas TRAILR4 has a truncated, nonfunctional DD.[359,360] OPG blocks TRAIL binding[361] and the lack of a functional DD indicates that TRAILR3-and R4 may antagonize TRAIL-induced apoptosis. Alternative splicing is a common feature among TRAIL and its receptors and also contributes to the regulation of programmed cell death.[362-364] The genetic organization of the TRAIL receptor loci in the mouse differs from that seen in humans. The mouse genome has TRAILR2, 3, and 4 homologues, but the TRAILR4 homologue does not bind TRAIL.

The TRAIL-TRAILR signaling network induces apoptosis and to a lesser degree NFκB activation. For example, the apoptotic function of the TRAIL system is necessary for regulating the response of memory CD8+ T cells to rechallenge with antigen. In the absence of CD4 T cell help, CD8 T cells undergo apoptosis mediated by TRAIL.[365] In much the same way as FAS, TRAILR DDs recruit FADD and initiator caspases 8 and/or 10 into the DISC to activate downstream effector caspases that carry out apoptosis.[366] Caspase 8 contribution to this process is well-defined, but caspase 10 involvement was less so until recent studies in ALPS II patients and some carcinoma cell lines revealed that mutant caspase 10 abolishes TRAIL-induced apoptosis.[359] In some cell types, the extrinsic mitochondrial pathway is important for amplifying DISC signals in TRAIL-induced apoptosis, which is analogous to mechanisms seen in for FAS-FASL signaling.[367]

TRAIL is involved in regulating immune homeostasis and in clearing virally infected and cancerous cells. IFNγ treatment of tumor cells increases sensitization to TRAIL-induced death. This has been observed in melanoma and ovarian carcinoma cells[359] and is analogous to FAS-FASL apoptotic signaling in the human colorectal carcinoma line HT-29. IFNα, which possesses powerful antiviral and antitumor properties, upregulates TRAIL via JAK-STAT pathways in stimulated human multiple myeloma *in vitro*.[368] The apoptosis-inducing activity of TRAIL has generated interest in the TRAIL network in the context of anticancer therapeutics.

TL1A-DR3 System

The TNF-like ligand 1A (TL1A; TNFSF15)-death receptor-3 (DR3) system provides essential signals that promote the expansion of effector T cells at the sites of inflammation.[274,369] The broad tissue expression profile of DR3 indicates its involvement in a range of cellular responses in multiple subsets of T cells, including Treg and ILC. TL1A is a paralog of FASL, LIGHT, and LTβ, and its receptor DR3 is a homolog of TNFR1. TL1A, like LIGHT and FASL, binds the decoy DcR3, potentially limiting signaling mediated by DR3. The TL1A-DR3 system has emerged as an important mechanism controlling mucosal homeostasis.[370] Gene deletions of TL1A or DR3 in mice result in poor T cell responses but no overt developmental abnormalities.[272,371-373] Deletion of TL1A or DR3 or blockade by anti-TL1A antibodies demonstrates a critical involvement of these elements in the pathogenesis of chronic mucosal inflammation. Excess TL1A beyond levels controlled by DcR3 leads to allergic gastrointestinal inflammation through stimulation of type 3 innate lymphoid cells.[374] TL1A and DR3 localize at sites of intestinal inflammation in patients with inflammatory bowel disease, phenomena also observed in mice with experimental ileitis colitis. *TNFSF15* polymorphic variants are identified as risk factors for Crohn disease,[375] and specific mutations in the ligand-binding domain of DR3 are linked to rheumatoid arthritis.[376]

TL1A-DR3 signaling maintains a homeostatic environment by eliminating inflammatory cells via cell death pathways, yet also promotes immunity by costimulation of T cell and ILC responses through TRAF2-NFκB. Blocking mAbs against TL1A are in clinical development.

TWEAK-FN14 SYSTEM

The TNF-like weak inducer of apoptosis (TWEAK)-fibroblast growth factor-inducible 14 (Fn14) system is emerging as an important mechanism of signaling between the immune system and epithelial and stromal tissues.[377-379] TWEAK is limited to a monogamous interaction with Fn14, its cognate receptor. Fn14 has one CRD and binds TWEAK homotrimers in a manner similar to BAFF, APRIL, and their receptors.[380,381] TWEAK, like BAFF and APRIL, is cleaved by furin protease to produce an active soluble form.[380] Human and mouse TWEAK share ≥90% sequence identity in the receptor-binding domain and manifest high-affinity cross-species interactions with respective Fn14 homologues.[307] The biological functions of the TWEAK-Fn14 system are broad and perhaps paradoxical, promoting both cell survival and death. TWEAK/Fn14 signaling induces apoptosis in some cancer lines (HT29, HSC3) and primary neurons.[382-384] TWEAK is produced primarily by leukocytes, whereas Fn14 expression is induced during inflammatory and tissue destructive processes and observed in lymph node endothelial cells, neutrophils, and alveolar macrophages. The TWEAK-Fn14 system may provide important homeostatic functions in promoting regenerative processes that conclude inflammatory responses. These homeostasis promoting functions include induction of survival and growth in endothelial cells, migration and wound closure in endothelial cells, and angiogenesis. Fn14 contributes to the activation of progenitor cell types in liver, skeletal muscle, and mesenchyme. These include osteoblast, chondrocyte, adipocyte, and neuronal progenitors.

TWEAK regulates the shift from innate to adaptive immune responsiveness by repressing inflammatory cytokines (IL-12, IFNγ) that are important in promoting Th1 immunity—a potential counter balance to the proinflammatory actions of TNF.[382] When dysregulated, TWEAK may also have a potentiating influence in chronic tissue inflammation, as evidenced by its role in excessive demyelination and ultimate progression to extensive autoimmune encephalomyelitis.[385] In a mouse model of collagen-induced arthritis, elevated levels of TWEAK in serum correlate with severe disease progression and enhanced expression of arthritogenic mediator molecules.[386] In obese patients with type 2 diabetes, TWEAK-Fn14 signaling augments proinflammatory cytokine release by adipocytes.[387] In rats, TWEAK may prevent excessive luteinization after gonadotrophin-induced ovulation by controlling progesterone production.[388]

The seemingly contradictory functions of TWEAK/Fn14 are less puzzling when TWEAK/Fn14 is viewed as an important signaling system that aids in achieving homeostasis following inflammation since cell death and survival are both essential for homeostasis.

EDA-EDAR CIRCUIT

The ligand *EDA*, *EDAR*, and the signaling protein EDAR-associated death domain (*EDARADD*) play key roles in embryonic development of the barrier tissues that contribute to host defense. These include skin eccrine sweat glands and submucosal glands in lungs and mammary tissue.[389] The EDA-EDAR system also cooperates with RANK, LTβ, and TNF pathways in controlling branching of the mammary gland, hair shaft formation, and hair follicle cycling.[4] Alternative splicing generates two isoforms EDA-A1 and EDA-A2 that differ in just two amino acids in the receptor-binding region.[4] This difference functions as a switch-controlling receptor specificity in which EDA-A1 binds only to EDAR, while EDA-A2 is specific for a second receptor, XEDAR.[15]

Mutations in EDA-A1, EDAR, or EDARADD genes result in hypohidrotic ectodermal dysplasias, a group of hereditary differentiation disorders in multiple ectodermal structures.[389]

The X-linked disorder hypohidrotic ectodermal dysplasia (XLHED) is a rare disease caused by mutations in EDA-A1 that result in the failure to activate EDAR signaling. The XLHED patient's lack of eccrine sweat glands can give rise hyperthermia, dry eczematous skin, dry eye problems, and respiratory issues such as frequent infections and allergic asthma. Ectodysplasin-A2 (EDA-A2) induces apoptosis in cultured human hair follicle cells and promotes the regression of hair follicles in mice.[390] Experimental models in mice indicate that agonists of EDAR could serve as a rescue therapy for correcting XLHED.[391] In humans, a recombinant form of human EDA-A1 containing the ligand-binding domain fused to the Fc region of IgG1 is an agonist, thus providing the opportunity for correcting XLHED.[392] The Fc-EDA protein is administered prenatally by injection into the amniotic fluid at the time of sweat gland formation. Amniotic fluid is swallowed by the fetus and transferred into the fetal circulation via the neonatal Fc receptor, which is already expressed in the intestine for the uptake of antibodies from breast milk. XLHED embryos receiving Fc-EDA *in utero* had skin-associated structures restored and the newborn patients did not experience episodes of eczematous skin, dry eye, or respiratory problems.[393] This remarkable pharmacological correction of a genetic defect underscores the success of fundamental research brought to the patient's bedside.

RANK LIGAND, RANK, AND OPG CIRCUIT

OPG, TNFRSF11B is a key regulator of bone density.[394] As a soluble receptor, OPG is a regulator of RANK (receptor activator of NFκB, *TNFRSF11A*) and RANK ligand (aka TRANCE) signaling pathways involved in the differentiation of bone-resorptive osteoclasts from hematopoietic progenitors (osteoclastogenesis).[395-398] However, this system is also crucial for development of mammary glands,[399] lymph nodes, and DC-T cell interactions.[400,401] Bone remodeling needed to sustain skeletal integrity and calcium homeostasis is regulated by a dynamic equilibrium between osteoblasts (bone forming) and osteoclasts (bone resorbing), a balance that is regulated by the RANKL-RANK-OPG system. Dysregulation of the RANKL-RANK-OPG system has consequences in inflammatory, osteological, and cancer pathophysiology. These include bone loss in inflammatory autoimmune diseases, preferential breast and prostate tumor metastasis to the bone, and gender bias in osteoporosis.

OPG functions as a decoy receptor that binds RANKL with high affinity. OPG also binds TRAIL, albeit at lower affinity in comparison to RANKL.[402,403] Despite this weak binding, *in vitro* studies show that OPG can block TRAIL-mediated apoptosis, suggesting that the OPG-TRAIL system contributes to osteoclastogenesis. RANK interacts with TRAF2, 5, and 6. The membrane-proximal TRAF site is highly specific for TRAF6, and several studies have shown TRAF6 to be essential for RANKL-RANK signaling in osteoclastogenesis and lymph node genesis.[404,405] Several "osteotropic" factors, including TGFβ, IL-1, TNF, estrogen, prostaglandin E2, glucocorticoids, and vitamin D3,[404,406] have been reported to regulate RANKL and OPG expression in osteoblast and osteoclast lineages. In a few cases, OPG and RANKL are differentially modulated by the same factor; for example, parathyroid-related protein increases RANKL while decreasing OPG mRNA expression in osteoclast-like odontoclasts.[407] The components of the RANK signaling triad are widely expressed. RANK is upregulated on CD40L-stimulated maturing DC, while RANKL is upregulated during T cell activation and is constitutively expressed in some tumors and mammary gland epithelial cells.

RANK−/− and TRAF6−/− mice lack NFκB activation in osteoclasts and exhibit abnormally high-bone density (osteopetrosis).[408,409] Osteoblasts and stromal cells stimulate differentiation via direct cell contact with osteoclast precursors. Macrophage colony-stimulating factor (M-CSF) cooperates with RANKL-expressing osteoblasts to transmit positive signals through c-Fms and RANK expressed on precursors, promoting the differentiation of these progenitors.[397] In contrast, OPG negatively regulates this process by competitively binding RANKL, thus ablating signaling through RANK. RANK−/− and RANKL−/− mice phenocopy each other, displaying severe osteopetrosis due to the lack of osteoclasts.[410] Complementing the observations in RANKL and RANK deficiencies, OPG blocks osteoclast differentiation in a dose-dependent manner. OPG overexpression in transgenic mice causes acute osteopetrosis, while OPG-deficient mice suffer from osteoporosis due to excessive osteoclastogenesis. In humans, OPG and RANK mutations are linked to Juvenile Paget disease, familial expansile osteolysis, and other osteolytic disorders.[396,404]

The RANKL-RANK system is proving to be an important enhancer of cell-mediated immune responses by promoting DC survival and naïve T cell proliferation. RANKL stimulates DC to produce proinflammatory cytokines like IL-12.[411,412] The RANK pathway may serve to bolster CD40-CD40L interaction between DC and T cells after TCR stimulation or to prolong survival of activated DC to ensure the establishment of T cell memory. For instance, in CD40L−/− mice, RANKL-RANK costimulation protects against *Leishmania* infection by inducing IL-12 secretion and consequent Th1 immune response.[413]

Bone loss in autoimmune and infectious models of inflammatory disease is mediated through slightly different mechanisms. In recent studies, activated CD4+RANKL+ T cells have been shown to support osteoclastogenesis and to mediate bone loss in humanized mouse models of *Actinobacillus*-specific periodontal disease suggesting a role for Th1 immunity in inflammatory bone destruction.[414,415] In HIV+ patients, the virion envelope gp120 induces disease-related osteoporosis via induction of RANKL on CD4+ T cells and augmentation of osteoclastogenesis.[416]

In mouse models of tumor metastasis, high OPG and RANK levels are risk factors for metastasis to the bone and subsequent osteolysis, especially in breast tumors.[417] OPG from human bone marrow stroma can protect prostate tumors from TRAIL-mediated apoptosis.[418] Treatment with anti-TRAIL neutralizing antibodies reduces osteoclastogenesis in mixed lymphocyte-bone marrow cultures from patients with multiple myeloma, presumably by interfering with OPG-TRAIL complexes formed by activated T cells expressing TRAIL, OPG, and RANKL.[419] The RANKL-RANK-OPG system has also been implicated in diabetes-associated osteopenia, as serum levels of OPG are elevated in patients with type I and II diabetes.[420] Interestingly these patients are at greater risk for developing atherosclerotic plaques, a phenomenon that is perhaps reflected in the phenotype of OPG-deficient mice. These mice suffer from arterial calcification,[421] presumably from unrestricted bone resorption and Ca^{2+} release. Although elevated endogenous OPG levels are considered a risk factor in certain disease manifestations, the complexity of the cytokine networks that regulate RANKL-RANK-OPG interactions and vice versa render the clinical data difficult to interpret.

RELT, IGFLR1, TROY, AND DR6 SYSTEMS

The TNFR superfamily members, RELT (receptor expressed in lymphoid tissues; TNFRSF19L), IGFLR1 (Insulin growth factor like receptor-1), TROY (TNFRSF19), and DR6 (death receptor 6) are implicated in inflammation and immune responses and in organ development but are largely unexplored systems. Interestingly, none of these TNFRSF members bind TNFSF ligands. RELT, a transmembrane protein, is expressed primarily in hematopoietic cells, macrophages, and neutrophils. The prominent expression of RELT in myeloid cells suggests a role in the immune system, which was confirmed in mice. RELT-deficient mice exhibit increased proliferation of CD4+ T helper cells and enhanced antitumor responses by CD8+ T cells, implicating RELT as negative regulator of T cells.[422] The signaling mechanisms of RELT remain incompletely understood. The ectodomain of RELT has a single atypical CRD; however, no TNFSF ligand has been identified as partner. RELT forms complexes with two RELT homologs, RELL1 and RELL2, that colocalize in the plasma membrane with RELT.[423] However, RELL1 and 2 have short ectodomains with the primary sequence homology in the cytoplasmic domains that likely influences RELT signaling.[424,425] Interestingly, RELL1 and 2 have functions independent of RELT including inflammation, host defense, and oncogenesis. Mutations in RELT cause autosomal recessive amelogenesis imperfecta, a developmental failure in tooth enamel formation in mice and humans.[426]

IGFLR1 (insulin growth factor like receptor-1) contains a p75NTR like DD. IGFL1 mRNA is expressed in psoriatic skin lesions and the receptor is detected in T cells suggesting a possible role in skin inflammation. Recent results show a strong presence of IGFLR1 expression in tumor infiltrating T cells.[427]

TROY is coexpressed with EDAR and functions in the epithelial enamel knot signaling pathway via TRAF6. TROY is involved in the regulation of cuspal morphogenesis during tooth development.[428] Additional evidence indicates that TROY functions more like p75NTR, capable of binding myelin inhibitory factors.[39]

DR6 functions as a regulator of oligodendrocyte-induced myelination.[429] In EAE, DR6-deficiency was shown to protect against CNS demyelination and leukocyte infiltration but also enhance overall CD4+ T cell proliferation and Th2 differentiation, underscoring a role for DR6 in mediating Th1-specific immune responses in EAE progression.[430]

TNF SUPERFAMILY IN THE THERAPY OF HUMAN DISEASE

Impact of TNF Blockade in Autoimmune Disease

TNF inhibitors. The TNFSF is associated with a wide range of human diseases, evidenced by genetic deficiencies or mutations in individual components of the TNFSF and in the therapeutic efficacy of TNF inhibitors in autoimmune diseases (Table 10.7). Although cancer therapy was the original motivation for developing TNF as a drug, the success of TNF inhibitors in limiting symptoms in several autoimmune diseases provides significant motivation to explore therapeutics targeting other members of the TNFSF. Research into the function of TNFSF in human diseases is providing strong rationale for testing both inhibitors and agonists of these cytokines (Table 10.8).

TABLE 10.7 Human Inherited Genetic Diseases Associated With the TNF Superfamily

System	Disease	Mutation
TNFR1	Familial periodic fever	Mutation in first cysteine-rich domain
CD40 ligand	Hyper IgM Syndrome	Multiple mutations affecting receptor binding and processing
EDA	X-linked hypohidrotic ectodermal dysplasia	Multiple mutations affecting receptor binding, trimerization, and secretion
RANK	Familial expansile osteolysis	Mutation in the signal peptide
FAS	Autoimmune lymphoproliferative syndrome (ALPS)	Multiple mutations/deletion in FAS
DcR3	Juvenile Crohn disease	Secretion and translation mutations
A20	Otulipenia	Ubiquitin dysregulation

TNF inhibitors include specific antibodies and receptor-Ig constructs, which are classified as biologics (protein-based drugs) (Fig. 10.11). These biologics are specific, stable, and provide better pharmacodynamics (serum half-life and bioavailability) when compared with the native ligand. Both antibody and receptor-based drugs function as competitive antagonists, sterically hindering ligand-receptor binding. The IgG1 Fc region of anti-TNF mAb may also act to induce ADCC of cells expressing membrane TNF. The FDA-approved antibody-based drugs include a partially humanized mouse monoclonal antibody to human TNF (infliximab), two fully human antibodies (adalimumab and golimumab), and a monovalent Fab fragment of a TNF antibody (certolizumab pegol). To create a soluble "decoy" receptor, the extracellular domain of TNFR-2 was genetically linked to the Fc region of human IgG1 to form a bivalent molecule (etanercept) that has

TABLE 10.8 TNF Superfamily Therapeutics

TNFSF Member	Drug	Therapeutic Designation	Indication	Status
TNF	Chimeric Ab	Remicade (infliximab)	RA, Ps, IBD	Approved
TNF LTα	TNFR2-Fc	Enbrel (etanercept)	RA Ps	Approved
TNF	Hu Fab-PEG	Cimzia (certolizumab)	RA, Ps, IBD	Approved
TNF	Hu mAb	Humira (adalimumab)	RA, Ps, IBD	Approved
TNF	Hu mAb	Simponi (golimumab)	RA, Ps, IBD	Approved
BAFF	Hu mAb	Benlysta (belimumab)	Lupus	Approved
RANK L	Hu mAb	Prolia (denosumab)	Osteoporosis	Approved
	Hu mAb	Xgeva (denosumab)	Bone metastasis	Approved
EDA1	Fc-EDA	ED004	XLHED	Completed
TRAIL	Trimer fusion	ABBV-621	Cancer	NCT03082209
TRAILR1	Agonist mAb	Mapatumumab	Cancer	Completed
TRAILR2	Agonist mAb	Conatumumab	Cancer	Completed
CD40L	Chimeric mAb	Ruplizumab	Lupus	Stopped
	Hu Fab-PEG	Dapirolizumab	Lupus	NCT04294667
LTβR	LTβR-Fc	Baminercept	RA	Concluded
LTα	Hu mAb	MLTA3698A	RA	Concluded
LIGHT	Hu mAb	CERC002	COVID ARDS	NCT04412057
OX40	Agonist mAb	INBRX-106	Cancer	NCT04198766
	PD1-Fc-OX40L	SL-279252	Cancer	NCT03894618
TWEAK	chimeric mAb	BIIB 023	SLE nephritis	Terminated

Fab-PEG, fragment of antibody polyethylene glycolated; IBD, inflammatory bowel disease; Ps, psoriasis; RA, rheumatoid arthritis; XLHED, X-linked hypohidrotic ectodermal dysplasia. NCT (National Clinical Trials), details at https://clinicaltrials.gov.

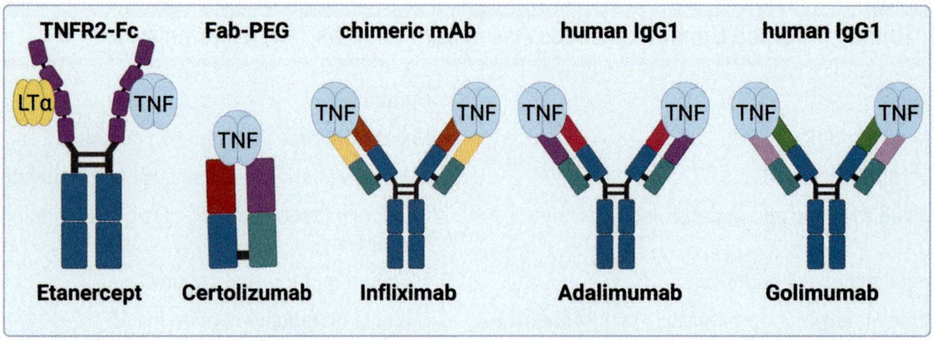

FIG. 10.11. Tumor necrosis factor (TNF) inhibitors in clinical use. The diagram depicts the structural features of the antibody and receptor-based drugs targeted at TNF that are approved for use in patients with autoimmune diseases. TNF decoy receptor (etanercept) engineered as a genetic fusion protein of the ectodomain of TNFR2 with the Fc region of human IgG1, which forms disulfide-linked dimer. Etanercept binds and blocks both TNF and LTα from activating TNFR1 and TNFR2. Monoclonal antibodies are specific for human TNF Infliximab is derived from mice and partially humanized; adalimumab and golimumab are fully human antibodies, and certolizumab is a monovalent engineered from the antigen-binding fragment (Fab) of a human anti-TNF antibody. Certolizumab is derivatized with polyethyleneglycol to extend the half-life *in vivo.* Both Ig and receptor-based inhibitors are approved for use in treating autoimmune diseases: rheumatoid arthritis, psoriasis, and inflammatory bowel diseases. (Figure generated using BioRender [BioRender.com].)

higher avidity than naturally occurring soluble receptors and increased pharmacodynamics. Etanercept binds both TNF and LTα, whereas anti-TNF antibodies are specific for TNF and do not cross-react with LTα. There are differences in the clinical efficacy of anti-TNF mAb and etanercept in certain inflammatory disease (Crohn disease and colitis)[431] and in eliciting antidrug antibody responses.[432]

TNF inhibitors are approved for use in the treatment of rheumatoid arthritis (RA), psoriatic arthritis, psoriasis, ankylosing spondylitis, juvenile rheumatoid arthritis, Crohn disease, and ulcerative colitis. Rheumatoid arthritis is a chronic autoimmune joint disease that occurs in genetically predisposed individuals. Treatment of RA is usually initiated with disease-modifying antirheumatic drugs (DMARDs) such as methotrexate or corticosteroids that alleviate symptoms and reduce joint damage. The American College of Rheumatology uses clinical and laboratory measures to assess improvement in response to therapy in RA. A patient with a 70% or greater improvement according to these parameters is designated as having an ACR-70 response. TNF inhibitors in combination with methotrexate (the most widely used DMARD) produce ACR-70 responses in 33% to 40% of patients with early RA and in 10% to 27% of patients with established RA. In patients receiving methotrexate alone, the ACR-70 responses occurred in 19% to 21% of early RA patients and fewer than 5% of patients with established disease.[433] In RA, joint damage begins early in the disease course as articular erosions occur in 40% of patients in the first year and 90% by the second year following diagnosis.[434] TNF inhibitors slow progression of joint erosion in RA.[435,436]

The inflammatory bowel diseases, ulcerative colitis (limited to colon and bowel), and Crohn disease (entire intestinal tract) are chronic inflammatory conditions in which TNF is disproportionately expressed during active disease. The formation of inflammatory fistula and granuloma is a pathologic feature of Crohn disease. Clinical trials have demonstrated an initial response in 59% to 69% of patients treated with anti-TNF mAb (adalimumab). When categorized by location, dose-dependent clinical remission ranged between 28% and 38% in nonfistulating Crohn disease and between 64% and 97% in patients with complete fistula disease.[437,438] In ulcerative colitis, controlled trials (800 patients with active UC) yielded an initial 8-week response in 65% of patients treated with anti-TNF (infliximab) versus 35% in the placebo group. Remission was achieved in 35% and 15% in infliximab and placebo-treated patients, respectively.[439] In controlled trials, anti-TNF mAb (adalimumab) was efficacious in patients with moderate to severe Crohn disease.[437] By contrast, soluble TNFR2-Fc (etanercept) at doses known to be effective in rheumatoid arthritis failed to produce benefit in a controlled trial for Crohn disease.[440] The clinical disparity between adalimumab and etanercept are not clear but may be due to differences in their ligand specificity. Soluble LTα which is also neutralized by etanercept is needed for gut-protecting IgA antibody,[441] perhaps accounting for the ineffectiveness of etanercept in inflammatory bowel disease.

Epidermal hypertrophy and erythrosquamous lesions containing infiltrating leukocytes characterize psoriasis, a chronic inflammatory skin disease. Skin lesions of psoriatic patients contain elevated levels of TNF. Following treatment with TNF inhibitors, decreased serum and skin levels of TNF are associated with remarkable clinical response.[442,443] A significant subset of patients with psoriasis develop inflammatory, erosive arthritis.[444] In clinical trials, both etanercept and infliximab significantly inhibited disease activity and improved the quality of life in patients with psoriatic arthritis.[445-449]

Serious adverse effects. TNF inhibitors are inherently safe because of their specificity. However, consequences that occur with the use of TNF inhibitors reflect the multiple biologic roles of TNF. Patients may exhibit lupus-like syndrome, aplastic anemia, hepatotoxicity, interstitial lung disease, optic neuritis, and exacerbations of quiescent multiple sclerosis. TNF inhibition is associated with diminished host defense, which is perhaps expected based on the role of TNF in innate and adaptive immune responses. The rate of serious skin and soft tissue infections is higher in treated patients, and there is an increased risk of bacterial infections, primary tuberculosis, and reactivation of latent *Mycobacterium tuberculosis* among patients treated with TNF inhibitors.[450,451] Risk due to infection can be minimized and controlled (eg, a tuberculosis skin test prior to treatment in patients that have not been previously vaccinated for *Mycobacterium*). The contribution of anti-TNF therapy to an increased risk of cancer is still unclear. A systematic review of randomized clinical trials using infliximab and adalimumab found a dose-related increase of malignancies in RA patients.[452] Some of the increased risk of malignancy (ie, lymphoma) is difficult to assess due to the known increased association of lymphoma with RA.[451] TNF has been implicated in the pathogenesis of cardiac dysfunction, in part due to negative ionotropic activities *in vitro* and *in vivo*. However, since TNF blockade significantly increased the incidence of cardiac death, it has been recommended that TNF inhibitor treatment be discontinued in patients with cardiac dysfunction.[453-455]

Adverse effects have also been observed with biologics targeting other TNFSF members. Although interference with CD40L ameliorates autoimmune disease and allograft rejection in animal models,[456,457] anti-CD40L treatment in humans had an unacceptable impact on platelet function in allotransplantation and autoimmune diseases. The use of monovalent Fab against anti-CD40L may avoid these problems.[458] OPG was dropped from clinical investigation because the recombinant protein induced antibody responses that limited its therapeutic effects. Thus, dose or mechanism-associated toxicities can impact physiology in unanticipated ways, limiting the success of clinical outcomes.

TNF inhibitor treatment failure. There is considerable variability in the response to TNF inhibitors. Approximately 60% of patients with rheumatic diseases fail to achieve clinical remission. These patients can be divided into several groups: those with unacceptable side effects, those who do not respond at all to a particular inhibitor, and those who exhibit loss of an initial response. Primary failure could be due to lack of responsiveness to a particular TNF inhibitor, differences in pathogenic mechanism contributed by different cytokines, and/or genetic polymorphisms. A common reason for loss of responsiveness includes the generation of neutralizing antibodies to the TNF antibody.[459,460] Substitution of a different TNF inhibitor is used to overcome antidrug antibodies.

Clinical trials with baminercept, a fusion protein between the ectodomain of the LTβR and the Fc of IgG1 targeting LTαβ and LIGHT, found decreased biomarkers associated with LTβR-HVEM signaling, such as IFN responsive genes, chemokines and lower circulating T and B cells levels. Baminercept treatment of patients with rheumatoid arthritis and Sjogren syndrome did not improve disease symptoms,[459,461] a lack of effect that was repeated with a neutralizing antibody to LTα that also blocks LTαβ complex.[462] The interpretation of these clinical results in terms of mechanism of drug action remains unclear due to the complexity and dynamics in the LTαβ and LIGHT pathways in mediating both proinflammatory and inhibitory signaling.

RANK Ligand Inhibitor

The central function of the RANK-RANKL system in osteoclastogenesis led to the development of anti-RANKL antibodies to treat bone-resorbing conditions. Human anti-RANKL monoclonal antibody (denosumab) as a means of disrupting RANKL-RANK signaling has been approved for the treatment of osteoporosis in postmenopausal women and for treatment of skeletal changes resulting from metastasis of solid tumors among other conditions.[463] Anti-RANKL therapy reduces osteoclastic bone resorption, thereby increasing bone mineral density in both the lumbar spine and the hip. Treatment with anti-RANKL for 36 months decreased the risk of vertebral, nonvertebral, and hip fractures. However, a rebound effect occurs in patients when denosumab treatment is discontinued.[464] There is no apparent increased risk of cancer or serious infection associated with anti-RANKL therapy in subjects with osteoporosis. Indeed, treatment with denosumab increases bone mass in bone fracture-susceptible women that are receiving adjuvant aromatase inhibitors. The same is true in men who are at high risk for fracture while undergoing androgen-deprivation therapy for prostate cancer.[465,466] In addition to its utility in prostate and breast cancer patients undergoing hormone deprivation therapy, anti-RANKL aids in prevention of skeletal-related events in patients with bone metastasis from solid tumors.[465]

BAFF Inhibitor

Systemic lupus erythematous (SLE) is characterized by hyperreactive B cells, production of anti-DNA antibodies, and overproduction of BAFF.[467] The anti-BAFF antibody (belimumab) was approved in 2011 for the treatment of SLE, becoming the first new drug to gain approval for treating this condition in over 50 years. In a phase III clinical trial, anti-BAFF decreased the rate of disease flares and increased the time between disease flares compared with placebo.[468] Anti-BAFF treatment increased SLE responder index (SRI; a reduction of SELENA-SLEDAI score of 4 or more points) compared to placebo.[468] However, only a minor subset of SLE patients respond to anti-BAFF treatment.[469]

TNFSF in Cancer Therapy

The TNFSF regulates several aspects of immune responses, including T cell costimulation and effects on lymphoid tissue microenvironments, providing the impetus to investigate the efficacy of these immunotherapeutics for cancer. Cancer models in mice exhibit striking enhancement of antitumor T cell responses when treated with TNFRSF agonists, especially in combination with the checkpoint inhibitors CTLA4 and PD1.[470-472] While mouse cancer models have limited capacity for predicting outcomes in human cancers, increasing knowledge of the physiological properties of the TNFSF will nevertheless continue to inspire clinical trials, especially when combined with patient-specific canceromics. Clinical trials in cancer patients that received recombinant TNF therapy have reported severe adverse effects characterized as a "septic shock-like syndrome" that occur at a dose limiting toxicity 10 times lower than the calculated antitumor dose extrapolated from animal studies.[473-475] The use of TNF as an antitumor agent in humans has been limited to treatment of localized tumors by isolated limb perfusion, which limits the systemic toxicity of TNF.[476]

TRAIL-induced apoptosis is under investigation for cancer therapy based on its ability to selectively target cancer cells while sparing normal cells. Recombinant TRAIL and antibody-based receptor agonists that induce apoptosis are being used as anti-cancer therapeutics.[477-479] Although recombinant TRAIL and agonist mAbs to TRAILR1 or R2 have exhibited no dose limiting toxicities,[480-483] the ability of some tumor cells to resist TRAIL-induced apoptosis may limit the usefulness of TRAIL in controlling cancer.[484] Several cancer-related clinical trials with TRAIL-targeted immunotherapeutics are ongoing in 2021 (clinicaltrials.gov).

Costimulation pathways are also being pursued in clinical trials. Agonist antibodies to OX40 appear promising for induction of antitumor immunity.[485,486] Targeting the LTβR-LIGHT-HVEM pathways based on their costimulatory action in T cells and lymphoid tissue microenvironments are also being pursued as potential cancer therapeutics.[223]

Cancer immunotherapy with chimeric antigen receptor (CAR)-T cells utilizes the TRAF signaling motif of 41BB to enhance T cell activation.[487,488] Patient-derived T cells are genetically modified using a retroviral vector encoding an anti-CD19 mAb that targets both malignant and normal B cells. The basic construct encodes the anti-CD19 mAb ectodomain fused to the cytosolic domains of CD3 and 41BB.[489] This construct confers T cells with the ability to recognize B cells and to activate the CD3 signaling pathway and cosignaling through 41BB. This combination acts to drive the differentiation of cytotoxic T effector cells that can kill malignant B cells. CAR–T cell immunotherapy is an FDA-approved treatment for B cell leukemia and lymphoma, with promising second generation modifications currently in clinical development.[490,491]

BCMA is primarily expressed by memory B cells and plasma cells, providing a key rationale for immunotherapeutic approaches using BCMA antagonism to treat multiple myeloma.[492] Anti-BCMA mAb is used for the treatment of patients with multiple myeloma that utilizes a novel mechanism of action.[493] The BCMA mAb (belantamab mafodotin) preferentially delivers a cytotoxic payload to the BCMA-expressing cancer as well as to normal B cells. The human anti-BCMA mAb is conjugated with mafodotin, the tubulin inhibitor monomethyl auristatin-F, via a stable maleimidocaproyl link. The mAb-toxin conjugate binds BCMA expressing cells and is internalized, resulting in release of the toxin to destabilize microtubules required for cell division. A second mechanism of killing occurs through altered glycosylation of the anti-BCMA mAb (afucosylated), which enhances ADCC of BCMA-expressing cells. The BCMA-directed immunotherapy approved for drug-resistant and recurrent multiple myeloma is under restricted use due to serious side effects.[494] CAR–T cell based immunotherapy targeting BCMA is also in clinical trials for multiple myeloma.[495]

The diversity of roles played by members of the TNF superfamily in the immune system predict that continued research will yield a bounty of new approaches for treating inflammatory, infectious, and malignant diseases.

ACKNOWLEDGMENTS

The authors gratefully recognize the assistance of Dr. Parham Ramezani-Rad, Dr. John Sedy, Dr. Jon Grist and editing assistance from Dr. Bill Stallcup and Theresa Villalpando. The authors gratefully acknowledge the consistent support from the National Institutes of Health in making this endeavor possible. Conflicts of Interest: NHR is a consultant to L2 Diagnostics; CFW consults for Coherus Biosciences and receives research support from Eli Lilly & Co and Boehringer Ingelheim.

This Chapter is dedicated to Professor Gale "Morrie" Granger and Professor Bryon H. Waksman for their founding contributions to the TNF Superfamily.

REFERENCES

1. Bodmer JL, Schneider P, Tschopp J. The molecular architecture of the TNF superfamily. *Trends Biochem Sci.* 2002;27:19-26.
2. Mukai Y, Nakamura T, Yoshikawa M, et al. Solution of the structure of the TNF-TNFR2 complex. *Sci Signal.* 2010;3:ra83.
3. Chattopadhyay K, Ramagopal UA, Nathenson SG, et al. 1.8 A structure of murine GITR ligand dimer expressed in *Drosophila melanogaster* S2 cells. *Acta Crystallogr D Biol Crystallogr.* 2009;65:434-439.
4. Hymowitz SG, Compaan DM, Yan M, et al. The crystal structures of EDA-A1 and EDA-A2: splice variants with distinct receptor specificity. *Structure.* 2003;11:1513-1520.
5. Gordon NC, Pan B, Hymowitz SG, et al. BAFF/BLyS receptor 3 comprises a minimal TNF receptor-like module that encodes a highly focused ligand-binding site. *Biochemistry.* 2003;42:5977-5983.
6. Cha SS, Kim MS, Choi YH, et al. 2.8 A resolution crystal structure of human trail, a cytokine with selective antitumor activity. *Immunity.* 1999;11:253-261.
7. Jones EY, Stuart DI, Walker NP. Crystal structure of TNF. *Immunol Ser.* 1992;56:93-127.
8. Eck MJ, Ultsch M, Rinderknecht E, et al. The structure of human lymphotoxin (tumor necrosis factor-β) at 1.9-A resolution. *J Biol Chem.* 1992;267:2119-2122.
9. Liu W, Ramagopal U, Cheng H, et al. Crystal structure of the complex of human FASL and its decoy receptor DcR3. *Structure.* 2016;24:2016-2023.
10. Liu W, Zhan C, Cheng H, et al. Mechanistic basis for functional promiscuity in the TNF and TNF receptor superfamilies: structure of the LIGHT:DcR3 assembly. *Structure.* 2014;22:1252-1262.
11. Zhan C, Patskovsky Y, Yan Q, et al. Decoy strategies: the structure of TL1A:DcR3 complex. *Structure.* 2011;19:162-171.
12. Browning JL, Ngam-ek A, Lawton P, et al. Lymphotoxin B, a novel member of the TNF family that forms a heteromeric complex with lymphotoxin on the cell surface. *Cell.* 1993;72:847-856.
13. Roschke V, Sosnovtseva S, Ward CD, et al. BLyS and APRIL form biologically active heterotrimers that are expressed in patients with systemic immune-based rheumatic diseases. *J Immunol.* 2002;169:4314-4321.
14. Pradet-Balade B, Medema JP, Lopez-Fraga M, et al. An endogenous hybrid mRNA encodes TWE-PRIL, a functional cell surface TWEAK-APRIL fusion protein. *EMBO J.* 2002;21:5711-5720.
15. Yan M, Wang LC, Hymowitz SG, et al. Two-amino acid molecular switch in an epithelial morphogen that regulates binding to two distinct receptors. *Science.* 2000;290:523-527.
16. Granger SW, Butrovich KD, Houshmand P, et al. Genomic characterization of LIGHT reveals linkage to an immune response locus on chromosome 19p13.3 and distinct isoforms generated by alternate splicing or proteolysis. *J Immunol.* 2001;167:5122-5128.
17. Muller U, Jongeneel CV, Nedospasov SA, et al. Tumour necrosis factor and lymphotoxin genes map close to H-2D in the mouse major histocompatibility complex. *Nature.* 1987;325:265-267.
18. Flajnik MF, Kasahara M. Comparative genomics of the MHC: glimpses into the evolution of the adaptive immune system. *Immunity.* 2001;15:351-362.
19. Banner DW, D'Arcy A, Janes W, et al. Crystal structure of the soluble human 55 kd TNF receptor-human TNFβ complex: implications for TNF receptor activation. *Cell.* 1993;73:431-445.

20. Carfi A, Willis SH, Whitbeck JC, et al. Herpes simplex virus glycoprotein D bound to the human receptor HVEA. *Mol Cell*. 2001;8:169-179.
21. Hymowitz SG, Patel DR, Wallweber HJ, et al. Structures of APRIL-receptor complexes: like BCMA, TACI employs only a single cysteine-rich domain for high affinity ligand binding. *J Biol Chem*. 2005;280:7218-7227.
22. Bossen C, Ingold K, Tardivel A, et al. Interactions of tumor necrosis factor (TNF) and TNF receptor family members in the mouse and human. *J Biol Chem*. 2006;281:13964-13971.
23. Boisvert J, Edmondson S, Krummel MF. Immunological synapse formation licenses CD40-CD40L accumulations at T-APC contact sites. *J Immunol*. 2004;173:3647-3652.
24. Saliba DG, Céspedes-Donoso PF, Bálint Š, et al. Composition and structure of synaptic ectosomes exporting antigen receptor linked to functional CD40 ligand from helper T cells. *Elife*. 2019;8:e47528.
25. Mintz MA, Felce JH, Chou MY, et al. The HVEM-BTLA axis restrains T cell help to germinal center B cells and functions as a cell-extrinsic suppressor in lymphomagenesis. *Immunity*. 2019;51:310.e7-323.e7.
26. Engelmann H, Holtmann H, Brakebusch C, et al. Antibodies to a soluble form of a tumor necrosis factor (TNF) receptor have TNF-like activity. *J Biol Chem*. 1990;265:14497-14504.
27. Li B, Russell SJ, Compaan DM, et al. Activation of the proapoptotic death receptor DR5 by oligomeric peptide and antibody agonists. *J Mol Biol*. 2006;361:522-536.
28. Wang F, Chau B, West SM, et al. Structures of mouse and human GITR-GITRL complexes reveal unique TNF superfamily interactions. *Nat Commun*. 2021;12:1378.
29. Karathanasis C, Medler J, Fricke F, et al. Single-molecule imaging reveals the oligomeric state of functional TNFα-induced plasma membrane TNFR1 clusters in cells. *Sci Signal*. 2020;13:eaax5647.
30. Liu Y, Hong X, Kappler J, et al. Ligand-receptor binding revealed by the TNF family member TALL-1. *Nature*. 2003;423:49-56.
31. Eslami M, Meinl E, Eibel H, et al. BAFF 60-mer, and differential BAFF 60-mer dissociating activities in human serum, cord blood and cerebrospinal fluid. *Front Cell Dev Biol*. 2020;8:577662.
32. Boldin MP, Mett IL, Varfolomeev EE, et al. Self-association of the "death domains" of the p55 tumor necrosis factor (TNF) receptor and FAS/APO1 prompts signaling for TNF and Fas/APO1 effects. *J Biol Chem*. 1995;270:387-391.
33. Chan FK. Three is better than one: pre-ligand receptor assembly in the regulation of TNF receptor signaling. *Cytokine*. 2007;37:101-107.
34. Kishore U, Gaboriaud C, Waters P, et al. C1q and tumor necrosis factor superfamily: Modularity and versatility. *Trends Immunol*. 2004;25:551-561.
35. Seldin MM, Tan SY, Wong GW. Metabolic function of the CTRP family of hormones. *Rev Endocr Metab Disord*. 2014;15:111-123.
36. Bothwell M. Recent advances in understanding context-dependent mechanisms controlling neurotrophin signaling and function. *F1000Res*. 2019;8:F1000.
37. He XL, Garcia KC. Structure of nerve growth factor complexed with the shared neurotrophin receptor p75. *Science*. 2004;304:870-875.
38. Sacchetti M, Bruscolini A, Lambiase A. Neurotrophic factors and nerve growth factor in ocular allergy. *Curr Opin Allergy Clin Immunol*. 2019;19:510-516.
39. Shao Z, Browning JL, Lee X, et al. TAJ/TROY, an orphan TNF receptor family member, binds Nogo-66 receptor 1 and regulates axonal regeneration. *Neuron*. 2005;45:353-359.
40. Park JB, Yiu G, Kaneko S, et al. A TNF receptor family member, TROY, is a coreceptor with Nogo receptor in mediating the inhibitory activity of myelin inhibitors. *Neuron*. 2005;45:345-351.
41. Collette Y, Gilles A, Pontarotti P, et al. A Co-evolution perspective of the TNFSF and TNFRSF families in the immune system. *Trends Immunol*. 2003;24:387-394.
42. Quistad SD, Stotland A, Barott KL, et al. Evolution of TNF-induced apoptosis reveals 550 My of functional conservation. *Proc Natl Acad Sci U S A*. 2014;111:9567-9572.
43. Igaki T, Miura M. The Drosophila TNF ortholog Eiger: emerging physiological roles and evolution of the TNF system. *Semin Immunol*. 2014;26:267-274.
44. Kato K, Losada-Perez M, Hidalgo A. Gene network underlying the glial regenerative response to central nervous system injury. *Dev Dyn*. 2018;247:85-93.
45. Quistad SD, Traylor-Knowles N. Precambrian origins of the TNFR superfamily. *Cell Death Dis*. 2016;2:16058.
46. Wiens GD, Glenney GW. Origin and evolution of TNF and TNF receptor superfamilies. *Dev Comp Immunol*. 2011;35:1324-1335.
47. Roca FJ, Mulero I, López-Muñoz A, et al. Evolution of the inflammatory response in vertebrates: fish TNFα is a powerful activator of endothelial cells but hardly activates phagocytes. *J Immunol*. 2008;181:5071-5081.
48. Barshis DJ, Ladner JT, Oliver TA, et al. Genomic basis for coral resilience to climate change. *Proc Natl Acad Sci U S A*. 2013;110:1387-1392.
49. Wenger Y, Buzgariu W, Reiter S, et al. Injury-induced immune responses in Hydra. *Semin Immunol*. 2014; 26:277-294.
50. Oliveira KC, Carvalho ML, Venancio TM, et al. Identification of the *Schistosoma mansoni* TNFα receptor gene and the effect of human TNFα on the parasite gene expression profile. *PLoS Negl Trop Dis*. 2009;3:e556.
51. Bertevello CR, Russo BRA, Tahira AC, et al. The evolution of TNF signaling in platyhelminths suggests the cooptation of TNF receptor in the host-parasite interplay. *Parasit Vectors*. 2020;13:491.
52. Benedict CA. Viruses and the TNF-related cytokines, an evolving battle. *Cytokine Growth Factor Rev*. 2003;14:349-357.
53. McFadden G. Poxvirus tropism. *Nat Rev Microbiol*. 2005;3:201-213.
54. Massung RF, Esposito JJ, Liu LI, et al. Potential virulence determinants in terminal regions of variola smallpox virus genome. *Nature*. 1993;366:748-751.
55. Upton C, Macen J, Schreiber M, et al. Myxoma virus expresses a secreted protein with homology to the tumor necrosis factor receptor gene family that contributes to viral virulence. *Virology*. 1991;184:370-382.
56. Rahman MM, McFadden G. Modulation of tumor necrosis factor by microbial pathogens. *PLoS Pathog*. 2006;2:e4.
57. Saraiva M, Alcami A. CrmE, a novel soluble tumor necrosis factor receptor encoded by poxviruses. *J Virol*. 2001;75:226-233.
58. Alejo A, Ruiz-Argüello MB, Pontejo SM, et al. Chemokines cooperate with TNF to provide protective anti-viral immunity and to enhance inflammation. *Nat Commun*. 2018;9:1790.
59. Montgomery RI, Warner MS, Lum B, et al. Herpes simplex virus 1 entry into cells mediated by a novel member of the TNF/NGF receptor family. *Cell*. 1996;87:427-436.
60. Compaan DM, Gonzalez LC, Tom I, et al. Attenuating lymphocyte activity: the crystal structure of the BTLA-HVEM complex. *J Biol Chem*. 2005;280:39553-39561.
61. Cheung TC, Humphreys IR, Potter KG, et al. Evolutionarily divergent herpesviruses modulate T cell activation by targeting the herpesvirus entry mediator cosignaling pathway. *Proc Natl Acad Sci U S A*. 2005;102:13218-13223.
62. Burn Aschner C, Loh LN, Galen B, et al. HVEM signaling promotes protective antibody-dependent cellular cytotoxicity (ADCC) vaccine responses to herpes simplex viruses. *Sci Immunol*. 2020;5:eaax2454.
63. Benedict CA, Butrovich KD, Lurain NS, et al. Cutting edge: a novel viral TNF receptor superfamily member in virulent strains of human cytomegalovirus. *J Immunol*. 1999;162:6967-6970.
64. Bitra A, Nemcovicova I, Picarda G, et al. Structure of human cytomegalovirus Ul144, an HVEM orthologue, bound to the B and T cell lymphocyte attenuator. *J Biol Chem*. 2019;294:10519-10529.
65. Poole E, King CA, Sinclair JH, et al. The Ul144 gene product of human cytomegalovirus activates NFκB via a TRAF6-dependent mechanism. *EMBO J*. 2006;25:4390-4399.
66. Picarda G, Ghosh R, McDonald B, et al. Cytomegalovirus evades TRAIL-mediated innate lymphoid cell 1 defenses. *J Virol*. 2019;93:e00617-e00619.
67. Nishimura M, Novita BD, Kato T, et al. Structural basis for the interaction of human herpesvirus 6b Tetrameric glycoprotein complex with the cellular receptor, human CD134. *PLoS Pathog*. 2020;16:e1008648.
68. Nishimura M, Mori Y. Entry of betaherpesviruses. *Adv Virus Res*. 2019;104:283-312.
69. Bolze PA, Mommert M, Mallet F. Contribution of syncytins and other endogenous retroviral envelopes to human placenta pathologies. *Prog Mol Biol Transl Sci*. 2017;145:111-162.
70. Zhang B, Sun C, Jin S, et al. Mapping of equine lentivirus receptor 1 residues critical for equine infectious anemia virus envelope binding. *J Virol*. 2008;82:1204-1213.
71. Zhang B, Jin S, Jin J, et al. A tumor necrosis factor receptor family protein serves as a cellular receptor for the macrophage-tTropic equine lentivirus. *Proc Natl Acad Sci U S A*. 2005;102:9918-9923.
72. Corbi-Botto CM, Sadaba SA, Zappa ME, et al. Nonsynonymous changes of equine lentivirus receptor-1 (ELR1) gene in amino acids involved in the interaction with equine infectious anemia virus (EIAV). *Res Vet Sci*. 2017;112:185-191.
73. Kieser A, Sterz KR. The latent membrane protein 1 (LMP1). *Curr Top Microbiol Immunol*. 2015;391:119-149.
74. Xie P, Kraus ZJ, Stunz LL, et al. Roles of TRAF molecules in B lymphocyte function. *Cytokine Growth Factor Rev*. 2008;19:199-207.
75. Bishop GA, Stunz LL, Hostager BS. TRAF3 as a multifaceted regulator of B lymphocyte survival and activation. *Front Immunol*. 2018;9:2161.
76. Perlin DS, Zafir-Lavie I, Roadcap L, et al. Levels of the TNF-related cytokine LIGHT increase in hospitalized COVID-19 patients with Cytokine Release Syndrome and ARDS. *mSphere*. 2020;5:e00699-e00720.
77. Arunachalam PS, Wimmers F, Mok CKP, et al. Systems biological assessment of immunity to mild versus severe COVID-19 infection in humans. *Science*. 2020;369:1210-1220.
78. Duivelshof BL, Murisier A, Camperi J, et al. Therapeutic Fc-fusion proteins: current analytical strategies. *J Separ Sci*. 2020;44:35-62.
79. Kerensky TA, Gottlieb AB, Yaniv S, et al. Etanercept: efficacy and safety for approved indications. *Expet Opin Drug Saf*. 2012;11:121-139.
80. Sedy JR, Balmert MO, Ware BC, et al. A herpesvirus entry mediator mutein with selective agonist action for the inhibitory receptor B and T lymphocyte attenuator. *J Biol Chem*. 2017;292:21060-21070.
81. Yaron JR, Zhang L, Guo Q, et al. Deriving immune modulating drugs from viruses-a new class of biologics. *J Clin Med*. 2020;9:972.
82. Gutierrez MJ, Nino G, Hong X, et al. Epigenetic dynamics of the infant immune system reveals a tumor necrosis factor superfamily signature in early human life. *Epigenomes*. 2020;4:12.
83. Kontoyiannis D, Pasparakis M, Pizarro TT, et al. Impaired on/off regulation of Tnf biosynthesis in mice lacking Tnf AU- rich elements: implications for joint and gut-associated immunopathologies. *Immunity*. 1999;10:387-398.
84. Kontoyiannis D, Boulougouris G, Manoloukos M, et al. Genetic dissection of the cellular pathways and signaling mechanisms in modeled tumor necrosis factor-induced Crohn's-like inflammatory bowel disease. *J Exp Med*. 2002;196:1563-1574.
85. Venkatesh D, Ernandez T, Rosetti F, et al. Endothelial TNF receptor 2 induces IRF1 transcription factor-dependent interferon-B autocrine signaling to promote monocyte recruitment. *Immunity*. 2013;38:1025-1037.
86. Webster JD, Vucic D. The balance of TNF mediated pathways regulates inflammatory cell death signaling in healthy and diseased tissues. *Front Cell Dev Biol*. 2020;8:365.
87. Häcker G. The morphology of apoptosis. *Cell Tissue Res*. 2000;301:5-17.
88. Boatright KM, Salvesen GS. Mechanisms of caspase activation. *Curr Opin Cell Biol*. 2003;15:725-731.
89. Zhang H, Ji J, Liu Q, et al. Muc1 downregulation promotes TNF-A-induced necroptosis in human bronchial epithelial cells via regulation of the RIPK1/RIPK3 pathway. *J Cell Physiol*. 2019;234:15080-15088.
90. O'Donnell MA, Perez-Jimenez E, Oberst A, et al. Caspase 8 inhibits programmed necrosis by processing CYLD. *Nat Cell Biol*. 2011;13:1437-1442.
91. Scott FL, Stec B, Pop C, et al. The FAS-FADD death domain complex structure unravels signalling by receptor clustering. *Nature*. 2009;457:1019-1022.
92. Cai Z, Jitkaew S, Zhao J, et al. Plasma membrane translocation of trimerized MLKL protein is required for TNF-induced necroptosis. *Nat Cell Biol*. 2014;16:55-65.

93. Chen J, Kos R, Garssen J, et al. Molecular insights into the mechanism of necroptosis: the necrosome as a potential therapeutic target. *Cells.* 2019;8:1486.
94. Shi JH, Sun SC. Tumor necrosis factor receptor-associated factor regulation of nuclear factor κB and mitogen-activated protein kinase pathways. *Front Immunol.* 2018;9:1849.
95. Pedros C, Altman A, Kong KF. Role of TRAFS in signaling pathways controlling T follicular helper cell differentiation and T cell-dependent antibody responses. *Front Immunol.* 2018;9:2412.
96. Walczak H. TNF and ubiquitin at the crossroads of gene activation, cell death, inflammation, and cancer. *Immunol Rev.* 2011;244:9-28.
97. Gough P, Myles IA. Tumor necrosis factor receptors: pleiotropic signaling complexes and their differential effects. *Front Immunol.* 2020;11:585880.
98. Belgnaoui SM, Paz S, Hiscott J. Orchestrating the interferon antiviral response through the mitochondrial antiviral signaling (MAVS) adapter. *Curr Opin Immunol.* 2011;23:564-572.
99. Sun SC. Non-canonical NF-κB signaling pathway. *Cell Res.* 2011;21:71-85.
100. Hacker H, Tseng PH, Karin M. Expanding TRAF function: TRAF3 as a tri-faced immune regulator. *Nat Rev Immunol.* 2011;11:457-468.
101. Li C, Norris PS, Ni CZ, et al. Structurally distinct recognition motifs in lymphotoxin-β receptor and CD40 for tumor necrosis factor receptor-associated factor (TRAF)-mediated signaling. *J Biol Chem.* 2003;278:50523-50529.
102. Varfolomeev E, Goncharov T, Maecker H, et al. Cellular inhibitors of apoptosis are global regulators of NF-κB and MAPK activation by members of the TNF family of receptors. *Sci Signal.* 2012;5:ra22.
103. Jarosz-Griffiths HH, Holbrook J, Lara-Reyna S, et al. TNF receptor signalling in autoinflammatory diseases. *Int Immunol.* 2019;31:639-648.
104. Ghosh S, Dass JFP. Study of pathway cross-talk interactions with NF-κB leading to its activation via ubiquitination or phosphorylation: a brief review. *Gene.* 2016;584:97-109.
105. Karin M, Ben-Neriah Y. Phosphorylation meets ubiquitination: the control of NF-κB activity. *Annu Rev Immunol.* 2000;18:621-663.
106. Schrofelbauer B, Hoffmann A. How do Pleiotropic kinase hubs mediate specific signaling by TNFR superfamily members? *Immunol Rev.* 2011;244:29-43.
107. Ghosh S, Karin M. Missing pieces in the NF-κB puzzle. *Cell.* 2002;109:S81-S96.
108. Sanjo H, Zajonc DM, Braden R, et al. Allosteric regulation of the ubiquitin:NIK and ubiquitin:TRAF3 E3 ligases by the lymphotoxin-β receptor. *J Biol Chem.* 2010;285:17148-17155.
109. Pomerantz JL, Baltimore D. Two pathways to NF-κB. *Mol Cell.* 2002;10:693-695.
110. Staudt LM. Oncogenic activation of NF-κB. *Cold Spring Harb Perspect Biol.* 2010;2:a000109.
111. Granger GA, Williams TW. Lymphocyte cytotoxicity in vitro: activation and release of a cytotoxic factor. *Nature.* 1968;218:1253-1254.
112. Ruddle NH, Waksman BH. Cytotoxicity mediated by soluble antigen and lymphocytes in delayed hypersensitivity. 3. Analysis of mechanism. *J Exp Med.* 1968;128:1267-1279.
113. Carswell EA, Old LJ, Kassel RL, et al. An Endotoxin-induced serum factor that causes necrosis of tumors. *Proc Natl Acad Sci U S A.* 1975;72:3666-3670.
114. Beutler B, Greenwald D, Hulmes JD, et al. Identity of tumour necrosis factor and the macrophage-secreted factor cachectin. *Nature.* 1985;316:552-554.
115. Ward-Kavanagh LK, Lin WW, Sedy JR, et al. The TNF receptor superfamily in Co-stimulating and Co-inhibitory responses. *Immunity.* 2016;44:1005-1019.
116. Wajant H, Siegmund D. TNFR1 and TNFR2 in the control of the life and death balance of macrophages. *Front Cell Dev Biol.* 2019;7:91.
117. Kulbe H, Thompson R, Wilson JL, et al. The inflammatory cytokine tumor necrosis factor-α generates an autocrine tumor-promoting network in epithelial ovarian cancer cells. *Cancer Res.* 2007;67:585-592.
118. Drayton DL, Liao S, Mounzer RH, et al. Lymphoid organ development: from ontogeny to neogenesis. *Nat Immunol.* 2006;7:344-353.
119. Rutschmann S, Hoebe K, Zalevsky J, et al. PanR1, a dominant negative missense allele of the gene encoding TNF-α (TNF), does not impair lymphoid development. *J Immunol.* 2006;176:7525-7532.
120. Neumann B, Luz A, Pfeffer K, et al. Defective Peyer's patch organogenesis in mice lacking the 55-kD receptor for tumor necrosis factor. *J Exp Med.* 1996;184:259-264.
121. Iizuka K, Chaplin DD, Wang Y, et al. Requirement for membrane lymphotoxin in natural killer cell development. *Proc Natl Acad Sci U S A.* 1999;96:6336-6340.
122. Elewaut D, Shaikh RB, Hammond KJ, et al. NIK-dependent RelB activation defines a unique signaling pathway for the development of Vα14i NKT cells. *J Exp Med.* 2003;197:1623-1633.
123. Kabashima K, Banks TA, Ansel KM, et al. Intrinsic lymphotoxin-β receptor requirement for homeostasis of lymphoid tissue dendritic cells. *Immunity.* 2005;22:439-450.
124. Tumanov AV, Grivennikov SI, Shakhov AN, et al. Dissecting the role of lymphotoxin in lymphoid organs by conditional targeting. *Immunol Rev.* 2003;195:106-116.
125. Ware CF. Network communications: Lymphotoxins, LIGHT, and Tnf. *Annu Rev Immunol.* 2005;23:787-819.
126. De Togni P, Goellner J, Ruddle NH, et al. Abnormal development of peripheral lymphoid organs in mice deficient in lymphotoxin. *Science.* 1994;264:703-707.
127. Cupedo T, Mebius RE. Cellular interactions in lymph node development. *J Immunol.* 2005;174:21-25.
128. Georgopoulos K, Bigby M, Wang J, et al. The Ikaros gene is required for the development of all lymphoid lineages. *Cell.* 1994;79:143-156.
129. Yokota Y, Mansouri A, Mori S, et al. Development of peripheral lymphoid organs and natural killer cells depends on the helix-loop-helix inhibitor Id2. *Nature.* 1999;397:702-706.
130. Sun Z, Unutmaz D, Zou YR, et al. Requirement for RORγ in thymocyte survival and lymphoid organ development. *Science.* 2000;288:2369-2373.
131. Kurebayashi S, Ueda E, Sakaue M, et al. Retinoid-related orphan receptor γ (RORγ) is essential for lymphoid organogenesis and controls apoptosis during thymopoiesis. *Proc Natl Acad Sci U S A.* 2000;97:10132-10137.
132. Eberl G, Marmon S, Sunshine MJ, et al. An essential function for the nuclear receptor RORγ(T) in the generation of fetal lymphoid tissue inducer cells. *Nat Immunol.* 2004;5:64-73.
133. Rennert PD, Browning JL, Mebius R, et al. Surface lymphotoxin α/β complex is required for the development of peripheral lymphoid organs. *J Exp Med.* 1996;184:1999-2006.
134. Rennert PD, James D, Mackay F, et al. Lymph node genesis is induced by signaling through the lymphotoxin β receptor. *Immunity.* 1998;9:71-80.
135. Ruddle NH. Basics of inducible lymphoid organs. *Curr Top Microbiol Immunol.* 2020;426:1-19.
136. Gago da Graca C, van Baarsen LGM, Mebius RE. Tertiary lymphoid structures: diversity in their development, composition, and role. *J Immunol.* 2021;206:273-281.
137. Nishikawa S, Honda K, Vieira P, et al. Organogenesis of peripheral lymphoid organs. *Immunol Rev.* 2003;195:72-80.
138. Lane PJ, Gaspal FM, Kim MY. Two sides of a cellular coin: CD4(+)CD3− cells regulate memory responses and lymph-node organization. *Nat Rev Immunol.* 2005;5:655-660.
139. Scandella E, Bolinger B, Lattmann E, et al. Restoration of lymphoid organ integrity through the interaction of lymphoid tissue-inducer cells with stroma of the T cell zone. *Nat Immunol.* 2008;9:667-675.
140. Mebius RE, Kraal G. Structure and function of the spleen. *Nat Rev Immunol.* 2005;5:606-616.
141. Cyster JG. Lymphoid organ development and cell migration. *Immunol Rev.* 2003;195:5-14.
142. Ngo VN, Korner H, Gunn MD, et al. Lymphotoxin α/β and tumor necrosis factor are required for stromal cell expression of homing chemokines in B and T cell areas of the spleen. *J Exp Med.* 1999;189:403-412.
143. Luther SA, Bidgol A, Hargreaves DC, et al. Differing activities of homeostatic chemokines CCL19, CCL21, and CXCL12 in lymphocyte and dendritic cell recruitment and lymphoid Neogenesis. *J Immunol.* 2002;169:424-433.
144. Ansel KM, Ngo VN, Hyman PL, et al. A chemokine-driven positive feedback loop organizes lymphoid follicles. *Nature.* 2000;406:309-314.
145. Katakai T, Hara T, Sugai M, et al. Lymph node fibroblastic reticular cells construct the stromal reticulum via contact with lymphocytes. *J Exp Med.* 2004;200:783-795.
146. Kindler V, Sappino AP, Grau GE, et al. The inducing role of tumor necrosis factor in the development of bactericidal granulomas during BCG infection. *Cell.* 1989;56:731-740.
147. Akirav EM, Xu Y, Ruddle NH. Resident B cells regulate thymic expression of myelin oligodendrocyte glycoprotein. *J Neuroimmunol.* 2011;235:33-39.
148. Silva-Santos B, Pennington DJ, Hayday AC. Lymphotoxin-mediated regulation of γδ cell differentiation by αβ T cell progenitors. *Science.* 2005;307:925-928.
149. Boehm T, Scheu S, Pfeffer K, et al. Thymic medullary epithelial cell differentiation, thymocyte emigration, and the control of autoimmunity require lympho-epithelial cross talk via LTβR. *J Exp Med.* 2003;198:757-769.
150. Elewaut D, Brossay L, Santee SM, et al. Membrane lymphotoxin is required early in ontogeny for development of different subpopulations of NK T cells. *J Immunol.* 2000;165:671-679.
151. Franki AS, Van Beneden K, Dewint P, et al. A unique lymphotoxin αβ-dependent pathway regulates thymic emigration of Vα14 invariant natural killer T cells. *Proc Natl Acad Sci U S A.* 2006;103:9160-9165.
152. Chin RK, Zhu M, Christiansen PA, et al. Lymphotoxin pathway-directed, autoimmune regulator-independent central tolerance to arthritogenic collagen. *J Immunol.* 2006;177:290-297.
153. Akirav EM, Ruddle NH, Herold KC. The role of AIRE in human autoimmune disease. *Nat Rev Endocrinol.* 2011;7:25-33.
154. Whitbeck JC, Peng C, Lou H, et al. Glycoprotein D of herpes simplex virus (HSV) binds directly to HVEM, a member of the tumor necrosis factor receptor superfamily and a mediator of HSV entry. *J Virol.* 1997;71:6083-6093.
155. Kuncewicz K, Spodzieja M, Sieradzan A, et al. A structural model of the immune checkpoint CD160-HVEM complex derived from HDX-mass spectrometry and molecular modeling. *Oncotarget.* 2019;10:536-550.
156. Connolly SA, Landsburg DJ, Carfi A, et al. Structure-based analysis of the herpes simplex virus glycoprotein D binding site present on herpesvirus entry mediator HVEA (HVEM). *J Virol.* 2002;76:10894-10904.
157. Shrestha R, Garrett-Thomson SC, Liu W, et al. Redesigning HVEM interface for selective binding to LIGHT, BTLA, and CD160. *Structure.* 2020;28(11):1197.e2-1205.e2.
158. Zhu Y, Yao S, Augustine MM, et al. Neuron-specific SALM5 limits inflammation in the CNS via its interaction with HVEM. *Sci Adv.* 2016;2:e1500637.
159. Cheung TC, Coppieters K, Sanjo H, et al. Polymorphic variants of LIGHT (TNF superfamily-14) alter receptor avidity and bioavailability. *J Immunol.* 2010;185:1949-1958.
160. Hsieh SL, Lin WW. Decoy receptor 3: an endogenous immunomodulator in cancer growth and inflammatory reactions. *J Biomed Sci.* 2017;24:39.
161. Sedy J, Bekiaris V, Ware CF. Tumor necrosis factor superfamily in innate immunity and inflammation. *Cold Spring Harb Perspect Biol.* 2014;7:a016279.
162. Cheung TC, Oborne LM, Steinberg MW, et al. T cell intrinsic heterodimeric complexes between HVEM and BTLA determine receptivity to the surrounding microenvironment. *J Immunol.* 2009;183:7286-7296.
163. Cheung TC, Steinberg MW, Oborne LM, et al. Unconventional ligand activation of herpesvirus entry mediator signals cell survival. *Proc Natl Acad Sci U S A.* 2009;106:6244-6249.
164. Sedy JR, Ramezani-Rad P. HVEM network signaling in cancer. *Adv Cancer Res.* 2019;142:145-186.
165. Sedy JR, Bjordahl RL, Bekiaris V, et al. CD160 activation by herpesvirus entry mediator augments inflammatory cytokine production and cytolytic function by NK cells. *J Immunol.* 2013;191:828-836.
166. Steinberg MW, Cheung TC, Ware CF. The signaling networks of the herpesvirus entry mediator (TNFRSF14) in immune regulation. *Immunol Rev.* 2011;244:169-187.
167. Krieg C, Boyman O, Fu YX, et al. B and T Lymphocyte attenuator regulates CD8(+) T cell-intrinsic homeostasis and memory cell generation. *Nat Immunol.* 2007;8:162-171.

168. Desai P, Abboud G, Stanfield J, et al. HVEM imprints memory potential on effector CD8 T cells required for protective mucosal immunity. *J Immunol.* 2017;199:2968-2975.
169. Flynn R, Hutchinson T, Murphy KM, et al. CD8 T cell memory to a viral pathogen requires trans cosignaling between HVEM and BTLA. *PLoS One.* 2013;8:e77991.
170. Kumaraguru U, Davis IA, Deshpande S, et al. Lymphotoxin α(−/−) mice develop functionally impaired CD8(+) T cell responses and fail to contain virus infection of the central nervous system. *J Immunol.* 2001;166:1066-1074.
171. Schneider K, Loewendorf A, De Trez C, et al. Lymphotoxin-mediated crosstalk between B cells and splenic stroma promotes the initial type I interferon response to cytomegalovirus. *Cell Host Microbe.* 2008;3:67-76.
172. Benedict CA, Banks TA, Senderowicz L, et al. Lymphotoxins and cytomegalovirus cooperatively induce interferon-β, establishing host-virus detente. *Immunity.* 2001;15:617-626.
173. Berger DP, Naniche D, Crowley MT, et al. Lymphotoxin-β-deficient mice show defective antiviral immunity. *Virol.* 1999;260:136-147.
174. Suresh M, Lanier G, Large MK, et al. Role of lymphotoxin α in T-cell responses during an acute viral infection. *J Virol.* 2002;76:3943-3951.
175. Junt T, Tumanov AV, Harris N, et al. Expression of lymphotoxin beta governs immunity at two distinct levels. *Eur J Immunol.* 2006;36:2061-2075.
176. Puglielli MT, Browning JL, Brewer AW, et al. Reversal of virus-induced systemic shock and respiratory failure by blockade of the lymphotoxin pathway. *Nat Med.* 1999;5:1370-1374.
177. Lin X, Ma X, Rodriguez M, et al. Membrane lymphotoxin is required for resistance to Theiler's virus infection. *Int Immunol.* 2003;15:955-962.
178. Lund FE, Partida-Sanchez S, Lee BO, et al. Lymphotoxin-α-deficient mice make delayed, but effective, T and B cell responses to influenza. *J Immunol.* 2002;169:5236-5243.
179. Ehlers S, Holscher C, Scheu S, et al. The lymphotoxin β receptor is critically involved in controlling infections with the intracellular pathogens *Mycobacterium tuberculosis* and *Listeria monocytogenes*. *J Immunol.* 2003;170:5210-5218.
180. Roach DR, Briscoe H, Saunders B, et al. Secreted lymphotoxin-α is essential for the control of an intracellular bacterial infection. *J Exp Med.* 2001;193:239-246.
181. Lucas R, Tacchini-Cottier F, Guler R, et al. A role for lymphotoxin β receptor in host defense against *Mycobacterium bovis* BCG infection. *Eur J Immunol.* 1999;29:4002-4010.
182. Kutsch S, Degrandi D, Pfeffer K. Immediate lymphotoxin beta receptor-mediated transcriptional response in host defense against *L. monocytogenes*. *Immunobiology.* 2008;213:353-366.
183. Wilhelm P, Riminton DS, Ritter U, et al. Membrane lymphotoxin contributes to anti-leishmanial immunity by controlling structural integrity of lymphoid organs. *Eur J Immunol.* 2002;32:1993-2003.
184. Stanley AC, de Labastida Rivera F, Haque A, et al. Critical roles for LIGHT and its receptors in generating T cell-mediated immunity during Leishmania donovani infection. *PLoS Pathog.* 2011;7:e1002279.
185. Tersteegen A, Sorg UR, Virgen-Slane R, et al. Lymphotoxin β receptor: a crucial role in innate and adaptive immune responses against Toxoplasma gondii. *Infect Immun.* 2021;89(6):e00026-e00121.
186. Schluter D, Kwok LY, Lutjen S, et al. Both lymphotoxin-α and TNF are crucial for control of *Toxoplasma gondii* in the central nervous system. *J Immunol.* 2003;170:6172-6182.
187. Engwerda CR, Mynott TL, Sawhney S, et al. Locally up-regulated lymphotoxin α, not systemic tumor necrosis factor α, is the principle mediator of murine cerebral malaria. *J Exp Med.* 2002;195:1371-1377.
188. Beutler B, Cerami A. The biology of cachectin/TNF – A primary mediator of the host response. *Annu Rev Immunol.* 1989;7:625-655.
189. Pfeffer K, Matsuyama T, Kundig TM, et al. Mice deficient for the 55 kd tumor necrosis factor receptor are resistant to endotoxic shock, yet succumb to *L. Monocytogenes* infection. *Cell.* 1993;73:457-467.
190. Wallis RS, Broder MS, Wong JY, et al. Granulomatous infectious diseases associated with tumor necrosis factor antagonists. *Clin Infect Dis.* 2004;38:1261-1265.
191. Gommerman JL, Browning JL, Ware CF. The Lymphotoxin Network: orchestrating a type I interferon response to optimize adaptive immunity. *Cytokine Growth Factor Rev.* 2014;25:139-145.
192. Li C, Lam E, Perez-Shibayama C, et al. Early-life programming of mesenteric lymph node stromal cell identity by the lymphotoxin pathway regulates adult mucosal immunity. *Sci Immunol.* 2019;4:eaax1027.
193. Koroleva EP, Fu YX, Tumanov AV. Lymphotoxin in physiology of lymphoid tissues - implication for antiviral defense. *Cytokine.* 2018;101:39-47.
194. Kashimura M. The human spleen as the center of the blood defense system. *Int J Hematol.* 2020;112:147-158.
195. Xu HC, Huang J, Khairnar V, et al. Deficiency of the B cell-activating factor receptor results in limited CD169+ macrophage function during viral infection. *J Virol.* 2015;89:4748-4759.
196. Reboldi A, Arnon TI, Rodda LB, et al. IgA production requires B cell interaction with subepithelial dendritic cells in Peyer's patches. *Science.* 2016;352:aaf4822.
197. Edwards RG, Longnecker R. Herpesvirus entry mediator and ocular herpesvirus infection: more than meets the eye. *J Virol.* 2017;91:e00115-e00117.
198. Park SJ, Riccio RE, Kopp SJ, et al. Herpesvirus entry mediator binding partners mediate immunopathogenesis of ocular herpes simplex virus 1 infection. *mBio.* 2020;11:e00790-e00820.
199. Aschner CB, Herold BC. Alphaherpesvirus vaccines. *Curr Issues Mol Biol.* 2020;41:469-508.
200. Haljasmagi L, Salumets A, Rumm AP, et al. Longitudinal proteomic profiling reveals increased early inflammation and sustained apoptosis proteins in severe COVID-19. *Sci Rep.* 2020;10:20533.
201. Keffer J, Probert L, Cazlaris H, et al. Transgenic mice expressing human tumour necrosis factor: a predictive genetic model of arthritis. *EMBO J.* 1991;10:4025-4031.
202. Shaikh RB, Santee S, Granger SW, et al. Constitutive expression of LIGHT on T cells leads to lymphocyte activation, inflammation, and tissue destruction. *J Immunol.* 2001;167:6330-6337.
203. Wang J, Lo JC, Foster A, et al. The regulation of T cell homeostasis and autoimmunity by T cell derived LIGHT. *J Clin Invest.* 2001;108:1771-1780.
204. Drayton DL, Ying X, Lee J, et al. Ectopic LTαβ directs lymphoid organ neogenesis with concomitant expression of peripheral node addressin and a HEV-restricted sulfotransferase. *J Exp Med.* 2003;197:1153-1163.
205. Tseng CC, Lin YZ, Lin CH, et al. Next-generation sequencing profiles of the methylome and transcriptome in peripheral blood mononuclear cells of rheumatoid arthritis. *J Clin Med.* 2019;8:128.
206. Yau AC, Tuncel J, Haag S, et al. Conserved 33-kb haplotype in the MHC class III region regulates chronic arthritis. *Proc Natl Acad Sci U S A.* 2016;113:E3716-E3724.
207. Mehta AK, Doherty T, Broide D, et al. Tumor necrosis factor family member LIGHT acts with IL-1β and TGF-β to promote airway remodeling during rhinovirus infection. *Allergy.* 2018;73:1415-1424.
208. Conlon TM, John-Schuster G, Heide D, et al. Inhibition of LTβR signalling activates WNT-induced regeneration in lung. *Nature.* 2020;588:151-156.
209. Cho JH, Feldman M. Heterogeneity of autoimmune diseases: Pathophysiologic insights from genetics and implications for new therapies. *Nat Med.* 2015;21:730-738.
210. Cardinale CJ, Wei Z, Panossian S, et al. Targeted resequencing identifies defective variants of decoy receptor 3 in pediatric-onset inflammatory bowel disease. *Gene Immun.* 2013;14(7):447-452.
211. Beecham AH, Patsopoulos NA, Xifara DK, et al. Analysis of immune-related loci identifies 48 new susceptibility variants for multiple sclerosis. *Nat Genet.* 2013;45:1353-1360.
212. Martinet L, Garrido I, Filleron T, et al. Human solid tumors contain high endothelial venules: association with T- and B-lymphocyte infiltration and favorable prognosis in breast cancer. *Cancer Res.* 2011;71:5678-5687.
213. Sautes-Fridman C, Petitprez F, Calderaro J, et al. Tertiary lymphoid structures in the era of cancer immunotherapy. *Nat Rev Canc.* 2019;19:307-325.
214. Helmink BA, Reddy SM, Gao J, et al. B cells and tertiary lymphoid structures promote immunotherapy response. *Nature.* 2020;577:549-555.
215. Cabrita R, Lauss M, Sanna A, et al. Tertiary lymphoid structures improve immunotherapy and survival in melanoma. *Nature.* 2020;577:561-565.
216. Haybaeck J, Zeller N, Wolf MJ, et al. A lymphotoxin-driven pathway to hepatocellular carcinoma. *Cancer Cell.* 2009;16:295-308.
217. Scarzello AJ, Jiang Q, Back T, et al. LTβR signalling preferentially accelerates oncogenic AKT-initiated liver tumours. *Gut.* 2016;65:1765-1775.
218. Behnke K, Zhuang Y, Xu HC, et al. B cell-mediated maintenance of cluster of differentiation 169-positive cells is critical for liver regeneration. *Hepatology.* 2018;68:2348-2361.
219. Kridel R, Sehn LH, Gascoyne RD. Pathogenesis of follicular lymphoma. *J Clin Invest.* 2012;122:3424-3431.
220. Bjordahl RL, Steidl C, Gascoyne RD, et al. Lymphotoxin network pathways shape the tumor microenvironment. *Curr Opin Immunol.* 2013;25:222-229.
221. Mintz MA, Cyster JG. T follicular helper cells in germinal center B cell selection and lymphomagenesis. *Immunol Rev.* 2020;296:48-61.
222. Yu P, Lee Y, Liu W, et al. Priming of naive T cells inside tumors leads to eradication of established tumors. *Nat Immunol.* 2004;5:141-149.
223. Skeate JG, Otsmaa ME, Prins R, et al. TNFSF14: lighting the way for effective cancer immunotherapy. *Front Immunol.* 2020;11:922.
224. Azuma M. Co-signal molecules in T-cell activation: historical overview and perspective. *Adv Exp Med Biol.* 2019;1189:3-23.
225. Granger SW, Ware CF. Commentary: turning on LIGHT. *J Clin Invest.* 2001;108:1741-1742.
226. Locksley RM, Killeen N, Lenardo MJ. The TNF and TNF receptor superfamilies: Integrating mammalian biology. *Cell.* 2001;104:487-501.
227. Croft M. Co-stimulatory members of the TNFR family: Keys to effective T-cell immunity? *Nat Rev Immunol.* 2003;3:609-620.
228. Croft M. Costimulation of T cells by OX40, 4-1BB, and CD27. *Cytokine Growth Factor Rev.* 2003;14:265-273.
229. Watts TH. TNF/TNFR family members in costimulation of T cell responses. *Annu Rev Immunol.* 2005;23:23-68.
230. Toubi E, Shoenfeld Y. The role of CD40-CD154 interactions in autoimmunity and the benefit of disrupting this pathway. *Autoimmunity.* 2004;37:457-464.
231. Schneider P. The role of APRIL and BAFF in lymphocyte activation. *Curr Opin Immunol.* 2005;17:282-289.
232. Vonderheide RH, June CH. A translational bridge to cancer immunotherapy: Exploiting costimulation and target antigens for active and passive T cell immunotherapy. *Immunol Res.* 2003;27:341-356.
233. Appleman LJ, Boussiotis VA. T cell anergy and costimulation. *Immunol Rev.* 2003;192:161-180.
234. Saito T, Yamasaki S. Negative feedback of T cell activation through inhibitory adapters and costimulatory receptors. *Immunol Rev.* 2003;192:143-160.
235. Riley JL, June CH. The CD28 family: a T-cell rheostat for therapeutic control of T-cell activation. *Blood.* 2005;105:13-21.
236. Subudhi SK, Alegre ML, Fu YX. The balance of immune responses: costimulation verse coinhibition. *J Mol Med.* 2005;83:193-202.
237. Chang YH, Wang KC, Chu KL, et al. Dichotomous expression of TNF superfamily ligands on antigen-presenting cells controls post-priming anti-viral CD4(+) T cell immunity. *Immunity.* 2017;47:943.e9-958.e9.
238. Clouthier DL, Zhou AC, Watts TH. Anti-GITR agonist therapy intrinsically enhances CD8 T cell responses to chronic lymphocytic choriomeningitis virus (LCMV), thereby circumventing LCMV-induced downregulation of costimulatory GITR ligand on APC. *J Immunol.* 2014;193:5033-5043.
239. Hirano T, Kikuchi T, Tode N, et al. OX40 ligand newly expressed on bronchiolar progenitors mediates influenza infection and further exacerbates pneumonia. *EMBO Mol Med.* 2016;8:422-436.
240. Elias G, Ogunjimi B, Van Tendeloo V. Activation-induced surface proteins in the identification of antigen-responsive CD4 T cells. *Immunol Lett.* 2020;219:1-7.

241. Hintzen RQ, Lens SM, Lammers K, et al. Engagement of CD27 with its ligand CD70 provides a second signal for T cell activation. *J Immunol.* 1995;154:2612-2623.
242. Hendriks J, Gravestein LA, Tesselaar K, et al. CD27 is required for generation and long-term maintenance of T cell immunity. *Nat Immunol.* 2000;1:433-440.
243. Bansal-Pakala P, Halteman BS, Cheng MH, et al. Costimulation of CD8 T cell responses by OX40. *J Immunol.* 2004;172:4821-4825.
244. Rogers PR, Song J, Gramaglia I, et al. OX40 promotes Bcl-xL and Bcl-2 expression and is essential for long-term survival of CD4 T cells. *Immunity.* 2001;15:445-455.
245. Cooper D, Bansal-Pakala P, Croft M. 4-1BB (CD137) controls the clonal expansion and survival of CD8 T cells *in vivo* but does not contribute to the development of cytotoxicity. *Eur J Immunol.* 2002;32:521-529.
246. Withers DR, Gaspal FM, Bekiaris V, et al. OX40 and CD30 signals in CD4(+) T-cell effector and memory function: a distinct role for lymphoid tissue inducer cells in maintaining CD4(+) T-cell memory but not effector function. *Immunol Rev.* 2011;244:134-148.
247. Hendriks J, Xiao Y, Rossen JW, et al. During viral infection of the respiratory tract, CD27, 4-1BB, and OX40 collectively determine formation of CD8+ memory T cells and their capacity for secondary expansion. *J Immunol.* 2005;175:1665-1676.
248. Dawicki W, Bertram EM, Sharpe AH, et al. 4-1BB and OX40 act independently to facilitate robust CD8 and CD4 recall responses. *J Immunol.* 2004;173:5944-5951.
249. DeBenedette MA, Wen T, Bachmann MF, et al. Analysis of 4-1BB ligand (4-1BBL)-deficient mice and of mice lacking both 4-1BBL and CD28 reveals a role for 4-1BBL in skin allograft rejection and in the cytotoxic T cell response to influenza virus. *J Immunol.* 1999;163:4833-4841.
250. Zhou AC, Snell LM, Wortzman ME, et al. CD30 is dispensable for T-cell responses to influenza virus and lymphocytic choriomeningitis virus clone 13 but contributes to age-associated T-cell expansion in mice. *Front Immunol.* 2017;8:1156.
251. Vashist N, Trittel S, Ebensen T, et al. Influenza-activated ILC1s contribute to antiviral immunity partially influenced by differential GITR expression. *Front Immunol.* 2018;9:505.
252. Jaeger-Ruckstuhl CA, Hinterbrandner M, Höpner S, et al. TNIK signaling Imprints CD8(+) T cell memory formation early after priming. *Nat Commun.* 2020;11:1632.
253. Kopf M, Ruedl C, Schmitz N, et al. OX40-Deficient mice are defective in Th cell proliferation but are competent in generating B cell and CTL responses after virus infection. *Immunity.* 1999;11:699-708.
254. Wang KC, Chu KL, Batista NV, et al. Conserved and differential features of TNF superfamily ligand expression on APC subsets over the course of a chronic viral infection in mice. *Immunohorizons.* 2018;2:407-417.
255. Teoh JJ, Gamache AE, Gillespie AL, et al. Acute virus control mediated by licensed NK cells sets primary CD8+ T cell dependence on CD27 costimulation. *J Immunol.* 2016;197:4360-4370.
256. Humphreys IR, de Trez C, Kinkade A, et al. Cytomegalovirus exploits IL-10-mediated immune regulation in the salivary glands. *J Exp Med.* 2007;204:1217-1225.
257. Humphreys IR, Loewendorf A, de Trez C, et al. OX40 costimulation promotes persistence of cytomegalovirus-specific CD8 T cells: a CD4-dependent mechanism. *J Immunol.* 2007;179:2195-2202.
258. Humphreys IR, Lee SW, Jones M, et al. Biphasic role of 4-1BB in the regulation of mouse cytomegalovirus-specific CD8(+) T cells. *Eur J Immunol.* 2010;40:2762-2768.
259. Bekiaris V, Gaspal F, Kim MY, et al. Synergistic OX40 and CD30 signals sustain CD8+ T cells during antigenic challenge. *Eur J Immunol.* 2009;39:2120-2125.
260. Soares H, Waechter H, Glaichenhaus N, et al. A subset of dendritic cells induces CD4+ T cells to produce IFN-γ by an IL-12-independent but CD70-dependent mechanism in vivo. *J Exp Med.* 2007;204:1095-1106.
261. Tuladhar R, Oghumu S, Dong R, et al. OX40l-OX40 pathway plays distinct roles in regulating Th2 responses but does not determine outcome of cutaneous leishmaniasis caused by *Leishmania Mexicana* and *Leishmania major*. *Exp Parasitol.* 2015;148:49-55.
262. Akiba H, Miyahira Y, Atsuta M, et al. Critical contribution of OX40 ligand to T helper cell type 2 differentiation in experimental leishmaniasis. *J Exp Med.* 2000;191:375-380.
263. Ekkens MJ, Liu Z, Liu Q, et al. The role of OX40 ligand interactions in the development of the Th2 response to the gastrointestinal nematode parasite *Heligmosomoides polygyrus*. *J Immunol.* 2003;170:384-393.
264. Sun Y, Blink SE, Liu W, et al. Inhibition of Th2-mediated allergic airway inflammatory disease by CD137 costimulation. *J Immunol.* 2006;177:814-821.
265. van der Werf N, Redpath SA, Phythian-Adams AT, et al. Th2 responses to helminth parasites can be therapeutically enhanced by, but are not dependent upon, GITR-GITR ligand costimulation in vivo. *J Immunol.* 2011;187:1411-1420.
266. Humphreys IR, Edwards L, Walzl G, et al. OX40 ligation on activated T cells enhances the control of *Cryptococcus neoformans* and reduces pulmonary Eosinophilia. *J Immunol.* 2003;170:6125-6132.
267. Florido M, Borges M, Yagita H, et al. Contribution of CD30/Cd153 but not of CD27/CD70, CD134/OX40l, or CD137/4-1BBL to the optimal induction of protective immunity to *Mycobacterium avium*. *J Leukoc Biol.* 2004;76:1039-1046.
268. Laouar A, Haridas V, Vargas D, et al. CD70+ antigen-presenting cells control the proliferation and differentiation of T cells in the intestinal mucosa. *Nat Immunol.* 2005;6:698-706.
269. Gajdasik DW, Gaspal F, Halford EE, et al. Th1 responses in vivo require cell-specific provision of OX40l dictated by environmental cues. *Nat Commun.* 2020;11:3421.
270. Shedlock DJ, Whitmire JK, Tan J, et al. Role of CD4 T cell help and costimulation in CD8 T cell responses during *Listeria monocytogenes* infection. *J Immunol.* 2003;170:2053-2063.
271. Nishimura H, Yajima T, Muta H, et al. A novel role of CD30/CD30 ligand signaling in the generation of long-lived memory CD8+ T cells. *J Immunol.* 2005;175:4627-4634.
272. Bamias G, Mishina M, Nyce M, et al. Role of TL1A and its receptor DR3 in two models of chronic murine ileitis. *Proc Natl Acad Sci U S A.* 2006;103:8441-8446.
273. Papadakis KA, Prehn JL, Landers C, et al. TL1A synergizes with IL-12 and IL-18 to enhance IFN-γ production in human T cells and NK cells. *J Immunol.* 2004;172:7002-7007.
274. Meylan F, Richard AC, Siegel RM. TL1A and DR3, a TNF family ligand-receptor pair that promotes lymphocyte costimulation, mucosal hyperplasia, and autoimmune inflammation. *Immunol Rev.* 2011;244:188-196.
275. Dubois PC, Trynka G, Franke L, et al. Multiple common variants for celiac disease influencing immune gene expression. *Nat Genet.* 2010;42:295-302.
276. Barrett JC, Hansoul S, Nicolae DL, et al. Genome-wide association defines more than 30 distinct susceptibility loci for Crohn's disease. *Nat Genet.* 2008;40:955-962.
277. Shimizu J, Yamazaki S, Takahashi T, et al. Stimulation of CD25+CD4+ regulatory T cells through GITR breaks immunological self-tolerance. *Nat Immunol.* 2002;3(2):135-142.
278. Ruddle NH, Bergman CM, McGrath KM, et al. An antibody to lymphotoxin and tumor necrosis factor prevents transfer of experimental allergic encephalomyelitis. *J Exp Med.* 1990;172:1193-1200.
279. Weinberg AD, Wegmann KW, Funatake C, et al. Blocking OX-40/OX-40 ligand interaction in vitro and in vivo leads to decreased T cell function and amelioration of experimental allergic encephalomyelitis. *J Immunol.* 1999;162:1818-1826.
280. Pakala SV, Bansal-Pakala P, Halteman BS, et al. Prevention of diabetes in NOD mice at a late stage by targeting OX40/OX40 ligand interactions. *Eur J Immunol.* 2004;34:3039-3046.
281. Nakajima A, Oshima H, Nohara C, et al. Involvement of CD70-CD27 interactions in the induction of experimental autoimmune encephalomyelitis. *J Neuroimmunol.* 2000;109:188-196.
282. Blazar BR, Kwon BS, Panoskaltsis-Mortari A, et al. Ligation of 4-1BB (CDw137) regulates graft-versus-host disease, graft-versus-leukemia, and graft rejection in allogeneic bone marrow transplant recipients. *J Immunol.* 2001;166:3174-3183.
283. Tsukada N, Akiba H, Kobata T, et al. Blockade of CD134 (OX40)-CD134L interaction ameliorates lethal acute graft-versus-host disease in a murine model of allogeneic bone marrow transplantation. *Blood.* 2000;95:2434-2439.
284. Blazar BR, Sharpe AH, Chen AI, et al. Ligation of OX40 (CD134) regulates graft-versus-host disease (GVHD) and graft rejection in allogeneic bone marrow transplant recipients. *Blood.* 2003;101:3741-3748.
285. Blazar BR, Levy RB, Mak TW, et al. CD30/CD30 ligand (CD153) interaction regulates CD4+ T cell-mediated graft-versus-host disease. *J Immunol.* 2004;173:2933-2941.
286. Tan JT, Whitmire JK, Murali-Krishna K, et al. 4-1BB costimulation is required for protective anti-viral immunity after peptide vaccination. *J Immunol.* 2000;164:2320-2325.
287. Soroosh P, Doherty TA, So T, et al. Herpesvirus entry mediator (TNFRSF14) regulates the persistence of T helper memory cell populations. *J Exp Med.* 2011;208:797-809.
288. Croft M. Control of immunity by the TNFR-related molecule OX40 (CD134). *Annu Rev Immunol.* 2010;28:57-78.
289. Yoshioka T, Nakajima A, Akiba H, et al. Contribution of OX40/OX40 ligand interaction to the pathogenesis of rheumatoid arthritis. *Eur J Immunol.* 2000;30:2815-2823.
290. Salek-Ardakani S, Song J, Halteman BS, et al. OX40 (CD134) controls memory T helper 2 cells that drive lung inflammation. *J Exp Med.* 2003;198:315-324.
291. Song J, So T, Cheng M, et al. Sustained survivin expression from OX40 costimulatory signals drives T cell clonal expansion. *Immunity.* 2005;22:621-631.
292. Kurts C, Carbone FR, Krummel MF, et al. Signalling through CD30 protects against autoimmune diabetes mediated by CD8 T cells. *Nature.* 1999;398:341-344.
293. Lee SW, Vella AT, Kwon BS, et al. Enhanced CD4 T cell responsiveness in the absence of 4-1BB. *J Immunol.* 2005;174:6803-6808.
294. Sun Y, Chen HM, Subudhi SK, et al. Costimulatory molecule-targeted antibody therapy of a spontaneous autoimmune disease. *Nat Med.* 2002;8:1405-1413.
295. Seo SK, Choi JH, Kim YH, et al. 4-1BB-Mediated immunotherapy of rheumatoid arthritis. *Nat Med.* 2004;10:1088-1094.
296. Clark EA, Craxton A. A CD40 bridge between innate and adaptive immunity. *Immunity.* 2003;18:724-725.
297. Quezada SA, Jarvinen LZ, Lind EF, et al. CD40/CD154 interactions at the interface of tolerance and immunity. *Annu Rev Immunol.* 2004;22:307-328.
298. Chen Z, Wang JH. Signaling control of antibody isotype switching. *Adv Immunol.* 2019;141:105-164.
299. Hildebrand JM, Yi Z, Buchta CM, et al. Roles of tumor necrosis factor receptor associated factor 3 (TRAF3) and TRAF5 in immune cell functions. *Immunol Rev.* 2011;244:55-74.
300. Craxton A, Otipoby KL, Jiang A, et al. Signal transduction pathways that regulate the fate of B lymphocytes. *Adv Immunol.* 1999;73:79-152.
301. Jhamnani RD, Nunes-Santos CJ, Bergerson J, et al. Class-switch recombination (CSR)/hyper-IgM (HIGM) syndromes and phosphoinositide 3-kinase (PI3K) defects. *Front Immunol.* 2018;9:2172.
302. Aruffo A, Farrington M, Hollenbaugh D, et al. The CD40 ligand, gp39, is defective in activated T cells from patients with X-linked hyper-IgM syndrome. *Cell.* 1993;72:291-300.
303. Allen RC, Armitage RJ, Conley ME, et al. CD40 ligand gene defects responsible for X-linked hyper-IgM syndrome. *Science.* 1993;259:990-993.
304. Durandy A, Revy P, Fischer A. Human models of inherited immunoglobulin class switch recombination and somatic hypermutation defects (hyper-IgM syndromes). *Adv Immunol.* 2004;82:295-330.
305. Litinskiy MB, Nardelli B, Hilbert DM, et al. DCs induce CD40-independent immunoglobulin class switching through BLyS and APRIL. *Nat Immunol.* 2002;3:822-829.
306. Mackay F, Leung H. The role of the BAFF/APRIL system on T cell function. *Semin Immunol.* 2006;18:284-289.
307. Bossen C, Schneider P. BAFF, APRIL and their receptors: structure, function and signaling. *Semin Immunol.* 2006;18:263-275.
308. Kolfschoten GM, Pradet-Balade B, Hahne M, et al. TWE-PRIL; a fusion protein of TWEAK and APRIL. *Biochem Pharmacol.* 2003;66:1427-1432.
309. Ingold K, Zumsteg A, Tardivel A, et al. Identification of proteoglycans as the APRIL-specific binding partners. *J Exp Med.* 2005;201:1375-1383.

310. Smulski CR, Kury P, Seidel LM, et al. BAFF- and TACI-dependent processing of BAFFR by ADAM proteases regulates the survival of B cells. *Cell Rep.* 2017;18:2189-2202.
311. O'Connor BP, Raman VS, Erickson LD, et al. BCMA is essential for the survival of long-lived bone marrow plasma cells. *J Exp Med.* 2004;199:91-98.
312. Sasaki Y, Casola S, Kutok JL, et al. TNF family member B cell-activating factor (BAFF) receptor-dependent and -independent roles for BAFF in B cell physiology. *J Immunol.* 2004;173:2245-2252.
313. Varfolomeev E, Kischkel F, Martin F, et al. APRIL-deficient mice have normal immune system development. *Mol Cell Biol.* 2004;24:997-1006.
314. Castigli E, Scott S, Dedeoglu F, et al. Impaired IgA class switching in APRIL-deficient mice. *Proc Natl Acad Sci U S A.* 2004;101:3903-3908.
315. Jung Y, Wen T, Mingler MK, et al. IL-1β in eosinophil-mediated small intestinal homeostasis and IgA production. *Mucosal Immunol.* 2015;8:930-942.
316. Yan M, Wang H, Chan B, et al. Activation and accumulation of B cells in TACI-deficient mice. *Nat Immunol.* 2001;2:638-643.
317. Castigli E, Wilson SA, Scott S, et al. TACI and BAFF-R mediate isotype switching in B cells. *J Exp Med.* 2005;201:35-39.
318. Ye Q, Wang L, Wells AD, et al. BAFF binding to T cell-expressed BAFF-R costimulates T cell proliferation and alloresponses. *Eur J Immunol.* 2004;34:2750-2759.
319. Lee JJ, Ozcan E, Rauter I, et al. Transmembrane activator and calcium-Modulator and cyclophilin ligand interactor mutations in common variable immunodeficiency. *Curr Opin Allergy Clin Immunol.* 2008;8:520-526.
320. Sutherland AP, Mackay F, Mackay CR. Targeting BAFF: Immunomodulation for autoimmune diseases and lymphomas. *Pharmacol Ther.* 2006;112:774-786.
321. Dillon SR, Gross JA, Ansell SM, et al. An APRIL to remember: novel TNF ligands as therapeutic targets. *Nat Rev Drug Discov.* 2006;5:235-246.
322. Moreaux J, Legouffe E, Jourdan E, et al. BAFF and APRIL protect myeloma cells from apoptosis induced by interleukin 6 deprivation and dexamethasone. *Blood.* 2004;103:3148-3157.
323. Golstein P, Griffiths GM. An early history of T cell-mediated cytotoxicity. *Nat Rev Immunol.* 2018;18:527-535.
324. Guégan JP, Legembre P. Nonapoptotic functions of FAS/CD95 in the immune response. *FEBS J.* 2018;285:809-827.
325. Watanabe-Fukunaga R, Brannan CI, Copeland NG, et al. Lymphoproliferation disorder in mice explained by defects in FAS antigen that mediates apoptosis. *Nature.* 1992;356:314-317.
326. Takahashi T, Tanaka M, Brannan CI, et al. Generalized lymphoproliferative disease in mice, caused by a point mutation in the FAS ligand. *Cell.* 1994;76:969-976.
327. Fisher GH, Rosenberg FJ, Straus SE, et al. Dominant interfering FAS gene mutations impair apoptosis in a human autoimmune lymphoproliferative syndrome. *Cell.* 1995;81:935-946.
328. Matson DR, Yang DT. Autoimmune lymphoproliferative syndrome: an overview. *Arch Pathol Lab Med.* 2020;144:245-251.
329. Rieux-Laucat F, Magérus-Chatinet A, Neven B. The autoimmune lymphoproliferative syndrome with defective FAS or FAS-ligand functions. *J Clin Immunol.* 2018;38:558-568.
330. Lee HO, Ferguson TA. Biology of FASL. *Cytokine Growth Factor Rev.* 2003;14:325-335.
331. Griffith TS, Yu X, Herndon JM, et al. CD95-induced apoptosis of lymphocytes in an immune privileged site induces immunological tolerance. *Immunity.* 1996;5:7-16.
332. Prager I, Watzl C. Mechanisms of natural killer cell-mediated cellular cytotoxicity. *J Leukoc Biol.* 2019;105:1319-1329.
333. Gregory MS, Repp AC, Holhbaum AM, et al. Membrane FAS ligand activates innate immunity and terminates ocular immune privilege. *J Immunol.* 2002;169:2727-2735.
334. Taylor AW, Ng TF. Negative regulators that mediate ocular immune privilege. *J Leukoc Biol.* 2018.
335. Schneider P, Holler N, Bodmer JL, et al. Conversion of membrane-bound FAS(CD95) ligand to its soluble form is associated with downregulation of its proapoptotic activity and loss of liver toxicity. *J Exp Med.* 1998;187:1205-1213.
336. Seyrek K, Lavrik IN. Modulation of CD95-mediated signaling by post-translational modifications: towards understanding CD95 signaling networks. *Apoptosis.* 2019;24:385-394.
337. Wajant H, Pfizenmaier K, Scheurich P. Non-apoptotic FAS signaling. *Cytokine Growth Factor Rev.* 2003;14:53-66.
338. Dosreis GA, Borges VM, Zin WA. The central role of FAS-ligand cell signaling in inflammatory lung diseases. *J Cell Mol Med.* 2004;8:285-293.
339. Houston A, O'Connell J. The FAS signalling pathway and its role in the pathogenesis of cancer. *Curr Opin Pharmacol.* 2004;4:321-326.
340. Green DR, Droin N, Pinkoski M. Activation-induced cell death in T cells. *Immunol Rev.* 2003;193:70-81.
341. Hayashi N, Mita E. Involvement of FAS system-mediated apoptosis in pathogenesis of viral hepatitis. *J Viral Hepat.* 1999;6:357-365.
342. Benedict CA, Banks TA, Ware CF. Death and survival: viral regulation of TNF signaling pathways. *Curr Opin Immunol.* 2003;15:59-65.
343. Benedict CA, Norris PS, Prigozy TI, et al. Three adenovirus E3 proteins cooperate to evade apoptosis by tumor necrosis factor-related apoptosis-inducing ligand receptor-1 and -2. *J Biol Chem.* 2001;276:3270-3278.
344. Thome M, Schneider P, Hofmann K, et al. Viral FLICE-inhibitory proteins (FLIPs) prevent apoptosis induced by death receptors. *Nature.* 1997;386:517-521.
345. Kim R, Emi M, Tanabe K, et al. Tumor-driven evolution of immunosuppressive networks during malignant progression. *Cancer Res.* 2006;66:5527-5536.
346. Ennaciri J, Menezes J, Proulx F, et al. Induction of apoptosis by herpes simplex virus-1 in neonatal, but not adult, neutrophils. *Pediatr Res.* 2006;59:7-12.
347. Niederkorn JY. See no evil, hear no evil, do no evil: the lessons of immune privilege. *Nat Immunol.* 2006;7:354-359.
348. Wang XY, Sun J, Wang C, et al. Effect of liver transplantation on islet allografts: up-regulation of FAS ligand and apoptosis of T lymphocytes are associated with islet graft tolerance. *Transplantation.* 2001;71:102-111.
349. Stuart PM, Yin X, Plambeck S, et al. The role of FAS ligand as an effector molecule in corneal graft rejection. *Eur J Immunol.* 2005;35:2591-2597.
350. Turvey SE, Gonzalez-Nicolini V, Kingsley CI, et al. FAS ligand-transfected myoblasts and islet cell transplantation. *Transplantation.* 2000;69:1972-1976.
351. Allison J, Georgiou HM, Strasser A, et al. Transgenic expression of CD95 ligand on islet beta cells induces a granulocytic infiltration but does not confer immune privilege upon islet allografts. *Proc Natl Acad Sci U S A.* 1997;94:3943-3947.
352. Yada S, Takamura N, Inagaki-Ohara K, et al. The role of p53 and FAS in a model of acute murine graft-versus-host disease. *J Immunol.* 2005;174:1291-1297.
353. Park H, Jung YK, Park OJ, et al. Interaction of FAS ligand and FAS expressed on osteoclast precursors increases osteoclastogenesis. *J Immunol.* 2005;175:7193-7201.
354. Wiley SR, Schooley K, Smolak PJ, et al. Identification and characterization of a new member of the TNF family that induces apoptosis. *Immunity.* 1995;3:673-682.
355. Zauli G, Secchiero P. The role of the TRAIL/TRAIL receptors system in hematopoiesis and endothelial cell biology. *Cytokine Growth Factor Rev.* 2006;17:245-257.
356. Degli-Esposti MA, Smolak PJ, Walczak H, et al. Cloning and characterization of TRAIL-R3, a novel member of the emerging TRAIL receptor family. *J Exp Med.* 1997;186:1165-1170.
357. Schneider P, Tschopp J. Apoptosis induced by death receptors. *Pharm Acta Helv.* 2000;74:281-286.
358. Kischkel FC, Lawrence DA, Chuntharapai A, et al. Apo2L/TRAIL-Dependent recruitment of endogenous FADD and caspase-8 to death receptors 4 and 5. *Immunity.* 2000;12:611-620.
359. Almasan A, Ashkenazi A. Apo2L/TRAIL: apoptosis signaling, biology, and potential for cancer therapy. *Cytokine Growth Factor Rev.* 2003;14:337-348.
360. Yang X, Wang J, Liu C, et al. Cleavage of p53-vimentin complex enhances tumor necrosis factor-related apoptosis-inducing ligand-mediated apoptosis of rheumatoid arthritis synovial fibroblasts. *Am J Pathol.* 2005;167:705-719.
361. Deligiorgi MV, Panayiotidis MI, Griniatsos J, et al. Harnessing the versatile role of OPG in bone oncology: counterbalancing RANKL and TRAIL signaling and beyond. *Clin Exp Metastasis.* 2020;37:13-30.
362. Krieg A, Krieg T, Wenzel M, et al. TRAIL-β and TRAIL-γ: two novel splice variants of the human Tnf-related apoptosis-inducing ligand (TRAIL) without apoptotic potential. *Br J Cancer.* 2003;88:918-927.
363. Kimberley FC, Screaton GR. Following a TRAIL: update on a ligand and its five receptors. *Cell Res.* 2004;14:359-372.
364. Krieg A, Schulte am Esch J II, Ramp U, et al. TRAIL-R4-β: a new splice variant of TRAIL-receptor 4 lacking the cysteine rich domain 1. *Biochem Biophys Res Commun.* 2006;349:115-121.
365. Hamilton SE, Wolkers MC, Schoenberger SP, et al. The generation of protective memory-like CD8+ T cells during homeostatic proliferation requires CD4+ T cells. *Nat Immunol.* 2006;7:475-481.
366. Füllsack S, Rosenthal A, Wajant H, et al. Redundant and receptor-specific activities of TRADD, RIPK1 and FADD in death receptor signaling. *Cell Death Dis.* 2019;10:122.
367. Barnhart BC, Alappat EC, Peter ME. The CD95 type I/type II model. *Semin Immunol.* 2003;15:185-193.
368. Sato K, Hida S, Takayanagi H, et al. Antiviral response by natural killer cells through TRAIL gene induction by IFN-α/β. *Eur J Immunol.* 2001;31:3138-3146.
369. Valatas V, Kolios G, Bamias G. TL1A (TNFSF15) and DR3 (TNFRSF25): a co-stimulatory system of cytokines with diverse functions in gut mucosal immunity. *Front Immunol.* 2019;10:583.
370. Richard AC, Ferdinand JR, Meylan F, et al. The Tnf-family cytokine TL1A: from lymphocyte costimulator to disease co-conspirator. *J Leukoc Biol.* 2015;98:333-345.
371. Papadakis KA, Zhu D, Prehn JL, et al. Dominant role for TL1A/DR3 pathway in IL-12 plus IL-18-induced IFN-γ production by peripheral blood and mucosal CCR9+ T lymphocytes. *J Immunol.* 2005;174:4985-4990.
372. Zhai Y, Ni J, Jiang GW, et al. VEGI, a novel cytokine of the tumor necrosis factor family, is an angiogenesis inhibitor that suppresses the growth of colon carcinomas *in vivo*. *FASEB J.* 1999;13:181-189.
373. Kim S, Zhang L. Identification of naturally secreted soluble form of TL1A, a TNF-like cytokine. *J Immunol Methods.* 2005;298:1-8.
374. Castellanos JG, Longman RS. Innate lymphoid cells link gut microbes with mucosal T cell immunity. *Gut Microb.* 2020;11:231-236.
375. Kamada N, Rogler G. The innate immune system: a trigger for many chronic inflammatory intestinal diseases. *Inflamm Intest Dis.* 2016;1:70-77.
376. Borysenko CW, Furey WF, Blair HC. Comparative modeling of TNFRSF25 (DR3) predicts receptor destabilization by a mutation linked to rheumatoid arthritis. *Biochem Biophys Res Commun.* 2005;328:794-799.
377. Burkly LC, Michaelson JS, Zheng TS. TWEAK/Fn14 pathway: an immunological switch for shaping tissue responses. *Immunol Rev.* 2011;244:99-114.
378. Boulamery A, Desplat-Jégo S. Regulation of neuroinflammation: what role for the tumor necrosis factor-like weak inducer of apoptosis/Fn14 pathway? *Front Immunol.* 2017;8:1534.
379. Burkly LC. TWEAK/Fn14 Axis: the current paradigm of tissue injury-inducible function in the midst of complexities. *Semin Immunol.* 2014;26:229-236.
380. Wiley SR, Winkles JA. TWEAK, a member of the TNF superfamily, is a multifunctional cytokine that binds the TWEAKR/Fn14 receptor. *Cytokine Growth Factor Rev.* 2003;14:241-249.
381. Vince JE, Silke J. TWEAK shall inherit the earth. *Cell Death Differ.* 2006;13:1842-1844.
382. Maecker H, Varfolomeev E, Kischkel F, et al. TWEAK attenuates the transition from innate to adaptive immunity. *Cell.* 2005;123:931-944.
383. Potrovita I, Zhang W, Burkly L, et al. Tumor necrosis factor-like weak inducer of apoptosis-induced neurodegeneration. *J Neurosci.* 2004;24:8237-8244.
384. Nakayama M, Ishidoh K, Kayagaki N, et al. Multiple pathways of TWEAK-induced cell death. *J Immunol.* 2002;168:734-743.
385. Aktas O, Prozorovski T, Zipp F. Death ligands and autoimmune demyelination. *Neuroscientist.* 2006;12:305-316.
386. Perper SJ, Browning B, Burkly LC, et al. TWEAK is a novel arthritogenic mediator. *J Immunol.* 2006;177:2610-2620.
387. Chacon MR, Richart C, Gomez JM, et al. Expression of TWEAK and its receptor Fn14 in human subcutaneous adipose tissue. Relationship with other inflammatory cytokines in obesity. *Cytokine.* 2006;33:129-137.

388. De A, Park JI, Kawamura K, et al. Intraovarian TWEAK/Fn14 ligand-receptor system limits ovarian preovulatory follicles from excessive luteinization. *Mol Endocrinol.* 2006.
389. Wright JT, Fete M, Schneider H, et al. Ectodermal dysplasias: classification and organization by phenotype, genotype and molecular pathway. *Am J Med Genet A.* 2019;179:442-447.
390. Kwack MH, Jun MS, Sung YK, et al. Ectodysplasin-A2 induces dickkopf 1 expression in human balding dermal papilla cells overexpressing the ectodysplasin A2 receptor. *Biochem Biophys Res Commun.* 2020;529: 766-772.
391. Jia S, Zhou J, Wee Y, et al. Anti-edar agonist antibody therapy resolves palate defects in Pax9(−/−) mice. *J Dent Res.* 2017;96:1282-1289.
392. Körber I, Klein OD, Morhart P, et al. Safety and immunogenicity of Fc-EDA, a recombinant ectodysplasin A1 replacement protein, in human subjects. *Br J Clin Pharmacol.* 2020;86:2063-2069.
393. Schneider H, Faschingbauer F, Schuepbach-Mallepell S, et al. Prenatal correction of X-linked hypohidrotic ectodermal dysplasia. *N Engl J Med.* 2018;378:1604-1610.
394. Simonet WS, Lacey DL, Dunstan CR, et al. Osteoprotegerin: a novel secreted protein involved in the regulation of bone density. *Cell.* 1997;89:309-319.
395. Walsh MC, Kim N, Kadono Y, et al. Osteoimmunology: interplay between the immune system and bone metabolism. *Annu Rev Immunol.* 2006;24:33-63.
396. Tanaka S, Nakamura K, Takahasi N, et al. Role of RANKL in physiological and pathological bone resorption and therapeutics targeting the RANKL-RANK signaling system. *Immunol Rev.* 2005;208:30-49.
397. Xing L, Schwarz EM, Boyce BF. Osteoclast precursors, RANKL/RANK, and immunology. *Immunol Rev.* 2005;208:19-29.
398. Yasuda H. Discovery of the RANKL/RANK/OPG system. *J Bone Miner Metab.* 2021;39:2-11.
399. Kong YY, Yoshida H, Sarosi I, et al. OPGL is a key regulator of osteoclastogenesis, lymphocyte development and lymph-node organogenesis. *Nature.* 1999;397:315-323.
400. Anderson DM, Maraskovsky E, Billingsley WL, et al. A homologue of the TNF receptor and its ligand enhance T-cell growth and dendritic-cell function. *Nature.* 1997;390:175-179.
401. Fata JE, Kong YY, Li J, et al. The osteoclast differentiation factor osteoprotegerin-ligand is essential for mammary gland development. *Cell.* 2000;103:41-50.
402. Emery JG, McDonnell P, Burke MB, et al. Osteoprotegerin is a receptor for the cytotoxic ligand TRAIL. *J Biol Chem.* 1998;273:14363-14367.
403. Miyashita T, Kawakami A, Nakashima T, et al. Osteoprotegerin (OPG) acts as an endogenous decoy receptor in tumour necrosis factor-related apoptosis-inducing ligand (TRAIL)-Mediated apoptosis of fibroblast-like synovial cells. *Clin Exp Immunol.* 2004;137:430-436.
404. Wada T, Nakashima T, Hiroshi N, et al. RANKL-RANK signaling in osteoclastogenesis and bone disease. *Trends Mol Med.* 2006;12:17-25.
405. Bharti AC, Aggarwal BB. Ranking the role of RANK ligand in apoptosis. *Apoptosis.* 2004;9:677-690.
406. Walsh MC, Choi Y. Biology of the TRANCE axis. *Cytokine Growth Factor Rev.* 2003;14:251-263.
407. Fukushima H, Jimi E, Kajiya H, et al. Parathyroid-hormone-related protein induces expression of receptor activator of Nf-κB ligand in human periodontal ligament cells via a cAMP/protein kinase a-independent pathway. *J Dent Res.* 2005;84:329-334.
408. Lomaga MA, Yeh WC, Sarosi I, et al. TRAF6 deficiency results in osteopetrosis and defective interleukin-1, CD40, and LPS signaling. *Genes Dev.* 1999;13:1015-1024.
409. Naito A, Azuma S, Tanaka S, et al. Severe osteopetrosis, defective interleukin-1 signalling and lymph node organogenesis in TRAF6-deficient mice. *Gene Cell.* 1999;4:353-362.
410. Dougall WC, Glaccum M, Charrier K, et al. RANK is essential for osteoclast and lymph node development. *Genes Dev.* 1999;13:2412-2424.
411. Josien R, Li HL, Ingulli E, et al. Trance, a tumor necrosis factor family member, enhances the longevity and adjuvant properties of dendritic cells in vivo. *J Exp Med.* 2000;191:495-502.
412. Josien R, Wong BR, Li HL, et al. TRANCE, a TNF family member, is differentially expressed on T cell subsets and induces cytokine production in dendritic cells. *J Immunol.* 1999;162:2562-2568.
413. Padigel UM, Kim N, Choi Y, et al. TRANCE-RANK costimulation is required for IL-12 production and the initiation of a Th1-type response to *Leishmania major* infection in CD40L-deficient mice. *J Immunol.* 2003;171:5437-5441.
414. Kotake S, Nanke Y, Mogi M, et al. IFN-γ-producing human T cells directly induce osteoclastogenesis from human monocytes via the expression of RANKL. *Eur J Immunol.* 2005;35:3353-3363.
415. Teng YT, Mahamed D, Singh B. Gamma interferon positively modulates *Actinobacillus actinomycetemcomitans*-specific RANKL+ CD4+ Th-cell-mediated alveolar bone destruction in vivo. *Infect Immun.* 2005;73:3453-3461.
416. Fakruddin JM, Laurence J. HIV envelope gp120-mediated regulation of osteoclastogenesis via receptor activator of nuclear factor κB ligand (RANKL) secretion and its modulation by certain HIV protease inhibitors through interferon-γ/RANKL cross-Talk. *J Biol Chem.* 2003;278:48251-48258.
417. Fisher JL, Thomas-Mudge RJ, Elliott J, et al. Osteoprotegerin overexpression by breast cancer cells enhances orthotopic and osseous tumor growth and contrasts with that delivered therapeutically. *Cancer Res.* 2006;66:3620-3628.
418. Nyambo R, Cross N, Lippitt J, et al. Human bone marrow stromal cells protect prostate cancer cells from TRAIL-induced apoptosis. *J Bone Miner Res.* 2004;19:1712-1721.
419. Colucci S, Brunetti G, Rizzi R, et al. T cells support osteoclastogenesis in an in vitro model derived from human multiple myeloma bone disease: the role of the OPG/TRAIL interaction. *Blood.* 2004;104:3722-3730.
420. Galluzzi F, Stagi S, Salti R, et al. Osteoprotegerin serum levels in children with type 1 diabetes: a potential modulating role in bone status. *Eur J Endocrinol.* 2005;153:879-885.
421. Bucay N, Sarosi I, Dunstan CR, et al. Osteoprotegerin-deficient mice develop early onset osteoporosis and arterial calcification. *Genes Dev.* 1998;12:1260-1268.
422. Choi BK, Kim SH, Kim YH, et al. RELT negatively regulates the early phase of the T-cell response in mice. *Eur J Immunol.* 2018;48:1739-1749.
423. Cusick JK, Xu LG, Bin LH, et al. Identification of RELT homologues that associate with RELT and are phosphorylated by OSR1. *Biochem Biophys Res Commun.* 2006;340:535-543.
424. Cusick JK, Alhomsy Y, Wong S, et al. RELT stains prominently in B-cell lymphomas and binds the hematopoietic transcription factor MDFIC. *Biochem Biophys Rep.* 2020;24:100868.
425. Moua P, Checketts M, Xu LG, et al. RELT family members activate p38 and induce apoptosis by a mechanism distinct from TNFR1. *Biochem Biophys Res Commun.* 2017;491:25-32.
426. Kim JW, Zhang H, Seymen F, et al. Mutations in RELT cause autosomal recessive amelogenesis imperfecta. *Clin Genet.* 2019;95:375-383.
427. Zhang L, Yu X, Zheng L, et al. Lineage tracking reveals dynamic relationships of T cells in colorectal cancer. *Nature.* 2018;564:268-272.
428. Morikawa Y, Hisaoka T, Kitamura T, et al. TROY, a novel member of the tumor necrosis factor receptor superfamily in the central nervous system. *Ann N Y Acad Sci.* 2008;1126:A1-A10.
429. Mi S, Lee X, Hu Y, et al. Death receptor 6 negatively regulates oligodendrocyte survival, maturation and myelination. *Nat Med.* 2011;17:816-821.
430. Schmidt CS, Zhao J, Chain J, et al. Resistance to myelin oligodendrocyte glycoprotein-induced experimental autoimmune encephalomyelitis by death receptor 6-deficient mice. *J Immunol.* 2005;175:2286-2292.
431. Rigby WF. Drug insight: different mechanisms of action of tumor necrosis factor antagonists-passive-aggressive behavior? *Nat Clin Pract Rheumatol.* 2007;3:227-233.
432. Strand V, Balsa A, Al-Saleh J, et al. Immunogenicity of biologics in chronic inflammatory diseases: a systematic review. *BioDrugs.* 2017;31:299-316.
433. Scott DL, Kingsley GH. Tumor necrosis factor inhibitors for rheumatoid arthritis. *N Engl J Med.* 2006;355: 704-712.
434. van der Heijde DM, van Leeuwen MA, van Riel PL, et al. Biannual radiographic assessments of hands and feet in a three-year prospective followup of patients with early rheumatoid arthritis. *Arthritis Rheum.* 1992;35:26-34.
435. van der Heijde DM. Overview of radiologic efficacy of new treatments. *Rheum Dis Clin North Am.* 2004;30:285-293, vi.
436. Lipsky PE, van der Heijde DM, St Clair EW, et al. Infliximab and methotrexate in the treatment of rheumatoid arthritis. Anti-tumor necrosis factor trial in rheumatoid arthritis with concomitant therapy study group. *N Engl J Med.* 2000;343:1594-1602.
437. Hanauer SB, Sandborn WJ, Rutgeerts P, et al. Human anti-tumor necrosis factor monoclonal antibody (adalimumab) in Crohn's disease: the CLASSIC-I trial. *Gastroenterology* 2006;130:323-333; quiz 591.
438. Sands BE, Anderson FH, Bernstein CN, et al. Infliximab maintenance therapy for fistulizing Crohn's disease. *N Engl J Med.* 2004;350:876-885.
439. Rutgeerts P, Sandborn WJ, Feagan BG, et al. Infliximab for induction and maintenance therapy for ulcerative colitis. *N Engl J Med.* 2005;353:2462-2476.
440. Sandborn WJ, Targan SR. Biologic therapy of inflammatory bowel disease. *Gastroenterology.* 2002;122:1592-1608.
441. Kruglov AA, Grivennikov SI, Kuprash DV, et al. Nonredundant function of soluble LTα3 produced by innate lymphoid cells in intestinal homeostasis. *Science.* 2013;342:1243-1246.
442. Bonifati C, Ameglio F. Cytokines in psoriasis. *Int J Dermatol.* 1999;38:241-251.
443. Schon MP, Boehncke WH. Psoriasis. *N Engl J Med.* 2005;352:1899-1912.
444. Mease PJ. Psoriatic arthritis therapy advances. *Curr Opin Rheumatol.* 2005;17:426-432.
445. Antoni CE, Kavanaugh A, Kirkham B, et al. Sustained benefits of infliximab therapy for dermatologic and articular manifestations of psoriatic arthritis: results from the infliximab multinational psoriatic arthritis controlled trial (impact). *Arthritis Rheum.* 2005;52:1227-1236.
446. Gordon KB, Langley RG, Leonardi C, et al. Clinical response to adalimumab treatment in patients with moderate to severe psoriasis: double-Blind, randomized controlled trial and open-label extension study. *J Am Acad Dermatol.* 2006;55:598-606.
447. Leonardi CL, Powers JL, Matheson RT, et al. Etanercept as monotherapy in patients with psoriasis. *N Engl J Med.* 2003;349:2014-2022.
448. Papp KA, Tyring S, Lahfa M, et al. A global phase III randomized controlled trial of etanercept in psoriasis: safety, efficacy, and effect of dose reduction. *Br J Dermatol.* 2005;152:1304-1312.
449. Tobin AM, Kirby B. TNFα inhibitors in the treatment of psoriasis and psoriatic arthritis. *BioDrugs.* 2005;19:47-57.
450. Dixon WG, Watson K, Lunt M, et al. Rates of serious infection, including site-specific and bacterial intracellular infection, in rheumatoid arthritis patients receiving anti-tumor necrosis factor therapy: results from the British Society for Rheumatology Biologics Register. *Arthritis Rheum.* 2006;54:2368-2376.
451. Hochberg MC, Lebwohl MG, Plevy SE, et al. The benefit/risk profile of Tnf-blocking agents: findings of a consensus panel. *Semin Arthritis Rheum.* 2005;34:819-836.
452. Bongartz T, Sutton AJ, Sweeting MJ, et al. Anti-TNF antibody therapy in rheumatoid arthritis and the risk of serious infections and malignancies: systematic review and meta-analysis of rare harmful effects in randomized controlled trials. *Jama.* 2006;295: 2275-2285.
453. von Haehling S, Jankowska EA, Anker SD. Tumour necrosis factor-α and the failing heart – Pathophysiology and therapeutic implications. *Basic Res Cardiol.* 2004;99:18-28.
454. Anker SD, Coats AJ. How to RECOVER from RENAISSANCE? The significance of the results of RECOVER, RENAISSANCE, RENEWAL and ATTACH. *Int J Cardiol.* 2002;86:123-130.
455. Chung ES, Packer M, Lo KH, et al. Randomized, double-bBlind, placebo-controlled, pilot trial of infliximab, a chimeric monoclonal antibody to tumor necrosis factor-α, in patients with moderate-to-severe heart failure: results of the anti-Tnf Therapy Against Congestive heart Failure (ATTACH) trial. *Circulation.* 2003;107:3133-3140.

456. Kalled SL, Cutler AH, Datta SK, et al. Anti-CD40 ligand antibody treatment of SNF1 mice with established nephritis: preservation of kidney function. *J Immunol.* 1998;160:2158-2165.
457. Larsen CP, Alexander DZ, Hollenbaugh D, et al. CD40-gp39 interactions play a critical role during allograft rejection. Suppression of allograft rejection by blockade of the CD40-gp39 pathway. *Transplantation.* 1996;61:4-9.
458. Shock A, Burkly L, Wakefield I, et al. CDP7657, an anti-CD40L antibody lacking an Fc domain, inhibits CD40L-dependent immune responses without thrombotic complications: an in vivo study. *Arthritis Res Ther.* 2015;17:234.
459. Kalden JR, Schulze-Koops H. Immunogenicity and loss of response to Tnf inhibitors: implications for rheumatoid arthritis treatment. *Nat Rev Rheumatol.* 2017;13:707-718.
460. Rubbert-Roth A, Szabo MZ, Kedves M, et al. Failure of anti-Tnf treatment in patients with rheumatoid arthritis: the pros and cons of the early use of alternative biological agents. *Autoimmun Rev.* 2019;18:102398.
461. Bienkowska J, Allaire N, Thai A, et al. Lymphotoxin-light pathway regulates the interferon signature in rheumatoid arthritis. *PLoS One.* 2014;9:e112545.
462. Kennedy WP, Simon JA, Offutt C, et al. Efficacy and safety of pateclizumab (anti-lymphotoxin-A) compared to adalimumab in rheumatoid arthritis: a head-to-head phase 2 randomized controlled study (the ALTARA study). *Arthritis Res Ther.* 2014;16:467.
463. Matsumoto T, Endo I. RANKL as a target for the treatment of osteoporosis. *J Bone Miner Metab.* 2021;39:91-105.
464. Anastasilakis AD, Makras P, Yavropoulou MP, et al. Denosumab discontinuation and the rebound phenomenon: a narrative review. *J Clin Med.* 2021;10:152.
465. Schwartz E, Reichert Z, Van Poznak C. Pharmacologic management of metastatic bone disease. *Bone.* 2020:115735.
466. Deligiorgi MV, Trafalis DT. The safety profile of denosumab in oncology beyond the safety of denosumab as an anti-osteoporotic agent: still more to learn. *Expet Opin Drug Saf.* 2021;20:191-213.
467. Crow MK, Ronnblom L. Type I interferons in host defence and inflammatory diseases. *Lupus Sci Med.* 2019;6:e000336.
468. Navarra SV, Guzman RM, Gallacher AE, et al. Efficacy and safety of belimumab in patients with active systemic lupus erythematosus: a randomised, placebo-controlled, phase 3 trial. *Lancet.* 2011;377:721-731.
469. Murphy G, Isenberg DA. New therapies for systemic lupus erythematosus - past imperfect, future tense. *Nat Rev Rheumatol.* 2019;15:403-412.
470. Tang H, Qiao J, Fu YX. Immunotherapy and tumor microenvironment. *Cancer Lett.* 2016;370:85-90.
471. Allen E, Jabouille A, Rivera LB, et al. Combined antiangiogenic and anti-PD-L1 therapy stimulates tumor immunity through HEV formation. *Sci Transl Med.* 2017;9:eaak9679.
472. He B, Jabouille A, Steri V, et al. Vascular targeting of LIGHT normalizes blood vessels in primary brain cancer and induces intratumoural high endothelial venules. *J Pathol.* 2018;245:209-221.
473. Taguchi T, Sohmura Y. Clinical studies with TNF. *Biotherapy.* 1991;3:177-186.
474. Skillings J, Wierzbicki R, Eisenhauer E, et al. A phase II study of recombinant tumor necrosis factor in renal cell carcinoma: a study of the National Cancer Institute of Canada Clinical Trials Group. *J Immunother.* 1992;11:67-70.
475. Tracey KJ, Lowry SF, Cerami A. The pathophysiologic role of cachectin/TNF in septic shock and cachexia. *Ann Inst Pasteur Immunol.* 1988;139:311-317.
476. Lejeune FJ, Lienard D, Matter M, et al. Efficiency of recombinant human TNF in human cancer therapy. *Cancer Immun.* 2006;6:6.
477. von Karstedt S, Walczak H. An unexpected turn of fortune: targeting TRAIL-Rs in KRAS-driven cancer. *Cell Death Discov.* 2020;6:14.
478. Wajant H. Molecular mode of action of TRAIL receptor agonists-common principles and their translational exploitation. *Cancers (Basel).* 2019;11:954.
479. Duiker EW, Mom CH, de Jong S, et al. The clinical trail of TRAIL. *Eur J Cancer.* 2006;42:2233-2240.
480. Younes A, Vose JM, Zelenetz AD, et al. A phase 1b/2 trial of mapatumumab in patients with relapsed/refractory non-Hodgkin's lymphoma. *Br J Cancer.* 2010;103:1783-1787.
481. Herbst RS, Eckhardt SG, Kurzrock R, et al. Phase I dose-escalation study of recombinant human Apo2L/TRAIL, a dual proapoptotic receptor agonist, in patients with advanced cancer. *J Clin Oncol.* 2010;28:2839-2846.
482. Forero-Torres A, Shah J, Wood T, et al. Phase I trial of weekly tigatuzumab, an agonistic humanized monoclonal antibody targeting death receptor 5 (DR5). *Cancer Biother Radiopharm.* 2010;25:13-19.
483. Rosevear HM, Lightfoot AJ, Griffith TS. Conatumumab, a fully human mAb against death receptor 5 for the treatment of cancer. *Curr Opin Investig Drugs.* 2010;11:688-698.
484. Wong SHM, Kong WY, Fang CM, et al. The TRAIL to cancer therapy: Hindrances and potential solutions. *Crit Rev Oncol Hematol.* 2019;143:81-94.
485. Weinberg AD, Morris NP, Kovacsovics-Bankowski M, et al. Science gone translational: the OX40 agonist story. *Immunol Rev.* 2011;244:218-231.
486. Aspeslagh S, Postel-Vinay S, Rusakiewicz S, et al. Rationale for anti-OX40 cancer immunotherapy. *Eur J Cancer.* 2016;52:50-66.
487. Kalos M, Levine BL, Porter DL, et al. T cells with chimeric antigen receptors have potent antitumor effects and can establish memory in patients with advanced leukemia. *Sci Transl Med.* 2011;3:95ra73.
488. Zhao Y, Moon E, Carpenito C, et al. Multiple injections of electroporated autologous T cells expressing a chimeric antigen receptor mediate regression of human disseminated tumor. *Cancer Res.* 2010;70:9053-9061.
489. Porter DL, Levine BL, Kalos M, et al. Chimeric antigen receptor-modified T cells in chronic lymphoid leukemia. *N Engl J Med.* 2011.
490. Boice M, Salloum D, Mourcin F, et al. Loss of the HVEM tumor suppressor in lymphoma and restoration by modified car-T cells. *Cell.* 2016;167:405.e13-418.e13.
491. Roselli E, Faramand R, Davila ML. Insight into next-generation CAR therapeutics: designing CAR T cells to improve clinical outcomes. *J Clin Invest.* 2021;131.
492. Shah N, Chari A, Scott E, et al. B-cell maturation antigen (BCMA) in multiple myeloma: rationale for targeting and current therapeutic approaches. *Leukemia.* 2020;34:985-1005.
493. Bera TK. Anti-BCMA immunotoxins: design, production, and preclinical evaluation. *Biomolecules.* 2020;10:1387.
494. Markham A. Belantamab mafodotin: first approval. *Drugs.* 2020;80:1607-1613.
495. Mikkilineni L, Kochenderfer JN. Car T cell therapies for patients with multiple myeloma. *Nat Rev Clin Oncol.* 2021;18:71-84.

CHAPTER 11

Chemokines and Chemokine Receptors

Philip M. Murphy, MD

INTRODUCTION

The immune system is a system of cells in motion. Leukocyte movement is passive in the blood, as determined by flow, but is active and energy requiring in other tissues, directed in part by gradients of chemoattractants signaling through G protein–coupled receptors (GPCRs). Before 1987, only a few leukocyte chemoattractants had been defined, including bacterial *N*-formylpeptides, C5a, leukotriene B4, and platelet-activating factor.[1,2] Although immunologically important, these chemoattractants mainly target myeloid cells, without discriminating between neutrophils and monocytes, and thus cannot by themselves explain the selective pathologic associations of neutrophils with acute inflammation and monocytes with chronic inflammation, let alone the complex trafficking itineraries of lymphocytes. In 1987 and 1988, in independent landmark studies by groups led by Marco Baggiolini at the Theodore Kocher Institute in Switzerland, Joost Oppenheim and Ed Leonard at the National Institutes of Health in the United States, Jo Van Damme at the Rega Institute in Leuven and Jens Schroder at the University of Kiel, a monocyte-derived neutrophil-specific chemoattractant and activating factor, renamed interleukin 8 in 1988, then CXCL8 in 2000, was discovered as the first of a large family of structurally related leukocyte chemoattractants named the chemokines, a contraction of "chemotactic cytokine", each of which has high selectivity for different types of leukocytes.[3-6]

A few similar molecules with unknown function had been discovered before CXCL8. The first, CXCL4, was described in the 1950s as a major constituent of platelet alpha granules and accordingly was first named platelet factor-4. Ironically, CXCL4 is one of the only chemokines never shown to have potent leukocyte chemotactic activity. The other chemokines, now numbering ~45 in both mouse and man, together coordinate most trafficking itineraries *in vivo* for all subsets of leukocytes, as determined by dynamic and differential spatial and temporal expression patterns of chemokines in source cells and of GPCRs on leukocytes.[7] Since the 1990s, chemokine-directed selective leukocyte trafficking mediated by specific GPCRs, a type of receptor that is in general highly druggable, has motivated many drug development programs, resulting in the discovery of hundreds of blocking agents and clinical trials aimed at creating a new generation of safer and better-targeted therapeutics, initially for inflammation but subsequently for many other indications, resulting so far in three approvals.[8]

Since chemokines are so numerous and since they share cell motility as a major function, the topic is usually covered separately from other cytokines in most textbooks of immunology, as it is here. As a measure of the scope of the field, a search of Titles and Abstracts in PubMed on September 23, 2020 using the query term "chemokine" returned 48,136 papers. Annual productivity in the field by this metric rose linearly from 1992, when the term chemokine was coined, to a plateau from 2015 to 2020 of ~3000 published papers per year. As a result, any overview of the chemokine system can only aspire to convey the basic principles of its organization and function, together with examples of the best-established clinical correlates and important exceptions to the "rules." Additional detailed information about individual molecules in specific immunologic contexts can be found in dedicated chapters throughout this textbook. Moreover, information from the primary scientific literature is continuously being aggregated and updated in multiple excellent databases, including the Immunological Genome Project for chemokine and chemokine receptor distribution in cells and tissues (www.immgen.org), the Mouse Genome Initiative for genetically modified alleles and phenotypes (www.informatics.jax.org), GenBank for sequence information (www.ncbi.nlm.nih.gov/genbank), the Protein Data Bank for structures (www.pdb.org), the International Union of Pharmacology/British Pharmacology Society Guide to PHARMACOLOGY for drugs (www.guidetopharmacology.org/), the Online Mendelian Inheritance in Man for human mutations and disease phenotypes (www.omim.org), and ClinicalTrials.gov for therapeutic interventions in humans (www.clinicaltrials.gov).

The amount of information currently available for each chemokine and chemokine receptor is highly heterogeneous across the system, reflecting in part how recently a molecule was identified, how broadly it is expressed, and how important it appears to be, with clinical connection as a major catalyst to funding, discovery, and, ultimately, published papers (Fig. 11.1A). From this perspective, the receptor CXCR4 still stands out as an extreme outlier, with ~2- to >100-fold more PubMed listings than for other chemokine receptors, reflecting (1) its expression by all subsets of leukocytes and some nonhematopoietic cell types, including many types of cancer cells; (2) its protean activities, including important roles in organ development, immunity, inflammation, and HIV pathogenesis; and (3) US Food and Drug Administration (FDA) approval of an antagonist for clinical application. Similarly, CCL2 is highly overrepresented in the literature compared with other chemokines, most likely reflecting its unusually broad roles in both innate and adaptive immunity.

In the immune system, chemokine stimulation may induce multiple cellular responses, including differentiation,

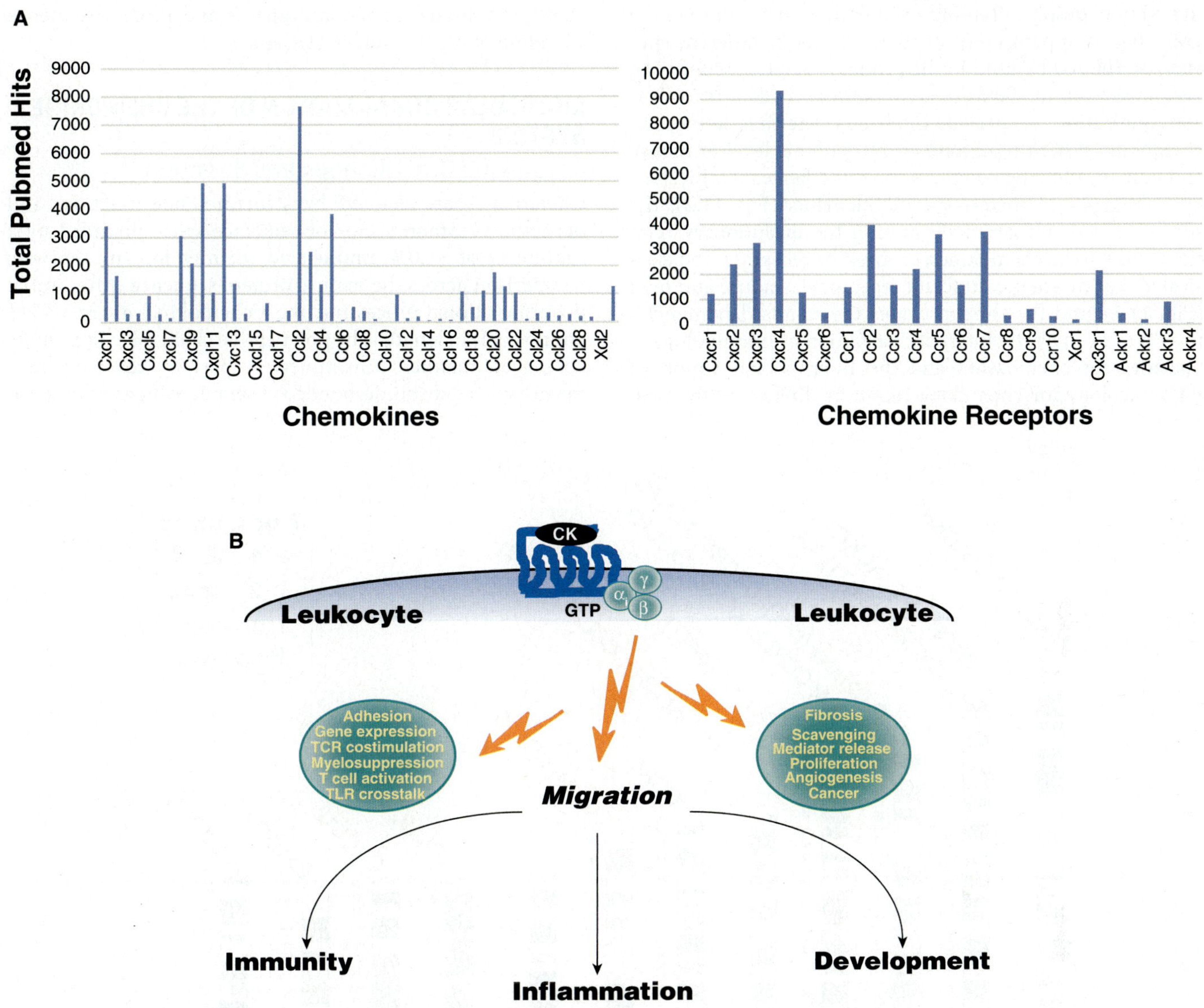

FIG. 11.1. The broad landscape of chemokine biology and research activity. A: Research activity is highly heterogeneous across the chemokine system. The graphs show the number of published papers returned when each of the molecules listed on the x-axis was used as the sole PubMed search term in April 2020. The searches for CXCR4 and CCR5 specifically excluded papers matching the search term "HIV." Results for all human chemokines are shown, but only the names of every other chemokine are shown because of space limitations. **B:** Chemokines multitask in development, immunity, and inflammation in large part by inducing leukocyte migration through activation of G protein–coupled receptors. Other functions regulated by chemokines are listed in the green spaces. CK, chemokine.

trafficking, tissue positioning, proliferation, and activation of effector functions (Fig. 11.1B). However, chemokines can behave like a double-edged sword, acting beneficially, for example, in host defense and tissue repair, or maladaptively and harmfully, for example, in cancer, autoimmunity, and infectious disease. In addition to mediating leukocyte accumulation and activation, chemokines may also promote inflammation through effects on gene expression, angiogenesis, and fibrosis. Many pathogens make broad-spectrum chemokine blocking agents, attesting to the importance of chemokines in antimicrobial host defense; and some pathogens, most notably HIV and *Plasmodium vivax*, exploit host chemokine receptors for target cell recognition and entry.

Chemokine and chemokine receptor genes are found in all vertebrates, where the system originated between 564 and 650 million years ago, expanding and diversifying by en bloc and tandem duplications.[9] Its major divisions were present even in unjawed vertebrates and preceded the appearance of RAG-based adaptive immunity. Paradoxically, there are more chemokine genes in fish species than in mammals. However, this chapter will be limited to information in mouse and human. In this regard, it is important to note that the chemokine system may differ substantially between these two species, starting at the genomic level where most genes have cross-species orthologs, whereas others do not, providing a cautionary note in the extrapolation of mouse studies to humans.

In addition, despite adopting a standard nomenclature in the year 2000, competing nonstandard aliases in different species continue to be used in the primary literature, confusing even experts in the field. To address these challenges to clear communication, we provide three aids. The first defines the chemokine system repertoire at the genomic level in mouse and human, allowing comprehensive and facile identification of orthologs and species-specific elements (Fig. 11.2); the second is a dendrogram illustrating the evolutionary relationships among chemokines, chemokine receptors, and classical leukocyte chemoattractant receptors in mouse and man (Fig. 11.3); and the third provides the standard nomenclatures approved by the International Union of Immunological Societies for chemokines and the International Union of Pharmacology for chemokine receptors, linked to the most commonly used nonstandard aliases and protein sequence accession numbers (Tables 11.1 and 11.2).

MOLECULAR ORGANIZATION OF THE CHEMOKINE SYSTEM

Chemokine Classification and Structure

Chemokines are defined by structure, not function, and occupy a common sector of sequence space bounded by a minimum of ~20% amino acid identity for any pairwise comparison across the mature protein sequence. After initial posttranslational processing, most chemokines are 66 to 111 amino acids in length. The polypeptide chain folds in a highly conserved manner constrained in part by relatively uniformly spaced disulfide-bonded cysteines, which are used for

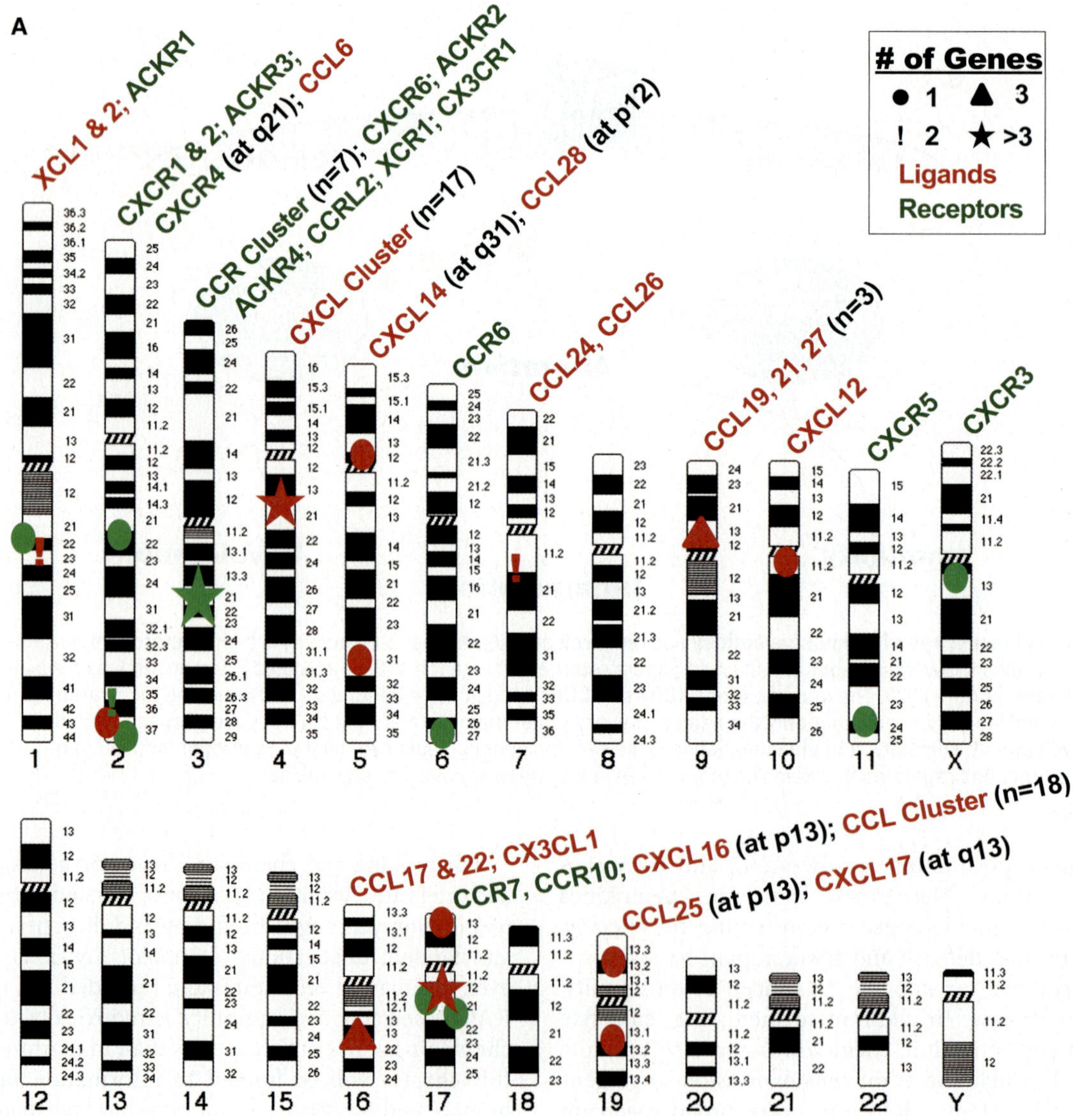

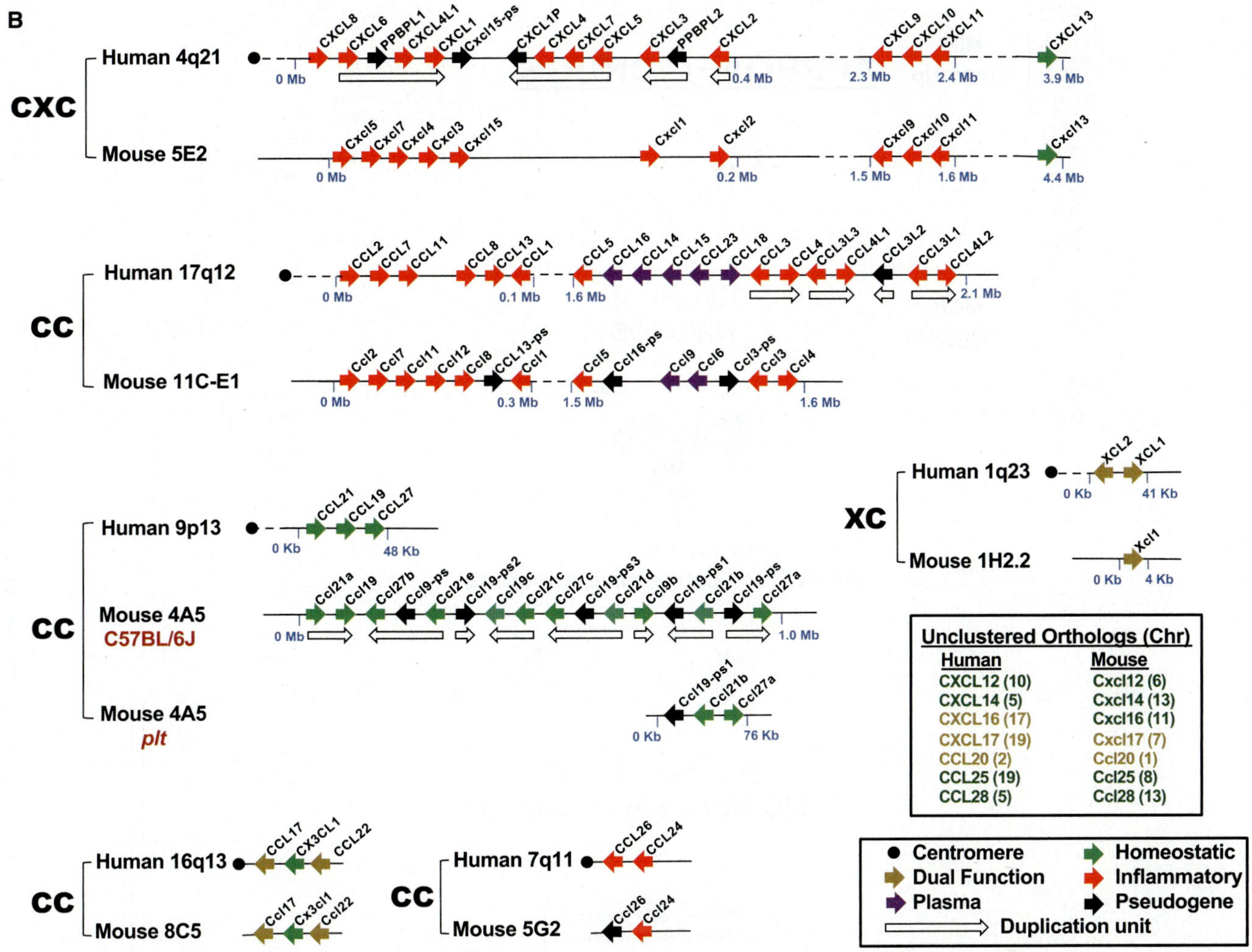

FIG. 11.2. Chemokine system genomics. A: Chromosomal location of human chemokines and chemokine receptor genes. The code for each symbol and color is given in the inset. **B:** Chemokine gene repertoires in mouse and human. Brackets pair regions of conserved synteny between the indicated mouse and human chromosomes. Bracket labels denote chemokine groups (see text). Gene names are from the HUGO Gene Nomenclature Committee assignments. Arrows, gene location and orientation relative to centromere; Mb, megabase; Kb, kilobase; ps, pseudogene; plasma, chemokines found at high concentrations constitutively in plasma. Other codes are in box at lower right. (B, Data are from http://www.ncbi.nlm.nih.gov/genome for *Homo sapiens* (Build 37.2) and *Mus musculus*. The figure layout is based on a previously published design and helpful advice from Hisayuki Nomiyama of Kumamoto University.)

subclassification. All chemokines have at least two cysteines, and all but two have at least four (Fig. 11.4A). In the four-cysteine group, the first two are either adjacent (CC motif, $n = 24$ in human) or separated by either one (CXC motif, $n = 16$ in human) or three (CX3C motif, $n = 1$) amino acids. C chemokines have only two cysteines, corresponding to the location of Cys-2 and Cys-4 in the other groups ($n =$ one in mouse and two in human). The group motifs are used as roots followed by the letter "L" for "ligand" and a number to create standard chemokine names (Fig. 11.4A). For human names, all letters are capitalized, whereas in the corresponding mouse names only the first letter is capitalized. Amino acid sequence identity is <30% between members of the four major groups, but it ranges from ~30% to 99% among members of the same group (Fig. 11.3).

Although chemokine genes are located on 10 different human chromosomes, most are found in two large clusters, one for CXC genes on chromosome 4q21 and one for CC genes on chromosome 17q12 (Fig. 11.2). These two clusters contain most of the chemokines that mainly mediate proinflammatory immune responses, whereas other chemokine genes, scattered across the genome either alone or in small clusters on chromosomes 1, 2, 5, 7, 9, 10, 16, 17, and 19, may preferentially regulate homeostatic immune functions. Most chemokine genes have clear orthologs in mouse and human, but there are notable exceptions (Fig. 11.2A), for example, *CXCL8*, *Cxcl15*, *CCL18*, *Ccl12*, and *CCL13*. Variation in gene copy number between mouse and human also occurs, for example, for CCL19 and CCL21, and even among individuals of the same species, for example, for *CCL3* in human (Fig. 11.2B).

Chemokine genes are relatively small and contain three and four exons each for members of the CC and CXC subgroups, respectively, as illustrated for *CXCL8* in Figure 11.4B. *CCL14* and *CCL15* are an exceptional example of neighboring genes that give rise to a family of mono- and bicistronic

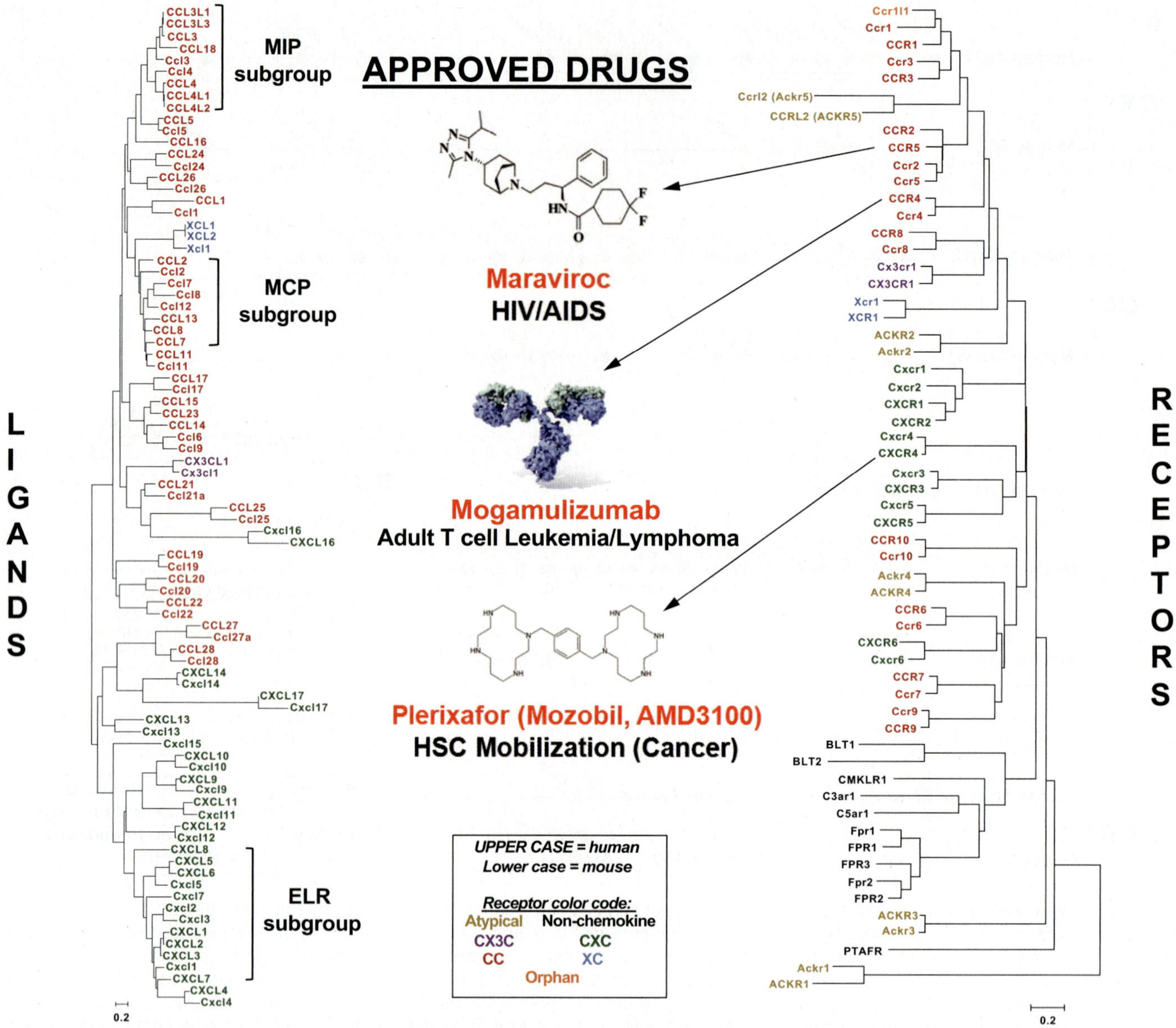

FIG. 11.3. **Chemokine and chemokine receptor family trees in mouse and human.** Branch distances are calibrated by the bar at the bottom of each tree. The dendrograms were created in collaboration with Steven Tsang. FDA-approved drugs targeting the indicated receptors are shown in the center.

transcripts. Chemokine gene promoters are usually immediately upstream of exon 1, which typically encodes a leader sequence and a few amino acids of the mature peptide. Alternative splicing has been reported for several chemokines. *CXCL12* is a particularly well-characterized example, generating six distinct differentially expressed isoforms by this mechanism (designated α, β, γ, δ, ε, and φ).[10] Most studies do not discriminate among the six forms or focus specifically on CXCL12α. An extended C terminus of the CXCL12γ isoform contains several basic residues, which contribute to binding to glycosaminoglycans (GAGs) with very high affinity, thereby promoting efficient formation of chemokine gradients, and may account for functional specialization, for example, for revascularization and infiltration of cells into ischemic tissue.

The strongest, although not invariant, functional correlate of the four major chemokine subclasses is receptor specificity: that is, most chemokine receptors have multiple chemokine ligands, but they are almost always restricted to one subclass (Fig. 11.5). With regard to leukocyte specificity, most neutrophils respond mainly to CXC chemokines, whereas most monocyte/macrophage-targeted chemokines are in the CC group. Major T and B cell–targeted chemokines can be found in all four groups.

CC and CXC chemokines can be subclassified further. The ELR subgroup of CXC chemokines is defined by the amino acid motif glu-leu-arg N-terminal to Cys-1 and is encoded by a cluster of seven genes on human chromosome 4q21 (Fig. 11.2). The ELR motif is required for function by these chemokines, which are all >40% identical at the amino acid

TABLE 11.1 Directory of Chemokine Nomenclature, Aliases, Sequence, Structure, Sources, and Function

Standard Name	Common Aliases	Accession Numbers			Cell Sources	Prominent Immunoregulatory Functions
		Protein		PDB		
		Human	Mouse			
CXCL1	GROα, MGSA (Mouse: KC)	P09341	P12850	1MGS	Induced in epithelial cells and ECs	Neutrophil trafficking and activation
CXCL2	Groβ; MIP-2α (Mouse: MIP-2)	P19875	P10889	1QNK	Similar to CXCL1	Similar to CXCL1
CXCL3	Groγ; MIP-2β,	P19876	Q6W5C0	NA	Similar to CXCL1	Similar to CXCL1
CXCL4	Platelet Factor-4	P02776	Q9Z126	1RHP	Preformed in platelets	Procoagulant
CXCL4L1	PF4V1	P10720	na	na	Similar to CXCL4	Similar to CXCL4
CXCL5	ENA-78 (Mouse: LIX)	P42830	P50228	2MGS	Induced in epithelial cells of gut and lung; N, Mo, Plts, EC	Neutrophil trafficking and activation
CXCL6	GCP-2	P80162	na	na	Induced in lung microvascular EC; Mo; alveolar epithelial cells, mesothelial cells, EC, and MΦ	Similar to CXCL1
CXCL7	NAP-2	P02775	Q9EQI5	1NAP	Preformed in platelets	Similar to CXCL1
CXCL8	IL-8	P10145	na	3IL8	Inducible in most cell types	Similar to CXCL1
CXCL9	Mig	Q07325	P18340	na	Induced in PMN, MΦ, T cells, astrocytes, microglial cells, hepatocytes, EC, fibroblasts, keratinocytes, thymic stromal cells	Type 1 immune response
CXCL10	γIP-10	P02778	P17515	1LV9	Similar to CXCL9	Similar to CXCL9
CXCL11	I-TAC	O14625	Q8R392	1RJT	Similar to CXCL9	Similar to CXCL9
CXCL12	SDF-1	P48061	P40224	3HP3	Constitutive in bone marrow stromal cells and many tissues	Hematopoiesis Organ development Homeostatic leukocyte maturation, survival, and trafficking Adaptive immunity
CXCL13	BLC	O43927	O55038	4ZAI	Constitutive in follicular HEV of secondary lymphoid tissue	B and T cell trafficking in 2° lymphoid tissue
CXCL14	BRAK	O95715	Q6AXC2	2HDL	Constitutive in most tissues, breast and kidney tumors	Macrophage positioning in tissue
Cxcl15	lungkine	na	Q9WVL7	na	Constitutive in lung epithelial cells	Neutrophil trafficking and innate immunity
CXCL16	SR-PSOX	Q9H2A7	Q8BSU2	na	DCs, activated T cells, endothelial cells, monocytes, macrophages, fibroblastic reticular cells	T cell and DC homing to spleen, NK and NKT cell trafficking, activation, and survival
CXCL17		Q6UXB2	Q8R3U6	na	Constitutive in lung	Mo and DC chemotaxis

(Continued)

TABLE 11.1 Directory of Chemokine Nomenclature, Aliases, Sequence, Structure, Sources, and Function (Continued)

Standard Name	Common Aliases	Accession Numbers			Cell Sources	Prominent Immunoregulatory Functions
		Protein		PDB		
		Human	Mouse			
CCL1	I-309	P22362	P10146	1ELO	Inducible in Mo and $CD4^+$ and $CD8^+$ $\alpha\beta$ and $CD4^-CD8^-$ $\gamma\delta$ T cells, dermal endothelial cells, epithelial APCs	Type 2 immune response, T_{reg} function
CCL2	MCP-1 (Mouse: JE)	P13500	P10148	1DOK	• Inducible in Mo, fibroblasts, keratinocytes, EC, neutrophils, synoviocytes, mesangial cells, astrocytes, lung epithelial cells, and MΦ • Constitutively made in splenic arteriolar lymphatic sheath and medullary region of lymph node and arterial plaque EC	Inflammatory Mo trafficking Tissue macrophage development Type 1 and 2 immune response
CCL3	MIP-1α	P10147	P10855	5D65	Inducible in Mo/MΦ, $CD8^+$ T cells, B cells, plts, PMN, Eo, Ba, DC, NK, mast cells, keratinocytes, fibroblasts, mesangial cells, astrocytes, microglial cells, epithelial cells	Innate immunity Type 1 and 2 immune response T/DC interaction HIV suppression
CCL3L1		P16619	P10855	na	Similar to CCL3	
CCL3L3		P16619	na	na	Similar to CCL3	Similar to CCL3
CCL4	MIP-1β	P13236	P14097	2X6L	Similar to CCL3	Similar to CCL3
CCL4L1		Q8NHW4	na	na	Similar to CCL3	Similar to CCL3
CCL4L2		Q8NHW4	na	na	Similar to CCL3	Similar to CCL3
CCL5	RANTES	P13501	P30882	5COY	• Inducible in EC, T cells, epithelial cells, Mo, fibroblasts, mesangial cells, NK cells Constitutively expressed and stored in plt and Eo granules	Similar to CCL3
Ccl6	C10, MRP-1	na	P27784	na	Inducible in bone marrow and peritoneal-derived MΦ	na
CCL7	MCP-3	P80098	Q03366	1NCV	Inducible in Mo, plts, fibroblasts, EC, skin, bronchial epithelial cells, astrocytes	Type 2 immune response
CCL8	MCP-2	P80075	Q9Z121	1ESR	• Inducible in fibroblasts, PMN, MΦ • Constitutively expressed in colon, small intest, heart, lung, thymus, pancreas, spinal cord, ovary, placenta	Type 2 immune response
Ccl9	MRP-2, MIP-1γ	na	P51670	na	• Constitutively expressed in all mouse organs except brain; highest in lung, liver and thymus • Induced in heart and lung	na

TABLE 11.1 Directory of Chemokine Nomenclature, Aliases, Sequence, Structure, Sources, and Function (Continued)

Standard Name	Common Aliases	Accession Numbers			Cell Sources	Prominent Immunoregulatory Functions
		Protein		PDB		
		Human	Mouse			
CCL10 (reserved)		na	na	na	na	na
CCL11	Eotaxin	P51671	P48298	2EOT	Epithelial cells, EC, smooth muscle, cardiac muscle, Eo, dermal fibroblasts, mast cells, MΦ, Reed-Sternberg cells	Homeostatic Eo positioning in gut and immune organs Type 2 immune response Eo, Ba, MC trafficking and activation
Ccl12	Mcp-5	na	Q62401	na	• Inducible in lung alveolar MΦ and smooth muscle cells; spinal cord • Constitutive expression in lymph node and thymic stromal cells	Type 2 immune response Mo, T lymphocyte, Eo, Ba, MC trafficking and activation
CCL13	MCP-4	Q99616	na	2RA4	• Inducible in nasal and bronchial epithelial cells; dermal fibroblasts; PBMC; atherosclerotic plaque EC and MΦ • Constitutively expressed in small intestine, colon, thymus, heart, and placenta	Type 2 immune response
CCL14	HCC-1	Q16627	na	2Q8T	Constitutively expressed in most organs; high plasma levels	na
CCL15	HCC-2	Q16663	na	2HCC	• Inducible in Mo and DC • Constitutive RNA expression in liver, gut, heart and skeletal muscle, adrenal gland, and lung leukocytes	na
CCL16	HCC-4	O15467	na	5LTL	• Constitutively expressed in liver, possibly many other organs. Also, Mo, T cells, and NK cells express mRNA	DC maturation factor
CCL17	TARC	Q92583	Q9WUZ6	1NR2	Constitutive in peripheral tissue mature $CD11b^+$ myeloid DCs	Type 2 immune response
CCL18	PARC	P55774	na	4MHE	• Constitutive in Mo/MΦ, germinal center DC • Induced by type 2 cytokines	DC attraction of naïve T cells Hematopoiesis Type 2 immune response
CCL19	ELC	Q99731	O70460	2MP1	Constitutive on interdigitating DC and fibroblastic reticular cells in secondary lymphoid tissue	Naïve and memory T cell and DC homing to lymph node

(Continued)

TABLE 11.1 Directory of Chemokine Nomenclature, Aliases, Sequence, Structure, Sources, and Function (Continued)

Standard Name	Common Aliases	Accession Numbers			Cell Sources	Prominent Immunoregulatory Functions
		Protein		PDB		
		Human	Mouse			
CCL20	MIP-3α, LARC	P78556	089093	2JY0	Epithelial cells and endothelial cells Osteoblasts Inducible in GALT	GALT development B and DC homing to GALT IgA Humoral response in gut Type 17 immune response
CCL21	SLC	000585	P84444	2L4N	Constitutive in lymphatic EC, HEV, and interdigitating DC in T areas of 2° lymphoid tissue, thymic medullary epith cells and EC	Naïve and memory T cell and DC homing to lymph node
CCL22	MDC	000626	088430	na	• Constitutive in DC and MΦ • Inducible in Mo, T, and B cells	Type 2 immune response
CCL23	MPIF-1	P55773	na	1G91	Constitutive in pancreas and skeletal muscle	na
CCL24	Eotaxin-2	000175	Q9JKC0	1EIH	Inducible by IL-13 in luminal airway Mf	Eo trafficking
CCL25	TECK	015444	035903	na	Constitutive in thymic stromal cells and small intestine	T cell progenitor homing to thymus Memory T cell homing to gut
CCL26	Eotaxin-3	Q9Y258	Q5C9Q0	1G2S	• Constitutive in heart and ovary • Inducible on dermal fibroblasts and EC	Type 2 immune response
CCL27	CTACK	Q9Y4X3	Q9Z1X0	2KUM	Constitutive in placenta, keratinocytes, testis, and brain	Lymphocyte homing to skin
CCL28	MEC	Q9NRJ3	Q9JIL2	6CWS	Constitutive in epith cells of gut, airway, salivary gland, mammary gland, and female reproductive tract	Lymphocyte homing to mucosa; IgA responses
XCL1 XCL2	Lymphotactin α Lymphotactin β	P47992 Q9UBD3	P47993 na	2N54 na	γδ epidermal T cells, NK, NK-T, activated CD8⁺ and Th1 CD4⁺ T cells	CD8⁺ T cell responses to Ag cross-presented by XCR1⁺CD8⁺ DCs
CX_3CL1	Fractalkine	P78423	035188	1F2L	EC, neurons, Mo, DC	NK, Mo, MΦ, DC and Type 1 T cell trafficking

Ag, antigen; Ba, basophil; CTACK, cutaneous T cell–associated chemokine; DC, dendritic cell; EC, endothelial cell; ELC, Epstein-Barr virus-induced receptor ligand chemokine; Eo, eosinophil; GALT, gut-associated lymphoid tissue; HCC, hemofiltrate CC chemokine; HEV, high endothelial venule; LARC, liver and activation related chemokine; MC, mast cell; MCP, monocyte chemoattractant protein; MDC, macrophage-derived chemokine; MEC, mucosa-associated epithelial cell chemokine; MIP, macrophage inflammatory protein; Mo, monocyte; MPC, myeloid progenitor cell; MPIF, myeloid progenitor inhibitory factor; MRP, MIP-related protein; MΦ, macrophage; na, not applicable; NK, natural killer cell; PARC, pulmonary and activation related chemokine; plt, platelet; PMN, neutrophil; RANTES, regulated upon activation normal T cell expressed and secreted; SLC, secondary lymphoid tissue chemokine; TARC, thymus and activation related chemokine; TECK, thymus expressed chemokine.

TABLE 11.2 Directory of Human Chemokine Receptor Systematic Nomenclature, Aliases, Sequence, Structure, and Immunologic Function. For a List of Cell Types Expressing Each Receptor, See Figure 11.5

Name	CD#	Previous Names	Accession Numbers		Prominent Immunoregulatory Functions
			Protein	PDB	
CXCR1	CD181	$IL8R_A$, IL-8R-I, IL-8Rα	P25024	2LNL	Neutrophil activation
CXCR2	CD182	$IL8R_B$, IL-8R-II, IL-8Rβ	P25025	na	B lymphopoiesis Neutrophil egress from bone marrow and trafficking in innate immunity
CXCR3	CD183	IP10/Mig R, GPR9, CKR-L2, CMKAR3, IP10-R, MigR	P49682	na	Type 1 adaptive immunity
CXCR4	CD184	HUMSTSR, LESTR, fusin, HM89, LCR1, NPYR, NPY3R, D2S201E, D2S201E, HSY3RR,	P61073	3OE9	Hematopoiesis Organ and vascular development Homeostatic leukocyte maturation, survival, and trafficking Adaptive immunity
CXCR5	CD185	BLR-1, MDR15	P32302	na	B and T cell trafficking in lymphoid tissue to B cell zone/follicles
CXCR6	CD186	BONZO, STRL33, TYMSTR	O00574	na	Innate lymphoid cell trafficking Adaptive immunity NKT cell trafficking
CCR1	CD191	CC CKR1, MIP-1α/RANTES receptor	P32246	na	Innate and adaptive immunity
CCR2	CD192	CKR2, CC CKR2, CMKBR2, MCP-1-R, CC-CKR-2	P41597	6GPX	Inflammatory monocyte trafficking Types 1 and 2 adaptive immunity
CCR3	CD193	CKR3, CC CKR3, Eotaxin receptor, CMKBR3	P51677	na	Type 2 adaptive immunity Eosinophil, basophil, and mast cell trafficking
CCR4	CD194	CKR4, CC CKR4, K5-5, CMKBR4, CHEMR1	P51679	na	Lymphocyte homing to skin Thymopoiesis; Type 2 immune response
CCR5	CD195	CKR5, CC CKR_5, ChemR13, CMKBR5, IDDM22	P51681	4MBS	Type 1 adaptive immunity Innate immunity
CCR6	CD196	GPR-CY4, CKR-L3, STRL22, DRY-6, DCR2, BN-1, GPR29, CMKBR6	P51684	na	DC trafficking GALT development Type 17 immune responses
CCR7	CD197	EBI-1, BLR-2, CMKBR7	P32248	na	mDC, and B and T cell trafficking in lymphoid tissue to T cell zones Egress of T cells from tissue
CCR8	CDw198	TER1, CKR-L1, GPR-CY6, CMKBR8, CKR-L1	P51685	na	Thymopoiesis, Treg function Immune surveillance in skin Type 2 adaptive immunity Allergic inflammation
CCR9	CD199	GPR28, GPR 9-6	P51686	5LWE	Thymopoiesis; memory T cell homing to gut GALT development and function
CCR10	na	GPR2	P46092	na	IgA Ab-secreting cell trafficking to mucosal sites; humoral immunity Lymphocyte homing to skin
XCR1	na	GPR5	P46094	na	$CD8^+$ T cell responses to Ag cross-presented by $XCR1^+CD8^+$ DCs
CX_3CR1	na	fractalkine receptor, GPR13, V28, CMKBRL1, CCRL1, CMKDR1	P49238	na	Patrolling monocytes in innate immunity Microglial cell, DC and NK cell migration Type 1 adaptive immunity Inflammatory conditions involving cells of the myeloid lineage

(Continued)

TABLE 11.2 Directory of Human Chemokine Receptor Systematic Nomenclature, Aliases, Sequence, Structure, and Immunologic Function. For a List of Cell Types Expressing Each Receptor, See Figure 11.5 (Continued)

Name	CD#	Previous Names	Accession Numbers		Prominent Immunoregulatory Functions
			Protein	PDB	
ACKR1	CD234	Duffy, DARC, glycoprotein D, CCBP1, Dfy, GPD, FY	Q16570	na	Chemokine transcytosis by endothelium Inflammatory chemokine scavenging by erythrocytes
ACKR2	na	D6, CCBP2, CCR9, CCR10	O00590	na	Inflammatory chemokine scavenging at tissue barriers
ACKR3	na	CXCR7, RDC1, GPR159, CMKOR1	P25106	na	Shaping chemokine gradients for CXCR4 Cardiovascular development
ACKR4	na	CCX-CKR, CCBP2, CCR11, PPR1, VSHK1	Q9NPB9	na	Chemokine scavenging at tissue barriers

Ba, basophil; DC, dendritic cell; EC, endothelial cell; Eo, eosinophil; GALT, gut-associated lymphoid tissue; HEV, high endothelial venule; MC, mast cell; Mo, monocyte; MPC, myeloid progenitor cell; MΦ, macrophage; na, not applicable; NK, natural killer cell; plt, platelet; PMN, neutrophil.

level (Fig. 11.3), attract neutrophils, bind the same receptor CXCR2 with similar high affinity, and have been reported to have angiogenic activity. Of the ELR-negative CXC chemokines, only CXCL12 is angiogenic and attracts neutrophils (among many other cell types), but by activating CXCR4. A second CXC subgroup, comprising CXCL9, 10, and 11, is encoded by a distinct cluster of genes on human chromosome 4q21. All three members share the receptor CXCR3 but have been reported to be angiostatic, not angiogenic. The C57BL/6 strain of mice uniquely and naturally lacks *Cxcl11*.

CC chemokines can be divided into multiple structurally related subgroups. The two largest are the MCP (monocyte chemoattractant protein) and MIP (macrophage inflammatory protein) subgroups clustered together on human chromosome 17q12, whose members typically chemoattract monocytes/macrophages, plus other differential granulocyte and lymphocyte targets (Fig. 11.3; Table 11.2). CC chemokines may also be subclassified by the presence or absence of additional cysteines (Fig. 11.4A). CXCL16 and CX3CL1 form a unique multimodular subgroup, characterized by a typical chemokine domain, as well as a mucin-like stalk, a transmembrane domain and a C-terminal cytoplasmic module. Both molecules can exist as a membrane-bound form or, after protease cleavage of the stalk, as a soluble shed form, mediating direct G protein–independent cell-cell adhesion and chemotaxis, respectively. The discovery of CXCL17 by sequence threading techniques, not by primary sequence homology, provides proof of principle that the boundaries of chemokine sequence space could expand as more secreted cytokine-like protein structures are determined or modeled.

Structures are now available for 35 of the ~45 human chemokines (Fig. 11.4B; Table 11.1). All fold into a highly conserved and compact three-dimensional structure. Disulfide bonds link Cys-1 to Cys-3 and Cys-2 to Cys-4. The core, which contains 3 β sheets arranged in the shape of a Greek key, is overlaid by a C-terminal α-helical domain and is flanked by a flexible N-terminal domain that lacks order (Fig. 11.4). The structures of four chemokines (CCL15, CCL23, CCL26, and XCL1) resolve as monomers, and the others are mainly dimers; CXCL4 is a tetramer. The dimer interface is completely different for CXC chemokines (apposed β1 strands from each monomer to form a compact six-stranded β sheet core) and CC chemokines (short interactions between the N termini to form an elongated structure) (Fig. 11.4B). The chemokine "interactome" also includes numerous examples of potential chemokine heterodimers,[11] both CC/CC and CC/CXC, and some may form preferentially over homodimers in a GAG-dependent manner. Native heterodimers have also been identified.

Chemokine Receptor Classification and Structure

Chemokine receptors are defined as proteins that activate cellular responses upon binding chemokines; that is, unlike chemokines they are defined by function not structure. Twenty-two chemokine receptors have been identified (Fig. 11.5). Together with the receptors for the classical lipid and peptide chemoattractant receptors, all are members of the rhodopsin-like 7-transmembrane (7TM) domain receptor gene superfamily (Fig. 11.3; www.gpcrdb.org). After the odorant receptor genes, the leukocyte chemoattractant receptor genes form the largest gene cluster in the 7TM superfamily, which itself is the largest gene family by far in any vertebrate genome.

Unlike chemokines, the chemokine receptor repertoire appears to be very similar in mouse and man. Differences are limited so far to *CXCR2P1* (originally called *IL8RP*), a CXCR2-related pseudogene found in human but not mouse, and *Ccr1l1*, a still orphan receptor gene selectively duplicated from *Ccr1* in the rodent lineage. Polymorphism is common among individuals for most chemokine and chemokine receptor genes. However, the degree of polymorphism varies greatly among different genes. An extreme and important example is the *CCR2-CCR5* locus in human, in which combinations of common dimorphic single nucleotide polymorphisms (SNPs) in the *CCR5* promoter with a SNP in the open reading frame (ORF) of *CCR2* named *CCR2 V64I* and a 32-base-pair (bp) deletion in the ORF of *CCR5* named *CCR5*

Δ32 together form multiple haplotypes, some of which dramatically affect HIV disease susceptibility, as detailed further in the section on disease.

Chemokine receptors are functionally subclassified into two main groups, the conventional G protein–coupled chemotactic chemokine receptors (GPCRs; n = 18) and atypical G protein–uncoupled chemokine receptors (ACKRs; n = 4). Standard chemokine GPCR names are based on the property of chemokine ligand-group restriction, as follows: receptor name = the ligand-group root + R (for "receptor") + number, in order of discovery.[7] An exception is the C chemokine receptor XCR1, where "X" distinguishes it from CR1, the previously assigned name for complement receptor 1. For consistency, the XCR1 ligands are named XCL1 and XCL2. Of note, GPR35 is a human orphan receptor that was originally named unofficially as CXCR8 after it was reported to signal in response

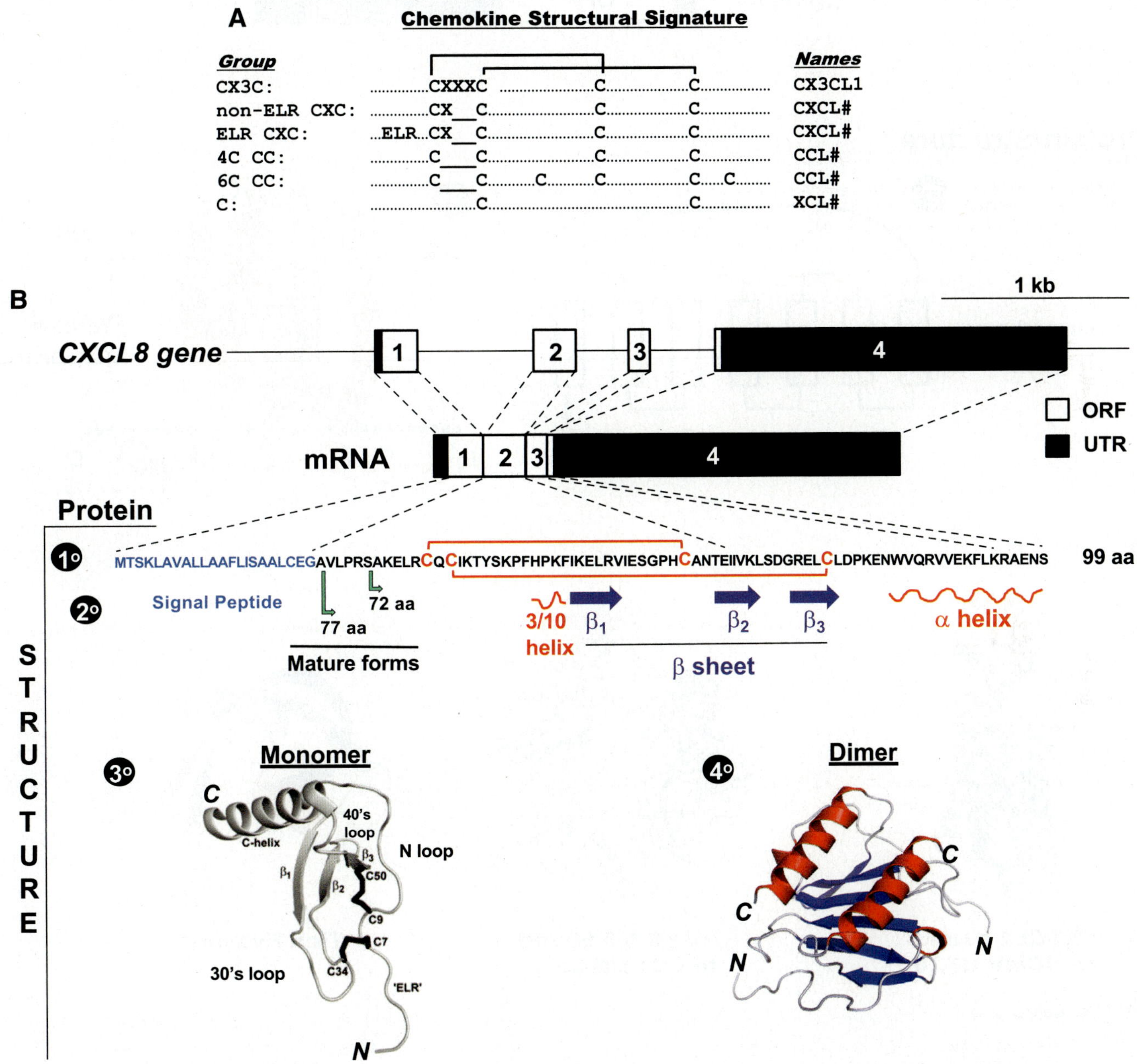

FIG. 11.4. **Molecular biology of the chemokine system. A:** Chemokine classification and standard nomenclature system. Chemokine groups are defined by the number and arrangement of conserved cysteines, as shown. Brackets link cysteines that form disulfide bonds. ELR, glu-leu-arg; X, an amino acid other than cysteine. The underscore is a spacer used to optimize the alignment. The N and C termini can vary considerably in length (not illustrated). For the molecules with four cysteines, there are approximately 24 amino acids between Cys-2 and Cys-3 and 15 amino acids between Cys-3 and Cys-4. At right is the nomenclature system. Examples of gene organization, mRNA splicing, posttranslational processing, and protein structure are shown for CXCL8 as a representative CXC chemokine **(B)** and its receptor CXCR1 **(C)**. **D:** Modes of chemokine binding to chemokine receptors (*left*) and to chemokine scavengers encoded by a virus (*middle*) and a tick (*right*). US28 is a G protein–coupled chemokine receptor encoded by human cytomegalovirus (HCMV). MHV68, gamma herpesvirus-68.

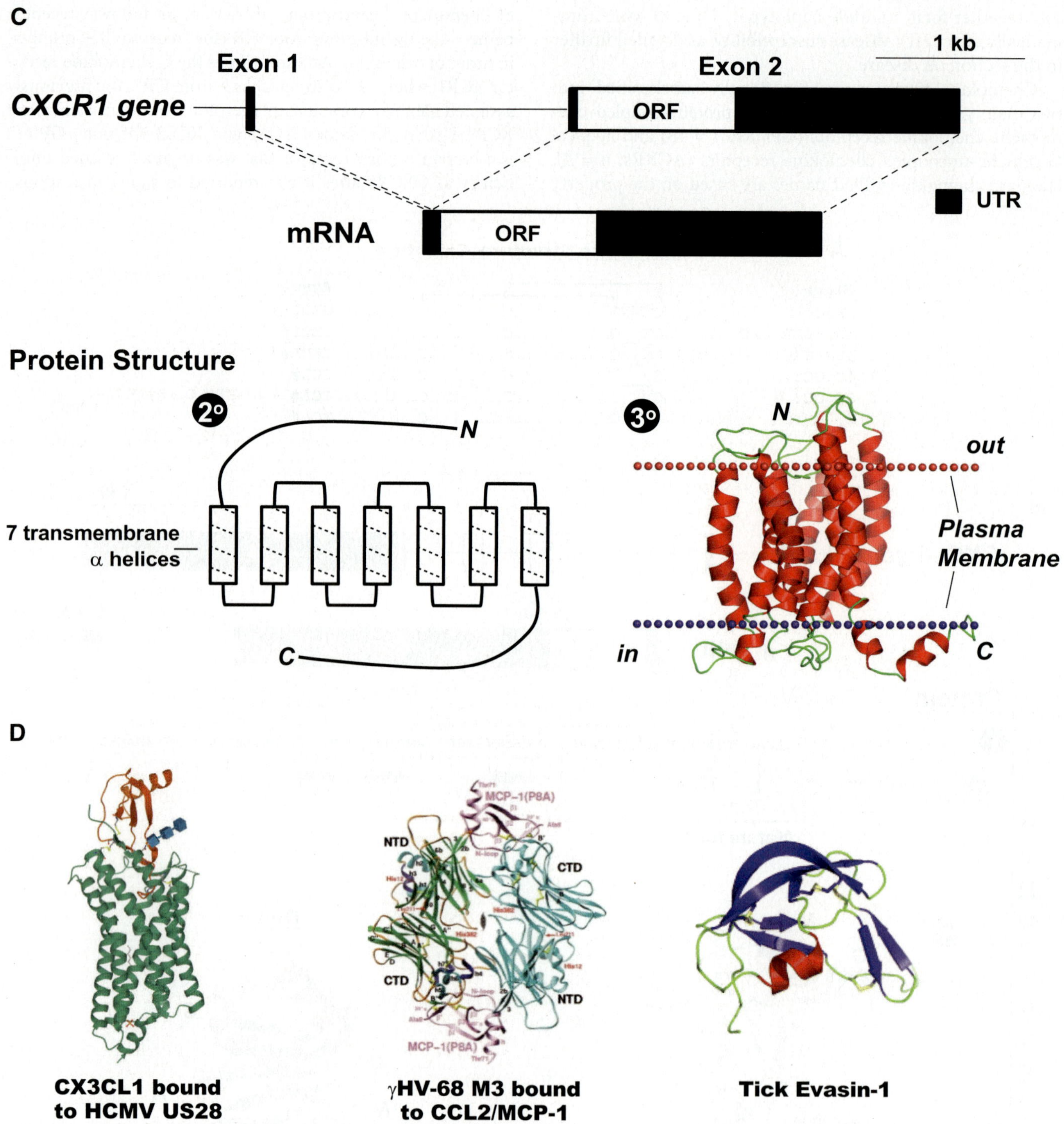

FIG. 11.4. Cont'd

to CXCL17; however, direct binding was not shown and other groups have not confirmed this function. CXCL17 remains the only human chemokine without at least one confirmed GPCR. Atypical chemokine receptors are named using the root ACKR followed by a number. ACKR3 was originally named RDC1 and then CXCR7, which is now reserved only as an alias, since it does not appear to signal through G proteins.

The 22 chemokine receptor genes can be divided by chromosomal location into three groups: a large cluster on human chromosome 3p21-23 that includes multiple *CCRs*, *CX3CR1*, *XCR1*, *CXCR6*, *ACKR2*, *ACKR4*, *ACKR5*, and the pseudogene *ACKR4P1*; *CXCR1*, *CXCR2*, and the pseudogene *CXCR2P1 (IL8RP)* clustered on 2q34-q35; and the rest unclustered. With the exception of *CCR9*, genes in the two clustered groups

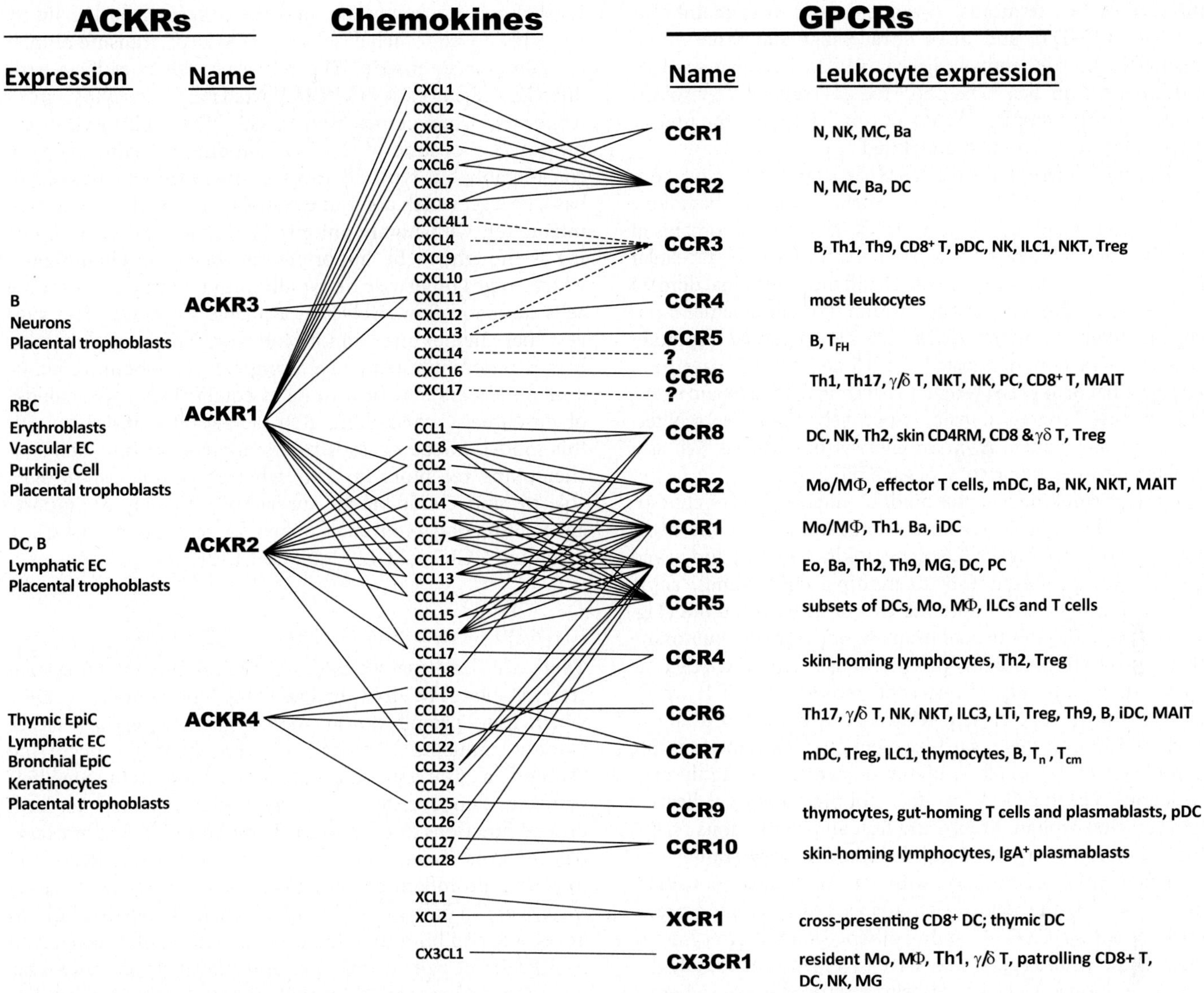

FIG. 11.5. Chemokine receptor specificity for chemokine ligands and cells. ACKR, atypical chemokine receptor; B, B lymphocyte; Ba, basophil; DC, dendritic cell; EC, endothelial cell; Eo, eosinophil; EpiC, epithelial cell; GPCR, G protein–coupled receptor; iDC, immature dendritic cells; ILC, innate lymphoid cell; MAIT, mucosal-associated invariant T cell; MG, microglial cells; Mo, monocyte; MΦ, macrophage; N, neutrophil; NK, natural killer cell; NKT, NKT cell; PC, plasma cells; pDC, plasmacytoid dendritic cells; RBC, red blood cell; T_{cm}, central memory T cells; T_{FH}, follicular help T cells; T_{mem}, memory T cells; T_{reg}, regulatory T cells; T_{RM}, resident memory T cells. Dashed lines indicate low affinity and unknown interactions.

lack introns in the ORF but have at least 1 and as many as 10 introns separating the promoter from the ORF (Fig. 11.4C). In contrast, an intron divides the N terminus of the majority of unclustered receptor genes (CXCR3, 4, and 5; CCR6 and 10). Several of these undergo alternative splicing, but the products appear to function similarly. CCR9A and B are coexpressed and have different EC_{50}s for activation, possibly extending the effective range of ligand concentration. CCR2 undergoes alternative splicing of a virtual intron in the C-terminal region of the ORF, but the two products have similar function and CCR2B appears to be the major expressed form.

In 2010, CXCR4 became the third mammalian 7TM protein to be crystallized and to have its structure determined.[12] This was followed by CXCR1, CCR2, CCR5, CCR9, and the human cytomegalovirus-encoded chemokine receptor US28.[13] Success involved conformational stabilization of the receptors by ligands, including small molecule antagonists in the case of CXCR4, CCR5, and CCR9 and CX3CL1 in the case of US28 (Fig. 11.4D, left panel). A structure of CXCR4 bound to the antagonistic viral chemokine vMIP-II has also been solved.[14] CXCR1 is the first 7TM receptor whose unmodified structure was solved in the absence of ligand or antibody and the first to be solved by nuclear magnetic resonance spectroscopy (Fig. 11.4C).[15] Several significant differences were observed for CXCR1 relative to the crystal structure of other chemokine receptors, including a monomeric structure (the others resolved as dimers), the presence of a short C-terminal domain α helix, and shifts of the transmembrane domains.

The ligand binding, membrane anchoring, and signaling domains of chemokine receptors are formed from a single polypeptide chain, ~340 to 370 aa long, stitched into the plasma membrane, creating an extracellular N terminus, an intracellular C terminus, seven α-helical transmembrane domains (TMDs), and three intracellular and three extracellular loops connecting adjacent TMDs. Evidence for heterodimerization has been reported for several chemokine receptors, for example, CXCR4 and ACKR3,[16] but the precise physiologic forms remain undefined.

The monomeric CX3CL1-US28 structural data have confirmed and extended the classic two-site model of chemokine binding to chemokine receptors, in which the core domains of the chemokine interact with the N-terminal and extracellular loop domains of the receptor (site 1) and the N-terminal domain of the chemokine intercalates within the intramembranous channel bounded by the TMDs (site 2). Additional important sites of interaction, designated site 0.5 and site 1.5, have been revealed that involve the receptor N-terminal domain and external juxtamembranous domain, respectively. The receptor N termini typically contain *N*-linked glycosylation sites as well as a sulfated tyrosine important for ligand binding.[17]

Each chemokine receptor binds a unique subset of chemokines. Of the GPCRs, six bind monogamously to one dedicated ligand, five bind to two chemokine ligands, and seven bind highly promiscuously to multiple chemokines, some of which may be shared among two or more receptors (Fig. 11.5). CXCL8 is an unusual example of a human chemokine that binds to both a highly promiscuous receptor, CXCR2, as well as to a more ligand-restricted receptor, CXCR1, which is structurally very closely related to CXCR2 (79% amino acid identity). In this regard, an even more extreme example is mouse Ccrl11, which is highly related to the highly promiscuous receptor Ccr1, yet does not bind any Ccr1 ligands and remains orphan. Mouse and human promiscuous receptor counterparts usually bind the same or very similar sets of orthologous chemokines, with several exceptions usually determined by the lack of a cross-species ortholog, as previously noted for CXCL8, among others. In addition, syntenic chemokine counterparts may have diverged functionally, for example, mouse Ccl8 is a potent agonist at Ccr8, whereas human CCL8 is inactive at human CCR8.[18]

In addition to the uncertainty about the correct ligand for "CXCR8/GPR35" noted previously, it is worth noting that the ligands for CXCR2, CCR3, CCR4, ACKR3, and ACKR4 were all originally misidentified in the literature. Thus, there may still be additional chemokine receptors that remain to be discovered among the many remaining orphan GPCRs, and additional ligand corrections may still be needed for known chemokine receptors as *in vitro* specificities defined using overexpressed recombinant receptors in cultured cell lines are further tested *in vivo* for the corresponding native receptors expressed at natural abundance.

Chemokines are typically found in very low concentrations under homeostatic conditions in plasma. CCL14, 15, and 16 are exceptional in this regard and are found in very high homeostatic concentrations in plasma, although the reason for this is not yet clear. Under conditions of systemic inflammation including during severe infection, concentrations of inflammatory chemokines may rise considerably and typically exceed those of most other types of cytokines. Under conditions of local infection and inflammation, inflammatory chemokines more typically become concentrated locally by tethering to endothelial cells via GAGs or transmembrane domains, or to matrix. The tethering cell may have produced the chemokine or else imported it by transcytosis from neighbors. Chemokines bind to GAGs as multimeric complexes. The location of the GAG-binding domain may vary considerably but typically involves discontinuous patches of basic charge, which are found in the majority of chemokines, interacting with highly anionic GAG surfaces. GAG binding is thought to be important for localizing chemokines, which might otherwise be rapidly washed away from source cells by blood flow. However, available structural data suggest that chemokines cannot bind to receptors as dimers or higher-order multimers, suggesting that GAG-bound chemokine multimers may instead be in equilibrium with "clouds" of monomeric chemokine at these sites.[19] Consistent with this model, forced chemokine monomers are fully active at chemokine receptors *in vitro*, whereas complex multimers able to bind GAGs on the surface of cells may be important for function *in vivo*, as shown for multiple chemokines, including CCL5, CXCL10, and CXCL12γ.[20-23]

Atypical Chemokine System Components

There are three main classes of atypical chemokine system components. The first includes the eponymous ACKRs, which bind promiscuously to multiple chemokines either without inducing a signal or with signaling occurring through a β-arrestin-dependent G protein–independent pathway. Functionally, they have been reported to have diverse functions distinct from chemokine GPCRs, for example, to shape chemokine gradients presented to GPCRs and to buffer proinflammatory effects of chemokines, keeping the system in balance. Although all chemokines that bind to an ACKR also bind to at least one GPCR, dually recognized ACKR/GPCR pairs are all distantly related structurally. This illustrates the general point that, although groups of similar chemokines may act in lockstep at the same receptor and have similar function, they may also diverge by activating dissimilar receptors and functions.

ACKRs may be subclassified by their degree of chemokine ligand promiscuity. ACKR1 and ACKR2 bind overlapping sets of many proinflammatory CC chemokines, and ACKR1 also binds multiple proinflammatory CXC chemokines, the most promiscuous of all known chemokine receptors. Prominent expression of both receptors on tissue barrier cells, including vascular and lymphatic endothelial cells and placental trophoblasts, and of ACKR1 on mature erythrocytes, which massively outnumber leukocytes in blood, has suggested they may function as chemokine sinks to shape more effective chemokine gradients or to dampen excessive inflammatory responses, depending on the context. As one example, inflammatory chemokine scavenging by placental trophoblast-expressed ACKRs might support immunologic tolerance during pregnancy.

ACKR1 may also present chemokines on endothelial cells either directly or indirectly by mediating transcytosis to the abluminal side of the cell.[24] Internalization of bound chemokine through a β-arrestin- and clathrin-dependent mechanism is a general property of almost all chemokine receptors; however, ACKR2 is particularly well adapted for this by constitutive association with β-arrestin allowing extremely rapid internalization of ligand destined for degradation and rapid recycling of receptor.[25]

In contrast to ACKR1 and 2, ACKR3 and 4 bind only two and two chemokines, respectively, and functionally coordinate chemokine gradient shaping with specific GPCRs with which they share ligand specificity. The clearest example of this has been demonstrated elegantly *in vivo* for zebrafish primordial germ cell migration during development,[26] which is mediated by CXCR4 activation by CXCL12 presented in gradients that may be formed in part as a result of a much slower off-rate and 10-fold higher binding affinity for CXCL12 at ACKR3,[27] which is expressed on stromal cells, than at CXCR4, which is expressed on germ cells. A similar process may occur for ACKR4 shaping of gradients for shared ligands signaling through CCR7 to facilitate B, T, and dendritic cell trafficking in mammalian lymph node. ACKR3 may also function independently of CXCR4, for example, to position marginal zone B cells in mouse spleen and during cardiac valve development.[28]

In addition to the four validated ACKRs, two other names, ACKR5 and ACKR6, are being reserved for CCRL2 and PITPNM3/Nir1 pending confirmation and further characterization of chemokine binding. PITPNM3/Nir1 is not a 7TM protein but in breast cancer cells was reported to activate a proliferative response to CCL18,[29] a primate-specific chemokine very closely related structurally to CCL3 (Fig. 11.3). Curiously, CCL18 is not an agonist at either of the two CCL3 receptors, CCR1 and CCR5, and PITPNM3/Nir1 is not expressed in leukocytes. Instead, CCL18 binds to CCR8 explaining its potent chemotactic activity for Th2 cells *in vitro*,[30] and has been reported to be an antagonist at CCR3. *In vivo* investigation of CCL18 is difficult in the absence of a mouse counterpart.

The second class of atypical chemokine system components includes endogenous nonchemokine ligands reported to bind to chemokine receptors (eg, macrophage inhibitory factor [MIF] at CXCR2, the antimicrobial peptide β-defensin-2 at CCR6, and the cytokine-like protein CYTL1 at CCR2). A variation on this theme is provided by HMGB1 (high mobility group protein B1), a nonhistone nucleoprotein released from damaged cells at inflammatory loci that is able to bind cooperatively to CXCL12 and augment its activity at CXCR4.[31] Several other examples of CXCL12-binding factors have been described, including MIF and ubiquitin. The extent to which these interactions occur and are functionally important *in vivo* is difficult to prove experimentally. As one particularly interesting biological example, in the adrenal gland intermediate opioid peptides were shown to act at ACKR3 in a paracrine manner resulting in augmentation of circadian glucocorticoid oscillations and modulation of anxiety in mice.[32,33]

The third class of atypical components includes microbe-encoded chemokine system mimics, including viral chemokines, 7TM chemokine receptors, structurally unique secreted chemokine scavengers, and nonchemokine chemokine receptor agonists and antagonists (Table 11.3).[7,34,35] These microbial elements may function to evade the immune system, recruit new target cells, reprogram gene expression for cell proliferation and angiogenesis, and mediate target cell entry and have been identified in most herpesviruses and poxviruses and some retroviruses, including HIV. A rich example of this centers on the host receptor CCR5, which can bind viral (HIV gp120), bacterial (*M. tuberculosis* Hsp70), and protozoan (*T. gondii* immunophilin) agonists that all lack chemokine structure. Broad-spectrum anti-inflammatory chemokine binding proteins are encoded by the mouse herpesvirus gHV-68 (M3 protein), as well as by *Schistosoma mansoni* and many ectoparasitic ticks.[36] The first four tick proteins described, named evasins 1 to 4, were purified from tick saliva, where they may contribute to the lack of inflammation associated with tick bites. *Staphylococcus aureus* produces the protein toxins LukED and H1gAB that bind to multiple cellular chemokine receptors, including CCR5 to LukED, CCR2 to HlgAB, and CXCR1, CXCR2, and ACKR1 to both toxins.[37] Binding to ACKR1 mediates red cell lysis with release of hemoglobin-bound iron, favoring bacterial survival, although *in vivo S. aureus* infection is not associated with hemolytic anemia. HCMV US28 is an example of a constitutively active viral chemokine receptor able to bind both multiple CC chemokines and CX3CL1, which enhance signaling, with evidence of biased agonism and activation of diverse functional outputs.[38]

Many atypical functions have been identified for mammalian chemokines. For example, CXCL16 was originally identified as a scavenger receptor and named SR-PSOX (scavenger receptor for phosphatidyl serine and oxidized LDL). As a second example, a truncated form of CCL23, unlike full-length CCL23, has not been shown to activate any chemokine receptors, yet it is an agonist at FPR2, a member of the classical chemoattractant fMet-Leu-Phe receptor subfamily.

Immunologic Classification of the Chemokine System

Apart from mature erythrocytes, which express only ACKR1, all other mature hematopoietic cells, including platelets, differentially express multiple chemokine receptors. In some cases, one of the receptors is sufficiently highly and uniformly expressed in a given subset to serve as a reliable signature (eg, CXCR1 and CXCR2 for neutrophils, CCR3 for eosinophils, CCR2 and CX3CR1 for inflammatory and patrolling monocytes, respectively, CXCR3 for Th1 cells, CCR4 for Th2 cells, CCR6 for Th17 cells, CXCR5 for B cells, CCR7 for naïve and central memory T cells and mature dendritic cells, among others), or in some cases even as a defining marker (eg, CXCR5 for follicular helper T cells and XCR1 for antigen-cross-presenting dendritic cells [DCs]).

Chemokines and chemokine receptors may be loosely divided into two main functional subsystems, designated *homeostatic* and *inflammatory*. Homeostatic chemokines are differentially and constitutively expressed primarily in specific microenvironments of primary and secondary immune organs and coordinate migration of hematopoietic precursor cells, mature DCs, and naïve and central memory lymphocyte subsets via constitutively expressed receptors. Inflammatory chemokines are induced by noxious stimuli in tissue cells

TABLE 11.3 Herpesvirus-Encoded Chemokine Components

Molecule	Virus	Biochemical Properties	Biological Roles
Chemokine-Binding Proteins (CKBP)			
gG1	HSV-1	Envelope structural protein	Binds CXCL12α and β, enhancing CXCR4 signaling
gG2	HSV-2	Envelope structural protein; soluble viral CKBP	Binds CXCL12α and β, enhancing CXCR4 signaling
gC	VZV	Envelope structural protein	Potentiates chemokine activity
UL21.5	HCMV	Binds CCL5	Blocks CCL5 binding to cellular receptors
M3	gHV68	Binds XC, CC, CXC, and CX3C chemokines	Important for latency. Blocks chemokine-GAG interaction and chemokine-receptor interaction
Chemokine Receptors and Chemokine Receptor Homologs			
US27	HCMV	Orphan	Enhances CXCR4 signaling Protects host cells against apoptosis
US28	HCMV	Constitutively coupled to G proteins. Binds CX3CL1 and several CC chemokines	Host cell proliferation and chemotaxis Angiogenesis Potentiates HCMV entry HIV-1 coreceptor activity
UL33	HCMV	Constitutively coupled to G proteins. Forms heterodimers with host and viral chemokine receptors	Viral dissemination Negative modulation of CCR5 and CXCR4 signaling
UL78	HCMV	Orphan	Promotes viral growth in endothelial and epithelial cells Negative modulation of CCR5 and CXCR4 signaling
U12	HHV-6	Activated by CCL2, CCL3, CCL4, and CCL5	Undefined
U12	HHV-7	Activated by CCL17, CCL19, CCL21 and CCL22. Forms heterodimers with CCR4 and CCR7	Undefined
U51	HHV-6	Constitutively coupled to IP production. Activated by several chemokines including vCCL2	Undefined
U51	HHV-7	Activated by CCL17, CCL19, CCL21 and CCL22	Undefined
BILF1	EBV	Constitutively active with no identified ligands. Forms heterodimers with CXCR4 and MHC-1	Promotes tumor growth Inhibits CXCR4 signaling and MHC-1 expression
ORF74[a]	KSHV[a]	Constitutively coupled to G proteins. Signaling modified CXC and CC chemokine agonists and inverse agonists Activation of ERK1/2, JNK, RhoA, and Rac1 signaling pathways linked to cytokine expression and angiogenesis	Kaposi sarcoma tumorigenesis
E1	EHV2	CCL11 binding	Unknown
Chemokines			
gL	HSV-1 and -2, VZV, EBV, KSHV	Forms dimers with gH, which determines viral tropism	Binds to integrins or ephrin type A receptor A2 promoting viral entry
vCXCL1	HCMV	Agonist at CXCR1, CXCR2, and CX3CR1	Promotes viral dissemination by recruiting leukocytes
vCXCL2	HCMV	Undefined	Undefined

TABLE 11.3 Herpesvirus-Encoded Chemokine Components (Continued)

Molecule	Virus	Biochemical Properties	Biological Roles
UL128 UL130	HCMV	Form a pentamer with gH/gL dimers and UL131A, which determines viral tropism	HCMV infection of epithelial, endothelial, and monocytic cells
MCK2 *ORF m131/m129*	MCMV	GAG binding Forms an entry complex with gH and gL	Macrophage entry Viral dissemination in salivary gland
vCCL4a	HHV-6A	Agonist at CCR1, CCR4, CCR5, CCR6, and CCR8	Promotes viral dissemination by recruiting leukocytes
vCCL4b	HHV-6B	Agonist at CCR2	Promotes viral dissemination by recruiting leukocytes
vCCL1	KSHV	Agonist at CCR8	Promotes host cell survival, angiogenesis, and vascular dissemination
vCCL2	KSHV	Antagonist of several chemokine receptors. Agonist at CCR3 and CCR8	Anti-inflammatory and angiogenic; immune evasion
vCCL3	KSHV	Agonist at XCR1 and CCR4	Recruits XCR1$^+$ DCs susceptible of latent infection

[a]Other γ herpesviruses, including Ateline herpesvirus, Herpesvirus saimiri, Equine herpesvirus 2, and Mouse herpesvirus 68 have ORF74 homologs. Chemokine modulators are also encoded by many poxviruses and retroviruses.

and leukocytes. The distinction between inflammatory and homeostatic chemokines may not be sharp since many constitutive chemokines may be further induced, and chemokines that are highly inducible in some cell types may be constitutively expressed in others. For example, CXCL8 is a highly inducible inflammatory chemokine important in acute neutrophil inflammation, but it is also made constitutively by sweat glands where it may help shape the microbiome through its direct antibiotic activity. In addition, the CXCL8 receptor CXCR2 is important in acute neutrophilic inflammation and also regulates homeostatic release of neutrophils from bone marrow to blood. Table 11.1 aggregates a small portion of this vast subject.

The five ligand-monogamous chemokine receptors all bind homeostatic chemokines, whereas the seven ligand-promiscuous chemokine receptors all bind inflammatory chemokines. Inflammatory chemokine receptors tend to be constitutively expressed on phagocytes, innate lymphoid cells, and effector but not naïve lymphocytes or immature DCs. Dynamic shifts in receptor expression, particularly in lymphocytes and DCs, occur during differentiation, maturation, and activation and may involve different receptors for the same cell type in different tissues.[39]

The Issue of Chemokine Redundancy

Almost all chemokines are chemotactic agonists, and a few may be agonists at one GPCR but antagonists at another (eg, CCL11 agonism at the Th2 receptor CCR3 and antagonism at the Th1 receptor CXCR3), in addition to binding to ACKRs. Differential receptor usage, differential regulation of expression, and biased agonism may all account for nonredundant function *in vivo* observed for chemokines acting at the same GPCR.[40] A good example of differential expression is CXCL7 and CXCL8. Both are agonists for neutrophil CXCR2, but CXCL7 is stored in platelet α granules and released immediately upon platelet activation,[41] whereas CXCL8 is made by many human cell types only upon gene induction by proinflammatory stimuli, together allowing for robust sequential amplification of neutrophil accumulation in tissue. Detailed study of CXCR3 has shown that its three chemokine ligands, CXCL9, 10, and 11, have nonredundant actions *in vivo*, the result of multiple factors including differential regulation and expression, differential binding to CXCR3, and possibly additional distinct binding sites. For example, CXCL11, but not CXCL9 and 10, binds to ACKR3.

There may be no inflammatory processes in which only one chemokine is involved. Instead, inflammatory chemokines may act cooperatively, synergistically, or antagonistically, on the same or separate cell types, but in a hierarchical manner through promiscuous chemokine receptors, both GPCRs and ACKRs, sometimes in concert with classical chemoattractant receptors.[31] The net result is that there may be enough functional redundancy in the whole system such that loss of a single inflammatory chemokine or chemokine receptor, for example, in a knockout mouse, does not cause altered susceptibility to naturally acquired infections or other diseases, yet there may not be sufficient redundancy to handle a stronger experimental or natural stress when one particularly important component is missing. In contrast, homeostatic chemokines act much less redundantly, and their loss typically results in major defects in development and leukocyte trafficking, as well as perinatal mortality in the case of CXCL14, ACKR3, and either CXCL12 or its receptor CXCR4. Of interest, these four essential genes are also unusual in being much more highly conserved in sequence throughout phylogeny than other chemokines and chemokine receptors.

LEUKOCYTE RESPONSES TO CHEMOKINES

Leukocytes respond to chemokine gradients by changing their shape to produce a front and a back and by crawling toward the front. Motion involves coordinated actin cytoskeleton remodeling at the front producing a leading edge or lamellipodium, myosin-based contraction at the back or uropod, release of the uropod from substrate, and treadmilling forward movement of membrane lipid. Navigation through tissue may require relays of chemokines and adhesion molecules. Fine navigation within tissue microenvironments has also been shown to be directed by specific chemokines up to the point of immunological synapse formation.[42] For example, CCR5 and CXCR4 may accumulate in the immunological synapse formed by CD4+ T cells and DCs *in vitro* and provide costimulatory signals.[43]

In vitro, leukocytes undergo random migration in the absence of chemoattractants. In the presence of a chemoattractant, migration may be chemokinetic (nondirectional motion in the presence of a homogenously distributed soluble chemoattractant), chemotactic (directed migration in a gradient of soluble chemoattractant), haptokinetic (nondirected migration induced by a homogeneously distributed chemoattractant tethered to a surface), or haptotactic (directed migration in a gradient of chemoattractant tethered to a surface). *In vivo*, most leukocyte trafficking appears to involve haptotactic mechanisms.

A major mode in both homeostatic and inflammatory leukocyte trafficking is transendothelial migration across a vascular barrier, a multistep process involving a combinatorial endothelium- and leukocyte-specific molecular code, in which chemokines regulate at least two steps.[44,45] In an initial chemokine-independent step, leukocytes roll on inflamed endothelium in a selectin-dependent manner. Then, chemokines presented by endothelium stimulate rolling leukocytes to activate β1 and β2 integrins, which mediate shear-resistant firm adhesion of leukocytes to endothelium by binding intercellular adhesion molecules (eg, VCAM, ICAM) and other ligands.

Important tissue-specific variations of this classic model may exist, for example, for neutrophil transendothelial migration in lung, which does not appear to involve selectins, and for patrolling CD8+ T cells, which may crawl on blood vessel endothelium against the direction of blood flow in a β2 integrin–independent manner. Moreover, detailed studies have revealed that multiple chemoattractant receptors may mediate distinct intermediate steps during transendothelial migration. For example, for mucosa-associated invariant T (MAIT) cell migration across TNF-activated human umbilical vein endothelial cells, at least three chemokine receptors are involved: CCR6 to mediate leukocyte arrest on endothelium, CCR5 to shorten the time for crawling across the abluminal endothelial surface, and CCR2 to mediate transendothelial migration (Fig. 11.6A).[46] Similarly, in an autoantibody-dependent arthritis mouse model, Ccr1, C5a receptor, and Cxcr2 were demonstrated to act sequentially at specific steps to drive neutrophil transendothelial migration *in vivo*.[47] Moreover, there is now evidence that multiple additional factors may regulate transendothelial migration, including sheer force, intracellular vesicle-localized chemokines (eg, Ccl2 attraction of T cells), and chemoattractants produced by pericytes (eg, Cxcl1 attraction of neutrophils).[48,49] Neutrophil transendothelial migration has been observed to be occasionally bidirectional, and fugetaxis has been defined as a process in which leukocytes move away from high concentrations of chemokines.[50] The concepts of specialized tissue-specific arrest chemokines, TEM chemokines, chemokine relays, and primary and secondary chemoattractants that compete for primacy in directing cell migration are all now supported by experimental evidence and continue to be advanced.[51]

Chemokines and their receptors have increasingly been implicated in other cell functions, including regulation of cell proliferation, quiescence and apoptosis (eg, hematopoietic stem cell [HSC] CXCR4), regulation of antigen sampling by immature DCs (eg, CCR7 and CX3CR1), and DC maturation (eg, CCR7).[52,53] Inflammatory chemokine activation of appropriate leukocyte subsets may induce mediator release (eg, defensins, proteases, perforins, histamine, eicosanoids) and reactive oxygen species production, resulting in cytotoxic or vasomotor responses. Some chemokines even have direct antibacterial activity, which could explain why some can be found constitutively at high concentrations in body fluids such as blood and sweat.[54] Angiostatic (eg, CXCL4) and angiogenic chemokines (eg, all ELR^+ CXC chemokines) and myelosuppressive chemokines have been described. The precise functions of the "plasma chemokines" CCL14 to 16, which are present at very high concentrations in plasma, have not been clearly defined. Thus, chemokine function is protean, not limited to chemotaxis, and may be extended in atypical ways.

CHEMOKINE SIGNALING PATHWAYS

G_i-dependent Effectors

Classically, agonist-bound chemokine receptors act as guanine nucleotide exchange factors (GEFs), especially for pertussis toxin–sensitive G_i-type heterotrimeric G proteins such as G_{i2} (Fig. 11.7).[55] Although the term G_i refers to the ability of this type of G protein to inhibit adenylyl cyclase, in leukocytes G_i is important in activating the cell and driving the chemotactic response. G protein activation presumably involves a conformational change in the transmembrane domains of the receptor and binding of the G protein to multiple intracellular domains, including the C tail, intracellular loop 3, and the DRY motif at the end of TMD2. The DRY motif is highly conserved in chemokine GPCRs but not ACKRs, which do not couple to G proteins. Activation of G proteins involves exchange of GDP for GTP by the α subunit, leading to dissociation of the trimer into α and βγ subunits. The duration of G protein signaling is regulated in part by intrinsic GTPase activity possessed by the α subunit, which can be markedly increased by GTPase activating proteins (GAPs) known as RGS proteins (regulators of G protein signaling).[55] Signaling appears to be mainly mediated by βγ subunits,[56] which activate diverse effectors, including phospholipases A2, C (subtypes β2 and β3), and D; phosphatidylinositol-3-kinase γ (PI3K γ); protein tyrosine kinases (PTKs) and phosphatases; low-molecular-weight GTPases; and mitogen-activated protein kinases. The requirements of these enzymes for chemotaxis may be both chemoattractant and cell type dependent, and the complex mechanistic details underlying specificity remain to be elucidated in most biological contexts.

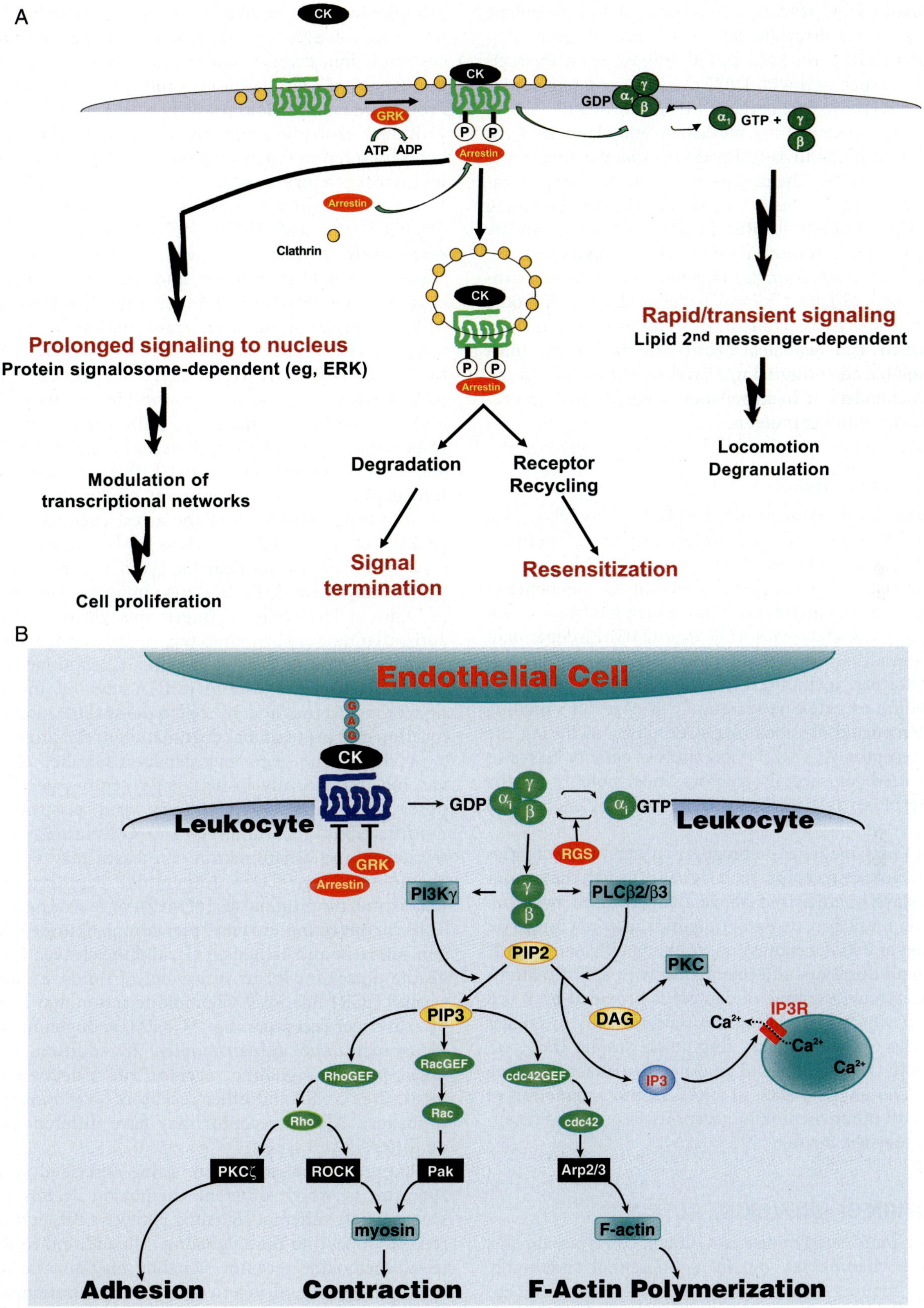

FIG. 11.6. Chemokine signal transduction. A: Chemokines signal through both G proteins and β-arrestin pathways resulting in diverse changes in cell state. **B:** Lipid mediators in the G protein signaling pathway. CK, chemokine; DAG, diacylglycerol; GAG, glycosaminoglycan; GEF, guanine nucleotide exchange factor; GRK, G protein–coupled receptor kinase; IP3, inositol trisphosphate; PI3K, phosphatidylinositol-3-kinase; PIP, phosphatidylinositol phosphate; PKC, protein kinase C; PLC, phospholipase C; RGS, regulator of G protein signaling. (B, Model is modified from the Alliance for Cell Signaling (www.signaling-gateway.org).)

Phospholipase C (PLC) hydrolyzes PI-4,5-bisphosphate (PIP_2) to form 1,2-diacylglycerol (DAG) and inositol-1,4,5-trisphosphate (IP_3). IP_3 induces Ca^{2+} release from intracellular stores, which acts with DAG to activate protein kinase C (PKC). PI3K γ phosphorylates PIP_2 to form PIP_3, which recruits proteins containing pleckstrin homology (PH) or Phox (PX) domains to lamellipodium, thereby converting shallow extracellular chemokine gradients to steep intracellular effector gradients.[57,58] Four PH domain-containing targets—Akt, and GEFs for Rac, Rho, and cdc42—modulate distinct phases of cell movement in various model systems. Rho regulates cell adhesion and chemotaxis and myosin contraction. Rac and cdc42 control lamellipodia and filipodia formation, respectively. Downstream targets of Rac include PAK1, which also regulates myosin contraction. Motility, phagocytosis, and immunologic synapse formation are all critically dependent on local polymerization of actin, coordinated by actin-binding proteins.

G_i-independent Effectors

Chemokine receptors may couple to G proteins other than G_i *in vitro*; however, roles *in vivo* are not as well defined. Chemokines may activate effectors such as MAP kinases and nonreceptor PTKs by a variety of G_i-independent mechanisms. An example is CXCR4, which has been shown to physically associate with TCR in a CXCL12-dependent manner, resulting in activation of ZAP-70 and downstream effectors, including ERK, calcium flux, AP-1, and costimulation of cytokine secretion.[59] β-Arrestin signaling has been reported for chemokine receptors, including the atypical receptor ACKR2.[60] Constitutive activity has also been reported for several receptors, most notably for the human herpesvirus 8 (HHV8) CXCR2 homolog encoded by ORF74.[61]

CXCR4 may mediate the broadest spectrum of activities of any chemokine receptor, yet it signals through the canonical pathways summarized above that are used by other chemokine receptors. Its exceptionalism may rely more on the context of CXCR4 expression, coupling efficiency, signal strength and duration, and interaction with specific modifiers. An extreme example of context is provided by B cell subsets, in which stimulating the receptor fails to trigger calcium flux or chemotactic responses, despite the presence of both the receptor and an otherwise intact signaling pathway. The physical state of CXCL12 and interaction of CXCR4 with other chemokine receptors in potential receptorsomes may also matter.

REGULATION OF CHEMOKINE ACTION

Virtually all nucleated mouse and human cell types are able to produce chemokines, but in a differential and highly regulated manner. Chemokine and chemokine receptor expression may be positively or negatively regulated at the transcriptional level by diverse factors, including proinflammatory cytokines, hypoxia, viruses, bacterial products, cell adhesion, shear stress, antigen uptake, T cell costimulation, diverse transcription factors, and cis-acting motifs, as well as microRNAs and long noncoding RNAs. Proinflammatory cytokines such as IL-1, IL-6, TNF, and IL-15 induce expression of many of the inflammatory chemokines involved in innate immunity such as CXCL8, whereas immunoregulatory cytokines such as IFN-γ, IL-4, and IL-17 are more tightly focused on Th1 (eg, CXCL9-11)-, Th2 (eg, CCL11)-, and Th17 (CCL20)-selective chemokines, respectively, through positive feedback loops involving tissue sources of chemokine, and leukocyte sources of the cognate chemokine receptor and signature cytokine (Fig. 11.8). Examples of important transcription factors that regulate the chemokine system include NF-κB for inflammatory chemokines; HIF-1α for oxygen-dependent expression of CXCR4; the signature transcription factors of Th1 (T-bet), Th2 (GATA3), and Th17 (RORγt) cells for their corresponding signature chemokine receptors CXCR3, CCR4, and CCR6, respectively; and CEBP–δ, which induces a full-service chemotactic program on human MAIT cells by activating expression of the arrest chemokine receptor CCR6 (but not CCR2) as well as *FUT7* and *ST3GAL4*, the glycosyltransferases critical for synthesis of the L-selectin ligand Sialyl-Lewis^X; PLZF is also important for expression of CCR6 in Th17 cells.[46,62] Interferons, glucocorticoids, and anti-inflammatory cytokines (eg, IL-10) may inhibit inflammatory chemokine gene expression. Chemokines may also be regulated at the level of mRNA stability, including by noncoding RNAs, and by cell type–specific polyubiquitinylation and proteasomal degradation of the proteins.

A chemokine gene may generate families of proteins varying dramatically in activity, receptor specificity and potency by alternative splicing and posttranslational modifications, including N- and C-terminal proteolytic trimming and citrullination. Proteases may target many chemokines (eg, CD26 [dipeptidyl peptidase IV] and matrix metalloproteinases [MMP]), or few or only one (eg, TNF α converting enzyme, plasmin, urokinase plasminogen activator and cathepsin G), and the cleaved forms may be the dominant forms in biological fluids, as shown for several CCR1 ligands.[63] Chemokine action may be blocked by scavenger receptors (eg, ACKRs), endogenous receptor antagonists, and autoantibodies. In addition, cytokines may convert a signaling receptor into a decoy (eg, IL-10 inactivates CCR2 and other receptors on monocytes/macrophages). Also, a receptor may have different functions on different cell types.

Receptor signalosomes are being described and biased agonism, in which different chemokine ligands preferentially activate different signaling pathways through the same receptor, may fine-tune signaling.[40] In addition to RGS proteins, chemokine receptor signaling may also be regulated by homologous and heterologous desensitization, which is a way of terminating signaling. Desensitization mechanisms involve phosphorylation of serine and threonine residues in

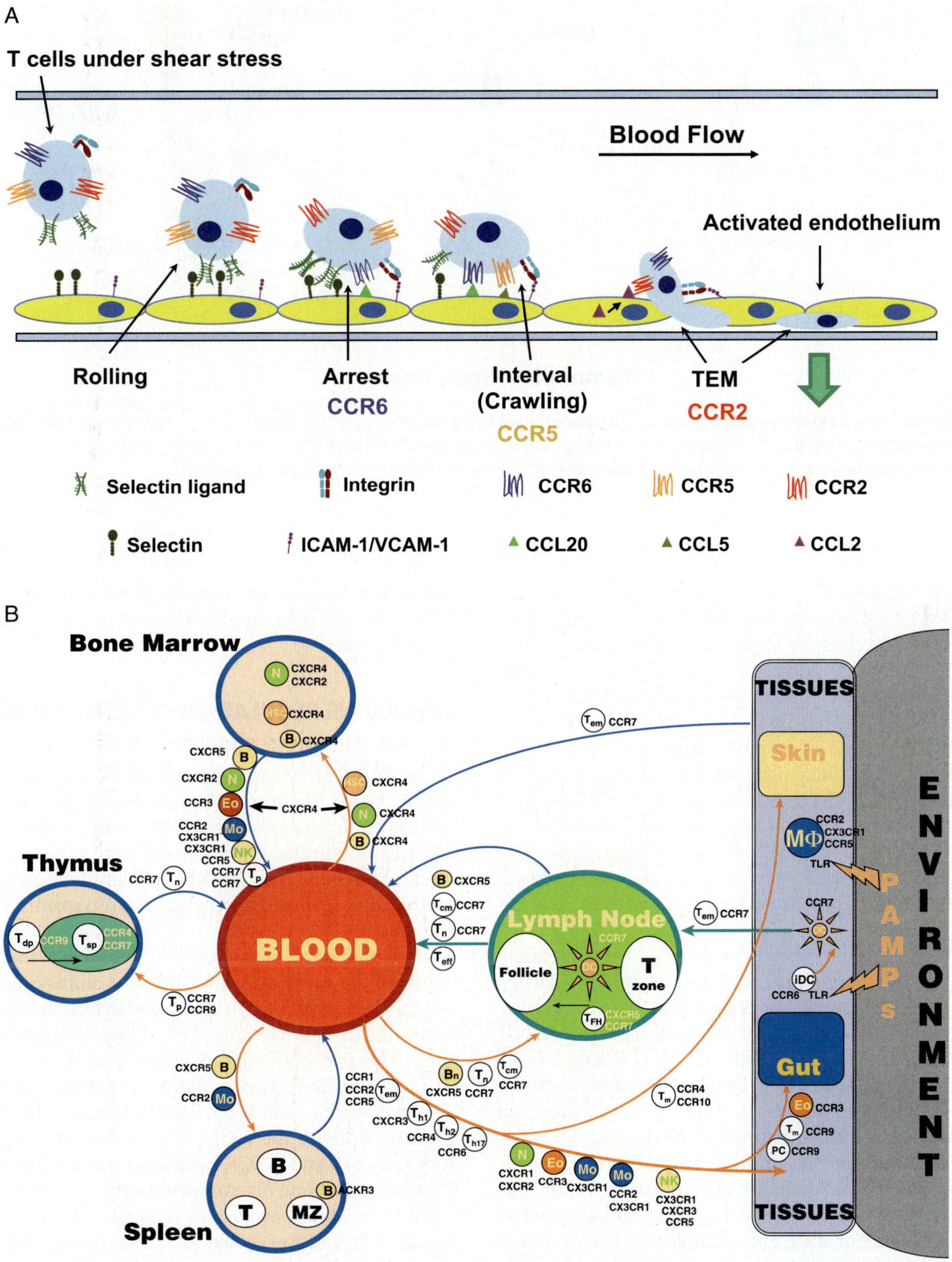

FIG. 11.7. Chemokine control of leukocyte trafficking. A: Expanded multistep model of leukocyte transendothelial migration (TEM). The figure was created and provided by Joshua Farber and Howard Zhang (NIAID) based on experimental analysis of multireceptor-expressing highly differentiated human MAIT cells migrating across TNF-activated HUVECs in flow chambers. CXCR3 involvement was not analyzed since its ligands are not produced in this system. **B:** Examples of homeostatic and inflammatory leukocyte trafficking itineraries. Shown are routes among primary and secondary immune organs and the periphery, leukocyte subtypes trafficking on those routes and some of the chemokine receptors that appear to be most important in regulating each route. Eo, eosinophil; GC, germinal center; HSC, hematopoietic stem cell; iDC, immature dendritic cells; Mo, monocyte; MZ, marginal zone; MΦ, macrophage; N, neutrophil; NK, natural killer cell; PAMP, pathogen-associated molecular patterns; PC, plasma cell; T_{cm}, central memory T cells; T_{em}, effector memory T cells; T_{FH}, follicular help T cells; TLR, Toll-like receptor; Tm, memory T cells; T_n, naïve T cells; T_p, precursor T cells. The model is based primarily on studies in knockout mice. (A, The figure is modified from a version available at https://elifesciences.org/articles/32532/figures)

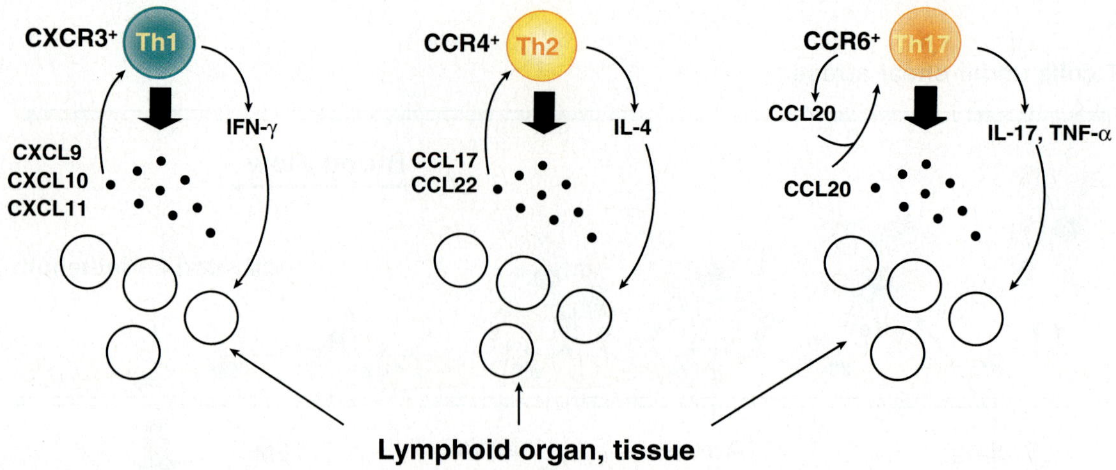

FIG. 11.8. Chemokine-cytokine feedback loops in T helper cell polarization. Dots denote gradients of the chemokines indicated to the *left*. *Thick arrows* denote the direction of T helper cell movement. *Thin arrows* denote release of the indicated factors from the indicated cell and action of the indicated factors on the indicated cell. (Courtesy of Joshua Farber and Howard Zhang (NIAID).)

the C tail by G protein-coupled receptor kinases or PKC and PKA, followed by β-arrestin binding to the tail and internalization of the receptor by clathrin-dependent and –independent mechanisms.[64] After internalization, the receptor may be degraded or stripped of ligand and recycled to the plasma membrane. The precise biologic importance of receptor downregulation has been most clearly revealed by the primary immunodeficiency disorder WHIM syndrome,[65] in which CXCR4 mutations that truncate the C tail and prevent receptor phosphorylation and internalization result in increased CXCL12-dependent signaling, distorted panleukocyte trafficking itineraries and distribution, and increased susceptibility to infections, as described in more detail in the section on disease.

Chemokine secretion mechanisms may vary dramatically for different chemokines. CXCL4 is stored in platelet α granules, whereas the highly-related CXCL4L1 is continuously produced and released. CCL5 is released from T cell granules preassociated with GAGs. CCL3 is released multidirectionally from activated T cells, presumably to create chemotactic gradients, unlike IL-2 and IFN-γ which are secreted directly into the immunological synapse. It is also important to point out that immune responses may represent a balance between mutually antagonistic effector and regulatory cells which may be colocalized and controlled in part by the same or different chemokines. For example, CXCR3, CCR6, and CCR8 are expressed on Tregs but can also be expressed on effector subsets of Th1, Th17, and Th2 cells, respectively.

For the next two sections, Figure 11.7B provides an overview of chemokine receptors that regulate movement of specific leukocyte subsets between compartments of the hematopoietic system during immune system development and during the immune response. Table 11.4 contains a representative list of phenotypes observed in chemokine receptor knockout mice.

CHEMOKINE REGULATION OF HEMATOPOIESIS

Regulation of Hematopoiesis in Bone Marrow

Originally identified as a stromal cell–derived pre-B cell stimulatory factor that also enhanced hematopoietic progenitor cell colony formation ex vivo, the homeostatic chemokine CXCL12 is expressed in many cell types, but is particularly abundant in bone marrow, where its production is under the control of G-CSF and adrenergic neurotransmission, among other factors.[66,67]

Loss of Cxcl12 or its receptor Cxcr4 in knockout mice results in the same dramatic and fatal multisystem developmental phenotype: ventricular septal defect, defective arterial patterning in multiple organs including stomach, kidney and lung, defective cerebellar development, defective myogenesis and limb innervation, and severe defects in bone marrow myelopoiesis and B cell lymphopoiesis.[68] The phenotype is consistent with the widespread constitutive expression of both CXCL12 and CXCR4 acting as a monogamous signaling system. In bone marrow, leukocyte CXCR4 activation by CXCL12 produced by various stromal cells, including eponymous CXCL12-associated reticular (CAR) cells, osterix-expressing cells and nestin-expressing cells, is important for development of distinct hematopoietic stem cell niches and both the B lymphoid and myeloid lineages.[69] It is also important for localizing HSCs to the niches and maintaining them in a quiescent state.

The balance of CXCR2 and CXCR4 signaling is thought to be particularly important in regulating neutrophil

TABLE 11.4 Major Immunological Phenotypes in Chemokine Receptor Knockout Mice

Receptor	Knockout Phenotypes
Cxcr1[a,b,c,d]	*Increased susceptibility*: systemic *Candida albicans* and pulmonary *Pseudomonas* infection
Cxcr2[a,b,c,d]	Abnormal *development:* Microbiome-dependent extramedullary myelopoiesis causing splenomegaly, lymphadenopathy, and massively increased PMN and B cell levels in blood; BM myeloid hyperplasia (grossly white marrow); oligodendrocyte precursor proliferation, migration, and positioning in white matter *Increased susceptibility: HSV* keratitis*; T. gondii; C. albicans;* brain abscess; *O. volvulus;* pulmonary infection by *Nocardia asteroides* and *Aspergillus* adenovirus; and *E. coli* pyelonephritis *Decreased:* wound healing; *A. fumigatus* and *RSV*-induced AHR*; both* urate crystal- and K/BxN serum transfer-induced synovitis; pI:pC-induced neutrophilic peritonitis; pulmonary TB immunopathology; CLP septic injury; autoantibody and Lyme-mediated arthritis; DSS-induced colitis; NKT cell trafficking to spleen and Treg induction in oral Ova tolerance protocol; atherosclerosis
Cxcr3[a,b,c,d]	*Abnormal development:* Impaired plasma cell differentiation in germinal centers *Increased susceptibility*: MTb; neurotropic mouse hepatitis virus; fatal pulmonary fibroproliferation after bleomycin injury related to deficient NK cell and $CD8^+$ T cell IFNγ *Decreased:* cardiac allograft rejection; Th1 and Th17 nephritis in MRL/lpr lupus model; TNF^+ T and Th17 cell trafficking from gut to bone marrow in hyperparathyroidism-induced bone loss; T cell–dependent postischemia neovascularization; protection from fatal memory response to CNS LCMV challenge; *Plasmodium berghei* cerebral malaria symptoms; *T. gondii* pathology; HSV-1 ocular pathology and encephalitis; HSV-2-induced vaginal immunopathology; IL-10 null-dependent inflammatory bowel disease; autoimmune diabetes; lupus immunopathology; contact hypersensitivity
Cxcr4[a,b,e]	*Abnormal development:* PERINATAL LETHAL; ventricular septal defect; gastric vascularization; cerebellar granule cell migration; bone marrow myelopoiesis, B cell lymphopoiesis, thymopoiesis
Cxcr5[a,b,c]	*Abnormal development:* Few Peyer patches; no inguinal LN; defective germinal center formation and B cell homing in secondary lymphoid tissue *Decreased:* chronic LCMV infection
Cxcr6[a,c]	*Abnormal development:* reduction of $CD1d^+$ NKT cells in the liver and lung *Increased susceptibility*: Metastasis of LLC and B16 tumor cells to liver *Decreased*: T cell–dependent hepatitis and fibrosis models
Ccr1[a]	*Increased susceptibility to: T. gondii,* Pneumonia virus of mice, Th1-dependent glomerular injury in nephrotoxic nephritis model *Decreased*: Neutrophilic alveolitis in pancreatitis-ARDS model*;* airway remodeling and Th2 response in *A. fumigatus* model; Th1-dependent EAE model*;* Th1 response to PPD*;* cardiac allograft rejection; Th2-dependent Schistosome egg antigen–induced granuloma formation in liver; serum transfer arthritis
Ccr2[a,b,c,d,e]	*Abnormal development:* Monocytopenia*;* enhanced myeloid progenitor cell cycling, and apoptosis *Increased susceptibility: Cryptococcus neoformans*; mouse hepatitis virus*; Listeria monocytogenes; Leishmania major; M. tuberculosis;* AHR to OVA and to *Aspergillus fumigatus;* glomerulonephritis in antiglomerular basement membrane antibody model *Decreased susceptibility: L. donovani, Influenza A;* EAE*;* DVT resolution*;* atherosclerosis*;* cardiac allograft rejection*;* AHR to CRA*;* dextran sulfate-mediated colitis*;* thioglycolate-induced peritonitis; FITC- and bleomycin-induced pulmonary fibrosis; neuropathic pain; DTH and granuloma formation to *M. bovis* PPD and Con A challenge in lung and liver, respectively*;* MRL/lpr lupus nephritis model; diabetes model; wound healing; ischemia-reperfusion injury; collagen-induced arthritis; osteoporosis; age-related macular degeneration
Ccr3[a]	*Increased:* AHR and airway mast cell accumulation with decreased airway eos after *ip OVA sensitization→OVA challenge* *Decreased:* allergic skin inflammation and AHR after epicutaneous OVA sensitization→OVA challenge (Eos and mast cells not recruited to skin or lung); IL-13-induced pulmonary eosinophilia and remodeling
Ccr4[a]	*Increased susceptibility:* lps; colitis in adoptive T cell transfer IBD model *Decreased:* Th2 response after lung challenge in primed mice with both iv PPD (Th1 response) and iv SEA-coated beads; contact hypersensitivity; tolerance induction
Ccr5[a,c,d]	*Increased susceptibility: L. monocytogenes*; encephalomyocarditis virus; genital herpes simplex; *C. neoformans; T. gondii;* West Nile virus*; Influenza* A*; L. major;* DTH; humoral response to T dependent Ag; ischemia-reperfusion injury *Decreased susceptibility: L. donovani*; hypoxia-induced pulmonary hypertension; dextran sulfate–mediated colitis; lps toxicity; mouse hepatitis virus–induced demyelination; cardiac allograft rejection; transplant-associated arteriosclerosis; atherosclerosis; acute GVHD; insulin resistance, glucose intolerance, and hepatic steatosis induced by high-fat feeding; cerebral ischemia challenge
Ccr6[a,c]	*Abnormal development:* Defective M cell differentiation over Peyer patches and of isolated lymphoid follicles in the intestine; increased $CD4^+$ and $CD8^+$ $\alpha\beta$-TCR T cells in intestinal mucosa; abnormal osteogenesis and decreased bone mass *Increased:* contact hypersensitivity to 2,4-dinitrofluorobenzene; IBD, EAE *Decreased:* production of IgA against cholera toxin and commensal bacteria; humoral immune response to oral- but not subQ-administered antigen and to rotavirus; CRA-induced allergic airway inflammation*;* DTH to allogeneic splenocytes; IL-23-induced psoriasoform lesions in skin; EAE; arthritis; acute gvhd

(Continued)

TABLE 11.4 Major Immunological Phenotypes in Chemokine Receptor Knockout Mice (Continued)

Receptor	Knockout Phenotypes
Ccr7[a,c,e]	*Abnormal development:* undeveloped T cell zones in 2° lymphoid tissue; abnormal negative selection in thymus *Increased:* autoimmunity *Decreased:* humoral responses; contact sensitivity; DTH; renal fibrosis and fibrocyte accumulation in Unilateral Ureteral Obstruction model; lymphocyte and DC migration to LN; normal EAE but T cell MOG responses occur in spleen, not lymph node
Ccr8[a,c]	*Decreased:* Th2-dependent SEA-induced granuloma formation and OVA- and CRA-induced allergic airway inflammation; GVHD and EAE suppression by Tregs; allergen-induced skin CD301b$^+$ DC migration from subcapsular sinus to LN parenchyma and IL-5-producing Th2 cell migration to skin in an atopic dermatitis model
Ccr9[a,c]	*Abnormal development:* preproB cells, but normal T and B cells; abnormal ratio of gut intraepithelial T-cell-to-epithelial cell ratio due to decreased γδ T cells *Increased:* Ventricular remodeling after myocardial infarction *Decreased:* CIA; OVA-induced acute allergic airway inflammation
Ccr10[a,c]	*Abnormal development:* Defective localization and accumulation of IgA antibody-secreting cells (ASC) to lactating mammary gland, and possibly colon; low IgA in milk and feces; aberrant skin-seeding and dermal localization by skin intraepithelial Vγ3$^+$ T lymphocytes (sIELs) precursors; anosmia *Increased:* susceptibility to *Salmonella enterica* serovar Typhimurium
Xcr1[a,c,f]	*Abnormal development:* deficient in intraepithelial and lamina propria T cells. *Increased* susceptibility: chemically induced colitis; *Listeria monocytogenes* infection *Decreased:* CD8$^+$ cytotoxic T cell responses to antigens cross-presented by CD8α+ DCs, including dsDNA and *L. monocytogenes*
Cx3cr1[a,b,c,d,e,f]	*Increased:* susceptibility to *Salmonella* and vaccinia infection; EAE severity; microglia-dependent neurotoxicity; age-related macular degeneration; IBD; central pain; allergic airway disease *Decreased:* cardiac allograft rejection in subtherapeutic cyclosporin A-treated animals; atherogenesis in ApoE ko mice; inflammation in a focal cerebral ischemia model and in a mouse model of Alzheimer disease; peripheral pain
Ackr1[a,b,c]	*Abnormal development:* defective balance and behavior; defective myelopoiesis *Increased:* lps-induced inflammatory infiltrates in lung and liver *Decreased:* neutrophil recruitment in acute inflammation models; macrophage infiltration in a bone fracture model; atherogenesis in the Apo E ko model; renal ischemia after lps challenge
Ackr2[a]	*Abnormal development:* inflammation-associated miscarriage *Increased:* gut, skin, liver, and lung inflammation in various models; granulomatous inflammatory responses; tissue damage after myocardial infarction; susceptibility to *Mycobacterium tuberculosis* infection
Ackr3[a,b,c]	*Abnormal development:* PERINATAL LETHAL; ventricular septal defects and semilunar heart valve malformation; less nerve-associated vessel branches in developing limb skin; abnormal positioning of cortical hem and Pallium-Subpallium boundary-derived Cajal-Retzius cells
Ackr4[a,c]	*Increased:* EAE severity; excessive Th17 responses

AHR, airway hyperreactivity; CRA, cockroach antigen; DC, dendritic cell; DTH, delayed type hypersensitivity; EAE, experimental autoimmune encephalomyelitis; LN, lymph node; OVA, ovalbumin; SEA, Schistosoma mansoni soluble egg antigen.
[a-f] Mice are available with genetically modified alleles for the indicated mouse receptor, including:
[a]straight knockout allele;
[b]floxed allele for cre-dependent conditional deletion;
[c]reporter allele (eg, LacZ, GFP);
[d]human wild-type gene knockin allele;
[e]human disease gene knockin allele; and;
[f]cre knockin fate mapping allele.
Details for each allele for each receptor are available at www.informatics.jax.org.

distribution between blood and bone marrow.[70] Immature neutrophils express high levels of CXCR4 and low levels of CXCR2. As they mature, there is a reciprocal switch in expression of these two receptors and the cells leave the marrow in response in part to reduced CXCR4 signaling in response to CXCL12 and increased CXCR2 signaling induced by oppositely oriented gradients of CXCL1 produced by bone marrow microvascular endothelial cells. Consistent with this, HSCs and neutrophils can both be rapidly mobilized pharmacologically by G-CSF, which acutely induces CXCL12 degradation by inducing release of neutrophil elastase, or by CXCR4 antagonists and CXCR2 agonists. Mice genetically deficient in either Cxcl12 or Cxcr4 both display defective bone marrow myelopoiesis, and *Cxcr2* knockout mice present with resting peripheral blood and bone marrow neutrophilia and impaired acute neutrophil migration into tissue under inflammatory conditions. Conversely, gain-of-function *CXCR4* mutations and loss-of-function *CXCR2* mutations, as occur naturally in patients with WHIM syndrome, both cause myelokathexis, a form of severe congenital neutropenia due to exaggerated retention in bone marrow of mature neutrophils. Cxcr2-dependent myelosuppression revealed by the phenotype of *Cxcr2* knockout mice may be important for opposing overstimulation of hematopoiesis induced by environmental flora, since this phenotype is absent in mice derived in germ-free conditions.[71] These mice also develop lymph node B cell hyperplasia, which may occur by an indirect mechanism. CXCR4 pairs antagonistically with CCR2 instead of CXCR2 signaling in control of monocyte release from bone marrow to blood.[72] Thus, both WHIM patients and Ccr2 or Ccl2 knockout mice may have severe monocytopenia.

CXCL12 binds to CXCR4 with high affinity but with even higher affinity to ACKR3,[27] yet these receptors are very distantly related structurally (Fig. 11.3) and their distribution and biological roles are quite different. Although ACKR3 may shape CXCL12 gradients for presentation to and activation of CXCR4 in select contexts, ACKR3 is expressed in a more restricted and often independent manner, including on marginal zone B cells, on some T cell subsets, and on endothelial cells of all four cardiac valves. The latter feature is dramatically revealed by the phenotype of *Ackr3* knockout mice, which display high perinatal mortality due to atresia of all four valves.[28] Mice deficient in the Ackr3 ligands Cxcl12 and Cxcl11 do not display the same cardiac phenotype, and Cxcr4 is not expressed on cardiac valves. Instead, there is indirect evidence for Ackr3 interaction with adrenomedullin, which is also important for cardiac valve development.[73] ACKR3 is also able to modulate cell survival, adhesion, and tumorigenesis[74] and may function as a chemokine scavenger *in vitro*.

Regulation of Thymopoiesis

During development T cell precursors must home from bone marrow to the thymus and then differentiate in sequential steps as they traffic from the thymic cortex to the medulla. CCR9 and its ligand CCL25 are highly expressed constitutively in thymus and appear to be important in thymic homing, yet Ccr9 knockout mice compensate at the level of differentiation and lack an overt defect in T cell development. CCL25 is expressed by medullary DCs and both cortical and medullary epithelial cells, and CCR9 is expressed on the majority of immature $CD4^+CD8^+$ thymocytes, but is downregulated during transition to the $CD4^+$ or $CD8^+$ single-positive stage. Just before thymic egress, thymocytes become CCR9 negative and upregulate L-selectin. Transition from $CD4^+CD8^+$ thymocytes in the cortex to CD4 or CD8 single positive thymocytes in the medulla is also associated with upregulation of CCR4 and CCR7, receptors for CCL22, and CCL19 and CCL21, respectively, which are expressed on endothelial cells of medullary venules. Studies with knockout mice have suggested that CCR7-dependent cortex to medulla migration of positively selected thymocytes may be essential for establishing central tolerance but not for maturation or export of thymocytes *in vivo*.[75]

CHEMOKINE REGULATION OF THE IMMUNE RESPONSE

Innate Immunity

Platelet-Derived Chemokines

Made primarily during platelet development, stored in platelet α granules, and rapidly released during platelet degranulation, CXCL4, CXCL4L1, and CXCL7 may be among the first chemokines to appear at sites of tissue injury and infection, particularly when there is hemorrhage and vascular damage, and may reach high concentrations.[76,77] CXCL4 aggregates to form tetramers critical for binding to chondroitin sulfate proteoglycans. In contrast, CXCL7 is activated by sequential proteolysis of its N terminus.

The prepropeptide form, named platelet basic protein (92 aa), is trimmed during platelet maturation to produce the 85-aa major stored form, named connective tissue activating peptide-III (CTAP-III), which is inactive on neutrophils. CTAP-III is further processed during degranulation to 81-aa β-thromboglobulin, also inactive. This is then cleaved by a cell surface–bound, cathepsin G–like enzyme on neutrophils to form 70-aa CXCL7, which has high homology to CXCL4 (70% aa identity). Thus, CXCL7 may function as an immediate-early mediator of neutrophil recruitment released from platelets at sites of inflammation.

Although it is not a prominent leukocyte chemoattractant and does not induce degranulation of neutrophil lysosomal enzymes, CXCL4 is able to induce secondary granule exocytosis and release matrix-degrading enzymes, which may facilitate neutrophil penetration of infected or injured tissues. Both CXCL4 and CXCL4L1, the product of the highly homologous nonallelic variant gene of CXCL4, *PF4V1*, are strong angiostatic factors.[78] In contrast, another platelet-derived chemokine, CXCL12, may be a major determinant of revascularization, as suggested by an ischemic hindlimb mouse model, where its release could be regulated by Kit ligand and thrombopoietin and it functioned to recruit $CXCR4^+$ $VEGFR^+$ hemangiocytes to the site of injury.[79] The ultimate response in inflammation and revascularization may depend on the relative concentration and activities of these counterregulatory chemokines.

Neutrophil-Targeted Chemokines

All seven ELR^+ CXC chemokines preferentially recruit neutrophils *in vitro* by binding to CXCR2. One of these, CXCL8, is an equipotent agonist at CXCR1, which is coexpressed at similar levels on neutrophils; CXCL6 is a CXCR1 agonist with lower potency. The ELR^+ CXC chemokines are rapidly inducible but may differ biologically owing to temporal and spatial differences in expression, which may provide a mechanism for graded navigation of neutrophils through tissue. Blocking studies in multiple animal models have demonstrated the importance of CXCL8 and CXCR2 in neutrophil accumulation in response to infectious and noninfectious stimuli. The ultimate outcome on survival depends on the organism. For example, in a high-dose pulmonary TB challenge model, there is increased survival in both *Cxcl5*$^{-/-}$ and *Cxcr2*$^{-/-}$ mice, the result of decreased neutrophil-mediated immunopathology driven by TLR2-induced airway epithelial cell production of Cxcl5 acting at Cxcr2 to recruit neutrophils, whereas with acute bacterial and fungal infections, mortality is increased in the absence of neutrophil Cxcr2.[80]

The role of CXCR1 *in vivo* has not yet been clearly defined and is difficult to analyze since its major human agonist, CXCL8, is not found in the mouse and its minor agonist Cxcl5 (previously named LIX, a mouse counterpart of human CXCL6) has very low potency. The discovery of mouse Cxcr1 was delayed for over a decade owing to its lower homology to Cxcr2 than the homology difference between human CXCR1 and CXCR2 (Fig. 11.3). Although CXCR1 can mediate chemotactic responses *in vitro*, there is no evidence to date that mouse Cxcr1 mediates leukocyte trafficking *in vivo*. Instead, a *Cxcr1* knockout mouse has revealed a role in neutrophil antimicrobial responses in models of *Candida albicans* infection in the kidney and *Pseudomonas aeruginosa* infection in the lung. CXCR1-specific antagonists have not been studied in clinical trials.[81,82]

In a human blister model, endogenous CXCL8 peaks at ~24 hours, whereas C5a and leukotriene B4, which also recruit neutrophils, appear earlier. Thus, the primary role of CXCL8 may be to amplify recruitment of neutrophils initiated by other chemoattractants. Intradermal injection of CXCL8 in man causes rapid (<30 minutes) and selective accumulation of large numbers of neutrophils in perivascular regions of the skin without causing edema. Tissue-specific transgenic overexpression of mouse CXCL8 paralogs Cxcl1 (KC) and Cxcl2 (MIP-2) results in a similar picture, suggesting that *in vivo* these factors may recruit cells but do not independently activate cytotoxic mechanisms. *In vitro*, the ability of CXC chemokines to activate neutrophil cytotoxic responses is modest. Instead, they may prolong survival of neutrophils in the setting of bacterial infection and paradoxically provide an intracellular niche in which at least some bacterial species, such as *S. aureus*, may persist.[83] Studies in mice have revealed that αβ TCR$^+$CD4$^+$ T cells expressing IL-17 are critical in providing CXC chemokines that foster neutrophil responses to bacterial infection. In addition, pathogens such as *Streptococci* may actively interfere with the action of CXC chemokines by degrading them with specific proteases.[84] Thus, although there is no question that neutrophils are critical in defense against bacterial and fungal pathogens and that CXCR2 agonists are critical for neutrophil trafficking in the setting of infectious disease, the ultimate outcome of infection may be determined by a balance between CXCR2-dependent neutrophil recruitment, niche development, and negative regulatory mechanisms.

CXCL1, 2, 3, 7, and 8 have also been reported to induce basophil chemotaxis and histamine release *in vitro*, which together with other factors such as complement-derived anaphylatoxins may promote vasodilatation during early stages of the innate immune response. Some CC chemokines can also activate human neutrophils, but this typically requires prestimulation of the cells with IFN-γ and other factors able to induce expression of appropriate receptors. Conversely, late apoptotic neutrophils have been reported to actively promote the resolution phase of acute inflammation by terminating chemokine signaling through scavenging by CCR5, which is expressed in a manner dependent on caspases and anti-inflammatory lipid mediators. Many other types of leukocytes and nonhematopoietic cells have also been reported to express CXCR1 and CXCR2 (endothelial cells, epithelial cells, neurons, mesenchymal stem cells), consistent with roles for CXCR2 in diverse pathologies, including infectious disease, cardiovascular disease, cancer, central nervous system pathologies, and pain regulation, as well as in morphogenesis. In mouse, Ccr1 is an important neutrophil chemotactic receptor and has been demonstrated to function in a sequential manner with the LTB4 receptor Blt1, the C5a receptor, and Cxcr2 for neutrophil recruitment in a mouse serum transfer model of acute arthritis.[47]

Monocyte/Macrophage-Targeted Chemokines

Circulating blood monocytes can be divided into two principal subsets with distinct migratory properties based on chemokine receptor expression.[85] Inflammatory monocytes are short-lived CX3CR1^{int}CCR2$^+$Ly6C$^+$ cells poised to be recruited to inflamed tissues in the mouse that correspond to human CD14$^+$CD16$^+$ and CD14$^+$CD16$^-$ monocytes. Patrolling monocytes are long-lived CX3CR1^{hi}CCR2$^-$Ly6C$^-$ cells in the mouse, characterized by CX3CR1-dependent engraftment and recruitment to noninflamed tissues by adhering to and crawling on the luminal surface of endothelium. These cells correspond to human CD14^{dim}CD16$^+$ monocytes, which are CX3CR1^{hi} but do not express inflammatory chemokine receptors. Inflammatory monocytes can return to bone marrow and convert into patrolling monocytes. Both subsets can differentiate into DCs *in vivo* that stimulate naïve T cells.

There are numerous examples of the functional importance of these receptors in the monocyte/macrophage lineage *in vivo*. As previously mentioned, monocyte Ccr2 is required for normal bone marrow egress of monocytes to blood under homeostatic conditions in mice, as well as after infection with pathogens such as *Listeria* or West Nile virus.[72,86,87] Monocytes express additional inflammatory chemokine receptors including CCR1 and CCR5, and in some contexts CXCR1 and CXCR2. Differentiation *in vitro* to macrophages is associated with downregulation of CCR2 and induction of CCR5 and CX3CR1 in specific subsets. Polarized M1 and M2 subsets of macrophages have been described in keeping with the Th1/Th2 polarization paradigm for T cells. In this regard, the classic anti-inflammatory Th2 cytokine Il-10 is able to downregulate signaling by inflammatory chemokine receptors on macrophages.[88] CXCL14 is a highly selective chemoattractant for monocytes and DCs and may be important for their positioning in lung and certain tumors.[89] Breeding programs designed to produce Cxcl14 knockout mice have been reported to result in lower-than-expected Mendelian proportions, small size and significant perinatal lethality for the knockouts; however, the mechanisms have not been established.

Innate Lymphoid Cell-Targeted Chemokines

Human NK cell subsets express unique repertoires of chemokine receptors.[90] The $CD56^{dim}CD16^{+}$ subset, which is associated with high cytotoxic capacity and low cytokine production, expresses primarily CXCR1 and CX3CR1, whereas the minor subset of $CD56^{bright}CD16^{dim}$ cells, which produce large amounts of cytokines but have low killing capacity, preferentially express CCR7. The exact profile of chemokine receptor expression can be modulated by adherence and stimulation *ex vivo* with IL-2. Cognate chemokines chemoattract NK cells and promote degranulation and killing. The importance of chemokines in NK cell function *in vivo* is well illustrated by mouse cytomegalovirus (MCMV) infection, a cause of hepatitis. MCMV induces CCL3 production in the liver, which is required for recruitment of NK cells. NK cells are the major source of IFN-γ in this model, and IFN-γ induces CXCL9, which is required for protection. Thus a cytokine to chemokine to cytokine cascade is required for NK cell–mediated host defense against this pathogen.[91] Precursors of type 1, 2, and 3 subsets of innate lymphoid cells typically express CCR7 but undergo a receptor switch to prominent expression of CXCR3, CCR4, and CCR6, respectively, aligning with chemokine receptor expression on the corresponding subsets of effector T cells. More specifically, CCR6 marks the LTi subset of ILC3 cells in the gut.

Dendritic Cells and Transition to the Adaptive Immune Response

Transition from innate to adaptive phases of the immune response involves antigen uptake by antigen-presenting cells, especially DCs, mediated by Fc and complement phagocytic receptors, as well as pattern recognition receptors (PRRs), including DC-SIGN and Toll-like receptors (TLRs). Through specific PRR ligands, pathogens may shape the nature and magnitude of the immune response in a specific direction, in part by inducing production of specific sets of chemokines with distinct leukocyte specificities.[92,93]

Chemokine receptor expression and function depend not only on the nature of the inflammatory stimulus but also on the type of DC. For example, blood-derived plasmacytoid and myeloid DCs express a similar repertoire of inflammatory chemoattractant receptors, but they may be functional only on myeloid DCs. CCL3, CCL4, and CCL5 may be particularly important for recruiting additional mononuclear phagocytes and DCs to sites of infection. This can amplify the late stage of the innate immune response and in the extreme may be involved in supporting the pathogenesis of endotoxic shock. Consistent with this idea, genetic disruption of the CCL3/CCL4/CCL5 receptor CCR5 renders mice relatively resistant to LPS-induced endotoxemia. $CCR4^{-/-}$ mice are also resistant to endotoxic shock, and this is associated with reduced macrophage extravasation to the peritoneum and production of the CCR4 ligand CCL17, which can be produced by DCs.[94] CX3CR1 also regulates location of myeloid DCs in Peyer patches and may promote dendrite protrusion into the lumen as a method of antigen sampling. Consistent with this, CX3CR1 is important for controlling *Salmonella* infection after oral challenge.[52]

Adaptive Immunity

Afferent Trafficking to Secondary Lymphoid Tissue

The homeostatic receptors CXCR5 and CCR7 and their ligands are important regulators of lymphoid development and the adaptive immune response.[95-98] DC maturation in peripheral tissues is associated with downregulation of inflammatory chemokine receptors such as CCR6, which may be important for recruitment, migration, and retention in the periphery, and reciprocal upregulation of CCR7, which mediates mature DC and T cell migration via afferent lymphatics to draining lymph nodes in response to CCL21 expressed on lymphatic endothelial cells. CCR7 in coordination with L-selectin also mediates both naïve and central memory T cell entry from blood into lymph node responding to CCL21 produced by endothelial cells in high endothelial venules.

Ccr7$^{-/-}$ mice and the *plt* mouse, which is naturally deficient in Ccl19 and the Ccl21 isoforms expressed in secondary lymphoid organs (see Fig. 11.2B for *plt* locus description), have similar phenotypes: atrophic T cell zones with few naïve T cells caused by defective homing. DC trafficking to draining lymph nodes is also defective. Phenotypes in these mice include impaired control of various pathogens, as well as impaired contact sensitivity, delayed type hypersensitivity, and antibody production. Yet, paradoxically, they also develop spontaneous autoimmune pathologies, which may be explained by impaired negative selection in the thymus, where Ccl21 is expressed selectively in the medulla.

CXCR5 is expressed on all peripheral blood and tonsillar B cells and a fraction of bone marrow B cells, but not plasma cells. Its ligand CXCL13 is also produced by HEV endothelial cells and is constitutively displayed on follicular HEV. The CXCL13/CXCR5 axis supports ~50% of the signaling required for B cell entry from blood to Peyer patch; CXCR4 may also contribute. In *Cxcr5*$^{-/-}$ mice B cells do not migrate to lymph nodes, which lack germinal centers. Peyer patches are also abnormal, and inguinal lymph nodes are completely absent. CXCL13 is also required for B1 cell homing, natural antibody production, body cavity immunity, and lymphoneogenesis in the setting of autoimmunity. *Cxcr5*$^{-/-}$ mice still can produce antibody, perhaps in part because B cells and follicular DCs are able to form ectopic germinal centers within T zones of the periarteriolar lymphocyte sheath of spleen. B cell entry into secondary lymphoid tissue is also dependent on CCR7 and CXCR4 signaling.

CXCR5 and CCR7 are probably not the only chemokine receptors responsible for afferent trafficking of leukocytes to lymph node. CCL9 has been reported to mediate monocyte homing. In addition, inflamed peripheral tissues may exert "remote control" for monocyte homing to draining lymph nodes from the blood by "projecting" their local chemokine profile to HEVs of the draining lymph node. Migration of T cells to splenic red pulp may involve local production of CXCL16. NK-T cells, and activated $CD4^+$ and $CD8^+$ T cells are found in this area and express the CXCL16 receptor CXCR6. CXCL16 is also made by DCs in the T zone, and CXCR6 is also found on intraepithelial lymphocytes. Thus, CXCL16 may function in T cell–DC interactions and in regulating movements of activated T cells in the splenic red pulp and in peripheral tissues. In addition, in allergen-induced inflammation in the skin, Ccl8 produced by macrophages in lymph node is important for $CD301b^+$ DC trafficking into the draining lymph node parenchyma by activating Ccr8 on DCs that have been prestaged in the subcapsular sinus by Ccr7 signaling.[99]

Migration Within Lymph Node Microenvironments

CXCR5 and CCR7 also appear to be important for lymphocyte movement within the lymphoid microenvironment. CXCR5 expression is common on, in addition to B cells, lymph node but not blood $CD4^+$ T cells. Follicular help T cells (T_{FH}), a $CD57^+$ subset of $CD4^+$ T cells that provides help for B cell maturation and antibody production, lack CCR7 but express CXCR5, which appears to facilitate the migration of these cells from the T cell zone to the follicles following activation by CXCL13 made in follicular stromal cells. Reciprocally, B cells activated by antigen in the follicles upregulate CCR7 and move toward the T cell zone in response to CCL21 produced by fibroblastic reticular cells. CCL21 is immobilized on the reticular network by GAG interaction with its extended C-terminal domain. The chemokine domain appears to be cleaved by a protease from the GAG-binding domain. CCL19, the other CCR7 ligand, is produced by adjacent stromal cells. It lacks a C-terminal domain extension and may preferentially function chemotactically as a soluble ligand, not haptotactically bound to surfaces. Thus, B-T_{FH} interaction may be facilitated by reciprocal movement of these cells, which may be determined in part by the balance of CXCR5 and CCR7 ligands made in adjacent lymphoid zones. DCs appear to swarm to T cell zones in lymph node by using immobilized CCL21 for adhesive random migration and soluble CCL21 and CCL19 for chemotactic steering, both acting through CCR7 activation.[100,101] There is also evidence that T cell zone chemokines such as CCL21 are bound to the surface of lymph node DCs *in vivo* and function to capture and prime naïve T cells for activation by peptide-MHC. Thus, T cells may be costimulated "in trans" and sequentially after initial engagement with their chemokine-rich environment.[102]

CXCR4 signaling may also be important in naïve and memory B cell trafficking to germinal centers. An active chemokine-dependent process has also been identified for $CD4^+$ T cell–dependent activation of rare antigen-specific naïve $CD8^+$ T cells.[103] After immunization in a TCR-transgenic mouse model, but before antigen recognition, naïve $CD8^+$ T cells in immunogen-draining lymph nodes upregulate CCR5 permitting the cells to home to sites of $CD4^+$ T cell–DC interaction in lymph nodes, where CCL3 and CCL4 are produced. Interference with this process may inhibit antigen-specific memory $CD8^+$ T cell generation.

CXCR3 signaling may also contribute to T cell trafficking within secondary lymphoid tissue. In mouse, Cxcr3 ligands are upregulated in lymph node in the context of both infection and immunization and promote trafficking of recently activated $CD4^+$ T cells and central memory $CD4^+$ T cells out of the T cell zone and into the interfollicular and medullary regions where they encounter antigen-activated innate immune cells. In human, CXCR3 ligands are expressed in activated tonsil and lymph node around and within germinal centers, where they colocalize with $CXCR3^+CD4^+$ T cells. In spleen, Cxcr3 promotes $CD8^+$ T cell encounter with antigen and inflammatory cytokines after both vaccinia and lymphocytic choriomeningitis infection, resulting in efficient generation of effector and memory $CD8^+$ T cells.[104]

Efferent Trafficking

Critical determinants of lymphocyte egress from lymph node are sphingosine-1-phosphate and its receptor S1P.[95,105] Naïve lymphocytes that do not encounter antigen continue to recirculate between the blood and secondary lymphoid tissue without acquiring any tissue-specific homing properties. In contrast, most antigen-stimulated T cells die by apoptosis. The survivors may be divided into functionally distinct subsets that preferentially express certain chemokine receptors. Within the $CD4^+$ subpopulation, multiple memory subsets and two primary effector subsets can be distinguished. The memory subsets include T_{FH} cells, described previously, and effector memory (T_{EM}) and central memory cells (T_{CM}).[106] Memory cells classically express CD45RO. In addition to this, classic T_{CM} cells also express L-selectin (CD62L) and CCR7, which are thought to mediate homing of the cells from blood to secondary lymphoid organs across HEV; these cells are not polarized and lack immediate effector function. Instead, they are thought to interact with DCs in lymph node and differentiate into T_{EM} upon secondary stimulation. T_{EM} cells are L-selectin^{lo} and lack CCR7 and were originally proposed to differentiate from T_{CM} and to traffic through peripheral tissues as immune surveillance cells, rapidly releasing cytokines in response to activation by recall antigens.

It is increasingly clear that the original T_{CM}/T_{EM} paradigm requires revision, in part because $CCR7^+$ effector memory cells have been detected in inflamed tissues and can produce effector cytokines and because CCR7 facilitates their exit into lymphatics.[107-110] Also, rather than a dualism with a single precursor-product relationship, analysis of freshly isolated human peripheral blood $CD4^+$ T cells has suggested that memory cell differentiation may be a continuum with multiple branch points arising probabilistically, giving rise to many subpopulations that preferentially express inflammatory chemokine receptors.[111]

Upon activation, the classic polarized effector subsets, Th1 and Th2, downregulate CCR7 and upregulate inflammatory chemokine receptors.[112] This switch facilitates exit from lymph node via efferent lymphatics and homing to inflamed sites. *In vivo*, Th1 cells, which by the simplest definition secrete IFN-γ but not IL-4 and control cellular and humoral immunity to intracellular pathogens, more frequently express CXCR3, CXCR6, CCR2, CCR5, and CX3CR1 than Th2 cells. In contrast, Th2 cells, which express IL-4 but not IFN-γ and are associated with cellular and humoral immunity to extracellular pathogens and allergic inflammation, more frequently express CCR3, CCR4, and CCR8 than Th1 cells. Furthermore, all Th1 cells express CXCR3, all Th2 cells express CCR4, and all Th17 cells express CCR6. These three relationships are unique among chemokine receptors and these subsets, justifying their status as signature chemokine receptors for these subsets. Chemokine receptor expression can be used to further divide effector T cells into additional subsets that differentially accumulate at pathologic sites depending on the irritant and context. For example, in a skin model of atopic dermatitis, Ccl8 but not Ccl1 signaling through Ccr8 is important for recruitment of $Ccr8^+$ Th2 cells preferentially producing large amounts of IL-5, whereas few $Ccr4^+$ Th2 cells preferentially producing IL-4 accumulate in the model.[18]

CXCR3 expression is dependent on T-bet expression and has been most consistently associated with type 1 immune responses and diseases. Consistent with this, its agonists CXCL9-11 are highly induced by IFN-γ but not by IL-4. Thus, in type 1 immunity, there is the potential for a positive feedback loop in which IFN-γ induces production of CXCL9-11, which then recruit $CXCR3^+$ Th1 cells that produce IFN-γ (Fig. 11.7). The "type 1 chemokines" may also help maintain type 1 dominance in part through their ability to block CCR3. Specific cytokines, microenvironments, and inflammatory stresses may differentially regulate CXCL9-11 expression, which may account for specialized biological roles. For example, although all three chemokines are highly inducible by IFN-γ, CXCL10 and CXCL11, but not CXCL9, are also inducible by type I interferon.

Conversely, research on chemokine expression and targets has led to a model of type 2 immunity in which IL-4 and IL-13 made at inflamed sites in the periphery may induce production of CCL7, CCL11, and other CCR3 ligands; the CCR4 ligands CCL17 and CCL22; and CCR8 ligands, including CCL1, Ccl8, and CCL18.[30] CCR3 is expressed on a subset of Th2-polarized T lymphocytes as well as on eosinophils and basophils, the three major cell types associated with Th2-type allergic inflammation, and Th2 cells are associated with CCR4 expression. The CCR3 ligand CCL11 but not CCL24 is also important during development for homeostatic trafficking of eosinophils to the gastrointestinal tract and spleen.[113]

Arrival of Th2 cells amplifies a positive feedback loop through secretion of additional IL-4. Moreover, CCL7 and CCL11 may block Th1 responses by antagonizing CCR2, CXCR3, and CCR5. Many chemokine receptors have been identified on $CD4^+$ regulatory T cells, including CCR4, CCR5, and CCR8, and differential expression of chemokine receptors may functionally distinguish T_{reg} subsets subspecialized to differentially suppress Th1, Th2, and Th17 responses. CCR5 has been shown to play an important role in homing of Tregs to skin during *Leishmania major* infection, favoring pathogen persistence.[114] CCR8 has been reported to be important in the maintenance of T_{regs} and to attenuate experimental autoimmune encephalomyelitis (EAE) pathology.[115]

All Th17 cells express CCR6 controlled by the signature Th17 transcription factor Rorγt, and CCR6 may also participate in a positive feedback loop with tissue cells and Th17 cells themselves expressing its ligand CCL20[116,117] (Fig. 11.8). The $CCR6^+$ subset of human $CD4^+$ T cells is much larger, however, and includes IL-22-producing IL-17-negative skin-homing memory T cells and IL-10-producing, autoreactive memory T cells with context-dependent regulatory function.[118,119]

T Cell Differentiation and Activation

Some chemokines appear to regulate not just trafficking but also T cell polarization and activation at least in certain contexts. CCL2 and its receptor CCR2 have been most extensively studied in this regard, but the results are complex and appear at first glance contradictory since *in vivo* CCR2 is strongly associated with Th1 immune responses and CCL2 may be associated with Th2 responses.[120] CCL2 appears to promote Th2 polarization directly, by inhibiting IL-12 production in monocytes and by enhancing IL-4 but not IFN-γ production in memory and activated T cells. Thus, CCL2 influences both innate immunity, through effects on monocyte trafficking, and adaptive immunity, through control of T helper cell trafficking and potentially T cell polarization. In the case of aerosol challenge with *M. tuberculosis* in $Ccr2^{-/-}$ mice, DC migration to draining lymph node is markedly impaired, which preempts any direct effects of CCL2 on T cells.

Xcr1 and its ligand Xcl1 provide an excellent example of how chemokines are able to regulate T cell activation, specifically, cytotoxic immune responses to cross-presented antigen, as revealed by studies using knockout, reporter, and floxed alleles.[121-123] Xcr1 has the narrowest known leukocyte distribution of any chemokine GPCR. After conjugate formation with antigen-cross-presenting $Xcr1^+CD8\alpha^+$ DCs, $CD8^+$ T cells secrete large amounts of Xcl1, which induces DC migration and survival and differentiation of antigen-specific $CD8^+$ T cells able to make IFN-γ. $Xcr1^+CD8\alpha^+$ DCs represent <1% of splenic DCs, yet they are required for optimal $CD8^+$ T cell responses to both dsRNA and *Listeria monocytogenes* infection. Intravital two-photon imaging showed that, after soluble antigen immunization, $CD8^+$ T cells preferred to form conjugates in skin draining lymph nodes with newly arriving migratory $Xcr1^+CD8\alpha^+$ DCs rather than resident lymph node cross-presenting $Xcr1^+$ DCs.

T cell–derived Xcl1 is also important for intestinal $Xcr1^+$ DC activation, migration, and support for T cell survival and function. In thymus, Xcl1 is expressed selectively by medullary thymic epithelial cells and Xcr1 is expressed by thymic DCs. Accordingly, Xcl1 knockout mice have reduced accumulation of thymic DCs in the medulla, which results in impaired thymic generation of naturally occurring regulatory T cells.

Consistent with the importance of XCL1 and XCR1 in cross-presentation, several viruses produce factors that modulate XCR1 signaling, possibly as an immune evasion mechanism. Examples include the broad-spectrum chemokine-binding protein M3 from mouse γ-herpesvirus MHV68 (Fig. 11.4D), which scavenges Xcl1, and two chemokines produced by HHV8, vCCL2, and vCCL3, which are nonselective and selective antagonists at XCR1, respectively.

Tissue-Specific Lymphocyte Homing

CLA+ T lymphocytes, which home to skin, preferentially express CCR4 and CCR10.[124] The CCR4 ligand CCL22 is made by resident dermal macrophages and DCs, whereas the CCR10 ligand CCL27 is made by keratinocytes. Blocking both of these pathways, but not either one alone, has been reported to inhibit lymphocyte recruitment to the skin in a DTH model, implying that, in this model, these two molecules act redundantly and independently of inflammatory chemokines.[125]

Homing to small intestine is determined in part by T lymphocyte expression of the integrin $\alpha_4\beta_7$ and CCR9. The $\alpha_4\beta_7$ ligand MAdCAM-1 and the CCR9 ligand CCL25 colocalize on normal and inflamed small intestinal endothelium, and most T cells in the intraepithelial and lamina propria zones of the small intestine express CCR9. Intraepithelial lymphocytes, which are mainly $\gamma\delta$ TCR$^+$ or CD8 $\alpha\alpha^+$ $\alpha\beta$TCR$^+$, are reduced in small intestine from *Ccr9*$^{-/-}$ mice.[126] CCL20 and its receptor CCR6 are also important in gut mucosal immune responses. CCL20 is constitutively expressed in follicle-associated epithelial cells (FAEs) overlying GALT inductive sites (Peyer patches and isolated lymphoid follicles), and it can also be induced in GALT under inflammatory conditions. It is important for GALT development and for B and DC homing to GALT. CCR6$^+$ DCs recruited to the dome region of Peyer patches upon invasion of FAE by an enteric pathogen have been shown to mediate rapid activation of pathogen-specific T cells.[127]

As B cells differentiate into plasma cells, they also downregulate CXCR5 and CCR7 and exit the lymph node. B immunoblasts expressing IgG coordinately upregulate CXCR4, which promotes homing to the bone marrow, whereas B immunoblasts expressing IgA specifically migrate to mucosal sites.[128] Like gut-homing T cells, gut-homing IgA-secreting B immunoblasts express $\alpha_4\beta_7$ integrin and CCR9 and respond to CCL25. Intestinal DCs may help shape mucosal immunity by generating gut-homing IgA-secreting B cells through the synergistic action of several mediators, including retinoic acid, IL-5, and IL-6.[129,130] Ccr10 knockout mice have defective localization and accumulation of IgA antibody–secreting cells to lactating mammary gland, but not to the gastrointestinal tract, as well as aberrant skin-seeding and dermal localization by skin intraepithelial Vγ3$^+$ T lymphocytes (sIELs) precursors.[131,132]

CHEMOKINES AND DISEASE

There is a vast literature on chemokine and chemokine receptor expression in disease, and the functional importance has been broadly investigated using knockout, reporter, and floxed mouse alleles as well as by administering specific blocking agents (Table 11.4). At least one knockout allele has been created for every chemokine receptor and for most chemokines (the full list and official allele designations are archived in the Mouse Genome Initiative at www.informatics.jax.org). This section will highlight diseases with some of the strongest evidence in humans for prominent chemokine involvement. (Table 11.5)

HIV Infection

HIV infection requires fusion of the viral envelope to a target cell membrane, a process that is initiated by direct binding of the HIV envelope glycoprotein gp120 to both CD4 and a coreceptor on the target cell.[133-135] Although many chemokine receptors and a few related GPCRs can function as coreceptors *in vitro*, CCR5 and CXCR4 appear to be the most relevant ones on primary cells and the most important *in vivo* (Fig. 11.9). Their interaction with CD4 may be constitutive and stabilized by gp120. The usage of these receptors is so important that HIV strains are now functionally classified and named according to their specificity for CXCR4 (X4 strains), CCR5 (R5 strains), or both (R5X4 strains).

The importance of CCR5 in clinical HIV/AIDS was established at the population level by the seminal discovery that *CCR5 Δ32*, a nonfunctional allele founded more than 5000 years ago in Caucasians and now present in ~20% of North American Caucasians, is a strong AIDS restriction factor, strongly protecting homozygotes from initial infection and substantially slowing the rate of progression to AIDS in infected heterozygotes. Additional genetic risk factors affecting the rate of HIV disease progression have been identified in the CCR5 signaling system, including SNPs in the promoters for *CCR5* and its ligand *CCL5*, variation in the gene copy number of the CCR5 ligand CCL3L1, and expression of a long noncoding CCR5-specific RNA.[136-138] Molecular epidemiologic studies do not support positive selective pressure as the mechanism responsible for fixation of *CCR5 Δ32* in Caucasians.[139] *CCR5 Δ32* homozygotes appear healthy, as do unstressed CCR5 knockout mice.

These observations provided proof of principle for the central role of CCR5 in HIV transmission and progression as well as for development and FDA approval in 2007 of the allosteric CCR5 modulator maraviroc (Selzentry, Pfizer) as a component of HAART therapy in HIV/AIDS.[140,141] Additional evidence was provided by two HIV$^+$ patients known as the Berlin patient and the London patient, who were both treated successfully with HAART but developed acute leukemia. Both patients were fortuitously transplanted after cytoreductive therapy with bone marrow from *CCR5 Δ32* homozygous donors and subsequently remained well off antiretroviral therapy with undetectable virus, interpreted as a functional cure.[142,143] The Berlin patient, Timothy Brown, unfortunately died of leukemia relapse but with continued undetectable HIV 13 years after the original transplant. This has fortified efforts to apply gene editing methods to inactivate CCR5 specifically in infected patient T cells as a gene therapy cure strategy. In this regard, major controversy erupted in 2019 over the announcement by Chinese investigators that germline gene editing using CRISPR-Cas9 was used to successfully generate humans in China with CCR5 mutations. None of the mutations succeeded in establishing a CCR5 null phenotype.

CCR5 is also important in the neuropathogenesis of West Nile virus (WNV) infection in humans and in a mouse model of the disease, but unlike HIV, it plays a protective role by facilitating antiviral leukocyte trafficking to the brain.[144,145] Pathogenesis in the mouse model is also regulated by neuronal CXCL10 through recruitment of CD8+ T cells to the brain.[146] CXCL10 also controls pathogenesis in a mouse model of the related flavivirus Dengue virus, both by regulating T cell recruitment through CXCR3 and by competitively inhibiting viral binding to cell surface heparan sulfate.[147]

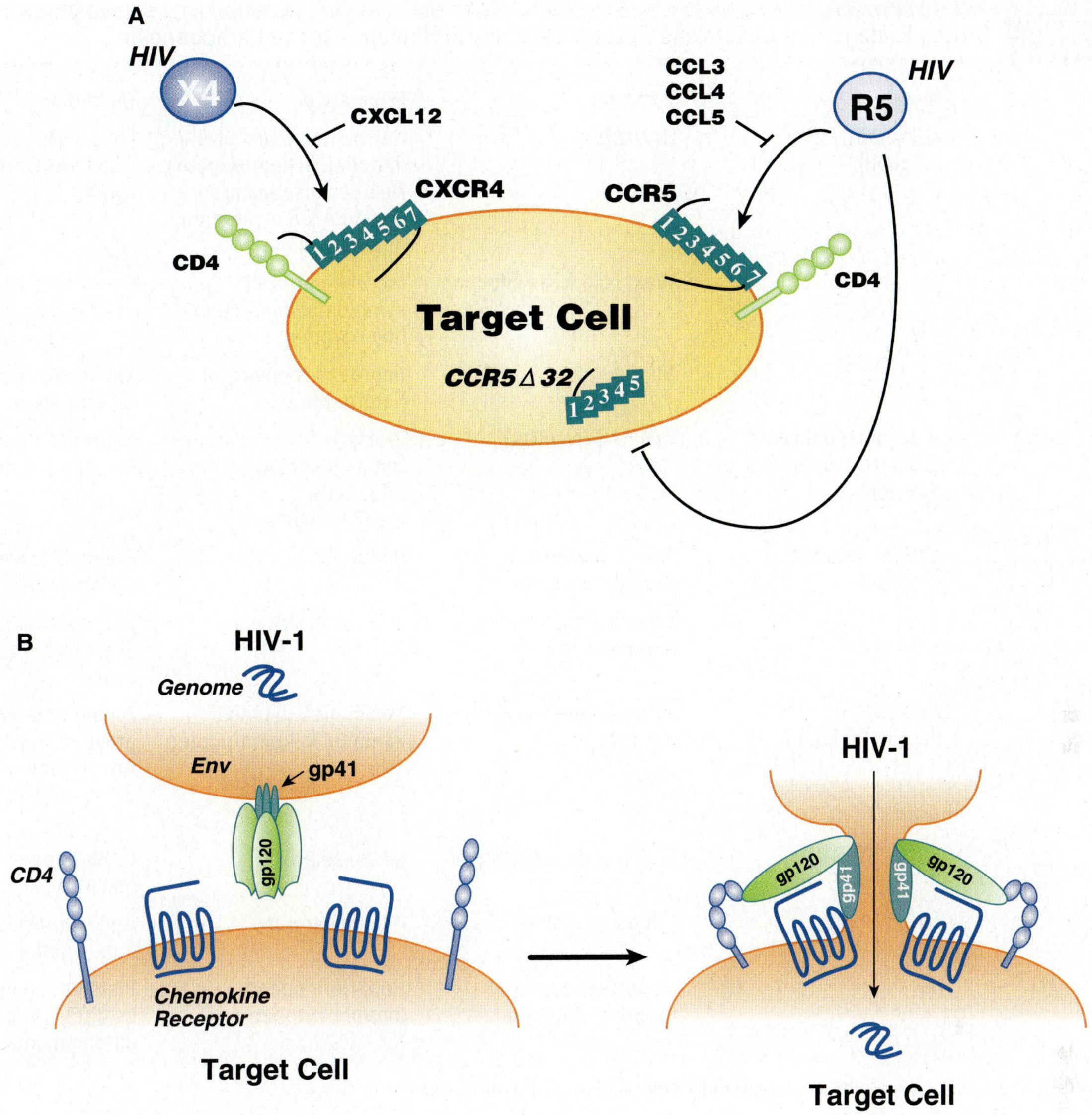

FIG. 11.9. Model of HIV coreceptor activity. A: HIV cytotropism is explained by differential usage of CXCR4 and CCR5 as coreceptors with CD4. Some viruses can use both CCR5 and CXCR4 (not shown). The mutant receptor CCR5Δ32 is not expressed on the cell surface and cannot be used for cell entry by HIV. **B:** Chemokine receptors cooperating with CD4 in a proposed mechanism of HIV entry.

X4 HIV strains typically do not transmit disease but instead appear to evolve from R5 quasispecies in a minority of patients late in infection during the transition to clinical disease. Consistent with this, these strains appear more virulent than R5 strains *in vitro*. The reason is not fully defined; however, it is important to note that the percentage of $CD4^+$ T cells expressing CXCR4 in vagina, blood, and rectum exceeds that expressing CCR5. Moreover, the X4 envelope protein expressed on $CD4^+$ T cells has been shown to induce autophagic programmed cell death of uninfected bystander cells by binding to CXCR4 on the cell membrane, signaling in a non-G_i-, non-caspase 3-dependent manner.[148] An unresolved question is why X4 strains do not readily transmit disease, given the broad expression of CXCR4 on macrophages and $CD4^+$ T cells.

The bicyclam plerixafor (Mozobil, AMD3100; Sanofi-Genzyme) was originally discovered by scientists at the Rega Institute in Leuven as an impurity in a cyclam chemical library that inhibited HIV replication.[149] It was later found to work at the level of virus entry by binding to CXCR4. Plerixafor was effective in phase 1 clinical trials for HIV but was abandoned because (1) X4 HIV strains are less prevalent than R5 strains and do not initiate infection, (2) the high doses needed for efficacy resulted in significant cardiac arrhythmias, and (3) the drug is not orally bioavailable and has a short half-life. Nevertheless, plerixafor was found to also have remarkable panleukocyte mobilization activity by reversibly blocking binding of the N terminus of CXCL12 to the receptor, preventing receptor activation, and was therefore repurposed in combination with G-CSF to enhance hematopoietic stem cell mobilization and collection for transplantation after cytoreductive therapy in multiple myeloma and non-Hodgkin lymphoma and was approved in 2008.[150]

TABLE 11.5 Strong Evidence for Chemokine System Causality in Human Disease Pathogenesis

Molecule	Mutation	Disease	Phenotype	Mechanism
CCR5	*CCR5Δ32* (loss of function)	HIV/AIDS	Resistance to initial HIV infection in homozygotes Delayed disease progression to AIDS in infected heterozygotes	Loss of HIV coreceptor CCR5 prevents cell entry by R5-tropic HIV
		West Nile virus infection	Increased risk of symptomatic infection in homozygotes	Reduced leukocyte trafficking to brain
		Stroke	Improved recovery in heterozygotes	Enhanced neuronal regeneration
CXCR4	Autosomal dominant C-tail mutations (gain of function)	WHIM SYNDROME	Warts Hypoγglobulinemia Infections Myelokathexis	Retention of mature leukocytes in bone marrow
CX3CR1	*CX3CR1-M280* (loss of function)	Atherosclerotic cardiovascular disease	Reduced risk	Reduced foam cell accumulation
		Age-related macular degeneration	Increased risk	Microglial cell accumulation in the subretinal space
ACKR1	*FYB(ES)* (loss of function)	*Plasmodium vivax* malaria	Resistance to initial infection in homozygotes	Promoter mutation prevents erythrocyte ACKR1 expression, which is required for cell entry by *P. vivax*
CCL26	CCL26 GG SNP	Eosinophilic esophagitis	Increased risk	CCR3-dependent eosinophil trafficking
HHV8 ORF74	na	Kaposi sarcoma	Kaposi sarcoma	Angiogenesis Inflammation
CXCL4	na	Heparin-induced thrombocytopenia	Heparin-induced thrombocytopenia	Platelet depletion by CXCL4-heparin autoantibodies

na, not applicable; HHV8, human herpesvirus 8; ORF, open reading frame; SNP, single nucleotide polymorphism.

Malaria

Just as CCR5 and CXCR4 are exploited by HIV to infect white blood cells, the promiscuous chemokine-binding and nonsignaling 7TM protein ACKR1 is exploited by *Plasmodium vivax* to infect red blood cells, causing a major form of malaria. ACKR1 is the only chemokine receptor expressed on red blood cells. It was originally discovered in the 1950s as a minor blood group antigen designated Duffy and was renamed DARC (Duffy antigen receptor for chemokines) in the 1990s before receiving a standardized ACKR designation by the International Union of Pharmacology in 2014.

The parasite ligand, which is named the *Plasmodium vivax* Duffy Binding Protein (PvDBP), is expressed in micronemes of merozoites and binds to the N-terminal domain of ACKR1 via a cysteine-rich domain.[151] This interaction is required for junction formation during invasion but not for initial binding or parasite orientation. ACKR1 deficiency, which is caused by an inherited homozygous single-nucleotide substitution named *FYB(ES)* of T to C at nucleotide *-67* in the promoter of the *FYB* allele of ACKR1, affecting the binding site of the erythroid-specific transcription factor GATA-1,[152] is fixed in sub-Saharan Africa but not in other malaria-endemic regions of the world. Accordingly, *P. vivax* malaria is rare in sub-Saharan Africa but common in Central and South America, India, and Southeast Asia. Fixation of the mutation in Africa presumably occurred because of positive selective pressure from malaria. Together with *CCR5 Δ32*, ACKR1 *FYB(ES)* is the strongest genetic resistance factor known for any infectious disease in man, although the identification of some infected ACKR1-negative individuals in several countries, including Madagascar and Papua New Guinea, indicates that it is not absolutely required for infection.[153]

Chemokine control of *P. falciparum* malaria is less well defined. *In vitro*, platelets are toxic to the parasite and the toxic factor is CXCL4, which can be internalized by binding to ACKR1 and thereby delivered to the parasite. Also *in vitro*, endothelial cell CX3CL1 is one of several endothelial surface molecules that can bind to parasite-infected erythrocytes.[154] Two 25-kDa parasite proteins bind to CX3CL1 and have been designated CBP1 and CBP2 (CX3CL1-binding proteins 1 and 2). These proteins have a unique structure but mimic CX3CR1 sufficiently to have been identified with anti-CX3CR1 antibodies.[155]

Conversely, the G glycoprotein of respiratory syncytial virus has been reported to function as a CX3CL1 mimic binding to CX3CR1 to induce chemotaxis and possibly to promote viral entry.[156] Extending this theme, human cytomegalovirus encodes US28, a promiscuous receptor for multiple inflammatory CC chemokines and CX3CL1, which the virus may use to mediate attachment to target cells.[157] Finally, two classical chemoattractant receptors, the platelet-activating factor receptor and FPR1, are exploited for target cell interaction by *Streptococcus pneumoniae* and the causative agent of plague, *Yersinia pestis*, respectively.[158,159] Together, these examples illustrate the rich dialogue between hosts and pathogens, using the chemoattractant system as a very popular method of communication.

ACKR1 is also expressed on vascular endothelial cells, where it mediates chemokine presentation and transcytosis, and on Purkinje cells in the cerebellum, where it affects balance, activity, and behavior in mice. *FYB(ES)* does not affect ACKR1 expression on these cells. Normal red cell expression of ACKR1 is thought to provide for inflammatory chemokine scavenging to negatively regulate inflammatory responses and to maintain tissue gradients of chemokines. A genome-wide association study (GWAS) has strongly linked ACKR1 deficiency to benign neutropenia, which is common in individuals of African ancestry.[160,161] The mechanism may involve ACKR1 expression on erythroblasts, which appears to be important for normal bone marrow myelopoiesis in mice.[162]

WHIM Syndrome

WHIM syndrome (*w*arts, *h*ypogammaglobulinemia, *i*nfection and *m*yelokathexis) is a rare combined primary immunodeficiency disorder almost always caused by autosomal dominant inheritance of gain-of-function mutations truncating various portions of the C tail of CXCR4, a domain that is 100% conserved from birds to man.[163,164] This causes exaggeration of CXCR4's normal neutrophil and HSC retention function in bone marrow niches resulting in myelokathexis (neutropenia without myeloid maturation arrest due to retention in the bone marrow of dysmorphic hyperlobulated and hypervacuolized neutrophils), which is almost pathognomonic for the disease. CXCR4 gain of function appears to result from enhanced receptor dimerization and G protein and β-arrestin signaling associated with impaired receptor desensitization and internalization.[65] One WHIM mutation, E343K, simply changes a single charge in this domain. Two patients with myelokathexis were found to have loss-of-function mutations in CXCR2, highlighting the reciprocal relationship between CXCR2 and CXCR4 in regulation of neutrophil release from bone marrow to blood.

Patients with WHIM syndrome often present with incomplete penetrance of wart and hypogammaglobulinemia phenotypes and may also present with cardiovascular and urogenital developmental abnormalities. Panleukopenia is common, but infections most commonly recur in sinopulmonary and skin sites without becoming invasive or life-threatening and are mainly caused by pyogenic bacteria typically controlled by neutrophils and antibody, not organisms typically associated with T cell deficiency, with the intriguing and eponymous exception of HPV, the cause of warts. This may be caused by virus-induced CXCR4 expression and signaling directly in keratinocytes.[165] Neutrophils may be mobilized in the setting of infection by unknown mechanisms, which may delay diagnosis for years or decades. Recurrent infections may result in chronic end organ damage, but childhood mortality is rare. The few reported deaths in adults have mostly been related to EBV^+ lymphomas and HPV^+ squamous cell cancer.

Administration of the CXCR4 antagonist plerixafor results in mobilization of almost all leukocyte subpopulations, but not erythrocytes or platelets, to the blood in healthy human subjects as well as in patients with WHIM syndrome, and in multiyear open-label studies of three patients who were intolerant of G-CSF has demonstrated preliminary evidence of safety and sustained clinical efficacy at very low doses given twice daily subcutaneously in reducing wart burden and infection frequency.[166] A phase 3 study of 19 patients in WHIM syndrome of plerixafor versus G-CSF, the standard of care for severe congenital neutropenia, was completed in 2020 with results pending. The goal of therapy with plerixafor is to restore signaling to normal, not to completely block CXCR4, which would heighten safety concerns. Its chronic use in WHIM syndrome is limited by a very short half-life, parenteral administration, high cost, and potential risks from partial agonist activity at the second CXCL12 receptor ACKR3. These concerns are addressed by the more potent and selective oral agent mavorixafor (X4-Pharma), which has demonstrated acceptable safety and preliminary evidence of efficacy in eight patients with WHIM syndrome in a phase 2 clinical trial and has now entered phase 3.[167]

Nine patients with WHIM syndrome have been cured by allotransplantation, and one other patient was spontaneously cured by a somatic mutation known as chromothripsis (chromosome shattering), which deleted the disease copy of *CXCR4* and 163 other genes on chromosome 2 in a single HSC.[168] Remarkably, this $CXCR4^{+/o}$ HSC evidently acquired a selective advantage and repopulated the bone marrow, resulting in durable correction of neutropenia and monocytopenia (but not lymphopenia or hypogammaglobulinemia), complete disappearance of warts, and cessation of recurrent bacterial infections. $CXCR4^{+/o}$ mouse HSCs have enhanced proliferative capacity *in vivo* and phenocopy the engraftment advantage in the patient, suggesting that selective WHIM allele silencing by gene editing in patient HSCs is a preferred gene therapy cure strategy for the disease.

Kaposi Sarcoma

A full discussion of the many virally encoded chemokines and chemokine receptors is beyond the scope of this overview but can be found in recent reviews.[34,169,170] The human γ herpesvirus HHV8 deserves special mention, however. HHV8 encodes three CC chemokines, vMIP-I, II, and III (systematically renamed vCCL1, vCCL2, and vCCL3), as well as a constitutively active CC/CXC chemokine receptor homolog of CXCR2 named vGPCR, encoded by ORF74.[61,171] All of these factors are angiogenic and may contribute to the pathogenesis of Kaposi sarcoma (KS), a highly vascular multicentric nonclonal tumor caused by HHV8 typically in the setting of immunosuppression such as in HIV/AIDS. Consistent with this, vGPCR induces KS-like tumors when expressed constitutively or inducibly on endothelial cells in transgenic mice.[172] The mechanism may involve signaling through diverse pathways including the nonconventional G protein Gq and phosphorylation of tuberin, which promotes activation of mTOR, a rapamycin sensitive target, through direct and paracrine mechanisms. The latter may include activation of NF-κB and induction of angiogenic factors such as VEGF and proinflammatory chemokines and cytokines. The receptor is constitutively active, and the activity can be further increased by its human chemokine agonists and decreased by inverse chemokine agonists. Thus, this virus appears to have converted a hijacked host receptor into a regulator of gene expression. The model identifies rapamycin as a candidate therapeutic in KS.

ORF74 is conserved in multiple other γ herpesviruses, including the nonhuman primate *Herpesvirus saimiri* where it has been shown to signal in response to multiple CXC chemokines. A second viral chemokine receptor, US28 of human cytomegalovirus, has also been extensively studied and shown to induce tumors when overexpressed in the colon. Remarkably, chimpanzee cytomegalovirus has five US28 homologs.

COVID-19

Although there is currently no strong evidence that specific chemokines and chemokine receptors determine outcomes in COVID-19 pathogenesis, some progress has been made in prioritizing candidates for detailed study. For example, chromosome 3p21.31, a region containing the genes *SLC6A20*, *LZTFL1*, and *FYCO1* along with the chemokine receptor genes *CCR9*, *CXCR6*, and *XCR1*, was identified as the major locus linked to severe COVID-19 disease by a GWAS.[173] Moreover, analysis of serum samples from patients with COVID-19 has revealed high levels of many inflammatory chemokines, including CXCL8, CXCL9, CXCL10, CCL2, CCL3, CCL5, CCL7, and CCL20. CXCL10 is particularly strongly associated with severe/critical disease. In patients presenting with mild to moderate disease severity, high CCL17 levels at presentation were predictive of a more benign outcome, whereas low levels were strongly predictive of progression to severe/critical disease.[174]

Inflammatory chemokine levels could be increased even in asymptomatic individuals infected with SARS-CoV-2, and in all patient groups levels typically returned to normal after convalescence and were only weakly correlated with viral load.[175] The CCR5 blocking monoclonal antibody leronlimab has been entered into clinical trials for the treatment of hospitalized patients with COVID-19.

Atherosclerosis

Multiple chemokine system components have been implicated in atherosclerosis. For CX3CR1, the evidence in man is based in part on consistent association of a defective receptor variant named CX3CR1 M280, which replaces threonine with methionine at position 280, with reduced risk of cardiovascular disease endpoints in multiple cohort studies. Mice lacking Cx3cr1 on the $ApoE^{-/-}$ background fed a Western diet are also at reduced risk of developing vascular lesions.[176,177] The mechanism whereby CX3CR1 and CX3CL1 contribute to atherosclerosis remains unclear but may involve induction of CX3CL1 on vascular smooth muscle cells and recruitment and retention of CX3CR1-positive monocytes to plaque through the dual chemotactic and adhesive functions of this ligand-receptor pair.[178] Treatment with a CX3CR1 antagonist was able to phenocopy the protective effect of *CX3CR1* gene knockout in atherosclerosis. There is also extensive evidence for a proatherogenic role for CCL2 and its receptor CCR2.

Adoptive transfer studies of bone marrow from knockout mice have also revealed a role for CCR5 and CXCR2 in mouse models of atherosclerosis. CXCR2 may work by promoting monocyte adhesion to early atherosclerotic endothelium, through interaction with its mouse ligand KC and activation of the VLA-4/VCAM-1 adhesion system. Immunoneutralization experiments have revealed a role for CCR7 on foam cells in a mouse model of atheroregression. Surprisingly, CCR1 and CXCL16 deficiency appeared to increase inflammation in atherosclerosis models. The atheroprotective role of CXCL16 may reflect a dominant feature of its oxidized LDL scavenging activity over its leukocyte activity, since knockout of its receptor CXCR6 is protective in atherosclerosis. CXCL10 has also been implicated in atherogenesis in a mouse model, possibly acting by recruitment of $CXCR3^+$ Th1 cells. Lack of CXCL10 was also associated with a relative increase of regulatory T cells in plaque. CCR6 and ACKR1 have also been implicated as proatherogenic factors in gene knockout models.

The magnitude of the change in atherosclerosis severity in mice with single chemokine or chemokine receptor knockouts is usually about 30% to 40%. Remarkably, $Apoe^{-/-}/Ccl2^{-/-}/Cx3cr1^{-/-}$ triple-knockout mice treated with the CCR5 antagonist AOP-RANTES (modified CCL5) and fed a Western diet were almost completely protected from atherosclerosis.[179] However, there has been little effort in clinical translation so far in this space. Bindarit, a selective CCL2 antagonist, was tested for 6 months at phase 2 in preventing coronary stent restenosis following angioplasty, but the results were negative.

Age-Related Macular Degeneration

CX3CR1 is particularly important in microglial cell migration and activation in the inflamed central nervous system. In the eye, mice lacking Cx3cr1 develop macular degeneration and human age-related macular degeneration is associated with the dysfunctional mutant allele *CX3CR1-M280*.[180] The mechanism appears to involve dysregulated activation of microglial cells in the retina. CCR3 has also been detected in human AMD, and targeting Ccr3 was effective in a mouse model of AMD.[181]

There is evidence for chemokine regulation of multiple other neuropsychoimmunologic pathologies, including multiple sclerosis, Parkinson disease, Alzheimer disease, pain, behavior, and stroke recovery. With regard to stroke, experimental cerebral ischemia in mice induced Ccr5 on cortical neurons, which normally do not express the receptor. Both genetic deletion and pharmacologic blockade of Ccr5 improved neuronal regeneration after the insult, and the *CCR5Δ32* allele was associated with better functional outcomes after stroke in humans, suggesting a potential utility for repurposing CCR5 blocking agents for stroke recovery.[182]

Autoimmunity

Heparin-induced thrombocytopenia (HIT) is a human autoimmune disease directly linked mechanistically to chemokines.[41] An established risk factor for thromboembolic complications of heparin therapy, HIT occurs in 1% to 5% of patients treated with heparin and is the result of autoantibodies that bind specifically to CXCL4-heparin complexes in plasma.

In general, T cell–dependent autoimmune diseases in man, such as psoriasis, multiple sclerosis (MS), rheumatoid arthritis (RA), and type I diabetes mellitus, are associated with inflammatory chemokine expression in disease tissue and tissue infiltration by T lymphocytes and monocyte-derived cells expressing inflammatory chemokine receptors.[183] In general, the complexity of chemokine-mediated pathogenesis in autoimmunity is revealed by the many inflammatory chemokines that appear to play a role and the partial phenotypic effects typically found in inbred mouse models of disease. For example, neutrophil trafficking to synovium in a mouse model of serum transfer arthritis revealed sequential contributions of Ccr1, Cxcr2, Blt1, and the C5a receptor for cell accumulation.[47] In humans, RA has been negatively associated with homozygous *CCR5 Δ32* suggesting that CCR5 is important in pathogenesis.[184]

A particularly well-studied example of chemokine regulation of immunologically mediated disease is CCR6. In an IL-23-induced psoriasiform dermatitis model in mice, Ccr6 deficiency was almost completely protective, markedly attenuating epidermal trafficking of γδ-T17 cells, which is a critical pathogenic cell type in this model. In epidermis and dermis, TNF-producing and IL-1β-producing monocyte-derived DCs are critical in the model, including a population of inflammatory Langerhans cells, and are dependent on Ccr6 expression for their accumulation in skin in the model.[185]

The eosinophil selective CCL26/CCR3 axis has been implicated in eosinophilic esophagitis, based on high expression of CCL26 in the esophagus from patients, association of a polymorphism in *CCL26* with human disease, and resistance of *Ccr3*$^{-/-}$ mice to experimental eosinophilic esophagitis.[186] The type 2 receptor CCR8 and its ligand CCL18 are also highly expressed in disease tissue.[30]

Nevertheless, clinical translation has been very disappointing so far in autoimmune disorders, with programs launched for inflammatory chemokine receptors in MS (CCR1 antagonist BX-471; CCR2 antagonist MLN1202 and MK-0812), RA (CCR1 antagonists CP-481715, CCX354, BMS-817399, and MLN3897; CCR2 antagonists MLN1202 and MK-0812; and CCR5 antagonist maraviroc), and psoriasis (CXCR1/2 antagonist SCH 527123), all returning negative results.[187] Only the CCR9 antagonist vercirnon and the dual CXCR1/CXCR2 inhibitor reparixin reached phase 3, in inflammatory bowel disease and type 2 diabetes mellitus, respectively, before being terminated.

Acute Neutrophil-Mediated Inflammatory Disorders

Many neutrophil-mediated human diseases have been associated with the presence of CXCL8, including psoriasis, gout, acute glomerulonephritis, chronic obstructive pulmonary disease (COPD), acute respiratory distress syndrome (ARDS), and ischemia-reperfusion injury. Studies of *Cxcr2* knockout mice as well as systemic administration of neutralizing anti-CXCL8 antibodies have shown protective phenotypes in diverse preclinical models of neutrophil-mediated acute inflammation, providing proof of concept that CXCL8 is a nonredundant mediator of innate immunity and acute pathologic inflammation in these settings.[188,189] Together the results identify CXCL8 and its receptors as candidate drug targets for diseases mediated by acute neutrophilic inflammation. However, clinical trials with CXCR1, CXCR2, CCR1, and CCR2 antagonists in COPD have not resulted in approvals.

Transplant Rejection

An advantage of transplant rejection over other animal models of human disease is that in both the human and animal situation the time of antigenic challenge is precisely known. The most extensive analysis of the role of chemokines in transplant rejection has been carried out in an MHC class I/II-mismatched cardiac allograft rejection model in the mouse, which is mediated by a Th1 immune response.[190] Similar sets of inflammatory chemokines are found in the mouse model as in the human disease. Analysis of knockout mice has demonstrated that, although multiple chemokine receptors contribute to rejection in this model, there is a marked rank order: CXCR3>>CCR5>CCR1 = CX3CR1 = CCR2. Most impressively, rejection and graft arteriosclerosis do not occur if the recipient mouse, treated with a brief, subtherapeutic course of cyclosporin A, is *Cxcr3*$^{-/-}$ or if the donor heart is *Cxcl10*$^{-/-}$, identifying this axis as a potential drug target.[191] Neutralization of Cxcl9, a Cxcr3 ligand that appears later than Cxcl10, can also prolong cardiac allograft survival and delay T cell infiltration and acute rejection in class II MHC-disparate skin allografts.[192] In man, CCR5 may be important in chronic kidney allograft rejection, since individuals homozygous for *CCR5 Δ32* are underrepresented among patients with this outcome in a large German kidney transplantation cohort.[193]

Allergic Airway Disease

Chemokine receptors associated with asthma include CXCR2, CXCR4, CCR3, CCR4, and CCR8. CCR3 is present on eosinophils, basophils, mast cells, and some Th2 T cells, and CCR4 and CCR8 identify airway T cells of allergen-challenged atopic asthmatics. Ccr8 knockout mice have reduced allergic airway inflammation in response to multiple different Th2-polarizing antigens, including *Schistosoma mansoni* soluble egg antigen, ovalbumin, and cockroach antigen.[194] A role for the CCR3 axis in asthma has been supported by Ccl11 neutralization in guinea pig and Ccr3 gene knockout in mouse. The net effect of Ccr3 knockout was expected to be more profound; however, the exact phenotype depends dramatically on the specific method of sensitization and challenge due to complex and opposite effects on eosinophil and mast cell trafficking.[195,196] CCR3 and its ligands have also been clearly shown to play a critical role in an *Aspergillus*-induced chronic allergic airway disease mouse model, in part due to failure of eosinophils to traffic to the lung.[197] CCR6 also appears to play a role, since *Ccr6*$^{-/-}$ mice have decreased allergic airway inflammation in response to sensitization and challenge with cockroach antigen, which is consistent with the induction of its ligand CCL20 in this model. The CXCL12/CXCR4 axis may be more than a homeostatic regulator since neutralization of either ligand or receptor resulted in decreased lung eosinophilia and airway hyperreactivity in a mouse model of allergic airway disease.[198] CXCR2, CCR3, and CCR4 antagonists entered into clinical trials in asthma have all failed so far.

Cancer

Diverse chemokines and chemokine receptors have been detected in many human cancers, produced by both cancer cells and stromal cells.[199] In fact, some chemokines were first discovered in tumors, and one, CXCL1, was originally named for its cancer association alternatively as MGSA, for melanoma growth-stimulatory activity, and Groα, for growth-related oncogene α. In 1993, CXCL10 was the first chemokine shown to inhibit tumorigenesis *in vivo* when injected into mice.[200] However, the paper that focused and accelerated research into chemokine regulation of cancer was not published until 2001, when Muller et al. reported that ectopic expression of CXCR4 on breast cancer cells was a key factor determining the organ-specific pattern of metastasis to lung, lymph node, bone marrow, brain, and liver, all of which are rich in the CXCR4 ligand CXCL12.[201] High expression of CXCR4 and other chemokine receptors, such as ACKR3, CCR4, CCR7, and CCR10, have been shown to correlate with poor prognosis across many cancers.

Chemokines have now been shown to have multiple functions in cancer through effects on both cancer cells and stromal cells, including at the levels of cancer stem cell biology, cancer cell proliferation and survival, immune evasion by fugetaxis, control of T cell polarization, recruitment of tumor-infiltrating leukocytes and endothelial progenitor cells, angiogenesis, epithelial-mesenchymal transition, fibrosis, and metastasis.[88] Chemokines may recruit both antitumor effector leukocytes and protumoral suppressive leukocytes, such as myeloid-derived suppressor cells, M2 macrophages, and regulatory T cells, so that the net effect of the chemokine system on tumorigenesis is complex and combinatorial, the sum of all these activities, and may differ for different tumors and different tumor stages. Thus, the effect of targeting a single chemokine or chemokine receptor in cancer is not readily predictable. The efficacy of immune checkpoint inhibitors in cancer may depend in part on chemokine recruitment of disinhibited tumor infiltrating lymphocytes. For example, CXCR4 blockade has shown preclinical promise in ovarian cancer and in a phase 2 trial in pancreatic cancer in combination with PD-1 blockade, possibly working in part by reducing Treg accumulation.[202,203] A complementary role of CXCR4 blockade may be to facilitate checkpoint inhibitors by favoring release of effector T cells from stroma to access the cancer cells.

Controlling the balance of angiogenic and angiostatic chemokines in cancer may be particularly important.[204] This has been shown in several instances, including human non-small cell lung carcinoma, in which the ratio of ELR to non-ELR CXC chemokine expression is high, and where in a *Scid* mouse model neutralization of endogenous tumor-derived CXCL8 (angiogenic) could inhibit tumor growth and metastasis by about 50% through a decrease in tumor-derived vessel density, without directly affecting tumor cell proliferation. ELR$^+$ CXC chemokines promote angiogenesis by affecting proliferation, survival, and migration of endothelial cells and the formation of capillary-like structures. Chemokine receptors on tumor cells have been shown to directly mediate chemokine-dependent proliferation, for example, in the case of CXCL1 in melanoma and metastasis; in the case of CXCR4 in a mouse model of breast cancer, prostate cancer, and glioblastoma; and in the case of CCR4 and CCR10 in melanoma.[205] Moreover, chemokines may function in cancer by reprogramming cancer-associated stromal fibroblasts to a senescent protumorigenic state, as shown in ovarian cancer for CXCL1.[206]

A large number of clinical trials have been launched in the past few years in cancer, including melanoma and pancreatic, breast, lung, and colorectal solid tumors, as well as leukemia/lymphoma, targeting various chemokine receptors (CXCR2, CXCR4, CCR2, CCR4, CCR5, and CCR7), resulting in one FDA approval so far, in 2018 for the humanized afucosylated cell-depleting anti-CCR4 monoclonal antibody Mogamulizumab-kpkc (Poteligeo; Kyowa-Kirin) on the basis of results from the pivotal phase 3 MAVORIC trial in cutaneous T cell lymphoma (CTCL) (Sezary syndrome and Mycosis fungoides), and in Japan in 2012 for adult T cell leukemia/lymphoma (ATL).[207,208] Response rates were ~20% to 37% in heavily pretreated patients, in whom the malignant T cells expressed CCR4 at high levels. The drug depletes CCR4-expressing cells by antibody-dependent cell-mediated cytotoxicity.

The most extensively studied chemokine receptor in cancer is CXCR4, which is expressed by over 20 types of cancer and has been implicated in multiple functions, including proliferation, anchorage, quiescence, radiation, and chemotherapy resistance of leukemic stem cells in bone marrow, and metastasis.[209,210] Of note, the second CXCL12 receptor ACKR3 is commonly coexpressed in CXCR4-expressing cancers and may collaborate functionally as it does in other settings. Clinical trials have been conducted to test plerixafor and other CXCR4 antagonists for mobilizing leukemia stem cells from bone marrow niches as a chemosensitization strategy in acute lymphocytic leukemia and other hematologic malignancies, so far without positive results despite preclinical promise.[211] This is an extension of plerixafor's licensed indication for combined single-dose administration with G-CSF to mobilize and collect large numbers of hematopoietic stem cells for eventual rescue transplantation after cytoreductive chemotherapy in patients with multiple myeloma and non-Hodgkin lymphoma. Balixafortide (Polyphor), a CXCR4 antagonist and derivative of a natural product isolated from horseshoe crabs, has reached phase 3 trials in metastatic breast cancer.

It should also be reemphasized here that the Mendelian disease WHIM syndrome, caused by congenital gain-of-function CXCR4 mutations, appears to increase the risk of virus-associated cancer, including HPV-associated squamous cell cancer and EBV-associated B cell lymphoma. Of interest, acquired CXCR4 mutations, including several of the same gain-of-function mutations that cause WHIM syndrome, have been found in many patients with Waldenstrom macroglobulinemia, a lymphoplasmacytic cell lymphoma in bone marrow, and have been associated with poor prognosis and poor treatment responses to ibrutinib.[212] Finally, CXCR4 ligands have been developed for positron emission tomography imaging of CXCR4-expressing cancers. One of the first examples is Cu^{64}-AMD3100 (plerixafor), which successfully imaged adrenocortical carcinoma metastasis in lung in a patient.[213]

THERAPEUTIC APPLICATIONS

Chemokines and Chemokine Receptors as Targets for Drug Development

The three FDA-approved drugs targeting the chemokine system, plerixafor for HSC mobilization, maraviroc for HIV/AIDS, and Mogamulizumab-kpkc for ATL and CTCL, are all used for indications that could not have been anticipated by the classic functional conception of the system's importance in leukocyte trafficking and inflammation. In contrast, despite major investment and 25 years of effort, there has not been success yet in therapeutically targeting the chemokine system for classic indications in acute and chronic inflammation. For these areas, despite the ability of the pharmaceutical industry to develop many orally available high potency small molecule chemokine receptor antagonists and other chemokine blocking agents, there are major obstacles to clinical application, including the potential issue of inflammatory chemokine redundancy, difficulties in optimizing bioavailability and timing in patients, and the unreliability of many preclinical disease models in predicting the best indications and targets.[187] The high target molecule promiscuity of chemokine scavengers produced naturally by viruses and ectoparasites attests to the likely need for polytargeting for chemokine blocking agents to be effective in full-blown inflammatory and autoimmune diseases where multiple chemokines and chemokine receptors are almost always in play. More tractable may be diseases like poststent arteriosclerosis, graft-versus-host disease, and transplant rejection, where the time the trigger is pulled is known exactly, allowing early intervention before the onset of full-blown inflammatory cascades.

The first wave of chemokine clinical trials focused on inflammation is ongoing but has been joined by a second wave involving a large number of trials in other areas, particularly in cancer, where chemokine receptors are now the most actively targeted class of GPCRs, which as a whole account for ~37% of all FDA-approved drugs.[209] In 2020, there were an astounding 677 registered and/or published clinical trials targeting the chemokine system, including 645 targeting chemokine receptors and only 32 trials targeting chemokines (Table 11.6). This reflects the fact that most chemokine receptors bind multiple chemokines, making it more efficient to target receptors. The only approved agent targeting a chemokine is Abcream (Anogen-Yes Biotech), a topical formulation marketed in China of a CXCL8-blocking monoclonal antibody reported to be effective in psoriasis. Of the 645 receptor trials, 496 are for only three of the receptors, CXCR4 (251), CCR5 (218), and CCR4,[27] mostly for the FDA-approved drugs and indications. Similarly, of the 59 phase 3 trials, 41 were for the approved drugs and indications, maraviroc in HIV,[30] plerixafor for stem cell mobilization,[9] and Mogamulizumab-kpkc for ATL and CTCL.[2] The rest include the CCR5 monoclonal antibody leronlimab (PRO140, Cytodyn) in HIV, CCR5+ triple-negative metastatic breast cancer, and COVID-19; the small molecule dual CCR2/CCR5 antagonist cenicriviroc in NASH (nonalcoholic steatohepatitis) and liver fibrosis, a complication of HAART treatment in HIV-infected individuals; the CXCR4 antagonists plerixafor and mavorixafor in WHIM syndrome; and plerixafor and balixafortide in advanced cancer and leukemia.

This scorecard reveals the many disappointments in targeting chronic inflammation, and also an overall success rate that exceeds the global 30% benchmark for drugs completing phase 3, and identifies many untargeted opportunities in the dark matter, perhaps the most obvious being ACKR1 in malaria. More than 20 new agents are now in trials for multiple indications, including cancer, HIV, RA, diabetes, asthma, and COPD. Attempts are also underway to repurpose the three approved drugs for multiple other indications in inflammation and cancer. However, there is still enormous unexplored space in humans, with no trials registered yet for CXCR5, CXCR6, CCR6, CCR8, CCR10, XCR1, CX3CR1, or for any of the ACKRs.

Although orally administered drugs are the main goal, other blocking strategies have also been tested, such as ribozymes, modified chemokines (eg, amino-terminally modified versions of CCL5), intrakines (which are modified forms of chemokines delivered by gene therapy that remain in the endoplasmic reticulum and block surface expression of newly synthesized receptors), monoclonal antibodies, and llama nanobodies,[214] as well as specific gene editing using zing finger nucleases, TALENs, and, most promising, CRISPR-Cas9. Another interesting strategy that takes account of the substantial redundancy built into the chemokine system involves multispecific agents. For example, dual antagonists have been developed for CXCR1/CXCR2 and CCR2/CCR5. These pairs each include receptors with very high homology, 79% and 70% identity at the amino acid level, respectively. Although broad-spectrum viral and tick-derived chemokine scavengers are biologically interesting, translation to the clinic involves major difficulties, including a limited market imposed by antibodies that are likely to be developed against these nonself proteins if administered in the setting of chronic disease. Finally, chemokine-based immunotoxins are under development, for example, in glioma using a CX3CL1 immunotoxin to target the HCMV-encoded CX3CL1 receptor US28 in virally infected tumor cells.[215,216]

TABLE 11.6 Clinical Trial Activity Targeting the Chemokine System

Target	Clinical Trial Registrations (n)				Phase 3	
	Phase			Total		
	1	2	3		Indications	Agent
Chemokine receptors						
CXCR4	127	112	12	251	HSC mobilization	Plerixafor (approved) BL-8040
					WHIM syndrome	Plerixafor Mavorixafor
					Cancer	Plerixafor Balixafortide
CCR5	121	57	39	217	HIV/AIDS	Maraviroc (approved) Leronlimab
					COVID-19	Leronlimab
					NASH liver fibrosis	Cenicriviroc
CCR4	13	12	2	27	Adult T cell leukemia/lymphoma Cutaneous T cell lymphoma	Mogamulizumab (approved)
Other CKRs	56	88	5	153		
CXCR1/2	1	7	1	9	Type 2 diabetes	Reparixin (terminated)
CCR9	2	3	4	9	Crohn disease	Vercirnon (terminated)
Receptor Trials Total	317	269	59	645		
Chemokine Trials Total	24	8	0	32	na	na
Total Registrations	341	277	59	677	9	10 (3 approvals)

Data were aggregated in July 2020 for (1) all unique registrations in www.clinicaltrials.gov returned by searching with each chemokine and chemokine receptor name and (2) published clinical trials returned by PubMed searching for each molecule and the term "clinical trial." See text for details about the status of Phase 1 and 2 clinical trials. NASH, nonalcoholic steatohepatitis. Agents highlighted in red are FDA-approved drugs.

Chemokines as Biological Response Modifiers

Both inflammatory and homeostatic chemokines are being evaluated for therapeutic potential as biological response modifiers, acting mainly as immunomodulators or as regulators of angiogenesis. Studies to date have not revealed major problems with toxicity, and efficacy has been observed in animal models of cancer, inflammation, and infection. Chemokines are also being developed as vaccine adjuvants, delivered either as pure protein, immunomodulatory plasmid, or as recombinant protein within antigen-pulsed DC. Chemokine gene administration has also been shown to induce neutralizing antibody against the encoded chemokine and has successfully blocked immune responses and improved EAE and arthritis scores in rodent models.

Many chemokines, when delivered pharmacologically as recombinant proteins or by plasmid DNA or in transfected tumor cells, are able to induce immunologically mediated antitumor effects in mouse models and could be clinically useful. Mechanisms may differ depending on the model but may involve recruitment of monocytes, NK cells, and $CD8^+$ cytotoxic T cells to tumor. Chemokines may also function as adjuvants in tumor antigen vaccines. Chemokine-tumor antigen fusion proteins represent a novel twist on this approach that facilitates uptake of tumor antigens by APCs via the normal process of ligand-receptor internalization. Non-ELR CXC chemokines such as CXCL4 also exert antitumor effects through angiostatic mechanisms.

CONCLUSION

The immune system is a system in motion, and the study of chemokines has provided major new insights into how order is maintained as cells enter a tissue, traffic to targets, and in some cases leave and recirculate. In addition, chemokines are increasingly appreciated as multifunctional cytokines, playing large roles in some cases such as CXCL12 in development, and perhaps smaller or less well-appreciated ones in immune activation processes. A major challenge for the future is to deepen basic knowledge of chemokine-mediated immunoregulation as a way to advance new therapeutics for immunologically mediated disease.

ACKNOWLEDGMENTS

This review was supported with funding from the Division of Intramural Research, National Institute of Allergy and Infectious Disease. I am grateful to Hisayuki Nomiyama of Kumamoto University for advice regarding Figure 11.2B, Steven Tsang of NIAID for creating the dendrogram in Figure 11.3, Joshua Farber and Howard Zhang of NIAID for providing Figures 11.7A and 11.8, and Joshua Farber for helpful discussions and critical review of the manuscript.

REFERENCES

1. Schiffmann E, Showell HV, Corcoran BA, Ward PA, Smith E, Becker EL. The isolation and partial characterization of neutrophil chemotactic factors from *Escherichia coli*. *J Immunol*. 1975;114(6):1831-1837.
2. Zhang M, Gao JL, Chen K, et al. A critical role of formyl peptide receptors in host defense against *Escherichia coli*. *J Immunol*. 2020;204(9):2464-2473.
3. Schroder JM, Mrowietz U, Morita E, Christophers E. Purification and partial biochemical characterization of a human monocyte-derived, neutrophil-activating peptide that lacks interleukin 1 activity. *J Immunol*. 1987;139(10):3474-3483.
4. Van Damme J, Van Beeumen J, Opdenakker G, Billiau A. A novel, NH2-terminal sequence-characterized human monokine possessing neutrophil chemotactic, skin-reactive, and granulocytosis-promoting activity. *J Exp Med*. 1988;167(4):1364-1376.
5. Walz A, Peveri P, Aschauer H, Baggiolini M. Purification and amino acid sequencing of NAF, a novel neutrophil- activating factor produced by monocytes. *Biochem Biophys Res Commun*. 1987;149(2):755-761.
6. Yoshimura T, Matsushima K, Tanaka S, et al. Purification of a human monocyte-derived neutrophil chemotactic factor that has peptide sequence similarity to other host defense cytokines. *Proc Natl Acad Sci U S A*. 1987;84(24):9233-9237.
7. Bachelerie F, Ben-Baruch A, Burkhardt AM, et al. International Union of Basic and Clinical Pharmacology. [corrected]. LXXXIX. Update on the extended family of chemokine receptors and introducing a new nomenclature for atypical chemokine receptors. *Pharmacol Rev*. 2014;66(1):1-79.
8. Miao M, De Clercq E, Li G. Clinical significance of chemokine receptor antagonists. *Expert Opin Drug Metab Toxicol*. 2020;16(1):11-30.
9. DeVries ME, Kelvin AA, Xu L, Ran L, Robinson J, Kelvin DJ. Defining the origins and evolution of the chemokine/chemokine receptor system. *J Immunol*. 2006;176(1):401-415.
10. Zhao S, Chang SL, Linderman JJ, Feng FY, Luker GD. A comprehensive analysis of CXCL12 isoforms in breast cancer(1,2). *Transl Oncol*. 2014;7(3):429-438.
11. von Hundelshausen P, Agten SM, Eckardt V, et al. Chemokine interactome mapping enables tailored intervention in acute and chronic inflammation. *Sci Transl Med*. 2017;9(384):eaah6650.
12. Wu B, Chien EY, Mol CD, et al. Structures of the CXCR4 chemokine GPCR with small-molecule and cyclic peptide antagonists. *Science*. 2010;330(6007):1066-1071.
13. Kufareva I, Gustavsson M, Zheng Y, Stephens BS, Handel TM. What do structures tell us about chemokine receptor function and antagonism? *Annu Rev Biophys*. 2017;46:175-198.
14. Qin L, Kufareva I, Holden LG, et al. Structural biology. Crystal structure of the chemokine receptor CXCR4 in complex with a viral chemokine. *Science*. 2015;347(6226):1117-1122.
15. Park SH, Das BB, Casagrande F, et al. Structure of the chemokine receptor CXCR1 in phospholipid bilayers. *Nature*. 2012;491(7426):779-783.
16. Levoye A, Balabanian K, Baleux F, Bachelerie F, Lagane B. CXCR7 heterodimerizes with CXCR4 and regulates CXCL12-mediated G protein signaling. *Blood*. 2009;113(24):6085-6093.
17. Ludeman JP, Stone MJ. The structural role of receptor tyrosine sulfation in chemokine recognition. *Br J Pharmacol*. 2014;171(5):1167-1179.
18. Islam SA, Chang DS, Colvin RA, et al. Mouse CCL8, a CCR8 agonist, promotes atopic dermatitis by recruiting IL-5+ T(H)2 cells. *Nat Immunol*. 2011;12(2):167-177.
19. Graham GJ, Handel TM, Proudfoot AEI. Leukocyte adhesion: reconceptualizing chemokine presentation by glycosaminoglycans. *Trends Immunol*. 2019;40(6):472-481.
20. Campanella GS, Grimm J, Manice LA, et al. Oligomerization of CXCL10 is necessary for endothelial cell presentation and in vivo activity. *J Immunol*. 2006;177(10):6991-6998.
21. Dyer DP, Salanga CL, Volkman BF, Kawamura T, Handel TM. The dependence of chemokine-glycosaminoglycan interactions on chemokine oligomerization. *Glycobiology*. 2016;26(3):312-326.
22. Handel TM, Johnson Z, Crown SE, Lau EK, Proudfoot AE. Regulation of protein function by glycosaminoglycans – as exemplified by chemokines. *Annu Rev Biochem*. 2005;74:385-410.
23. Jiang D, Liang J, Campanella GS, et al. Inhibition of pulmonary fibrosis in mice by CXCL10 requires glycosaminoglycan binding and syndecan-4. *J Clin Invest*. 2010;120(6):2049-2057.
24. Pruenster M, Mudde L, Bombosi P, et al. The Duffy antigen receptor for chemokines transports chemokines and supports their promigratory activity. *Nat Immunol*. 2009;10(1):101-108.
25. Graham GJ, Locati M, Mantovani A, Rot A, Thelen M. The biochemistry and biology of the atypical chemokine receptors. *Immunol Lett*. 2012;145(1-2):30-38.
26. Doitsidou M, Reichman-Fried M, Stebler J, et al. Guidance of primordial germ cell migration by the chemokine SDF-1. *Cell*. 2002;111(5):647-659.
27. Gustavsson M, Dyer DP, Zhao C, Handel TM. Kinetics of CXCL12 binding to atypical chemokine receptor 3 reveal a role for the receptor N terminus in chemokine binding. *Sci Signal*. 2019;12(598):eaaw3657.
28. Sierro F, Biben C, Martinez-Munoz L, et al. Disrupted cardiac development but normal hematopoiesis in mice deficient in the second CXCL12/SDF-1 receptor, CXCR7. *Proc Natl Acad Sci U S A*. 2007;104(37):14759-14764.
29. Chen J, Yao Y, Gong C, et al. CCL18 from tumor-associated macrophages promotes breast cancer metastasis via PITPNM3. *Cancer Cell*. 2011;19(4):541-555.
30. Islam SA, Ling MF, Leung J, Shreffler WG, Luster AD. Identification of human CCR8 as a CCL18 receptor. *J Exp Med*. 2013;210(10):1889-1898.
31. Proudfoot AE, Uguccioni M. Modulation of chemokine responses: synergy and cooperativity. *Front Immunol*. 2016;7:183.
32. Ikeda Y, Kumagai H, Skach A, Sato M, Yanagisawa M. Modulation of circadian glucocorticoid oscillation via adrenal opioid-CXCR7 signaling alters emotional behavior. *Cell*. 2013;155(6):1323-1336.
33. Meyrath M, Szpakowska M, Zeiner J, et al. The atypical chemokine receptor ACKR3/CXCR7 is a broad-spectrum scavenger for opioid peptides. *Nat Commun*. 2020;11(1):3033.
34. Pontejo SM, Murphy PM, Pease JE. Chemokine subversion by human herpesviruses. *J Innate Immun*. 2018;10(5-6):465-478.
35. Vischer HF, Siderius M, Leurs R, Smit MJ. Herpesvirus-encoded GPCRs: neglected players in inflammatory and proliferative diseases? *Nat Rev Drug Discov*. 2014;13(2):123-139.
36. Bhusal RP, Eaton JRO, Chowdhury ST, et al. Evasins: tick salivary proteins that inhibit mammalian chemokines. *Trends Biochem Sci*. 2020;45(2):108-122.
37. Spaan AN, Reyes-Robles T, Badiou C, et al. *Staphylococcus aureus* targets the duffy antigen receptor for chemokines (DARC) to lyse erythrocytes. *Cell Host Microbe*. 2015;18(3):363-370.
38. Miles TF, Spiess K, Jude KM, et al. Viral GPCR US28 can signal in response to chemokine agonists of nearly unlimited structural degeneracy. *Elife*. 2018;7:e35850.
39. Mantovani A, Garlanda C, Locati M. Macrophage diversity and polarization in atherosclerosis: a question of balance. *Arterioscler Thromb Vasc Biol*. 2009;29(10):1419-1423.
40. Jorgensen AS, Rosenkilde MM, Hjorto GM. Biased signaling of G protein-coupled receptors – from a chemokine receptor CCR7 perspective. *Gen Comp Endocrinol*. 2018;258:4-14.
41. Vandercappellen J, Van Damme J, Struyf S. The role of the CXC chemokines platelet factor-4 (CXCL4/PF-4) and its variant (CXCL4L1/PF-4var) in inflammation, angiogenesis and cancer. *Cytokine Growth Factor Rev*. 2011;22(1):1-18.
42. Castellino F, Germain RN. Chemokine-guided CD4+ T cell help enhances generation of IL-6RalphahighIL-7Ralpha high prememory CD8+ T cells. *J Immunol*. 2007;178(2):778-787.
43. Viola A, Contento RL, Molon B. T cells and their partners: the chemokine dating agency. *Trends Immunol*. 2006;27(9):421-427.
44. Butcher EC, Picker LJ. Lymphocyte homing and homeostasis. *Science*. 1996;272(5258):60-66.
45. Springer TA. Traffic signals for lymphocyte recirculation and leukocyte emigration: the multistep paradigm. *Cell*. 1994;76(2):301-314.
46. Lee CH, Zhang HH, Singh SP, et al. C/EBPdelta drives interactions between human MAIT cells and endothelial cells that are important for extravasation. *Elife*. 2018;7:e32532.
47. Chou RC, Kim ND, Sadik CD, et al. Lipid-cytokine-chemokine cascade drives neutrophil recruitment in a murine model of inflammatory arthritis. *Immunity*. 2010;33(2):266-278.
48. Cinamon G, Shinder V, Alon R. Shear forces promote lymphocyte migration across vascular endothelium bearing apical chemokines. *Nat Immunol*. 2001;2(6):515-522.
49. Girbl T, Lenn T, Perez L, et al. Distinct compartmentalization of the chemokines CXCL1 and CXCL2 and the atypical receptor ACKR1 determine discrete stages of neutrophil diapedesis. *Immunity*. 2018;49(6):1062-1076.e6.
50. Nourshargh S, Renshaw SA, Imhof BA. Reverse migration of neutrophils: where, when, how, and why? *Trends Immunol*. 2016;37(5):273-286.
51. Ley K. Arrest chemokines. *Microcirculation*. 2003;10(3-4):289-295.
52. Niess JH, Brand S, Gu X, et al. CX3CR1-mediated dendritic cell access to the intestinal lumen and bacterial clearance. *Science*. 2005;307(5707):254-258.
53. Sanchez-Sanchez N, Riol-Blanco L, Rodriguez-Fernandez JL. The multiple personalities of the chemokine receptor CCR7 in dendritic cells. *J Immunol*. 2006;176(9):5153-5159.
54. Forssmann U, Magert HJ, Adermann K, Escher SE, Forssmann WG. Hemofiltrate CC chemokines with unique biochemical properties: HCC-1/CCL14a and HCC-2/CCL15. *J Leukoc Biol*. 2001;70(3):357-366.
55. Kehrl JH. Chemoattractant receptor signaling and the control of lymphocyte migration. *Immunol Res*. 2006;34(3): 211-227.
56. Neptune ER, Iiri T, Bourne HR. Galphai is not required for chemotaxis mediated by Gi-coupled receptors. *J Biol Chem*. 1999;274(5):2824-2828.
57. Parent CA, Devreotes PN. A cell's sense of direction. *Science*. 1999;284(5415):765-770.
58. Wang F, Herzmark P, Weiner OD, Srinivasan S, Servant G, Bourne HR. Lipid products of PI(3)Ks maintain persistent cell polarity and directed motility in neutrophils. *Nat Cell Biol*. 2002;4(7):513-518.
59. Kumar A, Humphreys TD, Kremer KN, et al. CXCR4 physically associates with the T cell receptor to signal in T cells. *Immunity*. 2006;25(2):213-224.
60. Galliera E, Jala VR, Trent JO, et al. beta-Arrestin-dependent constitutive internalization of the human chemokine decoy receptor D6. *J Biol Chem*. 2004;279(24):25590-25597.
61. Arvanitakis L, Geras-Raaka E, Varma A, Gershengorn MC, Cesarman E. Human herpesvirus KSHV encodes a constitutively active G-protein- coupled receptor linked to cell proliferation. *Nature*. 1997;385(6614):347-350.
62. Singh SP, Zhang HH, Tsang H, et al. PLZF regulates CCR6 and is critical for the acquisition and maintenance of the Th17 phenotype in human cells. *J Immunol*. 2015;194(9):4350-4361.
63. Proost P, Struyf S, Van Damme J, Fiten P, Ugarte-Berzal E, Opdenakker G. Chemokine isoforms and processing in inflammation and immunity. *J Autoimmun*. 2017;85:45-57.
64. Neel NF, Schutyser E, Sai J, Fan GH, Richmond A. Chemokine receptor internalization and intracellular trafficking. *Cytokine Growth Factor Rev*. 2005;16(6):637-658.
65. Lagane B, Chow KY, Balabanian K, et al. CXCR4 dimerization and beta-arrestin-mediated signaling account for the enhanced chemotaxis to CXCL12 in WHIM syndrome. *Blood*. 2008;112(1):34-44.
66. Katayama Y, Battista M, Kao WM, et al. Signals from the sympathetic nervous system regulate hematopoietic stem cell egress from bone marrow. *Cell*. 2006;124(2):407-421.
67. Nagasawa T, Tachibana K, Kishimoto T. A novel CXC chemokine PBSF/SDF-1 and its receptor CXCR4: their functions in development, hematopoiesis and HIV infection. *Semin Immunol*. 1998;10(3):179-185.
68. Zou YR, Kottmann AH, Kuroda M, Taniuchi I, Littman DR. Function of the chemokine receptor CXCR4 in haematopoiesis and in cerebellar development. *Nature*. 1998;393(6685):595-599.
69. Hartmann TN, Grabovsky V, Pasvolsky R, et al. A crosstalk between intracellular CXCR7 and CXCR4 involved in rapid CXCL12-triggered integrin activation but not in chemokine-triggered motility of human T lymphocytes and CD34+ cells. *J Leukoc Biol*. 2008;84(4):1130-1140.
70. Murphy PM, Heusinkveld L. Multisystem multitasking by CXCL12 and its receptors CXCR4 and ACKR3. *Cytokine*. 2018;109:2-10.
71. Cacalano G, Lee J, Kikly K, et al. Neutrophil and B cell expansion in mice that lack the murine IL-8 receptor homolog. *Science*. 1994;265(5172):682-684.
72. Serbina NV, Jia T, Hohl TM, Pamer EG. Monocyte-mediated defense against microbial pathogens. *Annu Rev Immunol*. 2008;26:421-452.
73. Quinn KE, Mackie DI, Caron KM. Emerging roles of atypical chemokine receptor 3 (ACKR3) in normal development and physiology. *Cytokine*. 2018;109:17-23.
74. Burns JM, Summers BC, Wang Y, et al. A novel chemokine receptor for SDF-1 and I-TAC involved in cell survival, cell adhesion, and tumor development. *J Exp Med*. 2006;203(9):2201-2213.
75. Kurobe H, Liu C, Ueno T, et al. CCR7-dependent cortex-to-medulla migration of positively selected thymocytes is essential for establishing central tolerance. *Immunity*. 2006;24(2):165-177.
76. Brandt E, Petersen F, Ludwig A, Ehlert JE, Bock L, Flad HD. The beta-thromboglobulins and platelet factor 4: blood platelet-derived CXC chemokines with divergent roles in early neutrophil regulation. *J Leukoc Biol*. 2000;67(4):471-478.
77. Gear AR, Camerini D. Platelet chemokines and chemokine receptors: linking hemostasis, inflammation, and host defense. *Microcirculation*. 2003;10(3-4):335-350.
78. Struyf S, Burdick MD, Peeters E, et al. Platelet factor-4 variant chemokine CXCL4L1 inhibits melanoma and lung carcinoma growth and metastasis by preventing angiogenesis. *Cancer Res*. 2007;67(12):5940-5948.
79. Massberg S, Konrad I, Schurzinger K, et al. Platelets secrete stromal cell-derived factor 1alpha and recruit bone marrow-derived progenitor cells to arterial thrombi in vivo. *J Exp Med*. 2006;203(5):1221-1233.
80. Nouailles G, Dorhoi A, Koch M, et al. CXCL5-secreting pulmonary epithelial cells drive destructive neutrophilic inflammation in tuberculosis. *J Clin Invest*. 2014;124(3):1268-1282.
81. Carevic M, Oz H, Fuchs K, et al. CXCR1 regulates pulmonary anti-Pseudomonas host defense. *J Innate Immun*. 2016;8(4):362-373.
82. Swamydas M, Gao JL, Break TJ, et al. CXCR1-mediated neutrophil degranulation and fungal killing promote Candida clearance and host survival. *Sci Transl Med*. 2016;8(322):322ra10.
83. McLoughlin RM, Solinga RM, Rich J, et al. CD4+ T cells and CXC chemokines modulate the pathogenesis of *Staphylococcus aureus* wound infections. *Proc Natl Acad Sci U S A*. 2006;103(27):10408-10413.
84. Hidalgo-Grass C, Mishalian I, Dan-Goor M, et al. A streptococcal protease that degrades CXC chemokines and

impairs bacterial clearance from infected tissues. *EMBO J.* 2006;25(19):4628-4637.
85. Geissmann F, Jung S, Littman DR. Blood monocytes consist of two principal subsets with distinct migratory properties. *Immunity.* 2003;19(1):71-82.
86. Lim JK, Obara CJ, Rivollier A, Pletnev AG, Kelsall BL, Murphy PM. Chemokine receptor Ccr2 is critical for monocyte accumulation and survival in West Nile virus encephalitis. *J Immunol.* 2011;186(1):471-478.
87. Serbina NV, Pamer EG. Monocyte emigration from bone marrow during bacterial infection requires signals mediated by chemokine receptor CCR2. *Nat Immunol.* 2006;7(3):311-317.
88. Mantovani A, Allavena P, Sozzani S, Vecchi A, Locati M, Sica A. Chemokines in the recruitment and shaping of the leukocyte infiltrate of tumors. *Semin Cancer Biol.* 2004;14(3):155-160.
89. Shurin GV, Ferris RL, Tourkova IL, et al. Loss of new chemokine CXCL14 in tumor tissue is associated with low infiltration by dendritic cells (DC), while restoration of human CXCL14 expression in tumor cells causes attraction of DC both in vitro and in vivo. *J Immunol.* 2005;174(9):5490-5498.
90. Berahovich RD, Lai NL, Wei Z, Lanier LL, Schall TJ. Evidence for NK cell subsets based on chemokine receptor expression. *J Immunol.* 2006;177(11):7833-7840.
91. Salazar-Mather TP, Hamilton TA, Biron CA. A chemokine-to-cytokine-to-chemokine cascade critical in antiviral defense. *J Clin Invest.* 2000;105(7):985-993.
92. Luster AD. The role of chemokines in linking innate and adaptive immunity. *Curr Opin Immunol.* 2002;14(1):129-135.
93. Coelho AL, Hogaboam CM, Kunkel SL. Chemokines provide the sustained inflammatory bridge between innate and acquired immunity. *Cytokine Growth Factor Rev.* 2005;16(6):553-560.
94. Ness TL, Ewing JL, Hogaboam CM, Kunkel SL. CCR4 is a key modulator of innate immune responses. *J Immunol.* 2006;177(11):7531-7539.
95. Cyster JG. Chemokines, sphingosine-1-phosphate, and cell migration in secondary lymphoid organs. *Annu Rev Immunol.* 2005;23:127-159.
96. Moser K, Tokoyoda K, Radbruch A, MacLennan I, Manz RA. Stromal niches, plasma cell differentiation and survival. *Curr Opin Immunol.* 2006;18(3):265-270.
97. Schulz O, Hammerschmidt SI, Moschovakis GL, Forster R. Chemokines and chemokine receptors in lymphoid tissue dynamics. *Annu Rev Immunol.* 2016;34:203-242.
98. Griffith JW, Sokol CL, Luster AD. Chemokines and chemokine receptors: positioning cells for host defense and immunity. *Annu Rev Immunol.* 2014;32:659-702.
99. Sokol CL, Camire RB, Jones MC, Luster AD. The chemokine receptor CCR8 promotes the migration of dendritic cells into the lymph node parenchyma to initiate the allergic immune response. *Immunity.* 2018;49(3):449-463.e6.
100. Schumann K, Lammermann T, Bruckner M, Legler DF, Polleux J, Spatz JP, et al. Immobilized chemokine fields and soluble chemokine gradients cooperatively shape migration patterns of dendritic cells. *Immunity.* 2010;32(5):703-713.
101. Wendland M, Willenzon S, Kocks J, et al. Lymph node T cell homeostasis relies on steady state homing of dendritic cells. *Immunity.* 2011;35(6):945-957.
102. Friedman RS, Jacobelli J, Krummel MF. Surface-bound chemokines capture and prime T cells for synapse formation. *Nat Immunol.* 2006;7(10):1101-1108.
103. Castellino F, Huang AY, Altan-Bonnet G, Stoll S, Scheinecker C, Germain RN. Chemokines enhance immunity by guiding naive CD8+ T cells to sites of CD4+ T cell-dendritic cell interaction. *Nature.* 2006;440(7086):890-895.
104. Rahimi RA, Luster AD. Chemokines: critical regulators of memory T cell development, maintenance, and function. *Adv Immunol.* 2018;138:71-98.
105. Cyster JG. Shining a light on germinal center B cells. *Cell.* 2010;143(4):503-505.
106. Sallusto F, Lenig D, Forster R, Lipp M, Lanzavecchia A. Two subsets of memory T lymphocytes with distinct homing potentials and effector functions. *Nature.* 1999;401(6754):708-712.
107. Bromley SK, Thomas SY, Luster AD. Chemokine receptor CCR7 guides T cell exit from peripheral tissues and entry into afferent lymphatics. *Nat Immunol.* 2005;6(9):895-901.
108. Debes GF, Arnold CN, Young AJ, et al. Chemokine receptor CCR7 required for T lymphocyte exit from peripheral tissues. *Nat Immunol.* 2005;6(9):889-894.
109. Forster R, Davalos-Misslitz AC, Rot A. CCR7 and its ligands: balancing immunity and tolerance. *Nat Rev Immunol.* 2008;8(5):362-371.
110. Schneider MA, Meingassner JG, Lipp M, Moore HD, Rot A. CCR7 is required for the in vivo function of CD4+ CD25+ regulatory T cells. *J Exp Med.* 2007;204(4):735-745.
111. Song K, Rabin RL, Hill BJ, et al. Characterization of subsets of CD4+ memory T cells reveals early branched pathways of T cell differentiation in humans. *Proc Natl Acad Sci U S A.* 2005;102(22):7916-7921.
112. Sallusto F, Lenig D, Mackay CR, Lanzavecchia A. Flexible programs of chemokine receptor expression on human polarized T helper 1 and 2 lymphocytes. *J Exp Med.* 1998;187(6):875-883.
113. Rothenberg ME, Mishra A, Brandt EB, Hogan SP. Gastrointestinal eosinophils. *Immunol Rev.* 2001;179:139-155.
114. Yurchenko E, Tritt M, Hay V, Shevach EM, Belkaid Y, Piccirillo CA. CCR5-dependent homing of naturally occurring CD4+ regulatory T cells to sites of Leishmania major infection favors pathogen persistence. *J Exp Med.* 2006;203(11):2451-2460.
115. Barsheshet Y, Wildbaum G, Levy E, et al. CCR8(+)FOXp3(+) Treg cells as master drivers of immune regulation. *Proc Natl Acad Sci U S A.* 2017;114(23):6086-6091.
116. Acosta-Rodriguez EV, Rivino L, Geginat J, et al. Surface phenotype and antigenic specificity of human interleukin 17-producing T helper memory cells. *Nat Immunol.* 2007;8(6):639-646.
117. Singh SP, Zhang HH, Foley JF, Hedrick MN, Farber JM. Human T cells that are able to produce IL-17 express the chemokine receptor CCR6. *J Immunol.* 2008;180(1):214-221.
118. Duhen T, Geiger R, Jarrossay D, Lanzavecchia A, Sallusto F. Production of interleukin 22 but not interleukin 17 by a subset of human skin-homing memory T cells. *Nat Immunol.* 2009;10(8):857-863.
119. Rivino L, Gruarin P, Haringer B, et al. CCR6 is expressed on an IL-10-producing, autoreactive memory T cell population with context-dependent regulatory function. *J Exp Med.* 2010;207(3):565-577.
120. Gerard C, Rollins BJ. Chemokines and disease. *Nat Immunol.* 2001;2(2):108-115.
121. Dorner BG, Dorner MB, Zhou X, et al. Selective expression of the chemokine receptor XCR1 on cross-presenting dendritic cells determines cooperation with CD8+ T cells. *Immunity.* 2009;31(5):823-833.
122. Ohta T, Sugiyama M, Hemmi H, et al. Crucial roles of XCR1-expressing dendritic cells and the XCR1-XCL1 chemokine axis in intestinal immune homeostasis. *Sci Rep.* 2016;6:23505.
123. Yamazaki C, Sugiyama M, Ohta T, et al. Critical roles of a dendritic cell subset expressing a chemokine receptor, XCR1. *J Immunol.* 2013;190(12):6071-6082.
124. Kunkel EJ, Butcher EC. Chemokines and the tissue-specific migration of lymphocytes. *Immunity.* 2002;16(1):1-4.
125. Homey B, Alenius H, Muller A, et al. CCL27-CCR10 interactions regulate T cell-mediated skin inflammation. *Nat Med.* 2002;8(2):157-165.
126. Onai N, Kitabatake M, Zhang YY, Ishikawa H, Ishikawa S, Matsushima K. Pivotal role of CCL25 (TECK)-CCR9 in the formation of gut cryptopatches and consequent appearance of intestinal intraepithelial T lymphocytes. *Int Immunol.* 2002;14(7):687-694.
127. Salazar-Gonzalez RM, Niess JH, Zammit DJ, et al. CCR6-mediated dendritic cell activation of pathogen-specific T cells in Peyer's patches. *Immunity.* 2006;24(5):623-632.
128. Bowman EP, Kuklin NA, Youngman KR, et al. The intestinal chemokine thymus-expressed chemokine (CCL25) attracts IgA antibody-secreting cells. *J Exp Med.* 2002;195(2):269-275.
129. Mora JR, Bono MR, Manjunath N, et al. Selective imprinting of gut-homing T cells by Peyer's patch dendritic cells. *Nature.* 2003;424(6944):88-93.
130. Mora JR, von Andrian UH. Differentiation and homing of IgA-secreting cells. *Mucosal Immunol.* 2008;1(2):96-109.
131. Jin Y, Xia M, Sun A, Saylor CM, Xiong N. CCR10 is important for the development of skin-specific gammadeltaT cells by regulating their migration and location. *J Immunol.* 2010;185(10):5723-5731.
132. Morteau O, Gerard C, Lu B, et al. An indispensable role for the chemokine receptor CCR10 in IgA antibody-secreting cell accumulation. *J Immunol.* 2008;181(9):6309-6315.
133. Feng Y, Broder CC, Kennedy PE, Berger EA. HIV-1 entry cofactor: functional cDNA cloning of a seven-transmembrane, G protein-coupled receptor. *Science.* 1996;272(5263):872-877.
134. Berger EA, Murphy PM, Farber JM. Chemokine receptors as HIV-1 coreceptors: roles in viral entry, tropism, and disease. *Annu Rev Immunol.* 1999;17:657-700.
135. Lusso P. HIV and the chemokine system: 10 years later. *EMBO J.* 2006;25(3):447-456.
136. Arenzana-Seisdedos F, Parmentier M. Genetics of resistance to HIV infection: role of co-receptors and co-receptor ligands. *Semin Immunol.* 2006;18(6):387-403.
137. Gonzalez E, Kulkarni H, Bolivar H, et al. The influence of CCL3L1 gene-containing segmental duplications on HIV-1/AIDS susceptibility. *Science.* 2005;307(5714):1434-1440.
138. O'Brien SJ, Moore JP. The effect of genetic variation in chemokines and their receptors on HIV transmission and progression to AIDS. *Immunol Rev.* 2000;177:99-111.
139. Sabeti PC, Walsh E, Schaffner SF, et al. The case for selection at CCR5-Delta32. *PLoS Biol.* 2005;3(11):e378.
140. Meanwell NA, Kadow JF. Maraviroc, a chemokine CCR5 receptor antagonist for the treatment of HIV infection and AIDS. *Curr Opin Invest Drugs.* 2007;8(8):669-681.
141. Troia-Cancio P, Asmuth DM. Lessons from maraviroc clinical trials. *Expert Rev Anti Infect Ther.* 2011;9(6):649-651.
142. Allers K, Hutter G, Hofmann J, Loddenkemper C, Rieger K, Thiel E, et al. Evidence for the cure of HIV infection by CCR5Delta32/Delta32 stem cell transplantation. *Blood.* 2011;117(10):2791-2799.
143. Hutter G, Nowak D, Mossner M, et al. Long-term control of HIV by CCR5 Delta32/Delta32 stem-cell transplantation. *N Engl J Med.* 2009;360(7):692-698.
144. Glass WG, Lim JK, Cholera R, Pletnev AG, Gao JL, Murphy PM. Chemokine receptor CCR5 promotes leukocyte trafficking to the brain and survival in West Nile virus infection. *J Exp Med.* 2005;202(8):1087-1098.
145. Glass WG, McDermott DH, Lim JK, et al. CCR5 deficiency increases risk of symptomatic West Nile virus infection. *J Exp Med.* 2006;203(1):35-40.
146. Klein RS, Lin E, Zhang B, et al. Neuronal CXCL10 directs CD8+ T-cell recruitment and control of West Nile virus encephalitis. *J Virol.* 2005;79(17):11457-11466.
147. Hsieh MF, Lai SL, Chen JP, et al. Both CXCR3 and CXCL10/IFN-inducible protein 10 are required for resistance to primary infection by dengue virus. *J Immunol.* 2006;177(3):1855-1863.
148. Levine B, Sodora DL. HIV and CXCR4 in a kiss of autophagic death. *J Clin Invest.* 2006;116(8):2078-2080.
149. De Clercq E. The AMD3100 story: the path to the discovery of a stem cell mobilizer (Mozobil). *Biochem Pharmacol.* 2009;77(11):1655-1664.
150. DiPersio JF, Micallef IN, Stiff PJ, et al. Phase III prospective randomized double-blind placebo-controlled trial of plerixafor plus granulocyte colony-stimulating factor compared with placebo plus granulocyte colony-stimulating factor for autologous stem-cell mobilization and transplantation for patients with non-Hodgkin's lymphoma. *J Clin Oncol.* 2009;27(28):4767-4773.
151. Gaur D, Mayer DC, Miller LH. Parasite ligand-host receptor interactions during invasion of erythrocytes by Plasmodium merozoites. *Int J Parasitol.* 2004;34(13-14):1413-1429.
152. Tournamille C, Colin Y, Cartron JP, Le Van Kim C. Disruption of a GATA motif in the Duffy gene promoter abolishes erythroid gene expression in Duffy-negative individuals. *Nat Genet.* 1995;10(2):224-228.
153. Zimmerman PA, Woolley I, Masinde GL, et al. Emergence of FY*A(null) in a Plasmodium vivax-endemic region of Papua New Guinea. *Proc Natl Acad Sci U S A.* 1999;96(24):13973-13977.
154. Hatabu T, Kawazu S, Aikawa M, Kano S. Binding of Plasmodium falciparum-infected erythrocytes to the membrane-bound form of Fractalkine/CX3CL1. *Proc Natl Acad Sci U S A.* 2003;100(26):15942-15946.
155. Hermand P, Ciceron L, Pionneau C, Vaquero C, Combadiere C, Deterre P. Plasmodium falciparum proteins involved in cytoadherence of infected erythrocytes to chemokine CX3CL1. *Sci Rep.* 2016;6:33786.
156. Tripp RA, Jones LP, Haynes LM, Zheng H, Murphy PM, Anderson LJ. CX3C chemokine mimicry by respiratory syncytial virus G glycoprotein. *Nat Immunol.* 2001;2(8):732-738.
157. Noriega VM, Gardner TJ, Redmann V, Bongers G, Lira SA, Tortorella D. Human cytomegalovirus US28 facilitates cell-to-cell viral dissemination. *Viruses.* 2014;6(3):1202-1218.
158. Cundell DR, Gerard NP, Gerard C, Idanpaan-Heikkila I, Tuomanen EI. Streptococcus pneumoniae anchor to activated human cells by the receptor for platelet-activating factor. *Nature.* 1995;377(6548):435-438.
159. Osei-Owusu P, Charlton TM, Kim HK, Missiakas D, Schneewind O. FPR1 is the plague receptor on host immune cells. *Nature.* 2019;574(7776):57-62.
160. Charles BA, Hsieh MM, Adeyemo AA, et al. Analyses of genome wide association data, cytokines, and gene expression in African-Americans with benign ethnic neutropenia. *PLoS One.* 2018;13(3):e0194400.
161. Reich D, Nalls MA, Kao WH, et al. Reduced neutrophil count in people of African descent is due to a regulatory variant in the Duffy antigen receptor for chemokines gene. *PLoS Genet.* 2009;5(1):e1000360.
162. Duchene J, Novitzky-Basso I, Thiriot A, et al. Atypical chemokine receptor 1 on nucleated erythroid cells regulates hematopoiesis. *Nat Immunol.* 2017;18(7):753-761.
163. Diaz GA. CXCR4 mutations in WHIM syndrome: a misguided immune system? *Immunol Rev.* 2005;203:235-243.
164. Heusinkveld LE, Majumdar S, Gao JL, McDermott DH, Murphy PM. WHIM syndrome: from pathogenesis towards personalized medicine and cure. *J Clin Immunol.* 2019;39(6):532-556.
165. Chow KY, Brotin E, Ben Khalifa Y, et al. A pivotal role for CXCL12 signaling in HPV-mediated transformation of keratinocytes: clues to understanding HPV-pathogenesis in WHIM syndrome. *Cell Host Microbe.* 2010;8(6):523-533.
166. McDermott DH, Pastrana DV, Calvo KR, et al. Plerixafor for the treatment of WHIM syndrome. *N Engl J Med.* 2019;380(2):163-170.
167. Dale DC, Firkin FC, Bolyard AA, et al. Results of a phase 2 trial of an oral CXCR4 antagonist mavorixafor for treatment of WHIM syndrome. *Blood.* 2020;136(26):2994-3003.
168. McDermott DH, Gao JL, Liu Q, et al. Chromothriptic cure of WHIM syndrome. *Cell.* 2015;160(4):686-699.
169. Alcami A, Lira SA. Modulation of chemokine activity by viruses. *Curr Opin Immunol.* 2010;22(4):482-487.
170. Slinger E, Langemeijer E, Siderius M, Vischer HF, Smit MJ. Herpesvirus-encoded GPCRs rewire cellular signaling. *Mol Cell Endocrinol.* 2011;331(2):179-184.

171. Jensen KK, Lira SA. Chemokines and Kaposi's sarcoma. *Semin Canc Biol.* 2004;14(3):187-194.
172. Holst PJ, Rosenkilde MM, Manfra D, et al. Tumorigenesis induced by the HHV8-encoded chemokine receptor requires ligand modulation of high constitutive activity. *J Clin Invest.* 2001;108(12):1789-1796.
173. Severe Covid GG, Ellinghaus D, Degenhardt F, et al. Genomewide association study of severe covid-19 with respiratory failure. *N Engl J Med.* 2020;383(16):1522-1534.
174. Sugiyama M, Kinoshita N, Ide S, et al. Serum CCL17 level becomes a predictive marker to distinguish between mild/moderate and severe/critical disease in patients with COVID-19. *Gene.* 2021;766:145145.
175. Chi Y, Ge Y, Wu B, et al. Serum cytokine and chemokine profile in relation to the severity of coronavirus disease 2019 in China. *J Infect Dis.* 2020;222(5):746-754.
176. Combadiere C, Potteaux S, Gao JL, et al. Decreased atherosclerotic lesion formation in CX3CR1/apolipoprotein E double knockout mice. *Circulation.* 2003;107(7):1009-1016.
177. McDermott DH, Fong AM, Yang Q, et al. Chemokine receptor mutant CX3CR1-M280 has impaired adhesive function and correlates with protection from cardiovascular disease in humans. *J Clin Invest.* 2003;111(8):1241-1250.
178. Barlic J, Zhang Y, Foley JF, Murphy PM. Oxidized lipid-driven chemokine receptor switch, CCR2 to CX3CR1, mediates adhesion of human macrophages to coronary artery smooth muscle cells through a peroxisome proliferator-activated receptor gamma-dependent pathway. *Circulation.* 2006;114(8):807-819.
179. Combadiere C, Potteaux S, Rodero M, et al. Combined inhibition of CCL2, CX3CR1, and CCR5 abrogates Ly6C(hi) and Ly6C(lo) monocytosis and almost abolishes atherosclerosis in hypercholesterolemic mice. *Circulation.* 2008;117(13):1649-1657.
180. Combadiere C, Feumi C, Raoul W, et al. CX3CR1-dependent subretinal microglia cell accumulation is associated with cardinal features of age-related macular degeneration. *J Clin Invest.* 2007;117(10):2920-2928.
181. Takeda A, Baffi JZ, Kleinman ME, et al. CCR3 is a target for age-related macular degeneration diagnosis and therapy. *Nature.* 2009;460(7252):225-230.
182. Joy MT, Ben Assayag E, Shabashov-Stone D, et al. CCR5 is a therapeutic target for recovery after stroke and traumatic brain injury. *Cell.* 2019;176(5):1143-1157.e13.
183. Miyabe Y, Lian J, Miyabe C, Luster AD. Chemokines in rheumatic diseases: pathogenic role and therapeutic implications. *Nat Rev Rheumatol.* 2019;15(12):731-746.
184. Prahalad S. Negative association between the chemokine receptor CCR5-Delta32 polymorphism and rheumatoid arthritis: a meta-analysis. *Gene Immun.* 2006;7(3):264-268.
185. Singh TP, Zhang HH, Borek I, et al. Monocyte-derived inflammatory Langerhans cells and dermal dendritic cells mediate psoriasis-like inflammation. *Nat Commun.* 2016;7:13581.
186. Blanchard C, Wang N, Stringer KF, et al. Eotaxin-3 and a uniquely conserved gene-expression profile in eosinophilic esophagitis. *J Clin Invest.* 2006;116(2):536-547.
187. Schall TJ, Proudfoot AE. Overcoming hurdles in developing successful drugs targeting chemokine receptors. *Nat Rev Immunol.* 2011;11(5):355-363.
188. Harada A, Mukaida N, Matsushima K. Interleukin 8 as a novel target for intervention therapy in acute inflammatory diseases. *Mol Med Today.* 1996;2(11):482-489.
189. Harada A, Sekido N, Akahoshi T, Wada T, Mukaida N, Matsushima K. Essential involvement of interleukin-8 (IL-8) in acute inflammation. *J Leukoc Biol.* 1994;56(5):559-564.
190. Hancock WW. Chemokine receptor-dependent alloresponses. *Immunol Rev.* 2003;196:37-50.
191. Hancock WW, Lu B, Gao W, et al. Requirement of the chemokine receptor CXCR3 for acute allograft rejection. *J Exp Med.* 2000;192(10):1515-1520.
192. Miura M, Morita K, Kobayashi H, et al. Monokine induced by IFN-gamma is a dominant factor directing T cells into murine cardiac allografts during acute rejection. *J Immunol.* 2001;167(6):3494-3504.
193. Fischereder M, Luckow B, Hocher B, et al. CC chemokine receptor 5 and renal-transplant survival. *Lancet.* 2001;357(9270):1758-1761.
194. Lukacs NW, Miller AL, Hogaboam CM. Chemokine receptors in asthma: searching for the correct immune targets. *J Immunol.* 2003;171(1):11-15.
195. Humbles AA, Lu B, Friend DS, et al. The murine CCR3 receptor regulates both the role of eosinophils and mast cells in allergen-induced airway inflammation and hyperresponsiveness. *Proc Natl Acad Sci U S A.* 2002;99(3):1479-1484.
196. Ma W, Bryce PJ, Humbles AA, et al. CCR3 is essential for skin eosinophilia and airway hyperresponsiveness in a murine model of allergic skin inflammation. *J Clin Invest.* 2002;109(5):621-628.
197. Fulkerson PC, Fischetti CA, McBride ML, Hassman LM, Hogan SP, Rothenberg ME. A central regulatory role for eosinophils and the eotaxin/CCR3 axis in chronic experimental allergic airway inflammation. *Proc Natl Acad Sci U S A.* 2006;103(44):16418-16423.
198. Gonzalo JA, Lloyd CM, Peled A, Delaney T, Coyle AJ, Gutierrez-Ramos JC. Critical involvement of the chemotactic axis CXCR4/stromal cell-derived factor-1 alpha in the inflammatory component of allergic airway disease. *J Immunol.* 2000;165(1):499-508.
199. Karin N. Chemokines and cancer: new immune checkpoints for cancer therapy. *Curr Opin Immunol.* 2018;51:140-145.
200. Luster AD, Leder P. IP-10, a -C-X-C- chemokine, elicits a potent thymus-dependent antitumor response *in vivo*. *J Exp Med.* 1993;178(3):1057-1065.
201. Muller A, Homey B, Soto H, et al. Involvement of chemokine receptors in breast cancer metastasis. *Nature.* 2001;410(6824):50-56.
202. Bockorny B, Semenisty V, Macarulla T, et al. BL-8040, a CXCR4 antagonist, in combination with pembrolizumab and chemotherapy for pancreatic cancer: the COMBAT trial. *Nat Med.* 2020;26(6):878-885.
203. Zeng Y, Li B, Liang Y, et al. Dual blockade of CXCL12-CXCR4 and PD-1-PD-L1 pathways prolongs survival of ovarian tumor-bearing mice by prevention of immunosuppression in the tumor microenvironment. *FASEB J.* 2019;33(5):6596-6608.
204. Strieter RM, Burdick MD, Mestas J, Gomperts B, Keane MP, Belperio JA. Cancer CXC chemokine networks and tumour angiogenesis. *Eur J Cancer.* 2006;42(6):768-778.
205. Orimo A, Gupta PB, Sgroi DC, et al. Stromal fibroblasts present in invasive human breast carcinomas promote tumor growth and angiogenesis through elevated SDF-1/CXCL12 secretion. *Cell.* 2005;121(3):335-348.
206. Yang G, Rosen DG, Zhang Z, et al. The chemokine growth-regulated oncogene 1 (Gro-1) links RAS signaling to the senescence of stromal fibroblasts and ovarian tumorigenesis. *Proc Natl Acad Sci U S A.* 2006;103(44):16472-16477.
207. Kim YH, Bagot M, Pinter-Brown L, et al. Mogamulizumab versus vorinostat in previously treated cutaneous T-cell lymphoma (MAVORIC): an international, open-label, randomised, controlled phase 3 trial. *Lancet Oncol.* 2018;19(9):1192-1204.
208. Lewis DJ, Rook AH. Mogamulizumab in the treatment of advanced mycosis fungoides and Sezary syndrome: safety and efficacy. *Expert Rev Anticancer Ther.* 2020;20(6):447-452.
209. Hauser AS, Chavali S, Masuho I, et al. Pharmacogenomics of GPCR drug targets. *Cell.* 2018;172(1-2):41-54.e19.
210. Scala S, D'Alterio C, Milanesi S, et al. New insights on the emerging genomic landscape of CXCR4 in cancer: a lesson from WHIM. *Vaccines (Basel).* 2020;8(2):164.
211. Maganti H, Visram A, Shorr R, Fulcher J, Sabloff M, Allan DS. Plerixafor in combination with chemotherapy and/or hematopoietic cell transplantation to treat acute leukemia: a systematic review and metanalysis of preclinical and clinical studies. *Leuk Res.* 2020;97:106442.
212. Castillo JJ, Moreno DF, Arbelaez MI, Hunter ZR, Treon SP. CXCR4 mutations affect presentation and outcomes in patients with Waldenstrom macroglobulinemia: a systematic review. *Expert Rev Hematol.* 2019;12(10):873-881.
213. Weiss ID, Huff LM, Evbuomwan MO, et al. Screening of cancer tissue arrays identifies CXCR4 on adrenocortical carcinoma: correlates with expression and quantification on metastases using (64)Cu-plerixafor PET. *Oncotarget.* 2017;8(43):73387-73406.
214. Bobkov V, Zarca AM, Van Hout A, et al. Nanobody-Fc constructs targeting chemokine receptor CXCR4 potently inhibit signaling and CXCR4-mediated HIV-entry and induce antibody effector functions. *Biochem Pharmacol.* 2018;158:413-424.
215. Soroceanu L, Matlaf L, Bezrookove V, et al. Human cytomegalovirus US28 found in glioblastoma promotes an invasive and angiogenic phenotype. *Cancer Res.* 2011;71(21):6643-6653.
216. Spiess K, Jeppesen MG, Malmgaard-Clausen M, Krzywkowski K, Kledal TN, Rosenkilde MM. Novel chemokine-based immunotoxins for potent and selective targeting of cytomegalovirus infected cells. *J Immunol Res.* 2017;2017:4069260.

SECTION IV

Innate Immunity

CHAPTER 12

Pattern Recognition Receptors and the IL-1 Family

Rajendra Karki, PhD • Thirumala-Devi Kanneganti, PhD

INTRODUCTION

In 1989, Charles Janeway Jr proposed a microbial pattern recognition model that suggested a mechanism by which the innate immune system distinguishes "non-infectious self" from "infectious non-self." At that time, it was known that bacterial products were needed to serve as an adjuvant for the induction of robust T and B cell responses, but the reason for this was unclear. Janeway proposed an unconventional theory of microbial pattern recognition and suggested the principle of innate control of adaptive immunity. In a landmark essay published by Cold Spring Harbor Laboratory Press, Janeway predicted that innate immune cells instruct T and B cells to develop efficient adaptive responses.[1] Moreover, Janeway proposed the ability of innate immune cells (phagocytes) to identify microbes specifically and distinguish them from self-antigens.

The ability of the innate immune system to carry out this differentiating function relies on a series of germline-encoded receptors that have evolved to detect conserved molecular patterns produced by pathogens, but not by their metazoan hosts. Since this concept was proposed in 1989, many microbial ligands, ranging from structural components of bacteria, fungi, and viruses to biosynthetic molecules such as nucleic acids, have been identified that are recognized by specific receptors to induce the innate immune responses that protect the host from infectious diseases.

It is now known that innate immune cells are equipped with multiple pattern recognition receptors (PRRs) that recognize conserved molecular structures of microbial origin called pathogen-associated molecular patterns (PAMPs). Different types of microbes express different PAMPs. The PRRs also recognize endogenous molecules that are produced by or released from damaged and dying cells called damage-associated molecular patterns (DAMPs).[2] Recognition of molecular patterns by the receptors triggers multiple signaling events that are essential to initiate transcriptional and nontranscriptional cellular responses. One of the major transcriptional cellular responses is the upregulation of the genes encoding proinflammatory cytokines, chemokines, and interferons (IFNs). The induction of phagocytosis, autophagy, cell death, and cytokine processing are just some examples of nontranscriptional cellular responses. Additionally, PRR stimulation plays a crucial role in coordinating both the innate and adaptive immune systems by educating antigen-presenting cells, such as dendritic cells (DCs), to effectively elicit a T cell–mediated adaptive immune response against a particular antigen.

Based on protein domain homology, most PRRs are classified into six families: Toll-like receptors (TLRs); C-type lectin receptors (CLRs); scavenger receptors (SRs); nucleotide-binding domain leucine-rich repeat (LRR)–containing (or NOD-like) receptors (NLRs); absent in melanoma 2 (AIM2)–like receptors (ALRs); and retinoic acid–inducible gene (RIG) I–like receptors (RLRs). Based on their localization, these molecules can be divided into two major groups, membrane-bound receptors (TLRs, CLRs, and SRs) and cytosolic receptors (NLRs, ALRs, and RLRs) (Table 12.1). This array of PRRs ensures that a cell is able to recognize PAMPs from both extracellular and intracellular pathogens. Many cell types in the human body express PRRs, including myeloid white blood cells (monocytes, macrophages, DCs, neutrophils, eosinophils, mast cells, and basophils) and subsets of three types of lymphocytes (B cells, T cells, and natural killer [NK] cells). PRRs are also expressed by some other nonimmune cells such as keratinocytes, epithelial cells, endothelial cells, and fibroblasts.[3] Not all PRRs are expressed in a single cell.

PRR sensing of PAMPs and DAMPs leads to the activation of transcriptional programs producing proinflammatory cytokines. One such group of cytokines is the interleukin-1 (IL-1) family. IL-1 was discovered and identified in the early 1980s by Charles Dinarello.[4] Since then, the IL-1 family has grown to include three subfamilies (IL-1, IL-18, and IL-36)

TABLE 12.1 Pattern Recognition Receptor Families

Family	Members	Shared Domains	Localization
TLR	TLR 1-10 (Humans)	LRR, TIR	Cell surface (TLR1, 2, 4-6, 10) Endosomes (TLR3, 7-9, 11-13)
	TLR 1-9 (Mice)		
	TLR 11-13 (Mice)		
CLR	Dectin-2, MINCLE	ITAM	Cell surface
	Dectin-1	hemITAM	
	MICL, DCIR	ITIM	
SR	SR-A (SR-A1)	Collagen, SRCR	Cell surface
	SR-B (CD36)	CD36	
	SR-C (Insects)	Compliment control protein region, MAM motif	
	SR-D (CD68)	Mucin-like, LAMP	
	SR-E (LOX-1)	CLEC	
	SR-F (SCARF1, SCARF2, MEGF10)	EGF-like	
	SR-G (SR-PSOX/CXCL16)	CXC-chemokine	
	SR-H (FEEL1, FEEL2)	FEEL	
	SR-I (CD163A, CD163-L1, CD163c-α)	SRCR	
	SR-J (RAGE)	Ig-like regions	
	SR-K (CD44)	N/A	
	SR-L (LRP1, LRP2)	N/A	
NLR	NLRC1-5	LRR, NBD	Cytoplasm
	NLRP1-9, 11-14		
	NAIPs		
ALR	AIM2, IFI16	Pyrin, HIN-200	Cytoplasm
RLR	RIG-I, MDA5, LGP2	DExD/H helicase	Cytoplasm

based on their IL-1 consensus sequence and primary ligand binding receptor. The IL-1 family is primarily associated with acute and chronic inflammation. There are 11 members of the IL-1 family of cytokines and 10 members of the IL-1 family of receptors. Among IL-1 family cytokines, IL-1β has been widely studied because of its therapeutic potential for a number of autoinflammatory diseases. Although the IL-1 family provides beneficial functions to counteract infection, these molecules can also be detrimental to the host by causing acute or chronic inflammation. Therefore, several physiological mechanisms exist to inhibit IL-1 family signaling, including antagonists, cytokines, and decoy receptors. Each member of the IL-1 receptor family contains the functional cytoplasmic Toll-IL-1 receptor (TIR) domain, which is highly homologous to the TIR domains of the TLRs. Because of this homology, the fundamental inflammatory responses caused by the IL-1 and TLR families are indistinguishable.

PRR sensing of stimuli and the subsequent signaling and cytokine secretion are key parts of innate immunity. In this chapter, we describe PRR and IL-1 families and the signaling pathways that they activate for an effective innate immune response.

TOLL-LIKE RECEPTORS

The TLR family was the first family of PRRs to be discovered, and they are still the best characterized in terms of their structure, ligand specificity, and innate immune response induction. As of 2020, 13 TLRs have been identified in mice and humans. TLRs 1 to 10 are conserved between mice and humans (Table 12.2), although TLR10 is not functional in mice. TLRs 11 to 13 are expressed in mice but not in humans.

Discovery of Toll-Like Receptors

In the 1980s, researchers in Germany discovered that mutation of the *Toll* gene led to underdevelopment in the ventral portion of *Drosophila* fruit fly larva. The name "Toll" refers to the German slang for "weird," denoting the unusual phenotype of the Toll mutant flies. Thereafter, researchers focused on characterizing the *Toll* gene and identifying its role in embryonic development,[5,6] for which Christiane Nusslein-Volhard, Eric Wieschaus, and Edward B. Lewis were awarded the Nobel Prize in Physiology and Medicine in 1995. This was followed by a striking observation that the Toll mutant flies succumbed to fungal infection[7]; Jules Hoffman was awarded the Nobel

Prize in Physiology and Medicine for this discovery in 2011. This discovery advanced the concept that Toll and the related proteins played a role in the activation of innate immune responses in invertebrates.[8] Characterization of the Toll protein surprisingly revealed that its cytoplasmic signaling domain was homologous to that of the vertebrate receptor for the cytokine IL-1 (IL-1R). Based on this fact, Charles Janeway and Ruslan Medzhitov identified a human gene for a protein similar to Toll. This, along with other vertebrate Toll relatives discovered soon thereafter, was named a Toll-like receptor (TLR).[9]

Additional vertebrate TLRs were found by Bruce A. Beutler, Shizuo Akira, and others. In 1998, Beutler observed that mice homozygous for a mutant form of the *lps* gene were resistant to endotoxin shock.[10] He found that *lps* encoded the TLR4 protein and that mutation in this gene produced a version of the TLR4 protein that differed by a single amino acid so that it no longer was activated by lipopolysaccharide (LPS). For identifying TLR4 as an LPS-sensing receptor, Beutler earned a share of the 2011 Nobel Prize in Physiology and Medicine together with Jules Hoffman. In turn, the other TLR genes were largely discovered in the laboratory of Shizuo Akira and colleagues. Each TLR is now believed to have a distinct repertoire of specificities for conserved PAMPs.

TLRs and Their Ligands

Biochemical studies have revealed the structure of several TLRs and their binding specificity to ligands. TLRs are type I transmembrane receptors composed of an N-terminal ligand-binding domain, a single transmembrane helix, and a C-terminal signaling domain. Each TLR N-terminal domain (NTD) contains tandem copies of a motif known as the LRR, which is conserved in many proteins in animals, plants, and microorganisms. Multiple LRRs form a solenoid structure which is important for ligand binding and activation. The signaling domains of TLRs share homology with the signaling domains of the IL-1R family; therefore, they are called Toll/IL-1R (TIR) domains. TLRs exist both on the plasma membrane and in the membranes of endosomes to respond optimally to their cognate PAMPs and DAMPs. Broadly, TLRs 1, 2, 4 to 6, and 10 are expressed on the plasma membrane, and TLRs 3, 7 to 9, and 11 to 13 are confined to the endosomes. Both plasma membrane-bound and endosomal TLRs detect microbial ligands, such as LPS, polysaccharides, nucleic acids, hemozoin, lipopeptides, and flagellin. In addition to microbial ligands, TLRs also recognize products released by dead/dying cells or damaged tissues, leading to inflammatory responses similar to those triggered by microbial infection. The sterile inflammatory DAMPs include high-mobility group protein 1 (HMGB1), heat shock proteins such as Hsp90, chromatin proteins, fragments of extracellular matrix components such as fibronectin and hyaluronan, oxidized low-density lipoprotein (oxLDL), and amyloid-β.[11] While self-nucleic acids usually do not activate the intracellular PRRs, under certain circumstances they can be endocytosed and activate endosomal TLRs, contributing to disease pathogenesis. This phenomenon is particularly important in the context of autoimmune diseases such as systemic lupus erythematosus (SLE). A comprehensive list of known human TLRs and their ligands is shown in Table 12.2.

TLR Signaling Pathways

Recognition of PAMPs or DAMPs by TLRs induces dimerization of extracellular LRR domains, which in turn facilitates the localization and subsequent dimerization of intracellular TIR domains. This process is followed by the recruitment of TIR adaptor molecules. TIR domain interactions between receptors and adaptors are important for the downstream signaling required for the nuclear translocation of transcription factors such as nuclear factor kappa B (NF-κB) and interferon regulatory factors (IRFs) to induce the production of inflammatory cytokines and antimicrobial factors. To date, five TLR adaptor proteins have been identified: myeloid differentiation factor 88 (MyD88), TIR domain–containing adaptor-inducing IFN-β factor (TRIF, also known as TICAM1), MyD88 adaptor–like (MAL, also known as TIR domain–containing adaptor protein [TIRAP]), TRIF-related adaptor molecule (TRAM, also known as TICAM2), and sterile-alpha and armadillo-motif-containing protein (SARM) (Table 12.3).

The particular signaling pathways activated by a TLR are largely determined by the identity of the TLR and the initial adaptor molecule that is recruited to its cytoplasmic TIR domain. In general, TLRs transduce signals via two key adaptors, MyD88 and TRIF (Fig. 12.1). Almost all TLRs can signal via MyD88, whereas TLR3 signals exclusively via the TRIF axis. TLR4 is unique in that it binds both MyD88 and TRIF. Within the structure of the TLRs, the BB loop is critical for most TIR–TIR interactions, including those between the TLR and MyD88. With the exception of TLR3, all human TLRs contain a conserved proline within the BB loop of their TIR domain. TLR3 has an alanine at this position. Mutation of this alanine to proline switches TLR3 signaling from a TRIF- to a MyD88-directed pathway.[12] MyD88 signaling through TLR2[13,14] and cell surface-bound TLR4[15] requires an additional TIRAP adaptor protein. TIRAP acts as a bridging adaptor to facilitate MyD88 recruitment to the receptor's TIR domain, and a PIP2 binding domain allows for its localization to the plasma membrane.

MyD88-Dependent Signaling Pathway

MyD88 contains not only a TIR domain but also an N-terminal death domain (DD). This DD allows for homotypic interactions with other DD-containing proteins called IL-1 receptor–associated kinases (IRAKs), thereby forming a hetero-oligomeric structure called the Myddosome (Fig. 12.2). Among the four IRAKs that have been described in humans, IRAK1, IRAK2, and IRAK4 are involved in Myddosome formation. IRAK4 is the first to be recruited to MyD88. Once recruited, IRAK4 undergoes autophosphorylation, which in turn phosphorylates IRAK1 and IRAK2. Formation of this signaling platform then induces dissociation of the IRAKs from the complex, allowing IRAK1 or IRAK2 to interact with the E3 ubiquitin (Ub) ligase TNF receptor–associated factor 6 (TRAF6). Following its activation, TRAF6 is recruited to the TAK1 complex containing transforming growth factor β-activated kinase 1 (TAK1), TAK1-binding protein 1 (TAB1), and TAB2/3 to drive two major signaling pathways (Fig. 12.1A). The TAK1 complex

TABLE 12.2 Human TLRs and Their Microbial Ligands

TLR Family Member	Ligand(s)/PAMP(s)	Pathogen(s)	Expression in Cells/Tissues
TLR1	Triacyl lipopeptides	Mycobacteria Gram-negative bacteria	Myeloid cells Lymphocytes Epithelial cells
TLR2	Lipoproteins	Gram-positive bacteria Gram-negative bacteria	Leukocytes Myeloid cells Lymphocytes
	Lipoteichoic acid	Gram-positive bacteria	
	Lipoarabinomannan	Mycobacteria	
	Zymosan	Yeast and other fungi	
TLR3	dsRNA	Viruses	Dendritic cells Macrophages Epithelial cells Fibroblasts
TLR4	LPS Mannuronic acid	Gram-negative bacteria	Macrophages Dendritic cells
	Teichuronic acid	Gram-positive bacteria	
	F-protein	Respiratory syncytial virus	
TLR5	Flagellin	Bacteria	Epithelial cells Tissue-resident dendritic cells
TLR6	Diacyl lipopolypeptides	Mycobacteria Gram-positive bacteria	Leukocytes Myeloid cells Lymphocytes
	Zymosan	Yeasts and fungi	
TLR7	ssRNA	Viruses	pDCs B cells Eosinophils
TLR8	ssRNA	Viruses	Monocytes Macrophages
TLR9	Dinucleotides	Bacterial DNA	pDCs B cells
	Hemozoin	*Plasmodium*	
TLR10	HIV-1 gp41	HIV-1 virus	B cells Macrophages

TABLE 12.3 Key Innate Immunity Adaptor Proteins

Adaptor	Interacting Receptor Domain	Signaling Domain	Localization
TIRAP–MyD88	TIR domain of TLR2 and TLR4	Death domain	Cell surface
MyD88	TIR domain	Death domain	Endosomes
TRAM–TRIF	TLR4	TRAF binding	Cell surface
TRIF	TLR3	RHIM	Endosomes
ASC	NLR PYD	CARD	Cytoplasm
MAVS	RLR CARD	TRAF binding	Mitochondria
FcRγ DAP12	ITAM motif of CLRs	Phosphorylation by SYK	Cytoplasm Plasma membrane

stimulates activation of an IκB kinase (IKK) complex made up of IKKα, IKKβ, and IKKγ (NEMO); IKK activation induces phosphorylation and degradation of IκBα (allowing the p50 and p65 subunits of NF-κB to translocate to the nucleus, a phenomenon called canonical NF-κB activation). Additionally, TAK1 can activate mitogen-activated protein kinase (MAPK) cascades, leading to activation of the transcription factors AP1 and ATF3/CREB.

In addition to stimulating the NF-κB and MAPK pathways, MyD88 signaling promotes IRF1, IRF5, and IRF9

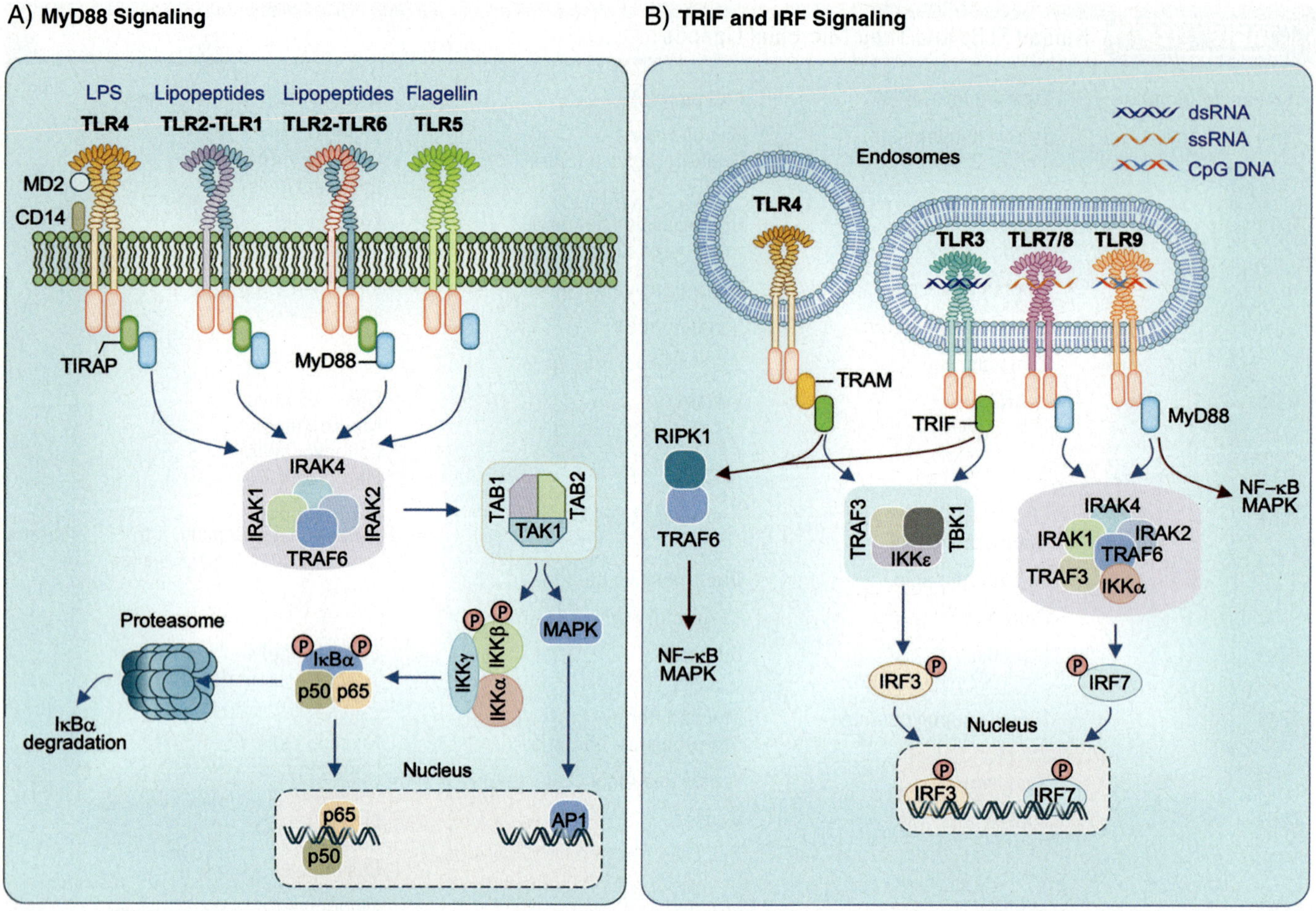

FIG. 12.1. TLR activation pathways. A: MyD88 signaling: Lipopolysaccharide (LPS)-binding protein (LBP) transfers LPS to the glycosylphosphatidylinositol-anchored membrane protein CD14. CD14 then delivers LPS to the MD2 protein associated with the extracellular domain of Toll-like receptor 4 (TLR4). Bacterial lipopeptides are recognized by TLR2-TLR1 or TLR2-TLR6 heterodimers, and flagellin is detected by TLR5. Toll/interleukin-1 receptor (TIR) domain-containing adaptor protein (TIRAP) acts as a bridge between TLRs and the adaptor protein MyD88, which subsequently leads to the activation of IRAK4 followed by IRAK1, IRAK2, and TRAF6. This TRAF6 complex promotes formation of the TAK1-TAB complex, which leads to the activation of NF-κB– and AP1–dependent gene expression through the NF-κB and MAPK pathways, respectively. In the NF-κB pathway, the IKK complex targets IκBα for phosphorylation and subsequent degradative polyubiquitination. This degradation releases the p50-p65 heterodimer from IκBα, allowing the heterodimer to translocate into the nucleus and induce the expression of various proinflammatory genes. **B:** TRIF and IRF signaling: Double-stranded RNA (dsRNA), single-stranded RNA (ssRNA), and CpG DNA are ligands for TLR3, TLR7 or TLR8, and TLR9, respectively. TLR4 can be activated by viral envelope glycoproteins. TRAM bridges TLR4 with TRIF. TRIF is recruited in response to TLR4 and TLR3 activation, leading to subsequent interferon regulatory factor 3 (IRF3) activation via the TRAF3 complex. After phosphorylation, IRF3 translocates to the nucleus. TRIF also recruits RIPK1 and TRAF6, which leads to NF-κB and MAPK-activating pathways. The TLR7-, TLR8-, and TLR9-dependent NF-κB, MAPK, and IRF7-acitivating pathways rely on the MyD88 adaptor protein to form the TRAF6 complex. Phosphorylated IRF7 translocates to the nucleus. Both IRF3 and IRF7 bind to interferon-sensitive response elements (ISREs) and induce the expression of interferon-dependent genes.

activation. MyD88 facilitates the nuclear localization of IRF1, enhancing the TLR-dependent gene induction program of IFNs.[16] This process is crucial in mounting a proper immune response during *Aspergillus* infection.[17] MyD88 also promotes IRF5 nuclear localization via a TAK1-IKKβ axis for the production of inflammatory cytokines and type I IFN.[18] In addition, MyD88 induces the expression of IFN-α exclusively in plasmacytoid DCs (pDCs) via IRAK4/IRAK1-mediated activation of IRF7 in response to TLR7 and TLR9 activation. Although the mechanisms involved in MyD88 signaling have been widely explored,[19] how posttranslational modifications of MyD88 affect its signaling activity has been understudied. MyD88 is known to be phosphorylated by SYK,[20] but the functional relevance of this modification is unknown. Future work will be required to fully describe these functions.

Because it is a proximal signaling adaptor for almost all TLRs, MyD88 plays a critical role in health and disease. While deficiency of MyD88 in mice leads to resistance against LPS-induced sepsis and lethality,[21] these mice also lack the ability to signal through IL-1R family members. As a result, MyD88-deficient mice are unable to mount an effective host defense response to multiple pathogens including bacteria, viruses, fungi, and parasites. In addition, MyD88 is pathogenic in a number of inflammatory disease models such as SLE,[22] psoriasis,[23] atherosclerosis,[24] and neutrophilic dermatosis.[20] In humans, MyD88 deficiency leads to abnormally

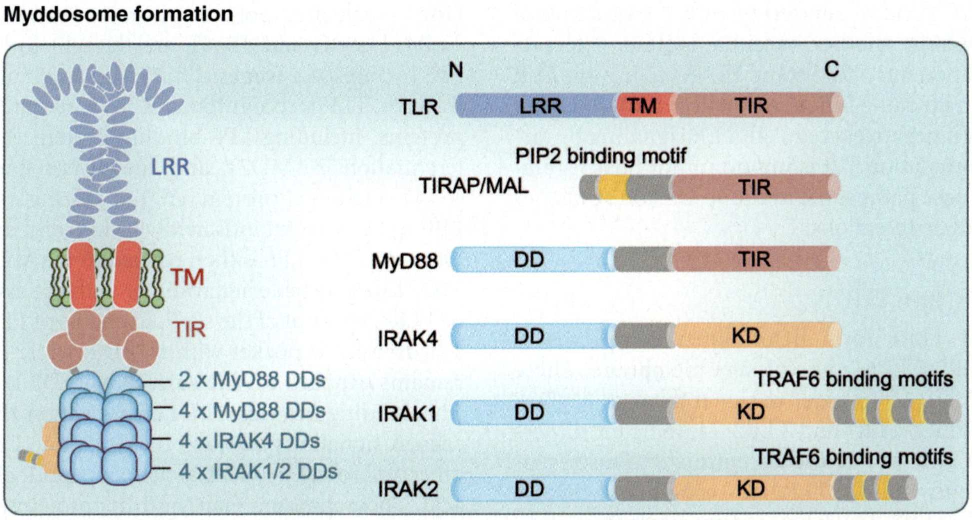

FIG. 12.2. Myddosome formation. Toll-like receptors (TLRs) contain an extracellular leucine-rich repeat (LRR) domain, a transmembrane (TM) domain, and a cytosolic Toll/interleukin-1 receptor (TIR) domain. Following TLR activation, MyD88 is recruited to the TLR via homotypic TIR domain interactions. TIR domain-containing adaptor protein (TIRAP), which is required for signaling by a subset of TLRs, contains a phosphoinositide (PIP2) binding region, allowing TIRAP to interact with the plasma membrane. Six MyD88 death domains (DDs) (arranged in two rings) provide a platform for the subsequent recruitment of 4 IRAK4 and 4 IRAK1/2 DDs, forming the complete Myddosome complex. The C-terminal regions of IRAK1 and IRAK2 contain putative binding motifs for the recruitment of TRAF6 and the activation of downstream signaling events.

frequent and severe infections by pyogenic bacteria such as *Streptococcus pneumoniae* and *Staphylococcus aureus* in children.[25] However, affected individuals have normal resistance to other microbes. Multiple patients with pyogenic bacterial infections have been found to express a variant of MyD88 with the inactivating mutation R196C (within the BB loop). Also, children with a homozygous in-frame deletion of the E52 residue of MyD88 suffer from invasive pneumococcal disease and do not induce cytokine responses typically driven by TLR agonists.[26] MyD88 pathway deficiencies are not evident during adulthood, likely due to the development of adaptive immunity or potential compensatory roles of other signaling pathways. However, a gain-of-function variant of MyD88 (L256P) has been described that has oncogenic potential in some classes of B cell lymphomas. This mutant can spontaneously lead to persistent NF-κB and STAT3 activation.[27] Overall, alterations in MyD88 signaling can have impacts in infectious and inflammatory diseases and even cancer.

TRIF-Dependent Signaling Pathway

TRIF is expressed by most cell types and localizes to the cytoplasm in resting cells. Signaling through TRIF is responsible for the activation of several TLR4- and TLR3-dependent transcription factors, including NF-κB, IRF3, and AP1[28,29] (Fig. 12.1B). This activation leads to the production of cytokines and type I IFNs[30] as well as maturation of myeloid DCs.[31] In contrast to TLR3, which recruits TRIF directly, TLR4 uses the bridging adaptor TRAM to associate with TRIF. TRAM has an N-terminal myristoylation motif that localizes it to the plasma membrane and a C-terminal TIR domain. Phosphorylation of TRAM by protein kinase C-epsilon translocates the TLR4/TRAM complex to the endosome.[32]

Structurally, TRIF contains an NTD for autoregulation, a consensus TRAF6-binding motif, a TIR domain, and a C-terminal receptor-interacting protein homotypic interaction motif (RHIM). TRIF interacts with TBK1 and IKKi/IKKε to induce IRF3 phosphorylation.[28]

TRIF-dependent TLR signaling causes type I IFN production and contributes to caspase activation, thereby leading to cell death. TRIF activates caspase-11 through type I IFN signaling specifically during infection with some gram-negative bacteria to induce a form of inflammatory cell death called pyroptosis.[33,34] TRIF can also induce cell death through caspase-independent mechanisms. TRIF can directly bind to receptor-interacting serine/threonine-protein kinase 1 (RIPK1) via homotypic RHIM–RHIM interactions to activate RIPK3.[35] Additionally, when caspase-8 is inhibited, which could occur as a result of pathogen immune evasion strategies, TLR-mediated TRIF signaling drives necroptosis.[36]

Through its functions in promoting downstream immune processes and cell death, TRIF-dependent signaling contributes to defense against several viral infections including influenza A virus (IAV), Sindbis virus, and poliovirus.[37] Additionally, TRIF polymorphisms are associated with pediatric HSV-induced encephalitis.[38]

Regulation of MyD88 and TRIF Signaling Pathways

Endogenous host molecules and virulence factors from pathogens can positively and negatively regulate TLR signaling downstream of MyD88 and TRIF. The endogenous TIR domain–containing adaptor protein SARM was originally identified as a negative regulator of TRIF-dependent signaling,[39] but it was later shown to be required for the optimal production of TLR-dependent CCL5 in macrophages.[40] Other endogenous factors that inhibit TLR signaling include microRNAs, decoy receptors, and some ubiquitin regulators. In addition to the endogenous host regulatory strategies, some pathogens encode virulence factors that target TLR signaling to facilitate immune evasion. For example, the A46R protein from vaccinia virus interacts with TRIF, TRAM,

MyD88, and TIRAP, thereby interfering with the function of these adaptors.[41] A52R, another vaccinia protein, interacts with signaling components TRAF6 and IRAKs to inhibit TLR signaling.[42] Human immunodeficiency virus (HIV)–encoded proteins Vpr and Vif negatively regulate TLR signaling by targeting IRF3 for degradation.[43] It is important for host regulation to overcome these pathogenic immune evasion strategies to effectively eliminate infections.

TLR1, TLR2, TLR6, and TLR10

TLR1, TLR2, and TLR6 form heterodimer complexes of TLR2-TLR1 or TLR2-TLR6 in the plasma membrane. These complexes require TIRAP in order to signal through MyD88 (Fig. 12.1A). The TLR2-TLR1 and TLR2-TLR6 heterodimers utilize CD14 and CD36 as major coreceptors.[44,45] This receptor complex flexibility allows TLR2 to recognize a diverse set of ligands. The TLR2-TLR1 heterodimer recognizes triacylated lipopeptides, whereas the TLR2-TLR6 heterodimer detects diacylated lipopeptides of mycoplasma or gram-negative bacteria. TLR1, TLR2, and TLR6 are expressed in macrophages, monocytes, DCs, as well as on regulatory T cells. TLR2 contributes to type I IFN production specifically in inflammatory monocytes in response to viral ligands in a MyD88-dependent, but TRIF-independent, manner.[46,47]

TLR10 shares sequence homology with TLR1 and TLR6. However, it lacks the downstream signaling typically associated with other TLR family members. It remains an orphan receptor without a known agonist, and it exists as a pseudogene in mice. TLR10 is unique among the TLR family in that it does not activate the immune system and instead suppresses inflammatory signaling in primary human cells.[48] It is primarily expressed in tissues rich in immune cells such as the lymph nodes, spleen, lungs, and tonsils. It senses HIV-1 gp41 and enhances HIV-1 infection.[49] Genetic polymorphisms in human TLR10 are linked to various diseases such as asthma,[50] bladder and nasopharyngeal carcinomas,[51] and Crohn disease.[52]

TLR5

TLR5 is also found in the plasma membrane, and it recognizes the bacterial structural protein flagellin, which is the only ligand that has been identified to date for TLR5. It is expressed in the lung and gut epithelial cells, as well as in tissue-resident DCs. TLR5 is not expressed in macrophages and conventional DCs in mice. TLR5 plays a crucial role in maintaining intestinal and lung homeostasis. Airway and gut epithelial TLR5 senses *Pseudomonas aeruginosa* and *Salmonella typhimurium* and initiates an early host inflammatory response to clear the invading pathogens. Polymorphisms in TLR5 are associated with susceptibility to infections[53-55] and autoimmune disease.[56]

TLR4

TLR4 is the most well-studied and best characterized TLR, and it was the first PRR to be discovered in mammalian innate immune cells. It is expressed predominantly on macrophages and DCs. The main ligand of TLR4 is LPS, a major component of the outer membrane of gram-negative bacteria. Single-nucleotide polymorphisms at two locations in human TLR4, D299G and T399I, are thought to be associated with LPS hyporesponsiveness.[57] LPS does not bind to TLR4 directly. Instead, TLR4 recognition of LPS requires three accessory proteins, including LPS-binding protein (LBP), myeloid differentiation 2 (MD2), and the glycosylphosphatidylinositol (GPI)–anchored protein CD14. During bacterial infection, LBP in the serum binds avidly to LPS and delivers LPS monomers to CD14. LPS is then relayed to the MD2-TLR4 complex. MD2 forms a stable heterodimer with the extracellular domain of TLR4. Five out of the six lipid chains of LPS submerge inside a hydrophobic pocket within MD2, while the last lipid chain remains exposed on the surface of MD2 and binds to TLR4. This binding induces TLR4 dimerization that activates intracellular signaling pathways. In addition to LPS, CD14 can bind other microbial products, such as peptidoglycan, lipoteichoic acid, lipoarabinomannan, and lipoproteins. Therefore, CD14 has broad ligand specificity and functions as a PRR.[58]

TLR4 also recognizes several other components of pathogens such as mannuronic acid polymers from gram-negative bacteria, teichuronic acid from gram-positive bacteria, and the F protein from respiratory syncytial virus. In addition to exogenous PAMPs, TLR4 also recognizes endogenous molecules such as heat shock proteins,[59] heme,[60] and lipoproteins.[61]

TLR3, TLR7, TLR8, and TLR9

TLR3, TLR7, TLR8, and TLR9 are intracellular TLRs capable of detecting nucleic acids (Fig. 12.1B). The trafficking of these intracellular TLRs from the endoplasmic reticulum to the endosome requires a specific protein, UNC93B1. Mice lacking UNC93B1 have defects in endosomal TLR signaling. UNC93B1 in mice is also associated with TLR13, which is located within endosomes, but its ligand remains unknown. These intracellular TLRs primarily sense viruses.

TLR3 is expressed constitutively in neurons, astrocytes, and microglia. In immune cells, only polymorphonuclear leukocytes, T cells, B cells, NK cells, myeloid DCs, and macrophages, but not monocytes, express TLR3. It is also expressed in fibroblasts and various epithelial cells including airway, corneal, cervical, biliary, and intestinal cells, which are the target sites of several viral infections. The expression of TLR3 is upregulated by type I IFN in a positive feedback loop during viral infection. TLR3 interacts with the ribose-phosphate backbone of dsRNA irrespective of specific sequence requirements. It recognizes dsRNA longer than 40 bp, which is generally absent under physiological conditions. TLR3 preferentially detects dsRNA from dsRNA viruses such as reovirus or the synthetic analog of dsRNA, polyinosinic-polycytidylic acid (poly I:C). Additionally, TLR3 recognizes some ssRNA viruses such as West Nile virus, respiratory syncytial virus, and encephalomyocarditis virus, and even dsDNA viruses such as herpes simplex virus and mouse cytomegalovirus.[62] The dsRNA produced during the replication of ssRNA or dsDNA viruses is also sensed by TLR3.[63]

TLR7 and TLR8 recognize ssRNA, imidazoquinoline derivatives such as resiquimod (R848) and imiquimod, and guanine analogs such as loxoribine. Some of the ssRNA viruses that TLR7 and TLR8 detect include HIV, influenza, and vesicular stomatitis virus.[62,64] TLR7 is expressed by pDCs, B cells, and eosinophils, while TLR8 is expressed primarily by monocytes and macrophages.

TLR9 is the DNA sensing TLR. It is predominantly expressed by pDCs and B cells, which produce the largest amount of type I IFN in response to viral infections.[65] It preferentially binds to hexamers containing unmethylated CpG motifs of DNA, which are commonly found in bacteria and viruses. CpG must be shuttled to endosomes in a controlled manner to encounter TLR9.[66] This shuttling is regulated by molecules such as granulin[67] and HMGB1.[68]

C-TYPE LECTIN RECEPTORS

Carbohydrate moieties present in certain microbes often act as PAMPs. For example, glucans present on the cell walls of fungi and mycobacteria; high-mannose structures expressed by some viruses, fungi, and bacteria; and fucose structures found on the surface of helminths and some bacteria are recognized by CLRs to mount immune responses. CLRs are glycosylated transmembrane proteins with a carbohydrate recognition domain (CBD) in their extracellular C-terminus. The CLR family contains more than 1000 proteins which are divided into 17 subgroups based on their structures and domain functions. The subgroups all utilize the same intracellular signaling pathways which involve recruitment and phosphorylation of spleen tyrosine kinase (SYK). SYK phosphorylation subsequently triggers its binding to caspase recruitment domain–containing protein 9 (CARD9), B cell lymphoma/leukemia 10 (BCL10), and mucosa-associated lymphoid tissue 1 (MALT1). SYK binding to these proteins leads to downstream signaling and cytokine production.

Based on their intracellular signaling motifs, CLRs can be classified largely in the following subgroups: immunoreceptor tyrosine–based activation motif (ITAM)–coupled CLRs, hemi-ITAM (hemITAM)–coupled CLRs, immunoreceptor tyrosine–based inhibitory motif (ITIM)–coupled CLRs, and a group of CLRs that lack typical signaling motifs[69] (Table 12.4). Examples of ITAM-coupled CLRs include Dectin-2 (CLEC6A), macrophage-inducible C-type lectin (MINCLE), and myeloid DAP12-associating lectin (MDL)-1. This subgroup interacts with ITAM-containing adaptor proteins such as Fc receptor γ (FcRγ) chain or DNAX activation protein 12 (DAP12).

TABLE 12.4 Mouse Membrane-Bound CLRs and Their Ligands

Associated Signaling Modules	CLR Family Member	Gene Symbol	Ligands/PAMPs	Functions
Non-ITAM, Non-ITIM	DC-SIGN (SIGN-R3)	*Cd209d* (Chr. 8)	Fucose, mannose, HIV-1gp120, ICAM-2, ICAM-3	DC migration, T cell interaction, antigen uptake
	Mannose receptor	*Mrc1* (Chr. 2)	Fucose, mannose, N-acetylglucosamine	Suppression of Th1, activation of Th2 differentiation
	CLEC1A	*Clec1a* (Chr. 6)	DHN-melanin	Immune response against *Aspergillus fumigatus*
ITAM	Dectin-2	*Clec4n* (Chr. 6)	Mannans	Defense against *Mycobacterium* and fungi
	MDL-1	*Clec5a* (Chr. 6)	Dengue virus particle	Macrophage survival, osteoclastogenesis
	DCAR2	*Clec4b1* (Chr. 6)	Phosphatidyl inositol mannosides (PIM)	T cell response against *Mycobacterium*
	MCL	*Clec4d* (Chr. 6)	TDM, glucuronoxylomannan	Defense against *Mycobacterium* and *Cryptococcus*
	MINCLE	*Clec4e* (Chr. 6)	TDM, SAP130	Defense against *Mycobacterium* and fungi
hemITAM	Dectin-1	*Clec7a* (Chr. 6)	β-glucans, galectin-9, vimentin	Defense against *Mycobacterium* and fungi, suppression of tumorigenesis
	DNGR-1	*Clec9a* (Chr. 6)	Mycobacteria, necrotic cells	Defense against *Mycobacterium*, necrotic cell antigen cross-presentation
	MGL-1	*Clec10a* (Chr. 11)	MUC1, siglec-1	Regulation of effector T cell, suppression of Treg
	CLEC-2	*Clec1b* (Chr. 6)	Podoplanin, rhodocytin	Lymphangiogenesis, regulation of tumor cell growth
	DC-SIGN	*Cd209d* (Chr. 8)		MHC-I upregulation
ITIM	DCIR	*Clec4a2* (Chr. 6)	Sulfated lactose, LacNAc	DC and osteoclast differentiation, immunity against tuberculosis
	MICL	*Clec12a* (Chr. 6)	Uric acid crystals	Recognition of apoptotic cells, antigen uptake, and cross-presentation
	CLEC128	*Clec12b* (Chr. 6)	GlcNAc, GalNAc, galactose	Inhibition of the NK receptor, NKG2D-mediated signaling

Dectin-2 and MINCLE use the FcRγ chain adaptor, while MDL-1 interacts with DAP12. Examples of hemITAM-coupled CLRs include Dectin-1 (CLEC7A), CLEC-2, and dendritic cell NK lectin group receptor-1 (DNGR-1). Both the ITAM and hemITAM CLRs are typically considered activating receptors that lead to downstream activation of NF-κB and MAPK pathways, resulting in the induction of proinflammatory cytokines. Conversely, ITIM-coupled CLRs suppress signaling initiated by heterologous receptors through the recruitment of tyrosine phosphatases, such as Src homology region 2 domain–containing phosphatase (SHP) 1 or 2[70] (Fig. 12.3). For example, activation of the ITIM-coupled CLR DC immunoreceptor (DCIR) leads to inhibition of TLR8-mediated IL-12 and TNF production and TLR9-induced IFN production.[71,72] CLRs that lack both ITAM and ITIM motifs such as DC-SIGN are not involved in activation or cytokine secretion. Instead, they modulate signaling by heterologous receptors or contribute to antigen processing/presentation to T cells.[69]

Dectin-1 and Dectin-2

Dectin-1 (a hemITAM-coupled CLR) was the first CLR discovered, and it was initially described as a DC-specific lectin that stimulates CD4+ T cell proliferation.[73] It is the best characterized CLR and is involved in antifungal immunity. Dectin-1 is predominantly expressed in myeloid cells, including monocytes, macrophages, DCs, and neutrophils. It binds to β-glucans, which are major fungal cell wall components.[74] Dectin-1 can also sense several endogenous ligands such as vimentin, galactosylated immunoglobulins, and galactins. Ligand recognition by Dectin-1 induces multiple downstream pathways. These pathways include those mediated by Raf-1 and SYK/CARD9, which regulate numerous cellular responses, such as phagocytosis, respiratory burst, neutrophil extracellular trap formation, autophagy, DC maturation, antigen presentation, and inflammasome activation.[17] Initial sensing of β-glucan from *Candida albicans* by Dectin-1 contributes to trained immunity and enhanced protection against subsequent microbial infection, a phenomenon similar to the immunological memory attributed to T and B cell function.[75,76] Additionally, polysaccharides from the fungal cell wall induce Th17 differentiation through the activation of Dectin-1 in DCs.[77] Th17 cell–derived IL-17A and IL-17F can then recruit neutrophils to clear fungi.[78] However, these Dectin-1–induced Th17 cytokines can also worsen asthma and allergic diseases.[79]

Dectin-1 deficiency leads to increased susceptibility to infections with fungal species such as *C. albicans* and *Aspergillus fumigatus*,[80,81] and mutations in human Dectin-1 are associated with familial chronic mucocutaneous candidiasis.[82] In patients with hematological malignancies undergoing hematopoietic stem cell transplantation, having an early stop codon polymorphism in CLEC7A (Y238X) is associated with susceptibility to invasive aspergillosis.[83] Similarly, polymorphisms in CLEC7A are strongly associated with inflammatory bowel disease.[84] Furthermore, the SYK-CARD9 signaling axis, the common downstream signalosome associated with most CLRs (Fig. 12.3), contributes to anticolorectal cancer immunity by promoting epithelial barrier restitution via inflammasome activation and IFN-γ production.[85]

CLR signaling

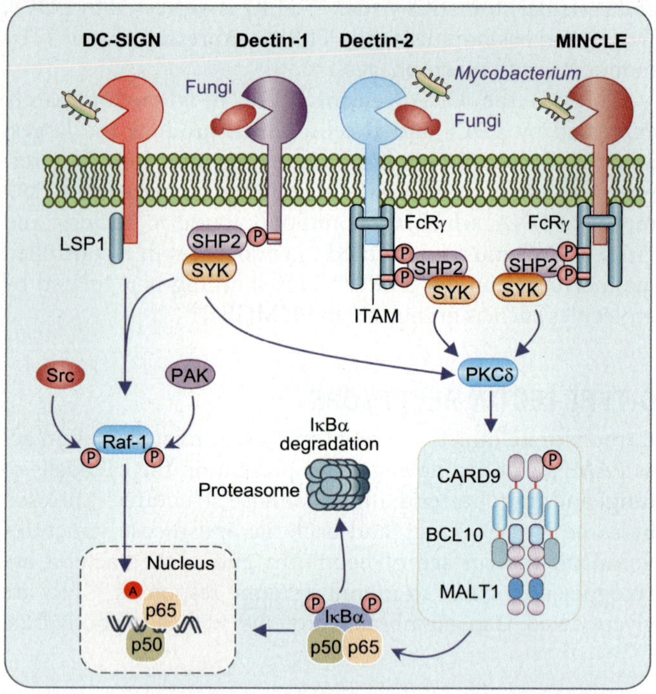

FIG. 12.3. CLR signaling. DC-SIGN is constitutively associated with the adapter molecule LSP1 in combination with Raf-1 under homeostatic conditions. Upon mannose or other ligand binding to DC-SIGN, the receptor undergoes conformational changes that allow for Raf-1 to dissociate from the complex and undergo subsequent phosphorylation by Src and PAK kinases. DC-SIGN–induced Raf-1 activation mediates the phosphorylation of NF-κB subunit p65, which in turn leads to p65 acetylation. Acetylation of p65 prolongs and increases IL-10 production. The recognition of fungi or fungal ligands by Dectin-1 induces phosphorylation of its immunoreceptor tyrosine-based activation motif (ITAM)–like sequence. Dectin-2 and MINCLE require a membrane-associated adaptor protein FcRγ that contains ITAMs to induce intracellular signaling. Phosphorylation of ITAMs subsequently recruits SHP2 and SYK to the receptor complex. Activation of SYK induces assembly of the CARD9-BCL10-MALT1 complex, which mediates NF-κB activation for the production of various inflammatory cytokines.

Dectin-2 (CLEC4N; an ITAM-coupled CLR) is a type 2 transmembrane receptor that was discovered as a DC-specific CLR along with Dectin-1.[86] Dectin-2 contains an EPN motif, a putative Ca^{2+}-dependent mannose-binding amino acid sequence, in its extracellular C-terminus.[87] It recognizes mannose structures present in a wide range of organisms including fungi, parasites, bacteria, and mammals.[88] Dectin-2–deficient mice have increased susceptibility to infection with *C. albicans*, and Dectin-2 is crucial for the IL-17 production that is essential for fungal clearance.[89]

MINCLE (CLEC4E)

MINCLE (an ITAM-coupled CLR) was first identified as an LPS-inducible gene under the control of nuclear factor for IL-6 (NF-IL6 or C/EBPβ).[90] The expression of MINCLE is strongly upregulated by TNF, IL-6, and IFNγ. The first ligand identified for MINCLE was a mycobacterial glycolipid.[91]

MINCLE can recognize glycolipids in the cell walls of both bacteria and fungi. A combination of sugar and lipid moieties is important for recognition by MINCLE. It can also sense self-glycolipids released from damaged cells such as soluble spliceosome-associated protein 130 (SAP130),[92] β-glucosylceramide,[93] or cholesterol sulfate.[94]

DC-SIGN

DC-SIGN (DC-specific intercellular adhesion molecule-3-grabbing non-integrin; a non-ITAM–, non-ITIM–coupled CLR) senses a broad spectrum of mannose and fucose moieties in polysaccharides present on various microbes, including viruses, parasites, fungi, and bacteria, and on self-receptors, such as ICAM-2, ICAM-3, and Mac-1.[95-97] DC-SIGN forms a ligand-specific signalosome that acts differently depending on the type of ligand bound. DC-SIGN promotes proinflammatory responses upon binding to mannosylated glucans such as those present on ICAM-3 or *Mycobacterium tuberculosis*, whereas it represses TLR-induced proinflammatory cytokine expression and promotes Th2 polarization upon recognition of fucosylated glucans such as those present on Mac-1 or *Helicobacter pylori*.[96,97] Under homeostatic conditions, DC-SIGN is associated with the adapter molecule LSP1 and Raf-1. Upon sensing of ligands or pathogens, such as *Mycobacterium* and *Candida*, the receptor undergoes conformational changes that allow Raf-1 to dissociate from the complex. Raf-1 is then phosphorylated by Src and PAK kinases. DC-SIGN–induced Raf-1 activation leads to p65 acetylation, which prolongs IL-10 production (Fig. 12.3). Polymorphisms in DC-SIGN have been associated with dengue infection and increased susceptibility to tuberculosis.[98]

Mannose Receptor

Mannose receptor (MR; a non-ITAM–, non-ITIM–coupled CLR) was one of the first CLRs to be discovered, and it has been shown to be involved in the clearance of glycoproteins. MR is a type I integral transmembrane receptor that contains a cysteine-rich domain, a fibronectin type II–like domain, and 8 C-type lectin domains (CTLDs) in the extracellular region. The fibronectin type II–like domain mediates collagen binding.[99] MR also has a hydrophilic internal cytoplasmic domain which lacks ITAM and ITIM motifs. The cytoplasmic domain is involved in receptor internalization and recycling. MR binds structures like L-fucose, D-mannose, and N-acetylglucosamine through a Ca^{2+}-dependent mechanisms.[100] In addition to binding microbial patterns, MR can also bind endogenous self-antigens, including ligands in the B cell follicles in the spleen and thyroid.[101] Upon ligand binding, the MR cytoplasmic tail interacts with FcRγ, recruiting another adaptor protein called Grb2 that subsequently activates the Rac-1, Cdc42, and PAK1 complex. The formation of this complex is important for phagocytosis.[102]

Mannan-Binding Lectin

Mannan-binding lectin (MBL), also known as mannose-binding lectin, is a soluble collagen-containing C-type lectin (collectin) with primary specificity for fucose, mannose, and N-acetylglucosamine, which are exposed on the surface of a number of pathogens.[103] MBL is produced mostly by the liver for secretion into the circulation. The serum level of MBL is increased during stress and infections.[104] A low MBL level is a risk factor for the development of neonatal infections.[105] MBL is involved in the phagocytosis of bacteria as well as in recognition and clearance of apoptotic cell bodies.[106] It also plays a role in complement activation. In particular, MBL co-operates with three MBL-associated serine proteases (MASPs-1, -2, and -3) to initiate lectin-dependent complement activation.[107] Because of the structural homology between MBL and complement C1q, MBL also binds to complement receptor 1/CD35.[108]

SCAVENGER RECEPTORS

SRs are membrane-bound PRRs that are structurally diverse and participate in a wide range of biological functions. Similar to other PRRs, SRs have a central role in innate immunity because of their diverse affinity for modified lipids and pathogens. SRs were originally defined by Goldstein and Brown in 1979 for their ability to recognize and remove modified lipoproteins.[109,110] Modified lipoproteins and other endogenous proteins can contain epitopes generated by peroxidation that resemble microbial structures; these epitopes are classified as DAMPs and are recognized by SRs.[111,112] Other biological functions of SRs include endocytosis, phagocytosis, adhesion, and signaling that ultimately lead to the elimination of degraded or harmful substances. These receptors are expressed predominantly by myeloid cells.

On the basis of sequence alignments and protein domain architecture, SRs are subdivided into eight classes, designated as A–H. In addition, there are proteins that exhibit SR activity, but have not been included in general SR nomenclature.

Class A SRs (SR-A)

Proteins in the SR-A class are expressed primarily on tissue macrophages and DCs.[113] SR-As contain an N-terminal cytoplasmic tail, transmembrane domain, spacer region, α-helical coiled coil domain, collagen domain, and a C-terminal scavenger receptor cysteine-rich (SRCR) domain. SCARA1/MSR1 was the first prototypical SR-A molecule to be cloned, and it is now referred to as SR-A1. A diverse array of ligands is recognized by members of this class. SR-A1 binds to β-amyloid,[114] heat shock proteins,[115] surface molecules of bacteria,[116] hepatitis C virus,[117] and modified LDL,[118] for example.

Class B SRs (SR-B)

Class B SRs contain a CD36 domain. CD36 is the prototypical SR-B receptor and is one of the most widely studied SRs. It was initially identified as a receptor for thrombospondin.[119] It is expressed in multiple cell types, including hematopoietic cells such as platelets, monocytes, and macrophages, endothelial cells, and some epithelial cells. It regulates various biological processes such as migration, inflammatory signaling, and foam cell formation in macrophages. CD36 has been characterized as a receptor for oxLDL.[120] It also cooperates with TLR family members to activate the innate immune response.

Class C SRs (SR-C)

Class C SRs have not been described in mammals.

Class D SRs (SR-D)

Class D SRs contain mucin-like and lysosome-associated membrane glycoprotein (LAMP) domains. CD68 is the only member of this class and is found on monocytes and tissue-specific macrophages. It serves as a SR for oxLDL.[121] It is used as a differentiation marker on hematopoietic cells of the monocyte/macrophage lineage.

Class E SRs (SR-E)

SR-E proteins are type 2 transmembrane proteins with C-type lectin-like (CLEC) domains. Lectin-like oxidized LDLR (LOX-1) and Dectin-1 belong to this family and have been shown to have scavenger activity. LOX-1 is expressed on macrophages, DCs, vascular endothelial cells, platelets, smooth muscle cells, and adipocytes. LOX-1 recognizes oxLDL and acute-phase protein C-reactive protein.[122]

Class F SRs (SR-F)

Class F SRs contain EGF-like domains. There are three members in this family: SCARF1, SCARF2, and MEGF10. SCARF1 was identified as an endothelial SR for modified LDL.[122] It is expressed in endothelial cells and macrophages as well. It binds calreticulin and associated peptides. It also recognizes fungal pathogens.[123] SCARF2 is expressed on endothelial cells and macrophages. SCARF2 is less well-characterized. MEGF10 is a receptor for the protein amyloid-β.[124] MEGF10 also recognizes complement protein C1Q. Mutations in MEGF10 in humans result in early-onset myopathy, respiratory distress, and dysphagia.[125]

Class G SRs (SR-G)

SR-G proteins contain a CXC-chemokine domain. SRs that bind phosphatidylserine and oxidized lipids (SR-PSOX) belong to this family.[126] The chemokine CXCL16 is identical to SR-PSOX in its amino acid sequence. SR-PSOX/CXCL16 exist in both membrane and soluble forms. The membrane-bound form acts as a SR for oxLDL and mediates phagocytosis of bacteria by antigen-presenting cells.[127] In addition, it regulates migration of immune cells. The soluble form of SR-PSOX/CXCL16 is produced by cleavage from the membrane domain and functions as a chemoattractant for activated T cells and bone marrow plasma cells via its receptor CXCR6.[127]

Class H SRs (SR-H)

Class H SRs consist of fasciclin, EGF-like, and lamin-type EGF-like (FEEL) domains. This class includes FEEL-1/stabilin-1/CLEVER-1 and FEEL-2/stabilin-2/HARE. FEEL-1 is predominantly expressed in macrophages, mononuclear cells, hematopoietic stem cells, and endothelial cells, whereas FEEL-2 is found primarily in sinusoidal endothelial cells. FEEL-1 and FEEL-2 are involved in lymphocyte adhesion, transmigration, angiogenesis, apoptotic cell clearance, and intracellular trafficking.[128,129]

Class I SRs (SR-I)

Class I SRs (CD163) contain multiple extracellular SRCR domains. SR-I1, also called CD163A, is composed of nine SRCR domains and is a prototypical class I SR. It is expressed exclusively by monocytes and macrophages and is also called "hemoglobin SR." It binds to haptoglobin-hemoglobin complexes to promote the clearance of plasma hemoglobin via endocytosis during intravascular hemolysis.[130] SR-I2 (CD163-L1, M160, CD163B) is composed of 12 SRCR domains and is expressed by cells of the myeloid lineage. It regulates the differentiation of monocytes to macrophages.[131] SR-I3 (CD163c-α) contains five SRCR domains. It is expressed by αβ and γδ T cells and in the small intestine and colon.[132]

Class J SRs (SR-J)

The only receptor in the SR-J class is the receptor for advanced glycation end products (RAGE). It belongs to the Ig superfamily of cell surface molecules. It interacts with a broad spectrum of ligands such as high mobility group protein box 1 (HMGB1) and the S-100 protein. RAGE recognizes endogenous molecules released during infection, physiological stress, or inflammation.[133] RAGE regulates various biological processes associated with atherosclerosis, diabetes, cancer, neurodegenerative disease, and stroke.[134] In addition to the membrane-bound form of RAGE, there is also soluble RAGE (sRAGE), the level of which increases in patients with myocarditis.[135]

Class K SRs (SR-K)

CD44, a cell surface glycoprotein, belongs to the SR-K class. CD44 is a receptor for hyaluronan.[136] In addition, it has a broad repertoire of ligands that include both host and microbial products. Like other SRs, it contributes to the clearance of its extracellular matrix ligands via endocytosis.[137]

Class L SRs (SR-L)

The SR-L class contains members of the LDLR gene family. LDLR is the principal clearance receptor for plasma cholesterol. SR-L1 is also known as LDLR-related protein 1 (LRP1), α_2-macroglobulin receptor (A2MR), CD91, or apolipoprotein E receptor (APOER). LRP1 is ubiquitously expressed in multiple tissues, and it is most abundant in vascular smooth muscle cells, hepatocytes, and neurons. In addition to its role in lipoprotein metabolism, it is also involved in the effective clearance of β-amyloid from the brain to the periphery across the blood–brain barrier.[138] Also, it functions in clearing proteases such as plasmin and metalloproteinases, which contributes to the prevention of cancer invasion. It is implicated in Alzheimer disease, cardiovascular diseases, and cancer.[139,140]

Megalin (SR-L2), also known as LRP2, is expressed on the apical membrane of several epithelial cells such as kidney proximal tubule cells, lung, thyroid, gallbladder, epididymis, prostate, ovaries, and uterus.[141] It internalizes a variety of ligands such as leptin, insulin, and β-amyloid. LRP2 is also involved in the uptake of albumin, vitamin D–binding protein, and retinol-binding protein.

NOD-LIKE AND INFLAMMASOME-FORMING RECEPTORS

The NLR family is a large family of proteins that are activated by intracellular PAMPs and DAMPs. Unlike TLRs and CLRs, which are membrane-associated PRRs, NLRs are cytosolic proteins which play major roles in driving inflammatory responses. The human and mouse genomes contain approximately 23 and 34 NLR genes, respectively. All NLRs share a typical domain structure containing (a) an N-terminal effector domain, which interacts with other proteins and mediates signal transduction; (b) a centrally located nucleotide-binding NACHT domain, which mediates self-oligomerization; and (c) a C-terminal domain, which contains LRR motifs involved in the recognition of PAMPs and DAMPs.[142] Based on the NTD structure, NLRs can be largely classified into four subfamilies: (1) acidic transactivation domain (AD)–containing (NLRA); (2) baculoviral inhibitory repeat–like domain (BIR)–containing (NLRB); (3) caspase activation and recruitment domain (CARD)–containing (NLRC); and (4) pyrin domain (PYD)–containing (NLRP) (Table 12.5; Fig. 12.4).

The NLRA subfamily contains only one member, called the class II major histocompatibility complex (MHC) transactivator (CIITA). In addition to an AD domain, CIITA also contains four LRRs, a GTP-binding domain, and three nuclear localization signals (NLSs).[143] CIITA has the ability to induce de novo transcription of the MHC class II gene, as well as to reinforce MHC class I gene expression.[144] Similar to the NLRA subfamily, the NLRB subfamily also only contains one member in humans, named the NLR family apoptosis inhibitory protein (NAIP), which is generally known to inhibit apoptotic caspases. The NLRC subfamily consists of six members: NLRC1, NLRC2, NLRC3, NLRC4, NLRC5, and NLRX1. The NLRP subfamily consists of 14 members. Upon activation, some members of the NLRC and NLRP subfamilies assemble into activation platforms for caspase-1 autoproteolysis. These activation platforms are called inflammasomes, and inflammasomes are essential for inducing cell death and for the secretion of bioactive IL-1β and IL-18. NLRP1, NLRP3, and NLRC4

TABLE 12.5 NLRs and Their Microbial Ligands and Associated Diseases

Subfamily	NLR	Ligands/Activating PAMPs	Functions	Associated Diseases
NLRA	CIITA	Unknown	Regulation of MHC-II expression	Bare lymphocyte syndrome, Hodgkin disease
NLRB	NAIPs	Flagellin, T3SS	NLRC4 inflammasome formation	Spinal muscular atrophy
NLRC	NLRC1	γLRCal muscular atDAP	RIP2-dependent NF-κB and MAPK activation	Asthma, inflammatory bowel disease
	NLRC2	Muramyl dipeptide	RIP2-dependent NF-κB and MAPK activation	Crohn disease, Blau syndrome, atopic dermatitis, eczema
	NLRC3	Unknown	Inhibition of TLR responses	Obesity, cancer
	NLRC4	Flagellin, T3SS	Inflammasome formation	Familial cold autoinflammatory syndrome, infantile enterocolitis, MAS
	NLRC5	Unknown	MHC-I upregulation	Unknown
	NLRX1	Unknown	ROS generation, viral-induced autophagy	Unknown
NLRP	NLRP1	Anthrax lethal toxin	Inflammasome formation	Vitiligo, celiac disease, SLE, rheumatoid arthritis, systemic sclerosis, type I diabetes
	NLRP2	Unknown	Negative regulation of NF-κB	Beckwith-Wiedemann syndrome
	NLRP3	PAMPs/DAMPs	Inflammasome formation	CAPS, inflammatory bowel disease, psoriasis, gout, diabetes
	NLRP4	Unknown	Inhibition of type I IFN production, autophagy	Unknown
	NLRP6	Lipoteichoic acid	Inflammasome formation	Colitis and colon cancer
	NLRP7	Acylated lipopeptide	Inflammasome formation	Hydatidiform mole, testicular seminoma, endometrial cancer
	NLRP12	Unknown	Negative regulation of NF-κB	Atopic dermatitis, periodic fever syndrome

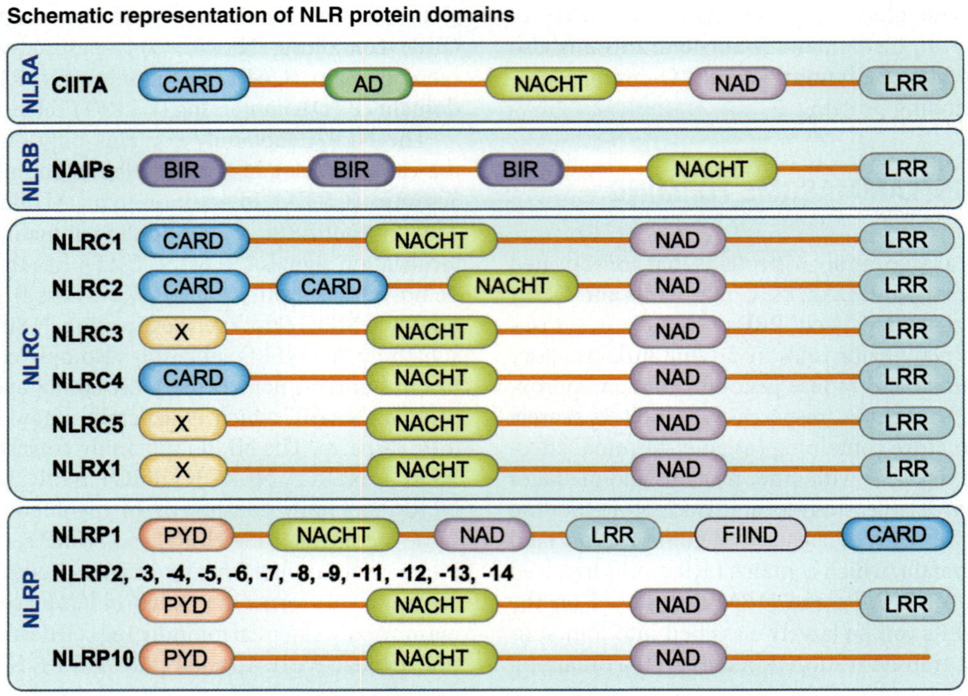

FIG. 12.4. Schematic representation of NLR protein domains. Human NOD-like receptors (NLRs) are subclassified into four categories: NLRA, NLRB, NLRC, and NLRP. All human NLRs contain a central NACHT domain, and all except NLRP10 contain a C-terminal ligand-sensing leucine-rich repeat (LRR) domain; most also contain an NBD-associated domain (NAD). The N-terminal domains ascribe functional properties to the NLRs. In the NLRC subfamily, the X displays an unknown domain that has no homology with sequences from the other NLR family members.

have well-established abilities to assemble inflammasomes. Additionally, the non-NLR sensors Pyrin and AIM2 are also known to assemble inflammasomes. Several other NLR sensors, including NLRP6, NLRP7, NLRP9, NLRP12, NLRC3, and NLRC5, may form inflammasomes in context-dependent manners.[145] Inflammasomes have largely been described in immune cells, such as macrophages and DCs, but they are also expressed and assembled in nonhematopoietic cells.

NLRC1 and NLRC2

NLRC1 and NLRC2, also known as NOD1 and NOD2, respectively, were the first intracellular sensors of bacterial peptidoglycan to be discovered, and they have a similar domain organization. NLRC1 contains a single N-terminal CARD domain, whereas there are two CARD domains in NLRC2 (Fig. 12.4). Peptidoglycan is a polymer comprised of amino acids and sugars with a backbone of alternating N-acetylglucosamine and N-acetylmuramic acid residues (MurNAc). While NLRC1 recognizes γ-D-glutamyl-meso-diaminopimelic acid, NLRC2 detects the muramyl dipeptide, both of which are found in peptidoglycans of bacteria.[146-148] Upon recognition of their respective cognate ligands, NLRC1 and NLRC2 undergo oligomerization and a conformational change which subsequently allows the recruitment of the scaffolding kinase RIPK2 (also known as RICK) through a homotypic CARD–CARD interaction.[149] Recruitment of RIPK2 leads to the activation of NF-κB, MAPK, and IFN signaling pathways. Interestingly, the protein YopJ produced by *Yersinia* species directly inhibits NLRC2 signaling by acetylating downstream mediators, such as TAK1.[150] In addition to their activation of signaling pathways, NLRC1 and NLRC2 also recruit the cellular autophagy machinery to counteract invading pathogens. Notably, NLRC1-RIPK2 signaling mediates mitophagy to inhibit inflammasome activation during viral infection.[151] However, NLRC1 and NLRC2 signaling via RIPK2 promotes inflammation and autoimmune disease.[152] Polymorphisms in human NOD2 have been associated with the development of leprosy,[153] susceptibility to tuberculosis,[154] *H. pylori*–associated gastric lymphoma,[155] and Crohn disease.[156]

NLRC3, NLRC5, and NLRX1

NLRC3, NLRC5, and NLRX1 have several atypical features that contribute to their complexity and uniqueness within the NLRC family. Unlike NLRC1 and NLRC2, they do not have specific ligands. These molecules have atypical NTDs (Fig. 12.4). They are largely involved in the suppression of NF-κB, MAPK, and IFN signaling pathways induced during microbial infection.

Despite its name, NLRC3 does not contain a CARD domain, and the function of its N-terminus is currently unknown. NLRC3 is expressed primarily in lymphoid progenitor–derived populations, such as CD4+ T cells, CD8+ T cells, and NK cells. The expression of NLRC3 in T cells decreases upon stimulation with the T cell receptor (TCR)–activating triggers anti-CD3/CD28 or phorbol myristate acetate (PMA)/ionomycin.[157] NLRC3 acts as a potential negative regulator of inflammation in macrophages. It inhibits TLR4 signaling–mediated TRAF6 activation to suppress inflammatory cytokine production.[158] In addition, NLRC3 inhibits the production of IFN-β by interfering with the STING-TBK1 interaction during DNA viral infection.[159,160] NLRC3 has been shown to be associated with obesity and cancer; specifically, it suppresses PI3K-mTOR signaling to prevent colon cancer.[161] In addition, NLRC3 acts as a rheostat that fine-tunes T cell responses in infection and autoimmunity.[162]

NLRC5 is phylogenetically closest to CIITA. Similar to CIITA, NLRC5 contains an NLS within its N-terminus and, therefore, can translocate into the nucleus. NLRC5 also contains an atypical CARD domain in its N-terminus. As an MHC class I transactivator, NLRC5 induces the expression of MHC class I genes.[163] NLRC5 is predominantly expressed in leukocytes including CD4+ T cells, CD8+ T cells, CD19+ B cells, NK cells, and NKT cells.

NLRX1 lacks a fully characterized N-terminus, hence the "X" used to define the gene/protein. It is ubiquitously expressed in immune cells and is the only NLR with a mitochondria-targeting sequence in its N-terminus.[164] NLRX1 shuttles from the cytosol to the mitochondria to regulate inflammation, apoptosis, and other mitochondrial functions.[165-167]

Inflammasome-Forming Receptors

NLRP1, NLRP3, and NLRC4 are widely studied inflammasome-forming NLR sensors. Non-NLR sensors, including Pyrin and AIM2, can also assemble inflammasomes. The core components of the inflammasome platform are the sensor, adaptor, and caspase-1 (Fig. 12.5). When a stimulus is sensed, the sensors generally recruit the common adaptor molecule apoptosis-associated speck-like protein containing a CARD (ASC) to form an activation platform for caspase-1. Recruitment and oligomerization of these components takes place through homotypic CARD–CARD and PYD–PYD interactions. ASC contains both a PYD and a CARD domain. The PYD-containing sensors NLRP3, Pyrin, and AIM2 interact with the CARD of caspase-1 via ASC, whereas the CARD-containing sensors NLRP1 and NLRC4 do not necessarily require ASC for their interaction with caspase-1. Assembly of the inflammasome leads to proximity-induced autoprocessing of caspase-1, resulting in the catalytically active form of caspase-1. Active caspase-1 proteolytically cleaves inactive pro–IL-1β and pro–IL-18 into bioactive IL-1β and IL-18. Similarly, active caspase-1 cleaves gasdermin D (GSDMD), which subsequently forms pores in the cell membrane, thereby allowing the release of bioactive IL-1β and IL-18 as well as other DAMPs. The cells undergo an inflammatory form of cell death referred to as pyroptosis following GSDMD activation (Fig. 12.5). The molecular mechanisms for the initiation of inflammasome assembly can be largely categorized based on the ligand specificity of the sensors. NLRP1, NLRP3, and Pyrin assemble inflammasomes through indirect sensing of cellular homeostasis, while AIM2 and NLRC4 assemble inflammasomes by direct ligand binding.

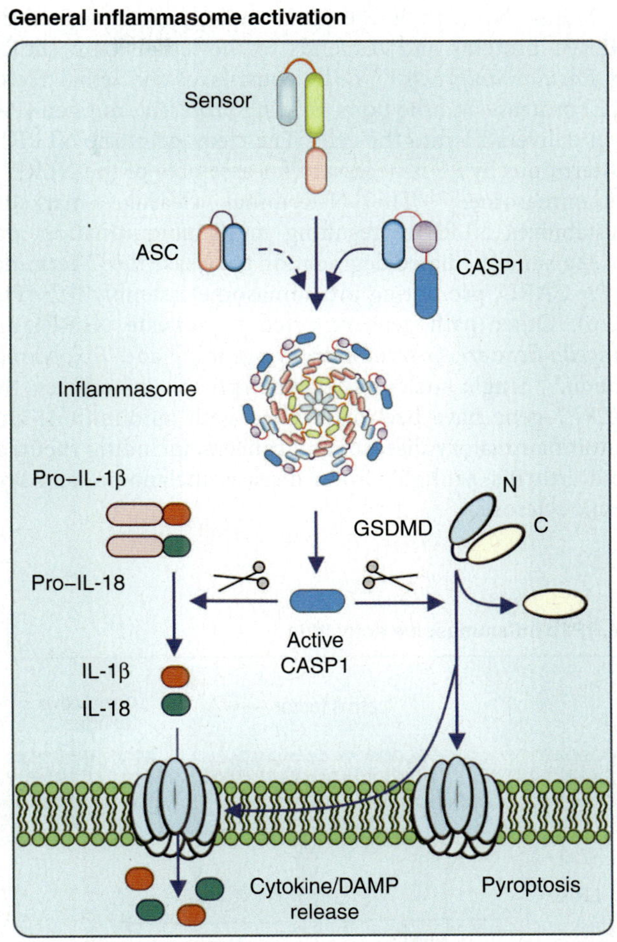

FIG. 12.5. General inflammasome activation. Inflammasome sensors can interact directly with their target ligand or respond to a variety of physiological changes. Inflammasome sensors which contain a pyrin domain may recruit apoptosis-associated speck-like protein containing a caspase activation and recruitment domain (CARD) (ASC) to mediate CARD–CARD interactions with pro–caspase-1. Alternatively, some sensors containing a CARD may directly bind to the CARD of caspase-1 and form an inflammasome complex without ASC. Caspase-1 (CASP1) then undergoes proximity-induced autocleavage to take on its active form. Activated CASP1 cleaves gasdermin D (GSDMD) to release the N-terminal domain to oligomerize in the plasma membrane and form pores, thereby inducing a form of cell death called pyroptosis. CASP1 also cleaves pro–interleukin (IL)-1β and pro–IL-18 into their active forms, which are released through the GSDMD pores.

NLRP1

NLRP1 was the first inflammasome-forming sensor to be discovered.[168] It is expressed in many cell types, including granulocytes, monocytes, DCs, B cells, T cells, and neurons. Human NLRP1 has three orthologs in mice, NLRP1a–c. Unlike human NLRP1, mouse NLRP1 orthologs lack the PYD. NLRP1 has two unusual features: (1) it has a CARD at the C-terminus and (2) it contains a FIIND (function-to-find domain) which undergoes self-cleavage resulting in two distinct polypeptides, NBD-LRR-ZU5 and UPA-CARD.[169] After stimulation, the UPA-CARD polypeptide then self-assembles and forms a platform for caspase-1 recruitment and activation.

Mouse NLRP1b is the most well-characterized of the NLRP1 proteins and responds to the lethal toxin (LeTx) of *Bacillus anthracis*.[170] LeTx consists of the lethal factor (LF) protease and the pore-forming protective antigen (PA) that delivers LF into the cells. The cleavage of the NLRP1b N-terminus by LF is necessary for assembly of the NLRP1b inflammasome.[171] The N-terminal cleavage markedly destabilizes NLRP1b, resulting in its ubiquitination and proteasome-mediated degradation to release the C-terminal UPA-CARD, promoting inflammasome assembly[169,172] (Fig. 12.6). Other pathogens reported to activate NLRP1 are *Shigella flexneri*, *Listeria monocytogenes*,[173] and *Toxoplasma gondii*.[174] Single nucleotide polymorphisms in and near the *NLRP1* gene have been associated with autoimmune and autoinflammatory diseases and cancers, including rheumatoid arthritis, vitiligo, Crohn disease, melanoma, and systemic sclerosis.[175]

NLRP1b inflammasome activation

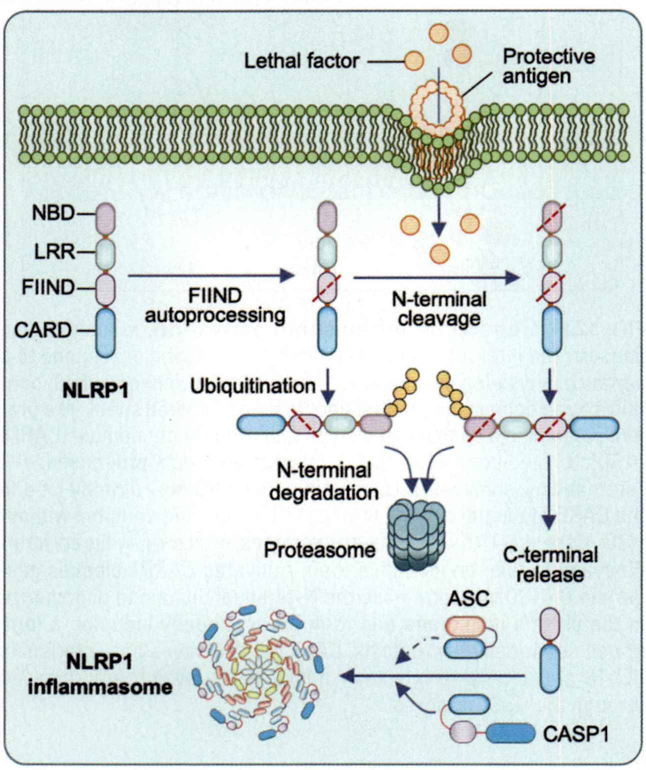

FIG. 12.6. NLRP1b inflammasome activation. Activation of murine NLRP1b involves a series of proteolytic cleavage events. First, the function-to-find domain (FIIND) of NLRP1b undergoes autoproteolytic cleavage. The protective antigen of anthrax lethal toxin delivers lethal factor across the cell membrane, allowing the lethal factor to directly cleave the N-terminal portion of the nucleotide-binding domain (NBD)-leucine-rich repeat domain (LRR)-FIIND fragment. The cleavage of NLRP1b is followed by ubiquitination, marking it for proteasomal degradation. This degradation frees the C-terminal fragment, which contains a caspase activation and recruitment domain (CARD) and can initiate inflammasome assembly with or without apoptosis-associated speck-like protein containing a CARD (ASC).

NLRP3

NLRP3, also known as cryopyrin or NALP3, was initially discovered due to its genetic association with autoinflammatory diseases such as familial cold autoinflammation syndrome (FCAS), Muckle-Wells syndrome (MWS), and neonatal-onset multisystem inflammatory disease/chronic infantile neurologic cutaneous articular syndrome (NOMID/CINCA), collectively referred to as cryopyrin-associated periodic syndromes (CAPS).[176-178] NLRP3 is expressed by innate immune cells, particularly macrophages. Canonical activation of the NLRP3 inflammasome requires a two-step mechanism. First, a priming step occurs that involves transcriptional upregulation of the essential components including NLRP3 as well as pro–IL-1β. Priming can be achieved by triggers that activate the NF-κB pathway (such as TLR ligands, cytokines, pathogens, DAMPs). Second, an activation step occurs, which is complex and not clearly understood to date (Fig. 12.7A). While two separate triggers are typically required to fulfill the priming and activation steps, LPS alone is sufficient to trigger NLRP3 inflammasome activation in human monocytes.[179]

Although most PRRs have limited specificity for PAMPs or DAMPs, NLRP3 is unique in that it is activated by a wide variety of PAMPs from bacteria, fungi, viruses, and parasites.[180-183] In addition to microbial components, NLRP3 can also be activated by sterile insults, including several DAMPs released by damaged tissues and cells.[184,185] There is no consensus model of NLRP3 activation. NLRP3 responds to cellular aberrations that result from these diverse stimuli, including the efflux of potassium ions (K^+) or chloride ions (Cl^-), flux of calcium ions (Ca^{2+}), generation of reactive oxygen species, lysosomal disruption, mitochondrial dysfunction, metabolic changes, trans-Golgi disassembly, and ribosomal disruption. Among these cellular aberrations, K^+ efflux is generally considered as a common upstream event for NLRP3 activation.[186]

In addition to this "canonical" pathway for activation, NLRP3 inflammasome activation can also be achieved as a result of mouse caspase-11 or the human analogues, caspase-4 and caspase-5, directly sensing LPS in the cytoplasm; this form of activation is termed "noncanonical NLRP3 inflammasome activation"[187-190] (Fig. 12.7B). The molecules involved in type I IFN production and signaling such as TRIF, IRF3, IRF8, IFNAR, IRF9, and STAT1 are important for NLRP3 inflammasome activation by gram-negative bacteria.[33,34,191] Type I IFNs induced during infection with gram-negative bacteria such as *Escherichia coli*, *Citrobacter rodentium*, and *S. typhimurium* upregulate the expression of caspase-11 along with a cluster of small GTPases, guanylate-binding proteins (GBPs), and their phylogenetic relative IRGB10.[192-194] GBPs and IRGB10 are recruited to the intracellular bacteria, where they lyse the bacteria and liberate LPS into the cytoplasm.[195] When caspase-11 senses the cytoplasmic LPS, it undergoes oligomerization and activation through autoproteolytic cleavage.[196] Activated caspase-11 subsequently cleaves GSDMD to induce pyroptosis, leading to an efflux of K^+ and ultimately triggering NLRP3 inflammasome activation[197,198] (Fig. 12.7B). FADD and caspase-8, which are involved in canonical NLRP3 inflammasome activation, also mediate activation of the noncanonical NLRP3 inflammasome.[193]

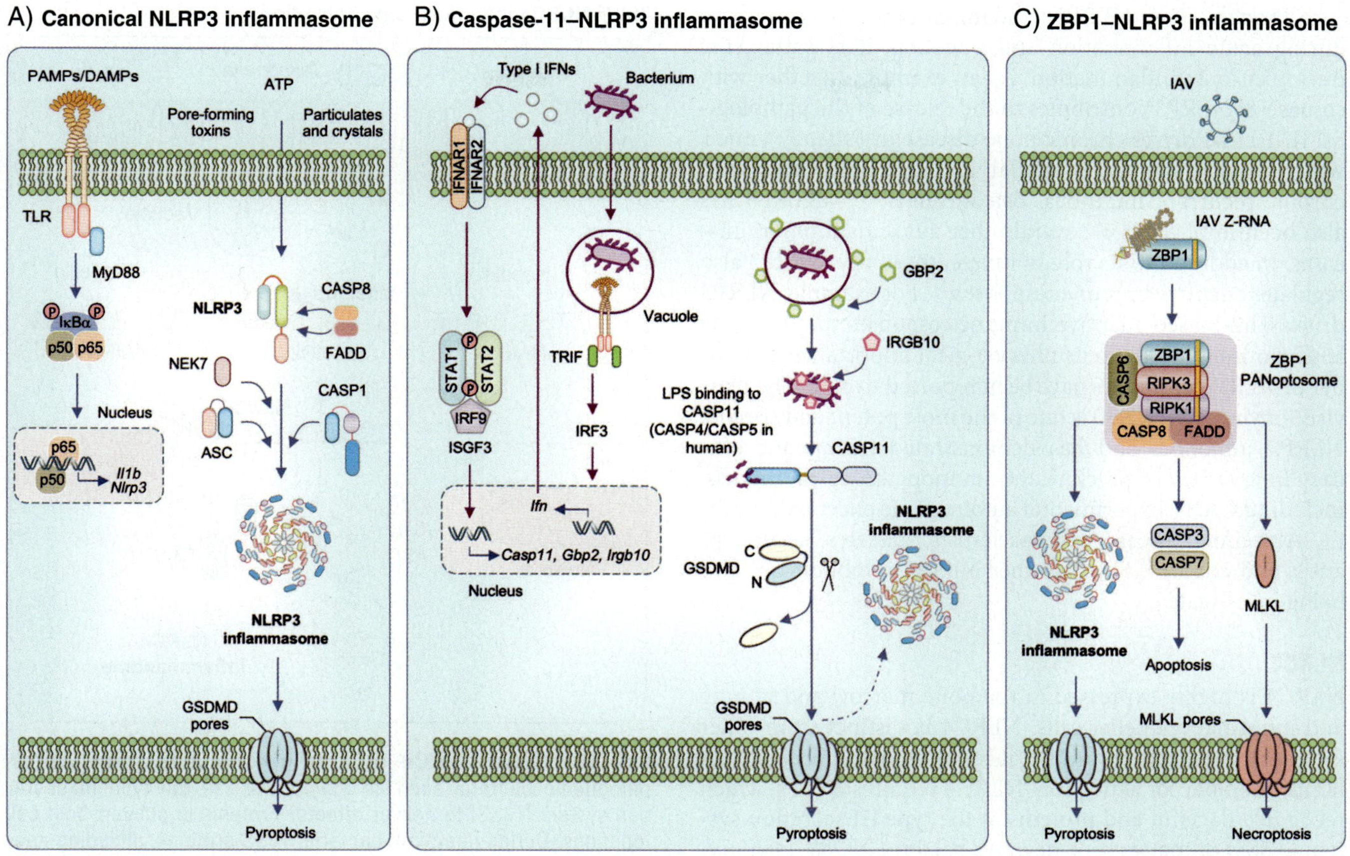

FIG. 12.7. NLRP3 inflammasome activation. A: Canonical NLRP3 inflammasome: Canonical NLRP3 inflammasome activation requires two signals: (1) priming and (2) activation. Signal 1, or the priming signal, is initiated by several classes of receptors including the Toll-like receptors (TLRs). Signaling through these receptors leads to the transcriptional upregulation of NLRP3 and pro–IL-1β. Signal 2, or the activation signal, is then induced by pathogen- or damage-associated molecular patterns (PAMPs or DAMPs) that activate multiple downstream signaling events. Sensing of physiological changes by NLRP3 is followed by the recruitment of apoptosis-associated speck-like protein containing a caspase activation and recruitment domain (ASC) and caspase-1 (CASP1) to form an inflammasome complex. Numerous other molecules such as CASP8, Fas-associated death domain (FADD), and NIMA-related kinase 7 (NEK7) are involved in NLRP3 activation in response to a variety of stimuli. **B:** Caspase-11–NLRP3 inflammasome: Gram-negative bacterial infections induce production of type I interferons (IFNs) via the Toll-like receptor 4 (TLR4)-Toll/interleukin-1 receptor domain–containing adaptor-inducing IFN-β factor (TRIF) signaling pathway. Type I IFN signaling upregulates expression of caspase-11 (CASP11), guanylate-binding proteins (GBPs), and IRGB10. GBPs and IRGB10 target bacterial and vacuolar membranes to facilitate the release of lipopolysaccharide (LPS), which binds to mouse CASP11 or human CASP4/5 in the cytosol. Activation of CASP11 (human CASP4/5) by LPS results in cleavage of gasdermin D (GSDMD). Potassium efflux triggered by GSDMD pores then induces the NLRP3 inflammasome. **C:** ZBP1–NLRP3 inflammasome: Z-DNA-binding protein 1 (ZBP1) is an innate immune receptor that senses Z-RNAs from influenza A virus (IAV) and other sources. ZBP1 activation leads to its interaction with receptor-interacting protein kinase 3 (RIPK3) and recruitment of FADD and CASP8 to form a cell death–inducing complex called the PANoptosome, which triggers (i) assembly of the NLRP3 inflammasome, leading to pyroptosis; (ii) CASP3 and CASP7 activation, leading to apoptosis; and (iii) MLKL phosphorylation and membrane localization, leading to necroptosis. CASP6 facilitates the interaction between ZBP1 and RIPK3. ZBP1 also induces RIPK1-driven NF-κB activation.

Another alternative form of NLRP3 inflammasome activation occurs in the context of IAV infection. Sensing of viral RNA genomes in the Z-conformation by Z-DNA-binding protein 1 (ZBP1) induces pyroptosis, apoptosis, and necroptosis (collectively PANoptosis),[199] which is executed by a single molecular platform called the PANoptosome.[200,201] ZBP1-mediated pyroptosis and caspase-1 cleavage require the assembly of a particular form of the NLRP3 inflammasome, called the ZBP1-NLRP3 inflammasome. After sensing the Z-RNA, ZBP1 recruits RIPK3 and caspase-8 to assemble the ZBP1-NLRP3 inflammasome (Fig. 12.7C). The assembly of the ZBP1-NLRP3 inflammasome in response to IAV is crucial for the host protection[202] and is regulated by several host proteins, including molecules associated with IFN signaling and caspase-6.[203]

Several regulators of NLRP3 activation have been discovered. For example, NIMA-related kinase 7 (NEK7), a serine-threonine kinase that is known to be involved in mitosis, is essential for NLRP3 activation.[204,205] In addition, TAK1 suppresses spontaneous NLRP3 inflammasome activation.[206] Furthermore, stress granules and the NLRP3 inflammasome compete for DDX3X to coordinate the activation of innate responses.[207]

Although the NLRP3 inflammasome is protective during acute inflammation and infection, it is pathogenic during chronic inflammation. As an example, together with caspase-8, NLRP3 contributes to the release of the pathological IL-1β that drives chronic bone disease in *Pstpip2*[cmo] mice, which display symptoms similar to those of patients with chronic recurrent multifocal osteomyelitis.[208,209] NLRP3 has also been implicated in several other autoinflammatory diseases. In addition to its role in innate immunity, NLRP3 also regulates adaptive immune responses.[210] For example, NLRP3 drives Th2-biased adaptive immune responses that promote leishmaniasis.[180] Due to its physiological importance, a number of NLRP3 inhibitors have been reported to date. The diarylsulfonylurea MCC950 is one of the most potent and specific NLRP3 inhibitors and has demonstrated therapeutic efficacy in a variety of preclinical immunopathological models, including CAPS, experimental autoimmune encephalomyelitis, Alzheimer disease, atherosclerosis, diabetes, steatohepatitis, and colitis.[186] Several other NLRP3 inhibitors are also being investigated.

NLRC4

NLRC4 is highly expressed in the bone marrow and in lung and intestinal epithelial cells. NLRC4 is distinct from other inflammasome sensors in that it requires an additional NLR family member for activation. NLRC4 requires NAIPs, which recognize flagellin and proteins of the type III secretion system (T3SS) of many gram-negative bacteria.[211,212] *S. typhimurium*, *Legionella pneumophila*, *P. aeruginosa*, *S. flexneri*, and *Burkholderia* spp. are known to activate the NAIP-NLRC4 inflammasome. There are seven NAIP proteins in mice with distinct ligand specificities.[213,214] NAIP1 and NAIP2 recognize the needle and rod proteins, respectively, of the bacterial T3SS, while NAIP5 and NAIP6 recognize flagellin.[215-217] Interestingly, the flagellins of *Campylobacter jejuni* and *H. pylori* do not activate the NLRC4 inflammasome.[218,219] NAIP3, NAIP4, and NAIP7 are not well studied compared with NAIP1, NAIP2, NAIP5, and NAIP6. In humans, there is only one NAIP which is capable of detecting the same set of ligands as all the murine NAIPs. After sensing their respective ligand, activated NAIPs subsequently bind NLRC4 to initiate inflammasome assembly (Fig. 12.8). IRF8 regulates the basal expression of several NAIPs and NLRC4.[220] Gain-of-function mutations in NLRC4 are associated with familial cold autoinflammatory syndrome (FCAS), infantile enterocolitis, and recurrent macrophage activation syndrome (MAS).[221-223]

PYRIN

In addition to the NLR proteins that form inflammasomes, some non-NLR proteins can also form these complexes. Rho-modifying proteins, such as the glycotransferase TcdB from *Clostridium difficile*, the adenyltransferase VopS from *Vibrio parahaemolyticus*, and the GTPase-activating protein YopE and cysteine protease YopT from *Yersinia pestis* are known to activate the Pyrin inflammasome.[224-226] Pyrin is mainly expressed by the cells of the innate immune system, particularly granulocytes, eosinophils, monocytes, and DCs. During homeostasis, Pyrin is normally phosphorylated by the

NAIP-NLRC4 inflammasome activation

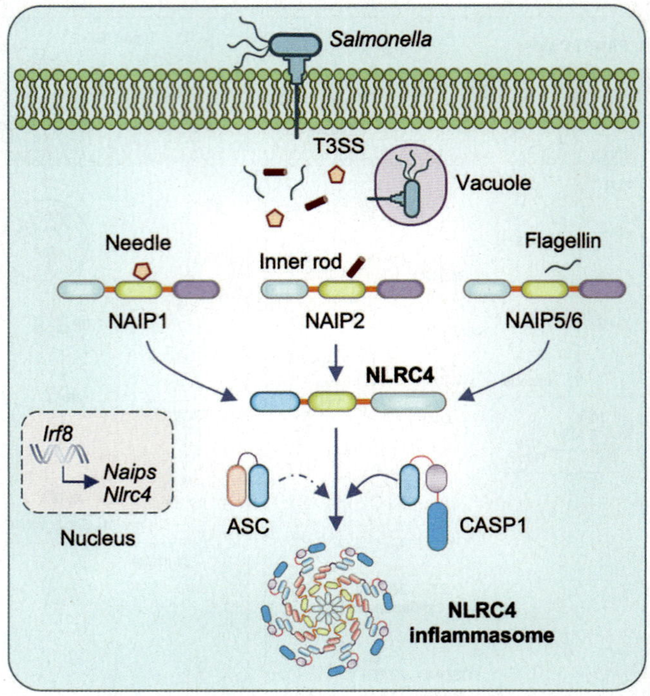

FIG. 12.8. **NAIP-NLRC4 inflammasome activation.** Certain pathogenic bacteria such as *Salmonella* use the type III secretion system (T3SS) to deliver effector proteins to subvert host cell functions. During infection, bacterial T3SS proteins, including needle and inner rod, and flagellin are injected into the host cell cytoplasm. Murine NOD-like receptor family apoptosis inhibitory protein (NAIP) 1, NAIP2, and NAIP5/6 detect needle, inner rod, and flagellin, respectively. NAIP sensing of ligands induces NLRC4 inflammasome assembly. Using its caspase activation and recruitment domain (CARD), NLRC4 may directly bind to the CARD of caspase-1 (CASP1) and form an inflammasome complex, or it may interact via apoptosis-associated speck-like protein containing a CARD (ASC). Interferon regulatory factor 8 (IRF8) regulates the transcription of the genes encoding NAIPs and NLRC4.

RhoA-activated serine-threonine kinases PKN1 and PKN2. This phosphorylation event allows the inhibitory protein 14-3-3 to bind to Pyrin. Inactivation of RhoA by bacterial toxins dephosphorylates Pyrin, thereby relieving inhibition by 14-3-3 and allowing Pyrin to assemble the inflammasome[227,228] (Fig. 12.9). Pyrin is encoded by the human *MEFV* gene. Gain-of-function mutations in *MEFV* are associated with the autoinflammatory condition familial Mediterranean fever (FMF).[229] Mice carrying mutations associated with FMF also develop autoinflammation, and this can be rescued by the genetic deletion of caspase-1 or IL-1β,[230] GSDMD,[231,232] or the specific blockade of TNFR1 signaling.[233] Pyrin signaling also promotes intestinal barrier integrity, prevents colon inflammation, and inhibits tumorigenesis.[234]

AIM2 and IFI16

AIM2 belongs to the ALR family of receptors. The number of ALR genes present varies widely among mammals, ranging from 4 in humans (*AIM2, IFI16, PYHIN1*, and *MNDA*) to 13 in mice. AIM2 and IFI16 (or the IFI204 mouse ortholog)

Pyrin inflammasome activation

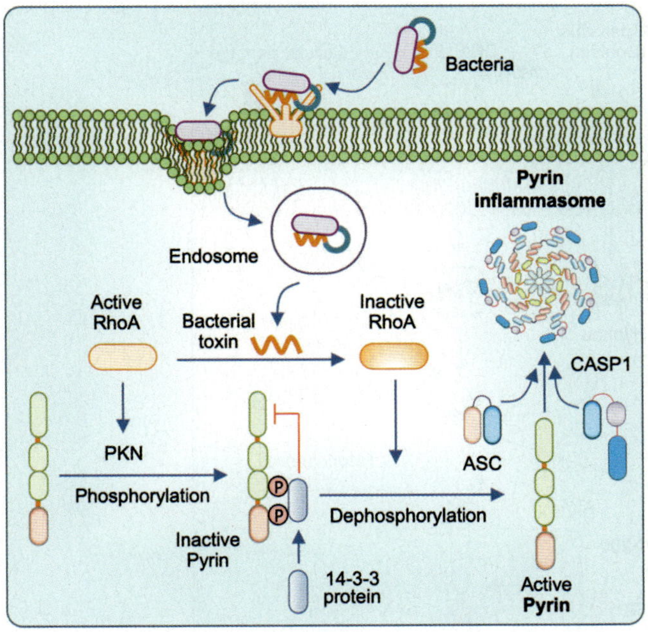

FIG. 12.9. **Pyrin inflammasome activation.** Under homeostatic conditions, Pyrin is kept in an inactive state by Ras homolog family member A (RhoA) effector kinases, protein kinase N1/2 (PKN1/2), and 14-3-3 proteins. Bacterial toxins such as *Clostridium difficile* TcdA and TcdB can inhibit RhoA activity and subsequent PKN1/2 phosphorylation, thereby leading to Pyrin dephosphorylation and 14-3-3 protein dissociation. Active Pyrin recruits apoptosis-associated speck-like protein containing a caspase activation and recruitment domain (ASC) and caspase-1 (CASP1) to form an active inflammasome complex.

are the two best studied ALR proteins. AIM2 is predominantly expressed in macrophages and epithelial cells. AIM2 contains a hematopoietic IFN-inducible nuclear protein (HIN) domain and a PYD, whereas IFI16 contains two HIN domains and a PYD. The HIN domain of AIM2 recognizes cytosolic double-stranded (ds)DNA of microbial or host origin.[235-240] The positively charged HIN domain of AIM2 forms an electrostatic attraction with the negatively charged sugar-phosphate backbone of the dsDNA.[238,240,241] The dsDNA needs to be at least 70 bp in length to be recognized by AIM2.[240,241]

The AIM2-activating pathogens such as *Francisella* induce type I IFN production, which in turn induces the expression of GBPs and IRGB10. The IFN-inducible GBPs and IRGB10 together target vacuolar and/or bacterial membranes for lysis, thereby promoting DNA release into the cytosol (Fig. 12.10). dsDNA from fungal hyphae and viruses can also be sensed. In the context of *Francisella novicida* infection, the GBPs and IRGB10 are induced in an IRF1-dependent manner.[192,242]

During homeostasis, the sensing of self-DNA by AIM2 is regulated by a conserved mechanism that has evolved to compartmentalize self-DNA to the nucleus or degrade mislocalized DNA. Interestingly, some DNA sequences, such as the TTAGGG repeat that is found most frequently in mammalian telomeric DNA, inhibit AIM2 inflammasome activation by competing with dsDNA for binding and preventing inflammasome complex formation.[243] Although the AIM2 inflammasome is generally protective in infectious diseases, it has been found to be detrimental in several inflammatory diseases including atherosclerosis, chronic kidney diseases, skin disease, liver disease, and neuroinflammation.[244] Additionally, the AIM2 inflammasome has recently been found to be a key component of the AIM2-PANoptosome to mediate inflammatory cell death (PANoptosis).[245] AIM2 can also have inflammasome-independent functions as observed in colorectal tumorigenesis.[246,247]

IFI16, another member of the ALR family, senses microbial DNA in both the cytoplasm and nucleus to initiate immune responses. It is predominantly expressed in lymphocytes, monocytes, and epithelial cells. There are limited studies regarding the role of IFI16 in inflammasome activation.[248] Nuclear IFI16 assembles an inflammasome during infection with Kaposi sarcoma–associated herpesvirus. Additionally, IFI16 senses incomplete DNA reverse transcripts that accumulate in abortive HIV-infected cells, resulting in pyroptosis and contributing to the debilitating CD4+ T cell depletion that underlies the progression of AIDS.[249] Cytoplasmic IFI16 and the murine ortholog IFI204 sense viral DNA to induce type I IFN expression during infection.[250] Both are upregulated by type I and type II IFNs. In addition to inflammasome activation and type I IFN expression, IFI16 has been associated with cell cycle regulation, apoptosis, tumor suppression, and DNA repair.

NLRP6 and NLRP12

NLRP6 is highly expressed in the liver, lung, kidney, and intestine of mice. Sensing of lipoteichoic acid, a major component of the cell wall of gram-positive bacteria such as *L. monocytogenes* and *S. aureus*, allows NLRP6 to recruit caspase-11 and caspase-1 via ASC to assemble the inflammasome.[251] In addition, NLRP6 plays a detrimental role in mice during infection with bacteria because it suppresses inflammatory responses mediated by NF-κB and MAPK signaling.[252] NLRP6 also functions cooperatively with DHX15 to drive the expression of IFN-inducible genes.[253]

NLRP12 is expressed in bone marrow neutrophils, granulocytes, macrophages, and DCs. It inhibits NF-κB and MAPK activation downstream of TLR signaling during various microbial infections. The inhibition of NF-κB signaling by NLRP12 suppresses colon inflammation and colitis-associated colorectal cancer in mice.[254] NLRP12 also interacts with the E3 ubiquitin ligase TRIM25 to disrupt its association with RIG-I, resulting in reduced type I IFN production.[255] Limited studies have shown NLRP12 to be an inflammasome sensor during infection with *Yersinia* and *Plasmodium*.[256,257]

NLRP9

Human NLRP9 has three orthologs, NLRP9a–c, in mice. NLRP9 is expressed in the ovaries and intestines.[258,259] Because of limited studies, the functions of NLRP9 in physiology are still not clear. Mouse NLRP9b senses short dsRNA stretches via the RNA helicase DEAH-box helicase DHX9 to assemble an inflammasome complex. NLRP9b specifically functions in the intestine to provide host defense against rotavirus infection in mice.[259]

AIM2 inflammasome activation

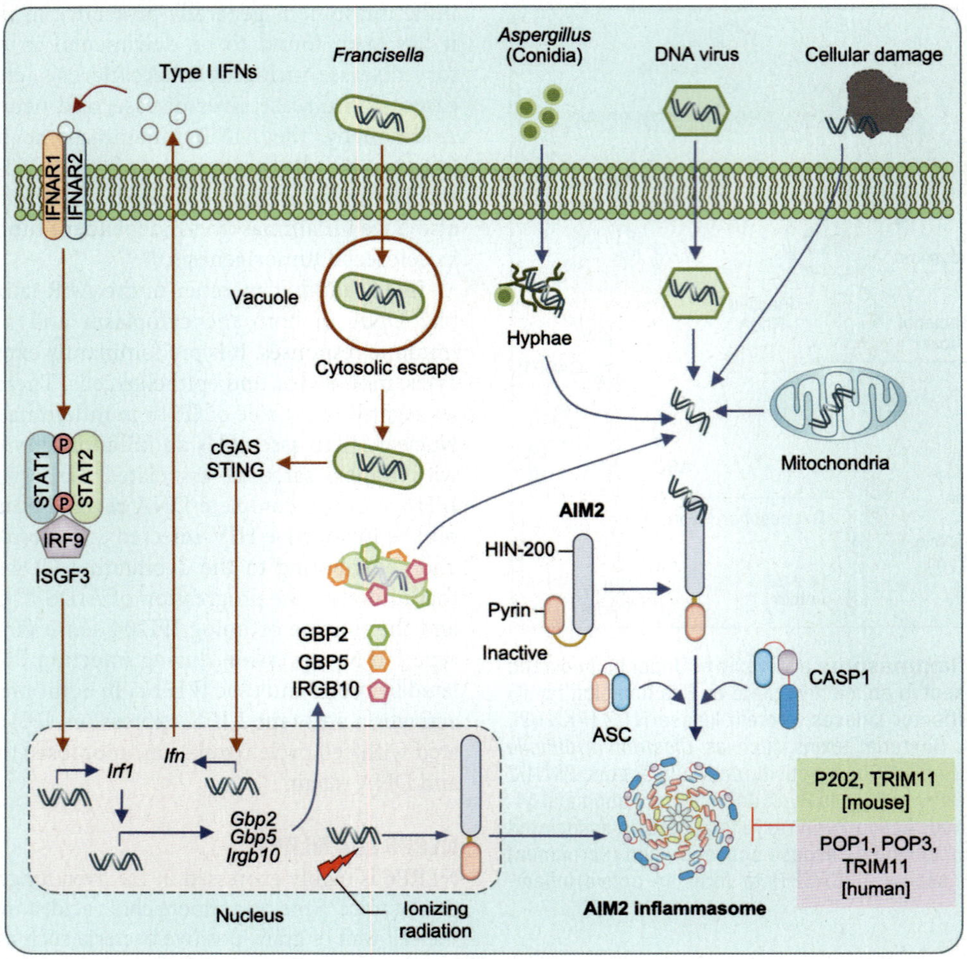

FIG. 12.10. AIM2 inflammasome activation. AIM2 is activated in response to microbial or self-DNA. The presence of cytosolic bacteria, such as *Francisella*, drives the production of type I interferons (IFNs) to induce the expression of IFN regulatory factor 1 (IRF1), which in turn activates the expression of guanylate-binding proteins (GBPs) and IRGB10. GBPs and IRGB10 target bacterial and vacuolar membranes for destruction, thereby releasing bacterial DNA in the cytosol where it can be sensed by AIM2. DNA viruses, such as mouse cytomegalovirus, transfected poly(dA:dT), host DNA from cellular damage, and fungal ligands from *Aspergillus* bypass the requirement for type I IFNs to activate the AIM2 inflammasome. AIM2 also detects radiation-induced host DNA damage in the nucleus or host DNA leaked into the cytoplasm following damage to the nucleus or mitochondria. AIM2 activation can be inhibited by several host factors, including P202 and tripartite motif-containing 11 (TRIM11) in mice and Popeye domain–containing (POP) 1 and 3 and TRIM11 in humans. (From Steele C, Rapaka RR, Metz A, Pop SM, Williams DL, Gordon S, et al. The beta-glucan receptor dectin-1 recognizes specific morphologies of Aspergillus fumigatus. *PLoS Pathog.* 2005;1(4):e42.)

RIG-I-LIKE RECEPTORS

RLRs are a family of cytoplasmic molecules that sense viruses. RLRs recognize dsRNA molecules that are produced by certain viruses and can induce antiviral immune responses.[260] The family consists of three DExD/H box RNA helicases, RIG-I (also known as DDX58), MDA5 (melanoma differentiation–associated factor 5, also known as IFIH1), and LGP2 (laboratory of genetics and physiology 2); activation of these DExD/H box RNA helicases produces type I IFN. RLRs are broadly expressed in most tissues. They are expressed constitutively at low levels in resting cells, but their expression increases in response to IFN or during viral infection.[261] RIG-I and MDA5 share a number of structural similarities, including their organization into three distinct domains: (1) an N-terminal CARD that facilitates signaling, (2) a central DExD/H box RNA helicase domain that facilitates ATP hydrolysis and RNA binding, and (3) a C-terminal regulatory domain (CTD), also known as regulatory domain (RD), that confers specificity to RNA ligand recognition and binding.[262]

RIG-I and MDA5 recognize different viral species. RIG-I generally responds potently to negative-strand RNA viruses, such as members of the Paramyxoviridae, Rhabdoviridae, and Orthomyxoviridae families, while MDA5 senses positive-strand RNA viruses such as members of the Picornaviridae and Arteriviridae families.[263] The preference for specific RNA species by RIG-I and MDA5 allows for them to distinguish between non-self and self-RNA. Adenosine-to-inosine (A-to-I) dsRNA editing by ADAR1 (adenosine deaminase acting on RNA) prevents RIG-I and MDA5 from sensing endogenous dsRNA.[264]

RIG-I and MDA5 can recognize diverse regions within viral genomes due to their differential preferences for RNA

TABLE 12.6 RNA Species for RIG-I and MDA5

RNA Species	RNA Name	Source(s)	Sensor
	5′ppp-dsRNA	Viral genomes, RNA pol III, and synthetic RNAs	RIG-I
	5′pp-dsRNA	Viral genomes and synthetic RNAs	RIG-I
	5′p-dsRNA 3′p-dsRNA	Short synthetic oligos, RNase III, and RNase L digested RNA	RIG-I
	5′ppp-ssRNA	Synthetic	RIG-I
	cRNA	Host cell during viral infection and synthetic	RIG-I
	SKIV2L products	Host cell	RIG-I
	Hairpin structures	Host cell	RIG-I
	Small hairpin structures	Host cell during viral infection	RIG-I
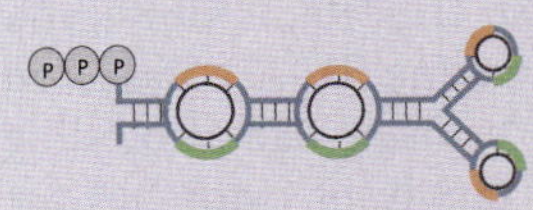	Vault RNAs, pseudogene ribosomal RNAs	Host cell during viral infection	RIG-I
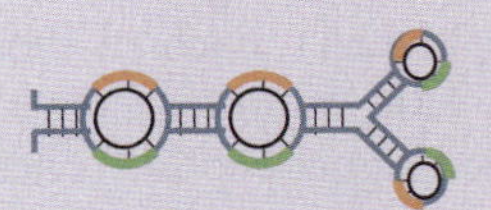	Retrotransposons	Host cell	RIG-I, MDA5
	Long dsRNA	Viral genomes and synthetic	MDA5

binding (Table 12.6). RIG-I senses RNA motifs such as 5′-ppp RNA, dsRNA structures, and ssRNA encoding polynucleotide runs of a specific length. RIG-I preferentially recognizes RNA sequences marked with 5′-triphosphorylated (5′-ppp) ends. Removal or modification of the 5′-ppp severely attenuates signaling.[265] RIG-I shows a preference for shorter RNA fragments, whereas MDA5 preferentially recognizes high molecular weight poly(I:C) fragments.[266] The sequence composition of an RNA ligand contributes to the activation of RIG-I–dependent signaling. For example, RIG-I activation occurs in response to poly-uridine motifs containing interspersed C nucleotides (poly-U/UC) that are present in the genome of hepatitis C virus (HCV), which has a ssRNA genome that is noncapped and includes a 5′-ppp.[267,268] Deletion of the poly-U/UC motif from the HCV genome completely abrogates RIG-I–dependent signaling despite the presence of a 5′-ppp.

Under homeostatic conditions, RIG-I and MDA5 RDs interact with their CARD to maintain an inactive state (Fig. 12.11). Phosphorylation of the CARDs by PKCα/β maintains the autoinhibited state of RIG-I. Following binding to nonself RNAs, the interaction between the RD and the CARD is disrupted, and these RLRs undergo posttranslational modifications.[269] The phosphate groups are removed by two specific phosphatases, PP1α and PP1γ, which are recruited to the sensors after RNA binds to them. In the case of RIG-I, the activating polyubiquitin chains are generated by the actions of two E3 ubiquitin ligases, TRIM25 and RNF135 (Riplet). TRIM25 and Riplet promote K63-polyubiquitination of the second CARD and RD of RIG-I, respectively. TRIM65 polyubiquitinates the helicase domain of MDA5.[270,271] Following these posttranslational modifications and activation, RIG-I and MDA5 translocate to the mitochondria where they interact with the adaptor protein mitochondria antiviral signaling (MAVS) via homotypic CARD–CARD interactions[272,273] to form a MAVS signalosome that drives IFN production. The key transcription factors involved in RIG-I and MDA5 signaling and MAVS signalosome formation include IRF3, IRF7, and NF-κB.

Although the third member of the RLR family, LGP2, is similarly organized and shares homology with RIG-I and MDA5 within its DExD/H helicase domain and CTD, it lacks the N-terminal CARD required for downstream antiviral signaling. LGP2 can bind a range of RNA ligands, including short dsRNA (<2 kb), 5′-ppp ssRNA (which is also a ligand for RIG-I), and long dsRNAs (>2 kb, which is also a ligand for MDA5). LGP2 is essential for producing antiviral responses against many viruses that are recognized by RIG-I and MDA5, and LGP2 also regulates RIG-I– and MDA5–mediated signaling. LGP2 interacts directly with RIG-I via its CTD, and the helicase activity of LGP2 positively regulates RIG-I signaling[274] (Fig. 12.11).

RIG-I and MDA5 signaling

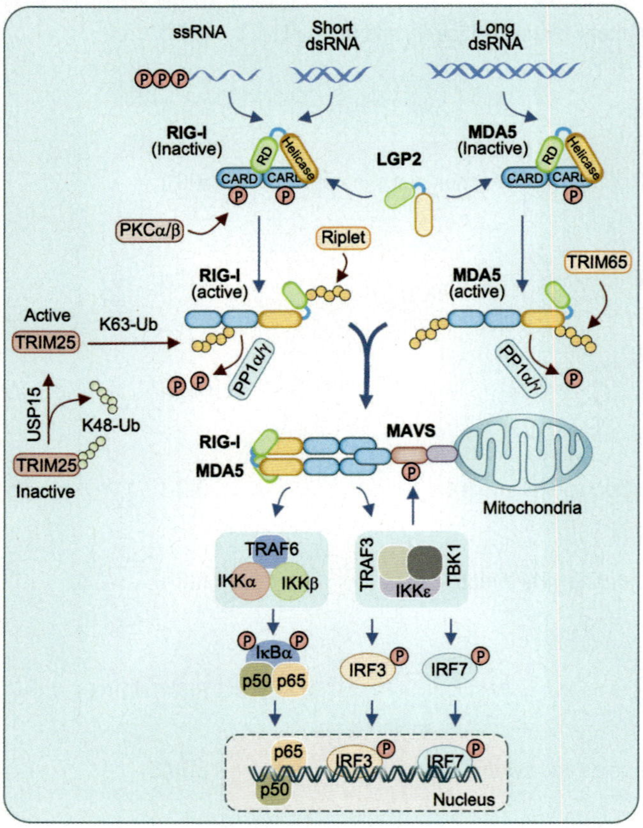

FIG. 12.11. **RIG-I and MDA5 signaling.** RIG-I binds specifically to single-stranded RNA (ssRNA) containing 5′-triphosphate, such as viral RNA, and short double-stranded RNA (dsRNA). MDA5 binds preferentially to long dsRNA. Although RIG-I and MDA5 recognize different ligands, they share common signaling features. The DExD/H box RNA helicase domain along with the adjacent C-terminal regulatory domain (RD) are required for RNA binding. These C-terminal domains keep the receptors in an inactive conformation until they are bound by an activating RNA. Both RIG-I and MDA5 also have phosphorylation sites in their caspase activation and recruitment domain (CARD) that suppress their function. Protein kinases Cα and β (PKCα and PKCβ) phosphorylate RIG-I to keep it in its inactive form. RNA binding to RIG-I and MDA5 is followed by the posttranslational modification of the receptors, which promotes their stabilization. PP1α/γ dephosphorylates the CARDs, and the E3 ubiquitin ligases TRIM25 and Riplet promote K63-polyubiquitionation of RIG-I, while TRIM65 K63 polyubiquitinates MDA5. Upon activation, the two N-terminal CARDs from both RIG-I and MDA5 mediate signaling by interacting with the CARD of mitochondrial antiviral signaling protein (MAVS). Another RNA-binding receptor called LGP2 shares homology with RIG-I and MDA5 but lacks the CARD and cannot activate MAVS. Homotypic CARD–CARD interactions between RIG-I and MDA5 and MAVS trigger the formation of a signaling synapse that includes the recruitment of adaptor proteins (TRAF3, TRAF6) and kinases (TBK1, IKKα, IKKβ, IKKε) leading to the activation of NF-κB, interferon regulatory factor 3 (IRF3), and IRF7.

Polymorphisms in RLRs have been associated with several autoimmune diseases. RIG-I mutations have been linked with atypical Singleton-Merten syndrome, while MDA5 mutations are associated with classical Singleton-Merten syndrome, Aicardi-Goutières syndrome, SLE, type 1 diabetes, and Graves disease.[275-277] Notably, RIG-I–deficient, but not MDA5-deficient, C57BL/6 mice are embryonically lethal.

IL-1 FAMILY CYTOKINES AND RECEPTORS

The IL-1 family of cytokines and their receptors are primarily associated with innate immune responses. The IL-1 family of receptors shares similar functions with TLRs. Each member of the IL-1 receptor and TLR families contains a cytoplasmic TIR domain. Some of the IL-1 family members, particularly IL-1α and IL-33, function as DAMPs. In addition to their roles in innate immunity, the IL-1 cytokines and their receptors can also play a role in adaptive immunity. The members of the IL-1 family are shown in Table 12.7 (including both their historic and new names) and Figure 12.12.

IL-1 Family of Cytokines

The IL-1 family of cytokines includes seven molecules with agonist activity (IL-1α, IL-1β, IL-18, IL-33, IL-36α, IL-36β, and IL-36γ), three receptor antagonists (IL-1Ra, IL-36Ra, and IL-38), and an anti-inflammatory cytokine (IL-37). Based on their IL-1 consensus sequence and primary ligand binding receptor, the members of the IL-1 family of cytokines are divided into three subfamilies (Fig. 12.12A). Each of these cytokines, with the exception of IL-1Ra, cannot be readily secreted by the cell because they lack a signal peptide. These proteins are located diffusely in the cytoplasm as precursors containing the consensus sequence AXD (any aliphatic amino acid [A] followed by any amino acid [X] and then aspartic acid [D]). The nine amino acids before the AXD sequence in the N-terminus allow for optimal folding to ensure receptor binding. However, truncation of the ninth amino acid before the AXD motif in IL-36 results in increased bioactivity by 1000-fold.[278] In the case of IL-1β and IL-18, the N-terminal regions of pro–IL-1β and pro–IL-18 act as the caspase-1 cleavage site to generate bioactive IL-1β and IL-18.

IL-1 Family of Receptors

There are 10 members of the IL-1 family of receptors (Fig. 12.12B), and they can come together in various combinations to form four signaling receptor complexes: the IL-1 receptor, the IL-33 receptor, the IL-36 receptor, and the IL-18 receptor. The receptor chains generally contain three extracellular Ig-like domains, with the exception of IL-18–binding protein (IL-18BP) and IL-1R8 (TIR8), which each has a single Ig-like domain. IL-1R1 binds IL-1α, IL-1β, and IL-1Ra. The binding of IL-1α or IL-1β to IL-1R1 recruits the additional co-receptor IL-1RAcP (IL-1R3), forming a trimeric signaling complex that brings the TIR domains of each receptor chain close to MyD88, triggering activation of proinflammatory signals such as NF-κB (Fig. 12.12B). IL-1RAcP also serves as a co-receptor for IL-33, IL-36α, IL-36β, and IL-36γ. Although

TABLE 12.7 IL-1 Family Cytokines

IL-1 Family Cytokines	New Name	Primary Receptor(s)	Co-receptor	Function
IL-1α	IL-1F1	IL-1R1	IL-1RAcP (IL-1R3)	Inflammatory
IL-1β	IL-1F2	IL-1R2		
IL-1Ra	IL-1F3	IL-1R1		Anti-inflammatory
IL-18	IL-1F4	IL-18Rα (IL-1R5)	IL-18Rβ (IL-1R7)	Inflammatory
IL-33	IL-1F11	ST2 (IL-1R4)	IL-1RAcP (IL-1R3)	Inflammatory
IL-36α	IL-1F6			
IL-36β	IL-1F7	IL-36R (IL-1R6)	IL-1RAcP (IL-1R3)	Inflammatory
IL-36γ	IL-1F8			
IL-36Ra	IL-1F5			Anti-inflammatory
IL-37	IL-1F7	IL-18Rα (IL-1R5)	IL-1R8	Anti-inflammatory
IL-38	IL-1F10	IL-36R (IL-1R6)	IL-1R9	Anti-inflammatory

IL-1RAcP exists as an integral membrane protein, its extracellular domain can be released from the cell surface after cleavage, allowing it to function as a soluble receptor (sIL-1RAcP). IL-1R2 acts as a decoy receptor for IL-1β, as it lacks the cytoplasmic domain required for signaling. A soluble form of IL-1R2 or IL-1RAcP can neutralize IL-1β activity by binding extracellular IL-1β. IL-1R2 also prevents the release of IL-1α from the cell and prevents the extracellular processing of IL-1α by calpain. IL-1R4, formerly referred to as ST2, is a binding receptor for IL-33, while IL-1R6 (formerly IL-36R) binds

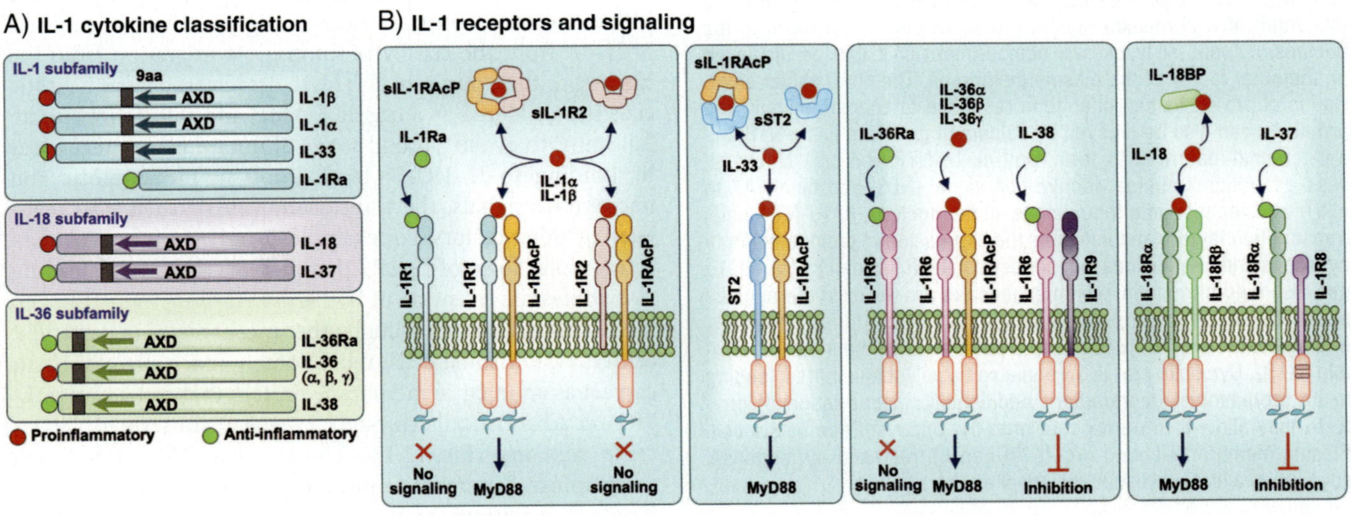

FIG. 12.12. **IL-1 superfamily and receptor signaling. A:** IL-1 cytokine classification: The interleukin (IL)-1 superfamily of cytokines can be divided into three subfamilies, the IL-1, IL-18, and IL-36 subfamilies. The presence of the IL-1 family consensus sequence (AXD) is indicated for each cytokine, and it is located 9 amino acids after the cleavage site (denoted by *gray vertical bars*). The red circles indicate cytokines with proinflammatory activities, whereas the green circles indicate cytokines with anti-inflammatory activities. IL-1Ra is readily secreted, and it does not contain the AXD sequence. **B:** IL-1 receptors and signaling: The biological activities of IL-1α and IL-1β are mediated by IL-1R1. Binding of IL-1α or IL-1β induces a conformational change in the extracellular domain of IL-1R1, enabling its interaction with IL-1RAcP to initiate downstream signaling. IL-1Ra competes with IL-1α and IL-1β for binding to IL-1R1 and does not recruit IL-1RAcP or activate intracellular signaling pathways. IL-1R2 acts as a decoy receptor on the cell surface or as a soluble form (sIL-1R2) in the extracellular environment. IL-1RAcP also exists in a soluble form (sIL-1RAcP) and can form a complex with sIL-1R2, increasing the inhibition of IL-1α– and IL-1β–mediated signaling by sIL-1R2. Binding of IL-33 to its receptor ST2 also recruits IL-1RAcP. Soluble ST2 (sST2) and sIL-1RAcP act as decoy receptors for IL-33. Also, binding of IL-36α, IL-36β, and IL-36γ to their receptor IL-1R6 (IL-36R) induces recruitment of IL-1RAcP. Similar to the function of IL-1Ra in the IL-1 subfamily, IL-36Ra competes with IL-36 for IL-1R6 binding, thereby acting as a natural IL-36 inhibitor. Binding of IL-38 to IL-1R6 induces recruitment of IL-1R9, which inhibits inflammatory signaling. Binding of IL-18 to its receptor IL-18Rα induces recruitment of IL-18Rβ. IL-18BP acts a soluble antagonist to inhibit IL-18 activity. IL-37 requires the receptors IL-18Rα and IL-1R8 to exert anti-inflammatory signal transduction. The tertiary complex formed by certain IL-1 cytokines and heterodimeric receptors then recruits intracellular signaling molecules, including MyD88 to activate NF-κB as indicated.

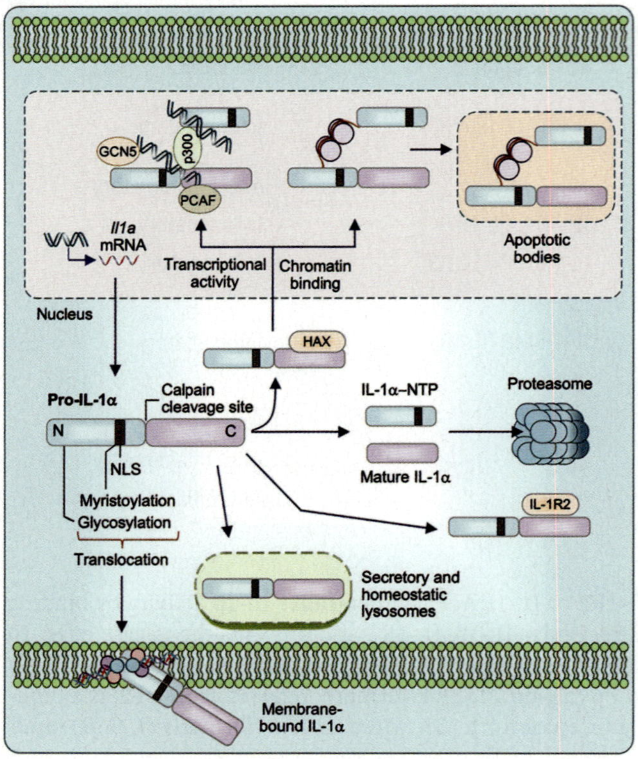

FIG. 12.13. IL-1α processing and functions. After synthesis of full-length pro–interleukin (IL)-1α, pro–IL-1α can be localized to the nucleus, cytosol, or lysosomal compartment or it can be displayed on the outer leaflet of the plasma membrane. The intracellular localization of pro–IL-1α can change in response to specific stimuli, and pro–IL-1α can also be cleaved by calpain to generate IL-1α-NTP and a C-terminal mature IL-1α form. Both IL-1α-NTP and pro–IL-1α possess a functional nuclear localization signal (NLS) and bind HAX to allow translocation into the nucleus. In the nucleus, IL-1α-NTP binds transcription factors and activates the expression of proinflammatory cytokines and chemokines independent of IL-1R1 signaling. Pro–IL-1α can also bind chromatin, which sequesters pro–IL-1α in the nucleus and limits aberrant IL-1α–dependent inflammation during apoptosis. In the cytosol, pro–IL-1α can form a complex with the inhibitory protein IL-1R2. Cytosolic pro–IL-1α can also be polyubiquitinated, leading to its proteasomal degradation. Additionally, glycosylation of pro–IL-1α may allow it to translocate onto the outer surface of the cell. Plasma membrane–bound pro–IL-1α can be released by mannose, suggesting a lectin-like mechanism of anchorage.

IL-36 and IL-38. Binding of IL-18 to its receptor IL-18Rα (IL-1R5) induces recruitment of IL-18Rβ (IL-1R7). Other members of the IL-1 family of receptors, IL-1R8 and IL-1R9, are considered as anti-inflammatory receptors.[279] IL-1R10 is an orphan receptor with unknown ligand and functions.

IL-1 Cytokines

IL-1α and IL-1β are potent proinflammatory cytokines that are involved in both innate and adaptive immunity. Although encoded by different genes, the primary receptor for both IL-1α and IL-1β is IL-1R1, and they have largely similar biological properties. However, producing and responding cell types, secretion mechanism, functions, and localization of these two cytokines are distinct.

IL-1α

The IL-1α cytokine is expressed constitutively in the epithelium of the gastrointestinal tract, lung, liver, and kidney and in endothelial cells and astrocytes. Expression of IL-1α can be induced in response to growth factors and proinflammatory or stress-associated stimuli. While constitutive IL-1α expression is regulated by Sp1-family transcription factors,[280] the inducible expression of IL-1α depends on the presence of binding sites for AP1 and NF-κB transcription factors.[281,282] IL-1α is produced as a pro-form that is cleaved by proteases, including calpain, chymase, elastase, and granzyme B, to yield mature IL-1α (Fig. 12.13). Both pro–IL-1α and mature IL-1α bind to IL-1R with similar kinetics.

IL-1α initiates signal transduction through its receptor, but it also acts as a transcription factor.[283] It has an NLS in the precursor region and can shuttle between the cytoplasm and the nucleus rapidly (Fig. 12.13). While IL-1α is localized to the nucleus of keratinocytes and fibroblasts under homeostatic conditions, this localization requires stimulation in other cell types such as macrophages. Notably, IL-1α behaves differently in response to apoptotic and necroptotic signals. Proapoptotic signals induce translocation of IL-1α from the cytoplasm to the nucleus where IL-1α binds tightly to chromatin, thereby limiting the availability of IL-1α to initiate inflammation. In contrast, necrotic signals trigger the release of IL-1α from the cell by promoting its migration from the nucleus to the cytoplasm.[284] The IL-1α released from necrotic cells is considered as a classic DAMP, and it functions as an "alarmin" to evoke a broad range of inflammatory responses by binding to IL-1R1.[285] In addition to proapoptotic and necroptotic signals, the translocation of IL-1α is also regulated by other factors. For example, pro–IL-1α is bound to an intracellular form of IL-1R2 that may mask the NLS, leading to cytoplasmic retention of IL-1α.

IL-1α can also act through other mechanisms in particular cell types. On some cells, most notably macrophages, IL-1α can act as an integral membrane protein that can activate IL-1R1 on adjacent cells through a mechanism termed "juxtacrine" signaling (Fig. 12.13). The ability of IL-1α to function as an integral membrane protein is unique within the IL-1 family.[286] Additionally, IL-1α released in membrane-bound apoptotic bodies in endothelial cells is highly inflammatory and contributes to vasculitis.[287]

Because both IL-1α and IL-1β use IL-1R1 for signal transduction, it can be difficult to determine whether a particular disease is caused by IL-1α specifically. However, IL-1α has been found to act as an apical driver in cutaneous inflammation,[288] colon inflammation and cancer,[289] cardiovascular disease,[290] and neural inflammation.[291] IL-1α is the primary cytokine that causes the development of autoinflammatory disease in *Ptpn6*^{spin} mice, an *in vivo* animal model for neutrophilic dermatoses, a heterogenous group of autoinflammatory skin disorders that include Sweet syndrome, pyoderma granulosum, subcorneal pustular dermatosis, and hidradenitis suppurativa.[288] Studies using *Ptpn6*^{spin} mice demonstrate that IL-1α production from nonhematopoietic cells acts on myeloid cells in a SYK-, MyD88-, RIPK1-, and TAK1-dependent manner.[20] Targeting IL-1α specifically is a promising treatment for patients with hidradenitis suppurativa.[292] Additionally, IL-1α from the intestinal epithelium drives inflammation and the associated colon cancer in mice.[293] In patients with head and neck squamous cell carcinoma, IL-1α expression has been considered as a prognostic marker of distant metastases. Patients with higher IL-1α expression in tumors have lower survival rates than patients with lower IL-1α expression.[294] Clinical trials with an IL-1α–blocking monoclonal antibody have shown promising outcomes in patients with end-stage cancers.[295]

IL-1β

IL-1β is produced by hematopoietic cells such as monocytes, macrophages, DCs, microglia, and neutrophils. The expression of IL-1β is rapidly induced in response to TLR activation by PAMPs or cytokines including IL-1 itself. Unlike the IL-1α precursor, the IL-1β precursor (pro–IL-1β) needs to be processed to become biologically active. Pro–IL-1β is cleaved by caspase-1 during inflammasome activation (Fig. 12.5). Pro–IL-1β is also cleaved by caspase-8 during fungal infection.[296] Pro–IL-1β cleavage by active caspase-1 can take place in the cytoplasm or in specialized secretory lysosomes. IL-1β release from the cell can be achieved through multiple pathways, including exocytosis of the secretory lysosomes,[297] shedding of plasma membrane microvesicles, direct release via transporters or multivesicular bodies containing exosomes,[298] or through the process of pyroptosis executed by GSDMD. Additionally, pro–IL-1β can be released from intracellular stores in dying neutrophils during the sterile inflammation associated with neutrophilic infiltration. Extracellular proteases such as proteinase-3, elastase, matrix metalloprotease 9, and granzyme A can then process pro–IL-1β extracellularly to yield the active form.[279]

IL-1β is the principle driver of dysregulated cellular homeostasis, extracellular matrix composition, proinflammatory cytokine production, and osteolysis in a diverse array of autoinflammatory, hematologic, and bone diseases including osteoarthritis and multiple myeloma.[299] Deletion of IL-1β, but not IL-1α, provides protection against bone disease in *Pstpip2*^{cmo} mice, an *in vivo* animal model of osteomyelitis that is driven by IL-1β produced by neutrophils specifically.[300,301] In *Pstpip2*^{cmo} mice, the expression of pro–IL-1β is dependent on SYK,[302] whereas bioactive IL-1β production is dependent on caspase-1 activation by the NLRP3 inflammasome and on caspase-8.[209] Additionally, the IL-1β produced during acute and chronic inflammation functions in both a paracrine and autocrine manner to induce cellular proliferation, immunosuppression, angiogenesis, and metastasis, all of which are central stages of tumorigenesis.[145] IL-1β suppresses tumor immune responses by recruiting myeloid-derived suppressor cells and inducing the production of proangiogenic factors, such as growth factors and chemokines. Also, IL-1β promotes cardiovascular diseases such as atherosclerosis in mice with bone marrow TET2 (ten-eleven-translocation 2) deficiency. The clinical trial Canakinumab Anti-inflammatory Thrombosis Outcomes Study (CANTOS) was designed to determine whether IL-1β inhibition could prevent cardiovascular events in patients who had high expression of inflammatory markers after a myocardial infarction. Results from this trial demonstrate that canakinumab, a specific IL-1β inhibitor, can reduce the incidence and mortality of lung cancer in these patients.[303]

IL-18

IL-18 was first discovered in 1989 and was described as the "IFN-γ–inducing factor" in endotoxemic mice. The name was changed to IL-18 in 1995.[304] Like IL-1β, IL-18 is initially produced as an inactive precursor (pro–IL-18) and typically requires caspase-1 cleavage to process it into its mature form[305] (Fig. 12.5). Pro–IL-18 is constitutively expressed in endothelial cells, keratinocytes, and epithelial cells, and it can be produced in response to stimulation in macrophages and DCs. In addition to its processing by caspase-1, IL-18 can also be processed by Fas ligand stimulation[306] and proteinase-3.[307]

IL-18 forms a low-affinity signaling complex by binding to its receptor IL-18Rα (IL-1R5) (Fig. 12.12B). When it is together with IL-18Rα and the co-receptor IL-18Rβ (IL-1R7), IL-18 can form a high affinity complex that induces a signaling cascade, leading to NF-κB activation.[4] One of the other major functions of IL-18 is for it to serve as an immunoregulatory molecule and induce the expression of IFN-γ in NK cells. IL-18 requires IL-12 or IL-15 to induce this IFN-γ expression.[308]

IL-18 is associated with several human autoimmune diseases, including SLE, rheumatoid arthritis, diabetes, Crohn disease, and psoriasis, where IFN-γ production is elevated.[279] IL-18 also induces IL-17 production from T cells and promotes autoimmune responses to specific antigens. Although IL-18 promotes gastric tumorigenesis and metastasis in melanoma, it is protective during colorectal cancer,[145,309] and IL-18 is a key cytokine for epithelial barrier regeneration during the early stages of colitis.[310] Notably, IL-18 can inhibit the expression of soluble IL-22 binding protein, largely in hematopoietic cells, which modulates the bioavailability of IL-22 and tumorigenesis.[311] Additionally, IL-18–deficient mice become insulin resistant and obese starting at 16 weeks of age, and this phenotype can be rescued by administering exogenous recombinant IL-18.[312]

IL-18BP is a constitutively secreted soluble protein which specifically neutralizes IL-18 activities (Fig. 12.12B). Some members of the Poxviridae family, notably *Molluscum contagiosum*, contain a gene that encodes an amino acid sequence similar to that of human IL-18BP.[313] The activity of IL-18 can be naturally neutralized by this viral IL-18BP during common viral infection without an inflammatory or immunological response being elicited. IL-18BP is also crucial in downregulating Th1 responses by binding to IL-18 and thus reducing the induction of IFN-γ.[308] Measuring the concentration of free IL-18 and IL-18BP in the serum is important in health and disease. The serum concentration of IL-18BP in a healthy human is 20-fold higher than the concentration of IL-18.[314] In patients with MAS, free IL-18 correlates with clinical status and biological markers of disease.[315]

IL-37

IL-37 is a recently discovered IL-1 family cytokine. It is expressed by human monocytes, macrophages, synovial cells, B cells, T cells, and epithelial cells, but is absent in mice. IL-37 is inducible in monocytes. Among the five IL-37 isoforms in humans, IL-37b is the most studied. It broadly suppresses innate inflammation and acquired immunity and limits the production of cytokines induced by IL-1 and TLRs.[316] It is a dual function cytokine that translocates to the nucleus upon caspase-1 cleavage.[317] Similar to IL-18, IL-37 can also be bound by IL-18BP, which inhibits the induction of IFN-γ by IL-18.[318] IL-37 can also bind to the IL-18 receptor (IL-18Rα), but with low affinity; this binding recruits an IL-1R–related molecule, IL-1R8 (formerly referred to as SIGIRR).[319,320] This is in contrast to IL-18 binding to IL-18Rα, which recruits IL-18Rβ instead of IL-1R8.[321] The TIR domain of IL-1R8 is mutated (termed TIRb); because of this mutation, MyD88 cannot be recruited to the complex to initiate proinflammatory signaling[322] (Fig. 12.12B). The IL-37 signaling complex suppresses the activation of several inflammatory kinases including mTOR.[316] Intracellular IL-37 binds Smad3, an important signaling molecule in the TGFβ-pathway. The production of IL-37 is often elevated in several human disease including inflammatory bowel disease, Grave disease, rheumatoid arthritis, psoriasis, and SLE. However, mice overexpressing human IL-37 (IL-37tg) are protected from LPS-induced sepsis and DSS-induced acute colitis and colon carcinogenesis.[316,323,324]

IL-36 and IL-38

The IL-36 group of cytokines is composed of four isoforms, IL-36α, IL-36β, IL-36γ, and IL-36 receptor antagonist (IL-36Ra). While the isoforms IL-36α, IL-36β, and IL-36γ act as receptor agonists for proinflammatory functions, IL-36Ra acts as an anti-inflammatory mediator (Fig. 12.12). IL-36 cytokines and their receptor, IL-1R6, are predominantly expressed in the skin. Keratinocytes produce a large amount of IL-36γ when stimulated with poly (I:C), and the release of this IL-36γ depends on caspase-1.[325] The signaling pathway downstream of the IL-1R6 agonists is similar to that activated by IL-1α and IL-1β at the IL-1 receptor. IL-36Ra also binds to IL-1R6 but does not recruit the IL-1RAcP co-receptor, preventing MyD88 recruitment and blocking downstream signaling. Another IL-1 family cytokine, IL-38, binds to IL-1R6 in a similar way that IL-36Ra does. Additionally, both IL-36Ra and IL-38 suppress IL-22 and IL-17 production by peripheral blood mononuclear cells stimulated with *Candida*.[326] Therefore, IL-36Ra and IL-38 are potent natural inhibitors of proinflammatory IL-36 agonists and have similar biological effects on immune activation. Notably, N-terminal processing of the IL-36 cytokines by neutrophil proteases increases their activity by 500-fold.[327]

IL-36 cytokines are associated with inflammatory diseases, particularly psoriasis, where the expression of IL-36α and IL-36γ, but not IL-36β, is observed.[328] The expression of IL-36α, IL-36β, and IL-36γ is also elevated in the lesioned skin of patients with hidradenitis suppurativa.[329] Additionally, IL-36 cytokines are associated with colitis, rheumatoid arthritis, and Sjögren syndrome.

IL-33

IL-33 is expressed constitutively in epithelial cells and reticular fibroblasts and can be induced in immune cells. The protein is composed of two evolutionarily conserved domains, the N-terminal nuclear domain and C-terminal IL-1–like cytokine domain, separated by a divergent central part. The N-terminal of IL-33 contains a chromatin-binding motif and an NLS. IL-33 is usually localized to the nucleus and is released upon cell death; therefore, IL-33 is considered an alarmin. IL-33 can be processed extracellularly by enzymes such as neutrophil elastase and cathepsins. The mature form of IL-33 that is processed by mast cell proteases is more potent than its precursor. IL-33 can be proteolytically inactivated by caspases including caspase-1, -3, and -7.[330,331] Similar to the sequestration of IL-1α, chromatin sequesters IL-33 in the nucleus to reduce inflammation. Transgenic mice expressing a form of IL-33 with a mutated NLS succumb to lethality that is associated with the constitutive release of IL-33 into the serum.[332] Thus, keeping IL-33 in the nucleus of the cells producing it is essential for avoiding unwanted cytokine activity and maintaining organismal survival.

IL-33 is bioactive in its full-length form. Binding of IL-33 to its receptor IL-1R4 (ST2) induces a conformational change which allows ST2 to form a heterodimer with IL-1RAcP, thereby initiating MyD88 signaling[333,334] (Fig. 12.12B). IL-33 can be sequestered by the soluble receptor sST2, which serves as a decoy receptor to limit the activity of IL-33.[335] The mRNA encoding sST2 is generated by alternative splicing and lacks the sequence encoding the transmembrane domain of full-length ST2.

IL-33 has been reported to have both anti-inflammatory and proinflammatory functions. It is protective in heart disease, obesity, and uveitis, which may be associated with increased circulating Th2 cytokines, decreased IFNγ production, and increased numbers of eosinophils.[336] IL-33 also serves as an amplifier of the type 2 cytokine response from Th2 cells, group 2 innate lymphoid cells (ILC2), mast cells, and alternatively activated macrophages. However, IL-33 is pathogenic in inflammatory bowel disease, colon cancer, asthma, and rheumatoid arthritis.[293,337] In humans, SNPs in the genes encoding ST2 and IL-33 are associated with asthma.[337]

IL-1 Receptor Antagonist (IL-1Ra)

IL-1Ra was first identified as a suppressor of IL-1–mediated thymocyte proliferation. It was isolated from the urine of patients with monoblastic leukemia and was shown to inhibit the binding of IL-1 to cells. Endogenous IL-1Ra, which is present at very low levels in the circulation, is produced rapidly after an infusion of endotoxin.[338] Also, endogenous IL-1Ra, but not recombinant IL-1Ra expressed in *E. coli*, is glycosylated. IL-1Ra specifically blocks IL-1R1 and therefore inhibits the activity of both IL-1α and IL-1β[279] (Fig. 12.12B). In addition to the secreted soluble form of IL-1Ra, there are two more intracellular isoforms which are released upon cell death, limiting the proinflammatory effects of cell death.[339] IL-1Ra deficiency results in systemic and local inflammation, including pustular skin eruptions, vasculitis, and osteomyelitis in children. Similarly, mice deficient in IL-1Ra develop spontaneous arthritis and psoriasis. A recombinant form of IL-1Ra called anakinra is used to treat a broad variety of diseases such as rheumatoid arthritis, gout, idiopathic pericarditis, and other hereditary diseases.[4]

CONCLUSION

The PRRs and IL-1 family members represent a diverse set of immune molecules that each play critical roles in regulating health and disease. Smooth operation of the sensors, cytokines, receptors, and their downstream pathways is essential to prevent disease, and aberrant activation of these mechanisms can have devastating consequences. While these molecules have been found to play significant roles in homeostasis as well as infectious and inflammatory diseases and cancer, new and important functions for these proteins and pathways are still being described. This active area of research will likely continue to shape our understanding of the balance between hosts and pathogens and between beneficial and detrimental inflammation.

REFERENCES

1. Janeway CA Jr. Approaching the asymptote? Evolution and revolution in immunology. *Cold Spring Harb Symp Quant Biol.* 1989;54(pt 1):1-13.
2. Bianchi ME. DAMPs, PAMPs and alarmins: all we need to know about danger. *J Leukoc Biol.* 2007;81(1):1-5.
3. Mogensen TH. Pathogen recognition and inflammatory signaling in innate immune defenses. *Clin Microbiol Rev.* 2009;22(2):240-273.
4. Dinarello CA. Immunological and inflammatory functions of the interleukin-1 family. *Annu Rev Immunol.* 2009;27:519-550.
5. Lewis EB. A gene complex controlling segmentation in *Drosophila. Nature.* 1978;276(5688):565-570.
6. Nüsslein-Volhard C, Wieschaus E. Mutations affecting segment number and polarity in *Drosophila. Nature.* 1980;287(5785):795-801.
7. Lemaitre B, Nicolas E, Michaut L, Reichhart JM, Hoffmann JA. The dorsoventral regulatory gene cassette spätzle/Toll/cactus controls the potent antifungal response in *Drosophila* adults. *Cell.* 1996;86(6):973-983.
8. Williams MJ, Rodriguez A, Kimbrell DA, Eldon ED. The 18-wheeler mutation reveals complex antibacterial gene regulation in *Drosophila* host defense. *EMBO J.* 1997;16(20):6120-6130.
9. Janeway CA Jr, Medzhitov R. Innate immune recognition. *Annu Rev Immunol.* 2002;20:197-216.
10. Poltorak A, He X, Smirnova I, et al. Defective LPS signaling in C3H/HeJ and C57BL/10ScCr mice: mutations in TLR4 gene. *Science.* 1998;282(5396):2085-2088.
11. Chen GY, Nuñez G. Sterile inflammation: sensing and reacting to damage. *Nat Rev Immunol.* 2010;10(12):826-837.
12. Verstak B, Arnot CJ, Gay NJ. An alanine-to-proline mutation in the BB-loop of TLR3 Toll/IL-1R domain switches signalling adaptor specificity from TRIF to MyD88. *J Immunol.* 2013;191(12):6101-6109.
13. Horng T, Barton GM, Flavell RA, Medzhitov R. The adaptor molecule TIRAP provides signalling specificity for Toll-like receptors. *Nature.* 2002;420(6913):329-333.
14. Yamamoto M, Sato S, Hemmi H, et al. Essential role for TIRAP in activation of the signalling cascade shared by TLR2 and TLR4. *Nature.* 2002;420(6913):324-329.
15. Fitzgerald KA, Palsson-McDermott EM, Bowie AG, et al. Mal (MyD88-adapter-like) is required for Toll-like receptor-4 signal transduction. *Nature.* 2001;413(6851):78-83.
16. Negishi H, Fujita Y, Yanai H, et al. Evidence for licensing of IFN-gamma-induced IFN regulatory factor 1 transcription factor by MyD88 in Toll-like receptor-dependent gene induction program. *Proc Natl Acad Sci U S A.* 2006;103(41): 15136-15141.
17. Briard B, Karki R, Malireddi RKS, et al. Fungal ligands released by innate immune effectors promote inflammasome activation during *Aspergillus fumigatus* infection. *Nat Microbiol.* 2019;4(2):316-327.
18. Bergstrom B, Aune MH, Awuh JA, et al. TLR8 senses *Staphylococcus aureus* RNA in human primary monocytes and macrophages and induces IFN-β production via a TAK1-IKKβ-IRF5 signaling pathway. *J Immunol.* 2015;195(3):1100-1111.
19. Deguine J, Barton GM. MyD88: a central player in innate immune signaling. *F1000Prime Rep.* 2014;6:97.
20. Gurung P, Fan G, Lukens JR, Vogel P, Tonks NK, Kanneganti TD. Tyrosine kinase SYK licenses MyD88 adaptor protein to instigate IL-1α-mediated inflammatory disease. *Immunity.* 2017;46(4):635-648.
21. Kawai T, Adachi O, Ogawa T, Takeda K, Akira S. Unresponsiveness of MyD88-deficient mice to endotoxin. *Immunity.* 1999;11(1):115-122.
22. Groom JR, Fletcher CA, Walters SN, et al. BAFF and MyD88 signals promote a lupuslike disease independent of T cells. *J Exp Med.* 2007;204(8):1959-1971.
23. Sharma BR, Karki R, Lee E, Zhu Q, Gurung P, Kanneganti TD. Innate immune adaptor MyD88 deficiency prevents skin inflammation in SHARPIN-deficient mice. *Cell Death Differ.* 2019;26(4):741-750.
24. Michelsen KS, Wong MH, Shah PK, et al. Lack of Toll-like receptor 4 or myeloid differentiation factor 88 reduces atherosclerosis and alters plaque phenotype in mice deficient in apolipoprotein E. *Proc Natl Acad Sci U S A.* 2004;101(29):10679-10684.
25. Picard C, Casanova JL, Puel A. Infectious diseases in patients with IRAK-4, MyD88, NEMO, or IκBα deficiency. *Clin Microbiol Rev.* 2011;24(3):490-497.
26. von Bernuth H, Picard C, Jin Z, et al. Pyogenic bacterial infections in humans with MyD88 deficiency. *Science.* 2008;321(5889):691-696.
27. Ngo VN, Young RM, Schmitz R, et al. Oncogenically active MyD88 mutations in human lymphoma. *Nature.* 2011;470(7332):115-119.
28. Fitzgerald KA, McWhirter SM, Faia KL, et al. IKKepsilon and TBK1 are essential components of the IRF3 signaling pathway. *Nat Immunol.* 2003;4(5):491-496.
29. Yamamoto M, Sato S, Hemmi H, et al. Role of adaptor TRIF in the MyD88-independent Toll-like receptor signaling pathway. *Science.* 2003;301(5633):640-643.
30. Oshiumi H, Matsumoto M, Funami K, Akazawa T, Seya T. TICAM-1, an adaptor molecule that participates in Toll-like receptor 3-mediated interferon-beta induction. *Nat Immunol.* 2003;4(2):161-167.
31. Shen H, Tesar BM, Walker WE, Goldstein DR. Dual signaling of MyD88 and TRIF is critical for maximal TLR4-induced dendritic cell maturation. *J Immunol.* 2008;181(3):1849-1858.
32. Rowe DC, McGettrick AF, Latz E, et al. The myristoylation of TRIF-related adaptor molecule is essential for Toll-like receptor 4 signal transduction. *Proc Natl Acad Sci U S A.* 2006;103(16):6299-6304.
33. Gurung P, Malireddi RK, Anand PK, et al. Toll or interleukin-1 receptor (TIR) domain-containing adaptor inducing interferon-β (TRIF)-mediated caspase-11 protease production integrates Toll-like receptor 4 (TLR4) protein-and NLRP3 inflammasome-mediated host defense against enteropathogens. *J Biol Chem.* 2012;287(41):34474-34483.
34. Rathinam VA, Vanaja SK, Waggoner L, et al. TRIF licenses caspase-11-dependent NLRP3 inflammasome activation by gram-negative bacteria. *Cell.* 2012;150(3):606-619.
35. He S, Liang Y, Shao F, Wang X. Toll-like receptors activate programmed necrosis in macrophages through a receptor-interacting kinase-3-mediated pathway. *Proc Natl Acad Sci U S A.* 2011;108(50):20054-20059.
36. Kaiser WJ, Sridharan H, Huang C, et al. Toll-like receptor 3-mediated necrosis via TRIF, RIP3, and MLKL. *J Biol Chem.* 2013;288(43):31268-31279.
37. Ullah MO, Sweet MJ, Mansell A, Kellie S, Kobe B. TRIF-dependent TLR signaling, its functions in host defense and inflammation, and its potential as a therapeutic target. *J Leukoc Biol.* 2016;100(1):27-45.
38. Sancho-Shimizu V, Perez de Diego R, Lorenzo L, et al. Herpes simplex encephalitis in children with autosomal recessive and dominant TRIF deficiency. *J Clin Invest.* 2011;121(12):4889-4902.
39. Carty M, Goodbody R, Schroder M, Stack J, Moynagh PN, Bowie AG. The human adaptor SARM negatively regulates adaptor protein TRIF-dependent Toll-like receptor signaling. *Nat Immunol.* 2006;7(10):1074-1081.
40. Gurtler C, Carty M, Kearney J, et al. SARM regulates CCL5 production in macrophages by promoting the recruitment of transcription factors and RNA polymerase II to the Ccl5 promoter. *J Immunol.* 2014;192(10):4821-4832.
41. Stack J, Haga IR, Schroder M, et al. Vaccinia virus protein A46R targets multiple Toll-like-interleukin-1 receptor adaptors and contributes to virulence. *J Exp Med.* 2005;201(6):1007-1018.
42. Harte MT, Haga IR, Maloney G, et al. The poxvirus protein A52R targets Toll-like receptor signaling complexes to suppress host defense. *J Exp Med.* 2003;197(3):343-351.
43. Okumura A, Alce T, Lubyova B, Ezelle H, Strebel K, Pitha PM. HIV-1 accessory proteins VPR and Vif modulate antiviral response by targeting IRF-3 for degradation. *Virology.* 2008;373(1):85-97.
44. Hoebe K, Georgel P, Rutschmann S, et al. CD36 is a sensor of diacylglycerides. *Nature.* 2005;433(7025):523-527.
45. Manukyan M, Triantafilou K, Triantafilou M, et al. Binding of lipopeptide to CD14 induces physical proximity of CD14, TLR2 and TLR1. *Eur J Immunol.* 2005;35(3):911-921.
46. Barbalat R, Lau L, Locksley RM, Barton GM. Toll-like receptor 2 on inflammatory monocytes induces type I interferon in response to viral but not bacterial ligands. *Nat Immunol.* 2009;10(11):1200-1207.
47. Dietrich N, Lienenklaus S, Weiss S, Gekara NO. Murine Toll-like receptor 2 activation induces type I interferon responses from endolysosomal compartments. *PLoS One.* 2010;5(4):e10250.
48. Jiang S, Li X, Hess NJ, Guan Y, Tapping RI. TLR10 is a negative regulator of both MyD88-dependent and-independent TLR signaling. *J Immunol.* 2016;196(9):3834-3841.
49. Henrick BM, Yao XD, Zahoor MA, Abimiku A, Osawe S, Rosenthal KL. TLR10 senses HIV-1 proteins and significantly enhances HIV-1 infection. *Front Immunol.* 2019;10:482.
50. Lazarus R, Raby BA, Lange C, et al. Toll-like receptor 10 genetic variation is associated with asthma in two independent samples. *Am J Respir Crit Care Med.* 2004;170(6):594-600.
51. Zhou XX, Jia WH, Shen GP, et al. Sequence variants in Toll-like receptor 10 are associated with nasopharyngeal carcinoma risk. *Cancer Epidemiol Biomarkers Prev.* 2006;15(5):862-866.
52. Morgan AR, Lam WJ, Han DY, Fraser AG, Ferguson LR. Genetic variation within TLR10 is associated with Crohn's disease in a New Zealand population. *Hum Immunol.* 2012;73(4):416-420.
53. Grube M, Loeffler J, Mezger M, et al. TLR5 stop codon polymorphism is associated with invasive aspergillosis after allogeneic stem cell transplantation. *Med Mycol.* 2013;51(8):818-825.
54. Hawn TR, Verbon A, Lettinga KD, et al. A common dominant TLR5 stop codon polymorphism abolishes flagellin signaling and is associated with susceptibility to legionnaires' disease. *J Exp Med.* 2003;198(10):1563-1572.
55. West TE, Chantratita N, Chierakul W, et al. Impaired TLR5 functionality is associated with survival in melioidosis. *J Immunol.* 2013;190(7):3373-3379.
56. Gewirtz AT, Vijay-Kumar M, Brant SR, Duerr RH, Nicolae DL, Cho JH. Dominant-negative TLR5 polymorphism reduces adaptive immune response to flagellin and negatively associates with Crohn's disease. *Am J Physiol Gastrointest Liver Physiol.* 2006;290(6):G1157-G1163.
57. Ohto U, Yamakawa N, Akashi-Takamura S, Miyake K, Shimizu T. Structural analyses of human Toll-like receptor 4 polymorphisms D299G and T399I. *J Biol Chem.* 2012;287(48):40611-40617.

58. Dziarski R. Recognition of bacterial peptidoglycan by the innate immune system. *Cell Mol Life Sci.* 2003;60(9):1793-1804.
59. Ohashi K, Burkart V, Flohe S, Kolb H. Cutting edge: heat shock protein 60 is a putative endogenous ligand of the Toll-like receptor-4 complex. *J Immunol.* 2000;164(2):558-561.
60. Figueiredo RT, Fernandez PL, Mourao-Sa DS, et al. Characterization of heme as activator of Toll-like receptor 4. *J Biol Chem.* 2007;282(28):20221-20229.
61. Choi SH, Harkewicz R, Lee JH, et al. Lipoprotein accumulation in macrophages via Toll-like receptor-4-dependent fluid phase uptake. *Circ Res.* 2009;104(12):1355-1363.
62. Blasius AL, Beutler B. Intracellular Toll-like receptors. *Immunity.* 2010;32(3):305-315.
63. Weber F, Wagner V, Rasmussen SB, Hartmann R, Paludan SR. Double-stranded RNA is produced by positive-strand RNA viruses and DNA viruses but not in detectable amounts by negative-strand RNA viruses. *J Virol.* 2006;80(10):5059-5064.
64. Jurk M, Heil F, Vollmer J, et al. Human TLR7 or TLR8 independently confer responsiveness to the antiviral compound R-848. *Nat Immunol.* 2002;3(6):499.
65. Swiecki M, Gilfillan S, Vermi W, Wang Y, Colonna M. Plasmacytoid dendritic cell ablation impacts early interferon responses and antiviral NK and CD8(+) T cell accrual. *Immunity.* 2010;33(6):955-966.
66. Latz E, Schoenemeyer A, Visintin A, et al. TLR9 signals after translocating from the ER to CpG DNA in the lysosome. *Nat Immunol.* 2004;5(2):190-198.
67. Moresco EM, Beutler B. Special delivery: granulin brings CpG DNA to Toll-like receptor 9. *Immunity.* 2011;34(4):453-455.
68. Ivanov S, Dragoi AM, Wang X, et al. A novel role for HMGB1 in TLR9-mediated inflammatory responses to CpG-DNA. *Blood.* 2007;110(6):1970-1981.
69. Geijtenbeek TB, Gringhuis SI. Signalling through C-type lectin receptors: shaping immune responses. *Nat Rev Immunol.* 2009;9(7):465-479.
70. Richard M, Thibault N, Veilleux P, Gareau-Page G, Beaulieu AD. Granulocyte macrophage-colony stimulating factor reduces the affinity of SHP-2 for the ITIM of CLECSF6 in neutrophils: a new mechanism of action for SHP-2. *Mol Immunol.* 2006;43(10):1716-1721.
71. Meyer-Wentrup F, Benitez-Ribas D, Tacken PJ, et al. Targeting DCIR on human plasmacytoid dendritic cells results in antigen presentation and inhibits IFN-alpha production. *Blood.* 2008;111(8):4245-4253.
72. Meyer-Wentrup F, Cambi A, Joosten B, et al. DCIR is endocytosed into human dendritic cells and inhibits TLR8-mediated cytokine production. *J Leukoc Biol.* 2009;85(3):518-525.
73. Ariizumi K, Shen GL, Shikano S, et al. Identification of a novel, dendritic cell-associated molecule, Dectin-1, by subtractive cDNA cloning. *J Biol Chem.* 2000;275(26):20157-20167.
74. Brown GD, Gordon S. Immune recognition. A new receptor for beta-glucans. *Nature.* 2001;413(6851):36-37.
75. Cheng SC, Quintin J, Cramer RA, et al. mTOR-and HIF-1α-mediated aerobic glycolysis as metabolic basis for trained immunity. *Science.* 2014;345(6204):1250684.
76. Quintin J, Saeed S, Martens JHA, et al. *Candida albicans* infection affords protection against reinfection via functional reprogramming of monocytes. *Cell Host Microbe.* 2012;12(2):223-232.
77. LeibundGut-Landmann S, Gross O, Robinson MJ, et al. Syk- and CARD9-dependent coupling of innate immunity to the induction of T helper cells that produce interleukin 17. *Nat Immunol.* 2007;8(6):630-638.
78. Iwakura Y, Nakae S, Saijo S, Ishigame H. The roles of IL-17A in inflammatory immune responses and host defense against pathogens. *Immunol Rev.* 2008;226:57-79.
79. Zhang Z, Biagini Myers JM, Brandt EB, et al. β-Glucan exacerbates allergic asthma independent of fungal sensitization and promotes steroid-resistant TH2/TH17 responses. *J Allergy Clin Immunol.* 2017;139(1):54.e8-65.e8.
80. Steele C, Rapaka RR, Metz A, et al. The beta-glucan receptor Dectin-1 recognizes specific morphologies of *Aspergillus fumigatus. PLoS Pathog.* 2005;1(4):e42.
81. Taylor PR, Tsoni SV, Willment JA, et al. Dectin-1 is required for beta-glucan recognition and control of fungal infection. *Nat Immunol.* 2007;8(1):31-38.
82. Ferwerda B, Ferwerda G, Plantinga TS, et al. Human Dectin-1 deficiency and mucocutaneous fungal infections. *N Engl J Med.* 2009;361(18):1760-1767.
83. Cunha C, Di Ianni M, Bozza S, et al. Dectin-1 Y238X polymorphism associates with susceptibility to invasive aspergillosis in hematopoietic transplantation through impairment of both recipient-and donor-dependent mechanisms of antifungal immunity. *Blood.* 2010;116(24):5394-5402.
84. Bonder MJ, Kurilshikov A, Tigchelaar EF, et al. The effect of host genetics on the gut microbiome. *Nat Genet.* 2016;48(11):1407-1412.
85. Malik A, Sharma D, Malireddi RKS, et al. SYK-CARD9 signaling axis promotes gut fungi-mediated inflammasome activation to restrict colitis and colon cancer. *Immunity.* 2018;49(3):515.e5-530.e5.
86. Ariizumi K, Shen GL, Shikano S, et al. Cloning of a second dendritic cell-associated C-type lectin (Dectin-2) and its alternatively spliced isoforms. *J Biol Chem.* 2000;275(16):11957-11963.
87. Sato K, Yang XL, Yudate T, et al. Dectin-2 is a pattern recognition receptor for fungi that couples with the Fc receptor gamma chain to induce innate immune responses. *J Biol Chem.* 2006;281(50):38854-38866.
88. McGreal EP, Rosas M, Brown GD, et al. The carbohydrate-recognition domain of Dectin-2 is a C-type lectin with specificity for high mannose. *Glycobiology.* 2006;16(5):422-430.
89. Saijo S, Ikeda S, Yamabe K, et al. Dectin-2 recognition of alpha-mannans and induction of Th17 cell differentiation is essential for host defense against *Candida albicans. Immunity.* 2010;32(5):681-691.
90. Matsumoto M, Tanaka T, Kaisho T, et al. A novel LPS-inducible C-type lectin is a transcriptional target of NF-IL6 in macrophages. *J Immunol.* 1999;163(9):5039-5048.
91. Ishikawa E, Ishikawa T, Morita YS, et al. Direct recognition of the mycobacterial glycolipid, trehalose dimycolate, by C-type lectin Mincle. *J Exp Med.* 2009;206(13):2879-2888.
92. Yamasaki S, Ishikawa E, Sakuma M, Hara H, Ogata K, Saito T. Mincle is an ITAM-coupled activating receptor that senses damaged cells. *Nat Immunol.* 2008;9(10):1179-1188.
93. Nagata M, Izumi Y, Ishikawa E, et al. Intracellular metabolite β-glucosylceramide is an endogenous Mincle ligand possessing immunostimulatory activity. *Proc Natl Acad Sci U S A.* 2017;114(16):E3285-E3294.
94. Kostarnoy AV, Gancheva PG, Lepenies B, et al. Receptor Mincle promotes skin allergies and is capable of recognizing cholesterol sulfate. *Proc Natl Acad Sci U S A.* 2017;114(13):E2758-E2765.
95. Geijtenbeek TB, Torensma R, van Vliet SJ, et al. Identification of DC-SIGN, a novel dendritic cell-specific ICAM-3 receptor that supports primary immune responses. *Cell.* 2000;100(5):575-585.
96. Gringhuis SI, Kaptein TM, Wevers BA, et al. Fucose-based PAMPs prime dendritic cells for follicular T helper cell polarization via DC-SIGN-dependent IL-27 production. *Nat Commun.* 2014;5:5074.
97. van Gisbergen KP, Sanchez-Hernandez M, Geijtenbeek TB, van Kooyk Y. Neutrophils mediate immune modulation of dendritic cells through glycosylation-dependent interactions between Mac-1 and DC-SIGN. *J Exp Med.* 2005;201(8):1281-1292.
98. Wang L, Chen RF, Liu JW, et al. DC-SIGN (CD209) Promoter-336 A/G polymorphism is associated with dengue hemorrhagic fever and correlated to DC-SIGN expression and immune augmentation. *PLoS Negl Trop Dis.* 2011;5(1):e934.
99. Taylor PR, Gordon S, Martinez-Pomares L. The mannose receptor: linking homeostasis and immunity through sugar recognition. *Trends Immunol.* 2005;26(2):104-110.
100. Taylor ME, Conary JT, Lennartz MR, Stahl PD, Drickamer K. Primary structure of the mannose receptor contains multiple motifs resembling carbohydrate-recognition domains. *J Biol Chem.* 1990;265(21):12156-12162.
101. Taylor PR, Zamze S, Stillion RJ, Wong SY, Gordon S, Martinez-Pomares L. Development of a specific system for targeting protein to metallophilic macrophages. *Proc Natl Acad Sci U S A.* 2004;101(7):1963-1968.
102. Rajaram MVS, Arnett E, Azad AK, et al. M. tuberculosis-initiated human mannose receptor signaling regulates macrophage recognition and vesicle trafficking by FcRγ-chain, Grb2, and SHP-1. *Cell Rep.* 2017;21(1):126-140.
103. Kawasaki N, Kawasaki T, Yamashina I. Isolation and characterization of a mannan-binding protein from human serum. *J Biochem.* 1983;94(3):937-947.
104. Dean MM, Minchinton RM, Heatley S, Eisen DP. Mannose binding lectin acute phase activity in patients with severe infection. *J Clin Immunol.* 2005;25(4):346-352.
105. de Benedetti F, Auriti C, D'Urbano LE, et al. Low serum levels of mannose binding lectin are a risk factor for neonatal sepsis. *Pediatr Res.* 2007;61(3):325-328.
106. Nauta AJ, Raaschou-Jensen N, Roos A, et al. Mannose-binding lectin engagement with late apoptotic and necrotic cells. *Eur J Immunol.* 2003;33(10):2853-2863.
107. Thiel S, Vorup-Jensen T, Stover CM, et al. A second serine protease associated with mannan-binding lectin that activates complement. *Nature.* 1997;386(6624):506-510.
108. Ghiran I, Barbashov SF, Klickstein LB, Tas SW, Jensenius JC, Nicholson-Weller A. Complement receptor 1/CD35 is a receptor for mannan-binding lectin. *J Exp Med.* 2000;192(12):1797-1808.
109. Brown MS, Goldstein JL. Receptor-mediated endocytosis: insights from the lipoprotein receptor system. *Proc Natl Acad Sci U S A.* 1979;76(7):3330-3337.
110. Brown MS, Goldstein JL, Krieger M, Ho YK, Anderson RG. Reversible accumulation of cholesteryl esters in macrophages incubated with acetylated lipoproteins. *J Cell Biol.* 1979;82(3):597-613.
111. Greaves DR, Gordon S. The macrophage scavenger receptor at 30 years of age: current knowledge and future challenges. *J Lipid Res.* 2009;50 suppl:S282-S286.
112. Miller YI, Choi SH, Wiesner P, et al. Oxidation-specific epitopes are danger-associated molecular patterns recognized by pattern recognition receptors of innate immunity. *Circ Res.* 2011;108(2):235-248.
113. Hughes DA, Fraser IP, Gordon S. Murine macrophage scavenger receptor: in vivo expression and function as receptor for macrophage adhesion in lymphoid and non-lymphoid organs. *Eur J Immunol.* 1995;25(2):466-473.
114. Frenkel D, Wilkinson K, Zhao L, et al. Scara1 deficiency impairs clearance of soluble amyloid-β by mononuclear phagocytes and accelerates Alzheimer's-like disease progression. *Nat Commun.* 2013;4:2030.
115. Berwin B, Hart JP, Rice S, et al. Scavenger receptor-A mediates gp96/GRP94 and calreticulin internalization by antigen-presenting cells. *EMBO J.* 2003;22(22):6127-6136.
116. Hampton RY, Golenbock DT, Penman M, Krieger M, Raetz CR. Recognition and plasma clearance of endotoxin by scavenger receptors. *Nature.* 1991;352(6333):342-344.
117. Labonte AC, Sung SJ, Jennelle LT, Dandekar AP, Hahn YS. Expression of scavenger receptor-AI promotes alternative activation of murine macrophages to limit hepatic inflammation and fibrosis. *Hepatology.* 2017;65(1):32-43.
118. Kodama T, Freeman M, Rohrer L, Zabrecky J, Matsudaira P, Krieger M. Type I macrophage scavenger receptor contains alpha-helical and collagen-like coiled coils. *Nature.* 1990;343(6258):531-535.
119. Asch AS, Barnwell J, Silverstein RL, Nachman RL. Isolation of the thrombospondin membrane receptor. *J Clin Invest.* 1987;79(4):1054-1061.
120. Endemann G, Stanton LW, Madden KS, Bryant CM, White RT, Protter AA. CD36 is a receptor for oxidized low density lipoprotein. *J Biol Chem.* 1993;268(16):11811-11816.
121. Ramprasad MP, Terpstra V, Kondratenko N, Quehenberger O, Steinberg D. Cell surface expression of mouse macrosialin and human CD68 and their role as macrophage receptors for oxidized low density lipoprotein. *Proc Natl Acad Sci U S A.* 1996;93(25):14833-14838.
122. Yoshimoto R, Fujita Y, Kakino A, Iwamoto S, Takaya T, Sawamura T. The discovery of LOX-1, its ligands and clinical significance. *Cardiovasc Drugs Ther.* 2011;25(5):379-391.
123. Means TK, Mylonakis E, Tampakakis E, et al. Evolutionarily conserved recognition and innate immunity to fungal pathogens by the scavenger receptors SCARF1 and CD36. *J Exp Med.* 2009;206(3):637-653.
124. Wilkinson K, El Khoury J. Microglial scavenger receptors and their roles in the pathogenesis of Alzheimer's disease. *Int J Alzheimers Dis.* 2012;2012:489456.
125. Iram T, Ramirez-Ortiz Z, Byrne MH, et al. Megf10 is a receptor for C1Q that mediates clearance of apoptotic cells by astrocytes. *J Neurosci.* 2016;36(19):5185-5192.
126. Shimaoka T, Kume N, Minami M, et al. Molecular cloning of a novel scavenger receptor for oxidized low density lipoprotein, SR-PSOX, on macrophages. *J Biol Chem.* 2000;275(52):40663-40666.
127. Shimaoka T, Nakayama T, Hieshima K, et al. Chemokines generally exhibit scavenger receptor activity through their receptor-binding domain. *J Biol Chem.* 2004;279(26):26807-26810.
128. Park SY, Jung MY, Kim HJ, et al. Rapid cell corpse clearance by stabilin-2, a membrane phosphatidylserine receptor. *Cell Death Differ.* 2008;15(1):192-201.
129. Park SY, Jung MY, Lee SJ, et al. Stabilin-1 mediates phosphatidylserine-dependent clearance of cell corpses in alternatively activated macrophages. *J Cell Sci.* 2009;122(pt 18):3365-3373.
130. Kristiansen M, Graversen JH, Jacobsen C, et al. Identification of the haemoglobin scavenger receptor. *Nature.* 2001;409(6817):198-201.
131. Moeller JB, Nielsen MJ, Reichhardt MP, et al. CD163-L1 is an endocytic macrophage protein strongly regulated by mediators in the inflammatory response. *J Immunol.* 2012;188(5):2399-2409.
132. Holm D, Fink DR, Steffensen MA, et al. Characterization of a novel human scavenger receptor cysteine-rich molecule SCART1 expressed by lymphocytes. *Immunobiology.* 2013;218(3):408-417.
133. Ibrahim ZA, Armour CL, Phipps S, Sukkar MB. RAGE and TLRs: relatives, friends or neighbours? *Mol Immunol.* 2013;56(4):739-744.
134. Lee EJ, Park JH. Receptor for advanced glycation endproducts (RAGE), its ligands, and soluble RAGE: potential biomarkers for diagnosis and therapeutic targets for human renal diseases. *Genomics Inform.* 2013;11(4):224-229.
135. Bangert A, Andrassy M, Müller AM, et al. Critical role of RAGE and HMGB1 in inflammatory heart disease. *Proc Natl Acad Sci U S A.* 2016;113(2):E155-E164.
136. Culty M, Nguyen HA, Underhill CB. The hyaluronan receptor (CD44) participates in the uptake and degradation of hyaluronan. *J Cell Biol.* 1992;116(4):1055-1062.
137. Thankamony SP, Knudson W. Acylation of CD44 and its association with lipid rafts are required for receptor and hyaluronan endocytosis. *J Biol Chem.* 2006;281(45):34601-34609.
138. Kanekiyo T, Bu G. The low-density lipoprotein receptor-related protein 1 and amyloid-β clearance in Alzheimer's disease. *Front Aging Neurosci.* 2014;6:93.
139. Herz J, Strickland DK. LRP: a multifunctional scavenger and signaling receptor. *J Clin Invest.* 2001;108(6):779-784.

140. Strickland DK, Au DT, Cunfer P, Muratoglu SC. Low-density lipoprotein receptor-related protein-1: role in the regulation of vascular integrity. *Arterioscler Thromb Vasc Biol.* 2014;34(3):487-498.
141. Oh SJ, Kim TH, Lim JM, Jeong JW. Progesterone induces expression of Lrp2 in the murine uterus. *Biochem Biophys Res Commun.* 2013;441(1):175-179.
142. Kanneganti TD, Lamkanfi M, Nunez G. Intracellular NOD-like receptors in host defense and disease. *Immunity.* 2007;27(4):549-559.
143. Raval A, Weissman JD, Howcroft TK, Singer DS. The GTP-binding domain of class II transactivator regulates its nuclear export. *J Immunol.* 2003;170(2):922-930.
144. Devaiah BN, Singer DS. CIITA and its dual roles in MHC gene transcription. *Front Immunol.* 2013;4:476.
145. Karki R, Kanneganti TD. Diverging inflammasome signals in tumorigenesis and potential targeting. *Nat Rev Cancer.* 2019;19(4):197-214.
146. Chamaillard M, Hashimoto M, Horie Y, et al. An essential role for NOD1 in host recognition of bacterial peptidoglycan containing diaminopimelic acid. *Nat Immunol.* 2003;4(7):702-707.
147. Girardin SE, Boneca IG, Carneiro LA, et al. Nod1 detects a unique muropeptide from gram-negative bacterial peptidoglycan. *Science.* 2003;300(5625):1584-1587.
148. Girardin SE, Boneca IG, Viala J, et al. Nod2 is a general sensor of peptidoglycan through muramyl dipeptide (MDP) detection. *J Biol Chem.* 2003;278(11):8869-8872.
149. Inohara N, Koseki T, del Peso L, et al. Nod1, an Apaf-1-like activator of caspase-9 and nuclear factor-κB. *J Biol Chem.* 1999;274(21):14560-14567.
150. Meinzer U, Barreau F, Esmiol-Welterlin S, et al. Yersinia pseudotuberculosis effector YopJ subverts the Nod2/RICK/TAK1 pathway and activates caspase-1 to induce intestinal barrier dysfunction. *Cell Host Microbe.* 2012;11(4):337-351.
151. Lupfer C, Thomas PG, Anand PK, et al. Receptor interacting protein kinase 2-mediated mitophagy regulates inflammasome activation during virus infection. *Nat Immunol.* 2013;14(5):480-488.
152. Shaw PJ, Barr MJ, Lukens JR, et al. Signaling via the RIP2 adaptor protein in central nervous system-infiltrating dendritic cells promotes inflammation and autoimmunity. *Immunity.* 2011;34(1):75-84.
153. Zhang FR, Huang W, Chen SM, et al. Genomewide association study of leprosy. *N Engl J Med.* 2009;361(27):2609-2618.
154. Austin CM, Ma X, Graviss EA. Common nonsynonymous polymorphisms in the NOD2 gene are associated with resistance or susceptibility to tuberculosis disease in African Americans. *J Infect Dis.* 2008;197(12):1713-1716.
155. Rosenstiel P, Hellmig S, Hampe J, et al. Influence of polymorphisms in the NOD1/CARD4 and NOD2/CARD15 genes on the clinical outcome of *Helicobacter pylori* infection. *Cell Microbiol.* 2006;8(7):1188-1198.
156. Hampe J, Cuthbert A, Croucher PJ, et al. Association between insertion mutation in NOD2 gene and Crohn's disease in German and British populations. *Lancet.* 2001;357(9272):1925-1928.
157. Conti BJ, Davis BK, Zhang J, O'Connor W Jr, Williams KL, Ting JP. CATERPILLER 16.2 (CLR16.2), a novel NBD/LRR family member that negatively regulates T cell function. *J Biol Chem.* 2005;280(18):18375-18385.
158. Schneider M, Zimmermann AG, Roberts RA, et al. The innate immune sensor NLRC3 attenuates Toll-like receptor signaling via modification of the signaling adaptor TRAF6 and transcription factor NF-κB. *Nat Immunol.* 2012;13(9):823-831.
159. Li X, Deng M, Petrucelli AS, et al. Viral DNA binding to NLRC3, an inhibitory nucleic acid sensor, unleashes STING, a cyclic dinucleotide receptor that activates type I interferon. *Immunity.* 2019;50(3):591.e6-599.e6.
160. Zhang L, Mo J, Swanson KV, et al. NLRC3, a member of the NLR family of proteins, is a negative regulator of innate immune signaling induced by the DNA sensor STING. *Immunity.* 2014;40(3):329-341.
161. Karki R, Man SM, Malireddi RKS, et al. NLRC3 is an inhibitory sensor of PI3K-mTOR pathways in cancer. *Nature.* 2016;540(7634):583-587.
162. Uchimura T, Oyama Y, Deng M, et al. The innate immune sensor NLRC3 acts as a rheostat that fine-tunes T cell responses in infection and autoimmunity. *Immunity.* 2018;49(6):1049.e6-1061.e6.
163. Meissner TB, Li A, Biswas A, et al. NLR family member NLRC5 is a transcriptional regulator of MHC class I genes. *Proc Natl Acad Sci U S A.* 2010;107(31):13794-13799.
164. Arnoult D, Soares F, Tattoli I, Castanier C, Philpott DJ, Girardin SE. An N-terminal addressing sequence targets NLRX1 to the mitochondrial matrix. *J Cell Sci.* 2009;122(pt 17):3161-3168.
165. Allen IC, Moore CB, Schneider M, et al. NLRX1 protein attenuates inflammatory responses to infection by interfering with the RIG-I-MAVS and TRAF6-NF-κB signaling pathways. *Immunity.* 2011;34(6):854-865.
166. Eitas TK, Chou WC, Wen H, et al. The nucleotide-binding leucine-rich repeat (NLR) family member NLRX1 mediates protection against experimental autoimmune encephalomyelitis and represses macrophage/microglia-induced inflammation. *J Biol Chem.* 2014;289(7):4173-4179.
167. Jaworska J, Coulombe F, Downey J, et al. NLRX1 prevents mitochondrial induced apoptosis and enhances macrophage antiviral immunity by interacting with influenza virus PB1-F2 protein. *Proc Natl Acad Sci U S A.* 2014;111(20):E2110-E2119.
168. Martinon F, Burns K, Tschopp J. The inflammasome: a molecular platform triggering activation of inflammatory caspases and processing of proIL-beta. *Mol Cell.* 2002;10(2):417-426.
169. Sandstrom A, Mitchell PS, Goers L, Mu EW, Lesser CF, Vance RE. Functional degradation: a mechanism of NLRP1 inflammasome activation by diverse pathogen enzymes. *Science.* 2019;364(6435):eaau1330.
170. Boyden ED, Dietrich WF. Nalp1b controls mouse macrophage susceptibility to anthrax lethal toxin. *Nat Genet.* 2006;38(2):240-244.
171. Chavarria-Smith J, Vance RE. Direct proteolytic cleavage of NLRP1B is necessary and sufficient for inflammasome activation by anthrax lethal factor. *PLoS Pathog.* 2013;9(6):e1003452.
172. Chui AJ, Okondo MC, Rao SD, et al. N-terminal degradation activates the NLRP1B inflammasome. *Science.* 2019;364(6435):82-85.
173. Neiman-Zenevich J, Stuart S, Abdel-Nour M, Girardin SE, Mogridge J. *Listeria monocytogenes* and *Shigella flexneri* activate the NLRP1B inflammasome. *Infect Immun.* 2017;85(11):e00338-e00417.
174. Ewald SE, Chavarria-Smith J, Boothroyd JC. NLRP1 is an inflammasome sensor for *Toxoplasma gondii*. *Infect Immun.* 2014;82(1):460-468.
175. Chavarria-Smith J, Vance RE. The NLRP1 inflammasomes. *Immunol Rev.* 2015;265(1):22-34.
176. Aganna E, Martinon F, Hawkins PN, et al. Association of mutations in the NALP3/CIAS1/PYPAF1 gene with a broad phenotype including recurrent fever, cold sensitivity, sensorineural deafness, and AA amyloidosis. *Arthritis Rheum.* 2002;46(9):2445-2452.
177. Aksentijevich I, Nowak M, Mallah M, et al. De novo CIAS1 mutations, cytokine activation, and evidence for genetic heterogeneity in patients with neonatal-onset multisystem inflammatory disease (NOMID): a new member of the expanding family of pyrin-associated autoinflammatory diseases. *Arthritis Rheum.* 2002;46(12):3340-3348.
178. Hoffman HM, Mueller JL, Broide DH, Wanderer AA, Kolodner RD. Mutation of a new gene encoding a putative pyrin-like protein causes familial cold autoinflammatory syndrome and Muckle-Wells syndrome. *Nat Genet.* 2001;29(3):301-305.
179. Gaidt MM, Ebert TS, Chauhan D, et al. Human monocytes engage an alternative inflammasome pathway. *Immunity.* 2016;44(4):833-846.
180. Gurung P, Karki R, Vogel P, et al. An NLRP3 inflammasome-triggered Th2-biased adaptive immune response promotes leishmaniasis. *J Clin Invest.* 2015;125(3):1329-1338.
181. Kanneganti TD, Body-Malapel M, Amer A, et al. Critical role for Cryopyrin/Nalp3 in activation of caspase-1 in response to viral infection and double-stranded RNA. *J Biol Chem.* 2006;281(48):36560-36568.
182. Kanneganti TD, Ozoren N, Body-Malapel M, et al. Bacterial RNA and small antiviral compounds activate caspase-1 through cryopyrin/NALP3. *Nature.* 2006;440(7081):233-236.
183. Lamkanfi M, Malireddi RK, Kanneganti TD. Fungal zymosan and mannan activate the cryopyrin inflammasome. *J Biol Chem.* 2009;284(31):20574-20581.
184. Mariathasan S, Weiss DS, Newton K, et al. Cryopyrin activates the inflammasome in response to toxins and ATP. *Nature.* 2006;440(7081):228-232.
185. Martinon F, Petrilli V, Mayor A, Tardivel A, Tschopp J. Gout-associated uric acid crystals activate the NALP3 inflammasome. *Nature.* 2006;440(7081):237-241.
186. Swanson KV, Deng M, Ting JP. The NLRP3 inflammasome: molecular activation and regulation to therapeutics. *Nat Rev Immunol.* 2019;19(8):477-489.
187. Aachoui Y, Leaf IA, Hagar JA, et al. Caspase-11 protects against bacteria that escape the vacuole. *Science.* 2013;339(6122):975-978.
188. Kayagaki N, Warming S, Lamkanfi M, et al. Non-canonical inflammasome activation targets caspase-11. *Nature.* 2011;479(7371):117-121.
189. Kayagaki N, Wong MT, Stowe IB, et al. Noncanonical inflammasome activation by intracellular LPS independent of TLR4. *Science.* 2013;341(6151):1246-1249.
190. Shi J, Zhao Y, Wang Y, et al. Inflammatory caspases are innate immune receptors for intracellular LPS. *Nature.* 2014;514(7521):187-192.
191. Karki R, Lee E, Sharma BR, Banoth B, Kanneganti TD. IRF8 regulates gram-negative bacteria-mediated NLRP3 inflammasome activation and cell death. *J Immunol.* 2020;204(9):2514-2522.
192. Man SM, Karki R, Sasai M, et al. IRGB10 liberates bacterial ligands for sensing by the AIM2 and caspase-11-NLRP3 inflammasomes. *Cell.* 2016;167(2):382.e17-396.e17.
193. Gurung P, Anand PK, Malireddi RK, et al. FADD and caspase-8 mediate priming and activation of the canonical and noncanonical NLRP3 inflammasomes. *J Immunol.* 2014;192(4):1835-1846.
194. Broz P, Ruby T, Belhocine K, et al. Caspase-11 increases susceptibility to *Salmonella* infection in the absence of caspase-1. *Nature.* 2012;490(7419):288-291.
195. Meunier E, Dick MS, Dreier RF, et al. Caspase-11 activation requires lysis of pathogen-containing vacuoles by IFN-induced GTPases. *Nature.* 2014;509(7500):366-370.
196. Lee BL, Stowe IB, Gupta A, et al. Caspase-11 auto-proteolysis is crucial for noncanonical inflammasome activation. *J Exp Med.* 2018;215(9):2279-2288.
197. Kayagaki N, Stowe IB, Lee BL, et al. Caspase-11 cleaves gasdermin D for non-canonical inflammasome signalling. *Nature.* 2015;526(7575):666-671.
198. Shi J, Zhao Y, Wang K, et al. Cleavage of GSDMD by inflammatory caspases determines pyroptotic cell death. *Nature.* 2015;526(7575):660-665.
199. Kuriakose T, Man SM, Malireddi RK, et al. ZBP1/DAI is an innate sensor of influenza virus triggering the NLRP3 inflammasome and programmed cell death pathways. *Sci Immunol.* 2016;1(2):aag2045.
200. Christgen S, Zheng M, Kesavardhana S, et al. Identification of the PANoptosome: a molecular platform triggering pyroptosis, apoptosis, and necroptosis (PANoptosis). *Front Cell Infect Microbiol.* 2020;10:237.
201. Samir PM, Malireddi RKS, Kanneganti TD. The PANoptosome: a deadly protein complex driving pyroptosis, apoptosis, and necroptosis (PANoptosis). *Front Cell Infect Microbiol.* 2020;10:238.
202. Thomas PG, Dash P, Aldridge JR Jr, et al. The intracellular sensor NLRP3 mediates key innate and healing responses to influenza A virus via the regulation of caspase-1. *Immunity.* 2009;30(4):566-575.
203. Zheng M, Karki R, Vogel P, Kanneganti TD. Caspase-6 is a key regulator of innate immunity, inflammasome activation, and host defense. *Cell.* 2020;181(3):674.e13-687.e13.
204. He Y, Zeng MY, Yang D, Motro B, Nunez G. NEK7 is an essential mediator of NLRP3 activation downstream of potassium efflux. *Nature.* 2016;530(7590):354-357.
205. Schmid-Burgk JL, Chauhan D, Schmidt T, et al. A genome-wide CRISPR (clustered regularly interspaced short palindromic repeats) screen identifies NEK7 as an essential component of NLRP3 inflammasome activation. *J Biol Chem.* 2016;291(1):103-109.
206. Malireddi RKS, Gurung P, Mavuluri J, et al. TAK1 restricts spontaneous NLRP3 activation and cell death to control myeloid proliferation. *J Exp Med.* 2018;215(4):1023-1034.
207. Samir P, Kesavardhana S, Patmore DM, et al. DDX3X acts as a live-or-die checkpoint in stressed cells by regulating NLRP3 inflammasome. *Nature.* 2019;573(7775):590-594.
208. Ferguson PJ, Bing X, Vasef MA, et al. A missense mutation in pstpip2 is associated with the murine autoinflammatory disorder chronic multifocal osteomyelitis. *Bone.* 2006;38(1):41-47.
209. Gurung P, Burton A, Kanneganti TD. NLRP3 inflammasome plays a redundant role with caspase 8 to promote IL-1β-mediated osteomyelitis. *Proc Natl Acad Sci U S A.* 2016;113(16):4452-4457.
210. Sutterwala FS, Ogura Y, Szczepanik M, et al. Critical role for NALP3/CIAS1/Cryopyrin in innate and adaptive immunity through its regulation of caspase-1. *Immunity.* 2006;24(3):317-327.
211. Franchi L, Amer A, Body-Malapel M, et al. Cytosolic flagellin requires Ipaf for activation of caspase-1 and interleukin 1beta in *salmonella*-infected macrophages. *Nat Immunol.* 2006;7(6):576-582.
212. Ren T, Zamboni DS, Roy CR, Dietrich WF, Vance RE. Flagellin-deficient *Legionella* mutants evade caspase-1-and Naip5-mediated macrophage immunity. *PLoS Pathog.* 2006;2(3):e18.
213. Endrizzi MG, Hadinoto V, Growney JD, Miller W, Dietrich WF. Genomic sequence analysis of the mouse *Naip* gene array. *Genome Res.* 2000;10(8):1095-1102.
214. Tenthorey JL, Kofoed EM, Daugherty MD, Malik HS, Vance RE. Molecular basis for specific recognition of bacterial ligands by NAIP/NLRC4 inflammasomes. *Mol Cell.* 2014;54(1):17-29.
215. Kofoed EM, Vance RE. Innate immune recognition of bacterial ligands by NAIPs determines inflammasome specificity. *Nature.* 2011;477(7366):592-595.
216. Lightfield KL, Persson J, Brubaker SW, et al. Critical function for Naip5 in inflammasome activation by a conserved carboxy-terminal domain of flagellin. *Nat Immunol.* 2008;9(10):1171-1178.
217. Zhao Y, Yang J, Shi J, et al. The NLRC4 inflammasome receptors for bacterial flagellin and type III secretion apparatus. *Nature.* 2011;477(7366):596-600.
218. Bouwman LI, de Zoete MR, Bleumink-Pluym NM, Flavell RA, van Putten JP. Inflammasome activation by *Campylobacter jejuni*. *J Immunol.* 2014;193(9):4548-4557.
219. Matusiak M, Van Opdenbosch N, Vande Walle L, Sirard JC, Kanneganti TD, Lamkanfi M. Flagellin-induced NLRC4 phosphorylation primes the inflammasome for activation by NAIP5. *Proc Natl Acad Sci U S A.* 2015;112(5):1541-1546.
220. Karki R, Lee E, Place D, et al. IRF8 regulates transcription of Naips for NLRC4 inflammasome activation. *Cell.* 2018;173(4):920-933 e13.
221. Canna SW, de Jesus AA, Gouni S, et al. An activating NLRC4 inflammasome mutation causes autoinflammation with recurrent macrophage activation syndrome. *Nat Genet.* 2014;46(10):1140-1146.

222. Kitamura A, Sasaki Y, Abe T, Kano H, Yasutomo K. An inherited mutation in NLRC4 causes autoinflammation in human and mice. *J Exp Med.* 2014;211(12):2385-2396.
223. Romberg N, Al Moussawi K, Nelson-Williams C, et al. Mutation of NLRC4 causes a syndrome of enterocolitis and autoinflammation. *Nat Genet.* 2014;46(10):1135-1139.
224. Chung LK, Park YH, Zheng Y, et al. The *Yersinia* virulence factor YopM hijacks host kinases to inhibit type III effector-triggered activation of the Pyrin inflammasome. *Cell Host Microbe.* 2016;20(3):296-306.
225. Ratner D, Orning MP, Proulx MK, et al. The *Yersinia pestis* effector YopM inhibits Pyrin inflammasome activation. *PLoS Pathog.* 2016;12(12):e1006035.
226. Xu H, Yang J, Gao W, et al. Innate immune sensing of bacterial modifications of Rho GTPases by the Pyrin inflammasome. *Nature.* 2014;513(7517):237-241.
227. Gao W, Yang J, Liu W, Wang Y, Shao F. Site-specific phosphorylation and microtubule dynamics control Pyrin inflammasome activation. *Proc Natl Acad Sci U S A.* 2016;113(33):E4857-E4866.
228. Park YH, Wood G, Kastner DL, Chae JJ. Pyrin inflammasome activation and RhoA signaling in the autoinflammatory diseases FMF and HIDS. *Nat Immunol.* 2016;17(8):914-921.
229. Chae JJ, Cho YH, Lee GS, et al. Gain-of-function Pyrin mutations induce NLRP3 protein-independent interleukin-1β activation and severe autoinflammation in mice. *Immunity.* 2011;34(5):755-768.
230. Sharma D, Sharma BR, Vogel P, Kanneganti TD. IL-1β and caspase-1 drive autoinflammatory disease independently of IL-1α or caspase-8 in a mouse model of familial Mediterranean fever. *Am J Pathol.* 2017;187(2):236-244.
231. Kanneganti A, Malireddi RKS, Saavedra PHV, et al. GSDMD is critical for autoinflammatory pathology in a mouse model of Familial Mediterranean Fever. *J Exp Med.* 2018;215(6):1519-1529.
232. Xiao J, Wang C, Yao JC, et al. Gasdermin D mediates the pathogenesis of neonatal-onset multisystem inflammatory disease in mice. *PLoS Biol.* 2018;16(11):e3000047.
233. Sharma D, Malik A, Guy C, Vogel P, Kanneganti TD. TNF/TNFR axis promotes pyrin inflammasome activation and distinctly modulates pyrin inflammasomopathy. *J Clin Invest.* 2019;129(1):150-162.
234. Sharma D, Malik A, Guy CS, Karki R, Vogel P, Kanneganti TD. Pyrin inflammasome regulates tight junction integrity to restrict colitis and tumorigenesis. *Gastroenterology.* 2018;154(4):948-964.e8.
235. Burckstummer T, Baumann C, Bluml S, et al. An orthogonal proteomic-genomic screen identifies AIM2 as a cytoplasmic DNA sensor for the inflammasome. *Nat Immunol.* 2009;10(3):266-272.
236. Fernandes-Alnemri T, Yu JW, Datta P, Wu J, Alnemri ES. AIM2 activates the inflammasome and cell death in response to cytoplasmic DNA. *Nature.* 2009;458(7237):509-513.
237. Fernandes-Alnemri T, Yu JW, Juliana C, et al. The AIM2 inflammasome is critical for innate immunity to *Francisella tularensis*. *Nat Immunol.* 2010;11(5):385-393.
238. Hornung V, Ablasser A, Charrel-Dennis M, et al. AIM2 recognizes cytosolic dsDNA and forms a caspase-1-activating inflammasome with ASC. *Nature.* 2009;458(7237):514-518.
239. Rathinam VA, Jiang Z, Waggoner SN, et al. The AIM2 inflammasome is essential for host defense against cytosolic bacteria and DNA viruses. *Nat Immunol.* 2010;11(5):395-402.
240. Roberts TL, Idris A, Dunn JA, et al. HIN-200 proteins regulate caspase activation in response to foreign cytoplasmic DNA. *Science.* 2009;323(5917):1057-1060.
241. Jin T, Perry A, Jiang J, et al. Structures of the HIN domain:DNA complexes reveal ligand binding and activation mechanisms of the AIM2 inflammasome and IFI16 receptor. *Immunity.* 2012;36(4):561-571.
242. Man SM, Karki R, Malireddi RK, et al. The transcription factor IRF1 and guanylate-binding proteins target activation of the AIM2 inflammasome by *Francisella* infection. *Nat Immunol.* 2015;16(5):467-475.
243. Kaminski JJ, Schattgen SA, Tzeng TC, Bode C, Klinman DM, Fitzgerald KA. Synthetic oligodeoxynucleotides containing suppressive TTAGGG motifs inhibit AIM2 inflammasome activation. *J Immunol.* 2013;191(7):3876-3883.
244. Sharma BR, Karki R, Kanneganti TD. Role of AIM2 inflammasome in inflammatory diseases, cancer and infection. *Eur J Immunol.* 2019;49(11):1998-2011.
245. Lee S, Karki R, Wang Y, Nguyen LN, Kalathur RC, Kanneganti TD. AIM2 forms a complex with pyrin and ZBP1 to drive PANoptosis and host defence. *Nature.* 2021;597(7876):415-419.
246. Man SM, Zhu Q, Zhu L, et al. Critical role for the DNA sensor AIM2 in stem cell proliferation and cancer. *Cell.* 2015;162(1):45-58.
247. Wilson JE, Petrucelli AS, Chen L, et al. Inflammasome-independent role of AIM2 in suppressing colon tumorigenesis via DNA-PK and Akt. *Nat Med.* 2015;21(8):906-913.
248. Kerur N, Veettil MV, Sharma-Walia N, et al. IFI16 acts as a nuclear pathogen sensor to induce the inflammasome in response to Kaposi Sarcoma-associated herpesvirus infection. *Cell Host Microbe.* 2011;9(5):363-375.
249. Monroe KM, Yang Z, Johnson JR, et al. IFI16 DNA sensor is required for death of lymphoid CD4 T cells abortively infected with HIV. *Science.* 2014;343(6169):428-432.
250. Chunfa L, Xin S, Qiang L, et al. The central role of IFI204 in IFN-beta release and autophagy activation during *Mycobacterium bovis* infection. *Front Cell Infect Microbiol.* 2017;7:169.
251. Hara H, Seregin SS, Yang D, et al. The NLRP6 inflammasome recognizes lipoteichoic acid and regulates gram-positive pathogen infection. *Cell.* 2018;175(6):1651-1664 e14.
252. Anand PK, Malireddi RK, Lukens JR, et al. NLRP6 negatively regulates innate immunity and host defence against bacterial pathogens. *Nature.* 2012;488(7411):389-393.
253. Wang P, Zhu S, Yang L, et al. NLRP6 regulates intestinal antiviral innate immunity. *Science.* 2015;350(6262):826-830.
254. Zaki MH, Vogel P, Malireddi RK, et al. The NOD-like receptor NLRP12 attenuates colon inflammation and tumorigenesis. *Cancer Cell.* 2011;20(5):649-660.
255. Chen ST, Chen L, Lin DS, et al. NLRP12 regulates anti-viral RIG-I activation via interaction with TRIM25. *Cell Host Microbe.* 2019;25(4):602.e7-616.e7.
256. Ataide MA, Andrade WA, Zamboni DS, et al. Malaria-induced NLRP12/NLRP3-dependent caspase-1 activation mediates inflammation and hypersensitivity to bacterial superinfection. *PLoS Pathog.* 2014;10(1):e1003885.
257. Vladimer GI, Weng D, Paquette SW, et al. The NLRP12 inflammasome recognizes *Yersinia pestis*. *Immunity.* 2012;37(1):96-107.
258. Peng H, Lin X, Liu F, Wang C, Zhang W. NLRP9B protein is dispensable for oocyte maturation and early embryonic development in the mouse. *J Reprod Dev.* 2015;61(6):559-564.
259. Zhu S, Ding S, Wang P, et al. Nlrp9b inflammasome restricts rotavirus infection in intestinal epithelial cells. *Nature.* 2017;546(7660):667-670.
260. Onoguchi K, Yoneyama M, Fujita T. Retinoic acid-inducible gene-I-like receptors. *J Interferon Cytokine Res.* 2011;31(1):27-31.
261. Loo YM, Gale M Jr. Immune signaling by RIG-I-like receptors. *Immunity.* 2011;34(5):680-692.
262. Yoneyama M, Kikuchi M, Matsumoto K, et al. Shared and unique functions of the DExD/H-box helicases RIG-I, MDA5, and LGP2 in antiviral innate immunity. *J Immunol.* 2005;175(5):2851-2858.
263. Kato H, Takeuchi O, Sato S, et al. Differential roles of MDA5 and RIG-I helicases in the recognition of RNA viruses. *Nature.* 2006;441(7089):101-105.
264. Liddicoat BJ, Piskol R, Chalk AM, et al. RNA editing by ADAR1 prevents MDA5 sensing of endogenous dsRNA as nonself. *Science.* 2015;349(6252):1115-1120.
265. Hornung V, Ellegast J, Kim S, et al. 5'-Triphosphate RNA is the ligand for RIG-I. *Science.* 2006;314(5801):994-997.
266. Kato H, Takeuchi O, Mikamo-Satoh E, et al. Length-dependent recognition of double-stranded ribonucleic acids by retinoic acid-inducible gene-I and melanoma differentiation-associated gene 5. *J Exp Med.* 2008;205(7):1601-1610.
267. Saito T, Owen DM, Jiang F, Marcotrigiano J, Gale M Jr. Innate immunity induced by composition-dependent RIG-I recognition of hepatitis C virus RNA. *Nature.* 2008;454(7203):523-527.
268. Uzri D, Gehrke L. Nucleotide sequences and modifications that determine RIG-I/RNA binding and signaling activities. *J Virol.* 2009;83(9):4174-4184.
269. Saito T, Hirai R, Loo YM, et al. Regulation of innate antiviral defenses through a shared repressor domain in RIG-I and LGP2. *Proc Natl Acad Sci U S A.* 2007;104(2):582-587.
270. Gack MU, Shin YC, Joo CH, et al. TRIM25 RING-finger E3 ubiquitin ligase is essential for RIG-I-mediated antiviral activity. *Nature.* 2007;446(7138):916-920.
271. Lang X, Tang T, Jin T, Ding C, Zhou R, Jiang W. TRIM65-catalized ubiquitination is essential for MDA5-mediated antiviral innate immunity. *J Exp Med.* 2017;214(2):459-473.
272. Horner SM, Liu HM, Park HS, Briley J, Gale M Jr. Mitochondrial-associated endoplasmic reticulum membranes (MAM) form innate immune synapses and are targeted by hepatitis C virus. *Proc Natl Acad Sci U S A.* 2011;108(35):14590-14595.
273. Liu HM, Loo YM, Horner SM, Zornetzer GA, Katze MG, Gale M Jr. The mitochondrial targeting chaperone 14-3-3ε regulates a RIG-I translocon that mediates membrane association and innate antiviral immunity. *Cell Host Microbe.* 2012;11(5):528-537.
274. Quicke KM, Diamond MS, Suthar MS. Negative regulators of the RIG-I-like receptor signaling pathway. *Eur J Immunol.* 2017;47(4):615-628.
275. Nejentsev S, Walker N, Riches D, Egholm M, Todd JA. Rare variants of IFIH1, a gene implicated in antiviral responses, protect against type 1 diabetes. *Science.* 2009;324(5925):387-389.
276. Onengut-Gumuscu S, Chen WM, Burren O, et al. Fine mapping of type 1 diabetes susceptibility loci and evidence for colocalization of causal variants with lymphoid gene enhancers. *Nat Genet.* 2015;47(4):381-386.
277. Willcox A, Richardson SJ, Bone AJ, Foulis AK, Morgan NG. Analysis of islet inflammation in human type 1 diabetes. *Clin Exp Immunol.* 2009;155(2):173-181.
278. Towne JE, Renshaw BR, Douangpanya J, et al. Interleukin-36 (IL-36) ligands require processing for full agonist (IL-36α, IL-36β, and IL-36γ) or antagonist (IL-36Ra) activity. *J Biol Chem.* 2011;286(49):42594-42602.
279. Dinarello CA. Overview of the IL-1 family in innate inflammation and acquired immunity. *Immunol Rev.* 2018;281(1):8-27.
280. McDowell TL, Symons JA, Duff GW. Human interleukin-1α gene expression is regulated by Sp1 and a transcriptional repressor. *Cytokine.* 2005;30(4):141-153.
281. Bailly S, Fay M, Israel N, Gougerot-Pocidalo MA. The transcription factor AP-1 binds to the human interleukin 1 alpha promoter. *Eur Cytokine Netw.* 1996;7(2):125-128.
282. Mori N, Prager D. Transactivation of the interleukin-1α promoter by human T-cell leukemia virus type I and type II Tax proteins. *Blood.* 1996;87(8):3410-3417.
283. Werman A, Werman-Venkert R, White R, et al. The precursor form of IL-1α is an intracrine proinflammatory activator of transcription. *Proc Natl Acad Sci U S A.* 2004;101(8):2434-2439.
284. Cohen I, Rider P, Carmi Y, et al. Differential release of chromatin-bound IL-1α discriminates between necrotic and apoptotic cell death by the ability to induce sterile inflammation. *Proc Natl Acad Sci U S A.* 2010;107(6):2574-2579.
285. Di Paolo NC, Shayakhmetov DM. Interleukin 1α and the inflammatory process. *Nat Immunol.* 2016;17(8):906-913.
286. Kurt-Jones EA, Beller DI, Mizel SB, Unanue ER. Identification of a membrane-associated interleukin 1 in macrophages. *Proc Natl Acad Sci U S A.* 1985;82(4):1204-1208.
287. Berda-Haddad Y, Robert S, Salers P, et al. Sterile inflammation of endothelial cell-derived apoptotic bodies is mediated by interleukin-1α. *Proc Natl Acad Sci U S A.* 2011;108(51):20684-20689.
288. Lukens JR, Vogel P, Johnson GR, et al. RIP1-driven autoinflammation targets IL-1α independently of inflammasomes and RIP3. *Nature.* 2013;498(7453):224-227.
289. Bersudsky M, Luski L, Fishman D, et al. Non-redundant properties of IL-1α and IL-1β during acute colon inflammation in mice. *Gut.* 2014;63(4):598-609.
290. Freigang S, Ampenberger F, Weiss A, et al. Fatty acid-induced mitochondrial uncoupling elicits inflammasome-independent IL-1α and sterile vascular inflammation in atherosclerosis. *Nat Immunol.* 2013;14(10):1045-1053.
291. Brough D, Denes A. Interleukin-1α and brain inflammation. *IUBMB Life.* 2015;67(5):323-330.
292. Kanni T, Argyropoulou M, Spyridopoulos T, et al. MABp1 targeting IL-1α for Moderate to severe hidradenitis suppurativa not eligible for adalimumab: a randomized study. *J Invest Dermatol.* 2018;138(4):795-801.
293. Malik A, Sharma D, Zhu Q, et al. IL-33 regulates the IgA-microbiota axis to restrain IL-1α-dependent colitis and tumorigenesis. *J Clin Invest.* 2016;126(12):4469-4481.
294. Leon X, Bothe C, Garcia J, et al. Expression of IL-1α correlates with distant metastasis in patients with head and neck squamous cell carcinoma. *Oncotarget.* 2015;6(35):37398-37409.
295. Dinarello CA. Interleukin-1α neutralisation in patients with cancer. *Lancet Oncol.* 2014;15(6):552-553.
296. Karki R, Man SM, Malireddi RKS, et al. Concerted activation of the AIM2 and NLRP3 inflammasomes orchestrates host protection against Aspergillus infection. *Cell Host Microbe.* 2015;17(3):357-368.
297. Andrei C, Dazzi C, Lotti L, Torrisi MR, Chimini G, Rubartelli A. The secretory route of the leaderless protein interleukin 1β involves exocytosis of endolysosome-related vesicles. *Mol Biol Cell.* 1999;10(5):1463-1475.
298. Qu Y, Franchi L, Nunez G, Dubyak GR. Nonclassical IL-1β secretion stimulated by P2X7 receptors is dependent on inflammasome activation and correlated with exosome release in murine macrophages. *J Immunol.* 2007;179(3):1913-1925.
299. Li Y, Li N, Yan Z, et al. Dysregulation of the NLRP3 inflammasome complex and related cytokines in patients with multiple myeloma. *Hematology.* 2016;21(3):144-151.
300. Lukens JR, Gross JM, Calabrese C, et al. Critical role for inflammasome-independent IL-1β production in osteomyelitis. *Proc Natl Acad Sci U S A.* 2014;111(3):1066-1071.
301. Lukens JR, Gurung P, Vogel P, et al. Dietary modulation of the microbiome affects autoinflammatory disease. *Nature.* 2014;516(7530):246-249.
302. Dasari TK, Geiger R, Karki R, et al. The nonreceptor tyrosine kinase SYK induces autoinflammatory osteomyelitis in a mouse model of chronic recurrent multifocal osteomyelitis. *J Biol Chem.* 2020;295(11):3394-3400.
303. Ridker PM, MacFadyen JG, Thuren T, et al. Effect of interleukin-1β inhibition with canakinumab on incident lung cancer in patients with atherosclerosis: exploratory results from a randomised, double-blind, placebo-controlled trial. *Lancet.* 2017;390(10105):1833-1842.
304. Okamura H, Nagata K, Komatsu T, et al. A novel costimulatory factor for gamma interferon induction found in the livers of mice causes endotoxic shock. *Infect Immun.* 1995;63(10):3966-3972.
305. Man SM, Karki R, Kanneganti TD. Molecular mechanisms and functions of pyroptosis, inflammatory caspases and inflammasomes in infectious diseases. *Immunol Rev.* 2017;277(1):61-75.

306. Tsutsui H, Kayagaki N, Kuida K, et al. Caspase-1-independent, Fas/Fas ligand-mediated IL-18 secretion from macrophages causes acute liver injury in mice. *Immunity.* 1999;11(3):359-367.
307. Sugawara S, Uehara A, Nochi T, et al. Neutrophil proteinase 3-mediated induction of bioactive IL-18 secretion by human oral epithelial cells. *J Immunol.* 2001;167(11):6568-6575.
308. Nakanishi K, Yoshimoto T, Tsutsui H, Okamura H. Interleukin-18 is a unique cytokine that stimulates both Th1 and Th2 responses depending on its cytokine milieu. *Cytokine Growth Factor Rev.* 2001;12(1):53-72.
309. Zaki MH, Vogel P, Body-Malapel M, Lamkanfi M, Kanneganti TD. IL-18 production downstream of the Nlrp3 inflammasome confers protection against colorectal tumor formation. *J Immunol.* 2010;185(8):4912-4920.
310. Zaki MH, Boyd KL, Vogel P, Kastan MB, Lamkanfi M, Kanneganti TD. The NLRP3 inflammasome protects against loss of epithelial integrity and mortality during experimental colitis. *Immunity.* 2010;32(3):379-391.
311. Huber S, Gagliani N, Zenewicz LA, et al. IL-22BP is regulated by the inflammasome and modulates tumorigenesis in the intestine. *Nature.* 2012;491(7423):259-263.
312. Netea MG, Joosten LA, Lewis E, et al. Deficiency of interleukin-18 in mice leads to hyperphagia, obesity and insulin resistance. *Nat Med.* 2006;12(6):650-656.
313. Xiang Y, Moss B. Correspondence of the functional epitopes of poxvirus and human interleukin-18-binding proteins. *J Virol.* 2001;75(20):9947-9954.
314. Novick D, Schwartsburd B, Pinkus R, et al. A novel IL-18BP ELISA shows elevated serum IL-18BP in sepsis and extensive decrease of free IL-18. *Cytokine.* 2001;14(6):334-342.
315. Mazodier K, Marin V, Novick D, et al. Severe imbalance of IL-18/IL-18BP in patients with secondary hemophagocytic syndrome. *Blood.* 2005;106(10):3483-3489.
316. Nold MF, Nold-Petry CA, Zepp JA, Palmer BE, Bufler P, Dinarello CA. IL-37 is a fundamental inhibitor of innate immunity. *Nat Immunol.* 2010;11(11):1014-1022.
317. Bulau AM, Nold MF, Li S, et al. Role of caspase-1 in nuclear translocation of IL-37, release of the cytokine, and IL-37 inhibition of innate immune responses. *Proc Natl Acad Sci U S A.* 2014;111(7):2650-2655.
318. Bufler P, Azam T, Gamboni-Robertson F, et al. A complex of the IL-1 homologue IL-1F7b and IL-18-binding protein reduces IL-18 activity. *Proc Natl Acad Sci U S A.* 2002;99(21):13723-13728.
319. Kumar S, Hanning CR, Brigham-Burke MR, et al. Interleukin-1F7B (IL-1H4/IL-1F7) is processed by caspase-1 and mature IL-1F7B binds to the IL-18 receptor but does not induce IFN-gamma production. *Cytokine.* 2002;18(2):61-71.
320. Nold-Petry CA, Lo CY, Rudloff I, et al. IL-37 requires the receptors IL-18Rα and IL-1R8 (SIGIRR) to carry out its multifaceted anti-inflammatory program upon innate signal transduction. *Nat Immunol.* 2015;16(4):354-365.
321. Tsutsumi N, Kimura T, Arita K, et al. The structural basis for receptor recognition of human interleukin-18. *Nat Commun.* 2014;5:5340.
322. Garlanda C, Riva F, Bonavita E, Gentile S, Mantovani A. Decoys and regulatory "receptors" of the IL-1/Toll-like receptor superfamily. *Front Immunol.* 2013;4:180.
323. McNamee EN, Masterson JC, Jedlicka P, et al. Interleukin 37 expression protects mice from colitis. *Proc Natl Acad Sci U S A.* 2011;108(40):16711-16716.
324. Mountford S, Ringleb A, Schwaiger R, et al. Interleukin-37 inhibits colon carcinogensis during chronic colitis. *Front Immunol.* 2019;10:2632.
325. Lian LH, Milora KA, Manupipatpong KK, Jensen LE. The double-stranded RNA analogue polyinosinic-polycytidylic acid induces keratinocyte pyroptosis and release of IL-36gamma. *J Invest Dermatol.* 2012;132(5):1346-1353.
326. van de Veerdonk FL, Stoeckman AK, Wu G, et al. IL-38 binds to the IL-36 receptor and has biological effects on immune cells similar to IL-36 receptor antagonist. *Proc Natl Acad Sci U S A.* 2012;109(8):3001-3005.
327. Henry CM, Sullivan GP, Clancy DM, Afonina IS, Kulms D, Martin SJ. Neutrophil-derived proteases escalate inflammation through activation of IL-36 family cytokines. *Cell Rep.* 2016;14(4):708-722.
328. Boutet MA, Bart G, Penhoat M, et al. Distinct expression of interleukin (IL)-36α, β and γ, their antagonist IL-36Ra and IL-38 in psoriasis, rheumatoid arthritis and Crohn's disease. *Clin Exp Immunol.* 2016;184(2):159-173.
329. Hessam S, Sand M, Gambichler T, Skrygan M, Ruddel I, Bechara FG. Interleukin-36 in hidradenitis suppurativa: evidence for a distinctive proinflammatory role and a key factor in the development of an inflammatory loop. *Br J Dermatol.* 2018;178(3):761-767.
330. Cayrol C, Girard JP. The IL-1-like cytokine IL-33 is inactivated after maturation by caspase-1. *Proc Natl Acad Sci U S A.* 2009;106(22):9021-9026.
331. Luthi AU, Cullen SP, McNeela EA, et al. Suppression of interleukin-33 bioactivity through proteolysis by apoptotic caspases. *Immunity.* 2009;31(1):84-98.
332. Bessa J, Meyer CA, de Vera Mudry MC, et al. Altered subcellular localization of IL-33 leads to non-resolving lethal inflammation. *J Autoimmun.* 2014;55:33-41.
333. Casadio R, Frigimelica E, Bossu P, et al. Model of interaction of the IL-1 receptor accessory protein IL-1RAcP with the IL-1beta/IL-1R(I) complex. *FEBS Lett.* 2001;499(1-2):65-68.
334. Lingel A, Weiss TM, Niebuhr M, et al. Structure of IL-33 and its interaction with the ST2 and IL-1RAcP receptors--insight into heterotrimeric IL-1 signaling complexes. *Structure.* 2009;17(10):1398-1410.
335. Hayakawa H, Hayakawa M, Kume A, Tominaga S. Soluble ST2 blocks interleukin-33 signaling in allergic airway inflammation. *J Biol Chem.* 2007;282(36):26369-26380.
336. Biton J, Khaleghparast Athari S, Thiolat A, et al. In vivo expansion of activated Foxp3+ regulatory T cells and establishment of a type 2 immune response upon IL-33 treatment protect against experimental arthritis. *J Immunol.* 2016;197(5):1708-1719.
337. Cayrol C, Girard JP. Interleukin-33 (IL-33): a nuclear cytokine from the IL-1 family. *Immunol Rev.* 2018;281(1):154-168.
338. Granowitz EV, Santos AA, Poutsiaka DD, et al. Production of interleukin-1-receptor antagonist during experimental endotoxaemia. *Lancet.* 1991;338(8780):1423-1424.
339. Arend WP, Guthridge CJ. Biological role of interleukin 1 receptor antagonist isoforms. *Ann Rheum Dis.* 2000;59(suppl 1):i60-i644.

CHAPTER 13 The Complement System

Erin E. West, PhD • Claudia Kemper, PhD

COMPLEMENT DISCOVERY AND KEY HISTORIC TURNING POINTS

The complement system is grouped within the *pathogen recognition receptor* (PRR) systems and is commonly recognized as one of the first lines of defense against invading pathogens and other noxious dangers to the host. It is a complex protein system with over 50 components of which some circulate in the blood, lymph, or interstitial fluids, and some are cell surface and membrane-bound or reside within cells. Complement was the first of the PRRs to be discovered and the initial publication mentioning this ancient system traces back to 1888 when the American scientist George Nuttal observed that human serum had bactericidal activity.[1] In 1891, the German bacteriologist Hans Buchner then found that the substance in blood capable of killing bacteria is heat-labile and gave it the name "alexin" (Greek, "to fend-off").[2] Jules Bordet, a Belgian clinical academic working in the laboratory of Eli Metchnikoff (the discoverer of phagocytosis and cellular immunity[3]) at the Pasteur Institute in Paris, noted in 1895 that the bactericidal activity of "alexin" required also the presence of a second, heat-stabile, blood component, which he termed "sensitizing substances" because it agglutinated bacteria and allowed for their subsequent lysis by alexin.[4] In 1899, Paul Ehrlich correctly theorized that the "sensitizing substances" are antibodies that bind to pathogens that are functionally complemented by alexin to kill microbes; he hence formally renamed alexin as "complement."[5,6] Bordet and Ehrlich are often considered among the founding fathers of immunology, and their early fundamental insights into protective humoral immunity were recognized with Nobel Prizes for Medicine, Metchnikoff and Ehrlich in 1908, and Bordet in 1919. Of note, even today, complement remains the oldest innate immune system that we can clearly trace back in evolution.[7,8]

Although Bordet and Ehrlich agreed on the requirement for cooperative action of antibodies and complement during pathogen destruction, they differed strongly in opinion about the nature of complement: Jules Bordet favored the existence of a single complement component that interacted nonspecifically and reversibly with antigens/antibodies, while Ehrlich argued that each antibody has its own matching complement and that their functional interaction was based on a nonreversible chemical reaction.[9]

The first wave of intense research over the next two decades on this new and critical immune mediator in the blood showed that neither Bordet nor Ehrlich was entirely correct: In 1907, Ferrata and Brand demonstrated in two independent studies that serum dialyzed against water separates into a globulin precipitate and an albumin "in solution" fraction and that neither of them retained complement (hemo)lytic activity on its own. Recombining those fractions, however, restored complement activity, strongly indicating that complement may be a multicomponent system. Indeed, they subsequently identified a complement component in the globulin fraction (later termed C1), which attaches to sensitized surfaces and a complement component in the albumin fraction (later termed C2) that interacts with the immobilized C1 and then fosters the lytic reaction.[10,11] Employing a battery of distinct early biochemical methods, several groups focused on better understanding this phenomenon known as "complement splitting," and by the mid-1920s, C3 and C4 were also discovered and the complement family had four members.[12-14] In the wake of these studies, Emil von Dungern and Arthur F. Coca noted that incubation of serum with yeasts inactivated complement[15,16] and Omorokow observed a similar outcome when serum was treated with cobra venom.[17] Today we know that these were the earliest observations of activation and subsequent consumption of the complement core component C3 (see below).

Although progress in the field of complement biology was clearly made, one crucial piece of information was still missing—the actual biochemical nature of the single components. Because complement activity cosegregated during the initial experiments with colloidal fractions,[12] it was considered by most to be of lipoid or "soap-like" nature. Two seminal studies by the American immunologist Louis Pillemer in the years 1941 and 1942 not only confirmed that the C1–C4 components are indeed proteins but also demonstrated that the components work in a sequential and interdependent fashion to achieve immune fixation and target lysis.[18,19] This observation, together with the development of more advanced protein chemistry techniques, sparked a second wave of research into complement, which was marked by the discovery of additional complement proteins and their assembly into the functional complement system as we know it today. For example, Ulf Nilsson and/or Hans Müller-Eberhard together with colleagues purified and characterized in detail human C4, C5, C6, C7, C8, and C9,[20-24] with the C1–C9 terminology for the discovered complement components set forth by the WHO in 1968.[25] Work during this productive period also revealed that complement is an enzymatic system, with its components circulating in blood in a proenzymatic form during the absence of infection.[26] Sensing of pathogens then activates this system in an elegant cascade-like fashion, beginning with the initiating component, C1,[26] which then cleavage-activates the next component and the "tripped-off" cascade finally culminating in the assembly of pore-forming C9 multimers into target surfaces and ultimately the lysis of the microbe; this C1-seeded activation sequence is today known as the classical pathway (see below).[6] Importantly, several groups made, in parallel, progress in identifying and purifying the corresponding complement components in guinea pigs and discovered a strain of rabbits deficient in functional C6 and provided the critical early animal models to begin studying complement *in vivo*.[27]

By the end of the 1960s, the prevailing opinion was that complement was fully understood and interest began to wane. This changed, however, dramatically, when Mark B. Pepys, a British scientist, discovered in 1972 that C3 plays a nonredundant role in T cell–dependent B cell responses,[28] a first suggestion that complement activity may be important for normal adaptive immunity. Over the next three decades, many groups collectively demonstrated that receptors for basically all complement activation fragments and breakdown products that are generated during complement activation exist (complement receptor 1 [CR1], CR2, CR3, and CR4; C3aR, C5aR1 and 2, etc.).[29-36] Further, it was found that most cells, including innate and adaptive immune cells, express a variety of complement receptors and that their engagement triggers specific signals which impact on the effector function of the respective signal-receiving cell.[36,37] Thus, the complement system not only directly kills invading microbes via lysis but also orchestrates the migration of scavenger cells to the site of infection, the phagocytosis of the pathogen "corpse," and the initiation and instruction of ensuing subsequent adaptive immune cell responses.[36,38,39] Moreover, it also became evident that complement is not just vigilant against pathogens but that this system, similar to other PRRs, is vital in the detection and clearance of self-derived danger, such as apoptotic cells, as elegantly demonstrated by Marina Botto in 1998.[40,41] Overall, by the 1990s, complement was becoming broadly recognized as central to the host's immunological health by occupying a key position at the interface of innate and adaptive immune activities.[37] This understanding was accompanied by the realization that complement is extremely potent and needed to be tightly regulated[42,43] as its uncontrolled or unwanted activity was associated with an increasing range of human diseases, from recurrent infections to autoimmunity and cancer.[44-48] Thus, a major research effort in the field focused, and still focuses, on understanding the intricate regulative mechanism that keeps this system at bay.[49]

In the wake of these studies, researchers made another exciting discovery that changed our view of complement substantially. It was long known that most complement components circulating in the blood are, with some exceptions, produced by the liver.[50,51] Although extrahepatic production of complement, particularly by immune cells, had been noted early on,[52,53] the functional significance of this observation remained obscure for a long time. Mike Carroll's group then demonstrated in 1998 that C3 produced by cells of the lymphoid compartment restores the impaired immune responses to T cell–dependent antigens in serum C3-deficient mice.[54] It is now understood that complement production by immune and nonimmune cells is a critical driver of many cellular responses, and that such local complement activity operates often independent of serum complement.[55] Further work aimed at better defining the roles of local complement ultimately also led to a departure from the generalized view of complement as a solely proinflammatory system. For example, from the 2000s on, a constantly growing body of work begun underpinning complement's equal involvement in the contraction of innate and adaptive immune responses and, thus, maintenance of general immune homeostasis.[56,57] Further, it turned out that complement not only partakes in processes underlying tissue repair and regeneration (including lens repair and healing of fractured bones),[58,59] but also its localized activity is a requirement for coordinated cell migration during tissue development.[60] Particularly, the unexpected findings that complement produced in the central nervous system[61] fosters normal brain development via controlling physiological synapse pruning[62] and that certain complement gene variants associate with increased risk for Alzheimer disease (AD)[63] have initiated an entirely new and ongoing research area, with frequent and exciting new publications about the physiological and pathophysiological roles of complement in neurology.[64] Another area of intense research focuses on expanding our understanding of how, and with what functional consequences, complement engages in crosstalk with other PRRs, such as the Toll-like receptors (TLRs), the inflammasomes, etc.[65]

The most recent paradigm shift in the field centers around the unexpected finding that complement activation is not confined, as always thought, to the extracellular space but can occur intracellularly.[66] The engagement of intracellular complement receptors by cell-intrinsic complement turned out to be an integral part of the signaling machinery that controls basic cellular processes, specifically including those of metabolic nature.[67] Moreover, deviations in intracellular complement activity fuel the hypermetabolic and -effector activity states of immune cells during autoimmunity, including rheumatoid arthritis (RA), systemic lupus erythematosus (SLE), and systemic sclerosis[68]; thus, much effort is currently being put toward defining the molecular mechanisms inducing and regulating intracellular complement activity. Overall, these are exciting times for complementologists: complement is shaking off its layer of dust and is emerging as an unanticipated central player in general cellular physiology. This realization, in combination with the increasing number of disease conditions that are found to involve aberrant complement activity,[69-71] and the first decade of clinical experience with the complement inhibitor eculizumab has reignited broad interest in complement biology.[72] This chapter aims at succinctly informing the reader about the current known core functions of complement in immunity and at providing a foundation for further reading into the many and ever-changing facets of this ancient system.

COMPLEMENT ACTIVATION AND REGULATION

The contemporary complement system has over 50 members (see Tables 13.1-13.3). While complement is considered a PRR system, it is particularly the lectin pathway (LP) with its ability to recognize molecular patterns found on pathogens and

TABLE 13.1 Human Complement Activation Pathway Components

Protein (*Gene*)	Structure	Function	Expression[a]
C1–C9 components			
C1q (*C1QA, C1QB, C1QC*)	Oligomer (460 kDa) of 18 C1qA, B, and C chains (~25 kDa each)	PRR of the CP, C1 complex component, activates CP	Endothelial cells, monocytes, microglia
C1r (*C1R*)	Single chain (85 kDa)	C1 complex component, activates C1s	Liver, gall bladder, other tissues
C1s (*C1S*)	Single chain (85 kDa)	C1 complex component, cleaves C4 and C2	Mostly liver, lung
C2 (*C2*)	Single chain (100 kDa)	C2b cleaves C3 as part of the CP C3/C5 convertase	Liver, lung, spleen
C3 (*C3*)	Thioester-containing protein (~184 kDa) composed of two covalently linked chains	Central to all pathways; C3b is a C3/C5 convertase component and major opsonin	Liver for circulation, most other cells in tissue, and immune cells for local activity
C4 (*C4A* and *C4B*)	Thioester-containing protein (~210 kDa) composed of three covalently linked chains	C4b is the receptor for C2; forms part of the CP C3/C5 convertase	Mostly liver, also lung, kidney, monocytes
C5 (*C5*)	Thioester-containing protein (~184 kDa) composed of two covalently linked chains	C5b initiates the TP, C5a is an anaphylatoxin	Mostly liver, lung, some activated immune cells, neutrophils
C6 (*C6*)	Single chain (110 kDa)	Binds C5b, C5b6 receives C7	Mostly liver, kidney, female reproductive tract
C7 (*C7*)	Single chain (100 kDa)	Binds C5b6, C5b–7 attaches to target cell	Mostly extrahepatic, several tissues, including adrenal glands, female reproductive tract, etc.
C8 (*C8A, C8B*, and *C8G*)	Protein composed of three single chains (~150 kDa overall MW)	Binds C5b–7, C5b–8 anchors C9	Mostly liver
C9 (*C9*)	Single chain (70 kDa)	Binds C5b–8/9 to form MAC	Mostly liver
Other components			
Factor B (*CFB*)	Single chain protease (110 kDa)	Bb part of the C3/C5 convertase, activates C3	Mostly liver, some kidney
Factor D (*CFD*)	Single chain protease (25 kDa)	Activates FB	Mostly adipose tissue
Collectin-10 (*COLEC10*)	Oligomer of varying sizes (250-460 kDa) composed of larger assemblies of homotrimeric single chains (~25 kDa each)	PRR of the LP, in complex with MASP-1/2, activates LP	Liver, placenta, thyroid glands
Collectin-11 (*COLEC11*)	Oligomer of varying sizes (250-460 kDa) composed of larger assemblies of homotrimeric single chains (~25 kDa each)	PRR of the LP, in complex with MASP-1/2, activates LP	Liver, gall bladder, kidney, ovary, adrenal gland
Ficolin-1 (*FCN1*)	Oligomer (~430 kDa) composed of assembly of four homotrimeric subunits from 36 kDa single chains	PRR of the LP, in complex with MASP-1/2, activates LP	Spleen, lung, leukocytes
Ficolin-2 (*FCN2*)	Oligomer (~420 kDa) composed of assembly of four homotrimeric subunits from 35 kDa single chains	PRR of the LP, in complex with MASP-1/2, activates LP	Mostly liver
Ficolin-3 (*FCN3*)	Oligomer (~590 kDa) composed of assembly of six homotrimeric subunits from 33 kDa single chains	PRR of the LP, in complex with MASP-1/2, activates LP	Mostly liver and lung

TABLE 13.1 Human Complement Activation Pathway Components (Continued)

Protein (*Gene*)	Structure	Function	Expression[a]
MBL (*MBL2*)	200-600 kDa collectin, two to six subunits each comprising three 32 kDa chains	PRR of the LP, in complex with MASP-1/2, binds mannan sugars on pathogens	Mostly liver
MASP-1 (*MASP1*)	Single chain (90 kDa)	Activates MASP-2	Mostly liver, also male/female reproductive tract, muscle tissue
MASP-2 (*MASP2*)	Single chain (75 kDa)	In complex with MBL, collectin, or ficolin, cleaves C2 and C4	Mostly liver, also male/female reproductive tract, muscle tissue
MASP-3 (*MASP1*)	Single chain (94 kDa)	Splice product from *MASP1* gene, activates zymogen FD	Mostly liver, also male/female reproductive tract, muscle tissue
MAP-1 (*MASP1*)	Single chain (44 kDa)	Splice product of *MASP1* gene, MASP-2 inhibitor	Mostly liver, expression may differ from MASP1
MAP-2 (*MASP2*)	Single chain (19 kDa)	Splice product of *MASP2* gene, MASP-2 inhibitor	Mostly liver, expression may differ from MASP2

AP, alternative pathway; CP, classical pathway; FB, Factor B; FD, Factor D; LP, lectin pathway; MAC, membrane attack complex; MAP, MASP-associated protein; MASP, mannan-binding lectin–associated serine protease; MBL, mannan-binding lectin; PRR, pattern recognition receptor; TP, terminal pathway.

[a]Expression profiles are based on entries in the "The Human Protein Atlas" at the time of writing.

altered self that drives categorization of complement as classic PRR. The classical pathway sensor C1q is more "specialized" and detects mostly altered/dangerous self, while the alternative pathway (AP) is triggered indiscriminately by hydroxyl- and amine groups which are common to most "living" surfaces.

The initiation and regulation of complement activation initially seems the most complex among the PRR systems and is, thus, often "off-putting" to students and newcomers to the field. However, albeit engaging distinct components initially, all complement activation pathways (and their regulation) function based on similar mechanistic principles. We will focus here on transporting these core principles as well as the wide-ranging impact of complement on host immunity and homeostasis in broader and more easily digestible strokes. For in-depth insights into the exact biochemical nature of complement components, the fine details of their respective activation and interaction mechanisms, and more comprehensive information about the single molecular pathways underlying the multitude of complement activities, we will refer to recent expert reviews and books in the pertinent sections. Complement activation occurs via three distinct pathways, the *classical pathway* (*CP*), the *LP*, and the *AP*. All three pathways share several common components and culminate in the activation of the pathogen-killing *terminal pathway* (*TP*) but differ in the nature of their respective initiation trigger and recognition molecules due to differences in their danger-sensing PRR components.[35,38,73-75] This is advantageous to the host as it allows for the detection of a broad range of pathogens and other threats. Furthermore, "alternative" C3 and C5 activation (CP-, LP-, and AP-independent) by specific proteases has emerged as important fourth driver of physiological complement activation and is now broadly acknowledged along the three traditional pathways.

Classical Pathway

The CP was the first complement activation pathway discovered and is indeed, as Ehrlich and contemporaries proposed, dependent on the presence of antibodies. Particularly, the pentameric immunoglobulin (Ig) M and the IgG1, 2, and 3 subtypes are considered complement-fixing and strong complement activators. When in free circulation, these IgGs are in a structural conformation that is usually not recognized by serum patrolling *C1*. However, structural changes induced by binding of these antibodies to targets, such as pathogens, cell surfaces, or immune complexes (ICs), render them visible to C1. For example, surface-bound IgM antibodies adopt a distinct dome-shaped staple conformation and five or six IgM protomers bound in close proximity create a structure that is recognizable to C1.[76] Similarly, IgG antibody binding to surfaces triggers noncovalent interactions between neighboring IgG antibodies, which induces the formation of ordered hexamers that are then recognized by C1.[77] C1 is a large multiprotein complex composed of one *C1q* sensor protein that assembles in a noncovalent but calcium-dependent fashion with two molecules of "enzymatic" *C1r* and *C1s* each ($C1qr_2s_2$).[78,79] C1q is the PRR component of C1 and mediates the recognition of surface-bound IgG and IgM antibodies. C1q is well known for its elegant tulip-like structure that assembles into an 18-unit protein from three distinct protein chains termed A, B, and C.[80] Each of these single chains is organized into an *N*-terminal collagen or stalk region and *C*-terminal globular head region.[81] A fully assembled C1q molecule consists of 6 A, B, and C chains. The distinct chains are initially arranged as single heterotrimeric A, B, and C subunits via interaction within their respective collagen/stalk domains. They then form first covalently linked subunit doublets and finally arrange three such doublets into the noncovalently linked hexameric C1q form[82] (Fig. 13.1). The

TABLE 13.2 Human Complement Regulatory Proteins

Protein (Gene)	Structure	Function	Expression[a]
Fluid phase			
C1 inhibitor (*SERPING1*)	Single chain (100 kDa)	Inhibits C1r, C1s, and MASP-2, in their PRR complexes	Mostly liver, lung, female reproductive tract
Factor I (*CFI*)	Heterodimer of two covalently linked protein chains (~88 kDa)	Cleaves C3b/C4b in presence of cofactor	Mostly liver, also several immune cells
Factor H (*CFH*)	Single chain (150 kDa), composed of 20 SCRs	AP convertase decay accelerator and FI cofactor	Mostly liver, immune cells
Factor H–like 1 (*CFH*)	Splice product of *CFH* gene, single chain (42 kDa), composed of seven SCRs	AP convertase decay accelerator and FI cofactor	Mostly liver, immune cells
Factor H–related protein 1 (*CFHR1*)	Single chain (35 kDa), composed of 5 SCRs	FH antagonist, competes with FH for binding to C3b/convertases	Mostly liver
Factor H–related protein 2 (*CFHR2*)	Single chain (26 kDa), composed of 54SCRs	FH antagonist, competes with FH for binding to C3b/convertases	Mostly liver
Factor H–related protein 3 (*CFHR3*)	Single chain (35 kDa), composed of 5 SCRs	Binds C3d and may regulate its interaction with CRs	Mostly liver
Factor H–related protein 4 (*CFHR4*)	Single chain (35 kDa), four SCRs, can assemble into two chains composed of nine SCRs total via splice variants	May bind to apoptotic cells to enhance opsonization	Mostly liver
Factor H–related protein 5 (*CFHR5*)	Single chain (62 kDa), composed of 9 SCRs	FH antagonist, competes with FH for binding to C3b/convertases	Mostly liver
C4b-binding protein (*C4BPA* and *C4BPB*)	Oligomer (480 kDa) composed of seven chains with eight SCRs and one chain with three SCRs	CP convertase decay accelerator and FI cofactor	Mostly liver, some lung and granulocytes
Carboxypeptidase B2 (*CPB2*)	Single protein chain (48 kDa)	Des-arginizes C3a and C5a	Mostly liver and lung
Carboxypeptidase N (*CPN1*)	Heterodimer (~280-330 kDa) of two protein chains	Des-arginizes C3a and C5a	Mostly liver and lung
Vitronectin (*VTN*)	Single chain (84 kDa)	Binds C5b–7 in fluid phase to inhibit MAC formation	Mostly liver
Clusterin (*CLU*)	Heterodimer (70 kDa) of two protein chains	Binds C5b–7 in fluid phase to inhibit MAC formation	Broadly expressed
Properdin (*CFP*)	Multiple different homo-oligomers of single chain (each 53 kDa)	Only positive complement regulator, stabilizes the AP C3 convertase	Liver, neutrophils, activated immune cells
Membrane-bound			
CD35, CR1 (*CR1*)	Single chain (220 kDa), transmembrane protein composed of 30 SCRs	CP and AP convertase decay accelerator, cofactor for FI, signaling capacity	RBC, leukocytes, kidney, brain, lung, intestine, others
CD46, MCP (*CD46*)	Single chain (55-70 kDa) transmembrane protein, composed of four SCRs	Cofactor for FI, signaling capacity	Broadly expressed, only absent from RBCs
CD55, DAF (*CD55*)	Single chain (70 kDa), GPI-linked and composed of four SCRs	CP and AP convertase decay accelerator	Broadly expressed
CD59 (*CD59*)	Single chain (~60 kDA), GPI-linked	Inhibits full MAC formation by binding to C5b–8	Broadly expressed

AP, alternative pathway; CP, classical pathway; CR, complement receptor; DAF, decay acceleration factor; FB, Factor B; FD, Factor D; FI, Factor I; GPI, glycosylphosphatidylinositol; LP, lectin pathway; MAC, membrane attack complex; MAP, MASP-associated protein; MASP, mannan-binding lectin–associated serine protease; MBL, mannan-binding lectin; MCP, membrane cofactor protein; PRR, pattern recognition receptor; RBC, red blood cell; SCR, short consensus repeat; TP, terminal pathway.

[a]Expression profiles are based on entries in the "The Human Protein Atlas" at the time of writing.

TABLE 13.3 Human Complement Receptors

Protein (*Gene*)	Structure	Function	Expression[a]
CD35 (CR1) (*CR1*)	Single chain (220 kDa) transmembrane protein composed of 30 SCRs	Binds C3b/C4b-opsonized targets for removal in the liver/spleen	RBC, leukocytes, kidney, brain, lung, intestine others
CD21 (CR2) (*CR2*)	Single chain (145 kDa) transmembrane protein composed of 15 or 16 SCRs	Binds iC3b/C3dg/C3d, functions as BCR costimulator	B cells, some T cells, FDC, small intestine
CD11b/CD18 (CR3, Mac1) (*ITGAM/ITGB2*)	Integrin heterodimer (265 kDa) transmembrane protein	Binds iC3b and regulates phagocytosis and migration; also binds ICAMs, FX, fibrinogen, heparin	Myeloid cells,, NK cells, activated T cells, brain
CD11c/CD18 (CR4) (*ITGAX/ITGB2*)	Integrin heterodimer (255 kDa) transmembrane protein	Binds iC3b and regulates phagocytosis and migration; also binds ICAMs, fibrinogen, heparin, and collagen	Myeloid cells, T cells, NK cells
CRIg (*VSIG4*)	45 kDa single chain Ig superfamily member	Binds C3b/iC3b and bacterial LTA, central phagocytic receptor for opsonized pathogens	Kupffer cells and several other tissue macrophages
CD46, MCP (*CD46*)	Single chain (55-70 kDa) transmembrane protein, composed of four SCRs	Functions as costimulator for TCR activation, drives Th1 induction and contraction	Broadly expressed, only absent from RBCs
C3a receptor (*C3AR1*)	Seven transmembrane spanning GPCR protein (54 kDa)	Binds C3a and can evoke anti-inflammatory or proinflammatory cellular responses	Broadly expressed
C5a receptor 1 (CD88) (*C5AR1*)	Seven transmembrane spanning GPCR protein (45 kDa)	Binds C5a/C5a-desArg and is generally considered proinflammatory	Broadly expressed
C5a receptor 2 (aka C5L2, GPR77) (*C5AR2*)	Seven transmembrane spanning protein (37 kDa); couples via arrestins, not G proteins	Binds C5a-desArg/C5a and is generally considered anti-inflammatory (may depend on context)	Broadly expressed
C1q receptors			
CD35, CR1 (*CR1*)	Single chain (220 kDa), transmembrane protein composed of 30 SCRs	C1q, controversial	RBCs, leukocytes, kidney, brain, others
CD91 (aka LRP1, A2MR) (*LRP1*)	Heterodimer (~600 kDa) transmembrane protein composed of different protein domains	C1q, induces noninflammatory phagocytosis; also binds A2M, calreticulin, HSP70/90, etc.	Broadly expressed
CD93 (aka C1qRp) (*CD93*)	Single chain (25 kDa) transmembrane protein	C1q (controversial), MBL, SP-A and induces phagocytosis	Broadly expressed
cC1qR (calreticulin) (*CALR*)	Single chain (46 kDa)	C1q-collagen region, induces phagocytosis and IL-12 in APCs	Broadly expressed (intra/extracellular)
gC1qR (aka C1qBP, p32/33) (*C1QBP*)	Single protein chains (33 kDa) form a tetramer	C1q-globular head region, phagocytic receptor and also modulates mitochondrial activity	Broadly expressed
CD305 (LAIR-1) (*LAIR1*)	Single protein chain (32 kDa)	C1q, induces inhibition of DCs; binds also collagens	Immune cells, lung, CNS
α2β1 integrin (*ITGA2/ITGB1*)	Integrin, heterodimer (~210 kDa)	C1q, induces general cell attachment and activation; binds also collagens, collectins, E-cadherin, etc.	Broadly expressed

A2MR, α-2-macroglobulin receptor; BCR, B cell receptor; C1qBP, C1q-binding protein; CR1, complement receptor 1; CR2, complement receptor 2; CR3, complement receptor 3; CR4, complement receptor 4; CRIg, complement receptor of the immunoglobulin superfamily; FDC, follicular dendritic cell; FX, Factor X; gC1qR, globular head C1q receptor; GPCR, G protein–coupled receptor; ICAM, intercellular adhesion molecule; LAIR-1, leukocyte-associated immunoglobulin-like receptor 1; LTA, lipoteichoic acid; LRP1, LDL receptor–related protein 1; MBL, mannan-binding lectin; MCP, membrane cofactor protein; NK, natural killer; RBCs, red blood cells; SCR, short consensus repeat; TCR, T cell receptor.

[a]Expression profiles are based on entries in the "The Human Protein Atlas" at the time of writing.

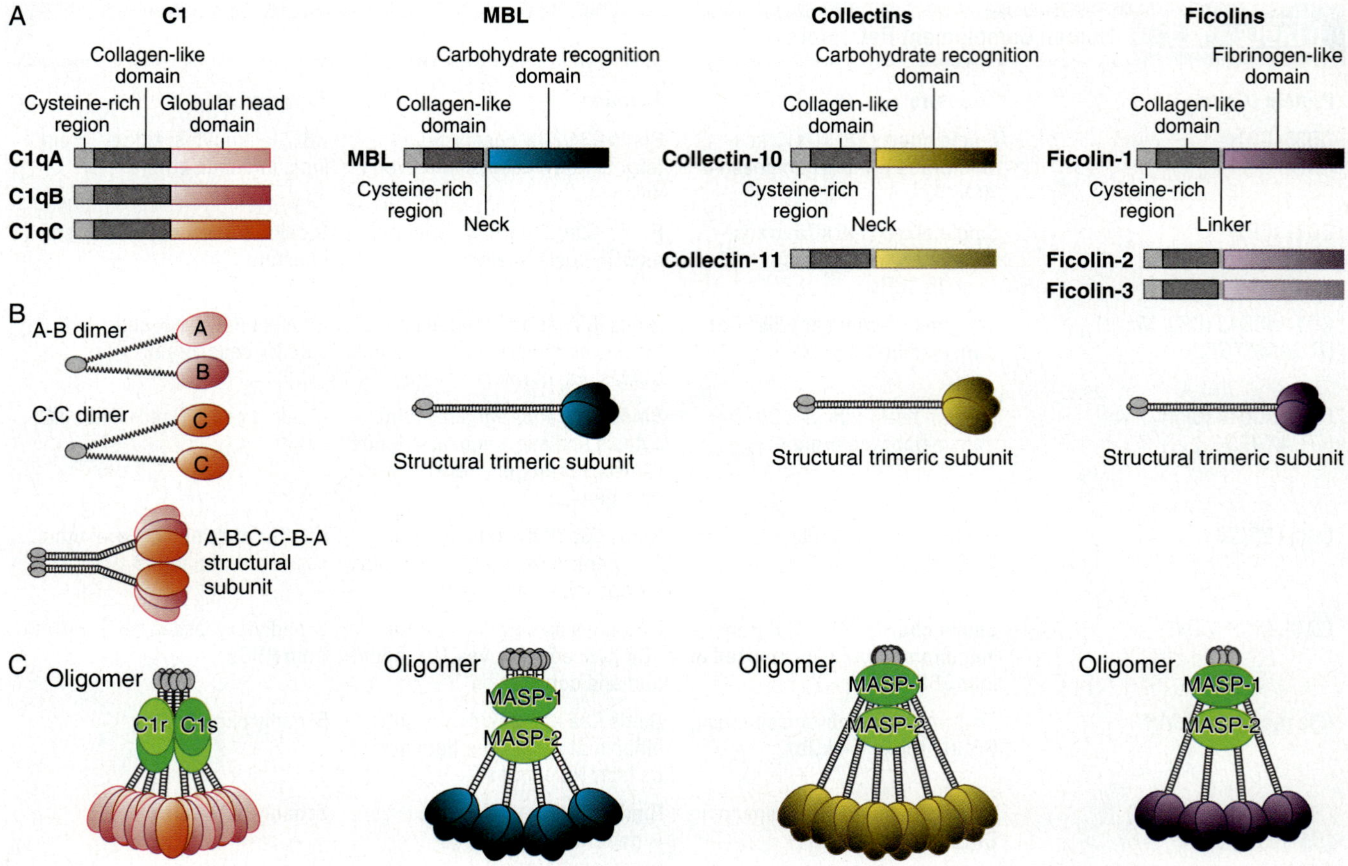

FIG. 13.1. Structures of the CP and LP-initiating complement PRRs. C1 triggers CP activation while MBL, Collectin-10 and -11, and the ficolins 1 to 3 activate the LP. **A:** Single protein chain domain structures that form the basis of each fully assembled PRR. **B:** Structural organization of the next-order subunits formed by single protein chains depicted under "A." MBL, collectins, and ficolins assemble into homotrimeric subunits. C1q initially forms A-B and C-C dimers, which then pair into a hexamer with a C-C dimer forming a covalent bridge between two A-B dimers. **C:** Subunits assemble into their final oligomeric structure together with their specific pathway-initiating serine proteases. Two molecules of C1r and C1s with the C1q oligomer form the C1 complex ($C1qC1r_2C1s_2$), while one molecule of MASP-1 and MASP-2 each interacts with MBL, collectin-10 or -11, or ficolin 1 to 3 oligomers. Of note, heterotrimeric subunits between collectin-10 and collectin-11 have recently been observed (not depicted). Also not shown are the MASP regulatory proteins MAP-1 and MAP-2 that can be part of the LP MBL, collectin, and ficolin PRR complexes. CP, classical pathway; LP, lectin pathway; MAP-1 or -2, MBL/ficolin-associated protein 1 or 2; MASP-1 or -2, mannose-binding lectin–associated serine protease 1 or 2; MBL, mannan-binding lectin; PRR, pattern recognition receptor.

tulip-like structure of C1q makes it one of the complement systems most powerful PRRs: it is highly flexible, has extensive reach, and can detect not only the altered structures in surface-bound IgM and IgG but also an additional broad range of other dangerous cellular alterations and noxious targets[80] (see below).

And in another ingenious twist, C1q also carries along the effector enzymes needed to kick-off the CP in full once indicated. The two serine proteases C1r and C1s within the C1 complex are inactive when C1 is not interacting with a target but undergo (auto)catalytic sequential activation once C1q is target-bound.[83] Enzymatically active C1s within C1 captures *C4* from the fluid phase and cleaves it into a smaller *C4a* and a larger *C4b* fragment[84] (Fig. 13.2). C4a, whose biological role is still not resolved, diffuses away from the initiating Ig–C1 complex. C4b, on the other hand, is highly reactive due to an exposed *thioester* seeking to interact with hydroxyl or amino groups, core components of all biological macromolecules, and hence "deposits" quickly and covalently onto the activating target. C4b within, or close to, Ig–C1 complexes recruits fluid phase *C2* and C1s then processes C2 into *C2a* (a protein with so far no known function) and the active serine protease *C2b*,[85] which remains associated with C4b to form the CP *C3 convertase C4bC2b* (Fig. 13.2). C4 is "oddly" placed before C2 in the CP activation sequence because it was discovered before C2. Of note, we use the complement nomenclature put forth in a recent update on the terminology of complement components.[74]

C3 convertases are at the heart of all complement activation pathways as they promote the activation of the central complement component *C3*,[35] which is highly homologous in structure to C4 and *C5* and the most abundant complement component in circulation.[7] The C3 convertase C4bC2b of the CP cleaves C3 into bioactive *C3a* and *C3b*. C3a is a so-called anaphylatoxin and mediates the recruitment and activation of innate immune cells to/at the side of infection.[86,87] C3b (just like antibodies) is among the host's major opsonins and tags pathogens and dangerous self for detection and removal by scavenger cells.[88-90] Some of the C3b generated remain within

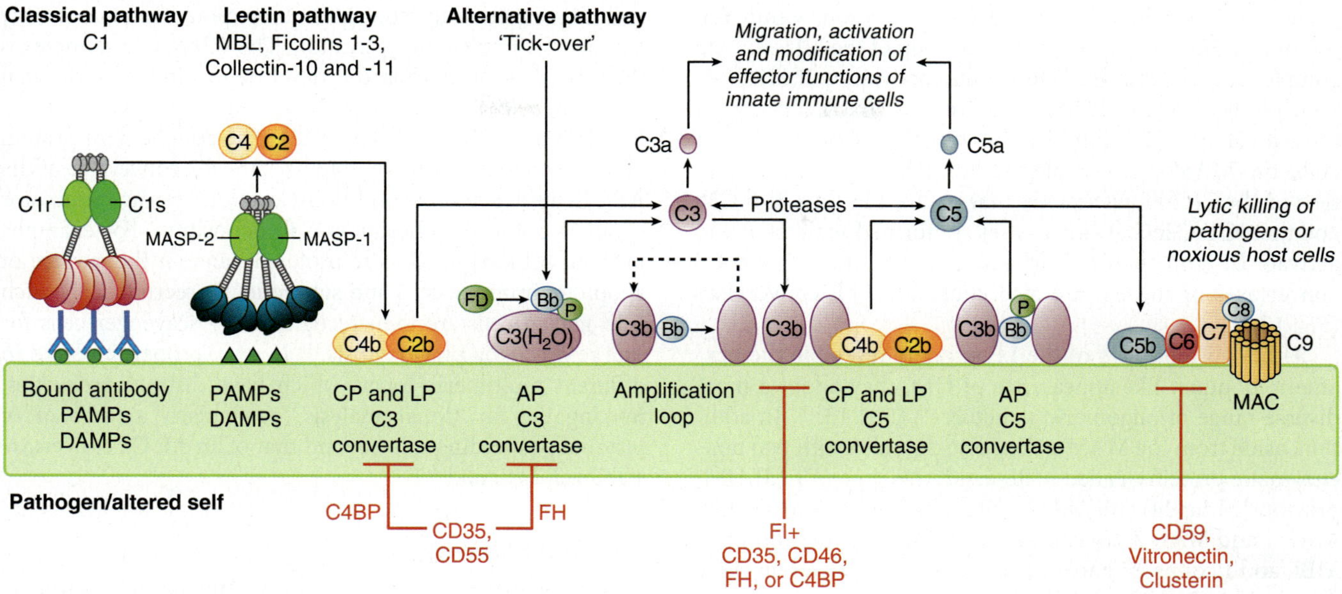

FIG. 13.2. Complement activation and regulation. Liver-derived, systemically circulating complement can be activated through three pathways: the classical (CP), lectin (LP), and alternative (AP) pathways with the initial deposition of C3b or C3b(H_2O) on a surface also initiating a feedback amplification loop. Via the formation of C3 convertases (C4bC2b for the CP and LP and C3bBb for the AP), all pathways lead to the cleavage activation of C3 into C3a and C3b. Subsequent C5 convertase formation (C4bC2bC3b for the CP and LP and C3bBbC3b for the AP) activates C5 into C5a and C5b, with surface-bound C5b triggering the formation and insertion of the MAC into pathogens or target membranes. Of note, a range of proteases can also activate C3 and/or C5 independent of convertase formation (see text for details). The inhibitory activity of regulators is indicated, and the main functional outcomes of complement activation during protective immune responses are listed. DAMP, danger-associated molecular pattern; FB, Factor B; FD, Factor D; FH, Factor H; FI, Factor I; MAC, membrane attack complex; MASP-1 or -2, mannose-binding lectin–associated serine protease 1 or 2; P, properdin; PAMP, pathogen-associated molecular pattern.

the C4bC2b complex, which leads to another critical event, the formation of the *C5 convertase of the CP* (*C4bC2bC3b*).[91,92] C5 convertases contain *C5* binding sites that allow C2b to cleave C5 recruited from the fluid phase into bioactive *C5a* and *C5b*, the final enzymatic step in the CP (Fig. 13.2). C5a synergizes with C3a to further mobilize immune cells and alerts the host about the need for an inflammatory response[86,87] and C5b mediates the activation of the TP[35] (see below).

Lectin Pathway

Although the LP was only discovered recently in 1987 by Ikeda and colleagues,[93] and therefore well after the CP and AP, LP proteins are among the evolutionary oldest complement components and arose together with or very shortly after the appearance of C3 and *Factor B* (*FB*) and thus much earlier than CP components around 600 million years ago.[94,95] It is therefore not surprising that LP activation is independent of antibodies (adaptive immunity evolved about 240 million years later[96]) and that this pathway shares several features with the CP and AP. In fact, the major difference between the LP and CP is that while C1 is the sole initiating PRR component of the CP, the LP can be triggered by at least three distinct soluble activation initiating PRRs. These include *mannose-binding lectin* (*MBL*), three *ficolins* (*Ficolin 1-3*, with only ficolin-2 and -3 found in serum), and two *collectins* (*Collectin-10 and -11*), together with their associated serine proteases providing enzymatic activity (see below).[97-100]

MBL, ficolins, and collectins share a similar collagen-like triple helix structure that is coupled to a specific recognition domain[101,102] (Fig. 13.1). The carbohydrate recognition domains of MBL and Collectin-10 and -11 places them into the C-type lectin protein family while the ficolins have a fibrinogen-like recognition domain and belong to the superfamily of fibrinogen-like proteins. Both protein families are specialized in recognizing distinct carbohydrate entities, such as mannose residues specific to pathogen surfaces (MBL) or N-acetyl glucosamine (GlcNAc) exposed by injured or dying cells (MBL and ficolins) or altered L-fucose and D-galactose patterns on stressed cells (ficolins and Collectin-10 and -11).[103] Of note, several recent studies indicate that the molecular pattern recognition repertoire of the PRRs of the LP is not confined to carbohydrates but extends, at minimum, to host DNA exposed on apoptotic cells[104,105] and most likely to additional targets. The recently discovered third, membrane-bound, collectin family member *Collectin-12* does not trigger the LP but may be able to activate the AP in its soluble form and in cooperation with properdin.[106] The LP has three associated serine proteases called *MBL-associated serine protease 1* (*MASP-1*), *MASP-2*, and *MASP-3*.[107-109] The common collagen region within MBL, ficolins, and collectins houses the binding sites for one molecule of MASP-1 and MASP-2 each, while MASP-3 activity has been only recently connected to AP activation through its proteolytic cleavage of Factor D (FD) (see below)[110] in addition to its other known complement-unrelated substrates, such as insulin growth factor–like protein 5.[111] Similar to C1r and C1s activation within C1, binding of the pattern recognition portion of the LP initiating complexes to a target leads to the autoactivation of MASP-1, which in turn cleaves MASP-2 to "free" its enzymatic activity toward C4 and C2 bound by the initiating MBL, ficolin, or collectin–MASP-1/2 complexes. From here on, the activation of the LP follows exactly that of the CP; thus, C4bC2b and then C4bC2bC3b convertases are formed sequentially to activate C3 and then C5. In consequence, the C3 and C5 convertases of the CP are also considered LP convertases (Fig. 13.2).

Although the PRRs of the LP strongly resemble the hexameric bouquet-like appearance of C1q, they adopt a more diverse range of oligomeric structures[98] (Fig. 13.1). In addition, aside from the MASPs, they associate also with two nonenzymatic proteins, called *MBL-associated protein 1* (*MAP-1*, previously Map44) and *MAP-2* (previously Map19 or sMAP). MAP-1 and MAP-2 can compete with MASP-2 for binding to MBL and ficolins and are hence thought to play a regulative role during LP activation.[112] MBL arises from a single gene in humans that gives rise to a single protein chain, which initially forms a homotrimeric subunit that then assembles into larger structures, from dimers to hexamers. The Collectin-10 and Collectin-11 also follow a similar scheme: a single peptide chain forms initially a homotrimeric structure via covalent bonds and these subunits then assembly into larger, oligomeric macrostructures (Fig. 13.1). Of note, Collectin-10 and -11 subunits can interact with each other and form "mixed" macrostructures, elegantly further expanding the pattern recognition range of the LP.[113]

Alternative Pathway

The AP of complement is listed in most textbooks (including here) as separate activation pathway. However, although it can clearly initiate complement activation on itself, it is appreciated by most in the field rather as an *amplification pathway/loop* that strongly enhances complement activation triggered by the CP and/or LP. An interesting feature of the AP is that it idles in serum constantly at a low level of activation: about 3% to 5% of C3 in circulation is continuously hydrolyzed into a *C3(H$_2$O)* form.[35,73,75,114] C3(H$_2$O) has an exposed thioester and can hence interact spontaneously with targets, such as microbes that entered the bloodstream. This constant "*tick-over*" of the AP in the blood, lymph, and interstitial fluids ensures a remarkably rapid protective response against pathogenic invaders as they become immediately tagged with active C3(H$_2$O). Such surface-bound C3(H$_2$O) as well as C3b deposited onto targets via the CP or LP is recognized and bound by the inactive serine protease FB. The *C3(H$_2$O)/C3bFB* complex creates a binding site for a second serine protease *FD*, which processes FB into *Ba* and *Bb* fragments. The smaller Ba fragment dissociates from the initiating C3(H$_2$O)/C3bFB complex, while Bb, which is now enzymatically active, remains bound to form *C3(H$_2$O)/C3bBb*, the *C3 convertase of the AP*. Bb within the AP C3 convertase activates C3 into C3a and C3b exactly as the CP/LP C3 convertases. C3b binds more FB, which in turn generates additional C3 convertases and this positive amplification loop potently and quickly augments complement activation. The *C3bBbC3b* complexes that are also formed via the rapid deposition of additional C3b molecules form the *AP C5 convertases*, which activate C5 into C5a and C5b[35,73,114] (Fig. 13.2). The complement protein *properdin*, which normally stabilizes the C3 convertase of the AP, can, under certain conditions, also function as an initiator or focus point for subsequent C3b deposition. For example, properdin binds to specific proteoglycans on the surface of apoptotic immune cells and subsequently recruits C3b. Such C3b-tagged cells are then recognized by scavenger cells for safe removal via phagocytosis.[115] *In vivo*, properdin exists in different multimeric forms which have distinct properties, making the functional analysis (particularly *ex vivo* or *in vitro*) of properdin's role beyond that of an AP C3 convertase stabilizer difficult.[116,117]

Terminal Pathway

All complement activation pathways and modes culminate at the level of the TP, the final and lytic part of complement activation that directly kills pathogens via the formation and insertion of the pore-forming *membrane attack complex* (*MAC*) into the microbes' surface[118,119] (Fig. 13.2). Five complement components *C5*, *C6*, *C7*, *C8*, and *C9* participate in the stepwise and sequential formation of the MAC with the components C6 to C9 belonging to the MAC/perforin/cholesterol-dependent cytolysin (MACPF/CDC) protein superfamily.[120] All family members share the same MACPF/CDC transmembrane domain that is driving the formation of the so-called β-barrel hairpins that initiate pore formation in target membranes.[118] Although the formation of the MAC has been studied since decades, recent advances specifically in cryogenic electron microscopy techniques initiated a "resolution revolution" in the field of structural biology and fostered unprecedented detailed insights into the structures and characteristics of the MAC at different stages of formation. MAC assembly is effectively initiated with the formation of C5 convertases and generation of C5b, no matter their activation pathway origin. The structural changes accompanying the processing of C5 into C5b uncovers a C6 binding site within C5b. Although there is high structural similarity between C3b and C5b, C5b is not covalently linked to the target membrane but rather dissipates away and is thus able to interact with C6 in solution to form a stable *C5b6 complex* that weakly binds to lipid bilayers in the vicinity.[121-123] The recruitment of C7 and generation of a *C5b–7 complex* allows a more stable interaction with the outer leaf of the lipid bilayer forming the target membrane. Next, C5b–7 recruits C8 which has hemolytic activity and initiates the protrusion of the *C5b–8 complex* deeper into the membrane, reaching through the inner leaf of its lipid bilayer.[121] The binding of C8 to C5b–7 also generates a novel C9 binding site which leads to the recruitment of a C9 molecule into the C5b–8 complex to form *C5b–9*.[124] C9 within this complex now has an exposed elongation surface that allows binding of the next C9 molecule(s). The number of additional C9 molecules further incorporated into the first C5b–9 complex varies[125] and the largest sized MACs are of about 10 nm diameter and most often contain 16 C9 molecules.[126] Detailed cryo-electron tomography analyses of membranes under MAC attack revealed large variations of inserted MAC compositions, indicating that the formation of MAC is a dynamic process and adapts to the nature of the biological target membrane.[126] For example, while the incorporation of a single MAC can famously induce osmotic *lysis* in sensitive cells with stiffer membranes such as erythrocytes,[125] the incorporation of a large(r) number of MACs is generally needed to lyse targets with more fluid membranes, as those are often able to actively reseal smaller pores.[127,128] Of interest, C9 has structural and functional similarities with two other pore-forming proteins that play central roles in the immune system, perforin 1 and 2.[129] Perforins are MACPF domain containing central components of granules in cytotoxic T cells (CTLs) and natural killer T (NKT) cells. CTL or NKT cell activation induces their degranulation, and released perforins oligomerize and insert into target cells which mediates target cell death via the diffusion influx of apoptosis-inducing granzymes.[129] With the increasing understanding that complement serves unexpected noncanonical roles, there is work currently underway to define whether sublytic MAC formation may also function as a similar "diffusion portal" driving cellular responses distinct from the induction of lytic cell death.[130]

Convertase-Independent Complement Activation

Although C3 and C5 activation via one or several of the canonical activating pathways is clearly central to complement activity during an infectious insult, it has been known since the late 1970s that C3 and/or C5 is also *activated by specific proteases in a convertase-independent fashion*[131] (Fig. 13.2). For example, C3 can be activated by limited digestion with the serine protease trypsin[132] or by elastase,[133] while the serine protease plasma kallikrein activates C5.[134] It is now becoming apparent that such alternative modes of C3 and C5 activation are of greater physiological importance than previously anticipated. Several proteases belonging to the coagglutination system cleavage activate C3 and/or C5. These include thrombin, plasmin, Factor VII (FVII), FIXa, FXa, and FXIa.[131,135-137] Of note, recent work performed using nonhuman primates failed to observe complement activation by plasmin, thrombin, or FXa *in vivo*.[138] It is currently unclear whether C3/C5 activation by clotting cascade factors is human/mouse-specific or whether experiments performed *ex vivo* with human samples may not have reflected the *in vivo* situation. The evolutionary old cysteine protease family of cathepsins and granzyme B has also been shown to activate C3 or C5 efficiently.[66,139,140] The biological significance of such convertase-independent complement activation has been explored in most detail in trauma and sepsis. A body of work using patient samples and appropriate mouse models connected the dramatic increase in C3- and C5-activating granzyme B, cathepsin D, and FVII functionally with augmented anaphylatoxin generation and increase in injury scores in patients.[140] Protease-mediated activation of complement components also occurs within cells and is of functional relevance. For example, the intrinsic activation of T cell–expressed C3 by cathepsin L in lysosomes and possibly the endoplasmic reticulum (ER) is an integral part of interferon (IFN)-γ expression induction by human $CD4^+$ T lymphocytes (see below).[66]

Compared to C3 and C5 convertase-driven C3/C5 activation, processing of C3/C5 by a single protease is more rapid and more efficient with regard to the cell's energy expenditure. Many of the complement processing proteases are evolutionary as old as complement itself[141,142]; it is therefore conceivable that protease-mediated C3/C5 activation preceded the convertase-dependent processes. This could also mean that this mechanism may have important currently undiscovered and new functions pertinent to complement biology. Indeed, cathepsin G, elastase, and protease 3 generated by activated neutrophils inactivate the intrinsically expressed C5a receptor C5aR1, thereby controlling neutrophil activity.[143] Finally, proteolytic activation may not be the only way to move C3 or C5 into an active state: phospholipids on the surface of liposomes and other cell and material interfaces induce hydrolysis of C3 into C3(H_2O) that can then bind to target (membranes).[144] Thus, research into convertase-independent complement activation is clearly a space to watch.

Complement Regulation

Efficient complement activation ensures the generation of the main complement effector molecules: the opsonins C3b and C4b, the anaphylatoxins C3a and C5a, and the MAC, a direct pathogen killer. Opsonization of pathogens also marks them for removal by phagocytic cells and the anaphylatoxins are potent chemoattractants and activators of immune cells that fight the infection[36,38] (see below). However, due to the highly reactive nature of the thioester in C3b and C4b, once complement is activated, it latches onto targets indiscriminately—including healthy host cells in the vicinity. Thus, to prevent detrimental damage to the host (for example, via the constant tick-over reaction of the AP in blood), complement is tightly controlled through several *fluid phase or surface-bound regulators* (Table 13.2).

Complement regulation seems not only as complex as its activation but can also be broken down into three major, staggered, strategies: (1) prevention of substantial C3b/C4b deposition on host cells in the first place, (2) increased decay of C3 and C5 convertases that nonetheless succeeded in forming on host cells, and (3) inhibition of MAC formation, where mechanisms 1 and 2 were not fully effective (Fig. 13.2). In combination, these protective principles are in most cases sufficient to shield the healthy host against unwarranted complement attack.[145,146] These major complement regulative strategies are, though, supported by some additional important regulative measures that do not fall into these categories and that we will summarize at the end of this section.

Prevention/removal of C3b and C4b surface deposition: The most "economic" way to halt complement activation on healthy host tissue is prevention or significant limitation of initial C3b and C4b deposition. It is important to understand that although activated C3b and C4b are highly reactive due to their exposed thioester, their window of opportunity to latch onto the targets is extremely short (in the range of milliseconds) as interaction of the thioester with H_2O renders the molecule inactive for further complement activities.[147] C3b and C4b molecules that are successfully deposited onto cell surfaces, on the other hand, become a target for complement regulation, as they undergo substantial structural changes that create a new binding site for the fluid phase serine protease *Factor I (FI)*[148] and circulating or cell membrane–bound complement regulators that function as *cofactor proteins* for FI. Such cofactors driving proteolytic degradation of C3b and/or C4b include the membrane-bound regulators *CD46* (*membrane cofactor protein, MCP*) and *CD35* (*CR1*).[33,149,150] CD46 cleaves C4 to release a large *C4c* fragment into the fluid phase while retaining a small inactive *C4d* fragment at the membrane (Fig. 13.3). CD46 and CR1 both assist cleavage of C3b into *inactivated C3b* (*iC3b*) and *C3f* (released from deposited C3b) fragments and CD35 can further support the processing of iC3b into a surface-bound *C3dg* fragment[145,151] (Fig. 13.3). This is important as iC3b and C3dg are ligands for (complement) receptors and mediate major and distinct biological activities (see below). Similar roles have, somewhat puzzling to many in the field, not been observed for C4c or C4g. It may be that the evolutionary older C3 is sufficient to fulfill the additional roles of complement activation breakdown products or that we simply have not yet discovered the functions of the C4b-derived fragments. Fluid phase regulators such as *Factor H* (*FH*) or *C4b-binding protein* (*C4BP*) also serve as cofactors to FI and inactivate C3b or C4b in a similar fashion[152-154] (Fig. 13.3). Of note, FH binds only C3b while C4BP interacts almost exclusively with C4b.[155] Normal circulating levels of FH are particularly important to control complement on host membranes as FH has a uniquely high affinity for sialic acid[156]: sialic acid is an essential component of all mammalian cell surfaces but not of pathogens, and FH hence forms a protective shield on most host surfaces even before C3b deposition can occur.[153]

All complement regulators are expressed in a common gene region termed "*regulators of complement activation*" (*RCA*) gene cluster on chromosome 1 in humans[42] and are built of varying numbers of evolutionary conserved and repetitive units, the so-called *short consensus repeats* (*SCRs*; also known as sushi domain or complement control protein [CCP] domain)[145,157] (Fig. 13.3), giving the proteins a highly flexible, "beads-on-a-string"–like, appearance. This structure allows these regulators, similar to the PRRs of the CP and LP, to efficiently cover larger regions of complement deposition on surfaces and also to continuously adopt to the growing sizes of C3b/C4b hot spots.

Acceleration of convertase decay: In cases where deposition of C3b/C4b on host cells cannot be prevented (for example, in the direct vicinity of pathogens), formation of C3 and/or C5 convertases will inevitably occur. In such cases, the second level of protection kicks in, which is mediated by complement regulators that significantly accelerate the natural decay rate of assembled C3 convertase enzyme complexes[145] (Fig. 13.3). Most prominent among those regulators is the glycosylphosphatidylinositol (GPI)–linked membrane regulator *CD55* (*decay acceleration factor*).[158-161] CD55 reduces the half-life of the membrane-associated C3 and C5 convertases of all three activation pathways, as it displaces C2b from the C3/C5 convertases of the CP and LP and the Bb fragment from the C3 and C5 convertases of the AP.[161] CD35 also has decay acceleration activity toward the AP C3 convertase and hence plays a dual role in complement regulation.[150] This also is the case for FH, which increases the decay rate of the C3/C5 convertases of the AP pathway, while C4BP reduces the half-life of the CP and LP convertases[162-164] (Fig. 13.3). FH and C4BP are particularly critical in regulating complement on a systemic level. Because they control convertases that may have formed in the fluid phase, they prevent detrimental serum complement consumption that would leave the host otherwise vulnerable to infections[49] (see below).

Inhibition of MAC formation: As a final save measure against complement-mediated host damage, activation of the TP and formation of the MAC are also regulated at multiple stages.[123,146,165] For example, non–membrane-attached MAC complexes in a range of states of assembly are highly susceptible to spontaneous hydrolysis, which renders them incapable of proceeding with productive MAC formation.[145] In addition, *vitronectin* (previously S-Protein or S40) and *clusterin*, two proteins that circulate in the blood, latch onto C5b–7, C5b–8, or even C5b–9 MAC complexes that have not yet inserted into cell membranes and prevent their further oligomerization[166,167] (Fig. 13.2). Also, the binding of the C8 component to non–membrane-associated C5b–7 is one of the most potent inhibitors of the MAC, as it prevents the subsequent interaction of the formed complex with the lipid bilayer.[168] Any C5b–8 MAC complexes that progress to insertion into the host membrane will encounter *CD59*, a small, compact GPI-anchored molecule with almost ubiquitous expression profile.[169] CD59 can bind to either C8 or C9 and prevent further recruitment of the first or additional C9 molecules, and with this the assembly of a lytic MAC pore[170,171] (Fig. 13.2). As an additional protective measure, nucleated cells have developed the ability to shed or internalize already-inserted MAC complexes and via this remove the dangerous pore formed and reseal their membrane.[118] The exact molecular events that underlie such MAC sensing and "deflection" are currently under investigation.

Additional regulative mechanisms: The initiating PRRs of the CP and LP are also subject to regulation: the plasma serine protease *C1 inhibitor* (*C1-INH*),[172] which belongs to the serpin superfamily, irreversibly binds to and inactivates $C1r_2$ and $C1s_2$ in activated C1 or MASP-2 in activated MBL/ficolin/collectin–MASP complexes.[173] C1-INH also inhibits enzymes of the coagulation and fibrinolytic systems and the kinin-forming pathways.[174] Its importance to human health was initially discovered through the characterization of patients with C1-INH deficiency that display dysregulated complement activation and suffer from life-threatening angioedema due to uncontrolled C1 activation.[174,175] As we are still discovering new and unexpected functions for complement (see below), the repertoire of complement regulators may also not yet be fully explored. For example, the membrane-bound *sushi domain-containing*

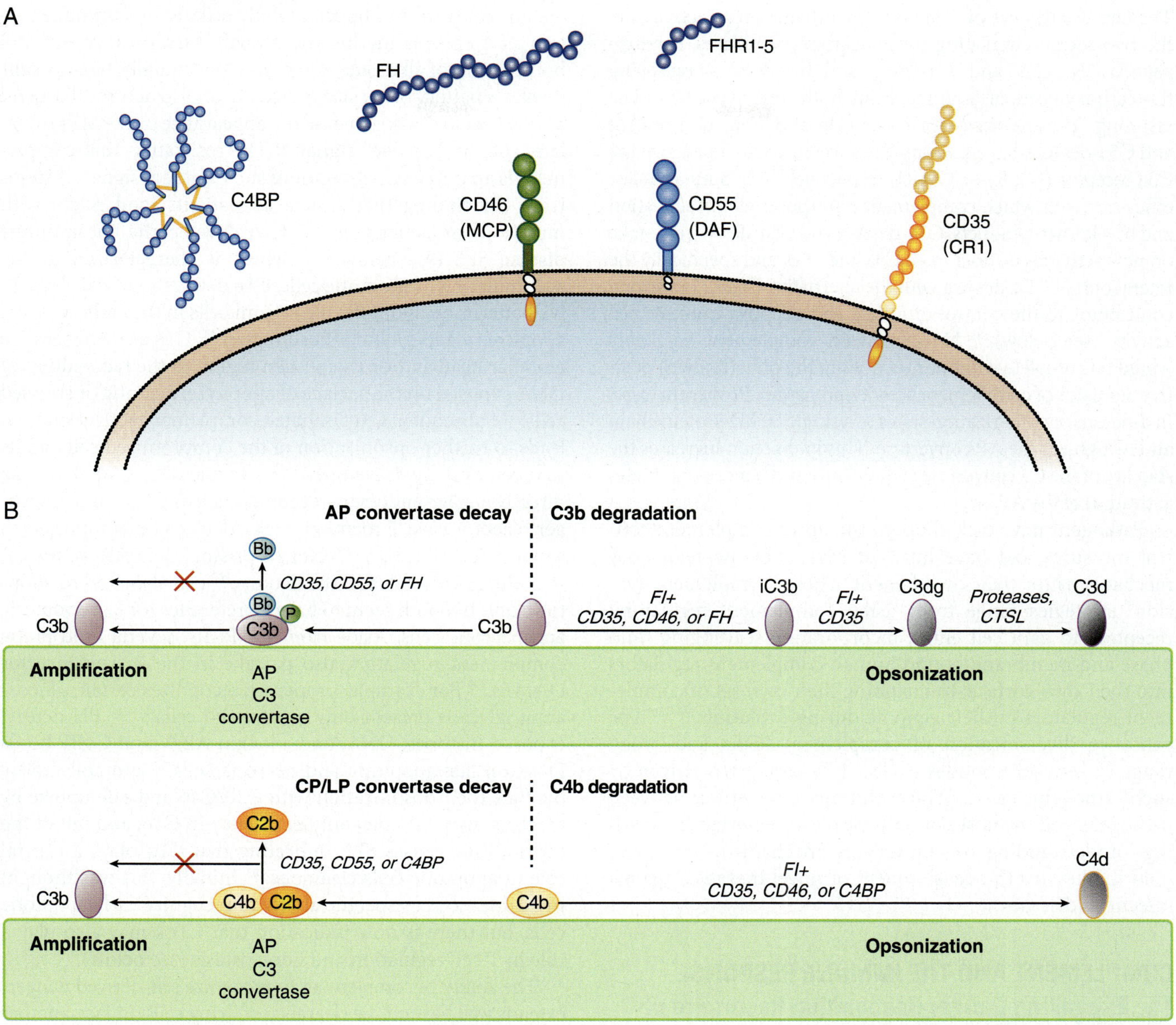

FIG. 13.3. C3b/C4b degradation and acceleration of convertase decay. A: Simplified schematic of fluid phase and cell membrane–bound complement regulators that are involved in C3b/C4b degradation and/or convertase decay. Single round units denote short consensus repeats (SCRs). **B:** The role of regulators in the decay acceleration of the AP pathway C3 convertases and degradation of cell surface deposited C3b (upper panel) and in the decay acceleration of the CP and LP C3 convertases and degradation of deposited C4b (lower panel). Note that indicated regulators also increase the decay of C5 convertases specific to their respective pathway (not shown). Also not shown is the ability of FHR proteins to negatively regulate FH activity. AP, alternative pathway; Bb, active fragment of Factor B; C4BP, C4b-binding protein; CP, classical pathway; CR1, complement receptor 1; DAF, decay acceleration factor; FH, Factor H; FHR, Factor H–related protein; FI, Factor I; LP, lectin pathway; MCP, membrane cofactor protein; P, properdin.

(*SUSD*) *protein* family and human *complement C1r/C1s Uegf, Bmp1* (*CUB*) *and sushi multiple domains* (*CSMD*) families contain SCRs and inhibit C3b and C4b membrane deposition by the CP and LP.[176] These are early findings and it will be interesting to see where and how these new regulatory members may function during host immunity. Similarly, we are still learning about the biology of the established complement regulatory pathways, as the recently developed idea of "regulating the regulators" demonstrates clearly. Under this concept, the FH splice variant, *FH-like protein 1* (*FHL-1*) and the *FH-related proteins 1 to 5* (*FHR-1 to FHR-5*) function as fine-tuners of the complement inhibitory role of FH. This is based on the ability of several FHRs to form homo- and heterodimers, which have a strong avidity for membrane-bound complement fragments and thus compete with FH for ligand binding. The unexpected finding that FHRs function as competitive antagonists of FH to modulate complement activation *in vivo* delivered a mechanistic explanation why FHR variations predispose to certain diseases.[177-180] One of the most powerful effectors generated during complement activation are the anaphylatoxins C3a and C5a. Without them, immune responses would not take place, and uncontrolled generation of C3a/C5a, however, can induce life-threatening conditions such as anaphylaxis (see below). It is hence not surprising that a mechanism to control their activity exists.

The key inactivators of C3a and C5a inflammatory activity are the two serum-circulating metallocarboxypeptidases, *carboxypeptidase N* (*CPN*) and *B2* (*CPB2*), and they do so by removing the carboxyterminal arginine from both anaphylatoxins. The resulting "*des-arginized*" forms of C3a and C5a, C3a-desArg and C5a-desArg, have a strongly reduced capacity to engage the C3a receptor (C3aR) or C5aR1, respectively.[181,182] Similar to the efficiency with which complement repurposes all C3 activation and breakdown products, C3a-desArg and C5a-desArg also take on new activities distinct from C3a and C5a, and specifically the interaction of C5a-desArg with the alternative C5aR2 is a strong contributor to the control of innate and adaptive immune cell activity (see below).[183] No section on complement regulators would be complete without mentioning the only (known) positive regulator of complement activity, *properdin.* Properdin exists in different oligomerization states of a single 50 kD protein chain and recognizes the AP convertase. This interaction increases the stability of the C3 convertase significantly and hence perpetuates activation of the AP.

Pathogens have picked up on the host's complement control measures and have hijacked several of the regulatory mechanisms for their own benefit. These "complement evasion" strategies range from using complement regulators/receptors to gain cell entry, incorporating/scavenging fluid phase and membrane-bound human complement regulators into their own surface, to including their own set of complement regulators in their genome during evolution.[184-186] The extensive diversification of complement PRRs and broad range in activation modes is clearly in large parts driven by such pathogenic tactics. The evolutionary tug of war between pathogens and the host not only makes for entertaining reading—understanding the underlying mechanisms may also hold the key for the development of novel therapies against infectious diseases.

COMPLEMENT AND THE IMMUNE RESPONSE

The Expanding Danger Recognition Repertoire of Complement

The host's ability to differentiate between self and non-self is a central paradigm in immunology and allows the immune system to clear infections while controlling unwanted injury to host tissues. Complement is central to this process. Further, although complement has, similar to all PRR systems, initially been discovered and defined as key effector in the fight against pathogens via recognition of *pathogen-associated molecular patterns* (*PAMPs*), it is now understood that it also plays important roles in the removal of harmful self-derived molecules, so-called *danger-associated molecular patterns* (*DAMPs*).[35,41,57] While the inflammasomes and the retinoic acid–inducible gene I (RIG-I) and melanoma differentiation–associated protein 5 (MDA5) (collectively known as the RIG-I–like receptors [RLRs])[187] protect mostly the cell cytosolic space, and the TLRs are membrane-operative DAMP sensors, complement is the prime fluid phase guardian of the host.[68] DAMPs are commonly generated when cells are infected, stressed or dying.[188] Particularly, the continuous safe and noninflammatory removal of *apoptotic and necrotic cells* are one of the most important daily tasks of our immune system as it ensures the healthy growth, turn-over, repair, and homeostasis of all tissues and cells. Conceptually, healthy cells display a battery of surface molecules and structures that send a "*don't eat me*" signal, while the appearance of DAMPs translates into an "*eat me*" signal.[189] The realization that complement is an important detector of such "eat me" signals[190] stems from the finding that almost all patients (and mice) with mutations or deficiencies in C1q develop the autoimmune disease *SLE* (see below)[40,191] which is characterized by the accumulation of apoptotic cells. C1q detects aberrant glycocalyx patterns on apoptotic blebs from cells in the early stages of apoptosis via its globular head region.[192] C1q also engages the globular head and/or its collagen region in the recognition of DNA exposed on the surface of apoptotic, necrotic or stressed cells.[193] Subsequent C1q-initiated complement activation then leads to further opsonization of the compromised cell and its safe removal by scavengers.[194,195] In parallel, C1q may also directly engage endocytic receptors such as CD91 and scavenger receptor class F member 1 (SCARF1) on macrophages to support this process.[196,197] Not surprising, the PRRs of the LP, including MBL[196,198] and the ficolins,[100] can also bind to apoptotic cells but both seem to have a preference for late apoptotic and necrotic cells. Aside from the PRRs, several fluid phase complement regulators also partake in the surveillance for DAMPs.[199] For example, properdin recognizes certain glucosaminoglycans present only on stressed cells.[115,200] FH detects exposed histones, DNA, and annexin A2,[201] and C4BP binds DNA on late apoptotic and necrotic cells[202] and collectively mediate their opsonization with C3b/C4b and safe uptake by macrophages. Of note, only deficiency in C1q, and not of the other PRRs, causes SLE indicating that C1q plays a central role in apoptotic cells clearance.[203] Initially, this was thought to be due to C1q's specific ability to recognize early apoptotic cells, but there is now indication that C1q's may also play a role in T cell regulation and homeostasis (see below).[204]

The ability of complement to recognize self-derived dangers extends well beyond the clearance of dying cells. In fact, another central "danger disposal" role of complement is the *removal of ICs* from the circulation.[205] The coating of ICs with CP-induced C3b, iC3b, and C3d modifies the IC lattice and produces smaller, more soluble, and easier to remove complexes. Furthermore, the transmembrane complement receptor CD35 (CR1) (Table 13.3) expressed on red blood cells (RBCs) captures opsonized ICs via interaction with C3b and delivers them to macrophages in the liver and spleen for destruction.[88] Abnormality in complement-mediated removal of ICs is a further hallmark of SLE and many other IC-mediated autoimmune diseases.[206] Complement also has antitumorigenic capacity as several cancers display changes in their membrane composition that trigger complement deposition on their surfaces via AP and/or CP activation.[47,48] Both lead to opsonization of the tumor and increased MAC-mediated attack as well as augmented "visibility" to patrolling phagocytes. Similar to pathogens, tumors respond most often with defensive measures (as part of their immune editing program[207]), such as the upregulation of surface complement regulators to evade complement attack. Moreover, there is accumulating evidence that the immunomodulatory activity of some complement activation fragments can be utilized by malignantly transformed cells to sustain an immunosuppressive micro-environment.[208] The complex relationship between complement and cancer biology is a rapidly expanding field (see below). The increasing accumulation of cellular debris and altered protein or lipid structures in tissues is a normal part of the aging process of multicellular organisms.[209] Such debris depots include, for example, the formation of atherosclerotic plaques in blood vessel walls or amyloid plaque accumulation in the brain. These structures are recognized as altered self and tagged by complement for their degradation and/or removal. For example, C1q, MBL and ficolins can bind to cholesterol crystals and the ensuing local complement activation attracts macrophages that ingest the complement-opsonized crystals.[210-212] Similarly, a role for complement in the clearance and/or reduction of amyloid plaque accumulation in the brain is supported by the finding that C3-deficient mice display accelerated plaque formation and neurodegeneration in a murine model of AD.[213] Although complement's vigilance in detecting stressed or dangerously altered self serves the healthy host well, it poses also one of the biggest hurdles to successful long-term organ transplantation. The cell surfaces changes in the donor graft endothelium introduced during ischemic reperfusion after grafting are a strong trigger of the recipient's complement system and a major cause for subsequent acute and/or chronic organ rejection.[214]

While PAMP sensing warrants the "full-blown" initiation of proinflammatory innate and adaptive responses to combat this immediate threat to the host and generate protective long-term memory responses (see below), the goal upon DAMP sensing is the removal of the trigger in an immunologically silent manner and with minimal collateral damage. Under both scenarios, the complement system achieves the desired response through the catered engagement of a variety of distinct *signaling-competent complement receptors* on immune cells (Table 13.3) and via crosstalk with other PRR systems (see below).

Complement, Innate Immunity, and the Inflammatory Reaction

Complement-mediated immune protection aims at directly killing an invader by heaping complement activation fragments onto its surface, ultimately leading to lytic MAC formation. This process is accompanied by two important additional events. Firstly, the anaphylatoxins generated are a danger signal which recruits immune cells and prepares the tissue for their arrival. Secondly, incoming cells which include neutrophils, macrophages, mast cells, and eosinophils express complement receptors that trigger the safe uptake and digestion of the opsonized microbe (cor pse)[36,38,215] (Fig. 13.4). Importantly, virtually all complement regulators/receptors transduce intracellular signals upon complement activation fragment binding and actively shape the response of the cell they are expressed on. The early stages of pathogen breach are initially confined locally to a wound or entry site that are typically located at the mucosal/environmental interfaces, such as skin, lung and intestinal system. Here, the microbe faces a localized inflammatory response mediated by a combination of alarmins and inflammatory cytokines produced primarily by tissue resident macrophages[216,217] and will simultaneously induce anaphylatoxin generation through activation of

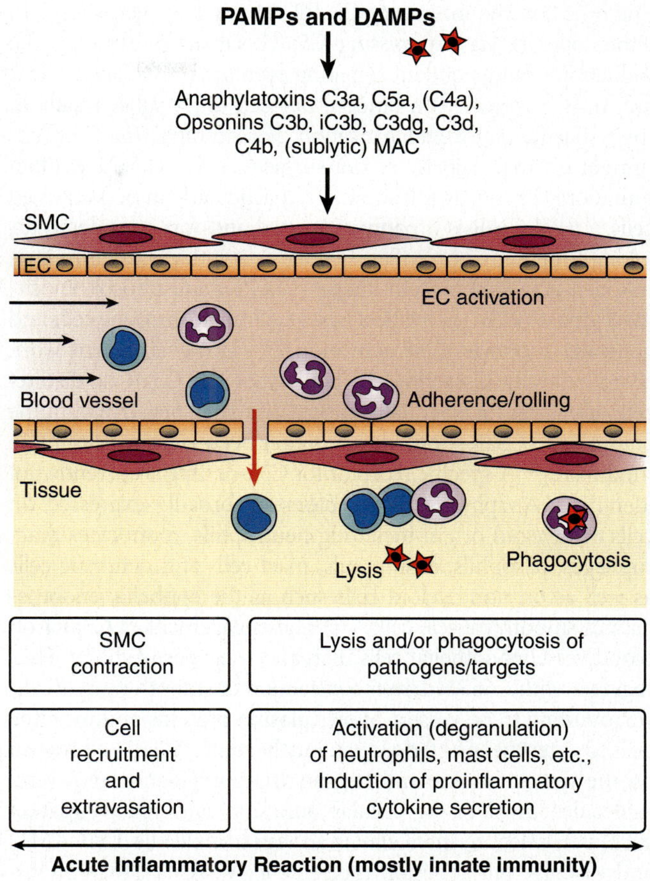

FIG. 13.4. Complement and the inflammatory reaction. During the initial phase of pathogen or danger sensing, complement activation fragments engage mostly tissue and innate immune cells to initiate the protective inflammatory reaction. The anaphylatoxins (and sublytic MAC) support SMC contraction and an increase in permeability of blood vessels, activate ECs to allow immune cell adherence and transmigration, and attract immune cells (monocytes, neutrophils, mast cells, eosinophils, etc.) toward the site of infection/danger. Incoming cells express complement receptors that mediate immune cell activation, trigger the safe phagocytosis of the C3b/C4b breakdown fragment opsonized microbe, while lytic MAC initiates direct killing. DAMP, danger-associated molecular pattern; EC, endothelial cell; MAC, membrane attack complex; PAMP, pathogen-associated molecular pattern; SMC, smooth muscle cell.

complement present in interstitial fluids. The *anaphylatoxins C3a and C5a* are small proteins (~10 kDa molecular weight) and have a range of pleiotropic effector functions central to initial innate immune cell activation and also help orchestrate the subsequent adaptive immune response.[87,218-220] They mediate their effects by binding to their specific seven transmembrane domain spanning *G protein-coupled receptors* (*GPCRs*), the *C3a receptor* (C3aR),[221] the *C5a receptor 1* (*C5aR1* or CD88)[32] or the "alternative" *C5aR2* (GPR77 or formerly C5L2).[222] The C5aR1 binds C5a while C5aR2 has higher affinity for the carboxypeptidase-processed and des-arginized form of C5a, C5a-desArg.[87] Although in reality the situation is more complex and there are exceptions to this rule, C5aR1 is generally considered to induce proinflammatory signals and responses while C5aR2 (which is uncoupled from G-proteins and engages intracellularly preferentially β-arrestins[87]) is an inhibitory receptor

(Table 13.3). The inhibitory C5aR2 activity has been shown to either operate via suppressing C5aR1 signals or through the induction of independent signaling events.[183,223,224] C5aR1 and 2 are mostly expressed in tandem but can differ substantially in their specific (sub)cellular location. For example, the C5aR1 is present on the majority of cell surfaces, while C5aR2 is often found only or mostly intracellularly, particularly in nonactivated cells.[183,223,224] C3aR is broadly expressed and was considered for a long time highly proinflammatory as it induces calcium flux, mitogen-activated protein kinase (MAPK) and phosphoinositide 3 kinase (PI3K) activation in several innate immune cells and generally supports their activation.[87,225] However, recent work more carefully dissecting the activity of the C3aR in distinct cells advocates for an equally important role in negative immune control, and C3aR should hence rather been viewed as immune modulator.[226] A specific receptor for C3a-desArg is currently not identified. Anaphylatoxin receptors are broadly expressed on cells of myeloid origin including neutrophils, monocytes/macrophages, basophils, eosinophils, mast cells and dendritic cells as well as on nonmyeloid cells such as the epithelia, endothelia and smooth muscle cells.[87,225] The engagement of C5aR1 on blood vessel endothelial cells increases their permeability. This, together with C5aR1-driven contraction of smooth muscle cells surrounding these vessels, allows plasma proteins and immune cells to enter the infected tissues. Furthermore, C5aR1 activation on the vessel endothelium triggers the upregulation of several molecules involved in cellular adhesion and transmigration such as E-selectin, intercellular adhesion molecule 1 (ICAM1) and vascular cell adhesion molecule 1 (VCAM1), thereby initiating immune cell adhesion and extravasation[87,218,219,225] (Fig. 13.4). Of note, *sublytic MAC* formation on endothelial cells may also support this process as it induces the expression of the neutrophil migration factor interleukin (IL)-8 as well as IL-1β and also fosters monocyte adherence.[227,228] In parallel to priming the vessels for immune cell transmigration, anaphylatoxin gradients also drive the migration of incoming innate immune cells deeper into tissues as they are the host's most potent *chemoattractant*.[87,218,219,225] C5a preferentially induces recruitment of neutrophils, macrophages, DCs and basophils while C3a gradients attract mast cells. Finally, C5a and C3a not only control immune cell tissue infiltration but also their activation upon arrival: C3a and C5a induce neutrophil degranulation and also the oxidative burst that mediates killing of intracellular bacteria[229,230] and C5aR1 signaling activates the lipoxygenase pathway, arachidonic acid metabolism and eicosanoid production in neutrophils and monocytes, and macrophages.[231] C3aR signaling on mast cells initiates histamine release by these cells.[232] Also, C3aR and C5aR1 are important mediators of proinflammatory cytokine production on monocytes. C3aR stimulation triggers the NOD-, LRR- and pyrin domain-containing protein 3 (NLRP3) inflammasome assembly and IL-1β production in monocytes via adenosine triphosphate (ATP) influx[233] and C5aR1 augments TLR4 induced IL-1β production as well as IL-6 secretion in these cells via the MAPK p38 pathway.[234] Together, these complement-driven events underly the early phases of the host's innate response to infection (Fig. 13.4) and the typical associated characteristic symptoms of fever, malaise and localized swelling, are collectively known as the *inflammatory reaction*.

Aside from the anaphylatoxins, complement activation triggered by pathogen or danger sensing also generates the *opsonins* C4b, C3b, iC3b, C3dg, and C3d. These latter fragments bind to the microbe/target surface and are subsequently recognized by specific complement fragment receptors expressed on phagocytes attracted and activated via anaphylatoxins.[36,38,46] For example, CD35 engagement on neutrophils and macrophages by C3b or C4b induces the phagocytic uptake of the opsonized pathogen/target[235]—an event that is synergistically supported by C5aR1 activation.[236] In addition, CD35 activation on macrophages induces secretion of the proinflammatory cytokine interleukin IL-12, a potent activator of T cells.[237] The *complement receptors 3* (*CR3*, CD11b-CD18) and *4* (*CR4*, CD11c-CD18) are also expressed by neutrophils, monocytes, macrophages, mast cells, and natural killer (NK) cells and their activation induces *phagocytosis of targets*, but their ligands are target-deposited iC3b and/or C3dg.[238,239] CR3 and 4 are both heterodimers composed of a specific α-integrin chain (CR3, CD11b, and CR4, CD11c) and a common β integrin chain (CD18).[36,38,46,239] They therefore are not dedicated complement receptors, but also belong to the β2 integrin family of adhesion molecules and, by definition, are engaged by a variety of additional ligands such as adhesion molecules, coagulation proteins, etc.[239,240] Similar to CD35, CR3 signals serve additional roles aside from inducing phagocytosis and are a strong driver of nuclear factor kappa-light-chain-enhancer of activated B cells (NF-kB) activation and the neutrophil oxidative burst.[241] Nonetheless, pathogens that may have been able to transition from the initial entry site into the bloodstream face a massive second onslaught by complement which circulates in high concentration. Here, the recently discovered complement receptor of the *immune globulin super family* (*CRIg*) is the dominant component of the phagocytic system removing complement opsonized targets.[242] CRIg binds C3b and iC3b and is abundantly expressed on Kupffer cells, the liver-resident macrophages that mediate removal of tagged antigens. CRIg-deficient animals display a defect in the clearance of C3b/iC3b-opsonized pathogens and succumb to infections, underpinning the important of CRIg in systemic immune surveillance.[242]

In sum, complement activation fragments are critical mediators of the immediate inflammatory reaction driven by complement receptors expressed by innate immune cells and aim to contain breaches by pathogens in their earliest stages (Fig. 13.4).

Complement and Adaptive Immunity

Many pathogens require the co-engagement of adaptive B and T lymphocyte effector activities for their complete clearance and, importantly, for the induction of immunological memory and long-lasting protection against reinfections.[37] The first proof that complement is also a central player in adaptive immunity was delivered by a seminal study from the British scientist Mark Pepys, in the 1970s[243] in which he showed that C3 is required for normal B cell activity—work that founded a completely new research area within the complement field that is still thriving. Today we know that complement, much like the TLRs, functions as a bridge between innate and adaptive immunity and controls B and T cell responses either directly, or via the modulation of antigen-presenting cell (APC) activities.

B cell responses: In 1974, Mark Pepys observed that murine B lymphocytes are able to bind C3 activation fragments and that mice depleted of serum-circulating C3 via cobra venom factor treatment cannot mount normal B cell responses against T cell–dependent antigens.[28,243] Subsequent studies by Pepys, Douglas Fearon and several other groups then established that the *complement receptor CD21* (*CR2*, binds iC3b, C3d, and C3dg) (Table 13.3) functions as a *costimulatory molecule* within a CD21-CD19-CD81 trimeric complex and supports B cell receptor (BCR) activation (Fig. 13.5). Mechanistically, co-engagement of CD21 by C3d-opsonized antigens amplifies BCR signaling in a PI3K-dependent fashion and reduces the threshold for normal antibody production by about 1000-fold.[244-247] Although BCR signaling without costimulatory help can eventually trigger antibody production, this scenario would require large amounts of antigen. CD21 hence enables important early and low-affinity antigen detection and with this, migration of B cells to the T cell–B cell boundaries in lymph nodes where B cells receive T cell help and then undergo expansion and somatic hypermutation in the dark zone of the germinal centers (GCs).[248] Indeed, serum C3 deficiency is often associated with common variable immunodeficiency (CVID), underpinning the role of C3 as a critical endogenous adjuvant during B cell responses.[249]

Today, we know that complement partakes in basically all phases of the B cell life cycle, that is, from the early initiation of the host's B cell repertoire to the formation of B cell memory (Fig. 13.5). B cells develop in the bone marrow and undergo, similar to T cells in the thymus, positive and negative selection processes that prevent autoreactive B cells from being released into the periphery to patrol for antigens. CD21 contributes to the negative selection of self-reactive B cells, as mice deficient either in the *C4* or *Cr2* gene fail to induce B cell anergy toward self-antigens.[250] The CD21-mediated mechanism preventing the production of autoantibodies is not fully understood, but may involve CD21 signals that lower the sensitivity of B cells to apoptosis during the negative selection process.[251] Furthermore, CD21 and C1q-mediated signals also have a say during normal positive selection, expansion and maintenance of the innate-like B-1 cell population taking place in the fetal liver. B1 cells are the key producers of

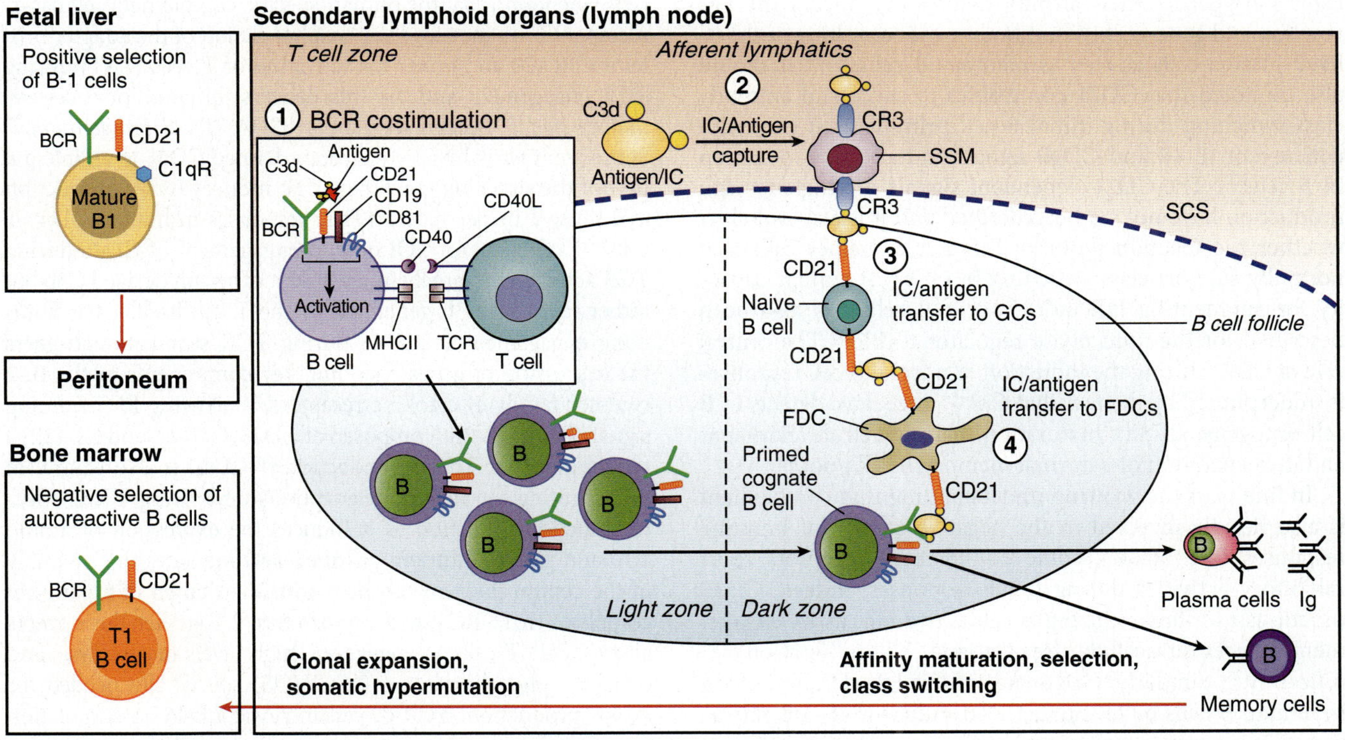

FIG. 13.5. Complement and B cell responses. CD21 (and C1qR) help shaping the B cell repertoire by controlling positive selection of B1 cells in the fetal liver and supporting negative selection of autoreactive B cells in the bone marrow. For induction of optimal B cell responses, naïve B cells rely on incoming CD21 signals (as part of a trimeric complex with CD19 and CD81) for normal BCR activation, antigen internalization, and proliferation in the B cell follicles (light zone) (1). SSM patrolling the SCS capture C3d-opsonized IC or antigens via CR3 (2) and transfer these through the SCS to CD21 expressed by naïve/noncognate B cells in the B cell follicles (3). These B cells transfer C3d-opsonized ICs/antigens to FDCs in the dark zone (4), which retain antigen long enough for (re)presentation to cognate primed B cells during the germinal center reaction which induces antibody-producing plasma cells and memory B cells. Complement may also coordinate the migration of formed memory B cells to the bone marrow and mature B1 cells to the peritoneum (indicated with red arrows). BCR, B cell receptor; C1qR, C1q receptor; FDC, follicular dendritic cell; GC, germinal center; IC, immune complex; Ig, immunoglobulin or antibody; MHCII, major histocompatibility complex class II; SCS, subcapsular sinus; SSM, subcapsular sinus macrophages; TCR, T cell receptor.

natural IgM antibodies, which are particularly important in the detection of pathogens that the host encounters for the first time.[252] Thus, complement is important in shaping the host's initial B cell repertoire (Fig. 13.5).

Protective humoral responses toward pathogens are triggered when B cells detect their cognate antigen in secondary lymphoid tissues (for example, lymph nodes) which are specialized in capturing antigens from circulating body fluids. Importantly, CD21 and CD35 play central roles in such antigen capture and subsequent presentation to B cells. For example, FcRIIb and CD35 expressed on the surface of subcapsular sinus macrophages (SSMs) in lymph nodes cooperatively trap C3d-opsinized IG complexes coming in from the afferent lymphatics.[253] Naïve or noncognate B cells within the B cell follicles then "grab" the C3 fragment-opsonized IC/antigen from the SSM in a CD21-dependent fashion and subsequently transport and transfer it to the surface of follicular dendritic cells (FDCs) in the light zone of the GC, in a process that also involves CD21.[254] FDCs are central to the humoral response as they have the unique capacity for long-term retention of the native antigen in GCs and (re)presentation of the antigen to previously primed B cells (see above). Such cognate primed B cells capture antigen directly from the surface of FDCs in a BCR/CD21 complex-dependent manner and this induces signals triggering their affinity maturation during the GC reaction and generation of antibody-producing long or short-lived plasma cells as well as memory B cells.[255,256] It should also be noted that CD21 contributes to successful antibody class-switching during initial B cell priming as it can, with concurrent IL-4R and CD40 signals, induce IgE production in B cells.[257] The CD21-dependent signals driving such IgE production have not yet been defined and it is also not clear whether the reaction requires T follicular helper cells that normally support class switching in GCs.[258] A similar capacity for augmenting IgE induction in B cells has also been described for the fluid phase regulator C4BP.[259] The central role of CD21 during the induction of normal B cell responses is underpinned by the fact that *Cr2*$^{-/-}$ mice have defects in B cell ontogeny, affinity maturation, plasma cell development and the generation of a normal memory B cell pool.[260]

In line with the growing understanding that complement is also heavily involved in the negative control of immune responses (see below), some complement receptors exert inhibitory activities during B cell responses. Indeed, CD35 functions as a brake during B cell activation, as its engagement reduces intracellular free Ca^+ and cell proliferation significantly.[261] Similarly, CD46 can inhibit IgE class switching in human B cells by blocking CD40-induced NF- κB activation.[262] It is important to note that *CR1* (encoding CD35) and *CR2* (encoding CD21) are two separate genes and proteins in humans, while in mice, the *Cr2* gene encodes a single protein that is a hybrid CD21/CD35 protein.[263] Similarly, CD46 is not expressed on somatic cells in mice, thus CD35 and CD46 represent human-specific regulators of B cell activity.[264]

T cell responses: Albeit the fact that CD21 is central to normal B cell function, it surprisingly does not seem to play a similar direct role during T cell activation. In fact, CD21 is only expressed by ~15 % of human CD4$^+$ and CD8$^+$ T cells in circulation and is upregulated upon (T cell receptor) TCR stimulation.[265,266] *In vitro* experiments in which CD21 was engaged during T cell stimulation showed that it inhibits proliferation and IL-2 production by T cells and suggest that CD21 is rather a negative regulator of T cell activity.[267] The role of CD35 in T cells biology also remains relatively unexplored. CD35 is found on activated T cells and ligation of CD35 on human CD4$^+$ T cells infected with the human deficiency virus (HIV) augments viral replication.[268] Interestingly, a current study showed that human recent thymic emigrants (RTEs) express high levels of CD21 and CD35, produce IL-8 predominantly upon activation, and are precursors of long-lived tissue-homing memory cells.[269] Thus, similar to their roles on B cells, CD21 and/or CD35 may also be involved in the initial shaping of certain T cell lineages. Mouse T cells express the CD21/CD35 hybrid protein but its function here is unexplored.[270] CR3 and CR4 are detected on ~5% of circulating CD4$^+$ and CD8$^+$ T cells and can be upregulated upon T cell stimulation. No T cell activity has been attributed to CR4 thus far, but early studies indicate that CR3 may be inhibitory as engagement of CR3 during T cell activation reduces IL-2 production and proliferation.[271]

A seminal study by the French immunologist Anne Astier published in 2000, however, demonstrated that the general concept for complement as a key provider of direct adjuvant signals during adaptive immune cell activation does indeed extend to T cells. She found that the human-specific complement regulator/receptor CD46 induces the phosphorylation of the adapter proteins CBL120 and linker for activation of T cells (LAT) during TCR engagement and via this delivers the additional *costimulatory signals* needed for human CD4$^+$ *T cell proliferation*.[272] Subsequent work by several groups defined CD46 as an integral part of the signaling machinery that induces *IFN-γ* production and thus, T helper type 1 (Th1) responses in human CD4$^+$ T cells[273-275] (Fig. 13.6A). Of note, coengagement of CD46 during TCR activation is largely independent of serum derived C3b but rather relies on C3b generated by the T cell itself.[66] The autocrine engagement of CD46 during TCR stimulation triggers the expression of genes encoding key components of the IL-2 receptor family of cytokine receptors,[274] allowing for incoming signals via the IL-2R (composed of CD25, CD122, and CD132) a requirement for Th1 cell induction.[276] CD46 is also central in orchestrating the metabolic adaptions underlying T cell effector function induction as it induces the expression of amino acid and glucose nutrient channels and mediates the assembly of the central nutrient sensor mammalian target of rapamycin complex 1 (mTORC1) to the lysosomes.[277] Together, these events allow CD4$^+$ T cells to engage the high levels of glycolysis and oxidative phosphorylation (OXPHOS) specifically needed for IFN-γ production. TCR-driven intrinsic CD46 co-stimulation also significantly augments fatty acid metabolism in activated human CD8$^+$ T cells and via this supports IFN-γ secretion, granzyme B production and effector activity in cytotoxic T lymphocytes (CTLs).[278] Thus, patients deficient in either CD46 or C3 are unable to mount productive Th1 responses and suffer from recurrent infections early in their life.[274,275]

The anaphylatoxins also play important and nonredundant direct roles during the initiation of T cell responses. This is exemplified by the finding that CD4$^+$ and CD8$^+$ T cells from mice deficient in either *C5ar1* or *C3ar1* genes have a significantly reduced ability to produce IFN-γ *in vitro* and *in vivo*.[279-281] T cells from

C5ar1-deficient mice also show reduced survival in the periphery as C5aR1-mediated signals counteract programmed cell death signals in activated T cells. In human T cells, C5aR1 stimulation induces the generation of reactive oxygen species (ROS) and subsequent NLRP3 inflammasome assembly and intrinsic IL-1β production, all supporting further IFN-γ generation, while C3aR activation further increases mTOR activation.[66,282] Thus, C3aR and C5aR1 on T cells seem to directly support the induction of their effector activity (Fig. 13.6A). This notion aligns with the observation that genetic ablation of pharmacological blockage of these receptors may induce a (spontaneous) regulatory phenotype in human and mouse T cells, characterized by Foxp3 expression and secretion of the immunosuppressive cytokines transforming growth factor (TGF)-β and/or IL-10.[283,284] However, while the expression of C3aR and C5aR1/2 on human T cells is broadly acknowledged, there is currently an ongoing debate in the field about their presence on mouse T cells; some groups observe anaphylatoxin receptor expression on murine T cells, while other groups fail to do so.[279-281,285-287] Because the levels of locally produced complement activation fragments have emerged as key controller of immune cell regulation (see below), it is not surprising that deviations in the expression of the GPI-linked cell surface convertase regulator CD55 on T cells has a strong impact. Indeed, activated $CD4^+$ T cells from *Cd55*$^{-/-}$ mice display increased Th1 induction due to unrestrained local anaphylatoxin generation.[288,289] Similarly, isolated T cells from a patient with *CD55*-deficiency showed substantially increased tumor necrosis factor production and a defect in normal Th1 contraction.[290] As to the GPI-linked MAC regulator CD59, early work showed that CD59 engagement during human T cell activation induces Ca^{2+} influx, IL-2 production and proliferation[291] possibly via moving into lipid rafts and activating LAT.[292] However, there are also studies demonstrating that CD59 deficiency enhances T cell activation in a complement independent manner,[289,293] suggesting that there may be another ligand and, thus, function for CD59 that remains to be discovered.[294] Further, CD55 and CD59 are encoded in humans each by a single gene (*CD55* and *CD59*). In mice, two genes (*Daf1* and *Daf2*) encode two distinct CD55 proteins and two genes (*Cd59a* and *Cd59b*) encode two distinct CD59 proteins.[295] Although *Daf2* and *CD59b* show more restricted expression, only *Daf1* and *Cd59a* knock-out animals have been generated and analyzed and no work on the other two proteins has been performed on immune cells.

Recently, C1q has emerged as an important controller of cellular behavior, with observed direct impacts on T cell function. For example, C1q-opsonized ICs can support IFN-γ secretion by T cells via interaction with a T cell–expressed C1qR with or without concurrent TCR stimulation.[296] The situation seems to be complex though as the responsiveness of T cells toward C1q depends on the timing of C1qR-mediated signals because preincubation of T cells with C1q-opsonized ICs prior to TCR stimulation rather reduces cytokine production.[297] The reason why direct complement receptor signaling on $CD4^+$ and $CD8^+$ T cells is specifically driving Th1/CTL induction but not Th2, Th9, or Th17 responses are currently unclear. References [275, 298, and 299] represent current and in-depth reviews about the direct role of complement in human T cell activation/contraction (Fig. 13.6A).

The previous paragraphs have focused on the T cell activating role of complement receptors. However, it is now becoming more broadly acknowledged that complement also contributes to the *contraction of T cell responses*, which is an important mechanism to prevent host tissue damage due to extended immune activation. After driving successful Th1 induction, CD46 engages in a crosstalk with the IL-2R and together they induce the co-expression of IL-10 in Th1 cells and the return of the cells to a more metabolic quiescent state.[273,275,277,300,301] The exact molecular mechanisms underlying the IL-2R and CD46 crosstalk mediating specifically human Th1 cell contraction are currently being explored. Similar to the C3b–CD46 interaction that directs Th1 induction but also contraction, T cell–derived C5a also plays a dual role. Activated Th1 cells have the ability to process C5a into C5a-desArg, which then engages the surface expressed C5aR2 in an autocrine fashion. C5aR2 activation exerts negative control over the activating C5aR1 signals and thus inhibits IFN-γ production and fosters IL-10 secretion.[282] The exact mechanisms by which C5aR2 negatively regulates Th1 function are not yet understood. CD59 may also restrain activated T cells, as cell from CD59a-deficient mice display a hyperactive phenotype with increased proliferation and proinflammatory cytokine production.[302] Finally, C1q does not always support T cell effector function but has been also shown to induce antiproliferative signals in these cells[298] and restrict CTL effector responses via modulation of $CD8^+$ T cells mitochondrial activity[204] (Fig. 13.6A).

Effects on APCs: Complement receptors also have a profound impact on APC activity, which in turn are master regulators of T cell behavior. For example, mouse studies employing *C3ar1* and/or *C5ar1*-deficient strains have demonstrated a strict requirement for anaphylatoxin-mediated signals for normal and efficient *dendritic cell (DC) maturation* (specifically the upregulation of CD40, CD54, and CD86) and the production of *polarizing cytokines* that subsequently direct T cell differentiation.[303,304] Particularly the central role for the anaphylatoxins in endowing APCs with Th2 or Th17 inducing capacity has gathered much attention,[305] since direct T cell effects of complement receptors seem to favor Th1 induction. C3aR and C5aR1 signaling mediated by cyclic adenosine monophosphate (cAMP), extracellular signal–regulated kinases (ERK), NF-kB, and/or induction of the NLRP3 inflammasome stimulate monocytes, macrophages or dendritic cells to secrete IL-1β, IL-12, IL-23, IL-6, and/or TGF-β which are instrumental in mediating either Th1, Th2, or Th17 responses[303-307] (Fig. 13.6B). Importantly, the timing of the incoming signal, the specific APC subtype and the nature of other co-engaged PRR systems are important for the net outcome of complement receptor activation on APCs. For example, while TLR2 and C5aR1 synergize in mouse bone-marrow derived DCs to induce IL-12 secretion, C5aR1-deficient DCs from the spleen generate more IL-6 and IL-23 compared to C5aR1-sufficient DCs when stimulated with the TLR1/2 ligand Pam3Cys.[308] Also, C5a inhibits house dust mite-induced IL-23 production from bone-marrow derived DCs in an asthma model, while C5aR1-deficient bone marrow-derived DCs do not produce IL-1β, IL-6, or IL-23 when exposed to combined ovalbumin (OVA) and LPS activation.[309] In humans, a protective Th17 response against *Candida albicans* relies on C5aR1-triggered IL-6 secretion by blood mononuclear cells (PBMC).[310] C1q is now also emerging as

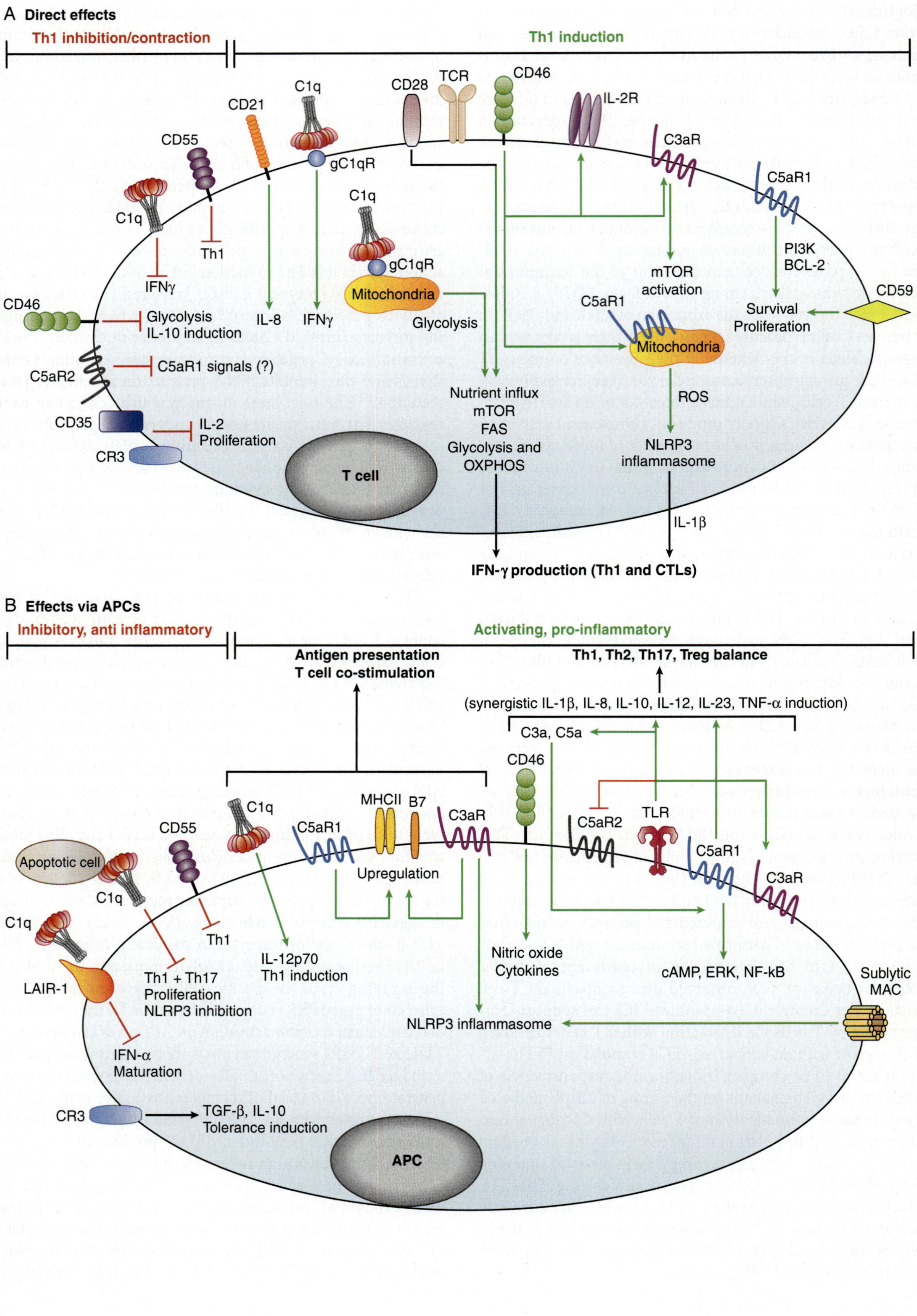
A Direct effects
Th1 inhibition/contraction
Th1 induction
CD21
C1q
gC1qR
CD28
TCR
CD46
IL-2R
C3aR
C5aR1
CD55
C1q
Th1
IFNγ
IL-8
IFNγ
C1q
gC1qR
Mitochondria
Glycolysis
mTOR activation
PI3K
BCL-2
CD59
Survival
Proliferation
C5aR1
Mitochondria
CD46
Glycolysis
IL-10 induction
C5aR2
C5aR1 signals (?)
CD35
IL-2
Proliferation
CR3
Nutrient influx
mTOR
FAS
Glycolysis and
OXPHOS
ROS
NLRP3
inflammasome
T cell
IL-1β
IFN-γ production (Th1 and CTLs)
B Effects via APCs
Inhibitory, anti inflammatory
Activating, pro-inflammatory
Antigen presentation
T cell co-stimulation
Th1, Th2, Th17, Treg balance
(synergistic IL-1β, IL-8, IL-10, IL-12, IL-23, TNF-α induction)
C3a, C5a
CD46
C1q
MHCII
B7
C3aR
C5aR1
CD55
C5aR2
TLR
C5aR1
C3aR
Apoptotic cell
Upregulation
C1q
Th1
C1q
LAIR-1
Th1 + Th17
Proliferation
NLRP3 inhibition
IL-12p70
Th1 induction
Nitric oxide
Cytokines
cAMP, ERK, NF-kB
Sublytic
MAC
NLRP3 inflammasome
IFN-α
Maturation
CR3
TGF-β, IL-10
Tolerance induction
APC

an important regulator of APC function and supports the expression of CD83, CD86, and HLA-DR, as well as IL-12 secretion during DC maturation.[311] Furthermore, C1q and C3 are also indicated in the expression regulation of CCR7, a chemokine receptor required for immune cell homing to the lymph nodes, and therefore also coordination of the trafficking and migration of APCs.[312] Also, sublytic MAC formation within membranes of host cells[313,314] can control key biological events such as inflammasome activation, cytokine secretion, and initiation of cell death versus proliferative pathways.[227,315-317] So far, these noncanonical activities of the MAC are most often viewed as anecdotal and simple defense reactions of cells under MAC attack. Given that most complement components now seem to contribute to normal cellular behavior in the healthy host (see below), looking at the MAC from a cell physiological angle during APC instruction may unearth unexpected new functions for his classic killer (Fig. 13.6B).

Although less explored, complement receptors also contribute to the generation of APCs with a more *tolerogenic phenotype*. Andrea Tenner's group recently found that primary human macrophages ingesting C1q-opsinized late apoptotic cells decrease allogenic and autologous Th1 and Th17 responses.[318] Exciting work by Betty Diamond's group also supports a key role for C1q in the negative control of APC function: they observed that C1q binding to the leukocyte-associated immunoglobulin-like receptor 1 (LAIR-1) reduces IFN-α production and limits DC maturation and activation.[319] Similarly, CR3 engagement on DCs renders these cells tolerogenic with reduced T cell activation potential[320] with the underlying mechanism not yet explored (Fig. 13.6B).

In sum, the combined balance of incoming complement receptors signals on both APCs and T/B cells during their cognate interaction are becoming recognized as key modulators of adaptive immunity (Figs. 13.5 and 13.6). However, this interplay is highly complex, involves several other PRRs (see below), and often engages species-specific pathways—we are therefore far away from understanding the exact role of complement in adaptive immunity.

COMPLEMENT IN TISSUE DEVELOPMENT, HOMEOSTASIS, AND REPAIR

For many years, complement was considered an exclusively proinflammatory system. This view has now changed, and it is becoming increasingly clear that complement is not only equally involved in the contraction phase of immune responses (see above) but also in the physiological pathways underlying cell/tissue development, normal function, and tissue regeneration after injury (Fig. 13.7). For example, complement contributes to the molecular processes underlying fertilization, implantation, embryogenesis, organogenesis and also reaches well into postnatal development.[321] Complement's contributions here involve such diverse actions as CD46 regulating the acrosome reaction,[322] CD59 ensuring survival of sperm in the testis,[323] and C3 activation fragments having embryotrophic activity, meaning they foster the attachment and subsequent growth of the blastocyst.[324] Further, deficiency in C1q causes defective placentation and predisposes to the development of preeclampsia in mice.[325] Complement also partakes in organogenesis in the developing embryo[321] and emerges as a central regulator for the coordinated migration and homing of embryonic stem cells needed to form specific tissues and organs. Early work using "classic" animal systems to study development, such as zebrafish, the aquatic frog *xenopus*, and mice hinted toward a particularly close relationship between complement and the neuronal system. In all of these species, most of the complement receptors (anaphylatoxin receptors, C1q receptors, CR1, etc.) and many core components (including C3, C5, and all complement PRRs) are particularly heavily expressed early on in the neuronal crest and coordinate the migration of neuronal crest cells, which will eventually form the host's (central) nervous system (CNS).[326] Any deviation in the expression of these receptors is accompanied by aberrant neuronal tube formation in the embryo with different degrees of severity.[321] One of the most exciting findings in the last decade is that complement is critical to brain development, not only prenatally but also postnatally. This occurs through the regulation of synaptic pruning, the process of eliminating "unnecessary" synapses, an important shaping process required for optimal brain function. The classical pathway labels, or tags, redundant synapses meant for elimination by microglial cells likely via CR3 engagement,[327] allowing for efficient synaptic pruning. Thus, *C1qa*- or *C3*-deficient mice display defects in synaptic pruning with abnormal brain development and behavior.[62] Further, C1q sustains neuronal survival in mice.[328] While much of this research is obviously not easily translated directly into humans, an emerging role for complement in brain development, neurodegenerative diseases and neuroinflammation in man is nonetheless becoming clearly evident: there is a strong association between complement system perturbations and diseases such as schizophrenia and epilepsy, disorders that

FIG. 13.6. Complement and T cell regulation. A: Direct effects of complement receptor engagement on T cells. Activation of complement sensors on a T cell drives cell survival and proliferation via actively reducing activation-induced cell death and allowing responsiveness to IL-2. CD46 and C1q control nutrient influx and key metabolic nodes that are particularly needed for the induction of IFN-γ in $CD4^+$ and $CD8^+$ T cells. Several complement receptors also partake in the negative control of Th1/CTL induction, which is important to prevent tissue pathology due to exuberant effector T cell activity. **B:** Complement receptor modulation of APCs affecting T cell lineage induction. The cytokine expression profile driven by complement receptor engagement on APCs is a critical determinant of Th1, Th2, and Th17 versus T_{reg} induction. In principle, anaphylatoxin receptors induce APC maturation and antigen presentation capacity. In crosstalk with diverse incoming TLR and/or inflammasome-mediated signals, complement receptors induce a specific cytokine (and chemokine, not shown) expression pattern that dictates the nature of the T cell response triggered by the cognate APC/T cell interaction (not shown). Similar to those in T cells, complement receptors also contribute to the negative control and/or induction of a tolerogenic phenotype in APCs. B7, B7 family member CD80 or CD86 costimulators; BCL-2, B cell lymphoma 2; CR3, complement receptor 3; CTL, cytotoxic T lymphocyte; FAS, fatty acid synthesis; IL-2R, interleukin 2 receptor; LAIR-1, leukocyte-associated immunoglobulin-like receptor 1; MAC, membrane attack complex; MHC, major histocompatibility complex; Mitoch., mitochondria; mTOR, mammalian target of rapamycin; NLRP3, NOD-, LRR-, and pyrin domain–containing protein 3; NO, nitric oxide; Oxphos, oxidative phosphorylation; PI3K, phosphoinositide 3-kinase; ROS, reactive oxygen species; TCR, T cell receptor; Th, T helper (cell); TLR, Toll-like receptor; T_{reg}, regulatory T cell (natural and induced).

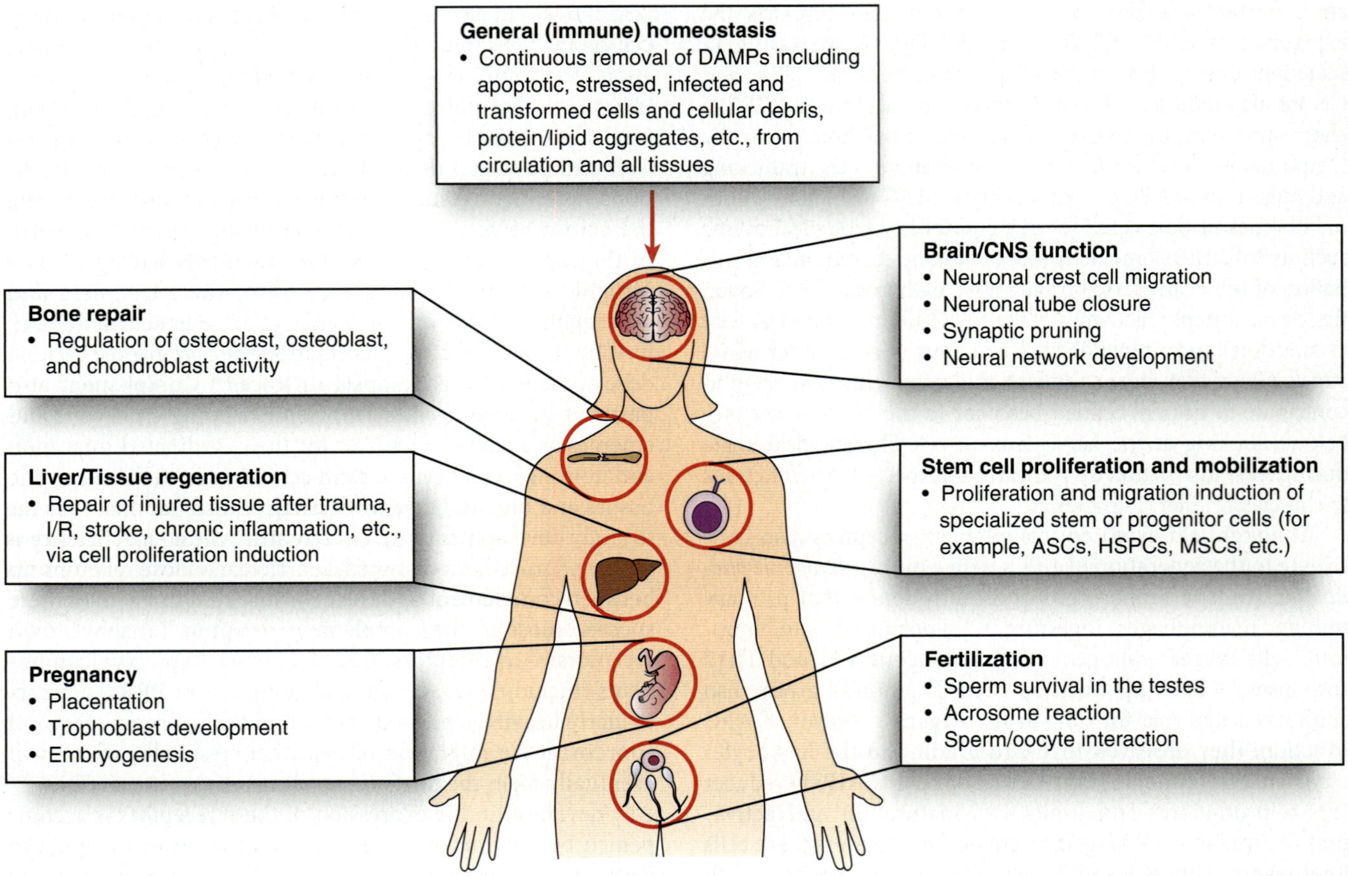

FIG. 13.7. Complement in tissue development, homeostasis, and repair. Constant surveillance by complement for DAMPs ensures general (immune) homeostasis and the host's health. However, complement is also required for the indicated diverse activities that range from fertilization to normal brain and central nervous system (CNS) development and function, to tissue repair after injury, etc.

are linked to altered brain development.[329] Furthermore, increased or dysregulated complement activation is seen in a spectrum of acute and chronic neurological disorders that are often characterized by neuroinflammation, including traumatic brain and spinal cord injuries, amyotrophic lateral sclerosis (ALS), multiple sclerosis (MS), neuromyelitis optica (NMO), stroke, and AD, Parkinson disease (PD), and Huntington disease (HD)[330] (see below). Thus, research on the role of complement in neurological diseases is an area of intense research and we should expect many exciting new insights over the next few years here.

Complement is not only a physiological housekeeper via the removal of apoptotic cells and cellular debris[57] (see above), it is also at the heart of the pathways that regulate tissue regeneration—in essence, complement induces repair pathways that will help to remedy the injuries and tissue pathologies it caused during induction of protective innate and adaptive immune responses.[331,332] In the liver, C3aR and C5aR1 are required to induce signal transducer and activator of transcription (STAT) 3–dependent pathways and cell cycle phases that induce hepatocyte proliferation and tissue repair after injury, thus, animals deficient in these receptors have impaired liver regeneration.[333] Similarly, mice that cannot produce C3 or C5 in sufficient amounts have a substantial defect in bone healing after fractures and here, the anaphylatoxin receptors but also the MAC and the PRRs MBL and C1q are indicated in mediating the needed repair signals in chondroblasts.[59,334] The role of complement activation in the healing of cutaneous (chronic) wounds is still currently thought to be rather detrimental as complement inhibition in animal models or topically applied complement inhibitors in patients generally promote wound healing.[335] However, there is indication that C1q has a protective role as it may promote wound healing via induction of angiogenesis,[336] warranting a future closer look into the role of distinct complement components in healing and/or scar formation. Similarly, although local complement activation and anaphylatoxin receptor activation on brain cells clearly contributes to the pathology of neuronal demyelinating diseases such MS,[337,338] C3aR and C5aR1 may play a dual role here as signals mediated by these receptors on astrocytes reduce cytokine production and aided in the remyelination process in a cuprizone-induced animal model of demyelinating disease.[339] Complement also directs the mobilization of a range of adult specialized progenitor or stem cell populations that can be found in a range of organs, such as liver, gut, bone marrow, heart, etc. Such cells are a reservoir for replenishing needed tissues and cells after injury[340,341] and include, for example, mesenchymal stem cells (MSCs), adipose-derived stem cells (ASCs), intestinal stem cells (ISTs), hematopoietic stem and progenitor cells (HSPCs), and many more.[341,342] In almost all cases, incoming signals from the anaphylatoxin receptors C3aR and C5aR1 are required to induce proliferation of stem cells and/or their migration via induction of chemokine receptors to the tissue site needed.[321] In several cases, C3aR and C5aR1 also induce the expression of specific differentiation markers, indicating that they also contribute to the differentiation of stem cells into desired, tissue-specific, lineages.[343]

As exciting as the new roles of the complement system in tissue repair are (Fig. 13.7),[344] they can also turn against the host, if normal complement activity is perturbed or if the system becomes chronically active. Successful repair of injury ideally occurs with minimal scarring and aims at limiting changes to the restored tissue that may affect its function.[344] The last few years have identified dysregulated complement as a central driver of fibrosis, the pathological thickening, and scarring of repaired tissues and wounds.[345-347] Most of the evidence has so far been obtained using kidney ischemic-reperfusion or lung injury models,[348,349] but the expectation is that complement may play a generalized role here and will contribute to additional fibrotic disease states.

COMPLEMENT-DRIVEN DISEASES AND SYNDROMES

Because the complement system plays such broad roles in immune cell and tissue biology, it is not surprising that aberrant complement activity is associated with a large, and in fact, increasing, number of human diseases.[350] Although clearly too simplified, consequences of complement perturbation fall grossly into two categories: deficiencies in PRRs or core components (primary or secondary) which cause *recurrent infections*, while changes in complement regulation are connected with a *broad range of (auto)inflammatory disorders*. The latter often involve trickle-down effects of unwanted or sustained complement activation on immune and tissue cells that perpetuate inflammation and tissue injury. Further, due to the multiple and distinct roles that most complement components serve in immunity, disease presentation is often complex with several parallel pathologies.

Complement Deficiencies

Most complete complement deficiencies are rare, only accounting for ~5% of all primary immunodeficiencies.[351] The majority of these present in early childhood with reoccurring bacterial infections and/or the autoimmune disease SLE. Diagnosis in patients with suspected complement deficiency occurs via measurement of serum complement activity using the *CH50* (assesses CP activity) and the *AH50* (measures AP activity) tests (Fig. 13.8) and measurement of complement components in serum and/or genetic screening. Total deficiencies in a complement protein result also in the loss of all cascade activity downstream of the affected component. These conditions are typically inherited in an autosomal recessive manner, thus requiring two abnormal alleles for complete deficiency (except for properdin [X-linked] and C1-INH [autosomal dominant] deficiency[350]) and are normally categorized based on the pathways or regulators they belong to: the CP, LP, AP, or TP, deficiencies in the regulators of these pathways, and deficiencies in complement receptors (Tables 13.4 and 13.5).

The most common complement deficiency is a deficiency in the PRR of the LP, MBL, with frequency of alleles that result in MBL deficiency reaching up to 30% in certain populations.[351,352] Although MBL deficiency can be connected with increased burden of bacterial, viral, fungal, and/or parasitic load, the LP appears to be somewhat redundant in protective immunity because prospective studies have failed to show that patients lacking functional serum MBL are more susceptible to infections. On the contrary, because enhanced LP activity may be associated with autoimmune disorders (see below), deficiencies in MBL could be advantageous to some individuals.[352] Ficolin-3 deficiency manifests with different pathologies, including membranous nephropathy, pulmonary infections and lung diseases, brain abscesses, and congenital heart disease[353,354] underpinning the fact that the PRRs serve a broad range of activities. MASP-2 deficiency is rare and in one patient resulted in pneumococcal pneumonia and immune manifestations such as ulcerative colitis and a hypersensitivity disorder involving the skin.[355] Homozygous deficiency in MASP-1 or in Collectin-11 results in 3MC syndrome (Carnevale, Mingarelli, Malpuech, and Michels syndrome), a disease that involves defects in both host immune defense and normal development.[356]

Total CP deficiencies result in undetectable CH50 activity along with low levels of the PRR C1qrs complex or C2 or C4 protein absence/strong reductions and are the driving cause for recurrent infections with *Streptococcus pneumoniae* and/or development of SLE. This combination is based on the fact that C1q is central not only in the detection and removal of apoptotic cells and control of T cell responses (thus, prevention of autoimmunity) but also in the initiation of the CP that protects against infections. Interestingly, 90% of patients deficient in C1q develop SLE but not all suffer from recurrent bacterial infections, indicating that LP and AP can provide substantial protection in absence of a fully functional CP. C2 deficiency also commonly results in SLE with and without recurrent pyogenic infections with encapsulated bacteria, and some patients present with a combination of multiple autoimmune disorders. Absence or strongly reduced expression of the CP core component C4 is rarely based on complete gene ablation but rather on changes in the complex *C4* locus. C4 is encoded by two genes, *C4A* and *C4B*, which are highly polymorphic and are often eliminated or duplicated, such that 35% to 50% of the population has 1 to 8 functional copies of *C4* genes rather than the normal two *C4A* and two *C4B* genes.[350] As functional copies of *C4* decrease in some individuals, predisposition to SLE increases and decreased C4 protein has also been associated with other immune disorders such as scleroderma, childhood diabetes, and immunoglobulin A nephritis and vasculitis.[350] Complete C3 deficiency exists but is also not observed often, likely because C3 absence affects all of the complement pathways. As C3 is the major complement opsonin, C3-deficient patients suffer from severe infections with encapsulated bacteria, particularly *S. pneumoniae*, beginning very early in life and later in life may develop conditions like glomerulonephritis due to inordinate

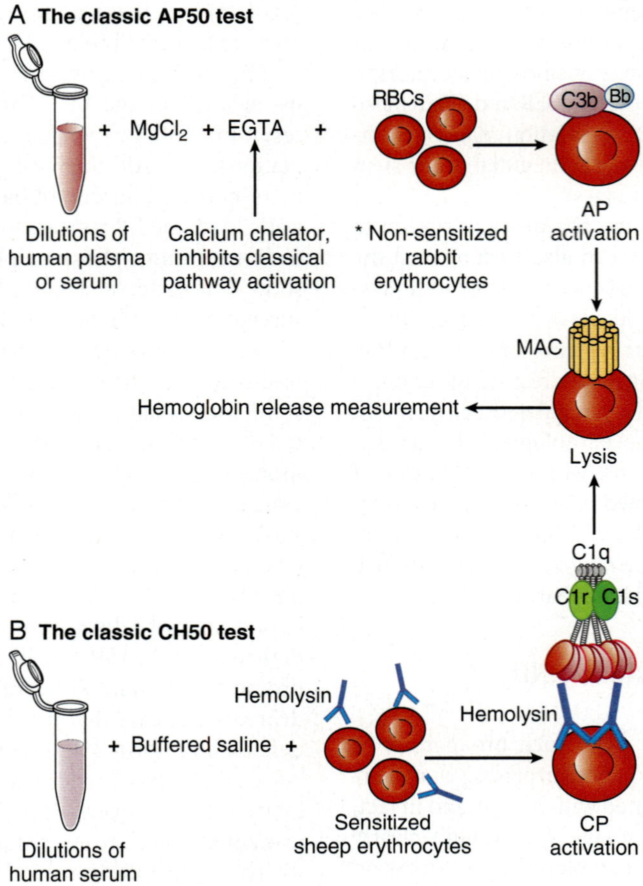

FIG. 13.8. Functional human classical and alternative pathway complement tests. A: The classic AP50 test detects alterations/deficiencies in the AP. Dilutions of human plasma + $MgCl_2$ + a calcium chelator (which inhibits CP activation) are mixed with nonsensitized rabbit erythrocytes (*, rabbit erythrocytes have a relative deficiency of surface sialic acid molecules and thus activate the human AP naturally and without need for antibody-mediated sensitization). The AP pathway convertase (C3bBb) is formed leading to the eventual formation of the MAC and erythrocyte lysis. Hemoglobin released by lysed erythrocytes is then measured utilizing a spectrophotometer. **B:** The classic CH50 test detects alterations/deficiencies in the CP. Here, dilutions of human serum + buffer are mixed with sheep erythrocytes sensitized with hemolysin (a specific rabbit anti-sheep red blood cell antibody). This results in C1q binding to the surface of the antibody-sensitized sheep erythrocytes and activating the CP, which in turn induces cell lysis by the MAC. The AP50 and CH50 values are each calculated as the dilution of the plasma or serum required to obtain 50% hemolysis of the erythrocytes. EGTA, ethyleneglycol-bis(β-aminoethyl ether)-N,N,N′,N′-tetraacetic acid; MAC, membrane attack complex. Of note, testing LP activity is more complex due to the many PRRs that can trigger its activation. However, MBL deficiency can be measured using an ELISA plate assay detecting cleaved C4 fragments generated in human serum diluted in hypertonic buffer, which inhibits C1q binding to antibodies but leaves MBL function intact (not shown). AP, alternative pathway; CP, classical pathway; LP, lectin pathway; MBL, mannan-binding lectin; PRR, pattern recognition receptor.

IC formation.[350] Similarly, FB or FD deficiency associates with dysfunctional AP activity (albeit normal C3 levels) and patients suffer from increased bacterial infections, often with meningococcal Neisseria.[351,357] Patients with deficiencies in the TP components C5, C6, C7, C8, or C9 are unable to form a normal MAC, which is critical in the protection against Neisseria infections. Deficiency in serum C5 has only been observed in a handful of patients and the meningococcal disease they suffer is combated by high doses of antibiotics. Interestingly, one of the most common genetic disorders in the Japanese population is heterozygous *C9* gene deficiency, and here, it is associated unexpectedly with decreased mortality and a reduction in infantile gastroenteritis incidences.[358] The reasons are unclear but may indicate that reduced C9 levels could have a selective evolutionary advantage in this population, protecting against sustained inflammation. Of note, MAC-specific bacterial killing can occur in the presence of reduced C9 amounts, just less efficiently.[128,359]

Complete loss of certain regulatory proteins can lead to the consumption of multiple components of a pathway, often resulting in an increased incidence of infections. The first deficiency in a regulator discovered was C1-INH deficiency. Lack of C1-INH activity results in uncontrolled activation of the LP (MASPs) and CP (C1r/s) and induces hereditary angioedema, in part because C1-INH also regulates enzymes in the coagulation, fibrinolytic, and kinin pathways, which also contribute to angioedema induction.[360-362] Complete deficiencies in FI or FH trigger AP serum activation, C3 consumption, and thus lead to *secondary or acquired C3 deficiency*. Again, disease pathologies differ between patient groups, with FI deficiency resulting in

TABLE 13.4 Common Disease Phenotypes Associated With Total Complement Deficiencies

Protein Affected	Associated Clinical Characteristics[a]
Lectin pathway	
MASP-1 or Collectin-11	3MC syndrome—defects in development and host defense
MASP-2	Pneumococcal pneumonia, skin hypersensitivity, ulcerative colitis
MBL	Increased susceptibility to infections (?)[b]
Ficolin-3	Membranous nephropathy, pulmonary infections and lung diseases, brain abscesses, congenital heart disease[c]
Alternative pathway	
Factor B	Increased bacterial infections (often Neisserial infections)
Factor D	Increased bacterial infections (often Neisserial infections)
Classical pathway	
C1q C1qr C1qs	Recurrent infections with pyogenic bacteria, SLE
C2	A combination of autoimmune disorders, recurrent pyogenic bacterial infections, SLE
Terminal pathway	
C5-C9	Autoimmune diseases, Neisserial infections
Core components	
C3	Glomerulonephritis, severe infections with encapsulated bacteria
C4	Childhood diabetes, IgA nephritis, recurrent infections with pyogenic bacteria, scleroderma, SLE, vasculitis
Regulatory proteins	
C1-INH	Hereditary angioedema
CD46 (MCP)	aHUS, recurrent infections, loss of Th1 responses
CD59	PNH
Factor H	Renal disease
Factor I	Increased infections
Properdin	Increased infections
Receptors	
CR1	SLE
CR2	Hypogammaglobulinemia, recurrent infections
CR3 CR4	LAD-1—delayed umbilical cord detachment, recurrent severe bacterial infections

3MC syndrome, Carnevale, Mingarelli, Malpuech, and Michels syndrome; aHUS, atypical hemolytic uremic syndrome; C1-INH, C1 esterase inhibitor; LAD-1, leukocyte adhesion deficiency type 1; MASP, mannan-binding lectin–associated serine protease; MBL, mannan-binding lectin; PNH, paroxysmal nocturnal hemoglobinuria; SLE, systemic lupus erythematosus; Th1, T helper type 1.

[a]Listed are the most common reported clinical characteristics only.

[b]Controversial.

[c]There are only few reported Ficolin-3 deficiencies and patients are heterogeneous in their clinical phenotype.

increased incidence of infections, while FH deficiency typically presents with renal disease because FH's functional repertoire also includes protection of the kidney basal membrane during inflammation. Individuals with complete lack of the C3 convertase stabilizer properdin are also prone to severe Neisserial infections, with males predominantly affected as the properdin gene (*CFP*) is located on the X chromosome. Total CD46 deficiency is also rare, but all patients suffer from hemolytic uremic syndrome (aHUS, see below) along with recurrent infections in early childhood likely due to altered adaptive immune responses caused by absence of CD46-mediated signals on T and B cells.[275] As expected, complete deficiency in the GPI-linked MAC regulator CD59 is not associated with increased infections. CD59 absence, however, causes the hematological disorder paroxysmal nocturnal hemoglobinuria (PNH), which is characterized by spontaneous complement attack of

TABLE 13.5 Complement-Based Therapeutics Currently in Clinical Development[a]

Target	Drug candidate(s)/Class(es)	Mode(s) of Action	Targeted Disease(s)
Lectin and classical pathway			
C1r/s, MASPs, other serine proteases	C1-INH protein	Inhibits serine proteases (CP and LP)	Hereditary angioedema[b], I/R, stroke, transplant rejection, trauma
C1s	Monoclonal antibody	Inhibits protease activity (CP)	Autoimmune hemolytic anemias
MASP-2	Monoclonal antibody	Blocks MASP-2 activity (LP)	aHUS, C3G, IgA nephropathy, transplant-associated TMAs
Alternative pathway			
Factor B	Small molecules	Inhibit AP C3 convertase	C3G, PNH
	Antisense oligonucleotides	Inhibit FB expression	AMD
Factor D	Small molecules	Inhibit AP C3 convertase	C3G, PNH
	Fab fragment of monoclonal antibody	Blocks AP C3 convertase formation	AMD
Properdin	Monoclonal antibody	Inhibits AP amplification	AMD
Core components			
C3	Pegylated and nonpegylated cyclic peptides (compstatins)	Inhibit C3 cleavage	AMD, autoimmune hemolytic anemias, C3G, hemodialysis-induced inflammation, periodontal disease, PNH
C5	Multiple monoclonal antibodies	Block C5 activation	aHUS[b], AMD, autoimmune neuromuscular disorders[b], neuromyelitis optica, PNH[b], transplantation
	Peptide macrocycle	Inhibits C5 activation	PNH
	RNA aptamer	Inhibits C5 expression	AMD
	RNAi	Inhibits hepatic C5 expression	aHUS, PNH
C3/C5 convertases	Recombinant CR1 protein or modified fragments of the CR1 protein	Accelerates C3/C5 convertase dissociation	Kidney I/R, myocardial infarction
Other components			
C5a	Monoclonal antibody	Blocks C5aR1 signaling (by preventing C5a binding)	Inflammatory skin diseases, multiorgan dysfunction, sepsis
C5aR1	Monoclonal antibody	Blocks C5aR1 signaling	Cancer
	Small molecule	C5aR1 antagonist	aHUS, ANCA-associated vasculitis, IgA nephropathy
DR5	Hexabody	Augments CDC of $DR5^+$ tumors	Cancer

aHUS, atypical hemolytic uremic syndrome; AMD, age-related macular degeneration; ANCA, antineutrophil cytoplasmic antibody; AP, alternative pathway; C1-INH, C1 esterase inhibitor; C3G, C3 glomerulopathy; CDC, complement-dependent cytotoxicity; CP, classical pathway; DR5, death receptor 5; Fab, antigen-binding fragment; I/R, ischemia-reperfusion injury; LP, lectin pathway; MASP, mannan-binding lectin–associated serine proteases; PNH, paroxysmal nocturnal hemoglobinuria.

[a]This is a current snapshot of some ongoing trials and does not represent a complete list of all complement-targeted therapeutics under evaluation.

[b]Designates indications already approved for the clinic.

erythrocytes due to the loss of GPI-linked proteins including the complement regulators CD59 and CD55.[363,364]

In general, there are less reported deficiencies in complement receptors. Loss of CD35 causes SLE,[365] while CD21-deficient individuals suffer from recurrent infections and hypogammaglobulinemia.[366] Deficiencies in the complement receptors CR3 and CR4 result in leukocyte adhesion deficiency type 1 (LAD-1), which is characterized by a gross functional dysregulation of monocyte, neutrophil and T cell effector activity,[367,368] failure of immune cell diapedesis, and recurrent severe bacterial infections.[365,369]

Overall, the repercussions of total deficiencies in complement proteins (Table 13.5) highlight the crucial role that complement plays not only as an initial innate barrier of defense against incoming pathogens but also as a coordination hub of the immune response where its absence can result in an increased prevalence of autoimmune diseases.

Complement Mutations and Polymorphisms

While complete genetic complement deficiencies are rare, polymorphism and mutations in genes encoding complement proteins are relatively common. An individual's unique combination of polymorphisms in all complement genes, the "complotype,"[370] in conjunction with environmental factors, dictate complement activation net outcome to PAMPs and DAMPs and a person's susceptibility to some diseases/disorders. Mutations or polymorphisms resulting ultimately in dramatically reduced complement PRR function or pathway(s) activation have, not surprisingly, many features of complete deficiencies, including recurrent infections, and/or reduced IC/DAMP clearance-mediated autoimmunity (mostly SLE). Alterations in complement proteins that result in more nuanced changes or even increases/sustenance in activity, on the other hand, are linked with a broad range of pathophysiological conditions spanning both acute and chronic disorders, including a range of arthritic diseases, asthma, allergies, etc. The specific pathological manifestations of such disorders are heavily impacted by the initial impetus for complement activation, as well as the nature, degree, duration, and localization of complement dysregulation. Due to space limitations, we will highlight here succinctly only three major disease states in which the underlying genetic changes in complement proteins have, so far, been relatively well defined: aHUS and renal disease, age-related macular degeneration (AMD), and pathologies of the CNS. For more in-depth reading on the roles of complement in these and also other human pathologies (including arthritis, asthma and allergies, etc.), we refer the reader to the many excellent reviews, including but not limited to Refs. 46,370-373.

Atypical HUS and renal disease: Atypical HUS, a rare renal thrombotic microangiopathy (TMA), is characterized by hemolytic anemia, thrombocytopenia, and kidney dysfunction and can be traced in the majority of cases to increased AP activation.[374] Over 60% of aHUS patients have genetic variations in AP regulators including FB, FH, FI, or in CD46 and ~10% of patients have autoantibodies recognizing FH.[375-377] Recent work also demonstrates an important role for the CFHR proteins (Table 13.2) in aHUS. These CFHR proteins are encoded by genes adjacent to *CFH* and are structurally related to FH. Gene deletions of *CFHR1* and *CFHR3* are associated with the development of autoantibodies specific for FH (via a mechanism that is currently not clear) in humans, which results in acquired FH deficiency and development of aHUS.[378,379] AP dysregulation can also lead to complement C3 glomerulopathy (C3G). C3G is a rare heterogeneous renal disease that includes two major subgroups, dense deposit disease (DDD) and C3 glomerulonephritis (C3GN).[380] About 25% of patients with C3G also have mutations in, or rare variants of, genes encoding FB, FH, FI, CFHR-5, or C3.[381,382] However, there must be subtle differences in the consequences of increased AP activation in C3G as patients display distinct renal pathology compared to individuals suffering from aHUS. Furthermore, specifically C3G is often driven by the appearance of complement convertase-stabilizing autoantibodies that lead to enhanced complement turnover and complement activation fragment deposition in the kidneys.[380] For reasons that are currently not well understood, the kidney is particularly vulnerable to augmented (AP) complement activity: aside from aHUS and C3G, uncontrolled C3 activation is the reason for, or at least a strong contributing factor of, many additional acute and chronic renal diseases including bacterial endocarditis, ANCA-associated vasculitis, IgA and membranous nephropathies, lupus nephritis, cryoglobulinemia, antiglomerular basement membrane diseases, postinfectious glomerulonephritis, and renal transplant rejection.[383,384] Although unrestrained C3 activation is the initial trigger in renal disease, the resulting increased activation of C5 is a defining factor in the perpetuation of disease and progression toward kidney failure.[385] Interestingly, pathologically increased C5 activation is classically connected with disorders of the vasculature, mostly because C5a activates endothelial cells lining the blood vessels and engages in heavy crosstalk with the coagulation system,[377,386,387] which, when perturbed, aggravates renal disease.[386,387] The humanized anti-C5 antibody eculizumab (see below) treats aHUS successfully and also prevents acute renal transplant rejection with high efficiency,[388,389] indicating that the blood vessels in the kidney (possibly due to their specialized "filter" function) require a particularly well-controlled complement milieu.[390] Attack of the vasculature by complement (in conjunction with subsequent immune cell activation) is also a driving force in the systemic inflammatory condition sepsis, underlies the ischemic reperfusion injury that can lead to strokes or myocardial infarctions,[391-393] and contributes to systemic autoimmunity as seen in arthritis and vasculitides.[394] Further, recent research insights suggest a causal connection also between augmented anaphylatoxin activity and the induction and/or progression of cardiovascular disease, such as atherosclerosis, via inflamed vasculature.[395]

Age-related macular degeneration: AMD is another complement-driven disease that garnered substantial interest. This is not surprising as AMD is the most common cause of irreversible blindness in the Western world. It is a chronic inflammatory disease of the retina that results in progressive loss of central vision. AMD is caused by the accumulation of waste products (proteins and lipids), called drusen, between the retinal pigment epithelium (RPE) and the Bruch's membrane, the gradual loss of photoreceptor activity, and invasion of choroidal vasculature into the retina.[396] RPEs support the function of photoreceptors, and the accumulation of drusen over time disturbs this activity. AMD develops in the elderly with aging, and with environment, and genetics all playing a role in the disease, with heritability estimated at 41% to 76%.[397] Understanding and treating this disease is a major research/clinical goal as the burden on the nations' health care systems is substantial and steadily increasing due to population aging.[398] Over a decade ago, the risk of AMD was first associated with a common gene variant of FH,[399-401] a major breakthrough in the field of ophthalmology as it allowed to finally explore molecular mechanisms underlying the condition. By now, we know that one-third of the common and rare gene variants identified to be associated with AMD are in or around complement-related genes, including *CFB*, *CFH*, and *CFI*, *CD46*, the core components *C2*, *C3*, *C9*, and *VTN*, indicating the dysregulation of all complement pathways in this disease.[49] In addition, it is now also accepted that many cells in the eye express a full range of complement components and that the source of dysregulated complement underlying AMD is local. Mutations or single nucleotide polymorphisms (SNPs) cause different effects on the affected complement proteins including reduced expression levels due to misfolding or degradation or reduced functional activity. Functionally, perturbations of normal complement regulator activities in the eye contribute to increased and sustained local complement activation, which then translates into tissue attack in combination with altered cell behavior. Indeed, deposition of complement activation fragments within drusen and on cells in the eye have been noted in AMD. For example, MAC-mediated RPE death has been reported[402] and C3a and C5a have been shown to aggravate the neovascularization of the eye, a hallmark of AMD, via vascular endothelial growth factor (VEGF) production.[403] Further, the increased production of anaphylatoxins locally recruits phagocytic cells into the eye, which then promote further loss of photoreceptors and RPEs. Mutations and SNPs in genes encoding FH and FHL-1 seem to play a particularly central role in this disease, possibly because FH not only regulates the AP but it is also required for clearing debris in the eye (Fig. 13.7) as it normally recognizes malondialdehyde (MDA), a lipid peroxidation product that accumulates in AMD.[404] FH binding to MDA blocks both its uptake by macrophages and the production of proinflammatory mediators by these macrophages. A major FH polymorphism, H402, reduces the ability of FH to bind to MDA and thus interrupts this protective measure. Interestingly, the H402 variant, on the other hand, provides increased resistance to pneumococcal invasive diseases.[405] This indicates that a highly potent complement system (or active complotype) may provide superior protection against infections in childhood but could come at the price of an increased risk for AMD, or other chronic inflammatory/age-related conditions, later in life.

Neuropathologies: One of the most exciting recent developments is the realization that normal brain function is dependent on complement-controlled processes. In consequence, perturbation in complement activity has emerged as a major driver of a range of neuropathologies and this has garnered the interest of researchers and clinicians well beyond those working directly in the complement field. Earlier, we had discussed the local complement-mediated events supporting the development of a normal CNS via control of neuronal stem cell migration, normal crest closure, etc. (Fig. 13.7). Here, we will summarize the effects of complement component mutations and polymorphisms not only on the postnatal brain with an eye on neuroinflammation driving degenerative conditions including AD and PD but also on psychiatric disorders such as schizophrenia.[64,406] Of note, similar to AMD in which complement produced in the eye drives disease, it is now broadly accepted that complement is produced locally in the brain by specialized cells such as glial cells, astrocytes, neurons, etc., and that perturbations in this local production contributes to neuroinflammation.[64,406]

Changes in complement component functions that affect neurodevelopment prenatally or early postnatally will also trickle down into deviations of adult brain activity. For example, mice that lack C3 or C1q expression and have a defect in synaptic pruning (a process that is needed to eliminate unwanted synaptic connections and allow development of a mature functional neural network) show altered behavior.[407] Further, blockade of C5aR1 during embryonic brain development results in microstructural brain changes that manifest in impaired short-term memory and coordination as well as increased anxiety in mice.[408] The long-term impact on mental wellbeing can be severe as new evidence links the complement system conclusively with schizophrenia in humans. Schizophrenia is a complex neurodevelopmental disorder in which patients interpret reality abnormally and can ultimately not function in society. Schizophrenia is rooted in the interaction of a multitude of combined genetic and environmental factors ultimately contributing to the onset and progression of this debilitating disease. Recent work has associated enrichment of SNPs in the *C3*, *C4*, or *CFH* genes with development of this condition. Changes in CNS C3 expression lead, similar to C1q, to alterations in synaptic pruning and normal neuronal connectivity.[64,406] However, there is a particularly tight connection between the C4 component and schizophrenia: *C4A* polymorphisms that result in enhanced C4A protein expression have been linked to onset or increased risk for developing schizophrenia[329]; in fact, *C4A* SNPs are currently the strongest risk predictor for schizophrenia in humans. Post-mortem examination of brains from patients with schizophrenia show indeed dramatically increased C4 protein levels when compared to brain sections derived from deceased donors that had no neuropathological changes. The exact underlying mechanisms are not clear but may be due to aberrant circuit wiring in the frontal cortex.[409] SNPs in the *CFH* gene are associated with increased local complement activation which contributes to the severity of schizophrenia symptoms rather than initial disease onset.[410] Evidence is mounting indicating that not just reduced but also excessive synaptic pruning can occur when CNS complement activity is pathologically increased, and that excessive pruning plays also a role in schizophrenia as it equally prevents the development of normal neural networks.[411] Lastly, dysregulation of maternal C1q may increase psychosis of adult offspring[412] and impaired developmental synaptic pruning may be involved in epilepsy as well, as C1q deficient mice display an increase in seizures.[413]

AD is a neurodegenerative disease and characterized by loss of memory and steady decline in cognitive ability that ultimately culminates in severe dementia. Similar to AMD, it is an age-related disease as it is in large parts triggered by the accumulation of insoluble plaques formed by aggregation of β-amyloid proteins over time.[330] These DAMPs induce local complement activation which normally leads to the safe removal of plaques via the actions of C1q, CR3 (opsonization and uptake), C5aR1 (recruitment of phagocytes) and microglia (brain macrophages) and maintenance of healthy brain homeostasis. However, with age plaque removal becomes inefficient and continuous presence of this DAMP then leads to chronic unwanted complement activation and neuroinflammation, the major driver of AD.[330] Previously protective complement activities now turn against the host and perpetuate disease: sustained complement activity leads to the reactivation of usually "developmentally confined" synapse pruning by complement and leads to synapse loss and cognitive decline. The deposition of the MAC has been observed on neurons[330] and neuronal death and changes in the synapse network attract more phagocytes/microglia. The pathologically increased C5a levels, however, induce a more activated and proinflammatory phenotype in these scavenger cells which contributes to further brain tissue damage.[330] Of note, local complement expression in the brain increases with age, further fueling this detrimental inflammatory fire.[414] The central role that complement plays in AD and dementia has mostly been demonstrated using animal models of AD in which mice deficient in, for example, C3, or C1q, or C5aR1 showed reduced disease markers and cognitive decline.[415,416] Moreover, treatment of mice with established disease with a C5aR1 antagonist reduced pathology and improved cognitive performance.[417] Evidence that complement perturbation is indeed causative for dementia in humans was then delivered in 2009, when Lambert and colleagues showed in a genome-wide gene association study that variants in the genes encoding clusterin (*CLU*) and CD35 (*CR1*)[63] are associated with AD. Interestingly, in both cases the SNPs resulting in reduced, and not increased, expression of the CD35 or clusterin proteins were associated with disease, indicating a protective role in AD. Subsequent work by several groups indeed then showed that CD35 expressed on RBCs mediates the clearance of β-amyloid from blood and reduced expression levels of CD35 impaired this function.[418] Similarly, clusterin (also known as apolipoprotein J) seems to act as an amyloid response agent and helps in β-amyloid removal, increases neuronal survival (like C1q), and negatively controls complement activation.[419] It should be noted that while the role of complement in neuronal diseases is clear in some animal models, employment of alternative *in vivo* models or mouse strains have delivered distinct/contradictory results exist. Further, complement clearly can exert protective and aggravating role in ADs, often dependent on the stage of disease progression.[330] Thus, there is a clear need for more research to pinpoint and clarify the exact involvement and timing of complement in these disease processes, particularly if the system is to be targeted successfully therapeutically in AD in the future. There is much excitement and momentum to do so, and larger research programs in academia and the pharmaceutical industry pour major resources into this field. This has resulted already in the development of some preclinical models and in progress surrounding the development of small molecule therapeutics that can cross the blood–brain barrier.

COMPLEMENT-BASED THERAPIES

The early realization that complement underlies a broad spectrum of human diseases characterized by immunodeficiency or immunohyperactivity, sparked traditionally strong interest in the therapeutic targeting of complement. Specifically, the last two decades have brought about a plethora of reagents that are in the pipeline for, or are undergoing, clinical testing in trials. Here, we give a snapshot only of the current "complement-targeted therapeutics landscape" (Table 13.5) as this is an extremely fast-moving field and by the time this book is published, there will likely be major advances and changes—which the reader should extract from then up-to-date reviews. Currently, only two types of complement-based therapies are approved for use in the clinic: two C5-targeting antibodies, eculizumab and ravulizumab, and a C1-INH protein (see below). These compounds are, at the moment, approved for only a few indications, their application range, however, is being extended to other disease conditions in which aberrant complement plays a role.

In the late 1980s pioneering work that led to the development of an C5 activation-blocking monoclonal antibody (mAb BB5.1)[420] showed therapeutic benefit in several animal inflammatory disease models, such as collagen-induced arthritis and a lupus-like autoimmune disease.[421,422] Mab BB5.1 formed the basis for the development of the first approved complement-based therapeutic, eculizumab. Eculizumab was first indicated for the hemolytic condition PNH due to successful inhibition of MAC formation on RBCs[423] but since, has been approved for several other C5-driven pathologies, including aHUS, myasthenia gravis (gMG), and neuromyelitis optica spectrum disorder[388,424,425] (Table 13.5). The relative clinical success of eculizumab has triggered efforts to develop additional C5 activation blocking antibodies, and several of those are in the process of being (or have been) tested in clinical trials for a range of disease states including AMD, organ transplant rejection and transplant-related TMAs. One of these antibodies, ravulizumab, a longer lasting version of eculizumab (but recognizes the same epitope of C5), has proven to be non-inferior to eculizumab in the treatment of PNH and now allows for prolonged intervals between treatment for patients.[426] Multiple other C5 antibodies, targeting distinct epitopes from those recognized by eculizumab and ravulizumab, have been (or are) in clinical trials, and may especially benefit patients that are refractory to eculizumab and ravulizumab therapy due to their specific disease inducing *C5* gene mutations. These are exciting developments, but they have also brought to light the limitations of antibody-based therapies for practical and broad use in many patients: for example, eculizumab treatment is extremely expensive (it can exceed ~$500,000 USD/year per patient),[427] needs to be applied frequently and intravenously in a hospital setting, and increases the patient's susceptibility to severe infections with encapsulated bacteria as it blocks complement activation broadly and not just at the desired site of inflammation.[71] Thus, to overcome these serious drawbacks, other strategies to therapeutically inhibit C5 are being investigated, including C5-binding peptides that inhibit C5 activation/cleavage for alternative treatments of PNH and aHUS and RNAi therapeutics targeting hepatic C5 production in PNH and aHUS, as well as an RNA aptamer that inhibits C5 expression and is tested currently in AMD.[428]

Blockage of C5 activation does not prevent C3 cleavage and C3 opsonization of surfaces which can be a disease driver as well. Thus, there has also been put much thought and effort into the generation of therapeutics targeting C3 or the C3/C5 convertases. Both pegylated and nonpegylated compstatins,[429] cyclic peptides that inhibit C3 cleavage independent of the initiating pathway,[430] are being clinically tested in a wide spectrum of disorders including PNH, hemodialysis-induced inflammation, periodontal disease, autoimmune hemolytic anemias, AMD, and C3G.[428] Moreover, soluble (partial) CD35 proteins that accelerate the decay of C3/C5 convertases are currently also lined up for clinical trials.[431]

While initial focus of the pharmaceutical industry was on targeting downstream C3 and C5 activation common to all activation pathways to reduce/modulate the net outcome of complement activation, there are now a growing number of therapeutics under development that inhibit or alter specific complement activation pathways. The advantage of a more nuanced targeting of complement is that some of its host protective activity can be retained through the other nondysregulated pathways and may reduce incidences of infection. The C1-INHs are the only currently clinically approved complement therapeutics in this group and are given to patients with hereditary angioedema. As they target specifically the CP and LP via blockage of C1 and MASPs, they are now also evaluated for efficacy in a wider range of disorders characterized by inappropriate CP/LP engagement, including stroke, trauma, antibody-mediated kidney transplant rejection, multiorgan dysfunction, and kidney ischemia-reperfusion injury. Importantly, C1-INHs also reduce activity of other serine proteases, for example, several belonging to the coagulation system's kinin and fibrinolytic pathways, and thus may have undesired side effects.[432] Therefore, more specific CP and LP inhibitors such as a monoclonal antibody only inhibiting C1s protease activity are under development. A major focus of therapeutic modulation has always been the AP, because an unleashed amplification loop fuels many/most complement-connected disorders.[35] As FB and FD are necessary for formation of the C3 convertase of the AP, clinical testing of small molecule inhibitors, antibodies, or antisense RNAs targeting these two proteins in C3G and aHUS are underway. Furthermore, inhibition of the AP C3 convertase stabilizer properdin is also an avenue that is currently explored by the pharmaceutical industry for treatment of AMD.[428]

In addition to targeting initiators and/or propagators of the complement cascades, a second range of therapeutic agents focuses on the controlled modulation of complement-induced signaling events. For example, antibodies that block binding of C5a to the C5aR receptors or small molecule antagonists of the C5aR1 receptor are being tested in a range of conditions including aHUS, IgA nephropathy, and sepsis or multiorgan dysfunction. Also, in solid tumors, an antibody against C5aR1 that blocks C5aR1 signaling is currently evaluated in combination with anti–programmed cell death ligand 1 (PDL-1) therapy[428] (Table 13.5). Of course, agonists to C5aR1 bear the hope to increase effector function of immune cells in patients with complement deficiencies and severe recurrent infections.[433] Similarly, new reagents that enhance complement activation/target fixation in an attempt to combat the increasing number of life-threatening hospital-acquired infections with *Pseudomonas aeruginosa* in immune compromised patients are a major research target.[428] Successful reagents increasing complement activity in a controlled fashion could also inform on the development of therapeutics that aim at augmenting complement-dependent cytotoxicity of tumor cells.[428]

While the development and testing of complement-based therapeutics have increased exponentially over the last decade and brought several successes, substantial challenges remain, including the high cost of many antibodies, frequent visits to hospitals for treatment, and the current inability to target reagents to desired tissues, for example, the brain to treat AD. Major leaps in technology development and recent advancements in suitable preclinical models allow tackling some of these problems. For example, the current development and testing of peptide and small molecule inhibitors, which are less expensive and easier to manufacture, will not only decrease the associated cost burden substantially but also may be simpler to administer and hence be more practical for the treatment of more patients across the globe. Furthermore, some of these smaller therapeutics cross the blood–brain barrier more easily and have opened the door to start thinking about targeting neurological disorders with complement-based therapeutics. Finally, as more as we learn about the different localizations and functions of complement activity and specifically its intense cross-talks with other PRR and cell effector systems (see below), we will be able to fine-tune complement medications for highest effect on disease phenotype with lowest adverse events.

EMERGING NEW AREAS OF COMPLEMENT RESEARCH

The prevailing opinion for many years was that complement biology was fully explored and that we know what is to know about this system. However, this is clearly not the case and there are still many surprising discoveries in store about this ancient system. Particularly, the discovery of new locations and modes of activation (intracellular) and novel unexpected noncanonical activities, a deeper look into the so-called complement hybrid proteins, and complement's emerging role as a central player in cancer and the crosstalk with other cell effector system has revived interest and created a new appreciation for the complement system as a whole. Here, we will briefly showcase some of the emerging novel areas of complement research (Fig. 13.9).

Intracellular Complement

It has long been appreciated that there are many extrahepatic sources of complement[434] and a large body of work now suggests that local complement production by immune cells is a critical modulator of their function. Particularly, the work aimed at understanding the impact of complement on T cell immunity has led to three surprising findings. Firstly, direct complement receptor-mediated signaling is critically required for normal induction of Th1 responses but appears to affect Th2 and Th17 responses only indirectly (see above). Secondly, Th1 induction requires complement activation fragments generated by the APCs and/or T cell itself and is largely independent of serum-derived complement.[435] This was elegantly delineated *in vivo* using mouse transplant model in which C3-deficient kidneys were protected from T cell–mediated graft rejection when transplanted into a wild-type host, despite a normal systemic complement compartment.[279] During the cognate T cells/APC interaction, TCR activation and CD28 costimulation induce secretion of complement C3, C5, and FB and FD in both cells leading to extracellular C3 and C5 convertase formation and

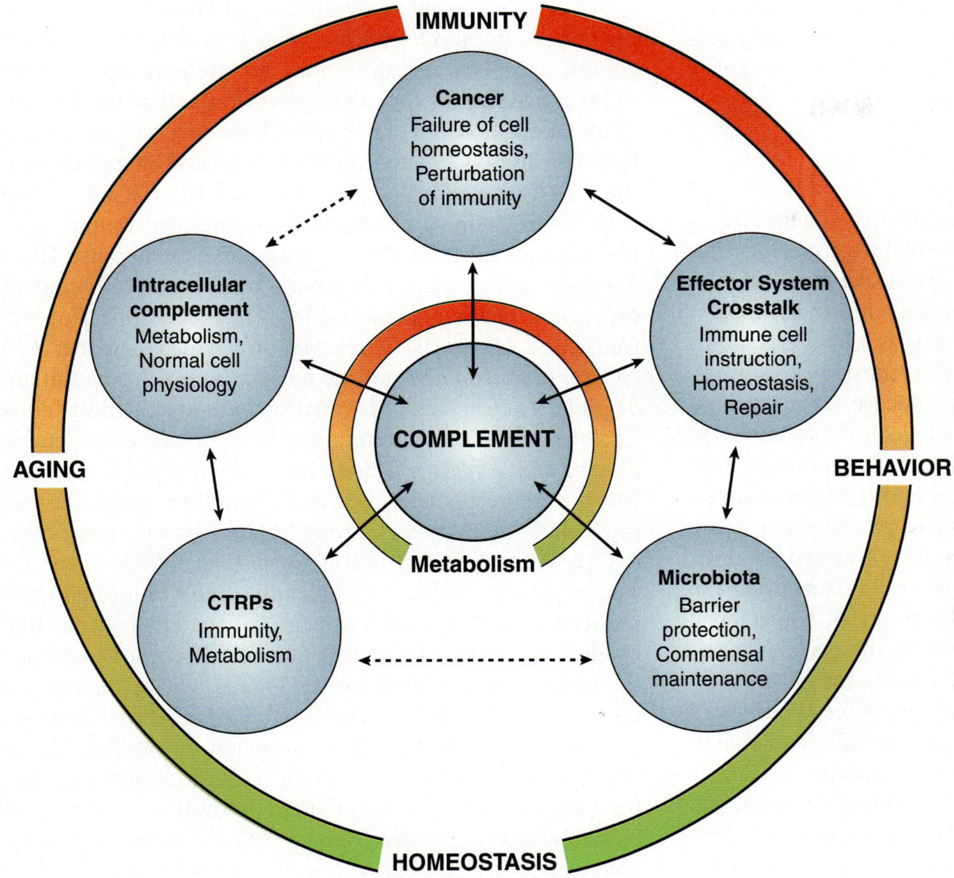

FIG. 13.9. Emerging areas in complement research. Shown are areas that have experienced a recent surge in interest from the complement field and beyond. The functional intersection between complement and other effector systems is multifactorial, bidirectional, and more extensive than previously thought. CTRPs, similar to intracellular complement, seem to mostly be involved in basic processes of the cell. The impact of complement in cancer biology is highly complex and can be tumorigenic or antitumorigenic and we are currently standing only at the beginning in our understanding of the complement and microbiota crosstalk. We suggest that the emerging role of complement in core physiological metabolic pathways may be one crucial functional intersection point in this network and, furthermore, that complement-mediated activities—via this network—contribute not only to normal immunity and host homeostasis but also to aging and behavior. CTRPs, C1q/tumor necrosis factor–related proteins.

the generation of the complement activation fragments. Thus, local immune cell–derived anaphylatoxins initiate C3aR- and C5aR1-driven signals for the specific production of IFN-γ and Th1 induction.[275] Importantly, in humans, normal Th1 biology is also reliant on intracellular complement activity. Human $CD4^+$ T cells express and store C3 intracellularly in the ER and the lysosomes continuously in the resting state. Portions of intracellular T cell–intrinsic C3 are cleaved by the ancient protease cathepsin L into C3a and C3b, and intracellular engagement of the C3aR expressed on lysosomes by C3a sustains homeostatic survival of circulating T cells through maintenance of basic cell metabolism (see next section).[66] TCR activation induces the shuttle of intracellularly generated C3a and C3b to the surface where they engage C3aR and CD46 on the cell surface to trigger Th1 lineage induction via metabolic reprogramming. Human $CD4^+$ T cells also express C5, generate intracellular C5a (the intrinsic intracellular C5a generating mechanism in T cells is not yet defined), and express intracellular C5aR1 and C5aR2.[282] During T cell stimulation, C5a engages intracellular C5aR1 which drives the ROS production that is needed in human $CD4^+$ T cells to support optimal IFN-γ secretion. Deviations in intracellular complement—coined the complosome—in T cells contribute to human disease states. For example, lack of C3 expression associates with reduced T cell survival[66] and Th1 induction in primary immune deficiencies[368,436] and augmented intracellular C3 activity contributes significantly to the hyperactive Th1 responses observed in RA, SLE, and scleroderma.[437-439] Excitingly, this system seems amendable to pharmaceutical intervention, as culturing T cells from the inflamed joints of RA patients with a cell permeable cathepsin L inhibitor normalize overactive Th1 responses *in vitro*.[66] Thus, an overall picture emerges in which the location of complement activation directs its activity: complement produced by the liver and released into (vascular) circulation protects against invading pathogens via direct tagging and lytic killing of microbes, while complement produced and secreted locally by immune cells drives their effector function, and intracellular complement emerges as direct regulator of biochemical reactions, currently mostly of metabolic nature.[68] The complosome is not confined to T cells but has been observed so far in all cells analyzed, indicating that it is of broad physiological importance.[66] The existence of the complosome is exciting, particularly because it may represent a novel target for complement-aimed therapeutics. However, this is a very recent discovery and therefore we currently only have limited understanding of the molecular mechanisms and pathways underlying its biology.[440] For example, we know little about the exact cellular subcompartments that "house" C3, C5, activating enzymes, and receptors, we do not know how the complosome is regulated, and whether it functionally engages with extracellular complement[441] and/or other PRR systems. Answering such key question will help us to better define its exact roles in health and to possibly, in the future, modulate complosome activity at will to ameliorate disease.

Single Cell Metabolism

The discovery of an intracellular complement system and its key role in cell metabolism is likely one of the reasons for complement's impressive and seemingly ever-expanding range of effector functions. The idea that complement activity impacts on the host's metabolic balance is not new as it has been shown already in the 1980s that serum complement C3a-desArg regulates whole body lipid metabolism, glucose transport, and energy usage.[442] In consequence, pathological deviations in plasma complement activities are associated with several metabolic disease states including diabetes and fatty liver disease, etc.[443] It was, however, the deeper insights gained when studying the activity of the complosome that demonstrated how intimately complement is involved in single cell metabolism. The metabolic machinery of immune cells comprises its metabolic enzymes and/or pathways (basal signature), the available nutrient sensors and metabolic checkpoint kinases (adaptation signature), and the epigenetic reprogramming of metabolic genes (memory transcriptional signature). Polarized differentiation of immune cells after stimulation by distinct activation or PAMP/DAMP signals drives cell-specific remodeling of metabolic pathways.[444,445] Recent years have given us particularly deep insights into the metabolic pathways underlying T cell lineage induction.[446] Because the human-specific complement regulator and receptor CD46 emerges here again as a central player,[277] it is important to understand that there are clear species-specific differences in complement-mediated signaling pathways underlying cell physiology, and that this should be taken into consideration when studying complement in animal models.[67] CD46's role in T cells lies in the direct coordination of a number of key metabolic events: CD46 drives gene transcription of *SLC2A1*, encoding the glucose transporter GLUT-1, of *SLC7A5*, encoding the amino acid transporter 1 (LAT-1) and therefore allows for the nutrient influx sustaining cell proliferation and induction of effector function.[277] At the same time, CD46 regulates the expression as well as activation state of enzymes (such as GAPDH or fatty acid synthase) that are needed to metabolize incoming nutrients and turn them into fuel for cell activity (Fig. 13.6A). Further, CD46 engages the oxygen metabolism machinery of the cell during activation via increasing intrinsic C5a generation and intracellular C5aR1 signaling which drives ROS production.[282] Finally, CD46 controls the assembly of the cell's prime nutrient sensor mTORC1 at the lysosomes via expression regulation of the mTOR scaffolding component LAMTOR5. Together, these events drive the high levels of glycolysis and OXPHOS particularly needed for Th1 induction. T cells from patients deficient in CD46 fail to induce these events and cannot produce IFN-γ.[277] Importantly, the Th1 contraction program is also initiated by CD46-controlled metabolic pathways: CD46-mediated signals induce changes in the T cell–intrinsic cholesterol biosynthesis pathway that allow ultimately for the co-induction of IL-10 and cessation of Th1 activity.[300]

An exciting recent study further solidifies the notion that the noncanonical roles of complement in controlling the single cell metabolic repertoire may be extensive and of broad significance. In this work, the authors found that uncontrolled C5aR1 activation is a driver of the lysosomal storage disease, Gaucher disease.[447] This condition is caused by mutations in the *GBA1* gene encoding glucocerebrosidase (GCase), which leads to excessive accumulation of glucosylceramide (GC) in immune cells and chronic inflammation because GC is not degraded. C5aR1 engagement on immune cells controls expression of the enzyme UDP-glucose ceramide glucosyltransferase (supports GC formation), which, together with GCase, maintains the normal balance of GC formation and degradation. Chronic inflammation and unwanted sustained C5aR1 stimulation on immune cells in patients with Gaucher disease hence fuels a vicious cycle of cellular GC accumulation and perpetuates disease.[447] Another molecule with intracellular activity and strong bonds with cell metabolism is C1q. Apart from its well-defined role as PRR of the CP, C1q is present within cells and the gC1qR is located on/in mitochondria.[448] Mitochondrial gC1qR activity controls ROS production by these organelles in cortical neurons[449] and also the balance of glycolysis versus OXPHPOS in human and mouse CD8$^+$ T cells.[204,440] The emergence of complement as a key player in metabolism also conspicuously parallels the rise of other ancient PRRs, such as the TLRs and inflammasome from mere pathogen detectors to orchestrators of normal cell physiology. Its heavy involvement in basic biochemical pathway regulation aligns also well with a new vantage point about the evolutionary roots of complement as an originally "single cell" system recognizing and rectifying metabolic stress and an evolutionary later development into a systemic immune intravascular guardian as we know it today.[68,95,450] It would be interesting to address if systemic and local metabolic changes in the host are integrated by complement-dependent mechanisms at the cellular level.

TNF/C1q-Related Proteins

Somewhat serendipitously, the family of C1q/TNF-related proteins (CTRPs) is now also receiving much attention. This is in large parts because CTRPs may form another novel (complement-controlled) link between immunity and metabolism as they represent proteins of dual function at this important interface. At least 15 CTRPs have been described to date,[451] of which many are secreted by adipose tissues. CTRPs assemble into homotrimeric forms and most members display metabolic functions. For example, CTRPs 2, 5, 6, 9, and 13 all activate the 5′ AMP-activated protein kinase (AMPK) pathway and/or induce acetyl coenzyme A carboxylase (ACC),[452] both essential to fatty acid oxidation, in several cell types. In addition, CTRPs 1, 3, and 9 also trigger MAPK and protein kinase B (PKB, aka AKT) activation which are key upstream events for normal mTOR activity. Intracellular expression of CTRP-3 drives OXPHOS and via this supports protein translation and mitochondrial ROS production in smooth muscle cells in an autocrine fashion.[453] Also, and similar to C1q, intracellular CTRP-3 can bind directly to mitochondrial protein cyclophilin

D and protect cells from the detrimental effects of oxidative stress.[453,454] CTRP-8 interacts with the relaxin/insulin-like family peptide receptor 1 (RXFP1) and induces signals that support proliferative expansion of glioblastoma cells.[455] Recent work indicates that CTRP-9 may play a role in the development of ischemic stroke, hypertension, and/or more general coronary atherosclerosis disease (CAD) because it is important in whole body energy metabolism regulation, control of platelet activation and foam cell generation, and protection of the endothelium via engagement with adiponectin receptor 1 (AdipoR1).[456] Of note, the underlying molecular pathways driving the biological effects of CTRPs are often not explored as this field is also only recently gaining momentum. In fact, the receptors for some but not all CTRPs have been defined yet. Also, most work on CTRPs has so far focused on the C1q portion, and it will therefore be important to also inquire about the effects of the TNF domain in the observations made and to dissect the contributions of each hybrid protein partner. In all, this is a young but important area to watch and we are expecting many exciting new insights with strong implications for human (metabolic) diseases in the future.

Cancer

A growing body of evidence demonstrates that complement plays a complex and important role in cancer biology and hence large research efforts focus on this important area. For many years, research centered on the fact that almost all cancers upregulate complement regulators, likely in an effort to evade complement attack triggered by their display of DAMPs.[48] Thus, with regards to therapeutic efforts, most strategies initially aimed at inducing complement-mediated tumor lysis by directing/increasing complement activation fragment deposition through antibody sensitization (antibody-mediated cytotoxicity).[457,458] In both humans and mice, almost all complement genes and proteins (not just regulators) display strongly altered expression within tumors compared to the healthy surrounding or "source" tissue.[208,459,460] Particularly, changes in the expression of genes encoding proteins of the CP and AP have been implicated in tumor progression. In principle, this makes sense as chronic inflammation is a major driver of malignant transformation and the role of complement in inflammation is undisputed.[461] In line with this, C3 deficiency in mice halts tumor progression in almost all malignancy models tested.[47] Importantly though, such protection is not simply due to absence of inflammation: C3a and/or C5a have been shown to alter the tumor microenvironment and the trafficking and function of immune cells to/within the tumor to the benefit of the cancer. For example, tumor-derived C5a recruits myeloid-derived suppressor cells (MDSCs) which dampen antitumor $CD4^+$ and $CD8^+$ T cell responses, increase the local generation of macrophages of the M2 or protumorigenic phenotype, which in turn slow the recruitment of NK cells, $CD4^+$ T cells, and neutrophils.[462] Similarly, tumor-derived C3a can modulate the activity of tumor-associated macrophages (TAMs) to promote the tumor microenvironment. Importantly, this effect could not be achieved by provision of exogenous C3a, demonstrating that, similar to the finding in human T cells, tumor-intrinsic and intracellularly active complement serves roles distinct from its serum-derived counterparts.[463] The mechanism by which intracellular C3a supports tumor expansion is unclear. However, given that intracellular C3a is a driver of T cell metabolism and metabolic changes are a hallmark of every cancer, it is tempting to speculate that altered complosome activity may underly the metabolic adaptions that fuel cancer growth at least in some cases. Overall, cancer is a prime example of how the nature/source of the transformed cell, the specific location of the ensuing tumor (solid vs leukemias), and its microenvironment greatly influence specific immune response and ultimately disease progression and outcome. Given the equally complex and diverse roles and effects of complement activation and specific locations (systemic, (immune) cell-derived, and intracellular), it is therefore not surprising that the effects of complement in a given cancer are highly diverse and range from antitumorigenic and/or antimetastatic (attempt of lysis, recruitment of tumor-fighting immune cells) to tumor-supportive (inducing neoangiogenesis, supporting tumor proliferation, immune cell suppression, etc). In fact, a recent study that analyzed several publicly available cancer databases revealed that some human cancers could possibly be stratified according to their distinct complement gene expression profiles and that these may correlate with disease outcome.[48] Thus, the in-depth analyses of data from large and well-defined patient cohorts with sophisticated and stringent bioinformatics methods may hold the key to identify complement-specific disease signatures in different cancers and ultimately inform which patients may benefit most from what type of complement-based therapeutics. Excitingly, complement seems indeed a viable therapeutic target here as the combinational treatment of mice with checkpoint inhibitors (anti–PD-1 antibody RMP-14) and an L-aptamer to block C5aR1 significantly improved efficacy of anti–PD-1 in a mouse model of lung adenocarcinoma.[464] Of course, big data analyses must be combined with a better understanding of the exact roles of systemic, local, and intracellular complement in a specific cancer setting. Thus, there is much to do, but this fact is met in the field with excitement and the awareness that understanding the complex complement and cancer relationship will likely translate into real impact on human health.

Crosstalk With Other Effector or PRR Systems

None of the PRR systems work in isolation in the host, and this is not different for complement. In fact, we are only now beginning to understand the extensiveness of the crosstalk between complement and other immune regulators. Functional intersections between complement and the coagulation system and cell surface–expressed TLRs have long been recognized and are somewhat intuitive as the coagulation cascade is the other "big" blood effector system and TLRs serve similar central functions during immunity. However, with the recent discovery of intracellular complement activity and the increasing number of noncanonical functions, unexpected novel communications with effector systems have emerged.

The functional relationship between the complement and *coagulation systems* is intimate but, due to the composite nature of each system alone already, extremely complex. Both systems are evolutionarily ancient, engage structurally similar serine proteases and associated inhibitors for their functions, and aim at the same outcome—pathogen removal and tissue repair. Crosstalk is bidirectional and is characterized by synergistic and/or antagonistic effects and intricate positive and negative feedback loops.[465-467] For example, many proteases specific to the coagulation system also activate C3 and/or C5, and these include plasmin, thrombin, and factors IX, X, and XI.[468,469] C3a and/or C5a, on the other hand, promote tissue factor expression and activate both platelets and endothelial cells, resulting in endothelial cell secretion of von Willebrand factor, a major coagulation factor.[469] Further, recently it was shown that C4a binds to and activates the protease-activated receptors (PARs) 1 and 4, which are expressed by platelets and endothelial cells and drive their activation during inflammation.[386,470] The disease-relevant crosstalk between complement and the clotting cascade is underpinned by the thrombotic diathesis associated with complement dysregulation in several conditions such as PNH, aHUS, and sepsis and the remarkable response of patients to eculizumab.[465,471] An increase of the insights into the molecular mechanisms underlying the functional complement and coagulation intersections will thus deliver additional anchor points for novel therapeutic interventions for thrombotic conditions. The relationship between complement and another "long-known" functional partner system—namely antibodies—is also currently being reevaluated in more depth. It is now understood that the complement and antibody relationship is not limited to complement-mediated B cell help during antibody production and antibody-provided triggers for CP activation.[472,473] The engagement of the Fc domain of an antibody Ig subtype with its specific activating or inhibitory *Fc receptor* on an immune cell induces signaling events that dictate the activity of the signal receiving cell. Unwanted engagement of Fc receptors by autoantibodies underly a wide range of IC diseases such as SLE, glomerulonephritis, etc.[474] Recent exciting work identified the C5aR1 as a general regulator of the expression of the proinflammatory Fcγ receptor.[472] The finding that Fcγ receptor signaling drives the increased generation of local C5a[472,474] explains the devastating synergistic impact that both systems have on the host when unleashed without control. Thus, checks and balances must exist during the complement–Fc receptor crosstalk; this is indeed the case as the inhibitory Fc receptor FcRIIB, together with the C-type lectin receptor dectin-1, suppresses C5aR1 proinflammatory signals[475] and provides a means for negative control of C5a activities in certain settings. Understanding the effects of Fc receptors on complement and vice versa better is therefore currently actively researched by several groups.

Naturally, complement's functional intersection with *other PRRs* has always been an intense research area and we are making good progress in understanding the crossconnections between complement and *TLRs*. Agonists to TLR-2, 4, or 9 induce not only specific known TLR-mediated signaling events but also activate complement C3 via a not-yet-understood mechanism.[476,477] Further, such TLR-driven complement activation in turn affects the production of several proinflammatory cytokines in response to TLR stimulation in myeloid cells via C3aR and/or C5aR1/2 engagement.[478-484] CD46-driven signals restrain TLR-4 induced IL-12 production in monocytes, while C1q increases those and CR3 dampens TLR2 activity.[484] Thus, the specific outcome of the complement-TLR dialogue on the host's immune response is clearly context-dependent. It will be worthwhile to detangle this crosstalk though because aberrant complement and TLR signals combine to harmful effects in, for example, sepsis which remains a major health burden worldwide.[485] Indeed, efforts into tackling both systems in parallel (inhibiting C5 and the TLR co-receptor CD14) in an *E. coli* animal model of sepsis decreased inflammation, proinflammatory cytokine secretion, and thrombus formation.[486,487] Additionally, in a polymicrobial sepsis model, while single inhibition of C5 or CD14 did not enhance survival, combined inhibition of CD14 and C5 both enhanced survival and decreased inflammation, demonstrating the benefit of a combinational therapeutic approach.[485,488,489] These are exciting developments particularly since complement and TLRs also cooperate during sterile inflammation[490] and may therefore also prove to be valuable combinational targets in cardiovascular disease.

A decade ago, few researchers would have argued that the *microbiota* is an immune effector system. Today, the microbiome is solidly acknowledged as an integral part of the development, education, and also sustained function of our immune system.[491,492] Although studies on functional interactions between complement and the microbiota are only beginning to appear, what we know so far strongly supports also a bidirectional crosstalk here: the inhibition of C5aR1 activity in the skin of mice reduces the diversity and alters the composition of their cutaneous microbiota. Conversely, the commensal skin microbiota regulates actively the expression of several complement genes in skin cells.[493] An interesting triangular crosstalk between TLRs, complement, and a pathogen that leads to mucosal dysbiosis has also recently been described. The periodontitis-associated bacterium *Porphyromonas gingivalis* manipulates TLR-2 and complement signaling in the oral cavity in such a way that *P. gingivalis* can thrive at the cost of the local healthy microbiota. This pathogenic evasion mechanism can be intercepted by a C5aR antagonist suggesting that complement may be a suitable target in the widespread human disease periodontitis.[494] Of note, the intestinal microbiota is central to the normal dietary fat absorption by gut epithelial cells and thus maintains the host's normal energy balance. Given that (intracellular) complement is shaping intestinal barrier function[495] and drives nutrient influx and usage at the single cell level,[67] we predict that local/intracellular complement may also partake in this process.

There is now also accumulating evidence that complement evolved a tight functional connection (with activating and inhibitory circuits) with intracellular PRRs, such as the *inflammasomes, the RLR systems, and the mitochondrial antiviral signaling protein (MAVS)*. For example, C1q binding to the gC1qR on myeloid cells triggers the MDA5 pathway and induces protective IFN-α production.[496] Further, optimal NLRP3-mediated IL-1β production by human monocytes and macrophages requires combined TLR and C3aR and C5aR1 signals[210,211,233] and human $CD4^+$ T cells rely on intracellular C5aR1 stimulation for intrinsic NLRP3 assembly and normal Th1 induction.[282,497] Similarly, during certain viral infections, sublytic MAC deposition induces IL-1β and IL-18 in a NLRP3 inflammasome-dependent manner in mouse dendritic and lung epithelial cells,[315,498] C1q can synergize with PTX3 to promote NLPR3 activation in monocytes and aggravate arthritis,[499] and CTRP9 inhibits NLRP3 on monocytes and may ameliorate atherosclerosis.[500] Sensing of cytosolic C3 activation fragments may also activate mitochondrial-associated PRRs: C3 fragments that have been introduced into the cell via a complement-opsonized pathogen activate the MAVS pathway and trigger proinflammatory cytokine production. Furthermore, complement can support cell-autonomous immunity also by inducing autophagy-dependent control of cyto-invasive bacteria.[501,502] Listed here are only few examples about the buoying field of complement and RPP crosstalk. Although the functional significance of the crosstalk between the complosome and other effector systems is evident for the protection against infection, these newly identified roles of complement in the regulation of basic cellular processes and its role as a nexus during PRR system communications (Fig. 13.9) suggest to reevaluate also a range of autoimmune and chronic inflammatory diseases for dysregulation of pathways and networks discussed here.

In summary, research on complement remains a vibrant field since its initial inception as canonical pathogen killer in the last century. This is largely due to the fact that complement's impact on the host's immune system is more far-reaching than previously thought and that it now emerges as a sophisticated and critical orchestrator of normal cell behavior. We, thus, expect future additional exciting and unexpected discoveries about this "old dog" of immunology and also invite the readers to keep an eye out for a complement/complosome signature in their own studies.

ACKNOWLEDGMENTS

We thank the countless complementologists whose work over the last century has contributed to our better understanding of the roles of complement in health and disease. In December 2020, a giant in our field, Sir Peter Lachmann (the co-discoverer of FI and also known as the 'father of the amplification loop') passed away. We dedicate this chapter to him. We apologize to those whose work we have not been able to display or cite due to space limitations. Work in the Kemper lab and CIRS is supported by the Intramural Research Program of the National Institutes of Health (project number zia/hl006223).

REFERENCES

1. Nuttal G. Experimente uber die bacterienfeindlichen Einfluss des thierischen Korpers. *Zeitschrift fur Hygiene und Infektionskrankheiten*. 1888;4:353-394.
2. Buchner H. Zur Nomenklatur der schutzenden Eiweisskorper. *Centr Bakteriol Parasitenk*. 1891;10:699-701.
3. Gordon S. Phagocytosis: the legacy of Metchnikoff. *Cell*. 2016;166:1065-1068.
4. Bordet J, Gengou O. Sur l'existence de substances sensibilisatrices dans la plupart des serums antimicrobiens. *Ann De l'Inst Pasteur*. 1901;15:289-303.
5. Ehrlich PM J. Zur theorie der Lysenwirkung. *Berlin Klin Woch*. 1899;36:6-9.
6. Nesargikar PN, Spiller B, Chavez R. The complement system: history, pathways, cascade and inhibitors. *Eur J Microbiol Immunol (Bp)*. 2012;2:103-111.
7. Shokal U, Eleftherianos I. Evolution and function of thioester-containing proteins and the complement system in the innate immune response. *Front Immunol*. 2017;8:759.
8. Merle NS, Singh P, Rahman J, Kemper C. Integrins meet complement: the evolutionary tip of an iceberg orchestrating metabolism and immunity. *Br J Pharmacol*. 2021;178(14):2754-2770.
9. Witebsky E. Ehrlich's side-chain theory in the light of present immunology. *Ann N Y Acad Sci*. 1954;59:168-181.
10. Ferrata A. Die Unwirksamkeit der komplexen Haemolysine in salzfeien Loesungen und ihre Ursache. *Berlin Klin Woch*. 1907;44:366.
11. Brand E. Ueber das Verhalten der Komplementr bei der Dialyse. *Berlin Klin Woch*. 1907;44:1075.
12. Bronfenbrenner J, Noguchi H. A biochemical study of the phenomena known as complement-splitting. First paper: splitting of the complement associated with globulin precipitation. *J Exp Med*. 1912;15:598-624.
13. Whitehead HR, Gordon J, Wormall A. The "third component" or heat-stable factor of complement. *Biochem J*. 1925;19:618-625.
14. Gordon J, Whitehead HR, Wormall A. The fourth component of complement and its relation to opsonin. *Biochem J*. 1926;20:1044-1045.
15. von Dungern E. Beitrage zur immunitatslehre. *Munchener Med Wochenschr*. 1900;47:677-680.
16. Coca AF. A study of the anticomplementary action of yeast, of certain bacteria, and of cobra venom. *Z Immunitätsforschung*. 1914;21:604-622.
17. Omorokow Z. über die Wirkung des Cobragiftes auf die Komplemente. II. Mittlg. *Z Immunitätsforschung*. 1911;11:710.
18. Pillemer L, Ecker EE. The terminology of the components of complement. *Science*. 1941;94:437.
19. Pillemer L, Seifter S, Ecker EE. The role of the components of complement in specific immune fixation. *J Exp Med*. 1942;75:421-435.
20. Mueller-Eberhard HJ, Biro CE. Isolation and description of the fourth component of human complement. *J Exp Med*. 1963;118:447-466.
21. Nilsson UR, Mueller-Eberhard HJ. Isolation of beta if-globulin from human serum and its characterization as the fifth component of complement. *J Exp Med*. 1965;122:277-298.
22. Nilsson U. Separation and partial purification of the sixth, seventh and eighth components of human haemolytic complement. *Acta Pathol Microbiol Scand*. 1967;70:469-480.
23. Manni JA, Muller-Eberhard HJ. The eighth component of human complement (C8): isolation, characterization, and hemolytic efficiency. *J Exp Med*. 1969;130:1145-1160.
24. Hadding U, Muller-Eberhard HJ. The ninth component of human complement: isolation, description and mode of action. *Immunology*. 1969;16:719-735.
25. World Health Organization. Nomenclature of complement. *Bull World Health Organ*. 1968;39(6):935-938.
26. Lepow IH, Naff GB, Todd EW, Pensky J, Hinz CF. Chromatographic resolution of the first component of human complement into three activities. *J Exp Med*. 1963;117:983-1008.
27. Nelson RA Jr, Jensen J, Gigli I, Tamura N. Methods for the separation, purification and measurement of nine components of hemolytic complement in guinea-pig serum. *Immunochemistry*. 1966;3:111-135.
28. Pepys MB. Role of complement in induction of the allergic response. *Nat New Biol*. 1972;237:157-159.
29. Weis JJ, Fearon DT, Klickstein LB, et al. Identification of a partial cDNA clone for the C3d/Epstein-Barr virus receptor of human B lymphocytes: homology with the receptor for fragments C3b and C4b of the third and fourth components of complement. *Proc Natl Acad Sci U S A*. 1986;83:5639-5643.
30. Ho MK, Springer TA. Mac-2, a novel 32,000 Mr mouse macrophage subpopulation-specific antigen defined by monoclonal antibodies. *J Immunol*. 1982;128:1221-1228.
31. Roglic A, Prossnitz ER, Cavanagh SL, Pan Z, Zou A, Ye RD. cDNA cloning of a novel G protein-coupled receptor with a large extracellular loop structure. *Biochim Biophys Acta*. 1996;1305:39-43.
32. Gerard NP, Gerard C. The chemotactic receptor for human C5a anaphylatoxin. *Nature*. 1991;349:614-617.
33. Seya T, Turner JR, Atkinson JP. Purification and characterization of a membrane protein (gp45-70) that is a cofactor for cleavage of C3b and C4b. *J Exp Med*. 1986;163:837-855.
34. Reis ES, Mastellos DC, Hajishengallis G, Lambris JD. New insights into the immune functions of complement. *Nat Rev Immunol*. 2019;19:503-516.
35. Merle NS, Church SE, Fremeaux-Bacchi V, Roumenina LT. Complement system part I – molecular mechanisms of activation and regulation. *Front Immunol*. 2015;6:262.
36. Merle NS, Noe R, Halbwachs-Mecarelli L, Fremeaux-Bacchi V, Roumenina LT. Complement system part II: role in immunity. *Front Immunol*. 2015;6:257.
37. Dunkelberger JR, Song WC. Complement and its role in innate and adaptive immune responses. *Cell Res*. 2010;20:34-50.
38. Walport MJ. Complement. Second of two parts. *N Engl J Med*. 2001;344:1140-1144.
39. Sim RB, Schwaeble W, Fujita T. Complement research in the 18th-21st centuries: progress comes with new technology. *Immunobiology*. 2016;221:1037-1045.
40. Botto M. C1q knock-out mice for the study of complement deficiency in autoimmune disease. *Exp Clin Immunogenet*. 1998;15:231-234.
41. Kohl J. Self, non-self, and danger: a complementary view. *Adv Exp Med Biol*. 2006;586:71-94.
42. Hourcade D, Holers VM, Atkinson JP. The regulators of complement activation (RCA) gene cluster. *Adv Immunol*. 1989;45:381-416.
43. Liszewski MK, Atkinson JP. Complement regulators in human disease: lessons from modern genetics. *J Intern Med*. 2015;277:294-305.
44. Daha MR. Role of complement in innate immunity and infections. *Crit Rev Immunol*. 2010;30:47-52.
45. Wong EKS, Kavanagh D. Diseases of complement dysregulation-an overview. *Semin Immunopathol*. 2018;40:49-64.
46. Holers VM. Complement and its receptors: new insights into human disease. *Annu Rev Immunol*. 2014;32:433-459.
47. Reis ES, Mastellos DC, Ricklin D, Mantovani A, Lambris JD. Complement in cancer: untangling an intricate relationship. *Nat Rev Immunol*. 2018;18:5-18.
48. Roumenina LT, Daugan MV, Petitprez F, Sautes-Fridman C, Fridman WH. Context-dependent roles of complement in cancer. *Nat Rev Cancer*. 2019;19:698-715.
49. Schroder-Braunstein J, Kirschfink M. Complement deficiencies and dysregulation: pathophysiological consequences, modern analysis, and clinical management. *Mol Immunol*. 2019;114:299-311.
50. Alper CA, Johnson AM, Birtch AG, Moore FD. Human C'3: evidence for the liver as the primary site of synthesis. *Science*. 1969;163:286-288.
51. Fey G, Colten HR. Biosynthesis of complement components. *Fed Proc*. 1981;40:2099-2104.
52. Tenner AJ, Volkin DB. Complement subcomponent C1q secreted by cultured human monocytes has subunit structure identical with that of serum C1q. *Biochem J*. 1986;233:451-458.
53. Morgan BP, Gasque P. Extrahepatic complement biosynthesis: where, when and why? *Clin Exp Immunol*. 1997;107:1-7.
54. Fischer MB, Ma M, Hsu NC, Carroll MC. Local synthesis of C3 within the splenic lymphoid compartment can reconstitute the impaired immune response in C3-deficient mice. *J Immunol*. 1998;160:2619-2625.
55. Kolev M, Le Friec G, Kemper C. Complement—Tapping into new sites and effector systems. *Nat Rev Immunol*. 2014;14:811-820.
56. Kemper C, Chan AC, Green JM, Brett KA, Murphy KM, Atkinson JP. Activation of human $CD4^+$ cells with CD3 and CD46 induces a T-regulatory cell 1 phenotype. *Nature*. 2003;421:388-392.
57. Ricklin D, Hajishengallis G, Yang K, Lambris JD. Complement: a key system for immune surveillance and homeostasis. *Nat Immunol*. 2010;11:785-797.
58. Kimura Y, Madhavan M, Call MK, et al. Expression of complement 3 and complement 5 in newt limb and lens regeneration. *J Immunol*. 2003;170:2331-2339.
59. Modinger Y, Loffler B, Huber-Lang M, Ignatius A. Complement involvement in bone homeostasis and bone disorders. *Semin Immunol*. 2018;37:53-65.
60. Carmona-Fontaine C, Theveneau E, Tzekou A, et al. Complement fragment C3a controls mutual cell attraction during collective cell migration. *Dev Cell*. 2011;21:1026-1037.
61. Veerhuis R, Nielsen HM, Tenner AJ. Complement in the brain. *Mol Immunol*. 2011;48:1592-1603.
62. Stevens B, Allen NJ, Vazquez LE, et al. The classical complement cascade mediates CNS synapse elimination. *Cell*. 2007;131:1164-1178.
63. Lambert JC, Heath S, Even G, et al. Genome-wide association study identifies variants at CLU and CR1 associated with Alzheimer's disease. *Nat Genet*. 2009;41:1094-1099.
64. Tenner AJ, Stevens B, Woodruff TM. New tricks for an ancient system: physiological and pathological roles of complement in the CNS. *Mol Immunol*. 2018;102:3-13.
65. Kumar V. The complement system, Toll-like receptors and inflammasomes in host defense: three musketeers' one target. *Int Rev Immunol*. 2019;38:131-156.
66. Liszewski MK, Kolev M, Le Friec G, et al. Intracellular complement activation sustains T cell homeostasis and mediates effector differentiation. *Immunity*. 2013;39:1143-1157.
67. Hess C, Kemper C. Complement-mediated regulation of metabolism and basic cellular processes. *Immunity*. 2016;45:240-254.
68. West EE, Kunz N, Kemper C. Complement and human T cell metabolism: location, location, location. *Immunol Rev*. 2020;295:68-81.

69. Morgan BP, Harris CL. Complement, a target for therapy in inflammatory and degenerative diseases. *Nat Rev Drug Discov.* 2015;14:857-877.
70. Harris CL, Pouw RB, Kavanagh D, Sun R, Ricklin D. Developments in anti-complement therapy; from disease to clinical trial. *Mol Immunol.* 2018;102:89-119.
71. Ricklin D, Mastellos DC, Reis ES, Lambris JD. The renaissance of complement therapeutics. *Nat Rev Nephrol.* 2018;14:26-47.
72. Patriquin CJ, Kuo KHM. Eculizumab and beyond: the past, present, and future of complement therapeutics. *Transfus Med Rev.* 2019;33:256-265.
73. Sarma JV, Ward PA. The complement system. *Cell Tissue Res.* 2011;343:227-235.
74. Bohlson SS, Garred P, Kemper C, Tenner AJ. Complement nomenclature-deconvoluted. *Front Immunol.* 2019;10:1308.
75. Barnum S, Schein T, eds. 2017. *The Complement Factsbook* 2nd ed. Academic Press.
76. Sharp TH, Boyle AL, Diebolder CA, Kros A, Koster AJ, Gros P. Insights into IgM-mediated complement activation based on *in situ* structures of IgM-C1-C4b. *Proc Natl Acad Sci U S A.* 2019;116:11900-11905.
77. Diebolder CA, Beurskens FJ, de Jong RN, et al. Complement is activated by IgG hexamers assembled at the cell surface. *Science.* 2014;343:1260-1263.
78. Naff GB, Pensky J, Lepow IH. The macromolecular nature of the first component of human complement. *J Exp Med.* 1964;119:593-613.
79. Lu J, Kishore U. C1 complex: an adaptable proteolytic module for complement and non-complement functions. *Front Immunol.* 2017;8:592.
80. Thielens NM, Tedesco F, Bohlson SS, Gaboriaud C, Tenner AJ. C1q: a fresh look upon an old molecule. *Mol Immunol.* 2017;89:73-83.
81. Kishore U, Reid KB. C1q: structure, function, and receptors. *Immunopharmacology.* 2000;49:159-170.
82. Ghebrehiwet B, Kandov E, Kishore U, Peerschke EIB. Is the A-chain the engine that drives the diversity of C1q functions? Revisiting its unique structure. *Front Immunol.* 2018;9:162.
83. Gaboriaud C, Thielens NM, Gregory LA, Rossi V, Fontecilla-Camps JC, Arlaud GJ. Structure and activation of the C1 complex of complement: unraveling the puzzle. *Trends Immunol.* 2004;25:368-373.
84. Reboul A, Thielens N, Villiers MB, Colomb MG. Structural changes in C4 produced by cleavage with C1-s. *FEBS Lett.* 1980;115:118-122.
85. Nagasawa S, Stroud RM. Cleavage of C2 by C1s into the antigenically distinct fragments C2a and C2b: demonstration of binding of C2b to C4b. *Proc Natl Acad Sci U S A.* 1977;74:2998-3001.
86. Haas PJ, van Strijp J. Anaphylatoxins: their role in bacterial infection and inflammation. *Immunol Res.* 2007;37:161-175.
87. Klos A, Tenner AJ, Johswich KO, Ager RR, Reis ES, Kohl J. The role of the anaphylatoxins in health and disease. *Mol Immunol.* 2009;46:2753-2766.
88. van Lookeren Campagne M, Wiesmann C, Brown EJ. Macrophage complement receptors and pathogen clearance. *Cell Microbiol.* 2007;9:2095-2102.
89. Roos D. Complement and phagocytes—A complicated interaction. *Mol Immunol.* 2015;68:31-34.
90. Janssen BJ, Christodoulidou A, McCarthy A, Lambris JD, Gros P. Structure of C3b reveals conformational changes that underlie complement activity. *Nature.* 2006;444:213-216.
91. Takata Y, Kinoshita T, Kozono H, et al. Covalent association of C3b with C4b within C5 convertase of the classical complement pathway. *J Exp Med.* 1987;165:1494-1507.
92. Vogt W, Schmidt G, Von Buttlar B, Dieminger L. A new function of the activated third component of complement: binding to C5, an essential step for C5 activation. *Immunology.* 1978;34:29-40.
93. Ikeda K, Sannoh T, Kawasaki N, Kawasaki T, Yamashina I. Serum lectin with known structure activates complement through the classical pathway. *J Biol Chem.* 1987;262:7451-7454.
94. Nonaka M. Evolution of the complement system. *Subcell Biochem.* 2014;80:31-43.
95. Kolev M, Kemper C. Keeping it all going-complement meets metabolism. *Front Immunol.* 2017;8:1.
96. Boehm T, Swann JB. Origin and evolution of adaptive immunity. *Annu Rev Anim Biosci.* 2014;2:259-283.
97. Kawasaki T, Etoh R, Yamashina I. Isolation and characterization of a mannan-binding protein from rabbit liver. *Biochem Biophys Res Commun.* 1978;81:1018-1024.
98. Garred P, Genster N, Pilely K, et al. A journey through the lectin pathway of complement-MBL and beyond. *Immunol Rev.* 2016;274:74-97.
99. Ma YJ, Skjoedt MO, Garred P. Collectin-11/MASP complex formation triggers activation of the lectin complement pathway—The fifth lectin pathway initiation complex. *J Innate Immun.* 2013;5:242-250.
100. Kuraya M, Ming Z, Liu X, Matsushita M, Fujita T. Specific binding of L-ficolin and H-ficolin to apoptotic cells leads to complement activation. *Immunobiology.* 2005;209:689-697.
101. Holmskov U, Thiel S, Jensenius JC. Collections and ficolins: humoral lectins of the innate immune defense. *Annu Rev Immunol.* 2003;21:547-578.
102. Genster N, Takahashi M, Sekine H, Endo Y, Garred P, Fujita T. Lessons learned from mice deficient in lectin complement pathway molecules. *Mol Immunol.* 2014;61:59-68.
103. Howard M, Farrar CA, Sacks SH. Structural and functional diversity of collectins and ficolins and their relationship to disease. *Semin Immunopathol.* 2018;40:75-85.
104. Jensen ML, Honore C, Hummelshoj T, Hansen BE, Madsen HO, Garred P. Ficolin-2 recognizes DNA and participates in the clearance of dying host cells. *Mol Immunol.* 2007;44:856-865.
105. Nakamura N, Nonaka M, Ma BY, et al. Characterization of the interaction between serum mannan-binding protein and nucleic acid ligands. *J Leukoc Biol.* 2009;86:737-748.
106. Ma YJ, Hein E, Munthe-Fog L, et al. Soluble collectin-12 (CL-12) is a pattern recognition molecule initiating complement activation via the alternative pathway. *J Immunol.* 2015;195:3365-3373.
107. Matsushita M, Fujita T. Activation of the classical complement pathway by mannose-binding protein in association with a novel C1s-like serine protease. *J Exp Med.* 1992;176:1497-1502.
108. Thiel S, Vorup-Jensen T, Stover CM, et al. A second serine protease associated with mannan-binding lectin that activates complement. *Nature.* 1997;386:506-510.
109. Terai I, Kobayashi K, Matsushita M, Fujita T. Human serum mannose-binding lectin (MBL)-associated serine protease-1 (MASP-1): determination of levels in body fluids and identification of two forms in serum. *Clin Exp Immunol.* 1997;110:317-323.
110. Iwaki D, Kanno K, Takahashi M, Endo Y, Matsushita M, Fujita T. The role of mannose-binding lectin-associated serine protease-3 in activation of the alternative complement pathway. *J Immunol.* 2011;187:3751-3758.
111. Cortesio CL, Jiang W. Mannan-binding lectin-associated serine protease 3 cleaves synthetic peptides and insulin-like growth factor-binding protein 5. *Arch Biochem Biophys.* 2006;449:164-170.
112. Degn SE, Hansen AG, Steffensen R, Jacobsen C, Jensenius JC, Thiel S. MAp44, a human protein associated with pattern recognition molecules of the complement system and regulating the lectin pathway of complement activation. *J Immunol.* 2009;183:7371-7378.
113. Henriksen ML, Brandt J, Andrieu JP, et al. Heteromeric complexes of native collectin kidney 1 and collectin liver 1 are found in the circulation with MASPs and activate the complement system. *J Immunol.* 2013;191:6117-6127.
114. Walport MJ. Complement. First of two parts. *N Engl J Med.* 2001;344:1058-1066.
115. Kemper C, Mitchell LM, Zhang L, Hourcade DE. The complement protein properdin binds apoptotic T cells and promotes complement activation and phagocytosis. *Proc Natl Acad Sci U S A.* 2008;105:9023-9028.
116. Lesher AM, Nilsson B, Song WC. Properdin in complement activation and tissue injury. *Mol Immunol.* 2013;56:191-198.
117. Chen JY, Cortes C, Ferreira VP. Properdin: a multifaceted molecule involved in inflammation and diseases. *Mol Immunol.* 2018;102:58-72.
118. Morgan BP, Boyd C, Bubeck D. Molecular cell biology of complement membrane attack. *Semin Cell Dev Biol.* 2017;72:124-132.
119. Tegla CA, Cudrici C, Patel S, et al. Membrane attack by complement: the assembly and biology of terminal complement complexes. *Immunol Res.* 2011;51:45-60.
120. Moreno-Hagelsieb G, Vitug B, Medrano-Soto A, Saier MH Jr. The membrane attack complex/perforin superfamily. *J Mol Microbiol Biotechnol.* 2017;27:252-267.
121. Menny A, Serna M, Boyd CM, et al. CryoEM reveals how the complement membrane attack complex ruptures lipid bilayers. *Nat Commun.* 2018;9:5316.
122. Lukoyanova N, Hoogenboom BW, Saibil HR. The membrane attack complex, perforin and cholesterol-dependent cytolysin superfamily of pore-forming proteins. *J Cell Sci.* 2016;129:2125-2133.
123. Hadders MA, Bubeck D, Roversi P, et al. Assembly and regulation of the membrane attack complex based on structures of C5b6 and sC5b9. *Cell Rep.* 2012;1:200-207.
124. Bubeck D, Roversi P, Donev R, Morgan BP, Llorca O, Lea SM. Structure of human complement C8, a precursor to membrane attack. *J Mol Biol.* 2011;405:325-330.
125. Tschopp J. Ultrastructure of the membrane attack complex of complement. Heterogeneity of the complex caused by different degree of C9 polymerization. *J Biol Chem.* 1984;259:7857-7863.
126. Sharp TH, Koster AJ, Gros P. Heterogeneous MAC initiator and pore structures in a lipid bilayer by phase-plate cryo-electron tomography. *Cell Rep.* 2016;15:1-8.
127. Lakkaraju A, Toops KA, Xu J. Should I stay or should I go? Trafficking of sub-lytic MAC in the retinal pigment epithelium. *Adv Exp Med Biol.* 2014;801:267-274.
128. Heesterbeek DA, Bardoel BW, Parsons ES, et al. Bacterial killing by complement requires membrane attack complex formation via surface-bound C5 convertases. *EMBO J.* 2019;38.
129. Voskoboinik I, Whisstock JC, Trapani JA. Perforin and granzymes: function, dysfunction and human pathology. *Nat Rev Immunol.* 2015;15:388-400.
130. Kumar-Singh R. The role of complement membrane attack complex in dry and wet AMD—From hypothesis to clinical trials. *Exp Eye Res.* 2019;184:266-277.
131. Kirschfink M, Borsos T. Binding and activation of C4 and C3 on the red cell surface by non-complement enzymes. *Mol Immunol.* 1988;25:505-512.
132. Minta JO, Man D, Movat HZ. Kinetic studies on the fragmentation of the third component of complement (C3) by trypsin. *J Immunol.* 1977;118:2192-2198.
133. Gardinali M, Padalino P, Vesconi S, et al. Complement activation and polymorphonuclear neutrophil leukocyte elastase in sepsis. Correlation with severity of disease. *Arch Surg.* 1992;127:1219-1224.
134. Wiggins RC, Giclas PC, Henson PM. Chemotactic activity generated from the fifth component of complement by plasma kallikrein of the rabbit. *J Exp Med.* 1981;153:1391-1404.
135. Kanse SM, Gallenmueller A, Zeerleder S, et al. Factor VII-activating protease is activated in multiple trauma patients and generates anaphylatoxin C5a. *J Immunol.* 2012;188:2858-2865.
136. Huber-Lang M, Sarma JV, Zetoune FS, et al. Generation of C5a in the absence of C3: a new complement activation pathway. *Nat Med.* 2006;12:682-687.
137. Lupu F, Keshari RS, Lambris JD, Coggeshall KM. Crosstalk between the coagulation and complement systems in sepsis. *Thromb Res.* 2014;133(suppl 1):S28-S31.
138. Keshari RS, Silasi R, Lupu C, Taylor FB Jr, Lupu F. *In vivo*-generated thrombin and plasmin do not activate the complement system in baboons. *Blood.* 2017;130:2678-2681.
139. Huber-Lang M, Denk S, Fulda S, et al. Cathepsin D is released after severe tissue trauma *in vivo* and is capable of generating C5a *in vitro*. *Mol Immunol.* 2012;50:60-65.
140. Perl M, Denk S, Kalbitz M, Huber-Lang M. Granzyme B: a new crossroad of complement and apoptosis. *Adv Exp Med Biol.* 2012;946:135-146.
141. Riesgo A, Maldonado M, Lopez-Legentil S, Giribet G. A proposal for the evolution of cathepsin and silicatein in sponges. *J Mol Evol.* 2015;80:278-291.
142. Hellman L, Thorpe M. Granule proteases of hematopoietic cells, a family of versatile inflammatory mediators—An update on their cleavage specificity, *in vivo* substrates, and evolution. *Biol Chem.* 2014;395:15-49.
143. van den Berg CW, Tambourgi DV, Clark HW, Hoong SJ, Spiller OB, McGreal EP. Mechanism of neutrophil dysfunction: neutrophil serine proteases cleave and inactivate the C5a receptor. *J Immunol.* 2014;192:1787-1795.
144. Bexborn F, Andersson PO, Chen H, Nilsson B, Ekdahl KN. The tick-over theory revisited: formation and regulation of the soluble alternative complement C3 convertase (C3(H2O)Bb). *Mol Immunol.* 2008;45:2370-2379.
145. Morgan BP, Harris CL, eds. *Complement Regulatory Proteins.* Academic Press; 1999.
146. Zipfel PF, Skerka C. Complement regulators and inhibitory proteins. *Nat Rev Immunol.* 2009;9:729-740.
147. Ricklin D, Reis ES, Mastellos DC, Gros P, Lambris JD. Complement component C3 – The "Swiss Army Knife" of innate immunity and host defense. *Immunol Rev.* 2016;274:33-58.
148. Lachmann PJ. The story of complement factor I. *Immunobiology.* 2019;224:511-517.
149. Medof ME, Nussenzweig V. Control of the function of substrate-bound C4b-C3b by the complement receptor Cr1. *J Exp Med.* 1984;159:1669-1685.
150. Fearon DT. Identification of the membrane glycoprotein that is the C3b receptor of the human erythrocyte, polymorphonuclear leukocyte, B lymphocyte, and monocyte. *J Exp Med.* 1980;152:20-30.
151. Fonseca MI, Chu S, Pierce AL, et al. Analysis of the putative role of CR1 in Alzheimer's disease: genetic association, expression and function. *PLoS One.* 2016;11:e0149792.
152. Kristensen T, Wetsel RA, Tack BF. Structural analysis of human complement protein H: homology with C4b binding protein, beta 2-glycoprotein I, and the Ba fragment of B2. *J Immunol.* 1986;136:3407-3411.
153. Parente R, Clark SJ, Inforzato A, Day AJ. Complement factor H in host defense and immune evasion. *Cell Mol Life Sci.* 2017;74:1605-1624.
154. Blom AM, Webb J, Villoutreix BO, Dahlback B. A cluster of positively charged amino acids in the C4BP alpha-chain is crucial for C4b binding and factor I cofactor function. *J Biol Chem.* 1999;274:19237-19245.
155. Thurman JM, Nester CM. All things complement. *Clin J Am Soc Nephrol.* 2016;11:1856-1866.
156. Meri S, Pangburn MK. Regulation of alternative pathway complement activation by glycosaminoglycans: specificity of the polyanion binding site on factor H. *Biochem Biophys Res Commun.* 1994;198:52-59.
157. Nonaka M, Kimura A. Genomic view of the evolution of the complement system. *Immunogenetics.* 2006;58:701-713.
158. Hoffmann EM. Inhibition of complement by a substance isolated from human erythrocytes. II. Studies on the site and mechanism of action. *Immunochemistry.* 1969;6:405-419.
159. Medof ME, Kinoshita T, Nussenzweig V. Inhibition of complement activation on the surface of cells after incorporation of decay-accelerating factor (DAF) into their membranes. *J Exp Med.* 1984;160:1558-1578.
160. Medof ME, Lublin DM, Holers VM, et al. Cloning and characterization of cDNAs encoding the complete sequence of decay-accelerating factor of human complement. *Proc Natl Acad Sci U S A.* 1987;84:2007-2011.
161. Dho SH, Lim JC, Kim LK. Beyond the role of CD55 as a complement component. *Immune Netw.* 2018;18:e11.
162. Skerka C, Chen Q, Fremeaux-Bacchi V, Roumenina LT. Complement factor H related proteins (CFHRs). *Mol Immunol.* 2013;56:170-180.

163. Ferreira VP, Pangburn MK, Cortes C. Complement control protein factor H: the good, the bad, and the inadequate. *Mol Immunol.* 2010;47:2187-2197.
164. Dahlback B. Vitamin K-dependent protein S: beyond the protein C pathway. *Semin Thromb Hemost.* 2018;44:176-184.
165. Bayly-Jones C, Bubeck D, Dunstone MA. The mystery behind membrane insertion: a review of the complement membrane attack complex. *Philos Trans R Soc Lond B Biol Sci.* 2017;372(1726):2016022.
166. Milis L, Morris CA, Sheehan MC, Charlesworth JA, Pussell BA. Vitronectin-mediated inhibition of complement: evidence for different binding sites for C5b-7 and C9. *Clin Exp Immunol.* 1993;92:114-119.
167. Jenne DE, Tschopp J. Clusterin: the intriguing guises of a widely expressed glycoprotein. *Trends Biochem Sci.* 1992;17:154-159.
168. Nemerow GR, Yamamoto KI, Lint TF. Restriction of complement-mediated membrane damage by the eighth component of complement: a dual role for C8 in the complement attack sequence. *J Immunol.* 1979;123:1245-1252.
169. Farkas I, Baranyi L, Ishikawa Y, et al. CD59 blocks not only the insertion of C9 into MAC but inhibits ion channel formation by homologous C5b-8 as well as C5b-9. *J Physiol.* 2002;539:537-545.
170. Huang Y, Fedarovich A, Tomlinson S, Davies C. Crystal structure of CD59: implications for molecular recognition of the complement proteins C8 and C9 in the membrane-attack complex. *Acta Crystallogr D Biol Crystallogr.* 2007;63:714-721.
171. Wickham SE, Hotze EM, Farrand AJ, et al. Mapping the intermedilysin-human CD59 receptor interface reveals a deep correspondence with the binding site on CD59 for complement binding proteins C8alpha and C9. *J Biol Chem.* 2011;286:20952-20962.
172. Carter PE, Duponchel C, Tosi M, Fothergill JE. Complete nucleotide sequence of the gene for human C1 inhibitor with an unusually high density of Alu elements. *Eur J Biochem.* 1991;197:301-308.
173. C1 esterase inhibitor (human). *P T.* 2010;35:2-3.
174. Zeerleder S, Levi M. Hereditary and acquired C1-inhibitor-dependent angioedema: from pathophysiology to treatment. *Ann Med.* 2016;48:256-267.
175. Frank MM, Gelfand JA, Atkinson JP. Hereditary angioedema: the clinical syndrome and its management. *Ann Intern Med.* 1976;84:580-593.
176. Gialeli C, Gungor B, Blom AM. Novel potential inhibitors of complement system and their roles in complement regulation and beyond. *Mol Immunol.* 2018;102:73-83.
177. Goicoechea de Jorge E, Tortajada A, Garcia SP, et al. Factor H competitor generated by gene conversion events associates with atypical hemolytic uremic syndrome. *J Am Soc Nephrol.* 2018;29:240-249.
178. Goicoechea de Jorge E, Caesar JJ, Malik TH, et al. Dimerization of complement factor H-related proteins modulates complement activation *in vivo. Proc Natl Acad Sci U S A.* 2013;110:4685-4690.
179. Medjeral-Thomas N, Pickering MC. The complement factor H-related proteins. *Immunol Rev.* 2016;274:191-201.
180. Cserhalmi M, Papp A, Brandus B, Uzonyi B, Jozsi M. Regulation of regulators: role of the complement factor H-related proteins. *Semin Immunol.* 2019;45:101341.
181. Mueller-Ortiz SL, Wang D, Morales JE, Li L, Chang JY, Wetsel RA. Targeted disruption of the gene encoding the murine small subunit of carboxypeptidase N (CPN1) causes susceptibility to C5a anaphylatoxin-mediated shock. *J Immunol.* 2009;182:6533-6539.
182. Morser J, Shao Z, Nishimura T, et al. Carboxypeptidase B2 and N play different roles in regulation of activated complements C3a and C5a in mice. *J Thromb Haemostasis.* 2018;16:991-1002.
183. Li XX, Lee JD, Kemper C, Woodruff TM. The complement receptor C5aR2: a powerful modulator of innate and adaptive immunity. *J Immunol.* 2019;202:3339-3348.
184. Laabei M, Ermert D. Catch me if you can: Streptococcus pyogenes complement evasion strategies. *J Innate Immun.* 2019;11:3-12.
185. Garcia BL, Zwarthoff SA, Rooijakkers SH, Geisbrecht BV. Novel evasion mechanisms of the classical complement pathway. *J Immunol.* 2016;197:2051-2060.
186. Rosbjerg A, Genster N, Pilely K, Garred P. Evasion mechanisms used by pathogens to escape the lectin complement pathway. *Front Microbiol.* 2017;8:868.
187. Brubaker SW, Bonham KS, Zanoni I, Kagan JC. Innate immune pattern recognition: a cell biological perspective. *Annu Rev Immunol.* 2015;33:257-290.
188. Venereau E, Ceriotti C, Bianchi ME. DAMPs from cell death to new life. *Front Immunol.* 2015;6:422.
189. Lemke G. How macrophages deal with death. *Nat Rev Immunol.* 2019;19:539-549.
190. Gasque P. Complement: a unique innate immune sensor for danger signals. *Mol Immunol.* 2004;41:1089-1098.
191. Botto M, Dell'Agnola C, Bygrave AE, et al. Homozygous C1q deficiency causes glomerulonephritis associated with multiple apoptotic bodies. *Nat Genet.* 1998;19:56-59.
192. Korb LC, Ahearn JM. C1q binds directly and specifically to surface blebs of apoptotic human keratinocytes: complement deficiency and systemic lupus erythematosus revisited. *J Immunol.* 1997;158:4525-4528.
193. Jiang H, Cooper B, Robey FA, Gewurz H. DNA binds and activates complement via residues 14-26 of the human C1q A chain. *J Biol Chem.* 1992;267:25597-25601.
194. Nauta AJ, Trouw LA, Daha MR, et al. Direct binding of C1q to apoptotic cells and cell blebs induces complement activation. *Eur J Immunol.* 2002;32:1726-1736.
195. Taylor PR, Carugati A, Fadok VA, et al. A hierarchical role for classical pathway complement proteins in the clearance of apoptotic cells *in vivo. J Exp Med.* 2000;192:359-366.
196. Ogden CA, deCathelineau A, Hoffmann PR, et al. C1q and mannose binding lectin engagement of cell surface calreticulin and CD91 initiates macropinocytosis and uptake of apoptotic cells. *J Exp Med.* 2001;194:781-795.
197. Ramirez-Ortiz ZG, Pendergraft WF III, Prasad A, et al. The scavenger receptor SCARF1 mediates the clearance of apoptotic cells and prevents autoimmunity. *Nat Immunol.* 2013;14:917-926.
198. Nauta AJ, Raaschou-Jensen N, Roos A, et al. Mannose-binding lectin engagement with late apoptotic and necrotic cells. *Eur J Immunol.* 2003;33:2853-2863.
199. Trouw LA, Bengtsson AA, Gelderman KA, Dahlback B, Sturfelt G, Blom AM. C4b-binding protein and factor H compensate for the loss of membrane-bound complement inhibitors to protect apoptotic cells against excessive complement attack. *J Biol Chem.* 2007;282:28540-28548.
200. Xu W, Berger SP, Trouw LA, et al. Properdin binds to late apoptotic and necrotic cells independently of C3b and regulates alternative pathway complement activation. *J Immunol.* 2008;180:7613-7621.
201. Leffler J, Herbert AP, Norstrom E, et al. Annexin-II, DNA, and histones serve as factor H ligands on the surface of apoptotic cells. *J Biol Chem.* 2010;285:3766-3776.
202. Webb JH, Blom AM, Dahlback B. Vitamin K-dependent protein S localizing complement regulator C4b-binding protein to the surface of apoptotic cells. *J Immunol.* 2002;169:2580-2586.
203. Truedsson L, Bengtsson AA, Sturfelt G. Complement deficiencies and systemic lupus erythematosus. *Autoimmunity.* 2007;40:560-566.
204. Ling GS, Crawford G, Buang N, et al. C1q restrains autoimmunity and viral infection by regulating CD8(+) T cell metabolism. *Science.* 2018;360:558-563.
205. Schifferli JA. Elimination of immune complexes: role of complement. Article in French. *Schweiz Med Wochenschr.* 1985;115:1422-1424.
206. Kavai M. Immune complex clearance by complement receptor type 1 in SLE. *Autoimmun Rev.* 2008;8:160-164.
207. Schreiber RD, Old LJ, Smyth MJ. Cancer immunoediting: integrating immunity's roles in cancer suppression and promotion. *Science.* 2011;331:1565-1570.
208. Mamidi S, Hone S, Kirschfink M. The complement system in cancer: Ambivalence between tumour destruction and promotion. *Immunobiology.* 2017;222:45-54.
209. Franceschi C, Garagnani P, Vitale G, Capri M, Salvioli S. Inflammaging and 'Garb-aging'. *Trends Endocrinol Metab.* 2017;28:199-212.
210. Samstad EO, Niyonzima N, Nymo S, et al. Cholesterol crystals induce complement-dependent inflammasome activation and cytokine release. *J Immunol.* 2014;192:2837-2845.
211. Niyonzima N, Halvorsen B, Sporsheim B, et al. Complement activation by cholesterol crystals triggers a subsequent cytokine response. *Mol Immunol.* 2017;84:43-50.
212. Pilely K, Rosbjerg A, Genster N, et al. Cholesterol crystals activate the lectin complement pathway via ficolin-2 and mannose-binding lectin: implications for the progression of atherosclerosis. *J Immunol.* 2016;196:5064-5074.
213. Maier M, Peng Y, Jiang L, Seabrook TJ, Carroll MC, Lemere CA. Complement C3 deficiency leads to accelerated amyloid beta plaque deposition and neurodegeneration and modulation of the microglia/macrophage phenotype in amyloid precursor protein transgenic mice. *J Neurosci.* 2008;28:6333-6341.
214. Grafals M, Thurman JM. The role of complement in organ transplantation. *Front Immunol.* 2019;10:2380.
215. Le Friec G, Kemper C. Complement: coming full circle. *Arch Immunol Ther Exp.* 2009;57:393-407.
216. Medzhitov R. Origin and physiological roles of inflammation. *Nature.* 2008;454:428-435.
217. Davies LC, Jenkins SJ, Allen JE, Taylor PR. Tissue-resident macrophages. *Nat Immunol.* 2013;14:986-995.
218. Wetsel RA. Structure, function and cellular expression of complement anaphylatoxin receptors. *Curr Opin Immunol.* 1995;7:48-53.
219. Guo RF, Ward PA. Role of C5a in inflammatory responses. *Annu Rev Immunol.* 2005;23:821-852.
220. Hawksworth OA, Li XX, Coulthard LG, Wolvetang EJ, Woodruff TM. New concepts on the therapeutic control of complement anaphylatoxin receptors. *Mol Immunol.* 2017;89:36-43.
221. Crass T, Raffetseder U, Martin U, et al. Expression cloning of the human C3a anaphylatoxin receptor (C3aR) from differentiated U-937 cells. *Eur J Immunol.* 1996;26:1944-1950.
222. Ohno M, Hirata T, Enomoto M, Araki T, Ishimaru H, Takahashi TA. A putative chemoattractant receptor, C5L2, is expressed in granulocyte and immature dendritic cells, but not in mature dendritic cells. *Mol Immunol.* 2000;37:407-412.
223. Li R, Coulthard LG, Wu MC, Taylor SM, Woodruff TM. C5L2: a controversial receptor of complement anaphylatoxin, C5a. *FASEB J.* 2013;27:855-864.
224. Lo MW, Woodruff TM. Complement: bridging the innate and adaptive immune systems in sterile inflammation. *J Leukoc Biol.* 2020;108(1):339-351.
225. Norgauer J, Dobos G, Kownatzki E, et al. Complement fragment C3a stimulates Ca2+ influx in neutrophils via a pertussis-toxin-sensitive G protein. *Eur J Biochem.* 1993;217:289-294.
226. Coulthard LG, Woodruff TM. Is the complement activation product C3a a proinflammatory molecule? Re-evaluating the evidence and the myth. *J Immunol.* 2015;194:3542-3548.
227. Kilgore KS, Flory CM, Miller BF, Evans VM, Warren JS. The membrane attack complex of complement induces interleukin-8 and monocyte chemoattractant protein-1 secretion from human umbilical vein endothelial cells. *Am J Pathol.* 1996;149:953-961.
228. Xie CB, Qin L, Li G, et al. Complement membrane attack complexes assemble NLRP3 inflammasomes triggering IL-1 activation of IFN-γ-primed human endothelium. *Circ Res.* 2019;124:1747-1759.
229. Brekke OL, Christiansen D, Fure H, Fung M, Mollnes TE. The role of complement C3 opsonization, C5a receptor, and CD14 in *E. coli*-induced up-regulation of granulocyte and monocyte CD11b/CD18 (CR3), phagocytosis, and oxidative burst in human whole blood. *J Leukoc Biol.* 2007;81:1404-1413.
230. Mollnes TE, Brekke OL, Fung M, et al. Essential role of the C5a receptor in *E coli*-induced oxidative burst and phagocytosis revealed by a novel lepirudin-based human whole blood model of inflammation. *Blood.* 2002;100:1869-1877.
231. Farnam J, Grant JA, Lett-Brown MA, Hunt C, Thueson DO, Giclas PC. Complement-and IgE-mediated release of histamine from basophils *in vitro*. V. Differential effects of drugs modulating arachidonic acid metabolism. *J Immunol.* 1985;134:541-547.
232. Johnson AR, Hugli TE, Muller-Eberhard HJ. Release of histamine from rat mast cells by the complement peptides C3a and C5a. *Immunology.* 1975;28:1067-1080.
233. Asgari E, Le Friec G, Yamamoto H, et al. C3a modulates IL-1β secretion in human monocytes by regulating ATP efflux and subsequent NLRP3 inflammasome activation. *Blood.* 2013;122:3473-3481.
234. Haggadone MD, Grailer JJ, Fattahi F, Zetoune FS, Ward PA. Bidirectional crosstalk between C5a receptors and the NLRP3 inflammasome in macrophages and monocytes. *Mediat Inflamm.* 2016;2016:1340156.
235. Holers VM, Kinoshita T, Molina H. The evolution of mouse and human complement C3-binding proteins: divergence of form but conservation of function. *Immunol Today.* 1992;13:231-236.
236. Fearon DT, Collins LA. Increased expression of C3b receptors on polymorphonuclear leukocytes induced by chemotactic factors and by purification procedures. *J Immunol.* 1983;130:370-375.
237. Foley JH, Peterson EA, Lei V, Wan LW, Krisinger MJ, Conway EM. Interplay between fibrinolysis and complement: plasmin cleavage of iC3b modulates immune responses. *J Thromb Haemost.* 2015;13:610-618.
238. Bilsland CA, Diamond MS, Springer TA. The leukocyte integrin p150,95 (CD11c/CD18) as a receptor for iC3b. Activation by a heterologous beta subunit and localization of a ligand recognition site to the I domain. *J Immunol.* 1994;152:4582-4589.
239. Vorup-Jensen T, Jensen RK. Structural immunology of complement receptors 3 and 4. *Front Immunol.* 2018;9:2716.
240. Erdei A, Lukacsi S, Macsik-Valent B, Nagy-Balo Z, Kurucz I, Bajtay Z. Non-identical twins: different faces of CR3 and CR4 in myeloid and lymphoid cells of mice and men. *Semin Cell Dev Biol.* 2019;85:110-121.
241. Ehlers MR. CR3: a general purpose adhesion-recognition receptor essential for innate immunity. *Microbes Infect.* 2000;2:289-294.
242. Helmy KY, Katschke KJ Jr, Gorgani NN, et al. CRIg: a macrophage complement receptor required for phagocytosis of circulating pathogens. *Cell.* 2006;124:915-927.
243. Pepys MB. Role of complement in induction of antibody production *in vivo*. Effect of cobra factor and other C3-reactive agents on thymus-dependent and thymus-independent antibody responses. *J Exp Med.* 1974;140:126-145.
244. Hebell T, Ahearn JM, Fearon DT. Suppression of the immune response by a soluble complement receptor of B lymphocytes. *Science.* 1991;254:102-105.
245. Molina H, Holers VM, Li B, et al. Markedly impaired humoral immune response in mice deficient in complement receptors 1 and 2. *Proc Natl Acad Sci U S A.* 1996;93:3357-3361.
246. Carter RH, Fearon DT. Pillars Article. CD19. Lowering the threshold for antigen receptor stimulation of B lymphocytes. Science, 1992. 256: 105-107. *J Immunol.* 2010;184:2233-2235.
247. Matsumoto AK, Martin DR, Carter RH, Klickstein LB, Ahearn JM, Fearon DT. Functional dissection of the CD21/CD19/TAPA-1/Leu-13 complex of B lymphocytes. *J Exp Med.* 1993;178:1407-1417.
248. Gonzalez SF, Lukacs-Kornek V, Kuligowski MP, et al. Complement-dependent transport of antigen into B cell follicles. *J Immunol.* 2010;185:2659-2664.
249. Bergmann-Leitner ES, Leitner WW, Tsokos GC. Complement 3d: from molecular adjuvant to target of immune escape mechanisms. *Clin Immunol.* 2006;121:177-185.

250. Prodeus AP, Goerg S, Shen LM, et al. A critical role for complement in maintenance of self-tolerance. *Immunity.* 1998;9:721-731.
251. Pappworth IY, Kulik L, Haluszczak C, Reuter JW, Holers VM, Marchbank KJ. Increased B cell deletion and significantly reduced auto-antibody titre due to premature expression of human complement receptor 2 (CR2, CD21). *Mol Immunol.* 2009;46:1042-1049.
252. Ferry H, Potter PK, Crockford TL, et al. Increased positive selection of B1 cells and reduced B cell tolerance to intracellular antigens in c1q-deficient mice. *J Immunol.* 2007;178:2916-2922.
253. Phan TG, Grigorova I, Okada T, Cyster JG. Subcapsular encounter and complement-dependent transport of immune complexes by lymph node B cells. *Nat Immunol.* 2007;8:992-1000.
254. Phan TG, Green JA, Gray EE, Xu Y, Cyster JG. Immune complex relay by subcapsular sinus macrophages and noncognate B cells drives antibody affinity maturation. *Nat Immunol.* 2009;10:786-793.
255. Carroll MC, Isenman DE. Regulation of humoral immunity by complement. *Immunity.* 2012;37:199-207.
256. Heesters BA, van der Poel CE, Das A, Carroll MC. Antigen presentation to B cells. *Trends Immunol.* 2016;37:844-854.
257. Aubry JP, Pochon S, Graber P, Jansen KU, Bonnefoy JY. CD21 is a ligand for CD23 and regulates IgE production. *Nature.* 1992;358:505-507.
258. Crotty S. T follicular helper cell differentiation, function, and roles in disease. *Immunity.* 2014;41:529-542.
259. Brodeur SR, Angelini F, Bacharier LB, et al. C4b-binding protein (C4BP) activates B cells through the CD40 receptor. *Immunity.* 2003;18:837-848.
260. Killick J, Morisse G, Sieger D, Astier AL. Complement as a regulator of adaptive immunity. *Semin Immunopathol.* 2018;40:37-48.
261. Jozsi M, Prechl J, Bajtay Z, Erdei A. Complement receptor type 1 (CD35) mediates inhibitory signals in human B lymphocytes. *J Immunol.* 2002;168:2782-2788.
262. Imani F, Proud D, Griffin DE. Measles virus infection synergizes with IL-4 in IgE class switching. *J Immunol.* 1999;162:1597-1602.
263. Aegerter-Shaw M, Cole JL, Klickstein LB, et al. Expansion of the complement receptor gene family. Identification in the mouse of two new genes related to the CR1 and CR2 gene family. *J Immunol.* 1987;138:3488-3494.
264. Nomura M, Tsujimura A, Shida K, et al. Membrane and secretory forms of mouse membrane cofactor protein (CD46) generated from a single gene through alternative splicing. *Immunogenetics.* 1999;50:245-254.
265. Wilson JG, Tedder TF, Fearon DT. Characterization of human T lymphocytes that express the C3b receptor. *J Immunol.* 1983;131:684-689.
266. Yaskanin DD, Thompson LF, Waxman FJ. Distribution and quantitative expression of the complement receptor type 1 (CR1) on human peripheral blood T lymphocytes. *Cell Immunol.* 1992;142:159-176.
267. Torok K, Dezso B, Bencsik A, Uzonyi B, Erdei A. Complement receptor type 1 (CR1/CD35) expressed on activated human CD4+ T cells contributes to generation of regulatory T cells. *Immunol Lett.* 2015;164:117-124.
268. Mouhoub A, Delibrias CC, Fischer E, Boyer V, Kazatchkine MD. Ligation of CR1 (C3b receptor, CD35) on CD4+ T lymphocytes enhances viral replication in HIV-infected cells. *Clin Exp Immunol.* 1996;106:297-303.
269. Pekalski ML, Garcia AR, Ferreira RC, et al. Neonatal and adult recent thymic emigrants produce IL-8 and express complement receptors CR1 and CR2. *JCI Insight.* 2017;2:e93739.
270. Kerekes K, Prechl J, Bajtay Z, Jozsi M, Erdei A. A further link between innate and adaptive immunity: C3 deposition on antigen-presenting cells enhances the proliferation of antigen-specific T cells. *Int Immunol.* 1998;10:1923-1930.
271. Torok K, Kremlitzka M, Sandor N, Toth EA, Bajtay Z, Erdei A. Human T cell derived, cell-bound complement iC3b is integrally involved in T cell activation. *Immunol Lett.* 2012;143:131-136.
272. Astier A, Trescol-Biemont MC, Azocar O, Lamouille B, Rabourdin-Combe C. Cutting edge: CD46, a new costimulatory molecule for T cells, that induces p120CBL and LAT phosphorylation. *J Immunol.* 2000;164:6091-6095.
273. Cardone J, Le Friec G, Vantourout P, et al. Complement regulator CD46 temporally regulates cytokine production by conventional and unconventional T cells. *Nat Immunol.* 2010;11:862-871.
274. Le Friec G, Sheppard D, Whiteman P, et al. The CD46-Jagged1 interaction is critical for human TH1 immunity. *Nat Immunol.* 2012;13:1213-1221.
275. West EE, Kolev M, Kemper C. Complement and the regulation of T cell responses. *Annu Rev Immunol.* 2018;36:309-338.
276. Liao W, Lin JX, Leonard WJ. IL-2 family cytokines: new insights into the complex roles of IL-2 as a broad regulator of T helper cell differentiation. *Curr Opin Immunol.* 2011;23:598-604.
277. Kolev M, Dimeloe S, Le Friec G, et al. Complement regulates nutrient influx and metabolic reprogramming during Th1 cell responses. *Immunity.* 2015;42:1033-1047.
278. Arbore G, West EE, Rahman J, et al. Complement receptor CD46 co-stimulates optimal human CD8(+) T cell effector function via fatty acid metabolism. *Nat Commun.* 2018;9:4186.
279. Lalli PN, Zhou W, Sacks S, Medof ME, Heeger PS. Locally produced and activated complement as a mediator of alloreactive T cells. *Front Biosci (Schol Ed).* 2009;1:117-124.
280. Liu J, Lin F, Strainic MG, et al. IFN-gamma and IL-17 production in experimental autoimmune encephalomyelitis depends on local APC-T cell complement production. *J Immunol.* 2008;180:5882-5889.
281. Vieyra M, Leisman S, Raedler H, et al. Complement regulates CD4 T-cell help to CD8 T cells required for murine allograft rejection. *Am J Pathol.* 2011;179:766-774.
282. Arbore G, West EE, Spolski R, et al. T helper 1 immunity requires complement-driven NLRP3 inflammasome activity in CD4(+) T cells. *Science.* 2016;352:aad1210.
283. Strainic MG, Shevach EM, An F, Lin F, Medof ME. Absence of signaling into CD4(+) cells via C3aR and C5aR enables autoinductive TGF-β1 signaling and induction of Foxp3(+) regulatory T cells. *Nat Immunol.* 2013;14:162-171.
284. Kwan WH, van der Touw W, Paz-Artal E, Li MO, Heeger PS. Signaling through C5a receptor and C3a receptor diminishes function of murine natural regulatory T cells. *J Exp Med.* 2013;210:257-268.
285. Dunkelberger J, Zhou L, Miwa T, Song WC. C5aR expression in a novel GFP reporter gene knockin mouse: implications for the mechanism of action of C5aR signaling in T cell immunity. *J Immunol.* 2012;188:4032-4042.
286. Quell KM, Karsten CM, Kordowski A, et al. Monitoring C3aR expression using a floxed tdTomato-C3aR reporter knock-in mouse. *J Immunol.* 2017;199:688-706.
287. Laumonnier Y, Karsten CM, Kohl J. Novel insights into the expression pattern of anaphylatoxin receptors in mice and men. *Mol Immunol.* 2017;89:44-58.
288. Heeger PS, Lalli PN, Lin F, et al. Decay-accelerating factor modulates induction of T cell immunity. *J Exp Med.* 2005;201:1523-1530.
289. Liu J, Miwa T, Hilliard B, et al. The complement inhibitory protein DAF (CD55) suppresses T cell immunity *in vivo. J Exp Med.* 2005;201:567-577.
290. Ozen A, Comrie WA, Ardy RC, et al. CD55 deficiency, early-onset protein-losing enteropathy, and thrombosis. *N Engl J Med.* 2017;377:52-61.
291. Korty PE, Brando C, Shevach EM. CD59 functions as a signal-transducing molecule for human T cell activation. *J Immunol.* 1991;146:4092-4098.
292. Wang LN, Gao MH, Wang B, Cong BB, Zhang SC. A role for GPI-CD59 in promoting T-cell signal transduction via LAT. *Oncol Lett.* 2018;15:4873-4881.
293. Longhi MP, Harris CL, Morgan BP, Gallimore A. Holding T cells in check – A new role for complement regulators? *Trends Immunol.* 2006;27:102-108.
294. Kimberley FC, Sivasankar B, Paul Morgan B. Alternative roles for CD59. *Mol Immunol.* 2007;44:73-81.
295. Zhang R, Liu Q, Liao Q, Zhao Y. CD59: a promising target for tumor immunotherapy. *Future Oncol.* 2018;14:781-791.
296. Chen A, Gaddipati S, Hong Y, Volkman DJ, Peerschke EI, Ghebrehiwet B. Human T cells express specific binding sites for C1q. Role in T cell activation and proliferation. *J Immunol.* 1994;153:1430-1440.
297. Jiang K, Chen Y, Jarvis JN. Cord blood and adult T cells show different responses to C1q-bearing immune complexes. *Cell Immunol.* 2004;229:62-67.
298. Clarke EV, Tenner AJ. Complement modulation of T cell immune responses during homeostasis and disease. *J Leukoc Biol.* 2014;96:745-756.
299. Kemper C, Kohl J. Novel roles for complement receptors in T cell regulation and beyond. *Mol Immunol.* 2013;56:181-190.
300. Perucha E, Melchiotti R, Bibby JA, et al. The cholesterol biosynthesis pathway regulates IL-10 expression in human Th1 cells. *Nat Commun.* 2019;10:498.
301. Barchet W, Price JD, Cella M, et al. Complement-induced regulatory T cells suppress T-cell responses but allow for dendritic-cell maturation. *Blood.* 2006;107:1497-1504.
302. Longhi MP, Sivasankar B, Omidvar N, Morgan BP, Gallimore A. Cutting edge: murine CD59a modulates antiviral CD4+ T cell activity in a complement-independent manner. *J Immunol.* 2005;175:7098-7102.
303. Li K, Fazekasova H, Wang N, et al. Functional modulation of human monocytes derived DCs by anaphylatoxins C3a and C5a. *Immunobiology.* 2012;217:65-73.
304. Li K, Fazekasova H, Wang N, et al. Expression of complement components, receptors and regulators by human dendritic cells. *Mol Immunol.* 2011;48:1121-1127.
305. Grailer JJ, Bosmann M, Ward PA. Regulatory effects of C5a on IL-17A, IL-17F, and IL-23. *Front Immunol.* 2012;3:387.
306. Hashimoto M, Hirota K, Yoshitomi H, et al. Complement drives Th17 cell differentiation and triggers autoimmune arthritis. *J Exp Med.* 2010;207:1135-1143.
307. Strainic MG, Liu J, Huang D, et al. Locally produced complement fragments C5a and C3a provide both costimulatory and survival signals to naive CD4+ T cells. *Immunity.* 2008;28:425-435.
308. Weaver DJ Jr, Reis ES, Pandey MK, et al. C5a receptor-deficient dendritic cells promote induction of T_{reg} and Th17 cells. *Eur J Immunol.* 2010;40:710-721.
309. Schmudde I, Strover HA, Vollbrandt T, et al. C5a receptor signalling in dendritic cells controls the development of maladaptive Th2 and Th17 immunity in experimental allergic asthma. *Mucosal Immunol.* 2013;6:807-825.
310. Cheng SC, Sprong T, Joosten LA, et al. Complement plays a central role in Candida albicans-induced cytokine production by human PBMCs. *Eur J Immunol.* 2012;42:993-1004.
311. Csomor E, Bajtay Z, Sandor N, et al. Complement protein C1q induces maturation of human dendritic cells. *Mol Immunol.* 2007;44:3389-3397.
312. Kandasamy M, Ying PC, Ho AW, et al. Complement mediated signaling on pulmonary CD103(+) dendritic cells is critical for their migratory function in response to influenza infection. *PLoS Pathog.* 2013;9:e1003115.
313. Lo TN, Boyle MD. Relationship between the intracellular cyclic adenosine 3':5'-monophosphate level of tumor cells and their sensitivity to killing by antibody and complement. *Cancer Res.* 1979;39:3156-3162.
314. Kraus S, Seger R, Fishelson Z. Involvement of the ERK mitogen-activated protein kinase in cell resistance to complement-mediated lysis. *Clin Exp Immunol.* 2001;123:366-374.
315. Triantafilou K, Hughes TR, Triantafilou M, Morgan BP. The complement membrane attack complex triggers intracellular Ca2+ fluxes leading to NLRP3 inflammasome activation. *J Cell Sci.* 2013;126:2903-2913.
316. Chauhan AK, Moore TL. T cell activation by terminal complex of complement and immune complexes. *J Biol Chem.* 2011;286:38627-38637.
317. Ziporen L, Donin N, Shmushkovich T, Gross A, Fishelson Z. Programmed necrotic cell death induced by complement involves a Bid-dependent pathway. *J Immunol.* 2009;182:515-521.
318. Clarke EV, Weist BM, Walsh CM, Tenner AJ. Complement protein C1q bound to apoptotic cells suppresses human macrophage and dendritic cell-mediated Th17 and Th1 T cell subset proliferation. *J Leukoc Biol.* 2015;97:147-160.
319. Son M, Santiago-Schwarz F, Al-Abed Y, Diamond B. C1q limits dendritic cell differentiation and activation by engaging LAIR-1. *Proc Natl Acad Sci U S A.* 2012;109:E3160-E3167.
320. Skoberne M, Somersan S, Almodovar W, et al. The apoptotic-cell receptor CR3, but not alphavbeta5, is a regulator of human dendritic-cell immunostimulatory function. *Blood.* 2006;108:947-955.
321. Hawksworth OA, Coulthard LG, Mantovani S, Woodruff TM. Complement in stem cells and development. *Semin Immunol.* 2018;37:74-84.
322. Riley RC, Tannenbaum PL, Abbott DH, Atkinson JP. Cutting edge. Inhibiting measles virus infection but promoting reproduction: an explanation for splicing and tissue-specific expression of CD46. *J Immunol.* 2002;169:5405-5409.
323. Qin X, Dobarro M, Bedford SJ, et al. Further characterization of reproductive abnormalities in mCd59b knockout mice: a potential new function of mCd59 in male reproduction. *J Immunol.* 2005;175:6294-6302.
324. Lee YL, Lee KF, Xu JS, et al. The embryotrophic activity of oviductal cell-derived complement C3b and iC3b, a novel function of complement protein in reproduction. *J Biol Chem.* 2004;279:12763-12768.
325. Singh J, Ahmed A, Girardi G. Role of complement component C1q in the onset of preeclampsia in mice. *Hypertension.* 2011;58:716-724.
326. McLin VA, Hu CH, Shah R, Jamrich M. Expression of complement components coincides with early patterning and organogenesis in *Xenopus laevis. Int J Dev Biol.* 2008;52:1123-1133.
327. Schafer DP, Lehrman EK, Kautzman AG, et al. Microglia sculpt postnatal neural circuits in an activity and complement-dependent manner. *Neuron.* 2012;74:691-705.
328. Benoit ME, Tenner AJ. Complement protein C1q-mediated neuroprotection is correlated with regulation of neuronal gene and microRNA expression. *J Neurosci.* 2011;31:3459-3469.
329. Sekar A, Bialas AR, de Rivera H, et al. Schizophrenia risk from complex variation of complement component 4. *Nature.* 2016;530:177-183.
330. Lee JD, Coulthard LG, Woodruff TM. Complement dysregulation in the central nervous system during development and disease. *Semin Immunol.* 2019;45:101340.
331. Mastellos DC, Deangelis RA, Lambris JD. Complement-triggered pathways orchestrate regenerative responses throughout phylogenesis. *Semin Immunol.* 2013;25:29-38.
332. Rutkowski MJ, Sughrue ME, Kane AJ, Ahn BJ, Fang S, Parsa AT. The complement cascade as a mediator of tissue growth and regeneration. *Inflamm Res.* 2010;59:897-905.
333. Strey CW, Markiewski M, Mastellos D, et al. The proinflammatory mediators C3a and C5a are essential for liver regeneration. *J Exp Med.* 2003;198:913-923.
334. Ehrnthaller C, Huber-Lang M, Nilsson P, et al. Complement C3 and C5 deficiency affects fracture healing. *PLoS One.* 2013;8:e81341.
335. Cazander G, Jukema GN, Nibbering PH. Complement activation and inhibition in wound healing. *Clin Dev Immunol.* 2012. 2012:534291.
336. Bossi F, Tripodo C, Rizzi L, et al. C1q as a unique player in angiogenesis with therapeutic implication in wound healing. *Proc Natl Acad Sci U S A.* 2014;111:4209-4214.
337. Liu Y, Given KS, Harlow DE, et al. Myelin-specific multiple sclerosis antibodies cause complement-dependent oligodendrocyte loss and demyelination. *Acta Neuropathol Commun.* 2017;5:25.
338. Carpanini SM, Torvell M, Morgan BP. Therapeutic inhibition of the complement system in diseases of the central nervous system. *Front Immunol.* 2019;10:362.

339. Ingersoll SA, Martin CB, Barnum SR, Martin BK. CNS-specific expression of C3a and C5a exacerbate demyelination severity in the cuprizone model. *Mol Immunol*. 2010;48:219-230.
340. Fu X, Liu G, Halim A, Ju Y, Luo Q, Song AG. Mesenchymal stem cell migration and tissue repair. *Cells*. 2019;8:784.
341. Galliot B, Crescenzi M, Jacinto A, Tajbakhsh S. Trends in tissue repair and regeneration. *Development*. 2017;144:357-364.
342. Dekoninck S, Blanpain C. Stem cell dynamics, migration and plasticity during wound healing. *Nat Cell Biol*. 2019;21:18-24.
343. Lara-Astiaso D, Izarra A, Estrada JC, et al. Complement anaphylatoxins C3a and C5a induce a failing regenerative program in cardiac resident cells. Evidence of a role for cardiac resident stem cells other than cardiomyocyte renewal. *Springerplus*. 2012;1:63.
344. Kulkarni OP, Lichtnekert J, Anders HJ, Mulay SR. The immune system in tissue environments regaining homeostasis after injury: is "inflammation" always inflammation? *Mediat Inflamm*. 2016;2016:2856213.
345. Mathern DR, Heeger PS. Molecules great and small: the complement system. *Clin J Am Soc Nephrol*. 2015;10:1636-1650.
346. Martin IV, Bohner A, Boor P, et al. Complement C5a receptors C5L2 and C5aR in renal fibrosis. *Am J Physiol Renal Physiol*. 2018;314:F35-F46.
347. Xavier S, Sahu RK, Landes SG, et al. Pericytes and immune cells contribute to complement activation in tubulointerstitial fibrosis. *Am J Physiol Renal Physiol*. 2017;312:F516-F32.
348. Fisher AJ, Cipolla E, Varre A, Gu H, Mickler EA, Vittal R. Potential mechanisms underlying TGF-beta-mediated complement activation in lung fibrosis. *Cell Mol Med Open Access*. 2017;3:14.
349. Gu H, Fisher AJ, Mickler EA, et al. Contribution of the anaphylatoxin receptors, C3aR and C5aR, to the pathogenesis of pulmonary fibrosis. *FASEB J*. 2016;30:2336-2350.
350. Liszewski MK, Atkinson JP. Inherited disorders of the complement system. In: Feldweg AM, ed. *UpToDate*. 2019.
351. El Sissy C, Rosain J, Vieira-Martins P, et al. Clinical and genetic spectrum of a large cohort with total and sub-total complement deficiencies. *Front Immunol*. 2019;10:1936.
352. Casanova JL, Abel L. Human mannose-binding lectin in immunity: friend, foe, or both? *J Exp Med*. 2004;199:1295-1299.
353. Munthe-Fog L, Hummelshoj T, Honore C, Madsen HO, Permin H, Garred P. Immunodeficiency associated with FCN3 mutation and ficolin-3 deficiency. *N Engl J Med*. 2009;360:2637-2644.
354. Michalski M, Swierzko AS, Pagowska-Klimek I, et al. Primary Ficolin-3 deficiency—Is it associated with increased susceptibility to infections? *Immunobiology*. 2015;220:711-713.
355. Stengaard-Pedersen K, Thiel S, Gadjeva M, et al. Inherited deficiency of mannan-binding lectin-associated serine protease 2. *N Engl J Med*. 2003;349:554-560.
356. Rooryck C, Diaz-Font A, Osborn DP, et al. Mutations in lectin complement pathway genes COLEC11 and MASP1 cause 3MC syndrome. *Nat Genet*. 2011;43:197-203.
357. Biesma DH, Hannema AJ, van Velzen-Blad H, et al. A family with complement factor D deficiency. *J Clin Invest*. 2001;108:233-240.
358. Lachmann PJ. Heberden oration 1986. Complement—Friend or foe? *Br J Rheumatol*. 1987;26:409-415.
359. O'Hara AM, Moran AP, Wurzner R, Orren A. Complement-mediated lipopolysaccharide release and outer membrane damage in Escherichia coli J5: requirement for C9. *Immunology*. 2001;102:365-372.
360. van Geffen M, Cugno M, Lap P, Loof A, Cicardi M, van Heerde W. Alterations of coagulation and fibrinolysis in patients with angioedema due to C1-inhibitor deficiency. *Clin Exp Immunol*. 2012;167:472-478.
361. Morgan BP. Hereditary angioedema—Therapies old and new. *N Engl J Med*. 2010;363:581-583.
362. Longhurst H, Cicardi M. Hereditary angio-oedema. *Lancet*. 2012;379:474-481.
363. Nicholson-Weller A, March JP, Rosenfeld SI, Austen KF. Affected erythrocytes of patients with paroxysmal nocturnal hemoglobinuria are deficient in the complement regulatory protein, decay accelerating factor. *Proc Natl Acad Sci U S A*. 1983;80:5066-5070.
364. Holguin MH, Fredrick LR, Bernshaw NJ, Wilcox LA, Parker CJ. Isolation and characterization of a membrane protein from normal human erythrocytes that inhibits reactive lysis of the erythrocytes of paroxysmal nocturnal hemoglobinuria. *J Clin Invest*. 1989;84:7-17.
365. Kazatchkine MD, Jouvin MH, Wilson JG, Fischer E, Fischer A. Human diseases associated with C3 receptor deficiencies. *Immunol Lett*. 1987;14:191-195.
366. Thiel J, Kimmig L, Salzer U, et al. Genetic CD21 deficiency is associated with hypogammaglobulinemia. *J Allergy Clin Immunol*. 2012;129:801.e6-810.e6.
367. Hajishengallis G, Moutsopoulos NM. Etiology of leukocyte adhesion deficiency-associated periodontitis revisited: not a raging infection but a raging inflammatory response. *Expert Rev Clin Immunol*. 2014;10:973-975.
368. Kolev M, West EE, Kunz N, et al. Diapedesis-induced integrin signaling via LFA-1 facilitates tissue immunity by inducing intrinsic complement C3 expression in immune cells. *Immunity*. 2020;52:513.e8-527.e8.
369. Mayilyan KR. Complement genetics, deficiencies, and disease associations. *Protein Cell*. 2012;3:487-496.
370. Harris CL, Heurich M, Rodriguez de Cordoba S, Morgan BP. The complotype: dictating risk for inflammation and infection. *Trends Immunol*. 2012;33:513-521.
371. Cedzynski M, Thielens NM, Mollnes TE, Vorup-Jensen T. Editorial: the role of complement in health and disease. *Front Immunol*. 2019;10:1869.
372. Ricklin D, Lambris JD. Complement in immune and inflammatory disorders: pathophysiological mechanisms. *J Immunol*. 2013;190:3831-3838.
373. Khan MA, Nicolls MR, Surguladze B, Saadoun I. Complement components as potential therapeutic targets for asthma treatment. *Respir Med*. 2014;108:543-549.
374. Jourde-Chiche N, Fakhouri F, Dou L, et al. Endothelium structure and function in kidney health and disease. *Nat Rev Nephrol*. 2019;15:87-108.
375. Durey MA, Sinha A, Togarsimalemath SK, Bagga A. Anti-complement-factor H-associated glomerulopathies. *Nat Rev Nephrol*. 2016;12:563-578.
376. Bu F, Zhang Y, Wang K, et al. Genetic analysis of 400 patients refines understanding and implicates a new gene in atypical hemolytic uremic syndrome. *J Am Soc Nephrol*. 2018;29:2809-2819.
377. Taylor RP, Lindorfer MA. Mechanisms of complement-mediated damage in hematological disorders. *Semin Hematol*. 2018;55:118-123.
378. Hofer J, Janecke AR, Zimmerhackl LB, et al. Complement factor H-related protein 1 deficiency and factor H antibodies in pediatric patients with atypical hemolytic uremic syndrome. *Clin J Am Soc Nephrol*. 2013;8:407-415.
379. Zipfel PF, Edey M, Heinen S, et al. Deletion of complement factor H-related genes CFHR1 and CFHR3 is associated with atypical hemolytic uremic syndrome. *PLoS Genet*. 2007;3:e41.
380. Smith RJH, Appel GB, Blom AM, et al. C3 glomerulopathy – Understanding a rare complement-driven renal disease. *Nat Rev Nephrol*. 2019;15:129-143.
381. Servais A, Noel LH, Roumenina LT, et al. Acquired and genetic complement abnormalities play a critical role in dense deposit disease and other C3 glomerulopathies. *Kidney Int*. 2012;82:454-464.
382. Gale DP, de Jorge EG, Cook HT, et al. Identification of a mutation in complement factor H-related protein 5 in patients of Cypriot origin with glomerulonephritis. *Lancet*. 2010;376:794-801.
383. Willows J, Brown M, Sheerin NS. The role of complement in kidney disease. *Clin Med*. 2020;20:156-160.
384. Lesher AM, Song WC. Review: complement and its regulatory proteins in kidney diseases. *Nephrology (Carlton)*. 2010;15:663-675.
385. Williams AL, Gullipalli D, Ueda Y, et al. C5 inhibition prevents renal failure in a mouse model of lethal C3 glomerulopathy. *Kidney Int*. 2017;91:1386-1397.
386. Roumenina LT, Rayes J, Frimat M, Fremeaux-Bacchi V. Endothelial cells: source, barrier, and target of defensive mediators. *Immunol Rev*. 2016;274:307-329.
387. Ekdahl KN, Teramura Y, Hamad OA, et al. Dangerous liaisons: complement, coagulation, and kallikrein/kinin cross-talk act as a linchpin in the events leading to thromboinflammation. *Immunol Rev*. 2016;274:245-269.
388. Legendre CM, Licht C, Muus P, et al. Terminal complement inhibitor eculizumab in atypical hemolytic-uremic syndrome. *N Engl J Med*. 2013;368:2169-2181.
389. Stegall MD, Diwan T, Raghavaiah S, et al. Terminal complement inhibition decreases antibody-mediated rejection in sensitized renal transplant recipients. *Am J Transplant*. 2011;11:2405-2413.
390. De Vriese AS, Sethi S, Van Praet J, Nath KA, Fervenza FC. Kidney disease caused by dysregulation of the complement alternative pathway: an etiologic approach. *J Am Soc Nephrol*. 2015;26:2917-2929.
391. Huang J, Kim LJ, Mealey R, et al. Neuronal protection in stroke by an sLex-glycosylated complement inhibitory protein. *Science*. 1999;285:595-599.
392. Clarke AR, Christophe BR, Khahera A, Sim JL, Connolly ES Jr. Therapeutic modulation of the complement cascade in stroke. *Front Immunol*. 2019;10:1723.
393. Lagrand WK, Nijmeijer R, Niessen HW, Visser CA, Hermens WT, Hack CE. C-reactive protein as a pro-inflammatory mediator in cardiovascular disease by its ability to activate complement: additional proof and hypothetical mechanisms. *Neth Heart J*. 2002;10:189-197.
394. Chen M, Daha MR, Kallenberg CG. The complement system in systemic autoimmune disease. *J Autoimmun*. 2010;34:J276-J286.
395. Hovland A, Jonasson L, Garred P, et al. The complement system and toll-like receptors as integrated players in the pathophysiology of atherosclerosis. *Atherosclerosis*. 2015;241:480-494.
396. Ambati J, Atkinson JP, Gelfand BD. Immunology of age-related macular degeneration. *Nat Rev Immunol*. 2013;13:438-451.
397. Geerlings MJ, de Jong EK, den Hollander AI. The complement system in age-related macular degeneration: a review of rare genetic variants and implications for personalized treatment. *Mol Immunol*. 2017;84:65-76.
398. Baumal CR. Wet age-related macular degeneration: treatment advances to reduce the injection burden. *Am J Manag Care*. 2020;26:S103-S11.
399. Hageman GS, Anderson DH, Johnson LV, et al. A common haplotype in the complement regulatory gene factor H (HF1/CFH) predisposes individuals to age-related macular degeneration. *Proc Natl Acad Sci U S A*. 2005;102:7227-7232.
400. Haines JL, Hauser MA, Schmidt S, et al. Complement factor H variant increases the risk of age-related macular degeneration. *Science*. 2005;308:419-421.
401. Klein RJ, Zeiss C, Chew EY, et al. Complement factor H polymorphism in age-related macular degeneration. *Science*. 2005;308:385-389.
402. Johnson LV, Leitner WP, Staples MK, Anderson DH. Complement activation and inflammatory processes in Drusen formation and age related macular degeneration. *Exp Eye Res*. 2001;73:887-896.
403. Nozaki M, Raisler BJ, Sakurai E, et al. Drusen complement components C3a and C5a promote choroidal neovascularization. *Proc Natl Acad Sci U S A*. 2006;103:2328-2333.
404. Weismann D, Hartvigsen K, Lauer N, et al. Complement factor H binds malondialdehyde epitopes and protects from oxidative stress. *Nature*. 2011;478:76-81.
405. van der Maten E, Westra D, van Selm S, et al. Complement factor H serum levels determine resistance to pneumococcal invasive disease. *J Infect Dis*. 2016;213:1820-1827.
406. Druart M, Le Magueresse C. Emerging roles of complement in psychiatric disorders. *Front Psychiatr*. 2019;10:573.
407. Perry VH, O'Connor V. C1q: the perfect complement for a synaptic feast? *Nat Rev Neurosci*. 2008;9:807-811.
408. Coulthard LG, Hawksworth OA, Li R, et al. Complement C5aR1 signaling promotes polarization and proliferation of embryonic neural progenitor cells through PKCζ. *J Neurosci*. 2017;37:5395-5407.
409. Comer ALJ T, Kretsge LN, Nguyen TPH, et al. Increased expression of schizophrenia-associated gene C4 leads to hypoconnectivity of prefrontal cortex and reduced social interaction. *PLoS Biol*. 2020;18(1):e3000604.
410. Zhang C, Lv Q, Fan W, Tang W, Yi Z. Influence of CFH gene on symptom severity of schizophrenia. *Neuropsychiatr Dis Treat*. 2017;13:697-706.
411. Howes OD, McCutcheon R. Inflammation and the neural diathesis-stress hypothesis of schizophrenia: a reconceptualization. *Transl Psychiatry*. 2017;7:e1024.
412. Severance EG, Gressitt KL, Buka SL, Cannon TD, Yolken RH. Maternal complement C1q and increased odds for psychosis in adult offspring. *Schizophr Res*. 2014;159:14-19.
413. Chu Y, Jin X, Parada I, et al. Enhanced synaptic connectivity and epilepsy in C1q knockout mice. *Proc Natl Acad Sci U S A*. 2010;107:7975-7980.
414. Cribbs DH, Berchtold NC, Perreau V, et al. Extensive innate immune gene activation accompanies brain aging, increasing vulnerability to cognitive decline and neurodegeneration: a microarray study. *J Neuroinflammation*. 2012;9:179.
415. Shi Q, Chowdhury S, Ma R, et al. Complement C3 deficiency protects against neurodegeneration in aged plaque-rich APP/PS1 mice. *Sci Transl Med*. 2017;9:eaaf6295.
416. Hernandez MX, Jiang S, Cole TA, et al. Prevention of C5aR1 signaling delays microglial inflammatory polarization, favors clearance pathways and suppresses cognitive loss. *Mol Neurodegener*. 2017;12:66.
417. Fonseca MI, Ager RR, Chu SH, et al. Treatment with a C5aR antagonist decreases pathology and enhances behavioral performance in murine models of Alzheimer's disease. *J Immunol*. 2009;183:1375-1383.
418. Brubaker WD, Crane A, Johansson JU, et al. Peripheral complement interactions with amyloid β peptide: erythrocyte clearance mechanisms. *Alzheimers Dement*. 2017;13:1397-1409.
419. Yu JT, Tan L. The role of clusterin in Alzheimer's disease: pathways, pathogenesis, and therapy. *Mol Neurobiol*. 2012;45:314-326.
420. Frei Y, Lambris JD, Stockinger B. Generation of a monoclonal antibody to mouse C5 application in an ELISA assay for detection of anti-C5 antibodies. *Mol Cell Probes*. 1987;1:141-149.
421. Wang Y, Rollins SA, Madri JA, Matis LA. Anti-C5 monoclonal antibody therapy prevents collagen-induced arthritis and ameliorates established disease. *Proc Natl Acad Sci U S A*. 1995;92:8955-8959.
422. Wang Y, Hu Q, Madri JA, Rollins SA, Chodera A, Matis LA. Amelioration of lupus-like autoimmune disease in NZB/WF1 mice after treatment with a blocking monoclonal antibody specific for complement component C5. *Proc Natl Acad Sci U S A*. 1996;93:8563-8568.
423. Hillmen P, Young NS, Schubert J, et al. The complement inhibitor eculizumab in paroxysmal nocturnal hemoglobinuria. *N Engl J Med*. 2006;355:1233-1243.
424. Howard JF Jr, Utsugisawa K, Benatar M, et al. Safety and efficacy of eculizumab in anti-acetylcholine receptor antibody-positive refractory generalised myasthenia gravis (REGAIN): a phase 3, randomised, double-blind, placebo-controlled, multicentre study. *Lancet Neurol*. 2017;16:976-986.
425. Pittock SJ, Berthele A, Fujihara K, et al. Eculizumab in aquaporin-4-positive neuromyelitis optica spectrum disorder. *N Engl J Med*. 2019;381:614-625.
426. Kulasekararaj AG, Hill A, Rottinghaus ST, et al. Ravulizumab (ALXN1210) vs eculizumab in C5-inhibitor-experienced adult patients with PNH: the 302 study. *Blood*. 2019;133:540-549.

427. Coyle D, Cheung MC, Evans GA. Opportunity cost of funding drugs for rare diseases: the cost-effectiveness of eculizumab in paroxysmal nocturnal hemoglobinuria. *Med Decis Making.* 2014;34:1016-1029.
428. Mastellos DC, Ricklin D, Lambris JD. Clinical promise of next-generation complement therapeutics. *Nat Rev Drug Discov.* 2019;18:707-729.
429. Ricklin D, Lambris JD. Compstatin: a complement inhibitor on its way to clinical application. *Adv Exp Med Biol.* 2008;632:273-292.
430. Janssen BJ, Halff EF, Lambris JD, Gros P. Structure of compstatin in complex with complement component C3c reveals a new mechanism of complement inhibition. *J Biol Chem.* 2007;282:29241-29247.
431. Zhang Y, Nester CM, Holanda DG, et al. Soluble CR1 therapy improves complement regulation in C3 glomerulopathy. *J Am Soc Nephrol.* 2013;24:1820-1829.
432. Varga L, Farkas H. rhC1INH: a new drug for the treatment of attacks in hereditary angioedema caused by C1-inhibitor deficiency. *Expert Rev Clin Immunol.* 2011;7:143-153.
433. Monk PN, Scola AM, Madala P, Fairlie DP. Function, structure and therapeutic potential of complement C5a receptors. *Br J Pharmacol.* 2007;152:429-448.
434. Lubbers R, van Essen MF, van Kooten C, Trouw LA. Production of complement components by cells of the immune system. *Clin Exp Immunol.* 2017;188:183-194.
435. Heeger PS, Kemper C. Novel roles of complement in T effector cell regulation. *Immunobiology.* 2012;217:216-224.
436. Ghannam A, Fauquert JL, Thomas C, Kemper C, Drouet C. Human complement C3 deficiency: Th1 induction requires T cell-derived complement C3a and CD46 activation. *Mol Immunol.* 2014;58:98-107.
437. Ellinghaus U, Cortini A, Pinder CL, Le Friec G, Kemper C, Vyse TJ. Dysregulated CD46 shedding interferes with Th1-contraction in systemic lupus erythematosus. *Eur J Immunol.* 2017;47:1200-1210.
438. Cope A, Le Friec G, Cardone J, Kemper C. The Th1 life cycle: molecular control of IFN-γ to IL-10 switching. *Trends Immunol.* 2011;32:278-286.
439. Arbore G, Ong VH, Costantini B, et al. Deep phenotyping detects a pathological CD4(+) T-cell complosome signature in systemic sclerosis. *Cell Mol Immunol.* 2020;17(9):1010-1013.
440. Rahman J, Singh P, Merle NS, Niyonzima N, Kemper C. Complement's favourite organelle—Mitochondria? *Br J Pharmacol.* 2021;178:2771-2785.
441. Elvington M, Liszewski MK, Bertram P, Kulkarni HS, Atkinson JP. A $C_3(H_2O)$ recycling pathway is a component of the intracellular complement system. *J Clin Invest.* 2017;127:970-981.
442. Cui W, Paglialunga S, Kalant D, et al. Acylation-stimulating protein/C5L2-neutralizing antibodies alter triglyceride metabolism *in vitro* and *in vivo*. *Am J Physiol Endocrinol Metab.* 2007;293:E1482-E1491.
443. Phieler J, Garcia-Martin R, Lambris JD, Chavakis T. The role of the complement system in metabolic organs and metabolic diseases. *Semin Immunol.* 2013;25:47-53.
444. Loftus RM, Finlay DK. Immunometabolism: cellular metabolism turns immune regulator. *J Biol Chem.* 2016;291:1-10.
445. Mills EL, Kelly B, O'Neill LAJ. Mitochondria are the powerhouses of immunity. *Nat Immunol.* 2017;18:488-498.
446. O'Neill LA, Kishton RJ, Rathmell J. A guide to immunometabolism for immunologists. *Nat Rev Immunol.* 2016;16:553-565.
447. Pandey MK, Burrow TA, Rani R, et al. Complement drives glucosylceramide accumulation and tissue inflammation in Gaucher disease. *Nature.* 2017;543:108-112.
448. Ghebrehiwet B, Hosszu KK, Valentino A, Peerschke EI. The C1q family of proteins: insights into the emerging non-traditional functions. *Front Immunol.* 2012;3.
449. Ten VS, Yao J, Ratner V, et al. Complement component c1q mediates mitochondria-driven oxidative stress in neonatal hypoxic-ischemic brain injury. *J Neurosci.* 2010;30:2077-2087.
450. Liszewski MK, Elvington M, Kulkarni HS, Atkinson JP. Complement's hidden arsenal: new insights and novel functions inside the cell. *Mol Immunol.* 2017;84:2-9.
451. Ghai R, Waters P, Roumenina LT, et al. C1q and its growing family. *Immunobiology.* 2007;212:253-266.
452. Schaffler A, Buechler C. CTRP family: linking immunity to metabolism. *Trends Endocrinol Metab.* 2012;23:194-204.
453. McGee AM, Baines CP. Complement 1q-binding protein inhibits the mitochondrial permeability transition pore and protects against oxidative stress-induced death. *Biochem J.* 2011;433:119-125.
454. Zolkiewicz J, Stochmal A, Rudnicka L. The role of adipokines in systemic sclerosis: a missing link? *Arch Dermatol Res.* 2019;311:251-263.
455. Klonisch T, Glogowska A, Thanasupawat T, et al. Structural commonality of C1q TNF-related proteins and their potential to activate relaxin/insulin-like family peptide receptor 1 signalling pathways in cancer cells. *Br J Pharmacol.* 2017;174:1025-1033.
456. Yu XH, Zhang DW, Zheng XL, Tang CK. C1q tumor necrosis factor-related protein 9 in atherosclerosis: mechanistic insights and therapeutic potential. *Atherosclerosis.* 2018;276:109-116.
457. Kolev M, Markiewski MM. Targeting complement-mediated immunoregulation for cancer immunotherapy. *Semin Immunol.* 2018;37:85-97.
458. Taylor RP, Lindorfer MA. Cytotoxic mechanisms of immunotherapy: harnessing complement in the action of anti-tumor monoclonal antibodies. *Semin Immunol.* 2016;28:309-316.
459. Hajishengallis G, Reis ES, Mastellos DC, Ricklin D, Lambris JD. Novel mechanisms and functions of complement. *Nat Immunol.* 2017;18:1288-1298.
460. Afshar-Kharghan V. The role of the complement system in cancer. *J Clin Invest.* 2017;127:780-789.
461. Mantovani A. Molecular pathways linking inflammation and cancer. *Curr Mol Med.* 2010;10:369-373.
462. Markiewski MM, DeAngelis RA, Benencia F, et al. Modulation of the antitumor immune response by complement. *Nat Immunol.* 2008;9:1225-1235.
463. Zha H, Wang X, Zhu Y, et al. Intracellular activation of complement C3 leads to PD-L1 antibody treatment resistance by modulating tumor-associated macrophages. *Cancer Immunol Res.* 2019;7:193-207.
464. Ajona D, Ortiz-Espinosa S, Moreno H, et al. A combined PD-1/C5a blockade synergistically protects against lung cancer growth and metastasis. *Cancer Discov.* 2017;7:694-703.
465. Conway EM. Complement-coagulation connections. *Blood Coagul Fibrinolysis.* 2018;29:243-251.
466. Foley JH. Examining coagulation-complement cross-talk: complement activation and thrombosis. *Thromb Res.* 2016;141(suppl 2):S50-S54.
467. Wiegner R, Chakraborty S, Huber-Lang M. Complement-coagulation crosstalk on cellular and artificial surfaces. *Immunobiology.* 2016;221:1073-1079.
468. Ward PA. A plasmin-split fragment of C'3 as a new chemotactic factor. *J Exp Med.* 1967;126:189-206.
469. Keragala CB, Draxler DF, McQuilten ZK, Medcalf RL. Haemostasis and innate immunity—A complementary relationship: a review of the intricate relationship between coagulation and complement pathways. *Br J Haematol.* 2018;180:782-798.
470. Wang H, Ricklin D, Lambris JD. Complement-activation fragment C4a mediates effector functions by binding as untethered agonist to protease-activated receptors 1 and 4. *Proc Natl Acad Sci U S A.* 2017;114:10948-10953.
471. Gruppo RA, Rother RP. Eculizumab for congenital atypical hemolytic-uremic syndrome. *N Engl J Med.* 2009;360:544-546.
472. Karsten CM, Kohl J. The immunoglobulin, IgG Fc receptor and complement triangle in autoimmune diseases. *Immunobiology.* 2012;217:1067-1079.
473. Mihai S, Nimmerjahn F. The role of Fc receptors and complement in autoimmunity. *Autoimmun Rev.* 2013;12:657-660.
474. Schmidt RE, Gessner JE. Fc receptors and their interaction with complement in autoimmunity. *Immunol Lett.* 2005;100:56-67.
475. Karsten CM, Pandey MK, Figge J, et al. Anti-inflammatory activity of IgG1 mediated by Fc galactosylation and association of FcγRIIB and Dectin-1. *Nat Med.* 2012;18:1401-1406.
476. Mangsbo SM, Sanchez J, Anger K, et al. Complement activation by CpG in a human whole blood loop system: mechanisms and immunomodulatory effects. *J Immunol.* 2009;183:6724-6732.
477. Zhang X, Kimura Y, Fang C, et al. Regulation of Toll-like receptor-mediated inflammatory response by complement *in vivo*. *Blood.* 2007;110:228-236.
478. Seow V, Lim J, Iyer A, et al. Inflammatory responses induced by lipopolysaccharide are amplified in primary human monocytes but suppressed in macrophages by complement protein C5a. *J Immunol.* 2013;191:4308-4316.
479. Fang C, Zhang X, Miwa T, Song WC. Complement promotes the development of inflammatory T-helper 17 cells through synergistic interaction with Toll-like receptor signaling and interleukin-6 production. *Blood.* 2009;114:1005-1015.
480. Maekawa T, Krauss JL, Abe T, et al. Porphyromonas gingivalis manipulates complement and TLR signaling to uncouple bacterial clearance from inflammation and promote dysbiosis. *Cell Host Microbe.* 2014;15:768-778.
481. Zaal A, Lissenberg-Thunnissen SN, van Schijndel G, Wouters D, van Ham SM, ten Brinke A. Crosstalk between Toll-like receptors and C5a receptor in human monocyte derived DCs suppress inflammatory cytokine production. *Immunobiology.* 2013;218:175-180.
482. Bosmann M, Sarma JV, Atefi G, Zetoune FS, Ward PA. Evidence for anti-inflammatory effects of C5a on the innate IL-17A/IL-23 axis. *FASEB J.* 2012;26:1640-1651.
483. Yamada M, Oritani K, Kaisho T, et al. Complement C1q regulates LPS-induced cytokine production in bone marrow-derived dendritic cells. *Eur J Immunol.* 2004;34:221-230.
484. Hajishengallis G, Lambris JD. Crosstalk pathways between Toll-like receptors and the complement system. *Trends Immunol.* 2010;31:154-163.
485. Barratt-Due A, Pischke SE, Nilsson PH, Espevik T, Mollnes TE. Dual inhibition of complement and Toll-like receptors as a novel approach to treat inflammatory diseases-C3 or C5 emerge together with CD14 as promising targets. *J Leukoc Biol.* 2017;101:193-204.
486. Barratt-Due A, Thorgersen EB, Egge K, et al. Combined inhibition of complement C5 and CD14 markedly attenuates inflammation, thrombogenicity, and hemodynamic changes in porcine sepsis. *J Immunol.* 2013;191:819-827.
487. Egge KH, Thorgersen EB, Pischke SE, et al. Organ inflammation in porcine *Escherichia coli* sepsis is markedly attenuated by combined inhibition of C5 and CD14. *Immunobiology.* 2015;220:999-1005.
488. Skjeflo EW, Sagatun C, Dybwik K, et al. Combined inhibition of complement and CD14 improved outcome in porcine polymicrobial sepsis. *Crit Care.* 2015;19:415.
489. Huber-Lang M, Barratt-Due A, Pischke SE, et al. Double blockade of CD14 and complement C5 abolishes the cytokine storm and improves morbidity and survival in polymicrobial sepsis in mice. *J Immunol.* 2014;192:5324-5331.
490. Barratt-Due A, Pischke SE, Brekke OL, et al. Bride and groom in systemic inflammation—the bells ring for complement and Toll in cooperation. *Immunobiology.* 2012;217:1047-1056.
491. Kau AL, Ahern PP, Griffin NW, Goodman AL, Gordon JI. Human nutrition, the gut microbiome and the immune system. *Nature.* 2011;474:327-336.
492. Blander JM, Longman RS, Iliev ID, Sonnenberg GF, Artis D. Regulation of inflammation by microbiota interactions with the host. *Nat Immunol.* 2017;18:851-860.
493. Chehoud C, Rafail S, Tyldsley AS, Seykora JT, Lambris JD, Grice EA. Complement modulates the cutaneous microbiome and inflammatory milieu. *Proc Natl Acad Sci U S A.* 2013;110:15061-15066.
494. Abe T, Hosur KB, Hajishengallis E, et al. Local complement-targeted intervention in periodontitis: proof-of-concept using a C5a receptor (CD88) antagonist. *J Immunol.* 2012;189:5442-5448.
495. Sina C, Kemper C, Derer S. The intestinal complement system in inflammatory bowel disease: shaping intestinal barrier function. *Semin Immunol.* 2018;37:66-73.
496. Wang Y, Tong X, Zhang J, Ye X. The complement C1qA enhances retinoic acid-inducible gene-I-mediated immune signalling. *Immunology.* 2012;136:78-85.
497. Arbore G, Kemper C, Kolev M. Intracellular complement - the complosome - in immune cell regulation. *Mol Immunol.* 2017;89:2-9.
498. Laudisi F, Spreafico R, Evrard M, et al. Cutting edge: the NLRP3 inflammasome links complement-mediated inflammation and IL-1β release. *J Immunol.* 2013;191:1006-1010.
499. Wu XY, Li KT, Yang HX, et al. Complement C1q synergizes with PTX3 in promoting NLRP3 inflammasome over-activation and pyroptosis in rheumatoid arthritis. *J Autoimmun.* 2020;106:102336.
500. Zhang H, Gong X, Ni S, Wang Y, Zhu L, Ji N. C1q/TNF-related protein-9 attenuates atherosclerosis through AMPK-NLRP3 inflammasome singling pathway. *Int Immunopharmacol.* 2019;77:105934.
501. Sorbara MT, Foerster EG, Tsalikis J, et al. Complement C3 drives autophagy-dependent restriction of cyto-invasive bacteria. *Cell Host Microbe.* 2018;23:644.e5-652.e5.
502. Viret C, Rozieres A, Duclaux-Loras R, Boschetti G, Nancey S, Faure M. Regulation of anti-microbial autophagy by factors of the complement system. *Microb Cell.* 2020;7:93-105.

CHAPTER 14

Neutrophils

Andres Hidalgo, PhD • Mariana J. Kaplan, MD

INTRODUCTION

Neutrophils are short-lived, granulocytic leukocytes that contain several types of granules in their cytoplasm in which they store adhesion receptors, enzymes, and antimicrobial peptides that promote their migration toward alarm signals, as well as prominent cytotoxic activities against invading pathogens. These represent the neutrophils' best characterized functions as first-line responders during infections, trauma, and other types of tissue stress. Beyond these canonical roles in immune defense, studies over the past decade have unveiled new aspects of their life cycle such as robust circadian oscillations in number and function, previously unknown mechanisms that mediate immune responses (such as the production of neutrophil extracellular traps or NETs), and functional properties that are not traditionally considered immune responses in various tissues. In addition to patrolling tissues by circulating in the blood, neutrophils enter tissues and can be even found for extended periods in certain organs. While neutrophils feature low transcriptional activity, they are transcriptionally more active than previously thought and manifest variable expression of multiple genes, some of which become prominently expressed depending on the context. Neutrophils are involved in many types of sterile inflammatory reactions and underlie several types of autoimmune disorders and can also contribute to cancer development, which has generated renewed interest in targeting their activity for therapeutic benefit. In this chapter, we describe the fundamental properties of these cells, the most abundant among myeloid leukocytes, with a description of their canonical and best understood functions, as well as recent developments in their biology in health and disease.

GRANULOPOIESIS

Granulopoiesis is the mechanism that gives rise to granulocytes, or granule-bearing leukocytes, within hematopoietic organs, which are the bone marrow and the spleen in healthy individuals. Besides neutrophilic granulocytes, this term includes the generation of eosinophils and basophils. Mast cells, which are also granulocytes, mature extramedullary and their generation is typically not under the definition of granulopoiesis. The main purpose of granulopoiesis is to maintain homeostatic levels of neutrophils. Given the potential toxicity of these cells, this process is under multiple layers of control, including specification of neutrophil progenitors from hematopoietic stem cells (HSCs), the synthesis of granule proteins, storage of mature neutrophils in the marrow, and their mobilization into the bloodstream (Fig. 14.1).

Because neutrophils are abundant in the human blood (representing 50%-70% of all circulating leukocytes, vs about 10% in mice) but are short-lived (see the section on the *Life cycle* below), they are continuously produced by the bone marrow, which devotes a significant fraction of its anabolic efforts toward the production of these cells. Like all other blood lineages, neutrophils originate from HSC. Classically, the several stages of granulocytic maturation have been classified by hematologists based on morphological features of the cells, including, by the order of maturation, myeloblasts, promyelocytes, myelocytes, metamyelocytes, band cells, and mature granulocytes.[1,2] It is important to highlight that, while certain myeloid lineages including tissue-resident macrophages and mast cells have embryonic or dual embryonic/postnatal origin, neutrophils are strictly derived from postnatal HSC.

Contrary to the classical hierarchy of hematopoiesis, recent studies have suggested that phenotypic short-term HSC and multipotent progenitors are at the apex of the hematopoietic tree under steady-state conditions, while long-term HSC are biased toward megakaryopoiesis and contribute to polyclonal hematopoiesis only under conditions of hematopoietic stress.[3,4] The next stage toward neutrophil production involves the common myeloid progenitor and finally the granulocytic/monocytic myeloid progenitor (GMP), which in humans are characterized by expression of IL3Ra and CD45RA markers or FcgR and CD34 in mice.[1,5] Beyond the use of these markers, which is often inaccurate and identifies heterogeneous populations of progenitor cells, new analytical tools that include single-cell sequencing and epigenetic analyses have been used to refine the different stages of myeloid progenitors and have allowed the identification of unipotent, proliferating precursors in mice and humans already committed to the production of neutrophils.[1,6] These studies in part relied on the expression of genes encoding for transcription factors, including *Cebpe* and *Gfi1*, which are important for granulopoiesis.[7] Gf1-1 is a key transcription factor for neutrophil differentiation,[8] as it blocks genes that maintain stemness and repress the monopoietic program in GMP (reviewed in Ref. 9). Further maturation of granulocytic progenitors proceeds under the control of the master granulopoietic cytokine granulocyte colony-stimulating factor (G-CSF)[10] and a family of transcription factors belonging to the CCAAT enhancer-binding protein (C/EBP). Studies in gene-deficient mice showed that G-CSF is needed (but not essential) for normal

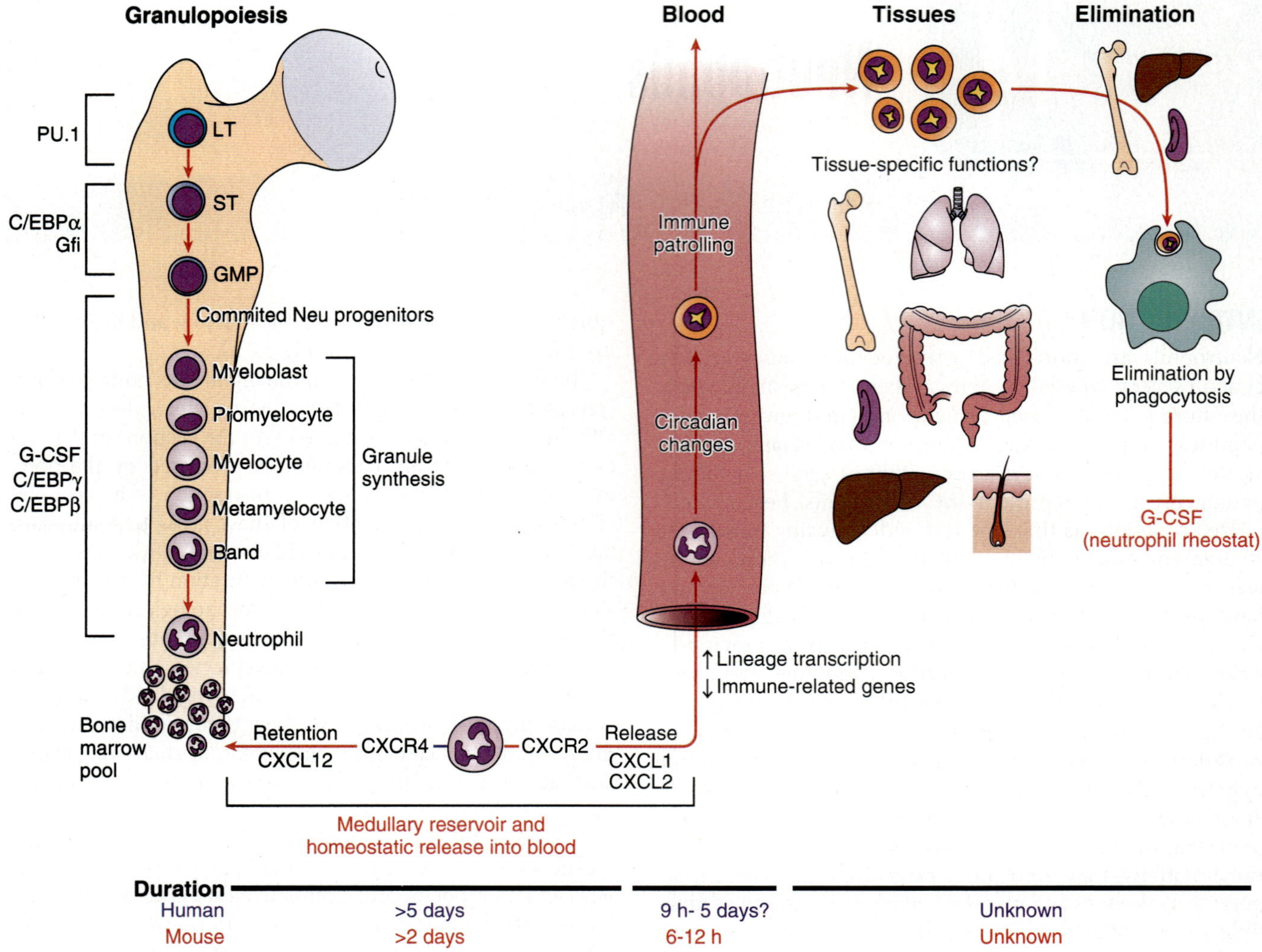

FIG. 14.1. Granulopoiesis and life cycle of neutrophils. Neutrophils are produced mainly in the bone marrow in a process known as granulopoiesis, through a number of intermediary stages that rely on specific transcription factors, and that generates a pool of mature neutrophils in the marrow. The mobilization of these neutrophils is regulated through the action of CXCR4 (retention in the marrow) and CXCR2 (release into the blood). Once in circulation, neutrophils undergo circadian-related changes in phenotype and function that determine the type of response at different times of day. At the end of their lifetime in the blood, neutrophils enter multiple tissues with unknown functions in most cases. The ultimate fate of neutrophils is believed to be phagocytosis by tissue macrophages mainly in the bone marrow, spleen, and liver. This process regulates the levels of granulocyte colony-stimulating factor (G-CSF) and controls granulopoiesis through a negative feedback loop (rheostat). Indicated at the bottom is the estimated duration for each step in human and mice.

neutrophil maturation and production[11] and that C/EBPα is an early regulator of steady-state granulopoiesis that induces myeloid bias in immature precursors, in part by controlling expression of the receptor for G-CSF.[12] In contrast, C/EBPε is involved in the final stages of maturation and is important for normal granule formation and full immune competence.[1,13,14]

A key process during granulopoiesis and functional maturation of neutrophils is the formation of granules and the synthesis of granule proteins, which occurs as the granulocytic progenitors mature in the bone marrow. Granule synthesis appears at the level of promyelocytes and continues until the segmented stage. The different types of granules are formed sequentially during granulocytic maturation through a model referred to as "targeting by timing of biosynthesis" that reflects the uptake of proteins as they are synthesized at different stages of maturation[9] and explains the different protein contents of primary, secondary, and tertiary granules (Table 14.1), which appears highly conserved between mice and humans.[1] An important notion is that biosynthetic processes are strongly compartmentalized during neutrophil maturation. Studies in mouse and human neutrophils have shown that genes encoding for maturation and cytotoxic granule proteins are predominantly transcribed in immature neutrophils in the bone marrow, and transition into the blood is marked by reduced transcriptional rates but preferred expression of genes involved in migration, production of reactive oxygen species (ROS), and vesicle transport.[1,15] The marked genetic regulation of neutrophils at different stages of maturation and anatomical locations has been proposed to confer protection of hematopoietic areas from the neutrophil's cytotoxic activity.

TABLE 14.1 Human Neutrophil Granules and Their Proteins

Type of Granule	Proteins Present	Time of Development
Primary (azurophilic)	Myeloperoxidase, defensins, cathepsin G, bactericidal/permeability-increasing protein, neutrophil elastase, proteinase 3, azurocidin, lysozyme, neutrophil serine protease 4, alkaline phosphatase, hydrolases, phospholipase A2, phospholipase C, phospholipase D, arginase-I, CD63, CD68	Promyelocyte
Secondary (specific)	Lysozyme, lactoferrin, fMLP receptor, cathelicidin, ficolin-1, CD11b/CD18, gelatinase-associated lipocalin (NGAL), OLFM-4, CD15, CD66, CD177, CD67, neutrophil collagenase (MMP8), laminin receptor, transcobalamin II, cytochrome b_{558}	Myelocyte
Tertiary (gelatinase)	fMLP receptor, MMP-9, heparanase, arginase-I, lysozyme, leukolysin, ADAM9 CD11b/CD18, CD45 receptor, laminin receptor, cytochrome b_{558}, β2-microglobulin, gelatinase, diamine oxidase, H+-ATPase, tetranectin, DAG lipase	Metamyelocyte
Secretory vesicles	fMLP receptor, alkaline phosphatase, CD13,C1q-R, CD11a/CD18, CD11b/CD18, complement receptor 1, cytochrome b_{588}	Neutrophil

The production of neutrophils in healthy organisms, which is described above, undergoes significant alterations under conditions of stress, such as infections, chronic inflammation, or cancer. Under these scenarios, a process of *emergency granulopoiesis* takes over to allow for enhanced production of neutrophils to efficiently combat pathogens and to replenish cells that are more rapidly "consumed" at the infection site.[5] During this emergency granulopoiesis, myeloid progenitors leave the bone marrow and production of neutrophils becomes prominent in extramedullary organs, such as the spleen and liver, and can even occur at the site of infection.[16,17] Another relevant developmental change is that the process becomes dependent on C/EBPβ, a transcription factor that allows simultaneous proliferation and differentiation of granulocytic progenitors.[18] The process is activated by the presence of pathogen-associated molecular patterns (PAMPs) recognized by dedicated sensors, such as Toll-like receptors (TLRs), which signal at the level of myeloid progenitors to favor granulocytic commitment.[19] Studies in tissue-specific deficient mice have additionally demonstrated that production of G-CSF by the vascular endothelium is an important early event driving emergency granulopoiesis.[20]

A final step toward effective granulopoiesis is the release of neutrophils to the bloodstream, where they perform their immune functions. Classical studies showed that this release, however, is not immediate when the granulocytic progenitors stop proliferating.[21,22] Indeed, a "postmitotic" pool is retained in the marrow for 5 to 6 days (2-3 days in rodents) for further maturation or for release under conditions of emergency.[1,2] This phenomenon of *mobilization* must be tightly regulated in order to prevent excessive accumulation of neutrophils in peripheral tissues and potential inflammatory damage. Illustrating this strict control, neutrophil numbers in the blood tightly oscillate during the day[23] and are kept at relatively low values despite the presence of the large reservoir of mature neutrophils in the bone marrow, which accounts for over 90% of the total neutrophil pool. Under homeostatic conditions, the release of neutrophils is based on opposing signals from two chemokine receptors: on the one hand, CXCR4 retains neutrophils (as well as other hematopoietic cells) in the marrow, which is rich in its ligand CXCL12. On the other hand, signals through CXCR2 promote mobilization.[24] The balance between the two chemokine-regulated signals determines whether neutrophils are retained in or released from the bone marrow, depending on the environmental demands. Inhibition of the CXCL12/CXCR4 axis under situations of infection or inflammation allows the *en masse* release of neutrophils into the blood.[25,26] G-CSF, the key granulopoietic cytokine, inhibits CXCR4 signaling and is important for the mobilization of neutrophils out of the bone marrow, particularly under emergency conditions.[10] For this reason, this cytokine is used in the clinic to increase the mobilization and neutrophil counts in the blood of immunocompromised patients.[27] Notably, low levels of neutrophils in peripheral tissues elicit the production of G-CSF, a phenomenon known as the "neutrophil rheostat" that is caused when tissue macrophages are deprived of target cells for phagocytosis and that helps to maintain physiological levels of neutrophils in circulation.[28]

LIFE CYCLE

The initial stages of the neutrophil life cycle, including their production from primitive HSC and more-committed myeloid progenitors and their release into the blood, have been described in the previous section and will not be further discussed here. This section rather focuses on other important stages in the neutrophil life cycle after they leave the bone marrow, as they get access to the circulation and disseminate in tissues (Fig. 14.1).

A key feature that endows neutrophils with a distinct life cycle among other hematopoietic cells is their short lifespan, compared with most other leukocyte subsets. Neutrophils present in the blood or in tissues are postmitotic and unable to divide. Illustrating this key feature, it has been shown that mitogens (proliferation-inducing agents), instead of inducing proliferation, elicit a type of "aborted mitosis" in neutrophils that causes the extrusion of their DNA into the milieu in the form of NETs (see more on this in *Function and Antimicrobial*

Strategies section).[29] Given this inability to proliferate and their essential antimicrobial functions, much effort has been invested in defining the actual lifetime of neutrophils in the blood and extramedullary tissues. By measuring the clearance time of neutrophils transferred into healthy individuals, classical studies inferred half-lives of 7 to 9 hours in the human blood.[22] More refined studies using isotype labeling of unmanipulated cells challenged these early findings to propose much longer lifespans of >5 days; however, these studies remain controversial because concurrent labeling by the same agents of myeloid progenitors, which feed new waves of labeled neutrophils in the blood, can make these calculations inaccurate.[30,31]

Contrary to studies in humans, measurements in mice consistently reveal shorter half-lives of 9 to 18 hours.[30] These values, shorter than 1 day, are relevant to understand temporal aspects of the neutrophil life cycle as they seem to adjust to circadian cycles,[2,32] as discussed in more detail in *Function and Antimicrobial Strategies* section. Neutrophils of mouse or human origin placed in culture also remain viable for short periods of time (again, less than 1 day), which suggested that built-in, cell-intrinsic mechanisms limit the viability of these cells. One of these mechanisms is regulation of the stability of the short-lived protein Mcl-1, a member of the Bcl2 family.[33] Mcl-1 prevents apoptosis in part by coupling and blocking proapoptotic factors, but when degraded over time leaves neutrophils subject to the action of these factors.[34] As a consequence, mice lacking the gene encoding Mcl-1 in myeloid cells display profound neutropenia.[35] Importantly, proinflammatory factors such as GM-CSF or lipopolysaccharide, as well as certain infection-associated factors, have been shown to extend the lifespan of neutrophils.[2] This life prolongation, which may be useful to combat pathogens, has been proposed to act by preventing degradation of Mcl-1.[34] In contrast, precipitating neutrophil apoptosis may be therapeutically beneficial in the context of chronic inflammatory diseases and can be triggered by inhibitors of cyclin-dependent kinases.[36] Illustrative of the physiological importance of keeping neutrophil numbers and susceptibility to apoptosis under check is the constant discovery of new regulatory mechanisms. For example, a long noncoding RNA called *Morrbid* has been shown to regulate neutrophil survival by controlling the transcription of Bim, a neighboring proapoptotic gene.[37]

The extension of the life cycle of neutrophils in tissues, beyond the circulation, remains poorly studied and there is no consensus as to their fate once they extravasate and reach organs under homeostatic, noninflammatory conditions. Studies focused on neutrophils that home back to the bone marrow, however, have demonstrated that their engulfment by resident macrophages elicits signals that can have relevant effects in the tissues. In the case of the marrow, infiltrating neutrophils cause the suppression of hematopoietic niche factors (such as CXCL12) and the mobilization of HSC into the circulation.[23] In other organs, uptake of neutrophils by local phagocytes suppresses the transcription of IL-23, an upstream regulator of G-CSF production, thereby contributing to the rheostat mechanism that regulates systemic neutrophil numbers.[28] These studies have helped to illustrate the need to further understand the role of neutrophils that enter tissues for their final elimination, even in infection-free scenarios.

NEUTROPHIL SUBSETS

Studies have identified a remarkable diversity of neutrophil phenotypes in specific contexts of infection, chronic inflammation, or cancer. However, the existence of bona fide neutrophils subsets remains controversial. There are several reasons behind this lack of consensus. One is the actual definition of a lineage, since major deviation from a "classical" phenotype raises concern about the actual lineage of the cells,[38] as illustrated, for example, by the identification of a hybrid population of neutrophils sharing features with dendritic cells (DCs).[39] A second concern is whether changes in a limited set of markers can define a true subset or rather an activation state of the cell.[40] Further, the short lifespan and limited transcriptional activity of neutrophils when compared with other leukocytes has been classically believed to limit their capacity to adopt distinct transcriptional profiles and functional properties.[41]

Despite these concerns, studies focused mainly on human blood neutrophils have identified small populations of neutrophils featuring distinct expression of antigens such as CD177 or granule proteins like olfactomedin 4.[41] The physiological significance of such populations remains, however, unknown. Changes in neutrophil phenotypes can be also associated with the time that neutrophils have spent in the circulation. Indeed, neutrophils that have spent longer times in circulation (referred to as "aged") express higher levels of the chemokine receptor CXCR4 on their surface and lower levels of CXCR2 and the selectin CD62L,[23,32] as discussed in more detail in the *Circadian Regulation of Neutrophil Function* section.

More physiologically relevant evidence that specialized neutrophil subsets exist, however, derives from studies under certain pathological contexts. For example, studies in ischemic tissues revealed populations of neutrophils expressing CXCR4, CD49d, and VGEFR1 that featured angiogenic properties, which depended in part in the metalloprotease MMP-9.[42,43] Cancer is another pathological state in which the phenotypic and functional divergence of neutrophils is prominent. Depending on the type and stage of the cancer, neutrophils acquire changes in surface markers, physical properties (ie, density), and functions (reviewed in Ref. 38). Early experiments by Fridlender and colleagues discovered that depletion of neutrophils in mice with lung adenocarcinoma normally caused delayed tumor growth; however, if the tumor-bearing mice were pretreated with an inhibitor of TGFβ signaling, then neutrophil depletion accelerated tumor growth.[44] These and additional experiments in murine mouse models and in tumor-bearing human patients revealed the existence of neutrophils that either promoted or inhibited tumor growth. These effects in cancer have been proposed, in some instances, to correlate with the capacity of various phenotypically different subsets of neutrophils to suppress T cell function. For example, studies in humans have found that neutrophils expressing CD10,[45] or high levels of CD16 and low levels for CD62L,[46] potently suppress T cell activation and appear in the circulation after inducing mobilization or endotoxemia. In contrast, a CD10-negative CD15^{high} population promotes antitumor T cell responses.[47] These tumor-associated neutrophils (TANs) with pro- or antitumoral and inflammatory properties are sometimes referred to as N1 and N2 neutrophils, respectively.[48] Related to the tumor-supportive N2 population, a type of immature neutrophil population with immune-suppressive properties can be found in the marrow and spleen of mice and humans. These cells are referred to as *granulocytic myeloid–derived suppressor cells* or G-MDSC[49,50]; however, because no specific markers define G-MDSC, it is unclear if they represent a true subset of neutrophils or rather a general feature of immature myeloid cells that mobilize to the periphery in the presence of a tumor.

A distinct subset of human proinflammatory neutrophils that are purified from the mononuclear cell fraction following density centrifugation has been described to be expanded in a variety of autoimmune diseases, and they are therefore called low-density granulocytes (LDGs).[51] Originally described in systemic lupus erythematosus (SLE),[51,52] LDGs have subsequently been detected in other autoimmune diseases including antineutrophil cytoplasmic antibody–associated vasculitis,[53] antiphospholipid syndrome,[54] idiopathic inflammatory myopathies,[55] and psoriasis.[56,57] LDGs synthesize higher levels of proinflammatory cytokines than normal-density neutrophils. Characteristically, LDGs have an enhanced ability to form immunogenic and vasculopathic NETs.[55,58,59] Indeed, LDGs have been associated with enhanced vascular damage *in vitro* and *in vivo*, at least in part through this enhanced ability to form NETs.[60,61] This will be discussed in more detail in *Function and Antimicrobial Strategies and Neutrophils in Disease States* sections. Overall, LDGs appear to significantly differ from normal-density neutrophils at the transcriptional, epigenetic, and functional levels, supporting the notion that they represent a distinct neutrophil subset.[52]

Contrasting with these emerging data in the context of disease, there is current controversy on whether neutrophil subsets exist in healthy tissues. This location-associated diversity in tissues is well known for macrophages[62] and other myeloid lineages;[63] however, this has not been yet reported for neutrophils beyond the bone marrow, where changes associate with maturational stages,[1] and for a type of B helper neutrophils localized in marginal zone of the spleen.[64] The advent of technologies that allow mass analysis of receptors and transcripts at the single-cell resolution[7,63] will allow establishing whether dedicated, tissue-specific populations of neutrophils exist (Fig. 14.1).

NEUTROPHIL TRAFFICKING INTO AND WITHIN TISSUES

A primary function of neutrophils is responding to infections and other types of tissue insults by rapidly migrating to the attacked tissues, which allows elimination of the infection and repair of the tissue. Trafficking into the affected tissues is achieved by the capacity of neutrophils to recognize cues on vessel-lining endothelial cells, which are activated upon recognition of a variety of PAMPs or damage-associated molecular patterns. The migration from the blood into the tissue parenchyma occurs in a series of steps that form part of the so-called neutrophil *recruitment cascade*. Although the modality of recruitment can vary from tissue to tissue, depending on the particular vascular architecture of each organ,[65] here we summarize the general features of the recruitment cascade, with only occasional reference to organ peculiarities.

Activation of endothelial cells generally induces the expression of receptors that allow interactions with neutrophils and other leukocyte types. In the vasculature of the cremaster muscle, where most studies of the recruitment cascade have been made due to its optimal optical properties for intravital microscopy, recruitment is initiated in postcapillary venules. In these initial stages, endothelial selectins (P-selectin and E-selectin) are essential to engage circulating neutrophils through high-affinity interactions with selectin ligands expressed on the leukocyte. Both selectins recognize sialylated and fucosylated carbohydrate moieties on only certain glycoproteins, including P-selectin glycoprotein ligand-1 (PSGL-1), E-selectin ligand 1 (ESL-1), and CD44,[66] but also glycolipids in human neutrophils.[67] Besides mediating the physical capture of the cells, selectin-mediated interactions initiate signals in the leukocyte that favors subsequent stages of recruitment, namely cellular polarization and integrin activation. Specifically, engagement of PSGL-1 with E-selectin triggers signals that activate the β2 integrin LFA-1 (CD11a/CD18 or αLβ2) to bind ICAM-1 expressed on the endothelium;[68] for this reason, this phenomenon is believed to coordinate the transition from slow rolling to firm arrest onto the vascular bed. Chemokines like CXCL1 and CXCL2, which are expressed by the endothelium, perivascular cells (eg, pericytes), or the neutrophils themselves,[69,70] are important for firm arrest and migration of the cells on the activated vessels. This process of intravascular *crawling* on the activated endothelium is mediated by Mac-1, another β2 integrin, which binds ICAM-1/2 on the vascular surface, and is believed to allow the migration to specific sites of extravasation.[71] Alternative to this canonical cascade, activation and migration of neutrophils can be initiated by alternative cues, such as hyaluronan in the liver,[72] or by passive entrapment, as occurs in the renal glomeruli.[73]

Another process of importance for efficient recruitment of neutrophils is early polarization, a process that entails the molecular redistribution of receptors when neutrophils and other leukocytes first attach to the vascular surface.[74] Although still poorly appreciated, this phenomenon has at least two major effects during recruitment; first, it allows the directional movement of neutrophils within the vessel;[75] second, it concentrates receptors in subcellular domains that may function as adhesive and signaling platforms during inflammation.[76] In the domain at the rear of the crawling neutrophil, or uropod, clustering of PSGL-1 mediates the capture of activated platelets (which express its ligand, P-selectin), and delivery of signals through this receptor promotes neutrophil crawling and extravasation.[77]

After crawling on the vascular surface, neutrophils eventually find appropriate sites for extravasation, typically through junctions between endothelial cells in a process referred to as *paracellular* diapedesis.[78] This step involves multiple heterotypic and homotypic interactions and appears to be guided in part by deposition of the chemokine CXCL2, produced by the neutrophils themselves, which is presented by the atypical chemokine receptor 1 (ACKR1) at the endothelial junctions.[70] Neutrophils that have crossed the endothelial lining into the perivascular space must, at least in some tissues, cross a barrier of perivascular cells and matrix to reach the tissue parenchyma. This final stage occurs in special areas of low matrix and cell density and is guided by CXCL1 produced by pericytes that surround the vessels.[70,79] However, the absence of perivascular cells in certain tissues and the presence of fenestrated vessels in others, such as the kidney, suggest alternative mechanisms for neutrophils to gain access to the tissue parenchyma. These various steps and alternative mechanisms illustrate the enormous diversity of vascular beds and pathological scenarios and the multiple adaptations of neutrophil recruitment under each condition.[78] In more recent years, a number of laboratories have identified the reverse migration of neutrophils, namely from the extravascular space back into the circulation. Initial studies of migration across endothelial layers *in vitro*[80] were later confirmed by imaging of neutrophils exiting wounds and returning to the circulation in the transparent zebrafish model.[81] In mouse models of inflammation induced by specific inflammatory stimuli, such as the lipid chemoattractant leukotriene B4 (LTB4), or by ischemia-reperfusion injury, reverse migration of ICAM-1+ neutrophils has been proposed as one mechanism that enables dissemination of activated cells and amplification of inflammation.[82]

Finally, neutrophils that have gained access to the tissue parenchyma must navigate a completely different environment rich in cellular and extracellular matrix constraints. This entails an entirely different modality of migration of polarized neutrophils, which now follow interstitial gradients of small chemotactic, mechanical, and even chemical signals.[83] These regional cues, together with integrin-independent propulsion by the cell's cytoskeleton,[84] as well as selection of micropores of sufficient size for their nuclei[85] drive the efficient navigation of inflammatory leukocytes in the tissue toward their final destination. Interestingly, neutrophils that have reached foci of danger, such as a dying cell or bacteria,[86] are able to amplify the recruitment of more neutrophils to effectively contain the affected area. This phenomenon, referred to as *swarming*, is a distinct strategy of neutrophils that is initiated by a small number of "scouting" cells in the tissue,[87] is mediated by the paracrine release of LTB4, and allows *en masse* recruitment of neighboring neutrophils to the foci of danger.[86]

FUNCTION AND ANTIMICROBIAL STRATEGIES (FIG. 14.2)

Degranulation/Reactive Oxygen Species

As an important antimicrobial strategy, neutrophils utilize preformed proteins stored in intracellular granules (Table 14.1). Neutrophil degranulation is the mechanism of deployment of the neutrophil armamentarium against microorganisms and can also induce robust inflammatory responses. Primary (azurophilic), secondary (specific), and tertiary (gelatinase) neutrophil granules contain distinct potent enzymes, adhesion molecules, receptors, and opsonins.[88] Primary granules contain myeloperoxidase (MPO), cathepsin G, neutrophil elastase, proteinase 3, various defensins, and lysozyme. Secondary granules (also known as specific granules) contain lactoferrin, cathelicidin, lysozyme, collagenase, gelatinase, gelatinase-associated lipocalin (NGAL), and OLFM-4, among others. Tertiary granules contain

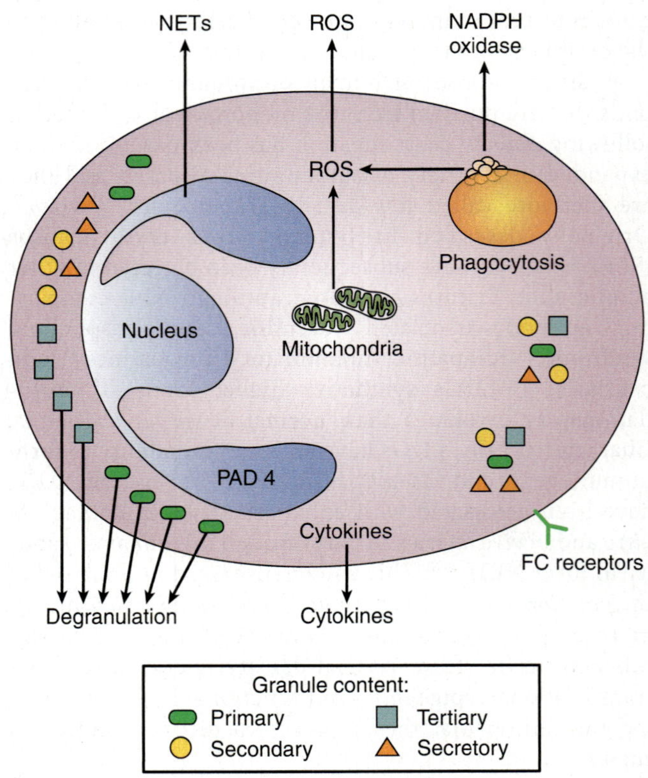

FIG. 14.2. Neutrophil functions. Neutrophils are multifunctional granulocytes with crucial roles in antimicrobial responses and inflammation. Activation of neutrophils can occur through a variety of specific receptors including pattern recognition receptors and Fc receptors. Reactive oxygen species (ROS) are produced intracellularly through multiple mechanisms including the nicotinamide adenine dinucleotide phosphate (NADPH) oxidase complex and mitochondria. Degranulation of the antimicrobial, cytotoxic contents within various types of neutrophil granules has antimicrobial properties but is also toxic to the local tissue environment. Secretion of cytokines/chemokines can promote broader immune responses. Formation of neutrophil extracellular traps (NETs) serves a variety of antibacterial, immunogenic, and immunosuppressive functions in human disease.

matrix metalloproteinase-9 (MMP-9), arginase 1, lysozyme, leukolysin, and ADAM9[88] (Table 14.1). During degranulation, in response to danger signals, neutrophil granules translocate to the phagosomal or plasma membranes where they dock and fuse with the membrane and release their content in a controlled and tightly regulated manner. The exocytosis of granules requires energy generation and reorganization of the actin cytoskeleton. Granule proteins, including neutrophil elastase, lysozymes, and cathepsin G, can harm bacterial walls and degrade virulence factors, while cathelicidin and defensins are microbicidal and can also activate innate and adaptive immune responses.[89] Secretory vesicles constitute a reservoir of membrane-associated receptors such as CD10, CD11b/CD18, CD15, and CD16, among others. They also contain actin and actin-binding proteins. The molecules contained in these vesicles are crucial for a neutrophil to establish contact with activated endothelial cells and undergo chemotaxis-driven migration within inflamed tissues.

Of all the human phagocytes, neutrophils are the prime producers of ROS. The generation of superoxide radicals and other ROS through an oxidative burst is a fundamental mechanism by which activated neutrophils display antimicrobial properties. The nicotinamide adenine dinucleotide phosphate (NADPH) oxidase complex synthesizes superoxide, a reactive oxygen intermediate. Superoxide dismutase converts the superoxide anion into hydrogen peroxide, followed by either production of hypochlorous acid in a reaction catalyzed by MPO or production of water that is catalyzed by catalase. ROS can either be released to the extracellular space at the site of infection or intracellularly in the phagolysosome, following microbe phagocytosis. ROS generation activates degranulation, promotes the formation of NETs, regulates autophagy and transcriptional activity, and stimulates the synthesis of various proinflammatory cytokines. ROS generated by neutrophils can cross microbial membranes and damage intracellular content.[90-92]

The NADPH oxidase is a multiprotein electron transfer system composed of five molecules and Rac2. The catalytic core (flavocytochrome b_{558}; $cytb_{588}$) is made up of two transmembrane proteins ($gp91^{phox}$ and $gp22^{phox}$). During the resting state, $cytb_{588}$ is localized in the phagosomal membrane, the plasma membrane, and the membranes of specific granules and secretory vesicles. This molecule catalyzes electron transfer from NADPH to oxygen, generating superoxide anions.[93] The regulatory subunits of the NADPH complex ($p40^{phox}$, $p47^{phox}$, and $p67^{phox}$) are localized to the cytoplasm of resting cells. NADPH components become activated by a variety of proinflammatory mediators and the activation of pattern recognition receptors. Following activation of neutrophils, $cytb_{588}$ translocates to the plasma membrane, where phosphorylation of $p47^{phox}$ occurs, with subsequent conformational changes. During neutrophil activation, Rac2 (a small GTPase protein) binds directly to $gp91^{phox}$ and $p67^{phox}$ and is required for ROS production.[94] Following activation, the formed NADPH oxidase complex promotes the transfer of electrons from the cytosol to oxygen, leading to the formation of superoxide anions.[93]

Neutrophils also express various membrane-bound and endocytic TLRs that sense a variety of danger signals and activate complex signaling pathways that include the activation of the NADPH oxidase machinery and ROS synthesis.[95]

Phagocytosis

When neutrophils phagocytose bacteria, this leads to direct bactericidal programs that evolve into complete degradation of the internalized microbe. Neutrophils also engulf other cells and cell fragments. Neutrophil phagocytosis is mediated by a variety of pattern recognition receptors, including C-type lectins (particularly Dectin-1) and Fcγ receptors.[9,96,97] When C-type lectin receptors bind to microbial surfaces, they facilitate their capture and ingestion. C-type lectin receptors can also promote ROS synthesis, trigger intracellular pathways that promote membrane reorganization and microbial phagocytosis, and influence the rate of phagosome maturation.[97] Neutrophil phagocytosis is enabled by microbe opsonization with immune complexes. Ligation of membrane-bound opsonin receptors initiates changes in the cytoskeleton that promotes the start of phagocytosis. Neutrophils express both high-affinity and low-affinity Fc receptors that recognize and phagocytose antibody-opsonized pathogens and promote ROS synthesis.[98] Human neutrophils express FcγRIIA and RIIIB, and signaling through these receptors is required to elicit ROS synthesis by IgG immune complexes.[99] Binding of IgG-opsonized particles to neutrophil Fcγ receptors leads to clustering of the receptor and activation of receptor-proximal Src family kinases (SFKs). These SFKs phosphorylate the tyrosine residues within the immunoreceptor tyrosine-based activation motif (ITAM) in the cytoplasmic tail of the Fc-gamma receptors. This, in turn, provides a docking site for Syk recruitment and activation. The activation of Syk leads to modification in lipids and remodeling of actin, promoting cell membrane protrusion and engulfment of the microbe.[100] In addition to immune complexes, the complement system contributes to efficient phagocytosis by neutrophils, for example, through the integrin αMβ2 (also known as C3R).

The phagocytic process is characterized by different phases: (1) the opsonized particle attaches to specific receptors; (2) there is formation of pseudopod extensions around the attached particle; and (3) completion of the engulfment with formation of a phagosome (outside-in compartment inside the cell). This is followed by mobilization and fusion of the phagosome with various granules, leading to degranulation and killing of the microorganism. This finally triggers an oxidative burst, ROS formation, and subsequent bacterial killing and digestion of the microbe.[96]

Bacteria can employ several strategies to evade phagocytosis, including anti-opsonic strategies such as the secretion of IgG-binding proteins, cleavage of IgG and complement 3 (C3) protein by secreted bacterial peptides, and complement inhibitors. Bacteria have also developed strategies to evade killing after being internalized by neutrophils. For example, they can secrete toxins that lyse the neutrophil from within following phagocytosis. Some bacteria can secrete scavenger enzymes like catalase to evade ROS-mediated killing.[101]

Neutrophil Extracellular Traps

NETs are extracellular web-like structures released by the neutrophils upon encountering danger signals from microbes or their products, crystals, immune complexes, cytokines, and chemical agonists.[59,102-104] NETs contain decondensed chromatin (nucleic acids, histones) bound through electrical charge to granule proteins. While the precise intracellular mechanisms leading to NET formation remain to be fully characterized and may vary according to the type of stimulus,[105] this process has been associated with NADPH oxidase assembly and ROS generation. ROS promotes activation and dissociation of a complex called "azurosome," formed by neutrophil elastase, MPO, azurocidin, and other granule proteins. This process leads to the translocation of neutrophil elastase and MPO into the nucleus, and these enzymes cause cleavage of histones and chromatin decondensation.[106,107] Peptidylarginine deiminase 4 (PAD4) is activated during NET formation and induces histone citrullination, further promoting chromatin decondensation.[108] Eventually, the nuclear membrane disintegrates and NETs are extruded to the extracellular space. NETs can immobilize microbes such as *Candida albicans*[109] and have been reported to have microbicidal activity against various bacteria, fungi, parasites, and viruses.[110-113] Beyond trapping and killing microorganisms, NETs can neutralize virulence factors, inhibit microbial growth, and promote functional opsonization of fungi.[114] NETs are degraded by the endonuclease DNAse1, which cleaves NET-bound DNA.[115] NETs can also be engulfed by phagocytes, such as macrophages and DCs, and defensins and nucleases promote this intracellular uptake and intracellular digestion of these traps.[116,117]

Microbes have acquired strategies to evade NETs. For example, some pathogens inhibit oxidative pathways that are important in NET formation, whereas other microbes degrade NETs through synthesis of DNAses. Certain pathogens produce biofilm or undergo electrochemical modifications in their surface that make them more resistant to the effects of NETs. Some bacterial virulence factors convey resistance to the effects of antimicrobial peptides present in NETs.[114]

Importantly, NETs also externalize a variety of autoantigens that are known to be primary targets of the immune system in systemic autoimmunity, including modified genomic and mitochondrial DNA and various proteins, such as MPO and citrullinated peptides.[53,58,59,103,118] Dysregulation in NET formation and/or in the clearance of NETs has been implicated in a variety of chronic disorders including autoimmune and autoinflammatory conditions, cancer, and atherothrombotic syndromes.[52,55-58,60,103,119-122] This is explored further in *Neutrophils in Disease States* section. In addition to lytic cell death through NET formation, it has been described that certain microbes like *Staphylococcus aureus* can trigger NET formation in a nonlytic, oxidant-independent, rapid manner, without causing damage to neutrophils' cell membrane integrity.[110]

Angiogenesis/Cytokines

Under conditions of hypoxia and tissue injury, sprouting vessels promote neutrophil migration and, in turn, neutrophils promote neoangiogenesis.[123] Neutrophils can release proangiogenic factors such as vascular endothelial growth factor A (VEGF-A) and MMP-9. NET formation can also have proangiogenic roles.[124] A distinct proangiogenic neutrophil subset has been reported in mice and humans. This subset exhibits high expression of the VEGF-A receptor VEGFR1 and CD49d and is recruited to ischemic tissues to promote local angiogenesis. These neutrophils interact with the endothelium in a CXCL12-dependent manner and modulate angiogenesis through matrix modulation and increasing the action of proangiogenic factors.[43] Neutrophils can also phagocytose cellular debris deriving from damaged blood vessels and create new paths for revascularization. Neutrophils induce proangiogenic T cells that appear to be important during pregnancy at the fetal-maternal interface.[125] Additionally, they can exert antiangiogenic roles through neutrophil elastase–dependent cleavage of VEGF and other proangiogenic factors.[126]

Besides the angiogenic factors mentioned above, neutrophils can synthesize and release a large number of cytokines and chemokines, including proinflammatory (TNF, RANK ligand, IL-1, IL-6, IL-7, BAFF, etc.), anti-inflammatory (TGFβ), colony-stimulating factors such as G-CSF and M-CSF, and many CC and CXC chemokines.[127,128]

Circadian Regulation of Neutrophil Function

As indicated above (see the section on the Life Cycle), the brief lifetime of neutrophils relative to other leukocytes creates unique temporal properties for these cells.[129] Because the half-life of neutrophils in the circulation is roughly half a day and these cells released from the marrow preferably during the night (in mice), the numbers of neutrophils follow strict 24-hour, diurnal cycles. These cycles are entrained by light and are maintained under constant darkness,[32] which indicates that neutrophil abundance is a bona fide circadian process. In addition to circadian oscillations in cell numbers, the neutrophil phenotype also varies predictably during the day. For example, certain markers change diurnally, with the surface levels of CD62L and the chemokine receptor CXCR2 being reduced, and those of CXCR4 increasing to reach a peak at zeitgeber time 5 (ZT5; or 5 hours after the onset of light) both in mice and humans.[23,32,130] Differences among species, however, may exist since oscillations in the capacity to produce ROS and to phagocytose bacteria are prominent in human neutrophils, but not in mice.[131] Transcriptomic and functional analyses have demonstrated that these circadian changes affect the transcriptional and migratory properties of murine neutrophils in the blood, such that nighttime neutrophils are more migratory than at daytime,[23,32] thus explaining the circadian infiltration of healthy tissues even in the steady state. Deletion of the core circadian gene *Arntl* (which encodes the circadian transcription factor Bmal1) from neutrophils demonstrated that this circadian-driven program in neutrophils is important for the differential response to infections and during sterile inflammatory reactions (such as myocardial infarction or vascular occlusion in sickle cell disease models).[32,130,132] Mechanistic studies showed that the underlying process behind these responses is the direct transcriptional regulation by Bmal1 of the gene encoding CXCL2, an agonist for the receptor CXCR2. Autocrine signaling through CXCR2 is then required to elicit the circadian changes seen in murine neutrophils. Because CXCR4 inhibits signaling through CXCR2,[32,133] this receptor acts as a negative regulator of the Bmal1- and CXCR2-dependent neutrophil circadian clock.

In addition to the transcriptional and migratory properties, the granule content and proteome composition of neutrophils is also regulated by this circadian clock. Using a model of acute pulmonary injury, studies have shown that diurnal loss of granule content renders blood neutrophils less toxic to tissues, both in mice and humans.[134] Overall, these studies suggest that the circadian properties intrinsic to neutrophils provide an important regulatory mechanism to balance the benefits of fighting off microbial infections with the risk of exaggerated damage to the host tissues due to inflammation.[135] An additional advantage of this intrinsic system, as for most circadian physiological processes, is the capacity to anticipate changes in the environment, for example, the predictable exposure of the organism to pathogens or risk of injury during the active period of the animal, which appears to be a general property of the immune and hematopoietic systems.[136]

Coagulation

Neutrophils play crucial roles in various steps of the clot formation process. NETs in particular can promote coagulation through pleiotropic effects on the blood cells, soluble plasma components, interactions with blood vessel wall, and disturbances in blood flow.[137-139] NETs can induce activation of factor XII, leading to further activation of the coagulation cascade. Furthermore, the release of DNA in NETs triggers thrombin generation in tissue factor-dependent and independent manner.[137] Histones can also promote activation of coagulation and thrombosis through platelet aggregation and thrombin generation, endothelial cell activation, vascular dysfunction, and direct effects on red blood cells.[140] Overall, NETs can act as scaffolds that promote platelet aggregation and activation and coagulation factor activation and also lead to the release of proteases that inactivate various anticoagulants.[141]This has implications not only for coagulation under homeostatic conditions but also in disease pathogenesis, including sepsis, autoimmune disorders, atherosclerosis, sickle cell disease, and cancer.[142]

Tissue Repair

Neutrophils play fundamental roles in wound healing and tissue repair.[143] Danger signals at sites of tissue injury promote the migration of neutrophils into these damaged areas. Using their canonical functions, neutrophils play an important role in preventing secondary infection at sites of tissue injury. In addition, there is evidence that neutrophils also have direct and indirect (ie, by modulating the phenotype and function of other inflammatory cells at sites of injury) tissue restorative functions.[144] As examples, tissue macrophages acquire tissue remodeling capabilities by phagocytosis of apoptotic neutrophils in tissues.[145] Neutrophils can directly induce epithelial cell proliferation and healing of colonic wounds.[146] The release of adenosine by neutrophils has prohealing effects in the intestinal wall[147] and neutrophils can also promote the release of resolvins, microRNAs, and TGFβ to further promote wound repair.[148,149] Conversely, neutrophils have been associated with the development of nonhealing wounds such as what has been observed in diabetics and other metabolic diseases. NET formation may interfere with proper wound healing.[150] Furthermore, neutrophils can release microparticles carrying microRNAs that promote accumulation of double-stranded DNA breaks, leading to genomic instability in epithelial cells.[151]

NEUTROPHILS IN DISEASE STATES

Immunodeficiencies

Defects in neutrophils' numbers and/or function frequently result in an enhanced predisposition to frequent and more severe infections, many times with opportunistic bacteria.[152] Some of these diseases are summarized below.

Severe Congenital Neutropenia

This condition results from mutations in various neutrophil genes, such as those encoding for HCLS1-associated protein X 1(HAX-1), adenylate kinase 2 (AK2), glucose-6-phosphate complex 3 (G6PC3), neutrophil elastase, CXCR4, CSF3R, etc. In this condition, neutrophil formation arrests at the promyelocytic stage and subjects display very low neutrophil counts and life-threatening infections that start early in life.[153] Subjects affected by this condition require treatment with G-CSF and, in some cases, bone marrow transplant.

Chronic Granulomatous Disease

This is an inherited heterogeneous immunodeficiency syndrome that affects 1 in every 200,000 to 250,000 live human births.[154] Chronic granulomatous disease (CGD) is caused by mutations in genes encoding for components of the NADPH oxidase complex, with subsequent defects in ROS synthesis by phagocytes, including neutrophils. Neutrophils then fail to mount a respiratory burst leading to a clinical phenotype characterized by recurrent and life-threatening fungal and bacterial infections, including *S. aureus*, *Serratia marcescens*, *Burkholderia cepacia*, *Nocardia* species, and *Aspergillus* species. Patients often develop abscesses in many organs. In addition to the immunodeficiency, some individuals affected by this condition present with large and diffuse granulomas in various areas of the body, fibrosis, and chronic inflammatory conditions. A percentage of CGD subjects also develop autoimmunity.[155] In addition to chronic use of antibiotics and antifungals, people affected by this condition may be treated with recombinant interferon (IFN) gamma, in order to improve function of phagocytes. Some individuals affected by CGD have been treated successfully with bone marrow transplantation.[156]

Myeloperoxidase Deficiency

This is the most common inherited disorder of phagocytes and in many cases is not associated with clinical symptoms. When clinically relevant, mutations in the MPO gene are associated with recurrent *Candida* and *Aspergillus* infections.[157]

Leukocyte Adhesion Deficiencies

These autosomal recessive diseases develop due to defects in neutrophil cell adhesion to extracellular matrix or to the endothelium, therefore impairing neutrophil recruitment. Mutations in genes encoding for specific integrins are associated with severe infections by *S. aureus* and gram-negative bacteria, particularly affecting the skin and oral mucosa. Leukocyte adhesion deficiency (LAD) type I is the most common type and affects CD18, the common chain of the β2 integrin family, which is required for the expression of CD11a/CD18, CD11b/CD18, and CD11c/CD18.[158] Neutrophil adhesion to the endothelium is severely compromised in LAD I. Patients develop frequent skin and periodontal infections, deep tissue abscesses, and delayed separation of the umbilical cord. Patients present with increased numbers of neutrophils in circulation but decreased numbers at inflamed or infected sites. In LAD II, the defect is in *SLC35C1*, a gene that encodes a GDP-fucose transporter 1 and therefore results in defective expression of fucosylated proteins, which are needed for functional expression of receptors including selectin ligands.[152,159] The clinical presentation is similar to LAD I but with less severe infections and includes developmental delay. LAD III results from a defect in *FERMT3*, which encodes for kindlin-3, resulting in defective integrin activation and impaired leukocyte and platelet adhesion.[159] The phenotype is similar to LAD I but there is also a bleeding diathesis. Given the severity of infections, allogeneic HSC transplantation is recommended for LAD I. In LAD II, some patients have benefitted from therapy with oral fucose.[158]

Papillon-Lefevre Syndrome

This is a rare, autosomal recessive disorder characterized by palmoplantar keratosis and severe periodontitis. The condition is caused by a mutation in the gene encoding for cathepsin C. Subjects display significant abnormalities in neutrophil function, including impaired chemotaxis and NET formation.[160] The disease becomes apparent in early childhood, and management requires extensive dental and skin treatment, including retinoids.[161]

Chediak-Higashi Syndrome

This is a rare, autosomal recessive disorder, due to a mutation in the *CHS1* gene that encodes for a lysosomal trafficking regulator protein. Chediak-Higashi syndrome (CHS) leads to impairments in microbicidal capacity due to failure of phagolysosome formation in neutrophils and other phagocytes. The condition then manifests as ineffective granulopoiesis, neutropenia, and incomplete degranulation. The neutrophils in CHS are characterized by giant peroxidase-positive granules, particularly prominent in the bone marrow. CHS is associated with enhanced prevalence of serious *S. aureus* infections in the skin and lung, periodontitis, and enhanced mortality. The disease is also characterized by oculocutaneous albinism, mild tendency to bleed, and neuropathy.[162] A proportion of CHS patients that survive beyond childhood develop inherited hemophagocytic histiocytosis.[159]

Warts, Hypogammaglobulinemia, Infections, and Myelokathexis Syndrome

Warts, hypogammaglobulinemia, infections, and myelokathexis (WHIM) syndrome is a rare genetic disorder characterized by hypogammaglobulinemia, warts, recurrent infections, and a failure of neutrophils to move from the bone marrow into the circulation. WHIM is caused by mutations in the *CXCR4* gene that results in excessive CXCR4 activity, thereby trapping neutrophils inside the bone marrow and preventing their ability to migrate into the blood and tissues.[163] Subjects affected by WHIM syndrome are treated with intravenous immunoglobulin and G-CSF. Plerixafor, a drug that blocks the activity of CXCR4, has shown efficacy in this condition.[164]

Blood Vessel Damage and Thrombosis (Fig. 14.3)

Atherosclerosis

Neutrophils play important roles during cardiovascular inflammation and repair.[165] Traditional and nontraditional risk factors for atherosclerosis alter neutrophil phenotype and function.[166] For example, hyperlipidemia and increased blood glucose enhance neutrophil synthesis in the bone marrow and at extramedullary sites, promoting cardiovascular inflammation.[166,167] Furthermore, neutrophils contribute to the development of atherosclerosis and its associated complications.[168] During atheroma formation, chemokines derived from platelets, such as CCL5, promote neutrophil activation and recruitment into the arterial wall.[169] The secretion of granule proteins by neutrophils at the level of the plaque further promotes recruitment of monocytes and activation of myeloid cells.[168] ROS synthesis by neutrophils activates endothelial cells and promotes leukocyte infiltration and low-density lipoprotein extravasation. MPO secreted by neutrophils can oxidize lipoproteins and perturb their function as well as the formation of foam cells.[170] In addition, neutrophils promote monocyte recruitment into the plaque and activation of myeloid cells. Dysregulated NET formation has been proposed to play important roles in the development of atherosclerosis and its associated complications that may lead to myocardial infarction and stroke. For example, through the externalization of histone H4, NETs induce lytic death of vascular smooth muscle cells and arterial plaque destabilization.[122,171] NETs can stimulate plasmacytoid DCs (pDCs) to synthesize type I IFNs and macrophages to secrete inflammasome-associated cytokines. Consequently, abrogation of NET formation and PAD4 activity in myeloid cells in animal models can mitigate atherosclerosis.[120] LDGs have been associated with enhanced development of coronary atherosclerotic plaque and arterial wall inflammation in SLE and

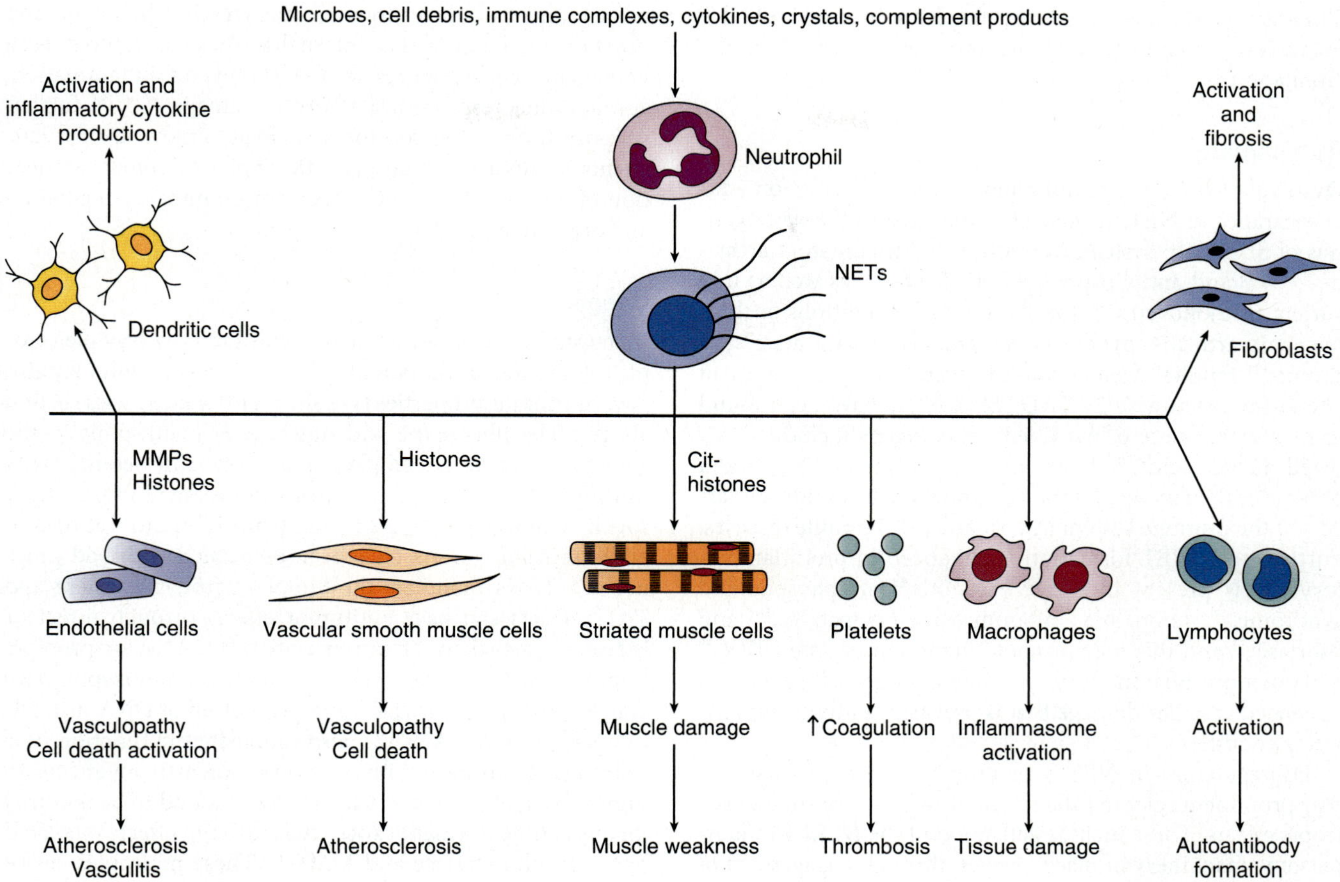

FIG. 14.3. Crosstalk of neutrophil extracellular traps (NETs) with immune and stromal cells. Upon exposure to various infectious and sterile proinflammatory stimuli, neutrophils release NETs. NETs externalize peptides present in granules in complex with nucleic acids and histones, which, in turn, activate a variety of immune and stromal cell types. Nucleic acid/granule protein complexes can activate plasmacytoid dendritic cells to synthetize type I interferons and also activate myeloid dendritic cells. NET peptides stimulate NLRP3 inflammasome activation in macrophages, leading to release of IL-1 and IL-18, which further promote neutrophil activation and NET formation. NETs can activate lymphocytes and promote autoantibody synthesis. NET-bound proteins, such as metalloproteinases (MMPs) and histones, promote vasculopathy through effects on endothelial cells and vascular smooth muscle cells. Histones can also harm striated muscle cells (myotubes), promoting muscle damage and weakness. NETs can be internalized by fibroblasts in various organs, promoting a proinflammatory and profibrotic phenotype. NETs have pleiotropic effects on coagulation system, including promoting platelet activation and thrombosis.

psoriasis.[52,57,61,134] Conversely, during neointima formation and myocardial infarction, neutrophils also display tissue repair capabilities and neoangiogenesis.

An aspect of particular interest in the activity of intravascular function of neutrophils is their capacity to bind platelets. Platelets are anucleated cellular fragments best known for their capacity to aggregate and to mediate thrombotic events; however, they are now known to be additionally endowed with potent immune regulatory functions, a reflection of the vast repertoire of receptors and cytokines stored in their granules and membrane.[172] For this reason, the conspicuous presence of platelets bound to neutrophils (and also to other circulating leukocytes) has been thoroughly studied in the context of vascular inflammation and thrombosis. Direct observation of inflamed vessels by intravital microscopy as well as flow cytometric analysis has revealed that interactions between neutrophils and platelets are frequent at the vessel wall interphase and that, at least under the flow forces present *in vivo*, they require the high-affinity binding properties of PSGL-1, a ligand for P-selectin in platelets, and the integrin Mac-1 ($\alpha M\beta 2$), respectively.[66,77]

The engagement of PSGL-1 at the neutrophil's uropod (the rear end of the polarized leukocyte) serves to trigger signals that are not yet fully characterized. These signals, in turn, elicit active crawling and activation of the intravascular neutrophil, resulting in vascular damage in the context of acute lung injury and stroke.[77] Reciprocally, platelets bound to the neutrophil surface or to the vessel wall release chemotactic cues that mediate integrin activation and adhesion to the endothelium and NETs, both of which contribute to exacerbating inflammation and organ injury, as shown in models of acute myocardial infarction.[173,174] In the context of atherosclerosis, leukocyte-bound platelets provide critical cues that enable accumulation of monocytes and neutrophils to the atherosclerotic plaque.[175] Interestingly, in the context of sepsis, activation of platelets through the endotoxin receptor

TLR4 is a sensor that triggers neutrophil activation and NET formation, which in turn allow removal of bacteria from the circulation.[176]

Autoimmunity

Dysregulation in neutrophil biology and the formation and/or clearance of NETs have been implicated in the pathogenesis of SLE,[58,59,115] systemic vasculitis,[53,177] rheumatoid arthritis,[103,118,178] and antiphospholipid syndrome,[179] as well as in a variety of monogenic autoinflammatory conditions such as pyogenic arthritis, pyoderma gangrenosum, and acne syndrome,[121] familial Mediterranean fever,[180] and deficiency of adenosine deaminase 2 (DADA2).[181] NETs have been found to be a rich source of modified autoantigens including oxidized nucleic acids,[59,182] citrullinated peptides,[103,118,178,183] and other modified molecules that are known to be important targets of the immune system in systemic autoimmune diseases. Furthermore, NET formation is enhanced in proinflammatory LDGs present in SLE,[58] vasculitis,[53] antiphospholipid syndrome,[179] idiopathic inflammatory myopathies,[55] and psoriasis,[56] and this may promote induction of type I IFNs[59] and other proinflammatory cytokines in target cells, as well as enhanced vascular damage that is prevalent in these autoimmune conditions.[60]

Dysregulation in NET formation has been proposed to play prominent roles in inflammatory diseases through pleiotropic effects. These include activation of the NLRP3 inflammasome machinery in macrophages, through engagement of P2X7 receptors,[184] leading to enhanced synthesis and secretion of IL-1β and IL-18. NETs and NETs/Ig immune complexes can also activate pDCs to synthesize higher levels of type I IFNs, with relevance to many autoimmune disorders.[58,185] NETs have been found to have the ability to activate T cells and autoreactive B cells and promote autoantibody generation in various autoimmune diseases.[118,186] In addition, NETs can be internalized by macrophages, fibroblasts, myeloid dendritic cells, and other target cells. This internalization has been found to induce dramatic changes in phenotype and function of the cells that internalize these structures, with pathogenic consequences in the context of inflammation, autoimmune responses, and organ damage,[118,187,188] further supporting an *in vivo* role for NETs in autoimmune disorders. In mouse models of lupus, genetic or pharmacologic abrogation of NET formation protects from immune dysregulation and organ damage.[59,189-192]

Cancer

Many studies suggest that neutrophils not only play pleiotropic roles in the promotion of cancer progression but may also have antitumor properties (see also *Neutrophil Subsets* section above). The phenotype and function of neutrophils in the context of cancer is complex, and neutrophil heterogeneity in different normal and cancerous tissues may impact their role in tumorigenesis. In general, neutrophil infiltration and NETs in solid cancers have been associated with bad prognosis.[193] TANs modulate the tumor microenvironment and TAN subsets can have antitumor effects through direct or indirect cytotoxicity of cancer cells (N1 TANs) or promote tumor growth and metastasis through immunosuppressive effects, proangiogenic roles, and promotion of DNA instability (N2 TANs).[44] TANs can also modulate the function of T cell subsets in tumors with important downstream effects on tumor size and growth. NETs have been found to be required for awakening dormant cancer cells through effects mediated by neutrophil elastase and MMP-9. These molecules cleave and remodel laminin, promoting cancer cell proliferation through integrin-mediated effects.[119] NETs can also directly alter the metabolic programming of neoplastic cells leading to increases in tumor growth.[194] Conversely, primary tumors have been reported to induce NETs, leading to enhance prometastatic effects.[195] Increased NET formation has been proposed to play roles not only in enhancing metastatic potential of tumors but also in promoting thrombosis in a variety of neoplasms.[196,197]

REFERENCES

1. Evrard M, Kwok IWH, Chong SZ, et al. Developmental analysis of bone marrow neutrophils reveals populations specialized in expansion, trafficking, and effector functions. *Immunity*. 2018;48(2):364-379.e8.
2. Hidalgo A, Chilvers ER, Summers C, Koenderman L. The neutrophil life cycle. *Trends Immunol*. 2019;40(7):584-597.
3. Busch K, Klapproth K, Barile M, et al. Fundamental properties of unperturbed haematopoiesis from stem cells in vivo. *Nature*. 2015;518(7540):542-546.
4. Rodriguez-Fraticelli AE, Wolock SL, Weinreb CS, et al. Clonal analysis of lineage fate in native haematopoiesis. *Nature*. 2018;553(7687):212-216.
5. Manz MG, Boettcher S. Emergency granulopoiesis. *Nat Rev Immunol*. 2014;14(5):302-314.
6. Zhu YP, Padgett L, Dinh HQ, et al. Identification of an early unipotent neutrophil progenitor with pro-tumoral activity in mouse and human bone marrow. *Cell Rep*. 2018;24(9):2329-2341.e8.
7. Paul F, Arkin Y, Giladi A, et al. Transcriptional heterogeneity and lineage commitment in myeloid progenitors. *Cell*. 2015;163(7):1663-1677.
8. Horman SR, Velu CS, Chaubey A, et al. Gfi1 integrates progenitor versus granulocytic transcriptional programming. *Blood*. 2009;113(22):5466-5475.
9. Borregaard N. Neutrophils, from marrow to microbes. *Immunity*. 2010;33(5):657-670.
10. Semerad CL, Liu F, Gregory AD, Stumpf K, Link DC. G-CSF is an essential regulator of neutrophil trafficking from the bone marrow to the blood. *Immunity*. 2002;17(4):413-423.
11. Lieschke G, Grail D, Hodgson G, et al. Mice lacking granulocyte colony-stimulating factor have chronic neutropenia, granulocyte and macrophage progenitor cell deficiency, and impaired neutrophil mobilization. *Blood*. 1994;84(6):1737-1746.
12. Zhang D-E, Zhang P, Wang ND, Hetherington CJ, Darlington GJ, Tenen DG. Absence of granulocyte colony-stimulating factor signaling and neutrophil development in CCAAT enhancer binding protein alpha-deficient mice. *Proc Natl Acad Sci U S A*. 1997;94(2):569-574.
13. Yamanaka R, Barlow C, Lekstrom-Himes J, et al. Impaired granulopoiesis, myelodysplasia, and early lethality in CCAAT/enhancer binding protein epsilon-deficient mice. *Proc Natl Acad Sci U S A*. 1997;94(24):13187-13192.
14. Lekstrom-Himes J, Xanthopoulos KG. CCAAT/enhancer binding protein epsilon is critical for effective neutrophil-mediated response to inflammatory challenge. *Blood*. 1999;93(9):3096-3105.
15. Grassi L, Pourfarzad F, Ullrich S, et al. Dynamics of transcription regulation in human bone marrow myeloid differentiation to mature blood neutrophils. *Cell Rep*. 2018;24(10):2784-2794.
16. Kim M-H, Granick JL, Kwok C, et al. Neutrophil survival and c-kit(+)-progenitor proliferation in *Staphylococcus aureus*-infected skin wounds promote resolution. *Blood*. 2011;117(12):3343-3352.
17. Granick JL, Falahee PC, Dahmubed D, Borjesson DL, Miller LS, Simon SI. *Staphylococcus aureus* recognition by hematopoietic stem and progenitor cells via TLR2/MyD88/PGE2 stimulates granulopoiesis in wounds. *Blood*. 2013;122(10):1770-1778.
18. Hirai H, Zhang P, Dayaram T, et al. C/EBPbeta is required for 'emergency' granulopoiesis. *Nat Immunol*. 2006;7(7):732-739.
19. Nagai Y, Garrett KP, Ohta S, et al. Toll-like receptors on hematopoietic progenitor cells stimulate innate immune system replenishment. *Immunity*. 2006;24(6):801-812.
20. Boettcher S, Gerosa RC, Radpour R, et al. Endothelial cells translate pathogen signals into G-CSF-driven emergency granulopoiesis. *Blood*. 2014;124(9):1393-1403.
21. Cronkite EP, Fliedner TM, Bond VP, Rubini JR. Dynamics of hemopoietic proliferation in man and mice studied by H3-thymidine incorporation into DNA. *Ann N Y Acad Sci*. 1959;77:803-820.
22. Dancey JT, Deubelbeiss KA, Harker LA, Finch CA. Neutrophil kinetics in man. *J Clin Invest*. 1976;58(3):705-715.
23. Casanova-Acebes M, Pitaval C, Weiss LA, et al. Rhythmic modulation of the hematopoietic niche through neutrophil clearance. *Cell*. 2013;153(5):1025-1035.
24. Eash KJ, Greenbaum AM, Gopalan PK, Link DC. CXCR2 and CXCR4 antagonistically regulate neutrophil trafficking from murine bone marrow. *J Clin Invest*. 2010;120(7):2423-2431.
25. Petit I, Szyper-Kravitz M, Nagler A, et al. G-CSF induces stem cell mobilization by decreasing bone marrow SDF-1 and up-regulating CXCR4. *Nat Immunol*. 2002;3(7):687-694.
26. Semerad CL, Christopher MJ, Liu F, et al. G-CSF potently inhibits osteoblast activity and CXCL12 mRNA expression in the bone marrow. *Blood*. 2005;106(9):3020-3027.
27. Geay J-F, Buet D, Zhang Y, et al. p210BCR-ABL inhibits SDF-1 chemotactic response via alteration of CXCR4 signaling and down-regulation of CXCR4 expression. *Cancer Res*. 2005;65(7):2676-2683.
28. Stark MA, Huo Y, Burcin TL, et al. Phagocytosis of apoptotic neutrophils regulates granulopoiesis via IL-23 and IL-17. *Immunity*. 2005;22(3):285-294.

29. Amulic B, Knackstedt SL, Abu Abed U, et al. Cell-cycle proteins control production of neutrophil extracellular traps. *Dev Cell.* 2017;43(4):449-462.e5.
30. Pillay J, den Braber I, Vrisekoop N, et al. In vivo labeling with 2H2O reveals a human neutrophil lifespan of 5.4 days. *Blood.* 2010;116(4):625-627.
31. Lahoz-Beneytez J, Elemans M, Zhang Y, et al. Human neutrophil kinetics: modeling of stable isotope labeling data supports short blood neutrophil half-lives. *Blood.* 2016;127(26):3431-3438.
32. Adrover JM, del Fresno C, Crainiciuc G, et al. A neutrophil timer coordinates immune defense and vascular protection. *Immunity.* 2019;50(2):390-402.e10.
33. Milot E, Filep JG. Regulation of neutrophil survival/apoptosis by Mcl-1. *ScientificWorldJournal.* 2011;11:1948-1962.
34. Akgul C, Moulding DA, Edwards SW. Molecular control of neutrophil apoptosis. *FEBS Lett.* 2001;487(3):318-322.
35. Dzhagalov I, StJohn A, He Y-W. The antiapoptotic protein Mcl-1 is essential for the survival of neutrophils but not macrophages. *Blood.* 2007;109(4):1620-1626.
36. Rossi AG, Sawatzky DA, Walker A, et al. Cyclin-dependent kinase inhibitors enhance the resolution of inflammation by promoting inflammatory cell apoptosis. *Nat Med.* 2006;12(9):1056-1064.
37. Kotzin JJ, Spencer SP, McCright SJ, et al. The long non-coding RNA Morrbid regulates Bim and short-lived myeloid cell lifespan. *Nature.* 2016;537(7619):239-243.
38. Ng LG, Ostuni R, Hidalgo A. Heterogeneity of neutrophils. *Nat Rev Immunol.* 2019;19(4):255-265.
39. Matsushima H, Geng S, Lu R, et al. Neutrophil differentiation into a unique hybrid population exhibiting dual phenotype and functionality of neutrophils and dendritic cells. *Blood.* 2013;121(10):1677-1689.
40. Deniset JF, Kubes P. Neutrophil heterogeneity: bona fide subsets or polarization states? *J Leukoc Biol.* 2018;103(5): 829-838.
41. Silvestre-Roig C, Hidalgo A, Soehnlein O. Neutrophil heterogeneity: implications for homeostasis and pathogenesis. *Blood.* 2016;127(18):2173-2181.
42. Christoffersson G, Vågesjö E, Vandooren J, et al. VEGF-A recruits a proangiogenic MMP-9-delivering neutrophil subset that induces angiogenesis in transplanted hypoxic tissue. *Blood.* 2012;120(23):4653-4662.
43. Massena S, Christoffersson G, Vågesjö E, et al. Identification and characterization of VEGF-A-responsive neutrophils expressing CD49d, VEGFR1, and CXCR4 in mice and humans. *Blood.* 2015;126(17):2016-2026.
44. Fridlender ZG, Sun J, Kim S, et al. Polarization of tumor-associated neutrophil phenotype by TGF-beta: "N1" versus "N2" TAN. *Canc Cell.* 2009;16(3):183-194.
45. Marini O, Costa S, Bevilacqua D, et al. Mature CD10(+) and immature CD10(−) neutrophils present in G-CSF-treated donors display opposite effects on T cells. *Blood.* 2017;129(10):1343-1356.
46. Pillay J, Kamp VM, van Hoffen E, et al. A subset of neutrophils in human systemic inflammation inhibits T cell responses through Mac-1. *J Clin Invest.* 2012;122(1):327-336.
47. Singhal S, Bhojnagarwala PS, O'Brien S, et al. Origin and role of a subset of tumor-associated neutrophils with antigen-presenting cell features in early-stage human lung cancer. *Canc Cell.* 2016;30(1):120-135.
48. Fridlender ZG, Albelda SM. Tumor-associated neutrophils: friend or foe? *Carcinogenesis.* 2012;33(5):949-955.
49. Gabrilovich DI, Nagaraj S. Myeloid-derived suppressor cells as regulators of the immune system. *Nat Rev Immunol.* 2009;9(3):162-174.
50. Bronte V, Brandau S, Chen S-H, et al. Recommendations for myeloid-derived suppressor cell nomenclature and characterization standards. *Nat Commun.* 2016;7:12150.
51. Denny MF, Yalavarthi S, Zhao W, et al. A distinct subset of proinflammatory neutrophils isolated from patients with systemic lupus erythematosus induces vascular damage and synthesizes type I IFNs. *J Immunol.* 2010;184(6):3284-3297.
52. Mistry P, Nakabo S, O'Neil L, et al. Transcriptomic, epigenetic, and functional analyses implicate neutrophil diversity in the pathogenesis of systemic lupus erythematosus. *Proc Natl Acad Sci U S A.* 2019;116(50):25222-25228.
53. Grayson PC, Carmona-Rivera C, Xu L, et al. Neutrophil-related gene expression and low-density granulocytes associated with disease activity and response to treatment in antineutrophil cytoplasmic antibody-associated vasculitis. *Arthritis Rheumatol.* 2015;67(7):1922-1932.
54. van den Hoogen LL, Fritsch-Stork RD, van Roon JA, Radstake TR. Low-density granulocytes are increased in antiphospholipid syndrome and are associated with anti-beta2-glycoprotein I antibodies: comment on the article by Yalavarthi et al. *Arthritis Rheumatol.* 2016;68(5):1320-1321.
55. Seto NL, Torres-Ruiz JJ, Carmona-Rivera C, et al. Neutrophil dysregulation is pathogenic in idiopathic inflammatory myopathies. *JCI Insight.* 2020;5:e134189.
56. Lin AM, Rubin CJ, Khandpur R, et al. Mast cells and neutrophils release IL-17 through extracellular trap formation in psoriasis. *J Immunol.* 2011;187(1):490-500.
57. Teague HL, Varghese NJ, Tsoi LC, et al. Neutrophil subsets, platelets, and vascular disease in psoriasis. *JACC Basic Transl Sci.* 2019;4(1):1-14.
58. Villanueva E, Yalavarthi S, Berthier CC, et al. Netting neutrophils induce endothelial damage, infiltrate tissues, and expose immunostimulatory molecules in systemic lupus erythematosus. *J Immunol.* 2011;187(1):538-552.
59. Lood C, Blanco LP, Purmalek MM, et al. Neutrophil extracellular traps enriched in oxidized mitochondrial DNA are interferogenic and contribute to lupus-like disease. *Nat Med.* 2016;22(2):146-153.
60. Carmona-Rivera C, Zhao W, Yalavarthi S, Kaplan MJ. Neutrophil extracellular traps induce endothelial dysfunction in systemic lupus erythematosus through the activation of matrix metalloproteinase-2. *Ann Rheum Dis.* 2015;74(7):1417-1424.
61. Carlucci PM, Purmalek MM, Dey AK, et al. Neutrophil subsets and their gene signature associate with vascular inflammation and coronary atherosclerosis in lupus. *JCI Insight.* 2018;3(8):e99276.
62. Wynn TA, Chawla A, Pollard JW. Macrophage biology in development, homeostasis and disease. *Nature.* 2013;496(7446):445-455.
63. Becher B, Schlitzer A, Chen J, et al. High-dimensional analysis of the murine myeloid cell system. *Nat Immunol.* 2014;15(12):1181-1189.
64. Puga I, Cols M, Barra CM, et al. B cell-helper neutrophils stimulate the diversification and production of immunoglobulin in the marginal zone of the spleen. *Nat Immunol.* 2011;13(2):170-180.
65. Maas SL, Soehnlein O, Viola JR. Organ-specific mechanisms of transendothelial neutrophil migration in the lung, liver, kidney, and aorta. *Front Immunol.* 2018;9:2739.
66. Hidalgo A, Peired AJ, Wild MK, Vestweber D, Frenette PS. Complete identification of E-selectin ligands on neutrophils reveals distinct functions of PSGL-1, ESL-1, and CD44. *Immunity.* 2007;26(4):477-489.
67. Zarbock A, Ley K, McEver RP, Hidalgo A. Leukocyte ligands for endothelial selectins: specialized glycoconjugates that mediate rolling and signaling under flow. *Blood.* 2011;118(26):6743-6751.
68. Zarbock A, Lowell CA, Ley K, et al. Spleen tyrosine kinase Syk is necessary for E-selectin-induced alpha(L)beta(2) integrin-mediated rolling on intercellular adhesion molecule-1. *Immunity.* 2007;26(6): 773-783.
69. Stark K, Eckart A, Haidari S, et al. Capillary and arteriolar pericytes attract innate leukocytes exiting through venules and 'instruct' them with pattern-recognition and motility programs. *Nat Immunol.* 2013;14(1):41-51.
70. Girbl T, Lenn T, Perez L, Rolas L, et al. Distinct compartmentalization of the chemokines CXCL1 and CXCL2 and the atypical receptor ACKR1 determine discrete stages of neutrophil diapedesis. *Immunity.* 2018;49(6):1062-1076 e1066.
71. Phillipson M, Heit B, Colarusso P, Liu L, Ballantyne CM, Kubes P. Intraluminal crawling of neutrophils to emigration sites: a molecularly distinct process from adhesion in the recruitment cascade. *J Exp Med.* 2006;203(12): 2569-2575.
72. McDonald B, McAvoy EF, Lam F, et al. Interaction of CD44 and hyaluronan is the dominant mechanism for neutrophil sequestration in inflamed liver sinusoids. *J Exp Med.* 2008;205(4):915-927.
73. Devi S, Li A, Westhorpe CLV, et al. Multiphoton imaging reveals a new leukocyte recruitment paradigm in the glomerulus. *Nat Med.* 2013;19(1):107-112.
74. Kinashi T. Intracellular signalling controlling integrin activation in lymphocytes. *Nat Rev Immunol.* 2005;5(7): 546-559.
75. Sanchez-Madrid F, del Pozo MA. Leukocyte polarization in cell migration and immune interactions. *EMBO J.* 1999;18(3):501-511.
76. Scheiermann C, Kunisaki Y, Jang J-E, Frenette PS. Neutrophil microdomains: linking heterocellular interactions with vascular injury. *Curr Opin Hematol.* 2010;17(1):25-30.
77. Sreeramkumar V, Adrover JM, Ballesteros I, et al. Neutrophils scan for activated platelets to initiate inflammation. *Science.* 2014;346(6214):1234-1238.
78. Nourshargh S, Alon R. Leukocyte migration into inflamed tissues. *Immunity.* 2014;41(5):694-707.
79. Proebstl D, Voisin M-B, Woodfin A, et al. Pericytes support neutrophil subendothelial cell crawling and breaching of venular walls in vivo. *J Exp Med.* 2012;209(6):1219-1234.
80. Buckley CD, Ross EA, McGettrick HM, et al. Identification of a phenotypically and functionally distinct population of long-lived neutrophils in a model of reverse endothelial migration. *J Leukoc Biol.* 2006;79(2):303-311.
81. Mathias JR, Perrin BJ, Liu T-X, et al. Resolution of inflammation by retrograde chemotaxis of neutrophils in transgenic zebrafish. *J Leukoc Biol.* 2006;80(6):1281-1288.
82. Woodfin A, Voisin M-B, Beyrau M, et al. The junctional adhesion molecule JAM-C regulates polarized transendothelial migration of neutrophils in vivo. *Nat Immunol.* 2011;12(8):761-769.
83. Sarris M, Sixt M. Navigating in tissue mazes: chemoattractant interpretation in complex environments. *Curr Opin Cell Biol.* 2015;36:93-102.
84. Lämmermann T, Bader BL, Monkley SJ, et al. Rapid leukocyte migration by integrin-independent flowing and squeezing. *Nature.* 2008;453(7191):51-55.
85. Renkawitz J, Kopf A, Stopp J, et al. Nuclear positioning facilitates amoeboid migration along the path of least resistance. *Nature.* 2019;568(7753):546-550.
86. Lämmermann T, Afonso PV, Angermann BR, et al. Neutrophil swarms require LTB4 and integrins at sites of cell death in vivo. *Nature.* 2013;498(7454):371-375.
87. Ng LG, Qin JS, Roediger B, et al. Visualizing the neutrophil response to sterile tissue injury in mouse dermis reveals a three-phase cascade of events. *J Invest Dermatol.* 2011;131(10):2058-2068.
88. Borregaard N, Sørensen OE, Theilgaard-Mönch K. Neutrophil granules: a library of innate immunity proteins. *Trends Immunol.* 2007;28(8):340-345.
89. Ramadass M, Catz SD. Molecular mechanisms regulating secretory organelles and endosomes in neutrophils and their implications for inflammation. *Immunol Rev.* 2016;273(1): 249-265.
90. Root RK, Metcalf JA. H2O2 release from human granulocytes during phagocytosis. Relationship to superoxide anion formation and cellular catabolism of H_2O_2: studies with normal and cytochalasin B-treated cells. *J Clin Invest.* 1977;60(6):1266-1279.
91. Robinson JM. Reactive oxygen species in phagocytic leukocytes. *Histochem Cell Biol.* 2008;130(2):281-297.
92. Dahlgren C, Karlsson A, Bylund J. Intracellular neutrophil oxidants: from laboratory curiosity to clinical reality. *J Immunol.* 2019;202(11):3127-3134.
93. Nguyen GT, Green ER, Mecsas J. Neutrophils to the ROScue: mechanisms of NADPH oxidase activation and bacterial resistance. *Front Cell Infect Microbiol.* 2017;7:373.
94. Miyano K, Sumimoto H. Assessment of the role for Rho family GTPases in NADPH oxidase activation. *Methods Mol Biol.* 2012;827:195-212.
95. Futosi K, Fodor S, Mócsai A. Reprint of Neutrophil cell surface receptors and their intracellular signal transduction pathways. *Int Immunopharmacol.* 2013;17(4):1185-1197.
96. Wang Y, Jonsson F. Expression, role, and regulation of neutrophil Fcgamma receptors. *Front Immunol.* 2019;10:1958.
97. Li K, Underhill DM. C-type lectin receptors in phagocytosis. *Curr Top Microbiol Immunol.* 2020;429:1-18.
98. Garcia-Garcia E, Rosales C. Signal transduction during Fc receptor-mediated phagocytosis. *J Leukoc Biol.* 2002;72(6):1092-1108.
99. Jakus Z, Simon E, Frommhold D, et al. Critical role of phospholipase Cgamma2 in integrin and Fc receptor-mediated neutrophil functions and the effector phase of autoimmune arthritis. *J Exp Med.* 2009;206(3):577-593.
100. Freeman SA, Grinstein S. Phagocytosis: receptors, signal integration, and the cytoskeleton. *Immunol Rev.* 2014;262(1):193-215.
101. van Kessel KP, Bestebroer J, van Strijp JA. Neutrophil-mediated phagocytosis of *Staphylococcus aureus. Front Immunol.* 2014;5:467.
102. Brinkmann V, Reichard U, Goosmann C, et al. Neutrophil extracellular traps kill bacteria. *Science.* 2004;303(5663):1532-1535.
103. Khandpur R, Carmona-Rivera C, Vivekanandan-Giri A, et al. NETs are a source of citrullinated autoantigens and stimulate inflammatory responses in rheumatoid arthritis. *Sci Transl Med.* 2013;5(178):178ra140.
104. Carmona-Rivera C, Purmalek MM, Moore E, et al. A role for muscarinic receptors in neutrophil extracellular trap formation and levamisole-induced autoimmunity. *JCI Insight.* 2017;2(3):e89780.
105. Kenny EF, Herzig A, Krüger R, et al. Diverse stimuli engage different neutrophil extracellular trap pathways. *Elife.* 2017;6:e24437.
106. Papayannopoulos V, Metzler KD, Hakkim A, Zychlinsky A. Neutrophil elastase and myeloperoxidase regulate the formation of neutrophil extracellular traps. *J Cell Biol.* 2010;191(3):677-691.
107. Metzler KD, Goosmann C, Lubojemska A, Zychlinsky A, Papayannopoulos V. A myeloperoxidase-containing complex regulates neutrophil elastase release and actin dynamics during NETosis. *Cell Rep.* 2014;8(3):883-896.
108. Wang Y, Li M, Stadler S, et al. Histone hypercitrullination mediates chromatin decondensation and neutrophil extracellular trap formation. *J Cell Biol.* 2009;184(2):205-213.
109. Urban CF, Reichard U, Brinkmann V, Zychlinsky A. Neutrophil extracellular traps capture and kill Candida albicans yeast and hyphal forms. *Cell Microbiol.* 2006;8(4):668-676.
110. Pilsczek FH, Salina D, Poon KKH, et al. A novel mechanism of rapid nuclear neutrophil extracellular trap formation in response to *Staphylococcus aureus. J Immunol.* 2010;185(12):7413-7425.
111. Menten-Dedoyart C, Faccinetto C, Golovchenko M, et al. Neutrophil extracellular traps entrap and kill Borrelia burgdorferi sensu stricto spirochetes and are not affected by Ixodes ricinus tick saliva. *J Immunol.* 2012;189(11):5393-5401.
112. Saitoh T, Komano J, Saitoh Y, et al. Neutrophil extracellular traps mediate a host defense response to human immunodeficiency virus-1. *Cell Host Microbe.* 2012;12(1):109-116.
113. Rochael NC, Guimarães-Costa AB, Nascimento MT, et al. Classical ROS-dependent and early/rapid ROS-independent release of Neutrophil Extracellular Traps triggered by Leishmania parasites. *Sci Rep.* 2015;5:18302.

114. Storisteanu DML, Pocock JM, Cowburn AS, et al. Evasion of neutrophil extracellular traps by respiratory pathogens. *Am J Respir Cell Mol Biol.* 2017;56(4):423-431.
115. Hakkim A, Furnrohr BG, Amann K, et al. Impairment of neutrophil extracellular trap degradation is associated with lupus nephritis. *Proc Natl Acad Sci U S A.* 2010;107(21):9813-9818.
116. Farrera C, Fadeel B. Macrophage clearance of neutrophil extracellular traps is a silent process. *J Immunol.* 2013;191(5):2647-2656.
117. Lazzaretto B, Fadeel B. Intra-and extracellular degradation of neutrophil extracellular traps by macrophages and dendritic cells. *J Immunol.* 2019;203(8):2276-2290.
118. Carmona-Rivera C, Carlucci PM, Moore E, et al. Synovial fibroblast-neutrophil interactions promote pathogenic adaptive immunity in rheumatoid arthritis. *Sci Immunol.* 2017;2(10):eaag3358.
119. Albrengues J, Shields MA, Ng D, et al. Neutrophil extracellular traps produced during inflammation awaken dormant cancer cells in mice. *Science.* 2018;361(6409):eaao4227.
120. Liu Y, Carmona-Rivera C, Moore E, et al. Myeloid-specific deletion of peptidylarginine deiminase 4 mitigates atherosclerosis. *Front Immunol.* 2018;9:1680.
121. Mistry P, Carmona-Rivera C, Ombrello AK, et al. Dysregulated neutrophil responses and neutrophil extracellular trap formation and degradation in PAPA syndrome. *Ann Rheum Dis.* 2018;77(12):1825-1833.
122. Silvestre-Roig C, Braster Q, Wichapong K, et al. Externalized histone H4 orchestrates chronic inflammation by inducing lytic cell death. *Nature.* 2019;569(7755):236-240.
123. Christoffersson G, Lomei J, O'Callaghan P, Kreuger J, Engblom S, Phillipson M. Vascular sprouts induce local attraction of proangiogenic neutrophils. *J Leukoc Biol.* 2017;102(3):741-751.
124. Aldabbous L, Abdul-Salam V, McKinnon T, et al. Neutrophil extracellular traps promote angiogenesis: evidence from vascular pathology in pulmonary hypertension. *Arterioscler Thromb Vasc Biol.* 2016;36(10):2078-2087.
125. Nadkarni S, Smith J, Sferruzzi-Perri AN, et al. Neutrophils induce proangiogenic T cells with a regulatory phenotype in pregnancy. *Proc Natl Acad Sci U S A.* 2016;113(52):E8415-E8424.
126. Ai S, Cheng XW, Inoue A, et al. Angiogenic activity of bFGF and VEGF suppressed by proteolytic cleavage by neutrophil elastase. *Biochem Biophys Res Commun.* 2007;364(2):395-401.
127. Tecchio C, Cassatella MA. Neutrophil-derived cytokines involved in physiological and pathological angiogenesis. *Chem Immunol Allergy.* 2014;99:123-137.
128. Braza MS, Conde P, Garcia M, et al. Neutrophil derived CSF1 induces macrophage polarization and promotes transplantation tolerance. *Am J Transplant.* 2018;18(5):1247-1255.
129. Adrover JM, Nicolás-Ávila JA, Hidalgo A, et al. Aging: a temporal dimension for neutrophils. *Trends Immunol.* 2016;37(5):334-345.
130. Zhang D, Chen G, Manwani D, et al. Neutrophil ageing is regulated by the microbiome. *Nature.* 2015;525(7570):528-532.
131. Mittal M, Siddiqui MR, Tran K, et al. Reactive oxygen species in inflammation and tissue injury. *Antioxid Redox Signal.* 2014;20(7):1126-1167.
132. Ella K, Csépányi-Kömi R, Káldi K. Circadian regulation of human peripheral neutrophils. *Brain Behav Immun.* 2016;57:209-221.
133. Martin C, Burdon PCE, Bridger G, Gutierrez-Ramos J-C, Williams TJ, Rankin SM. Chemokines acting via CXCR2 and CXCR4 control the release of neutrophils from the bone marrow and their return following senescence. *Immunity.* 2003;19(4): 583-593.
134. Adrover JM, Aroca-Crevillén A, Crainiciuc G, et al. Programmed 'disarming' of the neutrophil proteome reduces the magnitude of inflammation. *Nat Immunol.* 2020;21:135-144.
135. Scheiermann C, Gibbs J, Ince L, Loudon A. Clocking in to immunity. *Nat Rev Immunol.* 2018;18(7):423-437.
136. Man K, Loudon A, Chawla A. Immunity around the clock. *Science.* 2016;354(6315): 999-1003.
137. Gould TJ, Vu TT, Swystun LL, et al. Neutrophil extracellular traps promote thrombin generation through platelet-dependent and platelet-independent mechanisms. *Arterioscler Thromb Vasc Biol.* 2014;34(9):1977-1984.
138. Folco EJ, Mawson TL, Vromman A, et al. Neutrophil extracellular traps induce endothelial cell activation and tissue factor production through interleukin-1alpha and cathepsin G. *Arterioscler Thromb Vasc Biol.* 2018;38(8):1901-1912.
139. Wang Y, Jin Y, Laviña B, Jakobsson L, et al. Characterization of multi-cellular dynamics of angiogenesis and vascular remodelling by intravital imaging of the wounded mouse cornea. *Sci Rep.* 2018;8(1):10672.
140. Martinod K, Demers M, Fuchs TA, et al. Neutrophil histone modification by peptidylarginine deiminase 4 is critical for deep vein thrombosis in mice. *Proc Natl Acad Sci U S A.* 2013;110(21):8674-8679.
141. von Brühl M-L, Stark K, Steinhart A, et al. Monocytes, neutrophils, and platelets cooperate to initiate and propagate venous thrombosis in mice in vivo. *J Exp Med.* 2012;209(4): 819-835.
142. McDonald B, Davis RP, Kim S-J, et al. Platelets and neutrophil extracellular traps collaborate to promote intravascular coagulation during sepsis in mice. *Blood.* 2017;129(10): 1357-1367.
143. Brazil JC, Quiros M, Nusrat A, Parkos CA. Innate immune cell-epithelial crosstalk during wound repair. *J Clin Invest.* 2019;129(8):2983-2993.
144. Peiseler M, Kubes P. More friend than foe: the emerging role of neutrophils in tissue repair. *J Clin Invest.* 2019;129(7):2629-2639.
145. A-Gonzalez N, Quintana JA, García-Silva S, et al. Phagocytosis imprints heterogeneity in tissue-resident macrophages. *J Exp Med.* 2017;214(5):1281-1296.
146. Sumagin R, Brazil JC, Nava P, et al. Neutrophil interactions with epithelial-expressed ICAM-1 enhances intestinal mucosal wound healing. *Mucosal Immunol.* 2016;9(5):1151-1162.
147. Curtis VF, Cartwright IM, Lee JS, et al. Neutrophils as sources of dinucleotide polyphosphates and metabolism by epithelial ENPP1 to influence barrier function via adenosine signaling. *Mol Biol Cell.* 2018;29(22):2687-2699.
148. Calvente CJ, Tameda M, Johnson CD, et al. Neutrophils contribute to spontaneous resolution of liver inflammation and fibrosis via microRNA-223. *J Clin Invest.* 2019;130:4091-4109.
149. Yang W, Tao Y, Wu Y, et al. Neutrophils promote the development of reparative macrophages mediated by ROS to orchestrate liver repair. *Nat Commun.* 2019;10(1):1076.
150. Wong SL, Demers M, Martinod K, et al. Diabetes primes neutrophils to undergo NETosis, which impairs wound healing. *Nat Med.* 2015;21(7):815-819.
151. Butin-Israeli V, Bui TM, Wiesolek HL, et al. Neutrophil-induced genomic instability impedes resolution of inflammation and wound healing. *J Clin Invest.* 2019;129(2): 712-726.
152. Dinauer MC. Disorders of neutrophil function: an overview. *Methods Mol Biol.* 2014;1124:501-515.
153. Skokowa J, Dale DC, Touw IP, Zeidler C, Welte K. Severe congenital neutropenias. *Nat Rev Dis Primers.* 2017;3:17032.
154. Dinauer MC. Chronic granulomatous disease and other disorders of phagocyte function. *Hematology Am Soc Hematol Educ Program.* 2005;2005:89-95.
155. Schäppi MG, Jaquet V, Belli DC, Krause K-H. Hyperinflammation in chronic granulomatous disease and anti-inflammatory role of the phagocyte NADPH oxidase. *Semin Immunopathol.* 2008;30(3): 255-271.
156. Fox TA, Chakraverty R, Burns S, et al. Successful outcome following allogeneic hematopoietic stem cell transplantation in adults with primary immunodeficiency. *Blood.* 2018;131(8):917-931.
157. Klebanoff SJ, Kettle AJ, Rosen H, et al. Myeloperoxidase: a front-line defender against phagocytosed microorganisms. *J Leukoc Biol.* 2013;93(2):185-198.
158. van de Vijver E, Maddalena A, Sanal Ö, et al. Hematologically important mutations: leukocyte adhesion deficiency (first update). *Blood Cells Mol Dis.* 2012;48(1):53-61.
159. Dinauer MC. Neutrophil defects and diagnosis disorders of neutrophil function: an overview. *Methods Mol Biol.* 2020;2087:11-29.
160. Roberts H, White P, Dias I, et al. Characterization of neutrophil function in Papillon-Lefevre syndrome. *J Leukoc Biol.* 2016;100(2): 433-444.
161. Sreeramulu B, Shyam ND, Ajay P, Suman S. Papillon-Lefevre syndrome: clinical presentation and management options. *Clin Cosmet Investig Dent.* 2015;7:75-81.
162. Kaplan J, De Domenico I, Ward DM. Chediak-Higashi syndrome. *Curr Opin Hematol.* 2008;15(1):22-29.
163. Kawai T, Malech HL. WHIM syndrome: congenital immune deficiency disease. *Curr Opin Hematol.* 2009;16(1):20-26.
164. McDermott DH, Liu Q, Velez D, et al. A phase 1 clinical trial of long-term, low-dose treatment of WHIM syndrome with the CXCR4 antagonist plerixafor. *Blood.* 2014;123(15): 2308-2316.
165. Silvestre-Roig C, Braster Q, Ortega-Gomez A, Soehnlein O. Neutrophils as regulators of cardiovascular inflammation. *Nat Rev Cardiol.* 2020;17;327-340.
166. Gu Q, Yang X, Lv J, et al. AIBP-mediated cholesterol efflux instructs hematopoietic stem and progenitor cell fate. *Science.* 2019;363(6431):1085-1088.
167. Nagareddy PR, Murphy AJ, Stirzaker RA, et al. Hyperglycemia promotes myelopoiesis and impairs the resolution of atherosclerosis. *Cell Metab.* 2013;17(5):695-708.
168. Kraakman MJ, Lee MKS, Al-Sharea A, et al. Neutrophil-derived S100 calcium-binding proteins A8/A9 promote reticulated thrombocytosis and atherogenesis in diabetes. *J Clin Invest.* 2017;127(6):2133-2147.
169. Alard J-E, Ortega-Gomez A, Wichapong K, et al. Recruitment of classical monocytes can be inhibited by disturbing heteromers of neutrophil HNP1 and platelet CCL5. *Sci Transl Med.* 2015;7(317): 317ra196.
170. Smith CK, Vivekanandan-Giri A, Tang C, et al. Neutrophil extracellular trap-derived enzymes oxidize high-density lipoprotein: an additional proatherogenic mechanism in systemic lupus erythematosus. *Arthritis Rheumatol.* 2014;66(9): 2532-2544.
171. Warnatsch A, Ioannou M, Wang Q, Papayannopoulos V. Inflammation. Neutrophil extracellular traps license macrophages for cytokine production in atherosclerosis. *Science.* 2015;349(6245):316-320.
172. Li JL, Zarbock A, Hidalgo A, et al. Platelets as autonomous drones for hemostatic and immune surveillance. *J Exp Med.* 2017;214:2193-2204.
173. Mangold A, Alias S, Scherz T. Coronary neutrophil extracellular trap burden and deoxyribonuclease activity in ST-elevation acute coronary syndrome are predictors of ST-segment resolution and infarct size. *Circ Res.* 2015;116(7):1182-1192.
174. Stakos DA, Kambas K, Konstantinidis T, et al. Expression of functional tissue factor by neutrophil extracellular traps in culprit artery of acute myocardial infarction. *Eur Heart J.* 2015;36(22):1405-1414.
175. Huo Y, Schober A, Forlow SB, et al. Circulating activated platelets exacerbate atherosclerosis in mice deficient in apolipoprotein E. *Nat Med.* 2003;9(1):61-67.
176. Clark SR, Ma AC, Tavener SA, et al. Platelet TLR4 activates neutrophil extracellular traps to ensnare bacteria in septic blood. *Nat Med.* 2007;13(4):463-469.
177. Kessenbrock K, Krumbholz M, Schönermarck U, et al. Netting neutrophils in autoimmune small-vessel vasculitis. *Nat Med.* 2009;15(6):623-625.
178. Salinger AJ, Dubuke ML, Carmona-Rivera C, et al. Technical comment on "Synovial fibroblast-neutrophil interactions promote pathogenic adaptive immunity in rheumatoid arthritis". *Sci Immunol.* 2020;5(43):eaax5672.
179. Yalavarthi S, Gould TJ, Rao AN, et al. Release of neutrophil extracellular traps by neutrophils stimulated with antiphospholipid antibodies: a newly identified mechanism of thrombosis in the antiphospholipid syndrome. *Arthritis Rheumatol.* 2015;67(11):2990-3003.
180. Apostolidou E, Skendros P, Kambas K, et al. Neutrophil extracellular traps regulate IL-1beta-mediated inflammation in familial Mediterranean fever. *Ann Rheum Dis.* 2016;75(1):269-277.
181. Carmona-Rivera C, Khaznadar SS, Shwin KW, et al. Deficiency of adenosine deaminase 2 triggers adenosine-mediated NETosis and TNF production in patients with DADA2. *Blood.* 2019;134(4):395-406.
182. Gehrke N, Mertens C, Zillinger T, et al. Oxidative damage of DNA confers resistance to cytosolic nuclease TREX1 degradation and potentiates STING-dependent immune sensing. *Immunity.* 2013;39(3):482-495.
183. Carmona-Rivera C, Bicker KL, Thompson PR, et al. Response to comment on "Synovial fibroblast-neutrophil interactions promote pathogenic adaptive immunity in rheumatoid arthritis". *Sci Immunol.* 2018;3(21):eaar3701.
184. Kahlenberg JM, Carmona-Rivera C, Smith CK, Kaplan MJ. Neutrophil extracellular trap-associated protein activation of the NLRP3 inflammasome is enhanced in lupus macrophages. *J Immunol.* 2013;190(3):1217-1226.
185. Lande R, Ganguly D, Facchinetti V, et al. Neutrophils activate plasmacytoid dendritic cells by releasing self-DNA-peptide complexes in systemic lupus erythematosus. *Sci Transl Med.* 2011;3(73): 73ra19.
186. Gestermann N, Di Domizio J, Lande R, et al. Netting neutrophils activate autoreactive B cells in lupus. *J Immunol.* 2018;200(10):3364-3371.
187. Sangaletti S, Tripodo C, Chiodoni C, et al. Neutrophil extracellular traps mediate transfer of cytoplasmic neutrophil antigens to myeloid dendritic cells toward ANCA induction and associated autoimmunity. *Blood.* 2012;120(15): 3007-3018.
188. Barrera-Vargas A, Gómez-Martín D, Carmona-Rivera C, et al. Differential ubiquitination in NETs regulates macrophage responses in systemic lupus erythematosus. *Ann Rheum Dis.* 2018;77(6):944-950.
189. Knight JS, Zhao W, Luo W, et al. Peptidylarginine deiminase inhibition is immunomodulatory and vasculoprotective in murine lupus. *J Clin Invest.* 2013;123(7):2981-2993.
190. Knight JS, Luo W, O'Dell AA, et al. Peptidylarginine deiminase inhibition reduces vascular damage and modulates innate immune responses in murine models of atherosclerosis. *Circ Res.* 2014;114(6):947-956.
191. Knight JS, Subramanian V, O'Dell AA, et al. Peptidylarginine deiminase inhibition disrupts NET formation and protects against kidney, skin and vascular disease in lupus-prone MRL/lpr mice. *Ann Rheum Dis.* 2015;74(12): 2199-2206.
192. Liu Y, Lightfoot YL, Seto N, et al. Peptidylarginine deiminases 2 and 4 modulate innate and adaptive immune responses in TLR-7-dependent lupus. *JCI Insight.* 2018;3(23): e124729.
193. Li Y-W, Qiu S-J, Fan J, et al. Intratumoral neutrophils: a poor prognostic factor for hepatocellular carcinoma following resection. *J Hepatol.* 2011;54(3):497-505.
194. Yazdani HO, Roy E, Comerci AJ, et al. Neutrophil extracellular traps drive mitochondrial homeostasis in tumors to augment growth. *Cancer Res.* 2019;79(21):5626-5639.
195. Rayes RF, Mouhanna JG, Nicolau I, et al. Primary tumors induce neutrophil extracellular traps with targetable metastasis promoting effects. *JCI Insight.* 2019;5:e128008.
196. Park J, Wysocki RW, Amoozgar Z, et al. Cancer cells induce metastasis-supporting neutrophil extracellular DNA traps. *Sci Transl Med.* 2016;8(361):361ra138.
197. Wolach O, Sellar RS, Martinod K, et al. Increased neutrophil extracellular trap formation promotes thrombosis in myeloproliferative neoplasms. *Sci Transl Med.* 2018;10(436):eaan8292.

CHAPTER 15 Macrophages

Gwendalyn J. Randolph, PhD

DEFINING MACROPHAGES

Introducing Macrophages Through the Lens of History

Macrophages, with their dramatic capacity to engulf and destroy whole bacteria and other particulate objects through the process of phagocytosis, were the first cells of the immune system to be placed into functional context. They are quite aptly named from the Greek "makro" meaning "large" and "phagos" meaning "to eat."

In 1908, Elie Metchnikoff, who discovered and named macrophages and in so doing was considered to have established the innate immunity branch of immunology, shared the Nobel Prize with Paul Erlich, who demonstrated humoral immune protection conferred by antibodies. At the time, some regarded their concepts in immunology as competing, rather than cooperative. There is always a measure of irony in history, and, ironically, we now know that it would emerge that macrophages rely on antibodies as a major means to optimize and regulate their functions.

It was rather quickly understood that macrophages have critical roles in the removal of a wide range of tissue debris. Metchnikoff famously recorded using hand-drawn renderings of macrophages engulfing debris, including depictions of macrophages within starfish larvae that surround and appear to "attack" thorns that entered the starfish tissues. Figure 15.1 highlights the range of cargo that macrophages can interact with and remove within tissues.

Early on, the cell biology behind how macrophages removed and destroyed infectious or obstructive debris from wounds and developing organs homed in on the extraordinary capacity of macrophages to employ lysosomes to degrade cargo. James Hirsch at the Rockefeller University focused on the capacity of leukocytes to degrade bacteria and proteins that could be experimentally tracked. While pioneering the area of phagocytosis and the ability of phagocytes to use granule-associated enzymes to degrade engulfed cargo, Hirsch noted distinctions between neutrophils, eosinophils, and macrophages.[1] He highlighted one puzzling feature of macrophages that set them apart from the body's other major phagocyte in host defense, neutrophils. In contrast to neutrophils, macrophages displayed the ability to retain immunogenicity of antigens[2,3] foreshadowing a key function of macrophages beyond removal and degradation of debris, the presentation of antigen to lymphocytes. With antigen presentation, a field with its own fascinating history, we know that destruction of phagocytic cargo and the salvaging of antigens, that is, the salvaging of peptides from proteins, are both orchestrated by the macrophage's robust lysosomal compartment and the processes of phagocytosis and endocytosis, for display of appropriate peptides on MHC molecules.[4] By the mid-1970s, in part due to the research advances made by Samuel Silverstein and his group, a framework for uptake of soluble cargo via pinocytosis and uptake of particulate cargo via phagocytosis was established.[5] The clear outcome of uptake by macrophages was that all cargo was targeted to a degradative lysosomal compartment, contrasting with cells like endothelial cells that were capable of pinocytic uptake that allowed for cargo to move across the cells and be secreted through exocytic processes. At the same time, key roles for Fc receptors and complement receptors in the recognition of phagocytic cargo were established, while other "nonspecific" receptors were implicated.

PHAGOCYTIC ENGULFMENT AND ITS CONSEQUENCES

PHAGOCYTIC RECEPTORS. The cargo that macrophages may engulf is wide ranging. For instance, macrophages are responsible for engulfing and holding the pigment in place that is used for tattoos.[6] Since tattoos neither disappear nor typically set off an immune reaction, it is also clear that macrophages, with an impressive capacity to degrade cargo, do not degrade all cargo, nor should it be thought that engulfment leads inevitably to inflammation. Much depends upon the receptors that are engaged and the nature of the signaling that is triggered. There are many classes of receptors that macrophages use in the uptake of cargo during phagocytosis and these receptors often mediate other types of engulfment like pinocytosis or endocytosis under some circumstances. In fact, the list of relevant receptors that macrophages use to engage phagocytic cargo is so vast that it is not possible to cover the topic comprehensively here. However, the sections below classify some of the major groups of receptors and provide examples of key receptors in each group.

A recurring theme observed across these receptor types is that many receptors involved in phagocytosis rely on the identification of repeating **molecular patterns**, not just repeating patterns from pathogens that will be covered below but also repeating patterns that may recur in host molecules. Examples include carbohydrate structures that serve as ligands for many receptors or phosphatidyl serine that, as a common signature of apoptotic cells, can be recognized by a variety of receptors,

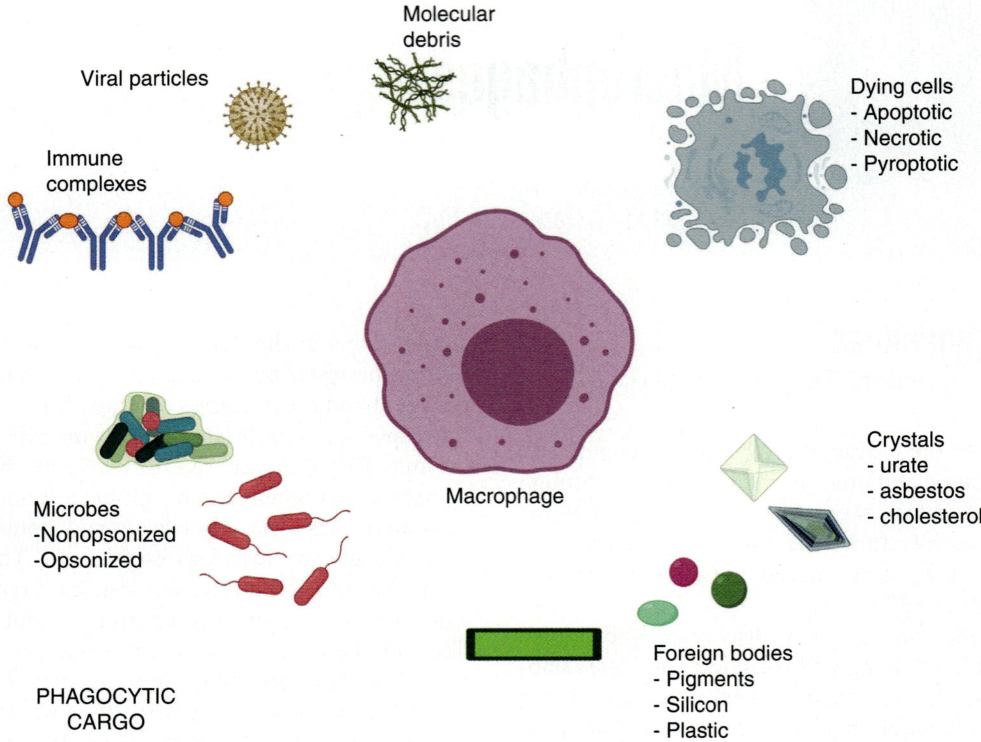

FIG. 15-1. Depiction of the range of phagocytic and pinocytic cargo engulfed by macrophage. (Created with BioRender.com.)

as presented below. As many molecules as there are to recognize pathogen-associated or host-derived phagocytic cargo, there is always a need to broaden recognition even further, to be ready to engulf pathogens possibly not encountered before. The macrophage achieves this readiness by expressing many receptor classes that bind not directly to cargo, but rather through adaptors. Adaptor molecules that promote phagocytic engulfment are called **opsonins**. Antibodies and complement are critical opsonins in immunity. Opsonins often engaged by macrophages include, among others, acute phase proteins (eg, mannose-binding protein and C-reactive protein), surfactants A and D, GAS6, protein S, and milk fat globule-EGF factor 8 (Mfge8) (Fig. 15.2). Table 15.1 lists a number of the most highly recognized phagocytic receptors used by macrophages and some of their properties.

Fc receptors and **complement receptors** both recognize cargo that have triggered humoral immune reactions, binding to immunoglobulin, especially IgG subtypes, or complement fragments, respectively. These antibodies or complement products, in turn, may be bound to antigens or to prokaryotic or eukaryotic cells that were permissive to complement activation on their surfaces. This feature makes antibodies and complement among the most important and well-known opsonins.

Fc receptors are so critical to the function of macrophages and in the link between humoral and cellular immunity that a whole chapter is devoted to them (Chapter 33). Briefly, Fc receptors are critical to phagocytosis because antibodies recognize foreign material and tag them for clearance by macrophages via binding to Fc receptors. There are three major classes of Fc gamma receptors (FcgR): Fcγr1 (CD64), Fcγr2b (CD32), FcγR3 (CD16), and Fcγr4 (CD16-2) in mice. Human homologs are FCGR1A (CD64), FCGR2B (CD32b), FCGR2A (CD32a), FCGR2C (CD32c), and FCGR3B (CD16a). FcγR1, FcγR3, and FcγR4 use the Fc receptor gamma chain (FcRγ) to signal through an immunoreceptor tyrosine-based *activation* motif. This activation signaling initiates cellular activation and phagocytosis, whereas the major isoform of FcγR2 has an immunoreceptor tyrosine-based *inhibitory* motif.

Complement receptors. The complement cascade is presented in Chapter 13. Complement fragments C3b or C4b react covalently with targets to create insoluble complexes and thereby serve as opsonins. Decades ago, major complement receptors that recognized these opsonins were named complement receptor 1 (CR1), CR2, CR3, and CR4. CR1 (also called CD35) is broadly expressed by many cell types. It recognizes C3b complement fragments and promotes their clearance, including clearance of dying cells bearing surface C3b. It is also the receptor on erythrocytes that takes in the malarial tachyzoite to propagate infection. CR2 (CD21) is not expressed on macrophages, but instead by B cells and follicular dendritic cells, where it serves as a platform to display antigen in B cell follicles and their germinal centers. The most relevant and widely expressed of complement receptors on macrophages, and on monocytes that serve as blood precursors for macrophages, are CR3 and CR4. Both CR3 and CR4 are integrins, a structurally conserved class of receptors

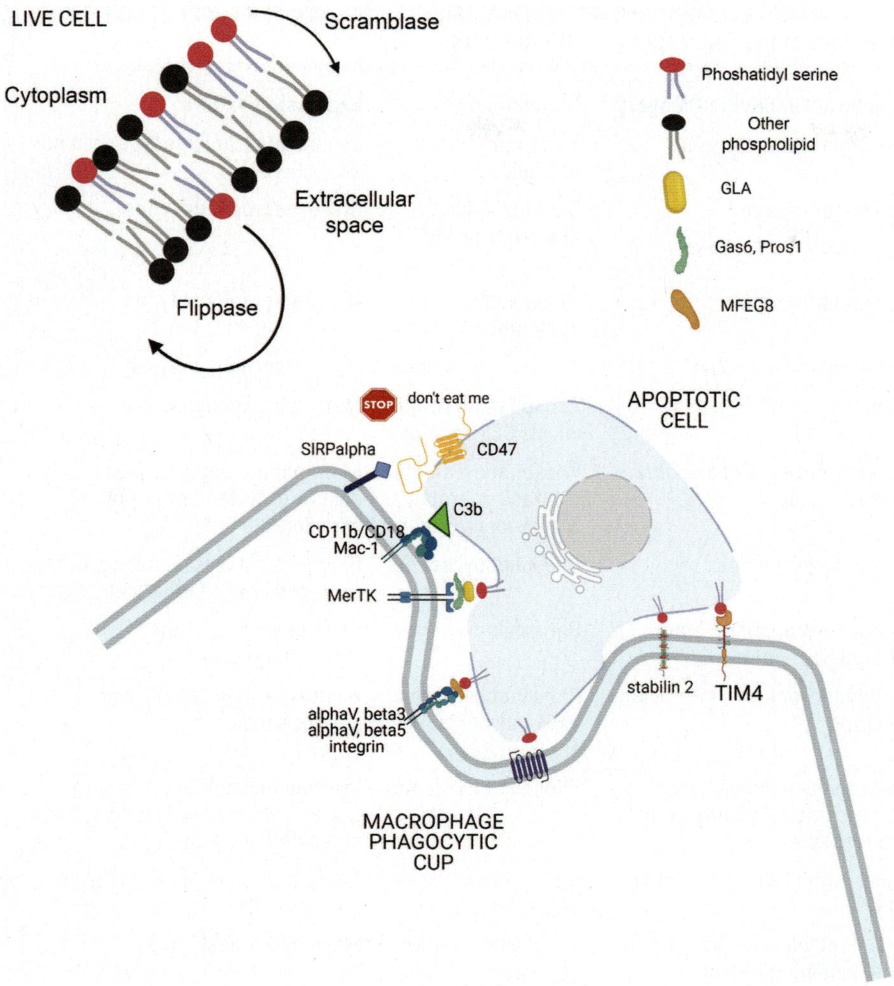

FIG. 15-2. **Engulfment of apoptotic cells by macrophages employs a range of different receptors.** Many of these receptors interact with the phospholipid phosphatidyl serine (PS), which is normally sequestered on the inner leaflet of the plasma membrane. However, extrusion of PS characterizes apoptotic cells, wherein scramblase that mobilizes PS to the outer leaflet dominates over the action of flippase that returns PS to the inner leaflet. Complement receptors that interact with complement iC3b deposited on the cell surface stands out as a receptor that does not engage PS. However, complement receptors like several other receptors do employ nontransmembrane adaptors to facilitate recognition of cargo by macrophages. CD47 engagement of SIRP1α stands out as a mechanism to prevent engulfment of target cells, rather than support it. (Created with BioRender.com.)

TABLE 15-1 Major Phagocytic Receptors of the Macrophage

Receptor Name	Gene Name	Structural/Functional Category	Ligands/Cargo	Expression pattern
FcgR1 (CD64)	Fcgr1	Fc receptor	High-affinity IgG, activating	Monocytes < most macrophages, not expressed by other cells
FcgRII (CD32)	Fcgr2	Fc receptor	Uptake, inhibit stimulation	Monocytes, macrophages, many cell types
FcgRIII (CD16)	Fcgr3	Fc receptor	Recognition of immune complexes, functional activation	Nonclassical monocytes > most tissue macrophages; NK cells
CD11b/Mac-1/CR3/amb2	Itgam	complement recetor, integrin	iC3b/ICAM-1	monocytes, most but not all macrophages, other myeloid cells
CD11c/CR4/axb2	Itgax	complement receptor, integrin	iC3b	some parenchymal macrophages, all mouse DCs; human monocytes, macrophages, DCs
CRIg	Vsig4	complement receptor	iC3b, LTA	Liver macrophages, peritoneal macrophages; many macrophages negative
CD36	Scarb3	scavenger receptor	Fatty acids, oxidized lipids, thrombospondin	Broad macrophage, some dendritic cells, nonhematopoietic cells (eg, intestinal epithelium)

(Continued)

TABLE 15-1 Major Phagocytic Receptors of the Macrophage (Continued)

Receptor Name	Gene Name	Structural/Functional Category	Ligands/Cargo	Expression pattern
MARCO	Scara2	scavenger receptor	Polyanions, modified lipids	Several macrophage types but not all
Macrophage scavenger receptor 1 (CD204, SR-A)	MSR1	scavenger receptor	Oxidized lipids, numerous ligands	Many macrophage types
CD163	CD163	scavenger receptor	Hemoglobin/ haptoglobin	Many macrophage types
TREM2	Trem2	immunoglobulin family	Aminophospholipids	Many macrophage types
avb3, avb5	itgav, itgb3, itgb5	Integrins	Phosphatidyl serine via MFGE8 adaptor	Many macrophage types
Mertk	Mertk	receptor tyrosine kinase, efferocytic receptor	Phosphatidyl serine via Gas6 or protein S adaptors	Mature macrophages, cleavable from surface by ADAM-17 metalloproteinase
CD300b	Cd300lb	CD300 family member	Phosphatidyl serine	High on most macrophages, except microglia; low on other cell types
Bai-1	Adgrb1	G-protein coupled receptor, efferocytic receptor	Phosphatidyl serine	Microglia, low on other macrophages
Stabilin 2	Stab2	scavenger receptor, efferocytic receptor	Phosphatidyl serine especially on aged erythrocytes	Highest on splenic and liver macrophages
TIM4	Timd4	T cell immunoglobulin and mucin-containing family, efferocytic receptor	Phosphatidyl serine	Multiple resident tissue macrophages; expressed after long period of residency
SIRPa	Sirpa	signal regulatory protein family (SIRP)	CD47, which blocks engulfment	Many macrophages and DC2 dendritic cells
CD169, sialoadhesin	Siglec1	immunoglobulin family, carbohydrate binding receptor	Sialic acid–containing ligands	Many macrophage types
CD206/mannose receptor	Mrc1	carbohydrate binding receptor	Terminal mannose residues in glycoproteins	Many macrophage types, including tumor macrophages, M2 polarized macrophages
Galectin 3	Lgals3	carbohydrate binding receptor	b-Galactosides in bacteria, fungi	Many macrophages, especially lung macrophages and inflammation-induced macrophages
Dectin/Clec7a	Clec7a	scavenger receptor/carbohydrate binding receptor	b1,3 or b1,6-Linked glucans enriched in fungi, plants	Many macrophages, very high in lung macrophages, peripheral nerve macrophages, activated microglia
Surfactant A	Sftpa1	Collectin	Secreted opsonin that promotes bacterial phagocytosis	Especially promotes phagocytic sampling of airway macrophages
Surfactant D	Sftpd	Collectin	Secreted opsonin, promotes bacterial phagocytosis	Especially promotes phagocytic sampling of airway macrophages

widely involved in cellular adhesion.[7] CR3 is also known as CD11b/CD18, $\alpha_m\beta_2$, or Mac-1 (macrophage receptor 1). CR4 is also known as CD11c/CD18 or $\alpha_x\beta_2$.

CD11b/Mac-1/CR3. The multifunctional role of Mac-1 in macrophage biology stems in part from the *different* ligands that it binds. First, like other integrins, CD11b/CD18 (CR3) binds to a counter-receptor from the Ig domain family called ICAM-1 that is expressed on the endothelial cells that line blood vessels and on T cells. This binding then permits Mac-1 to mediate transendothelial migration of monocytes, precursors of some macrophages (see below), out of the bloodstream. Transendothelial migration is a complex phenomenon requiring extensive and coordinated cytoskeletal rearrangement.[8] The role of Mac-1 in transendothelial migration is typically redundant with that of other integrins expressed by monocytes, including a_lb_2 (also called CD11a or LFA-1, lymphocyte function associated antigen) and a4b1 (also called CD49d/CD29 or VLA-4, very late antigen 4). The *inside-out signaling* that leads to integrin activation relies strongly on cytoskeletal rearrangement.[7]

Cytoskeletal rearrangement is, likewise, a critical element in phagocytic engulfment of cargo, so it is appropriate that the same signaling platform that mediates cell migration would be pivotal in phagocytosis. However, Mac-1 binds to a different ligand when serving as a complement receptor. Specifically, CR3 also binds to the opsonin and complement fragment iC3b that becomes covalently deposited onto surfaces that support complement activation. When Mac-1 engages surface-bound iC3b, cytoskeletal rearrangement permits the phagocytic cargo to be taken into the cell as a vacuole that coalesces with the lysosomal compartment for degradation. Thus, through the versatile molecule Mac-1, with distinct ligands in different scenarios, macrophages are able to sequentially carry out distinct functions with a single receptor and associated signaling pathway.

With its multiple functional roles, it is fitting that Mac-1 is widely expressed by monocytes and macrophages. In fact, it has frequently been regarded as a universal "macrophage marker." It is also expressed by other phagocytes, like neutrophils, where it too facilitates both migration and phagocytosis. In fact, it is so widely regarded as a universal macrophage marker in humans and mice that many cell identification strategies start off with the inclusion of Mac-1+ cells as a requirement for identifying macrophages in a heterogeneous mixture of cells, along with the exclusion of Mac-1-negative cells as "not containing macrophages." This strategy is successful in some contexts but can be incorrect. Mouse lung-resident macrophages and mouse spleen macrophages are two clear examples of macrophages that do not express Mac-1. Single, universal markers of macrophages may not exist. A molecule that more closely fits the definition of being broadly expressed on macrophages but not most other cells is CD64 (FcγR1).

CD11c/CR4. CD11c, or $\alpha_x\beta_2$, is the second integrin-type complement receptor on macrophages. It too plays a role in cell migration and phagocytosis for the same reasons as discussed for Mac-1. However, for many years in mice, the CD11c integrin was recognized for its more dominant expression on dendritic cells, and it became well known as a universal "dendritic cell marker," whereas Mac-1 become known as a "macrophage marker." This moniker prevailed despite clarity over decades that CD11c did not distinguish human macrophages and dendritic cells in many settings, particularly outside of secondary lymphoid tissues within organ parenchyma. CD11c is broadly expressed by dendritic cells in mice, but tissue-resident macrophages also express CD11c in mice. Thus, it is best to avoid strong reliance on CD11b and CD11c to define or distinguish macrophages from dendritic cells.

More recently, an additional receptor for C3b/iC3b called complement receptor Ig (CRIg; encoded by Vsig4 gene) was discovered to be expressed by Kupffer cells, the major resident macrophages of the liver located in liver sinusoids. CRIg is crucial to clearance of blood-borne bacteria, parasites, and complement-opsonized particulates[9,10] from the bloodstream, giving it a dominant role in host defense. Still further work revealed that it functions not only as a complement receptor but can directly recognize some pathogens, including gram-positive bacteria through binding of lipoteichoic acid.[11] The Immunological Genome Project, a database of gene expression by immune cells in mice (https://www.immgen.org), confirms that CRIg is most highly expressed among macrophages by Kupffer cells, with many other macrophage types being negative. Beyond Kupffer cells, interstitial macrophages of the lung, adipose tissue macrophages, and resident peritoneal macrophages express CRIg, albeit to a lower level than the Kupffer cells.

Scavenger receptors. Many textbooks, including Chapter 12 herein, and articles in the literature highlight a group of "scavenger receptors" that have historically been viewed as receptors on macrophages involved in phagocytic cargo recognition and engulfment. However, as the structural classification of receptors has matured over the decades, this term is losing relevance in favor of other receptor classifications. In particular, receptors with carbohydrate recognition domains called C-type lectin-like receptors (CLRs) account for several of the most relevant receptors also classified as scavenger receptors. These CLRs will be discussed below.

In 2017, a working group convened by the National Institutes of Health produced a consensus document for defining scavenger receptors.[12] The consensus definition is that "scavenger receptors are cell surface receptors that typically bind multiple ligands and promote the removal of nonself or altered-self targets. They often function by mechanisms that include endocytosis, phagocytosis, adhesion, and signaling that ultimately lead to the elimination of degraded or harmful substances." An example of a receptor that fits this definition and does not fall neatly into classification as another receptor type is CD36. CD36 has multiple ligands including the protein thrombospondin 1 and lipids including long-chain fatty acids and oxidized lipoproteins. The capacity to bind membranous lipids makes CD36 a relevant receptor for the uptake of lipoproteins (especially if oxidized or modified) or apoptotic cells. Additional scavenger receptors common on macrophages, such as macrophage receptor with collagenous structure (MARCO) or macrophage scavenger receptor 1 (MSR1; also called CD204, SR-A) are known for binding numerous different ligands. Others like CD163, which binds hemoglobin-haptoglobin complexes, have more well-defined roles in phagocytic clearance.

Receptors mediating uptake of apoptotic cells. Macrophages participate centrally in the recognition and uptake of apoptotic cells.[13] The engulfment-mediated removal of apoptotic cells is called **efferocytosis**. It is critical that uptake of apoptotic cells is not an inflammation-inducing response, because many physiological processes depend upon quiet removal of apoptotic cells. Remodeling of tissues during development is one such process, but in the adult, removal of aberrant sperm in the testis, clearance of apoptotic B cells in the germinal center or apoptotic T cells in the thymus, and removal of effete red blood cells in the spleen and liver are just a few examples of daily physiological processes that require macrophage-mediated phagocytosis and clearance. Of central importance is the display of the phospholipid phosphatidyl serine on the outer leaflet of the plasma membrane of apoptotic cells. This lipid is normally confined to the inner leaflet of the plasma membrane but scramblases activated during apoptosis allow this lipid to flip to the outer leaflet, while flippases that would return it to the inner leaflet are inactivated. A wide variety of different ligands for phosphatidyl serine can be expressed on macrophages, including TIM4 that directly binds to phosphatidyl serine for apoptotic cell clearance. Another receptor on macrophages called stabilin 2 binds directly to phosphatidyl serine and mediates clearance of effete red blood cells from the bloodstream in the spleen and liver. Yet other receptors have evolved to interact with adaptors that function much like opsonins that bridge these receptors to bind phosphatidyl serine. These receptors and their adaptors are shown in Figure 15.2.

Receptors interacting with carbohydrate domains. Receptors that recognize sugar residues (lectins) are often involved in interacting with and responding to phagocytic cargo.[14,15] Relevant classes of receptors include C-type lectin-like receptors (CLRs) and sialic acid binding Ig-like lectins (SIGLECs). Of the latter, CD169 (also called sialoadhesin) stands out as a prominent member of the SIGLEC family expressed on tissue-resident macrophages. Rather than promote uptake of phagocytic cargo, it mediates intercellular adhesion between macrophages and other immune cells. CLRs on macrophages that do mediate phagocytosis include CD206 (also called macrophage mannose receptor or MRC1) recognizing mannose-rich ligands, galectin 3 (Lgals3b), and Dectin1 (also called Clec7a) that recognizes β-glucan, a common component in the cell wall of fungi and present in some plants or bacteria. Soluble lectins are also involved in mediating phagocytic engulfment. Surfactant A and surfactant D are classified as collectins that through their recognition of sugars on bacteria can serve as critical opsonins in the airway.

PATTERN RECOGNITION RECEPTORS THAT COORDINATE SIGNALING

Chapter XXX is devoted to pattern recognition receptors including the families of toll-like receptors (TLRs) and NOD-like receptors (NLRs). These receptors are often engaged by macrophages during encounter with microbe-derived phagocytic cargo, alongside engagement of one or more of the receptors named above. In contrast to many of the receptors named above, the TLRs and NLRs are not considered triggers for phagocytic engulfment but rather critical triggers for transcriptional activation, leading, for example, to the activation of NF-κB, as discussed in Chapter 12. Among immune cells, TLR4 that binds lipopolysaccharide (LPS) from gram-negative bacteria stands out for its restricted expression on macrophages, at least among hematopoietic cells in the mouse.[16] Macrophages are also among the most potent producers of cytokines in the IL-1 family, especially IL-1β, the family of cytokines closely tied in evolution and function to TLRs.

THE ACT OF TAKING UP CARGO

After receptor engagement, uptake of cargo as large as bacteria through phagocytosis requires coordinated membrane remodeling. F-actin polymerization in response to receptor signaling allows for the formation of a *phagocytic cup* around the cargo, with eventual fusion of the membrane at the tips of the cup to form the *phagosome*. After the phagosome forms, F-actin remodeling is further required to allow its release and mobilization to the lysosome. Rho GTPases regulate F-actin assembly and key members of this group of molecules including RhoA, Rac1, and Cdc42.[17] Depending upon the cargo, there may be differential roles of one or more of these GTPases in phagocytosis of the relevant cargo. Guanidine nucleotide dissociation inhibitors (GDIs) prevent activation of Rho GTPases. *Yersinia* species of bacteria evolved a virulence factor YpkA that mimics the GDI to inhibit actin cytoskeletal dynamics and evade phagocytosis to avoid destruction in macrophages.[18]

From the earliest days of their isolation and study, macrophages were noted for their intensely active ruffling of the plasma membrane. This ruffling is associated with cytoskeletal dynamics that allow for the formation of micropinosomes and macropinosomes that can also mediate the uptake of cargo. Critically, pinocytosis stands out as a process that does not require receptor engagement to be triggered or to take in cargo. It is best described as a sampling of solutes in the environment, including some that might be as large as traditional phagocytic cargo. Like phagocytosis, cargo taken up via macropinocytosis can fuse with the lysosome and lead to degradation and possibly recycling of some acquired cargo. Highly active pinocytosis enables macrophages to thoroughly sample the surrounding environment. The panoply of receptors that can mediate phagocytosis, together with their adaptors, go a long way toward supporting broad sampling, but pinocytosis ensures that cargo can be sampled even when appropriate uptake receptors may be lacking.

LYSOSOMAL DESTRUCTION OF CARGO

Over the years, a defining molecule associated with macrophage identity has been the lysosomal protein CD68 (also called macrosialin). Dendritic cells also express CD68, but the number lysosomes in macrophages compared with dendritic cells is vastly greater. The cytoplasmic space of macrophages is typically filled with lysosomes throughout, whereas dendritic cells discretely position their lysosomes to a subdomain of the cytoplasm near the nucleus. This enrichment of lysosomes in macrophages is consistent with their role in killing micro-organisms that enter through phagocytosis, or pinocytosis, or in degradation of other cargo like apoptotic cells. Phagosomes and pinosomes fuse with lysosomes to generate a phagolysosome. *Mycobacterium tuberculosis* interferes with the generation of phagolysomes as part of its immunological resistance.

The role of the NADPH oxidase complex of enzymes that culminates in the generation of oxygen radicals in the phagolysosome is well described as a means for microbial killing, via the so-called oxidative burst. Mutations in various phagocyte oxidase (phox) enzymes lead to chronic granulomatous disease,[19] characterized by ongoing granulomas at sites where fungi or bacteria are poorly cleared within phagolysosomes owing to the mutations in one or more members of the enzyme complex. After ingestion of some bacteria, like *Salmonella typhimurium*, NADPH oxidase activity becomes extremely high in neutrophils but is far lower in macrophages. Macrophages, rather, rely on rapid and intense acidification of phagolysosomes through the action of vacuolar ATPase (V-ATPase) and potassium flux that collectively promotes high membrane potential[19] (Fig. 15.3). *M. tuberculosis* succeeds in intraphagosomal-lysosomal survival in part because of its ability to promote K^+ efflux from the lumen. This contrasts starkly with the phagolysosomes of neutrophils, which do not strongly acidify. A number of enzymes, like cathepsins, that are rich in the macrophage lysosomal proteome[20] rely on this acidification for optimal enzymatic activity. The macrophage lysosome is also rich in a variety of other proteases, lipases, and lysozyme, a glycoside hydrolase that degrades proteoglycans in the cell wall of gram-positive bacteria, along with ion channels and numerous molecules associated with vesicle fusion and remodeling.

A further mechanism of killing often emphasized in murine macrophages is the generation of the reactive nitrogen species (RNS) intermediate nitric oxide (NO) through expression and activity of the enzyme-inducible nitric oxide synthase (iNOS), which synthesizes NO from L-arginine and molecular oxygen with input of electrons from NADPH in the mitochondria. NO is a freely diffusible gas and can readily access the cytosol and cross membranes to enter organelles or enter the extracellular environment. It is also highly reactive and generates peroxynitrite, which often coincides with the generation of reactive oxygen species (ROS).[21,22] The generation of NO is necessary for optimal macrophage killing of *Mycobacterium bovis* and *M. tuberculosis*[23] and for killing of *Leishmania*.[24] The reaction producing NO drives the generation of ROS, including intermediates like peroxynitrite. RNS and ROS may directly kill pathogens, by providing toxic intermediates to the cytosol for the elimination of microbes that escape the phagolysosome or by entering the phagolysosome.[21] These species may also coordinate the killing of pathogens through activation of cellular processes like autophagy.[21] Macrophages continue to add lysosomes, and to upregulate lysosomal biogenesis, to accommodate organisms that survive and expand inside of phagolysosomes.[25] The eventual inability to keep up with the demand leads to phagolysosome disruption, with the spillage of K^+ ions driving the activation of the NLRP3 inflammasome.[25]

The sequential engulfment and destruction of apoptotic cell after apoptotic cell requires metabolic restructuring of macrophages.[26] The demand on a macrophage to export digested lipids, amino acids, and other metabolites after successful phagocytosis is great. The uptake of cargo by phagocytosis, micropinocytosis, or even receptor-mediated endocytosis can be so substantial as to position macrophages as the guardians of their own resistance to toxicity and that of their microenvironment. Lysosomal storage diseases result from the inability to clear cargo from the lysosome. An example is the loss of expression of a molecule called equilibrative nucleoside transporter 3 (ENT3) that allows for the clearance of nucleosides from the lysosome following degradation of cargo. In the absence of ENT3, nucleosides accumulate in the

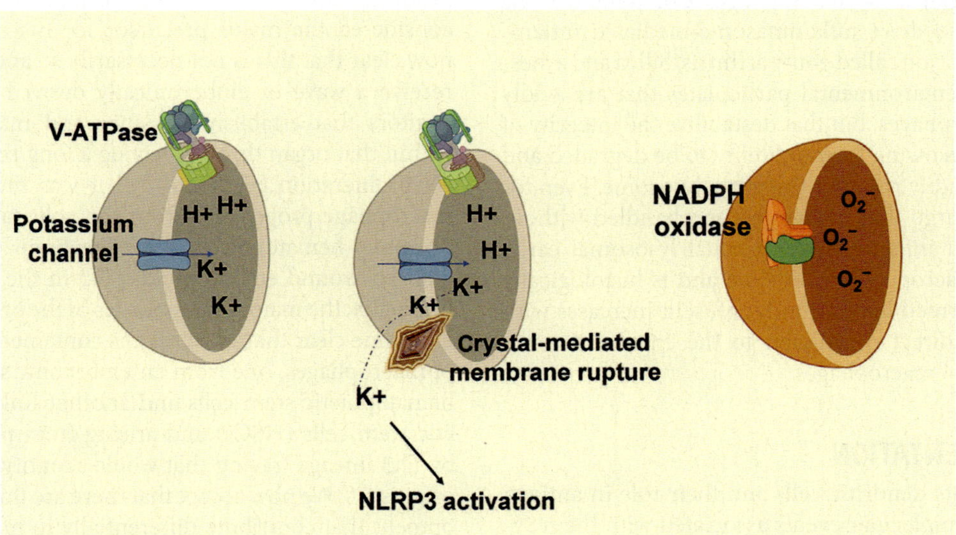

FIG. 15-3. Lysosomal acidification through the VATPase and potassium channels dominates as the means to activate degradation and pathogen killing in macrophages (*left*). This acidification and accumulation of potassium ions can be interrupted by crystals that form inside or outside the organelle that have the capacity to puncture the membrane. Spillage of K^+ into the cytosol serves as a means to activate the inflammasome. On the *right*, an alternative lysosomal mechanism of killing is the generation of oxygen radicals within the lysosome, via activation of the NADPH oxidase. This mechanism of lysosomal killing dominates in neutrophils over macrophages, but it also occurs in macrophages. (Created with BioRender.com.)

lysosomes, which alters their pH, and these changes in turn activate expression of the macrophage growth factor colony stimulating factor 1 (see below for more information).[27] This growth factor drives macrophage proliferation and leads to histiocytosis, linking lysosomal dysfunction to notable disease pathology beyond susceptibility to infection with a particular micro-organism.

Another classic example of the potential toxicity is accumulation of free cholesterol. Free cholesterol in the phospholipid membrane bilayer increases membrane fluidity and, if left unregulated, promotes loss of integrity and consequent toxicity. Cells have adapted to this problem by generating lipid droplets wherein cholesterol at the endoplasmic reticulum is esterified and packaged with stabilizing proteins into lipid droplets that are stored in the cytoplasm. Because of their robust sampling of the environment, macrophages are especially likely to become laden with lipid droplets. Their metabolic adaptation to activation furthers this likelihood.[22]

A heavily lipid droplet–containing macrophage is called a foam cell. It is thought that foam cells are designed to limit toxicity, because free cholesterol is sequestered away as cholesterol ester–rich lipid droplets. However, ongoing loading with lipid droplets is demanding for macrophages and renders them fragile, such that, although foam cells are not inherently proinflammatory,[28,29] they can promote inflammation. One possible way they do so is through crystallization of intracellular cholesterol[30] that either spatially impinges on adjacent lysosomes and activates the inflammasome through leak of K^+ ions or enters phagolysosomes through macrophage-mediated engulfment of cellular cargo bearing cholesterol crystals, such that the crystals disrupt the lysosomes of the efferocytic macrophage. In an analogous manner, sodium urate crystals engulfed by macrophages can puncture the phagolysosome and drive inflammasome-mediated inflammation in the condition called gouty arthritis. Silica and asbestos also generate environmental particulates that are avidly engulfed by macrophages, but that destabilize the integrity of the phagolysosome owing to poor ability to be degraded and a structure that readily ruptures the phagolysosome. Even for phagolysosomal cargo that is digested and handled without acute induction of inflammation, residual lysosomal cargo accumulates in macrophages over time and is histologically recognized and termed lipofuscin. Lipofuscin increases with age in presumed direct correlation to the engulfment and recycling activity of macrophages.

ANTIGEN PRESENTATION

Chapter 34 presents dendritic cells and their role in antigen presentation. The molecular events associated with the recycling of peptides from ingested foreign proteins for display on MHC molecules was worked out in macrophages and later reassessed and confirmed in dendritic cells. Many common features related to antigen presentation are shared between macrophages and dendritic cells. However, dendritic cells are capable of orchestrating CD8 T cell immunity, for instance, in a manner that is not duplicated by macrophages. Related to this issue, macrophages do not cross-present antigens, as observed for a subset of dendritic cells. Furthermore, a major advantage of dendritic cells over macrophages is the ability of dendritic cells to travel out of tissues to draining lymph nodes via lymphatic vessels. Macrophage membrane ruffling appears geared toward engulfment through pinocytosis and phagocytosis, rather than the types of cytoskeletal rearrangements associated with directed migration, as it has long been evident that such membrane ruffling activity is not a feature of dendritic cells.[31] Furthermore, macrophages generally do not upregulate the chemokine receptor CCR7, required for such migration.[32] A prominent exception to the paradigm is that Langerhans cells are now classified as macrophages, in part due to their dependence upon colony stimulating factor 1 (CSF-1) receptor[33] and their origin from yolk sac macrophages.[34] Prior to the more recent consideration as skin epidermal macrophages, Langerhans cells were classified as dendritic cells and display many dendritic cell features, including the trafficking of antigen from skin to draining lymph nodes.[35] Some immunologists may still classify Langerhans cells as dendritic cells for this reason.

Priming of immune responses often depends on antigen presentation in lymph nodes, such that dendritic cells are by far more likely than macrophages to initiate immunity to newly encountered peptides. However, it may be that tissue macrophages play a major role in activating memory T cells that have homed to organs where macrophages have sampled the microenvironment and displayed some of the acquired peptides on MHC II molecules.

EMBRYONIC ORIGIN OF MACROPHAGES AND MACROPHAGE DIVERSITY

From the late 1960s to recent years, blood monocytes were considered the major precursor for macrophages, but it is now clear that this is not necessarily so and that most organs receive a wave of embryonically derived macrophage progenitors that establish residence and maintain themselves within that organ through having a long cellular lifespan and via proliferation (Fig. 15.4).[36,37] It was first recognized that macrophage progenitors from the yolk sac, which supports primitive hematopoiesis, seed the brain of the developing embryo around embryonic day 7.5 in the mouse to become microglia, the major macrophages of the brain.[38] Shortly after, it became clear that most organs contained at least two types of macrophages, one from an embryonic source unlinked to hematopoietic stem cells and another linked to hematopoietic stem cells (HSCs) and arising from progenitors labeled by Flt3 lineage tracing that would identify their HSC-linked origins.[39,40] We now accept that there are three waves of development that contribute differentially to macrophages in different organs.[41] First, the only macrophages fully and truly derived from yolk sac progenitors are the microglia of the brain. Some Langerhans cells, macrophages bearing behaviors reminiscent of dendritic cells, are also derived from yolk sac progenitors. Second, between embryonic days 8.5 and 10.5, the fetal liver receives emigrated erythroid myeloid progenitors (EMPs) from the yolk sac. The EMPs, in turn, provide macrophage progenitors to developing organs like the

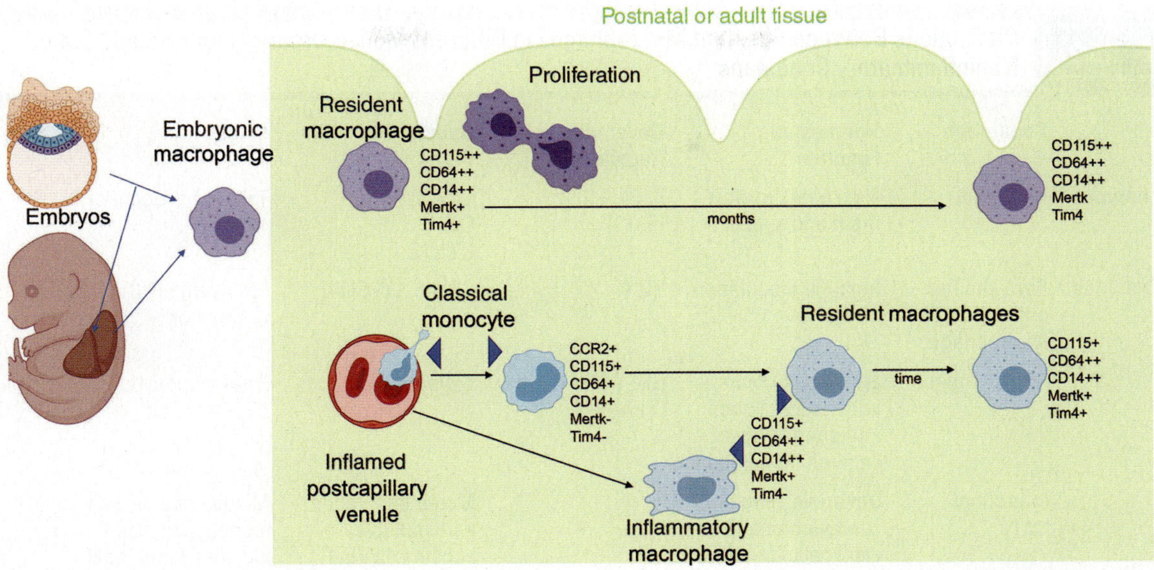

FIG. 15-4. A paradigm on the origin of macrophages. Macrophages in tissues arise from embryonic progenitors that, depending on the organ and time during development, arise from the yolk sac or fetal liver. Such macrophages are typically long lived and able to proliferate locally to sustain the organ macrophage pool. Another source of macrophages, typically more dominant during inflammation, is the blood monocyte. Macrophages that arise from monocytes can take on resident macrophage phenotypes but often first go through a differentiation phase that promotes inflammation in the organ. Tim-4 is a marker for the longevity of macrophages in tissues. Cell surface proteins that tend to be most highly expressed or restricted to macrophages include CD64, CD14, CD115, and MerTK. (Created with BioRender.com.)

liver, heart, pancreas, spleen, lymph nodes, peritoneal cavity, peripheral nerves, lung, and kidney. The embryonic source of macrophages establishes and forms a tissue-resident macrophage pool through self-renewal after what is often weeks, perhaps months, of survival in a relatively quiescent state. Finally, macrophages derived from HSC progenitors that first arise in the fetal liver but later mobilize to the bone marrow contribute to macrophages in the organs. These HSC-derived progenitors can add to the major resident macrophage population in some organs. Often, in mice, these major macrophages express high levels of F4/80 throughout their lifespan.[39] In some of these organs, the process of aging leads to a gradual replacement of part of the tissue-resident macrophage pool, suggesting an incapacity for local proliferation alone to meet the demands of homeostatic maintenance. An example of age-related replacement is observed in the peritoneal cavity for the major macrophages there,[42] now routinely referred to as large peritoneal macrophages.

Besides sometimes giving rise to some portion of the major resident macrophages in most organs, HSC-derived macrophages often give rise to a unique subset of macrophages in the organs (Table 15.2). The major HSC-derived cell type that serves as the precursor for these macrophages is the monocyte. It was noted early on that these macrophages expressed lower levels of F4/80.[39] Table 15.2 delineates the heterogeneity of macrophages found in a variety of organs in the steady state. Monocyte-derived macrophages include so-called small macrophages of the peritoneal cavity, interstitial macrophages of the lung, dermal macrophages of the skin, and CCR2+ macrophages of the heart (Table 15.2). The intestine stands out as an organ where embryonic macrophages are an extreme minority[124] and most resident macrophages are continuously replaced by monocytes.[125] The diversity of macrophages in a single organ is likely to become increasingly recognized and appreciated as single cell sequencing and other large-scale profiling efforts continue to contribute to the field.

MACROPHAGE MAINTENANCE VIA CYTOKINE SIGNALS

For survival, most macrophages depend at least partially upon CSF-1 receptor for development, survival, and proliferation.[126] This receptor tyrosine kinase is also called CD115 or macrophage colony stimulating factor receptor. It has two ligands, CSF-1 and IL-34. IL-34 promotes maintenance of microglia,[43,44] the resident macrophages of the brain. It also sustains Langerhans cells in the epidermis,[43] a population of macrophage-like cells that bears functional and molecular features of dendritic cells. Microglia arise from yolk sac progenitors,[38] and Langerhans cells partially depend upon yolk sac progenitors as well,[34] whereas most other macrophages arise from a later embryonic precursor.[40]

The study of mutations in CD115 across species ranging from mice to humans and beyond provides insight into the dependence of macrophages on CD115. Mutations generally manifest as skeletal defects that arise from the impact of mutations on bone macrophages called osteoclasts. Early research in this area identified a mouse line with osteopetrosis (abnormally dense bones prone to fracture) bearing a mutant allele (op) of the CD115 ligand CSF-1.[115,127] The op/op

TABLE 15-2 Distinctions Between Resident Macrophages in Different Mouse Organs Under Steady State, Noninflammatory Conditions

Organ	Population	Homeostatic Function	Developmental Regulators[a]	Specialized Markers	Origin	References
Central nervous system	Microglia	Neuronal development and support	IL-34, TGFβ; (Sall1)	CX3CR1^{hi} CD45^{lo} P2ry12	Yolk sac progenitor	43-48
Central nervous system	Perivascular and subdural macrophage	Immune modulation	PU.1	CD206, LYVE-1	Embryonic Flt3+ progenitor	49-51
Lung	Alveolar macs	Surfactant clearance, surveillance and phagocytosis, immune modulation	GM-CSF, TGFβ; (PPARγ)	Siglec F	Fetal monocytes	52-58
Lung	Interstitial macs	Unknown, immune surveillance and extracellular matrix remodeling?		Mixed population with multiple subsets; LYVE-1, CD206	Mixed, may include monocytes; also long lived and locally proliferative	59-62
Peritoneal cavity	Large peritoneal macrophages	Coagulation factor production to support phagocytic capture in fluid space	CSF-1, retinoic acid (Gata6, Bhlhe40, RXR)	ICAM2, CD93, P-selectin	EMP from fetal liver; slow input from monocytes during aging	16,42,63-68
Peritoneal cavity	Small peritoneal macrophages	T cell stimulation	(IRF4, AHR, MAFB)	CD206	HSC-derived monocytes	69-71
Spleen	Red pulp macrophages	Effete red blood cell removal; iron handling; host defense; cytokine secretion	(Spi-C, Bach1, IRF4 or IRF8)	VCAM-1, CD31, Stabilin 2, Heme oxidase 1	EMP from fetal liver, and HSC-derived monocytes	16,72-75
Spleen	Marginal zone macrophages	Bacterial, antigen, or apoptotic cell uptake from blood; support of B cell responses; immune suppression	CSF-1, lymphotoxin (LXRa, IRF8)	CD169, SIGNR1 (CD209b), MARCO	Postnatal development from monocytes	74,76-80
Spleen	Marginal metallophilic macrophages	Antigen uptake and transfer to other immune cells; viral clearance	CSF-1, lymphotoxin (LXRa)	CD169	Postnatal development from monocytes	74,76,81,82
Spleen	Tingible body macrophages	Located in germinal centers for uptake of apoptotic B cells		Poorly characterized: CD68+, MFG-E8-producing to support efferocytosis	Postnatal development from monocytes	74,83
Lymph node	Subcapsular sinus macrophages	Uptake of viral and phagocytic cargo; cytokine (e.g., IFN I) secretion; transfer of antigen to B cell follicle	Lymphatic endothelium-derived CSF-1; lymphotoxin	CD169, SIGNR1 (CD209b), F4/80-negative	Yolk sac progenitors, with replacement by HSC-derived monocytes postnatally; local proliferation	84-88
Lymph node	Medullary macrophages	Phagocytic clearance; support short-lived plasma cells; regulate lymphocyte egress	Lymphatic endothelium-derived CSF-1	CD169, SIGNR1 (CD209b), MARCO		84,88-90

TABLE 15-2 Distinctions Between Resident Macrophages in Different Mouse Organs Under Steady State, Noninflammatory Conditions (Continued)

Organ	Population	Homeostatic Function	Developmental Regulators[a]	Specialized Markers	Origin	References
Liver	Kupffer cells	Pathogen clearance; immune modulation; iron metabolism	(LXR-α, ID3)	Clec4F, Stabilin 2	EMP from fetal liver, and HSC-derived monocytes	91-93
Liver	Perivascular and capsular macrophages	Coordinate host defense and limit spread of infection		CXC3R1	Monocyte derived	94-96
Skin	Langerhans cell	Antigen transport to lymph nodes; adaptive immune regulation in skin and lymph nodes	IL-34, CSF1 receptor, TGFb; (ID2, RUNX3)	Langerin, capacity to express CCR7 unusual among macrophages (DC-like)	Yolk sac progenitors, EMP from fetal liver	32,38,97-99
Skin	Dermal macrophages	Scavengers, tattoo dye retention; nonmigratory and nonstimulatory of T cells	Partial dependence on CCR2	CD163, CX3CR1-negative, CD209f, PDGFc, C4b, a9 integrin	Progenitors from fetal liver, and HSC-derived monocytes that replenish throughout life	6
Intestine	Mucosal	Tissue quiescence, maintenance, and remodeling including secretion of immune modulating factors (e.g., IL-10, prostaglandin E2)	TGFb, IL-10	Expression of common macrophage markers with unique identifiers among macrophages presently unknown; CX3CR1^hi, CD11c+	Monocyte derived; mostly rapid turnover	100-103
Intestine	Muscularis	Neural cross talk, gut motility; vascular and tissue maintenance	CSF-1, norepinephrine signaling via b2 adrenergic receptor	CX3CR1^hi, CD11c^lo, RELMa^hi, TRPV4^+	Monocyte derived; slower turnover	100,104-106
Cardiac	Resident proliferating cardiac macrophages	Heart development; cardiomyocyte survival; dominant support of anti-inflammatory response; support electrical conduction	CSF1 receptor	TimD4, LYVE-1, Folate receptor 2, CD206, CD163, Factor XIIIA1, Connexin 43	Yolk sac progenitors, EMP from fetal liver	107-111
Cardiac	Recruited cardiac macrophages	Dominant support of inflammatory response	CCR2 and ligands	CCR2, MHC II	HSC-derived monocytes	37,107,112-114
Bone	Osteoclasts	Bone breakdown and remodeling; prevention of osteopetrosis	RANK-RANKL, osteoprotegerin, Blimp1-Bcl6, DC-STAMP (c-Fos, NFATC1, RXR)	Cathepsin K, ClC-7 chloride channel, multinucleated	EMP from fetal liver and ongoing HSC-derived monocyte recruitment	115-123

[a]Supporting growth factors listed first; supporting transcription factors in parentheses.

mouse survives only on some genetic backgrounds.[126] Without osteoclasts, the op/op mouse has growth impairments in part related to impaired skeletal remodeling and failed tooth eruption, functions normally supported out by degradative and tissue remodeling roles of osteoclasts during development. These bone defects tend to resolve with age.

Heterozygous mutations in CD115 in humans have been associated with skeletal defects. Furthermore, a distinct autosomal dominant condition arises in humans associated with mutations in CD115 called ALSP for adult-onset leukoencephalopathy with axonal spheroids and pigmented glia.[126] The condition results in adult-onset degenerative disease associated with impairments in cognition, behavior, or motor function. Mutations are typically observed in the highly conserved tyrosine kinase that appear to affect CD115 signaling, rather than generating null alleles that support a simple heterozygous state with one functional CD115 allele. Overall, bone and brain are the organs that manifest the greatest, or most obvious, functional impairments when CD115 is deleted or functionally modified.[126]

Aspects of how CD115 regulates macrophage homeostasis remain unknown. The ligand CSF-1 is produced in both a secreted form and a membrane form. CSF-1 concentrations in tissues may need to reach a certain threshold to sustain macrophage survival. Death of macrophages may result in reduced consumption of CSF-1 and thus a transient elevation in CSF-1, with the elevation possibly hitting a second threshold of a higher CSF-1 concentration that, in turn, promotes macrophage proliferation in a feedback loop.[128] This loop would ensure a balance between survival and proliferation to maintain a set number of macrophages in a given tissue microenvironment.[128]

An example of a resident population of macrophages only partially dependent upon CD115 signaling for maintenance is the Kupffer cell of the liver.[126] Alveolar macrophages resident in the lung do not depend upon CD115 or either of its ligands for maintenance, but instead rely on CSF-2, or granulocyte macrophage colony stimulating factor (GM-CSF), developing from fetal monocytes.[52] CSF-1 and CSF-2, respectively, support not only survival and proliferation of macrophages but also constitutive functional features of macrophages, including micropinocytosis.[129] Cytokines like TGFβ play a key role in supporting macrophage development in some tissues, including the development and homeostasis of alveolar macrophages[53] and brain microglia.[130]

RESIDENT TISSUE MACROPHAGES ARE SPECIALIZED TO THE ORGAN NICHE

Gene expression profiling of resident macrophages in different organs reveals a core signature of genes that defines all macrophages and also striking differences between tissue macrophages in different organs.[16,63,131,132] These distinct expression patterns reflect specializations that allow macrophages to carry out homeostatic functions of the organ.[128] Table 15.2 details features and functions of some of these macrophages and the transcription factors associated with programming these specific identities, factors that partner with pioneer, lineage-determining transcription factor PU.1 in macrophages.[131]

The structural architecture of an organ and the interaction of macrophages with organ parenchymal cells may drive the programming of tissue-specific macrophage identity. However, in many cases, the particular environmental signals that act to define the specialized tissue macrophage niche remain to be fully defined. In some cases, claims of key environmental signals have been made. For instance, retinoic acid is considered a critical environmental signal driving resident peritoneal macrophage identity.[63,64,131] However, it remains obscure how and why retinoic acid signaling is especially elevated in the peritoneal cavity as compared with other tissue spaces. Stromal cells in the serosal membrane are among those that can contribute to the production of retinoic acid,[133] but it is not known if this capacity is diminished in stromal cells in other organs. Understanding the specific nature of the tissue macrophage niche in relation to programming macrophage identity remains a quest for the future. Furthermore, recent data point to the possibility that differentiation trajectories of different resident macrophages in the same organ are predetermined before arrival in the organ, or rapidly following arrival, rather than solely by factors in the environment. That is, early progenitors from yolk sac that become microglia are entrained by TGFβ, but other embryonic progenitors not dependent upon TGFβ become brain-associated microglia (present in meninges and perivascular spaces).[130]

INFLAMMATION AND MACROPHAGES

Flexibility in activation states. It is often emphasized that macrophages are highly plastic cells, able to transform their identity to meet homeostatic needs in tissues, ranging from structural repair to host defense. For the last couple of decades, this emphasis on plasticity has usually been put forward as a preface to a concept called *macrophage polarization*.[134] Macrophage polarization connotes a distinct state of functional activation downstream of exposure to cytokines, immune complexes, or microbial products. The cytokines most extensively studied in relation to macrophage polarization are interferon gamma (IFN-γ) and IL-4, with IL-13 considered redundant with IL-4. The action of these cytokines on macrophages gives rise to states of activation that are, respectively, called M1 or M2, with M1 macrophages corresponding to type 1 immunity (including Th1 T cells producing IFN-γ) and M2 macrophages corresponding to type 2 immunity (including Th2 T cells producing IL-4 or IL-13) (Table 15.3). Other common terminology refers to IFN-γ-activated macrophages as classically activated macrophages, whereas IL-4-activated macrophages are often called alternatively activated macrophages. Regardless of the terminology used, these states should be regarded as paradigms rather than fixed identities, with a great deal of deviation from the ideal depending on the presence of other stimuli and the environment itself. The idealized paradigm is valuable in that macrophages bearing

TABLE 15-3 Macrophage Polarization by Type I or Type II Cytokines

Cytokine Effect on Macrophages	IFN-γ	IL-4 or IL-13
Function supported	Antibacterial Antiviral	Helminth responses; tissue remodeling
Selected effectors	iNOS/NO, inflammatory cytokines (TNFα, IL1β), Irg1, MHC, and antigen presentation	TGFb, IL-10, arginase

tendencies to these states are regularly, although not always, encountered and can be useful in interpreting the state of inflammation. Resident tissue macrophages rarely have basal states of activation that fits these paradigms, so the diversity in gene expression among macrophages described above does not typically reflect these states of activation.

IFN-γ is classified as type II interferon, as it was identified as a cytokine that promoted resistance to viral infection and tumor elimination. A major cellular target for its effector action is that of the macrophage.[135-137] Accordingly, IFN-γ improves antigen presentation by inducing higher expression of MHC I and MHC II molecules on antigen-presenting cells like macrophages and, in some cases, renders a macrophage lacking MHC II expression to become positive. It also promotes expression of new components of the proteosome that generates peptides from cytoplasmic proteins for presentation on MHC I. These induced components are called the immunoproteosome.[138] IFN-γ indeed is required to maintain latency of herpesvirus in infected macrophages.[139] By contrast, exposure to IL-4 reactivates the latent virus and suppresses the action of IFN-γ.[140] IL-4 expression is elevated during helminth infection, leading to the concept that viral reactivation in clinical scenarios might be in part related to the occurrences of coinfections with organisms that change the balance between M1 and M2 trajectories of activation.

IFN-γ action on macrophages is generally regarded to enhance both antiviral and antibacterial immunity. In the case of bacterial immunity, the mechanisms associated with bacterial destruction are greatly amplified in IFN-γ-conditioned macrophages. In mouse macrophages, this enhanced microbicidal activity is routinely tracked through monitoring the expression of iNOS or production of nitric oxide. M1 macrophages are known for the production of proinflammatory cytokines like TNFα, IL-1β, IL-12, and IL-6 with strong local and systemic effects marshalling endothelial cell activation for further immune cell recruitment and the acute phase response of the liver. IL-12 plays a major role in inducing Th1 T cells to produce IFN-γ that in turn acts on macrophages. Dendritic cells are considered a more relevant source of IL-12 than are macrophages. IFN-γ is also made by natural killer cells.

Cytokines that negatively regulate IFN-γ-driven macrophage activation include IL-10, IL-4, and IL-13.[141] These cytokines equip macrophages to facilitate defense against helminths. A clear challenge in immunity against worms is the large, multicellular nature of the pathogen. The major mode of killing bacteria through the engulfment and lysis within a phagolysosome cannot apply. Yet, macrophages are critical for optimizing defense against helminths, even though it is clear that the defense against helminths involves many other cell types. The overarching theme in natural immunity against helminths is to ensure that the host does not mete out such an immune response that the response itself puts survival of the host at risk. As death of helminths would require extracellular killing, one major risk is the dissemination of tissue from the dying worm that threatens tissue damage and outcomes like risk of vascular occlusion from embolism of worm material combined with activation of coagulation and complement in the vasculature. Rather than favoring this risk, the immune system has evolved a response that seals off the helminth within a mass of accumulated macrophages, other immune cells, and extracellular matrix that serves to encapsulate the helminth, walling it off and walling off the immune response surrounding it. M2 macrophages possess features that support this encapsulation, supporting extracellular matrix deposition through secretion of TGFβ, metabolic isolation of the parasite through the action of arginase-1 that reduces arginine accessibility, and through suppression of excessive inflammation via secretion of IL-10. IL-4 also promotes proliferation of macrophages,[142] allowing increased macrophage numbers to surround the organism without inciting a typical inflammatory response requiring monocyte infiltration from the blood.

Macrophages that tend toward an M2 polarization state have been associated with tumor microenvironments and other highly fibrotic environments. The same features of tissue remodeling and anti-inflammatory tendencies that optimize antihelminth responses seem adverse in these conditions, allowing for immune suppression that might facilitate killing of tumor cells or allowing for tissue destruction through excessive fibrosis.[143] Thus, control of macrophage activation is considered a target for the generation of a range of therapeutics.

The metabolic state of macrophages is closely coupled with activation status. Encounters with pathogen-derived signals like LPS or cytokines like IFN-γ or IL-4 can direct the metabolic state of macrophages as profoundly as the availability of nutrients or oxygen. It makes sense that the arrival of macrophages in a low-oxygen environment, which would include inflammatory environments rich in oxygen-consuming cells like neutrophils, would best rely on energy production that is independent of oxygen. Indeed, macrophages utilize glycolysis in hypoxic conditions. A break in the Krebs cycle such that the metabolic intermediate succinate accumulates facilitates stabilization and activation of the hypoxia induced factor (HIF) 1α (Fig. 15.5). M1 polarization of macrophages can also induce a state akin to hypoxia even in environments that are not low in oxygen. This comes about when the Krebs cycle is "broken" during M1 activation, raising the levels of succinate and citrate and inducing itaconate,[22] such that HIF1α is stabilized regardless of oxygen saturation. Glutamine metabolism also generates increased succinate and results in HIF1α activation. HIF1α in turn sustains expression of IL-1β by macrophages.

Further breaks in the Krebs cycle in these glycolytic macrophages generate citrate and itaconate. Citrate may drive lipid synthesis, but this interpretation has been challenged.[22] Itaconate is generated from the action of interferon induced regulatory gene (IRG) 1, a major gene product induced during M1 polarization (Table 15.3; Fig. 15.5). Although itaconate is induced during M1 polarization,[144] the action of itaconate is immunomodulatory, not one of enhancing type I inflammation. In fact, Irg1 activity is necessary to prevent hyperinflammation-related immunopathology and enhanced death in the context of murine *M. tuberculosis* infection.[145] The immunomodulatory properties of itaconate are under intense investigation.[146-149]

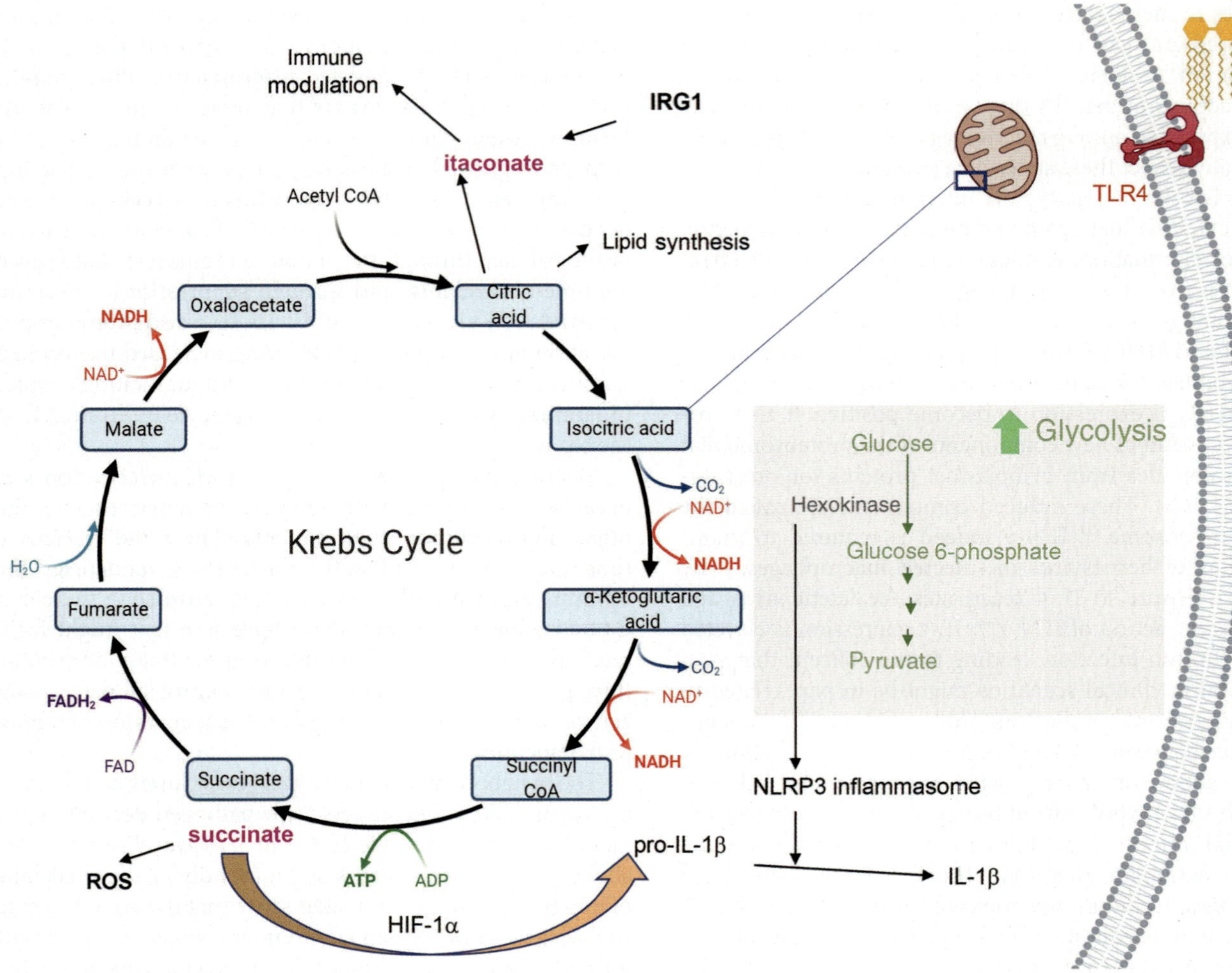

FIG. 15-5. Metabolic shifts that support macrophage phenotypes. Activation of macrophages via stimuli like lipopolysaccharide leads to metabolic shifts that include the generation of intermediates in the Krebs cycle that accumulate and act to effect processes like the stabilization and activation of hypoxia-induced factor, the generation of ROS, and the generation of novel intermediates like itaconate to fine-tune and modify the inflammatory response. Meanwhile, a reliance on glycolysis supports the energy demands of the macrophage and facilitates, along with accumulated succinate from the broken Krebs cycle, the generation of cytokines like IL-1β. (Created with BioRender.com.)

Accumulation of Macrophages at Sites of Inflammation

As discussed above, resident tissue macrophages can be self-sustained in adulthood by proliferation, and proliferation can under conditions of helminth infection allow for macrophage expansion. However, the need to expand the pool of macrophages at sites of inflammation or infection typically requires recruitment of macrophage precursors from the blood. Blood monocytes serve as macrophage precursors. There are two subsets of monocytes, classical and nonclassical.[150] Classical monocytes comprise greater than 90% of monocytes circulating in human blood and about 50% of monocytes in mouse blood. Classical monocytes are characterized by expression of CCR2 and CD14, among many other shared mRNA transcripts, in both species.[151] CCR2 is the chemokine receptor that engages the chemokine "monocyte chemoattractant protein" (also called JE historically) during monocyte recruitment to pulmonary granulomas.[152] Besides using CCR2 to gain access to tissues from the blood, classical monocytes emerge from the bone marrow in a manner dependent upon CCR2.[153] Other ligands for CCR2, CCL7, and CCL8, mediate egress from the marrow, rather than CCL2.[154] These newly emergent monocytes circulate for only a few days' time and get recruited to an inflammatory site, convert to nonclassical monocytes with the propensity to stay in the circulation for additional days, or turnover.[94] Nonclassical monocytes marginate within the vasculature and interact with vascular endothelium using CD11a/CD18 integrin.[155] This patrolling permits nonclassical monocytes to phagocytose cargo within the vasculature.[156,157]

The spleen serves as a major reservoir for blood monocytes. In contrast to the bone marrow monocytes, splenic monocytes are mature and identical to those in blood.[158] Once monocytes arrive at a site of inflammation, they begin to differentiate to mature macrophages. Even prior to leaving blood, monocytes are phagocytic, but their ability to kill phagocytic cargo, secrete cytokines, and coordinate local immune responses increases as they mature to macrophages. Extravasation from blood into tissue in the absence of inflammation does occur for a small fraction of CCR2+ monocytes; these monocytes do not readily mature to tissue-resident macrophages but traffic out of the tissue via the lymphatic vasculature.[159]

Likewise, some monocyte-derived cells that accumulate at sites of inflammation have the capacity to enter draining lymphatic vessels and thus travel to the draining lymph node,[160,161] where the cells may participate in organizing an ensuing adaptive immune response. This lymphatic vessel–mediated migration to lymph nodes does not quantitatively account for the removal of recruited macrophages from sites of resolving inflammation. Rather, local death accounts for the return of macrophage numbers to homeostatic levels.[161] Monocytes are also recruited to lymph nodes directly from the blood to facilitate the orchestration of immune responses.[162]

Resident Versus Recruited Monocytes in Inflammation and Beyond

Resident macrophages can participate in inflammation, but they are also able to dampen inflammation and keep hyperinflammatory reactions in check. This feature is relevant to the fact that macrophages resident in tissues are oriented toward maintenance of homeostasis. As indicated above, many resident macrophages arise from embryonic precursors. By contrast, inflammation brings in monocyte-derived macrophages from the bone marrow. In a variety of inflammatory settings, including ablation after irradiation, the original embryonically derived resident macrophages may die off. In such a case, resident macrophages are replaced by monocytes that develop into replacement macrophages. In some organs, like the heart, the macrophages arising from monocytes retain distinctive features, like expression of CCR2.[37,163,164] They may also retain distinct functions from the embryonic macrophages. For example, cardiac CCR2+ macrophages that once arose from monocytes are thought to respond to environmental stimuli by more avidly promoting further inflammation, as compared with the original embryonically derived macrophages. Thus, the relative increase of CCR2+ monocyte-derived macrophages over embryonic CCR2− cardiac macrophages is expected to have important implications in conditions like heart failure.

In other organs and in response to different inflammatory stimuli, monocyte-derived macrophages develop identical or nearly completely identical features to the original resident macrophages of embryonic origin,[165] yet they may nonetheless retain distinct responses to further insult.[166] However, while monocytes are in the transition phase toward becoming the specialized resident macrophage phenotype in that organ, the developing monocyte-derived macrophage transiently transitions through proinflammatory stages.[29,167,168] Thus, even if the outcome of monocyte recruitment to replace resident macrophages does not produce a more inflammation-prone resident macrophage, the tendency to be a less suppressive, more proinflammatory macrophage may characterize the intermediate stages of inflammation. There is also recent evidence that myeloid progenitors that bypass the classical monocyte stage may be recruited to sites of inflammation and exhibit extremely robust inflammatory capacity.[169]

During the course of inflammation, monocyte-derived macrophages appear to resist taking on the features of local resident macrophages. Instead, they may more closely resemble macrophages at other sites of inflammation, irrespective of the organ or niche, and they are most likely in the context of inflammation to closely resemble the M1 or M2 paradigm than at any other stage.[161] When recruited macrophages do develop identical features of embryonic resident macrophages within a given organ or niche, it takes time.[168] The effect of time may, in fact, be to wait out the inflammatory stimuli and reactions themselves.

One dominant model in the field has been that macrophages of distinct origin, resident versus recruited, would have different epigenetic marks and thus be incapable of responding to the same stimuli in precisely the same way. However, epigenetic landscapes shaped by niche-specific stimuli characteristic of different organ microenvironments are malleable. Indeed, it is becoming clear that even embryonic resident macrophages lose many of their niche-specific features during a strong inflammatory reaction.[170] This shift in programming is due to the redirection of niche-specific transcription factors to act on different target genes in response to inflammatory environmental cues that allow such transcription factors to partner with transcription factors induced during the inflammatory reaction itself. The outcome of this shift in the epigenetic landscape is that macrophages in inflamed organs lose much of their specialization; not only does the organ recruit monocyte-derived macrophages but the resident macrophages also become modified, and less distinctive, during an inflammatory reaction. The relative roles of embryonically derived and recruited monocytes, including those that take on resident macrophage features over time, is an intense subject of research with many open questions remaining.

CONCLUDING REMARKS

The intent of this chapter has been to primarily introduce the reader to the basics of macrophage biology, with an emphasis on macrophages from the mouse, presenting both established knowledge and developing frontiers in the field to lay the groundwork for further reading. The main text focuses on the general features that characterize most macrophages. Much activity in recent years has been focused on the delineation of tissue-resident macrophage functions and signals that drive their identity. Table 15.2 has been constructed to outline our understanding of the major functions and signals that drive the development of macrophages in a given organ. Continued growth in this area is anticipated in the near future, as new technologies like single cell sequencing and spatial imaging bring forward more information. It has not been possible to cover every relevant topic in relation to macrophage biology, nor to be in-depth in the topics covered. Furthermore, as the field evolves, some of the concepts presented here are likely to evolve. The reader should take away an understanding of the macrophage both as a crucial phagocyte and conductor of organ homeostasis and a cell that has the capacity to tune its inflammatory response to the nature of the pathogen. May the reader be inspired to build on this foundation going forward.

REFERENCES

1. Hirsch JG. Phagocytosis. *Annu Rev Microbiol.* 1965;19:339-350. doi:10.1146/annurev.mi.19.100165.002011
2. Cohn ZA. The fate of bacteria within phagocytic cells. 3. Destruction of an *Escherichia coli* agglutinogen within polymorphonuclear leucocytes and macrophages. *J Exp Med.* 1964;120(5):869-883.
3. Unanue ER, Askonas BA. Persistence of immunogenicity of antigen after uptake by macrophages. *J Exp Med.* 1968;127(5):915-926.
4. Unanue ER. From antigen processing to peptide-MHC binding. *Nat Immunol.* 2006;7(12):1277-1279. doi:10.1038/ni1206-1277
5. Silverstein, SC, Cohn ZA, Steinman RM. Endocytosis. *Annu Rev Biochem. Annu Rev Biochem.* 1977;46:669-722. doi:10.1146/annurev.bi.46.070177.003321
6. Baranska A, Shawket A, Jouve M, et al. Unveiling skin macrophage dynamics explains both tattoo persistence and strenuous removal. *J Exp Med.* 2018;215(4):1115-1133. doi:10.1084/jem.20171608
7. Hogg N, Patzak I, Willenbrock F. The insider's guide to leukocyte integrin signalling and function. *Nat Rev Immunol.* 2011;11(6):416-426. doi:10.1038/nri2986
8. Muller WA. Transendothelial migration: unifying principles from the endothelial perspective. *Immunol Rev.* 2016;273(1):61-75. doi:10.1111/imr.12443
9. Helmy KY, Katschke KJ, Gorgani NN, et al. CRIg: a macrophage complement receptor required for phagocytosis of circulating pathogens. *Cell.* 2006;124(5):915-927. doi:10.1016/j.cell.2005.12.039
10. Liu G, Fu Y, Yosri M, et al. CRIg plays an essential role in intravascular clearance of bloodborne parasites by interacting with complement. *Proc Natl Acad Sci U S A.* 2019;116(48):24214-24220. doi:10.1073/pnas.1913443116
11. Zeng Z, Surewaard BGJ, Wong CHY, Geoghegan JA, Jenne CN, Kubes P. CRIg functions as a macrophage pattern recognition receptor to directly bind and capture blood-borne gram-positive bacteria. *Cell Host Microbe.* 2016;20(1):99-106. doi:10.1016/j.chom.2016.06.002
12. PrabhuDas MR, Baldwin CL, Bollyky PL, et al. A consensus definitive classification of scavenger receptors and their roles in health and disease. *J Immunol.* 2017;198(10):3775-3789. doi:10.4049/jimmunol.1700373
13. Lemke G. How macrophages deal with death. *Nat Rev Immunol.* 2019;19(9):539-549. doi:10.1038/s41577-019-0167-y
14. Drouin M, Saenz J, Chiffoleau E. C-type lectin-like receptors: head or tail in cell death immunity. *Front Immunol.* 2020;11:251. doi:10.3389/fimmu.2020.00251
15. Li K, Underhill DM. *Current Topics in Microbiology and Immunology.* 2020.
16. Gautier EL, Shay T, Miller J, et al. Gene-expression profiles and transcriptional regulatory pathways that underlie the identity and diversity of mouse tissue macrophages. *Nat Immunol.* 2012;13(11):1118-1128.
17. Mao Y, Finnemann SC. Regulation of phagocytosis by Rho GTPases. *Small GTPases.* 2015;6(2):89-99. doi:10.4161/21541248.2014.989785
18. Prehna G, Ivanov MI, Bliska JB, Stebbins CE. Yersinia virulence depends on mimicry of host Rho-family nucleotide dissociation inhibitors. *Cell.* 2006;126(5):869-880. doi:10.1016/j.cell.2006.06.056
19. Nunes P, Demaurex N, Dinauer MC. Regulation of the NADPH oxidase and associated ion fluxes during phagocytosis. *Traffic.* 2013;14(11):1118-1131. doi:10.1111/tra.12115
20. Garin J, Diez R, Kieffer S, et al. The phagosome proteome: Insight into phagosome functions. *J Cell Biol.* 2001;152(1):165-80. doi:10.1083/jcb.152.1.165
21. Paiva CN, Bozza MT. Are reactive oxygen species always detrimental to pathogens? *Antioxid Redox Signal.* 2014;20(6):1000-1037. doi:10.1089/ars.2013.5447
22. Palmieri EM, McGinity C, Wink DA, McVicar DW. Nitric oxide in macrophage immunometabolism: hiding in plain sight. *Metabolites.* 2020;10(11):429. doi:10.3390/metabo10110429
23. Qualls JE, Subramanian C, Rafi W, et al. Sustained generation of nitric oxide and control of mycobacterial infection requires argininosuccinate synthase 1. *Cell Host Microbe.* 2012;12(3):313-323. doi:10.1016/j.chom.2012.07.012
24. Iniesta V, Gómez-Nieto LC, Corraliza I. The inhibition of arginase by Nω-hydroxy-L-arginine controls the growth of Leishmania inside macrophages. *J Exp Med.* 2001;193(6):777-784. doi:10.1084/jem.193.6.777
25. Westman J, Walpole GFW, Kasper L, et al. Lysosome fusion maintains phagosome integrity during fungal infection. *Cell Host Microbe.* 2020;28(6):798.e6-812.e6. doi:10.1016/j.chom.2020.09.004
26. Doran AC, Yurdagul A, Tabas I. Efferocytosis in health and disease. *Nat Rev Immunol.* 2020;20(4):254-267. doi:10.1038/s41577-019-0240-6
27. Hsu CL, Lin W, Seshasayee D, et al. Equilibrative nucleoside transporter 3 deficiency perturbs lysosome function and macrophage homeostasis. *Science.* 2012;335(6064):89-92. doi:10.1126/science.1213682
28. Spann NJ, Garmire LX, McDonald JG, et al. Regulated accumulation of desmosterol integrates macrophage lipid metabolism and inflammatory responses. *Cell.* 2012;151(1):138-152. doi:10.1016/j.cell.2012.06.054
29. Kim K, Shim D, Lee JS, et al. Transcriptome analysis reveals nonfoamy rather than foamy plaque macrophages are proinflammatory in atherosclerotic murine models. *Circ Res.* 2018;123(10):1127-1142. doi:10.1161/CIRCRESAHA.118.312804
30. Duewell P, Kono H, Rayner KJ, et al. NLRP3 inflammasomes are required for atherogenesis and activated by cholesterol crystals. *Nature.* 2010;464(7293):1357-1361. doi:10.1038/nature08938
31. Steinman RM, Cohn ZA. Identification of a novel cell type in peripheral lymphoid organs of mice: I. Morphology, quantitation, tissue distribution. *J Exp Med.* 1973;464(7293):1357-1361. doi:10.1084/jem.137.5.1142
32. Miller JC, Brown BD, Shay T, et al. Deciphering the transcriptional network of the dendritic cell lineage, *Nat Immunol.* 2012;13(9):888-899. doi:10.1038/ni.2370
33. Ginhoux F, Tacke F, Angeli V, et al. Langerhans cells arise from monocytes in vivo. *Nat Immunol.* 2006;7(3):265-273. doi:10.1038/ni1307
34. Hoeffel G, Wang Y, Greter M, et al. Adult Langerhans cells derive predominantly from embryonic fetal liver monocytes with a minor contribution of yolk sac-derived macrophages. *J Exp Med.* 2012;209(6):1167-1181. doi:10.1084/jem.20120340
35. Randolph GJ, Angeli V, Swartz MA. Dendritic-cell trafficking to lymph nodes through lymphatic vessels. *Nat Rev Immunol.* 2005;5(8):617-628. doi:10.1038/nri1670
36. Perdiguero EG, Geissmann F. The development and maintenance of resident macrophages. *Nat Immunol.* 2016;17(1):2-8. doi:10.1038/ni.3341
37. Epelman S, Lavine KJ, Beaudin AE, et al. Embryonic and adult-derived resident cardiac macrophages are maintained through distinct mechanisms at steady state and during inflammation. *Immunity.* 2014;40(1):91-104. doi:10.1016/j.immuni.2013.11.019
38. Ginhoux F, Greter M, Leboeuf M, et al. Fate mapping analysis reveals that adult microglia derive from primitive macrophages, *Science.* 2010;330(6005):841-845. doi:10.1126/science.1194637
39. Schulz C, Perdiguero EG, Chorro L, et al. A lineage of myeloid cells independent of Myb and hematopoietic stem cells, *Science.* 2012;336(6077):86-90. doi:10.1126/science.1219179
40. Gomez Perdiguero E, Klapproth K, Schulz C, et al. Tissue-resident macrophages originate from yolk-sac-derived erythro-myeloid progenitors. *Nature.* 2015;518(7540):547-551. doi:10.1038/nature13989
41. Mass E, Gentek R. Fetal-derived immune cells at the roots of lifelong pathophysiology. *Front Cell Dev Biol.* 2021;9:1-17.
42. Bain CC, Gibson DA, Steers NJ, et al. Rate of replenishment and microenvironment contribute to the sexually dimorphic phenotype and function of peritoneal macrophages. *Sci Immunol.* 2020;5(48):eabc4466. doi:10.1126/SCIIMMUNOL.ABC4466
43. Greter M, Lelios I, Pelczar P, et al. Stroma-derived interleukin-34 controls the development and maintenance of langerhans cells and the maintenance of microglia. *Immunity.* 2012;37(6):1050-1060. doi:10.1016/j.immuni.2012.11.001
44. Wang Y, Szretter KJ, Vermi W, et al. IL-34 is a tissue-restricted ligand of CSF1R required for the development of Langerhans cells and microglia. *Nat Immunol.* 2012;13(8):753-760. doi:10.1038/ni.2360
45. Bessis A, Béchade C, Bernard D, Roumier A. Microglial control of neuronal death and synaptic properties. *Glia.* 2007;55(3):233-238. doi:10.1002/glia.20459
46. Kana V, Desland FA, Casanova-Acebes M, et al. CSF-1 controls cerebellar microglia and is required for motor function and social interaction. *J Exp Med.* 2019;216(10):2265-2281. doi:10.1084/jem.20182037
47. Dror Michaelson M, Bieri PL, Mehler MF, et al. CSF-1 deficiency in mice results in abnormal brain development. *Development.* 1996;122(09):2661-2672.
48. Buttgereit A, Lelios I, Yu X, et al. Sall1 is a transcriptional regulator defining microglia identity and function. *Nat Immunol.* 2016;17(12):1397-1406. doi:10.1038/ni.3585
49. Goldmann T, Wieghofer P, Jordão MJC, et al. Origin, fate and dynamics of macrophages at central nervous system interfaces. *Nat Immunol.* 2016;17(7):797-805. doi:10.1038/ni.3423
50. Rua R, Lee JY, Silva AB, et al. Infection drives meningeal engraftment by inflammatory monocytes that impairs CNS immunity. *Nat Immunol.* 2019;20(4):407-419. doi:10.1038/s41590-019-0344-y
51. Brezovakova V, Jadhav S. Identification of Lyve-1 positive macrophages as resident cells in meninges of rats. *J Comp Neurol.* 2020;528(12):2021-2032. doi:10.1002/cne.24870
52. Guilliams M, De Kleer I, Henri S, et al. Alveolar macrophages develop from fetal monocytes that differentiate into long-lived cells in the first week of life via GM-CSF. *J Exp Med.* 2013;210(10):1977-1992. doi:10.1084/jem.20131199
53. Yu X, Buttgereit A, Lelios I, et al. The cytokine TGF-β promotes the development and homeostasis of alveolar macrophages. *Immunity.* 2017;47(5):903.e4-912.e4. doi:10.1016/j.immuni.2017.10.007
54. Yoshida M, Ikegami M, Reed JA, Chroneos ZC, Whitsett JA. GM-CSF regulates protein and lipid catabolism by alveolar macrophages. *Am J Physiol Lung Cell Mol Physiol.* 2001;280(3):L379-L386. doi:10.1152/ajplung.2001.280.3.l379
55. Huffman JA, Hull WM, Dranoff G, Mulligan RC, Whitsett JA. Pulmonary epithelial cell expression of GM-CSF corrects the alveolar proteinosis in GM-CSF-deficient mice. *J Clin Invest.* 1996;97(3):649-655. doi:10.1172/JCI118461
56. Shibata Y, Berclaz PY, Chroneos ZC, Yoshida M, Whitsett JA, Trapnell BC. GM-CSF regulates alveolar macrophage differentiation and innate immunity in the lung through PU.1. *Immunity.* 2001;15(4):557-567. doi:10.1016/S1074-7613(01)00218-7
57. Gautier EL, Chow A, Spanbroek R, et al. Systemic analysis of PPARγ in mouse macrophage populations reveals marked diversity in expression with critical roles in resolution of inflammation and airway immunity. *J Immunol.* 2012;189(5):2614-2624. doi:10.4049/jimmunol.1200495
58. Kopf M, Schneider C, Nobs SP. The development and function of lung-resident macrophages and dendritic cells. *Nat Immunol.* 2015;16(1):36-44. doi:10.1038/ni.3052
59. Gibbings SL, Thomas SM, Atif SM, et al. Three unique interstitial macrophages in the murine lung at steady state. *Am J Respir Cell Mol Biol.* 2017;57(1):66-76. doi:10.1165/rcmb.2016-0361OC
60. Chakarov S, Lim HY, Tan L, et al. Two distinct interstitial macrophage populations coexist across tissues in specific subtissular niches. *Science.* 2019;363(6432):eaau0964.doi:10.1126/science.aau0964
61. Schyns J, Bai Q, Ruscitti C, et al. Non-classical tissue monocytes and two functionally distinct populations of interstitial macrophages populate the mouse lung. *Nat Commun.* 2019;10(1):3964. doi:10.1038/s41467-019-11843-0
62. Shi T, Denney L, An H, Ho LP, Zheng Y. Alveolar and lung interstitial macrophages: Definitions, functions, and roles in lung fibrosis. *J Leukoc Biol.* 2020;110(1):107-114. doi:10.1002/JLB.3RU0720-418R
63. Okabe Y, Medzhitov R. Tissue-specific signals control reversible program of localization and functional polarization of macrophages. *Cell.* 2014;157(4):832-844. doi:10.1016/j.cell.2014.04.016
64. Casanova-Acebes M, Menéndez-Gutiérrez MP, Porcuna J, et al. RXRs control serous macrophage neonatal expansion and identity and contribute to ovarian cancer progression. *Nat Commun.* 2020;11(1):1655. doi:10.1038/s41467-020-15371-0
65. Rosas M, Davies LC, Giles PJ, et al. The transcription factor Gata6 links tissue macrophage phenotype and proliferative renewal, *Science.* 2014;344(6184):645-648. doi:10.1126/science.1251414
66. Gautier EL, Ivanov S, Williams JW, et al. Gata6 regulates aspartoacylase expression in resident peritoneal macrophages and controls their survival. *J Exp Med.* 2014;211:(8):1525-1531. doi:10.1084/jem.20140570
67. Zhang N, Czepielewski RS, Jarjour NN, et al. Expression of factor V by resident macrophages boosts host defense in the peritoneal cavity. *J Exp Med.* 2019;211(8):1525-1531. doi:10.1084/jem.20182024

68. Jarjour NN, Schwarzkopf EA, Bradstreet TR, et al. Bhlhe40 mediates tissue-specific control of macrophage proliferation in homeostasis and type 2 immunity. *Nat Immunol.* 2019;20(6):687-700. doi:10.1038/s41590-019-0382-5
69. Kim KW, Williams JW, Wang YT, et al. MHC II+ resident peritoneal and pleural macrophages rely on IRF4 for development from circulating monocytes. *J Exp Med.* 2016;213:(10):1951-1959. doi:10.1084/jem.20160486
70. Takenaka E, Van Vo A, Yamashita-Kanemaru Y, Shibuya A, Shibuya K. Selective DNAM-1 expression on small peritoneal macrophages contributes to CD4+ T cell costimulation. *Sci Rep.* 2018;8(1):15180. doi:10.1038/s41598-018-33437-4
71. Goudot C, Coillard A, Villani AC, et al. Aryl hydrocarbon receptor controls monocyte differentiation into dendritic cells versus macrophages. *Immunity.* 2017;47(3):582.e6-596.e6. doi:10.1016/j.immuni.2017.08.016
72. Kohyama M, Ise W, Edelson BT, et al. Role for Spi-C in the development of red pulp macrophages and splenic iron homeostasis. *Nature.* 2009;457(7227):318-321. doi:10.1038/nature07472
73. Theurl I, Hilgendorf I, Nairz M, et al. On-demand erythrocyte disposal and iron recycling requires transient macrophages in the liver. *Nat Med.* 2016;22(8):945-951. doi:10.1038/nm.4146
74. A-Gonzalez N, Castrillo A. Origin and specialization of splenic macrophages. *Cell Immunol.* 2018;330:151-158. doi:10.1016/j.cellimm.2018.05.005
75. Yamamoto M, Kato T, Hotta C, et al. Shared and distinct functions of the transcription factors IRF4 and IRF8 in myeloid cell development. *PLoS One.* 2011;6(10):e25812. doi:10.1371/journal.pone.0025812
76. A-Gonzalez N, Guillen JA, Gallardo G, et al. The nuclear receptor LXRα controls the functional specialization of splenic macrophages. *Nat Immunol.* 2013;14(8):831-839. doi:10.1038/ni.2622
77. Karlsson MCI, Guinamard R, Bolland S, Sankala M, Steinman RM, Ravetch JV. Macrophages control the retention and trafficking of B lymphocytes in the splenic marginal zone. *J Exp Med.* 2003;198(2):333-340. doi:10.1084/jem.20030684
78. Mcgaha TL, Karlsson MCI. Apoptotic cell responses in the splenic marginal zone: a paradigm for immunologic reactions to apoptotic antigens with implications for autoimmunity. *Immunol Rev.* 2016;269(1):26-43. doi:10.1111/imr.12382
79. Qi CF, Li Z, Raffeld M, Wang H, Kovalchuk AL, Morse HC. Differential expression of IRF8 in subsets of macrophages and dendritic cells and effects of IRF8 deficiency on splenic B cell and macrophage compartments. *Immunol Res.* 2009;45(1):62-74. doi:10.1007/s12026-008-8032-2
80. Pirgova G, Chauveau A, MacLean AJ, Cyster JG, Arnon TI. Marginal zone SIGN-R1+ macrophages are essential for the maturation of germinal center B cells in the spleen. *Proc Natl Acad Sci U S A.* 2020;117(22):12295-12305. doi:10.1073/pnas.1921673117
81. Backer R, Schwandt T, Greuter M, et al. Effective collaboration between marginal metallophilic macrophages and CD8+ dendritic cells in the generation of cytotoxic T cells. *Proc Natl Acad Sci U S A.* 2010;107(1):216-221. doi:10.1073/pnas.0909541107
82. Liu C, Yang XV, Wu J, et al. Oxysterols direct B-cell migration through EBI2. *Nature.* 2011;475(7357):519-523. doi:10.1038/nature10226
83. Hanayama R, Miyasaka K, Nakaya M, Nagata S. MFG-E8-dependent clearance of apoptotic cells, and autoimmunity caused by its failure. *Curr Dir Autoimmun.* 2006;9:162-172. doi:10.1159/000090780
84. Gray EE, Cyster JG. Lymph node macrophages. *J Innate Immun.* 2012;4(5-6):424-436. doi:10.1159/000337007
85. Roozendaal R, Mempel TR, Pitcher LA, et al. Conduits mediate transport of low-molecular-weight antigen to lymph node follicles. *Immunity.* 2009;30(2):264-276. doi:10.1016/j.immuni.2008.12.014
86. Phan TG, Green JA, Xu Y, Cyster JG. Immune complex relay by subcapsular sinus macrophages and noncognate B cells drives antibody affinity maturation. *Nat Immunol.* 2009;10(7):786-793. doi:10.1038/ni.1745
87. Iannacone M, Moseman EA, Tonti E, et al. Subcapsular sinus macrophages prevent CNS invasion on peripheral infection with a neurotropic virus. *Nature.* 2010;465(7301):1079-1083. doi:10.1038/nature09118
88. Mondor I, Baratin M, Lagueyrie M, et al. Lymphatic endothelial cells are essential components of the subcapsular sinus macrophage niche. *Immunity.* 2019;50(6):1453.e4-1466.e4. doi:10.1016/j.immuni.2019.04.002
89. Grande NR, Moreira de Sa C, Aguas AP, Carvalho E, Soares M. Time course and distribution of tungsten-laden macrophages in the hilar lymph nodes of the dog lung after experimental instillation of calcium tungstate into the left apical bronchus. *Lymphology.* 1990;23(4):171-182.
90. Grigorova IL, Schwab SR, Phan TG, Pham THM, Okada T, Cyster JG. Cortical sinus probing, S1P1-dependent entry and flow-based capture of egressing T cells. *Nat Immunol.* 2009;10(1):58-65. doi:10.1038/ni.1682
91. Bonnardel J, T'Jonck W, Gaublomme D, et al. Stellate cells, hepatocytes, and endothelial cells omprint the Kupffer cell identity on monocytes colonizing the liver macrophage niche. *Immunity.* 2019;51(4):638.e9-654.e9. doi:10.1016/j.immuni.2019.08.017
92. Scott CL, T'Jonck W, Martens L, et al. The transcription factor ZEB2 is required to maintain the tissue-specific identities of macrophages. *Immunity.* 2018;49(2):312.e5-325.e5. doi:10.1016/j.immuni.2018.07.004
93. Scott CL, Zheng F, De Baetselier P, et al. Bone marrow-derived monocytes give rise to self-renewing and fully differentiated Kupffer cells. *Nat Commun.* 2016;7:10321. doi:10.1038/ncomms10321
94. Yona S, Kim KW, Wolf Y, et al. Fate mapping reveals origins and dynamics of monocytes and tissue macrophages under homeostasis. *Immunity.* 2013;38(1):79-91. doi:10.1016/j.immuni.2012.12.001
95. David BA, Rezende RM, Antunes MM, et al. Combination of mass cytometry and imaging analysis reveals origin, location, and functional repopulation of liver myeloid cells in mice. *Gastroenterology.* 2016;151(6):1176-1191. doi:10.1053/j.gastro.2016.08.024
96. Sierro F, Evrard M, Rizzetto S, et al. A liver capsular network of monocyte-derived macrophages restricts hepatic dissemination of intraperitoneal bacteria by neutrophil recruitment. *Immunity.* 2017;47(2):374.e6-388.e6. doi:10.1016/j.immuni.2017.07.018
97. Merad M, Ginhoux F, Collin M. Origin, homeostasis and function of Langerhans cells and other langerin-expressing dendritic cells. *Nat Rev Immunol.* 2008;8(12):935-947. doi:10.1038/nri2455
98. Lavin Y, Mortha A, Rahman A, Merad M. Regulation of macrophage peripheral tissues. *Nat Rev Immunol.* 2015;15(12):731-744.
99. West HC, Bennett CL. Redefining the role of langerhans cells as immune regulators within the skin. *Front Immunol.* 2018;8:1941. doi:10.3389/fimmu.2017.01941
100. Bain CC, Schridde A. Origin, differentiation, and function of intestinal macrophages. *Front Immunol.* 2018;9:2733. doi:10.3389/fimmu.2018.02733
101. Schridde A, Bain CC, Mayer JU, et al. Tissue-specific differentiation of colonic macrophages requires TGFβ receptor-mediated signaling. *Mucosal Immunol.* 2017;10(6):1387-1399. doi:10.1038/mi.2016.142
102. Zigmond E, Bernshtein B, Friedlander G, et al. Macrophage-restricted interleukin-10 receptor deficiency, but not IL-10 deficiency, causes severe spontaneous colitis. *Immunity.* 2014;40(5):720-733. doi:10.1016/j.immuni.2014.03.012
103. Chikina AS, Nadalin F, Maurin M, et al. Macrophages maintain epithelium integrity by limiting fungal product absorption. *Cell.* 2020;183(2):411.e16-428.e16. doi:10.1016/j.cell.2020.08.048
104. Gabanyi I, Muller PA, Feighery L, Oliveira TY, Costa-Pinto FA, Mucida D. Neuro-immune interactions drive tissue programming in intestinal macrophages. *Cell.* 2016;164(3):378-391. doi:10.1016/j.cell.2015.12.023
105. Muller PA, Koscsó B, Rajani GM, et al. Crosstalk between muscularis macrophages and enteric neurons regulates gastrointestinal motility. *Cell.* 2014;158(2):300-313. doi:10.1016/j.cell.2014.04.050
106. Luo J, Qian A, Oetjen LK, et al. TRPV4 channel signaling in macrophages promotes gastrointestinal motility via direct effects on smooth muscle cells. *Immunity.* 2018;49(1):107.e4-119.e4. doi:10.1016/j.immuni.2018.04.021
107. Zaman R, Hamidzada H, Epelman S. Exploring cardiac macrophage heterogeneity in the healthy and diseased myocardium. *Curr Opin Immunol.* 2021;68:54-63. doi:10.1016/j.coi.2020.09.005
108. Hulsmans M, Clauss S, Xiao L, Milan DJ, Ellinor PT, Correspondence MN, Macrophages facilitate electrical conduction in the heart in brief heart-resident macrophages directly modulate the electrical properties of cardiomyocytes. *Cell.* 2017;169(3):510.e20-522.e20. doi:10.1016/j.cell.2017.03.050
109. Leid J, Carrelha J, Boukarabila H, Epelman S, Jacobsen SEW, Lavine KJ. Primitive embryonic macrophages are required for coronary development and maturation. *Circ Res.* 2016;118(10):1498-1511. doi:10.1161/CIRCRESAHA.115.308270
110. Leblond AL, Klinkert K, Martin K, et al. Systemic and cardiac depletion of M2 macrophage through CSF-1R signaling inhibition alters cardiac function post myocardial infarction. *PLoS One.* 2015;10(9):e0137515. doi:10.1371/journal.pone.0137515
111. Rojo R, Raper A, Ozdemir DD, et al. Deletion of a Csf1r enhancer selectively impacts CSF1R expression and development of tissue macrophage populations. *Nat Commun.* 2019;10(1):3215. doi:10.1038/s41467-019-11053-8
112. Epelman S, Lavine KJ, Randolph GJ. Origin and functions of tissue macrophages. *Immunity.* 2014;41:21-35. doi:10.1016/j.immuni.2014.06.013
113. Bajpai G, Bredemeyer A, Li W, et al. Tissue resident CCR2− and CCR2+ cardiac macrophages differentially orchestrate monocyte recruitment and fate specification following myocardial injury. *Circ Res.* 2019;124(2):263-278. doi:10.1161/CIRCRESAHA.118.314028
114. Rhee AJ, Lavine KJ. New approaches to target inflammation in heart failure: harnessing insights from studies of immune cell diversity. *Annu Rev Physiol.* 2020;82:1-20. doi:10.1146/annurev-physiol-021119-034412
115. Wiktor-Jedrzejczak W, Bartocci A, Ferrante AW, et al. Total absence of colony-stimulating factor 1 in the macrophage-deficient osteopetrotic (op/op) mouse. *Proc Natl Acad Sci U S A.* 1990;87(12):4828-4832. doi:10.1073/pnas.87.12.4828
116. Simonet WS, Lacey DL, Dunstan CR, et al. Osteoprotegerin: a novel secreted protein involved in the regulation of bone density. *Cell.* 1997;89(2):309-319. doi:10.1016/S0092-8674(00)80209-3
117. Dougall WC, Glaccum M, Charrier K, et al. RANK is essential for osteoclast and lymph node development. *Genes Dev.* 1999;13(18):2412-2424. doi:10.1101/gad.13.18.2412
118. Yagi M, Miyamoto T, Sawatani Y, et al. DC-STAMP is essential for cell-cell fusion in osteoclasts and foreign body giant cells. *J Exp Med.* 2005;207(4):751-762. doi:10.1084/jem.20050645
119. Miyauchi Y, Ninomiya K, Miyamoto H, et al. The Blimp1-Bcl6 axis is critical to regulate osteoclast differentiation and bone homeostasis. *J Exp Med.* 2010;207(4):751-762. doi:10.1084/jem.20091957
120. Gowen M, Lazner F, Dodds R, et al. Cathepsin K knockout mice develop osteopetrosis due to a deficit in matrix degradation but not demineralization. *J Bone Miner Res.* 1999;14(10):1654-1663. doi:10.1359/jbmr.1999.14.10.1654
121. Kornak U, Kasper D, Bösl MR, et al. Loss of the ClC-7 chloride channel leads to osteopetrosis in mice and man. *Cell.* 2001;104(2):205-215. doi:10.1016/S0092-8674(01)00206-9
122. Menéndez-Gutiérrez MP, Rőszer T, Fuentes L, et al. Retinoid X receptors orchestrate osteoclast differentiation and postnatal bone remodeling. *J Clin Invest.* 2015;125(2):809-823. doi:10.1172/JCI77186
123. Jacome-Galarza CE, Percin GI, Muller JT, et al. Developmental origin, functional maintenance and genetic rescue of osteoclasts. *Nature.* 2019;568(7753):541-545. doi:10.1038/s41586-019-1105-7
124. Shaw TN, Houston SA, Wemyss K, et al. Tissue-resident macrophages in the intestine are long lived and defined by Tim-4 and CD4 expression. *J Exp Med.* 2018;215(6):1507-1518. doi:10.1084/jem.20180019
125. Zigmond E, Jung S. Intestinal macrophages: Well educated exceptions from the rule. *Trends Immunol.* 2013;34(4):162-168. doi:10.1016/j.it.2013.02.001
126. Hume DA, Caruso M, Ferrari-Cestari M, Summers KM, Pridans C, Irvine KM. Phenotypic impacts of CSF1R deficiencies in humans and model organisms. *J Leukoc Biol.* 2020;107(2):205-219. doi:10.1002/JLB.MR0519-143R
127. Witmer-Pack MD, Hughes DA, Schuler G, et al. Identification of macrophages and dendritic cells in the osteopetrotic (op/op) mouse. *J Cell Sci.* 1993;104:1021-1029.
128. Guilliams M, Thierry GR, Bonnardel J, Bajenoff M. Establishment and maintenance of the macrophage niche. *Immunity.* 2020;52(3):434-451. doi:10.1016/j.immuni.2020.02.015
129. Anzinger JJ, Chang J, Xu Q, et al. Murine bone marrow-derived macrophages differentiated with GM-CSF become foam cells by PI3Kγ-dependent fluid-phase pinocytosis of native LDL. *J Lipid Res.* 2012;53(1):34-42. doi:10.1194/jlr.M018887
130. Utz SG, See P, Mildenberger W, et al. Early fate defines microglia and non-parenchymal brain macrophage development. *Cell.* 2020;181(3):557.e18-573.e18. doi:10.1016/j.cell.2020.03.021
131. Gosselin D, Link VM, Romanoski CE, et al. Environment drives selection and function of enhancers controlling tissue-specific macrophage identities. *Cell.* 2014;159(6):1327-1340. doi:10.1016/j.cell.2014.11.023
132. Lavin Y, Winter D, Blecher-Gonen R, et al. Tissue-resident macrophage enhancer landscapes are shaped by the local microenvironment. *Cell.* 2014;159(6):1312-1326. doi:10.1016/j.cell.2014.11.018
133. Buechler MB, Kim KW, Onufer EJ, et al. A stromal niche defined by expression of the transcription factor WT1 mediates programming and homeostasis of cavity-resident macrophages. *Immunity.* 2019;51:119.e5-130.e5. doi:10.1016/j.immuni.2019.05.010
134. Murray PJ, Allen JE, Biswas SK, et al. Macrophage activation and polarization: nomenclature and experimental guidelines. *Immunity.* 2014;41(1):14-20. doi:10.1016/j.immuni.2014.06.008
135. Celada A, Gray PW, Rinderknecht E, Schreiber RD. Evidence for a gamma-interferon receptor that regulates macrophage tumoricidal activity. *J Exp Med.* 1984;160(1):55-74. doi:10.1084/jem.160.1.55
136. Schreiber RD, Hicks LJ, Celada A, Buchmeier NA, Gray PW. Monoclonal antibodies to murine γ-interferon which differentially modulate macrophage activation and antiviral activity. *J Immunol.* 1985;134(3):1609-1618.
137. Alspach E, Lussier DM, Schreiber RD. Interferon γ and its important roles in promoting and inhibiting spontaneous and therapeutic cancer immunity. *Cold Spring Harb Perspect Biol.* 2019;11(3):a028480. doi:10.1101/cshperspect.a028480
138. Gaczynska M, Rock KL, Goldberg AL. γ-Interferon and expression of MHC genes regulate peptide hydrolysis by proteasomes. *Nature.* 1993;365(6445):264-267. doi:10.1038/365264a0
139. Steed AL, Barton ES, Tibbetts SA, et al. Gamma interferon blocks gammaherpesvirus reactivation from latency. *J Virol.* 2006;80(1):192-200. doi:10.1128/jvi.80.1.192-200.2006

140. Reese TA, Wakeman BS, Choi HS, et al. Helminth infection reactivates latent γ-herpesvirus via cytokine competition at a viral promoter, *Science*. 2014;345(6196):573-577. doi:10.1126/science.1254517
141. Gause WC, Wynn TA, Allen JE. Type 2 immunity and wound healing: Evolutionary refinement of adaptive immunity by helminths. *Nat Rev Immunol*. 2013;13(8):607-614. doi:10.1038/nri3476
142. Jenkins SJ, Ruckerl D, Cook PC, et al. Local macrophage proliferation, rather than recruitment from the blood, is a signature of TH2 inflammation. *Science*. 2011;332(6035):1284-1288. doi:10.1126/science.1204351
143. Gieseck RL, Wilson MS, Wynn TA. Type 2 immunity in tissue repair and fibrosis. *Nat Rev Immunol*. 2018;18(1):62-76. doi:10.1038/nri.2017.90
144. Jha AK, Huang SCC, Sergushichev A, et al. Network integration of parallel metabolic and transcriptional data reveals metabolic modules that regulate macrophage polarization. *Immunity*. 2015;42(3):419-430. doi:10.1016/j.immuni.2015.02.005
145. Nair S, Huynh JP, Lampropoulou V, et al. Irg1 expression in myeloid cells prevents immunopathology during *M. tuberculosis* infection. *J Exp Med*. 2018;215(4):1035-1045. doi:10.1084/jem.20180118
146. Lampropoulou V, Sergushichev A, Bambouskova M, et al. Itaconate links inhibition of succinate dehydrogenase with macrophage metabolic remodeling and regulation of inflammation. *Cell Metab*. 2016;24;158-166. doi:10.1016/j.cmet.2016.06.004
147. Bambouskova M, Gorvel L, Lampropoulou V, et al. Electrophilic properties of itaconate and derivatives regulate the IκBζ-ATF3 inflammatory axis. *Nature*. 2018;556:501-504. doi:10.1038/s41586-018-0052-z
148. O'Neill LAJ, Artyomov MN. Itaconate: the poster child of metabolic reprogramming in macrophage function. *Nat Rev Immunol*. 2019;19(5):273-281. doi:10.1038/s41577-019-0128-5
149. Swain A, Bambouskova M, Kim H, et al. Comparative evaluation of itaconate and its derivatives reveals divergent inflammasome and type I interferon regulation in macrophages. *Nat Metab*. 2020;2(7):594-602. doi:10.1038/s42255-020-0210-0
150. Jakubzick CV, Randolph GJ, Henson PM. Monocyte differentiation and antigen-presenting functions. *Nat Rev Immunol*. 2017;17(6):349-362. doi:10.1038/nri.2017.28
151. Ingersoll MA, Spanbroek R, Lottaz C, et al. Comparison of gene expression profiles between human and mouse monocyte subsets. *Blood*. 2010;115:e10-e19. doi:10.1182/blood-2009-07-235028
152. Flory CM, Jones ML, Warren JS. Pulmonary granuloma formation in the rat is partially dependent on monocyte chemoattractant protein 1. *Lab Investig*. 1993;69(4):396-404.
153. Serbina NV, Pamer EG. Monocyte emigration from bone marrow during bacterial infection requires signals mediated by chemokine receptor CCR2. *Nat Immunol*. 2006;7(3):311-317. doi:10.1038/ni1309
154. Tsou CL, Peters W, Si Y, et al. Critical roles for CCR2 and MCP-3 in monocyte mobilization from bone marrow and recruitment to inflammatory sites. *J Clin Invest*. 2007;117(4):902-909. doi:10.1172/JCI29919
155. Auffray C, Fogg D, Garfa M, et al. Monitoring of blood vessels and tissues by a population of monocytes with patrolling behavior. *Science*. 2007;317(5838):666-670. doi:10.1126/science.1142883
156. Carlin LM, Stamatiades EG, Auffray C, et al. Nr4a1-dependent Ly6Clow monocytes monitor endothelial cells and orchestrate their disposal. *Cell*. 2013;153(2):362-375. doi:10.1016/j.cell.2013.03.010
157. Evren E, Ringqvist E, Tripathi KP, et al. Distinct developmental pathways from blood monocytes generate human lung macrophage diversity. *Immunity*. 2020;54(2):259.e7-275.e7. doi:10.1016/j.immuni.2020.12.003
158. Swirski FK, Nahrendorf M, Etzrodt M, et al. Identification of splenic reservoir monocytes and their deployment to inflammatory sites. *Science*. 2009;325(5940):612-616. doi:10.1126/science.1175202
159. Jakubzick C, Gautier E, Gibbings S, et al. Minimal differentiation of classical monocytes as they survey steady-state tissues and transport antigen to lymph nodes. *Immunity*. 2013;39(3):599-610. doi:10.1016/j.immuni.2013.08.007
160. Bellingan GJ, Caldwell H, Howie SE, Dransfield I, Haslett C. In vivo fate of the inflammatory macrophage during the resolution of inflammation: inflammatory macrophages do not die locally, but emigrate to the draining lymph nodes. *J Immunol*. 1996;157(6):2577-2585.
161. Gautier EL, Ivanov S, Lesnik P, Randolph GJ. Local apoptosis mediates clearance of macrophages from resolving inflammation in mice. *Blood*. 2013;122:2714-2722. doi:10.1182/blood-2013-01-478206
162. Palframan RT, Jung S, Cheng G, et al. Inflammatory chemokine transport and presentation in HEV: A remote control mechanism for monocyte recruitment to lymph nodes in inflamed tissues. *J Exp Med*. 2001;194(9):1361-1373. doi:10.1084/jem.194.9.1361
163. Lavine KJ, Epelman S, Uchida K, et al. Distinct macrophage lineages contribute to disparate patterns of cardiac recovery and remodeling in the neonatal and adult heart. *Proc Natl Acad Sci U S A*. 2014;111:16029-16034. doi:10.1073/pnas.1406508111
164. Bajpai G, Schneider C, Wong N, et al. The human heart contains distinct macrophage subsets with divergent origins and functions. *Nat Med*. 2018;24(8):1234-1245. doi:10.1038/s41591-018-0059-x
165. Gibbings SL, Goyal R, Desch AN, et al. Transcriptome analysis highlights the conserved difference between embryonic and postnatal-derived alveolar macrophages. *Blood*. 2015;126(11):1357-1366. doi:10.1182/blood-2015-01-624809
166. Mould KJ, Barthel L, Mohning MP, et al. Cell origin dictates programming of resident versus recruited macrophages during acute lung injury. *Am J Respir Cell Mol Biol*. 2017;57(3):294-306. doi:10.1165/rcmb.2017-0061OC
167. Bain CC, Scott CL, Uronen-Hansson H, et al. Resident and pro-inflammatory macrophages in the colon represent alternative context-dependent fates of the same Ly6C hi monocyte precursors. *Mucosal Immunol*. 2013;6(3):498-510. doi:10.1038/mi.2012.89
168. Misharin AV, Morales-Nebreda L, Reyfman PA, et al. Monocyte-derived alveolar macrophages drive lung fibrosis and persist in the lung over the life span. *J Exp Med*. 2017;214(8):2387-2404. doi:10.1084/jem.20162152
169. Giladi A, Wagner LK, Li H, et al. Cxcl10+ monocytes define a pathogenic subset in the central nervous system during autoimmune neuroinflammation. *Nat Immunol*. 2020;21(5):525-534. doi:10.1038/s41590-020-0661-1
170. Seidman JS, Troutman TD, Sakai M, et al. Niche-specific reprogramming of epigenetic landscapes drives myeloid cell diversity in nonalcoholic steatohepatitis. *Immunity*. 2020;52(6):1057.e7-1074.e7. doi:10.1016/j.immuni.2020.04.001

CHAPTER 16

Mast Cells and Basophils

Mindy Tsai, DMSc • Stephen J. Galli, MD

INTRODUCTION

Mast cells and basophils are best known for their roles in IgE-associated allergic reactions. Basophils are the least frequent of the three granulocytes, typically accounting for less than 1% of blood leukocytes in humans and mice. Basophils ordinarily do not reside in tissues but circulate as mature cells and can migrate into tissues at sites of immunologic or inflammatory responses.[1-3] By contrast, mast cell progenitors arise in hematopoietic tissues and then migrate into and complete their maturation in most vascularized tissues.[4-7] Mast cells normally reside in the connective tissue and are particularly abundant beneath epithelial surfaces, around blood vessels and nerves, and, in some species, in serous cavities.[8,9] While blood basophils have a short life span of few days, mast cells can remain for long periods of time in the same tissue locations. Mast cell and basophil numbers are increased at sites of parasite infection or in association with certain chronic allergic diseases or other forms of pathology.

Human mast cells and basophils were described in the late 19th century.[10] For a long time, blood basophils were regarded as surrogates for tissue mast cells, given their similar biochemical and functional characteristics. The exact relationship of mast cells and basophils during hematopoiesis has been studied by several groups, but the data are not entirely consistent.[11-18] There is no convincing evidence that mature basophils can give rise to mast cells (or vice versa), and it seems best to regard them as distinct hematopoietic lineages with many overlapping functional features.[1,3,19,20]

In contrast to more extensive investigation of mast cells, there has been less progress in our understanding of basophils in health and disease. The recent development of mouse genetic approaches that enable deletion of basophils, or the manipulation of expression of basophil-specific products, has greatly facilitated our ability to study basophils in mice.[1,21,22] However, most of the findings we have learned from these mouse models remain to be validated in human studies.

It is generally accepted that mast cells and basophils can contribute importantly to the pathology associated with IgE-mediated allergic disorders. However, several lines of evidence indicate that mast cells and basophils also can contribute to protective host responses associated with IgE production, especially those directed against parasites, as well as to host defense in innate immune responses to certain pathogens. In mice, mast cells can enhance innate, and acquired IgE-dependent, defense against animal venoms. Mast cells and basophils also can participate in many other biological processes, with or without the involvement of IgE. In such settings, they can exhibit positive and/or negative effector or immunoregulatory functions through the production of cytokines/mediators and other mechanisms.

MAST CELLS AND BASOPHILS IN ANCIENT INNATE IMMUNITY

Mast cells and basophils are evolutionally conserved in many animal species.[1,23] Indeed, primitive mast cells and basophils are present in organisms which appeared more than 500 million years ago, before the development of immunoglobulins as components of adaptive immunity.[23-29] These ancient mast cell-/basophil-like cells primarily have been studied in urochordates. The "test cells" that surround the developing oocytes in the ascidian, *Styela plicata*,[24,25] and in the sea squirt, *Ciona intestinalis*,[28] are thought to represent the ancient counterparts of mammalian mast cells. These cells can degranulate and release heparin, histamine, and tryptase in response to compound 48/80,[24,28] which also can elicit degranulation of rodent and human mast cells.[30,31] Basophil-like granulocytes that contain heparin and histamine have also been found in the hemolymph of *S. plicata*.[26]

Given the abundance of test cells beneath the surface of the developing organism, test cells (and other heparin-containing cells, eg, in the intestinal lining and pharynx of *S. plicata*)[25] are thought to play a protective role in host defense.[24,25] While the proposed protective function of mast cell-/basophil-like cells in urochordates has not yet been verified, there is evidence that some of such innate abilities have been retained by mammalian mast cells and basophils. Thus, a variety of signals that can be generated during innate responses can induce (or, in some cases, suppress) the activation of mammalian mast cells and basophils. Whatever the functional repertoire of primitive mast cells and basophils, it is clear that, in mammals, mast cells and basophils have acquired the ability to bind IgE to their surface with high affinity.[29]

The first description of the mast cell was reported in 1863 by Friedrich von Recklinghausen, who described mast cells as highly granulated cells in the frog mesentery. Numerous subsequent studies have investigated the participation of mast cells and, to a lesser extent, basophils in inflammation and other pathology in nonmammalian vertebrates.[23,32] Mast cells or mast cell–like cells have been identified in many nonmammalian species, including fish, amphibians, reptiles, and birds. Like their mammalian counterparts, such nonmammalian mast cells contain heparin, histamine, proteases, and neuropeptides.[23] Mast cells in bony fishes can also produce antimicrobial peptides, suggesting that fish mast cells may participate in host defense.[33] There are fewer studies of basophils in lower vertebrates, but it is clear that basophils also exist in many animal species, including teleost fish and reptiles.[34] Snapping turtles have a large number of basophils (>50% of blood leukocytes), which can degranulate and release histamine after treatments with rabbit antiturtle Ig serum.[35]

RELATIONSHIP BETWEEN MAST CELLS AND BASOPHILS

Mature mast cells and basophils differ in morphology, natural history, tissue distribution, mediator production, cell surface phenotype, and growth factor requirements[1,3,20,36] (Table 16.1). Nevertheless, certain similarities of the two cells, taken together with evidence indicating that many tissue mast cells are derived from circulating bone marrow–derived precursors,[8,37] suggested to some investigators that basophils might be circulating precursors of mast cells.

However, the following evidence strongly favors the view that mature basophils represent terminally differentiated granulocytes, not circulating mast cell precursors: (1) no evidence has been presented in any species indicating that mature circulating basophils are capable of either mitosis or differentiation into mast cells; (2) rare reports of patients with hereditary or acquired abnormalities affecting basophil numbers or morphology indicate that eosinophils also may be affected in these disorders, but not mast cells[38-40]; (3) morphologically identifiable tissue mast cells in humans can exhibit mitoses,[41] showing that mast cells can replicate independently of a stage resembling basophils; and (4) in mice, evidence indicates that a mast cell–committed progenitor (MCP) present in the bone marrow is developmentally distinct during hematopoiesis from a Sca-1^{low} granulocyte-macrophage progenitor.[12,13] In studies of hematopoiesis in the mouse, evidence for a committed bipotential precursor of only mast cells and basophils has been reported in some studies[11,16] but not others.[12-14] (5) Mast cells and basophils exhibit distinct gene expression profiles by transcriptome analyses.[42-44]

DISTINGUISHING FEATURES OF MAST CELLS AND BASOPHILS

Mast Cells

Morphology

The morphological distinction between mast cells and basophils was first appreciated by Paul Ehrlich, who described the histochemical staining characteristics of human mast cells[45] and basophils[46] in the late 19th century. Tissue mast cells are poorly stained by routine histological methods but can readily be identified by basic dyes (eg, toluidine blue) which bind to granule-associated sulfated proteoglycans (chondroitin sulfates and heparin), revealing "metachromasia" (a shift in the color of the dye from blue to purple). Excellent visualization of tissue mast cells is achieved in appropriately prepared 1-μm sections (Fig. 16.1) or by electron microscopy (Fig. 16.2A and B). In tissue sections, mast cells typically appear as either round or spindle-shaped cells, ranging from 7 to 20 μm in diameter, usually with a round, nonsegmented nucleus with moderate condensation of nuclear chromatin, and contain prominent cytoplasmic granules. Mast cell granules are smaller, more numerous, and generally more variable in appearance than in basophils (Fig. 16.2A and C) and contain scroll-like structures, particles, and crystals, alone or in combination (Fig. 16.2A and B). In contrast to the irregularly spaced blunt surface projections of basophils, mast cells typically are covered by more uniformly distributed thin surface processes. Mast cells also differ from basophils in having many more cytoplasmic filaments and lacking cytoplasmic glycogen deposits. Human mast cells can contain numerous cytoplasmic lipid bodies (Figs. 16.2A, B and 16.3). Tissue mast cells can be localized by immunohistochemical staining of cell-associated tryptase or chymase.

Natural History and Regulation of Development

Mature mast cells reside near or within smooth muscle and mucus-producing glands and are particularly abundant near blood vessels and nerves and in connective tissues beneath surfaces that are exposed to the external environment, such as the skin, gastrointestinal and urogenital tracts, and respiratory system.[4,47,48] Some species, including murine rodents, have mast cells within mesothelial-lined cavities, such as the peritoneal or pulmonary cavities.[9] Mast cells can be long-lived in tissues, and at least, some mast cells can migrate and proliferate in the tissues during a variety of inflammatory or reparative processes.[4,47] Tissue mast cells increase by recruitment and differentiation/maturation of immature precursors and/or by local expansion of resident mast cells.[49-52]

Mast cells in lower vertebrates are derived from lymphatic organs, such as the thymus, and other hematopoietic tissues.[23] In mammals, except for a numerically minor population of mast cells residing in the bone marrow, many mast cell precursors travel through the circulation before they acquire more mature characteristics in peripheral tissues.[6,7,53-56] In mice, some adult mast cells are derived from progenitors that arise in the yolk sac and other hematopoietic tissues early in development.[12,13,37,53-57]

The original evidence for the bone marrow origin of mast cells was provided in a seminal study by Kitamura in 1977.[58] Committed mast cell progenitors named "promastocytes," which lack FcεRI and cytoplasmic granules, were first identified in fetal blood in mice.[37] Other mast cell progenitors were subsequently found in the bone marrow,[12,13,59] peripheral blood,[60] peritoneal cavity,[9] intestines,[5] white adipose tissue,[61] spleen,[11] and lungs[62] of adult mice. In agreement with the early report in 1983 that mouse yolk sac also contains mast cell progenitors,[57] data from recent studies have confirmed that most of the mast cells in mouse connective tissue originate from erythromyeloid progenitors in the yolk sac and that mast cells in some connective tissues are capable of self-renewal without the involvement of the bone marrow.[54,55] On the other hand, mouse mucosal mast cells originate from fetal

TABLE 16.1 Natural History, Major Mediators, and Surface Membrane Structures of Human Mast Cells and Basophils

Characteristics	Basophils	Mast Cells
Natural History		
Origin of precursor cells	Bone marrow	Bone marrow and yolk sac (in mice)
Site of maturation	Bone marrow	Connective tissue (a few in bone marrow)
Mature cells in circulation	Yes (usually <1% of blood leukocytes)	No
Mature cells recruited into tissues from circulation	Yes (during immunologic, inflammatory responses)	No
Mature cells normally residing in connective tissues	No (not detectable by microscopy)	Yes
Proliferative ability of morphologically mature cells	None reported	Yes (limited; under certain circumstances)
Life span	Days (like other granulocytes)	Weeks to months (according to studies in rodents)
Major growth factor	IL-3	SCF
Mediators		
Major mediators stored preformed in cytoplasmic granules	Histamine, chondroitin sulfates, tryptase,[a] chymase,[a] carboxypeptidase A,[a] neutral protease with bradykinin-generating activity, β-glucuronidase, elastase, cathepsin G-like enzyme, major basic protein, Charcot-Leyden crystal protein, granzyme B, and basogranulin (BB1) may contain low levels of eosinophil peroxidase (EPO), eosinophil-derived neurotoxin (EDN), and eosinophil cationic protein (ECP)	Histamine, heparin,[a] chondroitin sulfates,[a] chymase,[a] tryptase,[a] cathepsin G,[a] carboxypeptidase A, major basic protein, several acid hydrolases, peroxidase, phospholipases, granzyme B
Major lipid mediators produced on appropriate activation	Prostaglandin D_2, leukotriene C_4, platelet-activating factor	Prostaglandin D_2, leukotriene C_4, leukotriene B_2, platelet-activating factor, thromboxane A
Cytokines released on appropriate activation	IL-2, IL-3, IL-4, IL-5, IL-6, IL-9, IL-13, IL-15, IL-25, IL-31, GM-CSF, VEGF-A-B, TSLP, amphiregulin	TNF, TGF-β, IFN-α, VEGF-A-D, IL -1β, IL-2, IL-3, IL-4, IL-5, IL-6, IL-8, IL-9, IL-10, IL-11, IL-13, IL-15, IL-16, IL-17A, IL-18, IL-22, IL-24, IL-25, IL-33, amphiregulin, GM-CSF, FGF-2, NGF, PDGF, TSLP, SCF
Chemokines	CCL3 (MIP-1α), 4, 5 (RANTES), 11, 12, and 15; CXCL2, CXCL8 (IL-8), CXCL10 (IP-10)	CCL1-5, 7, 8, 13, 17–20, 22, 23, 25, 28; CXCLI-5, 7, 8, 10, 14, 16, 17; CX3CL1
Surface Structures		
Ig receptors	FcεRI, FcγRIIA, FcγRIIB, FcγRIIIB (very low levels)	FcεRI, FcγRI (after IFNγ exposure), FcγRIIA, FcγRIIB, FcγRIIIA
Cytokine or growth factor receptors	IL-1RIIb (CD121b), IL-2R (CD25), IL-3R, IL-4Rα, IL-5Rα, IL-6R, IL-8R, IL-18R, IL-25R, IL-33R (ST2), GM-CSFRα, TrkA, TSLPR; and receptors for IFNs and SCF (basophils express variable numbers of the SCF receptor, KIT)	KIT, IFN-γRα, IL-3R, IL-4Rα, IL-5Rα, IL-6R, IL-9R, IL-18R, IL-33R (ST2), thrombopoietin receptor (CD110), GM-CSFR, TrkA, TGF-β type-I and type-II β receptors
Chemokine receptors	CCR1, 2, 3, 5, 7; CXCR1, 2, 4, 8	CCR1, 3-5; CXCR1-4; CX3CR1
Additional GPCRs	C3aR, C5aR, CRTH2, MRGPRX2	C3aR, C5aR, EP3, BLT1, BLT2, cysLT1, platelet-activating factor receptor; MRGPRX2
TLRs	TLRs 1-8 (but lack CD14) and low levels of TLRs 9, 10	TLR1, 2-7, 9

IFN, interferon; Ig, immunoglobulin; IL, interleukin; GM-CSF, granulocyte-macrophage colony-stimulating factor; MIP, macrophage inflammatory protein; NGF, nerve growth factor; PDGF, platelet-derived growth factor; RANTES, regulated on activation, normal T-cell expressed, presumed secreted; SCF, stem cell factor; TGF, transforming growth factor; TLR, toll-like receptor; TNF, tumor necrosis factor; VEGF, vascular endothelial growth factor.

Note: Expression of these and other surface structures, including chemokine receptors, and production of individual cytokines and chemokines, can vary in different in vitro or in vivo derived basophil or mast cell populations.

[a]Basophil and mast cell content of these (and perhaps other) mediators vary, for example, in different subjects and tissues; and/or in association with certain inflammatory diseases.

Data Modified from Galli SJ, Dvorak AM, Dvorak HF. Basophils and mast cells: morphologic insights into their biology, secretory patterns, and function. *Prog Allergy.* 1984;34:1-141, Valent P. Immunophenotypic characterization of human basophils and mast cells. *Chem Immunol.* 1995;61:34-48, Boyce J, Fanning L. *Mast cells, Basophils, and Mastocytosis. Clinical Immunology: Principles and Practice.* 5th ed. Elsevier; 2019:335-347, and Galli SJ, Metcalfe DD, Arber DA. In: Kaushansky K, Lichtman MA, Beutler E, Kipps TJ, Prchal JT, eds. *Williams Hematology.* 10th ed. McGraw-Hill Medical Education; 2021:1093.

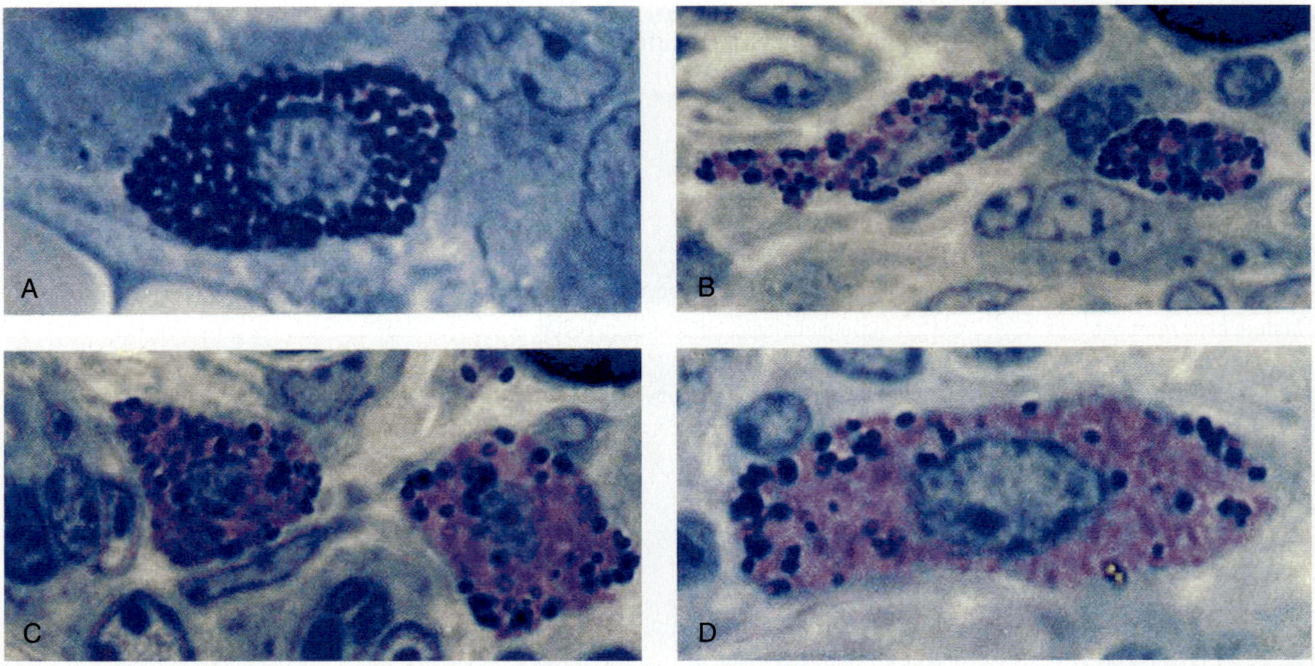

FIG. 16.1. **Light microscopy of alkaline-Giemsa-stained 1-μm sections of mouse tissue showing piecemeal degranulation of mast cells. A:** A mast cell with no altered granules. **B-D:** The mast cells undergoing piecemeal degranulation show many cytoplastic granules that stain pink, in contrast to the dark blue staining exhibited by the granules in the normal-appearing mast cell in **A.** However, the granules with altered staining remain within the cytoplasm. Original magnifications: **A:** ×2200, **B:** ×1500, **C:** ×1700, and **D:** ×2200. (Reprinted from Dvorak AM, Tepper RI, Weller PF, et al. Piecemeal degranulation of mast cells in the inflammatory eyelid lesions of interleukin-4 transgenic mice. Evidence of mast cell histamine release *in vivo* by diamine oxidase-gold enzyme-affinity ultrastructural cytochemistry. *Blood.* 1994;83:3600-3612, with permission from Elsevier.)

hematopoietic stem cells, and their replacement appears to depend on adult hematopoietic stem cells throughout life.[54,55]

These studies suggest that, in the steady state, mast cell numbers in the connective tissue are maintained by the expansion of local progenitors independent of new input from bone marrow progenitors, while bone marrow progenitors give rise to mucosal mast cells and, under various forms of tissue perturbance, also can contribute to mast cell expansion in connective tissues.[7,54-56] Thus, based on work in the mouse, there are multiple mast cell differentiation pathways and mast cell progenitors of different origins exhibit different tissue preferences.[7,54-56]

The extent to which this applies to other mammals, including humans, is not yet fully elucidated. Human mast cells can be generated from CD34+ progenitors in the bone marrow,[63,64] peripheral blood,[64-66] fetal liver,[67] and cord blood,[66,68] but the differentiation pathways of human mast cells are yet to be fully defined. Although human mast cells and basophils share multiple phenotypic and functional characteristics, there is no convincing evidence that they share a specific bipotent common precursor. A bipotent basophil/mast cell progenitor (BMCP) has been reported in mouse spleen[11] and bone marrow,[69] but the existence of BMCPs was confirmed in some but not all of the subsequent studies.[7,12-18,70] Thus, despite extensive research efforts, the developmental relationship between mast cells and basophils and other hematopoietic cell lineages are not yet fully delineated.

Studies in mice, nonhuman primates, and humans indicate that many aspects of mast cell development are critically regulated by stem cell factor (SCF), the ligand for the KIT tyrosine kinase receptor.[71,72] SCF is produced in membrane-associated and soluble forms, both of which are biologically active.[71,72] Besides promoting the migration, survival, proliferation, and maturation of cells in the mast cell lineage, SCF can directly promote mast cell mediator release[73-76] and, at even lower concentrations, can augment mast cell mediator release in response to stimulation by IgE and antigen.[73,74] Abnormalities affecting *KIT* (particularly gain-of-function mutations) are involved in the pathogenesis of many types of mastocytosis.[3] By contrast, alterations in the production of SCF by fibroblasts and other cells may contribute to the changes in mast cell numbers that occur during many chronic inflammatory conditions and other pathologic responses.[71,77,78]

Other cytokines that can influence mast cell proliferation, differentiation, maturation, and survival in humans and mice include IL-3, IL-4, IL-5, IL-6, IL-9, and IL-10. TGF-β inhibits mast cell proliferation[79,80] and promotes mast cell apoptosis[81] in mice. Nerve growth factor (NGF) induces increased numbers of tissue mast cells in neonatal rats[82] and in frog tadpoles.[83] In mice, mast cells can be generated in culture medium containing IL-3 or SCF, or SCF plus IL-3. The phenotypic characteristics and functional properties of mast cells are profoundly influenced by the cytokines/growth factors present in their microenvironment. SCF is regarded as the most critical growth factor for mast cell lineages in mice and humans, but, under certain settings, human mast cells can develop in IL-3 and IL-6–containing medium without SCF.[84] STAT5, GATA1, GATA2, FOG-1 (Friend of GATA 1), and MITF (microphthalmia-associated transcription factor) are the main transcription factors known to regulate mast cell differentiation.[85]

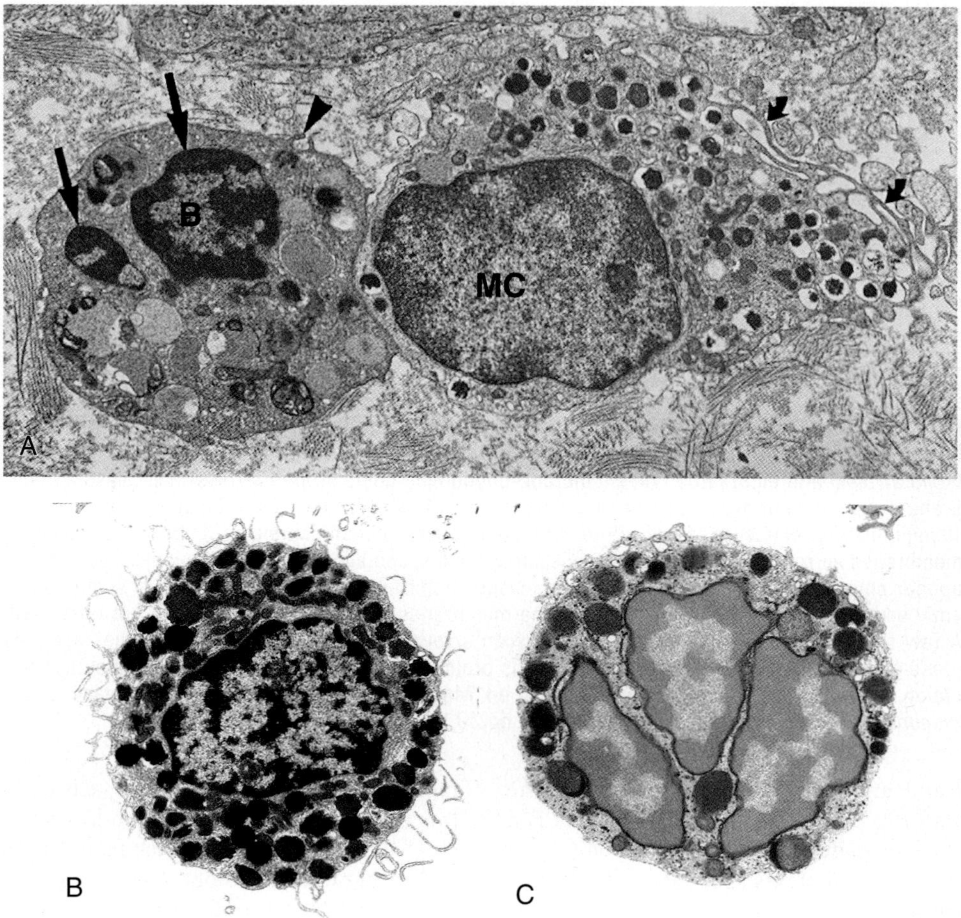

FIG. 16.2. Ultrastructure of human mast cells (MCs) and basophils. A: A basophil (B) adjacent to an **MC** in the ileal submucosa of a patient with Crohn disease. The basophil exhibits a bilobed nucleus (*solid arrows*) whose chromatin is strikingly condensed beneath the nuclear membrane. The basophil surface is relatively smooth with a few blunt process (*arrowhead*). The mast cell nucleus is larger and its chromatin less condensed than that of the basophil. The mast cell's granules are smaller, more numerous, and more variable in shape and content than those of the basophil. The mast cell surface has numerous elongated, thin folds (*curved arrows*). **B:** A human lung mast cell after isolation and culture for 36 hours shows electron-dense secretory granules, a monolobed nucleus, and numerous narrow surface folds. **C:** A human basophil isolated from the peripheral blood has electron-dense secretory granules and a polylobed nucleus. Original magnifications: **A:** ×9000; **B:** ×10,000; **C:** ×17,000. (**A**, Reprinted from Dvorak AM, Monahan RA, Osage JE, Dickersin GR. Crohn's disease: transmission electron microscopical studies. *Hum Pathol.* 1980;11:606-619; with permission from Elsevier and **B** and **C** are from Dvorak AM. Ultrastructural studies of human basophils and mast cells. *J Histochem Cytochem.* 2005;53:1043-1070. Reprinted by Permission of SAGE Publications.)

Regulation of Function: Positive and Negative

IgE/FcεRI-mediated activation is the most extensively studied and best characterized activation pathway for mast cells. The cross-linking of FcεRI by the binding of bivalent or multivalent antigens to antigen-specific IgE on the mast cell surface triggers multiple intracellular membrane events, resulting in fusion of cytoplasmic granules with the cell membrane, as well as initiation of a network of intracellular enzymatic cascades, resulting in activation of the PI3 kinase and MAP/ERK kinase signaling pathways. These complex signaling cascades and membrane events provoke mast cell degranulation (Fig. 16.3) and release of granule-associated mediators (eg, histamine, heparin), the *de novo* synthesis and secretion of lipid mediators (eg, prostaglandins, leukotrienes), and the transcription and translation of cytokines/chemokines/growth factors. See Chapter 33, "Fc receptor biology" (by Nimmerjah/Ravetch) and Chapter 47, "Allergic disorders" (by Nadeau/Scott/Milner), for a detailed description of the regulation of FcεRI signal transduction pathways and their roles in allergic disorders. In addition to antigen-specific IgE/FcεRI-dependent activation, mast cell functions can be regulated by a wide variety of receptor-ligand interactions which permit mast cells to respond to many non-IgE stimuli, some of which can also modulate IgE-dependent responses. Moreover, mast cells can undergo multiple rounds of activation by IgE and antigen or other stimuli.[86]

In certain mammals, including mice[87,88] and humans,[89-92] at least some mast cell populations also can bind immune

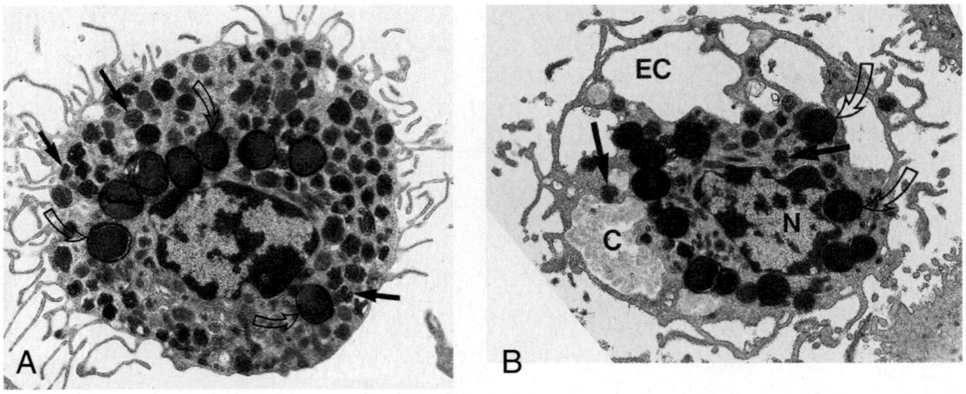

FIG. 16.3. Ultrastructural features of human mast cell degranulation. A: Mast cell purified from human lung. The cell contains many cytoplasmic granules with scroll-like substructural elements (*solid arrows*) and eight large non–membrane-bound lipid bodies (*open arrows*). The plasma membrane has prominent folds. (Original magnification ~ ×12,900). **B:** An isolated human lung mast cell 10 minutes after exposure to anti-IgE *in vitro.* Some degranulation channels (C), formed by fusion of membranes surrounding individual cytoplasmic granules, contain altered granule matrix; others (**EC**) appear empty. Cationized ferritin stains the plasma membrane and the membranes of some apparently empty degranulation channels (**EC**). The membranes lining other channels (C) are unstained. A few unaltered scroll-containing granules (*solid arrows*) remain. Numerous lipid bodies are also present (*open arrows*). The cytoplasm contains prominent filaments. **N**, nucleus. (Original magnification ~ ×9100). (Reproduced from Ishizaka K, ed. Mast cell activation and mediator release. *Chem Immunol Allergy.* Basel, Karger; 1984(34):1-141. doi:10.1159/000408370)

complexes of IgG and antigen. While it appears that the expression of high-affinity receptors for IgE (ie, FcεRI) is a feature of most if not all mature mast cells, their ability to bind IgG-antigen immune complexes may vary according to cell's adaptation to immune responses or diseases.[88,93-95] Thus, expression of FcγR on mast cells varies depending on species, anatomical location, and the local cytokine milieu.

Mouse mast cells express FcγRIIIA and FcγRIIB,[96] whereas human mast cells express FcγRIIA and IFN-γ–induced FcγRI.[93] Tissue mast cells in certain disease sites (eg, skin in psoriasis[93] and intestines in inflammatory bowel disease[97]) constitutively express FcγRI. FcγRIIB is present on *in vitro* derived cultured human and mouse mast cells and human mast cells in the gut, but not on skin mast cells.[98] Cross-linking of the ITAM-containing, activating Fc receptors, FcεRI, FcγRI, FcγRIIA, and FcγRIIIA, induces mast cell degranulation and secretion of chemical mediators and cytokines, whereas coligation of FcγRIIB with activating Fc receptors for IgE or IgG downregulates mast cell activation (reviewed in Ref. 96). Coligation of FcγRIIB mediated by antigen and antigen-specific IgG1 or IgG4 in humans, and IgG1 in mice, inhibits FcγR-mediated cell activation. Induction of allergen-specific IgG_4 commonly occurs during immunotherapy and is thought to be the primary mechanism that suppresses IgE-mediated mast cell and basophil activation.[99] Coaggregation of FcγRIIB with activating FcR receptors inhibits not only signals generated by the activating Fc receptors (cis-inhibition) but also signals generated by receptors that are not coligated with FcγRIIB (trans-inhibition).[100] There is no evidence that mast cells express IgA receptors or can be activated by IgA, but mast cells can bind IgD, suggesting that IgD ligation may also modulate mast cell functions.[101]

Many of the mast cells' innate functions are regulated by several pattern recognition receptors (PRRs) for pathogen-associated molecular patterns (PAMPs) and danger-associated molecular patterns. Mouse mast cells express toll-like receptors (TLRs) 1-9 at the mRNA and/or protein/functional levels.[102-104] TLRs-1/2/4/5/6 are expressed on the mast cell surface and recognize microbiota-derived PAMPs, whereas endosomal TLRs-3/7/8/9 recognize some virus-derived or host-derived nucleic acid, such as single-stranded RNA (ssRNA) and CpG-containing DNA.[105,106] Mast cells' responses to TLR ligands are influenced by their tissue locations and phenotypic maturation. Stimulation of the TLRs in mast cells leads to receptor-specific production of distinct sets of cytokines/chemokines and/or cysteinyl-leukotrienes and other eicosanoids, with or without degranulation.[104-106] Moreover, TLR activation may amplify or dampen the activation elicited by other stimuli (FcεRI, LL-37, substance P) in mast cells.[102-104]

In addition to PRRs, mast cells also can express other receptors which react to pathogen-derived products. For example, dectin-1, a member of the C-type lectin receptor family, can bind to zymosan and other β-glucan–rich components present in many pathogens.[107] Dectin-1 engagement on mast cells induces generation of reactive oxygen species, secretion of LTC_4, and the upregulation of cytokines and chemokines without degranulation.[107,108] Moreover, mast cells can phagocytose *Candida albicans* by both TLR-2 and dectin-1–dependent mechanisms.[109] The retinoic acid–inducible gene-1 (RIG-1)-like receptor and melanoma differentiation gene 5 are intracellular viral RNA sensors which can regulate cytokine and chemokine production by mast cells after viral infections, such as with dengue virus.[110]

Another important mast cell activation mechanism is via stimulation of G protein–coupled receptors (GPCRs), including MAS-related GPCR (MRGPR), complement receptors,[104,111,112] endothelin (ET) receptors,[113,114] neuropeptide and neurotransmitter receptors,[115] and chemokine receptors.[116] Many studies have shown that activated mast cells are sources of cytokines, growth factors, and chemokines,[117] and chemokine stimulation can also induce mast cell activation and chemotaxis.[116]

The Mrgprb2/MRGPRX2 receptor has been primarily analyzed for its participation in IgE-independent, drug-induced allergic reactions (ie, pseudoallergic reactions).[118-120] Recent work has emphasized the importance of the MRGPR family, specifically, Mrgprb2, in mast cell activation. Mrgprb2 is the mouse receptor for several cationic molecules and the ortholog of the human receptor MRGPRX2.[118] Natural endogenous ligands of Mrgprb2/MRGPRX2 have been reported, such as the neuropeptide substance P, antimicrobial peptides, β-defensin, cathelicidin, and PAMP9-20.[120-122] Synthetic compounds, like compound 48/48 and some cationic drugs, can activate MRGPRX2/Mrgprb2 receptors of human and mouse mast cells.[120] Furthermore, substance P stimulation of mast cells via Mrgprb2 but independently of its canonical receptor, the neurokinin-1 receptor (NK1R), can modulate mechanical and thermal hyperalgesia and recruit innate immune cells at sites of injury.[123] In a mouse model of atopic dermatitis, house dust mite extract, via its protease activity, activates nociceptors to release substance P, which in turn activates Mrgprb2 on mast cells, leading to exacerbation of allergic skin inflammation in mice.[124] This study suggests that activation of TRPV1+Tac1+ nociceptor/Mrgprb2+ mast cell sensory clusters may represent a key early event in the development of allergic skin reactions in humans.[124]

In addition to its effect on inflammation, in mice, Mrgprb2-mediated mast cell activation can excite itch-sensory neurons and nonhistaminergic itch distinct from that triggered by classical IgE/FcεRI-mediated histamine release.[125] Compared with classical IgE/antigen-mediated FcεRI activation, Mrgprb2/MRGPRX2 agonists (eg, substance P) and other innate signals (including the complement anaphylatoxins C3a [via C3R] and C5a [via C5R], and ET1 [via ET_A]) triggered more rapid degranulation of smaller mast cell granules, a process that was associated with a fast and transient form of inflammation[114] (Fig. 16.4). Mast cells also can respond to many other GPCR agonists, including sphingosine 1 phosphate (S1P), PGD_2, PGE_2, leukotrienes, and adenosine.[104,126-128]

The regulation of mast cell function by lipid mediators and allergen-associated lipids has been of considerable recent interest.[129] Many allergens contain lipid components, and allergen lipid-protein interactions can influence sensitization and development of Th2-type allergic responses.[130] Different lipid molecules can influence mast cell degranulation and activation,[129] but how mast cell responses can be modulated by allergen-associated lipids is not well understood.

Clearly, mast cells represent an important source of lipid mediators.[126] Activated mast cells produce three arachidonic acid–derived eicosanoids, PGD_2, LTB_4, and LTC_4. LTC_4 is the parent molecule of the cysteinyl leukotrienes (cysLTs), including LTC_4, LTD_4, and LTE_4. Mast cells express GPCRs specific for cysLTs, LTB_4, and PGE_2. These eicosanoid receptors can upregulate as well as downregulate mast cell functions in autocrine and paracrine manners.[126] For example, PGE_2 is a multifunctional eicosanoid, acting through G protein–coupled prostanoid receptors, EP_1, EP_2, EP_3, and EP_4. Human mast cells express EP_2, EP_3, and EP_4 receptors[131] and mouse mast cells express EP_1, EP_2, EP_3, and EP_4 receptors.[132] Stimulatory and inhibitory effects of PGE_2 have been reported and are mediated by different EP receptors in mast cells.[131-134] PGE_2/EP_3 interactions appear to represent a proinflammatory signaling pathway that potentiates IgE/antigen-stimulated degranulation and IL-6 cytokine production in mouse mast cells.[132]

In humans, PGE_2 causes exocytosis in IL-4–primed mast cells via stimulation of EP_3 receptors, whereas PGE_2/EP_2 stimulation reduces LTC_4 and PGD_2 generation and tumor necrosis factor (TNF) production upon FcεRI cross-linkage.[131] PGE_2 inhibition of hyperosmolar-induced degranulation of human mast cells is mediated through EP_2 and EP_4.[134] Thus, EP_3 receptor–dependent signaling induces or potentiates mast cell mediator release. By contrast, PGE_2 activation of EP_2 and EP_4 receptors on mast cells inhibits FcεRI-initiated mediator generation and may limit tissue pathology in mast cell–associated inflammation.[126,135] Mast cells also express BLT1 and BLT2 receptors for LTB_4, a potent chemoattractant, which can induce mast cell migration.[136,137]

In addition to proinflammatory lipid mediators (prostaglandins and leukotrienes), there is evidence that mast cell functions can be regulated by proresolving lipid mediators, including lipoxins, resolvins, protectins, and maresins.[129] The anti-inflammatory and cytoprotective effects of these lipids are considered important for the resolution of acute inflammation, including turning off mast cell activation.[129,138] The lipid derivatives, resolvin D_1 and D_2 and lipoxin A_4, inhibit histamine release from human lung mast cells.[139] In mice, lipoxin B_4 has been reported to decrease mast cell degranulation and cytokine release, as well as airway inflammation, mucus production, and airway hyperresponsiveness *in vivo*, in models of allergic airway inflammation.[140]

S1P is a sphingolipid metabolite that can be produced by mast cells and regulate mast cell function. Mast cells express at least two types of receptors for S1P, S1PR1 and S1PR2.[141] FcεRI cross-linking of mast cells activates sphingosine kinase isoforms (SphK1 and SphK2), leading to the production of S1P and upregulation of S1PR1 and 2.[141] S1P plays an important role in mediating mast cell degranulation, cytokine and lipid mediator production, and migration, by mechanisms that are both S1PR dependent and independent.[142] Chronic exposure to elevated S1P from a non–mast cell source can influence the differentiation and responsiveness of mast cells.[142]

Many endogenous peptides (vasoactive intestinal peptide [VIP],[115,143] substance P,[114,115,144] ET1,[145]) and cytokines (IL-1, IL-3, IL-33, SCF, IFNα2, and IFNγ) can trigger mast cell degranulation and/or cytokine release. Stimulation of mast cells with these cytokines synergizes with stimulation of other receptors, including FcεRI.[104] Mast cells can also be directly activated by components of reptile or insect venoms[143,146-148] (see below), and mast cell functions can be modulated by diverse receptors, including the aryl hydrocarbon receptor[149] and the vitamin D receptor.[104] There is evidence, from mouse studies, that mast cell function (including that stimulated by IgE) can be influenced by circadian rhythm.[150-154]

Because extensive mast cell degranulation can have catastrophic consequences, such as fatal anaphylaxis, it is critical that mast cell function be tightly regulated. There are a few inhibitory mechanisms identified that can downregulate mast

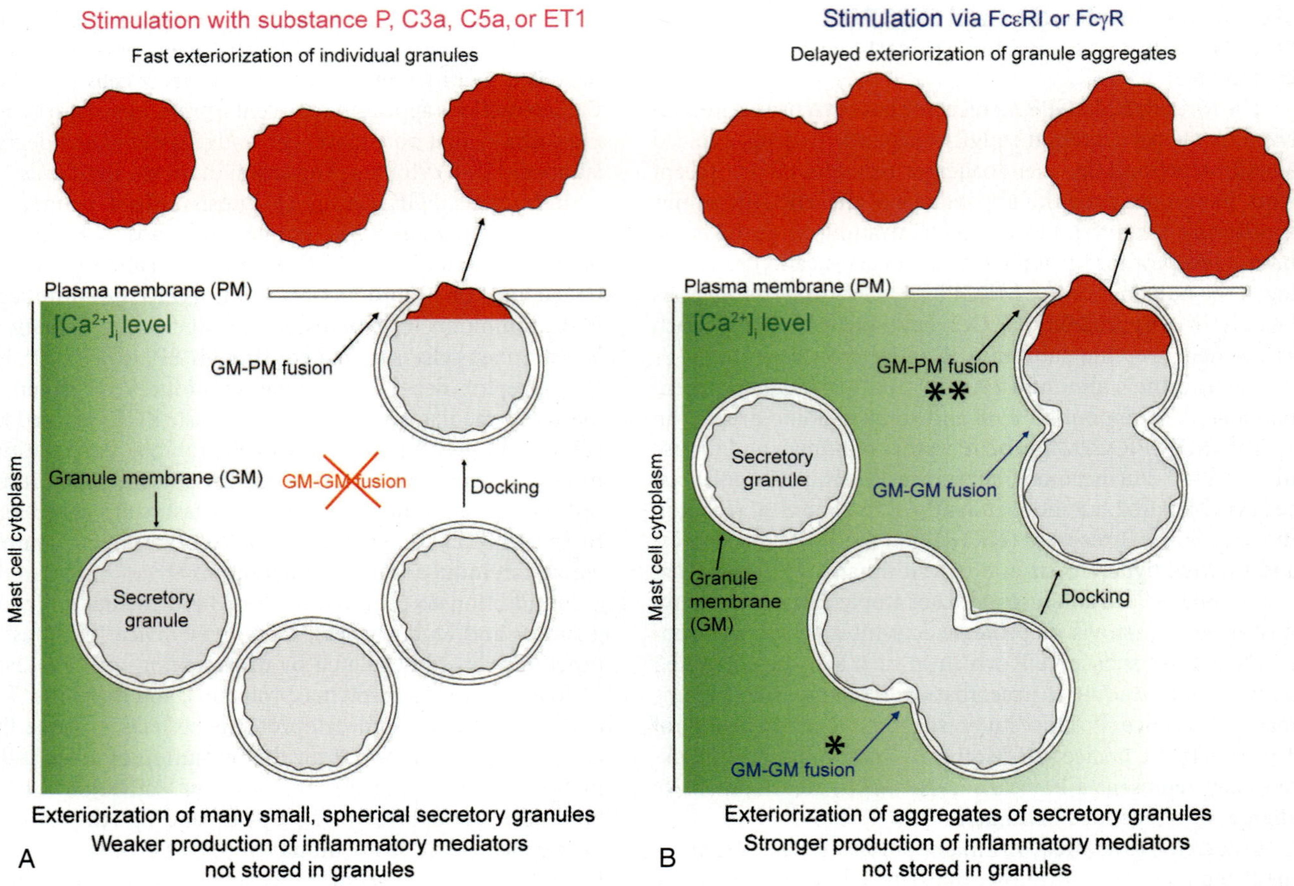

FIG. 16.4. Model of distinct patterns of human mast cell degranulation in response to substance P (SP), C3a, C5a, or endothelin-1 (ET1) versus anti-IgE or IgG immune complexes. A: Activation with SP, C3a, C5a, or ET1 induces rapid docking of discrete intracellular cytoplasmic granules with the plasma membrane, a process associated with the relatively transient elevation of levels of intracellular calcium ($[Ca^{2+}]_i$, indicated by a narrow green gradient) and resulting in the fast secretion of small spherical Av.SRho+ secretory granule structures that are likely to represent single granules. **B:** Activation via FcεRI or FcγR results in a longer time partition between signaling and secretion, associated with a period of sustained elevation of $[Ca^{2+}]_i$ levels (indicated by a wide green gradient). During this period, and before granule matrix material can be detected on the cell surface by binding of Av.SRho, some granules undergo deacidification, perhaps reflecting transient fusion of the granule membranes with the plasma membrane (not shown), favoring what may appear to represent, in static images, "intracellular" granule-granule fusion (*). The fusion of granules with each other and the plasma membrane that occurs during the process of compound exocytosis (**) results in the formation of heterogeneously shaped granule structures, likely representing aggregates of the matrices of multiple individual granules, that undergo exteriorization (as reflected by the binding of Av.SRho) over a more prolonged period than is seen with SP activation. Av.SRho: Fluorochrome-labeled avidin (sulforhodamine 101–coupled avidin), which labels externalized granule matrix on cell surface during mast cell degranulation, allowing monitoring of degranulation dynamics by high-resolution confocal microscopy. (With permission from Gaudenzio N, Sibilano R, Marichal T, et al. Different activation signals induce distinct mast cell degranulation strategies. *J Clin Invest.* 2016;126:3981-3998.)

cell activation.[104,155] The CD300 and tetraspanin (CD9, CD63, CD81, CD151) receptor families can positively or negatively regulate mast cell function, depending on the setting.[155] Stimulation of corticotropin-releasing factor receptors of subtype 1[156] and subtype 2[157] promotes and inhibits IgE-mediated mast cell activation, respectively. These findings highlight that immunological and psychologic stress can modulate mast cell function and mast cell–associated allergic disorders.

Siglec-7 and Siglec-8 inhibitory receptors are expressed by mast cells and basophils. Siglec-7 plays a significant role in inhibiting IgE-mediated mast cell activation but only moderately affects IgE-mediated activation in primary human basophils.[158] Siglec-8 is expressed by human eosinophils and mast cells, but weakly by basophils. Engaging Siglec-8 results in apoptosis of human eosinophils and inhibition of release of preformed and newly generated mediators from human mast cells but has no effect on their survival.[159] Recruitment and ligation of Siglec-3 (CD33) suppresses IgE/FcεRI signaling in human mast cells and dampens IgE-mediated anaphylaxis in transgenic mice expressing human CD33.[160]

Human mast cell activation is suppressed by the inhibitory cytokine IL-10.[161] TGF-β decreases FcεRI expression in mouse and human mast cells[162] and reduces KIT in human skin–derived mast cells.[163] TGF-β inhibits FcεRI-dependent degranulation and cytokine production[163] and IL-33 mediated cytokine release in human mast cells.[164] In mice, CD4+CD25+ T regulatory cells suppress mast cell degranulation in a cell-contact–dependent manner via OX40-OX40L interactions between T_{reg} cells and mast cells.[165] There are other intrinsic mechanisms that can suppress mast cell activation, such as those mediated by RabGEF1[166] and SHIP.[167] Recently, it has been shown that mast cell differentiation and IgE-mediated activation can be inhibited by proton pump inhibitors.[168]

Endocannabinoids are endogenous bioactive lipids that have anti-inflammatory activities. Cellular receptors for endocannabinoids include type-1 (CB1) and type-2 (CB2) G protein–coupled cannabinoid receptors, G protein–coupled receptor GPR55, transient receptor potential channel of the vanilloid subfamily 1 (TRPV1), and peroxisome proliferator-activated receptor γ.[169] Mast cells express cannabinoid receptors and respond to endocannabinoids,[170,171] which inhibit mast cell maturation[172] and mediator secretion.[171-173] Endocannabinoid anandamide (AEA)–dependent inhibition of FcεRI-induced mouse mast cell degranulation is mediated by a mechanism involving CB_2 and GPR55 receptor heteromers.[174]

Products and Mediators

The cytoplasmic granules of mast cells contain proteoglycans, consisting of sulfated glycosaminoglycans covalently linked to a protein core.[175-177] These substances can be stained metachromatically with basic dyes. In humans and murine species, individual mast cell populations can contain variable mixtures of heparin and chondroitin sulfate proteoglycans.[175-177] Although the biologic functions of mast cell proteoglycans are not fully understood, in mice, heparin is required for normal packaging of certain neutral proteases in mast cell cytoplasmic granules.[176,177] Activated mast cells expose negatively charged proteoglycans on their cell surface, allowing binding of the positively charged fluorescent tag avidin, which is being used to track mast cell activation *in vitro* and *in vivo*.[114,178]

Both human and mouse mast cells synthesize and store histamine. Although macrophages,[179] neutrophils,[180] and platelets,[181] as well as basophils,[20] can produce histamine, mast cells represent the source of virtually all the histamine stored in normal tissues in mice, with the notable exceptions of the glandular stomach and parts of the central nervous system.[182] In addition to histamine, murine and at least some human mast cells can store serotonin in their granules.[183,184]

Mast cells can be activated to secrete biologically active products not only by IgE and specific antigen but also by a long list of other stimuli as described above. Upon activation, mast cells release rapidly and/or secrete *de novo* a wide spectrum of biologically active mediators, many of which also can be produced by basophils and other cell types. During IgE-associated biologic responses, the cross-linking of antigen-specific IgE bound to FcεRI on the plasma membrane of mast cells induces the aggregation of FcεRI, thereby activating downstream signaling events that lead to the secretion of biologically active products implicated in allergic reactions.[48,126,185] Following activation, mast cells release into the extracellular space mediators prestored in their cytoplasmic granules, for example, vasoactive amines (histamine and serotonin), neutral proteases (tryptases, chymases, and carboxypeptidase A3 [CPA3]), proteoglycans (eg, heparin), and some cytokines and growth factors (eg, TNF) by a process called degranulation. A second wave of secreted products is generated by *de novo* synthesis of proinflammatory lipid mediators, such as prostaglandins and leukotrienes.

Finally, upon activation, mast cells have been reported to synthesize and secrete a long list of growth factors, cytokines, and chemokines, for example, IL-1β, IL-2, IL-3, IL-4, IL-5, IL-6, IL-9, IL-10, IL-11, IL-13, IL-16, IL-17A, IL-18, IL-22, IL-25, TNF, VEGF, TGF-β, NGF, SCF, and many others.[20,117] Much of this work has been done with mouse mast cells which have been generated and studied *in vitro* although several studies also have addressed cytokine production by isolated populations of human mast cells.

Many other stimuli can also induce mast cells to synthesize and secrete products with or without degranulation. The types and amounts of the products that are released by mast cells are influenced by factors, such as the type and species of origin of the mast cells, and the nature of the stimulus inducing mast cell activation.[48,186,187] In the case of IgE-dependent mast cell activation, whether the activation is by low- or high-affinity stimuli can influence FcεRI receptor signaling and the activation outcomes.[188]

Basophils

Morphology

Like mast cells, routine histological methods are poorly suited for demonstrating basophils. Optimal visualization of the basophil is best achieved by electron microscopy (Fig. 16.2A and C). By ultrastructure, human basophils are 5 to 7 μm in spherical diameter, exhibit a segmented or unsegmented nucleus with marked condensation of nuclear chromatin, and contain round or oval cytoplasmic granules surrounded by a membrane and containing a substructure of dense particles, less dense matrix, and, in some granules, membrane whorls and Charcot-Leyden crystals (Fig. 16.2A and C). A minor population of small, uniform granules is characteristically located near the nucleus. The cytoplasm of mature human basophils also contains glycogen particles, mitochondria, free ribosomes, and small membrane-bound vesicles. Lipid bodies are rarely present. Other organelles are inconspicuous. The basophil's

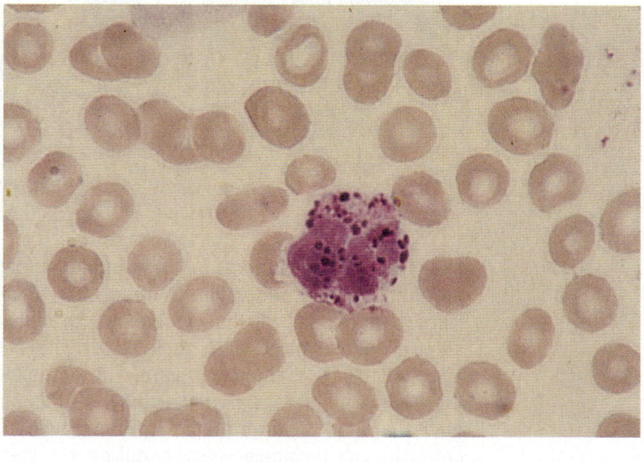

FIG. 16.5. A human basophil in peripheral blood. (From Anderson SC. *Anderson's Atlas of Hematology*. Wolters Kluwer Health/Lippincott Williams & Wilkins; 2003, Copyright 2003.)

prominent metachromatic cytoplasmic granules allow unmistakable identification in Wright-Giemsa–stained films of blood (Fig. 16.5) or bone marrow. Migration and tissue infiltration of basophils can occur during inflammation. Mouse basophils in tissue can be localized by the immunohistochemical staining of a protease largely restricted to basophils, mMCP8.[189]

Natural History and Regulation of Development

The basophil is the least common granulocyte in human blood, with a prevalence of approximately 0.5% to 0.6% of total leukocytes and approximately 0.3% of nucleated marrow cells. Basophils have a short life span of few days[1,190] and retain granulocytic features even after emigrating into tissues. Like other granulocytes, basophils appear to lack the ability to proliferate and are eliminated by apoptosis after participating in tissue responses.[191]

The differentiation trajectory of basophils and their relationship with that of other hematopoietic cell lineages have not been fully delineated. However, much evidence indicates that basophils share a common precursor with other granulocytes and monocytes.[1,3,36] Multiple developmental pathways have been reported for basophils, including from bipotent mast cell/basophil progenitors[11,16] of the mouse, and in humans, from common eosinophil-basophil progenitors.[192,193] Several transcription factors, STAT5,[194] P1-RUNX1,[14] GATA1,[195] GATA2,[194] and C/EBPα,[70] have been implicated in basophil differentiation,[85] whereas C/EBPα and MITF act antagonistically to determine basophil versus mast cell lineage development.[16]

Interleukin (IL)-3 is the major known growth factor for basophils. It promotes the production and survival of human and mouse basophils *in vitro* and induces basophilia in mice.[196-198] IL-3 is also important for the bone marrow and blood basophilia associated with the Th2-associated immunologic responses induced by infection with certain parasites, for example, *Nippostrongylus brasiliensis* or *Strongyloides venezuelensis*.[196,197,199,200] Thymic stromal lymphopoietin (TSLP) is another cytokine that can elicit basophilia associated with *Trichinella spiralis* infection[201] and can induce dermal basophil infiltration in atopic dermatitis-like skin lesions in mice.[202,203] Findings in IL-3−/−[199] and TSLPR−/−[204] mice indicate that neither IL-3 nor TSLP is essential for the baseline development of normal numbers of bone marrow or blood basophils, suggesting the involvement of additional growth factor(s) in basophil development.

Increased basophils are noted in settings with high-IgE levels in humans and mice, including in germ-free mice[205] and in mice treated with antibiotics.[206] In humans, increased frequencies of circulating basophils are detected in immunodeficient allergic subjects who exhibit high IgE levels.[206] IgE also can regulate basophil FcεRI levels in wild-type mice.[207]

Regulation of Function: Positive and Negative

Basophils are activated by a variety of stimuli, including immunoglobulins, complement components, cytokines/growth factors, products of pathogens, pathogen and allergens-derived proteases, etc. Similar to mast cells, the best studied pathway for basophil activation is via cross-linking of FcεRI by the binding of antigens to antigen-specific IgE, which triggers degranulation (Fig. 16.6) and release of granule-associated mediators (eg, histamine), *de novo* synthesis and secretion of lipid mediators (eg, LTC_4), and transcription and translation of cytokines/

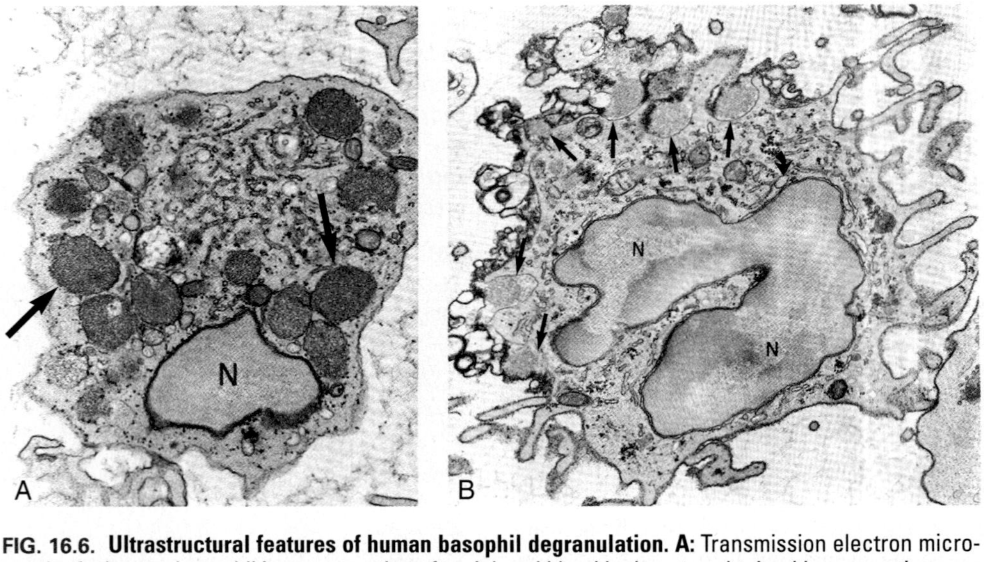

FIG. 16.6. Ultrastructural features of human basophil degranulation. A: Transmission electron micrograph of a human basophil in a preparation of peripheral blood leukocytes obtained by separation over Ficoll-Hypaque. All of the cytoplasmic granules (two indicated by *solid arrows*) contain particulate electron-dense material. **N**, nucleus. (Original magnification ~ ×19, 800). **B:** A human basophil 2 minutes after exposure to antigen *in vitro.* The cell exhibits extrusion of granules from six separate sites on the plasma membrane (*small arrows*). At this time after cell stimulation, particle-filled granules retain their shape and characteristic structure even after exposure to extracellular milieu. Cationized ferritin coats the cell surface and enters cul-de-sac that contain exteriorized granules. The cell exhibits no fully intracytoplasmic typical basophilic granules but one of a smaller kind of granule (*curved arrow*) can be observed in the perinuclear region. **N**, nucleus. (Original magnification ~ ×19, 200). (Reprinted by permission from Springer Nature: Dvorak AM, Newball HH, Dvorak HF, Lichtenstein LM. Antigen-induced IgE-mediated degranulation of human basophils. *Lab Invest.* 1980;43:126-139.)

chemokines/growth factors. Activation of basophils also triggers multiple intracellular membrane events and fusion of granules with the cell membrane, leading to elevated expression of activation markers (eg, CD11b, CD13, CD63, CD107a, CD203c, CD200R, and CD300a) on their surface. Because basophils are readily accessible in the peripheral blood, and given the rapid upregulation of activation markers following FcεRI cross-linking, basophil activation tests were developed in early 1990s as a noninvasive tool for the assessment of allergy severity and for monitoring the efficacy of allergen-specific immunotherapy.[208-211]

Basophils express several low-affinity IgG receptors. Human basophils express activating FcγRIIA, inhibitory FcγRIIB, and a very small amount of FcγRIIIB, whereas mouse basophils express FcγRIIIA and FcγRIIB.[96] It appears that IgG immune complexes minimally activate human basophils due to the dominant inhibitory effects of FcγRIIB,[96] but IgG-immune complexes induce basophil production of PAF and systemic anaphylaxis independent of IgE and mast cells in mice.[212] Thus, depending on the experimental conditions, allergen-IgG immune complex–stimulated basophils, macrophages, or neutrophils are the major producers of PAF.[88] IgG subclass might determine which type of leukocyte predominantly contributes to IgG-mediated systemic anaphylaxis.[87]

IgA receptors have not been detected on human basophils but secretory IgA can degranulate IL-3, IL-5, or GM-CSF primed human basophils.[213] IgD binds to basophils via galectin-9 and CD44, and cross-linking of IgD on basophil surface by anti-IgD induces production of B cell–activating factors (BAFF), proinflammatory cytokines, and antimicrobial peptides, but inhibits IgE-mediated basophil degranulation.[101,214] Basophils can be activated by GPCRs via the stimulation of C3a and C5a[111] or chemokines.[215] Human basophils also express MRGPRX2 and can be activated by anti-MRGPRX2 antibody and the MRGPRX2 agonist, ciprofloxacin.[216]

Many cytokines and growth factors, including IL-3,[217-219] IL-5,[220] GM-CSF,[218] IL-8,[221] IL-25,[222] IL-33,[222-226] NGF,[227] and TSLP,[202] can directly activate basophils and/or prime them for enhanced responses to IgE-dependent challenge and other stimuli. IL-3 is the main growth factor known to support human and mouse basophil development, expansion, and survival. IL-3 alone can upregulate CD203c and ST2 (IL-33 receptor) expression and synergizes with IgE-FcεRI activation to enhance cytokine production in basophils. Recent studies have shown the involvement of STIM1 and STIM2 in FcεRI- and IL-3 receptor (IL-3R)–dependent basophil responses, respectively.[219]

Basophils can amplify Th2 responses and other inflammatory responses by responding to innate cytokines of epithelial cells, such as IL-25 and IL-33, which are produced after exposure to allergens or infection with parasites or viruses.[222,228] Human and mouse basophils also express receptors for TSLP. TSLP is a critical cytokine in Th2 immunity and promotes basophil hematopoiesis.[202] Notably, basophils elicited by TSLP appear to be phenotypically and functionally different from those elicited by IL-3. IL-3–elicited basophils express higher levels of CD11b and CD62L and more degranulation upon FcεRI cross-linking, while TSLP-elicited basophils express more IL-33R, IL-18R, and IL-3R (CD123) and produce more cytokines and chemokines after IL-3, IL-18, or IL-33 stimulation.[202] IL-3- versus TSLP-elicited mouse basophils also exhibit distinct transcriptional profiles.[201]

Basophils from allergic subjects express more TSLP receptors than those of normal subjects, but the effects of TSLP on basophils are less consistent. One study showed that the effects of TSLP on basophils from allergic subjects were as potent as those induced by IL-3 or anti-IgE[229] and that TSLP stimulation of the basophils purified from patients with allergic asthma induced histamine release; upregulated CD203c, CCR3, and IL3R expression; increased intracellular IL-3, IL-4, IL-5, and IL-13; and primed basophils' responses to eotaxin.[229] But a subsequent study did not find that TSLP activated basophils from normal donors or allergic subjects.[230]

Pathogens and their products can activate basophils via TLRs, NOD-like receptors, and the formyl peptide receptor (FPR). TLRs 1 to 10 protein and/or mRNA are detected in human basophils.[20,231,232] Some TLR ligands (eg, peptidoglycan/TLR2) stimulate basophils to secrete cytokines (IL-4 and IL-13) without degranulation, while other TLR ligands synergistically enhance cytokine and chemokine release by basophils after FcεRI stimulation.[20,232] In addition to TLRs, bacterial products, like formyl-methionine-phenylalanine (fMLP), can stimulate basophils via FPRs[233] and muramyl dipeptides associated with bacterial peptidoglycans can activate basophils via intracellular nucleotide-binding oligomerization domain-containing protein 2 (NOD2) receptors.[234]

Basophils also can react directly (independently of IgE) to allergens containing protease activities (eg, papain and Der P 1 in house dust mite extract).[235] Leptin, a cytokine of the IL-6 family, can enhance survival of human basophils and induce their migration, degranulation, and cytokine production.[236] T_{reg} can upregulate basophil activation markers (CD69, CD203c, CD13) and production of cytokines (IL-13, IL-4, and IL-18).[237] Significant circadian variability of basophil responses to IgE-dependent and IgE-independent stimuli has been recently reported.[238]

A few signals have been reported to suppress basophil functions. Like in mast cells, cross-linking of FcγRIIB can suppress IgE-FcεRI–mediated basophil activation.[96,239,240] Several inhibitory receptors, such as CD300a,[241] CD200,[242] Siglec-7,[158] and Siglec-8,[243] are expressed by human basophils. Engagement of Siglec-7 inhibits IgE-mediated activation of human mast cells but only moderately suppresses IgE-mediated activation in primary human basophils.[158] Siglec-8 is uniquely expressed by human eosinophils and mast cells but is only weakly expressed by human basophils.[159] During venom immunotherapy, there is a rapid upregulation of the histamine receptor, H2R, on human basophils, and stimulation of H2R suppresses IgE/FcεRI-mediated degranulation and mediator release by basophils,[244] suggesting a mechanism by which basophil activation is inhibited by allergen-induced rapid desensitization.

Products and Mediators

Basophils are the source of most (if not all) histamine present in normal human blood.[245] The sulfated glycosaminoglycans of normal human blood basophils have not yet been characterized. However, two studies, that together analyzed five patients with myeloid leukemia containing 10% to 75% basophils, showed that such cells may produce solely chondroitin sulfates[246] or a mixture of chondroitin sulfates (50%-84%) and heparin (8%-43%).[247]

In addition to proteoglycans and histamine, basophils generate many other products that can influence the course of inflammatory processes. These substances are either preformed and granule associated (eg, histamine, neutral proteases, proteoglycans) or produced during activation of the cell (eg, LTC_4, PGE_2, PGD_2,[248] and platelet-activating factor [PAF]). The spectrum of basophil-derived cytokines appears to be more limited than mast cells but includes IL-3, IL-4, IL-13, IL-25, IL-31, vascular endothelial growth factor (VEGF),[249] certain chemokines, and at least in mice, IL-6, TNF, and TSLP.[1,2,19,20,250] Human basophils make IL-31 and are also activated by IL-31.[251] Basophils of chronic spontaneous urticaria (CSU) patients have been reported to produce substance P, express NK1R, and are degranulated by substance P.[252] In addition, basophils can produce retinoic acid, angiopoietin-1, basogranulin, granzyme B, chemokines (CCL5/RANTES, CXCL8/IL-8, CCL3/MIP-1α, CXCL10/IP-10), BAFF, and antimicrobial peptides.[20]

Basophils are a particularly important source of IL-4. They produce more IL-4 than CD4+ T cells on a per cell basis and produce IL-4 with a stimulation threshold lower than T cells.[253] In addition to FcεRI cross-linking, basophils produce IL-4 in response to other stimuli, such as IL-18, IL-33, TLR ligands, and proteases, which can be associated with helminth infection or exposure to allergens (house dust mite).[254,255]

IGE-DEPENDENT UPREGULATION OF FCεRI EXPRESSION AND FCεRI-DEPENDENT FUNCTION

As plasma levels of IgE increase (as often occurs in subjects with allergic diseases or parasite infections), levels of FcεRI expression on the surface of basophils and mast cells also increase.[207,256,257] Compared to cells with low "baseline" levels of FcεRI expression, such mast cells can bind more IgE, release mediators in response to lower concentrations of allergens, and produce significantly larger amounts of preformed and lipid mediators and cytokines.[258] Thus, basophils and mast cells in subjects with high levels of IgE may have significantly enhanced ability to express IgE-dependent and/or immunoregulatory functions.[48,258]

Notably, exposure of mouse mast cells to certain monoclonal IgE antibodies, in the absence of exposure to the antigen for which the IgE is known to have specificity, can enhance the survival of the cells and, in some cases, induce the cells to release all three classes of mediators (preformed, lipid, cytokines).[258] Exposure to IgE in the absence of known antigen also can induce enhanced survival and cytokine and chemokine release by *in vitro* derived human mast cells.[259] While the mechanisms responsible for these intriguing findings are not fully understood,[258,260] one monoclonal IgE antibody appears to induce aggregation of FcεRI by forming trace quantities of trimers in the absence of known antigen.[261] The clinical implications of these findings, if any, remain to be defined.

MAST CELL AND BASOPHIL HETEROGENEITY

Variation in the morphologic, biochemical, and/or functional characteristics of mast cells from different anatomic locations, or from the same organ or site, has been reported in several mammalian species, including humans.[53,187,262] Two subtypes of mast cells named "connective tissue mast cells" or "mucosal mast cells" were described in rodents by Enerback based on their anatomical location, metachromatic staining characteristics, and cytoplasmic granule proteoglycan content.[263,264] In humans, mast cell subtypes have been demonstrated based on their content of tryptase (MC_T) or tryptase and chymase (MC_{TC}).[175,177,187]

This phenomenon, often referred to as *mast cell heterogeneity*, raises the possibility that mast cells of different phenotype express different functions in health or disease and may exhibit different sensitivities to pharmacologic manipulation. At least four mechanisms may account for phenotypic variation in mast cell populations: (1) factors promoting irreversible branching within the mast cell lineage; (2) factors influencing differentiation and maturation (within single or, if they occur, within multiple pathways); (3) factors modulating mast cell function; and (4) factors influencing local concentrations of exogenous substances not derived from mast cells but taken up and stored in mast cell granules. Of these four mechanisms, experimental evidence has been obtained for all but the first.[53]

Human basophils can exhibit some variation in phenotypic characteristics also, such as immunoreactivity for tryptase, chymase, and carboxypeptidase A[265] or levels of expression of surface structures, including HLA-DR, CD32 (FcγRII), and receptors for cytokines. Such variation in basophil mediator content and/or cell surface phenotype may reflect individual differences among different subjects and/or the effects of disease processes[265] or the consequences of immunotherapy.[266] Basophil heterogeneity has also been noted in mice.[267] As discussed above, IL-3 and TSLP can expand mouse basophil populations that exhibit distinct patterns of gene expression, phenotype, and function. In comparison with IL-3–elicited basophils, TSLP-elicited basophils degranulate weakly to IgE-FcεRI cross-linking but produce more IL-4 by IL-3, IL-18, and IL-33 stimulation,[202] raising the possibility that basophils developed under different growth factor conditions may perform different functions in IgE versus non-IgE responses. However, the effects of IL-3 versus TSLP cytokine microenvironments on human basophils remain to be determined.

GENETIC APPROACHES FOR ANALYZING MAST CELL AND BASOPHIL FUNCTION

As described above, factors capable of inducing basophil infiltration, mast cell proliferation, and/or basophil or mast cell activation are generated during a wide variety of immunologic and pathologic processes.[1,4,20,21,48,104,268-272] As a result, there is considerable speculation that mast cells and basophils express critical roles in diverse biologic responses. However, the precise functions of mast cells and basophils in most of the biologic responses in which the cells have been implicated are obscure. Before the availability of genetically mast cell- and/or basophil-deficient mice, speculation about the roles of mast cells and basophils in biological or pathological settings was based on more indirect lines of evidence, such as assessing changes in the numbers of mast cells or basophils at sites of pathology (Tables 16.2 and 16.3) or by pharmacological approaches nonspecifically targeting mast cell or basophil mediators.

Studies of basophil function in guinea pigs[273] and mice[21,270,272] have employed antibodies to deplete basophils, and such

TABLE 16.2 Conditions Associated With Alterations in Numbers of Blood Basophils

I. Decreased Numbers (Basopenia)
  A. Hereditary absence of basophils (very rare)
  B. Elevated levels of glucocorticoids
  C. Hyperthyroidism or treatment with thyroid hormones
  D. Ovulation
  E. Hypersensitivity reactions
    1. Urticaria
    2. Anaphylaxis
    3. Drug-induced reactions
  F. Leukocytosis (in association with diverse disorders)
II. Increased Numbers (Basophilia)
  A. Allergy or inflammation
    1. Ulcerative colitis
    2. Drug, food, inhalant hypersensitivity
    3. Erythroderma, urticaria
    4. Juvenile rheumatoid arthritis
  B. Endocrinopathy
    1. Diabetes mellitus
    2. Estrogen administration
    3. Hypothyroidism (myxedema)
  C. Infection
    1. Chicken pox
    2. Influenza
    3. Smallpox
    4. Tuberculosis
  D. Iron deficiency
  E. Exposure to ionizing radiation
  F. Neoplasia
    1. "Basophilic leukemia"
  G. Myeloproliferative neoplasms (especially chronic myelogenous leukemia; also polycythemia vera, primary myelofibrosis, essential thrombocythemia)
  H. Carcinoma

From Galli SJ, Metcalfe DD, Arber DA. In: Kaushansky K, Lichtman MA, Beutler E, Kipps TJ, and Prchal JT, eds. *Williams Hematology.* 10th ed. McGraw-Hill Medical Education; 2021:1093.

approaches also have been used to deplete mast cells in mice[274]; however, the antibodies used may also influence other cell types.[1,2,21,197,270,275-277] Other approaches have employed pharmacological inhibitors to probe mast cell functions, but none of these drugs appear to target mast cells specifically. For example, cromolyn has been widely used as "mast cell stabilizer" to investigate biological functions of mast cells *in vivo*, but there is evidence that cromolyn does not selectively and effectively block mast cell activation, at least in mice.[278]

Various mutant or transgenic mice that exhibit constitutive or inducible depletion of some or all mast cell populations, as well as mast cell–deficient mice, which have been selectively engrafted with wild-type or genetically altered mast cells (so called "mast cell knockin mice"), have been used in efforts to define and quantify the contributions of mast cells to many different biologic responses.[275-277] Additional transgenic mice which lack certain mast cell–associated products (either globally or in selected cell populations) have been generated to interrogate the contribution of mast cell–specific products in biological or inflammatory responses.[275-277]

The contributions of mast cells and mast cell–associated products can be further investigated by a combination of genetic and *in vivo* imaging approaches in different disease models.[178,279,280] For example, a fluorescent imaging approach that enables the selective labeling and tracking of mast cell secretory granules by real-time intravital two-photon microscopy in living mice, and thereby permits the identification of such mast cells as a potential source of cytokines, has been used to sort out the controversial roles of mast cells in contact hypersensitivity (CHS) reactions in mice.[178,279] Genetically engineered mice which exhibit constitutive or inducible depletion of basophils or basophil-associated products are available for the investigation of basophil functions *in vivo*.[1,21,22,219,272]

As has been extensively reviewed,[1,4,21,48,271,272,275-277] the availability of such models has the potential to advance substantially our understanding of the roles of mast cells and basophils in health and disease in mice, including ascertaining the importance of strain background on some of the findings.[281-283] In evaluating the results of such studies, one should keep in mind that such genetic models can have various advantages and disadvantages; moreover, even very strong evidence that mast cells or basophils have particular roles in mice does not prove that the cells have identical functions in humans.[190,284]

No humans naturally devoid of mast cells have yet been reported. However, Ph+ chronic myeloid leukemia patients treated long-term with imatinib can have profound reductions in skin mast cells.[285] The findings in rare patients with a deficiency of basophils are not easy to interpret. One patient with profound basopenia experienced persistent and severe infestation with scabies,[38] a finding that might be viewed as consistent

TABLE 16.3 Conditions Associated With Secondary Changes in Mast Cell Numbers

I. Decreased Numbers
  A. Long-term treatment with glucocorticoids
  B. Primary or acquired immunodeficiency disorders (certain mast cell populations; see text)
II. Increased Numbers
  A. Immunoglobulin E–associated disorders
    1. Allergic rhinitis
    2. Asthma
    3. Urticaria
  B. Connective tissue disorders
    1. Rheumatoid arthritis
    2. Psoriatic arthritis
    3. Scleroderma
    4. Systemic lupus erythematosus
  C. Infectious diseases
    1. Tuberculosis
    2. Syphilis
    3. Parasitic diseases
  D. Neoplastic disorders
    1. Lymphoproliferative diseases[a] (lymphoplasmacytic lymphoma/Waldenström macroglobulinemia, lymphoma, chronic lymphocytic leukemia)
    2. Hematopoietic stem cell diseases[a] (acute or chronic myelogenous leukemias, myelodysplastic syndromes, idiopathic refractory sideroblastic anemia)
  E. Lymph nodes draining areas of tumor growth
  F. Osteoporosis[a]
  G. Chronic liver disease[a]
  H. Chronic renal disease[a]

[a]Can include increases in numbers of mast cells in the marrow.
From Galli SJ, Metcalfe DD, Arber DA. In: Kaushansky K, Lichtman MA, Beutler E, Kipps TJ, and Prchal JT, eds. *Williams Hematology*. 10th ed. McGraw-Hill Medical Education; 2021:1093.

with a role for basophils in resisting ectoparasites. However, that patient also had eosinopenia, IgA deficiency, and multiple other clinical problems.[38] A second basophil-deficient patient had recurrent bacterial and viral infections but also had a deficiency of eosinophils, hypogammaglobulinemia, abnormal suppressor T-cell function *in vitro*, and a thymoma.[40]

IMMUNOREGULATORY FUNCTIONS OF MAST CELLS AND BASOPHILS

Given the great capacity of mast cells and basophils to release a wide spectrum of immunoregulatory factors in their responses to exogenous and endogenous stimuli, these cells have the potential to influence many aspects of inflammatory and immunological processes. For example, many mast cell- and basophil-derived mediators/cytokines/chemokines/growth factors have been documented to positively or negatively regulate maturation, recruitment, and function of other innate and adaptive immune cells.[4,186,286-290]

Neutrophils are recruited by mast cells during innate[291,292] and adaptive[293-295] immune responses via mast cell–derived TNF,[293,294] macrophage inflammatory protein 2,[294] CXCL1/CXCL2,[292] LTB_4,[291] and other potent chemoattractants. Mast cell production of IL-1β has been implicated in the neutrophil infiltration and vascular leakage observed in chronic urticarial rashes of patients with cryopyrin-associated periodic syndrome, an autoinflammatory disease caused by inflammasome NLRP3 mutation.[296] Mast cells also express various components of inflammasomes, which can activate caspase-1-dependent cleavage of pro–IL-1β after stimulation of PPRs by internal danger signals or external signals from pathogens.[297] Mast cell inflammasomes and mast cell–derived IL-1β may participate in autoinflammatory diseases.[297]

Mast cells and ILC2s can be in close physical proximity, facilitating their functional interactions in tissues. In a model of intestinal helminth infection with *Heligmosomoides polygyrus*, mast cells are rapidly activated by ATP released from apoptotic intestinal epithelial cells and produce IL-33, which activates ILC2 cells.[298] Activated mast cells release the proteases, chymase and tryptase, that can cleave IL-33 to produce biologically active forms of IL-33 with increased potency on ILC2s.[299] Indeed, blocking of tryptase reduces IL-33–dependent allergic airway inflammation *in vivo*.[299] However, IL-33 degradation by mast cell chymase is protective in another model of allergic airway inflammation in mice.[300]

Although bioactive mediators, such as histamine, prostaglandins, and leukotrienes, were originally recognized for their effects on structural cells (eg, inducing vascular leakage and smooth muscle contractility), these mediators have additional functions in leukocyte migration and activation. Mast cells[301,302] and mast cell–associated histamine,[303-305] PGD_2,[306] PGE_2,[304] TNF,[307-310] IL-1β,[303,311] and IL-6[305] have been shown to regulate maturation, migration, and antigen presentation function of dendritic cells (DCs) or Langerhans cells *in vitro* and *in vivo*.

Another mechanism by which mast cells can regulate DC function is via exosomes, small vesicles implicated in the transfer of materials among cells. Mouse mast cell–derived exosomes are highly potent in inducing maturation (eg, CD40 and CD80 expression) and functional activation (secretion of IL12p70) of DCs and can potentiate the efficiency of antigen presentation by DCs.[312] In addition, mouse mast cells can form "immunological synapses" with immature DCs to transfer internalized mast cell–associated antigens to DCs; the transferred antigen is then processed and presented by DCs to activate T cells.[313] Conversely, mouse mast cells can acquire major histocompatibility complex class II (MHCII) proteins from DCs via direct synapse-like cell-cell contacts.[314] Furthermore, mouse mast cells can enhance antigen presentation indirectly by internalizing antigen bound to FcεRI-associated IgE; this mechanism is independent of mast-cell MHC class II expression but requires that such mast cells undergo apoptosis and then are phagocytosed by other antigen-presenting cells.[315]

Mast cells are normally anchored in tissues but mouse mast cells can migrate to lymph nodes upon activation.[316-319] This allows direct contacts of mast cells with T and B lymphocytes in the lymph node. Such close contacts of mast cells and lymphocytes can also be detected in peripheral tissue at sites of inflammation in mice and humans.[320] Mast cells and mast cell–derived inflammatory mediators/cytokines can influence polarization of naïve T cells to Th1, Th2, Th17, and T_{reg} cells and can modulate functional activation of these different T cell subsets.[320] Mouse mast cells can express several costimulatory molecules in the B7 family (ICOS-L, PD-L1, CD80, CD86) and the TNF/TNF receptor family (OX40L and CD40L), which also permits them to fine-tune antigen-specific T cell proliferation and activation.[321]

Mast cells may also influence T cell activation by their release of exosomes. Exosomes derived from mouse mast cells can induce lymphocyte proliferation and IL-2 and IFNγ production *in vitro*.[322] Injection of such mast cell–derived exosomes, which contain costimulatory molecules (such as MHC class II, CD86, CD40, CD40L, LFA-1, and intercellular adhesion molecule 1), can induce lymphocyte proliferation and cytokine production *in vivo*.[322]

Mouse mast cells can inhibit the suppressive function of T regulatory cells through both direct cell-cell contacts and IL-6 release.[323] On the other hand, mast cell IL-6 and direct mast cell/B cell contact via CD40 and CD40L can promote B cell proliferation, survival, and immunoglobulin production in mice.[324] Notably, mast cells can remotely influence immunopathology at distant sites.[280,317,325] Mast cell granules can travel via lymphatics[325] or be engulfed and carried by DCs[280] to draining lymph nodes to modulate adaptive immune responses.

In a mouse model of bacterial infection, mast cell granules release preformed TNF, which upregulates E-selectin on blood vessels, critical for DC recruitment and lymph node hypertrophy.[308] To explore such immunomodulatory features of mast cells, synthetic mast cell granules have been tested for their ability to promote and polarize immune responses.[326] Mast cell activators, such as compound 48/80, have been shown to enhance protective immunity and antibody production, against viral and bacterial infection in mice.[327-331] Given the mast cell's potent capacity to modulate the functions of DCs, T cells, and B cells in adaptive immunity, it is conceivable that "mast cell activators" may be used as mucosal adjuvants to enhance vaccination for infectious diseases.[31,332] However, such approaches need to be carefully developed, so as to avoid the possible anaphylaxis that can be induced by the rapid release of the mast cell's powerful mediators.

In other settings, activation of mast cells can be anti-inflammatory/immunosuppressive.[279,318,333-341] For example, mouse mast cells can function to reduce the intensity of models of severe[279,336] and/or chronic[338,341,342] contact dermatitis. In a model of severe contact dermatitis, mast cells suppressed the intensity of the response in part via their production of the immunosuppressive cytokine, IL-10.[279,336] By contrast, in a model in which mice were challenged with antigen multiple times, mast cells suppressed the intensity of the reactions via IL-2, which can promote T_{reg} expansion[338,342] (Fig. 16.7).

As described above, basophils have been reported to produce large amounts of stored and newly synthesized IL-4.[254,343] IL-4 in human[344] and mouse[345] basophils was first reported in the early 1990s. As a key driver of Th2 immunity in allergy and parasite infection, IL-4 arguably is one of the most important immunomodulatory molecules produced by basophils.[2,254,287,289,346] Basophils produce IL-4 by both IgE-dependent and independent mechanisms. TLR ligands, cytokines (eg, IL-3, IL-18, IL-33, TSLP), pathogen-derived molecules, and proteases associated with allergens (papain, house dust mite) and parasites activate basophils to produce IL-4 with or without concomitant FcεRI cross-linking.[254] Immunoregulatory effects of basophil-derived IL-4 potentially influence DCs, T cells, B cells, monocytes/macrophage, ILCs, and structural cells.[254] Several lines of evidence from *in vitro*[347,348] and *in vivo*[235,347-349] studies strongly support the critical involvement of basophil IL-4 in Th2 differentiation. The immediate release of IL-4 from basophils in response to danger signals (eg, observed with cell stress or tissue damage caused by allergens or pathogen invasion) is likely the initial signal that programs differentiation of naïve T cells to Th2 cells.[255] The later production of IL-4 and IL-13 by basophils after IgE/FcεRI cross-linking or other stimuli can further amplify Th2/allergic reactions.[350]

Basophil IL-4, together with IL-6, also can contribute to protective humoral memory responses against sepsis induced by *Streptococcus pneumoniae* in mice.[351] However, basophil-derived IL-4 can be detrimental in the pathogenesis of humoral autoimmunity by contributing to autoantibody production in mice susceptible to systemic lupus erythematosus (SLE)-like disease.[352] Basophils also can contribute to humoral responses via other mechanisms, such as by their production of BAFF after anti-IgD cross-linking of IgD[101] or TLR activation.[353] Activation of TLRs can enhance production of BAFF and IL-13 in basophils, leading to production of a large amount of IgG4 by B cells.[353]

Basophils can regulate other innate immune cells. Deletion of basophils in mice reduces infiltration of eosinophils and neutrophils in cutaneous IgE-dependent chronic allergic inflammation.[354,355] In another mouse model, irritant contact dermatitis, basophil-derived IL-4, and TNF induce eotaxin/CCL11 production by fibroblasts, leading to eosinophil recruitment.[356] Basophil-derived IL-4 promotes eosinophil

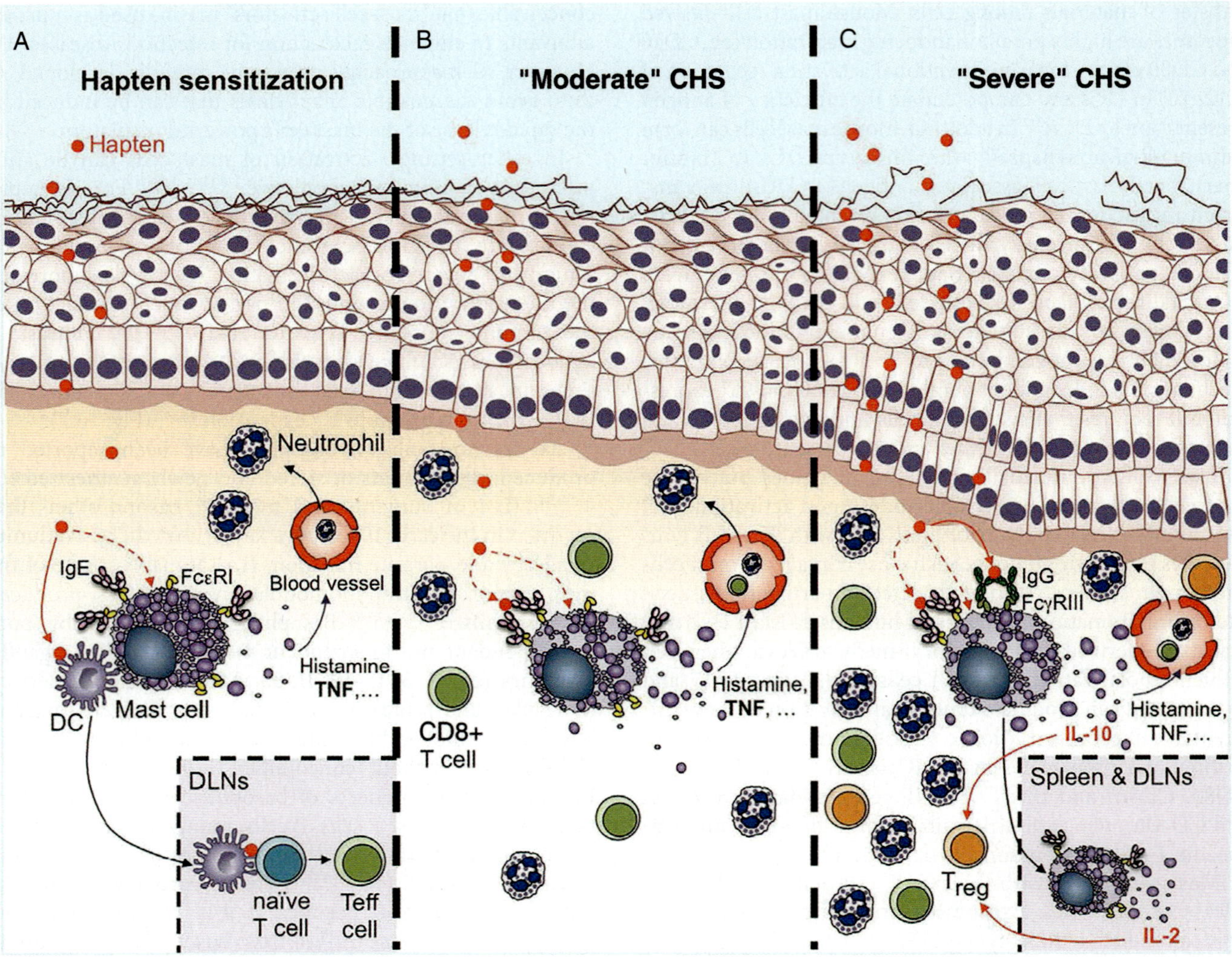

FIG. 16.7. **Roles of mast cells (MCs) in the sensitization and effector phases of contact hypersensitivity (CHS). A:** During the sensitization phase, MCs may be activated directly or indirectly by certain haptens to release a diverse spectrum of mediators, including histamine and tumor necrosis factor (TNF), which induce vasodilatation and recruitment of leukocytes, mainly neutrophils. Direct cell-to-cell contacts between MCs and dendritic cells (DCs), as well as MC-derived TNF, can amplify DC migration to the draining lymph nodes (DLNs), where these cells prime naïve T cells to become effector cells (T_{eff} cells) via antigen presentation. **B:** During "moderate" CHS responses, MCs and MC-derived TNF amplify tissue swelling, leukocyte recruitment (mainly neutrophils and CD8+ T cells), and epidermal hyperplasia. Evidence suggests that IgE and FcεRI can amplify the proinflammatory functions of MCs in moderate CHS. **C:** During more severe (and chronic) CHS responses, MCs represent an early source of IL-10 in the skin, which amplifies recruitment of regulatory T cells (T_{reg}) and helps to limit tissue swelling and epidermal hyperplasia. MC activation in the skin is amplified by the engagement of FcγRIII by hapten-IgG immune complexes. Additionally, MCs can migrate to the DLNs and the spleen in an IgE-dependent manner, where they produce IL-2. IL-2 helps in maintaining the T_{eff}:T_{reg} ratio at the site of inflammation, and thereby potentially contributes to limiting the severity of certain CHS responses. (From Gaudenzio N, Marichal T, Galli SJ, Reber LL. Genetic and imaging approaches reveal pro-inflammatory and immunoregulatory roles of mast cells in contact hypersensitivity. *Front Immunol.* 2018;9:1275.)

tissue migration by the induction of endothelial vascular cell adhesion molecule-1 (VCAM-1) on blood vessels in a mouse model of IgE-mediated skin inflammation.[357] Furthermore, by production of IL-4, basophils promote innate lymphoid cell responses in inflammation of the skin[358] and lung.[359]

In mice, it has been reported that TSLP and IL-3 generate two distinct populations of basophils with different phenotypic and functional properties.[267] This finding encouraged speculation that basophils influenced by TSLP versus IL-3 can regulate different aspects of immunity.[267] The importance of the TSLP/basophil axis in allergic inflammation is illustrated in a mouse model of food allergy elicited by epicutaneous sensitization with food allergens.[203,360] Atopic dermatitis with defective skin barrier function represents a significant risk for food allergy. Food allergen sensitization through an atopic dermatitis-like skin lesion in mice induced dermal eosinophil infiltration, TSLP-dependent expansion of basophils in the skin, and intestinal mast cell hyperplasia, with antigen-specific Th2 cytokine responses and increased serum levels of antigen-specific IgE. These sensitized mice went on to develop intestinal inflammation upon oral food challenge but disruption of TSLP responses or depletion of basophils or basophil-intrinsic IL-4, but not eosinophils, reduced skin inflammation and severity of intestinal food allergy.[203,360] *In vitro*, TSLP-elicited basophils promoted OX40L expression in DCs in a cell contact–dependent manner; also, TSLP-elicited basophils cocultured with DCs produced significantly more IL-4 than basophils cultured alone.[203] These findings indicate that, in mice, reciprocal interactions of DCs and basophils can contribute importantly to Th2 inflammation.

The production of IL-4 by basophils can have immunosuppressive effects. In a mouse model of IgE-dependent chronic allergic skin inflammation, basophil IL-4 can downregulate inflammation by converting inflammatory macrophages to M2-like macrophages with anti-inflammatory activity.[361] However, alternatively activated macrophages elicited by basophil-derived IL-4 can importantly promote host protection against the intestinal helminth *Nippostrongylus brasiliensis*.[362] In humans, basophil IL-4 and IL-13 can potentiate monocyte responses to IL-3 and the production of CCL17, promoting alternative differentiation of monocytes.[363] Human basophil IL-4 also can upregulate FcγRIIB and IL-10 production by monocytes, whereas basophil-specific histamine can inhibit LPS-induced production of the proinflammatory cytokine TNF and upregulation of the costimulatory molecule CD80 in monocytes.[364] Other anti-inflammatory effects of basophil IL-4 include suppression of Th1 responses in an experimental mouse model of colitis[365] and upregulation of IL-10 production by mouse CD8 T cells.[366]

Several reports suggest that mouse basophils can perform antigen presentation and promote Th2 cell differentiation via their expression of MHC class II and costimulatory molecules, CD40, CD80, and CD86.[367-369] However, such functions of basophils remained controversial, as some subsequent studies did not validate the antigen presentation function of basophils.[2,20,254] A possible explanation for certain antigen presentation activities detected in basophils is that mouse basophils, which express no or very low levels of MHC II at baseline, can acquire peptide-MHC II complexes from DCs by a cell-contact–dependent process called "trogocytosis."[370] Like mouse basophils, basophils in human peripheral blood normally express no or little HLA-DR but can be induced to express HLA-DR by cytokines (eg, IL-3, IFNγ, and GM-CSF).[371] Whether human basophils can perform antigen presentation, either by their HLA-DR expression or by trogocytosis, remains to be determined.

In summary, many mediators which can be produced by mast cells and basophils have been shown to have various positive or negative effects on the functions of diverse immune or structural cells. Such findings indicate that the immunoregulatory functions of mast cells and basophils at least have the potential to influence inflammation, hemostasis, tissue remodeling, cancer, metabolism, reproduction, behavior, sleep, and many other biological responses.[1,2,4,36,48,232,271,346]

BIOLOGICAL FUNCTIONS OF MAST CELLS AND BASOPHILS

Roles in IgE-Associated Allergic Responses

Mast cells and basophils express high-affinity receptors for immunoglobulin (Ig) E (FcεRI) on their surface. Both cell types can be triggered to release potent mediators in response to activation via FcεRI, when their cell-bound IgE recognizes bivalent or multivalent allergens. Accordingly, mast cells and basophils have long been regarded as important effector cells in allergic asthma, food allergies, hay fever, and other allergic disorders (Table 16.4). It is thought that the activation of these cells, resulting in the release of cytoplasmic granule-associated preformed mediators (including histamine and certain proteases), lipid mediators (such as prostaglandin D_2 [PGD_2] and leukotriene C_4), and cytokines, growth factors, and chemokines, contribute importantly to many of the signs and symptoms of these diseases.

Acute Phase

In IgE-associated immune responses, acute reactions typically occur within minutes of allergen exposure and primarily reflect the secretion of mediators by mast cells and basophils. The rapidity of such acute reactions has raised the problem of how allergens are so rapidly sensed by mast cells, which (unlike circulating basophils) reside outside of the circulation. Recently, it has been shown that a key mechanism accounting for the prompt mast cell activation during anaphylaxis is the rapid transmission of allergens to mast cells within microvesicles derived from perivascular DCs.[372]

However, it has recently become clear that certain IgE-dependent reactions may first occur 3 to 8 hours after consumption of the offending food.[373] Such reactions are produced when animal meats containing galactose-alpha-1,3,-galactose (alpha-gal) are consumed.[373] Notably, alpha-gal is found in all mammals tested, except for humans, apes, and old world monkeys.[374] Such reactions, that have been called alpha-gal allergy or mammalian meat allergy, are induced by the bites of certain ticks (present on six continents) which

TABLE 16.4 Mast Cells and Basophils in Health and Disease

- Mast cells and basophils are primary effector cells in atopic disorders (anaphylaxis, asthma, allergic rhinitis, atopic dermatitis).
- Antigen-specific, immunoglobulin E (IgE)–mediated mast cell activation results in an immediate clinical reaction and, in some cases, contributes to a late-phase reaction.
- Medications that block mast cell and basophil mediators are widely used for treatment of atopic diseases.
- Mast cells and/or basophils contribute to host defense against parasites, bacterial, and viral pathogens, and mast cells participate in defense against the toxicity of venoms (in mice).
- Mast cells and/or basophils may play a role in other human diseases, including certain autoimmune diseases and inflammatory arthritis.

Modified from Boyce J, Fanning L. Mast Cells, Basophils, and Mastocytosis. Clinical Immunology: Principles and Practice. 5th ed. Elsevier; 2019:335-347.

can transfer the alpha-gal carbohydrate to the victim.[373,375] These alpha-gal allergies, which are the first known food-induced allergy associated with a carbohydrate,[373] are also the first known food allergies that present the possibility of significantly delayed anaphylaxis.[373] Yet, for most acute IgE-dependent reactions, including those provoked by many other allergenic foods, there are only seconds to minutes between exposure and the onset of the reaction.

In acute IgE-dependent allergic reactions, cross-linking of adjacent IgE molecules by bivalent or multivalent allergen on mast cells and basophils results in the secretion of three classes of biologically active products: those stored in the cytoplasmic granules, lipid-derived mediators, and newly synthesized cytokines, chemokines, and growth factors, as well as other products. The secretion of preformed mediators occurs when the membrane of the mast cells' cytoplasmic granules fuses with the plasma membrane in a process called anaphylactic or compound degranulation, exposing the granule contents to the external environment. The released mediators include biogenic amines (histamine, serotonin), serglycin proteoglycans (heparin and chondroitin sulfate), serine proteases (such as tryptases and chymases) and carboxypeptidases, and various other enzymes, as well as certain cytokines and growth factors that have been reported to be associated with the granules (such as IL-4, TNF, and VEGFA[20,117]). Mast cells and basophils activated by the aggregation of FcεRI also release lipid-derived mediators (PGD_2), leukotriene B_4 (LTB_4), and cysteinyl leukotrienes (cys-LTs, particularly LTC_4).[1,20]

The sudden, massive release of mediators from basophils and mast cells is thought to provoke many of the clinical manifestations of acute immediate hypersensitivity reactions. Indeed, this is the critical underlying step in the evolution of the acute reactions observed in disorders such as certain forms of bronchial asthma (including fatal asthma, in which basophils can be prominent[376]); urticaria; allergic rhinitis; and anaphylaxis to foods, drugs, insect stings, and other antigens.[1,4,21,48,270,272] The critical signal in this event is the aggregation of FcεRI on the plasma membrane,[377,378] leading to the noncytolytic release of granule contents, such as histamine and other preformed mediators. See Chapter 33, "Fc receptor biology" (by Nimmerjah/Ravetch), and Chapter 47, "Allergic disorders" (by Nadeau/Scott/Milner) for the complex sequence of biochemical events associated with anaphylactic degranulation, the signaling mechanisms that positively and negatively regulate this response, and the rationale for the pharmacologic manipulation of these processes. In mice, both mast cells and basophils can bind IgG antibodies (as well as IgE), and the binding of IgG or IgG immune complexes can contribute to the activation of these cells (in certain mouse models of anaphylaxis)[88,96] or, in other circumstances, to the downregulation of their activation (eg, via FcγRIIB).[96]

Other stimuli, including certain complement fragments (anaphylatoxins), neutrophil lysosomal proteins, a variety of basic peptides and peptide hormones, components of insect or reptile venoms, radiocontrast solutions, cold, calcium ionophores, and certain drugs (such as narcotics and muscle relaxants), also may initiate rapid release of mediators from basophils and/or mast cells, independently of IgE.[20,104] For mast cells[379,380] and human basophils,[216] such signals include those which bind to and activate Mrgprs. Indeed, the clinical reactions provoked by these agents can closely mimic those of immediate hypersensitivity elicited by IgE cross-linking.[380]

Basophils activated via FcεRI or other mechanisms can exhibit increased surface levels of activation markers (eg, CD63, CD69, and CD203c), and the clinical value of using such findings to monitor basophil activation *in vivo* (eg, in the setting of allergic disorders or antigen-specific immunotherapy) is under investigation.[208,210,211]

Late Phase

Late-phase reactions occur when antigen challenge is followed, hours after initial IgE-dependent mast cell activation, by recurrence of signs (eg, cutaneous edema) and symptoms (eg, bronchoconstriction).[186,271] Much of the pathology associated with chronic allergic conditions, including allergic asthma, is widely believed to reflect the actions of leukocytes that are recruited to sites of late-phase reactions.[186,271] Late-phase reactions are thought to be coordinated in part by certain long-term consequences of the mediators released by activated mast cells during early-phase reactions, and in part, by antigen-stimulated T cells.

Mast cells respond to IgE and allergen to release a broad range of newly synthesized cytokines, chemokines, and growth factors, which are released more slowly than the preformed granule-associated mediators. Some mast-cell products have the potential to recruit other immune cells either directly or indirectly, to activate innate immune cells, and to affect many aspects of the biology of DCs, T cells, and B cells.[4,271] Certain mast cell–derived products can also influence the biology of structural cells, including vascular endothelial cells, epithelial cells, fibroblasts, smooth muscle cells, and nerve cells.[4,271] Other products that contribute to late-phase reactions can be derived from T cells that recognize allergen-derived peptides; such T cells may be either resident at or recruited to early-phase reactions at sites of allergen challenge.[271]

Studies in mast cell–deficient and mast cell knock in mice (genetically mast cell–deficient mice that have been selectively repaired of their mast cell deficiency) indicate that mast cells are responsible for virtually all of the vascular permeability changes and leukocyte infiltration associated with IgE-dependent cutaneous late-phase reactions[293,354] and that TNF can importantly contribute to these responses.[293] The extent to which mast cells (or TNF) contribute to late-phase reactions in humans, in which such reactions may have components that are either IgE or T-cell dependent (and in which it has been suggested that certain IgE-dependent mechanisms may not involve mast cells[92]), is not yet clear.[381,382] However, the lymphocytes, basophils, eosinophils, and other leukocytes that are recruited to these reactions likely produce cytokines and other mediators that regulate further development and, ultimately, resolution of these reactions.

Chronic Changes Associated With Allergic inflammation

When allergen exposure is continuous or repetitive, inflammation persists with the infiltration of many innate and adaptive immune cells at sites of allergen challenge. This is associated with changes in the structural cells at the affected sites, and in many cases, with markedly altered function of the affected organs. While acute reactions and late-phase reactions can easily be studied experimentally in humans, most investigations of chronic allergic inflammation involve either biopsy studies of patients with these disorders or experimental animal models of allergic disorders (none of which can be considered identical to the human diseases).

It is therefore not surprising that there is no clear understanding of how, after persistent and/or multiple exposures to allergen, local tissue inflammation changes from a series of early-phase and late-phase reactions to chronic allergic inflammation. However, studies in mice indicate that mast cells can contribute importantly to many of the features of asthma, as observed in certain mouse models of chronic allergic inflammation involving the lungs.[383,384] The features include the development of airway hyperreactivity to immunologically nonspecific agonists of bronchoconstriction (eg, methacholine); infiltration of the airways and lung interstitium with inflammatory cells, including eosinophils, neutrophils, and T cells; increased deposition of lung collagen and hyperplasia/hypertrophy of airway smooth muscle; and induction of increased numbers of mucus-producing goblet cells in the large airways.[271,383,384] Thus, such mouse models indicate that mast cells and their products can promote the development of much of the pathology and pathophysiologic changes observed in long-standing asthma in humans.[271,383,384] Studies in mice indicate that basophils also can enhance the development of IgE-dependent chronic inflammation of the skin, even in the absence of mast cells or T cells.[354]

Roles in Host Defense in Innate and Adaptive Immune Responses

It is well established that mast cells and basophils are major effector cells contributing to the inflammation, pathology, and tissue dysfunction associated with IgE-mediated allergic disorders. Yet, the evolutionary advantages which might be conferred by mast cells and basophils have been more difficult to define. Observational and epidemiological studies in humans,[385,386] as well as studies in experimental animals (including those employing mice genetically deficient in mast cells, basophils, or IgE) (reviewed in Ref. 387), strongly suggest that one beneficial role of IgE, mast cells, and basophils is to help to defend the host against ectoparasites such as ticks and to diminish the numbers of parasites and burden of disease in mammals infected with certain helminths. However, besides parasites, vertebrates also have been subjected to evolutionary pressure through millions of years of coevolution with venomous arthropods, reptiles, and other species. Moreover, there is now evidence that mast cells[143,146,147,388,389] and IgE-associated Th2 immunity[147,148,390] can enhance host defenses against animal venoms (Table 16.4).

As mast cells are particularly abundant along body surfaces near epithelium, they are well positioned to serve the first-line of defense against pathogen invasion and other environmental insults. Mast cells also can help to maintain tissue homeostasis by responding to endogenous danger signals produced by damaged tissues during inflammation. In contrast to circulating basophils and other immune cells, which are recruited to tissues and lymph nodes during inflammation and typically are present only transiently in the affected tissues, mast cells are long-lived tissue-dwelling cells, a property which permits them to participate in different phases of the immune responses, including inflammation resolution and tissue remodeling. In allergy, mast cell and basophil activation leads to pathology and tissue dysfunction. During infections, appropriate activation of mast cells and basophils can be beneficial, as it can enhance immunity and host defense. However, there is also evidence that excessive mast cell and basophil activation after infection can result in detrimental tissue damage. Potential protective versus pathological roles of mast cells in various bacterial,[391] viral,[392] and fungal[393] infections have been reviewed elsewhere. Thus, mast cells and basophils may have complex roles in innate immune responses, with some actions promoting host defense and survival and others enhancing the pathology associated with their activation.

Parasite Infection

A major "beneficial function" of the allergic effector mechanisms involving IgE, mast cells, and basophils is thought to be host defense against helminths and certain other parasites .[387,394-396] Increases in mucosal mast cell numbers and degranulation after helminth infection,[397] and basophil infiltration at sites of cutaneous infection with ticks,[398] have been documented, suggesting that basophils and mast cells may have critical roles in host resistance to certain parasites. However, it has been challenging to prove that mast cells or basophils dramatically influence the survival of parasite-infected animals. Abnormalities in host responses to certain parasites have been observed in mice that genetically lack IgE,[399,400] mast cells,[401-405] or basophils,[400,405] but such studies generally have not included an analysis of the effects of those deficiencies on the overall survival or reproductive success of the infected hosts. And some findings even suggest that, in certain settings, IgE or mast cells may have effects during host responses to parasites (eg, effects which result in increased parasite egg production) that may favor the parasite rather than the host.[406-408]

Nevertheless, a study employing basophil depleting antiserum provided evidence of a beneficial role of basophils in acquired immunity to the feeding of ixodid ticks, *Amblyomma americanum* ticks (the Lone star tick) in guinea pig.[273] Moreover, experiments in mice genetically deficient in mast cells or basophils showed that both mast cells and basophils, as well as IgE, can contribute to acquired immunity to the feeding of *Haemaphysalis longicornis* ticks in mice.[396,398] By contrast, it appears that basophils may have a more important role than mast cells in acquired resistance to the feeding of larval *Dermacentor variabilis* ticks in mice.[396]

On the other hand, mast cells have long been considered major sentinels in host defense against helminth infection.[397,409] Th2 responses and the production of certain specific cytokines (eg, SCF, IL-3, IL-4, IL-9, IL-10, and IL-18) associated with parasite infection can greatly expand intestinal mucosal mast cells. Such mucosal MMC hyperplasia and release of mast cell–specific chymase (in mice, eg, mMCP1),[410] tryptase (in mice, eg, mMCP6),[411] and other mediators contribute critically to host defense against certain intestinal nematodes, such as *T. spiralis*, *S. ratti*, or *S. venezuelensis*.[412-415] In addition, expulsion of *T. spiralis* is delayed by inhibition of mucosal mast cell proliferation.[416,417]

Also, mast cell–deficient and IL-3–deficient mice, which cannot undergo intestinal MMC hyperplasia, exhibit increased susceptibility to *S. venezuelensis* infection.[189,199] Basophils are also markedly increased in the jejunal mucosa during primary infections with *S. venezuelensis* but appear to be responsible for only a small, albeit significant, contribution to *S. venezuelensis* egg clearance.[189] Indeed, studies in mice deficient in various components of immune responses demonstrated the hierarchy of importance of the following immune effector mechanisms in *primary S. venezuelensis* infection: CD4 T cells/ILC2 cells, IgG, and FcRγ > mast cells > IgE and FcεRI > basophils.[189] Notably, in *secondary S. venezuelensis* infection, CD4 T cells are critically important, but mast cells, antibodies, and basophils have few or no nonredundant roles.[189]

The contributions of mast cells and basophils to antiparasite immunity has also been studied in mast cell- or basophil-deficient mice infected with *Strongyloides ratti*.[282,418-420] Depletion of T_{reg} can promote host resistance to infection with *S. ratti* by the increased production of IL-9 and accelerated mast cell degranulation.[282] While such IL-9–mediated mast cell activation is indispensable for the rapid expulsion of *S. ratti* during primary infection in T_{reg}-depleted BALB/c mice, other protective mechanisms can enhance host resistance to primary infection in T_{reg}-depleted C57BL/6 mice[282] and during secondary infection in both BALB/c and C57BL/6 mice.[420] Additional studies in mast cell–deficient mice that lack both mucosal and connective tissue mast cells ($Cpa3^{Cre}$ mice) or lack only connective tissue mast cells ($Mcpt5^{Cre}$ R-DTA mice), and in basophil–deficient $Mcpt8^{Cre}$ mice, showed that mast cells and basophils can contribute to the early expulsion of *S. ratti* adult worms from the intestine but the termination of *S. ratti* infection required mucosal mast cells.[418] While basophils can contribute to protection against primary infection with *S. ratti*,[418] they are not essential for the acquired antihelminth immunity expressed during secondary infection with *S. ratti*.[419]

Such findings support the notion that basophils and mast cells can express similar or complementary functions as effector cells in host defense against parasites and other agents.

Activated mast cells can potentially contribute to antiparasite immunity by multiple mechanisms. Mast cell–derived mediators (eg, proteases) can be toxic to parasites[421] and stimulate intestinal smooth muscle contraction, hastening worm expulsion.[422-424] They also can enhance mucosal permeability,[425,426] promoting the influx of antibodies into the gut lumen, thus contributing to a hostile environment for the parasites. Mast cells and mast cell–derived cytokines/chemokines/mediators also can modulate host-pathogen interactions by influencing the recruitment or function of other innate immune cells, and this could have positive or negative effects on the parasites.[387] Furthermore, mast cells have the potential to shape parasite-associated adaptive immune responses via production of soluble mediators or via cell-cell interactions with DCs and other antigen-presenting cells[103,409,427] (Fig. 16.8).

As discussed above, mast cell activation is likely to contribute to intestinal worm expulsion, at least in some settings. However, because host-defense mechanisms often employ redundant or partially overlapping elements, it has been

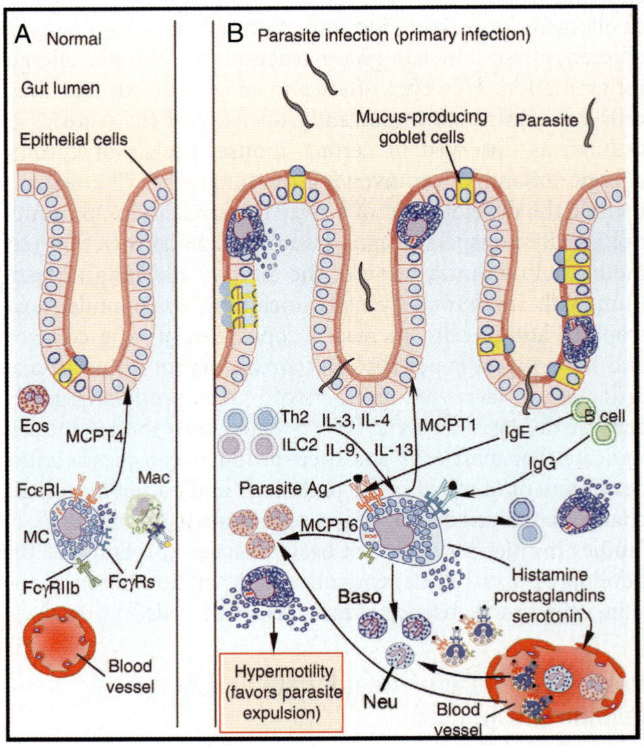

FIG. 16.8. The potential roles of mast cells in parasite infections. A: In the normal intestine, mast cells can contribute to the homeostatic regulation of the epithelial barrier through chymase (MCPT4, *a.k.a* mMCP4)-dependent mechanisms and few mast cells are present within the epithelium. **B:** During primary infections with parasites, the presence of IL-4 or IL-13, which may be derived from a variety of potential cellular sources, induces T cells to become Th2 cells. Th2 cells and ILC2 cells release IL-3 and IL-9, which promote the expansion of mucosal mast cells (MMCs), and some of these MMCs are found in the intestinal epithelium. Parasite antigen (Ag)-induced aggregation of IgE-bound FcεRI causes the prompt release of prestored mast cell mediators, including histamine, that can promote vasodilatation and increased vascular permeability. FcεRI activation also induces the production and release of *de novo*-synthesized compounds, such as leukotrienes, prostaglandins, and proinflammatory cytokines (such as IL-13) and chemokines. Such mast cell–derived products contribute to intestinal inflammation (including the recruitment and activation of neutrophils, basophils, and eosinophils and other leukocytes), increased intestinal permeability and motility, and worm expulsion. IgG–Ag immune complexes can potentially modulate mast cell activation through Fcγ receptors. Baso, basophils; Eos, eosinophils; FcεRI, high-affinity receptor for IgE; FcγRs: receptors for IgGs; IL, interleukin; ILC2, innate lymphoid cells type 2; Neu, neutrophils; Th2, T helper 2. (Reprinted by permission from Springer Nature: Reber LL, Sibilano R, Mukai K, Galli SJ. Potential effector and immunoregulatory functions of mast cells in mucosal immunity. *Mucosal Immunol.* 2015;8:444-463.)

challenging to identify conclusive evidence that mast cells, and especially IgE-dependent mast cell activation, confer host survival benefit during parasite infection. In the case of infection with *N. brasiliensis*, basophils appear to play a more critical role than mast cells in host resistance in mice.[400,405,428] Notably, it appears that basophils apparently do not contribute to intestinal expulsion of *N. brasiliensis*.[362] Instead, by promoting the generation of M2-type macrophages, they help to confine the worm larvae in the skin, therefore limiting larval migration and lung injury.[362] Depending on the basophil-deficient models used to analyze the responses, basophils either have been shown to promote defense against *N. brasiliensis*, especially during the secondary infection,[405,428] or to have no role in worm clearance or tissue eosinophilia during primary or secondary infections.[21]

Consistent with findings in c-*kit* mutant mast cell–deficient WBB6F$_1$-*Kit*$^{W/W-v}$ mice,[401,429] the contribution of mast cells to the immune response to *S. ratti* infection has recently been confirmed using c-*kit*–independent, mast cell–deficient mice.[282,418] However, findings in various mast cell–deficient models do not always yield consistent conclusions about the contributions of mast cells to other examples of parasite resistance (reviewed in Refs. 275-277). This may reflect differences in the other phenotypic abnormalities which occur in c-*kit* mutant versus alternative types of mast cell–deficient mice, differences in experimental design, and/or differences related to the animal facilities used for such work (eg, differences in mouse microbiomes) (reviewed in Refs. 275-277). For example, studies in c-*kit*–dependent or independent mast cell–deficient mice suggested that mast cells either make no contribution,[430,431] had effects that increased skin lesion size,[432] or enhanced protection against *Leishmania major*.[404] Furthermore, mast cell activation after parasite infection can have a negative impact on the host, as mast cell mediators can potentially cause pathology.[410] In some settings, mast cells can even have effects which promote parasite-induced immunosuppression[433] or otherwise favor the parasite.[406,408]

In light of the long-term coevolution of parasites and their hosts, with each attempting to probe weaknesses in or to coopt the defense mechanisms of the other, it should not be surprising that the engagement of mast cells and IgE during host/parasite interactions may not always produce results in favor of the host. Since the parasites' objectives may better be served by a viable rather than a dead host, it is not surprising that, depending on the parasite and the particular setting, effector mechanisms such as IgE and mast cells may be exploited by the parasites to their own advantage. For example, if the parasite elicits a Th2 response (that results in IgE-dependent mast cell activation and release of vasoactive mediators in response to parasite antigens at sites of infection), the parasite could regulate local blood flow and vascular permeability in ways that can help to enhance its ability to derive nutrition from the host.

Taken together, work in models of infection using helminths, Ixodid ticks, and other parasites suggests that there is redundancy in the protective mechanisms which can help to guard against the invasion or (for ticks and probably also other exoparasites) the feeding of these pathogens, the detrimental effects of the pathology they induce, or their ultimate clearance by the host. Whether basophils, mast cells, or both represent major effector cell types in these responses appears to vary according to factors such as species of parasite, species of host, and site of infection. Indeed, it is difficult to extend the results obtained in the analysis of the roles of IgE and mast cells (and basophils) in parasite immunity using experimental models in mice directly to humans. For example, humans exhibit a broader distribution of FcεRI among hematopoietic cells types than is observed in mice.[434] FcεRIs are primarily expressed by mast cells and basophils in mice under baseline conditions, whereas FcεRIs are detected in human eosinophils and platelets, which have been reported to be able to participate in parasite killing by IgE-mediated antibody-dependent cell-mediated cytotoxicity, a mechanism of defense against parasites that has not yet been reported in rodents.[435-438] Moreover, there are potentially important differences in the phenotype of mast cell populations in mice and humans.[439]

Bacterial Infection

Studies in mast cell–deficient and "mast cell knockin mice," or in mice that lack basophils, or certain mast cell- and basophil-associated products indicate that mast cells[103,275,277] or basophils[250] can contribute to innate host defense against some experimental bacterial infections. However, in other settings, mast cells may produce harmful levels of TNF and worsen survival.[440]

Depending on the mouse model system tested, the beneficial role of mast cells in models of innate immunity may partly reflect complement-dependent,[441] TLR 4-dependent,[442] ET1-dependent,[113] or neurotensin-dependent[443] activation of mast cells. Such mechanisms, by inducing the release of mast cell–derived mediators, in turn can contribute to enhanced local recruitment or activation of neutrophils and enhanced clearance of bacteria.[441,442] This also can result in the mast cell protease-dependent degradation of ET-1,[113,388] neurotensin[443], and perhaps other endogenous peptides that are produced during infections and which contribute to the associated pathology.[277] Studies in mice also indicate that mast cells can phagocytose bacteria,[103] and work in both mouse and human mast cells identifies them as sources of antimicrobial peptides (such as the cathelicidin, LL-37, in human mast cells[444] and lipocalin 2 in human and mouse mast cells[445]).

Topical treatment with mastoparan, a peptide toxin from wasp venom that can bind to MRGPRX2 in skin mast cells, has been shown to enhance the clearance of *Staphylococcus aureus* infection and accelerate the healing of lesions in infected mice.[446] Some secreted products of bacteria, such as Gram-positive bacterial quorum sensing molecules, can trigger Mrgprb2-and MRGPRX2-dependent mouse and human mast cell degranulation, resulting in direct killing of bacteria and prevention of biofilm formation.[447]

On the other hand, some mast cell functions that may be expressed during bacterial infections in mice may have detrimental consequences, such as the ability of mast cells to degrade IL-6 via dipeptidyl peptidase 1[448] or to produce IL-10 that can suppress acquired immunity.[449] Mast cell responses to bacterial products also can have detrimental effects in allergic diseases. *S. aureus* colonizes the skin of >90% of patients with atopic dermatitis. Mast cell degranulation in response to δ-toxin derived from *S. aureus* can exacerbate allergic skin inflammation in mice.[450]

Basophils can be activated by bacterial products via several mechanisms. Bacteria-derived fMLP binding to FPRs rapidly degranulates basophils and induces their chemotaxis,[233] but binding of *Helicobacter pylori*–derived peptides to FPR can induce chemotaxis without histamine release from human basophils.[451] Ligands for the intracellular PRR, NOD2, and various TLR receptors can also trigger basophil activation during bacterial infection.[234,452,453]

Also, human basophils and mast cells bind IgD, which can be produced during respiratory bacterial infections.[101] Anti-IgD cross-linking of basophils induces the production of BAFF, antimicrobial peptides, IL-1β, and TNF, suggesting a role for basophils in immune surveillance against bacterial infection.[101] In mice, basophil-derived TNF can confer survival benefit in the cecal ligation and puncture model of bacterial infection.[250]

Basophils and basophil IL-4 have been reported to contribute to liver homeostasis in a mouse model of *Listeria monocytogenes* infection.[454] In that model, IL-33 from infected hepatocytes induced IL-4 production from infiltrating basophils in the liver.[454] Basophil-derived IL-4 then recruits inflammatory monocytes and promotes their differentiation into alternatively activated macrophages that in turn replace dead hepatocytes and help in restoring liver homeostasis.[454] Human and mouse basophils also can exhibit antibacterial activity by forming extracellular DNA traps which contain mitochondrial DNA and granule proteins.[455]

Viral Infection

Mast cells and basophils can be directly infected and activated by viruses. While activation of these innate immune cells by viruses and viral products can enhance antivirus immunity, the excessive release of inflammatory mediators from these cells also can exacerbate pathological complications associated with the infection.[106,456] Some viruses such as Harvey sarcoma virus, Kirsten sarcoma virus, and Abelson murine leukemia virus can immortalize mast cells; by contrast, other (dengue) virus infections can induce mast cell apoptosis.[457]

Dengue virus infections are prevalent and are usually transmitted by the bites of mosquitoes and other insects. An extensive literature focuses on the protective versus harmful roles of mast cells in dengue virus infections.[106,456,458] These studies show that infection with dengue virus is associated with mast cell degranulation, apoptosis, and release of chemokines and cytokines.[459,460] Inflammation induced by the mast cells' responses to mosquito saliva has been proposed as constituting the first line of defense against this infection.[461] Mast cells are activated by dengue virus via their intracellular viral sensors, TLR3, melanoma differentiation associated protein (MDA) 5, and RIG-1.[110] The infection provokes secretion of mast cell granule-associated proteases, serotonin, prostaglandins, leukotrienes, cytokines (TNF, IFNα), and chemokines (CCL5, CXCL12, CX3CL1). These mast cell mediators increase vascular permeability and migration of inflammatory cells, including recruitment of NK, NKT, and γδ T cells to infected sites, that are thought to confine and clear the viruses.[110,462]

While local mast cell activation appears to be important for containment of the virus, systemic mast cell activation can be detrimental to hosts infected with dengue virus. During secondary infections, systemic mast cell activation can occur, possibly mediated by antibody/FcγRII- and complement/GPCR-dependent mechanisms,[459] and can contribute to dengue virus replication and systemic dengue disease severity. Some mast cell products have been associated with increased severity of various aspects of dengue-induced disease, including thrombocytopenia (serotonin)[463] and/or disruption of vascular integrity (VEGF, chymase, tryptase, leukotrienes).[464,465] Furthermore, human skin mast cells can be directly infected with dengue viruses, and mast cell granules containing the virus can spread experimental infections from the initial peripheral mosquito injection site.[460]

The roles of mast cells in viral immunity also have been investigated with other viruses, including respiratory syncytial virus, rhinovirus, reovirus, Sendai virus, Sindbis virus, HIV, Vaccinia virus, and influenza viruses (reviewed in Refs. 106,456). The beneficial versus detrimental effects of mast cells that have been observed in dengue infection may also occur during other viral infections. Moreover, the extent to which basophils may contribute to host defense or pathology during such viral infections is not clear.

For example, both human mast cell progenitors[466-468] and basophils[469] can become infected with "M tropic" strains of HIV. Although mature mast cells appear to be resistant to such infection, mast cells which matured from infected progenitors while harboring latent infection exhibited enhanced viral replication upon stimulation with ligands for TLR2, TLR4, or TLR9[467,468] or via an IgE-dependent mechanism.[468] At least one HIV-derived protein, gp120, can induce mast cell or basophil mediator release (histamine and, in basophils, IL-4 and IL-13) by binding to and cross-linking cell-surface–bound IgE.[469] Many patients infected with HIV develop high levels of IgE and can exhibit exacerbation of the symptoms and signs of their allergic disorders.[469] However, the clinical significance of such findings in HIV-infected patients remains to be determined.

Venom Toxicity

In contrast to parasites, bacteria, and viruses, most allergens do not pose a direct threat to the nonsensitized host. This is why type 2 immune responses associated with allergen exposures are widely considered to be "misdirected" or "maladaptive" immune responses. However, in 1974, Stebbings suggested that IgE-dependent "allergic" responses to insect bites and stings could confer immediate advantages to hosts, causing them to flee the attacks of such creatures.[470] In 1991, Margie Profet proposed a broader concept: that acute allergic reactions, such as sneezing, coughing, vomiting, and diarrhea, evolved as protective mechanisms allowing the sensitized host to respond rapidly to, and to expel, neutralize, and/or avoid diverse noxious substances which might be potentially life-threatening.[471] Profet's "toxin hypothesis" has recently been tested, and supported, by experiments in mice sensitized and challenged with animal venoms.[147,148]

The venoms of many animals contain substances that can activate mast cells via innate mechanisms and/or can induce specific IgE responses to venom components.[146-148,390] Many humans develop IgE antibodies against components of honeybee or wasp venom but only a small fraction of such "venom-sensitized" individuals have a history of anaphylaxis or other serious clinical response to such venoms.[472] Early work by Higginbotham and colleagues suggested that mast cells might be able to diminish the toxicity of certain venoms by degranulating and releasing heparin in response to venom exposure.[473,474]

Recent work in mast cell–deficient, mast cell knockin, and mast cell protease-deficient mice indicates that mast cells indeed can enhance the innate resistance of mice to diverse animal venoms and/or their toxic components. These include the whole venoms of four poisonous snakes (the Israeli mole viper [*Atractaspis engaddensis*], two North American pit vipers, the western diamondback rattlesnake [*Crotalus atrox*] and the southern copperhead [*Agkistrodon contortrix contortrix*][146] and Russel's viper [*Daboia russelii*][147,148]), the honey bee,[146,147] the Gila monster,[143] and two species of scorpions,[143] as well as individual venom toxins, sarafotoxin 6b (the major toxin in Israeli mole viper snake venom[388]) and helodermin (a toxin in Gila monster venom[143]).

Depending on the mammalian species, mast cells can contain several tryptases and chymases of distinct substrate specificity, as well as CPA3.[177,277,475] Notably, CPA3 or mMCP4 (the functional counterpart in the mouse to human chymase) appeared to account for much or all of the protective effects against various venoms that were attributable to mast cells.[143,146,147,388] CPA3 and mMCP4 can degrade both endogenous biologically active peptides (ET1[146,388] and VIP,[143] respectively) and similar peptides present in animal venoms (sarafotoxin 6b and helodermin, respectively), which are thought to act in mammals via the same receptors which bind the similar endogenous peptides.

By undergoing degranulation and releasing proteases that can inactivate potentially toxic endogenous peptides or peptides in venoms, mast cells can help to restore homeostasis, albeit while also enhancing features of the ensuing local and perhaps systemic inflammatory responses. This raises the possibility that one of the reasons that mast cells of various species contain several different proteases in their cytoplasmic granules is so that these cells, which are positioned in large numbers in the skin, the most common site of envenomation, are equipped to release a panel of proteases with the potential to degrade a variety of structurally distinct toxic compounds contained in animal venoms. Indeed, a study employing embryonic zebra fish to screen for venom toxicity shows that human mast cell tryptase, rather than human mast cell chymase or CPA3, is more capable of degrading several different snake venoms.[389] Mast cells might also contribute to resistance to venoms in other ways such as by increasing local vascular permeability and thereby favoring the interstitial access of circulating molecules that can antagonize the effects of venom and other toxins proteases.

Work in mice deficient in mast cells, IgE, or components of FcεRI suggests that mast cells also can contribute to the IgE antibody- and FcεRI-dependent enhanced resistance to challenge with potentially lethal amounts of honeybee[147] or *Daboia russelii* (Russell's viper)[148] venom that is observed in animals after an initial exposure to a sublethal amount of that venom (Fig. 16.9). Indeed, it has been hypothesized that the ability to participate in innate and acquired immune defenses against components of venoms and other toxins is an important protective function of mast cells and, in the case of acquired immunity, of IgE antibodies.[146,148,390,470,471,476]

Roles in Other Immunological Disorders

Increases in the numbers and/or the activation of mast cells and basophils at affected tissues have been documented in a variety of immunologic and pathological responses in both humans (Tables 16.2 and 16.3) and experimental animals.[1,21,48,186,270,272,275-277] However, despite years of study, the importance of the contributions of mast cells and basophils in many of these settings is still under debate. In part, this may reflect differences in the types of mast cell–deficient or basophil–deficient mice used to investigate such responses and/or the strain background of such mice,[281-283] and in part, this may reflect differences in the details of the experimental models and gender of the animals[477] used to probe the roles of mast cells and basophils in these settings and/or the species differences between mice and humans. For example, in some of the tested mouse models of T cell–dependent responses, authors have concluded that mast cell can either enhance or suppress features of the responses, and in some settings, evidence has been reported that mast cells can both *enhance* features of relatively weak responses and also *suppress* features of strong res ponses.[277,279,337]

Both mast cells and basophils have been implicated in autoimmune diseases. Circulating autoreactive IgEs that bind directly or indirectly to nucleic acids are highly prevalent in patients with autoimmune disorders such as SLE, bullous pemphigoid (BP), and CSU.[478] The importance of IgE in some of these settings is supported by the clinical efficacy of anti-IgE (omalizumab) treatments in patients with BP or CSU.[478] It is therefore reasonable to speculate that IgE-mediated activation of mast cells and basophils can contribute to the pathogenesis of these autoimmune disorders.

Systemic Lupus Erythematosus

Experiments in mast cell–deficient $Kit^{W/W-v}$ mice supported the idea that antibody-mediated mast cell activation can contribute to neutrophil recruitment in experimental BP.[479] In the case of SLE, depletion of mast cells in SLE-prone mice suggested that mast cells might be protective in this disease,[480] but the exact contribution of mast cells to the pathogenesis of SLE is not well characterized. By contrast, basophils and IgE have been strongly implicated in the pathogenesis of this systemic autoimmune disease.[352,478,481,482] This work included analyses of basophils in SLE patients, as well as in mice susceptible for

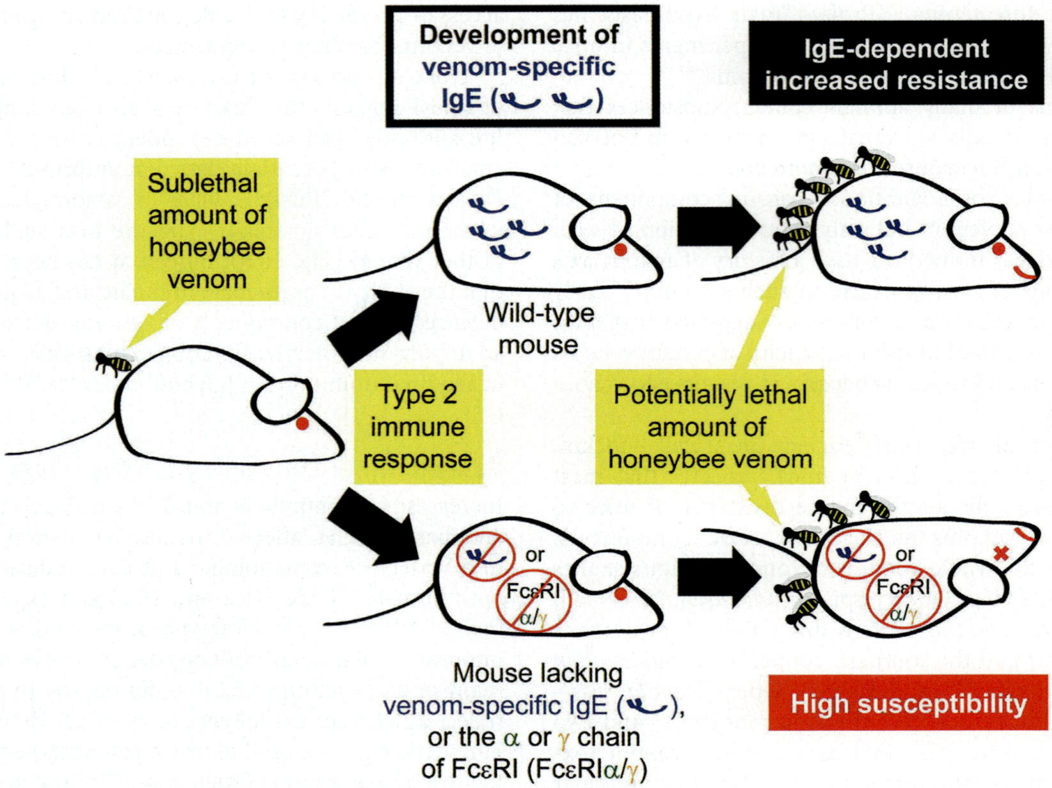

FIG. 16.9. Mast cells, IgE, and high-affinity IgE receptors (FcεRI) protect against lethal effects of honeybee venom in mice. (Reprinted from Marichal T, Starkl P, Reber LL, Kalesnikoff J, Oettgen HC, Tsai M, Metz M, Galli SJ. A beneficial role for immunoglobulin E in host defense against honeybee venom. *Immunity.* 2013;39:963-975, with permission from Elsevier.)

developing lupus-like disease.[352,478,482] SLE patients exhibited basophil infiltration in their lymph nodes and spleens, and basophils from these patients exhibited an activated phenotype and could be activated by a common SLE autoantigen.[478]

The mechanistic contribution of basophils in this disease was investigated by studies conducted in mouse models of a spontaneous lupus-like disease. In SLE-susceptible mice, autoreactive IgE activated basophils and induced their migration to lymph nodes, where basophils promoted Th2 cell differentiation and enhanced the production of self-reactive antibodies.[352] Moreover, basophil-depletion in susceptible mice reduced disease severity, while adoptive transfer of basophils aggravated the disease.[482] It is likely that basophils, by providing IL-4, BAFF, and other mediators in such settings, contribute to the production of autoantibodies, including self-reactive IgE, which in turn can activate basophils for more IL-4 production, thereby amplifying autoantibody production by B cells and disease progression.[352]

Inflammatory Arthritis

In contrast to SLE, evidence more strongly supports a role for mast cells, rather than basophils, in rheumatoid arthritis and other inflammatory joint diseases.[483] Increased mast cell numbers and activation are noted in the synovial tissue of rheumatic arthritis (RA),[484] osteoarthritis (OA),[485] spondyloarthritis,[486,487] and gouty arthritis.[488,489] In principle, many inflammatory stimuli in synovial tissues can potentially activate mast cells to release immunoregulatory mediators that can modulate the inflammation, cartilage/bone destruction, and resolution of such joint diseases. Thus, the contributions of mast cells in these diseases can be either protective or detrimental.

Mast cells are normally present in healthy synovial membranes, but their number significantly increases in RA synovitis.[490] Anti-inflammatory actions of mast cells in RA have been suggested,[340,491] based on the observations that serum levels of mast cell–derived tryptase are negatively correlated with C-reactive protein in early RA[491] and that human mast cell production of histamine and IL-10 after activation triggered by IL-33 and IgG can suppress monocyte activation, suggesting that synovial mast cell-monocyte interactions can inhibit the pathogenesis of RA.[340] On the other hand, mast cells have been associated with disease severity, and contributions to B cell autoantibody production, in early rheumatoid arthritis.[490]

The findings attributed to the roles of mast cells in experimental arthritis using mast cell–deficient mice have not been consistent.[492-496] This probably is due to phenotypic abnormalities unrelated to the mast cell deficiency in some of these mast cell–deficient mice, different experimental protocols that have been used to model human RA, and different stages of the disease that were analyzed in particular experiments. Thus, animal studies suggest that the contribution of mast cells to "models" of RA may be different at various disease stages, but exactly how mast cells contribute to RA in humans remains to be fully clarified.

Mast cells are aberrantly activated in the joint tissue of OA in humans and mice, and evidence for a central role for IgE-dependent mast cell activation in the pathogenesis of OA has recently been reported.[485] This study showed that a deficiency of mast cells (based on both c-*kit*–dependent and c-*kit*–independent mechanisms) alleviates a model of OA in mice. It also provided evidence that human mast cell–derived tryptase can induce inflammation, death of chondrocytes, and cartilage destruction, and that the IgE/FcεRI/Syk signaling axis is involved in the development of human OA.

Relatively little has been done to investigate the role of mast cells and their products in gouty arthritis. However, in a study of uric acid crystal–induced acute inflammation in the joints of mice, selective depletion of synovial mast cells was found to diminish the tissue swelling induced by uric acid crystals.[489] Moreover, evidence was reported that mast cell IL-1β production contributed to the development of early inflammation in this model, suggesting that mast cell cytokine secretion may play a role in gouty arthritis.[489]

Multiple Sclerosis /Experimental Autoimmune Encephalomyelitis

The role of mast cells in multiple sclerosis (MS), and particularly in mouse "models" of that disorder, is unsettled. Human MS is a T cell–mediated central nervous system (CNS) autoimmune disease. Mast cells are present and positive for FcεRI, around MS lesions.[497] Mast cells and mast cell–derived mediators and cytokines, such as histamine, tryptase, and TNF, can enhance vascular permeability and upregulate expression of vascular endothelial cell–associated molecules and have been implicated in the breakdown of the blood–brain barrier, allowing access of blood leukocytes into the brain parenchyma.[498] Moreover, mast cells secrete many immunomodulatory products which can directly induce migration and functional activation of other innate and adaptive immune cells involved in pathogenesis of this disease.[499] Direct myelinolysis by mast cell proteases has also been demonstrated, and mast cell inflammasome activity has been implicated in the pathogenesis of mouse experimental autoimmune encephalomyelitis (EAE).[500] On the other hand, activation of mast cells can exert a neuroprotective effect. For example, NGF can induce angiogenin production by human mast cells, and angiogenin can promote the survival and neuritogenesis of motor neurons.[499]

While these observations suggest that mast cells may play role(s) in the pathogenesis of MS, experiments in a mouse model of MS, EAE, have not yielded consistent findings supporting a role of mast cells in EAE. Early studies using mast cell–deficient $Kit^{W/W-v}$ mice engrafted with FcγRIII−/− or FcRγ−/− mast cells indicated that antigen- and antibody-dependent mast cell activation and mediator release can contribute to the expression of several features of these disorders.[501] While evidence for a detrimental role of mast cells in EAE has been independently confirmed in $Kit^{W-sh/W-sh}$ mice,[502] other studies reported that c-*kit* mutant $Kit^{W/W-v}$ and $Kit^{W-sh/W-sh}$, as well as c-*kit*–independent mast cell–deficient $Cpa3^{cre}$ mice, were fully susceptible to induction of EAE.[495] $Kit^{W-sh/W-sh}$ mice can even develop earlier and more severe EAE in comparison to the corresponding Kit+/+ mice under some experimental conditions.[503]

In addition to EAE and RA, disparate roles of mast cells have been reported in other experimental disease models, such as CHS and mouse models of asthma. Several factors could explain the experimental disparities in these disease models. As discussed, mast cells can reside in tissues for long periods of time and are likely to participate in different stages of the disease processes, from initiation to resolution. Functional plasticity of mast cells permits them to exert proinflammatory, as well as anti-inflammatory, activities, depending on the context and stage of the process. Except in the case of IgE-dependent responses (in which the role of mast cells is essentially always important), when experimental protocols induce strong reactions, the role of mast cells may be masked, as illustrated by some experiments in mouse models of asthma[504] and CHS.[279,337] Indeed, in some settings, mouse mast cells can actually downregulate the intensity of strong reactions to maintain tissue homeostasis and contribute to health.[336-338,341]

Finally, several factors can influence the outcome of mouse experiments, including housing conditions, strain background, and ages of the mice (and perhaps the composition of the animals' microbiota), as well as the strength of the immunization protocols used to induce the disease. Gender also can contribute to outcomes of the experiments that interrogate mast cell function. For example, recent data in an EAE model demonstrate that IL-33 is selectively induced in male mice and can activate ILC2s to promote and sustain a protective Th2 myelin-specific response.[505] In that setting, mast cells in male mice produce IL-33 induced by testosterone, and such mast cell–dependent IL-33 expression is protective against the development of a destructive immune response in male mice. By contrast, a deficiency in the IL-33 response in females generates a propathogenic Th17-dominant response and increases disease susceptibility.[505]

CONCLUSIONS

While mammalian mast cells and basophils share several notable biochemical and functional characteristics, they are not identical. In humans, basophils are the least frequent of the three granulocytes, typically accounting for <0.5% of blood leukocytes. Basophils circulate as mature cells and can be recruited into tissues, particularly at sites of immunologic or inflammatory responses, but they ordinarily do not reside in tissues. By contrast, mast cells are derived from yolk sac or blood precursors that lack many of the characteristic features of the mature cells and complete their maturation in the tissues.

Mature mast cells can reside in tissues for long periods of time and are particularly abundant near blood vessels and nerves and in connective tissues beneath surfaces that are exposed to the external environment, such as the skin, gastrointestinal and urogenital tracts, and respiratory system. Tissue mast cell numbers can increase at sites of parasite infection or in association with certain chronic allergic diseases or other forms of pathology. Mast cell expansion can occur by recruitment and local maturation of blood precursors and by proliferation of resident mast cells.

Mast cells and basophils express the high-affinity receptor for IgE (ie, the FcεRI) on their surface. Both cell types can be triggered to release potent mediators in response to activation via FcεRI, for example, when their cell-bound IgE recognizes bivalent or multivalent allergens. Accordingly, mast cells and basophils have long been regarded as important effector cells in allergic asthma, food allergies, hay fever, and other allergic disorders. It is thought that the cells' cytoplasmic granule-associated preformed mediators, including histamine and certain proteases, their lipid mediators (such as PGD_2 and leukotriene C_4), which are generated upon activation of the cells, and their cytokines, growth factors, and chemokines, contribute to many of the signs and symptoms of these diseases.

However, several lines of evidence indicate mast cells and basophils also can contribute to protective host responses associated with IgE production, especially those directed against certain parasites. Moreover, in mice, mast cells can enhance survival during innate and acquired (IgE-dependent) defense against some animal venoms and also can contribute to host defense in innate immune responses to certain bacterial infections. Mast cells and basophils also may express positive and negative immunoregulatory functions through cytokine production and other mechanisms. Indeed, many systemic disorders have been associated with changes in the numbers of blood basophils and many pathologic processes can be associated with changes in the numbers of tissue mast cells, but the exact nonredundant functions of mast cells and basophils in these conditions largely remains to be determined. Notably, patients with primary deficiencies in basophils appear to be exceedingly rare (if they exist at all). Patients with a primary deficiency of tissue mast cells have not been reported.

ACKNOWLEDGMENTS

We are grateful to all those whose work has been cited, and to all who have contributed to this field. The authors' studies are supported by NIH grants (U19 AI104209, R01 AR067145, R01 AI132494, and R01 AI125567), the United States-Israel Binational Science Foundation (Grant 2017182), and the Department of Pathology, Stanford University School of Medicine.

REFERENCES

1. Voehringer D. Protective and pathological roles of mast cells and basophils. *Nat Rev Immunol.* 2013;13(5):362-375.
2. Karasuyama H, Miyake K, Yoshikawa S, et al. Multifaceted roles of basophils in health and disease. *J Allergy Clin Immunol.* 2018;142(2):370-380.
3. Galli SJ, Metcalfe DD, Arber DA. Basophils and mast cells and their disorders. In: Kaushansky K, Lichtman MA, Beutler E, Kipps TJ, Prchal JT, eds. *Williams Hematology.* 10th ed. McGraw-Hill Medical Education; 2021:1093.
4. Galli SJ, Grimbaldeston M, Tsai M. Immunomodulatory mast cells: negative, as well as positive, regulators of immunity. *Nat Rev Immunol.* 2008;8(6):478-486.
5. Gurish MF, Austen KF. Developmental origin and functional specialization of mast cell subsets. *Immunity.* 2012;37(1):25-33.
6. Dahlin JS, Hallgren J. Mast cell progenitors: origin, development and migration to tissues. *Mol Immunol.* 2015;63(1):9-17.
7. Grootens J, Ungerstedt JS, Nilsson G, et al. Deciphering the differentiation trajectory from hematopoietic stem cells to mast cells. *Blood Adv.* 2018;2(17):2273-2281.
8. Kitamura Y. Heterogeneity of mast cells and phenotypic change between subpopulations. *Annu Rev Immunol.* 1989;7:59-76.
9. Jamur MC, Moreno AN, Mello LF, et al. Mast cell repopulation of the peritoneal cavity: contribution of mast cell progenitors versus bone marrow derived committed mast cell precursors. *BMC Immunol.* 2010;11:32.
10. Crivellato E, Nico B, Ribatti D. The history of the controversial relationship between mast cells and basophils. *Immunol Lett.* 2011;141(1):10-17.
11. Arinobu Y, Iwasaki H, Gurish MF, et al. Developmental checkpoints of the basophil/mast cell lineages in adult murine hematopoiesis. *Proc Natl Acad Sci U S A.* 2005;102(50):18105-18110.
12. Chen CC, Grimbaldeston MA, Tsai M, et al. Identification of mast cell progenitors in adult mice. *Proc Natl Acad Sci U S A.* 2005;102(32):11408-11413.
13. Franco CB, Chen CC, Drukker M, et al. Distinguishing mast cell and granulocyte differentiation at the single-cell level. *Cell Stem Cell.* 2010;6(4):361-368.
14. Mukai K, BenBarak MJ, Tachibana M, et al. Critical role of P1-Runx1 in mouse basophil development. *Blood.* 2012;120(1):76-85.
15. Metcalf D, Ng AP, Baldwin TM, et al. Concordant mast cell and basophil production by individual hematopoietic blast colony-forming cells. *Proc Natl Acad Sci U S A.* 2013;110(22):9031-9035.
16. Qi X, Hong J, Chaves L, et al. Antagonistic regulation by the transcription factors C/EBPalpha and MITF specifies basophil and mast cell fates. *Immunity.* 2013;39(1):97-110.
17. Sasaki H, Kurotaki D, Osato N, et al. Transcription factor IRF8 plays a critical role in the development of murine basophils and mast cells. *Blood.* 2015;125(2):358-369.
18. Huang H, Li Y, Liu B. Transcriptional regulation of mast cell and basophil lineage commitment. *Semin Immunopathol.* 2016;38(5):539-548.
19. Marone G, Borriello F, Varricchi G, et al. Basophils: historical reflections and perspectives. *Chem Immunol Allergy.* 2014;100:172-192.
20. Varricchi G, Raap U, Rivellese F, et al. Human mast cells and basophils-How are they similar how are they different?. *Immunol Rev.* 2018;282(1):8-34.
21. Sullivan BM, Liang HE, Bando JK, et al. Genetic analysis of basophil function *in vivo*. *Nat Immunol.* 2011;12(6):527-535.
22. Pellefigues C, Mehta P, Frout MS, et al. The Basoph8 mice enable an unbiased detection and a conditional depletion of basophils. *Front Immunol.* 2019;10:2143.
23. Baccari GC, Pinelli C, Santillo A, et al. Mast cells in nonmammalian vertebrates: an overview. *Int Rev Cell Mol Biol.* 2011;290:1-53.
24. Cavalcante MC, Allodi S, Valente AP, et al. Occurrence of heparin in the invertebrate styela plicata (Tunicata) is restricted to cell layers facing the outside environment. An ancient role in defense? *J Biol Chem.* 2000;275(46):36189-36196.
25. Cavalcante MC, de Andrade LR, Du Bocage Santos-Pinto C, et al. Colocalization of heparin and histamine in the intracellular granules of test cells from the invertebrate Styela plicata (Chordata-Tunicata). *J Struct Biol.* 2002;137(3):313-321.
26. de Barros CM, Andrade LR, Allodi S, et al. The Hemolymph of the ascidian Styela plicata (Chordata-Tunicata) contains heparin inside basophil-like cells and a unique sulfated galactoglucan in the plasma. *J Biol Chem.* 2007;282(3):1615-1626.
27. Crivellato E, Ribatti D. The mast cell: an evolutionary perspective. *Biol Rev Camb Philos Soc.* 2010;85(2):347-360.
28. Wong GW, Zhuo L, Kimata K, et al. Ancient origin of mast cells. *Biochem Biophys Res Commun.* 2014;451(2):314-318.
29. Hellman LT, Akula S, Thorpe M, et al. Tracing the origins of IgE, mast cells, and allergies by studies of wild animals. *Front Immunol.* 2017;8:1749.
30. Kashem SW, Subramanian H, Collington SJ, et al. G protein coupled receptor specificity for C3a and compound 48/80-induced degranulation in human mast cells: roles of Mas-related genes MrgX1 and MrgX2. *Eur J Pharmacol.* 2011;668(1-2):299-304.
31. Johnson-Weaver B, Choi HW, Abraham SN, et al. Mast cell activators as novel immune regulators. *Curr Opin Pharmacol.* 2018;41:89-95.
32. Selye H. *The Mast Cells.* Butterworth; 1965.
33. Silphaduang U, Noga EJ. Peptide antibiotics in mast cells of fish. *Nature.* 2001;414(6861):268-269.
34. Canfield PJ. Comparative cell morphology in the peripheral blood film from exotic and native animals. *Aust Vet J.* 1998;76(12):793-800.
35. Mead KF, Borysenko M, Findlay SR. Naturally abundant basophils in the snapping turtle, Chelydra serpentina, possess cytophilic surface antibody with reaginic function. *J Immunol.* 1983;130(1):334-340.
36. Metz M, Brockow K, Metcalfe D, et al. Mast cells, basophils and mastocytosis. In: Robert RR, Thomas AF, Shearer WT, Schroeder HW Jr, Frew AJ, Weyand CM, eds. *Clinical Immunology: Principles and Practice.* Elsevier Saunders; 2013:285.
37. Rodewald HR, Dessing M, Dvorak AM, et al. Identification of a committed precursor for the mast cell lineage. *Science.* 1996;271(5250):818-822.
38. Juhlin L, Michaelsson G. A new syndrome characterised by absence of eosinophils and basophils. *Lancet.* 1977;1(8024):1233-1235.
39. Tracey R, Smith H. An inherited anomaly of human eosinophils and basophils. *Blood Cell.* 1978;4(1-2):291-300.
40. Mitchell EB, Platts-Mills TA, Pereira RS, et al. Basophil and eosinophil deficiency in a patient with hypogammaglobulinemia associated with thymoma. *Birth Defects Orig Artic Ser.* 1983;19(3):331.
41. Dvorak AM, Mihm MC Jr, Dvorak HF. Morphology of delayed-type hypersensitivity reactions in man. II. Ultrastructural alterations affecting the microvasculature and the tissue mast cells. *Lab Invest.* 1976;34(2):179-191.
42. Motakis E, Guhl S, Ishizu Y, et al. Redefinition of the human mast cell transcriptome by deep-CAGE sequencing. *Blood.* 2014;123(17):e58-e67.
43. Dwyer DF, Barrett NA, Austen KF, et al. Expression profiling of constitutive mast cells reveals a unique identity within the immune system. *Nat Immunol.* 2016;17(7):878-887.
44. Plum T, Wang X, Rettel M, et al. Human mast cell proteome reveals unique lineage, putative functions, and structural basis for cell ablation. *Immunity.* 2020;52(2):404.e5-416.e5.
45. Ehrlich P. *Beitra¨ge zur Theorie und Praxis der histologiscen Fa¨rbung.* Thesis. Leipzig University; 1878.
46. Ehrlich P. Über die spezifischen Granulationen des Blutes. *Arch Anat Physiol (Leipzig).* 1879;3:517-579.
47. Metcalfe DD, Baram D, Mekori YA. Mast cells. *Physiol Rev.* 1997;77(4):1033-1079.
48. Galli SJ, Kalesnikoff J, Grimbaldeston MA, et al. Mast cells as "tunable" effector and immunoregulatory cells: recent advances. *Annu Rev Immunol.* 2005;23:749-786.
49. Sonoda T, Kanayama Y, Hara H, et al. Proliferation of peritoneal mast cells in the skin of W/Wv mice that genetically lack mast cells. *J Exp Med.* 1984;160(1):138-151.
50. Tsai M, Shih LS, Newlands GF, et al. The rat c-kit ligand, stem cell factor, induces the development of connective tissue-type and mucosal mast cells *in vivo*. Analysis by anatomical distribution, histochemistry, and protease phenotype. *J Exp Med.* 1991;174(1):125-131.
51. Tsai M, Takeishi T, Thompson H, et al. Induction of mast cell proliferation, maturation, and heparin synthesis by the rat c-kit ligand, stem cell factor. *Proc Natl Acad Sci U S A.* 1991;88(14):6382-6386.
52. Kambe N, Kambe M, Kochan JP, et al. Human skin-derived mast cells can proliferate while retaining their characteristic functional and protease phenotypes. *Blood.* 2001;97(7):2045-2052.

53. Galli SJ, Borregaard N, Wynn TA. Phenotypic and functional plasticity of cells of innate immunity: macrophages, mast cells and neutrophils. *Nat Immunol.* 2011;12(11):1035-1044.
54. Li Z, Liu S, Xu J, et al. Adult connective tissue-resident mast cells originate from late erythro-myeloid progenitors. *Immunity.* 2018;49(4):640.e5-653.e5.
55. Gentek R, Ghigo C, Hoeffel G, et al. Hemogenic endothelial fate mapping reveals dual developmental origin of mast cells. *Immunity.* 2018;48(6):1160.e5-1171.e5.
56. Nilsson G, Dahlin JS. New insights into the origin of mast cells. *Allergy.* 2019;74(4):844-845.
57. Sonoda T, Hayashi C, Kitamura Y. Presence of mast cell precursors in the yolk sac of mice. *Dev Biol.* 1983;97(1):89-94.
58. Kitamura Y, Shimada M, Hatanaka K, et al. Development of mast cells from grafted bone marrow cells in irradiated mice. *Nature.* 1977;268(5619):442-443.
59. Jamur MC, Grodzki AC, Berenstein EH, et al. Identification and characterization of undifferentiated mast cells in mouse bone marrow. *Blood.* 2005;105(11):4282-4289.
60. Dahlin JS, Heyman B, Hallgren J. Committed mast cell progenitors in mouse blood differ in maturity between Th1 and Th2 strains. *Allergy.* 2013;68(10):1333-1337.
61. Poglio S, De Toni-Costes F, Arnaud E, et al. Adipose tissue as a dedicated reservoir of functional mast cell progenitors. *Stem Cell.* 2010;28(11):2065-2072.
62. Abonia JP, Hallgren J, Jones T, et al. Alpha-4 integrins and VCAM-1, but not MAdCAM-1, are essential for recruitment of mast cell progenitors to the inflamed lung. *Blood.* 2006;108(5):1588-1594.
63. Kirshenbaum AS, Kessler SW, Goff JP, et al. Demonstration of the origin of human mast cells from CD34+ bone marrow progenitor cells. *J Immunol.* 1991;146(5):1410-1415.
64. Bandara G, Metcalfe DD, Kirshenbaum AS. Growth of human mast cells from bone marrow and peripheral blood-derived CD34(+) pluripotent hematopoietic cells. *Methods Mol Biol.* 2015;1220:155-162.
65. Kirshenbaum AS, Goff JP, Semere T, et al. Demonstration that human mast cells arise from a progenitor cell population that is CD34(+), c-kit(+), and expresses aminopeptidase N (CD13). *Blood.* 1999;94(7):2333-2342.
66. Radinger M, Jensen BM, Kuehn HS, et al. Generation, isolation, and maintenance of human mast cells and mast cell lines derived from peripheral blood or cord blood. *Curr Protoc Immunol.* 2010;Chapter 7:Unit 7.37.
67. Kambe N, Kambe M, Chang HW, et al. An improved procedure for the development of human mast cells from dispersed fetal liver cells in serum-free culture medium. *J Immunol Methods.* 2000;240(1-2):101-110.
68. Saito H, Ebisawa M, Tachimoto H, et al. Selective growth of human mast cells induced by Steel factor, IL-6, and prostaglandin E2 from cord blood mononuclear cells. *J Immunol.* 1996;157(1):343-350.
69. Dahlin JS, Hamey FK, Pijuan-Sala B, et al. A single-cell hematopoietic landscape resolves 8 lineage trajectories and defects in Kit mutant mice. *Blood.* 2018;131(21):e1-e11.
70. Iwasaki H, Mizuno S, Arinobu Y, et al. The order of expression of transcription factors directs hierarchical specification of hematopoietic lineages. *Genes Dev.* 2006;20(21):3010-3021.
71. Galli SJ, Zsebo KM, Geissler EN. The kit ligand, stem cell factor. *Adv Immunol.* 1994;55:1-96.
72. Smith MA, Court EL, Smith JG. Stem cell factor: laboratory and clinical aspects. *Blood Rev.* 2001;15(4):191-197.
73. Bischoff SC, Dahinden CA. c-kit ligand: a unique potentiator of mediator release by human lung mast cells. *J Exp Med.* 1992;175(1):237-244.
74. Columbo M, Horowitz EM, Botana LM, et al. The human recombinant c-kit receptor ligand, rhSCF, induces mediator release from human cutaneous mast cells and enhances IgE-dependent mediator release from both skin mast cells and peripheral blood basophils. *J Immunol.* 1992;149(2):599-608.
75. Wershil BK, Tsai M, Geissler EN, et al. The rat c-kit ligand, stem cell factor, induces c-kit receptor-dependent mouse mast cell activation *in vivo*. Evidence that signaling through the c-kit receptor can induce expression of cellular function. *J Exp Med.* 1992;175(1):245-255.
76. Costa JJ, Demetri GD, Harrist TJ, et al. Recombinant human stem cell factor (kit ligand) promotes human mast cell and melanocyte hyperplasia and functional activation *in vivo*. *J Exp Med.* 1996;183(6):2681-2686.
77. Galli SJ, Iemura A, Garlick DS, et al. Reversible expansion of primate mast cell populations *in vivo* by stem cell factor. *J Clin Invest.* 1993;91(1):148-152.
78. Finotto S, Mekori YA, Metcalfe DD. Glucocorticoids decrease tissue mast cell number by reducing the production of the c-kit ligand, stem cell factor, by resident cells: in vitro and in vivo evidence in murine systems. *J Clin Invest.* 1997;99(7):1721-1728.
79. Broide DH, Wasserman SI, Alvaro-Gracia J, et al. Transforming growth factor-beta 1 selectively inhibits IL-3-dependent mast cell proliferation without affecting mast cell function or differentiation. *J Immunol.* 1989;143(5):1591-1597.
80. Toyota N, Hashimoto Y, Matsuo S, et al. Transforming growth factor beta 1 inhibits IL-3-and IL-4-dependent mouse connective tissue-type mast cell proliferation. *Arch Dermatol Res.* 1995;287(2):198-201.
81. Mekori YA, Metcalfe DD. Transforming growth factor-beta prevents stem cell factor-mediated rescue of mast cells from apoptosis after IL-3 deprivation. *J Immunol.* 1994;153(5):2194-2203.
82. Aloe L, Levi-Montalcini R. Mast cells increase in tissues of neonatal rats injected with the nerve growth factor. *Brain Res.* 1977;133(2):358-366.
83. Baccari GC, Raucci F, Di Fiore MM, et al. Induced maturation of frog mast cells by nerve growth factor during ontogenesis. *Microsc Res Tech.* 2003;62(5):439-450.
84. Dahlin JS, Ekoff M, Grootens J, et al. KIT signaling is dispensable for human mast cell progenitor development. *Blood.* 2017;130(16):1785-1794.
85. Sasaki H, Kurotaki D, Tamura T. Regulation of basophil and mast cell development by transcription factors. *Allergol Int.* 2016;65(2):127-134.
86. Suurmond J, Habets KLL, Tatum Z, et al. Repeated FceRI triggering reveals modified mast cell function related to chronic allergic responses in tissue. *J Allergy Clin Immunol.* 2016;138(3):869-880.
87. Beutier H, Gillis CM, Iannascoli B, et al. IgG subclasses determine pathways of anaphylaxis in mice. *J Allergy Clin Immunol.* 2017;139(1):269.e7-280.e7.
88. Reber LL, Hernandez JD, Galli SJ. The pathophysiology of anaphylaxis. *J Allergy Clin Immunol.* 2017;140(2):335-348.
89. Okayama Y, Kirshenbaum AS, Metcalfe DD. Expression of a functional high-affinity IgG receptor, Fc gamma RI, on human mast cells: up-regulation by IFN-gamma. *J Immunol.* 2000;164(8):4332-4339.
90. Woolhiser MR, Okayama Y, Gilfillan AM, et al. IgG-dependent activation of human mast cells following up-regulation of FcgammaRI by IFN-gamma. *Eur J Immunol.* 2001;31(11):3298-3307.
91. Jonsson F, Mancardi DA, Zhao W, et al. Human FcgammaRIIA induces anaphylactic and allergic reactions. *Blood.* 2012;119(11):2533-2544.
92. Finkelman FD, Khodoun MV, Strait R. Human IgE-independent systemic anaphylaxis. *J Allergy Clin Immunol.* 2016;137(6):1674-1680.
93. Tkaczyk C, Okayama Y, Woolhiser MR, et al. Activation of human mast cells through the high affinity IgG receptor. *Mol Immunol.* 2002;38(16-18):1289-1293.
94. Malbec O, Daeron M. The mast cell IgG receptors and their roles in tissue inflammation. *Immunol Rev.* 2007;217:206-221.
95. Jonsson F, Daeron M. Mast cells and company. *Front Immunol.* 2012;3:16.
96. Daeron M. Innate myeloid cells under the control of adaptive immunity: the example of mast cells and basophils. *Curr Opin Immunol.* 2016;38:101-108.
97. Kobayashi R, Okamura S, Ohno T, et al. Hyperexpression of FcgammaRI and Toll-like receptor 4 in the intestinal mast cells of Crohn's disease patients. *Clin Immunol.* 2007;125(2):149-158.
98. Burton OT, Epp A, Fanny ME, et al. Tissue-specific expression of the low-affinity IgG receptor, FcgammaRIIB, on human mast cells. *Front Immunol.* 2018;9:1244.
99. Kulis MD, Patil SU, Wambre E, et al. Immune mechanisms of oral immunotherapy. *J Allergy Clin Immunol.* 2018;141(2):491-498.
100. Malbec O, Cassard L, Albanesi M, et al. Trans-inhibition of activation and proliferation signals by Fc receptors in mast cells and basophils. *Sci Signal.* 2016;9(459):ra126.
101. Chen K, Xu W, Wilson M, et al. Immunoglobulin D enhances immune surveillance by activating antimicrobial, proinflammatory and B cell-stimulating programs in basophils. *Nat Immunol.* 2009;10(8):889-898.
102. Marshall JS. Mast-cell responses to pathogens. *Nat Rev Immunol.* 2004;4(10):787-799.
103. Abraham SN, St John AL. Mast cell-orchestrated immunity to pathogens. *Nat Rev Immunol.* 2010;10(6):440-452.
104. Redegeld FA, Yu Y, Kumari S, et al. Non-IgE mediated mast cell activation. *Immunol Rev.* 2018;282(1):87-113.
105. Sandig H, Bulfone-Paus S. TLR signaling in mast cells: common and unique features. *Front Immunol.* 2012;3:185.
106. Marshall JS, Portales-Cervantes L, Leong E. Mast cell responses to viruses and pathogen products. *Int J Mol Sci.* 2019;20(17):4241.
107. Yang Z, Marshall JS. Zymosan treatment of mouse mast cells enhances dectin-1 expression and induces dectin-1-dependent reactive oxygen species (ROS) generation. *Immunobiology.* 2009;214(4):321-330.
108. Nieto-Patlan A, Campillo-Navarro M, Rodriguez-Cortes O, et al. Recognition of Candida albicans by Dectin-1 induces mast cell activation. *Immunobiology.* 2015;220(9):1093-1100.
109. Pinke KH, Lima HG, Cunha FQ, et al. Mast cells phagocyte Candida albicans and produce nitric oxide by mechanisms involving TLR2 and Dectin-1. *Immunobiology.* 2016;221(2):220-227.
110. St John AL, Rathore AP, Yap H, et al. Immune surveillance by mast cells during dengue infection promotes natural killer (NK) and NKT-cell recruitment and viral clearance. *Proc Natl Acad Sci U S A.* 2011;108(22):9190-9195.
111. Ali H. Regulation of human mast cell and basophil function by anaphylatoxins C3a and C5a. *Immunol Lett.* 2010;128(1):36-45.
112. Schafer B, Piliponsky AM, Oka T, et al. Mast cell anaphylatoxin receptor expression can enhance IgE-dependent skin inflammation in mice. *J Allergy Clin Immunol.* 2013;131(2):541-548.
113. Maurer M, Wedemeyer J, Metz M, et al. Mast cells promote homeostasis by limiting endothelin-1-induced toxicity. *Nature.* 2004;432(7016):512-516.
114. Gaudenzio N, Sibilano R, Marichal T, et al. Different activation signals induce distinct mast cell degranulation strategies. *J Clin Invest.* 2016;126(10):3981-3998.
115. Kulka M, Sheen CH, Tancowny BP, et al. Neuropeptides activate human mast cell degranulation and chemokine production. *Immunology.* 2008;123(3):398-410.
116. Migalovich-Sheikhet H, Friedman S, Mankuta D, et al. Novel identified receptors on mast cells. *Front Immunol.* 2012;3(238).
117. Mukai K, Tsai M, Saito H, et al. Mast cells as sources of cytokines, chemokines, and growth factors. *Immunol Rev.* 2018;282(1):121-150.
118. McNeil BD, Pundir P, Meeker S, et al. Identification of a mast-cell-specific receptor crucial for pseudo-allergic drug reactions. *Nature.* 2015;519(7542):237-241.
119. Subramanian H, Gupta K, Ali H. Roles of Mas-related G protein-coupled receptor X2 on mast cell-mediated host defense, pseudoallergic drug reactions, and chronic inflammatory diseases. *J Allergy Clin Immunol.* 2016;138(3):700-710.
120. Ali H. Emerging roles for Mas-related G protein-coupled receptor-X2 in host defense peptide, opioid, and neuropeptide-mediated inflammatory reactions. *Adv Immunol.* 2017;136:123-162.
121. Subramanian H, Gupta K, Guo Q, et al. Mas-related gene X2 (MrgX2) is a novel G protein-coupled receptor for the antimicrobial peptide LL-37 in human mast cells: resistance to receptor phosphorylation, desensitization, and internalization. *J Biol Chem.* 2011;286(52):44739-44749.
122. Meixiong J, Dong X. Mas-related G protein-coupled receptors and the biology of itch sensation. *Annu Rev Genet.* 2017;51:103-121.
123. Green DP, Limjunyawong N, Gour N, et al. A mast-cell-specific receptor mediates neurogenic inflammation and pain. *Neuron.* 2019;101(3):412.e3-420.e3.
124. Serhan N, Basso L, Sibilano R, et al. House dust mites activate nociceptor-mast cell clusters to drive type 2 skin inflammation. *Nat Immunol.* 2019;20(11):1435-1443.
125. Meixiong J, Anderson M, Limjunyawong N, et al. Activation of mast-cell-expressed Mas-related G-protein-coupled receptors drives non-histaminergic itch. *Immunity.* 2019;50(5):1163.e5-1171.e5.
126. Boyce JA. Mast cells and eicosanoid mediators: a system of reciprocal paracrine and autocrine regulation. *Immunol Rev.* 2007;217:168-185.
127. Olivera A, Mizugishi K, Tikhonova A, et al. The sphingosine kinase-sphingosine-1-phosphate axis is a determinant of mast cell function and anaphylaxis. *Immunity.* 2007;26(3):287-297.
128. Rudich N, Ravid K, Sagi-Eisenberg R. Mast cell adenosine receptors function: a focus on the a3 adenosine receptor and inflammation. *Front Immunol.* 2012;3:134.
129. Hagemann PM, Nsiah-Dosu S, Hundt JE, et al. Modulation of mast cell reactivity by lipids: the neglected side of allergic diseases. *Front Immunol.* 2019;10:1174.
130. Jappe U, Schwager C, Schromm AB, et al. Lipophilic allergens, different modes of allergen-lipid interaction and their impact on asthma and allergy. *Front Immunol.* 2019;10:122.
131. Feng C, Beller EM, Bagga S, et al. Human mast cells express multiple EP receptors for prostaglandin E2 that differentially modulate activation responses. *Blood.* 2006;107(8):3243-3250.
132. Nguyen M, Solle M, Audoly LP, et al. Receptors and signaling mechanisms required for prostaglandin E2-mediated regulation of mast cell degranulation and IL-6 production. *J Immunol.* 2002;169(8):4586-4593.
133. Kuehn HS, Gilfillan AM. G protein-coupled receptors and the modification of FcepsilonRI-mediated mast cell activation. *Immunol Lett.* 2007;113(2):59-69.
134. Torres-Atencio I, Ainsua-Enrich E, de Mora F, et al. Prostaglandin E2 prevents hyperosmolar-induced human mast cell activation through prostanoid receptors EP2 and EP4. *PLoS One.* 2014;9(10):e110870.
135. Torres R, Picado C, de Mora F. The PGE2-EP2-mast cell axis: an antiasthma mechanism. *Mol Immunol.* 2015;63(1):61-68.
136. Weller CL, Collington SJ, Brown JK, et al. Leukotriene B4, an activation product of mast cells, is a chemoattractant for their progenitors. *J Exp Med.* 2005;201(12):1961-1971.
137. Lundeen KA, Sun B, Karlsson L, et al. Leukotriene B4 receptors BLT1 and BLT2: expression and function in human and murine mast cells. *J Immunol.* 2006;177(5):3439-3447.
138. Serhan CN, Levy BD. Resolvins in inflammation: emergence of the pro-resolving superfamily of mediators. *J Clin Invest.* 2018;128(7):2657-2669.
139. Martin N, Ruddick A, Arthur GK, et al. Primary human airway epithelial cell-dependent inhibition of human lung mast cell degranulation. *PLoS One.* 2012;7(8):e43545.
140. Karra L, Haworth O, Priluck R, et al. Lipoxin B(4) promotes the resolution of allergic inflammation in the upper and lower airways of mice. *Mucosal Immunol.* 2015;8(4):852-862.
141. Jolly PS, Bektas M, Olivera A, et al. Transactivation of sphingosine-1-phosphate receptors by FcepsilonRI triggering

is required for normal mast cell degranulation and chemotaxis. *J Exp Med.* 2004;199(7):959-970.

142. Olivera A, Rivera J. An emerging role for the lipid mediator sphingosine-1-phosphate in mast cell effector function and allergic disease. *Adv Exp Med Biol.* 2011;716:123-142.
143. Akahoshi M, Song CH, Piliponsky AM, et al. Mast cell chymase reduces the toxicity of Gila monster venom, scorpion venom, and vasoactive intestinal polypeptide in mice. *J Clin Invest.* 2011;121(10):4180-4191.
144. Yano H, Wershil BK, Arizono N, et al. Substance P-induced augmentation of cutaneous vascular permeability and granulocyte infiltration in mice is mast cell dependent. *J Clin Invest.* 1989;84(4):1276-1286.
145. Boros M, Szalay L, Kaszaki J. Endothelin-1 induces mucosal mast cell degranulation and tissue injury via ETA receptors. *Clin Sci (Lond).* 2002;103(suppl 48):31S-4S.
146. Metz M, Piliponsky AM, Chen CC, et al. Mast cells can enhance resistance to snake and honeybee venoms. *Science.* 2006;313(5786):526-530.
147. Marichal T, Starkl P, Reber LL, et al. A beneficial role for immunoglobulin E in host defense against honeybee venom. *Immunity.* 2013;39(5):963-975.
148. Starkl P, Marichal T, Gaudenzio N, et al. IgE antibodies, FcεRIα, and IgE-mediated local anaphylaxis can limit snake venom toxicity. *J Allergy Clin Immunol.* 2016;137(1):246.e11-257.e11.
149. Sibilano R, Pucillo CE, Gri G. Allergic responses and aryl hydrocarbon receptor novel pathway of mast cell activation. *Mol Immunol.* 2015;63(1):69-73.
150. Baumann A, Gonnenwein S, Bischoff SC, et al. The circadian clock is functional in eosinophils and mast cells. *Immunology.* 2013;140(4):465-474.
151. Nakamura Y, Nakano N, Ishimaru K, et al. Circadian regulation of allergic reactions by the mast cell clock in mice. *J Allergy Clin Immunol.* 2014;133(2):568-575.
152. Nakamura Y, Ishimaru K, Shibata S, et al. Regulation of plasma histamine levels by the mast cell clock and its modulation by stress. *Sci Rep.* 2017;7:39934.
153. Kawauchi T, Ishimaru K, Nakamura Y, et al. Clock-dependent temporal regulation of IL-33/ST2-mediated mast cell response. *Allergol Int.* 2017;66(3):472-478.
154. Christ P, Sowa AS, Froy O, et al. The circadian clock drives mast cell functions in allergic reactions. *Front Immunol.* 2018;9:1526.
155. Bulfone-Paus S, Nilsson G, Draber P, et al. Positive and negative signals in mast cell activation. *Trends Immunol.* 2017;38(9):657-667.
156. Ayyadurai S, Gibson AJ, D'Costa S, et al. Frontline science: corticotropin-releasing factor receptor subtype 1 is a critical modulator of mast cell degranulation and stress-induced pathophysiology. *J Leukoc Biol.* 2017;102(6):1299-1312.
157. D'Costa S, Ayyadurai S, Gibson AJ, et al. Mast cell corticotropin-releasing factor subtype 2 suppresses mast cell degranulation and limits the severity of anaphylaxis and stress-induced intestinal permeability. *J Allergy Clin Immunol.* 2019;143(5):1865.e4-1877.e4.
158. Mizrahi S, Gibbs BF, Karra L, et al. Siglec-7 is an inhibitory receptor on human mast cells and basophils. *J Allergy Clin Immunol.* 2014;134(1):230-233.
159. Yokoi H, Choi OH, Hubbard W, et al. Inhibition of FcεRI-dependent mediator release and calcium flux from human mast cells by sialic acid-binding immunoglobulin-like lectin 8 engagement. *J Allergy Clin Immunol.* 2008;121(2):499.e1-505.e1.
160. Duan S, Koziol-White CJ, Jester WF Jr, et al. CD33 recruitment inhibits IgE-mediated anaphylaxis and desensitizes mast cells to allergen. *J Clin Invest.* 2019;129(3):1387-1401.
161. Royer B, Varadaradjalou S, Saas P, et al. Inhibition of IgE-induced activation of human mast cells by IL-10. *Clin Exp Allergy.* 2001;31(5):694-704.
162. Gomez G, Ramirez CD, Rivera J, et al. TGF-β1 inhibits mast cell Fc epsilon RI expression. *J Immunol.* 2005;174(10):5987-5993.
163. Zhao W, Gomez G, Yu SH, et al. TGF-beta1 attenuates mediator release and *de novo* Kit expression by human skin mast cells through a Smad-dependent pathway. *J Immunol.* 2008;181(10):7263-7272.
164. Ndaw VS, Abebayehu D, Spence AJ, et al. TGF-β1 suppresses IL-33-induced mast cell function. *J Immunol.* 2017;199(3):866-873.
165. Gri G, Piconese S, Frossi B, et al. CD4+CD25+ regulatory T cells suppress mast cell degranulation and allergic responses through OX40-OX40L interaction. *Immunity.* 2008;29(5):771-781.
166. Tam SY, Tsai M, Snouwaert JN, et al. RabGEF1 is a negative regulator of mast cell activation and skin inflammation. *Nat Immunol.* 2004;5(8):844-852.
167. Gimborn K, Lessmann E, Kuppig S, et al. SHIP down-regulates FcepsilonR1-induced degranulation at supraoptimal IgE or antigen levels. *J Immunol.* 2005;174(1):507-516.
168. Kanagaratham C, El Ansari YS, Sallis BF, et al. Omeprazole inhibits IgE-mediated mast cell activation and allergic inflammation induced by ingested allergen in mice. *J Allergy Clin Immunol.* 2020;146:884.e5-893.e5.
169. Maccarrone M, Bab I, Biro T, et al. Endocannabinoid signaling at the periphery: 50 years after THC. *Trends Pharmacol Sci.* 2015;36(5):277-296.
170. Lau AH, Chow SS. Effects of cannabinoid receptor agonists on immunologically induced histamine release from rat peritoneal mast cells. *Eur J Pharmacol.* 2003;464(2-3):229-235.
171. Samson MT, Small-Howard A, Shimoda LM, et al. Differential roles of CB1 and CB2 cannabinoid receptors in mast cells. *J Immunol.* 2003;170(10):4953-4962.
172. Sugawara K, Biro T, Tsuruta D, et al. Endocannabinoids limit excessive mast cell maturation and activation in human skin. *J Allergy Clin Immunol.* 2012;129(3):726.e8-738.e8.
173. Nam G, Jeong SK, Park BM, et al. Selective cannabinoid receptor-1 agonists regulate mast cell activation in an oxazolone-induced atopic dermatitis model. *Ann Dermatol.* 2016;28(1):22-29.
174. Cruz SL, Sanchez-Miranda E, Castillo-Arellano JI, et al. Anandamide inhibits FcepsilonRI-dependent degranulation and cytokine synthesis in mast cells through CB2 and GPR55 receptor activation. Possible involvement of CB2-GPR55 heteromers. *Int Immunopharmacol.* 2018;64:298-307.
175. Caughey GH. Mast cell proteases as protective and inflammatory mediators. *Adv Exp Med Biol.* 2011;716:212-234.
176. Ronnberg E, Melo FR, Pejler G. Mast cell proteoglycans. *J Histochem Cytochem.* 2012;60(12):950-962.
177. Wernersson S, Pejler G. Mast cell secretory granules: armed for battle. *Nat Rev Immunol.* 2014;14(7):478-494.
178. Gaudenzio N, Marichal T, Galli SJ, et al. Genetic and imaging approaches reveal pro-inflammatory and immunoregulatory roles of mast cells in contact hypersensitivity. *Front Immunol.* 2018;9:1275.
179. Oh C, Suzuki S, Nakashima I, et al. Histamine synthesis by non-mast cells through mitogen-dependent induction of histidine decarboxylase. *Immunology.* 1988;65(1):143-148.
180. Xu X, Zhang D, Zhang H, et al. Neutrophil histamine contributes to inflammation in mycoplasma pneumonia. *J Exp Med.* 2006;203(13):2907-2917.
181. Saxena SP, Brandes LJ, Becker AB, et al. Histamine is an intracellular messenger mediating platelet aggregation. *Science.* 1989;243(4898):1596-1599.
182. Galli SJ, Kitamura Y. Genetically mast-cell-deficient W/Wv and Sl/Sld mice. Their value for the analysis of the roles of mast cells in biologic responses *in vivo*. *Am J Pathol.* 1987;127(1):191-198.
183. Kushnir-Sukhov NM, Brown JM, Wu Y, et al. Human mast cells are capable of serotonin synthesis and release. *J Allergy Clin Immunol.* 2007;119(2):498-499.
184. Ringvall M, Ronnberg E, Wernersson S, et al. Serotonin and histamine storage in mast cell secretory granules is dependent on serglycin proteoglycan. *J Allergy Clin Immunol.* 2008;121(4):1020-1026.
185. Rivera J, Fierro NA, Olivera A, et al. New insights on mast cell activation via the high affinity receptor for IgE. *Adv Immunol.* 2008;98:85-120.
186. Galli SJ, Nakae S, Tsai M. Mast cells in the development of adaptive immune responses. *Nat Immunology.* 2005;6(2):135-142.
187. Moon TC, St Laurent CD, Morris KE, et al. Advances in mast cell biology: new understanding of heterogeneity and function. *Mucosal Immunol.* 2010;3(2):111-128.
188. Suzuki R, Leach S, Liu W, et al. Molecular editing of cellular responses by the high-affinity receptor for IgE. *Science.* 2014;343(6174):1021-1025.
189. Mukai K, Karasuyama H, Kabashima K, et al. Differences in the importance of mast cells, basophils, IgE, and IgG versus that of CD4(+) t cells and ILC2 cells in primary and secondary immunity to strongyloides venezuelensis. *Infect Immun.* 2017;85(5):e00053-e00117.
190. Siracusa MC, Kim BS, Spergel JM, et al. Basophils and allergic inflammation. *J Allergy Clin Immunol.* 2013;132(4):789-801; quiz 788.
191. Geering B, Stoeckle C, Conus S, et al. Living and dying for inflammation: neutrophils, eosinophils, basophils. *Trends Immunol.* 2013;34(8):398-409.
192. Grundstrom J, Reimer JM, Magnusson SE, et al. Human cord blood derived immature basophils show dual characteristics, expressing both basophil and eosinophil associated proteins. *PLoS One.* 2012;7(10):e48308.
193. Gorgens A, Radtke S, Mollmann M, et al. Revision of the human hematopoietic tree: granulocyte subtypes derive from distinct hematopoietic lineages. *Cell Rep.* 2013;3(5):1539-1552.
194. Li Y, Qi X, Liu B, et al. The STAT5-GATA2 pathway is critical in basophil and mast cell differentiation and maintenance. *J Immunol.* 2015;194(9):4328-4338.
195. Nei Y, Obata-Ninomiya K, Tsutsui H, et al. GATA-1 regulates the generation and function of basophils. *Proc Natl Acad Sci U S A.* 2013;110(46):18620-18625.
196. Lantz CS, Min B, Tsai M, et al. IL-3 is required for increases in blood basophils in nematode infection in mice and can enhance IgE-dependent IL-4 production by basophils *in vitro*. *Lab Invest.* 2008;88(11):1134-1142.
197. Shen T, Kim S, Do JS, et al. T cell-derived IL-3 plays key role in parasite infection-induced basophil production but is dispensable for *in vivo* basophil survival. *Int Immunol.* 2008;20(9):1201-1209.
198. Zellweger F, Buschor P, Hobi G, et al. IL-3 but not monomeric IgE regulates FcepsilonRI levels and cell survival in primary human basophils. *Cell Death Dis.* 2018;9(5):510.
199. Lantz CS, Boesiger J, Song CH, et al. Role for interleukin-3 in mast-cell and basophil development and in immunity to parasites. *Nature.* 1998;392(6671):90-93.
200. Ohmori K, Luo Y, Jia Y, et al. IL-3 induces basophil expansion *in vivo* by directing granulocyte-monocyte progenitors to differentiate into basophil lineage-restricted progenitors in the bone marrow and by increasing the number of basophil/mast cell progenitors in the spleen. *J Immunol.* 2009;182(5):2835-2841.
201. Giacomin PR, Siracusa MC, Walsh KP, et al. Thymic stromal lymphopoietin-dependent basophils promote Th2 cytokine responses following intestinal helminth infection. *J Immunol.* 2012;189(9):4371-4378.
202. Siracusa MC, Saenz SA, Hill DA, et al. TSLP promotes interleukin-3-independent basophil haematopoiesis and type 2 inflammation. *Nature.* 2011;477(7363):229-233.
203. Hussain M, Borcard L, Walsh KP, et al. Basophil-derived IL-4 promotes epicutaneous antigen sensitization concomitant with the development of food allergy. *J Allergy Clin Immunol.* 2018;141(1):223.e5-234.e5.
204. Muto T, Fukuoka A, Kabashima K, et al. The role of basophils and proallergic cytokines, TSLP and IL-33, in cutaneously sensitized food allergy. *Int Immunol.* 2014;26(10):539-549.
205. Herbst T, Sichelstiel A, Schar C, et al. Dysregulation of allergic airway inflammation in the absence of microbial colonization. *Am J Respir Crit Care Med.* 2011;184(2):198-205.
206. Hill DA, Siracusa MC, Abt MC, et al. Commensal bacteria-derived signals regulate basophil hematopoiesis and allergic inflammation. *Nat Med.* 2012;18(4):538-546.
207. Lantz CS, Yamaguchi M, Oettgen HC, et al. IgE regulates mouse basophil Fc epsilon RI expression *in vivo*. *J Immunol.* 1997;158(6):2517-2521.
208. Ebo DG, Bridts CH, Hagendorens MM, et al. Basophil activation test by flow cytometry: present and future applications in allergology. *Cytometry B Clin Cytom.* 2008;74(4):201-210.
209. MacGlashan DW Jr. Basophil activation testing. *J Allergy Clin Immunol.* 2013;132(4):777-787.
210. Hoffmann HJ, Santos AF, Mayorga C, et al. The clinical utility of basophil activation testing in diagnosis and monitoring of allergic disease. *Allergy.* 2015;70(11):1393-1405.
211. Hemmings O, Kwok M, McKendry R, et al. Basophil activation test: old and new applications in allergy. *Curr Allergy Asthma Rep.* 2018;18(12):77.
212. Tsujimura Y, Obata K, Mukai K, et al. Basophils play a pivotal role in immunoglobulin-G-mediated but not immunoglobulin-E-mediated systemic anaphylaxis. *Immunity.* 2008;28(4):581-589.
213. Iikura M, Yamaguchi M, Fujisawa T, et al. Secretory IgA induces degranulation of IL-3-primed basophils. *J Immunol.* 1998;161(3):1510-1515.
214. Shan M, Carrillo J, Yeste A, et al. Secreted IgD amplifies humoral T helper 2 cell responses by binding basophils via Galectin-9 and CD44. *Immunity.* 2018;49(4):709.e8-724.e8.
215. Bischoff SC, Krieger M, Brunner T, et al. RANTES and related chemokines activate human basophil granulocytes through different G protein-coupled receptors. *Eur J Immunol.* 1993;23(3):761-767.
216. Wedi B, Gehring M, Kapp A. The pseudoallergen receptor MRGPRX2 on peripheral blood basophils and eosinophils: expression and function. *Allergy.* 2020;75(9):2229-2242.
217. Valent P, Besemer J, Muhm M, et al. Interleukin 3 activates human blood basophils via high-affinity binding sites. *Proc Natl Acad Sci U S A.* 1989;86(14):5542-5546.
218. Bischoff SC, de Weck AL, Dahinden CA. Interleukin 3 and granulocyte/macrophage-colony-stimulating factor render human basophils responsive to low concentrations of complement component C3a. *Proc Natl Acad Sci U S A.* 1990;87(17):6813-6817.
219. Yoshikawa S, Oh-Hora M, Hashimoto R, et al. Pivotal role of STIM2, but not STIM1, in IL-4 production by IL-3-stimulated murine basophils. *Sci Signal.* 2019;12(576):eaav2060.
220. Bischoff SC, Brunner T, De Weck AL, et al. Interleukin 5 modifies histamine release and leukotriene generation by human basophils in response to diverse agonists. *J Exp Med.* 1990;172(6):1577-1582.
221. Krieger M, Brunner T, Bischoff SC, et al. Activation of human basophils through the IL-8 receptor. *J Immunol.* 1992;149(8):2662-2667.
222. Salter BM, Oliveria JP, Nusca G, et al. IL-25 and IL-33 induce Type 2 inflammation in basophils from subjects with allergic asthma. *Respir Res.* 2016;17:5.
223. Smithgall MD, Comeau MR, Yoon BR, et al. IL-33 amplifies both Th1-and Th2-type responses through its activity on human basophils, allergen-reactive Th2 cells, iNKT and NK cells. *Int Immunol.* 2008;20(8):1019-1030.
224. Valent P. Interleukin-33: a regulator of basophils. *Blood.* 2009;113(7):1396-1397.
225. Pecaric-Petkovic T, Didichenko SA, Kaempfer S, et al. Human basophils and eosinophils are the direct target leukocytes of the novel IL-1 family member IL-33. *Blood.* 2009;113(7):1526-1534.
226. Blom L, Poulsen BC, Jensen BM, et al. IL-33 induces IL-9 production in human CD4+ T cells and basophils. *PLoS One.* 2011;6(7):e21695.
227. Gibbs BF, Zillikens D, Grabbe J. Nerve growth factor influences IgE-mediated human basophil activation: functional

properties and intracellular mechanisms compared with IL-3. *Int Immunopharmacol.* 2005;5(4):735-747.
228. Hammad H, Lambrecht BN. Barrier epithelial cells and the control of type 2 immunity. *Immunity.* 2015;43(1):29-40.
229. Salter BM, Oliveria JP, Nusca G, et al. Thymic stromal lymphopoietin activation of basophils in patients with allergic asthma is IL-3 dependent. *J Allergy Clin Immunol.* 2015;136(6):1636-1644.
230. Salabert-Le Guen N, Hemont C, Delbove A, et al. Thymic stromal lymphopoietin does not activate human basophils. *J Allergy Clin Immunol.* 2018;141(4):1476.e6-1479.e6.
231. Suurmond J, Stoop JN, Rivellese F, et al. Activation of human basophils by combined toll-like receptor-and FcepsilonRI-triggering can promote Th2 skewing of naive T helper cells. *Eur J Immunol.* 2014;44(2):386-396.
232. Steiner M, Huber S, Harrer A, et al. The evolution of human basophil biology from neglect towards understanding of their immune functions. *BioMed Res Int.* 2016;2016:8232830.
233. de Paulis A, Montuori N, Prevete N, et al. Urokinase induces basophil chemotaxis through a urokinase receptor epitope that is an endogenous ligand for formyl peptide receptor-like 1 and-like 2. *J Immunol.* 2004;173(9):5739-5748.
234. Jiao D, Wong CK, Qiu HN, et al. NOD2 and TLR2 ligands trigger the activation of basophils and eosinophils by interacting with dermal fibroblasts in atopic dermatitis-like skin inflammation. *Cell Mol Immunol.* 2016;13(4):535-550.
235. Sokol CL, Barton GM, Farr AG, et al. A mechanism for the initiation of allergen-induced T helper type 2 responses. *Nat Immunol.* 2008;9(3):310-318.
236. Suzukawa M, Nagase H, Ogahara I, et al. Leptin enhances survival and induces migration, degranulation, and cytokine synthesis of human basophils. *J Immunol.* 2011;186(9):5254-5260.
237. Sharma M, Das M, Stephen-Victor E, et al. Regulatory T cells induce activation rather than suppression of human basophils. *Sci Immunol.* 2018;3(23):eaan0829.
238. Spadaro G, Giurato G, Stellato C, et al. Basophil degranulation in response to IgE ligation is controlled by a distinctive circadian clock in asthma. *Allergy.* 2020;75(1):158-168.
239. Kepley CL, Cambier JC, Morel PA, et al. Negative regulation of FcepsilonRI signaling by FcgammaRII costimulation in human blood basophils. *J Allergy Clin Immunol.* 2000;106(2):337-348.
240. Cady CT, Powell MS, Harbeck RJ, et al. IgG antibodies produced during subcutaneous allergen immunotherapy mediate inhibition of basophil activation via a mechanism involving both FcgammaRIIA and FcgammaRIIB. *Immunol Lett.* 2010;130(1-2):57-65.
241. Sabato V, Verweij MM, Bridts CH, et al. CD300a is expressed on human basophils and seems to inhibit IgE/FcepsilonRI-dependent anaphylactic degranulation. *Cytometry B Clin Cytom.* 2012;82(3):132-138.
242. Shiratori I, Yamaguchi M, Suzukawa M, et al. Down-regulation of basophil function by human CD200 and human herpesvirus-8 CD200. *J Immunol.* 2005;175(7):4441-4449.
243. Kikly KK, Bochner BS, Freeman SD, et al. Identification of SAF-2, a novel siglec expressed on eosinophils, mast cells, and basophils. *J Allergy Clin Immunol.* 2000;105(6 pt 1):1093-1100.
244. Novak N, Mete N, Bussmann C, et al. Early suppression of basophil activation during allergen-specific immunotherapy by histamine receptor 2. *J Allergy Clin Immunol.* 2012;130(5):1153-1158 e2.
245. MacGlashan D Jr. Histamine: a mediator of inflammation. *J Allergy Clin Immunol.* 2003;112(4 suppl):S53-S59.
246. Rothenberg ME, Caulfield JP, Austen KF, et al. Biochemical and morphological characterization of basophilic leukocytes from two patients with myelogenous leukemia. *J Immunol.* 1987;138(8):2616-2625.
247. Metcalfe DD, Bland CE, Wasserman SI. Biochemical and functional characterization of proteoglycans isolated from basophils of patients with chronic myelogenous leukemia. *J Immunol.* 1984;132(4):1943-1950.
248. Ugajin T, Satoh T, Kanamori T, et al. FcεRI, but not FcγR, signals induce prostaglandin D2 and E2 production from basophils. *Am J Pathol.* 2011;179(2):775-782.
249. de Paulis A, Prevete N, Fiorentino I, et al. Expression and functions of the vascular endothelial growth factors and their receptors in human basophils. *J Immunol.* 2006;177(10):7322-7331.
250. Piliponsky AM, Shubin NJ, Lahiri AK, et al. Basophil-derived tumor necrosis factor can enhance survival in a sepsis model in mice. *Nat Immunol.* 2019;20(2):129-140.
251. Raap U, Gehring M, Kleiner S, et al. Human basophils are a source of - and are differentially activated by - IL-31. *Clin Exp Allergy.* 2017;47(4):499-508.
252. Zheng W, Wang J, Zhu W, et al. Upregulated expression of substance P in basophils of the patients with chronic spontaneous urticaria: induction of histamine release and basophil accumulation by substance P. *Cell Biol Toxicol.* 2016;32(3):217-228.
253. Mitre E, Nutman TB. Basophils, basophilia and helminth infections. *Chem Immunol Allergy.* 2006;90:141-156.
254. Yamanishi Y, Karasuyama H. Basophil-derived IL-4 plays versatile roles in immunity. *Semin Immunopathol.* 2016;38(5):615-622.
255. Yoshimoto T. The hunt for the source of primary Interleukin-4: how we discovered that natural killer T cells and basophils determine T helper type 2 cell differentiation in vivo. *Front Immunol.* 2018;9:716.
256. MacGlashan DW Jr, Bochner BS, Adelman DC, et al. Down-regulation of FcεRI expression on human basophils during *in vivo* treatment of atopic patients with anti-IgE antibody. *J Immunol.* 1997;158(3):1438-1445.
257. Yamaguchi M, Lantz CS, Oettgen HC, et al. IgE enhances mouse mast cell FcεRI expression in vitro and in vivo: evidence for a novel amplification mechanism in IgE-dependent reactions. *J Exp Med.* 1997;185(4):663-672.
258. Kawakami T, Galli SJ. Regulation of mast-cell and basophil function and survival by IgE. *Nat Rev Immunol.* 2002;2(10):773-786.
259. Matsuda K, Piliponsky AM, Iikura M, et al. Monomeric IgE enhances human mast cell chemokine production: IL-4 augments and dexamethasone suppresses the response. *J Allergy Clin Immunol.* 2005;116(6):1357-1363.
260. James LC, Roversi P, Tawfik DS. Antibody multispecificity mediated by conformational diversity. *Science.* 2003;299(5611):1362-1367.
261. Bax HJ, Bowen H, Beavil RL, et al. IgE trimers drive SPE-7 cytokinergic activity. *Sci Rep.* 2017;7(1):8164.
262. Frossi B, Mion F, Sibilano R, et al. Is it time for a new classification of mast cells? What do we know about mast cell heterogeneity? *Immunol Rev.* 2018;282(1):35-46.
263. Enerback L. Mast cells in rat gastrointestinal mucosa. I. Effects of fixation. *Acta Pathol Microbiol Scand.* 1966;66(3):289-302.
264. Enerback L. Mast cells in rat gastrointestinal mucosa. 2. Dye-binding and metachromatic properties. *Acta Pathol Microbiol Scand.* 1966;66(3):303-312.
265. Li L, Li Y, Reddel SW, et al. Identification of basophilic cells that express mast cell granule proteases in the peripheral blood of asthma, allergy, and drug-reactive patients. *J Immunol.* 1998;161(9):5079-5086.
266. Siegmund R, Vogelsang H, Machnik A, et al. Surface membrane antigen alteration on blood basophils in patients with Hymenoptera venom allergy under immunotherapy. *J Allergy Clin Immunol.* 2000;106(6):1190-1195.
267. Siracusa MC, Wojno ED, Artis D. Functional heterogeneity in the basophil cell lineage. *Adv Immunol.* 2012;115:141-159.
268. da Silva EZ, Jamur MC, Oliver C. Mast cell function: a new vision of an old cell. *J Histochem Cytochem.* 2014;62(10):698-738.
269. Galli SJ. New insights into "the riddle of the mast cells": microenvironmental regulation of mast cell development and phenotypic heterogeneity. *Lab Invest.* 1990;62(1):5-33.
270. Min B. Basophils: what they 'can do' versus what they 'actually do'. *Nat Immunol.* 2008;9(12):1333-1339.
271. Galli SJ, Tsai M, Piliponsky AM. The development of allergic inflammation. *Nature.* 2008;454(7203):445-454.
272. Karasuyama H, Mukai K, Obata K, et al. Nonredundant roles of basophils in immunity. *Annu Rev Immunol.* 2011;29:45-69.
273. Brown SJ, Galli SJ, Gleich GJ, et al. Ablation of immunity to Amblyomma americanum by anti-basophil serum: cooperation between basophils and eosinophils in expression of immunity to ectoparasites (ticks) in Guinea pigs. *J Immunol.* 1982;129(2):790-796.
274. Brandt EB, Strait RT, Hershko D, et al. Mast cells are required for experimental oral allergen-induced diarrhea. *J Clin Invest.* 2003;112(11):1666-1677.
275. Reber LL, Marichal T, Galli SJ. New models for analyzing mast cell functions *in vivo*. *Trends Immunol.* 2012;33(12):613-625.
276. Rodewald HR, Feyerabend TB. Widespread immunological functions of mast cells: fact or fiction? *Immunity.* 2012;37(1):13-24.
277. Galli SJ, Tsai M, Marichal T, et al. Approaches for analyzing the roles of mast cells and their proteases *in vivo*. *Adv Immunol.* 2015;126:45-127.
278. Oka T, Kalesnikoff J, Starkl P, et al. Evidence questioning cromolyn's effectiveness and selectivity as a 'mast cell stabilizer' in mice. *Lab Invest.* 2012;92(10):1472-1482.
279. Reber LL, Sibilano R, Starkl P, et al. Imaging protective mast cells in living mice during severe contact hypersensitivity. *JCI Insight.* 2017;2(9):e9290.
280. Dudeck J, Froebel J, Kotrba J, et al. Engulfment of mast cell secretory granules on skin inflammation boosts dendritic cell migration and priming efficiency. *J Allergy Clin Immunol.* 2019;143(5):1849.e4-1864.e4.
281. Becker M, Reuter S, Friedrich P, et al. Genetic variation determines mast cell functions in experimental asthma. *J Immunol.* 2011;186(12):7225-7231.
282. Blankenhaus B, Reitz M, Brenz Y, et al. Foxp3(+) regulatory T cells delay expulsion of intestinal nematodes by suppression of IL-9-driven mast cell activation in BALB/c but not in C57BL/6 mice. *PLoS Pathog.* 2014;10(2):e1003913.
283. Hernandez JD, Yu M, Sibilano R, et al. Development of multiple features of antigen-induced asthma pathology in a new strain of mast cell deficient BALB/c-Kit(W-sh/W-sh) mice. *Lab Invest.* 2020;100(4):516-526.
284. Bradding P, Arthur G. Mast cells in asthma--state of the art. *Clin Exp Allergy.* 2016;46(2):194-263.
285. Cerny-Reiterer S, Rabenhorst A, Stefanzl G, et al. Long-term treatment with imatinib results in profound mast cell deficiency in Ph+ chronic myeloid leukemia. *Oncotarget.* 2015;6(5):3071-3084.
286. Sayed BA, Christy A, Quirion MR, et al. The master switch: the role of mast cells in autoimmunity and tolerance. *Annu Rev Immunol.* 2008;26:705-739.
287. Min B, Paul WE. Basophils and type 2 immunity. *Curr Opin Hematol.* 2008;15(1):59-63.
288. Tsai M, Grimbaldeston M, Galli SJ. Mast cells and immunoregulation/immunomodulation. *Adv Exp Med Biol.* 2011;716:186-211.
289. Yamanishi Y, Miyake K, Iki M, et al. Recent advances in understanding basophil-mediated Th2 immune responses. *Immunol Rev.* 2017;278(1):237-245.
290. Chirumbolo S, Bjorklund G, Sboarina A, et al. The role of basophils as innate immune regulatory cells in allergy and immunotherapy. *Hum Vaccin Immunother.* 2018;14(4):815-831.
291. Malaviya R, Abraham SN. Role of mast cell leukotrienes in neutrophil recruitment and bacterial clearance in infectious peritonitis. *J Leukoc Biol.* 2000;67(6):841-846.
292. De Filippo K, Dudeck A, Hasenberg M, et al. Mast cell and macrophage chemokines CXCL1/CXCL2 control the early stage of neutrophil recruitment during tissue inflammation. *Blood.* 2013;121(24):4930-4937.
293. Wershil BK, Wang ZS, Gordon JR, et al. Recruitment of neutrophils during IgE-dependent cutaneous late phase reactions in the mouse is mast cell-dependent. Partial inhibition of the reaction with antiserum against tumor necrosis factor-alpha. *J Clin Invest.* 1991;87(2):446-453.
294. Biedermann T, Kneilling M, Mailhammer R, et al. Mast cells control neutrophil recruitment during T cell-mediated delayed-type hypersensitivity reactions through tumor necrosis factor and macrophage inflammatory protein 2. *J Exp Med.* 2000;192(10):1441-1452.
295. Weber FC, Nemeth T, Csepregi JZ, et al. Neutrophils are required for both the sensitization and elicitation phase of contact hypersensitivity. *J Exp Med.* 2015;212(1):15-22.
296. Nakamura Y, Kambe N, Saito M, et al. Mast cells mediate neutrophil recruitment and vascular leakage through the NLRP3 inflammasome in histamine-independent urticaria. *J Exp Med.* 2009;206(5):1037-1046.
297. Bonnekoh H, Scheffel J, Kambe N, et al. The role of mast cells in autoinflammation. *Immunol Rev.* 2018;282(1):265-275.
298. Shimokawa C, Kanaya T, Hachisuka M, et al. Mast cells are crucial for induction of group 2 innate lymphoid cells and clearance of helminth infections. *Immunity.* 2017;46(5):863.e4-874.e4.
299. Lefrancais E, Duval A, Mirey E, et al. Central domain of IL-33 is cleaved by mast cell proteases for potent activation of group-2 innate lymphoid cells. *Proc Natl Acad Sci U S A.* 2014;111(43):15502-15507.
300. Waern I, Lundequist A, Pejler G, et al. Mast cell chymase modulates IL-33 levels and controls allergic sensitization in dust-mite induced airway inflammation. *Mucosal Immunol.* 2013;6(5):911-920.
301. Jawdat DM, Albert EJ, Rowden G, et al. IgE-mediated mast cell activation induces Langerhans cell migration *in vivo*. *J Immunol.* 2004;173(8):5275-5282.
302. Reuter S, Dehzad N, Martin H, et al. Mast cells induce migration of dendritic cells in a murine model of acute allergic airway disease. *Int Arch Allergy Immunol.* 2010;151(3):214-222.
303. Caron G, Delneste Y, Roelandts E, et al. Histamine polarizes human dendritic cells into Th2 cell-promoting effector dendritic cells. *J Immunol.* 2001;167(7):3682-3686.
304. McIlroy A, Caron G, Blanchard S, et al. Histamine and prostaglandin E up-regulate the production of Th2-attracting chemokines (CCL17 and CCL22) and down-regulate IFN-gamma-induced CXCL10 production by immature human dendritic cells. *Immunology.* 2006;117(4):507-516.
305. Dawicki W, Jawdat DW, Xu N, et al. Mast cells, histamine, and IL-6 regulate the selective influx of dendritic cell subsets into an inflamed lymph node. *J Immunol.* 2010;184(4):2116-2123.
306. Theiner G, Gessner A, Lutz MB. The mast cell mediator PGD2 suppresses IL-12 release by dendritic cells leading to Th2 polarized immune responses *in vivo*. *Immunobiology.* 2006;211(6-8):463-472.
307. Suto H, Nakae S, Kakurai M, et al. Mast cell-associated TNF promotes dendritic cell migration. *J Immunol.* 2006;176(7):4102-4112.
308. Shelburne CP, Nakano H, St John AL, et al. Mast cells augment adaptive immunity by orchestrating dendritic cell trafficking through infected tissues. *Cell Host Microbe.* 2009;6(4):331-342.
309. Otsuka A, Kubo M, Honda T, et al. Requirement of interaction between mast cells and skin dendritic cells to establish contact hypersensitivity. *PLoS One.* 2011;6(9):e25538.
310. Dudeck J, Ghouse SM, Lehmann CH, et al. Mast-cell-derived TNF amplifies CD8(+) dendritic cell functionality and CD8(+) T cell priming. *Cell Rep.* 2015;13(2):399-411.
311. Heib V, Becker M, Warger T, et al. Mast cells are crucial for early inflammation, migration of Langerhans cells, and CTL responses following topical application of TLR7 ligand in mice. *Blood.* 2007;110(3):946-953.
312. Skokos D, Botros HG, Demeure C, et al. Mast cell-derived exosomes induce phenotypic and functional maturation of dendritic cells and elicit specific immune responses *in vivo*. *J Immunol.* 2003;170(6):3037-3045.
313. Carroll-Portillo A, Cannon JL, te Riet J, et al. Mast cells and dendritic cells form synapses that facilitate antigen transfer for T cell activation. *J Cell Biol.* 2015;210(5):851-864.

314. Dudeck J, Medyukhina A, Frobel J, et al. Mast cells acquire MHCII from dendritic cells during skin inflammation. *J Exp Med*. 2017;214(12):3791-3811.
315. Kambayashi T, Baranski JD, Baker RG, et al. Indirect involvement of allergen-captured mast cells in antigen presentation. *Blood*. 2008;111(3):1489-1496.
316. Wang HW, Tedla N, Lloyd AR, et al. Mast cell activation and migration to lymph nodes during induction of an immune response in mice. *J Clin Invest*. 1998;102(8):1617-1626.
317. Tanzola MB, Robbie-Ryan M, Gutekunst CA, et al. Mast cells exert effects outside the central nervous system to influence experimental allergic encephalomyelitis disease course. *J Immunol*. 2003;171(8):4385-4391.
318. Hochegger K, Siebenhaar F, Vielhauer V, et al. Role of mast cells in experimental anti-glomerular basement membrane glomerulonephritis. *Eur J Immunol*. 2005;35(10):3074-3082.
319. Byrne SN, Limon-Flores AY, Ullrich SE. Mast cell migration from the skin to the draining lymph nodes upon ultraviolet irradiation represents a key step in the induction of immune suppression. *J Immunol*. 2008;180(7):4648-4655.
320. Mekori YA, Hershko AY, Frossi B, et al. Integrating innate and adaptive immune cells: mast cells as crossroads between regulatory and effector B and T cells. *Eur J Pharmacol*. 2016;778:84-89.
321. Nakae S, Suto H, Iikura M, et al. Mast cells enhance T cell activation: importance of mast cell costimulatory molecules and secreted TNF. *J Immunol*. 2006;176(4):2238-2248.
322. Skokos D, Le Panse S, Villa I, et al. Mast cell-dependent B and T lymphocyte activation is mediated by the secretion of immunologically active exosomes. *J Immunol*. 2001;166(2):868-876.
323. Piconese S, Gri G, Tripodo C, et al. Mast cells counteract regulatory T-cell suppression through interleukin-6 and OX40/OX40L axis toward Th17-cell differentiation. *Blood*. 2009;114(13):2639-2648.
324. Merluzzi S, Frossi B, Gri G, et al. Mast cells enhance proliferation of B lymphocytes and drive their differentiation toward IgA-secreting plasma cells. *Blood*. 2010;115(14):2810-2817.
325. McLachlan JB, Hart JP, Pizzo SV, et al. Mast cell-derived tumor necrosis factor induces hypertrophy of draining lymph nodes during infection. *Nat Immunol*. 2003;4(12):1199-1205.
326. St John AL, Chan CY, Staats HF, et al. Synthetic mast-cell granules as adjuvants to promote and polarize immunity in lymph nodes. *Nat Mater*. 2012;11(3):250-257.
327. McGowen AL, Hale LP, Shelburne CP, et al. The mast cell activator compound 48/80 is safe and effective when used as an adjuvant for intradermal immunization with Bacillus anthracis protective antigen. *Vaccine*. 2009;27(27):3544-3552.
328. Meng S, Liu Z, Xu L, et al. Intranasal immunization with recombinant HA and mast cell activator C48/80 elicits protective immunity against 2009 pandemic H1N1 influenza in mice. *PLoS One*. 2011;6(5):e19863.
329. Staats HF, Fielhauer JR, Thompson AL, et al. Mucosal targeting of a BoNT/A subunit vaccine adjuvanted with a mast cell activator enhances induction of BoNT/A neutralizing antibodies in rabbits. *PLoS One*. 2011;6(1):e16532.
330. Zeng L, Liu Y, Wang H, et al. Compound 48/80 acts as a potent mucosal adjuvant for vaccination against Streptococcus pneumoniae infection in young mice. *Vaccine*. 2015;33(8):1008-1016.
331. Bento D, Staats HF, Goncalves T, et al. Development of a novel adjuvanted nasal vaccine: C48/80 associated with chitosan nanoparticles as a path to enhance mucosal immunity. *Eur J Pharm Biopharm*. 2015;93:149-164.
332. Fang Y, Xiang Z. Roles and relevance of mast cells in infection and vaccination. *J Biomed Res*. 2016;30(4):253-263.
333. Hart PH, Grimbaldeston MA, Swift GJ, et al. Dermal mast cells determine susceptibility to ultraviolet B-induced systemic suppression of contact hypersensitivity responses in mice. *J Exp Med*. 1998;187(12):2045-2053.
334. Lu LF, Lind EF, Gondek DC, et al. Mast cells are essential intermediaries in regulatory T-cell tolerance. *Nature*. 2006;442(7106):997-1002.
335. Depinay N, Hacini F, Beghdadi W, et al. Mast cell-dependent down-regulation of antigen-specific immune responses by mosquito bites. *J Immunol*. 2006;176(7):4141-4146.
336. Grimbaldeston MA, Nakae S, Kalesnikoff J, et al. Mast cell-derived interleukin 10 limits skin pathology in contact dermatitis and chronic irradiation with ultraviolet B. *Nat Immunol*. 2007;8(10):1095-1104.
337. Norman MU, Hwang J, Hulliger S, et al. Mast cells regulate the magnitude and the cytokine microenvironment of the contact hypersensitivity response. *Am J Pathol*. 2008;172(6):1638-1649.
338. Hershko AY, Suzuki R, Charles N, et al. Mast cell interleukin-2 production contributes to suppression of chronic allergic dermatitis. *Immunity*. 2011;35(4):562-571.
339. Leveson-Gower DB, Sega EI, Kalesnikoff J, et al. Mast cells suppress murine GVHD in a mechanism independent of CD4+CD25+ regulatory T cells. *Blood*. 2013;122(22):3659-3665.
340. Rivellese F, Suurmond J, Habets K, et al. Ability of interleukin-33-and immune complex-triggered activation of human mast cells to down-regulate monocyte-mediated immune responses. *Arthritis Rheumatol*. 2015;67(9):2343-2353.
341. Gimenez-Rivera VA, Siebenhaar F, Zimmermann C, et al. Mast cells limit the exacerbation of chronic allergic contact dermatitis in response to repeated allergen exposure. *J Immunol*. 2016;197(11):4240-4246.
342. Salamon P, Shefler I, Moshkovits I, et al. IL-33 and IgE stimulate mast cell production of IL-2 and regulatory T cell expansion in allergic dermatitis. *Clin Exp Allergy*. 2017;47(11):1409-1416.
343. Gibbs BF, Haas H, Falcone FH, et al. Purified human peripheral blood basophils release interleukin-13 and preformed interleukin-4 following immunological activation. *Eur J Immunol*. 1996;26(10):2493-2498.
344. Piccinni MP, Macchia D, Parronchi P, et al. Human bone marrow non-B, non-T cells produce interleukin 4 in response to cross-linkage of Fc epsilon and Fc gamma receptors. *Proc Natl Acad Sci U S A*. 1991;88(19):8656-8660.
345. Seder RA, Paul WE, Dvorak AM, et al. Mouse splenic and bone marrow cell populations that express high-affinity Fc epsilon receptors and produce interleukin 4 are highly enriched in basophils. *Proc Natl Acad Sci U S A*. 1991;88(7):2835-2839.
346. Schroeder JT. Basophils beyond effector cells of allergic inflammation. *Adv Immunol*. 2009;101:123-161.
347. Hida S, Tadachi M, Saito T, et al. Negative control of basophil expansion by IRF-2 critical for the regulation of Th1/Th2 balance. *Blood*. 2005;106(6):2011-2017.
348. Oh K, Shen T, Le Gros G, et al. Induction of Th2 type immunity in a mouse system reveals a novel immunoregulatory role of basophils. *Blood*. 2007;109(7):2921-2927.
349. Charles N, Watford WT, Ramos HL, et al. Lyn kinase controls basophil GATA-3 transcription factor expression and induction of Th2 cell differentiation. *Immunity*. 2009;30(4):533-543.
350. Khodoun MV, Orekhova T, Potter C, et al. Basophils initiate IL-4 production during a memory T-dependent response. *J Exp Med*. 2004;200(7):857-870.
351. Denzel A, Maus UA, Rodriguez Gomez M, et al. Basophils enhance immunological memory responses. *Nat Immunol*. 2008;9(7):733-742.
352. Charles N, Hardwick D, Daugas E, et al. Basophils and the T helper 2 environment can promote the development of lupus nephritis. *Nat Med*. 2010;16(6):701-707.
353. Watanabe T, Yamashita K, Sakurai T, et al. Toll-like receptor activation in basophils contributes to the development of IgG4-related disease. *J Gastroenterol*. 2013;48(2):247-253.
354. Mukai K, Matsuoka K, Taya C, et al. Basophils play a critical role in the development of IgE-mediated chronic allergic inflammation independently of T cells and mast cells. *Immunity*. 2005;23(2):191-202.
355. Obata K, Mukai K, Tsujimura Y, et al. Basophils are essential initiators of a novel type of chronic allergic inflammation. *Blood*. 2007;110(3):913-920.
356. Nakashima C, Otsuka A, Kitoh A, et al. Basophils regulate the recruitment of eosinophils in a murine model of irritant contact dermatitis. *J Allergy Clin Immunol*. 2014;134(1):100-107.
357. Cheng LE, Sullivan BM, Retana LE, et al. IgE-activated basophils regulate eosinophil tissue entry by modulating endothelial function. *J Exp Med*. 2015;212(4):513-524.
358. Kim BS, Wang K, Siracusa MC, et al. Basophils promote innate lymphoid cell responses in inflamed skin. *J Immunol*. 2014;193(7):3717-3725.
359. Motomura Y, Morita H, Moro K, et al. Basophil-derived interleukin-4 controls the function of natural helper cells, a member of ILC2s, in lung inflammation. *Immunity*. 2014;40(5):758-771.
360. Noti M, Kim BS, Siracusa MC, et al. Exposure to food allergens through inflamed skin promotes intestinal food allergy through the thymic stromal lymphopoietin-basophil axis. *J Allergy Clin Immunol*. 2014;133(5):1390-1399, 1399.e1-1399.e6.
361. Egawa M, Mukai K, Yoshikawa S, et al. Inflammatory monocytes recruited to allergic skin acquire an anti-inflammatory M2 phenotype via basophil-derived interleukin-4. *Immunity*. 2013;38(3):570-580.
362. Obata-Ninomiya K, Ishiwata K, Tsutsui H, et al. The skin is an important bulwark of acquired immunity against intestinal helminths. *J Exp Med*. 2013;210(12):2583-2595.
363. Borriello F, Longo M, Spinelli R, et al. IL-3 synergises with basophil-derived IL-4 and IL-13 to promote the alternative activation of human monocytes. *Eur J Immunol*. 2015;45(7):2042-2051.
364. Rivellese F, Suurmond J, de Paulis A, et al. IgE and IL-33-mediated triggering of human basophils inhibits its TLR4-induced monocyte activation. *Eur J Immunol*. 2014;44(10):3045-3055.
365. Gomez MR, Talke Y, Hofmann C, et al. Basophils control T-cell responses and limit disease activity in experimental murine colitis. *Mucosal Immunol*. 2014;7(1):188-199.
366. Kim S, Shen T, Min B. Basophils can directly present or cross-present antigen to CD8 lymphocytes and alter CD8 T cell differentiation into IL-10-producing phenotypes. *J Immunol*. 2009;183(5):3033-3039.
367. Sokol CL, Chu NQ, Yu S, et al. Basophils function as antigen-presenting cells for an allergen-induced T helper type 2 response. *Nat Immunol*. 2009;10(7):713-720.
368. Yoshimoto T, Yasuda K, Tanaka H, et al. Basophils contribute to T(H)2-IgE responses *in vivo* via IL-4 production and presentation of peptide-MHC class II complexes to CD4+ T cells. *Nat Immunol*. 2009;10(7):706-712.
369. Perrigoue JG, Saenz SA, Siracusa MC, et al. MHC class II-dependent basophil-CD4+ T cell interactions promote T(H)2 cytokine-dependent immunity. *Nat Immunol*. 2009;10(7):697-705.
370. Miyake K, Shiozawa N, Nagao T, et al. Trogocytosis of peptide-MHC class II complexes from dendritic cells confers antigen-presenting ability on basophils. *Proc Natl Acad Sci U S A*. 2017;114(5):1111-1116.
371. Voskamp AL, Prickett SR, Mackay F, et al. MHC class II expression in human basophils: induction and lack of functional significance. *PLoS One*. 2013;8(12):e81777.
372. Choi HW, Suwanpradid J, Kim IH, et al. Perivascular dendritic cells elicit anaphylaxis by relaying allergens to mast cells via microvesicles. *Science*. 2018;362(6415):eaao0666.
373. Commins SP, Jerath MR, Cox K, et al. Delayed anaphylaxis to alpha-gal, an oligosaccharide in mammalian meat. *Allergol Int*. 2016;65(1):16-20.
374. Galili U. Evolution and pathophysiology of the human natural anti-alpha-galactosyl IgG (anti-Gal) antibody. *Springer Semin Immunopathol*. 1993;15(2-3):155-171.
375. Hamsten C, Starkhammar M, Tran TA, et al. Identification of galactose-alpha-1,3-galactose in the gastrointestinal tract of the tick Ixodes ricinus; possible relationship with red meat allergy. *Allergy*. 2013;68(4):549-552.
376. Kepley CL, McFeeley PJ, Oliver JM, et al. Immunohistochemical detection of human basophils in postmortem cases of fatal asthma. *Am J Respir Crit Care Med*. 2001;164(6):1053-1058.
377. Rivera J, Gilfillan AM. Molecular regulation of mast cell activation. *J Allergy Clin Immunol*. 2006;117(6):1214-1225; quiz 26.
378. Kraft S, Kinet JP. New developments in FcεRI regulation, function and inhibition. *Nat Rev Immunol*. 2007;7(5):365-378.
379. Varricchi G, Pecoraro A, Loffredo S, et al. Heterogeneity of human mast cells with respect to MRGPRX2 receptor expression and function. *Front Cell Neurosci*. 2019;13:299.
380. Babina M. The pseudo-allergic/neurogenic route of mast cell activation via MRGPRX2: discovery, functional programs, regulation, relevance to disease, and relation with allergic stimulation. *Itch*. 2020;5(2):e32.
381. Ong YE, Menzies-Gow A, Barkans J, et al. Anti-IgE (omalizumab) inhibits late-phase reactions and inflammatory cells after repeat skin allergen challenge. *J Allergy Clin Immunol*. 2005;116(3):558-564.
382. Conner E, Bochner BS, Brummet M, et al. The effect of etanercept on the human cutaneous allergic response. *J Allergy Clin Immunol*. 2008;121(1):258-260.
383. Yu M, Tsai M, Tam SY, et al. Mast cells can promote the development of multiple features of chronic asthma in mice. *J Clin Invest*. 2006;116(6):1633-1641.
384. Yu M, Eckart MR, Morgan AA, et al. Identification of an IFN-γ/mast cell axis in a mouse model of chronic asthma. *J Clin Invest*. 2011;121(8):3133-3143.
385. Rihet P, Demeure CE, Bourgois A, et al. Evidence for an association between human resistance to Schistosoma mansoni and high anti-larval IgE levels. *Eur J Immunol*. 1991;21(11):2679-2686.
386. Faulkner H, Turner J, Kamgno J, et al. Age-and infection intensity-dependent cytokine and antibody production in human trichuriasis: the importance of IgE. *J Infect Dis*. 2002;185(5):665-672.
387. Mukai K, Tsai M, Starkl P, et al. IgE and mast cells in host defense against parasites and venoms. *Semin Immunopathol*. 2016;38(5):581-603.
388. Schneider LA, Schlenner SM, Feyerabend TB, et al. Molecular mechanism of mast cell mediated innate defense against endothelin and snake venom sarafotoxin. *J Exp Med*. 2007;204(11):2629-2639.
389. Anderson E, Stavenhagen K, Kolarich D, et al. Human mast cell tryptase is a potential treatment for snakebite envenoming across multiple snake species. *Front Immunol*. 2018;9:1532.
390. Palm NW, Rosenstein RK, Yu S, et al. Bee venom phospholipase A2 induces a primary type 2 response that is dependent on the receptor ST2 and confers protective immunity. *Immunity*. 2013;39(5):976-985.
391. Johnzon CF, Ronnberg E, Pejler G. The role of mast cells in bacterial infection. *Am J Pathol*. 2016;186(1):4-14.
392. Graham AC, Temple RM, Obar JJ. Mast cells and influenza a virus: association with allergic responses and beyond. *Front Immunol*. 2015;6:238.
393. Saluja R, Metz M, Maurer M. Role and relevance of mast cells in fungal infections. *Front Immunol*. 2012;3:146.
394. Fitzsimmons CM, Dunne DW. Survival of the fittest: allergology or parasitology? *Trends Parasitol*. 2009;25(10):447-451.
395. Pulendran B, Artis D. New paradigms in type 2 immunity. *Science*. 2012;337(6093):431-435.
396. Karasuyama H, Tabakawa Y, Ohta T, et al. Crucial role for basophils in acquired protective immunity to tick infestation. *Front Physiol*. 2018;9:1769.
397. Pennock JL, Grencis RK. The mast cell and gut nematodes: damage and defence. *Chem Immunol Allergy*. 2006;90:128-140.
398. Wada T, Ishiwata K, Koseki H, et al. Selective ablation of basophils in mice reveals their nonredundant role in acquired immunity against ticks. *J Clin Invest*. 2010;120(8):2867-2875.

399. Spencer LA, Porte P, Zetoff C, et al. Mice genetically deficient in immunoglobulin E are more permissive hosts than wild-type mice to a primary, but not secondary, infection with the filarial nematode *Brugia malayi*. *Infect Immun.* 2003;71(5):2462-2467.
400. Schwartz C, Turqueti-Neves A, Hartmann S, et al. Basophil-mediated protection against gastrointestinal helminths requires IgE-induced cytokine secretion. *Proc Natl Acad Sci U S A.* 2014;111(48):E5169-E5177.
401. Nawa Y, Kiyota M, Korenaga M, et al. Defective protective capacity of W/Wv mice against Strongyloides ratti infection and its reconstitution with bone marrow cells. *Parasite Immunol.* 1985;7(4):429-438.
402. Knight PA, Wright SH, Lawrence CE, et al. Delayed expulsion of the nematode *Trichinella spiralis* in mice lacking the mucosal mast cell-specific granule chymase, mouse mast cell protease-1. *J Exp Med.* 2000;192(12):1849-1856.
403. Furuta T, Kikuchi T, Iwakura Y, et al. Protective roles of mast cells and mast cell-derived TNF in murine malaria. *J Immunol.* 2006;177(5):3294-3302.
404. Maurer M, Lopez Kostka S, Siebenhaar F, et al. Skin mast cells control T cell-dependent host defense in Leishmania major infections. *FASEB J.* 2006;20(14):2460-2467.
405. Ohnmacht C, Voehringer D. Basophils protect against reinfection with hookworms independently of mast cells and memory Th2 cells. *J Immunol.* 2010;184(1):344-350.
406. Arizono N, Kasugai T, Yamada M, et al. Infection of Nippostrongylus brasiliensis induces development of mucosal-type but not connective tissue-type mast cells in genetically mast cell-deficient Ws/Ws rats. *Blood.* 1993;81(10):2572-2578.
407. Amiri P, Haak-Frendscho M, Robbins K, et al. Anti-immunoglobulin E treatment decreases worm burden and egg production in Schistosoma mansoni-infected normal and interferon gamma knockout mice. *J Exp Med.* 1994;180(1):43-51.
408. Newlands GF, Miller HR, MacKellar A, et al. Stem cell factor contributes to intestinal mucosal mast cell hyperplasia in rats infected with Nippostrongylus brasiliensis or *Trichinella spiralis*, but anti-stem cell factor treatment decreases parasite egg production during N brasiliensis infection. *Blood.* 1995;86(5):1968-1976.
409. Hepworth MR, Maurer M, Hartmann S. Regulation of type 2 immunity to helminths by mast cells. *Gut Microb.* 2012;3(5):476-481.
410. Lawrence CE, Paterson YY, Wright SH, et al. Mouse mast cell protease-1 is required for the enteropathy induced by gastrointestinal helminth infection in the mouse. *Gastroenterology.* 2004;127(1):155-165.
411. Shin K, Watts GF, Oettgen HC, et al. Mouse mast cell tryptase mMCP-6 is a critical link between adaptive and innate immunity in the chronic phase of *Trichinella spiralis* infection. *J Immunol.* 2008;180(7):4885-4891.
412. Abe T, Nawa Y. Worm expulsion and mucosal mast cell response induced by repetitive IL-3 administration in Strongyloides ratti-infected nude mice. *Immunology.* 1988;63(2):181-185.
413. Abe T, Sugaya H, Ishida K, et al. Intestinal protection against Strongyloides ratti and mastocytosis induced by administration of interleukin-3 in mice. *Immunology.* 1993;80(1):116-121.
414. Korenaga M, Abe T, Hashiguchi Y. Injection of recombinant interleukin 3 hastens worm expulsion in mice infected with *Trichinella spiralis. Parasitol Res.* 1996;82(2):108-113.
415. Sasaki Y, Yoshimoto T, Maruyama H, et al. IL-18 with IL-2 protects against Strongyloides venezuelensis infection by activating mucosal mast cell-dependent type 2 innate immunity. *J Exp Med.* 2005;202(5):607-616.
416. Grencis RK, Else KJ, Huntley JF, et al. The *in vivo* role of stem cell factor (c-kit ligand) on mastocytosis and host protective immunity to the intestinal nematode *Trichinella spiralis* in mice. *Parasite Immunol.* 1993;15(1):55-59.
417. Donaldson LE, Schmitt E, Huntley JF, et al. A critical role for stem cell factor and c-kit in host protective immunity to an intestinal helminth. *Int Immunol.* 1996;8(4):559-567.
418. Reitz M, Brunn ML, Rodewald HR, et al. Mucosal mast cells are indispensable for the timely termination of *Strongyloides ratti* infection. *Mucosal Immunol.* 2017;10(2):481-492.
419. Reitz M, Brunn ML, Voehringer D, et al. Basophils are dispensable for the establishment of protective adaptive immunity against primary and challenge infection with the intestinal helminth parasite Strongyloides ratti. *PLoS Negl Trop Dis.* 2018;12(11):e0006992.
420. Reitz M, Hartmann W, Rudiger N, et al. Interleukin-9 promotes early mast cell-mediated expulsion of Strongyloides ratti but is dispensable for generation of protective memory. *Sci Rep.* 2018;8(1):8636.
421. McKean PG, Pritchard DI. The action of a mast cell protease on the cuticular collagens of Necator americanus. *Parasite Immunol.* 1989;11(3):293-297.
422. Vermillion DL, Ernst PB, Scicchitano R, et al. Antigen-induced contraction of jejunal smooth muscle in the sensitized rat. *Am J Physiol.* 1988;255(6 pt 1):G701-G708.
423. Marzio L, Blennerhassett P, Vermillion D, et al. Distribution of mast cells in intestinal muscle of nematode-sensitized rats. *Am J Physiol.* 1992;262(3 pt 1):G477-G482.
424. Bertaccini G, Morini G, Coruzzi G. Different mechanisms are responsible for the contractile effects of histaminergic compounds on isolated intestinal smooth muscle cells. *J Physiol Paris.* 1997;91(3-5):199-202.
425. Madden KB, Whitman L, Sullivan C, et al. Role of STAT6 and mast cells in IL-4-and IL-13-induced alterations in murine intestinal epithelial cell function. *J Immunol.* 2002;169(8):4417-4422.
426. McDermott JR, Bartram RE, Knight PA, et al. Mast cells disrupt epithelial barrier function during enteric nematode infection. *Proc Natl Acad Sci U S A.* 2003;100(13):7761-7766.
427. Vukman KV, Lalor R, Aldridge A, et al. Mast cells: new therapeutic target in helminth immune modulation. *Parasite Immunol.* 2016;38(1):45-52.
428. Ohnmacht C, Schwartz C, Panzer M, et al. Basophils orchestrate chronic allergic dermatitis and protective immunity against helminths. *Immunity.* 2010;33(3):364-374.
429. Abe T, Nawa Y. Localization of mucosal mast cells in W/Wv mice after reconstitution with bone marrow cells or cultured mast cells, and its relation to the protective capacity to Strongyloides ratti infection. *Parasite Immunol.* 1987;9(4):477-485.
430. Katakura K, Saito S, Hamada A, et al. Cutaneous leishmaniasis in mast cell-deficient W/Wv mice. *Infect Immun.* 1993;61(5):2242-2244.
431. Paul C, Wolff S, Zapf T, et al. Mast cells have no impact on cutaneous leishmaniasis severity and related Th2 differentiation in resistant and susceptible mice. *Eur J Immunol.* 2016;46(1):114-121.
432. Wershil BK, Theodos CM, Galli SJ, et al. Mast cells augment lesion size and persistence during experimental Leishmania major infection in the mouse. *J Immunol.* 1994;152(9):4563-4571.
433. Prendergast CT, Sanin DE, Mountford AP. CD4 T cell hyporesponsiveness induced by schistosome larvae is not dependent upon eosinophils but may involve connective tissue mast cells. *Parasite Immunol.* 2016;38(2):81-92.
434. Gould HJ, Sutton BJ. IgE in allergy and asthma today. *Nat Rev Immunol.* 2008;8(3):205-217.
435. Joseph M, Auriault C, Capron A, et al. A new function for platelets: IgE-dependent killing of schistosomes. *Nature.* 1983;303(5920):810-812.
436. Dunne DW, Butterworth AE, Fulford AJ, et al. Immunity after treatment of human schistosomiasis: association between IgE antibodies to adult worm antigens and resistance to reinfection. *Eur J Immunol.* 1992;22(6):1483-1494.
437. Capron M, Capron A. Immunoglobulin E and effector cells in schistosomiasis. *Science.* 1994;264(5167):1876-1877.
438. Gounni AS, Lamkhioued B, Ochiai K, et al. High-affinity IgE receptor on eosinophils is involved in defence against parasites. *Nature.* 1994;367(6459):183-186.
439. Bischoff SC. Role of mast cells in allergic and non-allergic immune responses: comparison of human and murine data. *Nat Rev Immunol.* 2007;7(2):93-104.
440. Piliponsky AM, Chen CC, Rios EJ, et al. The chymase mouse mast cell protease 4 degrades TNF, limits inflammation, and promotes survival in a model of sepsis. *Am J Pathol.* 2012;181(3):875-886.
441. Gommerman JL, Oh DY, Zhou X, et al. A role for CD21/CD35 and CD19 in responses to acute septic peritonitis: a potential mechanism for mast cell activation. *J Immunol.* 2000;165(12):6915-6921.
442. Supajatura V, Ushio H, Nakao A, et al. Protective roles of mast cells against enterobacterial infection are mediated by Toll-like receptor 4. *J Immunol.* 2001;167(4):2250-2256.
443. Piliponsky AM, Chen CC, Nishimura T, et al. Neurotensin increases mortality and mast cells reduce neurotensin levels in a mouse model of sepsis. *Nat Med.* 2008;14(4):392-398.
444. Di Nardo A, Vitiello A, Gallo RL. Cutting edge: mast cell antimicrobial activity is mediated by expression of cathelicidin antimicrobial peptide. *J Immunol.* 2003;170(5):2274-2278.
445. Chang YL, Wang Z, Igawa S, et al. Lipocalin 2: a new antimicrobial in mast cells. *Int J Mol Sci.* 2019;20(10):2380.
446. Arifuzzaman M, Mobley YR, Choi HW, et al. MRGPR-mediated activation of local mast cells clears cutaneous bacterial infection and protects against reinfection. *Sci Adv.* 2019;5(1):eaav0216.
447. Pundir P, Liu R, Vasavda C, et al. A connective tissue mast-cell-specific receptor detects bacterial quorum-sensing molecules and mediates antibacterial immunity. *Cell Host Microbe.* 2019;26(1):114.e8-122.e8.
448. Mallen-St Clair J, Pham CT, Villalta SA, et al. Mast cell dipeptidyl peptidase I mediates survival from sepsis. *J Clin Invest.* 2004;113(4):628-634.
449. Chan CY, St John AL, Abraham SN. Mast cell interleukin-10 drives localized tolerance in chronic bladder infection. *Immunity.* 2013;38(2):349-359.
450. Nakamura Y, Oscherwitz J, Cease KB, et al. Staphylococcus delta-toxin induces allergic skin disease by activating mast cells. *Nature.* 2013;503(7476):397-401.
451. de Paulis A, Prevete N, Fiorentino I, et al. Basophils infiltrate human gastric mucosa at sites of *Helicobacter pylori* infection, and exhibit chemotaxis in response to *H. pylori*-derived peptide Hp(2-20). *J Immunol.* 2004;172(12):7734-7743.
452. Bieneman AP, Chichester KL, Chen YH, et al. Toll-like receptor 2 ligands activate human basophils for both IgE-dependent and IgE-independent secretion. *J Allergy Clin Immunol.* 2005;115(2):295-301.
453. Komiya A, Nagase H, Okugawa S, et al. Expression and function of toll-like receptors in human basophils. *Int Arch Allergy Immunol.* 2006;140(suppl 1):23-27.
454. Bleriot C, Dupuis T, Jouvion G, et al. Liver-resident macrophage necroptosis orchestrates type 1 microbicidal inflammation and type-2-mediated tissue repair during bacterial infection. *Immunity.* 2015;42(1):145-158.
455. Yousefi S, Simon D, Stojkov D, et al. *In vivo* evidence for extracellular DNA trap formation. *Cell Death Dis.* 2020;11(4):300.
456. Piliponsky AM, Acharya M, Shubin NJ. Mast cells in viral, bacterial, and fungal infection immunity. *Int J Mol Sci.* 2019;20(12):2851.
457. Brown MG, Huang YY, Marshall JS, et al. Dramatic caspase-dependent apoptosis in antibody-enhanced dengue virus infection of human mast cells. *J Leukoc Biol.* 2009;85(1):71-80.
458. Wan SW, Wu-Hsieh BA, Lin YS, et al. The monocyte-macrophage-mast cell axis in dengue pathogenesis. *J Biomed Sci.* 2018;25(1):77.
459. St John AL. Influence of mast cells on dengue protective immunity and immune pathology. *PLoS Pathog.* 2013;9(12):e1003783.
460. Troupin A, Shirley D, Londono-Renteria B, et al. A role for human skin mast cells in dengue virus infection and systemic spread. *J Immunol.* 2016;197(11):4382-4391.
461. Demeure CE, Brahimi K, Hacini F, et al. Anopheles mosquito bites activate cutaneous mast cells leading to a local inflammatory response and lymph node hyperplasia. *J Immunol.* 2005;174(7):3932-3940.
462. Mantri CK, St John AL. Immune synapses between mast cells and γδ T cells limit viral infection. *J Clin Invest.* 2019;129(3):1094-1108.
463. Masri MFB, Mantri CK, Rathore APS, et al. Peripheral serotonin causes dengue virus-induced thrombocytopenia through 5HT2 receptors. *Blood.* 2019;133(21):2325-2337.
464. Furuta T, Murao LA, Lan NT, et al. Association of mast cell-derived VEGF and proteases in Dengue shock syndrome. *PLoS Negl Trop Dis.* 2012;6(2):e1505.
465. St John AL, Rathore AP, Raghavan B, et al. Contributions of mast cells and vasoactive products, leukotrienes and chymase, to dengue virus-induced vascular leakage. *Elife.* 2013;2:e00481.
466. Bannert N, Farzan M, Friend DS, et al. Human mast cell progenitors can be infected by macrophagetropic human immunodeficiency virus type 1 and retain virus with maturation *in vitro*. *J Virol.* 2001;75(22):10808-10814.
467. Sundstrom JB, Little DM, Villinger F, et al. Signaling through Toll-like receptors triggers HIV-1 replication in latently infected mast cells. *J Immunol.* 2004;172(7):4391-4401.
468. Sundstrom JB, Ellis JE, Hair GA, et al. Human tissue mast cells are an inducible reservoir of persistent HIV infection. *Blood.* 2007;109(12):5293-5300.
469. Li Y, Li LX, Wadley R, et al. Mast cells/basophils in the peripheral blood of allergic individuals who are HIV-1 susceptible due to their surface expression of CD4 and the chemokine receptors CCR3, CCR5, and CXCR4. *Blood.* 2001;97(11):3484-3490.
470. Stebbings JH Jr. Immediate hypersensitivity: a defense against arthropods? *Perspect Biol Med.* 1974;17(2):233-239.
471. Profet M. The function of allergy: immunological defense against toxins. *Q Rev Biol.* 1991;66(1):23-62.
472. Sahiner UM, Durham SR. Hymenoptera venom allergy: how does venom immunotherapy prevent anaphylaxis from bee and wasp stings? *Front Immunol.* 2019;10:1959.
473. Higginbotham RD. Mast cells and local resistance to Russell's viper venom. *J Immunol.* 1965;95(5):867-875.
474. Higginbotham RD, Karnella S. The significance of the mast cell response to bee venom. *J Immunol.* 1971;106(1):233-240.
475. Douaiher J, Succar J, Lancerotto L, et al. Development of mast cells and importance of their tryptase and chymase serine proteases in inflammation and wound healing. *Adv Immunol.* 2014;122:211-252.
476. Palm NW, Rosenstein RK, Medzhitov R. Allergic host defences. *Nature.* 2012;484(7395):465-472.
477. Brown MA. Studies of mast cells: adventures in serendipity. *Front Immunol.* 2018;9:520.
478. Sanjuan MA, Sagar D, Kolbeck R. Role of IgE in autoimmunity. *J Allergy Clin Immunol.* 2016;137(6):1651-1661.
479. Chen R, Ning G, Zhao ML, et al. Mast cells play a key role in neutrophil recruitment in experimental bullous pemphigoid. *J Clin Invest.* 2001;108(8):1151-1158.
480. Inaba Y, Kanazawa N, Yoshimasu T, et al. Severer lupus erythematosus-like skin lesions in MRL/lpr mice with homozygous Kit(wsh/wsh) mutation. *Mod Rheumatol.* 2018;28(2):319-326.
481. Pellefigues C, Charles N. The deleterious role of basophils in systemic lupus erythematosus. *Curr Opin Immunol.* 2013;25(6):704-711.
482. Pan Q, Gong L, Xiao H, et al. Basophil activation-dependent autoantibody and interleukin-17 production exacerbate systemic lupus erythematosus. *Front Immunol.* 2017;8:348.

483. Nigrovic PA, Lee DM. Mast cells in inflammatory arthritis. *Arthritis Res Ther.* 2005;7(1):1-11.
484. Nigrovic PA, Binstadt BA, Monach PA, et al. Mast cells contribute to initiation of autoantibody-mediated arthritis via IL-1. *Proc Natl Acad Sci U S A.* 2007;104(7):2325-2330.
485. Wang Q, Lepus CM, Raghu H, et al. IgE-mediated mast cell activation promotes inflammation and cartilage destruction in osteoarthritis. *Elife.* 2019;8:e39905.
486. Noordenbos T, Yeremenko N, Gofita I, et al. Interleukin-17-positive mast cells contribute to synovial inflammation in spondylarthritis. *Arthritis Rheum.* 2012;64(1):99-109.
487. Chen S, Noordenbos T, Blijdorp I, et al. Histologic evidence that mast cells contribute to local tissue inflammation in peripheral spondyloarthritis by regulating interleukin-17A content. *Rheumatology.* 2019;58(4):617-627.
488. Lee SJ, Nam KI, Jin HM, et al. Bone destruction by receptor activator of nuclear factor kappaB ligand-expressing T cells in chronic gouty arthritis. *Arthritis Res Ther.* 2011;13(5):R164.
489. Reber LL, Marichal T, Sokolove J, et al. Contribution of mast cell-derived interleukin-1beta to uric acid crystal-induced acute arthritis in mice. *Arthritis Rheumatol.* 2014;66(10):2881-2891.
490. Rivellese F, Mauro D, Nerviani A, et al. Mast cells in early rheumatoid arthritis associate with disease severity and support B cell autoantibody production. *Ann Rheum Dis.* 2018;77(12):1773-1781.
491. Rossini M, Viapiana O, Zanoni G, et al. Serum levels of tryptase suggest that mast cells might have an antiinflammatory role in rheumatoid arthritis: Comment on the article by Rivellese et al. *Arthritis Rheumatol.* 2016;68(3):769.
492. Lee DM, Friend DS, Gurish MF, et al. Mast cells: a cellular link between autoantibodies and inflammatory arthritis. *Science.* 2002;297(5587):1689-1692.
493. Zhou JS, Xing W, Friend DS, et al. Mast cell deficiency in Kit(W-sh) mice does not impair antibody-mediated arthritis. *J Exp Med.* 2007;204(12):2797-2802.
494. Pitman N, Asquith DL, Murphy G, et al. Collagen-induced arthritis is not impaired in mast cell-deficient mice. *Ann Rheum Dis.* 2011;70(6):1170-1171.
495. Feyerabend TB, Weiser A, Tietz A, et al. Cre-mediated cell ablation contests mast cell contribution in models of antibody-and T cell-mediated autoimmunity. *Immunity.* 2011;35(5):832-844.
496. Schubert N, Dudeck J, Liu P, et al. Mast cell promotion of T cell-driven antigen-induced arthritis despite being dispensable for antibody-induced arthritis in which T cells are bypassed. *Arthritis Rheumatol.* 2015;67(4):903-913.
497. Pedotti R, De Voss JJ, Steinman L, et al. Involvement of both 'allergic' and 'autoimmune' mechanisms in EAE, MS and other autoimmune diseases. *Trends Immunol.* 2003;24(9):479-484.
498. Brown M, Hatfield J. Mast cells are important modifiers of autoimmune disease: with so much evidence, why is there still controversy? *Front Immunol.* 2012;3:147.
499. Elieh-Ali-Komi D, Cao Y. Role of mast cells in the pathogenesis of multiple sclerosis and experimental autoimmune encephalomyelitis. *Clin Rev Allergy Immunol.* 2017;52(3):436-445.
500. Russi AE, Walker-Caulfield ME, Brown MA. Mast cell inflammasome activity in the meninges regulates EAE disease severity. *Clin Immunol.* 2018;189:14-22.
501. Robbie-Ryan M, Tanzola MB, Secor VH, et al. Cutting edge: both activating and inhibitory Fc receptors expressed on mast cells regulate experimental allergic encephalomyelitis disease severity. *J Immunol.* 2003;170(4):1630-1634.
502. Stelekati E, Bahri R, D'Orlando O, et al. Mast cell-mediated antigen presentation regulates CD8+ T cell effector functions. *Immunity.* 2009;31(4):665-676.
503. Piconese S, Costanza M, Musio S, et al. Exacerbated experimental autoimmune encephalomyelitis in mast-cell-deficient Kit W-sh/W-sh mice. *Lab Invest.* 2011;91(4):627-641.
504. Williams CM, Galli SJ. Mast cells can amplify airway reactivity and features of chronic inflammation in an asthma model in mice. *J Exp Med.* 2000;192(3):455-462.
505. Russi AE, Ebel ME, Yang Y, et al. Male-specific IL-33 expression regulates sex-dimorphic EAE susceptibility. *Proc Natl Acad Sci U S A.* 2018;115(7):E1520-E1529.

CHAPTER 17 Eosinophils

Elizabeth A. Jacobsen, PhD • Hirohito Kita, MD

INTRODUCTION

The eosinophil was first identified by Wharton Jones in 1846 in human peripheral blood. It was so named by Paul Ehrlich in 1879 because of the intense staining of its granules with the acidic dye eosin.[1] The eosinophil, or an eosinophil-like cell, has been evolutionarily maintained in vertebrates, including reptiles and fish,[2] suggesting that the eosinophil contributes favorable biology for the well-being of the animals. Indeed, the eosinophil has been thought to play a major role in host defense against parasite infection, in particular helminths.[3] On the other hand, the eosinophil is found to contribute to human disorders involving mucosal surfaces, such as asthma, chronic rhinosinusitis (CRS), and gastrointestinal disorders.[4-6] The eosinophil may be involved in the pathology of these diseases as a damaging inflammatory cell.[7] For the past 2 decades, our understanding of the immunobiology of eosinophils and their roles in health and diseases has advanced and evolved dramatically thanks to the availability of constitutive or conditional eosinophil-deficient mice.[8] These studies demonstrate that the eosinophil has immunoregulatory functions by promoting migration and functions of other immune cells and may play roles in maintenance of tissue homeostasis and repair in damaged organs. Furthermore, eosinophil infiltration has been recognized in a variety of disease conditions beyond helminth infection and allergic disorders. This chapter describes cell biology and immunologic aspects of eosinophils and mechanisms of eosinophilic disorders from a historical perspective to cutting-edge knowledge. The information is derived from *in vitro* and *in vivo* studies of both human and mouse eosinophils while the key differences in the eosinophil from these two species are noted.

EOSINOPHIL MORPHOLOGY, PRODUCTION, AND TISSUE DISTRIBUTION

Eosinophil Morphology

Both human and mouse eosinophils are characterized by their distinctive granules, as shown by their staining properties with acid dyes, such as eosin, and by their unique electron microscopic appearance[9] (Fig. 17.1). Human eosinophils contain granule proteins with higher cationic charges and stain more vibrantly to eosin than mouse eosinophils. The nuclei of circulating human eosinophils are bilobed or multilobed while those of mouse eosinophils show ring-like appearance[10] and can obtain multilobe appearance upon activation.

Human eosinophils contain four types of granules.[11] The "specific" granules or *secondary granules* that are reactive with acid dyes are composed of a crystalline electron-dense core and an electron-lucent matrix (Fig. 17.1). The core contains major basic protein (MBP) and the matrix contains eosinophil peroxidase (EPX), eosinophil cationic protein (ECP), and eosinophil-derived neurotoxin (EDN).[12] Certain cytokines, such as IL-4, are also localized in these specific granules.[13] Three other types of eosinophil granules have also been described: *Primary granules* are round, uniformly electron-dense, and characteristically seen in immature eosinophilic promyelocytes. *Small granules* contain acid phosphatase and arylsulfatase. *Secretory vesicles*, also referred to as microgranules or tubulovesicular with dumbbell-shaped structures, are the most abundant, and they contain membrane-bound cytochrome b558 (phox-22), CD11b (the α chain of a β_2 integrin), and albumin.[14] Eosinophils, in particular those with an activated phenotype, contain varying numbers of lipid bodies, which are nonmembrane-bound, lipid-rich inclusions, and serve a location for leukotriene synthesis.

Production

Eosinophils develop in the bone marrow from multipotent hematopoietic stem cells (HSCs) and are released into peripheral circulation as a mature leukocyte. Production of eosinophils involves a cascade of interdependent regulatory events of at least three classes of transcription factors, including GATA-1 (zinc family finger member), PU.1 (ETS family member), and C/EBP members (CCAAT/enhancer-binding protein family)[15,16] (Fig. 17.2). Expression of GATA-1 distinguishes a population of myeloid progenitors restricted to the mast cell, eosinophil, megakaryocyte, and erythroid lineages.[17] Specifically, human HSCs give rise directly to eosinophil/mast cell progenitors, from which committed eosinophil progenitors (EoPs) emerge that are recognized by the expression of the IL-5 receptor IL-5Rα (CD125) and the C-C chemokine receptor CCR3. In contrast, mouse EoPs are derived from granulocyte/macrophage progenitors.[18] PU.1 determines cell lineage fates, with relatively low levels inducing lymphocytic cells and high levels myeloid differentiation. C/EBPs are required for terminal differentiation of eosinophils.[19] Among these transcription factors, GATA-1 is likely the most critical as mice with a targeted deletion of the high-affinity GATA-1 binding site (ΔdblGATA-1) show a specific loss of eosinophils.[20] GATA-1 also regulates the expression of a number of eosinophil-associated genes, including those encoding MBP, Charcot-Leyden crystal (CLC) protein, CCR3, and IL-5Rα.[21] A number of unique genes, including

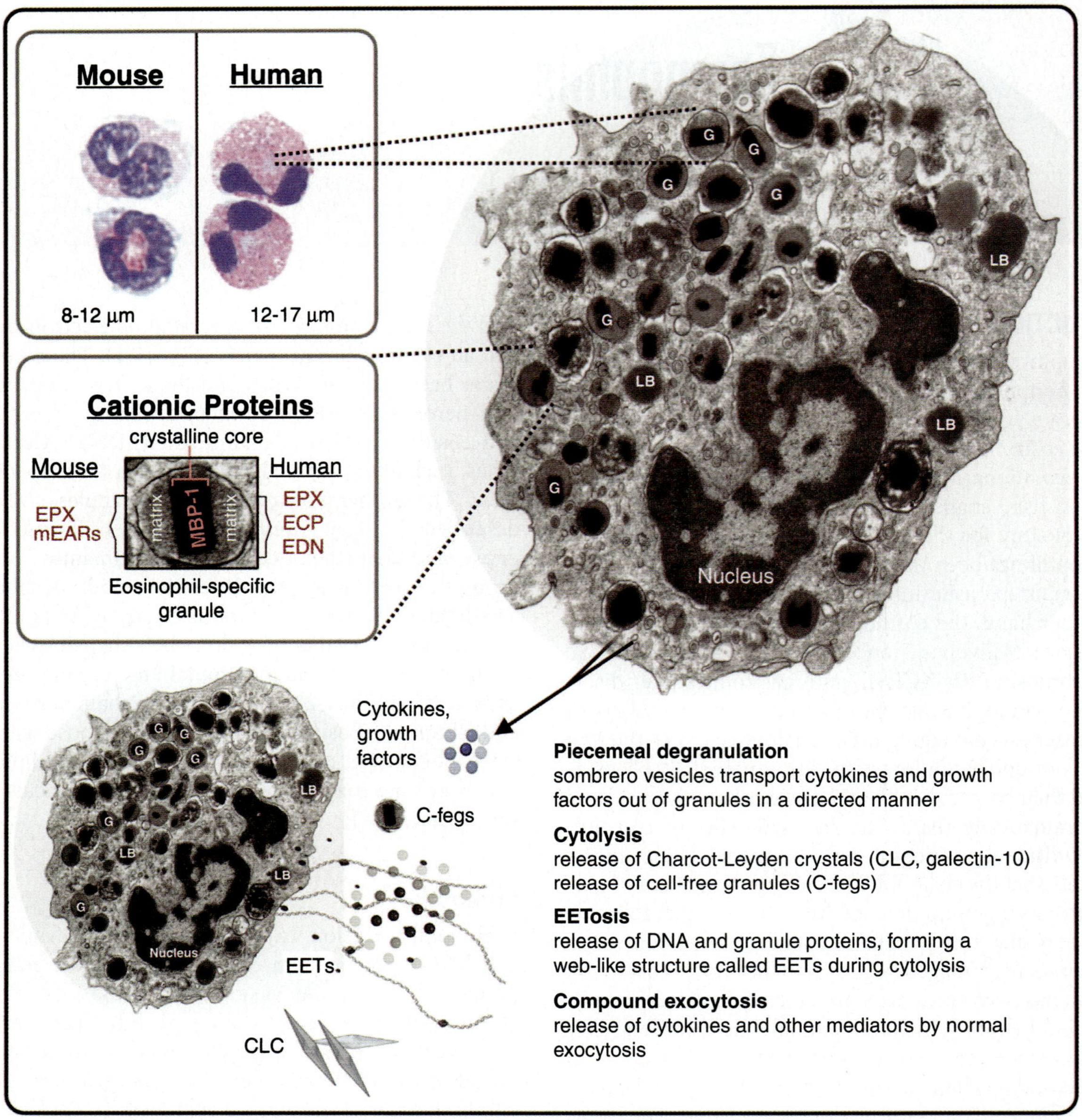

FIG. 17.1. **Eosinophil granule proteins and mechanisms of degranulation.** Light photomicrographs of mouse and human eosinophils stained with hematoxylin and eosin and an electron photomicrograph of a human eosinophil are shown. Location of granule proteins within the granule is indicated. Human eosinophils undergo three forms of granule release, including compound exocytosis (classical exocytosis), piecemeal degranulation, and cytolysis. Activated human eosinophils also release extracellular traps in a process called EETosis, whereby a net-like structure of DNA and granule proteins is released extracellularly. G, eosinophil-specific granule; LB, lipid body. (Created with BioRender.com.)

the Ikaros-family transcription factors Helios and Aiolos, are also expressed by eosinophils undergoing differentiation.[22]

Proliferation and differentiation of the eosinophil is induced by so-called eosinophil-activating cytokines and growth factors, including IL-3, granulocyte-macrophage colony-stimulating factor (GM-CSF), and IL-5. While IL-5 potently and specifically stimulates eosinophil production in the bone marrow, IL-3 and GM-CSF stimulate proliferation of neutrophils, basophils, and eosinophils.[23,24] Indeed, overproduction of IL-5 alone in transgenic mice results in profound eosinophilia.[25] Deletion of the IL-5 gene or the receptor for IL-5 (IL-5Rα) causes a eosinopenia,[26] yet not total elimination of eosinophils in the bone marrow and blood, providing evidence for a critical role for IL-5 in eosinophilopoiesis.

Eosinophilopoiesis is also regulated by several intrinsic factors. For example, the balance between paired immunoglobulin-like receptor A (PIR-A) and PIR-B plays an important role.[27] By recognizing major histocompatibility complex class I, PIR-B blocks the proapoptotic activity of PIR-A and allows IL-5–induced expansion of eosinophils. A long noncoding RNA *Morrbid* also prolongs the lifespan of myeloid cells, including eosinophils, by inhibiting the transcription of the proapoptotic gene *Bcl2l11* (*Bim*).[28] Importantly, successful biogenesis of their granules is

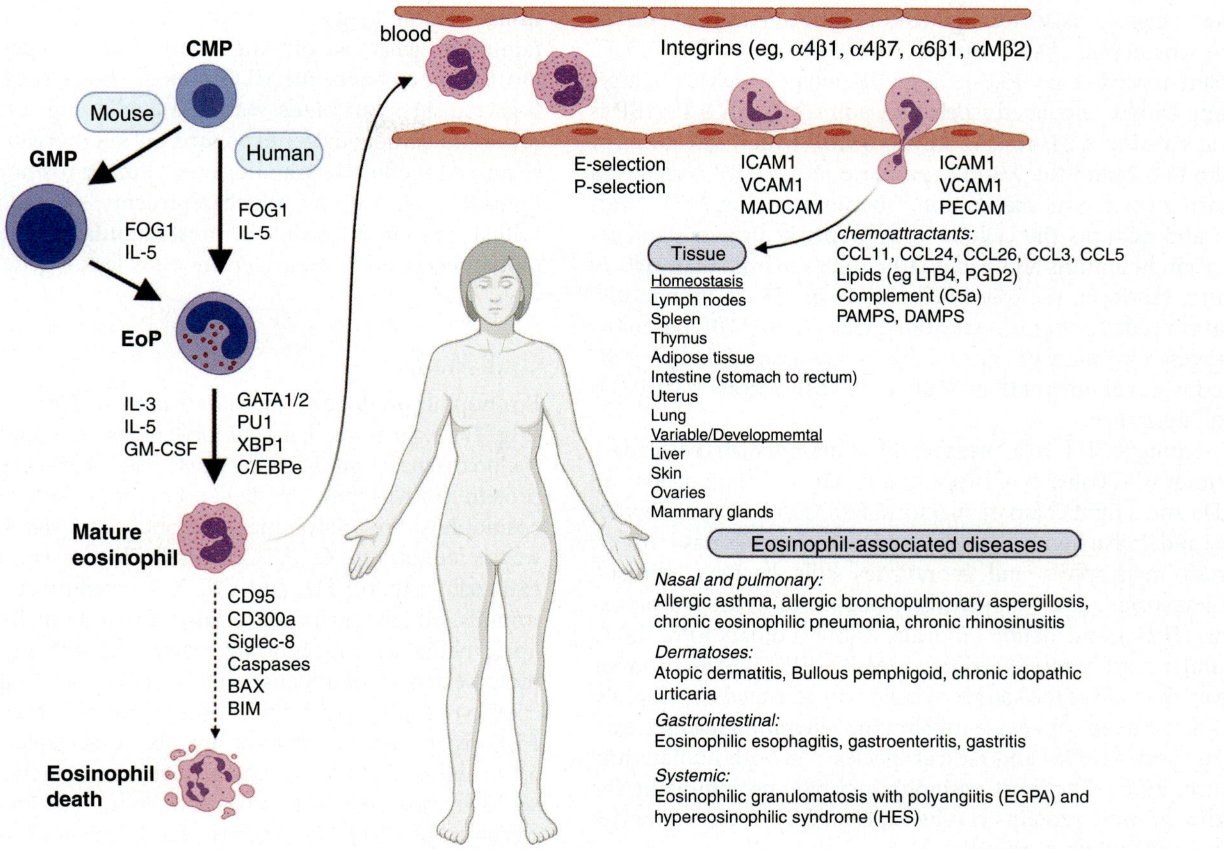

FIG. 17.2. Eosinophil development and recruitment to the tissues of eosinophil-associated disorders. The key transcription factors and cytokines involved in development of mouse and human eosinophils are described as well as the mechanisms leading to eosinophil cell death. Mature eosinophils enter peripheral circulation and are recruited to various tissues using selectin and integrin adhesion molecules and guided by gradients of various chemoattractants. Eosinophilic disorders involve various respiratory, cutaneous, and gastrointestinal organs. (Created with BioRender.com.)

indispensable for development of eosinophils. The transcription factor XBP1 regulates the unfolded protein response by promoting the transcription of genes encoding stress-response factors. Deletion of *Xbp1* in the hematopoietic lineage disrupts granule formation and specifically reduces both eosinophil precursors and mature eosinophils.[29] Similarly, loss of the endogenous cysteine protease inhibitor cystatin F impairs the integrity of granules and reduces the survival of eosinophils.[30] Finally, expression of the genes for eosinophil secondary granule proteins, including MBP and EPX, is required for successful completion of eosinophilopoiesis in the bone marrow in mice.[31]

Tissue Distribution

The eosinophil is primarily a tissue-dwelling cell. The life cycle of the eosinophil is divided into bone marrow, peripheral blood, and tissue phases (Fig. 17.2). The mean bone marrow maturation and storage time of the eosinophil is about 4.3 days.[32] Once the eosinophil has entered the blood, it has a short half-life of 8 to 18 hours.[33] The normal range of blood eosinophils is 0 to 0.5 × $10^3/mm^3$ and exhibits diurnal variation. In the homeostatic condition, eosinophils are present in thymus, mammary gland, uterus, secondary lymphoid tissues, adipose tissues, and the gastrointestinal tract, but not in the esophagus.[34-36] The localization of eosinophils to the gastrointestinal tract is likely mediated by constitutive tissue expression of the chemokine eotaxin-1 (CCL11).[37] The roles of eosinophils in these tissues and organs in healthy subjects are described later in this chapter.

The number of eosinophils in the mucosal tissues, such as upper and lower airways and the gastrointestinal tract, and in skin is greatly increased during type 2 immune responses. The lifespan of eosinophils in tissues ranges from 2 to 5 days,[38] depending partly on the tissue studied. However, cytokines, such as IL-5 or GM-CSF, increase eosinophil survival up to 14 days.[39] The role of tissue eosinophils and mediators that alter their lifespan and effector functions is a prominent area of current research, and some specifics are described below in the chapter.

EOSINOPHIL-DERIVED MEDIATORS

Granule Proteins

Eosinophils have a capacity to produce a variety of inflammatory and immunoregulatory molecules, including granule proteins, cytokines, lipid mediators, oxidative products, and enzymes.[40] The abundant presence of four highly cationic granule proteins, including MBP, ECP, EPX, and EDN, makes the eosinophil distinct from other leukocytes (Fig. 17.1). Human MBP-1 is a 13.8-kDa single-polypeptide rich in arginine with a calculated isoelectric point (pI) of 11.4. MBP is translated as a 23- to 25.2-kDa proMBP with a calculated pI of 6 to 6.2, and the 9.9-kDa propiece of proMBP is removed during eosinophil maturation.[41] Because human MBP binds to and disrupts the cellular membrane, it directly damages certain helminths and bacteria as well as mammalian cells *in vitro*. However, the role MBP-1 plays *in vivo* in human diseases needs further investigation. MBP-2, the MBP homolog, is present primarily in eosinophils. It has a calculated pI of 8.7 and is less abundant than MBP-1.[42] The function of MBP-2 is still unknown.

Human EPX is a member of a mammalian peroxidase family. EPX consists of two subunits, a heavy chain of 50 to 58 kDa and a light chain of 10.5 to 15.5 kDa in a 1:1 stoichiometry and is also highly basic (pI > 11). Myeloperoxidase (MPO) from neutrophils and monocytes kills bacteria, viruses, *Mycoplasma*, and fungi in the presence of hydrogen peroxide (H_2O_2) and halide chloride. While human EPX shows similar antimicrobial activity as MPO, EPX prefers bromide over chloride as the source of halide. In activated eosinophils, EPX is a major molecule involved in generating reactive oxygen species (ROS) and radical species.[43] In both humans and mice, EPX is the most eosinophil-specific protein while the other granule proteins can be expressed by other leukocytes (eg, neutrophils, mast cells).[44,45]

Human ECP and human EDN belong to the pancreatic RNase A superfamily.[46] ECP is highly basic (pI of 10.8) and consists of a single polypeptide chain of 15.5 kDa. Human ECP possesses RNase activity and is also known as RNase-3. Human EDN (RNase-2) is a single chain 18 to 21 kDa protein. It has marked amino acid sequence homology (67%) with ECP, yet has about 100 times more RNase activity than ECP. EDN is a powerful neurotoxin that can severely damage myelinated neurons when injected intrathecally into experimental animals.[47] EDN also enhances Th2-type immune responses through a toll-like receptor 2 (TLR2)–dependent mechanism.[48] mRNA for EDN, ECP, and CLC (see below) are abundantly expressed by mature peripheral blood eosinophils,[49] suggesting potential de novo synthesis of these proteins.

Although MBP-1 and MBP-2 and EPX are well-conserved between mouse and human eosinophils, mouse eosinophil-associated RNases represent a remarkably divergent and distant paralog of human EDN and ECP.[50] Mouse eosinophil granules contain seven types of ribonucleases that are also expressed by other myeloid cells in rodents.[50] The diversity of the RNase gene family, which is among the highest rapidly evolving lineages in the primate genome, ranking with gene families of immunoglobulins and T cell receptors,[51] suggests that evolutionary constraints are acting on the ECP/EDN superfamily.

The CLC protein is localized primarily in the cytoplasm[52] and forms hexagonal bipyramidal crystals when released extracellularly. In fact, the appearance of CLC crystals in the sputa has been regarded as a hallmark of eosinophilic inflammation in patients with asthma. CLC belongs to a family of galactose-binding lectins, named galectin-10.[53] On the other hand, the CLC protein shows no affinity for β-galactosides but binds mannose avidly, suggesting that it may bind carbohydrates expressed on microorganisms or be required for effective granulogenesis during eosinophil development.[54] While CLC, which represents ~7% to 10% total cellular protein, is one of the most abundant proteins in the human eosinophil, mouse eosinophils lack a gene encoding CLC.[12]

Lipid Mediators

Eosinophils produce a wide variety of lipid-derived mediators (Fig. 17.3). In general, arachidonic acids are metabolized by cyclooxygenase or 5-lipoxygenase (5-LO)/5-LO–activating protein (FLAP) into prostaglandins or leukotrienes.[55,56] In eosinophils, the predominant metabolite via the 5-LO pathway is leukotriene C_4 (LTC_4), which in turn is metabolized extracellularly to LTD_4 and LTE_4.[57] This contrasts with neutrophils, which can produce large amounts of LTB_4. These lipid mediators can contract airway smooth muscle, promote secretion of mucus, alter vascular permeability, and elicit eosinophil and neutrophil infiltration. In eosinophils, 15-lipoxygenase metabolites are also detectable, including 5,15- and 8,15-diHETEs and a novel chemotactic lipid, 5-oxo-15-hydroxy-6,8,11,13-eicosatetraenoic acid. Protectin D1, a product of 12/15-lipoxygenase, has been shown to enhance resolution of inflammation.[58] Finally, the eosinophil produces platelet-activating factor (PAF; 1-*O*-alkyl-2-acetyl-*sn*-glycero l-3-phosphocholine),[59] which is capable of activating platelets and inducing bronchoconstriction.

Enzymes and ROS

A variety of enzymes have been associated with the eosinophil, which may play a major role in its functions and biological activities. For example, eosinophils are a source of a 92-kDa matrix metalloproteinase (MMP-9).[60] MMP-9 can cleave type XVII collagen, a transmembrane molecule of the epidermal hemidesmosome, suggesting that production and release of the proteinase by eosinophils may contribute to tissue damage and remodeling. Furthermore, *in vitro* transmigration of eosinophils through the basement membrane is blocked by a neutralizing anti–MMP-9 antibody, suggesting that this enzyme is required during the migration process of eosinophils. Eosinophils also have lipases, serine proteases, cystatins, and other lipid- and protein-degrading enzymes.

Eosinophils are capable of producing reactive oxygen species and singlet oxygen. *Respiratory burst* is defined as the increase in cell metabolism and oxygen consumption, coupled with the release of ROS. The principal product of the respiratory burst in the eosinophil is superoxide (O_2^-), which undergoes dismutation into more ROS, including H_2O_2, the hydroxyl radical (OH^-), and hypohalous acids (eg, hypobromous acid). These eosinophils release a large amount of O_2^- extracellularly likely due to high levels of expression of the NADPH oxidase complex and preferred assembly of NADPH oxidase at the cell membrane.[61]

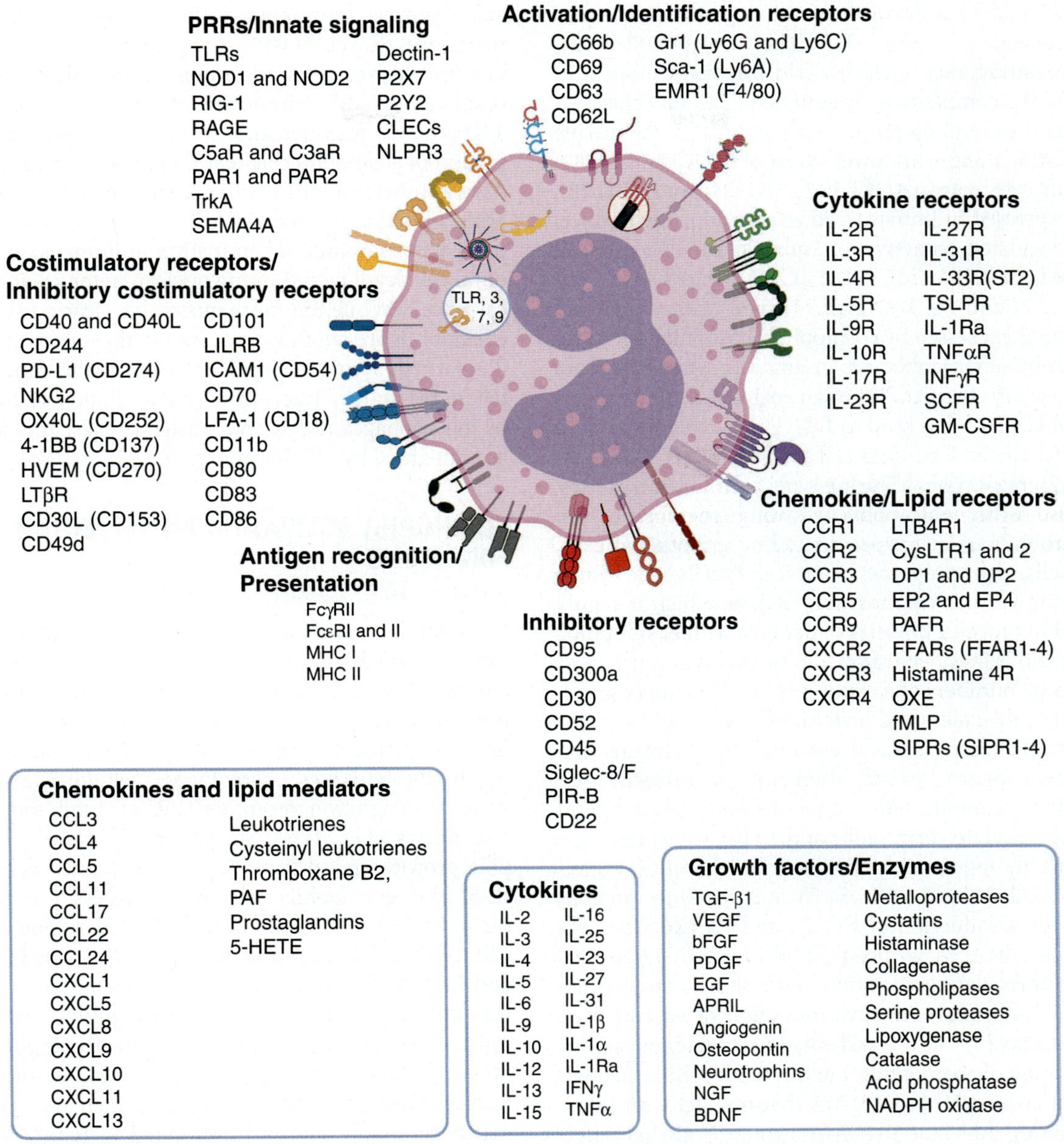

FIG. 17.3. Eosinophil receptors and mediators. Eosinophils express various types of receptors to receive signals for their environments. Note, some differences observed between mouse and human eosinophils are described in the text. Once activated, eosinophils produce and release chemokines, lipid mediators, cytokines, growth factors, and enzymes, which are likely important for proinflammatory, immunostimulatory, and immunoregulatory functions of eosinophils. (Created with BioRender.com.)

Cytokines

Eosinophils are a source of a number of cytokines and growth factors[35,40,62,63] (Fig. 17.3), which may contribute to immunoregulatory and tissue remodeling functions of the eosinophil. For example, human eosinophils infiltrating tissues express TGF-α and TGF-β1. Eosinophils can also produce osteopontin, vascular endothelial cell growth factor (VEGF), and nerve growth factor (NGF) that lead to tissue remodeling.[64,65] IL-13, which is associated with remodeling, has been localized to eosinophils in airway of asthmatics[66,67] and eosinophils in intestinal disease.[68] Eosinophils also produce cytokines, such as TNF-α and chemokines, such as MIP-1α (CCL3). Thus, the contribution of eosinophil-derived cytokines to the development and maintenance of inflammatory and immunologic responses may vary depending on the tissue and immune environment. Altogether, eosinophils can provide a strikingly wide variety of biologic molecules and may be involved in diverse biologic processes from tissue homeostasis, host defense, immune regulation, tissue pathology, and remodeling/repair as described in detail below.

EOSINOPHIL RECRUITMENT AND ACCUMULATION

Mechanisms of Recruitment

During inflammation, eosinophils are recruited across the vascular endothelium and into tissues through a selective process mediated by a network of chemotactic factors and adhesion molecules[62,69] (Fig. 17.2). The most potent eosinophil chemoattractants include lipid mediators, such as PAF and LTD_4, complement fragments C5a, and chemokine eotaxins. Eosinophils prominently express a chemokine receptor CCR3, along with lower levels of CCR1 and CCR2. CCR3 binds the eotaxins (CCL11, CCL24, and CCL26). CCL26 is expressed in humans and is a pseudogene in mice. RANTES (regulated on activation, normal T cells expressed and secreted, CCL5) binds CCR1, CCR3, and CCR5, and MCP-3 (CCL7) and MCP-4 (CCL13) bind both CCR1 and CCR3. CCR3 is expressed by eosinophils, basophils, and mast cells but not by other leukocytes in humans, while it is more eosinophil-specific in mice.[70] Human eosinophils also express CXCR3 and CXCR4 that bind to IFN-γ-inducible protein-10 (IP-10, CXCL10) and stromal cell–derived factor-1α (SDF-1α, CXCL12), respectively. During type 2 immune responses, eotaxins and other eosinophil-activating chemokines are produced by various cell types, including endothelial cells, epithelial cells, parasympathetic nerves, T cells, and fibroblasts. Among various chemokines, CCL26, which is highly upregulated in mucosal biopsies of patients with eosinophilic esophagitis (EoE), has been shown to be highly important as it recruits large numbers of eosinophils into the esophagus.[71]

Adhesion molecules and cytokines also play a major role in selective recruitment of eosinophils.[72] Selectins and their sialylated glycan ligands, along with α_4 integrins, are important for eosinophil tethering and rolling while α_4 and β_2 integrins mediate firm adhesion. The transmigration step is primarily regulated by β_2 integrins as well as eotaxins (CCL11, CCL24, and CCL26). Type 2 cytokines, such as IL-13, promote production of CCL24 and CCL26 from airway epithelial cells or macrophage.[73,74] IL-4 and IL-13 are also involved in upregulating eosinophil adhesion to endothelial cells, primarily through selective induction of vascular cell adhesion molecule 1.[75,76] IL-5 and eotaxin activate transendothelial migration of eosinophils, but not neutrophils, through activation of eosinophil β_2 integrins that interact with intercellular adhesion molecule 1 (ICAM-1) on endothelial cells.

Retention and Resolution

Once they enter target tissues, eosinophils do not return to the blood circulation, although studies in mice suggest that endobronchial eosinophils can travel to regional lymph nodes and act as antigen-presenting cells. Eosinophil survival and death are regulated by the tissue microenvironment,[77] and eosinophil numbers can remain high in tissues even when peripheral numbers are low. The cytokines that promote proliferation and differentiation of the eosinophil lineage, including IL-3, IL-5, and GM-CSF, also enhance survival of mature eosinophils. Activated eosinophils produce IL-3, IL-5, and GM-CSF and promote their own survival in an autocrine manner.[78] Withdrawal of these cytokines can induce apoptosis of eosinophils through activation of caspases. Caspase-3- and caspase-8-dependent apoptosis of eosinophils is also induced by ligation of Fas (CD95), particularly in the presence of TNF-α and IFN-γ.[79] Ligation of a sialic acid–binding immunoglobulin-like lectin (Siglec) family member-8 (Siglec-8) with endogenous sialylated keratin sulfate proteoglycans found in human lung also induces eosinophil cell death.[80] Additional cell surface receptors CD30, a TNFR family member, and CD300a are important in the regulation of eosinophil cell death. Extracellular matrix proteins have also been found to change the survival time of eosinophils.[81] Finally, glucocorticoids, which have been used widely to treat eosinophilic inflammation in humans, induce eosinophil cell death by inhibiting nuclear factor κB and activating caspases.[82] While the mechanisms for clearing the tissues of eosinophils are not fully understood, they likely include shedding of cells across mucosal surfaces into the lumen of the gut or the respiratory tract, engulfment of apoptotic eosinophils by macrophages and airway epithelial cells, and cellular lysis accompanied by cell degeneration.

EOSINOPHIL ACTIVATION AND EFFECTOR FUNCTIONS

Eosinophil Receptors

Eosinophils express a wide range of cell surface receptors through which they communicate with their surrounding environment and other cells in order to exert their effector functions. These include Fc receptors for immunoglobulins, complement receptors, receptors for soluble mediators (eg, lipids, cytokines, chemokines), cell adhesion molecules, pattern recognition receptors (PRRs), inhibitory receptors, and others (Fig. 17.3). As previously mentioned, eosinophil growth and differentiation are highly dependent on IL-5, and eosinophils selectively express the IL-5 receptor (CD125/CD131), along with related receptors for GM-CSF (CD116/CD131) and IL-3 (CD123/CD131). Both IL-5 and its receptor have been targeted using biologics to treat patients with asthma, CRS, and eosinophilia.[83] Although less cell-type selective, eosinophils can directly respond to many other cytokines, including IL-1, IL-4, TNF-α, and IFN-γ.[84,85] Furthermore, both TSLP and IL-33, which are produced primarily by tissue cells and implicated in type 2 immunity,[86] effectively activate eosinophil effector functions through the specific receptors TSLPR[87] and ST2 (IL1RL1), respectively.[85,88]

Eosinophils express various Fc receptors (FcR) that allow interaction with immunoglobulins (Igs). Conventionally, cross-linking of FcR on eosinophils by using ligands immobilized onto a solid surface, such as Sepharose beads, has been used to induce respiratory burst and degranulation in human eosinophils. Human eosinophils have functional IgA receptors (FcαR or CD89), and beads coated with IgA or secretory IgA (sIgA) potently induce degranulation of eosinophils. The hierarchy of effectiveness among Igs to induce eosinophil degranulation is in the order of sIgA ≥ IgA > IgG >> IgE.[89] Human eosinophils constitutively express low-affinity IgG receptor FcγRII (CD32), especially the activating isoform FcγRIIA, and cellular activation with cytokines, such as IFN-γ, induces the expression of high-affinity IgG receptors FcγRIII (CD16) and FcγRI (CD64). Human eosinophils synthesize detectable levels of α-chain protein of FcεRI, but this protein is usually not detectable on the cell surface. FcαRI and FcγRIIA and FcγRIIC are not identified in the mouse genome while mouse eosinophils express inhibitory isoform FcγRIIB.[90] In human eosinophils, these FcRs have been shown mediate phagocytosis, antigen-dependent cytotoxicity, oxygen metabolism, LTC_4 production, and degranulation *in vitro*. Nonetheless, the roles of these FcRs in host defense and disease *in vivo* are still an enigma.

Eosinophils express the complement receptors CR1 (CD35) and CR3 (Mac-1, CD11b/CD18, $\alpha_M\beta_2$), and their levels of expression are rapidly increased by many eosinophil-activating stimuli via mobilization to the cell surface from a preformed intracellular pool. The C3a receptor is constitutively expressed on human eosinophils but not on neutrophils while the C5a receptor (CD88) is expressed on both eosinophils and neutrophils.[91] Eosinophils express a variety of receptors for lipid mediators that they may also produce acting in a paracrine and autocrine manner to promote their functions. For example, cysteinyl leukotrienes (CysLTs; LTC_4, LTD_4, and LTE_4) interact with specific receptors, CysLT1 and CysLT2.[57] Both of these receptors are expressed by eosinophils, and their migration in response to CysLTs can be blocked by CysLT1 antagonists, which is used clinically to treat patients with asthma. LTB_4 induces several functional responses in eosinophils including chemotaxis through the high-affinity BLT1 and low-affinity BLT2 receptors. Human eosinophils also respond to PAF, one of their most potent and effective chemoattractants. PAF activation induces the release of granule proteins and production of ROS and LTC_4 from human eosinophils.[59] Eosinophils, along with basophils and Th2 cells, express a receptor for prostaglandin D_2, known as CRTH2 or DP2, which mediates eosinophil migration and activation.[56] Eosinophils also express sphingosine-1 phosphate receptors that promote recruitment out of lymphatic tissue and into the blood through sphingosine-1-phosphate, which can be inhibited by FTY720.[92]

Expression of a variety of PRRs suggests potential roles for the eosinophil in innate immunity. For example, human eosinophils express toll-like receptors TLR1 to 5, 7, and 9, NOD1, NOD2, and RAGE.[93] Functional examples include lipopolysaccharide (LPS) that activates cytokine-primed eosinophils to release DNA, which in turn kills bacteria,[94] and protease-activated receptors may contribute to the antifungal activities of eosinophils.[95] Additional description is given in the chapter below.

More recent studies suggest that eosinophils express four members of the Siglec family, including Siglec-3 (CD33), Siglec-7 (CD328), Siglec-8, and Siglec-10.[96] Siglec-8 is uniquely expressed on human eosinophils and mast cells and is actively being explored as a therapeutic target of biologics to treat eosinophilic inflammation.[97] Cross-linking of Siglec-8 with antibodies or a synthetic glycan ligand induces IL-5-primed human eosinophil death through a β2 integrin and ROS-dependent cell death pathway.[98] For example, the humanized nonfucosylated antibody to Siglec-8, lirentelimab, results in near-complete eosinophil depletion in patients with eosinophilic gastritis.[99] In contrast, a much more modest degree of mouse eosinophil death is seen *in vitro* and/or *in vivo* with antibodies to Siglec-F[100] or sialoglycans, such as those on airway mucins like Muc5b.[101]

A few other cell surface molecules should be mentioned as they have potential diagnostic and functional relevance. For example, cell surface molecule F4/80 is often used as marker of murine macrophages while it is also expressed by mouse eosinophils.[102] The human ortholog of F4/80, namely epidermal growth factor-like module containing mucin-like hormone receptor 1 (EMR1), is specifically expressed on human eosinophils although its function remains unknown.[103] Eosinophils also express type 4 histamine receptor (histamine H_4 receptor) and various purinergic receptors, such as P2X7 and P2Y2. While human cells do not express the Gr1 cell-surface marker, mouse eosinophils express variable levels of Ly6C and a low level of the Gr1 subtype Ly6-G, depending on activation state. Finally, eosinophils express several molecules that are potentially involved in suppression of their functions, including immunoreceptor tyrosine-based inhibitory motifs (ITIM)-containing molecules, such as leukocyte inhibitory receptor type 3 (LIR3, CD85a), killer cell Ig-like receptor, two domains, long cytoplasmic tail 3 (KIR2DL3), CD32, CD300a, and CD300f.[104]

Eosinophil Activation and Degranulation

The activation status of human eosinophils in peripheral blood differs among individuals. For example, eosinophils isolated from the blood of patients with asthma display enhanced proinflammatory properties as compared to those from normal individuals, including increased adhesion, chemotaxis, transendothelial migration, and cysteinyl leukotriene production.[72,105] These phenotypic and functional changes in human eosinophils are the result of enhanced cellular responsiveness termed *priming*, likely due to their prior exposure to proinflammatory mediators.[106] Indeed, incubation of eosinophils *in vitro* with a variety of cytokines, such as IL-3, IL-5, IL-4, IL-13, GM-CSF, and IFN-γ, mimics *in vivo* primed eosinophils.[107,108] The *priming* process may prepare blood eosinophils to respond vigorously to a variety of agonists, which they may encounter in the tissues.

Eosinophils also change their cell surface phenotype as they are activated. For example, *in vitro* stimulation of blood eosinophils with cytokines induces the expression of CD69, CD25, HLA-DR, and ICAM-1, which are not expressed by resting eosinophils, and upregulates the expression of the β_2 integrin, CD11b. Similarly, eosinophils activated *in vivo*, such as those in peripheral blood eosinophils from patients with helminth infection and bronchoalveolar lavage (BAL) fluid from patients with asthma, show increased expression of ICAM-1, HLA-DR, CD11b, CD11c, CD44, CD66, CD66b, CD69, and CD81.[105,109]

Extracellular release of granule contents has been considered to be one of the major effector functions of eosinophils.[11,35] The eosinophil exhibits three forms of granule release (Fig. 17.1). The first is *classical exocytosis* or *compound exocytosis* of crystalloid granules, in which their contents are released by granule fusion with the plasma membrane. This type of release is typically demonstrated *in vitro* using a patch-clamp technique. The second mode is by *piecemeal degranulation* (PMD). PMD was first reported as the appearance of numerous small vesicles in the cytoplasm coupled with the apparent loss of crystalloid granule core and matrix components, creating a "mottled" appearance on electron microscopy. Emptying of the granule contents is likely caused by budding of small vesicles from the larger secretory granules by PMD. The third mode is *cytolysis*. This process is likely regulated by the receptor-interacting protein kinase 3-mixed lineage kinase-like signaling pathway and may represent a pathologic event, resulting in the extracellular release of free granules (ie, cell-free granules) or granule proteins into surrounding tissues. Cytolytic degranulation is typically seen in the tissues from patients with atopic dermatitis, asthma, and EoE.[110] Finally, stimulated eosinophils can rapidly release extracellular traps in a process called *EETosis* whereby eosinophils release a net-like structure including chromatin DNA and granule proteins, a process usually associated with and similar to *NETosis* by neutrophils.[111]

Species-specific differences in the magnitudes and/or mechanisms of eosinophil degranulation likely exist between humans and mice. For example, cytolytic degranulation of eosinophils is not typically seen in most mouse models of allergen-induced airway inflammation.[112,113] PMD, however, has been reported in lung tissues of mice exposed to allergens or transgenic mice overexpressing IL-5 and CCL24.[114,115] The presence of preformed cytokines and growth factors in granules of human eosinophils that are released by PMD has not been fully demonstrated in mouse eosinophils.[13]

Several *in vitro* models have been used to investigate the immunologic mechanisms of degranulation. Notably, integrin adhesion molecule, Mac-1 (CD11b/CD18, $\alpha_M\beta_2$), likely plays a pivotal role for regulating human eosinophil degranulation.[116] Historically, receptor ligands immobilized to relatively large surfaces, such as IgG-coated Sepharose beads, and the large surface of parasites caused eosinophil degranulation[89] unlike soluble ligands such as aggregated IgG and antigens. Later, it was found that cellular adhesion to a surface through β_2 integrins, in particular $\alpha_M\beta_2$, plays a crucial role in the activation of eosinophils when IgG is immobilized onto beads. Similarly, the ability of eosinophils to release granule contents in response to lipid mediators or cytokines is greatly enhanced by cellular adhesion to a solid surface mediated by $\alpha_M\beta_2$.[116] This ability of human eosinophils to effectively engage their integrins and have a vigorous response to microbes/molecules with a large surface, such as parasites, likely distinguishes this leukocyte from other granulocytes.

Interaction With Other Immune Cells

Eosinophils have functions in both the innate and adaptive arms of the immune system (Fig. 17.4). Eosinophils show a bidirectional interaction with other myeloid cells, such as basophils, mast cells, dendritic cells, and macrophage. For example, eosinophils, basophils, and mast cells express many similar mediators upon activation, such as lipid mediators, neuropeptides, cytokines, and growth factors.[117-120] These cells tend to colocalize in the microenvironments of allergic responses, in particular skin and mucosal tissues, and may generate a coordinated and shared effector responses. For example, eosinophils release cytokines IL-5 and IL-9, chemokines CCL3 and CCL2, and lipids PGD2 and 15-hydroxy-eicosatetraenoic acid (HETE) that induce activation and chemotaxis of mast cells. Mast cells express factors that activate eosinophils, such as histamine that induces eosinophil chemotaxis and CysLTs that promote IL-4 production from eosinophils.[121] Basophils may also have similar functions to mast cells. In models of skin allergy, for example, basophil-derived IL-4 is an indirect mediator of eosinophil recruitment, promoting the production of eotaxins by stromal cells and macrophage.[122]

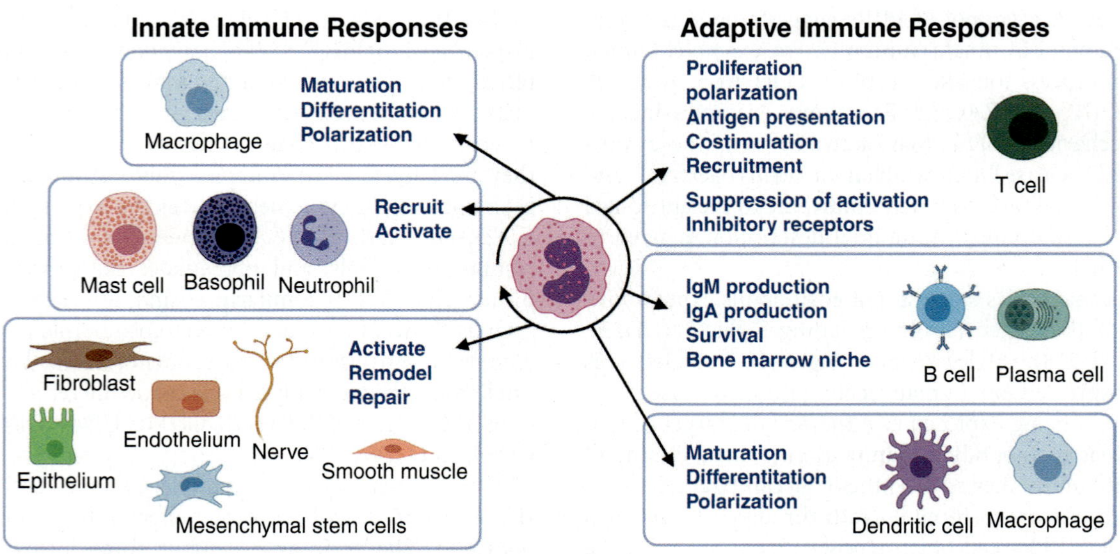

FIG. 17.4. Eosinophil interactions with the cells involved in innate and adaptive immunity. Eosinophils directly interact with a variety of inflammatory and immune cells that are involved in the innate and adaptive arms of immune responses. The biological outcomes may involve recruitment, proliferation, maturation, activation, and polarization of these cell types. (Created with BioRender.com.)

Eosinophils may also interact reciprocally with macrophages to promote their immune functions and tissue remodeling capacities. For example, eosinophil-derived IL-4 and IL-13 promote production of CCL24 from macrophage, which increase additional recruitment of eosinophils into tissues.[74,123] Eosinophil-derived IL-4 and IL-13 also promotes macrophage polarization to an M2 phenotype.[88,108,123] Macrophages that interact with eosinophils show enhanced effector functions, including release and activation of metalloproteases, lipid mediators, and growth factors, which may, in turn, activate eosinophils in a paracrine manner. Furthermore, eosinophil-derived IL-4, IL-13, TGF-β, and 12/15-lipoxygenase products aid macrophage to clear dead cells, such as dead neutrophils, and resolve inflammation as part of the return to physiological homeostasis.[58,124] Thus, eosinophils and macrophages may work together to contribute to tissue repair and remodeling to regain homeostasis during infection and tissue injury.[125]

Eosinophils' capacity to modulate DC function is likely dependent on the local cues of the immune environment. For example, eosinophils attenuate DC activation and T cell responses in allergen models[126,127] and in some nematode infections.[128] Conversely, eosinophil-derived IL-25[129] or EDN[48] induces upregulation of DC antigen presentation, costimulatory molecule expression, and polarization of Th2-type CD4+ T cells.[130] In particular, EDN acts as a TLR2 ligand that mediates chemotaxis of DCs and increases the production of IL-6 and TNF-α and expression of CD86 and MHC II. These cells colocalize in the lymph nodes[131] and the lung during allergic immune responses as visualized by live microscopy,[132] suggesting their close interactions.

Eosinophil interaction with other innate effector immune cells, such as neutrophils and NK cells, has also been reported. Eosinophils activated with TNF-α promote chemotaxis and activation of neutrophils through release of CXCL5 and production of ROS.[133] MBP induces neutrophils to produce IL-8, which promotes chemotaxis and activation of neutrophils in an autocrine and paracrine manner.[134] In contrast, eosinophils reduce neutrophil numbers by promoting macrophage to resolve inflammation in models of inflammatory bowel disease and peritonitis.[58,135] LPS-activated neutrophils promote the recruitment of eosinophils through release of lipid mediators, such as LTB4, and through release of elastase that aids in activating IL-1β and IL-33.[136] Eosinophils express some ligands for NK receptors (eg, 2B4) and engage with NK cells through the natural cytotoxicity receptors Nkp46 and NKp30 although the ligands on eosinophils remain to be defined.[137] In turn, NK cells induce either activation of eosinophils or cell death through caspase-independent mechanisms.[138]

Group 2 innate lymphocytes (ILC2s) have been identified recently as upstream mediators of eosinophil activation and recruitment in helminth infection and allergic immune responses.[139,140] ILC2s are potent producers of IL-5, which is involved in promoting eosinophilopoiesis in the bone marrow and survival and priming in tissues. In response to activating stimuli, such as IL-33 or TSLP, ILC2s promote accumulation of eosinophils into tissue, in part, by IL-13–mediated production of eotaxin. As both eosinophils and ILC2s are resident in mucosal and adipose tissues, their interactions are speculated to be bidirectional.[141,142] Multiplex imaging of nasal polyp tissues revealed colocalization of eosinophils and ILC2s.[143] In coculture with ILC2s *in vitro*, eosinophils increased their expression of CD69 and released IL-4, which when neutralized by antibody, reduced IL-5 and IL-13 production by ILC2s. Further studies are needed to clarify their bidirectional interactions as well as the contribution of eosinophils in regulating other ILCs, such as ILC1s or ILC3s.[144]

In addition to their involvement in innate immunity, eosinophils interact with adaptive immune cells, such as T cells and B cells (Fig. 17.4). Eosinophils have been shown to directly present antigen to T cells both through MHC I and MHC II molecules and to upregulate expression of costimulatory molecules such as CD86.[145] For example, eosinophils take up fluorescently tagged antigen followed by migration to the T cell zone of the lymph nodes in mouse models of asthma by both CCR7 and CysTLR activation.[146-148] In the intestine, MHC II expression on eosinophils is upregulated by antigen and required for T cell activation.[149] Yet, other studies show that adoptive transfer of MCH II-deficient eosinophils into eosinophil-deficient mice was sufficient to induce antigen-specific effector T cell activation in an asthma model,[148] suggesting that eosinophils have a minor role as antigen-presenting cells in asthma.[145,150] Human eosinophils also express MHC II, process antigen,[151] and express a Notch ligand, Jagged1, indicating a capacity to activate T cells.[152] MHC I-mediated antigen presentation by eosinophils to CD8+ T cells also occurs in models of cancer[153] and infection.[154] Eosinophils can suppress CD8+ T cell responses through inhibitory molecules, such as galectin-10 or PD-L1.[155,156]

One of the important functions of eosinophils may be recruitment of CD4+ and CD8+ T cells. Adoptive transfer of IL-33–stimulated eosinophils into lungs of eosinophil-deficient mice leads to eosinophil production of CCL17 and CCL22, which bind CCR4 on Th2 cells and promote their recruitment.[157] Alternatively, IFN-γ and TNF-α induce eosinophil production of CXCL9 and CXCL10 and recruitment of CD8+ T cells in models of cancer,[153,158] suggesting differential functions of eosinophils depending on the immune environment. Eosinophils may interact with unconventional mucosal associated invariant T cells (MAIT) cells or iNKT cells as well.[159]

Eosinophil, B cell, and plasma cell communication is only partially understood. Eosinophils aid plasma cell homing to bone marrow niches after immunizations although the mechanisms are unknown.[160] Once in the bone marrow, eosinophils release APRIL and IL-6 and promote plasma B cell survival. Similar eosinophil functions have been proposed for plasma cells in the intestine, particularly where IgA production is necessary for gut health.[161-163] Eosinophils also express IL-27 and CXCL13,[164] a B cell chemokine, suggesting a role for eosinophils in B cell recruitment or mucosal-associated lymphoid tissue formation. Finally, association between multiple myeloma and eosinophilia[165,166] indicates a potential for eosinophil-B cell interactions in humans. Altogether, as summarized in Figure 17.4, eosinophils interact with a variety of innate and adaptive immune cells and promote or regulate their functions. Considering the fact that eosinophils are resident in tissues, they may play a major role in optimizing the immune environment in homeostatic and pathologic conditions.

Eosinophil Subsets and Immunologic Functions

The immunologic functions of eosinophils are heterogeneous and likely are adapted to the milieu of the local immune microenvironment.[8,35,107,167] For example, earlier studies showed that IFN-γ or IL-4 are released by human eosinophils upon *in vitro* exposure to TNF-α or CCL11, respectively,[168,169] suggesting that distinct cytokines are released by eosinophils depending on the stimuli. Human eosinophils stimulated with IL-33 or IFN-γ also release either classic type 2 or type 1 associated chemokines, such as CCL17 or CXCL10, respectively.[84,85] The majority of eosinophil subset immune functions that have been identified *in vivo* in humans, though, are primarily identified type 2 immune responses, such as allergy. More, recently, though, through the use of mass cytometry, SARS-COV-2 patients were found to have unique eosinophil subtypes in their bloodstream that correlated with IFN-γ responses of acute respiratory distress syndrome.[170]

Generation of and studies with knockout/knockin and transgenic mice that target eosinophils have brought forth advances in understanding the diverse immunologic functions of eosinophils.[171,172] The first strains targeting eosinophils were the IL-5[173] and IL-5Rα–knockout mice.[174] While these mice fail to mount tissue eosinophilia in response to parasite or allergen challenge, the depletion of eosinophils is only moderately reduced in the bone marrow and blood.[26] These mouse strains were followed by generation of genetic knockouts for the eotaxin/CCR3 axis,[37] which further blunted eosinophil recruitment to tissues. In 2004, congenic eosinophil-deficient strains PHIL[175] and ΔdblGATA-1[176] provided further evidence to support the role of eosinophils in allergic immune responses. Soon thereafter, conditional eosinophil-deficient[32] and eosinophil-specific Cre[177] mice were created along with tools to derive eosinophils from bone marrow.[178]

These mouse models elucidated the role of eosinophils in regulation of type 2, type 1, and type 17 immune responses.[179] In general, eosinophil-deficient mice or mice that have eosinophil-specific knockout of type 2 cytokines or eotaxins showed impaired type 2 immune responses in mouse models of asthma.[171] The roles of eosinophil-derived IL-4 and IL-13 are relevant in nonallergic immune functions as well. Eosinophil IL-4 promotes M2 macrophage polarization in white adipose tissue[180] and in the skin lesions of leishmaniasis.[123] Eosinophil-derived IL-4 has also been found to induce differentiation of fibro-adipogenic progenitors,[181] enhance nerve regeneration,[182] and promote B cell functions.[183] Eosinophil-derived IL-13 is associated with tissue remodeling[125,184,185] and injury repair.[186]

Recently, the role of eosinophils in type 1 immune responses has expanded in models of cancer,[187] infection,[188,189] or transplant.[190] In these IFN-γ- and TNF-α-predominant environments, eosinophils produce CCL5, CXCL10, and CXCL9 that promote the recruitment of CD8+ T cells.[153] In addition, at the site of inflammation, eosinophils promote CD8+ T cell activation and present antigen through MHC I during respiratory virus infection[154] and cancer.[153,158] Conversely, in mice infected with with *Helicobacter pylori*, IFN-γ–stimulated eosinophils inhibit CD8+ T cell activation through the interaction of PD-L1 on eosinophils and PD-1 on T cells.[191] Similarly, in allograft lung-transplant models, eosinophil-derived iNOS and eosinophil upregulation of PD-L1 resulted in suppression of CD8 effector T cell activation, improving allograft survival.[156] In humans, eosinophils isolated from patients undergoing graft-versus-host disease have subpopulations of CD16+ eosinophils that release galectin-10 and express PD-L1 to inhibit T cell activation.[155] In addition to type 2 and type 1 responses, eosinophils modulate type 17 responses. Eosinophils may release IL-1β to promote IL-17 from other cells or suppress Th17 responses through release of IL-1Rα, which blocks IL-1β activity.[161,192] Eosinophils themselves have the capacity to produce IL-17, IL-23, and IL-6, particularly upon exposure to fungus *Aspergillus fumigatus*,[193,194] suggesting involvement of eosinophils in Th17-type response.

These observations led to hypothesize that eosinophil subsets may exist in polarized immune environments, similar to those for CD4+ T cells, ILCs, and macrophages. Recently, some additional features have been noted for eosinophils that may classify them into subsets. For example, eosinophils in type 2 immune environments show elevated cell surface expression of CD11b, CD69, MHC II, CD80, CD11c, and Siglec-F. These eosinophils also showed increased polymorphonuclear appearance and increased presence of vacuoles. In contrast, eosinophils found in type 1 immune environments, such as in infection or cancer, show increased ICAM-1, MHC I, and PD-L1 and less segmented nuclei and fewer vacuoles. In addition, eosinophils may have immune regulatory activity through expression of galectin-10, PD-L1, IL-10, protectin D1, and iNOS activity or production of a novel ontology of regulatory lung-resident eosinophils.[127] Future study into cytokines, small molecules, or extracellular matrix molecules[167] likely will help to characterize eosinophil subsets and to understand their regulatory mechanism. Cutting-edge technologies such as single-cell omics from isolated eosinophils from tissues in addition to multiplex digital spatial immune profiling of eosinophils *in situ*[195-197] will help to identify subtypes of eosinophils.

ROLE OF EOSINOPHILS IN HOST DEFENSE AND DISEASE

Eosinophils in Healthy Conditions

While earlier studies focused on the pathological roles of eosinophils, recent studies suggest that eosinophils play roles in maintenance of health and homeostatic conditions. The remodeling/repair (LIAR) hypothesis proposes that eosinophils act as immune cells involved in the remodeling and repair at sites with high cell turnover or stem cell activities.[8] This may include regions of developmental and physiological processes (eg, uterus) or regions with abnormal growth and cell death (eg, cancer).

As mentioned previously, eosinophils are present in all mucosal tissues, adipose tissue, bone marrow, thymus, secondary lymphatics, reproductive organs, and potentially at very low levels in liver, skin, and other organs.[34,36] The number of eosinophils in the thymus peaks at 2 weeks after birth in mice and within the first 6 months in humans. They migrate from the corticomedullary junction into the medulla where they remain during involution of the thymus. These eosinophils may be involved in MHC I–restricted negative T cell selection[198] and creation of Th2-type environment through production of kynurenine, which inhibits Th1 cytokine activation.[199] Yet, to date, no thymic developmental deficiencies or aberrations in T cell populations have been noted in eosinophil-deficient strains.

Secondary lymphatics, such as caudal mediastinal lymph nodes, have a resident population of eosinophils in the sinus region similar to the spatial-temporal location of sinus macrophage. Upon stimulation with antigen, eosinophils migrate to the T cell region.[146] While eosinophils express CCR7 and can chemoattract to CCL19, CCR7 does not appear to be the primary regulator of eosinophil lymph trafficking as it is for DCs.[147,148] Rather, LTC_4 expression by lymphoid cells appears to play a critical role in recruitment of eosinophils to lymph nodes. Furthermore, an absence of eosinophils reduces the size of Peyer patches as well as the numbers and activation of CD103+ DCs.[162] Intraepithelial eosinophils and those in lamina propria of the GI tract uniquely express elevated levels of CD11c and MHC II, form DC-like morphology, and acquire antigens from the lumen.[149] These eosinophils may participate in the maintenance of the overall mucosal health of the intestine by triggering IgA production. Eosinophils express free fatty acid receptor 2 (FFAR2) that recognizes microbiome-derived small chain fatty acids (SCFAs), such as butyrate,[200] suggesting that eosinophils might coordinate functions with the intestinal microbiome.[201] Similarly, intestinal eosinophils respond to incretins (eg, GLP-1[202]) and promote production of homeostatic factors related to lipogenesis.[203]

The homeostatic functions of eosinophils are also reported in adipose tissues. Eosinophils localize to crown-like structures in white adipose tissue,[204] in perivascular fat,[205] and have a role in its thermogenic activity.[180] Their potential homeostatic roles were demonstrated by stressing systemic metabolic conditions through either high-fat diet or cold exposure. In white adipose tissue, eosinophils release IL-4 to increase polarization of M2 macrophages, which are considered key players for glucose homeostasis and browning as a means of thermogenesis. While eosinophils are found in human fat,[206] their roles in metabolic health remains unknown.

Hemostasis and blood clotting are key features of maintaining normal physiology. Eosinophils control vasodilation through release of catecholamine that induces perivascular adipose tissue to release nitric oxide, which in turn promotes blood vessel relaxation.[205] Eosinophil-deficient mice showed impaired clot formation due to a deficiency in 12/15-lipoxygenase activity.[207] Eosinophils bind the endothelium through kindlin-3, an integrin found on most leukocytes, and interact with platelets to promote tissue factor activation and thrombin and fibrinogen coagulation.[208]

Crosstalk between neurons and eosinophils is evident in disease and likely has a role in maintaining neuronal homeostasis. Eosinophils release neuromodulating factors, such as NGF, BDNF, and neurotrophins[209] and express receptors for some of these factors.[210] Eosinophils have been found to localize around nerves in the lung[211] and intestine[212] and may play a role in development, regeneration,[182] or motility of smooth muscle.[213] In patients with tissue eosinophilia, eosinophils colocalize to nerves in diseases such as dermatoses,[214,215] neuromyelitis optica,[216] and spinal nerves in eosinophilic granulomatosis with polyangiitis (EGPA).[217]

Eosinophils aid in remodeling tissues during development and physiological changes. For example, in mice, eosinophils infiltrate the lung on day 7 after birth and resolve within 14 days.[218] This infiltration occurs in conjunction with other leukocytes, such as ILC2s, and type 2 immune responses associated with perinatal lung alveolar sac development.[219] It is unclear if a portion of these early eosinophils in the lung remain with a unique ontology[127] or if rather in adults lung eosinophils have a "patrolling" behavior in tissue after exiting the vasculature.[36] Remodeling of tissues also appears to occur in mammary gland budding during development and pregnancy, in which eosinophil infiltration is observed in a healthy condition.[35,220]

Altogether, a number of studies now suggest the potential roles for eosinophils in homeostatic conditions, in particular in mouse models. On the other hand, the importance of eosinophils in human health has been difficult to define as a complete absence of eosinophils from birth only occurs in scenarios such as rare mutations in granulocytes.[221] Treatment of patients with anti–IL-5Rα antibody, benralizaumab, which depletes eosinophils by antibody-dependent cytotoxicity (ADCC) has not revealed severe adverse events to date[6] nor have there been reports of adverse events for anti–Siglec-8 antibody, lirentelimab.[99] Eosinophil-deficient mice breed normally, have normal life spans, and do not appear to have apparent disruptions in overall health; albeit mice are in a regulated environment with congenic genotypes, unlike humans. Therefore, further studies are needed to precisely define the roles of eosinophils in maintaining health and potential redundancy with other immune cell types that may have similar homeostatic functions.

Infections

Helminth Infections

Infection with helminths has been recognized as a common cause of eosinophilia since discovery of the eosinophilic leukocyte in humans. The historic view that eosinophils promote host defense against helminths is derived largely from histological images where eosinophils and parasites are localized together and from *in vitro* studies. Indeed, eosinophils appear to respond vigorously to large, nonphagocytosable organisms, which other leukocytes, such as neutrophils and macrophages, have difficulty in handling. For example, human eosinophils can adhere to and kill helminth larvae through

ADCC, and deposition of eosinophil granule proteins on and around dead and dying parasites has been demonstrated in tissue specimens from infected patients.[222] Furthermore, eosinophil granule proteins MBP and ECP are toxic to both schistomules and the larvae of *Schistosoma mansoni*. In patients with schistosomiasis, the number of eosinophils correlated with resistance to reinfection.[223] Several *in vivo* mouse models also suggested protective roles for eosinophils in parasitic infection.[3,224]

However, some controversies still remain regarding the role of eosinophils in helminth infections.[107] For example, in humans, although treatment with anti–IL-5 reduced the number of circulating and tissue eosinophils, there was no evidence for change in the nature or extent of helminth infection. Future studies with IL-5Rα antibodies that nearly abolish eosinophils may provide more clear insights.[6] In mouse models, eosinophils are sometimes protective (eg, they protect from secondary infection by *Tricuris muris* and *Trichinella spiralis*), of no consequence (eg, they do not affect granuloma formation in schistosomiasis), or required for parasite survival (eg, they maintain *Trichinella* larvae encysted in muscle through their effects on nurse cells).[223] On the other hand, it is difficult to discern a precise role for eosinophils in mouse models as mice are generally not a natural host for helminths, and their roles may vary depending on the stages of infection. For example, studies with *Nippostrongylus brasiliensis*, often used to model of antigen-induced type 2 lung inflammation, have shown that although eosinophils may not be required on first infection, they have an important role in T cell–mediated expulsion of worms in secondary infection and early in infection in skin.[225]

Viral Immunity

The relationship between eosinophils and viruses likely depends on both the type of virus and the underlying host genetics. Eosinophils express TLR3, 7, and 9 and other PRRs (eg, RIG-I) that recognize viral RNA. Eosinophils are reportedly associated with three viruses, including rhinovirus (RV), respiratory syncytial virus (RSV), and influenza A (IVA).[226,227] RV and RSV infection leads to increased recruitment of eosinophils and type 2 inflammation in the lung and exacerbate clinical symptoms in subjects with asthma[228] while IVA promotes type 1 inflammation yet is potentially preventable with vaccination.[229]

Direct killing of viruses by eosinophils has been demonstrated in mice and in culture of human eosinophils. For example, EDN directly inhibits viability and replication of RSV,[51] and eosinophil-deficient mice develop poorer responses to RSV infection. RSV also induces EETosis, in which the DNA and granule protein webs directly capture viral particles.[230] Eosinophils can take up viral particles through receptors, such as ICAM1 for RV16, and present antigens to activate T cells.[231] Furthermore, eosinophils exposed to IVA express transcripts for *Ddx58* and *Ifih1*, genes encoding RIG-I and MDA-5, respectively, triggering antiviral pathways.[188] As such, IVA-exposed eosinophils present antigen to CD8+ T cells through their MHC I receptor, resulting in reduced viral load in the lungs of mice[154] in addition to promoting epithelial integrity.[232] Nonetheless, it should be noted that antiviral capacity of eosinophils varies depending on model, host susceptibility, and preconditioning of the immune microenvironment.

In humans, exposure of children to RSV early in their lives increases asthma susceptibility, likely due to the lung remodeling events the virus induces,[233] and eosinophils might play roles in the process. Interestingly, eosinophils from asthmatic patients with RV infection have fewer virus particles and treatment of patients with anti–IL-5 led to increased RV16 viral titers in the airways, suggesting that the eosinophils may function as a "sink" for viral particles. Yet eosinophils from severe asthmatics were less effective in reducing viral titer in airways, suggesting heterogenous population of eosinophils in asthmatics in their role in viral clearance.[234] Confounding results also exist for SARS-COV-2 patients.[235]

Bacterial Immunity

The capacity of eosinophils to phagocytize bacteria is limited as compared to neutrophils.[236] Nonetheless, eosinophils have some demonstrated activities against both gram-negative and gram-positive bacteria, such as *Staphylococcus aureus*, *Escherichia coli*, *Listeria monocytogenes*, and *Pseudomonas aeruginosa*. Eosinophil granule protein, ECP, has bactericidal activity to *E. coli* and *S. aureus* through physical disruption of the bacterial wall by binding glycolytic virulence factors or depolarizing the membrane wall.[237,238] Eosinophils also undergo EETosis to release DNA and granule protein webs that entrap and kill not only *E. coli* and *S. aureus* but also fungi (eg, *Aspergillus*).

Eosinophils aid in killing *Clostridium difficile*, a gram-positive anaerobic bacillus that can survive in a spore form.[239] *C. difficile* induces eosinophilia in the intestine, yet its binary toxin product triggers TLR2, resulting in eosinophil cell death. In contrast, TLR2-deficient eosinophils are able to mediate killing of *C. difficile*. *Mycobacterium* infection also induces increased eosinophil recruitment, through CCR3, to the lungs, resulting in eosinophilic pleurisy.[240] Similar to *C. difficile*, *M. tuberculosis* triggers eosinophils through TLR2.[241] Alternatively, gram-negative *H. pylori* produces peptide (Hp-2-20), which recruits eosinophils through formyl peptide receptors FPR2 and FPR3.[242] Thus, several pathways and molecules appear to be involved for recruitment and activation of eosinophils during bacterial infection.

Allergy

Asthma, Airway Remodeling, and Airway Hyperreactivity

Asthma is a chronic inflammatory condition resulting in reversible bronchoconstriction, immune cell infiltration into the airways, and lung tissue remodeling. Allergic asthma represents the majority of asthma phenotypes and is generally considered a type 2 inflammatory disease of the airways.[243] Eosinophils and lymphocytes are the predominant cells that infiltrate the lungs.[244] Earlier findings in patients with asthma demonstrated significant degranulation of eosinophils as seen in immunohistochemical imaging of biopsies[245] or from analysis of sputum samples that was associated with disease severity.[246] Studies demonstrated eosinophil granule proteins were sufficient to mediate cell killing and tissue damage in cultures and upon instillation into the lungs of animal models.[247-249] These observations led to a model that eosinophil toxic granule proteins play a pathologic role in asthma patients. Indeed, antibody-mediated depletion of eosinophils in patients with severe asthma improved collagen deposition and lung remodeling as well as incidence of asthma exacerbations.[250,251] Currently, three biologics, mepolizumab, reslizumab, and benralizumab, that target the IL-5/IL-5Rα pathway are approved by the FDA for the use to treat severe to mild asthma in patients with a blood eosinophilia greater than 150 cells/μL.[252]

More recently, research on the immune functions of eosinophils has provided new insight into the pathologic roles for eosinophils in asthma (Fig. 17.5). Mouse models using ovalbumin (OVA) antigen demonstrated an immune regulatory role for eosinophils that was independent of degranulation.[175,176,253,254] Transgenic mouse models of chronic asthma showed evidence for smooth muscle thickening, goblet cell

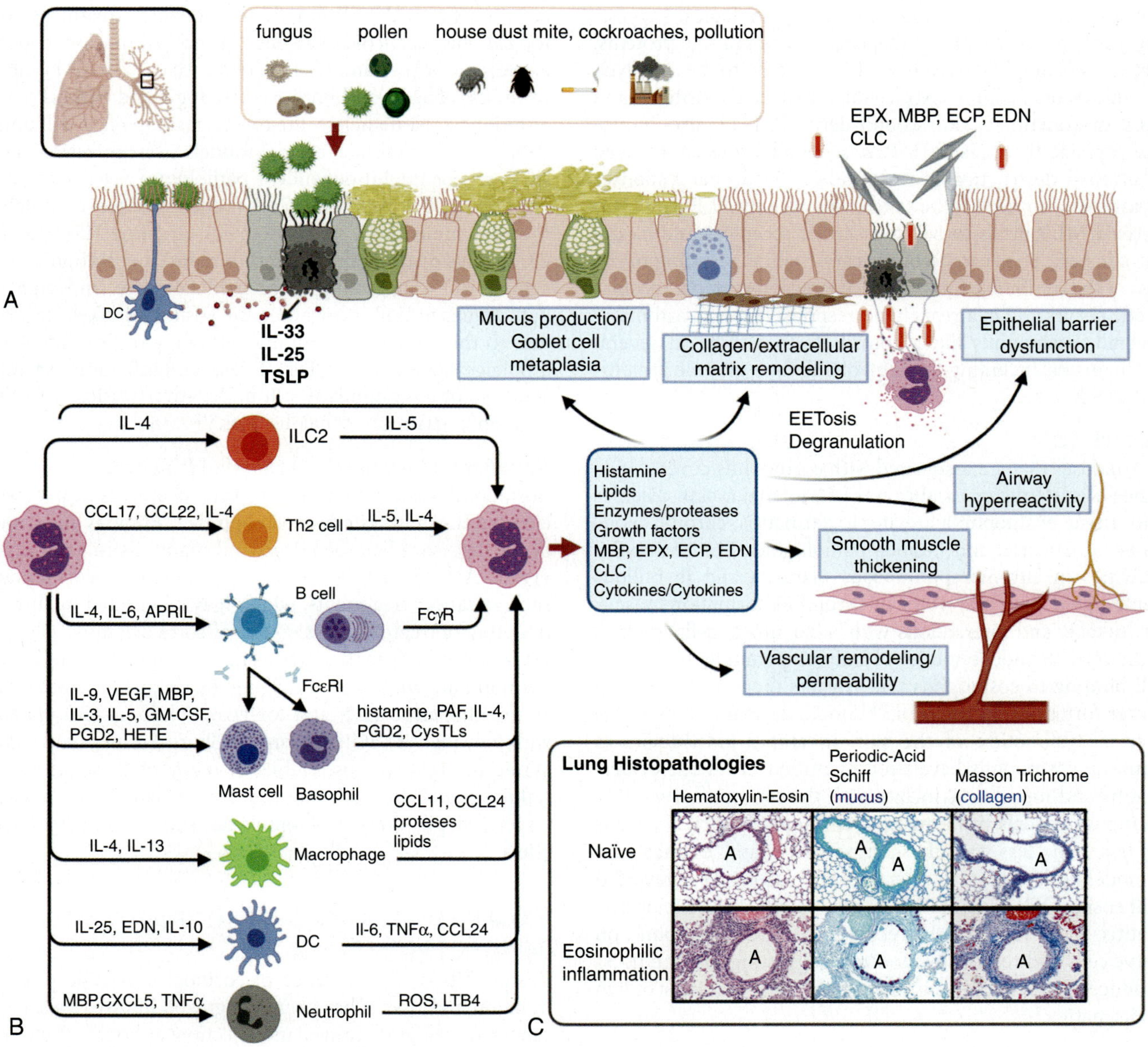

FIG. 17.5. Eosinophil functions in allergic asthma. A: Airborne particles such as fungal spores, pollen from grass and trees, house dust mites or cockroach cellular debris, or pollution can induce inflammation in the lung. These lead to production of epithelial cytokines IL-33, IL-25, and TSLP along with antigen uptake by antigen-presenting cells. **B:** The signaling cascade that follows is a network of interactions of eosinophils on other cells as well as other cells inducing eosinophil recruitment and activation. Ultimately, eosinophils enter the lungs and contribute to pathologies as noted in the blue boxes. **C:** Example staining of lung slices from either naïve mice or mice with severe eosinophilic inflammation. These mice show cellular infiltration in the parenchyma and around airways (A) and smooth muscle enlargement by hematoxylin-eosin staining, mucus production by periodic acid-Schiff assay, which shows as purple in goblet cells, and extracellular matrix remodeling around the airways as shown with Masson trichrome staining which stains collagen blue. (Created with BioRender.com.)

metaplasia, and collagen deposition that mimic pathologies of human asthma. Mice overexpressing IL-13 from club cells, which are nonmucous-producing columnar epithelial cells of the lung, had significant cellular infiltration and age-related lung remodeling, which were reduced upon crossing these mice with eosinophil-deficient ΔdblGATA-1 mice.[185] Furthermore, another mouse strain with increased expression of IL-5 from T cells and eotaxin-2 from club cells[114] resulted in eosinophil infiltration and degranulation. When crossed to eosinophil-deficient mice, the airway pathologies were completely ablated. Importantly crossing of these double-transgenic mice with either EPX or MBP-1 knockout mice failed to ameliorate all lung pathologies, suggesting a role for eosinophils independently of granule proteins. Rather, eosinophil-derived IL-13 is likely to be involved in mucus production, extracellular matrix deposition, and lung dysfunction.[255] Eosinophil-derived IL-13 also induce macrophage to release MMP12, a metalloprotease involved in alveolar destruction in this model.[184] Neuronal responses of asthma also need to be taken into account.[211] Eosinophil-derived MBP binds to M2 muscarinic receptors and blocks the allosteric structure to bind acetylcholine, resulting in the loss of the inhibitory function of M2 receptors in parasympathetic nerves and in airway hyperresponsiveness in asthma.[256] Overall, eosinophils likely contribute to asthma in several mechanisms, including those involving their granule proteins and cytokines.

Dermatoses

Dermatoses that are associated with eosinophils cover a wide range of systemic or local causes.[257] The major skin diseases with tissue eosinophilia are atopic dermatitis, chronic spontaneous urticaria, and bullous pemphigoid. The eosinophil activities in chronic spontaneous urticaria and in bullous pemphigoid may play roles in thrombus formation, vascular leakage, and interactions with other innate cells, such as mast cells. A mouse model of bullous pemphigoid showed IgE binding to eosinophils that triggers their activation and blister formation in the skin.[258] Atopic dermatitis represents barrier dysfunction of the skin in atopic individuals. In humans, eosinophils have been identified in lesions of dermatitis and found to associate with the sensory nerves.[214,215] In mouse models of dermatitis, eosinophil deficiency leads to less itch, less vascular permeability, and extracellular remodeling. Staining the nerves with substance P revealed that eosinophils promoted dendrite growth at the site of dermatitis. More recently, the receptor for IL-31 was found on nerve cells and responsible for itch, suggesting that the IL-31 produced by eosinophils is likely a potential mediator of itch in dermatitis.[259]

GI Disorders

Eosinophils are found in the GI tract of healthy individuals except for the esophagus. The prevalence of EoE, which was first reported in 1978, has increased for the past 4 decades. EoE is currently considered the most common eosinophil-associated GI diseases. EoE represents type 2 hypersensitivity responses to food antigens, resulting in marked infiltration of eosinophils and Th2-type CD4+ T cells producing IL-5 and IL-13 in the esophagus.[260,261] EoE is associated with polymorphisms in the genes encoding TSLP and epithelial barrier protein calpain 14 (CAPN14)[262] along with elevated levels of IL-13 and CCL26,[71] suggesting the roles for esophageal epithelium in promoting type 2 immunity and recruiting eosinophils.

Evidence for eosinophil activation is prominent in the biopsy specimens from patients with EoE, including the presence of eosinophil cell-free granules and deposition of granule proteins.[263] Treatment with anti–IL-5 antibody resulted in partial reduction in eosinophilia, yet improvement of clinical symptoms varied.[264,265] On the other hand, treatment with topical glucocorticoids has been successful to downregulate expression of IL-5 and CCL26, decrease eosinophil numbers in the esophagus, and improve histologic and clinical manifestations.[266] Nonetheless, the role of eosinophils is still under debate.[267] For example, animal models overexpressing IL-13 show an eosinophil-dependent pathology,[27] yet depletion of eosinophils has little effect on reducing tissue remodeling in other mouse models.[268] Biologics targeting IL-4 and IL-13, such as dupilumab, resulted in reduced eosinophilia and symptoms,[269] suggesting these mediators are important to pathologies in EoE. Altogether, the eosinophil likely contributes to the network of immunopathology of EoE, involving esophageal epithelium, CD4+ T cells, and inflammatory cells. The role of eosinophils is still under investigation for other eosinophil gastrointestinal diseases.[99,270]

Nonallergic Eosinophil-Associated Diseases

Eosinophil-associated diseases (EADs) are associated with elevated IL-5 and systemic eosinophilia.[4,5] Etiology and course of the disease is heterogeneous and manifests in a variety of organs. Acute EADs may occur in response to foreign stimuli, such as cancer treatments, which may be blunted by discontinuation of treatment or a course of corticosteroids. Chronic EADs are either genetic, such as in a leukemic form of hypereosinophilic syndrome (HES), or of unknown cause, such as for idiopathic HES and for eosinophilic granulomatosis with polyangiitis (EGPA, formerly Churg-Strauss syndrome). While the roles of eosinophils in many of these EADs are unknown and their relevance to disease pathologies unclear, targeted depletion of eosinophils has shown some beneficial effect.[6]

Novel Roles of Eosinophils in Transplant, Organ Injury, and Cancer

Eosinophils or the presence of eosinophil granule proteins in blood and tissues has been recognized in diseases such as cancer or allograft transplant rejection although their role has remained undefined. Speculations have been made that eosinophils are destructive and involved in ongoing tissue damage and inflammation characteristic of these diseases. For example, elevated levels of ECP and EDN in the blood have been suggested as an early indicator of acute graft rejection.[271] In lung and other organ transplants, eosinophils and degranulating eosinophils can be found *in situ* upon biopsy or in the BAL fluid.[272] Early studies in animal models demonstrated that in a type 2 environment, such as that occurs in mice deficient in IFNγ or CD8+ T cells, eosinophils infiltrate the transplanted organ and induce significant inflammation that ultimately results in organ rejection. Eosinophils are also suggested to aid in the activation of allograft-specific CD4+ Th2 cells.[273] Nonetheless, in lung allograft transplants[190] and human stem cell recipients,[155] eosinophils have shown to have an immune-suppressive effect on T cells and prevent organ rejection.

Recent studies in ischemia/reperfusion injury and repair models also have identified a potential protective role for eosinophils in transplanted organs. In models of liver ischemia/reperfusion, IL-33 was found elevated within hours, followed by recruitment of eosinophils and other type 2 immune cells, suggesting association between tissue eosinophilia and recovery from ischemic injury.[186] Similarly, IL-33 injection into mice with ischemic injury in the kidney improved clinical outcomes, which are correlated with both eosinophil and ILC2 recruitment into the tissue.[274] Altogether, these findings suggest eosinophils, and potentially IL-33,[275] are likely to be beneficial in ischemia/reperfusion injury.

Cancer progression and survival rates may correlate with the number of eosinophils and the levels of their granule proteins in the blood as well as in tissue.[187] Historically, eosinophils are found in a variety of cancers including oral carcinoma, glioblastoma, pancreatic, skin, bladder, uterine, lung, and colon cancer, for example. Mouse models show eosinophil infiltration in necrotic regions as well as in the stromal regions of the tumor microenvironment.[276] The role of eosinophils in cancer may vary dependent on the types of cancer and temporal-spatial relationship of eosinophils and tumor environment. For example, eosinophils have the potential for protumorigenic activities through release of growth factors (eg, VEGFA and TGFβ1), metalloproteases (eg, MMP9), and type 2 cytokines that aid the survival, expansion, evasion, and metastasis of the tumor. On the other hand, eosinophils also can kill tumors, which can be enhanced by cytokines, such as IFN-γ or IL-33. *In vivo*, IFN-γ was shown to be critical to promote antitumorigenic functions of eosinophils in colorectal cancer models.[277] In other models, GM-CSF activation of eosinophils enhanced CD8+ T cell responses to the cancer.[158] In a skin melanoma model, activation of eosinophils with IFN-γ upregulated their expression of CXCL10 and CXCL9 and promoted the CD8+ T cell–dependent tumor rejection.[153] Recent studies have identified IL-33 also promotes tumor rejection through distinct mechanisms. Administration of inhibitor of dipeptidyl peptidase 4 (DDP4) results in elevated levels of IL-33 and eosinophil chemotaxis to the tumor, reducing tumor growth in the several tumor cell lines.[278] These data highlight the complexity of antitumor activities of eosinophils.

CONCLUSION

The eosinophil remains an enigmatic cell that continues to intrigue biomedical scientists after more than a century. Eosinophils mainly reside in tissues and produce a variety of proinflammatory, immunostimulatory, and immunoregulatory factors. While eosinophils are classically involved in responses to parasites and allergic diseases, such as asthma, there is still much to learn about their roles in these responses. Potential roles for eosinophils in immune regulation, tissue homeostasis repair, and cancer have emerged. The constitutive and conditional eosinophil-deficient laboratory mice as well as deletion of specific molecules from the eosinophil will provide mechanistic understanding of the roles of eosinophils and their products in these conditions. Similarly, now that biologic agents that effectively and selectively deplete eosinophils in humans, including those present in a homeostatic condition, are available, one can gain more knowledge by studying the individuals treated with these agents. During the coming years, these clinical and research tools will greatly advance our understanding of the beneficial and pathologic roles for eosinophils in health and disease in humans. Finally, new technologies, such as single-cell "omics," which currently are challenging in eosinophils due to its ribonuclease activity and cellular fragility, will likely help bring forth specific understanding of the heterogeneity of eosinophils and their tissue- or organ-specific functions. It seems inevitable that a complex biologic system that includes eosinophils and their products may participate in a cascade of events leading to maintenance of health, inflammation, and disease.

ACKNOWLEDGMENTS

We wish to thank those who have mentored us, Dr. Gerald J. Gleich and Dr. James J. Lee, on our knowledge of eosinophil biology. We additionally thank our generous colleagues and collaborators that have provided the breadth of work presented in this chapter.

REFERENCES

1. Kay AB. The early history of the eosinophil. *Clin Exp Allergy*. 2015;45(3):575-582.
2. Stacy NI, Ackerman SJ. A tribute to eosinophils from a comparative and evolutionary perspective. *J Allergy Clin Immunol*. 2021;147(3):1115-1116.
3. Klion AD, Nutman TB. The role of eosinophils in host defense against helminth parasites. *J Allergy Clin Immunol*. 2004;113(1):30-37.
4. Roufosse F. Targeting the interleukin-5 pathway for treatment of eosinophilic conditions other than asthma. *Front Med*. 2018;5:49.
5. Klion AD, Ackerman SJ, Bochner BS. Contributions of eosinophils to human health and disease. *Annu Rev Pathol*. 2020;15:179-209.
6. Jacobsen EA, Jackson DJ, Heffler E, et al. Eosinophil knockout humans: Uncovering the role of eosinophils through eosinophil-directed biological therapies. *Annu Rev Immunol*. 2021;39:719-757.
7. Walsh GM, Al-Rabia M, Blaylock MG, et al. Control of eosinophil toxicity in the lung. *Curr Drug Targets Inflamm Allergy*. 2005;4(4):481-486.
8. Lee JJ, Jacobsen EA, McGarry MP, et al. Eosinophils in health and disease: the LIAR hypothesis. *Clin Exp Allergy*. 2010;40(4):563-575.
9. Dvorak AM, Weller PF. Ultrastructural analysis of human eosinophils. *Chem Immunol*. 2000;76:1-28.
10. Lee JJ, Jacobsen EA, Ochkur SI, et al. Human versus mouse eosinophils: "that which we call an eosinophil, by any other name would stain as red". *J Allergy Clin Immunol*. 2012;130(3):572-584.
11. Egesten A, Calafat J, Janssen H, et al. Granules of human eosinophilic leucocytes and their mobilization. *Clin Exp Allergy*. 2001;31(8):1173-1188.
12. Acharya KR, Ackerman SJ. Eosinophil granule proteins: form and function. *J Biol Chem*. 2014;289(25):17406-17415.
13. Spencer LA, Bonjour K, Melo RC, et al. Eosinophil secretion of granule-derived cytokines. *Front Immunol*. 2014;5:496.
14. Melo RC, Weller PF. Vesicular trafficking of immune mediators in human eosinophils revealed by immunoelectron microscopy. *Exp Cell Res*. 2016;347(2):385-390.
15. Fulkerson PC. Transcription factors in eosinophil development and as therapeutic targets. *Front Med*. 2017;4:115.
16. Ackerman SJ, Bochner BS. Mechanisms of eosinophilia in the pathogenesis of hypereosinophilic disorders. *Immunol Allergy Clin North Am*. 2007;27(3):357-375.
17. Drissen R, Buza-Vidas N, Woll P, et al. Distinct myeloid progenitor-differentiation pathways identified through single-cell RNA sequencing. *Nat Immunol*. 2016;17(6):666-676.
18. Mori Y, Iwasaki H, Kohno K, et al. Identification of the human eosinophil lineage-committed progenitor: revision of phenotypic definition of the human common myeloid progenitor. *J Exp Med*. 2009;206(1):183-193.
19. Bedi R, Du J, Sharma AK, et al. Human C/EBP-epsilon activator and repressor isoforms differentially reprogram myeloid lineage commitment and differentiation. *Blood*. 2009;113(2):317-327.

20. Yu C, Cantor AB, Yang H, et al. Targeted deletion of a high-affinity GATA-binding site in the GATA-1 promoter leads to selective loss of the eosinophil lineage in vivo. *J Exp Med.* 2002;195(11):1387-1395.
21. Uhm TG, Kim BS, Chung IY. Eosinophil development, regulation of eosinophil-specific genes, and role of eosinophils in the pathogenesis of asthma. *Allergy Asthma Immunol Res.* 2012;4(2):68-79.
22. Bouffi C, Kartashov AV, Schollaert KL, et al. Transcription factor repertoire of homeostatic eosinophilopoiesis. *J Immunol.* 2015;195(6):2683-2695.
23. Clutterbuck EJ, Hirst EM, Sanderson CJ. Human interleukin-5 (IL-5) regulates the production of eosinophils in human bone marrow cultures: comparison and interaction with IL-1, IL-3, IL-6, and GMCSF. *Blood.* 1989;73(6):1504-1512.
24. Ema H, Suda T, Nagayoshi K, et al. Target cells for granulocyte colony-stimulating factor, interleukin-3, and interleukin-5 in differentiation pathways of neutrophils and eosinophils. *Blood.* 1990;76(10):1956-1961.
25. Lee NA, McGarry MP, Larson KA, et al. Expression of IL-5 in thymocytes/T cells leads to the development of a massive eosinophilia, extramedullary eosinophilopoiesis, and unique histopathologies. *J Immunol.* 1997;158(3):1332-1344.
26. Matthaei KI, Foster P, Young IG. The role of interleukin-5 (IL-5) in vivo: studies with IL-5 deficient mice. *Mem Inst Oswaldo Cruz.* 1997;92(suppl 2):63-68.
27. Ben Baruch-Morgenstern N, Mingler MK, Stucke E, et al. Paired Ig-like receptor B inhibits IL-13-driven eosinophil accumulation and activation in the esophagus. *J Immunol.* 2016;197(3):707-714.
28. Kotzin JJ, Spencer SP, McCright SJ, et al. The long non-coding RNA Morrbid regulates Bim and short-lived myeloid cell lifespan. *Nature.* 2016;537(7619):239-243.
29. Bettigole SE, Lis R, Adoro S, et al. The transcription factor XBP1 is selectively required for eosinophil differentiation. *Nat Immunol.* 2015;16(8):829-837.
30. Matthews SP, McMillan SJ, Colbert JD, et al. Cystatin F ensures eosinophil survival by regulating granule biogenesis. *Immunity.* 2016;44(4):795-806.
31. Doyle AD, Jacobsen EA, Ochkur SI, et al. Expression of the secondary granule proteins major basic protein 1 (MBP-1) and eosinophil peroxidase (EPX) is required for eosinophilopoiesis in mice. *Blood.* 2013;122(5):781-790.
32. Jacobsen EA, Lesuer WE, Willetts L, et al. Eosinophil activities modulate the immune/inflammatory character of allergic respiratory responses in mice. *Allergy.* 2014;69(3):315-327.
33. Farahi N, Singh NR, Heard S, et al. Use of 111-Indium-labeled autologous eosinophils to establish the in vivo kinetics of human eosinophils in healthy subjects. *Blood.* 2012;120(19):4068-4071.
34. Kato M, Kephart GM, Talley NJ, et al. Eosinophil infiltration and degranulation in normal human tissue. *Anat Rec.* 1998;252(3):418-425.
35. Weller PF, Spencer LA. Functions of tissue-resident eosinophils. *Nat Rev Immunol.* 2017;17(12):746-760.
36. Chojnacki A, Wojcik K, Petri B, et al. Intravital imaging allows real-time characterization of tissue resident eosinophils. *Commun Biol.* 2019;2:181.
37. Matthews AN, Friend DS, Zimmermann N, et al. Eotaxin is required for the baseline level of tissue eosinophils. *Proc Natl Acad Sci U S A.* 1998;95(11):6273-6278.
38. Carlens J, Wahl B, Ballmaier M, et al. Common gamma-chain-dependent signals confer selective survival of eosinophils in the murine small intestine. *J Immunol.* 2009;183(9):5600-5607.
39. Wen T, Besse JA, Mingler MK, et al. Eosinophil adoptive transfer system to directly evaluate pulmonary eosinophil trafficking in vivo. *Proc Natl Acad Sci U S A.* 2013;110(15):6067-6072.
40. Giembycz MA, Lindsay MA. Pharmacology of the eosinophil. *Pharmacol Rev.* 1999;51(2):213-340.
41. Soragni A, Yousefi S, Stoeckle C, et al. Toxicity of eosinophil MBP is repressed by intracellular crystallization and promoted by extracellular aggregation. *Mol Cell.* 2015;57(6):1011-1021.
42. Plager DA, Loegering DA, Checkel JL, et al. Major basic protein homolog (MBP2): a specific human eosinophil marker. *J Immunol.* 2006;177(10):7340-7345.
43. Wu W, Samoszuk MK, Comhair SA, et al. Eosinophils generate brominating oxidants in allergen-induced asthma. *J Clin Invest.* 2000;105(10):1455-1463.
44. Abu-Ghazaleh RI, Dunnette SL, Loegering DA, et al. Eosinophil granule proteins in peripheral blood granulocytes. *J Leukoc Biol.* 1992;52(6):611-618.
45. Ochkur SI, Kim JD, Protheroe CA, et al. A sensitive high throughput ELISA for human eosinophil peroxidase: a specific assay to quantify eosinophil degranulation from patient-derived sources. *J Immunol Methods.* 2012;384(1-2):10-20.
46. Slifman NR, Loegering DA, McKean DJ, et al. Ribonuclease activity associated with human eosinophil-derived neurotoxin and eosinophil cationic protein. *J Immunol.* 1986;137(9):2913-2917.
47. Newton DL, Walbridge S, Mikulski SM, et al. Toxicity of an antitumor ribonuclease to Purkinje neurons. *J Neurosci.* 1994;14(2):538-544.
48. Yang D, Chen Q, Su SB, et al. Eosinophil-derived neurotoxin acts as an alarmin to activate the TLR2-MyD88 signal pathway in dendritic cells and enhances Th2 immune responses. *J Exp Med.* 2008;205(1):79-90.
49. Nakajima T, Matsumoto K, Suto H, et al. Gene expression screening of human mast cells and eosinophils using high-density oligonucleotide probe arrays: abundant expression of major basic protein in mast cells. *Blood.* 2001;98(4):1127-1134.
50. Zhang J, Dyer KD, Rosenberg HF. Evolution of the rodent eosinophil-associated RNase gene family by rapid gene sorting and positive selection. *Proc Natl Acad Sci U S A.* 2000;97(9):4701-4706.
51. Rosenberg HF. Eosinophil-derived neurotoxin (EDN/RNase 2) and the mouse eosinophil-associated RNases (mEars): expanding roles in promoting host defense. *Int J Mol Sci.* 2015;16(7):15442-15455.
52. Melo RCN, Wang H, Silva TP, et al. Galectin-10, the protein that forms Charcot-Leyden crystals, is not stored in granules but resides in the peripheral cytoplasm of human eosinophils. *J Leukoc Biol.* 2020;108(1):139-149.
53. Dyer KD, Handen JS, Rosenberg HF. The genomic structure of the human Charcot-Leyden crystal protein gene is analogous to those of the galectin genes. *Genomics.* 1997;40(2):217-221.
54. Grozdanovic MM, Doyle CB, Liu L, et al. Charcot-Leyden crystal protein/galectin-10 interacts with cationic ribonucleases and is required for eosinophil granulogenesis. *J Allergy Clin Immunol.* 2020;146(2):377-389 e10.
55. Bandeira-Melo C, Bozza PT, Weller PF. The cellular biology of eosinophil eicosanoid formation and function. *J Allergy Clin Immunol.* 2002;109(3):393-400.
56. Peinhaupt M, Sturm EM, Heinemann A. Prostaglandins and their receptors in eosinophil function and as therapeutic targets. *Front Med.* 2017;4:104.
57. Thompson-Souza GA, Gropillo I, Neves JS. Cysteinyl leukotrienes in eosinophil biology: functional roles and therapeutic perspectives in eosinophilic disorders. *Front Med (Lausanne).* 2017;4:106.
58. Isobe Y, Kato T, Arita M. Emerging roles of eosinophils and eosinophil-derived lipid mediators in the resolution of inflammation. *Front Immunol.* 2012;3:270.
59. Bartemes KR, McKinney S, Gleich GJ, et al. Endogenous platelet-activating factor is critically involved in effector functions of eosinophils stimulated with IL-5 or IgG. *J Immunol.* 1999;162(5):2982-2989.
60. Okada S, Kita H, George TJ, et al. Migration of eosinophils through basement membrane components in vitro: role of matrix metalloproteinase-9. *Am J Respir Cell Mol Biol.* 1997;17(4):519-528.
61. Kovacs I, Horvath M, Kovacs T, et al. Comparison of proton channel, phagocyte oxidase, and respiratory burst levels between human eosinophil and neutrophil granulocytes. *Free Radic Res.* 2014;48(10):1190-1199.
62. Davoine F, Lacy P. Eosinophil cytokines, chemokines, and growth factors: emerging roles in immunity. *Front Immunol.* 2014;5:570.
63. Spencer LA, Szela CT, Perez SA, et al. Human eosinophils constitutively express multiple Th1, Th2, and immunoregulatory cytokines that are secreted rapidly and differentially. *J Leukoc Biol.* 2009;85(1):117-123.
64. Aceves SS. Remodeling and fibrosis in chronic eosinophil inflammation. *Dig Dis.* 2014;32(1-2):15-21.
65. Morimoto Y, Hirahara K, Kiuchi M, et al. Amphiregulin-producing pathogenic memory T helper 2 cells instruct eosinophils to secrete osteopontin and facilitate airway fibrosis. *Immunity.* 2018;49(1):134.e6-150.e6.
66. Berry MA, Parker D, Neale N, et al. Sputum and bronchial submucosal IL-13 expression in asthma and eosinophilic bronchitis. *J Allergy Clin Immunol.* 2004;114(5):1106-1109.
67. Schmid-Grendelmeier P, Altznauer F, Fischer B, et al. Eosinophils express functional IL-13 in eosinophilic inflammatory diseases. *J Immunol.* 2002;169(2):1021-1027.
68. Masterson JC, Capocelli KE, Hosford L, et al. Eosinophils and IL-33 perpetuate chronic inflammation and fibrosis in a pediatric population with stricturing Crohn's ileitis. *Inflamm Bowel Dis.* 2015;21(10):2429-2440.
69. Rosenberg HF, Phipps S, Foster PS. Eosinophil trafficking in allergy and asthma. *J Allergy Clin Immunol.* 2007;119(6):1303-1310; quiz 1311-1312.
70. Pan ZZ, Parkyn L, Ray A, et al. Inducible lung-specific expression of RANTES: preferential recruitment of neutrophils. *Am J Physiol Lung Cell Mol Physiol.* 2000;279(4):L658-L666.
71. Blanchard C, Wang N, Stringer KF, et al. Eotaxin-3 and a uniquely conserved gene-expression profile in eosinophilic esophagitis. *J Clin Invest.* 2006;116(2):536-547.
72. Johansson MW. Activation states of blood eosinophils in asthma. *Clin Exp Allergy.* 2014;44(4):482-498.
73. Heiman AS, Abonyo BO, Darling-Reed SF, et al. Cytokine-stimulated human lung alveolar epithelial cells release eotaxin-2 (CCL24) and eotaxin-3 (CCL26). *J Interferon Cytokine Res.* 2005;25(2):82-91.
74. Pope SM, Fulkerson PC, Blanchard C, et al. Identification of a cooperative mechanism involving interleukin-13 and eotaxin-2 in experimental allergic lung inflammation. *J Biol Chem.* 2005;280(14):13952-13961.
75. Aulakh GK, Petri B, Wojcik KM, et al. Inhibiting focal adhesion kinase (FAK) blocks IL-4 induced VCAM-1 expression and eosinophil recruitment in vitro and in vivo. *J Leukoc Biol.* 2018;104(1):147-158.
76. Bochner BS. Adhesion molecules as therapeutic targets. *Immunol Allergy Clin North Am.* 2004;24(4):615-630, vi.
77. Park YM, Bochner BS. Eosinophil survival and apoptosis in health and disease. *Allergy Asthma Immunol Res.* 2010;2(2):87-101.
78. Kita H, Ohnishi T, Okubo Y, et al. Granulocyte/macrophage colony-stimulating factor and interleukin 3 release from human peripheral blood eosinophils and neutrophils. *J Exp Med.* 1991;174(3):745-748.
79. Letuve S, Druilhe A, Grandsaigne M, et al. Involvement of caspases and of mitochondria in Fas ligation-induced eosinophil apoptosis: modulation by interleukin-5 and interferon-gamma. *J Leukoc Biol.* 2001;70(5):767-775.
80. Gonzalez-Gil A, Porell RN, Fernandes SM, et al. Sialylated keratan sulfate proteoglycans are Siglec-8 ligands in human airways. *Glycobiology.* 2018;28(10):786-801.
81. Kita H, Horie S, Gleich GJ. Extracellular matrix proteins attenuate activation and degranulation of stimulated eosinophils. *J Immunol.* 1996;156(3):1174-1181.
82. Walsh GM, Sexton DW, Blaylock MG. Corticosteroids, eosinophils and bronchial epithelial cells: new insights into the resolution of inflammation in asthma. *J Endocrinol.* 2003;178(1):37-43.
83. Tabatabaian F, Ledford DK, Casale TB. Biologic and new therapies in asthma. *Immunol Allergy Clin North Am.* 2017;37(2):329-343.
84. Liu LY, Bates ME, Jarjour NN, et al. Generation of Th1 and Th2 chemokines by human eosinophils: evidence for a critical role of TNF-alpha. *J Immunol.* 2007;179(7):4840-4848.
85. Cherry WB, Yoon J, Bartemes KR, et al. A novel IL-1 family cytokine, IL-33, potently activates human eosinophils. *J Allergy Clin Immunol.* 2008;121(6):1484-1490.
86. Hammad H, Lambrecht BN. Barrier epithelial cells and the control of type 2 immunity. *Immunity.* 2015;43(1):29-40.
87. Wong CK, Hu S, Cheung PF, et al. Thymic stromal lymphopoietin induces chemotactic and prosurvival effects in eosinophils: implications in allergic inflammation. *Am J Respir Cell Mol Biol.* 2010;43(3):305-315.
88. Stolarski B, Kurowska-Stolarska M, Kewin P, et al. IL-33 exacerbates eosinophil-mediated airway inflammation. *J Immunol.* 2010;185(6):3472-3480.
89. Abu-Ghazaleh RI, Fujisawa T, Mestecky J, et al. IgA-induced eosinophil degranulation. *J Immunol.* 1989;142(7):2393-2400.
90. Bruhns P, Jonsson F. Mouse and human FcR effector functions. *Immunol Rev.* 2015;268(1):25-51.
91. Daffern PJ, Pfeifer PH, Ember JA, et al. C3a is a chemotaxin for human eosinophils but not for neutrophils. I. C3a stimulation of neutrophils is secondary to eosinophil activation. *J Exp Med.* 1995;181(6):2119-2127.
92. Park SJ, Im DS. Blockage of sphingosine-1-phosphate receptor 2 attenuates allergic asthma in mice. *Br J Pharmacol.* 2019;176(7):938-949.
93. Kvarnhammar AM, Cardell LO. Pattern-recognition receptors in human eosinophils. *Immunology.* 2012;136(1):11-20.
94. Yousefi S, Gold JA, Andina N, et al. Catapult-like release of mitochondrial DNA by eosinophils contributes to antibacterial defense. *Nat Med.* 2008;14(9):949-953.
95. Matsuwaki Y, Wada K, White TA, et al. Recognition of fungal protease activities induces cellular activation and eosinophil-derived neurotoxin release in human eosinophils. *J Immunol.* 2009;183(10):6708-6716.
96. Bochner BS. "Siglec"ting the allergic response for therapeutic targeting. *Glycobiology.* 2016;26(6):546-552.
97. Youngblood BA, Leung J, Falahati R, et al. Discovery, function, and therapeutic targeting of Siglec-8. *Cells.* 2020;10(1):19.
98. Carroll DJ, O'Sullivan JA, Nix DB, et al. Sialic acid-binding immunoglobulin-like lectin 8 (Siglec-8) is an activating receptor mediating beta2-integrin-dependent function in human eosinophils. *J Allergy Clin Immunol.* 2018;141(6):2196-2207.
99. Dellon ES, Peterson KA, Murray JA, et al. Anti-Siglec-8 antibody for eosinophilic gastritis and duodenitis. *N Engl J Med.* 2020;383(17):1624-1634.
100. Knuplez E, Krier-Burris R, Cao Y, et al. Frontline science. Superior mouse eosinophil depletion in vivo targeting transgenic Siglec-8 instead of endogenous Siglec-F: mechanisms and pitfalls. *J Leukoc Biol.* 2020;108(1):43-58.
101. O'Sullivan JA, Carroll DJ, Bochner BS. Glycobiology of eosinophilic inflammation: Contributions of Siglecs, glycans, and other glycan-binding proteins. *Front Med (Lausanne).* 2017;4:116.
102. McGarry MP, Stewart CC. Murine eosinophil granulocytes bind the murine macrophage-monocyte specific monoclonal antibody F4/80. *J Leukoc Biol.* 1991;50(5):471-478.
103. Legrand F, Tomasevic N, Simakova O, et al. The eosinophil surface receptor epidermal growth factor-like module containing mucin-like hormone receptor 1 (EMR1): a novel therapeutic target for eosinophilic disorders. *J Allergy Clin Immunol*, 2014; 133(5):1439-1447, 1447.e1-1447.e8.
104. Rozenberg P, Reichman H, Moshkovits I, et al. CD300 family receptors regulate eosinophil survival, chemotaxis, and effector functions. *J Leukoc Biol.* 2018;104(1):21-29.

105. Na HJ, Hamilton RG, Klion AD, et al. Biomarkers of eosinophil involvement in allergic and eosinophilic diseases: review of phenotypic and serum markers including a novel assay to quantify levels of soluble Siglec-8. *J Immunol Methods*. 2012;383(1-2):39-46.
106. Koenderman L, van der Bruggen T, Schweizer RC, et al. Eosinophil priming by cytokines: from cellular signal to in vivo modulation. *Eur Respir J Suppl*. 1996;22:119s-125s.
107. Kita H. Eosinophils: multifaceted biological properties and roles in health and disease. *Immunol Rev*. 2011;242(1):161-177.
108. Jacobsen EA, Doyle AD, Colbert DC, et al. Differential activation of airway eosinophils induces IL-13-mediated allergic Th2 pulmonary responses in mice. *Allergy*. 2015;70(9):1148-1159.
109. Johansson MW. Eosinophil activation status in separate compartments and association with asthma. *Front Med*. 2017;4:75.
110. Radonjic-Hoesli S, Wang X, de Graauw E, et al. Adhesion-induced eosinophil cytolysis requires the receptor-interacting protein kinase 3 (RIPK3)-mixed lineage kinase-like (MLKL) signaling pathway, which is counterregulated by autophagy. *J Allergy Clin Immunol*. 2017;140(6):1632-1642.
111. Ueki S, Tokunaga T, Melo RCN, et al. Charcot-Leyden crystal formation is closely associated with eosinophil extracellular trap cell death. *Blood*. 2018;132(20):2183-2187.
112. Malm-Erjefalt M, Persson CG, Erjefalt JS. Degranulation status of airway tissue eosinophils in mouse models of allergic airway inflammation. *Am J Respir Cell Mol Biol*. 2001;24(3):352-359.
113. Persson C, Uller L. Theirs but to die and do: primary lysis of eosinophils and free eosinophil granules in asthma. *Am J Respir Crit Care Med*. 2014;189(6):628-633.
114. Ochkur SI, Jacobsen EA, Protheroe CA, et al. Co-expression of IL-5 and eotaxin-2 in mice creates an eosinophil-dependent model of respiratory inflammation with characteristics of severe asthma. *J Immunol*. 2007;178(12):7879-7889.
115. Ochkur SI, Kim JD, Protheroe CA, et al. The development of a sensitive and specific ELISA for mouse eosinophil peroxidase: assessment of eosinophil degranulation ex vivo and in models of human disease. *J Immunol Methods*. 2012;375(1-2):138-147.
116. Kaneko M, Horie S, Kato M, et al. A crucial role for beta 2 integrin in the activation of eosinophils stimulated by IgG. *J Immunol*. 1995;155(5):2631-2641.
117. Lambrecht BN, Persson EK, Hammad H. Myeloid cells in asthma. *Microbiol Spectr*. 2017;5(1). doi:10.1128/microbiolspec.MCHD-0053-2016.
118. Robida PA, Puzzovio PG, Pahima H, et al. Human eosinophils and mast cells: Birds of a feather flock together. *Immunol Rev*. 2018;282(1):151-167.
119. Bochner BS, Schleimer RP. Mast cells, basophils, and eosinophils: distinct but overlapping pathways for recruitment. *Immunol Rev*. 2001;179:5-15.
120. Gessner A, Mohrs K, Mohrs M. Mast cells, basophils, and eosinophils acquire constitutive IL-4 and IL-13 transcripts during lineage differentiation that are sufficient for rapid cytokine production. *J Immunol*. 2005;174(2):1063-1072.
121. Schaper-Gerhardt K, Kother B, Wolff L, et al. The H4 R is highly expressed on eosinophils from AD patients and IL-4 upregulates expression and function via the JAK/STAT pathway. *Allergy*. 2021;76(4):1261-1264.
122. Miyake K, Karasuyama H. Emerging roles of basophils in allergic inflammation. *Allergol Int*. 2017;66(3):382-391.
123. Lee SH, Chaves MM, Kamenyeva O, et al. M2-like, dermal macrophages are maintained via IL-4/CCL24-mediated cooperative interaction with eosinophils in cutaneous leishmaniasis. *Sci Immunol*. 2020;5(46):eaaz4415.
124. Dalli J, Serhan C. Macrophage proresolving mediators-the when and where. *Microbiol Spectr*. 2016;4(3). doi:10.1128/microbiolspec.MCHD-0001-2014.
125. Gieseck RLIII, Wilson MS, Wynn TA. Type 2 immunity in tissue repair and fibrosis. *Nat Rev Immunol*. 2018;18(1):62-76.
126. Takeda K, Shiraishi Y, Ashino S, et al. Eosinophils contribute to the resolution of lung-allergic responses following repeated allergen challenge. *J Allergy Clin Immunol*. 2015;135(2):451-460.
127. Mesnil C, Raulier S, Paulissen G, et al. Lung-resident eosinophils represent a distinct regulatory eosinophil subset. *J Clin Invest*. 2016;126(9):3279-3295.
128. Huang L, Gebreselassie NG, Gagliardo LF, et al. Eosinophil-derived IL-10 supports chronic nematode infection. *J Immunol*. 2014;193(8):4178-4187.
129. Wang YH, Angkasekwinai P, Lu N, et al. IL-25 augments type 2 immune responses by enhancing the expansion and functions of TSLP-DC-activated Th2 memory cells. *J Exp Med*. 2007;204(8):1837-1847.
130. Wang YH, Liu YJ. Thymic stromal lymphopoietin, OX40-ligand, and interleukin-25 in allergic responses. *Clin Exp Allergy*. 2009;39(6):798-806.
131. Shi HZ, Humbles A, Gerard C, et al. Lymph node trafficking and antigen presentation by endobronchial eosinophils. *J Clin Invest*. 2000;105(7):945-953.
132. Veres TZ, Kopcsanyi T, van Panhuys N, et al. Allergen-induced CD4+ T cell cytokine production within airway mucosal dendritic cell-T cell clusters drives the local recruitment of myeloid effector cells. *J Immunol*. 2017;198(2):895-907.
133. Persson T, Monsef N, Andersson P, et al. Expression of the neutrophil-activating CXC chemokine ENA-78/CXCL5 by human eosinophils. *Clin Exp Allergy*. 2003;33(4):531-537.
134. Page SM, Gleich GJ, Roebuck KA, et al. Stimulation of neutrophil interleukin-8 production by eosinophil granule major basic protein. *Am J Respir Cell Mol Biol*. 1999;21(2):230-237.
135. Yamada T, Tani Y, Nakanishi H, et al. Eosinophils promote resolution of acute peritonitis by producing proresolving mediators in mice. *FASEB J*. 2011;25(2):561-568.
136. Clancy DM, Sullivan GP, Moran HBT, et al. Extracellular neutrophil proteases are efficient regulators of IL-1, IL-33, and IL-36 cytokine activity but poor effectors of microbial killing. *Cell Rep*. 2018;22(11):2937-2950.
137. Pesce S, Thoren FB, Cantoni C, et al. The innate immune cross talk between NK cells and eosinophils is regulated by the interaction of natural cytotoxicity receptors with eosinophil surface ligands. *Front Immunol*. 2017;8:510.
138. Awad A, Yassine H, Barrier M, et al. Natural killer cells induce eosinophil activation and apoptosis. *PLoS One*. 2014;9(4):e94492.
139. Doherty TA, Broide DH. Airway innate lymphoid cells in the induction and regulation of allergy. *Allergol Int*. 2019;68(1):9-16.
140. Martinez-Gonzalez I, Steer CA, Takei F. Lung ILC2s link innate and adaptive responses in allergic inflammation. *Trends Immunol*. 2015;36(3):189-195.
141. Lei A, He Y, Yang Q, et al. Role of myeloid cells in the regulation of group 2 innate lymphoid cell-mediated allergic inflammation. *Immunology*. 2020;161(1):18-24.
142. Mortha A, Burrows K. Cytokine networks between innate lymphoid cells and myeloid cells. *Front Immunol*. 2018;9:191.
143. Bal SM, Bernink JH, Nagasawa M, et al. IL-1beta, IL-4 and IL-12 control the fate of group 2 innate lymphoid cells in human airway inflammation in the lungs. *Nat Immunol*. 2016;17(6):636-645.
144. Ding X, Luo Y, Zhang X, et al. IL-33-driven ILC2/eosinophil axis in fat is induced by sympathetic tone and suppressed by obesity. *J Endocrinol*. 2016;231(1):35-48.
145. Schuijs MJ, Hammad H, Lambrecht BN. Professional and 'Amateur' antigen-presenting cells in type 2 immunity. *Trends Immunol*. 2019;40(1):22-34.
146. Wang HB, Ghiran I, Matthaei K, et al. Airway eosinophils: allergic inflammation recruited professional antigen-presenting cells. *J Immunol*. 2007;179(11):7585-7592.
147. Wang HB, Akuthota P, Kanaoka Y, et al. Airway eosinophil migration into lymph nodes in mice depends on leukotriene C4. *Allergy*. 2017;72(6):927-936.
148. Jacobsen EA, Zellner KR, Colbert D, et al. Eosinophils regulate dendritic cells and Th2 pulmonary immune responses following allergen provocation. *J Immunol*. 2011;187(11):6059-6068.
149. Xenakis JJ, Howard ED, Smith KM, et al. Resident intestinal eosinophils constitutively express antigen presentation markers and include two phenotypically distinct subsets of eosinophils. *Immunology*. 2018;154(2):298-308.
150. Kambayashi T, Laufer TM. Atypical MHC class II-expressing antigen-presenting cells: can anything replace a dendritic cell?. *Nat Rev Immunol*. 2014;14(11):719-730.
151. Weller PF, Rand TH, Barrett T, et al. Accessory cell function of human eosinophils. HLA-DR-dependent, MHC-restricted antigen-presentation and IL-1 alpha expression. *J Immunol*. 1993;150(6):2554-2562.
152. Radke AL, Reynolds LE, Melo RC, et al. Mature human eosinophils express functional Notch ligands mediating eosinophil autocrine regulation. *Blood*. 2009;113(13):3092-3101.
153. Carretero R, Sektioglu IM, Garbi N, et al. Eosinophils orchestrate cancer rejection by normalizing tumor vessels and enhancing infiltration of CD8(+) T cells. *Nat Immunol*. 2015;16(6):609-617.
154. Samarasinghe AE, Melo RC, Duan S, et al. Eosinophils promote antiviral immunity in mice infected with influenza A virus. *J Immunol*. 2017;198(8):3214-3226.
155. Andersson J, Cromvik J, Ingelsten M, et al. Eosinophils from hematopoietic stem cell recipients suppress allogeneic T cell proliferation. *Biol Blood Marrow Transplant*. 2014;20(12):1891-1898.
156. Onyema OO, G Y, Mahgoub B, et al. Eosinophils downregulate lung alloimmunity by decreasing TCR signal transduction. *JCI Insight*. 2019;4(11):e128241.
157. Jacobsen EA, Ochkur SI, Pero RS, et al. Allergic pulmonary inflammation in mice is dependent on eosinophil-induced recruitment of effector T cells. *J Exp Med*. 2008;205(3):699-710.
158. Arnold IC, Artola-Boran M, Gurtner A, et al. The GM-CSF-IRF5 signaling axis in eosinophils promotes antitumor immunity through activation of type 1 T cell responses. *J Exp Med*. 2020;217(12):e20190706.
159. Amini A, Pang D, Hackstein CP, et al. MAIT cells in barrier tissues: lessons from immediate neighbors. *Front Immunol*. 2020;11:584521.
160. Chu VT, Berek C. Immunization induces activation of bone marrow eosinophils required for plasma cell survival. *Eur J Immunol*. 2012;42(1):130-137.
161. Jung Y, Wen T, Mingler MK, et al. IL-1beta in eosinophil-mediated small intestinal homeostasis and IgA production. *Mucosal Immunol*. 2015;8(4):930-942.
162. Chu VT, Beller A, Rausch S, et al. Eosinophils promote generation and maintenance of immunoglobulin-A-expressing plasma cells and contribute to gut immune homeostasis. *Immunity*. 2014;40(4):582-593.
163. Coakley G, Wang H, Harris NL. Intestinal eosinophils: multifaceted roles in tissue homeostasis and disease. *Semin Immunopathol*. 2021;43(3):307-317.
164. Percopo CM, Dyer KD, Ochkur SI, et al. Activated mouse eosinophils protect against lethal respiratory virus infection. *Blood*. 2014;123(5):743-752.
165. Wong TW, Kita H, Hanson CA, et al. Induction of malignant plasma cell proliferation by eosinophils. *PLoS One*. 2013;8(7):e70554.
166. Calcinotto A, Brevi A, Chesi M, et al. Microbiota-driven interleukin-17-producing cells and eosinophils synergize to accelerate multiple myeloma progression. *Nat Commun*. 2018;9(1):4832.
167. Abdala-Valencia H, Coden ME, Chiarella SE, et al. Shaping eosinophil identity in the tissue contexts of development, homeostasis, and disease. *J Leukoc Biol*. 2018;104(1):95-108.
168. Carmo LAS, Bonjour K, Spencer LA, et al. Single-cell analyses of human eosinophils at high resolution to understand compartmentalization and vesicular trafficking of interferon-gamma. *Front Immunol*. 2018;9:1542.
169. Spencer LA, Melo RC, Perez SA, et al. Cytokine receptor-mediated trafficking of preformed IL-4 in eosinophils identifies an innate immune mechanism of cytokine secretion. *Proc Natl Acad Sci U S A*. 2006;103(9):3333-3338.
170. Rodriguez L, Pekkarinen PT, Lakshmikanth T, et al. Systems-level immunomonitoring from acute to recovery phase of severe COVID-19. *Cell Rep Med*. 2020;1(5):100078.
171. Jacobsen EA, Lee NA, Lee JJ. Re-defining the unique roles for eosinophils in allergic respiratory inflammation. *Clin Exp Allergy*. 2014;44(9):1119-1136.
172. Rosenberg HF, Dyer KD, Foster PS. Eosinophils: changing perspectives in health and disease. *Nat Rev Immunol*. 2013;13(1):9-22.
173. Kopf M, Brombacher F, Hodgkin PD, et al. IL-5-deficient mice have a developmental defect in CD5+ B-1 cells and lack eosinophilia but have normal antibody and cytotoxic T cell responses. *Immunity*. 1996;4(1):15-24.
174. Yoshida T, Ikuta K, Sugaya H, et al. Defective B-1 cell development and impaired immunity against Angiostrongylus cantonensis in IL-5R alpha-deficient mice. *Immunity*. 1996;4(5):483-494.
175. Lee JJ, Dimina D, Macias MP, et al. Defining a link with asthma in mice congenitally deficient in eosinophils. *Science*. 2004;305(5691):1773-1776.
176. Humbles AA, Lloyd CM, McMillan SJ, et al. A critical role for eosinophils in allergic airways remodeling. *Science*. 2004;305(5691):1776-1779.
177. Doyle AD, Jacobsen EA, Ochkur SI, et al. Homologous recombination into the eosinophil peroxidase locus generates a strain of mice expressing Cre recombinase exclusively in eosinophils. *J Leukoc Biol*. 2013;94(1):17-24.
178. Dyer KD, Moser JM, Czapiga M, et al. Functionally competent eosinophils differentiated ex vivo in high purity from normal mouse bone marrow. *J Immunol*. 2008;181(6):4004-4009.
179. Annunziato F, Romagnani C, Romagnani S. The 3 major types of innate and adaptive cell-mediated effector immunity. *J Allergy Clin Immunol*. 2015;135(3):626-635.
180. Qiu Y, Nguyen KD, Odegaard JI, et al. Eosinophils and type 2 cytokine signaling in macrophages orchestrate development of functional beige fat. *Cell*. 2014;157(6):1292-1308.
181. Heredia JE, Mukundan L, Chen FM, et al. Type 2 innate signals stimulate fibro/adipogenic progenitors to facilitate muscle regeneration. *Cell*. 2013;153(2):376-388.
182. Pan D, Hunter DA, Schellhardt L, et al. T cells modulate IL-4 expression by eosinophil recruitment within decellularized scaffolds to repair nerve defects. *Acta Biomater*. 2020;112:149-163.
183. Berek C. Eosinophils: important players in humoral immunity. *Clin Exp Immunol*. 2016;183(1):57-64.
184. Doyle AD, Mukherjee M, LeSuer WE, et al. Eosinophil-derived IL-13 promotes emphysema. *Eur Respir J*. 2019;53(5).
185. Fulkerson PC, Fischetti CA, Rothenberg ME. Eosinophils and CCR3 regulate interleukin-13 transgene-induced pulmonary remodeling. *Am J Pathol*. 2006;169(6):2117-2126.
186. Wang Y, Yang Y, Wang M, et al. Eosinophils attenuate hepatic ischemia-reperfusion injury in mice through ST2-dependent IL-13 production. *Sci Transl Med*. 2021;13(579):eabb6576.
187. Grisaru-Tal S, Itan M, Klion AD, et al. A new dawn for eosinophils in the tumour microenvironment. *Nat Rev Cancer*. 2020;20(10):594-607.
188. LeMessurier KS, Rooney R, Ghoneim HE, et al. Influenza A virus directly modulates mouse eosinophil responses. *J Leukoc Biol*. 2020;108(1):151-168.
189. Berek C. Eosinophils can more than kill. *J Exp Med*. 2018;215(8):1967-1969.
190. Onyema OO, Guo Y, Hata A, et al. Deciphering the role of eosinophils in solid organ transplantation. *Am J Transplant*. 2020;20(4):924-930.
191. Arnold IC, Artola-Boran M, Tallon de Lara P, et al. Eosinophils suppress Th1 responses and restrict bacterially induced gastrointestinal inflammation. *J Exp Med*. 2018;215(8):2055-2072.
192. Sugawara R, Lee EJ, Jang MS, et al. Small intestinal eosinophils regulate Th17 cells by producing IL-1 receptor antagonist. *J Exp Med*. 2016;213(4):555-567.

193. Guerra ES, Lee CK, Specht CA, et al. Central role of IL-23 and IL-17 producing eosinophils as Immunomodulatory effector cells in acute pulmonary Aspergillosis and allergic asthma. *PLoS Pathog.* 2017;13(1):e1006175.
194. Malacco N, Rachid MA, Gurgel I, et al. Eosinophil-associated innate IL-17 response promotes *Aspergillus fumigatus* lung pathology. *Front Cell Infect Microbiol.* 2018;8:453.
195. Merritt CR, Ong GT, Church SE, et al. Multiplex digital spatial profiling of proteins and RNA in fixed tissue. *Nat Biotechnol.* 2020;38(5):586-599.
196. Bingham GC, Lee F, Naba A, et al. Spatial-omics: novel approaches to probe cell heterogeneity and extracellular matrix biology. *Matrix Biol.* 2020;91-92:152-166.
197. Khoury P, Akuthota P, Ackerman SJ, et al. Revisiting the NIH Taskforce on the research needs of eosinophil-associated diseases (RE-TREAD). *J Leukoc Biol.* 2018;104(1):69-83.
198. Throsby M, Herbelin A, Pleau JM, et al. CD11c+ eosinophils in the murine thymus: developmental regulation and recruitment upon MHC class I-restricted thymocyte deletion. *J Immunol.* 2000;165(4):1965-1975.
199. Tulic MK, Sly PD, Andrews D, et al. Thymic indoleamine 2,3-dioxygenase-positive eosinophils in young children: potential role in maturation of the naive immune system. *Am J Pathol.* 2009;175(5):2043-2052.
200. Theiler A, Barnthaler T, Platzer W, et al. Butyrate ameliorates allergic airway inflammation by limiting eosinophil trafficking and survival. *J Allergy Clin Immunol.* 2019;144(3):764-776.
201. Yang H, Duan Z. The local defender and functional mediator: gut microbiome. *Digestion.* 2018;97(2):137-145.
202. Mitchell PD, Salter BM, Oliveria JP, et al. Glucagon-like peptide-1 receptor expression on human eosinophils and its regulation of eosinophil activation. *Clin Exp Allergy.* 2017;47(3):331-338.
203. Lee EH, Itan M, Jang J, et al. Eosinophils support adipocyte maturation and promote glucose tolerance in obesity. *Sci Rep.* 2018;8(1):9894.
204. Wu D, Molofsky AB, Liang HE, et al. Eosinophils sustain adipose alternatively activated macrophages associated with glucose homeostasis. *Science.* 2011;332(6026):243-247.
205. Withers SB, Forman R, Meza-Perez S, et al. Eosinophils are key regulators of perivascular adipose tissue and vascular functionality. *Sci Rep.* 2017;7:44571.
206. Hernandez JD, Tew BY, Li T, et al. A FACS-based approach to obtain viable eosinophils from human adipose tissue. *Sci Rep.* 2020;10(1):13210.
207. Uderhardt S, Ackermann JA, Fillep T, et al. Enzymatic lipid oxidation by eosinophils propagates coagulation, hemostasis, and thrombotic disease. *J Exp Med.* 2017;214(7):2121-2138.
208. Marx C, Novotny J, Salbeck D, et al. Eosinophil-platelet interactions promote atherosclerosis and stabilize thrombosis with eosinophil extracellular traps. *Blood.* 2019;134(21):1859-1872.
209. Kobayashi H, Gleich GJ, Butterfield JH, et al. Human eosinophils produce neurotrophins and secrete nerve growth factor on immunologic stimuli. *Blood.* 2002;99(6):2214-2220.
210. Dileepan M, Ge XN, Bastan I, et al. Regulation of eosinophil recruitment and allergic airway inflammation by Tropomyosin receptor kinase A. *J Immunol.* 2020;204(3):682-693.
211. Drake MG, Lebold KM, Roth-Carter QR, et al. Eosinophil and airway nerve interactions in asthma. *J Leukoc Biol.* 2018;104(1):61-67.
212. Nakanishi S, Mantani Y, Haruta T, et al. Three-dimensional analysis of neural connectivity with cells in rat ileal mucosa by serial block-face scanning electron microscopy. *J Vet Med Sci.* 2020;82(7):990-999.
213. Smyth CM, Akasheh N, Woods S, et al. Activated eosinophils in association with enteric nerves in inflammatory bowel disease. *PLoS One.* 2013;8(5):e64216.
214. Guseva D, Rudrich U, Kotnik N, et al. Neuronal branching of sensory neurons is associated with BDNF-positive eosinophils in atopic dermatitis. *Clin Exp Allergy.* 2020;50(5):577-584.
215. Foster EL, Simpson EL, Fredrikson LJ, et al. Eosinophils increase neuron branching in human and murine skin and in vitro. *PLoS One.* 2011;6(7):e22029.
216. Hengstman GJ, Wesseling P, Frenken CW, et al. Neuromyelitis optica with clinical and histopathological involvement of the brain. *Mult Scler.* 2007;13(5):679-682.
217. Nishi R, Koike H, Ohyama K, et al. Differential clinicopathologic features of EGPA-associated neuropathy with and without ANCA. *Neurology.* 2020;94(16):e1726-e1737.
218. Loffredo LF, Coden ME, Jeong BM, et al. Eosinophil accumulation in postnatal lung is specific to the primary septation phase of development. *Sci Rep.* 2020;10(1):4425.
219. de Kleer IM, Kool M, de Bruijn MJ, et al. Perinatal activation of the interleukin-33 pathway promotes type 2 immunity in the developing lung. *Immunity.* 2016;45(6):1285-1298.
220. Jacobsen EA, Helmers RA, Lee JJ, et al. The expanding role(s) of eosinophils in health and disease. *Blood.* 2012;120(19):3882-3890.
221. Gleich GJ, Klion AD, Lee JJ, et al. The consequences of not having eosinophils. *Allergy.* 2013;68(7):829-835.
222. Kephart GM, Gleich GJ, Connor DH, et al. Deposition of eosinophil granule major basic protein onto microfilariae of Onchocerca volvulus in the skin of patients treated with diethylcarbamazine. *Lab Invest.* 1984;50(1):51-61.
223. Huang L, Appleton JA. Eosinophils in helminth infection: Defenders and Dupes. *Trends Parasitol.* 2016;32(10):798-807.
224. Specht S, Saeftel M, Arndt M, et al. Lack of eosinophil peroxidase or major basic protein impairs defense against murine filarial infection. *Infect Immun.* 2006;74(9):5236-5243.
225. Voehringer D, Reese TA, Huang X, et al. Type 2 immunity is controlled by IL-4/IL-13 expression in hematopoietic non-eosinophil cells of the innate immune system. *J Exp Med.* 2006;203(6):1435-1446.
226. Flores-Torres AS, Salinas-Carmona MC, Salinas E, et al. Eosinophils and respiratory viruses. *Viral Immunol.* 2019;32(5):198-207.
227. Rodrigo-Munoz JM, Sastre B, Canas JA, et al. Eosinophil response against classical and emerging respiratory viruses: COVID-19. *J Investig Allergol Clin Immunol,* 2020;31(2):94-107.
228. Jackson DJ, Johnston SL. The role of viruses in acute exacerbations of asthma. *J Allergy Clin Immunol.* 2010;125(6):1178-1187; quiz 1188-1189.
229. Glezen WP. Asthma, influenza, and vaccination. *J Allergy Clin Immunol.* 2006;118(6):1199-1206; quiz 1207-1208.
230. Silveira JS, Antunes GL, Gassen RB, et al. Respiratory syncytial virus increases eosinophil extracellular traps in a murine model of asthma. *Asia Pac Allergy.* 2019;9(4):e32.
231. Handzel ZT, Busse WW, Sedgwick JB, et al. Eosinophils bind rhinovirus and activate virus-specific T cells. *J Immunol.* 1998;160(3):1279-1284.
232. Tiwary M, Rooney RJ, Liedmann S, et al. Eosinophil responses at the airway epithelial barrier during the early phase of influenza A virus infection in C57BL/6 mice. *Cells.* 2021;10(3):509.
233. Gern JE, Rosenthal LA, Sorkness RL, et al. Effects of viral respiratory infections on lung development and childhood asthma. *J Allergy Clin Immunol.* 2005;115(4):668-674; quiz 675.
234. Sabogal Pineros YS, Bal SM, Dijkhuis A, et al. Eosinophils capture viruses, a capacity that is defective in asthma. *Allergy.* 2019;74(10):1898-1909.
235. Skevaki C, Karsonova A, Karaulov A, et al. Asthma-associated risk for COVID-19 development. *J Allergy Clin Immunol.* 2020;146(6):1295-1301.
236. Hogan SP, Waddell A, Fulkerson PC. Eosinophils in infection and intestinal immunity. *Curr Opin Gastroenterol.* 2013;29(1):7-14.
237. Lehrer RI, Szklarek D, Barton A, et al. Antibacterial properties of eosinophil major basic protein and eosinophil cationic protein. *J Immunol.* 1989;142(12):4428-4434.
238. Boix E, Salazar VA, Torrent M, et al. Structural determinants of the eosinophil cationic protein antimicrobial activity. *Biol Chem.* 2012;393(8):801-815.
239. Cowardin CA, Buonomo EL, Saleh MM, et al. The binary toxin CDT enhances *Clostridium difficile* virulence by suppressing protective colonic eosinophilia. *Nat Microbiol.* 2016;1(8):16108.
240. Vijayan VK, Reetha AM, Jawahar MS, et al. Pulmonary eosinophilia in pulmonary tuberculosis. *Chest.* 1992;101(6):1708-1709.
241. D'Avila H, Almeida PE, Roque NR, et al. Toll-like receptor-2-mediated C-C chemokine receptor 3 and eotaxin-driven eosinophil influx induced by *Mycobacterium bovis* BCG pleurisy. *Infect Immun.* 2007;75(3):1507-1511.
242. Prevete N, Rossi FW, Rivellese F, et al. Helicobacter pylori HP(2-20) induces eosinophil activation and accumulation in superficial gastric mucosa and stimulates VEGF-alpha and TGF-beta release by interacting with formyl-peptide receptors. *Int J Immunopathol Pharmacol.* 2013;26(3):647-662.
243. Ray A, Camiolo M, Fitzpatrick A, et al. Are we meeting the promise of endotypes and precision medicine in asthma? *Physiol Rev.* 2020;100(3):983-1017.
244. Busse WW, Lemanske RFJr. Asthma. *N Engl J Med.* 2001;344(5):350-362.
245. Filley WV, Holley KE, Kephart GM, et al. Identification by immunofluorescence of eosinophil granule major basic protein in lung tissues of patients with bronchial asthma. *Lancet.* 1982;2(8288):11-16.
246. Frigas E, Loegering DA, Solley GO, et al. Elevated levels of the eosinophil granule major basic protein in the sputum of patients with bronchial asthma. *Mayo Clin Proc.* 1981;56(6):345-353.
247. Gundel RH, Letts LG, Gleich GJ. Human eosinophil major basic protein induces airway constriction and airway hyperresponsiveness in primates. *J Clin Invest.* 1991;87(4):1470-1473.
248. Coyle AJ, Uchida D, Ackerman SJ, et al. Role of cationic proteins in the airway. Hyperresponsiveness due to airway inflammation. *Am J Respir Crit Care Med.* 1994;150(5 pt 2):S63-S71.
249. Gleich GJ, Frigas E, Loegering DA, et al. Cytotoxic properties of the eosinophil major basic protein. *J Immunol.* 1979;123(6):2925-2927.
250. Haldar P, Brightling CE, Hargadon B, et al. Mepolizumab and exacerbations of refractory eosinophilic asthma. *N Engl J Med.* 2009;360(10):973-984.
251. Menzies-Gow A, Flood-Page P, Sehmi R, et al. Anti-IL-5 (mepolizumab) therapy induces bone marrow eosinophil maturational arrest and decreases eosinophil progenitors in the bronchial mucosa of atopic asthmatics. *J Allergy Clin Immunol.* 2003;111(4):714-719.
252. Walsh GM. Anti-IL-5 monoclonal antibodies for the treatment of asthma: an update. *Expert Opin Biol Ther.* 2020;20(10):1237-1244.
253. Cho JY, Miller M, Baek KJ, et al. Inhibition of airway remodeling in IL-5-deficient mice. *J Clin Invest.* 2004;113(4):551-560.
254. Mould AW, Ramsay AJ, Matthaei KI, et al. The effect of IL-5 and eotaxin expression in the lung on eosinophil trafficking and degranulation and the induction of bronchial hyperreactivity. *J Immunol.* 2000;164(4):2142-2150.
255. Jacobsen EA, Ochkur SI, Doyle AD, et al. Lung pathologies in a chronic inflammation mouse model are independent of eosinophil degranulation. *Am J Respir Crit Care Med.* 2017;195(10):1321-1332.
256. Fryer AD, Jacoby DB. Muscarinic receptors and control of airway smooth muscle. *Am J Respir Crit Care Med.* 1998;158(5 pt 3):S154-S160.
257. Simon D, Simon HU. Therapeutic strategies for eosinophilic dermatoses. *Curr Opin Pharmacol.* 2019;46:29-33.
258. Lin L, Hwang BJ, Culton DA, et al. Eosinophils mediate tissue injury in the Autoimmune skin disease bullous pemphigoid. *J Invest Dermatol.* 2018;138(5):1032-1043.
259. Rudrich U, Gehring M, Papakonstantinou E, et al. Eosinophils are a major source of interleukin-31 in bullous pemphigoid. *Acta Derm Venereol.* 2018;98(8):766-771.
260. Furuta GT. Eosinophilic esophagitis: update on clinicopathological manifestations and pathophysiology. *Curr Opin Gastroenterol.* 2011;27(4):383-388.
261. O'Shea KM, Aceves SS, Dellon ES, et al. Pathophysiology of eosinophilic esophagitis. *Gastroenterology.* 2018;154(2):333-345.
262. Kottyan LC, Parameswaran S, Weirauch MT, et al. The genetic etiology of eosinophilic esophagitis. *J Allergy Clin Immunol.* 2020;145(1):9-15.
263. Mueller S, Aigner T, Neureiter D, et al. Eosinophil infiltration and degranulation in oesophageal mucosa from adult patients with eosinophilic oesophagitis: a retrospective and comparative study on pathological biopsy. *J Clin Pathol.* 2006;59(11):1175-1180.
264. Straumann A, Conus S, Grzonka P, et al. Anti-interleukin-5 antibody treatment (mepolizumab) in active eosinophilic oesophagitis: a randomised, placebo-controlled, double-blind trial. *Gut.* 2010;59(1):21-30.
265. Spergel JM, Rothenberg ME, Collins MH, et al. Reslizumab in children and adolescents with eosinophilic esophagitis: results of a double-blind, randomized, placebo-controlled trial. *J Allergy Clin Immunol,* 2012;129(2):456-463; 463.e1-463.e3.
266. Butz BK, Wen T, Gleich GJ, et al. Efficacy, dose reduction, and resistance to high-dose fluticasone in patients with eosinophilic esophagitis. *Gastroenterology.* 2014;147(2):324.e5-333.e5.
267. Blanchard C, Simon D, Schoepfer A, et al. Eosinophilic esophagitis: unclear roles of IgE and eosinophils. *J Intern Med.* 2017;281(5):448-457.
268. Zuo L, Fulkerson PC, Finkelman FD, et al. IL-13 induces esophageal remodeling and gene expression by an eosinophil-independent, IL-13R alpha 2-inhibited pathway. *J Immunol.* 2010;185(1):660-669.
269. Hirano I, Dellon ES, Hamilton JD, et al. Efficacy of dupilumab in a phase 2 randomized trial of adults with active eosinophilic esophagitis. *Gastroenterology.* 2020;158(1):111.e10-122.e10.
270. Wechsler JB, Hirano I. Biological therapies for eosinophilic gastrointestinal diseases. *J Allergy Clin Immunol.* 2018;142(1):24-31 e2.
271. Goldman M, Le Moine A, Braun M, et al. A role for eosinophils in transplant rejection. *Trends Immunol.* 2001;22(5):247-251.
272. Trull A, Steel L, Cornelissen J, et al. Association between blood eosinophil counts and acute cardiac and pulmonary allograft rejection. *J Heart Lung Transplant.* 1998;17(5):517-524.
273. Braun MY, Desalle F, Le Moine A, et al. IL-5 and eosinophils mediate the rejection of fully histoincompatible vascularized cardiac allografts: regulatory role of alloreactive CD8(+) T lymphocytes and IFN-gamma. *Eur J Immunol.* 2000;30(5):1290-1296.
274. Cao Q, Wang Y, Niu Z, et al. Potentiating tissue-resident type 2 innate lymphoid cells by IL-33 to prevent renal ischemia-reperfusion injury. *J Am Soc Nephrol.* 2018;29(3):961-976.
275. Dwyer GK, Turnquist HR. Untangling local pro-inflammatory, reparative, and regulatory damage-associated molecular-patterns (DAMPs) pathways to improve transplant outcomes. *Front Immunol.* 2021;12:611910.
276. Grisaru-Tal S, Itan M, Grass DG, et al. Primary tumors from mucosal barrier organs drive unique eosinophil infiltration patterns and clinical associations. *OncoImmunology.* 2020;10(1):1859732.
277. Reichman H, Itan M, Rozenberg P, et al. Activated eosinophils exert Antitumorigenic activities in colorectal cancer. *Cancer Immunol Res.* 2019;7(3):388-400.
278. Hollande C, Boussier J, Ziai J, et al. Inhibition of the dipeptidyl peptidase DPP4 (CD26) reveals IL-33-dependent eosinophil-mediated control of tumor growth. *Nat Immunol.* 2019;20(3):257-264.

CHAPTER 18

The Biology of Natural Killer Cells and Innate Lymphoid Cells

Vincent Peng, BA, MD, PhD • Marco Colonna, MD

INTRODUCTION

The defining hallmark of adaptive immunity is the expression of somatically rearranged antigen receptors, which impart the exquisite diversity and specificity of T and B lymphocytes. For almost half a century, immunologists have described cells which possess lymphoid characteristics yet lack such antigen-specific receptors. Collectively, these populations are referred to as innate lymphoid cells (ILC). While there is significant diversity among ILC, they share (1) a lack of T- and B-cell antigen receptors, (2) a common progenitor that expresses the master transcription factor Id2, and (3) reliance on common gamma chain (γ_c) cytokines for their maturation and survival. **Natural killer (NK) cells** and **lymphoid tissue-inducing cells (LTi)** are the earliest described ILC. NK cells have remarkable cytotoxic potential and participate in antiviral and antitumor immunity. Two decades later, LTi were discovered to seed various tissues during embryonic development and instruct stromal cells to induce the formation of secondary lymphoid organs (SLO). Following the identification of these two populations, additional subsets of ILC have been described based on the signature cytokines they produce. **Group 1 ILC (ILC1)** produce IFNγ and are associated with early immune responses against viruses. **Group 2 ILC (ILC2)** produce IL-5 and IL-13 and are involved in type 2 immune responses that protect against parasitic helminths and contribute to allergic diseases. **Group 3 ILC (ILC3)** produce IL-22 and IL-17 and are largely involved in the control of extracellular microbes at barrier surfaces. The features of ILC1, ILC2, and ILC3 mirror those that define Th1, Th2, and Th17 cells, whereas NK cells are analogous to cytotoxic $CD8^+$ T cells (Fig. 18.1). In contrast to T cells which take several days to undergo clonal expansion and acquire effector functions, ILC secrete cytokines rapidly in response to activating signals in their tissue of residence. ILC also participate in immune responses by directly regulating adaptive lymphocytes through membrane-bound costimulatory molecules and soluble factors. Thus, ILC play important roles in both the initiation and amplification of the immune response. In this chapter, we will review current knowledge of ILC development and strategies used to identify these cells in tissue (Table 18.1). We will also review the diverse effector functions of ILC and their roles in disease (Fig. 18.2).

DEVELOPMENT OF ILC

Mouse ILC Development

A stepwise model has been proposed for the development of mouse ILC in bone marrow (BM) (Fig. 18.3).[2-4] The earliest step begins with the commitment of hematopoietic stem cells (HSC) to the common lymphoid progenitor (CLP) lineage, which gives rise to B cells, T cells, and ILC. Immediately downstream of the CLP is the alpha lymphoid progenitor (αLP), defined by expression of the integrin α4β7 and IL-7 receptor (CD127). The αLP lacks the potential to develop into T and B cells and is the earliest known progenitor selectively restricted to the ILC fate.[5-7] This heterogeneous population consists of a spectrum of multipotency. Along with the αLP, an early ILC progenitor (EILP) has been described that has the potential to generate all types of ILC, but not T or B lymphocytes. This population was identified based on the expression of TCF-1, a master transcription factor also involved in T cell fate specification.[8-10] EILP have been proposed to be downstream of αLP based on lineage tracing and *in vivo* transfer experiments.[4,11] Collectively, αLP and EILP are referred to as common innate lymphoid progenitors (CILP). CILP depend on the master transcription factors TOX, TCF-1, and NFIL3 (also called E4BP4), as genetic ablation of these factors leads to global deficiency of these progenitors and mature ILC.[5,9,10,12-17]

The next step constitutes a branchpoint where NK differentiation diverges from the development of other ILC. CILP contain populations of common helper ILC progenitors (CHILP), which are marked by high expression levels of Id2, and pre-NK progenitors (pre-NKP). High expression of Id2 is recognized to be a critical checkpoint factor in ILC development via the suppression of the E protein E2A, which drives T and B lymphocyte differentiation.[18-20] Furthermore, $Id2^{-/-}$ mice lack splenic NK cells, fetal LTi, lymph nodes (LN), and Peyer patches (PP).[21] Interestingly, while $Id2^{-/-}$ mice lack mature NK cells in peripheral organs, they retain normal NKP in BM, indicating a requirement for Id2 in the maturation, rather than the lineage specification, of NK cells.[19] Using Id2-reporter mice, $Id2^{hi}$ CHILP were shown to differentiate into ILC1, ILC2, ILC3, and LTi, but not NK *in vivo* and *in vitro*.[20] Within the CHILP, expression of the transcription factor PLZF marks the ILC progenitor (ILCP), a committed precursor to ILC1, ILC2, and ILC3, but not LTi.[22] In addition to PLZF and the aforementioned transcription factors, the development of ILCP is also dependent on the master transcriptional regulator GATA3.[23] In order to facilitate the identification of ILCP without using transcription factor reporter mice, programmed cell death protein-1 (PD-1) has been identified as a surrogate marker for ILCP within the BM.[24] The ILCP is the last known multipotent progenitor of ILCs, after which unipotent committed precursors for each ILC subset develop in the BM and seed the periphery.

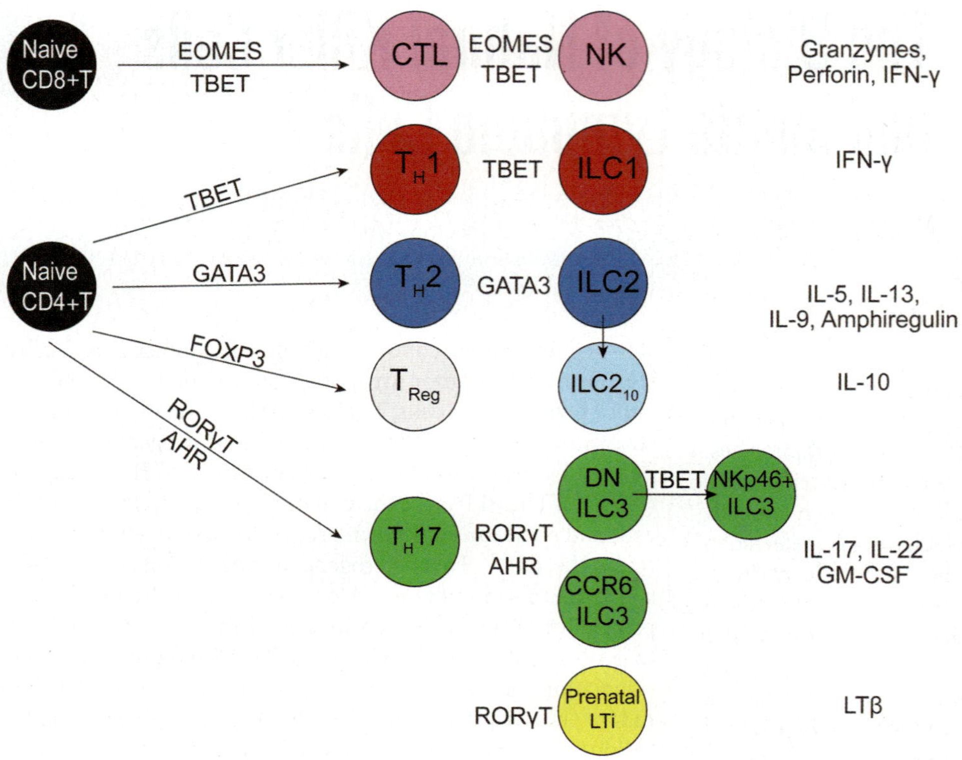

FIG. 18.1. Innate lymphoid cells (ILC) and T cells share similar effector modules. In the periphery, naive T cells (*left*) that recognize their specific antigen and are costimulated by antigen-presenting cells become activated and mature to effector cells. For CD4 T cells, antigen-presenting cell–derived cytokines drive transcription factor induction, leading to T helper polarization and the indicated effector functions. Meanwhile, ILC in the periphery are already mature cells that do not require antigen recognition or costimulation for their differentiation. The transcription factors required for T helper polarization and ILC development and the effector functions of cells in the same module are strikingly similar.

The most well-described of these precursors is the ILC2 progenitor (ILC2P). ILC2P are readily distinguished from multipotent progenitors by the expression of ILC2 markers such as IL-2–receptor alpha chain (CD25) and stem cells antigen-1 (Sca-1).[25] Specification and differentiation of ILC2P to mature ILC2 is dependent on the transcription factors Bcl11b, RORα, and ETS1.[26-31] Interestingly, the transcription factor Bcl11b appears to play a selective role in maintaining lineage fidelity, as Bcl11b-deficient ILC2s upregulate many ILC3-characteristic genes. These findings foreshadow a profound degree of plasticity among ILC that will be discussed later in the chapter.

We now return to the differentiation of NK cells from pre-NKP, which diverged at the level of the CILP. Pre-NKP have been identified as the earliest committed NK progenitor.[32] Transfer of these lineage negative (Lin^-) $cKit^-$ $Flk2^-$ $CD27^+$ $CD244^+$ $CD127^+$ $CD122^-$ BM progenitors to empty hosts gives rise to NK cells, but not other lymphocyte lineages, in the BM, blood, and spleen. Interestingly, the pre-NKP lacks expression of the IL-2/15Rβ (CD122) subunit, indicating that IL-15 signaling is not necessary for commitment to the NK fate. The transition from pre-NKP to the refined NK progenitor (rNKP) is marked by expression of CD122. CD122 transduces IL-15 signaling which acts through NFIL3 to facilitate rNKP differentiation into mature NK.[16] NFIL3-deficient mice have reduced numbers of both pre-NKP and rNKP.[33] Thus, NFIL3 is not only involved in the initial specification of pre-NKP but also the continued maturation of NKP in the BM.[16,17,33,34] Many developmental stages in the BM have been proposed and defined between the committed NKP and mature NK cells, and these definitions have been in flux as more sophisticated tools have been developed to address these questions. Rather than detail each of these stages, we will focus our attention on certain aspects of NK maturation that relate to their effector functions. The acquisition of NK cell receptors occurs in a stepwise manner throughout NK development, with NK1.1 and NKG2/CD94 among the first to be expressed. These receptors are followed by NKp46, then NKG2D, and then the Ly49s.[35] NK developmental stages are commonly described using the expression of CD27 and Mac-1 (or one of its subunits, CD11b).[36,37] As CD27 is expressed in early hematopoietic progenitors such as CLP, $CD27^{hi}$ $CD11b^-$ NK cells are the most immature, followed by $CD27^{hi}$ $CD11b^+$ cells, and then $CD27^{lo}CD11b^+$ cells, which are considered to be terminally differentiated. $CD27^{hi}$ $CD11b^+$ NK cells are more proliferative and more effective cytokine producers than $CD27^{lo}$ $CD11b^+$ NK but have a lower activation threshold for cytotoxicity due to reduced expression of inhibitory Ly49 receptors.[36]

The kinetics of ILC differentiation and distribution throughout the body are different from that of adaptive lymphocytes. Compared to T cells, B cells, and NK cells, almost all helper ILC (ILC1, ILC2, ILC3) are tissue-resident cells at steady-state. Studies using parabiotic mice demonstrated minimal (<5%) replacement of ILC from circulation after 130 days, whereas NK, T, and B cells rapidly equilibrated.[38,39] The minimal contribution of adult medullary hematopoiesis toward maintaining ILC in tissue suggests that the majority of helper ILC arise during fetal or neonatal life. The stepwise developmental scheme from CLP to lineage-restricted ILCP appears to apply during both embryonic and adult development. αLP, CHILP, and ILCP can be identified in primary hematopoietic organs, including the fetal liver, during both developmental timeframes.[6,7,40,41] In addition, multipotent

TABLE 18.1 Table Comparing the Phenotypic Markers of ILC in Mouse and Human

Mouse	Marker	NK	ILC1	ILC2	LTi	NKp46⁻ ILC3	NKp46⁺ ILC3
Cell-surface molecules	CD45	+	+	+	int	int	int
	CD127 (IL-7Ra)	+/–	+/–	+	++	++	+
	CD161 (NK1.1)	+	+	–	–	–	+/–
	ST2 (IL-33R)	–	–	+/–	–	–	–
	CD278 (ICOS)	W	nd	++	nd	W	W
	IL-17RB (IL-25R)	–	–	+	–	–	–
	CD294 (CRTH2)	–	–	+	–	–	–
	KLRG1	++	–	+/–	–	–	–
	CD117 (c-kit)	+/–	+/–	+/–	++	W	W
	CD69	–/*	+	nd	nd	nd	nd
	CD254 (RANKL)	nd	nd	nd	+	+	+
	CD196 (CCR6)	–	–	–	+	+/-	–
	CD4	–	–	–	+	+/–	–
	CD335 (NKp46)	+	+	–	–	–	+
	CD25 (IL-2Ra)	–	–/W	+	+/–	+/–	+/–
	MHC-II	–	–	+	+	+/–	–
	IL23R	–	–	–	+	+	+
	IL-1RB	–	+	nd	+	+	+
	CD122	+	+	W	–	–	–
	CD314 (NKG2D)	+	nd	–	–	–	–
	Ly49	+/–	+/–	–	–	–	–
	CD94	+/–	nd	+/–	nd	–	–
	Perforin	+	W	–	–	–	–
	CD253 (TRAIL)	–/*	+	nd	nd	nd	nd
	Sca-1 (Ly-6a)	–	+	+	–	nd	+
	CD49d (integrin α4β7)	nd	nd	–	nd	+	+
	CD49a (integrin α1β1)	–/*	+	nd	nd	nd	nd
	CD90 (Thy1)	+/–	+	+	++	++	++
	CD160	+/–	+	nd	nd	nd	nd
	CD103	–	–	nd	nd	nd	nd
	CD200R	–	+	nd	nd	nd	nd
	CD304 (NRP-1)	nd	nd	nd	+	+/–	–
Transcription factors	Tbet	+	+	–	–	+/–	+
	Eomes	+	+/–	–	–	–	–
	RORγT	–	–	–	+	+	+
	GATA3	W	W	+	W	W	W
	AhR	+/–	+	nd	++	++	++
	RORa	nd	+	++	nd	nd	nd

(Continued)

TABLE 18.1 Table Comparing the Phenotypic Markers of ILC in Mouse and Human (Continued)

Human	Marker	NK	ILC1	ILC2	LTi	NKp44⁻ ILC3	NKp44⁺ ILC3
Cell-surface molecules	CD45	+	+	+	int	+	+
	CD127 (IL-7Ra)	+/–	+/–	+	+	+	+
	CD161 (NK1.1)	+	+	+	+/–	+	+
	ST2 (IL-33R)	–	–	+	–	–	–
	CD278 (ICOS)	–	–	+	nd	nd	+
	IL-17RB (IL-25R)	–	–	+	–	nd	–
	CD294 (CRTH2)	–	–	+	–	–	–
	KLRG1	+	+/–	+	–	–	–
	CD117 (c-kit)	–/W	–	+/–	+	+	+
	CD69	W	+/–	nd	nd	nd	nd
	CD254 (RANKL)	–	–	nd	+	+	+
	CD196 (CCR6)	–	–	nd	+	+	+
	CD335 (NKp46)	+	–	–	–	–/W	+
	CD25 (IL-2Ra)	+/–	W	+	W	+/–	+
	MHC-II	+	+	+	+	+	+
	IL23R	–	–	–/W	+	+	+
	IL-1RB	+/–	+	W	+	+	+
	CD122	+	nd	nd	–/W	W	W
	CD314 (NKG2D)	+	nd	nd	–	–	–
	CD94	+/–	+/–	–	–	–	–
	Perforin	+	+/–	–	–	–	–
	IL-12RB	+	+	–	–	–	W
	CD194 (CCR4)	nd	+/–	+	nd	nd	nd
	CD56	+/++	++	–	–/W	+/–	+/–
	CD183 (CXCR3)	+/–	+	nd	nd	nd	nd
	CD337 (NKp30)	+	+	+	+/–	+/–	+/–
	CD336 (NKp44)	+/–	+	–	–	–	+
	CD103	–	+	nd	–	–	–
	CD16	+/–	–	–	–	–	–
	CD304 (NRP-1)	–	–	–	+	+	+
	NKp80	+	–	nd	nd	nd	nd
	CD300LF	–	–	nd	nd	–	+
Transcription factors	Tbet	+	+	–	–	–	–
	Eomes	+	+/–	–	–	–	–
	RORγT	–	–/W	–/W	+	+	+
	GATA3	–/W	–/W	+	–/W	–/W	–/W
	AhR	–/W	W	+	+	+	+

++, very high expression; +, high expression; int, intermediate expression; W, weak expression; +/–, expression detected in some, but not all, cells; *, induced by activation; nd, not determined.

Adapted from Vivier E, Artis D, Colonna M, et al. Innate lymphoid cells: 10 years on. *Cell.* 2018;174:1054-1066, with permission from Elsevier.

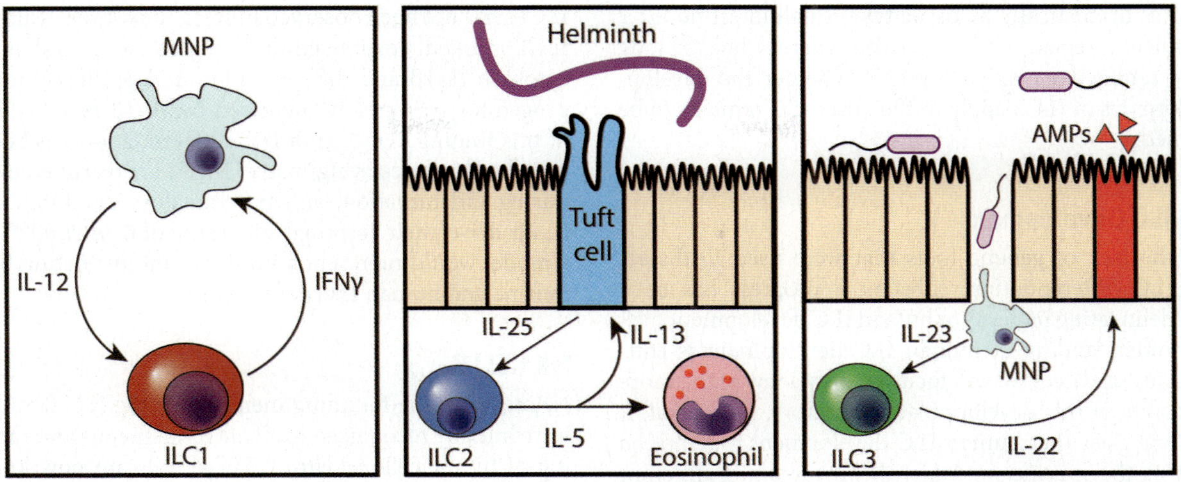

FIG. 18.2. Innate lymphoid cells (ILC) are involved in diverse immunological responses. (1) ILC1 are primarily responsive to IL-12 and produce IFNγ which can act back on dendritic cells to promote antigen presentation and clearance of intracellular pathogens. (2) In response to certain gastrointestinal nematode infections, small intestinal tuft cells produce IL-25 which activates ILC2 to promote production of IL-5 and IL-13. IL-5 induces eosinophilia, and IL-13 acts back on the small intestinal epithelium to induce goblet cell and tuft cell hyperplasia. (3) Intestinal bacterial infection is recognized by mononuclear phagocytes (MNPs) which produce IL-23. IL-23 acts on ILC3 to induce production of IL-22, which acts on the intestinal epithelium to stimulate the production of antimicrobial peptides (AMPs).

ILCPs have been identified in the fetal intestine as early as E13.5, based on the expression of the urea cycle enzyme arginase-1 (Arg1) and the absence of lineage markers.[42] Recent work using temporally controlled fate mapping of ILC2 has demonstrated two striking observations concerning the dynamics of ILC2 development: (1) while fetal ILC2s seed peripheral tissues early in embryonic development, the majority of adult ILC2 pools are derived from *de novo*–generated ILC2 during the first 8 weeks of life and (2) the replacement of these postnatal ILC2 pools by *de novo*–generated

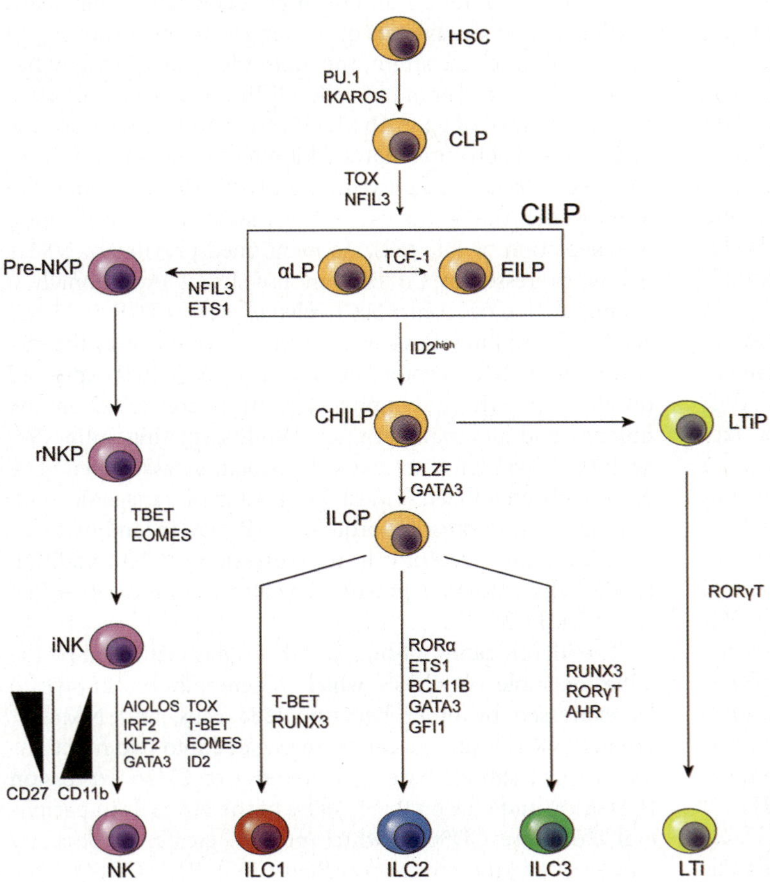

FIG. 18.3. Current paradigm of mouse innate lymphoid cell (ILC) development. Transcription factors involved at each developmental stage are indicated. αLP, alpha lymphoid progenitor; CHILP, common helper innate lymphoid progenitor; CILP, common innate lymphoid progenitor; CLP, common lymphoid progenitor; EILP, early innate lymphoid progenitor; HSC, hematopoietic stem cell; ILCP, innate lymphoid cell progenitor; iNK, immature NK; Pre-NKP, pre-NK progenitor; rNKP, refined NK progenitor.

ILC2 varies dramatically by tissue type; small intestine ILC2 are completely replaced by 32 weeks, whereas half of lung ILC2 are retained for up to 1 year.[43] Whether the developmental kinetics of ILC2 apply to the other ILC remains to be determined.

Human ILC Development

Despite the lack of genetic tools that were used to dissect murine ILC differentiation, substantial progress has been made in delineating patterns of human ILC development.[4,44,45] As our understanding of human ILC development is constantly being updated, we will focus on only a few of the more salient points in this developmental trajectory. In general, it has become clear that human ILC development may not be as linear as the murine model, with precursors at different developmental stages showing signs of commitment to specific ILC lineages.[46,47] Nevertheless, there are several comparisons that we can draw between species that help anchor this discussion. Similar to mouse, human ILC share their origins from a CLP-like precursor enriched in Lin$^-$ CD34$^+$ hematopoietic progenitor cells (HPC) that can be found in umbilical cord blood, thymus, BM, peripheral blood, or tonsils.[46,48-52] Broadly, they can be identified by a Lin$^-$ CD34$^+$ CD45RA$^+$ CD10$^+$ CD117$^-$ surface marker expression profile. However, the exact phenotype of this lymphoid-biased precursor is a matter of active debate, with multiple identification strategies proposed.[53-58] As opposed to the murine CLP, human CLP-like precursors still retain residual differentiation potential toward the myeloid lineage, further illustrating the heterogeneity of these progenitors.[55,59] Downstream of the CLP-like precursor, several committed ILCPs have been defined.

Putative committed NK precursors (NKP) have been identified in fetal liver, fetal BM, cord blood, and adult tonsil with a Lin$^-$CD34$^+$CD38$^+$CD123$^-$CD45RA$^+$CD7$^+$ CD10$^+$CD127$^-$ phenotype.[51] This population was shown to have a restricted capacity for NK cell differentiation and lacked the potential to produce T cells, B cells, myeloid cells, or other ILC *in vitro* and after transplantation into immunodeficient mice. In addition to the NKP, committed ILC3 precursors (ILC3P) were found in tonsil and intestinal lamina propria, suggesting that human ILC3 development could occur in nonlymphoid tissues.[50] Among CD34$^+$ HPC, ILC3P were defined by expression of the transcription factor RAR-related orphan receptor gamma (RORγt). A human equivalent of the CILP has been described in SLO but not in cord blood, peripheral blood, thymus, or BM with a CD34$^+$CD45RA$^+$CD117$^+$IL-1R1$^+$integrinβ7$^+$ phenotype.[46] *In vitro* cultures of these cells produced cytotoxic NK and ILC lineages with no sign of T cell, B cell, or dendritic cell (DC) potential. However, *in vivo* adoptive transfer studies demonstrated only NK potential, indicating a requirement for other factors in ILC differentiation. Up to this point, identification of putative multipotent ILC precursors had been limited to CD34$^+$ HPC and SLO. An important study identified multipotent ILCPs in peripheral blood with a Lin$^-$ CD34$^-$ CD117$^+$ phenotype that were able to give rise to NK cells, ILC1, ILC2, and ILC3 *in vitro* and *in vivo*.[47] The presence of circulating ILCPs has not been observed in mice. This circulating human ILCP markedly expressed the receptor for the inflammatory cytokine IL-1β and differentiation and proliferation of this progenitor was greatly increased by IL-1β *in vitro*. In light of this finding, a model of ILC differentiation has been proposed where circulating "naïve" ILCPs are recruited to tissues during inflammation, and inflammatory cytokines such as IL-1β drive their terminal differentiation *in situ*.[44,47,60] Such a model would represent a fundamental distinction between murine and human ILC development.

NK CELLS

NK cells are the founding members of the ILC family. They were initially recognized for their spontaneous and rapid killing of tumor cells *in vitro*.[61] These cells are now known to contribute to early immune responses to viruses and to modulate adaptive immunity. In this section, we will discuss the phenotype and regulation of NK cells as well as their effector functions and roles in the immunological response.

Phenotypic and Molecular Characterization of NK Cells

Before the identification of selective surface markers, NK cells were defined as large granular lymphocytes which lacked a T cell receptor (TCR) and which mediated perforin-dependent cytotoxic killing of target cells that lacked MHC-I.[62] However, monoclonal antibodies (mAbs) against certain NK surface receptors when used in conjunction with anti-CD3 or anti-CD19 to exclude T and B lymphocytes can reliably identify NK cells at steady-state or ongoing immune responses. In the C57BL/6 mouse strain, the molecule NK1.1 (*Klrb1c*) has been a useful marker of NK cells.[63-65] Intravenous administration of the anti-NK1.1 mAb PK136 efficiently depleted splenic NK cells and abrogated natural killing of tumors *in vivo* with no direct effects on adaptive immunity.[66,67] However, there are a few limitations in the use of this marker for identification and depletion of NK cells. As mentioned previously, NK1.1 is only expressed by C57BL/6 mouse strains. Additionally, a subpopulation of T cells (NKT cells) expresses TCR/CD3 and NK1.1. These limitations were partially overcome by the discovery of the NK receptor NKp46 (*Ncr1*). NKp46 is expressed on all CD3$^-$ NK1.1$^+$ lymphocytes and is conserved among humans and all strains of mice. While depleting antibodies against NKp46 have not been described, its associated gene *Ncr1* has been utilized toward the creation of transgenic mice, linking green fluorescent protein (GFP), cre recombinase, or diphtheria toxin receptor to its expression.[35,68-70] In addition to NK cells, NKp46 expression has also been described in a subset of ILC3.[71-74]

The identification of human NK cells is facilitated by the adhesion molecule CD56, which is generally not known to be expressed by other hematopoietic cells, and NKp46.[65] Human NK cell subsets can be segregated into two functionally distinct subsets based on the level of CD56 expression (CD56^{bright} and CD56^{dim}).[75,76] ~90% of the NK cells in peripheral blood are CD56^{dim}, which possess greater cytotoxicity and less cytokine production than CD56^{bright}.[77] CD56^{dim} NK cells also express the type III Fcγ receptor (CD16), enabling antibody-dependent cellular cytotoxicity (ADCC).[78] The CD56^{dim} population may be further divided into two subsets based on differential expression of CD57, with CD57^{+} NK cells considered to be more terminally differentiated and to more readily expand in response to human cytomegalovirus infection (CMV).[79,80]

In addition to surface molecules, multiple transcriptional mediators have been identified to regulate the unique identity of NK cells. While countless factors have been described to play a role in NK development or function, we will focus our discussion to two members of the T-box family of transcription factors, T-bet and Eomesodermin (Eomes), which play an essential role in NK development and maturation.[81,82] T-bet was first characterized by its essential role in the differentiation of Th1 cells and control of IFNγ production.[83] Subsequent work has demonstrated a requirement of T-bet for the terminal maturation of NK cells, in part by regulating sphingosine-1-phosphate receptor (S1PR) expression to facilitate NK trafficking from the BM and LN to peripheral tissues.[84,85] In conjunction with T-bet, another T-box family member Eomes has been found to play a critical role in NK development and maturation.[86] Eomes-deficient mice display a greater reduction in splenic NK cells than T-bet–deficient mice, and deletion of both Eomes and T-bet leads to near total loss of immature and mature NK cells.[86] Mechanistically, it has been shown in CD8^{+} T cells that T-bet and Eomes cooperate to drive expression of CD122, a component of the IL-15 receptor, which transduces IL-15 signaling that is critical for NK development.[87] The expression of T-bet and Eomes has also proven to be a valuable marker in the distinction between NK cells and ILC1, where NK are Tbet^{+}, Eomes^{+} and most ILC1 are T-bet^{+}, Eomes^{-}.[88-90]

Regulation of NK Cells by Common (γ_c) Cytokines

Named on the basis of their shared receptor subunit, the common cytokine receptor γ chain (γ_c) family of cytokines include IL-2, IL-4, IL-7, IL-9, IL-15, and IL-21.[91] These cytokines play essential roles in the development, differentiation, and homeostasis of all lymphoid cells and act as critical regulators of innate and adaptive immune responses. Historically, NK cells were discovered to display enhanced activity and proliferation when treated with high concentrations of the cytokine IL-2.[92-95] However, it is unlikely that NK cells are exposed to such concentrations of IL-2 *in vivo*. Indeed mice with genetic deficiency of either IL-2 or the alpha chain of its receptor (IL-2Rα) do not show any direct deficiency of NK cells.[96,97] Conversely, dramatic NK deficiency is observed with IL-2Rβ or IL-2Rγ (common cytokine receptor γ chain) deficiency in mice and humans.[98-101] This paradox is explained by the relationship between the IL-2 receptor and IL-15 receptor. The IL-2 receptor has different affinities depending on the composition of its receptor chains. The high-affinity IL-2R is composed of α (CD25), β (CD122), and γ (CD132) chains, whereas IL-2Rβγ has intermediate affinity. The high-affinity IL-15 receptor comprises IL-2Rβγ and a unique IL-15Rα chain that does not possess inherent signaling ability. Similar to its relationship with IL-2, the IL-2Rβγ complex alone can bind IL-15 with intermediate affinity and transduce signaling through the Jak1/Jak3 and Stat5/Stat3 pathways.[102,103] Resting NK cells express high amounts of IL-2Rβγ.[104,105] Thus, IL-15 signaling in NK cells is dependent on IL-2Rβγ and its associated downstream signaling components. IL-15 is profoundly consequential for NK cell biology.[106] Mice with genetic deficiency in either Il-15 or any component of the high-affinity trimeric IL-15 receptor lack NK cells.[98,105,107,108] Adoptive transfer of WT NK cells into Il15$^{-/-}$ hosts leads to abrupt loss of transferred NK cells, indicating that IL-15 is essential for their active maintenance and survival.[109] Intriguingly, transfer of WT NK cells into Il15ra$^{-/-}$ hosts also led to their abrupt disappearance.[110] These data proved that expression of the high-affinity IL-15Rα chain by accessory cells drives NK cell survival rather than NK-intrinsic expression of IL-15Rα. Subsequent studies showed that IL-15Rα expression by hematopoietic myeloid cells, such as DC, and radiation-resistant stromal cells is required for both the peripheral survival of NK cells during homeostasis and the priming of NK cells during inflammatory responses.[110-114] Moreover, *trans* presentation of IL-15 requires coordinate expression of IL-15 and IL-15Rα by the same accessory cell, suggesting intracellular endosomal assembly of IL-15/IL-15Rα complexes before presentation.[111,113] From these studies, we appreciate that NK survival and activation is highly dependent on IL-15 in complex with IL-15Rα that is presented in *trans* by accessory cells of hematopoietic and stromal origin.

Effector Functions of NK Cells

The hallmark of NK cell effector function is rapid killing of target cells, mediated by exocytosis of cytotoxic granules upon crosslinking of activating receptors.[115] Like cytotoxic CD8^{+} T cells, NK cell–mediated cytotoxicity is primarily mediated by the polarized extrusion of cytoplasmic granules containing perforin and granzymes.[116,117] Released perforin molecules then form pores on the target cell in a calcium-dependent manner, allowing entry of granzymes which cleave caspases and BH3-interacting domain death agonist (Bid), ultimately leading to target cell apoptosis.[118,119] While NK cells were previously thought to constitutively contain preformed cytotoxic granules, it was shown that priming with cytokines, such as IL-2 and IL-15, is needed to activate translation of perforin and granzymes and unleash full NK cytotoxicity.[120] NK cells also display perforin-independent modes via Fas/FasL binding and TRAIL/TRAIL-R interactions.[121-126] However, mice and humans with genetic defects in perforin-dependent pathways demonstrate substantial defects in NK-mediated cytotoxicity, indicating that granule exocytosis is the primary mode of cytotoxic killing.[127-130]

In addition to direct cytotoxicity, it has been appreciated that NK cells are significant sources of immunoregulatory cytokines which influence innate and adaptive responses. Upon activation, NK cells can produce a number of proinflammatory "type 1" cytokines such as IFNγ, TNFα, and GM-CSF.[131-133] In particular, NK-derived IFNγ has been shown to facilitate Th1 differentiation in LN.[134] Several studies have also described NK cell production of the immunosuppressive cytokine IL-10.[135-137] NK cells also secrete many chemokines, including CCL2 (MCP-1), CCL3 (MIP1-α), CCL4 (MIP1-β), CCL5 (RANTES), XCL1 (lymphotactin), and CXCL8 (IL-8).[138-140] IFNγ is a signature cytokine produced by activated NK cells and has been used commonly as a functional marker of activation. Exemplifying a general feature of ILC in amplifying inflammatory cascades, NK cytokine production is induced by myeloid-derived cytokines such as IL-12, IL-18, and type 1 interferon.[141-145]

General Principles of NK Cell Target Recognition

Although NK cells lack a somatically rearranged antigen receptor, target specificity is derived from a multitude of inhibitory and activating NK receptors expressed on each individual cell (Table 18.2). Contrary to T cell receptors, NK receptors are germline encoded and may bind to multiple, overlapping sets of ligands. Many of these receptors have either one of two general structures: (1) killer cell lectin-like receptors (KLRs) are C-type lectin-like receptors which are encoded in the NK gene complex (NKC) and (2) killer cell immunoglobulin-like receptors (KIRs) which are encoded in the leukocyte receptor complex (LRC). A major family of C-type lectin-like receptors (Ly49 receptors) are only present in mice while KIRs predominate in human. These receptors can be inhibitory or activating. All known inhibitory receptors possess cytoplasmic intracellular tyrosine-based inhibitory motifs (ITIMs) consisting of V/I/L/SxYxxL/V amino acid motif, where x is any amino acid.[146,147] Receptor-ligand binding leads to phosphorylation of ITIMs and recruitment of SH2 domain-containing phosphatases, such SHP-1 and SHP-2, to shut down activation. Several NK-activating receptors appear to share a common signaling mechanism with T and B cell receptors, forming complexes with adapter molecules containing immunoreceptor tyrosine-based activation motifs (ITAMs). The ITAM is defined by the prototypical amino acid sequence, YxxL/Ix6-8YxxL/I.[148,149] The major ITAM-containing molecules expressed by NK cells are FcεRIγ, CD3ζ, and DAP12 which contain one or three ITAMs. An alternative adapter molecule, DAP10 expresses an alternative tyrosine-based activation motif (YxxM) in lieu of the classical ITAM.[150-152] Upon receptor engagement, ITAMs are phosphorylated to recruit protein tyrosine kinases, such as Zap70 and Syk, to initiate multiple canonical signaling cascades. Alternatively, signaling through DAP10 recruits PI3K. NK cells must balance signals from inhibitory and activating receptors in order to specifically kill target cells while sparing healthy self (Fig. 18.4). Importantly, expression of certain NK receptors is stochastic such that any given NK cell only expresses a subset of receptors in its potential repertoire.[153] Thus, while NK cells may not be clonotypic in the sense of a unique antigen receptor, they do consist of "clones" with unique patterns of receptor expression and activation thresholds.

Inhibitory NK Receptors and the "Missing Self" Hypothesis

One of the first indications that NK cell activity is specific was that NK cells mediated *in vivo* rejection of MHC-I–deficient tumors.[154] Mutations in critical components of MHC-I antigen-processing machinery, such as β2-microglobulin (β2m) or transporter associated with processing (TAP) were sufficient to sensitize cells to NK cell killing *in vitro* and *in viv o*.[155-160] Separately, a phenomenon termed "hybrid resistance" had been observed where NK cells in irradiated F_1 hybrid mice efficiently reject parental BM.[161] This was proposed to be on account of the fact that the F_1 NK receptor repertoire is derived from both parental strains, so that BM from one homozygous parent essentially lacks half of the MHC-I molecules recognized by F_1 NK cells. Consistent with the notion that MHC-I are inhibitory ligands for NK activity, parental BM is swiftly rejected. In addition to evidence from mice, human NK cell alloreactivity was shown to be modulated by polymorphisms in HLA-B and HLA-C.[162-167] Collectively, these observations are summarized in the "missing self" hypothesis which postulates that, as opposed to cytotoxic T lymphocytes (CTL) which require MHC-I expression on target cells, NK kill target cells *because* they lack or have reduced expression of MHC-I.[168] Downregulation of MHC-I is a common method of immune evasion utilized by viruses and tumors.[169-172] Under the "missing self" hypothesis, the roles of CTL and NK complement each other to limit these mechanisms of escape from immune recognition.

The observation that MHC-I served to inhibit NK cell activation prompted the search for inhibitory NK cell receptors and their respective ligands in mice and humans. While the main inhibitory ligand of NK cells is MHC-I, a growing collection of inhibitory NK receptors have been found with alternative ligands. This section will discuss those inhibitory receptors that have been studied extensively in mouse and human.

Ly49 Receptors

The Ly49 family of receptors are KLRs encoded by the NKC on murine chromosome 6. They are most well-studied in rodents but essentially absent in humans.[173] The number of *Ly49* genes varies in different mouse strains and evidence shows extensive allelic polymorphism.[174] Ly49 receptors are best characterized in C57BL/6 and 129/J strains, where there are 15 genes.[62,175] The Ly49A receptor was the first inhibitory MHC-I–specific NK receptor to be functionally described.[176] Ly49A recognition of MHC-I is influenced by MHC haplotype, with Ly49A showing preferred binding to $H2D^d$, though Ly49A can also recognize other alleles.[176-180] Studies using blocking antibodies against different MHC-I domains as well as the crystal structure of Ly49A in complex with $H2D^d$ have identified Ly49A contact points in the α_1, α_2 domains of MHC-I, the peptide-binding groove, as well as a second common point that spans the α1, α2, α3 domains and β2-microglobulin.[176,177,179,181-185] The latter finding provided an explanation behind the species-specific requirement for β2-microglobulin in facilitating Ly49 recognition of MHC-I.[186] Despite binding of Ly49A to the MHC peptide-binding groove, interaction of Ly49A and $H2D^d$ appears to relatively independent of the specific peptide bound by $H2D^d$.[187,188] Functional and x-ray crystallography studies show that Ly49C binds $H2K^b$ in a similar manner to the interaction between Ly49A and $H2D^d$ but only the second contact point is observed.[180,189-191] Surprisingly, Ly49C recognition of $H2K^b$ may

TABLE 18.2 Table of NK Receptors and Their Known Ligands in Mouse and Human

Mouse	**Receptor**	**Other Names**	**Ligand**
Inhibitory	Ly49A	Ly49, Ly-49, Klra1	H2Dd, Dk, Dp, alleles in H2r, and H2q
	Ly49C	Klra3	H2Kb, numerous
	Ly49E	Klra5	Urokinase plasminogen activator
	Ly49G2	LGL-1, Klra7	H2Dd
	Ly49I129	Klra9	m157
	Ly49Q	Klra17	H2Kb
	Lilrb4	gp49	avb3 integrin
	CD94/NKG2A	Kp43	Qa-1
	Siglec-E		?
	Nkrp1b(d)	Klrb1b	Clrb
	Klrg1		Cadherins
Activating	FcγRIII	CD16	IgG Fc
	Ly49D		Chinese hamster MHC-I, H2Dd
	Ly49H		m157
	Ly49PMA/My		M04 + H2Dk
	CD94/NKG2C		Qa-1
	NKG2D	Klrk1	H60, Rae1, MULT1
	Nkrp1c	NK1.1	m12
	NKrp1f		Clrg
	CD2		CD48
	NTBA	SLAMF6, Ly108	NTBA
	Ly9	SLAMF3	Ly9
	CD84	SLAMF5	CD84
	CRACC	SLAMF7	CRACC
	NKp46	Ncr1	Hemagglutinin, complement factor P, Heparan sulphate
	Ly6		?
	CD226	Dnam-1	necl-5, nectin-2
	Crtam		necl-2, Cadm1
Human	**Receptor**	**Other Names**	**Ligand**
Inhibitory	KIR2DL1	CD158a, p58.1, EB6	HLA-C2 (HLA-Cw2, Cw4, -Cw5, -Cw6)
	KIR2DL2/KIR2DL3	CD158b, p58.2, GL183	HLA-C1 (HLA-Cw1, -Cw3, -Cw7, -Cw8)
	KIR2DL4	CD158d	HLA-G
	KIR2DL5	CD158f	PVR
	KIR3DL1	CD185e1, NKB1, p70, NKAT3	HLA-Bw4
	KIR3DL2	CD158k, p140, NKAT4	HLA-A3, -A11
	CD94/NKG2A	Kp43	HLA-E

(Continued)

TABLE 18.2 Table of NK Receptors and Their Known Ligands in Mouse and Human (Continued)

Human	Receptor	Other Names	Ligand
	LILRB1	CD85j, ILT2, LIR1	MHC-I, UL18
	LAIR-1		Collagen
	Siglec-7	P75, AIRM1	Carbohydrates
	Siglec-10		Carbohydrates?
	PILRa		CD99
	PILRb		CD99
	NKRP1A		LLT1
	CEACAM1	CD66a	CEA
	TIGIT		PVR, PVRL2
Activating	FcγRIII	CD16	IgG Fc
	KIR2DS1	CD158h, p50.1	HLA-Cw7
	KIR2DS2	CD158j, NKAT5, p50.2, clone 49	
	KIR2DS3	NKAT7	
	KIR2DS4	CD158i, NKAT8, clone 39	HLA-Cw4
	KIR2DS5	CD158g, NKAT9	
	KIR3DS1	CD158e2	
	CD94/NKG2C		HLA-E
	NKG2D	KLRK1	MICA, MICB, ULBP/RAET1
	NKp80		AICL
	NKp65	KLRF2	CLEC2A
	CD2		CD48, CD58
	NTBA	SLAMF6, Ly108	NTBA
	Ly9	SLAMF3	Ly9
	CD84	SLAMF5	CD84
	CRACC	SLAMF7	CRACC
	NKp46	Ncr1	Hemagglutinin, complement factor P, Heparan sulphate
	NKp44		PDGF-DD, HLA-DP, Heparan sulphate
	NKp30		B7-H6
	CD226	Dnam-1	necl-5, nectin-2
	CD96	Tactile	necl-5
	CRTAM		necl-2

Adapted from Yokoyama WM. *Paul's Fundametnal Immunology.* 7th ed.

be influenced by the type of MHC-bound peptide despite the lack of direct contact with the peptide binding groove.[192] Most but not all other Ly49 receptors are MHC-I–specific inhibitory receptors. In C57BL/6 mice, Ly49H and Ly49D are activating receptors. Ly49D recognizes H-D^d while Ly49H recognizes a viral MHC-I–like peptide.[193,194] Ly49H is especially well-studied due to its consequential role in viral infection, which will be discussed in later sections.

As mentioned earlier, NK cells often express multiple Ly49 receptors per cell in a stochastic manner. Regulation of Ly49 receptor expression is under several unique transcriptional controls. Several Ly49 genes show patterns of monoallelic expression and bidirectional, overlapping promoters appear to control the expression of some Ly49 genes.[195-197] The transcription factor TCF-1 has been implicated in the expression of some but not all Ly49 genes.[198-201] However, the mechanisms regulating Ly49 gene expression are still incompletely understood. Lastly, it has become clear that expression of Ly49 receptors is not restricted to NK cells. Notably, Ly49B and Ly49Q are not even expressed by NK cells but on myeloid

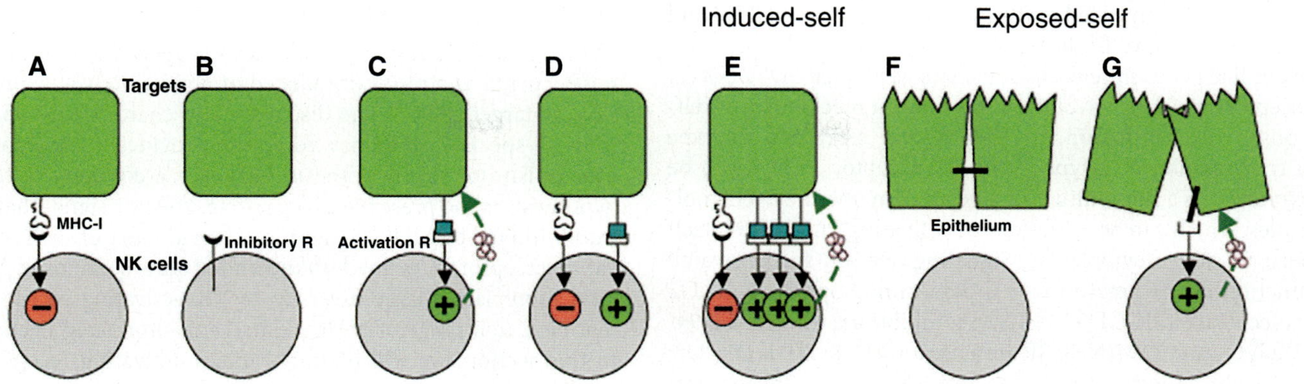

FIG. 18.4. Current principles of target recognition by natural killer (NK) cells. Pictured are several scenarios of interactions between ligands on targets (*top row*) and receptors on NK cells (*bottom row*). Successful activation of NK cells is shown by the dashed upward line. **A:** Targets expressing major histocompatibility complex (MHC)-I are resistant to lysis by NK cells because of MHC-I–specific inhibitory receptors. **B:** The absence of MHC-I (or lack of receptors specific for target MHC-I, not depicted) does not automatically result in target killing. **C:** Activation receptor engagement is required to trigger target killing in absence of MHC-I. **D:** In the situation where both inhibitory and activation receptors are engaged, the inhibitory receptor effect often dominates and no killing occurs. **E:** In the induced-self model, induced expression of NKG2D ligands can overcome the inhibitory influence of MHC-I, resulting in NK-cell activation. **F:** Normal epithelium masks ligands for NK-cell activation receptors at the tight junctions. **G:** Under pathologic situations, the epithelial architecture may be disrupted, leading to ligand exposure. (Not depicted are inhibitory ligands at the epithelial tight junctions that may inhibit NK cells when they transmigrate through epithelial barriers.) NK cells can also recognize pathogen-encoded ligands on infected cells (not shown, but similar to C or E). (Adapted from Yokoyama WM. *Paul's Fundametnal Immunology.* 7th ed.)

cells, where they appear to regulate TLR signaling.[202-204] Other inhibitory Ly49 receptors have been described and are listed in the associated table.[62,205]

KIR Receptors

While human NK cells do not express Ly49 receptors, KIRs predominately mediate inhibitory MHC-specific interactions involving these cells. Early work in identifying human NK receptors was made possible by two important technological advances: high-efficiency cloning of human lymphocytes and the generation of mAbs. Human NK cell clones were used for mouse immunization and the resulting monoclonal antibodies were selected for their ability to inhibit or enhance NK cytotoxic killing or cytokine production. Using this strategy, a series of studies identified two mAbs, GL183 and EB6, that recognized two homologous molecules expressed by partially overlapping NK subsets.[206-209] Treatment with these antibodies were found to strongly enhance the activity of human NK cells against certain tumor targets, and transfection of tumor cells with cDNA for HLA-C molecules could silence NK killing. Parallels with the "missing self" hypothesis and work on murine Ly49 receptors were clear; the molecules identified by these antibodies were inhibitory receptors belonging to the KIR family.

Compared to Ly49 receptors which are c-type lectin-like type II transmembrane receptors, KIRs are immunoglobulin-like type I transmembrane proteins. To date, 15 distinct KIR genes have been identified in the LRC on human chromosome 19q13 though the exact number varies between individuals due to extensive polymorphism in these loci. For the most up-to-date reference on KIR diversity, refer to the centralized repository of human KIR sequences.[163,210-213] It is now appreciated that significant structural diversity exists among KIRs and that KIRs can be activating as well as inhibitory.[214] A systematic nomenclature has been developed to classify all documented KIRs.[215] Inhibitory KIRs are identified by long (L) cytoplasmic tails, whereas activating KIRs have short (S) cytoplasmic tails. Furthermore, KIRs can either have two extracellular Ig-like domains (eg, KIR2D) or three Ig-like domains (eg, KIR3D). Finally, a single digit is added to the end to indicate a distinct receptor. Thus, "KIR2DL1" would describe an inhibitory KIR with two Ig-like domains.

KIRs recognize HLA-A, HLA-B, and HLA-C MHC proteins. There is substantial evidence that indicates KIR recognition is affected by HLA allotypes. KIR2DL1 and KIR2DL2/3 receptors recognize different polymorphisms at residues 77 and 80 on the α_1 chain of HLA-C.[164] KIR3DL1 recognizes the Bw4 epitope on HLA-B.[216,217] KIR interactions with HLA molecules have also been shown to exhibit some peptide selectivity in numerous studies though the physiological impact of these interactions remains unknown.[218-224] The crystal structure of KIR2DL2 in complex with HLA-Cw3 has provided mechanistic evidence for both allotypic and peptide specificity in KIR binding.[225] This relationship between polymorphic KIR alleles and HLA polymorphisms forms the basis of mismatching KIR and HLA repertoires to promote graft-versus-leukemia (GVL) effects in the context of hematopoietic stem cell transplant (HCST).[226-229]

Regulation of KIR expression appears to share several similarities with murine Ly49 expression. KIRs appear to be expressed stochastically and monoallelic expression of multiple KIRs has been described.[230-233] In addition to NK cells, KIRs are expressed on γδ T cells, memory/effector $CD8^+$ T cells, and some $CD4^+$ T cells.[234,235] Despite their different structures, the fact that murine Ly49s and human KIRs demonstrate remarkable homology in regulation, and function is a remarkable example of convergent evolution.

CD94/NKG2 Receptors

The last major group of inhibitory receptors we will discuss is the heterodimeric CD94/NKG2 family of receptors. CD94 and NKG2 genes are encoded in the NKC in mouse, human, and rat.[236-240] Compared to the Ly49 and KIR genes, CD94 and NKG2 genes have limited polymorphisms and the presence of similar genes in teleostean fish suggests that CD94/NKG2 receptors may be an evolutionarily ancient mechanism of self-nonself discrimination.[241] CD94/NKG2 receptors are type II transmembrane c-type lectin-like receptors. CD94 can be expressed as a homodimer or a heterodimer with NKG2 molecules, which can be inhibitory or activating.[242-246] CD94 itself lacks an active cytoplasmic signaling domain so the overall function of the heterodimer is determined by which NKG2 molecule it binds. CD94/NKG2A is inhibitory, whereas CD94/NKG2C and CD94/NKG2E are activating (NKG2B is an alternative spliceoform of NKG2A, and NKG2H is a spliceoform of NKG2E). The ligands of CD94/NKG2 were discovered to be nonclassical MHC-Ib ligands (HLA-E in humans, Qa-1 in mouse).[242,247-251] Interestingly, the antigens presented by nonclassical MHC-Ib molecules are the leader sequences derived from MHC-I molecules, and so the expression of HLA-E and Qa-1 are linked to the expression of classical MHC-I.[249,252-257] Thus, inhibitory CD94/NKG2A signaling mimics similar patterns of inhibitory MHC-I–specific receptors (Ly49 and KIR). However, this presents somewhat of a contradiction as the activating CD94/NKG2 receptors recognize the same ligand. Several pieces of evidence have helped explain these opposed functions. CD94/NKG2A has been shown to bind HLA-E with higher affinity than CD94/NKG2C.[258] The crystal structure of CD94/NKG2A in complex with HLA-E shows intimate contact with the leader peptide being presented, reminiscent of classical TCR-MHC-I interactions.[259,260] Consequently, it has been shown that CD94/NKG2 interactions are influenced by the peptide being presented by nonclassical MHC-Ib molecules.[258,261-263] Based on these premises, two examples have been described which demonstrate the dual inhibitory and activating functions of CD94/NKG2 receptors. It was shown that a stress-induced protein hsp60 is able to bind HLA-E and reduce the binding affinity of CD94/NKG2A, thus effectively reducing inhibitory NK signaling and prompting killing of distressed cells.[263] Alternatively, the expansion of NKG2C$^+$ NK cells is associated with a history of human cytomegalovirus (HCMV) infection and can be induced *in vitro* with HCMV-infected fibroblasts.[264-268] It was recently found that the HCMV-encoded protein UL40 is presented by HLA-E and preferentially expands NKG2C$^+$ NK cells.[269] Thus, the peptide selectivity exhibited by CD94/NKG2 receptors may be shaped by evolutionary adaptation to viral infections.

Like other NK receptors, CD94/NKG2 molecules are not restricted to NK cells. In particular, CD94/NKG2A is highly expressed by CTL and inhibits their cytotoxic activity during polyoma virus clearance and viral-induced oncogenesis.[270] However, in ectromelia virus infection, NKG2A enhances viral clearance by dampening CTL activation and reducing activation-induced cell death.[271] Lastly, recent studies have demonstrated a remarkable role for anti-NKG2A mAb as antitumor checkpoint therapy via NK- and CD8-dependent mechanisms.[272,273] These findings highlight a context-dependent role for NKG2A in regulating CTL responses. Thus, CD94/NKG2 interactions may be just as consequential in regulating T cells as NK cells, if not more.

NK-Activating Receptors

The original "missing self" hypothesis proposed that NK killing occurred as a result of reduced or absent MHC-I expression by target cells.[168] The discovery and characterization of MHC-I–specific inhibitory receptors provided a molecular mechanism for this hypothesis. However, a corollary to this hypothesis is the presence of a positive, effector signal that is thus inhibited by MHC recognition. The identity of such activating receptors remained unknown for quite some time. We now appreciate a large diversity of NK-activating receptors that bind to a range of MHC-related and -unrelated ligands. In this section, we will highlight major activating receptors which have been shown to have profound functional implications on NK cell responses.

FcγRIII (CD16)

Perhaps the first NK cell activation receptor identified and most extensively studied, CD16 is the low-affinity Fc γ receptor responsible for initiating antibody-dependent cellular cytotoxicity (ADCC).[274] Human CD16 can noncovalently associate with ITAM-containing FcεRIγ and/or CD3ζ, but mouse CD16 can only pair with FcεRIγ.[152,275,276] CD16 binds the Fc portion of IgG and crosslinking of the receptor results in rapid granule exocytosis and cytokine production. While CD16-deficient NK cells lack the ability to initiate ADCC, natural killing of target cells is preserved, indicating that these two cytotoxic mechanisms are functionally distinct. A more comprehensive discussion of CD16 can be found in Chapter 34.

NKG2D

The NKG2D receptor was first identified during the cloning of human NKG2 receptors.[236] Unlike Ly49 or KIR receptors, NKG2D displays little to no polymorphism in mice or humans. While the other NKG2 receptors are closely related to each other, it has very little homology with the other members of the NKG2 family. Thus, despite its naming, NKG2D should be considered distinct from other NKG2 receptors. NKG2D is a type II transmembrane glycoprotein, expressed as a disulfide-linked homodimer. Unlike other NKG2 receptors, it does not associate with CD94. Like other activating NK receptors, it has no intrinsic cytoplasmic signaling domain and relies on the association with adapter molecules for signal transduction. Mice express two isoforms of NKG2D (NKG2D-S and NKG2D-L) differing by the presence or absence of 13 amino acids at the N-terminus.[277,278] The longer isoform (NKG2D-L) associates with the adapter molecule DAP10, which does not contain ITAMs but a YxxM motif. While ITAMs bind Zap70 and Syk to initiate pathways evoked in TCR signaling, YxxM motifs bind to the p85 subunit of PI3K.[279] Interestingly, YxxM-containing proteins in T cells consist of costimulatory molecules such as CD28, suggesting a costimulatory role for NKG2D.[279] The shorter isoform (NKG2D-S) can associate with DAP12 or DAP10, and it was later revealed that the 13 amino acid segment in NKG2D-L specifically blocked association with DAP12.[280] In mice, resting NK cells primarily express NKG2D-L, whereas IL-2–activated NK cells predominately express NKG2D-S. However, humans only express NKG2D-L.[281] Debate exists as to whether NKG2D is primarily a costimulatory molecule or is sufficient to induce NK activation and killing by itself. NKG2D has been shown to function as a costimulatory molecule to ITAM-containing receptors in NK and T cells.[282-288] However, other reports have observed NKG2D-initiated cytotoxic killing that is independent of DAP12, Syk, and Zap70, suggesting that DAP10 can initiate NK activation without signaling from other traditional NK-activating receptors.[282,289,290] Thus, differential utilization of DAP12 and DAP10 represent pathways of functional diversification in NK cells.

NKG2D binds to a diverse family of ligands and exhibits cross-species reactivity where mouse ligands can bind to human NKG2D and vice-versa. The ligands for NKG2D in human are MHC-I chain-related proteins, A and B (MICA and MICB), and UL16-binding proteins (ULBP1-6).[291-297] The ligands for murine NKG2D are members of the minor histocompatibility antigen H-60 family (H60a, b, and c), members of the retinoic acid early inducible gene family (RAE-1) family (RAE-1α, RAE-1β, RAE-1γ, RAE-1δ, RAE-1ε), and murine ULBP-like transcript 1 (MULT1).[298-305] While all known ligands have structural homology with MHC-I, each family of ligands is quite distinct from each other (<25% homology). Despite high homology with MHC-I and presence of α_1 and α_2 domain, forming an MHC-I–like fold, NKG2D ligands do not bind peptide antigens and do not require β2-microglobulin to properly fold. Even though low homology exists among NKG2D ligands, crystal structures of NKG2D-ULBP3, NKG2D-MICA, and NKG2D-RAE-1β complexes have shown that receptor-ligand interfaces occur at similar points, despite different amino acid residues at these sites.[306-309] Human MIC and murine RAE-1 genes display extensive polymorphism, and it has been shown that allelic variants bind NKG2D with a wide range of affinities.[310,311]

The discovery of NKG2D led to the proposal of the "induced self" model of NK recognition. The initial experiments of the "missing self" hypothesis observed that NK cells tolerated MHC-I–expressing tumors (RMA), whereas MHC-I–deficient variants of the same tumor (RMA/S) were swiftly rejected.[154] RMA cells do not naturally express NKG2D ligands. Overexpression of the NKG2D ligands H60 or RAE-1 on MHC-I–expressing RMA cells induced NK killing of such tumors *in vivo* and *in vitro*.[298,299] These experiments demonstrated that in certain cases, expression of NKG2D ligands could override inhibitory signals derived from MHC-I recognition. While the regulation of NKG2D ligands is still poorly understood, it is appreciated that the expression of these proteins is fundamentally associated with cellular stress, such as DNA damage, inflammation, and oncogenesis.[312-317] MICA is negatively regulated by miRNAs which can be overexpressed by tumors in response to IFNγ signaling.[318-320] Furthermore, NKG2D-deficient mice display impaired tumor control and tumors in these mice had higher expression of NKG2D ligands than tumors in WT mice, suggesting tumor immunoediting of NKG2D ligands as a possible mechanism of evading immunosurveillance.[321] Interestingly, there is a profound difference in acute versus chronic stimulation of the NKG2D receptor. While acute induction of Rae-1 results in infiltration and activation of NKG2D$^+$ lymphocytes, sustained expression of Rae-1 or MICA leads to reduced NKG2D expression, NK hyporesponsiveness, and impaired tumor control.[322-325] These differences suggest that NK cell activation is regulated through a rheostat model whereby the basal availability of NKG2D ligands tune the sensitivity threshold of NKG2D receptor signaling; this concept will be addressed later in our discussion of NK education.

Natural Cytotoxicity Receptors

First identified in human, the natural cytotoxicity receptors (NCRs) were identified using an experimental technique known as "redirected lysis."[326] In CTL, it was found that cross-linking antibodies against certain cell surface receptors, such as the TCR, were capable of triggering receptor signaling. When the Fc components of these antibodies are bound by Fc receptor expressed by target cells, CTL can be redirected to lyse such target cells. Thus, "redirected lysis" assays were powerful tools in the discovery of novel-activating receptors.

The NCR family are comprised of three type I transmembrane immunoglobulin-like receptors, named NKp46 (NCR1), NKp44 (NCR2), and NKp30 (NCR3). One striking feature about the NCRs is that they recognize a diverse set of pathogen- and self-encoded ligands that are secreted, membrane-bound, nucleus-associated, or shed extracellularly. NCRs were initially identified by their roles in the killing of tumor cells.[327-330] Although blocking antibodies against only one NCR resulted in modest inhibition of target cell killing, blocking of multiple receptors at once showed a synergistic effect.[329] The cloning and molecular characterization of NKp46 led to the discovery that human NKp46 was homologous with a mouse cDNA, consistent with its ability to mediate killing of murine target cells by human NK cells.[327] To date, NKp46 is the most selective marker for NK cells in mice and human although it is also expressed on a subset of ILC3 and certain innate-like T cells. NKp46 binds influenza hemagglutinin proteins in a sialic acid–dependent manner, and mice deficient in NKp46 are more susceptible to influenza infection.[68,331,332] Recently, NKp46 has also been shown to bind complement factor P, indicating crosstalk between the alternative complement pathway and NK activation.[333] Using Ncr1$^{-/-}$ mice, NKp46 has been implicated in the control of several different types of tumors as well as promoting autoimmune diabetes.[334-338] However, a recent study has shown that NKp46 deficiency also ablates the surface expression of TRAIL (TNFSF10), raising the point that phenotypes due to NKp46 deficiency may also be TRAIL-mediated.[339]

Like NKp46, NKp44 plays an important role in facilitating NK cell activation, particularly against tumors.[340,341] In particular, NKp44 has been shown to bind the growth factor PDGF-DD, which is actively secreted by tumors to fuel their own growth.[341] Additionally, a recent study reported the ability of NKp44 to bind HLA-DP molecules, providing evidence of NK recognition of MHC-II.[342] While NKp46 and NKp44 are mainly activating receptors, NKp30 exists as three alternatively spliced isoforms (Nkp30A, NKp30B, and NKp30C). While the first two are clearly activating receptors, the third isoform has been implicated in IL-10 production and an immunosuppressive outcome.[343] Furthermore, binding of NKp30 to its ligand, B7-H6 has been reported to either promote or inhibit activation of human NK cells, suggesting additional complexity to NKp30 function.[344-347]

SLAM Family Receptors and Nectin-Binding Receptors

While we have discussed the most well-established NK activating and inhibitory receptors, there are some MHC-specific and many MHC-independent receptors that do not fit in these major categories. We will discuss two prominent families of receptors, SLAM receptors and nectin-binding receptors, due to their unique signaling and uncommon ligands.

2B4 and the SLAM Family of Receptors

2B4 (CD244, SLAMF4) was originally identified in mice as a type I transmembrane immunoglobulin-like receptor.[348] 2B4 belongs to a family of immunoglobulin-like receptors, including SLAM (CD150, SLAMF1), Ly-9 (CD229, SLAMF3), NK, T- and B-cell antigen (NTB-A, or Ly108 (SLAMF6), CD84 (SLAMF5), and CD2-like receptor activating cytotoxic cells (CRACC, SLAMF7).[349,350] SLAM-family receptors are only expressed on immune cells and most are typically self-ligands with the exception of 2B4, which interacts with CD48, an immunoglobulin superfamily member found on nearly all hematopoietic cells. SLAM family receptors can either be activating or inhibitory depending on the context in which they are engaged. Crosslinking of 2B4, NTB-A, and CRACC on NK cells *in vitro* promoted cytotoxic degranulation and IFNγ production, implying an activating role.[348,351-353] However, genetic deficiency of 2B4 in mice augmented NK cell cytotoxicity against target cells, though this effect may be partially dependent on sex.[354-356] On the other hand, genetic deficiency of CRACC impaired NK cytotoxicity toward CRACC-expressing targets, consistent with *in vitro* studies demonstrating CRACC to be an activating receptor.[357] SLAM family receptor signaling is distinct from that of other NK cell receptors. They associate with the SLAM-associated protein family of adaptor molecules which include, SAP, EAT-2, and ERT.[358] SAP mediates SLAM signaling through the recruitment of the Src-related protein tyrosine kinase FYNT (a hematopoietic-specific isoform of FYN).[359] The physiological importance of SAP was revealed by studies of patients with X-linked lymphoproliferative disease. 60% to 70% of these individuals have loss-of-function mutations in the SAP-encoding gene (SH2D1A).[358] While there are many clinical facets of X-linked lymphoproliferative diseases, one notable feature is that upon infection with Epstein-Barr virus, patients develop massive lymphoproliferation. Additionally, these patients have a heightened frequency of malignant lymphomas. In concordance with the observations from humans, triple knockout of SAP, EAT-2, and ERT in mice nearly abolished NK cytotoxic killing of target cells and impaired the ability of NK cells to reject hematopoietic tumor cells *in vivo*.[360] These observations suggest an essential role for SAP and SLAM family receptors in the initiation of NK responses against malignant or virally transformed hematopoietic cells.

Nectin-Binding Receptors: CD226 (DNAM-1), CD96 (TACTILE), TIGIT, and CRTAM

In addition to receptors for nonself or stressed self, NK cells express receptors for ligands expressed at cell-cell junctions in epithelial tissues. A prominent group of these receptors binds nectin or nectin-like (necl) ligands, which are often expressed at adherens junctions between cells. These receptors include CD226 (DNAM-1), CD96 (TACTILE), TIGIT, and CRTAM.[361-363] DNAM-1 was the first of these receptors to be identified, where it was observed to be expressed on human CTL and NK cells.[364] DNAM-1 was subsequently shown to be an activating receptor, and blocking it with a monoclonal antibody reduced CTL and NK cytotoxicity against a variety of tumor cell lines.[365] In contrast to other NK activation receptors, DNAM-1 does not associate with an ITAM-bearing molecule. While the specific intracellular signaling mechanisms are not well-defined, it was found that activation through DNAM-1 required physical association with the αLβ2 integrin (LFA-1).[366] The association between DNAM-1 and LFA-1 relies on phosphorylation of Ser_{329} in the cytoplasmic domain of DNAM-1, which is mediated by Protein Kinase C.[367] Through two independent studies, the ligands for DNAM-1 were shown to be CD155 (PVR; necl-5) and CD112 (nectin-2, PRR2).[368,369] Expression of Necl-5 and Nectin-2 is particularly pronounced in various tumors. Concordantly, DNAM-1 has been shown to mediate antitumor responses. DNAM1-deficient mice had reduced tumor rejection in models of fibrosarcoma and melanoma.[370,371] However, DNAM-1 may be partially redundant with other NK activating receptors as it was not required for the killing of tumors expressing NKG2D ligands.[372] In addition to DNAM-1, other receptors were identified to recognize nectins and nectin-like proteins. CD96 is strongly expressed by activated CTL and NK cells, and it recognizes necl-5 but not nectin-2.[373] Engagement of CD96 enhances NK cytotoxicity, though not as effectively as DNAM-1. CD96 has been shown to compete with DNAM-1 for binding to necl-5 and thus limit NK cell function by direct inhibition.[374] Interestingly, CD96-deficient mice not only displayed improved tumor rejection but also were more sensitive to LPS-induced endotoxicosis. TIGIT was identified as an immune cell-specific immunoglobulin superfamily receptor of the CD28 family.[375-377] Similar to DNAM-1, TIGIT binds necl-5 and nectin-2 but also binds CD113 (nectin-3, PVRL3). However, TIGIT contains an inhibitory ITIM motif and has been shown to restrain NK cell-mediated killing.[378,379] Since many tumors express nectins or nectin-like proteins, TIGIT is under current investigation for checkpoint blockade immunotherapy. Lastly, CRTAM is expressed by activated human NK, $CD8^+$ T cells, and a subset of $CD4^+$ T cells, though its expression declines in 72 hours. The ligand for CRTAM was identified to be Necl-2.[380-382] CRTAM promotes NK and CTL killing of necl-2–expressing tumors. In addition to its roles in cytotoxic killing, interactions between CRTAM on T cells and necl-2 on antigen-presenting cells may influence T cell priming and polarization.[383,384]

Though we have reviewed the major families of NK activating and inhibitory receptors, there are many more that were not discussed. To address these molecules, we provide an overview of NK receptors and their corresponding ligands in the accompanying table (Table 18.2) and refer the reader to a number of other excellent articles.[146,149,153,211,385,386]

NK Cell Education and the Licensing Hypothesis

With the multitude of NK receptors characterized, an overarching question in NK cell biology concerned how NK cells were able to integrate positive and inhibitory signals to selectively initiate natural killing. From the initial studies characterizing NK cells, it was clear that induced expression of MHC-I ligands completely ablated NK cell killing. This raised the question whether genetic manipulation of NK receptor ligands could induce NK cells to break tolerance and initiate autoimmunity. Broadly, two general approaches were employed to answer this question. One approach utilized models of transgenic expression of NKG2D ligands, such as RAE-1 and MIC, in all cells of the body.[322-324,387,388] Instead of initiating autoimmunity, NK cells become hypofunctional in these models through multiple mechanisms, including the downregulation of activating receptors. Surprisingly, this tolerogenic effect is not restricted to development as mature Ly49H+ NK cells exhibit functional defects when transferred into mice with transgenic expression of their cognate ligand.[387] The second approach examined whether somatic deletion of MHC-I, the main inhibitory ligand for NK cells, would be sufficient to enhance NK activation. However, NK cells from MHC-I–deficient mice are profoundly hypofunctional in response to target cells and NK1.1 crosslinking.[158,389,390] This effect was recapitulated in ITIM-mutated mice, demonstrating that inhibitory receptors were necessary to "license" functional NK cells. Interestingly, licensing is readily reversible as mature NK cells from MHC-I–deficient donors acquire functionality when transferred into MHC-I–sufficient hosts, and NK cells from MHC-I–sufficient donors become hypofunctional when transferred into MHC-I–deficient recipients.[391,392] Human NK cells appear to demonstrate features of licensing as NK subsets that do not express any inhibitory receptors were found to be hypofunctional.[393,394] To integrate these observations, two models have been proposed. The "licensing" or "arming" hypothesis posits that engagement of NK inhibitory receptors are necessary for proper maturation and effector function (Fig. 18.5). The "disarming" model proposes that in the absence of inhibitory

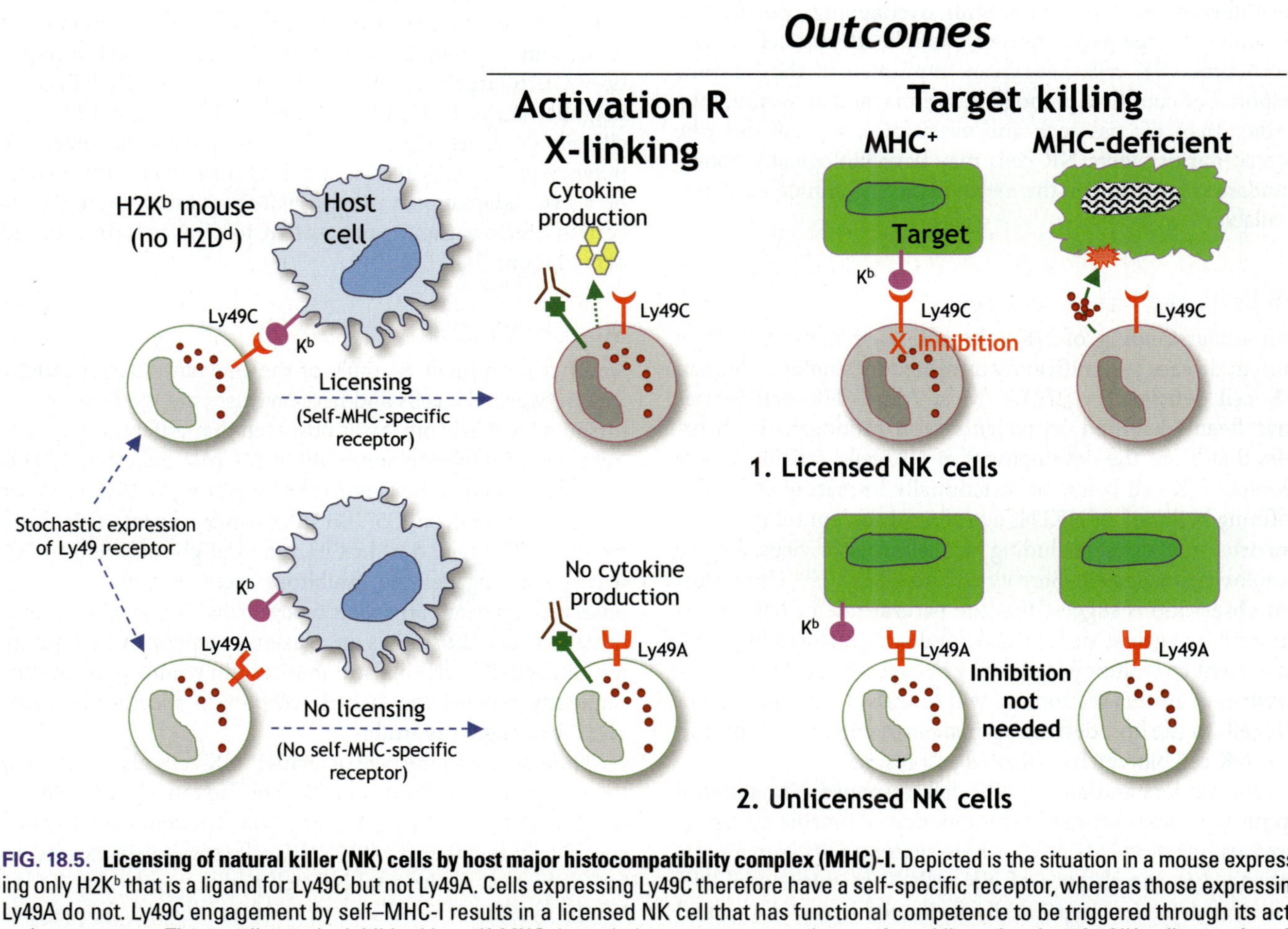

FIG. 18.5. Licensing of natural killer (NK) cells by host major histocompatibility complex (MHC)-I. Depicted is the situation in a mouse expressing only H2K^b that is a ligand for Ly49C but not Ly49A. Cells expressing Ly49C therefore have a self-specific receptor, whereas those expressing Ly49A do not. Ly49C engagement by self–MHC-I results in a licensed NK cell that has functional competence to be triggered through its activation receptors. These cells can be inhibited by self-MHC through the same receptor that conferred licensing. Ly49A+ NK cells remain self-tolerant because they are unlicensed. For clarity, only one MHC allele and cells expressing one Ly49 receptor are shown. Activation receptors are also not shown for the target killing depiction. Normally, the host has multiple MHC alleles, each of which could license different NK cells, and individual NK cells express multiple Ly49s simultaneously. Each NK cell is likely to be licensed separately, potentially by different MHC alleles, and through different Ly49s. (Adapted from Yokoyama WM. *Paul's Fundametnal Immunology.* 7th ed.)

signaling, unchecked activating signals induce NK anergy as shown in models of constitutive activating ligand expression. These models are not mutually exclusive, and it is possible that both mechanisms operate in parallel. Subsequent studies show that licensing or disarming are unlikely to be a binary phenomenon, as successive expression of inhibitory receptors confers increased effector function to NK cells.[395] Thus, NK are proposed to be regulated by a "rheostat" model whereby arming and disarming signals integrate to define an immunological set point that determines the activation threshold of every individual NK cell.[395] As few transcriptional features are observed between licensed and unlicensed cells, further research is warranted to understand the molecular processes behind NK cell education.[396]

NK Cells and Immune Responses

Despite the relative paucity of NK cells (~2.5% of splenic lymphocytes in C57/B6 mice), NK cells play a crucial role in the early stages of infection, when T cells are still undergoing clonal proliferation and differentiation. Furthermore, while T cells specific for a particular antigen are extremely rare during the onset of infection, NK cells constitutively express germline-encoded receptors with overlapping specificities, providing a large population of possible effector cells at any given time. NK cells have been implicated in the immune response of countless pathogens, tumors, and autoimmunity. Rather than discuss each and every case, we will describe two scenarios where NK cells may have biologically nonredundant roles based on the mechanistic and clinical evidence available.

NK Cells and Viral Infections

Our understanding of NK cells and their natural roles in human disease is significantly informed by studies of human NK cell deficiencies (NKD). Most "pure" NK deficiencies have been identified in patients with monogenetic disorders that blunt the development of NK cells, leading to low absolute NK cell count or functionally immature cells. The defining hallmark of NKD is a profound susceptibility to herpesvirus infections, including varicella zoster virus, herpes simplex virus, Epstein-Barr virus, and CMV.[397,398] These clinical observations suggest that the natural role of NK cells is the defense against such viruses, and such features highlight an ancient coevolutionary history between NK cells and herpesviruses. In this section, we will discuss the unique role of NK cells in the host defense against herpesvirus infection and how NK cell biology has adapted as a result.

Our understanding of NK cell responses in antiviral immunity comes primarily from models of murine cytomegalovirus infection (MCMV), where susceptibility is markedly enhanced in the absence of NK cells.[399] In this model, NK cell viral detection is primarily facilitated by cytokines and recognition by NK receptors. NK cell production of IFNγ is dependent on the myeloid-derived cytokine IL-12, and IL-12–deficient mice are extremely susceptible to infection with MCMV.[400-402] IL-18 plays a role in supporting NK production of IFNγ and perforin-dependent cytotoxicity, though IL-12 is more important in MCMV defense.[401,403] Lastly, type I interferons are produced in great quantities during viral infection, and they play important roles in the indirect and direct regulation of NK cell activity. In addition to promoting myeloid production of IL-12 and IL-18, type I interferons promote the production and presentation of IL-15, an important survival and priming factor for NK cells.[114,404] Type I interferons can act directly on NK to enhance or inhibit production of IFNγ through STAT4 and STAT1, respectively. While NK cells express high basal levels of STAT4, STAT1 is induced during infection, suggesting a regulatory feedback mechanism in restraining chronic NK activation.[405]

In addition to regulation by innate cytokines, NK cells respond to viral infection through a diverse set of activation receptors. NKG2D facilitates the killing of virus-infected cells through the recognition of stress-induced ligands. The importance of this recognition pathway is illustrated by the numerous viral immune evasion mechanisms that target it. MCMV possesses a number of viral proteins that bind to and inhibit the expression of NKG2D ligands: m152 inhibits RAE1; m155 and m138 downregulate expression of H60; and surface expression of MULT1 is inhibited by m145 and m138.[304,406-408] Similarly, HCMV has also devoted considerable amounts of its genomic resources toward inhibiting NKG2D recognition: UL16 targets MICB, ULBP1, ULBP2 and RAET1G for degradation; and a HCMV-encoded miRNA mir-UL112 specifically degrades MICB.[292,296,409] Consequently, the extensive polymorphism in MICA and MICB have been proposed to be a host adaptation to these HCMV evasion strategies, as certain allelic variants are resistant to viral protein-induced degradation.[410]

Memory NK Cells

Another prominent example of the coevolutionary relationship between NK cells and herpesviruses are the Ly49 receptors in mice. Early on, it was observed that different strains of mice had variable susceptibility to MCMV infection.[411] This was later found to be due to the activating Ly49H receptor which is present on C57/B6 mice while absent in BALB/c strains.[412] The ligand of Ly49H is a viral glycoprotein m157, which also engages the inhibitory receptor Ly49I.[194] These observations detail an evolutionary arms race where MCMV initially evolved *m157* as an evasion mechanism by binding inhibitory NK receptors and that certain strains of mice subsequently evolved the Ly49H receptor to specifically target *m157*-bearing virus strains.

Understanding the role of Ly49H$^+$ NK cells also facilitated the recognition of "memory-like" or "adaptive" NK cells. In the first 48 hours of MCMV infection, NK cells expand nonspecifically but then Ly49H$^+$ NK cells selectively expand to make up most of the total NK cell population by day 7.[413,414] While Ly49H$^+$ NK cells contracted to normal frequencies by day 37, adoptive transfer experiments showed that virus-experienced Ly49H$^+$ were preferentially maintained as long as 70 days.[414] These Ly49H$^+$ cells were found to be more responsive to cross-linking of NK receptors and more protective against MCMV infection in neonatal mice compared to naïve NK cells.[414] This pattern of immune response, expansion-contraction-recall, is reminiscent of adaptive immunity, and thus MCMV-induced Ly49H$^+$ NK cells were proposed to be "memory" NK cells. Similar populations of "adaptive" NK cells are broadly identified in HCMV-seropositive humans by a CD56^{dim} CD57$^+$ CD16$^+$ and display distinct transcriptional and epigenetic alterations.[415] However, evidence showing that cytokine activation alone is sufficient to induce an adaptive NK phenotype suggests that this phenomenon is not strictly antigen-specific.[416,417] Similar reports of memory-like NK cells have been reported to mediate hapten-specific responses in contact hypersensitivity and vaccination models, but the mechanisms behind these processes remain unknown.[418,419] Collectively, these findings suggest that NK cells possess the capacity to expand in response to specific antigens, self-maintain in the host for long periods of time, and are poised to mount stronger responses upon secondary activation. Blurring the lines between innate and adaptive immunity, the adaptive features of NK cells are the subject of active investigation, and it remains to be seen how this memory-like program relates to the concepts of "trained immunity" in myeloid cells.[420]

In addition to protection from herpesvirus infections, NK cells have been implicated in immune responses involving other human pathogens such as human immunodeficiency virus (HIV), hepatitis C virus (HCV), and influenza.[421] While NK cells are mainly thought to protect against viruses, they have been shown to play nonredundant roles in the defense against bacteria, parasites, and fungi.[422-424]

NK Cells and Cancer Immunotherapy

NK cells were originally identified by their exquisite ability to kill tumor cells.[61] Subsequent investigation of the mechanisms behind this ability have uncovered a multitude of strategies for recognizing cancer cells. We previously discussed the basis for tumor recognition as well as major receptors involved (NKG2D, NCRs). In this section, we will outline major translational applications of NK biology toward the treatment of human malignancies.

One of the early clinical observations of NK tumor recognition was in the context of allogeneic hematopoietic stem cell transplant (allo-HCST). In the case of liquid cancers, if donor and recipient are not HLA-matched, T cells from the graft eradicate malignant cells (GVL) but also initiate autoimmunity against normal cells (graft-versus-host [GVH]), which is a major factor behind transplant failure. If the donor and recipient are not HLA identical, T cell depletion of the BM graft can dramatically reduce GVH and improve transplant success.[425] Similarly, NK cells will engage in natural killing if KIR haplotypes do not recognize HLA haplotypes, and such NK clones are termed "alloreactive".[162] Transplant of alloreactive donor NK cells was found to potentiate GVL effects but have no effect, or even reduce, GVH pathology.[227,426] Several recent studies have identified particular activating KIRs that may play a more dominant role in GVL.[427,428] Such studies have established an argument for standard genotyping of KIR alleles for BM donors.

Early HCST studies established the remarkable potential for NK cells to selectively kill tumor cells while sparing patients from systemic autoimmunity. One of the underlying principles of allotypic NK transplant is that donor NK cells are released from the inhibitory actions of KIRs, thus tipping the balance toward activating receptors against tumor cells. A more direct approach to achieving this outcome is by blocking the KIRs and other inhibitory receptors with intravenously administered mAbs. The general strategy of blocking the checkpoints that inhibit the immune response (immune checkpoint blockade) has emerged as an exceptional breakthrough in the cases of anti-PD-1 and anti-CTLA-4 therapy. Currently, pan-KIR blocking antibodies and anti-NKG2A mAbs are under active investigation in liquid and solid tumors.[429]

LYMPHOID TISSUE INDUCER CELLS AND LYMPHOID TISSUE DEVELOPMENT

Mammalian immunity is intricately organized by SLO that are dispersed throughout the body. These structures serve as hubs for lymphocyte surveillance and differentiation and are discretely organized to efficiently amplify immune responses against antigens. SLO develop during embryogenesis and include the spleen, LN, and PP. Among these organs, LN and PP require a specialized type of ILC, lymphoid tissue inducer cells (LTi), for their development. LTi were initially observed as CD3$^-$ CD4$^+$ LTβ$^+$ cells that accumulated at lymph node anlagen during mouse embryogenesis.[430] These cells were further shown to express CD127 (IL-7Rα), c-kit, RANK, RANKL, CXCR5, and CCR6.[431] LTi orchestrate SLO formation through interactions with stromal cells; LTα1β2-expressing LTi activate lymphotoxin β receptor (LTβR)-expressing mesenchymal cells at the lymphoid tissue anlage.[432] As a result of these interactions, mesenchymal cells (stromal organizer cells), produce CXCL13, CCL19, and CCL21, chemokines which recruit additional CXCR5$^+$ and CCR7$^+$ LTi.[433,434] Mice with defects in LTi-stromal organizer cell interactions (LTα, LTβ, and LTβR), and mice lacking molecules required for normal LTi accumulation at the anlage (CXCR5/CXCL13 deficiency alone or CXCL13$^{-/-}$ in combination with CCR7 ligand deletion) exhibit defects in SLO development.

The necessity of LTI during SLO organogenesis was shown in multiple studies that demonstrated that mice with reduced LTi development and survival had defects in LN and/or PP development. LTi require the transcription factors RORγt and Id2 for their development.[21,435] Accordingly, mice with deficiencies in either transcription factor lack LN and PP.[21,436] Like other ILC, LTi survival depends on signaling through the common γ_c, and LTi-dependent SLO are absent in mice lacking the γ_c and its downstream signaling components.[100,437] LTi appear to be particularly reliant on signaling through IL-7 receptor (Il-7R), as deficiency of this receptor completely halts PP formation while reducing the number and size of various peripheral LN.[438-440] Consistent with these findings, transgenic overexpression of IL-7 induced an accumulation of LTi, and resulted in increased PP formation and the generation of ectopic LN and cecal patches.[441] However, deficiency of IL-7 modestly reduced total LTi and had variable effects on the number of peripheral LN, whereas development of mesenteric LN and PP were unaffected.[442] The discrepancy in PP phenotypes between $Il7^{-/-}$ and $Il7ra^{-/-}$ mice indicates that alternative ligands, such as TSLP, engage IL-7R in the absence of IL-7. Maternally derived retinoic acid has also been shown to directly regulate RORγt and drive the differentiation of LTi during embryogenesis, and mice with impaired retinoic acid receptor signaling have reduced LN size and fewer numbers of PP.[443]

In addition to SLO, postnatal LTi-like cells are required for the development of several tertiary lymphoid organs that develop postnatally. In naïve mice, the majority of these structures are in the intestine; these SILT include cryptopatches (CP) and isolated lymphoid follicles (ILF).[444,445] CP are characterized as clusters of LTi-like cells surrounded by dendritic cells, whereas ILF contain B cells and can mediate germinal center reactions. ILF are derived from CP, and their formation is dependent on innate sensing of commensal microbiota and CCR6-dependent recruitment of B cells.[446,447] Functionally, SILT may serve as adaptable reserve sites for induction of IgA responses.[448] Certain types of secondary and tertiary lymphoid organs have been found to be LTi-independent, as shown by RORγt-deficient mice, which still develop splenic white pulp, nasal-associated lymphoid tissue (NALT), bronchus-associated lymphoid tissue (BALT), tear duct-associated lymphoid tissue (TALT), and fat-associated lymphoid clusters (FALC).[436,449-451] These exceptions implicate RORγt-independent cell types in the organogenesis of those tissues.

LTi-like cells are present in adult C57BL/6 mice where they have a Lin^- $CD45^+$ $CCR6^+$ $CD4^+$ or Lin^- $CD45^+$ $CCR6^+$ $CD4^-$ phenotype in intestinal and SLO tissues.[431,452] Under the current nomenclature, adult LTi-like cells are considered to be part of the ILC3 family. In addition to their induction of SILT, LTi-like ILC3 play important roles in intestinal homeostasis, and we will discuss these functions in the corresponding section on ILC3.

LTi have been described in human fetal LN and spleen by a Lin^- $CD45^{int}$ $CD127^{hi}$ phenotype.[453] As opposed to their murine counterparts, human LTi do not express CD4, though they expressed RORγt, LTB, RANK, and RANKL, suggesting similar functional capacities as murine LTi. When cultured with mesenchymal stem cells, human LTi were able to induce expression of adhesion molecules on mesenchymal cells in a LTB- and TNF-receptor–dependent manner.[453] Furthermore, individuals with biallelic mutations in RORC (encoding RORγ and RORγT) lack palpable LN, though tonsils remained, suggesting that human SLO development is partially dependent on $ROR\gamma t^+$ LTi.[454]

ILC1 VERSUS NK: UNIQUE AND OVERLAPPING FEATURES

The identification of ILC1 as a discrete lineage from NK cells was hindered by the large degree of overlap in surface markers shared by these cells. ILC1 were first characterized as $TRAIL^+$ $NK1.1^+$ NK cells in the liver of adult mice, and it was first proposed that these were immature precursors to conventional NK cells based on their $CD127^+$ $CD27^+$ $CD11b^-$ phenotype and lack of expression of most Ly49 receptors, CD49b (DX5), and Eomes.[89] However, adoptive transfer studies using ILCPs have established that ILC1 and NK emerge from distinct committed precursors.[20,22,88] These developmental studies also established that ILC1 share a requirement for the transcription factor T-bet, but, unlike NK cells, develop independently of Eomes.[20,88] Parabiotic studies have established ILC1 to be primarily tissue-resident cells while NK circulate at steady-state.[38,455,456] Expression of integrin α_1 (CD49a), in particular, has been used widely for the identification of ILC1 in various tissues at steady-state, though NK cells have been observed to upregulate this marker during viral infection.[456] In mice, ILC1 have been identified in almost every tissue, such as liver, thymus, intestine, uterus, and salivary gland.[89,455,457-459] Contrary to NK cells, ILC1 are remarkably heterogeneous based upon their tissue of residence. For example, ILC1 in the liver and intestine do not express the transcription factor Eomes, while ILC1 in the uterus and salivary gland do. Additionally, the ILC master transcription factor NFIL3 is necessary for the development of liver and intestinal ILC1, whereas salivary gland, skin, kidney, and uterine NK develop independently of NFIL3.[5,13,14,455,457,458] Lastly, while NK and most ILC are largely depleted in IL15Rα-deficient mice, ILC1 in the intestinal intraepithelial lymphocyte (IEL) compartment were largely unaffected in $Il15ra^{-/-}$ mice.[458] The phenotypic characteristics and developmental requirements of ILC1 have been summarized in Table 18.3.

Functionally, ILC1 are similar to NK cells in that they rapidly produce IFNγ and are capable of cytotoxic killing. However, ILC1 have reduced ability to degranulate in response to stimulation with target cells, and ILC1 express lower amounts of perforin than NK.[20,456,460] Additionally, ILC1 produce greater amounts of TNFα than NK cells and exhibit higher responsiveness to IL-12 (Table 18.4).[20,455,461] ILC1 also express high amounts of the death receptor TRAIL at steady-state, though no study has yet investigated the contribution of this receptor in ILC1 function. ILC1 in human and mice lack expression of L-selectin (CD62L), a cell surface molecule which facilitates trafficking of lymphocytes from circulation to SLO, but express CD69, a surface antigen which inhibits sphingosine-1-phosphate receptor signaling to prevent migration of lymphocytes to the bloodstream.

In humans, ILC1 were first described as two distinct populations of IFNγ-producing cells in the tonsils and intestine.[458,462] One population has a $CD56^+$ $NKp44^+$ $CD103^+$ phenotype and is restricted to the IEL compartment.[458] The second has a $CD56^-$ $CD127^+$ $CD161^+$ $NKp44^-$ $c\text{-}kit^-$ phenotype and was most abundant in intestinal lamina propria, though subsequent studies have had difficulty in defining these cells due to their expression of certain T cell markers.[462,463] Both populations of ILC1 have been described to participate in ILC3 transdifferentiation but while $CD127^+$ ILC1 conversion can be bidirectional, $CD103^+$ ILC1 plasticity primarily occurs in the ILC3 → ILC1 direction *in vivo*.[464,465] This concept of ILC plasticity will be discussed in later sections.

TABLE 18.3 Table Showing the Phenotypic Markers and Developmental Requirements of Murine ILC1 and Conventional NK Cells in Various Tissues

Phenotypic Markers	Molecule	NK	Liver ILC1	IEL ILC1	SG ILC1
	CD127 (IL-7Ra)	–	+/–	+	–
	CD161 (NK1.1)	+	+	+	+
	KLRG1	+	–	–	–
	CD69	–/*	+	+	+
	CD335 (NKp46)	+	+	+	+
	CD226	+/–	+	+	nd
	CD253 (TRAIL)	–/*	+	nd	+
	CD49a (integrin α1β1)	–/*	+	+	+
	CD49b (DX5)	+	–	–	+
	CD160	+/–	+	+	+
	CD200R	–	+	nd	+
	CD62L	+	–	–	–
	CD11b	+	–	–	–
	Tbet	+	+	+	+
	Eomes	+	–	–	+
	CXCR6	–	+	+	+
Developmental requirements	IL-15	+	+	+	+
	IL-15RA	+	+	–	+
	NFIL3	+	+	+	–
	Tbet	–	+	+	+
	Eomes	+	–	–	nd
	TgfbrII	–	–	nd	+
	Hobit	–	+	–	–

+, high expression; +/–, expression detected in some, but not all, cells; *, induced by activation; nd, not determined.

Transcriptomic and epigenetic profiling of ILC1 and NK cells yielded several valuable insights into the unique regulatory programs present in these cell types.[460,466,467] Examination of human and mouse ILC1 transcriptomes revealed the enrichment of many genes associated with imprinting by transforming growth factor β (TGFβ), such as CD103, CD9, and NEDD9.[458,467] Conditional deletion of the common TGF-β-receptor type II receptor (TGF-βRII) on NK cells and ILC1 resulted in marked reduction of salivary gland ILC1 while sparing intestinal and liver ILC1.[467] Additionally, cells with similar phenotypes to salivary gland ILC1 have been identified in breast, prostate, and fibrosarcoma tumors.[468,469] Similarly, ablation of TGF-βRII on NK and ILC1 significantly reduced the number of tumor-associated ILC1. Intriguingly, this study implicated ILC1-derived TNFα as a tumorigenic factor, and thus depletion of tumor-associated ILC1 resulted in reduced tumor burden.[469] Collectively, these studies indicate a key role for tissue factors such as TGFβ in shaping the unique phenotypes and functions of ILC1. Epigenetic analysis of human ILC1 uncovered a significant enrichment of DNA binding motifs related to the master transcription factor PRDM1.[466] While ILC1 do not express PRDM1, they uniquely express high levels of the PRDM1 homolog, ZNF683 (Hobit).[470] Concordantly, Hobit-deficient mice exhibited a selective depletion of liver-resident ILC1, whereas splenic and circulating NK cells were unaffected.[471] Interestingly, TGFβ signaling and Hobit also control the differentiation of tissue-resident memory CD8$^+$ T cells (T_{RM}) and NKT cells.[471-474] Thus, it is likely that ILC1, T_{RM}, and NKT cells utilize similar transcriptional programs to establish tissue residency.

Due to the large overlap between ILC1 and NK molecular markers, it has been difficult to elucidate the unique contributions of ILC1 in immune responses. Many strategies that deplete NK cells such as anti-NK1.1, anti-asialo GM1, and *Ncr1-DTR* transgenic mice will also deplete ILC1. Hobit$^{-/-}$ mice have selective deletion of ILC1 in the liver, but ILC1 in other organs are unaffected.[471] Additionally, liver-resident NKT cells display altered effector function in Hobit$^{-/-}$ mice.[473] ILC1 are especially enriched in the liver, where they can comprise ~30% of NK1.1$^+$ cells.[461,475] One study has shown that ILC1-derived IFNγ protects mice in a model of acute liver injury.[475] This protective effect was mediated by ILC1

TABLE 18.4 Table of Known Regulators of ILC Function and Their Respective Signaling Pathways

Pathway	ILC1	ILC2	ILC3
JAK-STAT	IL-12 (+)	TSLP (+)	IL-23 (+)
		IL-2 (+)	IL-2 (+)
		IL-27 (−)	
		Type I/II interferon (−)	
NF-κB	IL-18 (+)	IL-25 (+)	RANKL (−)
		IL-33 (+)	TL1A (+)
		TL1A (+)	IL-1B (+)
GPCR		Lipoxin A4 (+)	VIP (+/−)
		PGI2 (−)	
		Leukotrienes (+)	
		PGD2 (+)	
		VIP (+)	
		NMU (+)	
		CGRP (−)	
		Catecholamines (−)	
Nuclear receptors		Retinoic acid (−)	Retinoic acid (+)
		AHR ligands (−)	AHR ligands (+)
		Androgens (−)	

+, indicates positive regulator; −, indicates negative regulator, +/−, indicates both positive and negative regulatory roles.

recognition of ATP released from damaged hepatocytes as well as costimulation through the activating receptor DNAM-1. Another study found that ILC1 conferred protection during the early phases of MCMV infection if the virus was delivered directly to the liver.[461] Both studies utilized Hobit$^{-/-}$ mice to demonstrate ILC1-specific effects in the liver. In addition to their role in the liver, numerous studies have implicated ILC1 in various gastrointestinal immune responses. Production of IFNγ by ILC1 has been proposed to mediate protection against the intracellular parasite *Toxoplasma gondii* as well as the bacteria *Clostridium difficile*.[20,476] However, these studies were performed using Tbet$^{-/-}$ or Ifng$^{-/-}$ mice, so the role of NK cells cannot be definitively excluded. In human, increases in CD127^{+} and CD103^{+} ILC1 have been observed in specimens from Crohn disease patients compared with non-IBD controls.[458,462] Collectively, these data show that ILC1 participate in a wide spectrum of immune responses, but further investigations are warranted to determine the exact contribution of ILC1 in these processes.

ILC2

ILC2 were first reported in three independent studies that described the existence of innate lymphocytes that produced the type 2 cytokines IL-5 and IL-13 in response to stimulation with IL-25 and/or IL-33.[477-479] In the first study, "natural helper cells" were identified in FALC, lymphoid structures located in mesenteric fat. These lineage-negative cells expressed c-Kit, Sca-1, IL-7R, and IL-33R, proliferated in response to IL-2, and produced large amounts of IL-5, IL-6, and IL-13 in response to combinations of IL-2, IL-25, and IL-33 *in vitro*.[477] The second study utilized IL-13 reporter mice to identify "nuocytes" in the MLN that robustly responded to IL-25 and IL-33 *in vivo*. "Nuocytes" were found to be the dominant early source of IL-13 in infection with the parasitic helminth *Nippostrongylus brasiliensis*.[478] The last study used IL-4 and IL-13 dual reporter mice to show that "innate type 2 helper cells" were systemically distributed among various tissues and robustly expanded during infection with parasitic helminths.[479] It is now understood that these three studies had described the same cell type, now referred to as ILC2.

ILC2 Phenotypic Characteristics

ILC2 were originally defined functionally by their capacity to produce type 2 cytokines. Additionally, parabiotic studies showed that ILC2 were mainly tissue-resident though they possess some capacity to recirculate during infection with the helminth *N. brasiliensis*.[38] Further studies have identified several surface markers that distinguish ILC2 among lineage-negative cells; these molecules include CD25, CD90, CD127 (IL-7Rα), SCA1, ICOS, IL-25R, ST2 (IL-33R), and KLRG1. However, as ILC2 are dispersed throughout most tissues of the body, the expression pattern of these markers varies between different tissues. In particular, ILC2 in the intestine are primarily KLRG1^{+} ST2^{-}, whereas ILC2 in the lung and fat are KLRG1^{-}ST2^{+}.[480] Intestinal ILC2 also express lower amounts of CD25 but a higher level of IL-25R.[480] A universal characteristic of ILC2 is markedly high expression of the transcription factor GATA3, which is necessary for mature ILC2 function and differentiation.[23,25,459,481-483] GATA3 is also partially required for differentiation of αLP and indispensable for differentiation of PLZF^{+} ILCP.[484] ILC2 can also be identified by expression of IL-5 and IL-13. A subset of ILC2 produce IL-5 at rest and reporter mice for this cytokine can be used to reliably track ILC2 *in vivo*.[485] Following the discovery of ILC2 in mice, human ILC2 were identified in the intestine, lungs, nasal tissues, and peripheral blood. Lin^{-} CD127^{+} human ILC2 were distinguished by the expression of chemoattractant receptor–homologous molecule expressed on TH2 lymphocytes (CRTH2) and the NK cell receptor CD161.[486] Like their mouse counterparts, CRTH2^{+} ILC2 produced IL-5 and IL-13 in response to stimulation with IL-25 and IL-33, and they also expressed high amounts of GATA-3.[483,486]

Regulation of ILC2 Effector Functions

As initially described, the primary effector function of ILC2 is the rapid and robust production of type 2 cytokines such as IL-4, IL-5, IL-9, and IL-13 (Table 18.4). In lymphoid tissues, IL-4 contributes toward IgE class-switching in B cells and Th2 differentiation. IL-5 mediates eosinophil development and survival and promotes B cell immunoglobulin production. IL-13 acts directly on intestinal stem cells (ISCs) to mediate goblet cell and tuft cell hyperplasia. IL-9 can further potentiate IL-5 and IL-13 secretion and promotes survival of ILC2 through autocrine signaling.[487-489] In addition to the production of type 2 cytokines, ILC2 also produce amphiregulin, an epidermal growth factor receptor (EGFR) ligand, which promotes wound healing and epithelial remodeling.[490,491] Lastly, ILC2 have immunoregulatory potential through production of the immunosuppressive cytokine IL-10.[492,493]

Unlike their Th2 counterparts, ILC2 constitutively produce IL-5 in the absence of infection.[485] During helminth infection, lung ILC2 produce large amounts of IL-13 but minimal amounts of IL-4 protein, despite the high expression of *Il4* transcript.[494] Comparison of reporter surrogates for *Il4* mRNA versus IL-4 protein showed that *Il4* mRNA abundance reflects the potential for IL4 protein production upon stimulation rather than the current state of IL4 protein abundance.[495] ILC2 can produce large amounts of IL-4 *in vitro* when stimulated with phorbol 12-myristate 13-acetate (PMA) and ionomycin or leukotriene D4.[477,496,497]

ILC2 activity is dynamically regulated by factors derived from the tissues in which they reside. Prior to the discovery of ILC2, it was known that epithelial-derived cytokines such as TSLP, IL-25, and IL-33 played a driving role in type 2 immunopathology.[498-501] Accordingly, the initial studies that described ILC2 showed IL-25 and IL-33 to be potent drivers of ILC2 proliferation and activation *in vitro* and *in vivo*, and mice with combined deficiency of TSLP receptor (TSLP-R), IL-25, and IL-33 receptor (ST2) contain hypoactive ILC2.[477-479,502] In certain situations, ILC2 effector functions are inhibited by IFNγ, type I interferons, and IL-27 via increased activity of Stat1.[39,503] ILC2 are also regulated by lipid mediators released during inflammation and tissue repair, such as prostaglandins and leukotrienes. Leukotrienes and prostaglandin D_2 (PGD_2) activate ILC2, whereas prostaglandin E_2 (PGE_2) inhibits ILC2 activity.[497,504-506] Interestingly, ILC2 production of IL-9 is not regulated by IL-25, IL-33, or TSLP but by IL-2 from adaptive lymphocytes.[489] This reveals a mechanism by which adaptive immunity can regulate ILC2 activity to properly calibrate immune responses. Similarly, it was shown that IL-2 induces the alternative activation of lung ILC2 to produce IL-10.[493] Further work in intestinal ILC2 supported these findings and also found multiple additional mediators, including IL-10 itself, to be positive regulators of ILC2 production of IL-10.[492] Lastly, the TNF superfamily member TL1A has been shown to be a potent activator of IL-5 and IL-13 production, whereas it completely ablates IL-2-induced production of IL-10.[492,507] Collectively, these studies show that ILC2 integrate multiple signals from stromal and hematopoietic origins to produce functionally diverse effector molecules.

ILC2 and Gastrointestinal Helminth Infections

Helminths are invertebrate parasites, of which nematodes are a major source of human parasitic infections. Many nematodes infect and reside within the GI tract of vertebrate host species.[508] Mouse models of nematode infections commonly utilize the model pathogen *Nippostrongylus* brasiliensis. This pathogen primarily infects the lung and intestine. Generally, *N. brasiliensis* infects the host as infectious larvae through subcutaneous invasion. Larvae then circulate to the lungs, enter the trachea, and are subsequently swallowed to facilitate entry to the intestine. Following the establishment of an intestinal niche, *N. brasiliensis* matures and reproduces, shedding eggs that produce larvae outside the host to restart the infectious cycle. Given their large size, helminths are difficult to eliminate using traditional cytolytic or antibacterial mechanisms. Type 2 immunity responses to helminths orchestrate profound changes in tissue physiology, inducing goblet cell hyperplasia and smooth muscle contractions. This process, termed "weep and sweep," relies on the type 2 cytokine IL-13. IL-13, along with IL-4, also promotes the differentiation of alternative activated macrophages (AAM) which contain parasites in granulomas. On account of these processes, IL-4 and IL-13 are required for expulsion of *N. brasiliensis*.[509] Transfer of wildtype T cells into T cell-deficient and IL-4/IL-13–deficient mice was not sufficient for expulsion of *N. brasiliensis*.[510] Furthermore, conditional deletion of IL-4 and IL-13 in T cells did not impact helminth expulsion.[511] Thus, IL-4/IL-13 from innate immune cells are necessary for defense from gastrointestinal nematodes. A series of studies found eosinophils and basophils to be dispensable during primary helminth infection.[510,512] Adoptive transfer of ILC2 to IL4/IL13-deficient mice was sufficient to restore helminth control.[478] Additionally, a genetic mouse model of selective ILC2 deficiency was sufficient to drastically impair helminth control despite a normal T cell compartment. Together, these findings show that ILC2 plays a significant and nonredundant role in expulsion of intestinal helminths.

Epithelial cytokines, especially IL-25, play a crucial role in inducing the production of type 2 cytokines and driving the expulsion of intestinal helminths.[478,513] While IL-25 was known to be necessary for proper helminth control, the source of IL-25 in the intestine was unknown. Three independent groups made the remarkable finding that tuft cells, a rare population of chemosensory epithelial cells in the small intestine, produce IL-25 to maintain ILC2 homeostasis and drive ILC2 expansion in response to helminth infection.[514-516] Tuft cells participate in a positive feedback system where tuft cell-derived IL-25 induces the production of IL-13 by ILC2. IL-13 then acts on ISCs to drive the differentiation of tuft and goblet cells, leading to more IL-25 production. Activation of this tuft cell-ILC2 module is not limited to helminth defense. Subsequent reports have described that tuft cell sensing of the protist-derived fermentative end-product, succinate, induces tuft cell hyperplasia and potentiates type 2 immune responses against protist pathosymbionts.[517,518]

ILC2 and Allergic Pathology

While type 2 immunity protects the host from parasitic infection, excessive or inappropriate responses result in allergic diseases. Increased ILC2 activity has been implicated in a host of human allergic pathologies involving the respiratory tract and skin. ILC2 were first observed to accumulate in the nasal mucosa of patients with chronic rhinosinusitis, a disease characterized by high IgE titers and eosinophilia.[486] Similarly, patients with severe asthma and chronic airway eosinophilia had significantly more ILC2 in blood and sputum samples, and ILC2 produced more IL-5 and IL-13.[519] Mouse models of allergic lung inflammation and airway hyper-reactivity have been a valuable resource in investigating the role of ILC2 in these processes. ILC2-deficient mice were shown to have impaired Th2 differentiation and reduced lung pathology in a model of protease allergen-induced airway inflammation.[31] Additionally, a model of influenza-induced airway hyper-reactivity demonstrated a significant contribution of ILC2-derived IL-13 for development of airway resistance, indicating a pathological role of ILC2.[520] However, later in the course of lung inflammation, ILC2 appear to play a beneficial role in mediating tissue repair through secretion of amphiregulin.[490] ILC2 also participate in cutaneous allergic pathology. Dermal ILC2 were identified in human skin by a Lin^- $CD25^+$ $IL\text{-}33R^+$ phenotype and were found to be enriched in skin lesions from atopic dermatitis (AD) patients compared with healthy control skin.[521,522] Similarly, mouse dermal ILC2 with the same phenotype expanded in a MC903-induced model of skin inflammation, concurrent with the observation of increased tissue-derived cytokines IL-25, IL-33, and TSLP in human AD lesions.[522] Studies using ILC2-deficient and ILC2-depleted mice demonstrated that ILC2 played a significant and nonredundant role in the development of skin inflammation.[521,522]

ILC2 and Metabolic Homeostasis

While a hallmark of type 2 immunity is defense against extracellular parasites, type 2 cytokines have been found to play important roles in maintaining metabolic homeostasis. ILC2 are present in visceral adipose tissue (VAT), where they produce IL-5 and IL-13 to sustain VAT-associated eosinophils and AAMs.[523] Eosinophils and AAMs have been demonstrated to mediate the development of beige fat, providing resistance against high-fat–induced obesity and metabolic dysregulation.[524,525] Concordantly, administration of IL-25 or IL-33 activates VAT ILC2 production of IL-13 and reduces weight gain and glucose intolerance in a model of diet-induced obesity.[526,527] ILC2 have also been shown to act independently of eosinophils and AAMs through secretion of IL-13 or methionine-enkephalin peptides that directly interact with adipocytes to promote beiging.[527,528] ILC2 have been identified in human subcutaneous white adipose tissue (WAT) and are depleted in WAT from obese donors relative to nonobese controls.[528] In addition to adipose tissue, ILC2 have also been identified in the pancreas, where they indirectly promote insulin secretion in an IL-33–dependent manner.[529] These findings highlight an unexpected role for ILC2 in the regulation of metabolic regulation, independent of pathogen recognition.

ILC3

When they were originally identified, ILC3 were identified based on the expression of NKp46 (in mouse) and NKp44 (in human).[71-74,453] ILC3 were further distinguished in mouse and human by restricted expression of the transcription factor RORγt and production of the cytokine IL-22. Unlike NK cells, ILC3 produced comparatively little IFNγ and had no cytotoxic potential. Despite the expression of NK cell receptors on subsets of ILC3, developmental studies have established ILC3 as a distinct lineage of ILC.[530-532]

Phenotypic Characterization, Effector Function, and Regulation

In mouse and human, ILC3 are identified as Lin^- $ROR\gamma t^+$ lymphocytes. They are distinguished from fetal LTi cells in that LTi are only found during embryogenesis, though LTi-like subsets of ILC3 may also contribute to formation of lymphoid tissue.[533,534] Compared to other ILC, ILC3 are especially enriched in the gastrointestinal tract, where they are the most abundant innate lymphocyte population in the small intestine.[38] ILC3 are relatively long-lived with a half-life of 22 to 26 days. The lack of ILC3 chimerism in parabiotic studies suggests that ILC3 may be replenished by tissue-resident progenitors.[38,452] In the steady-state intestine, ILC3 are a major source of IL-22 which binds to the heterodimeric receptor IL22RA1/IL10RB (IL-22R) expressed by non-hematopoietic cells, primarily intestinal epithelial cells.[535] IL-22 signaling maintains intestinal barrier function through multiple mechanisms: (1) production of antimicrobial peptides (AMPs) (ie, RegIIIβ, RegIIIγ, S100 proteins); (2) expression of tight junctions, such as claudin-2; and (3) fucosylation and glycosylation of the intestinal mucosa, which supports symbiotic intestinal commensals.[536-539] When the intestinal epithelium is damaged, IL-22 supports intestinal epithelial proliferation and survival during the wound healing process.[540,541] A fraction of ILC3 produce IL-17, which also acts on epithelial cells to mediate the recruitment of myeloid cells, and this function may be important for mediating the clearance of fungal pathogens.[542] Like IL-22, IL-17 also promotes the production of AMPs and synthesis of tight junctions.[543] Lastly, ILC3 have been observed to produce granulocyte-macrophage colony-stimulating factor (GM-CSF) in response to IL-1β from myeloid cells. ILC3 production of GM-CSF has been proposed to exacerbate autoimmune colitis through the enhanced recruitment of inflammatory myeloid cells.[544-546] Additionally, ILC3-derived GM-CSF maintains macrophage and dendritic cells in the intestine and facilitates their capacity to induce tolerogenic T_{reg} homeostasis in the large intestine.[547] Compared to ILC2 which are primarily regulated by epithelial-derived cytokines, ILC3 responses are classically regulated by IL-23 and IL1β, which are derived from hematopoietic cells (Table 18.4). These cytokines are mainly produced by myeloid cells, and Myd88-dependent TLR signaling has been shown to be a key pathway regulating their expression.[547,548] In addition to these canonical ILC3 regulatory cytokines, recent work has uncovered a role of TL1A and IL-2 in augmenting ILC3 production of IL-22.[549,550]

Human ILC3 were first identified in the tonsil and intestine by a $CD3^-CD56^+NKp44^+$ phenotype.[71] *In vitro*, human ILC3 uniquely produce IL-22 in response to IL-23, IL-1β, IL-2, IL-7, or TLR-activated monocytes. This response can be enhanced by TLR2 ligands or engagement of NKp44.[551,552] In addition to IL-22, ILC3 can produce IL-26, leukemia inhibitory factor (LIF), GM-CSF, B-cell activating factor (BAFF), and TNFα.[71,552,553] A second subset of ILC3 was also identified in human intestine with a Lin^- $NKp44^-$ c-Kit^+ $CD127^+$ phenotype.[453,551] Unlike $NKp44^+$ ILC3, this population can produce IL-17A/F. This subset is proposed to be a more immature subset of ILC3 due to its ability to differentiate into $NKp44^+$ ILC3 in culture.[453,551] Both populations of human ILC3 are proposed to be capable of interconversion to ILC1.[462,464,465]

Functional Heterogeneity of ILC3 in Intestinal Immunity

In mice, ILC3 can be divided into two distinct populations by expression of the NK cell receptor NKp46 ($NKp46^+$ ILC3 or NCR^+ ILC3) or the chemokine receptor CCR6 (LTi-like ILC3). While currently considered as part of the same family, $NKp46^+$ ILC3 and LTi-like ILC3 develop from distinct precursors.[22] Concordantly, they exhibit differences in effector functions, localization, and regulation. $NKp46^+$ ILC3 predominately produce IL-22 and are incapable of producing IL-17A/F, whereas LTi-like ILC3 can produce both IL-22 and IL-17A/F in addition to lymphotoxins.[72,554] $NKp46^+$ ILC3 are abundantly distributed throughout the small intestinal lamina propria but are markedly reduced in the colon and mesenteric LN.[555] In the small intestine, $NKp46^+$ ILC3 preferentially localize in the villi, close to the intestinal epithelium, in a CXCR6-dependent manner.[556] Interestingly, $NKp46^+$ ILC3 also express the ILC1/Th1 transcription factor T-bet, which is required for their development.[557-559] It has been proposed that graded acquisition of T-bet expression is necessary for the development of $NKp46^+$ ILC3 from $NKp46^-$ $CCR6^-$ ILC3 (DN ILC3) *in vitro* and *in vivo*.[558,559] $Tbet^+$ $NKp46^+$ ILC3 are able to produce limited amounts of IFNγ.[559] The expression T-bet among ILC3 is regulated by the presence of microbiota and IL-23. Fate-mapping studies using the *Rorc* locus have determined that ~30% of intestinal ILC1 ($Lin^-NKp46^+NK1.1^+ROR\gamma t^-Eomes^-$) are derived from $ROR\gamma t^+$ ILC3.[20,532] These "ex-ILC3" were demonstrated to be a source of IFNγ during infection with *Salmonella typhimurium*.[559]

Adult LTi-like ILC3 are the most frequent ILC3 subset in the colon and mesenteric LN.[555] In the intestine, $CCR6^+$ ILC3 are enriched in SILT near the base of ISC-containing crypts.[560] This positioning may allow for the efficient delivery of IL-22 to ISCs in order to facilitate epithelial regeneration and prevent genotoxic stress.[541,561,562] Like $NKp46^+$ ILC3, LTi-like ILC3 can participate directly in host defense against bacterial infection through secretion of IL-22.[563] Additionally, LTi-like ILC3 in PP and ILF produce lymphotoxins which are critical for the induction of IgA production by intestinal plasma cells through T cell–independent and –dependent mechanisms.[564-566] Aside from regulating other cells, LTi-like ILC3 have also been shown to regulate each other. Like fetal LTi, adult LTi-like ILC3 express receptor activator of nuclear factor kappa (RANK) and RANK ligand (RANKL). These molecules mediate ILC3-intrinsic quorum sensing, whereby engagement of RANK-RANKL on neighboring ILC3 limits their proliferation and activation.[567] IL-22 reporter mice have demonstrated differential expression of IL-22 between the two ILC3 subsets.[535] In healthy adult mouse intestine, ILC3 constitutively produced IL-22 and the most active IL-22 producers were LTi-like ILC3 concentrated in SILT. One possible explanation behind the high localization of IL-22 production in SILT is that lymphotoxin β from LTi-like ILC3 can stimulate IL-23 production from surrounding $LT\beta R^+$ DC.[568] While LTi-like ILC3 are present in SILT from birth, they are relatively quiescent until weaning, which significantly augments IL-22 and IL-17 production. This observation indicates that microbial colonization is necessary for constitutive ILC3 activity; accordingly, germ-free conditions or antibiotic treatment significantly ablate IL-22 production.[72,535] However, the relative abundances of ILC3 subsets and formation of CP appear to be programmed intrinsically and are not affected by microbiota depletion.[444,452,535,569] Thus, SILT may serve as discrete niches where LTi-like ILC3 cytokine production is amplified by lymphotoxin signaling and extrinsic microbial cues, but total cellularity and lymphoid structure is regulated via quorum-sensing mechanisms (Fig. 18.6).

Metabolic Regulation of ILC3

From their initial characterization, ILC3 were observed to express the aryl hydrocarbon receptor (AHR).[71] AHR is a ligand-dependent transcription factor belonging to the Per-Arnt-Sim (PAS) superfamily.[570,571] AHR is usually sequestered in the cytoplasm but ligand binding results in its nuclear translocation, heterodimerization with the AHR-nuclear translocator (ARNT), and binding to dioxin-responsive elements

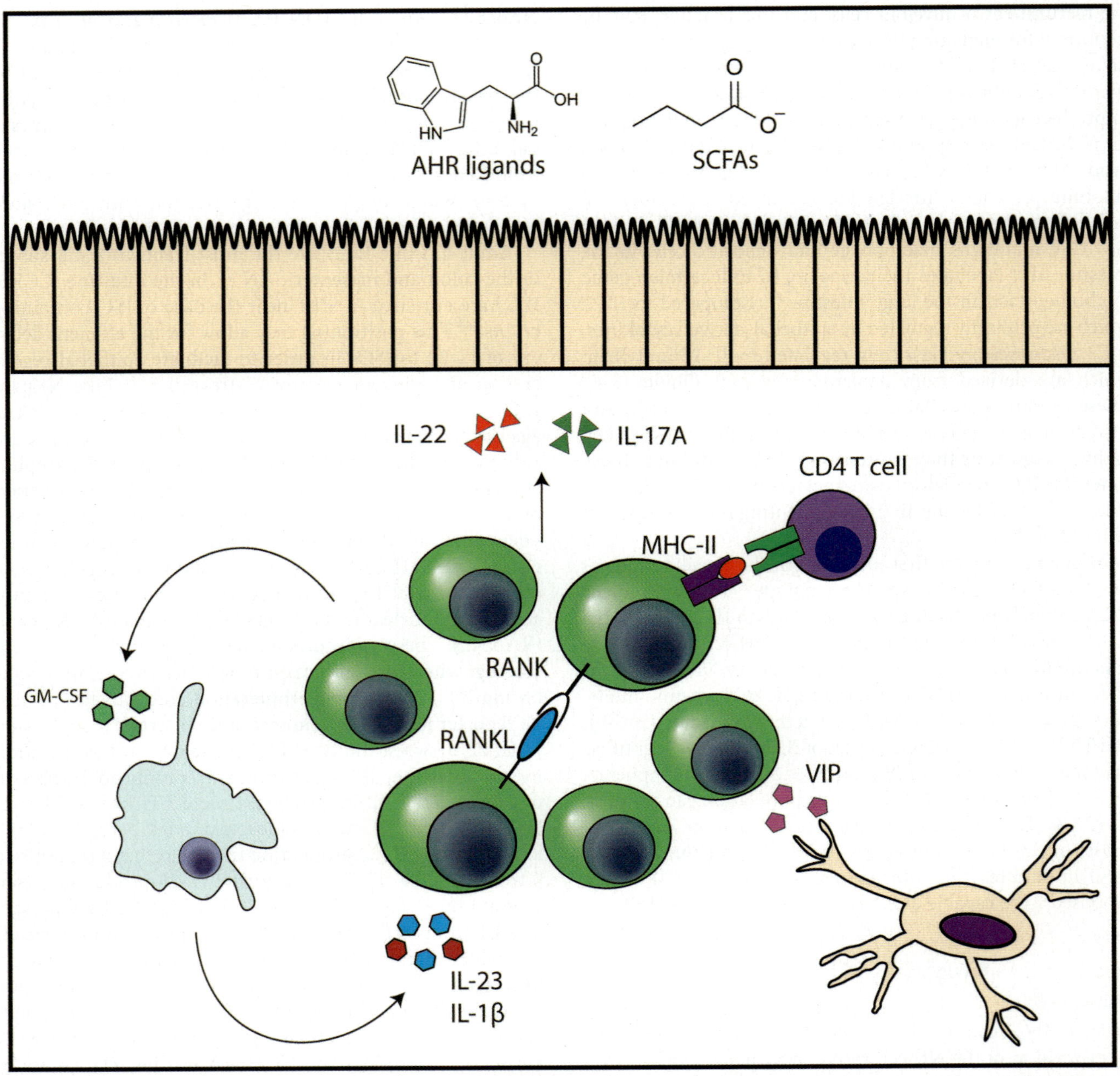

FIG. 18.6. Innate lymphoid cell (ILC3) integrate immunologic, dietary, microbial, and neuronal signals to coordinate intestinal immunity. ILC3 can be located in discrete lymphoid clusters throughout the lamina propria. ILC3 communicate with surrounding myeloid cells via lymphotoxin and granulocyte-macrophage colony-stimulating factor (GM-CSF) signaling. In response, myeloid cells enhance ILC3 activity through the secretion of IL-23 and IL-1β. ILC3 can also modulate T cell responses through major histocompatibility complex (MHC)-II. In addition to regulation by myeloid-derived cytokines, ILC3 activity can be modulated by neuronal signals, such as VIP. Dietary and microbial signals from the intestinal lumen can directly influence ILC3 differentiation and homeostasis. Lastly, ILC3 regulate each other through a quorum sensing mechanism by RANK-RANKL signaling.

(DRE) of various genes to initiate transcription. The ligands of AHR encompass a wide variety of compounds, including environmental pollutants, such as 2,3,7,8-Tetrachlorodiben zodioxin (TCDD), and products of tryptophan metabolism. The latter are abundantly derived from dietary components and some commensal bacteria, such as *Lactobacillus reuteri*.[572] AHR-deficient mice were found to contain a marked reduction of ILC3 and SILT.[533,573,574] Feeding colonies of mice with a diet deprived of plant-derived AHR ligands also reduced the number of ILC3 and CP. IL-22 transcription is induced by dual binding of AHR and RORγT, thus AHR-deficient ILC3 also produce less IL-22.[573,574] In human, silencing of AHR in ILC3 was shown to induce conversion into conventional CD56^{bright} NK cells.[575] AHR thus plays a central role in maintaining lineage fidelity and mediating the hallmark features of ILC3 (Fig. 18.6).

In addition to AHR, ILC3 and LTi express retinoic acid receptor alpha (RARα).[443,576] RARα is part of the steroid/thyroid/retinoid nuclear receptor family and binds to retinoic acid response elements (RARE).[577] Normally, retinoic acid receptors are localized in the nucleus and bind RAREs in complex with various corepressors. Upon binding retinoic acids, they undergo conformational change to dissociate from corepressors and bind transcriptional coactivators to initiate transcription. Retinoic acids are primarily derived from Vitamin A. Retinoic acids regulate LTi activity indirectly by affecting migration to SLO anlagen or directly by stabilizing RORγT expression.[433,443] Maternal vitamin A deficiency, genetic ablation of RARα, or pharmacological inhibition of retinoic acid receptor signaling during fetal development, reduced LTi and decreased the size of SLO.[443] Retinoids have also been found to positively regulate adult ILC3.[576,578] Depletion of retinoid signaling drastically reduced the number of ILC3 and their ability to produce IL-22. Conversely, supplementation with retinoic acid could induce ILC3 expression of RORγT, boost production of IL-22, and enhance proliferation. Similar to AHR ligands, retinoic acids have an inhibitory effect on ILC2, and depletion of vitamin A markedly promotes ILC2 proliferation and function, at the cost of ILC3.[578] Together, AHR and retinoic acid work to skew intestinal immunity toward more type 17 responses at the cost of dampening type 2 responses. It remains unknown if there is a physiological reason for why these ligands in particular trigger this immune switch.

The gastrointestinal tract is the primary site of lipid absorption. While lipids can influence the nutritional status of the host, bacteria-derived lipids also have profound roles in modulating immune responses. As opposed to the nuclear receptors described previously, lipid sensors expressed by ILC3 are G protein-coupled receptors. LTi-like ILC3 express GPR183, which binds the oxysterol 7α,25-OHC, which is derived from cholesterol. In the colon, fibroblastic stromal cells were found to produce 7α,25-OHC which facilitated the recruitment and organization of LTi-like ILC3.[579] Concordantly, GPR183-deficient mice lacked CP or ILF and produced less IL-22 secondary to the loss of these structures. There are conflicting reports about whether LTi-like ILC3 in the small intestine are also affected.[580] In addition to sensing oxysterols, ILC3 can also respond to short-chain fatty acids (SCFAs) through expression of free fatty acid receptor 2 (FFAR2).[581,582] SCFAs are produced by microbiota through the fermentation of dietary fiber. SCFAs can enhance ILC3 production of IL-22 when administered together with IL-23 and promote ILC3 proliferation in the tissue. Loss of FFAR2 led to selective defects in LTi-like ILC3 and reduction in IL-22 production by this subset. These data suggest that SCFAs are potent stimulatory ligands for all ILC3 but are primarily utilized by LTi-like ILC3 during steady-state (Fig. 18.6).

ILC3 in Intestinal Defense and Autoimmunity

Due to their localization to the GI tract and potent secretion of IL-22, ILC3 have been proposed to play crucial roles in maintaining the intestinal barrier against foreign pathogens. In mice, this role has primarily been interrogated using the model enteropathogen *Citrobacter rodentium*, due to its established sensitivity to IL-22–mediated immunity.[537]Both IL-22 and ILC3 have been shown to be necessary for proper defense against *Citrobacter rodentium*.[72,537] Additionally, depletion of ILC3 in Rag1-deficient mice results in peripheral dissemination of commensal bacteria and systemic inflammation.[583] However, the physiological relevance of NKp46$^+$ ILC3 is still a matter of active research. Tbet$^{-/-}$ mice, which lack NK cells, ILC1, and NKp46$^+$ ILC3, are susceptible to *Citrobacter rodentium* and the intracellular pathogen *Salmonella typhimurium*.[72,558,559] Subsequent studies have demonstrated that NKp46$^+$ ILC3 are redundant in defense against *Citrobacter rodentium*, whereas LTi-like ILC3 confer nonredundant advantages in host immunity.[544,584] ILC3 have also been implicated in immunity against rotavirus, *Clostridium difficile*, and *Candida albicans*.[15,476,542,582,585,586] Sensing of rotaviral infection by myeloid cells leads to the activation of ILC3 and production of IL-22. IL-22 was found to facilitate antiviral immunity and viral control through synergism with IFNλ.[585] *C. difficile* is a gram-positive, spore-forming anaerobe that causes colitis through the production of exotoxins that damage the colonic epithelium. ILC3-derived IL-22 is important in protection from *C. difficile* infection as neutralization of IL-22 or depletion of ILC3 exacerbates infection. *C. albicans* is a highly prevalent cause of yeast infections at multiple mucosal sites. In mouse models of oropharyngeal candidiasis, IL-17–producing ILC3 were found to play a significant role in fungal control.[542] LTi-like ILC3 also have a firmly established role in the induction of IgA responses that shape the commensal microbiome, but its relevance in disease has not yet been elucidated.[565,566,587]

When properly regulated, intestinal immunity maintains homeostasis and safeguards the health of the host; however, dysregulated immune responses drive intestinal pathology and impair host fitness. Intestinal autoimmunity in humans most commonly appears in the form of inflammatory bowel diseases (IBD). Although the drivers of IBD are multifaceted, genome-wide association studies have identified networks of genes that predispose individuals to IBD.[588] Many implicated genes are involved in ILC3-associated functions (ie, *RORC, NFIL3, IL22, IL23R*), and functional studies have implicated IL-23 signaling as a key driver in IBD pathogenesis.[589] The role of ILC3 in this paradigm is an area of active research. On one hand, ILC3 produce IL-22 which is generally considered to be important for epithelial healing and maintenance of intestinal homeostasis. On the other, ILC3 also produce IL-17A/F, GM-CSF, and IFNγ which have been shown to be potentially colitogenic in animal models.[545] IBD patient studies demonstrate a correlation between intestinal inflammation and reduced ILC3 frequency.[462,464,590,591] Furthermore, increased ratios of ILC1:ILC3 have been observed in inflamed human intestinal specimens.[458,462,590] One potential explanation for these observations is that intestinal inflammation is associated with conversion of ILC3 to ILC1. Thus, factors that promote ILC3 → ILC1 conversion may be attractive therapeutic targets for IBD.

EMERGING FIELDS IN ILC BIOLOGY

Neuroimmune Interactions

Recent studies have highlighted a role for neuronal signals in the regulation of ILC responses. Unlike naïve T cells, ILC are tissue-resident and often inhabit microniches innervated by peripheral nerves.[592] ILC have been primarily observed to be regulated by the autonomic nervous system, which contains sympathetic, parasympathetic, and enteric branches, and the sensory nervous system. Neuronal regulation of the immune system begins early in embryogenesis. The formation of PP coincides with the development of the enteric nervous system (ENS) when neural crest cells migrate to the developing intestine. The tyrosine kinase receptor RET is necessary for proper development of the ENS. RET is regulated by binding of its coreceptor, GFRα family members, to stromal-derived ligands and subsequent clustering of RET molecules. RET is also expressed by $CD45^+$ $CD4^-$ $CD3^-$ IL7Rα^- c-Kit^+ $CD11c^+$ hematopoietic cells that aggregate into the primordia of PP.[593] These $CD11c^+$ cells express lymphotoxin, which activate surrounding stromal organizer cells. Upon activation, stromal organizer cells express chemokines that attract LTi to developing PP. Consequently, deletion of RET or one of its coreceptors reduces the number of PP.[593] In adult mice, RET is expressed on mature ILC3, where it enhances IL-22 production.[594] Imaging of intestinal tissues revealed direct interaction between enteric glial cells and RET^+ ILC3 in CP. Glial cells sense microbial patterns through MYD88-dependent pathways to produce glial cell line–derived neurotrophic factor (GDNF) which elicits ILC3-derived IL-22 *in trans*. Thus, development of the ENS and ILC3 function rely on common receptor-ligand pairs.

One major feature of the sympathetic nervous system is the release of catecholamines which bind various adrenergic receptors to produce rapid and potent physiological changes at target sites throughout the body. Among ILC, NK cells and ILC2 also express adrenergic receptors. Catecholamines can increase circulating NK cells during acute stimulation, but chronic stimulation suppresses NK circulation and activity[595,596] ILC2 express β_2-adrenergic receptor (β_2AR), a Gα_s-protein–coupled receptor, in lung and intestine.[597] ILC2 in the intestine colocalize with adrenergic nerve fibers, and stimulation with β_2AR agonists inhibited ILC2 proliferation and production of IL-5 and IL-13 in intestine and lung.[597] Mice with genetic deficiency of β_2AR have hyperactive ILC2 and enhanced immunity against helminth infection.

In addition to catecholamine receptors, molecular profiling of ILC2 has also revealed the enrichment of several neuropeptide receptors, such as neuromedin U receptor (NMUR), calcitonin gene-related peptide (CGRP) receptor, and vasoactive intestinal peptide (VIP) receptor. Several groups reported the expression of NMUR, a Gα_q-coupled receptor, by intestinal and lung-resident ILC2.[598-600] In the lung, ILC2 were shown to be closely associated with sensory afferent nerves and the thoracic dorsal root ganglion, which houses the cell bodies of these nerves, expressed *Nmu* relative to the other ganglia.[598,599] In the intestine, $NMUR1^+$ ILC2 associate with NMU-expressing cholinergic neurons in the submucosa.[599,600] The immunomodulatory effect of NMU appears to differ between lung and intestinal ILC2. When administered alone, NMU does not stimulate production of IL-5 or IL-13 by lung ILC2 but can enhance IL-5 and IL-13 production driven by IL-25 or IL-33.[598,601] In the intestine, NMU appears to be capable of stimulating ILC2 proliferation and production of Il-5 and IL-13 by itself.[599,600] NMU is upregulated following worm infection and mice lacking NMUR1 were less protected against infection with *N. brasiliensis*.[599,600] The CGRP receptor complex is composed of two components: the Gα_s-coupled calcitonin receptor-like (CALCRL) and receptor activity-modifying protein 1 (RAMP1). Interestingly, both the components of the CGRP receptor and CGRP itself are expressed by ILC2.[601-603] In the lung, CGRP is secreted by rare pulmonary neuroendocrine cells which are innervated by sensory nerve fibers.[604] Costimulation of IL-33–treated ILC2 with CGRP augmented IL-5 production but inhibited IL-13 production.[601,604] Furthermore, CGRP potently inhibited cytokine-induced ILC2 proliferation, so total IL-5 production was reduced. CGRP also antagonized the costimulatory effects of NMU in lung ILC2. In the intestine, CGRP was expressed by cholinergic enteric neurons. Like lung ILC2, intestinal ILC2 production of IL-5 is enhanced by short-term stimulation with CGRP while IL-13 is inhibited, and their proliferation is profoundly inhibited as well.[602] Studies using a CGRP reporter mouse also showed that ILC2 were the dominant hematopoietic sources of CGRP in the lung and CGRP stimulated the production of CGRP from ILC2, indicating a potent role for feed-forward amplification of CGRP inhibitory signals. Additionally, NMU specifically increases the expression of CGRP receptor components in ILC2. Collectively, these observations show that ILC2 are under dynamic control by two competing neuronal impulses mediated by CGRP and NMU.

Circadian rhythms are significant to the physiology of the gastrointestinal tract, where they have been associated with feeding behavior, digestion, nutrient absorption, and gut motility. They are endogenously generated and are regulated by external rhythmically occurring cues but can also persist independently in static conditions. In mammals, there are two main clock mechanisms; the central clock, which is located in the suprachiasmatic nucleus (SCN) of the hypothalamus, and the peripheral clocks, which are thought to be present in most cells of the body. At the cellular level, the core mechanism for circadian oscillation involves transcriptional and translational feedback loops between a dedicated set of circadian clock genes. It was noted that intestinal ILC2 and, to an even greater extent, ILC3 had diurnal expression of key clock genes (including *Bmal1, Clock, Nfil3*, *Dbp* and *Per, Cry* and *Nr1d* family genes) with more readily apparent rhythmicity than did their progenitors, including the CLP and ILCP. These observations suggested that this upregulation of the circadian transcriptional program may be linked to the unique development and function of these cell types. By contrast, clock gene oscillations in intestinal ILC1 were less evident, with only the expression of *Per1* and *Cry2* reported to be statistically significant.[605] In addition to clock genes, rhythmic expression of a number of genes involved in diverse functional and signaling programs was evident in both ILC2 and ILC3.[606] A key, unifying aspect was the rhythmic production of cytokines. Under both constitutive and *ex vivo* stimulated conditions, IL-5 secretion from gut and lung ILC2 fluctuated over a 24-hour cycle.[606] ILC3 cytokine production also followed a rhythmic pattern. In purified ILC3, cyclic expression of *Il22* mRNA was detected *ex vivo*.[607] Similarly, studies using IL-22 reporter mice and intracellular cytokine staining indicated that IL-22 protein production in ILC3, both constitutively and following cell stimulation, followed a diurnal pattern.[591,606] Furthermore, *Il17a* expression also appeared to be rhythmic though cycling of IL-17 protein has not been conclusively defined.[591,607] This may be the result of poor temporal resolution compounded by the relatively lower levels of IL-17A secretion shown by ILC3.

The significance of clock genes in ILC development and function were interrogated in several studies. Genetic deletion of BMAL1 largely ablates circadian rhythm. Mice with genetic deletion of *Bmal1* in ILC3 had a large reduction in intestinal ILC3. Two mechanisms have been proposed to explain this reduction. ILC3 appear to rhythmically express the gut-homing chemokine receptor CCR9. Ablation of BMAL1 reduces the expression of CCR9 on ILC3 as well as a commensurate increase of ILC3 in mesenteric LN.[605] Thus, the reduction of intestinal ILC3 in BMAL1-deficient mice may be attributed to disrupted trafficking of ILC3 from mesenteric LN to intestinal lamina propria. A separate study reported that deficiency of BMAL1 resulted in hyperactivation of RORγT and increased apoptosis in ILC3.[591] BMAL1 deficiency abolished expression of the transcriptional repressors REV-ERBα/β, which act as competitive inhibitors of RORγT binding. Hyperactive RORγT binding was correlated with increased expression of proapoptotic proteins in ILC3s, including BCL2-interacting mediator of cell death (BIM), and blocking of RORγT with a small molecule inhibitor reduced expression of BIM. In addition to studies of BMAL1-deficient mice, mice lacking REV-ERBα were shown to have a dramatic cell-intrinsic reduction of NKp46$^+$ ILC3[607]. A significant accumulation of DN ILC3s, the direct progenitors of NKp46$^+$ ILC3, accompanied this reduction. Furthermore, the remaining NKp46$^+$ ILC3 expressed significantly reduced levels of RORγt due to increased inhibition by NFIL3.[607] It has been proposed that that due to REV-ERBα deficiency during NKp46$^+$ ILC3 development, these cells failed to maintain expression of RORγt, and hence halted development and accumulated as DN ILC3.[607] Collectively, these reports illustrate that core circadian clock genes play a central role in the differentiation and function of ILC3.

An important time-setting cue for the digestive system is the time of feeding. VIP is the major neuropeptide involved in synchronizing the circadian neuron network in the SCN.[608] It is also expressed by enteric neurons after food intake and regulates multiple aspects of intestinal physiology, including smooth muscle tone, epithelial cell secretion, and blood flow.[606,609] VIP signals through $G\alpha_s$ protein-coupled receptors $VPAC_1$ and $VPAC_2$, which are encoded by *Vipr1* and *Vipr2*, respectively. ILC2 in the lung and intestine express both *Vipr1* and *Vipr2*. Studies using IL-5 and Il-13 reporter mice demonstrated that fasting reduces IL-5 and IL-13 expression by intestinal ILC2.[485,606] Short-term culture of ILC2s in IL-7 with either VIP or a $VPAC_2$ agonist slightly increased IL-5 in the associated supernatant.[485] However, it is unknown whether VIP may prove to be inhibitory during longer treatments as was shown with CGRP, which acts on a similar signaling pathway. The significance of VIP signaling in ILC2 remains unclear. Circadian production of IL-5 in the serum has been proposed to regulate the circadian cycling of blood eosinophils.[485] Furthermore, release of VIP by lung nociceptors has been shown to promote allergic airway inflammation, though these effects are mediated through CD4$^+$ T cells as well.[610]

In addition to ILC2, CCR6$^+$ ILC3 were also reported to express $VPAC_2$ and were found situated adjacent to VIP-secreting neurons.[606,611] There are mixed reports concerning the effects of VIP on ILC3 function. One study found that 16-hour fasting in mice reduced IL-22 production by intestinal ILC3.[606] Furthermore, *in vivo* and *in vitro* treatment of bulk ILC3 with VIP or $VPAC_2$ agonists stimulated IL-22 secretion from ILC3 in a dose-dependent manner.[606] However, a separate study found the opposite, that VIP inhibited IL-22 and IL-17 secretion from CCR6$^+$ ILC3.[611] In a 6-hour fasting model, CCR6$^+$ ILC3 from fasted mice secreted higher levels of IL-22. *In vivo* activation of VIPergic neurons and *in vitro* treatment of purified CCR6$^+$ ILC3 with VIP corroborated the inhibitory effect of VIP.[611] Both studies agreed that these effects were mediated by $VPAC_2$ expression on CCR6$^+$ ILC3, as both constitutive and ILC3-specific *Vipr2* deletion abrogated the response to food-intake and VIP treatment. The conflicting findings may be a result of differences in fasting methods and *in vitro* treatment conditions. Although the two reports differed considerably with regard to the effects of VIP, both demonstrated that genetic deletion of *Vipr2* in ILC3 abrogated the cytokine response after food intake, resulting in abnormal homeostatic and inflammatory responses.[606,611]

ILC Plasticity

ILC were originally categorized by the expression of lineage-determining transcription factors and production of hallmark cytokines. These classifications were made with the assumption that ILC identity was immutable. However, recent studies have challenged this concept. Mature ILC are now recognized to have plasticity, or the capacity to transdifferentiate to other ILC types in response to changing environmental signals.[612] As ILC are resident cells, plasticity may allow a given population of ILC to tailor the immune response to the nature of the threat without having to rely on *de novo* differentiation of naïve cells recruited from circulation. The most well-established incidence of plasticity is the ability of ILC3 to transdifferentiate into ILC1. As previously discussed, a subset of ILC3 in mice and humans express the NK cell receptor NKp46 and NKp44, respectively. Murine NKp46$^+$ ILC3 coexpress T-bet and RORγT and possess the ability to produce both IL-22 and IFNγ.[532,559] These cells can transdifferentiate into RORγT⁻ ILC1 and lose the ability to produce IL-22.[559] Murine ILC3→ILC1 conversion is restrained by the transcription factor c-Maf.[613-615] Human NKp44⁺ ILC3, in contrast, do not coexpress T-bet at steady-state but can convert to CD103⁺ ILC1.[553] Single-cell analyses have shown that NKp44⁺ ILC3 and CD103⁺ ILC1 exist on opposite sides of a spectrum with multiple distinct intermediates that express both ILC3 and ILC1 features.[465] As shown by overexpression and pharmacological inhibition studies, NKp44⁺ ILC3→ILC1 conversion is mediated by the coordinated expression of T-bet and Aiolos.[465,616] Another population of CD117⁺ NKp44⁻ ILC3 have also been described to be plastic toward CD127⁺ NKp44⁻ ILC1.[462] ILC3→ILC1 conversion is mediated by IL-12 and IL-18. In addition, it has been reported that ILC1→ILC3 conversion can occur in culture with IL-23, IL-1β, and retinoic acid.[464]

While there is no evidence that Th2 cells can convert to Th17 cells, ILC2 have been observed to produce IL-17 in certain situations. In steady-state mice, lung ILC2 usually express IL-33 receptor (ST2) and do not express KLRG1, whereas intestinal ILC2 express IL-25 receptor (IL-17RB) and KLRG1. In mice injected with IL-25 or infected with helminths, KLRG1^{hi} ILC2 were observed to accumulate in the lung.[617] These "inflammatory ILC2" (iILC2) responded to IL-25 to produce IL-5, IL-13, and modest amounts of IL-17. Concordantly, iILC2 expressed low amounts of RORγT. These inflammatory features were enhanced by culture in Th17 polarizing conditions or Notch ligands.[617,618] Subsequent work has proposed that iILC2 are originally derived from the intestine and are mobilized by IL-25 during helminth infection, representing a significant break from the tissue-resident paradigm of ILC.[619] The generation of iILC2 is dependent on expression of the AP-1 transcription factor BATF, which plays an essential role in Th17 differentiation.[620,621] Human ILC2 have also been found to be highly plastic. Two subsets of CRTH2⁺ ILC2 have been identified in peripheral blood based on differential expression of CD117.[486,622,623] CD117⁺ ILC2 produced IL-17 in response to IL-23 and IL-β, whereas CD117⁻ ILC2 additionally required TGFβ.[623,624] Whether these findings constitute true ILC2→ILC3 plasticity is under debate, as there has not been evidence of ILC2 making IL-22 the signature cytokine of ILC3. In addition to IL-17, ILC2 have also been shown to produce IFNγ in mice and human.[625-628] Human ILC2 production of IFNγ is dependent on both IL-12 and IL-1β, as IL-1β is required for induction of the IL-12 receptor. Under these conditions, ILC2 lose the ability to produce IL-5 but still retain IL-13 production. IL-4 is able to reverse ILC2→ILC1 transition but is unable to induce plasticity toward ILC2 fate in any other ILC subsets. ILC2→ILC1 plasticity may be important in the context of chronic inflammatory lung diseases, such as COPD which have an enrichment of ILC1.[625,627] Overall, ILC2 exhibit some plasticity with other ILC types as well as the ability to enter recirculate from tissue. Human ILC2 are especially fluid; this may be due to the fact that ILC2 share many features with systemic circulating ILCP, including surface markers with the exception of CRTH2, enrichment in peripheral blood, and the requirement of IL-1β to differentiation events.[47] The discovery of multipotential ILC plasticity was an unanticipated and exciting addition to understanding of ILC biology. Further research is warranted to elucidate the molecular factors governing fate transitions and the relevance of these in disease.

CONCLUSION

Despite their overlapping functional programs with adaptive immunity, decades of research have defined unanticipated and unique roles for ILC in the immune system. Studies of selective NK deficiencies have shown a nonredundant role for ILC in primary responses against pathogens that otherwise subvert adaptive immunity. LTi play a unique and necessary role in establishing the framework of lymphoid tissues which organize and amplify immune responses. Entrenched in the tissue, ILC uniquely integrate the panoply of neuronal signals, inflammatory mediators, pathogen-associated patterns, and tissue-derived factors to orchestrate tissue immunity and steady-state homeostatic processes. ILC display unique epigenetic profiles that not only allow rapid deployment of effector functions but also remarkable plasticity. Understanding the diversity of ILCs and their unique functions will provide insights into the mechanisms of immune-mediated diseases and may designate the next generation of targets for therapeutic intervention.

ACKNOWLEDGMENTS

The authors thank Drs. Jennifer K. Bando, Marina Cella, and Qianli Wang for their review of the manuscript and thoughtful suggestions.

REFERENCES

1. Vivier E, Artis D, Colonna M, et al. Innate lymphoid cells: 10 years on. *Cell*. 2018;174:1054-1066.
2. Zook EC, Kee BL. Development of innate lymphoid cells. *Nat Immunol*. 2016;17:775-782.
3. Serafini N, Vosshenrich CAJ, Di Santo JP. Transcriptional regulation of innate lymphoid cell fate. *Nat Rev Immunol*. 2015;15:415-428.
4. Scoville SD, Freud AG, Caligiuri MA. Cellular pathways in the development of human and murine innate lymphoid cells. *Curr Opin Immunol*. 2019;56:100-106.
5. Yu X, Wang Y, Deng M, et al. The basic leucine zipper transcription factor NFIL3 directs the development of a common innate lymphoid cell precursor. *eLife*. 2014;3:e04406.
6. Possot C, Schmutz S, Chea S, et al. Notch signaling is necessary for adult, but not fetal, development of RORγt+ innate lymphoid cells. *Nat Immunol*. 2011;12:949-958.
7. Yoshida H, Kawamoto H, Santee SM, et al. Expression of $\alpha_4\beta_7$ integrin defines a distinct pathway of lymphoid progenitors committed to T cells, fetal intestinal lymphotoxin producer, NK, and dendritic cells. *J Immunol*. 2001;167:2511-2521.
8. Mielke LA, Groom JR, Rankin LC, et al. TCF-1 controls ILC2 and NKp46+RORγt+ innate lymphocyte differentiation and protection in intestinal inflammation. *J Immunol*. 2013;191:4383-4391.
9. Yang Q, Li F, Harly C, et al. TCF-1 upregulation identifies early innate lymphoid progenitors in the bone marrow. *Nat Immunol*. 2015;16:1044-1050.
10. Harly C, Kenney D, Ren G, et al. The transcription factor TCF-1 enforces commitment to the innate lymphoid cell lineage. *Nat Immunol*. 2019;20:1150-1160.
11. Harly C, Cam M, Kaye J, Bhandoola A. Development and differentiation of early innate lymphoid progenitors. *J Exp Med*. 2018;215:249-262.
12. Aliahmad P, de la Torre B, Kaye J. Shared dependence on the DNA-binding factor TOX for the development of lymphoid tissue–inducer cell and NK cell lineages. *Nat Immunol*. 2010;11:945-952.
13. Seillet C, Rankin LC, Groom JR, et al. Nfil3 is required for the development of all innate lymphoid cell subsets. *J Exp Med*. 2014;211:1733-1740.
14. Xu W, Domingues RG, Fonseca-Pereira D, et al. NFIL3 orchestrates the emergence of common helper innate lymphoid cell precursors. *Cell Rep*. 2015;10:2043-2054.
15. Geiger TL, Abt MC, Gasteiger G, et al. Nfil3 is crucial for development of innate lymphoid cells and host protection against intestinal pathogens. *J Exp Med*. 2014;211:1723-1731.
16. Gascoyne DM, Long E, Veiga-Fernandes H, et al. The basic leucine zipper transcription factor E4BP4 is essential for natural killer cell development. *Nat Immunol*. 2009;10:1118-1124.

17. Kamizono S, Duncan GS, Seidel MG, et al. Nfil3/E4bp4 is required for the development and maturation of NK cells *in vivo*. *J Exp Med*. 2009;206:2977-2986.
18. Zook EC, Li ZY, Xu Y, et al. Transcription factor ID2 prevents E proteins from enforcing a naïve T lymphocyte gene program during NK cell development. *Sci Immunol*. 2018;3:eaao2139. doi:10.1126/sciimmunol.aao2139
19. Boos MD, Yokota Y, Eberl G, Kee BL. Mature natural killer cell and lymphoid tissue–inducing cell development requires Id2-mediated suppression of E protein activity. *J Exp Med*. 2007;204:1119-1130.
20. Klose CSN, Flach M, Möhle L, et al. Differentiation of type 1 ILCs from a common progenitor to all helper-like innate lymphoid cell lineages. *Cell*. 2014;157:340-356.
21. Yokota Y, Mansouri A, Mori S, et al. Development of peripheral lymphoid organs and natural killer cells depends on the helix–loop–helix inhibitor Id2. *Nature*. 1999;397:702-706.
22. Constantinides MG, McDonald BD, Verhoef PA, Bendelac A. A committed precursor to innate lymphoid cells. *Nature*. 2014;508:397-401.
23. Yagi R, Zhong C, Northrup DL, et al. The transcription factor GATA3 is critical for the development of all IL-7Rα-expressing innate lymphoid cells. *Immunity*. 2014;40:378-388.
24. Yu Y, Tsang JCH, Wang C, et al. Single-cell RNA-seq identifies a PD-1^{hi} ILC progenitor and defines its development pathway. *Nature*. 2016;539:102-106.
25. Hoyler T, Klose CSN, Souabni A, et al. The transcription factor GATA3 controls cell fate and maintenance of type 2 innate lymphoid cells. *Immunity*. 2012;37:634-648.
26. Califano D, Cho JJ, Uddin MN, et al. Transcription factor Bcl11b controls identity and function of mature type 2 innate lymphoid cells. *Immunity*. 2015;43:354-368.
27. Walker JA, Oliphant CJ, Englezakis A, et al. Bcl11b is essential for group 2 innate lymphoid cell development. *J Exp Med*. 2015;212:875-882.
28. Yu Y, Wang C, Clare S, et al. The transcription factor Bcl11b is specifically expressed in group 2 innate lymphoid cells and is essential for their development. *J Exp Med*. 2015;212:865-874.
29. Zook EC, Ramirez K, Guo X, et al. The ETS1 transcription factor is required for the development and cytokine-induced expansion of ILC2. *J Exp Med*. 2016;213:687-696.
30. Wong SH, Walker JA, Jolin HE, et al. RORα is essential for nuocyte development. *Nat Immunol*. 2012;13:229-236.
31. Halim TYF, MacLaren A, Romanish MT, Gold MJ, McNagny KM, Takei F. Retinoic-acid-receptor-related orphan nuclear receptor alpha is required for natural helper cell development and allergic inflammation. *Immunity*. 2012;37:463-474.
32. Fathman JW, Bhattacharya D, Inlay MA, Seita J, Karsunky H, Weissman IL. Identification of the earliest natural killer cell–committed progenitor in murine bone marrow. *Blood*. 2011;118:5439-5447.
33. Male V, Nisoli I, Kostrzewski T, et al. The transcription factor E4bp4/Nfil3 controls commitment to the NK lineage and directly regulates Eomes and Id2 expression. *J Exp Med*. 2014;211:635-642.
34. Seillet C, Huntington ND, Gangatirkar P, et al. Differential requirement for Nfil3 during NK cell development. *J Immunol*. 2014;192:2667-2676.
35. Narni-Mancinelli E, Chaix J, Fenis A, et al. Fate mapping analysis of lymphoid cells expressing the NKp46 cell surface receptor. *Proc Natl Acad Sci U S A*. 2011;108:18324-18329.
36. Hayakawa Y, Smyth MJ. CD27 dissects mature NK cells into two subsets with distinct responsiveness and migratory capacity. *J Immunol*. 2006;176:1517-1524.
37. Chiossone L, Chaix J, Fuseri N, Roth C, Vivier E, Walzer T. Maturation of mouse NK cells is a 4-stage developmental program. *Blood*. 2009;113:5488-5496.
38. Gasteiger G, Fan X, Dikiy S, Lee SY, Rudensky AY. Tissue residency of innate lymphoid cells in lymphoid and nonlymphoid organs. *Science*. 2015;350:981-985.
39. Moro K, Kabata H, Tanabe M, et al. Interferon and IL-27 antagonize the function of group 2 innate lymphoid cells and type 2 innate immune responses. *Nat Immunol*. 2016;17:76-86.
40. Chea S, Schmutz S, Berthault C, et al. Single-cell gene expression analyses reveal heterogeneous responsiveness of fetal innate lymphoid progenitors to Notch signaling. *Cell Rep*. 2016;14:1500-1516.
41. Mebius RE, Miyamoto T, Christensen J, et al. The fetal liver counterpart of adult common lymphoid progenitors gives rise to all lymphoid lineages, CD45+CD4+CD3– cells, as well as macrophages. *J Immunol*. 2001;166:6593-6601.
42. Bando JK, Liang H-E, Locksley RM. Identification and distribution of developing innate lymphoid cells in the fetal mouse intestine. *Nat Immunol*. 2015;16:153-160.
43. Schneider C, Lee J, Koga S, et al. Tissue-resident group 2 innate lymphoid cells differentiate by layered ontogeny and *in situ* perinatal priming. *Immunity*. 2019;50:1425.e5-1438.e5.
44. Lim AI, Di Santo JP. ILC-poiesis: Ensuring tissue ILC differentiation at the right place and time. *Eur J Immunol*. 2019;49:11-18.
45. Juelke K, Romagnani C. Differentiation of human innate lymphoid cells (ILCs). *Curr Opin Immunol*. 2016;38:75-85.
46. Scoville SD, Mundy-Bosse BL, Zhang MH, et al. A progenitor cell expressing transcription factor RORγt generates all human innate lymphoid cell subsets. *Immunity*. 2016;44:1140-1150.
47. Lim AI, Li Y, Lopez-Lastra S, et al. Systemic human ILC precursors provide a substrate for tissue ILC differentiation. *Cell*. 2017;168:1086.e10-1100.e10.
48. Tang Q, Ahn Y-O, Southern P, Blazar BR, Miller JS, Verneris MR. Development of IL-22–producing NK lineage cells from umbilical cord blood hematopoietic stem cells in the absence of secondary lymphoid tissue. *Blood*. 2011;117:4052-4055.
49. Gentek R, Munneke JM, Helbig C, et al. Modulation of signal strength switches Notch from an inducer of T cells to an inducer of ILC2. *Front Immunol*. 2013;4:334. doi:10.3389/fimmu.2013.00334
50. Montaldo E, Teixeira-Alves LG, Glatzer T, et al. Human RORγt+CD34+ cells are lineage-specified progenitors of group 3 RORγt+ innate lymphoid cells. *Immunity*. 2014;41:988-1000.
51. Renoux VM, Zriwil A, Peitzsch C, et al. Identification of a human natural killer cell lineage-restricted progenitor in fetal and adult tissues. *Immunity*. 2015;43:394-407.
52. Res P, Martinez-Caceres E, Cristina Jaleco A, et al. CD34+CD38^{dim} cells in the human thymus can differentiate into T, natural killer, and dendritic cells but are distinct from pluripotent stem cells. *Blood*. 1996;87:5196-5206.
53. Galy A, Travis M, Cen D, Chen B. Human T, B, natural killer, and dendritic cells arise from a common bone marrow progenitor cell subset. *Immunity*. 1995;3:459-473.
54. Kohn LA, Hao Q-L, Sasidharan R, et al. Lymphoid priming in human bone marrow begins prior to CD10 expression with up-regulation of L-selectin. *Nat Immunol*. 2012;13:963-971.
55. Hoebeke I, De Smedt M, Stolz F, et al. T-, B- and NK-lymphoid, but not myeloid cells arise from human CD34+CD38−CD7+ common lymphoid progenitors expressing lymphoid-specific genes. *Leukemia*. 2007;21:311-319.
56. Hao Q-L, Zhu J, Price MA, Payne KJ, Barsky LW, Crooks GM. Identification of a novel, human multilymphoid progenitor in cord blood. *Blood*. 2001;97:3683-3690.
57. Six EM, Bonhomme D, Monteiro M, et al. A human postnatal lymphoid progenitor capable of circulating and seeding the thymus. *J Exp Med*. 2007;204:3085-3093.
58. Karamitros D, Stoilova B, Aboukhalil Z, et al. Heterogeneity of human lympho-myeloid progenitors at the single cell level. *Nat Immunol*. 2018;19:85-97.
59. Doulatov S, Notta F, Eppert K, Nguyen LT, Ohashi PS, Dick JE. Revised map of the human progenitor hierarchy shows the origin of macrophages and dendritic cells in early lymphoid development. *Nat Immunol*. 2010;11:585-593.
60. Di Santo JP, Lim AI, Yssel H. 'ILC-poiesis': generating tissue ILCs from naïve precursors. *Oncotarget*. 2017;8:81729-81730.
61. Kiessling R, Klein E, Wigzell H. "Natural" killer cells in the mouse. I. Cytotoxic cells with specificity for mouse Moloney leukemia cells. Specificity and distribution according to genotype. *Eur J Immunol*. 1975;5:112-117.
62. Yokoyama WM. *Paul's Fundametnal Immunology*. 7th ed. Cell Press; 2019.
63. Ryan JC, Turck J, Niemi EC, Yokoyama WM, Seaman WE. Molecular cloning of the NK1.1 antigen, a member of the NKR-P1 family of natural killer cell activation molecules. *J Immunol*. 1992;149:1631-1635.
64. Hackett J, Tutt M, Lipscomb M, Bennett M, Koo G, Kumar V. Origin and differentiation of natural killer cells. II. Functional and morphologic studies of purified NK-1.1+ cells. *J Immunol*. 1986;136:3124-3131.
65. Lanier LL, Phillips JH, Hackett J, Tutt M, Kumar V. Natural killer cells: definition of a cell type rather than a function. *J Immunol*. 1986;137:2735-2739.
66. Seaman WE, Sleisenger M, Eriksson E, Koo GC. Depletion of natural killer cells in mice by monoclonal antibody to NK-1.1. Reduction in host defense against malignancy without loss of cellular or humoral immunity. *J Immunol*. 1987;138:4539-4544.
67. Koo GC, Peppard JR. Establishment of monoclonal anti-Nk-1.1 antibody. *Hybridoma*. 1984;3:301-303.
68. Gazit R, Gruda R, Elboim M, et al. Lethal influenza infection in the absence of the natural killer cell receptor gene Ncr1. *Nat Immunol*. 2006;7:517-523.
69. Eckelhart E, Warsch W, Zebedin E, et al. A novel Ncr1-Cre mouse reveals the essential role of STAT5 for NK-cell survival and development. *Blood*. 2011;117:1565-1573.
70. Walzer T, Bléry M, Chaix J, et al. Identification, activation, and selective *in vivo* ablation of mouse NK cells via NKp46. *Proc Natl Acad Sci U S A*. 2007;104:3384-3389.
71. Cella M, Fuchs A, Vermi W, et al. A human NK cell subset provides an innate source of IL-22 for mucosal immunity. *Nature*. 2009;457:722-725.
72. Satoh-Takayama N, Vosshenrich CAJ, Lesjean-Pottier S, et al. Microbial flora drives interleukin 22 production in intestinal NKp46+ cells that provide innate mucosal immune defense. *Immunity*. 2008;29:958-970.
73. Sanos SL, Bui VL, Mortha A, et al. RORγt and commensal microflora are required for the differentiation of mucosal interleukin 22–producing NKp46+ cells. *Nat Immunol*. 2009;10:83-91.
74. Luci C, Reynders A, Ivanov II, et al. Influence of the transcription factor RORγt on the development of NKp46+ cell populations in gut and skin. *Nat Immunol*. 2009;10:75-82.
75. Freud AG, Mundy-Bosse BL, Yu J, Caligiuri MA. The broad spectrum of human natural killer cell diversity. *Immunity*. 2017;47:820-833.
76. Cooper MA, Fehniger TA, Caligiuri MA. The biology of human natural killer-cell subsets. *Trends Immunol*. 2001;22:633-640.
77. Lanier LL, Le AM, Civin CI, Loken MR, Phillips JH. The relationship of CD16 (Leu-11) and Leu-19 (NKH-1) antigen expression on human peripheral blood NK cells and cytotoxic T lymphocytes. *J Immunol*. 1986;136:4480-4486.
78. Nagler A, Lanier LL, Cwirla S, Phillips JH. Comparative studies of human FcRIII-positive and negative natural killer cells. *J Immunol*. 1989;143:3183-3191.
79. Lopez-Vergès S, Milush JM, Pandey S, et al. CD57 defines a functionally distinct population of mature NK cells in the human CD56dimCD16+ NK-cell subset. *Blood*. 2010;116:3865-3874.
80. Björkström NK, Riese P, Heuts F, et al. Expression patterns of NKG2A, KIR, and CD57 define a process of CD56dim NK-cell differentiation uncoupled from NK-cell education. *Blood*. 2010;116:3853-3864.
81. Sun JC. Natural killer cells. In: Vivier E, Di Santo J, Moretta A, eds. *Current Topics in Microbiology and Immunology*. Springer International Publishing; 2016:1-36. doi:10.1007/82_2015_452
82. Leong JW, Wagner JA, Ireland AR, Fehniger TA. Transcriptional and post-transcriptional regulation of NK cell development and function. *Clin Immunol*. 2017;177:60-69.
83. Szabo SJ, Kim ST, Costa GL, Zhang X, Fathman CG, Glimcher LH. A novel transcription factor, T-bet, directs Th1 lineage commitment. *Cell*. 2000;100:655-669.
84. Townsend MJ, Weinmann AS, Matsuda JL, et al. T-bet regulates the terminal maturation and homeostasis of NK and Vα14i NKT cells. *Immunity*. 2004;20:477-494.
85. Jenne CN, Enders A, Rivera R, et al. T-bet–dependent S1P5 expression in NK cells promotes egress from lymph nodes and bone marrow. *J Exp Med*. 2009;206:2469-2481.
86. Gordon SM, Chaix J, Rupp LJ, et al. The transcription factors T-bet and Eomes control key checkpoints of natural killer cell maturation. *Immunity*. 2012;36:55-67.
87. Intlekofer AM, Takemoto N, Wherry EJ, et al. Effector and memory CD8+ T cell fate coupled by T-bet and eomesodermin. *Nat Immunol*. 2005;6:1236-1244.
88. Daussy C, Faure F, Mayol K, et al. T-bet and Eomes instruct the development of two distinct natural killer cell lineages in the liver and in the bone marrow. *J Exp Med*. 2014;211:563-577.
89. Takeda K, Cretney E, Hayakawa Y, et al. TRAIL identifies immature natural killer cells in newborn mice and adult mouse liver. *Blood*. 2005;105:2082-2089.
90. Tayade C, Fang Y, Black GP, Va P, Erlebacher A, Croy BA. Differential transcription of Eomes and T-bet during maturation of mouse uterine natural killer cells. *J Leukoc Biol*. 2005;78:1347-1355.
91. Leonard WJ, Lin JX, O'Shea JJ. The γc family of cytokines: basic biology to therapeutic ramifications. *Immunity*. 2019;50:832-850.
92. Grimm EA, Mazumder A, Zhang HZ, Rosenberg SA. Lymphokine-activated killer cell phenomenon. Lysis of natural killer-resistant fresh solid tumor cells by interleukin 2-activated autologous human peripheral blood lymphocytes. *J Exp Med*. 1982;155:1823-1841.
93. Trinchieri G, Matsumoto-Kobayashi M, Clark SC, Seehra J, London L, Perussia B. Response of resting human peripheral blood natural killer cells to interleukin 2. *J Exp Med*. 1984;160:1147-1169.
94. Phillips JH, Lanier LL. Dissection of the lymphokine-activated killer phenomenon. Relative contribution of peripheral blood natural killer cells and T lymphocytes to cytolysis. *J Exp Med*. 1986;164:814-825.
95. Ortaldo JR, Mason A, Overton R. Lymphokine-activated killer cells. Analysis of progenitors and effectors. *J Exp Med*. 1986;164:1193-1205.
96. Willerford DM, Chen J, Ferry JA, Davidson L, Ma A, Alt FW. Interleukin-2 receptor α chain regulates the size and content of the peripheral lymphoid compartment. *Immunity*. 1995;3:521-530.
97. Schorle H, Holtschke T, Hünig T, Schimpl A, Horak I. Development and function of T cells in mice rendered interleukin-2 deficient by gene targeting. *Nature*. 1991;352:621-624.
98. Suzuki H, Duncan GS, Takimoto H, Mak TW. Abnormal development of intestinal intraepithelial lymphocytes and peripheral natural killer cells in mice lacking the IL-2 receptor beta chain. *J Exp Med*. 1997;185:499-505.
99. DiSanto JP, Müller W, Guy-Grand D, Fischer A, Rajewsky K. Lymphoid development in mice with a targeted deletion of the interleukin 2 receptor γ chain. *Proc Natl Acad Sci U S A*. 1995;92:377-381.
100. Cao X, Shores EW, Hu-Li J, et al. Defective lymphoid development in mice lacking expression of the common cytokine receptor gamma chain. *Immunity*. 1995;2:223-238.
101. Orange JS. Human natural killer cell deficiencies. *Curr Opin Allergy Clin Immunol*. 2006;6:399-409.

102. Giri JG, Ahdieh M, Eisenman J, et al. Utilization of the beta and gamma chains of the IL-2 receptor by the novel cytokine IL-15. *EMBO J.* 1994;13:2822-2830.
103. Johnston JA, Bacon CM, Finbloom DS, et al. Tyrosine phosphorylation and activation of STAT5, STAT3, and Janus kinases by interleukins 2 and 15. *Proc Natl Acad Sci U S A.* 1995;92:8705-8709.
104. Taniguchi T, Minami Y. The IL-2IL-2 receptor system: a current overview. *Cell.* 1993;73:5-8.
105. Nakarai T, Robertson MJ, Streuli M, et al. Interleukin 2 receptor gamma chain expression on resting and activated lymphoid cells. *J Exp Med.* 1994;180:241-251.
106. Ma A, Koka R, Burkett P. Diverse functions of IL-2, IL-15, and IL-7 in lymphoid homeostasis. *Annu Rev Immunol.* 2006;24:657-679.
107. Lodolce JP, Boone DL, Chai S, et al. IL-15 receptor maintains lymphoid homeostasis by supporting lymphocyte homing and proliferation. *Immunity.* 1998;9:669-676.
108. Kennedy MK, Glaccum M, Brown SN, et al. Reversible defects in natural killer and memory CD8 T cell lineages in interleukin 15–deficient mice. *J Exp Med.* 2000;191:771-780.
109. Cooper MA, Bush JE, Fehniger TA, et al. *In vivo* evidence for a dependence on interleukin 15 for survival of natural killer cells. *Blood.* 2002;100:3633-3638.
110. Koka R, Burkett PR, Chien M, et al. Interleukin (IL)-15Rα–deficient natural killer cells survive in normal but not IL-15Rα–deficient mice. *J Exp Med.* 2003;197:977-984.
111. Dubois S, Mariner J, Waldmann TA, Tagaya Y. IL-15Rα recycles and presents IL-15 in trans to neighboring cells. *Immunity.* 2002;17:537-547.
112. Koka R, Burkett P, Chien M, Chai S, Boone DL, Ma A. Cutting edge: murine dendritic cells require IL-15Rα to prime NK cells. *J Immunol.* 2004;173:3594-3598.
113. Burkett PR, Koka R, Chien M, Chai S, Boone DL, Ma A. Coordinate expression and trans presentation of interleukin (IL)-15Rα and IL-15 supports natural killer cell and memory CD8+ T cell homeostasis. *J Exp Med.* 2004;200:825-834.
114. Lucas M, Schachterle W, Oberle K, Aichele P, Diefenbach A. Dendritic cells prime natural killer cells by trans-presenting interleukin 15. *Immunity.* 2007;26:503-517.
115. Clark R, Griffiths GM. Lytic granules, secretory lysosomes and disease. *Curr Opin Immunol.* 2003;15:516-521.
116. Lieberman J. The ABCs of granule-mediated cytotoxicity: new weapons in the arsenal. *Nat Rev Immunol.* 2003;3:361-370.
117. Mentlik AN, Sanborn KB, Holzbaur EL, Orange JS. Rapid lytic granule convergence to the MTOC in natural killer cells is dependent on dynein but not cytolytic commitment. *Mol Biol Cell.* 2010;21:2241-2256.
118. Thiery J, Keefe D, Boulant S, et al. Perforin pores in the endosomal membrane trigger release of endocytosed granzyme B to the cytosol of target cells. *Nat Immunol.* 2011;12:770-777.
119. Afonina IS, Cullen SP, Martin SJ. Cytotoxic and non-cytotoxic roles of the CTL/NK protease granzyme B. *Immunol Rev.* 2010;235:105-116.
120. Fehniger TA, Cai SF, Cao X, et al. Acquisition of murine NK cell cytotoxicity requires the translation of a pre-existing pool of granzyme B and perforin mRNAs. *Immunity.* 2007;26:798-811.
121. Henkart PA. Lymphocyte-mediated cytotoxicity: two pathways and multiple effector molecules. *Immunity.* 1994;1:343-346.
122. Montel AH, Bochan MR, Hobbs JA, Lynch DH, Brahmi Z. Fas involvement in cytotoxicity mediated by human NK cells. *Cell Immunol.* 1995;166:236-246.
123. Arase H, Arase N, Saito T. Fas-mediated cytotoxicity by freshly isolated natural killer cells. *J Exp Med.* 1995;181:1235-1238.
124. Lee RK, Spielman J, Zhao DY, Olsen KJ, Podack ER. Perforin, Fas ligand, and tumor necrosis factor are the major cytotoxic molecules used by lymphokine-activated killer cells. *J Immunol.* 1996;157:1919-1925.
125. Oshimi Y, Oda S, Honda Y, Nagata S, Miyazaki S. Involvement of Fas ligand and Fas-mediated pathway in the cytotoxicity of human natural killer cells. *J Immunol.* 1996;157:2909-2915.
126. Screpanti V, Wallin RPA, Ljunggren H-G, Grandien A. A central role for death receptor-mediated apoptosis in the rejection of tumors by NK cells. *J Immunol.* 2001;167:2068-2073.
127. Kägi D, Ledermann B, Bürki K, et al. Cytotoxicity mediated by T cells and natural killer cells is greatly impaired in perforin-deficient mice. *Nature.* 1994;369:31-37.
128. Revell PA, Grossman WJ, Thomas DA, et al. Granzyme B and the downstream granzymes C and/or F are important for cytotoxic lymphocyte functions. *J Immunol.* 2005;174:2124-2131.
129. Pham CTN, Ley TJ. Dipeptidyl peptidase I is required for the processing and activation of granzymes A and B *in vivo. Proc Natl Acad Sci U S A.* 1999;96:8627-8632.
130. Crozat K, Hoebe K, Ugolini S, et al. Jinx, an MCMV susceptibility phenotype caused by disruption of Unc13d: a mouse model of type 3 familial hemophagocytic lymphohistiocytosis. *J Exp Med.* 2007;204:853-863.
131. Degliantoni G, Murphy M, Kobayashi M, Francis MK, Perussia B, Trinchieri G. Natural killer (NK) cell-derived hematopoietic colony-inhibiting activity and NK cytotoxic factor. Relationship with tumor necrosis factor and synergism with immune interferon. *J Exp Med.* 1985;162:1512-1530.
132. Anegón I, Cuturi MC, Trinchieri G, Perussia B. Interaction of Fc receptor (CD16) ligands induces transcription of interleukin 2 receptor (CD25) and lymphokine genes and expression of their products in human natural killer cells. *J Exp Med.* 1988;167:452-472.
133. Cuturi MC, Anegón I, Sherman F, et al. Production of hematopoietic colony-stimulating factors by human natural killer cells. *J Exp Med.* 1989;169:569-583.
134. Martín-Fontecha A, Thomsen LL, Brett S, et al. Induced recruitment of NK cells to lymph nodes provides IFN-γ for T(H)1 priming. *Nat Immunol.* 2004;5:1260-1265.
135. Perona-Wright G, Mohrs K, Szaba FM, et al. Systemic but not local infections elicit immunosuppressive IL-10 production by natural killer cells. *Cell Host Microbe.* 2009;6, 503-512.
136. Maroof A, Beattie L, Zubairi S, Svensson M, Stager S, Kaye PM. Posttranscriptional regulation of Il10 gene expression allows natural killer cells to express immunoregulatory function. *Immunity.* 2008;29:295-305.
137. Mehrotra PT, Donnelly RP, Wong S, et al. Production of IL-10 by human natural killer cells stimulated with IL-2 and/or IL-12. *J Immunol.* 1998;160:2637-2644.
138. Dorner BG, Scheffold A, Rolph MS, et al. MIP-1α, MIP-1β, RANTES, and ATAC/lymphotactin function together with IFN-γ as type 1 cytokines. *Proc Natl Acad Sci U S A.* 2002;99:6181-6186.
139. Walzer T, Dalod M, Robbins SH, Zitvogel L, Vivier E. Natural-killer cells and dendritic cells: "l'union fait la force". *Blood.* 2005;106:2252-2258.
140. Vivier E, Raulet DH, Moretta A, et al. Innate or adaptive immunity? The example of natural killer cells. *Science.* 2011;331:44-49.
141. Tripp CS, Wolf SF, Unanue ER. Interleukin 12 and tumor necrosis factor alpha are costimulators of interferon gamma production by natural killer cells in severe combined immunodeficiency mice with listeriosis, and interleukin 10 is a physiologic antagonist. *Proc Natl Acad Sci U S A.* 1993;90:3725-3729.
142. Guia S, Cognet C, de Beaucoudrey L, et al. A role for interleukin-12/23 in the maturation of human natural killer and CD56+ T cells *in vivo. Blood.* 2008;111:5008-5016.
143. Chaix J, Tessmer MS, Hoebe K, et al. Cutting edge: priming of NK cells by IL-18. *J Immunol.* 2008;181:1627-1631.
144. McCartney S, Vermi W, Gilfillan S, et al. Distinct and complementary functions of MDA5 and TLR3 in poly(I:C)-mediated activation of mouse NK cells. *J Exp Med.* 2009;206:2967-2976.
145. Biron CA, Sonnenfeld G, Welsh RM. Interferon induces natural killer cell blastogenesis *in vivo. J Leukoc Biol.* 1984;35:31-37.
146. Long EO. Regulation of immune responses through inhibitory receptors. *Annu Rev Immunol.* 1999;17:875-904.
147. Muta T, Kurosaki T, Misulovin Z, Sanchez M, Nussenzweig MC, Ravetch JV. A 13-amino-acid motif in the cytoplasmic domain of FcγRIIB modulates B-cell receptor signalling. *Nature.* 1994;368:70-73.
148. Reth M. Antigen receptor tail clue. *Nature.* 1989;338:383-384.
149. Vivier E, Nunès JA, Vély F. Natural killer cell signaling pathways. *Science.* 2004;306:1517-1519.
150. Anderson P, Caligiuri M, Ritz J, Schlossman SF. CD3-negative natural killer cells express ε TCR as part of a novel molecular complex. *Nature.* 1989;341:159-162.
151. Lanier LL, Corliss BC, Wu J, Leong C, Phillips JH. Immunoreceptor DAP12 bearing a tyrosine-based activation motif is involved in activating NK cells. *Nature.* 1998;391:703-707.
152. Hibbs ML, Selvaraj P, Carpen O, et al. Mechanisms for regulating expression of membrane isoforms of FcγRIII (CD16). *Science.* 1989;246:1608-1611.
153. Raulet DH, Vance RE, McMahon CW. Regulation of the natural killer cell receptor repertoire. *Annu Rev Immunol.* 2001;19:291-330.
154. Kärre K, Ljunggren HG, Piontek G, Kiessling R. Selective rejection of H-2-deficient lymphoma variants suggests alternative immune defence strategy. *Nature.* 1986;319:675-678.
155. Quillet A, Presse F, Marchiol-Fournigault C, et al. Increased resistance to non-MHC-restricted cytotoxicity related to HLA A, B expression. Direct demonstration using beta 2-microglobulin-transfected Daudi cells. *J Immunol.* 1988;141:17-20.
156. Franksson L, George E, Powis S, Butcher G, Howard J, Kärre K. Tumorigenicity conferred to lymphoma mutant by major histocompatibility complex-encoded transporter gene. *J Exp Med.* 1993;177:201-205.
157. Salcedo M, Momburg F, Hämmerling GJ, Ljunggren HG. Resistance to natural killer cell lysis conferred by TAP1/2 genes in human antigen-processing mutant cells. *J Immunol.* 1994;152:1702-1708.
158. Liao NS, Bix M, Zijlstra M, Jaenisch R, Raulet D. MHC class I deficiency: susceptibility to natural killer (NK) cells and impaired NK activity. *Science.* 1991;253:199-202.
159. Höglund P, Ohlén C, Carbone E, et al. Recognition of beta 2-microglobulin-negative (beta 2m-) T-cell blasts by natural killer cells from normal but not from beta 2m- mice: nonresponsiveness controlled by beta 2m- bone marrow in chimeric mice. *Proc Natl Acad Sci U S A.* 1991;88:10332-10336.
160. Storkus WJ, Alexander J, Payne JA, Dawson JR, Cresswell P. Reversal of natural killing susceptibility in target cells expressing transfected class I HLA genes. *Proc Natl Acad Sci U S A.* 1989;86:2361-2364.
161. Yu YYL, Kumar V, Bennett M. Murine natural killer cells and marrow graft rejection. *Annu Rev Immunol.* 1992;10:189-213.
162. Colonna M, Brooks EG, Falco M, Ferrara GB, Strominger JL. Generation of allospecific natural killer cells by stimulation across a polymorphism of HLA-C. *Science.* 1993;260:1121-1124.
163. Colonna M, Samaridis J. Cloning of immunoglobulin-superfamily members associated with HLA-C and HLA-B recognition by human natural killer cells. *Science.* 1995;268:405-408.
164. Colonna M, Spies T, Strominger JL, et al. Alloantigen recognition by two human natural killer cell clones is associated with HLA-C or a closely linked gene. *Proc Natl Acad Sci U S A.* 1992;89:7983-7985.
165. Litwin V, Gumperz J, Parham P, Phillips JH, Lanier LL. NKB1: a natural killer cell receptor involved in the recognition of polymorphic HLA-B molecules. *J Exp Med.* 1994;180:537-543.
166. Moretta A, Vitale M, Sivori S, et al. Human natural killer cell receptors for HLA-class I molecules. Evidence that the Kp43 (CD94) molecule functions as receptor for HLA-B alleles. *J Exp Med.* 1994;180:545-555.
167. Ciccone E, Pende D, Viale O, et al. Involvement of HLA class I alleles in natural killer (NK) cell-specific functions: expression of HLA-Cw3 confers selective protection from lysis by alloreactive NK clones displaying a defined specificity (specificity 2). *J Exp Med.* 1992;176:963-971.
168. Ljunggren HG, Kärre K. In search of the 'missing self': MHC molecules and NK cell recognition. *Immunol Today.* 1990;11:237-244.
169. Hansen TH, Bouvier M. MHC class I antigen presentation: learning from viral evasion strategies. *Nat Rev Immunol.* 2009;9:503-513.
170. Holtkamp B, Cramer M, Rajewsky K. Somatic variation of H-2Kk expression and structure in a T-cell lymphoma: instability, stabilization, high production and structural mutation. *EMBO J.* 1983;2:1943-1951.
171. Gattoni-Celli S, Kirsch K, Timpane R, Isselbacher KJ. β2-Microglobulin gene is mutated in a human colon cancer cell line (HCT) deficient in the expression of HLA class I antigens on the cell surface. *Cancer Res.* 1992;52:1201-1204.
172. Korkolopoulou P, Kaklamanis L, Pezzella F, Harris AL, Gatter KC. Loss of antigen-presenting molecules (MHC class I and TAP-1) in lung cancer. *Br J Cancer.* 1996;73:148-153.
173. Kelley J, Walter L, Trowsdale J. Comparative genomics of natural killer cell receptor gene clusters. *PLoS Genet.* 2005;1:e27.
174. Mehta IK, Wang J, Roland J, Margulies DH, Yokoyama WM. Ly49A allelic variation and MHC class I specificity. *Immunogenetics.* 2001;53:572-583.
175. Wilhelm BT, Gagnier L, Mager DL. Sequence analysis of the Ly49 cluster in C57BL/6 mice: a rapidly evolving multigene family in the immune system. *Genomics.* 2002;80:646-661.
176. Karlhofer FM, Ribaudo RK, Yokoyama WM. MHC class I alloantigen specificity of Ly-49+ IL-2-activated natural killer cells. *Nature.* 1992;358:66-70.
177. Karlhofer FM, Ribaudo RK, Yokoyama WM. The interaction of Ly-49 with H-2Dd globally inactivates natural killer cell cytolytic activity. *Trans Assoc Am Physicians.* 1992;105:72-85.
178. Correa I, Corral L, Raulet DH. Multiple natural killer cell-activating signals are inhibited by major histocompatibility complex class I expression in target cells. *Eur J Immunol.* 1994;24:1323-1331.
179. Karlhofer FM, Hunziker R, Reichlin A, Margulies DH, Yokoyama WM. Host MHC class I molecules modulate *in vivo* expression of a NK cell receptor. *J Immunol.* 1994;153:2407-2416.
180. Hanke T, Takizawa H, McMahon CW, et al. Direct assessment of MHC class I binding by seven Ly49 inhibitory NK cell receptors. *Immunity.* 1999;11:67-77.
181. Tormo J, Natarajan K, Margulies DH, Mariuzza RA. Crystal structure of a lectin-like natural killer cell receptor bound to its MHC class I ligand. *Nature.* 1999;402:623-631.
182. Matsumoto N, Mitsuki M, Tajima K, Yokoyama WM, Yamamoto K. The functional binding site for the C-type lectin-like natural killer cell receptor Ly49A spans three domains of its major histocompatibility complex class I ligand. *J Exp Med.* 2001;193:147-158.
183. Wang J, Whitman MC, Natarajan K, Tormo J, Mariuzza RA, Margulies DH. Binding of the natural killer cell inhibitory receptor Ly49A to its major histocompatibility complex class I ligand. Crucial contacts include both H-2D^d AND β2-microglobulin. *J Biol Chem.* 2002;277:1433-1442.
184. Kane KP. Ly-49 mediates EL4 lymphoma adhesion to isolated class I major histocompatibility complex molecules. *J Exp Med.* 1994;179:1011-1015.
185. Daniels BF, Karlhofer FM, Seaman WE, Yokoyama WM. A natural killer cell receptor specific for a major histocompatibility complex class I molecule. *J Exp Med.* 1994;180:687-692.

186. Mitsuki M, Matsumoto N, Yamamoto K. A species-specific determinant on β2-microglobulin required for Ly49A recognition of its MHC class I ligand. *Int Immunol.* 2004;16:197-204.
187. Correa I, Raulet DH. Binding of diverse peptides to MHC class I molecules inhibits target cell lysis by activated natural killer cells. *Immunity.* 1995;2:61-71.
188. Orihuela M, Margulies DH, Yokoyama WM. The natural killer cell receptor Ly-49A recognizes a peptide-induced conformational determinant on its major histocompatibility complex class I ligand. *Proc Natl Acad Sci U S A.* 1996;93:11792-11797.
189. Brennan J, Mager D, Jefferies W, Takei F. Expression of different members of the Ly-49 gene family defines distinct natural killer cell subsets and cell adhesion properties. *J Exp Med.* 1994;180:2287-2295.
190. Dam J, Guan R, Natarajan K, et al. Variable MHC class I engagement by Ly49 natural killer cell receptors demonstrated by the crystal structure of Ly49C bound to H-2K(b). *Nat Immunol.* 2003;4:1213-1222.
191. Deng L, Cho S, Malchiodi EL, Kerzic MC, Dam J, Mariuzza RA. Molecular architecture of the major histocompatibility complex class I-binding site of Ly49 natural killer cell receptors. *J Biol Chem.* 2008;283:16840-16849.
192. Franksson L, Sundbäck J, Achour A, Bernlind J, Glas R, Kärre K. Peptide dependency and selectivity of the NK cell inhibitory receptor Ly-49C. *Eur J Immunol.* 1999;29:2748-2758.
193. George TC, Mason LH, Ortaldo JR, Kumar V, Bennett M. Positive recognition of MHC class I molecules by the Ly49D receptor of murine NK cells. *J Immunol.* 1999;162:2035-2043.
194. Arase H, Mocarski ES, Campbell AE, Hill AB, Lanier LL. Direct recognition of cytomegalovirus by activating and inhibitory NK cell receptors. *Science.* 2002;296:1323-1326.
195. Held W, Roland J, Raulet DH. Allelic exclusion of Ly49-family genes encoding class I MHC-specific receptors on NK cells. *Nature.* 1995;376:355-358.
196. Held W, Raulet DH. Expression of the Ly49A gene in murine natural killer cell clones is predominantly but not exclusively mono-allelic. *Eur J Immunol.* 1997;27:2876-2884.
197. Saleh A, Davies GE, Pascal V, et al. Identification of probabilistic transcriptional switches in the Ly49 gene cluster: a eukaryotic mechanism for selective gene activation. *Immunity.* 2004;21:55-66.
198. Ioannidis V, Kunz B, Tanamachi DM, Scarpellino L, Held W. Initiation and limitation of Ly-49A NK cell receptor acquisition by T cell factor-1. *J Immunol.* 2003;171:769-775.
199. Held W, Kunz B, Lowin-Kropf B, van de Wetering M, Clevers H. Clonal acquisition of the Ly49A NK cell receptor is dependent on the trans-acting factor TCF-1. *Immunity.* 1999;11:433-442.
200. Kunz B, Held W. Positive and negative roles of the trans-acting T cell factor-1 for the acquisition of distinct Ly-49 MHC class I receptors by NK cells. *J Immunol.* 2001;166:6181-6187.
201. Held W, Clevers H, Grosschedl R. Redundant functions of TCF-1 and LEF-1 during T and NK cell development, but unique role of TCF-1 for Ly49 NK cell receptor acquisition. *Eur J Immunol.* 2003;33:1393-1398.
202. Gays F, Aust JG, Reid DM, et al. Ly49B is expressed on multiple subpopulations of myeloid cells. *J Immunol.* 2006;177:5840-5851.
203. Tai LH, Goulet ML, Belanger S, et al. Positive regulation of plasmacytoid dendritic cell function via Ly49Q recognition of class I MHC. *J Exp Med.* 2008;205:3187-3199.
204. Yoshizaki M, Tazawa A, Kasumi E, et al. Spatiotemporal regulation of intracellular trafficking of Toll-like receptor 9 by an inhibitory receptor, Ly49Q. *Blood.* 2009;114:1518-1527.
205. Orr MT, Lanier LL. Negative co-receptors and ligands. In: Ahmed R, Honjo T, eds. *Current Topics in Microbiology and Immunology.* Springer; 2011:67-87. doi:10.1007/82_2010_85
206. Moretta A, Tambussi G, Bottino C, et al. A novel surface antigen expressed by a subset of human CD3−CD16+ natural killer cells. Role in cell activation and regulation of cytolytic function. *J Exp Med.* 1990;171:695-714.
207. Moretta A, Bottino C, Pende D, et al. Identification of four subsets of human CD3−CD16+ natural killer (NK) cells by the expression of clonally distributed functional surface molecules: correlation between subset assignment of NK clones and ability to mediate specific alloantigen recognition. *J Exp Med.* 1990;172:1589-1598.
208. Moretta A, Vitale M, Bottino C, et al. P58 molecules as putative receptors for major histocompatibility complex (MHC) class I molecules in human natural killer (NK) cells. Anti-p58 antibodies reconstitute lysis of MHC class I-protected cells in NK clones displaying different specificities. *J Exp Med.* 1993;178:597-604.
209. Wagtmann N, Biassoni R, Cantoni C, et al. Molecular clones of the p58 NK cell receptor reveal immunoglobulin-related molecules with diversity in both the extra- and intracellular domains. *Immunity.* 1995;2:439-449.
210. IPD-KIR Home. IPD. EMBL-EBI. https://www.ebi.ac.uk/ipd/kir/index.html. Accessed May 14, 2020.
211. Lanier LL. NK cell recognition. *Annu Rev Immunol.* 2005;23:225-274.
212. Uhrberg M, Valiante NM, Shum BP, et al. Human diversity in killer cell inhibitory receptor genes. *Immunity.* 1997;7:753-763.
213. Middleton D, Gonzelez F. The extensive polymorphism of KIR genes. *Immunology.* 2010;129:8-19.
214. Moretta A, Sivori S, Vitale M, et al. Existence of both inhibitory (p58) and activatory (p50) receptors for HLA-C molecules in human natural killer cells. *J Exp Med.* 1995;182:875-884.
215. Marsh SGE, Parham P, Dupont B, et al. Killer-cell immunoglobulin-like receptor (KIR) nomenclature report, 2002. *Hum Immunol.* 2003;64:648-654.
216. Gumperz JE, Litwin V, Phillips JH, Lanier LL, Parham P. The Bw4 public epitope of HLA-B molecules confers reactivity with natural killer cell clones that express NKB1, a putative HLA receptor. *J Exp Med.* 1995;181:1133-1144.
217. Gumperz JE, Barber LD, Valiante NM, et al. Conserved and variable residues within the Bw4 motif of HLA-B make separable contributions to recognition by the NKB1 killer cell-inhibitory receptor. *J Immunol.* 1997;158:5237-5241.
218. Hansasuta P, Dong T, Thananchai H, et al. Recognition of HLA-A3 and HLA-A11 by KIR3DL2 is peptide-specific. *Eur J Immunol.* 2004;34:1673-1679.
219. Malnati MS, Peruzzi M, Parker KC, et al. Peptide specificity in the recognition of MHC class I by natural killer cell clones. *Science.* 1995;267:1016-1018.
220. Peruzzi M, Wagtmann N, Long EO. A p70 killer cell inhibitory receptor specific for several HLA-B allotypes discriminates among peptides bound to HLA-B*2705. *J Exp Med.* 1996;184:1585-1590.
221. Peruzzi M, Parker KC, Long EO, Malnati MS. Peptide sequence requirements for the recognition of HLA-B*2705 by specific natural killer cells. *J Immunol.* 1996;157:3350-3356.
222. Gavioli R, Zhang QJ, Masucci MG. HLA-A11-mediated protection from NK cell-mediated lysis: role of HLA-A11-presented peptides. *Hum Immunol.* 1996;49:1-12.
223. Rajagopalan S, Long EO. The direct binding of a p58 killer cell inhibitory receptor to human histocompatibility leukocyte antigen (HLA)-Cw4 exhibits peptide selectivity. *J Exp Med.* 1997;185:1523-1528.
224. Zappacosta F, Borrego F, Brooks AG, Parker KC, Coligan JE. Peptides isolated from HLA-Cw*0304 confer different degrees of protection from natural killer cell-mediated lysis. *Proc Natl Acad Sci U S A.* 1997;94:6313-6318.
225. Boyington JC, Motyka SA, Schuck P, Brooks AG, Sun PD. Crystal structure of an NK cell immunoglobulin-like receptor in complex with its class I MHC ligand. *Nature.* 2000;405:537-543.
226. Valiante NM, Parham P. Natural killer cells, HLA class I molecules, and marrow transplantation. *Biol Blood Marrow Transplant.* 1997;3:229-235.
227. Ruggeri L, Capanni M, Urbani E, et al. Effectiveness of donor natural killer cell alloreactivity in mismatched hematopoietic transplants. *Science.* 2002;295:2097-2100.
228. Pende D, Marcenaro S, Falco M, et al. Anti-leukemia activity of alloreactive NK cells in KIR ligand-mismatched haploidentical HSCT for pediatric patients: evaluation of the functional role of activating KIR and redefinition of inhibitory KIR specificity. *Blood.* 2009;113:3119-3129.
229. Chewning JH, Gudme CN, Hsu KC, Selvakumar A, Dupont B. KIR2DS1-positive NK cells mediate alloresponse against the C2 HLA-KIR ligand group *in vitro*. *J Immunol.* 2007;179:854-868.
230. Chan HW, Kurago ZB, Stewart CA, et al. DNA methylation maintains allele-specific KIR gene expression in human natural killer cells. *J Exp Med.* 2003;197:245-255.
231. Valiante NM, Uhrberg M, Shilling HG, et al. Functionally and structurally distinct NK cell receptor repertoires in the peripheral blood of two human donors. *Immunity.* 1997;7:739-751.
232. Santourlidis S, Trompeter HI, Weinhold S, et al. Crucial role of DNA methylation in determination of clonally distributed killer cell Ig-like receptor expression patterns in NK cells. *J Immunol.* 2002;169:4253-4261.
233. Mandelboim O, Davis DM, Reyburn HT, et al. Enhancement of class II-restricted T cell responses by costimulatory NK receptors for class I MHC proteins. *Science.* 1996;274:2097-2100.
234. Ferrini S, Cambiaggi A, Meazza R, et al. T cell clones expressing the natural killer cell-related p58 receptor molecule display heterogeneity in phenotypic properties and p58 function. *Eur J Immunol.* 1994;24:2294-2298.
235. Phillips JH, Gumperz JE, Parham P, Lanier LL. Superantigen-dependent, cell-mediated cytotoxicity inhibited by MHC class I receptors on T lymphocytes. *Science.* 1995;268:403-405.
236. Houchins JP, Yabe T, McSherry C, Bach FH. DNA sequence analysis of NKG2, a family of related cDNA clones encoding type II integral membrane proteins on human natural killer cells. *J Exp Med.* 1991;173:1017-1020.
237. Chang C, Rodríguez A, Carretero M, López-Botet M, Phillips JH, Lanier LL. Molecular characterization of human CD94: a type II membrane glycoprotein related to the C-type lectin superfamily. *Eur J Immunol.* 1995;25:2433-2437.
238. Dissen E, Berg SF, Westgaard IH, Fossum S. Molecular characterization of a gene in the rat homologous to human CD94. *Eur J Immunol.* 1997;27:2080-2086.
239. Berg SF, Dissen E, Westgaard IH, Fossum S. Two genes in the rat homologous to human NKG2. *Eur J Immunol.* 1998;28:444-450.
240. Lohwasser S, Hande P, Mager DL, Takei F. Cloning of murine NKG2A, B and C: second family of C-type lectin receptors on murine NK cells. *Eur J Immunol.* 1999;29:755-761.
241. Sato A, Mayer WE, Overath P, Klein J. Genes encoding putative natural killer cell C-type lectin receptors in teleostean fishes. *Proc Natl Acad Sci U S A.* 2003;100:7779-7784.
242. Vance RE, Jamieson AM, Raulet DH. Recognition of the Class Ib Molecule Qa-1b by putative activating receptors CD94/NKG2C and CD94/NKG2E on mouse natural killer cells. *J Exp Med.* 1999;190:1801-1812.
243. Lazetic S, Chang C, Houchins JP, Lanier LL, Phillips JH. Human natural killer cell receptors involved in MHC class I recognition are disulfide-linked heterodimers of CD94 and NKG2 subunits. *J Immunol.* 1996;157:4741-4745.
244. Brooks AG, Posch PE, Scorzelli CJ, Borrego F, Coligan JE. NKG2A complexed with CD94 defines a novel inhibitory natural killer cell receptor. *J Exp Med.* 1997;185:795-800.
245. Houchins JP, Lanier LL, Niemi EC, Phillips JH, Ryan JC. Natural killer cell cytolytic activity is inhibited by NKG2-A and activated by NKG2-C. *J Immunol.* 1997;158:3603-3609.
246. Aramburu J, Balboa MA, Izquierdo M, López-Botet M. A novel functional cell surface dimer (Kp43) expressed by natural killer cells and gamma/delta TCR+ T lymphocytes. II. Modulation of natural killer cytotoxicity by anti-Kp43 monoclonal antibody. *J Immunol.* 1991;147:714-721.
247. Vance RE, Kraft JR, Altman JD, Jensen PE, Raulet DH. Mouse CD94/NKG2A is a natural killer cell receptor for the nonclassical major histocompatibility complex (MHC) class I molecule Qa-1b. *J Exp Med.* 1998;188:1841-1848.
248. Braud VM, Allan DSJ, O'Callaghan CA, et al. HLA-E binds to natural killer cell receptors CD94/NKG2A, B and C. *Nature.* 1998;391:795-799.
249. Borrego F, Ulbrecht M, Weiss EH, Coligan JE, Brooks AG. Recognition of human histocompatibility leukocyte antigen (HLA)-E complexed with HLA class I signal sequence–derived peptides by CD94/NKG2 confers protection from natural killer cell–mediated lysis. *J Exp Med.* 1998;187:813-818.
250. Lee N, Llano M, Carretero M, et al. HLA-E is a major ligand for the natural killer inhibitory receptor CD94/NKG2A. *Proc Natl Acad Sci U S A.* 1998;95:5199-5204.
251. Salcedo M, Bousso P, Ljunggren HG, Kourilsky P, Abastado JP. The Qa-1b molecule binds to a large subpopulation of murine NK cells. *Eur J Immunol.* 1998;28:4356-4361.
252. Koller BH, Geraghty DE, Shimizu Y, DeMars R, Orr HT. HLA-E. A novel HLA class I gene expressed in resting T lymphocytes. *J Immunol.* 1988;141:897-904.
253. Ulbrecht M, Honka T, Person S, Johnson JP, Weiss EH. The HLA-E gene encodes two differentially regulated transcripts and a cell surface protein. *J Immunol.* 1992;149:2945-2953.
254. Soloski MJ, DeCloux A, Aldrich CJ, Forman J. Structural and functional characteristics of the class IB molecule, Qa-1. *Immunol Rev.* 1995;147:67-89.
255. Aldrich CJ, DeCloux A, Woods AS, Cotter RJ, Soloski MJ, Forman J. Identification of a Tap-dependent leader peptide recognized by alloreactive T cells specific for a class Ib antigen. *Cell.* 1994;79:649-658.
256. Braud V, Jones EY, McMichael A. The human major histocompatibility complex class Ib molecule HLA-E binds signal sequence-derived peptides with primary anchor residues at positions 2 and 9. *Eur J Immunol.* 1997;27:1164-1169.
257. Lee N, Goodlett DR, Ishitani A, Marquardt H, Geraghty DE. HLA-E surface expression depends on binding of TAP-dependent peptides derived from certain HLA class I signal sequences. *J Immunol.* 1998;160:4951-4960.
258. Valés-Gómez M, Reyburn HT, Erskine RA, López-Botet M, Strominger JL. Kinetics and peptide dependency of the binding of the inhibitory NK receptor CD94/NKG2-A and the activating receptor CD94/NKG2-C to HLA-E. *EMBO J.* 1999;18:4250-4260.
259. Petrie EJ, Clements CS, Lin J, et al. CD94-NKG2A recognition of human leukocyte antigen (HLA)-E bound to an HLA class I leader sequence. *J Exp Med.* 2008;205:725-735.
260. Kaiser BK, Pizarro JC, Kerns J, Strong RK. Structural basis for NKG2A/CD94 recognition of HLA-E. *Proc Natl Acad Sci U S A.* 2008;105:6696-6701.
261. Llano M, Lee N, Navarro F, et al. HLA-E-bound peptides influence recognition by inhibitory and triggering CD94/NKG2 receptors: preferential response to an HLA-G-derived nonamer. *Eur J Immunol.* 1998;28:2854-2863.
262. Kraft JR, Vance RE, Pohl J, Martin AM, Raulet DH, Jensen PE. Analysis of Qa-1^b Peptide binding specificity and the capacity of CD94/NKG2A to discriminate between Qa-1–peptide complexes. *J Exp Med.* 2000;192:613-624.
263. Michaëlsson J, Teixeira de Matos C, Achour A, Lanier LL, Kärre K, Söderström K. A signal peptide derived from hsp60 binds HLA-E and interferes with CD94/NKG2A recognition. *J Exp Med.* 2002;196:1403-1414.
264. Foley B, Cooley S, Verneris MR, et al. Human cytomegalovirus (CMV)-Induced memory-like NKG2C+ NK cells are transplantable and expand *in vivo* in response to recipient CMV antigen. *J Immunol.* 2012;189:5082-5088.
265. Lopez-Vergès S, Milush JM, Schwartz BS, et al. Expansion of a unique CD57+NKG2C^{hi} natural killer cell subset during acute

human cytomegalovirus infection. *Proc Natl Acad Sci U S A*. 2011;108:14725-14732.
266. Foley B, Cooley S, Verneris MR, et al. Cytomegalovirus reactivation after allogeneic transplantation promotes a lasting increase in educated NKG2C+ natural killer cells with potent function. *Blood*. 2012;119:2665-2674.
267. Béziat V, Dalgard O, Asselah T, et al. CMV drives clonal expansion of NKG2C+ NK cells expressing self-specific KIRs in chronic hepatitis patients. *Eur J Immunol*. 2012;42:447-457.
268. Gumá M, Budt M, Sáez A, et al. Expansion of CD94/NKG2C+ NK cells in response to human cytomegalovirus-infected fibroblasts. *Blood*. 2006;107:3624-3631.
269. Hammer Q, Rückert T, Borst EM, et al. Peptide-specific recognition of human cytomegalovirus strains controls adaptive natural killer cells. *Nat Immunol*. 2018;19:453-463.
270. Moser JM, Gibbs J, Jensen PE, Lukacher AE. CD94-NKG2A receptors regulate antiviral CD8+ T cell responses. *Nat Immunol*. 2002;3:189-195.
271. Rapaport AS, Schriewer J, Gilfillan S, et al. The inhibitory receptor NKG2A sustains virus-specific CD8+ T cells in response to a lethal poxvirus infection. *Immunity*. 2015;43:1112-1124.
272. André P, Denis C, Soulas C, et al. Anti-NKG2A mAb is a checkpoint inhibitor that promotes anti-tumor immunity by unleashing both T and NK cells. *Cell*. 2018;175:1731-1743.e13.
273. van Montfoort N, Borst L, Korrer MJ, et al. NKG2A blockade potentiates CD8 T cell immunity induced by cancer vaccines. *Cell*. 2018;175:1744-1755.e15.
274. Perussia B, Trinchieri G, Jackson A, et al. The Fc receptor for IgG on human natural killer cells: phenotypic, functional, and comparative studies with monoclonal antibodies. *J Immunol*. 1984;133:180-189.
275. Lanier LL, Yu G, Phillips JH. Co-association of CD3ζ with a receptor (CD16) for IgG Fc on human natural killer cells. *Nature*. 1989;342:803-805.
276. Kurosaki T, Gander I, Ravetch JV. A subunit common to an IgG Fc receptor and the T-cell receptor mediates assembly through different interactions. *Proc Natl Acad Sci U S A*. 1991;88:3837-3841.
277. Diefenbach A, Tomasello E, Lucas M, et al. Selective associations with signaling proteins determine stimulatory versus costimulatory activity of NKG2D. *Nat Immunol*. 2002;3:1142-1149.
278. Gilfillan S, Ho EL, Cella M, Yokoyama WM, Colonna M. NKG2D recruits two distinct adapters to trigger NK cell activation and costimulation. *Nat Immunol*. 2002;3:1150-1155.
279. Wu J, Song Y, Bakker ABH, et al. An activating immunoreceptor complex formed by NKG2D and DAP10. *Science*. 1999;285:730-732.
280. Rosen DB, Araki M, Hamerman JA, Chen T, Yamamura T, Lanier LL. A structural basis for the association of DAP12 with mouse, but not human, NKG2D. *J Immunol*. 2004;173:2470-2478.
281. André P, Castriconi R, Espéli M, et al. Comparative analysis of human NK cell activation induced by NKG2D and natural cytotoxicity receptors. *Eur J Immunol*. 2004;34:961-971.
282. Jamieson AM, Diefenbach A, McMahon CW, Xiong N, Carlyle JR, Raulet DH. The role of the NKG2D immunoreceptor in immune cell activation and natural killing. *Immunity*. 2002;17:19-29.
283. Roberts AI, Lee L, Schwarz E, et al. Cutting edge: NKG2D receptors induced by IL-15 costimulate CD28-negative effector CTL in the tissue microenvironment. *J Immunol*. 2001;167:5527-5530.
284. Ho EL, Carayannopoulos LN, Poursine-Laurent J, et al. Costimulation of multiple NK cell activation receptors by NKG2D. *J Immunol*. 2002;169:3667-3675.
285. Maasho K, Opoku-Anane J, Marusina AI, Coligan JE, Borrego F. Cutting edge: NKG2D is a costimulatory receptor for human naive CD8+ T cells. *J Immunol*. 2005;174:4480-4484.
286. Groh V, Rhinehart R, Randolph-Habecker J, Topp MS, Riddell SR, Spies T. Costimulation of CD8αβ T cells by NKG2D via engagement by MIC induced on virus-infected cells. *Nat Immunol*. 2001;2:255-260.
287. Bryceson YT, March ME, Ljunggren HG, Long EO. Synergy among receptors on resting NK cells for the activation of natural cytotoxicity and cytokine secretion. *Blood*. 2006;107:159-166.
288. Sáez-Borderías A, Gumá M, Angulo A, Bellosillo B, Pende D, López-Botet M. Expression and function of NKG2D in CD4+ T cells specific for human cytomegalovirus. *Eur J Immunol*. 2006;36:3198-3206.
289. Zompi S, Hamerman JA, Ogasawara K, et al. NKG2D triggers cytotoxicity in mouse NK cells lacking DAP12 or Syk family kinases. *Nat Immunol*. 2003;4:565-572.
290. Billadeau DD, Upshaw JL, Schoon RA, Dick CJ, Leibson PJ. NKG2D-DAP10 triggers human NK cell–mediated killing via a Syk-independent regulatory pathway. *Nat Immunol*. 2003;4:557-564.
291. Bauer S, Groh V, Wu J, et al. Activation of NK cells and T cells by NKG2D, a receptor for stress-inducible MICA. *Science*. 1999;285:727-729.
292. Cosman D, Müllberg J, Sutherland CL, et al. ULBPs, novel MHC class I–related molecules, bind to CMV glycoprotein UL16 and stimulate NK cytotoxicity through the NKG2D receptor. *Immunity*. 2001;14:123-133.
293. Jan Chalupny N, Sutherland CL, Lawrence WA, Rein-Weston A, Cosman D. ULBP4 is a novel ligand for human NKG2D. *Biochem Biophys Res Commun*. 2003;305:129-135.
294. Radosavljevic M, Cuillerier B, Wilson MJ, et al. A cluster of ten novel MHC class I related genes on human chromosome 6q24.2–q25.3. *Genomics*. 2002;79:114-123.
295. Onda H, Ohkubo S, Shintani Y, et al. A novel secreted tumor antigen with a glycosylphosphatidylinositol-anchored structure ubiquitously expressed in human cancers. *Biochem Biophys Res Commun*. 2001;285:235-243.
296. Bacon L, Eagle RA, Meyer M, Easom N, Young NT, Trowsdale J. Two human ULBP/RAET1 molecules with transmembrane regions are ligands for NKG2D. *J Immunol*. 2004;173:1078-1084.
297. Eagle RA, Traherne JA, Hair JR, Jafferji I, Trowsdale J. ULBP6/RAET1L is an additional human NKG2D ligand. *Eur J Immunol*. 2009;39:3207-3216.
298. Diefenbach A, Jensen ER, Jamieson AM, Raulet DH. Rae1 and H60 ligands of the NKG2D receptor stimulate tumour immunity. *Nature*. 2001;413:165-171.
299. Cerwenka A, Baron JL, Lanier LL. Ectopic expression of retinoic acid early inducible-1 gene (RAE-1) permits natural killer cell-mediated rejection of a MHC class I-bearing tumor *in vivo*. *Proc Natl Acad Sci U S A*. 2001;98:11521-11526.
300. Cerwenka A, Bakker ABH, McClanahan T, et al. Retinoic acid early inducible genes define a ligand family for the activating NKG2D receptor in mice. *Immunity*. 2000;12:721-727.
301. Carayannopoulos LN, Naidenko OV, Fremont DH, Yokoyama WM. Cutting edge: murine UL16-binding protein-like transcript 1. A newly described transcript encoding a high-affinity ligand for murine NKG2D. *J Immunol*. 2002;169:4079-4083.
302. Diefenbach A, Hsia JK, Hsiung MY, Raulet DH. A novel ligand for the NKG2D receptor activates NK cells and macrophages and induces tumor immunity. *Eur J Immunol*. 2003;33:381-391.
303. Diefenbach A, Jamieson AM, Liu SD, Shastri N, Raulet DH. Ligands for the murine NKG2D receptor: expression by tumor cells and activation of NK cells and macrophages. *Nat Immunol*. 2000;1:119-126.
304. Lodoen M, Ogasawara K, Hamerman JA, et al. NKG2D-mediated natural killer cell protection against cytomegalovirus is impaired by viral gp40 modulation of retinoic acid early inducible 1 gene molecules. *J Exp Med*. 2003;197:1245-1253.
305. Takada A, Yoshida S, Kajikawa M, et al. Two novel NKG2D ligands of the mouse H60 family with differential expression patterns and binding affinities to NKG2D. *J Immunol*. 2008;180:1678-1685.
306. Li P, Morris DL, Willcox BE, Steinle A, Spies T, Strong RK. Complex structure of the activating immunoreceptor NKG2D and its MHC class I–like ligand MICA. *Nat Immunol*. 2001;2:443-451.
307. Li P, McDermott G, Strong RK. Crystal structures of RAE-1β and its complex with the activating immunoreceptor NKG2D. *Immunity*. 2002;16:77-86.
308. Li P, Willie ST, Bauer S, Morris DL, Spies T, Strong RK. Crystal structure of the MHC class I homolog MIC-A, a γδ T cell ligand. *Immunity*. 1999;10:577-584.
309. Radaev S, Rostro B, Brooks AG, Colonna M, Sun PD. Conformational plasticity revealed by the cocrystal structure of NKG2D and its class I MHC-like ligand ULBP3. *Immunity*. 2001;15:1039-1049.
310. Li Z, Groh V, Strong RK, Spies T. A single amino acid substitution causes loss of expression of a MICA allele. *Immunogenetics*. 2000;51:246-248.
311. Steinle A, Li P, Morris DL, et al. Interactions of human NKG2D with its ligands MICA, MICB, and homologs of the mouse RAE-1 protein family. *Immunogenetics*. 2001;53:279-287.
312. Gasser S, Orsulic S, Brown EJ, Raulet DH. The DNA damage pathway regulates innate immune system ligands for the NKG2D receptor. *Nature*. 2005;436:1186-1190.
313. Groh V, Steinle A, Bauer S, Spies T. Recognition of stress-induced MHC molecules by intestinal epithelial γδ T cells. *Science*. 1998;279:1737-1740.
314. Hamerman JA, Ogasawara K, Lanier LL. Cutting edge: Toll-like receptor signaling in macrophages induces ligands for the NKG2D receptor. *J Immunol*. 2004;172:2001-2005.
315. Tokuyama M, Lorin C, Delebecque F, Jung H, Raulet DH, Coscoy L. Expression of the RAE-1 family of stimulatory NK-cell ligands requires activation of the PI3K pathway during viral infection and transformation. *PLoS Pathog*. 2011;7:e1002265.
316. Groh V, Bahram S, Bauer S, Herman A, Beauchamp M, Spies T. Cell stress-regulated human major histocompatibility complex class I gene expressed in gastrointestinal epithelium. *Proc Natl Acad Sci U S A*. 1996;93:12445-12450.
317. Raulet DH, Guerra N. Oncogenic stress sensed by the immune system: role of NK cell receptors. *Nat Rev Immunol*. 2009;9:568-580.
318. Stern-Ginossar N, Gur C, Biton M, et al. Human microRNAs regulate stress-induced immune responses mediated by the receptor NKG2D. *Nat Immunol*. 2008;9:1065-1073.
319. Bui JD, Carayannopoulos LN, Lanier LL, Yokoyama WM, Schreiber RD. IFN-dependent down-regulation of the NKG2D ligand H60 on tumors. *J Immunol*. 2006;176:905-913.
320. Yadav D, Ngolab J, Lim RSH, Krishnamurthy S, Bui JD. Cutting edge: down-regulation of MHC class I-related chain A on tumor cells by IFN-gamma-induced microRNA. *J Immunol*. 2009;182:39-43.
321. Guerra N, Tan YX, Joncker NT, et al. NKG2D-deficient mice are defective in tumor surveillance in models of spontaneous malignancy. *Immunity*. 2008;28:571-580.
322. Oppenheim DE, Roberts SJ, Clarke SL, et al. Sustained localized expression of ligand for the activating NKG2D receptor impairs natural cytotoxicity *in vivo* and reduces tumor immunosurveillance. *Nat Immunol*. 2005;6:928-937.
323. Wiemann K, Mittrücker HW, Feger U, et al. Systemic NKG2D down-regulation impairs NK and CD8 T cell responses *in vivo*. *J Immunol*. 2005;175:720-729.
324. Groh V, Wu J, Yee C, Spies T. Tumour-derived soluble MIC ligands impair expression of NKG2D and T-cell activation. *Nature*. 2002;419:734-738.
325. Strid J, Roberts SJ, Filler RB, et al. Acute upregulation of an NKG2D ligand promotes rapid reorganization of a local immune compartment with pleiotropic effects on carcinogenesis. *Nat Immunol*. 2008;9:146-154.
326. Barrow AD, Martin CJ, Colonna M. The natural cytotoxicity receptors in health and disease. *Front Immunol*. 2019;10:909. doi:10.3389/fimmu.2019.00909
327. Pessino A, Sivori S, Bottino C, et al. Molecular cloning of NKp46: a novel member of the immunoglobulin superfamily involved in triggering of natural cytotoxicity. *J Exp Med*. 1998;188:953-960.
328. Sivori S, Vitale M, Morelli L, et al. p46, a novel natural killer cell–specific surface molecule that mediates cell activation. *J Exp Med*. 1997;186:1129-1136.
329. Vitale M, Bottino C, Sivori S, et al. NKp44, a novel triggering surface molecule specifically expressed by activated natural killer cells, is involved in non-major histocompatibility complex-restricted tumor cell lysis. *J Exp Med*. 1998;187:2065-2072.
330. Pende D, Parolini S, Pessino A, et al. Identification and molecular characterization of NKp30, a novel triggering receptor involved in natural cytotoxicity mediated by human natural killer cells. *J Exp Med*. 1999;190:1505-1516.
331. Mandelboim O, Lieberman N, Lev M, et al. Recognition of haemagglutinins on virus-infected cells by NKp46 activates lysis by human NK cells. *Nature*. 2001;409:1055-1060.
332. Arnon TI, Achdout H, Lieberman N, et al. The mechanisms controlling the recognition of tumor- and virus-infected cells by NKp46. *Blood*. 2004;103:664-672.
333. Narni-Mancinelli E, Gauthier L, Baratin M, et al. Complement factor P is a ligand for the natural killer cell–activating receptor NKp46. *Science Immunology*. 2017;2:eaam9628. doi:10.1126/sciimmunol.aam9628
334. Glasner A, Ghadially H, Gur C, et al. Recognition and prevention of tumor metastasis by the NK receptor NKp46/NCR1. *J Immunol*. 2012;188:2509-2515.
335. Glasner A, Levi A, Enk J, et al. NKp46 receptor-mediated interferon-γ production by natural killer cells increases fibronectin 1 to alter tumor architecture and control metastasis. *Immunity*. 2018;48:107.e4-119.e4.
336. Gur C, Porgador A, Elboim M, et al. The activating receptor NKp46 is essential for the development of type 1 diabetes. *Nat Immunol*. 2010;11:121-128.
337. Halfteck GG, Elboim M, Gur C, Achdout H, Ghadially H, Mandelboim O. Enhanced *in vivo* growth of lymphoma tumors in the absence of the NK-activating receptor NKp46/NCR1. *J Immunol*. 2009;182:2221-2230.
338. Lakshmikanth T, Burke S, Ali TH, et al. NCRs and DNAM-1 mediate NK cell recognition and lysis of human and mouse melanoma cell lines *in vitro* and *in vivo*. *J Clin Invest*. 2009;119:1251-1263.
339. Sheppard S, Schuster IS, Andoniou CE, et al. The murine natural cytotoxic receptor NKp46/NCR1 controls TRAIL protein expression in NK cells and ILC1s. *Cell Rep*. 2018;22:3385-3392.
340. Baychelier F, Sennepin A, Ermonval M, Dorgham K, Debré P, Vieillard V. Identification of a cellular ligand for the natural cytotoxicity receptor NKp44. *Blood*. 2013;122:2935-2942.
341. Barrow AD, Edeling MA, Trifonov V, et al. Natural killer cells control tumor growth by sensing a growth factor. *Cell*. 2018;172:534.e19-548.e19.
342. Niehrs A, Garcia-Beltran WF, Norman PJ, et al. A subset of HLA-DP molecules serve as ligands for the natural cytotoxicity receptor NKp44. *Nat Immunol*. 2019;20:1129-1137.
343. Delahaye NF, Rusakiewicz S, Martins I, et al. Alternatively spliced NKp30 isoforms affect the prognosis of gastrointestinal stromal tumors. *Nat Med*. 2011;17:700-707.
344. Reiners KS, Topolar D, Henke A, et al. Soluble ligands for NK cell receptors promote evasion of chronic lymphocytic leukemia cells from NK cell anti-tumor activity. *Blood*. 2013;121:3658-3665.
345. Brandt CS, Baratin M, Yi EC, et al. The B7 family member B7-H6 is a tumor cell ligand for the activating natural killer cell receptor NKp30 in humans. *J Exp Med*. 2009;206:1495-1503.

346. Li Y, Wang Q, Mariuzza RA. Structure of the human activating natural cytotoxicity receptor NKp30 bound to its tumor cell ligand B7-H6. *J Exp Med.* 2011;208:703-714.
347. Binici J, Hartmann J, Herrmann J, et al. A soluble fragment of the tumor antigen BCL2-associated athanogene 6 (BAG-6) is essential and sufficient for inhibition of NKp30 receptor-dependent cytotoxicity of natural killer cells. *J Biol Chem.* 2013;288:34295-34303.
348. Garni-Wagner BA, Purohit A, Mathew PA, Bennett M, Kumar V. A novel function-associated molecule related to non-MHC-restricted cytotoxicity mediated by activated natural killer cells and T cells. *J Immunol.* 1993;151:60-70.
349. Veillette A. SLAM-family receptors: immune regulators with or without SAP-family adaptors. *Cold Spring Harb Perspect Biol.* 2010;2:a002469.
350. Schwartzberg PL, Mueller KL, Qi H, Cannons JL. SLAM receptors and SAP influence lymphocyte interactions, development and function. *Nat Rev Immunol.* 2009;9:39-46.
351. Nakajima H, Cella M, Langen H, Friedlein A, Colonna M. Activating interactions in human NK cell recognition: the role of 2B4-CD48. *Eur J Immunol.* 1999;29:1676-1683.
352. Bottino C, Falco M, Parolini S, et al. Gntb-A, a novel Sh2d1a-associated surface molecule contributing to the inability of natural killer cells to kill Epstein-Barr virus–infected B cells in X-linked lymphoproliferative disease. *J Exp Med.* 2001;194:235-246.
353. Bouchon A, Cella M, Grierson HL, Cohen JI, Colonna M. Cutting edge: activation of NK cell-mediated cytotoxicity by a SAP-independent receptor of the CD2 family. *J Immunol.* 2001;167:5517-5521.
354. Vaidya SV, Stepp SE, McNerney ME, et al. Targeted disruption of the 2B4 gene in mice reveals an in vivo role of 2B4 (CD244) in the rejection of B16 melanoma cells. *J Immunol.* 2005;174:800-807.
355. Lee K-M, McNerney ME, Stepp SE, et al. 2B4 acts as a non–major histocompatibility complex binding inhibitory receptor on mouse natural killer cells. *J Exp Med.* 2004;199: 1245-1254.
356. McNerney ME, Guzior D, Kumar V. 2B4 (CD244)-CD48 interactions provide a novel MHC class I-independent system for NK-cell self-tolerance in mice. *Blood.* 2005;106:1337-1340.
357. Cruz-Munoz ME, Dong Z, Shi X, Zhang S, Veillette A. Influence of CRACC, a SLAM family receptor coupled to the adaptor EAT-2, on natural killer cell function. *Nat Immunol.* 2009;10:297-305.
358. Sayos J, Wu C, Morra M, et al. The X-linked lymphoproliferative-disease gene product SAP regulates signals induced through the co-receptor SLAM. *Nature.* 1998;395: 462-469.
359. Latour S, Gish G, Helgason CD, Humphries RK, Pawson T, Veillette A. Regulation of SLAM-mediated signal transduction by SAP, the X-linked lymphoproliferative gene product. *Nat Immunol.* 2001;2:681-690.
360. Dong Z, Cruz-Munoz ME, Zhong MC, Chen R, Latour S, Veillette A. Essential function for SAP family adaptors in the surveillance of hematopoietic cells by natural killer cells. *Nat Immunol.* 2009;10:973-980.
361. de Andrade LF, Smyth MJ, Martinet L. DNAM-1 control of natural killer cells functions through nectin and nectin-like proteins. *Immunol Cell Biol.* 2014;92:237-244.
362. Fuchs A, Colonna M. The role of NK cell recognition of nectin and nectin-like proteins in tumor immunosurveillance. *Semin Cancer Biol.* 2006;16:359-366.
363. Martinet L, Smyth MJ. Balancing natural killer cell activation through paired receptors. *Nat Rev Immunol.* 2015;15:243-254.
364. Burns GF, Triglia T, Werkmeister JA, Begley CG, Boyd AW. TLiSA1, a human T lineage-specific activation antigen involved in the differentiation of cytotoxic T lymphocytes and anomalous killer cells from their precursors. *J Exp Med.* 1985;161:1063-1078.
365. Shibuya A, Campbell D, Hannum C, et al. DNAM-1, a novel adhesion molecule involved in the cytolytic function of T lymphocytes. *Immunity.* 1996;4:573-581.
366. Shibuya K, Lanier LL, Phillips JH, et al. Physical and functional association of LFA-1 with DNAM-1 adhesion molecule. *Immunity.* 1999;11:615-623.
367. Shibuya A, Lanier LL, Phillips JH. Protein kinase C is involved in the regulation of both signaling and adhesion mediated by DNAX accessory molecule-1 receptor. *J Immunol.* 1998;161:1671-1676.
368. Tahara-Hanaoka S, Shibuya K, Onoda Y, et al. Functional characterization of DNAM-1 (CD226) interaction with its ligands PVR (CD155) and nectin-2 (PRR-2/CD112). *Int Immunol.* 2004;16:533-538.
369. Bottino C, Castriconi R, Pende D, et al. Identification of PVR (CD155) and nectin-2 (CD112) as cell surface ligands for the human DNAM-1 (CD226) activating molecule. *J Exp Med.* 2003;198:557-567.
370. Iguchi-Manaka A, Kai H, Yamashita Y, et al. Accelerated tumor growth in mice deficient in DNAM-1 receptor. *J Exp Med.* 2008;205:2959-2964.
371. Gilfillan S, Chan CJ, Cella M, et al. DNAM-1 promotes activation of cytotoxic lymphocytes by nonprofessional antigen-presenting cells and tumors. *J Exp Med.* 2008;205:2965-2973.
372. Chan CJ, Andrews DM, McLaughlin NM, et al. DNAM-1/CD155 interactions promote cytokine and NK cell-mediated suppression of poorly immunogenic melanoma metastases. *J Immunol.* 2010;184:902-911.
373. Fuchs A, Cella M, Giurisato E, Shaw AS, Colonna M. Cutting edge: CD96 (tactile) promotes NK cell-target cell adhesion by interacting with the poliovirus receptor (CD155). *J Immunol.* 2004;172:3994-3998.
374. Chan CJ, Martinet L, Gilfillan S, et al. The receptors CD96 and CD226 oppose each other in the regulation of natural killer cell functions. *Nat Immunol.* 2014;15:431-438.
375. Yu X, Harden K, Gonzalez LC, et al. The surface protein TIGIT suppresses T cell activation by promoting the generation of mature immunoregulatory dendritic cells. *Nat Immunol.* 2009;10:48-57.
376. Levin SD, Taft DW, Brandt CS, et al. Vstm3 is a member of the CD28 family and an important modulator of T cell function. *Eur J Immunol.* 2011;41:902-915.
377. Boles KS, Vermi W, Facchetti F, et al. A novel molecular interaction for the adhesion of follicular CD4 T cells to follicular dendritic cells. *Eur J Immunol.* 2009;39:695-703.
378. Stanietsky N, Simic H, Arapovic J, et al. The interaction of TIGIT with PVR and PVRL2 inhibits human NK cell cytotoxicity. *Proc Natl Acad Sci U S A.* 2009;106:17858-17863.
379. Stanietsky N, Lenac T, Glasner A, et al. Mouse TIGIT inhibits NK-cell cytotoxicity upon interaction with PVR. *Eur J Immunol.* 2013;43:2138-2150.
380. Galibert L, Diemer GS, Liu Z, et al. Nectin-like protein 2 defines a subset of T-cell zone dendritic cells and is a ligand for class-I-restricted T-cell-associated molecule. *J Biol Chem.* 2005;280:21955-21964.
381. Arase N, Takeuchi A, Unno M, et al. Heterotypic interaction of CRTAM with Necl2 induces cell adhesion on activated NK cells and CD8+ T cells. *Int Immunol.* 2005;17:1227-1237.
382. Boles KS, Barchet W, Diacovo T, Cella M, Colonna M. The tumor suppressor TSLC1/NECL-2 triggers NK-cell and CD8+ T-cell responses through the cell-surface receptor CRTAM. *Blood.* 2005;106:779-786.
383. Takeuchi A, Itoh Y, Takumi A, et al. CRTAM confers late-stage activation of CD8+ T cells to regulate retention within lymph node. *J Immunol.* 2009;183:4220-4228.
384. Yeh JH, Sidhu SS, Chan AC. Regulation of a late phase of T cell polarity and effector functions by Crtam. *Cell.* 2008;132:846-859.
385. Moretta A, Bottino C, Vitale M, et al. Receptors for HLA class-I molecules in human natural killer cells. *Annu Rev Immunol.* 1996;14:619-648.
386. Sivori S, Vacca P, Del Zotto G, Munari E, Mingari MC, Moretta L, Human NK cells: surface receptors, inhibitory checkpoints, and translational applications. *Cell Mol Immunol.* 2019;16:430-441.
387. Tripathy SK, Keyel PA, Yang L, et al. Continuous engagement of a self-specific activation receptor induces NK cell tolerance. *J Exp Med.* 2008;205:1829-1841.
388. Sun JC, Lanier LL. Tolerance of NK cells encountering their viral ligand during development. *J Exp Med.* 2008;205:1819-1828.
389. Bix M, Liao NS, Zijlstra M, Loring J, Jaenisch R, Raulet D. Rejection of class I MHC-deficient haemopoietic cells by irradiated MHC-matched mice. *Nature.* 1991;349:329-331.
390. Kim S, Poursine-Laurent J, Truscott SM, et al. Licensing of natural killer cells by host major histocompatibility complex class I molecules. *Nature.* 2005;436:709-713.
391. Elliott JM, Wahle JA, Yokoyama WM. MHC class I–deficient natural killer cells acquire a licensed phenotype after transfer into an MHC class I–sufficient environment. *J Exp Med.* 2010;207:2073-2079.
392. Joncker NT, Shifrin N, Delebecque F, Raulet DH. Mature natural killer cells reset their responsiveness when exposed to an altered MHC environment. *J Exp Med.* 2010;207:2065-2072.
393. Fernandez NC, Treiner E, Vance RE, Jamieson AM, Lemieux S, Raulet DH. A subset of natural killer cells achieves self-tolerance without expressing inhibitory receptors specific for self-MHC molecules. *Blood.* 2005;105:4416-4423.
394. Anfossi N, André P, Guia S, et al. Human NK cell education by inhibitory receptors for MHC class I. *Immunity.* 2006;25:331-342.
395. Brodin P, Lakshmikanth T, Johansson S, Kärre K, Höglund P. The strength of inhibitory input during education quantitatively tunes the functional responsiveness of individual natural killer cells. *Blood.* 2009;113:2434-2441.
396. Guia S, Jaeger BN, Piatek S, et al. Confinement of activating receptors at the plasma membrane controls natural killer cell tolerance. *Sci Signal.* 2011;4:a21.
397. Mace EM, Orange JS. Emerging insights into human health and NK cell biology from the study of NK cell deficiencies. *Immunol Rev.* 2019;287:202-225.
398. Biron CA, Byron KS, Sullivan JL. Severe herpesvirus infections in an adolescent without natural killer cells. *N Engl J Med.* 1989;320:1731-1735.
399. Bukowski JF, Woda BA, Welsh RM. Pathogenesis of murine cytomegalovirus infection in natural killer cell-depleted mice. *J Virol.* 1984;52:119-128.
400. Orange JS, Biron CA. An absolute and restricted requirement for IL-12 in natural killer cell IFN-gamma production and antiviral defense. Studies of natural killer and T cell responses in contrasting viral infections. *J Immunol.* 1996;156:1138-1142.
401. Pien GC, Satoskar AR, Takeda K, Akira S, Biron CA. Cutting edge: selective IL-18 requirements for induction of compartmental IFN-γ responses during viral infection. *J Immunol.* 2000;165:4787-4791.
402. Orange JS, Wang B, Terhorst C, Biron CA. Requirement for natural killer cell-produced interferon gamma in defense against murine cytomegalovirus infection and enhancement of this defense pathway by interleukin 12 administration. *J Exp Med.* 1995;182:1045-1056.
403. Takeda K, Tsutsui H, Yoshimoto T, et al. Defective NK cell activity and Th1 response in IL-18–deficient mice. *Immunity.* 1998;8:383-390.
404. Paolini R, Bernardini G, Molfetta R, Santoni A. NK cells and interferons. *Cytokine Growth Factor Rev.* 2015;26:113-120.
405. Nguyen KB, Cousens LP, Doughty LA, Pien GC, Durbin JE, Biron CA. Interferon α/β-mediated inhibition and promotion of interferon γ: STAT1 resolves a paradox. *Nat Immunol.* 2000;1:70-76.
406. Lodoen MB, Abenes G, Umamoto S, Houchins JP, Liu F, Lanier LL. The cytomegalovirus m155 gene product subverts natural killer cell antiviral protection by disruption of H60–NKG2D interactions. *J Exp Med.* 2004;200:1075-1081.
407. Lenac T, Budt M, Arapovic J, et al. The herpesviral Fc receptor fcr-1 down-regulates the NKG2D ligands MULT-1 and H60. *J Exp Med.* 2006;203:1843-1850.
408. Krmpotic A, Hasan M, Loewendorf A, et al. NK cell activation through the NKG2D ligand MULT-1 is selectively prevented by the glycoprotein encoded by mouse cytomegalovirus gene m145. *J Exp Med.* 2005;201:211-220.
409. Stern-Ginossar N, Elefant N, Zimmermann A, et al. Host immune system gene targeting by a viral miRNA. *Science.* 2007;317:376-381.
410. Zou Y, Bresnahan W, Taylor RT, Stastny P. Effect of human cytomegalovirus on expression of MHC class I-related chains A. *J Immunol.* 2005;174:3098-3104.
411. Scalzo AA, Fitzgerald NA, Simmons A, La Vista AB, Shellam GR. Cmv-1, a genetic locus that controls murine cytomegalovirus replication in the spleen. *J Exp Med.* 1990;171:1469-1483.
412. Brown MG, Dokun AO, Heusel JW, et al. Vital involvement of a natural killer cell activation receptor in resistance to viral infection. *Science.* 2001;292:934-937.
413. Dokun AO, Kim S, Smith HRC, Kang HSP, Chu DT, Yokoyama WM. Specific and nonspecific NK cell activation during virus infection. *Nat Immunol.* 2001;2:951-956.
414. Sun JC, Beilke JN, Lanier LL. Adaptive immune features of natural killer cells. *Nature.* 2009;457:557-561.
415. Schlums H, Cichocki F, Tesi B, et al. Cytomegalovirus infection drives adaptive epigenetic diversification of NK cells with altered signaling and effector function. *Immunity.* 2015;42:443-456.
416. Cooper MA, Elliott JM, Keyel PA, Yang L, Carrero JA, Yokoyama WM. Cytokine-induced memory-like natural killer cells. *Proc Natl Acad Sci U S A.* 2009;106:1915-1919.
417. Romee R, Schneider SE, Leong JW, et al. Cytokine activation induces human memory-like NK cells. *Blood.* 2012;120:4751-4760.
418. O'Leary JG, Goodarzi M, Drayton DL, von Andrian UH. T cell– and B cell–independent adaptive immunity mediated by natural killer cells. *Nat Immunol.* 2006;7:507-516.
419. Paust S, Gill HS, Wang BZ, et al. Critical role for the chemokine receptor CXCR6 in NK cell-mediated antigen-specific memory of haptens and viruses. *Nat Immunol.* 2010;11:1127-1135.
420. Netea MG, Joosten LA, Latz E, et al. Trained immunity: a program of innate immune memory in health and disease. *Science.* 2016;352:aaf1098. doi:10.1126/science.aaf1098
421. Jost S, Altfeld M. Control of human viral infections by natural killer cells. *Annu Rev Immunol.* 2013;31:163-194.
422. Lodoen MB, Lanier LL. Natural killer cells as an initial defense against pathogens. *Curr Opin Immunol.* 2006;18:391-398.
423. Schmidt S, Tramsen L, Lehrnbecher T. Natural killer cells in antifungal immunity. *Front Immunol.* 2017;8:1623. doi:10.3389/fimmu.2017.01623
424. Goldszmid RS, Caspar P, Rivollier A, et al. NK cell-derived interferon-γ orchestrates the cellular dynamics and differentiation of monocytes into inflammatory dendritic cells at the site of infection. *Immunity.* 2012;36:1047-1059.
425. Moretta L, Locatelli F, Pende D, Marcenaro E, Mingari MC, Moretta A. Killer Ig–like receptor-mediated control of natural killer cell alloreactivity in haploidentical hematopoietic stem cell transplantation. *Blood.* 2011;117:764-771.
426. Ruggeri L, Capanni M, Casucci M, et al. Role of natural killer cell alloreactivity in HLA-mismatched hematopoietic stem cell transplantation. *Blood.* 1999;94:333-339.
427. Mancusi A, Ruggeri L, Urbani E, et al. Haploidentical hematopoietic transplantation from KIR ligand–mismatched donors with activating KIRs reduces nonrelapse mortality. *Blood.* 2015;125:3173-3182.
428. Oevermann L, Michaelis SU, Mezger M, et al. KIR B haplotype donors confer a reduced risk for relapse after

haploidentical transplantation in children with ALL. *Blood.* 2014;124:2744-2747.
429. Minetto P, Guolo F, Pesce S, et al. Harnessing NK cells for cancer treatment. *Front Immunol.* 2019;10:2836. doi:10.3389/fimmu.2019.02836
430. Mebius RE, Rennert P, Weissman IL. Developing lymph nodes collect CD4+CD3− LTβ+ cells that can differentiate to APC, NK cells, and follicular cells but not T or B cells. *Immunity.* 1997;7:493-504.
431. Schmutz S, Bosco N, Chappaz S, et al. Cutting edge: IL-7 regulates the peripheral pool of adult RORγ+ lymphoid tissue inducer cells. *J Immunol.* 2009;183:2217-2221.
432. Rennert PD, Browning JL, Mebius R, Mackay F, Hochman PS. Surface lymphotoxin α/β complex is required for the development of peripheral lymphoid organs. *J Exp Med.* 1996;184:1999-2006.
433. van de Pavert SA, Olivier BJ, Goverse G, et al. Chemokine CXCL13 is essential for lymph node initiation and is induced by retinoic acid and neuronal stimulation. *Nat Immunol.* 2009;10:1193-1199.
434. Finke D, Acha-Orbea H, Mattis A, Lipp M, Kraehenbuhl JP. CD4+CD3− cells induce Peyer's patch development: role of α4β1 integrin activation by CXCR5. *Immunity.* 2002;17:363-373.
435. Eberl G, Marmon S, Sunshine MJ, Rennert PD, Choi Y, Littman DR. An essential function for the nuclear receptor RORγt in the generation of fetal lymphoid tissue inducer cells. *Nat Immunol.* 2004;5:64-73.
436. Sun Z, Unutmaz D, Zou YR, et al. Requirement for RORγ in thymocyte survival and lymphoid organ development. *Science.* 2000;288:2369-2373.
437. Park SY, Saijo K, Takahashi T, et al. Developmental defects of lymphoid cells in Jak3 kinase-deficient mice. *Immunity.* 1995;3:771-782.
438. Luther SA, Ansel KM, Cyster JG. Overlapping roles of CXCL13, interleukin 7 receptor α, and CCR7 ligands in lymph node development. *J Exp Med.* 2003;197:1191-1198.
439. Adachi S, Yoshida H, Honda K, et al. Essential role of IL-7 receptor alpha in the formation of Peyer's patch anlage. *Int Immunol.* 1998;10:1-6.
440. Yoshida H, Honda K, Shinkura R, et al. IL-7 receptor α+ CD3− cells in the embryonic intestine induces the organizing center of Peyer's patches. *Int Immunol.* 1999;11:643-655.
441. Meier D, Bornmann C, Chappaz S, et al. Ectopic lymphoid-organ development occurs through interleukin 7-mediated enhanced survival of lymphoid-tissue-inducer cells. *Immunity.* 2007;26:643-654.
442. Chappaz S, Gärtner C, Rodewald HR, Finke D. Kit ligand and Il7 differentially regulate Peyer's patch and lymph node development. *J Immunol.* 2010;185:3514-3519.
443. van de Pavert SA, Ferreira M, Domingues RG, et al. Maternal retinoids control type 3 innate lymphoid cells and set the offspring immunity. *Nature.* 2014;508:123-127.
444. Kanamori Y, Ishimaru K, Nanno M, et al. Identification of novel lymphoid tissues in murine intestinal mucosa where clusters of c-kit⁺ IL-7R+ Thy1+ lympho-hemopoietic progenitors develop. *J Exp Med.* 1996;184:1449-1459.
445. Hamada H, Hiroi T, Nishiyama Y, et al. Identification of multiple isolated lymphoid follicles on the antimesenteric wall of the mouse small intestine. *J Immunol.* 2002;168:57-64.
446. Bouskra D, Brézillon C, Bérard M, et al. Lymphoid tissue genesis induced by commensals through NOD1 regulates intestinal homeostasis. *Nature.* 2008;456:507-510.
447. Pabst O, Herbrand H, Friedrichsen M, et al. Adaptation of solitary intestinal lymphoid tissue in response to microbiota and chemokine receptor CCR7 signaling. *J Immunol.* 2006;177:6824-6832.
448. Knoop KA, Newberry RD. Isolated lymphoid follicles are dynamic reservoirs for the induction of intestinal IgA. *Front Immunol.* 2012;3:84. doi:10.3389/fimmu.2012.00084
449. Nagatake T, Fukuyama S, Kim DY, et al. Id2-, RORγt-, and LTβR-independent initiation of lymphoid organogenesis in ocular immunity. *J Exp Med.* 2009;206:2351-2364.
450. Fukuyama S, Hiroi T, Yokota Y, et al. Initiation of NALT organogenesis is independent of the IL-7R, LTβR, and NIK signaling pathways but requires the Id2 gene and CD3−CD4+CD45+ cells. *Immunity.* 2002;17:31-40.
451. Rangel-Moreno J, Moyron-Quiroz JE, Carragher DM, et al. Omental milky spots develop in the absence of lymphoid tissue-inducer cells and support B and T cell responses to peritoneal antigens. *Immunity.* 2009;30:731-743.
452. Sawa S, Cherrier M, Lochner M, et al. Lineage relationship analysis of RORgt+ innate lymphoid cells. 2010;330:6.
453. Cupedo T, Crellin NK, Papazian N, et al. Human fetal lymphoid tissue–inducer cells are interleukin 17–producing precursors to RORC+ CD127+ natural killer–like cells. *Nat Immunol.* 2009;10:66-74.
454. Okada S, Markle JG, Deenick EK, et al. IMMUNODEFICIENCIES. Impairment of immunity to Candida and Mycobacterium in humans with bi-allelic RORC mutations. *Science.* 2015;349:606-613.
455. Sojka DK, Plougastel-Douglas B, Yang L, et al. Tissue-resident natural killer (NK) cells are cell lineages distinct from thymic and conventional splenic NK cells. *ELife.* 2014;3:e01659.
456. Peng H, Jiang X, Chen Y, et al. Liver-resident NK cells confer adaptive immunity in skin-contact inflammation. *J Clin Invest.* 2013;123:1444-1456.
457. Cortez VS, Fuchs A, Cella M, Gilfillan S, Colonna M. Cutting edge: salivary gland NK cells develop independently of Nfil3 in steady-state. *J Immunol.* 2014;192:4487-4491.
458. Fuchs A, Vermi W, Lee JS, et al. Intraepithelial type 1 innate lymphoid cells are a unique subset of cytokine responsive interferon-γ-producing cells. *Immunity.* 2013;38:769-781.
459. Vosshenrich CAJ, García-Ojeda ME, Samson-Villéger SI, et al. A thymic pathway of mouse natural killer cell development characterized by expression of GATA-3 and CD127. *Nat Immunol.* 2006;7:1217-1224.
460. Robinette ML, Fuchs A, Cortez VS, et al. Transcriptional programs define molecular characteristics of innate lymphoid cell classes and subsets. *Nat Immunol.* 2015;16:306-317.
461. Weizman OE, Adams NM, Schuster IS, et al. ILC1 confer early host protection at initial sites of viral infection. *Cell.* 2017;171:795-808.e12.
462. Bernink JH, Peters CP, Munneke M, et al. Human type 1 innate lymphoid cells accumulate in inflamed mucosal tissues. *Nat Immunol.* 2013;14:221-229.
463. Simoni Y, Fehlings M, Kløverpris HN, et al. Human innate lymphoid cell subsets possess tissue-type based heterogeneity in phenotype and frequency. *Immunity.* 2017;46:148-161.
464. Bernink JH, Krabbendam L, Germar K, et al. Interleukin-12 and -23 control plasticity of CD127(+) group 1 and group 3 innate lymphoid cells in the intestinal lamina propria. *Immunity.* 2015;43:146-160.
465. Cella M, Gamini R, Sécca C, et al. Subsets of ILC3-ILC1-like cells generate a diversity spectrum of innate lymphoid cells in human mucosal tissues. *Nat Immunol.* 2019;20:980-991.
466. Koues OI, Collins PL, Cella M, et al. Distinct gene regulatory pathways for human innate versus adaptive lymphoid cells. *Cell.* 2016;165:1134-1146.
467. Cortez VS, Cervantes-Barragan L, Robinette ML, et al. Transforming growth factor-β signaling guides the differentiation of innate lymphoid cells in salivary glands. *Immunity.* 2016;44:1127-1139.
468. Dadi S, Chhangawala S, Whitlock BM, et al. Cancer immunosurveillance by tissue-resident innate lymphoid cells and innate-like T cells. *Cell.* 2016;164:365-377.
469. Gao Y, Souza-Fonseca-Guimaraes F, Bald T, et al. Tumor immunoevasion by the conversion of effector NK cells into type 1 innate lymphoid cells. *Nat Immunol.* 2017;18:1004-1015.
470. Braga FAV, Hertoghs KML, Kragten NAM, et al. Blimp-1 homolog Hobit identifies effector-type lymphocytes in humans. *Eur J Immunol.* 2015;45:2945-2958.
471. Mackay LK, Minnich M, Kragten NAM, et al. Hobit and Blimp1 instruct a universal transcriptional program of tissue residency in lymphocytes. *Science.* 2016;352:459-463.
472. Mackay LK, Rahimpour A, Ma JZ, et al. The developmental pathway for CD103+ CD8+ tissue-resident memory T cells of skin. *Nat Immunol.* 2013;14:1294-1301.
473. van Gisbergen KP, Kragten NA, Hertoghs KM, et al. Mouse Hobit is a homolog of the transcriptional repressor Blimp-1 that regulates NKT cell effector differentiation. *Nat Immunol.* 2012;13:864-871.
474. Doisne JM, Bartholin L, Yan KP, et al. iNKT cell development is orchestrated by different branches of TGF-β signaling. *J Exp Med.* 2009;206:1365-1378.
475. Nabekura T, Riggan L, Hildreth AD, O'Sullivan TE, Shibuya A. Type 1 innate lymphoid cells protect mice from acute liver injury via interferon-γ secretion for upregulating Bcl-xL expression in hepatocytes. *Immunity.* 2020;52(1):96.e9-108.e9. doi:10.1016/j.immuni.2019.11.004
476. Abt MC, Lewis BB, Caballero S, et al. Innate immune defenses mediated by two ILC subsets are critical for protection against acute *Clostridium difficile* infection. *Cell Host Microbe.* 2015;18:27-37.
477. Moro K, Yamada T, Tanabe M, et al. Innate production of T(H)2 cytokines by adipose tissue-associated c-Kit⁺ Sca-1+ lymphoid cells. *Nature.* 2010;463:540-544.
478. Neill DR, Wong SH, Bellosi A, et al. Nuocytes represent a new innate effector leukocyte that mediates type-2 immunity. *Nature.* 2010;464:1367-1370.
479. Price AE, Liang HE, Sullivan BM, et al. Systemically dispersed innate IL-13–expressing cells in type 2 immunity. *Proc Natl Acad Sci U S A.* 2010;107:11489-11494.
480. Ricardo-Gonzalez RR, Van Dyken SJ, Schneider C, et al. Tissue signals imprint ILC2 identity with anticipatory function. *Nat Immunol.* 2018;19:1093-1099.
481. Wolterink RG, Serafini N, van Nimwegen M, et al. Essential, dose-dependent role for the transcription factor GATA3 in the development of IL-5+ and IL-13+ type 2 innate lymphoid cells. *Proc Natl Acad Sci U S A.* 2013;110:10240-10245.
482. Furusawa J, Moro K, Motomura Y, et al. Critical role of p38 and GATA3 in natural helper cell function. *J Immunol.* 2013;191:1818-1826.
483. Mjösberg J, Bernink J, Golebski K, et al. The transcription factor GATA3 is essential for the function of human type 2 innate lymphoid cells. *Immunity.* 2012;37:649-659.
484. Zhong C, Zheng M, Cui K, et al. Differential expression of the transcription factor GATA3 specifies lineage and functions of innate lymphoid cells. *Immunity.* 2019;52(1):83.e4-95.e4. doi:10.1016/j.immuni.2019.12.001
485. Nussbaum JC, Van Dyken SJ, von Moltke J, et al. Type 2 innate lymphoid cells control eosinophil homeostasis. *Nature.* 2013;502:245-248.
486. Mjösberg JM, Trifari S, Crellin NK, et al. Human IL-25- and IL-33-responsive type 2 innate lymphoid cells are defined by expression of CRTH2 and CD161. *Nat Immunol.* 2011;12:1055-1062.
487. Turner JE, Morrison PJ, Wilhelm C, et al. IL-9–mediated survival of type 2 innate lymphoid cells promotes damage control in helminth-induced lung inflammation. *J Exp Med.* 2013;210:2951-2965.
488. Licona-Limón P, Henao-Mejia J, Temann AU, et al. Th9 cells drive host immunity against gastrointestinal worm infection. *Immunity.* 2013;39:744-757.
489. Wilhelm C, Hirota K, Stieglitz B, et al. Interleukin 9 fate reporter reveals induction of innate IL-9 response in lung inflammation. *Nat Immunol.* 2011;12:1071-1077.
490. Monticelli LA, Sonnenberg GF, Abt MC, et al. Innate lymphoid cells promote lung-tissue homeostasis after infection with influenza virus. *Nat Immunol.* 2011;12:1045-1054.
491. Monticelli LA, Osborne LC, Noti M, Tran SV, Zaiss DMW, Artis D. IL-33 promotes an innate immune pathway of intestinal tissue protection dependent on amphiregulin–EGFR interactions. *Proc Natl Acad Sci U S A.* 2015;112:10762-10767.
492. Bando JK, Gilfillan S, Di Luccia B, et al. ILC2s are the predominant source of intestinal ILC-derived IL-10. *J Exp Med.* 2020;217:e20191520. doi:10.1084/jem.20191520
493. Seehus CR, Kadavallore A, de la Torre B, et al. Alternative activation generates IL-10 producing type 2 innate lymphoid cells. *Nat Commun.* 2017;8:1900.
494. Liang HE, Reinhardt RL, Bando JK, Sullivan BM, Ho IC, Locksley RM. Divergent expression patterns of IL-4 and IL-13 define unique functions in allergic immunity. *Nat Immunol.* 2012;13:58-66.
495. Mohrs K, Wakil AE, Killeen N, Locksley RM, Mohrs M. A two-step process for cytokine production revealed by IL-4 dual-reporter mice. *Immunity.* 2005;23:419-429.
496. Pelly VS, Kannan Y, Coomes SM, et al. IL-4-producing ILC2s are required for the differentiation of T(H)2 cells following Heligmosomoides polygyrus infection. *Mucosal Immunol.* 2016;9:1407-1417.
497. Doherty TA, Khorram N, Lund S, Mehta AK, Croft M, Broide DH. Lung type 2 innate lymphoid cells express cysteinyl leukotriene receptor 1, which regulates TH2 cytokine production. *J Allergy Clin Immunol.* 2013;132, 205-213.
498. Zhou B, Comeau MR, Smedt TD, et al. Thymic stromal lymphopoietin as a key initiator of allergic airway inflammation in mice. *Nat Immunol.* 2005;6:1047-1053.
499. Fort MM, Cheung J, Yen D, et al. IL-25 induces IL-4, IL-5, and IL-13 and Th2-associated pathologies *in vivo*. *Immunity.* 2001;15:985-995.
500. Schmitz J, Owyang A, Oldham E, et al. IL-33, an interleukin-1-like cytokine that signals via the IL-1 receptor-related protein ST2 and induces T helper type 2-associated cytokines. *Immunity.* 2005;23:479-490.
501. Townsend MJ, Fallon PG, Matthews DJ, Jolin HE, McKenzie ANJ. T1/ST2-Deficient mice demonstrate the importance of T1/ST2 in developing primary T helper cell type 2 responses. *J Exp Med.* 2000;191:1069-1076.
502. Van Dyken SJ, Mohapatra A, Nussbaum JC, et al. Chitin activates parallel immune modules that direct distinct inflammatory responses via innate lymphoid type 2 and γδ T cells. *Immunity.* 2014;40:414-424.
503. Molofsky AB, Van Gool F, Liang HE, et al. Interleukin-33 and interferon-γ counter-regulate group 2 innate lymphoid cell activation during immune perturbation. *Immunity.* 2015;43:161-174.
504. von Moltke J, O'Leary CE, Barrett NA, Kanaoka Y, Austen KF, Locksley RM. Leukotrienes provide an NFAT-dependent signal that synergizes with IL-33 to activate ILC2s. *J Exp Med.* 2017;214:27-37.
505. Xue L, Salimi M, Panse I, et al. Prostaglandin D2 activates group 2 innate lymphoid cells through chemoattractant receptor-homologous molecule expressed on TH2 cells. *J Allergy Clin Immunol.* 2014;133:1184-1194.e7.
506. Maric J, Ravindran A, Mazzurana L, et al. Prostaglandin E2 suppresses human group 2 innate lymphoid cell function. *J Allergy Clin Immunol.* 2018;141:1761-1773.e6.
507. Yu X, Pappu R, Ramirez-Carrozzi V, et al. TNF superfamily member TL1A elicits type 2 innate lymphoid cells at mucosal barriers. *Mucosal Immunol.* 2014;7:730-740.
508. Grencis RK. Immunity to helminths: resistance, regulation, and susceptibility to gastrointestinal nematodes. *Annu Rev Immunol.* 2015;33:201-225.
509. Urban JF, Noben-Trauth N, Donaldson DD, et al. IL-13, IL-4Rα, and Stat6 are required for the expulsion of the gastrointestinal nematode parasite *Nippostrongylus brasiliensis.* *Immunity.* 1998;8:255-264.
510. Voehringer D, Reese TA, Huang X, Shinkai K, Locksley RM. Type 2 immunity is controlled by IL-4/IL-13 expression in hematopoietic non-eosinophil cells of the innate immune system. *J Exp Med.* 2006;203:1435-1446.

511. Oeser K, Schwartz C, Voehringer D. Conditional IL-4/IL-13-deficient mice reveal a critical role of innate immune cells for protective immunity against gastrointestinal helminths. *Mucosal Immunol.* 2015;8:672-682.
512. Ohnmacht C, Schwartz C, Panzer M, Schiedewitz I, Naumann R, Voehringer D. Basophils orchestrate chronic allergic dermatitis and protective immunity against helminths. *Immunity.* 2010;33:364-374.
513. Fallon PG, Ballantyne SJ, Mangan NE, et al. Identification of an interleukin (IL)-25–dependent cell population that provides IL-4, IL-5, and IL-13 at the onset of helminth expulsion. *J Exp Med.* 2006;203:1105-1116.
514. von Moltke J, Ji M, Liang HE, Locksley RM. Tuft-cell-derived IL-25 regulates an intestinal ILC2–epithelial response circuit. *Nature.* 2016;529:221-225.
515. Gerbe F, Sidot E, Smyth DJ, et al. Intestinal epithelial tuft cells initiate type 2 mucosal immunity to helminth parasites. *Nature.* 2016;529:226-230.
516. Howitt MR, Lavoie S, Michaud M, et al. Tuft cells, taste-chemosensory cells, orchestrate parasite type 2 immunity in the gut. *Science.* 2016;351:1329-1333.
517. Schneider C, O'Leary CE, von Moltke J, et al. A metabolite-triggered tuft cell-ILC2 circuit drives small intestinal remodeling. *Cell.* 2018;174:271-284.e14.
518. Nadjsombati MS, McGinty JW, Lyons-Cohen MR, et al. Detection of succinate by intestinal tuft cells triggers a type 2 innate immune circuit. *Immunity.* 2018;49:33.e7-41.e7.
519. Smith SG, Chen R, Kjarsgaard M, et al. Increased numbers of activated group 2 innate lymphoid cells in the airways of patients with severe asthma and persistent airway eosinophilia. *J Allergy Clin Immunol.* 2016;137:75-86.e8.
520. Chang YJ, Kim HY, Albacker LA, et al. Innate lymphoid cells mediate influenza-induced airway hyper-reactivity independently of adaptive immunity. *Nat Immunol.* 2011;12:631-638.
521. Kim BS, Siracusa MC, Saenz SA, et al. TSLP elicits IL-33–independent innate lymphoid cell responses to promote skin inflammation. *Sci Transl Med.* 2013;5:170ra16.
522. Salimi M, Barlow JL, Saunders SP, et al. A role for IL-25 and IL-33–driven type-2 innate lymphoid cells in atopic dermatitis. *J Exp Med.* 2013;210:2939-2950.
523. Molofsky AB, Nussbaum JC, Liang HE, et al. Innate lymphoid type 2 cells sustain visceral adipose tissue eosinophils and alternatively activated macrophages. *J Exp Med.* 2013;210:535-549.
524. Qiu Y, Nguyen KD, Odegaard JI, et al. Eosinophils and type 2 cytokine signaling in macrophages orchestrate development of functional beige fat. *Cell.* 2014;157:1292-1308.
525. Wu D, Molofsky AB, Liang HE, et al. Eosinophils sustain adipose alternatively activated macrophages associated with glucose homeostasis. *Science.* 2011;332:243-247.
526. Hams E, Locksley RM, McKenzie ANJ, Fallon PG. IL-25 elicits innate lymphoid type 2 and type II natural killer T cells that regulate obesity in mice1. *J Immunol.* 2013;191:5349-5353. doi:10.4049/jimmunol.1301176
527. Lee MW, Odegaard JI, Mukundan L, et al. Activated type 2 innate lymphoid cells regulate beige fat biogenesis. *Cell.* 2015;160:74-87.
528. Brestoff JR, Kim BS, Saenz SA, et al. Group 2 innate lymphoid cells promote beiging of white adipose tissue and limit obesity. *Nature.* 2015;519:242-246.
529. Dalmas E, Lehmann FM, Dror E, et al. Interleukin-33-activated islet-resident innate lymphoid cells promote insulin secretion through myeloid cell retinoic acid production. *Immunity.* 2017;47:928.e7-942.e7.
530. Satoh-Takayama N, Lesjean-Pottier S, Vieira P, et al. IL-7 and IL-15 independently program the differentiation of intestinal CD3–NKp46+ cell subsets from Id2-dependent precursors. *J Exp Med.* 2010;207:273-280.
531. Crellin NK, Trifari S, Kaplan CD, Cupedo T, Spits H. Human NKp44+IL-22+ cells and LTi-like cells constitute a stable RORC+ lineage distinct from conventional natural killer cells. *J Exp Med.* 2010;207:281-290.
532. Vonarbourg C, Mortha A, Bui VL, et al. Regulated expression of nuclear receptor RORγt confers distinct functional fates to NK cell receptor-expressing RORγt(+) innate lymphocytes. *Immunity.* 2010;33:736-751.
533. Kiss EA, Vonarbourg C, Kopfmann S, et al. Natural aryl hydrocarbon receptor ligands control organogenesis of intestinal lymphoid follicles. *Science.* 2011;334:1561-1565.
534. Li S, Heller JJ, Bostick JW, et al. Ikaros inhibits group 3 innate lymphoid cell development and function by suppressing the aryl hydrocarbon receptor pathway. *Immunity.* 2016;45:185-197.
535. Savage AK, Liang HE, Locksley RM. The development of steady-state activation hubs between adult LTi ILC3s and primed macrophages in small intestine. *J Immunol.* 2017;199:1912-1922.
536. Goto Y, Obata T, Kunisawa J, et al. Innate lymphoid cells regulate intestinal epithelial cell glycosylation. *Science.* 2014;345:1254009.
537. Zheng Y, Valdez PA, Danilenko DM, et al. Interleukin-22 mediates early host defense against attaching and effacing bacterial pathogens. *Nat Med.* 2008;14:282-289.
538. Tsai PY, Zhang B, He WQ, et al. IL-22 upregulates epithelial claudin-2 to drive diarrhea and enteric pathogen clearance. *Cell Host Microbe.* 2017;21, 671.e4-681.e4.
539. Pickard JM, Maurice CF, Kinnebrew MA, et al. Rapid fucosylation of intestinal epithelium sustains host-commensal symbiosis in sickness. *Nature.* 2014;514:638-641.
540. Pickert G, Neufert C, Leppkes M, et al. STAT3 links IL-22 signaling in intestinal epithelial cells to mucosal wound healing. *J Exp Med.* 2009;206:1465-1472.
541. Hanash AM, Dudakov JA, Hua G, et al. Interleukin-22 protects intestinal stem cells from immune-mediated tissue damage and regulates sensitivity to graft versus host disease. *Immunity.* 2012;37:339-350.
542. Gladiator A, Wangler N, Trautwein-Weidner K, LeibundGut-Landmann S. Cutting edge: IL-17–secreting innate lymphoid cells are essential for host defense against fungal infection. *J Immunol.* 2013;190:521-525.
543. Pappu R, Rutz S, Ouyang W. Regulation of epithelial immunity by IL-17 family cytokines. *Trends Immunol.* 2012;33:343-349.
544. Song C, Lee JS, Gilfillan S, et al. Unique and redundant functions of NKp46+ ILC3s in models of intestinal inflammation. *J Exp Med.* 2015;212:1869-1882.
545. Buonocore S, Ahern PP, Uhlig HH, et al. Innate lymphoid cells drive interleukin-23-dependent innate intestinal pathology. *Nature.* 2010;464:1371-1375.
546. Pearson C, Thornton EE, McKenzie B, et al. ILC3 GM-CSF production and mobilisation orchestrate acute intestinal inflammation. *Elife.* 2016;5-e10066.
547. Mortha A, Chudnovskiy A, Hashimoto D, et al. Microbiota-dependent crosstalk between macrophages and ILC3 promotes intestinal homeostasis. *Science.* 2014;343:1249288. doi:10.1126/science.1249288
548. Longman RS, Diehl GE, Victorio DA, et al. CX3CR1+ mononuclear phagocytes support colitis-associated innate lymphoid cell production of IL-22. *J Exp Med.* 2014;211:1571-1583.
549. Castellanos JG, Woo V, Viladomiu M, et al. Microbiota-induced TNF-like ligand 1A drives group 3 innate lymphoid cell-mediated barrier protection and intestinal T cell activation during colitis. *Immunity.* 2018;49:1077.e5-1089.e5.
550. Bauché D, Joyce-Shaikh B, Fong J, et al. IL-23 and IL-2 activation of STAT5 is required for optimal IL-22 production in ILC3s during colitis. *Sci Immunol.* 2020;5:eaav1080. doi:10.1126/sciimmunol.aav1080
551. Crellin NK, Trifari S, Kaplan CD, Satoh-Takayama N, Di Santo JP, Spits H. Regulation of cytokine secretion in human CD127+ LTi-like innate lymphoid cells by Toll-like receptor 2. *Immunity.* 2010;33:752-764.
552. Glatzer T, Killig M, Meisig J, et al. RORγt+ innate lymphoid cells acquire a proinflammatory program upon engagement of the activating receptor NKp44. *Immunity.* 2013;38:1223-1235.
553. Cella M, Otero K, Colonna M. Expansion of human NK-22 cells with IL-7, IL-2, and IL-1β reveals intrinsic functional plasticity. *Proc Natl Acad Sci U S A.* 2010;107:10961-10966.
554. Takatori H, Kanno Y, Watford WT, et al. Lymphoid tissue inducer–like cells are an innate source of IL-17 and IL-22. *J Exp Med.* 2009;206:35-41.
555. Willinger T. Metabolic control of innate lymphoid cell migration. *Front Immunol.* 2019;10:2010.
556. Satoh-Takayama N, Serafini N, Verrier T, et al. The chemokine receptor CXCR6 controls the functional topography of interleukin-22 producing intestinal innate lymphoid cells. *Immunity.* 2014;41:776-788.
557. Sciumé G, Hirahara K, Takahashi H, et al. Distinct requirements for T-bet in gut innate lymphoid cells. *J Exp Med.* 2012;209:2331-2338.
558. Rankin LC, Groom JR, Chopin M, et al. The transcription factor T-bet is essential for the development of NKp46+ innate lymphocytes via the Notch pathway. *Nat Immunol.* 2013;14:389-395.
559. Klose CSN, Kiss EA, Schwierzeck V, et al. A T-bet gradient controls the fate and function of CCR6-RORγt+ innate lymphoid cells. *Nature.* 2013;494:261-265.
560. Lügering A, Ross M, Sieker M, et al. CCR6 identifies lymphoid tissue inducer cells within cryptopatches. *Clin Exp Immunol.* 2010;160:440-449.
561. Lindemans CA, Calafiore M, Mertelsmann AM, et al. Interleukin-22 promotes intestinal-stem-cell-mediated epithelial regeneration. *Nature.* 2015;528:560-564.
562. Gronke K, Hernández PP, Zimmermann J, et al. Interleukin-22 protects intestinal stem cells against genotoxic stress. *Nature.* 2019;566:249.
563. Sonnenberg GF, Monticelli LA, Elloso MM, Fouser LA, Artis D. CD4+ lymphoid tissue inducer cells promote innate immunity in the gut. *Immunity.* 2011;34:122-134.
564. Tsuji M, Suzuki K, Kitamura H, et al. Requirement for lymphoid tissue-inducer cells in isolated follicle formation and T cell-independent immunoglobulin A generation in the gut. *Immunity.* 2008;29:261-271.
565. Kruglov AA, Grivennikov SI, Kuprash DV, et al. Nonredundant function of soluble LTα3 produced by innate lymphoid cells in intestinal homeostasis. *Science.* 2013;342:1243-1246.
566. Reboldi A, Arnon TI, Rodda LB, Atakilit A, Sheppard D, Cyster JG. IgA production requires B cell interaction with subepithelial dendritic cells in Peyer's patches. *Science.* 2016;352:aaf4822.
567. Bando JK, Gilfillan S, Song C, et al. The tumor necrosis factor superfamily member RANKL suppresses effector cytokine production in group 3 innate lymphoid cells. *Immunity.* 2018;48:1208-1219.e4.
568. Tumanov AV, Koroleva EP, Guo X, et al. Lymphotoxin controls the IL-22 protection pathway in gut innate lymphoid cells during mucosal pathogen challenge. *Cell Host Microbe.* 2011;10:44-53.
569. Sawa S, Lochner M, Satoh-Takayama N, et al. RORγt+ innate lymphoid cells regulate intestinal homeostasis by integrating negative signals from the symbiotic microbiota. *Nat Immunol.* 2011;12:320-326.
570. Cervantes-Barragan L, Colonna M. AHR signaling in the development and function of intestinal immune cells and beyond. *Semin Immunopathol.* 2018;40:371-377.
571. Stockinger B, Meglio PD, Gialitakis M, Duarte JH. The aryl hydrocarbon receptor: multitasking in the immune system. *Annu Rev Immunol.* 2014;32:403-432.
572. Cervantes-Barragan L, Chai JN, Tianero MD, et al. Lactobacillus reuteri induces gut intraepithelial CD4+CD8αα+ T cells. *Science.* 2017;357:806-810.
573. Lee JS, Cella M, McDonald KG, et al. AHR drives the development of gut ILC22 cells and postnatal lymphoid tissues via pathways dependent on and independent of Notch. *Nat Immunol.* 2012;13:144-151.
574. Qiu J, Heller JJ, Guo X, et al. The aryl hydrocarbon receptor regulates gut immunity through modulation of innate lymphoid cells. *Immunity.* 2012;36:92-104.
575. Hughes T, Briercheck EL, Freud AG, et al. The transcription factor AHR prevents the differentiation of a stage 3 innate lymphoid cell subset to natural killer cells. *Cell Rep.* 2014;8:150-162.
576. Mielke LA, Jones SA, Raverdeau M, et al. Retinoic acid expression associates with enhanced IL-22 production by γδ T cells and innate lymphoid cells and attenuation of intestinal inflammation. *J Exp Med.* 2013;210:1117-1124.
577. Larange A, Cheroutre H. Retinoic acid and retinoic acid receptors as pleiotropic modulators of the immune system. *Annu Rev Immunol.* 2016;34:369-394.
578. Spencer SP, Wilhelm C, Yang Q, et al. Adaptation of innate lymphoid cells to a micronutrient deficiency promotes type 2 barrier immunity. *Science.* 2014;343:432-437.
579. Emgård J, Kammoun H, García-Cassani B, et al. Oxysterol sensing through the receptor GPR183 promotes the lymphoid-tissue-inducing function of innate lymphoid cells and colonic inflammation. *Immunity.* 2018;48:120-132.e8.
580. Chu C, Moriyama S, Li Z, et al. Anti-microbial functions of group 3 innate lymphoid cells in gut-associated lymphoid tissues are regulated by G-protein-coupled receptor 183. *Cell Rep.* 2018;23:3750-3758.
581. Chun E, Lavoie S, Fonseca-Pereira D, et al. Metabolite-sensing receptor Ffar2 regulates colonic group 3 innate lymphoid cells and gut immunity. *Immunity.* 2019;51:871-884.e6.
582. Fachi JL, Sécca C, Rodrigues PB, et al. Acetate coordinates neutrophil and ILC3 responses against *C. difficile* through FFAR2. *J Exp Med.* 2020;217:jem.20190489. doi:10.1084/jem.20190489
583. Sonnenberg GF, Monticelli LA, Alenghat T, et al. Innate lymphoid cells promote anatomical containment of lymphoid-resident commensal bacteria. *Science.* 2012;336:1321-1325.
584. Rankin LC, Girard-Madoux MJH, Seillet C, et al. Complementarity and redundancy of IL-22-producing innate lymphoid cells. *Nat Immunol.* 2016;17:179-186.
585. Hernández PP, Mahlakõiv T, Yang I, et al. Interferon-λ and interleukin 22 act synergistically for the induction of interferon-stimulated genes and control of rotavirus infection. *Nat Immunol.* 2015;16:698-707.
586. Zelante T, Iannitti RG, Cunha C, et al. Tryptophan catabolites from microbiota engage aryl hydrocarbon receptor and balance mucosal reactivity via interleukin-22. *Immunity.* 2013;39:372-385.
587. Melo-Gonzalez F, Kammoun H, Evren E, et al. Antigen-presenting ILC3 regulate T cell–dependent IgA responses to colonic mucosal bacteria. *J Exp Med.* 2019;216:728-742.
588. Khor B, Gardet A, Xavier RJ. Genetics and pathogenesis of inflammatory bowel disease. *Nature.* 2011;474:307-317.
589. Moschen AR, Tilg H, Raine T. IL-12, IL-23 and IL-17 in IBD: immunobiology and therapeutic targeting. *Nat Rev Gastroenterol Hepatol.* 2019;16:185-196.
590. Takayama T, Kamada N, Chinen H, et al. Imbalance of NKp44+NKp46– and NKp44–NKp46+ natural killer cells in the intestinal mucosa of patients with Crohn's disease. *Gastroenterology.* 2010;139:882-892.e3.
591. Teng F, Goc J, Zhou L, et al. A circadian clock is essential for homeostasis of group 3 innate lymphoid cells in the gut. *Sci Immunol.* 2019;4:eaax1215. doi:10.1126/sciimmunol.aax1215
592. Godinho-Silva C, Cardoso F, Veiga-Fernandes H. Neuro–immune cell units: a new paradigm in physiology. *Annu Rev Immunol.* 2019;37:19-46.
593. Veiga-Fernandes H, Coles MC, Foster KE, et al. Tyrosine kinase receptor RET is a key regulator of Peyer's patch organogenesis. *Nature.* 2007;446:547-551.
594. Ibiza S, García-Cassani B, Ribeiro H, et al. Glial-cell-derived neuroregulators control type 3 innate lymphoid cells and gut defence. *Nature.* 2016;535:440-443.

595. Benschop RJ, Rodriguez-Feuerhahn M, Schedlowski M. Catecholamine-induced leukocytosis: early observations, current research, and future directions. *Brain Behav Immun.* 1996;10:77-91.
596. Shakhar G, Ben-Eliyahu S. *In vivo* β-adrenergic stimulation suppresses natural killer activity and compromises resistance to tumor metastasis in rats. *J Immunol.* 1998;160:3251-3258.
597. Moriyama S, Brestoff JR, Flamar AL, et al. β2-adrenergic receptor–mediated negative regulation of group 2 innate lymphoid cell responses. *Science.* 2018;359:1056-1061.
598. Wallrapp A, Riesenfeld SJ, Burkett PR, et al. The neuropeptide NMU amplifies ILC2-driven allergic lung inflammation. *Nature.* 2017;549:351-356.
599. Cardoso V, Chesné J, Ribeiro H, et al. Neuronal regulation of type 2 innate lymphoid cells via neuromedin U. *Nature.* 2017;549:277-281.
600. Klose CSN, Mahlakõiv T, Moeller JB, et al. The neuropeptide neuromedin U stimulates innate lymphoid cells and type 2 inflammation. *Nature.* 2017;549:282-286.
601. Nagashima H, Mahlakõiv T, Shih HY, et al. Neuropeptide CGRP limits group 2 innate lymphoid cell responses and constrains type 2 inflammation. *Immunity.* 2019;51:682.e6-695.e6.
602. Xu H, Ding J, Porter CBM, et al. Transcriptional atlas of intestinal immune cells reveals that neuropeptide α-CGRP modulates group 2 innate lymphoid cell responses. *Immunity.* 2019;51:696-708.e9.
603. Wallrapp A, Burkett PR, Riesenfeld SJ, et al. Calcitonin gene-related peptide negatively regulates alarmin-driven type 2 innate lymphoid cell responses. *Immunity.* 2019;51:709-723.e6.
604. Sui P, Wiesner DL, Xu J, et al. Pulmonary neuroendocrine cells amplify allergic asthma responses. *Science.* 2018;360:eaan8546. doi:10.1126/science.aan8546
605. Godinho-Silva C, Domingues RG, Rendas M, et al. Light-entrained and brain-tuned circadian circuits regulate ILC3s and gut homeostasis. *Nature.* 2019;574:254-258.
606. Seillet C, Luong K, Tellier J, et al. The neuropeptide VIP confers anticipatory mucosal immunity by regulating ILC3 activity. *Nat Immunol.* 2020;21:168-177.
607. Wang Q, Robinette ML, Billon C, et al. Circadian rhythm–dependent and circadian rhythm–independent impacts of the molecular clock on type 3 innate lymphoid cells. *Sci Immunol.* 2019;4:eaay7501.
608. Aton SJ, Colwell CS, Harmar AJ, Waschek J, Herzog ED. Vasoactive intestinal polypeptide mediates circadian rhythmicity and synchrony in mammalian clock neurons. *Nat Neurosci.* 2005;8:476-483.
609. Iwasaki M, Akiba Y, Kaunitz JD. Recent advances in vasoactive intestinal peptide physiology and pathophysiology. Focus on the gastrointestinal system [version 1; peer review: 4 approved]. *F1000Res.* 2019;8:1-13.
610. Talbot S, Abdulnour REE, Burkett PR, et al. Silencing nociceptor neurons reduces allergic airway inflammation. *Neuron.* 2015;87:341-354.
611. Talbot J, Hahn P, Kroehling L, Nguyen H, Li D, Littman DR. Feeding-dependent VIP neuron–ILC3 circuit regulates the intestinal barrier. *Nature.* 2020;579:575-580.
612. Bal SM, Golebski K, Spits H. Plasticity of innate lymphoid cell subsets. *Nat Rev Immunol.* 2020:1-14.
613. Tizian C, Lahmann A, Hölsken O, et al. c-Maf restrains T-bet-driven programming of CCR6-negative group 3 innate lymphoid cells. *ELife.* 2020;9:e52549.
614. Parker ME, Barrera A, Wheaton JD, et al. c-Maf regulates the plasticity of group 3 innate lymphoid cells by restraining the type 1 program. *J Exp Med.* 2020;217:e20191030. doi:10.1084/jem.20191030
615. Pokrovskii M, Hall JA, Ochayon DE, et al. Characterization of transcriptional regulatory networks that promote and restrict identities and functions of intestinal innate lymphoid cells. *Immunity.* 2019;51:185.e6-197.e6.
616. Mazzurana L, Forkel M, Rao A, et al. Suppression of Aiolos and Ikaros expression by lenalidomide reduces human ILC3–ILC1/NK cell transdifferentiation. *Eur J Immunol.* 2019;49:1344-1355.
617. Huang Y, Guo L, Qiu J, et al. IL-25-responsive, lineage-negative KLRG1^{hi} cells are multipotential 'inflammatory' type 2 innate lymphoid cells. *Nat Immunol.* 2015;16:161-169.
618. Zhang K, Xu X, Pasha MA, et al. Cutting edge: Notch signaling promotes the plasticity of group-2 innate lymphoid cells. *J Immunol.* 2017;198:1798-1803.
619. Huang Y, Mao K, Chen X, et al. S1P-dependent interorgan trafficking of group 2 innate lymphoid cells supports host defense. *Science.* 2018;359:114-119.
620. Miller MM, Patel PS, Bao K, Danhorn T, O'Connor BP, Reinhardt RL. BATF acts as an essential regulator of IL-25–responsive migratory ILC2 cell fate and function. *Sci Immunol.* 2020;5(43):eaay3994. doi:10.1126/sciimmunol.aay3994
621. Schraml BU, Hildner K, Ise W, et al. The AP-1 transcription factor Batf controls T H 17 differentiation. *Nature.* 2009;460:405-409.
622. Bernink JH, Ohne Y, Teunissen MBM, et al. c-Kit-positive ILC2s exhibit an ILC3-like signature that may contribute to IL-17-mediated pathologies. *Nat Immunol.* 2019;20:992-1003.
623. Hochdörfer T, Winkler C, Pardali K, Mjösberg J. Expression of c-Kit discriminates between two functionally distinct subsets of human type 2 innate lymphoid cells. *Eur J Immunol.* 2019;49:884-893.
624. Golebski K, Ros XR, Nagasawa M, et al. IL-1β, IL-23, and TGF-β drive plasticity of human ILC2s towards IL-17-producing ILCs in nasal inflammation. *Nat Commun.* 2019;10:2162.
625. Silver JS, Kearley J, Copenhaver AM, et al. Inflammatory triggers associated with exacerbations of COPD orchestrate plasticity of group 2 innate lymphoid cells in the lungs. *Nat Immunol.* 2016;17:626-635.
626. Lim AI, Menegatti S, Bustamante J, et al. IL-12 drives functional plasticity of human group 2 innate lymphoid cells. *J Exp Med.* 2016;213:569-583.
627. Bal SM, Bernink JH, Nagasawa M, et al. IL-1β, IL-4 and IL-12 control the fate of group 2 innate lymphoid cells in human airway inflammation in the lungs. *Nat Immunol.* 2016;17:636-645.
628. Ohne Y, Silver JS, Thompson-Snipes L, et al. IL-1 is a critical regulator of group 2 innate lymphoid cell function and plasticity. *Nat Immunol.* 2016;17:646-655.

SECTION V Adaptive Immunity

CHAPTER 19 The Major Histocompatibility Complex

David H. Margulies, MD, PhD • Kannan Natarajan, PhD • Jamie Rossjohn, FRS • James McCluskey, FAA

INTRODUCTION

The major histocompatibility complex (MHC) is an extended genetic region found in jawed vertebrates that encodes, among a wide range of other molecules, the MHC glycoproteins, which include two major classes, MHC-I and MHC-II. MHC molecules are cell surface–expressed receptors that mediate and regulate many immune cellular interactions. The extended family of MHC and the related MHC-like molecules are critical components of intercellular recognition in both the innate and adaptive immune responses—particularly in their roles governing communication via cell surface receptors on T lymphocytes and on natural killer (NK) cells. The prototypical MHC-I and MHC-II molecules are obligate cell surface heterodimers that bind and display self or foreign peptides at the cell surface so that T cell receptors (TCRs) or NK cell receptors can interact with the molecular complexes in an MHC- and peptide-dependent manner.

Their name, *MHC* (italics denote the genes, regular font the proteins), derives from classic experiments that demonstrated the genetic control of one of the fundamental manifestations of an immune response, skin graft rejection.[1,2] Subsequent studies correlated control of other immune responses with genes of the *MHC*.[3-8] Contemporary immunologists, virologists, and tumor biologists rarely think about these molecules with allogeneic (ie, genetically distinct) skin transplantation in mind, but a detailed appreciation of the role of *MHC* genes and the functional contribution of MHC proteins in recognition of "non-self," of viral or other cellular parasitic infection, or of immunity to tumors is now well recognized. The name *MHC* is a general label, and specific immunological functions or molecular structures invoked when a particular *MHC* is mentioned are dependent on biological context. In recent years, our understanding of the structure, evolution, and molecular foundation of *MHC*-based interactions has blossomed, and also our ability to exploit engineered forms of these molecules for research, diagnostic, and vaccine applications has expanded enormously. Our blueprint for this chapter is to give an overview of *MHC*-regulated immunological function and then to summarize the explosive growth in understanding of the molecular structures of MHC molecules and their ligands. Of necessity, we discuss briefly some aspects of the cell biology and cellular maturation of MHC molecules and leave the detailed description of these essential aspects of the classic MHC molecules to the chapter on antigen presentation (Chapter 20). Basic aspects of the *MHC* genes, not only descriptions of their encoded molecules but also a view of their evolution, will also be touched upon.

Proteins are molecular machines, and their activities are governed by their shape, hydrophobicity, charge, and dynamics. These physicochemical properties dictate how these molecules interact with others, and how they change to convey a host of molecular signals. To understand proteins encoded by the *MHC* in a rational functional context, then, we consider it crucial to describe their overall structures and their structural interactions with bound ligands and cognate cell surface receptors. Despite a large and ever-increasing database of the structures of MHC and MHC-like molecules, a number of general questions remain unanswered, and we will present some of the many unsolved mysteries of the evolution and function of these molecules and their viral homologs.

The importance of the classical MHC molecules is underscored by two general characteristics—the enormous degree of polymorphism of genes of the *MHC* and the identification of a large number of human conditions and diseases that are profoundly influenced by either defective *MHC* expression or by polymorphic variants. Every student of immunology must become acquainted with the basic biology of *MHC*-encoded molecules. A comprehension of the regulation of the immune response, encompassing the inflammatory, NK, T, and B cell responses, is lacking without an understanding of the function and ongoing evolution of the *MHC*.

We stress that the molecular evolution of the MHC is ongoing, and for no other set of genetic markers is there such a large database cataloging genetic polymorphism. As a result, not only are new alleles regularly being identified but also many have arisen recently. Indeed, at the time of the writing of this chapter, more than 21,000 human *MHC-I* and 7800 *MHC-II* alleles have been listed in the Immuno Polymorphism Database (IPD-MHC) (https://www.ebi.ac.uk/ipd/index.html).[9,10] Dramatic examples of the ongoing conversation between MHC molecules and host pathogens are seen not only in new variants of the DNA sequences detected by those that perform the necessary typing of *MHC* genotypes for transplantation but also in surveys of the sequences of viruses, particularly the cytomegaloviruses (CMVs), that have purloined host *MHC* genes for immunoevasion.[11,12] In sequence gazing, we observe the interplay between host and pathogen as pathogenic viruses, exploiting their rapid generation times, develop new variations on the *MHC* structural theme for immunoevasion.

MHC molecules are crucial for both TCR- and NK-mediated interactions. Our efforts will be to preserve some sense of the historical development of this exciting field of study and also to focus on paradigmatic genetic, structural, and functional features that unify this extensive gene/protein family. Finally, the *MHC* provides a genetic link from immune responsiveness to autoimmune disease—those well-known strong associations of particular *MHC* genes with particular human conditions—and we will provide an outline of the molecular basis for such associations.

MHC Nomenclature

The names of the genes and proteins that are critical to understanding the *MHC* reflect the historical discovery of their functions and differ among different species. Often confusing to students of immunology, the nomenclature for different species must deal not only with a number of distinct, but usually linked, genetic loci but also with their encoded molecules, many of which are heterodimeric glycoproteins. Because the first genes of the *MHC* identified were those that encoded cell surface molecules that could be detected by antibodies or by transplantation responses, these are the ones that are referred to as "classical" *MHC* genes. Though technically genes that map to the extended *MHC* are all *MHC* genes, those encoding several complement components and related molecules also map here. Other *MHC*-encoded molecules with distinct structure and function are usually referred to by their specific names. MHC molecules are broadly denoted as MHC-I and MHC-II (also commonly called "class I" and "class II"), whose structural similarities and differences will be discussed below. Molecules that exhibit structural similarity with MHC-I molecules, whether they map to the *MHC* or not, may be called as a group "MHC-Ib" to distinguish them from the classical MHC-I molecules, also referred to as "MHC-Ia." Another subset of molecules that exhibit the MHC-I protein fold but that are expressed by viruses may be called "MHC-Iv" molecules.[11,13] Since many of the viral MHC-like molecules have been identified in their role as immunoevasins, or as viral regulators of antigen presentation, they are designated vRAPS.[14]

The mouse *MHC* is referred to as *H2* (previously called *H-2*) because it was the second genetic locus involved in control of expression of erythrocyte antigens reported by Gorer.[15,16] The *MHC* in the rat is known as *RT1*, the human locus is known as *HLA* (for **h**uman **l**eukocyte **a**ntigen), *DLA* for the dog, *GPLA* for the guinea pig, *SLA* for the swine, and *RLA* for the rabbit. For other species, based on a suggestion by Klein et al,[17] the taxonomic name forms the basis for the designation, contributing the first two letters of the genus and the first two of the species to name the locus. Thus, we have *Patr* for the chimpanzee, ***Pa**n **tr**oglodytes*; *Gogo* for the gorilla, ***Go**rilla **go**rilla*; *Mamu* for the rhesus macaque, ***Ma**caca **mu**latta*; *Mane* for the pig-tailed macaque, ***Ma**caca **ne**mestrina*; *Mafa* for the cynomolgus monkey, ***Ma**caca **fa**scicularis*; and *Papa* for the bonobo, ***Pa**n **pa**niscus*. A database for curated sequences of many species of MHC molecules is the IPD at http://www.ebi.ac.uk/ipd/mhc/.[9] The detailed rules for the "Nomenclature for Factors of the HLA System" are provided at http://hla.alleles.org. The extended HLA locus, which is in the 6p21.3 region of the short arm of human chromosome 6, contains more than 220 genes. New *HLA* genes, sequences, and quality control are determined by the WHO Nomenclature Committee for Factors of the HLA System (http://hla.alleles.org/nomenclature/committee.html). The human nomenclature is exemplified by "*HLA-A*02:101:01:02N*," in which the HLA prefix designates the species and the letter "*A*" the gene. Following the * separator, the first allele group is indicated here by "*02*" and the specific protein by "*101*." Synonymous DNA substitutions are shown by the third number ("*01*") and differences in coding region by the final number. Differences in level of expression are shown by the final letter—"N" is null, "L" is low, "S" for secreted molecules, and "Q" for questionable phenotype. Since MHC-I (eg, HLA-A, HLA-B, and HLA-C) proteins consist of a polymorphic heavy (α) chain and a monomorphic light chain (β_2 microglobulin [β_2m]), it is sufficient to denote MHC-I molecules by the heavy chain nomenclature described above.

Because the two protein chains (α and β) that assemble to form a particular MHC-II molecule may be polymorphic, it becomes necessary to specify both genes/chains of the molecule. Thus, for the *HLA-DR* locus, one should indicate both the α-encoding (eg, *HLA-DRA*, *HLA-DRA*01:01*) and the β-encoding (*HLA-DRB1*, *HLA-DRB1*01:01*) genes. For the *HLA-DR* locus, there are only two *DRA* alleles and numerous *HLA-DRB1* alleles, while at *HLA-DQ*, numerous *HLA-DQA1* and *HLA-DQB1* alleles have been identified. Although cumbersome, it is usually best to indicate exactly which α and β chain are of interest in a particular context. An example would be HLA-DQA1*01:01/HLA-DQB1*05:01 to denote a particular MHC-II heterodimer.

TABLE 19.1 Commonly Used Mouse Strains: *H2* Haplotypes[a]

Strain		H2 Loci/Regions									
		Class Ia	Class II				Class Ia		Class Ib		
	Haplotype	*K*	*Ab*	*Aa*	*Eb*	*Ea*	*D*	*L*	*Qa-2*	*T18*	*Qa-1*
Common strains											
129/J[b]	*Bc*	*b*	*b*	*b*	*b*	–	*b*	–	*a*	*f*	*b*
AKR/J	*k*	*k*	*k*	*k*	*k*	*k*	*k*	*k*	*b*	*b*	*b*
ASW/Sn	*s*	*s*	*s*	*s*	*s*	–	*s*		*a*	*b*	*b*
BALB/c	*d*	*d*	*d*	*d*	*d*	*d*	*d*		*a*	*d*	*b*
C3H/HeJ	*k*	*k*	*k*	*k*	*k*	*k*	*k*	*k*	*b*	*b*	*b*
CBA/J	*k*	*k*	*k*	*k*	*k*	*k*	*k*	*k*	*b*	*b*	*b*
C57BL/6	*b*	*b*	*b*	*b*	*b*	–	*b*	–	*a*	*b*	*b*
C57BL/10	*b*	*b*	*b*	*b*	*b*	–	*b*	–	*b*	*b*	*b*
C57BR	*k2*	*k*	*k*	*k*	*k*	*k*	*k*	*k*	*b*	*a*	*a*
DBA/2J	*d*	*d*	*d*	*d*	*d*	*d*	*d*	*d*			
NOD/ShiLtJ	*g7*	*d*	*g7*	*d*	–	–	*b*	–	*a*	*b*	*a*
NON/ShiLtJ	*nb1*	*b*	*nb1*	*nb1*	*k*	*k*	*b*				
Congenic Strains											
B10.BR	*k2*	*k*	*k*	*k*	*k*	*k*	*k*	*k*	*b*	*a*	*a*
B10.D2	*d*	*d*	*d*	*d*	*d*	*d*	*d*				
B10.S	*s*	*s*	*s*	*s*	*s*	–	*s*		*a*	*b*	*b*
BALB.B	*b*	*b*	*b*	*b*	*b*	–	*b*	–	*a*	*b*	*b*
BALB.K	*k*	*k*	*k*	*k*	*k*	*k*	*k*	*k*	*b*	*b*	*b*
C3H.SW	*b*	*b*	*b*	*b*	*b*	–	*b*	–	*a*	*b*	*b*
Recombinant Strains											
A	*a*	*k*	*k*	*k*	*k*	*k*	*d*	*d*	*a*	*a*	*a*
B10.A	*a*	*k*	*k*	*k*	*k*	*k*	*d*	*d*	*a*	*a*	*a*

[a]Adapted from Kruisbeek AM. Commonly used mouse strains. *Curr Protoc Immunol.* 2001;Appendix 1:Appendix 1C and http://jackson.jax.org/rs/444-BUH-304/images/MHC_h2_haplotypes.pdf. An en dash indicates abnormal gene expression, although precise mechanism may differ in different strains. Blanks indicate insufficient data for characterization.
[b]Different 129 strains are generally haplotype *b*, but some are *bc*.

The commonly used term "haplotype" refers to the linkage of alleles that occur as a group on a particular chromosome.[18] The concept of haplotype is important in typing the *HLA* loci in the human, where the linked *MHC* genes (those genes in "linkage disequilibrium") of one chromosome of one parent will generally segregate as a group to the children. Individual haplotypes of the *MHC* in the mouse are indicated by a lowercase letter superscript as $H2^b$, $H2^d$, or $H2^k$. Thus, the $H2^k$ haplotype refers to the full set of linked genes, such as $H2\text{-}K^k$, $H2\text{-}IA^k$, $H2\text{-}IE^k$, and $H2\text{-}D^k$, and extends to the genes of the *Q* and *T* regions as well.[19,20] Some haplotype designations, like $H2^a$, refer to natural *MHC* recombinants and thus have some of the linked genes from one haplotype and some from another. Congenic lines have been bred to preserve a specific *MHC* haplotype on a defined genomic background. Thus, B10.A preserves $H2^a$ on a C57BL10 (B10) background. Table 19.1 lists the genes at the *MHC* loci for some of the most commonly used mouse strains. A complication that demands the precise use of gene and encoded protein names is that the number of genes in a particular homologous genetic region may differ between strains or between individuals. In the mouse, while some strains have only a single gene at the *D* locus (for instance, $H2\text{-}D^b$), other strains may have as many as five genes in the homologous region ($H2\text{-}D^d$, $H2\text{-}D2^d$, $H2\text{-}D3^d$, $H2\text{-}D4^d$, and $H2\text{-}L^d$).[22,23]

Although detailed nucleotide sequences are available for the human and mouse, it is useful to have some schematic maps as a guide to general regions and relationships. Figures 19.1 (for human) and 19.2 (for mouse) are included here for that purpose. Although the focus of much current

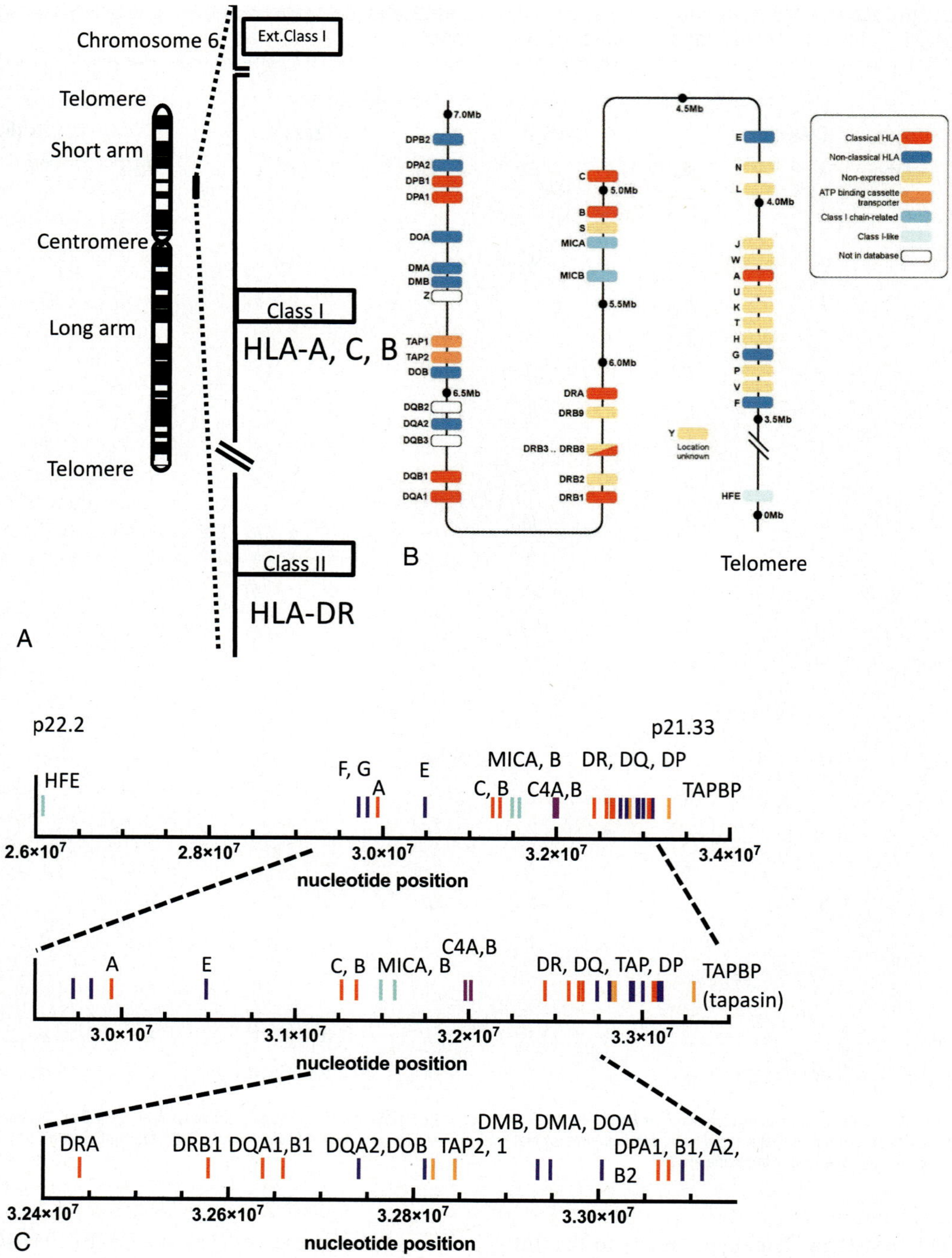

FIG. 19.1. Map of the human *MHC*. Schematic map of **(A)** human chromosome 6 and **(B)** the human *MHC* region. A blowup of the *MHC* region to scale based on the human genome sequence **(C)** is also shown. The ideogram of panel A is drawn from https://www.ncbi.nlm.nih.gov/genome/?term=homo+sapiens. Panel B is taken from http://hla.alleles.org/alleles/index.html.[24] C: Nucleotide positions are plotted as the first nucleotide of the indicated genes from https://www.ncbi.nlm.nih.gov/nuccore/NC_000006.12?report=graph, for GRCh38.p13 primary assembly.

immunology research has, understandably, been on human and primate models, the importance of mouse and humanized mouse studies cannot be overemphasized. The Jackson Laboratory is the recognized source for a host of inbred mouse strains and various congenic and recombinant inbred mouse lines. Description of numerous haplotypes may be obtained at (http://jackson.jax.org/rs/444-BUH-304/images/MHC_h2_haplotypes.pdf). Students of mouse genetics should be aware of the "Collaborative Cross," which has established a panel of mouse recombinant inbred lines derived from eight diverse mouse strains. These lines, similar to classical recombinant inbred lines, provide a resource to map quantitative

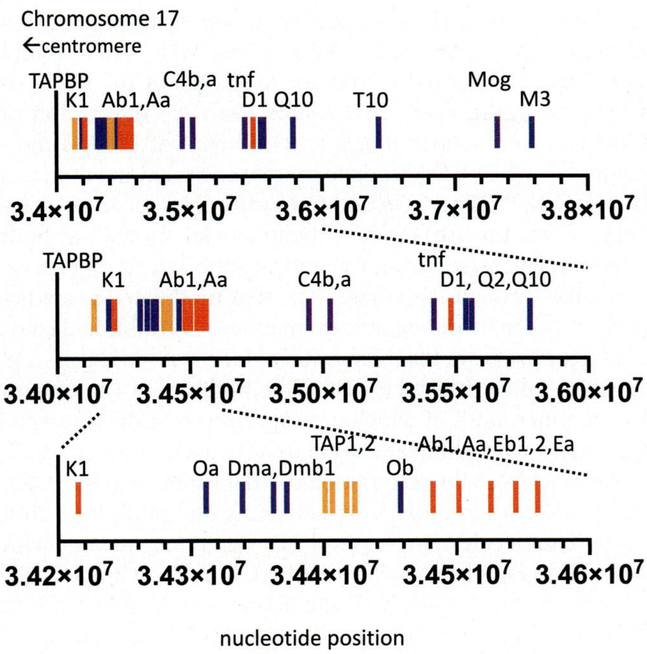

FIG. 19.2. Map of the mouse *MHC*. Positions of some of the most important genes of the mouse *MHC* are indicated, based on the first nucleotide of the annotated gene, taken from GenBank assembly of C57BL/6 genome chromosome 17, GRCm39, NC_000083.7.

trait loci to various immunological and disease phenotypes. Indeed, two recent examples demonstrate the power of analysis of such mouse strains in evaluating genetic contributions to the immune response to viral infection, including LCMV and SARS-CoV-1.[25-28]

THE IMMUNOLOGICAL FUNCTION OF MHC MOLECULES

MHC molecules are a molecular reflection of the health status of either the cell that synthesizes them (for MHC-I molecules) or of the local environment in which the cell resides (for MHC-II). The structure of the MHC molecule depends not only on the amino acid sequence of the two polypeptide chains, heavy (or α) and β_2m (for the light chain) for MHC-I, and α and β for MHC-II, that form the core of the complex but also on the variable-bound peptide that forms an integral part of the trimer. The MHC molecule must satisfy at least two distinct recognition functions: the binding of peptides (or nonpeptidic molecules for some MHC-Ib) and the interaction with either T or NK cells via their respective receptors. The TCR may augment its interaction with the MHC molecule by virtue of interaction with a T cell–expressed coreceptor (CD8 for MHC-I and CD4 for MHC-II). Some NK receptors, when expressed on T cells, may also provide distinct coreceptor function.[29] The binding of peptides by an MHC-I or MHC-II molecule is the initial selective event that permits the cell expressing the MHC molecule (the antigen-presenting cell [APC] or when this cell is to be the recipient of a cytolytic signal, the target cell which can be almost any cell type) to sample fragments derived either from proteins synthesized in the same cell (for MHC-I–restricted antigen presentation) or from proteins (self or nonself) ingested from the immediate extracellular environment (for the case of MHC-II). The biochemical steps involved in the production of antigen fragments from large molecules are collectively known as "antigen processing," while those that concern the binding of antigen fragments by MHC molecules and their display at the cell surface are known as "antigen presentation" (Chapter 20).

Specifically, MHC class I glycoproteins gather from the cell's biosynthetic pathway fragments of proteins derived from infecting viruses, intracellular parasites, or self-molecules. These peptides result from naturally expressed proteins of the cells, proteins that arise from the dysregulation that accompanies tumorigenesis, or abnormal sequences that may result from aberrant initiation of protein synthesis or protein splicing.[30-35] The cell-bound MHC-I/β_2m/peptide complex on the APC is exposed to the extracellular milieu and is available for interaction with either T cells or NK cells. The T cell bearing an αβ receptor recognizes the particular peptide/MHC (p/MHC) complex by virtue of a specific physical binding interaction. Each T cell is representative of a clonal population and bears a unique TCR encoded by somatically rearranged TCR genetic elements. A particular TCR can bind only a limited selection of p/MHC complexes. The recognition by T cells is considered "MHC restricted" in that a given TCR can bind only a limited set of p/MHC complexes. This discrimination is also termed "antigen specific" in that an individual T cell identifies only a limited set of peptides, usually related by sequence. For any given T cell clone, single amino acid substitutions of either the MHC or the peptide may severely diminish, obliterate, or rarely augment the functional interaction of the TCR with the p/MHC complex.

The MHC-I system draws its spectrum of peptides from proteins in the cytosol that are degraded by the proteolytic proteasome complex to peptides and transported from the cytosol to the endoplasmic reticulum (ER) with the aid of the intrinsic membrane peptide transporter, the transporter associated with antigen processing (TAP).[36] The imported peptides are then trimmed at their amino termini by an ER-associated amino terminal peptidase[37-39] and are cooperatively folded as an intrinsic component of the newly synthesized MHC-I molecule.[40] Interactions with the peptide loading complex (PLC) that includes TAP, the chaperones tapasin[41-44] (also designated TAP-binding protein [TAPBP], and ERp57),[44-46] and the lectin calreticulin[44] are crucial to MHC-I loading. The structural organization of the components of the PLC, originally deduced from biochemical pull-down experiments,[46] has recently been visualized in cryogenic electron microscopic (cryo-EM) image reconstruction of the complete PLC.[47] MHC-I molecules are unique among proteins in that their three-dimensional structure and thermal stability are exquisitely dependent on the heavy/light chain heterodimer being bound by an appropriate peptide.[48,49] In mutant cells that lack the necessary apparatus for generating and transporting peptides to the ER where peptide loading takes place, MHC-I molecules are expressed poorly at the cell surface and are inherently unstable[50] (see Chapter 20 for more detail).

MHC-II molecules, in contrast to MHC-I molecules that are found on virtually all somatic cells, are expressed on a more limited set—B cells, macrophages, dendritic cells, activated but not resting T cells in the human, and some unique T cell subsets in the mouse.[51] In general, they bind peptides derived from the degradation of proteins ingested by the APC, and they sort their MHC-II molecules into cellular compartments where the degraded peptides are generated and catalytically transferred to the binding site of the MHC-II. The MHC-II antigen presentation pathway is based on the initial assembly of the MHC-II αβ heterodimer with a multifunctional molecule, the invariant chain (Ii, CD74). CD74 guides the αβ heterodimer to an endosomal, acidic, protein processing location where it encounters antigenic peptides, and also protects the antigen-binding site of the MHC-II molecule from premature loading in the ER. The loading of the MHC-II molecule with antigenic peptide is dependent on the release of the Ii-derived class II–associated invariant chain peptide (CLIP), catalyzed by the MHC-II–like molecule, HLA-DM in the human[52-55] and H2-M in the mouse.[56-58] Finally, a stably loaded p/MHC-II complex is expressed at the cell surface.[59-63] The MHC-II–like molecules HLA-DO in the human and H2-O in the mouse regulate the activity of HLA-DM and H2-M, respectively, although details of the precise mechanism remain unclear.[64,65]

After p/MHC-II complexes arrive at the cell surface, circulating cognate T cells that encounter the appropriate complex may be activated to secrete lymphokines and/or be induced to proliferate or to undergo programmed cell death. Such MHC-restricted lymphokine production that facilitates and augments the recruitment of additional inflammatory cells as well as APCs and antibody-producing cells is a contemporary explanation for what was historically referred to as "T cell help." Under some physiological circumstances, particularly during viral infection, antigens incorporated into antigen-presenting dendritic cells via an "outside in" pathway may alternatively enter the MHC-I presentation pathway. This phenomenon is known as "cross-presentation."[66-69]

The crucial role that MHC molecules play in the thymus is to shape the TCR repertoire as T cell precursors mature into cells that eventually emigrate from the thymus and populate the spleen and other secondary lymphoid organs. In a complex, multistep process termed "thymic education,"[70,71] developing T cells expressing randomly rearranged TCRαβ receptors are first subjected to positive selection on MHC-I and MHC-II molecules expressed by cells of the thymic cortex for further maturation of only those cells capable of recognizing peptides in the context of self-MHC molecules (see Chapter 27). Positive selection may be viewed as setting a minimum threshold for T cell activation. T cells surviving positive selection then migrate to the thymic medulla where those T cells with higher-than-average affinity for self-p/MHC complexes are deleted by negative selection. Such negative selection protects against autoimmunity. MHC-I and MHC-II molecules thus play a critical role in shaping the peripheral TCR repertoire. Indeed, mice engineered to lack MHC-I or MHC-II expression are also deficient in corresponding T cell subsets.[57,72]

In addition to showing preference for distinct pathways of antigen presentation, the MHC-I and MHC-II molecules also show preferential restriction to T cells of the CD8- or CD4-bearing lineages. This results from the interaction of CD8 with the nonpolymorphic α3 domain of MHC-I molecules,[73-77] while CD4 binds to membrane proximal domains of MHC-II.[78-82] The CD8 and CD4 molecules serve as "coreceptors" on the surface of T lymphocytes, providing both adhesion (increase in avidity) and specific activating signals, mediated through the kinase, Lck, that modulate the avidity of the T cell in a time-dependent manner.[83-87] Additional complexities arise from the interaction of CD8 with MHC-I. CD8 is expressed as either the CD8αα homodimer or the CD8αβ heterodimer, both of which are expressed in a developmentally regulated fashion and seem to have distinct functions.[88]

Distinct from the recognition of MHC molecules by TCRs, many NK cell receptors, both activating and inhibitory, bind MHC-I molecules, and several NK receptors interact with MHC-I–like molecules encoded by CMVs[11,89] (Chapter 18). In general, the NK/MHC-I interaction, as compared to the TCR/MHC interaction, shows considerably less peptide specificity, although the interaction is peptide dependent, and in some cases may exhibit clear-cut peptide preference.[90-93] The functional purpose of the MHC-I molecule in NK cell recognition appears to be more subtle than that in T cell recognition. The NK cell is tuned to a balance of inhibitory and activating signals conveyed to it via MHC interaction, and in the steady state, the inhibitory signals predominate. MHC-I is a sensor of the biosynthetic and metabolic state of the cell in which it is synthesized. Healthy cells express levels of MHC-I to which the NK cell repertoire is "tuned," but when the level of MHC-I is downregulated by tumorigenesis or viral infection, the change can be detected by NK cells. This ability of the NK cell to sense reduced levels of surface MHC-I on target cells is the basis of the "missing self hypothesis," which explains that NK cells detect and lyse those cells defective in MHC-I expression due to the loss of the inhibitory signal.[94-96] The prototype NK receptor in the mouse is the inhibitory receptor, Ly49A, a C-type lectin-like molecule that signals its interaction with a normally expressed MHC-I molecule, such as H2-D^d.[97,98] MHC-I regulation of NK activity *in vivo* is complex. Distinct clones of NK cells differ in the combinatorial expression of receptors that have different MHC-I preferences. Thus, in the mouse, each distinct NK clone may express a different combination of inhibitory receptors such as Ly49A, Ly49C, Ly49G2, and Ly49I.[99] Since each inhibitory receptor may exhibit slight differences in its MHC-I and/or peptide preference and specificity, this kind of combinatorial expression of NK activity offers a breadth of reactivity toward different potential target cells. Recently it has been shown that developing NK cells are also subject to an education process, known as "licensing," that prevents improper activation of their cytolytic function in peripheral tissues.[100-105] NK cells that develop in MHC-I–deficient humans and mice fail to kill target cells that lack MHC-I expression. Such targets are readily killed by NK cells that develop in MHC-I–sufficient environments. Human NK cells seem to go through a selective/educating process similar to that of the mouse, but the relevant NK receptors on human cells are generally of the KIR (killer cell immunoreceptor) inhibitory receptor family.[106] The numerous interactions of MHC molecules with other cellular components, as well as with the wide variety of peptides and various immunological receptors, reflect the robust potential of the MHC structure as a molecular sensor and as a master regulator of immune responses.

Control of Expression of MHC Molecules

MHC molecules, synthesized in the ER and destined for cell surface expression, are controlled at many checkpoints before their final disposition as receptors available for interaction with either T cells or NK cells. The classical MHC-I molecules are trimers, consisting of the polymorphic H2 or HLA heavy chain, the light chain β_2m, and the assembled self-peptide. Thus, the cell surface expression of the assembled trimer may be reduced by a defect in the synthesis, steady state, or assembly of any of these requisite protein/peptide chains. The first level of MHC-I expression control is transcriptional, and IFN-γ–dependent regulation is particularly important.[107] For the most part, MHC-I molecules are ubiquitously expressed, and expression is dependent on a complex regulatory process that coordinately controls the transcription of both MHC-I and β_2m.[108,109]

Transcriptional and posttranscriptional control of MHC-I expression is a complex process dependent on several regions, primarily in the 5′ untranslated region of each MHC-I gene, which interact with a variety of regulatory molecules and molecular complexes.[110] NF-κB binds to a region designated "enhancer A," the interferon regulatory factor 1 interacts with an interferon-stimulated response element, and a multimolecular complex, now designated the "enhanceosome." This consists of NLRC5, an RFX-containing complex (RFX5/RFXAP/RFXA), ATF1/CREB, and the NFY complex (NFYa/b/c that binds an SXY module). MHC-I mRNA degradation is regulated in part by MEX-3C–mediated ubiquitination.[111] Clearly, the many contributions to MHC-I transcription and degradation offer opportunities for precise control over MHC-I expression levels under different physiological and developmental conditions.

The basis of the more limited tissue-specific expression of MHC-Ib molecules is of interest because of the potential importance in the role of some of the MHC-Ib molecules in tolerance to the placenta. HLA-E and HLA-G, expressed on placenta, and HLA-F, another MHC-Ib with limited tissue-specific expression, have been examined in considerable detail.[112-116] Another critical level of control of MHC-I expression is genetic, that is, the genes for a particular chain or key components of the PLC must be present for the trimer to be assembled and then expressed at the cell surface. A decrease in heavy chain synthesis as a result of a structural mutation in a coding region (such as HLA-A*30:14L) or in a noncoding region (such as HLA-A*24:02:01:02L) would reveal the simplest level of control. In addition, the light chain is absolutely required for cell surface expression of most MHC-I molecules, and deficiency of β_2m in various tumors is a relatively common occurrence.[117-119] Induced β_2m-defective animals (*$\beta_2m^{o/o}$*, "β_2m knockouts") lack normal levels of MHC-I expression because of this requirement for cell surface expression.[120,121] In addition, polymorphism of the MHC-I heavy chain may influence the efficiency with which the molecule can interact with the PLC or with which it binds self-peptides, and thus some alleles, because of a decreased ability to be loaded with self-peptides, may not be well-expressed at the surface. The rest of the MHC-I biosynthetic pathway is dependent on proper generation of cytosolic peptides by the proteasome and delivery to the ER by TAP, appropriate core glycosylation in the ER, transport through the Golgi, and arrival at the plasma membrane. A number of persistent viruses have evolved mechanisms for subverting this pathway of expression.[122,123] The herpes simplex virus encodes a protein, infected cell peptide 47,[124-126] that blocks the activity of the peptide transporter TAP.[127] Cowpox virus encodes two proteins, CPXV12 and CPXV203, that contribute to the downregulation of host MHC-I, CPXV12 inhibits TAP, and CPXV203 retains MHC-I in the ER.[128,129] Molluscum contagiosum virus encodes an MHC-I–like protein that impairs antigen presentation by blocking the PLC.[130] The human disease, bare lymphocyte syndrome, type I, manifested by chronic lung disease in late childhood, results from genetic lesions in either TAP1, TAP2, or tapasin, also known as TAPBP.[131-135] Several proteins encoded by the human cytomegalovirus (HCMV), unique short region proteins 2 and 11, cause rapid protein degradation of MHC-I molecules.[136] Murine CMV encodes an MHC-I–like protein, m152/gp40, that inhibits MHC-I expression by diverting MHC-I molecules to rapid degradation in the endosome.[137,138] Other molecules that assist the large DNA viruses in evading either the T cell or NK cell immune response include HCMV UL142.[139] Murine CMV[140] encodes a number of open reading frames considered MHC-I homologs that may function to deceive NK cells into the perception of normal MHC-I expression.[11,141-143] The adenovirus 2 protein, E3/19K, also functions to block the transfer of folded assembled MHC-I molecules from the ER to the Golgi.[144-147] Structural studies reveal the mechanism by which E3/19K impairs MHC-I expression.[148,149]

MHC-II molecules are also susceptible to regulation at multiple steps. The clear-cut tissue dependence of MHC-II expression—MHC-II molecules are generally found on cells that have specific antigen presentation functions such as macrophages, dendritic cells, Langerhans cells, thymic epithelial cells, and B cells, but can also be detected on activated T cells of the human and rat—suggests that transcriptional regulation plays an important role. Extensive studies of the promoter activities of *MHC-II* genes have defined a number of specific transcriptional regulatory sequences.[150] One transcription factor, MHC class II transcriptional activator (CIITA) is clearly crucial.[108,150-152] The expression of CIITA itself is regulated in a complex fashion, possessing at least three distinct promoters that function in a cell type–specific manner.[153] MHC-II deficiency diseases categorized as bare lymphocyte syndrome, type II, result from lesions in at least four different complementation groups, implicating defects in CTIIA, RFX5, RFXAP, or RFXANK, all encoding proteins regulating MHC-II transcription.[152,154-158] Polymorphisms in

MHC-II and the resulting differences in tissue- and lineage-specific expression of MHC-II suggest a role in the control of autoimmune disease.[159-161]

A unique aspect of MHC-II regulation is the need to protect its peptide-binding site from loading with self-peptides in the ER, and the requirement to traffic to an acidic endosomal compartment where antigenic peptides, the products of proteolytic digestion of exogenous proteins, can be obtained. These two functions are controlled by the type II membrane protein, invariant chain, Ii (CD74),[59,162-164] that forms a nine-subunit complex consisting of three Ii and three αβ MHC-II heterodimers. The region of Ii that protects the MHC-II peptide–binding groove, CLIP, is progressively trimmed from Ii and is ultimately released from the MHC-II by the catalytic action of HLA-DM in the endosome to allow exchange for peptides generated there. HLA-DO, another MHC-II–like molecules, regulates the repertoire of bound peptides, presumably by modulating the catalytic activity of the DM exchange reaction. The important role of Ii in regulating MHC-II expression has been emphasized by the behavior of induced mutant mice lacking normal Ii,[165-167] which exhibit a profound defect in MHC-II function and expression.

THE MAJOR HISTOCOMPATIBILITY COMPLEX

MHC Genetics

The *MHC* is an extended region of the genome that spans some 3.4 to 4 million base pairs (Mbp) on the short arm of human chromosome 6 in the region 6p21.3. The *MHC* in the human may be considered to map from *GABBR1* (genomic coordinates 29,602,230) to *KIFC1* (33,391,524) in human genome GRCh38.p13 assembly,[168] which includes *HLA-F* (29,723,434), to the gene encoding the MHC-I chaperone tapasin (33,299,694) (also designated as *TAPBP*[169-171]). In the mouse, the *MHC* encompasses a region on chromosome 17, extending from the *H2-K1* to the *H2-M2* gene, about 3.5 Mbp.[168] The *MHC* class I region is located at the telomeric end and the class II region is centromeric, with an intervening extended class III region. This class III region includes many genes related to stress response, complement components, some cytokine genes, and other regulatory genes. The class II region encodes proteasome genes and the peptide transporters TAP1 and TAP2. *TAPBP* is in the extended class II region. These maps are illustrated in Figures 19.1 and 19.2. Recent analysis of genomic sequences of the *MHC* of human, mouse, and macaque compares the human *HLA* region of 3.4 Mbp to the macaque, *Mamu-MHC* region of 4.7 Mbp, and the mouse *H2* region of 2.9 Mbp.[168] Although *MHC* genes were among the first to be mapped to this general region, it is now clear that many genes with function unrelated to immune recognition also reside here. The interested reader is referred to the regularly updated maps and linkages available at various websites including http://www.imgt.org. A schematic of the human *MHC* is shown in Figure 19.1. We refer the reader to the NCBI homology site (http://www.ncbi.nlm.nih.gov/projects/homology/maps/) for detailed sequences.

Further scrutiny of the human map reveals that *HLA-DP* genes (*DPA* encoding the α chain, and *DPB* encoding the β chain) are proximal to the centromere on the short arm of chromosome 6 and are linked to the genes encoding the related HLA-DM molecule (*DMB* and *DMA*). Between these and the *DQ* genes lie *LMP* (for **l**ow-**m**olecular-weight **p**roteins)[172-174] and *TAP*[175-179] genes that encode molecules that are involved in peptide generation in the cytosol and peptide transport across the ER membrane, respectively. LMPs are subunits of the catalytic proteolytic proteasome complex that regulate the specificity of cleavage of proteins, and thus modulate the repertoire of peptides available for MHC-I–restricted antigen presentation.[180-184] The *TAP* genes encode a two-chain intrinsic membrane protein that resides in the ER of all cells and functions as an ATP-dependent transporter that pumps peptides generated in the cytosol into the lumen of the ER.[185,186] Structural and mechanistic details of TAP function are now being revealed.[187,188] The selective transport of cytoplasmically generated peptides by different TAP proteins in the rat or the chicken demonstrates that the spectrum of p/MHC complexes expressed at the cell surface can be significantly altered by differences in the antigen presentation pathway,[189-191] and recent studies suggest that polymorphism of *TAP* genes may contribute to relative susceptibility to human immunodeficiency virus 1 (HIV-1)/acquired immunodeficiency syndrome (AIDS) infection.[192] TAP deficiency syndromes in humans[132,193-197] emphasize the importance of peptide delivery in the antigen presentation pathway for normal immune function.

The major *MHC-II* genes of the human are *HLA-DRA* and *HLA-DRB* that encode the α and β chains that form the HLA-DR molecule, a major antigen presentation element. Genetic mapping of the human *DRB* region indicates that several alternative arrangements of *DRB* loci explain the varied serotypes and genotypes observed among different individuals (Fig. 19.1). In particular, several HLA-DR "haplotypes," originally defined serologically as DR52, DR1, DR51, DR53, and DR8, differ not only in that they are encoded by particular allelic genes at the *DR* locus but also in the number and precise location of several of the *DRB* genes and pseudogenes. The *MHC-III* region is important in immunological terms for several reasons—the structural genes for several complement components map here, as well as the structural genes for 21-hydroxylase (CYP21A2),[198,199] an enzyme critical in the biosynthesis of glucocorticoids. Twenty-one hydroxylase deficiency, an *MHC*-linked genetic disease, is the cause of congenital adrenal hyperplasia. Also located in the *MHC-III* region are the structural genes for tumor necrosis factor α and lymphotoxin α and β, lymphokines made by activated T cells.[200-204] In both the human and mouse, a genetic region that includes representatives of primordial genes for the complement component C4, the 21-hydroxylase gene *CYP21A*, and the tenascin gene *TNX* are duplicated, generating two copies each of *C4* (*C4A* and *B*), of *CYP21A* (*CYP21A1* and *CYP21A2*), and of *TNX* (*A* and *B*). Remarkably, in the human, *CYP21A1* has been deactivated, and in the mouse *CYP21A1* is active and *CYP21A2* has been rendered inactive.[205]

The region of the *MHC* distal to the centromere encodes other MHC molecules. In humans the cluster of the major *MHC-I* genes lies here, including the genes encoding

HLA-B, HLA-C, HLA-E, and HLA-A, as well as HLA-H, HLA-G, and HLA-F. *HLA-A*, *HLA-B*, and *HLA-C* encode the major MHC-I molecules of man (Fig. 19.1). A recent example of HLA-C function in tumor immunity identifies HLA-C*08:02 presenting a G12D KRAS neoantigen to T cells.[206] HLA-C molecules may interact directly with NK receptors of the KIR2D family. HLA-C/KIR interactions contribute to long-term survival of HIV-1–infected patients.[207] Direct binding studies have analyzed the kinetics of the interaction of the KIR2 Ig-like NK receptors,[208-211] and three-dimensional structures of KIR2DL2 in complex with HLA-Cw3 and KIR2DL1 complexed with HLA-Cw4 have been published.[212,213] Recent structures of KIR3DL1 molecules in complex with HLA-B*27:05[214] or with HLA-B*57:01[215] emphasize the unique nature of different KIR/MHC interactions. The precise functions of HLA-E, HLA-F, and HLA-G are being explored. HLA-E and its murine analog Qa-1 bind hydrophobic leader peptides derived from some MHC-I molecules, forming a complex recognized by the C-type lectin-like NK receptor CD94/NKG2.[216-221] New evidence indicates that HLA-E can present pathogen-derived peptides to T cells.[222] This implies an important function for HLA-E since these molecules are expressed on placental trophoblast cells and would be expected to bind the inhibitory NK receptor, CD94/NKG2A, preventing NK-mediated rejection of the fetus.[223,224] Additional evidence now supports an antigen presentation function of HLA-F and HLA-G.[112,225-227] Tissue-restricted expression of *HLA-G* as well as the observation that a soluble form leads to apoptosis of CD8+ T cells suggests that this molecule may be involved in the mother's immunological tolerance of the fetus.[223,228] *HLA-H* is a pseudogene mapping to this region.[229] This should not be confused with the more distantly related *HLA-HFE*, an *MHC-Ib* gene erroneously called *HLA-H* by some authors,[229,230] that controls hereditary hemochromatosis by virtue of the interaction of its encoded protein with the transferrin receptor.[231-238] The observations that extravillous trophoblast cells express an unusual combination of HLA-C, HLA-E, and HLA-G molecules, that uterine NK cells express NK inhibitory receptor (NKIR) molecules known to interact with HLA-C allelic products, and that complex genotypes predispose to preeclampsia raise the possibility that important aspects of fetal rejection and clinical infertility may be related to MHC recognition by NK cells in the placenta[239-241] (Chapter 42).

Comparison of the human and mouse *MHC* maps (Figs. 19.1 and 19.2) reveals several interesting differences.[242-245] The mapping suggests that an intrachromosomal recombination event that occurred in some common rodent ancestor relocated some of the *MHC-I* genes from a distal location to the proximal site.[246] Inspection of the current human, mouse, and rat maps clearly indicates similarities in the relative locations and organization of *MHC-II, MHC-III,* and the distal *MHC-I* genes.[244] Various genetic expansions and contractions[168] are obvious as well. In particular, the mouse *Q* and *T* regions have expanded the pool of *MHC-I* genes, which are relatively few in the human and the rat.

Early studies of congenic mouse strains mapped multiple genes to the *Q* and *T* regions,[247-249] and recent evidence suggests significant differences in the number of genes of this region in different strains. The mouse has some *MHC-Ib* genes that possess relatively unique function. In particular, the *H2-M3* gene, which maps distal to the *Q* and *T* regions, encodes a protein that preferentially binds peptides that have N-formyl amino terminal modifications. This antigen presentation function may be geared to bacterial, protozoal, and mitochondrial antigens.[250-252] Rat homologs of the mouse *H2-M3* and *H2-M2* genes have also been identified,[244,253] though the human has none. The mouse vomeronasal organ expresses a family of *MHC-Ib* genes designated the *H2-Mv* family that co-expressed with *Vmn2r* G-protein–coupled receptor genes plays roles in detection of olfactory cues, which contribute to aggressive and sexual behavior.[254]

MHC Polymorphism

The *MHC*'s function in immune responsiveness is also reflected in its genetic polymorphism. Polymorphism is the presence at any given time of a larger-than-expected number of genetic variants in a population. As populations change and evolve, we expect that genetic variants should arise, but because of the selection exerted on most gene products, relatively few of these genetic variants will persist. A genetic locus that exhibits variant alleles at a frequency of more than one percent is considered "polymorphic."[242] A genetic locus that is relatively invariant may be referred to as "monomorphic," even if more than one allele is known. *HLA* genes exhibit a high degree of polymorphism and many different mechanisms may contribute to the generation and maintenance of polymorphism. Among these are the selective advantage of a heterozygous pool of antigen-presenting elements in any given individual that might allow the binding and presentation of antigenic peptides derived from a wide variety of environmental pathogens. Limited polymorphism would make the entire population susceptible to a chance infectious agent for which all individuals would be unable to respond, while widespread polymorphism would be expected to allow the APCs of at least a proportion of the population to bind and present effectively antigens derived from invading pathogens.[255] Although such a view was originally based on HLA molecules as presenting elements for pathogen-derived peptide fragments to T cells and their antigen-specific TCRs, studies suggest an additional role for *MHC-I*–related resistance to viral infection via NK cell–mediated recognition[256,257] and altered antigen presentation pathways.[258] Additionally, the presence of a high degree of polymorphism also implies that there is little or no selective disadvantage to the expression of new variant MHC molecules.

The human *MHC-I* and *MHC-II* genes are highly polymorphic, with thousands of alleles as noted above. In experimental inbred animals such as mice, it is more difficult to demonstrate polymorphism in terms of population genetics, although typing of wild mice has confirmed the existence of natural polymorphism predicted from the analysis of inbred strains and their mutants.[159,259] The polymorphism of *MHC-I* and *MHC-II* genes, so evident in man and mouse and all vertebrates, has been well documented in analyses of cichlid fishes, animals that diverged from the line leading to mammals at least several hundred million years ago.[260-263] Consideration of the peptide-binding repertoires of different human and avian MHC-I molecules leads to the hypothesis that the breadth of the repertoire governs the level of surface expression in an evolutionary balance between reactivity to pathogens and autoimmunity.[264]

Genetic Mechanisms for *MHC* Evolution

As both an extended genetic region and a group of genes with many belonging to the immunoglobulin supergene family,[249,265,266] the *MHC* has served as a prototype for elucidating mechanisms that contribute to the evolution of a multigene family and that add to the polymorphism that is such a dominant characteristic of the classical MHC molecules.[243] The analysis of mutations in the mouse and polymorphisms in the human, mostly of *MHC-I* genes, has led to an understanding of the genetic mechanisms that give rise to such variants. The regular identification of new human alleles, published in the journal *HLA, Immune Response Genetics* (https://onlinelibrary.wiley.com/journal/20592310?campaign=wolearlyview), and archived in the international ImMunoGeneTics information system® (IMGT)[9] contributes steadily to the knowledge base. In the mouse, both induced and spontaneous mutations affecting skin graft acceptance or rejection have been identified, and many of these have been mapped to the *MHC*. Gross recombinational events as well as more subtle mutations have been documented in the mouse.[267,268] Some mutants reveal multiple amino acid substitutions in a relatively small part of the gene that seem to derive from nonreciprocal recombinational events. Such recombination that occurs over short sequences is known as "gene conversion" because of its similarity to the phenomenon that occurs in yeast.[267-277] Convincing evidence for interlocus gene conversion has been documented for the *MHC* class II genes of teleost fish,[277] and by deep sequencing of *MHC* class I genes in recently founded bird populations[278] and in marmosets.[279] Some of the polymorphisms of *MHC* genes that have been identified clearly reflect point mutations. Structural studies have shown that the profound immunological effects of mutations of the *H2* genes *H2-K^bm1^* and *H2-K^bm8^*[275,280] result from minimal detectable changes that may affect thermostability.[281] In addition to such mouse mutants, a number of somatic cell variants and mutants, some due to major deletions or regulatory defects, others clearly point mutants of structural genes, have been described.[282]

The mechanisms that maintain *MHC* polymorphism have been the subject of considerable speculation for over half a century, and controversies continue.[283-286] In the early 1970s, Bodmer proposed that the evolutionary advantage associated with an *MHC* allele would be smaller, the higher its frequency in the population.[287] He reasoned that pathogens were poorly adapted to less common alleles having coevolved with the more common alleles to which adaptation (presumably through immune evasion) would be manifest. This concept is called negative frequency-dependent selection or "rare allele advantage" and has some observational[288] and experimental[289] basis. This model of maintaining polymorphism depends upon new alleles possessing a selective advantage, which is now thought to arise from selection of polymorphisms mainly occurring in the MHC antigen-binding cleft where they diversify peptide selection from pathogens to drive new T cell immunity. Selection of new alleles is evidenced by the preponderance of nonsynonymous mutations observed in the antigen-binding cleft of natural *MHC* alleles.[290]

Another mechanism believed to operate in maintaining MHC polymorphism is so-called "heterozygote advantage." This is thought to leverage the greater peptide-binding repertoire available from coexpression of distinct alleles from the same locus, thereby conferring broader and more effective T cell immunity associated with protective immunity to pathogens. This is sometimes called overdominance as *MHC* heterozygotes have greater "fitness" than homozygotes.[291] Fluctuating selection refers to maintenance of MHC polymorphism due to pathogen variation over multiple generations. Geography and migration of populations can also help maintain polymorphism through founder effects, different pathogen environments, and variable levels of racial admixture.

The *MHC* and Histocompatibility

The early description of the genes of the *MHC* was based on identification of loci involved in tumor and allograft rejection, leading to their designation as "histocompatibility" genes. Although these genes clearly play a role in such complex phenomena, a contemporary understanding of the function of *MHC* genes in immunology requires little understanding of the rules of transplantation. Although these genes are the *major* determinants of allograft rejection (but not the only determinants), it was also understood early on that these genes encode the molecules that determine antigen-specific immunity mediated by T cells, the so-called immune response genes. Hence, major histocompatibility genes and immune response genes are actually one and the same. Nonetheless, the role of *MHC* polymorphism and function in determining the rules of transplantation is very complex. The early history of transplantation is chronicled extensively in several books[2,242,292] and reflects a developing interest in tumor immunology and congenic mouse strains. The mouse has served as the model system for understanding the genetic underpinnings of skin, tumor, and organ transplantation, so a brief description of some relevant principles is in order. Comprehensive manuals and reviews are available.[293] Propagation of a mouse strain by repeated matings of brothers and sisters leads to the establishment of an inbred strain, a group of animals that is genetically identical at all loci. More complete descriptions of the process by which brother-sister mating leads to homozygosity at all loci are given elsewhere.[2,242]

"Congenic" mouse strains, also known as "congenic resistant" or "CR" strains, are those derived by first crossing two inbred strains that differ in a histocompatibility phenotype such as resistance to a transplantable tumor or ability to reject a skin graft. These are then successively backcrossed to one parental strain, and the resistance phenotype is preserved. Following at least 10 backcross generations (N10), a point at which $(1/2)^9 = 0.002 = 0.2\%$ of the genes of the selected strain should be present, the new strain is propagated by brother X sister mating. Several relevant inbred and congenic mouse strains are listed in Table 19.1 along with their *H2* designations.

The early rules of transplantation were determined by observation of the ability of either transplantable tumors or of allografts (usually from skin) to survive in a particular inbred mouse strain host. The graft rejection phenomenon is an extremely sensitive and specific bioassay that permits the detection of genetic differences as small as a single amino acid in an MHC protein. It has been particularly valuable in assessing spontaneous and induced mutants (see discussion above) and remains the absolute experimental discriminator of "histocompatibility."

In addition to the *MHC* genes, we should not overlook those that encode minor histocompatibility antigens. In the mouse, these were originally identified as genetic loci responsible for graft rejection after extended periods of time. The minor histocompatibility *H3* locus, originally identified by graft rejection to lie on chromosome 2 of the mouse, has been studied functionally and encodes several genes, including that for the MHC-I light chain, β_2m, which stimulate both MHC-I– and MHC-II–mediated responses.[294] More recently, several minor histocompatibility loci have been identified as those that encode polymorphic proteins that give rise to peptides presented by MHC molecules,[295-299] and we now can understand the complexities of transplantation tolerance in terms not only of *MHC* genes but in terms of numerous proteins that may give rise to variant peptides for T cell recognition. Not only can peptides derived from proteins encoded by polymorphic genes throughout the genome serve as minor histocompatibility antigens but also defective translation products or peptides resulting from transcription of introns or noncoding strands of DNA may also produce immunologically significant peptides that bound by self-MHC molecules may stimulate T cells.[34,300] Whether such defective, newly synthesized proteins serve as a major source of MHC-I–bound peptides remains controversial,[300] though refined proteomic methods are now identifying more examples.[301,302] Minor antigens that are confined to hematopoietic cells can be recognized as targets by antitumor cytolytic cells and may explain the lower incidence of relapse in hematopoietic stem cell (HSC) transplant recipients who experience graft-versus-host disease (GVHD).[303,304]

Mutations in the *H2* Region

The rapid pace of improvements in deriving mouse and other models for genetic disease based on the engineering of transgenic, targeted deletion and mutation, and conditional knockout mice emphasizes the importance of understanding the valuable reservoir of *MHC-I* and *MHC-II* mutants originally derived by laborious methods and still available for precise studies of the role of the *MHC* in immune responses. Mutations at the *H2* locus have been identified in animals screened by skin grafting in extensive experiments carried out over a 25-year period.[305,306] By grafting tail skin of siblings to and from each other, spontaneous or induced mutant animals that displayed either "gain," "loss," or "gain plus loss" transplantation phenotype were identified. Gain mutants are those that express a new transplantation antigen—thus their skin is rejected by their nonmutant siblings; loss mutants have lost a transplantation antigen—thus they recognize the skin of their siblings as foreign and reject that graft. Gain plus loss mutations give effects in both directions—they reject the skin of their siblings, and their skin is rejected by their siblings as well. In classic studies, Melvold, Kohn, and their colleagues screened a large number of mouse progeny. Homozygous inbred and F1 animals were examined, yielding a total of 25 *H2* mutations identified at *K, D, L,* and *Ab* loci, and an additional 80 mutations of non-*H2* histocompatibility genes. Although earlier studies suggested that all *H2* genes might be hypermutable, a more complete retrospective evaluation of the available data suggests that with the exception only of the *H2-K^b* gene, the spontaneous mutation rate for *H2* genes was comparable to that for non-*H2* genes. The characterization of these mutant animals, first based on peptide maps and amino acid sequences of the H2 proteins,[307-309] later based on the nucleotide sequences of the cloned cDNAs or genes,[275] provided some of the basic biochemical information on which further studies of structure and function and mechanism of gene evolution were based. X-ray structure determination of the H2-K^{bm1} and H2-K^{bm8} mutant proteins suggests explanations for the differences in T cell recognition that result from what might appear to be subtle amino acid substitutions.[281]

T Cell Alloreactivity

Processed foreign antigen complexed to MHC class I or class II molecules is recognized by a specific clonally distributed TCR for antigen on the surface of T lymphocytes. T cells bearing αβ receptors can recognize the unitary structure of the MHC molecule itself coordinately with the exposed parts of the peptide antigen. Corecognition of MHC and self-peptide antigen in the thymus where T cells develop results in culling of strongly self-reactive T cells (negative selection) to avoid autoimmunity, but persistence of weakly self-reactive T cells (positive selection) to create a self-reactive T cell pool. This process is called "thymic education" and results in a repertoire of developmentally selected T cells with TCRs that are highly specific and genetically restricted to recognizing thymic p/HLA complexes present in the individual from which they were derived (Chapter 27). These T cells are weakly reactive with self, just enough to maintain them in the periphery (secondary lymphoid tissues) where they are further selected during infection by pathogens expressing foreign antigens with higher reactivity than the original self-selecting peptides leading to T cell activation, proliferation, and acquisition of effector functions. Thus, a killer (cytotoxic) T cell raised against an influenza virus peptide in an individual expressing HLA-A2 (HLA-A*02) will not recognize influenza-infected HLA-A1 (HLA-A*01). This concept is known as "MHC restriction" and was described by Shevach and Rosenthal for recognition of alloantigens,[4] by Zinkernagel and Doherty[7,310] for recognition of viral antigens, and by Shearer et al[8] for recognition of altered self-ligands. Given that T cells are MHC restricted, it is difficult to understand why they should ever recognize any antigen presented by a foreign or allogeneic MHC type. However, in practice they do, and data indicate that such alloreactive T cells arise at high frequencies. Thus between 1/10 and 1/1000 of activated clonally distinct T cells are capable of responding to any random allogeneic HLA molecule.[311-314] Given the number of T cells in the human lymphoid system, this represents a striking tendency for T cells that are normally restricted to recognizing self-MHC molecules complexed to foreign peptides to cross-react on allogeneic MHC molecules. This cross-reaction can arise from direct recognition of the allogeneic p/MHC, which depends on TCR interaction with the peptide antigen as well as the allogeneic MHC molecule. The structural basis by which self-restricted TCRs cross-react with allogeneic MHC molecules can involve molecular mimicry between self-MHC complexed to cognate foreign antigen and allogeneic MHC and self-peptide antigen[315,316] or can result from disparate modes of TCR interaction with self and allogeneic p/MHC complexes.[317] Alternatively, allorecognition by T cells can occur indirectly.[318,319] In such cases, peptides derived from the allogeneic MHC molecules are presented as nominal antigen after processing by the host cells bearing self-MHC molecules. In the normal course of events, T cell alloreactivity may be considered an *in vitro* curiosity relevant to clinical transplantation. However, it also is manifest in the complexities of acceptance or rejection by the mother of the genetically distinct fetus. Although it is still not entirely clear why the fetal "allograft" does not stimulate the maternal immune system, current studies suggest that fetal success is influenced by cross talk between MHC molecules and NKIRs[320-323] (Chapter 42).

The *MHC* and Clinical Transplantation

Clinical transplantation of organs and HSCs across HLA compatibility barriers can result in graft rejection or GVHD due to T cell alloreactivity. Fully allogeneic transplants theoretically expose the recipient immune system to up to 12 non–self-HLA allele products expressed by the allograft. Moreover, the "self-peptides" constitutively presented by allogeneic HLA molecules are likely to be distinct from those presented by syngeneic HLA molecules because the polymorphisms of the peptide antigen-binding cleft of the MHC-I molecule that distinguish HLA alleles alter the spectrum of selected peptides. In the presence of appropriate costimuli, allogeneic p/HLA complexes evoke powerful T cell responses because of the high density of unusual determinants and the diversity of new peptide ligands presented. Because there are numerous *MHC*-linked genes encoding proteins lacking known immunological function, it is likely that polymorphisms in these molecules contribute significantly to the alloresponse (see earlier discussion of *MHC* genetic maps and *MHC* alleles). Renal graft survival improves with progressive matching of donor and recipient

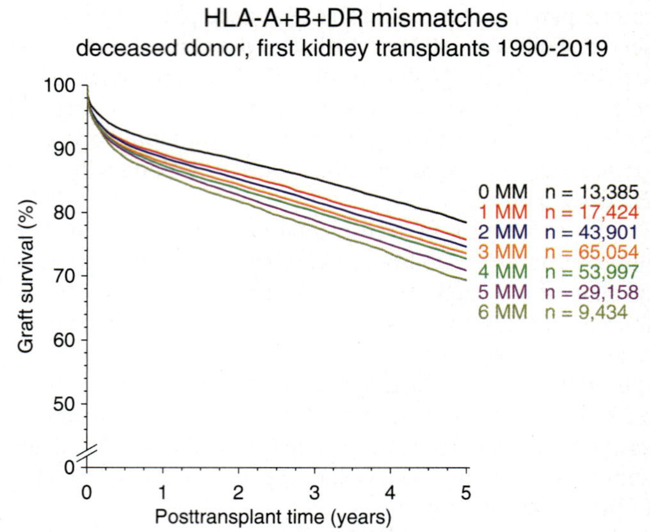

FIG. 19.3. Cumulative data for graft survival. The cumulative graft survival for the indicated number of HLA-A, HLA-B, and HLA-DR mismatches (MM) is plotted. Data are from the Collaborative Transplant Study (http://www.ctstransplant.org).

allotypes encoded at all HLA loci. In Figure 19.3, cumulative data for graft survival are plotted as a function of time. Curves represent those groups with the indicated number of mismatches. Therefore, for solid organ transplants, individuals are matched as closely as possible for their HLA types to avert cellular rejection mediated by alloreactive T cells and alloantibodies. The beneficial effects of HLA matching on solid organ allograft survival have been monitored by the Collaborative Transplant Study since 1982 (http://www.ctstransplant.org/). The degree of HLA matching based on family relationships usually required for renal transplantation is shown in Figure 19.4. Clearly, an HLA-identical sibling donor offers the best likelihood of long-term survival.

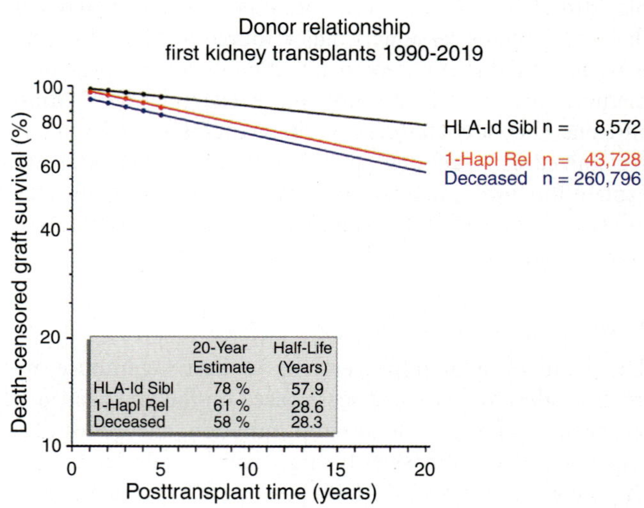

FIG. 19.4. Survival of renal transplants based on HLA or haploidentical matching. Data are taken from the Collaborative Transplant Study (http://www.ctstransplant.org).

As discussed below, it is also important to exclude preformed antidonor HLA antibodies in the host as these are associated with graft failure through rejection.

Alloantibodies and Transplantation

In addition to the allogeneic cellular response, the antibody response to HLA molecules and ABO blood groups can also cause rejection of certain grafts, especially where these antibodies are preformed and therefore present at the time of organ transplantation (Chapter 49). Antibodies to ABO blood group antigens react with these determinants on vascular endothelium and renal tubules, and therefore ABO-incompatible solid organs can be rapidly rejected by humoral mechanisms (hyperacute rejection). Historically, ABO incompatibility has been a contraindication to allogeneic renal transplantation, but desensitization protocols through plasma exchange, immunoadsorption, and anti–B cell immunotherapy have allowed successful ABO-incompatible transplantation to the point where they comprise between 10% and 20% of living related transplants in many centers. These protocols diminish isohemagglutinin antibody levels before and after transplantation in transplant recipients. A key advantage of this innovation is an estimated 30% increase in organ availability through acceptable HLA compatibility in otherwise ABO-incompatible donors.

Screening for Anti-HLA Alloantibodies

Preformed anti-HLA antibodies can lead to acute and hyperacute rejection of grafts expressing the particular HLA molecules recognized by these antibodies. The presence of preformed alloantibodies is most apparent in potential transplant recipients with a history of prior blood transfusion, previous organ allografts, or in multiparous females sensitized to allogeneic HLA through repeated childbirth. The presence of anti-HLA antibodies toward graft HLA molecules has been shown to be a good predictor of early allograft rejection.[324] Hence, for many years, it has been standard practice to test for antidonor HLA antibodies in the serum of patients immediately before transplantation. This test is known as the "crossmatch." Many laboratories crossmatch only for anti-HLA class I and HLA class II antibodies. Crossmatch compatibility to exclude anti-HLA antibodies is standard practice in solid organ transplantation including heart/lung transplantation. Crossmatching for liver transplantation is practiced at only some centers, and the evidence that a positive crossmatch predicts allogeneic liver graft rejection has not convinced everyone of its importance. Patients awaiting renal transplantation are usually monitored regularly for anti-HLA-I and anti-HLA-II antibodies because the level and specificity of these antibodies can change with time. Historically, this monitoring involved regular testing of patient serum antibodies for cytotoxicity against a panel of randomly selected cells bearing different HLA types. The percentage of positively reacting cells with different HLA types is known as the "panel reactivity." High panel reactivity in a recipient serum can make it difficult to find suitably matched donors as these alloantibodies predict poor graft outcome. Almost all centers now prescreen patients on transplant waiting lists for anti-HLA antibodies using single MHC-coated beads as a source of antigen and a highly sensitive Luminex or flow cytometry technology for their detection. Panel reactivity is now calculated using such assays and is termed CPRA (calculated panel reactive antibodies). The cytotoxic crossmatch assay has largely been replaced by flow cytometric crossmatching including the recently described Halifax or Halifaster methods where the parameters have been carefully optimized to shorten the assay.[325] This so-called "virtual crossmatching" assesses the compatibility of the recipient and prospective donor by analyzing the recipient's anti-HLA antibodies, without performing a physical donor-recipient crossmatch.[326,327] The virtual crossmatch is practiced widely and is very convenient as it does not require donor cells, a logistic efficiency in national cadaveric organ exchange programs where the donors and recipient are often located in different countries, states, or cities.[328]

Paired kidney exchange programs now exist at most transplant centers allowing groups of legal partners to enter into a "market" arrangement where live donor-recipient optimization is determined across the network of donor-recipient combinations, thereby increasing access to kidney allografts for patients with incompatible live partner donors.[329]

Prediction of HLA Alloantibody Reactivity and Immunogenicity *In Silico*

Alloantibody epitopes that are either shared or distinct between donor and recipient HLA molecules can be predicted with some accuracy *in silico* using the software HLAMatchmaker (http://www.epitopes.net).[330] HLAMatchmaker is a structurally based algorithm that defines triplet clusters of amino acids or "eplets" structurally separated by 3.0 to 3.5 Å (https://www.epregistry.com.br/). These eplets are common antibody epitopes recognized by HLA alloantibodies and are often shared by multiple HLA allotypes from different HLA loci such that eplets shared between donor and recipient HLA molecules are rarely immunogenic. Hence, an HLA mismatch can be interrogated for eplet mismatches across the complete set of HLA alleles in the recipient to determine the probability of any donor HLA mismatch being recognized by preformed recipient alloantibodies or likely to induce donor-specific alloantibodies following allografting. HLAMatchmaker can also be used as an adjunct to organ matching or in selection of the optimal HLA-matched platelets for transfusion in platelet refractory patients with anti-HLA class I antibodies. In highly sensitized patients with multiple host antidonor antibodies, these unwanted antibodies can sometimes be functionally eliminated using B cell ablation therapies (anti-CD20 mAb), plasmapheresis, and intravenous gamma-globulin infusions, thus allowing successful transplantation.

The role of antibody crossmatching in allogeneic HSC transplantation is less clear, and many clinical centers do not take the HLA class I or class II crossmatch into account when identifying a bone marrow transplant donor. However, some large centers place considerable importance on a positive crossmatch as a predictor of bone marrow or HSC rejection, and it is therefore advisable to crossmatch bone marrow donor-recipient pairs when there is a high risk of rejection (eg, aplastic anemia). Crossmatching is also used to detect anti-HLA antibodies that may cause refractoriness to platelet transfusion with random platelets.[331]

Family Studies and HSCT Matching

The most likely identical HLA match for a patient requiring HSCT is a sibling where HLA identity occurs for around 30% of donors. Sometimes another relative or even a parent can be closely matched or identical. HLA identity by descent can be confirmed in a family study where all relatives are HLA typed and the pedigree is constructed to confirm maximal likelihood of segregation of shared HLA alleles (Fig. 19.4). The linkage of HLA loci on chromosome 6 means that individuals will usually inherit a set of nonrecombined HLA alleles encoded at linked HLA loci from each parent. This set of genes is known as the haplotype, a term first coined by the Italian MHC biologist Ruggero Ceppellini[332] and now widely used to describe clusters of alleles at any linked loci, often clusters of SNPs used in genome-wide studies of disease association. In identifying donors for HSC transplantation, testing of family members is essential to determine haplotypes accurately.[333,334] This is because sharing of HLA antigens from different haplotypes is quite common in families, so mismatches in low resolution HLA subtypes (ie, allele-level mismatches) are easily overlooked as a result of mistaken haploidentity of siblings or other family members. Because HLA mismatching is poorly tolerated in HSC transplantation, high-resolution sequence-based DNA matching, or its equivalent, is required to avoid GVHD.[335-338] When a matched sibling donor does not exist for a patient requiring allogeneic HSC transplantation (70% of cases), searching of the extended family or unrelated bone marrow registries is indicated. The National Marrow Donor Panel (http://www.marrow.org) and other bone marrow registries across the world have 10 million potential donors suitable for unrelated stem cell transplantation, and these donors are used worldwide. Mobilization of stem cells in the blood following administration of hematopoietic growth factors is now widely used to avoid the need for marrow collections from donors.

The impact of HLA compatibility on the outcome of HSCT has been studied in large groups of unrelated donor-recipient pairs by a number of researchers.[339-341] These outcome data show that unrelated matched transplants can be as effective as sibling matched donor-recipient transplants in donor survival and risk of GVHD. The studies also show that mismatching at different HLA loci leads to distinct clinical risks including development of GVHD, graft failure, and mortality.[342] These effects are complex and not easily generalized across different diseases and ethnic groups.

Given the importance of evaluating all loci when selecting a donor, *HLA* typing is now done for *HLA-A*, *HLA-B*, *HLA-C*, *HLA-DRB1*, *HLA-DRB3*, *HLA-DRB4*, *HLA-DRB5*, *HLA-DQ*, and *HLA-DP* loci to optimize HSC transplant matching.[341,343] Testing is carried out by next-generation sequencing of the loci on both haplotypes, and resolution of genotypes is performed by informatic algorithms. The US National Marrow Donor Program guidelines for matching and selection of adult donors (bone marrow or PBSC) require six of eight loci match for *HLA-A*, *HLA-B*, *HLA-C*, and *HLA-DRB1*; each of these loci must be typed at high resolution by DNA-based methods.[207] Cord blood units require four of six loci matching for *HLA-A*, *HLA-B* (antigen level), and *HLA-DRB1* (allele level). Mismatching beyond a single, or at most two *HLA* loci, is associated with increased GVHD and inversely with a lower rate of leukemic relapse. Although allogeneic HSC transplantation can be carried out across single-locus mismatches, there is little correlation with the magnitude of a given mismatch (ie, number of polymorphisms between alleles) and the subsequent immune response. Donor/recipient MHC differences of just a single amino acid can provoke strong alloresponses comparable to reactions between vastly different MHC products.[336]

Matching for HLA-DP in allogeneic stem cell transplantation appears to improve graft success in HSCT in some studies[335] but is more commonly important in HSC when several donors are available and a rational choice of the best donor has to be made.[344-347] In addition, some HLA-DP mismatches provoke differential T cell alloreactivity such that some mismatches can be tolerated more than others.[348] An algorithm is available to calculate the permissiveness of a DP mismatch (website for calculator: https://www.ebi.ac.uk/ipd/imgt/hla/dpb.html). Survival of patients undergoing HSCT also varies according to the nature of the disease, disease stage, and age of the patient but can be >70% in some cases (see Center for International Blood and Marrow Transplant Research; http://www.cibmtr.org/pages/index.aspx).[349]

HLA-Haploidentical Donor Stem Cell Transplantation

HSC transplantation using HLA-haploidentical donor stem cells has become a mainstream alternative approach to sibling matched or unrelated HSC transplantation, with growing sophistication of pre- and posttransplant conditioning usually with cyclophosphamide. This intervention helps reduce severe GVHD and mortality unrelated to relapse. Logistically there are advantages too, as the donor is usually a sibling or parent and conditioning is low intensity without T cell depletion. These factors reduce cost and significantly increase probability of finding a donor. The success of HLA-haploidentical HSCT appears comparable to the outcome of HLA-matched HSCT. It is important to test for donor-specific antibodies which can lead to failure of engraftment and can be a factor in selection of donors where there are multiple possibilities. HLA-haploidentical HSCT has all but supplanted cord blood transplantation at many centers. Haplotype mismatching is generally unavoidable in renal transplantation from cadaveric donors. Fortunately, significant HLA mismatching is quite frequent in kidney transplantation, and while not preferable, it is clinically tolerated because of the potency of posttransplant immunosuppression.

In the family study shown in Figure 19.5, the mother and father are mismatched at both haplotypes. Jane's paternal haplotype is a recombinant involving a crossover event

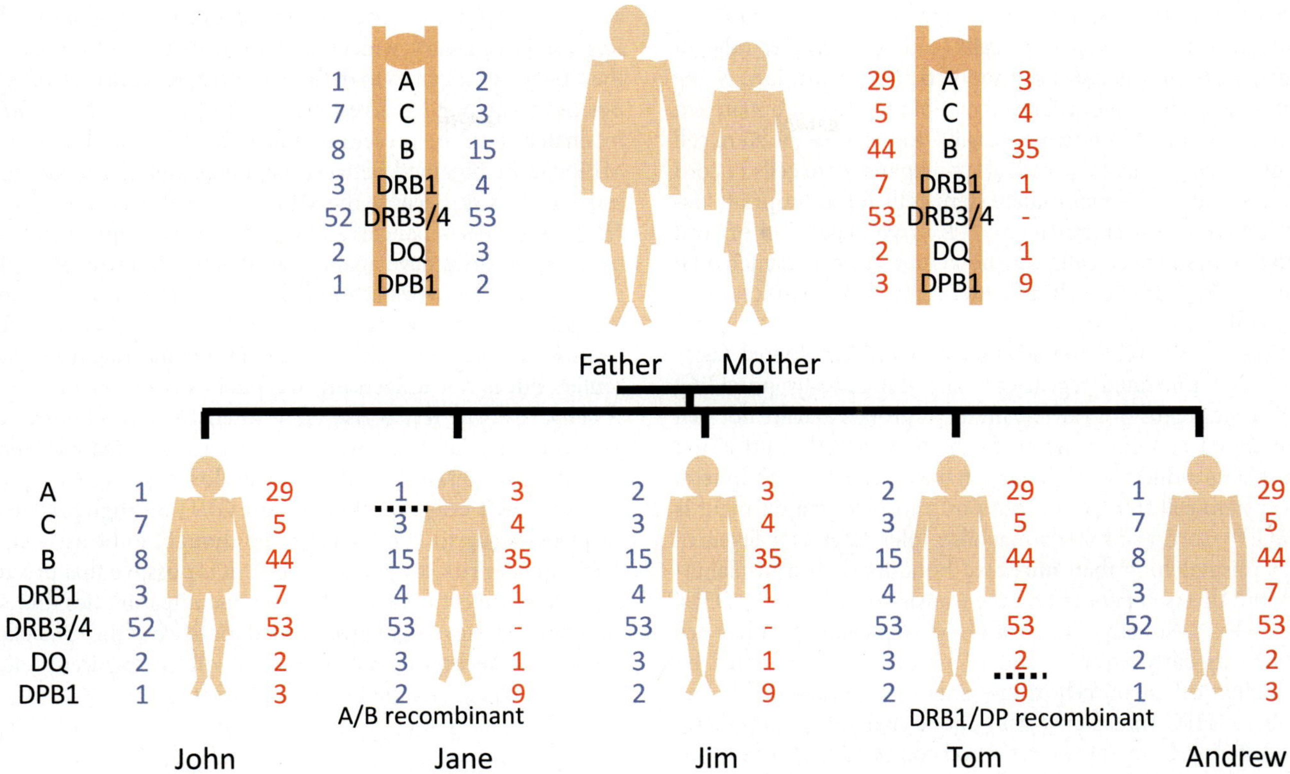

FIG. 19.5. Family study of linkage and recombination at the *HLA* locus. Father (*blue*) and mother (*red*) have the indicated HLA haplotypes that segregate Mendelianly in John, Jim, and Andrew. As described in the text, however, Jane reveals an A/B (in detail, A/C) paternal recombinant chromosome, and Tom has a DRB1/DP (DQ/DPB1) maternal recombinant.

between *HLA-A* and the *HLA-B* locus. Actually, the crossover is between *HLA-A* and *HLA-C*, noting the *HLA-B* and *HLA-C* loci are very closely linked and meiotic crossover between these loci is rare. Recombination is observed between *HLA-A* and *HLA-B*, and *HLA-B* and *HLA-DRB1* in about 1% of meiotic events. Recombination between *HLA-DRB1* and *HLA-DQ* and the *HLA-DP* locus is also relatively frequent and is evident in Tom's maternal haplotype. Among the other children, John and Andrew are *HLA* identical (and therefore phenotypically identical). Jane and Jim are almost *HLA* identical to each other except for Jane's *HLA-A1* created by the crossover event. Tom and Jim share one haplotype along with serendipitous *HLA-DP* identity. The implications of this family study are that Andrew and John would be ideal *HLA*-matched HSC donors for each other. Jane and Jim are relatively acceptable HSCT donors for each other with a single mismatch between *HLA-A1* and *HLA-A2*. None of the other siblings are an identical donor for John or Andrew or for Jane or Jim. However, there is sharing of a partial single haplotype between Tom and both John and Andrew with the exception of the *HLA-DP* mismatch on this haplotype due to a crossover (Tom's *HLA-DP9*), so that Tom is a potential haploidentical HSC donor if John or Andrew required such treatment and their *HLA*-identical sibling was not available. However, the predicted immunogenicity of the *DPB1* matching for Tom's *HLA-DP9* is "nonpermissive," making the parents plausible haploidentical donors for John or Andrew.

KIR Ligand Mismatching in HSC Transplantation

Controversial results suggest dramatically positive outcomes of HSC transplantation where deliberate KIR ligand mismatching leads to donor versus recipient NK cell alloreactivity toward cells of hematopoietic origin.[350-353] The Velardi group in Perugia has championed this approach providing evidence that in the absence of GVHD, a high frequency of donor NK cells in KIR ligand mismatched transplants can remove recipient's hemopoietic-derived tumor cells impacting on the frequency of relapse.[351] In particular, *HLA* haplotype mismatched hematopoietic transplants can be beneficial in adult acute myeloid leukemia and childhood acute lymphoblastic leukemia, leading to the development of adoptive NK cell transfers as an experimental therapy for graft-versus-leukemia treatment.[354] However, these transplants must be rigorously T cell depleted and variable results appear to reflect different transplant protocols, disease parameters, and KIR mismatching assignment.[352,353,355] Evaluation of KIR ligand mismatches to predict NK cell alloreactivity can be accomplished using a KIR ligand calculator available online (https://www.ebi.ac.uk/ipd/kir/ligand.html).

Umbilical Cord Blood Transplantation

Cord blood banks have also been established around the world.[356-358] However, cord blood donation of stem cells is often unsuitable for adult transplantation because of limitations in the volume of cord blood collections providing inadequate numbers of stem cells leading to higher rates of graft failure and delayed hemopoietic reconstitution. Many centers will combine different cord bloods that are matched with the recipient as much as feasible to increase the stem cell number. Experimental protocols preexpand stem cells in cord blood while others manipulate stem cells for appropriate tissue targeting and engraftment.[359] *Ex vivo* expansion of cord blood mononuclear cells can be achieved on mesenchymal stem cells that are thought to better reproduce the microenvironment of bone marrow.

Cord bloods offer the advantage of finding donors faster than adult unrelated registries[360] and theoretically providing banked stem cells from ethnic minority groups that are not well represented in bone marrow donor registries.[361] Cord blood transplants induce less GVHD than bone marrow or peripheral blood stem cell transplants, but posttransplant engraftment is slower.[362] Cord transplants may also tolerate greater levels of HLA mismatching than unrelated bone marrow transplantation, but this is dependent on the transfer of sufficient CD34+ stem cells.[363] Nonetheless, cord blood transplantation has not fulfilled its early promise and is increasingly supplanted by haploidentical stem cell transplantation worldwide.[364] The results of HSC transplants using bone marrow or peripheral blood stem cells are now considered comparable clinically but the impact of HLA matching in patients transplanted at early stage in their disease appears to be more marked.

Functional Tests of *HLA* Compatibility

Functional tests of *HLA* compatibility evaluating *HLA* identity at all *HLA* loci is a daunting task for most laboratories because of the very large number of alleles present in the population. Moreover, in renal transplantation, some mismatches appear to be well tolerated and are associated with long-term graft survival, whereas other mismatches of similar genetic disparity are poorly tolerated and are associated with early rejection.[356] Reliable methods for predicting these "taboo" mismatches are not readily available. Similarly, high-resolution HLA typing does not predict all GVHD when selecting suitable unrelated donors for HSC transplantation.[337,357] Therefore, there has been a great deal of interest in developing functional or *in vitro* cellular tests of overall donor-recipient compatibility. Unfortunately, none of the tests so far developed provides convincing predictability of impending graft rejection or, more important, GVHD. In addition, these tests are expensive and show poor reproducibility.

Among the tests used historically for assessing functional compatibility are the mixed lymphocyte reactivity (MLR) test and allogeneic T helper or cytotoxic T lymphocyte (CTL) precursor studies. The MLR or MLC (C = culture) involves measuring T cell proliferation of host T cells in response to donor lymphocytes and vice versa. In a one-way MLR, the stimulating lymphocytes are irradiated to prevent their proliferation, whereas in the two-way MLR, both stimulator and responder cells are allowed to proliferate. In MLR studies, it is necessary to include controls showing that responder cells can all respond and that stimulator cells can all stimulate across an appropriate barrier such as third-party donor cells. Responses can vary widely, and individual laboratories use their own cutoff values to define negative (ie, nonreactive) and positive (reactive) MLR results.[357] Unfortunately, known *HLA* mismatches can be present in a negative MLR, and a positive MLR can be obtained between phenotypically *HLA*-identical transplant pairs. Because the MLR is biased toward measuring *HLA* class II discrepancies and is notoriously irreproducible,[357] most laboratories have abandoned this test in favor of implementation of high-resolution PCR-based *HLA* class II typing. Measurement of allogeneic CTL or helper T cell precursor frequencies is carried out at specialized bone marrow transplant centers but is not universally accepted as being predictive of GVHD.[358,360] The test is also labor intensive and requires skill for reproducibility. Precursor frequencies are estimated by limiting dilution analysis or ELISPOT of donor-versus-host lymphocytes (ie, T cells expected to cause GVHD). High precursor frequencies (up to 1 in 10^4 cells) are thought to be associated with a greater risk of acute GVHD.[361] It is possible that precursor studies detect major and minor incompatibilities and so, theoretically, they might give a broad measure of the transplant barrier, but technical improvements will be required before this test is widely adopted in clinical practice.[365]

The *MHC* and Disease

In addition to the control of transplant acceptance or rejection and immune responsiveness, the *MHC* class I and II genes in the human play an important role in the etiology of a number of diseases, many of which are autoimmune or infectious in nature.[366-371] Several human diseases are associated with the *MHC* class III genes because some of the structural genes for enzymes involved in the adrenal steroid biosynthetic pathway (ie, 21-hydroxylase, CYP21A2) map to this region.[199,205] As many as 200 diseases have well-established genetic linkages to the *MHC*,[368,372-375] the most important of which are summarized in Table 19.2. As more information accumulates, other genetic loci and environmental factors clearly combine to influence the development and progression of *MHC*-linked disease.

Although the general mechanistic contributions of *MHC-I* and *MHC-II* genes/proteins are similar, and several diseases reveal contributions from both loci, it is convenient to describe their effects independently. Ankylosing spondylitis is so strongly associated with the *MHC-I* allele *HLA-B27* and the presence of some bacterial pathogens that it is a popular hypothesis that ankylosing spondylitis is due to the stimulation of particular T cells by *HLA-B*27*–presented bacterial antigens that cross-react on self-tissues. These T cells are then thought to initiate an inflammatory cascade. Despite the strong association of HLA with spondyloarthropathy, critical evaluation of the literature brings a postinfectious etiology into question, and certainly more studies are indicated.[403,404] Analysis of *HLA-B*27* subtypes reveals that disease association correlates with the fine structure of the HLA-B27 and that polymorphisms at individual amino acid positions contribute to susceptibility.[376,405,406] In particular, expression of *HLA-B*27:05* is strongly associated with ankylosing spondylitis,

TABLE 19.2 HLA/Disease Associations

Disease	MHC-I	Reference
Ankylosing spondylitis	HLA-B*27:05 (D116) (ERAP)	376-378
	HLA-B*27:09 (H116) (no association)	376,377
Reiter syndrome	HLA-B*27	379,380
Behcet disease	HLA-B*51:01 (ERAP1) HLA-B*27	381,382
Birdshot retinochoroidopathy	HLA-A*29 (ERAP1/2)	383,384
Graves disease	HLA-A*02:06	385
Hereditary hemochromatosis	HFE	238
Psoriasis	HLA-Cw1 HLA-Cw6 HLA-C*12 HLA-B*57:01 (ERAP1)	386-391
	MHC-II	
Celiac disease	HLA-DQ2 (DQA1*02:01/ DQB1*02:01) HLA-DQ8	392-394
Graves disease	HLA-DRB1*04:05 HLA-DRB1*14:03 HLA-DPB1*05:01	385,395
Hashimoto thyroiditis	HLA-DRB1*08:03 HLA-DRB1*09:01	395,396
Insulin-dependent diabetes mellitus (T1D)	HLA-DQ2 HLA-DQ8 HLA-DR2	397
Multiple sclerosis	HLA-DR2	398,399
Rheumatoid arthritis	HLA-DR4	400
Systemic lupus erythematosus	HLA-DRB1*15:01 HLA-DQB1 (Phe87)	401 402

Reprinted by permission from Springer Nature: de Bakker PI, McVean G, Sabeti PC, et al. A high-resolution HLA and SNP haplotype map for disease association studies in the extended human MHC. *Nat Genet.* 2006;38(10):1166-1172.

while *HLA-B*27:*09, which encodes a protein differing by only a single amino acid, H116 for D116, is not associated with disease. Recent genome-wide studies strongly indicate that an epistatic interaction between an enzyme involved in processing MHC-I–associated peptides, ERAP1, and HLA-B27 contributes to disease susceptibility.[407,408] The tendency of HLA-B27 to form disulfide-linked covalent dimers raises the question whether the cellular pathology resulting from poor cell surface expression of this MHC-I molecule may trigger the inflammation of joints that is characteristic of this disease.[409-412]

Hereditary hemochromatosis is one of the most common genetic disorders in Caucasian populations (with a prevalence of 1/300-1/400), and the gene controlling this condition (HFE) is *MHC* linked, mapping approximately 3 Mb telomeric to the *HLA-A* locus.[232] The HFE protein is an MHC-I–like molecule, the structure of which has been determined.[413,414] HFE assembles with β_2m and is expressed in the intestinal mucosa and placenta, where it plays a role in regulating iron uptake and transport.[415,416] Mice that are homozygous for an induced defect of β_2m as well as those with targeted inactivation of the HFE gene suffer from iron overload, hemochromatosis.[417-419] Mutations at loci other than β_2m or HFE also lead to the same disease phenotype.[420] HFE regulates the affinity of the transferrin receptor for transferrin, resulting in an alteration of the efficiency of iron transport. The most common molecular defect associated with hereditary hemochromatosis involves a point mutation that results in a Cys282Tyr substitution in the α3 domain of this class I–like molecule.[232] This mutation accounts for >80% of patients with hereditary hemochromatosis.[421] The disruption of the disulfide bond in the α3 domain at this site prevents efficient folding of the molecule and impairs assembly with β_2m, resulting in improper HFE protein expression. This leads to a failure to downregulate the affinity of the transferrin receptor for its ligand, transferrin, presumably causing increased iron uptake by cells and tissue damage as a result of iron overload. A second common HFE mutation, 187G, results in a His63Asp substitution and a very slight increase in susceptibility to developing hereditary hemochromatosis, depending on the genotype of the individual.[422] Incomplete penetrance of even the high-risk Cys282Tyr *HFE* genotypes can be partly explained by natural iron deficiency from limited dietary intake and menstrual losses in women.

Recent genome-wide association studies confirm the importance of the *HLA-DRB1* locus in rheumatoid arthritis and type 1 diabetes (T1D),[170,375,423,424] and fine mapping analysis suggests that a total of five amino acid polymorphisms (three in *HLA-DRB1*, one in *HLA-B*, and one in *HLA-DPB1*), all located in peptide-binding grooves, almost completely explain the *MHC* association with the risk of rheumatoid arthritis.[425] Recent evidence suggests a structural basis for joint inflammation found in rheumatoid arthritis.[400] In this hypothesis, the enzymatic conversion of arginine to citrulline in the joint, in the context of appropriate MHC-II molecules, generates a tightly binding peptide that stimulates reactive T cells. However, the precise mechanisms underlying the association of most of these diseases with the particular *MHC* haplotypes are unknown, but several models have been proposed, including the cross-reactivity of antimicrobial antibodies and the molecular mimicry of viral antigens that might induce T cell responses to self-antigens.[425-428] The very high incidence of some diseases associated with certain *HLA* genes assists in the diagnosis as well as the counseling of patients and their families. Several of these diseases are of particular note. Because virtually 100% of patients with narcolepsy have *HLA-DQB1*06:02* (associated with HLA-DR2),[428-430] HLA typing can be used as a test of disease exclusion. Thus, a diagnosis of narcolepsy can be excluded with reasonable certainty if the patient does not have *HLA-DQB1*06:02*. However, the presence of *HLA-DQB1*06:02* is of little predictive value in diagnosis of narcolepsy because this *HLA* type is relatively common in many populations and occurs frequently in the absence of disease.

As summarized in Table 19.2 and discussed earlier, there are a number of autoimmune diseases associated with particular alleles of *HLA* class II loci, especially with *HLA-DR* and *HLA-DQ*.[374] These diseases include T1D (insulin-dependent), rheumatoid arthritis, multiple sclerosis, systemic lupus erythematosus, thyrogastric autoimmunity, Sjögren syndrome, and many others. Rheumatoid arthritis is strongly associated with *HLA-DR4* subtypes that share a common sequence motif within the HLA-DRB chain,[431,432] suggesting preferential antigen presentation of self-epitopes by these molecules. The relative risk of severe rheumatoid arthritis is increased in *DR4* homozygotes, particularly compound heterozygotes with high-risk alleles,[433,434] indicating a gene dose effect in susceptibility to autoimmune inflammation.

Although the number of different *HLA* class I and class II alleles that are associated with T1D clearly indicates that this relatively common disease has a complex etiology, the identification of a novel *MHC-II* haplotype in the mutant nonobese diabetic mouse,[435-438] and the recognition that particular TCRs can mediate disease,[439] suggest that a cross-reactive response to a common self or environmental antigen may play an important role in the etiology of this disease as well. In human T1D, the incidence of disease is significantly increased in Caucasians with *HLA-DR3-DQ2* and *DR4-DQ8* haplotypes.[440,441] These haplotypes impart a synergistically increased relative risk when they occur as a heterozygous combination compared to the risk of disease conferred by either haplotype alone.[374,442,443] This raises the possibility of transcomplementation of *HLA-DQ* gene products producing new molecules involved in antigen presentation.[444] Analysis of large cohorts of patients in genome-wide marker studies offers a promising approach to identify additional genetic factors that act synergistically with *HLA* allelic linkages in raising the odds of disease susceptibility.[397] Online genomic maps and correlations of associations with particular loci further augment our ability to understand the relative contributions of MHC-linked and unlinked genes in autoimmune diseases such as T1D.

Celiac disease (CD) is an underdiagnosed inflammatory disorder affecting the small intestine[445] but resulting from a Th1 hypersensitivity to wheat gluten and related proteins present in barley and rye. Normally manifest in childhood as a failure to thrive, the disease may affect adults quite late in life leading to malabsorption and wasting.[446] CD can be subclinical, presenting with anemia, osteoporosis, or neurological symptoms.[445] This disease is strongly associated with *HLA-DQ2* and/or *HLA-DQ8* (DQ3)[398] as well as with autoantibodies to the enzyme transglutaminase (TG2). Deamidated gluten peptides are presented to CD4+ T cells by HLA-DQ2 or HLA-DQ8 molecules via macrophages, dendritic cells, and B cells.[447,448] HLA-DQ2–restricted peptides in general prefer glutamate at anchor positions P4 or P6, and sometimes P7.

Antigen presentation studies and epitope mapping of T cell responses in CD suggest that the interaction between the wild-type dominant gluten peptide and HLA-DQ2 is of low affinity but can be enhanced by deamidation of P6 glutamine.[447,448] HLA-DQ2 binds a dominant deamidated gliadin determinant through a network of hydrogen bonds.[449] This leads to enhanced Ag presentation and T cell recognition with intestinal inflammation in susceptible patients.[447,448,450] The protease resistance of the dominant DQ2-restricted gliadin peptides contrasts with DQ8-associated disease where the dominant gliadin peptides are protease sensitive. The HLA-DQ2–associated and HLA-DQ8–associated forms of CD have their own epitope patterns of deamidation dependence leading to T cell reactivity. DQ2-associated peptides possess single residue deamidation at P6 as compared with the preferred deamidation at positions P1 and P9 for DQ8-associated CD.[451,452] This dual deamidation of glutamine residues at P1 and P9 in gliadin peptides enables important salt-bridging interactions with HLA-DQ8.[452] While the mechanism of immunopathology of CD is well defined, the exact combination of factors that trigger celiac pathology is still mysterious.[453] Recently developed mouse models, reflecting the interaction among gluten, genetic background, and IL-15–driven tissue inflammation, offer new insights and promise improved approaches to rational therapy.[454]

HLA and Drug Reactions

Over the last decade, a number of adverse drug reactions (ADRs) have revealed strong associations with particular HLA types[455-457] (Table 19.3). The most notable HLA associations are *HLA-B*57:01* with abacavir hypersensitivity syndrome (AHS) in the Caucasian population,[458,474,475] *HLA-B*15:02* and *HLA-A*31:01* with carbamazepine-induced Stevens-Johnson syndrome (SJS) and toxic epidermal necrolysis (TEN) in Asian populations,[455,460,461,476] and *HLA-B*58:01* and allopurinol hypersensitivity syndrome and SJS/TEN.[459,477,478] Odds ratios or relative risks of these associations are greater than 500, 1000, and 800, respectively. These ADRs are all associated with *HLA* class I allotypes, with evidence of T cell involvement.

The kinetics of the syndromes are also consistent with T cell–mediated drug hypersensitivity.[475,479-482] The systemic features of many HLA-linked ADRs suggest that general mechanisms might mediate these reactions. Drug-specific CD8+ T cell responses can be generated in response to abacavir-treated APCs expressing HLA-B*57:01 but not closely related allotypes.[483] Indeed, unmodified abacavir specifically binds to HLA-B*57:01, lying across the bottom of the antigen-binding cleft and reaching into the F pocket where a carboxyl-terminal tryptophan typically anchors peptides bound to HLA-B*57:01.[484] As a result, abacavir changes the shape and chemistry of the antigen-binding cleft, thereby altering the repertoire of endogenous peptides that can bind HLA-B*57:01 while abacavir is also bound. In this way, abacavir guides selection of new endogenous peptides, inducing a dramatic alteration in "immunological self." The resultant peptide-focused "altered self" drives polyclonal CD8+ T cell activation resulting in a systemic reaction manifesting as AHS. The mode of abacavir binding provides a basis for understanding its exquisite specificity in binding to HLA-B*57:01

TABLE 19.3 HLA/Drug Hypersensitivity Associations

Drug Hypersensitivities	HLA Association	Adverse Drug Reaction	Reference
MHC-I Associations			
Abacavir	HLA-B*57:01	HSS	458
Allopurinol	HLA-B*58:01	SJS/TEN and HSS	459
Carbamazepine	HLA-B*1502	SJS/TEN	460,461
Feprazone	HLA-B*22	Fixed drug eruption	462
Flucloxacillin	HLA-B*57:01	Hepatitis	463,464
Levamisole	HLA-B*27	Agranulocytosis	465
Oxicam	HLA-A2 HLA-B12	SJS/TEN	466
Phenytoin	HLA-B*15:02	SJS/TEN	461,467
Sulfonamides	HLA-A*29 HLA-B*12	TEN	466
MHC-II Associations			
Aspirin	HLA-DPB1*03:01	Asthma	468
	DRB1*13:02-DQB1*06:09	Urticaria	469
Hydralazine	HLA-DR4	Systemic lupus erythematosus	470
Lapatinib	DRB1*07:01-DQA1*02:01-DQB1*0202/0203	Hepatotoxicity	471
NSAIDs	HLA-DR11	Anaphylactoid and cutaneous	472
Sulfonamides	HLA-DR7	TEN	466
Ximelagatran	DRB1*07, DQA1*02	Elevated alanine aminotransferase	473

HSS, hypersensitivity syndrome; MHC, major histocompatibility complex; NSAIDs, nonsteroidal anti-inflammatory drugs; SJS, Stevens-Johnson syndrome; TEN, toxic epidermal necrolysis.

Adapted from Illing PT, Purcell AW, McCluskey J. The role of HLA genes in pharmacogenomics: unravelling HLA associated adverse drug reactions. *Immunogenetics*. 2017;69(8-9):617-630 and Bharadwaj M, Illing P, Theodossis A, Purcell AW, Rossjohn J, McCluskey J. Drug hypersensitivity and human leukocyte antigens of the major histocompatibility complex. *Annu Rev Pharmacol Toxicol*. 2012;52:401-431.

and not to closely related allotypes with polymorphisms in the C, D, E, or F pockets of the MHC peptide–binding groove. It will be interesting to learn whether this represents a general mechanism for HLA-linked drug reactions applicable to other small molecule drugs that might bind specifically to HLA antigen–binding clefts. HLA testing before administration of these drugs is now recommended by many physicians and the Food and Drug Administration, in order to avoid life-threatening ADRs.[457,485] The very high relative risks of abacavir, carbamazepine, and allopurinol ADRs and *HLA* class I allotypes suggest a very high level of specificity in HLA binding, either directly or through a drug-peptide conjugate(s) that only binds the associated HLA allotype. Interestingly, such high relative risks are not commonplace in *HLA* allotypes and protective immunity, perhaps because it is unusual for any single determinant or *HLA* allotype to control protective phenotypes. In addition to recent progress in recognizing *MHC* associations with adverse reactions to small molecule drugs and a variety of autoimmune and immune deficiency conditions, further analysis has identified relationships of *MHC-I*–linked markers with susceptibility and prognosis in a number of infectious diseases.

HLA and Pathogen Immunity

The relationship of *HLA* with disease susceptibility is notably observed with respect to HIV where chronic infection allows the evolutionary arm wrestle between viral replication and immune control to play out in a manner that reveals the importance of host immunogenetics (especially *HLA* and *KIR* genotype) in controlling outcome. Among the most striking findings has been the recognition that *HLA-B*57:01* and *HLA-B27* both confer delayed progression of HIV infection to AIDS,[486,487] whereas certain subtypes of *HLA-B*35* are linked to rapid disease progression.[486,488] These HLA effects operate via the effectiveness of CD8+ T cell immunity elicited by these *HLA* allotypes. Moreover, HLA-mediated control of HIV can be further modulated by interactions between HLA and the NK cell receptors KIR3DL1 and KIR3DS1.

HIV-infected individuals who express both KIR3DS1 and HLA-B*04 alleles that express a molecule with isoleucine at position 80 have lower viral load set points and less risk of disease progression. This observation suggested that KIR3DS1 offered protection early in HIV infection for those patients bearing the appropriate Bw4 allele.[489] More recent observations indicate that patients with such a KIR3DS1/B*04 ile80 genotype also exhibit resistance to late opportunistic infections.[490]

The importance of HLA-mediated control of HIV is reflected in the complex pattern of HIV evasion of immunity mediated through viral regulation of expression of HLA-C.[491] HLA-C is expressed at lower levels than HLA-A and HLA-B, and yet is a key molecule in controlling viral set point load and CD4 T cell preservation. HLA-C mediates both CTL-mediated elimination of virus-infected cells and also interacts with different inhibitory receptors expressed by NK cells, thereby reducing the killing of virus-infected cells. Hence, HIV Nef-mediated downmodulation of HLA-A and HLA-B molecules spares HLA-C, so its expression continues to prevent NK-mediated killing of cells that fail to express MHC-I molecules on their surface. Nonetheless, HLA-C is downregulated by the HIV VPU protein in some HIV isolates underscoring the role of HLA-C as an important driver of CD8 CTL immunity in its own right.[491] Other studies now show that particular *HLA-DRB* alleles as well as *MHC-I* alleles significantly influence hepatitis B and C virus susceptibility, persistence, and response to treatment.[492-494]

STRUCTURE OF MHC AND MHC-LIKE MOLECULES

"There is nothing that living things do that cannot be understood from the point of view that they are made of atoms acting according to the laws of physics."

Richard Feynman[495]

"Love hides in molecular structures, yeah, Love is the answer."

Jim Morrison[496]

So central are *MHC* genes and their encoded molecules to both the regulation and the effector function of the immune system that it has been apparent almost since their discovery that an understanding of their structure and biochemical interactions would be fundamental to a comprehension of their physiological effects. Structural relationships of MHC molecules came first from understanding serological differences, then from analysis of cellular immune responses, and subsequently from biochemical studies of the MHC-I and MHC-II chains. Amino acid sequence comparisons suggested a domain structure for the MHC-I molecules, and an exon/domain correspondence was apparent with the initial identification of genomic and cDNA clones.[497] Shortly thereafter, a large number of sequences were determined leading to their alignment and comparison. Databases of these sequences are maintained and updated as indicated in previous sections of this chapter, and searches of standard sequence databases may find *MHC* gene and protein sequences of numerous species. Expression of recombinant clones encoding MHC molecules in several systems, mammalian cells, insect cells, and bacteria, then permitted the accumulation and purification of molecules for functional, binding, and structural studies by X-ray crystallography and nuclear magnetic resonance spectroscopy (NMR).

A current search of the Protein Data Bank (https://www.rcsb.org) for the keyword "histocompatibility" yields three-dimensional coordinates of more than 3000 MHC, p/MHC, p/MHC/TCR, p/MHC/coreceptor, and p/MHC/NK receptor complexes that allow an understanding of the function of these molecules in a detailed structural context.[498,499] However, these structures also pose a number of questions that may only be addressed by additional functional experiments *in vivo* complemented by biophysical methods applied *in vitro*. The molecular biological, functional, and structural studies have led to the development of the use of MHC multimers[500] as extremely powerful tools for imaging specific T cells or NK cells. The goal of this section of this chapter is to summarize these developments with an eye toward explanation of function by structure, and in hopes of revealing some of the current quandaries that continue to confound our understanding of the function of the *MHC*.

Amino Acid Sequences—Primary Structure

Before the cloning of *MHC* genes, the biochemical purification, and amino acid sequence determination of the MHC-I molecules, HLA-A2, HLA-B7, and H2-K^b [501,502] indicated that these had unique amino terminal "domains" and showed similarities to immunoglobulins in their membrane proximal regions. Early concerns were to identify the differences between allelic gene products as well as the differences between MHC proteins encoded at different loci. With the cloning of cDNAs and genomic clones for MHC-I molecules[274,503-505] and shortly thereafter for MHC-II molecules,[249,506-508] the encoded amino acid sequences of a large number of MHC molecules of a number of species quickly became available. The comparison of gene and cDNA sequences gave an indication of the exon/intron organization of the genes and explained the evolution of the MHC molecules as having been derived from primordial single-domain structures of a unit size of a single Ig domain (such as the light chain β_2m) and of two exons each encoding about 90 amino acids for coordinately forming a peptide-binding site.[243] The canonical MHC-I molecule has a heavy chain (also known as the α chain), which is an intrinsic type I integral membrane protein with amino terminal domains called α1, α2, and α3, is embedded in the cell membrane by a hydrophobic transmembrane domain, and extends into the cytoplasm of the cell with a short carboxyl-terminal tail. The light chain of the MHC-I molecule, β_2m, is a single-domain soluble molecule, encoded elsewhere in the genome in higher vertebrates.

The MHC-II molecule consists of two chains embedded in the membrane as type I proteins, an α chain and a β chain. Both α and β chains consist of two major extracellular domains, α1 and α2, and β1 and β2, respectively, each linked to a transmembrane domain and cytoplasmic sequences. An initial analysis of both MHC-I and MHC-II molecules suggests that they are noncovalently assembled heterodimers consisting of four extracellular domains, the two membrane proximal domains (α3 and β_2m for MHC-I and α2 and β2 for MHC-II) of each molecule are Ig like, while the two amino terminal domains (α1 and α2 of MHC-I and α1 and β1 of MHC-II) possess what is known as the MHC fold. The α1 domains of both MHC-I and MHC-II lack the intradomain disulfide bond characteristic of the other extracellular domains. The conserved cytoplasmic domain of MHC-I molecules can be regulated by alternative splicing, differential phosphorylation of conserved residues, or other modification and is likely to play a role in cell surface stability and cycling between the cell surface and other intracellular compartments.[509-515]

However, analysis of directed mutants of MHC-I in some systems indicates that the cytoplasmic domain is not required for cytoskeletal association or surface recycling.[516] (The cytoplasmic domain of MHC-I molecules is the target for the Nef protein of HIV, and the Nef/MHC interaction results in MHC-I downregulation in HIV-infected cells.)[517,518] The MHC-II transmembrane and cytoplasmic domains have clear effects on the level of cell surface expression, the efficiency of antigen presentation, and the rate of lateral diffusion of the molecules in the cell membrane.[519,520] Single-particle tracking of MHC-II molecules indicates that such lateral diffusion is governed, at least in part, by interaction with the actin-based membrane skeleton.[521] Amino acid sequence alignments of MHC-I and MHC-II proteins, particularly of the human molecules, are available in a number of databases such as http://www.ebi.ac.uk/imgt/hla/ and http://www.ebi.ac.uk/ipd/mhc/.

Identification of Peptides Bound by MHC Molecules

Although initial understanding of MHC molecules indicated that they are heterodimers, a complete understanding of their structure and resulting function requires the appreciation that they are, in fact, heterotrimers, in which the third component is a short peptide ranging from 8 to roughly 15 amino acids in length. This characteristic of mature MHC-I and MHC-II molecules makes them a unique class of proteins because their peptide composition and resulting structure derive from a heterogeneous array of peptides delivered to the assembling molecules during their maturation. Many different lines of evidence coalesced over a short period of time to demonstrate that MHC molecules function by binding peptides. From functional experiments, MHC-II–restricted T cell responses to protein antigens were shown to be dependent on peptide fragments.[522] The first direct evidence of p/MHC interactions was that purified MHC-II proteins could bind synthetic peptides in a specific, saturable and stable manner[523,524] with measurable affinity, and remarkably slow dissociation rate.[524] For MHC-I molecules, the results were at first less clear, but the realization that some cell lines defective in MHC-I surface expression could be induced to express higher levels of surface MHC-I molecules by exposure to the appropriate peptides[525] led the way for direct measurement of MHC-I peptide binding.[526]

Several laboratories succeeded in developing methods for the partial purification and identification of the peptides that copurified with MHC molecules. A viral antigenic peptide was identified bound to the MHC-I molecule H2-K^b, by recovering MHC molecules from virus-infected cells, fractionating the bound peptides chromatographically, identifying a peak of functional biological activity in a cytotoxic T cell assay, and determining the amino acid sequence of the recovered peptide by radiochemical techniques.[527] Another method that was applied to the identification of both virus-derived peptides and the "motif" of self-peptides by particular MHC-I molecules involved first the isolation of a large amount of detergent-solubilized MHC-I using appropriate antibodies, the elution of the bound peptides, their partial purification as pools by reverse phase high-pressure liquid chromatography (HPLC), and the determination of the amino acid sequence of the bound peptides by classic amino terminal sequencing.[528-530] The unpredicted and surprising result obtained from these studies of MHC-I–derived peptides was that specific amino acid residues were favored at particular positions of the sequence, depending on the MHC-I molecule from which the peptides were obtained, and that the length of the well-defined, short, bound peptides ranged from 8 to 10 amino acids. From such experiments, a number of peptide "motifs" of peptides bound to particular MHC-I molecules and allelic products were identified. Often, specific amino acids were identified at particular Edman degradation steps, indicating a common, highly preferred residue at the same spacing from the amino terminus of the peptide. Thus, the peptide "motif" could be determined even from heterogeneous pools of peptides eluted from particular MHC-I molecules. Further refinements in methodology have included the application of mass spectrometry to the identification and sequencing of individual peptides.[531,532] Alternative approaches for identifying peptide motifs include the use of soluble analogs of MHC-I molecules to ease the purification[533-535] or the use of peptide display libraries to identify those peptides that can bind the MHC.[536,537] With steady improvements in technology, including the use of transfected soluble versions of MHC-I molecules, immunoaffinity purification of the MHC-I molecules of interest, tandem reverse phase HPLC, and further identification and sequencing by tandem mass spectrometry, particular MHC-I motifs as well as tumor antigen peptides may now be identified routinely.[538,539] A summary of MHC-I peptide motifs is given in Table 19.4. It is clear that for some MHC-I molecules, dominant amino acids may be found at peptide position 2 (eg, for HLA-A*02, H2-L^d), position 5 (for H-D^b), and for the carboxyl-terminal residue (frequently hydrophobic). The carboxyl-terminal residue for MHC-I–bound peptides is frequently referred to as PΩ. Regularly maintained online databases of MHC-I peptide motifs may be found at several sites (http://www.imtech.res.in/raghava/mhcbn/links.html and http://www.syfpeithi.de).[541] Antigenic peptides that represent those observed as tumor antigens and available for recognition by T cells are summarized at http://www.cancerimmunity.org/peptidedatabase/differentiation.htm. In addition, the immune epitope database (http://www.immuneepitope.org)[542] offers a regularly updated and curated website that describes MHC motifs, allows evaluation of new proteins for MHC epitopes, and provides new computational tools for such evaluation. An algorithm that allows the prediction of candidates for MHC-I–restricted peptides based on the amino acid sequence of the protein of interest is also

TABLE 19.4 Peptide Motifs—MHC-I[a]

Position	1	2	3	4	5	6	7	8	9 (or Ω)
MHC-I Molecule									
HLA-A*01		(T)	D				(L)		Y
HLA-A*02:01		L/M				(V)			V/L
HLA-A*03		L/V/M	(F,Y)			(I,M,F,V,L)	(I,L,M,F)		K,Y,F
HLA-B*07		P	(R)						L,F
HLA-B*08:01			K		(K,R)		(E)		(L)
HLA-B*27:05		R							L,F (Y,R,H,K)
HLA-B*35:01		P							Y,F,M,L,I
HLA-B*53:01		P							W,F,L
HLA-B*57:01		A/T/S							F,W,Y
H2-K^b			(Y)		F/Y			L,M,I,V	8-mers
H2-D^b					N				M/I/L
H2-K^d		Y/F							I/L/V
H2-D^d		G	P		(R/K)				I/L/F
H2-L^d		P,S							F,L,M
H2-K^k		E						I/V	8-mers
H2-D^k	Uncertain	(R)							(L)
Qa-2		(M,L,Q)	(N,I,L)		(V,I)	(K,M,I)	H		L/I/F
H2-M3	N-formyl-M								
RT1.A1		(A,S,V)	F,Y						Y,F,L,M

[a]Peptide-binding motifs for the indicated MHC-I molecules are shown in the single amino acid code. Position refers to the amino acid position of the peptide from the amino terminus, or Ω for the C-terminal residue. Only the most common residues are shown. Assignments in parenthesis are less common than the others. These motifs are taken from the more extensive summary of http://www.syfpeithi.de/ and http://www.imtech.res.in/raghava/mhcbn/links.html and Biddison WE, Martin R. Peptide binding motifs for MHC class I and II molecules. *Curr Protoc Immunol*, 2001;Appendix 1:Appendix 1I. Full sequences of proteins may be queried for MHC peptides presented by particular alleles at the IEDB http://www.immuneepitope.org.

available.[543] The various predictive algorithms for HLA class I–binding peptides have been evaluated for their performance on a select group of HLA class I allotypes for which an independent data set was available.[544] The study found that MixMHCpred 2.0.1[545,546] was the best predictor for the HLA-I molecules tested among scoring function-based tools that used statistics to estimate the binding score for each peptide sequence. For the predictive tools relying on machine learning and consensus-based design, the best performers were NetMHCpan 4.0[547] and NetMHCcons 1.1.[548] None of the predictive algorithms are particularly capable of predicting immunogenicity of peptides even if they do bind HLA-I molecules and they do not efficiently accommodate prediction of longer peptides of 11 to 17 amino acids as have been identified in antiviral responses despite a priori predictions of shorter peptides expected from the same viral antigen.[549]

The distinction between "motif" residues of an MHC-restricted peptide and "anchor" residues is an important one. Motif refers to those amino acid residues that are identified based on the sequences of self or antigenic peptides that have been demonstrated to bind or copurify with a particular MHC molecule. Anchor implies a biophysical function of the particular amino acid residue as specifically interacting with a particular part of the MHC molecule itself.

The identification of MHC-II–bound self or antigenic peptides by biochemical methods similar to those employed for MHC-I molecules has proved more difficult because the MHC-II molecules do not have the rigorous requirements for a defined amino terminus or restricted length. Whereas MHC-I molecules bind peptides with a particular motif residue at a specific position as defined by the amino terminus, resulting in the ability to identify the dominant residue at a particular step in the Edman degradation even amid a pool of peptides, MHC-II molecules bind peptides with "ragged ends" and little information is obtained from the standard methods of sequencing of pools of peptides.[550-553] Thus, more precise fractionation of the peptides released from purified MHC-II molecules is required. Identification of MHC-II peptide–binding motifs by bacteriophage display is also possible.[554] In accord with the view that MHC-II molecules present peptides derived from an "outside in" pathway, many of the peptides that copurify with MHC-II molecules represent molecules derived from the extracellular milieu or the medium in which the cells were grown. Analysis of p/MHC-II complexes with cloned T cells and monoclonal antibodies with p/MHC specificity reveals that, in part because of the ability of MHC-II molecules to accommodate peptides with extensions at their amino and carboxyl termini and in part because of the smaller role that anchor residues seem to play in peptide binding by MHC-II molecules, occasionally even a unique peptide can bind a particular MHC-II molecule in more than one frame.[555-558] As a result of structural studies (summarized below) and compilation of sequences of those peptides bound by particular MHC-II molecules, the general conclusion is that MHC-II molecules all have binding pockets identifiable for the particular allelic product. For some alleles, such as HLA-DR1 and IA^d, these pockets are spaced at position P1, P4, P6, and P9 (or i, i + 3, i + 5, i + 8). (Some analyses, borne out in part by structural studies, suggest that a pocket at P7 [i + 6] may also contribute to the energy of binding.) Because of the complexity and potential degeneracy of peptide motifs for MHC-II molecules, more elaborate schemes have been devised for predicting those peptides that may bind to particular MHC-II molecules.[559] A summary of identified peptide-binding motifs for MHC-II molecules is in Table 19.5, but for some alleles the analysis of their binding peptides is not straightforward and demands sophisticated algorithms for identifying potential antigenic peptides from new proteins. The site http://www.imtech.res.in/raghava/mhcbn/links.html tabulates a number of links to various servers for predicting likely peptide epitopes for particular MHC-I and MHC-II molecules. Implicit in the description of MHC-I and MHC-II peptide motifs is that the bound peptides bind to the MHC molecule only in an amino to carboxyl-terminal orientation, that is, all peptides bound to MHC molecules bind with the amino terminus at the left-hand side of the peptide-binding groove, in the canonical structural representation that will be described in further detail below. However, a recent analysis employing both X-ray crystallography and nuclear magnetic resonance indicates that the CLIP peptide may undergo orientational inversion, binding dynamically in the carboxyl to amino as well as the amino to carboxyl direction.[560]

High-Resolution Structures

MHC-I Molecules

A most graphic description of the relationship of form to function of the MHC molecules was made by Bjorkman and colleagues who determined the three-dimensional structure of the human MHC-I molecule, HLA-A*02:01, by X-ray crystallography.[561,562] For these studies, the extracellular, soluble portion of the type I membrane–associated molecule was purified from protein cleaved by papain from the surface of tissue culture cells. At the time, there was not a clear appreciation of the role of peptide either in the assembly of the molecule or of the nature of the recognition of the MHC molecule by TCRs, or for that matter, NK receptors. Despite the fact that the first purified HLA-A2 molecules possessed a heterogeneous mixture of bound peptides, protein from these preparations crystallized readily, and electron density maps calculated from the diffraction data were interpretable, allowing modeling of the backbone molecular structure. The most important insight in the interpretation of the derived electron density map was that part of the electron density, and thus part of the structure, was due to this mixture of peptides bound tightly by the molecule and that this density could not be modeled based on the known amino acid sequence of HLA-A2.

This first MHC-I structure clarified several important aspects of the mechanism by which the MHC-I molecule carries out its peptide-binding function. The amino terminal domains (α1 and α2) form a unitary binding site for peptide. This domain unit (also called a "superdomain") consists of a floor of eight strands of antiparallel β-pleated sheet that supports two α-helices, one contributed from the α1 domain and one from the α2 domain, aligned in an antiparallel orientation. The membrane proximal α3 domain has an immunoglobulin C1-type fold and pairs asymmetrically with the other immunoglobulin domain of the molecule contributed by β_2m. The nature of recognition by T cells was suggested by comparing the location of those amino acid residues that had been characterized as being strong elements in T cell recognition, residues that distinguished closely related allelic gene products and amino acid residues that had been identified as those that were responsible for the transplant rejection of the mutants of the *H2-K*^{b} series.[562] Amino acid residues of the MHC-I molecule responsible for T cell recognition were clearly classified into one of two categories or an overlapping set: those residues that were "on the top of the molecule," exposed to solvent and available for direct interaction with the TCR, and those residues whose side chains pointed into the peptide-binding groove and might be considered crucial in the peptide-binding specificity of the particular MHC molecule. The original publications, based on a structure determined to a resolution of 3.5 Å, focused mainly on the general structural outline of the molecule. More recently, structures of a wide variety of MHC-I molecules of different species including human, mouse, rat, and chicken, complexed with specific peptides, have been determined. Ribbon diagrams of HLA-A2 as seen from the side (Fig. 19.6A) and from the top (Fig. 19.6B), complexed with a unique synthetic peptide, indicate the design of the entire molecule: the peptide-binding site is supported by the β sheet floor, and the floor in turn is supported by the two immunoglobulin-like domains, the α3 domain of the heavy chain and the β_2m light chain. (In the canonical view of the MHC-I HLA-A2 molecule shown in Figure 19.6B-D, the amino terminal residue of the bound peptide lies to the left, and the carboxyl-terminal residue lies to the right.)

TABLE 19.5 Peptide Motifs—MHC-II[a]

Position	i (P1)	i + 1	i + 2	i + 3 (P4)	i + 4	i + 5 (P6)	i + 6	i + 7	i + 8 (P9)
Allele									
DRB1*01:01	YVLFIAMW			LAIVMNQ		AGSTCP			LAIVNFY
DRB1*03:01	LIFMV			D		KREQN			YLF
DRB1*04:01	FYWILVM			PWILVADE		NSTQHR	DEHKNQR		DEHKNQR
DRB1*04:05	FYWVILM			VILMDE		NSTQKD			DEQ
DQA1*05:01/B1*03:01	WYAVM			A		AIVTS			QN
H2-IA^{b} (undefined)									
H2-IA^{d}	STYEVWMLI			VLIA		AV			
H2-IA^{g7}	KHSAV			L		VA			DSE
H2-IA^{k}	DN			IVLN		EQ			
H2-IA^{s} (undefined)									
H2-IE^{b}	WFYILV			LIFSA		QNASTHRE			KR
H2-IE^{d}	WYFILV			KRIV		ILVG			KR
H2-IE^{g7}	ILVFWM			DESMV		QNASTED			RKMF

[a]MHC-II peptide–binding motifs are summary of http://www.syfpeithi.de/. Full sequences of proteins may be queried for MHC peptides presented by particular alleles at the IEDB http://www.immuneepitope.org.

The comparison of this structure with that of the closely related human MHC-I molecule, HLA-A*68:01, suggested that surface depressions in the groove of the MHC-I molecule, now designated pockets A through F, would be available for interactions with some of the side chains of the bound peptide.[565,566] These six pockets are illustrated in Figure 19.6C and with bound peptide in Figure 19.6D. Concomitant with the determination of the X-ray structure of the human MHC-I molecule, HLA-B*27:05,[567,568] the motif of the peptides that were recovered from this molecule was determined, permitting the more precise modeling of the bound peptide in the cleft of the MHC-I. For HLA-B27, this was of particular interest since the bound peptides had a strong overrepresentation of arginine at position 2, and scrutiny of the HLA-B27 structure suggested that the amino acid residues lining the B pocket, particularly glutamic acid at position 45 as well as cysteine 67, were complementary to the long, positively charged arginine side chain of the peptide amino acid at that position. Structural studies supported a view of p/MHC-I binding in which the side chain of the carboxyl-terminal residue of the bound peptide sits deep in the F pocket. In addition, the amino terminal amino group of the peptide forms hydrogen bonds with the hydroxyl groups of conserved amino acids tyrosine 59 and tyrosine 171 that line the A pocket. A hydrogen bond from the amino group of conserved tryptophan 147 to the backbone carbonyl oxygen of the penultimate peptide amino acid (usually position 8) also seems important, as do charge interactions and hydrogen bonds of the free carboxyl group

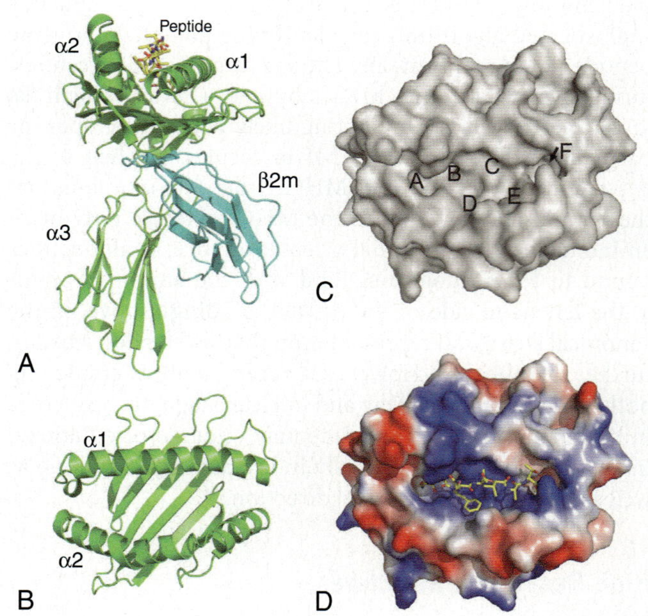

FIG. 19.6. Structure of the MHC-I molecule. Structure of HLA-A*02:01. **A:** Ribbon representation of HLA-A2 heavy chain (*green*), β_2m light chain (*cyan*), and bound peptide (*yellow*). **B:** Ribbon representation of the peptide-binding groove. **C:** Surface representation of the binding groove, with pockets labeled. **D:** Surface representation of binding groove colored by electrostatic charge (*red*, acidic; *blue*, basic) with peptide is shown in stick illustration. Figure generated from Protein Data Bank[563] structure 2BSV using PyMOL.[564] (Created with The PyMOL Molecular Graphics System, Version 2.0 Schrodinger, LLC.)

at the carboxyl terminus of the peptide with tyrosine 84, threonine 143, and lysine 146.

Other structures of MHC-I molecules were determined of complexes produced with homogeneous peptide, assembled either *in vitro* from bacterially expressed proteins with synthetic peptide[569] or exploiting MHC proteins expressed in insect cells.[569,570] The structures determined with homogeneous peptide confirmed the impression obtained from the structures obtained from molecules with heterogeneous self-peptides. Examination of the H2-K^b molecule complexed with synthetic, known antigenic peptides derived either from Sendai virus, vesicular stomatitis virus, or chicken ovalbumin revealed that the same MHC molecule can bind peptides of different sequence, length, and structure by virtue of their conserved motif residue side chains. Although small conformational changes of the MHC are detectable on binding the different peptides, the main distinction in the recognition of different peptides bound by the same MHC molecule is due to the location, context, size, and charge of amino acid side chains that are anchored in the MHC pockets. Conversely, solvent-exposed side chains contribute to the unique structure of the surface of the molecule available for interaction with TCR and other receptors.

The most consistent rule learned from the first X-ray structures and complemented by peptide recovery and early binding studies was that the carboxyl-terminal amino acid side chain of the MHC-I bound peptide was embedded in the F pocket. A recent study revealing that the MHC-I molecule H2-D^b binds a pentapeptide well anchored in the F pocket, but lacking amino terminal interactions, is consistent with this view.[571] However, other studies indicate that MHC-I molecules may bind longer peptides that extend beyond the residue anchored in the F pocket,[572] an observation confirmed by a crystallographic structure.[573] More recently, the lack of an absolute requirement for a free C-terminal amino acid has been exploited in the engineering of single-chain molecules in which peptide, β_2m, and the MHC-I heavy chain are covalently linked. These molecules have unusual thermal stability and are effective p/MHC immunogens.[574] An additional variation on the theme of MHC-I binding peptides based on particular anchor residues includes the demonstration that glycopeptides, bound to the MHC-I via amino acid side chains and termini, can expose their carbohydrate moieties to solvent and be available for TCR interaction.[571,575,576] An example of a 13-residue peptide bound to its MHC-I–presenting element, the rat MHC-I molecule, RT1-Aa, was crystallized and shown to produce a p/MHC complex with a large central bulge. Two different complexes consisting of the same MHC and peptide reveal significantly different conformations in this central bulge region.[577] "Bulged" viral peptides have also been characterized in complex with human MHC-I structures of HLA-B35 allotypes complexed with 13-mer and 11-mer peptides.[578-580] TCR recognition of such bulged peptides can involve conformational adjustments of the TCR in recognizing a fairly rigid peptide[578] or "crumplings" of the bound peptide by a largely rigid face of the TCR upon binding the p/MHC-I complex.[581] Another set of modified peptides that can be bound by MHC-I molecules and presented to TCR includes self-peptides modified by phosphorylation.[487,582] Although the fundamental topology of MHC-I structure and its interactions with antigenic peptide have been solved by the classic structures mentioned above, many more MHC-I complexes have been studied crystallographically in order to define the details of particular interactions with antigenic peptides.

A striking structural example illustrating the sensitivity of the p/MHC-I conformation to drugs is the characterization of the role that drugs may play in modifying the display of self-peptides.[484,583,584] Abacavir interacts with the peptide-binding groove of HLA-B*57:01 and thus alters the spectrum of peptides presented (Fig. 19.7).

A number of structures at a resolution of 1.40 Å or better have been deposited in the protein database. Structure determination of representative single-chain peptide-β_2m-MHC-I[585] and of MHC-I molecules containing photolabile peptides that can be exchanged with other peptides easily[586] reinforces our basic understanding of the structure of p/MHC-I complexes. MHC-I molecules with engineered intrachain disulfide bonds have recently been described. These are of value because they do not require bound peptide to preserve thermal stability.[587] Additionally, the use of various dipeptides in *in vitro* refolding reactions improves yield by stabilizing the F pocket during the process and enhancing peptide exchange.[588]

MHC-II Molecules

Before any MHC-II structure had been determined experimentally, a model was constructed based on the alignment of amino acid sequences and the available MHC-I three-dimensional structure.[589] This model made several valid predictions that were borne out by the subsequent X-ray structure determination of HLA-DR1.[590] MHC-II clearly

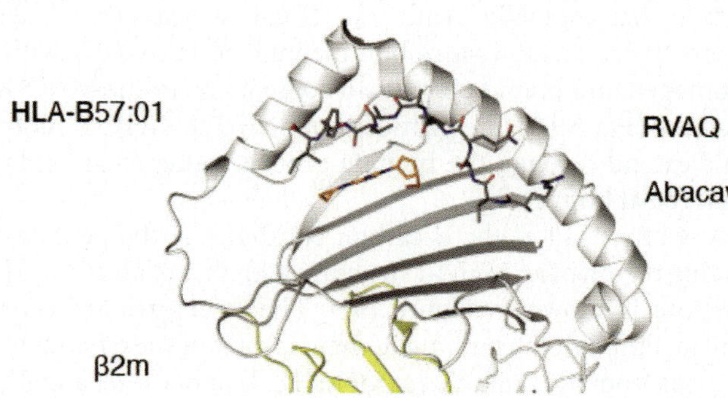

FIG. 19.7. Interaction of abacavir with HLA-B*57:01. Cartoon representation of the HLA-B*57:01-RVAQLENVYI in complex with abacavir (PDB:3VRI). The HLA-B*57:01 heavy chain is shown in *white*, β_2m is shown in yellow, the peptide is shown in *black sticks*, and the abacavir molecule in *orange*.

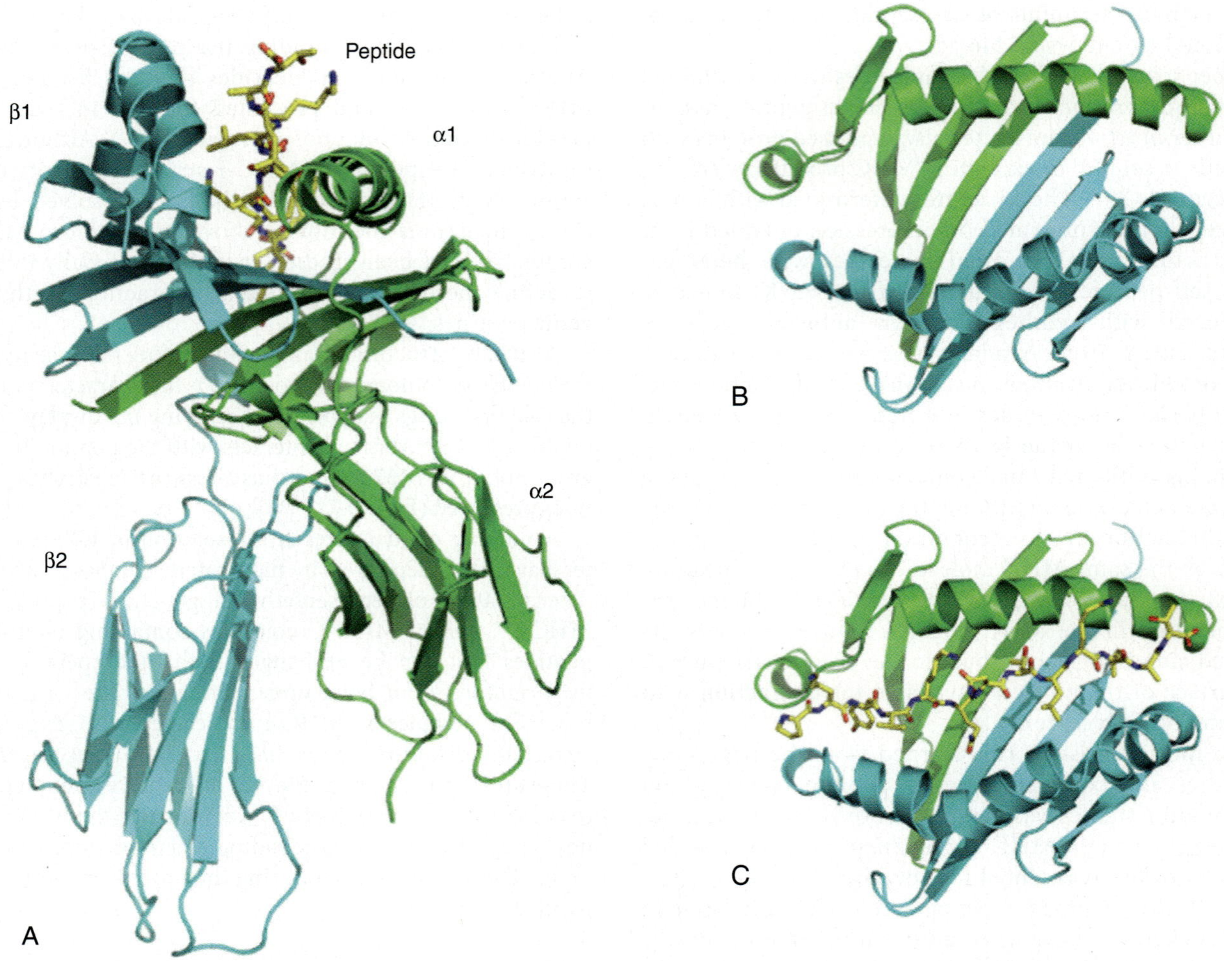

FIG. 19.8. **Structure of the MHC-II molecule. A:** HLA-DR1 in side view; α chain (*green*), β chain (*blue*), and peptide in stick representation (*yellow*). **B:** Top view. **C:** Top view with bound peptide (PKYVKQNTLKLAT) visualized. The illustration was made with PyMOL[564] on the Protein Data Bank[563] coordinates of 1DLH. (Created with The PyMOL Molecular Graphics System, Version 2.0 Schrodinger, LLC.)

showed similarity to MHC-I and formed its binding groove by the juxtaposition of the α1 and β1 domains. The position of the electron density representing the heterogeneous peptide that copurified with the HLA-DR1 was identified. Figure 19.8A, C shows a ribbon diagram of HLA-DR1 with a homogeneous bound peptide. In comparison to the MHC-I structure (Fig. 19.6), the peptides bound to the MHC-II molecule extend through the binding groove, rather than being anchored at both ends.

A comparison of the α-carbon backbone of the peptide-binding region of an MHC-I structure with that of an MHC-II structure is shown in Figure 19.9. The structures are very similar, although the binding domain is built of the α1 and α2 domains from the same chain (for MHC-I) or of the α1 and β1 domains that derive from two chains (for MHC-II). The location of polymorphic residues can be determined by variability plots based on multiple sequence alignments, as originally suggested by Kabat and Wu.[591] The current bioinformatics approach to variability prefers to calculate *Shannon entropy*, a function that describes the degree of unpredictability of an

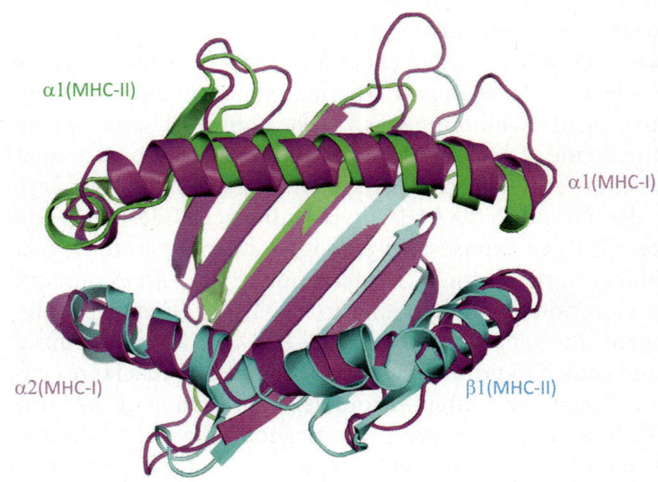

FIG. 19.9. **Similarity of MHC-I and MHC-II structures.** The α1α2 domains of MHC-I (HLA-A2, PDB 3HLA) and the α1 and β1 domains of MHC-II (HLA-DR1, PDB 1DLH) were superposed with PyMOL[564] and displayed as ribbon representations. MHC-I is *magenta,* the α1 domain of MHC-II is *green,* and the α2 domain of MHC-II is *cyan.* (Created with The PyMOL Molecular Graphics System, Version 2.0 Schrodinger, LLC.)

event.[592] For protein sequence alignments, Shannon entropy provides a sensitive measure of the diversity of particular amino acid positions.[593] Several online tools are available for implementing and displaying the results of such calculations: http://imed.med.ucm.es/PVS/ and https://consurfdb.tau.ac.il/index.php and http://weblogo.threeplusone.com/create.cgi. Comparative ribbon diagrams (Fig. 19.10), in which the location of the amino acid residues that are polymorphic in the human MHC-I and MHC-II chains are indicated with a color map, show that the bulk of MHC polymorphism derives from amino acid variability in regions that line the peptide-binding groove. This suggests that MHC polymorphism is required to allow the MHC molecules, and as a result, the organism and its species, to respond to a changing antigenic environment. The conservation of the Ig-like domains (α3 and β_2m for MHC-I and α2 and β2 for MHC-II) suggests the need to preserve function.

As with MHC-I, a further understanding of the details of the interactions of peptides with the MHC-II molecule came from crystallographic studies of molecules prepared with homogeneous peptide—in the first case HLA-DR1 complexed with an antigenic peptide derived from the hemagglutinin of influenza virus.[594] Based on this structure, a set of pockets was initially designated, numbered for the peptide position that is bound. For the influenza peptide studied in this example, the major interactions were from peptide positions 1, 4, 6, 7, and 9, which are indicated in Figure 19.11. The deep P1 pocket accommodates the tyrosine (the third position of the peptide PKYVKQNTLKLAT) and the pockets indicated by 4, 6, 7, and 9 fit the Q, T, L, and L residues, respectively. MHC-II exploits a similar mode of binding as compared with MHC-I, but there are key differences. MHC-II lacks the requirement for free amino and carboxyl of a type II polyproline helix, and the MHC-II

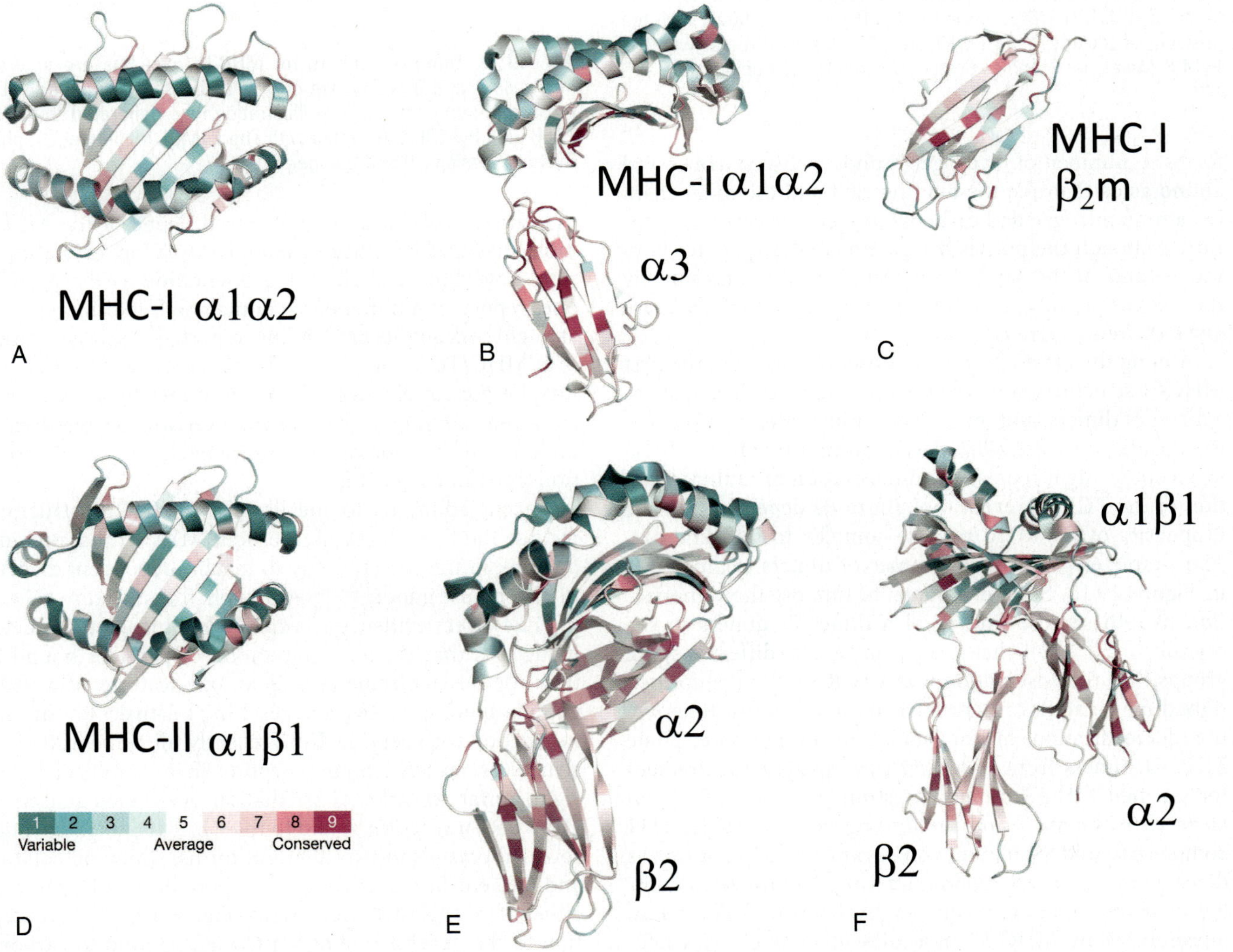

FIG. 19.10. Polymorphic amino acids predominate in the peptide-binding domains of both MHC-I and MHC-II. Amino acid sequence variability based on Shannon entropy of MHC-I **(A-C)** and MHC-II **(D-F)** was calculated and displayed with ConSurf. Color code for gradations of variable (*green*) to conserved (*maroon*) is provided. Clearly, α1α2 **(A)** and α1β1 **(D)** domains of MHC-I and MHC-II, respectively, are more variable than α3 and β_2m of MHC-I **(B, C)** and α2 and β2 of MHC-II **(E, F)**. (Data from https://consurfdb.tau.ac.il/index.php)

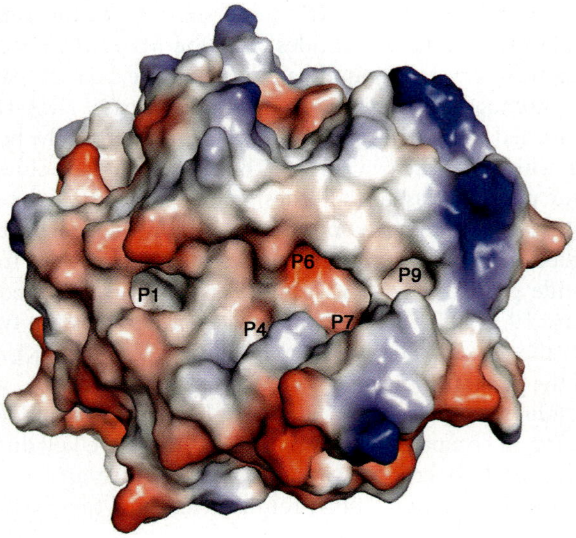

FIG. 19.11. Peptide-binding pockets of MHC-II. Location of pockets in HLA-DR1 is based on the cocrystal of HLA-DR1 with a peptide derived from the influenza hemagglutinin. An electrostatic surface representation of HLA-DR1 (1DLH) generated in PyMOL[564] is shown with the positions of pockets P1, P4, P6, P7, and P9 indicated. (Created with The PyMOL Molecular Graphics System, Version 2.0 Schrodinger, LLC.)

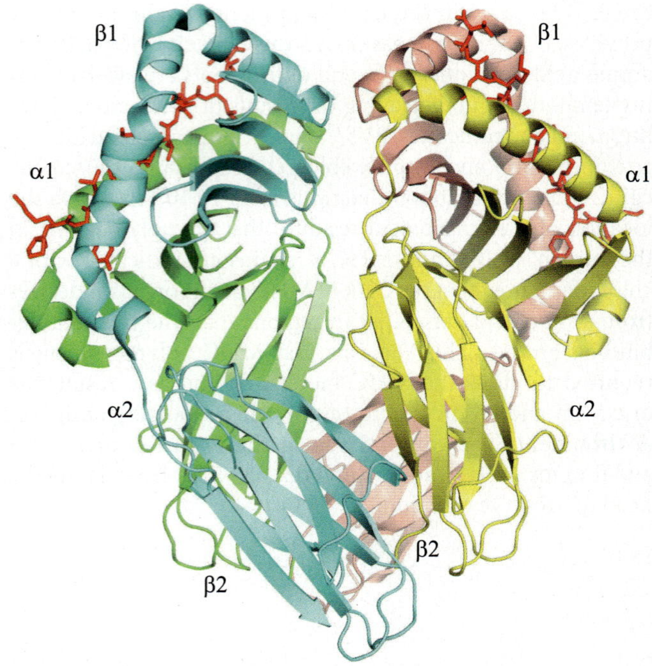

FIG. 19.12. Dimer of dimers for MHC-II. α chains are in *green* and *yellow*, and β chains are in *blue* and *salmon*. Peptide in stick representation is in *red*. The illustration was generated from PDB 1DLH with PyMOL.[564] (Created with The PyMOL Molecular Graphics System, Version 2.0 Schrodinger, LLC.)

forms a number of hydrogen bonds between conserved amino acids that line the binding cleft and the main chain (ie, amino nitrogen and carbonyl oxygen) atoms of the peptide. Although the pocket designations refer specifically to the features of the MHC-II molecule surface, occasionally the cognate peptide positions that reside in such pockets are given the same names.

Among the most provocative observations from the first MHC-II structures was that the molecule was visualized as a dimer of dimers, and this moved a number of investigators to consider the possibility that activation of the T cell via its receptor might require the dimerization or multimerization of the TCR, an event thought to be dependent on the propensity of the MHC/peptide complex to self-dimerize. The simple elegance of this dimer of dimers is illustrated in Figure 19.12. Several arguments support the dimerization hypothesis: the finding of a dimer of dimers in the crystals of HLA-DR that formed in several different space groups,[590,594] the observation that a TCR Vα domain formed tight dimers and its crystals formed dimers of the dimers,[595] the demonstration of the ability to immunoprecipitate MHC-II dimers from B cells,[596] the apparent requirement for purified MHC-I dimers for stimulation of a T cell in an *in vitro* system,[597] and the finding that p/MHC-II/TCR complexes could form higher order multimers in solution as detected by quasielastic light scattering.[598] However, a number of strong counter arguments draw this hypothesis into question. Many MHC-II molecules other than HLA-DR1 that have been crystallized do not seem to form the same kind of dimer of dimers in their crystals.[599,600] Modes of MHC-II dimerization with distinct topology that would preclude simultaneous interaction with multiple TCRs on the same T cell have been observed. None of the MHC-I molecules that have been examined by X-ray crystallography show dimers in the same orientation as the MHC-II ones reported. A different Vα domain fails to dimerize even at high concentration.[601] Many reported X-ray structures of p/MHC/TCR complexes[602-605] fail to show dimerization. Despite the simple elegance of the dimer hypothesis, it is clear that additional experimentation will be required to understand the topological requirements for T cell activation through the TCRαβ.

Recent additions to the library of MHC-II structures include the I-A^{g7} molecule, a unique MHC-II that provides one link in the susceptibility to insulin-dependent diabetes in the mouse model.[604,606] Although the structure fails to provide direct evidence to explain the linkage to diabetes, it suggests that the novel repertoire of peptides bound by this MHC-II molecule reveals unique features of a wider peptide-binding groove and resulting relatively low affinity interaction with peptide. Until recently, all of the p/MHC-II structures revealed bound peptide in a canonical left to right, amino to carboxyl orientation. A detailed analysis of HLA-DR1 molecules bound to the CLIP 106 to 120 peptide revealed crystals in two different forms, which on solution and refinement indicated that this peptide could interact in either of two distinct orientations (Fig. 19.13).[560] This exception to the general rule of left to right, amino to carboxyl orientation, may reflect the unique role that the CLIP peptide must play in protecting the MHC-II–binding site while simultaneously being available for DM-mediated peptide exchange in the endosome.

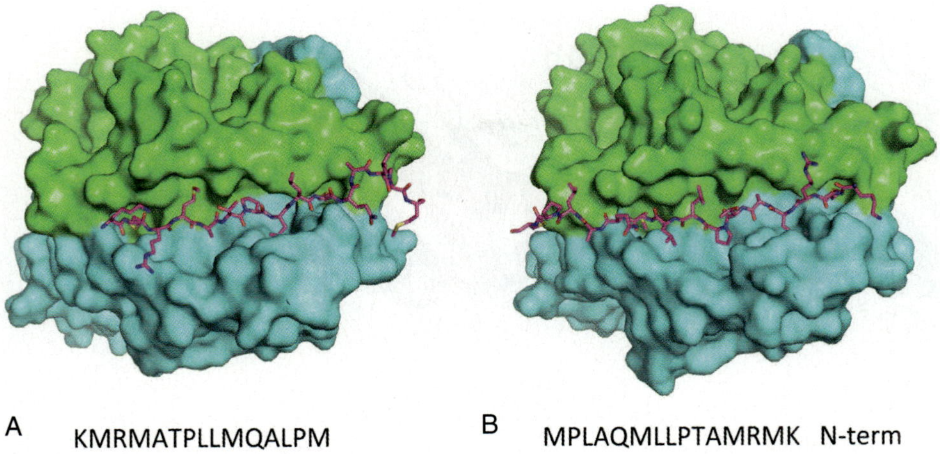

FIG. 19.13. MHC-II binds CLIP peptide in either N to C or C to N orientation. The structures of HLA-DR1 bound to the CLIP 106 to 120 peptide in either the N to C **(A)** (3PGD) or C to N **(B)** (3PGC) orientation are illustrated. α1 domain is in *green*, and β1 domain is in *cyan*. Peptide sequences are shown below.

MHC-Ib Molecules

H2-M3

To this point, our description of MHC-I molecules has focused on the classical MHC-Ia molecules, represented by HLA-A, HLA-B, and HLA-C in the human and by H2-K, H2-D, and H2-L in the mouse. Several MHC-Ib molecules for which three-dimensional structures have been determined are of particular interest, the CD1 molecules,[607-609] H2-M3,[252,610] MICA,[611,612] FcRn,[613] Rae-1,[614] and the mucosally expressed MR1 molecule that selects invariant T cells in the gut.[615]

Other molecules that are thought to show structural relatedness to the MHC-I fold but whose X-ray structures have not yet been reported include the stress-induced molecules H-60[299] and MULT1.[616] H2-M3 is of particular note because of its ability to bind and present peptide antigens that contain amino terminal N-formyl groups. H2-M3 was originally identified as the MHC-Ib molecule that presents an endogenous peptide derived from the mitochondrially encoded protein ND1 known as MTF (maternally transmitted factor).[252,617] Thus it was of interest to understand in structural terms how this molecule binds such N-formylated peptides.[610,618] The crystal structure of H2-M3 complexed with an N-formylated nonamer peptide, fMYFINILTL, revealed that the structure of the A pocket, highly conserved among MHC-Ia molecules, which have tyrosine 7, tyrosine 59, tyrosine 159, tryptophan 167, and tyrosine 171, is quite different, so that it can accommodate the N-formyl group in the A pocket. In particular, H2-M3 has an A pocket reduced in size and lined by hydrophobic residues, leucine at 167 and phenylalanine at 171, and leucine 16. These structural features cause the amino terminal nitrogen of the formylated peptide to be positioned where the peptide position 2 amino nitrogen would lie in an MHC-Ia molecule. Thus, the unique peptide selectivity of H2-M3 is explained in structural terms.

CD1

Another MHC-Ib molecule of great interest is CD1, representative of a class of MHC-I molecules that map outside of the MHC in higher vertebrates, that have limited tissue specific expression, and that are capable of interaction with both innate αβ and γδ T cells.[619] In the human, there are two clearly distinct groups of CD1 molecules: one consisting of CD1a, CD1b, CD1c, and CD1e; and another of CD1d alone.[620] In the mouse and rat, only CD1d is expressed.[621] As a group, these are β_2m-associated chains that bind hydrophobic antigens, primarily glycolipids, with the lipid moiety embedded in the CD1 heavy chain and the carbohydrate portion exposed to solvent (Chapter 36).

CD1a, CD1b, and CD1c are capable of binding and presenting various nonpeptidic mycobacterial cell wall components such as mycolic acid containing lipids and lipoarabinomannan lipoglycans.[622,623] NKT cells, a subset of TCRαβ-bearing T cells, of independent lineage and defined by the expression of the NK1.1 marker are restricted to CD1 recognition.[624] The crystal structures of both mouse and human CD1d1 have been determined,[608,625] revealing classic MHC-I structures with a basic α-carbon fold and β_2m association quite similar to that of the MHC-Ia molecules. Consistent with its apparent biological function of binding hydrophobic lipid containing molecules, its binding groove is somewhat narrower and deeper than that of either MHC-Ia molecules or MHC-II molecules. The backbone configuration of the α1α2 domain structure of CD1 is shown in Figure 19.14, where it is compared to the homologous region of H2-K^b, HLA-DR1, and another MHC-Ib molecule, FcRn, a neonatal Fc receptor (see below). The structures of a number of CD1-glycolipid complexes have been determined, and human and mouse CD1 molecules bound to different glycolipids have been compared.[621] The depth of the groove of CD1 results from the merging of pockets to form what have been termed the A′ and F′ pockets in place of the MHC-Ia A through F pockets. This A′ pocket is about the size of the binding site of a nonspecific lipid-binding protein.

In an effort to understand more precisely how CD1 molecules bind lipid antigens, Gadola et al crystallized human CD1b complexed with either phosphatidylinositol or ganglioside GM2 and determined their X-ray structures.[609] The structures were essentially identical for the CD1b heavy chain and β_2m in the two complexes and revealed a network of four

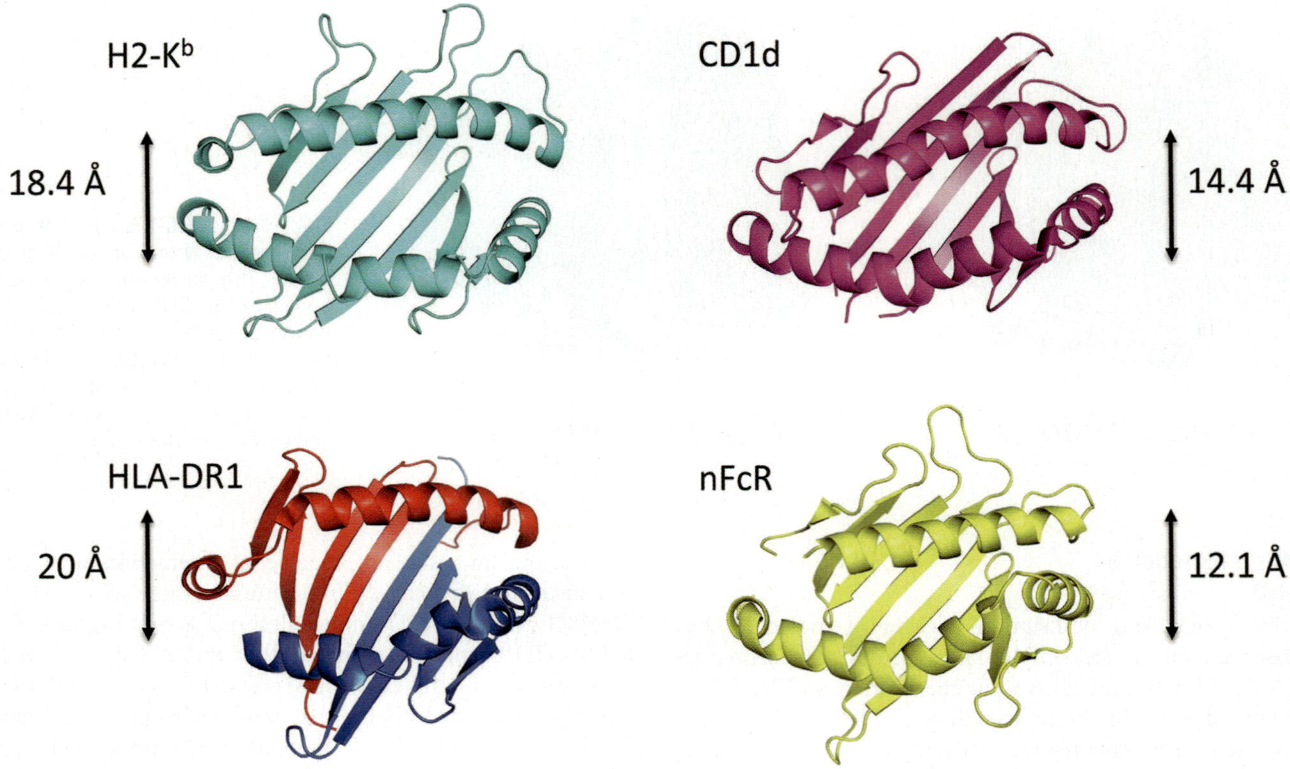

FIG. 19.14. Backbone structures of MHC-I-, MHC-II-, and MHC-I-like molecules. MHC-I molecule H2-K^b (2VAA), MHC-II HLA-DR1 (1DLH), MHC-Ib molecules CD1d (1Z5L), and FcRn (1EXU) are compared. Approximate maximal distances separating the two helices are indicated.

hydrophobic channels at the core of the α1α2 domain, which accommodate four hydrocarbon chains of length from 11 to 22 carbon atoms. These channels are called A′, C′, and F′ for the three pockets analogous to those of A, C, and F of the MHC-Ia molecules, and a fourth, termed T′, which is a distinct tunnel. Illustrations of the binding groove with the bound alkyl chains are shown in Figure 19.15, which shows views of the binding pockets of CD1a, b, and d with different ligands. These structures illustrate how the binding site of a classical MHC-I molecule may have evolved from (or to) the binding site of a molecule like CD1 to provide antigen selectivity for a distinct set of molecules that would be common to a set of important mycobacterial pathogens. Complexes of CD1 with their cognate NK TCR have been reported and will be discussed below.[626-629]

FcRn

Another example of an MHC-Ib molecule, noteworthy because it exemplifies a novel function of MHC molecules, is the neonatal Fc receptor. Originally described in the rat as a molecule of the intestinal epithelium that is involved in the transport of colostral IgG from the lumen to the bloodstream,[630,631] homologs in the mouse and human have also been described,[632-634] and the structure of the rat molecule has been determined crystallographically[613] (Fig. 19.16). As suggested by the amino acid sequence similarity of the FcRn to MHC-I proteins, the three-dimensional structure revealed considerable similarity to MHC-Ia molecules. Specifically, α1 and α2 domains have similar topology to the MHC-I molecule, although, as we discussed above, what would be the peptide-binding groove in the MHC-Ia molecules is closed tightly and lacks space sufficient for a ligand. The most provocative feature of the structure of the FcRn/Fc complex is that the MHC-I–like FcRn interacts with the Fc through contacts from the α2 and β_2m domains to interact with the Fc CH2-CH3 interface. As compared to the structure of the unliganded Fc, the complex reveals both conformational changes in the Fc and the presence of several titratable groups in the interface that must play a role in the pH-dependent binding and release of Ig molecules from the FcRn. The FcRn has taken the MHC-I fold and diverted its function for an interaction with the Fc of the immunoglobulin. Amino acids at what would classically be considered the “right-hand side” of the peptide-binding groove make contact with the Fc interface that lies between Cγ2 and Cγ3 domains. The FcRn serves as an excellent example of similar structures in the immune system being diverted for alternate purpose. The importance of the FcRn has been underscored by observations of differences in the serum half-life of immunoglobulin in animals that, as a result of an induced deletion of β_2m, lack the normal expression FcRn as well and seem to metabolize serum immunoglobulin aberrantly.[635] In recent years, the human FcRn has been the focus of structural and functional studies in efforts to modulate IgG transport and immune complex signaling in autoimmune disease and immune cytopenias.[636,637]

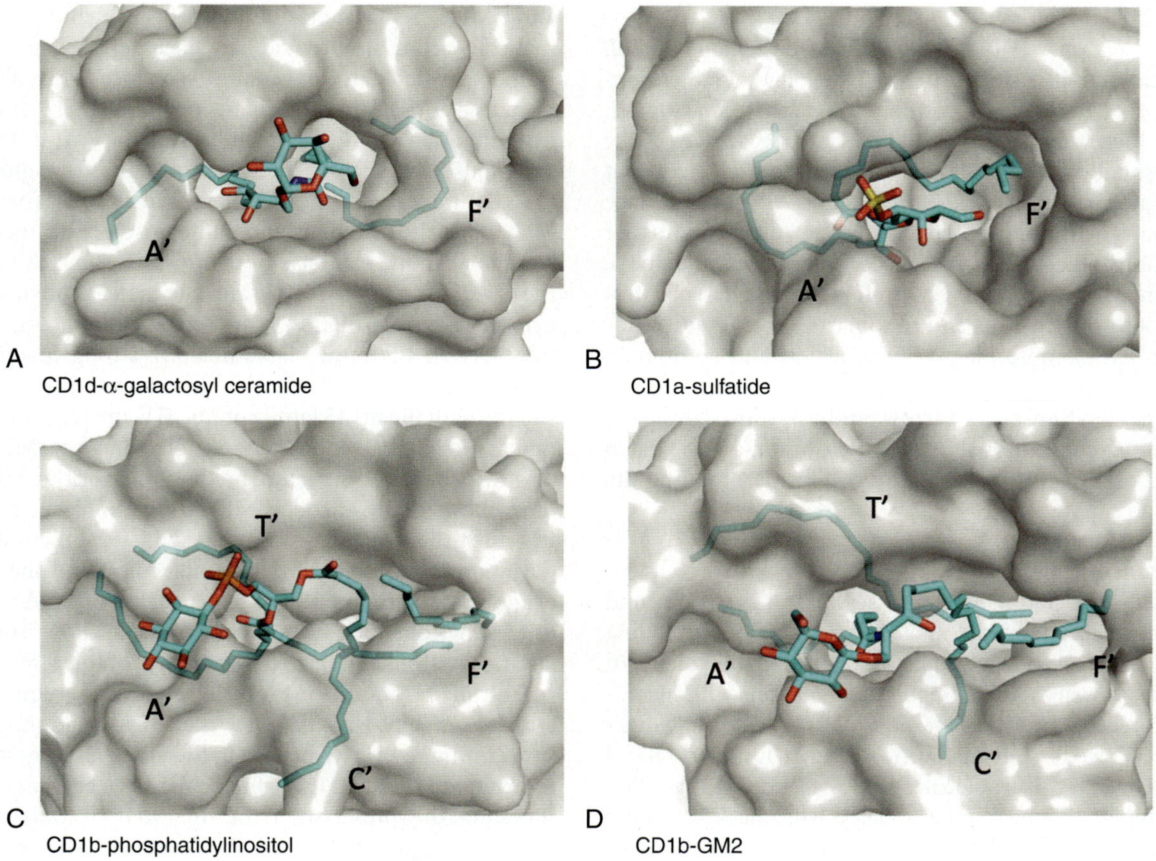

FIG. 19.15. Tunnels and pockets of CD1 molecules. Structures of **(A)** CD1d with α-galactosyl ceramide (1Z5L), **(B)** CD1a with sulfatide (1ONQ), **(C)** CD1b with phosphatidylinositol (1GZQ), and **(D)** CD1b with GM2 (1GZP) ganglioside are shown, with pockets A′, C′, and F′ indicated and tunnel T′.

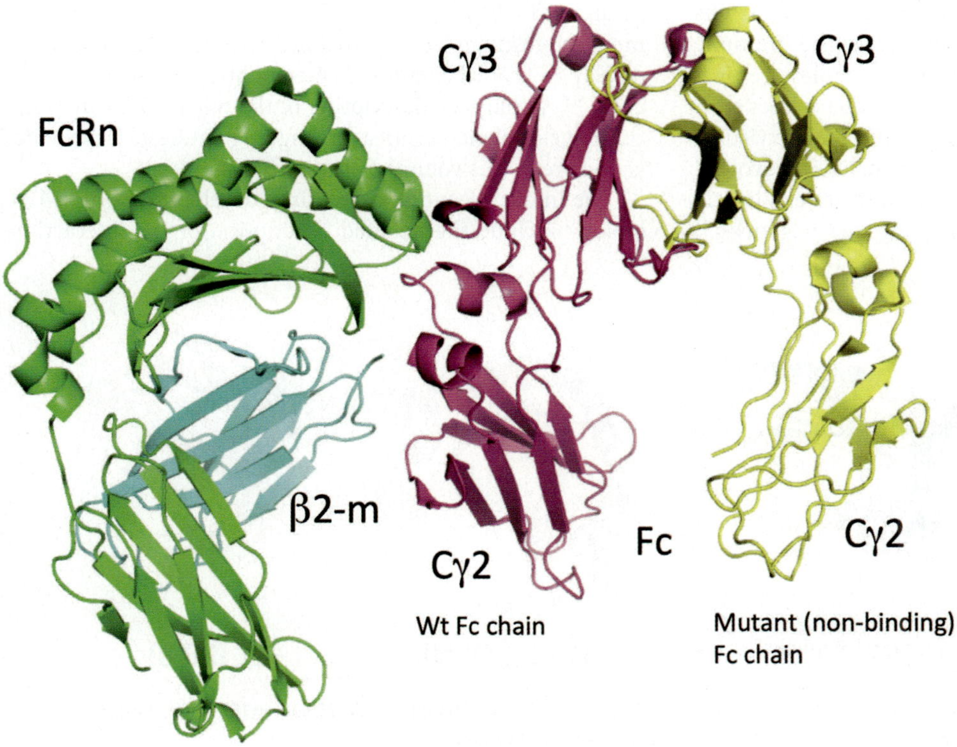

FIG. 19.16. Structure of the rat FcRn complexed with Fc. Ribbon diagram of rat FcRn complexes with the Fc heterodimer consisting of the wild-type (Wt) and nonbinding chains. This is from PDB 1I1A.

MILL MHC-Like Molecules

A novel set of MHC-Ib–like genes designated *MILL* (MHC class I–like located near the leukocyte receptor complex) has been identified in rodents, marsupials, and odd-toed ungulates.[638-641] These encode MICA/B-like glycophosphatidylinositol-linked cell surface molecules that are associated with β_2m and do not require TAP for cell surface expression. Their function remains unclear, although further examination indicates the presence of two family members, MILL1 and MILL2, observed on cycling thymocytes, proliferating smooth muscle cells and fibroblasts.[642] Structural studies of the MILL2 molecule reveal great flexibility in domain orientation of the $\alpha1\alpha2$ unit relative to $\alpha3$, plasticity in the association with β_2m, and a narrow binding groove consistent with lack of a conventional ligand.[643] However, a basic patch on the exposed aspect of the $\alpha3$ domain forms a binding site for heparan sulfate.

M10 proteins

The robust MHC fold seems to possess both the stability and flexibility requisite for the molecular function of binding ligands such as peptides, glycopeptides, phosphopeptides, and glycolipids, as well as for subsequent interactions with macromolecular receptors such as NK and TCRs. Likely, these qualities have allowed its adaptation for distinct function related to a class of MHC-Ib molecules, the M1 and M10 families, some members of which are associated with the V2R mouse pheromone receptors. The structure determination of M10.5 revealed a β_2m-associated MHC-I–like molecule with clear three-dimensional similarity to the MHC-I molecule H2-D^d.[644]

MHC-Iv Molecules

Since viruses are constantly engaged in a struggle for survival in their interplay with the immune system of their hosts, there is little surprise to note that pathogenic eukaryotic viruses can evolve numerous approaches to evade the immune response. MHC and structurally related molecules are produced not only by vertebrates but also by various viruses such as the CMVs of the β-herpesvirus family, and the poxviruses tanapox and molluscum contagiosum. These have coevolved with their vertebrate hosts over millennia and are exquisitely adapted to persist in the face of host immune responses.[11] Viral MHC-I–like molecules have been identified bioinformatically from amino acid sequences of the open reading frames predicted from genome sequence data. A number of putative MHC-Iv molecules, including UL18 and UL142 of the human CMV,[139,645] members of the "m145 family" of the mouse CMV,[13,646,647] the 2L molecule of tanapox,[648] and the U19 molecule of human herpesvirus 7,[649] as well as the MC80 protein of molluscum contagiosum,[130,650] may modulate the immune response of the host. In some cases such as the UL18 protein of HCMV and the m144 protein of mouse cytomegalovirus (MCMV), the sequence homology to bona fide MHC-I molecules is sufficiently strong to unambiguously identify them as MHC-I like. However, in most cases, such as with the m145 family of MCMV, the sequence homology to MHC-I is very weak and relatedness to MHC-I was originally based on structure prediction algorithms alone. We term these virally encoded MHC-I–like molecules MHC-Iv to reflect their distinct evolutionary history, their structural deviation from typical MHC-I molecules, and their role in evading host immune responses. A survey of DNA sequences of 11 genes obtained from 26 wild murine CMV isolates and laboratory strains indicated that several of the *MHC-Iv* genes (m144, m145, and m155) revealed significant sequence variation, consistent with the view that this variation may offer some immune-evasive benefit to the virus.[651] X-ray structures of several of these MHC-Iv molecules have been determined, revealing the preservation of the MHC-I fold, a narrow peptide-binding cleft, and variable assembly with the light chain, β_2m (Fig. 19.17).

MR1 and MAIT Cells

Originally identified as a monomorphic MHC-I–like molecule, MR1 has drawn increasing attention as it has been identified as the selecting element for mucosal-associated invariant T (MAIT) cells.[652] This molecule has proven even more interesting since the striking discovery that it serves to display microbial vitamin B metabolites and related molecules.[615] A complete description of the role of MR1 in MAIT cell development, an area of ongoing interest, is provided elsewhere in this volume (Chapter 37).

The most recent and exciting developments in MHC biology include the detailed description of the interactions

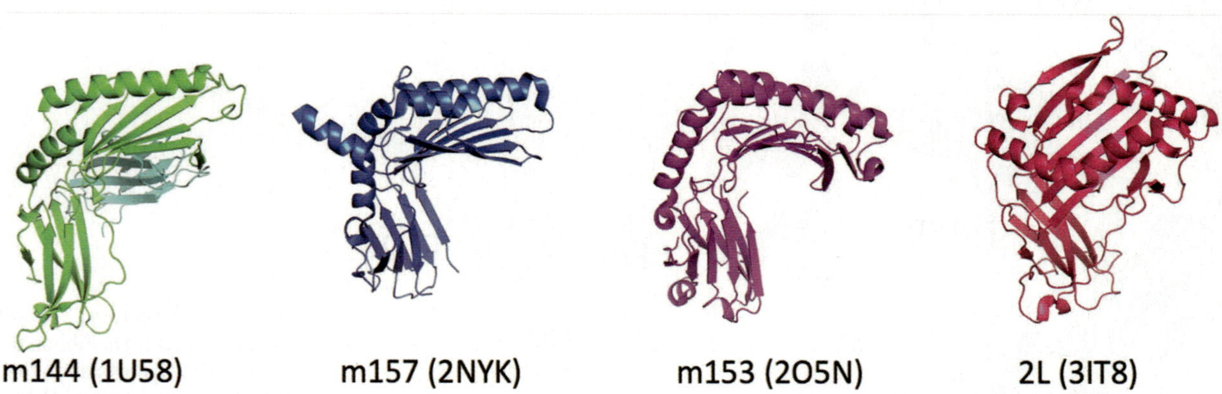

FIG. 19.17. Comparison of MHC-Iv molecules. Structures of MHC-Iv chains of m144 (1U58), m153 (2O5N), m157 (2NYK), and the 2L protein of tanapox virus (3IT8) were superposed and are illustrated as ribbon diagrams.

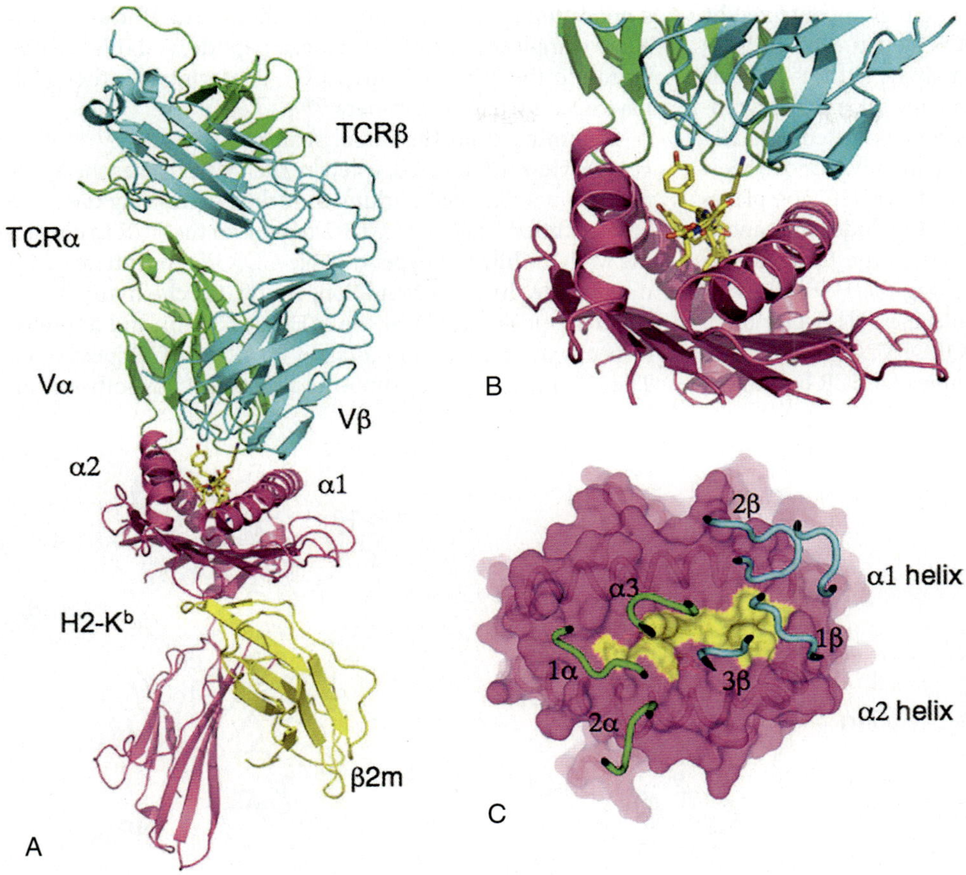

FIG. 19.18. Structure of a p/MHC-I/TCR complex and the TCR footprint on the p/MHC. The structure of the H2-K^b/dEV8/2C-TCR complex (2TCR) is displayed. **A:** Full complex, **(B)** a close-up of the p/MHC-I/TCR, and **(C)** the surface of the MHC (*magenta*)/peptide (*yellow*) complex with the CDR loops of the TCR Vα (*green*) and Vβ (*blue*) are shown and labeled.

of MHC-I and MHC-II molecules with their receptors and coreceptors on cells of the innate and adaptive immune system. Our structure/function survey will be completed by brief descriptions of the interactions of MHC-I and MHC-II molecules with TCRαβ, with the T cell coreceptors CD8 and CD4, and of MHC-I molecules with NK receptors. Details of the interactions of a number of CD1 molecules with their glycolipid antigens and specific NKT cell receptors shed light on the biology and evolution of these molecules. A brief description of interactions of MHC-II molecules with superantigen will follow. Each of these structural studies complements a host of biological experiments that have led up to an appreciation of the importance of understanding the structural basis of these immune reactions.

MHC/TCR Interactions

Second to the initial visualization of the structure of the MHC-I molecule HLA-A2 by crystallographic analysis, the images of p/MHC/TCR complexes initially reported some 15 years ago clarified the well-documented phenomenological details of T cell activation by APC such as MHC restriction, antigen specificity, TCR allo- and autoreactivity, the influence of HLA polymorphism, and biased patterns of TCR usage. The initial structures were determined in several systems,[602,604,653,654] and there remains ongoing interest in the details of particular MHC/TCR interactions. The first examples were of MHC-I–restricted TCR, one from the mouse and one from the human.[602] The mouse MHC-I molecule H2-K^b was analyzed in complex with a self-peptide, dEV8, and a TCR known as 2C,[653,655] and the human HLA-A2 complexed with the Tax peptide was studied with its cognate TCR derived from a cytolytic cell known as A6.[602] As a canonical example of the p/MHC/TCR complex, we include an illustration of the dEV8/H2-K^b/2C-TCR complex[655] (Figs. 19.18 and 19.19). These illustrations show that the complementarity-determining regions (CDRs) of the TCR (labeled 1α, 2α, and 3α for CDRs of Vα and 1β, 2β, and 3β for the CDRs of Vβ, respectively) sit symmetrically on the p/MHC complexes. For

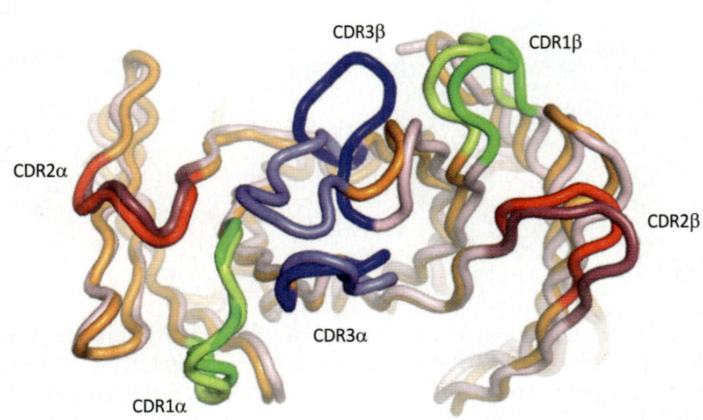

FIG. 19.19. CDR3 plasticity. Backbone tracings of the KB5-C20 TCR CDR loops free (1KB5) or bound to the p/MHC complex (1KJ2) are superposed. Unliganded forms are in darker colors. The largest conformational difference is seen in the CDR3β loop.

the dEV8/H2-K^b/2C-TCR complex, the region contacted by the CDRs of the Vα domain of the TCR lie to the left and that contacted by the CDRs of the Vβ domain lie to the right. The regions contacted by CDR3, labeled 3α and 3β, are at the center of the bound peptide, while the regions contacted by CDR1 and CDR2 of both Vα and Vβ lie peripherally. An overview comparing the range of interactions of p/MHC complexes with TCR is shown in Figure 19.20. This includes the canonical interaction of HLA-B*07:01 bound to the Epstein-Barr virus (EBV) peptide RPPIFIRRL binding the HD14 TCRαβ (panel A); the human MHC-II molecule (HLA-DQ*02.2) bound to a gluten peptide, APFSEQEQPVLG, binding the 594 TCRαβ (panel B); and as an example of a TCR binding in a noncanonical αβ orientation, the mouse H2-D^b class I molecule complexed with an influenza peptide, ASNENMETM, bound to the NP1 TCR (panel C). Footprints of other TCR have been reviewed elsewhere.[498]

Stemming from the initial structural reports,[602,604,653,654] a generic view was formed, whereby the TCR engages the MHC peptide in a diagonal docking modality, permitting the germ line–encoded CDR1 and CDR2 regions of the TCR to contact the MHC, while the hypervariable CDR3 loops interact with the peptide. In this orientation, the TCRα chain and TCRβ chain are positioned over the MHC-I α2-helix and α1-helix, respectively (β chain and α chain for MHC-II, respectively). These initial reports supported Jerne's[656] hypothesis on

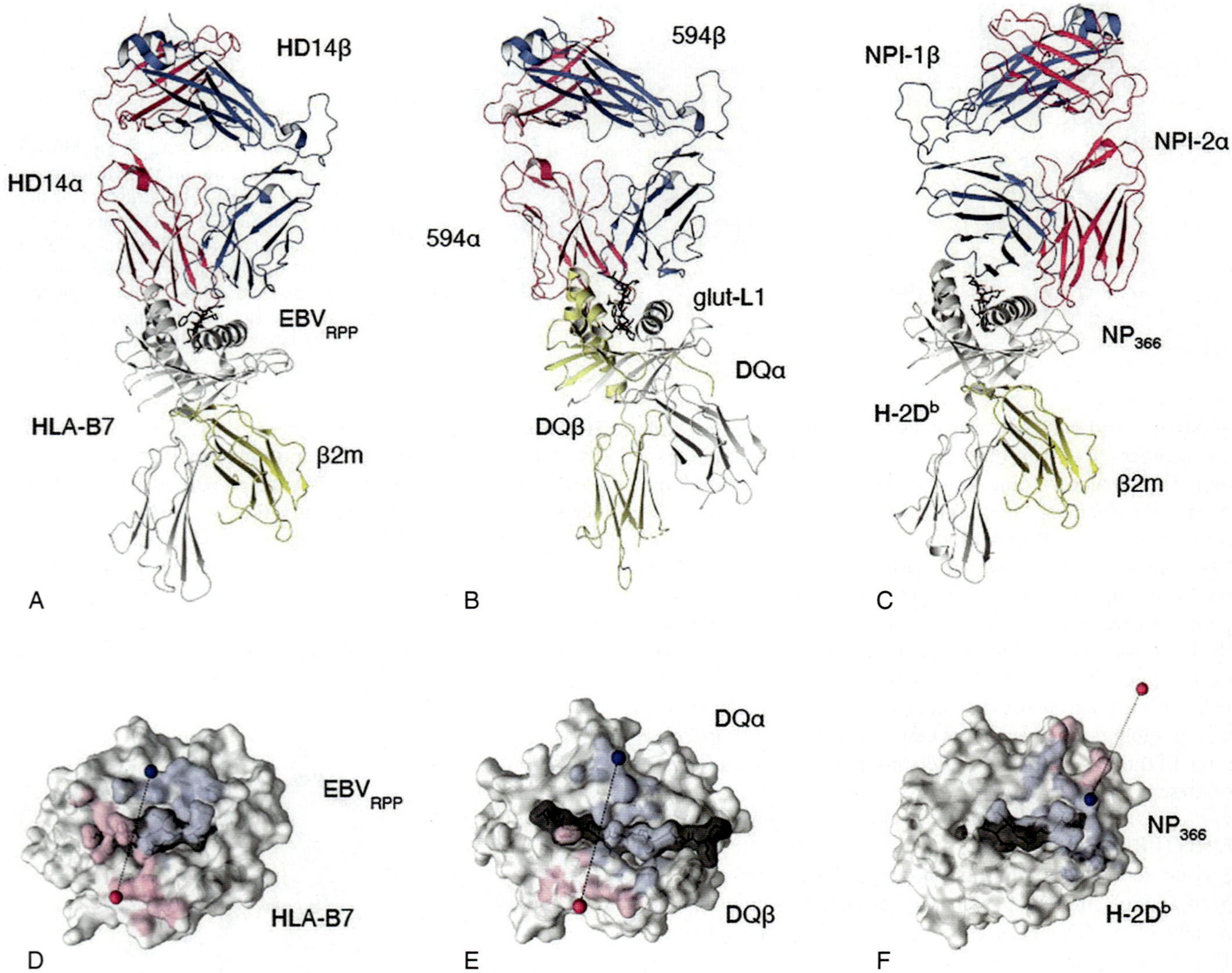

FIG. 19.20. Canonical and noncanonical p/MHC/TCR interactions. Overview of MHC molecule in complex with T cell receptors, **(A, D)** human MHC class I molecule HLA-B7 presenting an EBV epitope to HD14 TCRαβ (PDB:6VMX); **(B, E)** human MHC class II molecule HLA-DQ2.2 presenting a gluten epitope to TCR 594 (PDB:6PY2); **(C, F)** mouse MHC class I molecule H2-D^b presenting an influenza epitope to TCR NPI-1 (PDB: 5SWZ). A to C cartoon representation, the heavy chain (or class II α chain) is colored in *white*, $β_2$m (or class II β chain) is colored in *yellow*, peptide is shown in *black sticks*, the TCRα chain is shown in *pink*, and the TCRβ chain is shown in *blue*. **D-F:** Surface representation of the MHC-binding cleft colored in *white*, the peptide is shown in *dark gray*. Residues contacted by the different TCR chains are colored accordingly: *pink* for the TCRα chain and *blue* for the TCRβ chain. The center of mass of the TCR chains is represented as spheres and colored *pink* for α chain and *blue* for β chain.

TCR-MHC coevolution and offered a consistent first glimpse at the orientation of the TCR on the p/MHC complex. Since this time, however, numerous structural studies have shown that there is substantial variation in the nature of the p/MHC/TCR interaction.[657-659] Several additional points have emerged. (1) TCR recognition of unusually long and "bulged" MHC-I epitopes can show the TCR making limited contacts with the MHC molecule itself; (2) Often the germ line–encoded loops of the TCR were observed to make key interactions with the peptide, and conversely the CDR3 loops frequently engaged the MHC molecule itself; (3) Two recent examples, in the MHC-I[660] and MHC-II[661] settings, have reported "reversed TCR" docking polarity, where the TCR is flipped approximately 180° with respect to the MHC antigen-binding cleft. These "reversed" TCRs appear not to function so well however, suggesting that the consensus docking topology is associated with constraints that influence T cell signaling. (4) Finally, considerable plasticity in the conformation of the CDR loops of the TCR, particularly long CDR3 loops, is observed in the comparison of TCR free or bound to their cognate p/MHC ligands.[654,655]

Structural Insights Into Alloreactivity

Earlier in this chapter, we discussed the clinical approaches to remedy the practical problems raised by T cell alloreactivity. Here, we summarize recent experiments that elucidate the molecular underpinnings of this phenomenon. One approach to understanding the basis of such alloreactivity is to examine the structures of "allogeneic" p/MHC-I molecules complexed with TCR. Garcia and colleagues solved the structure of the 2C-TCR bound to its known allogeneic ligand, QL9/H2-L^d, and then compared it to the structure of 2C in complex with its positively selecting ligand, dEV8/H2-K^b.[315] These two TCR footprints were typical of known p/MHC/TCR structures but differed from each other. For instance, each chain of 2C made contacts with one MHC α-helix of QL9/H2-L^d, whereas both chains contacted both α-helices in dEV8/H2-K^b. The geometry of 2C adopted a more perpendicular orientation on QL9/H2-L^d and there was a relative rotation of dEV8/H2-K^b by 20°. The alloreactive complex reveals that both peptide-centric and MHC-centric interactions underpin direct T cell allorecognition by the 2C receptor, but with a heavy emphasis on MHC-centric interactions. Most surprising, however, was the small number of shared contacts between the two structures, implying a limited role for mimicry between cognate and allogeneic p/MHC. H2-K^b and H2-L^d have 31 amino acid differences and there is no sequence similarity between the H2-K^b–restricted octamer self-peptide, dEV8, and the H2-L^d-restricted nonamer, QL9.

Given these differences, it was a fair bet that the cross-reactivity of 2C on dEV8/H2-K^b and QL9/H2-L^d would depend upon plasticity in the CDR3 regions of the TCR as documented in comparisons of bound and free TCRs, including 2C. Surprisingly, this was not the case as the TCR actually adopted very similar conformations in the two structures. In contrast to the findings with the alloreactive murine, 2C-TCR, alloreactivity of the natural human LC13 TCR is based on striking molecular mimicry between the cognate and alloantigens.[316] LC13 recognizes an EBV peptide complexed with self-HLA-B*08:01 but also alloreacts with B44 allotypes (HLA-B*44:02 and 44:05) bound to an endogenous allopeptide from an ATP-binding cassette protein ABCD3. Despite significant polymorphism between HLA-B*08:01, HLA-B*44:02 and B*44:05, and the very different sequences of the virus and allopeptides, the LC13 TCR engaged the cognate and allogeneic p/HLA complexes identically. Moreover, there was mimicry of the viral peptide by the allopeptide. Thus, molecular mimicry involving HLA and peptide moieties is a mechanism for human T cell alloreactivity between disparate cognate and allogeneic p/HLA complexes. Interestingly, another HLA-B*08:01–restricted TCR (CF34) with the identical EBV peptide specificity as LC13 lacks B44 reactivity because it arises when HLA-B44 is coinherited *in trans* with HLA-B8. Notably, the CF34 TCR docked at the N-terminus of HLA-B8 peptide, coinciding with a polymorphic region between HLA-B8 and HLA-B44, whereas the alloreactive LC13 TCR docked at the C-terminus of the HLA-B8/peptide complex. The distinct footprints of the LC13 and CF34 TCRs provide a picture of how self-tolerance may guide the specificity of TCRs selected into the immune repertoire.[662]

The role of molecular mimicry and the relative importance of "peptide-centric" versus "MHC-centric" bias in T cell allorecognition is likely to vary in different systems. Hence, it is likely that the nature of the polymorphisms between cognate and allogeneic MHC allotypes will affect TCR focus. Thus, closely related MHC allotypes that differ by as little as one amino acid (eg, H2-K^b mutants in mice; HLA-A2, B44, and B27 families in humans) are set up for MHC mimicry to be a key component of T cell allorecognition where *specificity* is likely to be "peptide -centric." For example, HLA-B*44:02 and B*44:03 allotypes differ by only a single amino acid and yet stimulate strong mutual allogeneic T cell responses. Indeed, the potency of T cell alloresponses between closely related MHC allotypes probably occurs because positive selection of host T cells has evolved to create a repertoire responsive to subtle changes in peptide display. Therefore, closely related MHC allotypes, with differences in both peptide repertoire and p/MHC conformation of a shared repertoire, play straight into nature's design for T cell recognition.

Important questions concerning the nature of T cell signaling mediated by engagement of the TCR by p/MHC complexes remain. What are the molecular and biophysical determinants of agonist as compared with antagonist activity of particular p/MHC ligands? Although many parameters such as binding kinetics, half-life, and affinity (both two-dimensional and three-dimensional) as well as coreceptor contributions may play a role in the outcome of TCR engagement of p/MHC complexes, and the additional role of receptor clustering may contribute to such signaling, the precise contribution of each of these factors has not been established. A provocative approach to the question of the role of the geometry of docking of the TCR onto the p/MHC complex has been addressed in the analysis of the relationship of stimulatory parameters of four different peptides bound to H2-L^d with respect to the three-dimensional structure of the relevant p/MHC/TCR complex.[663] One of the four peptides failed to induce signaling, and remarkably, the TCR docked onto the p/MHC in a unique docking geometry. The authors interpret these results to demonstrate the relationship between p/MHC/TCR docking geometry, peptide cross-reactivity, and signaling, as well as explaining the basis for germ line bias in TCR/MHC interactions.

MHC/Coreceptor Complexes

The major coreceptors for recognition by TCRαβ are CD8, which interacts with MHC-I molecules, and CD4, which interacts with MHC-II. Coreceptor function plays a role in signaling the T cell in addition to contributions the MHC/coreceptor interaction may provide in increasing apparent avidity between the MHC/peptide and the TCR. CD8, the coreceptor on MHC-I–restricted αβ T cells, exists as a cell surface homodimer of two α chains or a heterodimer of α and β chains and plays an important role both in the activation of mature peripheral T cells and in the thymic development of MHC-I–restricted lymphocytes.[664-666] Mature peripheral CD8+ T cells express significant amounts of CD8αβ while other lymphocytes, such as intraepithelial lymphocytes in the gut, predominantly express the CD8αα homodimer. The CD8αβ heterodimer is considered the relevant functional coreceptor on peripheral T cells. The three-dimensional structures of human and mouse MHC-I/CD8αα complexes have been described,[667,668] and more recently a complex of mouse MHC-I/CD8αβ has been reported.[76] These structures localize the binding site of the CD8 immunoglobulin-like αα homodimer or CD8αβ heterodimer to a region on an exposed loop of the MHC-I α3 domain. This interaction is illustrated in Figure 19.21, which shows the flexible loop of residues 223-229 clamped by the CD8 combining site. This is one of relatively few examples of a non-Ig, Ig-like molecule exploiting an Ig-like heterodimer interface to bind another ligand. The structural identification of the CD8αα and CD8αβ on the MHC-I provides a context for understanding the contribution of CD8, which binds the kinase Lck through the cytoplasmic domain of its α chain, to T cell signaling. The orientation of the CD8αβ with respect to the MHC-I molecule and the role of this topology in signaling the T cell may be more complex than the X-ray structure suggests, as recent studies of the contribution of complex carbohydrates to CDαβ interactions have been analyzed[77] (Chapter 24).

The T cell coreceptor associated with cells restricted to MHC-II antigens, CD4, has also been the subject of detailed structural studies, in part because of its role as a receptor for attachment and entry of the HIV.[669] The X-ray structure determination of the complete extracellular portion of the molecule (domains D1 through D4) indicates a degree of segmental flexibility between domains D2 and D3, and both crystallographic and biochemical data suggest that dimerization of cell surface CD4 may occur.[670] These results have been interpreted to support a role for CD4-mediated MHC-II–dependent dimerization in facilitating TCR dimerization and signaling. Although a mixed species structure of human CD4 D1-D2 domains with a mouse MHC-II molecule was reported at low resolution some years ago,[80] only recently, employing an affinity-matured human CD4 molecule have diffraction quality crystals of human MHC-II/human CD4 D1-D2 molecules been obtained[81] (Fig. 19.22). This recently determined human MHC-II/hCD4 structure is remarkably similar to the low resolution mixed species structure determined more than 10 years earlier. A complete ternary complex of p/MHC-II/TCR/CD4 that reveals the relationship of the coreceptor to the MHC-II and the TCR has been reported.[82,671] These structures reveal the focus of the CD4 interaction on a stretch of amino acid residues on the Ig-like β2 domain of the MHC-II molecule. Such structural information has guided screening of site-directed HLA class II mutant libraries to identify high CD4 affinity variants to generate reliable staining reagents.[672]

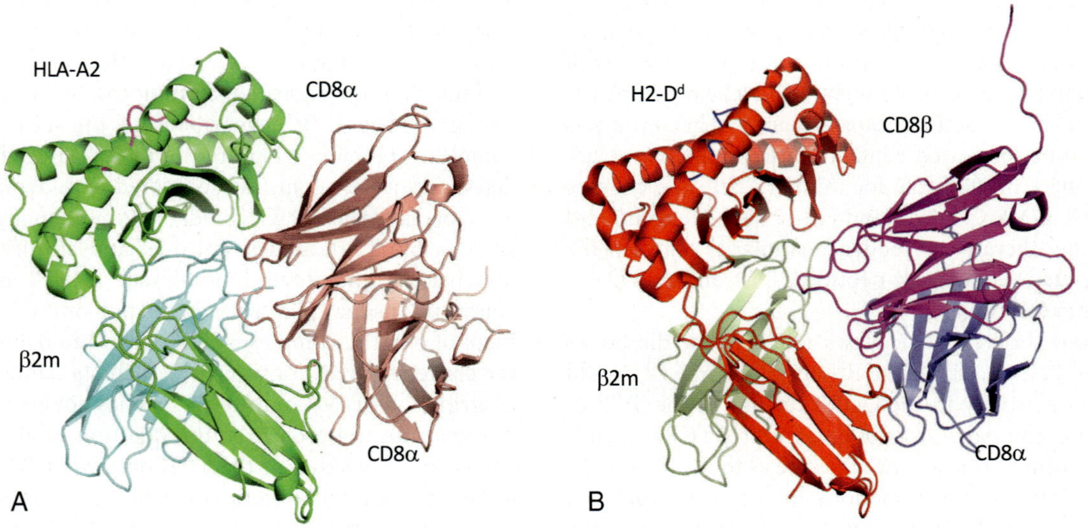

FIG. 19.21. CD8αα and CD8αβ bind the same site on the MHC-I α3 domain. The MHC-I interactions with the **(A)** CD8αα (PDB 1AKJ) homodimer and **(B)** CD8αβ (PDB 3DMM) heterodimer are shown.

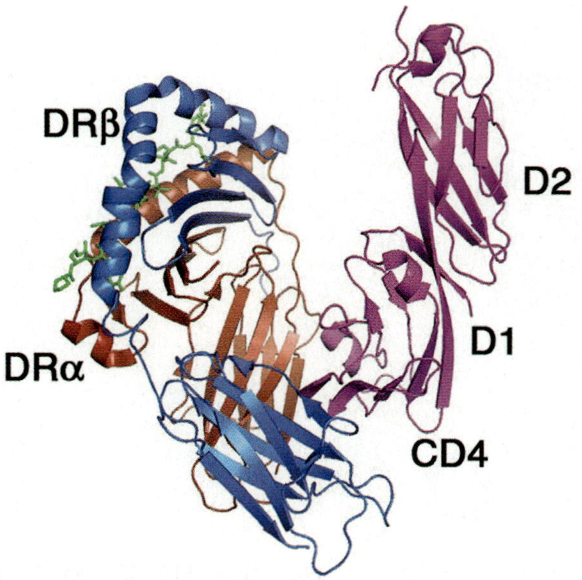

FIG. 19.22. CD4 interacts directly with the β2 domain of MHC-II. A ribbon illustration of a complex of human CD4 with HLA-DR1 (PDB 3S5L) is shown. HLA-DR1α chain is in firebrick, HLA-DR1β chain in *blue*, peptide in *green*, and CD4 in *magenta*.

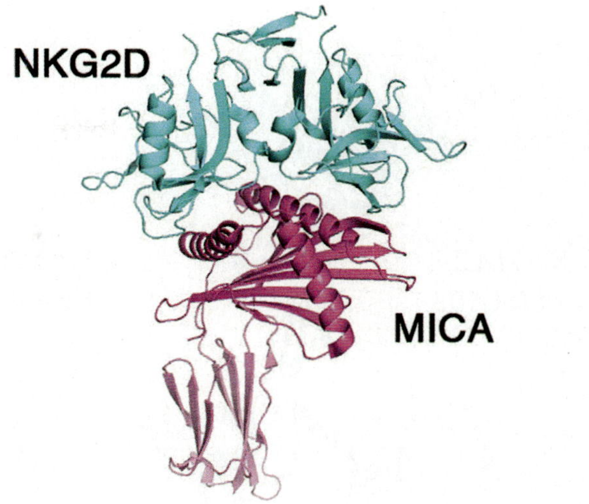

FIG. 19.23. NKG2D sees the MHC-Ib MICA α-helical superdomain. The structure of the human NKG2D/MICA complex (PDB 1HYR) reveals the homodimer NKG2D sitting asymmetrically astride the MICA α-helical superdomain. NKG2D domains are in *cyan* and MICA in *magenta*.

MHC/NK Receptor

NK cells provide a first line of defense against virus-infected and host cells, and their recognition of such targets is dependent on a balance of NK activating and inhibitory receptors.[89,673] In addition to their expression on NK cells, some of the NK receptors are also expressed on subsets of T cells and other hematopoietic cells. In general, each of these classes of NK receptors also falls into two different structural groups, the Ig-like receptors and the C-type lectin-like receptors. Because of their functional interactions with MHC-I– and MHC-I–like molecules, and because several different systems have evolved to recognize MHC-I molecules differently, it is worthwhile to examine the structures of several MHC-I/NK receptor complexes.

In the human, the major NK receptors fall into several categories: those of the NKG2 family, the NKp molecules, and the KIRs. The NKG2 family consists of type II membrane C-type lectin-like dimeric molecules that recognize ligands grouped broadly as the NKG2D ligands. These include the MHC-like molecules, MICA and MICB, as well as the ULBP family of stress-induced molecules. NKp molecules include NKp30, NKp44, and NKp46, all members of the immunoglobulin superfamily. The KIR molecules, also Ig superfamily members, bind either HLA-C or, for the KIR3 subfamily, HLA-B molecules. Because the focus of this chapter is members of the MHC protein family, we restrict our discussion of the human NK receptor/ligand interactions to the NKG2/MIC and ULBP complexes and to the KIR2D and KIR3D/MHC interactions.

The activating receptor, NKG2D, interacts with MICA, a stress-induced MHC-I–like molecule, allowing NK cells to eliminate virus-infected or tumor cells that are marked by the surface expression of MICA. The structure of the human NKG2D/MICA complex reveals that the homodimeric binding site of NKG2D recognizes the surface α-helical superdomain, in an orientation similar to that by which TCRs see MHC-I molecules[611] (Fig. 19.23). This structure shows remarkable similarity to that of the hNKG2D bound to another MHC-Ib protein, ULBP3.[674] The mouse NKG2D interacts in a similar orientation with the murine stress-induced molecule, RAE-1β.[614] Thus, each species appears to have evolved their own, but related specialized receptors for signaling stress in a general sense, and raising a flag for NK cell–activating receptors.

The short KIRs (eg, KIR3DS and KIR2DS molecules) are considered activating receptors because they have the potential to interact with the either the DAP10 or DAP12 signal-transducing molecules, and the long KIRs (eg, KIR3DL and KIR2DL) are inhibitory because they have cytoplasmic domains that contain ITIMs (immunoreceptor tyrosine-based inhibitory motifs). KIR2DL1 and KIR2DL2 have been studied extensively. They interact with the human MHC-I molecules HLA-C*04 and HLA-C*03, respectively, and show some preferences for MHC-I molecules complexed with particular peptides. Among the polymorphic amino acid residues that distinguish HLA-C*03 and HLA-C*04 are Asn80 of HLA-C*03 and Lys80 of HLA-C*04. Thus, it was of interest when the structures of KIR2DL2/HLA-C*03[213] and HLA-C*04/KIR2DL1 complexes were reported.[212] The KIR2DL2/HLA-C*03 interaction is illustrated in Figure 19.24. The important conclusions from this structure and those of similar KIR2DL complexes are that the recognition of the MHC-I is via amino acid residues of the elbow bend joining the two immunoglobulin-like domains of the KIR and that residues that vary among different KIRs determine the molecular specificity of the interaction with the particular allelic product of HLA-C. In addition, the interaction of the KIR with HLA-C is also modulated by the particular bound peptide,

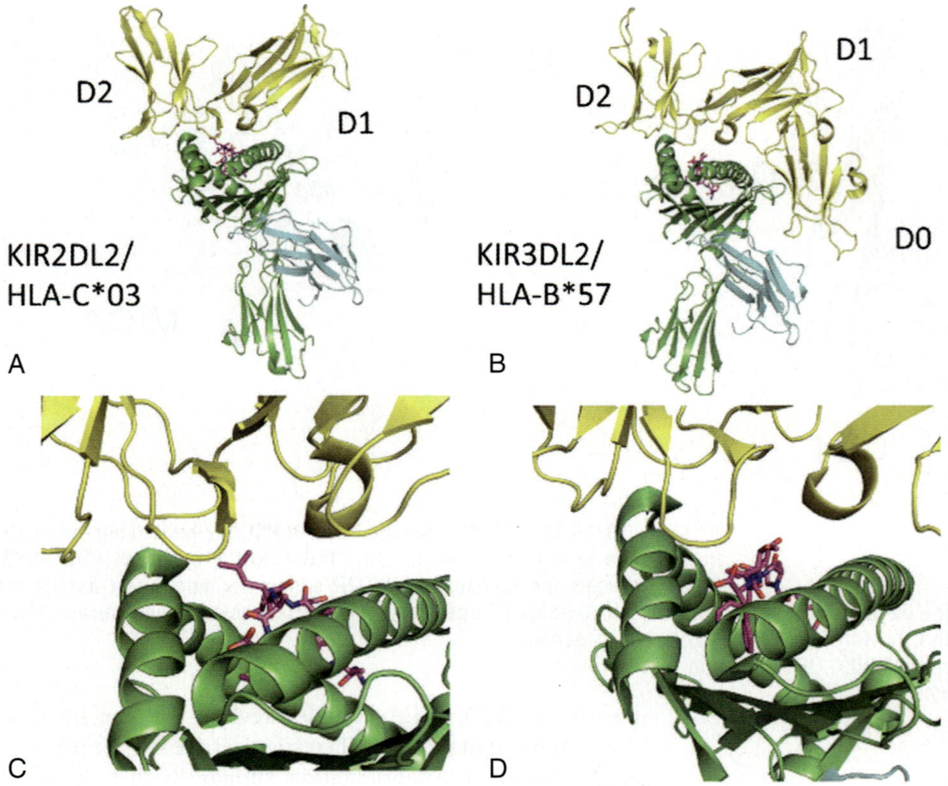

FIG. 19.24. KIR2DL2 and KIR3DL1 interact similarly with their p/MHC ligands. KIR2DL2 complexed with HLA-C*03 (PDB 1EFX) **(A, C)** and KIR3DL1 complexed with HLA-B*57:01 (PDB 3VH8) **(B, D)** are shown to illustrate the overall interactions **(A, B)** and close-ups of the contact with bound peptide **(C, D)**.

explaining the results of peptide-binding/specificity studies. Structural understanding of the KIR3 molecules that have three extracellular Ig-like domains eluded the field for many years. Molecular structures of complexes of HLA-B*57:01 with KIR3DL1 have been reported (Fig. 19.24).[215] These structures reveal the contribution of the D0 amino terminal domain of KIR3DL1 in recognizing a region of the MHC-I with limited polymorphism, the D1 domain in accommodating a region with sequence variation, and the D2 domain showing high complementarity. Recent discussions consider the polymorphisms of KIR3DL1 as they distinguish different HLA class I ligands.[675]

The mouse primarily exploits a set of NK receptors of the C-type lectin-like structural family to provide the same function. In particular, the predominant, and best studied, mouse NK receptors are those of the Ly49 family. Functional experiments had demonstrated that Ly49A interacts with the MHC-I molecule, H2-D^d, and also that for appropriate interaction, the H2-D^d needs to be complexed with a peptide. In contrast to human NK recognition, however, surveys of H2-D^d binding peptides revealed little if any peptide preference or specificity. This would explain the function of the NKIR in that by baseline engagement with normal p/MHC-I on somatic cells, they turn off the NK cell. When MHC-I is dysregulated by tumorigenesis or by pathogenic infection, the lower level of surface MHC-I diminishes the NKIR-mediated signal, and the NK cell is activated. The structure of the mouse Ly49A inhibitory receptor in complex with its MHC-I ligand, H2-D^d (Fig. 19.25), reveals several crucial features of the interaction: (1) the Ly49A C-type lectin-like molecule is a homodimer; (2) the Ly49A molecule makes no direct contact with residues of the MHC-bound peptide; and (3) in the X-ray structure, there are two potential sites for Ly49A interaction with the MHC molecule—site 1 at the end of the α1 and α2 helices, and site 2, an extensive region making contact with the floor of the peptide-binding groove, the α3 domain of the MHC-I molecule, and the β_2m domain as well. The ambiguity suggested by the X-ray structure has been resolved by extensive mutagenesis studies that are consistent with the

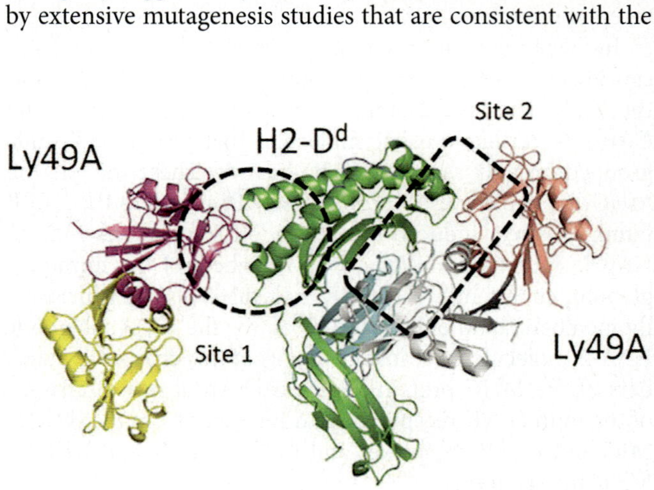

FIG. 19.25. Structure of the H2-D^d/Ly49A complex reveals an alternate site for MHC/NK receptor interaction. The structure of the complex (PDB 1QO3) is shown, illustrating both potential sites of interaction.

view that site 2 is functionally significant.[676,677] Recently, several other NK receptors have been studied structurally. These include the Ly49I inhibitory receptor, which interacts functionally with H2-K^b, which has been crystallized without a ligand, revealing a basic fold similar to that of Ly49A, but with a somewhat different dimeric arrangement.[678] Ly49C, another murine NKIR, has been examined in complex with its H2-K^b ligand.[679] Differences between H2-K^b/Ly49C and H2-D^d/Ly49A suggest different modes of NK receptor binding to MHC-I depending on whether the interaction is "cis" (ie, between the NK receptor and the MHC-I on the same NK cell) or "trans" (between the NK receptor on the NK cell and the MHC-I on its target).[680] Both modes of engagement take advantage of site 2.

MHC Interactions With Peptide Loading Chaperones

To understand the mechanisms of peptide loading, several studies have taken a structural approach to understand the conformational subtleties of interactions of the MHC-I molecule with the ER PLC. The PLC consists of the TAP peptide transporter, an MHC-binding chaperone tapasin, and additional molecules ERp57 and calreticulin. A cryo-EM study of the entire PLC,[47] complemented by high-resolution crystallographic studies of the direct interaction of a tapasin homolog, TAPBPR,[681,682] illustrates how interactions stabilize a peptide receptive form of the MHC molecule.

MHC-II Superantigen Complexes

Superantigens are molecules, frequently toxic products of bacteria, that bind MHC molecules on the cell surface and are then presented to a large subset of T cells, usually defined by the expression of a particular family of TCR V regions.[683] Most of the known superantigens bind MHC-II molecules, though one, the agglutinin from *Urtica dioica*, the stinging nettle, can be bound by both MHC-I and MHC-II molecules and presented to T cells of the Vβ8.3 family.[684] Its structure in complex with carbohydrate ligand has been reported.[685,686] MHC-II interactions with superantigens, such as those derived from pathogenic bacteria, are the first step in the presentation of the multivalent array of the APC-bound superantigen to T cells bearing receptors of the family or class that can bind the superantigen.[683] A number of structural studies examining the interaction of superantigen both with their MHC-II ligands and with TCR have been reported.[687] Structural analysis of crystals derived from staphylococcal enterotoxin B (SEB) complexed with HLA-DR1[688] and from toxic shock syndrome toxin-1 (TSST-1) complexed with either HLA-DR1[689] revealed that the two toxins bind to an overlapping site, primarily on the MHC-II α chain, and indicated that the SEB site would not be expected to be influenced by the specific peptide bound by the MHC, although the TSST-1 site would. These studies indicate that these superantigens can interact with the MHC-II β chain through a zinc-dependent site that includes superantigen contacts to bound peptide. Our understanding of superantigen interactions has been augmented by the determination of ternary structures of MHC-II/superantigen/TCR complexes for several examples, that of the *Mycoplasma arthritidis* mitogen in complex with HLA-DR1 and TCR,[690] and that of *Staphylococcus aureus* enterotoxin H complexed with HLA-DR1 and TCRβ.[691] In both of these cases, the superantigen interacts with the MHC-II and with both TCR Vα and Vβ domains.

Assays for Molecular Interactions

Whereas the crystal structures provide a vivid static illustration of the interactions of MHC molecules with their peptide, FcRn, CD8, CD4, superantigen, NK receptor, and TCR ligands, the dynamic aspects of these binding steps can be approached by a variety of biophysical methods.[692] It is important to note that affinities and kinetics of interaction of p/MHC complexes for TCR have been determined by several methods in a variety of systems.[693-698] In addition, MHC interactions with NK receptors[209,210,699] have been quantified by similar techniques. Although there are clear differences in the affinity and kinetics of binding of different TCR and NK receptors for their respective cognate MHC/peptide complexes, the generally consistent findings are that the affinities are low to moderate (ie, $K_D = 5 \times 10^{-5}$ to 10^{-7} M) and are characterized by relatively rapid dissociation rates (ie, $k_d = 10^{-1}$ to 10^{-3} s^{-1}). Attention has been drawn to differences in measuring interactions with one component in solution (three-dimensional affinity) as compared with measurements made between multivalent displays of the interactions on two surfaces, such as might mimic the apposition of two cells. Methods exploiting microscopic techniques[700] or a micropipette adhesion frequency assay[701-703] suggest that the biological readout of T cell activation may be more readily correlated with two-dimensional rather than three-dimensional measurements of MHC/TCR affinities.

Improved techniques for NMR, based on selective labeling of recombinant protein chains, have permitted the use of this sensitive, high-resolution, solution technique to be exploited in the analysis of macromolecular interactions. Recent studies from several laboratories have mapped MHC interaction sites of a viral immunoevasin,[704] of TCR,[605,705,706] and indicate allosteric changes in the TCR on binding of p/MHC.[605]

Multivalent MHC-Peptide Complexes

A major development in the study of antigen-specific MHC-restricted T cells, of both the MHC-I and MHC-II variety, has been the engineering of multivalent p/MHC complexes for identification, quantification, and purification.[500] Although several different formats have been developed, they all rely on the gain in avidity obtained by producing recombinant p/MHC complexes that are then multimerized as dimers[707] (in a recombinant Ig format), as the original tetramers[500] (using single biotin tags on the MHC to bind tightly to the tetravalent avidin), coupling to dextran,[708] or using other approaches. Soluble p/MHC complexes are generated in bacterial or eukaryotic expression systems and are conjugated to appropriate fluorophores. These reagents can be used in flow cytometric assays that permit the direct enumeration of p/MHC-specific T cells taken directly *ex vivo*. For MHC-I molecules, the technology has been reliable in producing multivalent (tetrameric) molecules loaded homogeneously with synthetic peptides, and a wide variety of specific p/MHC-I multimers are available either from a resource facility sponsored by the National Institute of Allergy and Infectious Diseases (https:/www.niaid.nih.gov/research/nih-tetramer-core-facility) or from commercial suppliers that offer either the tetramer, immunoglobulin multimer, pentamer, or other preparations. The MHC-I peptide multimers have also been exploited for identification of specific populations of NK cells and for assignment of various NK receptor specificities.[217,709] For many MHC-II molecules, similar success has been achieved in the production of such multimers, using insect cell or mammalian cell expression systems for molecules produced by either the tetramer or immunoglobulin chimera strategy.[710-714] p/MHC-II tetramers can be effectively used for a variety of CD4 T cell applications, but some TCRs (particularly those involved in autoimmune diseases) have an affinity for their cognate p/MHC-II that is beneath the threshold for detection.[701] The potential for peptides binding to a single MHC-II in several frames generates difficulties in engineering.[558]

SUMMARY

We have surveyed the *MHC* as a genetic region and a source for molecules crucial to immune regulation and immunological disease. These genes reflect the panoply of mechanisms involved in the evolution of complex systems and encode cell surface proteins that interact via complex orchestration with small molecules including peptides and glycolipids as well as with macromolecular receptors on T cells and NK cells. The MHC-I molecules provide the immune system with a window for viewing the biological health of the cell in which they are expressed, and MHC-II molecules function as scavengers to taste and display the remnants of the cellular environment. Viruses and bacterial pathogens contribute enormously to the genetic dance—they modulate and compete in the control of MHC expression, sometimes exploiting MHC mimics that they have acquired—and the host, by adjusting its T cell and NK cell repertoire on the time scale of both the individual organism and the species, resists the push to extinction. The immune system, dynamically, resourcefully, creatively, provides, through the concerted action of its MHC molecules, TCR, NK receptors, and antibodies as well as a host of other regulatory molecules, an organ system vital not only to the survival of the individual but to the success of the species. As we understand better the molecular functions of the MHC, we should better understand rational approaches to manipulating the immune system in the prevention, diagnosis, and treatment of immunological and infectious disease.

ACKNOWLEDGMENTS

We thank the members of our laboratories for their patience while we worked on this chapter. We take full responsibility for any errors and omissions in this chapter and apologize in advance to those whose work we overlooked in citation. DHM and KN are supported by the Intramural Research Program of the National Institute of Allergy and Infectious Diseases, NIH. The authors acknowledge the generous support of the National Health and Medical Research Council of Australia (program grants 1113293 and 1092262 and project grant 1120467), and National Institutes of Health grant R01AI148407. JR is also supported by an Australian Research Council, Australian Laureate Fellowship FT160100049. We thank Rhonda Holdsworth and Mary Diviney from the Victorian Transplantation and Immunogenetics Service for their generous advice. This chapter is dedicated to the memories of our colleagues and friends, Nilabh Shastri, Enzo Cerundolo, and of course, William E. Paul.

REFERENCES

1. Snell GD. Methods for the study of histocompatibility genes. *J Genet.* 1948;49(2):87-108.
2. Snell GD, Dausset J, Nathenson SG. *Histocompatibility.* Academic Press; 1976:xiv, 401.
3. Rosenthal AS, Shevach EM. Function of macrophages in antigen recognition by Guinea pig T lymphocytes. I. Requirement for histocompatible macrophages and lymphocytes. *J Exp Med.* 1973;138(5):1194-1212.
4. Shevach EM, Rosenthal AS. Function of macrophages in antigen recognition by Guinea pig T lymphocytes. II. Role of the macrophage in the regulation of genetic control of the immune response. *J Exp Med.* 1973;138(5):1213-1229.
5. Benacerraf B, McDevitt HO. Histocompatibility-linked immune response genes. *Science.* 1972;175(4019):273-279.
6. McDevitt HO, Tyan ML. Genetic control of the antibody response in inbred mice. Transfer of response by spleen cells and linkage to the major histocompatibility (H-2) locus. *J Exp Med.* 1968;128(1):1-11.
7. Zinkernagel RM, Doherty PC. Immunological surveillance against altered self components by sensitised T lymphocytes in lymphocytic choriomeningitis. *Nature.* 1974;251(5475):547-548.
8. Shearer GM, Rehn TG, Garbarino CA. Cell-mediated lympholysis of trinitrophenyl-modified autologous lymphocytes. Effector cell specificity to modified cell surface components controlled by H-2K and H-2D serological regions of the murine major histocompatibility complex. *J Exp Med.* 1975;141(6):1348-1364.
9. Robinson J, Barker DJ, Georgiou X, Cooper MA, Flicek P, Marsh SGE. IPD-IMGT/HLA database. *Nucleic Acids Res.* 2020;48(D1):D948-D955.
10. Robinson J, Halliwell JA, McWilliam H, Lopez R, Marsh SG. IPD – The Immuno polymorphism database. *Nucleic Acids Res.* 2013;41(Database issue):D1234-D1240.
11. Revilleza MJ, Wang R, Mans J, Hong M, Natarajan K, Margulies DH. How the virus outsmarts the host: function and structure of cytomegalovirus MHC-I-like molecules in the evasion of natural killer cell surveillance. *J Biomed Biotechnol.* 2011;2011:724607.
12. Tortorella D, Gewurz BE, Furman MH, Schust DJ, Ploegh HL. Viral subversion of the immune system. *Annu Rev Immunol.* 2000;18:861-926.
13. Mans J, Natarajan K, Balbo A, et al. Cellular expression and crystal structure of the murine cytomegalovirus major histocompatibility complex class I-like glycoprotein, m153. *J Biol Chem.* 2007;282(48):35247-35258.
14. Holtappels R, Freitag K, Renzaho A, Becker S, Lemmermann NAW, Reddehase MJ. Revisiting CD8 T-cell 'memory inflation': new insights with implications for cytomegaloviruses as vaccine vectors. *Vaccines (Basel).* 2020;8(3):402.
15. Gorer P. The detection of antigenic differences in mouse erythrocyte by the employment of immune sera. *Br J Exp Pathol.* 1936;17:42-50.
16. Gorer PA. Studies in antibody response of mice to tumour inoculation. *Br J Cancer.* 1950;4(4):372-379.
17. Klein J, Bontrop RE, Dawkins RL, et al. Nomenclature for the major histocompatibility complexes of different species: a proposal. *Immunogenetics.* 1990;31(4):217-219.
18. Ceppellini R. *Histocompatibility Testing, 1967.* Munksgaard; 1967.
19. Yoshino M, Xiao H, Jones EP, Kumánovics A, Amadou C, Fischer Lindahl K. Genomic evolution of the distal Mhc class I region on mouse Chr 17. *Hereditas.* 1997;127(1-2):141-148.
20. Yoshino M, Xiao H, Jones EP, Fischer Lindahl K. BAC/YAC contigs from the H2-M region of mouse Chr 17 define gene order as Znf173-Tctex5-Mog-D17Tu42-M3-M2. *Immunogenetics.* 1998;47(5):371-380.
21. Kruisbeek AM. Commonly used mouse strains. *Curr Protoc Immunol.* 2001;Appendix 1:Appendix 1C.
22. Duran LW, Horton RM, Birschbach CW, Chang-Miller A, Pease LR. Structural relationships among the H-2 D-regions of murine MHC haplotypes. *J Immunol.* 1989;142(1):288-296.
23. Hedley ML, Forman J, Tucker PW. Mutation of 3' splice sites in two different class I genes results in different usage of cryptic splice sites. *J Immunol.* 1989;143(3):1018-1025.
24. Robinson J, Halliwell JA, Hayhurst JD, Flicek P, Parham P, Marsh SG. The IPD and IMGT/HLA database: allele variant databases. *Nucleic Acids Res.* 2015;43(Database issue):D423-D431.
25. Graham JB, Swarts JL, Leist SR, et al. Baseline T cell immune phenotypes predict virologic and disease control upon SARS-CoV infection in Collaborative Cross mice. *PLoS Pathog.* 2021;17(1):e1009287.
26. Martin MD, Sompallae R, Winborn CS, Harty JT, Badovinac VP. Diverse CD8 T cell responses to viral infection revealed by the Collaborative Cross. *Cell Rep.* 2020;31(2):107508.
27. Gralinski LE, Ferris MT, Aylor DL, et al. Genome wide identification of SARS-CoV susceptibility loci using the Collaborative Cross. *PLoS Genet.* 2015;11(10):e1005504.
28. Aylor DL, Valdar W, Foulds-Mathes W, et al. Genetic analysis of complex traits in the emerging Collaborative Cross. *Genome Res.* 2011;21(8):1213-1222.
29. Jamieson AM, Diefenbach A, McMahon CW, Xiong N, Carlyle JR, Raulet DH. The role of the NKG2D immunoreceptor in immune cell activation and natural killing. *Immunity.* 2002;17(1):19-29.

30. Germain RN, Margulies DH. The biochemistry and cell biology of antigen processing and presentation. *Annu Rev Immunol.* 1993;11:403-450.
31. York IA, Rock KL. Antigen processing and presentation by the class I major histocompatibility complex. *Annu Rev Immunol.* 1996;14:369-396.
32. Yewdell JW, Bennink JR. Cell biology of antigen processing and presentation to major histocompatibility complex class I molecule-restricted T lymphocytes. *Adv Immunol.* 1992;52:1-123.
33. Yewdell JW, Anton LC, Bennink JR. Defective ribosomal products (DRiPs): a major source of antigenic peptides for MHC class I molecules? *J Immunol.* 1996;157(5):1823-1826.
34. Schwab SR, Li KC, Kang C, Shastri N. Constitutive display of cryptic translation products by MHC class I molecules. *Science.* 2003;301(5638):1367-1371.
35. Faridi P, Li C, Ramarathinam SH, et al. A subset of HLA-I peptides are not genomically templated: evidence for cis- and trans-spliced peptide ligands. *Sci Immunol.* 2018;3(28):eaar3947.
36. Procko E, Gaudet R. Antigen processing and presentation: TAPping into ABC transporters. *Curr Opin Immunol.* 2009;21(1):84-91.
37. Nguyen TT, Chang SC, Evnouchidou I, et al. Structural basis for antigenic peptide precursor processing by the endoplasmic reticulum aminopeptidase ERAP1. *Nat Struct Mol Biol.* 2011;18(5):604-613.
38. Li L, Batliwala M, Bouvier M. ERAP1 enzyme-mediated trimming and structural analyses of MHC I-bound precursor peptides yield novel insights into antigen processing and presentation. *J Biol Chem.* 2019;294(49):18534-18544.
39. Mavridis G, Arya R, Domnick A, et al. A systematic re-examination of processing of MHCI-bound antigenic peptide precursors by endoplasmic reticulum aminopeptidase 1. *J Biol Chem.* 2020;295(21):7193-7210.
40. Heemels MT, Ploegh H. Generation, translocation, and presentation of MHC class I-restricted peptides. *Annu Rev Biochem.* 1995;64:463-491.
41. Fisette O, Wingbermühle S, Tampé R, Schäfer LV. Molecular mechanism of peptide editing in the tapasin-MHC I complex. *Sci Rep.* 2016;6:19085.
42. Wearsch PA, Cresswell P. Selective loading of high-affinity peptides onto major histocompatibility complex class I molecules by the tapasin-ERp57 heterodimer. *Nat Immunol.* 2007;8(8):873-881.
43. Garbi N, Tiwari N, Momburg F, Hämmerling GJ. A major role for tapasin as a stabilizer of the TAP peptide transporter and consequences for MHC class I expression. *Eur J Immunol.* 2003;33(1):264-273.
44. Sadasivan B, Lehner PJ, Ortmann B, Spies T, Cresswell P. Roles for calreticulin and a novel glycoprotein, tapasin, in the interaction of MHC class I molecules with TAP. *Immunity.* 1996;5(2):103-114.
45. Dong G, Wearsch PA, Peaper DR, Cresswell P, Reinisch KM. Insights into MHC class I peptide loading from the structure of the tapasin-ERp57 thiol oxidoreductase heterodimer. *Immunity.* 2009;30(1):21-32.
46. Cresswell P, Bangia N, Dick T, Diedrich G. The nature of the MHC class I peptide loading complex. *Immunol Rev.* 1999;172:21-28.
47. Blees A, Januliene D, Hofmann T, et al. Structure of the human MHC-I peptide-loading complex. *Nature.* 2017;551(7681):525-528.
48. Bouvier M, Wiley DC. Structural characterization of a soluble and partially folded class I major histocompatibility heavy chain/beta 2m heterodimer. *Nat Struct Biol.* 1998;5(5):377-384.
49. Bouvier M, Wiley DC. Importance of peptide amino and carboxyl termini to the stability of MHC class I molecules. *Science.* 1994;265(5170):398-402.
50. Ljunggren HG, Stam NJ, Öhlén C, et al. Empty MHC class I molecules come out in the cold. *Nature.* 1990;346(6283):476-480.
51. Kiner E, Willie E, Vijaykumar B, et al. Gut CD4(+) T cell phenotypes are a continuum molded by microbes, not by TH archetypes. *Nat Immunol.* 2021;22(2):216-228.
52. Denzin LK, Hammond C, Cresswell P. HLA-DM interactions with intermediates in HLA-DR maturation and a role for HLA-DM in stabilizing empty HLA-DR molecules. *J Exp Med.* 1996;184(6):2153-2165.
53. Pierre P, Denzin LK, Hammond C, et al. HLA-DM is localized to conventional and unconventional MHC class II-containing endocytic compartments. *Immunity.* 1996;4(3):229-239.
54. Mellins ED, Stern LJ. HLA-DM and HLA-DO, key regulators of MHC-II processing and presentation. *Curr Opin Immunol.* 2014;26:115-122.
55. Yin L, Stern LJ. HLA-DM focuses on conformational flexibility around P1 pocket to catalyze peptide exchange. *Front Immunol.* 2013;4:336.
56. Fremont DH, Crawford F, Marrack P, Hendrickson WA, Kappler J. Crystal structure of mouse H2-M. *Immunity.* 1998;9(3):385-393.
57. Grusby MJ, Glimcher LH. Immune responses in MHC class II-deficient mice. *Annu Rev Immunol.* 1995;13:417-435.
58. Brocke P, Garbi N, Momburg F, Hämmerling GJ. HLA-DM, HLA-DO and tapasin: functional similarities and differences. *Curr Opin Immunol.* 2002;14(1):22-29.
59. Cresswell P. Invariant chain structure and MHC class II function. *Cell.* 1996;84(4):505-507.
60. Amigorena S, Webster P, Drake J, Newcomb J, Cresswell P, Mellman I. Invariant chain cleavage and peptide loading in major histocompatibility complex class II vesicles. *J Exp Med.* 1995;181(5):1729-1741.
61. Thery C, Amigorena S. The cell biology of antigen presentation in dendritic cells. *Curr Opin Immunol.* 2001;13(1):45-51.
62. Newcomb JR, Carboy-Newcomb C, Cresswell P. Trimeric interactions of the invariant chain and its association with major histocompatibility complex class II alpha beta dimers. *J Biol Chem.* 1996;271(39):24249-24256.
63. Sadegh-Nasseri S, Stern LJ, Wiley DC, Germain RN. MHC class II function preserved by low-affinity peptide interactions preceding stable binding. *Nature.* 1994;370(6491):647-650.
64. Denzin LK. Inhibition of HLA-DM mediated MHC class II peptide loading by HLA-DO promotes self tolerance. *Front Immunol.* 2013;4:465.
65. Sadegh-Nasseri S. A step-by-step overview of the dynamic process of epitope selection by major histocompatibility complex class II for presentation to helper T cells. *F1000Res.* 2016;5:F1000.
66. Moore MW, Carbone FR, Bevan MJ. Introduction of soluble protein into the class I pathway of antigen processing and presentation. *Cell.* 1988;54(6):777-785.
67. Cruz FM, Colbert JD, Merino E, Kriegsman BA, Rock KL. The biology and underlying mechanisms of cross-presentation of exogenous antigens on MHC-I molecules. *Annu Rev Immunol.* 2017;35:149-176.
68. Cresswell P, Ackerman AL, Giodini A, Peaper DR, Wearsch PA. Mechanisms of MHC class I-restricted antigen processing and cross-presentation. *Immunol Rev.* 2005;207:145-157.
69. Segura E, Villadangos JA. A modular and combinatorial view of the antigen cross-presentation pathway in dendritic cells. *Traffic.* 2011;12(12):1677-1685.
70. Gascoigne NR, Palmer E. Signaling in thymic selection. *Curr Opin Immunol.* 2011;23(2):207-212.
71. Lio CW, Hsieh CS. Becoming self-aware: the thymic education of regulatory T cells. *Curr Opin Immunol.* 2011;23(2):213-219.
72. Lemonnier FA. The utility of H-2 class I knockout mice. *Virus Res.* 2002;82(1-2):87-90.
73. Salter RD, Benjamin RJ, Wesley PK, et al. A binding site for the T-cell co-receptor CD8 on the alpha 3 domain of HLA-A2. *Nature.* 1990;345(6270):41-46.
74. Connolly JM, Potter TA, Wormstall EM, Hansen TH. The Lyt-2 molecule recognizes residues in the class I alpha 3 domain in allogeneic cytotoxic T cell responses. *J Exp Med.* 1988;168(1):325-341.
75. Norment AM, Salter RD, Parham P, Engelhard VH, Littman DR. Cell-cell adhesion mediated by CD8 and MHC class I molecules. *Nature.* 1988;336(6194):79-81.
76. Wang R, Natarajan K, Margulies DH. Structural basis of the CD8 alpha beta/MHC class I interaction: focused recognition orients CD8 beta to a T cell proximal position. *J Immunol.* 2009;183(4):2554-2564.
77. Liu Y, Cuendet MA, Goffin L, et al. CD8 binding of MHC-peptide complexes in cis or trans regulates CD8(+) T-cell responses. *J Mol Biol.* 2019;431(24):4941-4958.
78. Doyle C, Strominger JL. Interaction between CD4 and class II MHC molecules mediates cell adhesion. *Nature.* 1987;330(6145):256-259.
79. Doyle C, Shin J, Dunbrack RL, Strominger JL. Mutational analysis of the structure and function of the CD4 protein. *Immunol Rev.* 1989;109:17-37.
80. Wang JH, Meijers R, Xiong Y, et al. Crystal structure of the human CD4 N-terminal two-domain fragment complexed to a class II MHC molecule. *Proc Natl Acad Sci U S A.* 2001;98(19):10799-10804.
81. Wang XX, Li Y, Yin Y, et al. Affinity maturation of human CD4 by yeast surface display and crystal structure of a CD4-HLA-DR1 complex. *Proc Natl Acad Sci U S A.* 2011;108(38):15960-15965.
82. Yin Y, Wang XX, Mariuzza RA. Crystal structure of a complete ternary complex of T-cell receptor, peptide-MHC, and CD4. *Proc Natl Acad Sci U S A.* 2012;109(14):5405-5410.
83. Kittipatarin C, Tschammer N, Khaled AR. The interaction of LCK and the CD4 co-receptor alters the dose response of T-cells to interleukin-7. *Immunol Lett.* 2010;131(2):170-181.
84. Trobridge PA, Forbush KA, Levin SD. Positive and negative selection of thymocytes depends on Lck interaction with the CD4 and CD8 coreceptors. *J Immunol.* 2001;166(2):809-818.
85. Luescher IF, Vivier E, Layer A, et al. CD8 modulation of T-cell antigen receptor-ligand interactions on living cytotoxic T lymphocytes. *Nature.* 1995;373(6512):353-356.
86. Arcaro A, Grégoire C, Bakker TR, et al. CD8beta endows CD8 with efficient coreceptor function by coupling T cell receptor/CD3 to raft-associated CD8/p56(lck) complexes. *J Exp Med.* 2001;194(10):1485-1495.
87. Gilfillan CB, Wang C, Mohsen MO, et al. Murine CD8 T-cell functional avidity is stable in vivo but not in vitro: Independence from homologous prime/boost time interval and antigen density. *Eur J Immunol.* 2020;50(4):505-514.
88. Gangadharan D, Cheroutre H. The CD8 isoform CD8αα is not a functional homologue of the TCR co-receptor CD8αβ. *Curr Opin Immunol.* 2004;16(3):264-270.
89. Lanier LL. NK cell receptors. *Annu Rev Immunol.* 1998;16:359-393.
90. Correa I, Raulet DH. Binding of diverse peptides to MHC class I molecules inhibits target cell lysis by activated natural killer cells. *Immunity.* 1995;2(1):61-71.
91. Malnati MS, Peruzzi M, Parker KC, et al. Peptide specificity in the recognition of MHC class I by natural killer cell clones. *Science.* 1995;267(5200):1016-1018.
92. Orihuela M, Margulies D, Yokoyama W. The natural killer cell receptor Ly-49A recognizes a peptide-induced conformational determinant on its major histocompatibility complex class I ligand. *Proc Natl Acad Sci U S A.* 1996;93(21):11792-11797.
93. Chapel A, Garcia-Beltran WF, Hölzemer A, et al. Peptide-specific engagement of the activating NK cell receptor KIR2DS1. *Sci Rep.* 2017;7(1):2414.
94. Karre K. How to recognize a foreign submarine. *Immunol Rev.* 1997;155:5-9.
95. Karre K. NK cells, MHC class I molecules and the missing self. *Scand J Immunol.* 2002;55(3):221-228.
96. Ljunggren HG, Karre K. In search of the 'missing self': MHC molecules and NK cell recognition. *Immunol Today.* 1990;11(7):237-244.
97. Karlhofer FM, Ribaudo RK, Yokoyama WM. MHC class I alloantigen specificity of Ly-49+ IL-2-activated natural killer cells. *Nature.* 1992;358(6381):66-70.
98. Karlhofer FM, Hunziker R, Reichlin A, Margulies DH, Yokoyama WM. Host MHC class I molecules modulate *in vivo* expression of a NK cell receptor. *J Immunol.* 1994;153(6):2407-2416.
99. Dorfman JR, Raulet DH. Acquisition of Ly49 receptor expression by developing natural killer cells. *J Exp Med.* 1998;187(4):609-618.
100. Kim S, Poursine-Laurent J, Truscott SM, et al. Licensing of natural killer cells by host major histocompatibility complex class I molecules. *Nature.* 2005;436(7051):709-713.
101. Kadri N, Thanh TL, Hoglund P. Selection, tuning, and adaptation in mouse NK cell education. *Immunol Rev.* 2015;267(1):167-177.
102. Boudreau JE, Liu XR, Zhao Z, et al. Cell-extrinsic MHC class I molecule engagement augments human NK cell education programmed by cell-intrinsic MHC class I. *Immunity.* 2016;45(2):280-291.
103. Chen S, Li D, Wang Y, Li Q, Dong Z. Regulation of MHC class I-independent NK cell education by SLAM family receptors. *Adv Immunol.* 2020;145:159-185.
104. Brodin P, Karre K, Hoglund P. NK cell education: not an on-off switch but a tunable rheostat. *Trends Immunol.* 2009;30(4):143-149.
105. Orr MT, Lanier LL. Natural killer cell education and tolerance. *Cell.* 2010;142(6):847-856.
106. Anfossi N, André P, Guia S, et al. Human NK cell education by inhibitory receptors for MHC class I. *Immunity.* 2006;25(2):331-342.
107. Boehm U, Klamp T, Groot M, Howard JC. Cellular responses to interferon-gamma. *Annu Rev Immunol.* 1997;15:749-795.
108. Gobin SJ, van den Elsen PJ. Locus-specific regulation of HLA-A and HLA-B expression is not determined by nucleotide variation in the X2 box promoter element. *Blood.* 2001;97(5):1518-1521.
109. van den Elsen PJ, Gobin SJ. The common regulatory pathway of MHC class I and class II transactivation. *Microbes Infect.* 1999;1(11):887-892.
110. Jongsma MLM, Guarda G, Spaapen RM. The regulatory network behind MHC class I expression. *Mol Immunol.* 2019;113:16-21.
111. Cano F, Bye H, Duncan LM, et al. The RNA-binding E3 ubiquitin ligase MEX-3C links ubiquitination with MHC-I mRNA degradation. *EMBO J.* 2012;31(17):3596-3606.
112. Le Bouteiller P, Solier C, Pröll J, Aguerre-Girr M, Fournel S, Lenfant F. Placental HLA-G protein expression in vivo: where and what for? *Hum Reprod Update.* 1999;5(3):223-233.
113. Wainwright SD, Biro PA, Holmes CH. HLA-F is a predominantly empty, intracellular, TAP-associated MHC class Ib protein with a restricted expression pattern. *J Immunol.* 2000;164(1):319-328.
114. Houlihan JM, Biro PA, Harper HM, Jenkinson HJ, Holmes CH. The human amnion is a site of MHC class Ib expression: evidence for the expression of HLA-E and HLA-G. *J Immunol.* 1995;154(11):5665-5674.
115. McMaster MT, Librach CL, Zhou Y, et al. Human placental HLA-G expression is restricted to differentiated cytotrophoblasts. *J Immunol.* 1995;154(8):3771-3778.
116. Gobin SJ, van den Elsen PJ. Transcriptional regulation of the MHC class Ib genes HLA-E, HLA-F, and HLA-G. *Hum Immunol.* 2000;61(11):1102-1107.
117. Hicklin DJ, Wang Z, Arienti F, Rivoltini L, Parmiani G, Ferrone S. beta2-Microglobulin mutations, HLA class I antigen loss, and tumor progression in melanoma. *J Clin Invest.* 1998;101(12):2720-2729.

118. Hsieh CH, Hsu YJ, Chang CC, et al. Total HLA class I loss in a sarcomatoid renal carcinoma cell line caused by the coexistence of distinct mutations in the two encoding beta2-microglobulin genes. *Cancer Immunol Immunother.* 2009;58(3):395-408.
119. Chang CC, Ogino T, Mullins DW, et al. Defective human leukocyte antigen class I-associated antigen presentation caused by a novel beta2-microglobulin loss-of-function in melanoma cells. *J Biol Chem.* 2006;281(27):18763-18773.
120. Zijlstra M, Li E, Sajjadi F, Subramani S, Jaenisch R. Germ-line transmission of a disrupted beta 2-microglobulin gene produced by homologous recombination in embryonic stem cells. *Nature.* 1989;342(6248):435-438.
121. Koller BH, Smithies O. Inactivating the beta 2-microglobulin locus in mouse embryonic stem cells by homologous recombination. *Proc Natl Acad Sci U S A.* 1989;86(22):8932-8935.
122. Wiertz EJ, Mukherjee S, Ploegh HL. Viruses use stealth technology to escape from the host immune system. *Mol Med Today.* 1997;3(3):116-123.
123. Fruh K, Ahn K, Peterson PA. Inhibition of MHC class I antigen presentation by viral proteins. *J Mol Med.* 1997;75(1):18-27.
124. Easterfield AJ, Austen BM, Westwood OM. Inhibition of antigen transport by expression of infected cell peptide 47 (ICP47) prevents cell surface expression of HLA in choriocarcinoma cell lines. *J Reprod Immunol.* 2001;50(1):19-40.
125. Berger C, Xuereb S, Johnson DC, et al. Expression of herpes simplex virus ICP47 and human cytomegalovirus US11 prevents recognition of transgene products by CD8(+) cytotoxic T lymphocytes. *J Virol.* 2000;74(10):4465-4473.
126. Furukawa L, Brevetti LS, Brady SE, et al. Adenoviral-mediated gene transfer of ICP47 inhibits major histocompatibility complex class I expression on vascular cells *in vitro*. *J Vasc Surg.* 2000;31(3):558-566.
127. Jugovic P, Hill AM, Tomazin R, Ploegh H, Johnson DC. Inhibition of major histocompatibility complex class I antigen presentation in pig and primate cells by herpes simplex virus type 1 and 2 ICP47. *J Virol.* 1998;72(6):5076-5084.
128. Byun M, Verweij MC, Pickup DJ, Wiertz EJHJ, Hansen TH, Yokoyama WM. Two mechanistically distinct immune evasion proteins of cowpox virus combine to avoid antiviral CD8 T cells. *Cell Host Microbe.* 2009;6(5):422-432.
129. Alzhanova D, Edwards DM, Hammarlund E, et al. Cowpox virus inhibits the transporter associated with antigen processing to evade T cell recognition. *Cell Host Microbe.* 2009;6(5):433-445.
130. Harvey IB, Wang X, Fremont DH. Molluscum contagiosum virus MC80 sabotages MHC-I antigen presentation by targeting tapasin for ER-associated degradation. *PLoS Pathog.* 2019;15(4):e1007711.
131. de la Salle H, Hanau D, Fricker D, et al. Homozygous human TAP peptide transporter mutation in HLA class I deficiency. *Science.* 1994;265(5169):237-241.
132. Dogu F, Ikincioğullari A, Fricker D, et al. A novel mutation for TAP deficiency and its possible association with Toxoplasmosis. *Parasitol Int.* 2006;55(3):219-222.
133. Shrestha D, Szollosi J, Jenei A. Bare lymphocyte syndrome: an opportunity to discover our immune system. *Immunol Lett.* 2012;141(2):147-157.
134. de la Salle H, Saulquin X, Mansour I, et al. Asymptomatic deficiency in the peptide transporter associated to antigen processing (TAP). *Clin Exp Immunol.* 2002;128(3):525-531.
135. de la Salle H, Houssaint E, Peyrat MA, et al. Human peptide transporter deficiency: importance of HLA-B in the presentation of TAP-independent EBV antigens. *J Immunol.* 1997;158(10):4555-4563.
136. Rehm A, Engelsberg A, Tortorella D, et al. Human cytomegalovirus gene products US2 and US11 differ in their ability to attack major histocompatibility class I heavy chains in dendritic cells. *J Virol.* 2002;76(10):5043-5050.
137. Gold MC, Munks MW, Wagner M, Koszinowski UH, Hill AB, Fling SP. The murine cytomegalovirus immunomodulatory gene m152 prevents recognition of infected cells by M45-specific CTL but does not alter the immunodominance of the M45-specific CD8 T cell response *in vivo*. *J Immunol.* 2002;169(1):359-365.
138. Kavanagh DG, Gold MC, Wagner M, Koszinowski UH, Hill AB. The multiple immune-evasion genes of murine cytomegalovirus are not redundant: m4 and m152 inhibit antigen presentation in a complementary and cooperative fashion. *J Exp Med.* 2001;194(7):967-978.
139. Wills MR, Ashiru O, Reeves MB, et al. Human cytomegalovirus encodes an MHC class I-like molecule (UL142) that functions to inhibit NK cell lysis. *J Immunol.* 2005;175(11):7457-7465.
140. Koszinowski UH, Reddehase MJ, Del Val M. Principles of cytomegalovirus antigen presentation *in vitro* and *in vivo*. *Semin Immunol.* 1992;4(2):71-79.
141. Scalzo AA, Corbett AJ, Rawlinson WD, Scott GM, Degli-Esposti MA. The interplay between host and viral factors in shaping the outcome of cytomegalovirus infection. *Immunol Cell Biol.* 2007;85(1):46-54.
142. Rawlinson WD, Farrell HE, Barrell BG. Analysis of the complete DNA sequence of murine cytomegalovirus. *J Virol.* 1996;70(12):8833-8849.
143. Smith HR, Heusel JW, Mehta IK, et al. Recognition of a virus-encoded ligand by a natural killer cell activation receptor. *Proc Natl Acad Sci U S A.* 2002;99(13):8826-8831.
144. Burgert HG, Kvist S. An adenovirus type 2 glycoprotein blocks cell surface expression of human histocompatibility class I antigens. *Cell.* 1985;41(3):987-997.
145. Burgert HG, Kvist S. The E3/19K protein of adenovirus type 2 binds to the domains of histocompatibility antigens required for CTL recognition. *EMBO J.* 1987;6(7):2019-2026.
146. Sester M, Koebernick K, Owen D, et al. Conserved amino acids within the adenovirus 2 E3/19K protein differentially affect downregulation of MHC class I and MICA/B proteins. *J Immunol.* 2010;184(1):255-267.
147. Menz B, Sester M, Koebernick K, Schmid R, Burgert HG. Structural analysis of the adenovirus type 2 E3/19K protein using mutagenesis and a panel of conformation-sensitive monoclonal antibodies. *Mol Immunol.* 2008;46(1):16-26.
148. Li L, Muzahim Y, Bouvier M. Crystal structure of adenovirus E3-19K bound to HLA-A2 reveals mechanism for immunomodulation. *Nat Struct Mol Biol.* 2012;19(11):1176-1181.
149. Li L, Santarsiero BD, Bouvier M. Structure of the adenovirus type 4 (species E) E3-19K/HLA-A2 complex reveals species-specific features in MHC class I recognition. *J Immunol.* 2016;197(4):1399-1407.
150. Boss JM. Regulation of transcription of MHC class II genes. *Curr Opin Immunol.* 1997;9(1):107-113.
151. Fontes JD, Kanazawa S, Nekrep N, Peterlin BM. The class II transactivator CIITA is a transcriptional integrator. *Microbes Infect.* 1999;1(11):863-869.
152. Jabrane-Ferrat N, Nekrep N, Tosi G, Esserman LJ, Peterlin BM. Major histocompatibility complex class II transcriptional platform: assembly of nuclear factor Y and regulatory factor X (RFX) on DNA requires RFX5 dimers. *Mol Cell Biol.* 2002;22(15):5616-5625.
153. Zinzow-Kramer WM, Long AB, Youngblood BA, et al. CIITA promoter I CARD-deficient mice express functional MHC class II genes in myeloid and lymphoid compartments. *Gene Immun.* 2012;13(4):299-310.
154. Rohn WM, Lee YJ, Benveniste EN. Regulation of class II MHC expression. *Crit Rev Immunol.* 1996;16(3):311-330.
155. Reith W, Mach B. The bare lymphocyte syndrome and the regulation of MHC expression. *Annu Rev Immunol.* 2001;19:331-373.
156. Hume CR, Shookster LA, Collins N, O'Reilly R, Lee JS. Bare lymphocyte syndrome: altered HLA class II expression in B cell lines derived from two patients. *Hum Immunol.* 1989;25(1):1-11.
157. Mach B. MHC class II regulation – lessons from a disease. *N Engl J Med.* 1995;332(2):120-122.
158. Abdulkadir SA, Krishna S, Thanos D, Maniatis T, Strominger JL, Ono SJ. Functional roles of the transcription factor Oct-2A and the high mobility group protein I/Y in HLA-DRA gene expression. *J Exp Med.* 1995;182(2):487-500.
159. Mitchison NA, Mayer W. A survey of H2 gene sequences, including new wild-derived genes. *Int J Immunogenet.* 2007;34(1):3-12.
160. Guardiola J, Maffei A, Lauster R, Mitchison NA, Accolla RS, Sartoris S. Functional significance of polymorphism among MHC class II gene promoters. *Tissue Antigens.* 1996;48(6):615-625.
161. Mitchison NA, Schuhbauer D, Muller B. Natural and induced regulation of Th1/Th2 balance. *Springer Semin Immunopathol.* 1999;21(3):199-210.
162. Stumptner-Cuvelette P, Benaroch P. Multiple roles of the invariant chain in MHC class II function. *Biochim Biophys Acta.* 2002;1542(1-3):1-13.
163. Roche PA, Cresswell P. Invariant chain association with HLA-DR molecules inhibits immunogenic peptide binding. *Nature.* 1990;345(6276):615-618.
164. Roche PA, Marks MS, Cresswell P. Formation of a nine-subunit complex by HLA class-II glycoproteins and the invariant chain. *Nature.* 1991;354(6352):392-394.
165. Bikoff EK, Huang LY, Episkopou V, van Meerwijk J, Germain RN, Robertson EJ. Defective major histocompatibility complex class II assembly, transport, peptide acquisition, and CD4+ T cell selection in mice lacking invariant chain expression. *J Exp Med.* 1993;177(6):1699-1712.
166. Viville S, Neefjes J, Lotteau V, et al. Mice lacking the MHC class II-associated invariant chain. *Cell.* 1993;72(4):635-648.
167. Bonnerot C, Marks MS, Cosson P, et al. Association with BiP and aggregation of class II MHC molecules synthesized in the absence of invariant chain. *EMBO J.* 1994;13(4):934-944.
168. Shiina T, Blancher A, Inoko H, Kulski JK. Comparative genomics of the human, macaque and mouse major histocompatibility complex. *Immunology.* 2017;150(2):127-138.
169. Horton R, Wilming L, Rand V, et al. Gene map of the extended human MHC. *Nat Rev Genet.* 2004;5(12):889-899.
170. de Bakker PI, McVean G, Sabeti PC, et al. A high-resolution HLA and SNP haplotype map for disease association studies in the extended human MHC. *Nat Genet.* 2006;38(10):1166-1172.
171. Horton R, Gibson R, Coggill P, et al. Variation analysis and gene annotation of eight MHC haplotypes: the MHC Haplotype Project. *Immunogenetics.* 2008;60(1):1-18.
172. Monaco JJ, McDevitt HO. H-2-linked low-molecular weight polypeptide antigens assemble into an unusual macromolecular complex. *Nature.* 1984;309(5971):797-799.
173. Martinez CK, Monaco JJ. Post-translational processing of a major histocompatibility complex-encoded proteasome subunit, LMP-2. *Mol Immunol.* 1993;30(13):1177-1183.
174. Monaco JJ, McDevitt HO. The LMP antigens: a stable MHC-controlled multisubunit protein complex. *Hum Immunol.* 1986;15(4):416-426.
175. Suh WK, Mitchell EK, Yang Y, Peterson PA, Waneck GL, Williams DB. MHC class I molecules form ternary complexes with calnexin and TAP and undergo peptide-regulated interaction with TAP via their extracellular domains. *J Exp Med.* 1996;184(2):337-348.
176. Spies T, Cerundolo V, Colonna M, Cresswell P, Townsend A, DeMars R. Presentation of viral antigen by MHC class I molecules is dependent on a putative peptide transporter heterodimer. *Nature.* 1992;355(6361):644-646.
177. Powis SJ. Major histocompatibility complex class I molecules interact with both subunits of the transporter associated with antigen processing, TAP1 and TAP2. *Eur J Immunol.* 1997;27(10):2744-2747.
178. Hill A, Ploegh H. Getting the inside out: the transporter associated with antigen processing (TAP) and the presentation of viral antigen. *Proc Natl Acad Sci U S A.* 1995;92(2):341-343.
179. Yang Y, Sempe P, Peterson PA. Molecular mechanisms of class I major histocompatibility complex antigen processing and presentation. *Immunol Res.* 1996;15(3):208-233.
180. Collins GA, Goldberg AL. The logic of the 26S proteasome. *Cell.* 2017;169(5):792-806.
181. Driscoll J, Brown MG, Finley D, Monaco JJ. MHC-linked LMP gene products specifically alter peptidase activities of the proteasome. *Nature.* 1993;365(6443):262-264.
182. Nandi D, Jiang H, Monaco JJ. Identification of MECL-1 (LMP-10) as the third IFN-gamma-inducible proteasome subunit. *J Immunol.* 1996;156(7):2361-2364.
183. Kuckelkorn U, Frentzel S, Kraft R, Kostka S, Groettrup M, Kloetzel PM. Incorporation of major histocompatibility complex – Encoded subunits LMP2 and LMP7 changes the quality of the 20S proteasome polypeptide processing products independent of interferon-gamma. *Eur J Immunol.* 1995;25(9):2605-2611.
184. Ehring B, Meyer TH, Eckerskorn C, Lottspeich F, Tampé R. Effects of major-histocompatibility-complex-encoded subunits on the peptidase and proteolytic activities of human 20S proteasomes. Cleavage of proteins and antigenic peptides. *Eur J Biochem.* 1996;235(1-2):404-415.
185. Elliott T, Willis A, Cerundolo V, Townsend A. Processing of major histocompatibility class I-restricted antigens in the endoplasmic reticulum. *J Exp Med.* 1995;181(4):1481-1491.
186. Velarde G, Ford RC, Rosenberg MF, Powis SJ. Three-dimensional structure of transporter associated with antigen processing (TAP) obtained by single Particle image analysis. *J Biol Chem.* 2001;276(49):46054-46063.
187. Lehnert E, Tampe R. Structure and dynamics of antigenic peptides in complex with TAP. *Front Immunol.* 2017;8:10.
188. Noll A, Thomas C, Herbring V, et al. Crystal structure and mechanistic basis of a functional homolog of the antigen transporter TAP. *Proc Natl Acad Sci U S A.* 2017;114(4):E438-E447.
189. Powis SJ, Young LL, Joly E, et al. The rat cim effect: TAP allele-dependent changes in a class I MHC anchor motif and evidence against C-terminal trimming of peptides in the ER. *Immunity.* 1996;4(2):159-165.
190. Livingstone AM, Powis SJ, Diamond AG, Butcher GW, Howard JC. A trans-acting major histocompatibility complex-linked gene whose alleles determine gain and loss changes in the antigenic structure of a classical class I molecule. *J Exp Med.* 1989;170(3):777-795.
191. Livingstone AM, Powis SJ, Günther E, Cramer DV, Howard JC, Butcher GW. Cim: an MHC class II-linked allelism affecting the antigenicity of a classical class I molecule for T lymphocytes. *Immunogenetics.* 1991;34(3):157-163.
192. Abitew AM, Sobti RC, Sharma VL, Wanchu A. Analysis of transporter associated with antigen presentation (TAP) genes polymorphisms with HIV-1 infection. *Mol Cell Biochem.* 2020;464(1-2):65-71.
193. Caversaccio M, Bonél HM, Carter R, Williams AP, Gadola SD. TAP deficiency syndrome: chronic rhinosinusitis and conductive hearing loss. *Eur Arch Otorhinolaryngol.* 2008;265(10):1289-1292.
194. Parissiadis A, Dormoy A, Fricker D, et al. Unilateral necrotising toxoplasmic retinochoroiditis as the main clinical manifestation of a peptide transporter (TAP) deficiency. *Br J Ophthalmol.* 2005;89(12):1661-1662.
195. Hanna J, Mussaffi H, Steuer G, et al. Functional aberrant expression of CCR2 receptor on chronically activated NK cells in patients with TAP-2 deficiency. *Blood.* 2005;106(10):3465-3473.
196. Hanalioglu D, Ayvaz DC, Ozgur TT, van der Burg M, Sanal O, Tezcan I. A novel mutation in TAP1 gene leading to MHC class I deficiency: report of two cases and review of the literature. *Clin Immunol.* 2017;178:74-78.

197. Gadola SD, Moins-Teisserenc HT, Trowsdale J, Gross WL, Cerundolo V. TAP deficiency syndrome. *Clin Exp Immunol.* 2000;121(2):173-178.
198. White PC, Chaplin DD, Weis JH, Dupont B, New MI, Seidman JG. Two steroid 21-hydroxylase genes are located in the murine S region. *Nature.* 1984;312(5993):465-467.
199. Layrisse Z, White C, Gunczler P, et al. Sharing of MHC haplotypes among apparently unrelated patients with congenital adrenal hyperplasia due to 21-hydroxylase deficiency. *Immunogenetics.* 1987;25(2):99-103.
200. Lawton P, Nelson J, Tizard R, Browning JL. Characterization of the mouse lymphotoxin-beta gene. *J Immunol.* 1995;154(1):239-246.
201. Browning JL, Ngam-ek A, Lawton P, et al. Lymphotoxin beta, a novel member of the TNF family that forms a heteromeric complex with lymphotoxin on the cell surface. *Cell.* 1993;72(6):847-856.
202. Iraqi F, Teale A. Polymorphisms in the Tnfa gene of different inbred mouse strains. *Immunogenetics.* 1999;49(3):242-245.
203. Janer M, Geraghty DE. The human major histocompatibility complex: 42,221 bp of genomic sequence, high-density sequence-tagged site map, evolution, and polymorphism for HLA class I. *Genomics.* 1998;51(1):35-44.
204. Venditti CP, Harris JM, Geraghty DE, Chorney MJ. Mapping and characterization of non-HLA multigene assemblages in the human MHC class I region. *Genomics.* 1994;22(2):257-266.
205. White PC, New MI, Dupont B. Adrenal 21-hydroxylase cytochrome P-450 genes within the MHC class III region. *Immunol Rev.* 1985;87:123-150.
206. Sim MJW, Lu J, Spencer M, et al. High-affinity oligoclonal TCRs define effective adoptive T cell therapy targeting mutant KRAS-G12D. *Proc Natl Acad Sci U S A.* 2020;117(23):12826-12835.
207. Dehn J, Spellman S, Hurley CK, et al. Selection of unrelated donors and cord blood units for hematopoietic cell transplantation: guidelines from the NMDP/CIBMTR. *Blood.* 2019;134(12):924-934.
208. Kumar S, Sarkar P, Sim MJW, Rajagopalan S, Vogel SS, Long EO. A single amino acid change in inhibitory killer cell Ig-like receptor results in constitutive receptor self-association and phosphorylation. *J Immunol.* 2015;194(2):817-826.
209. Vales-Gomez M, Reyburn HT, Erskine RA, Lopez-Botet M, Strominger JL. Kinetics and peptide dependency of the binding of the inhibitory NK receptor CD94/NKG2-A and the activating receptor CD94/NKG2-C to HLA-E. *EMBO J.* 1999;18(15):4250-4260.
210. Vales-Gomez M, Reyburn HT, Mandelboim M, Strominger JL. Kinetics of interaction of HLA-C ligands with natural killer cell inhibitory receptors. *Immunity.* 1998;9(3):337-344.
211. Fan QR, Garboczi DN, Winter CC, Wagtmann N, Long EO, Wiley DC. Direct binding of a soluble natural killer cell inhibitory receptor to a soluble human leukocyte antigen-Cw4 class I major histocompatibility complex molecule. *Proc Natl Acad Scie U S A.* 1996;93(14):7178-7183.
212. Fan QR, Long EO, Wiley DC. Crystal structure of the human natural killer cell inhibitory receptor KIR2DL1-HLA-Cw4 complex. *Nat Immunol.* 2001;2(5):452-460.
213. Boyington JC, Motyka SA, Schuck P, Brooks AG, Sun PD. Crystal structure of an NK cell immunoglobulin-like receptor in complex with its class I MHC ligand. *Nature.* 2000;405(6786):537-543.
214. Stewart-Jones GB, di Gleria K, Kollnberger S, McMichael AJ, Jones EY, Bowness P. Crystal structures and KIR3DL1 recognition of three immunodominant viral peptides complexed to HLA-B*2705. *Eur J Immunol.* 2005;35(2):341-351.
215. Vivian JP, Duncan RC, Berry R, et al. Killer cell immunoglobulin-like receptor 3DL1-mediated recognition of human leukocyte antigen B. *Nature.* 2011;479(7373):401-405.
216. Braud VM, Allan DSJ, O'Callaghan CA, et al. HLA-E binds to natural killer cell receptors CD94/NKG2A, B and C. *Nature.* 1998;391(6669):795-799.
217. Braud V, Jones EY, McMichael A. The human major histocompatibility complex class Ib molecule HLA-E binds signal sequence-derived peptides with primary anchor residues at positions 2 and 9. *Eur J Immunol.* 1997;27(5):1164-1169.
218. Aldrich CJ, DeCloux A, Woods AS, Cotter RJ, Soloski MJ, Forman J. Identification of a Tap-dependent leader peptide recognized by alloreactive T cells specific for a class Ib antigen. *Cell.* 1994;79(4):649-658.
219. Soloski MJ, Decloux A, Aldrich CJ, Forman J. Structural and functional characteristics of the class IB molecule, Qa-1. *Immunol Rev.* 1995;147:67-89.
220. Kurepa Z, Hasemann CA, Forman J. Qa-1b binds conserved class I leader peptides derived from several mammalian species. *J Exp Med.* 1998;188(5):973-978.
221. Strong RK, Holmes MA, Li P, Braun L, Lee N, Geraghty DE. HLA-E allelic variants. Correlating differential expression, peptide affinities, crystal structures, and thermal stabilities. *J Biol Chem.* 2003;278(7):5082-5090.
222. Grant EJ, Nguyen AT, Lobos CA, Szeto C, Chatzileontiadou DSM, Gras S. The unconventional role of HLA-E: the road less traveled. *Mol Immunol.* 2020;120:101-112.
223. Hackmon R, Pinnaduwage L, Zhang J, Lye SJ, Geraghty DE, Dunk CE. Definitive class I human leukocyte antigen expression in gestational placentation: HLA-F, HLA-E, HLA-C, and HLA-G in extravillous trophoblast invasion on placentation, pregnancy, and parturition. *Am J Reprod Immunol.* 2017;77(6):e12643.
224. King A, Allan DSJ, Bowen M, et al. HLA-E is expressed on trophoblast and interacts with CD94/NKG2 receptors on decidual NK cells. *Eur J Immunol.* 2000;30(6):1623-1631.
225. Sim MJW, Sun PD. HLA-F: a new kid licensed for peptide presentation. *Immunity.* 2017;46(6):972-974.
226. Le Bouteiller P, Lenfant F. Antigen-presenting function(s) of the non-classical HLA-E, -F and -G class I molecules: the beginning of a story. *Res Immunol.* 1996;147(5):301-313.
227. Burian A, Wang KL, Finton KAK, et al. HLA-F and MHC-I open conformers bind natural killer cell Ig-like receptor KIR3DS1. *PLoS One.* 2016;11(9):e0163297.
228. Fournel S, Aguerre-Girr M, Huc X, et al. Cutting edge: soluble HLA-G1 triggers CD95/CD95 ligand-mediated apoptosis in activated CD8+ cells by interacting with CD8. *J Immunol.* 2000;164(12):6100-6104.
229. Mercier B, Mura C, Ferec C. Putting a hold on 'HLA-H'. *Nat Genet.* 1997;15(3):234.
230. Bodmer JG, Parham P, Albert ED, Marsh SGE. Putting a hold on 'HLA-H'. The WHO nomenclature committee for factors of the HLA system. *Nat Genet.* 1997;15(3):234-235.
231. Gerhard GS, Ten Elshof AE, Chorney MJ. Hereditary haemochromatosis as an immunological disease. *Br J Haematol.* 1998;100(2):247-255.
232. Gerhard GS, Levin KA, Goldstein JP, Wojnar MM, Chorney MJ, Belchis DA. Vibrio vulnificus septicemia in a patient with the hemochromatosis HFE C282Y mutation. *Arch Pathol Lab Med.* 2001;125(8):1107-1109.
233. Feder JN, Gnirke A, Thomas W, et al. A novel MHC class I-like gene is mutated in patients with hereditary haemochromatosis. *Nat Genet.* 1996;13(4):399-408.
234. Adams PD, Afonine PV, Bunkóczi G, et al. PHENIX: a comprehensive Python-based system for macromolecular structure solution. *Acta Crystallogr D Biol Crystallogr.* 2010;66(pt 2):213-221.
235. Hanson EH, Shue PM, Palm-Leis A, Rowley RK. An aviator with cardiomyopathy and genetic susceptibility to hereditary hemochromatosis: a case report. *Aviat Space Environ Med.* 2001;72(10):924-927.
236. Fleming RE, Ahmann JR, Migas MC, et al. Targeted mutagenesis of the murine transferrin receptor-2 gene produces hemochromatosis. *Proc Natl Acad Sci U S A.* 2002;99(16):10653-10658.
237. Cardoso CS, de Sousa M. HFE, the MHC and hemochromatosis: paradigm for an extended function for MHC class I. *Tissue Antigens.* 2003;61(4):263-275.
238. Press RD, Flora K, Gross C, Rabkin JM, Corless CL. Hepatic iron overload: direct HFE (HLA-H) mutation analysis vs quantitative iron assays for the diagnosis of hereditary hemochromatosis. *Am J Clin Pathol.* 1998;109(5):577-584.
239. Madeja Z, Yadi H, Apps R, et al. Paternal MHC expression on mouse trophoblast affects uterine vascularization and fetal growth. *Proc Natl Acad Sci U S A.* 2011;108(10):4012-4017.
240. Moffett A, Hiby S, Carrington M. The incidence of pre-eclampsia among couples consisting of Japanese women and Caucasian men'. *J Reprod Immunol.* 2006;71(2):132-133; author reply 134.
241. Moffett A, Hiby SE. How Does the maternal immune system contribute to the development of pre-eclampsia? *Placenta.* 2007;28(suppl A):S51-S56.
242. Klein J. *Natural History of the Major Histocompatibility Complex.* Wiley-Interscience; 1986.
243. Lawlor DA, Zemmour J, Ennis PD, Parham P. Evolution of class-I MHC genes and proteins: from natural selection to thymic selection. *Annu Rev Immunol.* 1990;8:23-63.
244. Gunther E, Walter L. The major histocompatibility complex of the rat (*Rattus norvegicus*). *Immunogenetics.* 2001;53(7):520-542.
245. Vincek V, Nizetić D, Golubić M, Figueroa F, Nevo E, Klein J. Evolutionary expansion of Mhc class I loci in the mole-rat, Spalax ehrenbergi. *Mol Biol Evol.* 1987;4(5):483-491.
246. Bodmer WF. HLA structure and function: a contemporary view. *Tissue Antigens.* 1981;17(1):9-20.
247. Pease LR, Nathenson SG, Leinwand LA. Mapping class I gene sequences in the major histocompatibility complex. *Nature.* 1982;298(5872):382-385.
248. Margulies DH, Evans GA, Flaherty L, Seidman JG. H-2-like genes in the Tla region of mouse chromosome 17. *Nature.* 1982;295(5845):168-170.
249. Hood L, Steinmetz M, Malissen B. Genes of the major histocompatibility complex of the mouse. *Annu Rev Immunol.* 1983;1:529-568.
250. Shawar SM, Vyas JM, Rodgers JR, Rich RR. Antigen presentation by major histocompatibility complex class I-B molecules. *Annu Rev Immunol.* 1994;12:839-880.
251. Kurlander RJ, Shawar SM, Brown ML, Rich RR. Specialized role for a murine class I-b MHC molecule in prokaryotic host defenses. *Science.* 1992;257(5070):678-679.
252. Lindahl KF, Byers DE, Dabhi VM, et al. H2-M3, a full-service class Ib histocompatibility antigen. *Annu Rev Immunol.* 1997;15:851-879.
253. Walter L, Gunther E. Physical mapping and evolution of the centromeric class I gene-containing region of the rat MHC. *Immunogenetics.* 2000;51(10):829-837.
254. Leinders-Zufall T, Ishii T, Chamero P, et al. A family of nonclassical class I MHC genes contributes to ultrasensitive chemodetection by mouse vomeronasal sensory neurons. *J Neurosci.* 2014;34(15):5121-5133.
255. Moore CB, John M, James IR, Christiansen FT, Witt CS, Mallal SA. Evidence of HIV-1 adaptation to HLA-restricted immune responses at a population level. *Science.* 2002;296(5572):1439-1443.
256. Michael NL. Host genetics and HIV--removing the mask. *Nat Med.* 2002;8(8):783-785.
257. Martin MP, Gao X, Lee JH, et al. Epistatic interaction between KIR3DS1 and HLA-B delays the progression to AIDS. *Nat Genet.* 2002;31(4):429-434.
258. Zernich D, Purcell AW, Macdonald WA, et al. Natural HLA class I polymorphism controls the pathway of antigen presentation and susceptibility to viral evasion. *J Exp Med.* 2004;200(1):13-24.
259. Figueroa F, Tichy H, Berry RJ, Klein J. MHC polymorphism in island populations of mice. *Curr Top Microbiol Immunol.* 1986;127:100-105.
260. Sultmann H, Mayer WE, Figueroa F, O'hUigin C, Klein J. Zebrafish Mhc class II alpha chain-encoding genes: polymorphism, expression, and function. *Immunogenetics.* 1993;38(6):408-420.
261. Sato A, Klein D, Sültmann H, O'hUigin C, Figueroa F, Klein J. Class I mhc genes of cichlid fishes: identification, expression, and polymorphism. *Immunogenetics.* 1997;46(1):63-72.
262. Sultmann H, Murray BW, Klein J. Identification of seven genes in the major histocompatibility complex class I region of the zebrafish. *Scand J Immunol.* 2000;51(6):577-585.
263. Sato A, Dongak R, Hao L, et al. Mhc class I genes of the cichlid fish *Oreochromis niloticus*. *Immunogenetics.* 2006;58(11):917-928.
264. Chappell P, Meziane el K, Harrison M, et al. Expression levels of MHC class I molecules are inversely correlated with promiscuity of peptide binding. *Elife.* 2015;4:e05345.
265. Turner SJ, Doherty PC, McCluskey J, Rossjohn J. Structural determinants of T-cell receptor bias in immunity. *Nat Rev Immunol.* 2006;6(12):883-894.
266. Williams AF. Immunoglobulin-related domains for cell surface recognition. *Nature.* 1985;314(6012):579-580.
267. Sun YH, Goodenow RS, Hood L. Molecular basis of the dm1 mutation in the major histocompatibility complex of the mouse: a D/L hybrid gene. *J Exp Med.* 1985;162(5):1588-1602.
268. Burnside SS, Hunt P, Ozato K, Sears DW. A molecular hybrid of the H-2Dd and H-2Ld genes expressed in the dm1 mutant. *Proc Natl Acad Sci U S A.* 1984;81(16):5204-5208.
269. Ohta T. Gene conversion vs point mutation in generating variability at the antigen recognition site of major histocompatibility complex loci. *J Mol Evol.* 1995;41(2):115-119.
270. Minias P, Bateson ZW, Whittingham LA, Johnson JA, Oyler-McCance S, Dunn PO. Contrasting evolutionary histories of MHC class I and class II loci in grouse – Effects of selection and gene conversion. *Heredity.* 2016;116(5):466-476.
271. Bahr A, Wilson AB. The evolution of MHC diversity: evidence of intralocus gene conversion and recombination in a single-locus system. *Gene.* 2012;497(1):52-57.
272. Geliebter J, Zeff RA, Schulze DH, et al. Interaction between Kb and Q4 gene sequences generates the Kbm6 mutation. *Mol Cell Biol.* 1986;6(2):645-652.
273. Schulze DH, Pease LR, Geier SS, et al. Comparison of the cloned H-2Kbm1 variant gene with the H-2Kb gene shows a cluster of seven nucleotide differences. *Proc Natl Acad Sci U S A.* 1983;80(7):2007-2011.
274. Evans GA, Margulies DH, Camerini-Otero RD, Ozato K, Seidman JG. Structure and expression of a mouse major histocompatibility antigen gene, H-2Ld. *Proc Natl Acad Sci U S A.* 1982;79(6):1994-1998.
275. Nathenson SG, Geliebter J, Pfaffenbach GM, Zeff RA. Murine major histocompatibility complex class-I mutants: molecular analysis and structure-function implications. *Annu Rev Immunol.* 1986;4:471-502.
276. Hasenkrug KJ, Nathenson SG. Nucleic acid sequences of the H-2Ks and H-2Ksm1 genes. *Immunogenetics.* 1991;34(1):60-61.
277. Reusch TB, Schaschl H, Wegner KM. Recent duplication and inter-locus gene conversion in major histocompatibility class II genes in a teleost, the three-spined stickleback. *Immunogenetics.* 2004;56(6):427-437.
278. Spurgin LG, Van Oosterhout C, Illera JC, et al. Gene conversion rapidly generates major histocompatibility complex diversity in recently founded bird populations. *Mol Ecol.* 2011;20(24):5213-5225.
279. Neehus AL, Wistuba J, Ladas N, Eiz-Vesper B, Schlatt S, Müller T. Gene conversion of the major histocompatibility complex class I Caja-G in common marmosets (*Callithrix jacchus*). *Immunology.* 2016;149(3):343-352.
280. Nathenson SG, Kesari K, Sheil JM, Ajitkumar P. Use of mutants to analyze regions on the H-2Kb molecule for interaction with immune receptors. *Cold Spring Harb Symp Quant Biol.* 1989;54(pt 1):521-528.

281. Rudolph MG, Speir JA, Brunmark A, et al. The crystal structures of K(bm1) and K(bm8) reveal that subtle changes in the peptide environment impact thermostability and alloreactivity. *Immunity.* 2001;14(3):231-242.
282. Rajan TV, Halay ED, Potter TA, Evans GA, Seidman JG, Margulies DH. H-2 hemizygous mutants from a heterozygous cell line: role of mitotic recombination. *EMBO J.* 1983;2(9):1537-1542.
283. Meyer D, Thomson G. How selection shapes variation of the human major histocompatibility complex: a review. *Ann Hum Genet.* 2001;65(pt 1):1-26.
284. Hedrick PW, Klitz W. Evolution of the human MHC: new haplotype frequency analysis is not informative. *Proc Natl Acad Sci U S A.* 2019;116(47):23386-23387.
285. Garrigan D, Hedrick PW. Perspective. Detecting adaptive molecular polymorphism: lessons from the MHC. *Evolution.* 2003;57(8):1707-1722.
286. Lobkovsky AE, Levi L, Wolf YI, et al. Multiplicative fitness, rapid haplotype discovery, and fitness decay explain evolution of human MHC. *Proc Natl Acad Sci U S A.* 2019;116(28):14098-14104.
287. Bodmer WF. Evolutionary significance of the HL-A system. *Nature.* 1972;237(5351):139-145 (passim).
288. Trachtenberg E, Korber B, Sollars C, et al. Advantage of rare HLA supertype in HIV disease progression. *Nat Med.* 2003;9(7):928-935.
289. Phillips KP, Cable J, Mohammed RS, et al. Immunogenetic novelty confers a selective advantage in host-pathogen coevolution. *Proc Natl Acad Sci U S A.* 2018;115(7):1552-1557.
290. Hughes AL, Nei M. Maintenance of MHC polymorphism. *Nature.* 1992;355(6359):402-403.
291. Penn DJ, Damjanovich K, Potts WK. MHC heterozygosity confers a selective advantage against multiple-strain infections. *Proc Natl Acad Sci U S A.* 2002;99(17):11260-11264.
292. Davis DM. *The Compatibility Gene.* Oxford; 2014:232.
293. Silver LM. *Mouse Genetics: Concepts and Applications.* Oxford University Press; 1995:362.
294. Roopenian DC, Davis AP. Responses against antigens encoded by the H-3 histocompatibility locus: antigens stimulating class I MHC- and class II MHC-restricted T cells are encoded by separate genes. *Immunogenetics.* 1989;30(5):335-343.
295. den Haan JM, Meadows LM, Wang W, et al. The minor histocompatibility antigen HA-1: a diallelic gene with a single amino acid polymorphism. *Science.* 1998;279(5353):1054-1057.
296. Simpson E. Minor transplantation antigens: mouse models for human host-versus-graft, graft-versus-host and graft-versus-leukemia reactions. *Arch Immunol Ther Exp.* 1998;46(6):331-339.
297. Simpson E, Scott D, James E, et al. Minor H antigens: genes and peptides. *Transpl Immunol.* 2002;10(2-3):115-123.
298. Mendoza LM, Villaflor G, Eden P, Roopenian D, Shastri N. Distinguishing self from nonself: immunogenicity of the murine H47 locus is determined by a single amino acid substitution in an unusual peptide. *J Immunol.* 2001;166(7):4438-4445.
299. Malarkannan S, Shih PP, Eden PA, et al. The molecular and functional characterization of a dominant minor H antigen, H60. *J Immunol.* 1998;161(7):3501-3509.
300. Eisenlohr LC, Huang L, Golovina TN. Rethinking peptide supply to MHC class I molecules. *Nat Rev Immunol.* 2007;7(5):403-410.
301. Wei J, Yewdell JW. Flu DRiPs in MHC class I immunosurveillance. *Virol Sin.* 2019;34(2):162-167.
302. Yewdell JW, Holly J. DRiPs get molecular. *Curr Opin Immunol.* 2020;64:130-136.
303. Oostvogels R, Lokhorst HM, Mutis T. Minor histocompatibility Ags: identification strategies, clinical results and translational perspectives. *Bone Marrow Transplant.* 2016;51(2):163-171.
304. Goulmy E. Minor histocompatibility antigens: from transplantation problems to therapy of cancer. *Hum Immunol.* 2006;67(6):433-438.
305. McKenzie IF, Morgan GM, Sandrin MS, Michaelides MM, Melvold RW, Kohn HI. B6.C-H-2bm12. A new H-2 mutation in the I region in the mouse. *J Exp Med.* 1979;150(6):1323-1338.
306. Kohn HI, Melvold RW. Spontaneous histocompatibility mutations detected by dermal grafts: significant changes in rate over a 10-year period in the mouse H-system. *Mutat Res.* 1974;24(2):163-169.
307. Yamaga KM, Pfaffenbach GM, Pease LR, et al. Biochemical studies of H-2K antigens from a group of related mutants. I. Identification of a shared mutation in B6-H-2bm5 and B6-H-2bm16. *Immunogenetics.* 1983;17(1):19-29.
308. Pease LR, Ewenstein BM, McGovern D, Melvold RW, Nisizawa T, Nathenso SG. Biochemical studies on the H-2K mutant B6.C-H-2bm10. *Immunogenetics.* 1983;17(1):7-17.
309. Ewenstein BM, Uehara H, Nisizawa T, Melvold RW, Kohn HI, Nathenson SG. Biochemical studies on the H-2K antigens of the MHC mutants bm3 and bm11. *Immunogenetics.* 1980;11(4):383-395.
310. Zinkernagel RM, Doherty PC. Restriction of *in vitro* T cell-mediated cytotoxicity in lymphocytic choriomeningitis within a syngeneic or semiallogeneic system. *Nature.* 1974;248(5450):701-702.
311. Lindahl KF, Wilson DB. Histocompatibility antigen-activated cytotoxic T lymphocytes. I. Estimates of the absolute frequency of killer cells generated *in vitro*. *J Exp Med.* 1977;145(3):500-507.
312. Matzinger P, Bevan MJ. Hypothesis: why do so many lymphocytes respond to major histocompatibility antigens? *Cell Immunol.* 1977;29(1):1-5.
313. Maryanski JL, MacDonald HR, Cerottini JC. Limiting dilution analysis of alloantigen-reactive T lymphocytes. IV. High frequency of cytolytic T lymphocyte precursor cells in MLC blasts separated by velocity sedimentation. *J Immunol.* 1980;124(1):42-47.
314. Dalchau R, Fangmann J, Fabre JW. Allorecognition of isolated, denatured chains of class I and class II major histocompatibility complex molecules. Evidence for an important role for indirect allorecognition in transplantation. *Eur J Immunol.* 1992;22(3):669-677.
315. Colf LA, Bankovich AJ, Hanick NA, et al. How a single T cell receptor recognizes both self and foreign MHC. *Cell.* 2007;129(1):135-146.
316. Macdonald WA, Chen Z, Gras S, et al. T cell allorecognition via molecular mimicry. *Immunity.* 2009;31(6):897-908.
317. Speir JA, Garcia KC, Brunmark A, et al. Structural basis of 2C TCR allorecognition of H-2Ld peptide complexes. *Immunity.* 1998;8(5):553-562.
318. Auchincloss H Jr, Sultan H. Antigen processing and presentation in transplantation. *Curr Opin Immunol.* 1996;8(5):681-687.
319. Tiercy JM. Molecular basis of HLA polymorphism: implications in clinical transplantation. *Transpl Immunol.* 2002;9(2-4):173-180.
320. Moffett A. NK cell allorecognition. *Nat Rev Immunol.* 2017;17(8):466.
321. Moffett A, Colucci F. Uterine NK cells: active regulators at the maternal-fetal interface. *J Clin Invest.* 2014;124(5):1872-1879.
322. Parham P, Moffett A. Variable NK cell receptors and their MHC class I ligands in immunity, reproduction and human evolution. *Nat Rev Immunol.* 2013;13(2):133-144.
323. Trowsdale J, Moffett A. NK receptor interactions with MHC class I molecules in pregnancy. *Semin Immunol.* 2008;20(6):317-320.
324. Patel R, Terasaki PI. Significance of the positive crossmatch test in kidney transplantation. *N Engl J Med.* 1969;280(14):735-739.
325. Liwski RS, Greenshields AL, Conrad DM, et al. Rapid optimized flow cytometric crossmatch (FCXM) assays: the Halifax and Halifaster protocols. *Hum Immunol.* 2018;79(1):28-38.
326. Tait BD. Detection of HLA antibodies in organ transplant recipients - Triumphs and challenges of the solid phase bead assay. *Front Immunol.* 2016;7:570.
327. Morris AB, Sullivan HC, Krummey SM, Gebel HM, Bray RA. Out with the old, in with the new: virtual versus physical crossmatching in the modern era. *HLA.* 2019;94(6):471-481.
328. Parsons RF, Locke JE, Redfield RR, Roll GR, Levine MH. Kidney transplantation of highly sensitized recipients under the new kidney allocation system: a reflection from five different transplant centers across the United States. *Hum Immunol.* 2017;78(1):30-36.
329. Murphey CL, Bingaman AW. Histocompatibility considerations for kidney paired donor exchange programs. *Curr Opin Organ Transplant.* 2012;17(4):427-432.
330. Duquesnoy RJ. HLA epitope based matching for transplantation. *Transpl Immunol.* 2014;31(1):1-6.
331. O'Rafferty C, Rooney G, Hagan R, Woolfson M, O'Donghaile D, Fitzgerald J. HLAMatchmaker is effective for selecting appropriate platelet units for alloimmunised thrombocytopaenic patients who are refractory to random donor platelets. *Transfus Med.* 2017;27(suppl 5):369-374.
332. Petersdorf EW. In celebration of Ruggero Ceppellini: HLA in transplantation. *HLA.* 2017;89(2):71-76.
333. Mickelson EM, Petersdorf E, Anasetti C, Martin P, Woolfrey A, Hansen JA. HLA matching in hematopoietic cell transplantation. *Hum Immunol.* 2000;61(2):92-100.
334. Hansen JA, Yamamoto K, Petersdorf E, Sasazuki T. The role of HLA matching in hematopoietic cell transplantation. *Rev Immunogenet.* 1999;1(3):359-373.
335. Varney MD, Lester S, McCluskey J, Gao X, Tait BD. Matching for HLA DPA1 and DPB1 alleles in unrelated bone marrow transplantation. *Hum Immunol.* 1999;60(6):532-538.
336. Petersdorf EW, Gooley T, Malkki M, et al. The biological significance of HLA-DP gene variation in haematopoietic cell transplantation. *Br J Haematol.* 2001;112(4):988-994.
337. Hansen JA, Petersdorf E, Martin PJ, Woolfry A, Anasetti C. Impact of HLA matching on hematopoietic cell transplants from unrelated donors. *Vox Sanguinis.* 2000;78(suppl 2):269-271.
338. Kollman C, Spellman SR, Zhang MJ, et al. The effect of donor characteristics on survival after unrelated donor transplantation for hematologic malignancy. *Blood.* 2016;127(2):260-267.
339. Morishima Y, Sasazuki T, Inoko H, et al. The clinical significance of human leukocyte antigen (HLA) allele compatibility in patients receiving a marrow transplant from serologically HLA-A, HLA-B, and HLA-DR matched unrelated donors. *Blood.* 2002;99(11):4200-4206.
340. Petersdorf EW, Hansen JA, Martin PJ, et al. Major-histocompatibility-complex class I alleles and antigens in hematopoietic-cell transplantation. *N Engl J Med.* 2001;345(25):1794-1800.
341. Petersdorf EW. Role of major histocompatibility complex variation in graft-versus-host disease after hematopoietic cell transplantation. *F1000Res.* 2017;6:617.
342. Petersdorf EW. Optimal HLA matching in hematopoietic cell transplantation. *Curr Opin Immunol.* 2008;20(5):588-593.
343. Gluckman E, Rocha V, Chevret S. Results of unrelated umbilical cord blood hematopoietic stem cell transplant. *Transfus Clin Biol.* 2001;8(3):146-154.
344. Rubinstein P. HLA matching for bone marrow transplantation--how much is enough? *N Engl J Med.* 2001;345(25):1842-1844.
345. Hakenberg P, Kogler G, Wernet P. NETCORD: a cord blood allocation network. *Bone Marrow Transplant.* 1998;22(suppl 1):S17-S18.
346. Kekre N, Mak KS, Stopsack KH, et al. Impact of HLA-mismatch in unrelated donor hematopoietic stem cell transplantation: a meta-analysis. *Am J Hematol.* 2016;91(6):551-555.
347. Fernandez-Vina MA, Klein JP, Haagenson M, et al. Multiple mismatches at the low expression HLA loci DP, DQ, and DRB3/4/5 associate with adverse outcomes in hematopoietic stem cell transplantation. *Blood.* 2013;121(22):4603-4610.
348. Fleischhauer K, Shaw BE, Gooley T, et al. Effect of T-cell-epitope matching at HLA-DPB1 in recipients of unrelated-donor haemopoietic-cell transplantation: a retrospective study. *Lancet Oncol.* 2012;13(4):366-374.
349. D'Souza A, Fretham C, Lee SJ, et al. Current use of and trends in hematopoietic cell transplantation in the United States. *Biol Blood Marrow Transplant.* 2020;26(8):e177-e182.
350. Ruggeri L, Capanni M, Urbani E, et al. Effectiveness of donor natural killer cell alloreactivity in mismatched hematopoietic transplants. *Science.* 2002;295(5562):2097-2100.
351. Ruggeri L, Capanni M, Casucci M, et al. Role of natural killer cell alloreactivity in HLA-mismatched hematopoietic stem cell transplantation. *Blood.* 1999;94(1):333-339.
352. Claas FH, Oudshoorn M. Role of NK cells in mismatched unrelated haematopoietic stem-cell transplantation: fact or fiction? *Tissue Antigens.* 2005;65(6):515-518.
353. Willemze R, Ruggeri A, Purtill D, Rodrigues CA, Gluckman E, Rocha V. Is there an impact of killer cell immunoglobulin-like receptors and KIR-ligand incompatibilities on outcomes after unrelated cord blood stem cell transplantation? *Best Pract Res Clin Haematol.* 2010;23(2):283-290.
354. Velardi A, Ruggeri L, Mancusi A, Aversa F, Christiansen FT. Natural killer cell allorecognition of missing self in allogeneic hematopoietic transplantation: a tool for immunotherapy of leukemia. *Curr Opin Immunol.* 2009;21(5):525-530.
355. Brunstein CG, Wagner JE, Weisdorf DJ, et al. Negative effect of KIR alloreactivity in recipients of umbilical cord blood transplant depends on transplantation conditioning intensity. *Blood.* 2009;113(22):5628-5634.
356. Doxiadis II, Smits JM, Th Schreuder GM, et al. Association between specific HLA combinations and probability of kidney allograft loss: the taboo concept. *Lancet.* 1996;348(9031):850-853.
357. Mickelson EM, Longton G, Anasetti C, et al. Evaluation of the mixed lymphocyte culture (MLC) assay as a method for selecting unrelated donors for marrow transplantation. *Tissue Antigens.* 1996;47(1):27-36.
358. Pei J, Akatsuka Y, Anasetti C, et al. Generation of HLA-C-specific cytotoxic T cells in association with marrow graft rejection: analysis of alloimmunity by T-cell cloning and testing of T-cell-receptor rearrangements. *Biol Blood Marrow Transplant.* 2001;7(7):378-383.
359. Maung KK, Horwitz ME. Current and future perspectives on allogeneic transplantation using *ex vivo* expansion or manipulation of umbilical cord blood cells. *Int J Hematol.* 2019;110(1):50-58.
360. Wang XN, Taylor PRA, Skinner R, et al. T-cell frequency analysis does not predict the incidence of graft-versus-host disease in HLA-matched sibling bone marrow transplantation. *Transplantation.* 2000;70(3):488-493.
361. Kaminski E, Hows J, Man S, et al. Prediction of graft versus host disease by frequency analysis of cytotoxic T cells after unrelated donor bone marrow transplantation. *Transplantation.* 1989;48(4):608-613.
362. Gluckman E, Rocha V, Chastang C. Peripheral stem cells in bone marrow transplantation. Cord blood stem cell transplantation. *Baillieres Best Pract Res Clin Haematol.* 1999;12(1-2):279-292.
363. Wagner JE, Barker JN, DeFor TE, et al. Transplantation of unrelated donor umbilical cord blood in 102 patients with malignant and nonmalignant diseases: influence of CD34 cell dose and HLA disparity on treatment-related mortality and survival. *Blood.* 2002;100(5):1611-1618.
364. Okeke C, Silas U, Okeke C, Chikwendu C. Current trends on hemopoietic stem cells. *Curr Stem Cell Res Ther.* 2021;16(2):199-208.
365. Oudshoorn M, Doxiadis IIN, van den Berg-Loonen PM, Voorter CEM, Verduyn W, Claas FHJ. Functional versus structural matching: can the CTLp test be replaced by HLA allele typing? *Hum Immunol.* 2002;63(3):176-184.

366. Nepom GT, Erlich H. MHC class-II molecules and autoimmunity. *Annu Rev Immunol.* 1991;9:493-525.
367. D'Antonio M, Reyna J, Jakubosky D, et al. Systematic genetic analysis of the MHC region reveals mechanistic underpinnings of HLA type associations with disease. *Elife.* 2019;8:e48476.
368. Thomson G. HLA disease associations: models for the study of complex human genetic disorders. *Crit Rev Clin Lab Sci.* 1995;32(2):183-219.
369. Winchester R, FitzGerald O. MHC class I associations beyond HLA-B27: the peptide binding hypothesis of psoriatic arthritis and its implications for disease pathogenesis. *Curr Opin Rheumatol.* 2020;32(4):330-336.
370. Unanue ER, Turk V, Neefjes J. Variations in MHC class II antigen processing and presentation in health and disease. *Annu Rev Immunol.* 2016;34:265-297.
371. de Sousa E, Ligeiro D, Lérias JR, et al. Mortality in COVID-19 disease patients: Correlating the association of major histocompatibility complex (MHC) with severe acute respiratory syndrome 2 (SARS-CoV-2) variants. *Int J Infect Dis.* 2020;98:454-459.
372. Malkki M, Single R, Carrington M, Thomson G, Petersdorf E. MHC microsatellite diversity and linkage disequilibrium among common HLA-A, HLA-B, DRB1 haplotypes: implications for unrelated donor hematopoietic transplantation and disease association studies. *Tissue Antigens.* 2005;66(2):114-124.
373. Shiina T, Inoko H, Kulski JK. An update of the HLA genomic region, locus information and disease associations. *Tissue Antigens.* 2004;64(6):631-649.
374. Lechler R, Warrens A. *HLA in Health and Disease.* 2nd ed. Academic Press; 2000.
375. Tiwari JL, Terasaki PI. *HLA and Disease Associations.* Springer-Verlag; 1985.
376. Loll B, Rückert C, Uchanska-Ziegler B, Ziegler A. Conformational plasticity of HLA-B27 molecules correlates inversely with efficiency of negative T cell selection. *Front Immunol.* 2020;11:179.
377. D'Amato M, Fiorillo MT, Carcassi C, et al. Relevance of residue 116 of HLA-B27 in determining susceptibility to ankylosing spondylitis. *Eur J Immunol.* 1995;25(11):3199-3201.
378. Pedersen SJ, Maksymowych WP. The pathogenesis of ankylosing spondylitis: an update. *Curr Rheumatol Rep.* 2019;21(10):58.
379. Khan MA, Khan MK. Diagnostic value of HLA-B27 testing ankylosing spondylitis and Reiter's syndrome. *Ann Intern Med.* 1982;96(1):70-76.
380. Rantapaa Dahlqvist S, Ström H, Bjelle A, Möller E. Clinical symptoms and HLA antigens in a family with Reiter's disease. *Scand J Rheumatol.* 1985;14(2):149-158.
381. Khabbazi A, Vahedi L, Ghojazadeh M, Pashazadeh F, Khameneh A. Association of HLA-B27 and Behcet's disease: a systematic review and meta-analysis. *Auto Immun Highlights.* 2019;10(1):2.
382. Guasp P, Alvarez-Navarro C, Gomez-Molina P, et al. The peptidome of Behcet's disease-associated HLA-B*51:01 includes two subpeptidomes differentially shaped by endoplasmic reticulum aminopeptidase 1. *Arthritis Rheumatol.* 2016;68(2):505-515.
383. Kuiper JJW, Venema WJ. HLA-A29 and birdshot uveitis: further down the rabbit hole. *Front Immunol.* 2020;11:599558.
384. Brezin AP, Monnet D, Cohen JHM, Levinson RD. HLA-A29 and birdshot chorioretinopathy. *Ocul Immunol Inflamm.* 2011;19(6):397-400.
385. Nakabayashi K, Tajima A, Yamamoto K, et al. Identification of independent risk loci for Graves' disease within the MHC in the Japanese population. *J Hum Genet.* 2011;56(11): 772-778.
386. Burlando M, Tajima A, Yamamoto K, et al. The HLA-Cw6 dilemma: is it really an outcome predictor in psoriasis patients under biologic therapy? A monocentric retrospective analysis. *J Clin Med.* 2020;9(10):3140.
387. Huang YW, Tsai TF. HLA-Cw1 and psoriasis. *Am J Clin Dermatol.* 2021;22(3):339-347.
388. Morelli M, Galluzzo M, Madonna S, et al. *HLA-Cw6* and other HLA-C alleles, as well as *MICB-DT*, *DDX58*, and *TYK2* genetic variants associate with optimal response to anti-IL-17A treatment in patients with psoriasis. *Expert Opin Biol Ther.* 2021;21(2):259-270.
389. Onsun N, Pirmit S, Ozkaya D, et al. The HLA-Cw12 allele is an important susceptibility allele for psoriasis and is associated with resistant psoriasis in the Turkish population. *Sci World J.* 2019;2019:7848314.
390. Stawczyk-Macieja M, Szczerkowska-Dobosz A, Rębała K, et al. *ERAP1* and *HLA-C*06* are strongly associated with the risk of psoriasis in the population of northern Poland. *Postepy Dermatol Alergol.* 2018;35(3):286-292.
391. Vasilca V, Vasilca A, Munteanu D, et al. HLA-B57 is significantly associated with psoriasis in Northeast Romania [corrected]. *Roum Arch Microbiol Immunol.* 2002;61(4): 259-265.
392. Dehghani SM, Dara N, Gharesifar B, Shahramian I, Dalili F, Salarzaei M. Prevalence of HLA DQ 2, 8 in children with celiac disease. *Hum Antibodies.* 2021;29:123-128.
393. Mansouri M, Dadfar M, Rostami-Nejad M, Ekhlasi G, Shahbazkhani A, Shahbazkhani B. The frequency of HLA-DQ2/DQ8 haplotypes and celiac disease among the first-degree relatives of patients with celiac disease. *Gastroenterol Hepatol Bed Bench.* 2021;14(1):36-43.
394. Siddiqui K, Uqaili AA, Rafiq M, Bhutto MA. Human leukocyte antigen (HLA)-DQ2 and -DQ8 haplotypes in celiac, celiac with type 1 diabetic, and celiac suspected pediatric cases. *Medicine (Baltimore).* 2021;100(11):e24954.
395. Katahira M, Ogata H, Takashima H, et al. Critical amino acid variants in HLA-DRB1 allotypes in the development of Graves' disease and Hashimoto's thyroiditis in the Japanese population. *Hum Immunol.* 2020;82(4):226-231.
396. Sasazuki T, Inoko H, Morishima S, Morishima Y. Gene map of the HLA region, Graves' disease and Hashimoto thyroiditis, and hematopoietic stem cell transplantation. *Adv Immunol.* 2016;129:175-249.
397. Brorsson C, Tue Hansen N, Bergholdt R, Brunak S, Pociot F. The type 1 diabetes - HLA susceptibility interactome – Identification of HLA genotype-specific disease genes for type 1 diabetes. *PLoS One.* 2010;5(3):e9576.
398. Spurkland A, Celius EG, Knutsen I, Belske A, Thorsby E, Vartdal F. The HLA-DQ(α1*0102, beta 1*0602) heterodimer may confer susceptibility to multiple sclerosis in the absence of the HLA-DR(α1*01, β1*1501) heterodimer. *Tissue Antigens.* 1997;50(1):15-22.
399. De Silvestri A, Capittini C, Mallucci G, et al. The involvement of HLA class II alleles in multiple sclerosis: a systematic review with meta-analysis. *Dis Markers.* 2019;2019:1409069.
400. Ting YT, Petersen J, Ramarathinam SH, et al. The interplay between citrullination and HLA-DRB1 polymorphism in shaping peptide binding hierarchies in rheumatoid arthritis. *J Biol Chem.* 2018;293(9):3236-3251.
401. Al-Motwee S, Jawdat D, Jehani GS, et al. Association of HLA-DRB1*15 and HLADQB1*06 with SLE in Saudis. *Ann Saudi Med.* 2013;33(3):229-234.
402. Sun J, Yang C, Fei W, et al. HLA-DQβ1 amino acid position 87 and DQB1*0301 are associated with Chinese Han SLE. *Mol Genet Genomic Med.* 2018;6(4):541-546.
403. Ringrose JH, Muijsers AO, Pannekoek Y, et al. Influence of infection of cells with bacteria associated with reactive arthritis on the peptide repertoire presented by HLA-B27. *J Med Microbiol.* 2001;50(4):385-389.
404. Ringrose JH. HLA-B27 associated spondyloarthropathy, an autoimmune disease based on crossreactivity between bacteria and HLA-B27? *Ann Rheum Dis.* 1999;58(10):598-610.
405. Loll B, Fabian H, Huser H, et al. Increased conformational flexibility of HLA-B*27 subtypes associated with ankylosing spondylitis. *Arthritis Rheumatol.* 2016;68(5):1172-1182.
406. Uchanska-Ziegler B, Loll B, Fabian H, Hee CS, Saenger W, Ziegler A. HLA class I-associated diseases with a suspected autoimmune etiology: HLA-B27 subtypes as a model system. *Eur J Cell Biol.* 2012;91(4):274-286.
407. Evans DM, Spencer CC, Pointon JJ, et al. Interaction between ERAP1 and HLA-B27 in ankylosing spondylitis implicates peptide handling in the mechanism for HLA-B27 in disease susceptibility. *Nat Genet.* 2011;43(8):761-767.
408. Lopez de Castro JA. How ERAP1 and ERAP2 shape the peptidomes of disease-associated MHC-I proteins. *Front Immunol.* 2018;9:2463.
409. Powis SJ, Colbert RA. Editorial. HLA-B27: the story continues to unfold. *Arthritis Rheumatol.* 2016;68(5):1057-1059.
410. Colbert RA, Tran TM, Layh-Schmitt G. HLA-B27 misfolding and ankylosing spondylitis. *Mol Immunol.* 2014;57(1): 44-51.
411. Bird LA, Peh CA, Kollnberger S, Elliott T, McMichael AJ, Bowness P. Lymphoblastoid cells express HLA-B27 homodimers both intracellularly and at the cell surface following endosomal recycling. *Eur J Immunol.* 2003;33(3):748-759.
412. Kollnberger S, Chan A, Sun MY, et al. Interaction of HLA-B27 homodimers with KIR3DL1 and KIR3DL2, unlike HLA-B27 heterotrimers, is independent of the sequence of bound peptide. *Eur J Immunol.* 2007;37(5):1313-1322.
413. Bennett MJ, Lebron JA, Bjorkman PJ. Crystal structure of the hereditary haemochromatosis protein HFE complexed with transferrin receptor. *Nature.* 2000;403(6765):46-53.
414. Lebron JA, Bennett MJ, Vaughn DE, et al. Crystal structure of the hemochromatosis protein HFE and characterization of its interaction with transferrin receptor. *Cell.* 1998;93(1):111-123.
415. Trinder D, Olynyk JK, Sly WS, Morgan EH. Iron uptake from plasma transferrin by the duodenum is impaired in the Hfe knockout mouse. *Proc Natl Acad Sci U S A.* 2002;99(8):5622-5626.
416. Waheed A, Grubb JH, Zhou XY, et al. Regulation of transferrin-mediated iron uptake by HFE, the protein defective in hereditary hemochromatosis. *Proc Natl Acad Sci U S A.* 2002;99(5):3117-3122.
417. Zhou XY, Tomatsu S, Fleming RE, et al. HFE gene knockout produces mouse model of hereditary hemochromatosis. *Proc Natl Acad Sci U S A.* 1998;95(5):2492-2497.
418. Santos M, Clevers HC, Marx JJ. Mutations of the hereditary hemochromatosis candidate gene HLA-H in porphyria cutanea tarda. *N Engl J Med.* 1997;336(18):1327-1328.
419. Rothenberg BE, Voland JR. β2 knockout mice develop parenchymal iron overload: a putative role for class I genes of the major histocompatibility complex in iron metabolism. *Proc Natl Acad Sci U S A.* 1996;93(4):1529-1534.
420. Fleming RE, Migas MC, Zhou X, et al. Mechanism of increased iron absorption in murine model of hereditary hemochromatosis: increased duodenal expression of the iron transporter DMT1. *Proc Natl Acad Sci U S A.* 1999;96(6):3143-3148.
421. The U. A simple genetic test identifies 90% of UK patients with haemochromatosis. The UK Haemochromatosis Consortium. *Gut.* 1997;41(6):841-844.
422. Lester S, Bardy P, McCluskey J. HFE genotypes and haemochromatosis: quantifying the risks of disease. *Tissue Antigens.* 1999;54(3):282-284.
423. Zhernakova A, Stahl EA, Trynka G, et al. Meta-analysis of genome-wide association studies in celiac disease and rheumatoid arthritis identifies fourteen non-HLA shared loci. *PLoS Genet.* 2011;7(2):e1002004.
424. Barrett JC, Clayton DG, Concannon P, et al. Genome-wide association study and meta-analysis find that over 40 loci affect risk of type 1 diabetes. *Nat Genet.* 2009;41(6):703-707.
425. Oldstone MB. Molecular mimicry as a mechanism for the cause and a probe uncovering etiologic agent(s) of autoimmune disease. *Curr Top Microbiol Immunol.* 1989;145:127-135.
426. Raychaudhuri S, Sandor C, Stahl EA, et al. Five amino acids in three HLA proteins explain most of the association between MHC and seropositive rheumatoid arthritis. *Nat Genet.* 2012;44(3):291-296.
427. Oldstone MB, von Herrath M, Evans CF, Horwitz MS. Virus-induced autoimmune disease: transgenic approach to mimic insulin-dependent diabetes mellitus and multiple sclerosis. *Curr Top Microbiol Immunol.* 1996;206:67-83.
428. Wucherpfennig KW. Structural basis of molecular mimicry. *J Autoimmun.* 2001;16(3):293-302.
429. Mignot E, Tafti M, Dement WC, Grumet FC. Narcolepsy and immunity. *Adv Neuroimmunol.* 1995;5(1):23-37.
430. Ellis MC, Hetisimer AH, Ruddy DA, et al. HLA class II haplotype and sequence analysis support a role for DQ in narcolepsy. *Immunogenetics.* 1997;46(5):410-417.
431. Newton JL, Harney S, Wordsworth B, Brown M. A review of the MHC genetics of rheumatoid arthritis. *Genes Immun.* 2004;5(3):151-157.
432. Wordsworth P. Rheumatoid arthritis. *Curr Opin Immunol.* 1992;4(6):766-769.
433. Wordsworth P, Pile KD, Buckely JD, et al. HLA heterozygosity contributes to susceptibility to rheumatoid arthritis. *Am J Hum Genet.* 1992;51(3):585-591.
434. Hall FC, Weeks DE, Camilleri JP, et al. Influence of the HLA-DRB1 locus on susceptibility and severity in rheumatoid arthritis. *QJM.* 1996;89(11):821-829.
435. Hattori M, Buse JB, Jackson RA, et al. The NOD mouse: recessive diabetogenic gene in the major histocompatibility complex. *Science.* 1986;231(4739):733-735.
436. Acha-Orbea H, McDevitt HO. The first external domain of the nonobese diabetic mouse class II I-A beta chain is unique. *Proc Natl Acad Sci U S A.* 1987;84(8):2435-2439.
437. Quartey-Papafio R, Lund T, Chandler P, et al. Aspartate at position 57 of nonobese diabetic I-Ag7 beta-chain diminishes the spontaneous incidence of insulin-dependent diabetes mellitus. *J Immunol.* 1995;154(10):5567-5575.
438. Atkinson MA, Leiter EH. The NOD mouse model of type 1 diabetes: as good as it gets? *Nat Med.* 1999;5(6):601-604.
439. Tisch R, McDevitt H. Insulin-dependent diabetes mellitus. *Cell.* 1996;85(3):291-297.
440. Noble JA, Johnson J, Lane JA, Valdes AM. Race-specific type 1 diabetes risk of HLA-DR7 haplotypes. *Tissue Antigens.* 2011;78(5):348-351.
441. Bell H, Berg JP, Undlien DE, et al. The clinical expression of hemochromatosis in Oslo, Norway. Excessive oral iron intake may lead to secondary hemochromatosis even in HFE C282Y mutation negative subjects. *Scand J Gastroenterol.* 2000;35(12):1301-1307.
442. Undlien DE, Lie BA, Thorsby E. HLA complex genes in type 1 diabetes and other autoimmune diseases. Which genes are involved? *Trends Genet.* 2001;17(2):93-100.
443. Undlien DE, Thorsby E. HLA associations in type 1 diabetes: merging genetics and immunology. *Trends Immunol.* 2001;22(9):467-469.
444. Morton NE, Green A, Dunsworth T, et al. Heterozygous expression of insulin-dependent diabetes mellitus (IDDM) determinants in the HLA system. *Am J Hum Genet.* 1983;35(2):201-213.
445. Green PH, Cellier C. Celiac disease. *N Engl J Med.* 2007;357(17):1731-1743.
446. Vilppula A, Kaukinen K, Luostarinen L, et al. Increasing prevalence and high incidence of celiac disease in elderly people: a population-based study. *BMC Gastroenterol.* 2009;9:49.
447. Arentz-Hansen H, Körner R, Molberg O, et al. The intestinal T cell response to alpha-gliadin in adult celiac disease is focused on a single deamidated glutamine targeted by tissue transglutaminase. *J Exp Med.* 2000;191(4):603-612.
448. Anderson RP, Degano P, Godkin AJ, Jewell DP, Hill AVS. *In vivo* antigen challenge in celiac disease identifies a single transglutaminase-modified peptide as the dominant A-gliadin T-cell epitope. *Nat Med.* 2000;6(3):337-342.

449. Kim CY, Quarsten H, Bergseng E, Khosla C, Sollid LM. Structural basis for HLA-DQ2-mediated presentation of gluten epitopes in celiac disease. *Proc Natl Acad Sci U S A.* 2004;101(12):4175-4179.
450. Tye-Din JA, Stewart JA, Dromey JA, et al. Comprehensive, quantitative mapping of T cell epitopes in gluten in celiac disease. *Sci Transl Med.* 2010;2(41):41ra51.
451. van de Wal Y, Kooy Y, van Veelen P, et al. Selective deamidation by tissue transglutaminase strongly enhances gliadin-specific T cell reactivity. *J Immunol.* 1998;161(4):1585-1588.
452. Henderson KN, Tye-Din JA, Reid HH, et al. A structural and immunological basis for the role of human leukocyte antigen DQ8 in celiac disease. *Immunity.* 2007;27(1):23-34.
453. Tye-Din J, Anderson R. Immunopathogenesis of celiac disease. *Curr Gastroenterol Rep.* 2008;10(5):458-465.
454. Abadie V, Kim SM, Lejeune T, et al. IL-15, gluten and HLA-DQ8 drive tissue destruction in coeliac disease. *Nature.* 2020;578(7796):600-604.
455. Hung SI, Chung WH, Jee SH, et al. Genetic susceptibility to carbamazepine-induced cutaneous adverse drug reactions. *Pharmacogenet Genomics.* 2006;16(4):297-306.
456. Illing PT, Purcell AW, McCluskey J. The role of HLA genes in pharmacogenomics: unravelling HLA associated adverse drug reactions. *Immunogenetics.* 2017;69(8-9):617-630.
457. Bharadwaj M, Illing P, Theodossis A, Purcell AW, Rossjohn J, McCluskey J. Drug hypersensitivity and human leukocyte antigens of the major histocompatibility complex. *Annu Rev Pharmacol Toxicol.* 2012;52:401-431.
458. Mallal S, Nolan D, Witt C, et al. Association between presence of HLA-B*5701, HLA-DR7, and HLA-DQ3 and hypersensitivity to HIV-1 reverse-transcriptase inhibitor abacavir. *Lancet.* 2002;359(9308):727-732.
459. Kaniwa N, Saito Y, Aihara M, et al. HLA-B locus in Japanese patients with anti-epileptics and allopurinol-related Stevens-Johnson syndrome and toxic epidermal necrolysis. *Pharmacogenomics.* 2008;9(11):1617-1622.
460. Simper GS, Gräser LS, Celik AA, et al. The mechanistic differences in HLA-associated carbamazepine hypersensitivity. *Pharmaceutics.* 2019;11(10):536.
461. Locharernkul C, Loplumlert J, Limotai C, et al. Carbamazepine and phenytoin induced Stevens-Johnson syndrome is associated with HLA-B*1502 allele in Thai population. *Epilepsia.* 2008;49(12):2087-2091.
462. Pellicano R, Lomuto M, Ciavarella G, Di Giorgio G, Gasparini P. Fixed drug eruptions with feprazone are linked to HLA-B22. *J Am Acad Dermatol.* 1997;36(5 pt 1):782-784.
463. Phillips EJ, Mallal SA. HLA-B*5701 and flucloxacillin associated drug-induced liver disease. *AIDS.* 2013;27(3):491-492.
464. Daly AK, Donaldson PT, Bhatnagar P, et al. HLA-B*5701 genotype is a major determinant of drug-induced liver injury due to flucloxacillin. *Nat Genet.* 2009;41(7):816-819.
465. Schmidt KL, Mueller-Eckhardt C. Agranulocytosis, levamisole, and HLA-B27. *Lancet.* 1977;2(8028):85.
466. Roujeau JC, Huynh TN, Bracq C, Guillaume JC. Genetic susceptibility to toxic epidermal necrolysis. *Arch Dermatol.* 1987;123(9):1171-1173.
467. Sabourirad S, Mortezaee R, Mojarad M, et al. Investigating the association of lamotrigine and phenytoin-induced Stevens-Johnson syndrome/toxic epidermal necrolysis with HLA-B*1502 in Iranian population. *Exp Dermatol.* 2021;30(2):284-287.
468. Dekker JW, Nizankowska E, Schmitz-schumann M, et al. Aspirin-induced asthma and HLA-DRB1 and HLA-DPB1 genotypes. *Clin Exp Allergy.* 1997;27(5):574-577.
469. Choi JH, Lee KW, Oh HB, et al. HLA association in aspirin-intolerant asthma: DPB1*0301 as a strong marker in a Korean population. *J Allergy Clin Immunol.* 2004;113(3):562-564.
470. Batchelor JR, Welsh KI, Mansilla Tinoco R, et al. Hydralazine-induced systemic lupus erythematosus: influence of HLA-DR and sex on susceptibility. *Lancet.* 1980;1(8178):1107-1109.
471. Spraggs CF, Budde LR, Briley LP, et al. HLA-DQA1*02:01 is a major risk factor for lapatinib-induced hepatotoxicity in women with advanced breast cancer. *J Clin Oncol.* 2011;29(6):667-673.
472. Quiralte J, Sánchez-García F, Torres MJ, et al. Association of HLA-DR11 with the anaphylactoid reaction caused by nonsteroidal anti-inflammatory drugs. *J Allergy Clin Immunol.* 1999;103(4):685-689.
473. Kindmark A, Jawaid A, Harbron CG, et al. Genome-wide pharmacogenetic investigation of a hepatic adverse event without clinical signs of immunopathology suggests an underlying immune pathogenesis. *Pharmacogenomics J.* 2008;8(3):186-195.
474. Hetherington S, Hughes AR, Mosteller M, et al. Genetic variations in HLA-B region and hypersensitivity reactions to abacavir. *Lancet.* 2002;359(9312):1121-1122.
475. Martin AM, Nolan D, Gaudieri S, et al. Predisposition to abacavir hypersensitivity conferred by HLA-B*5701 and a haplotypic Hsp70-Hom variant. *Proc Natl Acad Sci U S A.* 2004;101(12):4180-4185.
476. Chung WH, Hung SI, Hong HS, et al. Medical genetics: a marker for Stevens-Johnson syndrome. *Nature.* 2004;428(6982):486.
477. Hung SI, Chung WH, Liou LB, et al. HLA-B*5801 allele as a genetic marker for severe cutaneous adverse reactions caused by allopurinol. *Proc Natl Acad Sci U S A.* 2005;102(11):4134-4139.
478. Tassaneeyakul W, Jantararoungtong T, Chen P, et al. Strong association between HLA-B*5801 and allopurinol-induced Stevens-Johnson syndrome and toxic epidermal necrolysis in a Thai population. *Pharmacogenet Genomics.* 2009;19(9):704-709.
479. Phillips EJ, Sullivan JR, Knowles SR, Shear NH. Utility of patch testing in patients with hypersensitivity syndromes associated with abacavir. *AIDS.* 2002;16(16):2223-2225.
480. Nassif A, Bensussan A, Bachot N, et al. Drug specific cytotoxic T-cells in the skin lesions of a patient with toxic epidermal necrolysis. *J Invest Dermatol.* 2002;118(4):728-733.
481. Hari Y, Frutig-Schnyder K, Hurni M, et al. T cell involvement in cutaneous drug eruptions. *Clin Exp Allergy.* 2001;31(9):1398-1408.
482. Le Cleach L, Delaire S, Boumsell L, et al. Blister fluid T lymphocytes during toxic epidermal necrolysis are functional cytotoxic cells which express human natural killer (NK) inhibitory receptors. *Clin Exp Immunol.* 2000;119(1):225-230.
483. Chessman D, Kostenko L, Lethborg T, et al. Human leukocyte antigen class I-restricted activation of CD8+ T cells provides the immunogenetic basis of a systemic drug hypersensitivity. *Immunity.* 2008;28(6):822-832.
484. Illing PT, Vivian JP, Dudek NL, et al. Immune self-reactivity triggered by drug-modified HLA-peptide repertoire. *Nature.* 2012;486(7404):554-558.
485. Mallal S, Phillips E, Carosi G, et al. HLA-B*5701 screening for hypersensitivity to abacavir. *N Engl J Med.* 2008;358(6):568-579.
486. Kaslow RA, Carrington M, Apple R, et al. Influence of combinations of human major histocompatibility complex genes on the course of HIV-1 infection. *Nat Med.* 1996;2(4):405-411.
487. Pereyra F, Jia X, McLaren PJ, et al; International HIV Controllers Study. The major genetic determinants of HIV-1 control affect HLA class I peptide presentation. *Science.* 2010;330(6010):1551-1557.
488. Deeks SG, Walker BD. Human immunodeficiency virus controllers: mechanisms of durable virus control in the absence of antiretroviral therapy. *Immunity.* 2007;27(3):406-416.
489. Carrington M, Martin MP, van Bergen J. KIR-HLA intercourse in HIV disease. *Trends Microbiol.* 2008;16(12):620-627.
490. Qi Y, Martin MP, Gao X, et al. KIR/HLA pleiotropism: protection against both HIV and opportunistic infections. *PLoS Pathog.* 2006;2(8):e79.
491. Hopfensperger K, Richard J, Stürzel CM, et al. Convergent evolution of HLA-C downmodulation in HIV-1 and HIV-2. *mBio.* 2020;11(4):e00782-e00820.
492. Zeniya M, Takahashi H. Characteristics of autoimmune hepatitis in the Asia-Pacific Region: historical review. *Hepatol Int.* 2011;6(1):342-349.
493. Wawrzynowicz-Syczewska M, Underhill JA, Clare MA, Boron-Kaczmarska A, McFarlane IG, Donaldson PT. HLA class II genotypes associated with chronic hepatitis C virus infection and response to alpha-interferon treatment in Poland. *Liver.* 2000;20(3):234-239.
494. Elfaramawy AA, Elhossiny RM, Abbas AA, Aziz H. HLA-DRB1 as a risk factor in children with autoimmune hepatitis and its relation to hepatitis A infection. *Ital J Pediatr.* 2010;36:73.
495. Feynman R. The feynman lectures on physics. In: Feynman R, Leighton R, Sands M, eds. *The Feynman Lectures on Physics.* Addison-Wesley; 1963:5.
496. Morrison J. Love hides. In: Rothchild PA, ed. *Absolutely Live.* Elektra; 1970.
497. Margulies DH, McCluskey J. Exon shuffling: new genes from old. *Surv Immunol Res.* 1985;4(2):146-159.
498. Rudolph MG, Stanfield RL, Wilson IA. How TCRs bind MHCs, peptides, and coreceptors. *Annu Rev Immunol.* 2006;24:419-466.
499. Jiang J, Natarajan K, Margulies DH. MHC molecules, T cell receptors, natural killer cell receptors, and viral immunoevasins-key elements of adaptive and innate immunity. *Adv Exp Med Biol.* 2019;1172:21-62.
500. Altman JD, Moss PAH, Goulder PJR, et al. Phenotypic analysis of antigen-specific T lymphocytes. *Science.* 1996;274(5284):94-96.
501. Orr HT, Lopez de Castro JA, Parham P, Ploegh HL, Strominger JL. Comparison of amino acid sequences of two human histocompatibility antigens, HLA-A2 and HLA-B7: location of putative alloantigenic sites. *Proc Natl Acad Sci U S A.* 1979;76(9):4395-4399.
502. Coligan JE, Kindt TJ, Uehara H, Martinko J, Nathenson SG. Primary structure of a murine transplantation antigen. *Nature.* 1981;291(5810):35-39.
503. Steinmetz M, Frelinger JG, Fisher D, et al. Three cDNA clones encoding mouse transplantation antigens: homology to immunoglobulin genes. *Cell.* 1981;24(1):125-134.
504. Sood AK, Pereira D, Weissman SM. Isolation and partial nucleotide sequence of a cDNA clone for human histocompatibility antigen HLA-B by use of an oligodeoxynucleotide primer. *Proc Natl Acad Sci U S A.* 1981;78(1):616-620.
505. Ploegh HL, Orr HT, Strominger JL. Molecular cloning of a human histocompatibility antigen cDNA fragment. *Proc Natl Acad Sci U S A.* 1980;77(10):6081-6085.
506. Choi E, McIntyre K, Germain RN, Seidman JG. Murine I-A beta chain polymorphism: nucleotide sequences of three allelic I-A beta genes. *Science.* 1983;221(4607):283-286.
507. Mathis DJ, Benoist C, Williams VE, Kanter M, McDevitt HO. Several mechanisms can account for defective E alpha gene expression in different mouse haplotypes. *Proc Natl Acad Sci U S A.* 1983;80(1):273-277.
508. Davis MM, Cohen DI, Nielsen EA, Steinmetz M, Paul WE, Hood L. Cell-type-specific cDNA probes and the murine I region: the localization and orientation of Ad alpha. *Proc Natl Acad Sci U S A.* 1984;81(7):2194-2198.
509. Handy DE, McCluskey J, Lew AM, Coligan JE, Margulies DH. Signals controlling alternative splicing of major histocompatibility complex H-2 class I pre-mRNA. *Immunogenetics.* 1988;28(2):81-90.
510. Lew AM, Margulies DH, Maloy WL, Lillehoj EP, McCluskey J, Coligan JE. Alternative protein products with different carboxyl termini from a single class I gene, H-2Kb. *Proc Natl Acad Sci U S A.* 1986;83(16):6084-6088.
511. Vega MA, Strominger JL. Constitutive endocytosis of HLA class I antigens requires a specific portion of the intracytoplasmic tail that shares structural features with other endocytosed molecules. *Proc Natl Acad Sci U S A.* 1989;86(8):2688-2692.
512. Balomenos D, Poretz RD. An acidic modification of the cytoplasmic domain contributes to the charge heterogeneity of the MHC class I antigens. *Immunogenetics.* 1998;47(5):381-389.
513. Capps GG, Pine S, Edidin M, Zúñiga MC. Short class I major histocompatibility complex cytoplasmic tails differing in charge detect arbiters of lateral diffusion in the plasma membrane. *Biophys J.* 2004;86(5):2896-2909.
514. Fooksman DR, Edidin M, Barisas BG. Measuring rotational diffusion of MHC class I on live cells by polarized FPR. *Biophys Chem.* 2007;130(1-2):10-16.
515. Lew AM, McCluskey J, Maloy WL, Margulies DH, Coligan JE. Multiple class I molecules generated from single genes by alternative splicing of pre-mRNAs. *Immunol Res.* 1987;6(1-2):117-132.
516. Gur H, Geppert TD, Lipsky PE. Structural analysis of class I MHC molecules: the cytoplasmic domain is not required for cytoskeletal association, aggregation and internalization. *Mol Immunol.* 1997;34(2):125-132.
517. Singh RK, Lau D, Noviello CM, Ghosh P, Guatelli JC. An MHC-I cytoplasmic domain/HIV-1 Nef fusion protein binds directly to the mu subunit of the AP-1 endosomal coat complex. *PLoS One.* 2009;4(12):e8364.
518. Yi L, Rosales T, Rose JJ, Chaudhury B, Knutson JR, Venkatesan S. HIV-1 Nef binds a subpopulation of MHC-I throughout its trafficking itinerary and down-regulates MHC-I by perturbing both anterograde and retrograde trafficking. *J Biol Chem.* 2010;285(40):30884-30905.
519. St-Pierre Y, Nabavi N, Ghogawala Z, Glimcher LH, Watts TH. A functional role for signal transduction via the cytoplasmic domains of MHC class II proteins. *J Immunol.* 1989;143(3):808-812.
520. Nabavi N, Freeman GJ, Gault A, Godfrey D, Nadler LM, Glimcher LH. Signalling through the MHC class II cytoplasmic domain is required for antigen presentation and induces B7 expression. *Nature.* 1992;360(6401):266-268.
521. Umemura YM, Vrljic M, Nishimura SY, Fujiwara TK, Suzuki KGN, Kusumi A. Both MHC class II and its GPI-anchored form undergo hop diffusion as observed by single-molecule tracking. *Biophys J.* 2008;95(1):435-450.
522. Schwartz RH. Immune response (Ir) genes of the murine major histocompatibility complex. *Adv Immunol.* 1986;38:31-201.
523. Babbitt BP, Allen PM, Matsueda G, Haber E, Unanue ER. Binding of immunogenic peptides to Ia histocompatibility molecules. *Nature.* 1985;317(6035):359-361.
524. Buus S, Sette A, Colon SM, Jenis DM, Grey HM. Isolation and characterization of antigen-Ia complexes involved in T cell recognition. *Cell.* 1986;47(6):1071-1077.
525. Townsend A, Ohlën C, Foster L, Bastin J, Ljunggren HG, Kärre K. A mutant cell in which association of class I heavy and light chains is induced by viral peptides. *Cold Spring Harb Symp Quant Biol.* 1989;54(Pt 1):299-308.
526. Boyd LF, Kozlowski S, Margulies DH. Solution binding of an antigenic peptide to a major histocompatibility complex class I molecule and the role of beta 2-microglobulin. *Proc Natl Acad Sci U S A.* 1992;89(6):2242-2246.
527. Van Bleek GM, Nathenson SG. Isolation of an endogenously processed immunodominant viral peptide from the class I H-2Kb molecule. *Nature.* 1990;348(6298):213-216.
528. Rotzschke O, Falk K, Deres K, et al. Isolation and analysis of naturally processed viral peptides as recognized by cytotoxic T cells. *Nature.* 1990;348(6298):252-254.
529. Rotzschke O, Falk K, Wallny HJ, Faath S, Rammensee HG. Characterization of naturally occurring minor histocompatibility peptides including H-4 and H-Y. *Science.* 1990;249(4966):283-287.
530. Rammensee HG, Falk K, Rotzschke O. Peptides naturally presented by MHC class I molecules. *Annu Rev Immunol.* 1993;11:213-244.

531. Arnott D, Shabanowitz J, Hunt DF. Mass spectrometry of proteins and peptides: sensitive and accurate mass measurement and sequence analysis. *Clin Chem.* 1993;39(9):2005-2010.
532. Hunt DF, Henderson RA, Shabanowitz J, et al. Characterization of peptides bound to the class I MHC molecule HLA-A2.1 by mass spectrometry. *Science.* 1992;255(5049):1261-1263.
533. Corr M, Boyd LF, Frankel SR, Kozlowski S, Padlan EA, Margulies DH. Endogenous peptides of a soluble major histocompatibility complex class I molecule, H-2Lds: sequence motif, quantitative binding, and molecular modeling of the complex. *J Exp Med.* 1992;176(6):1681-1692.
534. Corr M, Boyd LF, Padlan EA, Margulies DH. H-2Dd exploits a four residue peptide binding motif. *J Exp Med.* 1993;178(6):1877-1892.
535. Barnea E, Beer I, Patoka R, et al. Analysis of endogenous peptides bound by soluble MHC class I molecules: a novel approach for identifying tumor-specific antigens. *Eur J Immunol.* 2002;32(1):213-222.
536. Gavin MA, Bevan MJ. Major histocompatibility complex allele-specific peptide libraries and identification of T-cell mimotopes. *Methods Mol Biol.* 1998;87:235-248.
537. Gavin MA, Dere B, Grandea AG, Hogquist KA, Bevan MJ. Major histocompatibility complex class I allele-specific peptide libraries: identification of peptides that mimic an H-Y T cell epitope. *Eur J Immunol.* 1994;24(9):2124-2133.
538. Hendrickson RC, Cicinnati VR, Albers A, et al. Identification of a 17β-hydroxysteroid dehydrogenase type 12 pseudogene as the source of a highly restricted BALB/c Meth A tumor rejection peptide. *Cancer Immunol Immunother.* 2010;59(1):113-124.
539. Kuznetsov A, Voronina A, Govorun V, Arapidi G. Critical review of existing MHC I Immunopeptidome isolation methods. *Molecules.* 2020;25(22):5409.
540. Biddison WE, Martin R. Peptide binding motifs for MHC class I and II molecules. *Curr Protoc Immunol,* 2001;Appendix 1:Appendix 1I.
541. Rammensee H, Bachmann J, Emmerich NP, Bachor OA, Stevanović S. SYFPEITHI: database for MHC ligands and peptide motifs. *Immunogenetics.* 1999;50(3-4):213-219.
542. Peters B, Sidney J, Bourne P, et al. The design and implementation of the immune epitope database and analysis resource. *Immunogenetics.* 2005;57(5):326-336.
543. Parker KC, Bednarek MA, Coligan JE. Scheme for ranking potential HLA-A2 binding peptides based on independent binding of individual peptide side-chains. *J Immunol.* 1994;152(1):163-175.
544. Mei S, Li F, Leier A, et al. A comprehensive review and performance evaluation of bioinformatics tools for HLA class I peptide-binding prediction. *Brief Bioinform.* 2020;21(4):1119-1135.
545. Bassani-Sternberg M, Chong C, Guillaume P, et al. Deciphering HLA-I motifs across HLA peptidomes improves neo-antigen predictions and identifies allostery regulating HLA specificity. *PLoS Comput Biol.* 2017;13(8):e1005725.
546. Gfeller D, Guillaume P, Michaux J, et al. The length distribution and multiple specificity of naturally presented HLA-I ligands. *J Immunol.* 2018;201(12):3705-3716.
547. Jurtz V, Paul S, Andreatta M, Marcatili P, Peters B, Nielsen M. NetMHCpan-4.0: improved peptide-MHC class I interaction predictions integrating eluted ligand and peptide binding affinity data. *J Immunol.* 2017;199(9):3360-3368.
548. Karosiene E, Lundegaard C, Lund O, Nielsen M. NetMHCcons: a consensus method for the major histocompatibility complex class I predictions. *Immunogenetics.* 2012;64(3):177-186.
549. Burrows SR, Rossjohn J, McCluskey J. Have we cut ourselves too short in mapping CTL epitopes? *Trends Immunol.* 2006;27(1):11-16.
550. Rudensky Y, Preston-Hurlburt P, Hong S.-C, Barlow A, Janeway CA. Sequence analysis of peptides bound to MHC class II molecules. *Nature.* 1991;353(6345):622-627.
551. Chicz RM, Urban RG, Lane WS, et al. Predominant naturally processed peptides bound to HLA-DR1 are derived from MHC-related molecules and are heterogeneous in size. *Nature.* 1992;358(6389):764-768.
552. Janeway CA Jr, Mamula MJ, Rudensky A. Rules for peptide presentation by MHC class II molecules. *Int Rev Immunol.* 1993;10(4):301-311.
553. Urban RG, Chicz RM, Vignali DAA, Strominger JL. The dichotomy of peptide presentation by class I and class II MHC proteins. *Chem Immunol.* 1993;57:197-234.
554. Hammer J, Takacs B, Sinigaglia F. Identification of a motif for HLA-DR1 binding peptides using M13 display libraries. *J Exp Med.* 1992;176(4):1007-1013.
555. Carrasco-Marin E, Petzold S, Unanue ER. Two structural states of complexes of peptide and class II major histocompatibility complex revealed by photoaffinity-labeled peptides. *J Biol Chem.* 1999;274(44):31333-31340.
556. Gugasyan R, Velazquez C, Vidavsky I, et al. Independent selection by I-Ak molecules of two epitopes found in tandem in an extended polypeptide antigen. *J Immunol.* 2000;165(6):3206-3213.
557. Cease KB, Berkower I, York-Jolley J, Berzofsky JA. T cell clones specific for an amphipathic alpha-helical region of sperm whale myoglobin show differing fine specificities for synthetic peptides. A multiview/single structure interpretation of immunodominance. *J Exp Med.* 1986;164(5):1779-1784.
558. Landais E, Romagnoli PA, Corper AL. New design of MHC class II tetramers to accommodate fundamental principles of antigen presentation. *J Immunol.* 2009;183(12):7949-7957.
559. Chang KY, Suri A, Unanue ER. Predicting peptides bound to I-Ag7 class II histocompatibility molecules using a novel expectation-maximization alignment algorithm. *Proteomics.* 2007;7(3):367-377.
560. Gunther S, Schlundt A, Sticht J, et al. Bidirectional binding of invariant chain peptides to an MHC class II molecule. *Proc Natl Acad Sci U S A.* 2010;107(51):22219-22224.
561. Bjorkman PJ, Saper MA, Samraoui B, Bennett WS, Strominger JL, Wiley DC. Structure of the human class I histocompatibility antigen, HLA-A2. *Nature.* 1987;329(6139):506-512.
562. Bjorkman PJ, Saper MA, Samraoui B, Bennett WS, Strominger JL, Wiley DC. The foreign antigen binding site and T cell recognition regions of class I histocompatibility antigens. *Nature.* 1987;329(6139):512-518.
563. Berman HM, Westbrook J, Feng Z, et al. The protein data bank. *Nucleic Acids Res.* 2000;28(1):235-242.
564. The PyMOL Molecular Graphics System, Version 2.4.2 Schrödinger, LLC.
565. Garrett TP, Saper MA, Bjorkman PJ, Strominger JL, Wiley DC. Specificity pockets for the side chains of peptide antigens in HLA-Aw68. *Nature.* 1989;342(6250):692-696.
566. Saper MA, Bjorkman PJ, Wiley DC. Refined structure of the human histocompatibility antigen HLA-A2 at 2.6 A resolution. *J Mol Biol.* 1991;219(2):277-319.
567. Madden DR, Gorga JC, Strominger JL, Wiley DC. The structure of HLA-B27 reveals nonamer self-peptides bound in an extended conformation. *Nature.* 1991;353(6342):321-325.
568. Jardetzky TS, Lane WS, Robinson RA, Madden DR, Wiley DC. Identification of self peptides bound to purified HLA-B27. *Nature.* 1991;353(6342):326-329.
569. Zhang W, Young AC, Imarai M, Nathenson SG, Sacchettini JC. Crystal structure of the major histocompatibility complex class I H-2Kb molecule containing a single viral peptide: implications for peptide binding and T-cell receptor recognition. *Proc Natl Acad Sci U S A.* 1992;89(17):8403-8407.
570. Fremont DH, Matsumura M, Stura EA, Peterson PA, Wilson IA. Crystal structures of two viral peptides in complex with murine MHC class I H-2Kb. *Science.* 1992;257(5072):919-927.
571. Glithero A, Tormo J, Haurum JS, et al. Crystal structures of two H-2Db/glycopeptide complexes suggest a molecular basis for CTL cross-reactivity. *Immunity.* 1999;10(1):63-74.
572. Joyce S, Kuzushima K, Kepecs G, Angeletti RH, Nathenson SG. Characterization of an incompletely assembled major histocompatibility class I molecule (H-2Kb) associated with unusually long peptides: implications for antigen processing and presentation. *Proc Natl Acad Sci U S A.* 1994;91(10):4145-4149.
573. Collins EJ, Garboczi DN, Wiley DC. Three-dimensional structure of a peptide extending from one end of a class I MHC binding site. *Nature.* 1994;371(6498):626-629.
574. Lybarger L, Yu YYL, Miley MJ, et al. Enhanced immune presentation of a single-chain major histocompatibility complex class I molecule engineered to optimize linkage of a C-terminally extended peptide. *J Biol Chem.* 2003;278(29):27105-27111.
575. Kastrup IB, Stevanovic S, Arsequell G, et al. Lectin purified human class I MHC-derived peptides: evidence for presentation of glycopeptides in vivo. *Tissue Antigens.* 2000;56(2):129-135.
576. Speir JA, Abdel-Motal UM, Jondal M, Wilson IA. Crystal structure of an MHC class I presented glycopeptide that generates carbohydrate-specific CTL. *Immunity.* 1999;10(1):51-61.
577. Speir JA, Stevens J, Joly E, Butcher GW, Wilson IA. Two different, highly exposed, bulged structures for an unusually long peptide bound to rat MHC class I RT1-Aa. *Immunity.* 2001;14(1):81-92.
578. Miles JJ, Elhassen D, Borg NA, et al. CTL recognition of a bulged viral peptide involves biased TCR selection. *J Immunol.* 2005;175(6):3826-3834.
579. Tynan FE, Borg NA, Miles JJ, et al. High resolution structures of highly bulged viral epitopes bound to major histocompatibility complex class I. Implications for T-cell receptor engagement and T-cell immunodominance. *J Biol Chem.* 2005;280(25):23900-23909.
580. Tynan FE, Burrows SR, Buckle AM, et al. T cell receptor recognition of a 'super-bulged' major histocompatibility complex class I-bound peptide. *Nat Immunol.* 2005;6(11):1114-1122.
581. Tynan FE, Reid HH, Kjer-Nielsen L, et al. A T cell receptor flattens a bulged antigenic peptide presented by a major histocompatibility complex class I molecule. *Nat Immunol.* 2007;8(3):268-276.
582. Petersen J, Wurzbacher SJ, Williamson NA, et al. Phosphorylated self-peptides alter human leukocyte antigen class I-restricted antigen presentation and generate tumor-specific epitopes. *Proc Natl Acad Sci U S A.* 2009;106(8):2776-2781.
583. Ostrov DA, Grant BJ, Pompeu YA, et al. Drug hypersensitivity caused by alteration of the MHC-presented self-peptide repertoire. *Proc Natl Acad Sci U S A.* 2012;109(25):9959-9964.
584. Norcross MA, Luo S, Lu L, et al. Abacavir induces loading of novel self-peptides into HLA-B*57: 01. An autoimmune model for HLA-associated drug hypersensitivity. *AIDS.* 2012;26(11):F21-F29.
585. Mitaksov V, Truscott SM, Lybarger L, Connolly JM, Hansen TH, Fremont DH. Structural engineering of pMHC reagents for T cell vaccines and diagnostics. *Chem Biol.* 2007;14(8):909-922.
586. Celie PH, Toebes M, Rodenko B, Ovaa H, Perrakis A, Schumacher TNM. UV-induced ligand exchange in MHC class I protein crystals. *J Am Chem Soc.* 2009;131(34):12298-12304.
587. Saini SK, Tamhane T, Anjanappa R, et al. Empty peptide-receptive MHC class I molecules for efficient detection of antigen-specific T cells. *Sci Immunol.* 2019;4(37):eaau9039.
588. Saini SK, Ostermeir K, Ramnarayan VR, Schuster H, Zacharias M, Springer S. Dipeptides promote folding and peptide binding of MHC class I molecules. *Proc Natl Acad Sci U S A.* 2013;110(38):15383-15388.
589. Brown JH, Jardetzky T, Saper MA, Samraoui B, Bjorkman PJ, Wiley DC. A hypothetical model of the foreign antigen binding site of class II histocompatibility molecules. *Nature.* 1988;332(6167):845-850.
590. Brown JH, Jardetzky TS, Gorga JC, et al. Three-dimensional structure of the human class II histocompatibility antigen HLA-DR1. *Nature.* 1993;364(6432):33-39.
591. Wu TT, Kabat EA. An analysis of the sequences of the variable regions of Bence Jones proteins and myeloma light chains and their implications for antibody complementarity. *J Exp Med.* 1970;132(2):211-250.
592. Shannon CE. The mathematical theory of communication. 1963. *MD Comput.* 1997;14(4):306-317.
593. Strait BJ, Dewey TG. The Shannon information entropy of protein sequences. *Biophys J.* 1996;71(1):148-155.
594. Stern LJ, Brown JH, Jardetzky TS, et al. Crystal structure of the human class II MHC protein HLA-DR1 complexed with an influenza virus peptide. *Nature.* 1994;368(6468):215-221.
595. Fields BA, Ober B, Malchiodi EL, et al. Crystal structure of the V alpha domain of a T cell antigen receptor. *Science.* 1995;270(5243):1821-1824.
596. Schafer PH, Malapati S, Hanfelt KK, Pierce SK. The assembly and stability of MHC class II-$(\alpha\beta)_2$ superdimers. *J Immunol.* 1998;161(5):2307-2316.
597. Abastado JP, Lone YC, Casrouge A, Boulot G, Kourilsky P. Dimerization of soluble major histocompatibility complex-peptide complexes is sufficient for activation of T cell hybridoma and induction of unresponsiveness. *J Exp Med.* 1995;182(2):439-447.
598. Reich Z, Boniface JJ, Lyons DS, Borochov N, Wachtel EJ, Davis MM. Ligand-specific oligomerization of T-cell receptor molecules. *Nature.* 1997;387(6633):617-620.
599. Fremont DH, Hendrickson WA, Marrack P, Kappler J. Structures of an MHC class II molecule with covalently bound single peptides. *Science.* 1996;272(5264):1001-1004.
600. Fremont DH, Dai S, Chiang H, Crawford F, Marrack P, Kappler J. Structural basis of cytochrome c presentation by IE(k). *J Exp Med.* 2002;195(8):1043-1052.
601. Plaksin D, Chacko S, McPhie P, Bax A, Padlan EA, Margulies DH. A T cell receptor V alpha domain expressed in bacteria: does it dimerize in solution? *J Exp Med.* 1996;184(4):1251-1258.
602. Garboczi DN, Ghosh P, Utz U, Fan QR, Biddison WE, Wiley DC. Structure of the complex between human T-cell receptor, viral peptide and HLA-A2. *Nature.* 1996;384(6605):134-141.
603. Reinherz EL, Tan K, Tang L, et al. The crystal structure of a T cell receptor in complex with peptide and MHC class II. *Science.* 1999;286(5446):1913-1921.
604. Hennecke J, Carfi A, Wiley DC. Structure of a covalently stabilized complex of a human αβ T-cell receptor, influenza HA peptide and MHC class II molecule, HLA-DR1. *EMBO J.* 2000;19(21):5611-5624.
605. Natarajan K, McShan AC, Jiang J, et al. An allosteric site in the T-cell receptor Cβ domain plays a critical signalling role. *Nat Commun.* 2017;8:15260.
606. Corper AL, Stratmann T, Apostolopoulos V, et al. A structural framework for deciphering the link between I-Ag7 and autoimmune diabetes. *Science.* 2000;288(5465):505-511.
607. Zajonc DM. The CD1 family: serving lipid antigens to T cells since the Mesozoic era. *Immunogenetics.* 2016;68(8):561-576.
608. Zeng Z, Castaño AR, Segelke BW, Stura EA, Peterson PA, Wilson IA. Crystal structure of mouse CD1: an MHC-like fold with a large hydrophobic binding groove. *Science.* 1997;277(5324):339-345.
609. Gadola SD, Zaccai NR, Harlos K, et al. Structure of human CD1b with bound ligands at 2.3 A, a maze for alkyl chains. *Nat Immunol.* 2002;3(8):721-726.
610. Wang CR, Lindahl KF, Deisenhofer J. Crystal structure of the MHC class Ib molecule H2-M3. *Res Immunol.* 1996;147(5):313-321.
611. Li P, Morris DL, Willcox BE, Steinle A, Spies T, Strong RK. Complex structure of the activating immunoreceptor NKG2D and its MHC class I-like ligand MICA. *Nat Immunol.* 2001;2(5):443-451.
612. Li P, Willie ST, Bauer S, Morris DL, Spies T, Strong RK. Crystal structure of the MHC class I homolog MIC-A, a γδ T cell ligand. *Immunity.* 1999;10(5):577-584.
613. Burmeister WP, Gastinel LN, Simister NE, Blum ML, Bjorkman PJ. Crystal structure at 2.2 A resolution of the MHC-related neonatal Fc receptor. *Nature.* 1994;372(6504):336-343.

614. Li P, McDermott G, Strong RK. Crystal structures of RAE-1β and its complex with the activating immunoreceptor NKG2D. *Immunity.* 2002;16(1):77-86.
615. Kjer-Nielsen L, Patel O, Corbett AJ, et al. MR1 presents microbial vitamin B metabolites to MAIT cells. *Nature.* 2012;491(7426):717-723.
616. Carayannopoulos LN, Naidenko OV, Kinder J, Ho EL, Fremont DH, Yokoyama WM. Ligands for murine NKG2D display heterogeneous binding behavior. *Eur J Immunol.* 2002;32(3):597-605.
617. Loveland B, Wang CR, Yonekawa H, Hermel E, Lindahl KF. Maternally transmitted histocompatibility antigen of mice: a hydrophobic peptide of a mitochondrially encoded protein. *Cell.* 1990;60(6):971-980.
618. Wang CR, Castaño AR, Peterson PA, Slaughter C, Fischer Lindahl K, Deisenhofer J. Nonclassical binding of formylated peptide in crystal structure of the MHC class Ib molecule H2-M3. *Cell.* 1995;82(4):655-664.
619. Moody DB, Zajonc DM, Wilson IA. Anatomy of CD1-lipid antigen complexes. *Nat Rev Immunol.* 2005;5(5):387-399.
620. Calabi F, Jarvis JM, Martin L, Milstein C. Two classes of CD1 genes. *Eur J Immunol.* 1989;19(2):285-292.
621. Zajonc DM, Kronenberg M. CD1 mediated T cell recognition of glycolipids. *Curr Opin Struct Biol.* 2007;17(5):521-529.
622. Shamshiev A, Donda A, Carena I, Mori L, Kappos L, De Libero G. Self glycolipids as T-cell autoantigens. *Eur J Immunol.* 1999;29(5):1667-1675.
623. Rosat JP, Grant EP, Beckman EM, et al. CD1-restricted microbial lipid antigen-specific recognition found in the CD8+ αβ T cell pool. *J Immunol.* 1999;162(1):366-371.
624. Bendelac A, Rivera MN, Park SH, Roark JH. Mouse CD1-specific NK1 T cells: development, specificity, and function. *Annu Rev Immunol.* 1997;15:535-562.
625. Koch M, Stronge VS, Shepherd D, et al. The crystal structure of human CD1d with and without alpha-galactosylceramide. *Nat Immunol.* 2005;6(8):819-826.
626. Wun KS, Cameron G, Patel O, et al. A molecular basis for the exquisite CD1d-restricted antigen specificity and functional responses of natural killer T cells. *Immunity.* 2011;34(3):327-339.
627. Borg NA, Wun KS, Kjer-Nielsen L, et al. CD1d-lipid-antigen recognition by the semi-invariant NKT T-cell receptor. *Nature.* 2007;448(7149):44-49.
628. Pellicci DG, Clarke AJ, Patel O, et al. Recognition of beta-linked self glycolipids mediated by natural killer T cell antigen receptors. *Nat Immunol.* 2011;12(9):827-833.
629. Girardi E, Zajonc DM. Molecular basis of lipid antigen presentation by CD1d and recognition by natural killer T cells. *Immunol Rev.* 2012;250(1):167-179.
630. Simister NE, Mostov KE. Cloning and expression of the neonatal rat intestinal Fc receptor, a major histocompatibility complex class I antigen homolog. *Cold Spring Harb Symp Quant Biol.* 1989;54(pt 1):571-580.
631. Simister NE, Rees AR. Isolation and characterization of an Fc receptor from neonatal rat small intestine. *Eur J Immunol.* 1985;15(7):733-738.
632. Simister NE. IgG Fc receptors that resemble class I major histocompatibility complex antigens. *Biochem Soc Trans.* 1993;21(4):973-976.
633. Ahouse JJ, Hagerman CL, Mittal P, et al. Mouse MHC class I-like Fc receptor encoded outside the MHC. *J Immunol.* 1993;151(11):6076-6088.
634. Story CM, Mikulska JE, Simister NE. A major histocompatibility complex class I-like Fc receptor cloned from human placenta: possible role in transfer of immunoglobulin G from mother to fetus. *J Exp Med.* 1994;180(6):2377-2381.
635. Israel EJ, Wilsker DF, Hayes KC, Schoenfeld D, Simister NE. Increased clearance of IgG in mice that lack beta 2-microglobulin: possible protective role of FcRn. *Immunology.* 1996;89(4):573-578.
636. Pyzik M, Rath T, Kuo TT, et al. Hepatic FcRn regulates albumin homeostasis and susceptibility to liver injury. *Proc Natl Acad Sci U S A.* 2017;114(14):E2862-E2871.
637. Wyckoff SL, Hudson KE. Targeting the neonatal Fc receptor (FcRn) to treat autoimmune diseases and maternal-fetal immune cytopenias. *Transfusion.* 2021;61(5):1350-1354.
638. Kasahara M, Watanabe Y, Sumasu M, Nagata T. A family of MHC class I-like genes located in the vicinity of the mouse leukocyte receptor complex. *Proc Natl Acad Sci U S A.* 2002;99(21):13687-13692.
639. Kondo M, Maruoka T, Otsuka N, et al. Comparative genomic analysis of mammalian NKG2D ligand family genes provides insights into their origin and evolution. *Immunogenetics.* 2010;62(7):441-450.
640. Watanabe Y, Maruoka T, Walter L, Kasahara M. Comparative genomics of the MILL family: a rapidly evolving MHC class I gene family. *Eur J Immunol.* 2004;34(6):1597-1607.
641. Kajikawa M, Baba T, Tomaru U, et al. MHC class I-like MILL molecules are β2-microglobulin-associated, GPI-anchored glycoproteins that do not require TAP for cell surface expression. *J Immunol.* 2006;177(5):3108-3115.
642. Rabinovich BA, Ketchem RR, Wolfson M, Goldstein L, Skelly M, Cosman D. A role for the MHC class I-like MILL molecules in nutrient metabolism and wound healing. *Immunol Cell Biol.* 2008;86(6):489-496.
643. Kajikawa M, Ose T, Fukunaga Y, et al. Structure of MHC class I-like MILL2 reveals heparan-sulfate binding and interdomain flexibility. *Nat Commun.* 2018;9(1):4330.
644. Olson R, Huey-Tubman KE, Dulac C, Bjorkman PJ. Structure of a pheromone receptor-associated MHC molecule with an open and empty groove. *PLoS Biol.* 2005;3(8):e257.
645. Yang Z, Bjorkman PJ. Structure of UL18, a peptide-binding viral MHC mimic, bound to a host inhibitory receptor. *Proc Natl Acad Sci U S A.* 2008;105(29):10095-10100.
646. Natarajan K, Hicks A, Mans J, et al. Crystal structure of the murine cytomegalovirus MHC-I homolog m144. *J Mol Biol.* 2006;358(1):157-171.
647. Zhi L, Mans J, Paskow MJ, et al. Direct interaction of the mouse cytomegalovirus m152/gp40 immunoevasin with RAE-1 isoforms. *Biochemistry.* 2010;49(11):2443-2453.
648. Yang Z, West AP Jr, Bjorkman PJ. Crystal structure of TNFalpha complexed with a poxvirus MHC-related TNF binding protein. *Nat Struct Mol Biol.* 2009;16(11):1189-1191.
649. Schneider CL, Hudson AW. The human herpesvirus-7 (HHV-7) U21 immunoevasin subverts NK-mediated cytoxicity through modulation of MICA and MICB. *PLoS Pathog.* 2011;7(11):e1002362.
650. Senkevich TG, Moss B. Domain structure, intracellular trafficking, and beta2-microglobulin binding of a major histocompatibility complex class I homolog encoded by molluscum contagiosum virus. *Virology.* 1998;250(2):397-407.
651. Smith LM, Shellam GR, Redwood AJ. Genes of murine cytomegalovirus exist as a number of distinct genotypes. *Virology.* 2006;352(2):450-465.
652. Treiner E, Duban L, Bahram S, et al. Selection of evolutionarily conserved mucosal-associated invariant T cells by MR1. *Nature.* 2003;422(6928):164-169.
653. Garcia KC, Degano M, Stanfield RL, et al. An αβ T cell receptor structure at 2.5 A and its orientation in the TCR-MHC complex. *Science.* 1996;274(5285):209-219.
654. Reiser JB, Darnault C, Guimezanes A, et al. Crystal structure of a T cell receptor bound to an allogeneic MHC molecule. *Nat Immunol.* 2000;1(4):291-297.
655. Garcia KC, Degano M, Pease LR, et al. Structural basis of plasticity in T cell receptor recognition of a self peptide-MHC antigen. *Science.* 1998;279(5354):1166-1172.
656. Jerne NK. The somatic generation of immune recognition. *Eur J Immunol.* 1971;1(1):1-9.
657. Marrack P, Scott-Browne JP, Dai S, Gapin L, Kappler JW. Evolutionarily conserved amino acids that control TCR-MHC interaction. *Annu Rev Immunol.* 2008;26:171-203.
658. Rossjohn J, Gras S, Miles JJ, Turner SJ, Godfrey DI, McCluskey J. T cell antigen receptor recognition of antigen-presenting molecules. *Annu Rev Immunol.* 2015;33:169-200.
659. Szeto C, Lobos CA, Nguyen AT, Gras S. TCR recognition of peptide-MHC-I: rule makers and breakers. *Int J Mol Sci.* 2020;22(1):68.
660. Gras S, Chadderton J, Del Campo CM, et al. Reversed T cell receptor docking on a major histocompatibility class I complex limits involvement in the immune response. *Immunity.* 2016;45(4):749-760.
661. Beringer DX, Kleijwegt FS, Wiede F, et al. T cell receptor reversed polarity recognition of a self-antigen major histocompatibility complex. *Nat Immunol.* 2015;16(11):1153-1161.
662. Gras S, Burrows SR, Kjer-Nielsen L, et al. The shaping of T cell receptor recognition by self-tolerance. *Immunity.* 2009;30(2):193-203.
663. Adams JJ, Narayanan S, Liu B, et al. T cell receptor signaling is limited by docking geometry to peptide-major histocompatibility complex. *Immunity.* 2011;35(5):681-693.
664. Crispe IN, Husmann LA, Bevan MJ. T cell receptor expression and receptor-mediated induction of clonal growth in the developing mouse thymus. High surface beta-chain density is a requirement for functional maturity. *Eur J Immunol.* 1986;16(10):1283-1288.
665. Devine L, Kavathas PB. Molecular analysis of protein interactions mediating the function of the cell surface protein CD8. *Immunol Res.* 1999;19(2-3):201-210.
666. Gao GF, Jakobsen BK. Molecular interactions of coreceptor CD8 and MHC class I: the molecular basis for functional coordination with the T-cell receptor. *Immunol Today.* 2000;21(12):630-636.
667. Gao GF, Tormo J, Gerth UC, et al. Crystal structure of the complex between human CD8α(α) and HLA-A2. *Nature.* 1997;387(6633):630-634.
668. Kern PS, Teng Mk, Smolyar A, et al. Structural basis of CD8 coreceptor function revealed by crystallographic analysis of a murine CD8αα ectodomain fragment in complex with H-2Kb. *Immunity.* 1998;9(4):519-530.
669. Maddon PJ, Dalgleish AG, McDougal JS, Clapham PR, Weiss RA, Axel R. The T4 gene encodes the AIDS virus receptor and is expressed in the immune system and the brain. *Cell.* 1986;47(3):333-348.
670. Wu H, Kwong PD, Hendrickson WA. Dimeric association and segmental variability in the structure of human CD4. *Nature.* 1997;387(6632):527-530.
671. Li Y, Yin Y, Mariuzza RA. Structural and biophysical insights into the role of CD4 and CD8 in T cell activation. *Front Immunol.* 2013;4:206.
672. Sugata K, Matsunaga Y, Yamashita Y, et al. Affinity-matured HLA class II dimers for robust staining of antigen-specific CD4(+) T cells. *Nat Biotechnol.* 2021;39(8):958-967.
673. Natarajan K, Dimasi N, Wang J, Mariuzza RA, Margulies DH. Structure and function of natural killer cell receptors: multiple molecular solutions to self, nonself discrimination. *Annu Rev Immunol.* 2002;20:853-885.
674. Radaev S, Rostro B, Brooks AG, Colonna M, Sun PD. Conformational plasticity revealed by the cocrystal structure of NKG2D and its class I MHC-like ligand ULBP3. *Immunity.* 2001;15(6):1039-1049.
675. Saunders PM, Pymm P, Pietra G, et al. Killer cell immunoglobulin-like receptor 3DL1 polymorphism defines distinct hierarchies of HLA class I recognition. *J Exp Med.* 2016;213(5):791-807.
676. Matsumoto N, Mitsuki M, Tajima K, Yokoyama WM, Yamamoto K. The functional binding site for the C-type lectin-like natural killer cell receptor Ly49A spans three domains of its major histocompatibility complex class I ligand. *J Exp Med.* 2001;193(2):147-158.
677. Wang J, Whitman MC, Natarajan K, Tormo J, Mariuzza RA, Margulies DH. Binding of the natural killer cell inhibitory receptor Ly49A to its major histocompatibility complex class I ligand - crucial contacts include both H-2D(d) and beta(2)-microglobulin. *J Biol Chem.* 2002;277(2):1433-1442.
678. Dimasi N, Sawicki MW, Reineck LA, et al. Crystal structure of the Ly49I natural killer cell receptor reveals variability in dimerization mode within the Ly49 family. *J Mol Biol.* 2002;320(3):573-585.
679. Dam J, Guan R, Natarajan K, et al. Variable MHC class I engagement by Ly49 natural killer cell receptors demonstrated by the crystal structure of Ly49C bound to H-2K(b). *Nat Immunol.* 2003;4(12):1213-1222.
680. Held W, Mariuzza RA. Cis interactions of immunoreceptors with MHC and non-MHC ligands. *Nat Rev Immunol.* 2008;8(4):269-278.
681. Jiang J, Natarajan K, Boyd LF, Morozov GI, Mage MG, Margulies DH. Crystal structure of a TAPBPR-MHC I complex reveals the mechanism of peptide editing in antigen presentation. *Science.* 2017;358(6366):1064-1068.
682. Thomas C, Tampe R. Structure of the TAPBPR-MHC I complex defines the mechanism of peptide loading and editing. *Science.* 2017;358(6366):1060-1064.
683. Herman A, Kappler JW, Marrack P, Pullen AM. Superantigens: mechanism of T-cell stimulation and role in immune responses. *Annu Rev Immunol.* 1991;9:745-772.
684. Rovira P, Buckle M, Abastado JP, Peumans WJ, Truffa-Bachi P. Major histocompatibility class I molecules present *Urtica dioica* agglutinin, a superantigen of vegetal origin, to T lymphocytes. *Eur J Immunol.* 1999;29(5):1571-1580.
685. Harata K, Muraki M. Crystal structures of *Urtica dioica* agglutinin and its complex with tri-N-acetylchitotriose. *J Mol Biol.* 2000;297(3):673-681.
686. Saul FA, Rovira P, Boulot G, et al. Crystal structure of *Urtica dioica* agglutinin, a superantigen presented by MHC molecules of class I and class II. *Structure.* 2000;8(6):593-603.
687. Papageorgiou AC, Acharya KR. Microbial superantigens: from structure to function. *Trends Microbiol.* 2000;8(8):369-375.
688. Jardetzky TS, Brown JH, Gorga JC, et al. Three-dimensional structure of a human class II histocompatibility molecule complexed with superantigen. *Nature.* 1994;368(6473):711-718.
689. Kim J, Urban RG, Strominger JL, Wiley DC. Toxic shock syndrome toxin-1 complexed with a class II major histocompatibility molecule HLA-DR1. *Science.* 1994;266(5192):1870-1874.
690. Wang L, Zhao Y, Li Z, et al. Crystal structure of a complete ternary complex of TCR, superantigen and peptide-MHC. *Nat Struct Mol Biol.* 2007;14(2):169-171.
691. Saline M, Rödström KEJ, Fischer G, Orekhov VY, Karlsson BG, Lindkvist-Petersson K. The structure of superantigen complexed with TCR and MHC reveals novel insights into superantigenic T cell activation. *Nat Commun.* 2010;1:119.
692. Fremont DH, Rees WA, Kozono H. Biophysical studies of T-cell receptors and their ligands. *Curr Opin Immunol.* 1996;8(1):93-100.
693. Matsui K, Boniface JJ, Reay PA, Schild H, de StGroth BF, Davis MM. Low affinity interaction of peptide-MHC complexes with T cell receptors. *Science.* 1991;254(5039):1788-1791.
694. Matsui K, Boniface JJ, Steffner P, Reay PA, Davis MM. Kinetics of T-cell receptor binding to peptide/I-Ek complexes: correlation of the dissociation rate with T-cell responsiveness. *Proc Natl Acad Sci U S A.* 1994;91(26):12862-12866.
695. Khilko SN, Corr M, Boyd LF, Lees A, Inman JK, Margulies DH. Direct detection of major histocompatibility complex class I binding to antigenic peptides using surface plasmon resonance. Peptide immobilization and characterization of binding specificity. *J Biol Chem.* 1993;268(21):15425-15434.
696. Corr M, Slanetz AE, Boyd LF, et al. T cell receptor-MHC class I peptide interactions: affinity, kinetics, and specificity. *Science.* 1994;265(5174):946-949.
697. Khilko SN, Jelonek MT, Corr M, Boyd LF, Bothwell AL, Margulies DH. Measuring interactions of MHC class I molecules using surface plasmon resonance. *J Immunol Methods.* 1995;183(1):77-94.

698. Margulies D, Plaksin D, Khilko S, Jelonek MT. Studying interactions involving the T-cell antigen receptor by surface plasmon resonance. *Curr Opin Immunol.* 1996;8(2):262-270.
699. Natarajan K, Boyd LF, Schuck P, Yokoyama WM, Eilat D, Margulies DH. Interaction of the NK cell inhibitory receptor Ly49A with H-2D(d): identification of a site distinct from the TCR site. *Immunity.* 1999;11(5):591-601.
700. Fooksman DR, Dustin ML. Affinity measured by microcluster. *J Exp Med.* 2010;207(5):907-909.
701. Sabatino JJ Jr, Huang J, Zhu C, Evavold BD. High prevalence of low affinity peptide-MHC II tetramer-negative effectors during polyclonal CD4+ T cell responses. *J Exp Med.* 2011;208(1):81-90.
702. Jiang N, Huang J, Edwards LJ, et al. Two-stage cooperative T cell receptor-peptide major histocompatibility complex-CD8 trimolecular interactions amplify antigen discrimination. *Immunity.* 2011;34(1):13-23.
703. Huang J, Zarnitsyna VI, Liu B, et al. The kinetics of two-dimensional TCR and pMHC interactions determine T-cell responsiveness. *Nature.* 2010;464(7290):932-936.
704. Sgourakis NG, May NA, Boyd LF, Ying J, Bax A, Margulies DH. A novel MHC-I surface targeted for binding by the MCMV m06 immunoevasin revealed by solution NMR. *J Biol Chem.* 2015;290(48):28857-28868.
705. Natarajan K, Jiang J, May NA, et al. The role of molecular flexibility in antigen presentation and T cell receptor-mediated signaling. *Front Immunol.* 2018;9:1657.
706. Varani L, Bankovich AJ, Liu CW, et al. Solution mapping of T cell receptor docking footprints on peptide-MHC. *Proc Natl Acad Sci U S A.* 2007;104(32):13080-13085.
707. Fahmy TM, Bieler JG, Schneck JP. Probing T cell membrane organization using dimeric MHC-Ig complexes. *J Immunol Methods.* 2002;268(1):93-106.
708. Bethune MT, Comin-Anduix B, Hwang Fu YH, Ribas A, Baltimore D. Preparation of peptide-MHC and T-cell receptor dextramers by biotinylated dextran doping. *Biotechniques.* 2017;62(3):123-130.
709. Hanke T, Takizawa H, McMahon CW, et al. Direct assessment of MHC class I binding by seven Ly49 inhibitory NK cell receptors. *Immunity.* 1999;11(1):67-77.
710. Cameron T, Norris PJ, Patel A, et al. Labeling antigen-specific CD4(+) T cells with class II MHC oligomers. *J Immunol Methods.* 2002;268(1):51.
711. Hugues S, Malherbe L, Filippi C, Glaichenhaus N. Generation and use of alternative multimers of peptide/MHC complexes. *J Immunol Methods.* 2002;268(1):83.
712. Malherbe L, Filippi C, Julia V, et al. Selective activation and expansion of high-affinity CD4+ T cells in resistant mice upon infection with Leishmania major. *Immunity.* 2000;13(6):771-782.
713. Lebowitz MS, O'Herrin SM, Hamad AR, et al. Soluble, high-affinity dimers of T-cell receptors and class II major histocompatibility complexes: biochemical probes for analysis and modulation of immune responses. *Cell Immunol.* 1999;192(2):175-184.
714. Crawford F, Kozono H, White J, Marrack P, Kappler J. Detection of antigen-specific T cells with multivalent soluble class II MHC covalent peptide complexes. *Immunity.* 1998;8(6):675-682.

CHAPTER 20

Classical Antigen Processing and Presentation

Jonathan W. Yewdell, MD, PhD • Paul Roche, PhD • Laurence C. Eisenlohr, VMD, PhD

ANTIGEN PRESENTATION PATHWAYS

T cell responses are a central element in host immunity to pathogens, abnormal cells, and toxins. Remarkably, T cells discriminate self from nonself by recognizing foreign peptides displayed on the cell surface by self–major histocompatibility complex (MHC) molecules. When naïve T cells encounter professional antigen-presenting cells (pAPCs) displaying foreign peptides, they become activated, leading to their rapid division and differentiation into effector and memory cells. T cells specific for MHC molecules presenting peptides from self-proteins are either eliminated during their ontogeny by negative selection or suppressed by peripheral tolerance mechanisms. At the molecular level, T cell recognition is based on the remarkable discriminatory power of the T cell receptor (TCR). The mechanisms by which peptide ligands for MHC molecules are generated are referred to as *antigen processing*, and the mechanisms that convey peptide/MHC complexes to the surface of APCs is called *antigen presentation*.

Although there are many exceptions, $CD8^+$ T cells typically recognize classical MHC class I molecules (class I) complexed with peptides generated from all types of proteins synthesized by the presenting cell's ribosomes. In contrast, $CD4^+$ T cells recognize MHC class II molecules with peptides generated from extracellular antigens in endocytic compartments. Further generalizing, peptide loading onto class I and class II molecules occurs, respectively, in the endoplasmic reticulum (ER) and endosomes/lysosomes. These are referred to as the classical class I and class II antigen processing pathways, which, respectively, generally present endogenous versus exogenous antigens. Differences in class I and class II molecules presentation are governed by their differential interaction with dedicated and general purpose molecular chaperones, with the system evolving to optimize T cell responses to different types of threats to the host. Note that a wide variety of *nonclassical* class I molecules are present in all vertebrate species and present a variety of self and nonself peptidic and nonpeptidic antigens, including lipids and metabolites.[1] In humans, the three classical class I molecules (HLA-A, HLA-B, and HLA-C) are outnumbered by the 16 nonclassical class I molecules. In other species, these numbers can vary considerably, with some species (eg, mice) expressing dozens of nonclassical class I molecules, and others (eg, cynomolgus monkeys) expressing at least 12 classical class I molecules.

MHC CLASS I ANTIGEN PROCESSING AND PRESENTATION

Overview of MHC Class I Processing and Presentation

Snell and Gorer discovered the class I system in the 1940s as polymorphic genes that govern the rejection of tissues and tumors transplanted between mice. From that time, it was generally assumed that tissue transplantation was a human artifact, and therefore that MHC-based evolution and polymorphism must have been driven by other selection factors. Be careful of your assumptions! In the past 15 years, natural tumor transmission between dogs[2] and Tasmanian devils[3] (at the brink of extinction due to the tumor) has been clearly established. This raises the possibility that the polymorphic family genes encoding class I, which is expressed on nearly all nucleated cells in jawed vertebrates, evolved, at least in part, to limit transmissible tumors.

This development is highly ironic since for decades, the relevance of class I–mediated tumor immunosurveillance was questioned, even ridiculed by mainstream immunologists. Rather, Doherty and Zinkernagel's discovery of class I–restricted $CD8^+$ T cell recognition[4] provided a facile explanation for the evolution of immunosurveillance based on TCR recognition of a self-molecule (class I) and a foreign (viral) antigen. The puzzling molecular nature of MHC restriction was vigorously debated until it was found, first for class II[5] and shortly thereafter for class I[6] that a clonally restricted TCR recognizes an MHC class I or II molecule that presents peptides derived from foreign or self-antigens.

Inasmuch as class I is ubiquitously expressed on nucleated cells, the class I antigen processing and presentation system is a basic feature of vertebrate cell biology. This system's essential element is to present peptides from "endogenous" antigens, proteins synthesized by the APC's own ribosomes. The system also functions to present extracellular antigens (exogenous antigens) delivered to the cytosol, as first shown for viral proteins that enter the cell during virus penetration.[7] Indeed, presentation of exogenous antigen is a feature of dendritic cells (DCs), which play a special role in activating naïve T cells in lymphoid tissues. Activation requires class I antigen complex-specific TCR signaling (Chapters 24 and 25) and signaling triggered by DC surface molecules (eg, CD86). Such presentation is termed *cross-presentation* and is based on the delivery of endocytosed antigens to the cytosol and the processing of antigens in the endosomes themselves. The cell biology of DC cross-presentation is detailed in Chapter 35.

Cross-presentation is the sole means of activating naïve T cells specific for tumor antigens and antigens from viruses that cannot infect DCs. For viruses that infect DCs, direct presentation[8] is augmented by cross-priming.[9,10] Importantly, cross-priming makes it extremely difficult for viruses to subvert the induction of $CD8^+$ T cell responses since the relevant viral gene products (*evasins*) would have to function in *trans* in cross-priming DC. Indeed, $CD8^+$ T cell response to herpes viruses, which possess an impressive armamentarium for blocking endogenous class I presentation, is often massive. Viral MHC class I and II evasins are described at the end of this chapter.

It is important to emphasize that what follows, like all textbook descriptions, is not written in stone, but is a condensed, consensus view on the central elements in class I biogenesis. As with any scientific field, the "facts" change as new evidence is presented, and it must be admitted, who is weighing the existing evidence. Progress is made when new experiments provide data inconsistent with existing models, prompting new hypotheses and exploration. It is, of course, impossible to know what you do not know, and young scientists are implored to question everything.[11]

Origin of Class I–Binding Peptides

As detailed in Chapter 20, class I proteins, which consist of a polymorphic heavy chain (HC) noncovalently bound to beta-2 microglobulin (β_2m), preferentially bind peptides of 8 to 11 amino acids.[12] Class I allele-specific polymorphic residues in the peptide-binding groove largely determine peptide specificity.[12,13] Each class I allomorph typically binds peptides with a consensus binding motif requiring a relatively specific amino acid central anchor and a C-terminal hydrophobic or—for some allomorphs—positively charged amino acid. Given the degeneracy in most peptide positions and high ratio of proteins synthesized by each cell (~3 million ribosomes each synthesizing one 400-residue protein per min, each with >400 potential class I–binding peptides of various lengths) relative to the rate of class I synthesis (hundreds/minute),[14] the selective processes controlling which peptides are presented are of considerable importance for the detection of pathogens and malignancies. The problem is exacerbated by ~250 housekeeping proteins accounting for 75% of the proteome by mass. Mass spectrometric characterization of the "immunopeptidome," the repertoire of peptide presented by MHC molecules, demonstrates that peptide presentation is not merely a molar representation of the proteome.[15,16]

CD8$^+$ T cell recognition of virus-infected cells would be of limited function in a strict law of mass action system of peptide presentation. Viruses can replicate in far less time than it takes for viral proteins to attain sufficient representation in the proteome to enable presentation. Yet, CD8$^+$ T cells can recognize virus-infected cells within an hour following infection[17] despite the viral source proteins being degraded with kinetics similar to host proteins (median of 46 hours across the proteome). This finding prompted the DRiP hypothesis: peptides are generated from translation products (**d**efective **r**ibosomal **p**roducts) that fail to fold due to errors in transcription, translation, or folding.[18,19] Initial supporting evidence for the DRiP hypothesis came from studies showing that, first, up to a-third of nascent translation products are degraded with a half-life of <10 minutes[20] and, second, cytosolic peptides transported by transporter associated with antigen presentation (TAP) are rapidly depleted upon blocking protein synthesis.[21] The rapid generation of peptides from ostensibly stable viral proteins now appears to the rule rather than the exception.[22] For host cell peptides, there appears to be a greater contribution of "retirees" proteins that have reached their expiration date and are degraded to maintain cellular homeostasis[16] (Fig. 20.1).

On top of this, peptides from noncanonical translation products are overrepresented in the immunopeptidome. This includes short open reading frames (ORFs) initiated on CUG and other non-AUG start codons, a process that is enhanced by cellular stress or exposure to proinflammatory cytokines and occurs independently of eIF2α is a preferred source of class I peptides.[28] Peptides generated from introns and 3′ and 5′ "nontranslated ORFs" in mature host mRNAs are also overrepresented in the immunopeptidome and are promising targets for immunotherapy of cancers, many of which exhibit translational dysregulation.[29] Indeed, unexpected translation from introns in nuclear pre-mRNAs[30] and the negative strand of negative strand viruses[31] appear to be a highly efficient source of antigenic peptides. Such translation may occur in the nucleus itself and may even generate antigenic peptides from exons.[32]

The many apparent non–mass action examples of peptide generation[33] led to the immunoribosome hypothesis[19] that a subset of ribosomes generates DRiPs whose peptide products have preferential access to TAP. Ribosomes functioning in an eIF2α-independent manner would meet this definition. There is also recent evidence that ribosomes lacking individual protein subunits (ribosomes have 80 structural proteins) vary in their ability to generate peptides from viral and cellular mRNA and, further, that tumor cells can avoid immunosurveillance by downregulating individual ribosome proteins.[34] DRiP-derived peptides appear to have preferential access to TAP, as their presentation by class I is not inhibited by excess amounts of minigene encoded peptides, which readily compete with others for class I binding in the ER.[35] This is consistent with a model[33] in which nascent proteins from ribosomes are channeled to proteasomes whose products are then channeled to TAP. Multiple rounds of translation/degradation from DRiPs from a single mRNA could account for the generation of clusters of class I molecules containing the same peptide,[26] which are maintained throughout the secretory pathway and plasma membrane, where their clustering enhances the sensitivity of T cell recognition.[26,36,37]

Mass action–independent peptide generation must also underlie the process of cross-presentation, in which minute amounts of viral,[7] bacterial,[38,39] or host proteins[40] introduced into cells generate peptides in the face of an overwhelming amount of degradation products from DRiPs and retirees. This is facilitated in DCs by sequestering nascent DRiPs into membrane-free proteinaceous depots[41] but also occurs in nonprofessional APCs.[42]

Antigenic peptides are commonly derived from all types of proteins, including proteins exported from cells via the ER and the Golgi complex (GC). The generation of peptides from *secreted* proteins (chicken ovalbumin and vaccinia virus B8R proteins are the poster children) provides a clear and absolute example of the contribution of DRiPs to the class I immunopeptidome, as the mature protein can hardly be a source of peptides, given its extracellular location. DRiPs are generated by multiple pathways, including mistranslation, misfolding, and mistargeting. A number of studies have examined the nature of DRiPs from ER-targeted proteins. Peptides arise from proteins that fail to be targeted to the ER[43] as well as proteins that misfold in the ER and are returned to the cytosol via the ERAD (ER-associated degradation) pathway.[44,45] As with cytosolic proteins,[46] misfolding in the ER does not always increase peptide supply.[47] The take-home message is that there is an inconstant relationship between the biochemical fate of a protein and its efficiency as a peptide source. As an extreme but telling example, upon removing a stabilizing

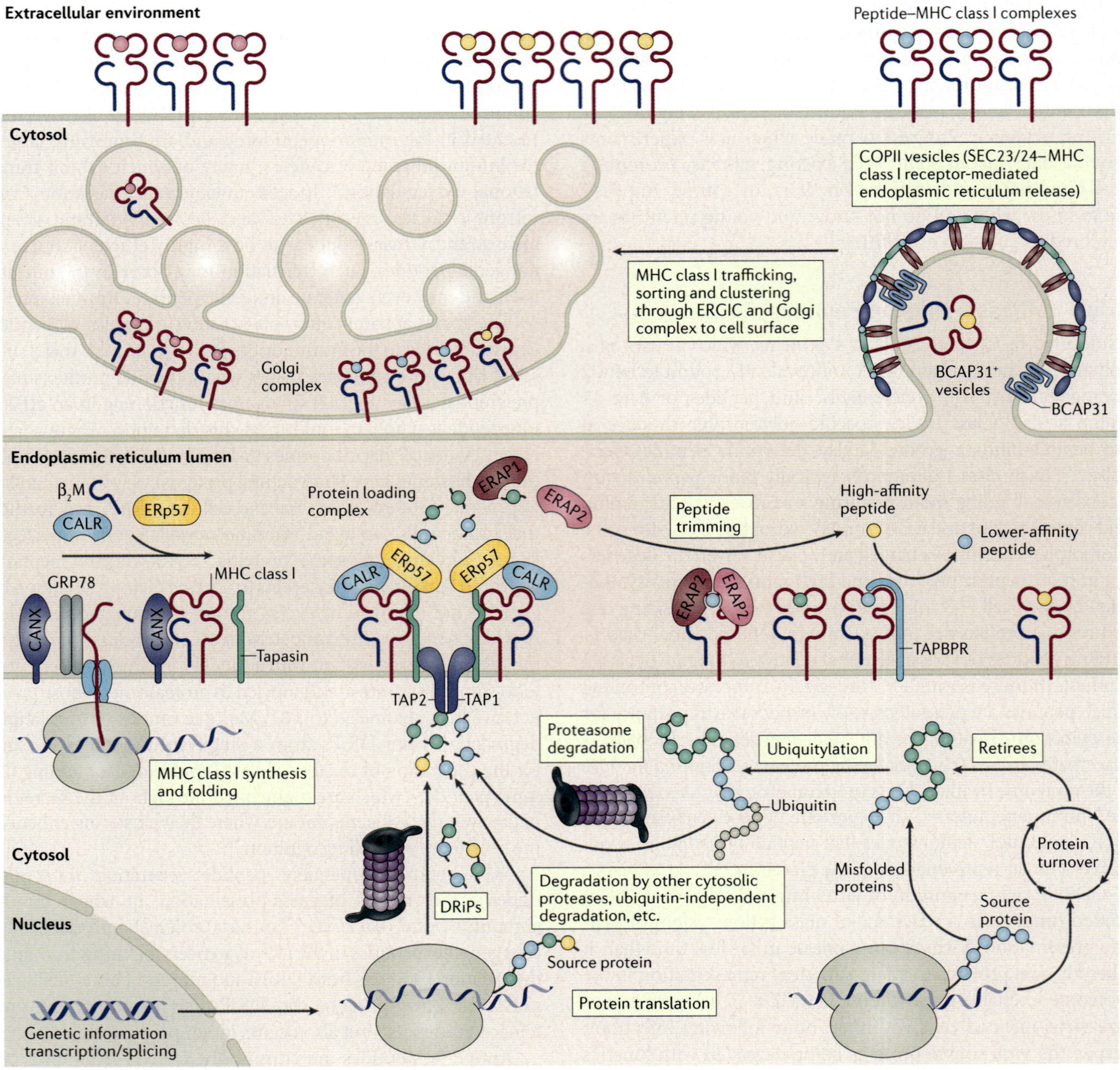

FIG. 20.1. Overview of MHC class I biogenesis, antigen processing, and presentation machinery. Peptides are generated from the degradation of source proteins, which can be categorized as "retirees" (natural turnover of functional proteins) or defective ribosomal products (DRiPs, rapidly degraded polypeptides not integrated into the proteome). Degradation is mediated by proteasomes via ubiquitin-dependent and ubiquitin-independent targeting and by nonproteasomal proteases. Peptides that escape destruction in the cytosol can be imported to the endoplasmic reticulum (ER) by the transporter associated with antigen presentation (TAP). Trimming of amino termini occurs via the activities of ER aminopeptidase 1 (ERAP1) and ERAP2, and the peptide loading complex assists in the loading and folding of MHC class I molecules with both peptide and β2-microglobulin (β2m). Folded MHC class I complexed with a high-affinity peptide is trafficked from the ER with the chaperone B cell receptor–associated protein 31 (BCAP31)[23,24] as an additional quality control step.[25] There is evidence that MHC class I molecules are clustered[24] based on their bound peptides,[26,27] perhaps due to localized peptide generation from individual mRNAs. Such clustering, which is detectable in the Golgi complex and cell surface, increases T cell sensitivity. MHC, major histocompatibility complex. (Reprinted by permission from Springer Nature: Dersh D, Holly J, Yewdell JW. A few good peptides: MHC class I-based cancer immunosurveillance and immunoevasion. *Nat Rev Immunol.* 2020;20(10):644.)

ligand, ostensibly identical native proteins generated hours previously from the same mRNA sequence demonstrated an approximately eightfold difference in generating peptide class I complexes depending on whether the mRNA was generated from a virus-encoded versus transfected gene.[48]

As DRiPs at the population level remain poorly characterized biochemically, it is not possible to estimate the efficiency of peptide generation from DRiPs relative to retirees. A general finding, however, is that targeting of protein for rapid generation typically has only a modest effect on increasing levels of peptide generation. As a practical point, this means that little is gained in using this strategy in vaccines meant to elicit CD8⁺ T cell responses. Indeed, rapid degradation of a vaccine antigen can be counterproductive inasmuch it will reduce cross-priming, which is usually based on DC acquisition of proteins and not protease-generated peptides.[49-51]

Proteolytic Liberation of Antigenic Peptides

A large fraction of cellular protein degradation is mediated by the proteasome, an abundant (~500,000 copies per mammalian cell) cylindrical protease formed by four stacked rings of seven subunits each (α7β7β7α7) that sediments at 20 Svedbergs (20S).[52] Proteasomes are present in eukaryotes, archaea, and actinobacteria. The inner two rings of the 20S proteasome are assembled from the seven β subunits, three of which are proteases with chymotryptic- (β5), tryptic- (β2), or caspase-like activity (β1). Thus, each proteasome has six active sites with the aggregate ability to cleave after most amino acid residues, although to differing efficiencies depending on the flanking residues. The catalytic sites are exposed on the interior of the central chamber of the 20S cylinder. The two outer rings of the 20S proteasome are assembled from seven different α subunits. The ends of the barrel are closed but can be gated open to generate a narrow channel that extended polypeptide chains can be threaded through. The gate is controlled by a 10-protein complex, termed the 19S regulatory particle, that can bind to each end of the 20S proteasome. The full assemblage of 19S-20S-19S subunits constitutes the 26S proteasome. The lid of the 19S regulator particle binds ubiquitylated proteins and deubiquitylates substrates before they gain entry to the 20S core; the base of the 19S regulatory particle contains ATPases that promote substrate unfolding and open the gate of the 20S proteasome to deliver denatured protein substrate to the proteolytic chamber.

Most proteasome substrates are targeted to proteasomes by ubiquitin (Ub), a highly abundant (up to 10^8 molecules/cell) 76-residue protein. Ub is typically conjugated to internal lysine residues of proteins substrates, but coupling can also occur at the N-terminus or on internal cysteine, serine, or threonine residues. In this highly choreographed process, activated Ub is transferred from Ub-activating enzyme E1 to one of 30 to 40 mammalian Ub conjugating enzymes (E2s) and then to the substrate that is bound by one of the hundreds of different ubiquitin protein ligases (E3s). The E3s are the major determinants of substrate specificity. This process can be repeated to form polyUb chains whereupon the next Ub moiety is added to one of the seven internal lysine residues of Ub. Proteins coupled with polyUb chains of four or more Ub moieties linked through Lys48 Ub residues are the prototypic signal for proteasome-mediated degradation.

The most important function of the proteasome is to recycle amino acids from unwanted proteins. However, in jawed vertebrates, modifications to the proteasome are made to promote antigen presentation during the inflammatory response.[53] Among the 1000+ genes induced by interferon gamma (IFN-γ) and TNF-α are alternative versions of the three catalytic proteasome subunits: β1i (LMP2), β2i (MECL-1), and β5i (LMP7), two of which (β1i and β5i) are encoded by MHC genes.[54] To facilitate their rapid induction, immunoproteasomes are assembled more rapidly than constitutive proteasomes. In immunoproteasomes, all six of the catalytic subunits are substituted by immunosubunits. In hybrid proteasomes, which are not unusual, up to five of the constitutive subunits are substituted. There is considerable overlap in the cleavage products of proteasomes versus hybrid proteasomes, while ~50% immunoproteasome peptides are generated at lower levels by proteasomes, and vice versa, as shown using mice lacking all three immunosubunit genes.[55] Thus, peptide generation may not be the only or even principal function of immunoproteasomes. Indeed, in many immune cell types, including lymphocytes, which are not considered pAPCs for CD8+ T cells, immunoproteasomes are the predominant form of proteasome in the absence of cytokine exposure. In mice lacking one or two immunosubunit genes, both B and T lymphocytes and DCs exhibit reduced proliferation and function associated with altered NF-κb activity.[56,57] While controversial,[58] there is compelling evidence that immunoproteasomes may have evolved to specialize in degrading oxidatively damaged proteins, whose generation is enhanced during inflammatory stress and rapid cell division.[59]

A third type of proteasome, present only in cortical thymic epithelial cells, consists of β1i, β2i, and a unique subunit β5t (thymus). Consistent with the function of these cells in positively selecting self-MHC restricted T cells, thymoproteasomes appear to selectively generate peptides that disfavor high-affinity TCR interactions, facilitating selection of self-restricted T cells that are tolerant to extrathymic self-peptides.[60] Like the catalytic immunoproteasome subunits, β5t may have additional functions, as its expression in thymocytes regulates nearly a thousand genes.[61]

In addition to βi proteasome subunits, proinflammatory cytokines also induce expression of the PA28α and PA28β subunits that associate to form the 11S proteasome regulator, which can replace 19S regulators. 11S regulators do not function in degrading ubiquitylated proteins, but they do induce conformational changes to open the 20S proteasome gate. This enhances the ability of peptides to enter and leave the catalytic chamber, potentially exerting opposite effects on generating antigenic peptides. The net effects of 11S regulators on antigen presentation published to date are modest, suggesting other biological functions. 11S regulators are also created by PA28γ subunits, which are not cytokine controlled, and are largely expressed in the nucleus. These regulators have several crucial cellular functions, but a recent report indicates that they may also enhance degradation of peptides generated in the nucleus.[62]

The involvement of proteasomes in creating the immunopeptidome is not as clear-cut as commonly concluded. Complete drug blockade of proteasome activity as confirmed by inhibition of total cellular or individual protein degradation can have complete to no discernible effect on the cell surface expression of individual peptide class I complexes or class I allomorphs.[63,64] Mass spectrometric analysis of the immunopeptidome proteasome inhibited cells revealed the ongoing presentation of a highly diverse repertoire of peptides.[65] Complete blockade of ubiquitylation using an E1-Ub ligase inhibitor has similar disparate effects on individual peptides and class I allomorphs.[66] Thus, while the Ub-26S proteasome pathway is a major source of antigenic peptides, peptides are likely also generated wholesale by 20S proteasomes in a Ub-independent manner, as well as by other cytosolic proteases. Complicating matters, until recently, proteasomes were believed to be exclusively present in the cytosol and nucleus. A substantial fraction (≈15%), however, also appears to be present in endosomes/phagosomes of DCs, along with Ub machinery, to enable their function in cross-presentation.[67]

Mammalian cells are rich in cytosolic aminopeptidases and nonproteasomal endopeptidases. While many cytosolic proteases (including insulin-degrading enzyme, tripeptidyl-peptidase II, leucine amino peptidase, bleomycin hydrolase, and puromycin-sensitive aminopeptidase) have been reported to favor the generation of individual peptides, all appear in general to destroy rather than create antigenic peptides.[68] This is consistent with current understanding that only the proteasome exhibits the universal ability to degrade large proteins, leaving the paradox of how peptides are liberated in the absence of active proteasomes. The paradox is deepened by the general absence of carboxypeptidase activity in the cytosol and secretory pathway of many cell types, leaving proteasomes to generate C-terminal anchor residues for binding to class I molecules,[69,70] though angiotensin-converting enzyme can serve this function[70-72] and is expressed in activated DCs.

Like all enzymes, proteases can work in reverse, in this case, to create rather than cleave peptide bonds. Peptide splicing requires that enzymatic products be present in sufficient concentrations to simultaneously occupy the catalytic site. Given rapid diffusion of the products and enormous dilution, this is normally an exceedingly rare event in cells, particularly since most proteasome products have an exceedingly short half-life (~10 seconds) due to aminopeptidases unless stabilized by binding to class I molecules (though there are exceptions for peptides that are stabilized by interacting with cell proteins[73,74]). But as a barrel with limited product egress, the proteasome can generate nongenetically templated novel peptides by splicing, even reversing the N- and C-ends of peptides[75] or linking peptides via the ε-amino group of lysine residues.[76] Peptide splicing was initially described for a handful of cancer peptides and was generally believed to be an oddity.[77-79] More recent studies concluded that up to 30% of the immunopeptidome consists of spliced peptides,[80] but there are questions regarding the accuracy of peptide identification by mass spectrometry,[79,81] and jury is out on the prevalence of spliced peptides and their potential clinical applications.

Peptide Transport Into the Endoplasmic Reticulum

Once it was clear that peptides from cytosolic/nuclear viral proteins are presented by MHC class I molecules,[82] the race was on to discover a transporter capable of delivering peptides from the cytosol to the ER. Several laboratories independently discovered two genes that, when co-expressed, restored the cell surface expression of class I molecules to cells that had been mutagenized and selected for low MHC expression.[83-86] The transporter, appropriately named TAP, consists of TAP1 and TAP2, members of the adenosine triphosphate (ATP)–binding cassette (ABC) family. TAP genes are located centrally in the MHC and are, as with all dedicated class I antigen processing genes, upregulated upon IFN-γ or TNF-α signaling. TAP predominantly localizes to the ER but can reach the cis-GC in some cells as well as phagosomes in DCs.[87]

TAP is a heterodimeric complex of TAP1 and TAP2 subunits; each encoded within the central region of the mouse and human MHC. The translocation pore of TAP is formed by six transmembrane domains of each subunit, whereas the remaining transmembrane domains (four for TAP1 and three for TAP2) interact with the peptide loading complex (PLC).[88] TAP transports peptides of 8 to 16 residues and longer peptides at lower efficiency. TAP demonstrates sequence selectivity, with specificity conferred by the side chains of the residues at each end of the peptide. For human TAP, Pro at position 2 and Glu and Asp at position 3 interfere with peptide binding, while N-terminal positively charged residues at P1 and P2 and aromatic residues at P3 and C-terminal hydrophobic or basic residues are favored, matching the preference of many class I allomorphs.[89] While peptide transport is ATP-dependent, peptide binding occurs independently. Peptide affinities vary widely, with many antigenic peptides binding with a K_d of ~500 nM.[90,91] Interestingly, in a typically sized cell, this corresponds to half-maximal binding with ~500,000 cytosolic peptides. To put this in perspective, cells may generate millions of peptides per second[14] with a predicted steady-state value of tens of millions (based on a 10-second half-life).[92] Yet the available evidence suggests that peptides are limiting for class I assembly,[21,93] consistent with the channeling hypothesis of peptide generation.[33]

In humans and mice, TAP polymorphism is limited and does not appear to have major biological effects, but in rats and birds, TAP polymorphism plays an important role in altering the peptide repertoire,[69,94,95] and TAP alleles are matched with MHC class I allomorphs that optimally bind to the TAP-provided repertoire. In both species, TAP genes are in close proximity to the relevant class I genes, which facilitates coevolution of binding specificities.[96] Presumably, diversity in peptide presentation at the population level is driven by selection for individuals resistant to $CD8^+$ T cell–mediated virus escape mutations.

If peptide affinity alone determined the loading of exported class I molecules, there would be no apparent selective advantage in matching TAP and class I peptide specificities, since class I molecules would ignore low-affinity peptides. That matching occurs suggests that lower-affinity peptides can compete with high-affinity peptides to the point of interfering with antigen presentation. Even with matching TAP, class I molecule presentation of low-affinity peptides (which broadens the immunopeptidome) likely accounts for the large fraction of peptide-receptive cell surface class I molecules (discussed in more detail below).

Peptide Trimming in the Endoplasmic Reticulum

Approximately half of proteasome-generated peptides of nine or more residues are too long to bind to class I molecules with high affinity. Many of these peptides with C-terminal residues compatible with high-affinity MHC binding have N-terminal extensions that must be cleaved to generate peptides of suitable length after TAP transport into the ER. Amino terminal peptide trimming in the ER was initially functionally inferred by the ability of TAP-deficient cells to present peptides or even proteins targeted to the ER by a signal sequence, but only when the peptide was located at the C-terminus of the inserted protein.[97,98] After identifying the relevant biochemical activity,[99] the aminopeptidase was discovered as a widely expressed, IFN-γ–inducible gene in the mouse[100] and man[101] and termed ERAAP (mouse)/ERAP1 (human) (ER aminopeptidase 1). ERAP1 was simultaneously discovered as an enzyme (ART1) that cleaves cytokine receptors, releasing them from cells.[102]

Unlike other aminopeptidases, ERAP1 preferentially cleaves peptides of 9 to 16 amino acids, matching the predominant peptide lengths transported by TAP. It is likely that most of the length preference in peptide trimming is intrinsic to ERAP1 itself, which can clearly function as a peptide "ruler."[103] The possibility that class I molecules guide trimming in the ER by binding to the peptide C-terminus as initially proposed[104] cannot be ruled out,[105] but it is appearing less likely due to steric constraints apparent in the recently revealed PLC structure (described below),[88] and the likelihood that ERAP1 in its active state cannot access class I–bound peptides.[106,107]

Although ERAP1 contributes to mouse and human immunopeptidomes, the effects of its absence on class I biogenesis, surface stability, or presentation of defined viral peptides range from moderate to mild, depending on the class I allomorph and cell line studied.[108-111] While humans have a second ERAP (ERAP2),[112] an allele present in ca. 50% of the population encodes a nonfunctional RNA that is rapidly degraded. Thus, 25% of humans are functionally ERAP2-deficient, demonstrating that its recent evolutionary selection is weak. ERAP2 generates an overlapping but distinct peptide repertoire and likely functions in cells as a dimer with ERAP1, which may increase the efficiency of generating a subset of peptides. There are ~10 alleles of ERAP2 at appreciable frequencies in humans, with a number of alleles being linked to a variety of autoimmune diseases, sparking considerable clinical interest in the enzyme.[113]

Chaperone-Assisted Peptide Loading in the Endoplasmic Reticulum

HC assembly with β_2m and peptide within the ER is orchestrated by general purpose and dedicated molecular chaperones that maintain folding intermediates in a conformation capable of attaining full assembly. Nascent HC transiently associates with membrane-anchored lectin calnexin (CXN), which, like its soluble paralog calreticulin (CRT), functions as a general chaperone for the assembly of oligomeric glycoproteins. CXN and CRT control glycoprotein biogenesis by monitoring terminal glucose residues on client N-linked glycans. The folding sensor UDP-glucose glycoprotein glucosyltransferase allows incompletely assembled substrates to reassociate with CXN or CRT. Cycles of glucose addition and removal by glucosidase II continue until substrates are either correctly folded and assembled or targeted for ERAD. After HC assembles with β_2m, the heterodimer exchanges CXN for CRT and enters the PLC, consisting of CRT, ERp57, TAP-binding protein (TAPBP), and TAP.[114,115]

Knockout cell lines, genetically deficient mouse strains, and mutagenesis studies have provided key insights into the cooperative roles of PLC components in class I assembly. CRT association with class I proteins is dependent upon the N-linked glycan at residue Asn86 of the HC. Interestingly, the location of the glycan is important for CRT but not CXN association with HCs, likely reflecting geometric constraints imposed by PLC assembly. CRT-deficient cells have suboptimal peptide loading resulting in reduced surface class I expression.[116] ERp57 is a thiol oxidoreductase that mediates disulfide bond formation of different substrates and is commonly associated with CXN and CRT. ERp57 forms a disulfide bond with TAPBP that is required for PLC assembly.[117,118] ERp57-deficient cells exhibit impaired peptide loading, surface expression, and class I antigen presentation.[119] It was originally believed that the ERp57 maintains the α_2 disulfide HC bond. It appears, however, that ERp57 reductase activity is irrelevant to class I assembly and that ERp57 functions strictly as a structural chaperone.[119,120]

TAPBP, a class I–dedicated molecular chaperone, is required to bridge β_2m-assembled HCs to TAP[121] and, along with other members of the PLC, limits class I with suboptimal peptides from dissociating from the PLC. TAPBP stabilizes nascent peptide-receptive class I and optimizes peptide loading to favor high-affinity peptides.[122-125] Presentation of full-length antigens can be reconstituted in insect cells (only jawed vertebrates have the MHC system of antigen presentation) by expressing MHC I, TAP, and TAPBP,[126] which represent the minimal number of specialized proteins required for MHC-based immunosurveillance.

TAPBP can be visualized in the assembled PLC by cryoelectron microscopy, but the resolution is insufficient to discern detailed interactions with class I.[88] Based on high-resolution structures of class I complexed with either TAPBPR[127,128] (described below) or a TAPBP-derived decapeptide, an attractive model posits that the tapasin loop containing this decapeptide binds the F peptide–binding pocket. This stabilizes class I molecules in the PLC until the peptide is displaced by TAP-transported peptides, binding initially from their N-termini.[129] Consistent with this model, F pocket polymorphism can have a profound effect of TAPBP dependencies of different class I alleles.[124,130] Further, it has long been known that mAbs can detect changes in this portion of class I based on peptide occupancy.[131] Recently, the critical missing link of the puzzle of class I assembly by the PLC was reported: the structure of peptide-receptive class I molecules.[132] Peptide binding results in subtle changes in the amino acids in the A and F pockets that reduce the mobility of the peptide-binding domain and immobilizes the C-terminus of the α1 helix and the A139 region through hydrogen bonds with the peptide N-terminus. The net result is locking class I into a stable conformation. With the peptide thus bound, TAPBP is displaced, and class I dissociates from the PLC.

But this is not the whole story of peptide loading in the secretory pathway. TAPBP-related (TAPBPR) is an IFN-γ–inducible, soluble ER/GC protein with 22% homology to TAPBP.[133] TAPBPR was found to interact with class I more than a decade following its discovery.[134] TAPBPR, like TAPBP, binds peptide-receptive class I and functions as a peptide editor,[135,136] though its effects on class I expression and peptide repertoire are more subtle, as shown by respective genetic knockout/knockdowns.[135] TAPBPR binds to UGT1,[137] which reglucosylates class I and restores its recognition by CRT, enabling retrieval by CRT[138] and another chance at peptide loading in the PLC.[139]

As fate would have it, unlike TAPBP, TAPBPR readily generates complexes with class I suitable for high-resolution X-ray crystallography.[127,128] These structures not only inform on TAPBPR function but also TAPBP, as discussed above. The "scoop loop" of TAPBR, consisting of residues 22 to 36, binds to class I near the F pocket, creating a competitor for peptide occupancy that functions as an editor favoring binding of higher affinity peptides. This model is supported by a growing number of studies.[140-143] As with TAPBP, TAPBPR demonstrates considerable heterogeneity in binding to different class I allomorphs. The function of the scoop loop can be mimicked by dipeptides, which at sufficient concentration destabilize bound peptides that can be replaced by standard peptides, a neat trick for reloading class I molecules on the cell surface.[144]

Generation of Peptide-Receptive Cell Surface Class I Molecules

A foundational finding for the central role of oligopeptides in class I–based immunosurveillance was the ability of synthetic peptides to sensitize cells for lysis by virus-specific $CD8^+$ T cells.[6] A considerable fraction of class I are in such a peptide-receptive state on the cell surface.[145] Although it remains widely believed that levels of peptide-receptive class I are enhanced in TAP-deficient cells and that QC mechanisms that retain class I molecules in the ER are overridden by incubating cells at 26°C,[146] later studies reversed these conclusions.[147-150] Low-temperature incubation actually decreases ER export of class I from TAP-deficient cells due to the general temperature dependence of membrane protein trafficking. The apparent increase in cell surface delivery at 26 °C is due to stabilizing peptide-receptive class I at the cell surface after the dissociation of weakly bound peptides.[151] This applies to TAP-expressing cells as well, but the effect *appears* to be less pronounced due to the constitutively high expression levels at 37°C.

That TAP-expressing cells express a substantial fraction of class I molecules with suboptimally loaded peptides is clearly shown by the ability of synthetic class I–binding peptides or class I–specific antibodies[147] to increase class I expression. The binding of synthetic peptides is almost wholly dependent on the presence of exogenous β_2m (though this can vary with class I allomorph[152]), which binds to peptide-receptive cell surface class I HCs,[153,154] preventing irreversible denaturation and proteolysis. As with peptide binding,[148] loss of TAP decreases β_2m binding by approximately fivefold. Cells lacking β_2m express minute levels of cell surface class I molecules and essentially cannot bind exogenous β_2m or present exogenous peptides[155] (though there can be limited presentation of endogenous peptides[156,157]), demonstrating that β_2m binds exclusively to cell surface class I molecules.

Robust binding of extracellular β_2m to cells has been demonstrated *in vivo*.[158] Interestingly, β_2m is released/secreted at high levels (0.13 mg/kg/h in humans), resulting in its presence in body fluids at concentrations (120 nM) well above its dissociation constant (10 nM). These findings are consistent with an important, yet-to-be-defined, physiological role for peptide-receptive class I generated from circulating β_2m. The extent to which serum β_2m derives from direct secretion versus release from cell surface class I remains to be established, though it appears that β_2m is synthesized in excess of HCs.[159] It is curious that class I molecules are not synthesized as a disulfide-linked dimers or with β_2m and HC as a fusion protein; such proteins are easy to engineer in functional form.[160,161] This argues that dissociation of β_2m from classical and nonclassical class I molecules has an important functional role. It would be surprising if none of the jawed vertebrates expressed a single chain class I molecule with a β_2m-like domain, given the variety in ecological niches and selection factors, which can even negate class I expression completely, as occurs in male anglerfish to enable their permanent attachment to females.[162]

Presumably, nearly all β_2m released from cells derive from class I molecules whose HCs are degraded. Class I molecules are internalized by Arf6-dependent clathrin-independent endocytosis[163] and sorted in early Rab5-EEA-1 positive endosomes to recycle via the endocytic recycling compartment and tubular recycling endosomes to the plasma membrane if conformed or late endosomes for destruction if unfolded.[164-167] It remains to be determined whether recycling of conformed class I molecules is truly stochastic and how many cycles each molecule experiences. Early findings suggested that nearly all complexes cycle several times per hour,[164] though the fraction is likely to vary with allomorph and cell type.[167] The slightly acidic pH of early endosomes (as low as 6.3) will favor peptide dissociation and possibly serve to winnow low-affinity peptide class I complexes.

MHC CLASS II ANTIGEN PROCESSING AND PRESENTATION

Like MHC-I, MHC-II molecules function as peptide "placeholders"; however, in this case they display antigenic peptides to CD4 helper T cells. For years, MHC-II was thought to be distinct from MHC-I in that the source of antigenic peptides for MHC-II was "exogenous" (internalized by the APC); however, it has become clear that the MHC-II antigen processing pathway includes presentation of epitopes generated by exogenous antigens as well as endogenous antigens generated within APCs themselves. In this part of our chapter we will review the sources of antigen capture by APCs, the mechanism of peptide generation and peptide binding to MHC-II, and the eventual expression of MHC-II-peptide complexes on the APC surface.

Sources of Antigen for MHC-II Presentation

As a general rule, the antigen processing machinery that is utilized by MHC-I is specialized for the degradation, import, and binding of peptides that are derived from antigens present in the cytosol of host cells. Although many viruses and bacteria thrive in the cytosol of infected cells, there are pathogens (such as *Toxoplasma gondii*) that reside in membrane-encapsulated intracellular compartments and do not access the cytosol.[168] Furthermore, almost any foreign material (including live or dead pathogens, fragments of apoptotic cells, or soluble proteins) can enter the cell by endocytosis and are therefore sequestered from the cytosol. It is for recognition of antigens such as these that MHC-II molecules exist; MHC-II binds peptides derived from antigens that gain access to intracellular compartments in cells by a variety of endocytic pathways (Fig. 20.2; Ref. 169). Fluid phase macropinocytosis is a process whereby extracellular fluid is taken up by plasma membrane protrusions that bring extracellular fluids, and soluble proteins, into the cell by endocytosis. Once internalized, macropinosomes fuse with early endosomes, thereby releasing their contents to the endocytic pathway. Immature (resting) DCs are particularly efficient in their capacity for macropinocytosis that enhances their significance as "sentinels" of the immune system.[170] Upon activation by encounter with foreign organisms, DCs greatly suppress macropinocytosis,[170-172] thereby limiting the ability of the activated DC to sample their microenvironment and restricting the repertoire of antigenic peptides presented to T lymphocytes.

As one can imagine, macropinocytosis is not a particularly efficient mechanism of antigen uptake. B cells possess antigen-specific receptors (immunoglobulins) that permit specific antigen internalization and presentation to T cells that can be 1000 times more efficient than fluid phase endocytosis.[173] Curiously, the function and fate of B cells that generate specific MHC-II-peptide complexes following immunoglobulin-mediated antigen uptake versus fluid phase uptake of the same antigen differ significantly,[174] highlighting the importance of the receptor-mediated antigen uptake pathway in B cells. Similarly, macrophages and DCs possess Fc receptors and complement receptors that allow efficient phagocytosis of large extracellular particles such as antibody-coated organisms or even apoptotic cells. Following endocytosis, the phagosome fuses with the membrane of a lysosome, forming a degradative phagolysosome that neutralizes and degrades internalized antigens into peptides.[175] Other surface receptors, such as mannose receptors[170,176,177] and DEC-205[178] on DCs, also allow for efficient antigen binding and uptake onto the endocytic pathway and these receptors have been exploited as a vehicle to efficiently deliver foreign vaccine antigens into the MHC-II processing pathway in APCs.[179]

A novel form of phagocytosis, termed LC3-associated phagocytosis (LAP), is stimulated by TLR signaling and involves regulation of phagocytosis by proteins classically associated with cellular autophagy.[180,181] LAP plays an important role as a source of antigen for MHC-II antigen presentation.[182,183] Curiously, in DCs, phagocytosis via LAP protects phagosomes from rapid fusion with lysosomes, thereby prolonging MHC-II antigen presentation.[183]

While antigens entering the endocytic pathway by phagocytosis, macropinocytosis, and receptor-mediated endocytosis give rise to many of the peptides that bind to MHC-II, there is increasing evidence that cytosolic proteins are capable of entering endosomal MHC-II processing compartments and are degraded there into MHC-II–binding peptides. Indeed, mass spectroscopy analysis of peptides eluted from MHC-II revealed that a large proportion of these peptides were derived from cytosolic or nuclear proteins.[184,185] Surprisingly, endogenous antigen processing drives the primary CD4 T cell response to influenza virus infection,[186] highlighting the importance of this pathway of antigen processing during *in vivo* immune responses to viral pathogens. The most likely cellular mechanism responsible for this phenomenon is a process termed autophagy.[187,188] The process of autophagy occurs by a variety of mechanistically distinct processes in APCs. Macroautophagy is constitutively active in DCs and is a process that leads to the "engulfment" of aggregated cytosolic protein complexes to form an autophagosome. Like phagosomes containing internalized extracellular antigens, the autophagosome fuses with a lysosome and results in the degradation of autophagosome contents into antigenic peptides.[189,190] A related process, termed chaperone-mediated autophagy, results from the direct import of cytosolic proteins into the lumen of endosomes/lysosomes and is facilitated by the chaperone protein hsc-70 and the lysosomal integral membrane protein Lamp2a.[191] Once inside this organelle, the imported protein is degraded into antigenic peptides just like "conventionally" internalized antigens. An additional type of autophagy, termed microautophagy, allows cytosolic contents to enter lysosomes through invaginations in the lysosomal membrane.[192] While the relative importance of autophagy during immune responses to pathogens remains to be fully elucidated, autophagy is known to be the predominant pathway of self-antigen delivery to antigen processing compartments in thymic epithelial cells that shape the CD4 T cell repertoire and is essential for self-tolerance.[193] Although not directly addressing how cytosolic antigens gain access to antigen processing compartments, Tewari et al have shown that at least for some viral antigens MHC-II–restricted antigen presentation is both proteasome- and TAP-dependent,[194] features normally attributed to the classical MHC-I antigen processing and presentation pathway.

Lastly, recent studies have shown that lymph can serve as a source of peptides available for MHC-II molecules.[195] Lymph contains a variety of proteins and peptides that can recirculate to all tissues of the body including, most importantly for MHC-II, secondary lymphoid organs. Peptide sequencing studies confirmed that many peptides eluted from MHC-II were identical to peptides found in lymph.[195] Although it remains to be determined how these peptides bind to MHC-II, it is possible that these peptides could bind "empty" MHC-II molecules[196] or MHC-II present in recycling early endosomes.[197] Whether lymph-derived peptide binding to MHC-II requires the MHC-II peptide editor HLA-DM (described below) remains to be determined.

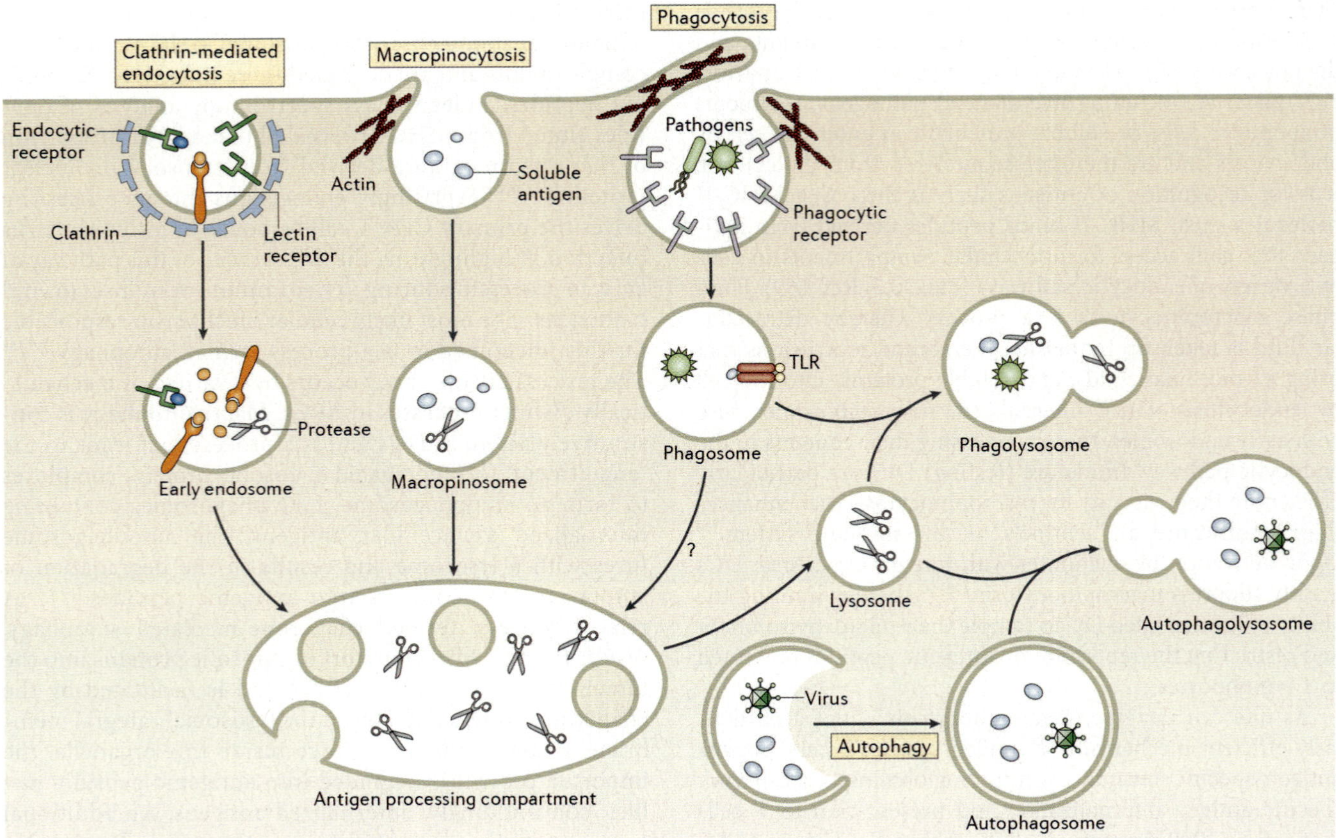

FIG. 20.2. Pathways of antigen endocytosis in antigen-presenting cells (APCs). Antigens can enter the endocytic pathway of APCs by a number of distinct mechanisms. Receptor-mediated endocytosis via clathrin-coated vesicles requires antigen binding to one of a variety of receptors on DCs, including lectin receptors, which result in their internalization into early endosomes. Macropinocytosis is an actin-dependent process leading to the uptake of soluble material into the cell in a so-called macropinosome. In each of these processes, internalized "early" endosomes eventually fuse with multivesicular late endosomal/lysosomal antigen processing compartments. Phagocytosis is an endocytic process in which opsonized particles bind to a variety of receptors and, following actin-dependent membrane reorganization around the particle, enter cells in membrane-derived phagosomes. Phagosomes are not particularly rich in proteinases or MHC-II, and after TLR-induced fusion with MHC-II+ late endosomal/lysosomal compartments, the resulting phagolysosome generates pMHC-II. Autophagy is a process in which membranes encapsulate cytosolic antigens to form an autophagosome. Upon autophagosome fusion with endosomal/lysosomal compartments, the resulting autophagolysosome generates pMHC-II. DCs, dendritic cells; MHC, major histocompatibility complex. (Courtesy of Roche and Furuta.)

MHC-II Assembly in the ER and Transport to the Endosomal Pathway

Since MHC-II is specialized to bind peptides generated in the endocytic pathway, there must be mechanisms in APCs to prevent the premature binding of peptides to MHC-II as the molecules assemble in the ER and traffic through the Golgi apparatus before their delivery to endosomal peptide loading compartments. In addition to the inability of MHC-II to bind to TAP/tapasin and other members of the MHC-I peptide-loading complex, MHC-II is unique in that it binds to a transmembrane chaperone protein termed the invariant chain (abbreviated here as Ii). It is the association of Ii with the MHC-II peptide-binding site that prevents MHC-II molecules in the early secretory pathway from binding peptides.[198,199] Indeed, failure of newly synthesized MHC-II to bind Ii allows for the presentation of ER-imported peptides to CD4 T cells by MHC-II.[200] A specific region of Ii, termed the **cl**ass II -associated **I**i **p**eptide, or CLIP, serves to block the peptide-binding grooves of nascent MHC-II molecules until they are ready for removal in endosomes/lysosomes. The MHC-II-CLIP complex has been crystallized, and as one might expect the structure of this complex is nearly identical to that of an MHC-II molecule containing an "antigenic" peptide.[201] As the name implies, unlike the highly polymorphic MHC-II molecule itself, Ii is nonpolymorphic. However, the human Ii mRNA can initiate protein synthesis using one of the two translation initiation sites (Fig. 20.3A), leading to generation of a major 33 kDa form of Ii (Ii-p33) as well as a minor population of larger Ii molecules whose translation is

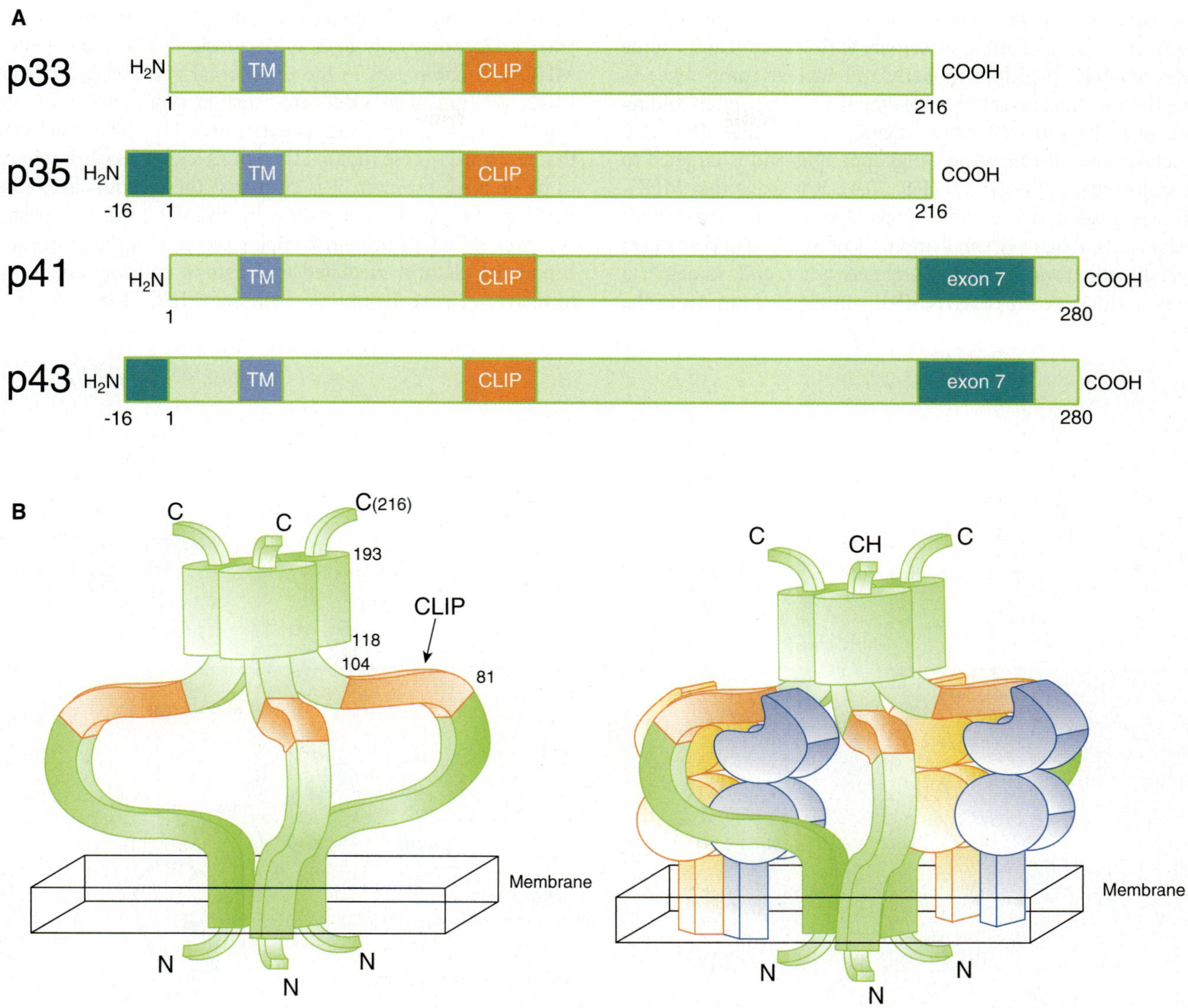

FIG. 20.3. Structure of Ii and MHC-II-Ii complexes. A: The major 216 amino acid form of human Ii is the 33 kDa Ii-p33 isoform. The use of an upstream initiation methionine results in the generation of Ii-p35 (which contains an ER-retention motif) and alternative splicing of each of these isoforms leads to the generation of Ii-p41 and Ii-p43 forms of Ii that possess proteinase inhibitor activity. The relative positions of the transmembrane domain TM and CLIP region of each isoform of Ii are indicated. **B:** Models of Ii trimer (*left panel*) and the MHC-II-Ii complex (*right panel*). Only two of the three pairs of MHC-II αβ dimers bound to the Ii trimer are shown. The model shows the peptide-binding groove of MHC-II αβ dimers occupied by the CLIP region of Ii. ER, endoplasmic reticulum; MHC, major histocompatibility complex. (Reprinted from Cresswell P. Invariant chain structure and MHC class II function. *Cell.* 1996;84:505-507, with permission from Elsevier.)

initiated using an upstream methionine (Ii-p35).[202] The use of alternative translation initiation sites appears to be unique to humans, as it is not seen in rodents.[203] Furthermore, alternative mRNA splicing in all species examined can give rise to an additional 8 kDa domain in each form of Ii, leading to the production of Ii-p41 and (in humans) Ii-p43.[204,205] This additional exon encodes a domain that serves as protease inhibitory domain (see below) that regulates the activity of proteases in endosomes/lysosomes.

The Ii molecule itself forms a trimer even in the absence of MHC-II, and each constituent subunit of this trimer contains Ii-p33, Ii-p35, Ii-p41, or Ii-p43 isoforms (Fig. 20.3B). Ii homotrimers and mixed heterotrimers are generated, and each trimer contains an Ii molecule whose presence in the trimer is proportional to its level of expression.[206-208] Onto this "scaffold" of an Ii trimer, pairs of MHC-II αβ heterodimers assemble, ultimately forming a mature, nine-chain MHC-II-Ii complex.[209] The partially formed MHC-II-Ii complex containing an Ii trimer and only one or two αβ heterodimers is retained in the ER by interacting with the chaperone CXN,[210] presumably allowing the MHC-II molecule additional time to form a fully functional MHC-II-Ii nonamer. In addition, the longer p35 and p43 forms of human Ii contain an ER retention motif, and free Ii trimers containing one of these longer Ii forms are retained in the ER.[211,212] Curiously, completely folded MHC-II-Ii nonamers containing one of these longer Ii isoforms are capable of exiting the ER and moving to antigen processing compartments,[208,213] demonstrating that assembly with MHC-II masks the ER retention motif on the Ii trimer.

Ii association does more for MHC-II than simply prevent premature peptide binding. Assembly with Ii is critical for some alleles of MHC-II to fold properly and achieve competence to leave the ER, thus revealing a role for Ii as an assembly/folding molecular chaperone.[214] After leaving the ER, the MHC-II-Ii complex moves through the Golgi apparatus and is targeted to the endosomal pathway[211,215] (Fig. 20.4). It is likely that MHC-II-Ii complexes arrive in the endocytic pathway by two distinct mechanisms. Relatively small amounts of MHC-II-Ii complexes target directly from the *trans*-Golgi network to endosomes[216,217]; however, the majority of MHC-II-Ii complexes move from the *trans*-Golgi network directly to the plasma membrane and enter endosomes only after endocytosis.[218] The movement of MHC-II-Ii complexes to the endosomal pathway is facilitated by recognition of two dileucine sorting motifs present in the cytosolic domain of every isoform of Ii.[219,220] The machinery that recognizes these motifs in the *trans*-Golgi network remains to be defined; however, it is clear that these motifs are recognized at the plasma membrane by the AP-2 adapter protein complex.[221,222] This protein complex serves as a linker between Ii and the clathrin-mediated endocytosis pathway and serves to deliver plasma membrane–localized MHC-II-Ii to the early

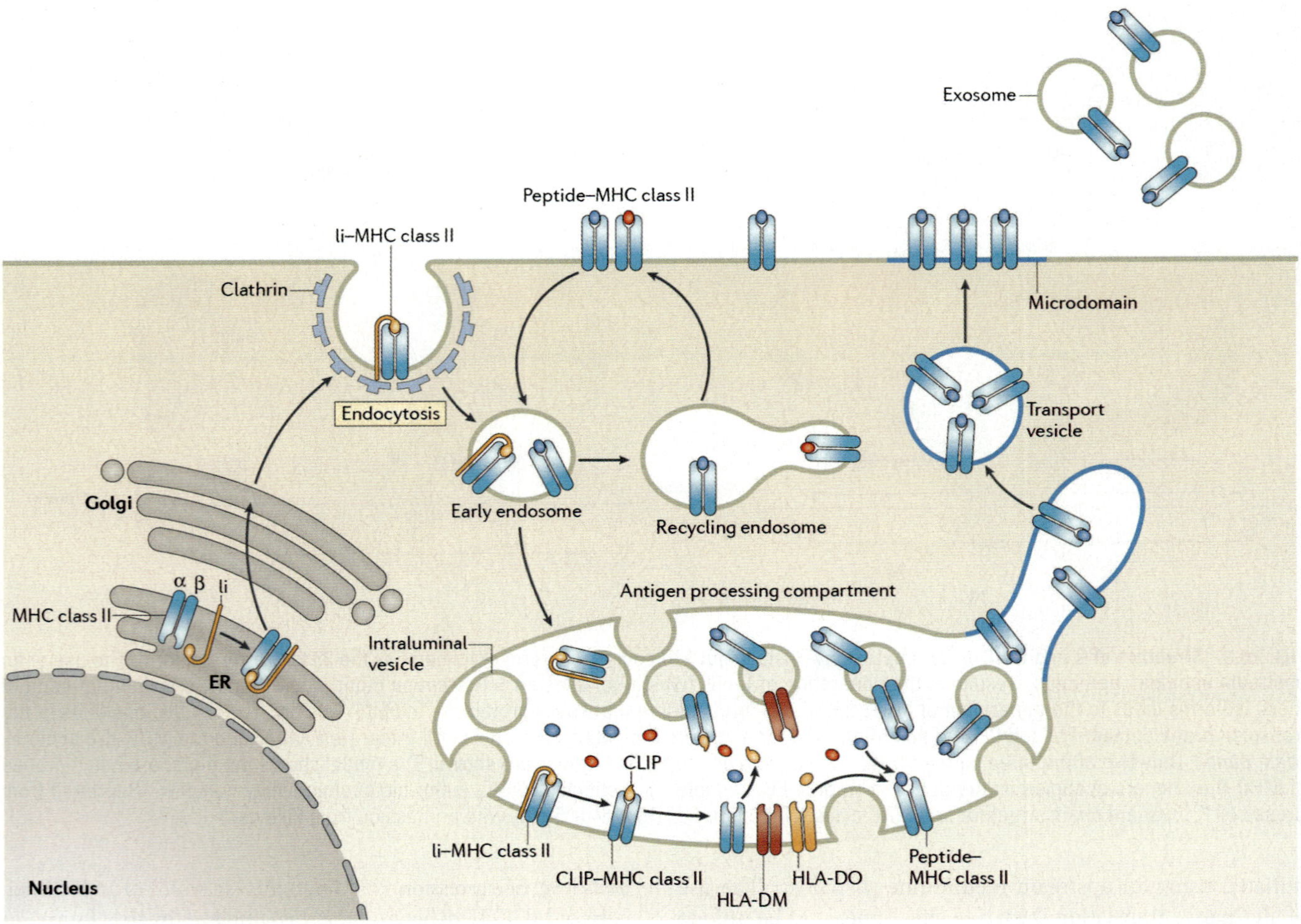

FIG. 20.4. Generation of MHC-II-peptide complexes in antigen-presenting cells (APCs). MHC-II αβ dimers associate with Ii in the ER, traffic through the Golgi apparatus, and are delivered to the plasma membrane. MHC-II-Ii complexes are internalized by clathrin-mediated endocytosis and traffic to multivesicular antigen processing compartments. Some of these complexes sort onto the ILV of MVB, where sequential Ii proteolysis leads to persistence of a fragment of Ii, termed CLIP, in the peptide-binding groove of MHC-II. CLIP is removed from MHC-II-CLIP complexes by the enzyme DM that is present on the MVB internal and limiting membranes, thereby allowing peptide binding onto nascent MHC-II. The activity of DM is regulated by DO. ILV-associated pMHC-II somehow "back-fuses" with the MVB limiting membrane, and endosomal/lysosomal tubules either directly fuse with or give rise to transport vesicles that fuse with the plasma membrane. MVB membranes are rich in the lipids that constitute lipid raft membrane microdomains (shown in *purple*), and fusion of pMHC-II from MVB-derived membranes with the plasma membrane leads to the deposition of lipid raft-associated pMHC-II directly into the plasma membrane. In circumstances in which an entire MVB fuses with the plasma membrane, the ILVs of MVB are released from the APC in the form of exosomes. Surface expressed pMHC-II can internalize using a clathrin-independent endocytosis pathway and is either targeted for lysosomal degradation or can recycle back to the plasma membrane. ER, endoplasmic reticulum; ILV, intralumenal vesicle; MHC, major histocompatibility complex; MVB, multivesicular body. (Courtesy of Roche and Furuta.)

endosomes. Recognition and internalization of MHC-II-Ii from the plasma membrane is very efficient and rapid,[218] and as a consequence of this, very little MHC-II-Ii is present on the plasma membrane of APCs at any given moment.

Removal of Invariant Chain in Antigen Processing Compartments

Regardless of the mechanism responsible for delivering MHC-II-Ii to endosomes, Ii must be removed in order to allow antigenic peptides to bind to nascent MHC-II. Ii is degraded in discrete steps by proteolysis in acidic endosomes, with proteolysis initiated at the luminal C-terminus of Ii in earlier endosomes and Ii degradation completed in more acidic late endosomal/prelysosomal structures. *In vitro* studies have shown that many different proteinases are capable of degrading Ii; however, *in vivo* the initial cleavage of Ii is initiated by a proteinase whose activity is insensitive to the protease inhibitor leupeptin while additional cleavage is strictly leupeptin-dependent.[223] Leupeptin treatment of living APCs results in the accumulation of a 21-kDa Ii degradation intermediate that retains the nonameric MHC-II-Ii stoichiometry and is unable to bind antigenic peptides.[213] After the initial cleavage of Ii, a variety of proteinases can cleave the remaining MHC-II–associated Ii molecule; however, the terminal stages of Ii degradation in B cells and DCs are dependent on the cysteine proteinase cathepsin S.[224] Inhibition of cathepsin S activity or analysis of DCs isolated from cathepsin S–deficient mice reveals the accumulation of a 10-kDa Ii degradation product, highlighting the importance of this proteinase in Ii degradation. Curiously, MHC-II–expressing epithelial cells in the thymus do not rely on cathepsin S for Ii degradation but instead require cathepsin L for terminal Ii degradation.[225] This thymic specialization of proteinase activity for Ii degradation for MHC-II is reminiscent of unique role of the thymoproteasome subunit β5t in peptide generation for MHC-I in the thymus.[60]

Proteolysis of Ii from both the luminal C-terminus and residues adjacent to the luminal Ii transmembrane domain ultimately leaves a small fragment of Ii bound to MHC-II. This class II–associated Ii polypeptide, termed CLIP, is bound to the peptide-binding groove of MHC-II and serves as a "placeholder" to occupy the peptide-binding site until antigenic peptides are available to bind (Fig. 20.1). While some MHC-II alleles have a very low affinity for CLIP and CLIP spontaneously dissociates from MHC-II,[226] in most cases CLIP must be actively removed from the peptide-binding site of MHC-II prior to antigenic peptide binding. A seminal finding in antigen processing and presentation came in 1994 with the discovery of HLA-DM.[227,228] HLA-DM (known as H-2M in mice) is a nonpolymorphic transmembrane heterodimer whose α- and β-subunits are encoded in the MHC gene locus. Like MHC-II, DM is expressed exclusively in APCs. An important function of DM is to facilitate the removal of CLIP from the MHC-II-CLIP complex.[229,230] DM functions by binding to a partially opened peptide binding groove on MHC-II, thereby resulting in a conformational change that disrupts hydrogen bonds between MHC-II and bound peptides, ultimately leading to their dissociation from MHC-II.[231,232] Not only does DM catalyze CLIP removal from MHC-II but also DM facilitates the dissociation of weakly bound antigenic peptides from MHC-II[229,233,234]; it is for this reason that DM has been termed a "peptide editor." It is differences in conformational flexibility of the MHC-II-peptide complex that is the primary determinant of whether a given MHC-II-peptide complex will be "edited out" by DM.[235] Unlike the MHC-II molecule itself, DM contains a tyrosine-based endocytic sorting motif in its cytosolic domain that directly interacts with clathrin-associated adapter proteins.[236] The presence of this signal allows for the entry of DM into the endocytic pathway after arrival at the cell surface. Most DM is localized in late endosomes/lysosomes that contain MHC-II-CLIP complexes,[237,238] and the activity of DM is enhanced by the low pH of these later endocytic compartments.[229,230,239] DM can also be found, albeit at lower amounts, in early endosomes[240] and the plasma membrane[241] where it could play a role as a peptide editor allowing high-affinity peptide binding in early recycling endosomes[242] or even at the plasma membrane.[241] DM editing activity in APCs can be regulated by its association with another MHC-encoded transmembrane protein termed HLA-DO (H-2O in mice).[243] *In vitro* studies have shown that DO binds to DM and inhibits the ability of DM to catalyze CLIP removal.[244,245] DO likely acts as a substrate mimic that competitively inhibits DM binding to DR, as crystallographic studies have shown that the co-crystal of DM with DO looks remarkably similar to the co-crystal of DM with DR.[246] Expression of DM and DO are relatively restricted to APCs; however, unlike DM, DO expression is downregulated upon APC activation.[247] Unlike MHC-II-Ii or DM, DO does not contain any intrinsic endosomal/lysosomal sorting motifs. Instead, DO associated with DM in the ER and DM serves to "escort" DO to antigen processing compartments.[248] It is in the low pH of these compartments that DO is thought to dissociate from DM, thereby enhancing DM activity.

Antigen Processing and Peptide Loading Onto MHC-II

While MHC-II in lysosome-like antigen processing compartments must be "prepared" for peptide binding (by degrading MHC-II–associated Ii and removing CLIP by DM), internalized foreign antigens must be similarly prepared for binding to the nascent MHC-II molecule. This task is initiated by proteolysis by any number of endocytic proteinases. Antigen processing compartments have a pH between 4.5 and 6.5, and most proteinases in these compartments have enzymatic activity optima at this pH. There is remarkable diversity in the proteinase pool in these compartments and great redundancy in the ability of distinct proteinases to degrade particular proteins. For this reason, it has been difficult to identify specific proteinases whose activity is required to generate specific antigenic peptides. There are exceptions to this, however, as asparaginyl cysteine endoprotease is required for the generation of an immunodominant epitope to tetanus toxoid[249] and both cathepsin L and cathepsin S affect quantitative and qualitative differences in the repertoire of peptides bound to MHC-II molecules.[250] It is also a known fact that many internalized foreign (and self) antigens contain intrachain disulfide bonds, and reduction of these bonds would alter protein conformation and render a protein more (or less) sensitive to proteolysis in antigen processing compartments. As its name implies, the expression of IFN-γ–induced thiol reductase GILT can be induced by exposure to IFN-γ.[251] This thiol reductase is expressed in antigen processing compartments in APCs, and the activity of GILT is important for the presentation of antigenic epitopes that are generated from proteins that contain many intrachain disulfide bonds.[252]

Just as important as generation of antigenic peptides is the ability of the antigen processing compartment to limit excessive proteolytic degradation of antigenic peptides. While asparaginyl cysteine endoprotease positively regulates expression of tetanus toxoid epitopes, this same proteinase destroys expression of the immunodominant epitope of myelin basic protein,[253] a protein thought to be involved in the demyelinating autoimmune disease multiple sclerosis. Clearly the proteinase activity of antigen processing compartments must be sufficient to generate antigenic peptides capable of binding to MHC-II but not so great as to completely destroy these same epitopes. Measurement of proteinase activity in various APC subtypes has revealed that macrophages have high proteinase activity, efficiently degrade proteins, but are relatively poor generators of antigenic peptides.[254] By contrast, DCs and B cells possess less proteinase activity, degrade proteins less efficiently, and are superb generators of antigen peptides.[254,255] This specialization of limiting proteinase activity helps DCs function as initiators of naïve T cell activation and B cells capacity for immunoglobulin class switching after encounter with antigen-specific T cells.

DCs are particularly interesting APCs in that they can regulate their proteinase activity upon cell activation. Immature (resting) DCs are highly phagocytic and have a remarkable capacity for macropinocytosis. Nevertheless, their relatively low proteinase activity makes them relatively inefficient generators of antigenic MHC-II-peptide complexes, potentially allowing them to serve as a "depot" for antigen internalized while the cell is in the quiescent state. It has been shown that internalized proteins can remain intact in immature DCs for days.[256,257] Activation of the immature DC by any of a number of mechanisms (either by TLR signaling or stimulation by activated T cells) leads to a number of changes in the cell, with perhaps the most important being the activation of the vacuolar proton ATPase.[258] This is mediated by activation of the transcription factor TFEB that regulates lysosome function.[259] Activation of this ATPase leads to the acidification of antigen processing compartments, enhances the activity of endosomal/lysosomal proteinases that have low pH optima, and leads to enhanced degradation of MHC-II–associated Ii as well as the degradation of internalized antigens to form antigenic peptides capable of binding to MHC-II. Curiously, DC activation also stimulates the LAP pathway of phagocytosis that actually delays phagosome–lysosome fusion, thereby prolonging antigen in LAP phagosomes.[183] Other changes also take place following activation of immature DCs, including downregulating expression of the endogenous cysteine proteinase inhibitor cystatin C[260] and reorganizing the internal membrane structure of the antigen processing compartments[261]; however, it is likely that acidification of these compartments is the major mechanism leading to enhanced antigen processing, peptide generation, and formation of MHC-II-peptide complexes upon DC activation.

The Ii molecule itself also can serve to limit proteinase activity of antigen processing compartments. The additional exon that is present in Ii isoforms that are products of alternative splicing (Ii-p41 and Ii-p43) encodes a domain that functions as an inhibitor of cathepsin L.[262,263] *In vitro* studies showed that co-expression with Ii-p41 inhibits the kinetics of Ii-p33 degradation,[264] a finding that is consistent with the role of Ii-p41 as a proteinase inhibitor. However, no defects in function have been identified using APCs isolated from mice expression only the Ii-p41,[265] leaving the importance of this domain in APC function an open issue.

It has been proposed that binding to the MHC-II molecules itself can serve as a mechanism to preserve the integrity of antigenic peptides. The peptide-binding groove of MHC-II is open and *in vitro* binding experiments have shown that MHC-II is capable of binding to unfolded (but otherwise intact) protein antigens.[266] The association of distinct regions within a full-length (unfolded) protein with the peptide-binding site of MHC-II could therefore protect these regions from proteolytic degradation by endosomal/lysosomal proteinases. This model of "bind first trim later" has been championed as a major mechanism of peptide loading onto MHC-II[267] and has been supported by experimental studies in B cells[268] and in cell-free systems[269]; however, the extent to which "bind first trim later" versus "trim first bind later" predominates in distinct APC subsets and for different antigens remains to be determined.

Delivery of MHC-II to the Cell Surface

The sites of antigenic protein degradation and subsequent binding of peptides onto nascent MHC-II are commonly referred to as antigen processing compartments. At one time it was thought that APCs possessed a "special" compartment that served as the major antigen processing compartment[270]; however, it is now clear that "antigen processing compartments" refer to a heterogeneous collection of intracellular (generally acidic) organelles that permit peptide binding onto MHC-II.[271] These organelles generally have characteristics of typical late endosomes/prelysosomes found in all cell types; however, in APCs, these organelles contain the specialized protein machinery required for antigen processing and presentation such as MHC-II, DM, DO, GILT, and Ii-processing enzymes. A "typical" antigen processing compartment is a late endosomal multivesicular body (MVB), an organelle that possesses intralumenal vesicles encapsulated in a limiting membrane.[272] These compartments can also have a multilamellar structure, containing numerous membrane sheets contained within an even more acidic lysosome-like organelles. Curiously, the majority of MHC-II is present on the intralumenal vesicles of these MVB, while the CLIP editor DM and MHC-II-peptide complexes are enriched on the MVB limiting membrane.[261] The location of these proteins in distinct MVB membranes has been revealed by immunoelectron microscopy; however, it is unclear on which membranes the process of Ii proteolysis, CLIP release, and peptide binding to MHC-II occurs. For example, while there is *in vitro* evidence that DM primarily interacts with MHC-II when these proteins are on the same membrane,[273,274] whether this is the intralumenal vesicle or the MVB limiting membrane remains to be determined.

The presence of MHC-II-peptide complexes on the intralumenal vesicles of MVB presents a topological problem for the APC, since direct fusion of the MVB with the plasma membrane would release the intralumenal vesicles from these compartments to the extracellular space. Such a process would effectively result in the secretion of MHC-II from the APC. This process does occur in DCs with some frequency, and these secreted intralumenal vesicles are termed exosomes that are capable of stimulating naïve T cells *in vivo*.[275] While APC-derived exosomes are being utilized as cell-free vaccines,[276] their physiological role remains to be unambiguously determined. The majority of MHC-II in antigen processing compartments is not released as exosomes but is instead either degraded following fusion of the MVB with lysosomes or inserted into the plasma membrane by "vesicle"-mediated protein transport. While transport vesicles containing intracellular MHC-II have been observed to fuse with the plasma membrane,[277] in DCs it is likely that these "vesicles" are actually tubules that bud from the MVB limiting membrane and subsequently fuse with the plasma membrane. The tubulation of these compartments is stimulated in DCs by TLR signaling.[261,278,279] The encounter of antigen-loaded, activated DCs with T cells leads to the fusion of these tubules directly to the plasma membrane region at the point of DC:T cell contact,[279] thereby delivering MHC-II peptides generated in DCs directly to the site of TCR concentration at the immunological synapse. Phagocytosis of TLR-containing pathogens "activates" individual phagosomes to fuse with lysosomes in DCs[280] and under these conditions MHC-II also traffics to the plasma membrane via elongated phagosomal tubules.[281]

The cellular processes that govern the movement of MHC-II to the surface of APCs are similar to those that govern the movement of any late endosomal protein to the surface of any cell. In some cases, the machinery regulating discrete transport steps remains a mystery while other steps are well characterized. For example, the mechanism leading to the delivery of intralumenal vesicle–bound MHC-II back to the MVB limiting membrane (from which tubules are formed) is termed back-fusion,[282] and the proteins and lipids that regulate this curious process remain to be identified. By contrast, it is known that the ability of MHC-II to insert into the plasma membrane is regulated by the small GTPase ARL14/ARF7 that couples the tubulovesicle membrane with the actin cytoskeleton.[283] Like other vesicle-mediated protein transport steps, microtubules and dynein/kinesin motors[284] as well as actin and myosin motors[285] are important for the locomotion of MHC-II into and out of antigen processing compartments.

Once at the plasma membrane, MHC-II-peptide complexes are poised to interact with antigen-specific TCRs. MHC-II-peptide complexes are clustered on the plasma membrane[257,286] and associate with distinct lipid raft[287] and tetraspanin[288] membrane microdomains, thereby locally concentrating MHC-II on the cell surface. Biochemical studies have revealed that MHC-II associates with these membrane microdomains while still present in antigen processing compartments even before antigenic peptide binding.[289] This finding is consistent with the observation that MVB contains large amounts of cholesterol, a major constituent of lipid raft membrane microdomains.[290] Indeed, it has been shown that following a specific pulse with antigen MHC-II-peptide complexes generated in antigen processing compartments actually arrive at the plasma membrane prepackaged in lipid rafts,[291] thereby increasing their concentration into distinct sites on the plasma membrane. While lipid raft- and tetraspanin microdomain associations are important for the T cell stimulatory function of APCs,[287,292] the molecular signals leading to membrane microdomain association of MHC-II remain to be determined.

MHC-II-peptide complexes are generated and reside in a relatively hostile (acidic, proteinase-rich) environment, and therefore it is not surprising that biochemical studies have confirmed that cell surface MHC-II molecules have a relatively long half-life.[293,294] While this is true on activated B cells and mature DCs, MHC-II on the surface of immature DCs has a significantly reduced half-life. This is likely due to rapid endocytosis and lysosomal degradation of surface MHC-II in resting (immature) DCs.[295,296] The dichotomy between the endocytosis and survival of MHC-II in immature and mature DCs can be explained in part by the fact that internalized MHC-II is ubiquitylated in immature DCs and targeted for lysosomal degradation while internalized MHC-II is not ubiquitinated in mature DCs and efficiently recycles back to the plasma membrane.[297,298] The E3 ubiquitin ligase MARCH1, whose expression appears to be mostly limited to DCs, B cells, and macrophages, ubiquitinates the cytosolic domain of the MHC-II β-chain in APCs,[299] thereby preventing the recycling of internalized MHC-II and targeting ubiquitylated MHC-II for lysosomal degradation.[300] This mechanism ensures that the kinetics of MHC-II synthesis and degradation in immature DCs is similar and provides a means to halt the degradation of relevant MHC-II-peptide complexes upon DC activation. In contrast to DCs, thymic epithelial cells do not express MARCH1 but instead express the related E3 ubiquitin ligase MARCH8 that ubiquitinates MHC-II and performs a (presumably) identical function in the thymus.[301,302]

Simply because MHC-II has a long half-life in B cells and DCs does not mean that these molecules do not internalize. In these APC subtypes, internalized MHC-II efficiently recycles from early (recycling) endosomes back to the plasma membrane,[295] thereby limiting their delivery to lysosomes and prolonging MHC-II half-life. This early endosomal recycling pathway is suppressed by ubiquitination of MHC-II and is regulated by the small GTPase Rab11.[303,304] The ability of internalized MHC-II to exchange one peptide for another in endosomes has been demonstrated *in vitro*[197,305]; however, it remains to be determined whether peptide exchange onto recycling MHC-II *in vivo* is a significant mechanism of peptide loading onto MHC-II. There is convincing evidence, however, that certain protein epitopes are primarily presented by recycling MHC-II.[306] Unlike presentation of epitopes from "classically processed" antigens, presentation of these epitopes does not require neosynthesis of MHC-II and is independent of Ii expression or DM activity. A particularly intriguing example of this pathway can be found in the analysis of an immunogenic epitope presentation of the influenza virus hemagglutinin,[194,304] in which presentation of one epitope is via the classical (late endosomal) processing and presentation pathway while another is presented via a nonclassical (early endosomal) recycling pathway.

EVADING ANTIGEN PRESENTATION

Predictably, pathogens employ a myriad of strategies to thwart peptide presentation by both MHC-I and MHC-II. Focus here will be on direct targeting of machinery within the antigen-presenting cell rather than strategies with an indirect effect such as manipulation of cytokine production.

EVADING MHC-I PRESENTATION

Viruses

The greatest number of immune evasion proteins ("immunoevasins") that have been described to date are expressed by viruses and target MHC-I–restricted antigen processing and presentation.[307] In fact, many aspects of MHC-I antigen processing have been elucidated or clarified through the study of viral immunoevasins. Most examples are encoded by DNA viruses with large genomes, particularly those viruses capable of latency or host coexistence. Indeed, virtually every step of MHC antigen processing and presentation is subject to viral interference (Fig. 20.5). For example, Epstein-Barr virus (EBV) and Kaposi sarcoma–associated herpesvirus (KSHV) express proteins during latency that escape CTL detection by containing sequences that inhibit proteasome processing.[308-312] Alternatively, blocking peptide transport by inhibiting TAP function is a commonly used immune evasion strategy. For example, ICP47 of HSV binds the cytosolic side of TAP and blocks peptide and not ATP binding,[313-318] whereas US6 of HCMV binds the luminal side of TAP and induces a conformational change resulting in inhibition of ATP hydrolysis and peptide translocation.[3,15,319-322] As a third mechanism of TAP blocking, UL49.5 of bovine herpesvirus induces TAP degradation.[323] There are also published examples of immune evasion proteins inhibiting Tpn function. For example, US3 of HCMV binds to Tpn and impairs optimization of peptide loading,[324,325] whereas E3-19K of AdV inhibits the ability of Tpn to bridge MHC-I with TAP.[326] The CPXV protein CPXV12 is a small ER-resident protein that interferes with both TAP function and peptide loading.[327,328]

Immune evasion proteins also employ strategies to misdirect trafficking of MHC-I. For example, the aforementioned E3-19K protein of AdV binds MHC-I proteins and blocks their transport out of the ER.[329-333] A second CPXV protein, CPXV203, returns fully assembled MHC-I proteins from the Golgi back to the ER.[327,328,334,335] A misdirection strategy employed by the Nef protein of HIV and gp48 protein of MCMV is shuttling of assembled MHC-I from the Golgi to the lysosome.[336-338]

A prevalent immunoevasion strategy is the targeting of MHC-I for degradation, accomplished via several distinct mechanisms. First are the viral ER ubiquitin ligases called the viral MARCH (**m**embrane-**a**ssociated **R**ING-**CH**) proteins, which share with their cellular homologs a transmembrane orientation and a highly conserved atypical RING domain that confers E3 ubiquitin ligase activity.[339-341] Viral proteins mK3 of MHV68 and proteins kK3 and kK5 of KSHV are the original members.[342-344] Mechanistically, mK3 binds to TAP, ubiquitinating the cytosolic tail of nascent MHC-I HCs, thereby inducing dislocation to the cytosol and subsequent degradation (ie, ERAD).[345-348] Of note, the US2 and US11 proteins of HCMV also mediate ERAD of MHC-I proteins by recruiting cellular E3 ligases.[349-354] In contrast to ERAD, the kK3 and kK5 MARCH ligases of KSHV induce endocytosis and lysosomal degradation of MHC-I proteins. Of interest, kK3 like mK3 appears to ubiquitinate only MHC-I proteins, whereas kK5 targets other surface receptors including the T cell costimulation molecule CD86 (B7.2) as well as natural killer (NK) cell ligands, AICL, MICA, and MICB.[355] Inhibition of NK responses is significant because downregulation of MHC-I proteins renders cells susceptible to NK cell lysis.

Bacteria

Many intracellular bacteria occupy the endosomal network but deliver to the host cell cytosol effector proteins that assist in establishing residency. These effector proteins can also serve as targets of MHC-I processing and presentation.[356] One such bacterium, *Shigella flexneri*, delivers via its type III secretion system (T3SS) the invasion plasmid antigen H 4.5, an E3 ubiquitin ligase that targets the RPN13 proteasomal subunit for degradation, resulting in reduced $CD8^+$ T cell activation.[357] Additionally, endosome-resident bacteria prevent endosomal acidification, thereby inhibiting acid-sensitive proteases that participate in endosomal cross-presentation, as exemplified by *Salmonella typhimurium*.[358]

EVADING MHC-II PRESENTATION

There are fewer examples of immunoevasins that interfere with MHC-II processing and presentation, although this may simply reflect fewer efforts in this area. As with MHC-I, there are many examples of pathogens that indirectly interfere with MHC-II presentation through manipulation of the cytokine environment, but focus here is on direct targeting.

Viruses

A comprehensive strategy, employed by many herpesviruses, is interference with transcription of CIITA, which upregulates

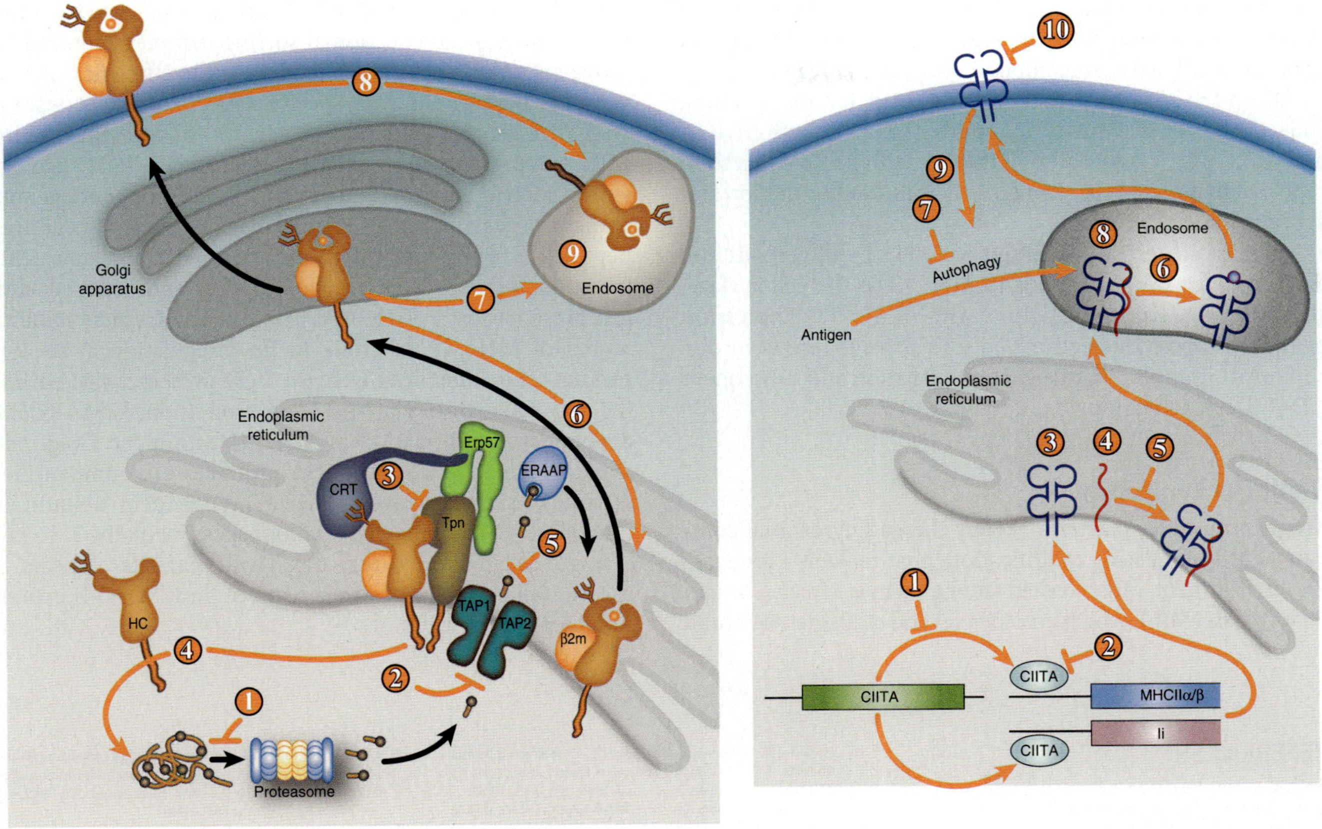

FIG. 20.5. **Immune evasion strategies used by pathogens to block antigen presentation by major histocompatibility complex (MHC)-I and II proteins.** *Black arrows* demarcate physiologic pathways for antigen presentation, whereas the numbered *red arrows* demarcate reported immune evasion strategies by virus proteins. Class I (*left panel*)—1: Blocking of antigen processing by the proteasome (EDNA1 protein of Epstein-Barr virus, the LANA1 protein of Kaposi sarcoma–associated herpesvirus, and the H 4.5 protein of *Shigella flexneri*, an E3 ubiquitin ligase that targets the RPN13 proteasomal subunit for degradation). 2: Blocking peptide transport by transporter associated with antigen presentation (ICP47 protein of herpes simplex virus, US6 protein of human cytomegalovirus, CPXV12 protein of cowpox virus, UL49.5 protein of bovine herpesvirus, and BNFL2a protein of Epstein-Barr virus). 3: Impairment of tapasin function (US3 of HCMV, which binds to Tpn and impairs optimization of peptide loading, and E3-19K of AdV, which inhibits the ability of Tpn to bridge MHC-I). 4: Dislocation of MHC-I proteins from the endoplasmic reticulum (ER) to the cytosol for proteasome-mediated degradation (US2 and US11 proteins of human cytomegalovirus and mK3 protein of MHV68). 5: Interference with peptide loading (CPXV12 of CPXV). 6: ER retention and retrieval of MHC-I proteins (E3-19K protein of AdV and the CPXV203 protein of cowpox virus). 7: Mistrafficking of MHC-I proteins from the Golgi to an endosomal compartment (gp48 protein of murine cytomegalovirus and Nef protein of human immunodeficiency virus 1). 8: Rapid internalization of MHC-I proteins from the cell surface to an endosomal/lysosomal compartment (kK3 and kK5 proteins of Kaposi sarcoma–associated herpesvirus). 9: Inhibition of endosomal acidification, thereby interfering with endosomal cross presentation (*Salmonella typhimurium*). Class II (*right panel*)—1: Inhibition of CIITA transcription (EBV proteins BZLF1, which binds to the CIITA promoter, preventing transcription, and LMP2A, which inhibits the positive regulators of CIITA transcription E47 and PU.1 in B cells). 2: Inhibition of MHC-II transcription (LANA protein of Kaposi sarcoma–associated herpesvirus, which associated enhansosome assembly and recruitment of CIITA to MHC-II promoters). 3: Posttranscriptional downregulation of MHC-II (US2 of HCMV and BDLF3 of EBV). 4: Posttranscriptional downregulation of Ii (BZLF1 protein of EBV). 5: Inhibition of nascent MHC-II-Ii association (US3 protein of HCMV). 6: Inhibition of Ii-peptide exchange in the endosome (NS5A of HCV, which inhibits cathepsin S, and the A35 protein of VACV). 7: Inhibition of autophagy-mediated delivery of cytosolic (endogenous) antigen to the endosomal network (PE_PGRS47 protein of Mtb). 8: Interference with endosomal sorting (EsxH protein of Mtb). 9: Polyubiquitinylation of MHC-II resulting in reduced expression at the cell surface (*Salmonella enterica* SteD, which targets MHC-II in the Golgi, and a *Francisella tularensis* soluble factor that drives MARCH1-dependent ubiquitinylation). 10: Inhibition of $CD4^+$ T cell recognition (gp42 of EBV, which binds to MHC-II at the cell surface, and ECTV C15, whose expression results in disruption of the immunological synapse).

many MHC-II processing components (xyz). The latency-associated nuclear antigen (LANA) of KSHV downregulates CIITA activity by interfering with CIITA-associated enhanceosome assembly and recruitment of CIITA to the MHC-II promoter.[359] The EBV protein BZLF1 binds to the CIITA promoter, repressing transcription.[360] Additionally, two positive regulators of CIITA transcription in B cells, E47 and PU.1, are inhibited by the EBV latency protein LMP2A.[361]

Ii function is compromised in several ways. HCMV US3 competes with Ii for binding to nascent αβ dimers, disrupting trafficking and peptide loading.[362] Through a mechanism that remains to be elucidated, EBV BZLF1 posttranscriptionally downregulates Ii.[363] The NS5A protein of hepatitis C virus indirectly targets Ii through inhibition of cathepsin S, which is critical for proteolysis of Ii that facilitates CLIP-peptide exchange.[364] Vaccinia virus (VACV) interferes with MHC-II function, potentially by targeting Ii.[365-368] A candidate VACV protein is A35, which locates to the endosomal compartment and enhances CLIP:MHC-II complex density at the cell surface.[365]

Like MHC-I, MHC-II is also targeted for degradation. Indeed, HCMV US2 has been reported to target both MHC-I and MHC-II.[369-371] Also, the BDLF3 protein of EBV has been reported to target MHC-I and MHC-II for ubiquitination and subsequent degradation.[372]

A final category is inhibition of CD4+ T cell recognition. The EBV protein gp42 binds to MHC-II at the cell surface, sterically hindering interaction with TCR.[373,374] Expression of the orthopoxvirus ectromelia C15 protein results in disruption of immunological synapse formation and subsequent CD4+ T cell activation.[375]

Bacterial Evasion

Many intracellular bacteria occupy the endolysosomal compartment, engineering modifications that make it a less hostile environment. Most studied among them is *Mycobacterium tuberculosis* (Mtb), which employs several strategies to inhibit MHC-II processing and presentation. The EsxH protein interferes with the endosomal sorting complex required for transport machinery, thereby reducing the efficiency of antigen presentation and Mtb-specific CD4+ T cell responses *in vivo*.[376] A genome-wide screen for loci of Mtb that inhibit MHC-II presentation revealed the PE_PGRS47 protein, which interferes with autophagy, implicating endogenous antigen processing as an important source of Mtb-derived peptides.[377] Finally, through unknown mechanisms, Mtb infection induces rapid DC maturation and reduced antigen presentation.[378] Two other bacteria induce polyubiquitination of MHC-II, resulting in downregulation at the cell surface. The *Salmonella enterica* SteD protein, injected into the host cell cytosol via type 3 secretion, induces MARCH8-dependent ubiquitination of MHC-II within the Golgi network.[379] *Francisella tularensis* produces an unidentified soluble factor that drives MARCH1-dependent ubiquitination of MHC-II in infected macrophages.[380] Finally, *T. gondii*, a single-celled intracellular eukaryotic parasite, appears to use multiple undefined mechanisms to limit MHC-II expression by infected APCs.[381]

REFERENCES

1. D'Souza MP, Adams E, Altman JD, et al. Casting a wider net: immunosurveillance by nonclassical MHC molecules. *PLoS Pathog*. 2019;15(2):e1007567.
2. Murgia C, Pritchard JK, Kim SY, Fassati A, Weiss RA. Clonal origin and evolution of a transmissible cancer. *Cell*. 2006;126(3):477-487.
3. Pearse AM, Swift K. Allograft theory: transmission of devil facial-tumour disease. *Nature*. 2006;439(7076):549.
4. Zinkernagel RM, Doherty PC. Restriction of *in vitro* T cell-mediated cytotoxicity in lymphocytic choriomeningitis within a syngeneic or semiallogeneic system. *Nature*. 1974;248:701-702.
5. Babbitt BP, Allen PM, Matsueda G, Haber E, Unanue ER. Binding of immunogenic peptides to Ia histocompatibility molecules. *Nature*. 1985;317:359-361.
6. Townsend AR, Rothbard J, Gotch FM, Bahadur G, Wraith D, McMichael AJ. The epitopes of influenza nucleoprotein recognized by cytotoxic T lymphocytes can be defined with short synthetic peptides. *Cell*. 1986;44:959-968.
7. Yewdell JW, Bennink JR, Hosaka Y. Cells process exogenous proteins for recognition by cytotoxic T lymphocytes. *Science*. 1988;239(4840):637-640.
8. Norbury CC, Malide D, Gibbs JS, Bennink JR, Yewdell JW. Visualizing priming of virus-specific CD8+ T cells by infected dendritic cells *in vivo*. *Nat Immunol*. 2002;3(3):265-271.
9. Sigal LJ, Crotty S, Andino R, Rock KL. Cytotoxic T-cell immunity to virus-infected non-haematopoietic cells requires presentation of exogenous antigen. *Nature*. 1999;398(6722):77-80.
10. Prasad SA, Norbury CC, Chen W, Bennink JR, Yewdell JW. Cutting edge: recombinant adenoviruses induce CD8 T cell responses to an inserted protein whose expression is limited to nonimmune cells. *J Immunol*. 2001;166(8):4809-4812.
11. Yewdell JW. *Truth Wins: A Practical Guide to Succeeding at Biomedical Research*. 2017.
12. Falk K, Rötzschke O, Stevanović S, Jung G, Rammensee HG. Allele-specific motifs revealed by sequencing of self-peptides eluted from MHC molecules. *Nature*. 1991;351(6324):290-296.
13. Bjorkman PJ, Saper MA, Samraoui B, Bennett WS, Strominger JL, Wiley DC. The foreign antigen binding site and T cell recognition regions of class I histocompatibility antigens. *Nature*. 1987;329:512-518.
14. Princiotta MF, Finzi D, Qian SB, Gibbs J. Quantitating protein synthesis, degradation, and endogenous antigen processing. *Immunity*. 2003;18(3):343-354.
15. Granados DP, Laumont CM, Thibault P, Perreault C. The nature of self for T cells-a systems-level perspective. *Curr Opin Immunol*. 2015;34:1-8.
16. Bourdetsky D, Schmelzer CEH, Admon A. The nature and extent of contributions by defective ribosome products to the HLA peptidome. *Proc Natl Acad Sci U S A*. 2014;111(16):E1591-E1599.
17. Esquivel F, Yewdell J, Bennink J. RMA/S cells present endogenously synthesized cytosolic proteins to class I-restricted cytotoxic T lymphocytes. *J Exp Med*. 1992;175:163-168.
18. Yewdell JW, Antón LC, Bennink JR. Defective ribosomal products (DRiPs): a major source of antigenic peptides for MHC class I molecules? *J Immunol*. 1996;157(5):1823-1826.
19. Yewdell JW, Nicchitta CV. The DRiP hypothesis decennial: support, controversy, refinement and extension. *Trends Immunol*. 2006;27(8):368-373.
20. Schubert U, Antón LC, Gibbs J, Norbury CC, Yewdell JW, Bennink JR. Rapid degradation of a large fraction of newly synthesized proteins by proteasomes. *Nature*. 2000;404(6779):770-774.
21. Reits EA, Vos JC, Grommé M, Neefjes J. The major substrates for TAP *in vivo* are derived from newly synthesized proteins. *Nature*. 2000;404(6779):774-778.
22. Purcell AW, Croft NP, Tscharke DC. Immunology by numbers: quantitation of antigen presentation completes the quantitative milieu of systems immunology! *Curr Opin Immunol*. 2016;40:88-95.
23. Spiliotis ET, Osorio M, Zúñiga MC, Edidin M. Selective export of MHC class I molecules from the ER after their dissociation from TAP. *Immunity*. 2000;13(6):841-851.
24. Pentcheva T, Edidin M. Clustering of peptide-loaded MHC class I molecules for endoplasmic reticulum export imaged by fluorescence resonance energy transfer. *J Immunol*. 2001;166(11):6625-6632.
25. Abe F, Van Prooyen N, Ladasky JJ, Edidin M. Interaction of Bap31 and MHC class I molecules and their traffic out of the endoplasmic reticulum. *J Immunol*. 2009;182(8):4776-4783.
26. Lu X, Gibbs JS, Hickman HD, et al. Endogenous viral antigen processing generates peptide-specific MHC class I cell-surface clusters. *Proc Natl Acad Sci U S A*. 2012;109(38):15407-15412.
27. Ferez M, Castro M, Alarcon B, van Santen HM. Cognate peptide-MHC complexes are expressed as tightly apposed nanoclusters in virus-infected cells to allow TCR crosslinking. *J Immunol*. 2013;192(1):52-58.
28. Starck SR, Shastri N. Nowhere to hide: unconventional translation yields cryptic peptides for immune surveillance. *Immunol Rev*. 2016;272(1):8-16.
29. Laumont CM, Vincent K, Hesnard L, et al. Noncoding regions are the main source of targetable tumor-specific antigens. *Sci Transl Med*. 2018;10(470):eaau5516.
30. Apcher S, Prado Martins R, Fåhraeus R. The source of MHC class I presented peptides and its implications. *Curr Opin Immunol*. 2016;40:117-122.
31. Wei J, Yewdell JW. Flu DRiPs in MHC class I immunosurveillance. *Virol Sin*. 2018;34(2):162-167.
32. Martins RP, Malbert-Colas L, Lista MJ, et al. Nuclear processing of nascent transcripts determines synthesis of full-length proteins and antigenic peptides. *Nucleic Acids Res*. 2019;47(6):3086-3100.
33. Yewdell JW, Dersh D, Fåhraeus R. Peptide channeling: the key to MHC class I immunosurveillance? *Trends Cell Biol*. 2019;29(12):929-939.
34. Wei J, Kishton RJ, Angel M, et al. Ribosomal proteins regulate MHC class I peptide generation for immunosurveillance. *Mol Cell*. 2019;73(6):1162.e5-1173.e5.
35. Lev A, Princiotta MF, Zanker D, et al. Compartmentalized MHC class I antigen processing enhances immunosurveillance by circumventing the law of mass action. *Proc Natl Acad Sci U S A*. 2010;107(15):6964-6969.
36. Fooksman DR, Grönvall GK, Tang Q, Edidin M. Clustering class I MHC modulates sensitivity of T cell recognition. *J Immunol*. 2006;176(11):6673-6680.
37. Ferez M, Castro M, Alarcon B, van Santen HM. Cognate peptide-MHC complexes are expressed as tightly apposed nanoclusters in virus-infected cells to allow TCR crosslinking. *J Immunol*. 2014;192(1):52-58.
38. Villanueva MS, Fischer P, Feen K, Pamer EG. Efficiency of MHC class I antigen processing: a quantitative analysis. *Immunity*. 1994;1:479-489.
39. Wolf BJ, Princiotta MF. Processing of recombinant *Listeria monocytogenes* proteins for MHC class I presentation follows a dedicated, high-efficiency pathway. *J Immunol*. 2013;190(6):2501-2509.
40. Moore MW, Carbone FR, Bevan MJ. Introduction of soluble protein into the class I pathway of antigen processing and presentation. *Cell*. 1988;54:777-785.
41. Lelouard H, Gatti E, Cappello F, Gresser O, Camosseto V, Pierre P. Transient aggregation of ubiquitinated proteins during dendritic cell maturation 3. *Nature*. 2002;417(6885):177-182.
42. Szeto J, Kaniuk NA, Canadien V, et al. ALIS are stress-induced protein storage compartments for substrates of the proteasome and autophagy. *Autophagy*. 2006;2(3):189-199.
43. Cosma GL, Lobby JL, Fay EJ, Siciliano NA, Langlois RA, Eisenlohr LC. Kinetically distinct processing pathways diversify the CD8(+) T cell response to a single viral epitope. *Proc Natl Acad Sci U S A*. 2020;117(32):19399-19407.
44. Mosse CA, Meadows L, Luckey CJ, et al. The class I antigen-processing pathway for the membrane protein tyrosinase involves translation in the endoplasmic reticulum and processing in the cytosol. *J Exp Med*. 1998;187(1):37-48.
45. Bacik I, Snyder HL, Antón LC, et al. Introduction of a glycosylation site into a secreted protein provides evidence for an alternative antigen processing pathway: transport of precursors of major histocompatibility complex class I-restricted peptides from the endoplasmic reticulum to the cytosol. *J Exp Med*. 1997;186(4):479-487.
46. Qian SB, Reits E, Neefjes J, Deslich JM, Bennink JR, Yewdell JW. Tight linkage between translation and MHC class I peptide ligand generation implies specialized antigen processing for defective ribosomal products. *J Immunol*. 2006;177(1):227-233.
47. Golovina TN, Morrison SE, Eisenlohr LC. The impact of misfolding versus targeted degradation on the efficiency of the MHC class I-restricted antigen processing. *J Immunol*. 2005;174(5):2763-2769.
48. Dolan BP, Sharma AA, Gibbs JS, Cunningham TJ, Bennink JR, Yewdell JW. MHC class I antigen processing distinguishes endogenous antigens based on their translation from cellular vs. viral mRNA. *Proc Natl Acad Sci U S A*. 2012;109(18):7025-7030.

49. Norbury CC, Basta S, Donohue KB, et al. CD8+ T cell cross-priming via transfer of proteasome substrates. *Science*. 2004;304(5675):1318-1321.
50. Wolkers MC, Brouwenstijn N, Bakker AH, Toebes M, Schumacher TN. Antigen bias in T cell cross-priming. *Science*. 2004;304(5675):1314-1317.
51. Shen L, Rock KL. Cellular protein is the source of cross-priming antigen *in vivo*. *Proc Natl Acad Sci U S A*. 2004;101(9):3035-3040.
52. Bard JAM, Goodall EA, Greene ER, Jonsson E, Dong KC, Martin A. Structure and function of the 26S proteasome. *Annu Rev Biochem*. 2018;87(1):697-724.
53. Kasahara M, Flajnik MF. Origin and evolution of the specialized forms of proteasomes involved in antigen presentation. *Immunogenetics*. 2019;71(3):251-261.
54. Murata S, Takahama Y, Kasahara M, Tanaka K. The immunoproteasome and thymoproteasome: functions, evolution and human disease. *Nat Immunol*. 2018;19(9):923-931.
55. Kincaid EZ, Che JW, York I, et al. Mice completely lacking immunoproteasomes show major changes in antigen presentation. *Nat Immunol*. 2012;13(2):129-135.
56. Hensley SE, Zanker D, Dolan BP, et al. Unexpected role for the immunoproteasome subunit LMP2 in antiviral humoral and innate immune responses. *J Immunol*. 2010;184(8):4115-4122.
57. Chen W, Norbury CC, Cho Y, Yewdell JW, Bennink JR. Immunoproteasomes shape immunodominance hierarchies of antiviral CD8(+) T cells at the levels of T cell repertoire and presentation of viral antigens. *J Exp Med*. 2001;193(11):1319-1326.
58. Nathan JA, Spinnenhirn V, Schmidtke G, Basler M, Groettrup M, Goldberg AL. Immuno and constitutive proteasomes do not differ in their abilities to degrade ubiquitinated proteins. *Cell*. 2013;152(5):1184-1194.
59. Seifert U, Bialy LP, Ebstein F, et al. Immunoproteasomes preserve protein homeostasis upon interferon-induced oxidative stress. *Cell*. 2010;142(4):613-624.
60. Sasaki K, Takada K, Ohte Y, et al. Thymoproteasomes produce unique peptide motifs for positive selection of CD8(+) T cells. *Nat Commun*. 2015;6:7484.
61. Apavaloaei A, Brochu S, Dong M, et al. PSMB11 orchestrates the development of CD4 and CD8 thymocytes via regulation of gene expression in cortical thymic epithelial cells. *J Immunol*. 2019;202(3):966-978.
62. Boulpicante M, Darrigrand R, Pierson A, et al. Tumors escape immunosurveillance by overexpressing the proteasome activator REGγ. *bioRxiv*. 2019:511873.
63. Vinitsky A, Antón LC, Snyder HL, Orlowski M, Bennink JR, Yewdell JW. The generation of MHC class I-associated peptides is only partially inhibited by proteasome inhibitors: involvement of nonproteasomal cytosolic proteases in antigen processing? *J Immunol*. 1997;159(2):554-564.
64. Benham AM, Grommé M, Neefjes J. Allelic differences in the relationship between proteasome activity and MHC class I peptide loading. *J Immunol*. 1998;161(1):83-89.
65. Milner E, Gutter-Kapon L, Bassani-Strenberg M, Barnea E, Beer I, Admon A. The effect of proteasome inhibition on the generation of the human leukocyte antigen (HLA) peptidome. *Mol Cell Proteomics*. 2013;12(7):1853-1864.
66. Wei J, Zanker D, Di Carluccio AR, et al. Varied role of ubiquitylation in generating MHC class I peptide ligands. *J Immunol*. 2017;198(10):3835-3845.
67. Sengupta D, Graham M, Liu X, Cresswell P. Proteasomal degradation within endocytic organelles mediates antigen cross-presentation. *EMBO J*. 2019;38(16):e99266.
68. Lázaro S, Gamarra D, Del Val M. Proteolytic enzymes involved in MHC class I antigen processing: a guerrilla army that partners with the proteasome. *Mol Immunol*. 2015;68(2 pt A):72-76.
69. Powis SJ, Young LL, Joly E, et al. The rat cim effect: TAP allele-dependent changes in a class I MHC anchor motif and evidence against C-terminal trimming of peptides in the ER. *Immunity*. 1996;4:159-165.
70. Eisenlohr LC, Bacik I, Bennink JR, Bernstein K, Yewdell JW. Expression of a membrane protease enhances presentation of endogenous antigens to MHC class I-restricted T lymphocytes. *Cell*. 1992;71(6):963-972.
71. Shen XZ, Lukacher AE, Billet S, Williams IR, Bernstein KE. Expression of angiotensin-converting enzyme changes major histocompatibility complex class I peptide presentation by modifying C termini of peptide precursors. *J Biol Chem*. 2008;283(15):9957-9965.
72. Shen XZ, Billet S, Lin C, et al. The carboxypeptidase ACE shapes the MHC class I peptide repertoire. *Nat Immunol*. 2011;12(11):1078-1085.
73. Lev A, Takeda K, Zanker D, et al. The exception that reinforces the rule: cross-priming by cytosolic peptides that escape degradation. *Immunity*. 2008;28(6):787-798.
74. Kunisawa J, Shastri N. Hsp90 alpha chaperones large C-terminally extended proteolytic intermediates in the MHC class I antigen processing pathway. *Immunity*. 2006;24(5):523-534.
75. Warren EH, Vigneron NJ, Gavin MA, et al. An antigen produced by splicing of noncontiguous peptides in the reverse order. *Science*. 2006;313(5792):1444-1447.
76. Berkers CR, de Jong A, Schuurman KG, et al. Peptide splicing in the proteasome creates a novel type of antigen with an isopeptide linkage. *J Immunol*. 2015;195(9):4075-4084.
77. Hanada K, Yewdell JW, Yang JC. Immune recognition of a human renal cancer antigen through post-translational protein splicing. *Nature*. 2004;427(6971):252-256.
78. Vigneron N, Stroobant V, Chapiro J, et al. An antigenic peptide produced by peptide splicing in the proteasome. *Science*. 2004;304(5670):587-590.
79. Vigneron N, Ferrari V, Stroobant V, Abi Habib J, Van den Eynde BJ. Peptide splicing by the proteasome. *J Biol Chem*. 2017;292(51):21170-21179.
80. Liepe J, Marino F, Sidney J, et al. A large fraction of HLA class I ligands are proteasome-generated spliced peptides. *Science*. 2016;354(6310):354-358.
81. Mylonas R, Beer I, Iseli C, et al. Estimating the contribution of proteasomal spliced peptides to the HLA-I ligandome. *Mol Cell Proteomics*. 2018;17(12):2347-2357.
82. Townsend AR, Gotch FM, Davey J. Cytotoxic T cells recognize fragments of the influenza nucleoprotein. *Cell*. 1985;42:457-467.
83. Deverson EV, Gow IR, Coadwell WJ, Monaco JJ, Butcher GW, Howard JC. MHC class II region encoding proteins related to the multidrug resistance family of transmembrane transporters. *Nature*. 1990;348:738-747.
84. Monaco JJ, Cho S, Attaya M. Transport protein genes in the murine MHC: possible implications for antigen processing. *Science*. 1990;250:1723-1726.
85. Spies T, Bresnahan M, Bahrain S, et al. A gene in the human major histocompatibility complex class II region controlling the class I antigen presentation pathway. *Nature*. 1990;348:744-747.
86. Trowsdale J, Hanson I, Mockridge I, Beck S, Townsendt A, Kelly A. Sequences encoded in the class II region of the MHC related to the "ABC" superfamily of transporters. *Nature*. 1990;348:741-744.
87. Ackerman AL, Kyritsis C, Tampe R, Cresswell P. Early phagosomes in dendritic cells form a cellular compartment sufficient for cross presentation of exogenous antigens. *Proc Natl Acad Sci U S A*. 2003;100(22):12889-12894.
88. Blees A, Januliene D, Hofmann T, et al. Structure of the human MHC-I peptide-loading complex. *Nature*. 2017;551(7681):525-528.
89. Abele R, Tampé R. Moving the cellular peptidome by transporters. *Front Cell Dev Biol*. 2018;6:43.
90. Herget M, Kreißig N, Kolbe C, Schölz C, Tampé R, Abele R. Purification and reconstitution of the antigen transport complex TAP: a prerequisite for determination of peptide stoichiometry and ATP hydrolysis. *J Biol Chem*. 2009;284(49):33740-33749.
91. Burgevin A, Saveanu L, Kim Y, et al. A detailed analysis of the murine TAP transporter substrate specificity. *PLoS One*. 2008;3(6):e2402.
92. Reits E, Griekspoor A, Neijssen J, et al. Peptide diffusion, protection, and degradation in nuclear and cytoplasmic compartments before antigen presentation by MHC class I. *Immunity*. 2003;18(1):97-108.
93. Reits EA, Hodge JW, Herberts CA, et al. Radiation modulates the peptide repertoire, enhances MHC class I expression, and induces successful antitumor immunotherapy. *J Exp Med*. 2006;203(5):1259-1271.
94. Joly E, Le Rolle AF, Gonzélez AL, et al. Co-evolution of rat TAP transporters and MHC class I RT1-A molecules. *Curr Biol*. 1998;8(3):169-172.
95. Tregaskes CA, Harrison M, Sowa AK, et al. Surface expression, peptide repertoire, and thermostability of chicken class I molecules correlate with peptide transporter specificity. *Proc Natl Acad Sci U S A*. 2016;113(3):692-697.
96. Kaufman J. Generalists and specialists: a new view of how MHC class I molecules fight infectious pathogens. *Trends Immunol*. 2018;39(5):367-379.
97. Snyder HL, Bacík I, Yewdell JW, Behrens TW, Bennink JR. Promiscuous liberation of MHC-class I-binding peptides from the C termini of membrane and soluble proteins in the secretory pathway. *Eur J Immunol*. 1998;28(4):1339-1346.
98. Snyder HL, Yewdell JW, Bennink JR. Trimming of antigenic peptides in an early secretory compartment. *J Exp Med*. 1994;180(6):2389-2394.
99. Serwold T, Gaw S, Shastri N. ER aminopeptidases generate a unique pool of peptides for MHC class I molecules. *Nat Immunol*. 2001;2(7):644-651.
100. Serwold T, Gonzalez F, Kim J, Jacob R, Shastri N. ERAAP customizes peptides for MHC class I molecules in the endoplasmic reticulum. *Nature*. 2002;419(6906):480-483.
101. Saric T, Chang SC, Hattori A, et al. An IFN-gamma-induced aminopeptidase in the ER, ERAP1, trims precursors to MHC class I-presented peptides. *Nat Immunol*. 2002;3(12):1169-1176.
102. Cui X, Hawari F, Alsaaty S, et al. Identification of ARTS-1 as a novel TNFR1-binding protein that promotes TNFR1 ectodomain shedding. *J Clin Invest*. 2002;110(4):515-526.
103. Chang SC, Momburg F, Bhutani N, Goldberg AL. The ER aminopeptidase, ERAN1, trims precursors to lengths of MHC class I peptides by a "molecular ruler" mechanism. *Proc Natl Acad Sci U S A*. 2005;102(47):17107-17112.
104. Falk K, Rötzschke O, Rammensee HG. Cellular peptide composition governed by major histocompatibility complex class I molecules. *Nature*. 1990;348:248-251.
105. Evnouchidou I, van Endert P. Peptide trimming by endoplasmic reticulum aminopeptidases: role of MHC class I binding and ERAP dimerization. *Hum Immunol*. 2019;80(5):290-295.
106. Nguyen TT, Chang SC, Evnouchidou I, et al. Structural basis for antigenic peptide precursor processing by the endoplasmic reticulum aminopeptidase ERAP1. *Nat Struct Mol Biol*. 2011;18(5):604-613.
107. Giastas P, Mpakali A, Papakyriakou A, et al. Mechanism for antigenic peptide selection by endoplasmic reticulum aminopeptidase 1. *Proc Natl Acad Sci U S A*. 2019;116(52):26709-26716.
108. Hammer GE, Gonzalez F, Champsaur M, Cado D, Shastri N. The aminopeptidase ERAAP shapes the peptide repertoire displayed by major histocompatibility complex class I molecules. *Nat Immunol*. 2006;7(1):103-112.
109. York IA, Brehm MA, Zendzian S, Towne CF, Rock KL. Endoplasmic reticulum aminopeptidase 1 (ERAP1) trims MHC class I-presented peptides *in vivo* and plays an important role in immunodominance. *Proc Natl Acad Sci U S A*. 2006;103(24):9202-9207.
110. Admon A. ERAP1 shapes just part of the immunopeptidome. *Hum Immunol*. 2019;80(5):296-301.
111. Nagarajan NA, de Verteuil DA, Sriranganadane D, et al. ERAAP shapes the peptidome associated with classical and nonclassical MHC class I molecules. *J Immunol*. 2016;197(4):1035-1043.
112. Saveanu L, Carroll O, Lindo V, et al. Concerted peptide trimming by human ERAP1 and ERAP2 aminopeptidase complexes in the endoplasmic reticulum. *Nat Immunol*. 2005;6(7):689-697.
113. Babaie F, Hosseinzadeh R, Ebrazeh M, et al. The roles of ERAP1 and ERAP2 in autoimmunity and cancer immunity: new insights and perspective. *Mol Immunol*. 2020;121:7-19.
114. Suh WK, Cohen-Doyle M, Fruh K, Wang K, Peterson P, Williams D. Interaction of MHC class I molecules with the transporter associated with antigen processing. *Science*. 1994;264:1322-1326.
115. Ortmann B, Androlewicz MJ, Cresswell P. MHC class I/β 2-microglobulin complexes associate with TAP transporters before peptide binding. *Nature*. 1994;368:864-867.
116. Gao B, Adhikari R, Howarth M, et al. Assembly and antigen-presenting function of MHC class I molecules in cells lacking the ER chaperone calreticulin. *Immunity*. 2002;16(1):99-109.
117. Peaper DR, Wearsch PA, Cresswell P. Tapasin and ERp57 form a stable disulfide-linked dimer within the MHC class I peptide-loading complex. *EMBO J*. 2005;24(20):3613-3623.
118. Dick TP, Bangia N, Peaper DR, Cresswell P. Disulfide bond isomerization and the assembly of MHC class I-peptide complexes. *Immunity*. 2002;16(1):87-98.
119. Garbi N, Tanaka S, Momburg F, Hämmerling GJ. Impaired assembly of the major histocompatibility complex class I peptide-loading complex in mice deficient in the oxidoreductase ERp57. *Nat Immunol*. 2006;7(1):93-102.
120. Dong G, Wearsch PA, Peaper DR, Cresswell P, Reinisch KM. Insights into MHC class I peptide loading from the structure of the tapasin-ERp57 thiol oxidoreductase heterodimer. *Immunity*. 2009;30(1):21-32.
121. Ortmann B, Copeman J, Lehner PJ, et al. A critical role for tapasin in the assembly and function of multimeric MHC class I-TAP complexes. *Science*. 1997;277:1306-1309.
122. Chen M, Bouvier M. Analysis of interactions in a tapasin/class I complex provides a mechanism for peptide selection. *EMBO J*. 2007;26(6):1681-1690.
123. Wearsch PA, Cresswell P. Selective loading of high-affinity peptides onto major histocompatibility complex class I molecules by the tapasin-ERp57 heterodimer. *Nat Immunol*. 2007;8(8):873-881.
124. Williams AP, Peh CA, Purcell AW, McCluskey J, Elliott T. Optimization of the MHC class I peptide cargo is dependent on tapasin. *Immunity*. 2002;16(4):509-520.
125. Howarth M, Williams A, Tolstrup AB, Elliott T. Tapasin enhances MHC class I peptide presentation according to peptide half-life. *Proc Natl Acad Sci U S A*. 2004;101(32):11737-11742.
126. Schoenhals GJ, Krishna RM, Grandea AG, et al. Retention of empty MHC class I molecules by tapasin is essential to reconstitute antigen presentation in invertebrate cells. *EMBO J*. 1999;18(3):743-753.
127. Thomas C, Tampe R. Structure of the TAPBPR-MHC I complex defines the mechanism of peptide loading and editing. *Science*. 2017;358(6366):1060-1064.
128. Jiang J, Natarajan K, Boyd LF, Morozov GI, Mage MG, Margulies DH. Crystal structure of a TAPBPR-MHC I complex reveals the mechanism of peptide editing in antigen presentation. *Science*. 2017;358(6366):1064-1068.
129. Hafstrand I, Sayitoglu EC, Apavaloaei A, et al. Successive crystal structure snapshots suggest the basis for MHC class I peptide loading and editing by tapasin. *Proc Natl Acad Sci U S A*. 2019;116(11):5055-5060.

130. Peh CA, Burrows SR, Barnden M, et al. HLA-B27-restricted antigen presentation in the absence of tapasin reveals polymorphism in mechanisms of HLA class I peptide loading. *Immunity*. 1998;8(5):531-542.
131. Smith JD, Myers NB, Gorka J, Hansen TH. Model for the *in vivo* assembly of nascent Ld class I molecules and for the expression of unfolded Ld molecules at the cell surface. *J Exp Med*. 1993;178:2035-2046.
132. Anjanappa R, Garcia-Alai M, Kopicki JD, et al. Structures of peptide-free and partially loaded MHC class I molecules reveal mechanisms of peptide selection. *Nat Commun*. 2020;11(1):1314.
133. Teng MS, Stephens R, Pasquier LD, Freeman T, Lindquist JA, Trowsdale J. A human TAPBP (TAPASIN)-related gene, TAPBP-R. *Eur J Immunol*. 2002;32(4):1059-1068.
134. Boyle LH, Hermann C, Boname JM, et al. Tapasin-related protein TAPBPR is an additional component of the MHC class I presentation pathway. *Proc Natl Acad Sci U S A*. 2013;110(9):3465-3470.
135. Hermann C, van Hateren A, Trautwein N, et al. TAPBPR alters MHC class I peptide presentation by functioning as a peptide exchange catalyst. *Elife*. 2015;4:e09617.
136. Morozov GI, Zhao H, Mage MG, et al. Interaction of TAPBPR, a tapasin homolog, with MHC-I molecules promotes peptide editing. *Proc Natl Acad Sci U S A*. 2016;113(8):E1006-E1015.
137. Neerincx A, Hermann C, Antrobus R, et al. TAPBPR bridges UDP-glucose: glycoprotein glucosyltransferase 1 onto MHC class I to provide quality control in the antigen presentation pathway. *Elife*. 2017;6:e23049.
138. Howe C, Garstka M, Al-Balushi M, et al. Calreticulin-dependent recycling in the early secretory pathway mediates optimal peptide loading of MHC class I molecules. *EMBO J*. 2009;28(23):3730-3744.
139. Zhang W, Wearsch PA, Zhu Y, Leonhardt RM, Cresswell P. A role for UDP-glucose glycoprotein glucosyltransferase in expression and quality control of MHC class I molecules. *Proc Natl Acad Sci U S A*. 2011;108(12):4956-4961.
140. McShan AC, Devlin CA, Overall SA, et al. Molecular determinants of chaperone interactions on MHC-I for folding and antigen repertoire selection. *Proc Natl Acad Sci U S A*. 2019;116(51):25602-25613.
141. McShan AC, Natarajan K, Kumirov VK, et al. Peptide exchange on MHC-I by TAPBPR is driven by a negative allostery release cycle. *Nat Chem Biol*. 2018;14(8):811-820.
142. Sagert L, Hennig F, Thomas C, Tampé R. A loop structure allows TAPBPR to exert its dual function as MHC I chaperone and peptide editor. *Elife*. 2020;9:e55326.
143. Ilca FT, Neerincx A, Hermann C, et al. TAPBPR mediates peptide dissociation from MHC class I using a leucine lever. *Elife*. 2018;7:e40126.
144. Saini SK, Schuster H, Ramnarayan VR, Rammensee HG, Stevanović S, Springer S. Dipeptides catalyze rapid peptide exchange on MHC class I molecules. *Proc Natl Acad Sci U S A*. 2015;112(1):202-207.
145. Christinck ER, Luscher MA, Barber BH, Williams DB. Peptide binding to class I MHC on living cells and quantitation of complexes required for CTL lysis. *Nature*. 1991;352(6330):67-70.
146. Ljunggren HG, Stam NJ, Öhlén C, et al. Empty MHC class I molecules come out in the cold. *Nature*. 1990;346:476-480.
147. Ortiz-Navarrete V, Hämmerling G. Surface appearance and instability of empty H-2 class I molecules under physiological conditions. *Proc Natl Acad Sci U S A*. 1991;88(9):3594-3597.
148. Day PM, Esquivel F, Lukszo J, Bennink JR, Yewdell JW. Effect of TAP on the generation and intracellular trafficking of peptide-receptive major histocompatibility complex class I molecules. *Immunity*. 1995;2(2):137-147.
149. Montealegre S, Venugopalan V, Fritzsche S, Kulicke C, Hein Z, Springer S. Dissociation of β2-microglobulin determines the surface quality control of major histocompatibility complex class I molecules. *FASEB J*. 2015;29(7):2780-2788.
150. Springer S. Transport and quality control of MHC class I molecules in the early secretory pathway. *Curr Opin Immunol*. 2015;34:83-90.
151. De Silva AD, Boesteanu A, Song R, et al. Thermolabile H-2Kb molecules expressed by transporter associated with antigen processing-deficient RMA-S cells are occupied by low-affinity peptides. *J Immunol*. 1999;163(8):4413-4420.
152. Smith JD, Lie WR, Gorka J, Myers NB, Hansen TH. Extensive peptide ligand exchange by surface class I major histocompatibility complex molecules independent of exogenous beta 2-microglobulin. *Proc Natl Acad Sci U S A*. 1992;89(16):7767-7771.
153. Rock KL, Rothstein LE, Gamble SR, Benacerraf B. Reassociation with beta 2-microglobulin is necessary for Kb class I major histocompatibility complex binding of exogenous peptides. *Proc Natl Acad Sci U S A*. 1990;87(19):7517-7521.
154. Rock KL, Gamble S, Rothstein L, Gramm C, Benacerraf B. Dissociation of beta 2-microglobulin leads to the accumulation of a substantial pool of inactive class I MHC heavy chains on the cell surface. *Cell*. 1991;65(4):611-620.
155. Vitiello A, Potter T, Sherman L. The role of beta 2-microglobulin in peptide binding by class I molecules. *Science*. 1990;250(4986):1423-1426.
156. Cook JR, Solheim JC, Connolly JM, Hansen TH. Induction of peptide-specific CD8+ CTL clones in beta 2-microglobulin-deficient mice. *J Immunol*. 1995;154(1):47-57.
157. Glas R, Franksson L, Ohlen C, et al. Major histocompatibility complex class I-specific and-restricted killing of beta 2-microglobulin-deficient cells by CD8+ cytotoxic T lymphocytes. *Proc Natl Acad Sci U S A*. 1992;89(23):11381-11385.
158. Jefferies WA, MacPherson GG. Expression of the W6/32 HLA epitope by cells of rat, mouse, human and other species: critical dependence on the interaction of specific MHC heavy chains with human or bovine beta 2-microglobulin. *Eur J Immunol*. 1987;17(9):1257-1263.
159. Krangel MS, Orr HT, Strominger JL. Assembly and maturation of HLA-A and HLA-B antigens *in vivo*. *Cell*. 1979;18:979-991.
160. Mottez E, Jaulin C, Godeau F, Choppin J, Levy JP, Kourilsky P. A single chain murine class I major transplantation antigen. *Eur J Immunol*. 1991;21:467-471.
161. Primeau T, Myers NB, Yu YY, et al. Applications of major histocompatibility complex class I molecules expressed as single chains. *Immunol Res*. 2005;32(1-3):109-121.
162. Swann JB, Holland SJ, Petersen M, Pietsch TW, Boehm T. The immunogenetics of sexual parasitism. *Science*. 2020:369(6511):1608-1615.
163. Radhakrishna H, Donaldson JG. ADP-ribosylation factor 6 regulates a novel plasma membrane recycling pathway. *J Cell Biol*. 1997;139(1):49-61.
164. Reid PA, Watts C. Cycling of cell-surface MHC glycoproteins through primaquine-sensitive intracellular compartments. *Nature*. 1990;346(6285):655-657.
165. Abdel Motal UM, Zhou X, Joki A, et al. Major histocompatibility complex class I-binding peptides are recycled to the cell surface after internalization. *Eur J Immunol*. 1993;23:3224-3229.
166. Mahmutefendić H, Blagojević G, Tomaš MI, Kučić N, Lučin P. Segregation of open Major Histocompatibility Class I conformers at the plasma membrane and during endosomal trafficking reveals conformation-based sorting in the endosomal system. *Int J Biochem Cell Biol*. 2011;43(4):504-515.
167. Montealegre S, van Endert PM. Endocytic recycling of MHC class I molecules in non-professional antigen presenting and dendritic cells. *Front Immunol*. 2018;9:3098.
168. Sinai AP, Joiner KA. Safe haven: the cell biology of nonfusogenic pathogen vacuoles. *Annu Rev Microbiol*. 1997;51:415-462.
169. Doherty GJ, McMahon HT. Mechanisms of endocytosis. *Annu Rev Biochem*. 2009;78:857-902.
170. Sallusto F, Cella M, Danieli C, Lanzavecchia A. Dendritic cells use macropinocytosis and the mannose receptor to concentrate macromolecules in the major histocompatibility complex class II compartment: downregulation by cytokines and bacterial products. *J Exp Med*. 1995;182(2):389-400.
171. Garrett WS, Chen LM, Kroschewski R, et al. Developmental control of endocytosis in dendritic cells by Cdc42. *Cell*. 2000;102(3):325-334.
172. West MA, Prescott AR, Eskelinen EL, Ridley AJ, Watts C. Rac is required for constitutive macropinocytosis by dendritic cells but does not control its downregulation. *Curr Biol*. 2000;10(14):839-848.
173. Lanzavecchia A. Antigen-specific interaction between T and B cells. *Nature*. 1985;314(6011):537-539.
174. Nashar TO, Drake JR. The pathway of antigen uptake and processing dictates MHC class II-mediated B cell survival and activation. *J Immunol*. 2005;174(3):1306-1316.
175. Trombetta ES, Mellman I. Cell biology of antigen processing *in vitro* and *in vivo*. *Annu Rev Immunol*. 2005;23:975-1028.
176. Tan MC, Mommaas AM, Drijfhout JW, et al. Mannose receptor-mediated uptake of antigens strongly enhances HLA class II-restricted antigen presentation by cultured dendritic cells. *Eur J Immunol*. 1997;27(9):2426-2435.
177. Engering AJ, Cella M, Fluitsma D, et al. The mannose receptor functions as a high capacity and broad specificity antigen receptor in human dendritic cells. *Eur J Immunol*. 1997;27(9):2417-2425.
178. Jiang W, Swiggard WJ, Heufler C, et al. The receptor DEC-205 expressed by dendritic cells and thymic epithelial cells is involved in antigen processing. *Nature*. 1995;375(6527):151-155.
179. Steinman RM, Banchereau J. Taking dendritic cells into medicine. *Nature*. 2007;449(7161):419-426.
180. Sanjuan MA, Dillon CP, Tait SW, et al. Toll-like receptor signalling in macrophages links the autophagy pathway to phagocytosis. *Nature*. 2007;450(7173):1253-1257.
181. Munz C. Autophagy proteins influence endocytosis for MHC restricted antigen presentation. *Semin Cancer Biol*. 2020;66:110-115.
182. Ma J, Becker C, Lowell CA, Underhill DM. Dectin-1-triggered recruitment of light chain 3 protein to phagosomes facilitates major histocompatibility complex class II presentation of fungal-derived antigens. *J Biol Chem*. 2012;287(41):34149-34156.
183. Romao S, Gasser N, Becker AC, et al. Autophagy proteins stabilize pathogen-containing phagosomes for prolonged MHC II antigen processing. *J Cell Biol*. 2013;203(5):757-766.
184. Mommen GP, Marino F, Meiring HD, et al. Sampling from the proteome to the human leukocyte antigen-DR (HLA-DR) ligandome Proceeds via high specificity. *Mol Cell Proteomics*. 2016;15(4):1412-1423.
185. Clement CC, Becerra A, Yin L, et al. The dendritic cell major histocompatibility complex II (MHC II) peptidome derives from a variety of processing pathways and includes peptides with a broad spectrum of HLA-DM sensitivity. *J Biol Chem*. 2016;291(11):5576-5595.
186. Miller MA, Ganesan APV, Luckashenak N, Mendonca M, Eisenlohr LC. Endogenous antigen processing drives the primary CD4+ T cell response to influenza. *Nat Med*. 2015;21(10):1216-1222.
187. Crotzer VL, Blum JS. Cytosol to lysosome transport of intracellular antigens during immune surveillance. *Traffic*. 2008;9(1):10-16.
188. Munz C. Enhancing immunity through autophagy. *Annu Rev Immunol*. 2009;27:423-449.
189. Paludan C, Schmid D, Landthaler M, et al. Endogenous MHC class II processing of a viral nuclear antigen after autophagy. *Science*. 2005;307(5709):593-596.
190. Schmid D, Pypaert M, Münz C. Antigen-loading compartments for major histocompatibility complex class II molecules continuously receive input from autophagosomes. *Immunity*. 2007;26(1):79-92.
191. Zhou D, Li P, Lin Y, et al. Lamp-2a facilitates MHC class II presentation of cytoplasmic antigens. *Immunity*. 2005;22(5):571-581.
192. Sahu R, Kaushik S, Clement CC, et al. Microautophagy of cytosolic proteins by late endosomes. *Dev Cell*. 2011;20(1):131-139.
193. Nedjic J, Aichinger M, Emmerich J, Mizushima N, Klein L. Autophagy in thymic epithelium shapes the T-cell repertoire and is essential for tolerance. *Nature*. 2008;455(7211):396-400.
194. Tewari MK, Sinnathamby G, Rajagopal D, Eisenlohr LC. A cytosolic pathway for MHC class II-restricted antigen processing that is proteasome and TAP dependent. *Nat Immunol*. 2005;6(3):287-294.
195. Clement CC, Rotzschke O, Santambrogio L. The lymph as a pool of self-antigens. *Trends Immunol*. 2011;32(1):6-11.
196. Santambrogio L, Sato AK, Fischer FR, Dorf ME, Stern LJ. Abundant empty class II MHC molecules on the surface of immature dendritic cells. *Proc Natl Acad Sci U S A*. 1999;96(26):15050-15055.
197. Pathak SS, Blum JS. Endocytic recycling is required for the presentation of an exogenous peptide via MHC class II molecules. *Traffic*. 2000;1(7):561-569.
198. Roche PA, Cresswell P. Invariant chain association with HLA-DR molecules inhibits immunogenic peptide binding. *Nature*. 1990;345:615-618.
199. Teyton L, O'Sullivan D, Dickson PW, et al. Invariant chain distinguishes between the exogenous and endogenous antigen presentation pathways. *Nature*. 1990;348:39-44.
200. Long EO, LaVaute T, Pinet V, Jaraquemada D. Invariant chain prevents the HLA-DR-restricted presentation of a cytosolic peptide. *J Immunol*. 1994;153(4):1487-1494.
201. Ghosh P, Amaya M, Mellins E, Wiley DC. The structure of an intermediate in class II MHC maturation: CLIP bound to HLA-DR3. *Nature*. 1995;378(6556):457-462.
202. Strubin M, Long EO, Mach B. Two forms of the Ia antigen-associated invariant chain result from alternative initiations at two in-phase AUGs. *Cell*. 1986;47:619-625.
203. Koch N, Lauer W, Habicht J, Dobberstein B. Primary structure of the gene for the murine Ia antigen-associated invariant chains (Ii). An alternatively spliced exon encodes a cysteine-rich domain highly homologous to a repetitive sequence of thyroglobulin. *EMBO J*. 1987;6(6):1677-1683.
204. Strubin M, Berte C, Mach B. Alternative splicing and alternative initiation of translation explain the four forms of the Ia antigen-associated invariant chain. *EMBO J*. 1986;5(13):3483-3488.
205. O'Sullivan DM, Noonan D, Quaranta V. Four Ia invariant chain forms derive from a single gene by alternate splicing and alternate initiation of transcription/translation. *J Exp Med*. 1987;166(2):444-460.
206. Marks MS, Blum JS, Cresswell P. Invariant chain trimers are sequestered in the rough endoplasmic reticulum in the absence of association with HLA class II antigens. *J Cell Biol*. 1990;111:839-855.
207. Lamb CA, Cresswell P. Assembly and transport properties of invariant chain trimers and HLA-DR-invariant chain complexes. *J Immunol*. 1992;148:3478-3482.
208. Arunachalam B, Lamb CA, Cresswell P. Transport properties of free and MHC class II-associated oligomers containing different isoforms of human invariant chain. *Int Immunol*. 1994;6:439-451.
209. Roche PA, Marks MS, Cresswell PJ. Formation of a nine-subunit complex by HLA class II glycoproteins and the invariant chain. *Nature*. 1991;354:392-394.
210. Anderson KS, Cresswell P. A role for calnexin (IP90) in the assembly of class II MHC molecules. *EMBO J*. 1994;13(3):675-682.
211. Lotteau V, Teyton L, Peleraux A, et al. Intracellular transport of class II MHC molecules directed by invariant chain. *Nature*. 1990;348:600-605.

212. Schutze MP, Peterson PA, Jackson MR. An N-terminal double-arginine motif maintains type II membrane proteins in the endoplasmic reticulum. *EMBO J.* 1994;13:1696-1705.
213. Newcomb JR, Cresswell P. Structural analysis of proteolytic products of MHC class II-invariant chain complexes generated *in vivo*. *J Immunol.* 1993;151:4153-4163.
214. Bikoff EK, Germain RN, Robertson EJ. Allelic differences affecting invariant chain dependency of MHC class II subunit assembly. *Immunity.* 1995;2(3):301-310.
215. Romagnoli P, Layet C, Yewdell J, Bakke O, Germain RN. Relationship between invariant chain expression and major histocompatibility complex class II transport into early and late endocytic compartments. *J Exp Med.* 1993;177:583-596.
216. Neefjes JJ, Stollorz V, Peters PJ, Geuze HJ, Ploegh HL. The biosynthetic pathway of MHC class II but not class I molecules intersects the endocytic route. *Cell.* 1990;61:171-183.
217. Benaroch P, Yilla M, Raposo G, et al. How MHC class II molecules reach the endocytic pathway. *EMBO J.* 1995;14:37-49.
218. Roche PA, Teletski CL, Stang E, Bakke O, Long EO. Cell surface HLA-DR-invariant chain complexes are targeted to endosomes by rapid internalization. *Proc Natl Acad Sci U S A.* 1993;90:8581-8585.
219. Bakke O, Dobberstein B. MHC class II-associated invariant chain contains a sorting signal for endosomal compartments. *Cell.* 1990;63:707-716.
220. Pieters J, Bakke O, Dobberstein B. The MHC class II-associated invariant chain contains two endosomal targeting signals within its cytoplasmic tail. *J Cell Sci.* 1993;106:831-846.
221. McCormick PJ, Martina JA, Bonifacino JS. Involvement of clathrin and AP-2 in the trafficking of MHC class II molecules to antigen-processing compartments. *Proc Natl Acad Sci U S A.* 2005;102(22):7910-7915.
222. Walseng E, Bakke O, Roche PA. Major histocompatibility complex class II-peptide complexes internalize using a clathrin-and dynamin-independent endocytosis pathway. *J Biol Chem.* 2008;283(21):14717-14727.
223. Blum JS, Cresswell P. Role for intracellular proteases in the processing and transport of class II HLA antigens. *Proc Natl Acad Sci U S A.* 1988;85:3975-3979.
224. Riese RJ, Wolf PR, Brömme D, et al. Essential role for cathepsin S in MHC class II-associated invariant chain processing and peptide loading. *Immunity.* 1996;4(4):357-366.
225. Shi GP, Villadangos JA, Dranoff G, et al. Cathepsin S required for normal MHC class II peptide loading and germinal center development. *Immunity.* 1999;10(2):197-206.
226. Sette A, Southwood S, Miller J, Appella E. Binding of major histocompatibility complex class II to the invariant chain-derived peptide, CLIP, is regulated by allelic polymorphism in class II. *J Exp Med.* 1995;181(2):677-683.
227. Morris P, Shaman J, Attaya M, et al. An essential role for HLA-DM in antigen presentation by class II major histocompatibility molecules. *Nature.* 1994;368(6471):551-554.
228. Fling SP, Arp B, Pious D. HLA-DMA and-DMB genes are both required for MHC class II/peptide complex formation in antigen-presenting cells. *Nature.* 1994;368(6471):554-558.
229. Sloan VS, Cameron P, Porter G, et al. Mediation by HLA-DM of dissociation of peptides from HLA-DR. *Nature.* 1995;375(6534):802-806.
230. Denzin LK, Cresswell P. HLA-DM induces CLIP dissociation from MHC class II alpha beta dimers and facilitates peptide loading. *Cell.* 1995;82:155-165.
231. Anders AK, Call MJ, Schulze MS, et al. HLA-DM captures partially empty HLA-DR molecules for catalyzed removal of peptide. *Nat Immunol.* 2011;12(1):54-61.
232. Painter CA, Negroni MP, Kellersberger KA, Zavala-Ruiz Z, Evans JE, Stern LJ. Conformational lability in the class II MHC 310 helix and adjacent extended strand dictate HLA-DM susceptibility and peptide exchange. *Proc Natl Acad Sci U S A.* 2011;108(48):19329-19334.
233. Kropshofer H, Vogt AB, Moldenhauer G, Hammer J, Blum JS, Hämmerling GJ. Editing of the HLA-DR-peptide repertoire by HLA-DM. *EMBO J.* 1996;15(22):6144-6154.
234. Weber DA, Evavold BD, Jensen PE. Enhanced dissociation of HLA-DR-bound peptides in the presence of HLA-DM. *Science.* 1996;274(5287):618-620.
235. Yin L, Trenh P, Guce A, et al. Susceptibility to HLA-DM protein is determined by a dynamic conformation of major histocompatibility complex class II molecule bound with peptide. *J Biol Chem.* 2014;289(34):23449-23464.
236. Marks MS, Roche PA, van Donselaar E, Woodruff L, Peters PJ, Bonifacino JS. A lysosomal targeting signal in the cytoplasmic tail of the beta chain directs HLA-DM to MHC class II compartments. *J Cell Biol.* 1995;131:351-369.
237. Sanderson F, Kleijmeer M, Kelly A, et al. Accumulation of HLA-DM, a regulator of antigen presentation, in MHC class II compartments. *Science.* 1994;266:1566-1569.
238. Stang E, Guerra CB, Amaya M, Paterson Y, Bakke O, Mellins ED. DR/CLIP (class II-associated invariant chain peptides) and DR/peptide complexes colocalize in prelysosomes in human B lymphoblastoid cells. *J Immunol.* 1998;160(10):4696-4707.
239. Sherman MA, Weber DA, Jensen PE. DM enhances peptide binding to class II MHC by release of invariant chain-derived peptide. *Immunity.* 1995;3(2):197-205.
240. Pierre P, Denzin LK, Hammond C, et al. HLA-DM is localized to conventional and unconventional MHC class II-containing endocytic compartments. *Immunity.* 1996;4:229-239.
241. Santambrogio L, Sato AK, Carven GJ, Belyanskaya SL, Strominger JL, Stern LJ. Extracellular antigen processing and presentation by immature dendritic cells. *Proc Natl Acad Sci U S A.* 1999;96(26):15056-15061.
242. Pathak SS, Lich JD, Blum JS. Cutting edge: editing of recycling class II:peptide complexes by HLA-DM. *J Immunol.* 2001;167(2):632-635.
243. Denzin LK, Fallas JL, Prendes M, Yi W. Right place, right time, right peptide: DO keeps DM focused. *Immunol Rev.* 2005;207:279-292.
244. Denzin LK, Sant'Angelo DB, Hammond C, Surman MJ, Cresswell P. Negative regulation by HLA-DO of MHC class II-restricted antigen processing. *Science.* 1997;278:106-109.
245. van Ham SM, Tjin EP, Lillemeier BF, et al. HLA-DO is a negative modulator of HLA-DM-mediated MHC class II peptide loading. *Curr Biol.* 1997;7(12):950-957.
246. Guce AI, Mortimer SE, Yoon T, et al. HLA-DO acts as a substrate mimic to inhibit HLA-DM by a competitive mechanism. *Nat Struct Mol Biol.* 2013;20(1):90-98.
247. Fallas JL, Tobin HM, Lou O, Guo D, Sant'Angelo DB, Denzin LK. Ectopic expression of HLA-DO in mouse dendritic cells diminishes MHC class II antigen presentation. *J Immunol.* 2004;173(3):1549-1560.
248. Liljedahl M, Kuwana T, Fung-Leung WP, Jackson MR, Peterson PA, Karlsson L. HLA-DO is a lysosomal resident which requires association with HLA-DM for efficient intracellular transport. *EMBO J.* 1996;15:4817-4824.
249. Manoury B, Hewitt EW, Morrice N, Dando PM, Barrett AJ, Watts C. An asparaginyl endopeptidase processes a microbial antigen for class II MHC presentation. *Nature.* 1998;396(6712):695-699.
250. Hsieh CS, deRoos P, Honey K, Beers C, Rudensky AY. A role for cathepsin L and cathepsin S in peptide generation for MHC class II presentation. *J Immunol.* 2002;168(6):2618-2625.
251. Arunachalam B, Phan UT, Geuze HJ, Cresswell P. Enzymatic reduction of disulfide bonds in lysosomes: characterization of a gamma-interferon-inducible lysosomal thiol reductase (GILT). *Proc Natl Acad Sci U S A.* 2000;97(2):745-750.
252. Maric M, Arunachalam B, Phan UT, et al. Defective antigen processing in GILT-free mice. *Science.* 2001;294(5545):1361-1365.
253. Manoury B, Mazzeo D, Fugger L, et al. Destructive processing by asparagine endopeptidase limits presentation of a dominant T cell epitope in MBP. *Nat Immunol.* 2002;3(2):169-174.
254. Delamarre L, Pack M, Chang H, Mellman I, Trombetta ES. Differential lysosomal proteolysis in antigen-presenting cells determines antigen fate. *Science.* 2005;307(5715):1630-1634.
255. Delamarre L, Couture R, Mellman I, Trombetta ES. Enhancing immunogenicity by limiting susceptibility to lysosomal proteolysis. *J Exp Med.* 2006;203(9):2049-2055.
256. Inaba K, Turley S, Iyoda T, et al. The formation of immunogenic major histocompatibility complex class II-peptide ligands in lysosomal compartments of dendritic cells is regulated by inflammatory stimuli. *J Exp Med.* 2000;191(6):927-936.
257. Turley SJ, Inaba K, Garrett WS, et al. Transport of peptide-MHC class II complexes in developing dendritic cells. *Science.* 2000;288(5465):522-527.
258. Trombetta ES, Ebersold M, Garrett W, Pypaert M, Mellman I. Activation of lysosomal function during dendritic cell maturation. *Science.* 2003;299(5611):1400-1403.
259. Samie M, Cresswell P. The transcription factor TFEB acts as a molecular switch that regulates exogenous antigen-presentation pathways. *Nat Immunol.* 2015;16(7):729-736.
260. Pierre P, Mellman I. Developmental regulation of invariant chain proteolysis controls MHC class II trafficking in mouse dendritic cells. *Cell.* 1998;93:1135-1145.
261. Kleijmeer M, Ramm G, Schuurhuis D, et al. Reorganization of multivesicular bodies regulates MHC class II antigen presentation by dendritic cells. *J Cell Biol.* 2001;155(1):53-63.
262. Bevec T, Stoka V, Pungercic G, Dolenc I, Turk V. Major histocompatibility complex class II-associated p41 invariant chain fragment is a strong inhibitor of lysosomal cathepsin L. *J Exp Med.* 1996;183(4):1331-1338.
263. Guncar G, Pungercic G, Klemencic I, Turk V, Turk D. Crystal structure of MHC class II-associated p41 Ii fragment bound to cathepsin L reveals the structural basis for differentiation between cathepsins L and S. *EMBO J.* 1999;18(4):793-803.
264. Fineschi B, Arneson LS, Naujokas MF, Miller J. Proteolysis of major histocompatibility complex class II-associated invariant chain is regulated by the alternatively spliced gene product, p41. *Proc Natl Acad Sci U S A.* 1995;92:10257-10261.
265. Takaesu NT, Lower JA, Yelon D, Robertson EJ, Bikoff EK. *In vivo* functions mediated by the p41 isoform of the MHC class II-associated invariant chain. *J Immunol.* 1997;158(1):187-199.
266. Sette A, Adorini L, Colon SM, Buus S, Grey HM. Capacity of intact proteins to bind to MHC class II molecules. *J Immunol.* 1989;143(4):1265-1267.
267. Sercarz EE, Maverakis E. Mhc-guided processing: binding of large antigen fragments. *Nat Rev Immunol.* 2003;3(8):621-629.
268. Moss CX, Tree TI, Watts C. Reconstruction of a pathway of antigen processing and class II MHC peptide capture. *EMBO J.* 2007;26(8):2137-2147.
269. Sadegh-Nasseri S, Kim A. Exogenous antigens bind MHC class II first, and are processed by cathepsins later. *Mol Immunol.* 2015;68(2 pt A):81-84.
270. Peters PJ, Neefjes JJ, Oorschot V, Ploegh HL, Geuze HJ. Segregation of MHC class II molecules from MHC class I molecules in the Golgi complex for transport to lysosomal compartments. *Nature.* 1991;349:669-676.
271. Kleijmeer MJ, Morkowski S, Griffith JM, Rudensky AY, Geuze HJ. Major histocompatibility complex class II compartments in human and mouse B lymphoblasts represent conventional endocytic compartments. *J Cell Biol.* 1997;139(3):639-649.
272. Geuze HJ. The role of endosomes and lysosomes in MHC class II functioning. *Immunol Today.* 1998;19:282-287.
273. Weber DA, Dao CT, Jun J, Wigal JL, Jensen PE. Transmembrane domain-mediated colocalization of HLA-DM and HLA-DR is required for optimal HLA-DM catalytic activity. *J Immunol.* 2001;167(9):5167-5174.
274. Zwart W, Griekspoor A, Kuijl C, et al. Spatial separation of HLA-DM/HLA-DR interactions within MIIC and phagosome-induced immune escape. *Immunity.* 2005;22(2):221-233.
275. Théry C, Duban L, Segura E, Véron P, Lantz O, Amigorena S. Indirect activation of naive CD4+ T cells by dendritic cell-derived exosomes. *Nat Immunol.* 2002;3(12):1156-1162.
276. Zitvogel L, Regnault A, Lozier A, et al. Eradication of established murine tumors using a novel cell-free vaccine: dendritic cell-derived exosomes. *Nat Med.* 1998;4(5):594-600.
277. Wubbolts R, Fernandez-Borja M, Oomen L, et al. Direct vesicular transport of MHC class II molecules from lysosomal structures to the cell surface. *J Cell Biol.* 1996;135:611-622.
278. Chow A, Toomre D, Garrett W, Mellman I. Dendritic cell maturation triggers retrograde MHC class II transport from lysosomes to the plasma membrane. *Nature.* 2002;418(6901):988-994.
279. Boes M, Cerny J, Massol R, et al. T-cell engagement of dendritic cells rapidly rearranges MHC class II transport. *Nature.* 2002;418(6901):983-988.
280. Blander JM, Medzhitov R. Toll-dependent selection of microbial antigens for presentation by dendritic cells. *Nature.* 2006;440(7085):808-812.
281. Mantegazza AR, Zajac AL, Twelvetrees A, Holzbaur ELF, Amigorena S, Marks MS. TLR-dependent phagosome tubulation in dendritic cells promotes phagosome cross-talk to optimize MHC-II antigen presentation. *Proc Natl Acad Sci U S A.* 2014;111(43):15508-15513.
282. Murk JL, Stoorvogel W, Kleijmeer MJ, Geuze HJ. The plasticity of multivesicular bodies and the regulation of antigen presentation. *Semin Cell Dev Biol.* 2002;13(4):303-311.
283. Paul P, van den Hoorn T, Jongsma ML, et al. A Genome-wide multidimensional RNAi screen reveals pathways controlling MHC class II antigen presentation. *Cell.* 2011;145(2):268-283.
284. Wubbolts R, Fernandez-Borja M, Jordens I, et al. Opposing motor activities of dynein and kinesin determine retention and transport of MHC class II-containing compartments. *J Cell Sci.* 1999;112(pt 6):785-795.
285. Vascotto F, Lankar D, Faure-André G, et al. The actin-based motor protein myosin II regulates MHC class II trafficking and BCR-driven antigen presentation. *J Cell Biol.* 2007;176(7):1007-1019.
286. Jenei A, Varga S, Bene L, et al. HLA class I and II antigens are partially co-clustered in the plasma membrane of human lymphoblastoid cells. *Proc Natl Acad Sci U S A.* 1997;94(14):7269-7274.
287. Anderson HA, Hiltbold EM, Roche PA. Concentration of MHC class II molecules in lipid rafts facilitates antigen presentation. *Nat Immunol.* 2000;1(2):156-162.
288. Kropshofer H, Spindeldreher S, Röhn TA, et al. Tetraspan microdomains distinct from lipid rafts enrich select peptide-MHC class II complexes. *Nat Immunol.* 2002;3(1):61-68.
289. Poloso NJ, Muntasell A, Roche PA. MHC class II molecules traffic into lipid rafts during intracellular transport. *J Immunol.* 2004;173(7):4539-4546.
290. Mobius W, Van Donselaar E, Ohno-Iwashita Y, et al. Recycling compartments and the internal vesicles of multivesicular bodies harbor most of the cholesterol found in the endocytic pathway. *Traffic.* 2003;4(4):222-231.
291. Bosch B, Heipertz EL, Drake JR, Roche PA. Major histocompatibility complex (MHC) class II-peptide complexes arrive at the plasma membrane in cholesterol-rich microclusters. *J Biol Chem.* 2013;288(19):13236-13242.
292. Eren E, Yates J, Cwynarski K, et al. Location of major histocompatibility complex class II molecules in rafts on dendritic cells enhances the efficiency of T-cell activation and proliferation. *Scand J Immunol.* 2006;63(1):7-16.
293. Davis JE, Cresswell P. Lack of detectable endocytosis of B lymphocyte MHC class II antigens using an antibody-independent technique. *J Immunol.* 1990;144(3):990-997.
294. Lanzavecchia A, Reid PA, Watts C. Irreversible association of peptides with class II MHC molecules in living cells. *Nature.* 1992;357(6375):249-252.
295. Cella M, Engering A, Pinet V, Pieters J, Lanzavecchia A. Inflammatory stimuli induce accumulation of MHC class II complexes on dendritic cells. *Nature.* 1997;388(6644):782-787.
296. Pierre P, Turley SJ, Gatti E, et al. Developmental regulation of MHC class II transport in mouse dendritic cells. *Nature.* 1997;388:787-792.

297. Shin JS, Ebersold M, Pypaert M, Delamarre L, Hartley A, Mellman I. Surface expression of MHC class II in dendritic cells is controlled by regulated ubiquitination. *Nature.* 2006;444(7115):115-118.
298. van Niel G, Wubbolts R, Ten Broeke T, et al. Dendritic cells regulate exposure of MHC class II at their plasma membrane by oligoubiquitination. *Immunity.* 2006;25(6):885-894.
299. Matsuki Y, Ohmura-Hoshino M, Goto E, et al. Novel regulation of MHC class II function in B cells. *EMBO J.* 2007;26(3):846-854.
300. Walseng E, Furuta K, Bosch B, et al. Ubiquitination regulates MHC class II-peptide complex retention and degradation in dendritic cells. *Proc Natl Acad Sci U S A.* 2010;107(47):20465-20470.
301. Liu H, Jain R, Guan J, et al. Ubiquitin ligase MARCH 8 cooperates with CD83 to control surface MHC II expression in thymic epithelium and CD4 T cell selection. *J Exp Med.* 2016;213(9):1695-1703.
302. von Rohrscheidt J, Petrozziello E, Nedjic J, et al. Thymic CD4 T cell selection requires attenuation of March8-mediated MHCII turnover in cortical epithelial cells through CD83. *J Exp Med.* 2016;213(9):1685-1694.
303. Cho KJ, Walseng E, Ishido S, Roche PA. Ubiquitination by March-I prevents MHC class II recycling and promotes MHC class II turnover in antigen-presenting cells. *Proc Natl Acad Sci U S A.* 2015;112(33):10449-10454.
304. Cho KJ, Ishido S, Eisenlohr LC, Roche PA. Activation of dendritic cells alters the mechanism of MHC class II antigen presentation to CD4 T cells. *J Immunol.* 2020;204(6):1621-1629.
305. Pinet V, Vergelli M, Martini R, Bakke O, Long EO. Antigen presentation mediated by recycling of surface HLA-DR molecules. *Nature.* 1995;375:603-606.
306. Sinnathamby G, Eisenlohr LC. Presentation by recycling MHC class II molecules of an influenza hemagglutinin-derived epitope that is revealed in the early endosome by acidification. *J Immunol.* 2003;170(7):3504-3513.
307. Hansen TH, Bouvier M. MHC class I antigen presentation: learning from viral evasion strategies. *Nat Rev Immunol.* 2009;9(7):503-513.
308. Levitskaya J, Coram M, Levitsky V, et al. Inhibiton of antigen processing by the internal repeat region of the Epstein-Barr virus nuclear antigen-1. *Nature.* 1995;375:685-688.
309. Dantuma NP, Heessen S, Lindsten K, Jellne M, Masucci MG. Inhibition of proteasomal degradation by the gly-Ala repeat of Epstein-Barr virus is influenced by the length of the repeat and the strength of the degradation signal. *Proc Natl Acad Sci U S A.* 2000;97(15):8381-8385.
310. Bennett NJ, May JS, Stevenson PG. Gamma-herpesvirus latency requires T cell evasion during episome maintenance. *PLoS Biol.* 2005;3(4):e120.
311. Zaldumbide A, Ossevoort M, Wiertz EJ, Hoeben RC. In cis inhibition of antigen processing by the latency-associated nuclear antigen I of Kaposi sarcoma herpes virus. *Mol Immunol.* 2007;44(6):1352-1360.
312. Kwun HJ, da Silva SR, Shah IM, Blake N, Moore PS, Chang Y. Kaposi's sarcoma-associated herpesvirus latency-associated nuclear antigen 1 mimics Epstein-Barr virus EBNA1 immune evasion through central repeat domain effects on protein processing. *J Virol.* 2007;81(15):8225-8235.
313. Früh K, Ahn K, Djaballah H, et al. A viral inhibitor of peptide transporters for antigen presentation. *Nature.* 1995;375:415-418.
314. Hill A, Jugovic P, York I, et al. Herpes simplex virus turns off the TAP to evade host immunity. *Nature.* 1995;375(6530):411-415.
315. Ahn K, Meyer TH, Uebel S, et al. Molecular mechanism and species specificity of TAP inhibition by herpes simplex virus ICP47. *EMBO J.* 1996;15(13):3247-3255.
316. Tomazin R, Hill AB, Jugovic P, et al. Stable binding of the herpes simplex virus ICP47 protein to the peptide binding site of TAP. *EMBO J.* 1996;15(13):3256-3266.
317. Lacaille VG, Androlewicz MJ. Herpes simplex virus inhibitor ICP47 destabilizes the transporter associated with antigen processing (TAP) heterodimer. *J Biol Chem.* 1998;273(28):17386-17390.
318. Gorbulev S, Abele R, Tampe R. Allosteric crosstalk between peptide-binding, transport, and ATP hydrolysis of the ABC transporter TAP. *Proc Natl Acad Sci U S A.* 2001;98(7):3732-3737.
319. Hengel H, Koopmann JO, Flohr T, et al. A viral ER-resident glycoprotein inactivates the MHC-encoded peptide transporter. *Immunity.* 1997;6(5):623-632.
320. Lehner PJ, Karttunen JT, Wilkinson GW, Cresswell P. The human cytomegalovirus US6 glycoprotein inhibits transporter associated with antigen processing-dependent peptide translocation. *Proc Natl Acad Sci U S A.* 1997;94(13):6904-6909.
321. Hewitt EW, Gupta SS, Lehner PJ. The human cytomegalovirus gene product US6 inhibits ATP binding by TAP. *EMBO J.* 2001;20(3):387-396.
322. Kyritsis C, Gorbulev S, Hutschenreiter S, Pawlitschko K, Abele R, Tampé R. Molecular mechanism and structural aspects of transporter associated with antigen processing inhibition by the cytomegalovirus protein US6. *J Biol Chem.* 2001;276(51):48031-48039.
323. Loch S, Klauschies F, Schölz C, et al. Signaling of a varicelloviral factor across the endoplasmic reticulum membrane induces destruction of the peptide-loading complex and immune evasion. *J Biol Chem.* 2008;283(19):13428-13436.
324. Lee S, Yoon J, Park B, et al. Structural and functional dissection of human cytomegalovirus US3 in binding major histocompatibility complex class I molecules. *J Virol.* 2000;74(23):11262-11269.
325. Park B, Kim Y, Shin J, et al. Human cytomegalovirus inhibits tapasin-dependent peptide loading and optimization of the MHC class I peptide cargo for immune evasion. *Immunity.* 2004;20(1):71-85.
326. Bennett EM, Bennink JR, Yewdell JW, Brodsky FM. Cutting edge: adenovirus E19 has two mechanisms for affecting class I MHC expression. *J Immunol.* 1999;162(9):5049-5052.
327. Byun M, Verweij MC, Pickup DJ, Wiertz EJ, Hansen TH, Yokoyama WM. Two mechanistically distinct immune evasion proteins of cowpox virus combine to avoid antiviral CD8 T cells. *Cell Host Microbe.* 2009;6(5):422-432.
328. Alzhanova D, Edwards DM, Hammarlund E, et al. Cowpox virus inhibits the transporter associated with antigen processing to evade T cell recognition. *Cell Host Microbe.* 2009;6(5):433-445.
329. Andersson M, Pääbo S, Nilsson T, Peterson PA. Impaired intracellular transport of class I MHC antigens as a possible means for adenovirus to evade immune surveillance. *Cell.* 1985;43:215-222.
330. Burgert HG, Kvist S. An adenovirus type 2 glycoprotein blocks cell surface expression of human histocompatibility class I antigens. *Cell.* 1985;41(3):987-997.
331. Pääbo S, Bhat BM, Wold WS, Peterson PA. A short sequence in the COOH-terminus makes an adenovirus membrane glycoprotein a resident of the endoplasmic reticulum. *Cell.* 1987;50:311-317.
332. Gabathuler R, Lévy F, Kvist S. Requirements for the association of adenovirus type 2 E3/19K wild-type and mutant proteins with HLA antigens. *J Virol.* 1990;64(8):3679-3685.
333. Liu H, Stafford WF, Bouvier M. The endoplasmic reticulum lumenal domain of the adenovirus type 2 E3-19K protein binds to peptide-filled and peptide-deficient HLA-A*1101 molecules. *J Virol.* 2005;79(21):13317-13325.
334. Dasgupta A, Hammarlund E, Slifka MK, Früh K. Cowpox virus evades CTL recognition and inhibits the intracellular transport of MHC class I molecules. *J Immunol.* 2007;178(3):1654-1661.
335. Byun M, Wang X, Pak M, Hansen TH, Yokoyama WM. Cowpox virus exploits the endoplasmic reticulum retention pathway to inhibit MHC class I transport to the cell surface. *Cell Host Microbe.* 2007;2(5):306-315.
336. Schwartz O, Maréchal V, Gall SL, Lemonnier F, Heard JM. Endocytosis of major histocompatibility complex class I molecules is induced by the HIV-1 Nef protein. *Nat Med.* 1996;2(3):338-342.
337. Collins KL, Chen BK, Kalams SA, Walker BD, Baltimore D. HIV-1 Nef protein protects infected primary cells against killing by cytotoxic T lymphocytes. *Nature.* 1998;391(6665):397-401.
338. Schaefer MR, Wonderlich ER, Roeth JF, Leonard JA, Collins KL. HIV-1 Nef targets MHC-I and CD4 for degradation via a final common beta-COP-dependent pathway in T cells. *PLoS Pathog.* 2008;4(8):e1000131.
339. Wang X, Herr RA, Hansen T. Viral and cellular MARCH ubiquitin ligases and cancer. *Semin Cancer Biol.* 2008;18(6):441-450.
340. Nathan JA, Lehner PJ. The trafficking and regulation of membrane receptors by the RING-CH ubiquitin E3 ligases. *Exp Cell Res.* 2009;315(9):1593-1600.
341. Ohmura-Hoshino M, Goto E, Matsuki Y, et al. A novel family of membrane-bound E3 ubiquitin ligases. *J Biochem.* 2006;140(2):147-154.
342. Stevenson PG, Efstathiou S, Doherty PC, Lehner PJ. Inhibition of MHC class I-restricted antigen presentation by gamma 2-herpesviruses. *Proc Natl Acad Sci U S A.* 2000;97(15):8455-8460.
343. Coscoy L, Ganem D. Kaposi's sarcoma-associated herpesvirus encodes two proteins that block cell surface display of MHC class I chains by enhancing their endocytosis. *Proc Natl Acad Sci U S A.* 2000;97(14):8051-8056.
344. Ishido S, Wang C, Lee BS, Cohen GB, Jung JU. Downregulation of major histocompatibility complex class I molecules by Kaposi's sarcoma-associated herpesvirus K3 and K5 proteins. *J Virol.* 2000;74(11):5300-5309.
345. Wang X, Connors R, Harris MR, Hansen TH, Lybarger L. Requirements for the selective degradation of endoplasmic reticulum-resident major histocompatibility complex class I proteins by the viral immune evasion molecule mK3. *J Virol.* 2005;79(7):4099-4108.
346. Boname JM, de Lima BD, Lehner PJ, Stevenson PG. Viral degradation of the MHC class I peptide loading complex. *Immunity.* 2004;20(3):305-317.
347. Wang X, Lybarger L, Connors R, Harris MR, Hansen TH. Model for the interaction of gammaherpesvirus 68 RING-CH finger protein mK3 with major histocompatibility complex class I and the peptide-loading complex. *J Virol.* 2004;78(16):8673-8686.
348. Lybarger L, Wang X, Harris MR, Virgin HW, Hansen TH. Virus subversion of the MHC class I peptide-loading complex. *Immunity.* 2003;18(1):121-130.
349. Wiertz EJ, Jones TR, Sun L, Bogyo M, Geuze HJ, Ploegh HL. The human cytomegalovirus US11 gene product dislocates MHC class I heavy chains from the endoplasmic reticulum to the cytosol. *Cell.* 1996;84(5):769-779.
350. Wiertz EJ, Tortorella D, Bogyo M, et al. Sec61-mediated transfer of a membrane protein from the endoplasmic reticulum to the proteasome for destruction. *Nature.* 1996;384(6608):432-438.
351. Loureiro J, Lilley BN, Spooner E, Noriega V, Tortorella D, Ploegh HL. Signal peptide peptidase is required for dislocation from the endoplasmic reticulum. *Nature.* 2006;441(7095):894-897.
352. Lilley BN, Ploegh HL. A membrane protein required for dislocation of misfolded proteins from the ER. *Nature.* 2004;429(6994):834-840.
353. Mueller B, Lilley BN, Ploegh HL. SEL1L, the homologue of yeast Hrd3p, is involved in protein dislocation from the mammalian ER. *J Cell Biol.* 2006;175(2):261-270.
354. Mueller B, Klemm EJ, Spooner E, Claessen JH, Ploegh HL. SEL1L nucleates a protein complex required for dislocation of misfolded glycoproteins. *Proc Natl Acad Sci U S A.* 2008;105(34):12325-12330.
355. Thomas M, Boname JM, Field S, et al. Down-regulation of NKG2D and NKp80 ligands by Kaposi's sarcoma-associated herpesvirus K5 protects against NK cell cytotoxicity. *Proc Natl Acad Sci U S A.* 2008;105(5):1656-1661.
356. Russmann H. Bacterial type III translocation: a unique mechanism for cytosolic display of heterologous antigens by attenuated Salmonella. *Int J Med Microbiol.* 2003;293(1):107-112.
357. Otsubo R, Mimuro H, Ashida H, Hamazaki J, Murata S, Sasakawa C. Shigella effector IpaH4.5 targets 19S regulatory particle subunit RPN13 in the 26S proteasome to dampen cytotoxic T lymphocyte activation. *Cell Microbiol.* 2019;21(3):e12974.
358. Tobar JA, González PA, Kalergis AM. Salmonella escape from antigen presentation can be overcome by targeting bacteria to Fc gamma receptors on dendritic cells. *J Immunol.* 2004;173(6):4058-4065.
359. Thakker S, Purushothaman P, Gupta N, Challa S, Cai Q, Verma SC. Kaposi's sarcoma-associated herpesvirus latency-associated nuclear antigen inhibits major histocompatibility complex class II expression by disrupting enhanceosome assembly through binding with the regulatory factor X complex. *J Virol.* 2015;89(10):5536-5556.
360. Li D, Qian L, Chen C, et al. Down-regulation of MHC class II expression through inhibition of CIITA transcription by lytic transactivator Zta during Epstein-Barr virus reactivation. *J Immunol.* 2009;182(4):1799-1809.
361. Lin JH, Lin JY, Chou YC, et al. Epstein-Barr virus LMP2A suppresses MHC class II expression by regulating the B-cell transcription factors E47 and PU.1. *Blood.* 2015;125(14):2228-2238.
362. Hegde NR, Tomazin RA, Wisner TW, et al. Inhibition of HLA-DR assembly, transport, and loading by human cytomegalovirus glycoprotein US3: a novel mechanism for evading major histocompatibility complex class II antigen presentation. *J Virol.* 2002;76(21):10929-10941.
363. Zuo J, Thomas WA, Haigh TA, et al. Epstein-Barr virus evades CD4+ T cell responses in lytic cycle through BZLF1-mediated downregulation of CD74 and the cooperation of vBcl-2. *PLoS Pathog.* 2011;7(12):e1002455.
364. Kim H, Mazumdar B, Bose SK, et al. Hepatitis C virus-mediated inhibition of cathepsin S increases invariant-chain expression on hepatocyte surface. *J Virol.* 2012;86(18):9919-9928.
365. Rehm KE, Jones GJ, Tripp AA, Metcalf MW, Roper RL. The poxvirus A35 protein is an immunoregulator. *J Virol.* 2010;84(1):418-425.
366. Li P, Wang N, Zhou D, et al. Disruption of MHC class II-restricted antigen presentation by vaccinia virus. *J Immunol.* 2005;175(10):6481-6488.
367. Yao Y, Li P, Singh P, et al. Vaccinia virus infection induces dendritic cell maturation but inhibits antigen presentation by MHC class II. *Cell Immunol.* 2007;246(2):92-102.
368. Wang N, Weber E, Blum JS. Diminished intracellular invariant chain expression after vaccinia virus infection. *J Immunol.* 2009;183(3):1542-1550.
369. Hegde NR, Johnson DC. Human cytomegalovirus US2 causes similar effects on both major histocompatibility complex class I and II proteins in epithelial and glial cells. *J Virol.* 2003;77(17):9287-9294.
370. Jones TR, Sun L. Human cytomegalovirus US2 destabilizes major histocompatibility complex class I heavy chains. *J Virol.* 1997;71(4):2970-2979.
371. Tomazin R, Boname J, Hegde NR, et al. Cytomegalovirus US2 destroys two components of the MHC class II pathway, preventing recognition by CD4+ T cells. *Nat Med.* 1999;5(9):1039-1043.
372. Quinn LL, Williams LR, White C, Forrest C, Zuo J, Rowe M. The missing link in Epstein-Barr virus immune evasion:

the BDLF3 gene induces ubiquitination and downregulation of major histocompatibility complex class I (MHC-I) and MHC-II. *J Virol.* 2016;90(1):356-367.
373. Ressing ME, van Leeuwen D, Verreck FAW, et al. Interference with T cell receptor-HLA-DR interactions by Epstein-Barr virus gp42 results in reduced T helper cell recognition. *Proc Natl Acad Sci U S A.* 2003;100(20):11583-11588.
374. Mullen MM, Haan KM, Longnecker R, Jardetzky TS. Structure of the Epstein-Barr virus gp42 protein bound to the MHC class II receptor HLA-DR1. *Mol Cell.* 2002;9(2):375-385.
375. Forsyth KS, Roy NH, Peauroi E, et al. Ectromelia-encoded virulence factor C15 specifically inhibits antigen presentation to CD4+ T cells post peptide loading. *PLoS Pathog.* 2020;16(8):e1008685.
376. Portal-Celhay C, Tufariello JM, Srivastava S, et al. *Mycobacterium tuberculosis* EsxH inhibits ESCRT-dependent CD4(+) T-cell activation. *Nat Microbiol.* 2016;2:16232.
377. Saini NK, Baena A, Ng TW, et al. Suppression of autophagy and antigen presentation by *Mycobacterium tuberculosis* PE_PGRS47. *Nat Microbiol.* 2016;1(9):16133.
378. Hava DL, van der Wel N, Cohen N, et al. Evasion of peptide, but not lipid antigen presentation, through pathogen-induced dendritic cell maturation. *Proc Natl Acad Sci U S A.* 2008;105(32):11281-11286.
379. Bayer-Santos E, Durkin CH, Rigano LA, et al. The Salmonella effector SteD mediates MARCH8-dependent ubiquitination of MHC II molecules and inhibits T cell activation. *Cell Host Microbe.* 2016;20(5):584-595.
380. Hunt D, Wilson JE, Weih KA, et al. Francisella tularensis elicits IL-10 via a PGE(2)-inducible factor, to drive macrophage MARCH1 expression and class II down-regulation. *PLoS One.* 2012;7(5):e37330.
381. Lang C, Algner M, Beinert N, Groß U, Lüder CGK. Diverse mechanisms employed by Toxoplasma gondii to inhibit IFN-gamma-induced major histocompatibility complex class II gene expression. *Microbes Infect.* 2006;8(8):1994-2005.

CHAPTER 21 Immunoglobulins: Structure and Function

Mohamed Khass, MD, PhD • David Wald, MD, PhD •
Neil S. Greenspan, MD, PhD • Harry W. Schroeder, Jr, MD, PhD

INTRODUCTION

Immunoglobulins (Igs) are marked by a duality of structure and function.[1] They provide a polyclonal set of receptors for B cells that permit them to produce a highly diverse range of ligand-binding sites in concert with a defined set of effector functions. Antigen-binding site diversity allows Igs to recognize an almost unlimited array of self- and nonself antigens, ranging from compounds as fundamental to life as DNA to man-made molecules that could not have played a role in the evolution of the immune system, while conserved effector function permits integration with other arms of the host defense armamentarium.

A HISTORICAL PERSPECTIVE

The identification of immunoglobulin as a key component of the immune response began in the 19th century. This section describes the history of the identification of immunoglobulin and introduces fundamental terminology.

Antibodies and Antigens

Aristotle and his contemporaries attributed disease to an imbalance of the four vital humors: the blood, the phlegm, and the yellow and black biles.[2] In 1890, Behring (later, von Behring) and Kitasato reported the existence of an activity in the blood that could neutralize diphtheria toxin.[3] They showed that sera containing this humoral antitoxin activity would protect other animals exposed to the same toxin. Ehrlich, who was the first to describe how diphtheria toxin and antitoxin interact,[4] made glancing reference to "Antikörper" in an 1891 paper describing discrimination between two immune bodies, or substances.[5] The term antigen was first introduced by Deutsch in 1899. He later explained that antigen is a contraction of "Antisomatogen + Immunkörperbildner," or that which induces the production of immune bodies (antibodies). Thus, the operational definition of antibody and antigen is a classic tautology.

Gamma Globulins

In 1939, Tiselius and Kabat immunized rabbits with ovalbumin and fractionated the immune serum by electrophoresis into albumin, alpha-globulin, beta-globulin, and gamma-globulin fractions.[6] Absorption of the serum against ovalbumin depleted the gamma-globulin fraction, hence the terms immunoglobulin and IgG. "Sizing" columns were used to separate immunoglobulins into those that were "heavy" (IgM), "regular" (IgA, IgE, IgD, IgG), and "light" (light chain dimers). Immunoelectrophoresis subsequently permitted identification of the various immunoglobulin classes and subclasses.

Fab and Fc

In 1949, Porter first used papain to digest IgG molecules into two types of fragments, termed Fragment antigen-binding (Fab) and Fragment crystallizable (Fc) (Table 21.1).[8] Papain digested IgG into two Fab fragments, each of which could bind antigen, and a single Fc fragment. Nisonoff developed the use of pepsin to split IgG into an Fc fragment and a single dimeric $F(ab)_2$ that could cross-link antigens.[9] Edelman broke disulfide bonds in IgG with reducing agents and was the first to show that IgG consisted of two H and two L chains.[10]

Two Genes, One Polypeptide

The two most C-terminal constant domains of the IgG H chain were the first portions of the antibody molecule to be subjected to amino acid sequence analysis. This fragment could be readily crystallized when chilled, hence the term Fc. The heterogeneity of the variable (V) domain precluded sequence and crystallographic analysis of an intact Ig chain until Bence-Jones myeloma proteins were identified as clonal, isolated Ig light chains. These intact chains could be purified and obtained in large quantities, which finally permitted rational analysis of antibody structure and function.[11] Recognition of the unique nature of a molecule consisting of one extremely variable V domain and one highly conserved C domain led to the then-heretical Dreyer-Bennett proposal of "two genes – one polypeptide,"[12] which was subsequently and spectacularly confirmed by Tonegawa.[13]

Immunoglobulins Are Heterodimers

Immunoglobulins are heterodimeric proteins, consisting of two heavy (H) and two light (L) chains (Figs. 21.1 and 21.2). The eponymous Ig domain serves as the basic building block for both chains. In mammals, each of the chains contains a single amino-terminal variable (V) Ig domain and one, three, or four carboxy-terminal constant (C) Ig domains. H chains contain three or four C domains, whereas L chains contain only one. H chains with three C domains tend to include a spacer *hinge* region between the first (C_H1) and second (C_H2).[15] The receptor and effector functions of each individual immunoglobulin can be localized to separate regions, or

TABLE 21.1 Definitions of Key Immunoglobulin Structure Nomenclature

Fc	A constant Region Dimer Lacking C_H1
Fab	A light chain dimerized to V_H-C_H1 resulting from papain cleavage; this is monomeric since papain cuts above the hinge disulfide bond(s)
$F(ab)'_2$	A dimer of Fab′ resulting from pepsin cleavage below the hinge disulfides; this is bivalent and can precipitate antigen
Fab′	A monomer resulting from mild reduction of $F(ab)'_2$: an Fab with part of the hinge
Fd	The heavy chain portion of Fab (V_H-C_H1) obtained following reductive denaturation of Fab
Fv	The variable part of Fab: a V_H-V_L dimer
Fb	The constant part of Fab: a C_H1-C_L dimer
pFc′	A C_H3 dimer

From Carayannopoulos L, Capra JD. Immunoglobulins: structure and function. In: Paul WE, ed. *Fundamental Immunology*. Raven Press; 1993:283-314, with permission.

domains. Each V or C domain consists of approximately 110 to 130 amino acids, averaging 12,000 to 13,000 kDa, which is typical of immunoglobulin superfamily (IgSF) domains.[16] A typical light (L) chain will thus mass approximately 25 kD, and a three C domain Cγ heavy chain, with its hinge, will mass approximately 55 kD.

At the primary structure level, immunoglobulins are marked by the interspersion of regions of impressive sequence variability with regions of equally impressive sequence conservation. The V domains, composed of V_H and V_L, demonstrate the greatest molecular heterogeneity, with some regions including non-germline-encoded variability and others exhibiting extensive germline conservation across 500 million years of evolution.[17] The molecular heterogeneity of the V domains permits the creation of binding sites, or *paratopes*, which can discriminate between antigens that may differ by as little as one atom. Thus, it is the V domains that encode receptor function and define the monovalent specificity of the antibody. The H chain C_H1 domain, which is immediately adjacent to the V, associates covalently with the single L chain C domain. Together, the C_H1 and C_L domains provide a stable platform for the paired set of V_H and V_L domains, which together create the antigen-binding site. The distal C_H2 and C_H3 domains, for those antibodies with a hinge; or the

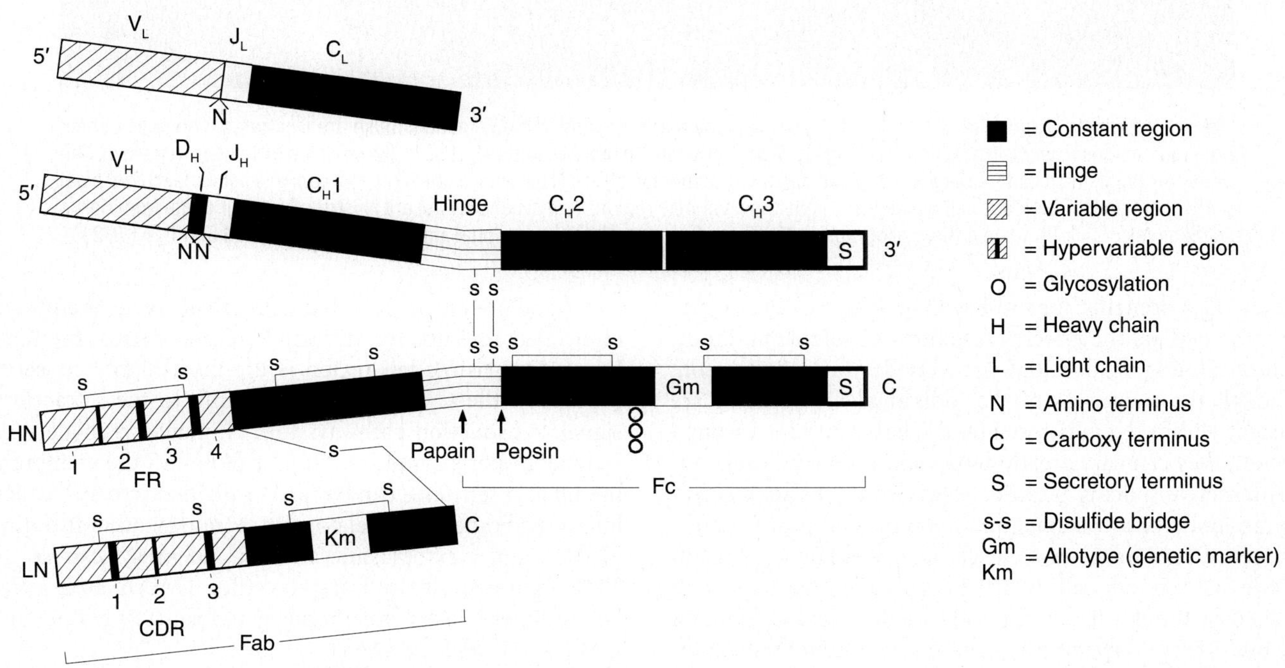

FIG. 21.1. Model of a secreted IgG. The germline exonic derivation of the sequence is shown at the top, and the protein structure is shown at the bottom. The location of the various cysteine residues that help hold both the individual domains and the various Ig subunits together are illustrated. Papain digests IgG molecules above the cysteine residues in the hinge that holds the two H chains together yielding two Fab molecules and an Fc, whereas pepsin digests below releasing an $(Fab)'_2$ fragment and an Fc fragment (which is typically degraded to smaller peptide fragments). The location of some allotypic variants is illustrated. C_H2 domains can be variably glycosylated, which also affects Ig structure and effector function.

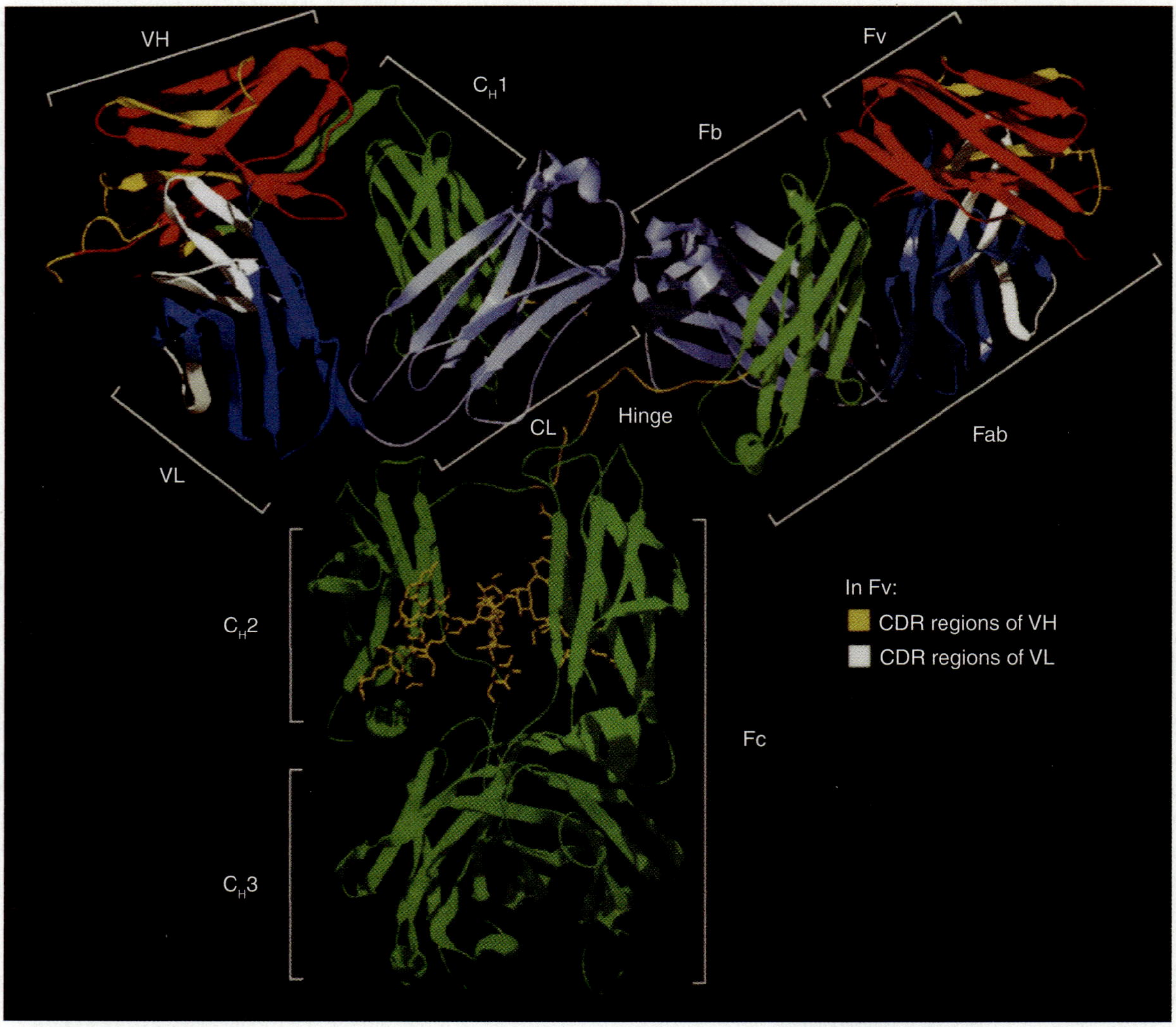

FIG. 21.2. Ribbon diagram of a complete IgG1 crystal. The major regions of the immunoglobulin are illustrated. The heavy chain constant regions (*green*) also include the hinge (*yellow*) between the first two domains. Cγ2 is glycosylated (also seen in *yellow*). The heavy and light chain variable regions (*red* and *dark blue*, respectively) are N terminal to the heavy (*green*) and light chain (*light blue*) constant regions. CDR loops in the heavy and light chain variable regions (*yellow* and *white*) are illustrated as well. (Data from Harris LJ, Skaletsky E, McPherson A. Crystallographic structure of an intact IgG1 monoclonal antibody. *J Mol Biol.* 1998;275(5):861-872.)

C_H3 and C_H4 domains, for those with an extra (C_H2) domain, typically mediate the effector functions of soluble antibody. Each of these Ig C_H domains is encoded by a separate exon. Although the sequences of the individual C_H domains are constant within the individual (and nearly constant within a species), they can vary greatly across species boundaries.

Alternative splicing permits production of either a secretory antibody or a transmembrane bound receptor for antigen. When read to the end, the carboxy terminal C_H domain encodes a secretory tail, which permits the antibody to be secreted by the B cell. Alternatively, the terminal C_H domain can use a cryptic splice site to access two membrane/cytoplasmic tail domain exons, termed M1 and M2. This alternative splicing event removes the secretory sequence typically encoded by the terminal C_H3 or C_H4 domain and replaces it with the peptides encoded by the M1 and M2 exons, converting a secretory immunoglobulin (sIg) that functions outside the cell as a soluble antibody to a membrane-bound immunoglobulin (mIg) that functions as a membrane-embedded receptor for antigen.[18] Across species, the membrane/cytoplasmic tail region is the most highly conserved portion of the C_H domains, which link to the intracellular signal transduction pathways that ultimately regulate B cell function.[19] For example, in the case of mIgM, the membrane-bound C_H4 terminus can be further divided into an extracellular membrane-proximal domain, a transmembrane domain (TMD), and a cytoplasmic domain. An mIgM can use the TMD to noncovalently associate with a TMD of an Ig-α/Ig-β heterodimer in a 1:1 stoichiometry to form a fully functional IgM B cell receptor complex.[20]

Paratopes and Epitopes

The Ig-antigen interaction takes place between the paratope, the site on the Ig at which the antigen binds, and the *epitope,* or antigenic determinant, is the site on the antigen that is bound. It is important to appreciate that antibodies do not recognize antigens, they recognize antigen epitopes.[21] This more localized feature of molecular recognition makes it possible for immunoglobulins to discriminate between two closely related antigens, each of which can be viewed as a collection of epitopes. Another consequence of this scenario is that the same antibody can bind different antigens that share equivalent epitopes, a phenomenon that represents one form of what is referred to as *cross-reactivity*.

It has been estimated that triggering of effector functions in solution typically requires aggregation of three or more effector domains and thus tends to involve the binding of three or more epitopes.[21] For antigens encoding repeating epitope structures, such as polysaccharides or antigen aggregates, binding of a single polymeric Ig molecule carrying multiple effector domains, such as pentameric IgM, can be sufficient to induce effector function. For antigens encoding diverse epitopes, which is more typical of monodisperse single-domain molecules in solution, triggering of inflammatory effector functions may require the binding of a diverse set of Ig molecules, all binding the same antigen, but at different epitopes.[22]

Membrane and Secretory Immunoglobulin

As mentioned, alternative splicing allows immunoglobulins to serve either as soluble antibodies or as membrane-bound antigen receptors. In their role as antibodies, Igs are released into the circulation from where they may traffic into the tissues and across mucosal surfaces. In their role as the B cell antigen receptor (BCR), they are anchored to the membrane by means of their M1:M2 transmembrane domain. Soluble antibodies can also be pressed into service as heterologous cell surface antigen receptors by means of their attachment to membrane-bound Fc receptors (Chapter 34).[23] This permits the power of antibody recognition to be extended to non-B cells, such as Fc receptor-expressing natural killer (NK) cells as well as nonlymphoid cells, such as dendritic cells, granulocytes, macrophages, and mast cells. The major difference between these two forms of cell surface receptors is that immunoglobulins as BCRs provide a monoclonal receptor for each B cell, whereas antibodies bound to Fc receptors endow the effector cell with a polyclonal set of antigen recognition molecules. This gives greater flexibility and increases the power of the effector cells to recognize antigens with multiple non-self-epitopes.

Isotypes and Idiotypes

Because of their great diversity, immunoglobulins can also serve as antigens for other immunoglobulins. Immunization of heterologous species with monoclonal antibodies (or a restricted set of immunoglobulins) has shown that immunoglobulins contain both common and individual antigenic determinants. In the context of immunoglobulin as antigen, the epitopes that are recognized within the V portion of the antibodies (Fig. 21.1) that have been used for immunization are termed *idiotypes*, whereas epitopes found in the constant (C) portion are termed *isotypes*. Recognition of isotypes first allowed grouping of immunoglobulins into recognized classes. Each class of immunoglobulin defines an individual set of C domains that corresponds to a single H chain constant region gene. For example, IgM utilizes μ H chain C domains, whereas IgE utilizes ε H chain C domains.

Some V domain epitopes derive from the germline sequence of V gene exons. These shared epitopes are commonly termed *public* or *cross-reactive idiotypes*. From a genetic perspective, however, they are isotypic because they can be found on many immunoglobulins of different antigen binding specificities that derive from the same germline V. Examples include cross-reactive idiotypes found on monoclonal IgM rheumatoid factors derived from individuals with mixed cryoglobulinemia, each of which can be linked to the use of specific V gene segments.[24]

Classes and Allotypes

Differences in the design and structure of the various Ig classes and subclasses enables each of them to play their own unique role in host defense (Fig. 21.3). After exposure to antigen, IgM is the first antibody to be made, and IgM pentamers facilitate aggregation and immune complex formation. IgG subclasses differ in their activation of the complement

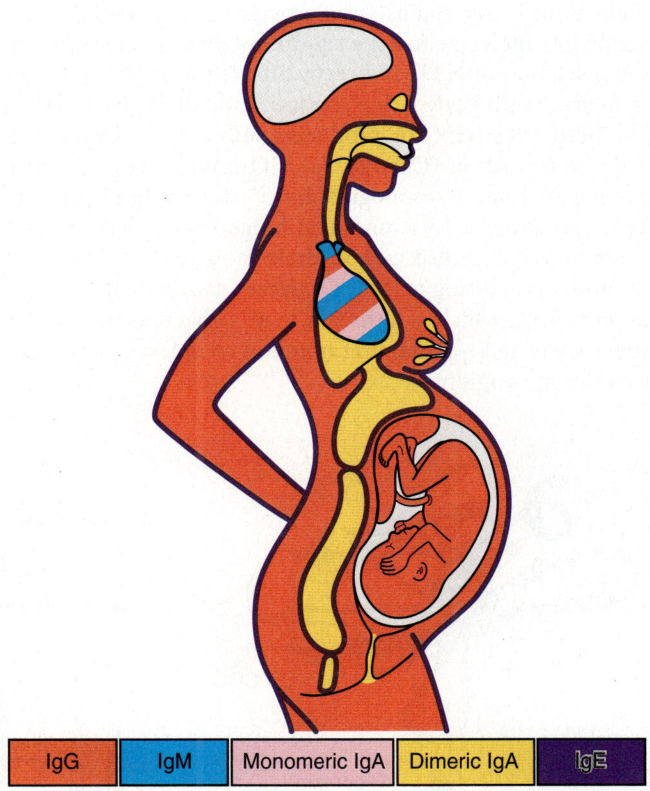

FIG. 21.3. Distribution of immunoglobulin classes in the body. IgG (*pink*), monomeric IgA (*pink*), and IgM (*blue*) are prominent in the circulation (depicted as a heart) in that order. IgG is prominent in the tissues. Dimeric IgA (*yellow*) is locally produced in the mucosal tissues, including the upper and lower respiratory tracts, the gastrointestinal system, and the vagina. IgA is also secreted into breast milk at a high level. IgE (*red*) is found at the interface between the mesoderm and both exodermal and endodermal tissues.

cascade and in their binding to the various Fcγ receptors and thus vary in effector function.[25] Dimeric IgA is the major class of immunoglobulin present in all external secretions. It is primarily responsible for protecting mucosal surfaces.

Determinants common to subsets of individuals within a species, yet differing between other members of that species, are termed *allotypes*. They define inherited polymorphisms that result from allelic forms of immunoglobulin C (less commonly, V) genes.[26]

Glycosylation

N-linked carbohydrates can be found in all constant domains as well as in some variable domains.[27] The structure of the attached N-linked carbohydrate can vary greatly, depending on the degree of processing. These carbohydrates can play a major role in Ig function.[28] For example, human IgG molecules contain a conserved glycosylation site at Asn 297 (N297), which is buried between the C_H2 domains[29] (Fig. 21.2). This oligosaccharide structure is almost as large as the C_H2 domain itself. The glycan present at N297 typically consists of two N-acetylglucosamine (GlcNAc), three mannose, and two more GlcNAc linked to the mannose to form a biantennary complex glycan. Additional fucose, galactose, sialic acid, and GlcNAc can be added to the core glycan structure. Since both heavy chains are glycosylated it is possible for a single IgG molecule to have significant glycan heterogeneity. Glycosylation of N297 indirectly affects the ability of the Fc to interact with FcγRs. Thus, optimization of Fc glycosylation has been important for producing antibody biologics with a desired function (Chapter 34).[30] O-linked sugars are also present in some immunoglobulins.[27] Human IgA1, but not IgA2, possesses a 13-amino-acid hinge region that contains three to five O-linked carbohydrate moieties.[31] A deficiency in proper processing of these O-glycans can contribute to IgA nephropathy, which is a disease that is characterized by the presence of IgA1-containing immune complexes in the glomerular mesangium.[32]

THE IG DOMAIN

As mentioned above, the immunoglobulin domain is the core unit that defines members of the immunoglobulin superfamily (IgSF) (reviewed in Williams and Barclay[16] and Harpaz and Chothia[33]). This section describes the Ig domain in detail.

The Immunoglobulin Superfamily

Each Ig domain consists of two sandwiched β pleated sheets "pinned" together by a disulfide bridge between two conserved cysteine residues (Fig. 21.4). The structure of the β pleated sheets in an Ig domain varies depending on the number and conformation of strands in each sheet. Two such structures, V and C1, are typically found in immunoglobulins. C1-type domains, which are the most compact, have seven antiparallel strands distributed as three strands in the first sheet and four strands in the second. Each of these strands has been given an alphabetical designation ranging from amino-terminal A to carboxy-terminal G. These strands are not contiguous in the primary sequence, where they are encoded in the following pattern: ABDE--CFG. Side chains positioned to lie sandwiched between the two strands tend to be nonpolar in nature. This hydrophobic core helps maintain the stability of the structure to the point that V domains engineered to replace the conserved cysteines with serine residues retain their ability to bind antigen. The residues that populate the external surface of the Ig domain and the residues that form the loops that link strands can vary greatly in sequence. These solvent-exposed residues offer multiple targets for docking with other molecules. This includes homotypic interactions with other antibodies, thereby improving antigen affinity and B cell activation.[34,35]

The V Domain

V-type domains add two additional antiparallel strands to the first sheet (C′ and C″), creating a five strand–four strand distribution. Domain stability results from the tight packing of

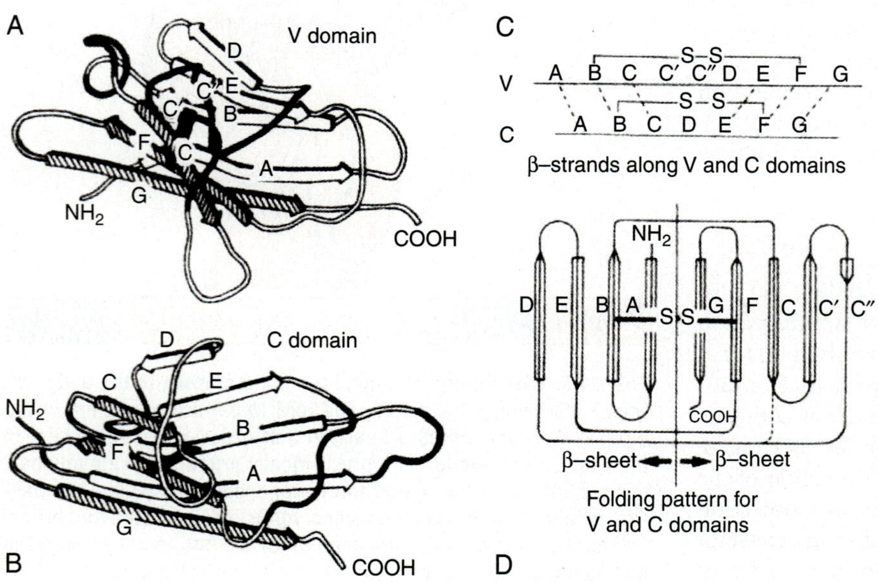

FIG. 21.4. The Ig domain. A: A typical V domain structure. Note the projection of the C-C′ strands and loop away from the core. **B:** A typical compact C domain structure. **C:** The cysteines used to pin the two β-sheets together are found in the B and F strands. **D:** The folding patterns for V and C domains. (Republished with permission of Annual Reviews, Inc. From Williams AF, Barclay AN. The immunoglobulin superfamily–domains for cell surface recognition. *Annu Rev Immunol.* 1988;6:381-405, permission conveyed through Copyright Clearance Center,Inc).

alternately inward-pointing residue side chains enriched for the presence of hydrophobic moieties to create a hydrophobic domain core. The H and L V domains are held together primarily through noncovalent interaction between the inner faces of the β sheets.[36,37]

Early comparisons of the primary sequences of the V domains of different antibodies identified four intervals of relative sequence stability, termed framework regions or FRs, which were separated by three hypervariable (HV) intervals, termed complementarity determining regions or CDRs (Figs. 21.5 and 21.6).[38] The exact location of these intervals has been adjusted over the years, first by a focus on the primary sequence,[38] then by a focus on the three-dimensional structure,[39] and, more recently, by a consensus integration of the two approaches by the international ImMunoGeneTics information system, or IMGT[42] (Fig. 21.5). (For students of the immunoglobulin repertoire, IMGT maintains an extremely useful web site, http://www.imgt.org, which contains a large database of immune-mediated sequences as well as a multiplicity of software tools.)

V domains differ from C domains owing to the inclusion of an additional framework loop, FR2. The C and C′ strands, which form FR2, project away from the core of the molecule (Figs. 21.4 and 21.6). They create a conserved structure that is parallel and opposite to the FR2 of the companion V and adjacent to the FR4 of the complementary chain. Approximately 50% of the interdomain contacts in the hydrophobic core of the V domain are formed by contacts between the FR2 of one chain and the FR4 of the complementary chain.[30] Another 30% to 45% is contributed by contacts between the CDR3 and the FR2 or CDR3 of the complementary domain. The overall interdomain contact includes between 12 and 21 residues from the L chain V domain and 16 to 22 residues from the H chain V domain, most of which derive from FR2, CDR3, and FR4.

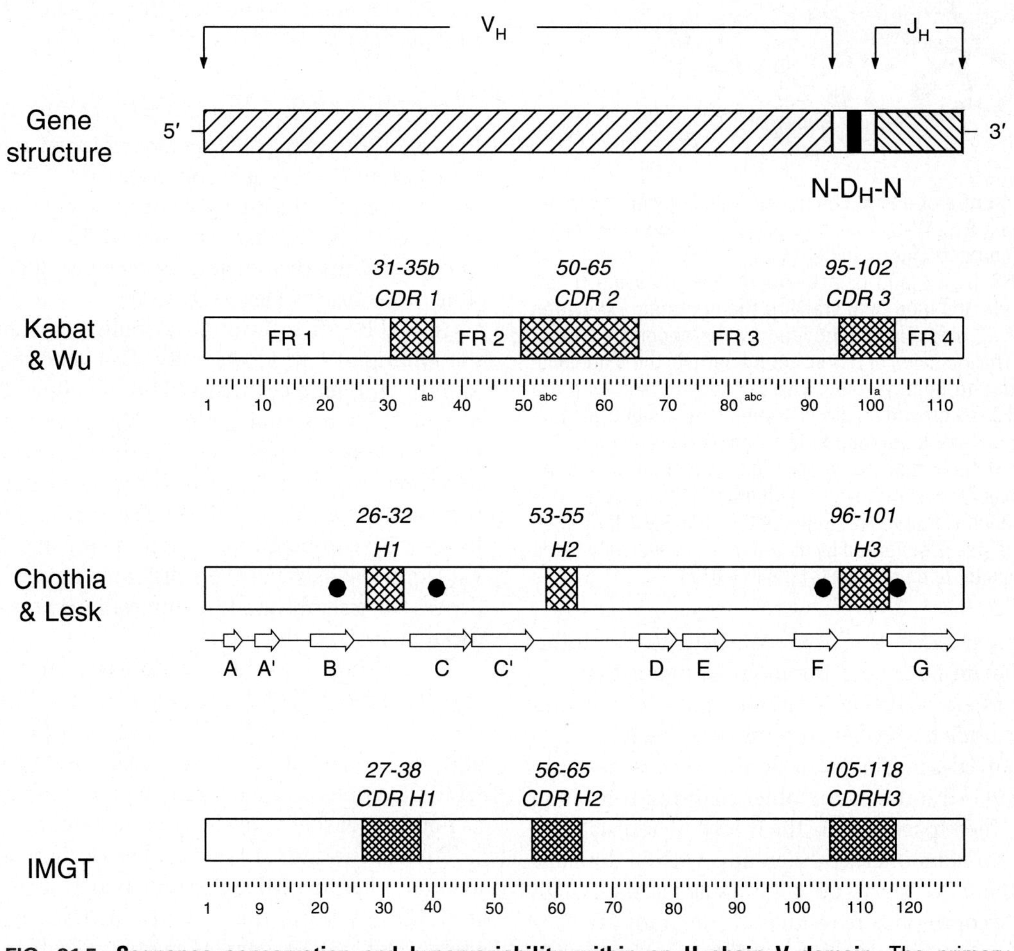

FIG. 21.5. Sequence conservation and hypervariability within an H chain V domain. The primary sequence the V domain can be divided into four regions of amino acid sequence conservation, termed framework regions or FRs, and three regions of sequence hypervariability, termed complementarity determining regions or CDRs. A schematic of the genomic origin of the variable domain is shown at the top of the figure. The classic separation of the sequence into FR and CDR by Kabat and Wu[38] is shown below the gene structure. The letter designation for individual β strands is given beneath the Chothia and Lesk nomenclature,[39] which focused more on structure. The positions of each of the four invariant residues of the V_H chain (FR1 Cys22, FR2 Trp36, FR3 Cys92, and FR4 Trp103) are shown as darkened circles on the Chothia and Lesk model. The IMGT designation has attempted to rationalize sequence variability with structure and is the current nomenclature of choice.[40]

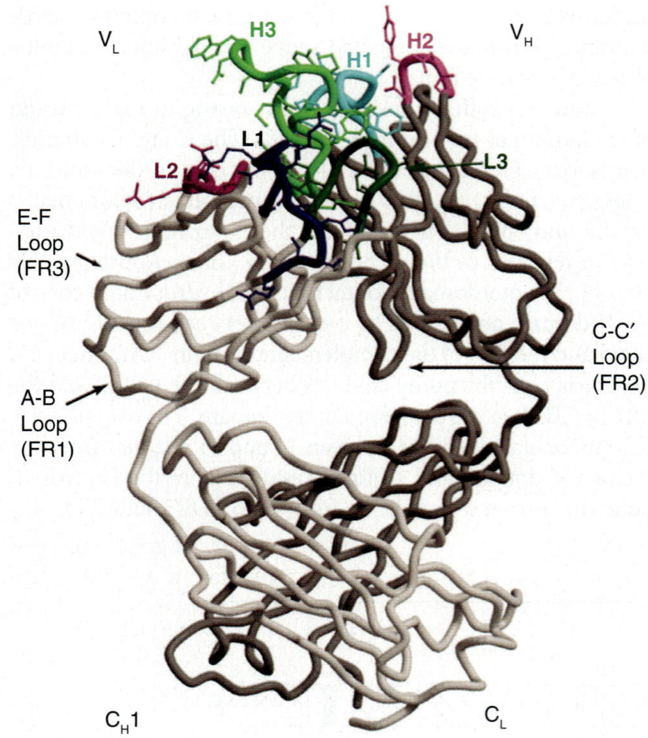

FIG. 21.6. The structure of a Fab. The antigen-binding site is formed by the H and L chain B-C, C′-C″, and F-G loops. Each loop encodes a separate complementarity determining region, or CDR. The locations of CDRs H1, H2, H3, L1, L2, and L3 are shown. The opposing H and L chain C-C′ strands and loop help stabilize the interaction between VH and VL. This C-C′ structure is encoded by the second V framework region, FR2. The inclusion of this structure permits the V domains to interact in a head-to-head fashion. The E-F strands and loop are encoded by the RF3 region and lie directly below the antigen-binding site. The A-B strands and loop encode FR1 and lie between the CH1 and CL domains and the rest of the versus. The β sheet strands of the CH1 and CL domains rest crosswise to each other. (The illustration is modified from Manivel V, Sahoo NC, Salunke DM, Rao KV. Maturation of an antibody response is governed by modulations in flexibility of the antigen-combining site. *Immunity.* 2000;13(5):611-620.)

There are approximately 40 crucial sequence sites that influence variable domain inter- and intradomain interactions.[39,43] Four of these sites are relatively invariant—the two cysteines that form the disulfide bridge between the beta sheets and two tryptophan (phenylalanine in Jκ) residues. One is near the beginning of the C strand and the other near the beginning of the G strand. These pack against the bridge to add stability. Thus, V and C domains interact differently, with V domains interacting via the CFG sheet and C domains via the ABDE sheet. Beyond the common core residues, Ig domains can vary widely in their primary amino acid sequences. However, a common secondary and tertiary structure characteristic of the core Ig V domain tends to be preserved.

FAB STRUCTURE AND FUNCTION

Introduction

The Fab region primarily allows immunoglobulins to discriminate between antigens. The Fab is individually manufactured to precise specifications by individual developing B cells. It shows an amazing array of binding capabilities while maintaining a highly homologous scaffold. This section describes the characteristics of the Fab domain, its component V domains, and the *paratope*, which is the part of the Fab that actually binds antigen.

Fab, Fv, and Fb

The antigen-binding fragment (Fab) is a heterodimer that contains a light chain in its entirety and the V and C_H1 portions of the H chain (Figs. 21.1 and 21.6). The Fab can be divided into a variable fragment (Fv) composed of the V_H and V_L domains and a constant fragment composed of the C_L and C_H1 domains (Table 21.1). Single Fv fragments can be produced in the laboratory through genetic engineering techniques.[44] They recapitulate the monovalent antigen binding characteristics of the original, parent antibody. As mentioned, other than minor allotypic differences, the sequences of the constant domains do not vary for a given H chain or L chain isotype.

Generation of Ig V Domains by Recombination

Ig variable domain genes are assembled in an ordered fashion by a series of recombination events.[13,45-48] The elegant mechanisms used for the assembly of these genes and the Fvs they create are fully discussed in Chapter 23. However, in order to understand the relationship between antibody structure and function, a brief review is in order.

In BALB/c mice, immunoglobulin V assembly begins with the joining of 1 of 13 diversity (D_H) gene segments to 1 of 4 joining (J_H) gene segments. This is followed by the joining of 1 of 110 functional variable (V_H) gene segments.[49-51] Each D_H gene segment has the potential to rearrange in any one of six reading frames (RFs), three by deletion and three by inversion. Thus, these 127 gene segments can come together in 3.4×10^4 combinations. The numbers of gene segments can vary quite widely between different mouse strains, but the process of assembly and the multiplicative effect of combinatorial diversity is the same.

In the final assembled V domain, the V_H gene segment encodes FRs H1-H3 and CDRs H1 and H2 in their entirety (Figs. 21.1 and 21.5); the J_H encodes FR-H4. CDR-H3 is created *de novo* in developing B cells by the joining process. CDR-H3 contains the D_H gene segment in its entirety, as well as portions of the V_H and J_H gene segments.

After a functional H chain has been created, light chain assembly begins. Mice contain two L chain loci, κ and λ. In C57BL/6 mice, the κ locus includes five Jκ gene segments and 140 Vκ gene segments, of which 4 and 73 have been shown to be functional, respectively.[52-54] This provides 292 combinations. There is only one Cκ.[55] The BALB/c λ locus contains 3 Vλ, three functional Jλ, and two or three functional Cλ chains.[56,57] The λ constant domains are functionally indistinguishable from each other. Owing to gene organization, the λ repertoire provides at most seven combinations. Each V_L encodes FRs L1-L3, CDR L1 and L2, and two-thirds of CDR-L3 (Fig. 21.1). Each J_L encodes one-third of CDR-L3 and FR4 in its entirety. Any one H chain can combine with any one L chain; thus, 211 V, D, and J gene segments can provide approximately 1×10^7 different H:L combinations.

At the V→D and D→J junctions, the potential for CDR-H3 diversity is amplified by imprecision in the site of joining, allowing exonucleolytic loss as well as palindromic (P junction) gain of terminal V_H, D_H, or J_H germline sequence. B cells that develop after birth express the enzyme terminal deoxynucleotidyl transferase (TdT) during the H chain rearrangement process.[13,45] TdT catalyzes the relatively random incorporation of non-germline-encoded nucleotides between V_H and D_H and between D_H and J_H. Each three nucleotides of N addition increase the potential diversity of CDR-H3 20-fold. Thus, sequences with nine nucleotides of N addition each between the V→D and D→J junctions would enhance the potential for diversity by $(20)^6$ or by 6×10^7; 6-fold greater than the potential diversity provided by VDJ gene segment combinations. These genomic gymnastics permit the length of CDR-H3 to vary from 5 to 20 amino acids among developing B cells in BALB/c bone marrow.[58] Together, imprecision in the site of VDJ joining and N addition provides the opportunity to create nearly random CDR-H3 sequence, potentially freeing the CDR-H3 repertoire from germline sequence constraints. Although a limited amount of N addition is observed between V_L and J_L in human,[59,60] N addition in murine L chains is distinctly uncommon. Moreover, the length of CDR-L3 appears to be under relatively strict control, greatly limiting the potential for somatic L chain junctional diversity.[57,59] Thus, CDR-H3 represents the greatest focus for the initial somatic diversification of the antibody repertoire.

Segmental Conservation and Diversity Within the V Domain

Although the large numbers of V gene segments might give the impression of a smooth incremental range of available diversity, multigene families are thought to evolve in concert through mechanisms of gene conversion and V gene segments are no exception. Sequence relationships allow grouping them into families and clans of sequences that share nucleotide homology,[61] as well as structural features. Close inspection of the V_H gene repertoire has shown that these family relationships reflect segmental gene conversion coupled with selection for function.[17,62,63]

Owing to the need to maintain a common secondary and tertiary core Ig V domain structure capable of associating randomly with a complementary V chain to form a stable Fv, the core sequence of FR2, which is encoded by the V_H gene segment, and the core sequence of FR4, which is encoded by the J_H gene segment, are highly conserved among all Ig V domains. Conversely, the need to generate a diverse repertoire of antigen-binding sites has led to extensive diversity in the CDR-1 and CDR-2 intervals. One might presume that the FR1 and FR3 intervals, which form the external surface of the antibody, would not be under any particular constraints, but sequence comparisons suggest otherwise.

Given the need to diversify the CDRs and the need to preserve FR2, it is not surprising that family identity, which might reflect ancestral relationships, can be assigned by the extent of FR 1 and FR3 similarity.[64] Of these, FR1 appears to be under the greatest constraints, with V_H gene segments belonging to different families both within and across species barriers exhibiting extensive similarities in FR1 sequence (Fig. 21.7). Sequence similarities in FR1 and, to a lesser extent, FR3 allow

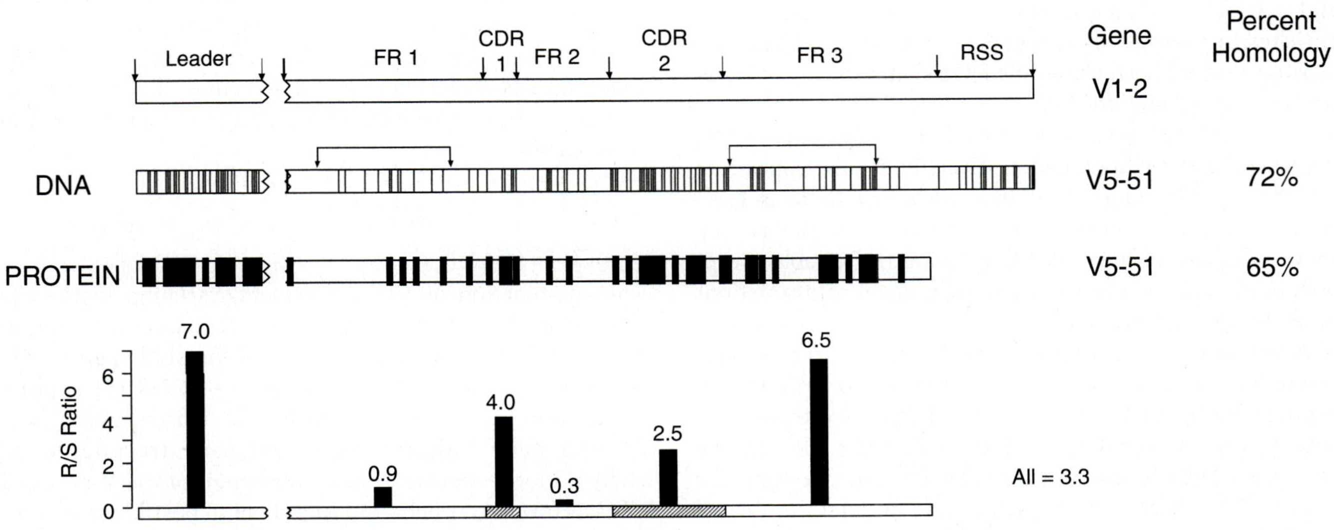

FIG. 21.7. A comparison of two human Clan I V_H sequences that belong to different V_H families. Shown is a comparison of the DNA and amino acid sequences of the V5-51 and V1-2 gene segments. Each line depicts a divergence in a nucleotide or amino acid at that position. Shown below is a replacement/silent site substitution analysis by interval. Random mutation tends to exhibit an R/S ratio of 2.9. The smaller the ratio, the greater the preservation of sequence. The intervals identified by the arrows predict the family and clan of origin. (Modified from Kirkham PM, Schroeder HW Jr. Antibody structure and the evolution of immunoglobulin V gene segments. *Semin Immunol.* 1994;6(6):347-60.)

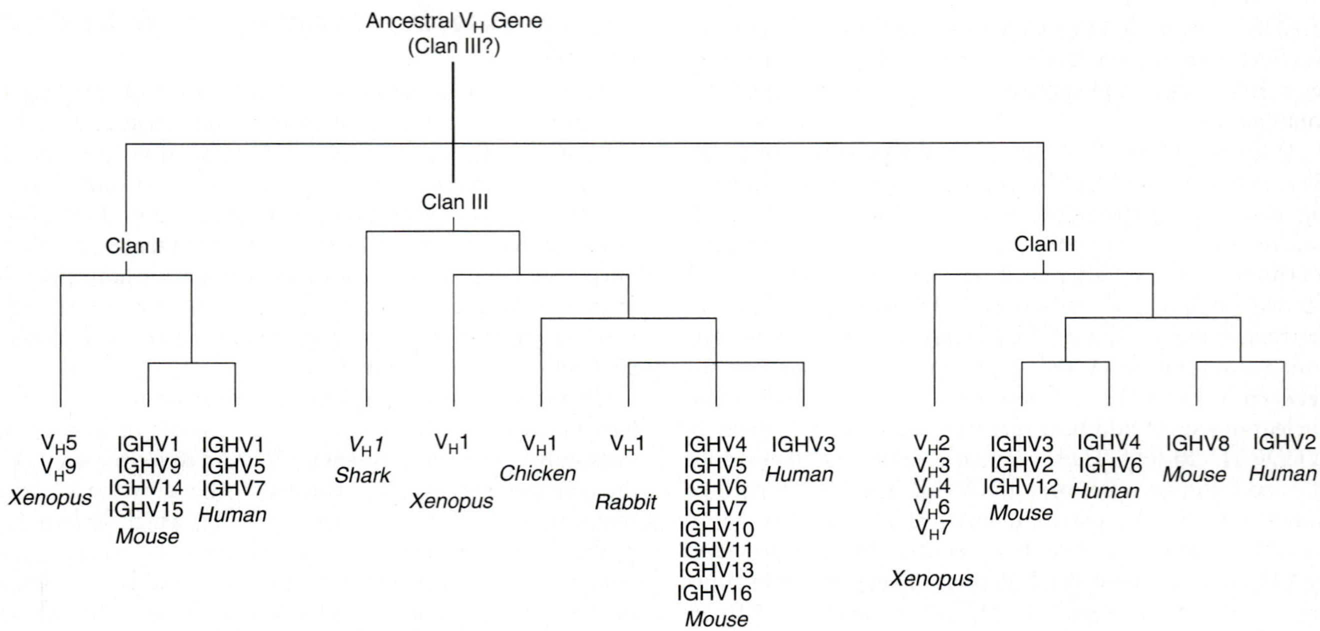

FIG. 21.8. Evolutionary relationships among selected vertebrate VH families. The sizes or relative placements of the evolutionary connecting lines are not to scale. (Modified from Kirkham PM, Schroeder HW Jr. Antibody structure and the evolution of immunoglobulin V gene segments. *Semin Immunol.* 1994;6(6):347-360.)

grouping of human and murine V_H families into three clans of related sequences (Fig. 21.8), presumably reflecting an early divergence in sequence from a primordial V_H gene sequence.

Constraints on the Sequence and Structure of V-Encoded CDRs

The antigen-binding site of an immunoglobulin is formed by the juxtaposition of the six hypervariable H and L chain V domain intervals: CDRs-H1, -H2, and -H3 and CDRs-L1, -L2, and -L3.[38] The CDR sequences of V gene segments tend to be enriched for codons where mutations maximize replacement substitutions.[65] This includes the DGYW motif that facilitates somatic hypermutation.[66-68]

Although evolution appears to favor CDR1 and CDR2 sequences that facilitate codon diversity, it also appears to preserve specific loop structures. Although there is great variation in the sequence and size of these CDRs, it has been shown that five of them, CDR-H3 being the notable exception, possess one of a small set of main-chain conformations termed canonical structures.[39,43,69-72] Each canonical structure is determined by the loop size and by the presence of certain residues at key positions in both the loop and framework regions. For example, three canonical structure types have been identified for CDR-H1, four types for CDR-H2, five types for CDR-L1, one type for CDR-L2, and five types for CDR-L3.[39,43,73] It can be calculated that the total number of possible combinations of canonical structures, or structure classes, is 300.[74] However, only 10 of these combinations, or classes, are sufficient to describe the majority of human and mouse Fab sequences. The lengths of CDR-H3 and -L1 appear to correlate with the type of recognized antigen. Antibodies with short loops in -H2 and -L1 appear to be preferentially specific for large antigens (proteins), whereas antibodies with long loops in -H2 and -L1 appear to be preferentially specific for small molecules (haptens).[74]

Given that the sequence and structure of the framework regions, which define families, influence the canonical structure of V_H-encoded CDRs, it is not surprising that the repertoire of canonical CDR structures is strongly associated with family and clan identity.[75] This implies restrictions to the random diversification of the hypervariable loop structures (canonical structures) and their combinations within the same V_H gene segment (canonical structure classes). It further suggests evolutionarily and structure-imposed restrictions operating to counteract the random diversification of these CDRs.

Diversity and Constraints on the Sequence and Structure of CDR-H3

The combination of VDJ assortment, variation in the site of gene segment rearrangement, and N nucleotide addition makes CDR-H3 the most variable of the six hypervariable regions. In some cases, the sequence of CDR-H3 appears designed to provide optimal flexibility.[76] Correspondingly, it has been more difficult to assign canonical structures to the CDR-H3 loops similar to those observed for the V-encoded CDRs. However, insight into a gradient of possible structures has been gained.

CDR-H3 can be separated into a base, which is adjacent to the frameworks, and a loop, which is positioned to interact with antigen. The base tends to be stabilized by two common residues, an arginine at Kabat position 94 (IMGT 106) and an

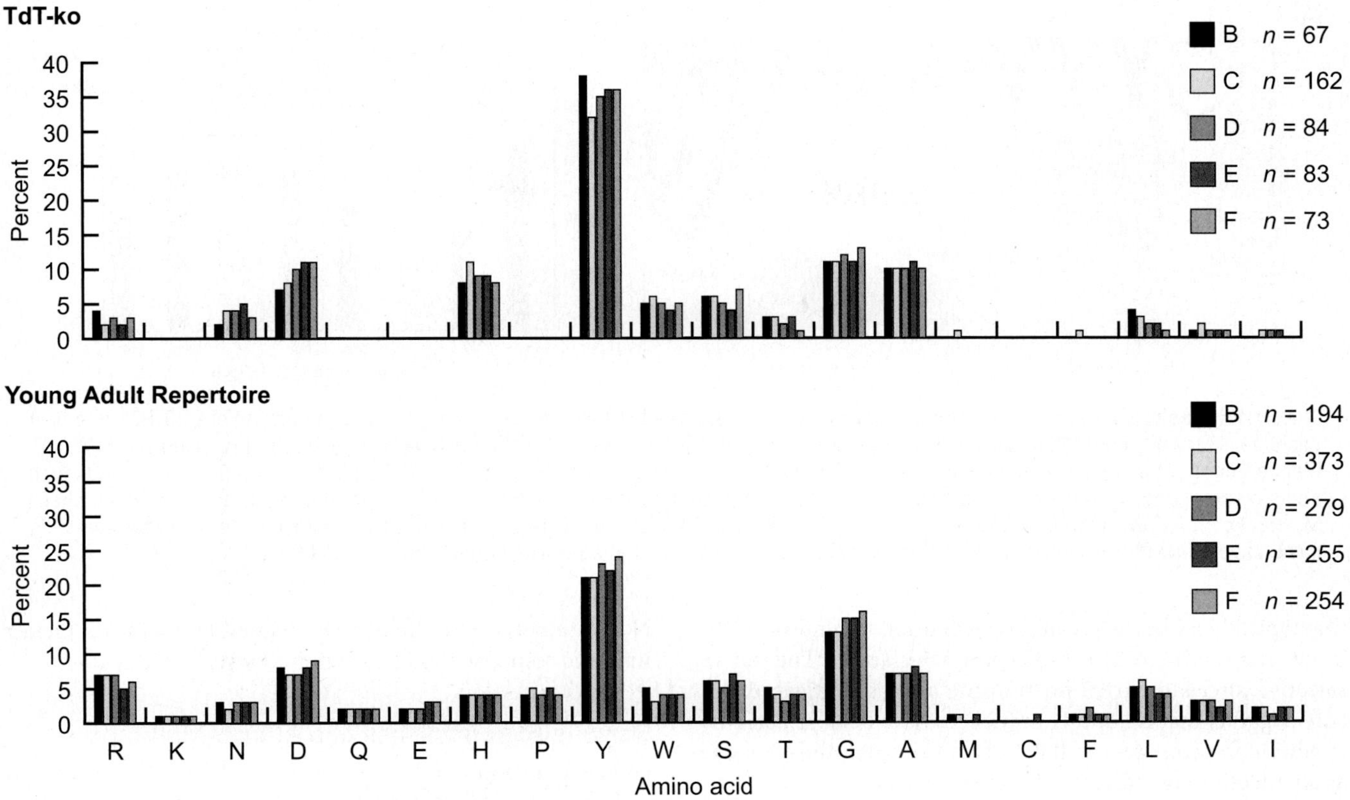

FIG. 21.9. Both in the absence and presence of N-addition, the preference for tyrosine and glycine in CDR-H3 begins early and intensifies with B cell development. V_H7183DJCμ transcripts were cloned and isolated from fractions B (pro–B cells) through F (mature B cells) from the bone marrow of 8- to 10-week-old TdT-sufficient (WT) and TdT-knockout BALB/c mice.[84] The amino acids are arranged by relative hydrophobicity, as assessed by a normalized Kyte-Doolittle scale.[85,86] Use is reported as the percent of the sequenced population. The number of unique sequences per fraction is shown.

aspartic acid at Kabat 101 (IMGT 116).[39,40] These form a salt bridge, which, together with the adjacent residues, tends to create one of three backbone conformations, termed kinked, extra-kinked, and extended.[77] For a minority of sequences with kinked or extra-kinked bases, it is possible to predict whether an intact hydrogen bond ladder may be formed within the loop of the CDR-H3 region or whether the hydrogen bond ladder is likely to be broken.[78,79]

The distribution of CDR-H3 lengths is regulated both as a function of B cell development and as a function of ontogeny.[58,80] In association with long V-encoded CDRs, short CDR-H3s create an antigen-binding cavity at the center of the antigen-binding site, and CDRs of intermediate length can create an antigen-binding groove. Each species appears to prefer a specific range of CDR-H3 lengths.[81] Long CDR-H3s, which can create "knobs" at the center of the antigen-binding site, are unevenly distributed between species and reflect both divergence in germline sequence and somatic selection.

In spite of the potential for totally random sequence provided by the introduction of N nucleotides, close inspection has shown that the distribution of amino acids in the CDR-H3 loop is enriched for tyrosine and glycine,[38,82] moderately depleted of highly polar (charged) amino acids, and highly depleted of nonpolar (hydrophobic) amino acids.[83] This pattern of amino acid utilization begins early in B cell development, prior to the expression of immunoglobulin on the surface of the cell (Fig. 21.9)[58,84,87,88] and reflects evolutionary conservation of J_H and D_H gene segment sequences. For example, although the absolute sequence of the D_H is not the same, the pattern of amino acid usage by reading frame is highly conserved. Of the six potential reading frames, RF1 by deletion is enriched for tyrosine and glycine. RF2 and RF3 by deletion are enriched for hydrophobic amino acids, as are RF2 and RF3 by inversion. RF1 by inversion tends to encode highly polar, often positively charged, amino acids.[82] Various species use different mechanisms to bias for use of RF1 by deletion, to limit use of hydrophobic reading frames, and to restrict or prevent use of reading frames enriched for charged amino acids. Forced rearrangement into reading frames with charged or hydrophobic amino acids yields an altered repertoire enriched for charge or hydrophobicity and depleted of tyrosine and glycine.[87]

Enrichment for tyrosine and underutilization of hydrophobic amino acids also reflects somatic selection processes that engage as B cells pass through developmental checkpoints in the bone marrow and the periphery.[89,90] The end effect is to enrich for tyrosine at positions within CDR3 where the structure permits direct interaction with surrogate light chain early in development and with antigen in the periphery (Fig. 21.10). Tyrosine is one of three amino acids

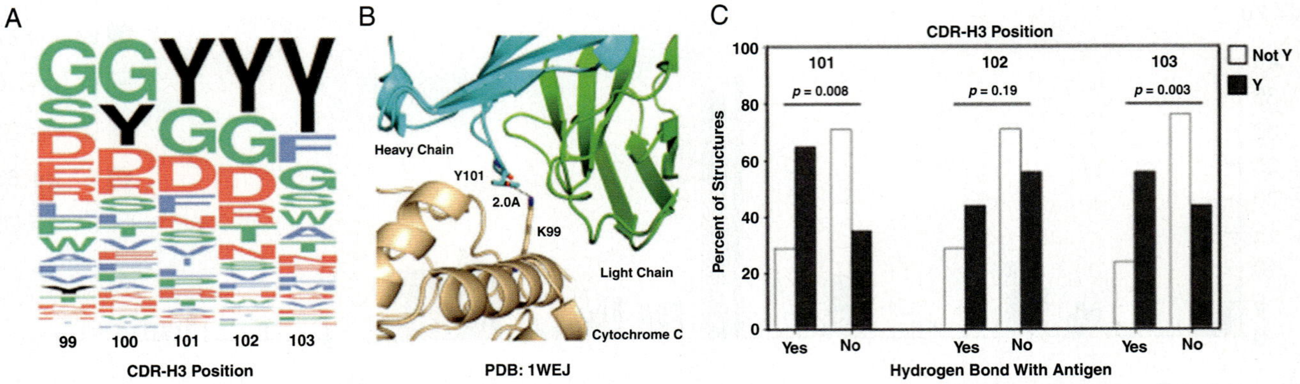

FIG. 21.10. Analysis of the CDR-H3 region in murine IgG Fab–antigen PDB structures. A: Amino acid composition of CDR-H3 regions at positions 99 to 103 in the 133 analyzed structures. **B:** Example of CDR-H3 Y101 interacting with antigen (cytochrome C) in PDB: 1WEJ. **C:** Fraction of structures with (*black*, Y) and without (*white*, Not Y) tyrosine in CDR-H3 positions 101 to 103 that form hydrogen bonds with antigen among those sequences where the tyrosine in the selected position lacks a flanking (one left, one right) tyrosine. (From Khass M, Blackburn T, Burrows PD, Walter MR, Capriotti E, Schroeder HW Jr. VpreB serves as an invariant surrogate antigen for selecting immunoglobulin antigen-binding sites. *Sci Immunol.* 2016;1(1):aaf6628. Reprinted with permission from AAAS.)

that typically make the greatest contribution to binding affinity at many protein ligand-receptor interfaces.[32] The use of selective forces to enrich for tyrosine prior to antigen exposure suggests that D_H and J_H sequences have been selected by evolution to promote the ability of immunoglobulins to serve as an effective and efficient binder of ligands.

The Antigen-Binding Site Is the Product of a Nested Gradient of Regulated Diversity

The tension between the need to conserve essential structure and the need to emphasize diversity in an environment subject to unpredictable antigen challenge appears to create a gradient of regulated diversity in the Fv. The most highly conserved components of the Fv are FR2 and FR4, which form the hydrophobic core of the V_H:V_L dimer (Fig. 21.6). FR1, which in the H chain presents with three conserved structures, helps form the ball and socket joint between the V_H and C_H1. FR3, which in the H chain defines the family and provides seven different structures in human versus 16 different structures in mouse, frames the antigen-binding site (Fig. 21.11). The V-encoded CDRs, -H1, -H2, -L1, -L2, and most of -L3, are programmed for diversity. However, conserved residues within these CDRs, which interact with V family-associated FR3 residues, constrain diversity within a preferred range of canonical structures. CDR-H3, the focus of junctional diversity, lies at the center of the antigen-binding site. The conformation of its base tends to fit within three basic structures. The loop varies greatly in sequence, yet still maintains a bias for the use of tyrosine and glycine. Thus, diversity increases with proximity to the tip of the antigen-binding site but appears to be held within regulated limits.

The extent and pattern of diversity in CDR-H3 can have a critical effect in the biologic function of immunoglobulin as a soluble effector molecule. Absence of N addition, with its enhancement of tyrosine and glycine usage (Fig. 21.9), facilitates B cell development, whereas enrichment for charged amino acids impairs it.[87,91,92] However, both the absence of N addition and enrichment for charged amino acids impair immune responses and protection *in vivo*.[87,91,93]

Somatic Hypermutation and Affinity Maturation

Following exposure to antigen and T cell help, the V domain genes of germinal center lymphocytes can undergo mutation at a rate of up to 10^{-3} changes per base pair per cell cycle,[94] a process termed *somatic hypermutation*. Somatic hypermutation followed by selection of the mutants on antigen in the germinal center allows *affinity maturation* of the antibody repertoire in response to repeated immunization or exposure to antigen.

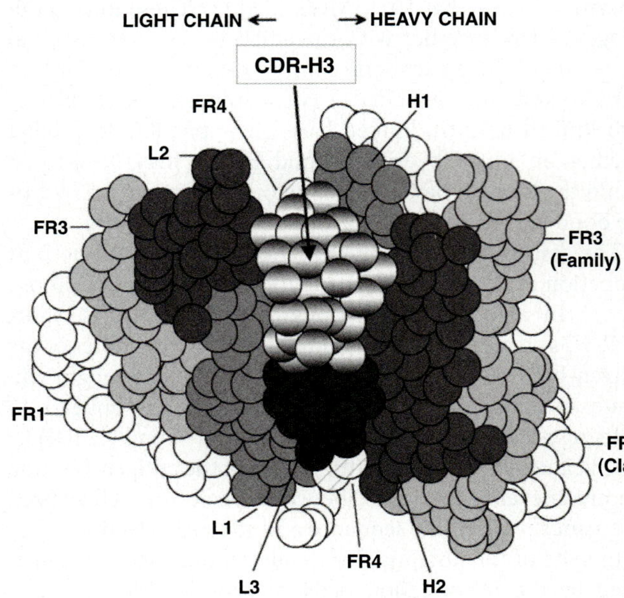

FIG. 21.11. Location and generation of CDR-H3. A cartoon of the classic antigen-binding site. Owing to its central location, most antigens bound to the antibody will interact with CDR-H3. (Modified from Kirkham PM, Schroeder HW Jr. Antibody structure and the evolution of immunoglobulin V gene segments. *Semin Immunol.* 1994;6(6):347-360.)

Although affinity maturation often preserves the canonical structure of the CDR loops, the distribution of diversity appears to differ between the primary and antigen-selected repertoire.[95] In the primary repertoire, diversity is focused at the center of the binding site in CDR-H3. With hypermutation, somatic diversity spreads to the V-encoded CDRs in the next ring of the binding site (Fig. 21.11), enabling a more "custom-tailored" fit.

Binding of "Superantigens" to Nonclassic V Domain Antigen-Binding Sites

Not all antigens bind to the paratope created by the classic antigen-binding site. Antigens that can bind to public idiotopes on V domain frameworks[96] and recognize large portions of the available repertoire are termed superantigens. There are indications that B cell superantigens influence the pathogenesis of some common infections, such as those caused by *Staphylococcus aureus*.

ANTIGEN-ANTIBODY INTERACTIONS

Technological advances in biomolecular structure determination, analysis of molecular dynamics, protein expression and mutagenesis, and biophysical investigation of receptor-ligand complex formation have facilitated significant advances in the understanding of antigen-antibody interactions. Particularly valuable has been the integration of high-resolution structural data with thermodynamic and kinetic analyses on a number of antigen–monoclonal antibody complexes. In this section, some of the key insights arising from these studies are reviewed.

Molecular Flexibility

Like many other protein domains, V domains exhibit varying degrees and modes of molecular flexibility. Evidence suggests that some V modules (ie, V_L-V_H pairs) can adopt two or more conformations with meaningful frequencies in the unbound state. Since these different conformational states can exhibit distinguishable binding proclivities, molecular flexibility provides monoclonal antibodies with a mechanism for polyspecificity.[97] Molecular flexibility can also play a role in binding a single ligand, which, in such instances, may be better understood as a process of binding rather than as a simple event.[98,99] The extent of conformational adjustment by antibody or antigen required for complex formation can influence both the thermodynamics and kinetics of that process.

Role of Water

A significant role for water molecules has become clear from the study of high-resolution structures and thermodynamic analyses of antigen-antibody complexes. Water molecules exhibit a broad range of association times with protein surfaces. Thus some of the more tightly protein-associated of these solvent molecules effectively behave as parts of the protein.[100] Water molecules found in the antigen-antibody interface, whether constitutively bound or newly recruited, can make important contributions to both the intrinsic (ie, monovalent) affinity of the complex and to the differential affinities for different ligands (ie, specificity).[101] Amino acid residues of antigen and antibody can interact, indirectly, through hydrogen bonds to one or more water molecules.[101]

Thermodynamics and Antigen-Antibody Interactions

Specific residues in the antibody V domains or the antigen can contribute to complex formation in different ways, and some residues can contribute in multiple ways.[102] In addition to making van der Waals contact, residues can be important owing to contributions to the free energy of complex formation or to the differential free energy of complex formation for two or more different ligands. Some residues may contribute primarily to modulation of the association rate, the "relaxation" of the forming complex, or the dissociation rate.[98,103] Other contact residues contribute minimally to the energetics of complex formation, and yet other noncontact residues can be significant thermodynamic contributors.[104,105]

Structural and thermodynamic/kinetic comparisons of antibodies with germline V domain sequences and with somatically mutated V domain sequences have provided new insights into the structural and energetic bases for affinity maturation.[104,105] Mutations in both contact and noncontact residues can have major consequences, positive or negative, for the affinity with which an antibody binds an antigen. They can also favor tighter binding by enhancing V domain rigidity, thereby reducing the entropic penalty associated with complex formation.[105] Antibodies derived from secondary or later responses that have incorporated somatic mutations have been shown to exhibit less than absolute specificity. Even these "mature" antibodies can bind multiple ligands when screened on libraries of peptides or proteins,[106,107] a lesson likely to be relevant to most biomolecules.

It has been routine to distinguish between interactions in which antibodies bind to protein antigens versus those in which antibodies bind to carbohydrate antigens. Studies of antibodies that bind to HIV-1 gp120 with high affinity and that exhibit potent and broad neutralizing activity suggest that antibodies can bind to epitopes composed of both peptide and glycan elements.[108,109]

IMMUNOGLOBULIN "ELBOW JOINTS" AND "HINGES"

The structure of the constant domains can affect antibody-antigen interactions by influencing the range of molecular flexibility permitted between the two Fabs. In this section, the role of the immunoglobulin hinge is discussed.

Elbow Joints

Individual Ig V and C domains tend to create rather rigid dimers. However, the antibody molecule as a whole, which consists of four or more such dimeric modules linked like beads on a string, can be viewed as a paradigm of molecular flexibility.[110] Flexibility begins between the Fv and the Fb of the Fab at what is termed the elbow bend or elbow angle.[111] This reflects both a ball and socket interaction between the

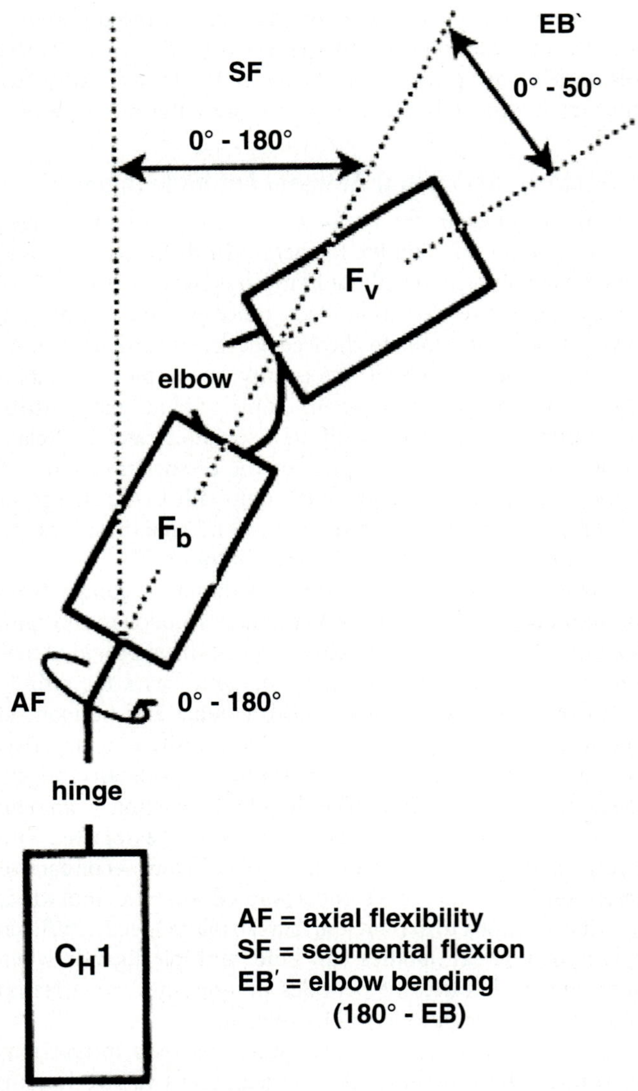

FIG. 21.12. **Illustration of the motions and flexibility of the immunoglobulin.** Axial and segmental flexibilities are determined by the hinge. The switch peptide (elbow) also contributes flexibility to the Fab. The measure of the elbow angle is defined with respect to the Fv and Fb axes of 2-fold symmetry. (From Carayannopoulos L, Capra JD. Immunoglobulins: structure and function. In: Paul WE, ed. *Fundamental Immunology*. Raven Press; 1993:283-314, with permission.)

FR1 of the H chain and the C_H1 domain[112] and the identity of the L chain.[111] Five residues (three in V_H and two in C_H1) that are highly conserved in both antibodies and T cell receptors make the key contacts that constitute this "joint." Elbow angles, assessed from crystal structures of homogeneous Fab fragments, range from 130° to 180° (Fig. 21.12). λ Light chains appear to permit an Fv to adopt a wider range of elbow angles than their κ chain counterparts.

Hinges

Some immunoglobulin isotypes contain a structural element that does not strictly correspond to the canonical structural motifs of IgSF V and C domains, which is termed the hinge. Where it occurs, it is located between the C terminus of the C_H1 and the N terminus of the C_H2 domains. In isotypes such as IgG and IgA, the hinge is encoded by one (or more) separate exon(s). In isotypes with four C_H domains, that is, IgM and IgE, the C_H2 domain serves in place of a classical hinge, although some might argue that the hinge is encoded by a degenerate C_H2 exon from primordial IgM.

The hinge, or the C_H2 domain in Igs that lack a hinge, permits a Fab arm to engage in motion relative both to the other Fab arm and to its Fc stem (Figs. 21.13 and 21.14). This permits the two Fab arms to cover a range of angles from maximal extension to an almost parallel alignment. The range of motion of the Fab arms reflects the nature of the hinge region, which in some C genes is relatively rigid and in others, such as human IgA1, functions more as a tether for each individual Fab than as a support. This flexibility has major implications for antibody function, because it enables a bivalent antibody molecule to bind epitopes in a variety of relative spatial arrangements.

Among human IgG subclasses, the most unusual hinge region is that of IgG3. Unlike other human IgG hinge regions, the IgG3 hinge is encoded by a quadruplicated hinge exon, making it the longest hinge (62 amino acids) by far (see Table 21.2). The primary structure of this hinge has been divided into upper, core (or middle), and lower hinge regions with somewhat different functional associations. The upper hinge in particular has been associated with the magnitude of segmental flexibility as assessed by fluorescence emission anisotropy kinetics[115] and with the magnitude of Fab-Fab flexibility by immunoelectron microscopy.[116] The core or middle hinge appears to serve, at least in part, a spacer function. The lower hinge functions primarily to facilitate C_H2-C_H2 interactions.

Segmental flexibility of the Ig molecule, conferred mainly by the hinge, permits or facilitates simultaneous binding through two or more Fab arms. Such monogamous bivalency or multivalency, which enhances overall binding,[117,118] is a crucial factor permitting biosynthetically feasible antibody concentrations to offer adequate immunity against replicating pathogens. Although it is more speculative why the Fab-Fc geometry needs to vary, it may have to do with optimizing effector function activation when antigen is bound, such as maintaining antigen binding when Fc receptors are simultaneously engaged.

The attribution of flexibility control to the hinge is supported by protein engineering studies in which V domain-identical IgGs of different subclasses were analyzed.[115,119] This basic conclusion is also supported by studies in which hinge regions have been selectively mutated or swapped among human or mouse IgG subclasses.[120-122] One of these studies indicated that structural variation among subclasses in the C_H1 domain also influenced segmental flexibility as assessed by nanosecond fluorescence polarization measurements.[120] Early suggestions that IgG subclass-related differences in activating the classical pathway of complement were explained by differences in segmental flexibility[115] were not confirmed by the studies in which mutant hinge regions were created.[120-122]

There are several types of molecular motion attributable to the hinge region that contribute to overall segmental

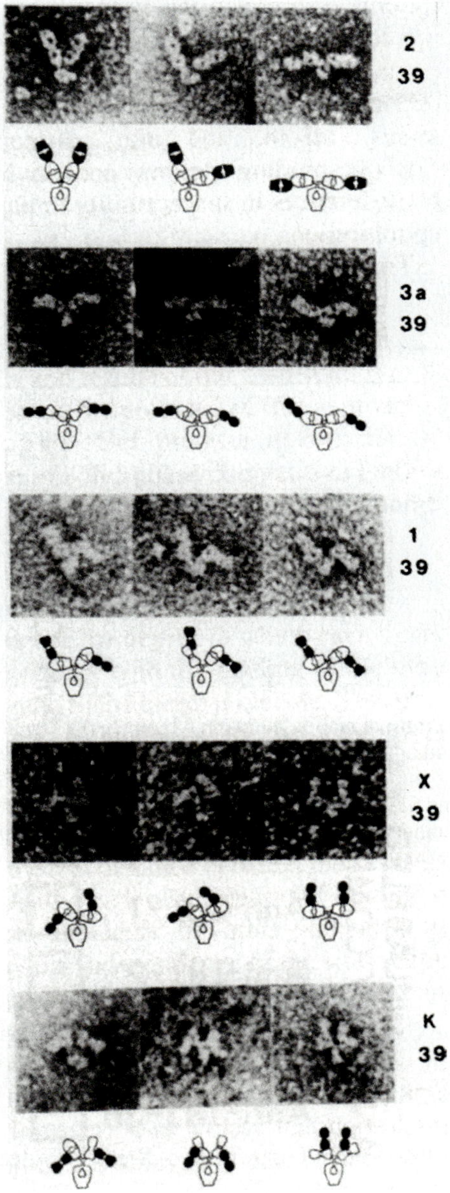

FIG. 21.13. Electron micrographs (*top*; ×350,000) and interpretive diagrams (*bottom*) of murine mAb HGAC 39 (specific for the cell wall polysaccharide of *Streptococcus pyogenes*) in complex with anti-idiotypic mAb Fab fragments. HGAC 39 is represented in the diagrams as an open figure, and the Fab anti-Id probes are represented as solid figures. The Fab arms of the antibody targets and probes are drawn to indicate their rotational orientation as planar (oval with open center) intermediate (bone shape with or without central opening), or perpendicular ("*dumbbell shaped*"). Different complexes illustrate the range of Fab-Fab angles made possible by segmental flexibility. Abbreviations: 39, HGAC 39; 2, anti-IdI-2 Fab; 3a, anti-IdI-3a Fab; 1, anti-IdI-1 Fab; X, anti-IdX Fab; K, anti-Cκ Fab. IdI designates an individual idiotope, and IdX, a cross-reactive idiotope. Antibody complexes were stained with 2% uranyl formate as described by Roux et al.[113] (Reproduced from Roux KH, Monafo WJ, Davie JM, Greenspan NS. Construction of an extended three-dimensional idiotope map by electron microscopic analysis of idiotope-anti-idiotope complexes. *Proc Natl Acad Sci U S A.* 1987;84(14):4984-4988, with permission.)

flexibility (see Fig. 21.12). These include flexing between Fab arms (motion toward or away from one another in the same plane), Fab arms moving in and out of the same plane, Fab arms rotating along their long axes, and Fab arms moving in or out of the same plane as the Fc region.[123,124] The inter-Fab angles observed by electron microscopy range from 0′ to 180′.[113,125] Similarly, Fab arm long-axis rotations can extend up to 180′.[126]

Another key role of the hinge is the maintenance of the C_H2-C_H2 interaction, ie, effectively constraining molecular mobility within the Fc region itself. The lower hinge stabilizes C_H2-C_H2 contacts by providing the key cysteine residues involved in inter–heavy chain disulfide bonds. Experiments in which IgG molecules were modified to eliminate the hinge region demonstrate that covalent linkage between the hinge regions just "upstream" of the two C_H2 domains is critical for the preservation of IgG effector function.[127]

HEAVY CHAIN STRUCTURE AND FUNCTION

What might be termed **the fundamental strategy of humoral immunity** is a two-step process that begins with the identification by antibodies (of appropriate binding specificity) of the molecules or molecular complexes that should be eliminated. Following such identification (ie, noncovalent complex formation), antibodies can then trigger other molecular systems (eg, complement) or cells (eg, phagocytes) to destroy or remove the antigenic material—guilt by association at the molecular level. Thus, antigen specificity, determined primarily by the V domains in the Fab arms, is physically and functionally linked to effector function, the activation of which is primarily attributable to the C domains of the Fc region. The effector functions associated with the humoral immune response primarily involve either complement or Fc receptor-bearing cells, such as natural killer cells, neutrophils, macrophages, and mast cells. As might be expected, therefore, the Fc contains sites for noncovalent interactions with complement components, such as C1q, and with Fc receptors. This section focuses on the structure and function of the Fc regions of the various Ig classes and subclasses. Further information can be found in Chapters 14 (complement) and 34 (Fc receptors).

Structure and Function of the Fc

The necessity for interacting effectively with relatively conserved molecules (eg, C1q) and Fc receptors provides a selective basis for maintaining the primary structures of the Fc region, at least where changes would undermine such intermolecular contacts. The degeneracy possible in noncovalent molecular recognition events permits selected primary sequence variations without catastrophic alterations in function. However, there are allotypic differences in the heavy chain constant domains among both human and mouse immunoglobulins. In some cases, these allotypic differences are associated with variation in function, at least *in vitro*. For example, two recombinant V domain-identical IgG3 antibodies, of different allotypes, exhibited differential abilities to

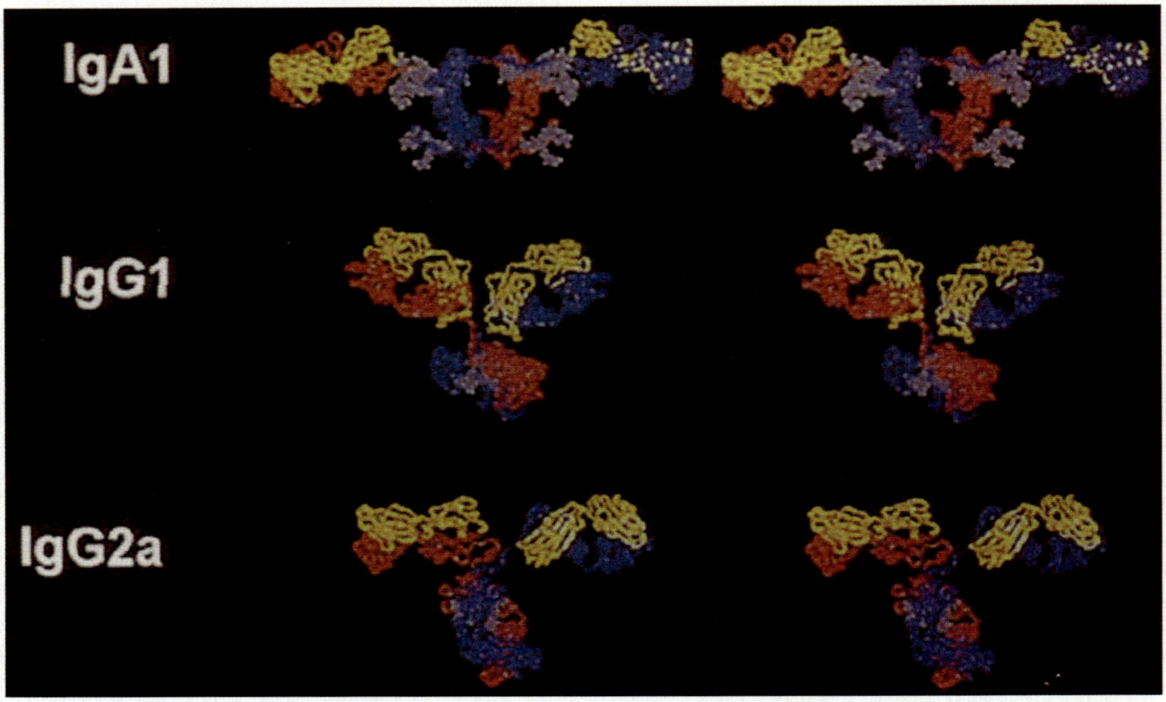

FIG. 21.14. A comparison of an x-ray and neutron-solution–scattering theoretical model (human IgA1) and x-ray crystal (murine IgG1 and IgG2a) structures. Light chains (*yellow*), heavy chains (red and dark blue), and glycosylation (*light blue*) are illustrated. The extended length of IgA1 over that of IgG can be seen along with extensive glycosylation that characterizes this isotype. (Reprinted from Boehm MK, Woof JM, Kerr MA, Perkins SJ. The Fab and Fc fragments of IgA1 exhibit a different arrangement from that in IgG: a study by X-ray and neutron solution scattering and homology modelling. *J Mol Biol.* 1999;286(5):1421-1447, with permission from Elsevier.)

bind C1q or initiate antibody-dependent cell-mediated cytotoxicity.[128] Nevertheless, the heavy chains of the human IgG subclasses are particularly well conserved, with >90% identity of amino acid sequences. In other mammalian species that have two or more IgG subclasses, they tend to exhibit less amino acid sequence identity than the human IgG subclasses.

By convention, the H chain constant domains are numbered from N terminus to C terminus, with C_H1 residing in the Fab arm and the remaining two (IgG, IgA, IgD) or three (IgM, IgE) C domains (C_H2, C_H3, and, if relevant, C_H4) residing in the Fc. In IgM and IgE, the C_H2 domain largely plays the role of the hinge region. The C domains of different isotypes and from different species share several key structural features. In distinction from V domains, which consist of four- and five-stranded ß-pleated sheets linked through an intrachain disulfide bond, C domains consist of three- and four-stranded ß-pleated sheets linked

TABLE 21.2 Properties of Hinges in IgG, IgA, and IgD

Ig Type	Upper Hinge Length	Middle Hinge Length	Lower Hinge Length	Genetic Hinge Configuration (Amino Acids/Exon)	Susceptibility to Proteolysis	Special Features
IgG1	4	10	6	15		
IgG2	3	8	6	12		
IgG3	12	49	6	17-15-15-15		
IgG4	7	4	6	12		
IgA1	1	23	2	19	High	Heavily O-linked glycosylation
IgA2	1	10	2	6		
IgD				34-24	High	Extensive charged amino acids: heavy O-linked glycosylation at N terminus

Note: Lengths represent amino acids.

through an intrachain disulfide bond (Fig. 21.4). Across isotypes, amino acid sequence identity for C_H domains is approximately 30%, while for subclasses (within an isotype) the amino acid sequence identity for C_H domains is in the range of 60% to 90%. Important physical and biological properties of the human Ig isotypes are summarized in Table 21.3.

As noted above, the traditional and accepted functional anatomy of immunoglobulins attributes antigen binding (both specificity and affinity) to the V modules in the Fab arms and effector function activation to the Fc region. Although this scheme is both well supported and appealing, structural variations in heavy chain domains (ie, C domains and hinge) can in some cases influence both discrimination among and the affinity of the antibody for antigens.[132-137] Although these instances of C domain influence on ligand binding (through the V domains) primarily involve multivalent antigens, there are also reports that suggest C domain influence in instances of monovalent recognition.[138-140] Mechanisms for these effects in the context of binding multivalent antigens include isotype-related differences in segmental flexibility as well as the tendency for self-association. Results from a study comparing resistance to pneumococcal infection for IgG3-deficient and IgG3-producing mice are consistent with the notion that the cooperative binding permitted by murine IgG3 antibodies contributes to the effectiveness of humoral immunity.[141]

Fc Glycosylation

Although the type and extent of glycosylation varies among isotypes, an N-linked oligosaccharide on Asn 297 in the C_H2 domain is conserved on all mammalian IgGs and homologous portions of IgM, IgD, and IgE. As the average serum IgG contains 2.8 oligosaccharides, there is often glycosylation present in the V domain as well.[142] The consensus sequence for the V domain N-linked oligosaccharides is not present in the germline, but it can be created during somatic hypermutation.[143] Glycosylation in the Fc region has been shown to be important for antibody half-life and effector functions.[144-146] Glycosylation of the Fc domain influences complement activation as Ig hypoglycosylation influences affinity for C1q as well as Ig binding to the FcR, possibly owing to its effects on Ig structure.[147] Differential sialylation of the core Fc polysaccharide has been shown to have dramatic effects on the proinflammatory versus anti-inflammatory activity of IgG[148,149] (Chapter 34). Further IgD N-linked glycans are necessary for IgD to bind to the IgD receptor on T cells.[150]

V domain glycosylation potentially affects the affinity for antigen, antibody half-life, antibody secretion, and organ targeting. Glycosylation has been shown to be capable of both positively and negatively affecting antigen binding.[151-154] The biologic significance of Ig glycosylation can be seen from studies demonstrating that IgG from patients with rheumatoid arthritis is galactosylated to a lesser extent (termed IgG G0) than IgG from normal controls. In some cases, hypogalactosylation correlates with disease activity.[155] Hypogalactosylation of IgG has also been found to occur in other chronic inflammatory diseases such as Crohn disease and systemic lupus erythematosus.[156]

Immunoglobulin M

IgM is an isotype of firsts. It is the first to be expressed on developing B lineage cells. It is the first to appear in the primary humoral immune response. It is probably, along with IgD, a phylogenetically primitive isotype in jawed vertebrates (an almost first) and may be the most phylogenetically stable isotype.[157]

IgM serves important immunological functions as a membrane-bound antigen receptor on the surfaces of B lymphocytes and as a soluble antibody in the blood and mucosal secretions. On the cell surface, IgM consists of two identical μ heavy chains and two identical light chains (μ_2L_2). μ H chains are initially expressed on developing pre–B cells bound to surrogate light chain to form pre–B cell receptors. Pre–B cells that undergo successful light chain gene rearrangement replace surrogate light chain with either a κ or λ light chain to form an immature B cell. On the B cell surface, IgM is noncovalently associated with two other polypeptide chains, Ig-α (CD79a) and Ig-β (CD79b).[158-160] These integral membrane proteins serve to transduce signals when surface IgM binds to and is cross-linked by cognate antigen.

In the secreted form, IgM can consist of either pentamers $(\mu_2L_2)_5$ or, less often, hexamers $(\mu_2L_2)_6$.[161] The μ_2L_2 monomers of the pentameric form are linked one to another by disulfide bonds in the C_H4 domains. Two of these monomers are, on one side, disulfide bonded not to another μ chain but to a 15,000-Da polypeptide, called J chain. The J chain is also found in dimeric IgA. There may be multiple patterns of such disulfide bonding, such that different cysteines participate in different monomeric units.[162]

The IgM-Fc pentamer is formed asymmetrically, resembling a hexagon with a missing triangle. The gap is filled by the J chain. The tailpieces, short C-terminal extensions of the IgM C_H4 domains, pack into an amyloid-like structure to stabilize the pentamer. The J chain caps the tailpiece assembly and bridges the interaction between IgM-Fc and the polymeric immunoglobulin receptor (pIgR), which undergoes a large conformational change to engage the IgM-J complex.[163] pIgR is a type I transmembrane protein that contains five extracellular immunoglobulin-like domains (D1-D5). It binds to J chain–containing secretory IgM and IgA at the basolateral surface of epithelial cells and escorts them to the apical side. There, the ectodomain, termed the secretory component (SC), is released by proteolysis and secreted together with IgM and IgA.

Application of electron microscopy to polymeric IgM molecules has suggested that IgM can adopt two different quaternary arrangements: star and staple.[164,165] The staple form can fix complement (Chapter 14), whereas star does not. All of the antigen-binding sites are arrayed in radial fashion, in the same plane as one another and the Fc regions, in the star arrangement. In the staple form, the Fab arms bend out of the plane of the Fc regions. It has been conjectured that the staple form is utilized in binding simultaneously to two or more epitopes on multivalent antigens, such as bacterial or viral surfaces.

A major pathway through which soluble IgM mediates immunity or immunopathology is the activation of the

TABLE 21.3 Properties of Immunoglobulin Isotypes

Class or Subclass Properties	IgM	IgD	IgG1	IgG2	IgG3	IgG4	IgA1	IgA2	IgE
Molecular weight of secreted form (kDa)[a]	950(p)	175	150	150	160	150	160(m), 300(d)	160(m), 350(d)	190
Sedimentation coefficient	19S	7S		6.6S			7S	11S	8S
Functional valency	5 or 10	2	2	2	2	2	2 or 4	2 or 4	2
Interheavy disulfide bonds per monomer	1	1	2	4	11	2	2	2	1
Membrane Ig cytoplasmic region	3	3	28	28	28	28	14	14	28
Secreted Ig tailpiece	20	9	2	2	2	2	20	20	2
Other chain	J chain (16 kDa)	—	—	—	—	—	J chain (16 kDa) Secretory component (70 kDa)		—
N-glycosylation sites	5	3	1	1	2	1	2	4	7
O-glycosylation sites	0	7	0	0	0	0	8	0	0
Carbohydrate average (%)	10-12	9-14	2-3	2-3	2-3	2-3	7-11	7-11	12-13
Adult level range (age 16-60 y) in serum (mg/ml)[b]	0.25-3.1	0.03-0.4	5-12	2-6	0.5-1	0.2-1	1.4-4.2	0.2-0 5	0.0001-0.0002
Approximate % total Ig in adult serum	10	0.2	45-53	11-15	3-6	1-4	11-14	1-4	0.004
Synthetic rate (mg/kg weight/day)	3.3	0.2	33	33	33	33	19-29	3.3-5.3	0.002
Biological half-life (days)	5-10	2-8	21-24	21-24	7-8	21-24	5-7	4-6	1-5
Transplacental transfer	0	0	++	+	++	++	0	0	0
Complement activation classical pathway (Clq)	++++	0	+++	+	++++	0	0	0	0
Complement activation alternative pathway	0	0	0	0	0	0	+	0	0
Reactivity with protein A via Fc	0	0	++	++	+/–	++	0	0	0
Allotypes	—	—	G1m	G2m	G3m	—	—	A2m	Em
Biological properties	Primary antibody response, some binding to pIgR, some binding to phagocytes	Mature B cell marker	Placental transfer, secondary antibody for most responses to pathogen, binds macrophages and other phagocytic cells by FcγR				Secretory Ig, binds pIgR		Allergy and parasite reactivity, binds FcεR on mast cells and basophiles

d, dimer; m, monomer; p, pentamer.
[a]Light chain molecular weight is 25 kDa.
[b]Total = 9.5 to 21.7 mg/mL.
Compiled from Carayannopoulos L, Capra JD. Immunoglobulins: structure and function. In: Paul WE, ed. *Fundamental Immunology.* Raven Press; 1993:283-314, Lefranc MP, Lefranc G. *The Immunoglobulin FactsBook.* Academic Press; 2001, Kuby J. *Immunology.* W.H. Freeman and Company; 1997, and Janeway CA Jr, Travers P, Walport M, Shlomchik M. *Immunobiology.* Garland Publishing; 1999, with permission.

classical pathway of complement. On a per molecule basis, relative to other isotypes, IgM is highly active in activating the classical pathway and can thereby effectively opsonize bacterial pathogens. In select cases (eg, *Neisseria meningitidis*), binding of IgM to bacterial surfaces, followed by complement activation, can cause direct lysis of the bacteria through the insertion of the membrane attack complex into the bacterial membrane.[166] IgM, like polymeric IgA, can reduce the effective number of colony- or plaque-forming units for, respectively, bacteria and viruses, through agglutination. Significant physical and biological properties of IgM and the other Ig isotypes are shown in Table 21.3.

Immunoglobulin D

IgD is primarily of interest in its membrane form, as the soluble form of IgD is found in relatively modest concentrations in the blood and other body fluids. The cell surface form of IgD is found along with IgM on all mature, naïve B cells, where it appears capable of transducing activating and tolerizing signals.[167] Simultaneous cell surface expression of two heavy chain isotypes expressing the same V_H domains and the same light chains occurs via differential RNA splicing.[168]

IgD appears optimized for responding to multivalent antigen in immune complexes as opposed to responding to monovalent antigen. This reflects, in part, differences in charge and O-linked glycosylation between the IgM and IgD hinge regions.[169] Higher IgD expression by anergic B cells helps keep them quiescent against monovalent autoantigens, reducing the risk of autoreactive responses. Absence of IgD leads to the retardation of T-dependent immune responses, antibody production, and affinity maturation.

B cell receptors reside as nanoclusters, or protein islands, on the surface of the cell. Just like IgM, the membrane form of IgD associates noncovalently with Ig-α (CD79a) and Ig-β (CD79b). However, on the B cell surface these two isotypes are functionally segregated and they can be stimulated independently from each other. In resting B cells, IgD-BCRs colocalize with CD19 and CXCR4, whereas IgM-BCRs gain proximity with CD19 only upon activation. IgD is required for CXCR4 signaling in mature B cells and thus influences the migration of mature B cells into secondary lymphoid tissues and induces actin cytoskeleton remodeling upon antigen binding.

In the serum, IgD exhibits greater sensitivity to proteolytic cleavage than IgM, which is consistent with a relatively short serum half-life of only 2.8 days. The relatively long hinge region is a primary target for proteolysis. The microbiota can help drive class switch recombination from IgM to IgD.[170] IgD-producing B cells in the upper respiratory mucosa can participate in host defense against pathogens relevant to this anatomical environment. Secreted IgD antibodies can bind to basophils and mast cells. IgD bound to basophils via galectin-9 and CD44 can amplify humoral Th2 responses.[171] When these antibodies are cross-linked by cognate antigen, the basophils release potent mediators that influence immune reactivity, inflammation, and pathogen viability.[172] Moreover, IgD activated mast cells can induce IgE synthesis in nasal polyps.[173]

Immunoglobulin G

IgG is the predominant isotype (approximately 70%-75% of the total Ig) in the blood and extravascular compartments. The four human IgG subclasses (IgG1, IgG2, IgG3, and IgG4) are named in order of their relative serum concentrations, with IgG1 the most prevalent and IgG4 the least. Although mice also have four IgG subclasses (Igl, IgG2a, IgG2b, and IgG3), shared numeric designations do not imply evolutionary relationships, since duplication events occurred separately in the two species. There are differences in effector functions (eg, complement activation and Fc receptor binding) (Table 21.4) and other biological properties (such as serum half-life) among these subclasses. However, there are also crucial functional commonalities, such as placental passage and complement activation (Tables 21.3 and 21.4). Although not exact matches, human IgG4 is most similar to mouse IgG1 and human IgG1 is most similar to mouse IgG2a in their function.

IgG antibodies are the hallmark of immunological memory in the humoral immune response. In addition to the isotype switch from IgM to IgG in a secondary antibody response, somatic hypermutation can lead to affinity maturation, a process by which the average affinity of antibody for the antigen eliciting the immune response can increase.

IgG antibodies of certain subclasses contribute to immunity directly and through the activation of complement or FcR-bearing cells. Important examples of immunity mediated directly through antibody binding include neutralization of toxins (eg, diphtheria toxin) and viruses (eg, poliovirus). Medically important examples of IgG-induced complement activation include immunity to encapsulated bacterial pathogens leading either to opsonization and destruction within phagocytes (eg, *Streptococcus pneumoniae*) or to direct complement-mediated lysis (eg, *Neisseria meningitidis*). Activation of FcR-bearing cells by IgG antibodies has also been implicated in immunity to pathogens (eg, *Cryptococcus neoformans*).[185]

The consensus view is that human IgG1 and IgG3 isotypes are effective activators of the classical complement pathway. Although some older sources state that IgG2 and IgG4 are weak or nonactivators of the classical complement pathway, more recent evidence suggests that, when epitope density is high, IgG2 is effective in activating complement through the classical pathway.[186,187] One possible source for the isotype-related variation in complement-activating ability is variation in affinity for C1q (IgG3>IgG1>IgG2>IgG4), the portion of the first component in the classical pathway that physically contacts the C_H2 domains of antibodies. However, isotype-associated differences in complement activation have also been found to occur at steps of the cascade subsequent to the binding of C1q to antibody.[128,179] For example, in one study of chimeric monoclonal antibodies engineered to express identical V domains and representing all four human IgG subclasses, the IgG3 antibody fixed C1q better than the IgG1 antibody but the IgG1 molecule was more effective in mediating complement-dependent cell lysis than the IgG3 molecule.[179] This may reflect the ability of hexameric assemblies of human IgG1, which are mediated by noncovalent Fc-Fc interactions, to maximally activate complement.[188,189] Human

TABLE 21.4 Comparative Properties of Human and Mouse IgG Subclasses

Species	Isotype	ADCC	C′	FcγRI Act	FcγRIIB Inh	FcγRIII	FcγRIIIA Act	FcRn Trans
Human	IgG1	++++	+++	+++	++	NA	++	+++
	IgG2	+/−	+	−	+	NA	+	+++
	IgG3	+++	++++	+++	++	NA	++	++
	IgG4	+	+/−	+++	++	NA	++	+++
Mouse	IgG1	++	+/−	−	++	++	NA	+++
	IgG2a	+++	++++	+++	++	++	NA	+++
	IgG2b	++++	+++	++	++	++	NA	+++
	IgG3	+/−	++	+	−	−	NA	+++

Act, activation; ADCC, antibody-dependent cell cytotoxicity; C′, C1q binding or complement fixation; Inh, inhibition; NA, not applicable (not present in species); Trans, transport.
Compiled from Damelang T, Rogerson SJ, Kent SJ, Chung AW. Role of IgG3 in infectious diseases. *Trends Immunol.* 2019;40(3):197-211, Bruhns P. Properties of mouse and human IgG receptors and their contribution to disease models. *Blood.* 2012;119(24):5640-5649, Klaus T, Bereta J. CH2 domain of mouse IgG3 governs antibody oligomerization, increases functional affinity to multivalent antigens and enhances hemagglutination. *Front Immunol.* 2018;9:1096, Leatherbarrow RJ, Dwek RA. Binding of complement subcomponent C1q to mouse IgG1, IgG2a and IgG2b: a novel C1q binding assay. *Mol Immunol.* 1984;21(4):321-327, Schumaker VN, Calcott MA, Spiegelberg HL, Muller-Eberhard HJ. Ultracentrifuge studies of the binding of IgG of different subclasses to the Clq subunit of the first component of complement. *Biochemistry.* 1976;15(23):517l5-5181, Bruggemann M, Williams GT, Bindon CI, et al. Comparison of the effector functions of human immunoglobulins using a matched set of chimeric antibodies. *J Exp Med.* 1987;166(5):1351-1361, Bindon CI, Hale G, Bruggemann M, Waldmann H. Human monoclonal IgG isotypes differ in complement activating function at the level of C4 as well as C1q. *J Exp Med.* 1988;168(1):127-142, Niwa R, Natsume A, Uehara A, et al. IgG subclass-independent improvement of antibody-dependent cellular cytotoxicity by fucose removal from Asn297-linked oligosaccharides. *J Immunol Methods.* 2005;306(1-2):151-160, Natsume A, In M, Takamura H, et al. Engineered antibodies of IgG1/IgG3 mixed isotype with enhanced cytotoxic activities. *Cancer Res.* 2008;68(10):3863-3872, Burton DR. Immunoglobulin G: functional sites. *Mol Immunol.* 1985;22:161-206, Unkeless JC, Scigliano E, Freedman VH. Structure and function of human and murine receptors for IgG. *Annu Rev Immunol.* 1988;6:251-281, and Collins AM. IgG subclass co-expression brings harmony to the quartet model of murine IgG function. *Immunol Cell Biol.* 2016;94(10):949-954.

IgG3 has been largely understudied. This reflects, in part, that the structure and enhanced functionality of human IgG3 is unique when compared with most mammals, including macaque species, which do not have an equivalent analog among their IgG subclasses. Prehuman clinical studies using macaque species can thus not be used to assess the human IgG3 response to vaccines.[174]

The affinities of IgG subclasses for Fc receptors vary from about 5×10^5 M^{-1} to about 10^8 M^{-1}. The relative contributions of IgG subclasses to various immunopathological processes depend on their relative affinities for the activating versus inhibiting isoforms of FcR.[190]

A remarkable attribute of IgG (for three of the four subclasses) is its serum half-life of about 23 days. This property, attributable to the Fc region and its interaction with the neonatal Fc receptor (FcRn), has been exploited for therapeutics through the genetic fusion of solubilized receptors (eg, CTLA4) to IgG Fc regions.[191]

IgG4 antibodies have the unique ability to spontaneously dissociate into half-molecules and to form antibodies composed of different half-molecules with distinct antigen specificities, ie, bispecific antibodies that cannot cross-link antigen and thus have anti-inflammatory activity.[192] This phenomenon may be of particular relevance to specific immunotherapy in the setting of allergy.[193]

Immunoglobulin A

Although IgG is the clearly predominant isotype in the blood, dimeric IgA is the dominant immunoglobulin isotype in the mucosal secretions as well as in breast milk and colostrum.[194]

In the blood, 10% to 15% of the Ig is IgA (vs 65%-75% IgG). Moreover, IgA has a shorter half-life than IgG in serum. The predominant form of IgA in human serum or plasma is monomeric (ie, α_2L_2), but there are small quantities of dimers $[(\alpha_2L_2)_2]$ and fewer still trimers and tetramers. Secretory IgA consists of dimers and lesser amounts of trimers and tetramers associated with one joining (J) chain (distinct from the J region in the heavy chain V domain) and one SC chain. As mentioned above, the latter is the extracellular portion of the polymeric Ig receptor (pIgR), which is expressed by mucosal epithelial cells and transfers polymeric IgA or IgM from basolateral to apical surfaces via transcytosis, thereby providing most of the Ig content of the mucosal secretions. The J chain is disulfide bonded to the tailpieces, short C-terminal extensions of the C_H3 domains, of the two IgA monomers of a dimer, whereas SC forms a disulfide bond to a cysteine in one C_H2 domain of one monomer.

The two α H chain C region genes correspond to two IgA subclasses, IgA1 and IgA2. IgA1 is the predominant (>80%) IgA subclass in the serum. Although IgA2 is the major form in some human mucosal secretions, such as those in the large intestine and the female genital tract, there is variation in the relative proportion of IgA1 and IgA2 in different secretions. The shorter hinge region of IgA2 confers increased resistance to bacterial proteases that might be encountered in the mucosal environment. The extended hinge region of IgA1 is believed to permit molecules of this isotype to accommodate variable epitope spacings on multivalent antigens. Although the long hinge region of IgA1 molecules might be expected to confer relatively high susceptibility to proteolysis (Fig. 21.14), relative protection against the activity of bacterial proteases is provided by heavy O-linked glycosylation in the hinge.

Nevertheless, the IgA1 hinge is uniquely susceptible to IgA proteases produced by certain pathogenic bacteria.[195]

Secretory dimeric IgA (sIgA1) Fcs are attached tail-to-tail in the same plane.[196] The two Fcs are held together at an ~110° angle by the J chain. The secretory component makes extensive contacts with both the Fcs and J chain, binding diagonally across an approximately 50° gap between the two monomers. sIgA2 tetramers and pentamers are assembled through incorporation of two or three additional Fcs in the same plane as sIgA1 and sIgA2 dimers. The pentamer adopts the same asymmetric pentagon structure seen in pentameric IgM,[163] complete with the J chain at the sixth hexagonal slice. As in pentameric IgM, SC is bound at the J chain–containing gap.

sIgA has been shown to participate in immunity against a range of viral, bacterial, and parasitic pathogens at mucosal surfaces. The relative absence of functional complement and phagocytes in mucosal secretions is consistent with a different mix of mechanisms for mediating immune effects associated with S-IgA versus, for example, serum IgG. Mechanisms associated with IgA-mediated immunity are less dependent on inflammation-producing molecules or cells, such as inhibition of microbial adherence through V module–mediated specific binding to microbial adhesins, agglutination of microbes, blocking of microbial receptors for cell surface carbohydrates with IgA-associated glycans, and mucus trapping (in which binding of S-IgA to bacteria makes them more adherent to host-generated mucus). There is also evidence that polymeric IgA can neutralize viruses in some circumstances by interfering with postattachment steps, such as internalization. In some cases, posttranslational modifications of viral surface molecules, related to proteolytic events associated with epithelial cell transit, lead to IgA antibodies expressing protection-related antigen specificities that have no parallel in the IgG pool.[197,198] Some of the contributions of IgA to immunity are mediated through binding to FcαRI (CD89) on human neutrophils, monocytes/macrophages, and eosinophils. For example, cells bearing FcαRI on their plasma membranes can phagocytose IgA-antigen complexes. FcαRI binds to the IgA Fc region between the Cα2 and the Cα3 domains.[199] Amino acid residues in the IgA Fc region critically involved in the interaction with CD89 are indicated in Figure 21.15.

IgA does not activate the classical complement pathway. Deposition of polymeric IgA1 in IgA nephropathy can, however, trigger the alternative and lectin complement pathways.[200] Only the latter appears dependent on mannose-binding lectin.[201]

Additional properties of the polymeric forms of IgM and IgA are considered below.

Immunoglobulin E

IgE is best known for its association with hypersensitivity reactions and allergy, but this isotype is also of interest in the context of immunity to parasites. In the blood, IgE is present at the lowest concentration of any of the immunoglobulin isotypes (with roughly five orders of magnitude less IgE than IgG) and has the shortest half-life. The unimpressive quantitative representation of IgE in the blood is related to the high affinity of IgE antibodies for FcεRI, often referred to as the high-affinity Fc receptor for IgE. FcεRI is expressed on mast cells, basophils, Langerhans cells, and eosinophils. Owing to the high affinity of FcεRI for its IgE ligand, mast cells and basophils are covered with relatively long-lived FcεRI-IgE complexes.

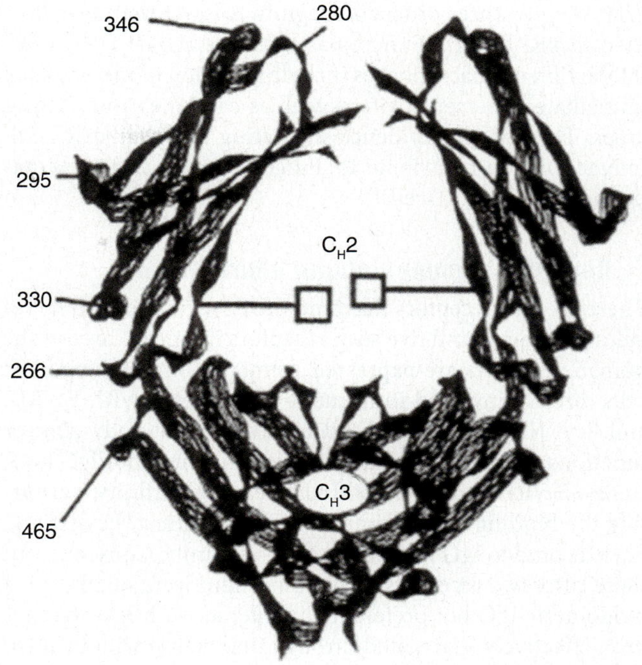

FIG. 21.15. Illustration of the residues essential for the binding of IgA to FcαR (CD89). Residues represent mutations made in IgA C_H regions as mapped on an Fcγ fragment. Of the residues mutated, L465 and L266 were found to be important for binding to CD89. (Republished with permission of Rockefeller University Press. From Carayannopoulos L, Hexham JM, Capra JD. Localization of the binding site for the monocyte immunoglobulin (Ig) A-Fc receptor (CD89) to the domain boundary between Calpha2 and Calpha3 in human IgA1. *J Exp Med.* 1996;183(4):1579-1586; permission conveyed through Copyright Clearance Center,Inc.)

The interaction between IgE and FcεRI has an affinity of ~10^{10} L/M. It primarily involves contacts between FcεRI and amino acids in the C_H3 domains, with some contributions from amino acids in the C_H2 domains. Although each IgE potentially has two sites for interacting with the FcεRI, the stoichiometry has been shown to be 1:1. Furthermore, it has been suggested that IgE binds to FcεRI in a kinked conformation.[202]

Upon ligation with bivalent or multivalent antigens specifically recognized by the bound IgE molecules, the FcεRI molecules transduce signals that activate the mast cells or basophils to secrete potent mediators of inflammation, such as histamine. These mediators are responsible for the symptoms associated with asthma, allergic rhinitis, and anaphylaxis.

There is a second receptor for IgE. FcεRII (CD23), a type II membrane protein, is expressed on monocytes/macrophages, B lymphocytes, natural killer (NK) cells, follicular dendritic cells, Langerhans cells, eosinophils, activated epithelial cells, and platelets. It binds monomeric IgE with an affinity of ~10^7

L/M, roughly three orders of magnitude lower than the affinity of FcεRI for IgE. Functional consequences of FcεRII-IgE interaction on macrophages include secretion of mediators of immediate hypersensitivity as well as cytokines and chemokines. There is also evidence suggesting that the FcεRII-IgE interaction can contribute to antigen capture and presentation to both B and T cells.[203]

Fc Receptor Immunoglobulin Interactions

There are four receptors that bind the Fc regions of IgG molecules in humans and five such receptors in mice. Three of the human receptors are expressed primarily on hematopoietic cells directly involved in immune responses: FcγRI, FcγRII, and FcγRIII. Among the FcγR involved in antibody effector functions, there is important variation in affinity for IgG molecules. FcγRI binds IgG with relatively high affinity, permitting the binding of monomeric IgG. In contrast, FcγRII and FcγRIII bind to IgG with relatively low affinity. Consequently, these latter two receptors do not bind significant quantities of monomeric IgG but preferentially interact with IgG that has been effectively aggregated through interaction with bivalent or multivalent antigens (ie, immune complexes). Key amino acid residues involved in the binding FcγRs and FcεR1 with their corresponding Ig can be seen in Figure 21.16. Allotypic variations in FcγRIIA can also influence affinity for IgG ligands and subsequent effector function.[205] A more extensive discussion of Fc receptor immunology can be found in Chapter 34.

There are different isoforms of FcγRII and FcγRIII. Of particular functional relevance, FcγRIIA is activating, whereas FcγRIIB is inhibiting. In mouse models, deficiency of FcγRIIB can be associated with an autoimmune syndrome similar to human lupus.[206] Studies in mice have also suggested that the effectiveness of antibodies of the various murine IgG subclasses in mediating FcγR-dependent effector functions is correlated with the ratio of affinities of antibodies of those subclasses for activating versus inhibiting FcγR.[190]

FcγRIIIA is expressed on NK cells while FcγRIIIB is expressed on neutrophils. The former receptor is attached to the membrane by a standard transmembrane polypeptide, whereas the latter is attached via a glycophospholipid tail.

The fourth human IgG Fc receptor, the neonatal FcγR, or FcRn (sometimes referred to as the Brambell receptor), transports IgG across the placenta. It also plays a crucial role in protecting IgG from proteolytic degradation, thereby prolonging serum half-life. Although FcγRI, FcγRII, and FcγRIII are members of the immunoglobulin superfamily, FcRn is structurally similar to MHC class I molecules, including noncovalent association with β_2-microglobulin. The interaction between FcRn and IgG involves amino acid residues in the C_H2-C_H3 interface[207] and is pH sensitive. This pH dependence is attributed to titration of histidine residues on Fc that interact with anionic pockets on the receptor. The FcRn dimer, which is required for high-affinity binding of IgG, is itself stabilized at acidic pH by histidine-mediated salt bridges and a side-chain rearrangement that creates a more favorable interaction with an anionic pocket at pH 6.5 relative to pH 8.[208] This latter property is consistent with the ability of FcRn to bind IgG in acidic vesicular compartments and then release it into the neutral-pH environment of the blood. A crystallographic structure of the rat FcRn in a 1:1 complex with a heterodimeric Fc (containing only one FcRn-binding site) reveals that there are conformational changes in the Fc on binding to Fcγn-b.[209] The investigators also identified three titratable salt bridges that confer pH-dependent binding of the IgG Fc to the FcRn.

There are other FcR that interact with the non-IgG isotypes: IgM, IgD, IgA, and IgE. The FcR for IgA and IgE are covered in the respective sections devoted to the corresponding isotypes. In addition to pIgR, two other IgM receptors exist in mammals: FcαμR and FcμR. They each contain a domain that is homologous to the immunoglobulin-like D1 domain of pIgR. Mice use an Fcα/μR to bind to both IgA and IgM, whereas humans use distinct FcαR and FcμR.[210,211] Galectin-9 and CD44 are used to bind IgD to the surface of basophils.[171]

There are functional as well as structural parallels among FcR for IgG and non-IgG isotypes. Some of the features of Fc-FcR interaction that are conserved across isotypes are illustrated in Figure 21.16.

Transmembrane and Cytoplasmic Domains

Igs are expressed in both membrane and secreted forms. In contrast to the secreted form, the membrane Ig contains a transmembrane and a cytoplasmic domain. The transmembrane domain is a typical single-pass polypeptide segment consisting of 26 hydrophobic residues. It extends from the C-terminal end of the C-terminal C_H domain and forms an alpha-helix followed by a variable number of basic amino acids. The cytoplasmic portion of the Ig H chain ranges in length from 3 amino acids for IgG to 30 amino acids for IgE.

Both membrane expression of Ig and the integrity of the cytoplasmic domain are important in antibody function. As the cytoplasmic domain is rather short, membrane-bound IgM does not "signal" directly, but through the associated Igα and Igβ molecules.[212] However, disruption of either membrane expression of IgG1 in mice or the cytoplasmic tail results in the failure to generate an effective IgG1 response and IgG1 memory.[213] Mice lacking IgE membrane expression due to deletion of the transmembrane and cytoplasmic domains exhibit significant impairment in IgE responses and have extremely low levels of secreted IgE.[214] Thus, for those Igs with longer membrane:cytoplasmic tails, membrane Ig affects signaling beyond its association with Igα and Igβ. Specific residues in the transmembrane domain have been identified that are crucial for signal transduction while having no effect on the association with Igα and Igβ.[215]

HIGHER-ORDER STRUCTURE

Many of the biological functions of IgA and IgM are dependent on their ability to form multimeric structures. This section discusses the role of multimeric immunoglobulin in immune function.

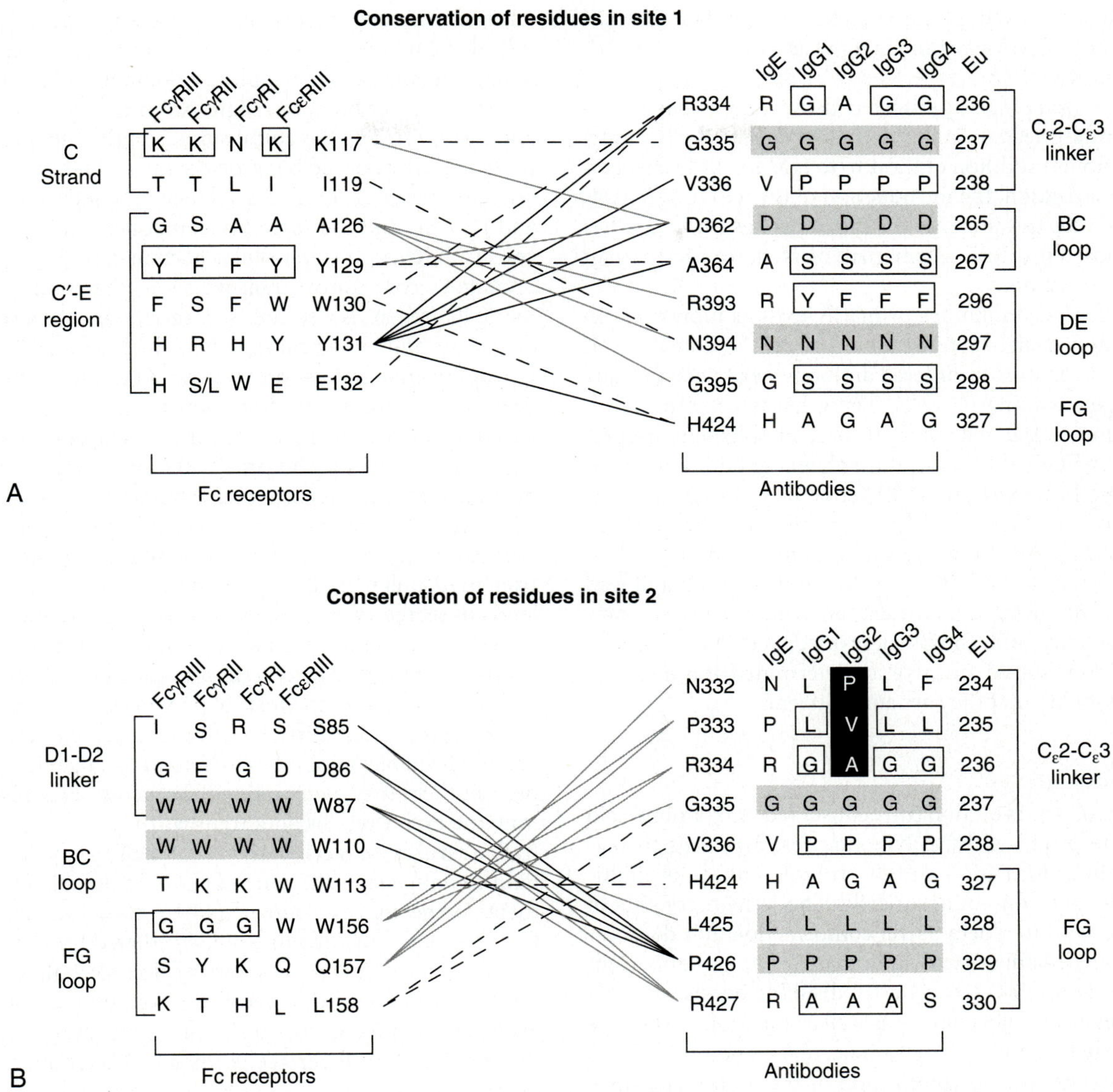

FIG. 21.16. **Certain residues are conserved between Fcγ Rs and FcεRI as well as between IgG and IgE that facilitate binding.** Two sites participate: site 1 in **(A)** and site 2 in **(B)**. Heavy lines indicate the highest number of contacts and dashed lines indicate the least. Of considerable note are residues W87 and W110 in site 2 of the receptors and P426 in the immunoglobulin that form a core "praline sandwich" in the interaction between immunoglobulin and receptor. (Reprinted by permission from Springer Nature: Garman SC, Wurzburg BA, Tarchevskaya SS, Kinet JP, Jardetzky TS. Structure of the Fc fragment of human IgE bound to its high-affinity receptor FcεRIα. *Nature.* 2000;406(6793):259-266.)

Dimers, Pentamers, and Hexamers

The majority of multimeric IgA exists as dimers and, less commonly, trimers and tetramers. IgM forms pentamers and occasionally hexamers. The polymeric structures of these antibodies enhance their functional affinity (avidity) for antigen, is essential for their active transport (both IgA and IgM) across epithelial cells to mucosal secretions, and, in the case of IgM, enhances the activation of the classical pathway of complement. Once multimerized, IgA or IgM in a complex with J chain can bind to pIgR and cross mucosal epithelial cells (Ref. 216, reviewed in Ref. 217). Although IgM can undergo transcytosis to the mucosal secretions, its principal actions are in the serum.

The ability of IgA and IgM to multimerize is due a tailpiece (tp), an additional C-terminal segment of 18 amino acids in the secreted forms of the μ and α heavy chains.[163,196] Tailpieces of both IgM and IgA contain a penultimate cysteine (residue 575 in IgM and 495 in IgA) that forms two different disulfide bonds important for multimer formation. In an Ig monomeric unit containing two identical heavy chains, one cysteine residue forms intermonomeric subunit bonds, whereas the remaining cysteine residue on the other heavy chain bonds to a cysteine on the J chain.[218-221]

Another cysteine residue, Cys 414, also forms intermonomeric subunit disulfide bonds in IgM, and this bond is important in hexamer formation.[222] Besides disulfide bonds, the highly conserved glycan linked to Asn 563 in IgM (and the homologous region in IgA) is also important for multimerization.[223]

Domain-swapping experiments demonstrate that the tp regulates multimerization in the context of the specific heavy chain. Although addition of the α tp to IgM has little effect on IgM polymerization, the introduction of the μ tp to IgA leads to higher-order IgA polymers.[224] Based on this finding, it has been proposed that IgM polymerization is more efficient than IgA polymerization.

Although IgG antibodies primarily exist as monomers in solution, noncovalent interactions between Fc domains can lead to the formation of ordered antibody hexamers typically after antigen binding on cells. These hexamers can recruit and activate C1, the first component of complement, triggering the complement cascade and enhancing killing of target cells.[168] The hexamerization efficiency of IgG3 and IgG1 is superior to that of IgG2 and IgG4, but all four subclasses can participate in these interactions. In chronic lymphocytic leukemia, however, hexamers can be found in solution and lead to chronic complement activation, and thus decreased complement activity and complement-mediated cytotoxicity.[225] The antigens involved are as yet undetermined, but there is a strong possibility that they are autoantigens.

The J Chain

The J chain, an evolutionarily conserved 137-amino-acid polypeptide produced by B lymphocytes, functions to regulate multimer formation and to promote linkage of multimeric Ig to pIgR on epithelial cells. The J chain consists of a single domain in a beta barrel conformation and does not show sequence similarity to Ig domains.[226] It contains eight cysteine residues that participate in disulfide bonds with two tp cysteines as described above as well as function to stabilize its own structure through intramolecular bonds.[224,227] The J chain influences the polymerization of the multimers, since in the absence of J chain, IgA forms fewer dimers and IgM forms fewer pentamers.[228]

The J chain exists in all polymeric forms of IgA and is important in IgA polymerization and secretion across the mucosa. The J chain is not required for IgM polymers but is required for external secretion. Although IgM pentamers contain J chain, hexamers typically do not.[163] The makeup of IgM is biologically significant since IgM hexamers have about 20-fold greater complement-activating activity than IgM pentamers. The presence of increased levels of hexameric IgM has been postulated to play a role in the pathogenesis of Waldenstrom macroglobulinemia, cold agglutinin disease, and recurrent urinary tract infections.[229]

Immunoglobulin Transport

Transport of dimeric IgA and pentameric IgM to the mucosal secretions occurs after binding to pIgR that is present on the basolateral surface of the lining epithelial cells. The J chain is essential for the secretion of IgA and IgM and as described above influences the polymeric structure of the Ig.[216] pIgR is a transmembrane receptor synthesized by mucosal epithelial cells that contains seven domains, including five extracellular V-like domains, a transmembrane domain, and a cytoplasmic domain.[230] Once bound to pIgR, polymeric Ig is endocytosed and transported to the apical surface of the cell. pIgR is then proteolytically cleaved between the fifth and sixth domains to release a complex (termed secretory Ig) containing the H, L, and J chains and the secretory component (SC), which represents the cleaved extracellular portion of pIgR.[231,232] As pIgR undergoes constitutive transcytosis in the absence of polymeric Ig, free SC is also released into the mucosal secretions. SC has several biological functions, including protecting the Ig from degradation by proteases and binding bacterial antigens such as the *Clostridium difficile* toxin A.[233,234] SC also functions to localize sIgA to the mucous layer to help protect against invasion by pathogens.[235] Results from pIgR-null mice demonstrate that alternate pathways exist to transport polymeric Ig to the mucosal secretions as some secretory Ig still crosses the epithelial cells in the absence of pIgR.[236] Results from pIgR-null mice also demonstrate the importance of high levels of secretory antibodies as these mice are more susceptible to mucosal infections with pathogens such as *Salmonella typhimurium* and *Streptococcus pneumoniae*.[237,238]

The epithelial transcytosis of polymeric Ig has several biological implications. First, by delivering the Ig to the mucosal surface, it enables antibodies to bind to pathogenic agents and prevent them from penetrating the mucosa, a process termed immune exclusion. Second, transcytosing antibody can neutralize viruses intracellularly.[239,240] Finally, polymeric Ig can bind to antigens in the mucosal lamina propria and excrete them to the mucosal lumen (where they can be removed from the body) by the same pIgR-mediated transcytosis process.[241,242] Some pathogens can exploit the pIgR-mediated transcytosis process in reverse to penetrate the mucosa. For example, the pneumococcal adhesin, CbpA, can bind pIgR at the epithelial apical surface, leading to bacterial penetration of the mucosa.[243]

THERAPEUTIC ANTIBODIES

The flexibility and biologic properties of immunoglobulins have made them a major focus of molecular engineering.[244] The US Food and Drug Administration approved the first monoclonal antibody for clinical use in 1986. The major benefit of these antibody drugs is that one can focus on a specific biologic effect with high specificity, and thus with a reduced risk of adverse events or side effects. As a result, antibodies have become the best-selling drugs in the pharmaceutical market. In 2018, 8 of the top 10 best-selling drugs worldwide were biologics. Immunoglobulins are being used as therapeutic agents for neoplastic, autoimmune, metabolic, and infectious diseases, as well as for biotechnology applications. These opportunities have led to a resurgence of interest in the structure-function aspects of antibodies as "designer antibodies" become more common. Both the variable and constant portions of these molecules represent substrates for engineering purposes, permitting alteration of both receptor and effector function.[245] The primacy of the immunoglobulin V domain even extends into therapeutic cell-mediated immunity. It is the general consensus that T cell receptors (Chapter 24) are designed to be polyreactive with low affinity, ie, to generally function as a sensor for various MHC:peptide combinations (Chapter 20), whereas Igs are designed to be effector molecules of high affinity and exquisite monospecificity. Thus, CAR-T cell therapy is increasingly focused on the use of T cell receptors with chimeric immunoglobulin V domain antigen-binding structures.[246]

Immunoglobulins are extremely versatile molecules that can carry out many biological activities at the same time. The need to be able to recognize unique antigen structures prior to any previous exposure coupled with the need to maintain host cell receptor or complement recognition properties presents a truly unique challenge for the system. As has been described, the system incorporates diversity within specific constraints. The precise biological niches may differ, but the overall design for these molecules is the same. The study of antibodies began with the need to understand how sera could neutralize toxins. It is likely that antibodies will continue to be a major focus for those who seek to take fundamental principles of protein chemistry to the bedside and for those who seek to develop new biologic therapeutic agents.

REFERENCES

1. Schroeder HW Jr, Wald D, Greenspan NS. Immunoglobulins: structure and function. In: Paul W, eds. *Fundamental Immunology*. Lippincott Wiliams & Wilkins; 2008:125-151.
2. Silverstein AM. *A History of Immunology*. Academic Press; 1989:1-422.
3. Behring E, Kitasato S. Ueber das Zustandekommen der Diphtherie-Immunitat und der Tetanus-Immunitat bei thieren. *Deutsche medizinsche Wochenschrift*. 1890;16:1145-1148.
4. Ehrlich P. Die Wertbemessung des Diphtherieheilserums. *Klinisches Jahrbuch*. 2006;6:299-326.
5. Lindenmann J. Origin of the terms "antibody" and "antigen." *Scand J Immunol*. 1984;19:281-285.
6. Tiselius A, Kabat EA. An electrophoretic study of immune sera and purified antibody preparations. *J Exp Med*. 1939;69:119-131.
7. Carayannopoulos L, Capra JD. Immunoglobulins: structure and function. In: Paul WE, ed. *Fundamental Immunology*. Raven Press; 1993:283-314.
8. Porter RR. Separation and isolation of fractions of rabbit gamma-globulin containing the antibody and antigenic combining sites. *Nature*. 1958;182(4636):670-671.
9. Nisonoff A, Wissler FC, Lipman LN. Properties of the major component of a peptic digest of rabbit antibody. *Science*. 1960;132:1770-1771.
10. Edelman GM, Poulik MD. Studies on the structural units of the gamma globulins. *J Exp Med*. 1961;113:861-884.
11. Putnam FW. Immunoglobulin structure: variability and homology. *Science*. 1969;163:633-644.
12. Dreyer WJ, Bennett JC. The molecular basis of antibody formation: a paradox. *Proc Natl Acad Sci U S A*. 1965;54:864-869.
13. Hozumi N, Tonegawa S. Evidence for somatic rearrangement of immunoglobulin genes coding for variable and constant regions. *Proc Natl Acad Sci U S A*. 1976;73(10):3628-3632.
14. Sarma VR, Davies DR, Labaw LW, Silverton EW, Terry WD. Crystal structure of an immunoglobulin molecule by x-ray diffraction and electron microscopy. *Cold Spring Harbor Symp Quant Biol*. 1972;36:413-419.
15. Sakano H, Rogers JH, Huppi K, et al. Domains and the hinge region of an immunoglobulin heavy chain are encoded in separate DNA segments. *Nature*. 1979;277(5698):627-633.
16. Williams AF, Barclay AN. The immunoglobulin superfamily–domains for cell surface recognition. *Annu Rev Immunol*. 1988;6:381-405.
17. Kirkham PM, Schroeder HW Jr. Antibody structure and the evolution of immunoglobulin V gene segments. *Semin Immunol*. 1994;6(6):347-360.
18. Rogers J, Early P, Carter C, et al. Two mRNAs with different 3' ends encode membrane-bound and secreted forms of immunoglobulin mu chain. *Cell*. 1980;20:303-312.
19. Geisberger R, Crameri R, Achatz G. Models of signal transduction through the B-cell antigen receptor. *Immunology*. 2003;110(4):401-410.
20. Friess MD, Pluhackova K, Bockmann RA. Structural model of the mIgM B-cell receptor transmembrane domain from self-association molecular dynamics simulations. *Front Immunol*. 2018;9:2947.
21. Cohn M. The immune system: a weapon of mass destruction invented by evolution to even the odds during the war of the DNAs. *Immunol Rev*. 2002;185:24-38.
22. Cohn M. What are the commonalities governing the behavior of humoral immune recognitive repertoires?. *Dev Comp Immunol*. 2006;30:19-42.
23. Ravetch JV. Fc receptors. *Curr Opin Immunol*. 1997;9(1):121-125.
24. Crowley JJ, Goldfien RD, Schrohenloher RE, et al. Incidence of three cross-reactive idiotypes on human rheumatoid factor paraproteins. *J Immunol*. 1988;140:3411-3418.
25. Nimmerjahn F, Ravetch JV. Divergent immunoglobulin g subclass activity through selective Fc receptor binding. *Science*. 2005;310(5753).
26. Sanchez-Mazas A, Fernandez-Vina M, Middleton D, et al. Immunogenetics as a tool in anthropological studies. *Immunology*. 2011;133(2):143-164.
27. Wright A, Morrison SL. Effect of glycosylation on antibody function: implications for genetic engineering. *Trends Biotechnol*. 1997;15(1):26-32.
28. Yagi H, Yanaka S, Kato K. Structure and dynamics of immunoglobulin G glycoproteins. *Adv Exp Med Biol*. 2018;1104:219-235.
29. Jefferis R, Lund J, Goodall M. Recognition sites on human IgG for Fc gamma receptors: the role of glycosylation. *Immunol Lett*. 1995;44(2-3):111-117.
30. Saunders KO. Conceptual approaches to modulating antibody effector functions and circulation half-life. *Front Immunol*. 2019;10:1296.
31. Yoo EM, Morrison SL. IgA: an immune glycoprotein. *Clin Immunol*. 2005;116(1):3-10.
32. Lai KN, Tang SC, Schena FP, et al. IgA nephropathy. *Nat Rev Dis Primers*. 2016;2:16001.
33. Harpaz Y, Chothia C. Many of the immunoglobulin superfamily domains in cell adhesion molecules and surface receptors belong to a new structural set which is close to that containing variable domains. *J Mol Biol*. 1994;238(4):528-539.
34. Imkeller K, Scally SW, Bosch A, et al. Antihomotypic affinity maturation improves human B cell responses against a repetitive epitope. *Science*. 2018;360(6395):1358-1362.
35. Rouge L, Chiang N, Steffek M, et al. Structure of CD20 in complex with the therapeutic monoclonal antibody rituximab. *Science*. 2020;367(6483):1224-1230.
36. Chothia C, Novotny J, Bruccoleri R, Karplus M. Domain association in immunoglobulin molecules. The packing of variable domains. *J Mol Biol*. 1985;186:651-663.
37. Padlan EA. Anatomy of the antibody molecule. *Mol Immunol*. 1994;31(3):169-217.
38. Kabat EA, Wu TT, Perry HM, Gottesman KS, Foeller C. *Sequences of Proteins of Immunological Interest*. U.S.Department of Health and Human Services; 1991:1-2387.
39. Chothia C, Lesk AM. Canonical structures for the hypervariable regions of immunoglobulins. *J Mol Biol*. 1987;196:901-917.
40. Ehrenmann F, Kaas Q, Lefranc MP. IMGT/3Dstructure-DB and IMGT/DomainGapAlign: a database and a tool for immunoglobulins or antibodies, T cell receptors, MHC, IgSF and MhcSF. *Nucleic Acids Res*. 2010;38(Database issue):D301-D307.
41. Manivel V, Sahoo NC, Salunke DM, Rao KV. Maturation of an antibody response is governed by modulations in flexibility of the antigen-combining site. *Immunity*. 2000;13(5):611-620.
42. Lefranc MP, Giudicelli V, Duroux P, et al. IMGT®, the international ImMunoGeneTics information system® 25 years on. *Nucleic Acids Res*. 2015;43(Database issue):D413-D422.
43. Chothia C, Lesk AM, Tramontano A, et al. Conformations of immunoglobulin hypervariable regions. *Nature*. 1989;342:877-883.
44. Huston JS, McCartney J, Tai MS, et al. Medical applications of single-chain antibodies. *Int Rev Immunol*. 1993;10(2-3):195-217.
45. Alt FW, Baltimore D. Joining of immunoglobulin heavy chain gene segments: implications from a chromosome with evidence of three D-J heavy fusions. *Proc Natl Acad Sci U S A*. 1982;79:4118-4122.
46. Rajewsky K. Clonal selection and learning in the antibody system. *Nature*. 1996;381(6585):751-758.
47. Hood L, Galas D. The digital code of DNA. *Nature*. 2003;421:444-448.
48. Nossal GJV. The double helix and immunology. *Nature*. 2003;421:440-444.
49. Ichihara Y, Hayashida H, Miyazawa S, Kurosawa Y. Only DFL16, DSP2, and DQ52 gene families exist in mouse immunoglobulin heavy chain diversity gene loci, of which DFL16 and DSP2 originate from the same primordial DH gene. *Eur J Immunol*. 1989;19:1849-1854.
50. Feeney AJ, Riblet R. DST4: a new, and probably the last, functional DH gene in the BALB/c mouse. *Immunogenetics*. 1993;37(3):217-221.
51. Johnston CM, Wood AL, Bolland DJ, Corcoran AE. Complete sequence assembly and characterization of the C57BL/6 mouse Ig heavy chain V region. *J Immunol*. 2006;176(7):4221-4234.
52. Max EE, Seidman JG, Leder P. Sequences of five potential recombination sites encoded close to an immunoglobulin kappa constant region gene. *Proc Natl Acad Sci U S A*. 1979;76(7):3450-3454.
53. Sakano H, Huppi K, Heinrich G, Tonegawa S. Sequences at the somatic recombination sites of immunoglobulin light-chain genes. *Nature*. 1979;280(5720):288-294.
54. Thiebe R, Schable KF, Bensch A, et al. The variable genes and gene families of the mouse immunoglobulin kappa locus. *Eur J Immunol*. 1999;29(7):2072-2081.
55. Seidman JG, Max EE, Leder P. A kappa-immunoglobulin gene is formed by site-specific recombination without further somatic mutation. *Nature*. 1979;280(5721):370-375.
56. Selsing E, Miller J, Wilson R, Storb U. Evolution of mouse immunoglobulin lambda genes. *Proc Natl Acad Sci U S A*. 1982;79(15):4681-4685.
57. Sanchez P, Marche PN, Rueff-Juy D, Cazenave PA. Mouse V lambda X gene sequence generates no junctional diversity and is conserved in mammalian species. *J Immunol*. 1990;144:2816-2820.
58. Ivanov II, Schelonka RL, Zhuang Y, Gartland GL, Zemlin M, Schroeder HW Jr. Development of the expressed immunoglobulin CDR-H3 repertoire is marked by focusing of constraints in length, amino acid utilization, and charge that are first established in early B cell progenitors. *J Immunol*. 2005;174:7773-7780.
59. Lee SK, Bridges SL Jr, Koopman WJ, Schroeder HW Jr. The immunoglobulin kappa light chain repertoire expressed in the synovium of a patient with rheumatoid arthritis. *Arthritis Rheum*. 1992;35:905-913.
60. Victor KD, Capra JD. An apparently common mechanism of generating antibody diversity: length variation of the VL-JL junction. *Mol Immunol*. 1994;31(1):39-46.
61. Brodeur PH, Riblet RJ. The immunoglobulin heavy chain variable region (IgH-V) locus in the mouse. I. One hundred Igh-V genes comprise seven families of homologous genes. *Eur J Immunol*. 1984;14:922-930.
62. Tutter A, Brodeur PH, Shlomchik MJ, Riblet RJ. Structure, map position, and evolution of two newly diverged mouse Ig VH gene families. *J Immunol*. 1991;147:3215-3223.
63. Perlmutter RM, Berson B, Griffin JA, Hood LE. Diversity in the germline antibody repertoire: molecular evolution of the T15 VH gene family. *J Exp Med*. 1985;162:1998-2016.
64. Kirkham PM, Mortari F, Newton JA, Schroeder HW Jr. Immunoglobulin VH clan and family identity predicts variable domain structure and may influence antigen binding. *EMBO J*. 1992;11:603-609.
65. Chang B, Casali P. The CDR1 sequences of a major proportion of human germline Ig VH genes are inherently susceptible to amino acid replacement. *Immunol Today*. 1994;15:367-373.
66. Rogozin IB, Kolchanov NA. Somatic hypermutagenesis in immunoglobulin genes. II. Influence of neighbouring base sequences on mutagenesis. *Biochim Biophys Acta*. 1992;1171(1):11-18.

67. Dorner T, Brezinschek HP, Brezinschek RI, Foster SJ, Domiati-Saad R, Lipsky PE. Analysis of the frequency and pattern of somatic mutations within nonproductively rearranged human variable heavy chain genes. *J Immunol.* 1997;158(6):2779-2789.
68. Rogozin IB, Diaz M. Cutting edge. DGYW/WRCH is a better predictor of mutability at G: C bases in Ig hypermutation than the widely accepted RGYW/WRCY motif and probably reflects a two-step activation-induced cytidine deaminase-triggered process. *J Immunol.* 2004;172(6):3382-3384.
69. Tramontano A, Chothia C, Lesk AM. Framework residue 71 is a major determinant of the position and conformation of the second hypervariable region in the VH domains of immunoglobulins. *J Mol Biol.* 1990;215:175-182.
70. Al Lazikani B, Lesk AM, Chothia C. Canonical structures for the hypervariable regions of T cell αβ receptors. *J Mol Biol.* 2000;295(4):979-995.
71. Kuroda D, Shirai H, Kobori M, Nakamura H. Systematic classification of CDR-L3 in antibodies: implications of the light chain subtypes and the VL-VH interface. *Proteins.* 2009;75(1):139-146.
72. North B, Lehmann A, Dunbrack RL Jr. A new clustering of antibody CDR loop conformations. *J Mol Biol.* 2011;406(2):228-256.
73. Chothia C, Lesk AM, Gherardi E, et al. Structural repertoire of the human VH segments. *J Mol Biol.* 1992;227:799-817.
74. Vargas-Madrazo E, Lara-Ochoa F, Almagro JC. Canonical structure repertoire of the antigen-binding site of immunoglobulins suggests strong geometrical restrictions associated to the mechanism of immune recognition.[erratum appears in *J Mol Biol* 1996 May 24;258(5):893]. *J Mol Biol.* 1995;254(3):497-504.
75. Almagro JC, Hernandez I, del Carmen RM, Vargas-Madrazo E. The differences between the structural repertoires of VH germ-line gene segments of mice and humans: implication for the molecular mechanism of the immune response. *Mol Immunol.* 1997;34(16-17):1199-1214.
76. Babor M, Kortemme T. Multi-constraint computational design suggests that native sequences of germline antibody H3 loops are nearly optimal for conformational flexibility. *Proteins.* 2009;75(4):846-858.
77. Shirai H, Kidera A, Nakamura H. Structural classification of CDR-H3 in antibodies. *FEBS Lett.* 1996;399(1-2):1-8.
78. Shirai H, Kidera A, Nakamura H. H3-rules: identification of CDR-H3 structures in antibodies. *FEBS Lett.* 1999;455(1-2):188-197.
79. Kuroda D, Shirai H, Kobori M, Nakamura H. Structural classification of CDR-H3 revisited: a lesson in antibody modeling. *Proteins.* 2008;73(3):608-620.
80. Schroeder HW Jr, Zhang L, Philips JB III. Slow, programmed maturation of the immunoglobulin HCDR3 repertoire during the third trimester of fetal life. *Blood.* 2001;98(9):2745-2751.
81. Link JM, Larson JE, Schroeder HW Jr. Despite extensive similarity in germline DH and JH sequence, the adult Rhesus macaque CDR-H3 repertoire differs from human. *Mol Immunol.* 2005;42:943-955.
82. Ivanov II, Link JM, Ippolito GC, Schroeder HW Jr. Constraints on hydropathicity and sequence composition of HCDR3 are conserved across evolution. In: Zanetti M, Capra JD, eds. *The Antibodies.* Vol 7. Taylor and Francis Group; 2002:43-67.
83. Zemlin M, Klinger M, Link J, et al. Expressed murine and human CDR-H3 intervals of equal length exhibit distinct repertoires that differ in their amino acid composition and predicted range of structures. *J Mol Biol.* 2003;334(4):733-749.
84. Schelonka RL, Ivanov II, Vale AM, et al. The CDR-H3 repertoire from TdT-deficient adult bone marrow is a close, but not exact, homologue of the CDR-H3 repertoire from perinatal liver. *J Immunol.* 2010;185(10):6075-6084.
85. Kyte J, Doolittle RF. A simple method for displaying the hydropathic character of a protein. *J Mol Biol.* 1982;157:105-132.
86. Eisenberg D. Three-dimensional structure of membrane and surface proteins. *Annu Rev Biochem.* 1984;53:595-623.
87. Ippolito GC, Schelonka RL, Zemlin M, et al. Forced usage of positively charged amino acids in immunoglobulin CDR-H3 impairs B cell development and antibody production. *J Exp Med.* 2006;203(6):1567-1578.
88. Schelonka RL, Ivanov II, Jung D, et al. A single DH gene segment is sufficient for B cell development and immune function. *J Immunol.* 2005;175:6624-6632.
89. Khass M, Blackburn T, Burrows PD, Walter MR, Capriotti E, Schroeder HW Jr. VpreB serves as an invariant surrogate antigen for selecting immunoglobulin antigen-binding sites. *Sci Immunol.* 2016;1(1):aaf6628.
90. Khass M, Blackburn T, Elgavish A, Burrows PD, Schroeder HW Jr. In the absence of central pre-B cell receptor selection, peripheral selection attempts to optimize the antibody repertoire by enriching for CDR-H3 Y101. *Front Immunol.* 2018;9:120.
91. Schelonka RL, Ivanov II, Vale AM, Dimmitt RA, Khaled M, Schroeder HW Jr. Absence of N addition facilitates B cell development, but impairs immune responses. *Immunogenetics.* 2011;63(9):599-609.
92. Carey JB, Moffatt-Blue CS, Watson LC, Gavin AL, Feeney AJ. Repertoire-based selection into the marginal zone compartment during B cell development. *J Exp Med.* 2008;205(9):2043-2052.
93. Nguyen HH, Zemlin M, Vu HL, et al. Heterosubtypic immunity to influenza A virus infection requires a properly diversified antibody repertoire. *J Virol.* 2007;81:9331-9338.
94. Shlomchik MJ, Marshak-Rothstein A, Wolfowicz CB, Rothstein TL, Weigert MG. The role of clonal selection and somatic mutation in autoimmunity. *Nature.* 1987;328:805-811.
95. Tomlinson IM, Walter G, Jones PT, Dear PH, Sonnhammer EL, Winter G. The imprint of somatic hypermutation on the repertoire of human germline V genes. *J Mol Biol.* 1996;256(5):813-817.
96. Goodyear CS, Silverman GJ. B cell superantigens: a microbe's answer to innate-like B cells and natural antibodies. *Springer Semin Immunopathol.* 2005;26(4):463-484.
97. James LC, Roversi P, Tawfik DS. Antibody multispecificity mediated by conformational diversity. *Science.* 2003;299(5611):1362-1367.
98. Li Y, Lipschultz CA, Mohan S, Smith-Gill SJ. Mutations of an epitope hot-spot residue alter rate limiting steps of antigen-antibody protein-protein associations. *Biochemistry.* 2001;40(7):2011-2022.
99. Lipschultz CA, Yee A, Mohan S, Li Y, Smith-Gill SJ. Temperature differentially affects encounter and docking thermodynamics of antibody–antigen association. *J Mol Recogn.* 2002;15(1):44-52.
100. Bhat TN, Bentley GA, Boulot G, et al. Bound water molecules and conformational stabilization help mediate an antigen-antibody association. *Proc Natl Acad Sci U S A.* 1994;91(3):1089-1093.
101. Levitt M, Park BH. Water: now you see it, now you don't. *Structure.* 1993;1(4):223-226.
102. Greenspan NS, Di Cera E. Defining epitopes: It's not as easy as it seems. *Nat Biotechnol.* 1999;17(10):936-937.
103. Li Y, Urrutia M, Smith-Gill SJ, Mariuzza RA. Dissection of binding interactions in the complex between the anti-lysozyme antibody HyHEL-63 and its antigen. *Biochemistry.* 2003;42(1):11-22.
104. Hawkins RE, Russell SJ, Baier M, Winter G. The contribution of contact and non-contact residues of antibody in the affinity of binding to antigen. The interaction of mutant D1.3 antibodies with lysozyme. *J Mol Biol.* 1993;234(4):958-964.
105. Patten PA, Gray NS, Yang PL, et al. The immunological evolution of catalysis. *Science.* 1996;271(5252):1086-1091.
106. Kramer A, Keitel T, Winkler K, Stocklein W, Hohne W, Schneider-Mergener J. Molecular basis for the binding promiscuity of an anti-p24 (HIV-1) monoclonal antibody. *Cell.* 1997;91(6):799-809.
107. Michaud GA, Salcius M, Zhou F, et al. Analyzing antibody specificity with whole proteome microarrays. *Nat Biotechnol.* 2003;21(12):1509-1512.
108. Pejchal R, Doores KJ, Walker LM, et al. A potent and broad neutralizing antibody recognizes and penetrates the HIV glycan shield. *Science.* 2011;334(6059):1097-1103.
109. McLellan JS, Pancera M, Carrico C, et al. Structure of HIV-1 gp120 V1/V2 domain with broadly neutralizing antibody PG9. *Nature.* 2011;480(7377):336-343.
110. Huber R, Bennett WS. Antibody-antigen flexibility. *Nature.* 1987;326(6111):334-335.
111. Stanfield RL, Zemla A, Wilson IA, Rupp B. Antibody elbow angles are influenced by their light chain class. *J Mol Biol.* 2006;357(5):1566-1574.
112. Lesk AM, Chothia C. Elbow motion in immunoglobulins involves a molecular ball-and-socket joint. *Nature.* 1988;335:188-190.
113. Roux KH, Monafo WJ, Davie JM, Greenspan NS. Construction of an extended three-dimensional idiotope map by electron microscopic analysis of idiotope-anti-idiotope complexes. *Proc Natl Acad Sci U S A.* 1987;84(14):4984-4988.
114. Boehm MK, Woof JM, Kerr MA, Perkins SJ. The Fab and Fc fragments of IgA1 exhibit a different arrangement from that in IgG: a study by X-ray and neutron solution scattering and homology modelling. *J Mol Biol.* 1999;286(5):1421-1447.
115. Dangl JL, Wensel TG, Morrison SL, Stryer L, Herzenberg LA, Oi VT. Segmental flexibility and complement fixation of genetically engineered chimeric human, rabbit and mouse antibodies. *EMBO J.* 1988;7:1989-1994.
116. Roux KH, Greenberg AS, Greene L, et al. Structural analysis of the nurse shark (new) antigen receptor (NAR): molecular convergence of NAR and unusual mammalian immunoglobulins. *Proc Natl Acad Sci U S A.* 1998;95(20):11804-11809.
117. Gopalakrishnan PV, Karush F. Antibody affinity. VI. Synthesis of bivalent lactosyl haptens and their interaction with anti-lactosyl antibodies. *Immunochemistry.* 1974;11(6):279-283.
118. Hornick CL, Karuch F. Antibody affinity. 3. The role of multivalance. *Immunochemistry.* 1972;9(3):325-340.
119. Phillips ML, Oi VT, Schumaker VN. Electron microscopic study of ring-shaped, bivalent hapten, bivalent antidansyl monoclonal antibody complexes with identical variable domains but IgG1, IgG2a and IgG2b constant domains. *Mol Immunol.* 1990;27(2):181-190.
120. Schneider WP, Wensel TG, Stryer L, Oi VT. Genetically engineered immunoglobulins reveal structural features controlling segmental flexibility. *Proc Natl Acad Sci U S A.* 1988;85(8):2509-2513.
121. Tan LK, Shopes RJ, Oi VT, Morrison SL. Influence of the hinge region on complement activation, C1q binding, and segmental flexibility in chimeric human immunoglobulins. *Proc Natl Acad Sci U S A.* 1990;87(1):162-166.
122. Shopes B. A genetically engineered human IgG with limited flexibility fully initiates cytolysis via complement. *Mol Immunol.* 1993;30(6):603-609.
123. Nezlin R. Internal movements in immunoglobulin molecules. *Adv Immunol.* 1990;48:1-40.
124. Schumaker VN, Phillips ML, Hanson DC. Dynamic aspects of antibody structure. *Mol Immunol.* 1991;28(12):1347-1360.
125. Valentine RC, Green NM. Electron microscopy of an antibody-hapten complex. *J Mol Biol.* 1967;27(3):615-617.
126. Wade RH, Taveau JC, Lamy JN. Concerning the axial rotational flexibility of the Fab regions of immunoglobulin G. *J Mol Biol.* 1989;206(2):349-356.
127. Brekke OH, Michaelsen TE, Sandin R, Sandlie I. Activation of complement by an IgG molecule without a genetic hinge. *Nature.* 1993;363(6430):628-630.
128. Bruggemann M, Williams GT, Bindon CI, et al. Comparison of the effector functions of human immunoglobulins using a matched set of chimeric antibodies. *J Exp Med.* 1987;166(5):1351-1361.
129. Lefranc MP, Lefranc G. *The Immunoglobulin FactsBook.* Academic Press; 2001.
130. Kuby J. *Immunology.* W.H. Freeman and Company; 1997.
131. Janeway CA Jr, Travers P, Walport M, Shlomchik M. *Immunobiology.* Garland Publishing; 1999.
132. Greenspan NS, Monafo WJ, Davie JM. Interaction of IgG3 anti-streptococcal group A carbohydrate (GAC) antibody with streptococcal group A vaccine: enhancing and inhibiting effects of anti-GAC, anti-isotypic, and anti-idiotypic antibodies. *J Immunol.* 1987;138(1):285-292.
133. Cooper LJ, Robertson D, Granzow R, Greenspan NS. Variable domain-identical antibodies exhibit IgG subclass-related differences in affinity and kinetic constants as determined by surface plasmon resonance. *Mol Immunol.* 1994;31(8):577-584.
134. Cooper LJ, Shikhman AR, Glass DD, Kangisser D, Cunningham MW, Greenspan NS. Role of heavy chain constant domains in antibody-antigen interaction. Apparent specificity differences among streptococcal IgG antibodies expressing identical variable domains. *J Immunol.* 1993;150(6):2231-2242.
135. Morelock MM, Rothlein R, Bright SM, et al. Isotype choice for chimeric antibodies affects binding properties. *J Biol Chem.* 1994;269(17):13048-13055.
136. Fulpius T, Spertini F, Reininger L, Izui S. Immunoglobulin heavy chain constant region determines the pathogenicity and the antigen-binding activity of rheumatoid factor. *Proc Natl Acad Sci U S A.* 1993;90(6):2345-2349.
137. Casadevall A, Scharff MD. The mouse antibody response to infection with Cryptococcus neoformans: VH and VL usage in polysaccharide binding antibodies. *J Exp Med.* 1991;174(1):151-160.
138. Pritsch O, Hudry-Clergeon G, Buckle M, et al. Can immunoglobulin C(H)1 constant region domain modulate antigen binding affinity of antibodies? *J Clin Invest.* 1996;98(10):2235-2243.
139. Pritsch O, Magnac C, Dumas G, Bouvet JP, Alzari P, Dighiero G. Can isotype switch modulate antigen-binding affinity and influence clonal selection? *Eur J Immunol.* 2000;30(12):3387-3395.
140. Torres M, May R, Scharff MD, Casadevall A. Variable-region-identical antibodies differing in isotype demonstrate differences in fine specificity and idiotype. *J Immunol.* 2005;174(4):2132-2142.
141. McLay J, Leonard E, Petersen S, Shapiro D, Greenspan NS, Schreiber JR. Gamma 3 gene-disrupted mice selectively deficient in the dominant IgG subclass made to bacterial polysaccharides. II. Increased susceptibility to fatal pneumococcal sepsis due to absence of anti-polysaccharide IgG3 is corrected by induction of anti-polysaccharide IgG1. *J Immunol.* 2002;168(7):3437-3443.
142. Kinoshita N, Ohno M, Nishiura T, et al. Glycosylation at the Fab portion of myeloma immunoglobulin G and increased fucosylated biantennary sugar chains: structural analysis by high-performance liquid chromatography and antibody-lectin enzyme immunoassay using Lens culinaris agglutinin. *Cancer Res.* 1991;51(21):5888-5892.
143. Dunn-Walters D, Boursier L, Spencer J. Effect of somatic hypermutation on potential N-glycosylation sites in human immunoglobulin heavy chain variable regions. *Mol Immunol.* 2000;37(3-4):107-113.
144. Tao MH, Morrison SL. Studies of aglycosylated chimeric mouse-human IgG. Role of carbohydrate in the structure and effector functions mediated by the human IgG constant region. *J Immunol.* 1989;143(8):2595-2601.
145. Wawrzynczak EJ, Cumber AJ, Parnell GD, Jones PT, Winter G. Blood clearance in the rat of a recombinant mouse monoclonal antibody lacking the N-linked oligosaccharide side chains of the CH2 domains. *Mol Immunol.* 1992;29(2):213-220.
146. Tsuchiya N, Endo T, Matsuta K, et al. Effects of galactose depletion from oligosaccharide chains on immunological activities of human IgG. *J Rheumatol.* 1989;16(3):285-290.

147. Krapp S, Mimura Y, Jefferis R, Huber R, Sondermann P. Structural analysis of human IgG-Fc glycoforms reveals a correlation between glycosylation and structural integrity. *J Mol Biol.* 2003;325(5):979-989.
148. Kaneko Y, Nimmerjahn F, Ravetch JV. Anti-inflammatory activity of immunoglobulin G resulting from Fc sialylation. *Science.* 2006;313(5787):670-673.
149. Anthony RM, Kobayashi T, Wermeling F, Ravetch JV. Intravenous gammaglobulin suppresses inflammation through a novel T(H)2 pathway. *Nature.* 2011;475(7354):110-113.
150. Amin AR, Tamma SM, Oppenheim JD, et al. Specificity of the murine IgD receptor on T cells is for N-linked glycans on IgD molecules. *Proc Natl Acad Sci U S A.* 1991;88(20):9238-9242.
151. Leibiger H, Wustner D, Stigler RD, Marx U. Variable domain-linked oligosaccharides of a human monoclonal IgG: structure and influence on antigen binding. *Biochem J.* 1999;338(2):529-538.
152. Wright A, Tao MH, Kabat EA, Morrison SL. Antibody variable region glycosylation: position effects on antigen binding and carbohydrate structure. *EMBO J.* 1991;10:2717-2723.
153. Coloma MJ, Trinh RK, Martinez AR, Morrison SL. Position effects of variable region carbohydrate on the affinity and in vivo behavior of an anti-(1-->6) dextran antibody. *J Immunol.* 1999;162(4):2162-2170.
154. Gala FA, Morrison SL. V region carbohydrate and antibody expression. *J Immunol.* 2004;172(9):5489-5494.
155. Parekh RB, Roitt IM, Isenberg DA, Dwek RA, Ansell BM, Rademacher TW. Galactosylation of IgG associated oligosaccharides: reduction in patients with adult and juvenile onset rheumatoid arthritis and relation to disease activity. *Lancet.* 1988;1(8592):966-969.
156. Tomana M, Schrohenloher RE, Koopman WJ, Alarcon GS, Paul WA. Abnormal glycosylation of serum IgG from patients with chronic inflammatory diseases. *Arthritis Rheum.* 1988;31:333-338.
157. Ohta Y, Flajnik M. IgD, like IgM, is a primordial immunoglobulin class perpetuated in most jawed vertebrates. *Proc Natl Acad Sci U S A.* 2006;103(28):10723-10728.
158. Hombach J, Tsubata T, Leclercq L, Stappert H, Reth MG. Molecular components of the B-cell antigen receptor complex of the IgM class. *Nature.* 1990;343:760-762.
159. Venkitaraman AR, Williams GT, Dariavach P, Neuberger MS. The B-cell antigen receptor of the five immunoglobulin classes. *Nature.* 1991;352(6338):777-781.
160. Campbell MA, Sefton BM. Protein tyrosine phosphorylation is induced in murine B lymphocytes in response to stimulation with anti-immunoglobulin. *EMBO J.* 1990;9(7):2519-2526.
161. Davis AC, Roux KH, Shulman MJ. On the structure of polymeric IgM. *Eur J Immunol.* 1988;18(7):1001-1008.
162. Davis AC, Roux KH, Pursey J, Shulman MJ. Intermolecular disulfide bonding in IgM: effects of replacing cysteine residues in the mu heavy chain. *EMBO J.* 1989;8(9):2125-2131.
163. Li Y, Wang G, Li N, et al. Structural insights into immunoglobulin M. *Science.* 2020;367(6481):1014-1017.
164. Feinstein A, Munn EA. Conformation of the free and antigen-bound IgM antibody molecules. *Nature.* 1969;224(226):1307-1309.
165. Svehag SE, Bloth B, Seligmann M. Ultrastructure of papain and pepsin digestion fragments of human IgM globulins. *J Exp Med.* 1969;130(4):691-705.
166. Griffiss JM, Jarvis GA, O'Brien JP, Eads MM, Schneider H. Lysis of Neisseria gonorrhoeae initiated by binding of normal human IgM to a hexosamine-containing lipooligosaccharide epitope(s) is augmented by strain-specific, properdin-binding-dependent alternative complement pathway activation. *J Immunol.* 1991;147(1):298-305.
167. Lutz C, Ledermann B, Kosco-Vilbois MH, et al. IgD can largely substitute for loss of IgM function in B cells. *Nature.* 1998;393(6687):797-801.
168. Cheng HL, Blattner FR, Fitzmaurice L, Mushinski JF, Tucker PW. Structure of genes for membrane and secreted murine IgD heavy chains. *Nature.* 1982;296(5856):410-415.
169. Maity PC, Datta M, Nicolo A, Jumaa H. Isotype specific assembly of B cell antigen receptors and synergism with chemokine receptor CXCR4. *Front Immunol.* 2018;9:2988.
170. Gutzeit C, Chen K, Cerutti A. The enigmatic function of IgD: some answers at last. *Eur J Immunol.* 2018;48(7):1101-1113.
171. Shan M, Carrillo J, Yeste A, et al. Secreted IgD amplifies humoral T helper 2 cell responses by binding basophils via galectin-9 and CD44. *Immunity.* 2018;49(4):709.e8-724.e8.
172. Chen K, Xu W, Wilson M, et al. Immunoglobulin D enhances immune surveillance by activating antimicrobial, proinflammatory and B cell-stimulating programs in basophils. *Nat Immunol.* 2009;10(8):889-898.
173. Zhai GT, Wang H, Li JX, et al. IgD-activated mast cells induce IgE synthesis in B cells in nasal polyps. *J Allergy Clin Immunol.* 2018;142(5):1489-1499 e23.
174. Damelang T, Rogerson SJ, Kent SJ, Chung AW. Role of IgG3 in infectious diseases. *Trends Immunol.* 2019;40(3):197-211.
175. Bruhns P. Properties of mouse and human IgG receptors and their contribution to disease models. *Blood.* 2012;119(24):5640-5649.
176. Klaus T, Bereta J. CH2 domain of mouse IgG3 governs antibody oligomerization, increases functional affinity to multivalent antigens and enhances hemagglutination. *Front Immunol.* 2018;9:1096.
177. Leatherbarrow RJ, Dwek RA. Binding of complement subcomponent C1q to mouse IgG1, IgG2a and IgG2b: a novel C1q binding assay. *Mol Immunol.* 1984;21(4):321-327.
178. Schumaker VN, Calcott MA, Spiegelberg HL, Muller-Eberhard HJ. Ultracentrifuge studies of the binding of IgG of different subclasses to the Clq subunit of the first component of complement. *Biochemistry.* 1976;15(23):5175-5181.
179. Bindon CI, Hale G, Bruggemann M, Waldmann H. Human monoclonal IgG isotypes differ in complement activating function at the level of C4 as well as C1q. *J Exp Med.* 1988;168(1):127-142.
180. Niwa R, Natsume A, Uehara A, et al. IgG subclass-independent improvement of antibody-dependent cellular cytotoxicity by fucose removal from Asn297-linked oligosaccharides. *J Immunol Methods.* 2005;306(1-2):151-160.
181. Natsume A, In M, Takamura H, et al. Engineered antibodies of IgG1/IgG3 mixed isotype with enhanced cytotoxic activities. *Cancer Res.* 2008;68(10):3863-3872.
182. Burton DR. Immunoglobulin G: functional sites. *Mol Immunol.* 1985;22:161-206.
183. Unkeless JC, Scigliano E, Freedman VH. Structure and function of human and murine receptors for IgG. *Annu Rev Immunol.* 1988;6:251-281.
184. Collins AM. IgG subclass co-expression brings harmony to the quartet model of murine IgG function. *Immunol Cell Biol.* 2016;94(10):949-954.
185. Yuan R, Clynes R, Oh J, Ravetch JV, Scharff MD. Antibody-mediated modulation of Cryptococcus neoformans infection is dependent on distinct Fc receptor functions and IgG subclasses. *J Exp Med.* 1998;187(4):641-648.
186. Garred P, Michaelsen TE, Aase A. The IgG subclass pattern of complement activation depends on epitope density and antibody and complement concentration. *Scand J Immunol.* 1989;30(3):379-382.
187. Michaelsen TE, Garred P, Aase A. Human IgG subclass pattern of inducing complement-mediated cytolysis depends on antigen concentration and to a lesser extent on epitope patchiness, antibody affinity and complement concentration. *Eur J Immunol.* 1991;21(1):11-16.
188. Diebolder CA, Beurskens FJ, de Jong RN, et al. Complement is activated by IgG hexamers assembled at the cell surface. *Science.* 2014;343(6176):1260-1263.
189. Uzgiris EE, Kornberg RD. Two-dimensional crystallization technique for imaging macromolecules, with application to antigen–antibody–complement complexes. *Nature.* 1983;301(5896):125-129.
190. Nimmerjahn F, Ravetch JV. Fcgamma receptors: old friends and new family members. *Immunity.* 2006;24(1):19-28.
191. Gribben JG, Guinan EC, Boussiotis VA, et al. Complete blockade of B7 family-mediated costimulation is necessary to induce human alloantigen-specific anergy: a method to ameliorate graft-versus-host disease and extend the donor pool. *Blood.* 1996;87(11):4887-4893.
192. van der Neut Kolfschoten M, Schuurman J, Losen M, et al. Anti-inflammatory activity of human IgG4 antibodies by dynamic Fab arm exchange. *Science.* 2007;317(5844):1554-1557.
193. Aalberse R. The role of IgG antibodies in allergy and immunotherapy. *Allergy.* 2011;66(suppl 95):28-30.
194. Mestecky J, Moro I, Kerr MA, Woof JM. Mucosal immunoglobulins. In: Mestecky J, Lamm ME, Strober W, eds. *Mucosal Immunol.* 3rd. Elsevier Academic Press; 2005:153-181.
195. Plaut AG, Wistar R Jr, Capra JD. Differential susceptibility of human IgA immunoglobulins to streptococcal IgA protease. *J Clin Invest.* 1974;54(6):1295-1300.
196. Kumar N, Arthur CP, Ciferri C, Matsumoto ML. Structure of the secretory immunoglobulin A core. *Science.* 2020;367(6481):1008-1014.
197. Sharpe AH, Fields BN. Pathogenesis of viral infections. Basic concepts derived from the reovirus model. *N Engl J Med.* 1985;312(8):486-497.
198. Zhaori G, Sun M, Faden HS, Ogra PL. Nasopharyngeal secretory antibody response to poliovirus type 3 virion proteins exhibit different specificities after immunization with live or inactivated poliovirus vaccines. *J Infect Dis.* 1989;159(6):1018-1024.
199. Carayannopoulos L, Hexham JM, Capra JD. Localization of the binding site for the monocyte immunoglobulin (Ig) A-Fc receptor (CD89) to the domain boundary between Calpha2 and Calpha3 in human IgA1. *J Exp Med.* 1996;183(4):1579-1586.
200. Dobo J, Kocsis A, Gal P. Be on target: Strategies of targeting alternative and lectin pathway components in complement-mediated diseases. *Front Immunol.* 2018;9:1851.
201. Roos A, Bouwman LH, van Gijlswijk-Janssen DJ, Faber-Krol MC, Stahl GL, Daha MR. Human IgA activates the complement system via the mannan-binding lectin pathway. *J Immunol.* 2001;167(5):2861-2868.
202. Zheng Y, Shopes B, Holowka D, Baird B. Conformations of IgE bound to its receptor Fc epsilon RI and in solution. *Biochemistry.* 1991;30(38):9125-9132.
203. Sutton BJ, Davies AM. Structure and dynamics of IgE-receptor interactions: FcεRI and CD23/FcεRII. *Immunol Rev.* 2015;268(1):222-235.
204. Garman SC, Wurzburg BA, Tarchevskaya SS, Kinet JP, Jardetzky TS. Structure of the Fc fragment of human IgE bound to its high-affinity receptor FcεRIα. *Nature.* 2000;406(6793):259-266.
205. Sanders LA, van de Winkel JG, Rijkers GT, et al. Fc gamma receptor IIa (CD32) heterogeneity in patients with recurrent bacterial respiratory tract infections. *J Infect Dis.* 1994;170(4):854-861.
206. Bolland S, Ravetch JV. Spontaneous autoimmune disease in FcγRIIB-deficient mice results from strain-specific epistasis. *Immunity.* 2000;13(2):277-285.
207. West AP Jr, Bjorkman PJ. Crystal structure and immunoglobulin G binding properties of the human major histocompatibility complex-related Fc receptor(,). *Biochemistry.* 2000;39(32):9698-9708.
208. Vaughn DE, Bjorkman PJ. Structural basis of pH-dependent antibody binding by the neonatal Fc receptor. *Structure.* 1998;6(1):63-73.
209. Martin WL, West AP Jr, Gan L, Bjorkman PJ. Crystal structure at 2.8 A of an FcRn/heterodimeric Fc complex: mechanism of pH-dependent binding. *Mol Cell.* 2001;7(4):867-877.
210. Kubagawa H, Oka S, Kubagawa Y, et al. Identity of the elusive IgM Fc receptor (FcμR) in humans. *J Exp Med.* 2009;206(12):2779-2793.
211. Maliszewski CR, March CJ, Schoenborn MA, Gimpel S, Shen L. Expression cloning of a human Fc receptor for IgA. *J Exp Med.* 1990;172(6):1665-1672.
212. Blum JH, Stevens TL, DeFranco AL. Role of the μ immunoglobulin heavy chain transmembrane and cytoplasmic domains in B cell antigen receptor expression and signal transduction. *J Biol Chem.* 1993;268(36):27236-27245.
213. Kaisho T, Schwenk F, Rajewsky K. The roles of γ1 heavy chain membrane expression and cytoplasmic tail in IgG1 responses. *Science.* 1997;276(5311):412-415.
214. Achatz G, Nitschke L, Lamers MC. Effect of transmembrane and cytoplasmic domains of IgE on the IgE response. *Science.* 1997;276(5311):409-411.
215. Pleiman CM, Chien NC, Cambier JC. Point mutations define a mIgM transmembrane region motif that determines intersubunit signal transduction in the antigen receptor. *J Immunol.* 1994;152(6):2837-2844.
216. Brandtzaeg P, Prydz H. Direct evidence for an integrated function of J chain and secretory component in epithelial transport of immunoglobulins. *Nature.* 1984;311(5981):71-73.
217. Rojas R, Apodaca G. Immunoglobulin transport across polarized epithelial cells. *Nat Rev Mol Cell Biol.* 2002;3(12):944-955.
218. Mestecky J, Schrohenloher RE. Site of attachment of J chain to human immunoglobulin M. *Nature.* 1974;249(458):650-652.
219. Mestecky J, Schrohenloher RE, Kulhavy R, Wright GP, Tomana M. Site of J chain attachment to human polymeric IgA. *Proc Natl Acad Sci U S A.* 1974;71(2):544-548.
220. Bastian A, Kratzin H, Fallgren-Gebauer E, Eckart K, Hilschmann N. Intra-and inter-chain disulfide bridges of J chain in human S-IgA. *Adv Exp Med Biol.* 1995;371A:581-583.
221. Frutiger S, Hughes GJ, Paquet N, Luthy R, Jaton JC. Disulfide bond assignment in human J chain and its covalent pairing with immunoglobulin M. *Biochemistry.* 1992;31(50):12643-12647.
222. Sorensen V, Sundvold V, Michaelsen TE, Sandlie I. Polymerization of IgA and IgM: roles of Cys309/Cys414 and the secretory tailpiece. *J Immunol.* 1999;162(6):3448-3455.
223. de Lalla C, Fagioli C, Cessi FS, Smilovich D, Sitia R. Biogenesis and function of IgM: the role of the conserved mu-chain tailpiece glycans. *Mol Immunol.* 1998;35(13):837-845.
224. Sorensen V, Rasmussen IB, Sundvold V, Michaelsen TE, Sandlie I. Structural requirements for incorporation of J chain into human IgM and IgA. *Int Immunol.* 2000;12(1):19-27.
225. Michelis R, Tadmor T, Aviv A, et al. Cell-free IgG-aggregates in plasma of patients with chronic lymphocytic leukemia cause chronic activation of the classical complement pathway. *PloS One.* 2020;15(3):e0230033.
226. Zikan J, Novotny J, Trapane TL, et al. Secondary structure of the immunoglobulin J chain. *Proc Natl Acad Sci U S A.* 1985;82(17):5905-5909.
227. Mole JE, Bhown AS, Bennett JC. Primary structure of human J chain: alignment of peptides from chemical and enzymatic hydrolyses. *Biochemistry.* 1977;16(16):3507-3513.
228. Randall TD, Brewer JW, Corley RB. Direct evidence that J chain regulates the polymeric structure of IgM in antibody-secreting B cells. *J Biol Chem.* 1992;267(25):18002-18007.
229. Petrusic V, Zivkovic I, Stojanovic M, Stojicevic I, Marinkovic E, Dimitrijevic L. Hexameric immunoglobulin M in humans: desired or unwanted? *Med Hypotheses.* 2011;77(6):959-961.
230. Mostov KE, Friedlander M, Blobel G. The receptor for transepithelial transport of IgA and IgM contains multiple immunoglobulin-like domains.. *Nature.* 1984;308:37-43.

231. Mostov KE, Blobel G. A transmembrane precursor of secretory component. The receptor for transcellular transport of polymeric immunoglobulins. *J Biol Chem.* 1982;257(19):11816-11821.
232. Kaetzel CS. The polymeric immunoglobulin receptor: bridging innate and adaptive immune responses at mucosal surfaces. *Immunol Rev.* 2005;206:83-99.
233. Perrier C, Sprenger N, Corthesy B. Glycans on secretory component participate in innate protection against mucosal pathogens. *J Biol Chem.* 2006;281(20):14280-14287.
234. Dallas SD, Rolfe RD. Binding of *Clostridium difficile* toxin A to human milk secretory component. *J Med Microbiol.* 1998;47(10):879-888.
235. Phalipon A, Cardona A, Kraehenbuhl JP, Edelman L, Sansonetti PJ, Corthesy B. Secretory component: a new role in secretory IgA-mediated immune exclusion in vivo. *Immunity.* 2002;17(1):107-115.
236. Shimada S, Kawaguchi-Miyashita M, Kushiro A, et al. Generation of polymeric immunoglobulin receptor-deficient mouse with marked reduction of secretory IgA. *J Immunol.* 1999;163(10):5367-5373.
237. Wijburg OL, Uren TK, Simpfendorfer K, Johansen FE, Brandtzaeg P, Strugnell RA. Innate secretory antibodies protect against natural *Salmonella typhimurium* infection. *J Exp Med.* 2006;203(1):21-26.
238. Sun K, Johansen FE, Eckmann L, Metzger DW. An important role for polymeric Ig receptor-mediated transport of IgA in protection against *Streptococcus pneumoniae* nasopharyngeal carriage. *J Immunol.* 2004;173(7):4576-4581.
239. Huang YT, Wright A, Gao X, Kulick L, Yan H, Lamm ME. Intraepithelial cell neutralization of HIV-1 replication by IgA. *J Immunol.* 2005;174(8):4828-4835.
240. Mazanec MB, Kaetzel CS, Lamm ME, Fletcher D, Nedrud JG. Intracellular neutralization of virus by immunoglobulin A antibodies. *Proc Natl Acad Sci U S A.* 1992;89(15):6901-6905.
241. Kaetzel CS, Robinson JK, Chintalacharuvu KR, Vaerman JP, Lamm ME. The polymeric immunoglobulin receptor (secretory component) mediates transport of immune complexes across epithelial cells: a local defense function for IgA. *Proc Natl Acad Sci U S A.* 1991;88(19):8796-8800.
242. Banks WA, Freed EO, Wolf KM, Robinson SM, Franko M, Kumar VB. Transport of human immunodeficiency virus type 1 pseudoviruses across the blood-brain barrier: role of envelope proteins and adsorptive endocytosis. *J Virol.* 2001;75(10):4681-4691.
243. Zhang JR, Mostov KE, Lamm ME, et al. The polymeric immunoglobulin receptor translocates pneumococci across human nasopharyngeal epithelial cells. *Cell.* 2000;102(6):827-837.
244. Lu RM, Hwang YC, Liu IJ, et al. Development of therapeutic antibodies for the treatment of diseases. *J Biomed Sci.* 2020;27(1):1.
245. Kang TH, Jung ST. Boosting therapeutic potency of antibodies by taming Fc domain functions. *Exp Mol Med.* 2019;51(11):1-9.
246. Dotti G, Gottschalk S, Savoldo B, Brenner MK. Design and development of therapies using chimeric antigen receptor-expressing T cells. *Immunol Rev.* 2014;257(1):107-126.

CHAPTER 22

Immunoglobulins: Molecular Genetics

Sebastian D. Fugmann, PhD

INTRODUCTION

Every mature naïve B cell in the circulation expresses a single unique immunoglobulin on the surface. Taking all unrelated mature naïve B cells in a human individual together, they represent a repertoire of antigen specificities that likely exceeds 1×10^{12}. Yet, it is obviously impossible to fit such a large number of separate immunoglobulin genes into the human genome with the approximate size being 3×10^9 bp. This conundrum was solved in the late 1970s by the discovery of V(D)J recombination by the Tonegawa laboratory,[1] a process in which the exon encoding the antigen-binding domain of each immunoglobulin chain is assembled from individual gene segments such that after transcription, splicing, and translation functional heavy and light chain proteins are produced that pair to form an immunoglobulin. This complex is either surface-bound or secreted, depending on the choice of splicing/polyadenylation in the last exon of the heavy chain locus in naïve or activated B cells. During immune responses, this immunoglobulin is further optimized by additional programmed DNA modification processes: class switch recombination (CSR) alters the Ig isotype, somatic hypermutation (SHM) facilitates optimizing its affinity for antigen, and in some select species, Ig gene conversion—a variant of SHM—diversifies Ig repertoires. Finally, it should not be surprising to anyone that the violation of genomic integrity that is integral to any of these endogenous DNA modification mechanisms brings with it the risk of genomic insults that are hallmarks of tumor cells. The central topics of this chapter are the human and mouse Ig gene loci, their overall structure and the nomenclature of the elements involved, the enzymatic machineries that modify these gene loci and their molecular mechanisms, the different transcripts that are arising from it, and the *cis*-regulatory elements and mechanisms that orchestrate these events.

Immunoglobulins consist of two different protein that are encoded by two gene loci, the immunoglobulin heavy chain (*IgH*) and immunoglobulin light chain (*IgL*) genes. While humans and mice carry only one *IGH* locus for all Ig isotypes (IgM, IgD, IgG, IgA, and IgE—see Chapter 21) and two alternatively used *IGL* loci, encoding Igκ and Igλ, these numbers differ in other vertebrates; every individual B cell, however, uses only one *IGL* gene locus at any given time to produce its immunoglobulins. The Ig genes are nonfunctional (ie, their transcripts do not encode a functional protein) in their germline configuration, as the single exons encoding the respective antigen-binding site of the IgH and IgL proteins are split up into numerous variable (V_H), diversity (D), and joining (J_H) segments, or V_L and J_L segments in case of the *IgH* and *IgL* genes, respectively. Each segment is flanked by conserved recombination signal sequences (RSSs) that direct the V(D)J recombination process. During B cell development in the bone marrow, a lymphocyte-specific enzyme complex, at the center of which are the products of the Recombination Activating Genes 1 and 2 (RAG1 and RAG2), catalyzes first the excision of the genomic DNA residing between each selected pair of gene segments (commonly referred to as the cleavage step). It subsequently also facilitates the joining of the DNA ends to restore genomic integrity. The stochastic choice of gene segments and the imprecision of the joining step (with nucleotides being added and deleted frequently) give rise to an enormous diversity of Ig gene sequences across the pool of immature B cells, with each individual B cell encoding its own unique Ig. This random repertoire of B cell specificities is subsequently shaped by selection processes to eliminate autoreactive clones (see Chapter 25).

B cells recognize antigen with their surface-bound Ig complexes, and with the help of follicular helper T (T_{FH}) cells become activated (for details see Chapter 26). One hallmark of this activation is turning on the expression of activation-induced cytidine deaminase (AID), an enzyme that initiates both, SHM and CSR, by replacing cytosines with uracils within transcribed *Ig* genes. The resulting U:G mismatches are subjected to nonfaithful (also called error-prone) repair—introducing point mutations in and around the original cytosine (in case of SHM) or DNA double-strand breaks (DSBs) in a pair of repetitive switch (S) regions within the *IgH* locus (in case of CSR). The latter concludes by the excision of the intervening DNA fragment and rejoining of the chromosomal DNA ends. This replaces the exons encoding the constant region of IgM with those of one of three other Ig isotypes, IgG, IgA, and IgE. Combined with selection for improved affinity, SHM and CSR results in the high affinity and isotype-switched immunoglobulins that are produced by plasma cells (terminally differentiated effector B cells). These secreted Igs are the key effectors during humoral immune responses and that provide long-term protection by neutralizing or collaborating with the innate immune system to destroy pathogens.

An individual B cell can produce two distinct forms of Igs, membrane-bound and secreted, with identical antigen recognition properties. The localization is solely determined by the nature of the IgH mRNA transcript. Alternative splicing of the primary transcript controls the inclusion of extra exons encoding a transmembrane domain and a short intracellular tail into the mature mRNA. Their presence anchors the IgH chain in the plasma membrane, and their absence results in Ig secretion, yet both exhibit the identical antigen specificity. The membrane-bound Igs are at the center of the B cell receptor (BCR) that transmits signals into B cells. Lastly, the secreted Igs (commonly named antibodies) are predominantly produced by plasma cells and are, as mentioned, key mediators of humoral immunity.

The endogenous damage that Ig genes incur during development and activation is reflected in frequent involvement in translocations and gene amplifications that result in malignant B cell phenotypes. In many cases, strong transcriptional control elements of Ig genes are brought in close proximity to oncogenes (eg, the t(8;14) *c-Myc*/*IgH* translocation in Burkitt lymphomas) and provide a growth advantage for such B cell clones. To limit the risk of malignancies, the programmed DNA modifications of Ig genes are controlled by intricate regulatory mechanisms, and the intensive studies of the respective *cis*-regulatory control DNA elements and their modes of action in the Ig loci shaped our understanding of gene and genome regulation in general. One central level of protection emerged as the RAG1/RAG2 transposase activity (discussed below) was lost because the excised Ig gene fragments are now disposed of as harmless excision circle instead of being integrated elsewhere in the genome.

In summary, Ig genes show enormous diversity that is generated both at the genomic and at the posttranscriptional level—this is in striking contrast to the fixed germline-encoded pattern recognition receptors of the innate immune system. Understanding this diversity is the key to the development of antibody drugs that are at the forefront of diagnosis and therapies for infectious diseases, cancers, and many other relevant diseases.

HISTORY

The starting point of the "molecular genetics of immunoglobulins" field was without doubt the discovery that somatic recombination alters the germline configuration of Ig genes to give rise to their rearranged functional state in B cells.[2] Subsequent cloning and DNA sequencing revealed that the exon encoding the antigen recognition site of Igs is initially split into individual gene segments—named variable (V), diversity (D), and joining (J)—and that these require assembly by a somatic gene diversification process thereafter called V(D)J recombination to give rise to functional Ig genes in mature B cells.

Shortly afterward, facilitated by the newly obtained Ig gene nucleotide sequence, the genetic basis of two additional Ig diversification processes, SHM and CSR, was solved. SHM had previously been characterized at the protein level by identifying differences in the amino acid sequences of clonally related Ig light chains.[3] Sequencing of the Ig genes of respective cell lines in the Tonegawa laboratory now confirmed that mutations in the nucleotide sequence of the VJ exon in the B cell genome were responsible for these amino acid changes.[4] A first model invoking a DNA recombination event as the molecular basis for class switching was proposed by the Honjo laboratory.[5] Subsequent competing studies in four laboratories led to the simultaneous discovery of the predicted DNA products in the IgH gene loci of cells that switch from expressing IgM to other isotypes.[6-9]

Ten years after the discovery of V(D)J recombination, the two genes encoding the lymphocyte-specific core of enzyme complex that catalyzes this process were cloned in David Baltimore's laboratory in 1988 as *RAG1* and *RAG2*.[10,11] It took, however, another 10 years until the initiator of both, CSR and SHM, was finally identified in 2000 as AID in Tasuku Honjo's and Anne Durandy's laboratories.[12,13]

Lastly, each of these five landmark discoveries initiated a flurry of follow-up studies whose numerous results culminate in the incredibly detailed molecular view and provide us now with a broad picture ranging from in-depth knowledge about their emergence in evolution, over intricate details of their mechanisms down to the atomic level, to the complex regulation in the living cells of all jawed vertebrates that are protected by these unique factors. In the following sections of this chapter, we do not trace the chronological order of discoveries but rather provide a comprehensive overview of the current state in the field.

IMMUNOGLOBULIN GENE LOCI

The most unique feature of Ig gene loci is the individual V, (D), and J elements that give rise to the exons encoding the antigen-binding site of the IgH and IgL subunits of each Ig molecule. Immediately after their discovery, numerous laboratories began to clone and sequence first parts, and then the entirety of the human and murine Ig loci using traditional molecular biology techniques prior to the genomics era, and the complete nucleotide sequence information of their germline configuration is now available as parts of the reference genomes at NCBI (and other reference genome sequence databases). There is a significant individual to individual variation in the Ig sequences, in terms of both the number and selection of gene elements. One example is the difference in the number of D elements (10 compared to 13) present in the two most commonly used laboratory mouse strains, C57BL/6 and BALBc, in the immunology community.

The international ImMunoGeneTics information system (http://imgt.org) includes the most comprehensive database for Ig and TCR genes from a large variety of species (currently 196,492 nucleotide sequence entries from 358 species) and includes maps, sequences, lists of chromosomal translocations, and numerous tools to assess and utilize the information.[14] Moreover, IgBLAST (www.ncbi.nlm.nih.gov/igblast/) is a nucleotide and amino acid sequence analysis online tool of the National Center for Biotechnology Information/National Institutes of Health (NCBI/NIH) that automatically identifies the V, D, and J elements of any mouse or human Ig transcript.[15]

The focus of the following sections is the structure and diversity of Ig loci in mice and humans. This includes not only the V, D, and J elements and their organization, and the exons encoding the constant region of the Igs but also noncoding elements that control both DNA recombination and expression of the Ig genes. For clarity, the description of murine IgH below is mainly focused on one standard inbred mouse strain for immunological studies, C57BL/6.

Immunoglobulin Heavy Chain Loci

Both mice and humans rely on a single gene locus to encode their IgH, and an overview of their structure and elements is shown in Figure 22.1. The murine *IgH* locus is located at the telomeric end of chromosome 12 and stretches over three megabases. The human *IGH* locus resides at the telomeric end of the long arm of chromosome 14 (14q32.33) and being 1.3 Mb in size is considerably smaller. Their overall arrangement including important *cis*-regulatory elements is similar, but relevant differences include the number and sequences of their germline V_H, D, and J_H elements and the variety of Ig isotypes they encode.

The Murine IgH Germline Gene Elements

V_H elements. The murine V_H region contains approximately 110 functional V_H elements and on the order of 85 nonfunctional V_H elements (also colloquially referred to V_H pseudogenes). All V_H elements are preceded by their own leader (L) exon encoding the first half of the signal peptide, carry a 23-RSS on their 3′ end, and their transcriptional orientation is identical to those of the J_H and C_H regions.[16] Based on sequence similarity, the murine V_H segments are classified into 16 distinct V_H families that can be further reduced into 3 ancient clans that already existed prior to the divergence of amphibians and reptiles.[17] This classification reflects a conservation primarily of their framework regions, forming the conserved structural backbone of the IgH variable region.

D and J_H elements. The murine D elements reside in a 100 kb region (depending on the mouse strain) located downstream of the most 3′ V_H element. The C57BL/6 mouse strain carries 10 D elements[18] (see Fig. 22.1) that are classified into four families (DSP2, DFL16, DST4, and DQ52). The D_H cluster contains a tandem array of seven 4.7kb repeats, with each repeat holding a single DFL16.1 or DSP element. All D elements are flanked by 12-RSSs on both sides (see Fig. 22.2), and thus the 12/23-rule ensures that all assembled V_HDJ_H exons (colloquially referred to as V exons) carry a single D within it. Although the usage of D segments in all three reading frames could theoretically increase Ig diversity, one of them (reading

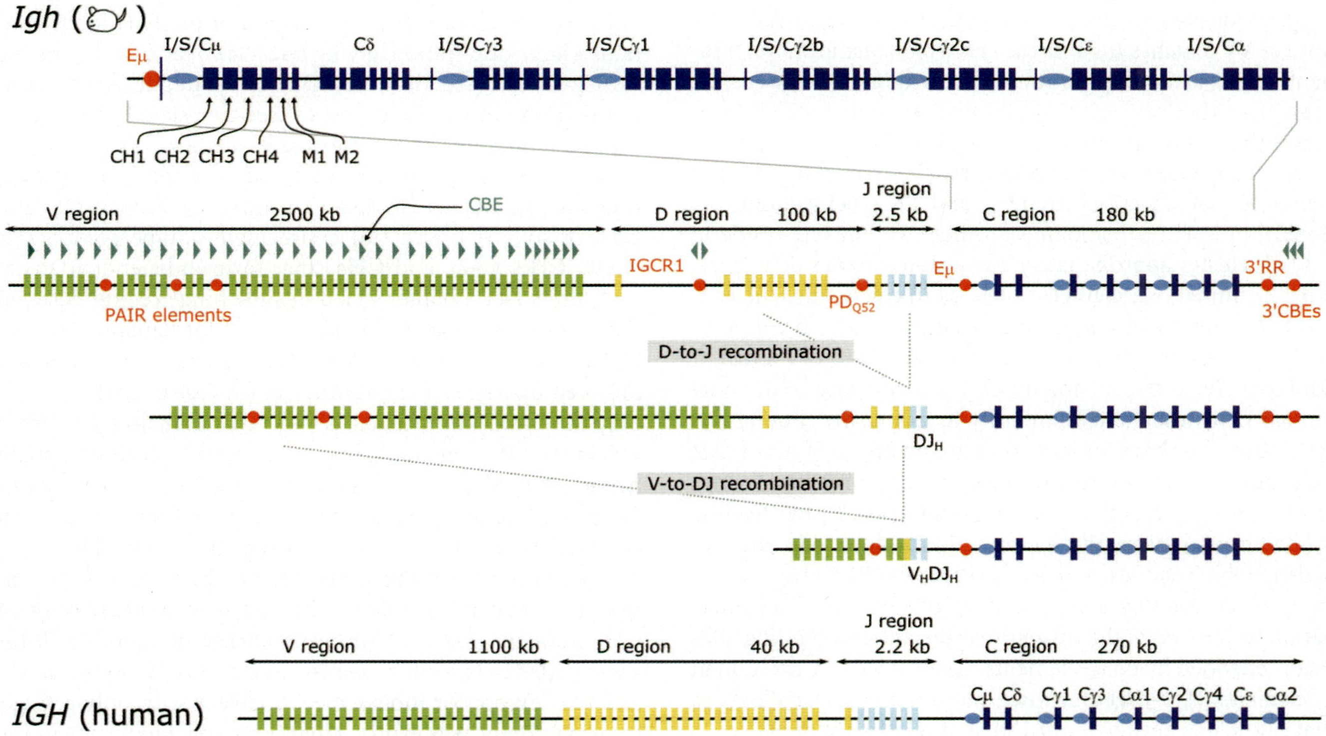

FIG. 22.1. The immunoglobulin heavy chain locus. This is a schematic overview of the structure and organization of the IgH locus of both mice and humans. The V region harbors numerous V_H elements (*green boxes*) and only some are shown here. While the number of J segments (*light blue boxes*) is comparable, human locus carries a much larger number of Ds (*yellow boxes*). The murine C region is shown enlarged above the locus to depict all exons (*blue boxes*, with systematic nomenclature shown for Cμ) encoding the different IgH isotypes, and this also includes the corresponding I exons (*blue lines*) and switch (S) regions (*blue ovals*) upstream of them that are essential for class switch recombination. V-to-J and the subsequent V-to-DJ recombination events excise large parts of the locus that end up as excision circles (not shown here). In the IgH transcripts, the V_HDJ_H exon gets spliced to the C exons to encode a functional IgH polypeptide. Important *cis*-regulatory elements that orchestrate this process in the mouse are shown as *red circles*, and CTCF-binding elements (CBEs; *green arrows*,) that organize the locus structure in three-dimensional (3D) space are distributed across the entire gene. (Note that this drawing is not to scale, and the sizes, locations, and numbers of many elements are only a schematic representation.)

		V		D		J		Combinatorial diversity
	Locus	RSS	Elements	RSS	Elements	RSS	Elements	
Human								
	Igκ	12▶	33			◀23	5	165
	Igλ	23▶	36			◀12	4-5	180
	IgH	23▶	45	◀12 12▶	25	◀23	6	6750
Mouse								
	Igκ	12▶	96			◀23	4	384
	Igλ	23▶	4			◀12	3	12
	IgH	23▶	25	◀12 12▶	10	◀23	4	1000

*Total combinatorial diversity: 2.3×10^6 for human Ig, 4×10^5 for mouse Ig

FIG. 22.2. **Diversity of immunoglobulin gene loci.** The recombination signal sequences (RSSs; *colored triangles*) coordinate recombination of the gene elements in the Ig loci of human and mouse. Randomly combining one of each elements within each locus produces a large amount of possible combinations. The total diversity is calculated by combining all possible IgHs with all possible light chains (Igκ + Igλ) in each species.

frame 2) is rarely found in B cells as its use gives rise to a nonfunctional truncated μ chain after D-J recombination.[19]

Lastly, the small 1.5 kb cluster of four J_H elements (J_{H1}-J_{H4}) resides about 700 bp downstream of the most 3′ D element, D_{Q52}. Each J_H carries a 23-RSS on its 5′ end and a splice donor site at its 3′ end to enable splicing of the recombined V_HDJ_H exon to the C region exons.

The Human IgH Germline Gene Elements

V_H elements. The human V_H region spans 1.1 Mb and contains approximately 45 V_H elements and a similar number of V_H pseudogenes. In terms of orientation and the 23-RSSs, the human V_H matches the murine elements. Furthermore, remnants of a partial duplication of the original *IGH* locus are present in the form of 24 additional nonfunctional V_H elements that reside on chromosomes 15 and 16.[20]

D and J_H elements. There is a much larger number of D elements, namely 27 (with $D_{1\text{-}14}$ and $D_{6\text{-}25}$ being nonfunctional), in humans compared to mice. All but one reside in a 40 kb cluster approximately 20 kb downstream of V_{H6}, the most 3′ human V_H element.[21] Just as in mice, this cluster is comprised of tandem duplications; here four 9.5 kb segments contain one representative of each of six distinct human D families. The 27th D element, D_{HQ52}, is not simply the only human D with sequence similarity to its murine counterpart (D_{Q52}) but also shares its location right upstream 5′ of the first J element, J_H1. In contrast to mice, the three reading frames of human D elements are more evenly used.[21] One reading frame encodes primarily hydrophilic residues, one encodes hydrophobic residues, and one includes frequent stop codons that are removed by end processing during V(D)J recombination to maximize the diversity of the antigen-binding sites that is encoded by these elements. Lastly, there are six human J_H elements (J_H1-J_H6) and three nonfunctional pseudogenes that reside in a cluster about 100 bp downstream of D_{HQ52}.

Heavy Chain Constant Regions

The IgH locus of C57BL/6 mice contains eight constant region (C_H) exon clusters (Cμ, Cδ, Cγ1, Cγ2b, Cγ2c, Cγ3, Cε, and Cα), each of which encodes the C_H of one of the corresponding Ig isotypes (IgM, IgD, IgG1, IgG2b, IgG2c, IgG3, IgE, and IgA) (see Fig. 22.1). They all share the same orientation and together span about 170 kb with the first Cμ exon residing just 5.7 kb downstream of the J_H4 element. In BALBc mice—the first mouse strain for which the C_H clusters were linked by contiguous phage clones in 1982[22]—the Cγ2c cluster is replaced by a different Cγ2 subtype, Cγ2a (encoding the constant region of IgG2a). Lastly, there are several nonfunctional Cγ clusters residing within the functional Cγ family cluster.[23]

The human IgH locus contains nine C_H exon clusters (Cμ, Cδ, Cγ3, Cγ1, Cα1, Cγ2, Cγ4, Cε, and Cα2) that were initially cloned from cosmid libraries,[24] and then eventually completely linked by the Human Genome Project. They are located 7 kb downstream of J_H6, the most 3′ J_H, and are spread out over 270 kb of DNA. This region of the *IGH* locus contains a large duplication, giving rise to two tandem copies of a Cγ-Cγ-Cε-Cα (see Fig. 22.1). One of the duplicated Cε clusters is also nonfunctional, and a sliced Cε-related pseudogene is present on human chromosome 9.

Exon structure. The constant region of the IgH polypeptide consists of two (mouse Cδ), three (γ, human Cδ, and Cα), or four (Cμ and Cε) consecutive C-type Ig domains (CH1, CH2, CH3, and CH4) that form independent structural units (see Chapter 21), a flexible hinge region between the first and second Ig domain (except for Cμ and Cε), and (for membrane-bound isoforms) a transmembrane domain followed by a short cytoplasmic tail (in case of IgM and IgD, only the three amino acids, KVK). This domain structure is reflected in their genomic organization, with each Ig domain being encoded by a separate exon, as are the hinge regions (with the human IgD and very long IgG3 hinge being even encoded by two and four exons, respectively) (see Fig. 22.1). Two exceptions are the hinge regions of human Cα1 and mouse Cα that are encoded contiguously with the subsequent CH2 domain. Lastly, the transmembrane domains of all Igs are encoded in two additional exons commonly labeled as M1 and M2 (except for mouse the Cα cluster with only a single transmembrane Mα exon). Alternative splicing results in the expression of either membrane-bound or secreted isoforms, and details are discussed in the penultimate section of this chapter.

I exons and switch repeat regions. As mentioned earlier in this chapter, changing the Ig isotype (with the exception of the reversible switch between IgM and IgD) requires CSR, a gene rearrangement process that removes the originally used Cμ/Cδ region exons and now places the target C region exons of either the Cγ, Cα, or Cε families downstream of the V_HDJ_H

exon. The recombination breakpoints typically reside within repetitive, G-rich DNA sequences known as switch (S) regions (also called repeats). They are located upstream of each C_H region (except for C_δ that lacks such region) and named after the C_H exons they precede (eg, Sμ upstream of the Cμ exons) (see Fig. 22.1). CSR requires transcription of both S regions involved, and hence there is a cytokine-inducible (I) promotor and I exon upstream of each S region. Together with the corresponding downstream C_H exon cluster, they form transcriptional units (frequently called Cμ, Cγ, Cα, and Cε genes). The respective Iμ, Iγ, Iα, and Iε transcripts originating from these units are the sterile (or "germline") switch transcripts with their the intronic S regions missing from the mRNAs.

The murine S regions are variable in size ranging from 1 kb (Sε) to up to 10 kb (Sγ1) and are composed of repeated core sequence elements. Sμ (4 kb) is built from tandem repeats of (GAGCT)n(GGGGT) with n ranging from 2 to 17. Moreover, the repeats of Sε and Sα (4.2 kb in length) are quite similar to that of Sμ but less perfect and contain numerous GGGGT and GAGCT motifs. The remaining S regions, Sγ3 (2.5 kb), Sγ1 (8 kb), Sγ2b (5.0 kb), and Sγ2a (2.5 kb), all share a different 49 bp core repeat containing AID hotspot motifs (AGCT) also present in Sμ, Sε, and Sα.

Important *Cis*-regulatory Elements

Transcription and V(D)J recombination of Ig gene loci are both controlled by *cis*-regulatory elements including classic promoters and enhancers. Here, we are focusing on murine elements as examples for these categories (see Fig. 22.1).

The IgH locus is rich in promoter sequences with individual promoters being present upstream of every leader exon preceding each V_H element, and these drive the transcription of functionally rearranged *IgH* genes. Additional elements with promoter activity (overlapping with the enhancers described below) are the PD_{Q52} element upstream of D_{Q52}, the Iμ promoter associated with the Eμ enhancer downstream of J_{H4}. Both promoters produce sterile transcripts that regulate V(D)J recombination. Lastly, there are the I promoters in front of each S region that orchestrate CSR in activated B cells.

The IgH locus contains three well-characterized enhancers: PD_{Q52} upstream of D_{Q52}, the intronic Eμ enhancer, and a cluster of hypersensitive sites (HS1-4), commonly referred to as the 3′ regulatory region (3′RR), at the very 3′ end of the IgH locus (see Fig. 22.1).

Apart from these classic control elements, there are novel classes of *cis*-regulatory sequences including CCCTC-binding factor (CTCF)-binding elements (CBEs). Two divergently oriented CBEs form the intergenic control region 1 (IGCR1), a cluster of ten CBEs, the 3′CBEs, resides within the HS5-8 region at the 3′ end of the locus, and more than a hundred CBEs are found throughout the V_H region. Moreover, 14 tandemly repeated PAX5-activated intergenic repeat (PAIR) elements are also located within the V_H region. All these elements act together with the enhancers imposing three-dimensional (3D) structures to the *IgH* locus that enable ordered gene rearrangements and prevent interference from neighboring genes.

Unusual IgH Loci and Gene Elements in Other Vertebrates

An increasing number of IgH loci in the genomes of other vertebrates have been sequenced, and these studies revealed a large variability in terms of their overall structure and the number and sequences of individual components (for details see Chapter 4). This reflects a remarkable level of genomic flexibility to implement receptor diversity enabled by the evolutionarily conserved somatic gene diversification machineries discussed here. The most remarkable structural divergences are (1) the ancient "cluster type" VDJ loci (V, D, and J elements are prealigned for recombination) in elasmobranchs that are dramatically divergent from the translocon organization (separate clusters for all Vs, Ds, and Js) in all other vertebrates including mice and humans,[25] (2) the emergence of the first (unusual) S regions in the ancestors of tetrapods, and (3) very unusual V_H elements and C_H exons that give rise to heavy chain-only IgH proteins in sharks (IgNAR) and HCAbs in camelids.[26,27] These unusual Igs are functional in the absence IgL chains, and this simplicity has a large appeal for biotechnological applications.[28]

Immunoglobulin Light Chain Gene Loci

In contrast to the single IgH locus, mice and humans both carry two Ig light chain loci, Igκ and Igλ (see Fig. 22.3), with no functional differences between these light chain types. Most mouse B cells express Igκ light chains (about 95%) while utilization of Igλ is much higher in human (about 40%) and other vertebrates. Igλ use largely correlates with the number of available Vλ elements.

Kappa Light Chain Genes

Murine Vκ elements. The murine Vκ locus shown in Figure 22.3 spans about 3.2 Mb on chromosome 6 and contains 96 functional Vκ and more than 40 nonfunctional Vκ elements.[29] They are each preceded by their own leader exon (encoding the signal peptide) and flanked by 12-RSSs. The Vκ elements can be grouped into 20 Vκ families, some of which are shared by humans and mice. Almost 60 functional Vκ elements lie in the opposite orientation to that of the Jκ and Cκ elements, and for these Vκ segments V(D)J recombination results in an inversion rather than deletion. A few Vκ sequences have been localized to chromosome 16 and 19 and are considered nonfunctional orphan elements.

Human Vκ elements. The human Vκ locus lies on the short arm of chromosome 2p11.2 spanning approximately 2 Mb of DNA. An exhaustive sequence analysis revealed that it harbors 132 Vκ sequences of which only 46 are functional with even less of these being used in mature B cells.[30] This locus includes a large inverted duplication, such that two copies (with >98% identity) of almost all Vκ sequences exist, with one copy lying in the cluster proximal to the Jκ elements and a second inverted copy in the distal cluster. This duplication occurred less than 5 million years ago as it is absent from the Igκ locus of chimpanzee (which diverged from the human lineage approximately 6 million years ago) and also

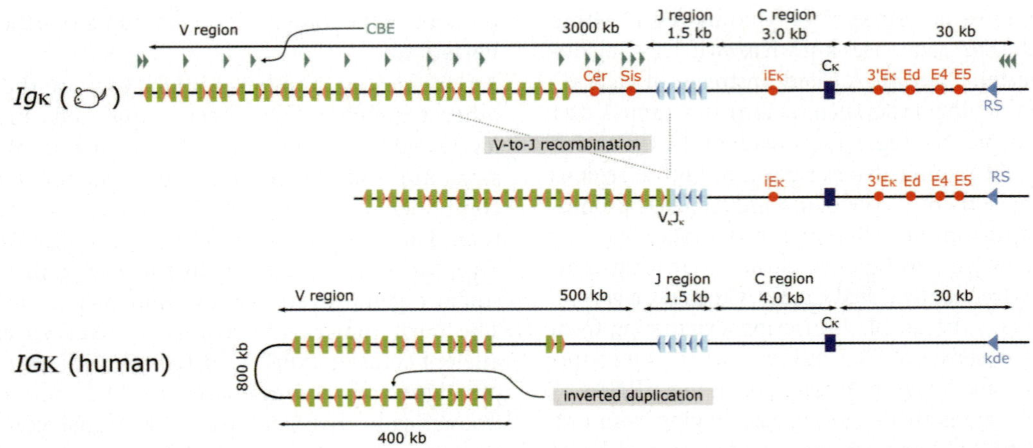

FIG. 22.3. **The Igκ gene loci.** The overall organization of the Igκ locus in mice and humans is shown. The gene elements and constant region exons are shown as boxes (V—*green*, J—*light blue*, C—*dark blue*) and many V elements are omitted for clarity. V-to-J recombination gives rise to a functional Igκ gene. Important *cis*-regulatory elements that orchestrate this process in the mouse are shown as red circles, and CTCF-binding elements (CBEs; *green arrows*) that organize the locus structure in 3D space are distributed across the entire gene. The V region of the human Igκ locus harbors a large inverted duplication. (Note that this drawing is not to scale, and the number and location of many elements are only a schematic representation.)

5% of human alleles. Lastly, at least 25 nonfunctional orphan Vκ elements reside in clusters on chromosomes 1, 2, and 22.

Jκ and Cκ elements. The mouse and human Igκ loci both contain a 1.5 kb cluster of five Jκ elements (flanked by 23-RSSs) residing 18 and 25 kb downstream of the Vκ region, respectively. Of these only mouse Jκ3 has not been observed in functional κ light chains. Compared to the IgH locus, the organization of the C region exons in the Igκ locus is extremely simple (see Fig. 22.3) with a single Cκ exon encoding the entire constant region.

Moreover, there is an isolated 23-RSS positioned downstream of the Cκ region that is named recombining sequence (RS) in the mouse[31] and kappa deleting element (kde) in humans.[32] Recombination involving RS or kde removes all Cκ exons and such events in both alleles frequently occur prior to the onset of V(D)J recombination in the Igλ locus.

Lambda Light Chain Genes

Murine Igλ locus. A complete sequence analysis[33] of the murine Igλ locus revealed two separate V-J-C clusters (Vλ2-Vλx-Jλ2-Cλ2-Jλ4-Cλ4 and Vλ1-Jλ3-Cλ3-Jλ1-Cλ1) about 110 kb apart (Fig. 22.4). Each Jλ element resides upstream its "own" Cλ region exon, but the Jλ4 segment is nonfunctional. V(D)J recombination occurs largely within each cluster, although light chains containing Vλ2 and Cλ1 sequences are occasionally observed. One of the three Vλ elements, Vλx, is unique in that it is more similar in sequence to Vκs. In contrast to the Vκ elements, all Vλ are flanked by 23-RSSs and the Jλ gene segments by 12-RSSs accordingly.

Human Igλ locus. The human Igλ locus is much bigger and covers about 1 Mb on chromosome 22q11,[34] and it contains about 70 Vλ elements (flanked by 23-RSSs) of which 36 are potentially functional. Each Jλ segment (flanked by a 12-RSSs) is paired with a Cλ exon, and seven such pairs are clustered within an approximately 33 kb region at the 3′ end of the locus (Fig. 22.4). The four major expressed human Igλ isotypes correspond to the functional Cλ1, Cλ2, Cλ3, and Cλ7 regions.[35] The Cλ4 and Cλ5 exons are inactive in all individuals, whereas a functional Cλ6 exon is present in a rare haplotype—the common variant has a 4 bp insertion leading to a prematurely truncated protein. Polymorphic variants of the human Jλ-Cλ region contain up to five additional Jλ-Cλ pairs as a result of gene duplication events.[36]

Important *Cis*-regulatory Elements

As Igλ-expressing B cells are rare in mice, we focus here on *cis*-regulatory elements in the *Igκ* locus (see Fig. 22.3). Just like in the *IgH* locus, individual promoters reside in front of each L exon-Vκ element pair, and they are responsible for the transcription in the functionally rearranged *Igκ* locus. In addition, a pair of germline promoters is located upstream of Jκ1 element which drive sterile κ0 transcripts to set up the locus for recombination at the pre-B cell stage. Moreover, five enhancers in the *Igκ* gene complete the group

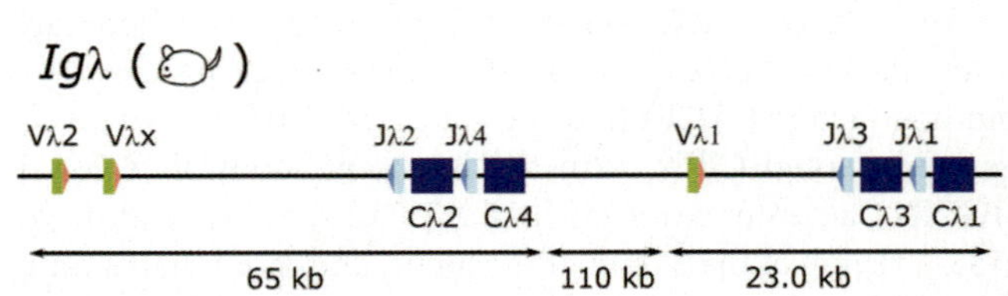

FIG. 22.4. **The murine Igλ gene locus.** This schematic overview depicts the location and organization of the V (*light green*) and J elements (*light blue*), as well as the constant region exons (*dark blue*) in the mouse Igλ locus. Recombination signal sequences (RSSs) are shown as *red* and *blue triangles*. Recombination occurs almost exclusively within each of the two clusters. (Note that this drawing is not to scale.)

of transcriptional control elements at the locus: the intronic enhancer (iEκ) residing between $J_{κ5}$ and the Cκ exon and four enhancers downstream of the Cκ, the 3′ enhancer (3′Eκ), the "distal" enhancer (Ed), and two recently discovered elements, E4 and E5.[37] While iEκ and 3′Eκ regulate V(D)J recombination, the downstream enhancers drive transcription and likely SHM in peripheral B cells.

The Igκ locus also contains approximately 60 CBEs that reside within the silencer in the intervening sequence (Sis—a recombination silencer in the intergenic region between Vκ and Jκ segments), in the newly described contracting element for recombination (Cer) immediately upstream of Sis, and in intergenic regions in the Vκ cluster. Furthermore, CBEs also act as insulators at the 5′ and 3′ boundaries of the locus.

Ig Light Chains in Other Species

In addition to the Igκ and Igλ isotypes originally identified in mice and humans, two additional Ig light chain families have been discovered in fish, Igσ and Igσ-cart. Comprehensive sequence analysis indicated that the four isotypes are ancient and existed as distinct genes for at least 450 million years.[38] Only elasmobranchs still carry gene for all four isotypes, while one or more isotypes have been lost from all other species.

Surrogate Light Chains

IgH chains can only be transported to the cell surface when paired with an appropriate IgL chain. Surprisingly pre-B cells, early B cell progenitors that have rearranged their IgH locus (see Chapter 25), express Igμ heavy chains on their surface despite their *IgL* loci being still in their germline configurations. Instead, two smaller proteins, VpreB1 (originally and commonly known as VpreB) and λ5, serve as a placeholder forming the "surrogate light chain" (SLC). They were identified and named based on their obvious similarities to V_L and C_L sequences.[39,40] Their function was demonstrated using an Ig-negative myeloma line in which Ig surface expression of a Igμ heavy chain was strictly dependent on simultaneously transfected λ5 and VpreB1 expression constructs.[41] In mouse, the SLC is encoded by closely linked genes (*Igll1* and *VpreB1*) on chromosome 16, and a second VpreB2 was discovered later.[42] In contrast, the human genome contains one functional λ5 (named 14.1) and two pseudogenes downstream of Cλ, and the single human VpreB homolog resides within the Vλ elements.[43] Importantly, the VpreB and λ5 genes are functional in their germline configuration and do not undergo V(D)J recombination.

Ig Gene Loci in 3D

In living cells, gene loci typically do not adopt a stretched out one-dimensional conformation shown in many schematic drawings (see eg, Fig. 22.1) but rather fold into complex 3D structures that are dictated by the limited space within the nucleus and interactions between *cis*-regulatory elements within each locus. CBEs and their *trans*-acting factor CTCF with its binding partner cohesin are main contributors to such structural arrangements. The chromosomal regions that form independent structural units are now referred to as topological-associated domains (TADs) and the *IgH* locus was one of the first gene loci for which this new concept has been intensively studied.

Central to these advances were improvement of both high-resolution imaging and high-throughput sequencing techniques named chromosome conformation capture (3C) and their 4C and 5C derivatives. The basis of the former is to simultaneously fluorescently tag multiple specific small regions of interest with the gene locus and to observe the spatial distribution of such marks in individual cells. Taking such a direct "image" of the locus conformation over many cells within a population can further inform about the locus dynamics. The basis of the latter is a population-based assessment of intra locus contacts that relies on fixing the locus confirmation by chemical crosslinking, subsequently breaking the DNA into small pieces, and finally ligation of DNA ends (which occurs preferentially for those in close spatial proximity). The identification of ligation products of sequences that are not collinear in the gene locus indicates the location of anchor regions that organize the folding of the gene locus. The current prevailing model is that each Ig locus forms an individual TAD, and that within these TADs, smaller subdomains exist that are altered during B cell development to allow for the controlled access of the somatic gene diversification machineries.

V(D)J RECOMBINATION

The mechanism by which germline V, D, and J gene segments are assembled to encode a functional Ig antigen-binding site has been pursued ever since Ig gene recombination was first discovered. In this section, we describe the substrates of the reaction, the components of the recombinase machinery, the molecular mechanism of the reaction, the topology of the recombination events, and the regulation of this process during B cell development.

To introduce important nomenclature that will be used, we begin with a brief overview of the molecular mechanism. V(D)J recombination is divided into two steps, cleavage and joining (see Fig. 22.5). First, the RAG1/2 complex binds to a pair of RSSs and cleaves the DNA precisely where the RSS are joined to their gene segments such that two hairpin-sealed coding ends (CEs; at the end of the gene segments) and two blunt signal ends (SEs; at the RSSs) arise. Subsequently, the two CEs are joined imprecisely (nucleotides are added and/or removed prior to connecting the DNA ends) to result in a coding joint (CJ), whereas the SEs are joined without further modification to emerge as a signal joint (SJ).

Recombination Signal Sequences

Upon the discovery of the V, D, and J elements themselves, it became immediately obvious that they are flanked by conserved short-sequence motifs in the germline configuration of the Ig loci and that such sequences were absent from rearranged V, D, and J elements. They were thus named recombination signal sequences and are the substrates of the V(D)J recombinase proteins RAG1 and RAG2. As the initial model for V(D)J recombination predicted, RSSs lie adjacent to each V, D, and J element.

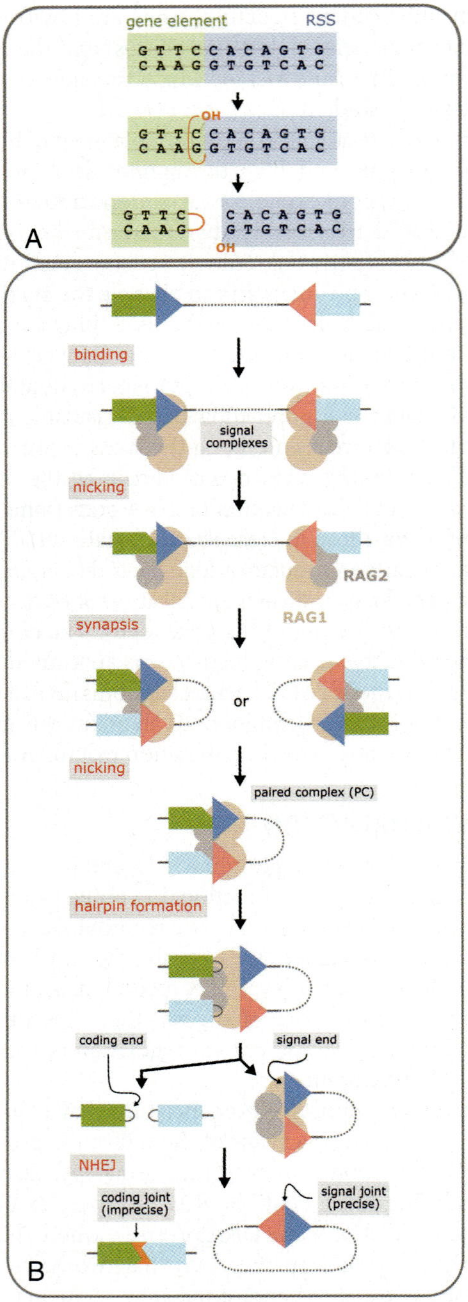

FIG. 22.5. The mechanism of V(D)J recombination. A: RAG1/2 cleave DNA in two steps. First one strand is nicked exposing a 3′-OH group, and this group subsequently performs a nucleophilic attack on the phosphodiester bond across from it. This results in a double-strand break with one DNA end showing a unique hairpin structure at its end. **B:** V(D)J recombination begins with binding of the RAG1/RAG2 complexes to their RSS substrates (*blue and red triangles*) and nicking occurs immediately afterward. Capturing a free recombination signal sequence (RSS) (following the 12/23-rule) results in the paired complex (PC). Nicking of the second RSS is followed by simultaneous hairpin formation of both RSSs, concluding the "Cleavage" phase. The hairpin-sealed coding ends (CEs) are rapidly released and end-processing and joining results in imprecise coding joints that harbor characteristic nucleotide additions and deletions. Signal ends (SEs) are ligated directly giving rise to precise signal joints (SJs) that typically reside on extrachromosomal excision circles. In both cases, the "Joining" phase is catalyzed by the NHEJ machinery.

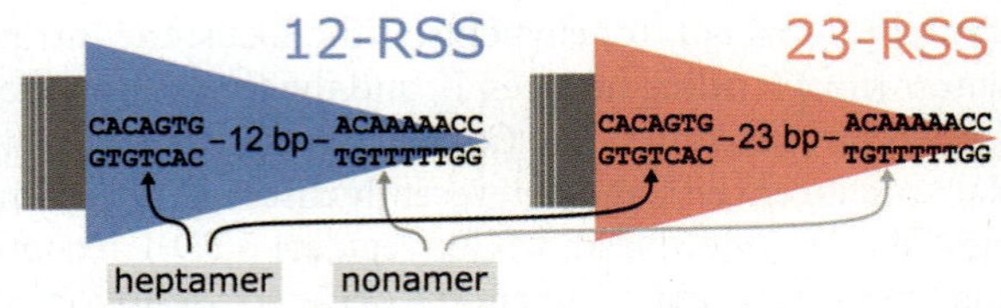

FIG. 22.6. The recombination signal sequence (RSS). RSSs are conserved recognition sites for the RAG1/RAG2 complex flanking each V and J segment at one end, and each D segment at both ends. They are commonly depicted as triangles with the blunt end facing the gene segment (*light gray box*). RSSs fall into two classes, 12-RSS (*red*) and 23-RSS (*blue*), depending on the length of the DNA spacer that separates the conserved heptamer and nonamer sequences.

RSSs consist of conserved seven base pairs (bp) long "heptamer" (consensus: CACAGTG) and nine bp long "nonamer" (consensus: ACAAAAACC) sequences that are separated by less well conserved spacers of either approximately 12 or 23 bp in length (see Fig. 22.6). Based on their spacer lengths, the RSSs are commonly referred to as 12-RSSs or 23-RSSs, respectively. Although almost all experiments to study the molecular mechanism employ artificial recombination substrates with "optimal" consensus RSSs, it is important to consider that no single naturally occurring RSS conforms to it. While the absolutely conserved nucleotides in the heptamer and nonamer are the major determinants of RSS function, there is an increasing body of evidence suggesting that spacer sequences can modulate recombination efficiencies and contribute to the nonrandom usage of human Vκ elements.[44] These observations led to the development of the RIC score, a single numerical value that can be calculated for a given sequence of 28 bp or 39 bp in length (corresponding to a 12-RSS or 23-RSS, respectively) that reflects their quality as a V(D)J recombination substrate.[45]

V(D)J recombination is directed by the spacer length of the RSSs and occurs only between RSSs of different spacer lengths, a requirement known as the "12/23-rule." Within each gene locus, all gene elements of one class (eg, all Vκs in the Igκ locus) carry RSSs with the same spacer length (see Fig. 22.2), largely preventing futile recombination products (eg, joining two Vκ or two Jκ elements). Thus, elements of different classes are flanked by RSSs of alternating spacer length to ensure the appropriate formation of functional V_HDJ_H and V_LJ_L exons.

In addition to RSSs, the RAG1/2 complex can also target heteroduplexes and unusual non-B form DNA structures. One example is a 150 bp region that resides within the major breakpoint region (Mbr) of the Bcl2 gene. *In vitro*, recombinant RAG proteins generate at least two independent nicks in a respective DNA substrate, thereby leading to a DNA DSB that *in vivo* could give rise to translocations between the IgH and the *Bcl2* locus occurring in most follicular lymphomas.[46]

The V(D)J Recombinase Enzyme Complex

Since the discovery of V(D)J recombination by the Tonegawa laboratory in 1976, a central question was the identity of the enzymatic machinery catalyzing this complex set of reaction steps (described in detail below, see Fig. 22.5). Genetic and biochemical work in a large number of laboratories led to the identification of a total of 13 different proteins that are directly involved in V(D)J recombination: RAG1, RAG2, HMG1, Ku70, Ku80, DNA-PKcs, Artemis, Pol μ, Pol λ, TdT, XLF (also named Cernunnos), XRCC4, and DNA ligase IV. The only lymphoid-specific factors are RAG1, RAG2, and TdT; all others are ubiquitously expressed in all cell types as general DSB repair factors, and this property allows investigators to study aspects of V(D)J recombination by ectopically expressing the RAG proteins in nonlymphoid cells. In 2008, a biochemical tour-de-force study demonstrated that cell-free CJ formation on artificial recombination substrates could be recapitulated with highly purified preparations of all 13 proteins.[47] The respective CJs showed all features observed *in vivo* (including the addition and deletion of nucleotides discussed in more detail below), suggesting that most, if not all, factors involved in CE processing have been identified. In contrast, SJ formation seems to require the removal of the RAG proteins after the cleavage reaction and this likely requires additional but as of yet unknown factors.

RAG Proteins: Mediators of Early Steps in V(D)J Recombination

Cloning. In pioneering experiments, David Schatz and his PhD thesis advisor David Baltimore conducted a genetic screen for the lymphoid-specific V(D)J recombinase enzyme.[48] The key tool was a nonlymphoid fibroblast reporter cell line that harbored an integrated artificial V(D)J recombination substrate that conferred drug resistance only upon V(D)J recombination. After the transfection of random libraries of human or murine genomic DNA fragments, only an extremely small number of cell clones grew under drug selection. One such murine genomic DNA fragment conferring recombinase activity was cloned and harbored two closely linked genes, *RAG1* and *RAG2*.[10,11] Both are indeed essential for recombination, and they would not have been discovered by this approach if they were not closely linked in the genome. First additional support for their importance in V(D)J recombination was provided by the striking evolutionary conservation of these genes and their structures. The gene locus was present in all jawed vertebrate species analyzed at that time and correlated perfectly with Ig- and TCR-mediated immunity. Recently, two deep sea fish species were discovered as the first species of vertebrates without functional RAG genes.[49] In these fish, the RAG genes were lost to allow for sexual parasitism by disabling adaptive immunity.

The key role of RAG1/RAG2 in the generation of functional Ig and TCR genes was validated in mouse strains lacking either gene (or carrying a catalytically inactive variant) which entirely lack mature B or T cells, as they are unable to initiate V(D)J recombination.[50-52] Similarly, a subset of human patients with null mutations in *RAG1/RAG2* suffer from a severe combined immunodeficiency (SCID) syndrome characterized by the complete absence of lymphocytes.[53] Moreover, hypomorphic alleles of RAG1/2 result in complex sets of clinical features (oligoclonal T cells, hepatosplenomegaly, eosinophilia, decreased serum immunoglobulin but elevated IgE) referred to as Omenn syndrome.[54]

Properties. Murine RAG1 consists of multiple modular domains including the sequence-specific DNA-binding domain (called the nonamer-binding domain, NBD) and the catalytic center of the recombinase complex. Murine RAG2, on the other hand, largely acts as a structural cofactor consisting of an N-terminal six-bladed β-propeller (also referred to as a WD40 repeat domain) that comprises the core of RAG2 essential for catalysis and a C-terminal plant homeo domain (PHD) zinc finger. Systematic deletion studies led to the "core" RAG proteins RAG1 (amino acids 384-1008) and RAG2 (amino acids 1-387), which display almost full catalytic activity on extrachromosomal substrates and can be readily purified as recombinant proteins. This culminated in the determination of a number of RAG1/RAG2 structures (with or without intact and pre-nicked RSSs) by Cryo-EM and X-ray crystallography[55-58] that largely confirmed a vast number of biochemical and exhaustive mutational studies. The recombinase complex consists of two RAG1/RAG2 heterodimers bound to a pair of RSSs. Each of the dimers acts largely in *cis*, with cleavage occurring at the RSS that is bound by the respective RAG1 NBD. The coordination of 12/23 couple cleavage is mediated by RAG1/RSS contacts in *trans* via the dimerization and DNA-binding domain in RAG1 that contacts the spacer of the opposite RSS. Lastly, to facilitate DNA cleavage, the dsDNA structure of the heptamer is severely distorted with first base pair of the being entirely flipped after the initial nicking to allow for hairpin formation (described below).

The N-terminal noncore region of RAG1 (amino acids 1-383) is required *in vivo* for optimal activity, and acts by modulating RAG1 stability,[59] by its interaction with chromatin loci,[60] and by promoting ubiquitination of histone H3.[61] Similarly, the noncore region of RAG2 (amino acids 388-527) is dispensable for catalysis *in vitro*, but acts as a key regulator for V(D)J recombination *in vivo*.[62] The PHD domain within this part of RAG2 binds specifically to the tails of histone H3 that are trimethylated at lysine 4 (H3K4me3),[63,64] and this is enhanced by the simultaneous presence of symmetrical dimethylation at arginine 2 (H3R2me2s).[65] In developing B cells, this interaction guides the RAG complex to Ig gene segments that are marked by this modification in a developmental stage–specific manner. It also alters the conformation of the RAG complex to boost its catalytic activity.[66] Finally, the noncore region of RAG2 ensures the proper fate of the SEs after cleavage, as its absence results in aberrant SJs,[67] and allows for RAG-mediated transposition.[68,69]

Transposition. *In vitro*, purified recombinant core RAG proteins can catalyze the excision of an RSSs-flanked DNA fragment and its subsequent insertion into an unrelated dsDNA target, acting as a transposase.[70,71] Even on using extremely sensitive reporter systems, such events can only rarely be detected in cell lines,[72] and recent cryo-EM structures of RAG proteins suggest that the arginine R848 evolved in vertebrates to suppress transposition.[73] Together these observations lend strong support for the early hypothesis that the V(D)J recombination system evolved from an ancient DNA transposon that integrated into a primordial antigen receptor exon, thereby generating a single pair of V and J gene segments which were flanked by the terminal inverted repeats of the transposon and now live on as RSSs.

The discovery of functional RAG1/RAG2-like gene pairs in echinoderms, however, suggests that the two RAG genes emerged far earlier than Ig/TCR-based adaptive immunity.[74] Importantly, an active transposon family encoding RAG1- and RAG2-like proteins is present in genomes of numerous invertebrates including some protostomes,[75-77] and studying these mobile DNA elements will likely bring us ever closer to the evolutionary origin of V(D)J recombination.

Regulation of RAG expression. RAG1 and RAG2 are expressed together in developing B and T cells, exclusively at the stages in which V(D)J recombination occurs. A comprehensive model for RAG gene expression would explain their lymphoid specificity, the two waves of RAG expression (during IgH and IgL rearrangements), and the autoantigen-induced upregulation associated with receptor editing. The *cis*-regulatory elements that regulate RAG expression have been characterized, and surprisingly, some of the elements regulating expression in B and T cells are distinct. Transcription of RAG genes is driven by individual promoters near their transcription start sites and occurs convergently in opposite directions. Five *cis*-regulatory B cell-specific enhancers have been identified, Erag (23 kb upstream of Rag2), R1B (5 kb upstream of Rag1), R2B (partially overlapping with Erag), D3, and Ep (residing in the RAG2 promoter region).[78,79] They are bound by an array of transcription factors (including Pax5, E2A, FoxP1, FoxO1, Ets1, IRF4, C/EBP, NFκB, c-myb, Lef1, and Ikaros) that mediate looping between these elements and thereby organize a locus structure and chromatin state that enables RAG expression.[80] Upon completion of IgL assembly at the end of the pre-B cell stage, RAG gene expression is terminated, but on occasion turned on again with the help of two transcription factors, NFκB and FoxO1, which facilitate receptor editing of autoreactive B cells.[81,82]

The protein levels of RAG2 across the cell cycle are regulated by cyclin-dependent kinase (cdk2)-mediated phosphorylation at T490 which induces ubiquitination and proteasomal degradation of RAG2 during S phase.[83] This finding is consistent with earlier observations that V(D)J recombination-associated DNA DSBs occur exclusively in the G0/G1 phase.[84] In contrast, RAG1 levels remain constant throughout the cell cycle, and modified mice expressing RAG2 T490A, a variant that cannot be phosphorylated and degraded, showed DSBs in Ig genes throughout the cell cycle, some of which resulted in chromosomal translocations.[85,86]

High Mobility Group Proteins

The search for RAG1/2 cofactors that stimulate cleavage activity on 23-RSSs *in vitro* led to the identification of high mobility group protein 1 (HMG1).[87] HMG1 (and the closely related HMG2) is an abundant ubiquitous protein that binds to dsDNA and bends it in a non-sequence-specific manner. The RAG complex associates stably with 12-RSSs, but efficient complex formation with a 23-RSS requires the addition of either HMG1 or HMG2.[88] In this scenario, HMG1/2 stabilizes the bending of the 23-RSS that is induced by the RAG proteins themselves.

The Nonhomologous End Joining Machinery

The RAG proteins are the essential lymphocyte-specific factors that mediate DNA cleavage at RSSs, but the central players in the subsequent joining phase are members of a DNA repair pathway known as nonhomologous end joining (NHEJ). This is the main pathway for repairing DNA DSBs in the G0/G1 phases of the cell cycle (this is the stage when RAG-mediated breaks are initiated), while homology-based repair pathways dominate in the S and G2 phases. Eight proteins represent the core components of NHEJ: Ku70, Ku80, DNA-PKcs, Artemis, DNA polymerase X family members, XLF, XRCC4, and DNA Ligase IV. In some scenarios, additional (and in part redundant) factors play a role in this repair pathway, but they are not essential for V(D)J recombination. Importantly, sensitivity to ionizing radiation, an exogenous source of DNA DSBs, is a property shared by mutants in any of the NHEJ factors and played a key role in their identification.

Ku70 and Ku80. The initial recognition of DNA DSBs is mediated by a heterodimer of Ku70 and Ku80 (also sometimes referred to as Ku86) which binds to ends of dsDNA irrespective of their nucleotide sequences. These proteins were originally identified as the autoantigens recognized by a patient antiserum and were named after the coded name of the patient (Ku) and their molecular weight of 70 kD and 80 to 86 kD, respectively. Once bound to DNA ends, the Ku complex forms the platform to recruit and assemble the NHEJ repair enzyme complexes. Ku70 and Ku80 mutant cell lines are defective in both SJ and CJ formation; and Ku70 and Ku80-deficient mice therefore exhibit a complete block in B and T cell development.[89,90]

DNA-PKcs. The DNA-dependent protein kinase catalytic subunit (DNA-PKcs) is a large protein (460 kD) that belongs to the phosphoinositide-3-kinase family. This kinase is only active when the protein is bound to dsDNA and a heterotrimer with Ku70/Ku80 is known as the DNA-PK complex. DNA-PKcs phosphorylates itself, Artemis (with which it forms a complex described below), and other proteins to orchestrate the DNA repair response. It was the first NHEJ component to be recognized as participating in V(D)J recombination when it was discovered that a mutation in DNA-PKcs causes the SCID in *scid* mice.[91-93] RAG-mediated cleavage and SJ formation occurs normally in lymphocytes with inactive DNA-PKcs,[94,95] but CJ formation and hence the generation of functional Ig and TCR genes is severely impaired. This emphasizes the role of DNA-PKcs in joining of DNA ends that required processing prior to joining—here the hairpin-sealed CEs at the ends of the gene elements.

Artemis. The Artemis protein is a nuclease that plays a key role in the processing and trimming of DNA ends prior to their ligation. It was discovered through positional cloning of the genetic defect in a group of human SCID patients with defects in V(D)J recombination and increased radiation sensitivity.[96] Hypomorphic Artemis mutations result in Omenn syndrome phenotypes similar to those caused by hypomorphic RAG1/2 alleles.[97] Artemis forms a complex with DNA-PKcs[98,99] and after phosphorylation it exhibits a single-strand endonuclease activity capable of opening the hairpin-sealed CEs.[100] Thus, the V(D)J recombination defects of Artemis null mutants former resemble that of DNA-PKcs mutants; SJs are formed normally while the ability to create regular CJs is impaired. Artemis (in complex with DNA-PKcs) can also cleave single-strand DNA near single-strand/double-strand transitions,[101] and as a purified recombinant protein has intrinsic exonuclease activity (independent of DNA-PKcs).

DNA Polymerase X family members: TdT, Polymerase μ, and Polymerase λ. Most DNA breaks repaired by NHEJ initially present as two noncompatible DNA ends (as is the case for CEs during V(D)J recombination), and thus, these DNA ends typically require processing before ligation can occur. In addition to nuclease digestion by Artemis (see above), this also involves the addition of low numbers of nucleotides by three members of the DNA polymerase X family: polymerase μ (Polm), polymerase λ (Poll), and Terminal deoxynucleotidyl Transferase (TdT). Common to all three enzymes is a BRCA1 C-terminal (BRCT) domain, which facilitates the efficient recruitment of these enzymes by the Ku complex that is bound to the DNA ends.[102]

TdT was originally purified as an enzyme with DNA polymerase activity from calf thymus. It is an unusual DNA polymerase as it catalyzes the nontemplated addition of random nucleotides (with a preference for Gs) to the 3′ end of DNA strands. This activity is exclusively directed toward CEs that are transiently generated during V(D)J recombination resulting in the appearance of random nontemplated nucleotides (named N nucleotides) in the CJs between the two gene segments. In the bone marrow, its expression is limited to the pro-B cell stage, and therefore, N region addition is common in IGH sequences (recombined in pro-B cells) but rare in murine Igκ and Igλ genes (recombined in pre-B cells). Importantly, TdT expression and thus also N nucleotide addition are characteristically absent from fetal lymphocytes.[103]

In a screen for novel DNA polymerases involved in Ig gene diversification, Polm and Poll were identified and cloned based on their sequence homologies to TdT and Polymerase β, respectively.[104] They are ubiquitously expressed DNA polymerases (with the highest levels of Polm observed in lymphoid tissues) that both readily fill in single-strand gaps in DNA, an activity important for DNA end processing and restoration of dsDNA prior to and after ligation of CEs during V(D)J recombination. In mice deficient in Polm, Igκ CJs show extended nucleotide deletion,[105] as is the case in the IgH genes of Poll-deficient mice.[106] The surprising selectivity reflects the relative transcript levels of these two polymerases at different stages of B cell development.

DNA Ligase IV, XRCC4, and XLF. The last step in V(D)J recombination (and NHEJ in general) is the joining of both, the (now compatible) CEs and the blunt SEs. The ligase essential for this step is DNA ligase IV,[107] and its cofactor XRCC4 that enhances the ligase activity significantly.[108] This DNA ligase IV-XRCC4 complex is recruited to the DNA ends by the Ku complex.[108,109] Deficiency in either of the two factors, XRCC4 or DNA ligase IV, is embryonically lethal, and this is associated with neuronal apoptosis. Human patients with ligase IV deficiency carry hypomorphic alleles and present with a severe clinical phenotype including chromosomal instability, developmental and growth retardation, microcephaly, radiosensitivity, and $T^-B^-NK^+$ immunodeficiency.[110,111] The V(D)J recombination defects vary in their magnitude, and the rare D_H-J_H junctions detected in one patient show extensive nucleotide deletion consistent with delayed ligation and prolonged exonuclease digestion.

Lastly, XLF, a small protein with weak sequence homology but clear structural similarity to XRCC4,[112] interacts with XRCC4 and thereby facilitates (but is not absolutely essential for) the synapsis of DNA ends to be joined by DNA ligase IV/XRCC4 complex. It was independently discovered by two laboratories in a yeast two-hybrid screen to search for XRCC4 interacting proteins[113] and (under its alternative name Cernnunos) in a cDNA expression screen for the gene causing a syndrome of T^+ B lymphocytopenia, increased radiosensitivity, and microcephaly in a Turkish family.[114] Testing V(D)J recombination on artificial extrachromosomal recombination substrates in XLF-deficient fibroblasts by ectopic expression of RAG1/2 revealed imprecise SJs reminiscent of those found in patients with hypomorphic DNA ligase IV mutations. Recently, another XRCC4 paralog, named PAXX, was discovered and subsequent experiments using genetically engineered mice suggest partial redundancy with XLF, with $XLF^{-/-}PAXX^{-/-}$ double-deficiency causing embryonic lethality[115] and lymphocytes with the same genotypes showing an inability to complete V(D)J recombination.[116]

Other Proteins Involved in V(D)J Recombination

In cell-free *in vitro* experiments, the classical core NHEJ proteins discussed above are necessary and sufficient to mediate the basic steps required for NHEJ. In living cells, however, a large number of additional DNA damage response (DDR) factors organize the cellular response to DNA breaks including the signals to halt cell division, to induce DNA repair, and in some cases, to trigger apoptosis. Some of these, including ATM (a kinase encoded by the gene mutated in the disease ataxia telangiectasia), Nbs1 (the product of the gene mutated in Nijmegen breakage syndrome), 53BP1 (p53 binding protein 1), and γH2AX, a phosphorylated form of the histone H2AX, are found to be associated with the DNA breaks generated during V(D)J recombination. Deficiency in any of these factors by themselves, however, does not lead to apparent defects in V(D)J recombination in humans or mice. It remains possible that they participate in backup DNA repair mechanisms to prevent aberrant V(D)J recombination and thus translocations.

The Mechanism

Recombination Model Overview

Several decades of work involving cell-free biochemistry with (recombinant) purified proteins, cell-based assays in both lymphoid and nonlymphoid cell lines, murine models, and material from human patients with SCIDs provides the basis for a comprehensive model of the molecular mechanism of V(D)J recombination (Fig. 22.5). Its perhaps most surprising feature is that the nucleotide sequences of the V, D, and J segments are completely irrelevant for any step of this process;

any RSS-flanked gene sequence will be acted upon, and this property is taken advantage of in commonly used artificial reporter substrates for V(D)J recombinase activities. Yet, their very sequences are what afterward provides the basis for the enormously diverse Ig repertoire of the human and mouse B cell pools.

Each recombination event begins with the assembly of a RAG1/RAG2 complex on a pair of RSSs that flank the two gene segments to be recombined. Next, the DNA is cleaved synchronously at both RSSs, precisely between the RSSs and the gene segments they flank. This results in four DNA ends, two SEs at the RSSs, and two CEs at the gene segments. Lastly, the precise joining of the two SEs gives rise to SJs while the two CEs must be processed before joining (except for homology-based joints), resulting in imprecise CJs which contribute enormously to Ig diversity.

The Cleavage Phase

The first step of the reaction is the binding of the RAG1/RAG2 complex to a compatible pair of RSSs. In cell-free systems, the complex settles first on one RSS, forming either a 12- or 23-signal complex and subsequently captures a matching partner RSS resulting in the paired complex (PC), the structure of which has been determined by Cryo-EM[55] and a X-ray crystal structure study.[56]

As shown in Figure 22.5, RAG-mediated cleavage occurs in two steps: first a nick is introduced on the top strand between each of the two gene segments and the heptamer of their flanking RSSs. This is followed by a direct transesterification reaction with the 3′-hydroxyl group serving as the nucleophile to attack the phosphodiester bond that still connects the bottom strands of the gene segment and the adjacent heptamer. This yields a hairpin structure on the CE and a blunt SE.[117] While nicking can occur asynchronously at the 12-RSS and 23-RSS, hairpin formation is "coupled" and occurs synchronously at both RSSs in the context of the PC, thereby establishing the 12/23-rule. This ensures the simultaneous formation of CEs on two compatible gene segments and thus contributes to the formation of undesired DNA recombination products including translocations. *In vitro* coupled cleavage requires only the RAG proteins, HMG1/2, and Mg^{2+} as the divalent metal ion in the reaction buffer, while Mn^{2+} enables single-site cleavage.[117]

Importantly, this mechanistic model for the cleavage phase established using cell-free assays largely reflects the order of events in developing B cells. Chromatin immunoprecipitation (ChIP) experiments in pro-B or pre-B cells of mice harboring a catalytic mutant of RAG1 show that RAG1/RAG2 complexes settle down first on an individual single RSS (the 23-RSS at the 5′ end of the J_H or Jκ elements, respectively) forming so-called "recombination centers."[52] Subsequently, appropriate partner 12-RSSs are captured involving a cohesin-mediated scanning process discussed below.[118] The RAG-mediated nicking on individual RSSs prior to the formation DSBs was demonstrated in a pre-B cell line using oligo capture and nick translation assays.[119] Why these nicks occur first on the Vκ 12-RSSs instead of the Jκ 23-RSS residing in the recombination centers is unclear. Lastly, given the prevalence of Mg^{2+} in cells, it likely also serves as the metal cofactor *in vivo*.

The Recombination Intermediates

Even prior to recombinant RAG proteins being available for biochemical studies of the cleavage phase, several laboratories employed ligation-mediated PCR (LM-PCR) strategies to study the nature of the transiently formed DNA ends as V(D)J recombination intermediates in living cells. Blunt SEs in the DNA obtained from cells undergoing V(D)J recombination were ligated to blunt double-stranded oligonucleotide linkers; the ligation junctions were amplified by PCR using a primer specific for the linker and a primer based on known genomic DNA sequences, and ultimately, cloned and sequenced. All SEs in the Ig and TCR loci identified with this technique mapped precisely to the heptamer end of RSSs,[84] but CEs failed to be detected with this approach.

Early Southern blot analyses had initially demonstrated two properties of CEs suggestive of a hairpin structure: (1) resistance to exonuclease treatments and (2) doubling of the apparent length of restriction fragments under denaturing electrophoresis conditions.[120] Hence, to successfully detect CEs by LM-PCR, samples require pretreatment with mung bean nuclease, a single-strand–specific endonuclease which recognizes and cleaves the distorted DNA structure at the tip of the hairpin. Sequences of respective LM-PCR products from CEs indicated that the hairpins are precisely at the end of the gene segment, usually without loss or gain of a single nucleotide.[121] The inability of detecting reaction intermediates in which the hairpin structures have been opened *in vivo* by other factors (described in the joining phase below) suggests that CJ formation occurs extremely rapidly after hairpin opening. Importantly, the properties of CEs and SEs formed *in vivo* are fully consistent with the reaction chemistry that was later on discovered in *in vitro* experiments with purified proteins and DNA substrates.

The Joining Phase

Immediately after DNA cleavage, the RAG proteins remain briefly in a complex together with all four DNA ends and subsequently facilitate aspects of the joining phase. This model is supported by the discovery of mutant forms of RAG1 or RAG2 that are competent for cleavage but show impairment in coding or signal joint formation.[68]

CEs. The first DNA ends to leave this postcleavage complex are the CEs, and they are rapidly bound by (or possibly handed over to) the ubiquitous Ku complex that binds to dsDNA ends independent of their sequence and structure. The Ku proteins then recruit the DNA-PKcs/Artemis complex, and DNA-PKcs is induced to phosphorylate itself[122] and Artemis. The autophosphorylation of DNA-PKcs induces conformational changes in both itself and Artemis, thereby activating this structure-specific nuclease. Next, Artemis nicks the DNA of both CEs at or near the DNA hairpin such that a 3′ overhang is generated.[100] If this reaction generates compatible ends (blunt or with complementary overhangs, like the ends generated by many restriction enzymes), these ends can be joined directly by the Xrcc4/DNA ligase IV complex without any additional processing.

Alternatively, and required when the ends are not compatible, end processing begins during which nucleotides are added by polymerases and deleted by Artemis from both CEs. This is initiated again by the Ku complex which now recruits additional template-independent (TdT) and template-dependent DNA polymerases (Polm and/or Poll). The importance of Ku proteins for this end-processing phase is supported by the sequence analysis of the extremely rare CJs formed in *Ku80*$^{-/-}$ mice with a characteristic reduced frequency of N nucleotides and high incidence of homology-mediated joining compared to wild-type mice.[123] Furthermore, analyses of CJs in TdT$^{-/-}$ mice revealed the complete absence of N nucleotides.[124] Conversely, when TdT was artificially expressed in cells undergoing Igκ or Igλ light chain rearrangement (a stage in development during which endogenous TdT is only present at low levels), the number of N nucleotides added to the V-J CJs was dramatically increased.[125] Lastly, the nucleotide composition of N nucleotides in CJs *in vivo* is biased toward G, reflecting the nucleotide preference observed for recombinant TdT in cell-free assays. Together, these reports support the notion that TdT is the primary enzyme for N nucleotide addition during V(D)J recombination.

The initial level of compatibility of most CEs is insufficient for direct ligation, but as few as one or two complementary nucleotides on the ssDNA overhangs (to whom TdT activity contributes) are sufficient to initiate the necessary additional nuclease processing and gap-filling by Polm or Poll to allow for ligation of the ends.

Lastly, Ku recruits the Xrcc4/DNA ligase IV complex via its BRCT domains to facilitate ATP-dependent end joining. XLF (and at times PAXX) can provide additional support at this step by bracing the two DNA ends together,[126] but the specific scenarios in which these factors are important to promote end synapsis are not understood. In most cases, one of the strands is joined first, and additional processing and filling in of potential gaps is required for ligation of the second strand to complete the CJ.

SEs. The SEs continue to form a stable RAG complex,[57] until the proteins are replaced by the Ku complex which directly recruits the Xrcc4/DNA ligase IV complex to ligate the compatible blunt ends. This occurs without any additional end processing and no N nucleotides are added to the SJs.

Signal Joints and Coding Joints

In the recombination products, SJs are almost exclusively direct ligation products of the SEs with the RSSs being joined precisely at the heptamers. In contrast, the hallmark of CJs is their "imprecise" nature with one or more of the following features:

1. Deletions—variable numbers of bases are deleted from one or both ends of the gene segments prior to joining.
2. Nontemplated (N) nucleotides—random nucleotides are added by TdT to the CEs with a slight bias toward G and C.
3. Palindromic (P) nucleotides—Artemis opens the hairpin structure of the CE by introducing a nick at some distance away from the tip. The resulting ssDNA overhang is filled in and now presents as a short DNA palindrome (one or two bp) mirroring the nucleotides at the end of the respective V, D, or J segment.

Topology of V(D)J Recombination

Deletion or Inversion

Each recombination event involves two gene segments and their flanking RSSs, and thus, there are two possible topologies of a standard joint formation, "deletion" or "inversion," that depend solely on the relative orientation of the two gene segments to each other. In the most common scenario, both gene segments share the same orientation (also most commonly depicted in schematic figures), resulting in a chromosomal CJ and an SJ located on an extrachromosomal excision circle—a deletional recombination event. If the gene segments, however, are in opposite orientation to each other like some Vκ and Jκ segments, the CJ and the SJ remain on the chromosome with the intervening sequence having been inverted—an inversional recombination event. The excision circles of deletional recombination events generally lack origins of replication, and these DNA molecules are thus diluted during subsequent B cell divisions. Importantly, they can be purified from cells[127] and serve as a telltale signs of recent V(D)J recombination events.

Nonstandard Joints

After RAG-mediated cleavage, two SEs and two CEs are joined to reinstate genomic integrity. The canonical outcomes are one SJ and one CJ, respectively. Alternatively, the CEs get joined either to their original partner SEs resulting in open-and-shut joints or the opposite SEs forming hybrid joints. While the latter can be easily identified as the gene segments are now flanked by opposing types of RSSs, the former are largely identical to the original state but can carry small nucleotide addition or deletions. Both alternative joints are present at low levels in the endogenous *Ig* loci.[128]

Regulation—Accessibility, Loop Extrusion, and Allelic Exclusion

V(D)J recombination is restricted to cells expressing RAG1/2, namely developing B and T lymphocytes. These proteins are the only lymphocyte-specific proteins required for this process, yet each set of gene elements largely recombines in the correct cell type—Ig elements in the B lineage and TCR elements in the T lineage (an exception is the D-J_H recombination in the IgH locus of early T cell progenitors)—and at the correct stage—IgH locus at pro-B cell stage followed by Igκ/Igλ in pre-B cells—and in the correct order—D to J_H precede V to DJ_H rearrangements (Fig. 22.7). The widely accepted explanation for this locus specificity was proposed more than three decades ago by the "accessibility" model.[129] This model states that only those gene segments programmed for recombination at a given stage of B and T cell development are "accessible" to the RAG recombinase. Strong experimental support of this model was provided by incubating purified nuclei from pro-B cells and pro-T cells with recombinant RAG proteins, resulting in DNA breaks at Ig J_H RSSs exclusively in the former and at TCRδ RSS exclusively in the latter, respectively.[130]

The main molecular correlate of accessibility is the chromatin state of the Ig and TCR gene loci. The smallest unit of chromatin is the nucleosome which consists of eight core

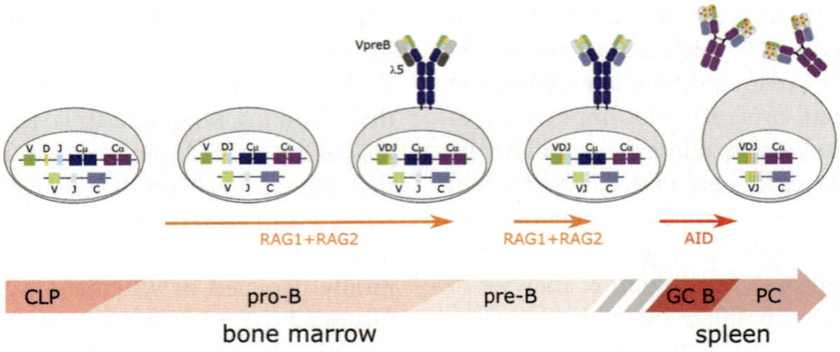

FIG. 22.7. Immunoglobulin diversification during B cell development. Starting from a common lymphoid progenitor (CLP), a B cell progresses through a series of developmental and activation stages to give rise to a terminally differentiated plasma cell (PC). The recombination status of the IgH locus and of one of the IgL loci are shown together with the immunoglobulin expressed at each stage. VpreB and λ5 are the components of the surrogate light chain. While RAG expression (*orange*) occurs in two waves in the bone marrow, a single wave of AID expression in activated B cells concludes in germinal center (GC) B cells. *Red lines* and *dots* indicate point mutations that increase the affinity of the antibody.

histones with two molecules of H2A, H2B, H3, and H4 forming a compact globular structure around which 146-bp DNA is wrapped. The tails of the histones are extending outward and can be decorated by a variety of posttranslational modifications (discussed below). *In vitro*, recombinant RAG proteins can readily bind and cleave RSS on nucleosome-free DNA, yet they are unable to execute the same actions when the DNA is wrapped around nucleosomes.[131,132] To allow for access, the nucleosomes are shifted or removed, in a process called chromatin remodeling. An additional layer of control is imposed by posttranslational modification of the histone tails, some of which regulate the tightness of DNA-nucleosome contacts. In the following section, we provide an overview about the mechanisms that regulate access to Ig gene loci and discuss some of the most relevant examples for each of them.

Transcription. Transcription typically occurs in "open" chromatin, and an emerging theme suggests that while initial locus "opening" precedes transcription, transcription per se also positively reinforces this chromatin state. As mentioned above briefly, transcription of individual elements within Ig gene loci correlates well with their availability for V(D)J recombination at that stage. In early pro-B cells when D to J_H recombination takes place, the sterile μ0 transcript starts upstream of D_{Q52} and proceeds all the way through the J elements,[133,134] while the promoters upstream of each V_H element are silent. After completion of D to J_H recombination, V_H sterile transcripts appear that correlate with the onset of V to DJ_H recombination events.[129] Similarly, sterile transcripts initiating upstream of the Vκ elements only start to appear at the pre-B cell stage once V(D)J recombination on the IgH locus has been completed. In addition to the different sterile IgH transcripts in the "sense" orientation (ie, in the same direction than the final IgH coding mRNA), there is also "antisense" transcription through the locus. This process is thought to represent the very first step of opening the chromatin in the IgH locus[135,136] but is also implicated in the repeat-mediated silencing of several D elements.[137] Lastly, it is still unknown whether transcription per se is essential for recombination or whether the correlation largely reflects the similar requirements for gaining access to chromatinized DNA.

Histone modifications. Posttranslational modifications of histone tails are epigenetic marks and important components of the chromatin state. Distinct modifications (and combinations thereof) are commonly referred to as the "histone code," and some such histone codes correlate well with actively transcribed and inactive (or repressed) gene loci.

These patterns hold true for Ig loci as well, active marks (eg, the acetylation of histone H3 and H4, H3Ac and H4Ac) and inactive marks (in particular the methylation of histones at selected lysines, H3K9me and H3K27me) correlate well with both, transcription and recombination accessibility.[138] The distribution of these histone mark in the Ig loci changes as B cell development progresses, thereby facilitating access selectively to the appropriate gene segments. Importantly, one particular mark for open chromatin, H3K4Me3, is directly linked to the RAG recombinase. As mentioned above, the PHD domain of RAG2 recognizes and binds to histone tails with this modification, and ChIP experiments reveal that the distribution of RAG2 in Ig loci matches exactly the pattern of H3K4Me3, a mark found on gene segments open for recombination.[52]

DNA methylation. Hypomethylation of CpG dinucleotides in mammalian DNA is associated with transcriptionally active genes, whereas hypermethylation at such sites is associated with transcriptional silence. Studies of the Igκ locus during B cell development revealed that DNA hypermethylation inhibits V(D)J recombination. At the transition from pro-B to pre-B cells, the Igκ locus becomes accessible for V(D)J recombination, and this correlates with a progression from being a hypermethylated, nontranscribed, and nonrearranging Igκ locus to one that is hypomethylated and transcribed.[139,140] In the IgH locus, CpG sequences surrounding the D (except for D_{Q52}) and J_H elements are hypermethylated and the DJ_H junction is selectively demethylated after recombination, thereby setting up the locus for the subsequent V-DJ rearrangements.[141] In contrast, a methylated artificial recombination substrate cannot undergo V(D)J recombination when it is transfected into B cells expressing the RAG1/RAG2 complex unlike its nonmethylated counterpart.[142] As the methylation of CpG sequences present within some RSS in many cases does not prevent recombination,[143] it appears that the closed chromatin status correlated with hypermethylation and it is not methylation per se what suppresses V(D)J recombination.

Chromatin remodeling. Beyond the modifications of DNA bases and histone tails that extend out from the nucleosomes around which the DNA is wrapped, the organization of the nucleosomes on the DNA is another layer of the chromatin state. The main enzymes that control this layer are the nucleosome remodeling complexes of the SWI-SNF family[144] which can promote accessibility by various mechanisms, including the removal or repositioning of nucleosomes. SWI-SNF complexes are bound to the IgH locus of murine pro-B cell lines, and depletion of SWI-SNF by RNAi led to a dramatical reduction in both, transcription and recombination.[136] Together with similar findings in TCR gene loci, these findings give rise to the model that chromatin remodeling enzymes are key regulators controlling the access of the RAG1/RAG2 recombinase to all Ig loci and facilitating their ordered assembly.

Subnuclear localization. In general, there is a correlation between the activity of a gene and its physical location in the nucleus. Inactive genes are found in the periphery of nuclei, while active genes reside in more central locations in so-called "transcription factories."[145] The position of the Ig loci changes during B cell development accordingly. Fluorescence in situ hybridization (FISH) studies with probes specific for Ig loci showed that the IgH and Igκ loci are located at the nuclear periphery in hematopoietic progenitors and pro-T cells but move to central areas of the nucleus in pro-B cells.[146] But whether the location per se contributes to accessibility is an open question.

The 3D structure of Ig loci. While accessibility allows the RAG proteins to selectively bind to appropriate sets of RSSs at each developmental stage, V(D)J recombination can only proceed if a pair of compatible RSSs comes into close physical proximity. This is facilitated by DNA looping, a process in which tethering of distant regulatory elements forms anchor points from which the intervening DNA emerges as loops, a structure often schematically depicted in 2D drawings as a rosette.[147] Such anchor points are distributed throughout the IgH locus and looping is mediated by a variety of DNA binding factors, especially CTCF that binds to pairs of convergently oriented CBEs[148] (see Fig. 22.8). These finding culminate in the RAG scanning and loop extrusion model shown in Figure 22.8.[118] After RAG1/2 forms a recombination center on the 23-RSS of a J_H, cohesion-mediated loop extrusion pulls the D region (beginning from its 3′ end) past the RAG complex until it latches onto a 12-RSS of a D segment. Loop extrusion is restricted to the D region by the tight tethering of a convergent CBE pair in the IGCR1 and the 3′CBE. Upon successful D to J_H rearrangement, the RAG complex now settles on the 12-RSS of the recombined DJ_H element and loop extrusion is reinitiated at various sites upstream of the IGCR1 to scanning for a matching 23-RSS on a V_H element. This model elegantly explains many aspects of V(D)J recombination *in vivo*, including a single preferred orientation of D elements during recombination (the 12-RSS on both ends theoretically allow for both orientations) and the order of recombination events in the IgH locus.

Mediators of accessibility and looping. All modes of regulation mentioned above are controlled by the *cis*-regulatory elements that are distributed throughout the Ig gene loci and respective *trans*-acting factors that bind to them. Enhancers play a major role in controlling the accessibility by two non-mutually exclusive mechanisms. First, they recruit transcription factors (some of which are lymphocyte-specific) that in turn recruit chromatin modifying and remodeling enzymes to open the chromatin structure of the Ig loci for transcription and recombination. Second, they serve as anchor points

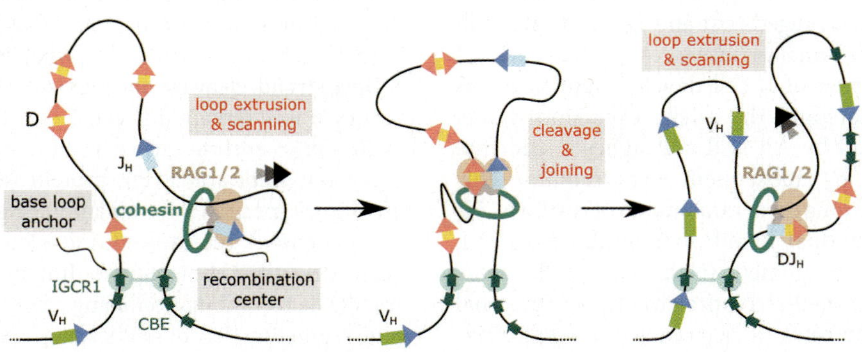

FIG. 22.8. The loop extrusion and scanning model of V(D)J recombination. First, a base loop anchor forms between convergently oriented CTCF-binding elements (CBEs) (*dark green arrows*) in the intergenic control region 1 (IGCR1) and downstream of the J_H region. Then, a RAG1/2 complex settles on a J_H 23-RSS (*blue triangle*), and cohesin pinches off a loop that is rapidly extruded, until a suitable 12-RSS (*red triangle*) is grabbed. Afterward the RAG complex settles on the 12-RSS at the 5′ end of the DJ_H segment and scanning for a V_H 23-RSS begins.

for loop formation with promoters and other *cis*-regulatory elements including CBEs of which are critical determinants of the 3D structure of Ig loci important for both V(D)J recombination and CSR.

Another important set of *cis*-regulatory elements (at times overlapping with enhancers), the most prominent of which are the numerous CBEs which can pair up over large distances when bound by CTCF, mediate looping. Loop formation is not only important for processes with in the Ig loci, but it is the basis of the TAD domains that insulate and uncouple Ig loci from neighboring genes. One prominent example for this is the cluster of CBEs downstream of the 3′RR.[149]

The Order of V(D)J Recombination and Allelic Exclusion

V(D)J recombination of Ig genes begins at the pro-B cell stage, with the first Ig gene rearrangements joining D to J_H segments (commonly occurring on both chromosomes). This is followed by V_H to DJ_H recombination, and events that produce a functional V_HDJ_H exon (ie, there is a contiguous open reading frame from the ATG in the leader exon until the stop codon in the last Cμ exon) result in the expression of the so-called preBCR, a functional μ heavy chain that is paired with the SLCs (see above). If the initial V_H to DJ_H rearrangement is nonfunctional (in cases where CE processing/joining introduced a frame-shift or a premature stop codon), a second V_H to DJ_H recombination occurs on the other allele. If the product of this event is also nonfunctional, then the cell has reached a dead end and is eliminated by apoptosis.[150]

The preBCR expression has two consequences. First, it shuts down RAG1/2 expression[151] and reduces the accessibility of unused V_H elements reflected by reduced transcription of these V_H.[152] In combination, this immediately prevents further IgH rearrangements and also renders the IgH locus inaccessible during subsequent rearrangement of the Igκ and Igλ loci at the pre-B cell stage. Second, it triggers the initiation of recombination of the IgL genes. Most B cells show isotypic exclusion of their Ig light chain; they express either Igκ or Igλ but not both. This is enforced at the level of V(D)J recombination; Igκ genes are rearranged first and Igλ only later, but what controls this order remains elusive.

The above description of B cell development provides the backdrop to understand the allelic exclusion model that was first proposed by Alt and colleagues[153] and that has been supported by subsequent experiments. This model states the functional rearrangement of an L (or H) chain gene in an individual B cell does inhibit further L (or H) chain gene rearrangement in the same cell. If this inhibition occurred immediately after the first functional rearrangement and the rates of rearrangement events are sufficiently low, then a single B cell would only encode a one IgH and one IgL chain and could never produce two immunoglobulins with different antigen specificities. An initial nonfunctional rearrangement has no inhibitory effect, so recombination continues until a functional immunoglobulin is produced or until the cell has used up all appropriate gene elements.

Secondary V(D)J Recombination

The imprecise joining of gene segments causes about two-thirds of all V(D)J recombination products to be out-of-frame (there is no contiguous open reading frame in the Ig mRNA), and hence, pre-B cells frequently end up with nonproductively rearranged Igκ genes. Subsequent secondary recombination events can now occur to obtain a functional $V_κJ_κ$ exon by (1) recombining unused upstream $V_κ$ and downstream $J_κ$ segments on the same allele, (2) $V_κJ_κ$ recombination on the other (allelic) copy of the Igκ locus (if it is still available), or (3) activating $V_λ$ to $J_λ$ recombination in the Igλ locus. These processes will be repeated until either the goal, a functional Igκ or Igλ gene, is achieved or all alternatives are exhausted. If the latter occurs or the IgL chain cannot pair with the IgH chain, the developing B cell will undergo apoptosis.

At times, despite successful V(D)J recombination, B cells expressing potentially deleterious autoreactive immunoglobulins arise frequently due to the unbiased gene segment choice. This is the second scenario in which secondary recombination occurs, and such events are referred to as V_H replacement (in the IgH locus) or receptor editing (in the IgL loci), respectively. Clear evidence for V_H replacement is found in about 5% of mature B cells in humans and mice,[154,155] while one estimate suggests that even large number of immunoglobulins, about 25%, are products of receptor editing.[156] Importantly, both processes require the reexpression of the RAG1/RAG2 gene pair, the mechanism of which remains unknown.

V_H replacement. In the IgH locus, secondary D-J rearrangements can occur but only before V_H–DJ_H recombination removes all unused upstream D_H segments. Once V_H to DJ_H recombination eliminates all cognate 12-RSSs (flanking D segments) from the IgH locus, there are no appropriate 12-RSSs present that could pair with the 23-RSSs flanking residual upstream V_H elements. In a process called V_H replacement, such V_H segments do recombine with an established V_HDJ_H unit, displacing most of the originally assembled V_H element.[157] This relies on cryptic RSSs present near the 3′ end of about 70% of all human and 90% of all murine V_H genes. An *in vitro* model suggests that in the context of V_H replacement, the RAG proteins nick on both DNA strands without forming a hairpin coding end,[158] but what controls such alternative double-strand cleavage mechanism and whether it actually occurs *in vivo* remain unclear.

Receptor editing. After production of an initial functional but undesirable Igκ protein (that when paired with IgH shows specificity for an auto-antigen or cannot pair with the IgH chain), receptor editing occurs by the same mechanisms described above for nonfunctional Igκ genes, secondary V(D)J recombination using as of yet unused $V_κJ_k$ or $V_λJ_λ$ gene segments. Recent studies of genetically engineered mice that harbor prerearranged IgH and Igκ sequences encoding autoantibodies in the respective Ig loci revealed complete replacement of the original $V_κJ_κ$ by de novo rearranged light chains, highlighting receptor editing as the major mechanism for B cell tolerance in these models.[159,160] Together with the data from V_H replacement studies, it now emerges that secondary recombination plays a larger role in "silencing" autoantibodies than the two alternatives, B cell anergy and B cell apoptosis, which are important later in the life of a B cell in the T1 and mature stages.

Ig Diversity Generated by VDJ Recombination

The outcome of V(D)J recombination is a highly diverse repertoire of immunoglobulins of the IgM isotype. The theoretical calculations of the maximal primary diversity first take into account the combinatorial diversity enabled by a random choice of gene elements and a random choice of the IgL type (Fig. 22.2) and reach about 4×10^5 in mice and 2.3×10^6 in humans. Including the junctional diversity of imperfect CJ formation, common estimates are in the range of 10^9 to 10^{11} possibilities in mice and humans, but these numbers are for obvious reasons far less accurate.

In vivo the diversity is much lower as additional forces shape the Ig repertoire: (1) some gene elements are rarely used, (2) not all possible IgH/IgL pairs can form a stable complex, (3) negative selection removes autoreactive clones from the repertoire, and (4) ongoing encounters with microorganisms support the outgrowth of individual B clones at the expense of clones with other specificities. The rapid development of high-throughput nucleotide sequencing technologies opened a new chapter in the field of immunoglobulin genetics as they begin to allow assessing the diversity of immunoglobulins encoded in the B lymphocytes of individuals. These experimental approaches are still far from being routine analyses; in particular, the bioinformatics strategies to process and transform the enormous number of nucleotide sequences into information that is easily accessible for physicians and scientists are still in their infancy. Yet, first important findings are already emerging including the size of immunoglobulin repertoires and public (ie, common to many individuals) responses against pathogens and allergens.

SOMATIC HYPERMUTATION AND CSR

After development in the bone marrow, B cells migrate to the follicles of secondary lymphoid tissues in which they may come into contact with antigen and helper factors from T cells. Several days after exposure to an antigen, these B cells begin additional maturation steps in newly emerging structures within the B cell follicles, the germinal centers (GCs) (see Chapter 26). During the GC response, antigen-specific B cells undergo cycles of proliferation and their Ig genes are again altered, this time by two unique processes: (1) SHM, a mutagenic process acting specifically on Ig gene loci, introduces random mutations in the V_HDJ_H and V_LJ_L exons and the surrounding regions, thereby altering the affinities of the encoded antibody for the antigen (Fig. 22.9). Combined with a selection for B cells producing antibodies with higher affinity for the respective antigen, this results in affinity maturation and ultimately the generation of plasma cells secreting high levels of antibodies (see Chapter 26) that produce high-affinity antibodies required for the clearance of pathogens. (2) CSR permanently alters the "class" (or isotype) of the antibody produced by an individual B cell from IgM to a new heavy chain isotype. It is a deletional DNA recombination process that proceeds via a pair of DNA DSBs in the

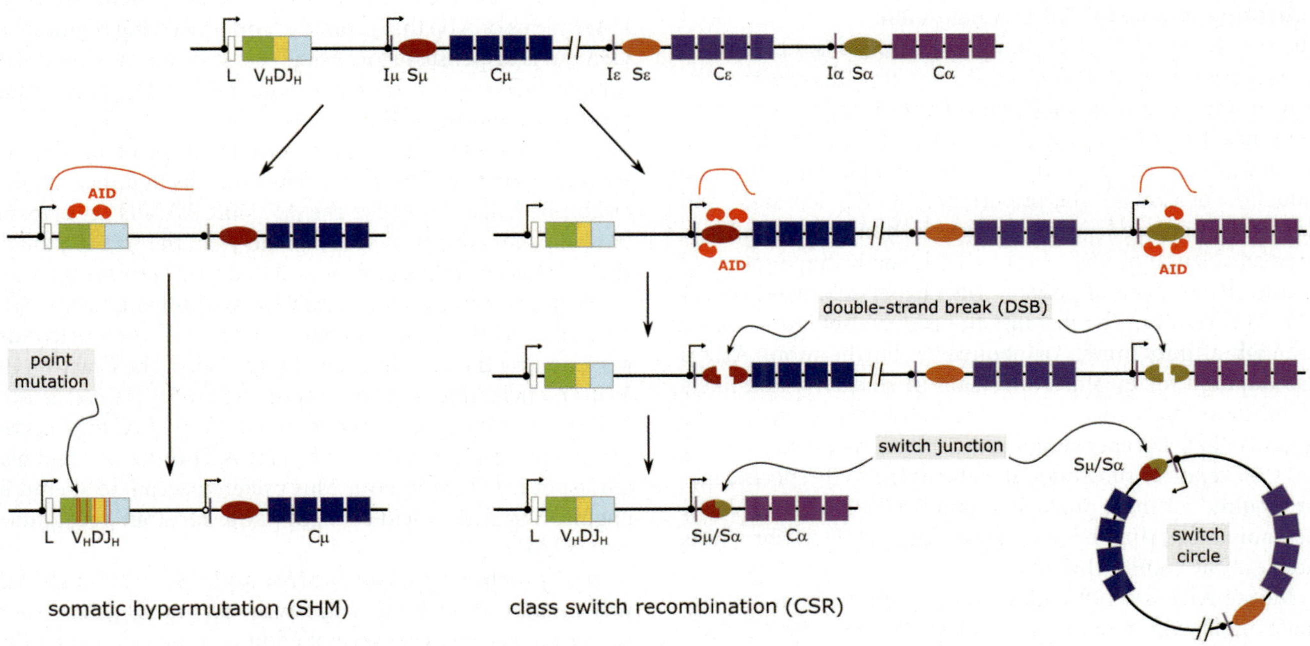

FIG. 22.9. Secondary immunoglobulin gene diversification. Different promoters (*black arrows*) upstream of the leader (L) and I exons are active and drive transcription through the VDJ exon (*green + orange + cyan box*) or selected S regions (*orange ovals*). AID activity is then focused on distinct transcribed regions (*red graphs*) to generate DNA lesions. In case of somatic hypermutation, error-prone DNA repair results in point mutations in the VDJ exon (*red lines*), while class switch recombination proceeds via DNA DSBs to give rise to two switch junctions, one in the IgH locus that now encodes IgA and one on an extrachromosomal switch circle.

Sμ "Switch" region, the repetitive noncoding DNA segment upstream of the C_{μ} exons, and a target S region upstream of the target C region. While the intervening DNA is lost from the chromosome, the chromosomal DNA ends are rejoined and the $V_{H}DJ_{H}$ exon now lies upstream of the Cx exons of the target isotype X (Fig. 22.9). All new *IgH* transcripts now encode antibodies with the same antigen specificity than before, but now of a new isotype. Typically, CSR occurs as a deletion within a single allele, and frequently, the nonfunctional *IgH* allele also switches to the same downstream isotype. On occasion, however, interallelic CSR takes place, linking the upstream Sμ region from one allele to a downstream S region on the other allele.[161]

For many years, CSR and SHM were considered as unrelated processes despite occurring in the same location, sharing the requirement of *Ig* gene transcription and an association with point mutations. The identification of a common enzyme, AID, as the sole initiator for both SHM and CSR and the subsequent delineation of their molecular mechanisms revealed surprising overlaps between them.

The Enzymatic Machinery

The mechanisms of SHM and CSR are closely related despite the distinct outcomes of these two processes. Hence, it is not surprising that they share numerous enzymes including a common initiator enzyme and DNA repair factors that handle mismatches in base pairing to alter the DNA sequences of immunoglobulin gene loci. This gives rise to a model in which the nature of the AID substrate, a $V_{H}DJ_{H}$/VLJL exon or a S region, governs the choice between these processes.

Activation-Induced Cytidine Deaminase

The very first step of SHM and CSR processes is mediated by the same enzyme, AID. The cDNA encoding this enzyme (the official gene symbol is *Aicda*, for activation-induced cytosine deaminase) was first cloned in a screen for transcripts that are strongly upregulated in CH12F3-2 lymphoma cells upon induction of CSR.[162] The importance of AID for SHM and CSR was discovered simultaneously in humans and mice: a targeted ablation of *Aicda* in mice and inactivating mutations in the *AICDA* gene of patients with hyper-IgM syndrome 2 (HIGM2) result in the inability to produce isotype-switched and high-affinity immunoglobulins.[12,13] Furthermore, AID is also required for Ig gene conversion in chicken[163] and likely all other animals that utilize this SHM-related process to diversify their Ig repertoire. Lastly, at least on artificial SHM and CSR reporter substrates, it is the only B cell–specific factor required for these processes (akin RAG1/RAG2 for V(D)J recombination) since ectopic expression of AID in fibroblasts induces CSR[164] and SHM.[165]

Mouse AID is a 198-amino-acid protein with an approximate molecular weight of 24 kDa. At the time of cloning, the only known protein with sequence similarity (34% amino acid identify) to AID was the apolipoprotein B (apoB) mRNA-editing enzyme catalytic polypeptide 1 (APOBEC1), a bona fide RNA editing enzyme which catalyzes the deamination of a cytosine residue to uracil in a specific position in the mRNA encoding apolipoprotein B (apoB).[166] Since then, three additional closely related proteins have been identified, APOBEC2, APOBEC3s, and APOBEC4, which together founded the AID/APOBEC cytidine deaminase superfamily, that now also includes related enzymes from jawless vertebrates[167] and invertebrates.[168,169]

The gene encoding AID resides on chromosome 6 in mice and at chromosome 12p13 in humans and is neighbored by the related *Apobec1* gene in both species, reflecting an ancient gene duplication event. Furthermore, AID is highly conserved in terms of size and amino acid sequence throughout the entire jawed vertebrate lineages, and its active site motif (H[A/V]E—$X_{(24-36)}$—PCXXC) is even shared with all other members of the vertebrate AID/APOBEC enzyme family.[170] Two crystal structures of truncated mouse AID with or without DNA[171,172] not only support these close relationships but also explain the lack of enzymatic activity in mutant AID HIGM2 patients. The structures also reveal two distinct nucleic acid-binding regions on opposing sides of AID, including one binding channel that positions the ssDNA substrate for the active site. Structure-function studies showed that the positively charged second polynucleotide binding site (named "assistant patch") has regulatory properties and is required for full enzymatic activity on DNA substrates with ssDNA/dsDNA transitions[172,173] like the G4 structures occurring during S region transcription. Notably, the assistant patch can bind both DNA and RNA, and thus explains early anecdotal reports about a presence and requirement of RNA in biochemical AID activity assays that have recently been experimentally substantiated.[174,175] Lastly, some HIGM2 patients carry defects in a small noncatalytic 17-amino-acid region (not covered by the X-ray crystal structures) at the C-terminus of AID that promotes end processing required for CSR but is dispensable for SHM,[176-178] a notion consistent with unique cofactor recruitment or regulation of DNA repair factors by AID during CSR.[179]

AID substrates. Studies in a wide range of model systems ranging from cell-free *in vitro systems* to mouse models established that AID (like the APOBEC3s) acts on cytosine bases in the context of ssDNA, and thus the initial thought that AID is an RNA editor was abandoned. Importantly, not every single C nucleotide in ssDNA is deaminated with equal efficiency, as the local sequence context (ie, the nucleotides surrounding the C) influences AID activity. The C within the WRCY nucleotide sequence motif—where W (Weak) is A or T, R (puRine) is A or G, and Y (pYrimidine) is C or T—is the preferred substrate for recombinant AID *in vitro*[180] and also a hotspot for SHM *in vivo*. This preference can be altered by engineering amino acid changes in the substrate recognition loop of AID.[181-183]

AID functions outside of SHM and CSR. Although AID was initially thought to exclusively act on *Ig* genes, more recent reports describe several additional beneficial and deleterious roles for this protein, even outside the B cell lineage.

1. ***Oncogenic mutations and translocations.*** Soon after AID was discovered, it was observed that tumors develop as a consequence of overexpressing this protein in respective

transgenic mice.[184] A series of studies revealed later that even the normal expression of AID in B cells contributes to tumorigenic translocations and oncogenic mutations,[185] and the most notable examples of this are the recurrent *c-Myc/IgH* translocations present in Burkitt lymphoma and murine myelomas.

2. ***Driver of drug resistance in tumors.*** Chronic myeloid leukemia (CML) is caused by the constitutively active BCR-ABL1 kinase fusion protein that arises from respective oncogenic t(9;22)(q34;q11) (BCR/ABL1) translocations. Patients are treated successfully with the kinase inhibitor imatinib (blocking the activity of the ABL1 kinase moiety) for years until drug resistance emerges. It was recently discovered that AID expression in CML cells is linked to the promotion of drug resistance, with AID driving genetic instability and hence mutability in respective CML cells.[186]
3. ***Tolerance.*** AID contributes to the establishment of B cell tolerance in both humans and mice.[187-190] In the absence of AID in mice or respective HIGM2 patients, autoantibodies emerge that are self-reactive in their germline configuration. In wild-type mice and healthy control patients, however, low levels of AID expression in immature and transitional B cells appear to suppress such autoantibodies. One plausible explanation is the higher sensitivity of these B cells to AID-induced genotoxic DNA lesions, and this culminates in apoptosis of autoreactive B cells.
4. ***DNA demethylation.*** Selective DNA methylation at cytosines is one epigenetic mechanism that regulates transcription, and DNA methylation patterns are normally maintained during cell division. During early embryonic development, a global erasure of methylation occurs in the primordial germ cells as part of the reprogramming to pluripotency, and AID apparently participates in this process.[191-193] However, a role for AID in active demethylation during zebrafish development[194,195] and for the generation of murine-induced pluripotent stem cells remains controversial.[196,197]

Regulation of AID activity. As a DNA mutator, AID needs to be tightly regulated to catalyze sufficiently high levels of cytosine deamination for efficient SHM and CSR, but at the same time limiting detrimental "off-target" alterations in non-Ig genes that could lead to tumor formation and apoptosis. Studies of *Aicda*$^{+/-}$ mice (expressing roughly 50% of AID compared to wild-type mice) show a reduction in SHM, CSR, and "off-target" events,[198,199] and this highlights that deamination is the rate-limiting step for all three outcomes.

Transcriptional regulation. *In vivo*, AID is expressed almost exclusively in GC B cells, and thus under normal conditions, other cell types are not subjected to its mutagenic activity. Its expression is induced by the same signals that trigger SHM and CSR including CD40L on the surface of Tfh cells. A genome-wide assessment of the *cis*-regulatory landscape in B cells showed that *Aicda* transcription is regulated by a complex array of at least six *cis*-regulatory elements, five enhancers (E1, E2, E3, E4, and E5), and one promoter that engage in long-range interaction with each other.[37] One of the upstream enhancers (E3) is responsive to T cell signals,[200] and numerous transcription factors that activate or repress *Aicda* transcription have been identified including C/EBP, E2F, cMyb, E proteins, and Hoxc4.[201-203]

Posttranscriptional regulation. Micro-RNAs are short (21-23 nt) RNAs which hybridize to complementary sequences in their target RNAs (typically a large set of different mRNAs), triggering their degradation or inhibiting their translation. At least one micro-RNA, miR-155, controls AID expression.[204,205] Mice engineered to lack miR-155 exhibited abnormally high AID levels that correlated with increased CSR and elevated levels of *c-Myc/IgH* translocations, while SHM remains normal. Importantly, mutating the miR-155 binding site in the 3′UTR of the *Aicda* mRNA leads to the same phenotype.

Posttranslational regulation. AID is phosphorylated *in vitro* and *in vivo* by the cyclic-AMP-dependent protein kinase A (PKA) at serine 38 (S38), a residue that conforms to the PKA target site motif [RRX(S/T)].[206,207] This modification is essential for AID to bind to replication protein A (RPA),[206,208] a trimeric ssDNA binding protein that can stabilize the ssDNA substrates of AID. Recombinant AID produced in *E. coli* or nonlymphoid eukaryotic cells is not phosphorylated at S38, and it can catalyze C-to-U deamination only on isolated ssDNA but not on ssDNA that is transiently generated during *in vitro* transcription. The activity on the latter substrate requires phosphorylation of at S38 and the presence of RPA. Moreover, this modification of AID also appears to play a role in interactions of AID with DNA repair factors after the deamination event.[209,210] Lastly, the roles of additional phosphorylation sites in AID, including S3, T27, T140, and Y184, in SHM and CSR are unclear.

Subcellular localization. AID shuttles between the cytoplasm and the nucleus.[211] In activated B cells, most of the protein is cytoplasmic and only a small fraction is in the nucleus where it targets Ig genes. This balance is maintained by the chaperone protein Hsp90 that stabilizes AID in the cytoplasm[212] and the counteracting ubiquitination and proteasomal degradation of AID in the nucleus.[213] The movement of AID between these compartments is controlled by at least four independent mechanisms: (1) active nuclear import mediated by an N-terminal noncontiguous nuclear localization signal,[214] (2) active nuclear export mediated by a C-terminal nuclear export signal,[211,215] (3) a passive import into the nucleus when the nuclear envelope breaks down during cell division,[216] (4) and a cytoplasmic retention mechanism.

Cell cycle. AID protein levels are modulated in a cell cycle–dependent manner. Nuclear AID acts on Ig genes to initiate SHM and CSR in the G1 phase, and during this time, it is degraded slowly. Once a B cell progresses to the S-G2/M phase of the cell cycle, AID gets more rapidly degraded to prevent undesired DNA damage during replication.[217]

Common DNA Repair Factors for SHM and CSR

Base excision repair (BER) factors. Uracil-N-glycosylase (UNG) is an evolutionarily conserved protein that removes uracil bases from DNA that is present in almost all prokaryotes and eukaryotes. Spontaneous deamination of cytosines to uracil occurs constantly at a considerable rate in the DNA genome of all living cells, and UNG is crucial for removing these inappropriate bases as the first step of BER, a ubiquitous DNA repair pathway. The UNG gene in mice and humans encodes two isoforms, UNG1 and UNG2, that differ in their N-termini and their subcellular localization with UNG1 residing in mitochondria and UNG2 (hereafter, simply referred to as UNG) in the nucleus.

The relevance of UNG for SHM and CSR became apparent during studies of cell lines, *Ung*$^{-/-}$ mice, and human patients. In the absence of UNG activity, SHM mutation patterns were drastically altered, CSR was suppressed, and uracils accumulated in the AID target regions of the Ig genes.[218-221] Together, these observations highlight that during CSR and SHM AID generates uracils in Ig genes and that UNG is one of the main players for removing this inappropriate base.

During canonical (or "short patch") BER, an apurinic/apyrimidinic (AP)-lyase, APEX1 or APEX2 (active in germinal center B cells), nicks the DNA right upstream of the abasic site, DNA polymerase β (POLB) subsequently fills this gap, and DNA ligase 1 or 3 reseals the DNA. In a variation of this process (named "long patch" BER), POLB is substituted by a complex of proliferating cell nuclear antigen PCNA and DNA polymerase δ (POLD) capable of displacement synthesis. The resulting single-stranded DNA flap is removed by flap structure-specific endonuclease 1 before ligation concludes the repair.

Mismatch repair (MMR) factors. MMR is an evolutionarily conserved DNA repair pathway that handles abnormalities in dsDNA structures, especially mispaired nucleotides. A U:G mismatch, the DNA lesion that initiates SHM and CSR, gets bound by the MutSα complex, a heterodimer of MSH2 and MSH6. This recruits the MutLα complex, a heterodimer of PMS1 and MLH1, that possesses endonuclease activity and generates a single-strand break immediately upstream of the mismatch. This nick is the entry point for the exonuclease EXO1 that creates a single-stranded gap that is filled by a replicative DNA polymerase. A noncanonical variant MMR (ncMMR) of this pathway acts during SHM and involves error-prone polymerases (discussed below).

While the absence of MSH2, MSH6, and Exo1 in mice leads to a consistent decrease in CSR and SHM at A:T base pairs (with a corresponding increase in the frequency of mutations from G:C base pairs),[222-224] the role of other MMR factors, PMS2 and MLH1, appears to be restricted to CSR.[225]

Error-prone polymerases. Although avoidance of errors is a central dogma for DNA replication, error-prone DNA synthesis plays an important role across all domains of life by facilitating replication over DNA lesions that would block progression of canonical stringent high-fidelity polymerases. The basis of tolerance toward abnormalities in DNA structure is the ability to allow for non–Watson-Crick base pairing during this translesion DNA synthesis (TLS) and the absence of proofreading activity. More than a dozen of TLS polymerases exist in eukaryotes, each acting preferentially on select lesions and showing distinct spectra of infidelity while replicating intact DNA. The distinct contributions of individual such polymerases to SHM were established by comparing their *in vitro* activity patterns with the changes in SHM patterns observed in mice or humans with mutations in the corresponding polymerase gene.

Polymerase η (POLH) incorporates mismatches opposite T nucleotides, and patients lacking this polymerase develop xeroderma pigmentosum variant (XPV) disease which predisposes them to sunlight-induced skin cancers. These patients and *POLH*$^{-/-}$ mice show severely reduced mutations at A:T base pairs.[226,227] The TLS polymerase REV1 almost exclusively inserts cytosine residues (especially opposite of abasic sites), and in its absence, C to G transversions were impaired.[228] Defects in additional TLS polymerases including REV3,[229,230] POLK (in XPV patients),[231] and POLQ[232,233] also show small but inconsistent alterations to SHM patterns.

A well-established mode of recruitment of TLS polymerases for DNA synthesis is the binding and activation by the sliding clamp protein proliferating cell nuclear protein (PCNA) when it is monoubiquitinated at the lysine residue K164. This mode is reflected in the loss of mutations at A:T sites in PCNA$^{K164R/K164R}$ mice mirroring the phenotype of POLH-deficient individuals.[234]

AID-Interacting Proteins

Despite the small size of AID, numerous putative interaction partners emerged from many biochemical and genetic screens, suggesting that AID is a component of many protein complexes in activated B cells. Surprisingly, only few of these interaction partners appear to be directly involved in Ig gene diversification as evidenced by robust, consistent and specific effects of their ablation on the outcome of CSR or SHM. The suppressor of Ty's 5 (Spt5) is a transcription-associated factor that is recruited to stalled RNA polymerase complexes, and it loads AID onto RNA polymerase complex to travel with them along Ig loci. Moreover, the RPA complex is a ubiquitous ssDNA binding protein that stabilizes and protects such structures that are key intermediates in both SHM and CSR. In fact, phosphorylated AID appears to recruit this complex to respective regions by interacting with the RPA32 subunit. In addition, the exosome is a multiprotein complex capable of RNA degradation and during CSR it interacts with AID to enable efficient deamination of the templated DNA strand.

The Molecular Mechanism of SHM

To initiate SHM, AID deaminates cytosines to uracils within ssDNA generated by transcription. After reannealing the complementary DNA strands, the resulting U:G mismatches are resolved by at least five distinct mechanisms to give rise to a variety of mutations at and around this initial lesion (Fig. 22.10).[235] This widely accepted model is derived from the detailed analysis of alterations of mutation patterns (ie, the spectrum and frequencies of mutations observed at A/G/C/T bases across a DNA substrate) that are observed when one or more DNA repair factors is missing. What governs the choice between these five (repair) pathways for a given mismatch remains unclear but is likely stochastic.

1. Direct replication places an A opposite of the U, which results in a U:A base pair in the daughter cell receiving the uracil-bearing DNA strand.[236-238] Faithful short-patch

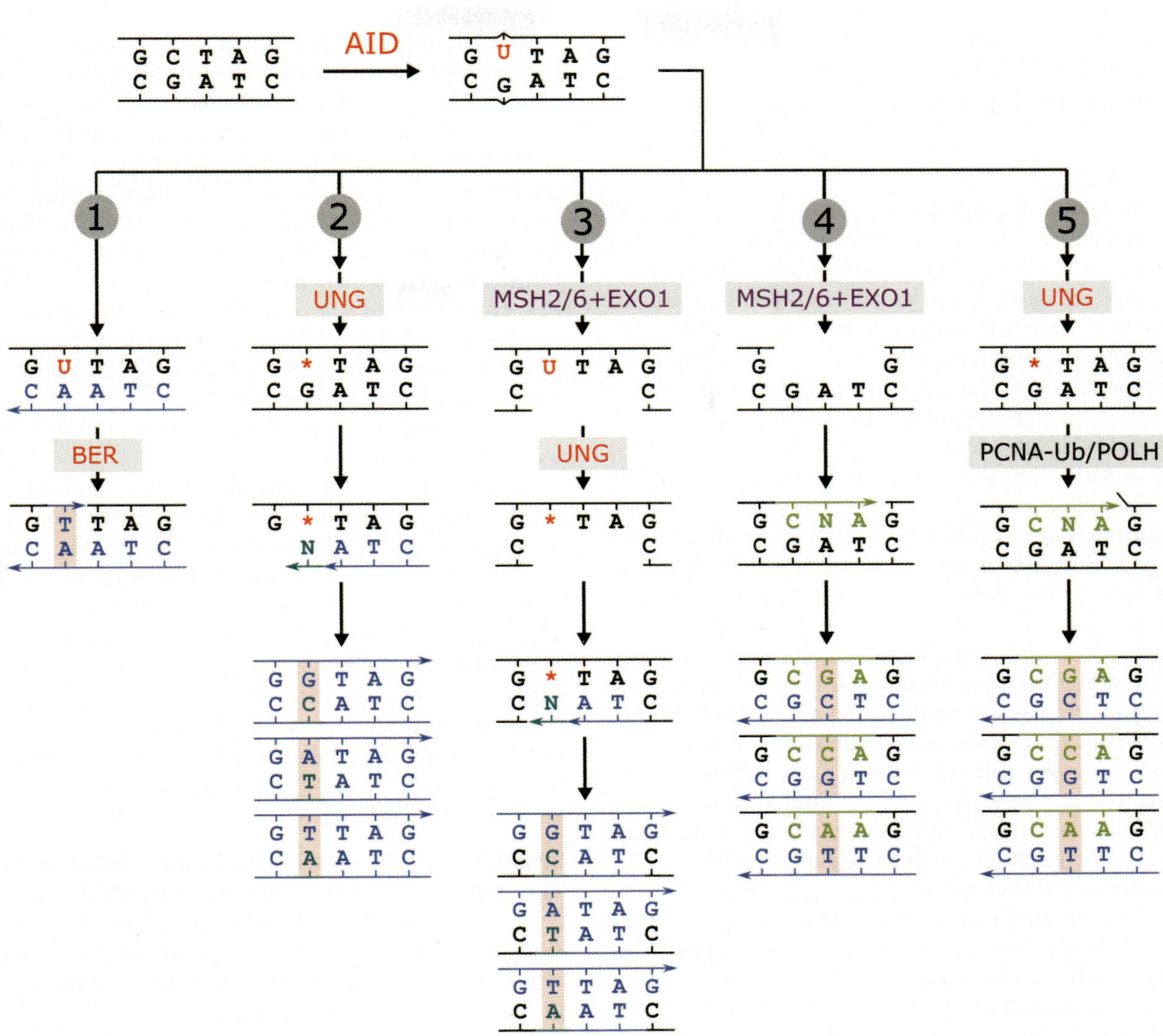

FIG. 22.10. The molecular mechanism of somatic hypermutation (SHM). AID converts cytosines (C) to uracils (*red U*) and the U:G mispair in the dsDNA is repaired by five different error-prone pathways. Some involve the generation of an abasic site (*red asterisk*). This results in mutations (*pink boxes*) at C:G and A:T base pairs at and around the initial site of deamination. Nucleotides are incorporated by replicative error-free DNA polymerases (*blue*), translesion synthesis (*teal*) mediated by REV1 and other enzymes, or error-prone DNA synthesis (*green*) by POLH and other polymerases. Additional enzymes and mechanisms involved in each pathway are labeled in red (uracil-N-glycosylase [UNG] and base excision repair [BER] factors), *purple* (mismatch repair [MMR] factors), and *black*. (Adapted from Pilzecker B, Jacobs H. Mutating for good: DNA damage responses during somatic hypermutation. *Front Immunol.* 2019;10:438.)

BER will replace this U with a T (short-patch BER), thereby ultimately giving rise to a C-to-T mutation.

2. UNG removes the uracil and a translesion DNA polymerase is recruited. One such enzyme, REV1, almost exclusively incorporates a C opposite the abasic site, resulting in C-to-G transversion mutations.[228,239] Other translesion polymerases are responsible for placing A or T opposite of C in this pathway, but their nature remains elusive.
3. ncMMR (described below) generates ssDNA, and within this single-stranded context, uracils are again removed by UNG followed by the same translesion synthesis pathway described above.[240]
4. During ncMMR, the main pathway for mutagenesis at A:T base pairs,[241] the MSH2/MSH6 complex, recognizes the initial U:G mismatch; a nick is subsequently introduced by the action of both UNG1 and PMS1 (the mechanism of which is unclear)[242]; and the Exo1 endonuclease then generates an ssDNA gap around the original U:G mismatch.[244] Finally, monoubiquitinated PCNA then recruits the error-prone DNA polymerase POLH which preferentially generates mismatches opposite of A.[243] Other error-prone polymerases with different mutagenic potential are also recruited albeit with lower efficiency.
5. UNG initiates long-patch BER and PCNA-Ub again recruits POLH for error-prone DNA synthesis opposite of an A in ssDNA.[240] This pathway accounts for 8% of all mutations at A/T residues.

Finally, despite countless in-depth studies to decipher the mechanism of SHM, one central question remains open: why is the repair of U:G mismatches in Ig genes particularly error-prone?

Targeting of SHM

The mutation rate of Ig genes in GC B cells undergoing SHM increases about 10^6-fold over the base line genomic mutation rate and reaches up to 10^{-3} mutations per base pair per division.[244] Such level of mutations would be lethal or tumorigenic if it were not specifically targeted to immunoglobulin genes, in which these mutations act to enhance antibody responses. Furthermore, mutagenic activity within Ig loci is directed to the V_HDJ_H/V_LJ_L exons and the S regions to initiate SHM and CSR (discussed below), respectively. Lastly, even within these regions, mutation/recombination events occur preferentially at and around hotspots.

Targeting to Ig gene loci. Although transcription is a prerequisite for AID-induced mutagenesis, the Ig genes are the only genes subject to SHM among thousands of transcribed genes in activated B cells. Early experiments indicated that neither the specific Ig coding sequence nor the Ig promoter was specifically required for SHM; any DNA sequences could be mutated and any active promoter was sufficient. Systematic genetic deletion studies of the IgL locus in the chicken DT40 B cell line finally led to the identification of a novel class of evolutionarily conserved *cis*-regulatory elements, called diversification activator (DIVAC)[245] or mutation enhancer element (MEE),[246] that is necessary and sufficient to render a gene locus susceptible to AID-mediated mutagenesis. Elements with DIVAC activity are also present in human and mouse Ig loci and overlapping with known enhancers. DIVACs are bound by recruiting transcription factors including Ikaros, E-box proteins, MEF2, NF-κB, and Ets family members,[247,248] but how these factors affect mutability remains a central question in the field.

Transcription. SHM events are distributed in a region that begins roughly 150 bp downstream of the promoter and extends for ~1.5 to 2 kb across the V_HDJ_H and V_LJ_L exons but ends long before the constant region exons. Moreover, shifting the positions of the Ig promoters in genetically engineered mice shifts the mutation distributions accordingly. These observations led to the model that first Spt5 mediates the loading of AID onto a paused RNA polymerase complex near the promoter,[249] and then AID goes along for a ride as the polymerase moves downstream. This transcription through the V_HDJ_H and V_LJ_L exons gives rise to ssDNA as determined using a bisulfite sequencing technique,[250] and this ssDNA is stabilized by RPA, an interaction partner of phosphorylated AID.[206] Now, AID deaminates individual cytosines within the ssDNA and eventually falls off the polymerase complex so that regions closer to the 3′ end of the Ig gene remain intact. Lastly, the window of mutations correlates with increased H3 and H4 acetylation levels,[251,252] but whether there is a distinct role for this chromatin state independent of its correlation with transcription is unclear.

Hotspot focusing. During immune responses, SHM is tightly coupled to selection of higher affinity antibody variants and hence studies of unbiased SHM patterns rely on analyzing the distribution of mutations in noncoding/nonselected DNA sequences, including Ig gene introns, nonfunctional V_HDJ_H alleles, and "passenger" transgenes. Just like *in vitro*, WRCY motifs are also hotspot for AID activity during SHM and CSR, and such motifs are concentrated in the complementarity determining regions of the V_HDJ_H/V_LJ_L exons[253,254] and S regions. The evolutionary driving force for this distribution is to enhance the potential for diversity generation in the parts of the protein critical for antigen contact. Lastly, the adenosines of WA motifs, despite not being directed substrates of AID, are hotspots for TLS polymerase–induced mutations.

Genome-wide SHM. AID also mutates a large number of transcribed genes and repetitive regions[255,256] but at much lower rates than its cognate substrate the Ig genes. Mutability of such "off-targets" of AID is controlled at two levels, the recruitment of AID to a gene and the propensity for mutagenic repair at this gene.[257] A wide array of molecular criteria that correlate mutability including chromatin and transcriptional state have been defined. AID activity at such loci is linked to genomic alterations resulting in B cell lymphoma,[258] but it is unclear how the underlying AID targeting relates to that active in the cognate substrates, the Ig loci, or why off-target genes are more effectively repaired.

Somatic Hypermutation During Immune Responses

The random generation of Ig variants by SHM is a required component of affinity maturation, the process by which specific high-affinity antibodies are generated during immune responses. Such antibodies are key for the clearance of pathogen and to prevent future infections.

Responses against pathogens are complex, and hence experimental models of SHM rely on small model hapten antigens with well-defined antibody responses, including (4-hydroxy-3-nitrophenyl)acetate (NP), p-azophenylarsonate (Ars), and phosphocholine. Unmutated V_HDJ_H and V_LJ_L sequences recognize a given antigen with low or moderate affinity, giving rise to an initial wave of IgM. The variants with mutations are observed 1 week after antigen injection, and the number of mutations continues to increase during the second week. These mutated antibodies (many of which also underwent CSR) exhibit higher affinity for the antigen. This "affinity maturation" phenomenon is based on a Darwinian mechanism linking iterative cycles of random SHM to a selection of higher affinity antibodies from the pool of random mutants. Most mutations obviously reduce the affinity for antigen, and some may even result in autoantibodies. The rare B cell clones with higher affinity preferentially proliferate and enter another round of SHM, while apoptosis is triggered in the lower affinity B cells. Continuous injection of antigen throughout the immune response interferes with affinity maturation, supporting a model in which a decrease of antigen concentration over time drives ever stricter selection. Booster injections given several weeks later, however, restart affinity maturation and yield even higher affinity antibodies.

The Mechanism of CSR

While CSR shares many features with SHM, it is more complex than SHM as it (1) requires simultaneous action on two S regions, (2) involves two DNA DSBs, and (3) relies on diverse end joining mechanisms for their repair.

Switch Regions as Substrates of CSR

During CSR, AID acts on both strands of S regions to initiate a cascade of events the results in the generation of DNA DSBs. As AID cannot access cytosines in dsDNA, transcription of the S regions leads to the generation of ssDNA structures, and within these, AID-mediated deamination gives rise to uracils. Transcription starts at the I promoters and extends through S region itself and the associated C exons (Fig. 22.11). These sterile transcripts are spliced (the actual S region RNA is excised), and splicing is in fact essential for CSR to occur. A deletion of the splice donor site at the 3′ end of the I exon completely abrogates CSR[259] but what links splicing and CSR is unknown.

R-loops. The most important consequence of transcription through the S region is the formation of R-loops, a structure in which the RNA transcript remains tightly bound to the template DNA in a stable hybrid RNA:DNA complex while the nontemplate DNA strand presents as a single-stranded loop (Fig. 22.11). This property of S regions is conferred by the nucleotide sequence itself, as cell-free transcription across S regions (but not in the reverse orientation) results in stable R-loop structures.[260] Moreover, bisulfite sequencing analysis has confirmed the presence of such accessible single-strand DNA regions in the S regions of activated B cells *in vivo*.[261] Lastly, the cytosines in the displaced single-stranded DNA are highly susceptible to deamination by AID.[262]

But how does AID get access to the template DNA strand that is held in a tight RNA:DNA hybrid? The incubation of recombinant AID with an S region R-loop substrate and a B cell protein extract provided a plausible answer to this question.[263] The RNA exosome, an evolutionarily conserved protein complex containing nine core proteins that can associate with RNA nucleases, was found to interact with AID. Degradation of the DNA-bound RNA by this complex exposes the template DNA strand in the form of ssDNA to AID prior to reannealing with the non template DNA strand.

Non-R-loop substrates. The clawed frog Sμ region is AT-rich and thus unable to form R-loops, yet still supports efficient CSR.[264] Most S regions, including those of frogs, are rich in AGCT motifs (a special version of the WRC motif, the preferred substrate of AID) on both strands as a self-complementary palindrome, and exposure to AID during transcription is likely sufficient to produce numerous uracils on both DNA strands. These U:G mismatches are readily converted into closely spaced nicks (as close as one nucleotide apart) on opposite strands leading to DSBs.[265] One implication of this model is that the ancestral CSR process did not involve R-loops and that these evolved to support efficient CSR in mammals.

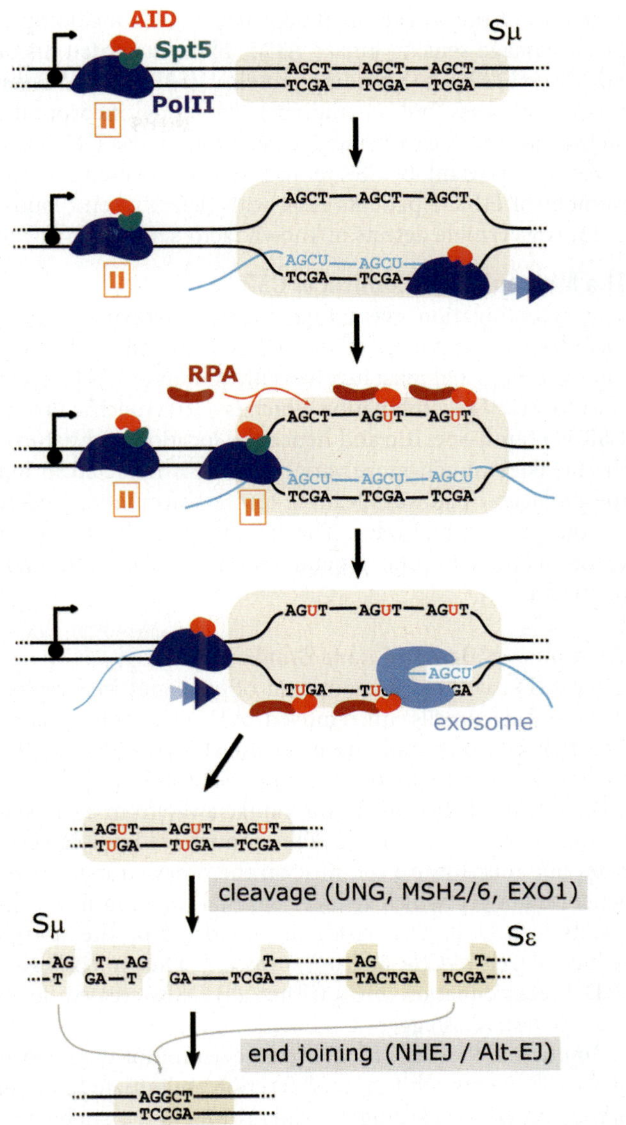

FIG. 22.11. The molecular mechanism of class switch recombination (CSR). The RNA polymerase complex pauses (*orange*) proximal to the I promoter, and Spt5 loads AID onto it. The polymerase travels through the Sμ region and a stable RNA:DNA hybrid is formed as part of an R-loop. AID interacts with the ssDNA-binding protein RPA to deaminate cytosines within AGCT motifs in the nontranscribed strand. Subsequent polymerase complexes are slowed down by the RNA:DNA hybrid and deliver more AID to the S region. The RNA transcript (*blue line*) is degraded by the exosome, and AID is recruited by it creating uracils on the transcribed strand. Base excision repair (BER) and mismatch repair (MMR) enzymes now recognize the numerous U:G mismatches on both DNA strands, and their combined action results in double-strand breaks (DSBs). One of two end-joining pathways now acts to join the upstream Sμ end with a downstream Sε end to create a switch junction.

DDR Proteins in CSR

The NHEJ machinery is one component of the DDR, the broad term for all processes that are involved in handling DNA damage caused by exogenous sources (eg, toxic chemical and radiation) and endogenous sources (eg, DNA damage during CSR).[266] During CSR, DBS repair occurs via NHEJ or by alternative-end-joining (Alt-EJ, described below). There are a large number of additional factors that orchestrate this DNA repair process by adjusting the chromatin state and 3D structure of the S regions in activated B cells including the DNA damage sensing kinase ATM, phosphorylated histone H2AX (γH2AX), p53 binding protein 1 (53BP1), and several ubiquitin ligases; but a comprehensive model incorporating all these factors has not emerged yet. Lastly, some DNA repair proteins important for CSR remain to be discovered as a small number of HIGM patients exist with defective end joining, but no discernible defects in known factors.[267]

The Molecular Mechanism of CSR

The recombination event that underlies isotype switching includes the formation of two DNA DSBs and subsequent joining events that must involve both strands of DNA. In contrast to V(D)J recombination which is a **site-specific** process, CSR is **region-specific** and hence the location and nature of the breaks is much more variable. This is reflected in both the molecular pathways leading to break formation and the various joining pathways. The following sections will thus focus on the main concepts and not rarely utilized alternative pathways.

Generation of Double-Strand Breaks

After AID-catalyzed deamination of cytosines in S regions, the resulting uracils are removed by UNG, creating abasic sites that are converted into nicks by APEX1/APEX2.[268] Two such nicks in close proximity on opposite strands give rise to a DNA DSB. If such nicks are too far away from each other to immediately create a DSB, the exonuclease Exo1 can chew from one nick toward the nick on the other strand, thereby generating a gap which reduces base-pairing and in the end results in a DSB. This model is supported by the strongly reduced levels of CSR in *Exo1*$^{-/-}$ mice.[224] Thus in contrast to V(D)J recombination, most (if not all) DSBs created during CSR are initially staggered.

In mice, UNG-dependent breaks account for the majority of all CSR events, while a small fraction are initiated by alternative pathways including the MSH2-/MSH6-dependent activation of MMR.[236] As human patients with UNG deficiency (due to mutations in both UNG alleles) exhibit HIGM2 syndrome with essentially no serum IgG, IgE, or IgA and no evidence for DSBs in Sμ,[220] it appears that DSB break formation in human S region occurs exclusively downstream of UNG.

DNA Break Intermediates in CSR

In contrast to V(D)J recombination, DNA breaks as CSR intermediates are not site-specific but occur over a wide region all across the S regions. Their presence and nature have been demonstrated directly by LM-PCR, or indirectly by the presence of chromatin states associated with DNA breaks, and by analyzing CSR products on artificial switch substrates.

Direct evidence. CSR-associated breaks were readily detected by LM-PCR protocols involving ligation of a blunt linker directly to blunt ends from genomic DNA or after pretreatment with T4 DNA polymerase which converts ssDNA overhangs (resulting from staggered nicks) into blunt ends. Importantly, the majority of breaks are staggered and not blunt, occur in the G1 phase of the cell cycle, and are dramatically reduced in the absence of AID.[269-272]

DNA repair foci. DNA breaks induce a DDR which involves the rapid accumulation of γH2AX at the break and in its vicinity, and these repair foci can be visualized by staining with anti-γH2AX antibodies. Simultaneous labeling of the IgH locus by FISH revealed the occurrence of AID-dependent γH2AX foci at the IgH locus in the G1 phase of B cells undergoing CSR.[273]

End Joining During CSR

The DNA DSBs in S regions result in DNA ends with single-stranded 3′ or 5′ overhangs that are mostly incompatible. Hence, extensive end processing occurs prior to end joining by canonical NHEJ and Alt-EJ (also called A-NHEJ). As many of the relevant DNA repair factors are also involved in V(D)J recombination and affect cell viability and proliferation in general, the accurate assessment of their roles during CSR is challenging. It typically requires new genetic mouse models with functional *Igh* and *Igl* knock-in alleles to bypass the need for V(D)J recombination during B cell development.

NHEJ. Exonuclease trimming of single-strand overhangs, filling in of recessed 3′ ends, or a combination of both processes results in blunt ends that are readily joined by NHEJ. If error-prone polymerases are involved in the filling in step, this would explain the mutations commonly observed around the switch junctions. The actual end joining relies on the core components of NHEJ (described in detail above): Ku70, Ku80, XLF, XRCC4, and DNA Ligase 4.[274-276] Their absence strongly reduces CSR, for example by 30% to 50% in *Ku80*$^{-/-}$ mice, and the residual switch junctions show increased junctional microhomologies (described below) compared to those in normal B cells. Unlike during V(D)J recombination in which the complex of DNA-PKcs and Artemis is essential for opening hairpin structures, the DNA ends in CSR are already "open" and therefore this complex only plays a very minor role in this process.

Alt-EJ. An additional pathway to repair DSBs—now commonly called alternative end joining (Alt-EJ)—joins the S region breaks using microhomology-facilitated ligation that relies on 1 to 16 nt of sequence homology flanking the DNA ends. It begins with end resection by the combined activities of the MRE11-RAD50-NBS1 (MRN) exonuclease and the CtIP endonuclease leading to short 3′ overhangs up to 20 nt in length. The ends are then loosely aligned by microhomology-mediated strand annealing of the upstream Sμ overhang to the complementary downstream partner at the target S region. PARP1 and POLQ are then recruited to fill in the remaining gaps, and the process concludes with resealing the DNA by the XRCC1/DNA ligase 3. Approximately, 50% of switch junctions show respective microhomology traces of one or more nucleotide in length, but only a small fraction of 3% present with five and more such nucleotides. Reduced levels or complete ablation of many Alt-EJ factors (including additional auxiliary factors not discussed above) alters the spectrum of switch junction sequences toward those lacking microhomologies.[277-279]

Switch Junctions and Switch Circles

Joining the four DNA ends in the two S regions gives rise to two switch junctions, one of which resides in the chromosome, whereas the other is located in a circular extrachromosomal DNA fragment called a "switch circle" (Fig. 22.9). Most switch junction sequences show mutations near the recombination breakpoint when compared to the corresponding germline switch sequences, but it is worth noting that the highly repetitive nature of the S region makes obtaining accurate DNA sequences enormously challenging.

Switch circles have been isolated from B cells actively undergoing CSR,[280,281] and their switch junctions are reciprocal to those formed on the chromosome. Switch circles are not linked to centromeres and may not contain origins of replication; hence, they are not efficiently replicated and eventually degraded. They are, however, a source of sterile transcripts that are driven by I promoters across the reciprocal switch junction, and RT-PCR detection of such transcripts is used as a telltale sign that a B cell recently underwent CSR.[282]

In rare instances, the DNA ends generated during CSR are joined to DNA ends on other chromosomes leading to translocations that are hallmarks of many hematopoietic tumors. Respective DNA breaks in other genes can be the outcome of "off target" AID action or of unrelated mechanisms. One example of the former is breaks in the *c-Myc* locus that give rise to *c-Myc*/IgH translocations in Burkitt lymphomas.

Regulation of CSR

CSR is controlled at multiple levels: (1) cellular proliferation, (2) AID expression and transcription of S regions, (3) selection and transcription of downstream target S regions, (4) recruitment of AID to S regions and, (5) synapsis of Sμ and the target S regions.

One central factor involved in regulating CSR is CD40L, a membrane-bound cytokine on the surface of T cells. It engages its receptor, CD40, on the surface of B cells to induce cell proliferation, AID expression, and germline S region transcription. The requirement of this factor for CSR is highlighted by the fact that genetic defects in numerous components in the CD40 signaling pathway (including CD40L, CD40, NEMO, NFκB proteins, and the JNK kinase) manifest as human HIGM syndromes.

Proliferation. When IgM$^+$ B cells are stimulated to undergo CSR *in vitro*, they need to divide several times before first B cells expressing other isotypes emerge. While the accessibility of S regions is rapidly established, AID expression correlates with cell division,[282] providing the first insight into the mechanistic basis of this phenomenon.

AID expression and sterile switch transcripts. As described earlier, the presence of cognate antigens and appropriate cytokines is a trigger of B cells to undergo SHM and CSR. Cytokine receptor signaling, BCR signaling, and crosstalk between both pathways activate a complex array of inducible transcription factors. These either directly or indirectly turn on AID expression and the transcription of sterile switch transcripts of the appropriate isotypes. What governs the choice between SHM and CSR in this context remains unclear.

Isotype choice. Different isotypes are known to be predominant in different immune responses. This depends on many factors including the nature of the antigen and the route by which the antigen is administered. These factors define the cytokine milieu in which CSR is initiated and thereby determines the choice of downstream target isotypes. As the CSR machinery is largely nonisotype-specific, it is these cytokine signals that render the target isotype S regions "open" and accessible while other S regions remain "closed." This is reflected in the observation that the accumulation of germline switch transcripts for a particular isotype correlates well with subsequent CSR to the same isotype. The I promoters harbor transcription factor binding sites for isotype specificity,[284] while the 3′RR, the main enhancer for transcription in mature B cells and plasma cells,[285] interacts with all these promoters with equal efficiently to drive high levels of transcription.

In vitro studies of CSR are based on stimuli that attempt to match the signals resting B cells would naturally encounter in secondary lymphoid organs, but an antigen-dependent BCR signal is commonly omitted. Treatment with anti-CD40 and IL-4 (a mimic of T-cell help) induces CSR to IgG1 and IgE; LPS alone (modeling a T-cell–independent engagement with bacteria) results in CSR to IgG2b and IgG3, while the combination of TGF-β (typically produced in the gut-associated Peyer patches), anti-CD40, and IL-4 promotes CSR to IgA. Surprisingly, responses of human B cells to these stimuli are different, and the mechanistic basis remains unclear.

Histone modifications. The act of transcription per se and the S region transcripts themselves play a role in CSR. Hence the chromatin state and the associated histone marks of S regions and their upstream I promoters are consistent with those present on active genes and active promoters. The acetylation of histone H3 (H3Ac) and its trimethylation at lysine 4 (H3K4me3) are examples for this. Accordingly, chromatin remodeling enzymes that are linked to the opening of chromatin also influence CSR in a similar manner. Surprisingly, one histone modification associated with silencing, trimethylation of H3 lysine 9 (H3K9me3), is present in active S regions targeted for CSR.[286,287] But a role of this mark in recruiting AID via the KRAB domain Associated Protein 1 (KAP1) that interacts with both AID and H3K9me3 appears unlikely, as the B cell–specific ablation of KAP1 only modestly reduced CSR.[288]

AID targeting. The mechanisms for recruiting AID to the IgH locus to initiate CSR and to focus AID activity within it toward its targets, the S regions, are very similar to those for SHM (see above). The Eμ enhancer acts as a DIVAC and hence licenses at least SHM in this locus,[247] but a likely equivalent role for the 3′RR that is critical for CSR and IgH transcription in mature B cells has not been experimentally tested yet.

Just like during SHM, AID loads onto an RNA polymerase complex, but this time at the I promoters driving transcription through Sμ and the downstream target S region. Its deamination footprint was recorded in *Ung*$^{-/-}$*Msh2*$^{-/-}$ B cells that are unable to undergo CSR[237] and indicates that AID begins to act 150 bp downstream of the I promoters and extends

4 to 5 kb covering the entire S region. As mentioned above, transcription of S regions facilitates the formation of stable RNA:DNA hybrids on the template strand, and these structures slow the progress of subsequent polymerase complexes moving through this region. This manifests in a high-density of stalled polymerase complexes enhancing AID dwell time and hence deamination rates.[289] These stalled complexes also associate with Spt5, an AID interacting protein, and this further focuses AID activity to these S regions. This model arose from experiments with *Spt5*$^{-/-}$mice which showed reduced AID occupancy of S regions and reduced CSR, while cellular levels of AID protein or sterile S region transcripts were unaffected.[249] AID activity on S regions is also strongly dependent on its phosphorylation at S38 by PKA, and this kinase is indeed present on transcribed S regions.[290] The reduced presence of RPA on S regions in the absence of AID phosphorylation indicates that AID facilitates the recruitment of RPA to stabilize ssDNA for efficient deamination. As mentioned above, AID also interacts with exosome components which degrade the RNA:DNA hybrid to provide access for AID to the template strand. Finally, all three AID interaction partners, Spt5, RPA, and the exosome contribute to promoting efficient cytosine deamination in S regions but whether they act in parallel or sequentially remains to be answered.

Switch Region Looping and Synapsis

CSR faces the same challenge than V(D)J recombination, namely how its substrates that lie great linear distances apart in the chromosome are brought into close spatial proximity before DNA cleavage. Not surprisingly, the mechanism by which this occurs is conserved between both processes and involves cohesin-mediated loop extrusion and scanning for substrates.[291] First, a small loop is established by cohesin at the 3′RR, and loop extrusion establishes a base loop which is anchored by a recombination center. This contains both active clusters of *cis*-regulatory elements the Eμ enhancer/Iμ promoter and the 3′RR/3′CBEs, and hence the Sμ region actively transcribed. Upon cytokine stimulation, transcription factors settle on the I promoter upstream of the target S region. Cohesion-mediated extrusion of a second loop now pulls the target I promoter into the recombination center where close proximity to the enhancers induces full-blown transcription. Lastly, the cohesin rings tightly maintain S region synapsis, while AID and DNA repair factors complete the CSR process.

Cellular Context of SHM and CSR

The question which of the two AID-dependent processes, SHM and CSR, occurs first during an immune response is as old as their discovery. A recent sophisticated analysis of their onset and completion used unique transcripts and protein products as markers and finally settled this question.[292]

CSR is predominantly initiated upon the first cognate B:T cell contact a full day before GC formation begins. GCs are then seeded by individual mostly class-switched B cells and emerge in secondary lymphoid organs where they provide a dedicated microenvironment for SHM. GCs promote the required contacts between B cells, follicular dendritic cells—which store, process, and present antigen—and Tfh cells, which activate B cells in part via CD40-CD40L interaction. GC B cells are frequently referred to as centroblasts and centrocytes. The former cells proliferate and diversify their Ig genes in the "dark zone" (DZ), and after migration appear as nondividing centrocytes in the "light zone" (LZ). Centrocytes are selected for higher antigen affinity as their default fate is apoptosis unless they are rescued by a strong survival signal driven by antigen and CD40L. Surviving centrocytes may return to the DZ to undergo several successive cycles of mutation and proliferation followed by selection. Ultimately plasma cells secreting class-switched high-affinity antibodies and memory B cells emerge from the GCs. Lastly, SHM and CSR on occasion occur independently of T cell help, driven by a combination of a strong BCR signal and strong innate immune signals. However, the microenvironment and structures that promote such events are less well-defined than those of GCs (see Chapter 26).

IMMUNOGLOBULIN DIVERSITY AND ALTERNATIVE SPLICING

Beyond the somatic Ig gene diversification processes described above, additional reversible/transient variations of Igs are established at the level of RNA processing: the expression of IgD and of membrane-bound or secreted Ig isoforms of all isotypes.

IgD expression. Mature IgM$^+$ B cells also express Igs of a second isotype, IgD. In contrast to CSR, this isotype switching occurs reversibly at the level of alternative RNA splicing. The V_HDJ_H exon can either be spliced to the C_H1 exon of the μ isotype or to the respective exon of the δ isotype that resides further downstream in IgH locus (Fig. 22.12). This choice is made for each primary IgH transcript, and hence a B cell can flip back and forth between making mRNAs encoding IgM and IgD. In fact, most IgM$^+$ B cells also express some levels of IgD and vice versa. Importantly, both isotypes within a single B cell always display identical

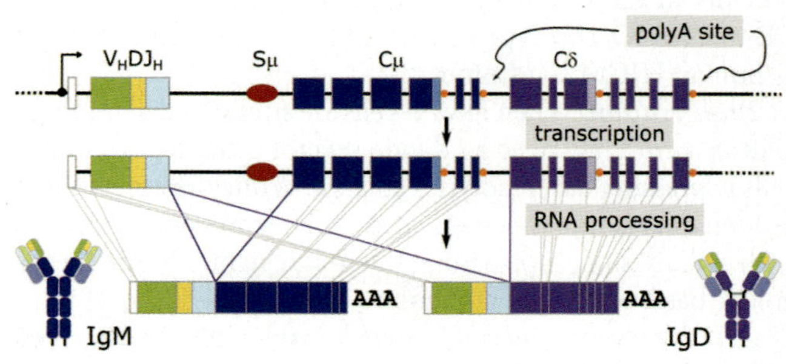

FIG. 22.12. Production of IgM and IgD. Transcription of the IgH locus in naïve B cells produces a long primary transcript containing exons (*colored boxes*) for the constant regions of both IgM (Cμ) and IgD (Cδ). Splicing and polyadenylation gives rise to the mature mRNA encoding either IgM or IgD—here the membrane-bound forms are shown—depending on alternative splicing (*blue and purple lines*) of the V_HDJ_H exon to the first constant region of each isotype.

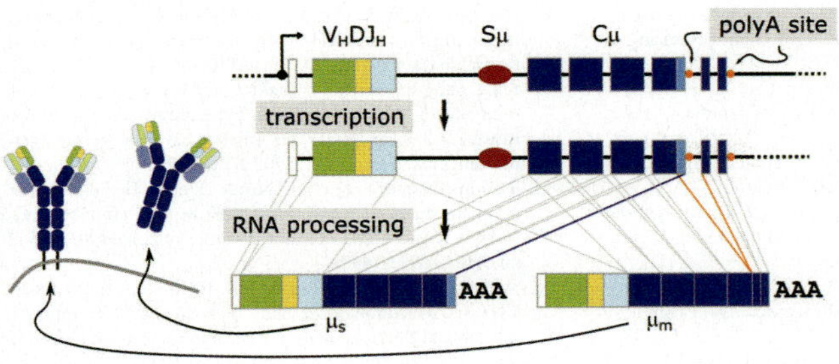

FIG. 22.13. Membrane-bound and secreted immunoglobulins. Transcription of the depicted IgH locus generates a primary *IgM* RNA transcript. Processing this RNA involves splicing of the exons (*colored boxes*) and polyadenylation. The choice of an alternative splice site in the CH3 exon (*orange lines*) and the corresponding polyadenylation (polyA) (*red circles*) determines whether the mature mRNA encodes membrane-bound (μm) or secreted (μs) IgM heavy chain.

antigen specificity and affinity. What governs the choice between the two isotypes in B1 cells, marginal zone B cells, and follicular B cells at the molecular level remains poorly understood.[293]

Membrane-bound and secreted isoforms. For each isotype, B cells produce two immunoglobulin isoforms with identical antigen specificity, a membrane-bound receptor, and secreted variant. The former serves as the antigen recognition module of the BCR (in a complex with Igα [CD79a] and Igβ [CD79b], see Chapter 21), whereas the latter is the central freely circulating effector molecule of humoral immunity. While the IgL polypeptide of both isoforms are identical, the distinct membrane (μ_m) and secreted (μ_s) heavy chains of IgM (and all other isotypes) represent the outcome of alternative splicing of the same primary IgH transcript. An internal splice site in the C_H4 exon is used to include two additional downstream Cμ exons encoding a C-terminal hydrophobic transmembrane domain that anchors the μ_m protein in the cell membrane lipid bilayer (see Fig. 22.13). The choice of this splice site and the utilization of the poly(A) site downstream of C_H4 is mutually exclusive (and occurs for all isotypes), but both transcript isoforms can exist in a B cell at the same time.

Upon differentiation into plasma cells, the balance of the two isoforms is almost entirely tilted toward the secreted variant at the transcript and hence also protein level. This results in high serum levels of Ig characteristic of ongoing immune responses. Thus, plasma cells suppress the use of the internal splice site in favor of polyadenylation at the μ_s poly(A) site. The balance is controlled by *cis*-regulatory elements in the primary RNA transcripts and corresponding *trans*-acting RNA binding proteins that recognize these elements. This includes a GU-rich element downstream of the μ_s poly(A) site[294] and similar elements in other isotypes,[295] the polyadenylation factor cleavage stimulator factor 64 (CstF-64),[296] and the U1A protein.[297] These factors likely function downstream of B lymphocyte–induced maturation protein-1 (Blimp-1; discussed in Chapter 26) whose expression in Ig-secreting plasma cells was also found to be critical for μ_s poly(A) site utilization.[298]

CONCLUSION

The discovery and cloning of Ig genes and their transcripts has revolutionized the study of the antibody response. Over the 40+ years since then, countless studies driven by increasingly sophisticated experimental technologies provided an evermore detailed view on the molecular details of the unique processes that diversify their sequence: V(D)J recombination, SHM, and CSR. This ranges from close-to-atomic-resolution views of the enzymatic machineries and their substrates to a comprehensive view of the cellular and organismal context in which these processes occur and how they are regulated.

Equally important, but out of the scope of this chapter, are the many practical applications involving Ig genes for the diagnosis and the treatment of challenging diseases in the clinical setting. The most relevant examples are the production of a wide range of recombinant monoclonal antibodies and the engineering of artificial Ig-fusion proteins that exploit either the exquisite specificity of antibodies to direct therapeutic agents including drugs and engineered immune cells to specific targets. Recombinant monoclonal antibodies against PD1/PD-L1 and anti-CTLA4 (commonly named "immune check point inhibitors" boosting the patients, anti-tumor T cell responses) are now commonly used in the immunotherapy treatment of tumors, and anti-CD20 antibodies are a well-established strategy to eliminate B cells in the context of cancer and autoimmune diseases. One recent family of Ig fusion proteins that gained a lot of attention are the chimeric antigen receptors (CARs),[299] an engineered Ig-TCRs chimera (the Ig fragment recognizes the tumor antigen), that is transduced into the T cells of cancer patients to facilitate the recognition and elimination of hematopoietic and solid tumors that are not responding to conventional cancer therapies. A final example for Ig gene technology in the context of tumor diagnosis is the use of patient-specific Ig gene rearrangements of leukemias or lymphomas to monitor their disease status by PCR.[300] Moreover, an enormous future potential lies in Ig repertoire sequencing experiments that are aimed at utilizing Ig gene sequences to define the health status of individuals including prior and current encounters with pathogens and allergens and to facilitate the discovery of antibodies with unique neutralizing properties for treatment of diseases where traditional vaccination strategies failed. In summary, both, further practical applications of immunoglobulin genes as well as a deeper scientific understanding of their biology, will keep scientists busy in the future.

ACKNOWLEDGMENTS

I would like to thank Dr. Shu Yuan Yang for her endless support and help with preparing the figure and Dr. Martin Flajnik for his infinite patience during this process. This work was funded by a MOST grant from Taiwan (108-2320-B-182-018-MY3) and Chang Gung Memorial Hospital Grants to S.D.F. (CMRPD1J0253).

REFERENCES

1. Tonegawa S. Somatic generation of antibody diversity. *Nature.* 1983;302:575-581.
2. Hozumi N, Tonegawa S. Evidence for somatic rearrangement of immunoglobulin genes coding for variable and constant regions. *Proc Natl Acad Sci U S A.* 1976;73:3628-3632.
3. Weigert MG, Cesari IM, Yonkovich SJ, et al. Variability in the lambda light chain sequences of mouse antibody. *Nature.* 1970;228(5276):1045-1047. https://pubmed.ncbi.nlm.nih.gov/5483159/
4. Bernard O, Hozumi N, Tonegawa S. Sequences of mouse immunoglobulin light chain genes before and after somatic changes. *Cell.* 1978;15:1133-1144.
5. Honjo T, Kataoka T. Organization of immunoglobulin heavy chain genes and allelic deletion model. *Proc Natl Acad Sci U S A.* 1978;75:2140-2144.
6. Yaoita Y, Honjo T. Deletion of immunoglobulin heavy chain genes from expressed allelic chromosome. *Nature.* 1980;286:850-853.
7. Kataoka T, Kawakami T, Takahashi N, et al. Rearrangement of immunoglobulin γ1-chain gene and mechanism for heavy-chain class switch. *Proc Natl Acad Sci U S A.* 1980;77:919-923.
8. Rabbitts TH, Forster A, Dunnick W, et al. The role of gene deletion in the immunoglobulin heavy chain switch. *Nature.* 1980;283:351-356.
9. Cory S, Adams JM. Deletions are associated with somatic rearrangement of immunoglobulin heavy chain genes. *Cell.* 1980;19:37-51.
10. Schatz DG, Oettinger MA, Baltimore D. The V(D)J recombination activating gene, RAG-1. *Cell.* 1989;59:1035-1048.
11. Oettinger MA, Schatz DG, Gorka C, et al. RAG-1 and RAG-2, adjacent genes that synergistically activate V(D)J recombination. *Science.* 1990;248:1517-1523.
12. Muramatsu M, Kinoshita K, Fagarasan S, et al. Class switch recombination and hypermutation require activation-induced cytidine deaminase (AID), a potential RNA editing enzyme. *Cell.* 2000;102:553-563.
13. Revy P, Muto T, Levy Y, et al. Activation-induced cytidine deaminase (AID) deficiency causes the autosomal recessive form of the Hyper-IgM syndrome (HIGM2). *Cell.* 2000;102:565-575.
14. Lefranc MP, Giudicelli V, Duroux P, et al. IMGT®, the international ImMunoGeneTics information system® 25 years on. *Nucleic Acids Res.* 2015;43:D413-D422.
15. Ye J, Ma N, Madden TL, et al. IgBLAST: an immunoglobulin variable domain sequence analysis tool. *Nucleic Acids Res.* 2013;41:W34-W40.
16. Johnston CM, Wood AL, Bolland DJ, et al. Complete sequence assembly and characterization of the C57BL/6 mouse Ig heavy chain V region. *J Immunol.* 2006;176:4221-4234.
17. Nei M, Gu X, Sitnikova T. Evolution by the birth-and-death process in multigene families of the vertebrate immune system. *Proc Natl Acad Sci U S A.* 1997;94:7799-7806.
18. Ye J. The immunoglobulin IGHD locus in C57BL/6 mice. *Immunogenetics.* 2004;56:399-404.
19. Zemlin M, Schelonka RL, Ippolito GC, et al. Regulation of repertoire development through genetic control of D H reading frame preference. *J Immunol.* 2008;181:8416-8424.
20. Tomlinson IM, Cook GP, Nigel PC, et al. Human immunoglobulin VH and D segments on chromosomes 15q11.2 and 16p11.2. *Hum Mol Genet.* 1994;3:853-860.
21. Corbett SJ, Tomlinson IM, Sonnhammer ELL, et al. Sequence of the human immunoglobulin diversity (D) segment locus: a systematic analysis provides no evidence for the use of DIR segments, inverted D segments, "Minor" D segments or D-D recombination. *J Mol Biol.* 1997;270:587-597.
22. Shimizu A, Takahashi N, Yaoita Y, et al. Organization of the constant-region gene family of the mouse immunoglobulin heavy chain. *Cell.* 1982;28:499-506.
23. Akahori Y, Kurosawa Y. Nucleotide sequences of all the γ gene loci of murine immunoglobulin heavy chains. *Genomics.* 1997;41:100-104.
24. Flanagan JG, Rabbitts TH. Arrangement of human immunoglobulin heavy chain constant region genes implies evolutionary duplication of a segment containing γ, ε and α genes. *Nature.* 1982;300:709-713.
25. Hinds KR, Litman GW. Major reorganization of immunoglobulin VH segmental elements during vertebrate evolution. *Nature.* 1986;320:546-549.
26. Greenberg AS, Avila D, Hughes M, et al. A new antigen receptor gene family that undergoes rearrangement and extensive somatic diversification in sharks. *Nature.* 1995;374:168-173.
27. Hamers-Casterman C, Atarhouch T, Muyldermans S, et al. Naturally occurring antibodies devoid of light chains. *Nature.* 1993;363:446-448.
28. Vincke C, Muyldermans S. Introduction to heavy chain antibodies and derived nanobodies. *Methods Mol Biol.* 2012;911:15-26. https://pubmed.ncbi.nlm.nih.gov/22886243/
29. Brekke KM, Garrard WT. Assembly and analysis of the mouse immunoglobulin kappa gene sequence. *Immunogenetics.* 2004;56:490-505.
30. Kawasaki K, Minoshima S, Nakato E, et al. Evolutionary dynamics of the human immunoglobulin κ locus and the germline repertoire of the Vκ genes. *Eur J Immunol.* 2001;31:1017-1028.
31. Durdik J, Moore MW, Selsing E. Novel kappa; light-chain gene rearrangements in mouse lambda; light chain-producing B lymphocytes. *Nature.* 1984;307:749-752.
32. Siminovitch KA, Bakhshi A, Goldman P, et al. A uniform deleting element mediates the loss of κ genes in human B cells. *Nature.* 1985;316:260-262.
33. Gerdes T, Wabl M. Physical map of the mouse λ light chain and related loci. *Immunogenetics.* 2002;54:62-65.
34. Kawasaki K, Minoshima S, Nakato E, et al. One-megabase sequence analysis of the human immunoglobulin λ gene locus. *Genome Res.* 1997;7:250-261.
35. Niewold TA, Murphy CL, Weiss DT, et al. Characterization of a light chain product of the human JC lambda 7 gene complex. *J Immunol.* 1996;157:4474-4477.
36. van der Burg M, Barendregt BH, van Gastel-Mol EJ, et al. Unraveling of the polymorphic Cλ2-Cλ3 amplification and the Ke+Oz– polymorphism in the human Igλ locus. *J Immunol.* 2002;169:271-276.
37. Kieffer-Kwon KR, Tang Z, Mathe E, et al. Interactome maps of mouse gene regulatory domains reveal basic principles of transcriptional regulation. *Cell.* 2013;155:1507-1520.
38. Criscitiello MF, Flajnik MF. Four primordial immunoglobulin light chain isotypes, including λ and κ, identified in the most primitive living jawed vertebrates. *Eur J Immunol.* 2007;37:2683-2694.
39. Kudo A, Sakaguchi N, Melchers F. Organization of the murine Ig-related lambda 5 gene transcribed selectively in pre-B lymphocytes. *EMBO J.* 1987;6:103-107.
40. Kudo A, Melchers F. A second gene, VpreB in the lambda 5 locus of the mouse, which appears to be selectively expressed in pre-B lymphocytes. *EMBO J.* 1987;6:2267-2272.
41. Tsubata T, Reth M. The products of pre-B cell-specific genes (λ5 and VpreB) and the immunoglobulin μ chain form a complex that is transported onto the cell surface. *J Exp Med.* 1990;172:973-976.
42. Dul JL, Argon Y, Winkler T, et al. The murine VpreB1 and VpreB2 genes both encode a protein of the surrogate light chain and are co-expressed during B cell development. *Eur J Immunol.* 1996;26:906-913.
43. Fippiat JP, Williams SC, Tomlinson LM, et al. Organization of the human immunoglobulin lambda light-chain locus on chromosome 22q11.2. *Hum Mol Genet.* 1995;4:983-991.
44. Nadel B, Tang A, Escuro G, et al. Sequence of the spacer in the recombination signal sequence affects V(D)J rearrangement frequency and correlates with nonrandom Vκ usage *in vivo.* *J Exp Med.* 1998;187:1495-1503.
45. Cowell LG, Davila M, Kepler TB, et al. Identification and utilization of arbitrary correlations in models of recombination signal sequences. *Genome Biol.* 2002;3:research0072.1–0072.20.
46. Raghavan SC, Swanson PC, Wu X, et al. A non-B-DNA structure at the Bcl-2 major breakpoint region is cleaved by the RAG complex. *Nature.* 2004;428:88-93.
47. Lu H, Shimazaki N, Raval P, et al. A biochemically defined system for coding joint formation in V(D)J recombination. *Mol Cell.* 2008;31:485-497.
48. Schatz DG, Baltimore D. Stable expression of immunoglobulin gene V(D)J recombinase activity by gene transfer into 3T3 fibroblasts. *Cell.* 1988;53:107-115.
49. Swann JB, Holland SJ, Petersen M, et al. The immunogenetics of sexual parasitism. *Science.* 2020;369:1608-1615.
50. Shinkai Y, Rathbun G, Lam KP, et al. RAG-2-deficient mice lack mature lymphocytes owing to inability to initiate V(D)J rearrangement. *Cell.* 1992;68:855-867.
51. Mombaerts P, Iacomini J, Johnson RS, et al. RAG-1-deficient mice have no mature B and T lymphocytes. *Cell.* 1992;68:869-877.
52. Ji Y, Resch W, Corbett E, et al. The *in vivo* pattern of binding of RAG1 and RAG2 to antigen receptor loci. *Cell.* 2010;141:419-431.
53. Schwarz K, Gauss GH, Ludwig L, et al. RAG mutations in human B cell-negative SCID. *Science.* 1996;274:97-99.
54. Villa A, Santagata S, Bozzi F, et al. Partial V(D)J recombination activity leads to Omenn syndrome. *Cell.* 1998;93:885-896.
55. Ru H, Chambers MG, Fu TM, et al. Molecular mechanism of V(D)J recombination from synaptic RAG1-RAG2 complex structures. *Cell.* 2015;163:1138-1152.
56. Kim MS, Chuenchor W, Chen X, et al. Cracking the DNA code for V(D)J recombination. *Mol Cell.* 2018;70:358-370.e4.
57. Ru H, Mi W, Zhang P, et al. DNA melting initiates the RAG catalytic pathway. *Nat Struct Mol Biol.* 2018;25:732-742.
58. Kim MS, Lapkouski M, Yang W, et al. Crystal structure of the V(D)J recombinase RAG1-RAG2. *Nature.* 2015;518:507-511.
59. Schabla NM, Perry GA, Palmer VL, et al. VprBP (DCAF1) regulates RAG1 expression independently of dicer by mediating RAG1 degradation. *J Immunol.* 2018;201:930-939.
60. Maman Y, Teng G, Seth R, et al. RAG1 targeting in the genome is dominated by chromatin interactions mediated by the non-core regions of RAG1 and RAG2. *Nucleic Acids Res.* 2016;44:9624-9637.
61. Grazini U, Zanardi F, Citterio E, et al. The RING domain of RAG1 ubiquitylates histone H3: a novel activity in chromatin-mediated regulation of V(D)J joining. *Mol Cell.* 2010;37:282-293.
62. Akamatsu Y, Monroe R, Dudley DD, et al. Deletion of the RAG2 C terminus leads to impaired lymphoid development in mice. *Proc Natl Acad Sci U S A.* 2003;100:1209-1214.
63. Matthews AGW, Kuo AJ, Ramón-Maiques S, et al. RAG2 PHD finger couples histone H3 lysine 4 trimethylation with V(D)J recombination. *Nature.* 2007;450:1106-1110.
64. Liu Y, Subrahmanyam R, Chakraborty T, et al. A plant homeodomain in RAG-2 that binds hypermethylated lysine 4 of histone H3 is necessary for efficient antigen-receptor-gene rearrangement. *Immunity.* 2007;27:561-571.
65. Ramón-Maiques S, Kuo AJ, Carney D, et al. The plant homeodomain finger of RAG2 recognizes histone H3 methylated at both lysine-4 and arginine-2. *Proc Natl Acad Sci U S A.* 2007;104:18993-18998.
66. Shimazaki N, Tsai AG, Lieber MR. H3K4me3 stimulates the V(D)J RAG complex for both nicking and hairpinning in trans in addition to tethering in cis: implications for translocations. *Mol Cell.* 2009;34:535-544.
67. Talukder SR, Dudley DD, Alt FW, et al. Increased frequency of aberrant V(D)J recombination products in core RAG-expressing mice. *Nucleic Acids Res.* 2004;32:4539-4549.
68. Tsai CL, Drejer AH, Schatz DG. Evidence of a critical architectural function for the RAG proteins in end processing, protection, and joining in V(D)J recombination. *Genes Dev.* 2002;16:1934-1949.
69. Elkin SK, Matthews AG, Oettinger MA. The C-terminal portion of RAG2 protects against transposition *in vitro.* *EMBO J.* 2003;22:1931-1938.
70. Agrawal A, Eastmant QM, Schatz DG. Transposition mediated by RAG1 and RAG2 and its implications for the evolution of the immune system. *Nature.* 1998;394:744-751.
71. Hiom K, Melek M, Gellert M. DNA transposition by the RAG1 and RAG2 proteins: a possible source of oncogenic translocations. *Cell.* 1998;94:463-470.
72. Reddy YVR, Perkins EJ, Ramsden DA. Genomic instability due to V(D)J recombination-associated transposition. *Genes Dev.* 2006;20:1575-1582.
73. Zhang Y, Corbett E, Wu S, et al. Structural basis for the activation and suppression of transposition during evolution of the RAG recombinase. *EMBO J.* 2020;39:e105857.
74. Fugmann SD, Messier C, Novack LA, et al. An ancient evolutionary origin of the Rag1/2 gene locus. *Proc Natl Acad Sci U S A.* 2006;103:3728-3733.
75. Huang S, Tao X, Yuan S, et al. Discovery of an active RAG transposon illuminates the origins of V(D)J recombination. *Cell.* 2016;166:102-114.
76. Morales Poole JR, Huang SF, Xu A, et al. The RAG transposon is active through the deuterostome evolution and domesticated in jawed vertebrates. *Immunogenetics.* 2017;69:391-400.
77. Martin EC, Vicari C, Tsakou-Ngouafo L, et al. Identification of RAG-like transposons in protostomes suggests their ancient bilaterian origin. *Mob DNA.* 2020;11:17.
78. Hsu LY, Lauring J, Liang HE, et al. A conserved transcriptional enhancer regulates RAG gene expression in developing B cells. *Immunity.* 2003;19:105-117.
79. Miyazaki K, Watanabe H, Yoshikawa G, et al. The transcription factor E2A activates multiple enhancers that drive Rag expression in developing T and B cells. *Sci Immunol.* 2020;5:eabb1455.
80. Miyazaki K, Miyazaki M. The interplay between chromatin architecture and lineage-specific transcription factors and the regulation of Rag gene expression. *Front Immunol.* 2021;12:659761.
81. Amin RH, Schlissel MS. Foxo1 directly regulates the transcription of recombination-activating genes during B cell development. *Nat Immunol.* 2008;9:613-622.
82. Verkoczy L, Aït-Azzouzene D, Skog P, et al. A role for nuclear factor kappa B/rel transcription factors in the regulation of the recombinase activator genes. *Immunity.* 2005;22:519-531.
83. Lin WC, Desiderio S. Regulation of V(D)J recombination activator protein RAG-2 by phosphorylation. *Science.* 1993;260:953-959.
84. Schlissel M, Constantinescu A, Morrow T, et al. Double-strand signal sequence breaks in V(D)J recombination are blunt, 5'- phosphorylated, RAG-dependent, and cell cycle regulated. *Genes Dev.* 1993;7:2520-2532.
85. Li Z, Dordai DI, Lee J. A conserved degradation signal regulates RAG-2 accumulation during cell division and links V(D)J recombination to the cell cycle. *Immunity.* 1996;5(6):575-589.
86. Zhang L, Reynolds TL, Shan X, et al. Coupling of V(D)J recombination to the cell cycle suppresses genomic instability and lymphoid tumorigenesis. *Immunity.* 2011;34:163-174.
87. van Gent DC, Hiom K, Paull TT, et al. Stimulation of V(D)J cleavage by high mobility group proteins. *EMBO J.* 1997;16:2665-2670.
88. Hiom K, Gellert M. Assembly of a 12/23 paired signal complex: a critical control point in V(D)J recombination. *Mol Cell.* 1998;1:1011-1019.
89. Zhu C, Bogue MA, Lim DS, et al. Ku86-deficient mice exhibit severe combined immunodeficiency and defective processing of V(D)J recombination intermediates. *Cell.* 1996;86:379-389.
90. Gu Y, Seidl KJ, Rathbun GA, et al. Growth retardation and leaky SCID phenotype of Ku70-deficient mice. *Immunity.* 1997;7:653-665.
91. Danska JS, Holland DP, Mariathasan S, et al. Biochemical and genetic defects in the DNA-dependent protein kinase in murine scid lymphocytes. *Mol Cell Biol.* 1996;16:5507-5517.

92. Blunt T, Gell D, Fox M, et al. Identification of a nonsense mutation in the carboxyl-terminal region of DNA-dependent protein kinase catalytic subunit in the scid mouse. *Proc Natl Acad Sci U S A*. 1996;93:10285-10290.
93. Araki R, Fujimori A, Hamatani K, et al. Nonsense mutation at Tyr-4046 in the DNA-dependent protein kinase catalytic subunit of severe combined immune deficiency mice. *Proc Natl Acad Sci U S A*. 1997;94:2438-2443.
94. Gao Y, Chaudhuri J, Zhu C, et al. A targeted DNA-PKcs-null mutation reveals DNA-PK-independent functions for KU in V(D)J recombination. *Immunity*. 1998;9:367-376.
95. Taccioli GE, Amatucci AG, Beamish HJ, et al. Targeted disruption of the catalytic subunit of the DNA-PK gene in mice confers severe combined immunodeficiency and radiosensitivity. *Immunity*. 1998;9:355-366.
96. Moshous D, Callebaut I, de Chasseval R, et al. Artemis, a novel DNA double-strand break repair/V(D)J recombination protein, is mutated in human severe combined immune deficiency. *Cell*. 2001;105:177-186.
97. Ege M, Ma Y, Manfras B, et al. Omenn syndrome due to ARTEMIS mutations. *Blood*. 2005;105:4179-4186.
98. Drouet J, Frit P, Delteil C, et al. Interplay between Ku, artemis, and the DNA-dependent protein kinase catalytic subunit at DNA ends. *J Biol Chem*. 2006;281:27784-27793.
99. Ma Y, Pannicke U, Lu H, et al. The DNA-dependent protein kinase catalytic subunit phosphorylation sites in human artemis. *J Biol Chem*. 2005;280:33839-33846.
100. Ma Y, Pannicke U, Schwarz K, et al. Hairpin opening and overhang processing by an Artemis/DNA-dependent protein kinase complex in nonhomologous end joining and V(D)J recombination. *Cell*. 2002;108:781-794.
101. Ma Y, Schwarz K, Lieber MR. The Artemis:DNA-PKcs endonuclease cleaves DNA loops, flaps, and gaps. *DNA Repair*. 2005;4:845-851.
102. Ma Y, Lu H, Tippin B, et al. A biochemically defined system for mammalian nonhomologous DNA end joining. *Mol Cell*. 2004;16:701-713.
103. Feeney AJ. Lack of N regions in fetal and neonatal mouse immunoglobulin VDJ functional sequences. *J Exp Med*. 1990;172:1377-1390.
104. Aoufouchi S, Flatter E, Dahan A, et al. Two novel human and mouse DNA polymerases of the polX family. *Nucleic Acids Res*. 2000;28:3684-3693.
105. Bertocci B, De Smet A, Berek C, et al. Immunoglobulin κ light chain gene rearrangement is impaired in mice deficient for DNA polymerase mu. *Immunity*. 2003;19:203-211.
106. Bertocci B, De Smet A, Weill JC, et al. Nonoverlapping functions of DNA polymerases mu, lambda, and terminal deoxynucleotidyltransferase during immunoglobulin V(D)J recombination *in vivo*. *Immunity*. 2006;25:31-41.
107. Grawunder U, Zimmer D, Fugmann S, et al. DNA ligase IV is essential for V(D)J recombination and DNA double-strand break repair in human precursor lymphocytes. *Mol Cell*. 1998;2:477-484.
108. Grawunder U, Wilm M, Wu X, et al. Activity of DNA ligase IV stimulated by complex formation with XRCC4 protein in mammalian cells. *Nature*. 1997;388:492-495.
109. Critchlow SE, Bowater RP, Jackson SP. Mammalian DNA double-strand break repair protein XRCC4 interacts with DNA ligase IV. *Curr Biol*. 1997;7:588-598.
110. O'Driscoll M, Cerosaletti KM, Girard PM, et al. DNA ligase IV mutations identified in patients exhibiting developmental delay and immunodeficiency. *Mol Cell*. 2001;8:1175-1185.
111. van der Burg M, van Veelen LR, Verkaik NS, et al. A new type of radiosensitive T-B-NK+ severe combined immunodeficiency caused by a LIG4 mutation. *J Clin Invest*. 2006;116:137-145.
112. Li Y, Chirgadze DY, Bolanos-Garcia VM, et al. Crystal structure of human XLF/Cernunnos reveals unexpected differences from XRCC4 with implications for NHEJ. *EMBO J*. 2008;27:290-300.
113. Ahnesorg P, Smith P, Jackson SP. XLF interacts with the XRCC4-DNA Ligase IV complex to promote DNA nonhomologous end-joining. *Cell*. 2006;124:301-313.
114. Buck D, Malivert L, De Chasseval R, et al. Cernunnos, a novel nonhomologous end-joining factor, is mutated in human immunodeficiency with microcephaly. *Cell*. 2006;124:287-299.
115. Balmus G, Barros AC, Wijnhoven PWG, et al. Synthetic lethality between PAXX and XLF in mammalian development. *Genes Dev*. 2016;30:2152-2157.
116. Lescale C, Lenden Hasse H, Blackford AN, et al. Specific roles of XRCC4 paralogs PAXX and XLF during V(D)J recombination. *Cell Rep*. 2016;16:2967-2979.
117. van Gent DC, Ramsden DA, Gellert M. The RAG1 and RAG2 proteins establish the 12/23 rule in V(D)J recombination. *Cell*. 1996;85:107-113.
118. Zhang Y, Zhang X, Ba Z, et al. The fundamental role of chromatin loop extrusion in physiological V(D)J recombination. *Nature*. 2019;573:600-604.
119. Curry JD, Geier JK, Schlissel MS. Single-strand recombination signal sequence nicks in vivo: evidence for a capture model of synapsis. *Nat Immunol*. 2005;6:1272-1279.
120. Roth DB, Nakajima PB, Menetski JP, et al. V(D)J recombination in mouse thymocytes: double-strand breaks near T cell receptor δ rearrangement signals. *Cell*. 1992;69:41-53.
121. Zhu C, Roth DB. Characterization of coding ends in thymocytes of scid mice: implications for the mechanism of V(D)J recombination. *Immunity*. 1995;2:101-112.
122. Meek K. Activation of DNA-PK by hairpinned DNA ends reveals a stepwise mechanism of kinase activation. *Nucleic Acids Res*. 2020;48:9098-9108.
123. Bogue MA, Wang C, Zhu C, et al. V(D)J recombination in Ku86-deficient mice: distinct effects on coding, signal, and hybrid joint formation. *Immunity*. 1997;7:37-47.
124. Komori T, Okada A, Stewart V, et al. Lack of N regions in antigen receptor variable region genes of TdT-deficient lymphocytes. *Science*. 1993;261:1171-1175.
125. Bentolila LA, Wu GE, Nourrit F, et al. Constitutive expression of terminal deoxynucleotidyl Transferase in transgenic mice is sufficient for N region diversity to occur at any Ig locus throughout B cell differentiation. *J Immunol*. 1997;158:715-723.
126. Zhao B, Watanabe G, Morten MJ, et al. The essential elements for the noncovalent association of two DNA ends during NHEJ synapsis. *Nat Commun*. 2019;10:3588.
127. Shimizu T, Iwasato T, Yamagishi H. Deletions of immunoglobulin Cκ region characterized by the circular excision products in mouse splenocytes. *J Exp Med*. 1991;173:1065-1072.
128. Lew S, Franco D, Chang Y. Activation of V(D)J recombination induces the formation of interlocus joints and hybrid joints in scid pre-B-cell lines. *Mol Cell Biol*. 2000;20:7170-7177.
129. Yancopoulos GD, Alt FW. Developmentally controlled and tissue-specific expression of unrearranged VH gene segments. *Cell*. 1985;40:271-281.
130. Stanhope-Baker P, Hudson KM, Shaffer AL, et al. Cell type-specific chromatin structure determines the targeting of V(D)J recombinase activity *in vitro*. *Cell*. 1996;85:887-897.
131. Golding A, Chandler S, Ballestar E, et al. Nucleosome structure completely inhibits *in vitro* cleavage by the V(D)J recombinase. *EMBO J*. 1999;18:3712-3723.
132. McBlane F, Boyes J. Stimulation of V(D)J recombination by histone acetylation. *Curr Biol*. 2000;10:483-486.
133. Alessandrini A, Desiderio SV. Coordination of immunoglobulin DJH transcription and D-to-JH rearrangement by promoter-enhancer approximation. *Mol Cell Biol*. 1991;11:2096-2107.
134. Schlissel MS, Corcoran LM, Baltimore D. Virus-transformed pre-B cells show ordered activation but not inactivation of immunoglobulin gene rearrangement and transcription. *J Exp Med*. 1991;173:711-720.
135. Bolland DJ, Wood AL, Afshar R, et al. Antisense intergenic transcription precedes Igh D-to-J recombination and is controlled by the intronic enhancer Eμ. *Mol Cell Biol*. 2007;27:5523-5533.
136. Osipovich OA, Subrahmanyam R, Pierce S, et al. Cutting edge: SWI/SNF mediates antisense Igh transcription and locus-wide accessibility in B cell precursors. *J Immunol*. 2009;183:1509-1513.
137. Chakraborty T, Chowdhury D, Keyes A, et al. Repeat organization and epigenetic regulation of the DH-cμ domain of the immunoglobulin heavy-chain gene locus. *Mol Cell*. 2007;27:842-850.
138. Osipovich O, Oltz EM. Regulation of antigen receptor gene assembly by genetic-epigenetic crosstalk. *Semin Immunol*. 2010;22:313-322.
139. Storb U, Arp B. Methylation patterns of immunoglobulin genes in lymphoid cells: correlation of expression and differentiation with undermethylation. *Proc Natl Acad Sci U S A*. 1983;80:6642-6646.
140. Goodhardt M, Cavelier P, Doyen N, et al. Methylation status of immunoglobulin × e segments correlates with their recombination potential. *Eur J Immunol*. 1993;23:1789-1795.
141. Selimyan R, Gerstein RM, Ivanova I, et al. Localized DNA demethylation at recombination intermediates during immunoglobulin heavy chain gene assembly. *PLoS Biol*. 2013;11:e1001475.
142. Hsieh CL, Lieber MR. CpG methylated minichromosomes become inaccessible for V(D)J recombination after undergoing replication. *EMBO J*. 1992;11:315-325.
143. Nakase H, Takahama Y, Akamatsu Y. Effect of CpG methylation on RAG1/RAG2 reactivity: implications of direct and indirect mechanisms for controlling V(D)J cleavage. *EMBO Rep*. 2003;4:774-780.
144. de la Serna IL, Ohkawa Y, Imbalzano AN. Chromatin remodelling in mammalian differentiation: Lessons from ATP-dependent remodellers. *Nat Rev Genet*. 2006;7(6):461-473. https://pubmed.ncbi.nlm.nih.gov/16708073/
145. Osborne CS, Chakalova L, Brown KE, et al. Active genes dynamically colocalize to shared sites of ongoing transcription. *Nat Genet*. 2004;36:1065-1071.
146. Kosak ST, Skok JA, Medina KL, et al. Subnuclear compartmentalization of immunoglobulin loci during lymphocyte development. *Science*. 2002;296:158-162.
147. Jhunjhunwala S, van Zelm MC, Peak MM, et al. The 3D structure of the immunoglobulin heavy-chain locus: implications for long-range genomic interactions. *Cell*. 2008;133:265-279.
148. Qiu X, Kumari G, Gerasimova T, et al. Sequential enhancer sequestration dysregulates recombination center formation at the IgH locus. *Mol Cell*. 2018;70:21-33.
149. Garrett FE, Emelyanov AV, Sepulveda MA, et al. Chromatin architecture near a potential 3′ end of the Igh locus involves modular regulation of histone modifications during B-cell development and *in vivo* occupancy at CTCF sites. *Mol Cell Biol*. 2005;25:1511-1525.
150. Fang W, Mueller DL, Pennell CA, et al. Frequent aberrant immunoglobulin gene rearrangements in Pro-B cells revealed by a bcl-xL transgene. *Immunity*. 1996;4:291-299.
151. Grawunder U, Leu TMJ, Schatz DG, et al. Down-regulation of RAG1 and RAG2 gene expression in PreB cells after functional immunoglobulin heavy chain rearrangement. *Immunity*. 1995;3:601-608.
152. Schlissel MS, Morrow T. Ig heavy chain protein controls B cell development by regulating germ- line transcription and retargeting V(D)J recombination. *J Immunol*. 1994;153:1645-1657.
153. Alt FW, Enea V, Bothwell ALM, et al. Activity of multiple light chain genes in murine myeloma cells producing a single, functional light chain. *Cell*. 1980;21:1-12.
154. Zhang Z, Zemlin M, Wang YH, et al. Contribution of VH gene replacement to the primary B cell repertoire. *Immunity*. 2003;19:21-31.
155. Huang L, Lange MD, Yu Y, et al. Contribution of VH replacement products in mouse antibody repertoire. *PLoS One*. 2013;8:e57877.
156. Casellas R, Shih TA, Kleinewietfeld M, et al. Contribution of receptor editing to the antibody repertoire. *Science*. 2001;291:1541-1544.
157. Reth M, Gehrmann P, Petrac E, et al. A novel VH to VHDJH joining mechanism in heavy-chain-negative (null) pre-B cells results in heavy-chain production. *Nature*. 1986;322:840-842.
158. Rahman NS, Godderz LAJ, Stray SJ, et al. DNA cleavage of a cryptic recombination signal sequence by RAG1 and RAG2: implications for partial VH gene replacement. *J Biol Chem*. 2006;281:12370-12380.
159. Hippen KL, Schram BR, Tze LE, et al. *In vivo* assessment of the relative contributions of deletion, anergy, and editing to B cell self-tolerance. *J Immunol*. 2005;175:909-916.
160. Halverson R, Torres RM, Pelanda R. Receptor editing is the main mechanism of B cell tolerance toward membrane antigens. *Nat Immunol*. 2004;5:645-650.
161. Reynaud S, Delpy L, Fleury L, et al. Interallelic class switch recombination contributes significantly to class switching in mouse B cells. *J Immunol*. 2005;174:6176-6183.
162. Muramatsu M, Sankaranand VS, Anant S, et al. Specific expression of activation-induced cytidine deaminase (AID), a novel member of the RNA-editing deaminase family in germinal center B cells. *J Biol Chem*. 1999;274:18470-18476.
163. Arakawa H, Hauschild J, Buerstedde JM. Requirement of the activation-induced deaminase (AID) gene for immunoglobulin gene conversion. *Science*. 2002;295:1301-1306.
164. Okazaki IM, Kinoshita K, Muramatsu M, et al. The AID enzyme induces class switch recombination in fibroblasts. *Nature*. 2002;416:340-345.
165. Yoshikawa K, Okazaki IM, Eto T, et al. AID enzyme-induced hypermutation in an actively transcribed gene in fibroblasts. *Science*. 2002;296:2033-2036.
166. Navaratnam N, Bhattacharya S, Fujino T, et al. Evolutionary origins of apoB mRNA editing: catalysis by a cytidine deaminase that has acquired a novel RNA-binding motif at its active site. *Cell*. 1995;81:187-195.
167. Rogozin IB, Iyer LM, Liang L, et al. Evolution and diversification of lamprey antigen receptors: evidence for involvement of an AID-APOBEC family cytosine deaminase. *Nat Immunol*. 2007;8:647-656.
168. Liu MC, Liao WY, Buckley KM, et al. AID/APOBEC-like cytidine deaminases are ancient innate immune mediators in invertebrates. *Nat Commun*. 2018;9:1948.
169. Krishnan A, Iyer LM, Holland SJ, et al. Diversification of AID/APOBEC-like deaminases in metazoa: multiplicity of clades and widespread roles in immunity. *Proc Natl Acad Sci U S A*. 2018;115:E3201-E3210.
170. Conticello SG. The AID/APOBEC family of nucleic acid mutators. *Genome Biol*. 2008;9:229.
171. Pham P, Afif SA, Shimoda M, et al. Structural analysis of the activation-induced deoxycytidine deaminase required in immunoglobulin diversification. *DNA Repair*. 2016;43:48-56.
172. Qiao Q, Wang L, Meng FL, et al. AID recognizes structured DNA for class switch recombination. *Mol Cell*. 2017;67:361-373.
173. Methot SP, Litzler LC, Subramani PG, et al. A licensing step links AID to transcription elongation for mutagenesis in B cells. *Nat Commun*. 2018;9:1248.
174. Abdouni HS, King JJ, Ghorbani A, et al. DNA/RNA hybrid substrates modulate the catalytic activity of purified AID. *Mol Immunol*. 2018;93:94-106.
175. Zheng S, Vuong BQ, Vaidyanathan B, et al. Non-coding RNA generated following lariat debranching mediates targeting of AID to DNA. *Cell*. 2015;161:762-773.
176. Ta VT, Nagaoka H, Catalan N, et al. AID mutant analyses indicate requirement for class-switch-specific cofactors. *Nat Immunol*. 2003;4:843-848.
177. Barreto V, Reina-San-Martin B, Ramiro AR, et al. C-terminal deletion of AID uncouples class switch recombination from somatic hypermutation and gene conversion. *Mol Cell*. 2003;12:501-508.

178. Zahn A, Eranki AK, Patenaude AM, et al. Activation induced deaminase C-terminal domain links DNA breaks to end protection and repair during class switch recombination. *Proc Natl Acad Sci U S A*. 2014;111:E988-E997.
179. Ranjit S, Khair L, Linehan EK, et al. AID binds cooperatively with UNG and msh2-msh6 to Ig switch regions dependent upon the AID C terminus. *J Immunol*. 2011;187:2464-2475.
180. Pham P, Bransteitter R, Petruska J, et al. Processive AID-catalysed cytosine deamination on single-stranded DNA simulates somatic hypermutation. *Nature*. 2003;424:103-107.
181. Wang M, Rada C, Neuberger MS. Altering the spectrum of immunoglobulin V gene somatic hypermutation by modifying the active site of AID. *J Exp Med*. 2010;207:141-153.
182. Kohli RM, Maul RW, Guminski AF, et al. Local sequence targeting in the AID/APOBEC family differentially impacts retroviral restriction and antibody diversification. *J Biol Chem*. 2010;285:40956-40964.
183. Carpenter MA, Rajagurubandara E, Wijesinghe P, et al. Determinants of sequence-specificity within human AID and APOBEC3G. *DNA Repair*. 2010;9:579-587.
184. Okazaki IM, Hiai H, Kakazu N, et al. Constitutive expression of AID leads to tumorigenesis. *J Exp Med*. 2003;197:1173-1181.
185. Ramiro AR, Jankovic M, Eisenreich T, et al. AID is required for c-myc/IgH chromosome translocations *in vivo*. *Cell*. 2004;118:431-438.
186. Klemm L, Duy C, Iacobucci I, et al. The B cell mutator AID promotes B lymphoid blast crisis and drug resistance in chronic myeloid leukemia. *Cancer Cell*. 2009;16:232-245.
187. Meyers G, Ng YS, Bannock JM, et al. Activation-induced cytidine deaminase (AID) is required for B-cell tolerance in humans. *Proc Natl Acad Sci U S A*. 2011;108:11554-11559.
188. Kuraoka M, Holl TM, Liao D, et al. Activation-induced cytidine deaminase mediates central tolerance in B cells. *Proc Natl Acad Sci U S A*. 2011;108:11560-11565.
189. Zaheen A, Boulianne B, Parsa JY, et al. AID constrains germinal center size by rendering B cells susceptible to apoptosis. *Blood*. 2009;114:547-554.
190. Cantaert T, Schickel JN, Bannock JM, et al. Activation-induced cytidine deaminase expression in human B cell precursors is essential for central B cell tolerance. *Immunity*. 2015;43:884-895.
191. Franchini DM, Chan CF, Morgan H, et al. Processive DNA demethylation via DNA deaminase-induced lesion resolution. *PLoS One*. 2014;9:e97754.
192. Popp C, Dean W, Feng S, et al. Genome-wide erasure of DNA methylation in mouse primordial germ cells is affected by AID deficiency. *Nature*. 2010;463:1101-1105.
193. Morgan HD, Dean W, Coker HA, et al. Activation-induced cytidine deaminase deaminates 5-methylcytosine in DNA and is expressed in pluripotent tissues: implications for epigenetic reprogramming. *J Biol Chem*. 2004;279:52353-52360.
194. Rai K, Huggins IJ, James SR, et al. DNA demethylation in zebrafish involves the coupling of a deaminase, a glycosylase, and gadd45. *Cell*. 2008;135:1201-1212.
195. Shimoda N, Hirose K, Kaneto R, et al. No evidence for AID/MBD4-coupled DNA demethylation in zebrafish embryos. *PLoS One*. 2014;9:e114816.
196. Habib O, Habib G, Do JT, et al. Activation-induced deaminase-coupled DNA demethylation is not crucial for the generation of induced pluripotent stem cells. *Stem Cell Dev*. 2014;23:209-218.
197. Bhutani N, Brady JJ, Damian M, et al. Reprogramming towards pluripotency requires AID-dependent DNA demethylation. *Nature*. 2010;463:1042-1047.
198. Sernández IV, de Yébenes VG, Dorsett Y, et al. Haploinsufficiency of activation-induced deaminase for antibody diversification and chromosome translocations both *in vitro* and *in vivo*. *PLoS One*. 2008;3:e3927.
199. Takizawa M, Á HT, Li Z, et al. AID expression levels determine the extent of cMyc oncogenic translocations and the incidence of B cell tumor development. *J Exp Med*. 2008;205:1949-1957.
200. Tran TH, Nakata M, Suzuki K, et al. B cell-specific and stimulation-responsive enhancers derepress Aicda by overcoming the effects of silencers. *Nat Immunol*. 2010;11:148-154.
201. Mai T, Zan H, Zhang J, et al. Estrogen receptors bind to and activate the HOXC4/HoxC4 promoter to potentiate HoxC4-mediated activation-induced cytosine deaminase induction, immunoglobulin class switch DNA recombination, and somatic hypermutation. *J Biol Chem*. 2010;285:37797-37810.
202. Sayegh CE, Quong MW, Agata Y, et al. E-proteins directly regulate expression of activation-induced deaminase in mature B cells. *Nat Immunol*. 2003;4:586-593.
203. Huong LT, Kobayashi M, Nakata M, et al. In vivo analysis of Aicda gene regulation: a critical balance between upstream enhancers and intronic silencers governs appropriate expression. *PLoS One*. 2013;8:e61433.
204. Teng G, Hakimpour P, Landgraf P, et al. MicroRNA-155 is a negative regulator of activation-induced cytidine deaminase. *Immunity*. 2008;28:621-629.
205. Dorsett Y, McBride KM, Jankovic M, et al. MicroRNA-155 suppresses activation-induced cytidine deaminase-mediated myc-igh translocation. *Immunity*. 2008;28:630-638.
206. Basu U, Chaudhuri J, Alpert C, et al. The AID antibody diversification enzyme is regulated by protein kinase a phosphorylation. *Nature*. 2005;438:508-511.
207. Pasqualucci L, Kitaura Y, Gu H, et al. PKA-mediated phosphorylation regulates the function of activation-induced deaminase (AID) in B cells. *Proc Natl Acad Sci U S A*. 2006;103:395-400.
208. Chaudhuri J, Khuong C, Alt FW. Replication protein A interacts with AID to promote deamination of somatic hypermutation targets. *Nature*. 2004;430:992-998.
209. Vuong BQ, Herrick-Reynolds K, Vaidyanathan B, et al. A DNA break- and phosphorylation-dependent positive feedback loop promotes immunoglobulin class-switch recombination. *Nat Immunol*. 2013;14:1183-1189.
210. Choi JE, Matthews AJ, Michel G, et al. AID phosphorylation regulates mismatch repair–dependent class switch recombination and affinity maturation. *J Immunol*. 2020;204:13-22.
211. Ito S, Nagaoka H, Shinkura R, et al. Activation-induced cytidine deaminase shuttles between nucleus and cytoplasm like apolipoprotein B mRNA editing catalytic polypeptide 1. *Proc Natl Acad Sci U S A*. 2004;101:1975-1980.
212. Orthwein A, Patenaude AM, Affar EB, et al. Regulation of activation-induced deaminase stability and antibody gene diversification by Hsp90. *J Exp Med*. 2010;207:2751-2765.
213. Aoufouchi S, Faili A, Zober C, et al. Proteasomal degradation restricts the nuclear lifespan of AID. *J Exp Med*. 2008;205:1357-1368.
214. Patenaude AM, Orthwein A, Hu Y, et al. Active nuclear import and cytoplasmic retention of activation-induced deaminase. *Nat Struct Mol Biol*. 2009;16:517-527.
215. McBride KM, Barreto V, Ramiro AR, et al. Somatic hypermutation is limited by CRM1-dependent nuclear export of activation-induced deaminase. *J Exp Med*. 2004;199:1235-1244.
216. Wang Q, Kieffer-Kwon KR, Oliveira TY, et al. The cell cycle restricts activation-induced cytidine deaminase activity to early G1. *J Exp Med*. 2017;214:49-58.
217. Le Q, Maizels N. Cell cycle regulates nuclear stability of AID and determines the cellular response to AID. *PLoS Genet*. 2015;11:e1005411.
218. Di Noia J, Neuberger MS. Altering the pathway of immunoglobulin hypermutation by inhibiting uracil-DNA glycosylase. *Nature*. 2002;419:43-48.
219. Rada C, Williams GT, Nilsen H, et al. Immunoglobulin isotype switching is inhibited and somatic hypermutation perturbed in UNG-deficient mice. *Curr Biol*. 2002;12:1748-1755.
220. Imai K, Slupphaug G, Lee WI, et al. Human uracil-DNA glycosylase deficiency associated with profoundly impaired immunoglobulin class-switch recombination. *Nat Immunol*. 2003;4:1023-1028.
221. Maul RW, Saribasak H, Martomo SA, et al. Uracil residues dependent on the deaminase AID in immunoglobulin gene variable and switch regions. *Nat Immunol*. 2011;12:70-76.
222. Wiesendanger M, Kneitz B, Edelmann W, et al. Somatic hypermutation in MutS homologue (MSH)3-, MSH6-, and MSH3/MSH6-deficient mice reveals a role for the MSH2-MSH6 heterodimer in modulating the base substitution pattern. *J Exp Med*. 2000;191:579-584.
223. Martomo SA, Yang WW, Gearhart PJ. A role for Msh6 but not Msh3 in somatic hypermutation and class switch recombination. *J Exp Med*. 2004;200:61-68.
224. Bardwell PD, Woo CJ, Wei K, et al. Altered somatic hypermutation and reduced class-switch recombination in exonuclease 1-mutant mice. *Nat Immunol*. 2004;5:224-229.
225. Péron S, Metin A, Gardès P, et al. Human PMS2 deficiency is associated with impaired immunoglobulin class switch recombination. *J Exp Med*. 2008;205:2465-2472.
226. Martomo SA, Yang WW, Wersto RP, et al. Different mutation signatures in DNA polymerase η- and MSH6-deficient mice suggest separate roles in antibody diversification. *Proc Natl Acad Sci U S A*. 2005;102:8656-8661.
227. Delbos F, De Smet A, Faili A, et al. Contribution of DNA polymerase η to immunoglobulin gene hypermutation in the mouse. *J Exp Med*. 2005;201:1191-1196.
228. Jansen JG, Langerak P, Tsaalbi-Shtylik A, et al. Strand-biased defect in C/G transversions in hypermutating immunoglobulin genes in Rev1-deficient mice. *J Exp Med*. 2006;203:319-323.
229. Zan H, Komori A, Li Z, et al. The translesion DNA polymerase ζ plays a major role in Ig and bcl-6 somatic hypermutation. *Immunity*. 2001;14:643-653.
230. Diaz M, Verkoczy LK, Flajnik MF, et al. Decreased frequency of somatic hypermutation and impaired affinity maturation but intact germinal center formation in mice expressing antisense RNA to DNA polymerase ζ. *J Immunol*. 2001;167:327-335.
231. Faili A, Stary A, Delbos F, et al. A backup role of DNA polymerase κ in Ig gene hypermutation only takes place in the complete absence of DNA polymerase η. *J Immunol*. 2009;182:6353-6359.
232. Zan H, Shima N, Xu Z, et al. The translesion DNA polymerase θ plays a dominant role in immunoglobulin gene somatic hypermutation. *EMBO J*. 2005;24:3757-3769.
233. Masuda K, Ouchida R, Hikida M, et al. Absence of DNA polymerase θ results in decreased somatic hypermutation frequency and altered mutation patterns in Ig genes. *DNA Repair*. 2006;5:1384-1391.
234. Langerak P, Nygren AOH, Krijger PHL, et al. A/T mutagenesis in hypermutated immunoglobulin genes strongly depends on PCNAK164 modification. *J Exp Med*. 2007;204:1989-1998.
235. Pilzecker B, Jacobs H. Mutating for good: DNA damage responses during somatic hypermutation. *Front Immunol*. 2019;10:438. https://pubmed.ncbi.nlm.nih.gov/30915081/
236. Rada C, Di Noia JM, Neuberger MS. Mismatch recognition and uracil excision provide complementary paths to both Ig switching and the A/T-focused phase of somatic mutation. *Mol Cell*. 2004;16:163-171.
237. Xue K, Rada C, Neuberger MS. The *in vivo* pattern of AID targeting to immunoglobulin switch regions deduced from mutation spectra in *msh2*$^{-/-}$ *ung*$^{-/-}$ mice. *J Exp Med*. 2006;203:2085-2094.
238. Shen HM, Tanaka A, Bozek G, et al. Somatic hypermutation and class switch recombination in *Msh6*$^{-/-}$*Ung*$^{-/-}$ double-knockout mice. *J Immunol*. 2006;177:5386-5392.
239. Krijger PH, Tsaalbi-Shtylik A, Wit N, et al. Rev1 is essential in generating G to C transversions downstream of the Ung2 pathway but not the Msh2+Ung2 hybrid pathway. *Eur J Immunol*. 2013;43:2765-2770.
240. Krijger PHL, Langerak P, van den Berk PCM, et al. Dependence of nucleotide substitutions on Ung2, Msh2, and PCNA-Ub during somatic hypermutation. *J Exp Med*. 2009;206:2603-2611.
241. Wilson TM, Vaisman A, Martomo SA, et al. MSH2-MSH6 stimulates DNA polymerase η, suggesting a role for A:T mutations in antibody genes. *J Exp Med*. 2005;201:637-645.
242. Zubani GG, Zivojnovic M, De Smet A, et al. Pms2 and uracil-DNA glycosylases act jointly in the mismatch repair pathway to generate Ig gene mutations at A-T base pairs. *J Exp Med*. 2017;214:1169-1180.
243. Pavlov YI, Rogozin IB, Galkin AP, et al. Correlation of somatic hypermutation specificity and A-T base pair substitution errors by DNA polymerase η during copying of a mouse immunoglobulin κ light chain transgene. *Proc Natl Acad Sci U S A*. 2002;99:9954-9959.
244. Kleinstein SH, Louzoun Y, Shlomchik MJ. Estimating hypermutation rates from clonal tree data. *J Immunol*. 2003;171:4639-4649.
245. Blagodatski A, Batrak V, Schmidl S, et al. A cis-acting diversification activator both necessary and sufficient for AID-mediated hypermutation. *PLoS Genet*. 2009;5:e1000332.
246. Kothapalli N, Norton DD, Fugmann SD. Cutting edge: a cis-acting DNA element targets AID-mediated sequence diversification to the chicken Ig light chain gene locus. *J Immunol*. 2008;180:2019-2023.
247. Buerstedde JM, Alinikula J, Arakawa H, et al. Targeting of somatic hypermutation by immunoglobulin enhancer and enhancer-like sequences. *PLoS Biol*. 2014;12:e1001831.
248. Dinesh RK, Barnhill B, Ilanges A, et al. Transcription factor binding at Ig enhancers is linked to somatic hypermutation targeting. *Eur J Immunol*. 2020;50:380-395.
249. Pavri R, Gazumyan A, Jankovic M, et al. Activation-induced cytidine deaminase targets DNA at sites of RNA polymerase II stalling by interaction with Spt5. *Cell*. 2010;143:122-133.
250. Ronai D, Iglesias-Ussel MD, Fan M, et al. Detection of chromatin-associated single-stranded DNA in regions targeted for somatic hypermutation. *J Exp Med*. 2007;204:181-190.
251. Woo CJ, Martin A, Scharff MD. Induction of somatic hypermutation is associated with modifications in immunoglobulin variable region chromatin. *Immunity*. 2003;19:479-489.
252. Odegard VH, Kim ST, Anderson SM, et al. Histone modifications associated with somatic hypermutation. *Immunity*. 2005;23:101-110.
253. Saini J, Hershberg U. B cell variable genes have evolved their codon usage to focus the targeted patterns of somatic mutation on the complementarity determining regions. *Mol Immunol*. 2015;65:157-167.
254. Shapiro GS, Aviszus K, Murphy J, et al. Evolution of Ig DNA sequence to target specific base positions within codons for somatic hypermutation. *J Immunol*. 2002;168:2302-2306.
255. Staszewski O, Baker RE, Ucher AJ, et al. Activation-induced cytidine deaminase induces reproducible DNA breaks at many non-ig loci in activated B cells. *Mol Cell*. 2011;41:232-242.
256. Álvarez-Prado ÁF, Pérez-Durán P, Pérez-García A, et al. A broad atlas of somatic hypermutation allows prediction of activation-induced deaminase targets. *J Exp Med*. 2018;215:761-771.
257. Liu M, Duke JL, Richter DJ, et al. Two levels of protection for the B cell genome during somatic hypermutation. *Nature*. 2008;451:841-845.
258. Ye X, Ren W, Liu D, et al. Genome-wide mutational signatures revealed distinct developmental paths for human B cell lymphomas. *J Exp Med*. 2021;218:e20200573.
259. Hein K, Lorenz MG, Siebenkotten G, et al. Processing of switch transcripts is required for targeting of antibody class switch recombination. *J Exp Med*. 1998;188:2369-2374.
260. Daniels GA, Lieber MR. RNA: DNA complex formation upon transcription of immunoglobulin switch regions. Implications for the mechanism and regulation of class switch recombination. *Nucleic Acids Res*. 1995;23:5006-5011.
261. Huang FT, Yu K, Hsieh CL, et al. Downstream boundary of chromosomal R-loops at murine switch regions: implications for the mechanism of switch recombination. *Proc Natl Acad Sci U S A*. 2006;103:5030-5035.
262. Chaudhuri J, Tian M, Khuong C, et al. Transcription-targeted DNA deamination by the AID antibody diversification enzyme. *Nature*. 2003;422:726-730.

263. Basu U, Meng FL, Keim C, et al. The RNA exosome targets the AID cytidine deaminase to both strands of transcribed duplex DNA substrates. *Cell.* 2011;144:353-363.
264. Zarrin AA, Alt FW, Chaudhuri J, et al. An evolutionarily conserved target motif for immunoglobulin class-switch recombination. *Nat Immunol.* 2004;5:1275-1281.
265. Han L, Masani S, Yu K. Overlapping activation-induced cytidine deaminase hotspot motifs in Ig class-switch recombination. *Proc Natl Acad Sci U S A.* 2011;108:11584-11589.
266. Polo SE, Jackson SP. Dynamics of DNA damage response proteins at DNA breaks: a focus on protein modifications. *Genes Dev.* 2011;25(5):409-433. https://pubmed.ncbi.nlm.nih.gov/21363960/
267. Imai K, Catalan N, Plebani A, et al. Hyper-IgM syndrome type 4 with a B lymphocyte-intrinsic selective deficiency in Ig class-switch recombination. *J Clin Invest.* 2003;112:136-142.
268. Schrader CE, Guikema JEJ, Wu X, et al. The roles of APE1, APE2, DNA polymerase β and mismatch repair in creating S region DNA breaks during antibody class switch. *Philos Trans R Soc B Biol Sci.* 2009;364:645-652.
269. Rush JS, Fugmann SD, Schatz DG. Staggered AID-dependent DNA double strand breaks are the predominant DNA lesions targeted to Sμ in Ig class switch recombination. *Int Immunol.* 2004;16:549-557.
270. Schrader CE, Linehan EK, Mochegova SN, et al. Inducible DNA breaks in Ig S regions are dependent on AID and UNG. *J Exp Med.* 2005;202:561-568.
271. Schrader CE, Guikema JEJ, Linehan EK, et al. Activation-induced cytidine deaminase-dependent DNA breaks in class switch recombination occur during G1 phase of the cell cycle and depend upon mismatch repair. *J Immunol.* 2007;179:6064-6071.
272. Catalan N, Selz F, Imai K, et al. The block in immunoglobulin class switch recombination caused by activation-induced cytidine deaminase deficiency occurs prior to the generation of DNA double strand breaks in switch μ region. *J Immunol.* 2003;171:2504-2509.
273. Petersen S, Casellas R, Reina-San-Martin B, et al. AID is required to initiate Nbs1/γ-H2AX focus formation and mutations at sites of class switching. *Nature.* 2001;414:660-665.
274. Yan CT, Boboila C, Souza EK, et al. IgH class switching and translocations use a robust non-classical end-joining pathway. *Nature.* 2007;449:478-482.
275. Li G, Alt FW, Cheng HL, et al. Lymphocyte-specific compensation for XLF/Cernunnos end-joining functions in V(D)J recombination. *Mol Cell.* 2008;31:631-640.
276. Boboila C, Yan C, Wesemann DR, et al. Alternative end-joining catalyzes class switch recombination in the absence of both Ku70 and DNA ligase. *J Exp Med.* 2010;207:417-427.
277. Dinkelmann M, Spehalski E, Stoneham T, et al. Multiple functions of MRN in end-joining pathways during isotype class switching. *Nat Struct Mol Biol.* 2009;16:808-813.
278. Lee-Theilen M, Matthews AJ, Kelly D, et al. CtIP promotes microhomology-mediated alternative end joining during class-switch recombination. *Nat Struct Mol Biol.* 2011;18:75-80.
279. Saribasak H, Maul RW, Cao Z, et al. XRCC1 suppresses somatic hypermutation and promotes alternative nonhomologous end joining in Igh genes. *J Exp Med.* 2011;208:2209-2216.
280. Iwasato T, Shimizu A, Honjo T, et al. Circular DNA is excised by immunoglobulin class switch recombination. *Cell.* 1990;62:143-149.
281. Matsuoka M, Yoshida K, Maeda T, et al. Switch circular DNA formed in cytokine-treated mouse splenocytes: evidence for intramolecular DNA deletion in immunoglobulin class switching. *Cell.* 1990;62:135-142.
282. Rush JS, Liu M, Odegard VH, et al. Expression of activation-induced cytidine deaminase is regulated by cell division, providing a mechanistic basis for division-linked class switch recombination. *Proc Natl Acad Sci U S A.* 2005;102:13242-13247.
283. Kinoshita K, Harigai M, Fagarasan S, et al. A hallmark of active class switch recombination: transcripts directed by I promoters on looped-out circular DNAs. *Proc Natl Acad Sci U S A.* 2001;98:12620-12623.
284. Mao C, Stavnezer J. Differential regulation of mouse germline Ig γ1 and ε promoters by IL-4 and CD40. *J Immunol.* 2001;167:1522-1534.
285. Vincent-Fabert C, Fiancette R, Pinaud E, et al. Genomic deletion of the whole IgH 3′ regulatory region (hs3a, hs1,2, hs3b, and hs4) dramatically affects class switch recombination and Ig secretion to all isotypes. *Blood.* 2010;116:1895-1898.
286. Kuang FL, Luo Z, Scharff MD. H3 trimethyl K9 and H3 acetyl K9 chromatin modifications are associated with class switch recombination. *Proc Natl Acad Sci U S A.* 2009;106:5288-5293.
287. Chowdhury M, Forouhi O, Dayal S, et al. Analysis of intergenic transcription and histone modification across the human immunoglobulin heavy-chain locus. *Proc Natl Acad Sci U S A.* 2008;105:15872-15877.
288. Jeevan-Raj BP, Robert I, Heyer V, et al. Epigenetic tethering of AID to the donor switch region during immunoglobulin class switch recombination. *J Exp Med.* 2011;208:1649-1660.
289. Rajagopal D, Maul RW, Ghosh A, et al. Immunoglobulin switch μ sequence causes RNA polymerase II accumulation and reduces dA hypermutation. *J Exp Med.* 2009;206:1237-1244.
290. Vuong BQ, Lee M, Kabir S, et al. Specific recruitment of protein kinase A to the immunoglobulin locus regulates class-switch recombination. *Nat Immunol.* 2009;10:420-426.
291. Zhang X, Zhang Y, Ba Z, et al. Fundamental roles of chromatin loop extrusion in antibody class switching. *Nature.* 2019;575:385-389.
292. Roco JA, Mesin L, Binder SC, et al. Class-switch recombination occurs infrequently in germinal centers. *Immunity.* 2019;51:337-350.
293. Geisberger R, Lamers M, Achatz G. The riddle of the dual expression of IgM and IgD. *Immunology.* 2006;118(4):429-437.
294. Phillips C, Kyriakopoulou CB, Virtanen A. Identification of a stem-loop structure important for polyadenylation at the murine IgM secretory poly(A) site. *Nucleic Acids Res.* 1999;27:429-438.
295. Coyle JH, Lebman DA. Correct immunoglobulin α mRNA processing depends on specific sequence in the C α 3-αM intron. *J Immunol.* 2000;164:3659-3665.
296. Takagaki Y, Seipelt RL, Peterson ML, et al. The polyadenylation factor CstF-64 regulates alternative processing of IgM heavy chain pre-mRNA during B cell differentiation. *Cell.* 1996;87:941-952.
297. Phillips C, Pachikara N, Gunderson SI. U1A inhibits cleavage at the immunoglobulin M heavy-chain secretory poly(A) site by binding between the two downstream GU-rich regions. *Mol Cell Biol.* 2004;24:6162-6171.
298. Shapiro-Shelef M, Lin KI, McHeyzer-Williams LJ, et al. Blimp-1 is required for the formation of immunoglobulin secreting plasma cells and pre-plasma memory B cells. *Immunity.* 2003;19:607-620.
299. June CH, O'Connor RS, Kawalekar OU, et al. CAR T cell immunotherapy for human cancer. *Science.* 2018;359(6382):1361-1365. https://pubmed.ncbi.nlm.nih.gov/29567707/
300. Garcia-Castillo H, Barros-Nunez P. Detection of clonal immunoglobulin and T-cell receptor gene recombination in hematological malignancies: monitoring minimal residual disease. *Cardiovasc Hematol Disord Targets.* 2009;9:124-135.

CHAPTER 23

T Cell Antigen Receptors and Antigen Recognition

Mark M. Davis, PhD • Yueh-Hsiu Chien, PhD

INTRODUCTION

T lymphocytes expressing αβ or γδ T cell antigen receptors together with B lymphocytes are the only cells that use combinatorial V, D, and J gene rearrangement to generate diverse antigen receptors. They are found together in all but the most primitive vertebrates, indicating that each type of cell has unique functions and all three are necessary for maintaining host immune competence. While all well-defined functions attributed to T cells in host immune defense and in autoimmunity are performed by αβ T cells, γδ T cells contribute uniquely to the initiation and regulation of the inflammatory response, the regulation of organismal immune metabolism, and in the neuroimmune interface. Yet, most αβ and γδ T cells produce similar sets of cytokines. Just like typical CD8+ αβ T cells, most γδ T cells also can mount cytotoxic responses. Moreover, other than murine skin epidermis, αβ T cells are always found alongside and usually in greater numbers than γδ T cells (including mucosal tissues), suggesting that how these two types of cells are triggered to act determines their unique functions. During the last 4 decades, there has been a great deal of progress in αβ and γδ T cell antigen recognition and the consequences of antigen recognition.

While T cell receptors (TCRs) share structural and functional similarities with B cell antigen receptors (immunoglobulins [Igs]), they also possess a number of unique features pertinent to their specific functions. The major difference between αβ T cells and Ig antigen recognition was first suggested by Zinkernagel and Doherty who found that cytotoxic cells specific for a viral antigen could only lyse infected cells that expressed a particular major histocompatibility complex (MHC) molecule.[1,2] This phenomenon of "MHC-restricted recognition" is in marked contrast to the recognition of intact antigens by antibodies.[3] Later work demonstrated that classical helper and cytotoxic T cells express αβ TCRs, which recognize peptides bound to a characteristic groove in a class I or class II MHC molecules.[4] The semi-invariant T cells, which express αβ TCRs with limited diversity, recognize lipid antigens or small organic molecules associated with nonclassical MHC molecules.[5,6]

The existence of γδ T cells had not been anticipated, and only came to light when an "extra" TCR chain was discovered.[7] Subsequent serological analysis uncovered a novel heterodimer on a subset of T cells.[8] Another surprise was that its antigen recognition is fundamentally different than that of αβ T cells was first suggested by the analysis of MHC recognition by specific γδ T cells.[9] This work showed that at least some γδ T cells do not require antigen processing and that the topology of γδ TCR interaction with the MHC (even in the rare cases when MHC is recognized) is distinct from that of αβ TCRs. Later work showed that most or all γδ T cells recognize antigens directly, like antibodies, and that the MHC molecules are not an obligatory component of γδ T cell antigens. And like antibodies, γδ T cells recognize molecules of different molecular natures, from small organic molecules to large multimeric proteins. Although both αβ and γδ T cells rearrange their TCR genes and develop in the thymus, there are differences in development of the antigen-specific repertoire, activation requirements, and effector fate of γδ T cells versus αβ T cells in the thymic maturation process. These features likely contribute to how these two types of T cells are able to play unique roles in maintaining host immune competence and will be discussed in this chapter.

TCR AND CD3 MOLECULES

Isolation and Characterization of αβ and γδ TCRs

The search for the molecules responsible for T cell recognition first focused on deriving antisera or monoclonal antibodies specific for molecules on T cell surfaces. Ultimately, a number of groups identified "clonotypic" sera[10] or monoclonal antibodies.[11-15] Several of these antibodies were able to block antigen-specific responses by the T cells they were raised against or, when coated on a surface, could activate the T cells they are specific for. They were also able to immunoprecipitate 85 to 90,000 MW disulfide-bonded heterodimers from different T cell clones or hybridomas consisting of two 40,000 to 50,000 MW glycosylated subunits referred to as the α and β TCR chains. Peptide mapping studies showed a striking degree of polymorphism between heterodimers isolated from T cells of differing specificity, thus suggesting that these antigen recognition molecules might be akin to Igs.[16,17]

Work in parallel to these serological studies exploited the small differences (~2%) observed between B and T cell gene expression[18] and isolated both a mouse[19,20] and a human[21] T cell–specific gene that had antibody-like V, J, and C region sequences and could rearrange in T lymphocytes.[20] This molecule was identified as TCRβ by partial sequence analysis of immunoprecipitated material.[22] Subsequent subtractive cloning work rapidly identified two other candidate TCRs cDNAs identified as TCRα[23,24] and TCRγ.[7] It was quickly established that all antigen-specific helper or cytotoxic T cell expressed TCRαβ heterodimers. Where TCRγ fit in remained a puzzle until work by Brenner et al[8] showed that it was expressed on a small (5%-10%) subset of peripheral T cells together with another polypeptide, TCRδ. The nature of TCRδ remained unknown until it was discovered within the TCRα locus, between the Vα and Jα regions.[25] The demonstration that expression of a specific αβ or γδ TCR can transfer antigen specificity provides formal proof that the TCRα and TCRβ subunits or TCRγ and TCRδ subunits were sufficient for antigen recognition.[26-28]

As shown in Figure 23.1, all TCR polypeptides have a similar primary structure, with distinct variable (V), diversity (D) in the case of TCRβ and TCRδ, joining (J), and constant (C) regions analogous to their Ig counterparts. They also share many of the amino acid residues thought to be important for the characteristic variable and constant domains of Igs.[30] Indeed, both αβ and γδ heterodimers have an antibody-like structure (discussed below). The TCR polypeptides all contain a single C region domain (vs up to four for Igs) followed by a connecting peptide, which usually contains the cysteine for the disulfide linkage that joins the two chains of the heterodimer (some human TCRγδ isoforms lack this cysteine and consequently are not disulfide linked).[31] *N*-linked glycosylation sites vary from two to four for each polypeptide with no indications of O-linked sugar addition. C-terminal to the connecting peptide sequences are the hydrophobic transmembrane regions. These have no similarity to those of the *IgH* genes but instead have one (TCRβ and TCRγ) or two (TCRα and TCRδ) positively charged residues. As discussed later, these charged residues are critical for the association of the ligand binding TCR polypeptides with the CD3 signaling polypeptides. This is important because the TCR polypeptides themselves have very short cytoplasmic regions with no known role in signaling.

Another member of the TCR polypeptide family is the pre-Tα chain,[32,33] as shown in Figure 23.1, which serves as a chaperone for TCRβ in early stage thymocytes, similar to the role of λ5 in pre-B and V pre-B cells. It consists of a single Ig constant region–like domain followed by a cysteine-containing connecting peptide, a transmembrane region containing two charged residues: an arginine and a lysine spaced identically to the TCRα transmembrane region. The cysteine in the connecting peptide is presumably what allows heterodimer formation with TCRβ. The similarity to TCRα in the transmembrane region is likely to accommodate the CD3

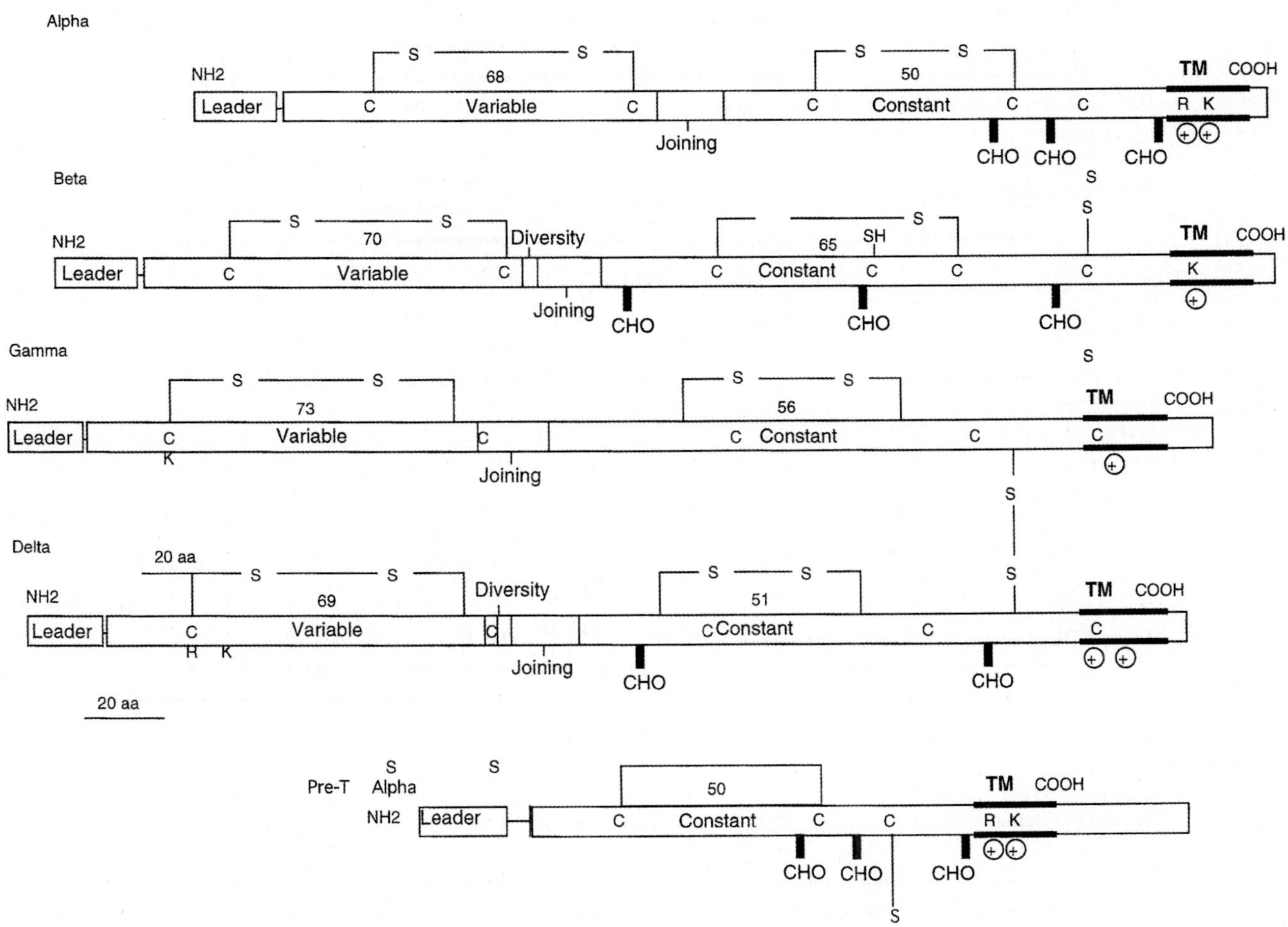

FIG. 23.1. Primary structure of T cell receptors and the pre-Tα polypeptide. Leader (L), variable (V), diversity (D), joining (J), and constant region (C) gene segments are indicated. TM and bold horizontal lines delineate the putative transmembrane regions; CHO indicates potential carbohydrate addition sites; C and S refer to cysteine residues that form interchain and intrachain disulfide bonds; R and K indicate the positively charged amino acids (arginine and lysine, respectively) that are found in the transmembrane regions. (Reprinted by permission from Springer Nature: Davis MM, Bjorkman PJ. T-cell antigen receptor genes and T-cell recognition. *Nature*. 1988;334:395-402.)

polypeptides. The structure of the pre-Tα chain is distinct from the constant (C) domain of the TCRα chain; nevertheless, the mode of association between pre-Tα and TCRβ mirrored that mediated by the Cα-Cβ domains of the αβ TCR.[34] Pre-Tα can dimerize in solution, which may enable the pre-Tα to interact with the variable (V) β domain through residues that are highly conserved across the Vβ and joining (J) β gene families, thus mimicking the interactions at the core of the αβ TCR's Vα-Vβ interface. Disruption of this pre–Tα-Vβ dimer interface abrogates pre-TCR dimerization in solution and impaired pre-TCR expression on the cell surface. These observations suggest a mechanism for pre-TCR self-association to simultaneously "sample" the correct folding of both the V and C domains of any TCRβ chain, regardless of its ultimate specificity,[34] which likely represents a critical checkpoint in T cell development.

CD3 Polypeptides

Immunoprecipitation of the human TCR with anti-idiotypic antibodies after membrane solubilization with detergents revealed four other proteins in addition to TCRα and TCRβ chains (as reviewed in Refs. 35-37). These are known as the CD3γ, δ, ε, and ζ. These CD3 subunits, in their various forms of spliced variants, are an integral part of TCR-mediated T cell activation because only these molecules, and not the cytoplasmic tails of the TCR chains, possess the immunoreceptor tyrosine-based activation motifs (ITAMs) that are necessary for cellular signaling when the TCR engages ligand.

Figure 23.2 illustrates the principal structural features of the CD3γ, CD3δ, CD3ε, and CD3ζ polypeptides as derived from gene cloning and sequencing (as reviewed in Refs. 36,38) and by crystal structures of the extracellular domains of γ, δ, and ε chains.[39-42] Each extracellular domain of the

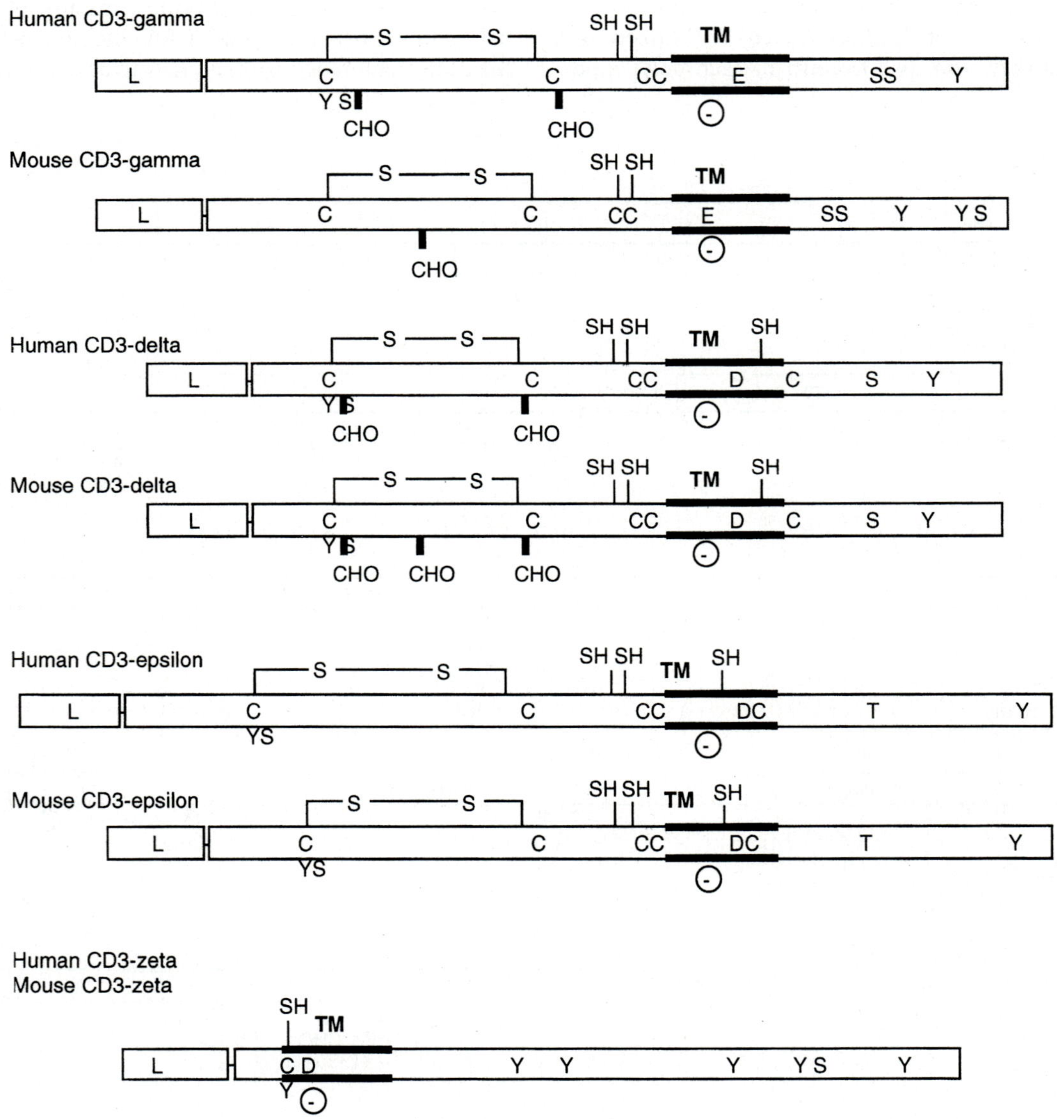

FIG. 23.2. Primary structure of the CD3 molecules. As in Figure 23.1, transmembrane region (TM), carbohydrate addition sites (CHO), and cysteine residues (C) are indicated. In addition, negatively charged membrane residues (D for aspartic acid and E for glutamic acid) and putative phosphorylation sites are shown.

CD3γ, CD3δ, and CD3ε chains consists of a single Ig superfamily domain. The compact Ig-fold is similar to a TCR constant region domain. The γ and δ subunits form distinct heterodimers with CD3ε via highly conserved residues at the dimerizing interface.[39-42] The connecting peptides of the CD3γ, CD3δ, and CD3ε chains all contain highly conserved, closely spaced cysteines just before the membrane-spanning regions. These residues form interchain disulfide bonds and appear to play a role in the assembly of the CD3 and TCR polypeptides.[39,43,52] The extracellular domain of the ζ chain consists of only nine amino acids and contains the only cysteine, which is responsible for the disulfide linkage of the ζζ homodimer or the ζ F_cεRIγ heterodimer. Each of the γ, δ, ε, and ζ polypeptides contains conserved, negatively charged amino acids in their transmembrane region complementary to the positively charged amino acids in the TCR membrane regions.[44-47]

The cytoplasmic regions of the γ, δ, ε, and ζ chains are the intracellular signaling "domains" of the TCR heterodimer. The intracellular sequences responsible for this activation are contained within an 18-amino-acid conserved ITAM motif[48] with the sequence $X_2YX_2L/IX_7YX_2L/I$. Both tyrosines in this motif are required to mediate signal transduction since mutation of either one completely prevents the mobilization of free Ca^{2+} or cytolytic activity.[49] This sequence occurs three times in the ζ chain and once in each of the CD3γ, δ, ε, and F_cεRIγ chains.

Assembly of the αβ TCR/CD3 Complex

The assembly of TCRα and TCRβ chains with the CD3γ, CD3δ, CD3ε, and CD3ζ chains and their intracellular fate has been studied in detail.[34,35,49-51] Mutagenesis studies suggests that the positive charges in each TCR transmembrane domain interacted with complementary negative charges in the transmembrane domains of the different CD3 components.[44] This has been confirmed by the recently determined structure of TCR-CD3 complexes[52] and will be discussed in more detail in a later section.

Pulse-chase experiments show that all six chains (TCRαβ, CD3γ, CD3δ, CD3ε, and CD3ζ) are assembled in the endoplasmic reticulum (ER), transported to the Golgi apparatus, and then transferred to the plasma membrane. It also appears that the amount of ζ chain is rate limiting, as it is synthesized at only 10% the level of the other chains. TCR/CD3 lacking CD3ζ chains migrate through the ER and Golgi intact but then are transported to and degraded in the lysosomes. This results in the vast majority of newly synthesized TCRα, TCRβ, or CD3 components being degraded within 4 hours of their synthesis. The remaining nondegraded chains are long lived due to the formation of complete TCR/CD3 complexes with the limiting ζ chain.[53] The significance of this pre-Golgi degradation pathway is most evident in CD4+ CD8+ thymocytes where, despite high levels of synthesis of both mRNA and protein for all the TCR, CD3, and ζ chains, surface expression is relatively low. The TCR chains in immature thymocytes seem to be selectively degraded.[53] Thus, posttranslation regulation appears to be an important means of controlling the cell surface expression of TCR heterodimers.

Assembly of the γδ TCR/CD3 Complex

Although γδ TCRs, like αβ TCRs, require the association with the CD3 molecules for cell surface expression and signaling, CD3 components associated with these two types of antigen receptors are different in composition, assembly geometry, and glycosylation pattern.[54] These differences are likely reflecting the structure differences in the constant regions of αβ and γδ TCRs (discussed in later sections) and strongly suggest that signal transduction through the TCR occurs differently in γδ versus αβ T cells.

As first demonstrated by Hayes and Love, the murine γδ TCR/CD3 complex lacks CD3δ chain, with the stoichiometry TCRγδCD3$ε_2γ_2ζ_2$.[55] This is in contrast to the murine and human αβ TCR/CD3 complex, which has the TCRαβCD3$ε_2$γδ$ζ_2$ stoichiometry. Interestingly, a differentially glycosylated form of CD3γ was found to be associated with γδ TCRs dependent on the activation state of the cells.[56] In addition, while CD3ζζ is incorporated into the complexes of naïve cells, T cell activation can result in the expression and incorporation of F_cεR1γ into the γδ TCR complex.[56] While human αβ and γδ TCR/CD3 have the same composition, TCRαβCD3$ε_2$γδ$ζ_2$ or TCRγδCD3$ε_2$γδ$ζ_2$, the glycosylation patterns of CD3γ and CD3δ chains in the αβ TCR/CD3 and γδ TCR/CD3 complexes are very different. Furthermore, αβ TCR/CD3 and γδ TCR/CD3 differ in the plasma membrane organization and their accessibility of signaling motifs in the CD3 intracellular tails.[54]

TCR Genes

As shown in Figure 23.3, TCR gene segments are organized similarly to those of Igs and the same recombination machinery is responsible for joining separate V and D segments to a particular J and C segments. This was initially indicated by the observation that the characteristic recombination signals (seven and nine nucleotide conserved sequences adjacent to the V, D, and J regions with the 12 or 23 nucleotide spacing between them), first described for Ig genes, are also present in TCRs.[57] The conclusive evidence of this common rearrangement mechanism is that both a naturally occurring recombination-deficient mouse strain (severe combined immunodeficiency; Ref. 58) and mice engineered to lack recombinase-activating genes 1[59] or 2[60] are unable to rearrange either *TCR* or Ig gene segments properly. Many of the other molecules involved in Ig gene rearrangement serve the same function in TCRs.[61] As with Igs, if the V region and J region gene segments are in the same transcriptional orientation, the intervening DNA is deleted during recombination. DNA circles of such material can be observed in the thymus,[62,63] the principal site of TCR recombination (see below). In the case of *TCRβ* and *TCRδ*, there is a single V region 3′ to the C in the opposite transcriptional orientation to J and C. Thus, rearrangement to these gene segments occurs via an inversion. Variable points of joining are seen along the *V*, *D*, and *J* gene segments. There are random nucleotide insertions (N regions) as well as the addition of nucleotides in an inverted repeat pattern, referred to a P element insertion in each recombination junction.[64]

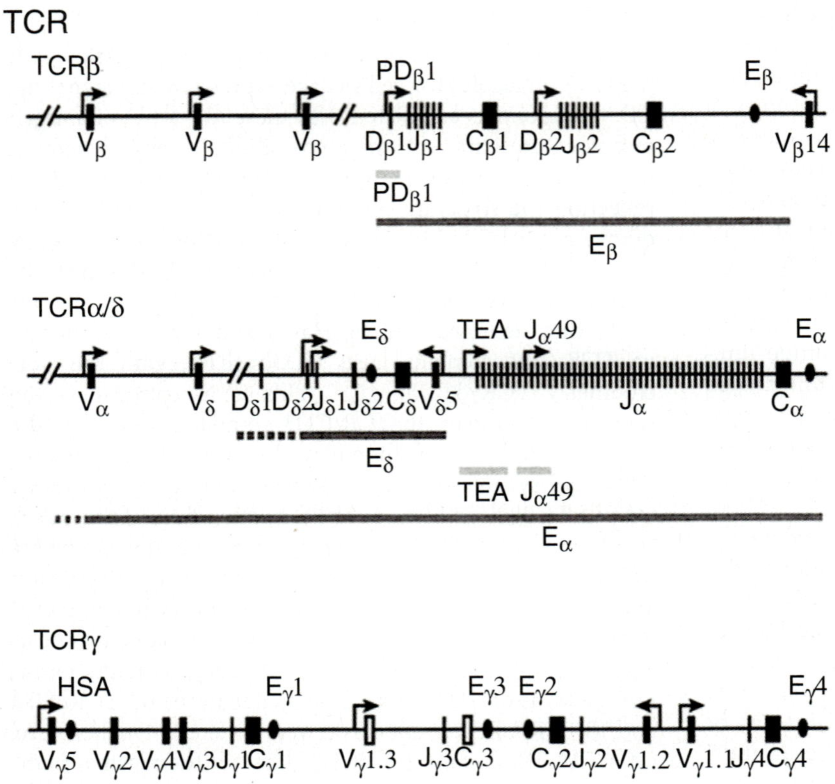

FIG. 23.3. T cell receptor gene organization. Schematic of V, D, J, and C elements of the T cell receptor genes. Transcriptional orientation is from left to right, except where noted. The overall size of each locus is as indicated to the right. E designates enhancer elements, and S designates silencer elements.

TCRα/δ Locus

In humans and in mice, there is a single α chain C region gene that is composed of four exons encoding (1) the first constant region domain, (2) the second domain that consists of 16 amino acids including the cysteine that forms the interchain disulfide bond, (3) the transmembrane and intracytoplasmic domains, and (4) the 3′ untranslated region (Fig. 23.3). The entire α/δ locus spans about 1.1 MB in both mice and humans. There are 50 different J region gene segments upstream of the C region of the TCRα chain in the murine locus. At least eight of these J regions are nonfunctional either because of inframe stop codons or having rearrangement and splicing signals that are likely to be defective. A similar number of α chain J regions are present in the human locus. This very large number of J regions compared to the Ig loci may indicate that the functional diversity contributed by the J segment of the TCR (which constitutes a major portion of the third complementarity-determining region [CDR3] loop) makes an important contribution to antigen recognition (see below).

In both the murine and human loci, the Cδ, Jδ, and two Dδ gene segments are located between the Vα and Jα gene segments. In the murine system, there are two Jδ and two Dδ gene segments on the 5′ side of Cδ and the Cδ exons are approximately 75 kb upstream of the Cα gene and approximately 8 kb upstream of the most 5′ known Jα gene segments. The human organization is similar, with three Dα gene segments and three Jδ segments. In both murine and human TCRα/δ locus, one of the Vδ gene segments—Vδ5 (murine), Vδ3 (human)—is located at the 3′ end of Cδ in the opposite transcriptional orientation and rearranges by inversion.

Surprisingly, all of the D elements of TCRδ can be joined to other D gene segments, which is not the case with TCRβ or IgH. This greatly increases the junctional or CDR3 diversity. With the potential for random nucleotide addition in between each jointed gene segment, TCRδs have the highest estimated diversity among of all the antigens receptor chains. The implications for this and comparisons with other antigen receptor genes are discussed below.

The location of Jδ and Cδ genes between Vα and Jα gene segments suggests that TCRδ and TCRα could share the same pool of V gene segments. But most TCRα and TCRδ chains use different V genes and even different members of the same gene family. Whereas approximately 50 V genes are associated with TCRα, TCRδ-associated V genes are only ~10 in mice and 8 in human. Nonetheless, a population of human δ/β T cells, expressing TCRs comprising a TCRδ variable gene (Vδ1) fused to α chain J and constant regions, paired with an array of TCRβ chains[65] have been reported. The mechanisms that account for the preferential usage of certain V gene segments to produce δ versus α chain are not known.

An implicit characteristic of the α/δ gene locus is that a rearrangement of Vα to Jα deletes the entire D-J-C core of the δ chain locus. Indeed, in many αβ T cells, the α chain locus is rearranged on both chromosomes and thus no TCRδ could be made. In most cases, this is due to Vα→Jα rearrangement but evidence suggesting an intermediate step in the deletion of TCRδ has been reported.[66] This involves rearrangements of an element termed TEA to a pseudo-Jα 3′ of Cδ. The rearrangement of TEA to this psuedo-Jα would eliminate the δ chain locus in αβ T cells. Gene targeting of the TEA element resulted in normal levels of αβ and γδ T cells, but usage of the most Jαs was severely restricted,[67] suggesting that its function is to govern the accessibility of the most proximal 5′ Jαs for recombination.

TCRβ Locus

The entire human 685 kb β chain gene locus was originally sequenced by Hood and coworkers.[68] One interesting feature is the tandem nature of JβCβ in the TCRβ locus. This arrangement is preserved in all higher vertebrate species that have been characterized thus far (mouse, human, chicken, frog). The two Cβ coding sequences are identical in the mouse and nearly so in humans and other species. Thus, it is unlikely that they represent two functionally distinct forms of Cβ. However, the Jβ clusters have relatively unique sequences and thus this may be a mechanism for increasing the number of Jβ gene segments. Together with the large number of Jα gene segments, there is far more potential diversity (Jα × Jβ = 50 × 12 = 600) provided by J regions in αβ TCRs than in Igs.

Most of the Vβ regions are located upstream of the joining and constant regions and in the same transcriptional orientation as the D and J gene element and rearrange to DβJβ gene via deletion. Similar to the case of Vδ5, a single Vβ gene, Vβ14, is located 3′ to C regions and in the opposite transcriptional orientation, thus rearrangements involving Vβ14 occur via inversion.

In the NZW strain of mouse, there is a deletion in the β chain locus that spans from Cβ1, suggesting a common ancestry with TCRδ, up to and including the Jβ2 cluster.[69] In SJL, C57BR, and C57L mice, there is a large deletion[70] in the V region locus from Vβ5 to Vβ9. These mice also express a V gene, Vβ17, that is not expressed in other strains of mice. Deletion of about half of the V genes (in SJL, C57BR, and C57L mice) does not seem to have any particular effect on the ability of these mice to mount immune responses, whereas mice that have deleted the $J_β2$ cluster show impaired responses.[71]

TCRγ Locus

The organization of the mouse and human γ chain loci are also shown in Figure 23.3. There is more than one nomenclature commonly used to describe the γ chain genes.[72-75] Here, we use that of Lefranc and Rabbitts[76] and Tonegawa and colleagues[72]—for the human and mouse γ chains, respectively. The relationship between this scheme and those of others is described in Raulet.[77]

The human γ chain locus consists of four subgroups; VγI includes Vγ2, Vγ3, Vγ4, Vγ5, and Vγ8. Among the three other Vγ subgroups, only Vγ9 (from the VγII group) is functional. The Vγs are located 5′ to two JγCγ clusters, and each of the V genes is potentially capable of rearranging to any of the five J regions. The sequences of the two human Cγ regions are very similar overall and only differ significantly in the second exon. In Cγ2, this exon is duplicated two to three times and the cysteine that forms in the interchain disulfide bond is absent. Thus, Cγ2-bearing human T cells have an extra-long γ chain (55,000 MW) that is not disulfide bonded to its δ chain partner.

The organization of the murine γ chain genes is different than that of the human genes in that there are three separate rearranging loci that span about 205 kb.[77,78] Of four murine Cγ genes, Cγ1 and Cγ2 are very similar in coding sequences. The major difference between these two genes is in the 5 amino acid deletion in the Cγ2 gene that is located in the C II exon at the amino acid terminal of the cysteine residue used for the disulfide formation with the δ chain. Cγ3 is apparently a pseudogene in BALB/c mice, and the Jγ3 Cγ3 region is deleted in several mouse strains including C57 BL/10. The Cγ4 gene differs significantly in sequences from the other Cγ genes (in 66% overall amino acid identity). In addition, when compared with Cγ2 gene, the Cγ4 gene has a 23 amino acid insertion preceding the cysteine residue.

The murine Vγ genes usually rearrange to the JγCγ gene that is most proximal and in the same transcriptional orientation. Thus, Vγ1 rearranges to Jγ4Cγ4; Vγ2 to Jγ2Cγ2; and Vγ4, Vγ5, Vγ6, and Vγ7 to Jγ1Cγ1. During murine fetal thymic ontogeny, the first two waves of T cells are γδ T cells that rearrange Vγ5 and Vγ6 in the γ locus and Vδ1 in the δ locus. The unique tissue distribution and function of the Vγ5Vδ1 and Vγ6Vδ1 γδ T cells will be discussed in later section. In the adult thymus, Vγ gene rearrangement switches to the production of γδ T cells expressing Vγ1, Vγ2, Vγ4, and Vγ7 gene segments.[72,77,79,80] The sequential gene rearrangement at the TCRγ loci to generate of subsets of γδ T cells at different stages of ontogeny is a fixed developmental program that cannot be easily reordered. Consistent with an intrinsic program of Vγ gene rearrangement, germline transcription of Vγ genes is found to correlate with their rearrangement pattern, suggesting a possible link between gene accessibility and rearrangement in this system.[81]

TCR Diversity

The basic V(D)J recombination machinery is shared between TCR loci and Igs. All three types of antigen receptors utilize the CDRs to bind specific antigens.[82-84] The CDR1 and CDR2 loops are encoded by the germline V regions. Different V gene combinations contribute to the diversity. The CDR3 loops are encoded by the junction of variable region gene segments joining to J (TCRα and TCRγ) or to D and J segments (TCRβ and TCRδ). Different gene segments, flexibility in the precise joining places, and the short N and P nucleotide insertions between the junctions of each gene segment (V and J, V and D, D and D, D and J) contribute to the diversity of the CDR3 regions. As discussed in Davis and Bjorkman,[29] while the overall estimated diversity of TCRs and antibodies is similar, it is clear that the antibodies have higher estimated diversities generated from V gene combination, but αβ and γδ TCRs have higher potential diversity in their CDR3 regions. In fact,

TCRδ chains have the highest potential diversity in the CDR3 region among all antigen receptor chains. This is because of the presence of multiple D gene segments (two in mice, three in human, and up to five in cattle) that can join together. Each D gene segment can be read in all three reading frames, and N nucleotides can be inserted into the junctions of the joining segments.

Initially it was thought that this concentration of diversity in the CDR3 regions of αβ TCRs made sense since αβ TCRs need to recognize a vast diversity of peptides in a more limited set of MHC molecules and that the peptide is displayed in a linear fashion that could be efficiently surveyed by the adjacent Vα and Vβ CDR3s.[29] However, as discussed in detail later, γδ TCRs recognize diverse set of ligands directly with no antigen processing and presentation requirement, suggesting that CDR3 diversity may simply be necessary for diverse antigen recognition. Indeed, immunizing transgenic mice expressing a human heavy chain minilocus, which contains 1 functional human Vh gene segment, 10 functional D segments, 6 Jh segments, and the Cμ and Cγ1 constant region on the *IgH−/−Igk−/−* background with various antigens, showed that these mice can generate specific antibody responses to diverse hapten and protein antigens and the antigen-specific antibodies utilizing identical heavy and lambda light chain V genes but differ in the Vh CDR3 regions,[85] indicating that diversity in the CDR3 region of Vh is sufficient for most antibody specificities. Thus, highly diverse CDR3 loops are the key determinant of specificity in antigen recognition in both TCRs and antibodies. Hence, despite having a limited number of V gene combination, γδ TCRs have the potential to recognize a large number of different antigens. It should be noted that these early estimates of TCR and Ig diversity were originally made with some very simple assumptions based on the number of gene segments, N region contributions, etc. Increasingly sophisticated sequencing technology has brought much more precise numbers of sequence diversity, at least for the TCRβ chain from Robins and colleagues,[86-88] analyzing tens of millions of human TCR sequences from peripheral blood T cells, and arriving at an estimate of ~10 million TCRβ, including both gene segment combinations and CDR3 diversity, only a fraction of which was expressed by a given individual. New analyses are just being reported for TCRβ and now TCRα, which are 5 to 10 times higher.[89-91] So the numbers are still somewhat in flux, but TCR sequences in humans are being generated at a great rate and so we should have firmer numbers in the not-too-distant future.

Allelic Exclusion

In Igs, normally only one allele of the heavy chain locus and one of the light chain alleles is productively rearranged and expressed, a phenomenon termed "allelic exclusion." With respect to αβ TCR expression, while TCRβ exhibits allelic exclusion,[92] TCRα seems much less constrained[93,94] and many mature T cells express two functional TCRα chains. As the chances of forming an inframe joint with any antigen receptor is only one in three, the probability that a T cell would have two productively rearranged TCRαs is only $1/3 \times 1/3 = 1/9$ or 11%. However, even when this happens, the two TCRα chains may not form heterodimers equally well with the single TCRβ that is expressed; thus, only one heterodimer may be expressed. But this simple calculation is complicated by the likelihood that only thymocytes that make at least one productive TCR rearrangement of each type will have a chance at maturation, which would eliminate almost half of the T cells (4/9ths, which is the product of a two-thirds chance of failure on one chromosome, followed by the same failure rate on the second). Secondly, it has been found that there is a mechanism called receptor editing, which means that a given rearrangement that is not productive, either because of an out-of-frame joint or for reasons of self-reactivity, can induce a V region further 5′ or "upstream" of the initial VJ joint to rearrange to one of the remaining Jαs.[95] In any event, it has been reported that one-third of human T cells express two TCRs.[93]

There also appears to be an important role for the pre-TCR heterodimer (eg, pre-Tα:TCRβ) in blocking further TCRβ rearrangement and thus ensuring allelic exclusion at that locus.[96,97] In particular, pre–Tα-deficient mice had a significant increase in the number of cells with two productive TCRβ rearrangements, compared with wild-type mice.[96]

Human γδ T cells that express two distinct TCRs utilizing two different γ chain alleles on cell surface have been reported.[98] This issue has not been systematically examined in γδ T cells from other animals.

αβ T CELL RECOGNITION

αβ TCR–CD3 Structure and Ligand Binding

Early efforts to derive x-ray crystal structures of TCR heterodimers and fragments of heterodimers presented many technical hurdles (as summarized in Ref. 99). One difficulty is that structure determination required engineering the molecules into a soluble form. A second problem is that many of the TCRs are heavily glycosylated, and it was necessary to eliminate most or all of the carbohydrates on each chain to achieve high-quality crystals. An alternative is to express soluble TCRs in insect cells, where they have compact N-linked sugars, or in *Escherichia coli*, where they are unglycosylated. The first successes in TCR crystallization came from the laboratory of Mariuzza and collaborators who solved the structure of first a Vβ Cβ polypeptide[100] and then a Vα fragment.[101] In the following year, the first complete αβ TCR structures were solved.[102,103] The structure of the 2C TCR by Garcia and colleagues is shown in Figure 23.4A.[102] In general, as predicted from sequence homologies, these domains are all Ig like, with the classical β-barrel structure in evidence in all three domains. At each end of the barrel in each V region domain, there are four loops between the β sheets, three of which form the CDRs of Igs. The fourth loop, between the D and E strands, has been implicated in superantigen binding. The six CDR loops from the two variable domains form the antigen-binding surface in both Igs and TCRs. The major anomaly in terms of similarity of TCRs to Igs is the structure of the Cα.[102] Cα consists of one half of the classical β-barrel, that is, one set (or "sheet") of β strands, while the rest of the partially truncated domain exhibits random coils. This type

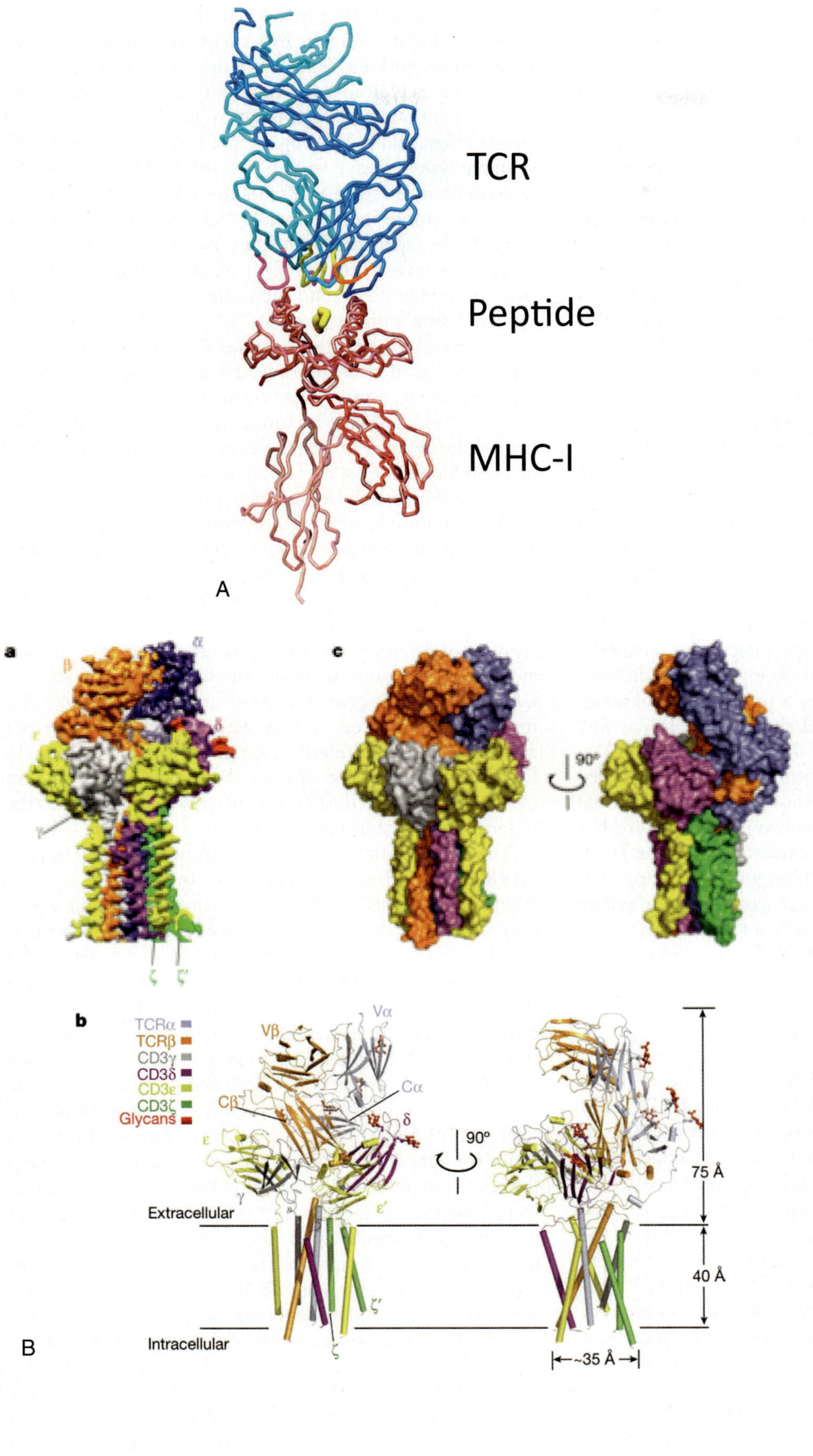

FIG. 23.4. A: T cell receptor (TCR)-peptide/major histocompatibility complex (MHC) crystal structure of a TCR-peptide/MHC complex. Peptide and complementarity-determining regions are portrayed in different colors.[102] **B:** CD3 molecules complexed to an αβ TCR. The recent solution of the first high-resolution structure of a CD3–αβ TCR complex by Dong and colleagues[52] both represents the culmination of many decades of effort and confirms some of the previous deductions from mutagenesis and assembly data, and also provides a wealth of molecular level detail. *a,* Represents rotations of a space filling picture of the complex. *b,* Ribbon diagrams of the TCR: CD3 structure. (A, Republished with permission of American Association for the Advancement of Science, from Garcia KC, Degano M, Stanfield RL, Brunmark A, Jackson MR, Peterson PA, Teyton L, Wilson IA. An αβ T cell receptor structure at 2.5 Å and its orientation in the TCR-MHC complex. *Science*. 1996;274:209-219; permission conveyed through Copyright Clearance Center, Inc. B, Reprinted by permission from Springer Nature: Dong D, Zheng L, Lin J, Zhang B, Zhu Y, Li N, Xie S, Wang Y, Gao N, Huang Z. Structural basis of assembly of the human T cell receptor–CD3 complex. *Nature*. 2019;573:546-552.)

of structure is unprecedented in the Ig gene family. The functional significance of such a variant structure in unknown, but it has been suggested that this incompletely formed Ig-like domain may be responsible for the observed lability of TCRα and this may allow greater flexibility in the regulation of its expression. Another possible explanation is that this configuration is designed to accommodate one or more of the CD3 molecules.

The now large number of solved αβ TCR structures can be compared to those of the Fab fragment of an antibody; there are several features that are unique to the αβ molecules. These include the following: (1) In one structure,[102] four out of seven N-linked sugars diffracted to high resolution, indicating that they are not free to move very much and thus may play a structural role, particularly in Cα-Cβ interactions. This correlates with mutagenesis data indicating that certain Cα sugars cannot be eliminated without abolishing protein expression[104] and the disordered state of a Cα domain in the structure of a TCR lacking glycosylation.[103] (2) There is significantly more contact between Vβ and Cβ and between Vα and Cα than in the equivalent regions of antibodies. (3) Between the CDR3 loops of Vα and Vβ, there is a pocket that can (and does in at least one case)[102] accommodate a large side chain from the peptide bound to an MHC.

Recently a complete structure of a TCR-CD3 complex has been solved[52] revealing the complex architecture of at least the unliganded form (Fig. 23.4B). This work is a tour de force of modern crystallography, adding a wealth of detail about the relationships between the TCR chains and CD3, as well as confirming the close packing of the TCRαβ, CD3δ, CD3ε, and CD3ζ chains and the complementary charge interactions in the membrane that were earlier shown to be critical for the assembly of the complex.[44,105] Not shown in the structure, but of importance is the interaction of cholesterol with the TCRβ chain as shown by Schamel and colleagues.[106] The apparent displacement of cholesterol by an excess of cholesterol sulfate inhibits T cell signaling, and the ratio of these two naturally occurring membrane lipids influences the sensitivity of thymocytes during thymic selection.[107]

αβ TCR Ligand Recognition

Although it has long been established that most αβ TCR–expressing T cells recognize peptides bound to an MHC molecule, except MAIT and NKT T cells, which use αβ TCRs to recognize lipids or other molecules bound to nonclassical class I molecules (see Chapters 36 and 37), a formal biochemical demonstration that this was due to TCR binding to a peptide/MHC complex took many years to establish. Part of the difficulty in obtaining measurements of this type has been the intrinsically membrane-bound nature of MHC and TCR molecules. Another major problem is that the affinities are very low, in the micromolar range, ~10,000-fold weaker than most antibodies, which is too unstable to measure by conventional means. This is consistent with the binding characteristics of most other membrane proteins that interact with proteins on other cell surfaces, since they involve direct cell-cell contact and considerable polyvalency, in most cases, in a largely two-dimensional (2D) environment. For this reason, it was not surprising that somatic hypermutation was not observed in the initial work on TCRs, since there was no need for the high affinities typically seen in affinity-matured (and highly mutated) antibodies, which typically need to work in a 3D environment. But reminding us that little is absolute in the immune system, recent work has shown that there is somatic hypermutation in some TCRs in some organisms.[108] Although here the purpose is not to create high affinity TCRs, but to diversify the repertoire for the purposes of thymic selection.

To some extent, the problem of measuring the interactions of membrane-bound molecules can be circumvented by expressing soluble forms of TCR and MHC, which is also essential for structural studies (see above). For TCRs, many successful strategies have been described, including replacing the transmembrane regions with signal sequences for glycolipid linkage,[109] expressing chains without transmembrane regions in either insect or mammalian cells,[110] or a combination of cysteine mutagenesis and *E. coli* expression.[103] Unfortunately, no one method seems to work for all TCR heterodimers, although the combination of insect cell expression and leucine zippers at the C-terminus to stabilize heterodimer expression has been successful in many cases.[111] The production of soluble forms of MHC molecules has a much longer history, starting with the enzymatic cleavage of detergent-solubilized native molecules,[112] as well as some of the same methods employed for TCR such as GPI linkage,[113] *E. coli* expression and refolding,[114,115] and insect cell expression of truncated (or leucine zippered) molecules.[116] One interesting variant that seems necessary for the stable expression of some class II MHC molecules in insect cells has been the addition of a covalent peptide to the N-terminus of the β chain.[117]

The first measurements of TCR affinities binding to peptide/MHC complexes were performed by Matsui et al[118] and Weber et al.[119] Matsui and colleagues used a high concentration of soluble peptide/MHC complexes to block the binding of a labeled anti-TCR Fab fragment to T cells specific for those complexes, obtaining an equilibrium-binding affinity (K_D) value of ~50 μM for several different T cells and two different cytochrome peptide/I-E^k complexes. Weber and colleagues used a soluble TCR to inhibit the recognition of a flu peptide/I-E^d complex by a T cell and obtained a K_D value of ~10 μM. While these measurements were an important start in TCR biochemistry, they gave no direct information about the kinetics of TCR-ligand interactions. Fortunately, the development of surface plasmon resonance (SPR) instruments, particularly the Biacore™ (Pharmacia Biosensor) with its remarkable sensitivity to weak macromolecular interactions,[120] has allowed rapid progress in this area. In this technique, one component is covalently cross-linked to a surface and then buffer containing the ligand is passed in solution over it. The binding of even ~5% of the surface-bound material is sufficient to cause a detectable change in the resonance state of gold electrons on the surface. This method allows the direct measurement of association and dissociation rates, that is, kinetic parameters, and also has the advantage of requiring neither cells nor radioactive labels.

Microcalorimetry has also been used to measure some TCR ligand affinities, and these analyses have confirmed the SPR values,[121] but do not allow kinetic measurements. These and other data (reviewed in Refs. 84,99) showed definitively that TCR and peptide-loaded MHC molecules alone are able to interact and also that expression in a soluble form has not altered their ability to bind to each other. SPR measurements show that while the on-rates of TCRs binding to peptide/MHC molecules vary from very slow (1000 M sec) to moderately fast (200,000 M sec), their off rates fall in a relatively narrow range (0.5-0.01 sec^{-1}) or a $t_{1/2}$ of 12 to 30 seconds at 25 °C. This is in the general range of other membrane-bound receptors, which recognize membrane molecules on other cells,[122] but it has been noted that most TCRs have very slow on-rates,[84] which reflects a flexibility in the binding site that might help to foster cross-reactivity (see below). In the case of a class I MHC–restricted TCR, 2C, this relatively fast off rate may be stabilized (10-fold) if soluble CD8 is introduced,[123] but this result is controversial.[124] CD8 stabilization of TCR binding has been seen by Luescher et al in cell-based TCR labeling assay[125]; however, no enhancement of TCR binding has been seen using soluble CD4[126] (see below for more discussion of CD4 and CD8).

Other biophysical methodologies have been utilized to measure TCR binding in its native state, that is, on a T cell surface and interacting with peptide-MHC ligands on either an artificial bilayer[127] or on the surface of a red blood cell "reporter." These types of measurements are important because studies of TCRs and peptide-MHC complexes binding in solution only partially reflects what is happening between the two membranes. This is because of the very severe constraints on binding that are present in this specialized environment. For one, it is likely that TCRs on T cells and pMHCs on antigen-presenting cells (APCs) are fixed into a fairly rigid alignment that maximizes their ability to bind to each other, as opposed to the same molecules in solution that are diffusing in all directions and for which only a fraction of collisions will result in a productive binding event. Secondly, binding is happening in a very small space/volume of liquid, making the effective concentration very high. While this environment is not strictly 2D, Dustin and others have used this as a simplifying assumption with relative success.[128] But a direct measurement of TCR affinity and kinetics has been made by Huppa and colleagues,[127] who used a fluorescent probe on an anti-TCR antibody to create a system in which this fluorophore could form a Förster or fluorescence resonance energy transfer (FRET) pair with a complementary label on the C-terminus of the peptide that could emit photons of the appropriate wavelength only when the TCR bound its peptide-MHC ligand. This enabled precise measurements of the dissociation rate (K_{off}) and affinity (K_D) within an immunological synapse, and these measurements could also be used to estimate an average association rate (K_{on}). Another advantage of these measurements is that they can be done at 37 °C, whereas most of the SPR measurements cited above were performed at 25 °C due to instrument limitations. Dissociation rates at higher temperatures are significantly faster. As expected from the discussion above, the association rate was greatly accelerated ~100× over the solution rate presumably because of the ideal orientation of the TCR and pMHC with respect to each other. Not expected was an accelerated dissociation rate of 4 to 10 faster. This turns out to be a function of actin polymerization and depolymerization activity within the T cell, as shown dramatically in the actin dynamics studies of Vale and Dustin.[129] Once this was inhibited, the dissociation rates within the synapse were quite close to the solution values.[130] Still, it is worth noting that TCR binding to pMHC is even more unstable within an active synapse than the solution measurements have suggested, with half-lives of less than 1 second in many cases.

A second experimental system that has been applied to the question of TCR binding in the context of cell-cell contact has been that of Zhu and colleagues, in which red blood cells coated with peptide-MHC complexes are brought into repeated contact with a T cell of the appropriate specificity. By visualizing the response of the pMHC coated red blood cell to different frequencies and durations of T cell contact, one can infer parameters of TCR affinity, at least in the 2D aspect.[128] While this system has worked well in describing integrin binding,[131] it is indirect, and in some ways contradicts results obtained with the FRET system of Huppa et al and even some of the basic properties determined in solution measurements. Nonetheless, it may be telling us important properties of cell-cell contact that are more difficult to discern with more directly molecular methods.[128,132-135]

Topology and Cross-Reactivity

As discussed above, TCR sequence diversity resides largely in the region between the V and J region gene segments, known as the CDR3 regions.[29] This led to early models in which the CDR3 loops of Vα and Vβ make the principal contacts with the antigenic peptide bound to the MHC.[29] Since multiple examples of unrearranged V-J–like sequences are known, it also led to speculation that TCRs arose from transposable elements inserting between V and J segment prototypes to create diversity that could then be repurposed to recognize different peptides bound to MHC molecules,[29] but this remains a speculation. Support for the model that V(D)J diversity was the primary factor in αβ TCR recognition has advanced much further and came well before any crystal structures were available (ca 1996) using a novel method developed by Jorgensen et al,[136] who made single amino acid changes in an antigenic peptide at positions that affect T cell recognition but not MHC binding. These variant peptides are then used to immunize mice that express either α or β chain of a TCR that recognizes the original peptide and the responding T cells are analyzed. Using these "hemitransgenic" mice allows the resulting T cells to keep one half of the receptor constant, while allowing considerable variation in the chain that pairs with it. Strikingly, in these experiments, changing a central lysine to an arginine in the peptide triggered a reciprocal change from an arginine in the Vα CDR3 to a lysine, suggesting a direct connection in the form of a "salt bridge," which was confirmed in the structural work of Newell et al.[137] In addition, Engel et al showed that a single CDR3 point mutation could alter the specificity of a TCR[138] and also a CDR3 "transplant" could confer the specificity of the donor TCR onto the recipient.[139]

These data were largely confirmed by the structures of TCR-pMHC complexes, where there are now over 50 structures of TCR-pMHC complexes.[99,140,141] These show that TCRs bind in roughly diagonal to 90° configurations, ranging over 30°. But in all cases, in these structures, the CDR3 loops are centrally located over the peptide, but what was unexpected is that the Vα CDR1 and the Vβ CDR1 are also in a position to contact the N-terminal and C-terminal peptide residues, respectively. The confined nature of TCR pMHC topology is a departure from the rotational flexibility of antibody-antigen interactions and may reflect a need to accommodate other molecules into a particular configuration that is optimal for signaling, such as CD4, CD8, and/or CD3 components, although other factors can also be critical for signaling (see below).

As αβ TCR heterodimers are first selected in the thymus for reactivity to self-peptides bound to MHC molecules, all foreign peptide-reactive TCRs could be considered to be inherently cross-reactive. Indeed, a number of T cells have reactivity to very different peptide sequences, as shown by Nanda and Sercarz.[142] It has also been argued by Mason[143] that the universe of peptides is so large that each T cell must on average be cross-reactive to ~10^6 different peptides (although many of the differences in peptide sequence in this calculation would not be accessible to the TCR, being buried in the MHC binding groove). Nevertheless, some degree of cross-reactivity to structurally distinct peptides would expand the effective repertoire. Several large-scale screens of a random 9-mer peptide library with different T cells did turn up a great many stimulatory peptides, but very few have changes in the 2-4 key TCR-sensitive residues.[144] Instead, the major requirement for cross-reactivity may be in thymic selection[145] and/or in the use of endogenous peptide MHCs to augment T cell sensitivity to agonist ligands.[146-148] Direct evidence of TCR plasticity was first obtained by Garcia et al[149] who compared the x-ray crystal structures of the same TCR bound to four different peptide-MHC ligands and found many ways in which the TCR binding site could adjust to recognize these different ligands (Fig. 23.5). Multiple other conformational changes were found in the CDR3β loop, particularly as it binds to a pMHC complex.[151] That each TCR may have many different conformations of its CDR3 loops is also suggested by the 2D nuclear magnetic resonance studies of Reinherz and Wagner and colleagues, who found that the CDR3 regions of a TCR in solution were significantly more mobile than the rest of the structure.[152] That this may be a general feature of most TCRs is supported by thermodynamic analyses of various TCRs binding to their peptide/MHC ligands, both class I and class II. Here the binding is often accompanied by a substantial loss of entropy (Table 23.1) and, in most cases, an "induced fit" mechanism.[121,170,171] This seems to be a situation where an inherently flexible binding site achieves greater order upon binding. This is a mechanism that is also employed by DNA recognition proteins, and Boniface et al[171] have suggested that it might represent a common mechanism of "scanning" an array of very similar molecular structures (MHCs or DNA)

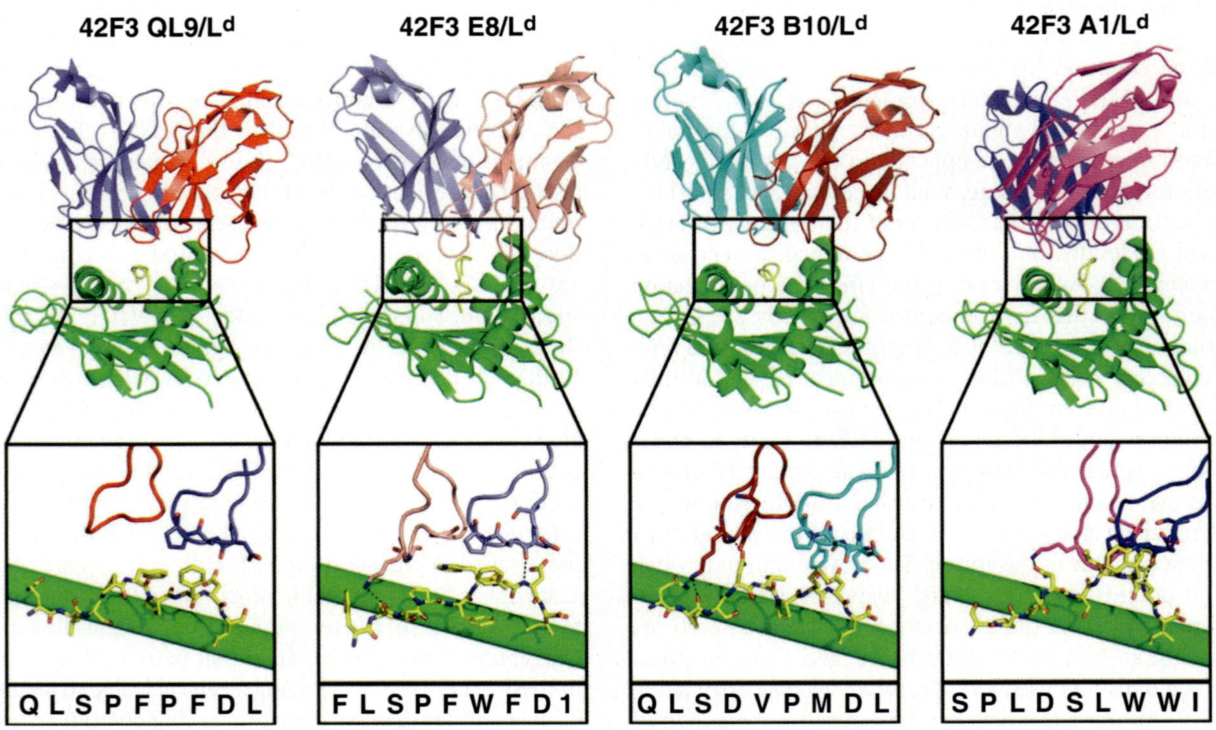

FIG. 23.5. αβ T cell receptor (TCR) binding plasticity. Structures of the same TCR bound four different peptide-MHC complexes highlight the plasticity of the binding interface of αβ TCRs generally and help to explain how the same receptor can recognize very different peptides. (This is from the work of Garcia and colleagues and is reproduced from Su LF, Davis MM. Antiviral memory phenotype T cells in unexposed adults. *Immunol Rev.* 2013;255:95-109.)

TABLE 23.1 Vβ Specificity of Exogenous and Endogenous Superantigens

Bacterial Superantigen	Human Vβ Specificity	Reference	Murine Vβ Specificity	Reference
				As reviewed in 153,154
SEA	ND		1,3,10,11,12,17	155
SEB	3,12,14,15,17,20	156,157	(3), 7,8.1,8.3, (11), (17)	155
SEC_1	12	157	7,8.2,8.3,11	155
SEC_2	12,13,14,15,17,20	156,157	8.2,10	155
SEC_3	5,12	157	(3), 7,8.2	155
SED	5,12	157	3,7, (8.2), 8.3,11,17	155
SEE	5.1,6.1-6.3,8,18	156,157	11,15,17	155
TSST-1	2	157	15,16	155
ExFT	2	156	10,11,15	155
Strep M	2,4,8	158	ND	159
Endogenous Proviruses	**Vβ Specificity**	**Mls Type[a]**	**Chromosome**	**Reference**
				As reviewed in 154,160
Mtv-1	3	c, 4a	7	
Mtv-2	14	NA	18	
Mtv-3	3,17	c	11	
Mtv-6	3,17	c, 3a	16	
Mtv-7	6,7,8.1,9	a, 1a	1	
Mtv-8	11,12	f, Dvbll.1	6	
Mtv-9	5,11,12	f, Etc-1	12	
Mtv-11	11,12	f, Dvbll.3	14	
Mtv-13	3	c, 2a	4	
Mtv-43	6,7,8.9,9	Mls like	ND	
Exogenous Viruses	**Vβ Specificity**	**Mls Type**	**Chromosome**	**Reference**
MMTV-C3H	14,15	NA		161
MMTV-SW	6,7,8.1,9	Mls like		162
Rabies	ND			163
EBV	ND	HERV-K18		158
CMV	ND			164
Herpes virus				165
Other Pathogens	**Vβ Specificity**	**Name**	**Chromosome**	**Reference**
Mycoplasma arthritidis	h17,6,8.1,8.3	MAM		166,167
Toxoplasma gondii	5			168
Yersinia enterocolitica	ND			164
Yersinia pseudotuberculosis	ND			123,169

Vβ in parentheses are reactive with commercial but not recombinant enterotoxins.
CMV, *Cytomegalovirus*; EBV, Epstein-Barr virus; MMTV, mouse mammary tumor virus; NA, not applicable; ND, not determined.
[a]The nomenclature in use before the discovery that the phenotype resulted from endogenous retroviruses.

rapidly for those few that "fit" properly. We have seen previously that the association rates are often remarkably slow, with association constants (K_as) ranging from 1000 to 10,000 $M^{-1}s^{-1}$. This indicates that either a multistep process is occurring before stable binding can be achieved or that only a fraction of the TCRs in solution have the correct conformation. Just how such a scanning mechanism might work for TCRs is seen in the analysis of Wu et al,[172] who found that a cytochrome c/MHC-II–specific TCR derived most of its stability of binding from antigenic peptide residues, but very little of its initial activation energy from these residues. In contrast, MHC residues contributed by far the most of the initial binding but had relatively modest effects on stability. This indicates that "scanning" may be a process that first involves contact with (and orientation by) the α-helices of the MHC and then a "fitting" process with and stabilization by peptide residues that involves a substantial loss of entropy. This model of TCR binding might help to explain the striking efficiency and sensitivity of T cell recognition with the MHC helices guiding the TCR into the correct orientation. It might also be the structural basis for cross-reactivity, in which structurally very different peptides bind to the same TCR, as the CDR3 regions of TCR could "fold" into the peptide in many possible configurations. While attractive, there remain caveats about this "two-step binding" model; one issue is the existence of some important human class I MHC antigens that "bulge" out of the binding groove, creating a barrier to "scanning" as a mechanism of binding to those antigens.[173] A second issue is that one would expect a reproducible "footprint" for MHC binding of particular TCR V regions, which until recently has been elusive. But Garcia and colleagues,[174] as well as Kappler and Marrack and their colleagues,[175] have documented such specific footprints that correlate to particular Vβ usage, although it does seem that there are multiple ways in which a particular V region can contact a particular MHC.

Renewed interest in TCR cross-reactivity and especially its possible biological implications has been triggered by the work of Su et al[150,176] and Jamieson[177] who found that in humans and mice, respectively, that there were T cells with an antigen-experienced phenotype (memory phenotype or virtual memory T cells) in organisms that had never experienced those pathogens. In the human work, this was particularly striking, since blood bank donors, who are rigorously excluded for HIV positivity, had HIV-specific CD4+ T cells, where between 10% and 80% had a memory phenotype and this was also true of several other viral specificities where the subjects were serologically negative, as shown in Figure 23.6. In contrast to these findings in adults, newborn cord blood samples had only naïve CD4+ T cells for the same HIV antigen. In this and subsequent work,[178] no difference could be found between these cells and authentic memory T cells. While this could be the result of homeostatic proliferation, as seen in CD8+ T cells,[177] a possibility is that these memory cells could be the product of cross-reactivity between these viral antigens and some of the myriad microbes that infect human beings. An illustration of this was made by showing that both cell lines and subjects vaccinated with a standard flu vaccine could result in cross-reactivity to various similar

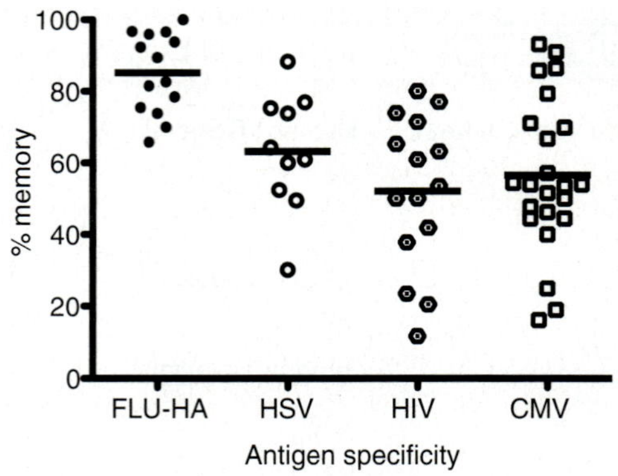

FIG. 23.6. Memory phenotype CD4+ T cells in unexposed individuals. Su and colleagues[150,176] found that even individuals who were serologically negative for exposure to common viruses, such as human immunodeficiency virus (HIV), herpes simplex virus (HSV), and *Cytomegalovirus* (CMV) (and others not shown here), nevertheless had an often large fraction of T cells specific for viral antigens that had a memory phenotype. Subsequent analysis further substantiated this shift toward a memory type of T cell,[178] and one explanation is that this is the result of cross-reactivity to other microbial antigens and may have protective value.

peptide from a variety of pathogens.[176] If these are authentic, functional memory T cells, they could play an important role in protection against pathogens with these cross-reactive antigens, as proposed many years ago by Welsh and colleagues,[179] and supported by recent work on SARS-CoV-2 specificities by Mallajosyula involving T cells specific for conserved coronavirus epitopes.[180]

Role of CD4 and CD8

What is the role of CD4 and CD8 with respect to the T cell response to agonist and antagonist peptides? Clearly their expression greatly augments activation and, in some cases, determines whether there is a response at all (as reviewed in Ref. 181). In addition, the results of Irvine et al[182] and Purbhoo et al[183] using a single-peptide labeling technique found an appreciable T cell response to even one agonist peptide in all four T cells analyzed, resulting in a "stop" signal for the T cell and a small, but detectable rise in intracellular calcium. In CD4+ T cells, these effects are attenuated by antibody blockade of CD4, such that many more[25,28,30,31] peptides are required in order to elicit a stop signal and a calcium flux.[182] Much of this effect likely comes from the recruitment of lck to the TCR/CD3 complexes via the cytoplasmic tails of either molecule. But in addition, there is also a significant positive effect even with CD4 molecules that are unable to bind lck and thus there appears to be an effect on TCR-ligand interaction as well. Nonetheless, while a weak binding of CD4 to class II MHC has been observed,[184] most recently in a complete crystal structure of CD4-TCR-pMHC,[185] there was no apparent stabilizing effect on TCR binding to peptide-MHC in measurements of binding within a synapse by Huppa et al.[127]

This is in contrast to the earlier data of Luescher and colleagues[186] who found that CD8 had a measureable stabilizing effect on TCR binding to ligand on T cells expressing that molecule. This later result is consistent with measurements of CD8 affinity for class I MHCs that are very similar, or even superior to that of many TCRs for pMHC.[124] So are CD4 and CD8 playing very different roles with respect to TCR binding to pMHC? Chakraborty and colleagues suggest not, in a model they have developed that indicates that the differences in CD4 and CD8 affinity for MHCs contribute at best marginally to activation and that delivering lck to productive TCR-pMHC complexes is their most important property.[187] In this view, the specificity of these molecules for their respective MHC molecules is more of a targeting mechanism than the stabilization function implicit in the ubiquitous illustrations originally proposed by Janeway[181] and featured in many subsequent models.

How could CD4 be facilitating the recognition of small numbers of peptides? Irvine et al[182] proposed a "pseudodimer" model, which suggests that a CD4 molecule associated with a TCR binding to an agonist peptide-MHC could bind laterally to an endogenous peptide-MHC complex that is also being bound by an adjacent TCR. This takes advantage of the apparent abundance of endogenous pMHCs that can be bound by a given TCR[188] and uses two weak interactions (CD4→MHC II and TCR→endogenous peptide-MHC) to help create a dimeric "trigger" for activation. This is supported by the work of Krogsgaard et al[147] who also showed that disabling CD4 binding to the endogenous pMHC in a heterodimer had a much a greater effect on activation than the same mutation on the agonist pMHC.

What about the timing of CD4 or CD8 binding with respect to TCR-pMHC? The first hints of this came from the work of Hampl et al[189] who found that antagonist pMHCs showed no apparent influence of CD4 versus agonists, and since these interactions are much briefer, they suggested that CD4 binding was subsequent to TCR engagement with a ligand. This is likely to be a general effect as shown in a very different binding system by Jiang et al,[135] who found that whereas very brief T cell contacts with pMHC loaded red blood cells involved TCR but not CD8, longer contact periods involved both molecules. Thus, it seems that the TCR must first bind to a high-quality (eg, agonist) ligand and then CD4 or CD8 comes in to deliver lck and dissociates (based on the lack of stabilization, at least in the case of CD4). This was the conclusion of modeling analysis by Chakraborty and colleagues.[187]

Very recent work[190] shows that whereas TCRs can dock in a variety of orientations around the same influenza peptide-MHC complex, only the more typical configuration was robustly amplified in a T cell response to that pathogen, arguing for this being the most favored in terms of signaling.

αβ T CELL RECOGNITION AND THE IMMUNOLOGICAL SYNAPSE

The complexity of T cell activation at the cellular level was first revealed by Kupfer and colleagues in 1998[184] who used high-resolution microscopy to show that there was a specific reorganization of cell surface, cytoskeletal and signaling molecules at the interface between CD4+ T cell and B cell lines bearing the appropriate peptide-MHC complexes, as shown in Figure 23.7A. Dustin subsequently dubbed these "immunological synapses,"[192] which is an apt description of the concentric layers of TCR-CD3 in the central "bull's eye" surrounded a circle of LFA-1 on the T cell side and its ligand ICAM-1. There is also a wholesale rearrangement of the actin cytoskeleton to orient toward the point of interaction, ending in a ring of actin just under the synapse. A storm of activity followed this, particularly using the then new high-speed video imaging technologies such as lipid bilayers with lipid-linked pMHCs to mimic the B cell APCs or sophisticated confocal microscopy systems that allowed microscopy of whole cell-cell interactions.[193-196] Later single-molecule imaging with TIRF or FRET or super-resolution methods were also added,[197] and most recently light sheet microscopy to gain ever higher resolution.[198] Another interesting feature of these synapses, and perhaps others, is that Huse and colleagues[191] found that molecules were secreted both into the synapse and also away from it, as shown in Figure 23.7B. This means that there is both a specific interaction between two cells, based at least in some cases on a shared antigen specificity, and also a more "public" activity of secretions designed to inflame the local environment or establish a chemokine gradient.

While these methods are now common throughout biology, T cell immunology was the first to take advantage of these new developments in imaging because the cells and molecules and specific questions were already in place. The work that emerged was also important because it was able to address and area of cell biology that was previously inaccessible to experimentation, which is the molecular nature of cell-cell interactions. This is because while tremendous progress has been made in mechanistic biology in systems that can be reduced to molecular components, such as those that can be purified and reassembled in solution, such as enzyme-substrate interactions or transcriptional regulation, this approach is impossible in cellular interactions that require intact cells and membranes and multiple molecules interacting on and between membranes. The planar bilayers were an effort in that direction, but only because most of what happens is occurring within the T cell and requires an intact T cell. The introduction of high-resolution video fluorescence microscopy has helped enormously to solve this problem because it has allowed us to track particular molecules or groups of molecules at all stages of T cell recognition.

So, what has emerged from this work?

T Cell Sensitivity

Early estimates of T cell sensitivity varied widely, from 1 to 400 peptides bound to a particular MHC allele on an average APC were needed to stimulate a particular T cell. But these experiments, based on quantitating the average number of pMHCs on a given APC,[199-203] were handicapped because it was not known how much of the APC surface was surveyed before a T cell gave up and moved on,

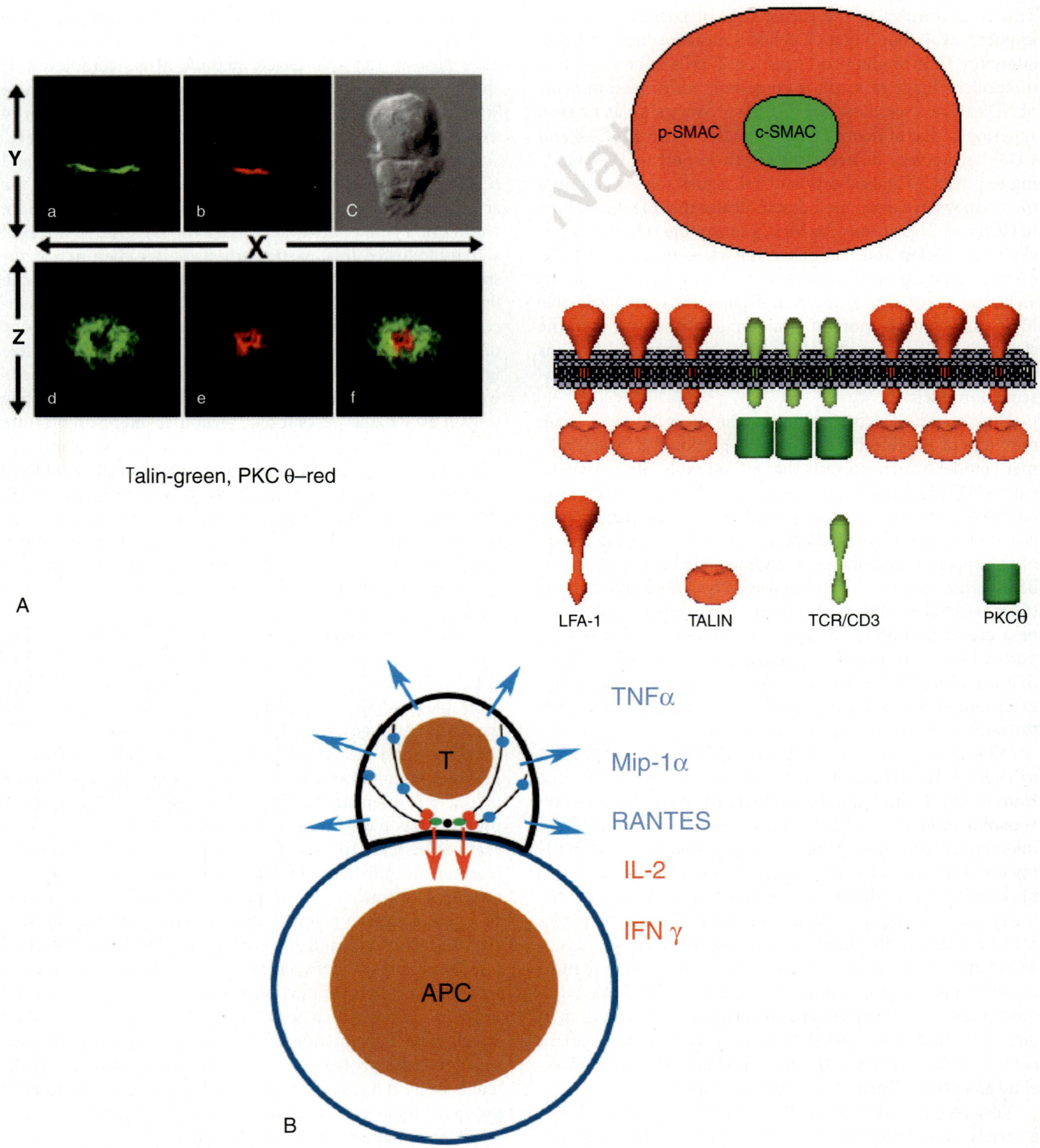

FIG. 23.7. Immunological Synapses. A: Kupfer and colleagues first showed that the antigen-specific interaction between a T cell and a B cell expressing the relevant peptide-MHC involved a complex rearrangement of both cytoskeletal and cell surface molecule components, with TCR-CD3 and the center opposite MHC molecules and LFA-1/1CAM-1 peripheral to that, forming a "bull's eye" pattern. They termed the central portion a C-SMAC for central supramolecular activation complex and the ring configuration as a P-SMAC for peripheral supramolecular activation complex. a + d show side and enface view of the T cell: B cell synapse, respectively. b + e show PKC θ staining and c show a Nomarski image. **B:** Bidirectional secretion. Huse et al[191] later showed that T cells in a "productive synapse" secreted cytokines in a bidirectional manner, with some, such as IL-2 and γIFN, being secreted into the synapse, while TNFα and the chemokines RANTES and Mip-1α are secreted away from the synapse. Thus there are both cell-specific and nonspecific consequences of specific T:B synapses and perhaps others as well. (A, Reprinted by permission from Springer Nature: Monks CRF, Freiberg BA, Kupfer H, Sciaky N, Kupfer A. Three-dimensional segregation of supramolecular activation clusters in T cells. *Nature*. 1998;395:82-86. B, Reprinted by permission from Springer Nature: Huse M, Lillemeier BF, Kuhns MS, Chen DS, Davis MM. T cells use two directionally distinct pathways for cytokine secretion. *Nat Immunol*. 2006;7:247-255.)

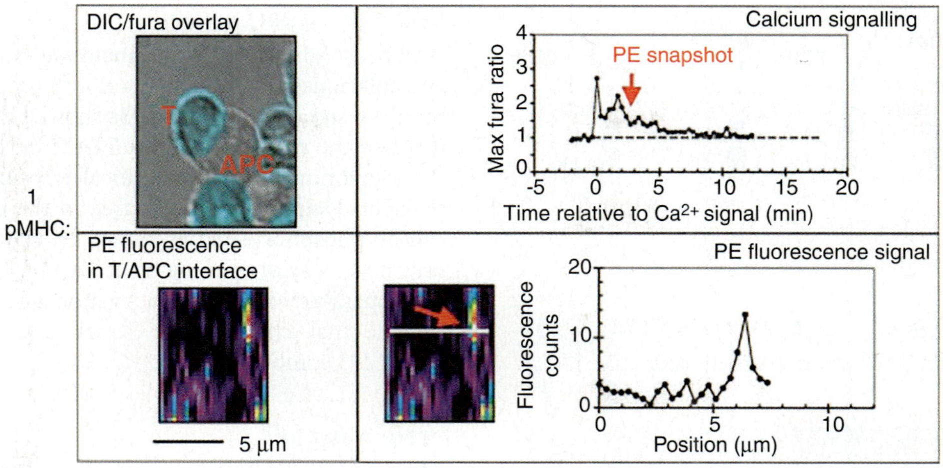

FIG. 23.8. T cells are able to detect even one pMHC ligand. Irvine and colleagues attached the very bright fluorophore, phycoerythrin to the E-terminus of an extended peptide antigen and used high-resolution video microscopy and calcium imaging to detect and quantitate the number of individual peptide-MHC complexes in a given T+B synapse.[182] As shown in the figure, even a single pMHC in the synapse could infiltrate a calcium flux. This was true for multiple T cell specificities, and later work with a more long-lived fluorescent tag showed that even one pMHC could result in both activation and cytokine secretion[204] regardless of what stage of maturation the CD4+ T cell was in, although memory stage cells responded much more quickly (a few hours) than naïve cells (~24 hours).

and this likely varies because APCs come in different sizes. Furthermore, the first T cells stimulated probably encountered "above average" pMHC densities, since there will be a distribution. So, this was a perfect case for imaging, particularly since Irvine et al[182,183] found they could label with phycoerythrin (PE) and image-labeled peptides within the synapse and show that even one agonist pMHC could trigger some T cell calcium flux and approximately 10 could produce the maximum elevation, as shown in Figure 23.8. Later, Huang et al[204] used the much more stable quantum dots to label the peptides and with longer incubation times could show that CD4+ T cells could secrete cytokines with only one pMHC, regardless of whether they were naïve, activated, or memory T cells, the difference being in how quickly this could happen, with naïve T cells taking almost a day, whereas activated and memory T cells took only a few hours. This study also found that exposure to 1-4 peptide MHCs in the synapse did not change the amounts of cytokine that were produced, suggesting an "all or none" activation program. So how is a T cell immune response "scaled"? Here the data suggested that the more of a particular antigen that was introduced, the more T cells with that specificity would be triggered.

Signaling Characteristics

Whereas initial speculation centered on the central TCR-CD3 complexes being where the key signaling activity occurred, subsequent work by Dustin and Saito and colleagues showed that most kinase activity was occurring on the periphery of that central feature and that those TCRs were bound to be internalized and degraded.[205] Work by Xie et al showed that engaging TCRs on the cell surface with a peptide-MHC complex caused them to be linked to the actin cytoskeleton and transported to the center of the synapse,[206] consistent with a peripheral activation and then receptor internalization once the T cell had committed to be activated. Additionally, Huang, et al[204] noted that there was no difference in the cytokine production whether one or four pMHCs were in the synapse, indicating that there are only two settings on or off. So how are T cell responses scaled? In that same paper, evidence is presented, indicating that increasing the antigen dose increases the number of T cells activated.

Displacing CD45

For many years, van der Merwe and Davis at Oxford argued that excluding the phosphatase CD45, which has a very large extracellular domain, from the synapse was necessary to allow for proper kinase activity and was facilitated by the fact that the extracellular domain of CD45 is very large compared to TCRs and pMHCs,[207] and it is illustrated in Figure 23.9. They termed this the "kinetic segregation model."[210] While the evidence for this was not clear for many years, some key experiments by that group[211] showed that increasing or decreasing the length of CD45 had the predicted effect, and recently Garcia and colleagues have shown that cross-linking CD45 to PD-1 with a bispecific antibody inhibits phosphorylation and activation.[212]

Role of CD4 and CD8 Molecules

In 1992, Janeway proposed that CD4 and CD8 acted as "coreceptors" to TCRs, binding simultaneously to conserved parts of class II and class I MHC molecules while the TCR bound to the pMHC. While this was a very popular idea,

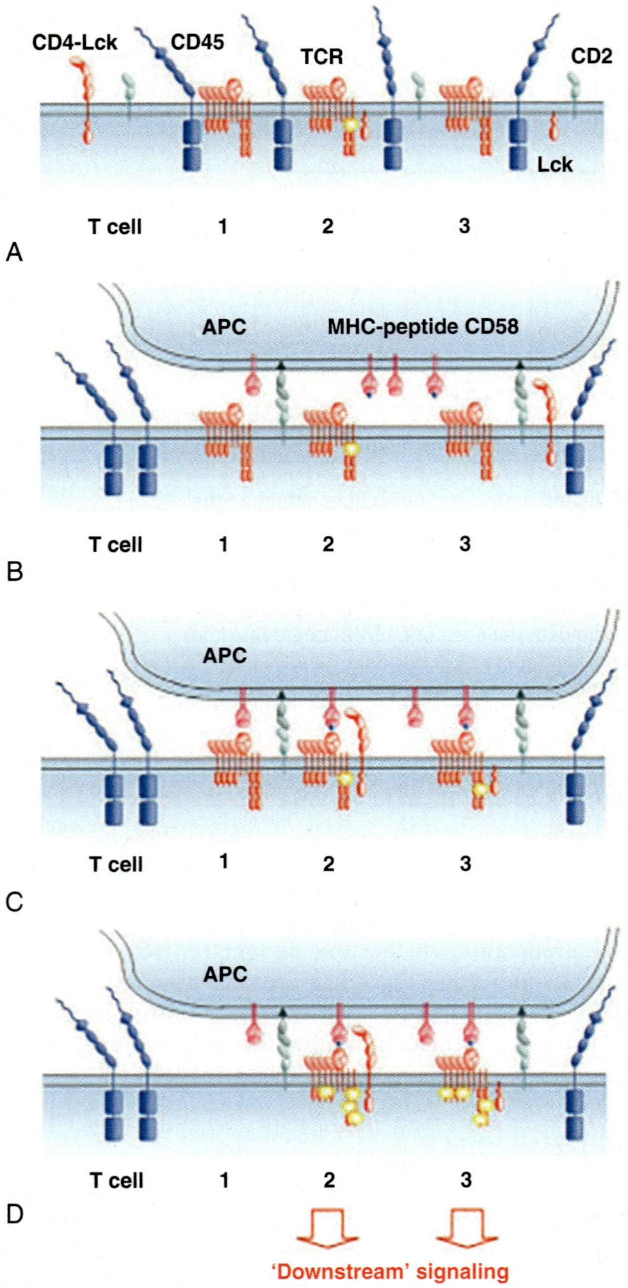

FIG. 23.9. Kinetic segregation model. In 1996, S. Davis and van der Merwe[208] and in subsequent papers[209] proposed that the phosphatase CD45 on T cells normally acts to dephosphorylate the many phosphorylation sites on the CD3 molecules (of which there are 10!), but that during synapse formation, the very large size of the CD45 extracellular domain causes it to be excluded for the synapse and thus facilitates phosphorylation and T cell activation.[210]

there was never any strong experimental evidence, and in 2010, Huppa et al showed that inhibiting CD4 had no effect on the stability of a TCR-pMHC between a T cell and a lipid bilayer using single molecule FRET as an assay, as shown in Figure 23.10. Other experiments suggest that these molecules act after TCR-pMHC engagement to bring lck to productive complexes.[187,189]

CD3 Movement

Another seminal FRET experiment was that of Wucherpfennig and colleagues who by using donor fluorophores in the membrane versus acceptors on CD3ε showed that T cell activation involves the movement of the ITAM on this molecule from the membrane, where the critical tyrosine is shielded from phosphorylation to be more free in the cytoplasm where it can be phosphorylated as shown in Figure 23.11. This is a very novel way of opening up at least part of the TCR signaling apparatus to promote activation and initiate the kinase cascade that characterizes T cell activation through the TCR-CD3 complex.[213]

Protein Islands

Lillemeier et al used the then nascent super-resolution microscopy method and electron microscopy to show that specific molecules like TCR-CD3s cluster on protein-rich structures termed "protein islands," and that separate islands containing the LAT molecules fuse with the TCR islands during T cell activation,[197] showing both a novel way to regulate activation and also that surface molecules are not randomly distributed but occur in distinct clusters. This has also been seen in B cell and NK cell work, and TCR clusters have been documented by Alarcon and Schamel as well.[214] An interesting property of protein islands is that they would concentrate TCR-CD3 complexes as much as 10-fold, which would facilitate the "serial engagement" of a rare peptide-MHC ligand with multiple TCR-CD3 molecules, as proposed by Valitutti and Lanzavecchia.[215,216] Since a typical TCR-pMHC interaction might have a half-life of only 1 second at 37°, it could engage hundreds of different TCRs on the same T cell over 5 to 10 minutes, which might be critical to stimulating enough phosphorylation events to activate the cell. Recent work of Lang and colleagues has disputed the relevance of serial engagement however, based on their finding that physically stimulating individual TCRs on T cells is sufficient to initiate activation.[217]

MOLECULAR MECHANICS OF αβ TCR-CD3 TRIGGERING

The mechanism by which TCR binds to pMHCs has long been a puzzle for T cell immunologists. A number of recent reviews provide more detail than space allows here.[153,218,219] Originally it was thought that standard binding kinetics would reveal a clear relationship between TCR-ligand interactions, although how that was communicated to the CD3 molecules, and where the actual phosphorylation occurs, was clearly an additional complication. But even after strenuous efforts to make TCRs and MHCs in a soluble form suitable for binding measurements, the first problem that occurred was that there was no straightforward way to measure such weak binding as it turned out to be. Some complex efforts[118,119] produced very low numbers for the affinity (~50 μM) that were so low as to be thought nonspecific binding by classical biochemists (but they were very specific), but these studies only really got going with the advent of SPR in the form of the Biacore instrument. This enabled many studies of the kinetics of binding of many

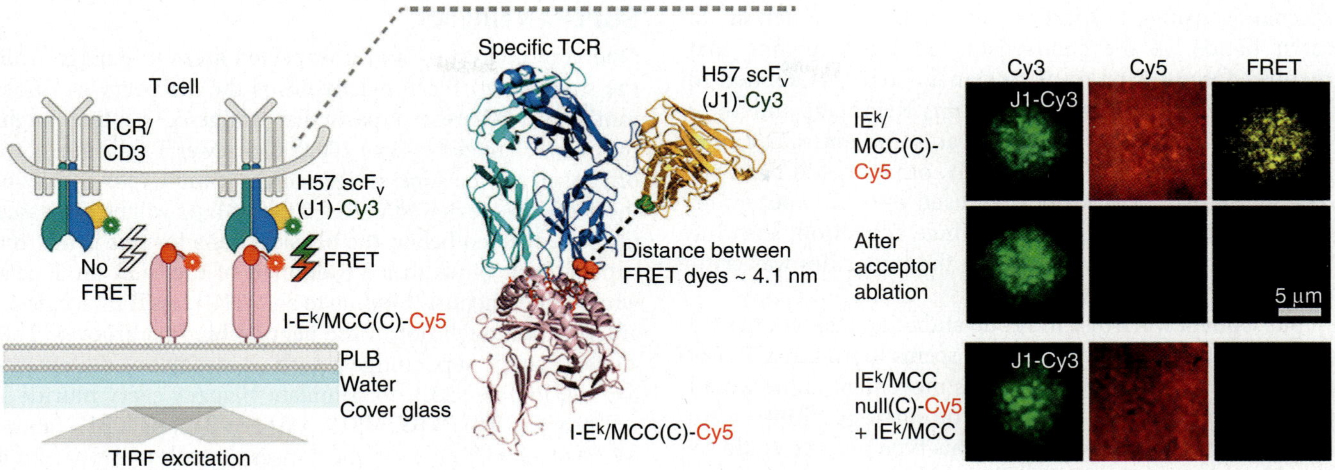

FIG. 23.10. TCR-pMHC kinetics. Actually measuring the kinetics of receptor-ligand interactions between cells is not an easy problem, but in 2010, Huppa and colleagues solved this for αβ TCR interactions between a T cell and an artificial bilayer by labeling an antibody probe that binds to the TCRβ constant region such that the fluorophore is close enough to a complementary fluorophore on the C-terminus of a peptide bound to a class II MHC molecule to allow FRET (Förster or fluorescence resonance energy transfer) when a single TCR binds to its ligand.[127] This enabled the precise measurement of the off rate and also showed that CD4 did not contribute to stability, indicating that it is not, strictly speaking, a "co-receptor." (Reprinted by permission from Springer Nature: Huppa JB, Axmann M, Mörtelmaier MA, et al. TCR-peptide-MHC interactions in situ show accelerated kinetics and increased affinity. *Nature*. 2010;463:963-967.)

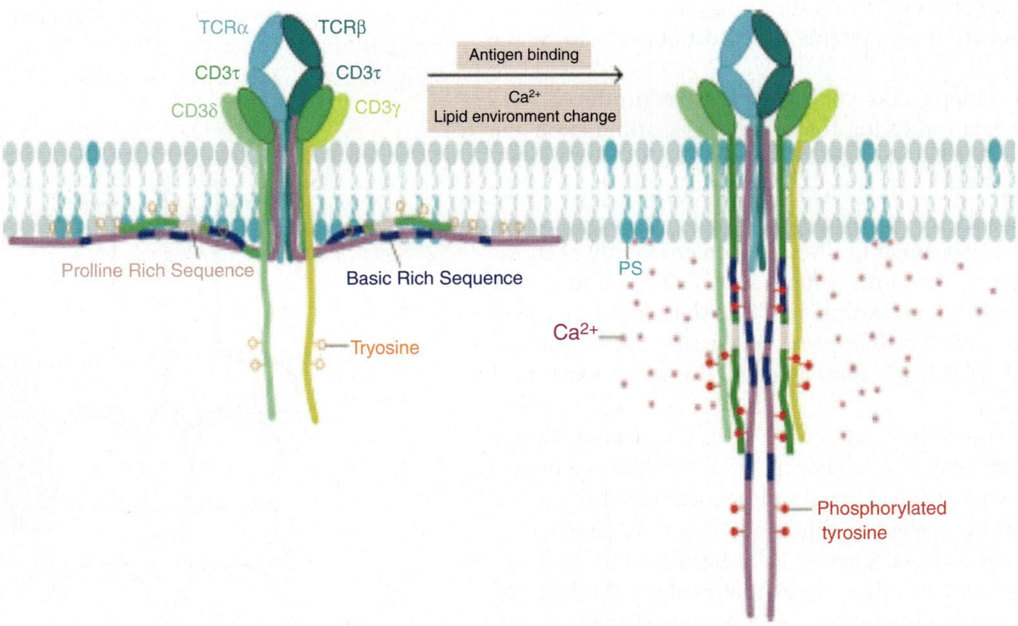

FIG. 23.11. Membrane shielding of CD3ε phosphorylation sites. Another important use of FRET was the work of Xu and colleagues, who showed that CD3ε and likely other CD3 molecules typically shielded their ITAM phosphorylation sites in the cell membrane, making them less accessible for phosphorylation. (Reprinted from Xu C, Gagnon E, Call ME, et al. Regulation of T Cell receptor activation by dynamic membrane binding of the CD3ε cytoplasmic tyrosine-based motif. *Cell*. 2008;135:702-713, with permission from Elsevier.)

TCRs to pMHCs, which were found to range from about 1 to 200 μM.[220] For comparison, a typical antibody that you might use for cell staining, for example, has an affinity of 2.0 to 0.02 nM, many orders of magnitude higher (here we are using affinities in K_D, which is defined as K_a/K_D, and is a useful metric as it is the state at which the concentrations of the receptor and the receptor-ligand complex are equal). From these numbers, you can see that there is a huge difference in the strength of an antibody-antigen complex versus a TCR-pMHC one, and the fact that both work and can be very specific is antibodies typically have to work in solution, whereas TCRs only have function at a cell-cell interface, essentially in two dimensions versus three, as quantified by Dustin and Springer.[130]

Recent work on TCR activation proximal to the TCR-CD3 complex has focused on the discovery of its mechanosensing properties and the phenomenon of catch bonds. A mechanosensory molecule is one that requires force acting on it to signal. That this applied to TCRs became apparent when Kim, Reinherz, and their colleagues discovered that whereas some anti-CD3 antibodies could activate T cells easily, others could bind but did not activate. Critically they used optical tweezers on the nonactivating antibody to induce activation, showing that applying mechanical force to the molecules was critical,[152,221] as reviewed by Huse.[222]

Subsequent work by this group and others has confirmed this,[131,132,223-228] and in retrospect, it seems logical that T cells need to scan cell surfaces that this movement alone would produce lateral forces if ligands of suitable affinity were engaged during this movement. This type of force could be considerable, not just by T cells brushing up against other cells, but in studies by Ueda et al[229] electron microscopy showed that T cell pseudopodia could probe deeply within an APC, almost to the nucleus. This must involve enormous force in cell biologic terms, and consequent shear stress on cell surface molecules. These results also speak indirectly to the long running controversy of whether or not dimerization of TCR-CD3 is required for activation. Activation with pMHC tetramers and antibodies suggested that this was true,[171,206] but in other systems this did not seem to be the case.[230]

Recently Huppa and colleagues[231] have produced very convincing data indicating that monomeric stimulation can be achieved. But how can this be reconciled with the previous findings? Here the evidence of mechanical force being needed provides an explanation because it suggests that monomeric engagement of pMHC, even covalently as in Xie and colleagues,[206] does not supply sufficient force to trigger activation, but that cross-linking TCRs does, not because of dimerization per se as interpreted originally but by "tilting" cross-linked TCR-CD3 complexes in the same way sheer stresses would.

Another interesting aspect of T cell activation is the increasing evidence of catch bonds.[131,154,160,232] This is a type of interaction where instead of a simple association-dissociation of two molecules, called a "slip bond," this type dissociation is a multi-step process. That is, as molecules start to dissociate, then "catch" in other places that prolong the half-life, and probably more importantly in the context of the mechanosensory data discussed above, this could induce a torque on the TCR-CD3 complex, as the T cell is brushing by (scanning) or burrowing into an APC. The importance of this likely process was made dramatically clear in by the work of Sibener et al,[223] who found peptide-MHC complexes that bound a cognate TCR with good affinity but could not activate T cells expressing that TCR. They then found a peptide variant that could activate, but x-ray structures revealed no discernible differences between the stimulating peptide and the nonstimulating one. But catch bond analysis showed that only the stimulating complex showed the characteristics of a catch bond, and modeling showed a plausible way in which the early dissociation of the molecules could lead to rebinding with other residues.

SUPERANTIGENS

One very interesting and unexpected areas to emerge from the study of αβ T cell reactivities is the discovery of "superantigens." Whereas a particular antigenic peptide might only be recognized by 1 in 100,000 or fewer T cells in a naïve organism, a given superantigen might stimulate 1% to 20% of the T cells (as reviewed in Ref. 233-240). As will be discussed in more detailed below, the physical basis for this is that the superantigen binds to a Vβ domain of the TCR on T cells while simultaneously binding to an MHC class II molecule on an APC (although not in the peptide-binding groove). This allows a single superantigen, such as staphylococcal enterotoxin A in Table 23.1, to stimulate virtually every murine T cell bearing Vβ1, Vβ3, Vβ10, Vβ11, Vβ12, or Vβ17 (about ~15% of all αβ T cells), in most cases regardless of what Vα it is paired with or what CDR3 sequence is expressed, as schematized in Figure 23.12. A fascinating recent development in this area has been the discovery of superantigens that have a similar effect on B cells, binding to specific VHs.[236] Clearly this is a very unique class of B and T lymphocyte stimulatory molecules whose function seems to be to distract the immune system, sometimes accompanied by serious toxicities.

The first indication of a superantigen effect was the discovery of minor lymphocyte stimulating determinants by Festenstein in the early 1970s.[241] Many years later, Kappler and Marrack and their colleagues characterized a mouse

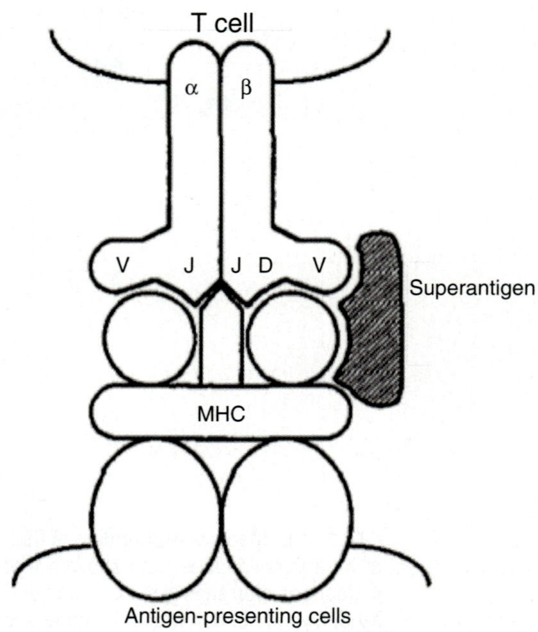

FIG. 23.12. Superantigens bind to TCRVβ's and class II MHCs in an antigen independent manner. The schematic shows the way in which many superantigens ligate TCRs and MHC class II molecules together and are able to activate T cells. MHC, major histocompatibility complex; TCR, T cell receptor. (Repulished with permission of Annual Reviews, Inc. from Herman A, Kappler JW, Marrack P, Pullen AM. Superantigens: mechanism of T-cell stimulation and role in immune responses. *Annu Rev Immunol.* 1991;9:745-772; permission conveyed through Copyright Clearance Center, Inc.)

strain–specific deletion of T cells expressing specific TCRVβs that were attributable to these loci.[242] It emerged that these effects were due to endogenous retroviruses of the mouse mammary tumor virus family.[71,155,243-245] Different family members bind different TCRVβ domains (as shown in Table 23.1) and stimulate T cells expressing them. Meanwhile, Janeway and colleagues[246] had shown earlier that staphylococcal enterotoxins could polyclonally active naïve T cells in a Vβ-specific manner without a requirement for antigen processing. Many of these enterotoxins have been characterized extensively.[233,234,247,248] Unlike the MMTV proteins, which are type II membrane proteins, the enterotoxins are secreted. Subsequently, proteins having similar properties have been isolated from other bacteria (*Yersinia pseudotuberculosis*,[123,169] *Streptococcus*,[159] and *Mycoplasma*).[166,167] There is also evidence of superantigen-like activities in other mammalian viruses such as rabies,[163] *Cytomegalovirus* (CMV),[164] herpes virus,[165] Epstein-Barr virus,[158] and also in *Toxoplasma gondii*.[168] Since so many pathogenic or parasitic organisms possess these molecules, apparently by convergent evolution, there must be some selective advantage, but in most cases, there is no conclusive evidence as to what this might be. The one exception is the case of the MMTV superantigens, where it has been shown that polyclonal T cell stimulation allows the virus to more efficiently infect the B lymphocytes that are activated by the T cells.[161,162] This may be a special case however and most authors have suggested that superantigens primarily serve to confuse and occupy the immune system while the pathogen escapes specific targeting and elimination. Large doses of superantigens have also been implicated in various "shock" syndromes, such as food poisoning or "toxic shock,"[233] but this is probably not their everyday purpose, as it would violate the general rule that the host and parasite should coexist.

It has also been suggested by Stauffer et al[249] that superantigens might be involved in triggering autoimmune diseases. Here the hypothesis is that a large number of some Vβ-bearing T cells are activated by a pathogenic superantigen and that subsequently self-reactive T cells within those activated cells are more easily stimulated by a particular tissue antigen. That this may occur in some cases is supported by their work on a human endogenous retrovirus, which specifically stimulates Vβ7 T cells and is implicated in the initiation of type 1 diabetes.[249] Another report implicates a superantigen in Crohn disease, another autoimmune disorder.[250]

While the biochemistry of superantigen binding to TCR and MHC is similar to that of TCR peptide/MHC interactions,[251] mutagenesis, and particularly x-ray structural data, has shown that the topology is both quite different and variable.[234] In particular, it has been found that Mls-la presentation to T cells is most affected by mutations on the "outside" surface of the Vβ domain, which do not affect peptide/MHC recognition.[101] In contrast, CDR1 and CDR2 of regions of Vβs are involved in bacterial superantigen reactivity.[252]

Why do all the many independently derived superantigens interact only with the TCRβ chain? One possibility is that the β chain offers the only accessible "face" of the TCR, perhaps because the CD4 molecules hinders access to the Vα side, as suggested by the antibody blocking studies of Janeway and colleagues.[253]

ANALYZING αβ T CELL RESPONSES

A longstanding problem in αβ T cell biology has been how to analyze a particular response. This has been so problematic that even now, there are no accepted correlates of protection in T cell responses to a vaccine, even in diseases that are largely dependent on these responses, such as TB or HIV. This is a serious problem since it means that you need to depend on large, hugely expensive and time-consuming efficacy studies to determine whether a vaccine will work or not. Recent results with SARS-CoV-2 vaccines are a bright spot, but in HIV, for example, there have been five distinct vaccine concepts tested, with seven efficacy trials completed at a cost of billions of dollars, all of which have failed. Other notable recent failures have involved TB, dengue, and malaria vaccines.[254-257] Very recent results for malaria and TB only report roughly a 50% efficacy. Hope springs eternal though, many new methods have been developed in the past decades to assess the magnitude, functional characteristics, and specificity of T cell responses in mice and humans, principally. This can range from stimulating peripheral blood mononuclear cells (PBMCs) from vaccinated subjects with whole pathogens/lysates/infected cells to antigen-specific assays such as ELISpots, peptide pools, or tetramers. In the case of the later, highly multiplexed methods like those of Newell and colleagues, who were able to survey over 400 different specificities[258-260] simultaneously with excellent sensitivity (0.01%-0.001%). ELISpots can also be used, and many groups have used them very effectively,[261] but the assay is not very sensitive (>0.1% threshold) and so may miss many responses. Oligo-tagged tetramers/multimers or nanoparticles have also been used[260,262]; however, since each multimers needs to be at a high enough concentration to bind cognate T cells effectively, there will be a point at which a staining solution will saturate and the proteins will precipitate. That threshold will be on the order of ~10 mg/mL or less. Still, multiple aliquots of different tetramer collections could extend this method into the thousands, if the input samples are large enough. Another route to antigen identification that has been successful is using tandem mass spectroscopy to sequence peptides or other ligands from isolated MHC molecules, and then bind them to the relevant MHC allele or use them to stimulate specific T cells. The first use of this method was by Hunt et al[263] More recent papers have expanded the scope and sensitivity of this method.[264-269]

Functionality and Phenotype of Responding T Cells

It is reasonable to assume that the functionality of the T cell response has an important bearing on the degree of protection, but what is the evidence that the particular phenotype a T cell has is effective? In *Mtb* infections, where the Th1* predominates, at least in the blood,[270] it is worth noting that ~90% of latent carriers are protected for life. Thus, it could be proposed that this is a fairly effective control mechanism, although it is perfectly likely that other T cell types, while less numerous, might also be making key contributions. Similarly, a robust antibody response was necessary.[271] Here labeled tetramers or other multiplex methods give the most flexibility since they can identify pathogen-specific T cells regardless of function. Nonspecific stimulation and then intracellular cytokine staining can then reveal the particular type or types of T cell with those characteristics or even that they are nonfunctional due to exhaustion or senescence. There is also evidence that T cells that make multiple cytokines indicate a more robust response.[272,273] Interestingly, Huang et al[274] found that highly expanded T cells specific for *Mtb* antigens or CMV showed a senescent phenotype. Thus, clonal expansion alone is no guarantee of protection. More recently, high-throughput, single-cell methods using next-generation sequencing for the analysis of T cell responses have become increasingly common. The first of these was developed by Han et al,[275] which focused on single-cell amplification and sequencing of both TCR chains and a set of cytokines and transcription factors that could designate the type of T cell that was responding. This typically yielded data on 500 to 1000 cells at a time and has been used extensively by my group, and also included primers that measure general transcriptional characteristics or epigenetic characteristics. Very recently, 10× Genomics has developed a single-cell transcriptional system and more recently added a TCR sequencing component. These methods can generate both chains of an αβ TCR (or γδ TCRs as well) and identify on the order of 2000 genes expressed in that cell. This falls short of the 5 to 10k genes that are expressed in most T cells, due to the fact that they tend to have much less RNA than most cell types. This means that the method will favor the most abundant genes and then give a fairly random sample of the many that are very rare (which is most). Thus, only a fraction of CD4+ T cells (in our experience) will "callout" the CD4 transcript. This seriously hinders the efficiency of the method, and so a very useful adjunct to this method is to use oligo-tagged antibodies (to CD4, among many others available) that are used to stain the cells first and then identified along with the RNAs in that cell during the amplification and sequencing steps.[276,277] In this way, one can at least know what the range of gene expression might be in a given major cell type. Data of this type have also been very useful in identifying novel subtypes within the "classical" T cell types, and this increased cellular resolution will undoubtedly be helpful in analyzing vaccine and pathogen responses.

Specific Targets

It is also reasonable to think that some potential targets of T cells might be more effective than others in controlling a pathogen. It has long been known, for example, that many specific antibodies are not effective in controlling a pathogen and may even represent a smokescreen the pathogen might use to divert the immune system. What these targets might be in terms of T cell responses has not been easy to discern, but recent progress in HIV analysis has yielded a new and potentially enabling strategy. This relies on earlier work of Chakraborty and colleagues who developed computational methods to define specific points of vulnerability, that is sequence variants that could not coexist in an HIV genome, presumably because they reduced viral fitness,[278] this was later expanded to develop "fitness landscapes" for HIV and other viruses.[279] But the recent paper of Gaiha and Walker[280] takes this further by refining this analysis and showing that such vulnerability regions are in fact targets of T cell responses in elite controllers—the small fraction of individuals infected with HIV who are able to control the virus indefinitely. Thus, analyses of this type may be an excellent way to pinpoint pathogen epitopes that would make particularly effective vaccines.

TCR Repertoire Analysis

Advances in DNA sequencing technology over the past decade have made possible what may be the ultimately broad way to analyze T cell responses. It is now relatively easy and inexpensive to obtain millions of TCR sequences using a variety of methods. But a major stumbling block has been how to extract meaning from raw sequence data. This is partly because αβ TCRs are so diverse, that even monozygotic twins share very few (1%-2%) identical sequences, even if they are making the same antigen responses using the same HLA alleles. There are typically hundreds or thousands of TCR sequences that can recognize the same peptide-MHC. This makes it nearly impossible to determine whether different individuals (or even inbred mice!) are responding to the same antigen. Thus, a recent set of analytical solutions to this problem have been welcome and have broad application to the analysis of T cell responses. Two methods to do this were published simultaneously just a few years ago, one is called Grouping of Lymphocyte Interactions by Paratope Hotspots or GLIPH, which looks for shared motifs within CDR3 sequences,[141] and the other is called TCRdist, which estimates sequence similarities.[281] Using the GLIPH method on ~5700 TCR sequences enriched for *Mtb* specificities, from 22 individuals latently infected with the bacterium identified, 5 commonly shared sequence groups, and most of the individual sequences in a cluster were shown to share a particular *Mtb* peptide-HLA class II specificity, as shown in Figure 23.13. This shows the similarities and differences in the TCRβ sequences GLIPH clustered, and also how individuals who shared sequence motifs also were enriched for particular class II MHC alleles. Using a large panel of Mtb peptides derived by Sette and colleagues, we were able to identify both the peptide antigen and the predicted MHC allele for each of the five specificity groups. The promise of this approach has advanced further very recently with the development of a new version of the GLIPH algorithm (GLIPH2), capable of analyzing millions of TCR sequences at much faster speeds, and without some of the artifacts encountered in earlier version.[282] It also incorporates some of the TCRdist features, such as analyzing global CDR3 sequence similarity and weighting N region creating conserved amino acids strongly. This chapter also demonstrates a whole genome screening method for CD4+ T cell epitopes for *Mtb*.[282] Since this is one of the largest genomes known for a pathogen, at 4.4 MB, this could be a useful strategy for any others. GLIPH2 has been applied to TCR datasets in the millions in cancer, influenza, and COVID-19 patients. It reduces these very large sequence collections into much smaller specificity groups, but since this can represent tens of thousands still, you can further reduce these into the most clinically significant. For example, in an analysis of almost 800K TCR sequences from 178 lung cancer

TCRID	Donor ID	CDR3b	Frequency	CDR3a	Frequency	DQA1		DQB1		DRB1		DRB3/4/5	
TCR001	01/0873	CASSFEETQYF	2/168	CIVKTNSGGSNYKLTF	2/158	*05:02	*01:02	*03:19	*06:02	*11:01	*15:03	DRB3*02:02	DRB3*02:02
TCR008	09/0018	CASSLEETQYF	2/400			*05:01	*01:02	*02:01	*06:03	*03:01	*15:03	DRB3*02:02	DRB5*01:01
TCR010	03/0492	CASSPEETQYF	1/112			*01:02	*01:02	*06:09	*06:02	*13:02	*15:03	DRB5*01:01	DRB3*03:01
TCR012	09/0217	CASSPEETQYF	49/166	CIVHTNSGGSNYKLTF	47/135	*01:03	*01:02	*06:04	*06:02	*13:01	*13:02	DRB3*03:01	DRB3*02:02
TCR003	01/0430	CASSLEETQYF	1/82	CGMSGNTGKLIF	1/70	*03:03	*01:05	*02:02	*05:01	*10:01	*09:01	DRB4*01:01	DRB4*01:01
TCR004	01/0873	CASSLEETQYF	21/168	CIEHTNSGGSNYKLTF	21/158	*05:02	*01:02	*03:19	*06:02	*11:01	[illegible]	DRB3*02:02	DRB3*02:02
TCR009	01/0873	CASSPEETQYF	2/304			*05:02	*01:02	*03:19	*06:02	*11:01		DRB3*02:02	DRB3*02:02
TCR011	09/0018	CASSPEETQYF	31/400	CAVPSGGANSKLTF	1/267	*05:01	*01:02	*02:01	*06:03	*03:01	15:03	DRB3*02:02	DRB5*01:01
TCR022	01/0873	CASSVALAGAEYF	1/69	CAVGGLSGANSKLTF	1/67	*05:02	*01:02	*03:19	*06:02	*11:01	*15:03	DRB3*02:02	DRB3*02:02
TCR023	02/0152	CASSVALASGANVLTF	2/41	CAGAGGGGFKTIF	2/28	*05:01	*01:02	*02:01	*06:01	*03:01	*15:01	DRB5*01:01	DRB3*01:01
TCR024	03/0492	CASSVALQGVHTQYF	2/112	CAGTNTGNQFYF	2/90	*01:02	*01:02	*06:09	*06:02	*13:02	*15:03	DRB5*01:01	DRB3*03:01
TCR026	09/0018	CASSVALYANEQFF	1/151	CAGPTTGYALNF	1/125	*05:01	*01:02	*02:01	*06:03	*03:01	*15:03	DRB3*02:02	DRB5*01:01
TCR036	09/0772	CASSVALLGETQYF	1/107	CAGAPTGNQFYF	1/98	*05:05	*01:02	*03:01	*06:02	*03:01	*15:03	DRB3*02:02	DRB5*01:01
TCR029	09/0328	CASSVALLGGEQYF	1/107	CAGLVGTSYGKLTF	1/73	*06:01	*04:01	*03:01	*04:02	*12:02	*03:02	DRB3*03:01	DRB3*01:01
TCR025	03/0492	CASSVALATGEQYF	1/112	CAGPTGGSYIPTF	1/90	*01:02	*01:02	*06:09	*06:02	*13:02	*15:03	DRB5*01:01	DRB3*03:01
TCR051	02/0152	CASSLIEGGTEAFF	1/41	CVVSAITNDYKLSF	1/28	*05:01	*01:02	*02:01	*06:01	*03:01	*15:01	DRB5*01:01	DRB3*01:01
TCR052	09/0772	CASSLIEGLEQYF	1/107	CAVQPGAGGFKTIF	1/98	*05:05	*01:02	*03:01	*06:02	*03:01	*15:03	DRB5*01:01	DRB3*02:02
TCR053	09/0018	CASSLIENTEAFF	1/151	CAVTIGATQGGSEKLVF	1/125	*05:01	*01:02	*02:01	*06:03	*03:01	*15:03	DRB5*01:01	DRB3*02:02
TCR054	02/0152	CASSLIEQQPQHF	1/41	CASQSNTGNQFYF	1/28	*05:01	*01:02	*02:01	*06:01	*03:01	*15:01	DRB5*01:01	DRB3*01:01
	03/0492	CASSSGQGHYNEQFF	1/162			*01:02	*01:02	*06:02	*06:09	*15:03	*13:02	DRB3*03:01	DRB5*01:01
	09/0328	CASSVGQGHYNEQFF	1/107	CAVISGGSNYKLTF	1/73	*06:01	*04:01	*03:01	*04:02	*12:02	*03:02	DRB3*03:01	DRB3*01:01
TCR098	03/0492	CASSLGQGHYNEQFF	3/162	CAVNGGGSNYKLTF	3/134	*01:02	*01:02	*06:02	*06:09	*15:03	*13:02	DRB3*03:01	DRB5*01:01
TCR099	09/0125	CASSPGQGHYNEQFF	4/56	CAVNSGGSNYKLTF	4/39	*06:01	*01:02	*03:01	*05:02	*12:02	*16:02	DRB3*03:01	DRB5*01:01
	01/0906	CSARSSGGEAKNIQYF	2/118			*02:01	*01:02	*02:02	*06:02	*07:01	*15:01	DRB4*01:03	DRB5*01:01
	09/0018	CSARKGGGEAKNIQYF	1/182			*05:01	*01:02	*02:01	*06:03	*03:01	*15:03	DRB3*02:02	DRB5*01:01
TCR087	03/0492	CSARAGGGEAKNIQYF	3/112	CAVSRAGAGSYQLTF	3/90	*01:02	*01:02	*06:09	*06:02	*13:02	*15:03	DRB3*03:01	DRB5*01:01
TCR088	01/0906	CSARASGGEAKNIQYF	1/106	CAVRDPGNTDKLIF	1/72	*02:01	*01:02	*02:02	*06:02	*07:01	*15:01	DRB4*01:03	DRB5*01:01

FIG. 23.13. Defining TCR specificity groups with GLIPH. αβ TCR sequences are highly diverse, with many possible solutions for the same pMHC ligand. So many in fact that even monozygotic twins share very few identical TCRs. Using the GLIPH algorithm (for grouping of lymphocyte interaction by paratope hotspots), Glanville and Huang were able to analyze 5700 TCR sequences from 22 individuals to find five TCRβ clusters that shared sequence motifs in their CDR3 regions. These shared motifs, from Mtb-specific CD4+ T cells, also enriched for specific class II alleles.[141] The subsequent development of GLIPH2[282] has allowed millions of TCR sequences to be analyzed in this way.

patients, GLIPH2 reduced this to 66K specificity groups, of which we determined that 435 were enriched in tumors versus adjacent tissue.[283] Thus, you can quickly focus in on what the most important T cell specificities might be, before you know what those specificities actually are (although it should be noted that the GLIPH algorithms have been very good predictors of the MHC restricting element).[141,282]

Infectious Disease History

This is typically not an issue in mice, but very definitely one in humans where even young children have typically been exposed to dozens if not hundreds of microbes, plus multiple vaccinations. Previous pathogen exposures to the other strains or even completely unrelated ones[176] can be very important in how a given individual, or group of individuals, may respond to a vaccine. In influenza work, this is a major topic, as it has been evident for many years that prior flu strain exposure contributes significantly to how an individual responds to a given vaccine. This goes back to the "original antigenic sin" observations of Thomas and others[281,284] and the recent work of Gostic,[285] indicating that the predominant strain exposures in different eras could be critical to the chances of a fatal infection in later years. It is also emerging as a possible explanation for why children are generally very resistant to SARS-CoV-2, possibly due to prior (and recent) infections with one of the four circulating related coronaviruses. As discussed above, T cell responses are also notoriously cross-reactive,[176] and thus we must also consider the possibility that exposure to unrelated pathogens might play an important role in vaccine responses. Here, advances in technology provide at least a possible route to this, with the development of array-based methods to broadly screen for pathogen exposure.[286,287]

NONCLASSICAL αβ T CELLS

A number of unique T cell types have been described, which have a role in infectious disease responses, but generally have not been targeted in vaccine development, but probably should be. The best known of these nonclassical TCRs are those of the γδ TCR type, which constitutes a minority of 3% to 20% of T cells in healthy humans, as mice deficient in these cells are more sensitive to infectious diseases. Interestingly, humans deficient in γδ T cells are yet to be described, and in evolution, organisms across a wide spectrum of vertebrate evolution that have been characterized this cell type along with αβ T cells and B cells.[288] Even in the most primitive vertebrates, such as hagfish or lampreys, which use completely different X genes for their antibody equivalents, there is a third cell type thought to be equivalent to γδ T cells.[289] Their TCRs recognize intact antigens directly, but their actual role in a pathogen response is generally not clear. An exception is in a murine form of Malaria (*Falciparum chabaldi*) where Mamedov et al[290] showed that a distinct species of γδ T cell arises late in the infection and blocks recurrent parasitemia through its secretion of MCSF, presumably through some action on monocytes or macrophages.

NKT cells utilize αβ TCRs, but typically have an invariant α TCR, and do not recognize their short chain fatty acid ligands in a way that makes use of TCR diversity the way that peptide-MHCs are recognized.[291] Their substrates are specific nonclassical MHCs that can bind these unique ligands. They seem chiefly to recognize bacterial antigens, and so should be considered in developing those vaccines. Similarly, MAIT cells have been shown to use a similar strategy to recognize metabolites, particular a vitamin (so far).[292,293] Single RNAseq analysis has shown that at least the NKT and MAIT cells that arise in the context of *Mtb* stimulation share virtually all of the same expressed genes,[274] suggesting that they are the same cell type that have been adapted to recognize different classes of ligands and do not require the diversity of classical αβ or γδ T cells.

γδ T CELL RECOGNITION

For many years, the recognition properties of γδ T cells remained a mystery. It was only in 1994 that things began to get somewhat clearer. Rock et al[294] found that an analysis of Vδ and Vγ CDR3 lengths showed that they resembled V_H and V_L distributions much more than Vα and Vβ in their asymmetry, as shown in Figure 23.14. Subsequent work has even shown that V_H-like gene segments can even substitute for Vδ in several organisms, as in Deiss et al, 2019.[295]

Even more important was the extensive work of Schild et al in the same year[9] who showed that γδ T cells specific for a nonclassical class MHC in the mouse did not require antigen processing, but seemed to recognize antigens directly, like antibodies, which was later confirmed by tetramer binding[296] and structural analysis,[297] as discussed in more detail below. Consistent with the Ig-like recognition properties of γδ TCRs, and the lack of constraints imposed by thymic maturation on the peripheral γδ T cell, antigen-specific repertoire γδ T cells recognize diverse antigens irrespective of their molecular or genetic nature and without antigen processing and presentation requirement. While γδ T cells can recognize classical and nonclassical MHC molecules, these molecules are not obligatory component of γδ T cell antigens. These aspects will be discussed in the following sections.

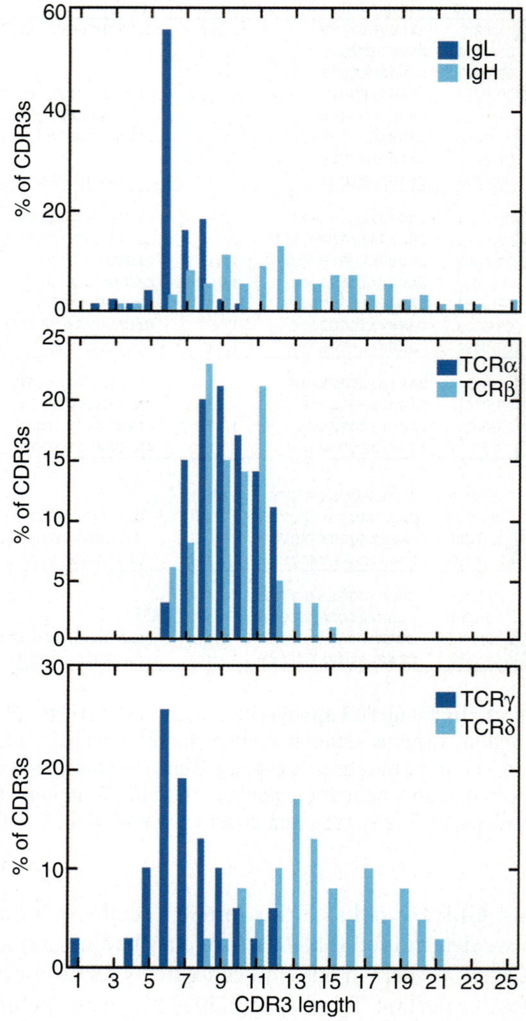

FIG. 23.14. CDR3 length analysis of T cell receptor (TCR) polypeptides versus immunoglobulin (Ig) heavy and light chains. These data, modified from Rock et al,[294] show that whereas TCRα and β CDR3 regions are relatively uniform with respect to each other, the other antigen receptor pairs show a marked asymmetry. Specifically, both Ig light chains (κ and λ) show very short CDR3s, as do TCRγ chains. In contrast, both IgH and TCRδ are quite heterogeneous and tend to be longer. These data suggest that γδ TCRs have a more antibody-like structure and binding properties. This has been borne out by subsequent analysis (see text).

Antigen-Specific Repertoire

γδ T cell antigens have been identified in many different ways: through stimulation, with targeted assays, and from pathological conditions.

Antigens Identified From γδ T Cells Responding to Stimulation

Immunization of mice with MHC-mismatched spleen cells leads to the identification of γδ T cells specific for MHC class II I-E and I-A and the nonclassical MHC class I molecules T10 and T22.[9,298] Stimulation of lymph node cells from HSV-infected C3H mouse with HSV-gI–transfected L cells generated γδ T cells specific for HSV-gI.[299]

In vitro exposure of peripheral blood lymphocytes (PBLs) with mycobacterial lysates or cancer cells induces human Vγ9Vδ2 γδ T cell expansion. These T cells recognize microbial as well as mammalian-derived phosphoantigen-induced butyrophilins (BTNs) to be expressed on target cells.[300] The human γδ T cell clone G115, which is derived from PBMC γδ T cells stimulated with irradiated PBLs and lymphoblastoid cells, recognizes ATP synthase and apolipoprotein A1 complex on tumor cells.[301] Stimulation of human Vδ2 γδ T cells with a broad panel of B cell lymphomas treated under different stress conditions generates annexin A2–specific γδ TCR.[302]

Other γδ T cell antigens identified from *in vitro* stimulation include those specific for cardiolipin associated with apolipoprotein H, cardiolipin associated with CD1d, polyGluTyr or polyGluTyr associated with Qa1, the human nonclassical MHC molecules MICA and MICB (summarized in Ref. 303), and HLA-A2 complexed with melanoma-derived peptides.[304]

γδ T Cells Recognize Some of the Same Antigens as αβ T and B Cells

Some γδ TCR ligands were identified by probing γδ T cells from naïve mice and/or human PBMCs with known αβ T cell targets. These include the nonclassical MHC CD1 family of proteins (CD1a, CD1b, CD1c, and CD1d in human and CD1d in murine) and MHC-related protein 1 (MR1).[305-309]

Known B cell antigens are also found to be both human and murine γδ T cell antigens.

These include the algae protein PE[310] and the haptens, cyanine 3, and 4-hydroxy-3-nitrophenylacetyl (NP).[311] Haptens are small organic molecules that when coupled to proteins induce strong γδ T cell–dependent, hapten-specific antibody response. The ability to recognize small organic molecules and chemical modifications of host molecules is an essential capability of the adaptive immune system. The observations that γδ T cells can directly recognize and respond to these small molecules represent a significant expansion in the scope of the γδ T cell antigen-specific repertoire (see Fig. 23.14).

Targets of γδ T Cells Associated With Pathological Conditions

A γδ T cell clone, M88, established from γδ T cells expanded in a case of autoimmune myositis recognizes tRNA synthetases and the translation initiation factor.[312] In this context, ~20% of patients with myositis also have autoantibodies against nuclear or cytoplasmic antigens, including tRNA synthetase. In fact, human histidyl-tRNA synthetase, a stimulatory antigen for M88, is also targeted by autoantibodies known as anti-Jo-1. This observation suggests a possible mechanism for γδ T cell involvement in this and other autoimmune diseases and support the notion that γδ T cell can recognize common B cell antigens.

Endothelial protein C receptor, a transmembrane protein with 20% identity to MHC class I and CD1 molecules, was shown to interact directly with the TCR from a γδ T cell clone isolated from a patient, which showed CMV activation after lung transplantation.[313] In the nonobese diabetic mouse model of type 1 diabetes, the insulin peptide B:9–23 is recognized by pancreatic islet infiltrating γδ T cells.[314] The same peptide is also recognized by B cells and by αβ T cells when complexed with IA^{g7}.

ANTIGEN RECOGNITION CHARACTERISTICS OF γδ T CELLS

Recognition of Peptide/MHC Complexes

The identification of γδ TCRs specific for classical and nonclassical MHC molecules allows a direct comparison of γδ and αβ T cell antigen recognition. As discussed below, compared to the "MHC restriction" type of αβ TCR/MHC complex interaction, the interactions between γδ TCR and the MHC complexes are much more diverse, both in whether and how the peptide or the lipid moieties associated with the MHC molecules confer the specificity of the recognition and in the TCR contact surfaces of the MHC molecules.

The LBK5 γδ TCR is specific for the classical MHC class II molecule I-E.[298] It was demonstrated that LBK5 recognizes IE/peptide complexes in a distinct way from αβ TCR recognition of peptide/MHC,[9] in that the bound peptide does not contribute to the specificity of the interaction, and mutations of residues located on the α helical regions of the IE α and β chains that affect αβ T cell recognition of peptide/I-E^k do not affect LBK5 activity. Instead, the epitope on I-E^k for LBK5 is centered on the polymorphic residues 67 and 70 of the I-E β chain.[315] This explains the fine specificity of LBK5, which is for I-$E^{b,k,s}$ but not I-E^d.

In addition to this mode of MHC/peptide complex recognition, γδ T cells recognize melanoma-associated antigens, MART-1, and gp100 peptide in association with HLA-A0201 in a peptide-specific, MHC-restricted manner. A crystal structure of a MART-1–specific γδ TCR, together with MHC/peptide, revealed distinctive but similar docking properties to those previously reported for αβ TCRs that recognize the same peptide/MHC complex.[304] In this context, it should be noted that while MHC-bound peptide often does not confer binding specificity of MHC-specific antibodies, peptide-specific, MHC restricted antibodies can be generated.[316]

Recognition of Nonclassical MHC Molecules

T10 and T22 are murine β_2M-associated molecules with a truncated MHC-like fold and lack a peptide-binding groove.[317] T10 and T22 do not bind peptide or any other moiety. They are recognized directly by the G8 γδ TCR with a dissociation constant (K_D) of 13 μM. T10/T22 are natural ligands of murine γδ T cells with ~0.1% to 1% of total spleen and intestinal intraepithelial lymphocyte (IEL) γδ T cells in naïve mice stained specifically with a T22 tetrameric staining reagent.[296] T10/T22-specific γδ T cells have TCRs that are diverse in both VγVδ gene usage and CDR3 sequences. And the antigen specificity is conferred by a somatic recombinant CDR3δ motif that consists of a Vδ- or Dδ1-encoded Trp (W), a Dδ2-encoded sequence of Ser, Glu, Gly, Tyr, and Glu (SEGYE); and a P-nucleotide–encoded Leu (L). The W-(S) EGYEL motif is necessary and sufficient for T10/T22 binding, and sequence variations in the CDR3 regions around this motif modulate the affinity and the kinetics of T22 binding.[318] That this CDR3 motif is important in antigen recognition is confirmed by the crystal structure of G8 TCR bound to T22,[297] as shown in Figure 23.15.

The nonpolymorphic MHC CD1 molecules have an MHC-like fold and display lipid antigens for the recognition by αβ T cells. In general, αβ TCRs interact with these molecules through binding on top of the α1 and α2 helices. The interactions between CD1 molecules and their specific γδ TCRs appear to be much more heterogeneous. Some of the γδ TCRs are specific for the lipid moiety associated with the CD1 molecules, whereas others appear to recognize CD1 with diverse bound ligands or do not show specificity to the bound ligands. Studies also show that in general, γδ TCR and αβ TCR interact differently with the same CD1 molecules.[307,311,321]

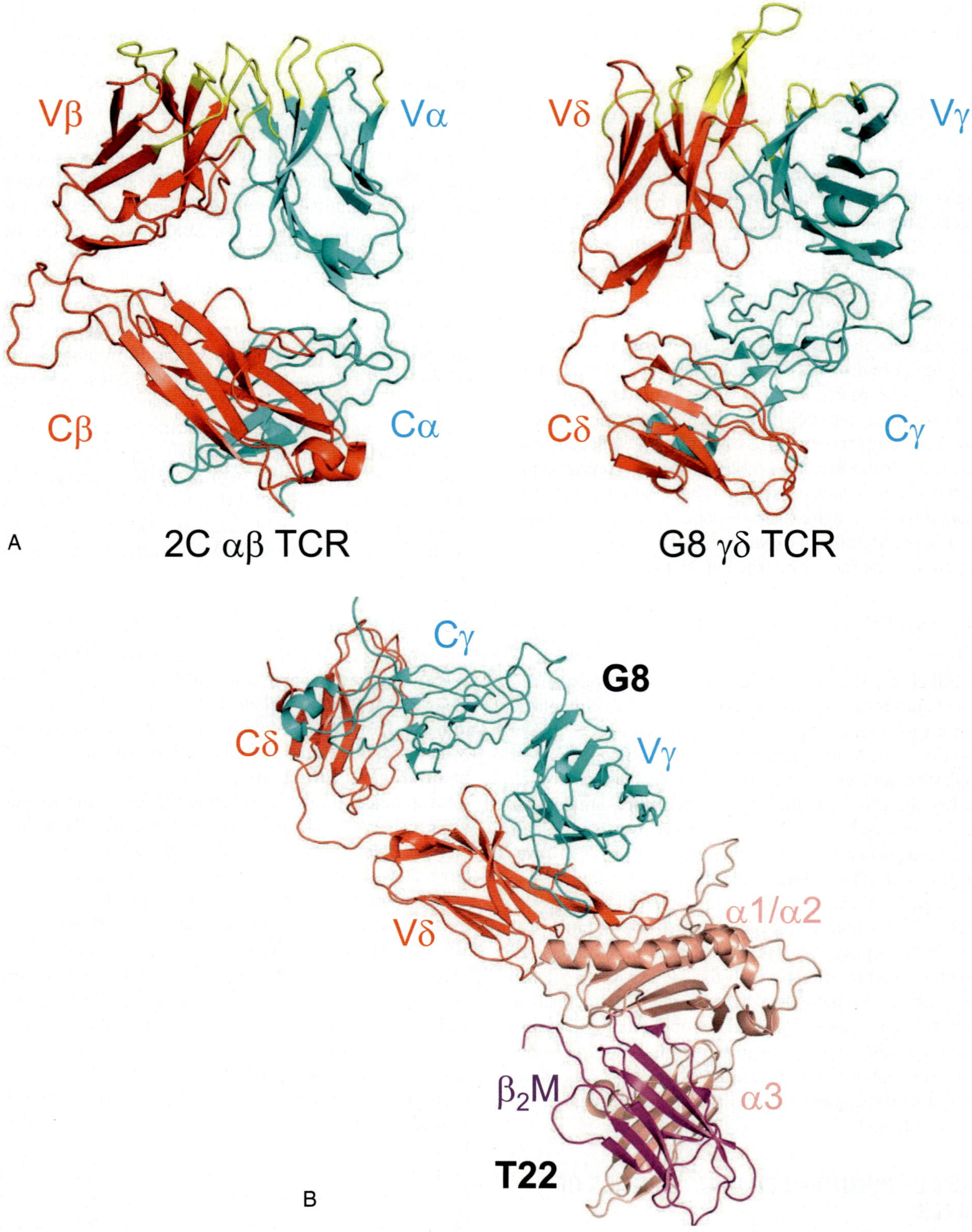

FIG. 23.15. **A:** Ribbon diagrams of the T cell receptor (TCR) structures. This shows the structures of the 2C αβ TCR[102] versus the G8 γδ TCR.[297] The TCRγ chains are in cyan, and the TCRα and γ chains are in vermillion. The CDRs of both are in yellow. The very long TCRδ CDR3 in G8, which binds the T10/T22 ligands, is very apparent here but is shorter in most other γδ TCRs. Note the different C region interactions with these TCRs and the deviations from the classic "β barrel" structure in both Cα and Cδ. The prominent Cβ loop to the left is also unusual and may mediate interactions with CD3 or other molecules on the T cell surface. **B:** The structure of the nonclassical MHC-I molecule T22 binding to the G8 γδ TCR, where the long TCRδ CDR3 accounts for almost all of the contact resides.[306] (Figures courtesy of Drs. X. Yang and K. C. Garcia.)

MR1 binds vitamin B precursors, small-molecule metabolites, which include derivatives of biosynthetic precursor to riboflavin, and folic acid degradation by-products. MR1-specific γδ TCRs utilize diverse VγVδ chains, and the bound small molecules confer little or no specificity to the TCR-MR1 interaction.[307] Other MR1-specific γδ TCRs interact with α3 domain of the molecule, with additional interactions to the α1α2 helical platform's underside. A similar pattern of reactivity was observed for diverse MR1-restricted γδ TCRs from multiple individuals.[308] This mode of interaction resembles CD8αα/class I MHC recognition, which is in the classic manner of an antibody-antigen interaction.

Recognition of PE

PE (MW of ~240 KD) is a fluorescent molecule located on the tip of the photosynthetic antenna of cyanobacteria and red algae. It is a hexameric protein with tetrapyrrole chromophores covalently linked to the cysteine residues. PE isolated from different algae has minor differences in amino acid sequences and the numbers, combinations, and positions of chromophores.[156] Individual human as well as murine γδ TCRs show difference in recognizing R-PE from *Porphyra tenera* and PE from red algae, P1 and P5. With respect to γδ T cells as a whole, R-PE as well as P1 stained twice as many murine splenic γδ T cells as P5 did. In one human sample, the percentages of P1- and P5-specific γδ T cells were similar. These observations indicate that γδ TCR PE recognition is highly specific.[310]

Mutagenesis studies indicated that the CDR3 regions of the TCRγ and δ chains encode PE-binding determinants, with contributions from the CDR1 and/or CDR2 encoded on the Vδ gene. Although there are common CDR3 features shared among murine PE-specific γδ TCRs, these features are not shared by human PE-specific γδ TCRs, and there is no apparent sequence similarity between the human and the murine PE-specific γδ TCR sequences.[310] Thus, the antigen specificity, but not the particular antigen TCR sequences, is conserved through evolution.

Recognition of Phosphoantigen-Induced BTN Molecules

γδ T cells express Vγ9 and Vδ2 TCRs are known to expand in a variety of infectious diseases, and after *in vitro* stimulation with either bacterial or mammalian phosphoantigens (pAgs) such as the microbially derived (E)-4-hydroxy-3-methyl-but-2-enyl pyrophosphate, and host-cell-derived isoprenyl pyrophosphate. Vγ9Vδ2-mediated PAg sensing requires cell-cell contact and depends on both Vγ and Vδ chains. Recent studies indicate that PAg induced γδ T cell activation requires the expression of BTN3A1 and BTN2A1 on the target cells.[157,300] BTNs are transmembrane proteins that are structurally related to the B7 family of costimulatory molecules. BTN2A1 associates closely with BTN3A1 on the surface of target cells and can bind directly to the side of the Vγ9 domain of the TCR. A putative second ligand-binding domain on a separate region of the TCR that incorporates Vδ2 has also been identified. Disruption of either of these binding sites abrogates the ability of Vγ9Vδ2 T cells to respond to pAg. The pAg likely modifies the BTN2A1-BTN3A1 complex through the interaction with the cytoplasmic region to make it stimulatory. These observations suggest a novel way of adaptive immune receptor antigen recognition.

Vγ-Specific Interaction With BTN-Like Molecules

In addition to the interaction of BTN2A1 with TCRγ chain of Vγ9Vδ2 TCRs, studies indicate that BTN-like family proteins, such as BTNL3 and BTNL8 can interact with human Vγ4 TCRs and BTNL1 and BTNL6 with murine intestine Vγ7 γδ TCRs.[320-322] The mode of interaction appears to be focusing on the germline-encoded V gene segments and may be similar to the superantigen binding to the Vβ chains. However, different studies disagree on whether the CDR3 region can modify the BTNL-Vγ interaction. One important issue is whether and how this Vγ-specific interaction would affect antigen recognition. As discussed above, both the T10/T22-specific and PE-specific γδ T cells in the spleen express Vγ1 and Vγ4 and those in the small intestine express Vγ7.[310,318] Along this line, human γδ T cells specific for CD1b utilize TCRs with different Vγ chains. And the CD1β-specific Vγ4+ T cell can also respond to BTNL3 and BTNL8 stimulation.[309]

γδ TCR Structure

The x-ray crystal structures of γδ TCRs and their ligands clearly indicate diverse modes of TCR-ligand interactions as compared to the rather constrained docking angles commonly observed of αβ TCRs on their peptide/MHC molecules, reflecting the differences in αβ TCR and γδ TCR antigen recognition as discussed above. For examples, the recognition of nonclassical MHC-like molecules (T22) by a specific γδ TCR (G8) is mediated exclusively through the TCRδ CDR3 region.[297] In contrast to the recognition of CD1d/lipid antigens, whereas both the γ and the δ chains of 9C2 TCR contribute to the recognition of CD1d/α2galactosyl ceramide, the DP10.7 γδ TCR interacted with CD1δ/sulfatide exclusively through the TCRδ chain. This mode of recognition is different than the recognition of CD1δ/lipid by the semi-invariant iNKT αβ TCR, where germline encoded CDR1, two loops appear to be restricted to the α1α2 platform of the CD1δ molecules.[305] Strikingly, a crystal structure of a γδ TCR–MR1 (the nonclassical MHC-related protein 1) antigen complex differs from all other TCR-MHC and TCR-MHC-I–like complex structures. In that, the γδ TCR binds underneath the MR1 antigen-binding cleft where contacts are dominated by the MR1 α3 domain.[308] Nonetheless, human γδ TCRs that recognize MART-1/HLA-A*0201 in a peptide-specific manner have distinctive but similar docking properties to those previously reported for αβ TCRs recognizing MART-1 on HLA-A*0201.[304]

There are also multiple γδ heterodimer structures: a γδ TCR from a human T cell clone G115[323] which can be activated by natural or synthetic pyrophosphomonoesters and a soluble form of G115 also interact with a soluble form of ATP synthase F1 and apolipoprotein A-1, and a human MIC (MHC class I chain-related)-reactive γδ TCR (δ1A/B-3), which was determined as a single-chain Fv construct.[324] In the case of G115, both Vδ and Vγ CDR3 loops protrude from the rest of the putative binding surface and create a cleft between them. Portions of the CDR1γ and CDR1δ and CDR2γ combine with the clefts between the CDR3 loops to form a pocket, which is surrounded by positively charged amino acid residues contributed by CDR2γ, CDR2δ, and CDR3γ. The jagged surface of this TCR resembles the surface of an antibody that binds a small molecule antigen. While the δ1A/B-3 TCR maintains an overall fold similar to the other γδ TCR structures, it was noted that unlike the G115 and G8 CDR3 regions, which are protruding out, the δ1A/B-3 CDR loops together generate a nearly flat surface on the combining site. This difference is anticipated, as the CDR3 length distribution of the TCRδ chains is quite variable as discussed above, and like antibodies should have a broad range of binding site shapes.

One distinctive feature of both the G115 and the G8 TCR, when compared with αβ TCRs and Igs, is that the C domains "swing out" from under the V domains. This unusual shape is highlighted by both a small elbow angle if 110°, defined as the angle between the pseudo twofold symmetry axes that relate V to V and C to C, and a small V-C interdomain angle. This contrasts with an average of 149° for αβ TCR structures (Adams and Garcia, unpublished communication). The small angle between the Vγ and Cγ domains shifts both Cδ and Cγ to one side. Moreover, the molecular surfaces of the constant domains are different than those of αβ TCRs with no clear similarities either in the shape or the nature of the CαCβ and CγCδ surfaces; there are only a few solvent-exposed residues that are conserved in both Cβ and Cγ domains as well. However, it is unclear where or how the extracellular domains of the CD3 subunits interact with the extracellular portions of γδ TCRs compared with αβ TCRs. This may explain why the CD3 components of αβ TCRs are different from those of γδ TCRs.

The γδ T Cell Repertoire

In unimmunized mice, ~0.1% to 2% and ~0.02% to 0.4% of total γδ T cells are T10/T22 and PE specific, respectively. In humans, ~0.025% of peripheral γδ T cells recognize PE, and the frequencies of CD1, MR1 tetramer–specific γδ T cells are in the similar range. The high initial frequencies of antigen-specific γδ T cells allow γδ T cells to mount sizable responses despite the paucity of the γδ T cells. In fact, the number of antigen-specific γδ T cells at the height of an immune response is comparable with those described for antigen-specific αβ T cell response.[325,326] However, if the frequencies of other antigen-specific γδ T cells are also in the similar range, then the numbers of distinct γδ T cell antigens would be ~10^3 to 10^4, which is considerably smaller than the estimated αβ T cell antigen–specific repertoire (~10^6). Along this line, the size of the B cell antigen–specific repertoire was originally estimated based on antigen-specific B cell frequencies: 0.004% to 0.007% of total B cells recognize nitrophenyl (NP), dinitrophenyl, and trinitrophenyl.[327,328] These numbers were obtained using antigen-specific spleen foci formation as a read-out and are likely to be underestimates, as this assay requires extensive proliferation of individual clones. In fact, fluorescence activated cell sorting analysis showed that in naïve mice, 0.1% of the B cells are PE specific and 0.02% are allophycocyanin (APC) specific.[310] Based on these values, the size of the antigen-specific B cell repertoire would be ~1000 to 5000 antigens, which is in the same range as that of γδ T cells. In this context, it has been proposed that the immune repertoire can offer efficient protection against about 10^3 to 10^4 distinct infections relevant for the survival of a given species.[329,330] Based on this argument, Zinkernagel and Hengartner proposed that "in mice…immunity is generated from a starting number of about 100 to 1000 antigen-specific precursor [αβ] T and B cells."[331] Although this range turns out to be much higher than the frequencies of naïve antigen-specific αβ T cells, it is surprisingly close to that of antigen-specific γδ T cells in naïve animals.

Shaping of γδ T Cell Antigen–Specific Repertoire and Function

γδ T cells, similar to αβ T cells, develop in the thymus before entering the periphery. For αβ T cells, this process determines what peripheral αβ T cells can recognize and whether they develop into CD4+ helper or CD8+ cytolytic T cells. Endogenous thymic ligand expression also plays an important role in γδ T cell development and function, but the regulations are fundamentally different from those of αβ T cells. In particular, encountering thymic ligand does little to constrain antigen specificities of the γδ T cell repertoire, and it endows γδ T cells with different effector functional programs.[332] However, encountering thymic antigen appears to have different effects on the development of the effector program of fetal and postnatal γδ T cells.[333]

Restricted TCR Repertoire on Fetal γδ Thymocytes, But Their Functional Characteristics Depend on Ligand Encounter

γδ T cells are the first T cells that appear during fetal thymic ontogeny. The first and second waves of murine γδ thymocytes express Vγ5Jγ1, Vδ1Dδ2Jδ2 and Vγ6Jγ1, Vδ1Dδ2Jδ2, respectively, with no N- or P-nucleotide addition at any of the V(D)J junctions of these TCRs. The same gene rearrangements are also present in the fetal thymus of TCRδ−/− mice,[334] indicating that these restricted γδ TCR rearrangements are not due to selection mediated through the TCR.

Vγ5Vδ1 cells appear as the dendritic epidermal T cells. In adult mice, their ability to react to keratinocytes and make IFNγ appear to require interaction with thymic antigen(s) during development.[335,336] Similarly, the presence of Vγ6Vδ1 γδ T cells in various immune and nonimmune tissues in adult animals and their ability to mount IL-17 response without explicit induction from an immune response (commonly referred to as the "innate like" or "natural" Tγδ17 cells) also appear to require strong signaling mediated through the TCR during development.[333,337] Even the antigens, which induce the development and function of either Vγ5Vδ1 or Vγ6Vδ1 γδ T cell responses, remain unclear.

Human fetal (17- to 19-week gestation) γδ thymocytes also express canonical TCR sequences (Vγ8JP1, Vδ2Jδ3, and Vδ1Jδ3) with no junctional diversity. Similar to that of murine fetal thymic γδ TCRs, the presence of these sequences is independent of functional TCR expression. These γδ thymocytes can mount IFNγ response.[338]

Thymic Maturation Determines γδ T Cell Effector Fate, But Not Their Antigen Specificity

γδ T cells generated postnatally express TCRs with various Vδs and Vγs and diverse CDR3 regions. These cells are found in all secondary lymphoid organs and mucosal tissues. As the MHC molecules are not obligatory components of γδ T cell ligands, and there is no apparent common structural motif shared by defined γδ T cell antigens, it is inconceivable that thymic γδ T cell development could be dependent on MHC-related molecules as αβ T cells do. Indeed, as discussed below, the β_2M associated nonclassical MHC T10/T22 are natural ligands for a sizable population (~0.2%-2%) of γδ T cells in normal mice of different T10/T22 expression patterns, for example, expressing T10 and T22 (C57BL/6), expressing only T10 (BALB/c), and mice lacking surface classical and nonclassical MHC class I T10/T22 expression (the β_2M−/− mice). A similar frequency of T10/T22-specific γδ T cells was also found in the thymuses and spleens of mice lacking both β_2M and class II MHC molecules.

Notably, T10/T22-specific γδ thymocytes from different genetic backgrounds show similar levels of signaling mediated through the TCR, indicating that γδ thymocytes can signal through the TCR in the absence of cognate ligand expression. It was demonstrated that γδ TCRs can dimerize through the interactions between the Vδs to induce signaling without cross-linking by ligands with the exception of Vδ1, which is expressed on the fetal γδ T cells as discussed above. Thus, TCR dimerization may be sufficient to drive the development of γδ thymocytes, and also enhance antigen-specific activation of peripheral γδ T cells, which are known to signal more efficiently than αβ T cells despite lacking CD4 or CD8 coreceptor expression.[55] Analysis of cell surface marker expressions and the turnover rate of total γδ T cells in the spleen and lymph nodes with those of T10/T22-specific γδ T cells developed in different genetic background suggests that a large fraction of both the thymic and the peripheral γδ T cells may not have encountered ligand during their development or in the periphery.[332] Along this line, human postnatal γδ thymocytes also have antigen-naïve phenotype.[339] With respect to γδ IELs in the small intestine, it was noted that Ag-naïve γδ thymocytes express high levels of the gut homing receptor CCR9.[340] This is in contrast to αβ IELs, which develop from thymocyte precursors that have interacted strongly with self-ligand in the thymus.[341]

Although encountering thymic ligand during thymic development does little to constrain the γδ T cell antigen–specific repertoire, it endows γδ T cells with different effector functions. In particular, murine postnatal γδ thymocytes that have not engaged ligand have the potential to develop into IL-17–producing cells in the periphery, and the encounter of cognate ligand drives these γδ thymocytes toward adopting the IFN-γ effector fate.[332] The exception is the γδ IELs; they make IFNγ. Whether the Vγ7-specific interaction with the BTNL molecules in the small intestine contribute to this alteration is unclear.

γδ TCRS WITH ANTIBODY-LIKE RECOGNITION PROPERTIES: AN EVOLUTIONARY PERSPECTIVE

In all mammals, bird, and amphibians examined, the genes encoding the α and δ chains are at a single (TCRα/δ) locus with the TCRβ and TCRγ chains each encoded at separate loci. Although some Vδ genes are in the Vα gene family, there are TCRδ isoforms that use V genes from Ig heavy chain V (VH) in both amphibians and birds. Such V genes have been designated as VHδ. In the chicken, *Gallus gallus*, genes encoding VHδs are located at a second TCRδ locus that is unlinked to the conventional TCRα/δ locus. In the frog, *Xenopus tropicalis*, and a passerine bird, the zebra finch *Taeniopygia guttata*, the VHδs are located within the TCRα/δ loci, where they coexist with conventional Vα and Vδ genes.[342,343]

Marsupials also have a TCR locus unlinked to TCRα/δ (TCRμ) that uses antibody-related V genes. TCR μ is related to TCRδ, but divergent in sequence and structure. TCR μ, which contains three extracellular domains, has evolved first in the ancestors of the cartilaginous fish in the form of NAR-TCR (new antigen receptor-TCR) and second in mammals as TCRδ. The NAR-TCR uses an N-terminal V domain related to the V domains found in Ig NAR (Immunoglobulin NAR) antibodies, which are unique to cartilaginous fish. The N-terminal V domains of TCR μ and NAR-TCR are unpaired, similar to the single antigen-binding domain, analogous to the V domains of heavy chain–only antibodies found in sharks and camelids. Other than the bony fish and placental mammals, the presence of TCR chains that use antibody-like V domains (such as TCRδ using VHδ, NAR-TCR, or TCRμ) are widely distributed in vertebrates, suggesting independent solutions for generating TCR chains with antibody V genes in different vertebrate lineages.[295,344-346]

They may reflect the selective pressure for TCR chains that could recognize antigen in ways similar to that of antibodies in different lineages of vertebrates.[295] See also Chapter 4 on vertebrate evolution in this volume.

CONCLUSION

Since TCR genes were first identified in the early 1980s, information about their genetics, biochemistry, structure, cell biology, and function has accumulated to become a broad field unto itself. Despite this very real progress, many issues still remain unsolved such as: What do γδ T cells normally "see" and what function do they serve in homeostasis? What do superantigens actually do during the course of a normal response and how is this of benefit to the pathogen/parasite? What is the structural/chemical basis of TCR specificity? What sort of rearrangements or conformational charges occur in the TCR/CD3 molecular ensemble upon ligand engagement? These and other questions will require more years of effort.

ACKNOWLEDGMENTS

We are grateful to C. Xu, J. Huppa, M. Kuhns, Z. Huang, Y. Yang, M. Call, K. Wucherpfennig, M. Krangel, and K. C. Garcia for providing some of the figures and tables used here. We also thank C. Tato, M. Kuhns, M. Krogsgaard, Qi-jing Li, and K. C. Garcia for their invaluable help with the manuscript, and R. Cuevas for expert assistance. This work was supported by grants from the National Institutes of Health (to MMD and YC) and from the Howard Hughes Medical Institute (to MMD).

REFERENCES

1. Zinkernagel RM, Doherty PC. Immunological surveillance against altered self components by sensitised T lymphocytes in lymphocytic choriomeningitis. *Nature*. 1974;251:547-548.
2. Zinkernagel RM, Doherty PC. H-2 compatability requirement for T-cell-mediated lysis of target cells infected with lymphocytic choriomeningitis virus. Different cytotoxic T-cell specificities are associated with structures coded for in H-2K or H-2D. *J Exp Med*. 1975;141:1427-1436.
3. Davies DR, Metzger H. Structural basis of antibody function. *Annu Rev Immunol*. 1983;1:87-117.
4. Bjorkman PJ, Saper MA, Samraoui B, Bennett WS, Strominger JL, Wiley DC. Structure of the human class I hitocompatibility antigen, HLA-A2. *Nature*. 1987;329:506-512.
5. Bendelac A, Savage PB, Teyton L. The biology of NKT cells. *Annu Rev Immunol*. 2007;25:297-336.
6. Cheroutre H. IELs: enforcing law and order in the court of the intestinal epithelium. *Immunol Rev*. 2005;206:114-131.
7. Saito H, Kranz DM, Takagaki Y, Hayday AC, Eisen HN, Tonegawa S. A third rearranged and expressed gene in a clone of cytotoxic T lymphocytes. *Nature*. 1984;312:36-40.
8. Brenner MB, McLean J, Dialynas DP, et al. Identification of a putative second T-cell receptor. *Nature*. 1986;322:145-149.
9. Schild H, Mavaddat N, Litzenberger C, et al. The nature of major histocompatibility complex recognition by γδ T cells. *Cell*. 1994;76:29-37.
10. Infante AJ, Infante PD, Gillis S, Fathman CG. Definition of T cell idiotypes using anti-idiotypic antisera produced by immunization with T cell clones. *J Exp Med*. 1982;155:1100-1107.
11. Allison JP, McIntyre BW, Bloch D. Tumor-specific antigen of murine T-lymphoma defined with monoclonal antibody. *J Immunol*. 1982;129:2293-2300.
12. Meuer SC, Fitzgerald KA, Hussey RE, Hodgdon JC, Schlossman SF, Reinherz EL. Clonotypic structures involved in antigen-specific human T cell function: relationship to the T3 molecular complex. *J Exp Med*. 1983;157:705-719.
13. Haskins K, Kubo R, White J, Pigeon M, Kappler J, Marrack P. The major histocompatibility complex-restricted antigen receptor on T cells: I. Isolation with a monoclonal antibody. *J Exp Med*. 1983;157:1149-1169.
14. Kaye J, Porcelli S, Tite J, Jones B, Janeway CA. Both a monoclonal antibody and antisera specific for determinants unique to individual cloned helper T cell lines can substitute for antigen and antigen-presenting cells in the activation of T cells. *J Exp Med*. 1983;158:836-856.
15. Samelson LE, Germain RN, Schwartz RH. Monoclonal antibodies against the antigen receptor on a cloned T-cell hybrid. *Proc Natl Acad Sci U S A*. 1983;80:6972-6976.
16. McIntyre BW, Allison JP. Biosynthesis and processing of murine T-cell antigen receptor. *Cell*. 1984;38:659-665.
17. Kappler J, Kubo R, Haskins K, et al. The major histocompatibility complex-restricted antigen receptor on T cells in mouse and man: identification of constant and variable peptides. *Cell*. 1983;35:295-302.
18. Davis MM, Cohen EA, Nielsen AL, DeFranco AL, Paul WE. *The Isolation of B and T Cell-Specific Genes*. Academic Press; 1982.
19. Hedrick SM, Cohen DI, Nielsen EA, Davis MM. Isolation of cDNA clones encoding T cell-specific membrane-associated proteins. *Nature*. 1984;308:149-153.
20. Hedrick SM, Nielsen EA, Kavaler J, Cohen DI, Davis MM. Sequence relationships between putative T-cell receptor polypeptides and immunoglobulins. *Nature*. 1984;308:153-158.
21. Yanagi Y, Yoshikai Y, Leggett K, Clark SP, Aleksander I, Mak TW. A human T cell-specific cDNA clone encodes a protein having extensive homology to immunoglobulin chains. *Nature*. 1984;308:145-149.
22. Acuto O, Fabbi M, Smart J. Purification and NH_2-terminal amino acid sequencing of the β subunit of a human T-cell antigen receptor. *Proc Natl Acad Sci U S A*. 1984;81:3851-3855.
23. Chien YH, Becker DM, Lindsten T, Okamura M, Cohen DI, Davis MM. A third type of murine T-cell receptor gene. *Nature*. 1984;312:31-35.
24. Saito H, Kranz DM, Takagaki Y, Hayday AC, Eisen HN, Tonegawa S. Complete primary structure of a heterodimeric T-cell receptor deduced from cDNA sequences. *Nature*. 1984;309:757-762.
25. Chien YH, Iwashima M, Kaplan KB, Elliott JF, Davis MM. A new T-cell receptor gene located within the alpha locus and expressed early in T-cell differentiation. *Nature*. 1987;327:677-682.
26. Dembić Z, Haas W, Weiss S, et al. Transfer of specificity by murine α and β T-cell receptor genes. *Nature*. 1986;320:232-238.
27. Saito T, Weiss A, Miller J, Norcross MA, Germain RN. Specific antigen-la activation of transfected human T cells expressing murine Ti αβ-human T3 receptor complexes. *Nature*. 1987;325:125-130.
28. Havran WL, Chien YH, Allison JP. Recognition of self antigens by skin-derived T cells with invariant γδ antigen receptors. *Science*. 1991;252:1430-1432.
29. Davis MM, Bjorkman PJ. T-cell antigen receptor genes and T-cell recognition. *Nature*. 1988;334:395-402.
30. Lefranc M-P, Lefranc G. *The T Cell Receptor: Facts Book*. Academic press, Ltd. Harcourt Science and Technology Company; 2001.
31. Brenner MB, McLean J, Scheft H, et al. Two forms of the T-cell receptor γ protein found on peripheral blood cytotoxic T lymphocytes. *Nature*. 1987;325:689-694.
32. Groettrup M, Ungewiss K, Azogui O, et al. A novel disulfide-linked heterodimer on pre-T cells consists of the T cell receptor β chain and a 33 kd glycoprotein. *Cell*. 1993;75:283-294.
33. Saint-Ruf C, Ungewiss K, Groettrup M, Bruno L, Fehling HJ, Von Boehmer H. Analysis and expression of a cloned pre-T cell receptor gene. *Science*. 1994;266:1208-1212.
34. Pang SS, Berry R, Chen Z, et al. The structural basis for autonomous dimerization of the pre-T-cell antigen receptor. *Nature*. 2010;467:844-848.
35. Call ME, Wucherpfennig KW. The T cell receptor: critical role of the membrane environment in receptor assembly and function. *Annu Rev Immunol*. 2005;23:101-125.
36. Terhorst C, Spits H, Stall F, Exley M. *T Lymphocyte Signal Transduction*. IRL Press; 1996.
37. Klausner RD, Lippincott-Schwartz J, Bonifacino JS. The T cell antigen receptor: insights into organelle biology. *Annu Rev Cell Biol*. 1990;6:403-431.
38. Clevers H, Alarcon B, Wileman T, Terhorst C. The T cell receptor/CD3 complex: a dynamic protein ensemble. *Annual Review of Immunology*. 1988;6:629-662.
39. Sun ZYJ, Kim KS, Wagner G, Reinherz EL. Mechanisms contributing to T cell receptor signaling and assembly revealed by the solution structure of an ectodomain fragment of the CD3εγ heterodimer. *Cell*. 2001;105:913-923.
40. Sun ZYJ, Sun TK, Il CK, Fahmy A, Reinherz EL, Wagner G. Solution structure of the CD3εδ ectodomain and comparison with CD3εγ as a basis for modeling T cell receptor topology and signaling. *Proc Natl Acad Sci U S A*. 2004;101:16867-16872.
41. Kjer-Nielsen L, Dunstone MA, Kostenko L, et al. Crystal structure of the human T cell receptor CD3εγ heterodimer complexed to the therapeutic mAb OKT3. *Proc Natl Acad of Sci U S A*. 2004;101:7675-7680.
42. Arnett KL, Harrison SC, Wiley DC. Crystal structure of a human CD3-ε/δ dimer in complex with a UCHT1 single-chain antibody fragment. *Proc Natl Acad Sci U S A*. 2004;101:16268-16273.
43. Xu C, Call ME, Wucherpfennig KW. A membrane-proximal tetracysteine motif contributes to assembly of CD3δε and CD3γε dimers with the T cell receptor. *J Biol Chem*. 2006;281:36977-36984.
44. Call ME, Pyrdol J, Wiedmann M, Wucherpfennig KW. The organizing principle in the formation of the T cell receptor-CD3 complex. *Cell*. 2002;111:967-979.
45. Cosson P, Lankford SP, Bonifacino JS, Klausner RD. Membrane protein association by potential intrarnembrane charge pairs. *Nature*. 1991;351:414-416.
46. Hall C, Berkhout B, Alarcon B, Sancho J, Wileman T, Terhorst C. Requirements for cell surface expression of the human TCR/CD3 complex in non-T cells. *Int Immunol*. 1991;3:359-368.
47. Alarcon B, Ley SC, Sanchez-Madrid F, et al. The CD3-γ and CD3-δ subunits of the T cell antigen receptor can be expressed within distinct functional TCR/CD3 complexes. *EMBO J*. 1991;10:903-912.
48. Reth M. Antigen receptor tail clue. *Nature*. 1989;338:383-384.
49. Ashwell JD, Klusner RD. Genetic and mutational analysis of the T-cell antigen receptor. *Annu Rev Immunol*. 1990;8:139-167.
50. Dave VP, Cao Z, Browne C, et al. CD3δ deficiency arrests development of the αβ but not the γδ T cell lineage. *EMBO J*. 1997;16:1360-1370.
51. Klausner RD, Samelson LE. T cell antigen receptor activation pathways: the tyrosine kinase connection. *Cell*. 1991;64:875-878.
52. Dong D, Zheng L, Lin J, et al. Structural basis of assembly of the human T cell receptor–CD3 complex. *Nature*. 2019;573:546-552.
53. Minami Y, Weissman AM, Samelson LE, Klausner RD. Building a multichain receptor: synthesis, degradation, an assembly of the T-cell antigen receptor. *Proc Natl Acad Sci U S A*. 1987;84:2688-2692.
54. Morath A, Schamel WW. αβ and γδ T cell receptors: similar but different. *J Leukoc Biol*. 2020;107:1045-1055.
55. Hayes SM, Love PE. Distinct structure and signaling potential of the γδTCR complex. *Immunity*. 2002;16:827-838.
56. Hayes SM, Laky K, El-Khoury D, Kappes DJ, Fowlkes BJ, Love PE. Activation-induced modification in the CD3 complex of the γδ T cell receptor. *J Exp Med*. 2002;196:1355-1361.
57. Chien YH, Gascoigne NRJ, Kavaler J, Lee NE, Davis MM. Somatic recombination in a murine T-cell receptor gene. *Nature*. 1984;309:322-326.
58. Bosma MJ. B and T cell leakiness in the scid mouse mutant. *Immunodefic Rev*. 1992;3:261-276.
59. Mombaerts P, Iacomini J, Johnson RS, Herrup K, Tonegawa S, Papaioannou VE. RAG-1-deficient mice have no mature B and T lymphocytes. *Cell*. 1992;68:869-877.
60. Shinkai Y, Rathbun G, Lam KP, et al. RAG-2-deficient mice lack mature lymphocytes owing to inability to initiate V(D)J rearrangement. *Cell*. 1992;68:855-867.
61. Krangel MS. Gene segment selection in V(D)J recombination: accessibility and beyond. *Nat Immunol*. 2003;4:624-630.
62. Fujimoto S, Yamagishi H. Isolation of an excision product of T-cell receptor α-chain gene rearrangements. *Nature*. 1987;327:242-243.
63. Okazaki K, Davis DD, Sakano H. T cell receptor beta gene sequences in the circular DNA of thymocyte nuclei: direct evidence for intramolecular DNA deletion in V-D-J joining. *Cell*. 1987;49:477-485.
64. Lafaille JJ, DeCloux A, Bonneville M, Takagaki Y, Tonegawa S. Junctional sequences of T cell receptor γδ genes: implications for γδ T cell lineages and for a novel intermediate of V-(D)-J joining. *Cell*. 1989;59:859-870.
65. Pellicci DG, Uldrich AP, Le Nours J, et al. The molecular bases of δ/αβ T cell-mediated antigen recognition. *J Exp Med*. 2014;211:2599-2615.
66. de Villartay JP, Lewis D, Hockett R, Waldmann TA, Korsmeyer SJ, Cohen DI. Deletional rearrangement in the human T-cell receptor alpha-chain locus. *Proc Natl Acad Sci U S A*. 1987;84:8608-8612.
67. Villey I, Caillol D, Selz F, Ferrier P, de Villartay JP. Defect in rearrangement of the most 5' TCR-J alpha following targeted deletion of T early alpha (TEA): implications for TCR alpha locus accessibility. *Immunity*. 1996;5:331-342.
68. Rowen L, Koop BF, Hood L. The complete 685-kilobase DNA sequence of the human β T cell receptor locus. *Science*. 1996;272:1755-1762.
69. Noonan DJ, Kofler R, Singer PA, Cardenas G, Dixon FJ, Theofilopoulos AN. Delineation of a defect in T cell receptor β genes of nzw mice predisposed to autoimmunity. *J Exp Med*. 1986;163:644-653.
70. Chou HS, Nelson CA, Godambe SA, Chaplin DD, Loh DY. Germline organization of the murine T cell receptor β-chain genes. *Science*. 1987;238:545-548.
71. Woodland DL, Happ MP, Gollob KJ, Palmer E. An endogenous retrovirus mediating deletion of αβ T cells? *Nature*. 1991;349:529-530.
72. Heilig JS, Tonegawa S. Diversity of murine gamma genes and expression in fetal and adult T lymphocytes. *Nature*. 1986;322:836-840.
73. Garman RD, Doherty PJ, Raulet DH. Diversity, rearrangement, and expression of murine T cell gamma genes. *Cell*. 1986;45:733-742.
74. LeFranc MP, Forster A, Baer R, Stinson MA, Rabbitts TH. Diversity and rearrangement of the human T cell rearranging gamma genes: nine germ-line variable genes belonging to two subgroups. *Cell*. 1986;45:237-246.
75. Concannon P, Robinson MA. Human T-cell receptor gene nomenclature. *Ann N Y Acad Sci*. 1995;756:124-129.
76. Lefranc MP, Forster A, Rabbitts TH. Genetic polymorphism and exon changes of the constant regions of the human T-cell rearranging gene γ. *Proc Natl Acad Sci U S A*. 1986;83:9596-9600.
77. Raulet DH. The structure, function, and molecular genetics of the γ/δ T cell receptor. *Annu Rev Immunol*. 1989;7:175-207.
78. Havran WL, Allison JP. Developmentally ordered appearance of thymocytes expressing different T-cell antigen receptors. *Nature*. 1988;335:443-445.
79. Korman AJ, Marusic-Galesic S, Spencer D, Kruisbeek AM, Raulet DH. Predominant variable region gene usage by γ/δ T cell receptor-bearing cells in the adult thymus. *J Exp Med*. 1988;168:1021-1104.
80. Takagaki Y, Nakanishi N, Ishida I, Kanagawa O, Tonegawa S. T cell receptor-γ and -δ genes preferentially utilized by adult thymocytes for the surface expression. *J Immunol*. 1989;142:2112-2121.
81. Goldman JP, Spencer DM, Raulet DH. Ordered rearrangement of variable region genes of the T cell receptor γ locus correlates with transcription of the unrearranged genes. *J Exp Med*. 1993;177:729-739.
82. Wu TT, Kabat EA. An analysis of the sequences of the variable regions of Bence Jones proteins and myeloma light chains and their implications for antibody complementarity. *J Exp Med*. 1970;132:211-250.
83. Davies DR, Padlan EA, Sheriff S. Antibody-antigen complexes. *Annu Rev Biochem*. 1990;59:439-473.
84. Davis MM, Jay Boniface J, Reich Z, et al. Ligand recognition by αβ T cell receptors. *Annu Rev Immunol*. 1998;16:523-544.
85. Xu JL, Davis MM. Diversity in the CDR3 region of V(H) is sufficient for most antibody specificities. *Immunity*. 2000;13:37-45.
86. Robins HS, Campregher PV, Srivastava SK, et al. Comprehensive assessment of T-cell receptor β-chain diversity in αβ T cells. *Blood*. 2009;114:4099-4107.
87. Robins HS, Srivastava SK, Campregher PV, et al. Overlap and effective size of the human CD8+ T cell receptor repertoire. *Sci Transl Med*. 2010;2.
88. Emerson RO, DeWitt WS, Vignali M, et al. Immunosequencing identifies signatures of cytomegalovirus exposure history and HLA-mediated effects on the T cell repertoire. *Nat Genet*. 2017;49:659-665.
89. Mora T, Walczak AM. *Quantifying Lymphocyte Receptor Diversity*. bioRxix; 2016.
90. Mora T, Walczak AM. *How many Different Clonotypes Do Immune Reprtoires Contain?* bioRxix; 2019.

91. Vanhanen R, Heikkilä N, Aggarwal K, et al. T cell receptor diversity in the human thymus. *Mol Immunol.* 2016;76:116-122.
92. Uematsu Y, Ryser S, Dembić Z, et al. In transgenic mice the introduced functional T cell receptor β gene prevents expression of endogenous β genes. *Cell.* 1988;52:831-841.
93. Padovan E, Casorati G, Dellabona P, Meyer S, Brockhaus M, Lanzavecchia A. Expression of two T cell receptor α chains: dual receptor T cells. *Science.* 1993;262:422-424.
94. Malissen M, Trucy J, Jouvin-Marche E, Cazenave PA, Scollay R, Malissen B. Regulation of TCR α and β gene allelic exclusion during T-cell development. *Immunol Today.* 1992;13:315-322.
95. Wagner DHJr. Re-shaping the T cell repertoire: TCR editing and TCR revision for good and for bad. *Clin Immunol.* 2007;123:1-6.
96. Aifantis I, Buer J, Von Boehmer H, Azogui O. Essential role of the pre-T cell receptor in allelic exclusion of the T cell receptor β locus. *Immunity.* 1997;7:601-607.
97. O'Shea CC, Thornell AP, Rosewell IR, Hayes B, Owen MJ. Exit of the pre-TCR from the ER/cis-Golgi is necessary for signaling differentiation, proliferation, and allelic exclusion in immature thymocytes. *Immunity.* 1997;7:591-599.
98. Davodeau F, Peyrat MA, Houde I, et al. Surface expression of two distinct functional antigen receptors on human γδ T cells. *Science.* 1993;260:1800-1802.
99. Rudolph MG, Stanfield RL, Wilson IA. How TCRs bind MHCs, peptides, and coreceptors. *Annu Rev Immunol.* 2006;24:419-466.
100. Bentley GA, Boulot G, Karjalainen K, Mariuzza RA. Crystal structure of the β chain of a T cell antigen receptor. *Science.* 1995;267:1984-1987.
101. Fields BA, Ober B, Malchiodi EL, et al. Crystal structure of the V_α domain of a T cell antigen receptor. *Science.* 1995; 270:1821.
102. Garcia KC, Degano M, Stanfield RL, et al. An αβ T cell receptor structure at 2.5 Å and its orientation in the TCR-MHC complex. *Science.* 1996;274:209-219.
103. Garboczi DN, Ghosh P, Utz U, Fan QR, Biddison WE, Wiley DC. Structure of the complex between human T-cell receptor, viral peptide and HLA-A2. *Nature.* 1996;384:134-141.
104. Strong RK, Penny DM, Feldman RMR, et al. Engineering and expression of a secreted murine TCR with reduced N-linked glycosylation. *J Immunol.* 1994;153:4111-4121.
105. Kuhns MS, Davis MM. Disruption of extracellular interactions impairs T cell receptor-CD3 complex stability and signaling. *Immunity.* 2007;26:357-369.
106. Martínez-Martín P, Carroza-García E, Frades-Payo B, Rodríguez-Blázquez C, Forjaz MJ, De Pedro-Cuesta J. Psychometric attributes of the scales for outcomes in Parkinson's disease-psychosocial (SCOPA-PS): validation in Spain and review. *Rev Neurol.* 2009;49:1-7.
107. Wang F, Beck-García K, Zorzin C, Schamel WWA, Davis MM. Inhibition of T cell receptor signaling by cholesterol sulfate, a naturally occurring derivative of membrane cholesterol. *Nat Immunol.* 2016;17:844-850.
108. Ott JA, Castro CD, Deiss TC, Ohta Y, Flajnik MF, Criscitiello MF. Somatic hypermutation of T cell receptor α chain contributes to selection in nurse shark thymus. *eLife.* 2018;7:e28477.
109. Lin AY, Devaux B, Green A, Sagerström C, Elliott JF, Davis MM. Expression of T cell antigen receptor heterodimers in a lipid-linked form. *Science.* 1990;249:677-679.
110. Grégoire C, Lin SY, Mazza G, Rebaï N, Luescher IF, Malissen B. Covalent assembly of a soluble T cell receptor-peptide-major histocompatibility class I complex. *Proc Natl Acad Sci U S A.* 1996;93:7184-7189.
111. Teyton L. Assembly and transport of major histocompatibility complex molecules. *Nouv Rev Fr Hematol.* 1994;36:S33-S36.
112. Parham P, Terhorst C, Herrmann H, Humphreys RE, Waterfield MD, Strominger JL. Immunological and chemical purity of papain solubilized HL A antigens. (Radiolabeling/immune complexes/gel filtration/N terminal amino acids). *Proc Natl Acad Sci U S A.* 1975;72:1594-1598.
113. Wettstein DA, Boniface JJ, Reay PA, Schild H, Davis MM. Expression of a class II major histocompatibility complex (MHC) heterodimer in a lipid-linked form with enhanced peptide/soluble MHC complex formation at low pH. *J Exp Med.* 1991;174:219-228.
114. Garboczi DN, Hung DT, Wiley DC. HLA-A2-peptide complexes: refolding and crystallization of molecules expressed in Escherichia coli and complexed with single antigenic peptides. *Proc Natl Acad Sci U S A.* 1992;89:3429-3433.
115. Altman JD, Reay PA, Davis MM. Formation of functional peptide complexes of class II major histocompatibility complex proteins from subunits produced in *Escherichia coli. Proc Natl Acad Sci U S A.* 1993;90:10330-10334.
116. Jackson MR, Song ES, Yang Y, Peterson PA. Empty and peptide-containing conformers of class I major histocompatibility complex molecules expressed in *Drosophila melanogaster* cells. *Proc Natl Acad Sci U S A.* 1992;89:12117-12121.
117. Kozono H, White J, Clements J, Marrack P, Kappler J. Production of soluble MHC class II proteins with covalently bound single peptides. *Nature.* 1994;369:151-154.
118. Matsui K, Boniface JJ, Reay PA, Schild H, Fazekas De St Groth B, Davis MM. Low affinity interaction of peptide-MHC complexes with T cell receptors. *Science.* 1991;254:1788-1791.
119. Weber S, Traunecker A, Oliveri F, Gerhard W, Karjalainen K. Specific low-affinity recognition of major histocompatibility complex plus peptide by soluble T-cell receptor. *Nature.* 1992;356:793-796.
120. Malmqvist M. Biospecific interaction analysis using biosensor technology. *Nature.* 1993;361:186-187.
121. Krogsgaard M, Prado N, Adams EJ, et al. Evidence that structural rearrangements and/or flexibility during TCR binding can contribute to T cell activation. *Mol Cell.* 2003;12:1367-1378.
122. Van Der Merwe PA, Barclay AN. Transient intercellular adhesion: the importance of weak protein-protein interactions. *Trends Biochem Sci.* 1994;19:354-358.
123. Stuart PM, Woodward JG. Yersinia enterocolitica produces superantigenic activity. *J Immunol.* 1992;148:225-233.
124. Wyer JR, Willcox BE, George FG, et al. T cell receptor and coreceptor CD8αα bind peptide-MHC independently and with distinct kinetics. *Immunity.* 1999;10:219-225.
125. Renard V, Romero P, Vivier E, Malissen B, Luescher IF. CD8β increases CD8 coreceptor function and participation in TCR-ligand binding. *J Exp Med.* 1996;184:2439-2444.
126. Xiong Y, Kern P, Chang HC, Reinherz EL. T cell receptor binding to a pMHCII ligand is kinetically distinct from and independent of CD4. *J Biol Chem.* 2001;276:5659-5667.
127. Huppa JB, Axmann M, Mörtelmaier MA, et al. TCR-peptide-MHC interactions in situ show accelerated kinetics and increased affinity. *Nature.* 2010;463:963-967.
128. Huang J, Zarnitsyna VI, Liu B, et al. The kinetics of two-dimensional TCR and pMHC interactions determine T-cell responsiveness. *Nature.* 2010;464:932-936.
129. Kaizuka Y, Douglass AD, Varma R, Dustin ML, Vale RD. Mechanisms for segregating T cell receptor and adhesion molecules during immunological synapse formation in Jurkat T cells. *Proc Natl Acad Sci U S A.* 2007;104:20296-20301.
130. Dustin ML, Ferguson LM, Chan PY, Springer TA, Golan DE. Visualization of CD2 interaction with LFA-3 and determination of the two-Dimensional dissociation constant for adhesion receptors in a contact area. *J Cell Biol.* 1996;132:465-474.
131. Marshall BT, Long M, Piper JW, Yago T, McEver RP, Zhu C. Direct observation of catch bonds involving cell-adhesion molecules. *Nature.* 2003;423:190-193.
132. Liu B, Chen W, Evavold BD, Zhu C. Accumulation of dynamic catch bonds between TCR and agonist peptide-MHC triggers T cell signaling. *Cell.* 2014;157:357-368.
133. Kolawole EM, Lamb TJ, Evavold BD. Relationship of 2d affinity to t cell functional outcomes. *Int J Mol Sci.* 2020;21:1-20.
134. Huang J, Edwards LJ, Evavold BD, Zhu C. Kinetics of MHC-CD8 interaction at the T cell membrane. *J Immunol.* 2007;179:7653-7662.
135. Jiang N, Huang J, Edwards LJ, et al. Two-stage cooperative T cell receptor-peptide major histocompatibility complex-CD8 trimolecular interactions amplify antigen discrimination. *Immunity.* 2011;34:13-23.
136. Jorgensen JL, Esser U, Fazekas de S., Groth B, Reay PA, Davis MM. Mapping T-cell receptor-peptide contacts by variant peptide immunization of single-chain transgenics. *Nature.* 1992;355:224-230.
137. Newell EW, Ely LK, Kruse AC, et al. Structural basis of specificity and cross-reactivity in T cell receptors specific for cytochrome c-I-Ek. *J Immunol.* 2011;186:5823-5832.
138. Engel I, Hedrick SM. Site-directed mutations in the VDJ junctional region of a T cell receptor β chain cause changes in antigenic peptide recognition. *Cell.* 1988;54:473-484.
139. Katayama CD, Eidelman FJ, Duncan A, Hooshmand F, Hedrick SM. Predicted complementarity determining regions of the T cell antigen receptor determine antigen specificity. *EMBO J.* 1995;14:927-938.
140. Hennecke J, Wiley DC. T cell receptor-MHC interactions up close. *Cell.* 2001;104:1-4.
141. Glanville J, Huang H, Nau A, et al. Identifying specificity groups in the T cell receptor repertoire. *Nature.* 2017;547:94-98.
142. Nanda NK, Arzoo KK, Mario Geysen H, Sette A, Sercarz EE. Recognition of multiple peptide cores by a single t cell receptor. *J Exp Med.* 1995;182:531-539.
143. Mason D. A very high level of crossreactivity is an essential feature of the T- cell receptor. *Immunol Today.* 1998;19: 395-404.
144. Wilson DB, Pinilla C, Wilson DH, et al. Immunogenicity. I. Use of peptide libraries to identify epitopes that activate clonotypic CD4+ T cells and induce T cell responses to native peptide ligands. *J Immunol.* 1999;163:6424-6434.
145. Huseby ES, Crawford F, White J, Marrack P, Kappler JW. Interface-disrupting amino acids establish specificity between T cell receptors and complexes of major histocompatibility complex and peptide. *Nat Immunol.* 2006;7:1191-1199.
146. Štefanová I, Hemmer B, Vergelli M, Martin R, Biddison WE, Germain RN. TCR ligand discrimination is enforced by competing ERK positive and SHP-I negative feedback pathways. *Nat Immunol.* 2003;4:248-254.
147. Krogsgaard M, Li QJ, Sumen C, Huppa JB, Huse M, Davis MM. Agonist/endogenous peptide-MHC heterodimers drive T cell activation and sensitivity. *Nature.* 2005;434:238-243.
148. Yachi PP, Ampudia J, Gascoigne NRJ, Zal T. Nonstimulatory peptides contribute to antigen-induced CD8-T cell receptor interaction at the immunological synapse. *Nat Immunol.* 2005;6:785-792.
149. Garcia KC, Degano M, Pease LR, et al. Structural basis of plasticity in T cell receptor recognition of a self peptide-MHC antigen. *Science.* 1998;279:1166-1172.
150. Su LF, Davis MM. Antiviral memory phenotype T cells in unexposed adults. *Immunol Rev.* 2013;255:95-109.
151. Reiser JB, Grégoire C, Darnault C, et al. A T cell receptor CDR3β loop undergoes conformational changes of unprecedented magnitude upon binding to a peptide/MHC class I complex. *Immunity.* 2002;16:345-354.
152. Reinherz EL, Tan K, Tang L, et al. The crystal structure of a T cell receptor in complex with peptide and MHC class II. *Science.* 1999;286:1913-1921.
153. Courtney AH, Lo WL, Weiss A. TCR signaling: mechanisms of initiation and propagation. *Trends Biochem Sci.* 2018;43:108-123.
154. Evans E, Leung A, Heinrich V, Zhu C. Mechanical switching and coupling between two dissociation pathways in a P-selectin adhesion bond. *Proc Natl Acad Sci U S A.* 2004;101:11281-11286.
155. Marrack P, Kushnir E, Kappler J. A maternally inherited superantigen encoded by a mammary tumour virus. *Nature.* 1991;349:524-526.
156. Apt KE, Collier JL, Grossman AR. Evolution of the phycobiliproteins. *J Mol Biol.* 1995;248:79-96.
157. Karunakaran MM, Willcox CR, Salim M, et al. Butyrophilin-2A1 directly binds germline-encoded regions of the Vγ9Vδ2 TCR and is essential for phosphoantigen sensing. *Immunity.* 2020;52:487-498.e6.
158. Sutkowski N, Conrad B, Thorley-Lawson DA, Huber BT. Epstein-Barr virus transactivates the human endogenous retrovirus HERV-K18 that encodes a superantigen. *Immunity.* 2001;15:579-589.
159. Tomai M, Kotb M, Majumdar G, Beachey EH. Superantigenicity of streptococcal M protein. *J Exp Med.* 1990;172:359-362.
160. Sarangapani KK, Yago T, Klopocki AG, et al. Low force decelerates L-selectin dissociation from P-selectin glycoprotein ligand-1 and endoglycan. *J Biol Chem.* 2004;279:2291-2298.
161. Held W, Waanders GA, Shakhov AN, Scarpellino L, Acha-Orbea H, MacDonald HR. Superantigen-induced immune stimulation amplifies mouse mammary tumor virus infection and allows virus transmission. *Cell.* 1993;74:529-540.
162. Golovkina TV, Chervonsky A, Dudley JP, Ross SR. Transgenic mouse mammary tumor virus superantigen expression prevents viral infection. *Cell.* 1992;69:637-645.
163. Lafon M, Scott-Algara D, Marche PN, Cazenave PA, Jouvin-Marche E. Neonatal deletion and selective expansion of mouse T cells by exposure to rabies virus nucleocapsid superantigen. *J Exp Med.* 1994;180:1207-1215.
164. Dobrescu D, Ursea B, Pope M, Ascht AS, Posnett DN. Enhanced HIV-1 replication in Vβ12 T cells due to human cytomegalovirus in monocytes: evidence for a putative herpesvirus superantigen. *Cell.* 1995;82:753-763.
165. Yao Z, Maraskovsky E, Spriggs MK, Cohen JI, Armitage RJ, Alderson MR. Herpesvirus saimiri open reading frame 14, a protein encoded by a T lymphotropic herpesvirus, binds to MHC class II molecules and stimulates T cell proliferation. *J Immunol.* 1996;156:3260-3266.
166. Cole BC, Kartchner DR, Wells DJ. Stimulation of mouse lymphocytes by a mitogen derived from Mycoplasma arthritidis. VII. Responsiveness is associated with expression of a product(s) of the V(β)8 gene family present on the T cell receptor α/β for antigen. *J Immunol.* 1989;142:4131-4137.
167. Friedman SM, Crow MK, Tumang JR, et al. Characterization of human T cells reactive with the mycoplasma arthritidis-derived superantigen (mam): generation of a monoclonal antibody against vβ17, the t cell receptor gene product expressed by a large fraction of mam-reactive human t cells. *J Exp Med.* 1991;174:891-900.
168. Denkers EY, Caspar P, Sher M. Toxoplasma gonclii possesses a superantigen activity that selectively expands murine T cell receptor Vβ-bearing CD8+ lymphocytes. *J Exp Med.* 1994;180:985-994.
169. Abe J, Takeda T, Watanabe Y, et al. Evidence for superantigen production by Yersinia pseudotuberculosis. *J Immunol.* 1993;148:4183-4188.
170. Willcox BE, Gao GF, Wyer JR, et al. TCR binding to peptide-MHC stabilizes a flexible recognition interface. *Immunity.* 1999;10:357-365.
171. Boniface JJ, Reich Z, Lyons DS, Davis MM. Thermodynamics of T cell receptor binding to peptide-MHC: evidence for a general mechanism of molecular scanning. *Proc Natl Acad Sci U S A.* 1999;96:11446-11451.
172. Wu LC, Tuot DS, Lyons DS, Garcia C, Davis MM. Two-step binding mechanism for T-cell receptor recognition of peptide-MHC. *Nature.* 2002;418:552-556.

173. Tynan FE, Burrows SR, Buckle AM, et al. T cell receptor recognition of a 'super-bulged' major histocompatibility complex class I-bound peptide. *Nat Immunol.* 2005;6:1114-1122.
174. Feng D, Bond CJ, Ely LK, Maynard J, Garcia KC. Structural evidence for a germline-encoded T cell receptor-major histocompatibility complex interaction 'codon'. *Nat Immunol.* 2007;8:975-983.
175. Dai S, Huseby ES, Rubtsova K, et al. Crossreactive T cells spotlight the germline rules for αβ T cell-receptor interactions with MHC molecules. *Immunity.* 2008;28:324-334.
176. Su LF, Kidd BA, Han A, Kotzin JJ, Davis MM. Virus-specific CD4(+) memory-phenotype T cells are abundant in unexposed adults. *Immunity.* 2013;38:373-383.
177. Akue AD, Lee JY, Jameson SC. Derivation and maintenance of virtual memory CD8 T cells. *J Immunol.* 2012;188:2516-2523.
178. Su LF, Del Alcazar D, Stelekati E, Wherry EJ, Davis MM. Antigen Exposure shapes the ratio between antigen-specific Tregs and conventional T cells in human peripheral blood. *Proc Natl Acad Sci U S A.* 2016;113:E6192-E6198.
179. Welsh RM, Che JW, Brehm MA, Selin LK. Heterologous immunity between viruses. *Immunol Rev.* 2010;235:244-266.
180. Mallajosyula V, Ganjavi C, Chakraborty S, et al. Improved staining reagent shows conserved SARS-CoV-2 pepetide-specific T cells correlate with milder disease. *Sci Immunol.* 2021;6(61):eabg5669.
181. Janeway CA. The T cell receptor as a multicomponent signalling machine: CD4/CD8 coreceptors and CD45 in T cell activation. *Ann Rev Immunol.* 1992;10:645-674.
182. Irvine DJ, Purbhoo MA, Krogsgaard M, Davis MM. Direct observation of ligand recognition by T cells. *Nature.* 2002;419:845-849.
183. Purbhoo MA, Irvine DJ, Huppa JB, Davis MM. T cell killing does not require the formation of a stable mature immunological synapse. *Nat Immunol.* 2004;5:524-530.
184. Monks CRF, Freiberg BA, Kupfer H, Sciaky N, Kupfer A. Three-dimensional segregation of supramolecular activation clusters in T cells. *Nature.* 1998;395:82-86.
185. Yin Y, Wang XX, Mariuzza RA. Crystal structure of a complete ternary complex of T-cell receptor, peptide-MHC, and CD4. *Proc Natl Acad Sci U S A.* 2012;109:5405-5410.
186. Luescher IF, Vivier E, Layer A, et al. CD8 modulation of t-cell antigen receptor–ligand interactions on living cytotoxic t lymphocytes. *Nature.* 1995;373:353-356.
187. Artyomov MN, Lis M, Devadas S, Davis MM, Chakraborty AK. CD4 and CD8 binding to MHC molecules primarily acts to enhance Lck delivery. *Proc Natl Acad Sci U S A.* 2010;107:16916-16921.
188. Wülfing C, Sumen C, Sjaastad MD, Wu LC, Dustin ML, Davis MM. Costimulation and endogenous mhc ligands contribute to t cell recognition. *Nat Immunol.* 2002;3:42-47.
189. Hampl J, Chien YH, Davis MM. CD4 augments the response of a T cell to agonist but not to antagonist ligands. *Immunity.* 1997;7:379-385.
190. Zareie P, Szeto C, Farenc C, et al. Canonical T cell receptor docking on peptide–MHC is essential for T cell signaling. *Science.* 2021;372.
191. Huse M, Lillemeier BF, Kuhns MS, Chen DS, Davis MM. T cells use two directionally distinct pathways for cytokine secretion. *Nat Immunol.* 2006;7:247-255.
192. Grakoui A, Bromley SK, Sumen C, et al. The immunological synapse: a molecular machine controlling T cell activation. *Science.* 1999;285:221-227.
193. Wülfing C, Sjaastad MD, Davis MM. Visualizing the dynamics of T cell activation: intracellular adhesion molecule 1 migrates rapidly to the T cell/B cell interface and acts to sustain calcium levels. *Proc Natl Acad Sci U S A.* 1998;95:6302-6307.
194. Krummel MF, Sjaastad MD, Wulfing CW, Davis MM. Differential clustering of CD4 and CD3ζ during T cell recognition. *Science.* 2000;289:1349-1352.
195. Ritter AT, Angus KL, Griffiths GM. The role of the cytoskeleton at the immunological synapse. *Immunol Rev.* 2013;256:107-117.
196. Stinchcombe JC, Griffiths GM. Secretory mechanisms in cell-mediated cytotoxicity. *Annu Rev Cell Dev Biol.* 2007;23:495-517.
197. Lillemeier BF, Mörtelmaier MA, Forstner MB, Huppa JB, Groves JT, Davis MM. TCR and Lat are expressed on separate protein islands on T cell membranes and concatenate during activation. *Nat Immunol.* 2010;11:90-96.
198. Gawden-Bone C, Griffiths GM. Phospholipids: pulling back the actin curtain for granule delivery to the immune synapse. *Front Immunol.* 2019;10:700.
199. Sykulev Y, Joo M, Vturina I, Tsomides TJ, Eisen HN. Evidence that a single peptide-MHC complex on a target cell can elicit a cytolytic T cell response. *Immunity.* 1996;4:565-571.
200. Harding CV, Unanue ER. Quantitation of antigen-presenting cell MHC class II/peptide complexes necessary for T-cell stimulation. *Nature.* 1990;346:574-576.
201. Demotz S, Grey HM, Sette A. The minimal number of class II MHC-antigen complexes needed for T cell activation. *Science.* 1990;249:1028-1030.
202. Kimachi K, Croft M, Grey HM. The minimal number of antigen-major histocompatibility complex class II complexes required for activation of naive and primed T cells. *Eur J Immunol.* 1997;27:3310-3317.
203. Reay PA, Matsui K, Haase K, Wulfing C, Chien YH, Davis MM. Determination of the relationship between T cell responsiveness and the number of MHC-peptide complexes using specific monoclonal antibodies. *J Immunol.* 2000;164:5626-5634.
204. Huang J, Brameshuber M, Zeng X, et al. A Single peptide-major histocompatibility complex ligand triggers digital cytokine secretion in CD4+ T Cells. *Immunity.* 2013;39:846-857.
205. Yokosuka T, Sakata-Sogawa K, Kobayashi W, et al. Newly generated T cell receptor microclusters initiate and sustain T cell activation by recruitment of Zap70 and SLP-76. *Nat Immunol.* 2005;6:1253-1262.
206. Xie J, Huppa JB, Newell EW, et al. Photocrosslinkable pMHC monomers stain T cells specifically and cause ligand-bound TCRs to be 'preferentially' transported to the cSMAC. *Nat Immunol.* 2012;13:674-680.
207. Van Der Merwe PA, Davis SJ. Molecular interactions mediating T cell antigen recognition. *Annu Rev Immunol.* 2003;21:659-684.
208. Davis SJ, Van Der Merwe PA. The structure and ligand interactions of CD2: implications for T-cell function. *Immunol Today.* 1996;17:177-187.
209. Van Der Merwe PA, Davis SJ, Shaw AS, Dustin ML. Cytoskeletal polarization and redistribution of cell-surface molecules during T cell antigen recognition. *Semin Immunol.* 2000;12:5-21.
210. Davis SJ, van der Merwe PA. The kinetic-segregation model: TCR triggering and beyond. *Nat Immunol.* 2006;7:803-809.
211. Choudhuri K, Wiseman D, Brown MH, Gould K, Van Der Merwe PA. T-cell receptor triggering is critically dependent on the dimensions of its peptide-MHC ligand. *Nature.* 2005;436:578-582.
212. Fernandes RA, Su L, Nishiga Y, et al. Immune receptor inhibition through enforced phosphatase recruitment. *Nature.* 2020;586:779-784.
213. Xu C, Gagnon E, Call ME, et al. Regulation of T Cell receptor activation by dynamic membrane binding of the CD3ε cytoplasmic tyrosine-based motif. *Cell.* 2008;135:702-713.
214. Schamel WW, Alarcon B, Minguet S. The TCR is an allosterically regulated macromolecular machinery changing its conformation while working. *Immunol Rev.* 2019;291:8-25.
215. Valitutti S, Miller S, Cella M, Padovan E, Lanzavecchia A. Serial triggering of many T-cell receptors by a few peptide-MHC complexes. *Nature.* 1995;375:148-151.
216. Viola A, Lanzavecchia A. T cell activation determined by T cell receptor number and tunable thresholds. *Science.* 1996;273:104-106.
217. Feng Y, Reinherz EL, Lang MJ. αβ T Cell Receptor Mechanosensing Forces out Serial Engagement. *Trends Immunol.* 2018;39(8):596-609. doi:10.1016/j.it.2018.05.005
218. Chakraborty AK, Weiss A. Insights into the initiation of TCR signaling. *Nat Immunol.* 2014;15:798-807.
219. Xu X, Li H, Xu C. Structural understanding of T cell receptor triggering. *Cell Mol Immunol.* 2020;17:193-202.
220. Davis MM, Krogsgaard M, Huppa JB, et al. Dynamics of cell surface molecules during T cell recognition. *Annu Rev Biochem.* 2003;72:717-742.
221. Kim ST, Takeuchi K, Sun ZYJ, et al. The αβ T cell receptor is an anisotropic mechanosensor. *J Biol Chem.* 2009;284:31028-31037.
222. Huse M. Mechanical forces in the immune system. *Nat Rev Immunol.* 2017;17:679-690.
223. Sibener LV, Fernandes RA, Kolawole EM, et al. Isolation of a structural mechanism for uncoupling T cell receptor signaling from peptide-MHC binding. *Cell.* 2018;174:672-687.e27.
224. Liu Y, Blanchfield L, Pui-Yan Ma V, et al. DNA-based nanoparticle tension sensors reveal that T-cell receptors transmit defined pN forces to their antigens for enhanced fidelity. *Proc Natl Acad Sci U S A.* 2016;113:5610-5615.
225. Das DK, Feng Y, Mallis RJ, et al. Force-dependent transition in the T-cell receptor β-subunit allosterically regulates peptide discrimination and pMHC bond lifetime. *Proc Natl Acad Sci U S A.* 2015;112:1517-1522.
226. Pryshchep S, Zarnitsyna VI, Hong J, Evavold BD, Zhu C. Accumulation of serial forces on TCR and CD8 frequently applied by agonist antigenic peptides embedded in MHC molecules triggers calcium in T cells. *J Immunol.* 2014;193:68-76.
227. Wu P, Zhang T, Liu B, et al. Mechano-regulation of peptide-MHC class I conformations determines TCR antigen recognition. *Mol Cell.* 2019;73:1015-1027.e7.
228. Suzuki T, Suzuki M, Ogino S, Umemoto R, Nishida N, Shimada I. Mechanical force effect on the two-state equilibrium of the hyaluronan-binding domain of CD44 in cell rolling. *Proc Natl Acad Sci U S A.* 2015;112:6991-6996.
229. Ueda H, Zhou J, Xie J, Davis MM. Distinct roles of cytoskeletal components in immunological synapse formation and directed secretion. *J Immunol.* 2015;195:4117-4125.
230. Delon J, Grégoire C, Malissen B, et al. CD8 expression allows T cell signaling by monomeric peptide-MHC complexes. *Immunity.* 1998;9(4):467-473. doi:10.1016/S1074-7613(00)80630-5
231. Brameshuber M, Kellner F, Rossboth BK, et al. Monomeric TCRs drive T cell antigen recognition article. *Nat Immunol.* 2018;19:487-496.
232. Dembo M, Torney DC, Saxman K, Hammer D. The reaction-limited kinetics of membrane-to-surface adhesion and detachment. *Proc Biol Sci.* 1988;234:55-83.
233. Marrack P, Kappler J. The staphylococcal enterotoxins and their relatives. *Science.* 1990;248:705-711.
234. Sundberg EJ, Li Y, Mariuzza RA. So many ways of getting in the way: diversity in the molecular architecture of superantigen-dependent T-cell signaling complexes. *Curr Opin Immunol.* 2002;14:36-44.
235. Imöhl M, Fitzner C, Perniciaro S, Van Der Linden M. Epidemiology and distribution of 10 superantigens among invasive Streptococcus pyogenes disease in Germany from 2009 to 2014. *PLoS One.* 2017;12:e0180757.
236. Bunker JJ, Drees C, Watson AR, et al. B cell superantigens in the human intestinal microbiota. *Sci Transl Med.* 2019;11(597):eaau9356.
237. Kulhankova K, Kinney KJ, Stach JM, et al. The superantigen toxic shock syndrome toxin 1 alters human aortic endothelial cell function. *Infect Immun.* 2018;86(3):e00848-e008417.
238. Xu R, Shears RK, Sharma R, et al. IL-35 is critical in suppressing superantigenic Staphylococcus aureus-driven inflammatory Th17 responses in human nasopharynx-associated lymphoid tissue. *Mucosal Immunol.* 2020;13:460-470.
239. Bennett MR, Thomsen IP. Epidemiological and clinical evidence for the role of toxins in *S. aureus* human disease. *Toxins.* 2020;12(6):408.
240. Altin F, Yasar H, Desrosiers M. *All Around the Nose.* Springer; 2020.
241. Festenstein H. Immunogenetic and biological aspects of in vitro lymphocyte allotransformation (MLR) in the mouse. *Immunol Rev.* 1973;15:62-88.
242. Kappler JW, Staerz U, White J, Marrack PC. Self-tolerance eliminates T cells specific for Mls-modified products of the major histocompatibility complex. *Nature.* 1988;332:35-40.
243. Choi Y, Kappler JW, Marrack P. A superantigen encoded in the open reading frame of the 3′ long terminal repeat of mouse mammary tumour virus. *Nature.* 1991;350:203-207.
244. Dyson PJ, Knight AM, Fairchild S, Simpson E, Tomonari K. Genes encoding ligands for deletion of Vβ11 T cells cosegregate with mammary tumour virus genomes. *Nature.* 1991;349:531-532.
245. Frankel WN, Rudy C, Coffin JM, Huber BT. Linkage of Mls genes to endogenous mammary tumour viruses of inbred mice. *Nature.* 1991;349:526-528.
246. Janeway CA, Yagi J, Conrad PJ, et al. T-Cell responses to mls and to bacterial proteins that mimic its behavior. *Immunol Rev.* 1989;107:61-88.
247. Acha-Orbea H, Robson MacDonald H. Superantigens of mouse mammary tumor virus. *Annu Rev Immunol.* 1995;13:459-486.
248. Li H, Llera A, Malchiodi EL, Mariuzza RA. The structural basis of T cell activation by superantigens. *Annu Rev Immunol.* 1999;17:435-466.
249. Stauffer Y, Marguerat S, Meylan F, et al. Interferon-α-induced endogenous superantigen: a model linking environment and autoimmunity. *Immunity.* 2001;15:591-601.
250. Dalwadi H, Wei B, Kronenberg M, Sutton CL, Braun J. The Crohn's disease-associated bacterial protein I2 is a novel enteric T cell superantigen. *Immunity.* 2001;15:149-158.
251. Malchiodi EL, Eisenstein E, Fields BA, et al. Superantigen binding to a T cell receptor R chain of known three-dimensional structure. *J Exp Med.* 1995;182:1833-1845.
252. Patten PA, Rock EP, Sonoda T, et al. Transfer of putative complementarity-determining region loops of T cell receptor V domains confers toxin reactivity but not peptide/MHC specificity. *J Immunol.* 1993;150:2281-2294.
253. Fields BA, Malchiodi EL, Li H, et al. Crystal structure of a T-cell receptor β-chain complexed with a superantigen. *Nature.* 1996;384:188-192.
254. Jameson SC, Masopust D. What is the predictive value of animal models for vaccine efficacy in humans? Reevaluating the potential of mouse models for the human immune system. *Cold Spring Harb Perspect Biol.* 2018;10(4):a029132.
255. Buchbinder SP, Mehrotra DV, Duerr A, et al. Efficacy assessment of a cell-mediated immunity HIV-1 vaccine (the Step Study): a double-blind, randomised, placebo-controlled, test-of-concept trial. *Lancet.* 2008;372:1881-1893.
256. Watkins DI, Burton DR, Kallas EG, Moore JP, Koff WC. Nonhuman primate models and the failure of the Merck HIV-1 vaccine in humans. *Nat Med.* 2008;14:617-621.
257. Tameris MD, Hatherill M, Landry BS, et al. Safety and efficacy of MVA85A, a new tuberculosis vaccine, in infants previously vaccinated with BCG: a randomised, placebo-controlled phase 2b trial. *Lancet.* 2013;381:1021-1028.
258. Newell EW, Klein LO, Yu W, Davis MM. Simultaneous detection of many T-cell specificities using combinatorial tetramer staining. *Nat Methods.* 2009;6:497-499.
259. Newell EW, Sigal N, Nair N, Kidd BA, Greenberg HB, Davis MM. Combinatorial tetramer staining and mass cytometry analysis facilitate T-cell epitope mapping and characterization. *Nat Biotechnol.* 2013;31:623-629.
260. Chng MHY, Lim MQ, Rouers A, et al. Large-scale HLA tetramer tracking of T cells during Dengue infection reveals broad acute activation and differentiation into two memory cell fates. *Immunity.* 2019;51:1119-1135.e5.

261. Tian Y, da Silva Antunes R, Sidney J, et al. A review on T cell epitopes identified using prediction and cell-mediated immune models for *Mycobacterium tuberculosis* and bordetella pertussis. *Front Immunol.* 2018;9:2778.
262. Kwong GA, Radu CG, Hwang K, et al. Modular nucleic acid assembled p/MHC microarrays for multiplexed sorting of antigen-specific T cells. *J Am Chem Soc.* 2009;131:9695-9703.
263. Hunt DF, Henderson RA, Shabanowitz J, et al. Characterization of peptides bound to the class I MHC molecule HLA-A2.1 by mass spectrometry. *Science.* 1992;255:1261-1263.
264. Sulzer D, Alcalay RN, Garretti F, et al. T cells from patients with Parkinson's disease recognize α-synuclein peptides. *Nature.* 2017;546:656-661.
265. Trolle T, McMurtrey CP, Sidney J, et al. The length distribution of class I-restricted T cell epitopes is determined by both peptide supply and MHC allele-specific binding preference. *J Immunol.* 2016;196:1480-1487.
266. Loffredo JT, Sidney J, Bean AT, et al. Two MHC class I molecules associated with elite control of immunodeficiency virus replication, Mamu-B*08 and HLA-B*2705, bind peptides with sequence similarity. *J Immunol.* 2009;182:7763-7775.
267. McMurtrey C, Trolle T, Sansom T, et al. Toxoplasma gondii peptide ligands open the gate of the HLA class I binding groove. *eLife.* 2016;5:e12556.
268. Khodadoust MS, Olsson N, Wagar LE, et al. Antigen presentation profiling reveals recognition of lymphoma immunoglobulin neoantigens. *Nature.* 2017;543:723-727.
269. Swaminathan K, Majzoub K, Lund PJ, et al. Immunoproteomic interrogation of antigen presentation during Dengue and Zika infections reveals novel targets for the design of broad acting T-cell vaccines. *J Immunol.* 2017;198:146.3.
270. Becattini S, Latorre D, Mele F, et al. T cell immunity. Functional heterogeneity of human memory CD4(+) T cell clones primed by pathogens or vaccines. *Science.* 2015;347:400-406.
271. Crotty S. Follicular helper CD4 T cells (TFH). *Annu Rev Immunol.* 2011;29:621-663.
272. Seder RA, Darrah PA, Roederer M. T-cell quality in memory and protection: implications for vaccine design. *Nat Rev Immunol.* 2008;8:247-258.
273. Darrah PA, Patel DT, De Luca PM, et al. Multifunctional TH1 cells define a correlate of vaccine-mediated protection against Leishmania major. *Nat Med.* 2007;13:843-850.
274. Huang H, Sikora MJ, Islam S, et al. Select sequencing of clonally expanded CD8(+) T cells reveals limits to clonal expansion. *Proc Natl Acad Sci U S A.* 2019;116:8995-9001.
275. Han A, Glanville J, Hansmann L, Davis MM. Linking T-cell receptor sequence to functional phenotype at the single-cell level. *Nat Biotechnol.* 2014;32:684-692.
276. Zemmour D, Zilionis R, Kiner E, Klein AM, Mathis D, Benoist C. Single-cell gene expression reveals a landscape of regulatory T cell phenotypes shaped by the TCR. *Nat Immunol.* 2018;19:291-301.
277. Zheng GX, Terry JM, Belgrader P, et al. Massively parallel digital transcriptional profiling of single cells. *Nat Commun.* 2017;8:14049.
278. Kosmrlj A, Read EL, Qi Y, et al. Effects of thymic selection of the T-cell repertoire on HLA class I-associated control of HIV infection. *Nature.* 2010;465:350-354.
279. Chakraborty AK. A perspective on the role of computational models in immunology. *Annu Rev Immunol.* 2017;35:403-439.
280. Gaiha GD, Rossin EJ, Urbach J, et al. Structural topology defines protective CD8(+) T cell epitopes in the HIV proteome. *Science.* 2019;364:480-484.
281. Dash P, Fiore-Gartland AJ, Hertz T, et al. Quantifiable predictive features define epitope-specific T cell receptor repertoires. *Nature.* 2017;547:89-93.
282. Huang H, Wang C, Rubelt F, Scriba TJ, Davis MM. Analyzing the *Mycobacterium tuberculosis* immune response by T-cell receptor clustering with GLIPH2 and genome-wide antigen screening. *Nat Biotechnol.* 2020;38:1194-1202.
283. Chiou SH, Tseng D, Reuben A, et al. Global analysis of shared T cell specificities in human non-small cell lung cancer enables HLA interference and antigen discovery. *Immunity.* 2021;54(3):586-602.e8.
284. Yewdell JW, Santos JJS. Original antigenic sin: how original? How sinful? *Cold Spring Harb Perspect Med.* 2021;11(5):a038786.
285. Gostic KM, Ambrose M, Worobey M, Lloyd-Smith JO. Potent protection against H5N1 and H7N9 influenza via childhood hemagglutinin imprinting. *Science.* 2016;354:722-726.
286. Rasmussen AL, Katze MG. Genomic signatures of emerging viruses: a new era of systems epidemiology. *Cell Host Microbe.* 2016;19:611-618.
287. Xu GJ, Kula T, Xu Q, et al. Viral immunology. Comprehensive serological profiling of human populations using a synthetic human virome. *Science.* 2015;348:aaa0698.
288. Lawand M, Dechanet-Merville J, Dieu-Nosjean MC. Key features of gamma-delta T-cell subsets in human diseases and their immunotherapeutic implications. *Front Immunol.* 2017;8:761.
289. Boehm T, Hirano M, Holland SJ, Das S, Schorpp M, Cooper MD. Evolution of alternative adaptive immune systems in vertebrates. *Annu Rev Immunol.* 2018;36:19-42.
290. Mamedov MR, Scholzen A, Nair RV, et al. A macrophage colony-stimulating-factor-producing gammadelta T cell subset prevents malarial parasitemic recurrence. *Immunity.* 2018;48:350-363.e7.
291. Crosby CM, Kronenberg M. Tissue-specific functions of invariant natural killer T cells. *Nat Rev Immunol.* 2018;18:559-574.
292. Provine NM, Klenerman P. MAIT cells in health and disease. *Annu Rev Immunol.* 2020;38:203-228.
293. Stacy A, Belkaid Y. Microbial guardians of skin health. *Science.* 2019;363:227-228.
294. Rock EP, Sibbald PR, Davis MM, Chien YH. CDR3 length in antigen-specific immune receptors. *J Exp Med.* 1994;179:323-328.
295. Deiss TC, Breaux B, Ott JA, et al. Ancient use of Ig variable domains contributes significantly to the TCRδ repertoire. *J Immunol.* 2019;203:1265-1275.
296. Crowley MP, Fahrer AM, Baumgarth N, et al. A population of murine γδ T cells that recognize an inducible MHC class Ib molecule. *Science.* 2000;287:314-316.
297. Adams EJ, Chien YH, Garcia KC. Structure of a γδ T cell receptor in complex with the nonclassical MHC T22. *Science.* 2005;308:227-231.
298. Matis LA, Fry AM, Cron RQ, Cotterman MM, Dick RF, Bluestone JA. Structure and specificity of a class II MHC alloreactive γδ T cell receptor heterodimer. *Science.* 1989;245:746-749.
299. Johnson RM, Lancki DW, Sperling AI, et al. A murine CD4−, CD8− T cell receptor-γδ T lymphocyte clone specific for herpes simplex virus glycoprotein I. *Journal of Immunology.* 1992;148:983-988.
300. Uldrich AP, Rigau M, Godfrey DI. Immune recognition of phosphoantigen-butyrophilin molecular complexes by γδ T cells. *Immunol Rev.* 2020;298:74-83.
301. Scotet E, Martinez LO, Grant E, et al. Tumor recognition following Vgamma9Vdelta2 T cell receptor interactions with a surface F1-ATPase-related structure and apolipoprotein A-I. *Immunity.* 2005;22:71-80.
302. Marlin R, Pappalardo A, Kaminski H, et al. Sensing of cell stress by human γδ TCR-dependent recognition of annexin A2. *Proc Natl Acad Sci U S A.* 2017;114:3163-3168.
303. Chien YH, Meyer C, Bonneville M. Gammadelta T cells: first line of defense and beyond. *Annu Rev Immunol.* 2014;32:121-155.
304. Benveniste PM, Roy S, Nakatsugawa M, et al. Generation and molecular recognition of melanoma-associated antigen-specific human γδ T cells. *Sci Immunol.* 2018;3(30):eaav4036.
305. Cotton RN, Shahine A, Rossjohn J, Moody DB. Lipids hide or step aside for CD1-autoreactive T cell receptors. *Curr Opin Immunol.* 2018;52:93-99.
306. Wun KS, Reijneveld JF, Cheng TY, et al. T cell autoreactivity directed toward CD1c itself rather than toward carried self lipids. *Nat Immunol.* 2018;19:397-406.
307. Gherardin NA, McCluskey J, Godfrey DI. The diverse family of MR1-restricted T cells. *J Immunol.* 2018;201:2862-2871.
308. Le Nours J, Gherardin NA, Ramarathinam SH, et al. A class of γδ T cell receptors recognize the underside of the antigen-presenting molecule MR1. *Science.* 2019;366:1522-1527.
309. Reijneveld JF, Ocampo TA, Shahine A, et al. Human γδ T cells recognize CD1b by two distinct mechanisms. *Proc Natl Acad Sci U S A.* 2020;117:22944-22952.
310. Zeng X, Wei YL, Huang J, et al. Gammadelta T cells recognize a microbial encoded B cell antigen to initiate a rapid antigen-specific interleukin-17 response. *Immunity.* 2012;37:524-534.
311. Zeng X, Meyer C, Huang J, et al. Gamma delta T cells recognize haptens and mount a hapten-specific response. *eLife.* 2014;3:e03609.
312. Bruder J, Siewert K, Obermeier B, et al. Target specificity of an autoreactive pathogenic human γδ-T cell receptor in myositis. *J Biol Chem.* 2012;287:20986-20995.
313. Willcox CR, Pitard V, Netzer S, et al. Cytomegalovirus and tumor stress surveillance by binding of a human γδ T cell antigen receptor to endothelial protein C receptor. *Nat Immunol.* 2012;13:872-879.
314. Zhang L, Jin N, Nakayama M, O'Brien RL, Eisenbarth GS, Born WK. Gamma delta T cell receptors confer autonomous responsiveness to the insulin-peptide B:9-23. *J Autoimmun.* 2010;34:478-484.
315. Hampl J, Schild H, Litzenberger C, Baron M, Crowley MP, Chien YH. The specificity of a weak γδ TCR interaction can be modulated by the glycosylation of the ligand. *J Immunol.* 1999;163:288-294.
316. Murphy DB, Rath S, Pizzo E, et al. Monoclonal antibody detection of a major self peptide: MHC class II complex. *J Immunol.* 1992;148:3483-3491.
317. Wingren C, Crowley MP, Degano M, Chien YH, Wilson IA. Crystal structure of a γδ T cell receptor ligand T22: a truncated MHC-like fold. *Science.* 2000;287:310-314.
318. Shin S, El-Diwany R, Schaffert S, et al. Antigen recognition determinants of gammadelta T cell receptors. *Science.* 2005;308:252-255.
319. Roy S, Ly D, Castro CD, et al. Molecular analysis of lipid-reactive Vδ1 γδ T cells identified by CD1c tetramers. *J Immunol.* 2016;196:1933-1942.
320. Di Marco Barros R, Roberts NA, Dart RJ, et al. Epithelia use butyrophilin-like molecules to shape organ-specific γδ T cell compartments. *Cell.* 2016;167:203-218.e17.
321. Willcox CR, Vantourout P, Salim M, et al. Butyrophilin-like 3 directly binds a human Vγ4+ T cell receptor using a modality distinct from clonally-restricted antigen. *Immunity.* 2019;51:813-825.e4.
322. Mayassi T, Ladell K, Gudjonson H, et al. Chronic inflammation permanently reshapes tissue-resident immunity in celiac disease. *Cell.* 2019;176:967-981.e19.
323. Allison TJ, Winter CC, Fournié JJ, Bonneville M, Garboczi DN. Structure of a human γδT-cell antigen receptor. *Nature.* 2001;411:820-824.
324. Xu B, Pizarro JC, Holmes MA, et al. Crystal structure of a gammadelta T-cell receptor specific for the human MHC class I homolog MICA. *Proc Natl Acad Sci U S A.* 2011;108:2414-2419.
325. Moon JJ, Chu HH, Pepper M, et al. Naive CD4(+) T cell frequency varies for different epitopes and predicts repertoire diversity and response magnitude. *Immunity.* 2007;27:203-213.
326. Kotturi MF, Scott I, Wolfe T, et al. Naive precursor frequencies and MHC binding rather than the degree of epitope diversity shape CD8+ T cell immunodominance. *J Immunol.* 2008;181:2124-2133.
327. Press JL, Klinman NR. Frequency of hapten-specific B cells in neonatal and adult murine spleens. *Eur J Immunol.* 1974;4:155-159.
328. Stashenko P, Klinman NR. Analysis of the primary anti-(4-hydroxy-3-nitrophenyl)acetyl (NP) responsive B cells in BALB/c and B10.D2 mice. *J Immunol.* 1980;125:531-537.
329. Mims CA. *Pathogenesis of Infectious Disease.* Academic Press; 1987.
330. Zinkernagel RM. Immunology taught by viruses. *Science.* 1996;271:348-350.
331. Zinkernagel RM, Hengartner H. Regulation of the immune response by antigen. *Science.* 2001;293:251-253.
332. Jensen KDC, Su X, Shin S, et al. Thymic selection determines γδ T cell effector fate: antigen-naive cells make interleukin-17 and antigen-experienced cells make interferon γ. *Immunity.* 2008;29:90-100.
333. Chien YH, Zeng X, Prinz I. The natural and the inducible: interleukin (IL)-17-producing γδ T cells. *Trends Immunol.* 2013;34:151-154.
334. Itohara S, Mombaerts P, Lafaille J, et al. T cell receptor δ gene mutant mice: independent generation of αβ T cells and programmed rearrangements of γδ TCR genes. *Cell.* 1993;72:337-348.
335. Barbee SD, Woodward MJ, Turchinovich G, et al. Skint-1 is a highly specific, unique selecting component for epidermal T cells. *Proc Natl Acad Sci U S A.* 2011;108:3330-3335.
336. Turchinovich G, Hayday AC. Skint-1 identifies a common molecular mechanism for the development of interferon-γ-secreting versus interleukin-17-secreting γδ T cells. *Immunity.* 2011;35:59-68.
337. Wencker M, Turchinovich G, Di Marco Barros R, et al. Innate-like T cells straddle innate and adaptive immunity by altering antigen-receptor responsiveness. *Nat Immunol.* 2014;15:80-87.
338. Tieppo P, Papadopoulou M, Gatti D, et al. The human fetal thymus generates invariant effector γδ T cells. *J Exp Med.* 2020;217(3):jem.20190580.
339. Ribot JC, Ribeiro ST, Correia DV, Sousa AE, Silva-Santos B. Human gd thymocytes are functionally immature and differentiate into cytotoxic type 1 effector T cells upon il-2/il-15 signaling. *J Immunol.* 2014;192:2237-2243.
340. Jensen KD, Shin S, Chien YH. Gammadelta intraepithelial lymphocytes of the small intestine are not biased toward thymic antigens. *J Immunol.* 2009;182:7348-7351.
341. Cheroutre H, Lambolez F. The thymus chapter in the life of gut-specific intra epithelial lymphocytes. *Curr Opin Immunol.* 2008;20:185-191.
342. Parra ZE, Ohta Y, Criscitiello MF, Flajnik MF, Miller RD. The dynamic TCRδ: TCRδ chains in the amphibian Xenopus tropicalis utilize antibody-like V genes. *Eur J Immunol.* 2010;40:2319-2329.
343. Parra ZE, Mitchell K, Dalloul RA, Miller RD. A second TCRδ locus in Galliformes uses antibody-like V domains: insight into the evolution of TCRδ and TCRμ genes in tetrapods. *J Immunol.* 2012;188:3912-3919.
344. Parra ZE, Baker ML, Schwarz RS, Deakin JE, Lindblad-Toh K, Miller RD. A unique T cell receptor discovered in marsupials. *Proc Natl Acad Sci U S A.* 2007;104:9776-9781.
345. Parra ZE, Baker ML, Hathaway J, et al. Comparative genomic analysis and evolution of the T cell receptor loci in the opossum Monodelphis domestica. *BMC Genomics.* 2008;9:111.
346. Criscitiello MF, Saltis M, Flajnik MF. An evolutionarily mobile antigen receptor variable region gene: doubly rearranging NAR-TcR genes in sharks. *Proc Natl Acad Sci U S A.* 2006;103:5036-5041.

CHAPTER 24 Antigen Receptor Signaling

Nevil J. Singh, PhD • Gideon Wolf, BA, MD, PhD

INTRODUCTION

The immune system is unique among all vertebrate organs since it is composed of a collection of sessile cells that can potentially migrate in and out of every nook and corner of the body. Each of these cells can determine their immediate course of action as well as long-term fate by sensing and responding to their immediate microenvironments. While multiple cellular receptors are involved in detecting the diversity of contextual information available, the critical ones for lymphocytes are those used to detect their targets. In the case of the adaptive immune system, these are the antigen receptors (AgR)—the B cell receptor (BCR) which is a membrane-anchored version of the antibody molecule and the T cell receptor (TCR).[1] The cellular consequences of AgR signaling depend on the developmental stage as well as the context in which the lymphocyte encounters antigen. In early development (as discussed in Chapter 27 for T cells and Chapter 25 for B cells) when genomic recombination successfully generates the functional AgR for the first time, signaling downstream of it is mostly coupled to cell survival. In the very next step of lymphocyte development known as negative selection, which is an integral part of central tolerance, strong signaling via the very same AgR receptor would lead to cell death, anergy, or receptor editing. Once the mature lymphocyte enters peripheral circulation, AgR signaling elicits one of several possible cellular responses. With the right inflammatory milieu, AgR signaling triggers extensive cell proliferation as well as differentiation that activates effector mechanisms. Alternately, if the antigen is detected without one of many different concomitant accessory signals or cellular help, the lymphocyte initiates programs of tolerance rather than continued effector function. In arriving at one of these many possible cell fates, the intracellular signaling machinery and the molecular genetic networks they control integrate a staggering array of information indicating antigen characteristics including dose, affinity, residence time, persistence, etc. Indeed, as discussed in this chapter, we understand significant details about different molecules and their biochemical chains of activities downstream of antigen receptors in lymphocytes. But the challenge for immunologists in the next decade is to decipher how these signaling networks detect multiple extrinsic parameters and then instigate a developmental stage and/or physiological context appropriate molecular genetic program that is essentially a cellular immune response. While we are used to thinking about DNA and protein sequences as the products of extensive evolutionary selection, it is important to remember that the molecular algorithm which controls the connectivity between receptors and cellular responses has also been extensively optimized by evolutionary pressures. For instance, the appropriate combination of molecular signals that trigger cell proliferation versus cell death must be consistently unique so that the cell does not confuse one for the other. Biochemically, this would include a combination of qualitative aspects (which particular signaling intermediates are being triggered by a proliferative vs apoptotic signal?) as well as quantitative ones (how much and how long each signaling flux must persist to drive a particular outcome?). The relationship of this combination of quantitative and qualitative signal flow to each particular cellular function can be thought of as a "signaling logic" with overlap across different cell types. To illustrate, hypothetically consider that expression of a particular phenotype (eg, upregulation of a homing receptor) requires signaling via the mitogen-activated protein kinase (MAPK) ERK. Typically ERK is activated by phosphorylation. The strength of ERK activation (roughly translated to be the number of molecules of ERK activated in the cell) is proportional to the amount of ligand and upstream signaling. Active ERK would have to phosphorylate a certain amount of targets and possibly for a certain amount of time, in order to successfully continue the chain of events leading to transcription of this homing receptor. Here the qualitative parameter is the identity of the signaling branch, ERK as opposed to P38 or JNK or other cascades. The quantitative parameters would be the extent (number of molecules of) ERK has to be activated and the time it needs to remain active. This would identify the combination of duration, extent, and identity of a signaling flow as the logical circuit which triggers this particular phenotype. Of the large number of environmental cues, including cytokine and chemokine signals, costimulatory and coinhibitory receptors, nutrient and trophic factor availability, etc., sensed by individual lymphocytes, it is increasingly clear that the signaling logic downstream of the AgR itself is critical to determine many cell fates. The dose and affinity of the antigen for T cells can influence the formation of different T cell subsets and differentiation trajectories.[2] Low doses of antigen have been shown to preferentially drive IL-4–producing effector cells instead of IFNγ producers.[3] Shorter dwell time of TCR with peptide-MHC (pMHC) ligands influences the formation of GC-TFH versus TH1 lineages.[4] Very short durations of TCR signaling result in abortive activation while overtly extended antigen availability leads to tolerance or exhaustion—with an optimal "goldilocks" duration of acute stimulation driving robust effector and memory differentiation. Even within this goldilocks window, shorter durations (with lower strength) favor the eventual formation of long-lived memory T cells while stronger stimulations promote terminally differentiating effector cells (reviewed in Ref. 5). While these data are still a limited discussion of the possible permutations and combinations of antigen affinity, dose, duration, and location—all of which affect T cell differentiation—it still raises important questions of how these antigen parameters elicit precise, unique, and reproducible signaling patterns—which deliver reproducible cellular fate outcomes each time.

Understanding the signaling logic employed by AgR signaling, to convert trigger-specific biochemical signals and relay them reproducibly into cellular functions is hardly a challenge limited to immunology, but one that immunologists have a unique vantage on. The sessile nature of lymphocytes makes them amenable to manipulate, transfer, and study at the single-cell level *in vivo*, using model systems such as in mice but also in humans. The diverse catalog of functional responses that lymphocytes can make also lend them to high-resolution mapping of signaling networks leading to cell fate decisions. At the same time, this is hardly an academic exercise, since cellular engineering of lymphocytes by modifying different aspects of AgR specificity as well as their downstream signaling parameters is already at the forefront of immunotherapy in the clinic. This is most dramatically illustrated in the case of CAR-T cells (Chapter 47) where precision engineering of signaling receptors and downstream pathways has allowed us to generate one of the most promising tumor treatment approaches in the past decade. But it is also evident in the clinical impact of drugs and biologicals targeting specific molecules in the lymphocyte AgR signaling cascades.

THE CELLULAR CIRCUITRY OF RECEPTOR SIGNALING

In the case of a typical membrane-bound receptor, biochemical changes closely associated (proximal) to the receptor are initiated after engaging the extracellular agonistic ligand. This triggering event alters the localization or activity of key enzymes (typically kinases) in the intracellular domains of the receptor, which proceed to activate downstream substrates. These substrates are often also enzymes themselves or adapters which can nucleate the activity of additional signaling proteins further (Fig. 24.12A). At the heart of such a signal transduction cascade, each step involving sequential protein interactions and enzyme activation, there is significant amplification of the overall flux—since each activated enzyme can modify multiple molecules of its substrate. The study of receptor signaling typically involves an iterative description of each of these molecular steps. Sensitive biochemical assays available today allow us to detect subtle alterations in the biochemical state (eg, phosphorylation) of a particular molecule and understand the regulation of that process in great detail. However, it is important for students to remember that understanding these precise molecular interactions must be in a context relevant to the functional outcomes they elicit at the cellular level. Such a holistic understanding of signaling, and its relationship to biological functions, is likely to require new approaches not only in research but also in teaching of the subject[6] that are currently being implemented.

In this context, a useful concept in thinking about (the otherwise overwhelmingly detailed) signaling networks in functional terms is a modular approach—wherein blocks of signaling networks are considered as discrete modules even when studying specific protein–protein interactions (Fig. 24.12A). The definition of what constitutes a signaling module can vary in each context, but very broadly, it is a collection of genes/proteins which participate in one aspect of a cellular function. Depending on the function, a module can include a complete set of molecular pathways or just a few interacting proteins. To be most useful in conceptualizing signaling (and studying it in typical reductionist models) however, each module needs to be self-contained. This means that there must be a particular input into the module which would be expected to produce a reproducible or predictable output—much like how the introduction of electricity into a circuit turns on a light (Fig. 24.12A). Inputs, depending on where the module is, can be the sensing of a ligand by a receptor or the phosphorylation of a particular kinase or the availability of a particular protein due to new gene expression. Similarly, output from a module can also span the spectrum of post-translational modification (phosphorylation, dephosphorylation, ubiquitinylation, etc.) to broad cellular functions such as initiation of cell cycle. Once a module is defined to have a predictable behavior of this sort, it allows us to visualize each of them as input→output devices and relate them to overall cellular functions within a cell. Different modules can be also connected together with the output of one module becoming the input of the next one (Fig. 24.12B). Therefore, cellular signaling can be studied from two different but complementary perspectives—the first at a classical biochemical level, where we endeavor to understand individual molecules and their molecular interactions; and the second, from the point of view of cellular circuits, considering the logic of how different input→output relationships can cause specific functional outcomes. Thinking about biochemical cascades being organized as discrete modules is not arbitrary. There is evidence for selection not only for the co-evolution of some modules in signaling and biosynthesis pathways, but also for the shunting of whole sets of proteins (modules) from different pathways to create new hybrid ones.

In this chapter, we discuss our current understanding of the signaling networks downstream of the TCR and BCR (Fig. 24.1A; *note that the legends of* Figures 24.1-24.11 *summarize key steps in AgR signaling and can be followed independent of the text below*). In both T and B cells, we can initially divide the signaling network into three broad groups of signaling modules based on their direct connection to the AgR and the resulting order in which the signaling in these stages initiate. The first group has a single module, herein referred to as the proximal module (Fig. 24.1A). This is composed of molecules that get recruited to the plasma membrane within the AgR's close proximity, very soon after the TCR or BCR is triggered by antigen binding on the extracellular side. This module is largely concerned with signal initiation, to trigger the signaling flux in lymphocytes, which is then amplified along multiple intracellular branches or a group of distal modules (Fig. 24.1A). Each of these distal signaling branches or modules is typically associated with eventually inducing a specific effector cellular function (eg, cytoskeletal rearrangement, transcription downstream of particular promoter, etc.). The connectivity between these two groups of modules (proximal and distal) is mediated by adapter modules, which serve as junction boxes in the cellular circuit or scaffolds to assemble complexes of multiple signaling molecules (signalosomes). This function in lymphocytes is carried out by relatively few scaffolding molecules. The output of each of the distal modules is linked to other modules that control transcriptional activation, specific effector function, etc., which are not discussed in this chapter. The docking and interaction of various modules (and the proteins within them) are facilitated by conserved peptide domains found in signaling proteins, some of which are summarized in Table 24.1.

An important characteristic of these distal modules is that most are not lymphocyte-specific. For example, proteins that make up the Ras-MAPK pathway in lymphocytes are also expressed in many other cell types and the module of PI3K linked to cell survival is also generally conserved. Since each of these distal modules has evolved relationships with "generic" cellular functions operating in multiple other cell types, lymphocytes merely re-use them to achieve the same goals rather than "reinvent the wheel." It should be noted that there is considerable crosstalk across modules and a particular biological function can be influenced by multiple distal modules. In contrast to their presence in multiple cell lineages, the proximal modules as well as the scaffolding (junctional) molecules of AgR receptor signaling are very lymphocyte-specific. They even differ between T cells and B cells, but are made up of homologous players. Finally, the kinetics of each of these modules are also important in determining functional outcomes (see the Control Mechanisms in the AgR Signaling Network section for a discussion). Figure 24.1B gives a flavor of some representative events in AgR signaling, relative to eventual cellular outcomes. Early proximal signaling is maximally triggered within the first few minutes of antigen engagement.[7] Transcription factors such as NFAT required for cytokine gene expression are also induced within the first few hours,[8] but even "rapid" cytokine production (eg, IL-2 for T cells) can take hours to days.[9] Cell proliferation and effector differentiation take much longer, but at least the latter requires additional signals than those provided by the AgR (Chapters 26, 28, and 30).

THE COMPOSITION OF THE ANTIGEN RECEPTOR SIGNALING COMPLEX

As with any cellular receptor, the TCR and BCR recognize their target by structural complementarity with the binding surfaces on their antigen. Yet, the two receptors differ significantly in their mode of antigen binding. The BCR recognizes native antigen directly (Chapter 5) by binding structural features (epitopes) exposed on these molecules while the TCR engages linear peptide antigens in combination with the presenting MHC surface (Chapter 11). Typically, the binding affinity of a TCR for its pMHC complex is low. Measured using conventional biochemical methods, it has a dissociation constant (K_D) in the range of ~50 μM and rather slow association/dissociation rates as well (see Chapter 23 Table 23.3). On the cell surface, the true binding affinities are harder to measure; but single-molecule fluorescence resonance energy transfer (FRET) experiments measuring the dissociation between labeled TCRs and pMHC on cell surface suggest that *in vivo* interactions can show a 4- to 12-fold increased affinity.[10] Even with this poorly understood *in vivo* affinity-enhancement process, TCR-pMHC affinity is considerably lower than that of a conventional receptor for its ligand. For comparison, the lower end of affinity for a BCR toward it's activating antigen tends to be in the nM range and many antibody molecules can bind with pM dissociation constants.[11] Despite these physicochemical differences in binding to specific ligands by their extracellular domains, both AgRs employ very similar strategies in signaling at their respective proximal signaling modules.

First, unlike many cell surface receptors, the intracellular domains of the TCR and BCR have no enzymatic activity (Fig. 24.2). Additionally, the transmembrane or intracellular regions do not directly recruit another active enzyme. Instead, signal transduction by lymphocyte AgRs relies on the structural proximity of key signaling facilitator molecules, whose cytoplasmic domains are in turn able to dock the signal-initiating kinases (Fig. 24.2). These AgR-associated molecules are often called coreceptors; but they do not really have a separate ligand (and coreceptors typically refer to CD4 and CD8 in T cells, as we discuss below). The TCR associates with the CD3 proteins (CD3-γ, ε, δ, and ζ—the ζ subunit is sometimes referred to as TCRζ due to historical antecedents[12,13]). Similarly, the BCR associates with Igα (CD79a) and Igβ (CD79b) and signals through them. All of these signaling facilitator/adapter molecules have an extracellular domain with an immunoglobulin domain, except for CD3ζ. Given the significance of these molecules for AgR function, stable association with the CD3 complex is necessary for the efficient trafficking and surface display of the TCR itself.[14] Early experiments suggested that introduction of cDNA for just the TCR chains into cells that lacked CD3 expression or vice versa led to accumulation of partially assembled complexes in the endoplasmic reticulum but no surface expression.[15-17] Similarly, although the BCR can be expressed on the surface of lymphoma lines by just transfecting in the IgH and IgL cDNA, in cell lines lacking B-lineage proteins such as the AtT20 pituitary line, surface display requires additional cotransfection of Igα and Igβ.[18] Igα and Igβ were originally immunoprecipitated as a 34-kDa heterodimer due to the formation of a disulphide link between them.[19] In this heterodimer, each Igα contacts the BCR IgH chain, allowing two heterodimers of Igα/β to be potentially associated with each BCR.

In 1989, Michael Reth drew attention to a conserved protein motif found in the cytoplasmic region of the CD3 molecules and also in the Ig signaling adapters (as well as

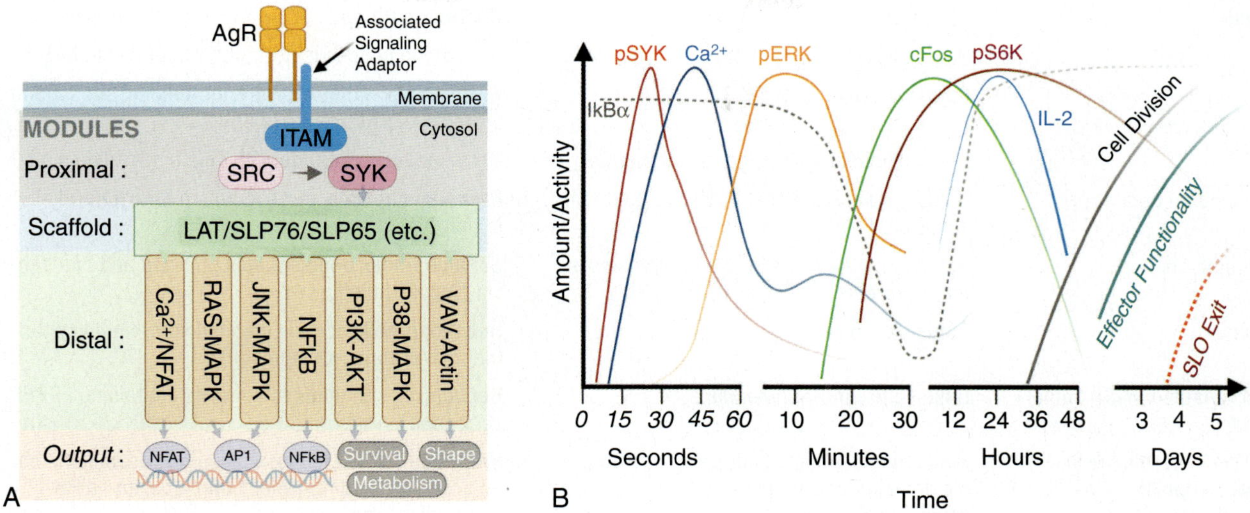

FIGURE 24.1. Overview of antigen receptor (AgR) signaling T and B cells. Figures 24.1-24.12 can be followed along, independent of the chapter text, to summarize key aspects of signaling in lymphocytes, downstream of their antigen receptors (TCR [T cell receptor] and BCR [B cell receptor]). Where the pathways (eg, the distal modules from Figs. 24.5-24.10) are common to both T and B cells, we discuss them together (and this is the default language); however, cases where the specifics differ (eg, the scaffolding molecules in Fig. 24.4) we discuss TCR and BCR signaling differently. **A:** The overall signaling network in T and B cells can be described as a collection of modules (see Fig. 24.12C). Although the AgRs do not have a dedicated cytoplasmic kinase domain or even docking sites for recruiting other signaling molecules, they are constitutively associated with accessory molecules (CD3 in T cells and Igα/β in B cells, see Fig. 24.2) that have intracellular ITAM domains which can be phosphorylated. The module closely associated with the surface ITAMs is the proximal module (Fig. 24.3), made up of SRC and SYK family kinases. The proximal module is connected with downstream signaling network by means of adaptors or scaffolds. These include LAT, SLP76, and SLP65 (see Fig. 24.4) as well as lipids such as PIP2 and DAG. Using the scaffolds as an intermediary, the proximal module phosphorylates several downstream signaling molecules. Each of these downstream enzymes helps initiate a distinct distal signaling module. These distal signaling modules are typically linked with a major output (which can be activation of transcription factors such as NFAT, AP1, NF-κB, or functional alterations in cell survival) (eg, by altering redox or mitochondrial function), metabolism (as discussed in Chapter 6), or cell shape (by altering the cytoskeleton), etc. Importantly, many of the distal modules are found in other (nonlymphocyte) cell lineages in the body but the proximal modules and scaffolds are more specific to lymphocytes immune system. **B:** This panel demonstrates the general timeline of signaling and cellular responses during T- and B-cell activation. The Y-axis shows an illustrative timeline starting with very rapid events (in seconds) and extending all the way to long-term changes (in days) separated into four segments. Typically, signaling changes in the proximal module (eg, *right red line*, shown as the phosphorylation of the SYK family [pSYK or ZAP70]) are not only quite rapid but also transient. Peak phosphorylation of this enzyme can be completed in minutes, and while low level of activity may persist with continued antigen engagement, the magnitude is considerably lower. Some of the distal modules are also activated rapidly—as shown by the cytosolic calcium flux (*blue line*, labeled Ca^{2+}). Not all signaling events dissipate completely. In the case of the Ca^{2+} flux, while the initial peak drops rather swiftly, a persistent (and potentially fluctuating) pulse is often maintained. Other modules, exemplified by phosphorylation of the MAPK ERK, are activated maximally within minutes; but significant levels of phosphor-ERK (pERK) can be maintained over the next few hours as the cells execute the functional programs associated with Ras-MAPK signaling. Not all signaling intermediates show a positive change. One example is the inhibitor of NF-κB, IκBα (gray dotted line), which is expressed highly in resting cells, degraded upon AgR signaling and then re-expressed after a while. Consistent with rapid induction of transcription factors, gene expression events are evident within 30 minutes, leading to the increase in early activation proteins such as cFos (*green line*), which can remain high for a few days after activation. Not all signaling events are rapidly triggered to be short-lasting. The phosphorylation of S6K (*purple line*), downstream of the mTOR pathway, is triggered relatively later, but persists for 2 to 4 days afterward, as it continues to regulate cellular metabolism and protein turnover. In the case of naïve T and B cells, the cellular responses canonically associated with activation take longer to appear than these signaling changes. In helper T cells, the secretion of the cytokine interleukin-2 (IL-2, light blue line) is an early marker for successful activation and can be made before cell division starts. In the case of T cells, cell division itself (black line) starts around 32 hours after initial activation, but then proceeds rapidly with a doubling time of 3 to 4 hours for the next few days. The acquisition of effector functions (*blue-green line*; eg, the ability to express cytokines like IFNγ in T cells and class-switched IgG in B cells) takes a few days to weeks to fully develop. In the case of a primary immune response, these events are largely occurring in the secondary lymphoid organs and it takes 3 to 4 days before the first wave of differentiated effector cells start migrating out (*dotted red line*, SLO exit). While these trendlines display general properties of naïve lymphocytes, multiple parameters including the dose and persistence of antigen, inflammation, etc., can influence the precise kinetics. (Created with BioRender.com)

TABLE 24.1 Key Signaling Domains

Domain	Function	Characteristics
14-3-3	Binds to pSer or pThr motifs	Nine alpha helices important for peptide binding
C1	Binds diacylglycerol (DAG)	Cysteine-rich domains with zinc-binding sites found in protein like PKC
C2	Binding lipid in a Ca^{2+}-dependent manner	Beta-sandwich strands with calcium-binding sites
CH (calponin homology)	Binds not only to actin but also to nonactin substrates	Four alpha helices, three of which form a parallel bundle
DD (death domain)	Binds to and recruits the death domains of other proteins	Six-alpha-helix bundle that is charged and can dimerize via electrostatic forces
EF hand	Binds Ca^{2+} ions	Helix-loop-helix domain with calcium-binding sites found in protein like calmodulin
EVH1 (enabled/VASP homology 1)	Binds to proline-rich domains	Recognizes and binds to domains containing FPPPP sequence
FERM (4.1 protein, ezrin, radixin, moesin)	Binds to phosphoinositides and hydrophobic domains and helps recruit molecules to plasma membrane	Has three distinct lobes: a ubiquitin-resembling lobe, an acyl-CoA binding-resembling lobe, and a PH-resembling lobe
FYVE	Binds phosphoinositides	Beta hairpins with an alpha-helix and zinc-binding sites, typically binds to PI(3)P
GBD (GTPase protein binding domain)	Binds to small GTPase molecules	Found in molecules like Wsp1
IQ (isoleucine glutamine)	Binds to calmodulin	Seven-turn alpha helix with: hydrophobic amino acid at position 1, conserved glutamine at position 2, basic charges at positions 6 and 11, variable glycine at position 7
ITAM (immunoreceptor tyrosine-based activation motif)	Conserved four amino acids important for initiating and transducing signal, and once phosphorylated recruits proteins containing SH2 domain	Found on cytoplasmic tails of proximal immune receptors (such as CD3 chains), conserved motif of YxxL/I
ITIM (immunoreceptor tyrosine-based inhibitory motif)	Conserved sequence important for recruiting immune signal–inhibitory components, such as phosphatases containing SH2 domains	Found on cytoplasmic tail of inhibitory molecules, such as PD-1, contain conserved motif of S/I/V/LxYxxI/V/L
PH (pleckstrin homology)	Binds phosphoinositides	Anti-parallel beta sheets which interact with phosphates of phosphoinositides
PX (phox homology)	Binds to phosphoinositides and can target them to cell membranes	Three-stranded beta sheet, four alpha helices, and a proline-rich region
RING (really interesting new gene)	Helps transfer ubiquitin from enzymes to substrates	Zinc finger structure with two zinc-binding sites
SAM (sterile-α-motif)	Involved in self-association	Five-helix bundle that can help form dimers and oligomers.
SH2 (Src homology 2)	Binds phosphotyrosine	Alpha helices and beta sheet with domain that binds at characteristic pYEEI motif
SH3 (Src homology 3)	Binds to proline-rich peptides	Anti-parallel beta sheets with a beta barrel fold
SH4 (Src homology 4)	Region on Src kinases that can be myristoylated or palmitoylated	Found at the N-terminal of Src kinases (such as Lck), often contains negative charge
UBA (ubiquitin-associated)	Ubiquitin-binding domain	Three-helix bundle and a hydrophobic core
VCA	Found in actin-binding proteins near the C-terminal	Three core sequences: one or two verpolin components that bind to actin, a connector sequence vital to nucleation, and an acidic component that interacts with Arp2/3
WH1 (WASP homology 1)	Found in actin-binding proteins near the N-terminal	Parallel beta-sandwich with a long alpha helix, contains conserved aromatic side chains for ligand binding
WW (tryptophan-tryptophan)	Binds to proline-rich domains	Two conserved tryptophans which fold into a beta sheet

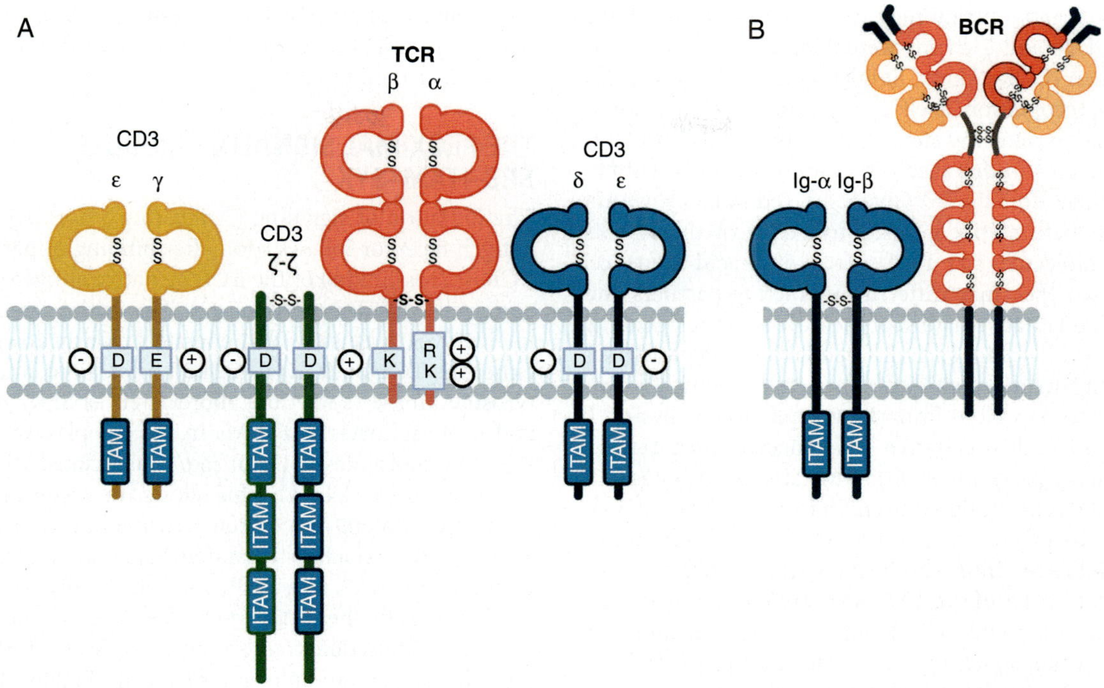

FIGURE 24.2. The components of the T cell receptor (TCR) and B cell receptor (BCR). A: The TCR is made up two chains (α and β, red polypeptides) which have characteristic extracellular Ig folds that are stabilized by disulfide bonds (see Chapter 23). The TCR does not have a cytosolic signaling domain. In order to couple extracellular ligand (peptide-MHC [pMHC] complexes) recognition to intracellular signaling, it constitutively associates with six polypeptides of the CD3 family—which exist as disulfide linked dimers of CD3ε-CD3γ, CD3ε-CD3δ, and CD3ζ-CD3ζ. The whole TCR-CD3 complex is stabilized by electrostatic interactions in the membrane-spanning region with the positive charges on lysine (K) and arginine (R) in the TCR α and β chains associating with the aspartic acid (D) on the CD3 molecules. In addition, glutamic acid residue (E) on the CD3g chain stabilizes the extended lattice-like interactions to form the holocomplex. The immunoreceptor tyrosine–based activation motif (ITAM) on the CD3 chains is marked as blue boxes. A total of 10 ITAMs are therefore associated with the TCR complex, the largest number for any receptor unit. The current consensus model for how pMHC can trigger the TCR is discussed in Chapter 23. **B:** A similar principle operates for the BCR (discussed in Chapter 21). Pairs of the heavy (*dark red*) and light (*orange color*) chains associate to form a tetrapeptide Ig molecule on the surface, which has a minimal cytoplasmic region with no known signaling properties. The association with Igα (CD79A) and Igβ (CD79B) allows the ITAMs in their cytoplasmic tail to recruit proximal signaling machinery. The BCR is typically triggered by crosslinking of the Ig when it binds to antigens which have multiple binding sites (epitopes) for it. (Created with BioRender.com)

Fc receptors and the gp30 protein of a virus)[20] but uniquely limited to receptors of the immune system. Over the next 4 years, mutational and biochemical analyses confirmed the critical role of this domain in AgR signals and it was eventually defined as the immunoreceptor tyrosine-based activation motif or ITAM (reviewed in Ref. 21). ITAMs have a consensus sequence of YxxL/I x6-8 YxxL, composed of two repeats of YxxL/I motif (Y; tyrosine, L/I; leucine/isoleucine) with a spacer of six to eight amino acids. CD3δ, γ, and ε have a single ITAM each and CD3ζ has three tandem ITAMs in their cytoplasmic tail. As a result, the TCRαβ $CD3\varepsilon^2\gamma\delta\zeta^2$ (the superscript of 2 indicates the presences of two copies of the protein in the complex) has a total of 10 ITAMs associated. The Igα and Igβ have one ITAM each, and accordingly, the BCR (IgH^2IgL^2) Ig TCR $(\alpha/\beta)^2$ has a total of four ITAMs available. In addition to the AgR, ITAMs are found in multiple immunoreceptors including various types of Fc receptors (FcRγ), adapter proteins associated with MHC-recognizing receptors, or paired receptors on natural killer (NK) cells (DAP12 and DAP10) and further in various pattern-recognizing receptors of innate cells such as macrophages and dendritic cells (DCs). Signaling molecules with ITAMs also contain characteristic negatively charged transmembrane residues (typically D or E in the +2 position relative to the first tyrosine [Y]) that allow constitutive association with the ligand-binding receptors, which (discussed in the TCR in Chapter 24) contain corresponding positively charged docking patches.[22] The critical nature of ITAM-bearing molecules in the initiation of AgR signaling was first highlighted by landmark experiments triggering T cell proliferation by just using crosslinking antibodies against CD3[23] in an accessory cell–independent fashion.[24] A similar outcome was also evident in B cells using antibodies against the extracellular region of Igβ.[25] Since signaling through the TCR and BCR is critical for different steps of positive selection during the development of these cells, loss of these signaling molecules also arrests lymphocyte development at key checkpoints—typically soon after the first VDJ recombination, where signaling via the newly generated receptors is tested. Intriguingly, using chimeric receptors where the CD3ε and ζ chains were connected to the extracellular human IL-2Rα chain, Shinkai and Alt demonstrated that antigen recognition itself can be completely bypassed for each of these signaling steps as long as the CD3 proteins were crosslinked—in this case with antibodies to the chimeric extracellular domain.[26,27] Given the redundancy within the ITAM signature (the x positions), each of the ITAMs in the CD3 molecules has a different amino acid sequence, suggesting that they may differ in interacting partners and therefore the qualitative nature of the downstream signaling. Biochemical evidence for such differential binding partners is based on using defined peptides from each ITAM with downstream enzymes as well as immunoprecipitation analysis.[28-30] In addition to such qualitative differences (which remain poorly defined), the number of ITAMs also contributes to a quantitative increase in signal strength in an additive fashion. Experiments using retroviral transduction of different numbers of CD3 ITAMs into lymphocytes from knockout mice suggest that at least 8 of the 10 ITAMs are required to transduce the correct amount of TCR signal, with lower numbers resulting in weaker signaling.[31] Consistent with this, TCRγδ-associated CD3 complexes lack CD3ζ, with a concomitant reduction in signaling intensity.[32] Nevertheless, the specificity of each ITAM and its precise contribution to the quality and quantity of TCR signaling in different contexts are not completely understood (reviewed in Ref. 33). ITAMs can be considered key integrators within the signaling networks of immune cells, allowing a wide diversity of external cues to be sensed by diverse receptors and then plugged-in to a limited set of conserved signaling modules.[34]

The critical residues within each ITAM from the point of AgR signal transduction are the tyrosines (Y). Once they are phosphorylated at a Y, proteins containing the SH2 (Src homology domain 2) domain can bind and assemble the proximal signalosome around them (see next section). SH2 domains are ubiquitous in tyrosine–kinase signaling pathways and contain a signature sequence of approximately 100 amino acids (reviewed in Ref. 35). The binding specificity of SH2 domains is due to conserved conformational domains rather than linear sequences. Therefore, these domains vary quite a lot between different pairs of interactors. In addition, increased avidity of interaction can often be afforded by two adjacent ITAM motifs which dock proteins containing tandem SH2 domains. Typically, ITAM–SH2 interaction circuits drive a positive signaling flux mediated by recruitment of activating kinases.

The converse function is anchored by ITIM (immunoreceptor tyrosine-based inhibitory motif) which were initially identified in the intracellular domain of FcγRIIB.[36] The ITIM consensus amino acid sequence is (I/V/L/S)-X-Y-X-X-(L/V), where X can be any amino acid. The function of the ITIM is similar to ITAMs in that they require a Src family PTK to phosphorylate the Y. Phosphorylated ITIMs recruit SH2 domain–containing phosphatases, typically the inositol phosphatase SHIP (Src homology 2–containing inositol polyphosphate 5-phosphatase) or the phosphatases SHP-1 (Src homology 2–containing protein tyrosine phosphatase-1) or SHP-2. As a result, ITIM-assembled signaling modifiers counteract the positive flux mediated by ITAM-recruited kinases.

THE PROXIMAL SIGNALING MODULE: SRC–ITAM–SYK

The initiation of signaling ("triggering") downstream of the antigen receptor follows soon after binding of pMHC by the TCR or crosslinking of the BCR by native antigen (TCR triggering discussed in Chapter 23). The initial signaling module involves three steps that operate similarly downstream of either the TCR or BCR (Fig. 24.3). First, a Src family protein tyrosine kinase (sometime abbreviated as SFK) is activated and it phosphorylates ITAMs in the cytoplasmic regions of signaling molecules (CD3 or Igα/β) associated with the AgR (see LCK in Fig. 24.3A). This allows the second step, which is the recruitment of a spleen tyrosine kinase (SYK) family kinase (with characteristic tandem SH2 domains) to the phosphorylated ITAMs (see ZAP70 in Fig. 24.3B). The recruited Syk kinase is further activated by the action of the Src as well as its own subsequent autocatalytic ability to further phosphorylate itself (autophosphorylation). Finally, the stimulation of the enzymatic activity of the activated Syk kinase allows it to phosphorylate multiple downstream substrates, which include the scaffolds (see next section) as well as the distal signaling branches assembling at these scaffolds. In the case of the B cells, the relevant Srcs are Lyn, Fyn, or Blk while the Syk family kinase is the prototypical Syk (Fig. 24.3D). T cells mostly use Lck or Fyn as the Src and Zeta-associated protein of 70 kDa (ZAP70) as the proximal Syk family kinase. Fyn is also expressed in multiple cell types, whereas Lck is relatively T-specific and NK-specific.

Activation of Src Family Kinases

The nonreceptor tyrosine kinases of the Src family bear the name of the original viral src gene identified as the cellular transforming factor from Rous sarcoma virus.[37,38] All members have a characteristic domain structure—including an N-terminal SH4 domain of ~16 amino acids (aa), a globular SH3 domain of ~65 aa which binds proline-rich stretches of proteins with a consensus PXXP motif, an SH2 domain of ~90 aa with a deep hydrophobic pocket capable of binding the phosphotyrosine (pTyr) residues (often within a conserved YEEI stretch in their targets[39]), and a catalytic kinase domain of ~250 aa which is involved in ATP binding as well as the phosphotransferase activity required for kinase function. In addition to these four domains, a stretch of intrinsically disordered 60 to 75 aa between the SH4 and SH3 domains (named the "unique" domain [UD]) plays an important role in regulating the activity of as well as conferring target specificity to each Src kinase (Fig. 24.3A). The N-terminal SH4 domain plays a key role in facilitating the membrane localization of Src kinases in large part due to a conserved glycine in position 2 (Gly-2), which can be myristoylated and a cysteine in position 3 (Cys-3) which can undergo palmitoylation. Typically, the addition of myristate to proteins can facilitate their membrane recruitment, but the binding of this

14-carbon saturated fatty acid to the inner membrane is quite weak. This has led to the suggestion of a two-signal process whereby either the presence of basic charged residues around the myristoylation site or the addition of palmitate to the second site strengthens their membrane docking.[40] In T cells, both Lck and Fyn show dual acylation status.[41-43] All known B cell Src kinases except Blk can be dually acylated; Blk has just the Gly-2 myristoylation motif without the N-terminal Cys sites for palmitoylation. It is important to note that acylation alone does not ensure cell membrane localization of Srcs; for instance, Lck in T cells is largely located to the plasma membrane while the closely related Src kinase Fyn is enriched on organelles and mitotic structures. Other domains, most notably structural features in the unique domain, also contribute to anchoring Src kinases subcellularly. Finally, the last domain in a Src kinase is a C-terminal stretch of about 15-20 aa which harbors a critical conserved Tyr residue that was found to be phosphorylated, counterintuitively, when the Src kinase was inactive in resting cells.[44] Structural studies revealed this residue (Tyr505 in Lck, Tyr531 in Fyn, Tyr507 in Lyn) helps restrain Src Kinase activity, by holding the protein in a structurally inactive conformation before antigenic stimulation.[45] In the case of Lck, this autoinhibitory phosphorylation on Tyr-505[46] allows the SH3 and SH2 domains to turn into the globular portion of Lck and make intramolecular contacts locking the catalytic domain in an inactive conformation. In contrast, phosphorylation of activating tyrosine sites such as Tyr394 in Lck leads to extension of the SH2 and SH3 domains with resultant enhancement in the kinase activity of the molecule. Analogous sites in Fyn (Tyr420) and Lyn (Tyr396) are also activating. While this reconfiguration has been demonstrated by structural analysis of c-Src, Lck, and Hck,[47] a similar mechanism is likely to contribute to the regulated activation of other Src kinases as well. This unique autoinhibitory state and regulated activation is thought to have evolved to constrain their activity very precisely, only after engaging their specific targets via the SH2/SH3 domains and in response to elaborate upstream input.

The maintenance of the inactive form of Src kinases therefore requires the constitutive activity of another kinase, known as the C-terminal Src kinase (Csk) in lymphocytes.[48,49] Genetic ablation of Csk leads to constitutive activation of multiple Src kinases causing embryonic lethality.[48] Conditional ablation of Csk in precursors to T cells (albeit using the inducible MX1-Cre system) results in constitutive Src- kinase activation, which was sufficient to drive even TCR-deficient T cells through multiple developmental checkpoints that normally require signaling via TCR engagement.[50] Csk is mainly a cytoplasmic kinase while Src kinases (as discussed earlier) function at the membrane (Fig. 24.3A). Recruitment of Csk to the membrane is facilitated by its constituent association with a phosphorylated ~90 kDa transmembrane protein that was accordingly named Csk-binding protein (Cbp).[51] With the help of a palmitoylation, Cbp accumulates in detergent-insoluble self-assembling cholesterol containing glycosphingolipid-enriched microdomain (GEMs—also known as rafts or detergent-resistant membranes [DRMs] for detergent-resistant microdomains) and was independently identified as a phosphoprotein associated with glycosphingolipid-enriched microdomains (PAG).[52,53] The regulation of Csk localization to Cbp/PAG maintains a dynamic feedback loop with Src kinases. Cbp/PAG has nine tyrosines capable of phosphorylation with four in the Src SH2 binding motifs (YEEI/V) and one in a Csk SH2 binding site (YSSV). Phosphorylation by Src allows Cbp/PAG to recruit more Csk into the GEMs and thereby negatively regulate Src activation. Although this transmembrane adapter protein has a complex domain structure and multiple phosphorylation sites (reviewed in Ref. 54), Cbp/PAG knockout mice have normal T cell development and function, despite altered Csk localization[55,56] leading to a lack of clarity on its precise roles in lymphocytes. GEMs are also suspected to be focal points for the initiation of antigen signaling in lymphocytes since dispersion of these structures (using antifungal agents such as filipin or nystatin) or modification of their composition (by cholesterol extraction or addition of exogenous gangliosides) affects T[57] and B[58] cell activation *in vitro*. Therefore, a critical step in initiating AgR signaling through Src kinase activation would be the inactivation of Csk—either by altering its location/activity or the removal of its phosphorylation from the C-terminal domain of Srcs. The latter is accomplished by plasma membrane–associated phosphatases in lymphocytes, most notably CD45.[59]

CD45 or protein tyrosine phosphatase, receptor type, C (PTPRC) is also known as the leukocyte common antigen since it is not only expressed on but is also the most abundant protein on the surface of all hematopoietically derived cells (besides, this is a fact that is commonly used in flow cytometry to identify such cells). CD45 is a type 1 single-transmembrane glycoprotein with two (tandem) protein tyrosine phosphatase (PTPase) domains (D1 and D2) in the cytoplasmic C-terminal tail. Although both domains are required for CD45 function, only D1—the first membrane proximal PTPase domain—is enzymatically active. The catalytically inert D2 domain regulates D1 by structural interactions[60] and this can also be regulated by phosphorylation by other enzymes including casein kinase 2.[61] The extracellular portion is quite large and has three conserved fibronectin type III (FNIII) domains closer to the membrane and a crucial N-terminal region which has an extended structure that is heavily O-linked glycosylated. The O-glycan tract and FNIII domains are separated by a globular cysteine-rich region with five cysteines that are conserved in all mammalian species that help form two intradomain disulfide bonds and one interdomain disulfide bond. Overall CD45 is heavily glycosylated with multiple N-glycosylation sites on the extracellular domain in addition to the O-glycosylation of the N-terminal domains. The N-terminal domains of CD45, however, have an intriguing biology, as they are encoded by variable exons (using different start sites) in different cell types in addition to modification by differential RNA splicing after activation—all of which results in surface expression of multiple lineage and activation state–dependent surface CD45 isoforms.[62] The isoform first identified in B cells (originally named T200 or B220—based on an epitope that binds a particular mAb) has a molecular weight of 220 kDa while thymocytes express a 180 kDa form, which can then switch to one of the four isoforms (190, 205, 220, or remain 180 kDa) in the periphery. The genomic locus consists of 34 exons spanning 130 kb with relatively conserved exon–intron architecture in humans as well.[63] Exon 1 is untranslated and followed by an unusually large intron (~50 kb), both contributing to the initiation of transcription of functional mRNA including all exons from exon 3 onwards. Most of the variant forms detected in lymphocytes arise because of differences in the use of exons 4, 5, and 6[62] (which are also referred to as exons A, B, and C), with the rest of the exons preserved in most cells. The 220 kDa B cell isoform (referred to as CD45R) includes

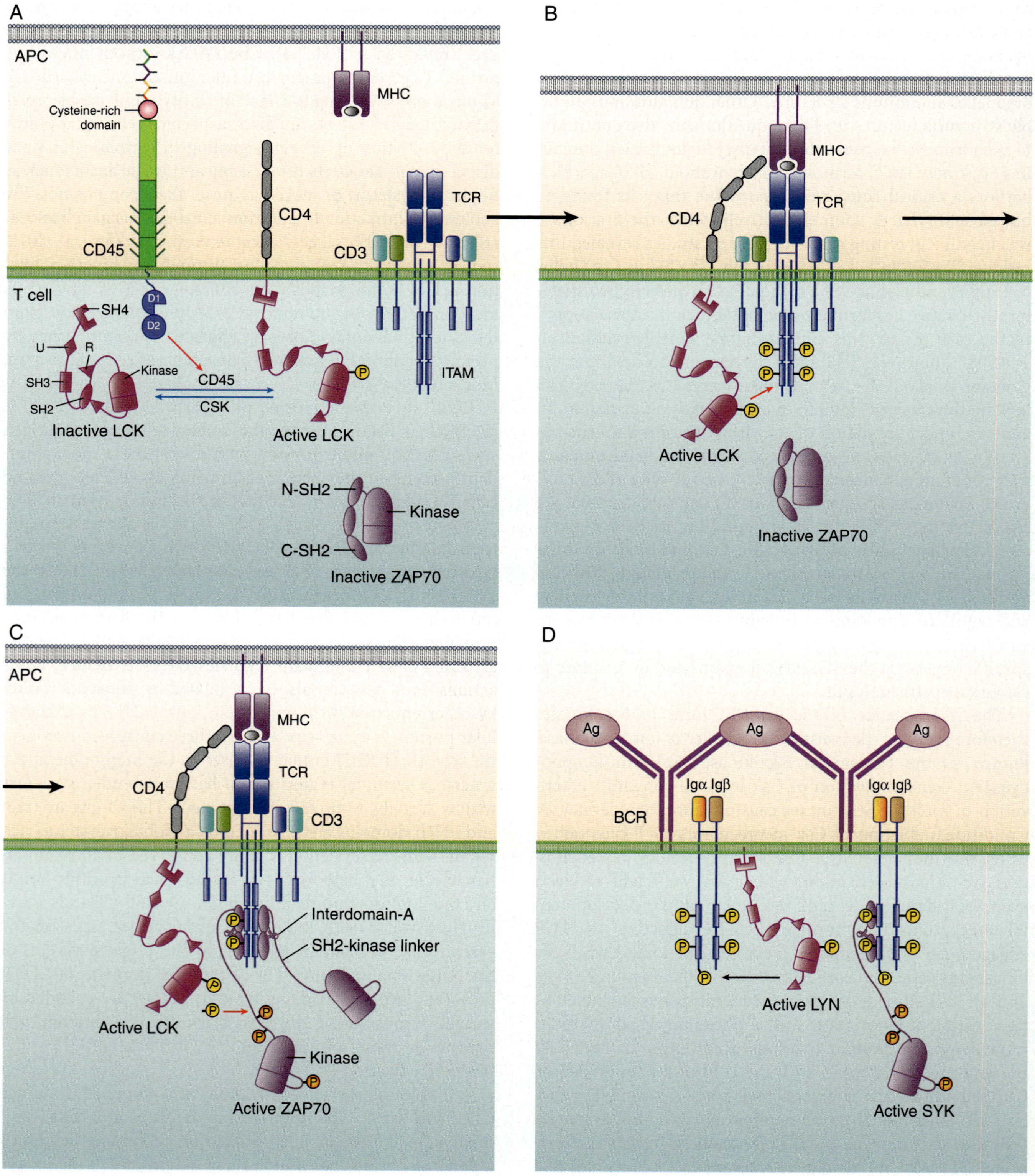

all the exons while naïve human T cell 180kD form typically splices out exons 4 and 5 (just using exon 3 from this group) leading to the designation CD45RA. In effector and memory T cells, the splicing excises exons 4, 5, and 6 leading to the surface expression of the CD45RO isoform. Importantly, the glycosylation and overall length of the extracellular domains of CD45 can therefore vary in different cell types. This has implications for the localization or exclusion of the linked intracellular phosphatase domains, when the TCR is closely associated with pMHC on the antigen-presenting cells (APCs), the large extracellular region of CD45 can cause it to be physically excluded from the junctional regions, which reduces the phosphatase in the signaling microdomains on the intracellular region.[64] A key target for the CD45 phosphatase itself is the site on Src kinases phosphorylated by Csk. Removal of this phosphate allows the unfolding of Src kinase and conversion to the active form. CD45-deficient cells have reduced Lyn and Fyn activity with corresponding increase in the amounts of the C-terminal phosphopeptide.[59] In CD45−/− mice, T cell development is significantly impaired at the positive selection stage, consistent with a failure to fully activate Lck.[65,66] However, B cell development in the bone marrow of these mice is relatively normal with defects evident only in the transitional stages in the spleen and in their subsequent activation. The activities of alternate phosphatases such as CD148 or PTPRJ have been suggested to compensate for lack of CD45 in early B cell development.[67] Finally, although the role of CD45 early in AgR signaling is a positive one, as a phosphatase it can subsequently also inhibit later stages in T cell activation by dephosphorylating other targets.

A critical piece of the puzzle in understanding the initial triggering of signaling downstream of AgR is the localization of the Src kinases themselves. In T cells, in addition to TCR binding the peptide-MHC surface, an invariant site on MHC molecules also binds the coreceptors CD4 or CD8. Because of this, most of the catalytically active Lck at the plasma membrane of T cells is bound to the cytoplasmic tail of these coreceptors[68,69] making Lck the first example of a protein-tyrosine kinase to constitutively dock with a surface receptor. The unique domain of Lck contains a conserved dicysteine motif (Cys-20 and Cys-23) that associates with a dual cysteine (Cys-445 and Cys-447) domain in the cytoplasmic tail of either CD4 or CD8α[70] using a unique zinc-dependent clasp domain[71] which facilitates very selective, tight, and fully reversible protein heterodimerization. The engagement of the coreceptor with MHC (ie, CD4 binding to MHC-II or CD8 to MHC-I) occurs after the TCR has engaged the peptide-MHC surface.[72] This binding contributes to the functional avidity and perhaps further stabilization of the pMHC–TCR–coreceptor complex.

As a result of this, the association of TCR with pMHC together with the binding of CD4/CD8 to MHC increases the local concentration of Lck intracellularly around the TCR. This redistribution effectively reduces Csk in the vicinity of Lck and shifts the equilibrium toward greater phosphatase activity on the C-terminus of the Lck. The enhanced kinase activity and specific recruitment to the ITAMs on CD3, resulting in their phosphorylation, sets the stage for the next step in the proximal module of TCR signaling involving Syk activation. Initial phosphorylation starts at the ITAMs of the CD3ζ chain.[73] The precise mechanisms of this early triggering event for TCRs are discussed in Chapter 23. It is also likely that a small pool of Lck bound to the coreceptor is constitutively active (due to proximal CD45 activity), allowing for rapid TCR activation.[74] Similarly, for BCRs, the aggregation of surface Ig colocalizes and concentrates Src kinases with their substrates including the ITAM tyrosines on Igαβ. In resting B cells, Lyn is weakly bound to the BCR as the N-terminal unique domain of Lyn engages the unphosphorylated Igα ITAM domain[75] in addition to the SH4 domain–dependent plasma membrane enrichment discussed earlier.

Activation of Syk Family Kinases

The ~70 KDa nonreceptor SYK and the related Zeta-associated protein of 70 kDa (ZAP70) are very similar

FIGURE 24.3. The proximal antigen receptor (AgR) signaling module. A: In resting T cells, the T cell receptor (TCR) is associated with the CD3 proteins as discussed in Figure 24.2A. The critical kinases required for T cell activation are the Src kinase LCK and the SYK family ZAP70. Both are maintained in inactive configurations, with their catalytic kinase domains folded inward and close to the SH2 and SH3 domains. In the case of LCK, the inactive state is maintained by a phosphorylation of its C-terminal domain by the kinase CSK. The intracellular phosphatase domain (D1) of the large surface molecule CD45 can remove this phosphorylation and make LCK more active. CD45 has a large extracellular domain, which is glycosylated, and this size allows it to be excluded from the immunological synapse (where it might otherwise inhibit critical positive signaling kinases). Activated LCK is recruited to the cytosolic tail of the coreceptors CD4 or CD8, as well as enriched to the membrane microdomains by lipid modifications of the SH4 domain. The unique domain (U) of LCK is critical for substrate specificity. The *blue boxes* in the cytosolic portion of CD3 molecules show the ITAM motifs (discussed in Fig. 24.2). **B:** Once the TCR engages a peptide-loaded MHC (pMHC) molecule on the relevant antigen-presenting cell (APC) surface, the coreceptor (CD4 shown here for MHC-II, but CD8 will operate similarly for MHC-I) binds to the invariant, non–peptide-loaded side of MHC. This brings active LCK to the proximity of the TCR complex, allowing it to phosphorylate (*yellow* P symbols) the ITAMs on CD3. **C:** Dually phosphorylated ITAMs can be bound by the tandem SH2 domains of ZAP70. This not only recruits ZAP70 to the TCR complex but also extends its active configuration. This also allows LCK to further phosphorylate and activate ZAP70, including in the interdomain region connecting the SH2 domains with the kinase domain. In some cases, phosphorylated ZAP70 can be further activated by autophosphorylation. Thus, while pMHC triggering of the TCR is the input for the proximal module, ZAP70 activity is the primary output that will now feed into all the other modules. **D:** In B cells, a similar sequence of events occurs following crosslinking of surface Ig (the BCR) by antigen. While B cells do not require dedicated APC for activation, since the BCR has to be crosslinked, the antigen must have repetitive epitopes or be already in a complex (eg, on the surface of an intact pathogen, associated with complement proteins, etc.). The ITAMs (*blue boxes*) are present in the cytosolic portion of the Igα/β molecules. The Src kinase recruited in B cells can be Lyn or Fyn and the active SYK at the end of this module operates like ZAP70 in **(B and C)**.

proteins which likely evolved from gene duplication events around the evolutionary appearance of adaptive immunity, discussed in Chapter 4. Syk (which is likely the more ancient copy, since it is found in invertebrates) is expressed widely and is the dominant Syk kinase in B cells while ZAP70 performs this role in T and NK cells. Structurally, both contain two tandem SH2 domains interrupted by a coiled coil region known as interdomain A (Fig. 24.3C). The C-terminal tyrosine kinase domain is connected to the tandem SH2s by an extended and flexible interdomain B. The kinase domain has the characteristically conserved bilobed catalytic core structure found in similar enzymes, with an ATP binding in a deep cleft located between the two lobes and a disordered activation loop following the cleft joining the two lobes.[76] The tandem SH2 domains of ZAP70 have been shown to directly bind dually phosphorylated (but not the unphosphorylated) CD3ζ chains.[28] In the same study, the N-terminal SH2 of ZAP70 was also shown to bind peptides from the ITAMs of CD3ζ chains but with 100-fold lower affinity than when the tandem SH2 was used. Crystal structures of ZAP70 domains in complex with ITAM regions revealed unique characteristics of the tandem SH2 function.[77] The pTyr binding pocket of the N-terminal SH2 domain is formed of residues from both the SH2 domains, implying that intact tandem SH2s as well as proper folding of interdomain A were crucial to allow this cross-domain binding interface to form de novo. The next SH2 domain (the C-term SH2) is more conventional since it forms a pTyr-binding pocket with just residues within the region. This allows ZAP70 to be recruited to the phosphorylated ITAM quickly with a single SH2 domain binding but eventually locking in with the tandem domains for a higher affinity interaction. Many resting T cells *in vivo* already show constitutive basal phosphorylation of one of the ITAMs associated with the CD3ζ chain (identified as a 21 kDa phosphoprotein or the p21 form of CD3ζ which is distinct from the ITAM dually phosphorylated version migrating at 23 kDa or the p23 form) which stems from engaging endogenous nonagonist pMHC molecules. Significant amounts of ZAP70 in the cell are already recruited to p21 CD3ζ in the plasma membrane.[78]

The general process described earlier also operates for Syk in B cells (Fig. 24.3D). A single tyrosine phosphorylation on the Igα ITAM is also sufficient to engage the SFK Lyn, while dual ITAM phosphorylation of the Igα chain is needed for recruiting and activating Syk.[79] The Syk kinases undergo very large conformational reorientations as they transition from resting (inactive in lymphocytes before antigen engagement) to p21-ITAM binding to fully active enzymatic states. The phosphorylation status of six tyrosines—one in interdomain A (127 in ZAP70 or 130 for Syk), three in interdomain B (291, 315, and 319 or 317, 342, and 346), and two in the activation loop of the kinase domain (492 and 493 or 519 and 520)—plays critical roles in these structural changes. In resting lymphocytes, Syk and Zap70 adopt the inactive configuration, wherein the noncatalytic face of the N and C lobes of the kinase domains snaps close to each other, oriented back to back with the tandem SH2 domains, and the interdomain linker regions get tucked in between these two ends of the protein. This positions the SH2 domains and the catalytic sites away from each other, but also does not allow the pTyr binding cleft of the N-term SH2 to form since it requires closer proximity to the C-term SH2 motif. In addition, the disordered activation loops in the kinase domain fall over the catalytic region, preventing ZAP70 enzymatic activity. Binding of the Syk or ZAP70 SH2 domains to phosphorylated ITAMs enhances their catalytic activity by two mechanisms. In the case of Syk, *in vitro* studies and reconstitution by overexpression in cell lines suggested that recruitment to ITAMs OR subsequent phosphorylation by the Src kinase Lyn was equally potent in activating Syk's own autophosphorylation (where it can phosphorylate its own activity and full signaling, leading to the suggestion that Syk operates via an "OR" switch[80]). ZAP70 differs from Syk[81] in that although a similar activation by recruitment is evident,[82] it is more reliant on subsequent Lck activity for full activation (operating more like an "AND" gate in signaling). The alignment of the SH2 domains with the ITAMs twists out the linker sandwich in the inactive ZAP70 configuration and accordingly extends the kinase domain, contributing to activation of its autocatalytic activity. In addition, the new proximity to active Lck allows tyrosines in the activation loop (Tyr-492/493) of ZAP70 to be phosphorylated which helps unfold this disordered domain, making the catalytic cleft available. As a result, the recruitment to the TCR complex and rapid enzymatic activation of Syk kinases are tightly coordinated. In addition to Tyr-492/493, Lck also phosphorylates the interdomain tyrosine residues 315 and 319, which are crucial for ZAP70 activity. While activated ZAP70 can autophosphorylate many of these residues subsequently, the need for Src activity to get the ball rolling implies that in the case of ZAP70 (unlike Syk), autophosphorylation is a necessary amplification mechanism.

As discussed in Chapters 27 and 25, early in T and B lymphocyte development, the appearance of fully rearranged receptors is validated by positive selection steps requiring AgR signaling. The nonredundant requirement for Syk family proteins (or for that matter the CD3 or Ig accessory receptor components) is highlighted by the complete absence of T cells in mice knocked out for ZAP70 or CD3 while B cell development is intact.[83] Since Syk is more broadly expressed than ZAP70, the knocked out mice have perinatal lethality. But Syk-deficient hematopoietic cells are also unable to generate B cells with lymphocyte development arrested at the pre–B cell stage.[84,85]

THE SCAFFOLDING OF AgR SIGNALOSOMES

Within seconds to minutes of antigen engagement, the proximal module is fully activated (Fig. 24.1B). The signaling flux emanating from this complex is then distributed and amplified through the distal modules that would eventually drive specific phenotypic consequences of lymphocyte activation (see The Distal Modules Operating in AgR Signaling section). These two modules are connected by specific linker/adapter molecules which lack intrinsic enzymatic activity. These act as scaffolds to facilitate the assembly and co-localization of multiple proteins (a signalosome) and thereby regulate the interconnectivity between different pathways. The key scaffolds in lymphocytes are the linker of activated T cells (LAT) and SH2 domain–containing leukocyte protein of 76 kDa (SLP76) in T cells (Fig. 24.4A) and SRC homology 2 (SH2) domain–containing leukocyte protein of 65 kDa (SLP-65 also known as BLNK, BASH, and BCA) in B cells (Fig. 24.4B). In addition to these general scaffolding molecules, additional and pathway-specific adapters are employed (eg, growth factor receptor-bound protein 2 [GRB-2] in the RAS-MAPK pathway (Fig. 24.6), which are discussed in those subsections). Furthermore, lipid products of signaling such as PIP2 and DAG (Fig. 24.5) can also nucleate different complexes to the membrane—although these are typically univalent and only scaffold due to their localization near similar membrane microdomains.

Scaffolding in T Cells

LAT is a type III transmembrane protein with a short (4 aa) N-terminal extracellular region, a single transmembrane spanning region followed by a critical and long cytosolic carboxyl terminus with no enzymatic activity (Fig. 24.4A). In Jurkat cells lacking LAT, the activation of TCR signaling up to ZAP70 activation is intact, but downstream modules such as phospholipase C γ1 (PLCγ1) or extracellular signal–regulated kinases (ERK) are inactive.[86] *In vivo*, Lat knockout mice are deficient in T cells, with development blocked at the similar (DN3) stage as TCR or ZAP70 knockout mice.[87]

Structurally, two conserved cysteine residues (Cys-26 and Cys-29 in human LAT) immediately following the transmembrane domain can attach to lipid groups by palmitoylation, which is critical for LAT function (Fig. 24.4A) and membrane mobility. The intracellular domain that follows provides key docking sites for multiple signaling molecules, which are recruited when one or more of the highly conserved tyrosines in this region are phosphorylated by activated ZAP70 from the proximal module.[88] For instance, phosphorylated Tyr-132 is sufficient for binding the SH2 domain of phospholipase C (PLCγ1) while multiple tyrosines (171, 191, and 226) must be phosphorylated to bind the Ras-MAPK adapter GRB-2 (TYR-171, 191, and 226) or the GRB2-related adapter protein 2 (GADs) (Tyr-171 and 191).[89] Binding of molecules such as PLCγ1 to their designated pTyr is also quantitatively affected by mutations in other sites, suggesting that there is cooperativity in the assembly of the signalosome on LAT (see Ref. 90 for an extensive review). Binding of GADs is also critical

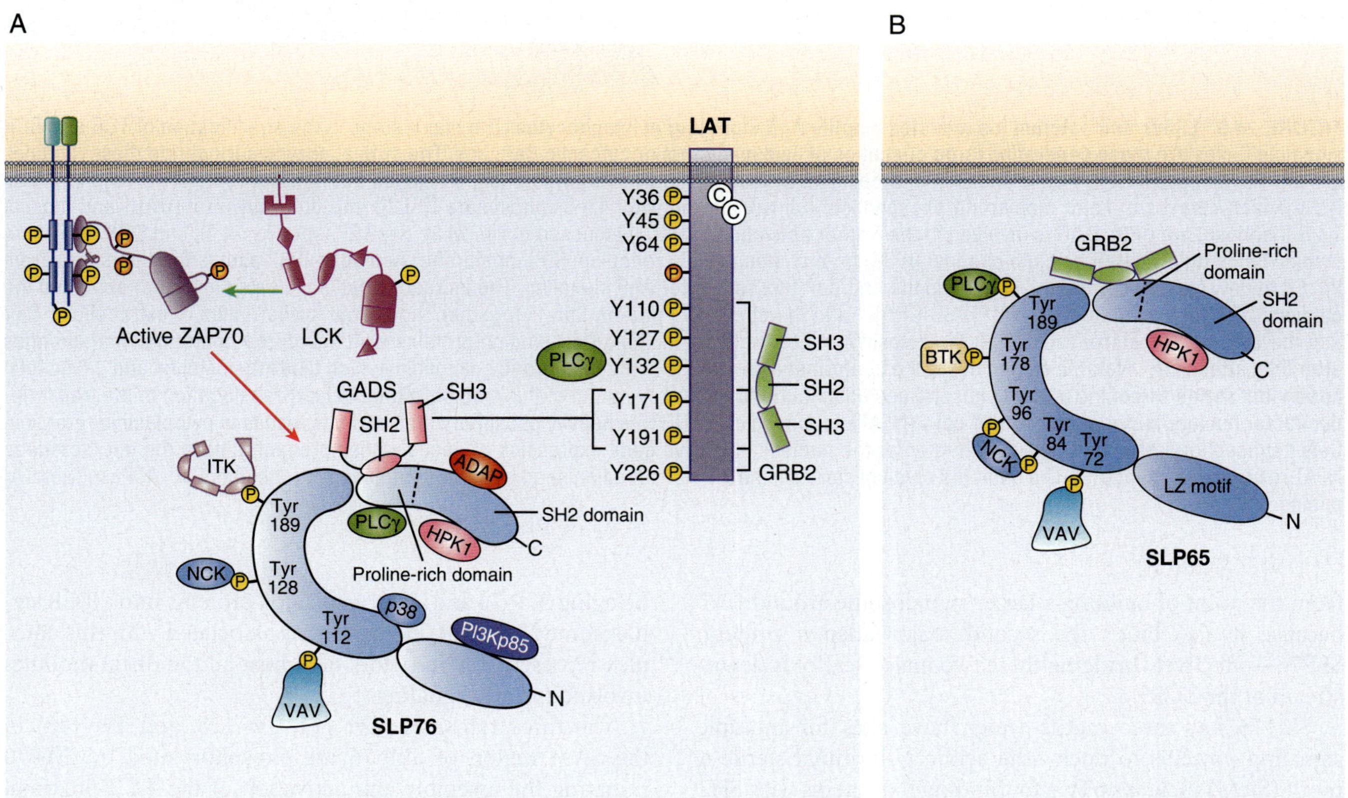

FIGURE 24.4. Scaffolding adaptors linking proximal AgR signaling to distal modules. A: Following activation of the proximal module, the output of that (essentially active ZAP70) phosphorylates multiple downstream targets, which are brought into proximity by the scaffolding molecules linker of activated T cells (LAT) and SH2 domain–containing leukocyte protein of 76 kDa (SLP76). Multiple tyrosines (marked as Y) on LAT and SLP76 are phosphorylated by ZAP70 and LCK. These phosphorylated tyrosines offer binding sites for key molecules in different distal signaling molecules, including ITK, GADS, NCK, PLCγ, VAV, P38, HPK1, ADAP, PI3K, and GRB2. **B:** In B cells, the major scaffolding protein is SH2 domain–containing leukocyte protein of 65 kDa (SLP65) which similarly binds BTK, NCK, PLCg, VAV, HPK1, PI3K, and GRB2. In this context, SLP65 fulfills the docking functions that are distributed among LAT and SLP76 in T cells.

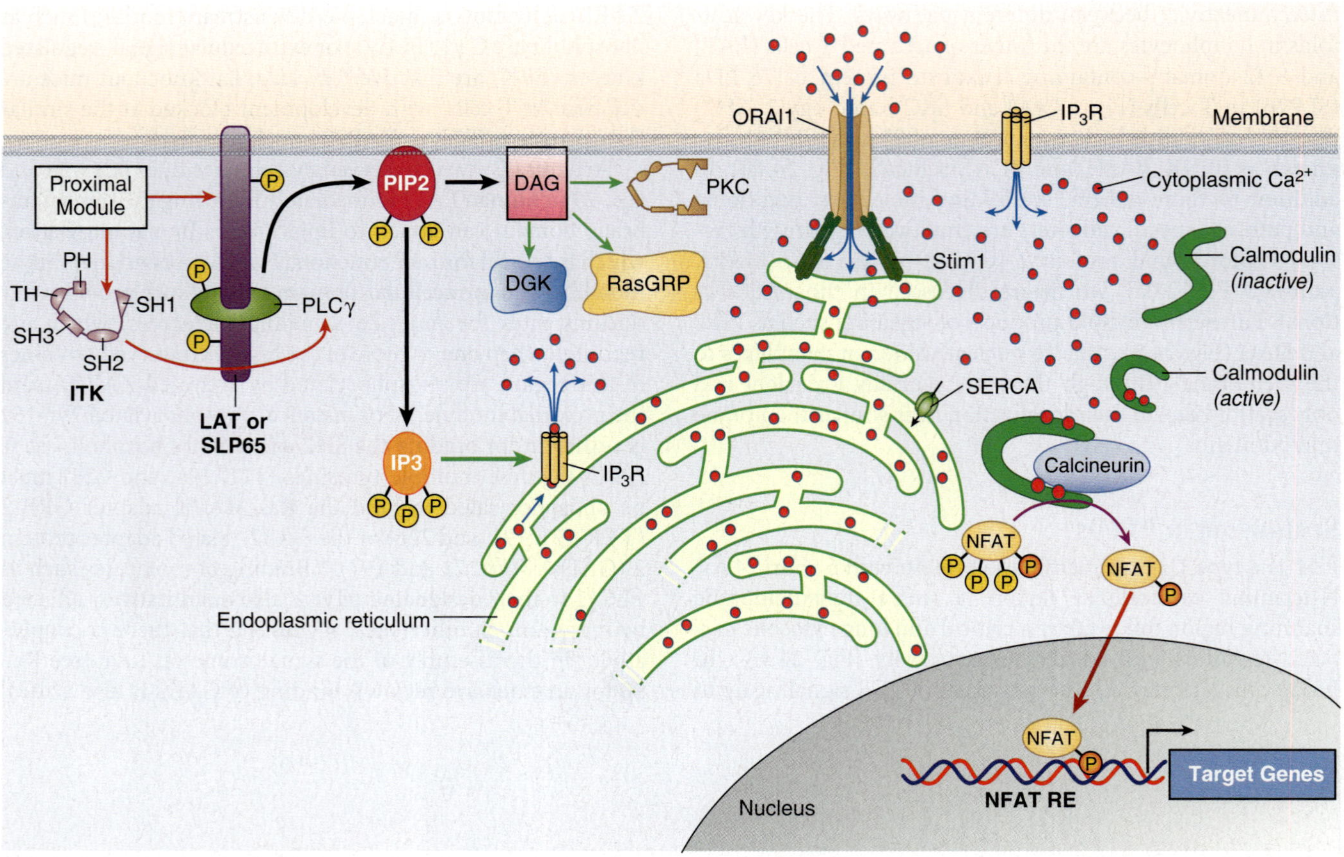

FIGURE 24.5. Lipids and calcium ion modules amplify AgR signaling in lymphocytes. The major steps in the amplification of TCR and BCR signaling fluxes are those generating large quantities of lipid mediators or cytosolic Ca^{2+} ions. The critical enzymes triggering these modules are the Tec kinases (ITK in T cells and BTK in B cells) as well as the lipolytic enzyme Phospho-Lipase C (PLC). Activated ITK phosphorylates PLCγ which cleaves a major membrane phospholipid (phosphatidylinositol 4,5-bisphosphate [PIP2]) into diacylglycerol (DAG) and inositol 1,4,5-trisphosphate (IP3). DAG is involved in the further activation of PKC (discussed in Fig. 24.8), RasGRP (see Fig. 24.5), and DGK. IP3 in the cytosol binds to IP3-gated calcium channel (IP3R) on the endoplasmic reticulum (ER) membrane. Activated IP3R allows Ca^{2+} ion stores from the ER to flow into the cytoplasm, setting up the first Ca^{2+} flux seen after AgR signaling. The increased cytosolic calcium further activated the calcium release–activated channel (CRAC) ORAI-1 and the accessory protein Stim1. Together, they allow further influx of extracellular Ca^{2+} into the cytosol. The increased Ca^{2+} in the cytosol can now activate multiple Ca^{2+}-binding proteins which undergo conformational changes after they attach Ca^{2+}. Notable among this is the conformational change in the chaperone calmodulin which binds the calcium and calmodulin dependent serine/threonine protein phosphatase calcineurin (CaN). A key output of this module is the nuclear translocation of the transcription factor nuclear factor of activated T cells (NFAT). In resting cells, NFAT is heavily phosphorylated and this results in cytoplasmic retention. CaN dephosphorylates NFAT, allowing it to enter the nucleus and drive gene expression of genes whose promoters have the characteristic NFAT response element (RE). The depleted calcium stores in the ER can be replenished by sarco/endoplasmic reticulum Ca^{2+}-ATPase (SERCA) pumps.

from the point of building a larger signalosome around LAT because it also binds the second major adapter protein, SLP76—effectively bridging these two major scaffolds downstream of the TCR.

SLP-76 has three motifs which have sites for multiple signaling proteins to dock—the acidic N-terminal sterile-α motif (SAM) includes pTyrs for binding to proteins with SH2 domains, a central proline-rich area for SH3 domains to associate and a C-terminal SH2 domain (Fig. 24.4A). The critical association of Gads happens via the SH3 domain binding to the central region of SLP-76, but intriguingly does not need the typical proline-rich (P-x-x-P-x-R) motif preferred by SH3 domains. Instead Gads binds primarily to a nonproline R-X-X-K motif with a very high affinity ($K_D = 240 \pm 45$ nM)[91] bringing SLP-76 and all its associated proteins into a TCR signalosome (Fig. 24.4). The proteins associated with this complex represent key initiators of almost all the distal modules involved in AgR signaling.

The three tyrosines (Tyr-112, Tyr-128, and Tyr-148) in the SAM region of SLP-76 are phosphorylated by ZAP70 initiating the assembly and activation of the TCR proximal signalosome.[92] Mutagenesis studies replacing tyrosines with phenylalanines (Y→F) suggest that although Tyr-112 and Tyr-128 allow for some redundancy,[93] Tyr-148 is critical for recruiting the TEC family kinase interleukin-2–inducible T cell kinase (ITK) which in turn activates PLCγ1[94] and the resulting calcium-NFAT signaling module, as discussed in the Lipids and Cations as Signal Amplifiers: the Ca^{2+} NFAT

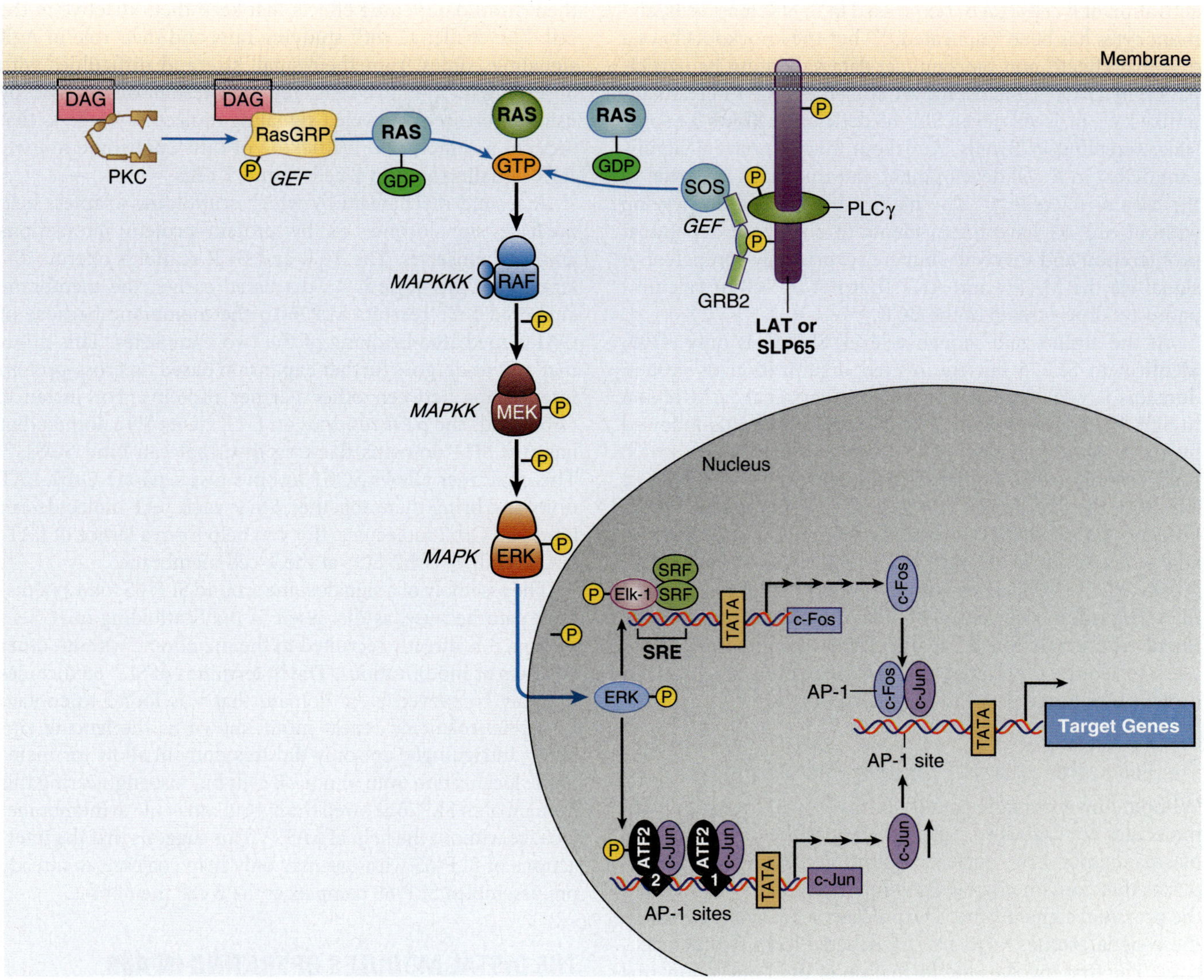

FIGURE 24.6. The Ras–mitogen-activated protein kinase (MAPK) signaling module. The Ras-MAPK pathway, like many distal signaling modules, is not unique to lymphocytes, but plays a pivotal role in activating lymphocyte. At the top of the pathway is the small 21 kDa protein Ras, which in its inactive state is bound to GDP. Activation of Ras-GDP requires guanine nucleotide exchange factor (GEF), of which there are two: RasGRP and Son of Sevenless (SOS). These GEFs transition Ras-GDP into the active Ras-GTP form. Upstream, DAG recruits protein kinase C (PKC), which can activate a recruited RasGRP, as shown on the *left* of RAS-GDP. Alternatively, the scaffold LAT, when phosphorylated, recruits Grb2, which can recruit and activate SOS (as shown on the *right*). Active Ras then functions to start the canonical MAPK cascade, beginning with the phosphorylation and activation of Raf (MAPKKK). Raf then phosphorylated and activates MEK (MAPKK), which in turn phosphorylates and activates ERK (MAPK). Once phosphorylated, ERK can translocate to the nucleus where it can phosphorylate ternary complex factors (TCFs) such as Elk-1. These TCFs associate with serum response factors (SRFs), promoting transcriptional activity by binding to promoters having serum response elements (SREs). Downstream of ERK signaling is the transcription of immediate early genes (IEGs); of note are c-Fos and c-Jun, two transcription factors shown in *blue* and *purple* (respectively). These two transcription factors are then known to form a heterodimer termed activator protein 1 (AP-1).

Pathway section. In addition, the N-terminal domain binds the guanine nucleotide exchange factor (GEF) VAV, the adapter noncatalytic region of tyrosine kinase (NCK), and the MAPK p38. PLCγ1 can also associate with SLP76 in addition to LAT, thereby further strengthening the dual-scaffolded signalosome. Finally, the C-terminal SH2 domain of SLP76 binds hematopoietic progenitor kinase 1 (HPK1)—which is upstream of the JNK and NF-κB pathways (discussed further in The Distal Modules Operating in AgR Signaling section) as well as the adhesion- and degranulation-promoting adapter protein (ADAP), a key regulator of cytoskeletal rearrangements in T cells.

Scaffolding in B Cells

Unlike T cells, the scaffolding function downstream of the BCR (Fig. 24.4B) is carried out mostly by SLP-65 (BLNK). A non–T cell activation linker (NTAL; or LAT2 or linker for activation of B cells [LAB]) expressed in B, NK, mast cells, and monocytes has been implicated,[95,96] but the knockouts have a minimal effect[97] and biochemical data validating its participation in a BCR signalosome are still emerging. Interestingly, unlike LAT in T cells, even SLP-65 deficiency allows for some leaky signaling in B cells. Knockout mice do have a significant defect in B cell development, but this block is partial in the case of B2 cells.[98,99] The mature B cells that do develop without SLP-65 have impairments in activation-dependent proliferation and survival—but are surprisingly competent to signal via the MAPK and AKT (but not NF-κB or calcium) pathways downstream of the BCR.[100]

At the amino acid sequence level, SLP-65 is only ~30% identical to SLP76 but the overall organization of protein domains is very similar. The acidic N-terminal SAM region in SLP-65 has five tyrosine phosphorylation motifs followed by the proline-rich and SH2 domains (Fig. 24.4B) which bind VAV and NCK similar to the role of SLP76 in T cells. In addition, the Tec family kinase (and ITK homolog) Bruton tyrosine kinase (BTK) also binds this region and plays an analogous role. In contrast to T cells, the key phospholipase in B cells, PLCγ2, associates with the N-terminal region, rather than the proline-rich central domain of SLP65. Unlike SLP76, the recruitment of SLP-65 to the plasma-membrane does not seem to require other adapters, but involves its SH2 domain directly binding to non-ITAM regions of Igα.[101,102]

The Formation of AgR-Proximal Signalosomes

Without any enzymatic activity of their own, the scaffolding molecules are crucial to AgR signaling as they translocate to plasma membrane to nucleate multimolecular signaling complexes that contain crucial signaling molecules, often termed the proximal signalosome. The precise mechanisms by which these signalosomes form are still debated from two perspectives. The first involves the formation of lipid microdomains in the plasma membrane (referred to as membrane rafts), which then recruits the scaffolded complexes to the membrane. The second model invokes the scaffolds being directly recruited to the membrane (as discussed earlier) leading to the membrane aggregation of associated binding proteins.

Rafts (also known as DRMs or glycosphingolipid-enriched membranes or simply liquid-ordered domains) are small membrane microdomains (~20 nM span) enriched for sphingolipids and cholesterol, which form dense structural units with the saturated hydrocarbons in sphingolipids closely intercalating cholesterol molecules.[103] Many transmembrane proteins that are enriched in these domains also help stabilize rafts. Importantly, lipid rafts can include or exclude particular proteins to different degrees. Glycosylphosphatidylinositol-anchored proteins, some doubly acylated proteins (as discussed earlier for Src family kinases), cholesterol-linked complexes, palmitoylated proteins, or distinct transmembrane domains (as in the case of LAT), are all thought to help mediate raft partitioning. This allows the raft to potentially segregate positive signaling molecules (kinases, etc.) from negative regulators (phosphatases, etc.). So, the act of recruiting key positive signaling molecules to rafts could theoretically free them from dampening effects that keep them inactive in the cell. The challenge with studying rafts and their role in AgR signaling comes from their small size and difficulties with observing them in live cells. Yet, high-resolution microscopy using fluorescently labeled signaling molecules suggests that ordered assembly and trafficking of rafts contribute to early AgR signaling both in B cells[104] and T cells.[105]

A second mechanism by which scaffolding adapters help nucleate signalosomes is by protein–protein interactions among themselves. The TCR and BCR scaffolds operate differently in this regard. As discussed earlier, the membrane anchored LAT recruits SLP76 to the membrane because of GADS mediated bridging of the two complexes. This minimal unit undergoes further expansion based on cross-protein interactions between other partner proteins. For instance, Grb2 binds the pTyr residues on LAT via its SH2 domain but has two SH3 domains flanking this, that can bind SOS1.[106] This multimer allows SOS1 to bind two separate Grb2-LAT units and bring them together. Since each LAT molecule can bind two Grb2 molecules, this can help form a lattice of LAT-GADs-SLP76-Grb2-SOS at the T cell membrane.

The assembly of a signalosome around SLP-65 follows a distinct path because, as discussed in the Scaffolding in B Cells section, it is directly recruited to the membrane without other adapters or modifications. The N-terminus of SLP-65 includes a highly conserved basic domain that was found to contain a leucine/isoleucine repeat motif known as the leucine zipper.[107] Intriguingly, not only did this domain allow for membrane localization even in non–B cells but just engineering this domain onto SLP76 allowed the hybrid molecule to membrane-localize without the help of LAT.[107] This suggests that the interactions of SLP-65 with Igα may only help coalesce an already pre-assembled SLP-65 complex at the B cell membrane.

THE DISTAL MODULES OPERATING IN AgR SIGNALING

Each of the distal pathways that plug-in to the scaffolding molecules helps execute one or more cellular functions (Fig. 24.1A). As discussed earlier, unlike the proximal and scaffolding modules which are mostly unique to lymphocytes, the distal branches are fairly ubiquitous and found in multiple cell types even outside of the immune system. Nevertheless, activating them is sufficient to trigger lymphocyte activation, as shown by experiments using phorbol esters and ionomycin, which bypass proximal signaling but trigger protein kinase C (PKC) as well as the calcium modules directly.[108] Typically, each distal module (especially the three MAPK modules discussed below) works as a phosphorylation relay system involving kinases (eg, MAPKKK) which activate downstream kinases (eg, MAPKK) and so forth. Each step amplifies the previous signal by allowing one enzyme to activate multiple copies of substrates. Although each of these modules does not necessarily operate in isolation within a lymphocyte, we will first consider how they are independently structured (for convenience) and then (in the Modifiers of AgR signaling-Costimulatory Receptors section) review mechanisms that interconnect these to form cellular signaling networks.

Lipids and Cations as Signal Amplifiers: the Ca^{2+} NFAT Pathway

A signature of AgR signaling in both B and T cells is a rapid rise in intracellular levels of phospholipids as well as Ca^{2+}.[109] The influx of calcium in particular is a routine measure of lymphocyte AgR activation, since it can be precisely quantified at the single-cell level using fluorescence-based assays with cell-permeable Ca^{2+}-binding dyes measured by microscopy or flow cytometry. This massive Ca^{2+} influx plays an important role in rapidly amplifying the proximal AgR signal to elicit a robust downstream transcriptional response. Typically, cytoplasmic concentration of calcium ions (Ca^{2+}) in lymphocytes is very low (<100 nM) compared to extracellular levels of 1 to 2 mM or the amounts in intracellular organelle stores (eg, in the ER ranging from 0.1 to 0.8 mM). This 10,000-fold differential across membranes is actively maintained by an elaborate series of calcium channels (reviewed in Ref. 110) and also creates a very strong driving force for Ca^{2+} entry into the cytoplasm if any of these channels are opened. The functional consequence of this flux is the result of the activation of multiple Ca^{2+}-binding signaling proteins in the cytoplasm, including ones facilitating the nuclear translocation of the transcription factor nuclear factor of activated T cells (NFAT,[111] more extensively discussed at the end of this section).

Briefly (see also legend for Fig. 24.5), the initiation of this overall module starts with recruitment and activation of phospholipase Cγ (PLCγ) to the AgR-proximal scaffolds. In addition to Zap70/Syk-dependent phosphorylation of PLCγ, Tec family kinases (ITK or BTK) are also critical amplifiers of PLCγ activity. Activated PLCγ catalyzes the hydrolysis of membrane lipid phosphatidylinositol 4,5-bisphosphate (PIP2) to produce two second messengers, diacylglycerol (DAG) and inositol 1,4,5-trisphosphate (IP3). As in the case of Ca^{2+}, the rapid enzymatic generation of large amounts of these small phospholipids greatly amplifies AgR-proximal signaling. IP3 is crucial for the cytoplasmic Ca^{2+} flux while DAG is discussed in The Canonical Growth Factor Module: Ras-MAPK and The Stress-Activated MAPK Module: SAPK/JNK sections for its role in the Ras and NF-κB pathways. Ca^{2+} influx in lymphocytes proceeds by a two-step mechanism linking levels of calcium in the ER stores with entry from outside, known as store-operated calcium entry (SOCE). First, IP3 acts on IP3R, a Ca^{2+} efflux channel present on the ER membranes, leading to rapid release of Ca^{2+} from the intraorganellar stores. The depletion of this ER calcium store triggers activation of plasma membrane Ca^{2+} release activated calcium (CRAC) channels. These not only help to maintain a higher Ca^{2+} ion concentration in the cytoplasm but also refill the initially depleted internal stores via the SERCA channels (see Fig. 24.5). As further discussed below, the primary CRAC channel in lymphocytes is Orai1, activated by the calcium sensor Stim1 in the ER. A major end point of this module involves NFAT-driven gene expression. NFAT is a transcription factor that is typically restrained in the cytoplasm due to multiple Ser/Thr phosphorylations which must be removed by the dedicated phosphatase calcineurin (CaN). The enzymatic activity of CaN requires association with the calcium-activated subunit calmodulin (CaM), linking Ca^{2+} influx to NFAT nuclear translocation and gene expression.

There are two isozymes of PLCγ found in the immune system, PLC-γ1 and PLC-γ2 which are 50% identical at the amino acid level. While PLC-γ1 knockouts are embryonic lethal (due to widespread expression and function outside of the immune system), PLC-γ2 knockouts are viable but notably B cell–deficient[112] emphasizing the T cell versus B cell dependence on PLC-γ1 versus PLC-γ2. Both enzymes have an N-terminal pleckstrin homology (PH) domain (which allows them to bind the membrane phosphoinositides, see Table 24.1), followed by a calcium binding EF hand, and a series of SH2 and SH3 domains and another lipid-binding C2 domain. The catalytic domains (just named X and Y) are separated by the central SH2–SH3 domains known as the regulatory core. This overall architecture allows PLCγ to bind both phosphorylated tyrosines on the scaffolding molecules as well as specific phospholipids on the plasma membrane. As discussed in The Scaffolding of AgR Signalosomes section, initial recruitment of PLCγ involved binding to pTyrs on LAT (in T cells) or SLP-65 (in B cells). Once recruited, the PH domains of PLCγ also bind phospholipids, particularly phosphatidylinositol (3,4,5)-trisphosphate (PIP3). This helps narrowly focus PLCγ localization to the site of AgR activation, since the production of PIP3 at the cell membrane requires active phosphoinositide 3-kinase (PI3K). This key enzyme (and therefore localized PIP3 production) is also activated by the proximal AgR module (discussed further in The PI3K-AKT Module section) as well.

In addition to PLCγ, PIP3 also provides a binding site for a Tec family kinase (mostly, ITK in T cells and BTK in B cells)[113] thus bringing in proximity a key amplifier of the overall module. In their inactive conformation, the N-terminal PH domain of Tec kinases, ITK or BTK, is folded over the C-terminal kinase domain, maintaining an autoinhibited state stabilized by additional intermolecular contacts.[114] Recruitment to the signalosome involves the binding of the SH2 domain to phosphorylated Tyr145 of SLP76 while the SH3 domain clasps the proline-rich region (PRR).[115] This open configuration, together with PIP3 binding by the PH domain, further allows activation of ITK/BTK. Intriguingly, although Zap70 and Syk are critical for phosphorylating many of the downstream targets at the head of each distal module directly, they seem dispensable for ITK/BTK activation. Instead this may primarily require Lck/FYN-mediated phosphorylation of Y511 in Tec kinase domain and Y180 in the ITK/BTK SH3 motif, such that all the active fraction of ITK molecules in T cells is SLP76-LAT associated.[116]

Following the general theme of latency to activation discussed earlier with other signaling intermediates, before activation by ITK-mediated phosphorylation of Tyr783 in PLCγ, the latter's regulatory core is positioned to block the catalytic core from gaining access to the membrane lipids that are its substrate; but the SH2 domain of the regulatory core has access to allow ITK to interact. This closed configuration is disrupted by phosphorylation of Tyr783 located within an exposed loop. The ensuing conformational rearrangement of PLCγ1 allows membrane binding as well as the hydrolysis of PIP2.[117] The IP3 that is released in this process, from the plasma membrane, associates with a tetrameric IP3-gated calcium channel on ER membranes to instigate the release of ER-stored calcium. There are three IP3R genes, all of which seem to be expressed in lymphocytes and can even form heterotetramers; but targeting of multiple forms using antisense transfection shows their critical role in T cell activation.[118] When all three are knocked out, T cell development is arrested at the late double negative stage underscoring a previously underappreciated role for Ca^{2+} signaling in thymocytes as well.[119] An intriguing Src kinase–dependent tyrosine phosphorylation of IP3R subunits has been described,[120] but its role in regulating channel function is not yet clear. Instead, these channels are primarily regulated by IP3. The emptying of the ER Ca^{2+} stores triggers a two-molecule pathway for opening CRAC channels which was identified after several years of hunting: the ER luminal Ca^{2+} sensor stromal interaction molecule 1 (STIM1) and the pore-forming ORAI1 proteins located in the plasma membrane. The single transmembrane span STIM proteins have an EF–SAM Ca^{2+} binding site in the ER luminal side and a C-terminal coiled coil region which extends into the cytosol. The Orai channel is composed of a hexamer of Orai proteins, of which lymphocytes primarily express Orai1. The mechanism by which Orai opening is regulated by a calcium-starved STIM1 is unique among vertebrate CRAC channels. In the presence of ER calcium, the EF hands in the ER lumen allow STIM1 to form a dimer which stretches out the transmembrane domain and collapse the cytosolic portion, such that it cannot bind the cytosolic tail of Orai. When calcium is lost in the ER, STIM1 now forms multimers, undergoes a conformational change exposing the Orai binding region in the cytosol, and migrates to the ER-PM junctions. STIM1 binding to Orai1 drives conformational changes that keep the channel open and allows Ca^{2+} influx into the lymphocyte. Eventually, as the ER stores are replenished, higher Ca^{2+} favors STIM1 dimers over multimers, leading it to release Orai1 that then folds back into a closed configuration.

In the cytoplasm, Ca^{2+} levels rapidly drive NFAT activation in a CaM-CaN dependent pathway. CaN (also known as protein phosphatase 3, and calcium-dependent serine-threonine phosphatase) is a heterodimer of subunit A which has the catalytic domain and B which is more structural. The catalytic domain of subunit A in the resting state is closed off by the auto-inhibitory domain (AID) attached to a flexible regulatory domain. This is the region to which CaM binds. In the cytoplasm, CaM molecules are present at very high concentrations, competing for available free Ca^{2+} ions, thereby making it a very sensitive sensor for Ca^{2+} influx. The binding of four such ions results in the activation of CaM, which then associates with the regulatory region of the CaN A subunit inducing a conformational change, releasing the AID from its inhibitory position over the catalytic domain, making it active. Active CaN binds its substrates on two sites on its phosphatase loop—named PxIxIT and LxVP sites based on the consensus motifs they bind to. While most substrates have both, even single site bearing substrates can be acted upon. Indeed a peptide inhibitor developed by the Anjana Rao Lab (VIVIT[121]) that dominantly competes for PxIxIT binding specifically prevents CaN activity on NFAT. The other binding sites, LxVP, are on the helix region of the A subunit which associates with the B subunit. The immunosuppressive drugs cyclosporin (brand name—Sandimmune) and FK506 (or Tacrolimus, sold as Prograf, etc.) bind in this region—making CaN activity one of the most relevant pharmacological targets for blocking lymphocyte activation, currently used in the clinic (see Chapter 47 for clinical applications in transplantation). Once activated, the CaM-CaN phosphatase acts on multiple cytoplasmic proteins including Akt,[122] transcription factor EB (TFEB[123]), and of course NFAT.

The NFAT family of transcription factors has four major members NFAT1 (also known as NFATc2 or NFATp), NFAT2 (NFATc1, NFATc), NFAT3 (NFATc4), and NFAT4 (NFATc3, NFATx) of which NFAT1 and NFAT2 are critical for AgR signaling (reviewed in Ref. 124). In the resting lymphocytes, NFAT proteins are phosphorylated on 18 different serine-rich motifs,[125] some of which help expose a nuclear export sequence (NES) that leads to active exclusion due to association with nuclear export complex of exportin-1 (XPO1, also known as chromosomal maintenance 1 or Crm1). Calcineurin-dependent dephosphorylation results in a conformational change that masks this NES while also exposing a nuclear localization sequence which then attracts binding to nuclear importins and helps translocate NFAT into the nucleus. The extensive phosphorylation of NFAT involves two kinds of kinases—maintenance kinases which act in the cytoplasm to keep NFAT in its phosphorylated state and nuclear export kinases which will rephosphorylate any NFAT that enters the nucleus. There are a wide variety of such kinases in lymphocytes; but they broadly fall into three families—glycogen synthase kinase 3 (GSK3), casein kinases (CK), and the dual specificity tyrosine phosphorylation regulated kinases (DYRK). CK1 can perform both maintenance and export phosphorylation; DYRK1A and GSK3 are mostly nucleus localized while DYRK2 is thought to be cytoplasmic. These compartmentalizations can change with activation status and additional structures involving long noncoding RNA and protein scaffolds may collaborate to regulate NFAT localization.[126]

These extensive enzymatic intermediates also serve as control mechanisms within the Ca^{2+}-NFAT module to precisely regulate lymphocyte function. The two phases of the initial AgR- activation–driven Ca^{2+} flux into lymphocytes can vary both in terms of the amount and duration. The different permutations of these two parameters on the store release versus extracellular influx can generate an elaborate syntax of Ca^{2+} signal variations, which can elicit different combinations of downstream transcriptional and cellular responses. For instance, different amplitudes and frequency of oscillations in the cellular calcium flux can dictate how much (and which) transcription factors get activated and thereby regulate the pattern of genes transcribed.[127] Combining this with the multiple sites available for NFAT phosphorylation, the balance of NFAT in the nucleus versus cytoplasm and the lifetime of different forms of NFAT in cells[128] allows for fine control over functions such as cytotoxicity, cell migration, etc., in T cells.

The Canonical Growth Factor Module: Ras-MAPK

The small, membrane tethered GTPase Ras which gained notoriety as an oncogene, is a major regulator of cellular growth and proliferation in multiple cell types and acts via a sequence of kinases under the MAPK pathway. As the name "mitogen" implies, these were originally described as key drivers of cell proliferation and oncogenic transformation but are now known to participate in multiple functions even in quiescent cells. The 21 kDa Ras proteins initiating this pathway (Fig. 24.6) are molecular switches because they can alternate between the inactive GDP-bound conformation and the active GTP-bound conformation. In resting cells, Ras is inactive and bound to guanosine diphosphate (GDP). Activation of Ras requires the GEFs or proteins that can exchange the GDP in Ras with guanosine triphosphate (GTP). GTP binding induces a conformational change in the Ras switch domains (aa 30-40 forming Switch 1 and aa 60-76 forming Switch 2) exposing the effector binding site. These downstream effectors include Raf (Fig. 24.4). Ras has an intrinsic GTPase activity that will return the protein to the inactive GDP-bound state. On its own, this process is rather slow, but GTPase activating proteins (GAPs) bind GTP-bound Ras and speed this up ≥1000-fold to ensure rapid inactivation.

In AgR signaling, there are two potential GEFs that can activate Ras—the Ras guanyl nucleotide releasing protein (RasGRP) or Son of Sevenless (SOS)—which are in turn activated by different upstream enzymes. One arm (Fig. 24.6, left arm) follows DAG production by activated PLCγ as discussed in the Lipids and Cations as Signal Amplifiers: the Ca^{2+} NFAT Pathway section. Binding to DAG recruits RasGRP as well the serine/threonine-specific kinase PKC to the membrane. In addition to recruitment, DAG also directly activates PKC which then proceeds to phosphorylate RasGRP to fully activate its GEF function. On the other side (Fig. 24.6, right arm) SOS is recruited and activated by GRB2 (which, as discussed in The PI3K-AKT Module section, is recruited to LAT or SLP-65).

Once Ras is in its active form, it can initiate several kinase cascades wherein one inactive kinase is activated by an upstream kinase, following which the activated substrate kinase now acts on another downstream substrate and so on. As shown in Figure 24.6, in the canonical MAPK pathway, the initial kinase is the MAPK-kinase-kinase (MAPKKK) Raf, which in turn activates a MAPK kinase (MEK) leading to its activation of extracellular signal–regulated protein kinase (ERK). Other than Raf1, the MAPK cascade can also be activated by the paralogue B-RAF, although the precise mechanisms upstream are still poorly understood. Being a critical node in the mitogen response, it seems that multiple cellular inputs feed into the Ras-MAPK module. The phosphorylation of the ERK isoforms (the 42 kDa proteins ERK1 and ERK2) is a signature feature of activation of the RAS-MAPK module and these enzymes regulate multiple cellular functions (see review in Ref. 129). The regulated entry of phosphorylated ERK1/2 to the nucleus (from which it is actively excluded in resting cells) allows it to access additional substrates. The canonical ERK substrates are the p90 ribosomal S6 kinase (RSK) family of proteins (reviewed in Ref. 130) as well as the MAP kinase-interacting serine/threonine-protein kinases (MNK).

A characteristic response to many mitogen receptors (including AgR) is the rapid expression of several mRNAs following ERK activation, commonly referred to as immediate early genes (IEGs). Importantly, the expression of IEGs does not require de novo protein synthesis, since the necessary signaling and transcriptional factors for this module are already present in resting cells. The serum response factor (SRF) is the DNA-binding transcription factor that promotes IEG transcription by binding to serum response elements (SREs) in the promoters of IEGs. SRFs are not directly activated by ERK but require RSK-mediated phosphorylation. To gain complete transcriptional activity, SRF must associate with three ternary complex factors (TCFs), a group of ETS domain-containing proteins are named Elk-1 (ELK1), ELK3, and ELK4. The TCFs are also phosphorylated by ERK on multiple sites within their transactivation domain. Key IEGs in lymphocytes responsive to ERK signaling include the early growth response, FOS, JUN, and MYC families of transcription factors. Fos combines with Jun (also discussed in The Stress-Activated MAPK Module: SAPK/JNK section) to form a heterodimeric transcription factor referred to as activator protein 1 (AP-1), one of the major transcriptional complexes critical for turning on multiple cytokine genes in T and B cells. The activation of Myc (in concert with mTOR signaling discussed below) is a key regulator of increase in cell size (or blastogenesis that accompanies lymphocyte activation) and subsequent cell division.

The basic Ras-MAPK module discussed earlier is found in multiple cell types and while all Ras isoforms (N-Ras, H-Ras, and K-Ras) are expressed by both T and B cells, their functions are not always redundant. H-Ras or N-Ras knockout mice have relatively normal T and B cell development but have specific peripheral activation defects in T cells.[131] Ras activation by SOS versus RasGRP may be significant here. SOS deletion impairs thymic development of T cells, past the double-negative stage, but RasGRP knockout mice do not have a significant impairment at this stage.[132] The Ras-MAPK pathway also plays a key role in discriminating between positive and negative selection peptide ligands during T cell development.

There are several negative regulators of the MAPK pathway, many of them acting as phosphatases that interfere with the signaling cascade. One of the most prominent negative regulators is Src homology region 2 domain–containing phosphatase-1 (SHP-1). When active, SHP-1 can dephosphorylate key proximal signaling molecules, for example, Lck and ZAP70 in T cells, thereby impacting downstream triggering of the MAPK cascade.[133,134] In a counter-balancing act, when MAPK triggering does occur, pERK can interfere with the ability of SHP-1 to dephosphorylate these molecules.[133] The exact targets of SHP-1 and the kinetics of how it binds to its partners during T or B cell signaling have yet to be fully determined. Additionally, SHP-2 and members of the dual-specificity phosphatases (DUSPs) are also important regulators of the MAPK pathway, particularly through dephosphorylation of tyrosine and serine/threonine residues. In several *in vivo* animal models where DUSP proteins have been knocked out or catalytically inactivated, there is a subsequent increase in the activity of the MAPK pathway.[135-137] The termination of the RAS-MAPK pathway at the proximal step relies on removal of membrane DAG by the action of diacylglycerol kinases (DGKs) which convert DAG to phosphatidic acid (PA). T cells express both DGKα and DGKζ[138] while regulation of MAPK intensity in B cells seems to be mostly reliant on DGKζ.

The Stress-Activated MAPK Module: SAPK/JNK

The c-Jun NH2-terminal kinases (JNK), also known as stress-activated MAP kinases (SAPK), are the second MAPK family operational in mammals, encoded by three genes of which JNK1 and JNK2 are the prominent ones in lymphocytes. Multiple proximal signaling molecules can help activate the JNK module, but in T and B cells, the most prominent one is downstream of the DBL family of RHO-GEF VAV,[139] which can be phosphorylated by both SYK and SRC family kinases (Fig. 24.7). Vav is recruited to SLP76 or SLP-65 and has been implicated in multiple aspects of AgR signaling—but is most critical for the JNK and cytoskeletal modules (actin polymerization, see Fig. 24.10) largely via its ability to exchange GDP with GTP on the GTPase Rac1 (Ras-related C3 botulinum toxin substrate 1). GTP-bound active Rac1 triggers key kinases including MAP3Ks of the mixed-lineage protein kinase (MLK) family, which in turn activates the MAP2K MKK4 (or MKK7, which is typically triggered by non-AgR signals), which phosphorylate JNK on Tyr and Thr residues. In addition to the shared scaffolds in lymphocytes, a JNK-specific scaffolding protein JIP1 also interacts with Src kinases, Vav, and the JNK-upstream kinases. It should be noted that there are multiple inputs to the JNK module that can even use alternate scaffolds (eg, POSH1 in B cells). Indeed the list of MAP3Ks activating JNK in response to different environmental triggers is quite extensive—including

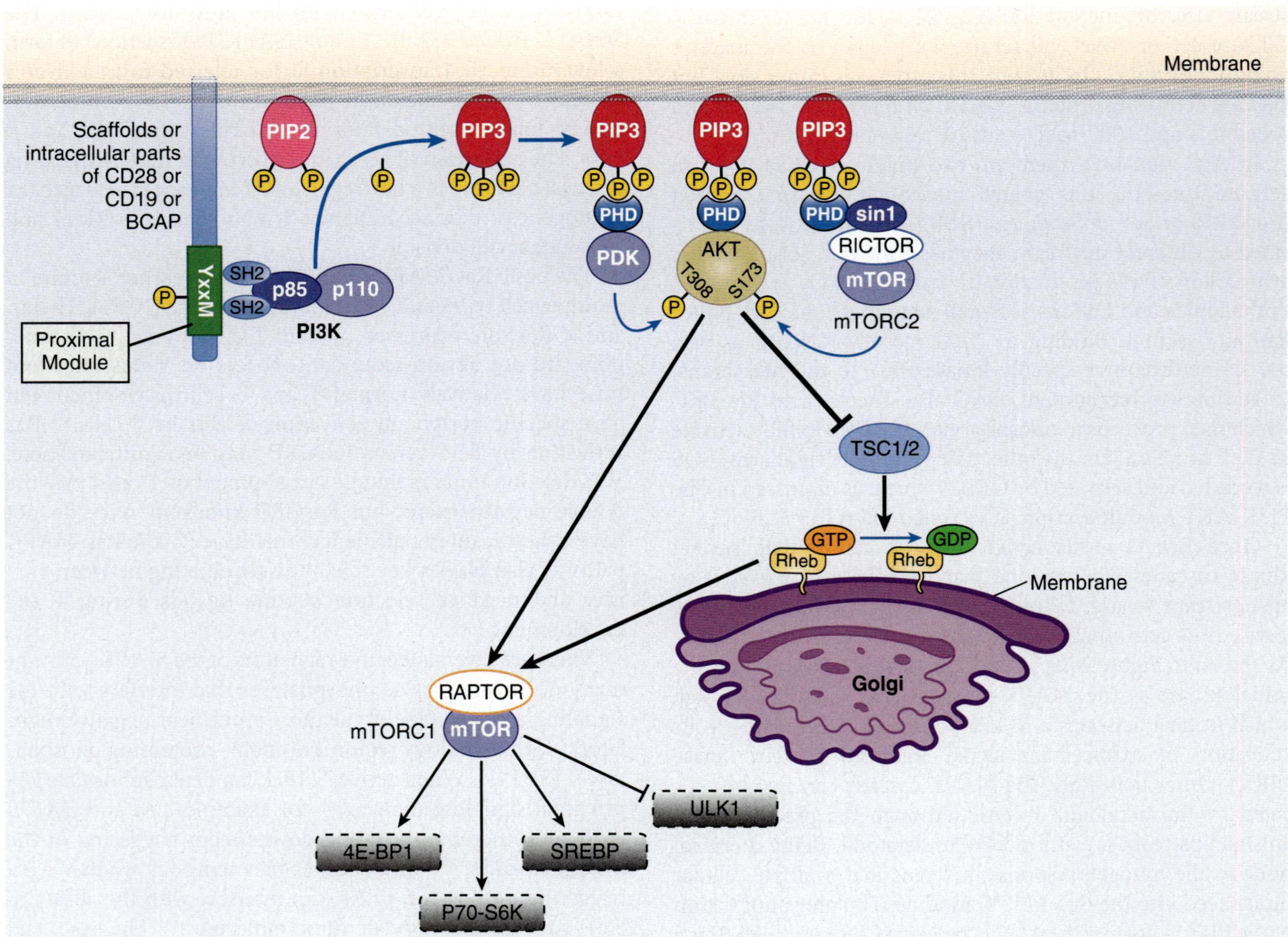

FIGURE 24.7. The Phosphoinositide 3-kinase (PI3K)–AKT module in lymphocytes. PI3K is composed of two subunits, typically referred to by their respective weight, 85 kDa and 110 kDa. The p85 subunit has multiple SH2 domains which allow it to bind to adaptor components of the proximal module, such as CD28, CD19, or BCAP. Activated PI3K is a lipid kinase phosphorylating PIP2 to generate phosphatidylinositol (3,4,5)-trisphosphate (PIP3). Membrane anchored PIP3 can recruit phosphoinositide-dependent protein kinase (PDK), as well as the complex mTORC2, via its SH domain. PDK phosphorylates Thr308 and mTORC2 phosphorylates Ser173 on the protein kinase B, AKT. Activated AKT has a PH domain (see Table 24.1) that allows it to bind PIP3. Akt can then activate the mTORC1 complex, which phosphorylates downstream effector molecules such as the ribosomal S6-kinase p70 (S6K or p70S6K) and regulators of translation such as the eukaryotic translation initiation factor 4E binding protein-1 (4E-BP1).

the MEKK, MLK, and ASK families (reviewed in Ref. 140). Activated JNK1 binds the N-terminal activation domain of c-Jun and phosphorylates it (on Ser-63 and Ser-73), sending it to the nucleus to form the AP1 transcriptional complex with c-Fos.

A critical variable in the JNK pathway relative to the other AgR signaling modules is that major components like JNK1, JNK2, and MKK4 are very poorly expressed in naïve T cells, if at all. Therefore, transcription and translation of these genes soon after antigen activation is a critical requirement for efficient JNK signaling (Fig. 24.8A), making this module more relevant at later stages of T cell activation (involved in T cell differentiation, cell survival, etc.). Once the T cell has gone through one round of antigen activation, it is no longer considered a naïve T cell and the basal expression of JNK pathway proteins remains higher, allowing the module to come online soon after activation of post-effector phase T cells. Of the three Jnk genes, Jnk1 and Jnk2 are found in T and B cells. There is evidence for both unique and redundant functions of these kinases in lymphocyte function. For instance, JNK1 was more critical for c-Jun phosphorylation while JNK2 is the preferred kinase for ATF2 (activating transcription factor 2). While both can form AP1 dimers by partnering with jun, fos, Maf, CREB, etc., the specific promoters each AP1 complex act on can vary; in addition, phosphorylated JNK targets have AP1-independent activities (eg, a histone-modifying

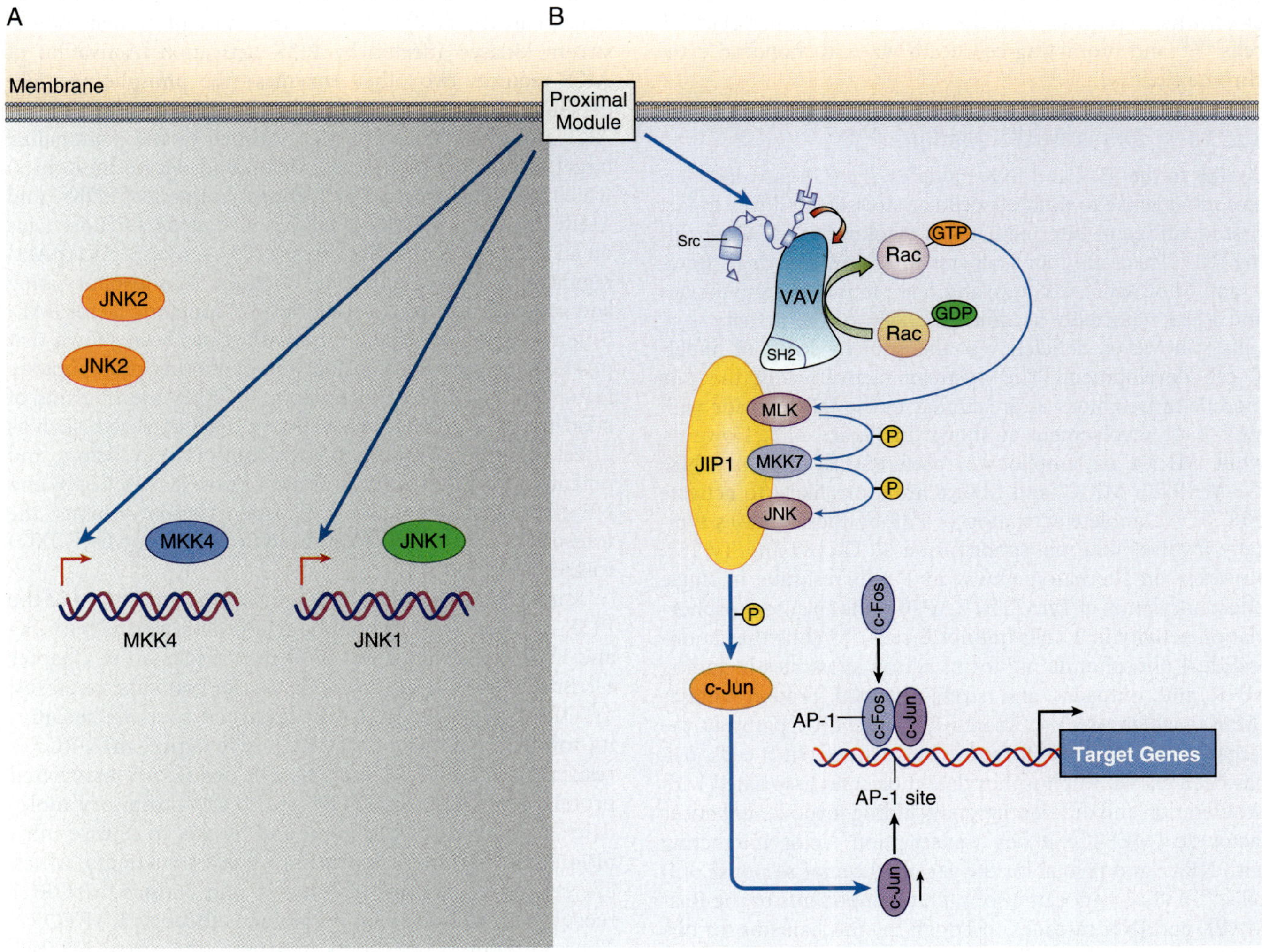

FIGURE 24.8. The c-Jun NHC-terminal kinase (JNK) MAPK module. In addition to the conventional Ras-MAPK pathway, there is another lymphocyte distal MAPK module termed the stress-activated MAPK pathway, involving the c-Jun NHC-terminal kinase (JNK). JNK1 and JNK2 are commonly found in lymphocytes but **(A)** while JNK2 protein is found constitutively in T and B cells, JNK1 (and some additional components of this pathway such as MKK4) have to be transcriptionally upregulated after activation. **B:** The JNK module is activated through phosphorylation of the GTP exchange factor (GEF) protein VAV by Syk and Src family kinases from the proximal module. VAV, as well as the subsequent downstream MAPK molecules, can be tethered to a pathway-specific scaffolding molecule Jnk-interacting protein-1 (JIP1), although this assembly is not essential. Although VAV has other functions, with regards to the JNK pathway, phosphorylated VAV can help transform the GTPase Rac1 from GDP-Rac1 to GTP-Rac1. GTP Rac1 can then initiate the JNK-MAPK cascade, beginning first with the phosphorylation of the mixed-lineage protein kinase (MLK), which then activates MKK4 or MKK7 (the latter is also driven by non-AgR signals), and finally MKK4 or MKK7 can phosphorylate JNK. Once active, JNK can bind to and phosphorylate the N-terminal of c-Jun, allowing it to translocate to the nucleus and form AP-1 with c-Fos.

activity of ATF2[141]). This pleiotropy is also evident in the phenotypes of JNK-deficient animals. JNK1-deficient mice generate T and B cells normally, but T cells are hyperproliferative and prone to greater TH2 differentiation.[142] JNK2 deficiencies do not show the hyperproliferative phenotype but have an intrinsic TH1 differentiation impairment that was rescued by exogenous IFNγ addition.[143] This is also cell type–specific; JNK1-deficient CD8 T cells are poorly proliferative while JNK2 knockout CD8s respond better to IL-2 and proliferate better to viral challenge.[144,145] Further upstream, although MKK4- and MKK7-deficient mice are embryonic lethals, complementation chimeras show that while neither is critical for T or B cell development, MKK4−/− T cells show poor IL-2 production.[146] Intriguingly, these studies also suggest a prosurvival function for JNK signals in DP thymocytes, which contrasts with their functioning in peripheral T cells[147,148] and promoting cell death via mitochondrial cytochrome c release.[149]

The Third Stress-MAPK Module: P38

Similar to the ERK and JNK modules, the P38 cascade operates in response to multiple cellular stressors and was in fact first identified as intermediates of cytokine and PRR signaling.[150-152] There are four isoforms of P38 in mammals (α, β, γ, and δ) of which P38α, β, and δ are found in lymphocytes and P38α reportedly accounts for 75% of the activity in T cells[153]; however, deficiency of this protein does not impair T cell development. The upstream activators of the P38 module in lymphocytes are similar to the JNK cascade with VAV-Rac1 involvement as shown in Figure 24.8. However, while MKK4 in lymphocytes preferentially acts on JNK, the MAP2Ks MKK7 and MKK6 are more likely to activate P38.[154,155] Complete activation of P38 by these kinases typically involves dual phosphorylation at Thr180 and Tyr182. However, an alternate pathway in T cells resulting in single phosphorylation at Tyr323 by ZAP70 gains an autophosphorylation activity in T cells but not B cells.[156] While the canonical dual phosphorylating forms act on substrates including MNK and mitogen- and stress-activated protein kinases (MSKs) which are also shared with the ERK pathway the targets of the alternate form are less defined. In B cells, p38 has been shown to phosphorylate molecules associated with proliferation and differentiation including myocyte-enhancer factor 2c (MEF2C), a key transcription factor interacting with NFAT and pivotal for the germinal center response of B cells.[157] P38 activity can also inhibit components of the Ras-MAPK and JNK cascades, although the mechanisms for this are still being explored.[158,159]

The PI3K-AKT Module

Downstream of both the BCR and TCR, PI3K activation and conversion of PIP2 into PIP3 leads to subsequent activation and phosphorylation of the kinase Akt (Fig. 24.7)—a module that has multiple functional roles including enhancing cell growth, metabolism, and survival.[160,161] Class I PI3Ks are heterodimers of two subunits named by their 85 kDa (p85) and 110 kDa (p110) molecular weights[162] with multiple isoforms of each expressed in different cell types. Of the multiple isoforms, lymphocytes predominantly express p110δ and p110γ together with p85; however, others are also expressed. The enzymatic activity is contained in the P110 catalytic subunit while P85 plays a regulatory role. A primary immunodeficiency syndrome in humans called activated PI3K-δ syndrome (APDS) is characterized by a gain-of-function mutation which increases the catalytic activity of the p110 subunit.[163,164] P85 has multiple SH2 domains, as well as an SH3 domain, all of which are important for PI3K docking to other adapter and kinase molecules. The SH2 domains of p85 recognize a p-YXXM motif[165] that is found in the cytoplasmic tail of membrane receptors such as CD28 and CD19.

The serine/threonine kinase AKT (also known as protein kinase B, PKB) is one of the most prominent downstream kinases affected by PI3K activation. Activation of AKT requires two other enzymes—the phosphoinositide-dependent kinase-1 (PDK-1) which phosphorylates T308 on AKT and the TORC2 complex (formed by the mammalian target of rapamycin [mTOR], rictor, and related molecules) which phosphorylates S473.[166] The recruitment of PDK-1 and TORC2 to the proximity of AKT is mediated by PH domains on all three molecules binding to PIP3. Activated Akt (pAkt) regulates multiple lymphocyte functions. By phosphorylating and subsequently deactivating the pro-apoptotic factor BAD, it increases cell survival.[167] Similarly, it has been shown that pAkt can prevent the translocation of another pro-apoptotic factor Bax into the mitochondria.[168] Other mechanisms of pAkt-induced cellular survival have been explored, such as direct inhibition of pro-apoptotic transcription factors, and potentiation of pro-survival transcription factors CREB and NF-κB.[169-171] AKT-mediated phosphorylation inhibits the transcriptional activity of Forkhead Box Subgroup O (FOXO) transcription factors.

Studies on the mTOR highlight an important arm of the PI3K-AKT module in lymphocyte function (this pathway and its role in cellular metabolism is discussed in Chapter 6). Briefly, mTOR acts via one of the two subcomplexes, mTORC1 and mTORC2, distinguished by their sensitivity toward rapamycin (mTORC1 is sensitive, mTORC2 is resistant).[172] mTORC1 includes the regulatory-associated protein of mTOR (RAPTOR) and a self-inhibitory molecule PRAS40 which can be bound by Akt to release inhibition.[173,174] mTORC1 activation requires nutrients, which are sensed as amino acid levels and induce mTORC1 recruitment to lysosomal membranes through RAPTOR2. There, mTORC1 meets its activator, the small GTPase RHEB (localized on organellar membranes), which conveys a second set of signals from environmental cues including energy, oxygen levels, and growth factors (Fig. 24.7). RHEB is also typically kept inactive (GDP-bound configuration) by the GAP tuberous sclerosis complex (TSC).[175] Activated AKT relieves this inhibitory loop by phosphorylating TSC.[176] mTORC2 is composed of multiple proteins including the rapamycin-insensitive companion of mTOR (RICTOR), as well as Sin1 a positive regulator of Akt phosphorylation.[177,178]

mTORC1 has been classically associated with the regulation of cell growth particularly through interactions with two proteins, S6 kinase (S6K) and 4E-BP1. mTORC1 can phosphorylate S6K at threonine 389 in its hydrophobic motif (HM) region.[179,180] S6K is a member of the AGC family of kinases (protein kinase A, protein kinase G, protein kinase C) and is involved in the ribosomal translation of proteins through phosphorylation of the 40S ribosomal subunit S6. 4E-BP1 is normally a repressor that keeps the eukaryotic translation initiation factor 4E (eIF4E) in an inactive state. Phosphorylation of 4E-BP1 releases eIF4E allowing the latter to function. eIF4E is a key effector in translation and a rate limiting step in protein synthesis. It binds to the 7-methylguanosine cap at the 5′ end of eukaryotic mRNA as a part of the eIF4F translation initiation complex and facilitates mRNA recruitment to ribosomes. Of course this increased protein translation is critical for newly activated lymphocytes to ramp up the cellular machinery for full functional responses. The phosphorylation of 4E-BP1 is regulated by mTORC1,[181,182] but the Ras-MAPK pathways also activate eIF4E activity. The role of mTORC2 is less well-understood, due to its more variable insensitivity to rapamycin. However, it (by way of RICTOR activity) increases the phosphorylation of Akt at the serine residue of 473.[183]

The overall connectivity of PI3K to multiple signal amplifying modules with the help of AKT therefore allows it to play an important role for integrating multiple inputs (including those from cytokine and costimulatory molecules). In T cells, costimulatory receptors such as CD28 and ICOS can also activate the Akt pathway through their enhancement of PI3K activation—though the precise mechanisms, and the importance of PI3K binding to these coreceptors, are not completely clear.[184-186] In B cells, some studies have reported a molecule, B cell adapter for PI3K (BCAP), as a potential recruiter and activator of PI3K/Akt downstream of CD19 ligation.[187,188] Thus the PI3K-Akt module is a key mechanism used by costimulatory or accessory signaling pathways to amplify AgR activation. The increased cellular PIP3 due to PI3K activity is also transient and is downregulated by the lipid phosphatase and tensin homolog (PTEN) which converts PIP3 back to PIP2. Accordingly, PTEN deficiency is associated with hyperactive AgR signaling, especially in T cells.

The NF-κB Signaling Module

In 1986, Sen and Baltimore identified a nuclear factor binding the promoter of the κ light-chain gene in B cells, or NF-κB,[189] which is now known to be one of a critical family of transcription factors involved in multiple biological pathways (see Ref. 190 for a historical review). These act as dimers, which in lymphocytes is typically a combination of RelA with either p50 or c-Rel. In mammals, there are a total of five NF-κB family members, namely RelA (or p65), RelB, c-Rel, p50 (NF-κB1), and p52 (NF-κB2) which can bind DNA either homodimers or heterodimers. Of these, p65, c-Rel, and RelB can directly activate transcription since they have transactivation domains (TADs) but p52 and p50 have to form heterodimers to be transcriptionally active. Therefore, in some cases, p50 or p52 homodimers can inhibit transcription in competition with other factors binding to κB sites but not being able to initiate. These NF-κB complexes are now known to regulate a large number of genes, binding to promoter/enhancer elements recognizing a 5′-GGGRNWYYCC-3′ motif (purines [R], adenine or thymine [W], pyrimidine [Y], or any base [N]). Before activation, all lymphocytes express significant amounts of multiple NF-κB family proteins; these are however retained in the cytoplasm by binding of the inhibitory proteins of the inhibitor of κB (IκB) family. Association with these proteins, typically IκBα and IκBβ in T and B cells, masks the nuclear localization signals of the NF-κB transcription factors. Accordingly, activation of NF-κB–mediated gene transcription requires upstream AgR signaling to effectively remove the IκB protein by phosphorylating it and targeting it for proteasomal degradation (Fig. 24.8).

The typical pathway for releasing IκB from NF-κB involves the triggering of the novel PKC family Ser/Thr kinase by the activated AgR proximal module.[191] As discussed in the Lipids and Cations as Signal Amplifiers: the Ca^{2+} NFAT Pathway section, the activity of PLCγ at this stage results in the generation of DAG at the plasma membrane, which in turn activates PKC (PKCθ in T cells and PKCβ in B cell) as a result of their cysteine-rich regulatory C2 domain binding DAG. PKCθ can also be activated by phorbol esters (eg, phorbol 12-myristate 13-acetate [PMA]). Indeed, this allows the combination of PMA with an ionophore such as ionomycin (which elicits the calcium flux) to sufficiently mimic TCR signaling to drive measurable T cell activation. PKC then recruits CARD11 (CARD11, also known as CARD-containing MAGUK protein 1 [CARMA1][192]) and phosphorylates it (together with other kinases such as casein kinase 1α), which initiates the formation of a complex with BCL10 and the paracaspase MALT1 on the inner leaflet of the membrane. This CARD11:Bcl10:MALT1 (CBM) complex is the central backbone for NF-κB activation in T and B cells. MALT1 recruits TRAF6 which catalyzes the formation of the cytosolic holoenzyme complex I κ kinase (IKK) that would eventually phosphorylate IκB, mark it for degradation, and thus free NF-κB to move to the nucleus. IKK also incorporates a noncatalytic regulatory partner, the NF-κB essential modifier (NEMO, also known as IKKγ) together with its two enzymatically active subunits IKKα and IKKβ. This adds another layer of regulation to the module, requiring polyubiquitination of NEMO by TRAF6 for full activation. Once phosphorylated by the activated IKKβ in the IKK holoenzyme, IκBα (phosphorylated on Ser-32 and Ser-36) or IκBβ (Ser-19 and Ser-23) are polyubiquitinated by the SCFβ–TrCP E3 ubiquitin ligases which sends them to the proteasome for degradation. This overall module (shown in Fig. 24.9) is known as the canonical pathway of NF-κB activation

An alternate pathway of NF-κB activation also operates via the NF-κB–inducing kinase (NIK) which is also stabilized by the removal of TRAF and then can phosphorylate IKKα. The dimeric IKKα in this pathway can phosphorylate NF-κB (in this case, it is usually composed of the p50 and p52 subunits) independent of IKKβ, IKKγ, or the classical IKK complex.

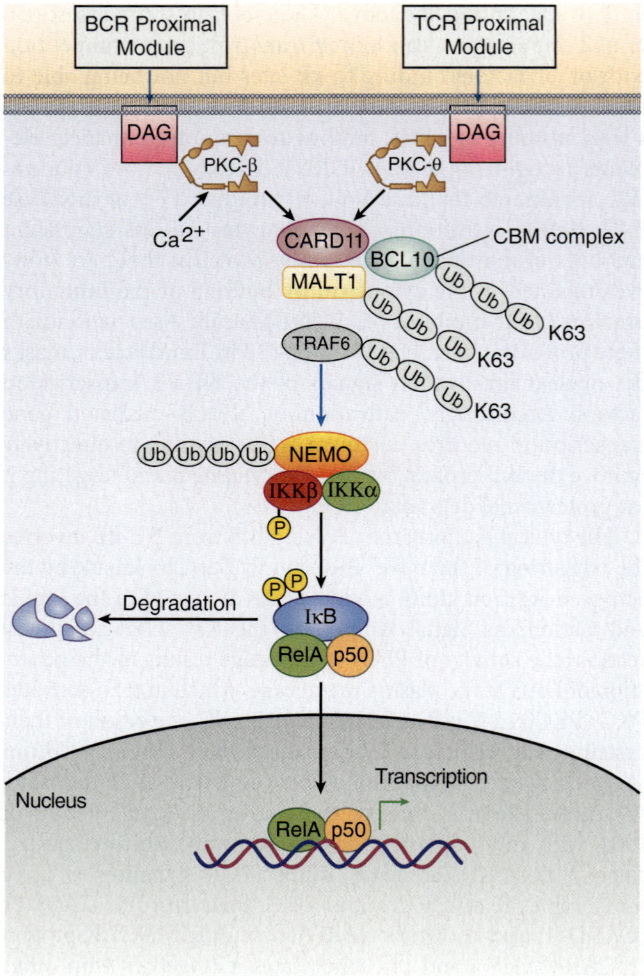

FIGURE 24.9. The canonical NF-κB signaling module in lymphocytes. The NF-κB pathway is another ubiquitous cell signaling module that is vital for full lymphocyte activation. There are five NF-κB protein family members in mammals: RelA/p65, RelB, c-Rel, p50, and p52, which can form dimeric transcription factors. Lymphocytes use mostly RelA+c-Rel or RelA+p50 dimers. We are using RelA and p50 to illustrate the canonical pathway of NF-κB signal transduction, but this applies to RelA+cRel as well. Prior to activation, both B and T cells sequester the transcriptional dimer of NF-κB (RelA+p50) in an inactive state by binding to the inhibitor of κB (Iκb). IκB masks the nuclear localization signals (NLS) on the NF-κB complexes. When AgR activation occurs, PKC (β in T cells, θ in B cells) is activated by DAG. This activated PKC phosphorylates and activates the molecule caspase recruitment domain family member 11 (CARD11, also known as CARD-containing MAGUK protein 1 [Carma 1]), which scaffolds a complex with BCL10 (B cell lymphoma/leukemia 10) and MALT1 (mucosa-associated lymphoid tissue lymphoma translocation 1). This complex is often referred to as the CBM (CARD11, BCL10, MALT1) complex. MALT1 is vital for the recruitment of TNF receptor–associated factor 6 (TRAF6), which recruits cytosolic I-kappa kinase (IKK). IKKα and IKKβ in turn recruit the molecule NF-κB essential modulator (NEMO) to form a complex. TRAF6 then is able to polyubiquitinate NEMO, eliciting its proteasomal degradation, and releases the active IKK. The active IKK can then phosphorylate IkB, triggering polyubiquitination of the IκBs, and releasing NFκB for nuclear translocation. BCR, B cell receptor; TCR, T cell receptor.

AgR signaling typically is more reliant on the classical pathway, with a variety of accessory receptors (eg, CD40, TNFR, etc.) capable of triggering the alternate pathway; but this is not absolute (see Sun, 2012[193] for a review of the alternate pathway). Indeed, overexpression of the CBM complex induces spontaneous NF-κB activation without T cell (or B cell) activation while mutations in members of either the CBM or IKK complexes lead to immunodeficiencies or hyperactive immunological disorders.[194,195]

The Cytoskeletal Remodeling Module and the Immunological Synapse

The cytoskeleton provides structural support and enables cytoplasmic transport of cargo, endocytosis, secretion, etc., for multiple cell types. The earliest indication that cytoskeletal remodeling was critical for coordinated signaling downstream of the AgR emerged from studies showing that pharmacologic disruption of the cytoskeleton impaired B cell activation.[196] The cellular consequences were subsequently elaborated in single-cell studies with T cells. We now know that immediately following TCR engagement of pMHC on APCs, rapid cytoskeletal rearrangement occurs, using at least two molecular complexes, F-actin and microtubules. The initial definition of a distinct contact zone or synapse (discussed in Fig. 24.11) between the APC and the T cell is driven by polymerization of actin monomers (formation of F-actin). At the same time, the secretory and protein transport apparatus of the T cell reorients to become polarized with an APC-proximal pole and a distal pole complex (on the opposite side) aided by the microtubule-organizing center. The ordered movement of cell surface receptors and associated signaling complexes (including those in membrane microdomains or rafts) eventually leads to the formation of a stable "immunological synapse (IS)"[197] between the T cell and APC after an initial burst of "presynapse" TCR signaling.[198] As it forms, the IS segregates critical molecules into specific regions known as the central, peripheral, and distal supramolecular activation clusters (cSMAC, pSMAC, and dSMAC)[199] which are illustrated in Figure 24.11. TCR, CD28, and key signaling mediators (active Lck, ZAP70, PKC, etc.) rapidly segregate to the central cSMAC, adhesion molecules (LFA1, talin, etc.) to the pSMAC, while molecules with large extracellular domains (CD44, CD45, etc.) move away from the center and into the d-SMAC. These segregations are not necessarily absolute, since molecules like CD2 can be found in multiple zones. Functionally, the organization of the IS, while not critical to initiate signal transduction, may help to maintain as well as terminate TCR signaling. The ordered trafficking of the TCR and signaling complexes into the cSMAC allows them to maximally engage the few available pMHC complexes on the APC (protected from phosphatases such as the cytoplasmic tail of CD45). The eventual endocytosis and downregulation of TCR at this site can also help terminate signaling. Similarly, B lymphocytes also form an IS after BCR engagement, with macrophages, DCs, or FDCs that display antigen,[200,201] and even use that process to acquire intact antigen for continued BCR signaling.[202] Intriguingly, FRET studies using labeled signaling molecules suggest that structures similar to an IS form at the antigen-contact site even as the BCR engages isolated antigen or antigen–opsonin complexes.[104]

Cytoskeletal remodeling downstream of AgR signaling (Fig. 24.10) requires the RHO family guanine nucleotide exchange factor (RHO-GEF) VAV, typically VAV1 in lymphocytes (but Vav2 is also involved in B cells). As a common theme in most of the signaling modules discussed earlier, in resting cells, VAV is held in an inactive configuration that is relieved by phosphorylation of Tyr-174[203] directly by the SRC or SYK kinases.[204,205] Vav is also recruited in a kinase-independent fashion to the membrane proximal signalosomes by binding to LAT-SLP76 (in T cells) or SLP65 (in B cells) where it is phosphorylated by the Tec kinases including ITK. In addition, an SH2 interaction is also thought to localize Vav to the cytoplasmic tails of CD19 which is critical for the signaling amplification afforded by this costimulatory molecule.[206] Once triggered by one or more of these recruitment mechanisms, the GEF activity of VAV (resulting in the inactive GDP-bound form turning into the GTP-bound form) turns on several GTPases in lymphocytes including CDC42, RAC1, RAC2, RHOG, etc. Each of these plays important and

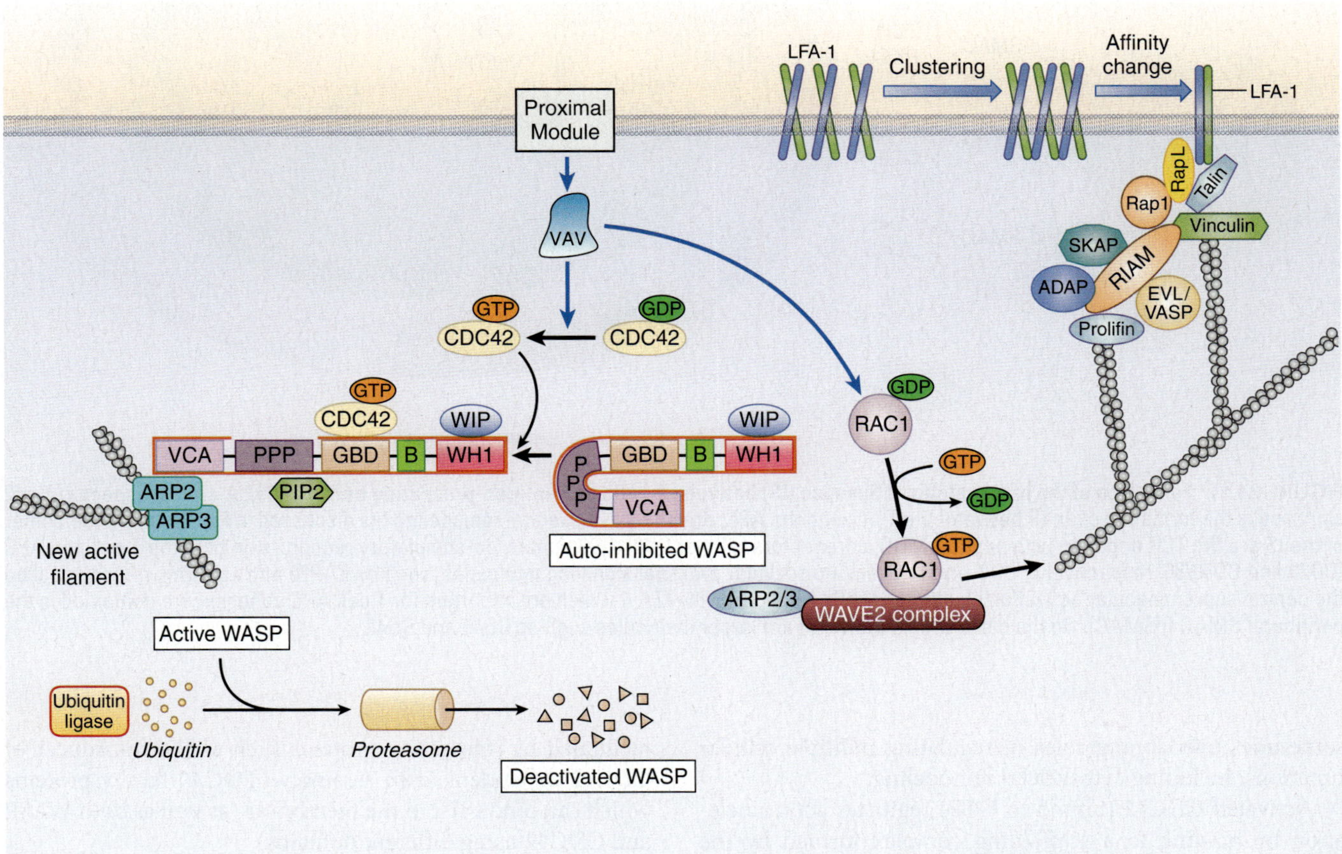

FIGURE 24.10. The F-actin polymerization module. The significant rearrangements of membrane microdomains, development of protrusions and microvilli, and movement of the T or B cells during activation involve cytoskeletal rearrangement, largely mediated by actin polymerization and microtubule remodeling. The RHO guanine nucleotide exchange fator (GEF) VAV (also discussed in Fig. 24.8) is a critical upstream connection between the proximal AgR module and this process. VAV is inactive until phosphorylation on Tyr-174 by a Src or SYK kinase. Active VAV can then use its GEF function to activate several GTPases, including CDC42 and Rac1. Once active, CDC42 binds to the Wiskott–Aldrich syndrome protein (WASP) and the WASP-interacting protein (WIP). This WASP/WIP complex can then reveal its C-terminal verpolin homology, cofilin homology, and acidic domain (VCA), in which the CA component can bind to the actin-related protein 2 and 3 complex (ARP2/3). ARP2/3 provides scaffolding for filamentous actin to form. WASP can be negatively regulated by ubiquitination and subsequent proteasomal degradation. The second network of forming actin is mediated by Rac1. Active Rac1 can bind to the WASP family verpolin homologous protein (WAVE) complex, which itself can bind to the ARP2/3 complex, steps important for inside-out signaling to increase T cell:APC adhesion. This adhesion is mediated by integrins such as LFA-1, which binds to ICAMs on APCs, and transitions from its inactive resting state to an active conformation, with the aid of the Rac1-WAVE complex mediated actin remodeling. Following TCR signaling, SLP76 binds to two molecules, ADAP and SKAP55, which form a core complex that transitions the activity of LFA-1. This complex also associates with several other proteins, namely Rap1-GTP interacting adaptor molecule (RIAM), and regulator of cell adhesion and polarization enriched in lymphoid tissue (RAPL). Another protein, the GTPase Rap1, also associates with RIAM and RAPL. ADAP/SKAP/RAPL binds to the α-chain of LFA-1, whereas ADAP/SKAP/RIAM binds to two FERM domain–containing proteins, talin and kindlin-3, which themselves bind to the β-chain of LFA-1. Talin and kindlin-3 also can bind directly to F-actin.

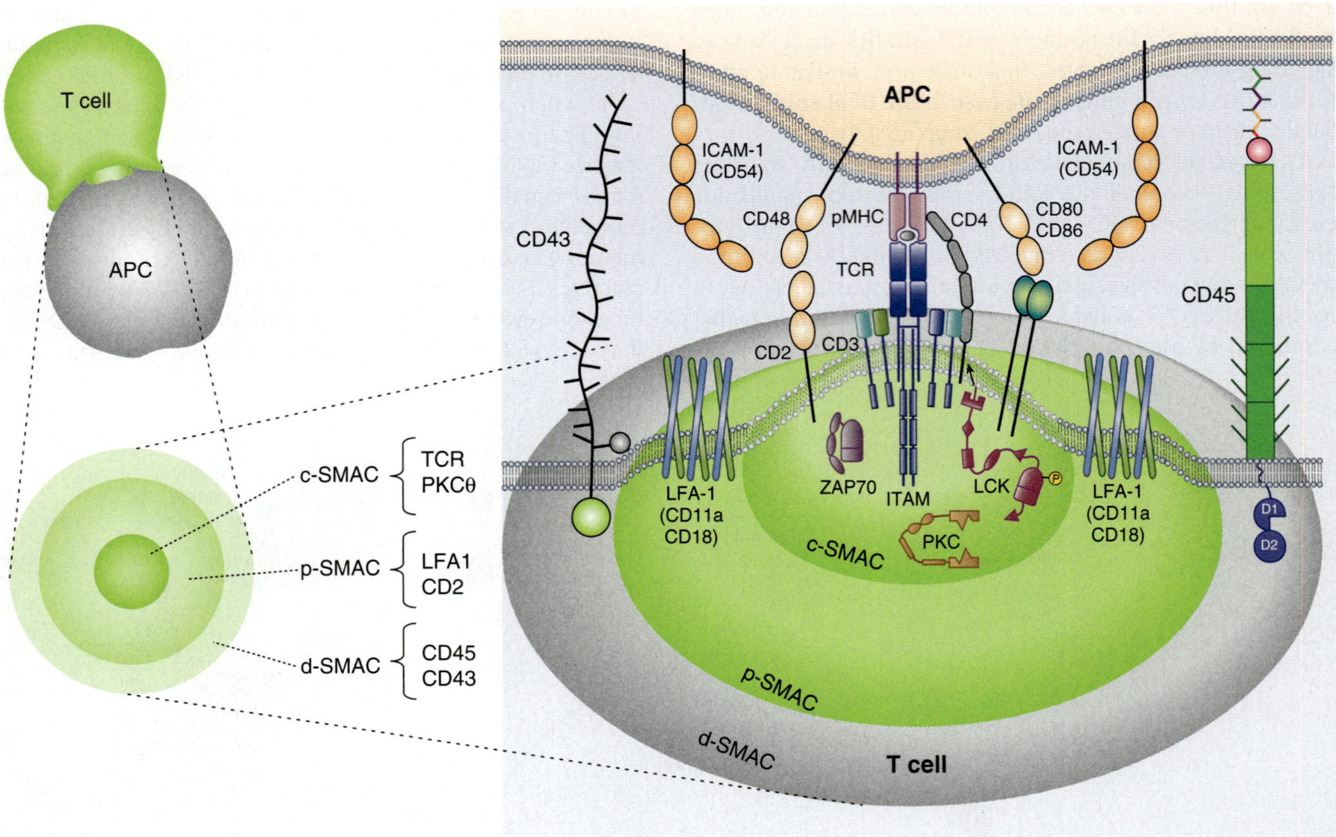

Immunological synapse

FIGURE 24.11. Formation of the Immunological Synapse (IS) between T cells and antigen-presenting cells (APCs). A crucial aspect to T cell signaling is the formation of an IS between the T cell and the APC, driven by cytoskeletal remodeling (as discussed in Fig. 24.10). At the center of the IS are the TCR engaged with peptide-MHC, coreceptors (CD4 or CD8), CD3 chains, co-stimulatory receptors on both the T cell and APC (CD28 and CD80/86, respectively), CD2, and relevant intracellular proximal signaling molecules, such as ZAP70 and Lck. This region is called the central supramolecular activation cluster (cSMAC). Molecules like LFA-1 which are important for T cell:APC adhesion are contained in the peripheral SMAC (pSMAC). On the distal SMAC (dSMAC) are larger molecules such as CD45 and CD43.

sometimes overlapping roles in regulating multiple cellular functions, including cytoskeletal remodeling.

Activated CDC42 (bound to GTP) regulates actin nucleation by binding to a scaffolding complex formed by the Wiskott–Aldrich syndrome protein (WASP) and its interacting partner WASP-interacting protein (WIP). WASP, encoded by an X-linked gene expressed largely in hematopoietic cells, is a multi-domain scaffolding protein with a critical GTPase-binding domain (GBD) recognized by CDC42. Activated WASP/WIP undergoes a conformational change that exposes the C-terminal verprolin homology, cofilin homology, and acidic (VCA) domain. In the VCA region, the CA motif binds the actin-related protein-2/3 (ARP2/3) complex while the V homology allows for binding G-actin monomers. The ARP2/3 complex formed of seven proteins has no catalytic activity of its own, but forms scaffolds allowing for the assembly of filamentous F-actin, extending it as well as branching it with the help of additional catalytically active nucleation promoting factors. In addition, the anchoring or recruitment of the branching actin filament to the plasma membrane is facilitated by interacting proteins such as the transducer of CDC42-dependent actin assembly (TOCA) family proteins which can bind PIP2 in the membrane (as well as both WASP and CDC42 using different domains).

Localization of the actin network to the membrane can also be mediated by the recruitment of the PIP3 binding WASP family verprolin-homologous protein (WAVE) family proteins by the second Rho family, GTPase Rac1 (also activated by VAV1). In lymphocytes, the WAVE2 complex also associates with ARP2/3 and is critical for the process of inside-out signaling which increased the adhesion of T cells to APC. T cell adhesion is mediated mainly by activation of integrins such as leukocyte function–associated antigen (LFA-1—a heterodimer of αL and β2 integrins), which can increase in affinity and avidity for their ligands (intercellular adhesion molecule [ICAM]-1 and 2) on APCs following TCR signaling. While classical ("outside-in") signaling involves a membrane receptor sensing an extracellular ligand and triggering changes in the cytosolic side of the receptor (or associated complexes), inside-out signaling reverses this by allowing

TCR signals ("inside" the cell) to alter the affinity of integrins outside the cell. LFA-1 can be found in three conformational states—resting, transitional, and active conformation, each requiring chemokine receptor (eg, CCR7) or TCR signals for the transitions. These changes involve extracellular structure conformational alterations of the head of αL and β2 driven by intracellular cytoskeleton extension. Following TCR-proximal module activation, SLP-76 binds a complex formed by ADAP (adhesion and degranulation promoting adapter protein) and SKAP55 (Src kinase–associated phosphoprotein of 55 kDa) binds a core module which scaffolds the assembly of two independent pools of LFA-1–activating signaling complexes.[207,208] These require either the Rap1 effector proteins RIAM (Rap1-GTP interacting adapter molecule) or RAPL (regulator of cell adhesion and polarization enriched in lymphoid tissue) which are constitutively associated with ADAP/SKAP1 module by interacting with SKAP1 (49-51). Both RAPL and RIAM also associate with the activated GTPase Rap1 (Ras proximity-1). These two activating complexes interact with different parts of the cytosolic regions of LFA-1. The ADAP/SKAP1/RAPL module (together with additional players such as the kinase Mst1) links up using RAPL binding to the α-chain (CD11a) of LFA-1. At the same time, the other LFA-1–activating complex (ADAP/SKAP1/RIAM) binds to the FERM (4.1 protein, ezrin, radixin, moesin) domain containing proteins talin and kindlin-3, which binds to the β1 (CD18) cytosolic region. Both talin and kindlin-3 also bind F-actin directly, completing the wiring between the integrins and the cytoskeleton.

Negative regulation of actin filament formation involves ubiquitylation and degradation of WASP complexes. But in resting lymphocytes, at least one regulator (filamin A) which binds with the cytoplasmic tail of the LFA-1 β-chain and keeps LFA-1 in the inactive conformation has been reported.[209] The release of filamin A from LFA-1, by poorly understood TCR-proximal signals, allows for the subsequent association with talin and kindlin-3 discussed earlier.

The regulation of F-actin also means that VAV1 can be considered a critical controller of the formation of the IS as well. In addition, the activation of RAC1 downstream of VAV also activated the ERM (ezrin, radixin, moesin) complex proteins. These proteins are typically associated with cellular submembrane structures such as filopodia, microvilli, furrow in mitotic cells, different poles, ruffling membranes, and adhesion sites, where the plasma membrane typically links up with F-actin. Typically, they have an N-terminal FERM domain and a C-terminal actin-binding domain, separated by a linker. Their structure undergoes changes from a closed to open conformation following AgR-mediated phosphorylation. In the open conformation, in the vicinity of the AgR, ERM proteins form a long connector that allows the plasma membrane receptors to detach from the cortical actin cytoskeleton. This favors a reduction in cellular rigidity around the junction with APC and promotes optimal rearrangement of the intercellular receptor–ligand interactions. The ERM proteins also help the formation of the distal pole complex after T cell activation. Thus, an extensive remodeling of the cellular cytoskeleton with specialized changes such as the synapse formation around the AgR itself sets up the lymphocyte for optimal activation.

MODIFIERS OF AgR SIGNALING–COSTIMULATORY RECEPTORS

The complete activation of lymphocytes requires more than just AgR signals. Although these accessory and costimulatory signals are discussed in other chapters (Chapter 28 for T cells and Chapter 26 for B cells), we briefly consider here the signaling pathways emanating from such receptors and how they synergize with TCR and BCR signals. In the case of T cells, the concept of costimulation came from studies showing that APCs presenting antigen, but unable to upregulate additional ligands typically induced tolerance.[210] Subsequently, it was clear that these additional (costimulatory) molecules are perceived by a second class of receptors on T cells typified by CD28. The ligands for CD28, namely CD80 (B7-1) and CD86 (B7-2), are found at very low levels on resting DCs but rapidly upregulated when DCs are activated by engaging PRRs (sensing DAMPs or PAMPs). This is the basis of the two-signal model for T cell activation (see Chapter 1), wherein the TCR signal is considered Signal 1 and the costimulatory signal from activated DC is Signal 2. The assumption is that when antigen is processed and presented to T cells by resting (not activated) DC, the T cell can only receive Signal 1—while fully activated DCs can concordantly provide Signal 2 as well. Accordingly, Signal 1 without Signal 2 is considered tolerogenic[210] but 1 + 2 would lead to activation. In cell culture experiments, mimicking both these signals can be accomplished by crosslinking either the TCR or CD3, together with antibodies to CD28.[211] The molecular basis by which CD28 signals and how that intersects with TCR signaling is significant.

A similar concept also exists for B cells, although they are very different in cellular and molecular terms. In B cells, the engagement of BCR with crosslinking antigen can give a Signal 1 which would lead to activation of the B cells. In the case of T-dependent antigens, subsequently, a Signal 2 (also known as help) must come from the B cell's interaction with fully activated CD4 T cells (likely a TFH) that are specific to the same antigen. Therefore, this Signal 2 or T cell help is considered in chapters dealing with B cell differentiation in the germinal centers. The molecular parallel for costimulation in B cells might be the influence of additional (non-BCR) receptors, including the complement receptors and TLRs, on augmenting initial B cell activation—although these are typically considered modifiers of the strength of BCR signals (ie, Signal 1). In the next segment of this chapter, we will explore how costimulatory pathways relate to the signaling modules we have previously examined.

Costimulatory Pathways in T Cells

After classic cellular experiments indicated that activated APC provide essential costimulatory signals for T cell activation,[210] the surface molecule previously identified as CD28[212] was characterized as the primary receptor for costimulation.

CD28 has a 202-residue extracellular region, a 27 amino acid hydrophobic transmembrane region, and a short cytoplasmic tail (41 amino acids).[212] The cytoplasmic portion of CD28 contains several tyrosines including an YxxM motif and two PxxP motifs. The YxxM motif, conserved in CD28 as well as the closely related (yet inhibitory) surface receptor CTLA-4 and another costimulatory molecule ICOS, is thought to be phosphorylated by the Src kinase Lck from the AgR-proximal module. Although *in vitro* analysis and an overexpression system indicated a critical function for these motifs in costimulation, *in vivo* analysis with specific knock-in mutations revealed that mice with mutated YMNM motifs had no overt phenotype. Other TCR-associated distal signaling modules can also interact with CD28. The V3 domain of PKCθ can associate with CD28 in an LCK-dependent fashion, amplifying NF-κB signaling. The phosphorylated YxxM is also a consensus motif for the binding of the p85 subunit of PI3K—altogether directly enhancing its activity and intersecting with the TCR signaling pathway. These principles, whereby CD28 is not linked to a qualitatively distinct signaling cascade from that triggered by the TCR, suggest that a major role of CD28 signaling is to bolster or amplify TCR signals, as we will discuss further. Similarly, the PxxP motif, which ICOS and CTLA-4 do not have, also binds Grb2/Gads. Grb2 recruits SOS and Vav, which in turn phosphorylates and activates Rac1 and Cdc42—as discussed earlier, leading to enhanced cytoskeletal remodeling. This also potentiated the MEKK1 cascade to activate the Jun kinase JNK. In this context, while the MAPK family members ERK1/2 are predominantly driven by TCR signals, CD28 engagement can augment some other MAPK modules including JNK. This altered balance between the activity of ERK and JNK as a consequence of CD28-mediated costimulation may regulate cell survival, which is a major functional consequence of CD28 costimulation of T cells. Mice with the mutant distal motif of the CD28 cytoplasmic tail had evidence of impaired CD28-mediated costimulation, reflected by reduced proliferation and IL-2 production of these T cells. The enhancement of IL-2 production by CD28 engagement is regulated at both transcriptional and posttranscriptional levels. A specific region responsible for CD28-mediated transcriptional activation (termed a CD28-response element, CD28RE) was found in the IL-2 promoter. The binding of transcription factors to CD28RE is dependent on CD28 signal. In addition to enhancement of the transcription of the IL-2 gene, CD28-mediated costimulation augments IL-2 production by increasing the stability of IL-2 mRNA. Many cytokine transcripts including IL-2 contain adenylate uridylate (AU)–rich elements within the 3′ untranslated region that stabilizes the message. Upon TCR stimulation, an AU-binding protein, TTP, binds to the AU-rich elements within the 3′ untranslated region and induces degradation of the mRNA. By contrast, CD28 costimulation induces other proteins, such as NF90, a transcription factor associated with NFAT, which may compete with TTP for binding at the AU-rich elements and enhance IL-2 mRNA stability.

A continuing debate is whether CD28 has unique downstream signaling that is distinct from that of the TCR, or it merely acts as an amplifier of signals along TCR-driven signaling pathways.[213] A large body of evidence supports the latter, but it is difficult to dissociate such quantitative enhancement effects from discrete or specific molecular signals. Typically signaling via CD28 alone does not have cellular consequences without concomitant TCR signals. An exception is the case of some superagonistic antibodies which can trigger polyclonal T cell activation in mice and humans. This method was particularly effective in stimulating regulatory T cell expansion,[214] but human trials identified a rather widespread effect which remains poorly understood. Classic experiments showing that that stimulation of purified T cells with anti-TCR or anti-CD3 leads to anergy—which is prevented by including antibodies to CD28 during stimulation—also suggested that CD28 may have a qualitatively distinct role.[215,216] Mechanistically however, these effects were attributed to an enhancement in the production of IL-2 when CD28 was co-engaged.[217,218] Early *in vivo* models in which CD28 was knocked out in mice also showed decreased (but not absent) IgG1 and IgG2b, as well as lowered T cell proliferation in response to TCR-mediated stimulation.[219] CD28 knockout mice when crossed to TCR transgenic strains show reduced ability to sustain T cell proliferation which could be rescued by PMA stimulation. As mentioned earlier, the cytoplasmic tail of CD28 contains an YXXM motif, which has been shown to directly recruit and activate the p85 subunit of PI3K.[220] Interestingly, in a study that used knock-in mutants of the proximal YXXM motif, as well as a distal proline-containing sequence PYAP, the authors found that mutating the distal site resulted in a decrease in IL-2 secretion, but a defect in PI3K binding and downstream PKB/Akt phosphorylation with a mutated proximal site.[221] Furthermore, it has been shown that the increase in IL-2 via CD28 is cyclosporin-A resistant, in contrast to the IL-2 induction as a result of signaling through the TCR, indicating that distinct pathways are triggered to converge on IL-2 transcription.[222] A general consensus is still that CD28 signaling can boost T cell proliferation, potentiate phosphorylation of Vav (and accordingly increase cytoskeletal rearrangement), increase T cell survival through BCL-X_L maintenance, and enhance metabolic fitness[223-226]—in large part through its ability to augment PI3K and mTOR signaling pathways. Yet, some of these effects can be replaced by stronger or prolonged antigenic stimulation.[227,228] Prolonging infections with high doses of virus compensate for the absence of CD28.[229] Consistent with this, increasing the duration of antigen presentation can also partially compensate for the strict requirement for adjuvants to drive potent T cell responses[230] and also allow extended stimulation to drive tolerance even in the presence of costimulation. Finally, the role of CD28 in human T cells is not well understood. A significant fraction of T cells in humans can be found to lose (or not have) CD28 expression. Interestingly congenital deficiency of CD28[231] does not lead to severe T cell defects, except for a quantitative decrease in immunity to some pathogens.

CD28 is closely linked to two other genes, inducible costimulatory (ICOS) and CTLA4, on the human chromosome 2q33 which are positive and negative regulators of costimulation, respectively. In addition, other TNF receptor family molecules including CD27, OX40 (CD134), 4-1BB (CD137), CD30, herpes virus entry mediator A (HVEM), and glucocorticoid-induced TNF receptor (GITR) have all been ascribed some degree of costimulatory function. Nevertheless, CD28 is unique among these since it is expressed by all naïve T cells while most other receptors are upregulated either after T cell activation or only in some specific subsets of differentiated T cells. Of these the TNF family receptors including OX40, 4-1BB, and CD40 have been extensively characterized. OX40, for instance, is expressed 2–3 days after T cell activation while 4-1BB is maximally present in subsets such as follicular helper T cells (TFH). CD40 is involved in specific interactions, such as the delivery of help to CD8 T cells by activated CD4s that upregulate CD40 L. Signaling via OX40, 4-1BB, and ICOS augments T cell activation and these are generally thought to operate via signaling pathways similar to CD28, augmenting the PI3K, NF-κB, and MAPK modules. ICOS and 4-1BB are also particularly relevant for costimulating the differentiation of TFH. CD80 and CD86 are primarily induced on professional APCs such as DCs and macrophages, whereas costimulation through ICOS, OX40, and 4-1BB can be induced by ICOS-L, OX40L, and 4-1BB-L expressed on nonhematopoietic cells in the periphery as well. The latter is critical for regulation of autoimmunity and infectious diseases.

Coreceptors for B Cell Activation

As discussed earlier, costimulation (Signal 2) for B cell activation involves T cell help, delivered by a combination of cytokine and CD40L-CD40 signals (discussed in Chapter 27). Also, unlike T cells, Signal 2 for B cells can be delivered a short period of time (up to a day) after Signal 1. The role of additional coreceptors that operate in synergy with the BCR, however, is largely to amplify Signal 1 and effectively lower the threshold for B cell activation.[232] The receptors CD19, CD21 (complement receptor 2 or CR2), and CD81 or TAPA (target of the antiproliferative antibody 1, also known as tetraspanin-28) are the key players in this role and often operate as part of the TAPA complex associated with the BCR. CD19 is an Ig superfamily surface receptor with nine tyrosines in its cytosolic domain that can be phosphorylated to allow SH2 domain–containing proteins to be recruited after BCR engagement. This has led to the suggestion that CD19 plays the role of a critical scaffold on the B cell membrane, perhaps similar to the role played by LAT in T cells.[232] This tail has been shown to co-immunoprecipitate with PLCγ1, VAV, and PI3K.[233-235] The initial presynaptic signaling by the BCR requires CD19 when the antigens are cell or membrane-bound but not in cases where they arrive to the B cell as soluble complexes.[236] On the other hand CD81 is a four-pass transmembrane protein that does not seem signal directly but it helps to transport and stabilize CD19 levels on the surface.[237] CD21 (CR2) is a 145 kDa transmembrane glycoprotein that can bind complement C3–derived fragments C3d, C3dg, and iC3b. So antigens covalently bound by these complement proteins can be concentrated on the BCR surface and allowed to interact with the BCR with better avidity. In addition to this avidity enhancement, signaling by the 34-aa intracellular tail of CD21 may also participate in amplifying AgR signaling. It should be noted that the coreceptor function of CR2 has been demonstrated in mouse B cells, but remains controversial in human cells.

Coinhibitory Receptors and Modifiers of AgR Signaling

In addition to positive regulation of signaling and lymphocyte activation, specific receptors are also tasked with recruiting intracellular components that modify the proximal signaling modules of antigen receptors. The case of CD45, which possesses phosphatase activity, was discussed earlier. Another example is the scavenger receptor cysteine-rich (SRCR) superfamily protein CD5 which is expressed on all T cells and subsets of B cells.[238-241] CD5 is initially upregulated on developing T cells, soon after positive selection[239] and operates in a ligand-independent fashion to negatively regulate TCR signaling[242] to endogenous (self-peptide) TCR ligands. Recent proteomic studies have validated multiple molecules (including the phosphatase SHP-1, the ubiquitin ligase cbl-b, the kinase casein kinase II, etc.) which bind to the cytoplasmic tail of CD5 and then recruited to the IS.[243] The ability to recruit such "dampeners" of AgR signaling may be a shared mechanism of action for multiple inhibitory receptors on the cell surface. Yet, CD5 is not strictly an inhibitor; it was originally reported to have costimulatory functions[244] and may also promote lymphocyte survival by modifying the NF-κB levels in peripheral T cells.[245]

Another group of molecules which have recently gained significance as the targets of checkpoint-blockade treatments (reviewed in Ref. 246) are surface molecules such as CTLA4, PD1, LAG3, etc., collectively referred to as coinhibitory receptors. CTLA4 is closely related to CD28 but has higher affinity for its ligands (CD80/86) allowing it to compete better for potential sources of Signal 2 to T cells. Further, while the intracellular tail of CTLA4 can bind PI3K similar to CD28, it also recruits the phosphatases SHP-2 and PP2A—effectively canceling out the positive flux. Similar mechanisms are thought to operate downstream of PD1 and LAG3; but these are not completely understood.

CONTROL MECHANISMS IN THE AgR SIGNALING NETWORK

We have discussed earlier some of the mechanisms which regulate the flow of biochemical signaling downstream of the AgR. In this section we review the control of this flux more broadly and discuss how these pathways form an interconnected AgR signaling network (Fig. 24.1A). Our understanding of both, but especially the latter, is in its early stages but current advances in technology are driving the rapid accumulation of data relevant to these areas. An area of growth has been the mapping of protein–protein interactions that form the physical basis of the signaling networks, with the help of large-scale quantitative proteomics. Introducing perturbations (such as mTOR inhibitors) allowed for measuring the dynamic alterations in the signaling proteins associated with specific functions such as nutrient sensing in lymphocytes.[247] Creative introduction of specific tags to signaling molecules in knock-in mice allowed for isolating multiple signaling complexes as they form in T cells *in vivo*.[248] Such studies have identified new scaffolding functions for key receptors such as CD5, CD6, BTLA, etc., in T cells[249] or adapters like Grb2 in B cells.[250] Proteomic studies have also highlighted the large-scale remodeling of the signaling molecules and networks in lymphocytes accompanied with a dramatic increase in total protein content and associated nutrient metabolism.[251] These and other new data are still being incorporated into the modular landscape of AgR signaling network, we discussed so far. As discussed in The Cellular Circuitry of Receptor Signaling section, the connectivity between modules in the network involves the distinctive proximal modules feeding to multiple distal modules via scaffolds (Fig. 24.1A). The final outputs of these modules then converge at the functional end point, whether to drive gene expression or alter cell shape, migration, etc. The operation of this network is, however, more complex than that depicted in such a two-dimensional map. For instance, subcellular localization of many of the players in the network is also dynamically regulated. The relevance of membrane rafts and synapses in positioning signaling complexes during lymphocyte activation has been discussed

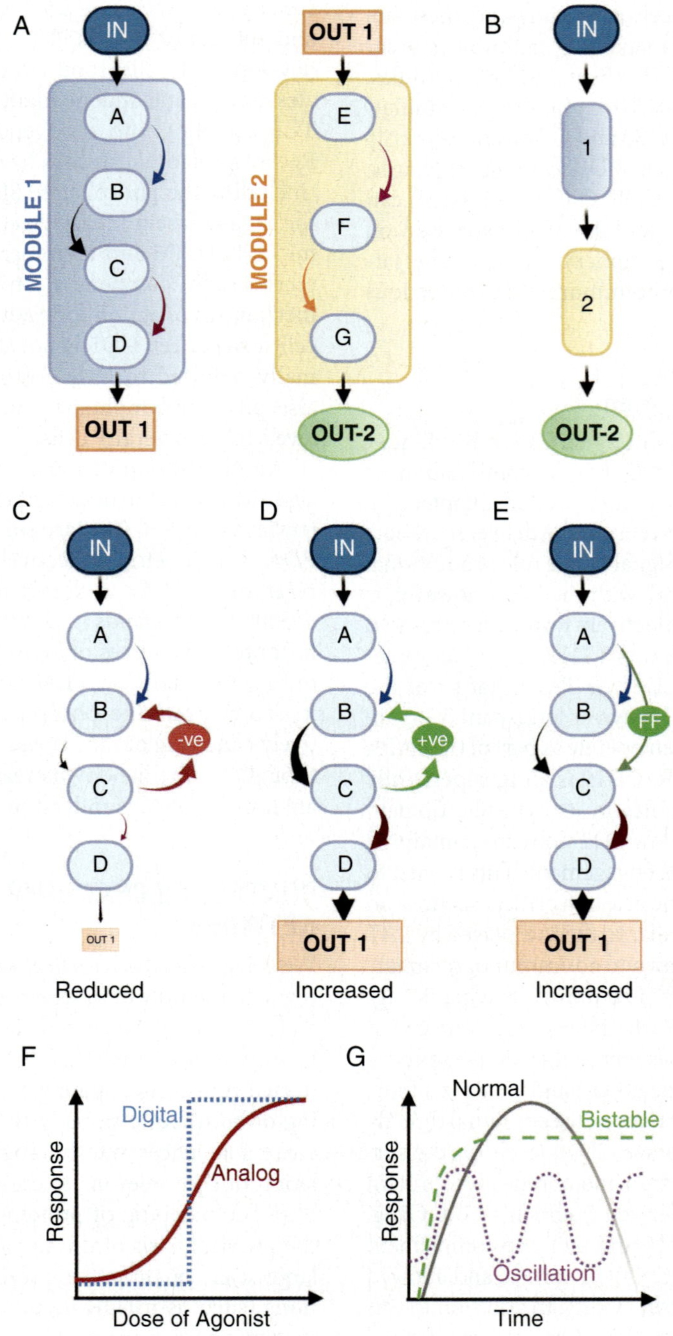

earlier. But the key molecules also localize to other distinct subcellular compartments prior to and during activation, allowing for even the same pathway to be geographically segregated, with potentially dissimilar outcomes. Ras isoforms, discussed in The Stress-Activated MAPK Module: SAPK/JNK section, are perhaps the best studied in this regard. In addition to palmitoylation-dependent recruitment to the plasma membrane, Ras can also localize to the ER or Golgi membranes with the help of specific scaffolding molecules such as KSR and β-arrestins, but these remain to be elucidated. Interestingly, activation of Ras in different subcellular compartments has been correlated with the ability of MAPK to signal different affinities of TCR ligands, contributing to positive versus negative selection in the thymus.[252] The partitioning of signaling complexes to discrete locations can also amplify cell fates quite significantly. The first cell division following activation of naïve T cells asymmetrically distributes AgR-associated signaling complexes to both daughter cells.[253] Intriguingly the daughter cell which inherits the distal pole complex (farther from the IS) has greater propensity to eventually become memory cells.

In addition to these factors, the overall flow of signals from the AgR is regulated by multiple mechanisms, which further amplify, dampen, or even generate new characteristics that are significant for eliciting appropriate functional outcomes (Fig. 24.12C-E)). Amplification of signaling is implicit in the nature of most cascades discussed earlier, eg, in the ability of one kinase molecule to phosphorylate multiple molecules of its substrates (ie, downstream kinases, scaffolds, or transcription factors). On the other hand, negative feedback along a pathway (also discussed earlier) can dampen subsequent signaling flux (Fig. 24.12C). In addition to intramodule feedback processes, the connectivity between different pathways allows for more complex regulatory mechanisms to emerge.

Negative and Positive Feedback

A linear signaling pathway is said to be controlled by feedback, when molecules that are activated later in that sequential cascade act back on upstream steps or substrates in order to regulate the overall signaling flux, either negatively or positively (see Gaud et al[254] for a review on feedback mechanisms regulating TCR signaling, and Reth and Brummer[255] for similar processes in B cells). At a cellular level, feedback can be regulated biochemically or transcriptionally. The latter is well described in the case of the NF-κB module[256] where genes encoding the inhibitory IκB proteins (see Fig. 24.9) are also under the control of NF-κB transcriptional activation. Increased NF-κB activity also generates more of the (upstream) negative regulatory proteins. Because the increased expression of IκB requires the positive action of NF-κB to first play out, feedback control of this nature incorporates a time delay into the signaling circuitie, in this case, the feedback mediated by increased IκB proteins occurs only after a certain duration of forward signaling proceeds. Such delay allows cellular signaling pathways to "measure" time or the duration of signaling. Since the amount of feedback accumulates proportional to the strength of initial signal, the cell

FIGURE 24.12. Concepts in modeling signaling networks and their emergent properties. A: The basic biochemistry of AgR signaling networks has been largely mapped by studying the enzymatic activity and biology of individual proteins as well as their immediate interactors. Illustrated here is a signaling input (IN, blue box)—which can be an antigen triggering the AgR or a proximal kinase acting on a distal module trigger which sets in motion a simple cascade of biochemical processes. Active A acts on B, activating it to then act on C, etc. The end point of this linear cascade can be the production of the first output, Out1. In biochemical terms, this could be, for example, the generation of DAG at the end of the module in Figure 24.5 or the active ERK in Figure 24.6. If a certain value of IN always results in a certain OUT1, then these steps are considered a module. Apart from their value in modeling, treating closely linked steps in a biochemical cascade associated as a module allows us to keep the biology of the pathways in perspective during reductionistic studies. This principle extends to multiple modules. If OUT1 works as an input to the next cascade, that lends itself to being treated as another module. **B:** The operation of two modules can then be combined both in interpreting biological data and computational modeling of signaling. The figure also makes the point that modules are not rigid. In some cases, considering module 1 and module 2 as the part of one module (with input IN and output OUT2) may be more useful. However, if OUT1 is subject to regulation by other factors in the cell, then treating these two separately may be more appropriate. Signaling pathways are regulated by a variety of control mechanisms, some of which are briefly discussed in the chapter. **C: Negative feedback** involves later steps (eg, enzyme C) initiating a process that results in turning down some of the activity of an upstream step (in this case enzyme B). The process may involve enzyme C activating another phosphatase, ubiquitin ligase, or such that de-activates enzyme B. This reduces the output of the module by reducing the subsequent flux (as indicated by the decreased size of *arrows* from B to C, C to D; compare to Fig. 24.12A). **D: Positive feedback** is similar to above, except that the later step (eg, enzyme C) helps further activate an upstream step (in this case enzyme B). The process may involve enzyme C–activating enzyme B itself or using another kinase, etc. This increases the output of the module by enhancing the subsequent flux (as indicated by the bigger size of *arrows* from B to C, C to D; compare to Fig. 24.12A). **E: Feedforward loops** can manifest in various ways; but a simple concept is that the active enzyme in an early step (eg, enzyme A in this case) enhances a later step (eg, enzyme C). This increases the output of the module by amplifying the activity of the downstream steps. Interactions between modules in a signaling pathway can cause unexpected changes in the overall response. Sometime these are called emergent properties. Some examples are shown in **F and G. F:** Typical dose–response curves (*red line*) involve the response from cells increasing with increasing dose of the stimulus. In some signaling pathways, operation of control loops leads to altered flux. One instance is when the response is essentially kept at baseline until a certain (threshold) value, at which point the signaling flux jumps to its near maximal value. This form of an "ultrasensitive" or "switch-like" response is referred to as a digital transition (*blue dotted line*). **G:** Typically, most signaling pathways turn on for a duration of time and then dissipate—because of loss of ligand, loss of receptor, or negative feedback; this would cause a bell-shaped (label, normal) curve when plotting the response over time. However, alterations in signaling interactions can result in different states—including a cyclical pattern of increase and decrease in signaling or a transition to a stable "on" state. The latter is the basis of some bistable states where once the signaling apparatus transitions into a new steady state it is hard to disrupt this new pattern.

can use feedback to calculate the rate of change of signaling as well, as we discuss further in the section below on tuning. Therefore, in addition to calibrating the lymphocyte response to strength of the ligand–receptor interactions, dose of stimuli, etc. (which are "counted" by the amounts of AgR complexes that are triggered), this allows different responses to antigens which are early versus late or acute versus persisting (see Grossman and Paul[257-259] for more extensive discussions). The overall perspective is that the signaling logic (see The Cellular Circuitry of Receptor signaling section) is encoded not only in the biochemical concentrations of antigen but also in the time and space variables of how the antigen is available to the cell.

In biochemical cascades, feedback regulation can proceed without new gene expression and this typically involves the posttranslational modification of upstream elements by phosphorylation, dephosphorylation, ubiquitination, etc. The recruitment of the phosphatase SHP1 upstream of MAPK discussed in The Canonical Growth Factor Module: Ras-MAPK section illustrates one such negative regulatory feedback. In other cell types, activity of ERK also triggers multiple such feedback processes, which are not very well described in lymphocytes yet. One example is the feedback inhibition of SOS and GAB2 by activated ribosomal protein S6 kinase 1 (RSK1). Furthermore, similar to the upregulation of IκB by NF-κB, transcriptional activation by ERK also results in the upregulation of the DUSPs as well as the Sprouty (SPRY) family proteins. DUSPs selectively dephosphorylate and inactivate ERK, whereas the SPRY family recruits RAS GAP proteins that reduce the amount of active RAS upstream (as a canonical negative feedback loop). A few members of the DUSP family are specific for MAPKs due to the presence of a MAPK-binding domain (MKB) and are also referred to as MAPK phosphatases (MKP). In T cells, DUSP6/MKP-3 is a highly specific regulator of ERK1/2 and a classic model of direct feedback interactions between a kinase and a phosphatase. Once upregulated in T cells by TCR signaling, DUSP6 can bind and dephosphorylate ERK1/2. However, strong activation of ERK1/2 allows them to directly bind and phosphorylate DUSP6 on serines 159 and 197. This primes DUSP6 for proteasomal targeting and degradation setting in motion competing positive and negative loops that control the MAPK signaling flux.

In addition to such protein phosphatases, the lipid phosphatase SHIP can degrade PIP3 to PIP2 (see the Lipids and Cations as Signal Amplifiers: the Ca^{2+} NFAT Pathway section), after being recruited to ITIM containing receptors on the cell surface. In this case, AgR-dependent upregulation of ITIM-bearing receptors offers a mechanism of negatively regulating the PIP3-activated modules of distal signaling. In T cells, ITIM-bearing receptors include the checkpoint molecule PD1, while in B cells FcγRIIB is a major docking site. At the level of the proximal module, Src and Syk can be inactivated by feedback-induced ubiquitination. Ubiquitination is accomplished by a series of enzymes (E1, E2, and E3) which cyclically transfer ubiquitins (Ub) to mark a protein for degradation by the 26S proteasome. The E3 ubiquitin ligases of the Cbl family, notably Cbl-b and c-Cbl, are key regulators of the process in lymphocytes which help target it to specific proteins. In addition to a characteristic RING finger domain critical for ubiquitin ligase activity, these proteins also have TKB and proline-rich domains that bind to Syk and Srcs as well as an incomplete SH2 domain that binds activated protein kinases. The expressions of Cbl proteins themselves are regulated by AgR signaling by a combination of transcriptional upregulation and proteolytic degradation. Accordingly, recently activated lymphocytes express low levels of Cbl-b (especially if appropriate costimulatory signals are provided), but AgR-proximal signals without strong costimulation or in other tolerogenic contexts (eg, prolonged stimulation) lead to further upregulation of these ubiquitin ligases which dampen AgR proximal signaling. Cbl can also be regulated at the level of recruitment, with surface receptors such as CD5 upregulated by TCR signaling subsequently serving as docking sites for accumulating them close to the AgR. In this sense, the upregulation of surface molecules such as CTLA4, PD1, CD5, etc., in T cells, with a time delay after initial AgR signaling, serves the role of a feedback process, attenuating the initial T cell trigger unless stronger TCR signaling allows the response to continue.[260]

Aspects of positive feedback and feedforward amplification (Fig. 24.12D and E) have been discussed in The Distal Modules Operating in AgR Signaling section. Although the definitions of these two often overlap, an example of positive feedback is in the case of the Ca^{2+} influx, where the initial IP3-mediated release of Ca^{2+} further enhances the previous steps, eg, ITK and PLCγ activity. An example of feedforward would be the additional opening of Ca^{2+} channels as well as Ca^{2+}-dependent enzymes further downstream of IP3 release. These amplification mechanisms contribute to the surprisingly high sensitivity of lymphocytes to antigen, where the TCR can be triggered by a single pMHC complex (discussed in Chapter 23). The earliest positive feedback loop is in the phosphorylation of proximal ITAMs. While the Src kinases kindle the process by phosphorylating ITAMs and Syk, the active Syk proteins recruited to the p-ITAMs are themselves very potent ITAM kinases[261] which rapidly increase the density of phosphorylated ITAMs and active Src around the AgR. Another example of a feedforward is the autophosphorylation of Syk (at Tyr-518/519) soon after activation which enhances its ability to activate downstream modules further. Although several mechanisms involved remain to be parsed out, the activities of Src and Syk offer some of the best evidence for how canonical feedforward amplifies AgR signaling. In T cells, the activation of proximal modules is linked to multiple distal modules at the scaffolding steps. Several of the enzymes at the apex of these distal modules are activated by preceding kinases; however, once recruited to the signalosome, enzymes in this distal module can also be regulated by Src and Syk. In the case of T cells, these included Lck-mediated phosphorylation of PI3K and ITK. While these are some narrow biochemical examples of forward loops in signaling, the responsiveness of peripheral cells can be programmed by feedforward loops that are set in place months in advance—as we discuss in thymic tuning and self-peptide sensitivity of naïve T cells, where calibration of signaling during development affects eventual responses to target antigen (agonist).

The Interactive and Emergent Properties in AgR Signaling Networks

The operation of multiple control elements, even connecting different modules, has the potential to generate new kinds of signaling outputs that would not be expected from just the linear steps in each biochemical cascade. In the study of cellular signaling, such states are collectively referred to as "emergent properties"—defined as biological activities that cannot be directly ascribed to one component of a pathway or network (discussed in Ref. 262), but arise reproducibly as a result of interactions between multiple different parts of a module (or different modules). While these concepts are yet to be widely elucidated in the AgR signaling network, there is evidence for such emergence in modulating several key functions starting with lymphocyte activation itself to cell division, differentiation, etc. Emergent properties in a signaling network can extend beyond merely additive (eg, as in the collaboration between fos and jun to make the transcription factor AP1) or synergistic (eg, with the feedforward amplification of the consequences of calcium signaling) outputs.

Typically, most enzymatic reactions, and by extension cellular responses to ligands, proceed along a sigmoidal dose response curve (Fig. 24.12F, red line) where increasing the ligand (or agonist) concentration elicits small increases in cellular response until concentration levels, or even a unique threshold concentration, are reached. At this point, the response increases sharply or abruptly and then proceeds to increase until at another concentration, it plateaus out. This pattern is seen in many T and B cell responses whether it be measuring the upregulation of surface markers, cytokine production, or cell proliferation in response to varying doses of antigen. Yet in the case of some responses, the transition from a basal state of no measurable response to the maximal response (the saturation or plateau phase) is so rapid that it almost looks like the cell switched the response from an off to on state in a digital manner (Fig. 24.12F, blue dotted line). This switching or switch-like response is also called "digital" or biphasic response to contrast with the smoother sigmoidal transitions observed in the other cases (discussed above) called "analog" (Fig. 24.12F). The digital response is not trivially explained by steady linear increases in input leading to nonlinear amplification of output as might be expected in a simple biochemical cascade. Instead, this is a result of an emergent property of cellular systems termed ultrasensitivity. The underlying mechanisms vary but often involve cooperativity, positive feedback, and bistability. Detecting ultrasensitivity and the resulting digital, biphasic response may require single-cell assays of signaling. This is because averaging the response of multiple cells (with digital responses on their own) can often convey the impression of an analog response, especially if the population of cells are activated asynchronously. Indeed, this typical pattern of concentration dependence, with varying rates of increase (slopes) during the exponential switching phase, is seen in many T and B cell responses whether it be measuring the upregulation of surface markers corresponding with activation, cytokine production or cell proliferation. In T cells, the expression of early activation markers such as CD69 or cytokines like interleukin-2 has been reported to be biphasic while the upregulation of CD44 or the production of IFNγ shows analog distributions. At the signaling level, both early TCR triggering (the activation of ZAP70) and the phosphorylation of the MAPK and ERK[263] have digital tendencies (at least in naïve T cells). The mechanisms, however, may differ. The former (discussed in Chapter 23) involves kinetic cooperativity in early signaling. The latter is thought to reflect the effect of interactions between a positive loop via Raf on the one hand and the negative feedback mediated by SHP-1 on the other, which buffers against any slow increase in phosphorylation of ERK until a critical threshold is met. This switch-like transition in MAPK signaling is considered significant at multiple stages in T cell development, including the sharp discrimination between affinities of peptides driving positive versus negative selection of thymocytes.

The relationship between patterns of cellular signaling and functional outcomes can be quite diverse as exemplified by the oscillation properties in the Ca^{2+} signaling module (Fig. 24.12G). Early studies suggested that quantitative parameters (amplitude, frequency, etc.) associated with the oscillation of intracellular Ca^{2+} can specify which transcription factors are turned on.[127,264] This in turn was employed by B cells to encode positive versus negative cellular outcomes; activating agonist BCR signals in peripheral B cells evoked strong spikes of Ca^{2+} while tolerogenic signals driven by self-antigens in immature B cells elicited slow oscillations of Ca^{2+} and induced anergy or deletion.[265] Similarly in double-positive thymocytes undergoing positive selection in thymic slice cultures, altering the precise frequency of calcium oscillations affected cellular fate.[266] The generation of sophisticated patterns of calcium signals is proposed to involve a large set of (unrelated) cellular proteins loosely called the "calcium toolkit."[263] At a minimum, the generation of repeated spikes requires feedback interactions between IP3Rs and cytosolic Ca^{2+} ATPases. Once IP3R activation opens, the store operated channels to allow a rapid flow of Ca^{2+} out of the ER, the increased cytosolic concentration turns on Ca^{2+} ATPases drive Ca^{2+} back into the ER. While this could generate cyclical fluctuations in the cytosolic Ca^{2+} levels, it is not clear how the amplitude and frequency of these oscillations are determined. This core cycle is regulated by input from other signaling modules (some of which are not yet known) including the regulation of IP3R expression by MAPK signaling and activation of Ets family transcription factors.[267] Even in simple model systems such as zebrafish, the calcium toolkit is thought to include hundreds of proteins; and the precise algorithm in lymphocytes for eliciting different patterns of Ca^{2+} fluxes remains to be clarified.

Conceptual Frameworks for the Signaling Logic in Lymphocytes

As ongoing studies accumulate the data necessary to decipher the signaling logic of lymphocytes, a conceptual framework to synthesize and develop models incorporating these new findings is instrumental. Two-signal models, where AgR signaling in combination with costimulation (for T cells) or help (for B cells) drives the initial activation of naïve lymphocytes, have been helpful in not only parsing out the early events in an immune response but also in practical terms toward the development of adjuvants, carriers, etc., for vaccines and immunotherapy. As discussed earlier, it is increasingly clear that the AgR signals (or Signal 1 in the two-signal models) alone can encode information that drives different fate programs, at least in T cells. For instance how long the antigen persists in the body can have profound consequences for T cell fate. Along the "goldilocks" scheme for T cell activation (see the Introduction section), prolonged antigenic stimulation leads to T cell dysfunction or exhaustion. In 1992, Zvi Grossman and the late William E Paul (the former editor of this book) proposed an operational framework, the tunable activation threshold (TAT) model, which has proven useful in trying to understand these dynamics.[257] Central to the TAT or "Tuning" model is that the signaling networks in lymphocytes adapt to the ambient antigen levels by adjusting their activation threshold dynamically. The framework considers the continuum of T cell behavior from the thymus to the periphery. As discussed in Chapter 27, T cells developing in the thymus adjust their signaling properties soon after positive selection. This is consistent with the process of tuning in the TAT model. Such naïve T cells who do not have any stimulating cognate (the intended agonist, usually a foreign-derived) antigen in the peripheral environment would keep the receptor calibration that they acquired in the thymus and remain maximally sensitive to the agonist. This would allow T cells to be activated by as low as one pMHC (as discussed in Chapter 23) and develop an effector response (see Chapter 28). If, however, the agonist-antigen persists *at a fairly constant level* in the system for prolonged period of time, the signaling network recalibrates such that the new level is nonstimulatory. This now becomes the new baseline for that particular T cell. What this would accomplish is that antigens which tend to persist in the body for a long time would be unable to sustain immune responses against them, minimizing the risk of deleterious T cell responses to self-antigens and commensals.[268] On the flipside, this also accounts for the failure of T cells to effectively sustain responses against chronic infections and tumors which persist beyond an initial acute window of opportunity. Importantly, the TAT model (supported by heuristic mathematical modeling) predicted that the driver of robust T cell activation would essentially be the rate of change of the stimulation strength. Accordingly, T cells act as change detectors. If cognate antigen levels change quickly (as would be expected from a replicating pathogen), T cell activation would be sustained. As discussed earlier, since costimulatory molecules can modify the signaling downstream of the AgR as well, inflammatory contexts can affect the "change detection" system as well. Mechanistically, the tuning process might be expected to proceed by a balancing act between positive AgR signals (kinases, etc.) and time-delayed negative signals as discussed in the Clinical Consequence of Altering AgR Signaling section. When the TCR engages an agonist, a rapid increase in positive signals (reinforced beyond a certain threshold by positive loops discussed earlier) can lead to T cell activation. But the delayed increase in negative molecules in the same cell also means that with the passage of time, the T cell accumulates more negative regulators as well. Both the positive and negative feedback are proportional to the strength of antigenic signals. As a result, after a time delay, these would cancel each other, leaving the T cell refractory to activation. Thus, adjustable, dynamic accumulation of recruitable negative signaling molecules renders the cell resistant to activation in response to moderate fluctuations in self-antigens or to persisting pathogens. In this framework, the cellular memory of ambient stimulation and the rapidity with which the stimulus increases relative to baseline would determine whether the cell can be activated.

While the molecular underpinnings of such signaling adaptations in T cells (or for that matter B cells) remains to be understood, there is evidence that T cells can dynamically adjust their signaling network to match ambient stimulus, ie, tune to antigen. The tuning process in the thymus (see above and Chapter 27) involves the downregulation of microRNA such as mir181[269] and the upregulation of potentially inhibitory receptors such as CD5.[238,241] In the periphery, tolerance, which was once (extending Burnet's clonal selection hypothesis) largely interpreted to be a binary choice between activation and cell death (or anergy), is now appreciated to be a tunable process with graded responsiveness in tolerant lymphocytes.[270,271] The clearest evidence for peripheral tuning of T cells comes from studies where monoclonal T cells exposed to persistent "self"-antigens in model systems adjust their TCR responsiveness to match the environmental antigen levels[272] and even recover some of their TCR sensitivity when removed from constant stimulation.[273-277] Beyond the context of tolerance, antigen responsiveness of a T cell changes dynamically in different cell fate contexts (immature, naïve, effector, tolerant, memory, etc.) even though the actual TCR stays the same.[278-280] This suggests that the TAT framework can be helpful in relating the biochemistry of cell signaling to functional consequences, albeit with modification to incorporate new findings as they emerge. For instance, the sensing of constant levels of self-peptide–MHC complexes does not always just dampen T cell's antigen sensitivity. Even after the thymic TCR tuning process discussed earlier, there is evidence that the T cells continue to engage these ligands and the signaling network continues to be aware of the presence of these "low or moderate affinity" ligands (reviewed by Hogquist & Jameson,[281] Morris & Allen[282]). This recognition also allows for greater cell survival, co-agonism with foreign antigen, and other functional alterations, which in the TAT model were covered as "subthreshold" responses[258,283] that were insufficient to drive robust T cell activation but assist in other biological roles. At a biochemical level, the machinery involved in sensing these subthreshold interactions and how they translate to cellular responses are not understood. There is evidence that the TCR proximal module is partially activated by interactions with self-pMHC to result in the phosphorylation of few ITAMs on the CD3ζ chain—although whether this favors or inhibits subsequent T cell activation remains to be clarified.[284] Interestingly, depriving T cells of this tonic signal were reported to render them less responsive to an agonist stimulation.[285] This suggested that in routine immune responses, endogenous self-peptide recognition may promote rather than dampen naïve T cell activation.

Consistent with these studies, T cells that sensed positively selecting self-peptides more strongly in the thymus (and were marked subsequently by high levels of CD5 expression) were better able to contribute to unrelated foreign agonist responses as well.[286-288] Mechanistically, the ability of upregulated CD5 to modulate NF-κB signaling could also contribute to these effects.[245] While all these data suggest the dynamic tunability of the TCR signaling network, these studies are also discussing this in the context of two different families of TCR ligands. The behavior of naïve T cells with tonic self-ligands is generally examining natural peptides akin to those involved in thymic positive selection. The tonic stimulation that causes exhaustion and dysfunction later in a T cell's life is considering a strong agonistic pMHC derived from pathogens, tumors, etc. Indeed, one would expect that the second set of pMHC would be negatively selecting, if they were presented to T cells during development. Importantly, how the first set of self-peptides signals, whether it be for acting as positively selecting ligands in the thymus or modifying T cell behavior in the periphery, is still being elucidated.

CLINICAL CONSEQUENCE OF ALTERING AgR SIGNALING

Given the central role of these pathways in regulating lymphocyte development and function at all stages, it is not surprising that they are of critical significance in a variety of scenarios. These fall into two major categories—first in the use of specific drugs or antibodies that target AgR signaling and second in understanding the clinical consequence of genetic variations in molecules involved in these pathways. The major pharmacological modulators relevant here are those used to dampen lymphocyte activation, of which rapamycin (inhibitor of mTOR) and cyclosporin (inhibitor of CaN) affect pathways discussed in The Third Stress-MAPK Module: P38 and The PI3K-AKT Module sections. Some genetic variations in signaling molecules affecting lymphocyte function are discussed in Chapter 50. The most apparent consequence would be in the loss of specific cells; loss of AgR signaling can prevent the development of T and B cell precursors leading to immunodeficiency. For instance, Zap70 deficiency, or inactivating mutations, cause severe immunodeficiencies specifically lacking T cells.[289,290] However, mutations that partially affect Zap70 function have intriguing and often unexpected functional phenotypes. A frameshift mutation causing a truncated Zap70 82-residues shorter than the wild-type caused a near total deficiency of peripheral CD8+ T cells, with an intact, albeit defective-in-signaling peripheral CD4+ T cell pool.[290] Equally surprising, several mutations of Zap70 manifest as autoimmune syndromes. Children that were identified with blistering skin, autoantibody production, and renal impairment in a 2016 study were found to have two mutations in Zap70: R192W and R360P. While R192W was associated with a decrease in the ability of Zap70 to bind to CD3ζ, the R360P mutation, in the kinase domain of Zap70, results in a decrease in autoinhibition and increased T cell activation.[291] Insight into the potential mechanisms of these defects comes from mouse models. For instance, in the SKG strain, a Balb/c mouse which develops spontaneous arthritis, a Zap70 W163C substitution alters the SH2 domain of the protein and impairs recruitment of Zap70 to its substrates.[292] While this mutation would seem to result in immunodeficiency rather than autoimmunity, it appears that the W163C mutation increases thymic T cells' passing rate through positive selection on self-peptide MHC ligands—this could skew the T cell repertoire to autoreactivity in the periphery.[292]

As discussed in The Scaffolding of AgR Signalosomes section, Slp76 is an important scaffolding molecule within the proximal TCR signaling pathway that is responsible for the recruitment of ITK, PLCγ, ADAP, and VAV. In 2020, there was a patient identified with a mutation in SLP76, resulting in the deletion of the C-terminal SH2 domain of the protein, and presented with a clinical condition of repeated dermatological abnormalities, splenomegaly, and fungal brain abscesses.[293] Peripheral CD4+ T cells were diminished in contrast to CD8+ T cells, which acquired a terminally differentiated state.[293] This corresponded with a reduced ability of the peripheral CD4+ T cells to produce inflammatory cytokines such as IFNγ and an excessive cytokine production from CD8+ T cells when both groups of cells were stimulated *ex vivo*.[293] Similarly, rare mutations in LAT that lead to a truncated form causing rapid degradation manifest as SCID.[294] The patients presented clinically with repeated infection and failure to thrive. In contrast to the mutations of SLP76 and Zap70, this defect in LAT resulted in a lack of peripheral T cells.[294] Another mutation results in a truncated LAT, resulting in the abrogation of major phosphorylation sites, such as Y132, Y171, Y191, and Y226; interestingly, peripheral T cells are still present with this mutation, and some downstream signaling, particularly the flux of Ca^{2+}, was still functional, though the authors did note a shift in cytokine production toward a Th2 response.[295] Three patients who had this truncated LAT presented with an array of symptoms, including autoimmune forms of anemia, lymphadenopathy, splenomegaly, chronic lung disease, fungal infections, and rashes.[295] Mutations in the phosphorylation sites of LAT causing an autoimmune Th2 skewing of T cells have been observed in targeted mutant mouse models and remain a topic of further exploration.[296-298] Although rare, a patient homozygous for mutations in the Lck kinase domain which impacted the structural stability of the alpha helix portion of the kinase domain significantly reduces the ability of Lck to phosphorylate downstream molecules.[299] The patient presented with severe immunodeficiency, with failure to thrive, lung infections, skin lesions, autoimmune forms of anemia, and recurrent diarrhea. Interestingly, a few CD4+ and CD8+ T cells displaying a memory or exhausted phenotype were found in the periphery.[299] These cells could still activate MAPK signaling, presumably due to some redundancy with Fyn. Lck mutations have also been implicated as oncogenic and have been found in human T cell leukemia lines.[300,301] CD45 mutations have been described in humans as resulting in a SCID clinical phenotype. A 1997 case report discussed a child with almost negligible detection of CD45, along with profound lymphopenia and a history of repeated infections.[302] A follow-up study in 2001 found that this patient had a homozygous 6 base pair deletion resulting in two missing amino acids from the extracellular CD45 portion, preventing its surface expression.[303] Another report found that a child lacked a significant portion of the CD45 gene resulting from chromosomal deletion, as well as a mutation on the other allele impacting proper CD45 splicing; these defects resulted not only in profound lymphopenia, but also a tremendously reduced ability of any developed lymphocytes to be activated.[304]

The interaction between ICOS on T cells and ICOSL on other cell types such as B cells is important for T cell differentiation and B cell germinal responses. In some patients diagnosed with common variable immunodeficiency (CVID), a condition which serologically presents with low immunoglobulin levels and susceptibility to bacterial infections, ICOS has been shown to be deficient.[305] Furthermore, a recent case study identified a mutation in a conserved amino acid N219 of ICOSL that prevents ICOSL from being expressed on the cell surface; this results in a clinical phenotype similar to that seen with CVID.[306]

BTK, as discussed in this chapter and elsewhere, is an important member of the Tec family of kinases involved in BCR signaling and is indispensable for proper B cell development and signaling. Mutations in BTK have been associated with a clinical syndrome called X-linked agammaglobulinemia, in which B cell isotype diversity is tremendously limited, and instead there is a prevalence of immature IgM-hi B cells.[307,308] This disease is further discussed in Chapter 5. PKC-δ is considered a novel protein kinase C family member and is important for intracellular signaling in B cells; though the precise mechanisms of signaling remain unclear, PKC-δ may be involved in inhibitory cell signaling, particularly in triggering apoptosis in self-reactive B cells during negative selection.[309] It has been shown in mouse models that PKC-δ deficiency results in massive B cell/germinal center expansion, with autoantibody production.[310,311] Similarly, a human mutation that prevented nuclear import of PKC-δ resulted in aberrant B cell production and an autoimmune clinical syndrome; the clinical syndrome presented as lymphadenopathy, hepatosplenomegaly, recurrent fevers, and autoantibody production.[312] BLNK, or B cell linker protein, also known as Slp-65, is an important adapter molecule in bridging the upstream protein tyrosine kinases, such as Syk, with downstream signaling cascades of PLCγ and MAPK involved in BCR activation.[313] BLNK's specific role in B cell activation and development has been explored through several mouse and human knockout models, with evidence highlighting that BLNK−/− mice had reduced number of mature B cells, and the mature B cells that survived had decreased IgM production, as well as intracellular Ca^{2+} mobilization.[98,314] A few studies in humans with mutations in BLNK have illustrated a drastic decrease in the number of mature circulating B cells, with a corresponding increase of immature B cells.[315,316]

As discussed earlier, Lyn is an important proximal Src kinase member involved in B cell activation and signaling. While it is postulated both through *in vitro* studies and *in vivo* mouse models that loss-of-function mutations in Lyn may result in immunodeficiency, there has recently been described a human gain-of-function mutation that results in an autoimmune phenotype. In this patient, there was a premature stop codon around the site of Y508; this site is normally dephosphorylated by CD45 upon B cell activation, but this particular mutation rendered it constitutively active.[317] Similar to the gain-of-function mutations of Lyn, there have been several observations of overactive Syk. A recent study reported two unrelated patients presenting with autoimmune syndromes including arthritis, colitis, and vasculitis. It was postulated that the mutations later identified in Syk were relevant to autoinhibition of Syk, and when mutated resulted in constitutive activation of B cell signaling.[318]

IN ANTICIPATION

While it could be helpful to conclude this chapter, pointing students to the excitement of anticipated advances in the coming years, that is a daunting task for this field. The sheer volume of data and rapid pace of new discoveries in antigen receptor signaling can be overwhelming. Yet, based on trends in the recent literature, we expect that major breakthroughs in the near future would fall under three main categories.

First, higher resolution tools are now allowing us to map multiple interactions between signaling molecules in primary cells and even in live animals. A large fraction of the foundational work discussed in this chapter was derived from necessary and painstaking work by numerous laboratories over the past half a century, characterizing the molecular steps in each biochemical cascade, one at a time. Over time, the connections between these steps have been pieced together and validated using conventional biochemistry and molecular biology techniques. Typically, these approaches relied heavily on cell lines, cell culture models, and recombinant proteins. While this represents a central framework of the overall signaling network, many pieces of the puzzle remained missing, especially relating to activating T and B cells *in vivo*. As discussed in the Control Mechanisms in the AgR Signaling Network section, newer approaches that allow the study of primary cells under different physiological conditions are now mapping a fuller spectrum of these interactions.[247-251] These additional data will be crucial in resolving multiple outstanding questions ranging from the precise mechanics of how the BCR or TCR are triggered to how crosstalk between different pathways is regulated (eg, between different JNK, P38, and Ras MAPKs) to how different AP1 or NFAT complexes are activated. Similarly, despite their importance in immunotherapy, the mechanisms by which checkpoint regulators like PD1, LAG3, or other regulators such as CD5 and CD6 modulate AgR signaling are still being elucidated. In addition, the use of functional screens using CRISPR libraries[319] as well as characterization of human mutations[320] has revealed new players such as Mg^{2+} channels, in AgR signaling.

Secondly, new tools have allowed recent studies to examine the kinetics and dynamics of signaling pathways in T and B cells after antigens encounter under physiological activation conditions *in vivo*. Previously, stimulations used (eg, antibodies to CD3 and CD28 or pharmacological activators like PMA/ionomycin or even isolated APCs co-cultured *in vitro*) were chosen to robustly and synchronously activate signaling pathways *in vitro*. Imaging studies have since revealed that rare antigen-specific lymphocytes find APCs, engage antigen, and activate differently under various tissue contexts *in vivo*.[321-324] Studying signaling under these conditions has been made possible by advances in multiparameter phospho-flow cytometry and single-cell proteomics.[325,326] Even further, the use of reporter strains relying on signaling sensors[327] or the expression of molecules like Nur77,[328,329] cFos, etc., now allow signaling to be read out without additional manipulation or processing. Additional clarity is expected with broader palettes of such reagents and advances in live imaging techniques.

Finally, as discussed extensively in this chapter, we expect significant advances in the coming years in deciphering the overall signaling logic evolved to operate lymphocytes as sensory organs.[1] The impact of dose, duration, periodicity, quality, and other variables associated with antigenic ligand, in determining different outcomes for B and T cell activation, is well-documented. Yet how these variables translate to patterns of signaling and altered gene expression or phenotype is not yet clear. Given that AgR signaling involves multiple distal modules, the possible quantitative and qualitative combinations of signaling that can map to a particular functional output can be quite vast. There is some evidence that this might mask a degree of redundancy; indeed, the list of knockouts in signaling molecules with minimal impact on particular T or B cell functions is long. Empirical parsing of such redundancies to decipher the underlying logic is also likely to require advances in computational modeling.

The impact of these advances will surely be felt in multiple translational arenas. At the time of writing the last edition of this chapter, the obvious clinical prospects would have been in developing drugs targeting key nodes in the AgR signaling network or in gene therapy approaches to correct immunodeficiencies. These remain key applications. But the explosion of interest in CAR-T cells and newer generations of hybrid AgRs have opened up a new frontier in this regard. These creative applications of basic science, combining different modules to precisely tailor signaling entities not previously found in nature and applying them to revolutionary cures, make it an even more exciting time to study lymphocyte signaling mechanisms.

ACKNOWLEDGMENTS

We would like to thank Dr Zvi Grossman, Elizabeth Hill, and Zachary Fasana for critical discussions and helpful comments that shaped the final form of this chapter. Limitations in the number of references we could reasonably cite in the chapter have forced us to leave out many seminal papers from this storied field and we apologize for those omissions.

REFERENCES

1. Davis MM, Krogsgaard M, Huse M, Huppa J, Lillemeier BF, Li QJ. T cells as a self-referential, sensory organ. *Annu Rev Immunol.* 2007;25:681-695. doi:10.1146/annurev.immunol.24.021605.090600
2. Valitutti S, Muller S, Dessing M, Lanzavecchia A. Different responses are elicited in cytotoxic T lymphocytes by different levels of T cell receptor occupancy. *J Exp Med.* 1996;183:1917-1921.
3. Constant S, Pfeiffer C, Woodard A, Pasqualini T, Bottomly K. Extent of T cell receptor ligation can determine the functional differentiation of naive CD4+ T cells. *J Exp Med.* 1995;182:1591-1596.
4. Tubo NJ, Pagan AJ, Taylor JJ, et al. Single naive CD4(+) T cells from a diverse repertoire produce different effector cell types during infection. *Cell.* 2013;153:785-796. doi:10.1016/j.cell.2013.04.007
5. Matson CA, Singh NJ. Manipulating the TCR signaling network for cellular immunotherapy: challenges & opportunities. *Mol Immunol.* 2020;123:64-73, doi:10.1016/j.molimm.2020.04.007
6. Lazebnik Y. Can a biologist fix a radio? – Or, what I learned while studying apoptosis. *Cancer Cell.* 2002;2:179-182.
7. Huse M, Klein LO, Girvin AT, et al. Spatial and temporal dynamics of T cell receptor signaling with a photoactivatable agonist. *Immunity.* 2007;27:76-88.
8. Chow CW, Rincon M, Davis RJ. Requirement for transcription factor NFAT in interleukin-2 expression. *Mol Cell Biol.* 1999;19:2300-2307.
9. Sojka DK, Bruniquel D, Schwartz RH, Singh NJ. IL-2 secretion by CD4+ T cells *in vivo* is rapid, transient, and influenced by TCR-specific competition. *J Immunol.* 2004;172:6136-6143.
10. Huppa JB, Axmann M, Mortelmaier MA, et al. TCR-peptide-MHC interactions *in situ* show accelerated kinetics and increased affinity. *Nature.* 2010;463:963-967. doi:10.1038/nature08746
11. Batista FD, Neuberger MS. Affinity dependence of the B cell response to antigen: a threshold, a ceiling, and the importance of off-rate. *Immunity.* 1998;8:751-759.
12. Samelson LE, Harford JB, Klausner RD. Identification of the components of the murine T cell antigen receptor complex. *Cell.* 1985;43:223-231. doi:10.1016/0092-8674(85)90027-3
13. Klausner RD, Weissman AM, Baniyash M, Bonifacino JS, Samelson LE. The role of the zeta chain in the expression, structure and function of the T cell receptor. *Adv Exp Med Biol.* 1989;254:21-24. doi:10.1007/978-1-4757-5803-0_3
14. Weiss A, Stobo JD. Requirement for the coexpression of T3 and the T cell antigen receptor on a malignant human T cell line. *J Exp Med.* 1984;160:1284-1299. doi:10.1084/jem.160.5.1284
15. Alarcon B, Berkhout B, Breitmeyer J, Terhorst C. Assembly of the human T cell receptor-CD3 complex takes place in the endoplasmic reticulum and involves intermediary complexes between the CD3-γ.δ.ε core and single T cell receptor α or β chains. *J Biol Chem.* 1988;263:2953-2961.
16. Saito T, Hochstenbach F, Marusic-Galesic S, Kruisbeek AM, Brenner M, Germain RN. Surface expression of only gamma delta and/or alpha beta T cell receptor heterodimers by cells with four (alpha, beta, gamma, delta) functional receptor chains. *J Exp Med.* 1988;168:1003-1020. doi:10.1084/jem.168.3.1003
17. Berkhout B, Alarcon B, Terhorst C. Transfection of genes encoding the T cell receptor-associated CD3 complex into COS cells results in assembly of the macromolecular structure. *J Biol Chem.* 1988;263:8528-8536.
18. Matsuuchi L, Gold MR, Travis A, Grosschedl R, DeFranco AL, Kelly RB. The membrane IgM-associated proteins MB-1 and Ig-beta are sufficient to promote surface expression of a partially functional B-cell antigen receptor in a nonlymphoid cell line. *Proc Natl Acad Sci U S A.* 1992;89:3404-3408. doi:10.1073/pnas.89.8.3404
19. Hombach J, Tsubata T, Leclercq L, Stappert H, Reth M. Molecular components of the B-cell antigen receptor complex of the IgM class. *Nature.* 1990;343:760-762. doi:10.1038/343760a0
20. Reth M. Antigen receptor tail clue. *Nature.* 1989;338:383-384. doi:10.1038/338383b0
21. Flaswinkel H, Barner M, Reth M. The tyrosine activation motif as a target of protein tyrosine kinases and SH2 domains. *Semin Immunol.* 1995;7:21-27. doi:10.1016/1044-5323(95)90004-7
22. Cosson P, Lankford SP, Bonifacino JS, Klausner RD. Membrane protein association by potential intramembrane charge pairs. *Nature.* 1991;351:414-416. doi:10.1038/351414a0
23. Van Wauwe JP, De Mey JR, Goossens JG. OKT3: a monoclonal anti-human T lymphocyte antibody with potent mitogenic properties. *J Immunol.* 1980;124:2708-2713.
24. Gerpert TD, Lipsky PE. Accessory cell independent proliferation of human T4 cells stimulated by immobilized monoclonal antibodies to CD3. *J Immunol.* 1987;138:1660-1666.
25. Van Kooten C, Galibert L, Seon BK, Garrone P, Liu YJ, Banchereau J. Cross-linking of antigen receptor via Ig-beta (B29, CD79b) can induce both positive and negative signals in CD40-activated human B cells. *Clin Exp Immunol.* 1997;110:509-515. doi:10.1046/j.1365-2249.1997.4201436.x
26. Shinkai Y, Alt FW. CD3 epsilon-mediated signals rescue the development of CD4+CD8+ thymocytes in RAG-2−/− mice in the absence of TCR beta chain expression. *Int Immunol.* 1994;6:995-1001. doi:10.1093/intimm/6.7.995
27. Shinkai Y, Ma A, Cheng HL, Alt FW. CD3 epsilon and CD3 zeta cytoplasmic domains can independently generate signals for T cell development and function. *Immunity.* 1995;2:401-411. doi:10.1016/1074-7613(95)90148-5
28. Isakov N, Wange RL, Burgess WH, Watts JD, Aebersold R, Samelson LE. ZAP-70 binding specificity to T cell receptor tyrosine-based activation motifs: the tandem SH2 domains of ZAP-70 bind distinct tyrosine-based activation motifs with varying affinity. *J Exp Med.* 1995;181:375-380.
29. Zenner G, Vorherr T, Mustelin T, Burn P. Differential and multiple binding of signal transducing molecules to the ITAMs of the TCR-zeta chain. *J Cell Biochem.* 1996;63:94-103. doi:10.1002/(sici)1097-4644(199610)63:1<94::aid-jcb8>3.0.co;2-v
30. Osman N, Turner H, Lucas S, Reif K, Cantrell DA. The protein interactions of the immunoglobulin receptor family tyrosine-based activation motifs present in the T cell receptor zeta subunits and the CD3 gamma, delta and epsilon chains. *Eur J Immunol.* 1996;26:1063-1068. doi:10.1002/eji.1830260516
31. Holst J, Wang H, Eder KD, et al. Scalable signaling mediated by T cell antigen receptor-CD3 ITAMs ensures effective negative selection and prevents autoimmunity. *Nat Immunol.* 2008;9:658-666. doi:10.1038/ni.1611
32. Hayes SM, Love PE. Distinct structure and signaling potential of the gamma delta TCR complex. *Immunity.* 2002;16:827-838. doi:10.1016/s1074-7613(02)00320-5
33. Love PE, Hayes SM. ITAM-mediated signaling by the T-cell antigen receptor. *Cold Spring Harb Perspect Biol.* 2010;2:a002485. doi:10.1101/cshperspect.a002485
34. Bezbradica JS, Rosenstein RK, Demarco RA, Brodsky I, Medzhitov R. A role for the ITAM signaling module in specifying cytokine-receptor functions. *Nat Immunol.* 2014;15(4):333-342. doi:10.1038/ni.2845
35. Pawson T, Gish GD, Nash P. SH2 domains, interaction modules and cellular wiring. *Trends Cell Biol.* 2001;11:504-511. doi:10.1016/s0962-8924(01)02154-7
36. Daëron M, Latour S, Malbec O, et al. The same tyrosine-based inhibition motif, in the intracytoplasmic domain of Fc gamma RIIB, regulates negatively BCR-, TCR-, and FcR-dependent cell activation. *Immunity.* 1995;3:635-646. doi:10.1016/1074-7613(95)90134-5
37. Martin GS. The hunting of the Src. *Nat Rev Mol Cell Biol.* 2001;2:467-475. doi:10.1038/35073094
38. Parsons SJ, Parsons JT. Src family kinases, key regulators of signal transduction. *Oncogene.* 2004;23:7906-7909. doi:10.1038/sj.onc.1208160
39. Songyang Z, Shoelson SE, Chaudhuri M, et al. SH2 domains recognize specific phosphopeptide sequences. *Cell.* 1993;72:767-778. doi:10.1016/0092-8674(93)90404-e
40. Resh MD. Fatty acylation of proteins: new insights into membrane targeting of myristoylated and palmitoylated proteins. *Biochim Biophys Acta.* 1999;1451:1-16. doi:10.1016/s0167-4889(99)00075-0
41. Yurchak LK, Sefton BM. Palmitoylation of either Cys-3 or Cys-5 is required for the biological activity of the Lck tyrosine protein kinase. *Mol Cell Biol.* 1995;15:6914-6922. doi:10.1128/mcb.15.12.6914

42. van't Hof W, Resh MD. Rapid plasma membrane anchoring of newly synthesized p59fyn: selective requirement for NH2-terminal myristoylation and palmitoylation at cysteine-3. *J Cell Biol.* 1997;136:1023-1035. doi:10.1083/jcb.136.5.1023
43. Zlatkine P, Mehul B, Magee AI. Retargeting of cytosolic proteins to the plasma membrane by the Lck protein tyrosine kinase dual acylation motif. *J Cell Sci.* 1997;110(pt 5):673-679.
44. Courtneidge SA. Activation of the pp60c-src kinase by middle T antigen binding or by dephosphorylation. *EMBO J.* 1985;4:1471-1477.
45. Xu W, Doshi A, Lei M, Eck MJ, Harrison SC. Crystal structures of c-Src reveal features of its autoinhibitory mechanism. *Mol Cell.* 1999;3:629-638. doi:10.1016/s1097-2765(00)80356-1
46. Bergman M, Mustelin T, Oetken C, et al. The human p50csk tyrosine kinase phosphorylates p56lck at Tyr-505 and down regulates its catalytic activity. *EMBO J.* 1992;11:2919-2924.
47. Sicheri F, Moarefi I, Kuriyan J. Crystal structure of the Src family tyrosine kinase Hck. *Nature.* 1997;385:602-609. doi:10.1038/385602a0
48. Nada S, Yagi T, Takeda H, et al. Constitutive activation of Src family kinases in mouse embryos that lack Csk. *Cell.* 1993;73:1125-1135. doi:10.1016/0092-8674(93)90642-4
49. Nada S, Okada M, MacAuley A, Cooper JA, Nakagawa H. Cloning of a complementary DNA for a protein-tyrosine kinase that specifically phosphorylates a negative regulatory site of p60c-src. *Nature.* 1991;351:69-72. doi:10.1038/351069a0
50. Schmedt C, aijo K, Niidome T, Kuhn R, Aizawa S, Tarakhovsky A. Csk controls antigen receptor-mediated development and selection of T-lineage cells. *Nature.* 1998;394:901-904. doi:10.1038/29802
51. Kawabuchi M, Satomi Y, Takao T, et al. Transmembrane phosphoprotein Cbp regulates the activities of Src-family tyrosine kinases. *Nature.* 2000;404:999-1003. doi:10.1038/35010121
52. Brdicka T, Pavlistova D, Leo A, et al. Phosphoprotein associated with glycosphingolipid-enriched microdomains (PAG), a novel ubiquitously expressed transmembrane adaptor protein, binds the protein tyrosine kinase csk and is involved in regulation of T cell activation. *J Exp Med.* 2000;191:1591-1604. doi:10.1084/jem.191.9.1591
53. Douglass AD, Vale RD. Single-molecule microscopy reveals plasma membrane microdomains created by protein-protein networks that exclude or trap signaling molecules in T cells. *Cell.* 2005;121:937-950. doi:10.1016/j.cell.2005.04.009
54. Hrdinka M, Horejsi V. PAG—A multipurpose transmembrane adaptor protein. *Oncogene.* 2014;33:4881-4892. doi:10.1038/onc.2013.485
55. Dobenecker MW, Schmedt C, Okada M, Tarakhovsky A. The ubiquitously expressed Csk adaptor protein Cbp is dispensable for embryogenesis and T-cell development and function. *Mol Cell Biol.* 2005;25:10533-10542. doi:10.1128/mcb.25.23.10533-10542.2005
56. Xu S, Huo J, Tan JE, Lam KP. Cbp deficiency alters Csk localization in lipid rafts but does not affect T-cell development. *Mol Cell Biol.* 2005;25:8486-8495. doi:10.1128/mcb.25.19.8486-8495.2005
57. Xavier R, Brennan T, Li Q, McCormack C, Seed B. Membrane compartmentation is required for efficient T cell activation. *Immunity.* 1998;8:723-732. doi:10.1016/s1074-7613(00)80577-4
58. Cheng PC, Dykstra ML, Mitchell RN, Pierce SK. A role for lipid rafts in B cell antigen receptor signaling and antigen targeting. *J Exp Med.* 1999;190:1549-1560. doi:10.1084/jem.190.11.1549
59. Cahir McFarland ED, Hurley TR, Pingel JT, Sefton BM, Shaw A, Thomas ML. Correlation between Src family member regulation by the protein-tyrosine-phosphatase CD45 and transmembrane signaling through the T-cell receptor. *Proc Natl Acad Sci U S A.* 1993;90:1402-1406. doi:10.1073/pnas.90.4.1402
60. Felberg J, Johnson P. Stable interdomain interaction within the cytoplasmic domain of CD45 increases enzyme stability. *Biochem Biophys Res Commun.* 2000;271:292-298. doi:10.1006/bbrc.2000.2623
61. Wang Y, Guo W, Liang L, Esselman WJ. Phosphorylation of CD45 by casein kinase 2. Modulation of activity and mutational analysis. *J Biol Chem.* 1999;274:7454-7461. doi:10.1074/jbc.274.11.7454
62. Streuli M, Hall LR, Saga Y, Schlossman SF, Saito H. Differential usage of three exons generates at least five different mRNAs encoding human leukocyte common antigens. *J Exp Med.* 1987;166:1548-1566. doi:10.1084/jem.166.5.1548
63. Saga Y, Tung JS, Shen FW, Pancoast TC, Boyse EA. Organization of the Ly-5 gene. *Mol Cell Biol.* 1988;8:4889-4895. doi:10.1128/mcb.8.11.4889-4895.1988
64. Cordoba SP, Choudhuri K, Zhang H, et al. The large ectodomains of CD45 and CD148 regulate their segregation from and inhibition of ligated T-cell receptor. *Blood.* 2013;121:4295-4302. doi:10.1182/blood-2012-07-442251
65. Byth KF, Conroy LA, Howlett S, et al. CD45-null transgenic mice reveal a positive regulatory role for CD45 in early thymocyte development, in the selection of CD4+CD8+ thymocytes, and B cell maturation. *J Exp Med.* 1996;183:1707-1718. doi:10.1084/jem.183.4.1707
66. Kishihara K, Penninger J, Wallace VA, et al. Normal B lymphocyte development but impaired T cell maturation in CD45-exon6 protein tyrosine phosphatase-deficient mice. *Cell.* 1993;74:143-156. doi:10.1016/0092-8674(93)90302-7
67. Zhu JW, Brdicka T, Katsumoto TR, Lin J, Weiss A. Structurally distinct phosphatases CD45 and CD148 both regulate B cell and macrophage immunoreceptor signaling. *Immunity.* 2008;28:183-196. doi:10.1016/j.immuni.2007.11.024
68. Barber EK, Dasgupta JD, Schlossman SF, Trevillyan JM, Rudd CE. The CD4 and CD8 antigens are coupled to a protein-tyrosine kinase (p56lck) that phosphorylates the CD3 complex. *Proc Natl Acad Sci U S A.* 1989;86:3277-3281. doi:10.1073/pnas.86.9.3277
69. Veillette A, Bookman MA, Horak EM, Bolen JB. The CD4 and CD8 T cell surface antigens are associated with the internal membrane tyrosine-protein kinase p56lck. *Cell.* 1988;55:301-308. doi:10.1016/0092-8674(88)90053-0
70. Turner JM, Brodsky MH, Irving BA, Levin SD, Perlmutter RM, Littman DR. Interaction of the unique N-terminal region of tyrosine kinase p56lck with cytoplasmic domains of CD4 and CD8 is mediated by cysteine motifs. *Cell.* 1990;60:755-765. doi:10.1016/0092-8674(90)90090-2
71. Kim PW, Sun ZY, Blacklow SC, Wagner G, Eck MJ. A zinc clasp structure tethers Lck to T cell coreceptors CD4 and CD8. *Science.* 2003;301:1725-1728. doi:10.1126/science.1085643
72. Yachi PP, Ampudia J, Zal T, Gascoigne NR. Altered peptide ligands induce delayed CD8-T cell receptor interaction – A role for CD8 in distinguishing antigen quality. *Immunity.* 2006;25:203-211. doi:10.1016/j.immuni.2006.05.015
73. Baniyash M, Garcia-Morales P, Luong E, Samelson LE, Klausner RD. The T cell antigen receptor zeta chain is tyrosine phosphorylated upon activation. *J Biol Chem.* 1988;263:18225-18230.
74. Nika K, Soldani C, Salek M, et al. Constitutively active Lck kinase in T cells drives antigen receptor signal transduction. *Immunity.* 2010;32:766-777. doi:10.1016/j.immuni.2010.05.011
75. Yamanashi Y, Kakiuchi T, Mizuguchi J, Yamamoto T, Toyoshima K. Association of B cell antigen receptor with protein tyrosine kinase Lyn. *Science.* 1991;251:192-194. doi:10.1126/science.1702903
76. Jin L, Pluskey S, Petrella EC, et al. The three-dimensional structure of the ZAP-70 kinase domain in complex with staurosporine: implications for the design of selective inhibitors. *J Biol Chem.* 2004;279(41):42818-42825. doi:10.1074/jbc.M407096200. PMID: 15292186
77. Hatada MH, Lu X, Laird ER, et al. Molecular basis for interaction of the protein tyrosine kinase ZAP-70 with the T-cell receptor. *Nature.* 1995;377:32-38. doi:10.1038/377032a0
78. van Oers NS, Killeen N, Weiss A. ZAP-70 is constitutively associated with tyrosine-phosphorylated TCR zeta in murine thymocytes and lymph node T cells. *Immunity.* 1994;1:675-685. doi:10.1016/1074-7613(94)90038-
79. Johnson SA, Pleiman CM, Pao L, Schneringer J, Hippen K, Cambier JC. Phosphorylated immunoreceptor signaling motifs (ITAMs) exhibit unique abilities to bind and activate Lyn and Syk tyrosine kinases. *J Immunol.* 1995;155:4596-4603.
80. Tsang E, Giannetti AM, Shaw D, et al. Molecular mechanism of the Syk activation switch. *J Biol Chem.* 2008;283:32650-32659. doi:10.1074/jbc.M806340200
81. Zoller KE, MacNeil IA, Brugge JS. Protein tyrosine kinases Syk and ZAP-70 display distinct requirements for Src family kinases in immune response receptor signal transduction. *J Immunol.* 1997;158:1650-1659.
82. Neumeister EN, Zhu Y, Richard S, Terhorst C, Chan AC, Shaw AS. Binding of ZAP-70 to phosphorylated T-cell receptor zeta and eta enhances its autophosphorylation and generates specific binding sites for SH2 domain-containing proteins. *Mol Cell Biol.* 1995;15:3171-3178. doi:10.1128/mcb.15.6.3171
83. Negishi I, Motoyama N, Nakayama K, et al. Essential role for ZAP-70 in both positive and negative selection of thymocytes. *Nature.* 1995;376:435-438. doi:10.1038/376435a0
84. Cheng AM, Rowley B, Pao W, Hayday A, Bolen JB, Pawson T. Syk tyrosine kinase required for mouse viability and B-cell development. *Nature.* 1995;378:303-306. doi:10.1038/378303a0
85. Turner M, Mee PJ, Costello PS, et al. Perinatal lethality and blocked B-cell development in mice lacking the tyrosine kinase Syk. *Nature.* 1995;378:298-302. doi:10.1038/378298a0
86. Finco TS, Kadlecek T, Zhang W, Samelson LE, Weiss A. LAT is required for TCR-mediated activation of PLCγ1 and the Ras pathway. *Immunity.* 1998;9:617-626. doi:10.1016/s1074-7613(00)80659-7
87. Zhang W, Sommers CL, Burshtyn DN, et al. Essential role of LAT in T cell development. *Immunity.* 1999;10:323-332. doi:10.1016/s1074-7613(00)80032-1
88. Zhang W, Sloan-Lancaster J, Kitchen J, Trible RP, Samelson LE. LAT: the ZAP-70 tyrosine kinase substrate that links T cell receptor to cellular activation. *Cell.* 1998;92:83-92. doi:10.1016/s0092-8674(00)80901-0
89. Zhang W, Trible RP, Zhu M, Liu SK, McGlade CJ, Samelson LE. Association of Grb2, Gads, and phospholipase C-gamma 1 with phosphorylated LAT tyrosine residues. Effect of LAT tyrosine mutations on T cell angigen receptor-mediated signaling. *J Biol Chem.* 2000;275:23355-23361. doi:10.1074/jbc.M000404200
90. Balagopalan L, Coussens NP, Sherman E, Samelson LE, Sommers CL. The LAT story: a tale of cooperativity, coordination, and choreography. *Cold Spring Harb Perspect Biol.* 2010;2:a005512. doi:10.1101/cshperspect.a005512
91. Berry DM, Nash P, Liu SK, Pawson T, McGlade CJ. A high-affinity Arg-X-X-Lys SH3 binding motif confers specificity for the interaction between Gads and SLP-76 in T cell signaling. *Curr Biol.* 2002;12:1336-1341. doi:10.1016/s0960-9822(02)01038-2
92. Bubeck Wardenburg J, Fu C, Jackman JK, et al. Phosphorylation of SLP-76 by the ZAP-70 protein-tyrosine kinase is required for T-cell receptor function. *J Biol Chem.* 1996;271:19641-19644.
93. Jordan MS, Sadler J, Austin JE, et al. Functional hierarchy of the N-terminal tyrosines of SLP-76. *J Immunol.* 2006;176:2430-2438. doi:10.4049/jimmunol.176.4.2430
94. Fang N, Motto DG, Ross SE, Koretzky GA. Tyrosines 113, 128, and 145 of SLP-76 are required for optimal augmentation of NFAT promoter activity. *J Immunol.* 1996;157:3769-3773.
95. Brdicka T, Imrich M, Angelisova P, et al. Non-T cell activation linker (NTAL): a transmembrane adaptor protein involved in immunoreceptor signaling. *J Exp Med.* 2002;196:1617-1626. doi:10.1084/jem.20021405
96. Janssen E, Zhu M, Zhang W, Koonpaew S, Zhang W. LAB: a new membrane-associated adaptor molecule in B cell activation. *Nat Immunol.* 2003;4:117-123. doi:10.1038/ni882
97. Wang Y, Horvath O, Hamm-Baarke A, et al. Single and combined deletions of the NTAL/LAB and LAT adaptors minimally affect B-cell development and function. *Mol Cell Biol.* 2005;25:4455-4465. doi:10.1128/mcb.25.11.4455-4465.2005
98. Jumaa H, Wollscheid B, Mitterer M, Wienands J, Reth M, Nielsen PJ. Abnormal development and function of B lymphocytes in mice deficient for the signaling adaptor protein SLP-65. *Immunity.* 1999;11:547-554. doi:10.1016/s1074-7613(00)80130-2
99. Pappu R, Cheng AM, Li B, et al. Requirement for B cell linker protein (BLNK) in B cell development. *Science.* 1999;286:1949-1954. doi:10.1126/science.286.5446.1949
100. Tan JE, Wong SC, Gan SK, Xu S, Lam KP. The adaptor protein BLNK is required for b cell antigen receptor-induced activation of nuclear factor-kappa B and cell cycle entry and survival of B lymphocytes. *J Biol Chem.* 2001;276:20055-20063. doi:10.1074/jbc.M010800200
101. Engels N, Wollscheid B, Wienands J. Association of SLP-65/BLNK with the B cell antigen receptor through a non-ITAM tyrosine of Ig-alpha. *Eur J Immunol.* 2001;31:2126-2134. doi:10.1002/1521-4141(200107)31:7<2126::aid-immu2126>3.0.co;2-o
102. Kabak S, Skaggs BJ, Gold MR, et al. The direct recruitment of BLNK to immunoglobulin alpha couples the B-cell antigen receptor to distal signaling pathways. *Mol Cell Biol.* 2002;22:2524-2535. doi:10.1128/mcb.22.8.2524-2535.2002
103. Simons K, Ikonen E. Functional rafts in cell membranes. *Nature.* 1997;387:569-572. doi:10.1038/42408
104. Sohn HW, Tolar P, Jin T, Pierce SK. Fluorescence resonance energy transfer in living cells reveals dynamic membrane changes in the initiation of B cell signaling. *Proc Natl Acad Sci U S A.* 2006;103:8143-8148. doi:10.1073/pnas.0509858103
105. Yokosuka T, Sakata-Sogawa K, Kobayashi W, et al. Newly generated T cell receptor microclusters initiate and sustain T cell activation by recruitment of Zap70 and SLP-76. *Nat Immunol.* 2005;6:1253-1262. doi:10.1038/ni1272
106. Kortum RL, Balagopalan L, Alexander CP, et al. The ability of Sos1 to oligomerize the adaptor protein LAT is separable from its guanine nucleotide exchange activity *in vivo*. *Sci Signal.* 2013;6:ra99. doi:10.1126/scisignal.2004494
107. Köhler F, Storch B, Kulathu Y, et al. A leucine zipper in the N terminus confers membrane association to SLP-65. *Nat Immunol.* 2005;6:204-210. doi:10.1038/ni1163
108. Truneh A, Albert F, Golstein P, Schmitt-Verhulst AM. Early steps of lymphocyte activation bypassed by synergy between calcium ionophores and phorbol ester. *Nature.* 1985;313:318-320. doi:10.1038/313318a0
109. Imboden JB, Stobo JD. Transmembrane signalling by the T cell antigen receptor. Perturbation of the T3-antigen receptor complex generates inositol phosphates and releases calcium ions from intracellular stores. *J Exp Med.* 1985;161:446-456. doi:10.1084/jem.161.3.446
110. Trebak M, Kinet JP. Calcium signalling in T cells. *Nat Rev Immunol.* 2019;19:154-169. doi:10.1038/s41577-018-0110-7
111. Northrop JP, Ho SN, Chen L, et al. NF-AT components define a family of transcription factors targeted in T-cell activation. *Nature.* 1994;369:497-502. doi:10.1038/369497a0
112. Wang D, Feng J, Wen R, et al. Phospholipase Cγ2 is essential in the functions of B cell and several Fc receptors. *Immunity.* 2000;13:25-35. doi:10.1016/s1074-7613(00)00005-4
113. August A, Sadra A, Dupont B, Hanafusa H. Src-induced activation of inducible T cell kinase (ITK) requires phosphatidylinositol 3-kinase activity and the Pleckstrin homology domain of inducible T cell kinase. *Proc Natl Acad Sci U S A.* 1997;94:11227-11232. doi:10.1073/pnas.94.21.11227
114. Andreotti AH, Bunnell SC, Feng S, Berg LJ, Schreiber SL. Regulatory intramolecular association in a tyrosine kinase of the Tec family. *Nature.* 1997;385:93-97. doi:10.1038/385093a0
115. Bunnell SC, Diehn M, Yaffe MB, Findell PR, Cantley LC, Berg LJ. Biochemical interactions integrating Itk with the T cell receptor-initiated signaling cascade. *J Biol Chem.* 2000;275:2219-2230. doi:10.1074/jbc.275.3.2219

116. Bogin Y, Ainey C, Beach D, Yablonski D. SLP-76 mediates and maintains activation of the Tec family kinase ITK via the T cell antigen receptor-induced association between SLP-76 and ITK. *Proc Natl Acad Sci U S A*. 2007;104:6638-6643. doi:10.1073/pnas.0609771104
117. Hajicek N, Keith NC, Siraliev-Perez E, et al. Structural basis for the activation of PLC-γ isozymes by phosphorylation and cancer-associated mutations. *Elife*. 2019;8:e51700. doi:10.7554/eLife.51700
118. Jayaraman T, Ondriasová E, Ondrias K, Harnick DJ, Marks AR. The inositol 1,4,5-trisphosphate receptor is essential for T-cell receptor signaling. *Proc Natl Acad Sci U S A*. 1995;92:6007-6011. doi:10.1073/pnas.92.13.6007
119. Ouyang K, Leandro Gomez-Amaro R, Stachura DL, et al. Loss of IP3R-dependent Ca^{2+} signalling in thymocytes leads to aberrant development and acute lymphoblastic leukemia. *Nat Commun*. 2014;5:4814. doi:10.1038/ncomms5814
120. Jayaraman T, Ondrias K, Ondriasová E, Marks AR. Regulation of the inositol 1,4,5-trisphosphate receptor by tyrosine phosphorylation. *Science*. 1996;272:1492-1494. doi:10.1126/science.272.5267.1492
121. Aramburu J, Yaffe MB, Lopez-Rodriguez C, Cantley LC, Hogan PG, Rao A. Affinity-driven peptide selection of an NFAT inhibitor more selective than cyclosporin A. *Science*. 1999;285:2129-2133. doi:10.1126/science.285.5436.2129
122. Ni YG, Wang N, Cao DJ, et al. FoxO transcription factors activate Akt and attenuate insulin signaling in heart by inhibiting protein phosphatases. *Proc Natl Acad Sci U S A*. 2007;104:20517-20522.
123. Medina DL, Di Paola S, Peluso I, et al. Lysosomal calcium signalling regulates autophagy through calcineurin and TFEB. *Nat Cell Biol*. 2015;17:288-299. doi:10.1038/ncb3114
124. Muller MR, Rao A. NFAT, immunity and cancer: a transcription factor comes of age. *Nat Rev Immunol*. 2010;10:645-656. doi:10.1038/nri2818
125. Okamura H, Aramburu J, Garcia-Rodriguez C, et al. Concerted dephosphorylation of the transcription factor NFAT1 induces a conformational switch that regulates transcriptional activity. *Mol Cell*. 2000;6:539-550. doi:10.1016/s1097-2765(00)00053-8
126. Sharma S, Findlay GM, Bandukwala HS, et al. Dephosphorylation of the nuclear factor of activated T cells (NFAT) transcription factor is regulated by an RNA-protein scaffold complex. *Proc Natl Acad Sci U S A*. 2011;108:11381-11386. doi:10.1073/pnas.1019711108
127. Dolmetsch RE, Xu K, Lewis RS. Calcium oscillations increase the efficiency and specificity of gene expression. *Nature*. 1998;392:933-936. doi:10.1038/31960
128. Tomida T, Hirose K, Takizawa A, Shibasaki F, Iino M. NFAT functions as a working memory of Ca^{2+} signals in decoding Ca^{2+} oscillation. *EMBO J*. 2003;22:3825-3832. doi:10.1093/emboj/cdg381
129. Lavoie H, Gagnon J, Therrien M. ERK signalling: a master regulator of cell behaviour, life and fate. *Nat Rev Mol Cell Biol*. 2020;21:607-632. doi:10.1038/s41580-020-0255-7
130. Anjum R, Blenis J. The RSK family of kinases: emerging roles in cellular signalling. *Nat Rev Mol Cell Biol*. 2008;9:747-758. doi:10.1038/nrm2509
131. Iborra S, Soto M, Stark-Aroeira L, et al. H-ras and N-ras are dispensable for T-cell development and activation but critical for protective Th1 immunity. *Blood*. 2011;117:5102-5111. doi:10.1182/blood-2010-10-315770
132. Kortum RL, Sommers CL, Alexander CP, et al. Targeted Sos1 deletion reveals its critical role in early T-cell development. *Proc Natl Acad Sci U S A*. 2011;108:12407-12412. doi:10.1073/pnas.1104295108
133. Stefanová I, Hemmer B, Vergelli M, Martin R, Biddison WE, Germain RN. TCR ligand discrimination is enforced by competing ERK positive and SHP-1 negative feedback pathways. *Nat Immunol*. 2003;4:248-254. doi:10.1038/ni895
134. Brockdorff J, Williams S, Couture C, Mustelin T. Dephosphorylation of ZAP-70 and inhibition of T cell activation by activated SHP1. *Eur J Immunol*. 1999;29:2539-2550, doi:10.1002/(sici)1521-4141(199908)29:08<2539::Aid-immu2539>3.0.Co;2-m
135. Maillet M, Purcell NH, Sargent MA, York AJ, Bueno OF, Molkentin JD. DUSP6 (MKP3) null mice show enhanced ERK1/2 phosphorylation at baseline and increased myocyte proliferation in the heart affecting disease susceptibility. *J Biol Chem*. 2008;283:31246-31255. doi:10.1074/jbc.M806085200
136. Zhao Q, Wang X, Nelin LD, et al. MAP kinase phosphatase 1 controls innate immune responses and suppresses endotoxic shock. *J Exp Med*. 2006;203:131-140. doi:10.1084/jem.20051794
137. Mandl M, Slack DN, Keyse SM. Specific inactivation and nuclear anchoring of extracellular signal-regulated kinase 2 by the inducible dual-specificity protein phosphatase DUSP5. *Mol Cell Biol*. 2005;25:1830-1845. doi:10.1128/MCB.25.5.1830-1845.2005
138. Zhong XP, Hainey EA, Olenchock BA, et al. Enhanced T cell responses due to diacylglycerol kinase zeta deficiency. *Nat Immunol*. 2003;4:882-890. doi:10.1038/ni958
139. Crespo P, Bustelo XR, Aaronson DS, et al. Rac-1 dependent stimulation of the JNK/SAPK signaling pathway by Vav. *Oncogene*. 1996;13:455-460.
140. Davis RJ. Signal transduction by the JNK group of MAP kinases. *Cell*. 2000;103:239-252. doi:10.1016/s0092-8674(00)00116-1
141. Kawasaki H, Schiltz L, Chiu R, et al. ATF-2 has intrinsic histone acetyltransferase activity which is modulated by phosphorylation. *Nature*. 2000;405:195-200. doi:10.1038/35012097
142. Dong C, Yang DD, Wysk M, Whitmarsh AJ, Davis RJ, Flavell RA. Defective T cell differentiation in the absence of Jnk1. *Science*. 1998;282:2092-2095.
143. Yang DD, Conze D, Whitmarsh AJ, et al. Differentiation of CD4+ T cells to Th1 cells requires MAP kinase JNK2. *Immunity*. 1998;9:575-585.
144. Conze D, Krahl T, Kennedy N, et al. c-Jun NH(2)-terminal kinase (JNK)1 and JNK2 have distinct roles in CD8(+) T cell activation. *J Exp Med*. 2002;195:811-823. doi:10.1084/jem.20011508
145. Arbour N, Naniche D, Homann D, Davis RJ, Flavell RA, Oldstone MBA. c-Jun NH(2)-terminal kinase (JNK)1 and JNK2 signaling pathways have divergent roles in CD8(+) T cell-mediated antiviral immunity. *J Exp Med*. 2002;195:801-810. doi:10.1084/jem.20011481
146. Nishina H, Fischer KD, Radvanyi L, et al. Stress-signalling kinase Sek1 protects thymocytes from apoptosis mediated by CD95 and CD3. *Nature*. 1997;385:350-353. doi:10.1038/385350a0
147. Rincón M, Whitmarsh A, Yang DD, et al. The JNK pathway regulates the *in vivo* deletion of immature CD4(+)CD8(+) thymocytes. *J Exp Med*. 1998;188:1817-1830. doi:10.1084/jem.188.10.1817
148. Sabapathy K, Hu Y, Kallunki T, et al. JNK2 is required for efficient T-cell activation and apoptosis but not for normal lymphocyte development. *Curr Biol*. 1999;9:116-125.
149. Tournier C, Hess P, Yang DD, et al. Requirement of JNK for stress-induced activation of the cytochrome c-mediated death pathway. *Science*. 2000;288:870-874. doi:10.1126/science.288.5467.870
150. Han J, Lee JD, Bibbs L, Ulevitch RJ. A MAP kinase targeted by endotoxin and hyperosmolarity in mammalian cells. *Science*. 1994;265:808-811. doi:10.1126/science.7914033
151. Price MA, Cruzalegui FH, Treisman R. The p38 and ERK MAP kinase pathways cooperate to activate ternary complex factors and c-fos transcription in response to UV light. *EMBO J*. 1996;15:6552-6563.
152. Lee JC, Laydon JT, McDonnell PC, et al. A protein kinase involved in the regulation of inflammatory cytokine biosynthesis. *Nature*. 1994;372:739-746. doi:10.1038/372739a0
153. Jirmanova L, Giardino Torchia ML, Sarma ND, Mittelstadt PR, Ashwell JD. Lack of the T cell-specific alternative p38 activation pathway reduces autoimmunity and inflammation. *Blood*. 2011;118:3280-3289. doi:10.1182/blood-2011-01-333039
154. Schafer PH, Wang L, Wadsworth SA, Davis JE, Siekierka JJ. T cell activation signals up-regulate p38 mitogen-activated protein kinase activity and induce TNF-α production in a manner distinct from LPS activation of monocytes. *J Immunol*. 1999;162:659.
155. Tanaka N, Kamanaka M, Enslen H, et al. Differential involvement of p38 mitogen-activated protein kinase kinases MKK3 and MKK6 in T-cell apoptosis. *EMBO Rep*. 2002;3:785-791. doi:10.1093/embo-reports/kvf153
156. Salvador JM, Mittelstadt PR, Guszczynski T, et al. Alternative p38 activation pathway mediated by T cell receptor-proximal tyrosine kinases. *Nat Immunol*. 2005;6:390-395. doi:10.1038/ni1177
157. Khiem D, Cyster JG, Schwarz JJ, Black BL. A p38 MAPK-MEF2C pathway regulates B-cell proliferation. *Proc Natl Acad Sci U S A*. 2008;105:17067-17072. doi:10.1073/pnas.0804868105
158. Chen G, Hitomi M, Han J, Stacey DW. The p38 pathway provides negative feedback for ras proliferative signaling. *J Biol Chem*. 2000;275:38973-38980. doi:10.1074/jbc.M002856200
159. Giardino Torchia ML, Dutta D, Mittelstadt PR, et al. Intensity and duration of TCR signaling is limited by p38 phosphorylation of ZAP-70. *Proc Natl Acad Sci U S A*. 2018;115:2174-2179. doi:10.1073/pnas.1713301115
160. Khwaja A, Rodriguez-Viciana P, Wennström S, Warne PH, Downward J. Matrix adhesion and Ras transformation both activate a phosphoinositide 3-OH kinase and protein kinase B/Akt cellular survival pathway. *EMBO J*. 1997;16:2783-2793. doi:10.1093/emboj/16.10.2783
161. Kulik G, Klippel A, Weber MJ. Antiapoptotic signalling by the insulin-like growth factor I receptor, phosphatidylinositol 3-kinase, and Akt. *Mol Cell Biol*. 1997;17:1595-1606. doi:10.1128/MCB.17.3.1595
162. Wymann MP, Pirola L. Structure and function of phosphoinositide 3-kinases. *Biochim Biophys Acta*. 1998;1436:127-150. doi:10.1016/S0005-2760(98)00139-8
163. Angulo I, Vadas O, Garcon F, et al. Phosphoinositide 3-kinase δ gene mutation predisposes to respiratory infection and airway damage. *Science*. 2013;342:866. doi:10.1126/science.1243292
164. Lucas CL, Kuehn HS, Zhao F, et al. Dominant-activating germline mutations in the gene encoding the PI(3)K catalytic subunit p110δ result in T cell senescence and human immunodeficiency. *Nat Immunol*. 2014;15:88-97. doi:10.1038/ni.2771
165. Songyang Z, Shoelson SE, McGlade J, et al. Specific motifs recognized by the SH2 domains of Csk, 3BP2, fps/fes, GRB-2, HCP, SHC, Syk, and Vav. *Mol Cell Biol*. 1994;14:2777-2785. doi:10.1128/mcb.14.4.2777-2785.1994
166. Alessi DR, James SR, Downes CP, et al. Characterization of a 3-phosphoinositide-dependent protein kinase which phosphorylates and activates protein kinase Bα. *Curr Biol*. 1997;7:261-269. doi:10.1016/S0960-9822(06)00122-9
167. Datta SR, Dudek H, Tao X, et al. Akt phosphorylation of BAD couples survival signals to the cell-intrinsic death machinery. *Cell*. 1997;91:231-241. doi:10.1016/s0092-8674(00)80405-5
168. Yamaguchi H, Wang HG. The protein kinase PKB/Akt regulates cell survival and apoptosis by inhibiting Bax conformational change. *Oncogene*. 2001;20:7779-7786. doi:10.1038/sj.onc.1204984
169. Brunet A, Bonni A, Zigmond MJ, et al. Akt promotes cell survival by phosphorylating and inhibiting a Forkhead transcription factor. *Cell*. 1999;96:857-868.
170. Du K, Montminy M. CREB is a regulatory target for the protein kinase Akt/PKB. *J Biol Chem*. 1998;273:32377-32379. doi:10.1074/jbc.273.49.32377
171. Madrid LV, Mayo MW, Reuther JY, Baldwin AS. Akt stimulates the transactivation potential of the RelA/p65 subunit of NF-κB through utilization of the IκB kinase and activation of the mitogen-activated protein kinase p38. *J Biol Chem*. 2001;276:18934-18940. doi:10.1074/jbc.M101103200
172. Loewith R, Jacinto E, Wullschleger S, et al. Two TOR complexes, only one of which is rapamycin sensitive, have distinct roles in cell growth control. *Mol Cell*. 2002;10:457-468. doi:10.1016/s1097-2765(02)00636-6
173. Sancak Y, Thoreen CC, Peterson TR, et al. PRAS40 is an insulin-regulated inhibitor of the mTORC1 protein kinase. *Mol Cell*. 2007;25:903-915. doi:10.1016/j.molcel.2007.03.003
174. Hara K, Maruki Y, Long X, et al. Raptor, a binding partner of target of rapamycin (TOR), mediates TOR action. *Cell*. 2002;110:177-189. doi:10.1016/s0092-8674(02)00833-4
175. Tee AR, Fingar DC, Manning BD, Kwiatkowski DJ, Cantley LC, Blenis J. Tuberous sclerosis complex-1 and -2 gene products function together to inhibit mammalian target of rapamycin (mTOR)-mediated downstream signaling. *Proc Natl Acad Sci U S A*. 2002;99:13571-13576. doi:10.1073/pnas.202476899
176. Inoki K, Li Y, Zhu T, Wu J, Guan KL. TSC2 is phosphorylated and inhibited by Akt and suppresses mTOR signalling. *Nat Cell Biol*. 2002;4:648-657. doi:10.1038/ncb839
177. Jacinto E, Loewith R, Schmidt A, et al. Mammalian TOR complex 2 controls the actin cytoskeleton and is rapamycin insensitive. *Nat Cell Biol*. 2004;6:1122-1128. doi:10.1038/ncb1183
178. Yang Q, Inoki K, Ikenoue T, Guan KL. Identification of Sin1 as an essential TORC2 component required for complex formation and kinase activity. *Genes Dev*. 2006;20:2820-2832. doi:10.1101/gad.1461206
179. Hornberger TA, Sukhija KB, Wang XR, Chien S. mTOR is the rapamycin-sensitive kinase that confers mechanically-induced phosphorylation of the hydrophobic motif site Thr(389) in p70(S6k). *FEBS Lett*. 2007;581:4562-4566. doi:10.1016/j.febslet.2007.08.045
180. Weng QP, Kozlowski M, Belham C, Zhang A, Comb MJ, Avruch J. Regulation of the p70 S6 kinase by phosphorylation *in vivo*. Analysis using site-specific anti-phosphopeptide antibodies. *J Biol Chem*. 1998;273:16621-16629. doi:10.1074/jbc.273.26.16621
181. von Manteuffel SR, Gingras AC, Ming XF, Sonenberg N, Thomas G. 4E-BP1 phosphorylation is mediated by the FRAP-p70s6k pathway and is independent of mitogen-activated protein kinase. *Proc Natl Acad Sci U S A*. 1996;93:4076-4080. doi:10.1073/pnas.93.9.4076
182. Gingras AC, Kennedy SG, O'Leary MA, Sonenberg N, Hay N. 4E-BP1, a repressor of mRNA translation, is phosphorylated and inactivated by the Akt(PKB) signaling pathway. *Genes Dev*. 1998;12:502-513. doi:10.1101/gad.12.4.502
183. Sarbassov DD, Guertin DA, Ali SM, Sabatini DM. Phosphorylation and regulation of Akt/PKB by the rictor-mTOR complex. *Science*. 2005;307:1098-1101. doi:10.1126/science.1106148
184. Céfaï D, Schneider H, Matangkasombut O, Kang H, Brody J, Rudd CE. CD28 receptor endocytosis is targeted by mutations that disrupt phosphatidylinositol 3-kinase binding and costimulation. *J Immunol*. 1998;160:2223-2230.
185. Garçon F, Patton DT, Emery JL, et al. CD28 provides T-cell costimulation and enhances PI3K activity at the immune synapse independently of its capacity to interact with the p85/p110 heterodimer. *Blood*. 2008;111:1464-1471. doi:10.1182/blood-2007-08-108050
186. Coyle AJ, Lehar S, Lloyd C, et al. The CD28-related molecule ICOS is required for effective T cell-dependent immune responses. *Immunity*. 2000;13:95-105. doi:10.1016/s1074-7613(00)00011-x
187. Inabe K, Kurosaki T. Tyrosine phosphorylation of B-cell adaptor for phosphoinositide 3-kinase is required for Akt activation in response to CD19 engagement. *Blood*. 2002;99:584-589. doi:10.1182/blood.v99.2.584
188. Castello A, Gaya M, Tucholski J, et al. Nck-mediated recruitment of BCAP to the BCR regulates the PI(3)K-Akt pathway in B cells. *Nat Immunol*. 2013;14:966-975. doi:10.1038/ni.2685

189. Sen R, Baltimore D. Multiple nuclear factors interact with the immunoglobulin enhancer sequences. *Cell.* 1986;46:705-716. doi:10.1016/0092-8674(86)90346-6
190. Zhang Q, Lenardo MJ, Baltimore D. 30 years of NF-κB: a Blossoming of relevance to human pathobiology. *Cell.* 2017;168:37-57. doi:10.1016/j.cell.2016.12.012
191. Isakov N, Mally MI, Scholz W, Altman A. T-lymphocyte activation: the role of protein kinase C and the bifurcating inositol phospholipid signal transduction pathway. *Immunol Rev.* 1987;95:89-111. doi:10.1111/j.1600-065x.1987.tb00501.x
192. Gaide O, Favier B, Legler DF, et al. CARMA1 is a critical lipid raft-associated regulator of TCR-induced NF-κB activation. *Nat Immunol.* 2002;3:836-843. doi:10.1038/ni830
193. Sun SC. The noncanonical NF-κB pathway. *Immunol Rev.* 2012;246:125-140. doi:10.1111/j.1600-065X.2011.01088.x
194. Lu HY, Biggs CM, Blanchard-Rohner G, Fung SY, Sharma M, Turvey SE. Germline CBM-opathies: from immunodeficiency to atopy. *J Allergy Clin Immunol.* 2019;143:1661-1673. doi:10.1016/j.jaci.2019.03.009
195. Senegas A, Gautheron J, Maurin AG, Courtois G. IKK-related genetic diseases: probing NF-κB functions in humans and other matters. *Cell Mol Life Sci.* 2015;72:1275-1287. doi:10.1007/s00018-014-1793-y
196. de Petris S, Raff MC. Normal distribution, patching and capping of lymphocyte surface immunoglobulin studied by electron microscopy. *Nat New Biol.* 1973;241:257-259. doi:10.1038/newbio241257a0
197. Grakoui A, Bromley SK, Sumen C, et al. The immunological synapse: a molecular machine controlling T cell activation. *Science.* 1999;285:221-227.
198. Lee KH, Holdorf AD, Dustin ML, Chan AC, Allen PM, Shaw AS. T cell receptor signaling precedes immunological synapse formation. *Science.* 2002;295:1539-1542.
199. Monks CR, Freiberg BA, Kupfer H, Sciaky N, Kupfer A. Three-dimensional segregation of supramolecular activation clusters in T cells. *Nature.* 1998;395:82-86.
200. Fleire SJ, Goldman JP, Carrasco YR, Weber M, Bray D, Batista FD. B cell ligand discrimination through a spreading and contraction response. *Science.* 2006;312:738-741.
201. Qi H, Egen JG, Huang AY, Germain RN. Extrafollicular activation of lymph node B cells by antigen-bearing dendritic cells. *Science.* 2006;312:1672-1676.
202. Batista FD, Iber D, Neuberger MS. B cells acquire antigen from target cells after synapse formation. *Nature.* 2001;411:489-494.
203. Aghazadeh B, Lowry WE, Huang XY, Rosen MK. Structural basis for relief of autoinhibition of the Dbl homology domain of proto-oncogene Vav by tyrosine phosphorylation. *Cell.* 2000;102:625-633. doi:10.1016/s0092-8674(00)00085-4
204. Han J, Das B, Wei W, et al. Lck regulates Vav activation of members of the Rho family of GTPases. *Mol Cell Biol.* 1997;17:1346-1353. doi:10.1128/mcb.17.3.1346
205. Deckert M, Tartare-Deckert S, Couture C, Mustelin T, Altman A. Functional and physical interactions of Syk family kinases with the Vav proto-oncogene product. *Immunity.* 1996;5:591-604. doi:10.1016/s1074-7613(00)80273-3
206. O'Rourke LM, Tooze R, Turner M, et al. CD19 as a membrane-anchored adaptor protein of B lymphocytes: costimulation of lipid and protein kinases by recruitment of Vav. *Immunity.* 1998;8:635-645. doi:10.1016/s1074-7613(00)80568-3
207. Kliche S, Worbs T, Wang X, et al. CCR7-mediated LFA-1 functions in T cells are regulated by 2 independent ADAP/SKAP55 modules. *Blood.* 2012;119:777-785. doi:10.1182/blood-2011-06-362269
208. Ménasché G, Kliche S, Chen EJH, Stradal TEB, Schraven B, Koretzky G. RIAM links the ADAP/SKAP-55 signaling module to Rap1, facilitating T-cell-receptor-mediated integrin activation. *Mol Cell Biol.* 2007;27:4070-4081. doi:10.1128/mcb.02011-06
209. Takala H, Nurminen E, Nurmi SM, et al. Beta2 integrin phosphorylation on Thr758 acts as a molecular switch to regulate 14-3-3 and filamin binding. *Blood.* 2008;112:1853-1862. doi:10.1182/blood-2007-12-127795
210. Jenkins MK, Schwartz RH. Antigen presentation by chemically modified splenocytes induces antigen- specific T cell unresponsiveness *in vitro* and *in vivo*. *J Exp Med.* 1987;165:302-319.
211. Baroja ML, Lorre K, Van Vaeck F, Ceuppens JL. The anti-T cell monoclonal antibody 9.3 (anti-CD28) provides a helper signal and bypasses the need for accessory cells in T cell activation with immobilized anti-CD3 and mitogens. *Cell Immunol.* 1989;120:205-217. doi:10.1016/0008-8749(89)90188-3
212. Aruffo A, Seed B. Molecular cloning of a CD28 cDNA by a high-efficiency COS cell expression system. *Proc Natl Acad Sci U S A.* 1987;84:8573-8577. doi:10.1073/pnas.84.23.8573
213. Michel F, Attal-Bonnefoy G, Mangino G, Mise-Omata S, Acuto O. CD28 as a molecular amplifier extending TCR ligation and signaling capabilities. *Immunity.* 2001;15:935-945.
214. Lin CH, Hunig T. Efficient expansion of regulatory T cells *in vitro* and *in vivo* with a CD28 superagonist. *Eur J Immunol.* 2003;33:626-638.
215. Jenkins MK, Taylor PS, Norton SD, Urdahl KB. CD28 delivers a costimulatory signal involved in antigen-specific IL-2 production by human T cells. *J Immunol.* 1991;147:2461-2466.
216. Harding FA, McArthur JG, Gross JA, Raulet DH, Allison JP. CD28-mediated signalling co-stimulates murine T cells and prevents induction of anergy in T-cell clones. *Nature.* 1992;356:607-609.
217. Fraser JD, Irving BA, Crabtree GR, Weiss A. Regulation of interleukin-2 gene enhancer activity by the T cell accessory molecule CD28. *Science.* 1991;251:313-316. doi:10.1126/science.1846244
218. Khoruts A, Mondino A, Pape KA, Reiner SL, Jenkins MK. A natural immunological adjuvant enhances T cell clonal expansion through a CD28-dependent, interleukin (IL)-2-independent mechanism. *J Exp Med.* 1998;187:225-236.
219. Shahinian A, Pfeffer K, Lee KP, et al. Differential T cell costimulatory requirements in CD28-deficient mice. *Science.* 1993;261:609-612.
220. Prasad KV, Cai YC, Raab M, et al. T-cell antigen CD28 interacts with the lipid kinase phosphatidylinositol 3-kinase by a cytoplasmic Tyr(P)-Met-Xaa-Met motif. *Proc Natl Acad Sci U S A.* 1994;91:2834-2838. doi:10.1073/pnas.91.7.2834
221. Dodson LF, Boomer JS, Deppong CM, et al. Targeted knock-in mice expressing mutations of CD28 reveal an essential pathway for costimulation. *Mol Cell Biol.* 2009;29:3710-3721. doi:10.1128/MCB.01869-08
222. June CH, Ledbetter JA, Gillespie MM, Lindsten T, Thompson CB. T-cell proliferation involving the CD28 pathway is associated with cyclosporine-resistant interleukin 2 gene expression. *Mol Cell Biol.* 1987;7:4472-4481. doi:10.1128/mcb.7.12.4472-4481.1987
223. Michel F, Mangino G, Attal-Bonnefoy G, et al. CD28 utilizes Vav-1 to enhance TCR-proximal signaling and NF-AT activation. *J Immunol.* 2000;165:3820-3829. doi:10.4049/jimmunol.165.7.3820
224. Boise LH, Minn AJ, Noel PJ, et al. CD28 costimulation can promote T cell survival by enhancing the expression of Bcl-XL. *Immunity.* 1995;3:87-98. doi:10.1016/1074-7613(95)90161-2
225. Kearney ER, Walunas TL, Karr RW, et al. Antigen-dependent clonal expansion of a trace population of antigen-specific CD4+ T cells *in vivo* is dependent on CD28 costimulation and inhibited by CTLA-4. *J Immunol.* 1995;155:1032-1036.
226. Pagès F, Ragueneau M, Rottapel R, et al. Binding of phosphatidylinositol-3-OH kinase to CD28 is required for T-cell signalling. *Nature.* 1994;369:327-329. doi:10.1038/369327a0
227. Iezzi G, Karjalainen K, Lanzavecchia A. The duration of antigenic stimulation determines the fate of naive and effector T cells. *Immunity.* 1998;8:89-95.
228. Marchingo JM, Kan A, Sutherland RM, et al. Antigen affinity, costimulation, and cytokine inputs sum linearly to amplify T cell expansion. *Science.* 2014;346:1123-1127. doi:10.1126/science.1260044
229. Kundig TM, Shahinian A, Kawai K, et al. Duration of TCR stimulation determines costimulatory requirement of T cells. *Immunity.* 1996;5:41-52.
230. Singh NJ, Cox M, Schwartz RH. TLR ligands differentially modulate T cell responses to acute and chronic antigen presentation. *J Immunol.* 2007;179:7999-8008.
231. Béziat V, Rapaport F, Hu J, et al. Humans with inherited T cell CD28 deficiency are susceptible to skin papillomaviruses but are otherwise healthy. *Cell.* 2021;184:3812.e30-3828.e30. doi:10.1016/j.cell.2021.06.004
232. Carter RH, Fearon DT. CD19: lowering the threshold for antigen receptor stimulation of B lymphocytes. *Science.* 1992;256:105-107. doi:10.1126/science.1373518
233. Fujimoto M, Poe JC, Jansen PJ, Sato S, Tedder TF. CD19 amplifies B lymphocyte signal transduction by regulating Src-family protein tyrosine kinase activation. *J Immunol.* 1999;162:7088-7094.
234. Li X, Sandoval D, Freeberg L, Carter RH. Role of CD19 tyrosine 391 in synergistic activation of B lymphocytes by coligation of CD19 and membrane Ig. *J Immunol.* 1997;158:5649-5657.
235. Tuveson DA, Carter RH, Soltoff SP, Fearon DT. CD19 of B cells as a surrogate kinase insert region to bind phosphatidylinositol 3-kinase. *Science.* 1993;260:986-989. doi:10.1126/science.7684160
236. Depoil D, Fleire S, Treanor BL, et al. CD19 is essential for B cell activation by promoting B cell receptor-antigen microcluster formation in response to membrane-bound ligand. *Nat Immunol.* 2008;9:63-72. doi:10.1038/ni1547
237. Shoham T, Rajapaksa R, Boucheix C, et al. The tetraspanin CD81 regulates the expression of CD19 during B cell development in a postendoplasmic reticulum compartment. *J Immunol.* 2003;171:4062-4072.
238. Tarakhovsky A, Kanner SB, Hombach J, et al. A role for CD5 in TCR-mediated signal-transduction and thymocyte selection. *Science.* 1995;269:535-537. doi:10.1126/science.7542801
239. Azzam HS, Grinberg A, Lui K, Shen H, Shores EW, Love PE. CD5 expression is developmentally regulated by T cell receptor (TCR) signals and TCR avidity. *J Exp Med.* 1998;188:2301-2311.
240. Gary-Gouy H, Bruhns P, Schmitt C, Dalloul A, Daeron M, Bismuth G. The pseudo-immunoreceptor tyrosine-based activation motif of CD5 mediates its inhibitory action on B-cell receptor signaling. *J Biol Chem.* 2000;275:548-556. doi:10.1074/jbc.275.1.548
241. Azzam HS, DeJarnette JB, Huang K, et al. Fine tuning of TCR signaling by CD5. *J Immunol.* 2001;166:5464-5472.
242. Bhandoola A, Bosselut R, Yu Q, et al. CD5-mediated inhibition of TCR signaling during intrathymic selection and development does not require the CD5 extracellular domain. *Eur J Immunol.* 2002;32:1811-1817. doi:10.1002/1521-4141(200206)32:6<1811::Aid-immu1811>3.0.Co;2-g
243. Voisinne G, Garcia-Blesa A, Chaoui K, et al. Co-recruitment analysis of the CBL and CBLB signalosomes in primary T cells identifies CD5 as a key regulator of TCR-induced ubiquitylation. *Mol Syst Biol.* 2016;12:876. doi:10.15252/msb.20166837
244. Ceuppens JL, Baroja ML. Monoclonal antibodies to the CD5 antigen can provide the necessary second signal for activation of isolated resting T cells by solid-phase-bound OKT3. *J Immunol.* 1986;137:1816-1821.
245. Matson CA, Choi S, Livak F, et al. CD5 dynamically calibrates basal NF-κB signaling in T cells during thymic development and peripheral activation. *Proc Natl Acad Sci U S A.* 2020;117:14342-14353. doi:10.1073/pnas.1922525117
246. Topalian SL, Drake CG, Pardoll DM. Immune checkpoint blockade: a common denominator approach to cancer therapy. *Cancer Cell.* 2015;27:450-461. doi:10.1016/j.ccell.2015.03.001
247. Howden AJM, Hukelmann JL, Brenes A, et al. Quantitative analysis of T cell proteomes and environmental sensors during T cell differentiation. *Nat Immunol.* 2019;20:1542-1554. doi:10.1038/s41590-019-0495-x
248. Voisinne G, Kersse K, Chaoui K, et al. Quantitative interactomics in primary T cells unveils TCR signal diversification extent and dynamics. *Nat Immunol.* 2019;20:1530-1541. doi:10.1038/s41590-019-0489-8
249. Roncagalli R, Hauri S, Fiore F, et al. Quantitative proteomics analysis of signalosome dynamics in primary T cells identifies the surface receptor CD6 as a Lat adaptor-independent TCR signaling hub. *Nat Immunol.* 2014;15(4):384-392. doi:10.1038/ni.2843
250. Neumann K, Oellerich T, Urlaub H, Wienands J. The B-lymphoid Grb2 interaction code. *Immunol Rev.* 2009;232:135-149. doi:10.1111/j.1600-065X.2009.00845.x
251. Tan H, Yang K, Li Y, et al. Integrative proteomics and phosphoproteomics profiling reveals dynamic signaling networks and bioenergetics pathways underlying T cell activation. *Immunity.* 2017;46:488-503. doi:10.1016/j.immuni.2017.02.010
252. Daniels MA, Teixeiro E, Gill J, et al. Thymic selection threshold defined by compartmentalization of Ras/MAPK signalling. *Nature.* 2006;444:724-729. doi:10.1038/nature05269
253. Chang JT, Palanivel VR, Kinjyo I, et al. Asymmetric T lymphocyte division in the initiation of adaptive immune responses. *Science.* 2007;315:1687-1691.
254. Gaud G, Lesourne R, Love PE. Regulatory mechanisms in T cell receptor signalling. *Nat Rev Immunol.* 2018;18:485-497. doi:10.1038/s41577-018-0020-8
255. Reth M, Brummer T. Feedback regulation of lymphocyte signalling. *Nat Rev Immunol.* 2004;4:269-277. doi:10.1038/nri1335
256. Hoffmann A, Levchenko A, Scott ML, Baltimore D. The IkappaB-NF-kappaB signaling module: temporal control and selective gene activation. *Science.* 2002;298:1241-1245. doi:10.1126/science.1071914
257. Grossman Z, Paul WE. Adaptive cellular interactions in the immune system: the tunable activation threshold and the significance of subthreshold responses. *Proc Natl Acad Sci U S A.* 1992;89:10365-10369.
258. Grossman Z, Paul WE. Autoreactivity, dynamic tuning and selectivity. *Curr Opin Immunol.* 2001;13:687-698.
259. Grossman Z, Paul WE. Dynamic tuning of lymphocytes: physiological basis, mechanisms, and function. *Annu Rev Immunol.* 2015;33:677-713. doi:10.1146/annurev-immunol-032712-100027
260. Thompson CB, Allison JP. The emerging role of CTLA-4 as an immune attenuator. *Immunity.* 1997;7:445-450.
261. Rolli V, Gallwitz M, Wossning T, et al. Amplification of B cell antigen receptor signaling by a Syk/ITAM positive feedback loop. *Mol cell.* 2002;10:1057-1069. doi:10.1016/s1097-2765(02)00739-6
262. Azeloglu EU, Iyengar R. Signaling networks: information flow, computation, and decision making. *Cold Spring Harb Perspect Biol.* 2015;7:a005934. doi:10.1101/cshperspect.a005934
263. Bootman MD, Collins TJ, Peppiatt CM, et al. Calcium signalling--an overview. *Semin Cell Dev Biol.* 2001;12, 3-10. doi:10.1006/scdb.2000.0211
264. Dolmetsch RE, Lewis RS, Goodnow CC, Healy JI. Differential activation of transcription factors induced by Ca^{2+} response amplitude and duration. *Nature.* 1997;386:855-858.
265. Healy JI, Dolmetsch RE, Timmerman LA, et al. Different nuclear signals are activated by the B cell receptor during positive versus negative signaling. *Immunity.* 1997;6:419-428.
266. Bhakta NR, Oh DY, Lewis RS. Calcium oscillations regulate thymocyte motility during positive selection in the three-dimensional thymic environment. *Nat Immunol.* 2005;6:143-151. doi:10.1038/ni1161
267. Nagaleekar VK, Diehl SA, Juncadella I, et al. IP3 receptor-mediated Ca^{2+} release in naive CD4 T cells dictates their cytokine program. *J Immunol.* 2008;181:8315-8322. doi:10.4049/jimmunol.181.12.8315
268. Grossman Z, Paul WE. Self-tolerance: context dependent tuning of T cell antigen recognition. *Semin Immunol.* 2000;12:197-203.

269. Li QJ, Chau J, Ebert PJR, et al. miR-181a is an intrinsic modulator of T cell sensitivity and selection. *Cell.* 2007;129:147-161. doi:10.1016/j.cell.2007.03.008
270. Akkaraju S, Ho WY, Leong D, Canaan K, Davis MM, Goodnow CC. A range of CD4 T cell tolerance: partial inactivation to organ-specific antigen allows nondestructive thyroiditis or insulitis. *Immunity.* 1997;7:255-271.
271. Wherry EJ, Blattman JN, Murali-Krishna K, van der MR, Ahmed R. Viral persistence alters CD8 T-cell immunodominance and tissue distribution and results in distinct stages of functional impairment. *J Virol.* 2003;77:4911-4927.
272. Singh NJ, Schwartz RH. The strength of persistent antigenic stimulation modulates adaptive tolerance in peripheral CD4+ T cells. *J Exp Med.* 2003;198:1107-1117.
273. Rocha B, Tanchot C, Von Boehmer H. Clonal anergy blocks *in vivo* growth of mature T cells and can be reversed in the absence of antigen. *J Exp Med.* 1993;177:1517-1521.
274. Tanchot C, Barber DL, Chiodetti L, Schwartz RH. Adaptive tolerance of CD4+ T cells in vivo: multiple thresholds in response to a constant level of antigen presentation. *J Immunol.* 2001;167:2030-2039.
275. Wong P, Barton GM, Forbush KA, Rudensky AY. Dynamic tuning of T cell reactivity by self-peptide-major histocompatibility complex ligands. *J Exp Med.* 2001;193:1179-1187.
276. Stamou P, de Jersey J, Carmignac D, Mamalaki C, Kioussis D, Stockinger B. Chronic exposure to low levels of antigen in the periphery causes reversible functional impairment correlating with changes in CD5 levels in monoclonal CD8 T cells. *J Immunol.* 2003;171:1278-1284.
277. Marquez ME, Ellmeier W, Sanchez-Guajardo V, Freitas AA, Acuto O, Di Bartolo V. CD8 T cell sensory adaptation dependent on TCR avidity for self-antigens. *J Immunol.* 2005;175:7388-7397.
278. Farber DL, Acuto O, Bottomly K. Differential T cell receptor-mediated signaling in naive and memory CD4 T cells. *Eur J Immunol.* 1997;27:2094-2101.
279. Kumar R, Ferez M, Swamy M, et al. Increased sensitivity of antigen-experienced T cells through the enrichment of oligomeric T cell receptor complexes. *Immunity.* 2011;35:375-387. doi:10.1016/j.immuni.2011.08.010
280. Huang J, Brameshuber M, Zeng X, et al. A single peptide-major histocompatibility complex ligand triggers digital cytokine secretion in CD4(+) T cells. *Immunity.* 2013;39:846-857. doi:10.1016/j.immuni.2013.08.036
281. Hogquist KA, Jameson SC. The self-obsession of T cells: how TCR signaling thresholds affect fate "decisions" and effector function. *Nat Immunol.* 2014;15:815-823. doi:10.1038/ni.2938
282. Morris GP, Allen PM. How the TCR balances sensitivity and specificity for the recognition of self and pathogens. *Nat Immunol.* 2012;13:121-128. doi:10.1038/ni.2190
283. Grossman Z, Min B, Meier-Schellersheim M, Paul WE. Concomitant regulation of T-cell activation and homeostasis. *Nat Rev Immunol.* 2004;4:387-395.
284. Kersh EN, Kersh GJ, Allen PM. Partially phosphorylated T cell receptor zeta molecules can inhibit T cell activation. *J Exp Med.* 1999;190:1627-1636.
285. Stefanova I, Dorfman JR, Germain RN. Self-recognition promotes the foreign antigen sensitivity of naive T lymphocytes. *Nature.* 2002;420:429-434.
286. Fulton RB, Hamilton SE, Xing Y, et al. The TCR's sensitivity to self peptide-MHC dictates the ability of naive CD8 T cells to respond to foreign antigens. *Nat Immunol.* 2014;16(1):107-117. doi:10.1038/ni.3043
287. Mandl JN, Monteiro JP, Vrisekoop N, Germain RN. T cell-positive selection uses self-ligand binding strength to optimize repertoire recognition of foreign antigens. *Immunity.* 2013;38:263-274. doi:10.1016/j.immuni.2012.09.011
288. Persaud SP, Parker CR, Lo WL, Weber KS, Allen PM. Intrinsic CD4 T cell sensitivity and response to a pathogen are set and sustained by avidity for thymic and peripheral complexes of self peptide and MHC. *Nat Immunol.* 2014;15(3):266-274. doi:10.1038/ni.2822
289. Chan AC, Kadlecek TA, Elder ME, et al. ZAP-70 deficiency in an autosomal recessive form of severe combined immunodeficiency. *Science.* 1994;264:1599-1601. doi:10.1126/science.8202713
290. Elder ME, Lin D, Clever J, et al. Human severe combined immunodeficiency due to a defect in ZAP-70, a T cell tyrosine kinase. *Science.* 1994;264:1596-1599. doi:10.1126/science.8202712
291. Chan AY, Punwani D, Kadlecek TA, et al. A novel human autoimmune syndrome caused by combined hypomorphic and activating mutations in ZAP-70. *J Exp Med.* 2016;213:155-165. doi:10.1084/jem.20150888
292. Sakaguchi N, Takahashi T, Hata H, et al. Altered thymic T-cell selection due to a mutation of the ZAP-70 gene causes autoimmune arthritis in mice. *Nature.* 2003;426:454-460. doi:10.1038/nature02119
293. Lev A, Lee YN, Sun G, et al. Inherited SLP76 deficiency in humans causes severe combined immunodeficiency, neutrophil and platelet defects. *J Exp Med.* 2021;218. doi:10.1084/jem.20201062
294. Bacchelli C, Moretti FA, Carmo M, et al. Mutations in linker for activation of T cells (LAT) lead to a novel form of severe combined immunodeficiency. *J Allergy Clin Immunol.* 2017;139:634.e5-642.e5. doi:10.1016/j.jaci.2016.05.036
295. Keller B, Zaidman I, Yousefi OS, et al. Early onset combined immunodeficiency and autoimmunity in patients with loss-of-function mutation in LAT. *J Exp Med.* 2016;213:1185-1199. doi:10.1084/jem.20151110
296. Sommers CL, Park CS, Lee J, et al. A LAT mutation that inhibits T cell development yet induces lymphoproliferation. *Science.* 2002;296:2040-2043. doi:10.1126/science.1069066
297. Weiss A. Human LAT mutation results in immune deficiency and autoimmunity but also raises questions about signaling pathways. *J Exp Med.* 2016;213:1114. doi:10.1084/jem.2137insight1
298. Aguado E, Richelme S, Nunez-Cruz S, et al. Induction of T helper type 2 immunity by a point mutation in the LAT adaptor. *Science.* 2002;296:2036-2040. doi:10.1126/science.1069057
299. Hauck F, Randriamampita C, Martin E, et al. Primary T-cell immunodeficiency with immunodysregulation caused by autosomal recessive LCK deficiency. *J Allergy Clin Immunol.* 2012;130:1144.e11-1152.e11. doi:10.1016/j.jaci.2012.07.029
300. Marth JD, Overell RW, Meier KE, Krebs EG, Perlmutter RM. Translational activation of the lck proto-oncogene. *Nature.* 1988;332:171-173. doi:10.1038/332171a0
301. Wright DD, Sefton BM, Kamps MP. Oncogenic activation of the Lck protein accompanies translocation of the LCK gene in the human HSB2 T-cell leukemia. *Mol Cell Biol.* 1994;14:2429-2437. doi:10.1128/mcb.14.4.2429-2437.1994
302. Cale CM, Klein NJ, Novelli V, Veys P, Jones AM, Morgan G. Severe combined immunodeficiency with abnormalities in expression of the common leucocyte antigen, CD45. *Arch Dis Child.* 1997;76(2):163-164.
303. Tchilian EZ, Wallace DL, Wells RS, Flower DR, Morgan G, Beverley PC. A deletion in the gene encoding the CD45 antigen in a patient with SCID. *J Immunol.* 2001;166:1308. doi:10.4049/jimmunol.166.2.1308
304. Kung C, Pingel JT, Heikinheimo M, et al. Mutations in the tyrosine phosphatase CD45 gene in a child with severe combined immunodeficiency disease. *Nat Med.* 2000;6(3):343-345.
305. Grimbacher B, Hutloff A, Schlesier M, et al. Homozygous loss of ICOS is associated with adult-onset common variable immunodeficiency. *Nat Immunol.* 2003;4:261-268. doi:10.1038/ni902
306. Roussel L, Landekic M, Golizeh M, et al. Loss of human ICOSL results in combined immunodeficiency. *J Exp Med.* 2018;215:3151-3164. doi:10.1084/jem.20180668
307. Vetrie D, Vorechovsky I, Sideras P, et al. The gene involved in X-linked agammaglobulinaemia is a member of the src family of protein-tyrosine kinases. *Nature.* 1993;361:226-233. doi:10.1038/361226a0
308. Tsukada S, Saffran DC, Rawlings DJ, et al. Deficient expression of a B cell cytoplasmic tyrosine kinase in human X-linked agammaglobulinemia. *Cell.* 1993;72:279-290. doi:10.1016/0092-8674(93)90667-f
309. DeVries TA, Neville MC, Reyland ME. Nuclear import of PKCδ is required for apoptosis: identification of a novel nuclear import sequence. *EMBO J.* 2002;21:6050-6060. doi:10.1093/emboj/cdf606
310. Miyamoto A, Nakayama K, Imaki H, et al. Increased proliferation of B cells and auto-immunity in mice lacking protein kinase Cdelta. *Nature.* 2002;416:865-869. doi:10.1038/416865a
311. Mecklenbräuker I, Saijo K, Zheng NY, Leitges M, Tarakhovsky A. Protein kinase Cdelta controls self-antigen-induced B-cell tolerance. *Nature.* 2002;416(6883):860-865.
312. Kuehn HS, Niemela JE, Rangel-Santos A, et al. Loss-of-function of the protein kinase C δ (PKCδ) causes a B-cell lymphoproliferative syndrome in humans. *Blood.* 2013;121:3117-3125. doi:10.1182/blood-2012-12-469544
313. Fu C, Turck CW, Kurosaki T, Chan AC. BLNK: a central linker protein in B cell activation. *Immunity.* 1998;9:93-103. doi:10.1016/s1074-7613(00)80591-9
314. Taguchi T, Kiyokawa N, Takenouch H, et al. Deficiency of BLNK hampers PLC-gamma2 phosphorylation and Ca^{2+} influx induced by the pre-B-cell receptor in human pre-B cells. *Immunology.* 2004;112:575-582. doi:10.1111/j.1365-2567.2004.01918.x
315. Lagresle-Peyrou C, Millili M, Luce S, et al. The BLNK adaptor protein has a nonredundant role in human B-cell differentiation. *J Allergy Clin Immunol.* 2014;134:145-154. doi:10.1016/j.jaci.2013.12.1083
316. Minegishi Y, Rohrer J, Coustan-Smith E, et al. An essential role for BLNK in human B cell development. *Science.* 1999;286:1954-1957. doi:10.1126/science.286.5446.1954
317. De Jesus AA, Montealegre G, Liu Y, et al. A de novo nonsense mutation in the tyrosine kinase lyn in a patient with an early onset autoinflammatory phenotype. *Pediatr Rheumatol Online J.* 2014;12:O25. doi:10.1186/1546-0096-12-S1-O25
318. Wang L, Aschenbrenner D, Zeng Z, et al. Gain-of-function variants in SYK cause immune dysregulation and systemic inflammation in humans and mice. *Nat Genet.* 2021;53:500-510. doi:10.1038/s41588-021-00803-4
319. Henriksson J, Chen X, Gomes T, et al. Genome-wide CRISPR screens in T helper cells reveal Pervasive crosstalk between activation and differentiation. *Cell.* 2019;176:882.e18-896.e18. doi:10.1016/j.cell.2018.11.044
320. Li FY, Chaigne-Delalande B, Kanellopoulou C, et al. Second messenger role for Mg^{2+} revealed by human T-cell immunodeficiency. *Nature.* 2011;475:471-U463. doi:10.1038/nature10246
321. Bousso P, Robey E. Dynamics of CD8+ T cell priming by dendritic cells in intact lymph nodes. *Nat Immunol.* 2003;4:579-585.
322. Qi H, Cannons JL, Klauschen F, Schwartzberg PL, Germain RN. SAP-controlled T-B cell interactions underlie germinal centre formation. *Nature.* 2008;455:764-769.
323. Egen JG, Rothfuchs AG, Feng CG, Horwitz MA, Sher A, Germain RN. Intravital imaging reveals limited antigen presentation and T cell effector function in mycobacterial granulomas. *Immunity.* 2011;34:807-819. doi:10.1016/j.immuni.2011.03.022
324. Celli S, Garcia Z, Bousso P. CD4 T cells integrate signals delivered during successive DC encounters *in vivo*. *J Exp Med.* 2005;202:1271-1278. doi:10.1084/jem.20051018
325. Zell T, Khoruts A, Ingulli E, Bonnevier JL, Mueller DL, Jenkins MK. Single-cell analysis of signal transduction in CD4 T cells stimulated by antigen *in vivo*. *Proc Natl Acad Sci U S A.* 2001;98:10805-10810. doi:10.1073/pnas.191567898
326. Krishnaswamy S, Spitzer MH, Mingueneau M, et al. Conditional density-based analysis of T cell signaling in single-cell data. *Science.* 2014;346(6213):1250689. doi:10.1126/science.1250689
327. Dong TX, Othy S, Jairaman A, et al. T-cell calcium dynamics visualized in a ratiometric tdTomato-GCaMP6f transgenic reporter mouse. *Elife.* 2017;6:e32417. doi:10.7554/eLife.32417
328. Zikherman J, Parameswaran R, Weiss A. Endogenous antigen tunes the responsiveness of naive B cells but not T cells. *Nature.* 2012;489:160-164. doi:10.1038/nature11311
329. Elliot TAE, Jennings EK, Lecky DAJ, et al. Antigen and checkpoint receptor engagement recalibrates T cell receptor signal strength. *Immunity.* 2021;54:2481.e6-2496.e6. doi:10.1016/j.immuni.2021.08.020

CHAPTER 25

B Cell Development in Mice

Nicole Baumgarth, DVM, PhD

INTRODUCTION

The pool of peripheral B lymphocytes found in mammalian hosts express antigen receptors, or B cell receptors (BCRs), that can bind to the vast array of pathogens, commensals, and other foreign antigens, while being largely unresponsive to self-antigens. This feat is achieved through a complex process, in which large-scale epigenetic and transcriptional programs, initiated and guided by environmental cues in the fetal liver before birth and the bone marrow after birth, drive hematopoietic stem cells (HSC) to commitment to the B lymphocyte lineage. Committed B cell precursors then undergo a stepwise process, in which "pro-BCR," "pre-BCR," and then mature BCRs are assembled through multiple immunoglobulin gene rearrangement processes, leading to expression of this antigen receptor that together with its signaling chains CD79a and CD79b and various coreceptors form the BCR signalosome. Successful gene rearrangement triggers critical signaling cascades inducing cell survival, proliferation at each step and induces the gene regulatory networks that support further differentiation, whereas failure to do so triggers death by neglect via apoptosis. Negative selection steps remove not only nonfunctional B cells but once the BCR is fully rearranged also purge cells that express strongly self-reactive BCR. Negative selection occurs both in the bone marrow and subsequently in the blood and secondary lymphoid tissues, inducing central and peripheral tolerance, respectively. Dysfunction of negative selection events enhances the likelihood of antibody-mediated autoimmunity.

In mice and humans B cell development from HSC is a highly ordered and complex process of differentiation that has been defined over the last 5 decades. The use of flow cytometry allowed identification of a changing set of cell surface receptors and internal proteins that help to delineate intermediate developmental stages that cells transit to develop into B lymphocytes. During B cell development, HSCs, with the capacity to generate all leukocytes, generate progeny with an increasingly more restricted ability to generate cell types other than B cells. Critical branching points have been identified in which common myeloid progenitors diverge from common lymphoid progenitors (CLPs), followed by further differentiation into common innate lymphocytes, pro-T and pro-B cells. A simplified example of such a developmental path is presented in Fig. 25.1.

This pathway of differentiation was confirmed by isolation and short-term culture of intermediate developmental stages, allowing developmental progression to occur, which helped to define the order. Separating B cell development into discrete stages and mapping the critical rearrangement of the BCR through V-D-J rearrangement and immunoglobulin (Ig) heavy and light chain expression allowed for the analysis and identification of key transcription factors, microenvironmental interactions, cytokines, and natural or engineered mutations on the path toward mature B cells. Checkpoints have been identified that lead to selection of mature B cells that express a highly diverse repertoire of antigen specificities and can bind to and help remove foreign antigens/pathogens, as well altered or damaged "self" antigens, without causing autoimmune disease.

The following sections will outline the sites of B cell lineage development at different stages of ontogeny, focusing on B cell development in the bone marrow of adult mice and the critical paths and checkpoints leading to the development of mature B-2 cells. Differences in the development of fetal/neonatal-derived B cells will be highlighted. Later sections will consider the stepwise differentiation process of cells from HSCs to the emergence of various specialized peripheral populations and the transcriptional and epigenetic mechanisms underlying the processes. For more detailed reading on antibody structure and function and the molecular genetics of VDJ recombination the reader is referred to Chapters 21 and 22, respectively.

EARLY DEVELOPMENT

Sites of B Lymphopoiesis During Ontogeny

In the mouse, hematopoiesis in which HSCs give rise to erythrocytes and leukocytes occurs predominantly in the fetal liver prior to birth, in the spleen just prior to and shortly after birth, and in the bone marrow thereafter. In addition, and prior to fetal liver hematopoiesis, the blood islands of the yolk sac (YS) contain hemogenic endothelium from which the first identifiable hematopoietic cells, nucleated erythrocytes with embryonic forms of hemoglobin, arise as early as embryonic day (E)7.25 prior to the emergence of HSC.[1,2] Several waves of hematopoiesis take place in the YS before the emergence of the first HSC at E11.5. These include the first wave of definitive hematopoiesis, in which precursors of erythroid and myeloid cells are generated starting around day E8.5. This leads to the development of some tissue-resident macrophage populations, including microglia in the brain and Kupffer cells in the liver, which are self-replenishing and retained into adulthood.[2] Although initial studies suggested

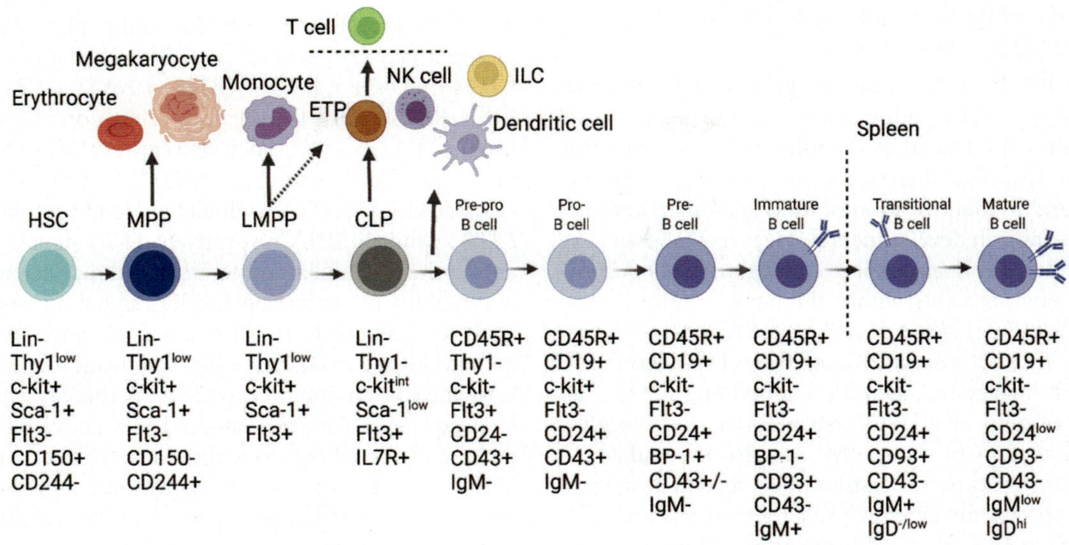

FIG. 25.1. Development of B cells from hematopoietic stem cells. Shown are the major steps in hematopoiesis in the bone marrow and spleen that lead to the development of mature B cells in the mouse. Phenotypic characteristics are described below each cell subtype. CLP, common lymphoid progenitor; ETP, early T cell progenitor; HSC, hematopoietic stem cell; ILC, innate-like lymphocytes; LMPP, lymphocyte and myeloid progenitor; MPP, multipotent progenitor.

lymphocyte development, and thus development of B cells, only occurs in subsequent waves of leukocyte development that arise from HSC precursors, increasing evidence now suggests that an early wave of non-HSC-derived lymphocyte precursors arises also directly from hemogenic endothelium in the YS and areas of the embryo, specifically the aortic-gonadomesonephros (AGM) region.[2,3] Whether there are differences in the hemogenic endothelial cells that give rise to these waves of non-HSC- and HSC-derived precursors, respectively, remains an important open question. Evidence for the presence of early, hemogenic endothelial-derived B cell precursors is consistent with an early study demonstrating that the earliest B cell precursors were found in the embryonic AGM region.[4] The first B cells emerging from the YS and AGM region independent of HSCs were shown to depend in development on the Sfpi1 regulatory element upstream of the transcriptional regulator PU.1[5] and to develop into innate-like B cells, giving them their name as the first-appearing B cells, that is, B-1 cells. Splenic marginal zone (MZ) B cells can also appear very early,[3,4] while most conventional, follicular or "B-2" cells emerge later in development.

HSCs capable of developing into all the blood cell types migrate to the fetal liver at about E10. Further waves of B cell hematopoiesis are observed from these long-term reconstituting "definitive" fetal liver HSCs until birth. They are generated by two coexisting but distinct types of HSCs in the fetal liver. One is a population of developmentally restricted (dr) HSCs with multilineage potential, which like the YS and AGM precursors predominantly gives rise to B-1 cells, whereas the other, adult-type HSC, can generate both B-1 and B-2 cells.[6] However, differences exist also between fetal and adult bone marrow B-2 cell development, as demonstrated recently by their differential dependence on Sfpi1/PU.1 as assessed in PU.1 hypomorphic mice.[5] Together, the data suggest the presence of at least three to four waves of B-1 cell development (pre-HCS, fetal-liver drHSC and HSC, and bone marrow HSC) as well as at least two waves of HCS-dependent B-2 cell development (fetal liver and bone marrow[5,6]).

The progression of B cell development from precursors to mature B cells can be delineated by flow cytometry, showing early pro-B cells appearing around E14 and then increasing rapidly in frequency and maturing further to pre-B and immature B cells until birth. Early precursors can also be found in the fetal omentum.[7] At birth, B cell development occurs also in the spleen for a period of 2 to 4 weeks.[8] The bone marrow then takes over rapidly as site of exclusively HSC-derived hematopoiesis. Elegant studies on B cell output from precursors demonstrated a rapid decline in the output of B-1 and an emergence of predominant B-2 cell development from bone marrow precursors, such that by about 6 weeks of age, B-2 cells are the nearly exclusive outcome of B cell development.[9] The regulation of this molecular switch from fetal-type, mostly B-1 generating, to adult-type, mostly B-2 cell generating, hematopoiesis has recently been shown to require suppression of Lin28b, a DNA-binding protein that regulates the Let7 family of microRNAs ([10,11], see also later sections). Ectopic overexpression of Lin28b resulted in enhanced B-1 generation from bone marrow HSC.[10,11] Nonetheless, in situations of hematopoietic crisis, such as that induced following whole-body irradiation of mice, reconstituted bone marrow precursors appear to generate an increasing number of (mostly CD5−) B-1 cells.[12,13] As mice age, B cell output is diminished, likely contributing to immune dysfunction in the aged.[14]

Stem Cells, Commitment, and Early B Cell Progenitors in Bone Marrow

Conventional B cells are continually generated from bone marrow–resident HSCs in adult mice. Analysis of the bone marrow niche is complicated by the difficulty in imaging the bone histologically. Nonetheless, recent progress in imaging technologies and in identification of individual markers that can cleanly distinguish developmental stages have shown that multipotent HSC and lineage-specific daughter cells occupy distinct microenvironments within the bone.[15,16] Flow cytometric identification of HSCs described them as Thy-1^{lo} Sca-1^{+} Lineage^{-} c-kit $^{+}$.[17,18] About 20% of these HSCs establish long-term multilineage reconstitution, a defining functional characteristic of definitive HSCs that requires them to give rise to cells in a recipient mouse that can also repopulate all the blood cell lineages upon retransfer into a second host, thus indicating a capacity for extensive self-renewal without differentiation into more restricted progenitors. More recently, differential gene expression analysis between this population of bone marrow cells and the multipotent progenitors (MPPs), which arise from HSC as Thy-1^{lo} Sca-1^{+} Lineage^{-}c-kit^{+} B220- cells,[18,19] identified expression of various SLAM family surface receptors as able to distinguish HSCs using a small number of surface markers, suitable for visualization by immunohistochemistry.[19] In those studies, HSCs were identified as CD150+ CD48− CD244−, which distinguished them from the CD150− CD48− CD244+ MPP and other bone marrow cells.[19] Scal+ c-kit+ CD150+ CD48− CD244− cells represented about 0.0084% of bone marrow cells, and nearly half of those were providing multilineage long-term reconstitution after single-cell transfer into adult mice. Further differentiation of MPPs leads to the branching of precursors that are restricted in their development to either cells of the myeloid or the lymphoid lineage, ie, common myeloid and common lymphoid progenitors.[20-22] The developmental paths that result in these major differential fates of leukocytes are increasingly identified and have shown the critical role of hematopoietic cytokines in selectively inducing and then reinforcing lineage commitment choices.[23,24]

CLPs give rise to all lymphocytes, including B cells, T cells, and the family of innate-like lymphocytes (ILCs) that includes classical NK cells and innate-like helper cells, now categorized as ILC1, 2, and 3,[25] as well as some dendritic cells (DCs). B cells, NK cells, and ILCs, as well as lymphoid-derived DCs continue their development in the bone marrow, while T cells develop in the thymus following the migration of precursors from bone marrow to the thymus.[26] Recent advances in the search for the thymus-seeding precursors have identified populations of CD34+ CD10+ CD44+ CD7− cells that overlap in their transcriptional profile with that of bone marrow–resident precursors as well as significant overlaps and some differences between these populations in humans and mice.[27,28] Cells with the phenotype of CLP constitute about 1/3000 of bone marrow cells. This precursor fraction was identified by lack of staining with a panel of "lineage markers (Lin-)" and of Thy-1, low-level expression of Scal (Ly6A/E), and distinctive, intermediate levels of c-Kit (CD117), compared with higher levels on HSCs.[20] In another phenotypic profiling of B cell precursors, CLPs were identified among the CD4− CD24− CD43+ CD45R− (B220−) AA4.1+ "Hardy Fraction A0" cells, following the naming convention of Hardy and colleagues (Fig. 25.2)[29,30]. They also express two critical cytokine receptors, the IL-7Rα chain (CD127) and CD135 (Fms-like tyrosine kinase 3, FLT3), which bind to the cytokines IL-7 and FLT3 ligand (FLT3L), respectively. FLT3 signaling is important for the continued proliferation and maintenance of CLP.[31] It functions by activating the Ras signaling cascade, which leads to increases in IL-7Rα expression and the suppression of Janus kinase 2 (Jak2)-mediated activation of suppressors of cytokine stimulation (SOCS) 2 and 3, thereby removing their blockade of IL-7Ra-mediated STAT5 activation.[32] Together with FLT3, IL-7R acts to maintain survival and proliferation of CLP.[33] In addition, it serves critical and nonredundant functions in supporting bone marrow lymphocyte growth, proliferation, and survival of B cell developmental stages from the CLP to the pre-B/large pre-BII stage.[34-36] Human patients with various IL7Ra mutations have been shown to retain some B lymphocyte development, and thus it appears that B cell development in humans may utilize signals other than those provided through the IL7R. However, those patients nonetheless suffer from hypogammaglobulinemia (in addition to loss of T cell development), suggesting that even in humans IL7R signaling is required for normal B cell development.[37]

Rather than a progression of strict branching points at which a precursor would irrevocably commit the B cell (or other) lineage, it appears that, instead, CLPs give rise to daughter cells that are less and less likely to develop into cells other than B cells. Analysis of the LIN− IL7Rα+ cKIT+ CD45/B220− CLP demonstrated that, although CLPs fail to efficiently generate myeloid cells upon transfer into irradiated hosts, they nevertheless retained significant capacity to produce such cells in short-term cultures, likely due to continued (albeit reduced) expression of receptors for myeloid growth factors. In contrast, this myeloid capacity is greatly reduced as cells begin to express CD45R/B220, concomitant with reduced expression of receptors for myeloid growth factor receptors. Although these B220+ cells poorly reconstituted T cells in cell transfer assays, they nonetheless retained the capacity to generate T-lineage cells in culture, mediated by engagement of Notch by its ligand DL1.[38,39] In addition, cells considered to be progressing toward a lymphoid/B-lineage path can be suppressed and redirected also to become DC by toll-like receptor (TLR) 9 signaling. IFN-γ has also been shown to affect lymphoid cell development. The data thus suggest that infections or other severe proinflammatory states can alter early stages of lymphoid hematopoiesis through direct action on early progenitors[40,41] in a manner well appreciated for myeloid cell development.[42,43]

The potential for alternate hematopoietic lineages thus appears to be lost gradually and perhaps later in progression down the B-lineage pathway in mouse bone marrow than previously thought. On the other hand, it appears that Ig rearrangement is initiated earlier than some studies have indicated. Determination of the extent of germline deoxyribonucleic

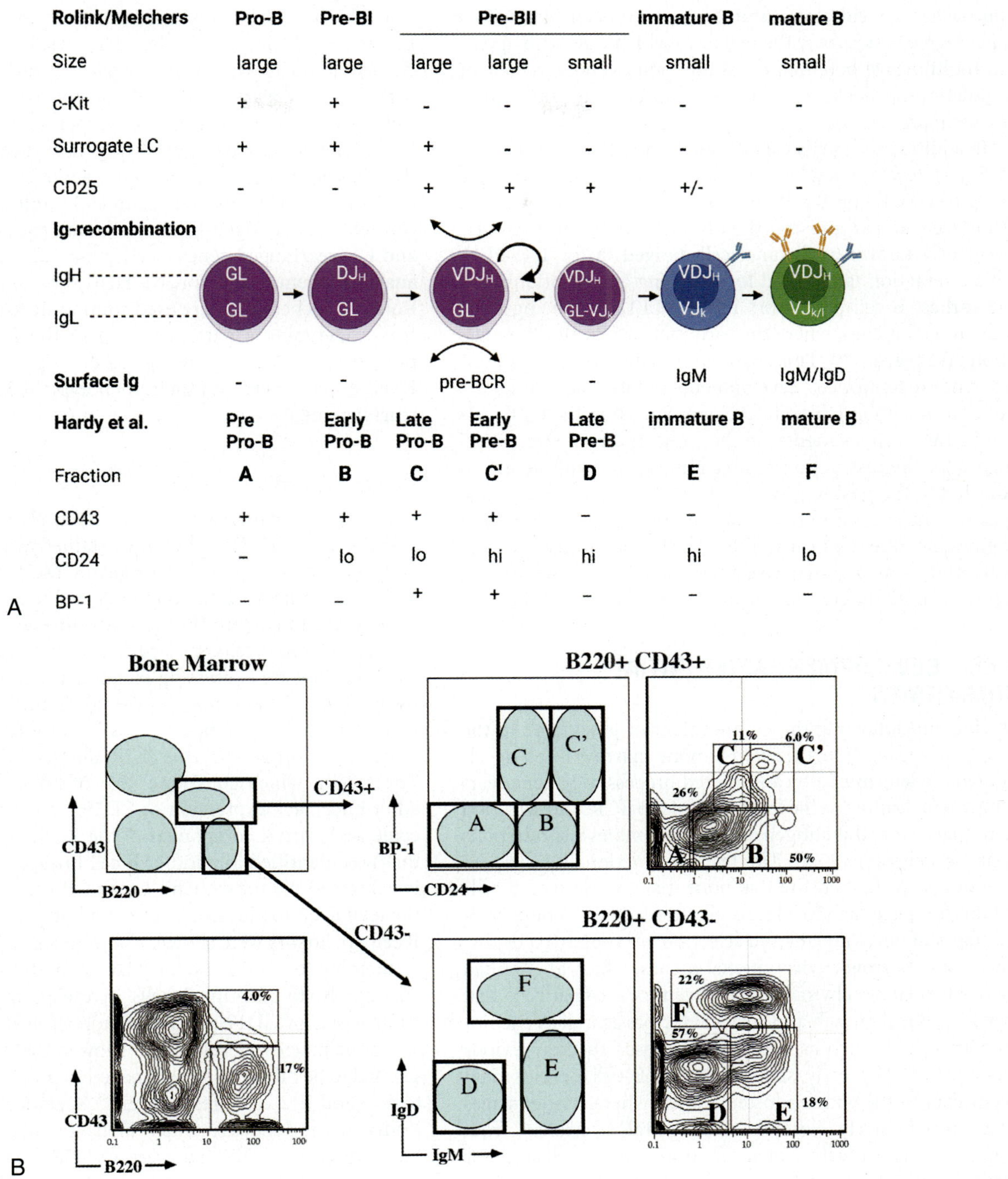

FIG. 25.2. **B cell precursor identification. A:** Shown are the phenotypic characteristics of indicated B cell developmental stages using two widely used cell subset definitions developed by the laboratories of Rolink and Melchers (top) and Hardy and colleagues (bottom). Note that each change in surface marker expression has a corresponding change in Ig heavy and/or light chain rearrangement. **B:** Shown is a flow cytometric analysis of bone marrow from C57BL6 mice using the Hardy scheme. GL, germline.

acid (DNA) segments lost upon D-J rearrangement and the formation of such D-J segments in individual cells isolated by flow cytometry showed that 30% to 50% of cells in CLPs and more than 80% of the B220+ cells contained a D-J rearrangement on at least one chromosome.[44] This is consistent with high-level expression of genes important in Ig rearrangement, including TdT, Rag-1, and Rag-2 in these cells.[45] Additional issues remain to be identified regarding the lineage restriction of cells at these early stages in B cell development. For example, there is evidence that cells restricted to generating B and myeloid/macrophage (but not T) lineage may exist in the fetal liver[46] and in bone marrow.[47] The availability of sensitive single-cell RNA sequencing approaches, as well as the combination of cellular barcoding and adoptive transfer approaches, provides opportunities now to determine with a high degree of accuracy the extent to which cells with apparent multilineage potential constitute homogenous precursor populations rather than mixed populations of distinct unidirectional precursors.

In addition to signals via Flt3 and IL7Rα, the earliest stages of B cell development require expression of the chemokine receptor CXCR4 on B cell precursors. On these cells, CXCR4 is induced by IL7 and can then bind to its ligand CXCL12, previously known as Stromal cell-derived factor 1 (SDF1). This interaction is required for retaining and positioning of the earliest B cell precursors into CXCL12-producing bone marrow cell niches, where the early steps of B cell differentiation take place.[24,48,49] This chemokine is dispensable for multipotent progenitor cell development, and its absence reduces but does not abolish T cell development. Thus, the CXCL12-producing compartments of the bone marrow are critical niches for the instruction of B cell lineage commitment.[49] In addition to the positioning of cells into the appropriate bone marrow microenvironments, signals by these cytokine and chemokine receptors cause the induction of transcriptional regulators, which instruct self-reenforced pathways toward stepwise B cell differentiation, as discussed further below.

B CELL DEVELOPMENT AFTER INITIAL COMMITMENT

During and following the commitment of precursors to the B cell lineage, each further step in bone marrow B cell development is determined by an ordered process of Ig gene rearrangement, leading to the formation of the BCR, and selection steps that ensure this antigen receptor assembles and functions appropriately (Fig. 25.2). The test for functionality includes selection steps that ensure that bone marrow–derived B cells exhibit no more than low levels of autoreactivity. These critical stages of development and selection are supported by distinct bone marrow environmental niches, which are defined by resident mesenchymal stem/progenitor stromal cells that provide critical growth and survival factors as well as chemokine attractants such as CXCL12, discussed in the previous section.[24,50,51] The early stages of adult B cell development occur nearest the bone endosteum layer, whereas latter stages occur progressively closer to the central arteriole,[50] from which B cells eventually exit at the "immature" B cell stage for further maturation and final selection in the spleen. Various stromal cell culture systems had been established initially to systematically characterize the process of B cell development and the molecular and phenotypic changes that occur as bone marrow and fetal liver precursor B cells undergo differentiation.[52,53] The combination of phenotypic characterization coupled with analysis of growth and differentiation in culture provided a powerful approach for delineating B cell development, and for the identification of critical cytokines and some of the cell adhesion molecules important for in vivo development of these cells.

From these studies, a stepwise path of B cell development from early progenitors, CLPs to pre-pro B cells, pro-B cells, pre-B cells, and then immature B cells can now be described (Fig. 25.2). Using flow cytometry, characteristic changes in surface phenotype can be identified that correspond to changes in the Ig-locus that take place, in order for the cells to express a fully rearranged BCR that consists of two identical, fully rearranged Igμ-heavy chains together with two identical kappa or two lambda-light chains. Each heavy chain of the BCR has one antigen-binding site encoded by three genes: the (V_H) variable, (D_H) diversity, and (J_H) joining genes, whereas the light chain antigen-binding sites are encoded by two ($V_{\kappa/\lambda}$ and $J_{\kappa/\lambda}$) rearranged genes (see Chapter 22). Two phenotypic analyses schemes, one from the Hardy lab[29] and the other from Rolink/Melchers lab,[54] are used commonly to identify precursors undergoing a particular step in the Ig-rearrangement process (Fig. 25.2). Following is a description of the steps of B cell development and their phenotypic delineation and Ig-rearrangement state.

Pre-pro B Cells

Phenotype Bone marrow cells that are surface CD24– (HSA) AA4.1+ and CD43$^{lo/+}$, the latter stained with the mAb S7, and in addition express the isoform CD45R of the common leukocyte antigen, CD45, often referred to as B220 in mice,[55] were found to contain the earliest B lineage precursors, also termed pre-pro B cells or Hardy fraction A_1, A_2 depending on whether they are expressing or not expressing CD4, respectively, (Fig. 25.2 and not shown), as defined by their ability to preferentially generate B cells in stromal cell cultures.[30] They also express c-Kit, the IL7R, and lack CD25 (IL2Rα).[54] These cells, which constitute <3% of bone marrow cells of adult mice, lack expression of CD19,[56] a costimulatory molecule and critical component of the BCR signaling complex that was identified as another B cell lineage marker and that is expressed on subsequent steps of B cell development. In the fetal liver and juvenile, but not adult, bone marrow, these B cell precursors were shown also to retain flt3 expression.[57]

Staining for CD19 and for B220 are both used, often interchangeably, to identify B cells in the periphery. However, comparison of CD19 and B220 expression during fetal liver and bone marrow development showed that these molecules seemed to be differentially expressed in precursors giving rise to B-1 and B-2 cells, respectively.[58] Specifically, Montecino-Rodriguez and colleagues identified a population of Lin- cells that expressed CD19 but were CD45R$^{lo/-}$ among the Lin- AA4.1$^+$ adult bone marrow populations. This population represented about 0.1%-0.4% of adult bone marrow cells, but significantly higher levels in the fetal liver and bone marrow environment[58,59]; was responsive to TLSP in vitro; and gave rise to B cells and macrophages in vitro, and peritoneal cavity B-1 cells in vivo after adoptive transfer.[47,58] Together these studies suggested that pre-pro B-1 precursors express CD19 but not B220, whereas the same precursors giving rise to B-2 cells have the opposite phenotype. This is consistent with the roughly 3-fold higher level expression of CD19 on B-1 cells and the strict requirement for CD19 expression for the generation of CD5+ B-1 cells,[60] as well as the reduced expression of B220 on mature B-1 cells in spleen and body cavities, when compared with that of mature B-2 cells.[61]

Thus, B-1 and B-2 cells appear to show subtle differences in BCR coreceptor expression as early as the pro-B cell stage of development.

Ig-DNA rearrangement At the molecular level, pre-pro B cells begin to express Igμ-heavy chain "sterile transcripts," signaling transcriptional accessibility of the Ig heavy chain (IgH)-gene locus and the beginning of the rearrangement process, further identifying these cells as B cell precursors. They also express the recombinase activating genes Rag-1 and Rag-2, which together form the double-stranded breaks in DNA required for Ig rearrangement.[62,63] Rag-1 binds in a highly specific fashion to discrete sites within the IgH locus, recombination signal sequences, defining "recombination centers."[64] Furthermore, in the adult bone marrow, but not the fetal liver, these cell express high levels of terminal deoxytransferase (TdT), the enzyme responsible for adding non-templated nucleotides at the D-J and V-D junctions of the heavy chain, further enhancing the variability and thus the repertoire of the B cells emerging from the bone marrow.[65,66] Indeed, most cells at this stage of development have undergone the first rearrangement process, the annealing of a IgH D and a J-gene (DJ_H) on either one or both Ig alleles. This process is enabled by epigenetic DNA and chromatin changes that provide access of the recombination machinery to this region of the Ig-locus.[67,68] In contrast to other Ig-rearrangement steps D_H to J_H rearrangement can proceed simultaneously on both alleles. B cell development is arrested at this stage, if neither allele productively rearranges, for example, due to the insertion of a stop codon. A more detailed description of the processes can be found in Chapter 22.

BCR signaling complex: Pre-pro-B cells also begin to express the two signaling chains, CD79a (Igα, mb–1) and CD79b (Igβ, B29), that are noncovalently associated with the pre-BCR and the BCR. They also begin to express the "surrogate light chains," lamda5 (CD179b) and VpreB (CD179a), following successful DJ_H rearrangement.[69] Lamda5 is similar to the lambda L chain constant region, and VpreB is similar to the V domain of the L chain. Together, these peptides assemble into what constitutes a pseudo- or surrogate Ig-light chain. These chains do not require rearrangement and associate with fully rearranged μ-heavy chains to form the pre-BCR.[69]

Pro B Cells/Pre-BI

This step in B cell development is considered the B cell commitment stage, as pro B/Pre-BI cells have little potential to develop into cells other than B cells. This is facilitated in large part through the induction of the transcriptional repressor and master regulator PAX-5, which promotes B cell commitment by suppressing other fate choices. PAX-5 is responsible also for the loss of flt3 expression (see below)[70] and the upregulation of CD19. Transcriptional regulation is discussed in more detail in later parts of this chapter.

Phenotype As pre-pro B cells develop further, these AA4.1+ CD4− CD43+ B220+ cells first upregulate surface expression of CD24 (HSA, Hardy Fraction B) and then BP-1 (Hardy Fraction C), a zinc-dependent cell surface metallopeptidase also known as aminopeptidase A (136), and they are CD19+. They do not express CD25 (IL2Rα) or flt3, identifying them as pre-BI according to the Melchers naming convention,[54] whereas both fractions B and C are considered pro B cells according to the Hardy naming convention,[29] to signify the stepwise developmental process prior to pre-BCR expression. A diagram summarizing this type of phenotypic subdivision and relating different nomenclatures is shown in Fig. 25.2. Pro/pre-BI cells continue to express the IL7Rα, a critical STAT-5-mediated signal that supports pro-B cell survival and suppresses light chain rearrangement.[71,72] The interaction of pro-B cells with cytokine-producing stromal cells is supported through their high expression of focal adhesion kinase and very late antigen 4 (VLA4) that promotes binding to vascular adhesion molecule 1 (VCAM-1) on the stroma.[24]

Ig-DNA rearrangement Pro-B cells in Hardy fraction B have completed DJ_H recombination and begin to undergo V_H–DJ_H rearrangement as they upregulate BP-1, thus move into fraction C. The rearrangement process is dependent not only on expression of the two Rag proteins but also on expression of the DNA-dependent protein kinase (DNA-Pk) that facilitates DNA double-stranded break repair during VDJ recombination.[73] Mice with mutations in the subunits of DNA-PK (Ku-70, Ku-80, and p350) show B cell developmental arrest at the pre-pro to pro-B cell stage, expressing very few B220+ CD43+ bone marrow precursors. The spontaneous mutation that is present in the T and B cell–deficient severe combined immunodeficient (*SCID*) mouse strain is caused by mutations in the DNA-PK catalytic subunit p350.[74] In addition to the prosurvival signals provided by IL7R signaling, recent studies have demonstrated a critical role for IL7R signaling in regulating access to distal 5′ V_H genes for recombination. In pro-B cells that lacked IL7R signaling, both DJ_H and VDJ_H rearrangement used only proximal D_H and V_H genes, demonstrating the importance of IL7 not only for the survival and expansion of B cell precursors but also for the breadth of the developing repertoire.[75]

BCR signaling complex Pro-B cells continue to express the surrogate light chains as well as the BCR signaling chains, CD79α and CD79β. Existing studies suggest that the surrogate light chains form a "pro-BCR" together with an adhesion molecule, cadherin-17.[76,77] Expression of this cadherin is downregulated after the pro-B cell stage, before being upregulated again at the immature B cell stage, suggesting its involvement in regulating the location of the developing B cells, possibly supporting a close association between these cells and stromal cells, which may also provide unknown additional stimuli.[78] Neither the constant region of the Igμ-heavy chain nor the BCR signaling chains seem to assemble with this "pro-BCR." It is unresolved whether the presence of CD79a and CD79b in the cell membrane of pro B cells transmit any signals at this stage of development.

Pre-B Cells/Large Pre-BII

The pre-B/Large pre-BII (called pre-B cells for short) complete the V_H-DJ_H rearrangement process of the μ-heavy chain and assemble a pre-BCR with the surrogate light chains and the CD79a and b signaling chains. Pre-BCR signaling of a functional receptor will induce rapid clonal expansion in an IL-7-dependent manner. The assembly and signaling of the pre-BCR is a critical checkpoint in B cell development. It is interesting to note that fetal/B-1 cell hematopoiesis did not appear to have the same dependence on IL7R signaling. Initial studies suggested that they instead were responsive to thymic stromal-derived lymphopoietin (TSLP;[79,80]), which implied that this cytokine might regulate fetal B cell development. However, TSLP receptor-deficient mice show no abnormality in fetal or adult B cell development. More recent studies indicated that signaling through FLT3 might be the main driver of fetal-derived IL-7R-independent B cell development.[81] Using *Flt3l*$^{-/-}$ or *Il7*$^{-/-}$ mice, it was shown that, although TSLP receptor deficiency failed to have any further effect on postnatal B cell development, a combined loss of FLT3L and IL-7 completely abolished all neonatal and adult B lymphopoiesis.[82] However, it remains possible that specific fetal liver and/or bone marrow niches provide preferential cytokine signals that support B cell development in a redundant manner.

Phenotype Pre-B cells begin to express the cytokine receptor IL-2Rα and rapidly become larger in size, as they undergo rapid clonal expansion (Melchers nomenclature, large pre-BII; Fig. 25.2).[83] They also increase expression of CD24 and remain BP-1$^+$ and CD43+ (Hardy Fraction C′).

Ig-DNA rearrangement Pre-B cells have mostly completed the VDJ$_H$ gene rearrangement process, which occurs at one of the two Ig heavy chain alleles at a time. Important, productive rearrangement that results in a functional Igμ-heavy chain protein triggers a rapid shutdown of any further rearrangement,[84] due to the rapid downregulation of Rag-1 and Rag-2 as well as other epigenetic changes that make the heavy chain locus inaccessible for further recombination.[85,86] This process, termed "allelic exclusion," ensures that each B cell expresses a BCR that is encoded by only one heavy chain allele and thus that it expresses BCRs of only one antigen binding specificity.[69,87] Furthermore, it ensures that the genomic integrity of the cells is not compromised by double-stranded DNA breaks while these cells are undergoing rapid proliferation, as gene rearrangement is confined to the G_0 and G_1 stages of the cell cycle.[88] As outlined in the next section, pre-BCR assembly and signaling ensures also the continued suppression of light chain rearrangement at this stage of B cell development.

The requirement for pre-BCR signaling at this stage of B cell development importantly shapes the peripheral BCR repertoire by selecting VDJ$_H$ rearrangements that can form a pre-BCR (see next section). An example of VDJ$_H$-rearrangement selection is the D-proximal V$_H$ gene, V$_H$81X, which is overrepresented among pro B cells due to its preferential rearrangement.[89] Regulatory sequences interspersed throughout the distal VH region, termed PAX5-activated intergenic repeat elements, appear to enhance utilization of distal J558 family V$_H$ genes.[90] Furthermore, a recombination regulatory region, the intergenic control region 1, which lies between the V$_H$ and D$_H$ clusters, was shown to inhibit proximal and promote distal VH gene utilization.[91,92] Although abundant in early B cell precursors, V$_H$81X is rare in the mature B cell compartment. Such a decrease in representation of cells with V$_H$81X rearrangements at the bone marrow pre-B clonal expansion phase[93,94] was explained by the frequent failure of heavy chains utilizing VH81X to associate with the surrogate light chains, and thus to form a pre-BCRs.[95-97] Thus, pre-BCR formation and signaling establishes a critical selection step during bone marrow B cell development.

BCR signaling complex Completion of VDJ$_H$ recombination leads to the rapid expression of the fully mature Igμ heavy chain protein. Two Igμ heavy chains assemble on the cell surface together with two surrogate light chains. The surrogate light chains are formed by VpreB and Lamda5, in lieu of light chain variable and constant regions, respectively. Together these four chains assemble in what is called the pre-BCR.[69,98] The μ heavy chain has a very short, three-amino-acid cytoplasmic region, and signaling of the pre-BCR, similar to the BCR, is facilitated through assembly of this complex with CD79a and CD79b. Consistent with a requirement for pre-BCR signaling in support for continued B cell development, inactivation of CD79a or CD79b resulted in a block before the pre-B cell/large pre BII stage in mouse bone marrow.[99] In further support, deletion of the Syk tyrosine kinase, which binds to the ITAM motives of the CD79 chains, caused B cell arrest at the pre-B cell stage.[100,101] Thus, mutations that affect the signaling capacity of the pre-BCR complex preclude efficient progression of B cell precursors past pre-BI or Hardy fraction C and prevents the clonal expansion of these precursors.

The precise nature of pre-BCR signaling has been the subject of intense investigation. It did not lead to the identification of a defined pre-BCR ligand. Instead, the fact that development past the pre-BCR checkpoint can occur in vitro in the absence of other cells,[102] and that kappa light chains can substitute for the surrogate light chains to support B cell development past the pre-BCR stage,[103] has led to the model that pre-BCR signaling is similar to "tonic" mature BCR signaling; that is, pre-BCR signaling might be initiated simply as the complex itself is assembled. Alternatively, increased pre-BCR expression may lead to homodimerization and associated conformational changes that affect signaling, which is supported by glycosylation of a conserved asparagine residue in the first μ constant region domain,[104] although other models of pre-BCR signaling remain to be explored.

In addition to inducing 4 to 5 rounds of cell division,[83,102] pre-BCR signaling induces the downregulation of TdT[105] and of the surrogate light chain genes,[106] and increases expression of prosurvival factors Mcl1 and Bcl2, and suppresses the proapoptotic factors Bim and Bad.[107] By inducing proliferation, the pre-BCR suppresses kappa light chain rearrangement, in part through epigenetic changes enforced by the Myc-dependent induction of cyclin D3, which inhibits DNA recombination.[108,109] Although initial studies suggested that pre-BCR signaling also initiates the rapid downregulation of the Rag genes through the activation of the PI3K-Akt signaling pathway[85] and suppression of FoxO1, a transcriptional regulator of Rag gene expression,[110,111] it has now been shown that the induction of the PI3K pathway in pre-B cells is facilitated through IL7R signaling.[112] The interplay and differential signaling through the pre-BCR and the IL7R explain how pre-BCR signaling can be required for the induction of cell cycle entry, clonal expansion, and repression of DNA recombination at the light chain locus, on the one hand, but then on the other hand for the induction of light chain rearrangement via enhanced kappa light chain accessibility and upregulation of Rag gene expression,[107,113-115] that is, the initiation of the next step in B cell development.

The eventual downregulation of the pre-BCR, which is initiated by the suppression of surrogate light chain genes, does not drive the exit of pre-B cells from the cell cycle, as the constitutive overexpression of surrogate light chains did not affect cell cycle exit.[114] Instead, activation of BLNK/Syk phosphorylation via pre-BCR signaling itself initiates cell cycle exit.[114] BLNK/Syk phosphorylation antagonizes PI3K/Akt-driven cell proliferation by inducing the mitogen-activated protein kinase p38. p38 then represses (IL7R-mediated) PI3K-Akt activation, causing the derepression of FoxO1 as well as PAX5, which leads to induction of IRF4 and IRF8.[112,116] IRF4 suppresses cell cycle entry by restraining the pre-BCR-induced expression of c-Myc and induces the cell cycle inhibitor p27. p27-mediated downregulation of cyclin D3 leads to the exit from the cell cycle. Cyclin D3 suppression also causes the derepression of light chain rearrangement, which together with IRF4-mediated changes to light chain enhancer activity initiates kappa light chain rearrangement.[107,116,117]

Thus, in addition to supporting B lineage commitment and development prior to the pre-B stage, IL7R signaling, via activation of the STAT5, PI3K, and the MAP/Erk signaling pathways, complements signals from a functional pre-BCR. Together these signals enhance cell survival, initiate proliferation, and restrain light chain recombination.[107,118] The overlapping induction of these signaling pathways by pre-BCR and IL7R has made it difficult to cleanly distinguish their individual contributions. The extent of pre-BCR and IL7R-mediated signaling not only supports pre-B cell survival and proliferation, but also regulates the further development of the pre-B cells by regulating cell cycle exit. It is the continued exposure of pre-B cells to IL7 that retains these cells in cell cycle and suppresses Ig-kappa recombination.[71]

In addition to the eventual repression of IL7R-mediated PI3K activation by pre-BCR signaling, repositioning of cycling pre-B cells in the bone marrow seems to cause their withdrawal from the stromal cell sources of IL7, and thus PI3K-signaling.[119] The process of repositioning appears to be facilitated by the IRF4-mediated induction of CXCR4, the chemokine receptor for CXCL12.[51,112,116] In support of that model, recent single-cell RNA sequencing of bone marrow stromal cells identified distinct stromal cell subsets differing in CXCR4 and IL7 expression. Certain stromal cells were found to express both CXCR4 and IL7 at high levels, others expressed one factor higher than the other, or one exclusively.[120] The data are consistent with earlier models that proposed that pre-B cells, supported by $IL7^{hi}$ $CXCL12^{lo}$ stromal cells, migrate toward $IL7^{lo}$ $CXCL12^{hi}$ cells as pre-B cells become $CXCR4^{hi}$. Reduced exposure to IL7 then initiates the above outlined changes that cause the critical next steps in further B cell development.[34,116,121]

The outlined steps guiding bone marrow B cell development of mammals are not seen in other vertebrates, such as sharks, which lack surrogate light chains.[122] Even in some mammals the processes are distinct from those outlined here. One example are swine, which rearrange their light chains prior to the μ-heavy chains, and B cell development seems to occur independent of surrogate light chains.[123] In mice, accumulating evidence now suggests that the development of at least some B-1 cells also proceeds independent of surrogate light chains, as outlined in the next section (Fig. 25.3).

Pre-BCR-Independent Alternative Pathway of B Cell Development

The earliest studies exploring the function and impact of the surrogate light chains on B cell development noted that the deletion of Lambda5 or the VpreB genes, and thus the lack of a pre-BCR, led to the arrest of most but not all B cells past the pre-B cell stage.[124,125] A small population of peripheral B cells remained present in these mice, many of which expressed CD5.[124-126] CD5 is a negative regulator of BCR signaling that in mice is expressed by fetal-derived, autoreactive B-1 cells as well as autoreactive, anergic B cells.[127] Indeed, V_H usage that can result in autoreactive B cell development was shown to disfavor assembly of a pre-BCR complex, and B cells that developed in the absence of surrogate light chains were enriched for autoreactive B cells[126] as well as B-1 cells.[128] Thus, pre-BCR-mediated signaling acts as both a checkpoint for the functionality of the rearranged μ-heavy chain and as a positive selection signal for the survival and expansion of B cells utilizing nonautoreactive V_H genes.

The presence of a distinct developmental pathway for B-1 cells, capable of generating self-reactive natural antibodies that can facilitate housekeeping functions such as removal of dead and dying cells[61] is highlighted by the above outlined findings. Consistent with studies suggesting that surrogate light chains might be dispensable for at least some B-1 cell development,[124,125] an early study from the Hardy lab showed that CD5+ B-1 cells that utilize VH11, which encodes binding to phosphatidylcholine-containing liposomes considered mimics of self-antigens, were unable to associate effectively with surrogate light chains. Yet, strong VH11+ B cell expansion and B-1 cell development still occurred in the absence of pre-BCR signaling.[129] Given that VH11 expression is limited to CD5+ B-1 cells, the data suggested that a different, pre-BCR-independent pathway supported their development (Fig. 25.3). Indeed, the differential requirement for pre-BCR signaling was recently shown to be connected to the reduced levels of IL7R expression[128] and STAT-5 phosphorylation[130] by fetal-derived pro-B cells compared with their counterparts in the bone marrow. This is consistent also with data showing that IL7Rα signaling supports, but is not absolutely required, for B-1 cell development in mice.[131] Reduction in IL7R signaling in fetal liver pro-B cells seems to result in a rapid derepression of kappa light chain transcription and initiation of kappa light chain rearrangement.[128] Thus, instead of utilizing surrogate light chains to form a pre-BCR that provides survival and expansion signals, the rapidly rearranged kappa light chains

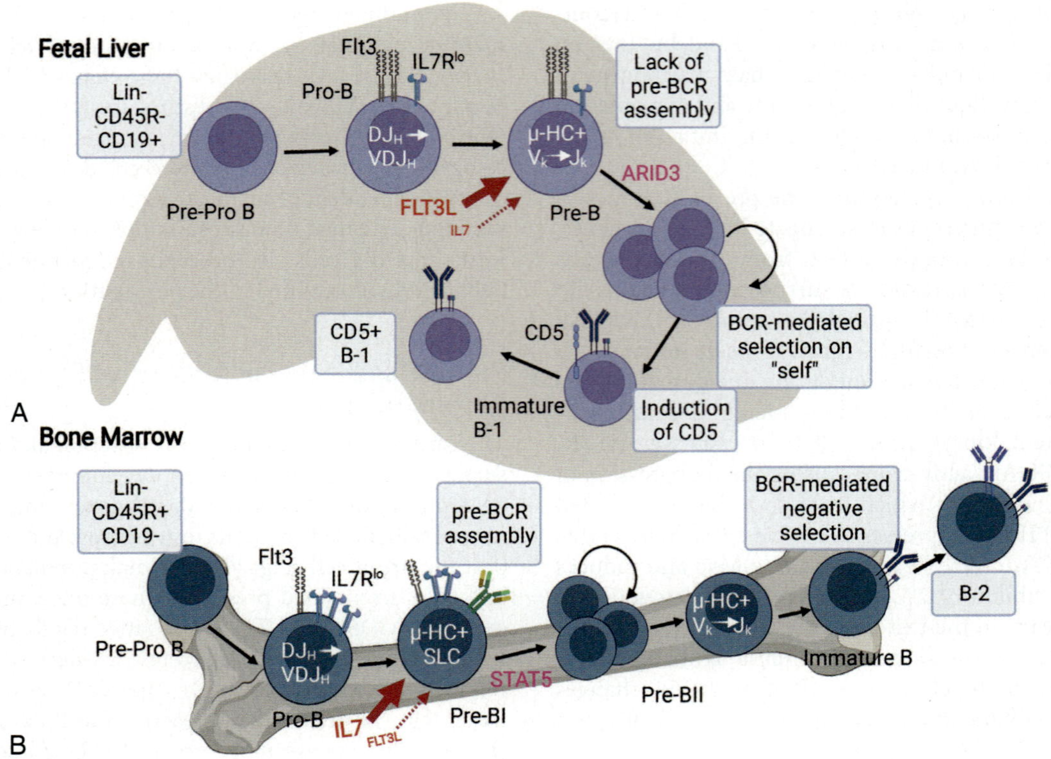

FIG. 25.3. B cell development in fetal liver and bone marrow. Shown are the major steps and characteristics of B cell development leading to the emergence of CD5+ B-1 cells from the fetal liver (top) and mature B-2 cells from the bone marrow (bottom), respectively.

assemble with the μ-heavy chains to form a BCR.[128] The regulation of this alternative, pre-BCR-independent development of CD5+ B-1 cells appears to be mediated through the Let7-induced upregulation of the transcription factor Arid3A.[132] Increases in Arid3A expression enhanced expression of Myc and Bhle41, the latter required for normal CD5+ B-1 cell development,[133] and reduced expression of Siglec-G and CD72, two known inhibitors of B-1 cell expansion.[134-136]

Thus, circumvention of pre-BCR signaling appears to support the development of CD5+ B-1 cells that carry a unique, autoreactive repertoire that would otherwise be deleted during the development of B cells in the bone marrow. These studies explain why V_H gene families that assemble poorly with the surrogate light chains, particularly belonging to the 7183 and Q52 families, are overrepresented in the CD5+ B-1 cell compartment.[137] Indeed, in apparent contrast to pre-BCR signaling at this stage of bone marrow B-2 development, which appears to occur absent of a particular ligand, B-1 cell development requires a ligand-mediated positive selection step, since B-1 cells assemble a mature BCR during the pro- to immature B cell transition and such a ligand would usually be a self-antigen. Indeed, the need for a self-ligand that can drive B-1 cell development is well described[138,139] and BCR autoreactivity can replace the need for pre-BCR signaling.[140] Given that the differential expression of IL7R in B cell precursors is dependent on the environment of the fetal liver and bone marrow, respectively, this alternative pathway model is consistent with the findings that B-1 cell development is largely restricted to the fetal liver, and/or early ontogeny, and consistent also with recent advances in our understanding of the epigenetic and transcriptional regulation of fetal and adult hematopoiesis (please see section on regulation of B cell development by micro-RNA, below).

In summary, driven by differential signaling through the IL7R, B cells that utilize V_H genes that poorly assemble a pre-BCR are rescued during fetal development by rapid kappa light chain rearrangement and expression of a mature BCR (Fig. 25.3). Positive selection at this stage of development drives their expansion and also drives the phenotypic and functional changes that define and differentiate CD5+ B-1 cells from mostly bone marrow–derived conventional B cells.[141] In so doing, this alternative pathway of B cell development ensures the maintenance of a restricted natural repertoire of BCR specificities that support homeostasis and early immune defense.[142]

Immature B Cells

B cell precursors that pass the pre-B cell checkpoint will exit the cell cycle and begin pre-BCR-mediated initiation of light chain rearrangement. In contrast to Ig heavy chain rearrangement, B cells can utilize one of two light chain constant regions, named kappa (κ) and lambda (λ), respectively. Furthermore, the light chain variable region is the product of only two, not three, genes, V and J. Assembly of a rearranged Ig light chain with the μ-heavy chain marks the first time B cells express a complete IgM-BCR on the cell surface that consists of two μ-heavy chains, two light chains, and the associated signaling molecules CD79a and b. Both positive and negative selection steps seem to occur at this point, where strong BCR-mediated antigen binding leads to the deletion of immature B cells, greatly reducing B cell bone marrow output of strongly autoreactive B cells. Alternatively, nonproductive or autoreactive heavy-light chain pairing can induce additional rounds of light chain rearrangements, a process termed "receptor editing" or "receptor re-rearrangement," either on the nonrearranged κ-allele and/or the λ-light chains, followed by a new round of selection. Successful completion of this developmental checkpoint termed "central tolerance induction" allows the immature B cells to leave the bone marrow.

Phenotype Having exited the cell cycle, these pre-B cells are small (small pre-BII) and initially remain surface BCR negative. They also lose surface expression of CD43, remain negative for CD23, and highly express CD24 ([143]; Hardy fraction D). Following successful rearrangement of the light chain genes, the cells begin to express surface IgM but remain $IgD^{lo/-}$ and CD23− (Hardy fraction E, immature B cells, Fig. 25.2). These newly formed cells also express CD93 (also termed AA4.1), recognized by mAb clone 493,[144] a type I glycoprotein that is expressed also in hematopoietic precursors but not on the surface of Ig+ mature B cells.

Ig-DNA rearrangement Epigenetic and transcriptional changes at the small pre-BII stage lead to reexpression of Rag1 and Rag2 in cells in which the IgH locus is now inaccessible for further rearrangement, while monoallelic activation and transcription of the κ-light chain locus is initiated. These pre-BCR signaling-driven alterations of Ig-locus accessibility are dependent on enhanced expression of two transcription factors, E2A and IRF4, which drive increased 3′ κ-enhancer activity.[116,145] Although exceptions have been reported, V-Jκ light chain rearrangement in mice and humans usually precedes rearrangement of the Vλ light chain genes, which are located on a different chromosome from the Vκ locus.[146,147] Over 100 Vκ genes in mice, and about half that number in humans, are present and can rearrange in both transcriptional orientations, enabling V-Jκ rearrangement either by deletion (generating an extrachromosomal excision circle) or by inversion.[146] The absence of intervening D segments also means that it is possible for upstream Vκ genes to rearrange to downstream Jκ segments.

In contrast to rearrangement at the IgH locus, unproductive rearrangement of the light chains, or the induction of a strongly self-reactive BCR, does not automatically lead to deletion. Instead, it can initiate receptor editing, ie, the initiation of additional rearrangement of either the second κ-allele and/or one and then the other λ-light chain locus.[148-150] Thus, a developing B cell may attempt up to four rounds of BCR expression, selection, and receptor editing. In addition to eliminating strongly autoreactive B cells from the repertoire, more extensive light chain rearrangement adds to the diversification of the B cell repertoire through utilization of a greater number of Vκ and Vλ genes. BCR editing instead of deletion or anergy induction has been suggested to be the main mechanism by which central tolerance is enforced in the B cell compartment.[151,152] Although defining the exact numbers of B cells undergoing re-rearrangement during development is challenging, one study calculated that about 20% of B cells undergo re-rearrangement at any given time.[153]

BCR signaling complex The signaling outcomes of BCR engagement, ie, antigen binding by immature B cells, is distinct from that of mature B cells. The former induces apoptosis, anergy, or receptor editing, whereas the latter induces cell activation and clonal expansion. The molecular mechanisms facilitating these vastly different outcomes remain insufficiently understood but are likely multifactorial. Studies by Pierce and colleagues first demonstrated that the organization of the antigen-engaged BCR signalosome in the cell membrane differs between immature and mature B cells.[154,155] Antigen binding by mature B cells causes the translocation of the BCR into glycosphingolipid- and cholesterol-rich plasma membrane microdomains, termed lipid rafts. Instead, the BCR of immature B cells appears to be excluded from lipid rafts.[154] Lipid rafts are critical in supporting and modulating BCR signaling,[156,157] BCR internalization, the targeting of BCR-bound antigens to the peptide loading compartment of the MHCII,[155] as well as the anchoring of the BCR to the cytoskeleton.[158] Thus, BCR exclusion from these rafts alone can fundamentally change the signaling outcomes of these cells in response to antigen.

A positive selection step is required for the cessation of gene recombination at the kappa and/or lambda light chain locus once a productive rearrangement has occurred. Similar to pre-BCR signaling, immature BCR signaling-mediated positive selection seems to be ligand independent but dependent on tonic BCR signaling, requiring expression of CD19, a coreceptor critically involved in PI3K induction.[159,160] Indeed, it was shown that the failure to induce appropriate signaling at this developmental stage enhances receptor editing.[159,161] This was due likely through dysregulated PI3K/Akt p110δ signaling and a failure to inactivate FoxO1 and thereby to repress Rag1/2.[162-165] Incomplete silencing of additional light chain alleles during the receptor editing step may result in "allelic inclusion," ie, the development of B cells expressing BCR with more than one light chain, which may be a factor in the development of autoreactive B cells.[166] Recent studies suggested that constitutive high PI3K signaling in autoreactive immature B cells, mimicking levels usually expressed by nonautoreactive immature B cells, allows their escape from central tolerance induction and their exit from the bone marrow.[163,167] Thus, dose-dependent regulation of PI3K signaling seems to facilitate positive selection of immature B cells.[168] This conclusion is further supported by a study that showed that overexpression of miR148a resulted in a lack of central tolerance induction,[169] likely due to suppression of the autoimmune suppressor Gadd45α, the tumor suppressor PTEN, a negative regulator of PI3K signaling, and the proapoptotic protein Bim.

Modifications in pro- and antisurvival factors such as Bcl2 and Bim are critical in negative selection outcomes,[169] a process that is required for reducing self-reactive, immature B cells. Estimates from human studies suggested that up to 55% to 75% of immature B cells are autoreactive and require elimination, through apoptosis or anergy,[170-175] or modification through receptor editing.[176] Of interest, TLR-mediated activation can overcome BCR signaling–mediated apoptosis induction in immature B cells.[177] Given the extensive overlap and integration of BCR and TLR signaling pathways,[178,179] these results suggest that the differential immature B cell fates are tightly linked to the quality/signaling strengths of BCR-mediated signaling in the steady state.[168] However, inflammatory signals and perhaps chronic inflammation may reduce central tolerance thresholds for BCR-mediated deletion, with long-term consequences for autoimmune response regulation.

PERIPHERAL MATURATION STAGES AND B CELL SUBSETS

Transitional B Cells

More than 75% of immature B cells that successfully pass through the above outlined checkpoints will enter the bloodstream and exit into the spleen, whereas the remainder seem to mature in the bone marrow.[180,181] The reader is referred to Chapter 8 for a detailed description of spleen tissue architecture. Cells that entered the spleen have been classified as "transitional" B cells, as they are transitioning from developing to mature B cells. Transitional B cells are short lived, phenotypically distinct from mature B cells, and contain an estimated 20% to 35% autoreactive cells,[182] which are purged from the repertoire through selection events that mark a so-called second or peripheral window of tolerance induction. Antigen-reactive transitional B cells are deleted through induction of apoptosis unless rescued at the T-2 stage by provision of CD4 T cell help in form of CD40-CD40L interaction.[183-186] Alternatively, these cells may be rendered unresponsive (anergic).[175,184] Successful passage through this final developmental checkpoint leads to their maturation into either follicular or MZ B cells.

Phenotype Transitional B cells in the spleen and PBMC can be subdivided based on differential expression of several surface proteins, including CD21, CD23, CD24, and CD93 (AA4.1), and expression of IgM and IgD (Fig. 25.4)[187,188]. Staining for CD93 (mAb 493) might be used to identify transitional B cells, as its expression differentiates these cells from all mature B cell subsets present in the spleen. However, distinction of CD93+ from CD93− cells can be challenging, as CD93 expression is rapidly lost as immature B cells leave the bone marrow. More useful, perhaps, transitional B cells are also CD24^{hi} and express high levels of IgM (Fig. 25.4), whereas mature B cells are CD24$^{int/lo}$. Within the splenic transitional B cell compartment two distinct developmental subsets of transitional B cells are differentiated: Transitional (T)1

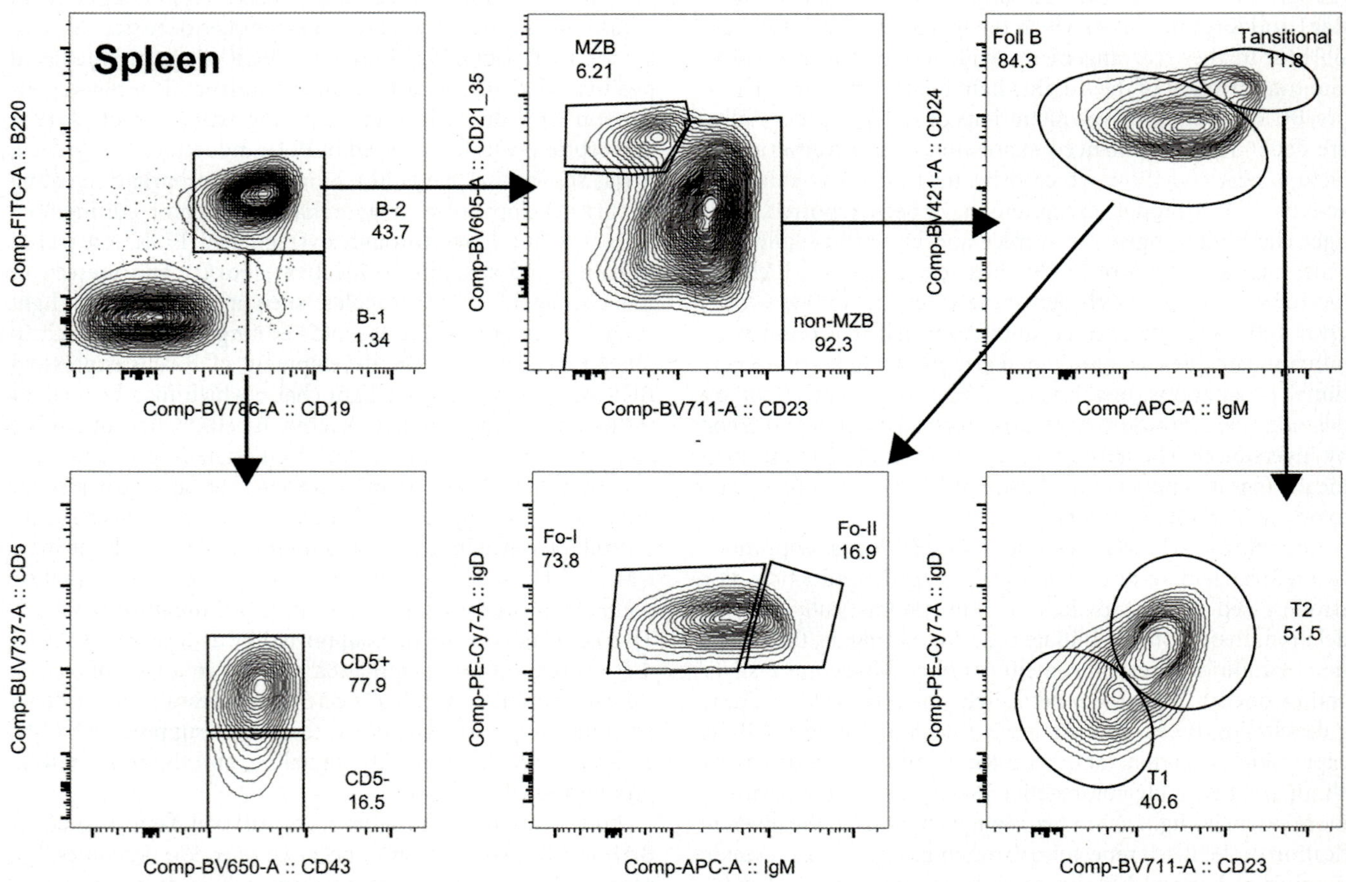

FIG. 25.4. Flow cytometric identification of spleen B cell subsets. Shown are contour plots of a flow cytometric analysis of spleen cells from C57BL/6 mice that identify the indicated B cell subsets: CD5+ and CD5− B-1, MZ B, T1 and T2 transitional, and follicular I and II.

and T-2 cells. Newly arriving T-1 cells are CD21− CD23− IgM^{hi} and IgD$^{lo/-}$ and are excluded from entering B cell follicles (Fig. 25.4 and not shown). Separating the IgD− from the IgD^{lo} cells may identify the earliest (T0) transitional B cells, that enter the spleen from the marginal sinus into the red pulp, and prior to their migration into the white pulp, which occurs in a Rac GTPase-dependent manner.[189] Within 2 days T-1 cells can become T-2, which are phenotypically defined as CD21$^{lo/-}$ CD23+, IgM^{hi} and IgD^{hi} (Fig. 25.4)[188,190,191]. Although early studies suggested a progression from T-2 to an additional stage, T3, identified as B cells with the phenotype CD23+, IgD^{hi} and IgM^{lo},[188] more recent work has suggested that the "T3" cells constitute a population of autoreactive anergic cells.[175,192]

Following transition from T-1 to T-2, B cells strongly upregulate the receptor for the tumor necrosis factor ligand superfamily member *B cell Activating Factor* (BAFF, also termed BlyS), encoded by the gene tnfsf13b, and become highly dependent on BAFFR signaling for survival.[193,194] However, BAFFR signaling may enhance survival already earlier in development.[195] BAFFR signaling drives increased expression of CD21 and CD23 and thus is responsible for the phenotypic changes that characterize T-2 cells.[196] Maturation into the mature follicular B cell pool leads to the loss of CD93 and continued high expression of IgD and downregulation of IgM.[188] Of note, two CD93− and CD24$^{int/lo}$ mature B cell subsets exist in the spleen that, similar to T-1 cells, lack CD23 expression. Both are IgM^{hi} IgD$^{lo/-}$ and either are CD21$^{lo/-}$ CD43+ CD5$^{+/-}$ B-1 cells or CD21^{hi} CD43− CD5− MZ B cells, respectively (Fig. 25.4).

Ig-DNA rearrangement In contrast to bone marrow immature B cells, transitional B cells are unable to undergo further rearrangement processes.[186,197] This appears to be regulated at the epigenetic level since T-1 cells can still express Rag 1/2.[198]

BCR signaling complex T-1 cells behave like bone marrow immature B cells following BCR engagement, in that "strong" BCR stimulation results in induction of apoptosis or anergy. However, they do not undergo receptor editing. In order to further mature, transitional B cells have to migrate from the blood via the marginal sinus and the red pulp to the white pulp,[189] where they undergo a negative selection step between the T-1 to T-2 stage.[143] At this step, autoreactive B cells are removed from the repertoire,[186] unless rescued by provision of T cell help in form of CD40L or IL4 signaling.[185] Interesting, both immature B cells in the bone marrow and T-1 B cells in spleen express the enzyme AID, which in peripheral mature B cells is required for support of class-switch recombination and affinity maturation in germinal centers.[199] Expression of AID was shown to be critical for both central and peripheral tolerance induction, by mechanisms that remain to be fully explored, but seem to be connected to their relative resistance to BCR-mediated apoptosis.[199,200] The presence of endosomal MyD88, an adaptor downstream of TLRs, was also shown to be critical for tolerance induction. Together, published studies suggest that the antigen-induced endocytosis of the BCRs allows the BCR to utilize MyD88 to increase BCR signaling.[201] Removing MyD88 reduced BCR-mediated signals, which led to the escape of autoreactive immature and transitional B cells from negative selection.

The molecular mechanisms that regulate BCR signaling-induced apoptosis at the T-1 stage of development and proliferation at the T-2 stage are a focus of ongoing research. They do not appear to be regulated by the increased expression of the IgD-BCR,[202] but rather depend on a complex set of epigenetic and transcriptional regulation and BCR and coreceptor signaling, which integrate to determine the distinct cell fates. Comparative next-generation sequencing showed that most prominent among the transcriptional changes occurring between T-1 and T-2 is the acquisition of c-Myc expression, which supports mature B cell proliferation following BCR stimulation.[198,203] A second major difference in gene regulation is the increased activation of the PI3K-Akt signaling pathway in T-2 (see also below)[198]. At this critical transition point, surviving T-2 cells gain the ability to enter the B cell follicles and increase expression of the BAFFR in a BCR signaling–dependent manner. BAFFR signaling imbues T-2 cells with survival signals in form of the prosurvival protein Bcl2 and via inhibition of the apoptosis inducer Bim. Thus, BAFFR signaling protects cells from cell death induced by tonic BCR signaling alone.[204,205] In a positive feedback loop, tonic BCR signaling induces p100, the substrate for BAFFR-mediated activation of the alternate pathway of NF-κB, further supporting cell survival.[206]

T-2 cells undergo some clonal expansion as they enter the B cell follicles and proliferate in response to antigen encounter, suggesting a positive selection event at this step, although the mechanisms are unresolved.[207] As discussed above, the cytokine BAFF, acting through the BAFFR, provides nonredundant signals to T-2 cells that integrate with tonic BCR signals to support further differentiation and support cell survival. This integration seems to occur at least in part through the BAFFR-induced phosphorylation of Syk[208] and the downstream induction of the PI3K/Akt and MAP/Erk signaling pathways. Of interest, BAFFR-mediated activation of Syk requires binding of Syk to the ITAM-motive in CD79a.[208] A similar integration of BCR via the phosphorylation of Syk by TLR-mediated signals has been reported to occur in mature B cells[179] and might explain the effects of TLR signaling on B cell maturation outcomes.[177,201]

Selection of T-2 cells into the mature B cell pool likely also involves interaction with CD4 T cells, as defects in CD40 signaling as well as MHCII expression were shown to affect peripheral tolerance induction in humans,[209] inhibition of CD4 T cell or CD40-CD40L interaction affects the repertoire of the mature B cell compartment,[210] and provision of CD40 signaling increased B cell maturation in Btk-deficient B cells.[211] Furthermore, CD4 T cells raised in B cell–deficient mice eliminated developing B cells at the T-1 to T-2 transition, suggesting that functional T-B interaction is a critical positive selection step for entering the mature follicular B cell pool.[212] Rescuing transitional B cells from cell death after antigen encounter either by TLR and/or T cell stimulation is important for the ability of newly developing antigen-specific B cells to participate rapidly with activation and differentiation, rather than elimination, during an ongoing infection.

Indeed, some have proposed that these newly emerging B cells, which would include autoreactive cells, are particularly sensitive to TLR-mediated signaling and rapidly differentiate into broadly reactive, class-switched, antibody-secreting cells that can also contribute to cytokine production and antigen presentation and thus provide a rapid and broad first line of immune defense (summarized in[213]).

Tonic BCR signaling together with signaling through the BAFFR, TLR and/or T cell interaction via CD40 and MHCII modulate peripheral tolerance induction of transitional B cells through negative and positive selection. Chronic infections and chronic inflammation may alter these processes, specifically by affecting circulating BAFF levels, the levels of PAMPS and DAMPS, ie, TLR ligands and/or T cell activation, potentially altering the effectiveness of peripheral tolerance induction.[214]

Final Maturation of Transitional B Cells

Following peripheral selection of bone marrow–derived transitional B cells, these cells enter the mature B cell pool as *follicular B cells* or as *MZ B cells*. The selection of cells into the follicular or MZ B cell pool constitutes the final developmental step for bone marrow–derived B cells that lack self-reactivity above a certain threshold.[215,216] Current data suggest that the relative strengths with which the BCR provides either tonic and/or antigen-stimulated signals determines whether these cells become follicular or MZ B cells.[217] Relatively stronger BCR signaling is thought to drive differentiation of cells into the follicular B cell pool, with simultaneous BAFFR signaling supporting cell survival. Relatively weaker BCR-mediated signaling plus signals provided by the interaction of the ligand Delta-like 1 (Dll1) expressed on splenic venules with Notch2 expressed by transitional B cells instead drives the development of MZ B cells.[217-219] Recent studies suggest that MZ B cells may arise not only from T-2 cells but also directly from T-1 or even follicular (type II) B cells.[198,220] In addition, the Follicular/MZ B cell fate decision might be reversible, as Notch2 signaling induced follicular B cells to acquire a MZ B cell phenotype and transcriptional profile,[218] providing a possible explanation for the multiple pathways that have been shown to result in MZ B cell development.

Follicular B Cells

Follicular B cells constitute over 75% of the mature B cell pool in mice. They are located in B cell follicles, hence the term follicular B cells, and are characterized as CD19+, B220+ $CD21^{int}$, CD23+, $IgM\text{-}BCR^{lo}$, and $IgD\text{-}BCR^{hi}$. Some studies further separate follicular (Fo)-I from Fo-II cells, based on their levels of IgM expression, with Fo-I expressing low levels and Fo-II expressing high levels IgM-BCR in addition to both expressing high levels of IgD (Fig. 25.4). Fo-II cells might be a developmental intermediate between T-2 and either Fo-I or MZ B cells.[198] Follicular B cells recirculate between lymphoid follicles and the blood, regulated by expression of the follicular homing chemokine receptor CXCR5 and its stromal-derived ligand, CXCL13.[221]

Expression of the mature BCR is required for B cell survival, as it provides tonic signals in the form of PI3K pathway stimulation that together with BAFFR signaling drives B cell survival.[222-225] Follicular B cells are noncycling cells with a half-life of about 4 to 5 months in the mouse.[226,227] Continuous generation of newly emerging follicular B cells replaces senescent cells. Little is known about the triggers supporting natural turnover of mature resting B cells. Overexpression of BAFF causes increases in peripheral mature B cells, suggesting that the B cell follicular pool is constraint in part through limiting access to this critical survival factor.[228] In the steady state, various surface molecules provide positive and negative signals that modulate the responsiveness of follicular B cells to BCR-mediated stimuli. These include CD19, CD21, CD22, CD72, and TLRs among others. Thus, the BCR on mature follicular B cells appears to be not only a receptor that recognizes antigen. It acts also as a cell surface–expressed "hub" that integrates signaling pathways from B cells in response to continuous homeostatic and antigen-induced stimulation. Dysregulation in BCR signaling, such as demonstrated in mice and humans lacking Bruton tyrosine kinase (Btk), the kinase Syk, or other critical signaling molecules that associate with the BCR or its signaling chains, negatively impact not only B cell development but also the responsiveness of the B cells following immunization and infection. Similarly, the lack or dysfunction of costimulatory molecules such as CD19, and of BAFFR and endosomal TLRs, as well as signaling through CD40 by helper T cells alters the ability of B cells to respond to antigen encounter, as their effects integrate in various signaling pathways to modify BCR-mediated signals. In contrast, signals from inhibitory receptors such as CD22, FcγRIIb, and CD72, when deleted or mutated alter B cell homeostasis, increasing follicular B cell responsiveness and/or peripheral expansion of follicular B cells and often result in antibody-mediated autoimmune diseases.[60,136,205,211,229-234]

Marginal Zone B Cells

MZ B cells in the mouse are confined to the spleen, where they constitute about 5% to 10% of B cells. They are situated along the marginal sinus that carries circulating blood as it enters the spleen and filters through the red pulp for removal of dead and dying erythrocytes, and from where leukocytes emigrate into the white pulp, the lymphoid compartment of the spleen[235] (see also Chapter 8). Although originally thought of as "sessile," MZ B continuously shuttle between the marginal zone and the splenic B cell follicles, carrying antigens acquired from the circulating blood for loading onto follicular dendritic cells in a process that involves antigen binding to complement receptor CD21 (CR2).[236,237] Activation of MZ B cells is associated with rapid differentiation into antibody-producing plasmablasts. Although MZ B cells have been shown to undergo class-switch recombination to IgG,[238-240] they are particularly important in providing immediate/early IgM antibodies in response to blood-derived pathogens.[235,241] In contrast to follicular B cells, MZ B cells have the potential to self-renew.[227] Thus, fetal-derived MZ B cell populations can persist long-term in adult mice.

Spleen MZ B cells are easily identified by flow cytometry as CD19+ B220+ CD23− $CD21^{hi}$ $CD1^{hi}$ $CD9^{hi}$ IgD^{lo} IgM^{hi} IL9R+ (Fig. 24.4 and not shown). In addition to high-level expression of the complement receptor CD21 and the nonclassical MHC molecule CD1, MZ B cells also express high levels of additional innate receptors, such as the LPS-binding receptor RP105, as well as high levels of endosomal TLRs 3, 7, and 9,[198] explaining their rapid and strong responsiveness to innate signals.[235]

MZ B cell maturation from transitional B cells requires Notch2 signaling.[217-219] This process depends on the BCR signaling–induced upregulation of the serine-threonine-kinase Taok3, which enables the relocation of the metalloproteinase Adam10 from intracellular vesicles to the cell surface. On the cell surface Adam10 then cleaves the intracellular domain of Notch2 to support its nuclear translocation in response to Notch2 binding by Dll1, a ligand found in splenic marginal sinus venules.[242] The need for BCR-induced upregulation of Taok3 may explain why mice with gene deletions that reduce BCR signaling strengths, such as Btk (xid)−/− and CD79a-deficient mice, show reduced MZ B cell populations.[217,241] In addition, tyrosine kinase Pyk-2 expression by MZ B cells is indispensable for the development of these cells, possibly through the regulation of MZ B cell positioning within the splenic marginal zone.[243] The cannabinoid receptor 2 has been shown to regulate the localization of MZ B cell precursors to the vicinity of Notch2 signals.[244]

MZ B cells are generated in mice even when B cell development is blocked shortly after birth by conditional deletion of Rag-2,[227] showing that a significant portion of MZ B cells is produced during fetal development. This is further supported by data showing that the developmentally restricted fetal-only HSC population, described in the beginning of this chapter, contributes strongly to the MZ B cell pool of adult mice,[6] as well as by the analysis of Ig heavy chains of MZ B cells, that contain significant numbers of cells having little or no TdT-mediated N-region insertions at their V-D and D-J junctions.[220,241] Of interest, irradiated mice repopulated with bone marrow precursors still accumulate significant numbers of MZ B cells with little N-addition.[245,246]

B-1 Cells

As discussed earlier in this chapter, B-1 cells are a population of mainly fetal/neonatally derived B cells, favored in development by early-arising and transient fetal liver HSCs.[6] As outlined further below, they require the Lin28/Let7-mediated fetal program of hematopoiesis for normal development.[10,11,61] Their emergence from multiple early waves of distinct fetal precursor sources is not exclusive to that stage in ontogeny, however, as adult bone marrow can reconstitute some B-1 cells, preferentially those lacking CD5 expression, after bone marrow ablation through irradiation and reconstitution of adult mice.[12,13] Similar to MZ B cells, but in contrast to follicular B cells, B-1 cells have the capacity for self-renewal and are maintained into adulthood through this process.[226,227,247] Numerous adoptive transfer and precursor studies have demonstrated the mostly fetal origins of B-1 cells.[247,248] Those studies are consistent with studies that demonstrated the greatly limited ability of bone marrow B cell precursors to give rise to B-1 cells after 6 weeks of age.[8,9,58] However, some studies have suggested that a continuous influx of B-1 cells can occur from adult bone marrow and that this de novo influx over time can skew the repertoire of these cells in the aging mouse, with important functional consequences.[13,249] The extent to which the pool of fetal/newborn-derived B-1 cells is reconstituted through de novo B-1 development after birth in nonmanipulated mice continues to be an area of active investigation.

Mature B-1 cells are present in blood, secondary lymphoid tissues, as well as the pleural and peritoneal cavities of mice and are distinguished by their phenotype $CD19^{hi}$ $B220^{lo}$ $CD21^{lo}$ CD23− CD43+ IgM^{hi} IgD^{lo}[142] (Fig. 25.4 and not shown). These cells have been further differentiated based on their expression of CD5 into CD5+ "B-1a" and CD5− "B-1b" cells. Although earlier reports suggested that B-1a and B-1b cells represent distinct B cell lineages,[248] more recent studies have shown that individual precursors can give rise to both B-1a and B-1b. Furthermore, activation of CD5+ B-1 cells via TLR stimulation causes a rapid loss of CD5 expression, showing that CD5− B-1 cells can arise from CD5+ B-1 cells.[250] The data suggest that, at a minimum, not all CD5− B-1 cells emerge during development from distinct precursors.

B-1 cells constitute 30% to 60% of B cells, or around 1 to 2 $\times 10^6$ total cells, found in the peritoneal and pleural cavities, where they also express CD11b, a feature that seems restricted to the body cavity.[142] The same total number of B-1 cells is also present in the spleen, but because of the much higher cellularity of the spleen, they constitute only about 1% to 2% of all B cells in that organ. B-1 cells in resting lymph nodes of adult mice make roughly 0.1% to 0.3% of B cells, whereas their frequencies in the blood is about 0.5%.[142] Their relative abundance in the body cavities might be related to their ability to rapidly respond to microbial insults with relocation to draining secondary lymphoid tissues, where they can differentiate into antibody or cytokine-secreting cells.[61,142] Another important function of B-1 cells is their ability to differentiate into spleen and bone marrow natural antibody-secreting cells, mostly IgM and IgG3, in a process that appears to be independent of microbial exposure.[251,252] Given their relatively strong self-reactivity, it is tempting to speculate that continuous exposure to these antigens drives their differentiation. However, BCR stimulation via anti-IgM cannot induce proliferation of CD5+ B-1 cells and their differentiation into antibody-secreting cells, suggesting other signals, perhaps provided through DAMPS or PAMPS, might be responsible.

As outlined above, CD5+ B-1 cells seem to preferentially develop through an alternate developmental path that does not form a classical pre-BCR but rather uses kappa light chains instead of the surrogate light chains to assemble a BCR (see: Pre-BCR independent alternative pathway of B cell development, Fig. 25.3). This difference in development critically shapes the defining feature of B-1 cells, namely, its unique BCR repertoire, which in turn shapes the transcriptional profile and the function of B-1 cells.[61,128,141] The repertoire of B-1 cells encodes for self-reactive specificities, notably for branched carbohydrates, glycolipids, and glycoproteins, including phosphorylcholine, phosphatidylcholine (PtC), oxidized lipids, the Thy-1 glycoprotein, and bacterial cell wall constituents, serving as a kind of "hardwired" immunological memory that regulates the microbiota, self-antigen removal and inflammation, and acute responses to pathogens.[253-255] Among other modulators of BCR signaling strengths, CD5 acts to downmodulate the strengths of BCR (and TCR) signaling.[127] Induction of CD5 expression, rather than deletion of re-rearrangement, occurs at the immature B cell stage, when relatively strong BCR signaling through self-antigen recognition appears to be required for selection, expansion, and/or maintenance of B-1 cells.[138] This alternative developmental process that appears to select self-reactive cells and subsequently "silences" BCR signaling through provision of CD5 is promoted by the epigenetic regulators of fetal hematopoiesis, Lin28/Let 7.[256]

Consistent with the importance of antigen selection in the generation and/or maintenance of B-1 cells, these cells are severely affected by mutations that alter BCR signaling intensity. The loss of negative mediators such as PTP1C/SHP-1 in "moth-eaten" mice,[257] of CD72 that recruits this phosphatase,[257] and of the CD22 coreceptor[233,234] as well as Siglec G[134,232] all result in an increased frequency of CD5+ B-1 cells relative to follicular B cells. On the other hand, the loss of critical BCR signaling components or positive mediators of this pathway such as an Ig-α tail mutant,[258] Btk deficiency,[231] CD19 null mice,[230,259] CD21 null mice,[229] and vav null mice[260] all negatively impact B-1 cell numbers.

TRANSCRIPTIONAL REGULATION OF B CELL DEVELOPMENT

Great progress has been made over the last decade in defining the major transcriptional regulators of hematopoiesis in general, and B cell development in particular. Cell type–specific patterns of gene expression have been shown to be controlled by the binding of transcription factors to enhancer regions of the genome. So-called super-enhancers, which are exceptionally large enhancer regions to which transcriptional regulators bind together with the mediator complex that recruits the RNA polymerase II to the promoter of hundreds of target genes, have been used to assess the critical transcriptional regulators that define HSC differentiation and B cell identity.[261-263] PU.1, E2A, EBF1, and FOX01 transcription factors were identified to strongly bind to pro-B cell super-enhancer sites[261]; all four were previously identified as key transcriptional regulators of B cell lineage commitment. For peripheral blood-derived B cells the transcription factors identified were Ikaros, Aiolos, BSAP, Bach-2, OCT-2, and IRF4 and IRF8,[262] also all previously identified as regulators of the B cell fate. Following is a brief description of these transcription factors and their known effects on B cell development (Fig. 25.5).

Purine Box Factor 1 (PU.1) PU.1, an Ets family transcription factor encoded by *sfpi1*, is critical for HSC development/maintenance in fetal liver[264] and bone marrow[265] and for B cell lineage commitment at the myeloid to lymphoid branching point.[266] Both the presence and levels of expression of PU.1 were shown to modulate precursor fate: low-level expression in HSC induced in otherwise PU.1 null mice promoted B-lineage development, whereas high-level expression supported myeloid lineage differentiation.[266] A complex regulatory circuit consisting of PU.1 and the zinc finger transcriptional repressor growth factor independent (GFI)-1 regulates PU.1 expression. GFI-1 expression is required to restrain PU.1 expression and in so doing GFI-1 supports the preferential development of B cells.[267] Induction of IL-7Rα seems to be a critical downstream target of this regulatory circuit, as forced expression of the IL-7Rα chain in PU.1 null bone marrow HPCs complemented defective B lymphopoiesis.[268] In contrast, cultures of fetal PU.1-deficient HCS showed outgrowth of IL-7Rα+ CD45R/B220− B cell precursors, a

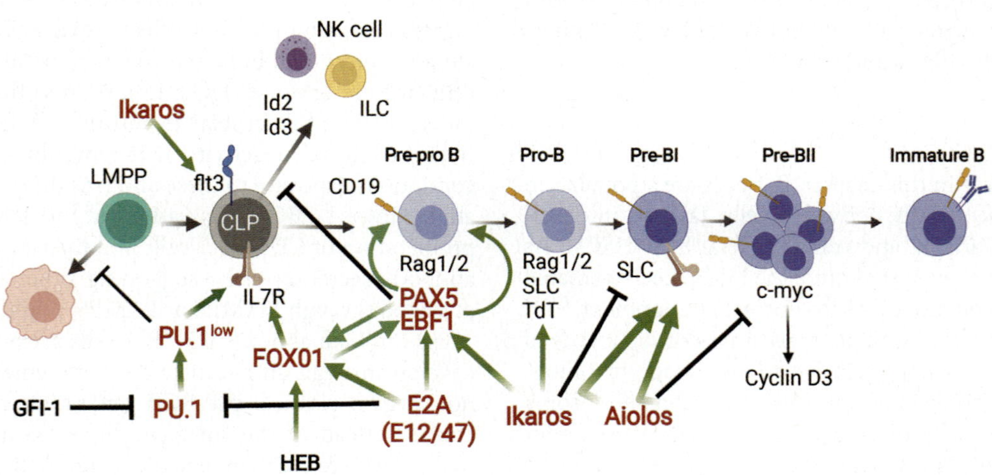

FIG. 25.5. Transcriptional regulation of early B cell development. Shown are the effects of some key transcriptional regulators (red) and key gene targets (black) that regulate the B cell fate at indicated stages of B cell development. Green arrows indicate gene induction and enhancement; black lines indicate inhibition. LMPP, lymphoid/myeloid progenitor; CLP, common lymphoid progenitor; ILC, innate-like lymphocytes; SLC, surrogate light chains (VpreB and lambda 5).

phenotype of B-1 cell precursors.[58] Consistent with that, conditional deletion of floxed PU.1 via CD19 re-expression forced a preferential outgrowth of B-1-like over B-2 cells in vivo,[269] suggesting a differential requirement of PU.1 in B-1 versus B-2 cell development, or, as suggested by the authors, a potential conversion of B-2 to B-1 or B-1-like cells.[269] Although expression of PU.1 was reported not to be required beyond the pre-B cell stage,[270] its expression was shown to modulate Vκ gene usage during kappa light chain rearrangement and thus to affect B cell repertoire development.[271,272] Given the importance of BCR specificity for B-1 cell development,[138,141] the PU.1-mediated effects on the BCR repertoire could have been in part responsible for such an apparent "conversion" from B-2 to B-1 cells after their emergence from the bone marrow and following (self-antigen) recognition in the periphery. This would be similar to reports by others, which showed a "conversion" of B-2 into B-1 cells when follicular B cells were induced to express a BCR usually only expressed on B-1 cells.[141]

E box binding protein 2A (E2A) E2A (*Tcf3*) encodes two proteins, E12 and E47, through alternative splicing. Both are members of the basic helix-loop-helix family of transcription factors that form homo- and heterodimers and function during early B cell commitment.[273] Induction of E2A is indispensable for B-lineage development and Ig gene recombination. B cell development is blocked at the CD19+ pre-pro-B cell stage in E2A null mice with precursors demonstrating a complete loss of D-J rearrangements.[274,275] This is explained by recent studies demonstrating that E2A functions by activating Rag-1 gene expression.[276] In addition, E2A has been shown to support B cell development through direct binding of E47 to PU.1, causing PU.1 repression. PU.1 repression then inhibits activation of the myeloid cell fate and drives B cell development.[277] Additional functions for E2A relate to its direct involvement in Igk light chain rearrangement,[145,278] a function antagonized by induction of the family of another helix-loop-helix family of transcriptional regulators, *id2 and id3*. These are regulators of an alternative, NK and innate-like lymphocyte fate,[279,280] which compete with E47 for DNA binding. Of importance, E12 induces expression of early B cell factor 1 (EBF1) at the pre-pro-B cell stage, critically supporting B cell lineage specification.[281,282]

Early B cell factor 1 (EBF1) EBF1, a member of the O/E protein transcription factor family, is required for progression of pre-pro B-lineage progenitors to the Ig heavy chain D-J rearranged pro-B stage (Hardy Fraction B).[283] The expression of EBF1 is induced by action of the epigenetic histone H2A deubiquitinase MYSM1.[284] EBF1 is distinct from the other lineage-promoting transcriptional regulators, as its forced expression in precursors lacking PU.1 or Ikaros can overcome the lineage block.[285] EBF1 acts as a transcriptional regulator that orchestrates a complex sequence of epigenetic and transcriptional events in early B cell development. Recent elegant studies provided a detailed analysis of the transcriptional and epigenetic changes that are orchestrated by EBF1 and that accompany the progression from multilineage progenitors to B cells.[286,287] These studies also indicated the presence of additional epigenetic regulators whose function requires further exploration, and they revealed apparent waves of transcriptional and epigenetic regulation and identified an intermediate stage between multilineage progenitor and committed B cell precursor.[287] EBF1 acts as both a linage instructor and as enforcer of B cell lineage commitment and identity. It does so in part by repressing ID2 and TCF1 to inhibit innate-like lymphocyte/NK and T cell lineage choices, respectively.[288] This latter function is shared with PAX5, the second lineage enforcing transcription factor that appears to form an autoregulatory loop with EBF1.[289,290]

Forkhead box protein O1 (FoxO1) FoxO1 is a member of a large group of transcription factors with a forkhead DNA-binding domain that plays a critical role at several stages of B cell development. This is consistent with its binding to super-enhancer sites that regulate B cell development. In addition, Fox01 is critically involved in B cell activation and class-switch recombination.[261,291,292] FoxO1 has multiple gene targets and functions in a regulatory network with PU.1 E2A and EBF1.[282] Its early induction by E2A/*tcf3* and a second transcription factor HEB[293] is subsequently reenforced through a positive feedback loop with EBF1, thus supporting the earliest steps of B cell commitment directly and indirectly through EBF1.[294] Through interaction with PAX5, it acts in concert to induce expression of critical PAX5 target genes such as CD79a. Deletion of FoxO1 was shown to arrest B cell development at the CLP stage, blocking pro-B cell development. At that stage it was shown to be necessary, among other functions, for the regulation of Rag-1 and Rag-2 expression and for the induction of IL7R.[111,295] FoxO1 activity is regulated through the phosphatidylinositol-3-OH kinase (PI3K)-Akt axis, thereby critically regulating survival and proliferation of pro- and pre-B cells.[291] Phosphorylation of Akt in response to BCR signaling leads to the removal of phosphorylated FoxO1 from the nucleus into the cytoplasm, derepressing various FoxO1 transcriptional targets, including genes regulating B cell proliferation (p27, cyclin D2), survival (Bim, CD95L), as well as migration (CD62L).[295]

Ikaros family of transcription factors The founding member of the family of Ikaros zinc finger transcriptional regulators, Ikaros (*ikzf1*), is expressed early in hematopoietic precursors and was shown to be required for all lymphocyte development.[296-299] Ikaros-null mice lack all fetal and adult B lymphocytes and their precursors.[300] In contrast, gene targeted mice lacking functional Aiolos (*ikzf3*), another member of the same transcription factor family that binds to the same DNA-binding sites as Ikaros, showed that Aiolos is not required for early commitment toward B cell differentiation. Instead, B cell development was abrogated at the pre-B cell stage as well as later in B cell maturation, consistent with its expression profile.[301] Patients with primary immunodeficiencies due to mutations in genes encoding Ikaros or Aiolos have shown similar severe reductions in B cell development as well as hypogammaglobulinemia as seen in gene-targeted mice.[302,303]

E2A and EBF1 direct CLPs into the B cell developmental pathway, consistent with their strong occupancy of pro-B cell super-enhancer sites.[261] Ikaros is required for induction of EBF1 expression, functioning together with PAX5 to maintain B-lineage commitment.[304] The absence of Ikaros in CLP results in increased NK cells at the expense of T and B cell development.[305] However, Ikaros functions at multiple additional stages of B cell commitment, B cell differentiation, including through regulation of Ig-gene recombination, which are explained only in part by its induction of EBF1. This is highlighted by earlier findings that rescue of EBF1 expression in Ikaros−/− hematopoietic progenitor cells restored the generation of CD19-expressing cells, but these cells were not committed to the B cell fate and failed to rearrange IgH genes.[304] Conditional deletion of *ikzf1* in pre-pro-B cells via *mb-1*-regulated expression of the Cre recombinase also identified Ikaros as a regulator for the transition from pro- to pre-B cells.[299] Ikaros activates numerous early B-lineage genes, including *Dntt* (encoding the TdT), *Rag-1*, and the surrogate light chains λ5 (*Igll1*) and VpreB (*vpre-B1*). It is critical in supporting Ig heavy chain and kappa light chain rearrangement through the regulation of Rag expression, as well as by interacting with chromatin modifiers and its effects on the epigenetic control of the Ig heavy and kappa light chain genes.[271,306] Furthermore, at the pre-B cell stage, Ikaros expression and DNA binding was shown to facilitate appropriate signaling through IL-7Rα and thus for enabling the transition of pre-BI to clonal expansion and further differentiation.[299,307] Conditional deletion of Ikaros in CD19+ or CD21+ B cells has been shown to enhance the presence of bone marrow and splenic, but not body cavity, B-1 cell populations.[308] The mechanisms underlying the suppressive effect of Ikaros on B-1 cell development or expansion remain to be explored.

Expression of *ikzf3,* encoding Aiolos, is induced at the pro-B cell stage in both mice and humans, and its expression increases throughout B cell maturation.[303] Homo- and heterodimers of Aiolos with Ikaros are critical regulators of the pro- to pre-B cell transition.[303] The transcriptional regulators interferon-regulatory factor (IRF) 4 and IRF8 support Aiolos expression at the pre-B cell stage, where continued expression of Aiolos depends on pre-BCR signaling–induced SLP-65 expression.[309] Although Ikaros and EBF1 act together at the pro-B cell stage to induce expression of *Igll1*, the gene encoding the λ5 surrogate light chain, Ikaros together with Aiolos facilitates *Igll1* silencing at the pre-BII cell stage, a process dependent on IRF4.[309,310] In addition, in cycling pre-B cells, Ikaros and Aiolos repress c-Myc, causing the downregulation of cyclin D3, critical for the transition of the cycling pre-BI to the noncycling pre-BII stage, that is, initiation of light chain rearrangement.[117,279] Thus, the combined actions of these two transcriptional regulators play key roles in B cell fate determination as well as the induction and regulation of downstream developmental steps, including at the pre-B cell stage, a critical checkpoint in B cell development.

B cell lineage-specific activator protein (BSAP, PAX5) PAX5 is a member of the paired box family of transcriptional regulator genes and encodes BSAP. PAX5/BSAP is a critical B cell lineage fate determinant that is expressed throughout B cell development.[311] Although usually lost at the terminal plasma cell stage, its forced expression still allows for robust plasma cell development, with only modest reductions in IgG-secreting cells.[312] PAX5 transcriptional targets include CD19 and BLNK, critical components of the BCR signalosome that further enforce lineage commitment. Furthermore, PAX5 acts to upregulate V to D-J heavy chain rearrangement by inducing V to D-J locus contraction and chromatin remodeling.[306,313] The absence of these critical BCR-signaling components together with the lack of V to D-J heavy chain rearrangement explains the B cell developmental arrest at the early pro-B stage (Hardy Fraction B, large pro-B) in PAX5-deficient mice.[290,314] As discussed, lineage commitment is as much about the induction of genes identifying a certain lineage, as it is about repressing alternate cell fates. In that regard, PAX5 is critical also, as its expression causes suppression of myeloid and T cell fates through repression of the myeloid growth factor receptor gene *c-fms*[315] and of the Notch1 signaling pathway,[316] respectively. Of note, in contrast to B cell development in the bone marrow, the absence of PAX5 in fetal liver precursors was shown to arrest B cell development prior to the B220+ stage.[314] More recent data showed the differential expression of the PAX5-controlled BCR coreceptor CD19 in B-1 and B-2 precursors, with early B-1 cell precursors expressing CD19 in the absence of B220, whereas B-2 cell precursors express B220 prior to induction of CD19.[58] Moreover, deletion of CD19 was shown to completely inhibit the development of CD5+ B-1 cells,[317] suggesting a crucial difference in the early dependence of CD19 signaling, and thus PAX5, for fetal versus adult development and/or maintenance.[314] CD19-deficient fetal liver precursors were able to reconstitute bone marrow hematopoiesis following their adoptive transfer into adult recipients, indicating that the bone marrow niche can overcome the strict dependency of developing B cell precursors for PAX5, and thus perhaps CD19 signaling.[314] Conditional targeting of PAX5 in more mature B cell stages through CD23-Cre-mediated deletion of PAX5 showed that its continued expression was necessary for the maintenance and function of MZ B and CD5+ B-1 cells, and to a lesser extent follicular B cells.[318] This seemed to be regulated at least in part through loss of the BCR- or TLR-mediated activation of the PI3K-AKT signaling axis, through the miRNA-regulated overexpression of the inhibitor of PI3K phosphorylation, the lipid phosphatase PTEN.[318]

Octamer binding protein 2 (OcT-2) OcT-2 binding to its octamer DNA motive induces a conformational change in its POU domain that enables the recruitment of the coactivator OCA-B (OBF-1/BOB-1), encoded by *POU2AF1*, which stabilizes the complex and further enhances transcriptional activation. Identification of OcT-2, a largely B cell–restricted transcription factor encoded by the gene *POU2F2*, as one of the super-enhancers of B cells in the periphery[262] is consistent with earlier work showing largely normal B cell development in gene-targeted oct2−/− mice up to the immature B cell stage, but severe defects in B cell responsiveness. However, these studies were complicated by the fact that global deletion of OcT-2 expression resulted in lethality shortly after birth.[319] Overexpression studies of OBF-1 suggested the need for tight regulation of this transcriptional regulator for effective early B cell development also.[320] Following adoptive transfer of fetal liver hematopoietic precursors from oct2−/− mice, the mature follicular B cell pool was reconstituted at reduced numbers, whereas these precursors completely failed to reconstitute the CD5+ B-1 cell compartment.[321] OcT-2 was recently shown to induce the micro-RNA (miR) 210, which supports the expansion of B-1 cells,[322] providing a potential mechanistic explanation for the failure to reconstitute B-1 cells by OcT-2-deficient precursors.

Gene targeting of the Oct-binding factor, OCA-B, resulted in viable offspring whose B cell compartment showed normal immunoglobulin rearrangement and similar defects in mature follicular B cells in some[323,324] but not all studies[325] and no apparent defect in B-1 cell development.[324] All reports, however, demonstrated a profound reduction in class-switched serum antibody levels. Similarly, double gene-targeted OcT-2/OCA-B-deficient mice showed a reduction in peripheral transitional splenic B cells in addition to severe reductions of the follicular B cell compartment.[326] Multiple studies also reported reductions in T-independent and T-dependent B cell responses, including a failure to form germinal centers, following the immunization of OcT-2-deficient mice.[323,326] This failure to induce robust germinal centers may be explained by the altered signaling following BCR engagement in OcT-2-deficient mature B cells, which results in reduced B cell proliferation.[322]

BTB and CNC homology 1 basic leucine zipper transcription factor 2 (BACH2) The transcriptional regulator Bach2 is induced in a PAX5-dependent manner at the pro-B cell stage and is active during the critical pre-B cell transition checkpoint.[327] Bach2 enhances V-DJ recombination by enhancing expression of Rag-1 and Rag-2. But at the same time Bach2 expression directs negative selection, that is, the elimination of pro-B cells that are not successfully undergoing productive V-DJ recombination through induction of Arf and the tumor suppressor gene p53 among other targets, inducing pre-B cell apoptosis. These cells can be rescued, however, if they undergo productive V-DJ_H recombination and assemble a pre-BCR. Pre-BCR signaling then induces expression of the transcriptional regulator, Bcl6, that supports not only B cell survival but also the exit of pre-B cells from the cell cycle through suppression of c-Myc.[327,328]

Bach2 competes with Bcl6 for binding to overlapping promoter-binding sites, differentially affecting target gene expression. Bcl6 induction was shown to drive positive selection of pre-B cells that express a functional pre-BCR by suppressing the function of Bach2. Thus, the balance of the transcriptional regulators Bach2 and Bcl6 determines pre-B cell developmental fates. Bach2 has critical additional functions in mature B cells, where it is required for class switch recombination and somatic hypermutation of immunoglobulin genes in activated B cells. Bach2 also regulates terminal B cell differentiation to memory B cells and plasma cells by antagonizing IRF4-mediated induction of Blimp-1, the transcriptional regulator of plasma cell differentiation.[329,330] Bach2 expression itself seems to be controlled by two transcriptional regulators, SpiB and SpiC.[331]

Interferon regulatory factors (IRF)4 and IRF8 Several factors have been identified that suppress proliferation in late pre-B cells and enable Ig light chain gene rearrangement. Among those, IRF4 and IRF8 were shown to be essential. Targeting of both transcriptional regulators arrested B cell development at the cycling (large) pre-B cell stage.[332] IRF4 and 8 support clonal expansion of pre-B cells by suppressing surrogate light chain expression and further development by downregulating expression of the pre-BCR, a process that depends on their induction of Ikaros and Aiolos and that eventually suppresses c-Myc-induced cell cycling.[117,310] In addition, IRF4 induces expression of CXCR4, which allows cycling pre-B cells to move away from IL-7-producing stromal cells toward a CXCL12-producing cell niche. Removal from IL-7-supported pre-B cell proliferation and V-DJ_H recombination enforces allelic exclusion at the heavy chain locus.[75,121] IRF4 and 8 are essential also for the regulation of B cell differentiation during mature B cell activation through modulation of BCR downstream signaling.[333] Finally, high ratios of IRF4:IRF8 expression support the rapid differentiation of B cells into plasmablasts and plasma cells, whereas lower ratios instead promote the entry of B cells into germinal center responses.[334]

Taken together, transcriptional regulatory networks that regulate the development of B cells consist of positive and negative feedback loops, in which a relatively small number of nonredundant transcriptional master regulators induce and then reinforce lineage commitment, selection, and maturation (Fig. 25.5). Some of these same regulators remain involved in the coordination of mature B cell activation in response to antigen stimulation, and in the terminal differentiation of B cells to the plasma cell stage. Recent data, exemplified by studies on IRF4 and IRF8, demonstrate that it is not just the presence or absence of these regulators that control B cell development. Control of their relative expression levels as well as the expression levels of counterregulators provide an extraordinary level of fine-tuning.[335] Having identified the major functions of these super-enhancers and further exploring their functional gene regulatory circuits, these data will aid in the understanding of processes leading to the malignant transformation of B cells and other leukocytes, which in turn can aid the development of precision therapies that interfere with these aberrant processes.[336,337]

Epigenetic Regulation of B Cell Development

Epigenetic and posttranscriptional regulations, such as chromatin remodeling through differential histone acetylation and ubiquitination, DNA methylation, and the expression of noncoding RNAs, are additional levels of regulation in B cell development that are currently being explored.[338-341] Given the above outlined important functions of transcription factors in the regulation of B cell development, it is not surprising that enhancement of access to transcription factor–binding sites via demethylation of these regions was a particularly prominent finding in the analysis of the global gene methylation status of B cell precursors.[342-344]

Noncoding (nc)RNAs, either short (snc)RNAs, also called micro (mi)RNAs, of about 18 to 22 nucleotides in length, or long (l)ncRNAs, >200 nucleotides in length, are transcribed from the 3′ noncoding region of genes. Both facilitate posttranscriptional regulation of gene expression and translation, not only through accelerated mRNA degradation and inhibition of mRNA translation but also through chromatin-remodeling.[337,345-347] A comprehensive analysis of lncRNAs recently suggested the differential expression of over 4500 lncRNAs at each step of B cell commitment and differentiation that bound to either promoters or enhancer regions.[348] Of interest, PAX5 appeared strongly involved in the regulation of lncRNA expression, revealing another gene regulatory pathway through which this critical B cell fate determinant affects B cell development.[348]

Below are just a few examples by which miRNAs are known to affect B cell development. The generation of mature functional miRNAs requires action of two enzymes, Dicer and Drosha. Drosha cleaves pre-miRNA in the nucleus, whereas Dicer is required for further processing of pre-miRNAs in the cytoplasm and for the recruitment of Agonaute proteins to form protein-containing miRNA-induced silencing complexes (RISCs). These RSICs bind to the 3′ UTR of mRNAs, which are targeted via imperfect complementary base pairing. ncRNAs can simultaneously affect hundreds of genes in a gene network, broadly altering gene expression. Recent large-scale analysis has comprehensively assessed the regulation of miRNA expression in lymphopoiesis.[349] The overall effects of miRNAs on B cell development were demonstrated through ablation of Dicer and the Drosher-associated molecule DiGeorge Critical Region 8 (DGCR8) in early B-lineage progenitors, respectively.[350,351] The studies demonstrated a near-complete block of B cell development at the pro- to pre-B cell transitional stage, likely due to upregulation of the proapoptotic molecule Bim, as development was rescued by gene targeting Bim or by overexpression of Bcl-2. Counteracting Bim function via forced Bcl-2 transgene expression revealed other targets of miRNA action in normal B cell development. This included the regulation of nontemplated nucleotide addition (N-regions) at the Ig light chain V-J junction, which is not usually seen, which was due to aberrant expression of the terminal deoxynucleotidyl transferase (TdT), whose expression is usually extinguished at the pre-B cell stage.[350,351]

Transgenic overexpression of the miR-17/92, a miRNA often found to be amplified in lymphoma, resulted in a lymphoproliferative syndrome and autoimmune disease.[352] One target of miR-17/92 was shown to be the aforementioned regulator of apoptosis, Bim, as well as the tumor suppressor PTEN.[352] However, ablation of miR-17/92 later in B cell development suggested a critical role for this miRNA also in the development of plasma cells through antigen-mediated stimulation of B cells and the migration of plasma cells from sites of induction to bone marrow niches for long-term survival.[353]

Experimental evidence recently demonstrated that c-Myc expression, the critical transcriptional regulator of B cell proliferation, is controlled by miRNA 144/451, whose loss can cause lymphoma generation via derepression of c-Myc.[354] Given the importance of c-Myc in B cell development, further studies on the role of miRNA 144/451 in B cell development and malignant transformation are warranted.[355] miRNAs can also influence lineage choice or act at key checkpoints during hematopoietic development. Retroviral provision of miR-34a in bone marrow hematopoietic progenitors blocked B cell development at the pro-B to pre-B checkpoint by repressing FoxP1, resulting in reduced numbers of mature B cells in mice repopulated with such precursors.[356]

Already discussed above, miRNA148a is a critical regulator of the immature B cell tolerance checkpoint.[169] miRNA148a was shown to regulate expression of the critical transcriptional regulator IRF4, likely by impacting IRF4 gene accessibility by the histone acetyltransferase GCN5, which enhances IRF4 transcription.[357] It also ensures the silencing of the light chain loci. Indeed, through their large-scale actions on networks of transcriptional regulators, key miRNAs are likely involved in the regulation of B cell development and selection at each step.

As mentioned in the section on alternative B cell developmental path, regulation of the Let-7 family of miRNA by the DNA-binding protein Lin28b was identified as critical for regulating the molecular switch from fetal to adult-type B cell development. Transcription of eight Let-7 miRNA family members was shown to be either absent or largely reduced in CD19− B220+ CD24+ CD43+ IgM− pro-B cells in the fetal liver, compared with their counterparts in the bone marrow, in a mechanism that required the presence of Lin28b.[11] Let-7 was shown to act, at least in part, by inducing expression of the transcription factor Arid3a, whose ectopic expression was able to drive fetal-like B-1 cell development.[10] Consistent with a role for Let-7 in regulating fetal versus adult hematopoiesis are studies that observed B-1/B-2 switching effects of Lin28b/Let-7 expression in IL-7Rα-deficient HSCs. IL-7Rα-deficient HSCs had been shown previously not to be able to generate adult bone marrow–derived B cells, because of their inability to respond to IL-7 with clonal expansion at the pre-B cell stage, whereas fetal B-1 cell development is less affected by the absence of IL-7Rα signaling.[131,358] Indeed, although adult HSCs from IL7Rα-deficient donor cells were unable to give rise to mature $CD19^+B220^+$ B cells, sizable populations of B cells with overrepresentation of B-1 and MZ B cells were detected in the peritoneal cavities and spleens, respectively, of BM chimeras reconstituted with Lin28b-transfected adult-derived HSCs.[10,11] The regulation of B-1 cell development through Lin28 has been linked to the need for positive selection of B-1 cells through binding to self-antigens.[359] Thus, Lin28 provides a critical link between the specificity of the BCR and its ability to drive a B-1 cell phenotype.[141] Expression of Lin 28b was observed also in fetal-derived thymic T cell precursors, where it conferred fetal-like expansions of certain TCRγδ-expressing cells,[11] further confirming that Let-7 miRNA is critical for regulating the switch from fetal to adult lymphopoiesis. Together these studies highlight the importance of epigenetic regulation in the ordered process of hematopoiesis and B cell development. They also support the long-standing concept that B cell development in the fetal liver and the adult bone marrow underlie fundamentally different regulatory mechanisms, resulting in the emergence of mature B cells with distinct repertoires and functions.

CONCLUSION

The development of B lymphocytes represents a fascinating example of the complexity by which gene regulatory networks guide the development of individual cell lineages from multipotent progenitors. B cell development is supported and initiated by environmental cues, such as from cytokines and chemokines produced by stromal cell networks. These signals induce critical gene regulatory circuits that themselves induce expression of transcriptional regulators and lead to epigenetic changes that guide first lineage commitment and later immunoglobulin gene rearrangement, selection, and BCR expression. Through a series of positive and negative selection steps that result in the removal of nonfunctional and strongly self-reactive B cells, and following the positive selection and clonal expansion of those that are deemed "appropriate," a repertoire of B cell specificities emerges during adult B cell development that, although not free of self-reactive cells, is free of strongly self-reactivity.

An increased understanding of the distinct processes that regulate B cell development during the fetal/neonatal versus adult period seems to have revealed an elegant solution to a long-standing question for the field, namely, whether B-1 and B-2 cells are part of the same or a distinct B cell developmental path and emerge from the same or distinct HSCs. Based on the existing data, it appears that it is not that HSCs exist that can give rise to only one or the other, but rather that fetal and adult B lymphopoiesis, through differential epigenetic and transcriptional regulation, results in the development of B cells that express distinct BCR repertoires, which leads to fundamental differences in the function of B-1 and also MZ B cells compared to B-2 cells.

The development of fetal-derived B-1 cells expressing a relatively strong self-reactive BCR appears to be generated through an alternative developmental path that often "skips" pre-BCR selection (Fig. 25.3), thus avoiding their elimination at this step of B cell development. This process is linked to the epigenetic regulation of fetal hematopoiesis via Lin 28/Let7 and thus cannot usually occur in adult lymphopoiesis. It is also reminiscent of B cell development in nonmammalian vertebrates. The subsequent expression of CD5 and other BCR coreceptors modulate and silence these self-reactive BCRs on B-1 cells to suppress inappropriate activation. This critical step not only imprints the distinct repertoire and phenotype that distinguishes B-1 from B-2 cells but also determines their functionality. This is because tonic BCR signaling has to be fundamentally different between B-1 and B-2 cells, and the integration of BCR tonic signaling with differential signaling from cytokine receptors, such as the BAFFR and perhaps also TLRs, results in the distinct functions and transcriptional profile of B-1 and B-2 cells. Such an interpretation would be consistent with the many studies that have shown the predisposition of B-1 cell development in the fetus, as well as the importance of the BCR in defining the distinct phenotype and transcriptional profiles of B-1 and B-2 cells.

ACKNOWLEDGMENTS

Randy R. Hardy wrote the chapter on B cell development for the previous edition of this book. That excellent chapter provided a basis and a guide for the writing of the chapter as it is now presented. Zheng Luo (UC Davis) generated the flow cytometry data shown in this chapter. Some figures were created using BioRender.com. The author gratefully acknowledges recent research support through grants from the National Institutes of Health, NIH/NIAID R01AI117890, RO1AI148652, R01AI157007, and R21AI151995.

REFERENCES

1. Moore MA, Metcalf D. Ontogeny of the haemopoietic system: yolk sac origin of in vivo and in vitro colony forming cells in the developing mouse embryo. *Br J Haematol.* 1970;18:279-296.
2. Dzierzak E, Bigas A. Blood development: hematopoietic stem cell dependence and independence. *Cell Stem Cell.* 2018;22:639-651.
3. Yoshimoto M, Montecino-Rodriguez E, Ferkowicz MJ, et al. Embryonic day 9 yolk sac and intra-embryonic hemogenic endothelium independently generate a B-1 and marginal zone progenitor lacking B-2 potential. *Proc Natl Acad Sci U S A.* 2011;108:1468-1473.
4. Godin IE, Garcia-Porrero JA, Coutinho A, Dieterlen-Lievre F, Marcos MA. Para-aortic splanchnopleura from early mouse embryos contains B1a cell progenitors. *Nature.* 1993;364:67-70.
5. Montecino-Rodriguez E, Fice M, Casero D, Berent-Maoz B, Barber CL, Dorshkind K. Distinct genetic networks orchestrate the emergence of specific waves of fetal and adult B-1 and B-2 development. *Immunity.* 2016;45:527-539.
6. Beaudin AE, Boyer SW, Perez-Cunningham J, et al. A transient developmental hematopoietic stem cell gives rise to innate-like B and T cells. *Cell Stem Cell.* 2016;19:768-783.
7. Solvason N, Lehuen A, Kearney JF. An embryonic source of Ly1 but not conventional B cells. *Int Immunol.* 1991;3:543-550.
8. Hardy RR, Hayakawa K. A developmental switch in B lymphopoiesis. *Proc Natl Acad Sci U S A.* 1991;88:11550-11554.
9. Barber CL, Montecino-Rodriguez E, Dorshkind K. Reduced production of B-1-specified common lymphoid progenitors results in diminished potential of adult marrow to generate B-1 cells. *Proc Natl Acad Sci U S A.* 2011;108:13700-13704.
10. Zhou Y, Li YS, Bandi SR, et al. Lin28b promotes fetal B lymphopoiesis through the transcription factor Arid3a. *J Exp Med.* 2015;212:569-580.
11. Yuan J, Nguyen CK, Liu X, Kanellopoulou C, Muljo SA. Lin28b reprograms adult bone marrow hematopoietic progenitors to mediate fetal-like lymphopoiesis. *Science.* 2012;335:1195-1200.
12. Brennecke AM, Düber S, Roy B, et al. Induced B cell development in adult mice. *Front Immunol.* 2018;9:2483.
13. Holodick NE, Repetny K, Zhong X, Rothstein TL. Adult BM generates CD5+ B1 cells containing abundant N-region additions. *Eur J Immunol.* 2009;39:2383-2394.
14. Ma S, Wang C, Mao X, Hao Y. B cell dysfunction associated with aging and autoimmune diseases. *Front Immunol.* 2019;10:318.
15. Morrison SJ, Scadden DT. The bone marrow niche for haematopoietic stem cells. *Nature.* 2014;505:327-334.
16. Shen B, Tasdogan A, Ubellacker JM, et al. A mechanosensitive peri-arteriolar niche for osteogenesis and lymphopoiesis. *Nature.* 2021;591:438-444.
17. Morrison SJ, Weissman IL. The long-term repopulating subset of hematopoietic stem cells is deterministic and isolatable by phenotype. *Immunity.* 1994;1:661-673.
18. Morrison SJ, Wandycz AM, Hemmati HD, Wright DE, Weissman IL. Identification of a lineage of multipotent hematopoietic progenitors. *Development.* 1997;124:1929-1939.
19. Kiel MJ, Yilmaz OH, Iwashita T, Yilmaz OH, Terhorst C, Morrison SJ. SLAM family receptors distinguish hematopoietic stem and progenitor cells and reveal endothelial niches for stem cells. *Cell.* 2005;121:1109-1121.
20. Kondo M, Weissman IL, Akashi K. Identification of clonogenic common lymphoid progenitors in mouse bone marrow. *Cell.* 1997;91:661-672.
21. Akashi K, Traver D, Miyamoto T, Weissman IL. A clonogenic common myeloid progenitor that gives rise to all myeloid lineages. *Nature.* 2000;404:193-197.
22. Kondo M, Scherer DC, Miyamoto T, et al. Cell-fate conversion of lymphoid-committed progenitors by instructive actions of cytokines. *Nature.* 2000;407:383-386.
23. Brown G. Towards a new understanding of decision-making by hematopoietic stem cells. *Int J Mol Sci.* 2020;21:2362.
24. Zehentmeier S, Pereira JP. Cell circuits and niches controlling B cell development. *Immunol Rev.* 2019;289:142-157.
25. Vivier E, Artis D, Colonna M, et al. Innate lymphoid cells: 10 Years on. *Cell.* 2018;174:1054-1066.
26. Boehm T, Bleul CC. Thymus-homing precursors and the thymic microenvironment. *Trends Immunol.* 2006;27:477-484.
27. Lavaert M, Liang KL, Vandamme N, et al. Integrated scRNA-seq identifies human postnatal thymus seeding progenitors and regulatory dynamics of differentiating immature thymocytes. *Immunity.* 2020;52:1088-1104.e1086.
28. Rothenberg EV. Single-cell insights into the hematopoietic generation of T-lymphocyte precursors in mouse and human. *Exp Hematol.* 2021;95:1-12.
29. Hardy RR, Carmack CE, Shinton SA, Kemp JD, Hayakawa K. Resolution and characterization of pro-B and pre-pro-B cell stages in normal mouse bone marrow. *J Exp Med.* 1991;173:1213-1225.
30. Li YS, Wasserman R, Hayakawa K, Hardy RR. Identification of the earliest B lineage stage in mouse bone marrow. *Immunity.* 1996;5:527-535.
31. Sitnicka E, Bryder D, Theilgaard-Mönch K, Buza-Vidas N, Adolfsson J, Jacobsen SE. Key role of flt3 ligand in regulation of the common lymphoid progenitor but not in maintenance of the hematopoietic stem cell pool. *Immunity.* 2002;17:463-472.
32. Li LX, Goetz CA, Katerndahl CD, Sakaguchi N, Farrar MA. A Flt3- and Ras-dependent pathway primes B cell development by inducing a state of IL-7 responsiveness. *J Immunol.* 2010;184:1728-1736.
33. Sitnicka E, Brakebusch C, Martensson IL, et al. Complementary signaling through flt3 and interleukin-7 receptor alpha is indispensable for fetal and adult B cell genesis. *J Exp Med.* 2003;198:1495-1506.

34. Fistonich C, Zehentmeier S, Bednarski JJ, et al. Cell circuits between B cell progenitors and IL-7(+) mesenchymal progenitor cells control B cell development. *J Exp Med.* 2018;215:2586-2599.
35. Grabstein KH, Waldschmidt TJ, Finkelman FD, et al. Inhibition of murine B and T lymphopoiesis in vivo by an anti-interleukin 7 monoclonal antibody. *J Exp Med.* 1993;178:257-264.
36. Carvalho TL, Mota-Santos T, Cumano A, Demengeot J, Vieira P. Arrested B lymphopoiesis and persistence of activated B cells in adult interleukin 7(-/)- mice. *J Exp Med.* 2001;194:1141-1150.
37. Clark MR, Mandal M, Ochiai K, Singh H. Orchestrating B cell lymphopoiesis through interplay of IL-7 receptor and pre-B cell receptor signalling. *Nat Rev Immunol.* 2014;14:69-80.
38. Hosokawa H, Rothenberg EV. How transcription factors drive choice of the T cell fate. *Nat Rev Immunol.* 2021;21:162-176.
39. Vanderbeck A, Maillard I. Notch signaling at the crossroads of innate and adaptive immunity. *J Leukoc Biol.* 2021;109:535-548.
40. Welner RS, Pelayo R, Nagai Y, et al. Lymphoid precursors are directed to produce dendritic cells as a result of TLR9 ligation during herpes infection. *Blood.* 2008;112:3753-3761.
41. Baratono SR, Chu N, Richman LP, Behrens EM. Toll-like receptor 9 and interferon-γ receptor signaling suppress the B-cell fate of uncommitted progenitors in mice. *Eur J Immunol.* 2015;45:1313-1325.
42. Rieger MA, Hoppe PS, Smejkal BM, Eitelhuber AC, Schroeder T. Hematopoietic cytokines can instruct lineage choice. *Science.* 2009;325:217-218.
43. Takizawa H, Boettcher S, Manz MG. Demand-adapted regulation of early hematopoiesis in infection and inflammation. *Blood.* 2012;119:2991-3002.
44. Rumfelt LL, Zhou Y, Rowley BM, Shinton SA, Hardy RR. Lineage specification and plasticity in CD19− early B cell precursors. *J Exp Med.* 2006;203:675-687.
45. Borghesi L, Hsu LY, Miller JP, et al. B lineage-specific regulation of V(D)J recombinase activity is established in common lymphoid progenitors. *J Exp Med.* 2004;199:491-502.
46. Cumano A, Paige CJ, Iscove NN, Brady G. Bipotential precursors of B cells and macrophages in murine fetal liver. *Nature.* 1992;356:612-615.
47. Montecino-Rodriguez E, Leathers H, Dorshkind K. Bipotential B-macrophage progenitors are present in adult bone marrow. *Nat Immunol.* 2001;2:83-88.
48. McHeik S, Van Eeckhout N, De Poorter C, Galés C, Parmentier M, Springael JY. Coexpression of CCR7 and CXCR4 during B cell development controls CXCR4 responsiveness and bone marrow homing. *Front Immunol.* 2019;10:2970.
49. Egawa T, Kawabata K, Kawamoto H, et al. The earliest stages of B cell development require a chemokine stromal cell-derived factor/pre-B cell growth-stimulating factor. *Immunity.* 2001;15:323-334.
50. Jacobsen K, Osmond DG. Microenvironmental organization and stromal cell associations of B lymphocyte precursor cells in mouse bone marrow. *Eur J Immunol.* 1990;20:2395-2404.
51. Lim VY, Zehentmeier S, Fistonich C, Pereira JP. A chemoattractant-guided walk through lymphopoiesis: from hematopoietic stem cells to mature B lymphocytes. *Adv Immunol.* 2017;134:47-88.
52. Whitlock C, Denis K, Robertson D, Witte O. In vitro analysis of murine B-cell development. *Annu Rev Immunol.* 1985;3:213-235.
53. Grawunder U, Melchers F, Rolink A. Interferon-gamma arrests proliferation and causes apoptosis in stromal cell/interleukin-7-dependent normal murine pre-B cell lines and clones in vitro, but does not induce differentiation to surface immunoglobulin-positive B cells. *Eur J Immunol.* 1993;23:544-551.
54. ten Boekel E, Melchers F, Rolink A. The status of Ig loci rearrangements in single cells from different stages of B cell development. *Int Immunol.* 1995;7:1013-1019.
55. Coffman RL. Surface antigen expression and immunoglobulin gene rearrangement during mouse pre-B cell development. *Immunol Rev.* 1982;69:5-23.
56. Nadler LM, Anderson KC, Marti G, et al. B4, a human B lymphocyte-associated antigen expressed on normal, mitogen-activated, and malignant B lymphocytes. *J Immunol.* 1983;131:244-250.
57. Ogawa M, ten Boekel E, Melchers F. Identification of CD19(−)B220(+)c-Kit(+)Flt3/Flk-2(+)cells as early B lymphoid precursors before pre-B-I cells in juvenile mouse bone marrow. *Int Immunol.* 2000;12:313-324.
58. Montecino-Rodriguez E, Leathers H, Dorshkind K. Identification of a B-1 B cell-specified progenitor. *Nat Immunol.* 2006;7:293-301.
59. de Andrés B, Gonzalo P, Minguet S, et al. The first 3 days of B-cell development in the mouse embryo. *Blood.* 2002;100:4074-4081.
60. Sato S, Ono N, Steeber DA, Pisetsky DS, Tedder TF. CD19 regulates B lymphocyte signaling thresholds critical for the development of B-1 lineage cells and autoimmunity. *J Immunol.* 1996;157:4371-4378.
61. Baumgarth N. The double life of a B-1 cell: self-reactivity selects for protective effector functions. *Nat Rev Immunol.* 2011;11:34-46.
62. Oettinger MA, Schatz DG, Gorka C, Baltimore D. RAG-1 and RAG-2, adjacent genes that synergistically activate V(D)J recombination. *Science.* 1990;248:1517-1523.
63. Schatz DG, Oettinger MA, Baltimore D. The V(D)J recombination activating gene, RAG-1. *Cell.* 1989;59:1035-1048.
64. Ji Y, Resch W, Corbett E, Yamane A, Casellas R, Schatz DG. The in vivo pattern of binding of RAG1 and RAG2 to antigen receptor loci. *Cell.* 2010;141:419-431.
65. Gilfillan S, Dierich A, Lemeur M, Benoist C, Mathis D. Mice lacking TdT: mature animals with an immature lymphocyte repertoire. *Science.* 1993;261:1175-1178.
66. Feeney AJ. Lack of N regions in fetal and neonatal mouse immunoglobulin V-D-J junctional sequences. *J Exp Med.* 1990;172:1377-1390.
67. Bossen C, Mansson R, Murre C. Chromatin topology and the regulation of antigen receptor assembly. *Annu Rev Immunol.* 2012;30:337-356.
68. Rother MB, Palstra RJ, Jhunjhunwala S, et al. Nuclear positioning rather than contraction controls ordered rearrangements of immunoglobulin loci. *Nucleic Acids Res.* 2016;44:175-186.
69. Winkler TH, Mårtensson IL. The role of the pre-B cell receptor in B cell development, repertoire selection, and tolerance. *Front Immunol.* 2018;9:2423.
70. Holmes ML, Carotta S, Corcoran LM, Nutt SL. Repression of Flt3 by Pax5 is crucial for B-cell lineage commitment. *Genes Dev.* 2006;20:933-938.
71. Malin S, McManus S, Cobaleda C, et al. Role of STAT5 in controlling cell survival and immunoglobulin gene recombination during pro-B cell development. *Nat Immunol.* 2010;11:171-179.
72. Giles AJ, Bender TP, Ravichandran KS. The adaptor protein Shc plays a key role during early B cell development. *J Immunol.* 2009;183:5468-5476.
73. Blunt T, Finnie NJ, Taccioli GE, et al. Defective DNA-dependent protein kinase activity is linked to V(D)J recombination and DNA repair defects associated with the murine scid mutation. *Cell.* 1995;80:813-823.
74. Peterson SR, Kurimasa A, Oshimura M, Dynan WS, Bradbury EM, Chen DJ. Loss of the catalytic subunit of the DNA-dependent protein kinase in DNA double-strand-break-repair mutant mammalian cells. *Proc Natl Acad Sci U S A.* 1995;92:3171-3174.
75. Baizan-Edge A, Stubbs BA, Stubbington MJT, et al. IL-7R signaling activates widespread V(H) and D(H) gene usage to drive antibody diversity in bone marrow B cells. *Cell Rep.* 2021;36:109349.
76. Karasuyama H, Rolink A, Melchers F. A complex of glycoproteins is associated with VpreB/lambda 5 surrogate light chain on the surface of mu heavy chain-negative early precursor B cell lines. *J Exp Med.* 1993;178:469-478.
77. Meffre E, Fougereau M, Argenson JN, Aubaniac JM, Schiff C. Cell surface expression of surrogate light chain (psi L) in the absence of mu on human pro-B cell lines and normal pro-B cells. *Eur J Immunol.* 1996;26:2172-2180.
78. Ohnishi K, Melchers F, Shimizu T. Lymphocyte-expressed BILL-cadherin/cadherin-17 contributes to the development of B cells at two stages. *Eur J Immunol.* 2005;35:957-963.
79. Vosshenrich CA, Cumano A, Müller W, Di Santo JP, Vieira P. Thymic stromal-derived lymphopoietin distinguishes fetal from adult B cell development. *Nat Immunol.* 2003;4:773-779.
80. Vosshenrich CA, Cumano A, Müller W, Di Santo JP, Vieira P. Pre-B cell receptor expression is necessary for thymic stromal lymphopoietin responsiveness in the bone marrow but not in the liver environment. *Proc Natl Acad Sci U S A.* 2004;101:11070-11075.
81. Jensen CT, Kharazi S, Böiers C, Liuba K, and Jacobsen SE. 2007. TSLP-mediated fetal B lymphopoiesis? *Nat Immunol* 8:897; author reply 898.
82. Jensen CT, Kharazi S, Böiers C, et al. FLT3 ligand and not TSLP is the key regulator of IL-7-independent B-1 and B-2 B lymphopoiesis. *Blood.* 2008;112:2297-2304.
83. Hess J, Werner A, Wirth T, Melchers F, Jäck HM, Winkler TH. Induction of pre-B cell proliferation after de novo synthesis of the pre-B cell receptor. *Proc Natl Acad Sci U S A.* 2001;98:1745-1750.
84. Aslam MA, Alemdehy MF, Hao B, et al. The Ig heavy chain protein but not its message controls early B cell development. *Proc Natl Acad Sci U S A.* 2020;117:31343-31352.
85. Grawunder U, Leu TM, Schatz DG, et al. Down-regulation of RAG1 and RAG2 gene expression in preB cells after functional immunoglobulin heavy chain rearrangement. *Immunity.* 1995;3:601-608.
86. Skok JA, Brown KE, Azuara V, et al. Nonequivalent nuclear location of immunoglobulin alleles in B lymphocytes. *Nat Immunol.* 2001;2:848-854.
87. Vettermann C, Schlissel MS. Allelic exclusion of immunoglobulin genes: models and mechanisms. *Immunol Rev.* 2010;237:22-42.
88. Bassing CH, Alt FW. The cellular response to general and programmed DNA double strand breaks. *DNA Repair.* 2004;3:781-796.
89. Decker DJ, Boyle NE, Koziol JA, Klinman NR. The expression of the Ig H chain repertoire in developing bone marrow B lineage cells. *J Immunol.* 1991;146:350-361.
90. Ebert A, McManus S, Tagoh H, et al. The distal V(H) gene cluster of the Igh locus contains distinct regulatory elements with Pax5 transcription factor-dependent activity in pro-B cells. *Immunity.* 2011;34:175-187.
91. Guo C, Yoon HS, Franklin A, et al. CTCF-binding elements mediate control of V(D)J recombination. *Nature.* 2011;477:424-430.
92. Lin SG, Guo C, Su A, Zhang Y, Alt FW. CTCF-binding elements 1 and 2 in the Igh intergenic control region cooperatively regulate V(D)J recombination. *Proc Natl Acad Sci U S A.* 2015;112:1815-1820.
93. Decker DJ, Boyle NE, Klinman NR. Predominance of nonproductive rearrangements of VH81X gene segments evidences a dependence of B cell clonal maturation on the structure of nascent H chains. *J Immunol.* 1991;147:1406-1411.
94. Decker DJ, Kline GH, Hayden TA, Zaharevitz SN, Klinman NR. Heavy chain V gene-specific elimination of B cells during the pre-B cell to B cell transition. *J Immunol.* 1995;154:4924-4935.
95. Keyna U, Beck-Engeser GB, Jongstra J, Applequist SE, Jäck HM. Surrogate light chain-dependent selection of Ig heavy chain V regions. *J Immunol.* 1995;155:5536-5542.
96. Kline GH, Hartwell L, Beck-Engeser GB, et al. Pre-B cell receptor-mediated selection of pre-B cells synthesizing functional mu heavy chains. *J Immunol.* 1998;161:1608-1618.
97. Keyna U, Applequist SE, Jongstra J, Beck-Engeser GB, Jäck HM. Ig mu heavy chains with VH81X variable regions do not associate with lambda 5. *Ann N Y Acad Sci.* 1995;764:39-42.
98. Tsubata T, Reth M. The products of pre-B cell-specific genes (lambda 5 and VpreB) and the immunoglobulin mu chain form a complex that is transported onto the cell surface. *J Exp Med.* 1990;172:973-976.
99. Pelanda R, Braun U, Hobeika E, Nussenzweig MC, Reth M. B cell progenitors are arrested in maturation but have intact VDJ recombination in the absence of Ig-alpha and Ig-beta. *J Immunol.* 2002;169:865-872.
100. Cheng AM, Rowley B, Pao W, Hayday A, Bolen JB, Pawson T. Syk tyrosine kinase required for mouse viability and B-cell development. *Nature.* 1995;378:303-306.
101. Turner M, Mee PJ, Costello PS, et al. Perinatal lethality and blocked B-cell development in mice lacking the tyrosine kinase Syk. *Nature.* 1995;378:298-302.
102. Rolink AG, Winkler T, Melchers F, Andersson J. Precursor B cell receptor-dependent B cell proliferation and differentiation does not require the bone marrow or fetal liver environment. *J Exp Med.* 2000;191:23-32.
103. Papavasiliou F, Jankovic M, Nussenzweig MC. Surrogate or conventional light chains are required for membrane immunoglobulin mu to activate the precursor B cell transition. *J Exp Med.* 1996;184:2025-2030.
104. Ubelhart R, Bach MP, Eschbach C, Wossning T, Reth M, Jumaa H. N-linked glycosylation selectively regulates autonomous precursor BCR function. *Nat Immunol.* 2010;11:759-765.
105. Wasserman R, Li YS, Hardy RR. Down-regulation of terminal deoxynucleotidyl transferase by Ig heavy chain in B lineage cells. *J Immunol.* 1997;158:1133-1138.
106. Parker MJ, Licence S, Erlandsson L, et al. The pre-B-cell receptor induces silencing of VpreB and lambda5 transcription. *EMBO J.* 2005;24:3895-3905.
107. McLean KC, Mandal M. It takes three receptors to raise a B cell. *Trends Immunol.* 2020;41:629-642.
108. Powers SE, Mandal M, Matsuda S, et al. Subnuclear cyclin D3 compartments and the coordinated regulation of proliferation and immunoglobulin variable gene repression. *J Exp Med.* 2012;209:2199-2213.
109. Karki S, Kennedy DE, McLean K, et al. Regulated capture of Vκ gene topologically associating domains by transcription factories. *Cell Rep.* 2018;24:2443-2456.
110. Herzog S, Hug E, Meixlsperger S, et al. SLP-65 regulates immunoglobulin light chain gene recombination through the PI(3)K-PKB-Foxo pathway. *Nat Immunol.* 2008;9:623-631.
111. Amin RH, Schlissel MS. Foxo1 directly regulates the transcription of recombination-activating genes during B cell development. *Nat Immunol.* 2008;9:613-622.
112. Ochiai K, Maienschein-Cline M, Mandal M, et al. A self-reinforcing regulatory network triggered by limiting IL-7 activates pre-BCR signaling and differentiation. *Nat Immunol.* 2012;13:300-307.
113. Reth M, Petrac E, Wiese P, Lobel L, Alt FW. Activation of V kappa gene rearrangement in pre-B cells follows the expression of membrane-bound immunoglobulin heavy chains. *EMBO J.* 1987;6:3299-3305.
114. van Loo PF, Dingjan GM, Maas A, Hendriks RW. Surrogate-light-chain silencing is not critical for the limitation of pre-B cell expansion but is for the termination of constitutive signaling. *Immunity.* 2007;27:468-480.
115. Sakamoto S, Wakae K, Anzai Y, et al. E2A and CBP/p300 act in synergy to promote chromatin accessibility of the immunoglobulin κ locus. *J Immunol.* 2012;188:5547-5560.

116. Johnson K, Hashimshony T, Sawai CM, et al. Regulation of immunoglobulin light-chain recombination by the transcription factor IRF-4 and the attenuation of interleukin-7 signaling. *Immunity*. 2008;28:335-345.
117. Ma S, Pathak S, Mandal M, Trinh L, Clark MR, Lu R. Ikaros and Aiolos inhibit pre-B-cell proliferation by directly suppressing c-Myc expression. *Mol Cell Biol*. 2010;30:4149-4158.
118. Erlandsson L, Licence S, Gaspal F, Lane P, Corcoran AE, Mårtensson IL. Both the pre-BCR and the IL-7Ralpha are essential for expansion at the pre-BII cell stage in vivo. *Eur J Immunol*. 2005;35:1969-1976.
119. Schlissel MS, Baltimore D. Activation of immunoglobulin kappa gene rearrangement correlates with induction of germline kappa gene transcription. *Cell*. 1989;58:1001-1007.
120. Mandal M, Okoreeh MK, Kennedy DE, et al. CXCR4 signaling directs Igk recombination and the molecular mechanisms of late B lymphopoiesis. *Nat Immunol*. 2019;20:1393-1403.
121. Tokoyoda K, Egawa T, Sugiyama T, Choi BI, Nagasawa T. Cellular niches controlling B lymphocyte behavior within bone marrow during development. *Immunity*. 2004;20:707-718.
122. Flajnik MF. A cold-blooded view of adaptive immunity. *Nat Rev Immunol*. 2018;18:438-453.
123. Sinkora M, Sinkorova J, Stepanova K. Ig light chain precedes heavy chain gene rearrangement during development of B cells in swine. *J Immunol*. 2017;198:1543-1552.
124. Kitamura D, Kudo A, Schaal S, Müller W, Melchers F, Rajewsky K. A critical role of lambda 5 protein in B cell development. *Cell*. 1992;69:823-831.
125. Mundt C, Licence S, Shimizu T, Melchers F, Mårtensson IL. Loss of precursor B cell expansion but not allelic exclusion in VpreB1/VpreB2 double-deficient mice. *J Exp Med*. 2001;193:435-445.
126. Keenan RA, De Riva A, Corleis B, et al. Censoring of autoreactive B cell development by the pre-B cell receptor. *Science*. 2008;321:696-699.
127. Burgueño-Bucio E, Mier-Aguilar CA, Soldevila G. The multiple faces of CD5. *J Leukoc Biol*. 2019;105:891-904.
128. Wong JB, Hewitt SL, Heltemes-Harris LM, et al. B-1a cells acquire their unique characteristics by bypassing the pre-BCR selection stage. *Nat Commun*. 2019;10:4768.
129. Wasserman R, Li YS, Shinton SA, et al. A novel mechanism for B cell repertoire maturation based on response by B cell precursors to pre-B receptor assembly. *J Exp Med*. 1998;187:259-264.
130. Johnson K, Chaumeil J, Micsinai M, et al. IL-7 functionally segregates the pro-B cell stage by regulating transcription of recombination mediators across cell cycle. *J Immunol*. 2012;188:6084-6092.
131. Erlandsson L, Licence S, Gaspal F, et al. Impaired B-1 and B-2 B cell development and atypical splenic B cell structures in IL-7 receptor-deficient mice. *Eur J Immunol*. 2004;34:3595-3603.
132. Hayakawa K, Li YS, Shinton SA, et al. Crucial role of increased Arid3a at the pre-B and immature B cell stages for B1a cell generation. *Front Immunol*. 2019;10:457.
133. Kreslavsky T, Vilagos B, Tagoh H, et al. Essential role for the transcription factor Bhlhe41 in regulating the development, self-renewal and BCR repertoire of B-1a cells. *Nat Immunol*. 2017;18:442-455.
134. Hoffmann A, Kerr S, Jellusova J, et al. Siglec-G is a B1 cell-inhibitory receptor that controls expansion and calcium signaling of the B1 cell population. *Nat Immunol*. 2007;8:695-704.
135. Hutzler S, Özgör L, Naito-Matsui Y, et al. The ligand-binding domain of Siglec-G is crucial for its selective inhibitory function on B1 cells. *J Immunol*. 2014;192:5406-5414.
136. Pan C, Baumgarth N, Parnes JR. CD72-deficient mice reveal nonredundant roles of CD72 in B cell development and activation. *Immunity*. 1999;11:495-506.
137. ten Boekel E, Melchers F, Rolink AG. Changes in the V(H) gene repertoire of developing precursor B lymphocytes in mouse bone marrow mediated by the pre-B cell receptor. *Immunity*. 1997;7:357-368.
138. Hayakawa K, Asano M, Shinton SA, et al. Positive selection of natural autoreactive B cells. *Science*. 1999;285:113-116.
139. Ferry H, Jones M, Vaux DJ, Roberts IS, Cornall RJ. The cellular location of self-antigen determines the positive and negative selection of autoreactive B cells. *J Exp Med*. 2003;198:1415-1425.
140. Köhler F, Hug E, Eschbach C, et al. Autoreactive B cell receptors mimic autonomous pre-B cell receptor signaling and induce proliferation of early B cells. *Immunity*. 2008;29:912-921.
141. Graf R, Seagal J, Otipoby KL, et al. BCR-dependent lineage plasticity in mature B cells. *Science*. 2019;363:748-753.
142. Baumgarth N, Waffarn EE, Nguyen TT. Natural and induced B-1 cell immunity to infections raises questions of nature versus nurture. *Ann N Y Acad Sci*. 2015;1362:188-199.
143. Allman DM, Ferguson SE, Cancro MP. Peripheral B cell maturation. I. Immature peripheral B cells in adults are heat-stable antigenhi and exhibit unique signaling characteristics. *J Immunol*. 1992;149:2533-2540.
144. Rolink AG, Andersson J, Melchers F. Characterization of immature B cells by a novel monoclonal antibody, by turnover and by mitogen reactivity. *Eur J Immunol*. 1998;28:3738-3748.
145. Lazorchak AS, Schlissel MS, Zhuang Y. E2A and IRF-4/Pip promote chromatin modification and transcription of the immunoglobulin kappa locus in pre-B cells. *Mol Cell Biol*. 2006;26:810-821.
146. Collins AM, Watson CT. Immunoglobulin light chain gene rearrangements, receptor editing and the development of a self-tolerant antibody repertoire. *Front Immunol*. 2018;9:2249.
147. Arakawa H, Shimizu T, Takeda S. Re-evaluation of the probabilities for productive arrangements on the kappa and lambda loci. *Int Immunol*. 1996;8:91-99.
148. Tiegs SL, Russell DM, Nemazee D. Receptor editing in self-reactive bone marrow B cells. *J Exp Med*. 1993;177:1009-1020.
149. Prak EL, Weigert M. Light chain replacement: a new model for antibody gene rearrangement. *J Exp Med*. 1995;182:541-548.
150. Gay D, Saunders T, Camper S, Weigert M. Receptor editing: an approach by autoreactive B cells to escape tolerance. *J Exp Med*. 1993;177:999-1008.
151. Halverson R, Torres RM, Pelanda R. Receptor editing is the main mechanism of B cell tolerance toward membrane antigens. *Nat Immunol*. 2004;5:645-650.
152. Nemazee D. Receptor editing in lymphocyte development and central tolerance. *Nat Rev Immunol*. 2006;6:728-740.
153. Duong BH, Ota T, Aoki-Ota M, et al. Negative selection by IgM superantigen defines a B cell central tolerance compartment and reveals mutations allowing escape. *J Immunol*. 2011;187:5596-5605.
154. Sproul TW, Malapati S, Kim J, Pierce SK. Cutting edge: B cell antigen receptor signaling occurs outside lipid rafts in immature B cells. *J Immunol*. 2000;165:6020-6023.
155. Cheng PC, Dykstra ML, Mitchell RN, Pierce SK. A role for lipid rafts in B cell antigen receptor signaling and antigen targeting. *J Exp Med*. 1999;190:1549-1560.
156. Schmidt C, Kim D, Ippolito GC, et al. Signalling of the BCR is regulated by a lipid rafts-localised transcription factor, Bright. *EMBO J*. 2009;28:711-724.
157. Harwood NE, Batista FD. New insights into the early molecular events underlying B cell activation. *Immunity*. 2008;28:609-619.
158. Schmidt C, Christian L, Smith TA, Tidwell J, Kim D, Tucker HO. Lipid rafts interaction of the ARID3A transcription factor with EZRIN and G-Actin regulates B-cell receptor signaling. *Diseases*. 2021;9.
159. Keren Z, Diamant E, Ostrovsky O, Bengal E, Melamed D. Modification of ligand-independent B cell receptor tonic signals activates receptor editing in immature B lymphocytes. *J Biol Chem*. 2004;279:13418-13424.
160. Bannish G, Fuentes-Pananá EM, Cambier JC, Pear WS, Monroe JG. Ligand-independent signaling functions for the B lymphocyte antigen receptor and their role in positive selection during B lymphopoiesis. *J Exp Med*. 2001;194:1583-1596.
161. Schram BR, Tze LE, Ramsey LB, et al. B cell receptor basal signaling regulates antigen-induced Ig light chain rearrangements. *J Immunol*. 2008;180:4728-4741.
162. Benhamou D, Labi V, Getahun A, et al. The c-Myc/miR17-92/PTEN axis tunes PI3K activity to control expression of recombination activating genes in early B cell development. *Front Immunol*. 2018;9:2715.
163. Greaves SA, Peterson JN, Strauch P, Torres RM, Pelanda R. Active PI3K abrogates central tolerance in high-avidity autoreactive B cells. *J Exp Med*. 2019;216:1135-1153.
164. Verkoczy L, Duong B, Skog P, et al. Basal B cell receptor-directed phosphatidylinositol 3-kinase signaling turns off RAGs and promotes B cell-positive selection. *J Immunol*. 2007;178:6332-6341.
165. Llorian M, Stamataki Z, Hill S, Turner M, Mårtensson IL. The PI3K p110delta is required for down-regulation of RAG expression in immature B cells. *J Immunol*. 2007;178:1981-1985.
166. Pelanda R. Dual immunoglobulin light chain B cells: Trojan horses of autoimmunity? *Curr Opin Immunol*. 2014;27:53-59.
167. Lau A, Avery DT, Jackson K, et al. Activated PI3Kδ breaches multiple B cell tolerance checkpoints and causes autoantibody production. *J Exp Med*. 2020;217:e20191336.
168. Jani PK, Kubagawa H, Melchers F. A rheostat sets B-cell receptor repertoire selection to distinguish self from non-self. *Curr Opin Immunol*. 2020;67:42-49.
169. Gonzalez-Martin A, Adams BD, Lai M, et al. The microRNA miR-148a functions as a critical regulator of B cell tolerance and autoimmunity. *Nat Immunol*. 2016;17:433-440.
170. Norvell A, Mandik L, Monroe JG. Engagement of the antigen-receptor on immature murine B lymphocytes results in death by apoptosis. *J Immunol*. 1995;154:4404-4413.
171. Hippen KL, Schram BR, Tze LE, Pape KA, Jenkins MK, Behrens TW. In vivo assessment of the relative contributions of deletion, anergy, and editing to B cell self-tolerance. *J Immunol*. 2005;175:909-916.
172. Goodnow CC, Crosbie J, Adelstein S, et al. Altered immunoglobulin expression and functional silencing of self-reactive B lymphocytes in transgenic mice. *Nature*. 1988;334:676-682.
173. Hartley SB, Crosbie J, Brink R, Kantor AB, Basten A, Goodnow CC. Elimination from peripheral lymphoid tissues of self-reactive B lymphocytes recognizing membrane-bound antigens. *Nature*. 1991;353:765-769.
174. Nemazee DA, Bürki K. Clonal deletion of B lymphocytes in a transgenic mouse bearing anti-MHC class I antibody genes. *Nature*. 1989;337:562-566.
175. Merrell KT, Benschop RJ, Gauld SB, et al. Identification of anergic B cells within a wild-type repertoire. *Immunity*. 2006;25:953-962.
176. Wardemann H, Yurasov S, Schaefer A, Young JW, Meffre E, Nussenzweig MC. Predominant autoantibody production by early human B cell precursors. *Science*. 2003;301:1374-1377.
177. Wechsler-Reya RJ, Monroe JG. Lipopolysaccharide prevents apoptosis and induces responsiveness to antigen receptor cross-linking in immature B cells. *Immunology*. 1996;89:356-362.
178. Rawlings DJ, Schwartz MA, Jackson SW, Meyer-Bahlburg A. Integration of B cell responses through toll-like receptors and antigen receptors. *Nat Rev Immunol*. 2012;12:282-294.
179. Schweighoffer E, Nys J, Vanes L, Smithers N, Tybulewicz VLJ. TLR4 signals in B lymphocytes are transduced via the B cell antigen receptor and SYK. *J Exp Med*. 2017;214:1269-1280.
180. Cariappa A, Chase C, Liu H, Russell P, Pillai S. Naive recirculating B cells mature simultaneously in the spleen and bone marrow. *Blood*. 2007;109:2339-2345.
181. Lindsley RC, Thomas M, Srivastava B, Allman D. Generation of peripheral B cells occurs via two spatially and temporally distinct pathways. *Blood*. 2007;109:2521-2528.
182. Meffre E, Wardemann H. B-cell tolerance checkpoints in health and autoimmunity. *Curr Opin Immunol*. 2008;20:632-638.
183. Chung JB, Sater RA, Fields ML, Erikson J, Monroe JG. CD23 defines two distinct subsets of immature B cells which differ in their responses to T cell help signals. *Int Immunol*. 2002;14:157-166.
184. Cambier JC, Gauld SB, Merrell KT, Vilen BJ. B-cell anergy: from transgenic models to naturally occurring anergic B cells? *Nat Rev Immunol*. 2007;7:633-643.
185. Sater RA, Sandel PC, Monroe JG. B cell receptor-induced apoptosis in primary transitional murine B cells: signaling requirements and modulation by T cell help. *Int Immunol*. 1998;10:1673-1682.
186. Russell DM, Dembić Z, Morahan G, Miller JF, Bürki K, Nemazee D. Peripheral deletion of self-reactive B cells. *Nature*. 1991;354:308-311.
187. Carsetti R, Köhler G, Lamers MC. Transitional B cells are the target of negative selection in the B cell compartment. *J Exp Med*. 1995;181:2129-2140.
188. Allman D, Lindsley RC, DeMuth W, Rudd K, Shinton SA, Hardy RR. Resolution of three nonproliferative immature splenic B cell subsets reveals multiple selection points during peripheral B cell maturation. *J Immunol*. 2001;167:6834-6840.
189. Henderson RB, Grys K, Vehlow A, et al. A novel Rac-dependent checkpoint in B cell development controls entry into the splenic white pulp and cell survival. *J Exp Med*. 2010;207:837-853.
190. Loder F, Mutschler B, Ray RJ, et al. B cell development in the spleen takes place in discrete steps and is determined by the quality of B cell receptor-derived signals. *J Exp Med*. 1999;190:75-89.
191. Su TT, Guo B, Wei B, Braun J, Rawlings DJ. Signaling in transitional type 2 B cells is critical for peripheral B-cell development. *Immunol Rev*. 2004;197:161-178.
192. Nojima T, Reynolds AE, Kitamura D, Kelsoe G, Kuraoka M. Tracing self-reactive B cells in normal mice. *J Immunol*. 2020;205:90-101.
193. Gross JA, Dillon SR, Mudri S, et al. TACI-Ig neutralizes molecules critical for B cell development and autoimmune disease. impaired B cell maturation in mice lacking BLyS. *Immunity*. 2001;15:289-302.
194. Schiemann B, Gommerman JL, Vora K, et al. An essential role for BAFF in the normal development of B cells through a BCMA-independent pathway. *Science*. 2001;293:2111-2114.
195. Rowland SL, Leahy KF, Halverson R, Torres RM, Pelanda R. BAFF receptor signaling aids the differentiation of immature B cells into transitional B cells following tonic BCR signaling. *J Immunol*. 2010;185:4570-4581.
196. Gorelik L, Cutler AH, Thill G, et al. Cutting edge: BAFF regulates CD21/35 and CD23 expression independent of its B cell survival function. *J Immunol*. 2004;172:762-766.
197. Wang H, Feng J, Qi CF, Li Z, Morse HC3rd, Clarke SH. Transitional B cells lose their ability to receptor edit but retain their potential for positive and negative selection. *J Immunol*. 2007;179:7544-7552.
198. Kleiman E, Salyakina D, De Heusch M, et al. Distinct transcriptomic features are associated with transitional and mature B-cell populations in the mouse spleen. *Front Immunol*. 2015;6:30.
199. Kuraoka M, Holl TM, Liao D, et al. Activation-induced cytidine deaminase mediates central tolerance in B cells. *Proc Natl Acad Sci U S A*. 2011;108:11560-11565.
200. Meyers G, Ng YS, Bannock JM, et al. Activation-induced cytidine deaminase (AID) is required for B-cell tolerance in humans. *Proc Natl Acad Sci U S A*. 2011;108:11554-11559.
201. Kuraoka M, Snowden PB, Nojima T, et al. BCR and endosomal TLR signals synergize to increase AID expression and establish central B cell tolerance. *Cell Rep*. 2017;18:1627-1635.

202. Norvell A, Monroe JG. Acquisition of surface IgD fails to protect from tolerance-induction. Both surface IgM- and surface IgD-mediated signals induce apoptosis of immature murine B lymphocytes. *J Immunol.* 1996;156:1328-1332.
203. Carman JA, Wechsler-Reya RJ, Monroe JG. Immature stage B cells enter but do not progress beyond the early G1 phase of the cell cycle in response to antigen receptor signaling. *J Immunol.* 1996;156:4562-4569.
204. Zikherman J, Parameswaran R, Weiss A. Endogenous antigen tunes the responsiveness of naive B cells but not T cells. *Nature.* 2012;489:160-164.
205. Khan WN, Nilsson A, Mizoguchi E, et al. Impaired B cell maturation in mice lacking Bruton's tyrosine kinase (Btk) and CD40. *Int Immunol.* 1997;9:395-405.
206. Stadanlick JE, Kaileh M, Karnell FG, et al. Tonic B cell antigen receptor signals supply an NF-kappaB substrate for prosurvival BLyS signaling. *Nat Immunol.* 2008;9:1379-1387.
207. Meyer-Bahlburg A, Andrews SF, Yu KO, Porcelli SA, Rawlings DJ. Characterization of a late transitional B cell population highly sensitive to BAFF-mediated homeostatic proliferation. *J Exp Med.* 2008;205:155-168.
208. Schweighoffer E, Vanes L, Nys J, et al. The BAFF receptor transduces survival signals by co-opting the B cell receptor signaling pathway. *Immunity.* 2013;38:475-488.
209. Hervé M, Isnardi I, Ng YS, et al. CD40 ligand and MHC class II expression are essential for human peripheral B cell tolerance. *J Exp Med.* 2007;204:1583-1593.
210. Schwartz MA, Kolhatkar NS, Thouvenel C, Khim S, Rawlings DJ. CD4+ T cells and CD40 participate in selection and homeostasis of peripheral B cells. *J Immunol.* 2014;193:3492-3502.
211. Tanwar S, Dhar A, Varanasi V, et al. Mediation of transitional B cell maturation in the absence of functional Bruton's tyrosine kinase. *Sci Rep.* 2017;7:46029.
212. Baumgarth N, Jager GC, Herman OC, Herzenberg LA. CD4+ T cells derived from B cell-deficient mice inhibit the establishment of peripheral B cell pools. *Proc Natl Acad Sci U S A.* 2000;97:4766-4771.
213. Giltiay NV, Giordano D, Clark EA. The plasticity of newly formed B cells. *J Immunol.* 2019;203:3095-3104.
214. Metzler G, Kolhatkar NS, Rawlings DJ. BCR and co-receptor crosstalk facilitate the positive selection of self-reactive transitional B cells. *Curr Opin Immunol.* 2015;37:46-53.
215. Cyster JG, Hartley SB, Goodnow CC. Competition for follicular niches excludes self-reactive cells from the recirculating B-cell repertoire. *Nature.* 1994;371:389-395.
216. Cyster JG, Goodnow CC. Antigen-induced exclusion from follicles and anergy are separate and complementary processes that influence peripheral B cell fate. *Immunity.* 1995;3:691-701.
217. Pillai S, Cariappa A. The follicular versus marginal zone B lymphocyte cell fate decision. *Nat Rev Immunol.* 2009;9:767-777.
218. Lechner M, Engleitner T, Babushku T, et al. Notch2-mediated plasticity between marginal zone and follicular B cells. *Nat Commun.* 2021;12:1111.
219. Saito T, Chiba S, Ichikawa M, et al. Notch2 is preferentially expressed in mature B cells and indispensable for marginal zone B lineage development. *Immunity.* 2003;18:675-685.
220. Carey JB, Moffatt-Blue CS, Watson LC, Gavin AL, Feeney AJ. Repertoire-based selection into the marginal zone compartment during B cell development. *J Exp Med.* 2008;205:2043-2052.
221. Ansel KM, Ngo VN, Hyman PL, et al. A chemokine-driven positive feedback loop organizes lymphoid follicles. *Nature.* 2000;406:309-314.
222. Srinivasan L, Sasaki Y, Calado DP, et al. PI3 kinase signals BCR-dependent mature B cell survival. *Cell.* 2009;139:573-586.
223. McAllister E, Jellusova J. BAFF signaling in B cell metabolism. *Curr Opin Immunol.* 2021;71:69-74.
224. Schweighoffer E, Tybulewicz VL. BAFF signaling in health and disease. *Curr Opin Immunol.* 2021;71:124-131.
225. Lam KP, Kühn R, Rajewsky K. In vivo ablation of surface immunoglobulin on mature B cells by inducible gene targeting results in rapid cell death. *Cell.* 1997;90:1073-1083.
226. Förster I, Rajewsky K. The bulk of the peripheral B-cell pool in mice is stable and not rapidly renewed from the bone marrow. *Proc Natl Acad Sci U S A.* 1990;87:4781-4784.
227. Hao Z, Rajewsky K. Homeostasis of peripheral B cells in the absence of B cell influx from the bone marrow. *J Exp Med.* 2001;194:1151-1164.
228. Mackay F, Schneider P. Cracking the BAFF code. *Nat Rev Immunol.* 2009;9:491-502.
229. Ahearn JM, Fischer MB, Croix D, et al. Disruption of the Cr2 locus results in a reduction in B-1a cells and in an impaired B cell response to T-dependent antigen. *Immunity.* 1996;4:251-262.
230. Engel P, Zhou LJ, Ord DC, Sato S, Koller B, Tedder TF. Abnormal B lymphocyte development, activation, and differentiation in mice that lack or overexpress the CD19 signal transduction molecule. *Immunity.* 1995;3:39-50.
231. Khan WN, Alt FW, Gerstein RM, et al. Defective B cell development and function in Btk-deficient mice. *Immunity.* 1995;3:283-299.
232. Meyer SJ, Linder AT, Brandl C, Nitschke L. B cell Siglecs-News on signaling and its interplay with ligand binding. *Front Immunol.* 2018;9:2820.
233. O'Keefe TL, Williams GT, Davies SL, Neuberger MS. Hyperresponsive B cells in CD22-deficient mice. *Science.* 1996;274:798-801.
234. Sato S, Miller AS, Inaoki M, et al. CD22 is both a positive and negative regulator of B lymphocyte antigen receptor signal transduction: altered signaling in CD22-deficient mice. *Immunity.* 1996;5:551-562.
235. Cerutti A, Cols M, Puga I. Marginal zone B cells: virtues of innate-like antibody-producing lymphocytes. *Nat Rev Immunol.* 2013;13:118-132.
236. Cinamon G, Zachariah MA, Lam OM, Foss FW Jr, Cyster JG. Follicular shuttling of marginal zone B cells facilitates antigen transport. *Nat Immunol.* 2008;9:54-62.
237. Arnon TI, Horton RM, Grigorova IL, Cyster JG. Visualization of splenic marginal zone B-cell shuttling and follicular B-cell egress. *Nature.* 2013;493:684-688.
238. Song H, Cerny J. Functional heterogeneity of marginal zone B cells revealed by their ability to generate both early antibody-forming cells and germinal centers with hypermutation and memory in response to a T-dependent antigen. *J Exp Med.* 2003;198:1923-1935.
239. Phan TG, Gardam S, Basten A, Brink R. Altered migration, recruitment, and somatic hypermutation in the early response of marginal zone B cells to T cell-dependent antigen. *J Immunol.* 2005;174:4567-4578.
240. Snapper CM, Yamada H, Smoot D, Sneed R, Lees A, Mond JJ. Comparative in vitro analysis of proliferation, Ig secretion, and Ig class switching by murine marginal zone and follicular B cells. *J Immunol.* 1993;150:2737-2745.
241. Martin F, Kearney JF. Positive selection from newly formed to marginal zone B cells depends on the rate of clonal production, CD19, and btk. *Immunity.* 2000;12:39-49.
242. Hammad H, Vanderkerken M, Pouliot P, et al. Transitional B cells commit to marginal zone B cell fate by Taok3-mediated surface expression of ADAM10. *Nat Immunol.* 2017;18:313-320.
243. Guinamard R, Okigaki M, Schlessinger J, Ravetch JV. Absence of marginal zone B cells in Pyk-2-deficient mice defines their role in the humoral response. *Nat Immunol.* 2000;1:31-36.
244. Muppidi JR, Arnon TI, Bronevetsky Y, et al. Cannabinoid receptor 2 positions and retains marginal zone B cells within the splenic marginal zone. *J Exp Med.* 2011;208:1941-1948.
245. Pujanauski LM, Janoff EN, McCarter MD, Pelanda R, Torres RM. Mouse marginal zone B cells harbor specificities similar to human broadly neutralizing HIV antibodies. *Proc Natl Acad Sci U S A.* 2013;110:1422-1427.
246. Chen X, Martin F, Forbush KA, Perlmutter RM, Kearney JF. Evidence for selection of a population of multi-reactive B cells into the splenic marginal zone. *Int Immunol.* 1997;9:27-41.
247. Hayakawa K, Hardy RR, Stall AM, Herzenberg LA, Herzenberg LA. Immunoglobulin-bearing B cells reconstitute and maintain the murine Ly-1 B cell lineage. *Eur J Immunol.* 1986;16:1313-1316.
248. Kantor AB, Stall AM, Adams S, Herzenberg LA, Herzenberg LA. Differential development of progenitor activity for three B-cell lineages. *Proc Natl Acad Sci U S A.* 1992;89:3320-3324.
249. Holodick NE, Rothstein TL. B cells in the aging immune system: time to consider B-1 cells. *Ann N Y Acad Sci.* 2015;1362:176-187.
250. Savage HP, Kläsener K, Smith FL, Luo Z, Reth M, Baumgarth N. TLR induces reorganization of the IgM-BCR complex regulating murine B-1 cell responses to infections. *Elife.* 2019;8:e46997.
251. Choi YS, Dieter JA, Rothaeusler K, Luo Z, Baumgarth N. B-1 cells in the bone marrow are a significant source of natural IgM. *Eur J Immunol.* 2012;42:120-129.
252. Savage HP, Yenson VM, Sawhney SS, Mousseau BJ, Lund FE, Baumgarth N. Blimp-1-dependent and -independent natural antibody production by B-1 and B-1-derived plasma cells. *J Exp Med.* 2017;214:2777-2794.
253. Baumgarth N, Tung JW, Herzenberg LA. Inherent specificities in natural antibodies: a key to immune defense against pathogen invasion. *Springer Semin Immunopathol.* 2005;26:347-362.
254. Grönwall C, Silverman GJ. Natural IgM: beneficial autoantibodies for the control of inflammatory and autoimmune disease. *J Clin Immunol.* 2014;34(Suppl 1):S12-S21.
255. Binder CJ, Papac-Milicevic N, Witztum JL. Innate sensing of oxidation-specific epitopes in health and disease. *Nat Rev Immunol.* 2016;16:485-497.
256. Vanhee S, Åkerstrand H, Kristiansen TA, et al. Lin28b controls a neonatal to adult switch in B cell positive selection. *Sci Immunol.* 2019;4:eaax4453.
257. Sidman CL, Shultz LD, Hardy RR, Hayakawa K, Herzenberg LA. Production of immunoglobulin isotypes by Ly-1+ B cells in viable motheaten and normal mice. *Science.* 1986;232:1423-1425.
258. Torres RM, Flaswinkel H, Reth M, Rajewsky K. Aberrant B cell development and immune response in mice with a compromised BCR complex. *Science.* 1996;272:1804-1808.
259. Rickert RC, Rajewsky K, Roes J. Impairment of T-cell-dependent B-cell responses and B-1 cell development in CD19-deficient mice. *Nature.* 1995;376:352-355.
260. Tarakhovsky A, Turner M, Schaal S, et al. Defective antigen receptor-mediated proliferation of B and T cells in the absence of Vav. *Nature.* 1995;374:467-470.
261. Whyte WA, Orlando DA, Hnisz D, et al. Master transcription factors and mediator establish super-enhancers at key cell identity genes. *Cell.* 2013;153:307-319.
262. Hnisz D, Abraham BJ, Lee TI, et al. Super-enhancers in the control of cell identity and disease. *Cell.* 2013;155:934-947.
263. Pongubala JMR, Murre C. Spatial organization of chromatin: transcriptional control of adaptive immune cell development. *Front Immunol.* 2021;12:633825.
264. Kim HG, de Guzman CG, Swindle CS, et al. The ETS family transcription factor PU.1 is necessary for the maintenance of fetal liver hematopoietic stem cells. *Blood.* 2004;104:3894-3900.
265. Iwasaki H, Somoza C, Shigematsu H, et al. Distinctive and indispensable roles of PU.1 in maintenance of hematopoietic stem cells and their differentiation. *Blood.* 2005;106:1590-1600.
266. DeKoter RP, Singh H. Regulation of B lymphocyte and macrophage development by graded expression of PU.1. *Science.* 2000;288:1439-1441.
267. Spooner CJ, Cheng JX, Pujadas E, Laslo P, Singh H. A recurrent network involving the transcription factors PU.1 and Gfi1 orchestrates innate and adaptive immune cell fates. *Immunity.* 2009;31:576-586.
268. DeKoter RP, Lee HJ, Singh H. PU.1 regulates expression of the interleukin-7 receptor in lymphoid progenitors. *Immunity.* 2002;16:297-309.
269. Ye M, Ermakova O, Graf T. PU.1 is not strictly required for B cell development and its absence induces a B-2 to B-1 cell switch. *J Exp Med.* 2005;202:1411-1422.
270. Polli M, Dakic A, Light A, Wu L, Tarlinton DM, Nutt SL. The development of functional B lymphocytes in conditional PU.1 knock-out mice. *Blood.* 2005;106:2083-2090.
271. Matheson LS, Bolland DJ, Chovanec P, et al. Local chromatin features including PU.1 and IKAROS binding and H3K4 methylation shape the repertoire of immunoglobulin kappa genes chosen for V(D)J recombination. *Front Immunol.* 2017;8:1550.
272. Kleiman E, Loguercio S, Feeney AJ. Epigenetic enhancer marks and transcription factor binding influence Vκ gene rearrangement in pre-B cells and pro-B cells. *Front Immunol.* 2018;9:2074.
273. Murre C. Helix-loop-helix proteins and lymphocyte development. *Nat Immunol.* 2005;6:1079-1086.
274. Bain G, Maandag EC, Izon DJ, et al. E2A proteins are required for proper B cell development and initiation of immunoglobulin gene rearrangements. *Cell.* 1994;79:885-892.
275. Zhuang Y, Soriano P, Weintraub H. The helix-loop-helix gene E2A is required for B cell formation. *Cell.* 1994;79:875-884.
276. Miyazaki K, Watanabe H, Yoshikawa G, et al. The transcription factor E2A activates multiple enhancers that drive Rag expression in developing T and B cells. *Sci Immunol.* 2020;5:eabb1455.
277. Rogers JH, Owens KS, Kurkewich J, et al. E2A antagonizes PU.1 activity through inhibition of DNA binding. *BioMed Res Int.* 2016;2016:3983686.
278. Inlay MA, Tian H, Lin T, Xu Y. Important roles for E protein binding sites within the immunoglobulin kappa chain intronic enhancer in activating Vkappa Jkappa rearrangement. *J Exp Med.* 2004;200:1205-1211.
279. Mandal M, Powers SE, Ochiai K, et al. Ras orchestrates exit from the cell cycle and light-chain recombination during early B cell development. *Nat Immunol.* 2009;10:1110-1117.
280. Teachenor R, Beck K, Wright LY, Shen Z, Briggs SP, Murre C. Biochemical and phosphoproteomic analysis of the helix-loop-helix protein E47. *Mol Cell Biol.* 2012;32:1671-1682.
281. Kee BL, Murre C. Induction of early B cell factor (EBF) and multiple B lineage genes by the basic helix-loop-helix transcription factor E12. *J Exp Med.* 1998;188:699-713.
282. Lin YC, Jhunjhunwala S, Benner C, et al. A global network of transcription factors, involving E2A, EBF1 and Foxo1, that orchestrates B cell fate. *Nat Immunol.* 2010;11:635-643.
283. Lin H, Grosschedl R. Failure of B-cell differentiation in mice lacking the transcription factor EBF. *Nature.* 1995;376:263-267.
284. Jiang XX, Nguyen Q, Chou Y, et al. Control of B cell development by the histone H2A deubiquitinase MYSM1. *Immunity.* 2011;35:883-896.
285. Medina KL, Pongubala JM, Reddy KL, et al. Assembling a gene regulatory network for specification of the B cell fate. *Dev Cell.* 2004;7:607-617.
286. Li R, Cauchy P, Ramamoorthy S, Boller S, Chavez L, Grosschedl R. Dynamic EBF1 occupancy directs sequential epigenetic and transcriptional events in B-cell programming. *Genes Dev.* 2018;32:96-111.
287. Miyai T, Takano J, Endo TA, et al. Three-step transcriptional priming that drives the commitment of multipotent progenitors toward B cells. *Genes Dev.* 2018;32:112-126.
288. Nechanitzky R, Akbas D, Scherer S, et al. Transcription factor EBF1 is essential for the maintenance of B cell identity and prevention of alternative fates in committed cells. *Nat Immunol.* 2013;14:867-875.

289. Ungerbäck J, Åhsberg J, Strid T, Somasundaram R, Sigvardsson M. Combined heterozygous loss of Ebf1 and Pax5 allows for T-lineage conversion of B cell progenitors. *J Exp Med.* 2015;212:1109-1123.
290. Urbánek P, Wang ZQ, Fetka I, Wagner EF, Busslinger M. Complete block of early B cell differentiation and altered patterning of the posterior midbrain in mice lacking Pax5/BSAP. *Cell.* 1994;79:901-912.
291. Ushmorov A, Wirth T. FOXO in B-cell lymphopoiesis and B cell neoplasia. *Semin Cancer Biol.* 2018;50:132-141.
292. Omori SA, Cato MH, Anzelon-Mills A, et al. Regulation of class-switch recombination and plasma cell differentiation by phosphatidylinositol 3-kinase signaling. *Immunity.* 2006;25:545-557.
293. Welinder E, Mansson R, Mercer EM, Bryder D, Sigvardsson M, Murre C. The transcription factors E2A and HEB act in concert to induce the expression of FOXO1 in the common lymphoid progenitor. *Proc Natl Acad Sci U S A.* 2011;108:17402-17407.
294. Mansson R, Welinder E, Åhsberg J, et al. Positive intergenic feedback circuitry, involving EBF1 and FOXO1, orchestrates B-cell fate. *Proc Natl Acad Sci U S A.* 2012;109:21028-21033.
295. Dengler HS, Baracho GV, Omori SA, et al. Distinct functions for the transcription factor Foxo1 at various stages of B cell differentiation. *Nat Immunol.* 2008;9:1388-1398.
296. Georgopoulos K, Bigby M, Wang JH, et al. The Ikaros gene is required for the development of all lymphoid lineages. *Cell.* 1994;79:143-156.
297. Heizmann B, Kastner P, Chan S. The Ikaros family in lymphocyte development. *Curr Opin Immunol.* 2018;51:14-23.
298. Yoshida T, Ng SY, Zuniga-Pflucker JC, Georgopoulos K. Early hematopoietic lineage restrictions directed by Ikaros. *Nat Immunol.* 2006;7:382-391.
299. Schwickert TA, Tagoh H, Gültekin S, et al. Stage-specific control of early B cell development by the transcription factor Ikaros. *Nat Immunol.* 2014;15:283-293.
300. Wang JH, Nichogiannopoulou A, Wu L, et al. Selective defects in the development of the fetal and adult lymphoid system in mice with an Ikaros null mutation. *Immunity.* 1996;5:537-549.
301. Wang JH, Avitahl N, Cariappa A, et al. Aiolos regulates B cell activation and maturation to effector state. *Immunity.* 1998;9:543-553.
302. Yamashita M, Morio T. Inborn errors of IKAROS and AIOLOS. *Curr Opin Immunol.* 2021;72:239-248.
303. Yamashita M, Kuehn HS, Okuyama K, et al. A variant in human AIOLOS impairs adaptive immunity by interfering with IKAROS. *Nat Immunol.* 2021;22:893-903.
304. Reynaud D, Demarco IA, Reddy KL, et al. Regulation of B cell fate commitment and immunoglobulin heavy-chain gene rearrangements by Ikaros. *Nat Immunol.* 2008;9:927-936.
305. Papathanasiou P, Attema JL, Karsunky H, et al. Self-renewal of the long-term reconstituting subset of hematopoietic stem cells is regulated by Ikaros. *Stem Cell.* 2009;27:3082-3092.
306. Rogers CH, Mielczarek O, Corcoran AE. Dynamic 3D locus organization and its drivers underpin immunoglobulin recombination. *Front Immunol.* 2020;11:633705.
307. Heizmann B, Kastner P, Chan S. Ikaros is absolutely required for pre-B cell differentiation by attenuating IL-7 signals. *J Exp Med.* 2013;210:2823-2832.
308. Macias-Garcia A, Heizmann B, Sellars M, et al. Ikaros is a negative regulator of B1 cell development and function. *J Biol Chem.* 2016;291:9073-9086.
309. Thompson EC, Cobb BS, Sabbattini P, et al. Ikaros DNA-binding proteins as integral components of B cell developmental-stage-specific regulatory circuits. *Immunity.* 2007;26:335-344.
310. Ma S, Pathak S, Trinh L, Lu R. Interferon regulatory factors 4 and 8 induce the expression of Ikaros and Aiolos to down-regulate pre-B-cell receptor and promote cell-cycle withdrawal in pre-B-cell development. *Blood.* 2008;111:1396-1403.
311. Cobaleda C, Schebesta A, Delogu A, Busslinger M. Pax5: the guardian of B cell identity and function. *Nat Immunol.* 2007;8:463-470.
312. Liu GJ, Jaritz M, Wöhner M, et al. Repression of the B cell identity factor Pax5 is not required for plasma cell development. *J Exp Med.* 2020;217.
313. Fuxa M, Skok J, Souabni A, Salvagiotto G, Roldan E, Busslinger M. Pax5 induces V-to-DJ rearrangements and locus contraction of the immunoglobulin heavy-chain gene. *Genes Dev.* 2004;18:411-422.
314. Nutt SL, Urbánek P, Rolink A, Busslinger M. Essential functions of Pax5 (BSAP) in pro-B cell development: difference between fetal and adult B lymphopoiesis and reduced V-to-DJ recombination at the IgH locus. *Genes Dev.* 1997;11:476-491.
315. Tagoh H, Ingram R, Wilson N, et al. The mechanism of repression of the myeloid-specific c-fms gene by Pax5 during B lineage restriction. *EMBO J.* 2006;25:1070-1080.
316. Souabni A, Cobaleda C, Schebesta M, Busslinger M. Pax5 promotes B lymphopoiesis and blocks T cell development by repressing Notch1. *Immunity.* 2002;17:781-793.
317. Haas KM, Poe JC, Steeber DA, Tedder TF. B-1a and B-1b cells exhibit distinct developmental requirements and have unique functional roles in innate and adaptive immunity to S. pneumoniae. *Immunity.* 2005;23:7-18.
318. Calderón L, Schindler K, Malin SG, et al. Pax5 regulates B cell immunity by promoting PI3K signaling via PTEN down-regulation. *Sci Immunol.* 2021;6.
319. Corcoran LM, Karvelas M, Nossal GJ, Ye ZS, Jacks T, Baltimore D. Oct-2, although not required for early B-cell development, is critical for later B-cell maturation and for postnatal survival. *Genes Dev.* 1993;7:570-582.
320. Bordon A, Bosco N, Du Roure C, et al. Enforced expression of the transcriptional coactivator OBF1 impairs B cell differentiation at the earliest stage of development. *PLoS One.* 2008;3:e4007.
321. Humbert PO, Corcoran LM. oct-2 gene disruption eliminates the peritoneal B-1 lymphocyte lineage and attenuates B-2 cell maturation and function. *J Immunol.* 1997;159:5273-5284.
322. Hodson DJ, Shaffer AL, Xiao W, et al. Regulation of normal B-cell differentiation and malignant B-cell survival by OCT2. *Proc Natl Acad Sci U S A.* 2016;113:E2039-E2046.
323. Kim U, Qin XF, Gong S, et al. The B-cell-specific transcription coactivator OCA-B/OBF-1/Bob-1 is essential for normal production of immunoglobulin isotypes. *Nature.* 1996;383:542-547.
324. Nielsen PJ, Georgiev O, Lorenz B, Schaffner W. B lymphocytes are impaired in mice lacking the transcriptional co-activator Bob1/OCA-B/OBF1. *Eur J Immunol.* 1996;26:3214-3218.
325. Schubart DB, Rolink A, Kosco-Vilbois MH, Botteri F, Matthias P. B-cell-specific coactivator OBF-1/OCA-B/Bob1 required for immune response and germinal centre formation. *Nature.* 1996;383:538-542.
326. Schubart K, Massa S, Schubart D, Corcoran LM, Rolink AG, Matthias P. B cell development and immunoglobulin gene transcription in the absence of Oct-2 and OBF-1. *Nat Immunol.* 2001;2:69-74.
327. Swaminathan S, Huang C, Geng H, et al. BACH2 mediates negative selection and p53-dependent tumor suppression at the pre-B cell receptor checkpoint. *Nat Med.* 2013;19:1014-1022.
328. Nahar R, Ramezani-Rad P, Mossner M, et al. Pre-B cell receptor-mediated activation of BCL6 induces pre-B cell quiescence through transcriptional repression of MYC. *Blood.* 2011;118:4174-4178.
329. Shinnakasu R, Kurosaki T. Regulation of memory B and plasma cell differentiation. *Curr Opin Immunol.* 2017;45:126-131.
330. Igarashi K, Ochiai K, Itoh-Nakadai A, Muto A. Orchestration of plasma cell differentiation by Bach2 and its gene regulatory network. *Immunol Rev.* 2014;261:116-125.
331. Laramée AS, Raczkowski H, Shao P, et al. Opposing roles for the related ETS-family transcription factors Spi-B and Spi-C in regulating B cell differentiation and function. *Front Immunol.* 2020;11:841.
332. Lu R, Medina KL, Lancki DW, Singh H. IRF-4,8 orchestrate the pre-B-to-B transition in lymphocyte development. *Genes Dev.* 2003;17:1703-1708.
333. Budzyńska PM, Niemelä M, Sarapulov AV, et al. IRF4 deficiency leads to altered BCR signalling revealed by enhanced PI3K pathway, decreased SHIP expression and defected cytoskeletal responses. *Scand J Immunol.* 2015;82:418-428.
334. Cook SL, Franke MC, Sievert EP, Sciammas R. A synchronous IRF4-dependent gene regulatory network in B and helper T cells orchestrating the antibody response. *Trends Immunol.* 2020;41:614-628.
335. Xu H, Chaudhri VK, Wu Z, et al. Regulation of bifurcating B cell trajectories by mutual antagonism between transcription factors IRF4 and IRF8. *Nat Immunol.* 2015;16:1274-1281.
336. Pasqualucci L, Klein U. Mouse models in the study of mature B-cell malignancies. *Cold Spring Harb Perspect Med.* 2021;11:a034827.
337. Katsaraki K, Karousi P, Artemaki PI, et al. MicroRNAs: Tiny regulators of gene expression with Pivotal roles in normal B-cell development and B-cell chronic lymphocytic leukemia. *Cancers.* 2021;13:593.
338. Lin YH, Liang Y, Wang H, et al. Regulation of B Lymphocyte development by histone H2A deubiquitinase BAP1. *Front Immunol.* 2021;12:626418.
339. Bao Y, Cao X. Epigenetic control of B cell development and B-cell-related immune disorders. *Clin Rev Allergy Immunol.* 2016;50:301-311.
340. Ikawa T. Genetic and epigenetic control of early lymphocyte development. *Curr Top Microbiol Immunol.* 2014;381:1-20.
341. Andrews JM, Payton JE. Epigenetic dynamics in normal and malignant B cells: die a hero or live to become a villain. *Curr Opin Immunol.* 2019;57:15-22.
342. Lee ST, Xiao Y, Muench MO, et al. A global DNA methylation and gene expression analysis of early human B-cell development reveals a demethylation signature and transcription factor network. *Nucleic Acids Res.* 2012;40:11339-11351.
343. Kulis M, Merkel A, Heath S, et al. Whole-genome fingerprint of the DNA methylome during human B cell differentiation. *Nat Genet.* 2015;47:746-756.
344. Almamun M, Levinson BT, Gater ST, et al. Genome-wide DNA methylation analysis in precursor B-cells. *Epigenetics.* 2014;9:1588-1595.
345. Zan H, Casali P. Epigenetics of peripheral B-cell differentiation and the antibody response. *Front Immunol.* 2015;6:631.
346. Wigton EJ, Ansel KM. Noncoding RNAs in B cell responses. *RNA Biol.* 2021;18:633-639.
347. Guttman M, Amit I, Garber M, et al. Chromatin signature reveals over a thousand highly conserved large non-coding RNAs in mammals. *Nature.* 2009;458:223-227.
348. Brazão TF, Johnson JS, Müller J, Heger A, Ponting CP, Tybulewicz VL. Long noncoding RNAs in B-cell development and activation. *Blood.* 2016;128:e10-19.
349. Kuchen S, Resch W, Yamane A, et al. Regulation of microRNA expression and abundance during lymphopoiesis. *Immunity.* 2010;32:828-839.
350. Koralov SB, Muljo SA, Galler GR, et al. Dicer ablation affects antibody diversity and cell survival in the B lymphocyte lineage. *Cell.* 2008;132:860-874.
351. Brandl A, Daum P, Brenner S, et al. The microprocessor component, DGCR8, is essential for early B-cell development in mice. *Eur J Immunol.* 2016;46:2710-2718.
352. Xiao C, Srinivasan L, Calado DP, et al. Lymphoproliferative disease and autoimmunity in mice with increased miR-17-92 expression in lymphocytes. *Nat Immunol.* 2008;9:405-414.
353. Xu S, Ou X, Huo J, et al. Mir-17-92 regulates bone marrow homing of plasma cells and production of immunoglobulin G2c. *Nat Commun.* 2015;6:6764.
354. Ding L, Zhang Y, Han L, et al. Activating and sustaining c-Myc by depletion of miR-144/451 gene locus contributes to B-lymphomagenesis. *Oncogene.* 2018;37:1293-1307.
355. Arman K, Möröy T. Crosstalk between MYC and lncRNAs in hematological malignancies. *Front Oncol.* 2020;10:579940.
356. Rao DS, O'Connell RM, Chaudhuri AA, Garcia-Flores Y, Geiger TL, Baltimore D. MicroRNA-34a perturbs B lymphocyte development by repressing the forkhead box transcription factor Foxp1. *Immunity.* 2010;33:48-59.
357. Kikuchi H, Nakayama M, Kuribayashi F, et al. GCN5 is essential for IRF-4 gene expression followed by transcriptional activation of Blimp-1 in immature B cells. *J Leukoc Biol.* 2014;95:399-404.
358. Kikuchi K, Kondo M. Developmental switch of mouse hematopoietic stem cells from fetal to adult type occurs in bone marrow after birth. *Proc Natl Acad Sci U S A.* 2006;103:17852-17857.
359. Xu X, Deobagkar-Lele M, Bull KR, et al. An ontogenetic switch drives the positive and negative selection of B cells. *Proc Natl Acad Sci U S A.* 2020;117:3718-3727.

CHAPTER 26

B Cell Responses

Claude-Agnès Reynaud, PhD • Jean-Claude Weill, PhD

INTRODUCTION

B cells are key actors of adaptive immune responses, and the antibodies they secrete upon activation are the hallmark of successful vaccine procedures.

Multiple signals activate B cell responses, and the diversity of effector and memory B cell profiles they induce has only recently begun to emerge. The notion of subsets generated by distinct activation conditions somehow differs in B cells compared to T cells as it does not systematically rely on the expression of a master transcriptional factor, with the exception of antibody-secreting cells (ASCs) that represent a specific entity and a terminally differentiated product of B cell activation.

Describing how the characteristics of activated/memory B cell subpopulations translate into specific effector functions is the focus of most recent studies, with the aim of understanding how the specific handling of different types of infections, notably through the immunoglobulin isotype these subsets express, may contribute to reduce pathogen escape and resistance.

We will describe in this chapter the different layers of the B cell response, which ranges from a rapid, immediate protection through an extrafollicular reaction leading to the secretion of low-affinity, broadly reactive antibodies, to the formation of an enduring memory, encoded within different types of long-lasting subsets that emerge following affinity maturation of their B cell receptor (BCR) against the eliciting antigen.

GERMINAL CENTERS AND T-DEPENDENT RESPONSES

The first surprise for immunologists was the discovery, around 60 years ago, that the B cell immune response against an antigen was improving within a period of weeks. Thereafter, it was shown that this phenomenon was due to a mutation process targeting the V genes involved in the response and finally, several decades later, that a cytidine deaminase, together with translesion DNA polymerases which are error prone when copying normal DNA, was coopted by this process. This mechanism of hypermutation, which operates during several weeks on an antibody gene target at a mutation rate a million times higher than the normal mutation rate of the genome, is unique in the living world. It takes place in specialized structures within lymphoid organs of warm-blooded vertebrates called germinal centers (GCs), allowing for the selection of cells harboring high-affinity antibody variants, which will be described in this section.

Initiation of the Immune Response

The immune system is composed of primary and secondary lymphoid organs (SLO). Primary lymphoid organs are the bone marrow and the thymus, which, respectively, maintain the production of newly formed B and T cells. SLO are the spleen, lymph nodes, and lymphoid tissues associated with mucosal surfaces, notably tonsils, Peyer's patches, and gut-associated and bronchial-associated lymphoid tissues. SLO filter foreign antigens as they are entering the body from the blood, the skin, or mucosal surfaces. The anatomical structures of these different SLO are unique, but they share common B and T cell areas. B cells are present in B cell follicles close to T cell zones, thus allowing the encounter of T and B cells activated by the same antigen. How a foreign antigen precisely enters SLO and triggers cognate B cells has been uncovered recently and will be detailed for lymph nodes and spleen.

Initiation of the Immune Response in Lymph Nodes

Lymph nodes are fed by two vascular systems: the lymph, which delivers antigen and antigen carriers from peripheral tissues, and the blood, which transfers lymphocytes to the node. The afferent lymph flows into the marginal sinus below the node capsule. Under the marginal sinus, the cortex is constituted by an outer region composed of B cell follicles and an inner region, the paracortex, composed of T lymphocytes. B and T cells enter the node from the blood by crossing the specialized high endothelial venules located in the paracortex. If they are not activated by an antigen, they return to the circulation via the efferent lymphatics and the thoracic duct.[1]

When lymph flows from peripheral tissues to the draining lymph nodes, it goes through sinuses where antigen-presenting cells (APCs) can get access to particulate antigens or pathogens. Lymph is first in contact with the subcapsular sinus lined with a specific subset of CD169$^+$ macrophages that can relay captured antigen to B cell follicles.[2-5] Small soluble antigens in the lymph go through the subcapsular sinus and enter small channels called conduits and reach directly the B cell follicles where they can activate B cells.[6-8] In the form of immune complexes, these antigens can also be retained by follicular dendritic cells (FDCs) through their complement receptor CR1/2 (CD35) or their Fc receptor, FcγRIIB (CD32). FDCs recycle unprocessed antigens through a nondegradative endocytic pathway that allows them to retain and reexpress them over a long period of time.[9] Subcapsular sinus macrophages can also transfer immune complexes to noncognate B cells, which then transfer them to FDCs.[4] SIGN-R1$^+$ dendritic cells (DCs) can also carry antigens to B cell follicles.[10]

Initiation of the Immune Response in Spleen

The spleen is separated into two regions, the white pulp and the red pulp. The red pulp consists of a reticular network containing stromal cells, macrophages, and plasma cells (PCs). The white pulp contains the lymphoid compartment. Antigens and lymphocytes enter the spleen from blood via the splenic artery that branches into central arterioles. Around the central arterioles is a T cell–rich zone designated the periarteriolar lymphoid sheath, surrounded by B cell follicles. The white pulp is separated from the red pulp by the marginal sinus that is embedded in a population of marginal zone (MZ) lymphocytes.[1]

Blood-borne antigens are captured by MZ B cells that express high amounts of the CR2 complement receptor (CD21) and therefore can bind circulating opsonized antigens. The location of B cells in the MZ depends on their expression of the sphingosine-1-phosphate receptor 1 (S1pr1) and on the integrins LFA-1 and VL4-4, binding to ICAM-1 and VCAM-1, respectively.[11] It has been reported that, instead of being a sessile population, MZ B cells constantly shuttle between the MZ and the B cell follicles, with this shuttling being driven by the expression of Cxcr5 by MZ B cells and the high abundance of Cxcl13 in the follicles.[12] After capture of the opsonized antigens by FDCs, MZ B cells slowly reexpress S1pr1 and migrate back to the MZ.

At the same time, the main role of MZ B cells is to respond rapidly to T-independent antigens. During an infection with encapsulated bacteria such as *Streptococcus pneumoniae*, neutrophils, monocytes, and DCs capture and transport the bacteria to the splenic MZ, where MZ B cells are activated and differentiate rapidly into ASCs. These cells then migrate to the red pulp where antibodies are secreted and pass in the general circulation.[13] Two types of macrophages present in the MZ, MARCO$^+$ MZ macrophages and metallophilic macrophages, are essentially playing a role in the elimination of bacteria from the circulation and not in antigen presentation.[14] Nevertheless, the microarchitecture of the MZ is perturbed in their absence and type II T-independent responses against polysaccharides are impaired.[15]

Interaction between B cells and pathogens in the spleen depends, therefore, on the type of the incoming antigen and induces either a rapid MZ B cell response or delivery of the antigen to FDC within B follicles. It is not clear at this stage which type of signals triggers either of these outcomes.

Triggering the Naïve B Cell Response

Using human immunodeficiency virus (HIV)–neutralizing antibodies as a model, S. Crotty and his colleagues analyzed the role of the frequency and affinity of antigen-specific naïve B cell in triggering a B cell immune response. They used transgenic mice expressing the unmutated form of the VRC01 broadly neutralizing antibody observed in some HIV-infected patients and, as immunogen, the cognate HIV envelop construct. With precursor frequencies as high as 1 in 10^3 B cells, an immunogen with any affinity could induce naïve cognate B cells to proliferate and enter GCs. When the precursor frequency was reduced to physiological values (1 in 10^5 or 1 in 10^6), antigen affinity became the limiting factor.[16] The same observation was made with the valence of antigen, a high antigen avidity being able to compensate for low precursor frequency. These observations on the role of precursor frequency in B cell responses were observed in other immunization settings.[17] Recall responses may show the same bias, since it was reported that reseeding of GCs by rare GC-derived memory B cells favored high-affinity clones.[18]

Entering a Germinal Center

W. Fleming in 1885 observed mitotic figures in lymphoid tissues that he called GCs, hypothesizing that they were the source of lymphocytes.[19] GCs offer a unique strategy for B cells engaged in a T-dependent immune response to improve their BCR. The B cell primary repertoire is unique by its diversity and may contain in human over 10^{15} variants of the BCRs.[20] Each BCR produced in the preimmune repertoire is of rather low affinity for foreign antigens, thus avoiding that too many holes are produced in the repertoire due to negative selection induced by self-antigens. When a B cell is engaged in a GC reaction, the affinity for the antigen can be rapidly improved by several orders of magnitude through the process of hypermutation. Affinity maturation was first demonstrated by H. Eisen and G. Siskind,[21] while thereafter M. Weigert and colleagues showed that this improvement in the binding capacity of the BCR was caused by mutations taking place in the variable region of the immunoglobulin gene.[22] Ig gene somatic hypermutation (SHM) is a Lamarckian/Darwinian process unique in the living world, Lamarckian because, upon signaling of a cell surface receptor (the BCR), the gene coding for this receptor will hypermutate without affecting, at least not at the same level, the rest of the genome, and Darwinian because, among B cells expressing variant BCRs generated by SHM, the one displaying the most fit mutations will be selected for proliferation by the immunizing antigen. The GC response is therefore responsible for the installation of an affinity-matured memory through the generation of long-lived B cells and PCs.[23,24]

A breakthrough in the field was the discovery of the activation-induced cytidine deaminase (AID) enzyme as a key factor for the molecular transactions occurring in mature B cells[25,26] (see Chapter 22). AID is triggered upon B cell activation, during the extrafollicular response and the GC reaction, and initiates class switch recombination (CSR) and SHM processes. At the molecular level, AID converts cytosines into uracils in the DNA, which are then processed in different manners. For SHM, processing mobilizes uracil glycosylase, mismatch repair, and translesional DNA polymerases to perform error-prone repair and to fix mutations at the initial deamination site and its vicinity.[23,27] For CSR, the high density of AID hotspots in switch repeats allows for deamination on both DNA strands at nearby sites and the formation of double-strand breaks in two different switch regions. Their joining leads to the expression of the same VDJ gene with a different constant region.[28]

Initiation of the GC response occurs in lymphoid tissues whose function is to filter blood, lymph, or mucosal fluids for the presence of foreign antigens. All these organs have B cell follicles surrounded by T cells. After BCR triggering, B cells upregulate Ccr7 and Ebi2 (encoded by the *Gpr183* gene), which guide them at the T-B interface of the follicle through a gradient of Ccl21 and oxysterol (7alpha25-HC) produced by stromal cells. After activation by DC presenting the same antigen in the T cell zone, follicular helper CD4 T cells also upregulate Ebi2 and move to the B cell follicle–T cell zone interface where rare antigen-specific T and B cells can meet (Fig. 26.1). If this T-B encounter does not occur within 12 to 14 hours, B cells enter into a Bim-mediated apoptotic program. This also occurs if the B cell is triggered by a self-antigen in the absence of T cell help. T-B interaction induces upregulation of CD86 in B cells, a ligand of the CD28 T cell coreceptor, leading to the secretion of IL-4 and IL-12 cytokines. These pre-GC B cells also start to express Bcl6, the master regulator of GC B cell differentiation. At this stage, B cells move to the center of the follicle within the FDC network. Ebi2 and Ccr7 are downregulated while the expression of Cxcr5 is maintained, thus keeping these B cells at the center of the follicle. Preexisting reticular cells expressing Cxcl12 help to organize the GC, which becomes divided into a dark zone near the T cell zone and a light zone facing the outer follicle.[30,31] Integrins reinforce the T-B collaboration, with LFA1 on helper T cells engaging ICAM1 and ICAM2 on B cells.[32,33] When GCs are formed at the center of the follicle, recirculating naïve B cells are positioned at the outer edge of the FDC network and form the follicular mantle zone.

Class Switch Recombination

CSR is the AID-mediated intrachromosomal DNA recombination process that allows replacement of the expressed heavy chain isotype.[25,26] Since AID is required for the initiation of CSR, it has been assumed that it takes place along with SHM during the GC reaction, most authors describing this process as being restricted to the light zone selection step.[34] By using transgenic as well as normal mice, C. Vinuesa and colleagues elegantly showed that heavy chain germline transcripts, which mark the access to the constant region locus targeted for CSR, are essentially expressed in B cells just after their interaction with cognate T cells, prior to their entry into GC. Accordingly, pre-GC B cells express the APE1 endonuclease required for CSR, while GC B cells downregulate APE1 and express APE2 involved in SHM.[35] Moreover, in silico modeling suggested that, if ongoing CSR were taking place in GCs, the probability of finding IgM-dominated GCs and consequently antigen-experienced IgM⁺ memory B cells, which are observed in most immunization models, would be negligible.[36]

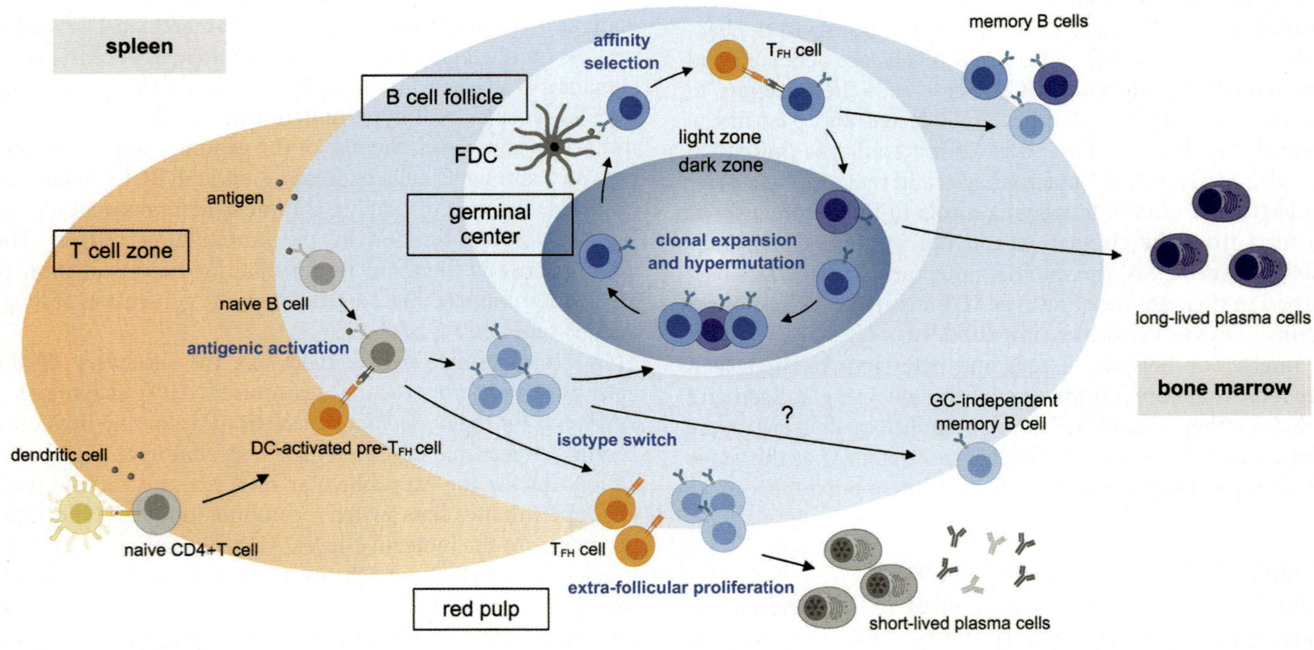

FIG. 26.1. T-dependent germinal center B cell responses.[29] Antigen-activated B cells and CD4⁺ T helper cells previously activated by the same antigen presented by dendritic cells interact at the T-B border of splenic B cell follicles. This interaction triggers either extrafollicular B cell proliferation and differentiation into short-lived plasma cells or migration to the center of the follicle and formation of a germinal center structure. Clonal amplification of responding B cells leading to the formation of a germinal center–independent memory pool may arise from this pre–GC-proliferating population. Germinal center B cells undergo high proliferation and mutation of their Ig V genes in the dark zone and migrate to the light zone where selection for high-affinity variants of the B cell receptor takes place, a selection mediated by competition for capture of antigens deposited on follicular dendritic cells (FDC) (in the form of antigen-antibody complexes) and presentation to germinal center T_{FH} cells. Selected B cells can either proceed to further rounds of hypermutation in the dark zone or exit from the germinal center after differentiation into long-lived memory or plasma cells (detailed in Fig. 26.2). Darker shades of blue depict increasing B cell affinities. (Adapted by permission from Springer Nature: Kurosaki T, Kometani K, Ise W. Memory B cells. *Nat Rev Immunol.* 2015;15:149-159.)

Negative Selection

In many cases, autoimmune antibodies that emerge from GCs lose their autoimmune specificities when reverted to their unmutated germline sequence, as shown, for example, for antidesmoglein antibodies in pemphigus vulgaris.[37] This implies that autoimmunity arising through SHM in GC can escape negative selection. However, the precise events allowing this escape are still not perfectly understood. Effectively, B cells arriving in GCs have been cleared for self-recognition by several processes: receptor editing occurring in bone marrow[38-40] (see Chapter 25), clonal deletion occurring in bone marrow and in peripheral lymphoid organs,[41-43] clonal ignorance often observed for B cells recognizing intracellular antigens,[44] and finally clonal anergy allowing B cells to survive with a desensitized BCR.[45] Anergic B cells lose their BCR signaling properties because they are chronically exposed to self-antigens, but if they receive an appropriate T cell help, they can differentiate in autoreactive ASCs.[46] However, as a safeguard, they express the Fas marker upon CD40-CD40L interactions. Without a strong signal through the BCR, they will not upregulate the activation marker CD86 that, through its interaction with CD28 on T cells, prevents their death by apoptosis.[47,48] Fas mutations resulting in autoimmune lymphoproliferative syndrome and autoantibody production in humans argue in favor of this scenario.[49-53] However, if these anergic B cells encounter a highly cross-linking antigen, they can enter into a GC reaction and mutate away from self and toward foreign recognition.[54,55] The observation by the group of C. Goodnow that anergic cells are still able under certain conditions to participate in an immune response, a phenomenon that they call "clonal redemption of autoantibodies," provided for the first time a functional role for the maintenance of anergic B cells in the periphery. However, the purging system cannot be perfect and if an ignorant B cell recognizes an intracellular self-antigen that cross-reacts with a foreign antigen, it will enter a GC reaction and be able to differentiate into high-affinity anti-self B cells, as occurs in Sjögren syndrome and systemic lupus erythematosus autoimmune conditions.[43,56]

Shuttling Between Germinal Center Dark Zone and Light Zone

The basis of the GC reaction is the generation of B cells with an improved affinity against the immunizing antigen. GCs are organized in light zone and dark zone regions (Fig. 26.2). The light zone contains B cells, FDCs, and T follicular helper (T_{FH}) cells, while the dark zone contains B cells and Cxcl12-expressing reticular cells.[24,57] Light zone B cells express more CD83 and CD86 than dark zone B cells.[58] The latter show stronger expression of AID, which triggers hypermutation, and of the chemokine receptor Cxcr4. Circulation between the two zones is controlled by the downregulation of Cxcr4, which takes place when dark zone B cells move back to the light zone, allowing Cxcr5-expressing GC B cells to migrate

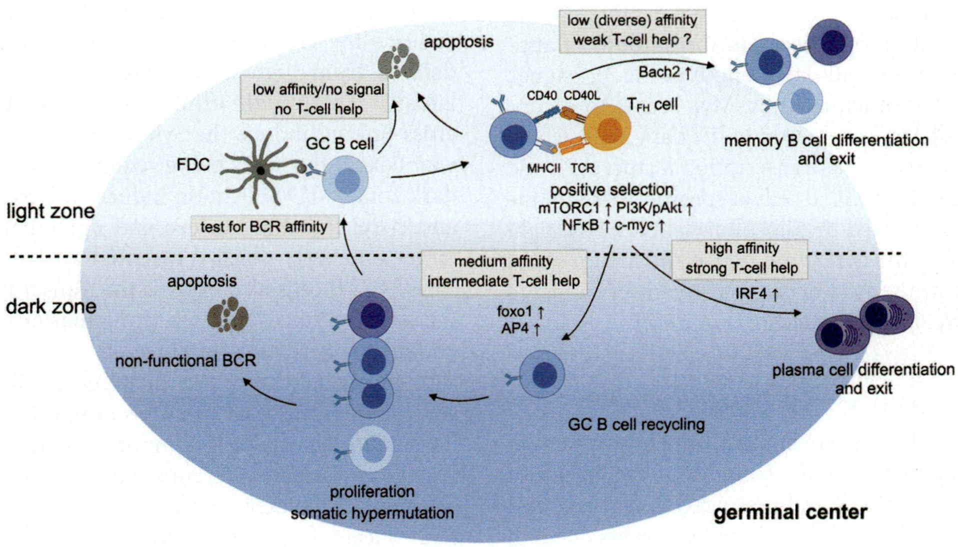

FIG. 26.2. B cell selection processes in the light zone and dark zone of germinal centers. Light zone B cells can undergo different fates following antigen capture deposited on follicular dendritic cell (FDC) and antigen presentation to T_{FH} cells, according to their affinity for the antigen and the amount of T cell help received, through both TCR/MHCII-peptide and CD40-CD40L interactions (other interactions, involving ICOS-ICOSL, LFA1-ICAM1, or SLAM receptors are not depicted). High-affinity/strong T_{FH} signals result in positive selection, activation of PI3K and mTORC1, and upregulation of NFκB and its c-Myc target; strong signaling through CD40 also upregulates IRF4, promoting plasma cell differentiation and exit. Lower CD40 signals allow the maintenance of Bcl6 expression in T_{FH}-activated B cells and their recycling to the dark zone, with Foxo1 and AP4 expression promoting intense cell proliferation and hypermutation. B cells with damaging mutations in their B cell receptor (BCR) die by apoptosis, while those having modified their antigenic specificity migrate back to the light zone for a new round of selection. Entry in the memory pool is generally described as a selection process favoring low affinity GC B cells in the light zone, and involving upregulation of Bach2, but, along this process, high-affinity B cells are equally recruited in the memory pool. The type of selection operating at this stage is therefore indicated with a question mark. Light zone B cells with very low-affinity BCRs or that failed to receive survival signals die by apoptosis. Darker shades of blue depict increasing B cell affinities. (Created with BioRender.com)

toward Cxcl13 secreted by FDC.[30,59] Fifty to 200 B cell clones can colonize a GC.[60] While affinity maturation enriches for certain clones, this does not obligatorily narrow clonal diversity, which can be maintained by low and moderate affinity B cells (see below). Proliferation and SHM take place in the dark zone, while selection occurs in the light zone. It is estimated that the cell cycle of dark zone centroblasts is of 6 to 7 hours.[61] At the same time, there is a high death rate in GCs, with the apoptotic cells being constantly scavenged by tingible body macrophages present in the dark zone.[62,63] One unresolved issue is to more precisely understand what are the signals controlling the behavior of B cells in the light zone, which, depending on their capacity to receive signals from FDCs and T_{FH} cells, can either return to the dark zone for a new round of mutations, leave the GC and give rise to either PCs or memory B cells, or otherwise die in situ.

Cell surface receptors and transcription factors coordinate the GC choreography. Engagement of ICOS on T_{FH} cell by B cell ICOSL leads to increased CD40L expression and thus increased T cell help. T cell help activates the mechanistic target of rapamycin complex 1 (mTORC1), which triggers the anabolic program that sustains proliferation in the dark zone.[64] Transcription factors are also involved, such as c-Myc and Foxo1, which are important for positive selection.[65] c-Myc is expressed shortly after antigen activation and establishes a cyclin D2-dependent proliferation program in GC precursor B cells. After a few days, c-Myc is repressed in proliferating B cell blasts and is only reexpressed in a subpopulation of light zone B cells that are dividing and selected for migration back to the dark zone.[66-68] c-Myc expression and cell growth are directly proportional to the amount of antigen captured by the B cell and the strength of the T_{FH} signal. Moreover, the amount of expressed c-Myc seems to dictate the number of cell divisions achieved in the dark zone, which take place every 6 hours.[68] The transcription factors AP4 and Foxo1 are also expressed in GC B cells at specific phases of the process. Foxo1 is inactivated by phosphorylation in the light zone but is required at the dark zone step, downstream of the PI3K/Akt pathway, while AP4 acts downstream of c-Myc to consolidate the dark zone program[69,70] (Fig. 26.2).

Positive Selection in the Light Zone

The light zone is the place where antigen, which is present in its native form as antigen-antibody complex at the surface of FDCs, is captured by the BCR and presented in its processed form to cognate T_{FH} cells. Which of these two events drives B cells to the next step of the selection process is still debated. G. Victora and colleagues could follow the fate of transgenic B cells harboring the same high-affinity anti-NP (nitrophenyl hapten) BCR after their photoactivation in GCs. By loading antigen in a BCR-independent mode to B cells engaged in a response to their cognate antigen in the light zone, they could observe their proliferation at the expense of cells not able to load the antigen. The loaded B cells at first proliferated in the light zone and, after 24 hours, appeared in the dark zone. Strikingly, if, in wild-type mice immunized with NP-OVA (NP coupled to ovalbumin), OVA was targeted to all GC B cells through the same BCR-independent mode irrespectively of their affinity, there was an increase in the titer of anti-NP antibodies but no maturation of affinity. The authors concluded that under conditions of equal T cell help, differential BCR cross-linking was not sufficient to mediate affinity-based selection and that, therefore, T cell help was the limiting step for cell expansion and further SHM in the dark zone.[58]

In another experimental approach, R. Brink and colleagues, by using transgenic B cells carrying high- or low-affinity BCR, showed that high-affinity B cells in the light zone were selectively engaged in PC differentiation, underlining the role of BCR affinity. Based on these and other observations, it was proposed that signals delivered during engagement of the BCR by intact antigen were necessary to initiate the process, while T_{FH} signals were required to direct migration of B cells to the dark zone and to complete their differentiation into PC.[65,71] Along the same line, it was shown that reducing by half MHC class II (MHCII) expression at the surface of GC B cells did not change the selection process within GCs, thus indicating that the amount of T cell help was not sufficient by itself to account for the positive selection of GC B cells.[72] In contrast, haploinsufficiency of MHCII played a role at an earlier step, the intensity of T cell help being critical for GC entry.[72,73]

Leaving the GC

One of the central questions concerning this topic is when do B cells leave the GC and what signals control their differentiation into either memory B cells or long-lived plasma cells (LLPCs). It has been shown that memory B cells start to leave the GC at early time points while PCs are mainly derived from high-affinity GC B cells, formed essentially later.[74,75] These high-affinity GC B cells under strong BCR cross-linking induce the Syk-PI3K-Akt pathway and inactivate Foxo1, the latter being required for the reentry into the dark zone.[69] CD40 ligation induces the translocation of RelA and c-Rel into the nucleus and upregulation of the NF-κB target gene *Irf4*, which is essential for PC differentiation.[76-78] These GC B cells also express the transcription factor *Prdm1* (encoding Blimp-1), which is the master gene for PC differentiation (see section on PCs).

Concerning the signals dictating the memory B cell fate, several scenarios have been proposed. H. Qi and colleagues have reported that GC memory precursors expressing the IL-9 receptor may differentiate into memory B cells upon the action of IL-9 secreted by T_{FH} cells.[79] Conversely, D. Kitamura and colleagues reported a normal primary response in IL-9R KO mice using NP-CGG (NP coupled to chicken gammaglobulin) as an antigen. Nevertheless, the antibody response in their mice after a boost was much diminished, with IL-$9R^{-/-}$ memory B cells going back preferentially to GC.[80] Their proposition was that IL-9 induced a downregulation of ICOSL on memory B cells, thus preventing their return to GC and favoring their proliferation and differentiation into PCs. R. Brink and colleagues identified B cell memory precursors in the light zone differently, based on their expression of Ccr6. Interestingly, these $Ccr6^+$ memory precursors within the light zone belonged to both the low and high-affinity pool.[81]

S1pr2 is expressed by light zone and dark zone GC B cells. Based on an S1pr2-Cre-ERT2 fate mapping mouse model, most $CD38^-GL7^+$ GC B cells were labeled after immunization with NP-CGG and ingestion of tamoxifen at day 9 or 18. Low-affinity B cells filled the memory compartment with the highest production at day 9, while B cells bearing a BCR with 10- to 100-fold higher affinity were found in GC. Gene profiling of low and high-affinity GC B cells showed that higher *Irf4* and *Prdm1* expression in the latter fostered PC differentiation, while *Bach2*, *Pax5*, *Bcl6*, and *Irf8* expression in the former blocked this process. The authors proposed that a weak help from T cells inducing a lower CD40 signature keeps high expression of Bach2 in low-affinity light zone B cells, leading to their entry into the memory compartment, while high-affinity light zone B cells expressing low amount of Bach2 are poised to enter the cell cycle or to undergo PC differentiation[82] (Fig. 26.2).

Germinal Center T_{FH} Cells

Interaction of naïve CD4 T cells with a professional APC gives rise to Th1, Th2, Th17, Th9, or T_{FH} cells[83,84] (see Chapter 28). T_{FH} cells are T helper cells whose property is to help B cells. Many factors influence CD4 decision, such as APC expression of costimulatory factors, the innate cytokine milieu, the amount of antigen presented but also the strength of T cell receptor (TCR) signaling that depends on TCR affinity, the amount of MHCII on APC, and the length of time of their interaction.[85]

CD4 T effector cells will express different master transcription factors depending on their polarization: Gata3 in Th2,[86] T-bet in Th1,[87] and RORγt in Th17.[88] T_{FH} cells express *Bcl6* that inhibits the Th1, Th2, and Th17 programs via direct repression of *Tbx21* (encoding T-bet),[89] *Gata3*, and *RorC* (encoding RORγt), respectively.[90-93] One should recall that *Bcl6* is also expressed after activation of most T cells by TCR ligation,[94-96] but that it is repressed by IL-2, which thus prevents T_{FH} differentiation.[97] T_{FH} will develop and produce the cytokines of their corresponding T effector subsets.[89,98-100] During a Th1 inflammatory response, a naïve CD4 T cell clone that receives a strong TCR stimulation in the presence of IFNγ and IL-12 will become a Th1 effector cell expressing high level of T-bet, Stat4, and Blimp-1.[101-104] They then migrate to various tissues, depending on their chemokines receptors, ready to secrete IFNγ in order to control an acute infection induced by intracellular pathogens. Pathogen clearance will also be mediated by various effector cells including NK cells and CD8 cytotoxic T cells.[105,106] In the same priming milieu, a CD4 T cell displaying a lower affinity for the antigen will express Cxcr5 and migrate to the T-B border as an IFNγ-secreting T_{FH} cell.[98] It can interact there with cognate B cells, thus reenforcing the T_{FH} program while inducing B cells to secrete IgG2a/2b (IgG1 in humans) in response to intracellular pathogens.

In a Th2 response, effector T cells home to tissues and express the eosinophil-recruiting chemokine IL-5, while $T_{FH}2$ cells thereafter secrete IL-4, 5, or 13 and induce B cells to secrete IgG1 and IgE for clearance of helminths (together with IgG4 in humans), with the combined action of epithelial cells, mast cells, eosinophils, and basophils.[99,107-109]

Priming by APC induces a pre-T_{FH} phenotype with the expression of Bcl6, as well as Cxcr5, PD-1, and ICOS, which constitute the classical T_{FH} markers.[110-113] In mice, IL-6 produced by DC induces the T_{FH} program and activates IL-21 production by T_{FH} cell, which signals in an autocrine manner to reenforce T_{FH} differentiation.[114-116] In humans, IL-6, IL-12, and TGF-β are involved in T_{FH} differentiation.[117,118] Full maturation of pre-T_{FH} cells depends on their interaction with cognate B cells, which takes place at the T-B border. Once activated, B cells proliferate and upregulate MHCII, CD80, and CD86, thus becoming more potent APC for T_{FH} cells. Such reciprocal signaling drives pre-T_{FH} into T_{FH} and their recruitment within the follicle. T_{FH} cells upregulate Cxcr5, and repress Psgl-1 and Ebi2, and can thus migrate toward the center of the follicle through a Cxcl13 gradient.[119,120] They are then engaged in a noncognate interaction with bystander B cells, which express ICOSL. These bystander B cells also express PD-L1, which prevents the recruitment of T_{FH} that express PD1 and are ICOS negative.[121]

T_{FH} deliver contact-dependent and as well soluble factor signals to cognate B cells such as CD40L, ICOS, IL-21, IL-4, and IFNγ. Synaptic delivery of CD40L occurs as antigen-presenting B cells form large areas of contact with GC T_{FH} cells.[122-124] Such interaction induces a calcium flux and an upregulation of CD40L expression in T_{FH} cells, with a concomitant upregulation of ICOSL in cognate B cells. This calcium flux also increases the expression of IL-21 and IL-4 by T_{FH}, which, respectively, maintains Bcl6 expression and Stat6 induction of the prosurvival factor Bcl-xL.[123,125,126] PD1 is highly expressed on GC T_{FH} as a negative regulator controlling the interaction with PD-L1 and PD-L2 on B cells.[127,128] IL-21 expressing T_{FH} regulate Bcl6 expression in GC B cells and therefore drive their proliferation and differentiation within GCs.[129,130] The dopamine neurotransmitter is expressed by human GC T_{FH} cells and enhances the expression of ICOSL by human GC B cells, but is surprisingly absent in mouse T_{FH} cells.[131]

Follicular Regulatory T Cells

Follicular regulatory T (T_{FR}) cells negatively control T_{FH} cells whose dysregulation can lead to autoimmune pathologies. T_{FR} cells originate from T_{reg} precursors and differentiate upon interaction with DCs and B cells. T_{FR} cells are $Cxcr5^+$ and $Foxp3^+$ CD4 T cells and coexpress Blimp-1 and Bcl-6, as well as ICOS and IL-10. They also express the coinhibitor receptor CTLA4, which seems to be important for their suppressive function.[132,133] Although T_{FR} cells have been reported almost a decade ago,[134-136] their precise mechanism of action to suppress B cell, dendritic cell, and T_{FH}-cell responses is still not completely understood.

GC B Cells

B cells that fail to upregulate Bcl6 do not repress Ebi2 and S1pr1 and cannot enter the follicle. Several transcription factors are mandatory for GC B cell differentiation, such as NF-κB at the initiation of the GC reaction,[137] Irf8 that contributes to Bcl6 expression,[137,138] E2a,[139] Ebf1,[140] and Bach2 whose deletion suppresses GC formation.[141] The antiapoptotic factor Mcl1 was also shown to be indispensable for the formation and maintenance of GC B cells.[142]

At the epigenetic level, Ezh2, which is the catalytic component of the Prc2 polycomb repressive complex responsible for the trimethylation of lysine 27 on histone H3 (H3K27me3 modification), is highly expressed in GC B cells.[143,144] Its ablation impairs the GC reaction and, consequently, affinity maturation and the formation of high-affinity antibodies. The role of Ezh2 consists in repressing genes involved in proliferation checkpoints, such as *Cdkn1a,* and genes promoting terminal differentiation, such as *Irf4* and *Prdm1*. Promoters of many genes involved in the exit and terminal differentiation of GC B cells display H3K4me3 activation marks together with the H3K27me3 repressive ones. Such bivalent chromatin marks allow the poising of GC B cell genes, leading to their repression or activation at different time points.[145] Surprisingly, GC B cells, which divide actively, display low glycolytic activity and generate their energy essentially by oxidation of fatty acids.[146]

Bcl6

Bcl6 is a transcription factor that belongs to the BTB-POZ (bric-a-brac, tramtrack, broad complex-poxvirus zinc finger) family of zinc finger–containing transcription factors. Acting as a transcriptional repressor, Bcl6 helps to coordinate enhancer and promoter pausing through the recruitment of corepressor proteins in GC B cells and T_{FH} cells. Its absence in either cell type abrogates GC formation.

Bcl6 in GC B Cells: A transcriptional Repressor

Bcl6 was first identified as a potential protooncogene from a chromosome translocation observed in diffuse large cell B cell lymphomas (DLBCLs).[147] Bcl6 represses DNA damage sensors such as *Tp53* (tumor protein p53) and *ATR* (ataxia telangiectasia and Rad3 related),[148,149] thus rendering GC B cells permissive to the DNA transactions taking place during SHM. It also represses the cell cycle inhibitor *Cdkn1a*,[150] as well as genes involved in the differentiation of GC B cells into PCs such as *Prdm1*.[151,152] It must therefore be switched off for this step to occur. Accordingly, a deregulation of Bcl6 in GC B cells is often observed in DLBCLs or follicular lymphomas (see Chapter 49).

At the molecular level, Bcl6 forms a ternary complex with BCOR (Bcl6 corepressor), SMRT (silencing mediator of retinoic and thyroid receptor), and NCOR (nuclear receptor corepressor), which repress target genes at their promoters.[153-155] SMRT/NCOR in complex with HDAC3 also induces H3K27 deacetylation, allowing for switching of enhancers between off and on states, and thus controlling gene expression alternate changes in the GC.[156,157] TBL1XR1 is a component of the SMRT/NCOR complex, which is often mutated in DLBCLs.[158,159] Such a mutation of the TBL1XR1 gene, implemented in transgenic mice, biases GC B cell fate toward memory and away from the PC lineage, causing a Bcl6 to Bach2 molecular switch. Interestingly, the memory B cells produced are mainly IgM$^+$ and CD80/PDL2 negative and tend to reenter GCs upon a boost.[160] As reported recently, the histone demethylase Lsd1 is directly recruited by Bcl6 to repress enhancers of genes involved in GC exit, terminal differentiation, and proliferation.[161]

Bcl6 in T_{FH} Cell: A Repressor of Repressors

As mentioned above, Bcl6 is also indispensable for the differentiation of T_{FH} cells.[91,92,152] Tcf1 and Lef-1 are involved in early induction of Bcl6,[162,163] which then mediates the induction of T_{FH} genes by repressing repressor genes. Bcl6 represses the expression of not only *Id2*[164] and *Prdm1*[152] but also *Runx2, Runx3, Gata-3, Klf2*, and *miR17-92*, which are repressors of PD-1, ICOS, CD200, IL-6Ra, IL-21, IL-4, and Cxcr5.[92,93,165] Bcl6 also represses genes that control T cell migration such as those coding for Ebi2, Ccr7, Ccr6, S1pr1, and Psgl-1.[93]

Germinal Center Responses to Complex Antigens

The B cell response after immunization with *Bacillus anthracis* protective antigen or recombinant influenza hemagglutinin (HA) was analyzed at the single cell level.[166] In the preimmune repertoire, B cells that bound these antigens were present at the frequency of 10^{-4} (0.02% of naïve B cells). After immunization, there was a 4000-fold expansion, reaching 30% to 50% of GC B cells. Affinity against the antigens increased up to 20-fold during 2 weeks of proliferation in GCs. However, 20% to 30% of GC B cells bound the eliciting antigen with much lower affinity. This surprising result suggested that selection forces operating in GCs allow for substantial differences in the affinities for antigen (10- to 100-fold differences) despite the open circulation of B cells, whether of high or low affinity, between GCs[167] and the presence of newly generated circulating antibodies driving affinity maturation within GCs.[168] Even more surprising was the presence of B cell clones with undetectable affinity for the native immunogen, the authors calling these clones "dark antigen"-responsive B cells. Their proposition was that such B cells may recognize a non-native form of the antigen exposing cryptic epitopes.[166,169] One should also mention that the SHM process generates nonfunctional BCR in mutating B cells, such cells which amount to 10% to 25% of the GC B cell population dying rapidly *in situ* in the dark zone.[63]

The clonal diversity of GCs was also described by G. Victora and colleagues, using "confetti" mice immunized with a protein antigen in which B cells engaged in a GC response can stochastically acquire 10 different color combinations, allowing for enumeration of GC founder clones.[60] Single GCs were found to be seeded by numerous clones, when analyzed shortly after initiation of the response. While the frequency of GC dominated by a single clone increased with time, a number of GCs retained substantial clonal diversity, 3 weeks after immunization, demonstrating more complex clonal dynamics than previously reported.[24,170]

Another type of GC structures, residual or persistent ones, can be observed in the spleen of nonimmunized mice, and is the result of spontaneous responses taking place in animals kept in pathogen-free conditions.[171] Such continuous activation may be driven by constant toll-like receptor 7 (TLR7) signaling and appears to constitute a reservoir of pathogenic B cells in autoimmune conditions.[172,173]

These results tend to indicate that many different parameters beyond affinity may control the entry and maintenance into GCs, such as the precursor frequency and the avidity of the BCR in the preimmune naïve compartment, as well as bystander activation signals. Understanding precisely how selection takes place in GCs remains therefore a challenging issue.

Extrafollicular Responses

Extrafollicular responses are of paramount importance since they provide the host with a rapid early protective antibody response.[174] Antigen-activated B cells, once in contact with cognate Bcl6$^+$ pre-T_{FH} cells, will on one hand give rise to GCs within follicles and on the other hand give rise to PCs in the extrafollicular zone[175] (Fig. 26.1). B cells engaged into follicles express Bcl6, which represses Blimp-1 and PC differentiation. Conversely, B cells engaged in the extrafollicular response express Blimp-1, which suppresses Pax5, a transcription factor required for the B cell lineage program and for GC formation, and induces differentiation into plasmablasts.[176,177] These plasmablasts express CD138 and become Cxcr4^{high} and Cxcr5/Ccr7^{low}, thus allowing their migration to extrafollicular foci in the red pulp of the spleen or in medullary cords of lymph nodes, which both produce Cxcl12, the Cxcr4 ligand.[178,179] When entering in contact with CD11c^{high} DCs, they differentiate into PCs and switch to other isotypes.[180] These PCs produce class-switched antibodies during a few days before dying by apoptosis.

Which signals trigger either a GC or an extrafollicular response? One proposition is that high- and low-affinity B cells can both enter stochastically the extrafollicular or the GC pathway.[181,182] However, earlier work showed that the primary antibody response to various T-dependent antigens was of a relatively high affinity.[183-185] Based on a transgenic mouse model in which the affinity of the BCR toward the antigen varied over 4 orders of magnitude, Brink and colleagues reported that a lower affinity was driving B cells toward a GC reaction, while, conversely, higher affinity B cells were more present in the extrafollicular response. As mentioned by the authors, there was no sharp threshold of interaction between the BCR and the antigen that would dictate an early PC differentiation, but rather a direct relation between the size of the extrafollicular response and the strength of the initial BCR trigger. In fact, lower affinity B cells could also be driven in the extrafollicular response when the density of antigen was increased.[186]

Major advances have been achieved in the last decades, allowing the detailed molecular, cellular, and biochemical description of the GC reaction. Nevertheless, there are still many unanswered questions concerning the signals dictating whether B cells should remain within GCs and go on diversifying their BCRs or exit this structure to become PCs or memory B cells. One would also like to understand which factors control, during repeated antigenic challenges, the reentry of memory B cells into GCs for further rounds of diversification. These are central basic issues with obvious direct translational applications.

T CELL–INDEPENDENT, EXTRAFOLLICULAR, AND INNATE-LIKE B CELL RESPONSES AND B CELL REGULATORY FUNCTIONS

T-Independent Responses to Nonpeptidic Antigens

T cell independent responses have in common to be triggered by antigens that cannot be processed and presented by T cells on class II MHC molecules, that is, essentially nonprotein antigens. They nevertheless cover a large panel of antigenic structures, from model antigens, like haptens coupled to dextran or Ficoll, to polysaccharides constituting the capsule of bacteria, like *S. pneumoniae,* or whole pathogens displaying repeated antigenic structures, as well as costimulatory molecules interacting with pattern recognition receptors. This diversity of antigenic stimuli may explain the complex picture that emerges from various experimental approaches or vaccine procedures.

Historically, T-independent responses have been classified as type I and type II, with type I responses representing responses to LPS challenge and type II to nonprotein antigens.[187] This terminology should now be abandoned, since LPS stimulation triggers a polyclonal B cell activation through TLR and not through BCR signals. Antigen-specific T-independent responses are thus restricted to type II responses. Type I and type II responses were distinguished by the fact that only type II responses were abrogated in XID mice, later shown to be deficient for the Bruton tyrosine kinase, a kinase activated by BCR cross-linking.[188,189]

T-independent responses indeed require a strong aggregation of the BCR, allowing for downstream signaling and activation, and, while being independent of cognate T cell help, they can benefit from additional stimuli from noncognate T cell help, TLR agonists, or cytokines secreted by accessory cells like DC[13,190,191] (Fig. 26.3).

B Cell Subsets and PC Differentiation

Strong BCR cross-linking induces rapid B cell activation, proliferation, and plasmablast differentiation. Isotype switch takes place in a fraction of proliferating cells,[193] allowing for secretion of IgM as well as of switched isotypes, mainly IgG3 in the mouse and IgG2 in humans.[194,195]

While cross-linking of multiple BCR molecules can induce activation of all B cell subsets *in vitro*, it is generally considered that MZ B cells and B1b cells are the main actors of T-independent responses in the mouse,[196] due to their specific preactivated status, reflected notably in a higher basal expression of the *Prdm1* gene.[197] Moreover, the localization of MZ B cells at the site of incoming blood flow in the spleen make them excellent candidates to be responders to blood-borne antigens.

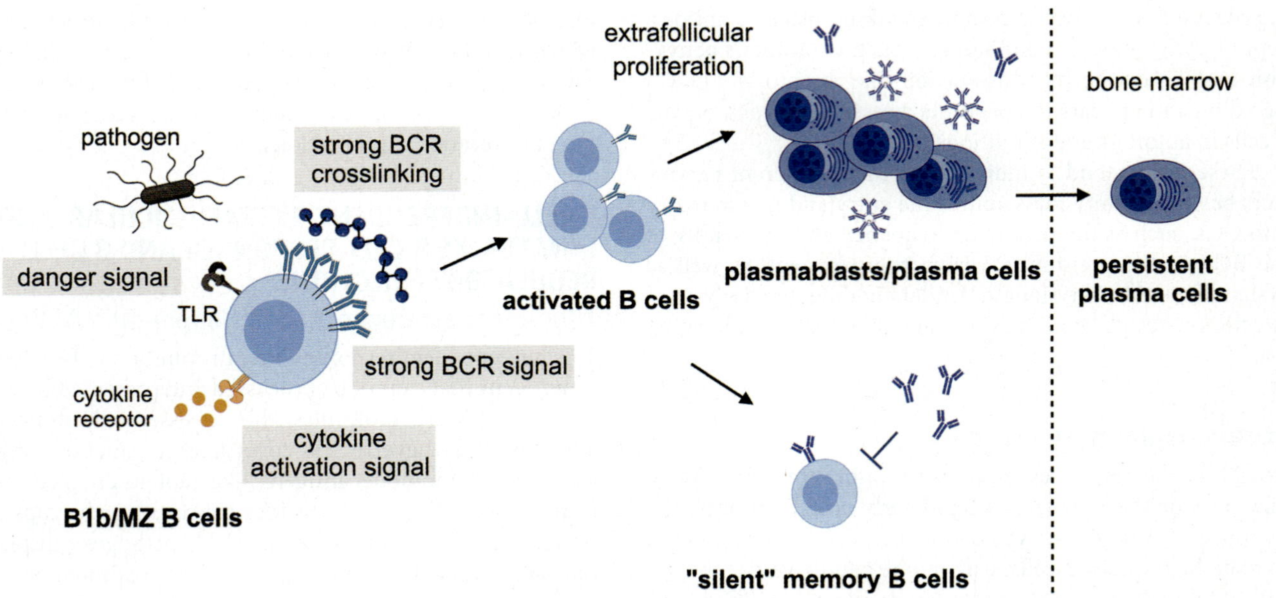

FIG. 26.3. **T-independent and extrafollicular responses.**[192] Initiation of T-independent responses requires a strong cross-linking of the B cell receptor (BCR) and may include toll-like receptor (TLR) and/or cytokine cosignals. While these responses are considered as essentially short-lived, strong stimulations can give rise to bone marrow plasma cells persisting up to 6 months in the mouse. A memory B cell response, whose duration has not been determined, has also been demonstrated in some T-independent responses, only revealed by cell transfer into adoptive hosts, thus suggesting a negative feedback on memory B cell activation by circulating antigen-specific antibodies. A strong and long-lasting extrafollicular response, T-independent or involving noncognate T cell help (eg, CD40-CD40L interaction), occurs in some bacterial infections, causing a delay of several weeks in the eventual initiation of germinal center responses against T-dependent epitopes. (Adapted from Defrance T, Taillardet M, Genestier L. T cell-independent B cell memory. *Curr Opin Immunol.* 2011;23:330-336, with permission from Elsevier.)

Persistence of T-independent Responses

T-independent antigenic stimulations are considered to be unable to generate an anamnestic response, that is, a B cell intrinsic enhanced and faster response to antigen rechallenge displaying affinity maturation. The massive decay observed after the initial PC burst also suggested that the vast majority of ASC produced with T-independent stimuli could be short-lived.[198]

The overall picture is nevertheless more complex (Fig. 26.3). PCs with extended life span (surviving at least for several months in the mouse) have been described after T-independent immunizations, this profile being markedly enhanced by costimulation with TLR ligands (NP hapten coupled to LPS or CpG administered shortly after *S. pneumoniae* PS3 injection), a configuration that may mimic responses to polysaccharides in the context of whole pathogens.[199-201] Several reports have also described a persistent extrafollicular PC response (largely IgM), notably following *Ehrlichia muris* infection, which may be driven by prolonged stimulation of nonpeptidic antigens that better resist processing and degradation.[202] It is interesting to note that this response, which can take place in T cell deficient mice, can be followed by a delayed T-dependent GC reaction, making the dichotomy between T-independent and T-dependent responses tenuous. Moreover, this delay was shown, in the case of malaria, to be caused by the vast metabolic requirements of the extrafollicular response, which inhibits the differentiation of GC B cells.[203]

Some forms of B cell memory have also been described, both in response to model antigens and to an infection with a polysaccharide-coated pathogen, *Borrelia hermsii*.[204,205] In both cases, the existence of an antigen-experienced population could only be revealed upon adoptive transfer, identifying B1b cells as the responding subset in the case of *B. hermsii*. To account for this unexpected observation, it was proposed that antigen-specific IgG antibodies are inhibiting recall responses, thus possibly avoiding recurrent stimulation by antigens with poor biodegradability. The "recall" response against *B. hermsii* involved differentiation into IgM-secreting cells, without mutation, and relied on the clonal amplification of cells issued from the first challenge, but not on their maturation into a memory phenotype. The duration of this memory was however not followed beyond a few months after challenge, and longevity of these antigen-experienced cells was similar to the one of naïve B cells.[204]

A last exception to the general characteristics of extrafollicular responses also concerns Ig gene somatic mutation. Some bacteria, like *Salmonella typhimurium*, induce an intense and prolonged extrafollicular reaction, while GC formation is delayed by several weeks.[206] One report described that this response, which involves isotype switching, mobilizes clones with very low initial affinity but whose specificity was eventually enhanced by V gene mutagenesis at these extrafollicular sites of proliferation.[207] This unusual process taking place in activated B cells/PCs did not however generate an affinity-matured memory response.

Anti-polysaccharide Vaccines in Humans

It should also be mentioned that anti-polysaccharide responses in humans generate serum responses lasting for a few years, which, although obviously shorter than antiviral responses, may imply the formation of PCs with some persistence. One epidemiologic study suggested that Pneumovax (a vaccine containing 23 serotypes of the *S. pneumoniae* capsule) may provide protection against infection for up to 5 years.[208] Along the same line, the follow-up of anti-PS serum titers during the few months of Rituximab-mediated B cell depletion did not report a significant decay in anti-PS titers, suggesting again some persistence of PCs in these T-independent responses.[209] Secondary challenges with polysaccharide vaccines do not generate an enhanced response and seem, on the contrary, to face a state of hyporesponsiveness and/or of exhaustion of the pool of responding B cells.[210-212] It should nevertheless be mentioned (as discussed in the last section) that anti-PS responses may mobilize different B cell subsets in humans, with different kinetics and outcomes.

Systemic Antibacterial/Antiglycan B Cell Subsets of Mucosal Origin

Mucosal immune responses are the topic of a specific chapter in this book (see chapter 38). What will be developed here are recent results indicating that responses at mucosal barriers are not completely compartmentalized even in homeostatic conditions.

Multiple pieces of evidence have been brought, indicating that elements from the gut flora can be translocated to peripheral lymphoid tissues and may result in systemic immune activation. Bacterial metabolites or cell wall components can modulate, globally or specifically, IgM, IgA, and even IgG present in the general circulation.[213,214]

But distinct activated/memory B cell subsets are also generated at the systemic level through mucosal activation. B1 cells are strong responders against bacterial glycans and contribute a large fraction of natural IgM present in the circulation.[215,216] Ongoing immune reactions in Peyer's patches also seed IgM and IgA memory subsets in spleen, as well as IgM and IgA PCs in spleen and bone marrow, these cells showing distinct antibacterial specificities and protective function.[171,217] A large fraction of these systemic memory B cells and PCs recognize cross-reactive bacterial epitopes, mainly glycans, indicating that their emergence from Peyer's patch GC reactions may proceed through distinct T-B interactions, involving different types of activation and selection signals, and may have multiple T-dependent and T-independent origins.[218]

One important contribution of these memory or PC subsets raised against mucosal antigens would be their capacity to control the presence at the systemic level of commensals, including pathobionts, arising from breaches of the mucosal barrier, as well as to provide protection and rapidly react against pathogens presenting cross-reactive antigenic structures.

B Cells as Cytokine Producers: B and PC Regulatory Functions

This last section will deal with B cell functions that are not strictly related to their antigenic specificity, that is, their capacity to secrete cytokines that may modulate positively or negatively immune responses.

A protective role for B cells has been described in several autoimmune, allergic, or inflammatory conditions, notably mouse models of ulcerative colitis, arthritis, or multiple sclerosis.[219-221] This has been attributed to their capacity to secrete IL-10 and to inhibit Th1 or Th17 responses as well as to promote T_{regs} function. The term of regulatory B cells (B_{regs}) has been coined for such B cell effectors, but there is no consensus in the definition of B_{regs}, the observation being that B cells able to upregulate IL-10 in response to inflammatory signals are enriched in the immature (T2/MZP) or MZ subset.[222,223] B_{regs} furthermore do not seem to constitute a distinct lineage, but rather to represent an activation stage resulting from stimulation through TLR or via proinflammatory cytokines.

Other studies have described that the major source of IL-10 or IL-35 secretion was in fact a subset of IgM-secreting nonproliferating PCs,[224,225] with, by analogy with T_{regs}, a possible implication of IRF4 or Blimp-1 in the regulation of cytokine expression.[226] These observations therefore led to propose the concept of "regulatory PCs."[227]

A positive regulatory function has also been described for B cells, linked with a distinct set of secreted cytokines: GM-CSF secretion by B1 B cells in microbial sepsis, IL-17 secretion in antiparasite immunity, or IL-6 secretion during influenza virus infection.[228-230] IL-6 secreted by B cells can also be detrimental and contribute to the progression of autoimmune diseases.[231]

Altogether, the various pro- or anti-inflammatory roles of B cells in distinct activation conditions may explain some unanticipated outcomes in B cell–depleting therapies, leading either to disease control or disease exacerbation, likely depending on the cytokine secretion profile of the B cells involved.

T-independent and other types of extrafollicular responses develop more rapidly than GC-dependent ones and thus represent a crucial immune layer against fast-replicating bacterial pathogens, reducing the bacterial burden at the early steps of infections. IgM plays a major role in the clearance of complex pathogens, and the various outcomes observed in distinct infectious or T-independent vaccine settings may reflect the contribution of accessory cells or stimuli, notably TLR ligands, as well as the diversity of antigenic structures.

PLASMA CELLS: SHORT-LIVED AND LONG-LIVED EFFECTORS OF SEROLOGICAL RESPONSES

PCs are terminally differentiated B cells, whose main contribution to immune function is the secretion of antibodies. Antibodies are key effectors of natural or adaptive immunity, including at mucosal surfaces, as well as contributors of several autoimmune pathologies. The formation of PCs takes place in different contexts: extrafollicular responses following T-dependent or T-independent stimulation of naïve B cells; PC differentiation from memory B cells during recall responses (see section on "Memory recall responses"); and PC formation following high-affinity selection in the GC, this last pathway generating the majority of long-lived effectors and being the hallmark of successful vaccines. One should also mention the steady-state activation of B cells leading to basal IgM secretion ("natural antibodies"), contributed in part, but not exclusively, by B1 cells in the mouse,[215,232] as well as the formation of IgM and IgA PCs generated from mucosal responses but residing at nonmucosal sites.[171,217] This section will briefly discuss these different pathways and will focus on the more specific issue of PC maintenance, addressing how intrinsic and extrinsic factors contribute to the persistence of PCs over a lifetime.

Formation of Antibody-Secreting Cells

ASCs are terminally differentiated B cells that undergo marked morphological changes, which led to the identification of their Ig secreting function by A. Fagraeus.[233] They harbor a large cytoplasmic content with expanded endoplasmic reticulum and Golgi compartments. These changes are linked to a major transcriptional reprogramming, allowing them to cope with the extraordinary metabolic load of immunoglobulin secretion: secretion rates may vary over several logs, up to 5000 to 10,000 Ig molecules per second.[234,235] Naïve B cells can be efficiently stimulated *in vitro* to differentiate into short-lived ASC, using T cell-derived stimuli (CD40L), cytokines, or through TLR activation (LPS in the mouse, or CpG). The term ASC includes both plasmablasts, undergoing rapid proliferation, and PCs, resting cells with various fates and life spans.

ASC as Long-lived Effectors: LLPCs

The proliferative burst observed after antigenic stimulation, followed by antibody decay in serum, as well as the rapid and transient differentiation of naïve B cells upon in vitro stimulation, has led to consider, for a long time, ASC as short-lived effectors of immune responses. Moreover, the large quantity of secreted antibodies was viewed as a metabolic disequilibrium, incompatible with long-term cell homeostasis. However, two seminal studies established that, after homing to the bone marrow, ASC can acquire a long-lived phenotype.[236,237] Studies based on B cell depletion, as well as splenectomy in macaques, further demonstrated that this longevity was cell intrinsic, thus excluding that ongoing memory B cell differentiation constitutes the main supply ensuring long-term persistence of the PC pool.[238-240]

A longitudinal study performed in humans to follow the serological response to various vaccines or infectious challenges revealed the remarkable longevity of some immune responses, with half-lives of several pathogen-specific antibodies exceeding a century, clearly establishing the bone marrow PC subset, together with memory B cells, as a major component of B cell–based memory[241,242] (Table 26.1).

TABLE 26.1 Duration of Humoral Immunity Against Viruses and Vaccines (Serum Antibody Half-Lives)[a]

Natural infections (acute)	
Measles	3014 y
Mumps	542 y
Rubella	110 y
Natural infections (chronic/latent)	
Epstein-Barr virus	No decay
Varicella-zoster virus	50 y
Vaccines	
Vaccinia	92 y
Tetanus	11 y[b]
Diphtheria	19 y[b]

[a]Amanna et al. (2007) Ref. 241.
[b]Durable protection ensured by high initial antibody titers, Antia et al. (2018) Ref. 242.

Transcriptional Rewiring: Facing the Challenge of Ig Secretion

An *Irf4/Prdm1*(encoding Blimp-1)/*Xbp1* transcription axis controls the differentiation of PCs. Irf4 functions as a rheostat, acting in a dose-dependent manner[243] (Fig. 26.4). While

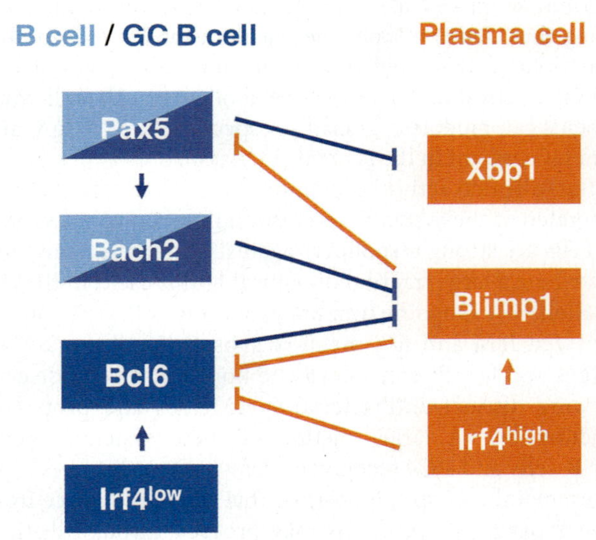

FIG. 26.4. Transcription factors regulating the germinal center/plasma cell fate.[244] Major transcription factors expressed at both B cell and germinal center stages or in GC B cells only (mixed *light/dark blue or dark blue* boxes, respectively) and at the plasma cell stage (*orange* boxes) are represented with arrows indicating transcriptional activation and lines with bar transcriptional repression. The Irf4 gene functions as a rheostat, its expression level conditioning its association with other transcription factors as well as its gene targets. (Adapted from Nutt SL, Taubenheim N, Hasbold J, Corcoran LM, Hodgkin PD. The genetic network controlling plasma cell differentiation. *Semin Immunol.* 2011;23:341-349, with permission from Elsevier.)

low Irf4 expression occurs in naïve and GC B cells, high Irf4 expression activates *Prdm1* transcription. Irf4/Blimp-1 will, on one side, repress the Pax5-controlled B cell program, thus relieving the repressing activity of the Pax5-activated gene *Bach2* on *Prdm1*.[176,245] It will also downregulate *Bcl6* and *Aicda* expression in GC B cells, thus turning off the GC program as well.[246]

Blimp-1 promotes Xbp1 expression, and a set of genes involved in the endoplasmic reticulum stress response, notably Ire-1 and ATF6, which enhance the Xbp1 controlled unfolded protein response in a feed-forward loop[248,249]: this step indeed involves an unusual molecular event promoting cytoplasmic splicing of Xbp1 mRNA to a shorter, more efficient form, Xbp1s, through endonucleolytic cleavage and ligation mediated, respectively, by the Ire-1 nuclease and tRNA ligase complex activities.[250] The ER stress response is a complex process activated in response to misfolded protein accumulation, with one arm triggering cell apoptosis or reduction in protein synthesis. In the case of PCs, minimal activation of this downstream apoptotic pathway takes place, the stress response allowing in contrast PCs to survive and cope with its intense protein synthesis activity. It was also recently reported that an early and independent activation of UPR genes via the mTOR pathway takes place during B cell activation, thus anticipating further Ig synthesis and cooperating with Xbp1s-mediated transcriptional activation at later stages.[251]

Irf4 inactivation at the GC stage severely impairs PC formation, while *Prdm1* deletion still allows the formation of a preplasmablast population, but no further maturation.[252,253] Xbp1 deficiency does not prevent PC formation and may be partially complemented by the mTOR pathway, but markedly reduced Ig secretion is observed in mice with a B cell–specific Xbp1 deficiency.[254-256]

PC Life Span: Migration Cues and Intrinsic and Extrinsic Factors of the Survival Niche

The bone marrow hosts the majority of LLPCs, but persisting PCs can also be found in spleen as well as, importantly, at mucosal locations.[257] PCs that emerge from GC reactions, or from recall responses, will leave the spleen or lymph node via the blood, and, through expression of Cxcr4, will migrate toward Cxcl12, secreted by resident stromal and reticular cells in the bone marrow.[179,258] Alternatively, in inflammatory or autoimmune responses, expression of Cxcr3 by PCs will allow their settlement in inflamed tissues where Cxcl9, Cxcl10, or Cxcl11 secretion occurs.[259,260]

Factors that control PC longevity are complex, and the relative contribution of stochastic versus programmed events, or of intrinsic versus extrinsic factors, is still debated. One view posits that bone marrow niches are in finite numbers and that each new wave of PC migration dislodges a small fraction of the cells formed upon previous immune responses, thus only marginally affecting previously stored memories.[261] Other propositions favor an intrinsic survival advantage imprinted during the immune responses, since distinct vaccines or infectious diseases endow the PCs they induce with different longevities.[262] Such specific survival capacity would be provided by the nature of the immunizing agent, for example, by the strength of multivalent antigens or by some adjuvants.[263,264] Such specific imprint may favor the capacity of newly formed PCs either to survive up to their relocation in their resident niche or to compete more efficiently for their settlement.

Whatever the model, PCs only survive for a short time when dislodged from their bone marrow niche, which clearly indicates the intricate roles played by intrinsic and extrinsic factors in PC survival, and we will briefly review these different elements.

Phenotype and Metabolic Profile of LLPC

Transition from cycling plasmablasts and from immature PCs involves further upregulation of *Prdm1* expression and distinct transcriptional changes.[265,266] PCs in bone marrow can be characterized by a set of markers, differing in mice and humans. In the mouse, bone marrow PCs are identified as B220^{neg}, CD138+, Sca1^{+}, TACI^{+}, and CD93+, while none of these markers marks the entire population.[267-269] In humans, PCs are identified as CD20^{neg}, CD138+, CD38^{high}, and CD27^{high}, and the PC fraction lacking expression of CD19 has been proposed as being the most long-lived one, harboring antigen specificities linked with childhood vaccines.[270,271] Nevertheless, gene expression profiling so far failed to identify clear long-lived signatures among bone marrow PC subsets, suggesting that posttranscriptional regulations and/or metabolic alterations might determine PC longevity.

PCs express a set of genes and surface markers that allow them to interact with accessory cells in the bone marrow niche and promote their local retention. Cxcr4 not only guides PC migration toward Cxcl12-rich sites but also strengthens their residence, together with VLA-4 and LFA-1 integrins, which promote interaction with cells expressing their ligand, VCAM-1 and ICAM-1, respectively, as well as colocalization of accessory cells expressing similar integrins[258,272] (Fig. 26.5).

BCMA is a key marker of bone marrow PC, allowing, upon binding of APRIL, upregulation of Mcl1, a member of the antiapoptotic Bcl2 family of proteins, and a major factor for PC longevity.[274,275] Mcl1 is a short-lived molecule, and thus constant stimulation through BCMA is required to prevent apoptosis.

PCs profoundly remodel their metabolism, which allows them to cope with constant antibody secretion and ensures their longevity. Increased autophagy enables the recycling of nutrients and may be sustained by the hypoxic environment of the bone marrow.[276,277] Higher respiratory capacity and glucose uptake is observed in LLPC, with glucose metabolism diverted in part for antibody glycosylation.[278] The ATP-degrading enzyme, ENPP1, was recently shown to impact PC longevity through the control of the glucose-driven metabolic activity.[279]

While a main emphasis has been put on IgG-secreting cells that constitute the major part of long-lasting serological responses, the bone marrow also hosts large numbers of persistent IgM- and IgA-secreting cells.[171,217,232,280] Such PCs still harbor IgM or IgA at their surface and, while not sensitive to mitogenic stimuli, they can be sensitive to their cognate

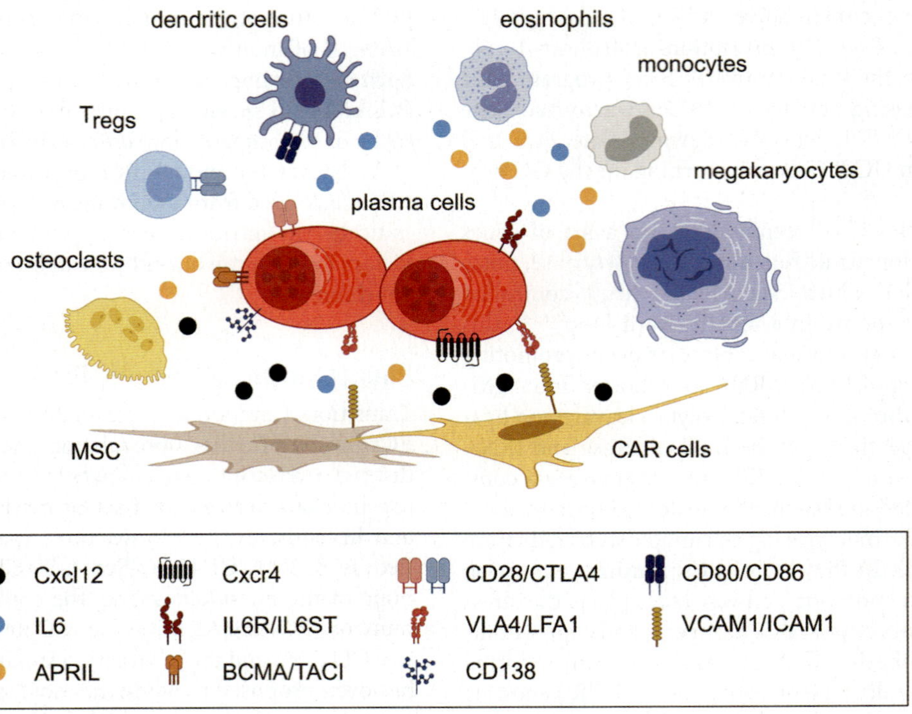

FIG. 26.5. The bone marrow plasma cell niche.[273] Multiple sessile and motile cell types interact with plasma cells in the bone marrow. Mesenchymal stromal cells (MSCs) and Cxcl12-abundant reticular (CAR) cells structure the niche, attracting and retaining plasma cells through Cxcl12 secretion. The cellular environment of the bone marrow provides diverse types of factors: factors promoting migration toward the niche (Cxcl12/Cxcr4); factors ensuring plasma cell retention (integrin pairs LFA1/ICAM1, VLA4/VCAM1) as well as colocalization of other cell types, like neutrophils; and factors promoting survival, essentially APRIL and IL-6, inducing signals through BCMA/TACI and IL-6R/IL-6ST, respectively. CD138, a heparan sulfate proteoglycan, can bind APRIL and trigger BCMA-mediated survival. MSCs and eosinophils, together with dendritic cells, are the main producers of IL-6, while numerous cell types can secrete APRIL, notably megakaryocytes, eosinophils, osteoclasts, and monocytes. A three-partner interaction between plasma cells, dendritic cells, and T_{regs} seems to fine-tune the inflammatory milieu and to regulate access to the niche. (Adapted from Tangye SG. Staying alive: regulation of plasma cell survival. *Trends Immunol.* 2011;32:595-602, with permission from Elsevier.)

antigen, and notably be induced to secrete cytokines like IL-10 or possibly other factors allowing them to modulate their own niche.[281,282]

Cellular Components Providing Surviving Signals in the Bone Marrow PC Niche

Imaging has identified several cell components located in close proximity to resident PCs, like megakaryocytes, eosinophils, or T cells, with two-photon microscopy further documenting long-lasting contacts. These different components contribute redundantly to the provision of key survival factors, APRIL and IL-6, as well as Cxcl12, required for migration and retention in the bone marrow niche (Fig. 26.5).

Mesenchymal stromal cells and fibroblastic reticular cells structure the bone marrow niche and secrete abundant Cxcl12. They also express integrin ligands, VCAM-1 and ICAM-1, that tether PCs to the extracellular matrix and allow their coclustering with other hematopoietic cells like neutrophils that also express the VLA-4 or LFA-1 integrins.[272]

Many cells types can secrete APRIL (or BAFF) in the vicinity of bone marrow PCs: eosinophils, osteoclasts, megakaryocytes, but also granulocytes, and macrophages.[283-286] Interestingly, CD138 (syndecan-1), expressed at the PC surface, is a heparan sulfate proteoglycan that can bind APRIL and leads to its oligomerization, thus triggering BCMA-mediated survival.[287]

A second cytokine with a major survival function, IL-6, is secreted mainly by mesenchymal stromal cells and eosinophils.[285,288] Since mature PC express low levels of IL-6R,[270] IL-6 signaling in LLPC may possibly proceed through trans-signaling, a pathway whereby IL-6 bound to soluble IL-6R subsequently interacts with IL-6ST (IL-6 signal transducer) expressed at the cell surface and triggers Stat3 activation.[289] DCs interact with CD28 expressed on PC through CD80/CD86, with CD28 signaling to DC triggering IL-6 secretion and sustaining PC survival.[290] A tripartite repressive interaction with T_{regs} has also been described, whereby CTLA4 on T_{regs} would downmodulate CD28-derived signals and DC activation, and thus cytokine secretion[291] (Fig. 26.5).

It should be noted that several components that are obligate requirements for bone marrow PC survival, like BCMA, are in fact dispensable for those PCs residing in spleen, clearly indicating that multiple factors can substitute for each other, the interaction BCMA/APRIL being replaced by TACI and/or BAFF-R/BAFF to provide surviving signals.[275] Similarly, the supportive role of IL-6–secreting basophils *in vivo* has been mainly documented in the splenic context.[292]

Altogether, both sessile stromal cells and short-lived, motile hematopoietic cells of various lineage origins appear to contribute to the long-term residency of PCs in the bone marrow niche, indicating its highly dynamic aspect.[293]

LLPCs represent the second arm of immune memory, secreting protective antibodies that can readily control a new pathogenic attack. This powerful and frontline immune defense is nevertheless constitutive, being unable to adapt quantitatively or qualitatively to pathogen load or changes. Like memory B cells, the key aspect of their effector function is their longevity, being able to survive for decades in the human bone marrow. The overall picture that emerges when considering the survival factors, accessory cells in the survival niche or metabolic pathways involved, is a global redundancy of the signals involved, allowing cells facing a changing environment of metabolic supplies, as well as inflammatory or infectious episodes, to survive in the long term while maintaining their high secretory profile.

MEMORY B CELL SUBSETS AND MEMORY RECALL RESPONSES

Memory B cells are essential components of immune memory, showing, for some infections or immunization conditions, lifelong persistence and being able to quickly respond not only to a new encounter with the same pathogen/immunogen but also to adapt to drifted variants by the selective activation and amplification of memory B cells with better-fitted specificities. Their response potential comes from an altered activation profile allowing them to respond faster to new stimuli, with an increased affinity for their eliciting antigen and a higher frequency due to the expansion of antigen-specific B cell clones. They also have the capacity to adopt diverse activation profiles: proliferation, which maintains the memory B pool and prevents its attrition, PC differentiation from proliferating and amplified memory B cells allowing for massive antibody secretion and rapid antigen control, or recruitment into new GC reactions for further affinity maturation.

Memory B Cell Subsets in the Mouse

The expression of a switched isotype (mainly IgG, but also IgA) has been used for a long time as a surrogate marker for B cell memory. Surface IgG expression indeed endows memory B cells with unique signaling properties, through the IgG intracytoplasmic tail that harbors a phosphorylation motif responsive to antigen binding. Such specific signaling accounts to a large extent for the clonal burst and biased potential toward PC differentiation observed in recall responses.[294-296] However, there are now clear indications that IgM^+ B cells represent an integral component of immune memory, their prevalence and specific function depending largely on the nature of the immunogen. IgM memory B cells are, for example, an important part of the response against sheep red blood cells, a model particulate antigen in the mouse, but also of some infections with pathogens like *E. muris* or *Plasmodium chabaudi*.[297-299]

The hallmark of memory B cells is affinity maturation, meaning that these cells experienced mutation of their BCR and selection for improved affinity, even though the selection in GCs seems to favor the output of memory B cells with an intermediate affinity for the eliciting antigen, while those with the highest affinity are selected in the LLPC pool, as described in the first section of this chapter.

A set of surface markers identifies GC-derived memory B cells, among which CD73 and CD80, and, to a more variable extent, PD-L2. Expression of these three markers appears to accompany a gradient of memory B cell maturation, as well as an increased proportion of IgG over IgM memory B cells in a transgenic mouse setting.[300] However, absence of CD73 and CD80 expression marks to a large extent an unusual memory B cell subset, named "GC independent," which emerges by clonal expansion of naïve B cells without their further maturation in GCs, and keeps a major IgM expression profile.[301,302] It is nevertheless unclear whether clonal expansion without affinity maturation can allow GC-independent memory B cells to compete with GC-derived B cells upon recall stimulations, and what is the physiological role of this memory entity.

Another memory entity was initially described as an age-associated subset, expressing mainly IgM and being refractory to BCR activation, but responding to TLR signals.[303,304] Several markers further associated with this subset, CD11c or expression of the T-bet transcription factor, revealed that they mark in fact an antigen-responsive entity produced in the context of Th1 cytokines and notably during viral or bacterial activation.[305] They can be either IgM^+ or IgG^+ and often harbor negative signaling receptors like FcRL4/FcRL5.[298,306] Such cells also accumulate during chronic inflammation and stimulation, likely explaining their progressive accumulation with age. This point is further developed in the last section, as numerous conditions in humans (chronic viral infection, autoimmune inflammatory processes) also lead to the accumulation of T-bet expressing antigen-experienced B cells.

The issue of tissue residency is a field that is still in its infancy for memory B cells. While it is clear that memory B cells recirculate and patrol in blood and lymph for antigen encounter, some memory B cells remains as sentinels in lymphoid tissues, lymph node, or spleen, where the initial response was elicited. Such tissue residency and repositioning may facilitate rapid interaction between memory B and T_{FH} cells.[307] Residency is also observed at sites of infection like pulmonary tissues or mucosal epithelia, with such tissue-resident memory B cells harboring functional characteristics distinct from recirculating cells.[308,309]

Memory B Cells in Recall Responses

Affinity maturation is an iterative process among GC B cells recycling between dark zone and light zones, and, analogously, the seminal studies of C. Berek and C. Milstein have laid the grounds for the widely held notion that repeated antigen stimulation also leads to an iterative process of improvement of the memory response[310] (Fig. 26.6). This work, based on the response to the phenyl oxazolone hapten, reported the successive accumulation of mutations and the increased affinity in antigen binding of B cells identified by hybridoma fusion and sequencing of the canonical V_HOx/VκOx genes that are recurrently mobilized in such anti-hapten responses. These mutations were followed during the course of a primary, secondary, and even a tertiary stimulation, at 1 year from the first challenge. This study also reported that, in addition to the iterative mobilization of V_HOx/VκOx expressing memory B cells, new V_H/V_L combinations, less frequent but with higher initial affinity, were also recruited during successive challenges.[310]

Nevertheless, other approaches based on adoptive transfer of IgG memory B cells only reported a PC fate with no further GC differentiation upon an antigen boost.[311] From the many recent studies that addressed the question of the behavior of memory B cells during recall responses, the consensus that emerges is that the default pathway for memory B cells is proliferation and PC differentiation. The mobilization in GCs is in contrast highly context-dependent, being affected by the nature of the immunogen, the preexisting immune level (notably circulating antibodies), and intrinsic characteristics of the memory B cells themselves (Fig. 26.7).

It has been proposed that IgM memory B cells might preferentially undergo migration and activation in GCs, while IgG memory B cells would be more prone to differentiation into ACSs.[297,312] This distinct behavior could be dictated by intrinsic gene expression profile, IgM memory B cells in humans expressing higher levels of FoxP1[313] and IgG memory cells expressing lower levels of Bach2,[314] two transcription factors involved in the repression of the *IRF4/Prdm1/Xbp1* axis, and thus affecting the Bcl6/Blimp-1 equilibrium. Such isotype preference matches also the PD-L2/CD80 profile in some mouse models,[315] but, as discussed above, the issue is rendered more complex by the fact that the PD-L2/CD80-negative population is essentially generated by a GC-independent pathway, that is, outside affinity maturation. Nevertheless, clear cases of IgM PC activation (in a malaria recall response) and of IgG1 memory B cell recruitment in new GC reactions (in a boost against the NP hapten) have also been reported.[299,316]

A major factor regulating the recall response includes the previous immune status and notably the level of circulating, antigen-specific IgG antibodies.[312] The impact of IgG feedback on the recall responses can take place at multiple levels, either through direct inhibition of GC responses via FcγRIIB-dependent or FcγRIIB-independent mechanisms or through epitope masking. High antibody levels may even inhibit the induction of Ig secretion, if they are potent enough to lead to antigen clearance.

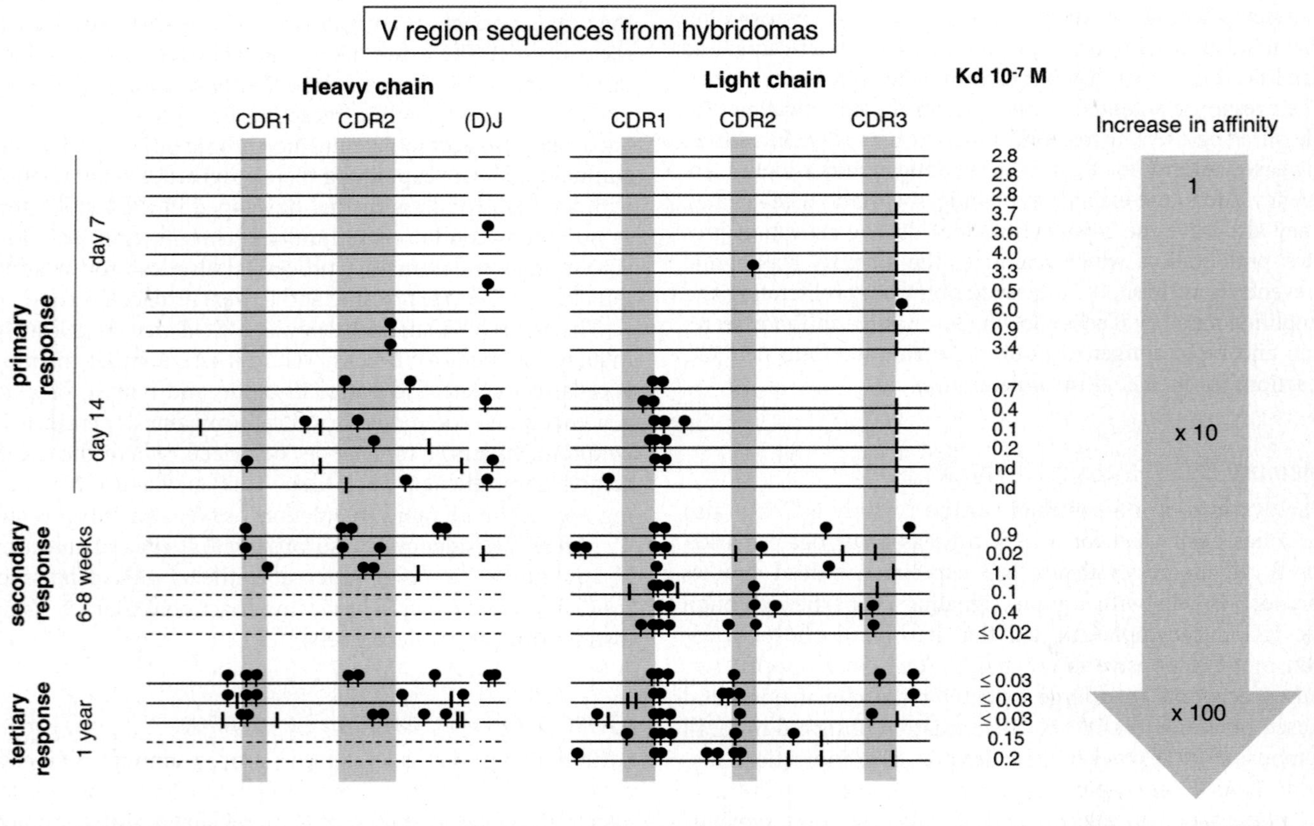

FIG. 26.6. Affinity maturation as an iterative process upon successive antigenic challenges.[310] Anti–oxazolone-specific hybridomas were obtained after primary, secondary, and tertiary immunization with 2-phenyl-5-oxazolone coupled to chicken serum albumin, and the heavy and light chain mRNA sequenced. Somatic mutations compared to the canonical V_H-Ox1 ("heavy chain") and Vκ-Ox1 ("light chain") genes are indicated, with *bars and black circles* for silent and replacement mutations, respectively. *Dotted lines* represent unrelated V_H or V_L genes. Dissociation constants (molarity) of the secreted antibodies are indicated. (Adapted from Berek C, Milstein C. Mutation drift and repertoire shift in the maturation of the immune response. *Immunol Rev.* 1987;96:23-41.)

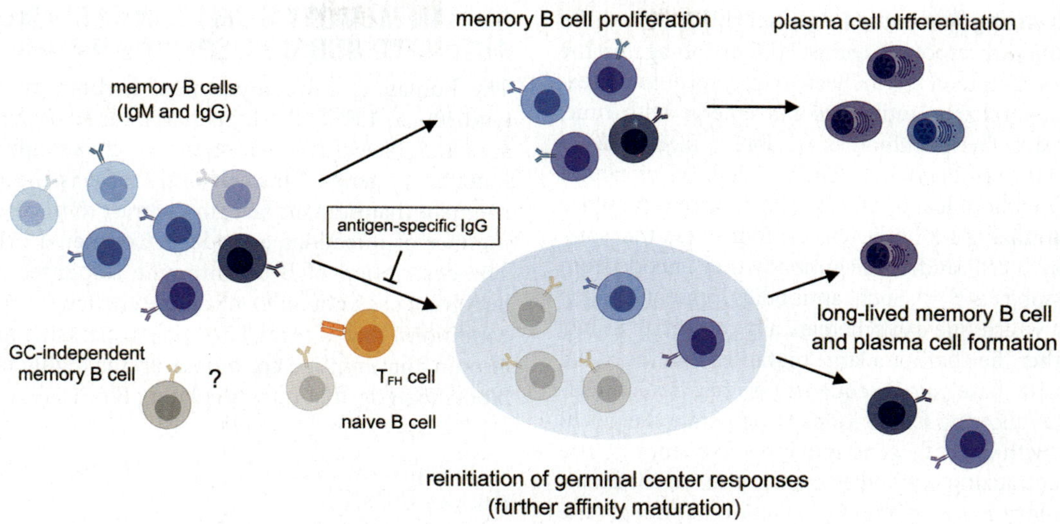

FIG. 26.7. Memory recall responses: plasma cell versus germinal center B cell differentiation. The default pathway for memory B cells upon secondary stimulations is plasma cell differentiation, while reactivation in germinal centers appears to depend on multiple factors: the concentration of circulating, antigen-specific antibodies, the presence of persistent germinal centers, the memory B cell isotype, the site of immunization and recall, the type of adjuvant and Th bias. Upon recall, naïve and memory B cells can be recruited in germinal centers, both benefiting from the help of memory T_{FH} cells. Restimulation with an identical antigen activates memory B cells with higher affinity, while restimulation with a drifted version of the antigen can activate different memory B cells present in the memory pool due to a broader affinity selection process (see Fig. 26.2). The behavior of germinal center–independent memory B cells is unsettled and indicated by a question mark. Darker shades of blue depict increasing B cell affinities. (Created with BioRender.com)

A very recent study, based on the tracing of cells previously engaged in GC reactions, reported that the majority of newly recruited GC B cells are of naïve origin, likely driven by the rapid availability of T cell help, with no contribution from a GC-independent subset.[18] This study further established that the fraction of memory B cells that is also mobilized in GC, albeit numerically less abundant, presents strong selection features for mutation and affinity and is clonally highly restricted. Such selection and clonal restriction is also observed in the antibody-secreting plasmablast pool, highlighting the fact that, in spite of the moderate selection operating on memory B cells exiting GC reactions, boosting achieves a strong discriminating step by only activating a small, highly fitted memory subset.

As just mentioned, a large part of the memory B cell pool has been shown to emerge at earlier time points of the immune response than LLPCs, and to harbor a less mutated repertoire with hallmarks of a less stringent selection. Nevertheless, output of memory B cells takes place during the several weeks of the GC reaction, even though these selected cells are less numerous with time. While such early export could appear counterintuitive for a B cell subset involved in long-term protection, this broader repertoire appears well adapted for coping with viral or bacterial escape mutants, allowing through the strong proliferative capacity of the memory pool, the emergence of rare memory clones whose specificity presents a better fit with the drifted form of the pathogen.[317]

Two Examples of Memory Recall From the Human Vaccine/Infection Field

Original Antigenic Sin in Memory Responses to Influenza

The fact that mobilization of memory B cells versus recruitment of naïve B cells depends heavily on the nature of antigenic challenge is illustrated by the concept of "original antigenic sin" in the response to diverse influenza strains,[318] a term which can be replaced by the more neutral one of "antigen imprinting"[319] or "antigenic seniority."[320] Antigen imprinting relates to the critical role of primary exposure to flu infection or vaccines in shaping subsequent responses to drifted variants or more distant strains. It corresponds to the fact that differentiation of preexisting memory B cells against cross-reactive epitopes ("back boosting") predominates over the recruitment of new naïve B cells with more adapted specificities. This bias can be either beneficial, triggering the induction of broadly neutralizing antibodies against conserved structures like the stalk of the HA antigen when there is no homology between the HA head parts of different influenza strains,[247] or it can be detrimental when cross-reactivity between antigenically drifted HA domains triggers nonprotective responses.[321] Responses to flu clearly highlight the persistence of memory B cells and their capacity to self-maintain and to be recurrently mobilized upon successive challenges to generate antibody responses. Recent data nevertheless suggest that some forms of adjuvanted flu vaccines may enhance the recruitment of naïve B cells upon successive challenges.[322]

Broadly Neutralizing Antibodies Against HIV

While initial antibody responses against HIV, mainly against the gp41 surface glycoprotein, are ineffective to control the infection, broadly neutralizing antibodies can emerge with time, usually after years. These antibodies are able to target sites of vulnerability in the gp120 glycoprotein, which the virus cannot further modify without losing fitness. The recurrent presence of broadly neutralizing antibodies in serum fostered the isolation of memory B cells and of the antibody they encode from HIV-infected subjects.[323,324] Such antibodies present unique characteristics, which one can schematically categorize in two main types: either they harbor a large mutation load (over 100 mutations in a 300 base pair V sequence) or they have unusually long CDR3 (allowing CDR3 loops to penetrate the gp120 glycan shield) with lower V gene mutation frequency.[325] The first type of neutralizing antibodies evokes a recurrent mobilization of memory B cells within GC reactions, although one cannot ascertain whether it represents B cells persisting within GCs and experiencing ongoing mutation and selection or successive waves of memory B cell recruitment in new responses against virus variants.[326] The second type of antibodies likely corresponds to the recruitment of very rare naïve B cells with unique CDR3 specificities, often deleted because of their self-reactivity,[327] which join the antiviral response and acquire more rapidly their neutralization capacity. Broadly neutralizing antibodies against HIV thus illustrate the recurrent/ongoing mobilization of antigen-experienced B cells as well as naïve B cells in GC reactions.

The default differentiation of memory B cells in recall responses thus consists in a strong proliferative phase, preventing the "burnout" of the memory B cell pool, followed by differentiation into ASCs. The capacity of the incoming antigen to selectively trigger the stimulation of the most-fitted memory B cells ensures a response adapted to the pathogen, in contrast to the constant Ab secretion of LLPCs that do not have such adaptability. Recruitment of memory B cells into new GC reactions to further improve antigen recognition and cope with drifted antigen profiles follows a more complex scheme, which strongly depends upon previous immune experience and immune status, thus precluding any simple scheme that could apply to each vaccine or infectious condition. Understanding and controlling these parameters is obviously at the core of vaccine design strategies.

HUMAN MEMORY B CELLS, INNATE-LIKE AND ACTIVATED SUBSETS: SPECIFIC ISSUES

The human B cell compartment harbors distinct features, with 30% to 40% of B cells, either in blood or in lymphoid tissues like spleen, having a memory/activated phenotype with mutated Ig genes. This obviously reflects the natural environment that humans face, in contrast to mice, as well as the number of infectious episodes encountered over a lifetime. The description of B cell immune responses in humans is largely focused not only on vaccine challenges or pathological conditions, that is, viral, bacterial, or parasitic infections, but also on autoimmune conditions and is highly biased toward blood analysis, that is, recirculating B cell subsets.

Genotypic and Phenotypic Characterization of Human B Cells

The locus encoding the human Ig constant region genes emerged through a duplication event, harboring two clusters encoding two IgG isotypes, one IgE (the first one being a pseudogene) and one IgA. As a consequence, isotype switch to the first IgA isotype, IgA1, does not preclude further evolution toward expression of the IgG2 or IgG4 isotype (Fig. 26.8). The four IgG isotypes, IgG1, 2, 3, and 4, were initially named according to their decreasing abundance in human serum.[328]

Memory B cells in humans are identified by the expression of CD27, which marks switched, as well as IgM^+IgD^+ and IgM-only B cells, all of them harboring mutated Ig genes, hallmark of antigen experience.[329] $CD27^+$ switched memory B cells constitute the main reservoir of long-lived memory B cells against most vaccines or infections, while the role of the IgM-positive subset appears more complex, and possibly includes different functional entities (discussed below). A set of other markers, among them CD21, CD11c, CD69, CD95, CD73, and $CD45RB^{MEM}$ (a glycosylated variant of CD45RB), are now extensively studied using multiparametric flow or mass cytometry to further refine this classification and identify subpopulations of physiological relevance.[330] This is a fast progressing field, and it is clearly premature to present a consensus overview of these studies. We will nevertheless discuss several memory-like or atypical subsets that have been recurrently described, even though their detailed description and identification may differ between disease conditions.

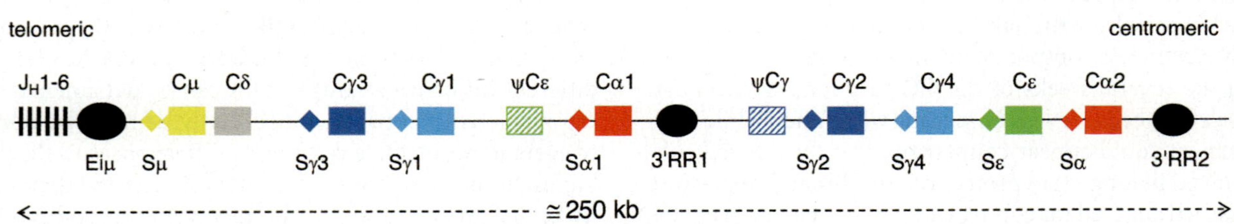

FIG. 26.8. The human heavy chain constant region gene locus. The different constant region genes present on human chromosome 12 are represented (not at scale), to highlight the duplication event of the Cγ-Cγ-Cε-Cα gene cluster, with an additional Cγ pseudogene in-between (pseudogenes are depicted as *hatched* boxes). Intronic (Eiμ) and 3′ (3′RR1 and 3′RR2) enhancers are represented as *black ovals*, and switch repeat regions as *diamonds*.

"Memory-like" Subsets: Multiple Phenotypes, Names, and Functions

At steady state, the blood B cell compartment comprises transitional B cells, naïve B cells, memory B cells, $IgM^+IgD^+CD27^+$ B cells, and a limited number of plasmablasts, which can all be easily discriminated by the use of CD24, CD38, IgD, CD27, and CD19 or CD20. A small "double-negative" subset, IgD^-CD27^-, is also present, which becomes amplified in a number of activation or pathological conditions[331] (Table 26.2).

A memory $CD27^-$ subset was initially identified in tonsils and named "tissue-based" memory: it was described as harboring FcRL4, a surface receptor with strong inhibitory potential, as well as CD11c and low levels of CD21, but its physiological function was not clearly established.[332,333] An expanded subset with similar characteristics ("tissue-like") was later observed in blood of HIV-infected patients and proposed to represent exhausted B cells produced by chronic activation, by analogy with the exhausted state of CD8 T cells in chronic viral infections.[334] Indeed, in such cells, an additional set of inhibitory receptors was described, notably LAIR1, LILRB1 (CD85j), LILRB4 (CD85k), or CD22, strengthening the proposition of their dysfunctional state. Malaria infection represents another disease state where double-negative cells are expanded, named in this case "atypical" memory B cells.[335] Again, the presence of inhibitory receptors, notably FcRL4

TABLE 26.2 B Cell Subsets in Human Blood[a]

B Cell Subset ($CD19^+$)	Core Markers (IgD, IgM, CD24, CD38, CD27, CD21)	Additional Markers	Function	Comments
Immature				
T1/T2	$IgM^{high}IgD^{low}CD27^-CD24^{high}CD38^{high}$	$CD10^+ABCB1^-$	Follicular B cell precursor	Decreased IgM expression with maturation
MZP/T3	$IgM^{high}IgD^{low}CD27^-CD24^{high}CD38^{high}$	$CD45RB^{MEM55}$	MZ B cell precursor/ late transitional stage	ABCB1 marks MZP
Naïve				
Resting	$IgD^{high}IgM^{low}CD27^-CD21^+$	$ABCB1^+CD23^+$	Naïve follicular B cells	
Activated (act)	$IgD^{high}IgM^{low}CD27^-CD21^{int}$	$ABCB1^-CD23^-CD11c^+CD71^+$	DN2/ASC precursor	
Memory (mem)				
MZB	$IgD^{low}IgM^{high}CD27^+CD24^+CD38^{low}$	$CD1c^{high}$	Mouse MZB cell equivalent	IgM and MZB subsets likely overlapping
IgM	$IgD^{neg}IgM^+CD27^+CD24^+CD38^{low}$		IgM memory	
Switched	$IgD^{low}IgM^{high}CD27^+CD24^+CD38^{low}$	IgG^+ or IgA^+	Switched memory	
Switched activated	$IgD^-CD27^+CD24^-CD38^-CD21^{low}$	$CD80/CD86^+CD95^+CD71^+$	Replenishment of memory—DN1/ASC precursor	
Atypical	IgD^-CD21^{low}	$CD11c^{high}T\text{-}bet^{high}CD19^{high}FcRL4^+$	Product of recurrent activation (functional/ exhausted)	Described as either $CD27^+$ or $CD27^-$
Double-negative (DN)				
DN1	$IgD^-CD27^-CD21^+$	$CD11c^-$	Extrafollicular ASC precursors (from memB cells)	
DN2	$IgD^-CD27^-CD21^{low}$	$CD11c^+T\text{-}bet^+$	Extrafollicular ASC precursors (from act. naïve)	
Antibody-secreting cells (ASCs)				
PB	$CD38^{high}CD27^{high}$	$Ki\text{-}67^+HLA\text{-}DR^+$	Proliferating ASC	CD138 present on some newly formed PB
PC	$CD38^{high}CD27^{high}$	$CD138^+CD20^-$	Mature ASC	

[a]Adapted from Sanz et al., 2019 Ref. 331.

and FcRL5, was considered to likely account for the ineffective protection offered by B cells against malaria and the multiple infection waves observed before an immune control of the disease takes place. Their expression of the transcription T-bet further linked them to the aged B cell population described in the mouse.

Lastly, autoimmune conditions and in particular lupus represent another pathological setting in which expansion of double-negative (DN, $CD27^-IgD^-$) B cells, expressing T-bet and FcRL4, is observed (and named DN2, to distinguish for non–T-bet expressing DN cells): such cells however differ from the ones generated in infectious contexts, which are related to memory B cells, as they emerge from the naïve B cell pool through an "activated-naïve" intermediate, identified through the CD71 activation/proliferation marker.[336] In the case of lupus, strong TLR7-mediated activation appears to bypass the control of inhibitory receptors.[337]

$T\text{-bet}^+$, $CD11c^+$, and $CD21^{low}$ DN cells are nevertheless observed in many Th1-biased, IFNγ-rich infectious or vaccine contexts, in which they are probably not dysfunctional. Transcriptional analyses have also shown that such cells are poised toward PC differentiation, which has led to novel proposals reinterpreting their inhibitory profile: such propositions indeed favor the notion that, in some cases, the set of inhibitory receptors expressed might in fact constrain, rather than inhibit, their activation, restricting it to strong, possibly surface-embedded antigen signaling and preventing their activation by abundant soluble antigens or bystander stimuli.[338] Inhibitory receptors on $T\text{-bet}^+$ cells would thus be the hallmark of potent activation controls, rather than of exhaustion or paralysis.

$T\text{-bet}^+$, $CD11c^+$, $CD21^{low}$, and DN B cells are thus observed in a multiplicity of contexts, with diverse phenotypes, origins, and functions, beneficial or detrimental, and represent an emerging and fast-evolving field.

MZ B Cells and T-independent Responses: A Still Controversial Issue

The nature of T-independent responses in humans, notably responses against the polysaccharides of encapsulated bacteria, as well as the B cell subset involved is still a debated question. There is indeed no clear equivalent in humans of the mouse B1b subset, and cells that harbor characteristics similar to mouse MZ B cells also display marked differences. Indeed, cells expressing IgM, IgD, and CD27 are present in human spleens around the corona of naïve B cells, that is, in a location analogous to MZ B cells in the mouse. Splenic $IgM^+IgD^+CD27^+$ B cells are also $CD21^{high}$ and express more strongly the invariant class I MHC molecule CD1c, similar to mouse MZ B cells that express the only CD1 molecule of the mouse genome, CD1d.[339] But $IgM^+IgD^+CD27^+$ B cells harbor mutated Ig genes, questioning the origin of such diversification. Another distinct feature, in contrast to the mouse, is their capacity to recirculate, harboring clonal relationships between blood and spleen.[339,340]

Arguments in favor of a distinct MZ lineage in humans, and of their diversification outside classical T-dependent responses, come from their presence, albeit at reduced frequencies but with mutated Ig genes, in patients with genetic deficiencies compromising T-B collaboration, that is, inactivation of CD40 or CD40L genes.[341] Identification of a distinct precursor, capable of differentiating into MZ-like B cells upon NOTCH2-triggering signals, further strengthened the analogy with the NOTCH2-directed differentiation pathway of mouse MZ B cells.[342]

However, mutations in the *Bcl6* gene, as well as clonal relationships between blood $IgM^+IgD^+CD27^+$ and switched B cells, were proposed as strong evidences for a GC origin, and thus of an IgM memory function.[343,344] A small subset of IgM-only B cells also exist, but it seems clear that IgD expression appears insufficient by itself to discriminate IgM memory B cells involved in T-dependent responses from those activated by polysaccharides and that both subsets overlap with the set of markers commonly used to phenotype human B cells.[340]

From a functional point of view, the evidence for a role of $IgM^+IgD^+CD27^+$ B cells in T-independent responses is largely indirect and is mainly based on the correlation of their reduced presence in the blood of immune-compromised patients showing susceptibility to bacterial infections, notably asplenic or splenectomized patients, or patients deficient in TLR effectors, Myd88, or IRAK4.[345,346] Isolation of PCs activated during pneumococcal or meningococcal vaccinations have confirmed that B cells with mutated Ig genes and switched isotypes are activated in such antigenic challenges, obviously questioning their origin.[347,348]

Very recently, a high-dimensional analysis has documented the distinct differentiation path of MZ B cells compared to memory B cells, their possible diversification in GC of gut-associated lymphoid tissues, and their dissemination to widely distant sites via blood recirculation.[349] Such observations are obviously reminiscent of species, like chicken or sheep, that diversify their immune repertoire in gut-associated lymphoid tissues,[350] but a global picture of the role of MZ B cells in humans still awaits a more thorough understanding of the development of their repertoire and of their role in T-independent responses.

Human T_{FH} Cells: Insight From Circulating Subsets

T_{FH} cells, contrarily to GC B cells, recirculate, and it is thus possible to identify, in human blood, T cells that likely originate from recent or ongoing responses and harbor a GC-like T_{FH} profile (CXCR5, PD1, ICOS)[351] or which may be derived from active T_{FH} cells after having downmodulated the PD1 and ICOS markers as well as Bcl6 expression (circulating T_{FH} or cT_{FH}).[352] Among $CXCR5^+$ circulating T cells, several subsets differing by chemokine receptor expression have been further identified, among them CXCR3, CCR6, and CCR7, defining different subsets resembling Th1 ($CXCR3^+CCR6^-$), Th2 ($CXCR3^-CCR6^-$), and Th17 cells ($CXCR3^-CCR6^+$).[353] CCR7 further discriminates activated from quiescent cells.[354] While $CXCR3^+CCR6^-$ cT_{FH} are unable to activate naïve B cells *in vitro*, their presence correlates with strong antibody responses following influenza vaccination, where their contribution could be to stimulate high-affinity memory B cells toward extrafollicular responses.

Circulating memory T_{FH} cells can thus be classified by their Th1-, Th2-, or Th17-like profile, which obviously raises the question of the identity of the T_{FH} lineage. It may suggest that the B cell activation function of T_{FH} (notably IL-21 secretion) may superimpose to their initial polarization, which would reemerge when immune reactions wane.[100] Additional profiles have also been described, like a new $PD1^{high}CXCR3^{+}$ subset lacking CXCR5 expression but triggering a strong B cell activation and abundant IL-21 secretion upon *in vitro* stimulation: it likely represents the recirculating fraction of a T_{FH} subset present in inflamed tissues, notably within tertiary lymphoid structures, and is named T_{PH}, for peripheral helper T cells.[355] These cells, mainly described in autoimmune conditions, but possibly existing as well in other infectious or malignant inflammatory contexts, appear to provide help mainly to memory B cells and to codevelop, in variable proportions, with cT_{FH} cells.[356]

Human studies of B cell responses are progressing at a rapid pace, notably through the use of high-dimensional analyses. They can only rarely address the question of immune reactions taking place in lymphoid or mucosal tissues, even though biopsies or surgical samples can be accessible in some infectious or pathological contexts and are increasingly studied. Even if restricted to the blood compartment, studies performed in humans allow us to challenge models drawn from immunization schemes used in the mouse that may sometimes be very distant from physiology, notably by the use of restricted B cell repertoires, immunization routes, antigen dose, or structure. Such mouse studies nevertheless laid the conceptual ground to confront and contrast human-based observations. There are also many aspects of the immune response that may be unique to humans, like gene activation networks, lymphoid organ structures, and, obviously, more complex individual infectious trajectories, which the emerging wealth of studies performed in humans aims to address.

CONCLUSION

Model systems elaborated in the mouse have identified major B cell effector pathways, which provide rationale for translation into immune interventions. The complexities and differences observed for the human immune system compared to the one of laboratory animals notably lie in the multiple infections the former encounters with time, shaping a B cell compartment in which a large fraction of B cells have acquired antigen experience, even in a young adult. Moreover, most pathogens are complex antigenic structures, mobilizing more than one type of B cell response, with such responses occurring in settings affected by previous antigen imprinting within and outside the B cell as well as the T cell lineage. The next challenge will be to better integrate these multiple, time-layered antigenic triggers into a comprehensive scheme of the adaptive immune response.

ACKNOWLEDGMENT

The authors wish to gratefully thank the Fondation Princesse Grace de Monaco for its long-term support.

REFERENCES

1. Fu YX, Chaplin DD. Development and maturation of secondary lymphoid tissues. *Annu Rev Immunol.* 1999;17:399-433.
2. Junt T, Moseman EA, Iannacone M, et al. Subcapsular sinus macrophages in lymph nodes clear lymph-borne viruses and present them to antiviral B cells. *Nature.* 2007;450:110-114.
3. Phan TG, Grigorova I, Okada T, Cyster JG. Subcapsular encounter and complement-dependent transport of immune complexes by lymph node B cells. *Nat Immunol.* 2007;8:992-1000.
4. Phan TG, Green JA, Gray EE, Xu Y, Cyster JG. Immune complex relay by subcapsular sinus macrophages and noncognate B cells drives antibody affinity maturation. *Nat Immunol.* 2009;10:786-793.
5. Carrasco YR, Batista FD. B cells acquire particulate antigen in a macrophage-rich area at the boundary between the follicle and the subcapsular sinus of the lymph node. *Immunity.* 2007;27:160-171.
6. Pape KA, Catron DM, Itano AA, Jenkins MK. The humoral immune response is initiated in lymph nodes by B cells that acquire soluble antigen directly in the follicles. *Immunity.* 2007;26:491-502.
7. Bajenoff M, Egen JG, Koo LY, et al. Stromal cell networks regulate lymphocyte entry, migration, and territoriality in lymph nodes. *Immunity.* 2006;25:989-1001.
8. Roozendaal R, Mempel TR, Pitcher LA, et al. Conduits mediate transport of low-molecular-weight antigen to lymph node follicles. *Immunity.* 2009;30:264-276.
9. Heesters BA, Chatterjee P, Kim YA, et al. Endocytosis and recycling of immune complexes by follicular dendritic cells enhances B cell antigen binding and activation. *Immunity.* 2013;38:1164-1175.
10. Gonzalez SF, Lukacs-Kornek V, Kuligowski MP, et al. Capture of influenza by medullary dendritic cells via SIGN-R1 is essential for humoral immunity in draining lymph nodes. *Nat Immunol.* 2010;11:427-434.
11. Cinamon G, Matloubian M, Lesneski MJ, et al. Sphingosine 1-phosphate receptor 1 promotes B cell localization in the splenic marginal zone. *Nat Immunol.* 2004;5:713-720.
12. Cinamon G, Zachariah MA, Lam OM, Foss FW Jr, Cyster JG. Follicular shuttling of marginal zone B cells facilitates antigen transport. *Nat Immunol.* 2008;9:54-62.
13. Balazs M, Martin F, Zhou T, Kearney J. Blood dendritic cells interact with splenic marginal zone B cells to initiate T-independent immune responses. *Immunity.* 2002;17:341-352.
14. Borges da Silva H, Fonseca R, Pereira RM, Cassado Ados A, Alvarez JM, D'Imperio Lima MR. Splenic macrophage subsets and their function during blood-borne infections. *Front Immunol.* 2015;6:480.
15. Chen Y, Pikkarainen T, Elomaa O, et al. Defective microarchitecture of the spleen marginal zone and impaired response to a thymus-independent type 2 antigen in mice lacking scavenger receptors MARCO and SR-A. *J Immunol.* 2005;175:8173-8180.
16. Abbott RK, Lee JH, Menis S, et al. Precursor frequency and affinity determine B cell competitive fitness in germinal centers, tested with germline-targeting HIV vaccine immunogens. *Immunity.* 2018;48:133-146 e136.
17. Abbott RK, Crotty S. Factors in B cell competition and immunodominance. *Immunol Rev.* 2020;296:120-131.
18. Mesin L, Schiepers A, Ersching J, et al. Restricted clonality and limited germinal center reentry characterize memory B cell reactivation by boosting. *Cell.* 2020;180:92-106.e11.
19. Flemming W. Studien uber regeneration der Gewebe. *Archiv fur mikroskopische Anatomie.* 1884;24:50-91.
20. Briney B, Inderbitzin A, Joyce C, Burton DR. Commonality despite exceptional diversity in the baseline human antibody repertoire. *Nature.* 2019;566:393-397.
21. Eisen HN, Siskind GW. Variations in affinities of antibodies during the immune response. *Biochemistry.* 1964;3:996-1008.
22. Weigert MG, Cesari IM, Yonkovich SJ, Cohn M. Variability in the lambda light chain sequences of mouse antibody. *Nature.* 1970;228:1045-1047.
23. Di Noia JM, Neuberger MS. Molecular mechanisms of antibody somatic hypermutation. *Annu Rev Biochem.* 2007;76:1-22.
24. MacLennan IC. Germinal centers. *Annu Rev Immunol.* 1994;12:117-139.
25. Revy P, Muto T, Levy Y, et al. Activation-induced cytidine deaminase (AID) deficiency causes the autosomal recessive form of the Hyper-IgM syndrome (HIGM2). *Cell.* 2000;102:565-575.
26. Muramatsu M, Kinoshita K, Fagarasan S, Yamada S, Shinkai Y, Honjo T. Class switch recombination and hypermutation require activation-induced cytidine deaminase (AID), a potential RNA editing enzyme. *Cell.* 2000;102:553-563.
27. Weill JC, Reynaud CA. DNA polymerases in adaptive immunity. *Nat Rev Immunol.* 2008;8:302-312.
28. Stavnezer J. Complex regulation and function of activation-induced cytidine deaminase. *Trends Immunol.* 2011;32:194-201.
29. Kurosaki T, Kometani K, Ise W. Memory B cells. *Nat Rev Immunol.* 2015;15:149-159.
30. Allen CD, Ansel KM, Low C, et al. Germinal center dark and light zone organization is mediated by CXCR4 and CXCR5. *Nat Immunol.* 2004;5:943-952.
31. Cyster JG, Dang EV, Reboldi A, Yi T. 25-Hydroxycholesterols in innate and adaptive immunity. *Nat Rev Immunol.* 2014;14:731-743.

32. Zaretsky I, Atrakchi O, Mazor RD, et al. ICAMs support B cell interactions with T follicular helper cells and promote clonal selection. *J Exp Med.* 2017;214:3435-3448.
33. Wang X, Rodda LB, Bannard O, Cyster JG. Integrin-mediated interactions between B cells and follicular dendritic cells influence germinal center B cell fitness. *J Immunol.* 2014;192:4601-4609.
34. Klein U, Dalla-Favera R. Germinal centres: role in B-cell physiology and malignancy. *Nat Rev Immunol.* 2008;8:22-33.
35. Stavnezer J, Linehan EK, Thompson MR, et al. Differential expression of APE1 and APE2 in germinal centers promotes error-prone repair and A:T mutations during somatic hypermutation. *Proc Natl Acad Sci U S A.* 2014;111:9217-9222.
36. Roco JA, Mesin L, Binder SC, et al. Class-switch recombination occurs infrequently in germinal centers. *Immunity.* 2019;51:337-350.e7.
37. Di Zenzo G, Di Lullo G, Corti D, et al. Pemphigus autoantibodies generated through somatic mutations target the desmoglein-3 cis-interface. *J Clin Invest.* 2012;122:3781-3790.
38. Gay D, Saunders T, Camper S, Weigert M. Receptor editing: an approach by autoreactive B cells to escape tolerance. *J Exp Med.* 1993;177:999-1008.
39. Tiegs SL, Russell DM, Nemazee D. Receptor editing in self-reactive bone marrow B cells. *J Exp Med.* 1993;177:1009-1020.
40. Wardemann H, Yurasov S, Schaefer A, Young JW, Meffre E, Nussenzweig MC. Predominant autoantibody production by early human B cell precursors. *Science.* 2003;301:1374-1377.
41. Hartley SB, Cooke MP, Fulcher DA, et al. Elimination of self-reactive B lymphocytes proceeds in two stages: arrested development and cell death. *Cell.* 1993;72:325-335.
42. Russell DM, Dembic Z, Morahan G, Miller JF, Burki K, Nemazee D. Peripheral deletion of self-reactive B cells. *Nature.* 1991;354:308-311.
43. Chan TD, Wood K, Hermes JR, et al. Elimination of germinal-center-derived self-reactive B cells is governed by the location and concentration of self-antigen. *Immunity.* 2012;37:893-904.
44. Aplin BD, Keech CL, de Kauwe AL, Gordon TP, Cavill D, McCluskey J. Tolerance through indifference: autoreactive B cells to the nuclear antigen La show no evidence of tolerance in a transgenic model. *J Immunol.* 2003;171:5890-5900.
45. Goodnow CC, Crosbie J, Jorgensen H, Brink RA, Basten A. Induction of self-tolerance in mature peripheral B lymphocytes. *Nature.* 1989;342:385-391.
46. Eris JM, Basten A, Brink R, Doherty K, Kehry MR, Hodgkin PD. Anergic self-reactive B cells present self antigen and respond normally to CD40-dependent T-cell signals but are defective in antigen-receptor-mediated functions. *Proc Natl Acad Sci U S A.* 1994;91:4392-4396.
47. Rathmell JC, Cooke MP, Ho WY, et al. CD95 (Fas)-dependent elimination of self-reactive B cells upon interaction with CD4+ T cells. *Nature.* 1995;376:181-184.
48. Rathmell JC, Townsend SE, Xu JC, Flavell RA, Goodnow CC. Expansion or elimination of B cells in vivo: dual roles for CD40- and Fas (CD95)-ligands modulated by the B cell antigen receptor. *Cell.* 1996;87:319-329.
49. Fisher GH, Rosenberg FJ, Straus SE, et al. Dominant interfering Fas gene mutations impair apoptosis in a human autoimmune lymphoproliferative syndrome. *Cell.* 1995;81:935-946.
50. Rieux-Laucat F, Le Deist F, Hivroz C, et al. Mutations in Fas associated with human lymphoproliferative syndrome and autoimmunity. *Science.* 1995;268:1347-1349.
51. Holzelova E, Vonarbourg C, Stolzenberg MC, et al. Autoimmune lymphoproliferative syndrome with somatic Fas mutations. *N Engl J Med.* 2004;351:1409-1418.
52. Dowdell KC, Niemela JE, Price S, et al. Somatic FAS mutations are common in patients with genetically undefined autoimmune lymphoproliferative syndrome. *Blood.* 2010;115:5164-5169.
53. Del-Rey M, Ruiz-Contreras J, Bosque A, et al. A homozygous Fas ligand gene mutation in a patient causes a new type of autoimmune lymphoproliferative syndrome. *Blood.* 2006;108:1306-1312.
54. Sabouri Z, Schofield P, Horikawa K, et al. Redemption of autoantibodies on anergic B cells by variable-region glycosylation and mutation away from self-reactivity. *Proc Natl Acad Sci U S A.* 2014;111:E2567-E2575.
55. Burnett DL, Langley DB, Schofield P, et al. Germinal center antibody mutation trajectories are determined by rapid self/foreign discrimination. *Science.* 2018;360:223-226.
56. Brink R, Phan TG. Self-reactive B cells in the germinal center reaction. *Annu Rev Immunol.* 2018;36:339-357.
57. Bannard O, Horton RM, Allen CD, An J, Nagasawa T, Cyster JG. Germinal center centroblasts transition to a centrocyte phenotype according to a timed program and depend on the dark zone for effective selection. *Immunity.* 2013;39:912-924.
58. Victora GD, Schwickert TA, Fooksman DR, et al. Germinal center dynamics revealed by multiphoton microscopy with a photoactivatable fluorescent reporter. *Cell.* 2010;143:592-605.
59. Caron G, Le Gallou S, Lamy T, Tarte K, Fest T. CXCR4 expression functionally discriminates centroblasts versus centrocytes within human germinal center B cells. *J Immunol.* 2009;182:7595-7602.
60. Tas JM, Mesin L, Pasqual G, et al. Visualizing antibody affinity maturation in germinal centers. *Science.* 2016;351:1048-1054.
61. Mesin L, Ersching J, Victora GD. Germinal center B cell dynamics. *Immunity.* 2016;45:471-482.
62. Mayer CT, Gazumyan A, Kara EE, et al. The microanatomic segregation of selection by apoptosis in the germinal center. *Science.* 2017;358:eaao2602.
63. Stewart I, Radtke D, Phillips B, McGowan SJ, Bannard O. Germinal center B cells replace their antigen receptors in dark zones and fail light zone entry when immunoglobulin gene mutations are damaging. *Immunity.* 2018;49:477-489.e7.
64. Ersching J, Efeyan A, Mesin L, et al. Germinal center selection and affinity maturation require dynamic regulation of mTORC1 kinase. *Immunity.* 2017;46:1045-1058.e6.
65. Luo W, Weisel F, Shlomchik MJ. B cell receptor and CD40 signaling are rewired for synergistic induction of the c-myc transcription factor in germinal center B cells. *Immunity.* 2018;48:313-326.e5.
66. Dominguez-Sola D, Victora GD, Ying CY, et al. The proto-oncogene MYC is required for selection in the germinal center and cyclic reentry. *Nat Immunol.* 2012;13:1083-1091.
67. Calado DP, Sasaki Y, Godinho SA, et al. The cell-cycle regulator c-Myc is essential for the formation and maintenance of germinal centers. *Nat Immunol.* 2012;13:1092-1100.
68. Finkin S, Hartweger H, Oliveira TY, Kara EE, Nussenzweig MC. Protein amounts of the MYC transcription factor determine germinal center B cell division capacity. *Immunity.* 2019;51:324-336.e5.
69. Sander S, Chu VT, Yasuda T, et al. PI3 kinase and FOXO1 transcription factor Activity differentially control B cells in the germinal center light and dark zones. *Immunity.* 2015;43:1075-1086.
70. Chou C, Verbaro DJ, Tonc E, et al. The transcription factor AP4 mediates resolution of chronic viral infection through amplification of germinal center B cell responses. *Immunity.* 2016;45:570-582.
71. Krautler NJ, Suan D, Butt D, et al. Differentiation of germinal center B cells into plasma cells is initiated by high-affinity antigen and completed by Tfh cells. *J Exp Med.* 2017;214:1259-1267.
72. Yeh CH, Nojima T, Kuraoka M, Kelsoe G. Germinal center entry not selection of B cells is controlled by peptide-MHCII complex density. *Nat Commun.* 2018;9:928.
73. Schwickert TA, Victora GD, Fooksman DR, et al. A dynamic T cell-limited checkpoint regulates affinity-dependent B cell entry into the germinal center. *J Exp Med.* 2011;208:1243-1252.
74. Smith KG, Light A, Nossal GJ, Tarlinton DM. The extent of affinity maturation differs between the memory and antibody-forming cell compartments in the primary immune response. *EMBO J.* 1997;16:2996-3006.
75. Weisel FJ, Zuccarino-Catania GV, Chikina M, Shlomchik MJ. A temporal switch in the germinal center determines differential output of memory B and plasma cells. *Immunity.* 2016;44:116-130.
76. Koike T, Harada K, Horiuchi S, Kitamura D. The quantity of CD40 signaling determines the differentiation of B cells into functionally distinct memory cell subsets. *Elife.* 2019;8:e44245.
77. Heise N, De Silva NS, Silva K, et al. Germinal center B cell maintenance and differentiation are controlled by distinct NF-kappaB transcription factor subunits. *J Exp Med.* 2014;211:2103-2118.
78. Roy K, Mitchell S, Liu Y, et al. A regulatory circuit controlling the dynamics of NFkappaB cRel transitions B cells from proliferation to plasma cell differentiation. *Immunity.* 2019;50:616-628.e6.
79. Wang Y, Shi J, Yan J, et al. Germinal-center development of memory B cells driven by IL-9 from follicular helper T cells. *Nat Immunol.* 2017;18:921-930.
80. Takatsuka S, Yamada H, Haniuda K, et al. IL-9 receptor signaling in memory B cells regulates humoral recall responses. *Nat Immunol.* 2018;19:1025-1034.
81. Suan D, Krautler NJ, Maag JLV, et al. CCR6 defines memory B cell precursors in mouse and human germinal centers, revealing light-zone location and predominant low antigen affinity. *Immunity.* 2017;47:1142-1153.e4.
82. Shinnakasu R, Inoue T, Kometani K, et al. Regulated selection of germinal-center cells into the memory B cell compartment. *Nat Immunol.* 2016;17:861-869.
83. Mosmann TR, Cherwinski H, Bond MW, Giedlin MA, Coffman RL. Two types of murine helper T cell clone. I. Definition according to profiles of lymphokine activities and secreted proteins. *J Immunol.* 1986;136:2348-2357.
84. Zhou L, Chong MM, Littman DR. Plasticity of CD4+ T cell lineage differentiation. *Immunity.* 2009;30:646-655.
85. Amsen D, Antov A, Flavell RA. The different faces of Notch in T-helper-cell differentiation. *Nat Rev Immunol.* 2009;9:116-124.
86. Zheng W, Flavell RA. The transcription factor GATA-3 is necessary and sufficient for Th2 cytokine gene expression in CD4 T cells. *Cell.* 1997;89:587-596.
87. Szabo SJ, Kim ST, Costa GL, Zhang X, Fathman CG, Glimcher LH. A novel transcription factor, T-bet, directs Th1 lineage commitment. *Cell.* 2000;100:655-669.
88. Ivanov II, McKenzie BS, Zhou L, et al. The orphan nuclear receptor RORgammat directs the differentiation program of proinflammatory IL-17+ T helper cells. *Cell.* 2006;126:1121-1133.
89. Reinhardt RL, Liang HE, Locksley RM. Cytokine-secreting follicular T cells shape the antibody repertoire. *Nat Immunol.* 2009;10:385-393.
90. Kusam S, Toney LM, Sato H, Dent AL. Inhibition of Th2 differentiation and GATA-3 expression by BCL-6. *J Immunol.* 2003;170:2435-2441.
91. Nurieva RI, Chung Y, Martinez GJ, et al. Bcl6 mediates the development of T follicular helper cells. *Science.* 2009;325:1001-1005.
92. Yu D, Rao S, Tsai LM, et al. The transcriptional repressor Bcl-6 directs T follicular helper cell lineage commitment. *Immunity.* 2009;31:457-468.
93. Hatzi K, Nance JP, Kroenke MA, et al. BCL6 orchestrates Tfh cell differentiation via multiple distinct mechanisms. *J Exp Med.* 2015;212:539-553.
94. Baumjohann D, Okada T, Ansel KM. Cutting edge: distinct waves of BCL6 expression during T follicular helper cell development. *J Immunol.* 2011;187:2089-2092.
95. Lund R, Ahlfors H, Kainonen E, Lahesmaa AM, Dixon C, Lahesmaa R. Identification of genes involved in the initiation of human Th1 or Th2 cell commitment. *Eur J Immunol.* 2005;35:3307-3319.
96. Nakayamada S, Kanno Y, Takahashi H, et al. Early Th1 cell differentiation is marked by a Tfh cell-like transition. *Immunity.* 2011;35:919-931.
97. Ballesteros-Tato A, Leon B, Graf BA, et al. Interleukin-2 inhibits germinal center formation by limiting T follicular helper cell differentiation. *Immunity.* 2012;36:847-856.
98. Fazilleau N, Mark L, McHeyzer-Williams LJ, McHeyzer-Williams MG. Follicular helper T cells: lineage and location. *Immunity.* 2009;30:324-335.
99. Lu KT, Kanno Y, Cannons JL, et al. Functional and epigenetic studies reveal multistep differentiation and plasticity of in vitro-generated and in vivo-derived follicular T helper cells. *Immunity.* 2011;35:622-632.
100. Ruterbusch M, Pruner KB, Shehata L, Pepper M. In vivo CD4(+) T cell differentiation and function: revisiting the Th1/Th2 paradigm. *Annu Rev Immunol.* 2020;38:705-725.
101. Trinchieri G, Pflanz S, Kastelein RA. The IL-12 family of heterodimeric cytokines: new players in the regulation of T cell responses. *Immunity.* 2003;19:641-644.

102. Thieu VT, Yu Q, Chang HC, et al. Signal transducer and activator of transcription 4 is required for the transcription factor T-bet to promote T helper 1 cell-fate determination. *Immunity*. 2008;29:679-690.
103. Zhu J, Yamane H, Paul WE. Differentiation of effector CD4 T cell populations. *Annu Rev Immunol*. 2010;28:445-489.
104. Oestreich KJ, Huang AC, Weinmann AS. The lineage-defining factors T-bet and Bcl-6 collaborate to regulate Th1 gene expression patterns. *J Exp Med*. 2011;208:1001-1013.
105. Lazarevic V, Glimcher LH. T-bet in disease. *Nat Immunol*. 2011;12:597-606.
106. Crotty S. T follicular helper cell differentiation, function, and roles in disease. *Immunity*. 2014;41:529-542.
107. Hogan SP, Mould AW, Young JM, et al. Cellular and molecular regulation of eosinophil trafficking to the lung. *Immunol Cell Biol*. 1998;76:454-460.
108. Kobayashi T, Iijima K, Dent AL, Kita H. Follicular helper T cells mediate IgE antibody response to airborne allergens. *J Allergy Clin Immunol*. 2017;139:300-313.e7.
109. Turner H, Kinet JP. Signalling through the high-affinity IgE receptor Fc epsilonRI. *Nature*. 1999;402:B24-B30.
110. Choi YS, Kageyama R, Eto D, et al. ICOS receptor instructs T follicular helper cell versus effector cell differentiation via induction of the transcriptional repressor Bcl6. *Immunity*. 2011;34:932-946.
111. Kato Y, Zaid A, Davey GM, et al. Targeting antigen to Clec9A primes follicular Th cell memory responses capable of robust recall. *J Immunol*. 2015;195:1006-1014.
112. Krishnaswamy JK, Gowthaman U, Zhang B, et al. Migratory CD11b(+) conventional dendritic cells induce T follicular helper cell-dependent antibody responses. *Sci Immunol*. 2017;2:eaam9169.
113. DiToro D, Winstead CJ, Pham D, et al. Differential IL-2 expression defines developmental fates of follicular versus nonfollicular helper T cells. *Science*. 2018;361:eaao2933.
114. Vogelzang A, McGuire HM, Yu D, Sprent J, Mackay CR, King C. A fundamental role for interleukin-21 in the generation of T follicular helper cells. *Immunity*. 2008;29:127-137.
115. Nurieva RI, Chung Y, Hwang D, et al. Generation of T follicular helper cells is mediated by interleukin-21 but independent of T helper 1, 2, or 17 cell lineages. *Immunity*. 2008;29:138-149.
116. Suto A, Kashiwakuma D, Kagami S, et al. Development and characterization of IL-21-producing CD4+ T cells. *J Exp Med*. 2008;205:1369-1379.
117. Ma CS, Tangye SG. Helping the helpers! *Immunity*. 2009;31:12-14.
118. Schmitt N, Morita R, Bourdery L, et al. Human dendritic cells induce the differentiation of interleukin-21-producing T follicular helper-like cells through interleukin-12. *Immunity*. 2009;31:158-169.
119. Meli AP, Fontes G, Avery DT, et al. The integrin LFA-1 controls T follicular helper cell generation and maintenance. *Immunity*. 2016;45:831-846.
120. Vinuesa CG, Linterman MA, Yu D, MacLennan IC. Follicular helper T cells. *Annu Rev Immunol*. 2016;34:335-368.
121. Xu H, Li X, Liu D, et al. Follicular T-helper cell recruitment governed by bystander B cells and ICOS-driven motility. *Nature*. 2013;496:523-527.
122. Foy TM, Laman JD, Ledbetter JA, Aruffo A, Claassen E, Noelle RJ. gp39-CD40 interactions are essential for germinal center formation and the development of B cell memory. *J Exp Med*. 1994;180:157-163.
123. Shulman Z, Gitlin AD, Weinstein JS, et al. Dynamic signaling by T follicular helper cells during germinal center B cell selection. *Science*. 2014;345:1058-1062.
124. Liu D, Xu H, Shih C, et al. T-B-cell entanglement and ICOSL-driven feed-forward regulation of germinal centre reaction. *Nature*. 2015;517:214-218.
125. Kim HP, Korn LL, Gamero AM, Leonard WJ. Calcium-dependent activation of interleukin-21 gene expression in T cells. *J Biol Chem*. 2005;280:25291-25297.
126. Ray JP, Staron MM, Shyer JA, et al. The interleukin-2-mTORc1 kinase Axis defines the signaling, differentiation, and metabolism of T helper 1 and follicular B helper T cells. *Immunity*. 2015;43:690-702.
127. Good-Jacobson KL, Szumilas CG, Chen L, Sharpe AH, Tomayko MM, Shlomchik MJ. PD-1 regulates germinal center B cell survival and the formation and affinity of long-lived plasma cells. *Nat Immunol*. 2010;11:535-542.
128. Kawamoto S, Tran TH, Maruya M, et al. The inhibitory receptor PD-1 regulates IgA selection and bacterial composition in the gut. *Science*. 2012;336:485-489.
129. Linterman MA, Beaton L, Yu D, et al. IL-21 acts directly on B cells to regulate Bcl-6 expression and germinal center responses. *J Exp Med*. 2010;207:353-363.
130. Zotos D, Coquet JM, Zhang Y, et al. IL-21 regulates germinal center B cell differentiation and proliferation through a B cell-intrinsic mechanism. *J Exp Med*. 2010;207:365-378.
131. Papa I, Saliba D, Ponzoni M, et al. TFH-derived dopamine accelerates productive synapses in germinal centres. *Nature*. 2017;547:318-323.
132. Sage PT, Paterson AM, Lovitch SB, Sharpe AH. The coinhibitory receptor CTLA-4 controls B cell responses by modulating T follicular helper, T follicular regulatory, and T regulatory cells. *Immunity*. 2014;41:1026-1039.
133. Wing JB, Ise W, Kurosaki T, Sakaguchi S. Regulatory T cells control antigen-specific expansion of Tfh cell number and humoral immune responses via the coreceptor CTLA-4. *Immunity*. 2014;41:1013-1025.
134. Chung Y, Tanaka S, Chu F, et al. Follicular regulatory T cells expressing Foxp3 and Bcl-6 suppress germinal center reactions. *Nat Med*. 2011;17:983-988.
135. Linterman MA, Pierson W, Lee SK, et al. Foxp3+ follicular regulatory T cells control the germinal center response. *Nat Med*. 2011;17:975-982.
136. Wollenberg I, Agua-Doce A, Hernandez A, et al. Regulation of the germinal center reaction by Foxp3+ follicular regulatory T cells. *J Immunol*. 2011;187:4553-4560.
137. Jacque E, Schweighoffer E, Visekruna A, et al. IKK-induced NF-kappaB1 p105 proteolysis is critical for B cell antibody responses to T cell-dependent antigen. *J Exp Med*. 2014;211:2085-2101.
138. Lee CH, Melchers M, Wang H, et al. Regulation of the germinal center gene program by interferon (IFN) regulatory factor 8/IFN consensus sequence-binding protein. *J Exp Med*. 2006;203:63-72.
139. Kwon K, Hutter C, Sun Q, et al. Instructive role of the transcription factor E2A in early B lymphopoiesis and germinal center B cell development. *Immunity*. 2008;28:751-762.
140. Vilagos B, Hoffmann M, Souabni A, et al. Essential role of EBF1 in the generation and function of distinct mature B cell types. *J Exp Med*. 2012;209:775-792.
141. Muto A, Tashiro S, Nakajima O, et al. The transcriptional programme of antibody class switching involves the repressor Bach2. *Nature*. 2004;429:566-571.
142. Vikstrom I, Carotta S, Luthje K, et al. Mcl-1 is essential for germinal center formation and B cell memory. *Science*. 2010;330:1095-1099.
143. Raaphorst FM, van Kemenade FJ, Fieret E, et al. Cutting edge: polycomb gene expression patterns reflect distinct B cell differentiation stages in human germinal centers. *J Immunol*. 2000;164:1-4.
144. Velichutina I, Shaknovich R, Geng H, et al. EZH2-mediated epigenetic silencing in germinal center B cells contributes to proliferation and lymphomagenesis. *Blood*. 2010;116:5247-5255.
145. Beguelin W, Popovic R, Teater M, et al. EZH2 is required for germinal center formation and somatic EZH2 mutations promote lymphoid transformation. *Cancer Cell*. 2013;23:677-692.
146. Weisel FJ, Mullett SJ, Elsner RA, et al. Germinal center B cells selectively oxidize fatty acids for energy while conducting minimal glycolysis. *Nat Immunol*. 2020;21:331-342.
147. Offit K, Lo Coco F, Louie DC, et al. Rearrangement of the bcl-6 gene as a prognostic marker in diffuse large-cell lymphoma. *N Engl J Med*. 1994;331:74-80.
148. Phan RT, Dalla-Favera R. The BCL6 proto-oncogene suppresses p53 expression in germinal-centre B cells. *Nature*. 2004;432:635-639.
149. Ranuncolo SM, Polo JM, Dierov J, et al. Bcl-6 mediates the germinal center B cell phenotype and lymphomagenesis through transcriptional repression of the DNA-damage sensor ATR. *Nat Immunol*. 2007;8:705-714.
150. Ranuncolo SM, Polo JM, Melnick A. BCL6 represses CHEK1 and suppresses DNA damage pathways in normal and malignant B-cells. *Blood Cells Mol Dis*. 2008;41:95-99.
151. Tunyaplin C, Shaffer AL, Angelin-Duclos CD, Yu X, Staudt LM, Calame KL. Direct repression of prdm1 by Bcl-6 inhibits plasmacytic differentiation. *J Immunol*. 2004;173:1158-1165.
152. Johnston RJ, Poholek AC, DiToro D, et al. Bcl6 and Blimp-1 are reciprocal and antagonistic regulators of T follicular helper cell differentiation. *Science*. 2009;325:1006-1010.
153. Hatzi K, Jiang Y, Huang C, et al. A hybrid mechanism of action for BCL6 in B cells defined by formation of functionally distinct complexes at enhancers and promoters. *Cell Rep*. 2013;4:578-588.
154. Ghetu AF, Corcoran CM, Cerchietti L, Bardwell VJ, Melnick A, Prive GG. Structure of a BCOR corepressor peptide in complex with the BCL6 BTB domain dimer. *Mol Cell*. 2008;29:384-391.
155. Gearhart MD, Corcoran CM, Wamstad JA, Bardwell VJ. Polycomb group and SCF ubiquitin ligases are found in a novel BCOR complex that is recruited to BCL6 targets. *Mol Cell Biol*. 2006;26:6880-6889.
156. Karagianni P, Wong J. HDAC3: taking the SMRT-N-CoRrect road to repression. *Oncogene*. 2007;26:5439-5449.
157. Beguelin W, Teater M, Gearhart MD, et al. EZH2 and BCL6 cooperate to assemble CBX8-BCOR complex to repress bivalent promoters, mediate germinal center formation and lymphomagenesis. *Cancer Cell*. 2016;30:197-213.
158. Chapuy B, Stewart C, Dunford AJ, et al. Molecular subtypes of diffuse large B cell lymphoma are associated with distinct pathogenic mechanisms and outcomes. *Nat Med*. 2018;24:679-690.
159. Schmitz R, Wright GW, Huang DW, et al. Genetics and pathogenesis of diffuse large B-cell lymphoma. *N Engl J Med*. 2018;378:1396-1407.
160. Venturutti L, Teater M, Zhai A, et al. TBL1XR1 mutations drive Extranodal lymphoma by inducing a pro-tumorigenic memory fate. *Cell*. 2020;182:297-316.e27.
161. Hatzi K, Geng H, Doane AS, et al. Histone demethylase LSD1 is required for germinal center formation and BCL6-driven lymphomagenesis. *Nat Immunol*. 2019;20:86-96.
162. Choi YS, Gullicksrud JA, Xing S, et al. LEF-1 and TCF-1 orchestrate T(FH) differentiation by regulating differentiation circuits upstream of the transcriptional repressor Bcl6. *Nat Immunol*. 2015;16:980-990.
163. Xu L, Cao Y, Xie Z, et al. The transcription factor TCF-1 initiates the differentiation of T(FH) cells during acute viral infection. *Nat Immunol*. 2015;16:991-999.
164. Shaw LA, Belanger S, Omilusik KD, et al. Id2 reinforces TH1 differentiation and inhibits E2A to repress TFH differentiation. *Nat Immunol*. 2016;17:834-843.
165. Choi J, Diao H, Faliti CE, et al. Bcl-6 is the nexus transcription factor of T follicular helper cells via repressor-of-repressor circuits. *Nat Immunol*. 2020;21:777-789.
166. Kuraoka M, Schmidt AG, Nojima T, et al. Complex antigens drive permissive clonal selection in germinal centers. *Immunity*. 2016;44:542-552.
167. Schwickert TA, Lindquist RL, Shakhar G, et al. In vivo imaging of germinal centres reveals a dynamic open structure. *Nature*. 2007;446:83-87.
168. Zhang Y, Meyer-Hermann M, George LA, et al. Germinal center B cells govern their own fate via antibody feedback. *J Exp Med*. 2013;210:457-464.
169. Finney J, Yeh CH, Kelsoe G, Kuraoka M. Germinal center responses to complex antigens. *Immunol Rev*. 2018;284:42-50.
170. Liu YJ, Zhang J, Lane PJ, Chan EY, MacLennan IC. Sites of specific B cell activation in primary and secondary responses to T cell-dependent and T cell-independent antigens. *Eur J Immunol*. 1991;21:2951-2962.
171. Le Gallou S, Zhou Z, Thai LH, et al. A splenic IgM memory subset with antibacterial specificities is sustained from persistent mucosal responses. *J Exp Med*. 2018;215:2035-2053.
172. Soni C, Wong EB, Domeier PP, et al. B cell-intrinsic TLR7 signaling is essential for the development of spontaneous germinal centers. *J Immunol*. 2014;193:4400-4414.
173. Domeier PP, Chodisetti SB, Soni C, et al. IFN-gamma receptor and STAT1 signaling in B cells are central to spontaneous germinal center formation and autoimmunity. *J Exp Med*. 2016;213:715-732.
174. MacLennan IC, Toellner KM, Cunningham AF, et al. Extrafollicular antibody responses. *Immunol Rev*. 2003;194:8-18.
175. Lee SK, Rigby RJ, Zotos D, et al. B cell priming for extrafollicular antibody responses requires Bcl-6 expression by T cells. *J Exp Med*. 2011;208:1377-1388.

176. Shaffer AL, Lin KI, Kuo TC, et al. Blimp-1 orchestrates plasma cell differentiation by extinguishing the mature B cell gene expression program. *Immunity*. 2002;17:51-62.
177. Horcher M, Souabni A, Busslinger M. Pax5/BSAP maintains the identity of B cells in late B lymphopoiesis. *Immunity*. 2001;14:779-790.
178. Wehrli N, Legler DF, Finke D, et al. Changing responsiveness to chemokines allows medullary plasmablasts to leave lymph nodes. *Eur J Immunol*. 2001;31:609-616.
179. Hargreaves DC, Hyman PL, Lu TT, et al. A coordinated change in chemokine responsiveness guides plasma cell movements. *J Exp Med*. 2001;194:45-56.
180. Garcia De Vinuesa C, Gulbranson-Judge A, Khan M, et al. Dendritic cells associated with plasmablast survival. *Eur J Immunol*. 1999;29:3712-3721.
181. Dal Porto JM, Haberman AM, Shlomchik MJ, Kelsoe G. Antigen drives very low affinity B cells to become plasmacytes and enter germinal centers. *J Immunol*. 1998;161:5373-5381.
182. Blink EJ, Light A, Kallies A, Nutt SL, Hodgkin PD, Tarlinton DM. Early appearance of germinal center-derived memory B cells and plasma cells in blood after primary immunization. *J Exp Med*. 2005;201:545-554.
183. Newman MA, Mainhart CR, Mallett CP, Lavoie TB, Smith-Gill SJ. Patterns of antibody specificity during the BALB/c immune response to hen eggwhite lysozyme. *J Immunol*. 1992;149:3260-3272.
184. Roost HP, Bachmann MF, Haag A, et al. Early high-affinity neutralizing anti-viral IgG responses without further overall improvements of affinity. *Proc Natl Acad Sci U S A*. 1995;92:1257-1261.
185. Foote J, Eisen HN. Kinetic and affinity limits on antibodies produced during immune responses. *Proc Natl Acad Sci U S A*. 1995;92:1254-1256.
186. Paus D, Phan TG, Chan TD, Gardam S, Basten A, Brink R. Antigen recognition strength regulates the choice between extrafollicular plasma cell and germinal center B cell differentiation. *J Exp Med*. 2006;203:1081-1091.
187. Mosier DE, Mond JJ, Goldings EA. The ontogeny of thymic independent antibody responses in vitro in normal mice and mice with an X-linked B cell defect. *J Immunol*. 1977;119:1874-1878.
188. Thomas JD, Sideras P, Smith CI, Vorechovsky I, Chapman V, Paul WE. Colocalization of X-linked agammaglobulinemia and X-linked immunodeficiency genes. *Science*. 1993;261:355-358.
189. Rawlings DJ, Saffran DC, Tsukada S, et al. Mutation of unique region of Bruton's tyrosine kinase in immunodeficient XID mice. *Science*. 1993;261:358-361.
190. Garcia de Vinuesa C, MacLennan IC, Holman M, Klaus GG. Anti-CD40 antibody enhances responses to polysaccharide without mimicking T cell help. *Eur J Immunol*. 1999;29:3216-3224.
191. Vos Q, Lees A, Wu ZQ, Snapper CM, Mond JJ. B-cell activation by T-cell-independent type 2 antigens as an integral part of the humoral immune response to pathogenic microorganisms. *Immunol Rev*. 2000;176:154-170.
192. Defrance T, Taillardet M, Genestier L. T cell-independent B cell memory. *Curr Opin Immunol*. 2011;23:330-336.
193. Marshall JL, Zhang Y, Pallan L, et al. Early B blasts acquire a capacity for Ig class switch recombination that is lost as they become plasmablasts. *Eur J Immunol*. 2011;41:3506-3512.
194. Perlmutter RM, Hansburg D, Briles DE, Nicolotti RA, Davie JM. Subclass restriction of murine anti-carbohydrate antibodies. *J Immunol*. 1978;121:566-572.
195. Barrett DJ, Ayoub EM. IgG2 subclass restriction of antibody to pneumococcal polysaccharides. *Clin Exp Immunol*. 1986;63:127-134.
196. Martin F, Oliver AM, Kearney JF. Marginal zone and B1 B cells unite in the early response against T-independent blood-borne particulate antigens. *Immunity*. 2001;14:617-629.
197. Fairfax KA, Corcoran LM, Pridans C, et al. Different kinetics of blimp-1 induction in B cell subsets revealed by reporter gene. *J Immunol*. 2007;178:4104-4111.
198. Smith KG, Hewitson TD, Nossal GJ, Tarlinton DM. The phenotype and fate of the antibody-forming cells of the splenic foci. *Eur J Immunol*. 1996;26:444-448.
199. Taillardet M, Haffar G, Mondiere P, et al. The thymus-independent immunity conferred by a pneumococcal polysaccharide is mediated by long-lived plasma cells. *Blood*. 2009;114:4432-4440.
200. Bortnick A, Chernova I, Quinn WJ III, Mugnier M, Cancro MP, Allman D. Long-lived bone marrow plasma cells are induced early in response to T cell-independent or T cell-dependent antigens. *J Immunol*. 2012;188:5389-5396.
201. Foote JB, Mahmoud TI, Vale AM, Kearney JF. Long-term maintenance of polysaccharide-specific antibodies by IgM-secreting cells. *J Immunol*. 2012;188:57-67.
202. Racine R, Chatterjee M, Winslow GM. CD11c expression identifies a population of extrafollicular antigen-specific splenic plasmablasts responsible for CD4 T-independent antibody responses during intracellular bacterial infection. *J Immunol*. 2008;181:1375-1385.
203. Vijay R, Guthmiller JJ, Sturtz AJ, et al. Infection-induced plasmablasts are a nutrient sink that impairs humoral immunity to malaria. *Nat Immunol*. 2020;21:790-801.
204. Obukhanych TV, Nussenzweig MC. T-independent type II immune responses generate memory B cells. *J Exp Med*. 2006;203:305-310.
205. Alugupalli KR, Leong JM, Woodland RT, Muramatsu M, Honjo T, Gerstein RM. B1b lymphocytes confer T cell-independent long-lasting immunity. *Immunity*. 2004;21:379-390.
206. Cunningham AF, Gaspal F, Serre K, et al. Salmonella induces a switched antibody response without germinal centers that impedes the extracellular spread of infection. *J Immunol*. 2007;178:6200-6207.
207. Di Niro R, Lee SJ, Vander Heiden JA, et al. Salmonella infection drives promiscuous B cell activation followed by extrafollicular affinity maturation. *Immunity*. 2015;43:120-131.
208. Shapiro ED, Berg AT, Austrian R, et al. The protective efficacy of polyvalent pneumococcal polysaccharide vaccine. *N Engl J Med*. 1991;325:1453-1460.
209. Ferraro AJ, Drayson MT, Savage CO, MacLennan IC. Levels of autoantibodies, unlike antibodies to all extrinsic antigen groups, fall following B cell depletion with Rituximab. *Eur J Immunol*. 2008;38:292-298.
210. Granoff DM, Gupta RK, Belshe RB, Anderson EL. Induction of immunologic refractoriness in adults by meningococcal C polysaccharide vaccination. *J Infect Dis*. 1998;178:870-874.
211. Granoff DM, Pollard AJ. Reconsideration of the use of meningococcal polysaccharide vaccine. *Pediatr Infect Dis J*. 2007;26:716-722.
212. O'Brien KL, Hochman M, Goldblatt D. Combined schedules of pneumococcal conjugate and polysaccharide vaccines: is hyporesponsiveness an issue? *Lancet Infect Dis*. 2007;7:597-606.
213. Kim M, Qie Y, Park J, Kim CH. Gut microbial metabolites fuel host antibody responses. *Cell Host Microbe*. 2016;20:202-214.
214. Zeng MY, Cisalpino D, Varadarajan S, et al. Gut microbiota-induced immunoglobulin G controls systemic infection by symbiotic bacteria and pathogens. *Immunity*. 2016;44:647-658.
215. Thurnheer MC, Zuercher AW, Cebra JJ, Bos NA. B1 cells contribute to serum IgM, but not to intestinal IgA, production in gnotobiotic Ig allotype chimeric mice. *J Immunol*. 2003;170:4564-4571.
216. New JS, Dizon BLP, Fucile CF, Rosenberg AF, Kearney JF, King RG. Neonatal exposure to commensal-bacteria-derived antigens directs polysaccharide-specific B-1 B cell repertoire development. *Immunity*. 2020;53:172-186.e6.
217. Wilmore JR, Gaudette BT, Gomez Atria D, et al. Commensal microbes induce serum IgA responses that protect against polymicrobial sepsis. *Cell Host Microbe*. 2018;23:302-311.e3.
218. Biram A, Shulman Z. T cell help to B cells: cognate and atypical interactions in peripheral and intestinal lymphoid tissues. *Immunol Rev*. 2020;296:36-47.
219. Mizoguchi A, Mizoguchi E, Takedatsu H, Blumberg RS, Bhan AK. Chronic intestinal inflammatory condition generates IL-10-producing regulatory B cell subset characterized by CD1d upregulation. *Immunity*. 2002;16:219-230.
220. Mauri C, Gray D, Mushtaq N, Londei M. Prevention of arthritis by interleukin 10-producing B cells. *J Exp Med*. 2003;197:489-501.
221. Fillatreau S, Sweenie CH, McGeachy MJ, Gray D, Anderton SM. B cells regulate autoimmunity by provision of IL-10. *Nat Immunol*. 2002;3:944-950.
222. Yanaba K, Bouaziz JD, Haas KM, Poe JC, Fujimoto M, Tedder TF. A regulatory B cell subset with a unique CD1dhiCD5+ phenotype controls T cell-dependent inflammatory responses. *Immunity*. 2008;28:639-650.
223. Blair PA, Norena LY, Flores-Borja F, et al. CD19(+)CD24(hi) CD38(hi) B cells exhibit regulatory capacity in healthy individuals but are functionally impaired in systemic Lupus Erythematosus patients. *Immunity*. 2010;32:129-140.
224. Matsumoto M, Baba A, Yokota T, et al. Interleukin-10-producing plasmablasts exert regulatory function in autoimmune inflammation. *Immunity*. 2014;41:1040-1051.
225. Shen P, Roch T, Lampropoulou V, et al. IL-35-producing B cells are critical regulators of immunity during autoimmune and infectious diseases. *Nature*. 2014;507:366-370.
226. Cretney E, Xin A, Shi W, et al. The transcription factors Blimp-1 and IRF4 jointly control the differentiation and function of effector regulatory T cells. *Nat Immunol*. 2011;12:304-311.
227. Fillatreau S. Regulatory plasma cells. *Curr Opin Pharmacol*. 2015;23:1-5.
228. Rauch PJ, Chudnovskiy A, Robbins CS, et al. Innate response activator B cells protect against microbial sepsis. *Science*. 2012;335:597-601.
229. Bermejo DA, Jackson SW, Gorosito-Serran M, et al. Trypanosoma cruzi trans-sialidase initiates a program independent of the transcription factors RORgammat and Ahr that leads to IL-17 production by activated B cells. *Nat Immunol*. 2013;14:514-522.
230. Karnowski A, Chevrier S, Belz GT, et al. B and T cells collaborate in antiviral responses via IL-6, IL-21, and transcriptional activator and coactivator, Oct2 and OBF-1. *J Exp Med*. 2012;209:2049-2064.
231. Barr TA, Shen P, Brown S, et al. B cell depletion therapy ameliorates autoimmune disease through ablation of IL-6-producing B cells. *J Exp Med*. 2012;209:1001-1010.
232. Reynolds AE, Kuraoka M, Kelsoe G. Natural IgM is produced by CD5- plasma cells that occupy a distinct survival niche in bone marrow. *J Immunol*. 2015;194:231-242.
233. Fagraeus A. The plasma cellular reaction and its relation to the formation of antibodies in vitro. *J Immunol*. 1948;58:1-13.
234. Hibi T, Dosch HM. Limiting dilution analysis of the B cell compartment in human bone marrow. *Eur J Immunol*. 1986;16:139-145.
235. Eyer K, Doineau RCL, Castrillon CE, et al. Single-cell deep phenotyping of IgG-secreting cells for high-resolution immune monitoring. *Nat Biotechnol*. 2017;35:977-982.
236. Manz RA, Thiel A, Radbruch A. Lifetime of plasma cells in the bone marrow. *Nature*. 1997;388:133-134.
237. Slifka MK, Antia R, Whitmire JK, Ahmed R. Humoral immunity due to long-lived plasma cells. *Immunity*. 1998;8:363-372.
238. Ahuja A, Anderson SM, Khalil A, Shlomchik MJ. Maintenance of the plasma cell pool is independent of memory B cells. *Proc Natl Acad Sci U S A*. 2008;105:4802-4807.
239. Hammarlund E, Thomas A, Amanna IJ, et al. Plasma cell survival in the absence of B cell memory. *Nat Commun*. 2017;8:1781.
240. DiLillo DJ, Hamaguchi Y, Ueda Y, et al. Maintenance of long-lived plasma cells and serological memory despite mature and memory B cell depletion during CD20 immunotherapy in mice. *J Immunol*. 2008;180:361-371.
241. Amanna IJ, Carlson NE, Slifka MK. Duration of humoral immunity to common viral and vaccine antigens. *N Engl J Med*. 2007;357:1903-1915.
242. Antia A, Ahmed H, Handel A, et al. Heterogeneity and longevity of antibody memory to viruses and vaccines. *PLoS Biol*. 2018;16:e2006601.
243. Sciammas R, Shaffer AL, Schatz JH, Zhao H, Staudt LM, Singh H. Graded expression of interferon regulatory factor-4 coordinates isotype switching with plasma cell differentiation. *Immunity*. 2006;25:225-236.
244. Nutt SL, Taubenheim N, Hasbold J, Corcoran LM, Hodgkin PD. The genetic network controlling plasma cell differentiation. *Semin Immunol*. 2011;23:341-349.
245. Lin KI, Angelin-Duclos C, Kuo TC, Calame K. Blimp-1-dependent repression of Pax-5 is required for differentiation of B cells to immunoglobulin M-secreting plasma cells. *Mol Cell Biol*. 2002;22:4771-4780.
246. Ochiai K, Maienschein-Cline M, Simonetti G, et al. Transcriptional regulation of germinal center B and plasma cell fates by dynamical control of IRF4. *Immunity*. 2013;38:918-929.

247. Wrammert J, Koutsonanos D, Li GM, et al. Broadly cross-reactive antibodies dominate the human B cell response against 2009 pandemic H1N1 influenza virus infection. *J Exp Med.* 2011;208:181-193.
248. Yoshida H, Matsui T, Yamamoto A, Okada T, Mori K. XBP1 mRNA is induced by ATF6 and spliced by IRE1 in response to ER stress to produce a highly active transcription factor. *Cell.* 2001;107:881-891.
249. Iwakoshi NN, Lee AH, Vallabhajosyula P, Otipoby KL, Rajewsky K, Glimcher LH. Plasma cell differentiation and the unfolded protein response intersect at the transcription factor XBP-1. *Nat Immunol.* 2003;4:321-329.
250. Jurkin J, Henkel T, Nielsen AF, et al. The mammalian tRNA ligase complex mediates splicing of XBP1 mRNA and controls antibody secretion in plasma cells. *EMBO J.* 2014;33:2922-2936.
251. Gaudette BT, Jones DD, Bortnick A, Argon Y, Allman D. mTORC1 coordinates an immediate unfolded protein response-related transcriptome in activated B cells preceding antibody secretion. *Nat Commun.* 2020;11:723.
252. Klein U, Casola S, Cattoretti G, et al. Transcription factor IRF4 controls plasma cell differentiation and class-switch recombination. *Nat Immunol.* 2006;7:773-782.
253. Kallies A, Hasbold J, Fairfax K, et al. Initiation of plasma-cell differentiation is independent of the transcription factor Blimp-1. *Immunity.* 2007;26:555-566.
254. Todd DJ, McHeyzer-Williams LJ, Kowal C, et al. XBP1 governs late events in plasma cell differentiation and is not required for antigen-specific memory B cell development. *J Exp Med.* 2009;206:2151-2159.
255. Taubenheim N, Tarlinton DM, Crawford S, Corcoran LM, Hodgkin PD, Nutt SL. High rate of antibody secretion is not integral to plasma cell differentiation as revealed by XBP-1 deficiency. *J Immunol.* 2012;189:3328-3338.
256. Benhamron S, Pattanayak SP, Berger M, Tirosh B. mTOR activation promotes plasma cell differentiation and bypasses XBP-1 for immunoglobulin secretion. *Mol Cell Biol.* 2015;35:153-166.
257. Landsverk OJ, Snir O, Casado RB, et al. Antibody-secreting plasma cells persist for decades in human intestine. *J Exp Med.* 2017;214:309-317.
258. Tokoyoda K, Egawa T, Sugiyama T, Choi BI, Nagasawa T. Cellular niches controlling B lymphocyte behavior within bone marrow during development. *Immunity.* 2004;20:707-718.
259. Hauser AE, Debes GF, Arce S, et al. Chemotactic responsiveness toward ligands for CXCR3 and CXCR4 is regulated on plasma blasts during the time course of a memory immune response. *J Immunol.* 2002;169:1277-1282.
260. Muehlinghaus G, Cigliano L, Huehn S, et al. Regulation of CXCR3 and CXCR4 expression during terminal differentiation of memory B cells into plasma cells. *Blood.* 2005;105:3965-3971.
261. Radbruch A, Muehlinghaus G, Luger EO, et al. Competence and competition: the challenge of becoming a long-lived plasma cell. *Nat Rev Immunol.* 2006;6:741-750.
262. Amanna IJ, Slifka MK. Mechanisms that determine plasma cell lifespan and the duration of humoral immunity. *Immunol Rev.* 2010;236:125-138.
263. Wang Y, Bhattacharya D. Adjuvant-specific regulation of long-term antibody responses by ZBTB20. *J Exp Med.* 2014;211:841-856.
264. Slifka MK, Amanna IJ. Role of multivalency and antigenic threshold in generating protective antibody responses. *Front Immunol.* 2019;10:956.
265. Kallies A, Hasbold J, Tarlinton DM, et al. Plasma cell ontogeny defined by quantitative changes in blimp-1 expression. *J Exp Med.* 2004;200:967-977.
266. Shi W, Liao Y, Willis SN, et al. Transcriptional profiling of mouse B cell terminal differentiation defines a signature for antibody-secreting plasma cells. *Nat Immunol.* 2015;16:663-673.
267. Chevrier S, Genton C, Kallies A, et al. CD93 is required for maintenance of antibody secretion and persistence of plasma cells in the bone marrow niche. *Proc Natl Acad Sci U S A.* 2009;106:3895-3900.
268. Pracht K, Meinzinger J, Daum P, et al. A new staining protocol for detection of murine antibody-secreting plasma cell subsets by flow cytometry. *Eur J Immunol.* 2017;47:1389-1392.
269. Wilmore JR, Jones DD, Allman D. Protocol for improved resolution of plasma cell subpopulations by flow cytometry. *Eur J Immunol.* 2017;47:1386-1388.
270. Halliley JL, Tipton CM, Liesveld J, et al. Long-lived plasma cells are contained within the CD19(−)CD38(hi) CD138(+) subset in human bone marrow. *Immunity.* 2015;43:132-145.
271. Mei HE, Wirries I, Frolich D, et al. A unique population of IgG-expressing plasma cells lacking CD19 is enriched in human bone marrow. *Blood.* 2015;125:1739-1748.
272. Belnoue E, Tougne C, Rochat AF, Lambert PH, Pinschewer DD, Siegrist CA. Homing and adhesion patterns determine the cellular composition of the bone marrow plasma cell niche. *J Immunol.* 2012;188:1283-1291.
273. Tangye SG. Staying alive: regulation of plasma cell survival. *Trends Immunol.* 2011;32:595-602.
274. O'Connor BP, Raman VS, Erickson LD, et al. BCMA is essential for the survival of long-lived bone marrow plasma cells. *J Exp Med.* 2004;199:91-98.
275. Peperzak V, Vikstrom I, Walker J, et al. Mcl-1 is essential for the survival of plasma cells. *Nat Immunol.* 2013;14:290-297.
276. Pengo N, Scolari M, Oliva L, et al. Plasma cells require autophagy for sustainable immunoglobulin production. *Nat Immunol.* 2013;14:298-305.
277. Nguyen DC, Garimalla S, Xiao H, et al. Factors of the bone marrow microniche that support human plasma cell survival and immunoglobulin secretion. *Nat Commun.* 2018;9:3698.
278. Lam WY, Becker AM, Kennerly KM, et al. Mitochondrial pyruvate import promotes long-term survival of antibody-secreting plasma cells. *Immunity.* 2016;45:60-73.
279. Wang H, Gonzalez-Garcia I, Traba J, et al. ATP-degrading ENPP1 is required for survival (or persistence) of long-lived plasma cells. *Sci Rep.* 2017;7:17867.
280. Lemke A, Kraft M, Roth K, Riedel R, Lammerding D, Hauser AE. Long-lived plasma cells are generated in mucosal immune responses and contribute to the bone marrow plasma cell pool in mice. *Mucosal Immunol.* 2016;9:83-97.
281. Blanc P, Moro-Sibilot L, Barthly L, et al. Mature IgM-expressing plasma cells sense antigen and develop competence for cytokine production upon antigenic challenge. *Nat Commun.* 2016;7:13600.
282. Pinto D, Montani E, Bolli M, et al. A functional BCR in human IgA and IgM plasma cells. *Blood.* 2013;121:4110-4114.
283. Belnoue E, Pihlgren M, McGaha TL, et al. APRIL is critical for plasmablast survival in the bone marrow and poorly expressed by early-life bone marrow stromal cells. *Blood.* 2008;111:2755-2764.
284. Winter O, Moser K, Mohr E, et al. Megakaryocytes constitute a functional component of a plasma cell niche in the bone marrow. *Blood.* 2010;116:1867-1875.
285. Chu VT, Frohlich A, Steinhauser G, et al. Eosinophils are required for the maintenance of plasma cells in the bone marrow. *Nat Immunol.* 2011;12:151-159.
286. Abe M, Kido S, Hiasa M, et al. BAFF and APRIL as osteoclast-derived survival factors for myeloma cells: a rationale for TACI-Fc treatment in patients with multiple myeloma. *Leukemia.* 2006;20:1313-1315.
287. McCarron MJ, Park PW, Fooksman DR. CD138 mediates selection of mature plasma cells by regulating their survival. *Blood.* 2017;129:2749-2759.
288. Minges Wols HA, Underhill GH, Kansas GS, Witte PL. The role of bone marrow-derived stromal cells in the maintenance of plasma cell longevity. *J Immunol.* 2002;169:4213-4221.
289. Rose-John S. Coordination of interleukin-6 biology by membrane bound and soluble receptors. *Adv Exp Med Biol.* 2001;495:145-151.
290. Rozanski CH, Arens R, Carlson LM, et al. Sustained antibody responses depend on CD28 function in bone marrow-resident plasma cells. *J Exp Med.* 2011;208:1435-1446.
291. Glatman Zaretsky A, Konradt C, Depis F, et al. T regulatory cells support plasma cell populations in the bone marrow. *Cell Rep.* 2017;18:1906-1916.
292. Rodriguez Gomez M, Talke Y, Goebel N, Hermann F, Reich B, Mack M. Basophils support the survival of plasma cells in mice. *J Immunol.* 2010;185:7180-7185.
293. Lindquist RL, Niesner RA, Hauser AE. In the right place, at the right time: spatiotemporal conditions determining plasma cell survival and function. *Front Immunol.* 2019;10:788.
294. Martin SW, Goodnow CC. Burst-enhancing role of the IgG membrane tail as a molecular determinant of memory. *Nat Immunol.* 2002;3:182-188.
295. Liu W, Meckel T, Tolar P, Sohn HW, Pierce SK. Intrinsic properties of immunoglobulin IgG1 isotype-switched B cell receptors promote microclustering and the initiation of signaling. *Immunity.* 2010;32:778-789.
296. Engels N, Konig LM, Heemann C, et al. Recruitment of the cytoplasmic adaptor Grb2 to surface IgG and IgE provides antigen receptor-intrinsic costimulation to class-switched B cells. *Nat Immunol.* 2009;10:1018-1025.
297. Dogan I, Bertocci B, Vilmont V, et al. Multiple layers of B cell memory with different effector functions. *Nat Immunol.* 2009;10:1292-1299.
298. Kenderes KJ, Levack RC, Papillion AM, Cabrera-Martinez B, Dishaw LM, Winslow GM. T-Bet(+) IgM memory cells generate multi-lineage effector B cells. *Cell Rep.* 2018;24:824-837 e823.
299. Krishnamurty AT, Thouvenel CD, Portugal S, et al. Somatically hypermutated plasmodium-specific IgM(+) memory B cells are rapid, plastic, early responders upon malaria rechallenge. *Immunity.* 2016;45:402-414.
300. Tomayko MM, Steinel NC, Anderson SM, Shlomchik MJ. Cutting edge: hierarchy of maturity of murine memory B cell subsets. *J Immunol.* 2010;185:7146-7150.
301. Taylor JJ, Pape KA, Jenkins MK. A germinal center-independent pathway generates unswitched memory B cells early in the primary response. *J Exp Med.* 2012;209:597-606.
302. Kaji T, Ishige A, Hikida M, et al. Distinct cellular pathways select germline-encoded and somatically mutated antibodies into immunological memory. *J Exp Med.* 2012;209:2079-2097.
303. Hao Y, O'Neill P, Naradikian MS, Scholz JL, Cancro MP. A B-cell subset uniquely responsive to innate stimuli accumulates in aged mice. *Blood.* 2011;118:1294-1304.
304. Rubtsov AV, Rubtsova K, Fischer A, et al. Toll-like receptor 7 (TLR7)-driven accumulation of a novel CD11c(+) B-cell population is important for the development of autoimmunity. *Blood.* 2011;118:1305-1315.
305. Rubtsova K, Rubtsov AV, van Dyk LF, Kappler JW, Marrack P. T-box transcription factor T-bet, a key player in a unique type of B-cell activation essential for effective viral clearance. *Proc Natl Acad Sci U S A.* 2013;110:E3216-E3224.
306. Kim CC, Baccarella AM, Bayat A, Pepper M, Fontana MF. FCRL5(+) memory B cells exhibit robust recall responses. *Cell Rep.* 2019;27:1446-1460.e4.
307. Moran I, Nguyen A, Khoo WH, et al. Memory B cells are reactivated in subcapsular proliferative foci of lymph nodes. *Nat Commun.* 2018;9:3372.
308. Adachi Y, Onodera T, Yamada Y, et al. Distinct germinal center selection at local sites shapes memory B cell response to viral escape. *J Exp Med.* 2015;212:1709-1723.
309. Allie SR, Bradley JE, Mudunuru U, et al. The establishment of resident memory B cells in the lung requires local antigen encounter. *Nat Immunol.* 2019;20:97-108.
310. Berek C, Milstein C. Mutation drift and repertoire shift in the maturation of the immune response. *Immunol Rev.* 1987;96:23-41.
311. Siekevitz M, Kocks C, Rajewsky K, Dildrop R. Analysis of somatic mutation and class switching in naive and memory B cells generating adoptive primary and secondary responses. *Cell.* 1987;48:757-770.
312. Pape KA, Taylor JJ, Maul RW, Gearhart PJ, Jenkins MK. Different B cell populations mediate early and late memory during an endogenous immune response. *Science.* 2011;331:1203-1207.
313. van Keimpema M, Gruneberg LJ, Mokry M, et al. The forkhead transcription factor FOXP1 represses human plasma cell differentiation. *Blood.* 2015;126:2098-2109.
314. Kometani K, Nakagawa R, Shinnakasu R, et al. Repression of the transcription factor Bach2 contributes to predisposition of IgG1 memory B cells toward plasma cell differentiation. *Immunity.* 2013;39:136-147.
315. Zuccarino-Catania GV, Sadanand S, Weisel FJ, et al. CD80 and PD-L2 define functionally distinct memory B cell subsets that are independent of antibody isotype. *Nat Immunol.* 2014;15:631-637.
316. McHeyzer-Williams LJ, Milpied PJ, Okitsu SL, McHeyzer-Williams MG. Class-switched memory B cells remodel BCRs within secondary germinal centers. *Nat Immunol.* 2015;16:296-305.
317. Purtha WE, Tedder TF, Johnson S, Bhattacharya D, Diamond MS. Memory B cells, but not long-lived plasma cells, possess antigen specificities for viral escape mutants. *J Exp Med.* 2011;208:2599-2606.

318. Francis T. On the doctrine of original antigenic sin. *Proc Am Philos Soc.* 1960;104:572-578.
319. Ma J, Dushoff J, Earn DJ. Age-specific mortality risk from pandemic influenza. *J Theor Biol.* 2011;288:29-34.
320. Lessler J, Riley S, Read JM, et al. Evidence for antigenic seniority in influenza A (H3N2) antibody responses in southern China. *PLoS Pathog.* 2012;8:e1002802.
321. Petrie JG, Parkhouse K, Ohmit SE, Malosh RE, Monto AS, Hensley SE. Antibodies against the current influenza A(H1N1) vaccine strain do not protect some individuals from infection with contemporary circulating influenza A(H1N1) virus strains. *J Infect Dis.* 2016;214:1947-1951.
322. Ellebedy AH, Nachbagauer R, Jackson KJL, et al. Adjuvanted H5N1 influenza vaccine enhances both cross-reactive memory B cell and strain-specific naive B cell responses in humans. *Proc Natl Acad Sci U S A.* 2020;117:17957-17964.
323. Scheid JF, Mouquet H, Feldhahn N, et al. Broad diversity of neutralizing antibodies isolated from memory B cells in HIV-infected individuals. *Nature.* 2009;458:636-640.
324. Corti D, Langedijk JP, Hinz A, et al. Analysis of memory B cell responses and isolation of novel monoclonal antibodies with neutralizing breadth from HIV-1-infected individuals. *PLoS One.* 2010;5:e8805.
325. Kwong PD, Mascola JR. Human antibodies that neutralize HIV-1: identification, structures, and B cell ontogenies. *Immunity.* 2012;37:412-425.
326. Liao HX, Lynch R, Zhou T, et al. Co-evolution of a broadly neutralizing HIV-1 antibody and founder virus. *Nature.* 2013;496:469-476.
327. Bonsignori M, Wiehe K, Grimm SK, et al. An autoreactive antibody from an SLE/HIV-1 individual broadly neutralizes HIV-1. *J Clin Invest.* 2014;124:1835-1843.
328. Schur PH. Human gamma-g subclasses. *Prog Clin Immunol.* 1972;1:71-104.
329. Klein U, Rajewsky K, Kuppers R. Human immunoglobulin (Ig)M+IgD+ peripheral blood B cells expressing the CD27 cell surface antigen carry somatically mutated variable region genes: CD27 as a general marker for somatically mutated (memory) B cells. *J Exp Med.* 1998;188:1679-1689.
330. Glass DR, Tsai AG, Oliveria JP, et al. An integrated multi-omic single-cell Atlas of human B cell identity. *Immunity.* 2020;53:217-232.e5.
331. Sanz I, Wei C, Jenks SA, et al. Challenges and opportunities for consistent classification of human B cell and plasma cell populations. *Front Immunol.* 2019;10:2458.
332. Ehrhardt GR, Hsu JT, Gartland L, et al. Expression of the immunoregulatory molecule FcRH4 defines a distinctive tissue-based population of memory B cells. *J Exp Med.* 2005;202:783-791.
333. Ehrhardt GR, Hijikata A, Kitamura H, Ohara O, Wang JY, Cooper MD. Discriminating gene expression profiles of memory B cell subpopulations. *J Exp Med.* 2008;205:1807-1817.
334. Moir S, Ho J, Malaspina A, et al. Evidence for HIV-associated B cell exhaustion in a dysfunctional memory B cell compartment in HIV-infected viremic individuals. *J Exp Med.* 2008;205:1797-1805.
335. Portugal S, Tipton CM, Sohn H, et al. Malaria-associated atypical memory B cells exhibit markedly reduced B cell receptor signaling and effector function. *Elife.* 2015;4:e07218.
336. Ellebedy AH, Jackson KJ, Kissick HT, et al. Defining antigen-specific plasmablast and memory B cell subsets in human blood after viral infection or vaccination. *Nat Immunol.* 2016;17:1226-1234.
337. Jenks SA, Cashman KS, Zumaquero E, et al. Distinct effector B cells induced by unregulated Toll-like receptor 7 contribute to pathogenic responses in systemic lupus erythematosus. *Immunity.* 2018;49:725-739.e6.
338. Holla P, Ambegaonkar A, Sohn H, Pierce SK. Exhaustion may not be in the human B cell vocabulary, at least not in malaria. *Immunol Rev.* 2019;292:139-148.
339. Weller S, Braun MC, Tan BK, et al. Human blood IgM "memory" B cells are circulating splenic marginal zone B cells harboring a prediversified immunoglobulin repertoire. *Blood.* 2004;104:3647-3654.
340. Bagnara D, Squillario M, Kipling D, et al. A reassessment of IgM memory subsets in humans. *J Immunol.* 2015;195:3716-3724.
341. Weller S, Faili A, Garcia C, et al. CD40-CD40L independent Ig gene hypermutation suggests a second B cell diversification pathway in humans. *Proc Natl Acad Sci U S A.* 2001;98:1166-1170.
342. Descatoire M, Weller S, Irtan S, et al. Identification of a human splenic marginal zone B cell precursor with NOTCH2-dependent differentiation properties. *J Exp Med.* 2014;211:987-1000.
343. Seifert M, Kuppers R. Molecular footprints of a germinal center derivation of human IgM+(IgD+)CD27+ B cells and the dynamics of memory B cell generation. *J Exp Med.* 2009;206:2659-2669.
344. Budeus B, Schweigle de Reynoso S, Przekopowitz M, Hoffmann D, Seifert M, Kuppers R. Complexity of the human memory B-cell compartment is determined by the versatility of clonal diversification in germinal centers. *Proc Natl Acad Sci U S A.* 2015;112:E5281-E5289.
345. Weller S, Bonnet M, Delagreverie H, et al. IgM+IgD+CD27+ B cells are markedly reduced in IRAK-4-, MyD88-, and TIRAP- but not UNC-93B-deficient patients. *Blood.* 2012;120:4992-5001.
346. Kruetzmann S, Rosado MM, Weber H, et al. Human immunoglobulin M memory B cells controlling *Streptococcus pneumoniae* infections are generated in the spleen. *J Exp Med.* 2003;197:939-945.
347. Lucas AH, Moulton KD, Tang VR, Reason DC. Combinatorial library cloning of human antibodies to Streptococcus pneumoniae capsular polysaccharides: variable region primary structures and evidence for somatic mutation of Fab fragments specific for capsular serotypes 6B, 14, and 23F. *Infect Immun.* 2001;69:853-864.
348. Smith K, Muther JJ, Duke AL, et al. Fully human monoclonal antibodies from antibody secreting cells after vaccination with Pneumovax(R)23 are serotype specific and facilitate opsonophagocytosis. *Immunobiology.* 2013;218:745-754.
349. Zhao Y, Uduman M, Siu JHY, et al. Spatiotemporal segregation of human marginal zone and memory B cell populations in lymphoid tissue. *Nat Commun.* 2018;9:3857.
350. Weill JC, Weller S, Reynaud CA. A bird's eye view on human B cells. *Semin Immunol.* 2004;16:277-281.
351. Heit A, Schmitz F, Gerdts S, et al. Vaccination establishes clonal relatives of germinal center T cells in the blood of humans. *J Exp Med.* 2017;214:2139-2152.
352. Locci M, Havenar-Daughton C, Landais E, et al. Human circulating PD-1+CXCR3-CXCR5+ memory Tfh cells are highly functional and correlate with broadly neutralizing HIV antibody responses. *Immunity.* 2013;39:758-769.
353. Morita R, Schmitt N, Bentebibel SE, et al. Human blood CXCR5(+)CD4(+) T cells are counterparts of T follicular cells and contain specific subsets that differentially support antibody secretion. *Immunity.* 2011;34:108-121.
354. He J, Tsai LM, Leong YA, et al. Circulating precursor CCR7(lo)PD-1(hi) CXCR5(+) CD4(+) T cells indicate Tfh cell activity and promote antibody responses upon antigen reexposure. *Immunity.* 2013;39:770-781.
355. Rao DA, Gurish MF, Marshall JL, et al. Pathologically expanded peripheral T helper cell subset drives B cells in rheumatoid arthritis. *Nature.* 2017;542:110-114.
356. Rao DA. T cells that help B cells in chronically inflamed tissues. *Front Immunol.* 2018;9:1924.

CHAPTER 27

T Cell Development in the Thymus

Avinash Bhandoola, MB, BS, PhD • Remy Bosselut, MD, PhD • Yousuke Takahama, PhD

OVERVIEW OF T CELL DEVELOPMENT

T cells, etymologically thymus-derived cells, play essential roles in immune responses. They express T cell receptors (TCRs), which consist of invariant CD3γ, δ, ε, and ζ subunits complexed with highly variable TCRαβ or γδ heterodimers. During their development, immature T cell precursors and TCR-expressing cells are often referred to as thymocytes, reflecting their residence in the thymus.

Like other blood cells, T cells develop from hematopoietic stem cells (HSCs) that reside in the bone marrow of adult mice or humans. Unlike other blood cells, however, precursors of T cells must migrate to the thymus for their differentiation. This chapter will present our current knowledge of intrathymic T cell development and repertoire selection. The mechanisms by which T cell development is largely constrained to the thymus will also be considered.

The requirement for cell migration to the thymus imposes a bottleneck, and only small numbers of precursors are thought to undertake this journey. Multiple precursor types are thought to contribute to T cell development, including T lineage–restricted precursors and multipotential lymphoid precursors that can generate many distinct types of blood cells. Thus, T cell development must begin with the imposition of T cell identity on multipotential intrathymic precursors, along with enormous proliferation to generate the numbers of T lineage–committed cells needed to generate a diverse T cell repertoire. T cell precursors in the αβ lineage must also gain expression of CD4 and CD8 coreceptors, which play key roles in T cell recognition of peptides on MHC II and MHC I molecules, respectively. Such CD4 and CD8 co-expressing cells subsequently finish assembling the T cell antigen receptor (TCR), and those that express a functional αβ TCR subsequently undergo intensive assessment and screening of their TCR specificities via processes termed positive and negative selection. It is only after these quality control steps are successfully completed that T cells are permitted to leave the thymus, to perform their critical immune functions in secondary lymphoid organs and tissues (Fig. 27.1).

In this chapter, we divide T cell development into discrete steps: **thymus settling; lineage commitment and receptor assembly**; and **selection** (Fig. 27.1). Thymus settling encompasses the mechanisms by which T competent progenitors leave their sites of generation, to migrate to and settle within the thymus. Lineage commitment and receptor assembly are distinct but overlapping processes that concern the mechanisms by which precursors express T lineage genes (specification), lose the ability to generate cell types other than T cells (commitment), and elaborate the T cell–specific gene network that includes expression of TCR gene loci and rearrangement of TCR gene segments, followed by assembly of the multisubunit TCR. Selection comprises processes that promote that survival of thymocytes that assemble functioning TCRs (TCRβ selection and positive selection) and eliminate thymocytes that assemble autoreactive (negative selection) or nonfunctional TCRs (death by neglect). Additionally, some lineages of T cells require strong signals for maturation, and the term "agonist selection" is used for the selection of regulatory T cells (T_{regs}) and innate-like T cell lineages. The development of T_{regs}, invariant natural killer T (iNKT) cells, mucosal-associated invariant T (MAIT) cells, and γδ T cells, the DNA rearrangement events critical for antigen receptor assembly, and the signaling events essential for TCR function are presented in separate sections of this textbook.

T cell development is critically dependent on multiple thymic microenvironments composed of the epithelial cells of the thymus and stepwise relocation of developing thymocytes through the thymus. To perform this essential function, thymic epithelial cells (TECs) manufacture key molecules that support T cell precursors and also display the peptide-MHC ligands involved in selection and thus shape and restrict the T cell repertoire. Because of the many distinct steps in T cell development, a diversity of TECs, as well as additional cells such as dendritic cells (DCs), is thought to be required. Initial steps of T cell development occur in the thymic cortex and are supported by cortical TECs (cTECs), whereas subsequent events occur in the medulla and are directed by medullary TECs (mTECs). Indeed, the thymus is properly considered a series of linked microenvironments made of distinct types of TECs that support T cell survival and selection at corresponding distinct stages of development. This chapter will commence with an **outline of thymus architecture** and **thymus development** to orient the process of T cell development and will later provide a discussion of TEC subsets and factors influencing subset differentiation and function, focusing on the **thymic medulla**.

Defects in TECs underlie many perturbations in T cell development that can result in immunodeficiency and/or autoimmunity. Further, age-related alterations in the thymus have profound consequences for TCR repertoire and immunity in aged humans; and damage to TECs from infection, radiotherapy, and chemotherapeutic drugs can also profoundly impact T cell development. In recent years, significant progress has been made in understanding the diversity of TECs, particularly mTECs, and we have expanded the section on TECs to include these advances.

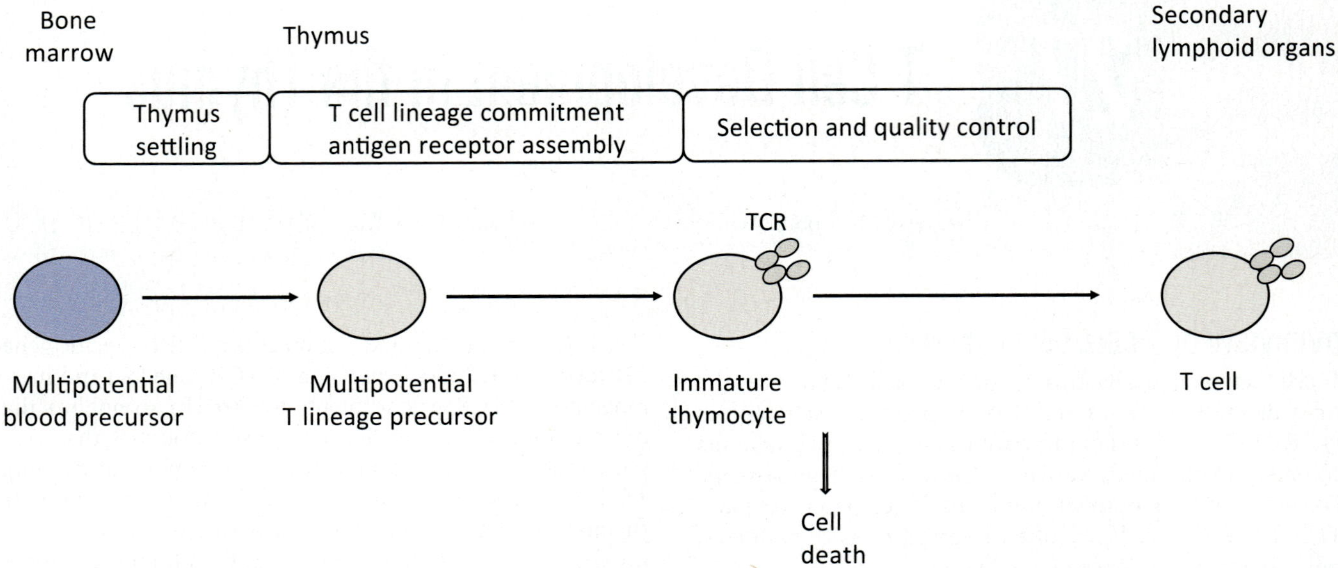

FIG. 27.1. Steps of T cell development, which are thymus settling; lineage commitment and receptor assembly; and selection. The first steps occur in the outer thymic cortex, whereas the last steps of selection take place in the inner medulla.

OUTLINE OF THYMUS ARCHITECTURE

In many mammalian species including human and mouse, the thymus consists of two lobes and is located in the anterosuperior mediastinum, on the pericardium and the great vessels, and behind the sternum. At birth, the human thymus measures approximately 5 cm in length, 4 cm in width, and 1 cm in thickness and weighs 10 to 15 g. The thymus grows in size during childhood to reach a maximum weight of 20 to 30 g during adolescence. With the onset of puberty, the thymus begins to diminish its size and its parenchyma is gradually replaced with adipose tissue, a process called thymic involution. By the age of 70, the human thymus weighs as little as 5 g and is hardly distinguishable from the surrounding fatty tissue.[1,2]

In addition to the thoracic thymus, small pieces of the thymic parenchyma can be detected in the cervical region. The cervical thymus is similar in microscopic morphology to the thoracic thymus and is thought to originate from thymic epithelial progenitor cells during their downward migration from the third pharyngeal pouch to the thoracic cavity.[3-5]

The thymic lobe is composed of many lobules surrounded by a mesenchymal capsule (Fig. 27.2A). The parenchyma of postnatal thymus contains two major microenvironments, the cortex and the medulla (Fig. 27.2B). The thymic cortex contains a high density of small lymphocytes, that is, thymocytes, surrounded by a network of cTECs, which appear as large, stretched, and extended cellular bodies (Fig. 27.3). A fraction of cTECs, called thymic nurse cells, envelop many thymocytes. Macrophages and DCs are also found in the cortex.

The thymic medulla is generally localized to the middle of the thymic parenchyma and often surrounded by the thymic cortex. mTECs are the primary component of the thymic medulla. A fraction of mTECs form concentric structures called Hassall corpuscles (Fig. 27.4). DCs and mesenchymal fibroblasts are enriched in the medullary region, whereas thymocytes are at a lower density in the medulla than the cortex. Macrophages are also found in the thymic medulla. Most thymocytes in the medulla are more mature than the majority of thymocytes in the cortex, in agreement with the notion that thymocytes migrate from the cortex to the medulla during T cell development.

Blood vessels are detected throughout the thymic parenchyma in the cortex and the medulla but are enriched in the medullary region and the area between the cortex and the medulla, called the corticomedullary junction. Some vascular structures in the postnatal thymus are surrounded by a perivascular space between endothelial cells and mesenchymal cells. The perivascular space is thought to allow the entrance and/or the exit of developing T cells between the thymic parenchyma and the circulation.[6] The thymus is also supplied with lymphatic vessels and neurons. Thus, the thymus contains hematopoietic cells, including developing T cells, DCs, and macrophages, and nonhematopoietic cells, including cTECs, mTECs, mesenchymal cells, endothelial cells, and nerve cells.

The corticomedullary structure of the thymic parenchyma constitutes the majority of the thymus during newborn and adolescent stages, but is replaced with adipose tissue postpuberty. Although the adult human thymus still contains thymic parenchyma, its corticomedullary structure tends to be unclear (Fig. 27.2C).

THYMUS DEVELOPMENT

Thymus Organogenesis

The thymus is derived from the third pharyngeal pouch, which forms during embryonic development of vertebrates. During embryogenesis, the thymic primordium is formed from endodermal epithelial cells at the posterior-ventral

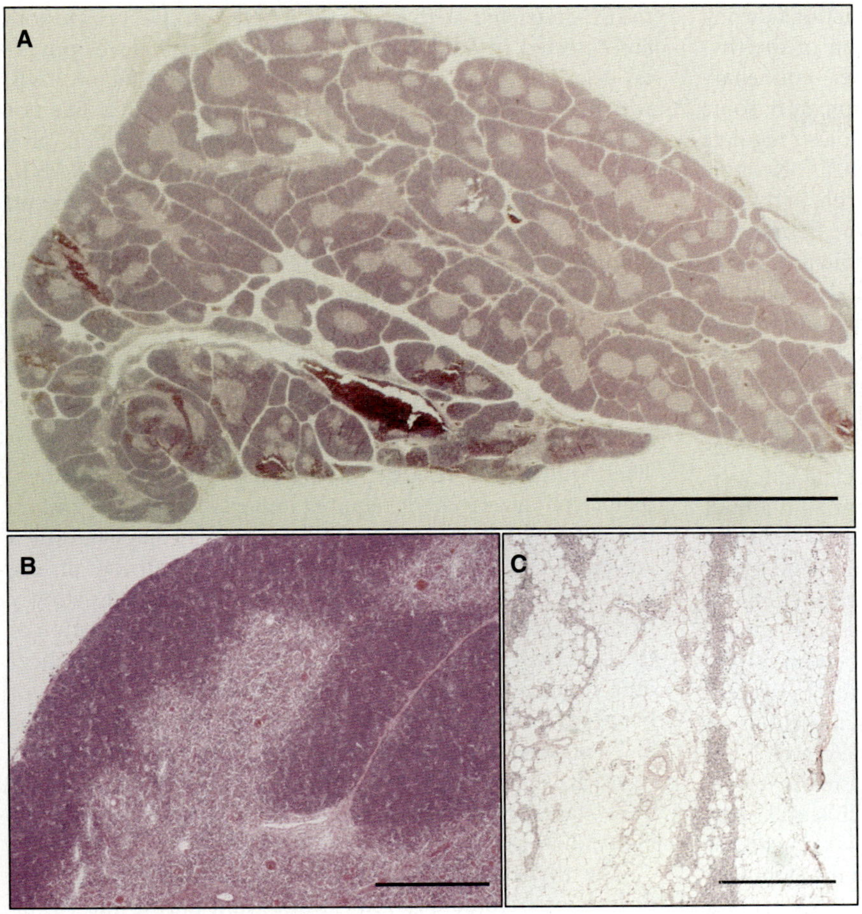

FIG. 27.2. Thymus architecture and involution in human. A: Thymic section of a 13-day-old infant, showing lobular structure. **B:** Thymic section of a 2-day-old infant, showing corticomedullary structure. **C:** Thymic section of a 55-year-old man, showing thymic involution. Hematoxylin and eosin stain. Scale bars, 5 mm **(A)**, 0.5 mm **(B and C)**. (Photos taken by Yasushi Nakagawa and Naoko Matsui.)

aspect of the third pharyngeal pouch, through interactions with mesenchymal cells of neural crest origin.[7,8] The primordium of the thymus is detectable by embryonic day 11 in mice and the 5th or 6th week of human embryogenesis. The transcription factor Tbx1 is necessary for the development of pharyngeal organs including the thymus, so that a defect in Tbx1 is associated with the onset of the DiGeorge syndrome, a congenital immunodeficiency often characterized with T cell deficiency.[9,10] Other transcription factors, including Hoxa3, Eya1, Six1, Pax1, and Pax9, also play a role in thymus primordium formation.[11-15] The thymic primordium expresses BMP4, a bone morphogenetic protein family member that

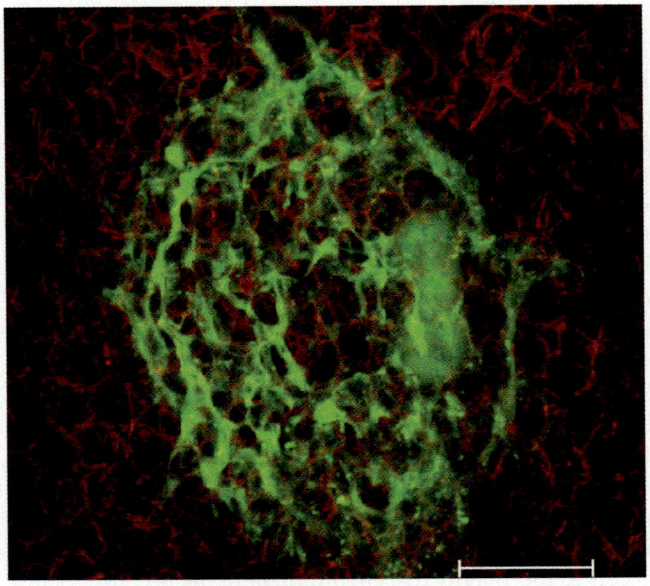

FIG. 27.3. Cortical thymic epithelial cell (*green*) and thymocytes (*red*) in mouse thymic cortex. Scale bar, 17.3 μm.

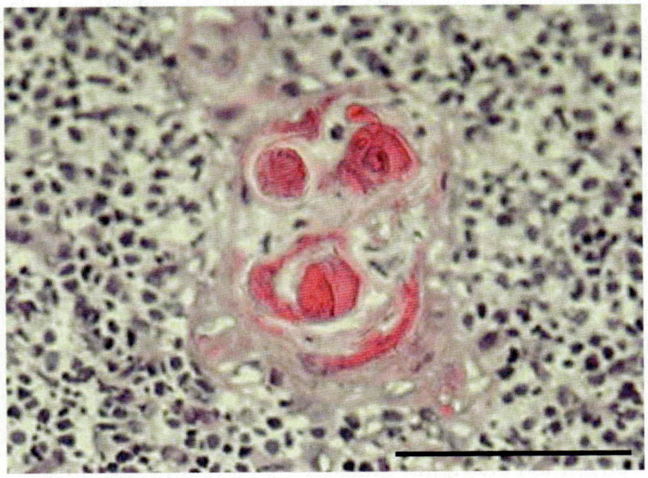

FIG. 27.4. Hassall corpuscle detected in human thymic medulla. Hematoxylin and eosin stained section of a 2-week-old infant. Scale bar, 0.1 mm. (Photo taken by Hirotsugu Kurobe.)

is important for epithelial–mesenchymal interactions during early thymus morphogenesis and the migration of the thymus to the thoracic mediastinum.[16,17] Other epithelial–mesenchymal signaling molecules, including ephrin B, sonic hedgehog, Wnts, and fibroblast growth factors, also regulate thymus organogenesis and thoracic migration.[18-22] Retinoic acid produced by neural crest cells also plays a role in thymus primordium formation.[23,24] The formation of the thymic primordium is followed by the seeding of T lymphoid progenitor cells derived from HSCs. Endodermal epithelial cells of the thymic primordium but not neighboring parathyroid primordium express the forkhead family transcription factor Foxn1, which is required for the differentiation and proliferation of TECs.[25,26] Congenital deficiency in Foxn1 results in a hairless, or "nude," phenotype and a severe defect in thymus development, which causes a severe combined immunodeficiency.[27,28] Foxn1 in TECs binds to a variety of genes essential to attract T cell precursors to the thymus, foster their early development, and select the T cell repertoire.[29] Mutation of a Foxn1-binding site in the vicinity of one of these genes (*Psmb11*, encoding for the proteosomal β5t subunit that is selectively expressed in cTECs) greatly decreased its expression, establishing that *Psmb11* is directly regulated by Foxn1.[30] A recent study reported that compound heterozygous mutations in Foxn1 cause selective thymic hypoplasia and defective T cell production with normal hair development in humans.[31]

In jawless vertebrates such as lampreys, which lack T cells, lymphoepithelial structures termed the thymoids, located at the tip of the gill filaments, are the site for the development of T-like VLRA$^+$ and VLRC$^+$ cells (see Chapter 4). Like TECs in jawed vertebrates, epithelial cells in the thymoid express Foxn1.[32,33]

Embryonic and Adult TEC Progenitors

Both cTECs and mTECs are derived from common endodermal progenitor cells, which are generated in a Foxn1-dependent manner during embryogenesis.[34] The common TEC progenitors progress through a developmental stage that expresses cTEC-associated molecules, including β5t, CD205, and high levels of IL-7.[35-37] Thus, bipotent TEC progenitors acquire hallmarks of the cTEC lineage, before their commitment to the mTEC lineage in a stepwise manner during initial thymus cortex and medulla formation.[38]

Bipotent TEC progenitors that give rise to cTECs and mTECs are detected in the embryonic and postnatal thymus.[39] However, more than 90% of mTECs in the adult thymus are derived from embryonic and neonatal β5t$^+$ progenitors, whereas adult β5t$^+$ progenitors play only a minor role in the maintenance of mTECs in the adult thymus. In contrast, more than 30% of cTECs in the adult thymus appear newly generated in adult life.[40]

The thymic medulla originates from islets of mTEC progenitors.[41] mTECs in the adult thymus exhibit active proliferation and rapid turnover,[42,43] so that mTEC lineage-committed cells likely maintain adult mTEC subpopulations via active development and/or proliferation throughout life. Indeed, SSEA-1$^+$ (stage-specific embryonic antigen 1, CD15) mTEC-restricted stem cells and RANK$^+$ mTEC progenitors are detected in adult mice,[44,45] although their developmental relationship is unclear. The transcription factor RelB, which is essential for the development of mTECs but not cTECs, is crucial for the generation of RANK$^+$ mTEC progenitors.[45] TNF family cytokine signals via RANKL-RANK, CD40L-CD40, and LT-LTβR promote further development, proliferation, and diversification of mTEC subpopulations, as discussed later in this chapter. Aire$^+$ mTECs are highly dependent on RANKL-RANK signals, whereas CCL21$^+$ mTECs are dependent on LT-LTβR signals.[46-49]

THYMUS SETTLING BY T CELL PRECURSORS

Prethymic Precursors of T Cells

The thymus lacks resident T lymphoid stem cells, and normal T cell development requires importation of progenitors throughout life.[50-52] Thymus settling is mediated by two different pathways: the vasculature-independent pathway during early embryogenesis and the vasculature-dependent pathway in late embryogenesis and postnatally, after vascularization of the thymus. Vasculature-independent colonization of embryonic thymus is regulated by the chemotactic attraction of lymphoid progenitor cells to the thymic primordium. Steps of T cell development and the cells participating in these processes are diagrammed in Figure 27.5, and the relationship of these stages to the anatomy of the thymus is diagrammed in Figure 27.6.

The thymus is settled by lymphoid progenitor cells as early as embryonic day 11.5 in mice and during the 8th or 9th week of gestation in humans.[53,54] Progenitors competent to migrate to the thymus are generated downstream of HSCs and multipotent progenitors (MPPs) in the bone marrow of adult mice or liver of fetal mice.[55] They include cells (1) within the lymphoid-primed multipotent progenitor (LMPP) compartment that possess myeloid and DC potential but not erythroid potential in addition to their potent lymphoid potential[56,57] and (2) within the common lymphoid progenitor (CLP) compartment[58] that possess efficient lymphoid and DC potential and retain residual myeloid potential.[59]

Within the thymus, the earliest recognized T cell progenitor populations are transcriptionally CD4$^-$ CD8$^-$ and so are termed double-negative (DN) cells[55,60] (Fig. 27.5). The earliest known T cell precursors residing within the thymus are termed early thymic progenitors (ETPs); these correspond to the small (<10%) Kit^{hi} fraction of the DN1 (CD44$^+$ CD25$^-$) compartment. In adult mice, ETPs arise by proliferation of the very small numbers of thymus settling progenitors that respond to the inductive environment of the thymus.[61] The source and nature of these thymus settling progenitors is not known with certainty.[55,62,63] Many progenitors in addition to LMPPs and CLPs can make T cells if placed in a T-inductive environment—this includes HSCs and MPPs, as well as some myeloid–erythroid precursors (common myeloid progenitors). *In vivo*, however, most precursors are unable to gain access to the T-inductive environment of the thymus because they lack expression of some or all of the molecules needed for migration to the organ.[55]

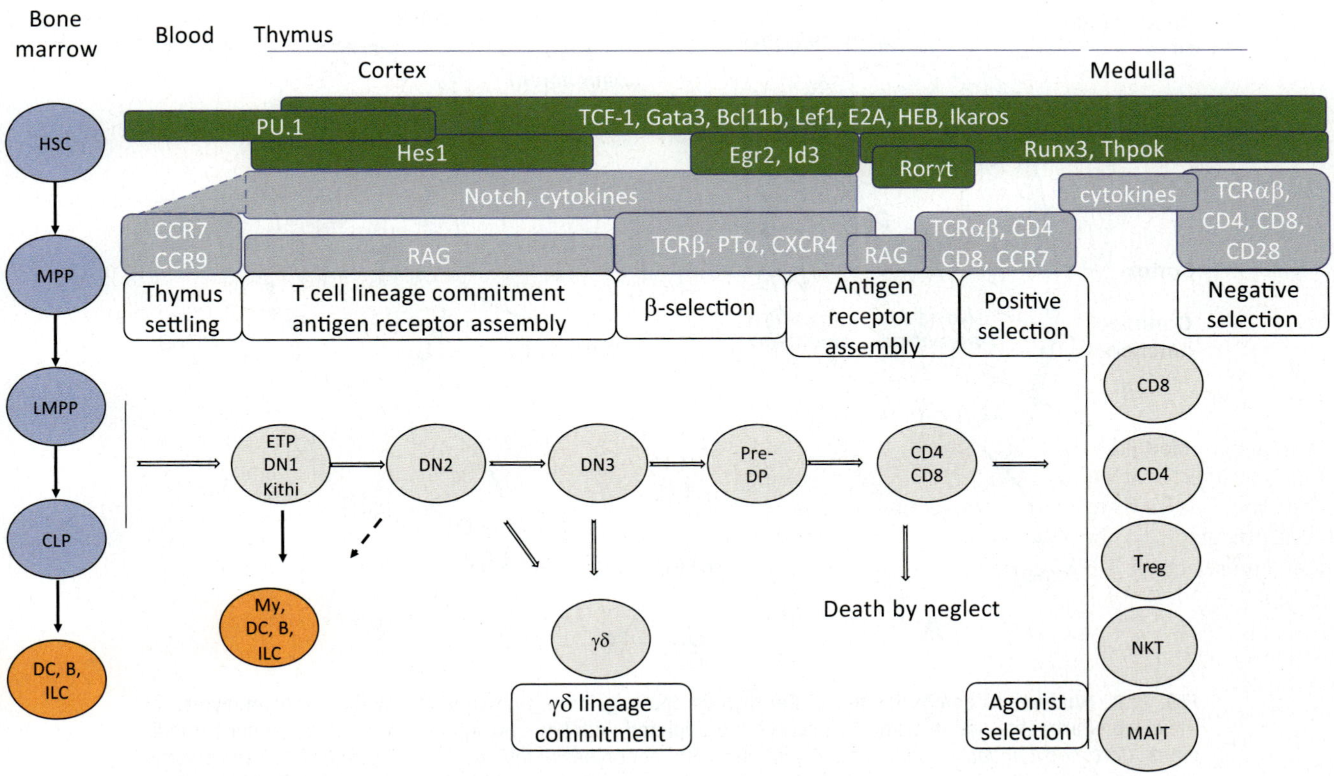

FIG. 27.5. Cartoon of the major types of thymocytes, critical structural molecules in *gray*, and major transcription factors in *green*. B, B cell; CLP, common lymphoid progenitor; DC, dendritic cell; DN, double-negative or CD4⁻ CD8⁻; ILC, innate lymphoid cell; HSC, hematopoietic stem cell; LMPP, lymphoid-primed multipotential progenitor; MAIT, mucosal-associated invariant T cell; MPP, multipotential progenitor; My, Myeloid cell; NKT, natural killer T cell; pre-DP, pre–double-positive ($CD4^+$ $CD8^+$); RAG, recombination-activating gene; T_{reg}, regulatory T cell; TCR, T cell receptor. This figure focuses on early phases of thymocyte development. As discussed in the text, negative selection is not limited to the medulla, but occurs in the cortex for some developing thymocytes.

Migration to the Thymus

To accomplish thymic entry, progenitors must leave the generative site, which includes the liver in embryos and the bone marrow in adults in humans and mice. Whether mobilization into the blood from these sites of origin is physiologically regulated *in vivo* is unknown. In fetal mice, relatively high frequencies of circulating progenitors are evident,[64] but circulating progenitors are very rare in adult mice.[65-67] After formation of the vasculature, precursors must enter through thymic blood vessels; progenitor entry has been suggested to occur from relatively large blood vessels near the corticomedullary junction.[68] Progenitors that circulate into these thymic vessels are thought to be recruited into the thymus in a manner analogous to the multistep adhesion cascade utilized by lymphocytes to access tissues, which is rolling on endothelial cells using selectin and selectin ligands; chemokine and chemokine receptor signaling resulting in rapid activation of integrins and arrest of cells; and extravasation.[69,70]

The entry of lymphoid progenitor cells into the thymus is an intermittent and gated event that occurs in waves during embryogenesis and in adulthood.[71,72] In fetal as well as adult mice, progenitors must express either of the chemokine receptors CCR7 or CCR9 to settle the thymus.[73-75] These chemokine receptors are expressed on subsets of LMPPs and CLPs but not on HSCs or MPPs in adult mice.[74-77] Additionally, the majority of ETPs express low levels of *Il7r* but are lineage-traced in *IL7rCre* mice (in which Cre is under the control of the *Il7r* gene that encodes IL-7Rα and permanently marks progeny in suitable experimental systems). This result indicates that ETPs likely derive from IL-7Rα⁺ CLPs or from subsets of LMPPs that express IL-7Rα.[58,59,78-80] Accordingly, cells within these precursor populations are shown to settle within the thymus in short-term homing assays, indicating that these two closely related progenitor populations likely contain the physiological thymus settling progenitors.[62,77,81,82]

After vascularization of the developing thymus, thymic entry additionally requires functional P-selectin glycoprotein ligand PSGL-1, highly expressed on LMPPs and CLPs.[83] PSGL-1 is thought to mediate the initial rolling of precursors on thymic endothelium expressing P-selectin.[84] After the initiation of rolling, CCR7 is thought to be activated by chemokine ligands CCL21 and CCL19 primarily manufactured by mTECs, whereas CCR9 is activated by CCL25 produced mainly by cTECs.[85-87] In fetal life, CCL21 is instead manufactured by the parathyroid primordium, and CCL25 is manufactured by the thymic primordium.[73] Additionally, a role for CXCL12 binding CXCR4 on progenitors is also established.[88] After chemokine signaling, integrins $\alpha_4\beta_1$ and $\alpha_L\beta_2$, expressed on precursors, and their respective ligands VCAM-1 and ICAM-1, are implicated in firm adhesion of progenitors to vascular endothelium, followed by diapedesis.[69]

Although many of these thymus settling requirements are shared between fetal and adult prethymic progenitors, the

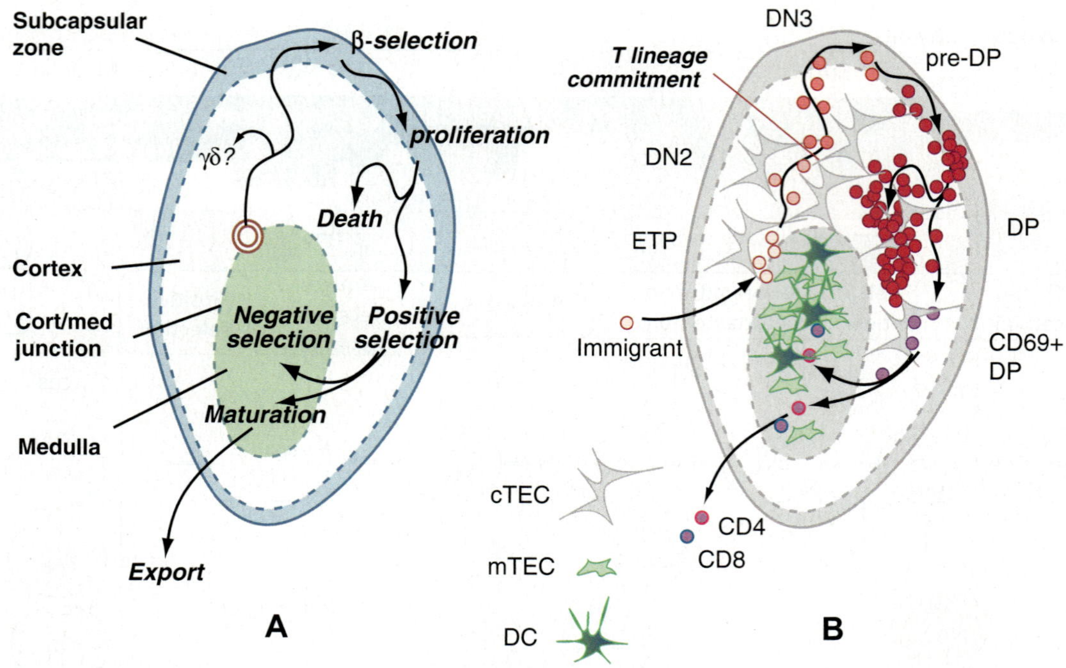

FIG. 27.6. Anatomical compartments of the thymus and migration pathways of developing thymocytes. A: Major compartments and migration pathways are depicted with key checkpoints in T cell development indicated. *Cort/med junction*, corticomedullary junction. **B:** Phenotypes of thymocytes (*pink*, *red*, *violet*) and stromal cell types in different compartments. cTEC, cortical thymic epithelial cell; DC, dendritic cell; mTEC, medullary thymic epithelial cell.

functional properties of precursors for T cells differ during ontogeny. In particular, there is a shift in developmental potential to give rise to different V-region–expressing γδ T cells during ontogeny, as described in Chapter 36; and some fetal progenitors are committed to the T and NK cell lineages prior to entry within the thymus[64,89] (although lineage potential for additional innate lymphoid cell types could not be addressed in these early experiments). Whether this commitment to T-NK lineages occurs by mechanisms similar to those by which T lineage commitment occurs intrathymically (discussed later in this chapter) is not known with certainty.[90,91]

Development of Lymphocyte Progenitors

The development of lymphoid progenitors competent to settle the thymus from HSCs is poorly understood. Labeling of HSCs in the bone marrow niche in adult mice (using an inducible Cre recombinase) reveals development of labeled erythroid and myeloid cells several weeks before the generation of lymphoid progenitors and subsequently B cells and T cells.[92] The reason for the long delay in generation of lymphoid progenitors from HSCs is unknown. The generation of CLPs depends on transcription factors including Runx factors, Ikaros, PU.1, and E2A; and CLPs fail to develop or develop poorly, when any of these transcription factors are experimentally mutated.[93-98] The generation of T-competent lymphoid progenitors also requires RBPJ, which is the transcription factor involved in Notch signaling. RBPJ is required to generate subsets of LMPPs capable of adopting the T cell fate, suggesting requirements for Notch signaling prethymically, in the adult bone marrow,[91] in addition to the well-established requirements for Notch signaling in intrathymic T cell development discussed below.

LINEAGE COMMITMENT AND RECEPTOR ASSEMBLY

Intrathymic Early T Cell Precursors

The DN1 term refers to CD4⁻ CD8⁻ (DN), CD44⁺ CD25⁻ cells. Although some researchers use the ETP and DN1 terms interchangeably, we use ETP here to specifically refer to the Kit^{hi} subset of DN1 cells.[99] ETPs are the most efficient progenitors of T cells within the mouse thymus, as they make the greatest number of T lineage progeny when injected into the thymus of recipient mice, whereas the remaining DN1 cells do not contain efficient T cell progenitors but rather are a heterogeneous mixture of mature and immature cells.[98,100]

ETPs arise by proliferation of small numbers of precursors settling the thymus. The chief inductive signals driving this considerable proliferation are Notch signals and cytokine receptor signals provided by IL-7R and Kit. Although there are four mammalian Notch genes and five Notch ligands, only Notch1 and the Notch ligand Delta-like 4 (Dll4) are essential for T cell development. Dll4, as well as cytokines IL-7 and stem cell factor (SCF, Kit ligand), are expressed by vascular endothelial cells and cTECs;[101,102] whereas Notch1 is expressed by multiple primitive hematopoietic precursors, including LMPPs and CLPs.[95]

Notch1 and Dll4 are essential at the earliest steps of intrathymic T cell development, and in the absence of either, thymus settling precursors adopt B cell fates and other non–T cell fates including differentiation to DCs.[103-108] Indeed, ETPs have the ability to generate non–T cell lineages (Fig. 27.7), including B

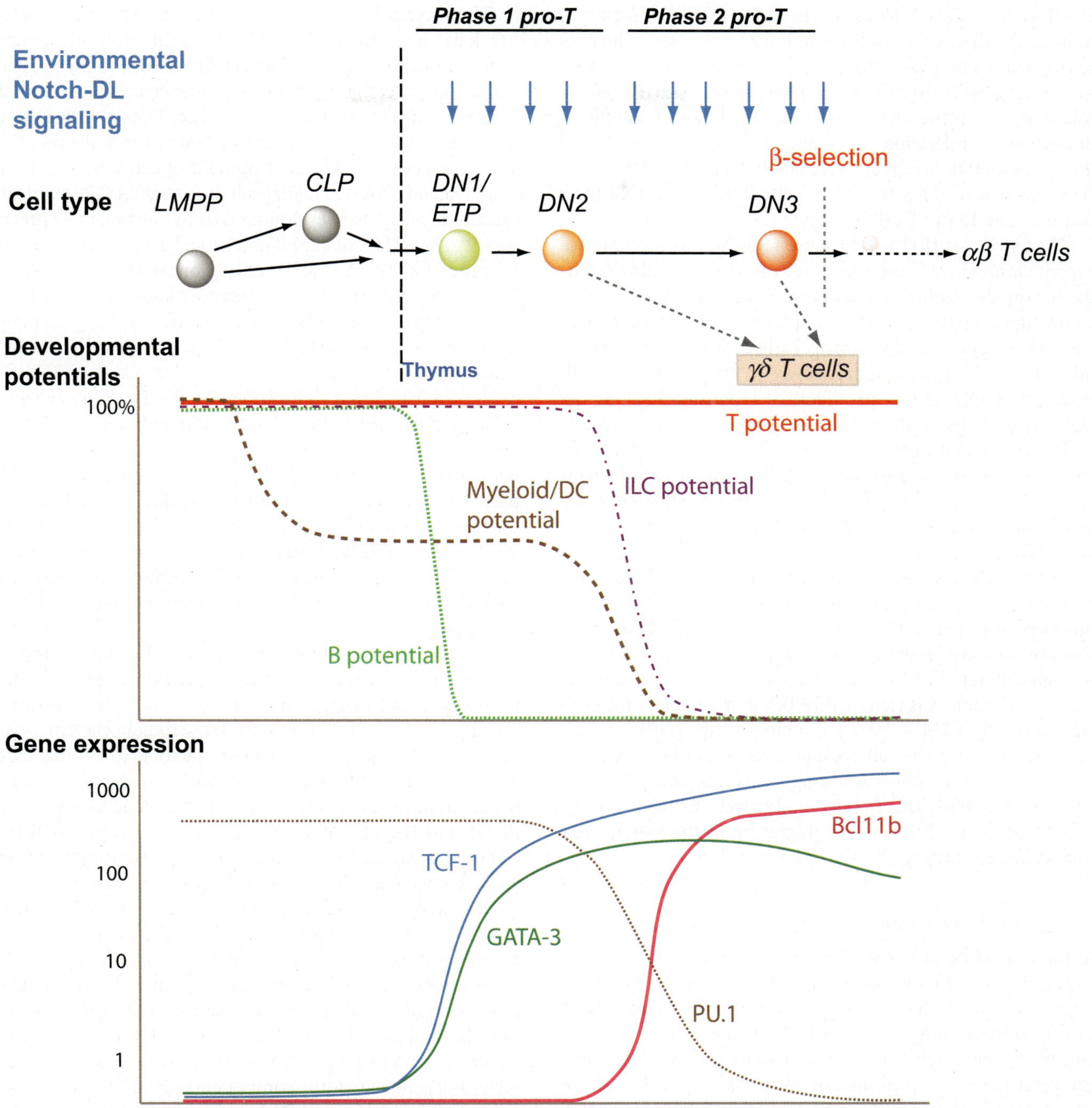

FIG. 27.7. T lineage commitment and expression patterns of contributing factors. Top panel: Prethymic and intrathymic stages involved in T lineage commitment. The thymus can be seeded by cells that retain myeloid potential (lymphoid-primed multipotent precursor) as well as by cells that have already reduced their access to this option (common lymphoid precursor). As shown in the figure, Notch-Delta signals are delivered to the cells repeatedly from their time of entry into the thymus until β-selection. **Middle panel:** Approximate times at which distinct cell-lineage options for T cell precursors are excluded. Y axis represents remaining ability of T cell precursor cells to enter alternative pathway at the indicated stages (relative units). **Bottom panel:** Expression patterns of key regulatory genes noted in the text. Factors used in the T cell program are depicted with *solid lines*, while PU.1, a factor used in multipotent progenitor, myeloid, and B cell programs, is shown with a *dotted line*. (Top and middle adapted from figures in Rothenberg EV. T cell lineage commitment: identity and renunciation. *J Immunol.* 2011;186(12):6649-6655; Copyright 2011. The American Association of Immunologists, Inc. and Rothenberg EV, Zhang J, Li L. Multilayered specification of the T-cell lineage fate. *Immunol Rev.* 2010;238(1):150-168.)

cells, DCs, myeloid cell types such as macrophages and granulocytes, and innate lymphoid cells including NK cells and ILC1, ILC2, and ILC3.[98,112-121] The non–T cell lineage potentials are readily evident when ETPs are placed in *in vitro* culture or if Notch signals are withdrawn experimentally,[120,122-124] indicating that Notch signaling constrains non–T cell lineage potentials during early T cell development.

ETPs generate myeloid cells in the adult thymus; however, the majority of DCs within the thymus do not arise from ETPs, and their origin is poorly understood.[79,119,120,125,126]

Whether non–T cell lineages arising from ETPs have functions in the thymus is not known. Functions of such lineages in the thymus may be distinct from their canonical extrathymic functions; in particular, it has been suggested that the role of B cells in the thymus is to provide B-specific antigens important for tolerizing developing T cells.[127-129] Non–T cell lineage potentials are greatly reduced in DN2 ($CD44^+$ $CD25^+$) cells and lost in DN3 ($CD44^{lo}$ $CD25^+$) cells, reflecting their commitment to the T cell lineage.[130]

The cytokines IL-7 and SCF are required for proliferative expansion of early T cell precursors, and IL-7 is also essential for the development of some γδ lineage T cells.[101,102,131-136] Upon signaling by IL-7, SCF, and Dll4, ETPs generate DN2 and DN3 stages rapidly during early fetal development in mice, but considerably more slowly in the adult, and with correspondingly greater proliferative expansion.[116,137] At the DN3 stage, T-committed cells undertake rearrangement of TCRβ, TCRγ, and TCRδ loci (see Chapter 24).[138,139] DN3 cells also downregulate expression of CCR7, which is important because re-expression of CCR7 guides corticomedullary migration at subsequent stages of T cell development, as discussed later in this chapter.

Thymocyte development from the ETP to the DN3 stage promotes the development of cTECs to form the cortical microenvironment.[140,141] Concomitantly, DN thymocytes relocate outward from the corticomedullary junction to the subcapsular region of the thymic cortex.[68] The subcapsular region of the cortex is rich in TGF-β, which regulates the generation of DP ($CD4^+$ $CD8^+$ or double-positive [DP]) thymocytes by impeding the cell cycle progression of DN4 ($CD44^{lo}$ $CD25^-$) cells and immature single-positive (mostly $CD8^+$ $CD4^-$ in C57BL/6 mice) cells, collectively termed pre-DP cells.[142] However, this outward migration to the subcapsular zone is not essential for T cell development.[143]

Mechanisms of T Lineage Specification

Notch Signaling in Early T Cell Precursors

Experimentally, loss of function of Notch1 results in a very early block in T cell development, with the concomitant accumulation of large numbers of B lineage cells in the thymus.[105,106] Conversely, forced Notch signaling in hematopoietic precursors, accomplished by retroviral expression of the active intracellular portion of Notch1, allows early steps of T cell development to proceed outside the thymus and at the expense of B lineage cells.[144] These early experiments established that signaling mediated by Notch1 is necessary and sufficient to impose the T cell fate on primitive hematopoietic precursors.

Notch signaling persists from thymic entry through commitment to the T cell lineage, which occurs between the DN2 and DN3 stages. Notch signaling continues in T-committed DN3 cells and then is muted at the DP stage concomitant with downregulated expression of Notch1 in DP thymocytes. Thus, Notch signals are not persistent in developing T cells; instead, Notch signaling triggers a pattern of T lineage gene expression that persists after Notch signals are withdrawn.[145]

The mechanisms by which Notch1 imposes T cell identity have been intensely studied. Notch molecules recruit coactivators to the transcription factor RBPJ; these coactivators include the p300 histone acetyltransferase and the Mastermind-like proteins.[146,147] Some Notch target genes including *Myc*, require dimers of Notch for activation.[148] A key innovation was the development of cell lines expressing Notch ligands. Mesenchymal cell lines such as OP9 support B cell and myeloid cell development from bone marrow precursors.[149] The enforced expression of Delta-like 1 or Delta-like 4 ligands for Notch in OP9 cells allowed such OP9-DL (for OP9-Delta) cells to guide hematopoietic precursors into the T cell lineage.[150] The OP9/OP9-DL system is efficient enough to allow the *ex vivo* study of single precursor cells. An early study of single precursors from fetal liver, plated on OP9-DL cells, established that commitment to the T cell lineage was a prolonged process, because precursors cultured on OP9-DL stromal cells for several days could yet commit to non–T cell fates if they were removed from these cultures and instead placed in cultures with OP9 cells lacking Notch ligands.[123] Further, the intensity of Notch signals needed to inhibit different non–T cell fates was different between fates, because the addition of low doses of Notch inhibitors still permitted OP9-DL cells to inhibit B cell development but not NK cell development.[130]

The transcriptional program imposed by Notch signaling has been elucidated in recent years, providing further insight into why T cell lineage commitment can be so protracted. T cell precursors placed on OP9-DL stromal cells upregulate expression of many transcription factors, including Hes1, TCF-1, Gata3, and Bcl11b.[123] In addition, Notch signaling results in upregulated expression of TCR gene loci; genes for RAG1 and RAG2 that are necessary to rearrange TCR loci; CD3 subunits that are essential components of the αβTCR and γδTCR; together with signaling molecules such as Lck that are required for transducing TCR signals.[145] Among the transcription factors upregulated by Notch1, TCF-1 is sufficient to impose the early T cell program when forcibly expressed in mouse bone marrow progenitors. This includes upregulated expression of the other T cell transcription factors Gata3 and Bcl11b; Lef-1, another sequence-specific HMG transcription factor in the same family as TCF-1; and structural T cell genes including genes encoding components of the TCR and related signaling molecules.[151-153] Of these factors, Gata3 enforces loss of B cell lineage potential early in T cell development.[154] The ability of TCF-1 to activate expression of Bcl11b remains evident when it is ectopically expressed in mouse fibroblasts, indicating that the ability of TCF-1 to mediate some aspects of T lineage specification does not strictly require a chromatin context specific for T cell precursors.[155] However, TCF-1 normally collaborates with other factors expressed in T cell precursors including Gata3, Runx, and Notch1, using a chromatin landscape derived from upstream lymphoid progenitors to activate expression of Bcl11b.[156]

T Cell Lineage Commitment

Bcl11b expression occurs at the DN2 to DN3 transition. Prior to Bcl11b expression, a noncoding RNA termed ThymoD that is transcribed from a region downstream of the Bcl11b gene repositions the Bcl11b locus away from the nuclear lamina into a transcriptionally permissive environment and brings the Bcl11b promoter and enhancer into the same chromosome loop.[157] Bcl11b induction correlates well with commitment to the T cell lineage and with large-scale reorganization of genomes in T cell precursors.[158] Expression of Bcl11b requires prolonged Notch signaling over many days in adult progenitors (Fig. 27.7) and, further, has a stochastic component, because the timing of initial expression can vary between the two alleles of Bcl11b in the same cell.[159] Because T cell lineage commitment requires Bcl11b,[160,161] the requirement for prolonged Notch signaling in Bcl11b upregulation can explain why prolonged Notch signaling is needed for T cell lineage commitment. Further, the requirement for prolonged Notch signaling provides a potential explanation for the observation that development of T lineage committed precursors is largely restricted to the thymus, even though Notch ligands can be abundant at extrathymic sites. Thus, precursors may not persist at extrathymic sites of Notch ligand expression for long enough to receive the needed duration of Notch signaling, as they do within the thymus.

Together with the upregulated expression of Hes1, TCF-1, Gata3, and Bcl11b, development from the ETP to DN3 stages is accompanied by eventual downregulated expression of transcription factors highly expressed in stem cells and myeloid progenitors, such as PU.1, Lmo2, Lyl1, Hhex, and SCL (Tal1) (Fig. 27.7). This constrained expression of stem and progenitor genes is likely to be important for normal T cell development, as sustained expression of these factors is associated with leukemic transformation.[162-164] Recent work indicates that after thymic entry, expression of some transcription factors like PU.1 may be initially further increased prior to eventual downregulation. The reasons for this increase are unknown, but the consequence may be lineage instability in early precursors toward non–T cell fates such as myeloid and DC lineages.[165,166] A role for Hes1 in constraining myeloid transcriptional programs is evident, because requirements for Hes1 in T cell development can be overcome by removing the myeloid transcription factor C/EBPα.[167] Other transcription factors important at this transition include Runx1, Ets-1, Myb, Gfi1, Lyl-1, and the E proteins E2A and HEB.[94,145,168-174] E2A is important for expression of *Rag* genes in prethymic T cell as well as B cell precursors, among other roles played by this versatile transcription factor.[95,175,176]

Expression of TCF-1, Gata3, and Bcl11b is shared not only between mouse and human T cell precursors[177] but also between T cell precursors and precursors of ILCs.[93] These latter observations suggest the establishment of a shared chromatin landscape in early development that may underlie some of the functional similarities between T cells and ILCs,[178] but they also raise the question of how T cell and ILC lineages diverge. One insight comes from the expression of Id2 in ILCs and ILC precursors, but not in early T cell precursors. Because E proteins bind DNA as dimers, the transcriptional regulatory activity of E proteins is effectively antagonized by Id proteins that heterodimerize with E proteins but lack DNA binding domains.[175,176] Expression of Id2 is normally antagonized by Bcl11b in T cell precursors, providing one insight into how Bcl11b acts to enforce T cell lineage commitment.[161] Experimental removal of Bcl11b from committed T cell precursors allows upregulated expression of Id2 and access to NK cell fates, which is a type of innate lymphocyte.[160] These results had initially suggested that the choice between innate and adaptive lymphocyte fates might be mediated by the expression of Id factors. Consistently, forced expression of Id proteins or removal of E proteins even in committed T cell precursors allows them to become innate lymphoid cells.[179,180] However, the converse experiment does not give the predicted result, that is, loss of Id2 is not sufficient to allow innate lymphocyte precursors to access adaptive lymphoid cell fates, even when they are exposed to T-inductive Notch signaling.[181,182] Additionally, Bcl11b and Id2 are co-expressed in another innate lymphoid cell type, ILC2, indicating that Bcl11b does not always repress Id2. Interestingly, Bcl11b binds to distinct chromatin regions and regulates distinct gene targets in T cells and ILC2, indicating that T cell precursors and ILC precursors diverge independently of the expression of Bcl11b.[183] The differences in early transcriptional regulatory events between T cell and ILC precursors that enable these distinct functions for Bcl11b are an active area of research.

αβ LINEAGE COMMITMENT

The TCRβ Selection Checkpoint

The Pre-TCR

Assembly of the αβ TCR is accomplished in stages. Although limited Dβ-Jβ TCR gene rearrangements are evident at the ETP stage,[184] these partial rearrangements do not encode functional antigen receptors, which require V to DJ rearrangement. By the DN3 stage, however, efficient recombination of Vβ segments to the joined Dβ-Jβ products begins, and in-frame rearrangement products now encode functional TCRβ chains. At the DN3 stage, assessment of whether TCRβ chains are productively rearranged occurs and is termed "TCRβ selection" (Fig. 27.5). The success of TCRβ rearrangement is sensed by the cell when the β-chain is assembled into a complex at the cell surface. This pre-TCR complex must form prior to TCRα rearrangement, which is limited at this early developmental stage;[185] a surrogate for TCRα is instead used. The *Ptcra* gene encoding the invariant pTα chain does not require rearrangement and serves as a surrogate for TCRα.[186] Following successful assembly of the pre-TCR, it signals from the cell surface; however, ligand recognition may not be needed.[187,188] Instead, the pre-TCR undergoes spontaneous clustering that is thought to result in signaling.[189,190] However, it is also suggested that binding of the pre-TCR to peptide-MHC ligands affects thymocyte development.[191,192] The pre-TCR signal largely requires the same second messengers, enzymatic mechanisms, and adapters used in signaling by the αβTCR on mature T cells,[193] with the exception that CD3δ is not required.[194] Thus, TCRβ selection assesses for functional assembly of nearly all of the αβTCR, except for TCRα and CD3δ. The pre-TCR expressed during T cell development resembles pre-BCR, comprising of IgH and surrogate light chains, VpreB and λ5, expressed during B cell development. In mice engineered to lack the RAG recombinase, or other components of the pre-TCR, T cell development is arrested at the DN3 stage.[195,196]

Consequences of Pre-TCR Signaling

DN3 cells that fail to receive a pre-TCR signal, because they fail to productively rearrange TCRβ, or for any other reason including failure to express necessary signaling molecules, undergo apoptosis. Conversely, successful pre-TCR assembly and signaling triggers a dramatic set of events including allelic exclusion, rapid proliferation, and differentiation,[197] resulting in a further 100-fold proliferative expansion from the already extensive proliferation that occurred at earlier stages of development.[198,199] This proliferation contributes to TCR repertoire diversity, because it expands clones with identical TCRβ rearrangements that subsequently can rearrange different TCRα chains. Proliferation at this stage is not only important to generate necessary numbers of thymocytes but also is required for differentiation.[200,201] DN3 cells that instead productively rearrange TCRγ and TCRδ genes to develop into γδ T cells do not undergo this proliferative burst, which may be the reason mice deficient in the ribosomal protein L22 (Rpl22) lack αβ but not γδ T cells.[202,203]

Successful TCRβ selection requires Notch, IL-7R, and CXCR4 signaling in addition to an expressed pre-TCR.[204-207] Both IL-7R and CXCR4 are required for proliferation. Interestingly, the function of CXCR4 is distinct from its canonical role as a chemotactic receptor, but instead involves signaling through the phosphatidylinositol 3-kinase (PI3K) pathway.[205] Notch signaling at the TCRβ selection checkpoint plays a predominantly metabolic or trophic role, and DN3 cells with Notch1 signaling experimentally blocked are unable to undertake normal proliferation.[204,208] Conversely, forced proliferation by genetic manipulation bypasses the Notch requirement at this stage.[201]

During the proliferative expansion at the TCRβ selection checkpoint, cells downregulate transcription of *Rag1* and *Rag2* genes and degrade RAG2 protein.[209] Cells further activate expression of CD4, CD8α, and CD8β, to eventually become $CD4^+$ $CD8^+$ (DP) thymocytes that cease proliferation. DP cells downregulate expression of Notch1 transcripts as they cease proliferation and subsequently initiate transcription of the TCRα locus, and re-express *Rag1* and *Rag2*, focusing rearrangement on TCRα genes.[197] Initial rearrangement events delete the TCRδ locus; and DP cells further transcriptionally silence TCRγ genes, thus permanently losing the ability to access γδ T cell fates and so becoming committed to the αβ T cell lineage.[210]

Mechanisms of Transcriptional Control

Notch signaling induces the expression of the transcriptional repressor Hes1, which at this stage acts to lower PTEN expression levels to further enable PI3K-dependent programs.[211] In addition, active Notch binds a unique enhancer in Myc to mediate its effects at TCRβ selection.[200,212] Recent work established that Notch signals induced the transcription of the F-box protein Fbxl1, whereas pre-TCR signals induced the transcription of Fbxl12; these are substrate recognition subunits of SCF (Skip-Cullin-F-box) ubiquitin ligase complexes that target the cell cycle inhibitor Cdkn1b for polyubiquitination and degradation in DN3 cells. This mechanism can explain why coordinated Notch and pre-TCR signals are required for proliferation in β-selected thymocytes.[213] Thus, the requirement for Notch1 at TCRβ selection is likely to be distinct from the Notch requirement in T cell lineage specification and commitment. In addition to Hes1, Myc, and Fbxl1, Notch1 also directly controls expression of the *Ptcra* gene encoding for pTα, perhaps ensuring that this step of T cell development is restricted to the thymus and does not normally occur extrathymically.[214] Nevertheless, exceptions are known; in particular, athymic mice, and irradiated mice given bone marrow transplants, generate DP immature TCR-expressing cells selectively in mesenteric lymph nodes.[215-217]

Transcriptional circuits downstream of the pre-TCR signal at the TCRβ checkpoint have been identified. TCF-1 is required, together with Lef-1, and in mice lacking both these factors T cell development is arrested prior to the DP stage.[218] TCF-1 acts upstream of HEB to promote development to the DP stage.[153,219] Imposition of the TCRβ selection checkpoint requires Ikaros and E proteins E2A and perhaps HEB, and cells can transit this checkpoint without the pre-TCR when these transcription factors are experimentally disrupted.[175,220,221] Pre-TCR signaling is thought to downregulate or alter the activity of these factors via mechanisms that are becoming understood. Pre-TCR signaling results in the upregulated expression of the immediate-early gene Egr2, followed by upregulated expression of Id3 and also Id2,[222] thus antagonizing the activity of E proteins.[175,176] This transcriptional circuit when activated by pre-TCR signaling thus allows T cell precursors to transit through the TCRβ selection checkpoint imposed by high activity of E proteins.

αβ and γδ T Cell lineage Determination

The transcriptional circuit just described for pre-TCR signaling is important for the choice between αβ and γδ T cell lineages in adult mice. γδ T cells themselves comprise many distinct subtypes characterized by their distinct γδ TCRs that develop at different times and adopt distinct effector programs and tissue tropism (see Chapter 36). The γδ TCR is expressed at considerably higher levels than the pre-TCR, and some γδ T cells are also thought to be selected by endogenous ligands. Perhaps for these reasons, TCRγδ signaling is stronger and accompanied by greater expression of Egr2 and Id3 than signaling downstream of the pre-TCR.[223,224] The increased Id3 expression, and increased antagonism of E proteins, has been proposed to direct cells to adopt γδ fates instead of αβ fates. Consistently, genetic manipulations that decrease TCR signal intensity allow γδ TCRs to instead transduce signals resembling those from the pre-TCR, directing cells bearing γδ TCRs to become DP thymocytes.[222,225,226] Conversely, experimental expression of the αβ TCR at the DN3 stage can deliver stronger signals than the pre-TCR when peptide-MHC ligands are present, directing cells bearing these αβ TCRs to remain DN and adopt a γδ gene expression program.[227] These results support a model where the intensity of signaling downstream of the γδ TCR directs cells that successfully rearrange γδ TCRs to the γδ fate; whereas αβ fates instead require weak signals transmitted by the pre-TCR. Thus, γδ versus αβ fates in this model are matched to the receptor that the cells rearrange and express. This model can accommodate the additional possibility that the rearrangement of TCR genes may not be random, but instead be under transcriptional control that may differ in different precursors. Thus, inherent biases in the selection of gene loci for rearrangement may exist, particularly in early development. Rearrangement and expression of distinct TCR genes then results in distinct signaling outcomes and so distinct cell fate choices.

Distinct from this model, transcriptional preprogramming for γδ fates and IL-17 production independent of TCR signaling is described for a population of Vγ2 cells (Tγδ17) that makes IL-17.[228,229] A reporter allele for transcription factor Sox13 established that Tγδ17 cells could be shown to adopt IL-17 fates prior to upregulation of CD73, which occurs after TCR signaling. Single-cell transcriptional profiling established that the Tγδ17 transcriptional program was established in precursors of Tγδ17 cells that lacked RAG proteins and so were unable to assemble a functional TCR. Additional work is needed to decipher the distinct developmental mechanisms and transcriptional programs that are likely to underlie the diversity of γδ T cell types.

SELECTION AND DIFFERENTIATION OF αβ T CELLS

Key Concepts of αβ T Cell Development

Most T cells in secondary lymphoid organs are conventional αβ T cells, which recognize peptide antigens associated to classical MHC I and MHC II molecules and which express either CD8 or CD4, respectively. These conventional T cells derive from $CD4^+$ $CD8^+$ DP thymocytes, which differentiate in the thymus into $CD4^+$ or $CD8^+$ single-positive thymocytes and T cells. Some DP thymocytes, carrying specific αβ TCR clonotypes, recognize nonpeptide antigens bound to MHC-like molecules, differentiate into innate-like αβ T cells, including iNKT and MAIT cells. The development of these cells is described in Chapters 37 and 38, and only a few specific issues will be discussed in this section.

The differentiation of DP thymocytes into T cells can be analyzed from two perspectives, developmental and immunological. As in all developmental processes, immature αβ lineage thymocytes respond to environmental cues that engage specific receptors, generate intracellular signals, and result in changes in transcription factor expression and activity. These affect the transcriptomic and epigenetic landscape of the cell, eventually resulting in the acquisition of new functional or survival properties and trafficking patterns. Notably absent from that list is cell proliferation: differentiating DP thymocytes mostly remain nondividing during their differentiation in mature SP thymocytes and T cells.[230,231] From an immunological perspective, the completion of TCRα rearrangement and the surface expression of TCRαβ-CD3 complexes make DP thymocytes the first antigen-recognizing cells in the αβ developmental pathway; the purpose of DP differentiation is to generate a set of clonotypic T cells (the "repertoire") that will appropriately serve immune responses. This involves the "random" (that is, not directed by antigen recognition) generation of a highly diverse set of clonotypic receptors by V(D)J rearrangement of TCR genes (see Chapter 24); these receptors differ by (1) their combination of V, D, and J segments and (2) the clonotype-specific amino acid sequence generated by the rearrangement process at the junction of V, D, and J segments, forming the "complementarity determining region (CDR) 3." Importantly, properties of T cell antigen recognition mandate that this initial repertoire be heavily culled by selection so that it can properly serve its function.

Antigen recognition by conventional αβ T cells involves interactions between TCR chains and a complex ligand formed by amino acid residues from both the antigenic peptide and the bordering MHC helices[232] (Chapters 20 and 24). While the peptide contacts ensure specificity of antigen recognition, the MHC contacts are needed for efficient TCR-ligand interactions; such a requirement provides the mechanistic basis for MHC restriction, which we define as the fact that T cell recognition of a particular peptide antigen is dependent on the presentation of that peptide by a specific MHC molecule. The MHC contacts by TCRs have major consequences for αβ T cell development because the allelic diversity of MHC molecules, at the species level, makes most αβ TCRs unable to efficiently dock on any given MHC molecule. Only a small subset of "useful" DP thymocytes carry TCRαβ pairs able to interact with the allelic variants of MHC molecules carried by any given individual; the other "useless" DP cells die in the thymic cortex by programmed cell death (by "neglect") (Fig. 27.8). Thus, a critical feature of T cell development is that engagement of the αβ TCR by complexes of MHC and self-peptide ligands is needed to rescue DP thymocytes from death by neglect, a process called "positive selection."[233] An essential consequence of positive selection is that the mature T cell repertoire, aimed at recognizing foreign peptides, is actually selected on self-peptides and necessarily carries self-reactivity. Several intrathymic processes contribute to make this repertoire self-tolerant (ie, not generate an immune response against self antigens), including deleting thymocytes with excessive self-reactivity (negative selection) or diverting them into T-lineages with regulatory activity. Together, these processes generate a T cell repertoire that is both self-MHC-restricted and self-tolerant[234,235] (Fig. 27.8).

Positively selected thymocytes are subject to several differentiation processes. They undergo functional preprogramming to match their transcriptional circuitry and CD4 or CD8 coreceptor expression to their MHC specificity, they migrate from the cortex to the medulla, and, before exiting the thymus, they adopt transcriptomic programs characteristic of mature T cells, which we refer to as late-stage thymic maturation. None of these processes occurs in thymocytes that have not undergone TCR signaling. However, not all these processes are individually triggered by TCR signals; as discussed later in this chapter, only thymocytes that have ceased TCR signaling are cleared to leave the thymus. There is no nomenclature consensus in the field as to whether these differentiation processes are part of, or distinct from, positive selection. Here, we adopt a more restrictive definition and consider positive selection as including the fixation of TCR

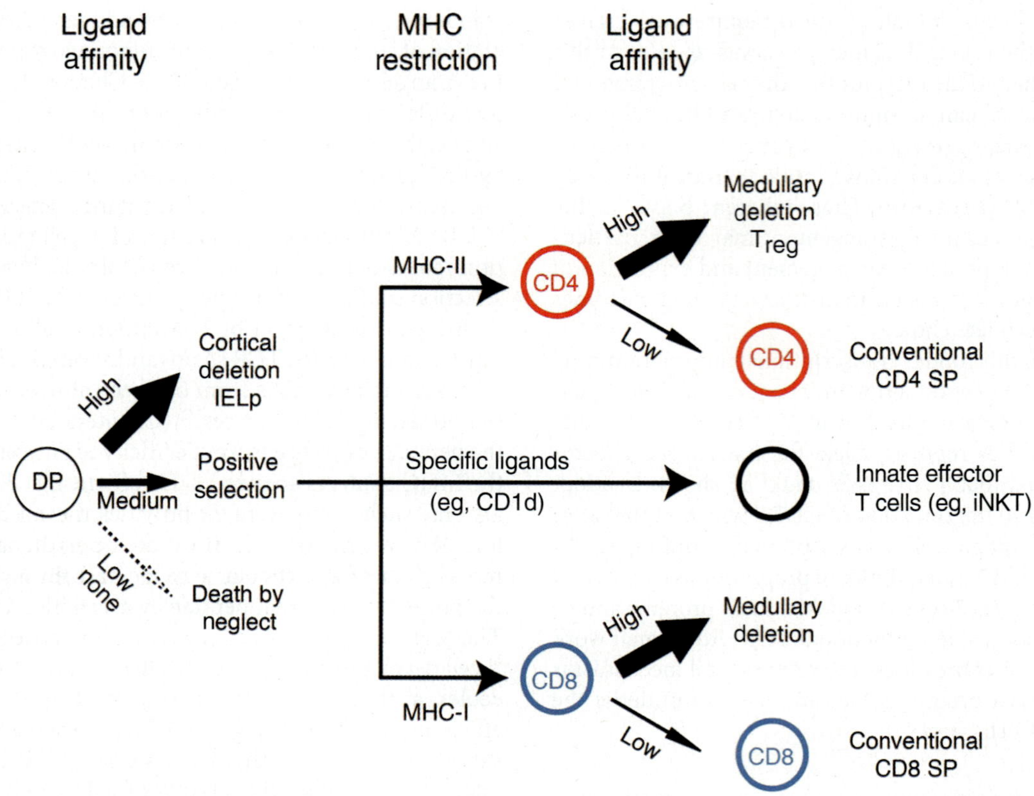

FIG. 27.8. Selection and differentiation of αβ lineage T cell precursors. The affinity of TCR for MHC ligands causes DP thymocytes to undergo TCR-induced cells death (high affinity ligands), rescue from programmed cell death (positive selection, intermediate affinity ligands), or to undergo programmed cell death (death by neglect, low affinity ligands). Surviving thymocytes undergo differentiation into $CD4^+$ or $CD8^+$-lineage depending on their MHC restriction, or into specific innate effector fates if restricted by nonclassical MHC molecules. $CD4^+$ and $CD8^+$-lineage cells migrate to the medulla where their affinity to peptide-MHC ligands determines whether they undergo TCR-induced cell death (medullary negative selection, high affinity ligands), or differentiation into conventional (naïve) $CD4^+$ or $CD8^+$ T cell (low affinity ligands). In addition, a subset of MHC II–restricted cells, enriched in cells with high-self reactivity differentiate into Foxp3-expressing T_{reg} cells.

specificity in addition to the rescue from programmed cell death. Accordingly, we consider $CD4^+$-$CD8^+$ lineage differentiation, thymocyte migration, and late-stage thymic maturation of thymocytes as separate from positive selection.

In experimental mouse models, up to 95% of "preselection" DP thymocytes (ie, those generated by the proliferation boost that follows pre-TCR signaling) fail to become mature T cells[236]; it has been speculated that most cells rescued from death by neglect differentiate into mature T cells, so that only a small number are subject to negative selection. This scenario fits with the idea that the preselection TCR repertoire is entirely unbiased, with an inverse relationship between cell frequency and TCR affinity for MHC ligands (Fig. 27.9); it is in line with the "random" nature of V(D)J rearrangement, that is that its outcome—the combination of V, D, and J segments and the CDR3 sequence—is neither predetermined nor dependent on the functionality of the rearrangement product. However, several independent assessments indicate that the frequency of pre-selection DP thymocytes signaled by MHC ligands is much higher than 5%, therefore implying that greater numbers of DP thymocytes are eliminated by negative selection than are allowed to differentiate into mature T cells.[237-242] These findings agree with the idea that the pre-selection TCR repertoire is evolutionarily biased toward recognition of MHC ligands[243-246]; such bias would not affect the randomly generated CDR3 regions, mostly involved in peptide binding, but the germline (V segment) encoded CDR1 and CDR2 regions,

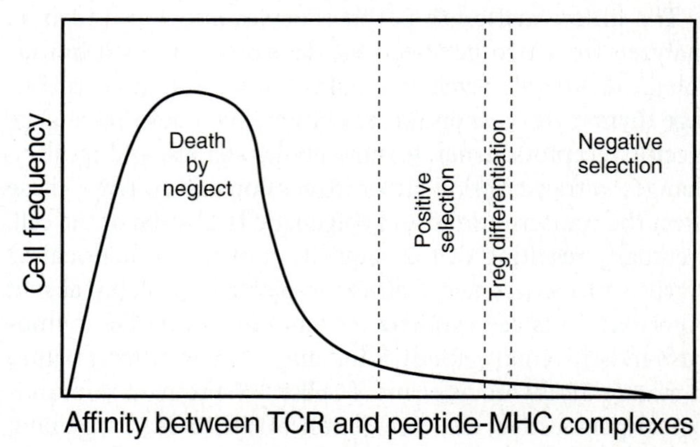

FIG. 27.9. Schematics representing the distribution of cells frequency (vertical axis) against TCR affinity for peptide-MHC ligands (horizontal axis) according to the conventional unimodal cell distribution with continuously decreasing cell frequency as TCR-ligand affinity increases, and fewer cells undergoing negative selection than positive selection.

mostly involved in MHC contacts. However, it is also evident that MHC-independent TCR specificities generated by V(D)J rearrangement are present within the preselection TCR repertoire of DP thymocytes[247-250]; as discussed later in this chapter, CD4 and CD8 coreceptor molecules notably serve to prevent the selection of such MHC-independent specificities.

Biology of DP Thymocytes

DP thymocytes, which differentiate from DN precursors upon completion of a productive TCRβ gene rearrangement and signaling by the pre-TCR, are the most abundant thymocyte population. DP thymocytes reside in the thymic cortex, where they interact with its other components, chiefly cTECs. Except for a small subset that just initiated expression of CD4 and CD8, DP cells are of small size and do not proliferate; they are programmed to undergo cell death by apoptosis, in a manner requiring both pro-apoptotic proteins Bak and Bax.[251] Accordingly, DP thymocytes are highly sensitive to a variety of agents promoting cell death, including natural and synthetic glucocorticoid hormones. In the mouse thymus, their half-life averages 3 days in the absence of TCR engagement by intrathymic ligands[230,252]; triggering of the apoptotic program rapidly results in engulfment and clearance by resident macrophages.[253] Even though limited, DP thymocyte survival requires the anti-apoptotic Bcl2-family protein Bcl-$_{XL}$, whose expression in turn depends on the transcription factor RORγt.[254] DP thymocytes are essentially unresponsive to most cytokines because they fail to express many cytokine receptors (notably for IL-7), and because of their high expression of Socs1, an intra-cellular inhibitor of cytokine signaling.[255,256]

A critical function of DP thymocytes is to rearrange the gene encoding TCRα. E-proteins E2A and HEB are needed for this process, and also serve as gate-keepers to prevent inappropriate differentiation of unsignaled DP thymocytes into mature T cells,[257,258] a dual activity similar to their function in β-selection.[175] Two features of TCRα gene rearrangement have important biological consequences.[259] First, it is not subject to allelic exclusion and proceeds on both alleles in a more or less synchronous manner. As a result, many T cells productively rearrange and express two TCRα chains. Nonetheless, even if they synthesize two TCRα chains, most αβ T cells predominantly express one TCRαβ pair at their surface, as a result of incompletely understood post-transcriptional and post-translational mechanisms.[260] Second, the absence of D segments allows each TCRα locus to undergo sequential rearrangement, initially involving V and J segments proximal to the center of the locus (occupied in the germline configuration by the TCRδ gene locus) and progressively involving segments in more distal locations. Accordingly, RORγt-deficient DP thymocytes, which have a reduced half-life, show a strong bias toward proximal V and J chain usage in the TCRα locus, a defect rescued by enforced expression of Bcl$_{XL}$.[261] Such sequential rearrangement, a form of receptor editing, continues unless DP cells receive TCR signals; this allows each DP thymocyte to "test" multiple TCRα chains for their ability to pair with the previously rearranged TCRβ chain and for the ability of the resulting TCRαβ to recognize self peptide-MHC complexes and therefore initiate positive selection. Many DP thymocytes are enveloped within thymic nurse cells, which are a subpopulation of cTECs (amounting to approximately 10% of cTECs) and are morphologically and functionally specialized for optimizing positive selection of thymocytes.[262-264] Thymic nurse cell engulfment is more frequently observed for DP thymocytes that undergo secondary TCRα-chain rearrangement,[263] suggesting that thymic nurse cells provide a microenvironment for optimal TCR repertoire selection by nurturing DP thymocytes to express the secondary rearranged TCRα chains, thereby enabling additional chances of positive selection. Together with the β-selection-induced proliferation burst that results in the generation of large numbers of DP cells sharing the same TCRβ rearrangement, sequential TCRα gene rearrangement contributes to leverage the potential for α − β combinatorial diversity that is theoretically allowed by the number of TCRα and TCRβ V, D, and J gene segments. DP cells are highly mobile, with velocities of approximately 3 to 8 μm/min, but lack directional persistence.[265,266] These motile DP thymocytes pause to interact through their TCRs with peptide-MHC complexes expressed by cTECs and other antigen-presenting cells in the cortex, thereby optimizing the potential for productive interactions between TCR and peptide-MHC complexes.

Positive Selection of the αβ T Cell Repertoire

Biology of Positive Selection

Cells signaled to undergo positive selection can be identified by their elevated expression of surface markers CD5 and CD69 (Fig. 27.10), and of the transcription factor Nur77[267-269]; however, all three markers are up-regulated by TCR signaling, and therefore do not unambiguously distinguish positive from negative selection signals. Experimentally, the concept of positive selection emerged from bone marrow chimera and thymic graft experiments demonstrating that, for T cells to respond to a given antigen, they must have developed in a thymic epithelial environment expressing the MHC restriction element that supports the antigenic response.[270-272] Placed into the current knowledge context, these observations have two essential implications: first, the thymus is essential to select a TCRαβ repertoire allowing responsiveness to foreign peptides presented by self MHC; second, the thymic epithelium (and specifically the cortical thymic epithelium, as discussed later in this chapter) provides signals for such selection.

Much effort has been dedicated to characterizing which ligands promote positive selection. Initial experiments with TCR transgenic mice, in which thymocytes are driven to express a specific TCRαβ clonotype, demonstrated that TCR recognition of peptide-MHC ligands is needed for positive selection, and confirmed the requirement for the restricting MHC molecules.[234,273-275] However, these initial experiments did not determine whether specific peptides were needed to generate positive selection signals. Despite the absence of an appropriate two-dimensional cell culture model, unlike for earlier stages of T cell development, addressing that question was facilitated by three-dimensional cultures of thymic organs from TCR transgenic mice.[276-278] Thymic organ culture allows the selection of thymocytes carrying defined TCR clonotypes in intact fetal thymic lobes cultured *in vitro* in defined conditions,

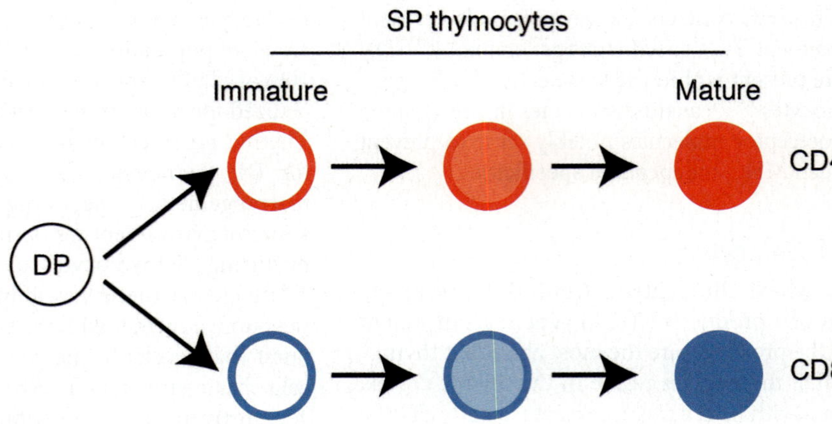

CD3	±	++	++	++
TCRβ	±	++	++	++
CD5	+	+++	+++	+++
CD69	−	++	++	±
CCR7	−	++	++	++
CD24	+++	+++	++	±
MHC-I	−	−	++	++
CD62L	−	−	−	+
S1pr1	−	−	−	+

FIG. 27.10. Markers commonly used to distinguish stages during the late-stage thymic differentiation of αβ lineage T cells.

including the addition of specific peptide ligands.[279,280] This combined approach has notably been used to study the selection of MHC I–restricted thymocytes expressing a transgenic TCR that recognizes an ovalbumin-derived peptide presented by H-2K^b MHC I molecules.[278,279,281] Three key concepts have emerged from fetal thymus organ culture experiments with this system. First, peptide sequence (and by implication TCR-peptide contacts) is critical for positive selection: for a given TCR, most peptides are unable to promote selection, causing DP thymocytes to die by neglect; accordingly, selection of a diverse TCR repertoire requires a diverse set of peptides. Second, positive selection is driven by "low-affinity" interactions,[282] which, for the ovalbumin-reactive TCR, can be provided by weak agonist analogs of the nominal peptide.[276,281] Third, the "nominal" antigenic peptide (ie, the peptide that activates mature T cells expressing the TCR), does not promote positive selection; rather, it causes negative selection or the diversion of cells to alternate regulatory lineages (discussed later in this chapter). However, additional studies supported the idea that ligand abundancy was important to determine the outcome of TCR engagement by peptide-MHC complexes, as they showed that a single peptide could promote negative selection at high concentration, but positive selection at low concentrations.[277,278,281,283] The role of peptide affinity has not been as extensively studied in the selection of MHC II–restricted thymocytes as it has been for their MHC I–restricted counterparts. However, *in vitro* and *in vivo* studies support the conclusion that low-affinity analogs of the nominal peptide promote selection of MHC II–restricted thymocytes.[284,285] Altogether, these findings favor a model whereby positive selection signals come from low-affinity interactions between TCR and peptide-MHC ligands, or from higher affinity ligands if displayed at low effective concentrations. Identifying naturally occurring positively selecting peptides would provide further insight into these questions and is a major objective of current studies; the current evidence suggests that such peptides are not necessarily related in sequence to the nominal peptide.[279,286-288] Such experiments would also examine whether the peptide-TCR interactions that promote positive selection are as specific as those promoting antigen recognition by mature T cells. Addressing this question, studies using MHC II molecules engineered to present a single peptide *in vivo* have reiterated the importance of peptide diversity for the positive selection of a rich repertoire, supporting the idea of nondegenerate interactions.[289] However, these same studies have revealed the potential for a single peptide to select multiple TCR specificities, in a sequence-specific manner.[290,291]

Signals and Intra-Cellular Mechanisms Driving Positive Selection

TCR signaling causes the positive selection and further differentiation of DP thymocytes. Signaling is initiated by co-engagement of the TCR and either CD4 or CD8 coreceptor by MHC II– or MHC I–peptide complexes, respectively. CD8 is required for the selection of MHC I–restricted thymocytes, whereas CD4 is important but not strictly required for that of MHC II–restricted thymocytes.[292-295] CD4 and CD8 bind nonpolymorphic regions of MHC II and MHC I molecules, respectively. Additionally, the cytosolic domains of CD4 and CD8 bind Lck, a tyrosine kinase needed for TCR signaling.[296-299] Thus, peptide-MHC co-engagement of CD4 or CD8 with TCR brings Lck in the immediate vicinity of the signaling chains of the TCR complex, thereby promoting phosphorylation of their Immunoreceptor Tyrosine-based Activation Motifs (ITAM), and initiation of TCR signaling.[300] Accordingly, Lck binding is important, even though not strictly required, for the function of CD4 and CD8 in positive selection.[301,302] Lck binds preferentially to CD4 relative to CD8,[303-305] which has functional consequences for signaling in DP thymocytes (discussed later in this chapter). The ability of CD4 and CD8 to both promote TCR-MHC interactions and initiate TCR signaling has important consequences for positive selection. Indeed, only MHC molecules, and not other ligands, can co-engage TCR and CD4 or CD8; this provides a signaling advantage to peptide-MHC-reactive TCRs, compared to TCRs reactive against non-MHC ligands that are generated during V(D)J TCR gene rearrangement.[306]

Membrane-proximal TCR signal transduction uses similar intermediates in thymocytes as it does in mature T cells; the pathways and mechanisms of TCR signal transduction are discussed in Chapter 24 and will not be addressed in this chapter. TCR signaling induces multiple transcriptomic changes in DP thymocytes, of which some are specific of positively versus negatively selected cells, whereas others may be common to both outcomes.[279,307] Of major biological significance are the termination of *Rag1* and *Rag2* gene expression and therefore the fixation of TCR specificity,[308] which is followed by the increase in surface TCR expression to the level characteristic of mature T cells. Inhibition of E-protein activity, largely through TCR-induced expression of their inhibitors Id2 and Id3, is important for the shut-down of the rearrangement machinery and the proper timing of CCR7 and IL-7R expression.[175,257,309] A critical question is which anti-apoptotic molecule promotes the survival of positively selected thymocytes,[310] as TCR signaling terminates expression of RORγt and its target Bcl_{XL}. Bcl2 appears important for this function. It is up-regulated in intrathymically signaled DP thymocytes and required for the survival of mature T cells.[311,312] Although Bcl2 is not strictly required for positive selection, as SP thymocytes and T cells develop in neonate Bcl2-deficient mice, the survival of these cells may rely on the functional overlap between Bcl2 and the related molecule Mcl1,[313-316] and on the low level of endogenous glucocorticoids in neonate mice (because thymocytes that do not express Bcl2 are highly sensitive to glucocorticoids).[317] Importantly, although TCR signaling up-regulates Bcl2 during positive selection,[318] signaling by intrathymic cytokines, including IL-7, is probably essential for sustained Bcl2 expression in and survival of more mature $CD8^+$-lineage thymocytes.[319,320] How TCR-induced transcriptomic changes are implemented has been difficult to address, in large part because of the lack of cell culture models suitable for biochemical and genetic studies. Initial research assessed the role of signal transduction pathways involved in TCR signaling.[279] These studies demonstrated that the Ras-Erk kinase pathway, but not store-operated calcium mobilization,[321-323] is required for positive selection and T cell differentiation.[322] Downstream effector mechanisms, including transcriptional circuitries and effector molecules, are incompletely understood.[324]

Positive Selection and the Paradox of Self-Reactivity

Although TCR engagement by self peptide-MHC contributes to the long-term survival and responsiveness of peripheral T cells,[325] mature T cells are not broadly self-reactive. Because TCR signaling in thymocytes and mature T cells use highly similar, even if not identical, mechanisms and pathways, this suggests that DP thymocytes and mature T cells are presented with distinct sets of self peptides, or that they differ in their reactivity. The unique features of cTECs described below, including expression of β5t, Cathepsin L and TSSP, ensure that many positively selecting peptides are in fact generated only in the thymus. However, there is no evidence that all positively selecting peptides are cTEC-specific, and low affinity analogs of nominal peptides, which promote positive selection, do not cause detectable activation when presented to mature T cells.[276] This suggests that, despite their low surface TCR expression, DP thymocytes show higher reactivity to low-affinity ligands than mature SP thymocytes and T cells do. This is indeed the case, and contrasts with unchanged responsiveness to high-affinity ligands.[326,327] Several complementary mechanisms have emerged to account for this difference.[328,329] The first includes "tuning" molecules that are induced by and provide negative feedback to TCR signaling. The best characterized of these molecules is CD5, although the mechanistic bases of its function remain to be elucidated.[329] An appealing idea is that the dynamics of expression of such molecules allows thymocytes to broaden the range of ligand affinities that are conducive to survival, between death by neglect and activation-induced deletion.[329]

A distinct group of mechanisms involves molecules expressed in DP thymocytes but not (or at reduced levels) in mature T cells, and that increase sensitivity to low-affinity ligands by promoting TCR signal transduction. Multiple proteins controlling components of TCR signal transduction, including some involved in calcium signaling, fit in this category.[329-331] An additional molecule, Themis, the prototype of a new class of signal transduction regulators, is needed for positive selection.[332-336] Ongoing studies point to the role of Themis in regulating the activity of the tyrosine phosphatase Shp1, and thereby TCR signaling in thymocytes.[329,337,338] Much interest has been raised by observations that the miR181 family of micro RNAs, abundantly expressed in DP thymocytes, promotes TCR signaling and acts by post-transcriptionally reducing expression of tyrosine phosphatases antagonizing TCR signal transduction[339]; these included PTPN22, which was independently shown to impede TCR signaling by weak agonist ligands.[340] This suggested that miR181 expression promotes TCR signaling in DP thymocytes in response to low-affinity peptide ligands, and thereby increases the number of cells undergoing positive selection. However, the role of miR181 in the thymic selection of conventional T cells remains to be clarified[341]; in contrast, miR181 has been shown to be needed for the generation of nonconventional and some T_{regs} cells.[341]

Processing and Presentation of Self-Peptides for Positive Selection in the Thymic Cortex

Antigen-Presenting Cells for Positive Selection

cTECs, expressing abundant MHC class I and class II molecules, are the key physiological antigen-presenting cells that induce positive selection of cortical thymocytes that differentiate into conventional CD4+ and CD8+ SP cells.[342,343] In contrast, for innate PLZF+ iNKT and MAIT cells, DP thymocytes play a primary role in inducing their positive selection by expressing the class Ib MHC molecule (CD1d and MR1) that serves as the recognizing ligand for iNKT and MAIT cells, respectively.[344-346] In addition to abundant expression of MHC class I and class II molecules, cTECs express a unique set of molecules for protein degradation that serves important roles in the processing of MHC-associated peptides and the turnover of peptide-MHC complexes (Fig. 27.11).

MHC Class II-Associated Self-Peptides in cTECs

Peptides associated with MHC II molecules are generated by proteolysis within the lumen of the endosomal-lysosomal system. Cathepsin L, a lysosomal endopeptidase abundant in cTECs, is involved in the processing of MHC II invariant chain and the generation of MHC II–associated self-peptides that promote cTEC-mediated positive selection of CD4+ T cells.[347,348] Another endosomal peptidase, thymus-specific serine protease (TSSP, encoded by *Prss16*) is abundant in cTECs and contributes to the production of MHC II–associated self-peptides in cTECs.[349,350] Unlike most cells, in which autophagy is activated by nutrient starvation, cTECs are constitutively active in autophagy-mediated degradation of cytoplasmic proteins.[351] Autophagy in cTECs is thought to provide another source of MHC II–associated self-peptides, including those derived from cytoplasmic protein-derived peptides, thereby optimizing the diversity of peptides required for the positive selection of CD4+ T cells.[352]

March family ubiquitin-protein ligases enhance degradation and turnover of MHC II molecules. cTECs predominantly express March 8, the activity of which is attenuated by the cell-surface protein CD83, which is also expressed in

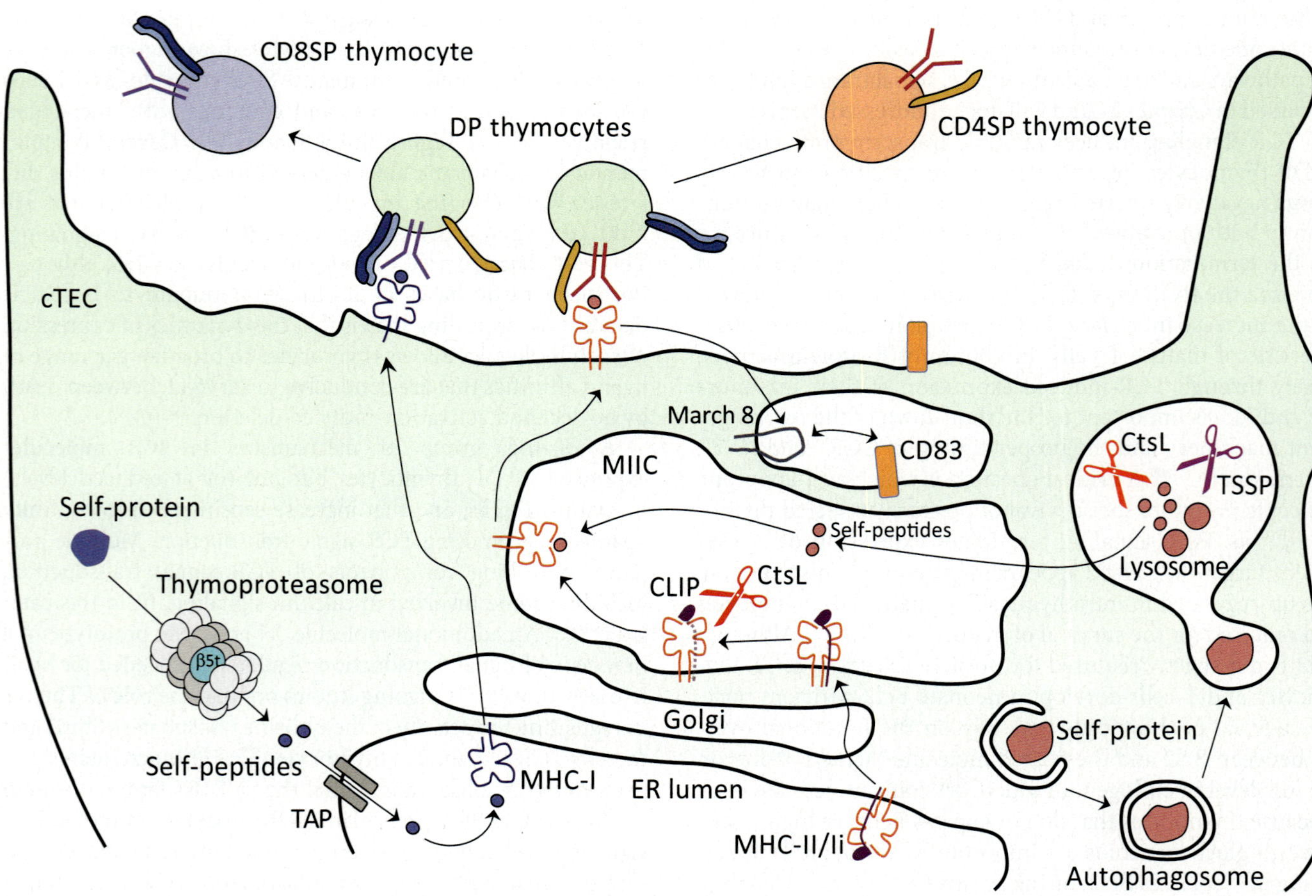

FIG. 27.11. Self-antigen processing and presentation in cTECs. Peptides derived from cytoplasmic proteins are degraded by β5t-containing thymoproteasomes, transported to endoplasmic reticulum (ER) through TAP, loaded to MHC class I (MHC I) molecules, and transported to the cell surface. Proteins derived from lysosomes and autophagosomes are delivered to MHC class II (MHC II) molecules. Autophagy is constitutively active in many cTECs. Cathepsin L (CtsL) is involved in invariant chain (Ii) degradation in endosomal MHC class II compartment (MIIC) and peptide processing in lysosomes. TSSP (Prss16) also contributes to the processing of MHC II–associated self-peptides in cTECs. Degradation and turnover of MHC II molecules in cTECs are regulated by March 8 and CD83.

cTECs. CD83 stabilizes surface MHC II molecules in cTECs by slowing March 8-mediated MHC II turnover. CD83-deficient mice are defective in cTEC MHC II stabilization and thymic CD4+ T cell generation, whereas these defects are rescued by the deficiency in March 8.[353-355] CD83-mediated stabilization of surface MHC II molecules in cTECs may maintain persistent TCR engagement, which is important for positive selection of CD4+ T cells.

Thymoproteasome and MHC Class I-Associated Self-Peptides in cTECs

Peptides associated with MHC I molecules are generated through proteasome-mediated degradation of cytoplasmic and nuclear proteins. cTECs express a unique form of the proteasome termed thymoproteasome, which is characterized by the inclusion of the catalytic subunit β5t, encoded by *Psmb11*.[356] The expression of β5t is promoted by the transcription factor Foxn1 and is highly specific to cTECs, so that the β5t-containing thymoproteasome is specifically abundant in cTECs.[30] In comparison with other forms of proteasome, β5t-containing thymoproteasomes exhibit altered proteolytic specificity, producing a unique set of MHC I–associated peptides.[357] In β5t-deficient mice, CD8+ T cells are reduced in number, altered in TCR repertoire, and defective in TCR responsiveness.[358-360] Thus, thymoproteasome-dependent positive selection in the thymic cortex is important for promoting optimal production of CD8+ T cells (Fig. 27.11).

For humans more than 150 genomic variations at the locus encoding β5t, including single nucleotide polymorphisms in the β5t-coding sequence, have been registered in the National Center for Biotechnology Information public database. Among those variants, the rs34457782 single nucleotide polymorphism that alters the 49th amino acid of human β5t proprotein from glycine to serine (G49S) is detected at an appreciable allele frequency (up to 5%) in various human populations. The 49th amino acid is located at the carboxyl terminus of the propeptide that is cleaved to generate catalytically active β5t protein, the amino terminus of which begins at the 50th residue (threonine). The G49S variant in either human or mouse β5t has inefficient processing of the propeptide to generate mature β5t protein.[361] The impact of this allele to human health is still unclear.[361,362]

Migration of Positively Selected Thymocytes From the Cortex to the Medulla

DP thymocytes that have received positively selecting TCR signals begin relocating from the cortex to the medulla. Upon TCR engagement, cortical DP thymocytes increase cell-surface expression of the chemokine receptor CCR7,[363] whereas a subpopulation of mTECs constitutively produce CCL21, a chemokine ligand for CCR7. Thus, cortical DP thymocytes that have received TCR-mediated signals are selectively attracted toward the medulla via CCR7/CCL21-mediated chemotaxis. In mice deficient in CCR7 or CCL21, positively selected mature CD4+ and CD8+ SP thymocytes are detected in the cortex because they fail to migrate into the medulla[363-365] (Fig. 27.6).

In contrast to the increase in CCR7 expression, TCR signals in DP thymocytes decrease the expression of other chemokine receptors CCR9 and CXCR4.[366,367] cTECs abundantly express the ligands for these receptors, chemokines CCL25 and CXCL12, respectively. These chemokine signals are thought to contribute to the retention of DP thymocytes in the cortex prior to TCR engagement and the release of TCR-signaled DP thymocytes from the cortex.[365-368]

In addition to chemotactic regulation, there is an inward flow of interstitial fluid within the thymic parenchyma. Intraperitoneal administration of labeled proteins or small particles indicated that liquid flow is present in the post-natal thymus from the capsular and subcapsular areas toward the medullary area.[369,370] Whether passive inward flow contributes to the migration of developing thymocytes from the cortex to the medulla is unclear.

CD4+-CD8+ Lineage Differentiation and Functional Programming

MHC Restriction, Coreceptor Expression, and Functional Programming

Unlike DP thymocytes, most mature αβ T cells express either CD4 or CD8 but not both. Three mouse genes (and their orthologs in other vertebrate species) encode components of CD4 and CD8: *Cd4*, and the closely linked *Cd8a* and *Cd8b1*, encoding CD8α and CD8β subunits of CD8. Most CD8+ T cells express CD8αβ heterodimers, although specific subsets, including MAIT cells and some intra-epithelial lymphocytes (IEL), express CD8α only as a homodimer (CD8β requires binding to CD8α for cell surface expression). Expression of the coreceptor is a stable cell attribute, thereby defining CD4+ and CD8+ lineages; it is highly dependent on MHC specificity, so that MHC II– and MHC I–restricted T cells typically are CD4+ and CD8+, respectively.[273,371-375] The current view is that CD4+ and CD8+ lineages diverge in thymocytes that have been rescued from cell death (ie, they have undergone positive selection), before their late-stage thymic maturation; this divergence does not require thymocyte migration to the medulla.[363] Most humans and most mouse strains have more CD4+ than CD8+ T cells; it is likely that several mechanisms contribute to account for this difference, including the greater association of tyrosine kinase Lck with CD4 than with CD8.[376,377] Coreceptor expression is associated with functional properties displayed by T cells when responding to antigen. CD8+ T cells typically become cytotoxic killer cells; in contrast, CD4+ T cells, depending on additional signals at the time of antigen encounter, differentiate into any of multiple effector types that control the function of other cells involved in immune responses, and are often referred to as "helper."[324,378] Programming for such killer or helper functions is initiated in the thymus, before thymocytes acquire the capability to mount functional immune responses. This programming is coupled with coreceptor gene expression rather than with MHC specificity, as demonstrated using mutant mice engineered to disrupt the normal MHC-coreceptor matching.[379,380] This suggested that the same transcriptional circuitries control coreceptor genes and functional differentiation, a prediction that has been verified since.

Connecting MHC Restriction to Lineage Commitment

How lineage differentiation is matched to MHC restriction has attracted considerable attention and generated much discussion, because of its own importance and of its paradigmatic value in studies of cell differentiation. The question was initially viewed in the light of opposing instructive and stochastic-selective perspectives.[381] Instruction posited that MHC I versus MHC II molecules send lineage-specific differentiating signals to thymocytes (eg, through co-engagement of CD8 vs CD4 coreceptors with TCR); in this instructive perspective, all selected thymocytes express the coreceptor matched to their MHC specificity. The selection hypothesis presumed no such signal. It stated that rescue from programmed cell death occurs after lineage commitment and requires co-engagement of TCR and coreceptor by MHC molecules. The selection hypothesis predicts that thymocytes with coreceptor expression mismatched to their MHC specificity transiently develop (and would be half the number of thymocytes signaled to undergo positive selection if commitment to either lineage was purely stochastic), but fail to complete positive selection. Several lines of evidence argue against a purely selective perspective, most notably observations that the efficiency of positive selection for cells carrying a given TCR clonotype can be too high to accommodate the selection idea.[382] However, efforts to identify MHC-driven lineage-specific signals have long been unfruitful, challenging the instructive hypothesis.

An important insight[383] was that "instruction" could come from the greater association of the tyrosine kinase Lck with CD4 than with CD8,[303,304] resulting in greater TCR signaling by TCR and CD4 co-engagement than by TCR and CD8 co-engagement. However, further assessment of this idea gave rise to distinct interpretations, in large part because of the pleiotropic effects of TCR signaling in thymocytes.[378] Subsequent research explored whether lineage differentiation was controlled by the duration, rather than the intensity, of TCR signaling (where duration refers not necessarily to individual TCR-peptide-MHC interactions, but also to the time it would take for repeated signaling events to result in lineage differentiation). Although differing on specific aspects, two studies proposed that CD4$^+$-lineage commitment required signals of greater duration than CD8$^+$-lineage commitment,[384,385] and this idea garnered support from *in vivo* analyses.[386,387] An elegant explanation to why MHC II–restricted thymocytes would undergo longer signaling than MHC I–restricted thymocytes came from two sets of observations.[378,385] First, the CD4$^+$ commitment of MHC II–restricted thymocytes follows positive selection and CD8 down-regulation, occurring in CD4$^+$ thymocytes expressing intermediate CD8 levels. Second, down-regulation of CD8 is not specific of MHC II signaling and can take place in MHC I–signaled thymocytes as well. Together, these observations form the basis for the "kinetic signaling" hypothesis,[378,385] which is currently the prevailing model for CD4$^+$-CD8$^+$-lineage differentiation (Fig. 27.12). It proposes that (1) CD8 down-regulation occurs in response to positively

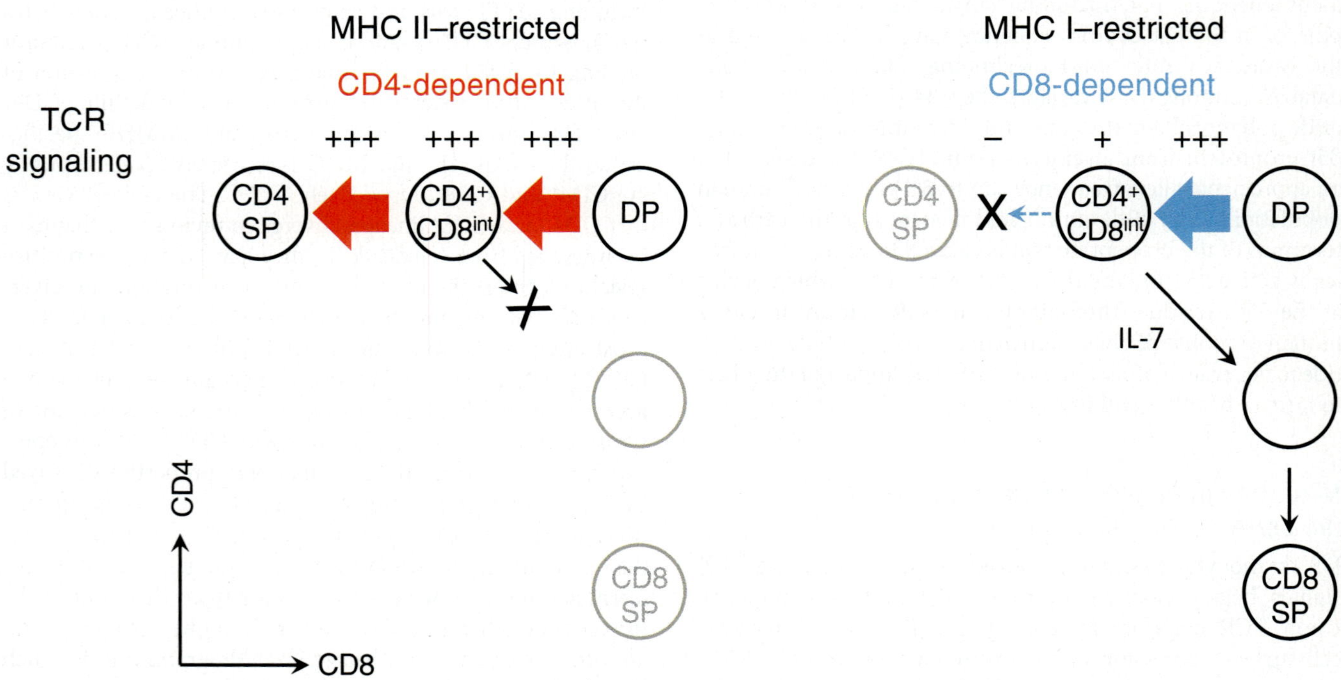

FIG. 27.12. The kinetic signaling model of CD4$^+$-CD8$^+$ lineage commitment. The developmental trajectories of MHC II– and MHC I–restricted thymocytes are schematically depicted on a virtual flow cytometry plot of CD4 and CD8 expression. The thickness of colored arrows indicates the intensity of TCR signaling; black arrows show TCR-independent developmental steps, Xs indicate differentiation paths (*gray-colored*) made unavailable by attributes of TCR signaling. The model first proposes that initial signaling in DP thymocytes reduces expression of *Cd8* genes, which (1) causes MHC I–induced signals to decrease because of the reduced CD8 protein expression, but (2) leaves MHC II–induced signals unaffected. The model further proposes that persistence of TCR signaling in CD4$^+$ CD8^{int} cells causes Thpok expression and CD4$^+$-lineage commitment, whereas cessation of TCR signaling in CD4$^+$ CD8^{int} cells makes them receptive to IL-7 signaling and results in CD8$^+$-lineage commitment.

selecting TCR signals, regardless of their MHC specificity, (2) CD8 down-regulation in turn compromises MHC I–induced but not MHC II–induced TCR signaling, because TCR signaling is coreceptor-dependent, and (3) the resulting asymmetry in the kinetics of MHC I– versus MHC II–induced TCR signaling determines lineage commitment, with persistent signals causing CD4+ commitment, and transient signals CD8+ commitment. The key concept of kinetic signaling is that TCR signaling in DP thymocytes generates a signaling asymmetry between MHC I- and MHC II–restricted cells at the time of lineage commitment; it does not require (nor does it exclude) that such an asymmetry already exists in pre-selection DP thymocytes as a consequence of CD4 or CD8 functional properties. Contrasting with the role of TCR signaling for CD4+-lineage commitment, the kinetic signaling model proposes that cytokine signaling, principally by IL-7, operates CD8+-lineage commitment, in line with the requirement of cytokine signaling for the differentiation of CD8+ SP thymocytes and T cells.[320,378,385,388]

A Dual Regulatory Loop Controls CD4+-CD8+-Lineage Differentiation

Another line of investigations has identified transcription factors controlling CD4+-CD8+ lineage commitment. These studies have built a model, by which expression of two transcription factors, Thpok and Runx3, determines the differentiation into CD4+ and CD8+ lineages, respectively. The Runx-family transcription factor Runx3 is expressed in CD8+-lineage thymocytes and T cells, whereas its paralog Runx1 is expressed in virtually all other thymocytes subsets and in naïve CD4+ T cells.[389-391] The role of Runx molecules in CD4+-CD8+ differentiation was uncovered from their ability to repress expression of *Cd4*, by binding a cis-regulatory silencer element located in the first *Cd4* intron[389,392-394] (Fig. 27.13). This silencer ensures the lineage-specificity of *Cd4* expression because the activity of *Cd4* enhancers is not specific of CD4+-lineage cells. Silencer activity (and therefore repression of *Cd4*) depends on its recruitment of Runx1, in DN thymocytes, or Runx3, in CD8+-lineage thymocytes. Runx3 is also important for the expression of CD8 and binds multiple enhancers located within or near *Cd8a* and *Cd8b1*, some of which show developmental-stage specific activity.[396-401] Runx3 also inhibits expression of *Runx1*, as shown by the expression of *Runx1* in CD8+ T cells from mice carrying disrupted *Runx3* genes. This has important experimental consequences because Runx1 and Runx3 are at least in part functionally interchangeable[402]: thus, limited numbers of CD8+ T cells develop in *Runx3*-deficient mice, and disruption of both *Runx1* and *Runx3* (or of their common obligatory cofactor Cbfβ) is necessary to completely arrest CD8+ T cell differentiation in the thymus.[391,403] In addition to its effect on *Cd4* and *Cd8* genes, Runx3 is important for the functional programming of CD8+ T cells and their cytotoxic activity.[389,390,404]

Mirroring in part the role of Runx proteins in CD8+-lineage differentiation, the zinc finger transcription factor Thpok,[405,406] encoded by *Zbtb7b* (which we will refer to as *Thpok* here for simplicity), is essential for CD4+-lineage differentiation. Among conventional αβ thymocytes, *Thpok* expression is specific of MHC II–signaled, CD4+-lineage cells, and its expression in the thymus is mutually exclusive with that of Runx3.[403] *Thpok* is also expressed in all mature CD4+ T cells (with the notable exception of CD4+CD8α+CD8β- IELs[407]). Thpok is not necessary for development of MHC II–restricted thymocytes but specifically supports their CD4+ lineage differentiation: Thpok-deficient MHC II–restricted thymocytes become CD8+ instead of CD4+ T cells.[405] This observation, actually made from a "helper-deficient" mouse line subsequently shown to harbor an inactivating *Thpok* mutation,[405,408] indicated that distinct mechanisms underpin CD4+-lineage commitment and positive selection, even though both depend on TCR signaling. Thpok inhibits CD8+-lineage differentiation,[405,406] and the expression of many CD8+-lineage genes, including *Cd8* genes and *Runx3*.[403,409,410] Similar to Runx3 for CD8+ T cells, Thpok is needed for the appropriate programming of helper functions in CD4+ T cells.[405,406,411-413] The control of *Thpok* expression involves multiple positive regulatory elements, with at least in part stage-specific activity.[409,414-416] However, similar to *Cd4* enhancers, these elements are active in both CD4+- and CD8+-lineage cells, and the CD4+-lineage specificity of *Thpok* expression depends on a silencer, located upstream of the *Thpok* gene itself.[414,415] Recruitment of Runx1 or Runx3 to specific binding sites activates this silencer,[415,417] and therefore inhibits *Thpok* expression in CD8+-lineage thymocytes and T cells. The zinc finger transcription factor Mazr is also important for silencer activity and the proper control of *Thpok* expression.[418] Thus, a simple model is that the repressive activity of Thpok and Runx3 on each other's gene, whether direct or not, accounts for their mutually exclusive and lineage-specific expression (Fig. 27.13), whereas Runx1 contributes to prevent premature *Thpok* expression in DP thymocytes.

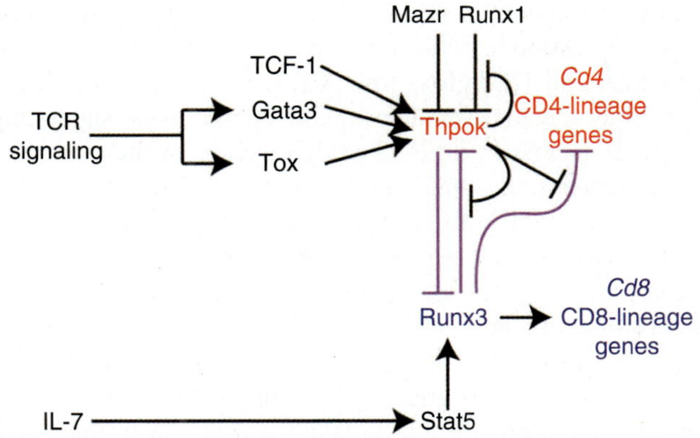

FIG. 27.13. Schematic and simplified representation of the transcriptional control of CD4+-CD8+ lineage commitment. The lineage commitment factors Thpok and Runx3 are engaged in a dual negative regulatory loop resulting in their mutually exclusive expression in SP thymocytes. In addition, Runx3 promotes expression of CD8+ lineage genes and represses that of CD4+ lineage genes, and Thpok antagonizes these two functions of Runx3. In pre selection DP thymocytes and TCR signaled precommitment cells, transcription factors Gata3 and Tox (both TCR-induced), and TCF-1 promote, and transcription factors Runx1 and Mazr inhibit, expression of *Thpok*. Reciprocally, there is evidence that Stat5 (activated by IL-7) contributes to *Runx3* expression, through mechanisms that remain to be determined.[257,395] Please see text for references.

This dual control mechanism is asymmetric in at least three respects. First, transgenic expression experiments have shown that Thpok is functionally dominant over Runx1 or Runx3 when co-expressed with either factor. That is, Thpok molecules inhibit the Runx-mediated repression of *Thpok* and of *Cd4*, in addition to inhibiting *Runx3* expression.[409,419] In line with this observation, CD4+ SP thymocytes and resting CD4+ T cells express both Thpok and Runx1. Thpok molecules bind both the *Thpok* and *Cd4* silencer, supporting the idea that Thpok directly affects silencer activity.[409] Second, CD8+ T cell differentiation is accompanied by DNA methylation-mediated epigenetic silencing of the *Cd4* locus, such that disruption of the *Cd4* silencer in post-thymic CD8+ T cells does not restore *Cd4* expression.[420,421] In contrast, repression of *Cd8* (and other CD8+-lineage genes) in mature CD4+ T cells is actively enforced by Thpok.[410,411] Third, whereas Runx activity is strictly required for the differentiation of CD8+ T cells,[391] some CD4+ SP thymocytes and T cells can differentiate from DP thymocytes fully defective for both Thpok and Runx activities.[403] This supports the idea that Thpok serves in the thymus mostly by antagonizing Runx functions rather than by inducing expression of CD4+-lineage genes. However, functional redundancies within the Runx family and between Thpok and its paralog LRF[411] complicate these analyses, and ongoing research addresses how CD4+-lineage gene expression is initiated in the thymus.

Because neither *Thpok* nor *Runx3* is expressed in DP thymocytes, the signals and transcription factors that control their expression have been intensively investigated. In line with the idea that strong or persistent TCR signals are needed for CD4+-lineage commitment, there is evidence that *Thpok* is induced by TCR signaling,[414] although not in DP thymocytes.[406] Ongoing efforts are directed at deciphering how TCR signaling induces Thpok expression. Several transcription factors promoting *Thpok* expression have been identified, including Gata3, E proteins E2A and HEB, Tox, and TCF-1.[422-427] Gata3 and TCF-1 both bind specific elements in the *Thpok* locus, suggesting that these factors could be directly involved in the control of *Thpok* expression. Gata3 and Tox have attracted much attention because they both are required for the differentiation of CD4+ T cells but at least in part dispensable for that of CD8+ T cells, and they are induced by TCR signaling. In particular, Gata3 expression is higher in MHC II– than in MHC I–signaled thymocytes, suggesting that it serves as a relay between TCR signals and *Thpok* upregulation. While ongoing research addresses that possibility, it was also noted that enforced *Thpok* expression fails to rescue the differentiation of Gata3-deficient thymocytes into CD4+ T cells, indicating that Gata3 serves additional roles in this process.[424] Other factors, including Bcl11b and Satb1, have complex functions in the regulation of *Thpok* and *Runx3* expression.[428,429] Determining how the combinatorial activity of these (and presumably other) factors contributes to *Thpok* expression and to the expression of other CD4+-lineage genes is obviously a complex task and an important objective of current research.

The control of *Runx3* expression involves distinct enhancers, stage-specific promoter usage and so far incompletely understood post-transcriptional mechanisms.[391,416,429] In line with the role of cytokines (principally IL-7) in the differentiation of CD8+-lineage thymocytes, there is evidence that cytokine signals, mediated by partly overlapping activities of transcription factors Stat5 and Stat6, promote expression of *Runx3*, in addition to having *Runx3*-independent eff ects.[320,388,395] Conversely, Thpok is the key factor preventing *Runx3* induction in MHC II selected thymocytes, because *Runx3* is similarly induced in MHC I-restricted and in Thpok-deficient MHC II–restricted thymocytes.[403] Thpok molecules bind several cis-regulatory elements important for *Runx3* expression, supporting the idea of a direct control.[429] Additionally, there is evidence for an indirect mechanism, whereby Thpok proteins enhance expression of Socs1, an inhibitor of cytokine signaling, and thereby prevent cytokine-induced *Runx3* expression.[430]

Role of the Thymus in Immune Tolerance

Evolving Concepts of Negative Selection

The concept of negative selection is a basic tenet of the clonal selection theory and common to B and T cells. For T cells, it posits that the thymic contribution to T cell tolerance is established by intrathymic clonal deletion, (ie, active cell death) of thymocytes that are overly self-reactive. Indeed, classical experiments addressing tissue transplantation tolerance or tolerance to superantigens (which bind MHC II outside of the peptide binding groove and are recognized by nonclonotypic determinants of specific TCR Vβ chains) provided evidence for active clonal deletion in the thymus.[233,431-433] These experiments further suggested that negatively selecting antigens are expressed by bone marrow–derived components located in the thymic medulla, including B cells and DCs.[434] This contrasted with the role of the cortical thymic epithelium in delivering positively selecting signals for the generation of conventional T cells.[342]

The simple picture of cortical positive selection and medullary deletion has become more complex over time[242] (Fig. 27.6). First, deletion is incomplete. *In vivo* models have consistently shown that large numbers of clonotypic T cells escape deletion, and antigen-specific self-reactive mature T cells can be detected in mice and humans, although there is evidence that they show lower affinity for their nominal antigen, and reduced functional responsiveness.[435,436] Second, in addition to deletion, the thymus diverts self-reactive cells toward lineages with immunosuppressive or regulatory activity. These include MHC II–restricted regulatory T (T_{reg}) cells expressing the transcription factor Foxp3, which fulfill an essential function of the thymus in immune tolerance.[437,438] The generation and function of T_{reg} cells is covered in detail in Chapter 30 and only a few specific aspects, limited to thymic T_{reg} differentiation, will be addressed below. Recent work points to the existence of a corresponding, although not Foxp3-based, lineage diversion pathway for thymocytes with high affinity for MHC I–presented peptides.[439] Third, evidence has accumulated that both cortical and medullary thymocytes are subject to deletion. Last, as discussed below, tolerance to tissue-restricted self-antigens (TRAs) requires their "promiscuous" expression by the medullary thymic epithelium.[440]

The Biology of Negative Selection: Current Perspectives

There is now decisive evidence that the thymus is essential to establish immune tolerance (referred to as "central tolerance"), and this function relies on both cell deleting and non–cell deleting mechanisms. Central tolerance is currently thought to comprise three main components: (1) cortical deletion, (2) medullary deletion, and (3) "lineage diversion," referring to the differentiation of self-reactive cells into lineages with immunosuppressive functions (Fig. 27.8). We will use negative selection to refer to thymocyte deletion by TCR-induced death.

Direct evidence for the cortical deletion of self-reactive clones comes from experiments enforcing expression of specific TCR clonotypes in thymocytes but avoiding premature expression before the DP stage.[239,441,442] Such experiments identified multiple TCRs triggering the death of DP (and thus cortical) thymocytes; many of these TCRs showed cross-reactivity to multiple MHC molecules.[239] Independent approaches have provided additional evidence for cortical deletion.[240,241] These experiments tracked expression of two transcription factors (Nur77 or Helios, whose high-level expression is indicative of negative selection signals) by thymocytes made unable to express the pro-apoptotic molecule Bim, a Bcl2-family effector of clonal deletion. Bim inactivation impairs programmed cell death, and therefore allows the survival of thymocytes targeted for clonal deletion, which are otherwise rapidly cleared from the thymus.[253,443] Both studies using this approach[240,241] reached the conclusion that most thymocyte TCR-induced death occurs in the cortex. Which cells provide the cortical deleting signals has yet to be determined. In addition to cTECs, self-antigen presenting cells in the thymic cortex include DCs, macrophages, thymocytes, and fibroblasts. However, the efficiency in antigen presentation varies among cell types, which differently express MHC molecules and costimulatory molecules. Although cortical DCs are clear candidates to provide negative selection signals, a potential role for macrophages or cTECs has not yet been excluded.[441,444] This early deletion is incomplete, and a (presumably) small subset of cells carrying TCRs that normally promote cortical deletion escape programmed cell death[239]; unexpectedly, these "escapee" cells are diverted toward a lineage characteristic of the epithelial layer of the small intestine mucosa and expressing the CD8α but not the CD8β subunit of CD8, and no CD4.[442,445,446] These "CD8αα" cells, for which a human equivalent has yet to be identified,[447] are thought to contribute to immune homeostasis in the intestine, which current research is evaluating.

Thymocytes reactive with TRAs are typically not deleted in the cortex, in which these antigens are not expressed; instead, such thymocytes can undergo deletion once they have reached the medulla, at the SP stage, where they encounter mTEC-expressed TRAs.[448,449] Evidence for medullary deletion comes from experimental systems combining thymocyte expression of a transgenic TCR and mTEC expression of its nominal antigen using cis-regulatory elements of a TRA, for example the rat insulin promoter that directs expression both in pancreatic islet cells and in mTECs.[450-452]

Signals and Intra-cellular Mechanisms of Negative Selection

The many facets of negative selection and the fact that the same TCR clonotype can promote clonal deletion and diversion to regulatory lineages have complicated the identification of the signals and intra-cellular mechanisms driving negative selection. Pro-apoptotic Bcl2-family molecules Bim and Puma serves redundantly to eliminate self-reactive cells,[453,454] and counteract the anti-apoptotic functions of Mcl1 and Bcl2. Transcription factors of the Nur77 family reach their highest expression in cells targeted for negative selection and promote cell death; they may serve not only as transcriptional regulators but also as direct apoptotic effectors through mitochondrial translocation and interactions with Bcl2.[455] They also are needed for the differentiation of T_{reg} cells.[456,457] How TCR signals by high affinity ligands promote expression of these genes is not fully understood, but there is evidence that it relies on MAP kinases of the p38 and Jnk family.[458] Genetic disruption of the anchoring proteins Stim1 and Stim2 has established that store-operated calcium mobilization contributes to negative selection,[323] although that function of calcium signaling does not appear to be mediated by calcineurin.[459]

Because both positive and negative selection are triggered by TCR signaling, the question arises of how and where signal transduction pathways diverge. Two complementary ideas have been explored as solutions to this question. The first is that positively and negatively selecting TCR signals are intrinsically distinct because they are delivered by low and high affinity peptide ligands. The second is that these signals are delivered by distinct APCs, that notably express separate sets of co-stimulatory molecules. There is ample support for the first idea and the concept of an intermediate affinity range resulting in signals conducive to positive but not negative selection.[281,460] Imaging studies have shown that, unlike low-affinity positively selecting ligands, signaling by high affinity ligands causes contacts of extended duration between thymocytes and APCs, and that this results in rapid commitment to cell death.[443,460,461] Such durable contacts and the signals they generate (including calcium entry) are thought to result in transcriptomic changes that cannot be achieved by the repetition of short-term signaling events triggered by positively selecting signals, which fits the earlier concept of kinetic proofreading.[462] How the TCR signal transduction machinery converts these distinct stimuli into survival versus death outcomes has attracted much interest. Ongoing work explores how the multiplicity of ITAM motifs in TCR-associated CD3 chains contribute to the discrimination between positive- and negative selection signals.[463-466] *In vitro* analyses have provided evidence that high-affinity peptides give rise to intense but short-lived activation of the Erk kinase pathway, in contrast to a durable activation achieved by low-affinity peptides. Even though Erk molecules themselves are not needed for negative selection,[467] there is evidence that their upstream activators, Ras-Raf complexes, are differentially activated by negatively and positively selecting signals.[281,468] Additional mechanisms contribute to set the border between positive and negative selection outcomes. These include the transcription factor Schnurri2, which is needed to protect TCR-signaled thymocytes from TCR-induced deletion.[469] Research has also focused on the physiological role of glucocorticoid hormones in setting the thresholds that distinguish positive from negative selection TCR signals.[317] Recent work has shown that these hormones inhibit expression of negative selection effector genes, including Helios and Nur77, and that their paracrine synthesis in TECs contributes to prevent negative selection of useful antigen-specific thymocytes and to allow their positive selection.[470,471]

Support is equally strong for a separation of tasks among thymic APCs, whereby cTEC-presented peptide-MHC complexes provide positively selecting signals and other types of APCs in the thymus, including DCs, mTECs and B cells, deliver negatively selecting signals.[440,444] This reflects several properties of these cells. mTECs express TRAs that promote the deletion of antigen-specific clones. Additionally, as discussed above, differences in antigen processing machineries cause cTECs and other APCs to present distinct sets of MHC-bound peptides; such differences may be important to avoid the negative selection of cells that received positive selection signals.[472-474] Last but not least, DCs and mTECs express co-stimulatory ligands for T cells, including CD80 and CD86 that engage CD28, which is expressed on all αβ lineage thymocytes. While the role of CD28 co-stimulation in negative selection has been debated,[475] recent studies have shown that it is an important contributor to active cell deletion and notably contributes to prevent diversion to the IEL lineage.[445,476]

Distinguishing Negative Selection From Lineage Diversion

Both clonal deletion and T_{reg} differentiation of MHC II–restricted thymocytes contribute to thymic tolerance. The conventional view is that while either fate awaits self-reactive thymocytes, T_{reg} differentiation occupies an intermediate niche in the TCR-ligand affinity spectrum, between highest affinity (deleting) and low affinity ligands (positively selecting) ligands[477-480] (Fig. 27.9). Indeed, many signaling mechanisms involved in clonal deletion also promote thymic T_{reg} differentiation, including signaling by CD28 or nuclear receptors of the Nur77 family.[456,457,481-483] However, although high-affinity ligands can promote T_{reg} differentiation and clonal deletion, not all T_{reg} cells carry TCRs with high affinity for self-ligands, and there is substantial overlap between the repertoires of conventional and T_{reg} $CD4^+$ T cells. Several factors, associated with or independent from TCR-ligand affinity, have been reported to affect the balance between deletion, T_{reg} differentiation and positive selection.[480] Studies in mice expressing pseudo self-antigens (eg, GFP) showed the importance of the expression pattern of self-antigens in the thymus. Expression driven by cis-regulatory elements with broad (or ubiquitous) activity favors TCR-induced death, whereas expression from tissue-restricted cis-regulatory elements, which cause antigen expression in thymic mTECs, preferentially drives thymocytes toward T_{reg} differentiation.[484,485] Expression of Aire in mTECs is needed for the generation of a normal T_{reg} repertoire,[486] at least in part because it diverts self-reactive clonotypes from a conventional to a T_{reg} fate.[487] As discussed later in this chapter, there is no evidence for a strict correspondence between APC and developmental fate: both direct presentation by mTECs and indirect DC-mediated presentation seem conducive to Aire-dependent T_{reg} differentiation and clonal deletion.[479,480] T_{reg} differentiation requires signaling by various ligands and costimulatory molecules,[488] including IL-2 (mainly produced by conventional thymocytes and provided in spatial association with the APC)[480,489-491] and receptors for TNF family members, including Ox40, Gitr, and TNFR2[492]; there is evidence that T_{reg} differentiation is favored by intrathymic production of TGFβ as a by-product of intrathymic cell death.[493]

In summary, considerable progress has been made in our understanding of the contribution of the thymus to immune tolerance, which we now see as having both deletional and nondeletional components and distinct mechanisms to target ubiquitous and tissue specific self-antigens. Furthermore, through the generation of cells with immuno-suppressive capacity, the thymus probably controls T cell tolerance to a variety of molecules, not only limited to genomic components but also including commensal microbiota and food-derived antigens.

Convergence and Divergence Among Agonist-Selected αβ T Cell Lineages

T_{reg} cells and IEL precursors are often grouped in a broad category of "agonist-selected" cells,[494] that also comprises lineages that acquire effector functions while in the thymus, iNKT and MAIT cells.[495,496] The development of these cells is examined in detail in other chapters, but it is interesting to highlight commonalities and differences among these subsets. The grouping of these cells within an "agonist-selected" class is justified by the evidence that they all recognize their selecting antigen with high affinity, generating strong TCR signals as assessed by expression of the TCR signaling target Nur77.[269] Accordingly, the development of T_{reg}, iNKT and IEL precursors depends on store-operated calcium entry, unlike that of conventional T cells.[323,442] However, important features differentiate iNKT and MAIT cells from T_{reg} and IEL precursors cells. First and foremost, iNKT and MAIT cells have low-diversity TCR repertoires, dominated by one or a few TCRα and TCRβ chains, unlike both T_{reg} and IEL precursors.[497] Accordingly, iNKT and MAIT cells recognize non peptide ligands of limited diversity presented by nonclassical MHC I (class Ib MHC) molecules (CD1d and MR1, respectively, expressed by DP thymocytes), whereas T_{reg} cells are restricted by classical class II MHC molecules, and IEL precursors appear to be selected by a broad array of classical and nonclassical MHC ligands.[442,446] Last, the selection of iNKT is followed by induction of the transcription factor PLZF, which promotes their intrathymic proliferation and acquisition of effector properties, including cytokine production[498]; the same is true for most MAIT cells,[496] whereas neither T_{reg} nor IEL precursors express PLZF, proliferate or produce intrathymic cytokines.

Thymic Medulla for Central Tolerance

Promiscuous Gene Expression in mTECs

At a population level, mTECs exhibit the expression of nearly 90% of all genes, including TRAs.[499] This characteristic unique to mTECs is termed promiscuous gene expression.[449,500] A nuclear protein, Aire (autoimmune regulator), is expressed in a subpopulation of mTECs and is a major contributor to the promiscuous gene expression, which in turn contributes to the establishment of self-tolerance of T cells to the entire genomic products of the body components.[501] Aire-deficiency causes a failure in establishing central tolerance to TRAs, resulting in autoimmune polyendocrinopathy-candidiasis-ectodermal dystrophy (APECED), also known as autoimmune polyendocrine syndrome 1 (APS-1), in humans and similar autoimmune diseases in mice.[501-503] In the nuclei of Aire$^+$ mTECs, Aire binds to many proteins that contribute to the enhancement of gene transcription, mRNA processing, and protein translation.[504-506] Many promiscuously expressed genes are dependent on Aire and are detected in the Aire$^+$ subpopulation of mTECs. However, a considerable number of promiscuously expressed genes are detectable in the Aire$^-$ subpopulation of mTECs, and are not diminished in mTECs in Aire-deficient mice.[507] A role for the transcription factor Fezf2 in promoting the Aire-independent promiscuous gene expression in mTECs has been reported.[508]

At a single-cell level, promiscuous expression of individual TRAs is detected at expression frequencies between 1% and 3% of mTECs.[509,510] A single mTEC co-expresses multiple promiscuously expressed genes, and a single Aire$^+$ mTEC co-expresses both Aire-dependent and Aire-independent genes.[509] The promiscuous expression of a particular gene is transient in single mTECs, and mTECs shift through heterogeneous patterns of promiscuous gene expression during ontogeny, which perhaps allows them to cover a diverse set of self-antigens.[511,512] Although mTECs express TRAs, actual antigen presentation to medullary SP thymocytes can be carried out directly by mTECs or indirectly, after antigen capture, by medullary DCs.[440] As with cortical deletion, not all medullary thymocytes reactive with self-TRAs are deleted, as a subset of them are diverted toward the Foxp3$^+$ T_{reg} cell lineage, in an Aire-dependent manner.[486,487,513,514]

Hematopoietic Antigen-Presenting Cells in Thymic Medulla

DCs in the thymus are enriched in the medullary region and play an important role in self-antigen presentation to induce the negative selection of self-reactive T cells and the generation of T_{regs}.[515,516] Thymic DCs include the three main DC lineages identified across tissues and species[517]: CD11b$^-$ CD8α$^+$ classical DC (cDC)1, CD11b$^+$CD8α$^-$ cDC2, and plasmacytoid DC (pDC).[444,480,518,519] While cDC1 are traditionally considered of intrathymic origin (derived from T cell precursors), recent evidence indicates that most of them are the progeny from myeloid precursors that migrate from the bone marrow into the thymus.[125,520] cDC2 and pDC also are of extrathymic origin; chemokine receptors CCR2 and CCR9, respectively, contribute to their intrathymic entry.[126,521,522] All types of thymic DCs can induce thymic deletion, whereas the complex interactions of cDC1 and cDC2 with mTECs for T_{reg} differentiation is being investigated.[479,523,524] After intrathymic maturation,[525] cDC1 can present antigens from mTECs, and are uniquely competent for cross-presentation; their localization in the thymic medulla is regulated by XCR1-XCL1-mediated chemokine signals. Thymic cDC1 express XCR1, whereas mTECs produce XCL1, the ligand for XCR1, in an Aire-dependent manner. Mice deficient in XCL1 are defective in the accumulation of DCs in the thymic medulla and the generation of T_{regs} in the thymus. T cells generated in the XCL1-deficient thymus are responsible for causing autoimmune disease in mice.[526] Migratory cDC2 and pDCs are competent for intrathymic presentation of, and tolerization to, antigens acquired extrathymically.[521,527,528] cDC2 can migrate to perivascular areas of the thymic cortex, and could therefore contribute to cortical negative selection.[441]

B cells in the thymus are reported to mediate the negative selection of SP thymocytes reactive to endogenous mammary tumor virus self-encoded superantigens. The maintenance of the cellularity of thymic B cells is dependent on the presence of mature SP thymocytes and the B-cell-autonomous expression of CD40, whereas thymic B cells reciprocally function in the negative selection of self-reactive thymocytes.[128] The CD40 signal in thymic B cells, by the interaction with autoreactive CD4SP thymocytes, enables thymic B cells to express Aire and present self-antigens to induce self-tolerance in T cells.[129]

Heterogeneity in mTECs

Epithelial cells localized within the medullary region of the thymus are highly heterogenous and comprise functionally and morphologically diverse cell populations.[38,529] In addition to the Aire$^+$ subpopulation, which exhibits promiscuous gene expression, mTECs contain a CCL21$^+$ subpopulation, which contributes to the recruitment of positively selected thymocytes from the cortex to the medullary region.[365] Like Aire$^+$ mTECs, CCL21$^+$ mTECs are essential for the establishment of self-tolerance in T cells. Most CCL21$^+$ mTECs belong to the MHC II^{low} population and are distinct from Aire$^+$ mTECs, which largely belong to the MHC II^{high} population.[49] CCL21$^+$ mTECs accumulate during postnatal ontogeny, whereas Aire$^+$ mTECs dominate during the perinatal period, when Aire plays an essential role in establishing long-lasting T cell tolerance.[486,530] Thus, CCL21$^+$ mTECs represent a functionally mature mTEC subpopulation.

MHC II^{low} mTECs also contain developmentally diverse subpopulations including precursors for Aire$^+$ mTECs and "post-Aire" mTECs. The MHC II^{low} mTECs include developmental potential to give rise to Aire$^+$ MHC II^{high} mTECs,[47] although the identity of mTEC progenitors is still poorly defined. Post-Aire mTECs are derived from Aire$^+$ MHC II^{high} mTECs and are generated in an Aire-dependent manner. Post-Aire mTECs exhibit promiscuous gene expression at a spectrum different from the gene expression profiles in Aire$^+$ mTECs.[46,48]

MHC II^{low} mTECs further contain morphologically diverse subpopulations including cystic ciliated cells, neurosecretory cells, concentric Hassall corpuscles and cholinergic chemosensory cells.[43,529,531] Recent studies have highlighted a subpopulation of mTECs that resemble tuft cells of the small intestine. Like tuft cells in the gut, thymic tuft cells are dependent on the transcription factor Pou2f3 for their generation, and express chemosensory receptors and the pro-inflammatory cytokine IL-25. Unlike intestinal tuft cells, however, thymic tuft cells are generated as a post-Aire mTEC subpopulation and express both MHC I and MHC

II molecules. The functions of thymic tuft cells are unclear, although reports suggest their roles in supporting type 2 invariant natural killer T cell (iNKT2) development by formation of an IL-4–rich microenvironment, and/or in controlling type 2 innate lymphoid cells (ILC2s) in the thymic medulla.[532,533] A recent study showed that CD104⁺CCL21⁺ MHC II^{low} mTECs are capable of IL-15 transpresentation for regulating iNKT1 and iNKT17 cells, while thymic tuft cells regulate iNKT2s via IL-25.[534] In addition, a subset of MHC II^{low} mTECs are involucrin^{+} and resemble terminally differentiated corneocytes; these cells constitute Hassall corpuscles that are detectable clearly in human thymus (Fig. 27.4). A study described that Hassall corpuscles contribute to the development of T_{reg} cells in the human thymus.[535]

Late-Stage Thymic Differentiation and Thymic Egress

Lineage-committed conventional CD4^{+} and CD8^{+} SP thymocytes are not yet fully mature.[536] They are functionally incompetent and susceptible to various apoptotic signals including glucocorticoids, whereas mature SP thymocytes are glucocorticoid-resistant, functional (eg, they have gained cytokine production potential), and able to enter cell cycle in response to TCR signals. SP thymocytes also initiate expression of trafficking molecules appropriate for their migration to peripheral lymphoid organs, including CD62L (L-selectin). Mouse SP thymocytes spend approximately 4 to 5 days in the medulla before being exported from the thymus,[537] as shown by (1) pulse-chase experiments of nucleoside analogs (eg, BrdU), which are incorporated in proliferating pre-DP cells and thus time-stamp cohorts of developing cells, and (2) by tracking the decay of fluorescence from GFP expressed from the *Rag* gene promoter, which is silenced upon TCR signaling at the DP stage. Although late-stage thymic maturation takes place in the thymic medulla, it is unclear whether it requires unique signals from the medullary environment, unlike tolerance to TRAs and T_{reg} differentiation.[538,539] A variety of surface markers, including CD69, CD24 (heat-stable antigen), and MHC I molecules (Fig. 27.10), tracks the late-stage thymic maturation of mouse SP thymocytes, although recent studies suggest that such phenotypic differences track continued developmental progression toward maturation rather than truly separate developmental stages.[476,540]

Thymocyte egress is an active process[541]; it depends on S1pr1, a G-protein-coupled surface receptor for sphingosine-1-phosphate (S1P), a lipid mediator abundant in the bloodstream.[542,543] Expression of S1pr1 is regulated at both the transcriptional and post-transcriptional levels. Transcription of *S1pr1* starts in SP thymocytes and is highest in the most mature SP cells. It requires the transcription factor Klf2, itself preferentially expressed in mature SP thymocytes.[544] Expression of *S1pr1* also requires two histone demethylases, Jmjd3 and Utx, suggesting that it is under control by epigenetic mechanisms in addition to transcription factors.[545] Surface expression of S1pr1 protein is controlled post-translationally by CD69: S1pr1 and CD69 associate at the cell surface, forming a complex that is rapidly internalized.[546] As a result, there is little if any surface expression of S1pr1 in immature SP thymocytes, which actively transcribe the *Cd69* gene. In contrast, in mature cells in which *Cd69* transcription subsides, increasing *S1pr1* transcription results in increased S1pr1 protein expression and elimination of CD69 from the cell surface. In addition to the S1P-mediated chemotaxis, CCR7 signals contribute to thymocyte emigration in newborn mice.[85] The export of mature thymocytes from the thymic parenchyma to the circulation is thought to occur through the perivascular space, which is channeled to postcapillary venules, arterioles, and lymphatics. The recirculation of mature T cells back to the thymic medulla has also been described.[547]

Two apparently distinct transcriptional circuits contribute to thymocyte maturation and survival. One, involving the serine-threonine kinases Tak1 and Ripk1 and the IκB kinase complex promotes thymocyte survival.[540,548] The second involves Foxo1, a Forkhead box family transcription factor that serves in part redundantly with Foxo3.[549,550] Foxo1 directs expression of genes involved in T cell trafficking, including *S1pr1*, *Klf2*, which are needed for thymic egress, and *Ccr7* (encoding CCR7) and *Sell* (encoding CD62L), which both promote homing to secondary lymphoid organs. Foxo1 also supports naïve T cell survival by enhancing expression of IL-7Rα. Current evidence suggests that two levels of controls prevent premature activation of this circuitry and thymocyte egress. In DP thymocytes, the transcription factor Gfi1 represses Foxo1 expression.[551] Additionally, TCR signaling enhances the activity of the PI3-kinase Akt signaling pathway, which post-translationally inhibits Foxo1 activity[552] and controls thymic egress.[553,554] Thus, it has been proposed that the dependence of late-stage thymic maturation genes on Foxo1 constitutes a gate-keeping mechanism preventing the maturation and egress of self-reactive medullary thymocytes[324,555]; such a mechanism would serve in addition to the post-translational control of S1pr1 expression by the TCR signaling target CD69.[546]

Thymic Crosstalk

Bidirectional Crosstalk Between Thymocytes and TECs

The thymus provides the microenvironment for T cell development, in which thymic nonhematopoietic cells critically regulate the fate determination and differentiation of thymocytes. The initial suggestion that developing thymocytes also affect the formation of the thymic microenvironment emerged from the analysis of the thymus in animals treated with the immunosuppressant cyclosporin A.[556] When mature thymocytes were eliminated by cyclosporin A treatment, the thymic medulla showed a reduction in size, which was recovered after cessation of the treatment. It was further shown genetically in mice that thymocyte development into mature TCRαβ^{+} T cells is required for the growth and/or maturation of mTECs.[557] Those results led to the concept of thymic crosstalk, in which lymphoid and stromal cells are symbiotic and influence each other during T cell development in the thymus.[558] Bidirectional crosstalk signals between developing thymocytes and mTECs that are important to establish self-tolerance in T cells include TNF family cytokines produced by positively selected thymocytes, which act on mTECs to mediate their growth and maturation, and CCR7 ligand chemokines produced by mTECs, which act on positively selected thymocytes to induce their migration into the thymic medulla.[559]

TNF Family Cytokines for Thymic Crosstalk

TCR signals that induce positive selection elevate the expression of many TNF family cytokines in DP thymocytes, whereas mTECs express the receptors for those TNF family cytokines, including RANK, CD40, and LTβR.[560] TNF family cytokines derived from positively selected thymocytes mediate the growth and maturation of mTECs to form the thymic medulla, which is essential for the establishment of self-tolerance in T cells. These TNF family cytokines affect mTECs through the activation of classical and nonclassical NF-κB signal pathways. TNF receptor associated factor 3 (TRAF3), an inhibitor of nonclassical NF-κB signaling, imposes crosstalk signal requirements for mTEC development. Indeed, the lack of TRAF3 in TECs restores mTEC development in the absence of thymocyte-mTEC crosstalk.[561]

In the thymus, RANK is predominantly expressed by mTECs.[47,560] The expression of RANK ligand (RANKL) in the thymus is detectable not only in positively selected thymocytes but also in innate immunity-associated cells, including $CD4^+$ $CD3^-$ lymphoid tissue inducer (LTi) cells, invariant Vγ5-TCR^+ thymocytes, and iNKT cells.[47,560,562,563] LTi cells and Vγ5-TCR^+ thymocytes are detectable in the fetal thymus, and RANKL produced by those cells is mainly involved in mTEC development during embryogenesis. On the other hand, the majority of positively selected thymocytes and iNKT cells are postnatally generated, and RANKL produced by those cells is predominantly involved in mTEC proliferation, and possibly maturation, after birth. Mice deficient in RANK or RANKL exhibit reductions in mTEC cellularity and medullary size. Many T cells generated in RANK- or RANKL-deficient thymus are overtly self-reactive and induce autoimmune diseases in mice. On the other hand, the overexpression of RANKL in thymocytes of TCRα-deficient mice, in which the positive selection of thymocytes is arrested and thereby the development of thymic medulla is impaired, restores mTEC cellularity and thymic medullary formation.[560]

Osteoprotegerin (OPG), the decoy RANKL receptor that inhibits RANK signaling, is expressed by a subpopulation in $Aire^+$ mTECs. As the cellularity of mTECs is elevated and the thymic medulla is enlarged in OPG-deficient mice, OPG regulates the formation of the thymic medulla by a negative feedback mechanism that fine-tunes mTEC cellularity.[560,564] Thus, the RANK-RANKL-OPG axis is essential for mediating thymic crosstalk signals to promote the proliferation of mTECs for the formation of the thymic medullary microenvironment and to establish self-tolerance in T cells.

CD40 is also detectable in mTECs, whereas CD40L is predominantly produced by CD4SP thymocytes.[560,565] While RANKL-expressing cells are primarily composed of newly positively selected CD4SP thymocytes and T_{regs}, CD40L-expressing cells include CD4SP thymocytes in the later stage of maturation, as defined by the reduced expression of CD69.[566] In contrast to their expression of RANKL, thymic LTi cells and Vγ5-TCR^+ thymocytes express a limited amount of CD40L.[566] Mice deficient in CD40 or CD40L exhibit only a minor, if any, decrease in the cellularity of mTECs, and T cells generated in the thymus of CD40-deficient mice do not induce autoimmune diseases. However, when the loss of either CD40 or CD40L was introduced in mice deficient in either RANK or RANKL, mTEC cellularity was severely diminished and severe autoimmune disease was detected in the mice.[567] Thus, the CD40-CD40L axis is involved in thymocyte-dependent thymic medulla formation in a manner synergistic with RANK-RANKL signaling.

LTβR is expressed comparably by both mTECs and cTECs.[560] However, the deficiency of LTβR affects the development of mTECs rather than cTECs.[568] Among the two ligands for LTβR, LTαβ is highly expressed in CD4SP and CD8SP thymocytes, whereas LIGHT expression is not abundant in thymocytes.[560] Mice deficient in LTβR or LTαβ exhibit a decrease in mTEC cellularity and the onset of autoimmune disease, whereas mTEC development is not diminished by the deficiency of LIGHT.[568] Thus, LTβR-LTαβ axis is important for mTEC development and self-tolerance establishment in T cells. The development of $Aire^+$ mTECs is independent of LTβR, whereas the LTβR signal promotes the development of $CCL21^+$ mTECs.[49] LTβR signals also regulate the development of the post-Aire mTECs, including the involucrin+ mTEC subpopulation.[569]

Costimulatory Interactions in Thymic Crosstalk

CD28 is widely expressed in thymocytes and at the highest level in DP thymocytes,[570] whereas the CD28 ligands, CD80 and CD86, are highly expressed in mTECs. CD28 provides a co-stimulatory signal to T cells and is important for the negative selection of self-reactive thymocytes and the generation of T_{regs}.[482,483,571,572] The CD28-CD80/CD86 co-stimulatory interaction up-regulates the expression of LTα and LTβ in thymocytes and of RANK in mTECs.[573] The absence of both CD28-CD80/CD86 and CD40-CD40L results in a profound deficiency in mTEC cellularity, and CD4SP thymocytes generated in CD40/CD80/CD86-deficient mice are highly autoreactive and are lethal upon adaptive transfer in mice. Thus, in cooperation with the CD40-CD40L pathway, the CD28-CD80/86 pathway plays an important role in the generation of the thymic medullary microenvironment and the establishment of self-tolerance in T cells.[573]

Involution and Regeneration of the Thymus

The thymus is one of the most susceptible organs to age-dependent atrophy, or involution (Fig. 27.2C). Thymic involution is associated with a decline in de novo T cell production and in the diversity of T cell repertoires, resulting in the deterioration of the immune system.[574] Age-dependent thymic involution is correlated with alterations in the morphology and global gene expression profiles, including micro-RNAs, of TECs.[575-577] Reduced expression and activity of transcription factors, Foxn1 and Myc, and antioxidant enzyme catalase, have each been reported.[577-579] Thymic involution is induced not only by aging but also by other stresses, including irradiation, infection, chemotherapeutic drugs, and sex hormone perturbation. In contrast to age-dependent thymic involution, stress-induced thymic injuries are transient and regenerable.[580] The involution of the thymus is associated with the reduction in the production of T cells in humans and mice.[581,582]

Mouse TECs populations are highly dynamic after birth and exhibit continuous turnover. It is estimated that approximately 10% of TECs in the adult thymus are newly supplied daily, and TECs turnover occurs every 10 to 14 days.[42] During the postinjury thymic repair, cTECs and mTECs are regenerated from injury-resistant cells that could be either bipotent progenitors or lineage-restricted cells. Cytokines, including FGF7 and IL-22, and sex hormone ablation help thymic regeneration and are potentially useful to enhance T cell reconstitution in immunocompromised situations.[580,583-587]

CONCLUDING REMARKS

T cell development is unique among immune cells, requiring migration of progenitors from their source of generation to a highly specialized organ that coordinates lineage specification and commitment, controls repertoire formation and imposition of self-recognition through selection, and development of functionally distinct T cell subsets (including T_{regs}, IELs, MAIT cells). This complex process relies on multiple choreographed cellular interactions and factors, as well as tightly regulated cell migration and cross-talk between epithelial cells and developing thymocytes.

This chapter addressed how T cell development is constrained to occur within the thymus. We suggest that the thymus provides a specialized environment that allows persistent engagement of Notch1 by Dll4, which is abundantly provided by TECs. We speculate that the need for a thymus arises from the requirement to generate unique peptides for positive selection. This is performed by highly specialized cTECs, which provide peptides distinct from other self-peptides, and that possibly serve to positively select TCRs that can subsequently avoid negative selection by mTECs and other antigen-presenting cells. We further speculate that the thymic medulla is needed to provide a specialized environment for the establishment of self-tolerance in T cells, by inducing negative selection and generation of regulatory lineages. These requirements are met by the thymus, with anatomically separate cortical and medullary regions, that display different sets of MHC-associated peptides sequentially to developing thymocytes. Positive selection ultimately serves to select the repertoire of conventional T cells, able to recognize foreign peptides in association with self-MHC, as well as T_{regs} able to extinguish pathological immune responses. Understanding how repertoire selection fine-tunes these distinct classes of response remains an important question for future experimental work.

The generation of the αβ T cell repertoire occurs early in life, as thymic activity in mammals declines quickly after puberty. Combined with constraints on naive T cell survival, the near cessation of T cell generation in adults results in a reduction of αβ T cell diversity in elderly individuals, that can be magnified by disease and therapeutic intervention, notably after myeloablative chemotherapy for cancer or bone marrow transplantation. Such a reduction in T cell diversity is associated with impaired immune responses, including to vaccines. Regenerative approaches focused on TECs have promise, but much work will be needed before such approaches can be routinely used in clinics. Understanding the diversity of TEC types and how they control distinct steps of T cell development is an important area of research. Generating T cells *in vitro*, or re-starting T cell development *in vivo*, are thus important practical objectives of T cell development research. The diversity of MHC molecules and its implication for functionality and tolerance of the T cell repertoire are formidable challenges to such endeavors, and the generation of appropriate *in vitro* systems would be an important step for progress. Since the discovery of the thymus as an immune organ,[588-592] research in the past 60 years has built a view of αβ lineage T cell development as driven by a sequence of transcriptional circuitries that control diverse processes, including TCR gene rearrangement, proliferation, lineage differentiation, survival, and acquisition of functional properties. Critical tasks for the generation of tractable *in vitro* models of αβ lineage cell development include deciphering the relationships between these elementary circuits and how they build the transcriptomic landscape of mature T cells, and understanding how they are articulated with signaling cascades triggered by cell surface signals. High throughput approaches, including at the single-cell level, and genetic models to track the development of antigen-specific T cells will be instrumental to fulfill the objective of understanding and reconstituting the developmental sequence for T cells.

ACKNOWLEDGMENTS

We apologize to authors whose work we were unable to cite for reasons of space. We thank Ellen Rothenberg for generous permission to use text, figures and concepts from the previous version of this chapter. Graham Anderson, Jonathan Ashwell, Laura Chopp, Thomas Ciucci, Wilfried Ellmeier, Christelle Harly, Joonsoo Kang, Ferenc Livak, Paul Love, Jia Nie, Ellen Robey, Susannah Shissler, Alfred Singer, Melanie Vacchio, David Wiest, and Juan-Carlos Zuniga-Pflucker provided essential insight and critical comments. The authors were supported by the Intramural Research Program of the Center for Cancer Research at the National Cancer Institute.

REFERENCES

1. Standring S. *Gray's Anatomy*. 40th ed. Elsevier; 2008.
2. Simpson JG, Gray ES, Beck JS. Age involution in the normal human adult thymus. *Clin Exp Immunol*. 1975;19(2):261-265.
3. Tovi F, Mares AJ. The aberrant cervical thymus. Embryology, pathology, and clinical implications. *Am J Surg*. 1978;136(5):631-637.
4. Terszowski G, Muller SM, Bleul CC, et al. Evidence for a functional second thymus in mice. *Science*. 2006;312(5771):284-287.
5. Dooley J, Erickson M, Gillard GO, Farr AG. Cervical thymus in the mouse. *J Immunol*. 2006;176(11):6484-6490.
6. Mori K, Itoi M, Tsukamoto N, Kubo H, Amagai T. The perivascular space as a path of hematopoietic progenitor cells and mature T cells between the blood circulation and the thymic parenchyma. *Int Immunol*. 2007;19(6):745-753.
7. Gordon J, Wilson VA, Blair NF, et al. Functional evidence for a single endodermal origin for the thymic epithelium. *Nat Immunol*. 2004;5(5):546-553.
8. Blackburn CC, Manley NR. Developing a new paradigm for thymus organogenesis. *Nat Rev Immunol*. 2004;4(4):278-289.
9. Merscher S, Funke B, Epstein JA, et al. TBX1 is responsible for cardiovascular defects in velo-cardio-facial/DiGeorge syndrome. *Cell*. 2001;104(4):619-629.
10. Jerome LA, Papaioannou VE. DiGeorge syndrome phenotype in mice mutant for the T-box gene, Tbx1. *Nat Genet*. 2001;27(3):286-291.
11. Zou D, Silvius D, Davenport J, Grifone R, Maire P, Xu PX. Patterning of the third pharyngeal pouch into thymus/parathyroid by Six and Eya1. *Dev Biol*. 2006;293(2):499-512.
12. Xu PX, Zheng W, Laclef C, et al. Eya1 is required for the morphogenesis of mammalian thymus, parathyroid and thyroid. *Development*. 2002;129(13):3033-3044.
13. Wallin J, Eibel H, Neubuser A, Wilting J, Koseki H, Balling R. Pax1 is expressed during development of the thymus epithelium and is required for normal T-cell maturation. *Development*. 1996;122(1):23-30.
14. Manley NR, Capecchi MR. Hox group 3 paralogs regulate the development and migration of the thymus, thyroid, and parathyroid glands. *Dev Biol*. 1998;195(1):1-15.
15. Hetzer-Egger C, Schorpp M, Haas-Assenbaum A, Balling R, Peters H, Boehm T. Thymopoiesis requires Pax9 function in thymic epithelial cells. *Eur J Immunol*. 2002;32(4):1175-1181.
16. Gordon J, Patel SR, Mishina Y, Manley NR. Evidence for an early role for BMP4 signaling in thymus and parathyroid morphogenesis. *Dev Biol*. 2010;339(1):141-154.
17. Bleul CC, Boehm T. BMP signaling is required for normal thymus development. *J Immunol*. 2005;175(8):5213-5221.
18. Revest JM, Suniara RK, Kerr K, Owen JJ, Dickson C. Development of the thymus requires signaling through the fibroblast growth factor receptor R2-IIIb. *J Immunol*. 2001;167(4):1954-1961.
19. Moore-Scott BA, Manley NR. Differential expression of Sonic hedgehog along the anterior-posterior axis regulates patterning of pharyngeal pouch endoderm and pharyngeal endoderm-derived organs. *Dev Biol*. 2005;278(2):323-335.
20. Frank DU, Fotheringham LK, Brewer JA, et al. An Fgf8 mouse mutant phenocopies human 22q11 deletion syndrome. *Development*. 2002;129(19):4591-4603.
21. Foster KE, Gordon J, Cardenas K, et al. EphB-ephrin-B2 interactions are required for thymus migration during organogenesis. *Proc Natl Acad Sci U S A*. 2010;107(30):13414-13419.
22. Balciunaite G, Keller MP, Balciunaite E, et al. Wnt glycoproteins regulate the expression of FoxN1, the gene defective in nude mice. *Nat Immunol*. 2002;3(11):1102-1108.
23. Sitnik KM, Kotarsky K, White AJ, Jenkinson WE, Anderson G, Agace WW. Mesenchymal cells regulate retinoic acid receptor-dependent cortical thymic epithelial cell homeostasis. *J Immunol*. 2012;188(10):4801-4809.
24. Dupe V, Ghyselinck NB, Wendling O, Chambon P, Mark M. Key roles of retinoic acid receptors alpha and beta in the patterning of the caudal hindbrain, pharyngeal arches and otocyst in the mouse. *Development*. 1999;126(22):5051-5059.
25. Nehls M, Pfeifer D, Schorpp M, Hedrich H, Boehm T. New member of the winged-helix protein family disrupted in mouse and rat nude mutations. *Nature*. 1994;372(6501):103-107.
26. Nehls M, Kyewski B, Messerle M, et al. Two genetically separable steps in the differentiation of thymic epithelium. *Science*. 1996;272(5263):886-889.
27. Pantelouris EM. Absence of thymus in a mouse mutant. *Nature*. 1968;217(5126):370-371.
28. Adriani M, Martinez-Mir A, Fusco F, et al. Ancestral founder mutation of the nude (FOXN1) gene in congenital severe combined immunodeficiency associated with alopecia in southern Italy population. *Ann Hum Genet*. 2004;68(pt 3):265-268.
29. Zuklys S, Handel A, Zhanybekova S, et al. Foxn1 regulates key target genes essential for T cell development in postnatal thymic epithelial cells. *Nat Immunol*. 2016;17:1206-1215.
30. Uddin MM, Ohigashi I, Motosugi R, et al. Foxn1-beta5t transcriptional axis controls CD8(+) T-cell production in the thymus. *Nat Commun*. 2017;8:14419.
31. Du Q, Huynh LK, Coskun F, et al. FOXN1 compound heterozygous mutations cause selective thymic hypoplasia in humans. *J Clin Invest*. 2019;129(11):4724-4738.
32. Bajoghli B, Guo P, Aghaallaei N, et al. A thymus candidate in lampreys. *Nature*. 2011;470(7332):90-94.
33. Hirano M, Guo P, McCurley N, et al. Evolutionary implications of a third lymphocyte lineage in lampreys. *Nature*. 2013;501(7467):435-438.
34. Rossi SW, Jenkinson WE, Anderson G, Jenkinson EJ. Clonal analysis reveals a common progenitor for thymic cortical and medullary epithelium. *Nature*. 2006;441(7096):988-991.
35. Ohigashi I, Zuklys S, Sakata M, et al. Aire-expressing thymic medullary epithelial cells originate from beta5t-expressing progenitor cells. *Proc Natl Acad Sci U S A*. 2013;110(24):9885-9890.
36. Baik S, Jenkinson EJ, Lane PJ, Anderson G, Jenkinson WE. Generation of both cortical and Aire(+) medullary thymic epithelial compartments from CD205(+) progenitors. *Eur J Immunol*. 2013;43(3):589-594.
37. Alves NL, Takahama Y, Ohigashi I, et al. Serial progression of cortical and medullary thymic epithelial microenvironments. *Eur J Immunol*. 2014;44(1):16-22.
38. Takahama Y, Ohigashi I, Baik S, Anderson G. Generation of diversity in thymic epithelial cells. *Nat Rev Immunol*. 2017;17(5):295-305.
39. Bleul CC, Corbeaux T, Reuter A, Fisch P, Monting JS, Boehm T. Formation of a functional thymus initiated by a postnatal epithelial progenitor cell. *Nature*. 2006;441(7096):992-996.
40. Ohigashi I, Zuklys S, Sakata M, et al. Adult thymic medullary epithelium is maintained and regenerated by lineage-restricted cells rather than bipotent progenitors. *Cell Rep*. 2015;13(7):1432-1443.
41. Rodewald HR, Paul S, Haller C, Bluethmann H, Blum C. Thymus medulla consisting of epithelial islets each derived from a single progenitor. *Nature*. 2001;414(6865):763-768.
42. Gray DH, Seach N, Ueno T, et al. Developmental kinetics, turnover, and stimulatory capacity of thymic epithelial cells. *Blood*. 2006;108(12):3777-3785.
43. Farr AG, Rudensky A. Medullary thymic epithelium: a mosaic of epithelial "self"? *J Exp Med*. 1998;188(1):1-4.
44. Sekai M, Hamazaki Y, Minato N. Medullary thymic epithelial stem cells maintain a functional thymus to ensure lifelong central T cell tolerance. *Immunity*. 2014;41(5):753-761.
45. Baik S, Sekai M, Hamazaki Y, Jenkinson WE, Anderson G. Relb acts downstream of medullary thymic epithelial stem cells and is essential for the emergence of RANK(+) medullary epithelial progenitors. *Eur J Immunol*. 2016;46(4):857-862.
46. Wang X, Laan M, Bichele R, Kisand K, Scott HS, Peterson P. Post-Aire maturation of thymic medullary epithelial cells involves selective expression of keratinocyte-specific autoantigens. *Front Immunol*. 2012;3(March):19.
47. Rossi SW, Kim MY, Leibbrandt A, et al. RANK signals from CD4(+)3(−) inducer cells regulate development of Aire-expressing epithelial cells in the thymic medulla. *J Exp Med*. 2007;204(6):1267-1272.
48. Metzger TC, Khan IS, Gardner JM, et al. Lineage tracing and cell ablation identify a post-Aire-expressing thymic epithelial cell population. *Cell Rep*. 2013;5(1):166-179.
49. Lkhagvasuren E, Sakata M, Ohigashi I, Takahama Y. Lymphotoxin beta receptor regulates the development of CCL21-expressing subset of postnatal medullary thymic epithelial cells. *J Immunol*. 2013;190(10):5110-5117.
50. Donskoy E, Goldschneider I. Thymocytopoiesis is maintained by blood-borne precursors throughout postnatal life. A study in parabiotic mice. *J Immunol*. 1992;148(6):1604-1612.
51. Martins VC, Ruggiero E, Schlenner SM, et al. Thymus-autonomous T cell development in the absence of progenitor import. *J Exp Med*. 2012;209(8):1409-1417.
52. Martins VC, Busch K, Juraeva D, et al. Cell competition is a tumour suppressor mechanism in the thymus. *Nature*. 2014;509(7501):465-470.
53. Owen JJ, Ritter MA. Tissue interaction in the development of thymus lymphocytes. *J Exp Med*. 1969;129(2):431-442.
54. Haynes BF, Heinly CS. Early human T cell development: analysis of the human thymus at the time of initial entry of hematopoietic stem cells into the fetal thymic microenvironment. *J Exp Med*. 1995;181(4):1445-1458.
55. Bhandoola A, von Boehmer H, Petrie HT, Zuniga-Pflucker JC. Commitment and developmental potential of extrathymic and intrathymic T cell precursors: plenty to choose from. *Immunity*. 2007;26(6):678-689.
56. Adolfsson J, Mansson R, Buza-Vidas N, et al. Identification of Flt3+ lympho-myeloid stem cells lacking erythro-megakaryocytic potential a revised road map for adult blood lineage commitment. *Cell*. 2005;121(2):295-306.
57. Adolfsson J, Borge OJ, Bryder D, et al. Upregulation of Flt3 expression within the bone marrow Lin(−)Sca1(+)c- kit(+) stem cell compartment is accompanied by loss of self-renewal capacity. *Immunity*. 2001;15(4):659-669.
58. Kondo M, Weissman IL, Akashi K. Identification of clonogenic common lymphoid progenitors in mouse bone marrow. *Cell*. 1997;91(5):661-672.
59. Inlay MA, Bhattacharya D, Sahoo D, et al. Ly6d marks the earliest stage of B-cell specification and identifies the branchpoint between B-cell and T-cell development. *Genes Dev*. 2009;23(20):2376-2381.
60. Petrie HT, Zuniga-Pflucker JC. Zoned out: functional mapping of stromal signaling microenvironments in the thymus. *Annu Rev Immunol*. 2007;25:649-679.
61. Zietara N, Lyszkiewicz M, Puchalka J, et al. Multicongenic fate mapping quantification of dynamics of thymus colonization. *J Exp Med*. 2015;212(10):1589-1601.
62. Saran N, Lyszkiewicz M, Pommerencke J, et al. Multiple extrathymic precursors contribute to T-cell development with different kinetics. *Blood*. 2010;115(6):1137-1144.
63. Krueger A. Thymus colonization: who, how, how many? *Arch Immunol Ther Exp*. 2018;66(2):81-88.
64. Rodewald HR, Kretzschmar K, Takeda S, Hohl C, Dessing M. Identification of pro-thymocytes in murine fetal blood: T lineage commitment can precede thymus colonization. *EMBO J*. 1994;13(18):4229-4240.
65. Wright DE, Wagers AJ, Gulati AP, Johnson FL, Weissman IL. Physiological migration of hematopoietic stem and progenitor cells. *Science*. 2001;294(5548):1933-1936.
66. Schwarz BA, Bhandoola A. Circulating hematopoietic progenitors with T lineage potential. *Nat Immunol*. 2004;5(9):953-960.
67. Umland O, Mwangi WN, Anderson BM, Walker JC, Petrie HT. The blood contains multiple distinct progenitor populations with clonogenic B and T lineage potential. *J Immunol*. 2007;178(7):4147-4152.
68. Lind EF, Prockop SE, Porritt HE, Petrie HT. Mapping precursor movement through the postnatal thymus reveals specific microenvironments supporting defined stages of early lymphoid development. *J Exp Med*. 2001;194(2):127-134.
69. Scimone ML, Aifantis I, Apostolou I, von Boehmer H, von Andrian UH. A multistep adhesion cascade for lymphoid progenitor cell homing to the thymus. *Proc Natl Acad Sci U S A*. 2006;103(18):7006-7011.
70. von Andrian UH, Mackay CR. T-cell function and migration. Two sides of the same coin. *N Engl J Med*. 2000;343(14):1020-1034.
71. Le Douarin NM, Jotereau FV. Tracing of cells of the avian thymus through embryonic life in interspecific chimeras. *J Exp Med*. 1975;142(1):17-40.
72. Foss DL, Donskoy E, Goldschneider I. The importation of hematogenous precursors by the thymus is a gated phenomenon in normal adult mice. *J Exp Med*. 2001;193(3):365-374.
73. Liu C, Saito F, Liu Z, et al. Coordination between CCR7- and CCR9-mediated chemokine signals in prevascular fetal thymus colonization. *Blood*. 2006;108(8):2531-2539.
74. Zlotoff DA, Sambandam A, Logan TD, Bell JJ, Schwarz BA, Bhandoola A. CCR7 and CCR9 together recruit hematopoietic progenitors to the adult thymus. *Blood*. 2010;115(10):1897-1905.
75. Krueger A, Willenzon S, Lyszkiewicz M, Kremmer E, Forster R. CC chemokine receptor 7 and 9 double-deficient hematopoietic progenitors are severely impaired in seeding the adult thymus. *Blood*. 2010;115(10):1906-1912.
76. Benz C, Bleul CC. A multipotent precursor in the thymus maps to the branching point of the T versus B lineage decision. *J Exp Med*. 2005;202(1):21-31.
77. Lai AY, Kondo M. Identification of a bone marrow precursor of the earliest thymocytes in adult mouse. *Proc Natl Acad Sci U S A*. 2007;104(15):6311-6316.
78. Ding L, Morrison SJ. Haematopoietic stem cells and early lymphoid progenitors occupy distinct bone marrow niches. *Nature*. 2013;495(7440):231-235.
79. Schlenner SM, Madan V, Busch K, et al. Fate mapping reveals separate origins of T cells and myeloid lineages in the thymus. *Immunity*. 2010;32(3):426-436.
80. Lai AY, Lin SM, Kondo M. Heterogeneity of Flt3-expressing multipotent progenitors in mouse bone marrow. *J Immunol*. 2005;175(8):5016-5023.
81. Serwold T, Ehrlich LI, Weissman IL. Reductive isolation from bone marrow and blood implicates common lymphoid progenitors as the major source of thymopoiesis. *Blood*. 2009;113(4):807-815.
82. Zhang SL, Wang X, Manna S, et al. Chemokine treatment rescues profound T-lineage progenitor homing defect after bone marrow transplant conditioning in mice. *Blood*. 2014;124(2):296-304.
83. Sultana DA, Zhang SL, Todd SP, Bhandoola A. Expression of functional P-selectin glycoprotein ligand 1 on hematopoietic progenitors is developmentally regulated. *J Immunol*. 2012;188(9):4385-4393.
84. Rossi FM, Corbel SY, Merzaban JS, et al. Recruitment of adult thymic progenitors is regulated by P-selectin and its ligand PSGL-1. *Nat Immunol*. 2005;6(6):626-634.
85. Ueno T, Hara K, Willis MS, et al. Role for CCR7 ligands in the emigration of newly generated T lymphocytes from the neonatal thymus. *Immunity*. 2002;16(2):205-218.
86. Uehara S, Song K, Farber JM, Love PE. Characterization of CCR9 expression and CCL25/thymus-expressed chemokine responsiveness during T cell development: CD3(high)CD69+ thymocytes and gammadeltaTCR+ thymocytes preferentially respond to CCL25. *J Immunol*. 2002;168(1):134-142.
87. Wurbel MA, Philippe JM, Nguyen C, et al. The chemokine TECK is expressed by thymic and intestinal epithelial cells and

attracts double- and single-positive thymocytes expressing the TECK receptor CCR9. *Eur J Immunol.* 2000;30(1):262-271.

88. Calderon L, Boehm T. Three chemokine receptors cooperatively regulate homing of hematopoietic progenitors to the embryonic mouse thymus. *Proc Natl Acad Sci U S A.* 2011;108(18):7517-7522.
89. Kawamoto H, Ohmura K, Fujimoto S, Katsura Y. Emergence of T cell progenitors without B cell or myeloid differentiation potential at the earliest stage of hematopoiesis in the murine fetal liver. *J Immunol.* 1999;162(5):2725-2731.
90. Harman BC, Jenkinson WE, Parnell SM, Rossi SW, Jenkinson EJ, Anderson G. T/B lineage choice occurs prior to intrathymic Notch signaling. *Blood.* 2005;106(3):886-892.
91. Chen ELY, Thompson PK, Zuniga-Pflucker JC. RBPJ-dependent Notch signaling initiates the T cell program in a subset of thymus-seeding progenitors. *Nat Immunol.* 2019;20(11):1456-1468.
92. Upadhaya S, Sawai CM, Papalexi E, et al. Kinetics of adult hematopoietic stem cell differentiation in vivo. *J Exp Med.* 2018;215(11):2815-2832.
93. Harly C, Cam M, Kaye J, Bhandoola A. Development and differentiation of early innate lymphoid progenitors. *J Exp Med.* 2018;215(1):249-262.
94. Guo Y, Maillard I, Chakraborti S, Rothenberg EV, Speck NA. Core binding factors are necessary for natural killer cell development and cooperate with Notch signaling during T-cell specification. *Blood.* 2008;112(3):480-492.
95. Dias S, Mansson R, Gurbuxani S, Sigvardsson M, Kee BL. E2A proteins promote development of lymphoid-primed multipotent progenitors. *Immunity.* 2008;29(2):217-227.
96. Yoshida T, Ng SY, Zuniga-Pflucker JC, Georgopoulos K. Early hematopoietic lineage restrictions directed by Ikaros. *Nat Immunol.* 2006;7(4):382-391.
97. Arinobu Y, Mizuno S, Chong Y, et al. Reciprocal activation of GATA-1 and PU.1 marks initial specification of hematopoietic stem cells into myeloerythroid and myelolymphoid lineages. *Cell Stem Cell.* 2007;1(4):416-427.
98. Allman D, Sambandam A, Kim S, et al. Thymopoiesis independent of common lymphoid progenitors. *Nat Immunol.* 2003;4(2):168-174.
99. Godfrey DI, Zlotnik A, Suda T. Phenotypic and functional characterization of c-kit expression during intrathymic T cell development. *J Immunol.* 1992;149(7):2281-2285.
100. Porritt HE, Rumfelt LL, Tabrizifard S, Schmitt TM, Zuniga-Pflucker JC, Petrie HT. Heterogeneity among DN1 prothymocytes reveals multiple progenitors with different capacities to generate T cell and non-T cell lineages. *Immunity.* 2004;20(6):735-745.
101. Buono M, Facchini R, Matsuoka S, et al. A dynamic niche provides Kit ligand in a stage-specific manner to the earliest thymocyte progenitors. *Nat Cell Biol.* 2016;18(2):157-167.
102. Rodewald HR, Kretzschmar K, Swat W, Takeda S. Intrathymically expressed c-kit ligand (stem cell factor) is a major factor driving expansion of very immature thymocytes in vivo. *Immunity.* 1995;3(3):313-319.
103. Izon DJ, Punt JA, Xu L, et al. Notch1 regulates maturation of CD4+ and CD8+ thymocytes by modulating TCR signal strength. *Immunity.* 2001;14(3):253-264.
104. Koch U, Fiorini E, Benedito R, et al. Delta-like 4 is the essential, nonredundant ligand for Notch1 during thymic T cell lineage commitment. *J Exp Med.* 2008;205(11):2515-2523.
105. Wilson A, MacDonald HR, Radtke F. Notch 1-deficient common lymphoid precursors adopt a b cell fate in the thymus. *J Exp Med.* 2001;194(7):1003-1012.
106. Radtke F, Wilson A, Stark G, et al. Deficient T cell fate specification in mice with an induced inactivation of Notch1. *Immunity.* 1999;10(5):547-558.
107. Hozumi K, Mailhos C, Negishi N, et al. Delta-like 4 is indispensable in thymic environment specific for T cell development. *J Exp Med.* 2008;205(11):2507-2513.
108. Alves NL, Richard-Le Goff O, Huntington ND, et al. Characterization of the thymic IL-7 niche in vivo. *Proc Natl Acad Sci U S A.* 2009;106(5):1512-1517.
109. Rothenberg EV. T cell lineage commitment: identity and renunciation. *J Immunol.* 2011;186(12):6649-6655.
110. Rothenberg EV, Zhang J, Li L. Multilayered specification of the T-cell lineage fate. *Immunol Rev.* 2010;238(1):150-168.
111. Rothenberg EV, Moore JE, Yui MA. Launching the T-cell-lineage developmental programme. *Nat Rev Immunol.* 2008;8(1):9-21.
112. Wu L, Scollay R, Egerton M, Pearse M, Spangrude GJ, Shortman K. CD4 expressed on earliest T-lineage precursor cells in the adult murine thymus. *Nature.* 1991;349(6304):71-74.
113. Carlyle JR, Michie AM, Furlonger C, et al. Identification of a novel developmental stage marking lineage commitment of progenitor thymocytes. *J Exp Med.* 1997;186(2):173-182.
114. Wu L, Antica M, Johnson GR, Scollay R, Shortman K. Developmental potential of the earliest precursor cells from the adult mouse thymus. *J Exp Med.* 1991;174(6):1617-1627.
115. Wong SH, Walker JA, Jolin HE, et al. Transcription factor RORalpha is critical for nuocyte development. *Nat Immunol.* 2012;13(3):229-236.
116. Ramond C, Berthault C, Burlen-Defranoux O, et al. Two waves of distinct hematopoietic progenitor cells colonize the fetal thymus. *Nat Immunol.* 2014;15(1):27-35.
117. Shortman K, Wu L. Early T lymphocyte progenitors. *Annu Rev Immunol.* 1996;14:29-47.
118. Matsuzaki Y, Gyotoku J, Ogawa M, et al. Characterization of c-kit positive intrathymic stem cells that are restricted to lymphoid differentiation. *J Exp Med.* 1993;178(4):1283-1292.
119. Wada H, Masuda K, Satoh R, et al. Adult T-cell progenitors retain myeloid potential. *Nature.* 2008;452(7188):768-772.
120. Bell JJ, Bhandoola A. The earliest thymic progenitors for T cells possess myeloid lineage potential. *Nature.* 2008;452(7188):764-767.
121. Vosshenrich CA, Garcia-Ojeda ME, Samson-Villeger SI, et al. A thymic pathway of mouse natural killer cell development characterized by expression of GATA-3 and CD127. *Nat Immunol.* 2006;7(11):1217-1224.
122. Franco CB, Scripture-Adams DD, Proekt I, et al. Notch/Delta signaling constrains reengineering of pro-T cells by PU.1. *Proc Natl Acad Sci U S A.* 2006;103(32):11993-11998.
123. Taghon TN, David ES, Zuniga-Pflucker JC, Rothenberg EV. Delayed, asynchronous, and reversible T-lineage specification induced by Notch/Delta signaling. *Genes Dev.* 2005;19(8):965-978.
124. Feyerabend TB, Terszowski G, Tietz A, et al. Deletion of Notch1 converts pro-T cells to dendritic cells and promotes thymic B cells by cell-extrinsic and cell-intrinsic mechanisms. *Immunity.* 2009;30(1):67-79.
125. De Obaldia ME, Bell JJ, Bhandoola A. Early T-cell progenitors are the major granulocyte precursors in the adult mouse thymus. *Blood.* 2013;121(1):64-71.
126. Li J, Park J, Foss D, Goldschneider I. Thymus-homing peripheral dendritic cells constitute two of the three major subsets of dendritic cells in the steady-state thymus. *J Exp Med.* 2009;206(3):607-622.
127. Perera J, Meng L, Meng F, Huang H. Autoreactive thymic B cells are efficient antigen-presenting cells of cognate self-antigens for T cell negative selection. *Proc Natl Acad Sci U S A.* 2013;110(42):17011-17016.
128. Fujihara C, Williams JA, Watanabe M, Jeon H, Sharrow SO, Hodes RJ. T cell-B cell thymic cross-talk: maintenance and function of thymic B cells requires cognate CD40-CD40 ligand interaction. *J Immunol.* 2014;193(11):5534-5544.
129. Yamano T, Nedjic J, Hinterberger M, et al. Thymic B cells are licensed to present self antigens for central T cell tolerance induction. *Immunity.* 2015;42(6):1048-1061.
130. Schmitt TM, Ciofani M, Petrie HT, Zuniga-Pflucker JC. Maintenance of T cell specification and differentiation requires recurrent notch receptor-ligand interactions. *J Exp Med.* 2004;200(4):469-479.
131. von Freeden-Jeffry U, Vieira P, Lucian LA, McNeil T, Burdach SE, Murray R. Lymphopenia in interleukin (IL)-7 gene-deleted mice identifies IL-7 as a nonredundant cytokine. *J Exp Med.* 1995;181(4):1519-1526.
132. Maki K, Sunaga S, Komagata Y, et al. Interleukin 7 receptor-deficient mice lack gammadelta T cells. *Proc Natl Acad Sci U S A.* 1996;93(14):7172-7177.
133. Cao X, Shores EW, Hu-Li J, et al. Defective lymphoid development in mice lacking expression of the common cytokine receptor gamma chain. *Immunity.* 1995;2(3):223-238.
134. He YW, Malek TR. Interleukin-7 receptor alpha is essential for the development of gamma delta + T cells, but not natural killer cells. *J Exp Med.* 1996;184(1):289-293.
135. Moore TA, von Freeden-Jeffry U, Murray R, Zlotnik A. Inhibition of gamma delta T cell development and early thymocyte maturation in IL-7 −/− mice. *J Immunol.* 1996;157(6):2366-2373.
136. Malissen M, Pereira P, Gerber DJ, Malissen B, DiSanto JP. The common cytokine receptor gamma chain controls survival of gamma/delta T cells. *J Exp Med.* 1997;186(8):1277-1285.
137. Lu M, Tayu R, Ikawa T, et al. The earliest thymic progenitors in adults are restricted to T, NK, and dendritic cell lineage and have a potential to form more diverse TCRbeta chains than fetal progenitors. *J Immunol.* 2005;175(9):5848-5856.
138. Livak F, Tourigny M, Schatz DG, Petrie HT. Characterization of TCR gene rearrangements during adult murine T cell development. *J Immunol.* 1999;162(5):2575-2580.
139. Tourigny MR, Mazel S, Burtrum DB, Petrie HT. T cell receptor (TCR)-beta gene recombination: dissociation from cell cycle regulation and developmental progression during T cell ontogeny. *J Exp Med.* 1997;185(9):1549-1556.
140. Klug DB, Carter C, Crouch E, Roop D, Conti CJ, Richie ER. Interdependence of cortical thymic epithelial cell differentiation and T-lineage commitment. *Proc Natl Acad Sci U S A.* 1998;95(20):11822-11827.
141. Hollander GA, Wang B, Nichogiannopoulou A, et al. Developmental control point in induction of thymic cortex regulated by a subpopulation of prothymocytes. *Nature.* 1995;373(6512):350-353.
142. Takahama Y, Letterio JJ, Suzuki H, Farr AG, Singer A. Early progression of thymocytes along the CD4/CD8 developmental pathway is regulated by a subset of thymic epithelial cells expressing transforming growth factor beta. *J Exp Med.* 1994;179(5):1495-1506.
143. Benz C, Heinzel K, Bleul CC. Homing of immature thymocytes to the subcapsular microenvironment within the thymus is not an absolute requirement for T cell development. *Eur J Immunol.* 2004;34(12):3652-3663.
144. Pui JC, Allman D, Xu L, et al. Notch1 expression in early lymphopoiesis influences B versus T lineage determination. *Immunity.* 1999;11(3):299-308.
145. Rothenberg EV. Programming for T-lymphocyte fates: modularity and mechanisms. *Genes Dev.* 2019;33(17-18):1117-1135.
146. Maillard I, Adler SH, Pear WS. Notch and the immune system. *Immunity.* 2003;19(6):781-791.
147. Thompson PK, Zuniga-Pflucker JC. On becoming a T cell, a convergence of factors kick it up a Notch along the way. *Semin Immunol.* 2011;23(5):350-359.
148. Liu H, Chi AW, Arnett KL, et al. Notch dimerization is required for leukemogenesis and T-cell development. *Genes Dev.* 2010;24(21):2395-2407.
149. Kodama H, Nose M, Niida S, Nishikawa S, Nishikawa S. Involvement of the c-kit receptor in the adhesion of hematopoietic stem cells to stromal cells. *Exp Hematol.* 1994;22(10):979-984.
150. Schmitt TM, Zuniga-Pflucker JC. Induction of T cell development from hematopoietic progenitor cells by delta-like-1 in vitro. *Immunity.* 2002;17(6):749-756.
151. Weber BN, Chi AW, Chavez A, et al. A critical role for TCF-1 in T-lineage specification and differentiation. *Nature.* 2011;476(7358):63-68.
152. Germar K, Dose M, Konstantinou T, et al. T-cell factor 1 is a gatekeeper for T-cell specification in response to Notch signaling. *Proc Natl Acad Sci U S A.* 2011;108(50):20060-20065.
153. Emmanuel AO, Arnovitz S, Haghi L, et al. TCF-1 and HEB cooperate to establish the epigenetic and transcription profiles of CD4(+)CD8(+) thymocytes. *Nat Immunol.* 2018;19(12):1366-1378.
154. Garcia-Ojeda ME, Klein Wolterink RG, Lemaitre F, et al. GATA-3 promotes T-cell specification by repressing B-cell potential in pro-T cells in mice. *Blood.* 2013;121(10):1749-1759.
155. Johnson JL, Georgakilas G, Petrovic J, et al. Lineage-determining transcription factor TCF-1 initiates the epigenetic identity of T cells. *Immunity.* 2018;48(2):243-257.e10.
156. Kueh HY, Yui MA, Ng KK, et al. Asynchronous combinatorial action of four regulatory factors activates Bcl11b for T cell commitment. *Nat Immunol.* 2016;17(8):956-965.
157. Isoda T, Moore AJ, He Z, et al. Non-coding transcription instructs chromatin folding and compartmentalization to dictate enhancer-promoter communication and T cell fate. *Cell.* 2017;171(1):103-119 e118.
158. Hu G, Cui K, Fang D, et al. Transformation of accessible chromatin and 3D nucleome underlies lineage commitment of early T cells. *Immunity.* 2018;48(2):227-242.e8.
159. Ng KK, Yui MA, Mehta A, et al. A stochastic epigenetic switch controls the dynamics of T-cell lineage commitment. *Elife.* 20;2018:7.
160. Li P, Burke S, Wang J, et al. Reprogramming of T cells to natural killer-like cells upon Bcl11b deletion. *Science.* 2010;329(5987):85-89.
161. Hosokawa H, Romero-Wolf M, Yui MA, et al. Bcl11b sets pro-T cell fate by site-specific cofactor recruitment and by repressing Id2 and Zbtb16. *Nat Immunol.* 2018;19(12):1427-1440.
162. McCormack MP, Shields BJ, Jackson JT, et al. Requirement for Lyl1 in a model of Lmo2-driven early T-cell precursor ALL. *Blood.* 2013;122(12):2093-2103.
163. McCormack MP, Young LF, Vasudevan S, et al. The Lmo2 oncogene initiates leukemia in mice by inducing thymocyte self-renewal. *Science.* 2010;327(5967):879-883.
164. Smith S, Tripathi R, Goodings C, et al. LIM domain only-2 (LMO2) induces T-cell leukemia by two distinct pathways. *PLoS One.* 2014;9(1):e85883.
165. Ungerback J, Hosokawa H, Wang X, et al. Pioneering, chromatin remodeling, and epigenetic constraint in early T-cell gene regulation by SPI1 (PU.1). *Genome Res.* 2018;28(10):1508-1519.
166. Zhou W, Yui MA, Williams BA, et al. Single-cell analysis reveals regulatory gene expression dynamics leading to lineage commitment in early T cell development. *Cell Syst.* 2019;9(4):321-337.e9.
167. De Obaldia ME, Bell JJ, Wang X, et al. T cell development requires constraint of the myeloid regulator C/EBP-alpha by the Notch target and transcriptional repressor Hes1. *Nat Immunol.* 2013;14(12):1277-1284.
168. Bain G, Engel I, Robanus Maandag EC, et al. E2A deficiency leads to abnormalities in alphabeta T-cell development and to rapid development of T-cell lymphomas. *Mol Cell Biol.* 1997;17(8):4782-4791.
169. Anderson MK, Hernandez-Hoyos G, Diamond RA, Rothenberg EV. Precise developmental regulation of Ets family transcription factors during specification and commitment to the T cell lineage. *Development.* 1999;126(14):3131-3148.
170. Bender TP, Kremer CS, Kraus M, Buch T, Rajewsky K. Critical functions for c-Myb at three checkpoints during thymocyte development. *Nat Immunol.* 2004;5(7):721-729.
171. Xu W, Carr T, Ramirez K, McGregor S, Sigvardsson M, Kee BL. E2A transcription factors limit expression of Gata3 to facilitate T lymphocyte lineage commitment. *Blood.* 2013;121(9):1534-1542.
172. Braunstein M, Anderson MK. HEB-deficient T-cell precursors lose T-cell potential and adopt an alternative pathway of differentiation. *Mol Cell Biol.* 2011;31(5):971-982.

173. Phelan JD, Saba I, Zeng H, et al. Growth factor independent-1 maintains Notch1-dependent transcriptional programming of lymphoid precursors. *PLoS Genet.* 2013;9(9):e1003713.
174. Zohren F, Souroullas GP, Luo M, et al. The transcription factor Lyl-1 regulates lymphoid specification and the maintenance of early T lineage progenitors. *Nat Immunol.* 2012;13(8):761-769.
175. Murre C. Helix-loop-helix proteins and lymphocyte development. *Nat Immunol.* 2005;6(11):1079-1086.
176. Kee BL. E and ID proteins branch out. *Nat Rev Immunol.* 2009;9(3):175-184.
177. Lavaert M, Liang KL, Vandamme N, et al. Integrated scRNA-Seq identifies human postnatal thymus seeding progenitors and regulatory dynamics of differentiating immature thymocytes. *Immunity.* 2020;52:1088-1104.
178. Shih HY, Sciume G, Mikami Y, et al. Developmental acquisition of Regulomes underlies innate lymphoid cell functionality. *Cell.* 2016;165(5):1120-1133.
179. Miyazaki M, Miyazaki K, Chen K, et al. The E-Id protein axis specifies adaptive lymphoid cell identity and suppresses thymic innate lymphoid cell development. *Immunity.* 2017;46(5):818-834.e4.
180. Wang HC, Qian L, Zhao Y, et al. Downregulation of E Protein activity augments an ILC2 differentiation program in the thymus. *J Immunol.* 2017;198(8):3149-3156.
181. Yang Q, Li F, Harly C, et al. TCF-1 upregulation identifies early innate lymphoid progenitors in the bone marrow. *Nat Immunol.* 2015;16(10):1044-1050.
182. Zook EC, Kee BL. Development of innate lymphoid cells. *Nat Immunol.* 2016;17(7):775-782.
183. Hosokawa H, Romero-Wolf M, Yang Q, et al. Cell type-specific actions of Bcl11b in early T-lineage and group 2 innate lymphoid cells. *J Exp Med.* 2020;217(1):e20190972.
184. King AG, Kondo M, Scherer DC, Weissman IL. Lineage infidelity in myeloid cells with TCR gene rearrangement: a latent developmental potential of proT cells revealed by ectopic cytokine receptor signaling. *Proc Natl Acad Sci U S A.* 2002;99(7):4508-4513.
185. Yannoutsos N, Wilson P, Yu W, et al. The role of recombination activating gene (RAG) reinduction in thymocyte development in vivo. *J Exp Med.* 2001;194(4):471-480.
186. Fehling HJ, Krotkova A, Saint-Ruf C, von Boehmer H. Crucial role of the pre-T-cell receptor alpha gene in development of alpha beta but not gamma delta T cells. *Nature.* 1995;375(6534):795-798.
187. Irving BA, Alt FW, Killeen N. Thymocyte development in the absence of pre-T cell receptor extracellular immunoglobulin domains. *Science.* 1998;280(5365):905-908.
188. O'Shea CC, Thornell AP, Rosewell IR, Hayes B, Owen MJ. Exit of the pre-TCR from the ER/cis-Golgi is necessary for signaling differentiation, proliferation, and allelic exclusion in immature thymocytes. *Immunity.* 1997;7(5):591-599.
189. Pang SS, Berry R, Chen Z, et al. The structural basis for autonomous dimerization of the pre-T-cell antigen receptor. *Nature.* 2010;467(7317):844-848.
190. Yamasaki S, Ishikawa E, Sakuma M, et al. Mechanistic basis of pre-T cell receptor-mediated autonomous signaling critical for thymocyte development. *Nat Immunol.* 2006;7(1):67-75.
191. Li X, Mizsei R, Tan K, et al. Pre-T cell receptors topologically sample self-ligands during thymocyte beta-selection. *Science.* 2021;371:181-185.
192. Mallis RJ, Bai K, Arthanari H, et al. Pre-TCR ligand binding impacts thymocyte development before alphabetaTCR expression. *Proc Natl Acad Sci U S A.* 2015;112(27):8373-8378.
193. von Boehmer H. Unique features of the pre-T-cell receptor alpha-chain: not just a surrogate. *Nat Rev Immunol.* 2005;5(7):571-577.
194. Berger MA, Dave V, Rhodes MR, et al. Subunit composition of pre-T cell receptor complexes expressed by primary thymocytes: CD3 delta is physically associated but not functionally required. *J Exp Med.* 1997;186(9):1461-1467.
195. Mombaerts P, Iacomini J, Johnson RS, Herrup K, Tonegawa S, Papaioannou VE. RAG-1-deficient mice have no mature B and T lymphocytes. *Cell.* 1992;68(5):869-877.
196. Shinkai Y, Rathbun G, Lam KP, et al. RAG-2-deficient mice lack mature lymphocytes owing to inability to initiate V(D)J rearrangement. *Cell.* 1992;68(5):855-867.
197. Kruisbeek AM, Haks MC, Carleton M, Michie AM, Zuniga-Pflucker JC, Wiest DL. Branching out to gain control: how the pre-TCR is linked to multiple functions. *Immunol Today.* 2000;21(12):637-644.
198. Borowski C, Li X, Aifantis I, Gounari F, von Boehmer H. Pre-TCRalpha and TCRalpha are not interchangeable partners of TCRbeta during T lymphocyte development. *J Exp Med.* 2004;199(5):607-615.
199. Aifantis I, Borowski C, Gounari F, Lacorazza HD, Nikolich-Zugich J, von Boehmer H. A critical role for the cytoplasmic tail of pTalpha in T lymphocyte development. *Nat Immunol.* 2002;3(5):483-488.
200. Weng AP, Millholland JM, Yashiro-Ohtani Y, et al. c-Myc is an important direct target of Notch1 in T-cell acute lymphoblastic leukemia/lymphoma. *Genes Dev.* 2006;20(15):2096-2109.
201. Kreslavsky T, Gleimer M, Miyazaki M, et al. β-selection-induced proliferation is required for αβ T cell differentiation. *Immunity.* 2012;37(5):840-853.
202. Solanki NR, Stadanlick JE, Zhang Y, et al. Rpl22 loss selectively impairs αβ T cell development by dysregulating endoplasmic reticulum stress signaling. *J Immunol.* 2016;197(6):2280-2289.
203. Stadanlick JE, Zhang Z, Lee SY, et al. Developmental arrest of T cells in Rpl22-deficient mice is dependent upon multiple p53 effectors. *J Immunol.* 2011;187(2):664-675.
204. Ciofani M, Schmitt TM, Ciofani A, et al. Obligatory role for cooperative signaling by pre-TCR and Notch during thymocyte differentiation. *J Immunol.* 2004;172(9):5230-5239.
205. Janas ML, Varano G, Gudmundsson K, Noda M, Nagasawa T, Turner M. Thymic development beyond beta-selection requires phosphatidylinositol 3-kinase activation by CXCR4. *J Exp Med.* 2010;207(1):247-261.
206. Trampont PC, Tosello-Trampont AC, Shen Y, et al. CXCR4 acts as a costimulator during thymic beta-selection. *Nat Immunol.* 2010;11(2):162-170.
207. Boudil A, Matei IR, Shih HY, et al. IL-7 coordinates proliferation, differentiation and Tcra recombination during thymocyte beta-selection. *Nat Immunol.* 2015;16(4):397-405.
208. Ciofani M, Zuniga-Pflucker JC. Notch promotes survival of pre-T cells at the beta-selection checkpoint by regulating cellular metabolism. *Nat Immunol.* 2005;6(9):881-888.
209. Lin WC, Desiderio S. Cell cycle regulation of V(D)J recombination-activating protein RAG-2. *Proc Natl Acad Sci U S A.* 1994;91(7):2733-2737.
210. Ishida I, Verbeek S, Bonneville M, Itohara S, Berns A, Tonegawa S. T-cell receptor gamma delta and gamma transgenic mice suggest a role of a gamma gene silencer in the generation of alpha beta T cells. *Proc Natl Acad Sci U S A.* 1990;87(8):3067-3071.
211. Wong GW, Knowles GC, Mak TW, Ferrando AA, Zuniga-Pflucker JC. HES1 opposes a PTEN-dependent check on survival, differentiation, and proliferation of TCRbeta-selected mouse thymocytes. *Blood.* 2012;120(7):1439-1448.
212. Palomero T, Lim WK, Odom DT, et al. NOTCH1 directly regulates c-MYC and activates a feed-forward-loop transcriptional network promoting leukemic cell growth. *Proc Natl Acad Sci U S A.* 2006;103(48):18261-18266.
213. Zhao B, Yoganathan K, Li L, Lee JY, Zuniga-Pflucker JC, Love PE. Notch and the pre-TCR coordinate thymocyte proliferation by induction of the SCF subunits Fbxl1 and Fbxl12. *Nat Immunol.* 2019;20(10):1381-1392.
214. Reizis B, Leder P. Direct induction of T lymphocyte-specific gene expression by the mammalian Notch signaling pathway. *Genes Dev.* 2002;16(3):295-300.
215. Guy-Grand D, Azogui O, Celli S, et al. Extrathymic T cell lymphopoiesis: ontogeny and contribution to gut intraepithelial lymphocytes in athymic and euthymic mice. *J Exp Med.* 2003;197(3):333-341.
216. Maillard I, Schwarz BA, Sambandam A, et al. Notch-dependent T-lineage commitment occurs at extrathymic sites following bone marrow transplantation. *Blood.* 2006;107(9):3511-3519.
217. Besseyrias V, Fiorini E, Strobl LJ, et al. Hierarchy of Notch-Delta interactions promoting T cell lineage commitment and maturation. *J Exp Med.* 2007;204(2):331-343.
218. Okamura RM, Sigvardsson M, Galceran J, Verbeek S, Clevers H, Grosschedl R. Redundant regulation of T cell differentiation and TCRalpha gene expression by the transcription factors LEF-1 and TCF-1. *Immunity.* 1998;8(1):11-20.
219. Barndt R, Dai MF, Zhuang Y. A novel role for HEB downstream or parallel to the pre-TCR signaling pathway during alpha beta thymopoiesis. *J Immunol.* 1999;163(6):3331-3343.
220. Engel I, Johns C, Bain G, Rivera RR, Murre C. Early thymocyte development is regulated by modulation of E2A protein activity. *J Exp Med.* 2001;194(6):733-745.
221. Winandy S, Wu L, Wang JH, Georgopoulos K. Pre-T cell receptor (TCR) and TCR-controlled checkpoints in T cell differentiation are set by Ikaros. *J Exp Med.* 1999;190(8):1039-1048.
222. Haks MC, Lefebvre JM, Lauritsen JP, et al. Attenuation of gammadeltaTCR signaling efficiently diverts thymocytes to the alphabeta lineage. *Immunity.* 2005;22(5):595-606.
223. Hayes SM, Love PE. Distinct structure and signaling potential of the gamma delta TCR complex. *Immunity.* 2002;16(6):827-838.
224. Lee SY, Stadanlick J, Kappes DJ, Wiest DL. Towards a molecular understanding of the differential signals regulating alphabeta/gammadelta T lineage choice. *Semin Immunol.* 2010;22(4):237-246.
225. Lee SY, Coffey F, Fahl SP, et al. Noncanonical mode of ERK action controls alternative alphabeta and gammadelta T cell lineage fates. *Immunity.* 2014;41(6):934-946.
226. Hayes SM, Li L, Love PE. TCR signal strength influences alphabeta/gammadelta lineage fate. *Immunity.* 2005;22(5):583-593.
227. Bruno L, Fehling HJ, von Boehmer H. The alpha beta T cell receptor can replace the gamma delta receptor in the development of gamma delta lineage cells. *Immunity.* 1996;5(4):343-352.
228. Malhotra N, Narayan K, Cho OH, et al. A network of high-mobility group box transcription factors programs innate interleukin-17 production. *Immunity.* 2013;38(4):681-693.
229. Spidale NA, Sylvia K, Narayan K, et al. Interleukin-17-Producing gammadelta T Cells originate from SOX13(+) progenitors that are independent of gammadeltaTCR signaling. *Immunity.* 2018;49(5):857-872.e5.
230. Huesmann M, Scott B, Kisielow P, von Boehmer H. Kinetics and efficacy of positive selection in the thymus of normal and T cell receptor transgenic mice. *Cell.* 1991;66(3):533-540.
231. Ernst B, Surh CD, Sprent J. Thymic selection and cell division. *J Exp Med.* 1995;182(4):961-971.
232. Garcia KC, Adams EJ. How the T cell receptor sees antigen—a structural view. *Cell.* 2005;122(3):333-336.
233. Sprent J, Kishimoto H. The thymus and central tolerance. *Philos Trans R Soc Lond Series B Biol Sci.* 2001;356(1409):609-616.
234. Kisielow P, Teh HS, Bluthmann H, von Boehmer H. Positive selection of antigen-specific T cells in thymus by restricting MHC molecules. *Nature.* 1988;335(6192):730-733.
235. Jameson SC, Hogquist KA, Bevan MJ. Positive selection of thymocytes. *Annu Rev Immunol.* 1995;13:93-126.
236. Shortman K, Vremec D, Egerton M. The kinetics of T cell antigen receptor expression by subgroups of CD4+8+ thymocytes: delineation of CD4+8+3(2+) thymocytes as post-selection intermediates leading to mature T cells. *J Exp Med.* 1991;173(2):323-332.
237. Merkenschlager M, Graf D, Lovatt M, Bommhardt U, Zamoyska R, Fisher AG. How many thymocytes audition for selection? *J Exp Med.* 1997;186(7):1149-1158.
238. Zerrahn J, Held W, Raulet DH. The MHC reactivity of the T cell repertoire prior to positive and negative selection. *Cell.* 1997;88(5):627-636.
239. McDonald BD, Bunker JJ, Erickson SA, Oh-Hora M, Bendelac A. Crossreactive alphabeta T Cell receptors are the predominant targets of thymocyte negative selection. *Immunity.* 2015;43(5):859-869.
240. Daley SR, Hu DY, Goodnow CC. Helios marks strongly autoreactive CD4+ T cells in two major waves of thymic deletion distinguished by induction of PD-1 or NF-kappaB. *J Exp Med.* 2013;210(2):269-285.
241. Stritesky GL, Xing Y, Erickson JR, et al. Murine thymic selection quantified using a unique method to capture deleted T cells. *Proc Natl Acad Sci U S A.* 2013;110(12):4679-4684.
242. Hogquist KA, Jameson SC. The self-obsession of T cells: how TCR signaling thresholds affect fate 'decisions' and effector function. *Nat Immunol.* 2014;15(9):815-823.
243. Huseby ES, White J, Crawford F, et al. How the T cell repertoire becomes peptide and MHC specific. *Cell.* 2005;122(2):247-260.
244. Marrack P, Scott-Browne JP, Dai S, Gapin L, Kappler JW. Evolutionarily conserved amino acids that control TCR-MHC interaction. *Annu Rev Immunol.* 2008;26:171-203.
245. Garcia KC. Reconciling views on T cell receptor germline bias for MHC. *Trends Immunol.* 2012;33(9):429-436.
246. Krovi SH, Kappler JW, Marrack P, Gapin L. Inherent reactivity of unselected TCR repertoires to peptide-MHC molecules. *Proc Natl Acad Sci U S A.* 2019;116(44):22252-22261.
247. Van Laethem F, Sarafova SD, Park JH, et al. Deletion of CD4 and CD8 coreceptors permits generation of alphabetaT cells that recognize antigens independently of the MHC. *Immunity.* 2007;27(5):735-750.
248. Van Laethem F, Tikhonova AN, Singer A. MHC restriction is imposed on a diverse T cell receptor repertoire by CD4 and CD8 co-receptors during thymic selection. *Trends Immunol.* 2012;33(9):437-441.
249. Lu J, Van Laethem F, Bhattacharya A, et al. Molecular constraints on CDR3 for thymic selection of MHC-restricted TCRs from a random pre-selection repertoire. *Nat Commun.* 2019;10(1):1019.
250. Lu J, Van Laethem F, Saba I, et al. Structure of MHC-independent TCRs and their recognition of native antigen CD155. *J Immunol.* 2020;204:3351-3359.
251. Rathmell JC, Lindsten T, Zong WX, Cinalli RM, Thompson CB. Deficiency in Bak and Bax perturbs thymic selection and lymphoid homeostasis. *Nat Immunol.* 2002;3(10):932-939.
252. Egerton M, Scollay R, Shortman K. Kinetics of mature T-cell development in the thymus. *Proc Natl Acad Sci U S A.* 1990;87(7):2579-2582.
253. Surh CD, Sprent J. T-cell apoptosis detected in situ during positive and negative selection in the thymus. *Nature.* 1994;372(6501):100-103.
254. Sun Z, Unutmaz D, Zou YR, et al. Requirement for RORgamma in thymocyte survival and lymphoid organ development. *Science.* 2000;288(5475):2369-2373.
255. Chong MMW, Cornish AL, Darwiche R, et al. Suppressor of cytokine signaling-1 is a critical regulator of interleukin-7-dependent CD8+ T cell differentiation. *Immunity.* 2003;18(4):475-487.
256. Yu Q, Park JH, Doan LL, Erman B, Feigenbaum L, Singer A. Cytokine signal transduction is suppressed in preselection double-positive thymocytes and restored by positive selection. *J Exp Med.* 2006;203(1):165-175.
257. Jones ME, Zhuang Y. Acquisition of a functional T cell receptor during T lymphocyte development is enforced by HEB and E2A transcription factors. *Immunity.* 2007;27(6):860-870.
258. Belle I, Zhuang Y. Chapter four—E proteins in lymphocyte development and lymphoid diseases. In: Taneja R, ed. *Current Topics in Developmental Biology.* Vol. 110. Academic Press; 2014:153-187.
259. Carico Z, Krangel MS. Chapter seven—chromatin dynamics and the development of the TCRα and TCRδ repertoires. In:

Murre C, ed. *Advances in Immunology*. Vol. 128. Academic Press; 2015:307-361.

260. Schuldt NJ, Binstadt BA. Dual TCR T cells: identity crisis or multitaskers? *J Immunol*. 2019;202(3):637-644.
261. Guo J, Hawwari A, Li H, et al. Regulation of the TCRalpha repertoire by the survival window of CD4(+)CD8(+) thymocytes. *Nat Immunol*. 2002;3(5):469-476.
262. Wekerle H, Ketelsen UP. Thymic nurse cells—Ia-bearing epithelium involved in T-lymphocyte differentiation? *Nature*. 1980;283(5745):402-404.
263. Nakagawa Y, Ohigashi I, Nitta T, et al. Thymic nurse cells provide microenvironment for secondary T cell receptor alpha rearrangement in cortical thymocytes. *Proc Natl Acad Sci U S A*. 2012;109(50):20572-20577.
264. Kyewski BA, Kaplan HS. Lymphoepithelial interactions in the mouse thymus: phenotypic and kinetic studies on thymic nurse cells. *J Immunol*. 1982;128(5):2287-2294.
265. Witt CM, Raychaudhuri S, Schaefer B, Chakraborty AK, Robey EA. Directed migration of positively selected thymocytes visualized in real time. *PLoS Biol*. 2005;3(6):e160.
266. Bousso P, Bhakta NR, Lewis RS, Robey E. Dynamics of thymocyte-stromal cell interactions visualized by two-photon microscopy. *Science*. 2002;296(5574):1876-1880.
267. Azzam HS, Grinberg A, Lui K, Shen H, Shores EW, Love PE. CD5 expression is developmentally regulated by T cell receptor (TCR) signals and TCR avidity. *J Exp Med*. 1998;188(12):2301-2311.
268. Swat W, Dessing M, von Boehmer H, Kisielow P. CD69 expression during selection and maturation of CD4+8+ thymocytes. *Eur J Immunol*. 1993;23(3):739-746.
269. Moran AE, Holzapfel KL, Xing Y, et al. T cell receptor signal strength in Treg and iNKT cell development demonstrated by a novel fluorescent reporter mouse. *J Exp Med*. 2011;208(6):1279-1289.
270. Bevan MJ. In a radiation chimaera, host H-2 antigens determine immune responsiveness of donor cytotoxic cells. *Nature*. 1977;269(5627):417-418.
271. Zinkernagel RM, Callahan GN, Althage A, Cooper S, Klein PA, Klein J. On the thymus in the differentiation of "H-2 self-recognition" by T cells: evidence for dual recognition? *J Exp Med*. 1978;147(3):882-896.
272. Singer A, Hathcock KS, Hodes RJ. Self recognition in allogeneic thymic chimeras. Self recognition by T helper cells from thymus-engrafted nude mice is restricted to the thymic H-2 haplotype. *J Exp Med*. 1982;155(1):339-344.
273. Kaye J, Hsu ML, Sauron ME, Jameson SC, Gascoigne NR, Hedrick SM. Selective development of CD4+ T cells in transgenic mice expressing a class II MHC-restricted antigen receptor. *Nature*. 1989;341(6244):746-749.
274. Sha WC, Nelson CA, Newberry RD, Kranz DM, Russell JH, Loh DY. Positive and negative selection of an antigen receptor on T cells in transgenic mice. *Nature*. 1988;336(6194):73-76.
275. Berg LJ, Pullen AM, Fazekas de St Groth B, Mathis D, Benoist C, Davis MM. Antigen/MHC-specific T cells are preferentially exported from the thymus in the presence of their MHC ligand. *Cell*. 1989;58(6):1035-1046.
276. Hogquist KA, Jameson SC, Heath WR, Howard JL, Bevan MJ, Carbone FR. T cell receptor antagonist peptides induce positive selection. *Cell*. 1994;76(1):17-27.
277. Ashton-Rickardt PG, Bandeira A, Delaney JR, et al. Evidence for a differential avidity model of T cell selection in the thymus. *Cell*. 1994;76(4):651-663.
278. Sebzda E, Wallace VA, Mayer J, Yeung RS, Mak TW, Ohashi PS. Positive and negative thymocyte selection induced by different concentrations of a single peptide. *Science*. 1994;263(5153):1615-1618.
279. Starr TK, Jameson SC, Hogquist KA. Positive and negative selection of T cells. *Annu Rev Immunol*. 2003;21:139-176.
280. Hedrick SM. Positive selection in the thymus: an enigma wrapped in a mystery. *J Immunol*. 2012;188(5):2043-2045.
281. Daniels MA, Teixeiro E, Gill J, et al. Thymic selection threshold defined by compartmentalization of Ras/MAPK signalling. *Nature*. 2006;444(7120):724-729.
282. Alam SM, Travers PJ, Wung JL, et al. T-cell-receptor affinity and thymocyte positive selection. *Nature*. 1996;381(6583):616-620.
283. Yamagata T, Mathis D, Benoist C. Self-reactivity in thymic double-positive cells commits cells to a CD8 alpha alpha lineage with characteristics of innate immune cells. *Nat Immunol*. 2004;5(6):597-605.
284. Williams CB, Engle DL, Kersh GJ, Michael White J, Allen PM. A kinetic threshold between negative and positive selection based on the longevity of the T cell receptor-ligand complex. *J Exp Med*. 1999;189(10):1531-1544.
285. Rubin RL, Hermanson TM. Plasticity in the positive selection of T cells: affinity of the selecting antigen and IL-7 affect T cell responsiveness. *Int Immunol*. 2005;17(7):959-971.
286. Sasada T, Ghendler Y, Neveu JM, Lane WS, Reinherz EL. A naturally processed mitochondrial self-peptide in complex with thymic MHC molecules functions as a selecting ligand for a viral-specific T cell receptor. *J Exp Med*. 2001;194(7):883-892.
287. Santori FR, Kieper WC, Brown SM, et al. Rare, structurally homologous self-peptides promote thymocyte positive selection. *Immunity*. 2002;17(2):131-142.
288. Takada K, Kondo K, Takahama Y. Generation of peptides that promote positive selection in the thymus. *J Immunol*. 2017;198(6):2215-2222.
289. Barton GM, Rudensky AY. Requirement for diverse, low-abundance peptides in positive selection of T cells. *Science*. 1999;283(5398):67-70.
290. Pacholczyk R, Kraj P, Ignatowicz L. An incremental increase in the complexity of peptides bound to class II MHC changes the diversity of positively selected alpha beta TCRs. *J Immunol*. 2001;166(4):2357-2363.
291. Barton GM, Beers C, deRoos P, et al. Positive selection of self-MHC-reactive T cells by individual peptide-MHC class II complexes. *Proc Natl Acad Sci U S A*. 2002;99(10):6937-6942.
292. Fung-Leung WP, Wallace VA, Gray D, et al. CD8 is needed for positive selection but differentially required for negative selection of T cells during thymic ontogeny. *Eur J Immunol*. 1993;23(1):212-216.
293. Rahemtulla A, Fung-Leung WP, Schilham MW, et al. Normal development and function of CD8+ cells but markedly decreased helper cell activity in mice lacking CD4. *Nature*. 1991;353(6340):180-184.
294. Bendelac A, Killeen N, Littman DR, Schwartz RH. A subset of CD4+ thymocytes selected by MHC class I molecules. *Science*. 1994;263(5154):1774-1778.
295. Tyznik AJ, Sun JC, Bevan MJ. The CD8 population in CD4-deficient mice is heavily contaminated with MHC class II-restricted T cells. *J Exp Med*. 2004;199(4):559-565.
296. Veillette A, Bookman MA, Horak EM, Bolen JB. The CD4 and CD8 T cell surface antigens are associated with the internal membrane tyrosine-protein kinase p56lck. *Cell*. 1988;55(2):301-308.
297. Barber EK, Dasgupta JD, Schlossman SF, Trevillyan JM, Rudd CE. The CD4 and CD8 antigens are coupled to a protein-tyrosine kinase (p56lck) that phosphorylates the CD3 complex. *Proc Natl Acad Sci U S A*. 1989;86(9):3277-3281.
298. Shaw AS, Amrein KE, Hammond C, Stern DF, Sefton BM, Rose JK. The lck tyrosine protein kinase interacts with the cytoplasmic tail of the CD4 glycoprotein through its unique amino-terminal domain. *Cell*. 1989;59(4):627-636.
299. Turner JM, Brodsky MH, Irving BA, Levin SD, Perlmutter RM, Littman DR. Interaction of the unique N-terminal region of tyrosine kinase p56lck with cytoplasmic domains of CD4 and CD8 is mediated by cysteine motifs. *Cell*. 1990;60(5):755-765.
300. Weiss A, Littman DR. Signal transduction by lymphocyte antigen receptors. *Cell*. 1994;76(2):263-274.
301. Chan IT, Limmer A, Louie MC, et al. Thymic selection of cytotoxic T cells independent of CD8 alpha-Lck association. *Science*. 1993;261(5128):1581-1584.
302. Killeen N, Littman DR. Helper T-cell development in the absence of CD4-p56lck association. *Nature*. 1993;364(6439):729-732.
303. Veillette A, Zuniga-Pflucker JC, Bolen JB, Kruisbeek AM. Engagement of CD4 and CD8 expressed on immature thymocytes induces activation of intracellular tyrosine phosphorylation pathways. *J Exp Med*. 1989;170(5):1671-1680.
304. Wiest DL, Yuan L, Jefferson J, et al. Regulation of T cell receptor expression in immature CD4+CD8+ thymocytes by p56lck tyrosine kinase: basis for differential signaling by CD4 and CD8 in immature thymocytes expressing both coreceptor molecules. *J Exp Med*. 1993;178(5):1701-1712.
305. Ravichandran KS, Burakoff SJ. Evidence for differential intracellular signaling via CD4 and CD8 molecules. *J Exp Med*. 1994;179(2):727-732.
306. Van Laethem F, Tikhonova AN, Pobezinsky LA, et al. Lck availability during thymic selection determines the recognition specificity of the T cell repertoire. *Cell*. 2013;154(6):1326-1341.
307. Mingueneau M, Kreslavsky T, Gray D, et al. The transcriptional landscape of alphabeta T cell differentiation. *Nat Immunol*. 2013;14(6):619-632.
308. Turka LA, Schatz DG, Oettinger MA, et al. Thymocyte expression of RAG-1 and RAG-2: termination by T cell receptor cross-linking. *Science*. 1991;253(5021):778-781.
309. Bain G, Cravatt CB, Loomans C, Alberola-Ila J, Hedrick SM, Murre C. Regulation of the helix-loop-helix proteins, E2A and Id3, by the Ras-ERK MAPK cascade. *Nat Immunol*. 2001;2(2):165-171.
310. Marsden VS, Strasser A. Control of apoptosis in the immune system: Bcl-2, BH3-only proteins and more. *Annu Rev Immunol*. 2003;21:71-105.
311. Veis DJ, Sorenson CM, Shutter JR, Korsmeyer SJ. Bcl-2-deficient mice demonstrate fulminant lymphoid apoptosis, polycystic kidneys, and hypopigmented hair. *Cell*. 1993;75(2):229-240.
312. Nakayama K, Nakayama K, Negishi I, et al. Disappearance of the lymphoid system in Bcl-2 homozygous mutant chimeric mice. *Science*. 1993;261(5128):1584-1588.
313. Dzhagalov I, Dunkle A, He Y-W. The anti-apoptotic Bcl-2 family member Mcl-1 promotes T lymphocyte survival at multiple stages. *J Immunol*. 2008;181(1):521-528.
314. Li KP, Shanmuganad S, Carroll K, Katz JD, Jordan MB, Hildeman DA. Dying to protect: cell death and the control of T-cell homeostasis. *Immunol Rev*. 2017;277(1):21-43.
315. Zhang N, Hartig H, Dzhagalov I, Draper D, He YW. The role of apoptosis in the development and function of T lymphocytes. *Cell Res*. 2005;15(10):749-769.
316. Yang C-Y, Lin N-H, Lee J-M, et al. Promoter knock-in mutations reveal a role of Mcl-1 in thymocyte-positive selection and tissue or cell lineage-specific regulation of Mcl-1 expression. *J Immunol*. 2009;182(5):2959-2968.
317. Ashwell JD, Lu FW, Vacchio MS. Glucocorticoids in T cell development and function. *Annu Rev Immunol*. 2000;18:309-345.
318. Groves T, Parsons M, Miyamoto NG, Guidos CJ. TCR engagement of CD4+CD8+ thymocytes in vitro induces early aspects of positive selection, but not apoptosis. *J Immunol*. 1997;158(1):65-75.
319. Yu Q, Erman B, Bhandoola A, Sharrow SO, Singer A. In vitro evidence that cytokine receptor signals are required for differentiation of double positive thymocytes into functionally mature CD8+ T cells. *J Exp Med*. 2003;197(4):475-487.
320. McCaughtry TM, Etzensperger R, Alag A, et al. Conditional deletion of cytokine receptor chains reveals that IL-7 and IL-15 specify CD8 cytotoxic lineage fate in the thymus. *J Exp Med*. 2012;209(12):2263-2276.
321. Alberola-Ila J, Forbush KA, Seger R, Krebs EG, Perlmutter RM. Selective requirement for MAP kinase activation in thymocyte differentiation. *Nature*. 1995;373(6515):620-623.
322. Fischer AM, Katayama CD, Pages G, Pouyssegur J, Hedrick SM. The role of erk1 and erk2 in multiple stages of T cell development. *Immunity*. 2005;23(4):431-443.
323. Oh-Hora M, Komatsu N, Pishyareh M, et al. Agonist-selected T cell development requires strong T cell receptor signaling and store-operated calcium entry. *Immunity*. 2013;38(5):881-895.
324. Carpenter AC, Bosselut R. Decision checkpoints in the thymus. *Nat Immunol*. 2010;11(8):666-673.
325. Cho JH, Sprent J. TCR tuning of T cell subsets. *Immunol Rev*. 2018;283(1):129-137.
326. Davey GM, Schober SL, Endrizzi BT, Dutcher AK, Jameson SC, Hogquist KA. Preselection thymocytes are more sensitive to T cell receptor stimulation than mature T cells. *J Exp Med*. 1998;188(10):1867-1874.
327. Lucas B, Stefanova I, Yasutomo K, Dautigny N, Germain RN. Divergent changes in the sensitivity of maturing T cells to structurally related ligands underlies formation of a useful T cell repertoire. *Immunity*. 1999;10(3):367-376.
328. Gascoigne NR, Rybakin V, Acuto O, Brzostek J. TCR signal strength and T cell development. *Annu Rev Cell Dev Biol*. 2016;32:327-348.
329. Gaud G, Lesourne R, Love PE. Regulatory mechanisms in T cell receptor signalling. *Nat Rev Immunol*. 2018;18(8):485-497.
330. Lo WL, Donermeyer DL, Allen PM. A voltage-gated sodium channel is essential for the positive selection of CD4(+) T cells. *Nat Immunol*. 2012;13(9):880-887.
331. Wang D, Zheng M, Lei L, et al. Tespa1 is involved in late thymocyte development through the regulation of TCR-mediated signaling. *Nat Immunol*. 2012;13(6):560-568.
332. Fu G, Vallée S, Rybakin V, et al. Themis controls thymocyte selection through regulation of T cell antigen receptor-mediated signaling. *Nat Immunol*. 2009;10(8):848-856.
333. Lesourne R, Uehara S, Lee J, et al. Themis, a T cell-specific protein important for late thymocyte development. *Nat Immunol*. 2009;10(8):840-847.
334. Johnson AL, Aravind L, Shulzhenko N, et al. Themis is a member of a new metazoan gene family and is required for the completion of thymocyte positive selection. *Nat Immunol*. 2009;10(8):831-839.
335. Patrick MS, Oda H, Hayakawa K, et al. Gasp, a Grb2-associating protein, is critical for positive selection of thymocytes. *Proc Natl Acad Sci U S A*. 2009;106(38):16345-16350.
336. Kakugawa K, Yasuda T, Miura I, et al. A novel gene essential for the development of single positive thymocytes. *Mol Cell Biol*. 2009;29(18):5128-5135.
337. Choi S, Warzecha C, Zvezdova E, et al. THEMIS enhances TCR signaling and enables positive selection by selective inhibition of the phosphatase SHP-1. *Nat Immunol*. 2017;18(4):433-441.
338. Mehta M, Brzostek J, Chen EW, et al. Themis-associated phosphatase activity controls signaling in T cell development. *Proc Natl Acad Sci U S A*. 2018;115(48):E11331-E11340.
339. Li Q-J, Chau J, Ebert PJR, et al. miR-181a is an intrinsic modulator of T cell sensitivity and selection. *Cell*. 2007;129(1):147-161.
340. Salmond RJ, Brownlie RJ, Morrison VL, Zamoyska R. The tyrosine phosphatase PTPN22 discriminates weak self peptides from strong agonist TCR signals. *Nat Immunol*. 2014;15(9):875-883.
341. Winter SJ, Krueger A. Development of unconventional T cells controlled by MicroRNA. *Front Immunol*. 2019;10:2520.
342. Laufer TM, DeKoning J, Markowitz JS, Lo D, Glimcher LH. Unopposed positive selection and autoreactivity in mice expressing class II MHC only on thymic cortex. *Nature*. 1996;383(6595):81-85.
343. Anderson G, Owen JJ, Moore NC, Jenkinson EJ. Thymic epithelial cells provide unique signals for positive selection of CD4+CD8+ thymocytes in vitro. *J Exp Med*. 1994;179(6):2027-2031.
344. Wei DG, Lee H, Park SH, et al. Expansion and long-range differentiation of the NKT cell lineage in mice expressing CD1d exclusively on cortical thymocytes. *J Exp Med*. 2005;202(2):239-248.

345. Gapin L, Matsuda JL, Surh CD, Kronenberg M. NKT cells derive from double-positive thymocytes that are positively selected by CD1d. *Nat Immunol.* 2001;2(10):971-978.
346. Seach N, Guerri L, Le Bourhis L, et al. Double-positive thymocytes select mucosal-associated invariant T cells. *J Immunol.* 2013;191(12):6002-6009.
347. Nakagawa T, Roth W, Wong P, et al. Cathepsin L: critical role in Ii degradation and CD4 T cell selection in the thymus. *Science.* 1998;280(5362):450-453.
348. Honey K, Nakagawa T, Peters C, Rudensky A. Cathepsin L regulates CD4+ T cell selection independently of its effect on invariant chain: a role in the generation of positively selecting peptide ligands. *J Exp Med.* 2002;195(10):1349-1358.
349. Viret C, Lamare C, Guiraud M, et al. Thymus-specific serine protease contributes to the diversification of the functional endogenous CD4 T cell receptor repertoire. *J Exp Med.* 2011;208(1):3-11.
350. Gommeaux J, Gregoire C, Nguessan P, et al. Thymus-specific serine protease regulates positive selection of a subset of CD4+ thymocytes. *Eur J Immunol.* 2009;39(4):956-964.
351. Mizushima N, Yamamoto A, Matsui M, Yoshimori T, Ohsumi Y. In vivo analysis of autophagy in response to nutrient starvation using transgenic mice expressing a fluorescent autophagosome marker. *Mol Biol Cell.* 2004;15(3):1101-1111.
352. Nedjic J, Aichinger M, Emmerich J, Mizushima N, Klein L. Autophagy in thymic epithelium shapes the T-cell repertoire and is essential for tolerance. *Nature.* 2008;455(7211):396-400.
353. von Rohrscheidt J, Petrozziello E, Nedjic J, et al. Thymic CD4 T cell selection requires attenuation of March8-mediated MHCII turnover in cortical epithelial cells through CD83. *J Exp Med.* 2016;213(9):1685-1694.
354. Liu H, Jain R, Guan J, et al. Ubiquitin ligase MARCH 8 cooperates with CD83 to control surface MHC II expression in thymic epithelium and CD4 T cell selection. *J Exp Med.* 2016;213(9):1695-1703.
355. Fujimoto Y, Tu L, Miller AS, et al. CD83 expression influences CD4+ T cell development in the thymus. *Cell.* 2002;108(6):755-767.
356. Murata S, Sasaki K, Kishimoto T, et al. Regulation of CD8+ T cell development by thymus-specific proteasomes. *Science.* 2007;316(5829):1349-1353.
357. Sasaki K, Takada K, Ohte Y, et al. Thymoproteasomes produce unique peptide motifs for positive selection of CD8(+) T cells. *Nat Commun.* 2015;6:7484.
358. Xing Y, Jameson SC, Hogquist KA. Thymoproteasome subunit-beta5T generates peptide-MHC complexes specialized for positive selection. *Proc Natl Acad Sci U S A.* 2013;110(17):6979-6984.
359. Takada K, Van Laethem F, Xing Y, et al. TCR affinity for thymoproteasome-dependent positively selecting peptides conditions antigen responsiveness in CD8(+) T cells. *Nat Immunol.* 2015;16(10):1069-1076.
360. Nitta T, Murata S, Sasaki K, et al. Thymoproteasome shapes immunocompetent repertoire of CD8+ T cells. *Immunity.* 2010;32(1):29-40.
361. Ohigashi I, Ohte Y, Setoh K, et al. A human PSMB11 variant affects thymoproteasome processing and CD8+ T cell production. *JCI Insight.* 2017;2(10).
362. Nitta T, Kochi Y, Muro R, et al. Human thymoproteasome variations influence CD8 T cell selection. *Sci Immunol.* 2017;2(12):eaan5165.
363. Ueno T, Saito F, Gray DH, et al. CCR7 signals are essential for cortex-medulla migration of developing thymocytes. *J Exp Med.* 2004;200(4):493-505.
364. Nitta T, Nitta S, Lei Y, Lipp M, Takahama Y. CCR7-mediated migration of developing thymocytes to the medulla is essential for negative selection to tissue-restricted antigens. *Proc Natl Acad Sci U S A.* 2009;106(40):17129-17133.
365. Kozai M, Kubo Y, Katakai T, et al. Essential role of CCL21 in establishment of central self-tolerance in T cells. *J Exp Med.* 2017;214(7):1925-1935.
366. Kadakia T, Tai X, Kruhlak M, et al. E-protein-regulated expression of CXCR4 adheres preselection thymocytes to the thymic cortex. *J Exp Med.* 2019;216(8):1749-1761.
367. Choi YI, Duke-Cohan JS, Ahmed WB, et al. PlexinD1 glycoprotein controls migration of positively selected thymocytes into the medulla. *Immunity.* 2008;29(6):888-898.
368. Halkias J, Melichar HJ, Taylor KT, et al. Opposing chemokine gradients control human thymocyte migration in situ. *J Clin Invest.* 2013;123(5):2131-2142.
369. Nieuwenhuis P, Stet RJ, Wagenaar JP, Wubbena AS, Kampinga J, Karrenbeld A. The transcapsular route: a new way for (self-) antigens to by-pass the blood-thymus barrier? *Immunol Today.* 1988;9(12):372-375.
370. Eggli P, Schaffner T, Gerber HA, Hess MW, Cottier H. Accessibility of thymic cortical lymphocytes to particles translocated from the peritoneal cavity to parathymic lymph nodes. *Thymus.* 1986;8(3):129-139.
371. Cosgrove D, Gray D, Dierich A, et al. Mice lacking MHC class II molecules. *Cell.* 1991;66(5):1051-1066.
372. Grusby MJ, Johnson RS, Papaioannou VE, Glimcher LH. Depletion of CD4+ T cells in major histocompatibility complex class II-deficient mice. *Science.* 1991;253(5026):1417-1420.
373. Zijlstra M, Bix M, Simister NE, Loring JM, Raulet DH, Jaenisch R. Beta 2-microglobulin deficient mice lack CD4-8+ cytolytic T cells. *Nature.* 1990;344(6268):742-746.
374. Teh HS, Kisielow P, Scott B, et al. Thymic major histocompatibility complex antigens and the alpha beta T-cell receptor determine the CD4/CD8 phenotype of T cells. *Nature.* 1988;335(6187):229-233.
375. Sha WC, Nelson CA, Newberry RD, Kranz DM, Russell JH, Loh DY. Selective expression of an antigen receptor on CD8-bearing T lymphocytes in transgenic mice. *Nature.* 1988;335(6187):271-274.
376. Bosselut R, Feigenbaum L, Sharrow SO, Singer A. Strength of signaling by CD4 and CD8 coreceptor tails determines the number but not the lineage direction of positively selected thymocytes. *Immunity.* 2001;14(4):483-494.
377. Myrick C, DiGuisto R, DeWolfe J, et al. Linkage analysis of variations in CD4:CD8 T cell subsets between C57BL/6 and DBA/2. *Genes Immun.* 2002;3(3):144-150.
378. Singer A, Adoro S, Park JH. Lineage fate and intense debate: myths, models and mechanisms of CD4- versus CD8-lineage choice. *Nat Rev Immunol.* 2008;8(10):788-801.
379. Corbella P, Moskophidis D, Spanopoulou E, et al. Functional commitment to helper T cell lineage precedes positive selection and is independent of T cell receptor MHC specificity. *Immunity.* 1994;1(4):269-276.
380. Matechak EO, Killeen N, Hedrick SM, Fowlkes BJ. MHC class II-specific T cells can develop in the CD8 lineage when CD4 is absent. *Immunity.* 1996;4(4):337-347.
381. Janeway CA Jr. T-cell development. Accessories or coreceptors? *Nature.* 1988;335(6187):208-210.
382. Itano A, Robey E. Highly efficient selection of CD4 and CD8 lineage thymocytes supports an instructive model of lineage commitment. *Immunity.* 2000;12(4):383-389.
383. Itano A, Salmon P, Kioussis D, Tolaini M, Corbella P, Robey E. The cytoplasmic domain of CD4 promotes the development of CD4 lineage T cells. *J Exp Med.* 1996;183(3):731-741.
384. Yasutomo K, Doyle C, Miele L, Fuchs C, Germain RN. The duration of antigen receptor signalling determines CD4+ versus CD8+ T-cell lineage fate. *Nature.* 2000;404(6777):506-510.
385. Brugnera E, Bhandoola A, Cibotti R, et al. Coreceptor reversal in the thymus: signaled CD4+8+ thymocytes initially terminate CD8 transcription even when differentiating into CD8+ T cells. *Immunity.* 2000;13(1):59-71.
386. Liu X, Bosselut R. Duration of TCR signaling controls CD4-CD8 lineage differentiation in vivo. *Nat Immunol.* 2004;5(3):280-288.
387. Adoro S, McCaughtry T, Erman B, et al. Coreceptor gene imprinting governs thymocyte lineage fate. *EMBO J.* 2012;31(2):366-377.
388. Etzensperger R, Kadakia T, Tai X, et al. Identification of lineage-specifying cytokines that signal all CD8+-cytotoxic-lineage-fate 'decisions' in the thymus. *Nat Immunol.* 2017;18(11):1218-1227.
389. Taniuchi I, Osato M, Egawa T, et al. Differential requirements for Runx proteins in CD4 repression and epigenetic silencing during T lymphocyte development. *Cell.* 2002;111(5):621-633.
390. Woolf E, Xiao C, Fainaru O, et al. Runx3 and Runx1 are required for CD8 T cell development during thymopoiesis. *Proc Natl Acad Sci U S A.* 2003;100(13):7731-7736.
391. Egawa T, Tillman RE, Naoe Y, Taniuchi I, Littman DR. The role of the Runx transcription factors in thymocyte differentiation and in homeostasis of naive T cells. *J Exp Med.* 2007;204(8):1945-1957.
392. Sawada S, Scarborough JD, Killeen N, Littman DR. A lineage-specific transcriptional silencer regulates CD4 gene expression during T lymphocyte development. *Cell.* 1994;77(6):917-929.
393. Siu G, Wurster AL, Duncan DD, Soliman TM, Hedrick SM. A transcriptional silencer controls the developmental expression of the CD4 gene. *EMBO J.* 1994;13(15):3570-3579.
394. Issuree PD, Ng CP, Littman DR. Heritable gene regulation in the CD4:CD8 T cell lineage choice. *Front Immunol.* 2017;8:291.
395. Park JH, Adoro S, Guinter T, et al. Signaling by intrathymic cytokines, not T cell antigen receptors, specifies CD8 lineage choice and promotes the differentiation of cytotoxic-lineage T cells. *Nat Immunol.* 2010;11(3):257-264.
396. Ellmeier W, Sunshine MJ, Losos K, Hatam F, Littman DR. An enhancer that directs lineage-specific expression of CD8 in positively selected thymocytes and mature T cells. *Immunity.* 1997;7(4):537-547.
397. Hostert A, Tolaini M, Roderick K, Harker N, Norton T, Kioussis D. A region in the CD8 gene locus that directs expression to the mature CD8 T cell subset in transgenic mice. *Immunity.* 1997;7(4):525-536.
398. Ellmeier W, Sunshine MJ, Losos K, Littman DR. Multiple developmental stage-specific enhancers regulate CD8 expression in developing thymocytes and in thymus-independent T cells. *Immunity.* 1998;9(4):485-496.
399. Hostert A, Garefalaki A, Mavria G, et al. Hierarchical interactions of control elements determine CD8alpha gene expression in subsets of thymocytes and peripheral T cells. *Immunity.* 1998;9(4):497-508.
400. Sato T, Ohno S, Hayashi T, et al. Dual functions of Runx proteins for reactivating CD8 and silencing CD4 at the commitment process into CD8 thymocytes. *Immunity.* 2005;22(3):317-328.
401. Hassan H, Sakaguchi S, Tenno M, et al. Cd8 enhancer E8I and Runx factors regulate CD8α expression in activated CD8+ T cells. *Proc Natl Acad Sci U S A.* 2011;108(45):18330-18335.
402. Collins A, Littman DR, Taniuchi I. RUNX proteins in transcription factor networks that regulate T-cell lineage choice. *Nat Rev Immunol.* 2009;9(2):106-115.
403. Egawa T, Littman DR. ThPOK acts late in specification of the helper T cell lineage and suppresses Runx-mediated commitment to the cytotoxic T cell lineage. *Nat Immunol.* 2008;9(10):1131-1139.
404. Cruz-Guilloty F, Pipkin ME, Djuretic IM, et al. Runx3 and T-box proteins cooperate to establish the transcriptional program of effector CTLs. *J Exp Med.* 2009;206(1):51-59.
405. He X, Dave VP, Zhang Y, et al. The zinc finger transcription factor Th-POK regulates CD4 versus CD8 T-cell lineage commitment. *Nature.* 2005;433(7028):826-833.
406. Sun G, Liu X, Mercado P, et al. The zinc finger protein cKrox directs CD4 lineage differentiation during intrathymic T cell positive selection. *Nat Immunol.* 2005;6(4):373-381.
407. Mucida D, Husain MM, Muroi S, et al. Transcriptional reprogramming of mature CD4(+) helper T cells generates distinct MHC class II-restricted cytotoxic T lymphocytes. *Nat Immunol.* 2013;14(3):281-289.
408. Keefe R, Dave V, Allman D, Wiest D, Kappes DJ. Regulation of lineage commitment distinct from positive selection. *Science.* 1999;286(5442):1149-1153.
409. Muroi S, Naoe Y, Miyamoto C, et al. Cascading suppression of transcriptional silencers by ThPOK seals helper T cell fate. *Nat Immunol.* 2008;9(10):1113-1121.
410. Wang L, Wildt KF, Castro E, et al. The zinc finger transcription factor Zbtb7b represses CD8-lineage gene expression in peripheral CD4+ T cells. *Immunity.* 2008;29(6):876-887.
411. Vacchio MS, Bosselut R. What happens in the thymus does not stay in the thymus: how T cells recycle the CD4+-CD8+ lineage commitment transcriptional circuitry to control their function. *J Immunol.* 2016;196(12):4848-4856.
412. Ciucci T, Vacchio MS, Gao Y, et al. The emergence and functional Fitness of memory CD4(+) T cells require the transcription factor Thpok. *Immunity.* 2019;50(1):91-105.e4.
413. Vacchio MS, Ciucci T, Gao Y, et al. A Thpok-directed transcriptional circuitry promotes Bcl6 and Maf expression to orchestrate T follicular helper differentiation. *Immunity.* 2019;51(3):465-478.e6.
414. He X, Park K, Wang H, et al. CD4-CD8 lineage commitment is regulated by a silencer element at the ThPOK transcription-factor locus. *Immunity.* 2008;28(3):346-358.
415. Setoguchi R, Tachibana M, Naoe Y, et al. Repression of the transcription factor Th-POK by Runx complexes in cytotoxic T cell development. *Science.* 2008;319(5864):822-825.
416. Taniuchi I. CD4 helper and CD8 cytotoxic T cell differentiation. *Annu Rev Immunol.* 2018;36:579-601.
417. Tanaka H, Naito T, Muroi S, et al. Epigenetic Thpok silencing limits the time window to choose CD4(+) helper-lineage fate in the thymus. *EMBO J.* 2013;32(8):1183-1194.
418. Sakaguchi S, Hombauer M, Bilic I, et al. The zinc-finger protein MAZR is part of the transcription factor network that controls the CD4 versus CD8 lineage fate of double-positive thymocytes. *Nat Immunol.* 2010;11(5):442-448.
419. Wildt KF, Sun G, Grueter B, et al. The transcription factor Zbtb7b promotes CD4 expression by antagonizing Runx-mediated activation of the CD4 silencer. *J Immunol.* 2007;179(7):4405-4414.
420. Zou YR, Sunshine MJ, Taniuchi I, Hatam F, Killeen N, Littman DR. Epigenetic silencing of CD4 in T cells committed to the cytotoxic lineage. *Nat Genet.* 2001;29(3):332-336.
421. Sellars M, Huh JR, Day K, et al. Regulation of DNA methylation dictates Cd4 expression during the development of helper and cytotoxic T cell lineages. *Nat Immunol.* 2015;16(7):746-754.
422. Hernandez-Hoyos G, Anderson MK, Wang C, Rothenberg EV, Alberola-Ila J. GATA-3 expression is controlled by TCR signals and regulates CD4/CD8 differentiation. *Immunity.* 2003;19(1):83-94.
423. Pai SY, Truitt ML, Ting CN, Leiden JM, Glimcher LH, Ho IC. Critical roles for transcription factor GATA-3 in thymocyte development. *Immunity.* 2003;19(6):863-875.
424. Wang L, Wildt KF, Zhu J, et al. Distinct functions for the transcription factors GATA-3 and ThPOK during intrathymic differentiation of CD4(+) T cells. *Nat Immunol.* 2008;9(10):1122-1130.
425. Jones-Mason ME, Zhao X, Kappes D, Lasorella A, Iavarone A, Zhuang Y. E protein transcription factors are required for the development of CD4(+) lineage T cells. *Immunity.* 2012;36(3):348-361.
426. Steinke FC, Yu S, Zhou X, et al. TCF-1 and LEF-1 act upstream of Th-POK to promote the CD4(+) T cell fate and interact with Runx3 to silence Cd4 in CD8(+) T cells. *Nat Immunol.* 2014;15(7):646-656.
427. Aliahmad P, Kadavallore A, de la Torre B, Kappes D, Kaye J. TOX is required for development of the CD4 T cell lineage gene program. *J Immunol.* 2011;187(11):5931-5940.
428. Kakugawa K, Kojo S, Tanaka H, et al. Essential roles of SATB1 in specifying T lymphocyte subsets. *Cell Rep.* 2017;19(6):1176-1188.

429. Kojo S, Tanaka H, Endo TA, et al. Priming of lineage-specifying genes by Bcl11b is required for lineage choice in post-selection thymocytes. *Nat Commun.* 2017;8(1):702.
430. Luckey MA, Kimura MY, Waickman AT, Feigenbaum L, Singer A, Park JH. The transcription factor ThPOK suppresses Runx3 and imposes CD4(+) lineage fate by inducing the SOCS suppressors of cytokine signaling. *Nat Immunol.* 2014;15(7):638-645.
431. Kappler JW, Wade T, White J, et al. A T cell receptor V beta segment that imparts reactivity to a class II major histocompatibility complex product. *Cell.* 1987;49(2):263-271.
432. Kappler JW, Staerz U, White J, Marrack PC. Self-tolerance eliminates T cells specific for Mls-modified products of the major histocompatibility complex. *Nature.* 1988;332(6159):35-40.
433. MacDonald HR, Schneider R, Lees RK, et al. T-cell receptor V beta use predicts reactivity and tolerance to Mlsa-encoded antigens. *Nature.* 1988;332(6159):40-45.
434. Palmer E. Negative selection – clearing out the bad apples from the T-cell repertoire. *Nat Rev Immunol.* 2003;3(5):383-391.
435. Yu W, Jiang N, Ebert PJ, et al. Clonal deletion prunes but does not eliminate self-specific alphabeta CD8(+) T lymphocytes. *Immunity.* 2015;42(5):929-941.
436. Hassler T, Urmann E, Teschner S, et al. Inventories of naive and tolerant mouse CD4 T cell repertoires reveal a hierarchy of deleted and diverted T cell receptors. *Proc Natl Acad Sci U S A.* 2019;116(37):18537-18543.
437. Asano M, Toda M, Sakaguchi N, Sakaguchi S. Autoimmune disease as a consequence of developmental abnormality of a T cell subpopulation. *J Exp Med.* 1996;184(2):387-396.
438. Sakaguchi S, Sakaguchi N, Asano M, Itoh M, Toda M. Immunologic self-tolerance maintained by activated T cells expressing IL-2 receptor alpha-chains (CD25). Breakdown of a single mechanism of self-tolerance causes various autoimmune diseases. *J Immunol.* 1995;155(3):1151-1164.
439. Miller CH, Klawon DEJ, Zeng S, Lee V, Socci ND, Savage PA. Eomes identifies thymic precursors of self-specific memory-phenotype CD8(+) T cells. *Nat Immunol.* 2020;21(5):567-577.
440. Klein L, Kyewski B, Allen PM, Hogquist KA. Positive and negative selection of the T cell repertoire: what thymocytes see (and don't see). *Nat Rev Immunol.* 2014;14(6):377-391.
441. McCaughtry TM, Baldwin TA, Wilken MS, Hogquist KA. Clonal deletion of thymocytes can occur in the cortex with no involvement of the medulla. *J Exp Med.* 2008;205(11):2575-2584.
442. McDonald BD, Bunker JJ, Ishizuka IE, Jabri B, Bendelac A. Elevated T cell receptor signaling identifies a thymic precursor to the TCRαβ(+)CD4(−)CD8β(−) intraepithelial lymphocyte lineage. *Immunity* 2014;41(2):219-229.
443. Dzhagalov IL, Chen KG, Herzmark P, Robey EA. Elimination of self-reactive T cells in the thymus: a timeline for negative selection. *PLoS Biol.* 2013;11(5):e1001566.
444. Breed ER, Lee ST, Hogquist KA. Directing T cell fate: how thymic antigen presenting cells coordinate thymocyte selection. *Semin Cell Dev Biol.* 2018;84:2-10.
445. Pobezinsky LA, Angelov GS, Tai X, et al. Clonal deletion and the fate of autoreactive thymocytes that survive negative selection. *Nat Immunol.* 2012;13(6):569-578.
446. Mayans S, Stepniak D, Palida S, et al. αβT cell receptors expressed by CD4(−)CD8αβ(−) intraepithelial T cells drive their fate into a unique lineage with unusual MHC reactivities. *Immunity.* 2014;41(2):207-218.
447. Mayassi T, Jabri B. Human intraepithelial lymphocytes. *Mucosal Immunol.* 2018;11(5):1281-1289.
448. Smith KM, Olson DC, Hirose R, Hanahan D. Pancreatic gene expression in rare cells of thymic medulla: evidence for functional contribution to T cell tolerance. *Int Immunol.* 1997;9(9):1355-1365.
449. Derbinski J, Schulte A, Kyewski B, Klein L. Promiscuous gene expression in medullary thymic epithelial cells mirrors the peripheral self. *Nat Immunol.* 2001;2(11):1032-1039.
450. Akkaraju S, Ho WY, Leong D, Canaan K, Davis MM, Goodnow CC. A range of CD4 T cell tolerance: partial inactivation to organ-specific antigen allows nondestructive thyroiditis or insulitis. *Immunity.* 1997;7(2):255-271.
451. Liston A, Lesage S, Wilson J, Peltonen L, Goodnow CC. Aire regulates negative selection of organ-specific T cells. *Nat Immunol.* 2003;4(4):350-354.
452. Anderson MS, Venanzi ES, Chen Z, Berzins SP, Benoist C, Mathis D. The cellular mechanism of Aire control of T cell tolerance. *Immunity.* 2005;23(2):227-239.
453. Bouillet P, Purton JF, Godfrey DI, et al. BH3-only Bcl-2 family member Bim is required for apoptosis of autoreactive thymocytes. *Nature.* 2002;415(6874):922-926.
454. Gray DH, Kupresanin F, Berzins SP, et al. The BH3-only proteins Bim and Puma cooperate to impose deletional tolerance of organ-specific antigens. *Immunity.* 2012;37(3):451-462.
455. Banta KL, Wang X, Das P, Winoto A. B cell lymphoma 2 (Bcl-2) residues essential for Bcl-2's apoptosis-inducing interaction with Nur77/Nor-1 orphan steroid receptors. *J Biol Chem.* 2018;293(13):4724-4734.
456. Sekiya T, Kashiwagi I, Yoshida R, et al. Nr4a receptors are essential for thymic regulatory T cell development and immune homeostasis. *Nat Immunol.* 2013;14(3):230-237.
457. Fassett MS, Jiang W, D'Alise AM, Mathis D, Benoist C. Nuclear receptor Nr4a1 modulates both regulatory T-cell (Treg) differentiation and clonal deletion. *Proc Natl Acad Sci U S A.* 2012;109(10):3891-3896.
458. Hernandez JB, Newton RH, Walsh CM. Life and death in the thymus–cell death signaling during T cell development. *Curr Opin Cell Biol.* 2010;22(6):865-871.
459. Neilson JR, Winslow MM, Hur EM, Crabtree GR. Calcineurin B1 is essential for positive but not negative selection during thymocyte development. *Immunity.* 2004;20(3):255-266.
460. Melichar HJ, Ross JO, Herzmark P, Hogquist KA, Robey EA. Distinct temporal patterns of T cell receptor signaling during positive versus negative selection in situ. *Sci Signal.* 2013;6(297):ra92.
461. Au-Yeung BB, Melichar HJ, Ross JO, et al. Quantitative and temporal requirements revealed for Zap70 catalytic activity during T cell development. *Nat Immunol.* 2014;15(7):687-694.
462. McKeithan TW. Kinetic proofreading in T-cell receptor signal transduction. *Proc Natl Acad Sci U S A.* 1995;92(11):5042-5046.
463. Ardouin L, Boyer C, Gillet A, et al. Crippling of CD3-zeta ITAMs does not impair T cell receptor signaling. *Immunity.* 1999;10(4):409-420.
464. Holst J, Wang H, Eder KD, et al. Scalable signaling mediated by T cell antigen receptor-CD3 ITAMs ensures effective negative selection and prevents autoimmunity. *Nat Immunol.* 2008;9(6):658-666.
465. Shores EW, Tran T, Grinberg A, Sommers CL, Shen H, Love PE. Role of the multiple T cell receptor (TCR)-zeta chain signaling motifs in selection of the T cell repertoire. *J Exp Med.* 1997;185(5):893-900.
466. Hwang S, Song KD, Lesourne R, et al. Reduced TCR signaling potential impairs negative selection but does not result in autoimmune disease. *J Exp Med.* 2012;209(10):1781-1795.
467. McGargill MA, Ch'en IL, Katayama CD, Pages G, Pouyssegur J, Hedrick SM. Cutting edge: extracellular signal-related kinase is not required for negative selection of developing T cells. *J Immunol.* 2009;183(8):4838-4842.
468. Kortum RL, Rouquette-Jazdanian AK, Samelson LE. Ras and extracellular signal-regulated kinase signaling in thymocytes and T cells. *Trends Immunol.* 2013;34(6):259-268.
469. Staton TL, Lazarevic V, Jones DC, et al. Dampening of death pathways by schnurri-2 is essential for T-cell development. *Nature.* 2011;472(7341):105-109.
470. Mittelstadt PR, Taves MD, Ashwell JD. Cutting edge: de novo glucocorticoid synthesis by thymic epithelial cells regulates antigen-specific thymocyte selection. *J Immunol.* 2018;200(6):1988-1994.
471. Mittelstadt PR, Taves MD, Ashwell JD. Glucocorticoids oppose thymocyte negative selection by inhibiting Helios and Nur77. *J Immunol.* 2019;203(8):2163-2170.
472. Marrack P, Kappler J. The T cell receptor. *Science.* 1987;238(4830):1073-1079.
473. Kourilsky P, Claverie JM. MHC restriction, alloreactivity, and thymic education: a common link? *Cell.* 1989;56(3):327-329.
474. Kincaid EZ, Murata S, Tanaka K, Rock KL. Specialized proteasome subunits have an essential role in the thymic selection of CD8(+) T cells. *Nat Immunol.* 2016;17(8):938-945.
475. Sebzda E, Mariathasan S, Ohteki T, Jones R, Bachmann MF, Ohashi PS. Selection of the T cell repertoire. *Annu Rev Immunol.* 1999;17:829-874.
476. Breed ER, Watanabe M, Hogquist KA. Measuring thymic clonal deletion at the population level. *J Immunol.* 2019;202(11):3226-3233.
477. Ohkura N, Kitagawa Y, Sakaguchi S. Development and maintenance of regulatory T cells. *Immunity.* 2013;38(3):414-423.
478. Li MO, Rudensky AY. T cell receptor signalling in the control of regulatory T cell differentiation and function. *Nat Rev Immunol.* 2016;16(4):220-233.
479. Owen DL, Sjaastad LE, Farrar MA. Regulatory T cell development in the thymus. *J Immunol.* 2019;203(8):2031-2041.
480. Klein L, Robey EA, Hsieh CS. Central CD4(+) T cell tolerance: deletion versus regulatory T cell differentiation. *Nat Rev Immunol.* 2019;19(1):7-18.
481. Salomon B, Lenschow DJ, Rhee L, et al. B7/CD28 costimulation is essential for the homeostasis of the CD4+CD25+ immunoregulatory T cells that control autoimmune diabetes. *Immunity.* 2000;12(4):431-440.
482. Tang Q, Henriksen KJ, Boden EK, et al. Cutting edge: CD28 controls peripheral homeostasis of CD4+CD25+ regulatory T cells. *J Immunol.* 2003;171(7):3348-3352.
483. Tai X, Cowan M, Feigenbaum L, Singer A. CD28 costimulation of developing thymocytes induces Foxp3 expression and regulatory T cell differentiation independently of interleukin 2. *Nat Immunol.* 2005;6(2):152-162.
484. Malhotra D, Linehan JL, Dileepan T, et al. Tolerance is established in polyclonal CD4(+) T cells by distinct mechanisms, according to self-peptide expression patterns. *Nat Immunol.* 2016;17(2):187-195.
485. Legoux FP, Lim JB, Cauley AW, et al. CD4+ T cell tolerance to tissue-restricted self antigens is mediated by antigen-specific regulatory T cells rather than deletion. *Immunity.* 2015;43(5):896-908.
486. Yang S, Fujikado N, Kolodin D, Benoist C, Mathis D. Immune tolerance. Regulatory T cells generated early in life play a distinct role in maintaining self-tolerance. *Science.* 2015;348(6234):589-594.
487. Malchow S, Leventhal Daniel S, Lee V, Nishi S, Socci Nicholas D, Savage Peter A. Aire enforces immune tolerance by directing autoreactive T cells into the regulatory T cell lineage. *Immunity.* 2016;44(5):1102-1113.
488. Wing JB, Tay C, Sakaguchi S. Control of regulatory T cells by Co-signal molecules. *Adv Exp Med Biol.* 2019;1189:179-210.
489. Weist BM, Kurd N, Boussier J, Chan SW, Robey EA. Thymic regulatory T cell niche size is dictated by limiting IL-2 from antigen-bearing dendritic cells and feedback competition. *Nat Immunol.* 2015;16(6):635-641.
490. Owen DL, Mahmud SA, Vang KB, et al. Identification of cellular sources of IL-2 needed for regulatory T cell development and homeostasis. *J Immunol.* 2018;200(12):3926-3933.
491. Hemmers S, Schizas M, Azizi E, et al. IL-2 production by self-reactive CD4 thymocytes scales regulatory T cell generation in the thymus. *J Exp Med.* 2019;216(11):2466-2478.
492. Mahmud SA, Manlove LS, Schmitz HM, et al. Costimulation via the tumor-necrosis factor receptor superfamily couples TCR signal strength to the thymic differentiation of regulatory T cells. *Nat Immunol.* 2014;15(5):473-481.
493. Konkel JE, Jin W, Abbatiello B, Grainger JR, Chen W. Thymocyte apoptosis drives the intrathymic generation of regulatory T cells. *Proc Natl Acad Sci U S A.* 2014;111(4):E465-E473.
494. Stritesky GL, Jameson SC, Hogquist KA. Selection of self-reactive T cells in the thymus. *Annu Rev Immunol.* 2012;30:95-114.
495. Godfrey DI, Uldrich AP, McCluskey J, Rossjohn J, Moody DB. The burgeoning family of unconventional T cells. *Nat Immunol.* 2015;16(11):1114-1123.
496. Godfrey DI, Koay HF, McCluskey J, Gherardin NA. The biology and functional importance of MAIT cells. *Nat Immunol.* 2019;20(9):1110-1128.
497. McDonald BD, Jabri B, Bendelac A. Diverse developmental pathways of intestinal intraepithelial lymphocytes. *Nat Rev Immunol.* 2018;18(8):514-525.
498. Constantinides MG, Bendelac A. Transcriptional regulation of the NKT cell lineage. *Curr Opin Immunol.* 2013;25(2):161-167.
499. Sansom SN, Shikama-Dorn N, Zhanybekova S, et al. Population and single-cell genomics reveal the Aire dependency, relief from Polycomb silencing, and distribution of self-antigen expression in thymic epithelia. *Genome Res.* 2014;24(12):1918-1931.
500. Jolicoeur C, Hanahan D, Smith KM. T-cell tolerance toward a transgenic beta-cell antigen and transcription of endogenous pancreatic genes in thymus. *Proc Natl Acad Sci U S A.* 1994;91(14):6707-6711.
501. Anderson MS, Venanzi ES, Klein L, et al. Projection of an immunological self shadow within the thymus by the aire protein. *Science.* 2002;298(5597):1395-1401.
502. Nagamine K, Peterson P, Scott HS, et al. Positional cloning of the APECED gene. *Nat Genet.* 1997;17(4):393-398.
503. Aaltonen J, Bjorses P, Perheentupa J, et al. An autoimmune disease, APECED, caused by mutations in a novel gene featuring two PHD-type zinc-finger domains. *Nat Genet.* 1997;17(4):399-403.
504. Abramson J, Giraud M, Benoist C, Mathis D. Aire's partners in the molecular control of immunological tolerance. *Cell.* 2010;140(1):123-135.
505. Waterfield M, Khan IS, Cortez JT, et al. The transcriptional regulator Aire coopts the repressive ATF7ip-MBD1 complex for the induction of immunotolerance. *Nat Immunol.* 2014;15(3):258-265.
506. Bansal K, Yoshida H, Benoist C, Mathis D. The transcriptional regulator Aire binds to and activates super-enhancers. *Nat Immunol.* 2017;18(3):263-273.
507. Derbinski J, Gabler J, Brors B, et al. Promiscuous gene expression in thymic epithelial cells is regulated at multiple levels. *J Exp Med.* 2005;202(1):33-45.
508. Takaba H, Morishita Y, Tomofuji Y, et al. Fezf2 orchestrates a thymic program of self-antigen expression for immune tolerance. *Cell.* 2015;163(4):975-987.
509. Derbinski J, Pinto S, Rosch S, Hexel K, Kyewski B. Promiscuous gene expression patterns in single medullary thymic epithelial cells argue for a stochastic mechanism. *Proc Natl Acad Sci U S A.* 2008;105(2):657-662.
510. Cloosen S, Arnold J, Thio M, Bos GM, Kyewski B, Germeraad WT. Expression of tumor-associated differentiation antigens, MUC1 glycoforms and CEA, in human thymic epithelial cells: implications for self-tolerance and tumor therapy. *Cancer Res.* 2007;67(8):3919-3926.
511. Tykocinski LO, Sinemus A, Rezavandy E, et al. Epigenetic regulation of promiscuous gene expression in thymic medullary epithelial cells. *Proc Natl Acad Sci U S A.* 2010;107(45):19426-19431.
512. Pinto S, Michel C, Schmidt-Glenewinkel H, et al. Overlapping gene coexpression patterns in human medullary thymic epithelial cells generate self-antigen diversity. *Proc Natl Acad Sci U S A.* 2013;110(37):E3497-E3505.
513. Kieback E, Hilgenberg E, Stervbo U, et al. Thymus-derived regulatory T cells are positively selected on natural self-antigen through cognate interactions of high functional avidity. *Immunity.* 2016;44(5):1114-1126.
514. Aschenbrenner K, D'Cruz LM, Vollmann EH, et al. Selection of Foxp3+ regulatory T cells specific for self antigen expressed

and presented by Aire+ medullary thymic epithelial cells. *Nat Immunol.* 2007;8(4):351-358.
515. Perry JSA, Lio CJ, Kau AL, et al. Distinct contributions of Aire and antigen-presenting-cell subsets to the generation of self-tolerance in the thymus. *Immunity.* 2014;41(3):414-426.
516. Gallegos AM, Bevan MJ. Central tolerance to tissue-specific antigens mediated by direct and indirect antigen presentation. *J Exp Med.* 2004;200(8):1039-1049.
517. Guilliams M, Dutertre C-A, Scott Charlotte L, et al. Unsupervised high-dimensional analysis aligns dendritic cells across tissues and species. *Immunity.* 2016;45(3):669-684.
518. Wu L, Shortman K. Heterogeneity of thymic dendritic cells. *Semin Immunol.* 2005;17(4):304-312.
519. Audiger C, Rahman MJ, Yun TJ, Tarbell KV, Lesage S. The importance of dendritic cells in maintaining immune tolerance. *J Immunol.* 2017;198(6):2223-2231.
520. Luche H, Ardouin L, Teo P, et al. The earliest intrathymic precursors of CD8alpha(+) thymic dendritic cells correspond to myeloid-type double-negative 1c cells. *Eur J Immunol.* 2011;41(8):2165-2175.
521. Baba T, Nakamoto Y, Mukaida N. Crucial contribution of thymic Sirpα+ conventional dendritic cells to central tolerance against blood-borne antigens in a CCR2-dependent manner. *J Immunol.* 2009;183(5):3053-3063.
522. Hadeiba H, Sato T, Habtezion A, Oderup C, Pan J, Butcher EC. CCR9 expression defines tolerogenic plasmacytoid dendritic cells able to suppress acute graft-versus-host disease. *Nat Immunol.* 2008;9(11):1253-1260.
523. Leventhal DS, Gilmore DC, Berger JM, et al. Dendritic cells coordinate the development and homeostasis of organ-specific regulatory T cells. *Immunity.* 2016;44(4):847-859.
524. Perry JSA, Russler-Germain EV, Zhou YW, et al. Transfer of cell-surface antigens by scavenger receptor CD36 promotes thymic regulatory T cell receptor repertoire development and allo-tolerance. *Immunity.* 2018;48(5):923-936.e924.
525. Ardouin L, Luche H, Chelbi R, et al. Broad and largely concordant molecular changes characterize tolerogenic and immunogenic dendritic cell maturation in thymus and periphery. *Immunity.* 2016;45(2):305-318.
526. Lei Y, Ripen AM, Ishimaru N, et al. Aire-dependent production of XCL1 mediates medullary accumulation of thymic dendritic cells and contributes to regulatory T cell development. *J Exp Med.* 2011;208(2):383-394.
527. Bonasio R, Scimone ML, Schaerli P, Grabie N, Lichtman AH, von Andrian UH. Clonal deletion of thymocytes by circulating dendritic cells homing to the thymus. *Nat Immunol.* 2006;7(10):1092-1100.
528. Hadeiba H, Lahl K, Edalati A, et al. Plasmacytoid dendritic cells transport peripheral antigens to the thymus to promote central tolerance. *Immunity.* 2012;36(3):438-450.
529. Kadouri N, Nevo S, Goldfarb Y, Abramson J. Thymic epithelial cell heterogeneity: TEC by TEC. *Nat Rev Immunol.* 2020;20(4):239-253.
530. Guerau-de-Arellano M, Martinic M, Benoist C, Mathis D. Neonatal tolerance revisited: a perinatal window for Aire control of autoimmunity. *J Exp Med.* 2009;206(6):1245-1252.
531. Boyd RL, Tucek CL, Godfrey DI, et al. The thymic microenvironment. *Immunol Today.* 1993;14(9):445-459.
532. Miller CN, Proekt I, von Moltke J, et al. Thymic tuft cells promote an IL-4-enriched medulla and shape thymocyte development. *Nature.* 2018;559(7715):627-631.
533. Bornstein C, Nevo S, Giladi A, et al. Single-cell mapping of the thymic stroma identifies IL-25-producing tuft epithelial cells. *Nature.* 2018;559(7715):622-626.
534. Lucas B, White AJ, Cosway EJ, et al. Diversity in medullary thymic epithelial cells controls the activity and availability of iNKT cells. *Nat Commun.* 2020;11(1):2198.
535. Watanabe N, Wang YH, Lee HK, et al. Hassall's corpuscles instruct dendritic cells to induce CD4+CD25+ regulatory T cells in human thymus. *Nature.* 2005;436(7054):1181-1185.
536. Ramsdell F, Jenkins M, Dinh Q, Fowlkes BJ. The majority of CD4+8- thymocytes are functionally immature. *J Immunol.* 1991;147(6):1779-1785.
537. McCaughtry TM, Wilken MS, Hogquist KA. Thymic emigration revisited. *J Exp Med.* 2007;204(11):2513-2520.
538. Kurobe H, Liu C, Ueno T, et al. CCR7-dependent cortex-to-medulla migration of positively selected thymocytes is essential for establishing central tolerance. *Immunity.* 2006;24(2):165-177.
539. Cowan JE, Parnell SM, Nakamura K, et al. The thymic medulla is required for Foxp3+ regulatory but not conventional CD4+ thymocyte development. *J Exp Med.* 2013;210(4):675-681.
540. Xing Y, Wang X, Jameson SC, Hogquist KA. Late stages of T cell maturation in the thymus involve NF-kappaB and tonic type I interferon signaling. *Nat Immunol.* 2016;17(5):565-573.
541. Scollay R, Godfrey DI. Thymic emigration: conveyor belts or lucky dips? *Immunol Today* 1995;16(6):268-273; discussion 273-264.
542. Matloubian M, Lo CG, Cinamon G, et al. Lymphocyte egress from thymus and peripheral lymphoid organs is dependent on S1P receptor 1. *Nature.* 2004;427(6972):355-360.
543. Allende ML, Dreier JL, Mandala S, Proia RL. Expression of the sphingosine 1-phosphate receptor, S1P1, on T-cells controls thymic emigration. *J Biol Chem.* 2004;279(15):15396-15401.
544. Carlson CM, Endrizzi BT, Wu J, et al. Kruppel-like factor 2 regulates thymocyte and T-cell migration. *Nature.* 2006;442(7100):299-302.
545. Manna S, Kim JK, Bauge C, et al. Histone H3 Lysine 27 demethylases Jmjd3 and Utx are required for T-cell differentiation. *Nat Commun.* 2015;6:8152.
546. Shiow LR, Rosen DB, Brdickova N, et al. CD69 acts downstream of interferon-alpha/beta to inhibit S1P1 and lymphocyte egress from lymphoid organs. *Nature.* 2006;440(7083):540-544.
547. Agus DB, Surh CD, Sprent J. Reentry of T cells to the adult thymus is restricted to activated T cells. *J Exp Med.* 1991;173(5):1039-1046.
548. Webb LV, Barbarulo A, Huysentruyt J, et al. Survival of single positive thymocytes depends upon developmental control of RIPK1 kinase signaling by the IKK complex independent of NF-kappaB. *Immunity.* 2019;50(2):348-361.e4.
549. Hedrick SM, Hess Michelini R, Doedens AL, Goldrath AW, Stone EL. FOXO transcription factors throughout T cell biology. *Nat Rev Immunol.* 2012;12(9):649-661.
550. Luo CT, Li MO. Foxo transcription factors in T cell biology and tumor immunity. *Semin Canc Biol.* 2018;50:13-20.
551. Shi LZ, Saravia J, Zeng H, et al. Gfi1-Foxo1 axis controls the fidelity of effector gene expression and developmental maturation of thymocytes. *Proc Natl Acad Sci U S A.* 2017;114(1):E67-e74.
552. Navarro MN, Cantrell DA. Serine-threonine kinases in TCR signaling. *Nat Immunol.* 2014;15(9):808-814.
553. Sinclair LV, Finlay D, Feijoo C, et al. Phosphatidylinositol-3-OH kinase and nutrient-sensing mTOR pathways control T lymphocyte trafficking. *Nat Immunol.* 2008;9(5):513-521.
554. Barbee SD, Alberola-Ila J. Phosphatidylinositol 3-kinase regulates thymic exit. *J Immunol.* 2005;174(3):1230-1238.
555. Love PE, Bhandoola A. Signal integration and crosstalk during thymocyte migration and emigration. *Nat Rev Immunol.* 2011;11(7):469-477.
556. Kanariou M, Huby R, Ladyman H, et al. Immunosuppression with cyclosporin A alters the thymic microenvironment. *Clin Exp Immunol.* 1989;78(2):263-270.
557. Shores EW, Van Ewijk W, Singer A. Disorganization and restoration of thymic medullary epithelial cells in T cell receptor-negative scid mice: evidence that receptor-bearing lymphocytes influence maturation of the thymic microenvironment. *Eur J Immunol.* 1991;21(7):1657-1661.
558. van Ewijk W, Shores EW, Singer A. Crosstalk in the mouse thymus. *Immunol Today.* 1994;15(5):214-217.
559. Nitta T, Ohigashi I, Nakagawa Y, Takahama Y. Cytokine crosstalk for thymic medulla formation. *Curr Opin Immunol.* 2011;23(2):190-197.
560. Hikosaka Y, Nitta T, Ohigashi I, et al. The cytokine RANKL produced by positively selected thymocytes fosters medullary thymic epithelial cells that express autoimmune regulator. *Immunity.* 2008;29(3):438-450.
561. Jenkinson SR, Williams JA, Jeon H, et al. TRAF3 enforces the requirement for T cell cross-talk in thymic medullary epithelial development. *Proc Natl Acad Sci U S A.* 2013;110(52):21107-21112.
562. White AJ, Jenkinson WE, Cowan JE, et al. An essential role for medullary thymic epithelial cells during the intrathymic development of invariant NKT cells. *J Immunol.* 2014;192(6):2659-2666.
563. Roberts NA, White AJ, Jenkinson WE, et al. Rank signaling links the development of invariant gammadelta T cell progenitors and Aire(+) medullary epithelium. *Immunity.* 2012;36(3):427-437.
564. McCarthy NI, Cowan JE, Nakamura K, et al. Osteoprotegerin-mediated homeostasis of Rank+ thymic epithelial cells does not limit Foxp3+ regulatory T cell development. *J Immunol.* 2015;195(6):2675-2682.
565. Foy TM, Page DM, Waldschmidt TJ, et al. An essential role for gp39, the ligand for CD40, in thymic selection. *J Exp Med.* 1995;182(5):1377-1388.
566. Desanti GE, Cowan JE, Baik S, et al. Developmentally regulated availability of RANKL and CD40 ligand reveals distinct mechanisms of fetal and adult cross-talk in the thymus medulla. *J Immunol.* 2012;189(12):5519-5526.
567. Akiyama T, Shimo Y, Yanai H, et al. The tumor necrosis factor family receptors RANK and CD40 cooperatively establish the thymic medullary microenvironment and self-tolerance. *Immunity.* 2008;29(3):423-437.
568. Boehm T, Scheu S, Pfeffer K, Bleul CC. Thymic medullary epithelial cell differentiation, thymocyte emigration, and the control of autoimmunity require lympho-epithelial cross talk via LTbetaR. *J Exp Med.* 2003;198(5):757-769.
569. White AJ, Nakamura K, Jenkinson WE, et al. Lymphotoxin signals from positively selected thymocytes regulate the terminal differentiation of medullary thymic epithelial cells. *J Immunol.* 2010;185(8):4769-4776.
570. Williams JA, Hathcock KS, Klug D, et al. Regulated costimulation in the thymus is critical for T cell development: dysregulated CD28 costimulation can bypass the pre-TCR checkpoint. *J Immunol.* 2005;175(7):4199-4207.
571. Punt JA, Osborne BA, Takahama Y, Sharrow SO, Singer A. Negative selection of CD4+CD8+ thymocytes by T cell receptor-induced apoptosis requires a costimulatory signal that can be provided by CD28. *J Exp Med.* 1994;179(2):709-713.
572. Kishimoto H, Sprent J. Several different cell surface molecules control negative selection of medullary thymocytes. *J Exp Med.* 1999;190(1):65-73.
573. Williams JA, Zhang J, Jeon H, et al. Thymic medullary epithelium and thymocyte self-tolerance require cooperation between CD28-CD80/86 and CD40-CD40L costimulatory pathways. *J Immunol.* 2014;192(2):630-640.
574. Dooley J, Liston A. Molecular control over thymic involution: from cytokines and microRNA to aging and adipose tissue. *Eur J Immunol.* 2012;42(5):1073-1079.
575. Venables T, Griffith AV, DeAraujo A, Petrie HT. Dynamic changes in epithelial cell morphology control thymic organ size during atrophy and regeneration. *Nat Commun.* 2019;10(1):4402.
576. Papadopoulou AS, Dooley J, Linterman MA, et al. The thymic epithelial microRNA network elevates the threshold for infection-associated thymic involution via miR-29a mediated suppression of the IFN-alpha receptor. *Nat Immunol.* 2011;13(2):181-187.
577. Cowan JE, Malin J, Zhao Y, et al. Myc controls a distinct transcriptional program in fetal thymic epithelial cells that determines thymus growth. *Nat Commun.* 2019;10(1):5498.
578. Griffith AV, Venables T, Shi J, et al. Metabolic damage and premature thymus aging caused by stromal catalase deficiency. *Cell Rep.* 2015;12(7):1071-1079.
579. Cheng L, Guo J, Sun L, et al. Postnatal tissue-specific disruption of transcription factor FoxN1 triggers acute thymic atrophy. *J Biol Chem.* 2010;285(8):5836-5847.
580. Chaudhry MS, Velardi E, Dudakov JA, van den Brink MR. Thymus: the next (re)generation. *Immunol Rev.* 2016;271(1):56-71.
581. Douek DC, McFarland RD, Keiser PH, et al. Changes in thymic function with age and during the treatment of HIV infection. *Nature.* 1998;396(6712):690-695.
582. Hale JS, Boursalian TE, Turk GL, Fink PJ. Thymic output in aged mice. *Proc Natl Acad Sci U S A.* 2006;103(22):8447-8452.
583. Min D, Taylor PA, Panoskaltsis-Mortari A, et al. Protection from thymic epithelial cell injury by keratinocyte growth factor: a new approach to improve thymic and peripheral T-cell reconstitution after bone marrow transplantation. *Blood.* 2002;99(12):4592-4600.
584. Sutherland JS, Goldberg GL, Hammett MV, et al. Activation of thymic regeneration in mice and humans following androgen blockade. *J Immunol.* 2005;175(4):2741-2753.
585. Dudakov JA, Hanash AM, Jenq RR, et al. Interleukin-22 drives endogenous thymic regeneration in mice. *Science.* 2012;336(6077):91-95.
586. Rossi S, Blazar BR, Farrell CL, et al. Keratinocyte growth factor preserves normal thymopoiesis and thymic microenvironment during experimental graft-versus-host disease. *Blood.* 2002;100(2):682-691.
587. Erickson M, Morkowski S, Lehar S, et al. Regulation of thymic epithelium by keratinocyte growth factor. *Blood.* 2002;100(9):3269-3278.
588. Fichtelius KE, Laurell G, Philipsson L. The influence of thymectomy on antibody formation. *Acta Pathol Microbiol Scand.* 1961;51:81-86.
589. Miller JF. Immunological function of the thymus. *Lancet.* 1961;2(7205):748-749.
590. Arnason BG, Jankovic BD, Waksman BH. Effect of thymectomy on 'delayed' hypersensitive reactions. *Nature.* 1962;194:99-100.
591. Parrott DM, East J. Role of the thymus in neonatal life. *Nature.* 1962;195:347-348.
592. Archer OK, Pierce JC. Role of the thymus in development of the immune response. *Fed Proc.* 1961;20:26.

CHAPTER 28 CD4+ Helper T Cells: Differentiation, Flexibility, and Heterogeneity

John J. O'Shea, MD • Marion Pepper, PhD • Marc K. Jenkins, PhD • Danielle A. Chisolm, PhD

INTRODUCTION

CD4+ T cells are critical for host defense and the elimination of microbial pathogens and their products, as vividly documented by the consequences of their loss due to genetic and infectious causes. Increasingly, the role of CD4+ T cells in elimination of cancer cells is recognized and capitalized upon therapeutically.[1] To effectively combat a new microbial infection, the few naïve CD4+ T cells expressing T cell antigen receptors (TCRs) specific for major histocompatibility complex class II (MHCII)–bound peptides from the microbe must differentiate to form effector cells with functions that clear the infection and generate memory cells that provide immunity to future infections. Different classes of microbes stimulate different cytokines from the innate immune system, which in turn drive the differentiation of the major subsets of CD4+ effector T cells, which we know as T helper 1 (Th), Th2, Th17, and follicular helper (Tfh) cells. Each of these subsets selectively produce different cytokines that activate myeloid cells, granulocytes, B cells, or NK cells to eliminate particular types of microorganisms. CD4+ T cells also support the activation of CD8+ T cells and their differentiation to memory cells.[2] CD4+ T cells are thus classically referred to as "helper cells" because they foster immunity by enhancing the microbicidal actions of other cells. In this chapter, we will discuss the cellular processes and molecular mechanisms that control CD4+ helper T cell subset specification. We do this with the recognition that the understanding of these critical cells will continue to evolve and new models will emerge.

LIFE OF NAÏVE T CELLS

The process of T cell development and selection (Chapter 27) creates a large and diverse set of self-tolerant naïve CD4+ T cells expressing TCRs with low affinities for MHCII-bound host peptides.[3] This repertoire will contain a few cells that by chance have TCRs with high affinity for any MHCII-bound "foreign" peptides that are not present in the host's peptidome, for example, peptides from microbes. On average, about one naive CD4+ T cell in 10^5 to 10^6 total CD4+ T cells expresses a TCR capable of strong binding to any given MHCII-bound foreign peptide. About 10% of the cells in each foreign epitope-specific naïve CD4+ T cell repertoire express the Foxp3 transcription factor and are regulatory T (T_{reg}) cells,[4] while 90% lack Foxp3 and are conventional T cells.[5] This chapter will focus on conventional cells, the precursors of the effector and memory T cells that arise in response to any foreign peptide; detailed information about T_{reg} cells can be found in Chapter 29.

Conventional naïve CD4+ T cells search the body for the MHCII-bound foreign peptide that their TCR is specific for. After leaving the thymus, these cells migrate into the blood and circulate through the secondary lymphoid tissues including the spleen, lymph nodes, and mucosal-associated lymphoid tissues such as Peyer patches.[6-8] T cells enter lymph nodes and Peyer patches via specialized blood vessels termed high endothelial venules (HEVs), which display GlyCAM-1 (and/or MadCAM-1), ICAM-1, and CCL21. Naive T cells express the receptors for these ligands, CD62L, LFA-1, and CCR7, which promote the extravasation of the T cells across the HEV and into the lymphoid tissue. Naïve T cells enter the spleen through terminal arterioles, not HEVs, as a result, expression of CD62L is not required. However, naïve T cell migration to the splenic white pulp requires integrin and chemokine receptor signaling.[9]

After entering a secondary lymphoid organ, naïve T cells concentrate in central zones called the paracortical regions of lymph nodes[10] and Peyer patches[11] or the periarteriolar lymphoid sheaths of the splenic white pulp[12] (Chapter 8). These T cell–rich areas contain fibroblastic reticular cells that produce CCL21 and CCL19, which attract the CCR7-expressing naïve T cells, as well as IL-7, which is important for naïve T cell survival.[13,14] Fibroblastic reticular cells also serve as scaffolds to which numerous dendritic cells (DCs) adhere.[15] Naïve T cells spend about a day in a secondary lymphoid organ migrating along the fibroblastic reticular cells and interacting with many DCs and in the process recognize MHCII-bound host peptides[16] and receive tonic survival signals through their TCRs and IL-7 receptors. Naïve CD4+ T cells remain in the G_0 phase of the cell cycle throughout this process due to the action of quiescence factors Foxp1 and miR-125b.[17,18]

If a naïve CD4+ T cell does not encounter its MHCII-bound foreign peptide, then it leaves the lymph nodes via a process requiring sphingosine 1-phosphate (S1P) receptor 1,[19,20] which like other elements of the naïve state is controlled by the transcription factor KLF2.[19,21,22] The concentration of S1P is higher in efferent lymph than in the T cell–rich areas of lymph nodes.[20] Lymphocytes sense this concentration gradient with S1P receptor 1 and exit into the efferent lymph, then the thoracic duct, and then the blood. Exit from the spleen is thought to occur by migration from white pulp to red pulp with subsequent entry into blood via the splenic vein. Naïve T cells then enter another secondary lymphoid organ and repeat the process. The immunomodulatory drugs, fingolimod, siponimod and ozanimod,[23] target S1P receptor 1 and block egress of T cells from the thymus and secondary lymphoid organs.

ENCOUNTERING ANTIGEN; NAÏVETE LOST

The behavior of a naïve T cell changes dramatically if it encounters its cognate MHCII-bound foreign peptide on a DC in a secondary lymphoid organ (Chapter 20). DCs can generate foreign peptide-MHCII complexes in several different ways,[24] which can be illustrated by a bacterial infection in the skin. Proteins that are secreted by the bacteria or cleaved from their surfaces at the infection site can be carried to a draining lymph node through afferent lymphatic vessels.[25,26] The proteins then enter thin conduits formed by the fibroblastic reticular cells in the T cell–rich zones[10,27] and are taken up by associated DCs into late endosomes where the proteins are cleaved by proteases into short peptides,[28] some of which have core sequences[29] capable of binding to MHCII molecules that are also in this compartment. The MHCII-bound peptides are then trafficked to the antigen-presenting cell surface. DCs in the skin infection site also take up secreted proteins and the bacteria themselves and produce MHCII-bound peptides.[25] These DCs also sense bacterial products using pattern recognition receptors, turn on CD28 ligands and CCR7, and migrate to the draining lymph node in a CCL21-dependent manner.[30] The few helper T cells that are in this lymph node and express TCRs specific for these particular MHCII-bound peptides, then detect them on the resident and migrated DCs, and receive signals from the TCR and CD28.[25] In this phase of activation, T cells upregulate the affinity of integrins such as LFA-1,[31] allowing stable interactions with DCs that last for several hours.[32-34] T cell interactions trigger DCs to produce the chemokines CCL3 and CCL4 that recruit more T cells to the area, thereby increasing the probability of CD4+ T cells encountering an appropriate antigen-presenting cell with specific MHCII-bound foreign peptides.[35]

The stage is then set for T cell proliferation. TCR and CD28 signals activate the transcription factors NF-κB, NFAT, and AP-1 to enter the nucleus[36-39] (Chapters 23 and 24) and initiate transcription of many genes including those encoding IL-2 and CD25, which pairs with CD122 and the common gamma chain to form the high-affinity IL-2 receptor.[40,41] The activated T cells secrete IL-2, which binds to the IL-2 receptor and activates STAT5[42,43] (Chapter 9), which causes the T cells to start dividing about 30 hours after initial antigen presentation.[44] Naïve CD4+ T cells exposed to MHCII-bound microbial peptides in the context of infection continue to proliferate rapidly in the secondary lymphoid organs for about a week, increasing about 1000-fold to produce a large population of lymphoblasts known as effector cells.[45]

EFFECTOR CELL BIFURCATION

It became clear in the late 1980s that CD4 effector T populations were not homogenous. Pioneering studies by Mosmann and Coffman[46] identified two types of T cell lines—one type called Th1 that produced IL-2, TNF, IFN-γ, and another called Th2 that produced IL-4, IL-5, and IL-13[47-49] (Fig. 28.1). This discovery led to an intense search for the conditions that cause naive CD4+ T cells to become Th1 or Th2 cells. The breakthrough came from the reductionist approach of culturing naïve TCR transgenic T cells with antigen-presenting cells and cognate peptide.[50] This work lead to the theory that different cytokines from the innate immune system cause proliferating effector T cells to turn on distinct transcription factors that cause the T cells to become different lineages that produce different cytokines.[47-49] It was recognized that intracellular microbes stimulate type 1 innate cytokines IL-12 and type I interferons, which cause CD4+ T cells to express T-bet, the master regulator of Th1 cell differentiation,[51] while allergens and helminths stimulate type 2 innate cytokines IL-4, IL-25, and IL-33, which cause CD4+ T cells to express GATA3, the master regulator of Th2 cell differentiation.[52] Later work identified IL-17–producing Th17 cells and type 17 innate cytokines IL-1, IL-6, IL-23, and TGF-beta, and the Th17 lineage-defining transcription factor RORγt.[53-55] *In vitro* culture systems, however, failed to identify B cell–helping Tfh cells, which were instead discovered from *in vivo* studies.[56-61] Unlike Th1, Th2, and Th17 cells, which were defined by selective cytokine production, Tfh cells were recognized by selective expression of the chemokine receptor CXCR5 that allowed localization within B cell–rich follicles and germinal centers, thereby positioning Tfh cells to "help" B cells.[62] Tfh cells can be considered to be a distinct subset like Th1, Th2, and Th17 cells because their formation is controlled by a master transcription factor, BCL6.[49,59,60]

Later studies revealed that CD4+ effector T cell populations generated *in vivo* initially contain a mixture of Tfh and non-Tfh lineage cells with the non-Tfh cells being Th1, Th2, or Th17 cells depending on the innate immune system context[63-68] (Fig. 28.1). For example, Tfh and Th1 cells are generated during infections that elicit type 1 innate immune responses,[63] while Tfh and Th2 cells are generated in type 2 innate immune responses.[69] These observations indicate that primary CD4+ T cell responses have a feature that ensures that a mixture of B cell–helping Tfh cells and granulocyte and myeloid cell–helping non-Tfh effector cells are generated at the same time. The process is beneficial to the host by simultaneously augmenting humoral and cellular immunity.

The mechanism that drives contemporaneous development of Tfh and non-Tfh effector cells has been the topic of much research. Evidence points to IL-2 receptor signaling and STAT5 activation (Fig. 28.1). Naïve CD4+ T cells produce IL-2 and induce CD25 early after receipt of TCR and CD28 signals.[40] STAT5 activation downstream of IL-2 receptor signaling[42,43] induces the Blimp1 transcription factor, which is a potent repressor of BCL6.[63] This repression of BCL6 is significant as evidenced by the observations that Th1 or Th2, cells do not form in CD25-deficient animals during infections that trigger type 1 or 2 innate immune responses, while Tfh cells form normally.[66,68,69] Moreover, genetic reduction in STAT5A/B levels result in increased BCL6 expression and expansion of Tfh cells.[70] These results indicate that IL-2 receptor signaling is required to prevent early effector cells from becoming Tfh cells. It is possible that TCR- and CD28-triggered naïve CD4+ T cells vary in the amount of IL-2 receptor signaling that they receive either because of variations in the amount of IL-2 or CD25 that they produce or

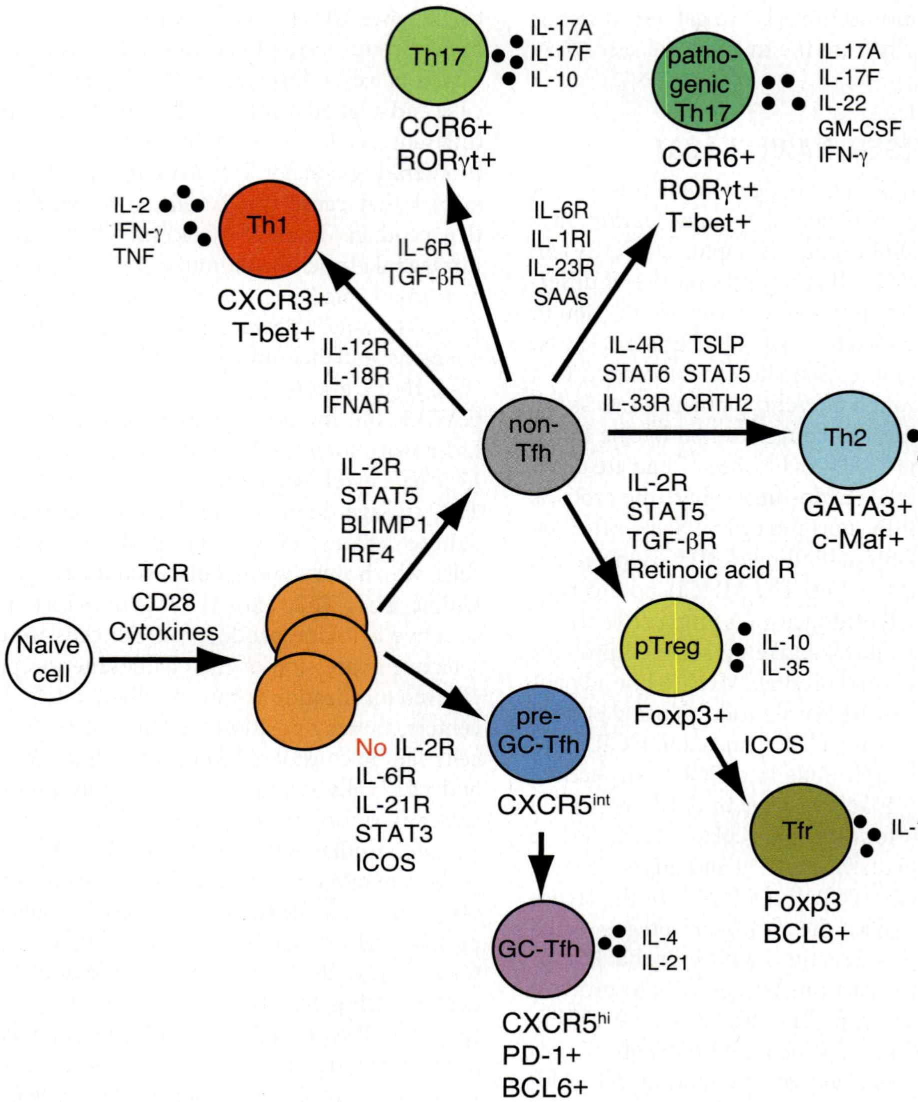

FIG. 28.1. Diverse fates of CD4+ T cells. Naïve CD4+ T cells are activated via signals from their T cell receptors, costimulatory receptors, and cytokines, and proliferate to form an uncommitted effector cell population. Cells from this population that do not receive or maintain signals from IL-2R then become CXCR5^{int} cells that further differentiate into CXCR5^{hi} GC-Tfh cells expressing the transcription factor BCL6. By contrast, T cells that receive signals that activate STAT5 (eg, IL-2) downregulate BCL6, express Blimp1, and become Th1, Th2, Th17, pathogenic Th17, or peripheral T_{reg} cells depending on innate immune system cytokines. The figure also shows typical cytokines produced by each subset.

their proximity to other IL-2–producing cells. It then follows that CD4+ T cells that receive strong IL-2 receptor-mediated STAT5 signals induce Blimp1 and are prevented from expressing BCL6 or becoming Tfh cells; conversely, CD4+ T cells that receive weak IL-2 receptor-mediated STAT5 signals do not induce Blimp1 and express BCL6 and become Tfh cells. It is important to note, however, that although BCL6 represses T-bet and RORγt, this repression is not complete. Thus, GC-Tfh cells that form in type 1 immune responses express low amounts of T-bet,[45] while GC-Tfh cells that form in type 17 immune responses express low amounts of RORγt.[71] In addition, the GC-Tfh cells that form in type 2 immune responses can be more potent IL-4 producers than Th2 cells.[69,72] Tfh cells may express aspects of the Th1 and Th2 programs to produce IFN-γ or IL-4 needed for immunogloblulin class switch recombination by the B cells they help. Tfh cells expressing chemokine receptors associated with Th1, Th2, or Th17 cells have also been identified in humans with autoimmune diseases.[73-75] Thus, effector cell bifurcation leads to Th1, Th17, or Th2 cells that lack BCL6 and Tfh cells that express BCL6 and lesser amounts of the transcription factor expressed by their non-Tfh siblings.

The preceding model raises the question of what could cause CD4+ T cells to vary in the amount of IL-2 receptor-mediated STAT5 activation that they receive. TCR signaling is one possibility[76,77] because it acts as a rheostat downstream of signaling molecules such as the IRF4 transcription factor and the IL-2 receptor.[71,78,79] High levels of IRF4 and IL-2 receptor signaling favor induction of Blimp1 over BCL6[79] and evidence in the literature supports the idea that strong TCR,

IRF4, and IL-2 receptor signaling favors non-Tfh (Th1 in this case) formation,[71,79-82] although it has been reported that IRF4 favors CXCR5+ cells in humans.[83] Thus, naïve murine T cells that have TCRs with high affinity for MHCII-bound peptides or happen to engage with antigen-presenting cells displaying large amounts of these ligands may strongly induce IRF4 and Blimp1, sustain IL-2 receptor expression, and become non-Tfh effector cells. Other naïve T cells that have TCRs with lower affinity for the MHCII-bound peptide or happen to engage with DCs displaying lower amounts of this ligand may not sustain the IL-2 receptor or Blimp1, express BCL6, and generate Tfh effector cells.

It should be noted, however, that some evidence does not support the concept that strong TCR signaling favors non-Tfh formation. Tfh effector cells responding to immunization with a protein and adjuvant bound more peptide: MHCII tetramer, a correlate of high TCR affinity, than other effector cells.[84] Although increasing the antigen dose and presumably TCR signaling in a bacterial infection increased Th1 cell formation up to a point, the largest antigen doses favored Tfh formation.[80] Likewise, CD4+ T cells receiving the strongest TCR signals produced IL-2 but did not respond to it and became Tfh cells.[85] These apparently conflicting results may be explained by the fact that TCR affinity and antigen dose, which might be thought to influence TCR signaling in the same way, can have opposing effects. It has been reported, for example, that although increasing TCR affinity increases non-Tfh formation, increasing antigen dose increases Tfh formation.[86] Another potentially confounding factor is that the amount of tonic signaling that a naïve T cell receives can vary depending on the affinity of its TCR for the relevant MHCII-bound *host* peptide. This variation can influence the probability that the naïve CD4+ T cell will produce Tfh or non-Tfh progeny when stimulated by its MHCII-bound *foreign* peptide.[87] Thus, the separate influences of antigen dose and TCR affinity for MHCII-bound self and foreign peptides have made it difficult to resolve the role that TCR plays Tfh and non-Tfh cell bifurcation during the primary response. This issue is not a trivial question as its answer has important implications for the development of effective vaccines.

Finally, the bifurcation may result from individual progeny of naïve cells migrating into different microniches in the secondary lymphoid organs containing different cytokines that favor Tfh or non-Tfh differentiation.[88] The cytokines that have this potential for Tfh, Th1, Th2, and Th17 differentiation are described in separate sections below.

TFH CELLS

Tfh populations generated during *in vivo* immune responses contain subsets that express intermediate (int) or large (hi) amounts of CXCR5 (Fig. 28.1). The CXCR5^{hi} cells also express PD-1 and folate receptor 4 (FR4) and are known as germinal center Tfh (GC-Tfh) cells[45,61,63,89-91] based on uniform expression of BCL6 and residence in germinal centers.[62,92] CXCR5^{int} Tfh cells reside in mantle regions of the follicles[92] where they provide CD40 ligand and IL-21 to CD40+ B cells that present MHCII-bound peptides to them.[93,94] These Tfh-derived signals cause the B cells to proliferate and become plasmablasts or germinal center cells and promote the survival of germinal center B cells with somatic mutations that improve antigen binding.[95] The CXCR5^{int} Tfh cells then differentiate into GC-Tfh cells,[90,96-98] which reside in germinal centers and drive the process of affinity maturation.[62,92]

Although it was thought that Tfh cells are confined to follicles and germinal centers in the secondary lymphoid organs, recent insights from single cell technologies and assessment of immune cell subsets in tissue and blood samples from humans with autoimmune disease reveal the existence of another CD4 subset with similarities to Tfh cells. These cells are designated T peripheral helper cell (Tph) and are found in blood and inflammed joints of patients with rheumatoid arthritis.[99,100] Tph cells express PD-1 and produce IL-21 but display features distinct from those of Tfh cells including relatively less BCL6, more Blimp1, and different chemokine receptors, CCR2, CX_3CR1, and CCR5. These cells may be relevant factors in driving autoimmune disease but require considerable additional investigation will be needed to understand these interesting cells.

As described above, a differentiating CD4+ effector T cell needs to receive minimal IL-2 receptor signaling to become a Tfh cell. Additional cytokines, however, are also needed for Tfh differentiation. Formation of Tfh cells depends on IL-6 and IL-21 receptor-mediated induction of STAT3, inducible costimulatory receptor (ICOS)-mediated induction of c-Maf,[45,68,89,101,102] and the transcription factor Thpok,[103] which cooperate to induce BCL6 (Fig. 28.1). The LEF1 and TCF-1 transcription factors coordinate this process by sustaining expression of IL-6 receptor and enhancing expression of ICOS.[104-106] ICOS signaling in Tfh cells is triggered by ICOS ligand on B cells and most efficiently by antigen-specific B cells that directly present MHCII-bound peptides to the Tfh cells.[45,68]

BCL6 then turns on the gene products that define the Tfh lineage, paradoxically by acting as a transcriptional repressor, in part by interaction via its BTB domain with the BCL corepressor.[107,108] BCL6 represses Blimp1, a promoter of non-Tfh options,[109] thereby leaving the door open to the Tfh fate. Equally important, however, BCL6 also represses repressors of the genes that positively regulate Tfh formation.[110] For example, BCL6 represses Id2, a repressor of *Cxcr5*, and other transcription factors that repress other important Tfh genes including those encoding PD-1, IL-6 receptor, and ICOS. The CXCR5^{int} Tfh precursors of GC-Tfh cells express less BCL6 than GC-Tfh cells,[45,65] and thus have repressed some but not all of the Tfh repressors, thereby allowing intermediate expression of CXCR5. Receipt of additional signals from IL-6 and IL-21 receptors and ICOS could increase BCL6 in these cells and cause their further progression to become GC-Tfh cells. This model is supported by the observation that CXCR5^{int} and GC-Tfh cells disappear in mice lacking both copies of the *Bcl6* gene.[45,90] CXCR5^{int} effector cells, however, appear in normal numbers in mice lacking one copy of the *Bcl6* gene,[66] while GC-Tfh cells are greatly reduced in this instance. The fact that CXCR5^{int} cells can form under conditions of suboptimal BCL6 induction is consistent with partial silencing of Tfh repressors and initiation but not completion of the full GC-Tfh program.

We will now turn our attention to the factors that influence the differentiation of the non-Tfh effector cells that form during the bifurcation process and summarize what has been learned from the *in vitro* approach and later *in vivo* studies for each of the major non-Th subsets.

TH1 CELLS

The early work on T cell clones identified IFN-γ as the signature cytokine produced by Th1 cells[46] although these cells also make IL-2 and TNF (Fig. 28.1). Th1 cells also express high levels of IL-12 receptor β1 and β2 subunits, IL-18 receptor subunits, E-selectin, P-selectin, CXCR3, and CCR5.[51] IFN-γ from Th1 cells activates macrophages to kill phagocytosed microbes and is critical for this process as evidenced by the dramatic sensitivity of CD4+ T cell- or IFN-γ-deficient people to infections of the macrophage phagosome.[111] IFN-γ also enhances antigen processing and presentation via upregulation of MHCI and MHCII molecules and thereby amplifies T cell–dependent immune responses. IFN-γ promotes B cell class switching to IgG1 and IgG3 (IgG2a or IgG3 in mice) and inhibits switching to IL-4–dependent isotypes (eg, IgE). Th1 cells or T-bet^{low} Tfh cells that develop in the presence of type 1 innate cytokines could be the sources of IFN-γ that promotes B cell class switching.

Th1 differentiation is stimulated by intracellular bacteria, viruses, and adjuvants that elicit production of IL-12, IL-18, IFN-γ, and/or type I IFN[51] (Fig. 28.1). Beyond infection, commensal and dysbiotic microbiota contribute to immune cell homeostasis.[112,113] In the case of inflammatory bowel disease, the oral cavity may serve as a reservoir for potential intestinal pathobionts including Klebsiella spp. that can drive Th1 responses and exacerbate intestinal disease.[51,113]

In vitro experiments indicate a step-wise model in which TCR, IFN-γ receptor, and STAT1 signaling causes T cells to express T-bet (encoded by the *Tbx21* gene),[114,115] the master regulator of Th1 cells[116-123] that directly regulates all the canonical Th1 genes including *Ifng*.[124-126] As TCR signaling wanes,[123] T-bet induces the IL-12 receptor and IL-12 receptor signaling activates STAT4, which drives proliferation of the Th1 cells,[121,127] enhances their T-bet expression and IFN-γ production capacity. This second step of the process causes epigenetic modifications discussed below that fully commit cells to the Th1 fate.[121,123,128] STAT4 also activates the transcriptional repressor TWIST1, which limits the expression of cytokines including IFN-γ, IL-2, and TNF.[129] Similarly, the protease Furin, which processes the anti-inflammatory cytokine TGF-β, is also regulated by STAT4.[127,130] Thus, STAT4 simultaneously promotes Th1-mediated immunity and constrains it, presumably to prevent immunopathology.

T-bet also promotes Th1 cell formation by inhibiting alternative fates. It inhibits expression of RORγt and BCL6, the master regulators for Th17 and Tfh cells, and blocks Th2 formation by inhibiting the action of Runx1,[117] GATA3, and IL-4, in a Runx3-dependent manner.[124,131] In addition to T-bet, Th1 cells express the Hlx, Runx3, and Ets family transcription factors that promote IFN-γ production while repressing *Il4* transcription.[132,133]

Many Th1 effector cells leave the secondary lymphoid organs where they are generated and migrate to nonlymphoid sites of inflammation.[7,134] T-bet regulates P-selectin glycoprotein ligand-1, CXCR3, and CCR5, which bind P-selectin, CXCL9, CXCL10, CXCL11, and RANTES, all of which are displayed by endothelial cells in inflamed tissues.[51] Th1 cells also accumulate in microbe-containing granulomas that form during persistent infections of macrophages, in part due to CXCR3 sensing of CXCL9 and CXCL10.[135] Concentration of Th1 cells in granulomas is likely key to Th1-based immunity since the infections that CD4+ T cell–deficient hosts are hyper susceptible to are controlled in granulomas.[111]

TH2 CELLS

The Th2 signature cytokines, IL-4, IL-13, and IL-5 (Fig. 28.1), are important for barrier defense at mucosal and epithelial surfaces. IL-4 stimulates the development of Th2 cells and functions as an autocrine growth factor for differentiated Th2 cells.[136] It also stimulates B cell growth and immunoglobulin class switching to IgG4 in humans (IgG1 in mice) and IgE isotypes. Conversely, IL-4 inhibits switching to the IgG2a and IgG3 isotypes in mice, both of which are stimulated by IFN-γ. IL-4, together with IL-13, promotes the formation of "alternatively activated" macrophages, suppresses IFN-γ–mediated classical macrophage activation, and thus inhibits defense against intracellular microbes. IL-4, IL-13, and IL-5 mobilize eosinophils, basophils, and mast cells and IL-13 increases mucus secretion from airway and gut epithelial cells. These actions contribute to host defense and repair at epithelial surfaces. In this manner, Th2 cells are important for protection against helminths and other parasites but are also the major drivers of allergy and asthma. The monoclonal antibody dupilumab targets the shared receptor for IL-4 and IL-13 and is approved for treatment of atopic dermatitis.[137]

In vivo, Th2 differentiation depends upon a relay circuit in which a variety of environmental triggers that induce epithelial cell damage or sensory neuron activation lead to the release of soluble factors. Unique subsets of DCs respond to these soluble cues and migrate to the lymph nodes where they prime Th2 cells that can migrate back to the tissues. Th2 cells gain their full effector capacity after returning to the tissue where they interact with innate lymphoid cells and other Th2-skewing cell populations. Understanding what fate-directing cues Th2 cells receive early in the lymphoid organs versus later in the tissues is an active area of research.

Many Th2-inducing immunogens including allergens, secreted helminth products, and venoms commonly possess enzymatic activities that are required for type 2 immune activation.[138] For example, house dust mite allergens contain cysteine proteases that can disrupt epithelial junctions, release and activate innate alarmins, and elicit the release of innate cell activating neuropeptides including substance P from activated TRPV1+ neurons.[139-141] Both alarmins and neuropeptides can activate a population of CD301+ CD11b+ DCs to migrate back to the lymph node to induce Th2 skewing.[142,143]

CD301+ CD11b+ DCs, also called cDC2, are required for the differentiation of Th2 cells and depend upon the transcription factors IRF4 and KLF4 for their formation.[144-146] Alarmins and neuropeptides can also activate other innate cell populations including ILC2s to indirectly induce DC migration to the lymph node through the production of cytokines[147] and promote Th2 responses by activating other innate cell populations.[148-153] ILC2s produce IL-13 and IL-5 but typically do not produce high levels of IL4.[152,154,155] ILC2s can also express MHCII molecules and may promote Th2 differentiation[156] via presentation of p:MHCII complexes either in the secondary lymphoid organs or later in nonlymphoid tissues.[157]

As stated above, IL-4 itself is a major driver of Th2 differentiation (Fig. 28.1). IL-4 can be produced by an array of innate cells including NK T cells, mast, cells and basophils.[52,153,158-160] IL-25, an IL-17-family cytokine (also termed IL-17e), IL-33, an IL-1–related cytokine, and thymic stromal lymphopoietin (TSLP) also promote Th2 responses in large measure by activating innate cells.[148-153]

IL-4 stimulates Th2 development by activating STAT6, which together with TCR-dependent signals, induces expression of the Th2 master regulator, GATA-3 (Fig. 28.1). STAT6 also regulates thousands of other genes in Th2 cells including pivotal transcription factors, cytokines and cytokine receptors, chemokines and chemokine receptors.[127,161-163] IL-2 and TSLP also contribute to Th2 differentiation by activating STAT5, which positively regulates expression of the IL-4 receptor.[164-167] STAT5 directly targets the *Il4r* gene in Th2 cells.[167] IL-2 signaling has also been shown to be essential for the formation of Th2 tissue-resident memory cells in a murine model of allergic airway inflammation. The lack of IL-2 signaling leads to diminished lung-resident Th2 cells and an increase in BCL6-expressing Tfh cells.[168] Additionally, IL-6 acting through STAT3 has been reported to contribute to Th2 differentiation.[169] IL-10–induced Blimp1 expression can also suppress BCL-6 and is also required for Th2 development, suggesting Blimp1 as an important factor in Th2 development.[170]

GATA3 is a critical factor for lineage commitment of Th2 cells, although it also plays essential roles in thymic differentiation. GATA3 binds to *Il4, Il5, Il13, Gata3,* and other Th2 genes.[171] GATA3 also actively represses RUNX3, which in turn leads to reduced IFN-γ production[172] and induces the transcription factor c-Maf, which can aid in Th2 differentiation. However, c-Maf deficiency does not abrogate production of Th2 cytokines[173] and c-Maf has functions beyond regulation of the Th2 program. With respect to its role in thymic development, GATA3 regulates key factors including Th-POK, Notch, and TCR subunits.[171]

Notch, IRF-4, and growth factor independent-1 (Gfi-1) also aid in Th2 differentiation. Notch promotes *Gata3* expression in a STAT6-independent manner.[174] IRF-4 contributes to Th2 differentiation, but it is also important for other subsets as described above.[175] Repressor of GATA3 (*Rog* or *Zbtb32*), as the name implies antagonizes the actions of GATA3 and limits Th2 differentiation.[176,177] Mina, a member of the jumonji C (JmjC) protein family, binds and represses the *Il4* promoter. Conversely, reducing levels of Mina in T cells leads to *Il4* derepression.[178] Eomesodermin (*Eomes*), a key regulator of IFN-γ production in CD8+ T cells, is important for limiting IL-5 production in Th2 memory cells[179] and acts by limiting GATA3 binding to the *Il5* promoter. The transcription factor Bcl11b is a GATA3 binding partner. Bcl11b binds Th2 genes but also binds and represses Th1 genes, limiting expression of both transcriptomic programs.[180,181]

Although CD4+ T cells primed under Th2 conditions in the lymph node can express transcripts for IL-4, these cells are Tfh cells. CXCR5– Th2 effector cells do not appear to express IL-4, IL-13, or IL-15 proteins until they have reached the tissues.[182-184] Much like ILC2s, tissue-resident Th2 cells express the receptor for IL-33 (ST2) and can also express the chemoattractant receptor-homologous molecule expressed on Th2 cells (CRTH2) (a receptor for the bioactive lipid prostaglandin D_2 [PGD_2]). PGD2 is an eicosanoid derived from dietary arachidonic acid that is produced in response to a variety of cues including the expression of IL-33 by epithelial cells. Both the IL-33 and CRTH2 receptors are associated with pathological memory Th2 cells that express high amounts of IL-5 and IL-13 and can be found in the blood of subjects with ongoing allergic responses.[185] Memory Th2 cells can express cytokines in response to innate cytokines such as IL33, suggesting that these cells become rapidly responsive sentinel cells.[186,187]

TH17 CELLS

The signature cytokines for Th17 cells are IL-17A, IL-17F, IL-21, and IL-22[53,54,188-196] (Fig. 28.1). IL-17 enhances neutrophil generation by increasing G-CSF production and promotes neutrophil recruitment by production of chemokines[197-199]; in this respect, IL-17 represents an important link between adaptive and innate immunity. Because neutrophils are a major defense mechanism against extracellular bacteria and fungi, Th17 cells play an especially important role in defense in such infections including those caused by *S. aureus, K. pneumoniae,* and *C. albicans.*[198,200,201] Additionally, IL-17 stimulates the production of antimicrobial substances, including defensins. The IL-17 axis also contributes to host defense against *M. tuberculosis*[202] and contributes to germinal centers and formation of lymphoid structures during inflammation.[202,203] Th17 cells are important in the pathogenesis of a variety of autoimmune and immune-mediated diseases including psoriasis and spondyloarthritis.[204] Th17 cells typically express the chemokine receptor CCR6, along with its ligand, CCL20,[205] which allow Th17 cells to migrate into inflamed tissues and recruit neutrophils.

Cytokines that promote Th17 differentiation include IL-6, IL-23, IL-1, serum amyloid A family members, and TGF-β[191,192,194,205-209] (Fig. 28.1). *In vivo*, segmented filamentous bacteria are major inducers of Th17 cells and promote autoimmunity.[210-212]

IL-6 and IL-23 activate STAT3,[213,214] which turns on *Rorc*, the gene encoding RORγt, the master regulator for Th17 cells and other transcription factors expressed by this T cell subset.[207,215] RORγt acts directly on many genes that contribute to the phenotype of Th17 cells including *Il17a/f*, *Il23r*, and *Il21*.[216] In addition to its role in Th17 cells, RORγt, like GATA3, has important functions in thymic development, as well as development of lymphoid tissue inducer cells and IL-22–producing NK cells.[217-219] SOCS3 is an important negative regulator of Th17 cell differentiation.[213] In addition to RORγt, RORa is also a contributor to Th17 differentiation.[220] BATF and IRF4 cooperate with STAT3 to promote chromatin accessibility and the Th17 transcriptional program.[221]

Other transcription factors expressed by Th17 cells include aryl hydrocarbon receptor (AHR), Batf, I kappaB zeta, IRF-4, and Runx1; absence of any of these molecules impairs IL-17 expression.[117,222-226] The key metabolic sensor Hypoxia-inducible factor 1 (HIF-1)[227] is also a positive regulator of Th17 differentiation. HIF-1 associates with RORγt and recruits the acetyl transferase p300 to the *Il17* gene.[227] TLR2 signaling[228] and mTor signaling via mTorc1 also promote Th17 cell differentiation[229] as does BCL6 corepressor, which represses Lef1, Runx2, and Dusp4 that otherwise suppress the pathway.[71] Th17 cell formation is inhibited by IL-4 and IFN-γ[53] and the transcription factors Gfi-1, Ets-1, T-bet, and Tcf-1.[230-232]

In vivo experiments in mice point to several modes of Th17 differentiation. CD4+ T cells that receive signals from TCR, IL-6 receptor, and TGF-β receptor activate STAT3, induce RORγt and AHR, and become Th17 cells that participate in helpful immune responses to infection.[54,190] Experiments in mice with natural T cell repertoires and intranasal *Streptococcus pyogenes* infection indicate that Th17 cell formation in this mode conforms to the effector cell bifurcation model in which Tfh and non-Tfh cells form at the same time with the non-Tfh cells (Th17 cells in this case) augmented by IL-2 receptor signaling.[71] It should be noted, however, that *in vitro* experiments with added cytokines or *in vivo* experiments with TCR transgenic T cells and vaccination with complete Fruend adjuvant show that IL-2 acting via STAT5 *negatively* regulates Th17 cell formation[233] by binding the *Il17a/f* locus and displacing STAT3.[214] Other studies have shown that T_{reg} cells may promote Th17 cell differentiation by limiting IL-2.[234,235] Thus, the role that the IL-2 receptor signaling plays in Th17 cell differentiation may vary depending on the context.

A pathogenic mode of Th17 cell differentiation has also been described. In certain situations, usually involving abundant or prolonged production of IL-23, CD4+ T cells receive signals from TCR, IL-6 receptor, IL-1 receptor, and IL-23 receptor, activate STAT3, induce RORγt and T-bet, and become pathogenic Th17 cells that can produce IL-17 and IFN-γ and participate in harmful responses to gut bacteria or host tissues[236] (Fig. 28.1). The combination of IL-6 and serum amyloid A1, A2, and A3 can also induce pathogenic Th17 cells.[237]

Although Th17 cells are generated in the secondary lymphoid organs, they exert their functions in nonlymphoid organs. Th17 cells typically express the chemokine receptor CCR6, along with its ligand, CCL20,[205] which allow Th17 cells to migrate into inflamed tissues and recruit neutrophils.

TH9 CELLS

Adding TGF-β to cultures of T cells with IL-4 generates a population of cells that produce IL-9 but not IL-4. Such cells have been referred to as Th9 cells.[238-241] IL-2 promotes IL-9 production,[242] but in addition, Th9 cells express the receptor for IL-25 (IL-17RB), which also promotes IL-9 production.[243] TNF family members also induce IL-9,[244] whereas retinoic acid inhibits IL-9 production.[245]

IL-9 in turn acts on mucosal cells to enhance mucus production.[228,246] IL-9 has also been reported to have variable effects on Th17 differentiation and Th17-dependent pathology.[247-249] The transcription factors PU.1, IRF-4, and STAT5 are important for generation of IL-9–producing T cells and allergic inflammatio n.[244,249-251] Innate cells are major sources of IL-9.[252]

TH22 CELLS

IL-22 is an IL-10–related cytokine located in proximity to the *Ifng* gene but regulated in a distinctive manner.[253] It acts on epithelial barrier cells, having both pro- and anti-inflammatory effects.[254-256] It is important for host defense against *K. pneumoniae*.[200] Although IL-22 is produced by Th17 cells, some cells produce IL-22 and not IL-17.[257-260] This is especially prominent in skin-homing cells that express the chemokine receptors CCR10, CCR6, and CCR4.[261,262] IL-22 is also produced by mucosal innate immune cells especially ILC3.[148,219,263-266]

Production of IL-6 and TNF by plasmacytoid DCs promotes IL-22 production without inducing IL-17 and IFN-γ. Unlike other Th subsets, no clear master regulator that directs this phenotype has been identified, but the AHR and RORγt are important for IL-22 production.[223,239,267] Notch drives IL-22 production by enhancing AHR expression.[268] Conversely, TGF-β inhibits IL-22; it does so by inducing Maf, which directly represses *Il22*.[259]

HELPER T CELLS: GRADIENTS, HETEROGENEITY, AND FLEXIBILITY

The sections described above may leave the impression that helper T cell subsets are distinct lineages in the same way that we think of CD4+ and CD8+ T cells as being distinct and phenotypically stable. On the contrary, recent work including single cell approaches indicate that it is more appropriate to think of effector T cells as resulting from a continuum or gradient of transcriptional programs; this is especially true for Th1 and Th17 cells with Th2 cells being more distinct; this has become more clear as more emphasis is placed on *in vivo* differentiation in response to diverse challenges. Different types of infections provoke distinct responses but IFN-γ and IL-17 production often overlaps; discrete clusters are not evident, whereas a gradient of activation is more obvious[269,270] (Fig. 28.2). As discussed above, effector programs are the product of commensal microbiota as well infectious microorganisms.[272] There are also circumstances in which phenotypic flexibility is seen.

To begin, although we have focused on selective production, some cytokines are broadly expressed. IL-10 was initially thought to be a Th2 cytokine, it is clear that it can also be made by Th1, Th2, IL-17, and T_{reg} cells, and a variety of innate immune cells.[253] Likewise, TGF-β is made by both effector cells and regulatory cells. Conversely, granulocyte macrophage colony-stimulating factor (GM-CSF) is made by Th1, Th2, and Th17 cells and is an important driver of immune-mediated disease.[273] IL-21, the "signature cytokine" of Tfh and Tph cells, can also be made by Th17 and Th1 cells.[136]

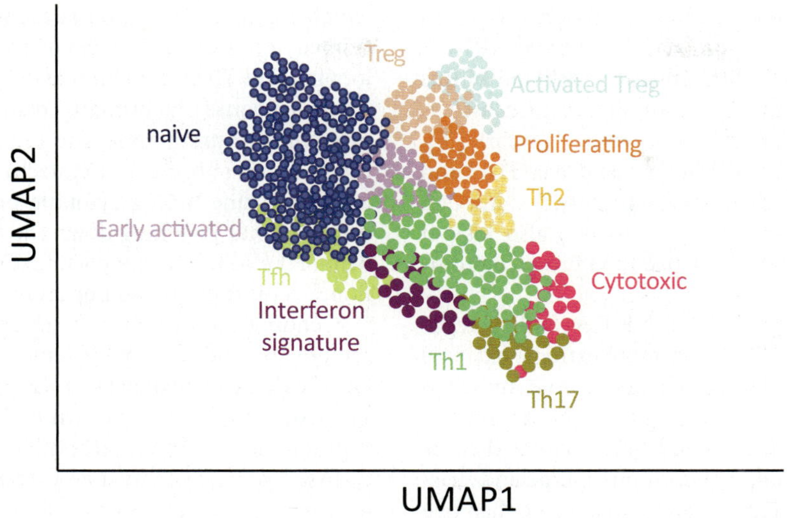

FIG. 28.2. Heterogeneity of CD4 responses *in vivo*. The recognition of distinct subsets of CD4 T cells was recognized nearly half a century ago and many key insights have been made by reductionist, *in vitro* approaches. However, as techniques have continued to improve enabling comprehensive study of small numbers of *ex vivo* derived cells—and even single cells—our views of helper T cells have evolved. Rather than conceptualizing fates of activated T cells as "lineages" defined by a limited number of transcription factors, surface markers, and products, the programs of CD4 T cells are probably better visualized as multidimensional gradients.[128,271] Aside from classic subsets ("Tfh, Th1, Th17, and Th2") defined by conventional markers, responses and differentiation of CD4 T cells reflect the amalgamation of levels of activation (eg, clusters denoted "early activation," "activated T_{regs}," "proliferating" cells) with effects of intrinsic factors including expression of lineage defining and induced transcription factors including "master regulator" transcription factors, extrinsic signals (eg, interferon signature) and signal-dependent transcription factors, and residence within tissues along with the impact of local tissue factors including neural-derived signals, as well as nutrients and metabolic state. The net result is palette of transcriptomic responses, and in this respect, helper cell "lineages" represent multidimensional continua with a degree of flexible outcomes driven by a myriad of stimuli. While scRNAseq data and evolving clustering algorithms provide new windows into the complexity of helper T cells, technological and computational improvements will continue and our concepts of CD4 T cells will also change over time.

Moreover, some suppressive cells can express IL-17.[195,274] In other words, cytokine production can be selective to a degree but is also quite flexible.

Although IL-17–secreting Th cells were suggested to represent a new lineage because they do not make other lineage-defining cytokines like IFN-γ and IL-4, it is now well-recognized that Th17 cells can become pathogenic IFN-γ producers,[209,275-279] especially when reinforced by IL-23.[280] Th17 cell populations can also be heterogenous with respect to IL-10 production. IL-27 can drive IL-10 expression in Th17 and other cells.[281-283] Reduced production of IL-10 in Th17 correlates with autoimmune disease in humans.[284] Th17 cells can acquire a Tfh phenotype in mucosal secondary lymphoid organs and can become Th1 cells in some model systems.[285,286] Thus, Th17 cells may be the least stable of the major helper T cell subsets compounded by variable IL-10 production.

Implicit in the notion of lineage-defining master regulators is that their expression dictates a distinctive phenotype. In the strictest sense, expression of a master regulator in a given subset would be expected to be distinct; the expectation being that specific lineages express one lineage-defining master regulator whose expression equates with the distinctive programs that do not overlap among the different subsets. However, this is not the case for CD4+ T cells.

For example, Foxp3+ cells can express T-bet, GATA3, STAT3, IRF4, RORγt, and BCL6—all of which are functionally relevant. Expression of T-bet in Foxp3+ cells is important for trafficking of the cells to sites of Th1-associated inflammation.[287] GATA3 is expressed in T_{reg} cells that reside in barrier sites such as the gastrointestinal tract and the skin but is also induced by TCR and IL-2 stimulation.[288,289] Although not required for basal T_{reg} cell homeostasis and function, T_{reg} cell expression of GATA3 is critical during inflammatory responses where it maintains Foxp3 expression and limits T_{reg} cell conversion to an effector T cell phenotype. Considerable numbers of human CD4+ T cells coexpress Foxp3 and RORγt. These T cells express CCR6 and produce IL-17 upon activation but also have suppressive activity.[290]

Foxp3 can be induced in conventional Foxp3 cells by factors including TGFβ and retinoic acid[291,292] (Fig. 28.1) to produce so-called peripheral T_{reg} cells, which may further differentiate into BCL6+ T follicular regulatory cells (Tfr) in the presence of the ICOS signals to constrain B cell responses.[293,294] The differentiation and function of several intestinal T_{reg} cell populations, including RORγt+ T_{reg} cells and Tfr cells are dependent upon Maf.[295] Although IRF4 is important for Th2 and Th17 differentiation, deletion of this factor in T_{reg} cells leads to autoimmunity disease associated with overproduction of Th2 cytokines.[296] Similarly, T_{reg} cell–specific deletion of STAT3, a factor that drives Th17 and Tfh differentiation, also leads to autoimmunity.[297] Although the subject is far from resolved, T_{reg} cells can also lose of Foxp3 expression and acquire the capacity to produce proinflammatory cytokines.[17,298]

In Th2 cells, the transcription factor NF-IL3 (E4BP4) regulates IL-13 independently of IL-4.[299] A more extreme example is that committed GATA-3+ Th2 cells can be reprogrammed to express T-bet+ and IFN-γ+ in the setting of viral infection.[300]

From all these examples, it should be clear that there is often not a simple relationship between phenotype and master regulator expression and that expression of more than one master regulator is by no means an infrequent event.[128] Clearly, the reality of CD4+ T cells is more nuanced. Collectively, the data indicate that although helper cells may exhibit features of phenotypic stability, it is very clear that they also have the capacity for considerable plasticity. This then raises the broader question of what factors preserve the phenotype of specialized cells.

EPIGENETIC REGULATION OF HELPER CELL DIFFERENTIATION

In considering genomic regulation of helper T cell differentiation, it is worth remembering that less than 2% of the genome represents conventional, protein-coding genes and that much of the genome is nonetheless transcribed and modified in a cell-specific manner. In this respect, exogenous and endogenous factors—including cytokines—are not only acting conventional genes but also broadly acting on the genome. Genome-wide epigenetic modifications that influence the transcriptome are referred to as the epigenome or regulome—an area that is intensively studied, especially as single cell technologies improve.

A major determinant in gene expression is whether the gene is accessible or not. DNA is associated with octameric histone molecules forming nucleosomes, which may be in a condensed (heterochromatin) or open (euchromatin) configuration (Fig. 28.3). Genes residing within regions of condensed chromatin are inaccessible and are therefore silenced, whereas genes that reside in open chromatin are readily expressed; however, there are multiple mechanisms involved in regulating accessibility and multiple techniques used to measure these changes. The 3 billion pairs of nucleotides forming our genome are compacted within a nucleus that is approximately 6 μm. Compaction is not random with active and inactive regions often residing within different nuclear territories.

Transcriptionally poised genes reside within loops of DNA that are formed via an extrusion process mediated by cohesin and blocked by two convergently orientated CCCTC-binding factor (CTCF) targeting sites.[301] CTCF is a multifunctional zinc finger TF that functions as an activator, a repressor or an insulator, and thus is a chromatin organizer that dominates higher-order chromatin architecture. These chromatin loops, which are measured genomewide by techniques like HiC, are termed topological-associated domains (TAD). Chromatin looping and TAD formation allows modulation of transcriptomic response by bringing distal regulatory elements in proximity to promoters. The mechanisms by which three-dimensional chromatin architecture coordinates gene regulation and fine tuning cytokine production remains largely unknown but knocking down the expression of one component of the cohesion complex, RAD21, reduces *Ifng* transcription.[302] This is clearly an important area for research.

Genomic accessibility is influenced by ATP-dependent remodeling and chromatin remodeling complexes that contain Brahma-related gene 1 (Brg1) displace nucleosomes. Consequently, knocking down *Brg1* results in decreased IFN-γ production.[303] While accessibility has been measured by DNAse hypersensitivity, newer techniques are now available in which small numbers of cells and even single cells can be assessed (eg, Assay for Transposase-Accessible Chromatin using sequencing, ATACseq). Measuring changes in accessibility may be more rapid and direct consequences of cell signaling and differentiation since measurement of mRNA can be influenced by posttranscriptional events.

DNA methylation is a classic correlate of genomic inaccessibility,[304] which can persist in the absence of continued exogenous signals or intrinsic factors that initially induced this modification. This allows a cell to "remember" its distinctive transcriptional profile and, by extension, its cellular identity. The conserved CpG rich region upstream of the *Foxp3* promoter is a good example—this region is fully demethylated in natural T_{reg} cells but exhibits residual methylation in induced T_{reg} cells.[305-307] This correlates with the finding that natural T_{reg} cells tend to be more stable in their phenotype than their induced cousins. When naïve T cells leave the thymus, the *Ifng* and *Il4* loci are hypermethylated primarily through DNA methyltransferase, DNMT1. Conditional ablation of DNMT1 or MbD2 (methyl-CpG binding domain protein 2) leads to increased expression of IFN-γ and Th2 cytokines and inability of Th1 or Th2 cells to properly silence the expression of opposing cytokine genes.[304,308,309] Loss of DMT3a also results in a failure to methylate the *Ifng* locus in alternative T helper cell fates.[310]

The 10-11 translocation (Tet) proteins mediate DNA demethylation by oxidizing DNA methylation (5 mC) through several confirmations including DNA hydroxymethylation (5 hmC).[311,312] In Th1 and Th17 cells, Tet2 is important for the expression of the lineage-specific cytokines. Tet2/Tet3 stabilizes *Foxp3* expression and vitamin C promotes Foxp3 expression by influencing Tet2/Tet3 enzymatic activity.[313] Exactly how changes in DNA methylation and hydromethylation occur during lymphocyte differentiation is still incompletely understood.

Accessibility of genes, including key loci in helper T cells, is associated with covalent modifications of histone tails[314-319] and deletion or reduction of factors that catalyze histone modifications impact helper T cell differentiation.[320] Histone 3 lysine 4 trimethylation (H3K4me3)—a mark of accessible promoters—on lineage-appropriate genes in the corresponding lineages (eg, *Ifng* in Th1 cells) is typically accompanied by repressive histone 3 lysine 27 trimethylation (H3K27me3) marks on lineage inappropriate genes (eg, *Il4* in Th1 cells).

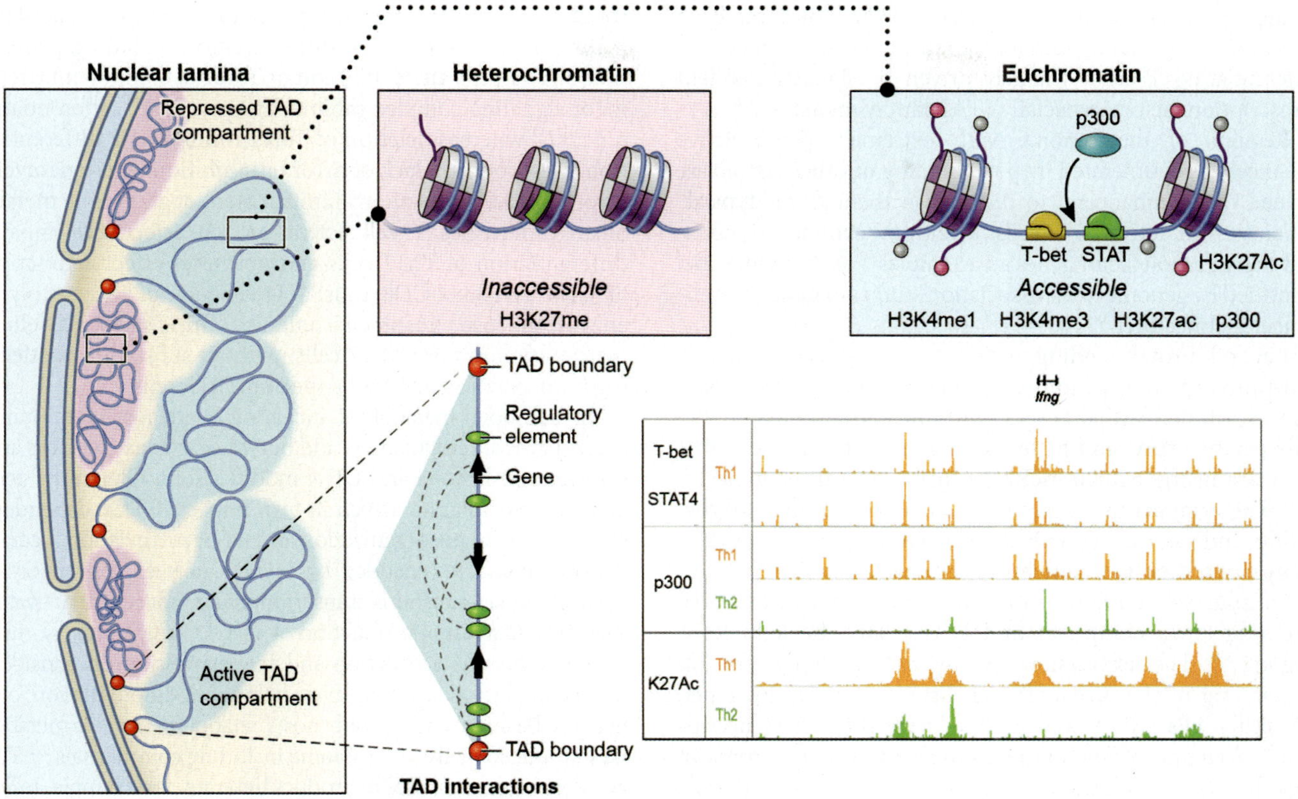

FIG. 28.3. Genomic views of and influences on helper cell differentiation. In consideration of actions and behavior of helper T cells, much effort has been directed at identifying products of helper T cells from proteins to mRNAs; however, the differentiation of cells encompasses far more than transcriptomic programs. In considering genomic correlates with cell specification, it needs to be borne in mind that a tiny fraction of the genome represents conventional, protein-coding genes (<2%) and much of the genome is transcribed and modified in cell- and activation-dependent manners. The behavior and stability of subsets are influenced by the regions of the genome that are accessible (euchromatin—blue arrows; regions marked by histone 3 lysine 4 monomethylation, H3K4me1; histone 3 lysine 4 monomethylation, H3K4me3; and histone 3 lysine 27 acetylation, H3K27Ac) and responsive to the action of signal-dependent transcription factors (eg, STATs). Many immune cell effector loci (eg, cytokines and cytokine receptors) reside within regions associated with high density of active enhancers termed superenhancers, which themselves are enriched for the presence of long noncoding RNAs. These regions can also be subject to higher order chromatin looping, termed topologically associated domains that encompass superenhancer loci. Chromatin accessibility, formation of superenhancers, regions with dense deposition of p300 and resultant H3K27Ac, and lncRNAs are few examples of genomic factors that comprise additional layers to effect regulation of cell behavior. The extended *Ifng* locus is illustrated in the right panel as an example of a superenhancer locus, with deposition of signal-dependent transcription factors such as STAT4 and the lineage defining transcription factor, T-bet, from Chip-seq data, along with p300 deposition and H3K27Ac. Increasingly, epigenomic changes are measured on a single cell level, thus allowing a better appreciation of the complexity of CD4+ T cells. Interactions between regulatory elements including enhancers and promoters (left panel) can be measured by techniques like HiC. Interactions between regions of chromatin are referred to as topological-associated domains (TADs).

Impaired expression of constituents of the Trithorax complex responsible for H3K4 trimethylation is associated with aberrant Th2 differentiation.[321,322] Mice that lack components of polycomb repressor complexes responsible for H3K27 trimethylation also have impaired helper cell differentiation.[323-325] Genomewide profiling of histone marks in CD4+ T cells generally confirms expectations regarding accessible and repressive marks of signature cytokine genes.[326-328] However, this simple scenario is not always the case; Th17 cells are intrinsically unstable[275,276] and the *Ifng* locus can be remodeled when Th17 cells are exposed to IL-12.[329]

CD4+ T cells can express more than one master regulator and an explanation for the flexibility of expression of key transcription factors is that they may exhibit both repressive and accessible marks—even when these transcription factors are not expressed.[327] For example, while *Tbx21* exhibits H3K27me3 marks in Th2 and Th17 cells, H3K4me3 marks are also present. Such regions are termed bivalent domains and are indicative of genes poised for expression.[330,331] This then helps explain how T-bet might be induced in T_{reg}, Th17, and even Th2 cells. Similarly, the *Bcl6* locus has H3K4me3 marks in polarized Th1, Th2, and Th17 cells and conversely, the *Tbx21*, *RORc*, and *Gata3* genes have accessible marks in Tfh cells.[332]

Loci in which key signature genes reside, including those encoding cytokine, cytokine receptor, and transcription factor genes, are very complex and encompass broad swaths of the genome. The *IFNG/IL26/IL22* (*Ifng/Iltifb/Il22 in mice*) gene cluster (>400 kb)[315,333-335] and *IL4/IL13/IL5* (nearly 120 kb), have numerous regulatory elements.[336-339] Poised enhancers are marked by histone 3 lysine 4 monomethylation (H3K4me1), whereas active enhancers are marked by histone 3 lysine 27 acetylation (H3K27Ac). Active enhancers are also revealed by deposition of the histone acetylase, p300, which catalyzes H3K27Ac. Importantly, the formation of active enhancers in differentiating T cells is largely driven by signal-dependent transcription factors, especially STAT family members.[340]

Regions of the genome with extremely dense active enhancers as enumerated by p300 binding or other factors are termed "superenhancers" to distinguish them from "typical" enhancers. In activated T cells, cytokines, cytokine receptor, and transcription factor genes are examples[341] and genomic loci identified by genome-wide association studies exhibit superenhancer architecture.[342]

Beyond protein-coding genes, transcripts cover nearly two-third of the genome.[343-346] Long noncoding RNAs (lncRNAs), defined as at least 200 nucleotides, have highly cell-specific expression patterns. The present estimate is that there are nearly 60,000 lncRNAs in the human genome.[343,347-349] They can act in *cis* and in *trans*, acting locally and distalling, and interacting with DNA, RNA, and proteins, but in many cases, their mechanisms of action are opaque.

Of note, superenhancer loci are enriched for lncRNAs.[341] The *Ifng* locus comprises and is regulated by a lncRNA *Ifng-as1*.[350] The expression of *Ifng-as1* is dependent on remodeling of the proximal and distal enhancers by T-bet, recruiting TFs NF-κB and Ets1 to drive *Ifng-as1* transcription.[351] Deletion of *Ifng-as1* leads to disruption of chromatin organization and reduced *Ifng* expression, indicating its role in maintenance of the chromatin architecture of the *Ifng*-extended locus. In part, this may be due to the deletion of a critical CTCF site in this locus. Inhibiting generation of *Ifng-as1* also increases susceptibility to *T. gondii*.

Similarly, the *Il4/Il13/Il5* locus is also associated with a lncRNA as is *Gata3* (*Gata-as1*). Another example is lncRNA Rroid (RNA-demarcated Regulatory region of Id2), which promotes the expression of Id2, which in turn regulates lymphocyte fates including ILC1 development.[352] Indeed, numerous cytokine and effector genes are associated with lncRNAs but few have been studied great in detail.

MicroRNAs (miRNAs or miRs) are small RNAs that bind to target messenger RNAs (mRNAs) and lead to inhibition of translation or mRNA degradation.[353-356] Relatively, more work has been done on the role of these RNAs and helper cell differentiation. Drosha and Dicer are two key components of the machinery responsible for miRNA generation and loss of these factors is associated with autoimmune diseases and loss of T_{reg} cell stability.[357,358] Examples of pertinent miRNAs include miR-146a, which is important for suppressive function of T_{reg},[359] miR-155, which influences Th1, Th2 differentiation, and T_{reg} cell development,[360-363] miR-326, which enhances Th17 differentiation, and miR-29, which regulates Th1 differentiation.[364] By contrast, miR221/222 targets *Il23 R* and *Maf* in Th17 and limits intestinal inflammation. Bcl-6 represses expression of many microRNAs (miRNAs) predicted to control the Tfh cell signature, including miR-17-92, which represses CXCR5 expression.[91] Yet another miR, miR-125b contributes to maintaining the naïve state of helper T cells and miR-182 promotes clonal expansion.[18,365]

Beyond cytokines and other immune stimuli, metabolic perturbation is an important aspect of T cell activation that impacts epigenomic status; importantly, helper cell specification can be impacted differentially by metabolic state. mTor inhibits effector cell differentiation including generation of Th1 cells and inhibition of Torc1, a component of the mTor signaling complex inhibits Th1 and Th17 differentiation,[229,366-368] whereas deletion of Torc2 inhibits Th2 differentiation.[229] By contrast, lack of mTor or inhibition with rapamycin favors T_{reg} cell generation.[366] Impairment of glutamine metabolism diminishes T cell activation but selectively impairs differentiation of Th17 cells and promotes effector function of CD4 Th1 and CD8 cells.[369] Deficiency of the glycolytic enzyme glucose phosphate isomerase (Gpi1) selectively eliminates inflammatory Th17 cells, without substantially affecting homeostatic microbiota-specific Th17 cells.[370]

In addition, metabolites including S-adenosylmethionine, acetyl-CoA, a-ketoglutarate, and butyrate serve as cofactors and substrates in histone and DNA modifications. Glutamine generates a-ketoglutarate, which in turn impacts CTCF-dependent changes in genome organization and DNA methylation. Lactate dehydrogenase A (encoded by *Ldha*) catalyzes conversion of pyruvate to lactate and is transcriptionally induced in activated Th1 cells, increasing production of acetyl CoA.[371] This, in turn, enhances histone acetylation and *Ifng* expression. In considering the impact of metabolism on helper cell differentiation, one needs to bear in mind not only host metabolites but also metabolites produced by the microbiome including commensals.[372] The gut-derived commensal byproduct butyrate, for example, influences acetylation of regulatory regions of *Foxp3*, stabilizing the expression of this transcription factor.

CD4+ T CELL MEMORY

Memory cells arise from naïve T cells during primary immune responses.[373] CD4+ effector T cell populations generated by acute infections start to contract after the peak of expansion when proliferation has ceased about a week after the infection begins.[45,374,375] The contraction phase continues for about 2 weeks with about 20% of the cells in the population lost per day such that about 10% of the cells that were present at the peak of the expansion phase survive as memory cells at the end of the contraction phase. The memory cells are G_1 cells[376] that retain the $CD44^{high}$ (mice) or CD45RO+ (humans) phenotype of antigen-experienced cells and persist for months but decline slowly with a half-life of 80 days in mice or three-five months in humans,[45,374,375,377] which equates to about 0.05 life times in both species. CD4+ memory T cells are maintained in interphase by IL-7 receptor signals and slow IL-15–driven homeostatic proliferation[375,378] but evidently not at a high enough rate to achieve the numerical stability typical of CD8+ memory T cell populations.[379]

Much of the current understanding of CD4+ memory T cell formation is based on studies of mice with acute infections that are cleared in about a week.[64,65,67,380] These infections induce the contemporaneous production of Tfh and non-Tfh effector cells as described in Figure 28.1, and in this case, the Tfh cells are $CXCR5^{int}$ $PD\text{-}1^-$ $BCL6^{lo}$ GC-Tfh precursors and $CXCR5^{hi}$ $PD\text{-}1^+$ $BCL6^{hi}$ GC-Tfh cells that express FR4, and the non-Tfh cells are T-bet+ Th1 cells.

The CD4+ memory T cell population that survives after these acute infections are cleared contains subsets that likely correspond to each of the earlier effector cell subsets. It contains CXCR5−T-bet^{hi} memory cells that are the progeny of CXCR5−T-bet^{hi} effector cells[45,380] and are located in the secondary lymphoid organs[66] where they are generated but also migrate into nonlymphoid organs[80] where a CD69+ fraction permanently resides.[381-383] Thus, some of the CXCR5− T-bet^{hi} effector cells survive the contraction phase to become Th1 memory cells, some expressing Ly6C,[64,65,380] some of the effector memory type[384] recirculating between lymphoid and nonlymphoid organs via the blood, and some of the resident memory type[385] locked mainly in nonlymphoid organs.

The CD4+ memory T cell population also contains CXCR5+ cells, all with the CXCR5^{int} phenotype but some FR4^{high} (and probably PD-1^{high} in humans)[380] and some FR4^{low}.[386] Humans also contain CXCR5+ memory T cell populations including a BCL6^{hi} CCR7^{lo} PD-1^{hi} subset resembling murine GC-Tfh cells and a BCL6^{lo} CCR7^{hi} PD-1^{lo} subset resembling the CXCR5^{int} memory cells.[387] CXCR5+ memory cells are found in the blood and secondary lymphoid organs of both species.[84,383,387] The origins of the CXCR5+ memory T cells subsets and their progeny during secondary responses have been hard to pin down because of a lack of definitive stable cell surface markers. A reasonable hypothesis, however, is that both CXCR5^{int} Tfh cells and GC-Tfh effector cells contribute cells to the memory cell pool.[65,380,386] These memory cells are more flexible than Th1 memory cells and can generate Th1 and Tfh effector cells during secondary responses.[45,65,380]

The appearance of all memory T cells is preceded by the loss of the majority of effector cells during the contraction phase. This loss is due to apoptosis as the process is blunted in mice lacking the proapoptotic factors BIM or p27(Kip1).[388-390] In addition, about 10% of TCR transgenic effector T cells express the apoptosis-inducing enzyme CASPASE3 and die rapidly after transfer into adoptive hosts.[391] Mice with a defect in autophagy contain CD8+ T cells that undergo an accelerated contraction phase associated with impaired mitochondrial function.[392] This results together with the facts that BIM and CASPASE3 act on mitochondria[393] suggest that mitochondrial damage contributes to effector cell death.

The mitochondrial damage that triggers apoptosis during the contraction phase may be induced by reactive oxygen species (ROS). Effector T cells use anaerobic glycolysis to produce the small molecules needed for the macromolecular synthesis associated with cell growth and division.[394] These cells also turn up the mitochondrial respiratory chain to generate ATP via oxidative phosphorylation—a process that produces ROS that could damage macromolecules and trigger apoptosis.[395,396] Reports that effector cells are susceptible to death in mice lacking mitochondrial superoxide dismutase 2,[397] which reduces mitochondrial ROS, or have reduced intracellular glutathione,[398] which buffers intracellular ROS, are consistent with this possibility. The finding that a superoxide dismutase inhibitor blocks *in vitro* death of T cells induced to die *in vivo* by superantigen injection[395] also fits the idea that ROS could trigger effector cell death by damaging mitochondria.

This model raises the question of what is special about the effector cells that survive to become memory cells? One possibility is that some effector T cells retain the IL-7 receptor allowing resistance to cell death and transition to the memory cell state. As attractive as this model is, it is contradicted by cell sorting experiments indicating that IL-7 receptor+ CD4+ effector cells do not have an advantage over IL-7 receptor− cells,[64] that the effector cell population undergoing contraction contains CASPASE3+ IL-7 receptor+ cells,[391] and that forced expression of IL-7 receptor does not prevent the contraction of CD8+ effector cells.[399]

Asymmetric division is also a possibility. Microscopy studies have shown that effector T cells that divide while attached to an antigen-presenting cell can partition different molecules between the proximal and distal daughter cells in such a way that affects T cell fate.[400] For example, it has also been demonstrated that proximal daughter cells acquire more of the amino acid transporter SLC1A5, which increases the activities of mTORC1 and c-Myc,[401] have higher metabolic activities than distal daughters, and are more likely to produce short-lived effector cells. Thus, the progeny of naïve T cells may be fated at the first cell division to become death-prone effector cells or long-lived memory cells depending on their relationship to the antigen-presenting cell.

Finally, the metabolic activities of effector cells in the population may vary for stochastic reasons. Indeed, the cells in early effector cell populations are very heterogenous in size, a feature that is proportional to c-Myc levels at least in B cells.[402] This heterogeneity could simply arise from variance in TCR, costimulatory, or cytokine receptor signals and downstream metabolic activities based simply on type or location of the antigen-presenting cells, the T cells happen to interact with. In this case, the effector cells that happen to achieve the highest metabolic activity could damage their macromolecules via ROS and die while cells that happen to achieve lower metabolic activity could survive to become memory cells. The observation that large clonally derived effector cell populations undergo more contraction than small ones[386,403] is consistent with this model.

SUMMARY

The 1980s and 1990s were a golden age in the CD4+ T cell field due to the discovery of Th1 and Th2 cells with different cytokine production profiles and helper functions. The following decades witnessed resolution of the molecular pathways that generate Th1 and Th2 cells from naïve precursors, the discovery of Th17, Tfh, and pT_{reg} cells, and the pathways that lead to their formation. *In vivo* experiments in this era revealed the phenomenon whereby Tfh and noneffector cells form at the same time during primary responses and provided glimpses of the mechanisms that underlie this bifurcation. Future challenges will be to work out how single T cells decide whether to follow one fork of the bifurcated path or the other, how fixed the different fates really are, and how to manipulate the bifurcation process to produce more effective vaccines.

ACKNOWLEDGMENTS

MKJ and JO'S were extremely fortunate to be mentored by Bill Paul during a most exciting time in immunology. We are very grateful for Bill's many insights and generosity; he was a true mensch—simply an extraordinary human being.

REFERENCES

1. Borst J, Ahrends T, Babala N, Melief CJM, Kastenmuller W. CD4(+) T cell help in cancer immunology and immunotherapy. *Nat Rev Immunol.* 2018;18(10):635-647. doi:10.1038/s41577-018-0044-0
2. Laidlaw BJ, Craft JE, Kaech SM. The multifaceted role of CD4(+) T cells in CD8(+) T cell memory. *Nat Rev Immunol.* 2016;16(2):102-111. doi:10.1038/nri.2015.10
3. Jenkins MK, Chu HH, McLachlan JB, Moon JJ. On the composition of the preimmune repertoire of T cells specific for Peptide-major histocompatibility complex ligands. *Annu Rev Immunol.* 2010;28:275-294. doi:10.1146/annurev-immunol-030409-101253
4. Fontenot JD, Gavin MA, Rudensky AY. Foxp3 programs the development and function of CD4+CD25+ regulatory T cells. *Nat Immunol.* 2003;4(4):330-336. doi:10.1038/ni904
5. Moon JJ, Dash P, Oguin THIII, et al. Quantitative impact of thymic selection on Foxp3+ and Foxp3− subsets of self-peptide/MHC class II-specific CD4+ T cells. *Proc Natl Acad Sci U S A.* 2011;108(35):14602-14607. doi:10.1073/pnas.1109806108
6. Cyster JG. Lymphoid organ development and cell migration. *Immunol Rev.* 2003;195:5-14. doi:10.1034/j.1600-065x.2003.00075.x
7. von Andrian UH, Mackay CR. T-cell function and migration. Two sides of the same coin. *N Engl J Med.* 2000;343(14):1020-1034. doi:10.1056/NEJM200010053431407
8. Girard JP, Moussion C, Forster R. HEVs, lymphatics and homeostatic immune cell trafficking in lymph nodes. *Nat Rev Immunol.* 2012;12(11):762-773. doi:10.1038/nri3298
9. Cyster JG, Goodnow CC. Pertussis toxin inhibits migration of B and T lymphocytes into splenic white pulp cords. *J Exp Med.* 1995;182(2):581-586. doi:10.1084/jem.182.2.581
10. Gretz JE, Anderson AO, Shaw S. Cords, channels, corridors and conduits: critical architectural elements facilitating cell interactions in the lymph node cortex. *Immunol Rev.* 1997;156:11-24. doi:10.1111/j.1600-065x.1997.tb00955.x
11. Reboldi A, Cyster JG. Peyer's patches: organizing B-cell responses at the intestinal frontier. *Immunol Rev.* 2016;271(1):230-245. doi:10.1111/imr.12400
12. Lewis SM, Williams A, Eisenbarth SC. Structure and function of the immune system in the spleen. *Sci Immunol.* 2019;4(33):eaau6085. doi:10.1126/sciimmunol.aau6085
13. Link A, Vogt TK, Favre S, et al. Fibroblastic reticular cells in lymph nodes regulate the homeostasis of naive T cells. *Nat Immunol.* 2007;8(11):1255-1265. doi:10.1038/ni1513
14. Mueller SN, Ahmed R. Lymphoid stroma in the initiation and control of immune responses. *Immunol Rev.* 2008;224:284-294. doi:10.1111/j.1600-065X.2008.00657.x
15. Bajenoff M, Egen JG, Koo LY, et al. Stromal cell networks regulate lymphocyte entry, migration, and territoriality in lymph nodes. *Immunity.* 2006;25(6):989-1001. doi:10.1016/j.immuni.2006.10.011
16. Feuillet V, Lucas B, Di Santo JP, Bismuth G, Trautmann A. Multiple survival signals are delivered by dendritic cells to naive CD4+ T cells. *Eur J Immunol.* 2005;35(9):2563-2572. doi:10.1002/eji.200526127
17. Feng X, Wang H, Takata H, Day TJ, Willen J, Hu H. Transcription factor Foxp1 exerts essential cell-intrinsic regulation of the quiescence of naive T cells. *Nat Immunol.* 2011;12(6):544-550. doi:10.1038/ni.2034
18. Rossi RL, Rossetti G, Wenandy L, et al. Distinct microRNA signatures in human lymphocyte subsets and enforcement of the naive state in CD4+ T cells by the microRNA miR-125b. *Nat Immunol.* 2011;12(8):796-803. doi:10.1038/ni.2057
19. Carlson CM, Endrizzi BT, Wu J, et al. Kruppel-like factor 2 regulates thymocyte and T-cell migration. *Nature.* 2006;442(7100):299-302. doi:10.1038/nature04882
20. Schwab SR, Cyster JG. Finding a way out: lymphocyte egress from lymphoid organs. *Nat Immunol.* 2007;8(12):1295-1301. doi:10.1038/ni1545
21. Weinreich MA, Takada K, Skon C, Reiner SL, Jameson SC, Hogquist KA. KLF2 transcription-factor deficiency in T cells results in unrestrained cytokine production and upregulation of bystander chemokine receptors. *Immunity.* 2009;31(1):122-130. doi:10.1016/j.immuni.2009.05.011
22. Takada K, Wang X, Hart GT, et al. Kruppel-like factor 2 is required for trafficking but not quiescence in postactivated T cells. *J Immunol.* 2011;186(2):775-783. doi:10.4049/jimmunol.1000094
23. Cartier A, Hla T. Sphingosine 1-phosphate: lipid signaling in pathology and therapy. *Science.* 2019;366(6463):eaar5551. doi:10.1126/science.aar5551
24. Dudziak D, Kamphorst AO, Heidkamp GF, et al. Differential antigen processing by dendritic cell subsets *in vivo*. *Science.* 2007;315(5808):107-111. doi:10.1126/science.1136080
25. Itano AA, Jenkins MK. Antigen presentation to naive CD4 T cells in the lymph node. *Nat Immunol.* 2003;4(8):733-739. doi:10.1038/ni957
26. Grant SM, Lou M, Yao L, Germain RN, Radtke AJ. The lymph node at a glance - how spatial organization optimizes the immune response. *J Cell Sci.* 2020;133(5):jcs241828. doi:10.1242/jcs.241828
27. Lammermann T, Sixt M. The microanatomy of T-cell responses. *Immunol Rev.* 2008;221:26-43. doi:10.1111/j.1600-065X.2008.00592.x
28. ten Broeke T, Wubbolts R, Stoorvogel W. MHC class II antigen presentation by dendritic cells regulated through endosomal sorting. *Cold Spring Harb Perspect Biol.* 2013;5(12):a016873. doi:10.1101/cshperspect.a016873
29. Wang P, Sidney J, Dow C, Mothe B, Sette A, Peters B. A systematic assessment of MHC class II peptide binding predictions and evaluation of a consensus approach. *PLoS Comput Biol.* 2008;4(4):e1000048. doi:10.1371/journal.pcbi.1000048
30. Forster R, Davalos-Misslitz AC, Rot A. CCR7 and its ligands: balancing immunity and tolerance. *Nat Rev Immunol.* 2008;8(5):362-371. doi:10.1038/nri2297
31. Dustin ML, Springer TA. T-cell receptor cross-linking transiently stimulates adhesiveness through LFA-1. *Nature.* 1989;341(6243):619-624. doi:10.1038/341619a0
32. Henrickson SE, von Andrian UH. Single-cell dynamics of T-cell priming. *Curr Opin Immunol.* 2007;19(3):249-258. doi:10.1016/j.coi.2007.04.013
33. Huang AY, Qi H, Germain RN. Illuminating the landscape of in vivo immunity: insights from dynamic in situ imaging of secondary lymphoid tissues. *Immunity.* 2004;21(3):331-339. doi:10.1016/j.immuni.2004.08.001
34. Germain RN, Jenkins MK. *In vivo* antigen presentation. *Curr Opin Immunol.* 2004;16(1):120-125. doi:10.1016/j.coi.2003.11.001
35. Castellino F, Germain RN. Cooperation between CD4+ and CD8+ T cells: when, where, and how. *Annu Rev Immunol.* 2006;24:519-540. doi:10.1146/annurev.immunol.23.021704.115825
36. Courtney AH, Lo WL, Weiss A. TCR signaling: mechanisms of initiation and propagation. *Trends Biochem Sci.* 2018;43(2):108-123. doi:10.1016/j.tibs.2017.11.008
37. Esensten JH, Helou YA, Chopra G, Weiss A, Bluestone JA. CD28 costimulation: from mechanism to therapy. *Immunity.* 2016;44(5):973-988. doi:10.1016/j.immuni.2016.04.020
38. Macian F NFAT proteins: key regulators of T-cell development and function. *Nat Rev Immunol.* 2005;5(6):472-484. doi:10.1038/nri1632
39. Rao A, Luo C, Hogan PG. Transcription factors of the NFAT family: regulation and function. *Annu Rev Immunol.* 1997;15:707-747. doi:10.1146/annurev.immunol.15.1.707
40. Malek TR. The biology of interleukin-2. *Annu Rev Immunol.* 2008;26:453-479. doi:10.1146/annurev.immunol.26.021607.090357
41. Spolski R, Li P, Leonard WJ. Biology and regulation of IL-2: from molecular mechanisms to human therapy. *Nat Rev Immunol.* 2018;18(10):648-659. doi:10.1038/s41577-018-0046-y
42. Lin JX, Migone TS, Tsang M, et al. The role of shared receptor motifs and common Stat proteins in the generation of cytokine pleiotropy and redundancy by IL-2, IL-4, IL-7, IL-13, and IL-15. *Immunity.* 1995;2(4):331-339. doi:10.1016/1074-7613(95)90141-8
43. Beadling C, Guschin D, Witthuhn BA, et al. Activation of JAK kinases and STAT proteins by interleukin-2 and interferon alpha, but not the T cell antigen receptor, in human T lymphocytes. *EMBO J.* 1994;13(23):5605-5615.
44. Gudmundsdottir H, Turka LA. A closer look at homeostatic proliferation of CD4+ T cells: costimulatory requirements and role in memory formation. *J Immunol.* 2001;167(7):3699-3707. doi:10.4049/jimmunol.167.7.3699
45. Pepper M, Jenkins MK. Origins of CD4(+) effector and central memory T cells. *Nat Immunol.* 2011;12(6):467-471. doi:10.1038/ni.2038
46. Mosmann TR, Cherwinski H, Bond MW, Giedlin MA, Coffman RL. Two types of murine helper T cell clone. I. Definition according to profiles of lymphokine activities and secreted proteins. *J Immunol.* 1986;136(7):2348-2357.
47. Paul WE, Seder RA. Lymphocyte responses and cytokines. *Cell.* 1994;76(2):241-251. doi:10.1016/0092-8674(94)90332-8
48. Seder RA, Paul WE. Acquisition of lymphokine-producing phenotype by CD4+ T cells. *Annu Rev Immunol.* 1994;12:635-673. doi:10.1146/annurev.iy.12.040194.003223
49. Abbas AK, Murphy KM, Sher A. Functional diversity of helper T lymphocytes. *Nature.* 1996;383(6603):787-793. doi:10.1038/383787a0
50. Hsieh CS, Macatonia SE, Tripp CS, Wolf SF, O'Garra A, Murphy KM. Development of TH1 CD4+ T cells through IL-12 produced by Listeria-induced macrophages. *Science.* 1993;260(5107):547-549. doi:10.1126/science.8097338
51. Zhang Y, Zhang Y, Gu W, Sun B. TH1/TH2 cell differentiation and molecular signals. *Adv Exp Med Biol.* 2014;841:15-44. doi:10.1007/978-94-017-9487-9_2
52. Paul WE. What determines Th2 differentiation, *in vitro* and *in vivo*? *Immunol Cell Biol.* 2010;88(3):236-239. doi:10.1038/icb.2010.2
53. Harrington LE, Hatton RD, Mangan PR, et al. Interleukin 17-producing CD4+ effector T cells develop via a lineage distinct from the T helper type 1 and 2 lineages. *Nat Immunol.* 2005;6(11):1123-1132. doi:10.1038/ni1254
54. Bettelli E, Korn T, Oukka M, Kuchroo VK. Induction and effector functions of T(H)17 cells. *Nature.* 2008;453(7198):1051-1057. doi:10.1038/nature07036
55. Veldhoen M, Hocking RJ, Flavell RA, Stockinger B. Signals mediated by transforming growth factor-beta initiate autoimmune encephalomyelitis, but chronic inflammation is needed to sustain disease. *Nat Immunol.* 2006;7(11):1151-1156. doi:10.1038/ni1391
56. Breitfeld D, Ohl L, Kremmer E, et al. Follicular B helper T cells express CXC chemokine receptor 5, localize to B cell follicles, and support immunoglobulin production. *J Exp Med.* 2000;192(11):1545-1552. doi:10.1084/jem.192.11.1545
57. Schaerli P, Willimann K, Lang AB, Lipp M, Loetscher P, Moser B. CXC chemokine receptor 5 expression defines follicular homing T cells with B cell helper function. *J Exp Med.* 2000;192(11):1553-1562. doi:10.1084/jem.192.11.1553
58. Kim CH, Rott LS, Clark-Lewis I, Campbell DJ, Wu L, Butcher EC. Subspecialization of CXCR5+ T cells: B helper activity is focused in a germinal center-localized subset of CXCR5+ T cells. *J Exp Med.* 2001;193(12):1373-1381. doi:10.1084/jem.193.12.1373
59. Crotty ST Follicular helper cell biology: a decade of discovery and diseases. *Immunity.* 2019;50(5):1132-1148. doi:10.1016/j.immuni.2019.04.011
60. Vinuesa CG, Linterman MA, Yu D, MacLennan IC. Follicular helper T cells. *Annu Rev Immunol.* 2016;34:335-368. doi:10.1146/annurev-immunol-041015-055605
61. Nurieva RI, Chung Y, Martinez GJ, et al. Bcl6 mediates the development of T follicular helper cells. *Science.* 2009;325(5943):1001-1005. doi:10.1126/science.1176676
62. Crotty S. Follicular helper CD4 T cells (TFH). *Annu Rev Immunol.* 2011;29:621-663.
63. Johnston RJ, Poholek AC, DiToro D, et al. Bcl6 and Blimp-1 are reciprocal and antagonistic regulators of T follicular helper cell differentiation. *Science.* 2009;325(5943):1006-1010. doi:10.1126/science.1175870
64. Marshall HD, Chandele A, Jung YW, et al. Differential expression of Ly6C and T-bet distinguish effector and memory Th1 CD4(+) cell properties during viral infection. *Immunity.* 2011;35(4):633-646. doi:10.1016/j.immuni.2011.08.016
65. Hale JS, Youngblood B, Latner DR, et al. Distinct memory CD4+ T cells with commitment to T follicular helper- and T helper 1-cell lineages are generated after acute viral infection. *Immunity.* 2013;38(4):805-817. doi:10.1016/j.immuni.2013.02.020
66. Pepper M, Pagan AJ, Igyarto BZ, Taylor JJ, Jenkins MK. Opposing signals from the Bcl6 transcription factor and the interleukin-2 receptor generate T helper 1 central and effector memory cells. *Immunity.* 2011;35(4):583-595. doi:10.1016/j.immuni.2011.09.009
67. Leon B, Bradley JE, Lund FE, Randall TD, Ballesteros-Tato A. FoxP3+ regulatory T cells promote influenza-specific Tfh responses by controlling IL-2 availability. *Nat Commun.* 2014;5:3495. doi:10.1038/ncomms4495
68. Choi YS, Kageyama R, Eto D, et al. ICOS receptor instructs T follicular helper cell versus effector cell differentiation via induction of the transcriptional repressor Bcl6. *Immunity.* 2011;34(6):932-946. doi:10.1016/j.immuni.2011.03.023
69. Hondowicz BD, Kim KS, Ruterbusch MJ, Keitany GJ, Pepper M. IL-2 is required for the generation of viral-specific CD4(+) Th1 tissue-resident memory cells and B cells are essential for maintenance in the lung. *Eur J Immunol.* 2018;48(1):80-86. doi:10.1002/eji.201746928
70. Villarino A, Laurence A, Robinson GW, et al. Signal transducer and activator of transcription 5 (STAT5) paralog dose governs T cell effector and regulatory functions. *Elife.* 2016;5:e08384. doi:10.7554/eLife.08384
71. Kotov DI, Mitchell JS, Pengo T, et al. TCR affinity biases Th Cell differentiation by regulating CD25, Eef1e1, and Gbp2. *J Immunol.* 2019;202(9):2535-2545. doi:10.4049/jimmunol.1801609
72. Reinhardt RL, Liang HE, Locksley RM. Cytokine-secreting follicular T cells shape the antibody repertoire. *Nat Immunol.* 2009;10(4):385-393. doi:10.1038/ni.1715
73. Le Coz C, Joublin A, Pasquali JL, Korganow AS, Dumortier H, Monneaux F. Circulating TFH subset distribution is strongly affected in lupus patients with an active disease. *PLoS One.* 2013;8(9):e75319. doi:10.1371/journal.pone.0075319
74. Espinosa-Ortega F, Gomez-Martin D, Santana-De Anda K, Romo-Tena J, Villasenor-Ovies P, Alcocer-Varela J. Quantitative T cell subsets profile in peripheral blood from patients with idiopathic inflammatory myopathies: tilting the balance towards proinflammatory and pro-apoptotic subsets. *Clin Exp Immunol.* 2015;179(3):520-528. doi:10.1111/cei.12475
75. Buck MD, O'Sullivan D, Klein Geltink RI, et al. Mitochondrial dynamics controls T cell fate through metabolic programming. *Cell.* 2016;166(1):63-76. doi:10.1016/j.cell.2016.05.035
76. Constant S, Pfeiffer C, Woodard A, Pasqualini T, Bottomly K. Extent of T cell receptor ligation can determine the functional differentiation of naive CD4+ T cells. *J Exp Med.* 1995;182(5):1591-1596. doi:10.1084/jem.182.5.1591
77. Brogdon JL, Leitenberg D, Bottomly K. The potency of TCR signaling differentially regulates NFATc/p activity and

early IL-4 transcription in naive CD4+ T cells. *J Immunol.* 2002;168(8):3825-3832. doi:10.4049/jimmunol.168.8.3825
78. Nayar R, Schutten E, Bautista B, et al. Graded levels of IRF4 regulate CD8+ T cell differentiation and expansion, but not attrition, in response to acute virus infection. *J Immunol.* 2014;192(12):5881-5893. doi:10.4049/jimmunol.1303187
79. Krishnamoorthy V, Kannanganat S, Maienschein-Cline M, et al. The IRF4 gene regulatory module functions as a read-write integrator to dynamically coordinate T helper cell fate. *Immunity.* 2017;47(3):481.e7-497.e7. doi:10.1016/j.immuni.2017.09.001
80. Tubo NJ, Pagan AJ, Taylor JJ, et al. Single naive CD4+ T cells from a diverse repertoire produce different effector cell types during infection. *Cell.* 2013;153(4):785-796. doi:10.1016/j.cell.2013.04.007
81. Kim C, Wilson T, Fischer KF, Williams MA. Sustained interactions between T cell receptors and antigens promote the differentiation of CD4(+) memory T cells. *Immunity.* 2013;39(3):508-520. doi:10.1016/j.immuni.2013.08.033
82. Snook JP, Kim C, Williams MA. TCR signal strength controls the differentiation of CD4(+) effector and memory T cells. *Sci Immunol.* 2018;3(25):eaas9103. doi:10.1126/sciimmunol.aas9103
83. Schmitt N, Liu Y, Bentebibel SE, Ueno H. Molecular mechanisms regulating T helper 1 versus T follicular helper cell differentiation in humans. *Cell Rep.* 2016;16(4):1082-1095. doi:10.1016/j.celrep.2016.06.063
84. Fazilleau N, McHeyzer-Williams LJ, Rosen H, McHeyzer-Williams MG. The function of follicular helper T cells is regulated by the strength of T cell antigen receptor binding. *Nat Immunol.* 2009;10(4):375-384. doi:10.1038/ni.1704
85. DiToro D, Winstead CJ, Pham D, et al. Differential IL-2 expression defines developmental fates of follicular versus nonfollicular helper T cells. *Science.* 2018;361(6407):eaao2933. doi:10.1126/science.aao2933
86. Keck S, Schmaler M, Ganter S, et al. Antigen affinity and antigen dose exert distinct influences on CD4 T-cell differentiation. *Proc Natl Acad Sci U S A.* 2014;111(41):14852-14857. doi:10.1073/pnas.1403271111
87. Bartleson JM, Viehmann Milam AA, Donermeyer DL, et al. Strength of tonic T cell receptor signaling instructs T follicular helper cell-fate decisions. *Nat Immunol.* 2020;21(11):1384-1396. doi:10.1038/s41590-020-0781-7
88. De Giovanni M, Cutillo V, Giladi A, et al. Spatiotemporal regulation of type I interferon expression determines the antiviral polarization of CD4(+) T cells. *Nat Immunol.* 2020;21(3):321-330. doi:10.1038/s41590-020-0596-6
89. Eto D, Lao C, DiToro D, et al. IL-21 and IL-6 are critical for different aspects of B cell immunity and redundantly induce optimal follicular helper CD4 T cell (Tfh) differentiation. *PLoS One.* 2011;6(3):e17739. doi:10.1371/journal.pone.0017739
90. Liu X, Yan X, Zhong B, et al. Bcl6 expression specifies the T follicular helper cell program *in vivo*. *J Exp Med.* 2012;209(10):1841-1852, S1-S24. doi:10.1084/jem.20120219
91. Yu D, Rao S, Tsai LM, et al. The transcriptional repressor Bcl-6 directs T follicular helper cell lineage commitment. *Immunity.* 2009;31(3):457-468. doi:10.1016/j.immuni.2009.07.002
92. Shulman Z, Gitlin AD, Targ S, et al. T follicular helper cell dynamics in germinal centers. *Science.* 2013;341(6146):673-677. doi:10.1126/science.1241680
93. Weinstein JS, Herman EI, Lainez B, et al. TFH cells progressively differentiate to regulate the germinal center response. *Nat Immunol.* 2016;17(10):1197-1205. doi:10.1038/ni.3554
94. Gonzalez DG, Cote CM, Patel JR, et al. Nonredundant roles of IL-21 and IL-4 in the phased initiation of germinal center B cells and subsequent self-renewal transitions. *J Immunol.* 2018;201(12):3569-3579. doi:10.4049/jimmunol.1500497
95. Mesin L, Ersching J, Victora GD. Germinal center B cell dynamics. *Immunity.* 2016;45(3):471-482. doi:10.1016/j.immuni.2016.09.001
96. Choi YH, Burdick MD, Strieter BA, Mehrad B, Strieter RM. CXCR4, but not CXCR7, discriminates metastatic behavior in non-small cell lung cancer cells. *Mol Cancer Res.* 2014;12(1):38-47. doi:10.1158/1541-7786.MCR-12-0334
97. Kitano M, Moriyama S, Ando Y, et al. Bcl6 protein expression shapes pre-germinal center B cell dynamics and follicular helper T cell heterogeneity. *Immunity.* 2011;34(6):961-972. doi:10.1016/j.immuni.2011.03.025
98. Takebe T, Sakamoto K, Higami Y, Harada Y. A novel mouse model for tracking the fate of CXCR5-expressing T cells. *Biochem Biophys Res Commun.* 2018;495(2):1642-1647. doi:10.1016/j.bbrc.2017.12.029
99. Bocharnikov AV, Keegan J, Wacleche VS, et al. PD-1hiCXCR5- T peripheral helper cells promote B cell responses in lupus via MAF and IL-21. *JCI Insight.* 2019;4(20):e130062. doi:10.1172/jci.insight.130062
100. Rao DA, Gurish MF, Marshall JL, et al. Pathologically expanded peripheral T helper cell subset drives B cells in rheumatoid arthritis. *Nature.* 2017;542(7639):110-114. doi:10.1038/nature20810
101. Bauquet AT, Jin H, Paterson AM, et al. The costimulatory molecule ICOS regulates the expression of c-Maf and IL-21 in the development of follicular T helper cells and TH-17 cells. *Nat Immunol.* 2009;10(2):167-175. doi:10.1038/ni.1690
102. Xu H, Li X, Liu D, et al. Follicular T-helper cell recruitment governed by bystander B cells and ICOS-driven motility. *Nature.* 2013;496(7446):523-527. doi:10.1038/nature12058
103. Vacchio MS, Ciucci T, Gao Y, et al. A Thpok-directed transcriptional circuitry promotes Bcl6 and Maf expression to orchestrate T follicular helper differentiation. *Immunity.* 2019;51(3):465.e6-478.e6. doi:10.1016/j.immuni.2019.06.023
104. Wu T, Shin HM, Moseman EA, et al. TCF1 is required for the T follicular helper cell response to viral infection. *Cell Rep.* 2015;12(12):2099-2110. doi:10.1016/j.celrep.2015.08.049
105. Choi YS, Gullicksrud JA, Xing S, et al. LEF-1 and TCF-1 orchestrate T(FH) differentiation by regulating differentiation circuits upstream of the transcriptional repressor Bcl6. *Nat Immunol.* 2015;16(9):980-990. doi:10.1038/ni.3226
106. Gullicksrud JA, Li F, Xing S, et al. Differential requirements for Tcf1 long isoforms in CD8(+) and CD4(+) T cell responses to acute viral infection. *J Immunol.* 2017;199(3):911-919. doi:10.4049/jimmunol.1700595
107. Yang JA, Tubo NJ, Gearhart MD, Bardwell VJ, Jenkins MK. Cutting edge: Bcl6-interacting corepressor contributes to germinal center T follicular helper cell formation and B cell helper function. *J Immunol.* 2015;194(12):5604-5608. doi:10.4049/jimmunol.1500201
108. Nance JP, Belanger S, Johnston RJ, Hu JK, Takemori T, Crotty S. Bcl6 middle domain repressor function is required for T follicular helper cell differentiation and utilizes the corepressor MTA3. *Proc Natl Acad Sci U S A.* 2015;112(43):13324-13329. doi:10.1073/pnas.1507312112
109. Crotty S. T follicular helper cell differentiation, function, and roles in disease. *Immunity.* 2014;41(4):529-542. doi:10.1016/j.immuni.2014.10.004
110. Choi J, Diao H, Faliti CE, et al. Bcl-6 is the nexus transcription factor of T follicular helper cells via repressor-of-repressor circuits. *Nat Immunol.* 2020;21(7):777-789. doi:10.1038/s41590-020-0706-5
111. Tubo NJ, Jenkins MK. CD4+ T Cells: guardians of the phagosome. *Clin Microbiol Rev.* 2014;27(2):200-213. doi:10.1128/CMR.00097-13
112. Honda K, Littman DR. The microbiota in adaptive immune homeostasis and disease. *Nature.* 2016;535(7610):75-84. doi:10.1038/nature18848
113. Brown EM, Kenny DJ, Xavier RJ. Gut microbiota regulation of T cells during inflammation and autoimmunity. *Annu Rev Immunol.* 2019;37:599-624. doi:10.1146/annurev-immunol-042718-041841
114. Lighvani AA, Frucht DM, Jankovic D, et al. T-bet is rapidly induced by interferon-Œ≥ in lymphoid and myeloid cells. *Proc Natl Acad Sci U S A.* 2001;98(26):15137-15142.
115. Thieu VT, Yu Q, Chang HC, et al. Signal transducer and activator of transcription 4 is required for the transcription factor T-bet to promote T helper 1 cell-fate determination. *Immunity.* 2008;29(5):679-690.
116. Szabo SJ, Sullivan BM, Peng SL, Glimcher LH. Molecular mechanisms regulating Th1 immune responses. *Annu Rev Immunol.* 2003;21:713-758.
117. Lazarevic V, Chen X, Shim JH, et al. T-bet represses TH 17 differentiation by preventing Runx1-mediated activation of the gene encoding RORγt. *Nat Immunol.* 2011;12(1):96-104.
118. Oestreich KJ, Weinmann AS. Transcriptional mechanisms that regulate T helper 1 cell differentiation. *Curr Opin Immunol.* 2012;24(2):191-195. doi:10.1016/j.coi.2011.12.004
119. Beima KM, Miazgowicz MM, Lewis MD, Yan PS, Huang TH, Weinmann AS. T-bet binding to newly identified target gene promoters is cell type-independent but results in variable context-dependent functional effects. *J Biol Chem.* 2006;281(17):11992-12000. doi:10.1074/jbc.M513613200
120. Miller SA, Weinmann AS. Molecular mechanisms by which T-bet regulates T-helper cell commitment. *Immunol Rev.* 2010;238(1):233-246.
121. Mullen AC, High FA, Hutchins AS, et al. Role of T-bet in commitment of TH1 cells before IL-12-dependent selection. *Science.* 2001;292(5523):1907-1910. doi:10.1126/science.1059835
122. Afkarian M, Sedy JR, Yang J, et al. T-bet is a STAT1-induced regulator of IL-12R expression in naive CD4+ T cells. *Nat Immunol.* 2002;3(6):549-557. doi:10.1038/ni794
123. Schulz EG, Mariani L, Radbruch A, Hofer T. Sequential polarization and imprinting of type 1 T helper lymphocytes by interferon-gamma and interleukin-12. *Immunity.* 2009;30(5):673-683. doi:10.1016/j.immuni.2009.03.013
124. Djuretic IM, Levanon D, Negreanu V, Groner Y, Rao A, Ansel KM. Transcription factors T-bet and Runx3 cooperate to activate Ifng and silence Il4 in T helper type 1 cells. *Nat Immunol.* 2007;8(2):145-153. doi:10.1038/ni1424
125. Sekimata M, Perez-Melgosa M, Miller SA, et al. CCCTC-binding factor and the transcription factor T-bet orchestrate T helper 1 cell-specific structure and function at the interferon-gamma locus. *Immunity.* 2009;31(4):551-564. doi:10.1016/j.immuni.2009.08.021
126. Kao C, Oestreich KJ, Paley MA, et al. Transcription factor T-bet represses expression of the inhibitory receptor PD-1 and sustains virus-specific CD8+ T cell responses during chronic infection. *Nat Immunol.* 2011;12(7):663-671. doi:10.1038/ni.2046
127. Wei L, Vahedi G, Sun HW, et al. Discrete roles of STAT4 and STAT6 transcription factors in tuning epigenetic modifications and transcription during T helper cell differentiation. *Immunity.* 2010;32(6):840-851. doi:10.1016/j.immuni.2010.06.003
128. O'Shea J, Paul WE. Mechanisms underlying lineage commitment and plasticity of helper CD4 + T cells. *Science.* 2010;327(5969):1098-1102.
129. Niesner U, Albrecht I, Janke M, et al. Autoregulation of Th1-mediated inflammation by twist1. *J Exp Med.* 2008;205(8):1889-1901. doi:10.1084/jem.20072468
130. Pesu M, Watford WT, Wei L, et al. T-cell-expressed proprotein convertase furin is essential for maintenance of peripheral immune tolerance. *Nature.* 2008;455(7210):246-250. doi:10.1038/nature07210
131. Hwang ES, Szabo SJ, Schwartzberg PL, Glimcher LH. T helper cell fate specified by kinase-mediated interaction of T-bet with GATA-3. *Science.* 2005;307(5708):430-433. doi:10.1126/science.1103336
132. Mullen AC, Hutchins AS, High FA, et al. Hlx is induced by and genetically interacts with T-bet to promote heritable T(H)1 gene induction. *Nat Immunol.* 2002;3(7):652-658. doi:10.1038/ni807
133. Zheng WP, Zhao Q, Zhao X, et al. Up-regulation of Hlx in immature Th cells induces IFN-γ expression. *J Immunol.* 2004;172(1):114-122.
134. Reinhardt RL, Khoruts A, Merica R, Zell T, Jenkins MK. Visualizing the generation of memory CD4 T cells in the whole body. *Nature.* 2001;410(6824):101-105. doi:10.1038/35065111
135. Goldberg MF, Roeske EK, Ward LN, et al. Salmonella persist in activated macrophages in T cell-sparse granulomas but are contained by surrounding CXCR3 ligand-positioned Th1 cells. *Immunity.* 2018;49(6):1090.e7-1102.e7. doi:10.1016/j.immuni.2018.10.009
136. Leonard WJ, Lin JX, O'Shea JJ. The γc family of cytokines: Basic biology to therapeutic ramifications. *Immunity.* 2019;50(4):832-850. doi:10.1016/j.immuni.2019.03.028
137. Beck LA, Thaci D, Hamilton JD, et al. Dupilumab treatment in adults with moderate-to-severe atopic dermatitis. *N Engl J Med.* 2014;371(2):130-139. doi:10.1056/NEJMoa1314768
138. Palm NW, Rosenstein RK, Medzhitov R. Allergic host defences. *Nature.* 2012;484(7395):465-472. doi:10.1038/nature11047
139. Wan H, Winton HL, Soeller C, et al. Der p 1 facilitates transepithelial allergen delivery by disruption of tight junctions. *J Clin Invest.* 1999;104(1):123-133. doi:10.1172/jci5844
140. Cayrol C, Duval A, Schmitt P, et al. Environmental allergens induce allergic inflammation through proteolytic maturation of IL-33. *Nat Immunol.* 2018;19(4):375-385. doi:10.1038/s41590-018-0067-5
141. Serhan N, Basso L, Sibilano R, et al. House dust mites activate nociceptor-mast cell clusters to drive type 2 skin inflammation. *Nat Immunol.* 2019;20(11):1435-1443. doi:10.1038/s41590-019-0493-z
142. Rank MA, Kobayashi T, Kozaki H, Bartemes KR, Squillace DL, Kita H. IL-33-activated dendritic cells induce an atypical TH2-type response. *J Allergy Clin Immunol.* 2009;123(5):1047-1054. doi:10.1016/j.jaci.2009.02.026
143. Perner C, Flayer CH, Zhu X, et al. Substance P release by sensory neurons triggers dendritic cell migration and initiates the type-2 immune response to allergens. *Immunity.* 2020;53(5):1063.e7-1077.e7. doi:10.1016/j.immuni.2020.10.001
144. Gao Y, Nish SA, Jiang R, et al. Control of T helper 2 responses by transcription factor IRF4-dependent dendritic cells. *Immunity.* 2013;39(4):722-732. doi:10.1016/j.immuni.2013.08.028
145. Kumamoto Y, Linehan M, Weinstein JS, Laidlaw BJ, Craft JE, Iwasaki A. CD301b(+) dermal dendritic cells drive T helper 2 cell-mediated immunity. *Immunity.* 2013;39(4):733-743. doi:10.1016/j.immuni.2013.08.029
146. Tussiwand R, Everts B, Grajales-Reyes GE, et al. Klf4 expression in conventional dendritic cells is required for T helper 2 cell responses. *Immunity.* 2015;42(5):916-928. doi:10.1016/j.immuni.2015.04.017
147. Halim TY, Steer CA, Matha L, et al. Group 2 innate lymphoid cells are critical for the initiation of adaptive T helper 2 cell-mediated allergic lung inflammation. *Immunity.* 2014;40(3):425-435. doi:10.1016/j.immuni.2014.01.011
148. Mjosberg JM, Trifari S, Crellin NK, et al. Human IL-25- and IL-33-responsive type 2 innate lymphoid cells are defined by expression of CRTH2 and CD161. *Nat Immunol.* 2011;12(11):1055-1062. doi:10.1038/ni.2104
149. Schmitz J, Owyang A, Oldham E, et al. IL-33, an interleukin-1-like cytokine that signals via the IL-1 receptor-related protein ST2 and induces T helper type 2-associated cytokines. *Immunity.* 2005;23(5):479-490. doi:10.1016/j.immuni.2005.09.015
150. Arend WP, Palmer G, Gabay C. IL-1, IL-18, and IL-33 families of cytokines. *Immunol Rev.* 2008;223:20-38. doi:10.1111/j.1600-065X.2008.00624.x
151. Palmer G, Gabay C. Interleukin-33 biology with potential insights into human diseases. *Nat Rev Rheumatol.* 2011;7(6):321-329. doi:10.1038/nrrheum.2011.53

152. Saenz SA, Siracusa MC, Perrigoue JG, et al. IL25 elicits a multipotent progenitor cell population that promotes T(H)2 cytokine responses. *Nature.* 2010;464(7293):1362-1366. doi:10.1038/nature08901
153. Siracusa MC, Saenz SA, Hill DA, et al. TSLP promotes interleukin-3-independent basophil haematopoiesis and type 2 inflammation. *Nature.* 2011;477(7363):229-233. doi:10.1038/nature10329
154. Moro K, Yamada T, Tanabe M, et al. Innate production of T(H)2 cytokines by adipose tissue-associated c-Kit(+) Sca-1(+) lymphoid cells. *Nature.* 2010;463(7280):540-544. doi:10.1038/nature08636
155. Barlow JL, McKenzie AN. Nuocytes: expanding the innate cell repertoire in type-2 immunity. *J Leukoc Biol.* 2011;90(5):867-874. doi:10.1189/jlb.0311160
156. Oliphant CJ, Hwang YY, Walker JA, et al. MHCII-mediated dialog between group 2 innate lymphoid cells and CD4(+) T cells potentiates type 2 immunity and promotes parasitic helminth expulsion. *Immunity.* 2014;41(2):283-295. doi:10.1016/j.immuni.2014.06.016
157. Van Dyken SJ, Nussbaum JC, Lee J, et al. A tissue checkpoint regulates type 2 immunity. *Nat Immunol.* 2016;17(12):1381-1387. doi:10.1038/ni.3582
158. Karasuyama H, Mukai K, Obata K, Tsujimura Y, Wada T. Nonredundant roles of basophils in immunity. *Annu Rev Immunol.* 2011;29:45-69. doi:10.1146/annurev-immunol-031210-101257
159. Ziegler SF, Artis D. Sensing the outside world: TSLP regulates barrier immunity. *Nat Immunol.* 2010;11(4):289-293. doi:10.1038/ni.1852
160. Sokol CL, Medzhitov R. Emerging functions of basophils in protective and allergic immune responses. *Mucosal Immunol.* 2010;3(2):129-137. doi:10.1038/mi.2009.137
161. Lund R, Ahlfors H, Kainonen E, Lahesmaa AM, Dixon C, Lahesmaa R. Identification of genes involved in the initiation of human Th1 or Th2 cell commitment. *Eur J Immunol.* 2005;35(11):3307-3319. doi:10.1002/eji.200526079
162. Elo LL, Jarvenpaa H, Tuomela S, et al. Genome-wide profiling of interleukin-4 and STAT6 transcription factor regulation of human Th2 cell programming. *Immunity.* 2010;32(6):852-862. doi:10.1016/j.immuni.2010.06.011
163. O'Shea JJ, Lahesmaa R, Vahedi G, Laurence A, Kanno Y. Genomic views of STAT function in CD4+ T helper cell differentiation. *Nat Rev Immunol.* 2011;11(4):239-250. doi:10.1038/nri2958
164. Cote-Sierra J, Foucras G, Guo L, et al. Interleukin 2 plays a central role in Th2 differentiation. *Proc Natl Acad Sci U S Am.* 2004;101(11):3880-3885.
165. Zhu J, Cote-Sierra J, Guo L, Paul WE. Stat5 activation plays a critical role in Th2 differentiation. *Immunity.* 2003;19(5):739-748.
166. Yamane H, Zhu J, Paul WE. Independent roles for IL-2 and GATA-3 in stimulating naive CD4+ T cells to generate a Th2-inducing cytokine environment. *J Exp Med.* 2005;202(6):793-804. doi:10.1084/jem.20051304
167. Liao W, Schones DE, Oh J, et al. Priming for T helper type 2 differentiation by interleukin 2-mediated induction of interleukin 4 receptor α-chain expression. *Nat Immunol.* 2008;9(11):1288-1296.
168. Hondowicz BD, An D, Schenkel JM, et al. Interleukin-2-dependent allergen-specific tissue-resident memory cells drive asthma. *Immunity.* 2016;44(1):155-166. doi:10.1016/j.immuni.2015.11.004
169. Stritesky GL, Muthukrishnan R, Sehra S, et al. The transcription factor STAT3 is required for T helper 2 cell development. *Immunity.* 2011;34(1):39-49.
170. He K, Hettinga A, Kale SL, et al. Blimp-1 is essential for allergen-induced asthma and Th2 cell development in the lung. *J Exp Med.* 2020;217(7):e20190742. doi:10.1084/jem.20190742.
171. Wei G, Abraham BJ, Yagi R, et al. Genome-wide analyses of transcription factor GATA3-mediated gene regulation in distinct T cell types. *Immunity.* 2011;35(2):299-311. doi:10.1016/j.immuni.2011.08.007
172. Yagi R, Junttila IS, Wei G, et al. The transcription factor GATA3 actively represses RUNX3 protein-regulated production of interferon-gamma. *Immunity.* 2010;32(4):507-517. doi:10.1016/j.immuni.2010.04.004
173. Kim JI, Ho IC, Grusby MJ, Glimcher LH. The transcription factor c-Maf controls the production of interleukin-4 but not other Th2 cytokines. *Immunity.* 1999;10(6):745-751. doi:S1074-7613(00)80073-4
174. Amsen D, Blander JM, Lee GR, Tanigaki K, Honjo T, Flavell RA. Instruction of distinct CD4 T helper cell fates by different notch ligands on antigen-presenting cells. *Cell.* 2004;117(4):515-526.
175. Lohoff M, Mittrucker HW, Prechtl S, et al. Dysregulated T helper cell differentiation in the absence of interferon regulatory factor 4. *Proc Natl Acad Sci U S A.* 2002;99(18):11808-11812. doi:10.1073/pnas.182425099 182425099
176. Miaw SC, Choi A, Yu E, Kishikawa H, Ho IC. ROG, repressor of GATA, regulates the expression of cytokine genes. *Immunity.* 2000;12(3):323-333. doi:10.1016/s1074-7613(00)80185-5
177. Hirahara K, Yamashita M, Iwamura C, et al. Repressor of GATA regulates TH2-driven allergic airway inflammation and airway hyperresponsiveness. *J Allergy Clin Immunol.* 2008;122(3):512.e11-520.e11. doi:10.1016/j.jaci.2008.06.004
178. Okamoto M, Van Stry M, Chung L, et al. Mina, an Il4 repressor, controls T helper type 2 bias. *Nat Immunol.* 2009;10(8):872-879. doi:10.1038/ni.1747
179. Endo Y, Iwamura C, Kuwahara M, et al. Eomesodermin controls interleukin-5 production in memory T helper 2 cells through inhibition of activity of the transcription factor GATA3. *Immunity.* 2011;35(5):733-745. doi:10.1016/j.immuni.2011.08.017
180. Fang D, Cui K, Hu G, et al. Bcl11b, a novel GATA3-interacting protein, suppresses Th1 while limiting Th2 cell differentiation. *J Exp Med.* 2018;215(5):1449-1462. doi:10.1084/jem.20171127
181. Lorentsen KJ, Cho JJ, Luo X, et al. Bcl11b is essential for licensing Th2 differentiation during helminth infection and allergic asthma. *Nat Commun.* 2018;9(1):1679. doi:10.1038/s41467-018-04111-0
182. King IL, Mohrs M. IL-4-producing CD4+ T cells in reactive lymph nodes during helminth infection are T follicular helper cells. *J Exp Med.* 2009;206(5):1001-1007. doi:10.1084/jem.20090313
183. Glatman Zaretsky A, Taylor JJ, King IL, Marshall FA, Mohrs M, Pearce EJ. T follicular helper cells differentiate from Th2 cells in response to helminth antigens. *J Exp Med.* 2009;206(5):991-999. doi:10.1084/jem.20090303
184. Liang HE, Reinhardt RL, Bando JK, Sullivan BM, Ho IC, Locksley RM. Divergent expression patterns of IL-4 and IL-13 define unique functions in allergic immunity. *Nat Immunol.* 2011;13(1):58-66. doi:10.1038/ni.2182
185. Wambre E, Bajzik V, DeLong JH, et al. A phenotypically and functionally distinct human T(H)2 cell subpopulation is associated with allergic disorders. *Sci Transl Med.* 2017;9(401):eaam9171. doi:10.1126/scitranslmed.aam9171
186. Endo Y, Hirahara K, Iinuma T, et al. The interleukin-33-p38 kinase axis confers memory T helper 2 cell pathogenicity in the airway. *Immunity.* 2015;42(2):294-308. doi:10.1016/j.immuni.2015.01.016
187. Guo L, Wei G, Zhu J, et al. IL-1 family members and STAT activators induce cytokine production by Th2, Th17, and Th1 cells. *Proc Natl Acad Sci U S A.* 2009;106(32):13463-13468. doi:10.1073/pnas.0906988106
188. Park H, Li Z, Yang XO, et al. A distinct lineage of CD4 T cells regulates tissue inflammation by producing interleukin 17. *Nat Immunol.* 2005;6(11):1133-1141. doi:10.1038/ni1261
189. Weaver CT, Harrington LE, Mangan PR, Gavrieli M, Murphy KM. Th17: an effector CD4 T cell lineage with regulatory T cell ties. *Immunity.* 2006;24(6):677-688.
190. Weaver CT, Hatton RD, Mangan PR, Harrington LE. IL-17 family cytokines and the expanding diversity of effector T cell lineages. *Annu Rev Immunol.* 2007;25:821-852. doi:10.1146/annurev.immunol.25.022106.141557
191. Stockinger B, Veldhoen M, Martin B. Th17 T cells: linking innate and adaptive immunity. *Semin Immunol.* 2007;19(6):353-361. doi:10.1016/j.smim.2007.10.008
192. Mangan PR, Harrington LE, O'Quinn DB, et al. Transforming growth factor-beta induces development of the T(H)17 lineage. *Nature.* 2006;441(7090):231-234. doi:10.1038/nature04754
193. Miossec P. IL-17 and Th17 cells in human inflammatory diseases. *Microbes Infect.* 2009;11(5):625-630. doi:10.1016/j.micinf.2009.04.003
194. Annunziato F, Cosmi L, Liotta F, Maggi E, Romagnani S. Type 17 T helper cells-origins, features and possible roles in rheumatic disease. *Nat Rev Rheumatol.* 2009;5(6):325-331. doi:10.1038/nrrheum.2009.80
195. Peters A, Lee Y, Kuchroo VK. The many faces of Th17 cells. *Curr Opin Immunol.* 2011;23(6):702-706. doi:10.1016/j.coi.2011.08.007
196. McGeachy MJ, Cua DJ, Gaffen SL. The IL-17 family of cytokines in health and disease. *Immunity.* 2019;50(4):892-906. doi:10.1016/j.immuni.2019.03.021
197. Kolls JK, Linden A. Interleukin-17 family members and inflammation. *Immunity.* 2004;21(4):467-476. doi:10.1016/j.immuni.2004.08.018
198. Kolls JK. Th17 cells in mucosal immunity and tissue inflammation. *Semin Immunopathol.* 2010;32(1):1-2. doi:10.1007/s00281-010-0198-8
199. Onishi RM, Gaffen SL. Interleukin-17 and its target genes: mechanisms of interleukin-17 function in disease. *Immunology.* 2010;129(3):311-321. doi:10.1111/j.1365-2567.2009.03240.x
200. Aujla SJ, Chan YR, Zheng M, et al. IL-22 mediates mucosal host defense against gram-negative bacterial pneumonia. *Nat Med.* 2008;14(3):275-281. doi:10.1038/nm1710
201. Puel A, Cypowyj S, Marodi L, Abel L, Picard C, Casanova JL. Inborn errors of human IL-17 immunity underlie chronic mucocutaneous candidiasis. *Curr Opin Allergy Clin Immunol.* 2012;12(6):616-622. doi:10.1097/ACI.0b013e328358cc0b
202. Rangel-Moreno J, Carragher DM, de la Luz Garcia-Hernandez M, et al. The development of inducible bronchus-associated lymphoid tissue depends on IL-17. *Nat Immunol.* 2011;12(7):639-646. doi:10.1038/ni.2053
203. Hsu HC, Yang P, Wang J, et al. Interleukin 17-producing T helper cells and interleukin 17 orchestrate autoreactive germinal center development in autoimmune BXD2 mice. *Nat Immunol.* 2008;9(2):166-175. doi:10.1038/ni1552
204. Tesmer LA, Lundy SK, Sarkar S, Fox DA. Th17 cells in human disease. *Immunol Rev.* 2008;223:87-113. doi:10.1111/j.1600-065X.2008.00628.x
205. Acosta-Rodriguez EV, Rivino L, Geginat J, et al. Surface phenotype and antigenic specificity of human interleukin 17-producing T helper memory cells. *Nat Immunol.* 2007;8(6):639-646. doi:10.1038/ni1467
206. Coombes JL, Siddiqui KR, Arancibia-Carcamo CV, et al. A functionally specialized population of mucosal CD103+ DCs induces Foxp3+ regulatory T cells via a TGF-beta and retinoic acid-dependent mechanism. *J Exp Med.* 2007;204(8):1757-1764. doi:10.1084/jem.20070590
207. Zhou L, Ivanov II, Spolski R, et al. IL-6 programs T(H)-17 cell differentiation by promoting sequential engagement of the IL-21 and IL-23 pathways. *Nat Immunol.* 2007;8(9):967-974. doi:10.1038/ni1488
208. Manel N, Unutmaz D, Littman DR. The differentiation of human T(H)-17 cells requires transforming growth factor-beta and induction of the nuclear receptor RORgammat. *Nat Immunol.* 2008;9(6):641-649. doi:10.1038/ni.1610
209. Ghoreschi K, Laurence A, Yang XP, et al. Generation of pathogenic T(H)17 cells in the absence of TGF-beta signalling. *Nature.* 2010;467(7318):967-971. doi:10.1038/nature09447
210. Yang Y, Torchinsky MB, Gobert M, et al. Focused specificity of intestinal TH17 cells towards commensal bacterial antigens. *Nature.* 2014;510(7503):152-156. doi:10.1038/nature13279
211. Wu HJ, Ivanov II, Darce J, et al. Gut-residing segmented filamentous bacteria drive autoimmune arthritis via T helper 17 cells. *Immunity.* 2010;32(6):815-827. doi:10.1016/j.immuni.2010.06.001
212. Ivanov II, Atarashi K, Manel N, et al. Induction of intestinal Th17 cells by segmented filamentous bacteria. *Cell.* 2009;139(3):485-498. doi:10.1016/j.cell.2009.09.033
213. Chen Z, Laurence A, Kanno Y, et al. Selective regulatory function of Socs3 in the formation of IL-17-secreting T cells. *Proc Natl Acad Sci U S A.* 2006;103(21):8137-8142. doi:10.1073/pnas.0600666103
214. Yang XP, Ghoreschi K, Steward-Tharp SM, et al. Opposing regulation of the locus encoding IL-17 through direct, reciprocal actions of STAT3 and STAT5. *Nat Immunol.* 2011;12(3):247-254.
215. Ivanov II, McKenzie BS, Zhou L, et al. The orphan nuclear receptor RORgammat directs the differentiation program of proinflammatory IL-17+ T helper cells. *Cell.* 2006;126(6):1121-1133.
216. Durant L, Watford WT, Ramos HL, et al. Diverse targets of the transcription factor STAT3 contribute to T cell pathogenicity and homeostasis. *Immunity.* 2010;32(5):605-615.
217. Luci C, Reynders A, Ivanov II, et al. Influence of the transcription factor RORgammat on the development of NKp46+ cell populations in gut and skin. *Nat Immunol.* 2009;10(1):75-82. doi:10.1038/ni.1681
218. Sawa S, Lochner M, Satoh-Takayama N, et al. RORgammat+ innate lymphoid cells regulate intestinal homeostasis by integrating negative signals from the symbiotic microbiota. *Nat Immunol.* 2011;12(4):320-326. doi:10.1038/ni.2002
219. Takatori H, Kanno Y, Watford WT, et al. Lymphoid tissue inducer-like cells are an innate source of IL-17 and IL-22. *J Exp Med.* 2009;206(1):35-41. doi:10.1084/jem.20072713
220. Yang XO, Pappu BP, Nurieva R, et al. T helper 17 lineage differentiation is programmed by orphan nuclear receptors ROR alpha and ROR gamma. *Immunity.* 2008;28(1):29-39.
221. Ciofani M, Madar A, Galan C, et al. A validated regulatory network for Th17 cell specification. *Cell.* 2012;151(2):289-303. doi:10.1016/j.cell.2012.09.016
222. Quintana FJ, Basso AS, Iglesias AH, et al. Control of T(reg) and T(H)17 cell differentiation by the aryl hydrocarbon receptor. *Nature.* 2008;453(7191):65-71. doi:10.1038/nature06880
223. Veldhoen M, Hirota K, Christensen J, O'Garra A, Stockinger B. Natural agonists for aryl hydrocarbon receptor in culture medium are essential for optimal differentiation of Th17 T cells. *J Exp Med.* 2009;206(1):43-49. doi:10.1084/jem.20081438
224. Schraml BU, Hildner K, Ise W, et al. The AP-1 transcription factor Batf controls T(H)17 differentiation. *Nature.* 2009;460(7253):405-409. doi:10.1038/nature08114
225. Okamoto K, Iwai Y, Oh-Hora M, et al. IkappaBzeta regulates T(H)17 development by cooperating with ROR nuclear receptors. *Nature.* 2010;464(7293):1381-1385. doi:10.1038/nature08922
226. Huber M, Brustle A, Reinhard K, et al. IRF4 is essential for IL-21-mediated induction, amplification, and stabilization of the Th17 phenotype. *Proc Natl Acad Sci U S A.* 2008;105(52):20846-20851. doi:10.1073/pnas.0809077106
227. Dang EV, Barbi J, Yang HY, et al. Control of T(H)17/T(reg) balance by hypoxia-inducible factor 1. *Cell.* 2011;146(5):772-784. doi:10.1016/j.cell.2011.07.033
228. Reynolds JM, Pappu BP, Peng J, et al. Toll-like receptor 2 signaling in CD4(+) T lymphocytes promotes T helper 17 responses and regulates the pathogenesis of autoimmune disease. *Immunity.* 2010;32(5):692-702. doi:10.1016/j.immuni.2010.04.010

229. Delgoffe GM, Pollizzi KN, Waickman AT, et al. The kinase mTOR regulates the differentiation of helper T cells through the selective activation of signaling by mTORC1 and mTORC2. *Nat Immunol.* 2011;12(4):295-303. doi:10.1038/ni.2005
230. Moisan J, Grenningloh R, Bettelli E, Oukka M, Ho IC. Ets-1 is a negative regulator of Th17 differentiation. *J Exp Med.* 2007;204(12):2825-2835. doi:10.1084/jem.20070994
231. Zhu J, Davidson TS, Wei G, et al. Down-regulation of Gfi-1 expression by TGF-beta is important for differentiation of Th17 and CD103+ inducible regulatory T cells. *J Exp Med.* 2009;206(2):329-341. doi:10.1084/jem.20081666
232. Ma J, Wang R, Fang X, Ding Y, Sun Z. Critical role of TCF-1 in repression of the IL-17 gene. *PLoS One.* 2011;6(9):e24768. doi:10.1371/journal.pone.0024768
233. Laurence A, Tato CM, Davidson TS, et al. Interleukin-2 signaling via STAT5 constrains T helper 17 cell generation. *Immunity.* 2007;26(3):371-381. doi:10.1016/j.immuni.2007.02.009
234. Pandiyan P, Conti HR, Zheng L, et al. CD4(+)CD25(+)Foxp3(+) regulatory T cells promote Th17 cells *in vitro* and enhance host resistance in mouse Candida albicans Th17 cell infection model. *Immunity.* 2011;34(3):422-434. doi:10.1016/j.immuni.2011.03.002
235. Chen Y, Haines CJ, Gutcher I, et al. Foxp3(+) regulatory T cells promote T helper 17 cell development *in vivo* through regulation of interleukin-2. *Immunity.* 2011;34(3):409-421. doi:10.1016/j.immuni.2011.02.011
236. Korn T, Bettelli E, Oukka M, Kuchroo VK. IL-17 and Th17 cells. *Annu Rev Immunol.* 2009;27:485-517. doi:10.1146/annurev.immunol.021908.132710
237. Lee JY, Hall JA, Kroehling L, et al. Serum amyloid A proteins induce pathogenic Th17 cells and promote inflammatory disease. *Cell.* 2020;183(7):2036-2039. doi:10.1016/j.cell.2020.12.008
238. Dardalhon V, Awasthi A, Kwon H, et al. IL-4 inhibits TGF-beta-induced Foxp3+ T cells and, together with TGF-beta, generates IL-9+ IL-10+ Foxp3(-) effector T cells. *Nat Immunol.* 2008;9(12):1347-1355. doi:10.1038/ni.1677
239. Veldhoen M, Uyttenhove C, van Snick J, et al. Transforming growth factor-beta 'reprograms' the differentiation of T helper 2 cells and promotes an interleukin 9-producing subset. *Nat Immunol.* 2008;9(12):1341-1346. doi:10.1038/ni.1659
240. Beriou G, Bradshaw EM, Lozano E, et al. TGF-beta induces IL-9 production from human Th17 cells. *J Immunol.* 2010;185(1):46-54. doi:10.4049/jimmunol.1000356
241. Perumal NB, Kaplan MH. Regulating Il9 transcription in T helper cells. *Trends Immunol.* 2011;32(4):146-150. doi:10.1016/j.it.2011.01.006
242. Schmitt E, Germann T, Goedert S, et al. IL-9 production of naive CD4+ T cells depends on IL-2, is synergistically enhanced by a combination of TGF-beta and IL-4, and is inhibited by IFN-gamma. *J Immunol.* 1994;153(9):3989-3996.
243. Angkasekwinai P, Chang SH, Thapa M, Watarai H, Dong C. Regulation of IL-9 expression by IL-25 signaling. *Nat Immunol.* 2010;11(3):250-256. doi:10.1038/ni.1846
244. Meylan F, Siegel RM. TNF superfamily cytokines in the promotion of Th9 differentiation and immunopathology. *Semin Immunopathol.* 2017;39(1):21-28. doi:10.1007/s00281-016-0612-y
245. Schwartz DM, Farley TK, Richoz N, et al. Retinoic acid receptor alpha represses a Th9 transcriptional and epigenomic program to reduce allergic pathology. *Immunity.* 2019;50(1):106.e10-120.e10. doi:10.1016/j.immuni.2018.12.014
246. Goswami R, Kaplan MH. A brief history of IL-9. *J Immunol.* 2011;186(6):3283-3288. doi:10.4049/jimmunol.1003049
247. Nowak EC, Weaver CT, Turner H, et al. IL-9 as a mediator of Th17-driven inflammatory disease. *J Exp Med.* 2009;206(8):1653-1660. doi:10.1084/jem.20090246
248. Elyaman W, Bradshaw EM, Uyttenhove C, et al. IL-9 induces differentiation of TH17 cells and enhances function of FoxP3+ natural regulatory T cells. *Proc Natl Acad Sci U S A.* 2009;106(31):12885-12890. doi:10.1073/pnas.0812530106
249. Staudt V, Bothur E, Klein M, et al. Interferon-regulatory factor 4 is essential for the developmental program of T helper 9 cells. *Immunity.* 2010;33(2):192-202. doi:10.1016/j.immuni.2010.07.014
250. Chang HC, Sehra S, Goswami R, et al. The transcription factor PU.1 is required for the development of IL-9-producing T cells and allergic inflammation. *Nat Immunol.* 2010;11(6):527-534. doi:10.1038/ni.1867
251. Richard AC, Tan C, Hawley ET, et al. The TNF-family ligand TL1A and its receptor DR3 promote T cell-mediated allergic immunopathology by enhancing differentiation and pathogenicity of IL-9-producing T cells. *J Immunol.* 2015;194(8):3567-3582. doi:10.4049/jimmunol.1401220
252. Wilhelm C, Hirota K, Stieglitz B, et al. An IL-9 fate reporter demonstrates the induction of an innate IL-9 response in lung inflammation. *Nat Immunol.* 2011;12(11):1071-1077. doi:10.1038/ni.2133
253. Ouyang W, O'Garra A. IL-10 family cytokines IL-10 and IL-22: from basic science to clinical translation. *Immunity.* 2019;50(4):871-891. doi:10.1016/j.immuni.2019.03.020
254. Zenewicz LA, Yancopoulos GD, Valenzuela DM, Murphy AJ, Stevens S, Flavell RA. Innate and adaptive interleukin-22 protects mice from inflammatory bowel disease. *Immunity.* 2008;29(6):947-957. doi:10.1016/j.immuni.2008.11.003
255. Zenewicz LA, Yancopoulos GD, Valenzuela DM, Murphy AJ, Karow M, Flavell RA. Interleukin-22 but not interleukin-17 provides protection to hepatocytes during acute liver inflammation. *Immunity.* 2007;27(4):647-659. doi:10.1016/j.immuni.2007.07.023
256. Zenewicz LA, Flavell RA. Recent advances in IL-22 biology. *Int Immunol.* 2011;23(3):159-163. doi:10.1093/intimm/dxr001
257. Zheng Y, Danilenko DM, Valdez P, et al. Interleukin-22, a T(H)17 cytokine, mediates IL-23-induced dermal inflammation and acanthosis. *Nature.* 2007;445(7128):648-651. doi:10.1038/nature05505
258. Zheng Y, Valdez PA, Danilenko DM, et al. Interleukin-22 mediates early host defense against attaching and effacing bacterial pathogens. *Nat Med.* 2008;14(3):282-289. doi:10.1038/nm1720
259. Rutz S, Noubade R, Eidenschenk C, et al. Transcription factor c-Maf mediates the TGF-beta-dependent suppression of IL-22 production in T(H)17 cells. *Nat Immunol.* 2011;12(12):1238-1245. doi:10.1038/ni.2134
260. Eyerich S, Eyerich K, Pennino D, et al. Th22 cells represent a distinct human T cell subset involved in epidermal immunity and remodeling. *J Clin Invest.* 2009;119(12):3573-3585. doi:10.1172/JCI40202
261. Duhen T, Geiger R, Jarrossay D, Lanzavecchia A, Sallusto F. Production of interleukin 22 but not interleukin 17 by a subset of human skin-homing memory T cells. *Nat Immunol.* 2009;10(8):857-863. doi:10.1038/ni.1767
262. Trifari S, Kaplan CD, Tran EH, Crellin NK, Spits H. Identification of a human helper T cell population that has abundant production of interleukin 22 and is distinct from T(H)-17, T(H)1 and T(H)2 cells. *Nat Immunol.* 2009;10(8):864-871. doi:10.1038/ni.1770
263. Crellin NK, Trifari S, Kaplan CD, Cupedo T, Spits H. Human NKp44+IL-22+ cells and LTi-like cells constitute a stable RORC+ lineage distinct from conventional natural killer cells. *J Exp Med.* 2010;207(2):281-290. doi:10.1084/jem.20091509
264. Cella M, Fuchs A, Vermi W, et al. A human natural killer cell subset provides an innate source of IL-22 for mucosal immunity. *Nature.* 2009;457(7230):722-725. doi:10.1038/nature07537
265. Vivier E, Spits H, Cupedo T. Interleukin-22-producing innate immune cells: new players in mucosal immunity and tissue repair? *Nat Rev Immunol.* 2009;9(4):229-234. doi:10.1038/nri2522
266. Spits H, Di Santo JP. The expanding family of innate lymphoid cells: regulators and effectors of immunity and tissue remodeling. *Nat Immunol.* 2011;12(1):21-27. doi:10.1038/ni.1962
267. Sanos SL, Bui VL, Mortha A, et al. RORgammat and commensal microflora are required for the differentiation of mucosal interleukin 22-producing NKp46+ cells. *Nat Immunol.* 2009;10(1):83-91. doi:10.1038/ni.1684
268. Alam MS, Maekawa Y, Kitamura A, et al. Notch signaling drives IL-22 secretion in CD4+ T cells by stimulating the aryl hydrocarbon receptor. *Proc Natl Acad Sci U S A.* 2010;107(13):5943-5948. doi:10.1073/pnas.0911755107
269. Cano-Gamez E, Soskic B, Roumeliotis TI, et al. Single-cell transcriptomics identifies an effectorness gradient shaping the response of CD4(+) T cells to cytokines. *Nat Commun.* 2020;11(1):1801. doi:10.1038/s41467-020-15543-y
270. Kiner E, Willie E, Vijaykumar B, et al. Gut CD4(+) T cell phenotypes are a continuum molded by microbes, not by TH archetypes. *Nat Immunol.* 2021;22(2):216-228. doi:10.1038/s41590-020-00836-7
271. Zemmour D, Kiner E, Benoist C. CD4(+) teff cell heterogeneity: the perspective from single-cell transcriptomics. *Curr Opin Immunol.* 2020;63:61-67. doi:10.1016/j.coi.2020.02.004
272. Ladinsky MS, Araujo LP, Zhang X, et al. Endocytosis of commensal antigens by intestinal epithelial cells regulates mucosal T cell homeostasis. *Science.* 2019;363(6431):eaat4042. doi:10.1126/science.aat4042
273. Batlle E, Massague J. Transforming growth factor-beta signaling in immunity and cancer. *Immunity.* 2019;50(4):924-940. doi:10.1016/j.immuni.2019.03.024
274. Esplugues E, Huber S, Gagliani N, et al. Control of TH17 cells occurs in the small intestine. *Nature.* 2011;475(7357):514-518. doi:10.1038/nature10228
275. Lee YK, Turner H, Maynard CL, et al. Late developmental plasticity in the T helper 17 lineage. *Immunity.* 2009;30(1):92-107. doi:10.1016/j.immuni.2008.11.005.
276. Shi G, Cox CA, Vistica BP, Tan C, Wawrousek EF, Gery I. Phenotype switching by inflammation-inducing polarized Th17 cells, but not by Th1 cells. *J Immunol.* 2008;181(10):7205-7213. doi:10.4049/jimmunol.181.10.7205
277. Muranski P, Boni A, Antony PA, et al. Tumor-specific Th17-polarized cells eradicate large established melanoma. *Blood.* 2008;112(2):362-373. doi:10.1182/blood-2007-11-120998
278. Hirota K, Duarte JH, Veldhoen M, et al. Fate mapping of IL-17-producing T cells in inflammatory responses. *Nat Immunol.* 2011;12(3):255-263. doi:10.1038/ni.1993
279. Lee Y, Awasthi A, Yosef N, et al. Induction and molecular signature of pathogenic TH17 cells. *Nat Immunol.* 2012;13(10):991-999. doi:10.1038/ni.2416.
280. Jain R, Chen Y, Kanno Y, et al. Interleukin-23-Induced transcription factor Blimp-1 promotes pathogenicity of T helper 17 cells. *Immunity.* 2016;44(1):131-142. doi:10.1016/j.immuni.2015.11.009
281. Stumhofer JS, Silver JS, Laurence A, et al. Interleukins 27 and 6 induce STAT3-mediated T cell production of interleukin 10. *Nat Immunol.* 2007;8(12):1363-1371. doi:10.1038/ni1537
282. Awasthi A, Carrier Y, Peron JP, et al. A dominant function for interleukin 27 in generating interleukin 10-producing anti-inflammatory T cells. *Nat Immunol.* 2007;8(12):1380-1389. doi:10.1038/ni1541
283. Tait Wojno ED, Hunter CA, Stumhofer JS. The immunobiology of the interleukin-12 family: Room for discovery. *Immunity.* 2019;50(4):851-870. doi:10.1016/j.immuni.2019.03.011
284. Hu D, Notarbartolo S, Croonenborghs T, et al. Transcriptional signature of human pro-inflammatory TH17 cells identifies reduced IL10 gene expression in multiple sclerosis. *Nat Commun.* 2017;8(1):1600. doi:10.1038/s41467-017-01571-8
285. Hirota K, Turner JE, Villa M, et al. Plasticity of Th17 cells in Peyer's patches is responsible for the induction of T cell-dependent IgA responses. *Nat Immunol.* 2013;14(4):372-379. doi:10.1038/ni.2552
286. Bending D, De la Pena H, Veldhoen M, et al. Highly purified Th17 cells from BDC2.5NOD mice convert into Th1-like cells in NOD/SCID recipient mice. *J Clin Invest.* 2009;119(3):565-572. doi:10.1172/JCI37865
287. Koch MA, Tucker-Heard G, Perdue NR, Killebrew JR, Urdahl KB, Campbell DJ. The transcription factor T-bet controls regulatory T cell homeostasis and function during type 1 inflammation. *Nat Immunol.* 2009;10(6):595-602. doi:10.1038/ni.1731
288. Wohlfert EA, Grainger JR, Bouladoux N, et al. GATA3 controls Foxp3(+) regulatory T cell fate during inflammation in mice. *J Clin Invest.* 2011;121(11):4503-4515. doi:10.1172/JCI57456
289. Wang Y, Su MA, Wan YY. An essential role of the transcription factor GATA-3 for the function of regulatory T cells. *Immunity.* 2011;35(3):337-348. doi:10.1016/j.immuni.2011.08.012
290. Voo KS, Wang YH, Santori FR, et al. Identification of IL-17-producing FOXP3+ regulatory T cells in humans. *Proc Natl Acad Sci U S A.* 2009;106(12):4793-4798. doi:10.1073/pnas.0900408106.
291. Kanamori M, Nakatsukasa H, Okada M, Lu Q, Yoshimura A. Induced regulatory T cells: their development, stability, and applications. *Trends Immunol.* 2016;37(11):803-811. doi:10.1016/j.it.2016.08.012
292. Bilate AM, Lafaille JJ. Induced CD4+Foxp3+ regulatory T cells in immune tolerance. *Annu Rev Immunol.* 2012;30:733-758. doi:10.1146/annurev-immunol-020711-075043
293. Linterman MA, Pierson W, Lee SK, et al. Foxp3+ follicular regulatory T cells control the germinal center response. *Nat Med.* 2011;17(8):975-982. doi:10.1038/nm.2425
294. Chung Y, Tanaka S, Chu F, et al. Follicular regulatory T cells expressing Foxp3 and Bcl-6 suppress germinal center reactions. *Nat Med.* 2011;17(8):983-988. doi:10.1038/nm.2426
295. Neumann C, Blume J, Roy U, et al. c-Maf-dependent T_{reg} cell control of intestinal TH17 cells and IgA establishes host-microbiota homeostasis. *Nat Immunol.* 2019;20(4):471-481. doi:10.1038/s41590-019-0316-2
296. Zheng Y, Chaudhry A, Kas A, et al. Regulatory T-cell suppressor program co-opts transcription factor IRF4 to control T(H)2 responses. *Nature.* 2009;458(7236):351-356. doi:10.1038/nature07674
297. Chaudhry A, Rudra D, Treuting P, et al. CD4+ regulatory T cells control TH17 responses in a Stat3-dependent manner. *Science.* 2009;326(5955):986-991. doi:10.1126/science.1172702
298. Oldenhove G, Bouladoux N, Wohlfert EA, et al. Decrease of Foxp3+ T_{reg} cell number and acquisition of effector cell phenotype during lethal infection. *Immunity.* 2009;31(5):772-786. doi:10.1016/j.immuni.2009.10.001
299. Motomura Y, Kitamura H, Hijikata A, et al. The transcription factor E4BP4 regulates the production of IL-10 and IL-13 in CD4+ T cells. *Nat Immunol.* 2011;12(5):450-459. doi:10.1038/ni.2020
300. Hegazy AN, Peine M, Helmstetter C, et al. Interferons direct Th2 cell reprogramming to generate a stable GATA-3+T-bet+ cell subset with combined Th2 and Th1 cell functions. *Immunity.* 2010;32(1):116-128.
301. Stadhouders R, Filion GJ, Graf T. Transcription factors and 3D genome conformation in cell-fate decisions. *Nature.* 2019;569(7756):345-354. doi:10.1038/s41586-019-1182-7
302. Hadjur S, Williams LM, Ryan NK, et al. Cohesins form chromosomal cis-interactions at the developmentally regulated IFNG locus. *Nature.* 2009;460(7253):410-413. doi:10.1038/nature08079
303. Zhang F, Boothby M. T helper type 1-specific Brg1 recruitment and remodeling of nucleosomes positioned at the IFN-gamma promoter are Stat4 dependent. *J Exp Med.* 2006;203(6):1493-1505.

304. Lee PP, Fitzpatrick DR, Beard C, et al. A critical role for Dnmt1 and DNA methylation in T cell development, function, and survival. *Immunity*. 2001;15(5):763-774.
305. Floess S, Freyer J, Siewert C, et al. Epigenetic control of the foxp3 locus in regulatory T cells. *PLoS Biol*. 2007;5(2):0169-0178.
306. Huehn J, Polansky JK, Hamann A. Epigenetic control of FOXP3 expression: the key to a stable regulatory T-cell lineage? *Nat Rev Immunol*. 2009;9(2):83-89. doi:10.1038/nri2474
307. Kim HP, Leonard WJ. CREB/ATF-dependent T cell receptor-induced FoxP3 gene expression: a role for DNA methylation. *J Exp Med*. 2007;204(7):1543-1551. doi:10.1084/jem.20070109
308. Makar KW, Perez-Melgosa M, Shnyreva M, Weaver WM, Fitzpatrick DR, Wilson CB. Active recruitment of DNA methyltransferases regulates interleukin 4 in thymocytes and T cells. *Nat Immunol*. 2003;4(12):1183-1190. doi:10.1038/ni1004
309. Hutchins AS, Mullen AC, Lee HW, et al. Gene silencing quantitatively controls the function of a developmental trans-activator. *Mol Cell*. 2002;10(1):81-91.
310. Thomas RM, Gamper CJ, Ladle BH, Powell JD, Wells AD. De novo DNA methylation is required to restrict T helper lineage plasticity. *J Biol Chem*. 2012;287(27):22900-22909. doi:10.1074/jbc.M111.312785
311. Yue X, Lio CJ, Samaniego-Castruita D, Li X, Rao A. Loss of TET2 and TET3 in regulatory T cells unleashes effector function. *Nat Commun*. 2019;10(1):2011. doi:10.1038/s41467-019-09541-y
312. Lio CJ, Rao A. TET enzymes and 5hmC in adaptive and innate immune systems. *Front Immunol*. 2019;10:210. doi:10.3389/fimmu.2019.00210
313. Sasidharan Nair V, Song MH, Oh KI. Vitamin C facilitates demethylation of the Foxp3 enhancer in a Tet-dependent manner. *J Immunol*. 2016;196(5):2119-2131. doi:10.4049/jimmunol.1502352
314. Amsen D, Spilianakis CG, Flavell RA. How are T(H)1 and T(H)2 effector cells made? *Curr Opin Immunol*. 2009;21(2):153-160. doi:10.1016/j.coi.2009.03.010
315. Aune TM, Collins PL, Chang S. Epigenetics and T helper 1 differentiation. *Immunology*. 2009;126(3):299-305. doi:10.1111/j.1365-2567.2008.03026.x
316. Ansel KM, Lee DU, Rao A. An epigenetic view of helper T cell differentiation. *Nat Immunol*. 2003;4(7):616-623. doi:10.1038/ni0703-616
317. Shih HY, Sciume G, Poholek AC, et al. Transcriptional and epigenetic networks of helper T and innate lymphoid cells. *Immunol Rev*. 2014;261(1):23-49. doi:10.1111/imr.12208
318. Li P, Spolski R, Liao W, Leonard WJ. Complex interactions of transcription factors in mediating cytokine biology in T cells. *Immunol Rev*. 2014;261(1):141-156. doi:10.1111/imr.12199
319. Li P, Leonard WJ. Chromatin accessibility and interactions in the transcriptional regulation of T cells. *Front Immunol*. 2018;9:2738. doi:10.3389/fimmu.2018.02738
320. Nakayama T, Yamashita M. Initiation and maintenance of Th2 cell identity. *Curr Opin Immunol*. 2008;20(3):265-271. doi:10.1016/j.coi.2008.03.011
321. Yamashita M, Hirahara K, Shinnakasu R, et al. Crucial role of MLL for the maintenance of memory T helper type 2 cell responses. *Immunity*. 2006;24(5):611-622. doi:10.1016/j.immuni.2006.03.017
322. Onodera A, Yamashita M, Endo Y, et al. STAT6-mediated displacement of polycomb by trithorax complex establishes long-term maintenance of GATA3 expression in T helper type 2 cells. *J Exp Med*. 2010;207(11):2493-2506. doi:10.1084/jem.20100760
323. Kimura M, Koseki Y, Yamashita M, et al. Regulation of Th2 cell differentiation by mel-18, a mammalian Polycomb group gene. *Immunity*. 2001;15(2):275-287.
324. Yamashita M, Kuwahara M, Suzuki A, et al. Bmi1 regulates memory CD4 T cell survival via repression of the Noxa gene. *J Exp Med*. 2008;205(5):1109-1120. doi:10.1084/jem.20072000
325. Koyanagi M, Baguet A, Martens J, Margueron R, Jenuwein T, Bix M. EZH2 and histone 3 trimethyl lysine 27 associated with Il4 and Il13 gene silencing in TH1 cells. *J Biol Chem*. 2005;280(36):31470-31477.
326. Barski A, Cuddapah S, Cui K, et al. High-resolution profiling of histone methylations in the human genome. *Cell*. 2007;129(4):823-837. doi:10.1016/j.cell.2007.05.009
327. Wei G, Wei L, Zhu J, et al. Global mapping of H3K4me3 and H3K27me3 reveals specificity and plasticity in lineage fate determination of differentiating CD4+ T cells. *Immunity*. 2009;30(1):155-167. doi:10.1016/j.immuni.2008.12.009
328. Cuddapah S, Barski A, Zhao K. Epigenomics of T cell activation, differentiation, and memory. *Curr Opin Immunol*. 2010;22(3):341-347. doi:10.1016/j.coi.2010.02.007
329. Mukasa R, Balasubramani A, Lee YK, et al. Epigenetic instability of cytokine and transcription factor gene loci underlies plasticity of the T helper 17 cell lineage. *Immunity*. 2010;32(5):616-627.
330. Bernstein BE, Kamal M, Lindblad-Toh K, et al. Genomic maps and comparative analysis of histone modifications in human and mouse. *Cell*. 2005;120(2):169-181.
331. Zhou VW, Goren A, Bernstein BE. Charting histone modifications and the functional organization of mammalian genomes. *Nat Rev Genet*. 2011;12(1):7-18. doi:10.1038/nrg2905
332. Lu KT, Kanno Y, Cannons JL, et al. Functional and epigenetic studies reveal multistep differentiation and plasticity of *in vitro*-generated and *in vivo*-derived follicular T helper cells. *Immunity*. 2011;35(4):622-632. doi:10.1016/j.immuni.2011.07.015
333. Balasubramani A, Mukasa R, Hatton RD, Weaver CT. Regulation of the Ifng locus in the context of T-lineage specification and plasticity. *Immunol Rev*. 2010;238(1):216-232. doi:10.1111/j.1600-065X.2010.00961.x
334. Soutto M, Zhang F, Enerson B, Tong Y, Boothby M, Aune TM. A minimal IFN-gamma promoter confers Th1 selective expression. *J Immunol*. 2002;169(8):4205-4212. doi:10.4049/jimmunol.169.8.4205
335. Collins PL, Chang S, Henderson M, et al. Distal regions of the human IFNG locus direct cell type-specific expression. *J Immunol*. 2010;185(3):1492-1501. doi:10.4049/jimmunol.1000124
336. Lee DU, Agarwal S, Rao A. Th2 lineage commitment and efficient IL-4 production involves extended demethylation of the IL-4 gene. *Immunity*. 2002;16(5):649-660. doi:10.1016/s1074-7613(02)00314-x
337. Ansel KM, Djuretic I, Tanasa B, Rao A. Regulation of Th2 differentiation and Il4 locus accessibility. *Annu Rev Immunol*. 2006;24:607-656. doi:10.1146/annurev.immunol.23.021704.115821
338. Ansel KM, Greenwald RJ, Agarwal S, et al. Deletion of a conserved Il4 silencer impairs T helper type 1-mediated immunity. *Nat Immunol*. 2004;5(12):1251-1259. doi:10.1038/ni1135
339. Naoe Y, Setoguchi R, Akiyama K, et al. Repression of interleukin-4 in T helper type 1 cells by Runx/Cbf beta binding to the Il4 silencer. *J Exp Med*. 2007;204(8):1749-1755. doi:10.1084/jem.20062456
340. Vahedi G, Takahashi H, Nakayamada S, et al. STATs shape the active enhancer landscape of T cell populations. *Cell*. 2012;151(5):981-993. doi:10.1016/j.cell.2012.09.044
341. Vahedi G, Kanno Y, Furumoto Y, et al. Super-enhancers delineate disease-associated regulatory nodes in T cells. *Nature*. 2015;520(7548):558-562. doi:10.1038/nature14154
342. Whyte WA, Orlando DA, Hnisz D, et al. Master transcription factors and mediator establish super-enhancers at key cell identity genes. *Cell*. 2013;153(2):307-319. doi:10.1016/j.cell.2013.03.035
343. Chen YG, Satpathy AT, Chang HY. Gene regulation in the immune system by long noncoding RNAs. *Nat Immunol*. 2017;18(9):962-972. doi:10.1038/ni.3771
344. Ponting CP, Oliver PL, Reik W. Evolution and functions of long noncoding RNAs. *Cell*. 2009;136(4):629-641. doi:10.1016/j.cell.2009.02.006
345. Lander ES, Linton LM, Birren B, et al. Initial sequencing and analysis of the human genome. *Nature*. 2001;409(6822):860-921. doi:10.1038/35057062
346. Djebali S, Davis CA, Merkel A, et al. Landscape of transcription in human cells. *Nature*. 2012;489(7414):101-108. doi:10.1038/nature11233
347. Uszczynska-Ratajczak B, Lagarde J, Frankish A, Guigo R, Johnson R. Towards a complete map of the human long non-coding RNA transcriptome. *Nat Rev Genet*. 2018;19(9):535-548. doi:10.1038/s41576-018-0017-y
348. West KA, Lagos D. Long non-coding RNA function in CD4(+) T cells: what we know and what next? *Noncoding RNA*. 2019;5(3):43. doi:10.3390/ncrna5030043
349. Iyer MK, Niknafs YS, Malik R, et al. The landscape of long noncoding RNAs in the human transcriptome. *Nat Genet*. 2015;47(3):199-208. doi:10.1038/ng.3192
350. Petermann F, Pekowska A, Johnson CA, et al. The magnitude of IFN-gamma responses is fine-tuned by DNA architecture and the non-coding transcript of Ifng-as1. *Mol Cell*. 2019;75(6):1229.e5-1242.e5. doi:10.1016/j.molcel.2019.06.025
351. Collier SP, Henderson MA, Tossberg JT, Aune TM. Regulation of the Th1 genomic locus from Ifng through Tmevpg1 by T-bet. *J Immunol*. 2014;193(8):3959-3965. doi:10.4049/jimmunol.1401099
352. Mowel WK, McCright SJ, Kotzin JJ, et al. Group 1 innate lymphoid cell lineage identity is determined by a cis-regulatory element marked by a long non-coding RNA. *Immunity*. 2017;47(3):435.e8-449.e8. doi:10.1016/j.immuni.2017.08.012
353. Baltimore D, Boldin MP, O'Connell RM, Rao DS, Taganov KD. MicroRNAs: new regulators of immune cell development and function. *Nat Immunol*. 2008;9(8):839-845. doi:10.1038/ni.f.209
354. O'Connell RM, Rao DS, Chaudhuri AA, Baltimore D. Physiological and pathological roles for microRNAs in the immune system. *Nat Rev Immunol*. 2010;10(2):111-122. doi:10.1038/nri2708
355. Monticelli S. MicroRNAs in T helper cell differentiation and plasticity. *Semin Immunol*. 2013;25(4):291-298. doi:10.1016/j.smim.2013.10.015
356. Simpson LJ, Ansel KM. MicroRNA regulation of lymphocyte tolerance and autoimmunity. *J Clin Invest*. 2015;125(6):2242-2249. doi:10.1172/JCI78090
357. Muljo SA, Ansel KM, Kanellopoulou C, Livingston DM, Rao A, Rajewsky K. Aberrant T cell differentiation in the absence of Dicer. *J Exp Med*. 2005;202(2):261-269. doi:10.1084/jem.20050678
358. Zhou X, Jeker LT, Fife BT, et al. Selective miRNA disruption in T_{reg} cells leads to uncontrolled autoimmunity. *J Exp Med*. 2008;205(9):1983-1991. doi:10.1084/jem.20080707
359. Lu LF, Boldin MP, Chaudhry A, et al. Function of miR-146a in controlling T_{reg} cell-mediated regulation of Th1 responses. *Cell*. 2010;142(6):914-929. doi:10.1016/j.cell.2010.08.012
360. Maul J, Alterauge D, Baumjohann D. MicroRNA-mediated regulation of T follicular helper and T follicular regulatory cell identity. *Immunol Rev*. 2019;288(1):97-111. doi:10.1111/imr.12735
361. Kohlhaas S, Garden OA, Scudamore C, Turner M, Okkenhaug K, Vigorito E. Cutting edge: the Foxp3 target miR-155 contributes to the development of regulatory T cells. *J Immunol*. 2009;182(5):2578-2582. doi:10.4049/jimmunol.0803162
362. Banerjee A, Schambach F, DeJong CS, Hammond SM, Reiner SL. Micro-RNA-155 inhibits IFN-gamma signaling in CD4+ T cells. *Eur J Immunol*. 2010;40(1):225-231. doi:10.1002/eji.200939381
363. O'Connell RM, Kahn D, Gibson WS, et al. MicroRNA-155 promotes autoimmune inflammation by enhancing inflammatory T cell development. *Immunity*. 2010;33(4):607-619. doi:10.1016/j.immuni.2010.09.009
364. Steiner DF, Thomas MF, Hu JK, et al. MicroRNA-29 regulates T-box transcription factors and interferon-gamma production in helper T cells. *Immunity*. 2011;35(2):169-181. doi:10.1016/j.immuni.2011.07.009
365. Stittrich AB, Haftmann C, Sgouroudis E, et al. The microRNA miR-182 is induced by IL-2 and promotes clonal expansion of activated helper T lymphocytes. *Nat Immunol*. 2010;11(11):1057-1062. doi:10.1038/ni.1945
366. Delgoffe GM, Powell JD. mTOR: taking cues from the immune microenvironment. *Immunology*. 2009;127(4):459-465. doi:10.1111/j.1365-2567.2009.03125.x
367. Lee K, Gudapati P, Dragovic S, et al. Mammalian target of rapamycin protein complex 2 regulates differentiation of Th1 and Th2 cell subsets via distinct signaling pathways. *Immunity*. 2010;32(6):743-753. doi:10.1016/j.immuni.2010.06.002
368. Patel CH, Powell JD. Targeting T cell metabolism to regulate T cell activation, differentiation and function in disease. *Curr Opin Immunol*. 2017;46:82-88. doi:10.1016/j.coi.2017.04.006
369. Johnson MO, Wolf MM, Madden MZ, et al. Distinct regulation of Th17 and Th1 cell differentiation by glutaminase-dependent metabolism. *Cell*. 2018;175(7):1780.e19-1795.e19. doi:10.1016/j.cell.2018.10.001
370. Wu L, Hollinshead KER, Hao Y, et al. Niche-selective inhibition of pathogenic Th17 cells by targeting metabolic redundancy. *Cell*. 2020;182(3):641.e20-654.e20. doi:10.1016/j.cell.2020.06.014
371. Peng M, Yin N, Chhangawala S, Xu K, Leslie CS, Li MO. Aerobic glycolysis promotes T helper 1 cell differentiation through an epigenetic mechanism. *Science*. 2016;354(6311):481-484. doi:10.1126/science.aaf6284
372. Skelly AN, Sato Y, Kearney S, Honda K. Mining the microbiota for microbial and metabolite-based immunotherapies. *Nat Rev Immunol*. 2019;19(5):305-323. doi:10.1038/s41577-019-0144-5
373. Jameson SC, Masopust D. Understanding subset diversity in T cell memory. *Immunity*. 2018;48(2):214-226. doi:10.1016/j.immuni.2018.02.010
374. Homann D, Teyton L, Oldstone MB. Differential regulation of antiviral T-cell immunity results in stable CD8+ but declining CD4+ T-cell memory. *Nat Med*. 2001;7(8):913-919. doi:10.1038/90950
375. Pepper M, Linehan JL, Pagan AJ, et al. Different routes of bacterial infection induce long-lived TH1 memory cells and short-lived TH17 cells. *Nat Immunol*. 2010;11(1):83-89. doi:10.1038/ni.1826
376. Munitic I, Ryan PE, Ashwell JD. T cells in G1 provide a memory-like response to secondary stimulation. *J Immunol*. 2005;174(7):4010-4018. doi:10.4049/jimmunol.174.7.4010
377. Dan JM, Mateus J, Kato Y, et al. Immunological memory to SARS-CoV-2 assessed for up to 8 months after infection. *Science*. 2021;371(6529):eabf4063. doi:10.1126/science.abf4063
378. Purton JF, Tan JT, Rubinstein MP, Kim DM, Sprent J, Surh CD. Antiviral CD4+ memory T cells are IL-15 dependent. *J Exp Med*. 2007;204(4):951-961. doi:10.1084/jem.20061805
379. Murali-Krishna K, Lau LL, Sambhara S, Lemonnier F, Altman J, Ahmed R. Persistence of memory CD8 T cells in MHC class I-deficient mice. *Science*. 1999;286(5443):1377-1381. doi:10.1126/science.286.5443.1377
380. Kunzli M, Schreiner D, Pereboom TC, et al. Long-lived T follicular helper cells retain plasticity and help sustain humoral immunity. *Sci Immunol*. 2020;5(45):eaay5552. doi:10.1126/sciimmunol.aay5552
381. Romagnoli PA, Fu HH, Qiu Z, et al. Differentiation of distinct long-lived memory CD4 T cells in intestinal tissues after oral Listeria monocytogenes infection. *Mucosal Immunol*. 2017;10(2):520-530. doi:10.1038/mi.2016.66
382. Beura LK, Fares-Frederickson NJ, Steinert EM, et al. CD4(+) resident memory T cells dominate immunosurveillance and orchestrate local recall responses. *J Exp Med*. 2019;216(5):1214-1229. doi:10.1084/jem.20181365

383. Malhotra D, Burrack KS, Jenkins MK, Frosch AE. Antigen-specific CD4(+) T cells exhibit distinct kinetic and phenotypic patterns during primary and secondary responses to infection. *Front Immunol.* 2020;11:2125. doi:10.3389/fimmu.2020.02125
384. Sallusto F, Lenig D, Forster R, Lipp M, Lanzavecchia A. Two subsets of memory T lymphocytes with distinct homing potentials and effector functions. *Nature.* 1999;401(6754):708-712. doi:10.1038/44385
385. Schenkel JM, Masopust D. Tissue-resident memory T cells. *Immunity.* 2014;41(6):886-897. doi:10.1016/j.immuni.2014.12.007
386. Tubo NJ, Fife BT, Pagan AJ, Kotov DI, Goldberg MF, Jenkins MK. Most microbe-specific naive CD4(+) T cells produce memory cells during infection. *Science.* 2016;351(6272):511-514. doi:10.1126/science.aad0483
387. He J, Tsai LM, Leong YA, et al. Circulating precursor CCR7(lo)PD-1(hi) CXCR5(+) CD4(+) T cells indicate Tfh cell activity and promote antibody responses upon antigen reexposure. *Immunity.* 2013;39(4):770-781. doi:10.1016/j.immuni.2013.09.007
388. Wojciechowski S, Jordan MB, Zhu Y, White J, Zajac AJ, Hildeman DA. Bim mediates apoptosis of CD127(lo) effector T cells and limits T cell memory. *Eur J Immunol.* 2006;36(7):1694-1706. doi:10.1002/eji.200635897
389. Prlic M, Bevan MJ. Exploring regulatory mechanisms of CD8+ T cell contraction. *Proc Natl Acad Sci U S A.* 2008;105(43):16689-16694. doi:10.1073/pnas.0808997105
390. Jatzek A, Tejera MM, Singh A, Sullivan JA, Plisch EH, Suresh M. p27(Kip1) negatively regulates the magnitude and persistence of CD4 T cell memory. *J Immunol.* 2012;189(11):5119-5128. doi:10.4049/jimmunol.1201482
391. Garrod KR, Moreau HD, Garcia Z, et al. Dissecting T cell contraction *in vivo* using a genetically encoded reporter of apoptosis. *Cell Rep.* 2012;2(5):1438-1447. doi:10.1016/j.celrep.2012.10.015
392. Xu X, Araki K, Li S, et al. Autophagy is essential for effector CD8(+) T cell survival and memory formation. *Nat Immunol.* 2014;15(12):1152-1161. doi:10.1038/ni.3025
393. Harada H, Grant S. Apoptosis regulators. *Rev Clin Exp Hematol.* 2003;7(2):117-138.
394. Buck MD, Sowell RT, Kaech SM, Pearce EL. Metabolic instruction of immunity. *Cell.* 2017;169(4):570-586. doi:10.1016/j.cell.2017.04.004
395. Hildeman DA, Mitchell T, Teague TK, et al. Reactive oxygen species regulate activation-induced T cell apoptosis. *Immunity.* 1999;10(6):735-744. doi:10.1016/s1074-7613(00)80072-2
396. Belikov AV, Schraven B, Simeoni L. T cells and reactive oxygen species. *J Biomed Sci.* 2015;22:85. doi:10.1186/s12929-015-0194-3
397. Case AJ, McGill JL, Tygrett LT, et al. Elevated mitochondrial superoxide disrupts normal T cell development, impairing adaptive immune responses to an influenza challenge. *Free Radic Biol Med.* 2011;50(3):448-458. doi:10.1016/j.freeradbiomed.2010.11.025
398. Mak TW, Grusdat M, Duncan GS, et al. Glutathione primes T cell metabolism for inflammation. *Immunity.* 2017;46(6):1089-1090. doi:10.1016/j.immuni.2017.06.009
399. Hand TW, Morre M, Kaech SM. Expression of IL-7 receptor alpha is necessary but not sufficient for the formation of memory CD8 T cells during viral infection. *Proc Natl Acad Sci U S A.* 2007;104(28):11730-11735. doi:10.1073/pnas.0705007104
400. Reiner SL, Adams WC. Lymphocyte fate specification as a deterministic but highly plastic process. *Nat Rev Immunol.* 2014;14(10):699-704. doi:10.1038/nri3734
401. Verbist KC, Guy CS, Milasta S, et al. Metabolic maintenance of cell asymmetry following division in activated T lymphocytes. *Nature.* 2016;532(7599):389-393. doi:10.1038/nature17442
402. Finkin S, Hartweger H, Oliveira TY, Kara EE, Nussenzweig MC. Protein amounts of the MYC transcription factor determine germinal center B cell division capacity. *Immunity.* 2019;51(2):324.e5-336.e5. doi:10.1016/j.immuni.2019.06.013
403. Buchholz VR, Flossdorf M, Hensel I, et al. Disparate individual fates compose robust CD8+ T cell immunity. *Science.* 2013;340(6132):630-635. doi:10.1126/science.1235454

CHAPTER 29

Regulatory T Cells

Ethan M. Shevach, MD

INTRODUCTION

T cells are crucial in the immune response because they can function as both effector cells in cell-mediated responses and as helper cells in both humoral and cell-mediated responses. Most biologic systems are subject to complex regulatory controls, and the immune system is not an exception. In addition to T cells that upregulate (help), other populations that downregulate (suppress) the immune response must exist. Once a normal immune response is initiated by antigenic stimulation, mechanisms must be in place to control the magnitude of that response and to terminate it over time. Downregulation should contribute to the homeostatic control of all immune responses serving to limit clonal expansion and effector cell activity in response to any antigenic stimulus. An active mechanism of T cell suppression is also needed to control potentially pathogenic autoreactive T cells. The primary mechanism that leads to tolerance (See Chapter 33 for in-depth discussion of Tolerance) to self-antigens is thymic deletion of autoreactive T cells but some autoreactive T cells may escape thymic deletion or recognize antigens that are expressed only extrathymically. T cell anergy[1] and T cell ignorance/indifference[2] have been proposed as the primary mechanisms used to control autoreactive T cells in the periphery, although these "passive" mechanisms for self-tolerance may not be sufficient to control potentially pathogenic T cells.

It was proposed more than 50 years ago that a distinct subset of T cells is responsible for immune suppression.[3] A suppressor T cell is functionally defined as a T cell that inhibits an immune response by influencing the activity of another cell type. Although a strong theoretical basis exists for T cell-mediated suppression, this area of immunologic research for many years was plagued by controversy. The past 2 decades have seen a tremendous resurgence of interest in the concept of T cell suppression mediated by a distinct subset of T cells that are uniquely equipped to mediate suppressor activity and are characterized primarily by expression of the transcription factor, Foxp3.

HISTORICAL PERSPECTIVE

Gershon and Kondo[3,4] first identified suppressor T cells during studies designed to understand the process of "high-zone" tolerance. Injection of supraoptimal doses of an antigen including sheep red blood cells (SRBC) resulted in specific tolerance or nonresponsiveness to subsequent challenge with that antigen. It was believed at that time that the antibody-producing B cell was rendered nonresponsive by exposure to the high concentration of antigen. To investigate whether B cell tolerance was dependent on the presence of T cells, Gershon and Kondo injected high doses (2.5×10^{10}) of SRBC into thymectomized (Tx), irradiated, bone-marrow reconstituted mice, and then assayed the functional status of B cells from these mice by a secondary challenge with SRBC in the presence of added thymocytes as a source of T cell help. Surprisingly, nonresponsiveness as measured by deficient antibody production was only induced in the B cells of animals that had received thymocytes as well as bone marrow cells during the initial exposure to high dose antigen but not in mice that received bone marrow alone (Fig. 29.1). This result fulfilled Gershon's prediction that under certain conditions antigen seen by T cells can induce not only helper and effector cells but also cells that are able to suppress immune responses. Furthermore, when spleen cells from tolerized animals were transferred into secondary recipients together with normal thymocytes and bone marrow cells, they were capable of suppressing the otherwise competent response of these animals to SRBC. This suppression, or "infectious tolerance," as it was originally termed, was antigen-specific, as the immune response to an unrelated antigen, horse RBC, was not inhibited.[4] T cells were necessary for the induction of B cell tolerance, and these T suppressor cells were assumed to be a distinct cell population with a fully differentiated gene program that allowed them to perform a very specialized function. Other studies[5-7] during the 1970s supported the existence of T cell-mediated suppression. Most of the early studies in these models demonstrated that the T cells mediating suppression were distinct from T cells mediating help because the former were CD8+, while the latter were CD8−. The finding that T suppressor effector cells were CD8+ distinguished them from helper cells but did not allow them to be distinguished from cytotoxic cells, which are also CD8+. It remained possible that suppressor T cells were actually cytotoxic cells that killed the helper or effector T cells. CD8+ suppressor cells were also claimed to bind antigen in the absence of major histocompatibility complex (MHC) molecules.

Research in this area rapidly shifted from studies of the function of intact T cells to studies of their soluble products.[8] By the late 1970s, soluble factors produced by T suppressor cells were described by several groups and cloned T cell hybridomas that produced such factors were generated. T cell suppression was regarded as being mediated by numerous soluble antigen-specific and nonspecific factors that comprised a functionally unique network.[9,10] These elaborate highly convoluted suppressor cell pathways and circuits fell out of favor in the mid-1980s. No studies were ever performed that convincingly

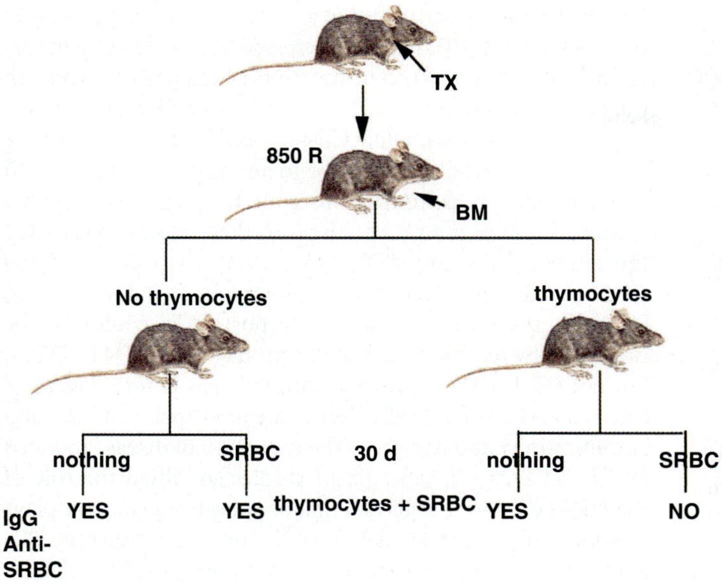

FIG. 29.1. Original experimental protocol used to identify suppressor T cells. See text for details. (Adapted from Gershon RK, Kondo K. Cell interactions in the induction of tolerance: the role of thymic lymphocytes. *Immunology*. 1970;18:723-735.)

characterized at the molecular level any of the suppressor T cell factors. These studies, together with the inability to identify a marker specific for suppressor T cells and the inability to purify suppressor T cells, raised considerable doubts about the existence of a distinct functional lineage of suppressor T cells.

Two completely different approaches to the demonstration of the importance of T regulatory (T_{reg}) or suppressor T cells in the prevention of organ-specific autoimmunity were also developed in the 1970s. In one, mice that were Tx on the third day of life (d3Tx) were shown to develop organ-specific autoimmune diseases (Fig. 29.2). The specific disease that developed varied with the strain of mouse under study; more than one organ could be involved in a given mouse. Most importantly, autoimmunity was not seen if the mouse was Tx on day 1 or day 7 of life, and disease could be completely prevented if the d3Tx mouse received a thymus transplant between days 10 and 15 of life.[11] These observations led to the hypothesis that autoreactive T cells were exported from the thymus during the first 3 days of life and that somewhat later in ontogeny, a population of suppressor cells emigrated from the thymus that controlled the autoreactive T cells. Removal of the thymus before the suppressor cells reached the periphery resulted in autoimmune disease. A number of other protocols (Table 29.1) that induced a lymphopenic state also resulted in the development of organ-specific autoimmunity. It was believed that these procedures resulted in a selective depletion of suppressor T cells while leaving the autoreactive effector populations intact. Subsequent studies demonstrated that the effector cells in this model were CD4+ T cells. The suppressor T cells were also CD4+ cells, and the development of autoimmune disease could be prevented by reconstitution of the d3Tx animals with peripheral CD4+ T cells from normal adult mice.[12] A second approach to define the role of T regulatory cells (T_{reg}) in the control of autoimmunity was described by Penhale and coworkers in the 1970s.[13,14] They devised a procedure to deplete T_{reg} from adult animals while leaving the

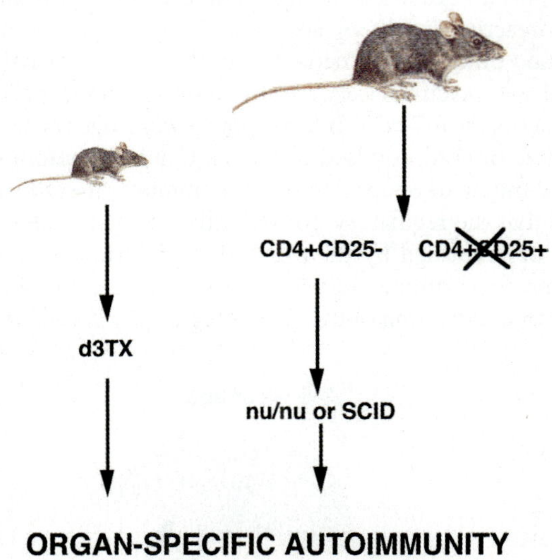

FIG. 29.2. Depletion of CD4+ CD25+ T cells results in organ-specific autoimmunity.[11,17]

helper population responsible for autoantibody production intact. Spontaneous thyroiditis and circulating IgG autoantibodies developed in 60% of rats following the selective depletion of T cells by adult Tx followed by irradiation. Tx was performed between 3 and 5 weeks of age, and the rats were then given 4 to 5 repeated doses of 200 rad at 14-day intervals (Fig. 29.3). The active role of T cells in preventing the development of autoimmunity in this model was confirmed by reconstituting, shortly after the final dose of irradiation, the Tx-irradiated mice with lymphoid cells from normal donors. Penhale and colleagues[15] also demonstrated that autoimmune diabetes would develop following the Tx-irradiation protocol in a strain of rats that was normally not susceptible to this disease. Taken together, the d3Tx model in the mouse and the Tx-irradiation model in the rat demonstrated that normally

TABLE 29.1 Organ-specific Autoimmune Disease and Lymphopenia

- 3dTx
- Neonatal administration of cyclosporine
- Thymectomy + repeated low-dose irradiation
- High-dose fractionated total lymphoid irradiation
- Adult thymectomy + cyclophosphamide
- Single TCR α-chain mice
- Transfer of T cells to T cell–deficient mice

Data from Sakaguchi S, Sakaguchi N. *Thymus, T cells, and autoimmunity: various causes but a common mechanism of autoimmune disease.* In: *Autoimmunity: Physiology and Disease.* Wiley-Liss, Inc.; 1994:203-227.

autoreactive helper and suppressor cells may coexist and that certain autoimmune responses are held in check by the equilibrium favoring suppressor activity.

FOXP3+ T_{REG}

Identification of CD4+CD25+ T_{reg}

An important extension of this hypothesis was that inhibition of autoreactive T cells by suppressor T cells was not a phenomenon unique to the neonate, but that in the normal adult animal, autoreactive T cells are also under the constant control of the suppressor T cells. If the suppressor lineage was deleted, damaged, or compromised in the adult animal, autoimmune disease might develop. Although a number of studies suggested that the regulatory T cells in the normal adult animal might be identified by the expression of certain membrane antigens, for example, high levels of CD5,[16] a major advance in our understanding of the role of regulatory T cells was the demonstration by Sakaguchi et al,[17,18] that a minor population of CD4+ T cells (10%) that coexpressed the CD25 antigen (the IL-2R α-chain) appeared to function as regulatory T cells in the normal adult. When CD25+ T cells were depleted from a population of normal adult CD4+ T cells and the remaining CD4+CD25− T cells transferred to an immunocompromised recipient such as a *nu/nu* mouse, the recipients developed a spectrum of autoimmune diseases that closely resembled those seen following d3Tx (Fig. 29.2). Cotransfer of the CD25+ cells prevented the development of autoimmunity. Similarly, the induction of disease post-d3Tx could also be prevented by reconstitution of the animals with CD4+CD25+, but not CD4+CD25−, normal adult T cells before day 14 of life.[19] CD8+CD25− T cells alone were not capable of inducing autoimmunity and enhanced the induction of disease induced by CD4+CD25− T cells. These studies solidified the role of the CD4+CD25+ T cells as a major subset of cells that plays a unique role in the regulation of the immune response. The autoimmune diseases induced by depletion of CD4+CD25+ T cells are uniformly accompanied by the development of organ-specific autoantibodies, suggesting that this mode of loss of T cell tolerance also results in the breakdown of B cell tolerance as well. It is likely that the activated self-reactive T helper cells provide signals to self-reactive B cells, rescue them from apoptosis, and stimulate autoantibody production.

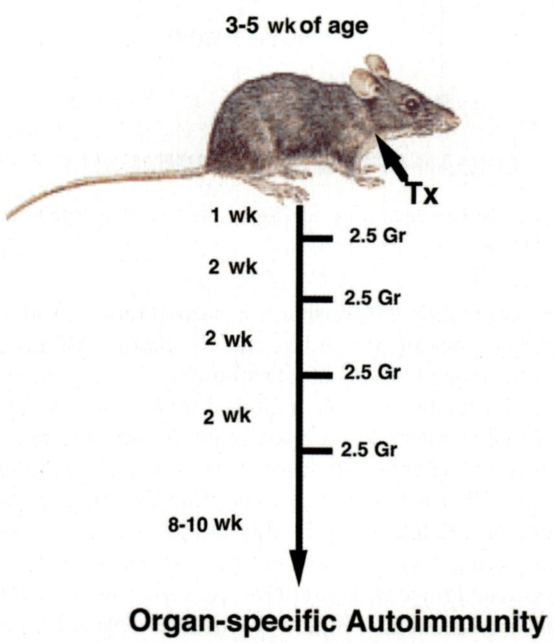

FIG. 29.3. Induction of organ-specific autoimmunity in rats by thymectomy and irradiation. (From Penhale WJ, Farmer A, Irvine WJ. Thyroiditis in T cell-depleted rats: influence of strain, radiation dose, adjuvants and antilymphocyte serum. *Clin Exp Immunol.* 1975;21:362-375.)

Powrie and Mason[20] were the first to identify cell surface markers that distinguished between regulatory and effector T cell populations in the rat. When athymic rats were reconstituted with small numbers of CD4+CD45RC^{high} T cells, they developed a severe wasting disease characterized by extensive mononuclear cell infiltration in the lungs, liver, thyroid, stomach, and pancreas 6 to 10 weeks later. No pathology developed in rats that received unseparated CD4+ T cells or CD45RC^{low} cells. It seemed likely from these studies that the CD45RC^{low} subset controlled the capacity of the RC^{high} subset to mediate the wasting disease. Fowell and Mason[21] directly demonstrated the suppressive effect of the RC^{low} subset in the Tx-irradiation model developed by Penhale. Transfer of RC^{low} CD4+ T cells completely inhibited the development of diabetes and insulitis. RC^{low} T cell from long-term Tx donors could protect as efficiently as cells from normal donors, demonstrating that the regulatory T cell is long-lived in the periphery. Subsequently, CD4+CD45RB^{low} subset in the mouse was shown to have regulatory properties similar to the CD4+CD45RC^{low} subset of the rat.[22] Taken together, these studies in mouse and rat model systems demonstrated for the first time that two well-characterized cell-surface antigens (CD25 and CD45RC^{low}/CD45RB^{low}) could be used to identify suppressor CD4+ T cell subpopulations present in normal animals. In later studies,[23] it was shown that the CD25+ T within the CD4+CD45RB^{low} population mediated their suppressor activity.

CD4+CD25+ T cells typically represent 5% to 8% of the total population of T cells in the normal mouse lymph node (LN), or 10% to 15% of mouse CD4+ T cells (Fig. 29.4). CD4+CD25+ T cells can also be found in the thymus where they also represent about 5% to 10% of the mature CD4+CD8− population, or 0.5% of mouse thymocytes.[17,24] A population with an identical phenotype has been identified in the rat, in man,[25-32] and many other species.[33] When compared directly to CD4+CD25− T

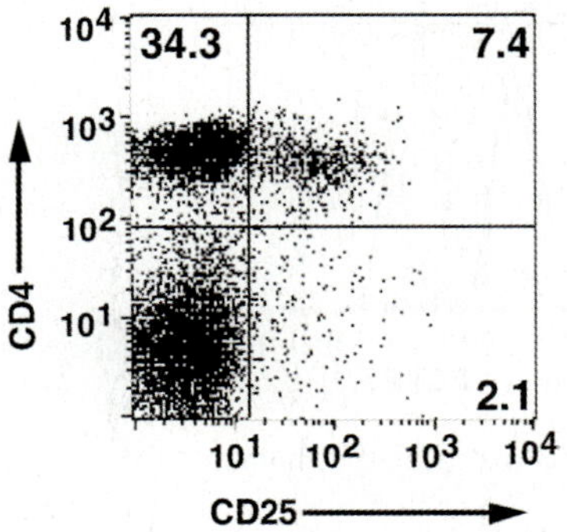

FIG. 29.4. Identification of CD4+ CD25+ T cells in normal mouse lymph node. (Adapted with permission of Rockefeller University Press, from Thornton AM, Shevach EM. CD4+CD25+ immunoregulatory T cells suppress polyclonal T cell activation *in vitro* by inhibiting interleukin 2 production. *J Exp Med.* 1998;188:287-296; permission conveyed through Copyright Clearance Center, Inc.)

cells, CD4+CD25+ T cells express slightly higher levels of CD5, have a slightly higher proportion of CD62L^{low} cells, and have a higher proportion of CD69+ cells. They express both intermediate and low levels of CD45RB and are completely absent from the CD45RB^{high} population.[34] All of the CD4+CD25+ T cells express the T cell receptor (TCR αβ) receptor and are NK 1.1 negative. One other unique property of CD25+ T cells in both mouse and man is that they are the only nonactivated T cell population that expresses high levels of the CTLA-4 antigen intracellularly.[23,35,36] The glucocorticoid-induced TNF-like receptor (GITR, TNFRSF18) has also been shown to be expressed on the majority of resting CD4+CD25+ T cells and to be expressed at very low levels on CD4+CD25− T cells.[37,38] The GITR is upregulated on CD4+CD25− T cells following TCR-mediated activation.

Foxp3 as a T_{reg} Marker

A major advance in the study of T_{reg} was derived from two related experiments of nature. Although CD4+CD25+ T cells were shown to display suppressive properties in multiple disease models, the value of CD25 as a marker is limited as CD25 is highly expressed on both activated CD4 and CD8 T cells, compromising its usefulness in studying T_{reg} cells in settings of immune activation. The X-chromosome–encoded forkhead transcription factor, Foxp3, was identified as the key player in the biology of CD4+CD25+ T_{reg}.[39-42] Young males with the IPEX syndrome (immune dysregulation, polyendocrinopathy, enteropathy, X-linked) or a mutant mouse strain, scurfy, both succumb to similar autoimmune and inflammatory diseases as a result of uncontrolled activation and expansion of CD4+ T cells. Both IPEX patients and scurfy mice have mutations in a common gene, *Foxp3*, which encodes a forkhead-winged-helix transcription factor. There is an excellent correlation between expression of Foxp3 and CD25, but a minor population (~10%) of Foxp3+ cells is CD25− and about 10% of CD25+ cells are Foxp3− (Fig. 29.5); the latter represent activated effector cells. Expression of Foxp3 only in the thymus using a proximal lck-driven transgene does not prevent disease in scurfy mice. Foxp3 expression in peripheral T cells is thus required for maintenance of T_{reg} function. Retroviral-mediated transduction of Foxp3 expression in naïve T cells can convert these cells to a regulatory phenotype functioning *in vitro* in a manner similar to CD4+CD25+ T_{reg}. Furthermore, Foxp3-transduced naïve CD25− T cells also manifest suppressor activity *in vivo* and can inhibit the weight loss, diarrhea, and histological development of colitis induced by transfer of CD25−CD45RB^{high} cells as effectively as T_{reg}.

Fontenot et al[43] generated Foxp3-deficient (−/−) mice and demonstrated that Foxp3 is specifically required for the thymic development of T_{reg}. Mixed bone marrow chimeras using wild type (WT) and Foxp3−/− bone marrow demonstrated that Foxp3−/− T cells behaved normally in the presence of WT Foxp3+ T_{reg}. Thus, the lethal autoimmune syndrome in Foxp3−/− mice results from a deficiency of T_{reg} and not from a cell-intrinsic defect of Foxp3− T cells. Further study of the role of Foxp3 in T_{reg} function has been greatly facilitated by the development of strains of mice in which a fluorescent protein such as GFP has been "knocked in" to the Foxp3 locus, permitting ready identification and isolation of Foxp3+ T_{reg} by cell sorting.[44] Studies of T_{reg} development in the thymus of the Foxp3^{gfp} mice demonstrated that less than 0.1% of the CD4+CD8− thymocytes expressed Foxp3 within 12 hours after birth and that the percentage of Foxp3+ CD4 SP thymocytes increased slowly over the following days and reached a plateau of ~4% at ~21 days after birth. The largest single-day change in the percentage of Foxp3 expressing SP occurred between days 3 and 4. 80% of the Foxp3+ T cells on day 1 were CD4+CD8− SP cells. This result correlates with the early studies on the d3Tx mice that suggested that the generation of T_{reg} cells is delayed relative to generation of nonregulatory CD4+ T cells. Foxp3 was not expressed in nonhematopoietic tissues.

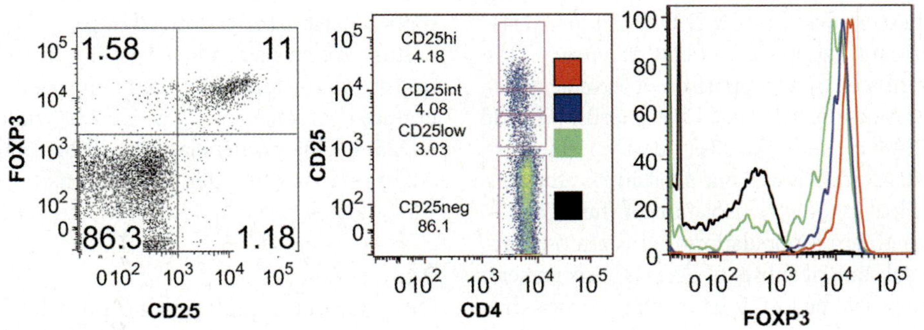

FIG. 29.5. Correlation between CD25 expression and Foxp3 expression in mouse CD4+ T cells.

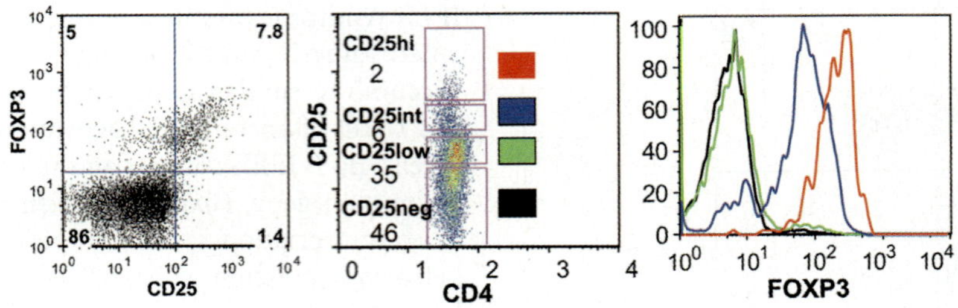

FIG. 29.6. Correlation between CD25 expression and Foxp3 expression in human CD4+ T cells.

Foxp3 expression in peripheral TCR αβ⁺ T cells irrespective of CD25 expression correlates with suppressor activity.

Direct proof that Foxp3+ T cells maintain control of autoreactive T cells in the periphery has been shown by using a targeting construct encoding the human diphtheria toxin receptor fused to sequences encoding GFP and equipped with an internal ribosome entry site into the 3′ untranslated region of Foxp3.[45] Complete elimination of Foxp3+ cells was achieved after 7d of treatment with diphtheria toxin. Foxp3 elimination at birth led to a syndrome very similar to that seen in Foxp3−/− mice. T_{reg} cell elimination in adult nonlymphopenic mice resulted in an even more rapid development of terminal autoimmune disease than in neonates. These studies demonstrate that T cells, probably expressing self-reactive TCRs, are targets of continuous T_{reg}-mediated suppression. After T_{reg} cell ablation, those T cells become activated, produced secreted and membrane-bound cytokines, and facilitated dendritic cell (DC) maturation. At day 7 after ablation, increased numbers of B cells, macrophages, Gr-1⁺ granulocytes, natural killer (NK) cells, and a 10-fold increase in the absolute numbers of DC. Thus, DC dysregulation is a general phenomenon resulting from the absence of T_{reg} cells. Furthermore, Foxp3-independent recessive and dominant tolerance mechanisms established in adult mice are not sufficient to protect mice from fatal autoimmunity after elimination of T_{reg}. Even a partial decreased level of Foxp3 protein in T_{reg} cells as a result of induced dysregulation of Foxp3 gene transcription results in marked attenuation of T_{reg} suppressor function.[46]

While Foxp3 is a robust marker for T_{reg} in the mouse, thymic-derived and -induced T_{reg} cells, human CD4+ T cells transiently express Foxp3 during activation.[47-49] Isolation of human T_{reg} continues to be a problematical as one must rely on expression of cell surface antigens. Even after 2 decades of study, no single cell surface antigen has been identified which uniquely identifies Foxp3+ T_{reg} in either mouse or man. It was reported that among T cells, only the highest expressors of CD25 (~2% of CD4+ T cells) exerted significant suppressive effects *in vitro*.[25] This observation was substantiated by a comparison of Foxp3 expression with CD25 expression on human CD4+ T cells. Almost all human CD4+CD25^{high} cells are Foxp3+, but a variable percentage of CD25^{int} cells express lower but substantial amounts of Foxp3 (Fig. 29.6). Flow cytometric isolation of the CD25^{high} cells does allow one to obtain a population that is almost uniformly Foxp3+, but a substantial subset of Foxp3+ cells present in the CD25^{int} population will be lost. It has been proposed that almost all of human Foxp3+ T cells can be identified as expressing low levels of the IL-7 receptor (CD127) and that CD127^{low} expression can be used to isolate human T_{reg}.[50] However, CD127 is downregulated early in the course of T cell activation, so the CD127^{low} phenotype is also unlikely to be T_{reg}-specific during an ongoing immune or inflammatory response. In addition, only about 40% of the CD127^{low} population is Foxp3+ and even purified CD4+CD127^{low}CD25+ cells were only 85% to 90% Foxp3+. Nevertheless, the differential expression of CD127 has proven useful for the isolation of human T_{reg} (Fig. 29.7). Miyara et al[51] have proposed that human T_{regs} can be divided into three main fractions (Fig. 29.8): Fr I, CD45RA+, Foxp3^{lo}/CD25^{lo} resting or naïve T_{regs}; Fr II, CD45RA−FOXP3^{high}CD25^{high} effector T_{regs}; and Fr III, CD45RA−Foxp3^{low}/CD25^{low} cells, the majority of which are not T_{reg}. Fr I T_{regs} are stable T_{regs} express a demethylated TSDR (Section IV. A), modest suppressive capacity, express various naive T cell–associated cell surface markers such as CCR7 and CD62L, and are relatively enriched in the thymic emigrant marker CD31. Upon TCR stimulation, Fr. I naïve T_{regs} differentiate into Fr. II effector T_{regs} with downregulation of CD45RA and upregulation of CD45RO, upregulation of Foxp3, CD25, and CTLA-4, completely demethylated TSDR, and potent suppressor activity. The major problem with this model is that it is difficult to clearly distinguish between Fr. II and Fr. III. While it has been proposed that the majority of Fr II T_{reg} are activated T_{conv}, this fraction may also contain bona fide T_{reg} with demethylated TSDR[52] as well as CXCR5+ T follicular regulatory (Tfr) cells.[53]

Since the number of Foxp3+ T cells is remarkably constant in normal animals in the absence of perturbation of the immune system, they have been frequently referred to as thymic-derived, "naturally occurring" T_{reg},[54] while Foxp3+ T cells that are generated extrathymically are termed "adaptive or induced" T_{reg}. We recommend the nomenclature proposed by a group of investigators in the field.[55] T_{reg} that have been definitively demonstrated to be generated in the thymus should be referred to as thymic-derived or tT_{reg}, while T_{reg} that have been shown to be generated at peripheral sites should be named peripherally induced or pT_{reg}. T_{reg} that are generated in culture are named induced or iT_{reg}. If the site of generation is unknown, the term "T_{reg}" might be used. Some of the older literature characterizes T_{reg} as CD25+ while non-T_{reg} are termed CD25−. This nomenclature will be retained in this Chapter when referring to older publications.

pT_{reg} and iT_{reg}

One important question that must be addressed is whether the thymus is the only site for the generation of Foxp3+ T_{reg}. A number of studies performed both *in vivo* and *in vitro*

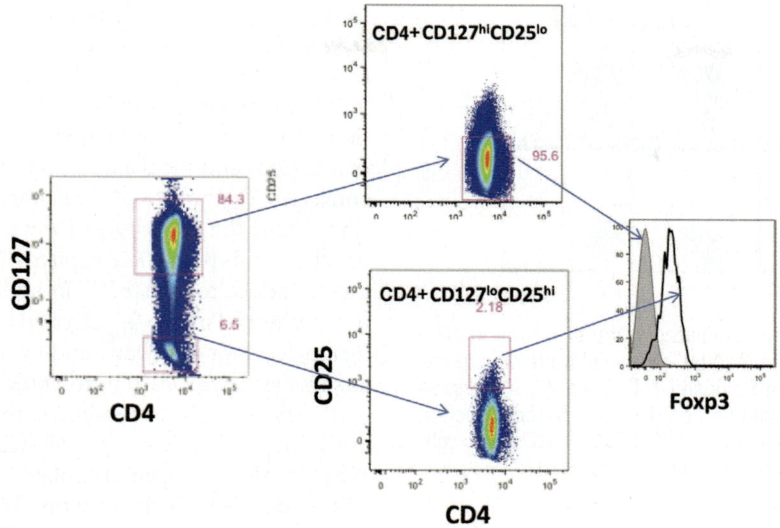

FIG. 29.7. The combination of $CD25^{hi}$ and $CD127^{lo}$ is the most practical for isolation of human T_{reg}.

have demonstrated that under certain conditions Foxp3+ T_{reg} can be generated extrathymically. One highly effective method for the induction of T_{reg} with antigen was to expose TCR-transgenic T cells on a RAG−/− background *in vivo* to a continuous supply of subimmunogenic doses of agonist peptides delivered via a miniosmotic pump. Many of the T cells became Foxp3+ and resembled thymic-derived T_{reg} in all their phenotypic and functional properties.[56,57] It was postulated from these studies that low doses of peptide presented over relatively long periods of time on nonactivated DC represented an important mechanism by which T_{reg} could be generated in the periphery particularly to self-components that do not lead to tolerance in the thymus. Foxp3+ T cells were also induced when mice were treated with peptide agonist ligands targeted to DC in minute doses by coupling them to antibodies directed to DC antigens under conditions of suboptimal DC activation.[58] T_{reg} induced with subimmunogenic doses could subsequently be expanded by delivery of antigen in immunogenic conditions. DC-targeted de novo generation

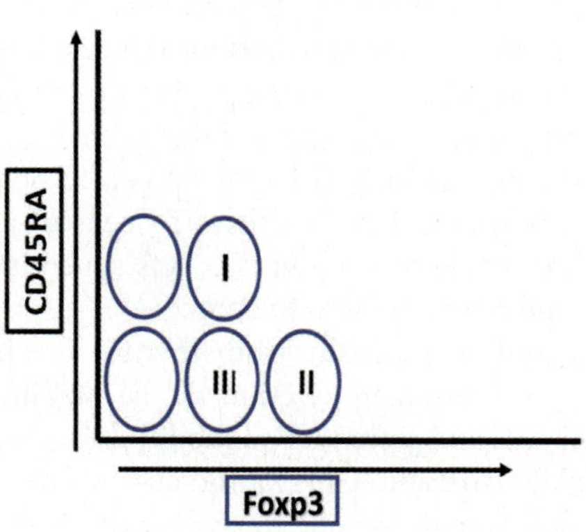

FIG. 29.8. Subsets of human T_{reg}. (Adapted from Miyara M, Yoshioka Y, Kitoj A, et al. Functional delineation and differentiation dynamics of human CD4(+) T cells expressing the FoxP3 transcription factor. *Immunity.* 2009;30:899-911, with permission from Elsevier.)

in vivo results in efficient demethylation of conserved CpG motifs within the TSDR (conserved noncoding sequences [CNS2]) region (Section IV. A) of the Foxp3 locus, a predictive parameter for long-term stability of induced Foxp3 expression.[59]

One interpretation of the studies described above is that weak TCR stimulation is favorable for the peripheral induction of Foxp3 and that robust TCR signaling during T_{reg} cell generation stimulated the Akt-PI3K-mTOR pathway that antagonizes the induction of Foxp3[60] and results in cell cycle–dependent maintenance of a silenced state of the Foxp3 locus.[61] Several studies have examined the role of the strength of the antigenic signal required for the induction of Foxp3+ T cells in the periphery.[62] Foxp3+ cells induced by weak agonist peptides did not persist compared to cells generated by a higher affinity ligand. There appear to be qualitative differences between weak and strong signals whereby a few strong signals may be different from many weaker signals. The few strong signals may specifically induce survival signals, while the many weaker signals may induce proapoptotic signals. There also appears to be unique subpopulation of recent thymic emigrants that are $CD25^{hi}CD62L^{int}$ CD69+ that are capable of becoming Foxp3+ upon exposure to IL-2 alone in the absence of TGF-β.[63]

A number of groups[64-67] have presented studies that demonstrate a pivotal role for TGF-β in the generation of Foxp3+ T_{reg} *in vitro*. TCR activation in the presence of TGF-β converted naïve T cells into Foxp3+ (Fig. 29.9) cells with an anergic/suppressor phenotype. IL-2 plays a nonredundant role in TGF-β–mediated induction of Foxp3 expression in mouse CD4+ T cells. Once induced, Foxp3 expression was maintained both *in vitro* and *in vivo* in the absence of IL-2. Other cytokines utilizing the common γ-chain as part of their receptors were unable to induce Foxp3 expression in IL-2−/− cells. When T cells are activated through the TCR in the presence of TGF-β, both RORγt and Foxp3 are transcribed and interact with each other. In the presence of IL-2, Foxp3 expression dominates leading to suppression of RORγt

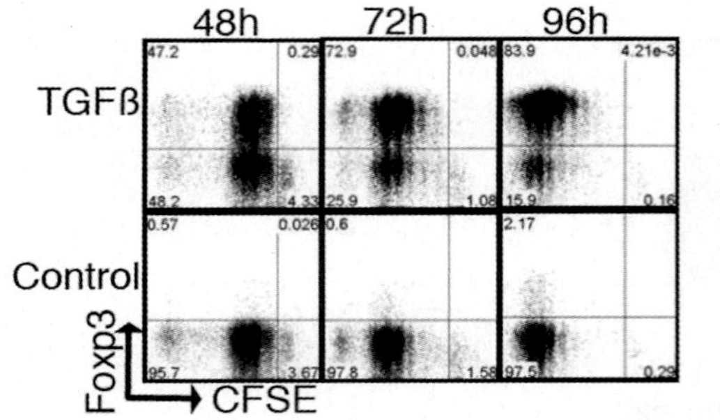

FIG. 29.9. Foxp3− T cells were stimulated in the presence of TGF-β. Foxp3 expression was determined by intracellular staining at the indicated times. (Adapted from Davidson T, DiPaolo RJ, Andersson J, et al. Cutting edge: IL-2 is essential for TGF-β-mediated induction of Foxp3+ regulatory cells. *J Immunol.* 2007;178:4027-4032. Copyright 2007. The American Association of Immunologists, Inc.)

transcription promoting T_{reg} development. In contrast, in the presence of IL-6 and low levels of IL-2, the transcription of RORγt is favored, which in turn represses Foxp3 transcription with resultant Th17 development.[68]

TGF-β–induced Foxp3+ T cells not only exhibit unresponsiveness to TCR stimulation but also suppress normal CD4+ T cell activation and Th1 and Th2 cytokine production *in vitro*. The induction of Foxp3 results in downregulation of Smad7 and renders T cells highly susceptible to the regulatory effects of TGF-β signaling via Smad3/4.[65] Foxp3+ T cells induced *in vitro* can suppress antigen-specific proliferation of CD4+ T cells *in vivo* and prevent house dust mite-induced mouse asthma.[64] One of the major factors preventing induction of Foxp3 expression *in vitro* is the presence of IFN-γ with subsequent STAT1 signaling.[69]

Mouse iT_{reg} exhibits potent suppressor activity *in vitro* and *in vivo*.[70] It remains controversial if human T cells induced to express Foxp3 by activation in the presence of TGF-β acquired regulatory function.[49] It has recently been shown that culture of naïve human T cells (CD4+CD25−CD45RA+) in the presence of TGF-β and rapamycin[71] or in the presence of TGF-β and retinoic acid (RA)[72] results in the generation of functional Foxp3+ human T_{regs} with increased suppressor activity *in vitro*. However, iT_{regs} generated by these protocols did not stably retain Foxp3 expression and did not suppress *in vivo* in a humanized graft-versus-host disease mouse model.[73]

The gut appears to be one of the major sites for the generation of T_{regs} in the periphery, and the gut contains a large number of Foxp3+ T_{reg}. Mucida et al[74] first demonstrated that repeated antigen-feeding induced antigen-specific Foxp3+ T_{reg} cells in the absence of thymic-derived T_{regs}. Converted cells were rapidly found in the mesenteric LN (MLN), Peyer patches (PP), and in the small intestine lamina propria (LP). Gut-associated lymphoid tissue has a number of primarily unique properties that facilitate the generation of Foxp3+ T cells. A vitamin A metabolite, RA, produced by gut DCs can selectively induce molecules such as CCR9 and the integrin, α4β7, on both conventional T cells and T_{reg} cells, and these cell surface antigens are involved in directing homing of cells to the gut. DCs from the LP of the small intestine have the unique ability to generate T_{reg} cells *in vitro* via a mechanism that involves the synergistic action of TGF-β and RA. A subpopulation (25%) of LP DCs express CD103, and this CD103+ subpopulation was responsible for the induction of Foxp3 in the absence of exogenously added TGF-β. Endogenous TGF-β was responsible for T_{reg} conversion, as anti–TGF-β largely abrogated T_{reg} induction. Addition of RA to splenic DC cultures enhanced T_{reg} conversion, and blockade of nuclear RA signaling decreased converted T_{reg}, but RA alone is not sufficient for Foxp3 induction. Thus, the capacity of the intestinal immune system to produce TGF-β and RA explains the unique capacity of the gut to induce tolerance.[75,76] In addition to the TGF-β induction pathway, Foxp3+ T_{reg} can also be induced via PD-L1-PD-1 ligand-receptor pathway[77] and by environmental factors such as some aryl hydrocarbon receptor agonists.[78]

It has recently been shown that chemical inhibition of cyclin-dependent kinases 8/19 (CDK 8/19) is able to induce Foxp3 both in antigen-stimulated effector cells and in naïve CD4+ and CD8+ cells *in vitro*. Administration of an inhibitor to CDK8/19 together with TCR-stimulation generated functionally stable Foxp3+ T cells that also expressed T_{reg} signature antigens including CTLA-4, PD-1, and GITR.[79,80] In animal models *in vivo*, these iT_{regs} ameliorated encephalomyelitis (EAE). The induction of Foxp3 expression in this model was independent of TGF-β. In contrast to TGF-β induction of Foxp3, inhibition of CDK8/19 could induce Foxp3 expression in activated/memory T cells. CDK8/19 inhibitors were also effective when administered *in vivo*, and the induced Foxp3+ T cells had stable T_{reg}-specific demethylation at the Foxp3 locus (Section IV. A).

One major issue that remains to be addressed is the relative contribution of thymic-derived versus peripherally induced T_{reg} to the establishment of immune tolerance. Although T_{regs} are induced in the colitis model from Foxp3− T cells, the induced T_{reg} are not sufficient to prevent disease.[81] Some studies suggest that only the combination of tT_{reg} and *in vitro* generated T_{reg} cells could fully protect mice from developing IBD induced by naïve T cells from Foxp3−/− donors. When scurfy mice were reconstituted with tT_{reg} alone, a marked inhibition of lethality was observed, but the treated mice still displayed signs of inflammation.[82] Surprisingly, the combined treatment with tT_{reg} and conventional Foxp3− T cells that could differentiate into Foxp3+ T cells decreased inflammation. *In vitro* generated T_{reg} cells could substitute for T_{reg} generated *in vivo* from Foxp3− conventional cells. One explanation for this result was that the TCR repertoires of the tT_{reg} and the pT_{reg} induced in the periphery were largely nonoverlapping and that each T_{reg} cell subset is responsible for recognizing different antigens. The major conclusions drawn from this study is that pT_{reg} cells are an essential, nonredundant regulatory subset that acts synergistically with tT_{reg} cells to enforce peripheral tolerance.

pT_{regs} may differ markedly from thymic T_{regs} in terms of TCR repertoire expression (Section III. E), stability, suppressor mechanisms, and potential suitability for *in vitro* expansion and cellular therapy. Thus far, no cell surface antigen has been identified that allows one to reliably distinguish thymic-derived T_{regs} from T_{regs} induced in peripheral sites. Some studies have suggested that expression of Helios, a member of the Ikaros transcription factor family, distinguishes thymic T_{regs} from peripheral T_{regs}.[83] Helios was expressed intracellulary in

100% of Foxp3+ mouse thymocytes but is only expressed by ~70% of Foxp3+ T cells in peripheral lymphoid tissues in both mouse and man (Fig. 29.10). Foxp3+ T cells induced *in vivo* by oral tolerance, following transfer of Foxp3− T cells to RAG−/− mice, or following induction by reconstitution of germ-free mice with bacteria[84,85] were primarily Helios−. These studies suggest that a significant proportion of Foxp3+ T_{regs} are generated extrathymically. Further studies are needed to validate the use of Helios expression as a marker for thymic-derived T_{regs}. One major drawback in the use of Helios to distinguish T_{reg} subpopulations is that Helios is expressed intracellularly and cannot be used for cell isolation. Some studies have suggested that the cell surface antigen neuropilin 1 (Nrp1) distinguishes tT_{regs} from pT_{regs} or iT_{regs}.[86,87] Nrp1 was found to be expressed on Foxp3+ cells from unmanipulated mice, but not on pT_{regs} induced *in vivo* by oral tolerance similar to what had been observed with the expression of Helios. Nrp1 and Helios expression correlated well in that Nrp1+ cells were also Helios+, but Foxp3+Helios− T cells contained a significant fraction of of Nrp1+ cells.[88] The expression of Nrp1 is also positively regulated by TGF-β and Nrp1 is also expressed on TGF-β–induced iT_{regs}.[87] Furthermore, under inflammatory conditions, Nrp1 is expressed at high levels on pT_{regs}. Overall, the ideal marker that distinguishes tT_{regs} and pT_{regs} has not been identified.

One possible mechanism to increase the suppressor capacity of Foxp3+ T_{reg} is that T_{reg} induces anergy or suppressor activity in the Foxp3− responders as a form of infectious immunological tolerance or that anergic T cells become T_{reg}.[89,90] It has been shown that human CD4+CD25+ T cells induce CD4+CD25− T cells to become T_{reg} that are capable of suppressing by producing IL-10[91] or TGF-β[92] and thereby mediating suppression of naïve CD4+ T cells. The responder cells remained Foxp3− and exerted suppressor activity in a cell contact-dependent manner that was partially abrogated by anti–TGF-β. One problem with the interpretation of these early studies is that it was difficult to determine if the

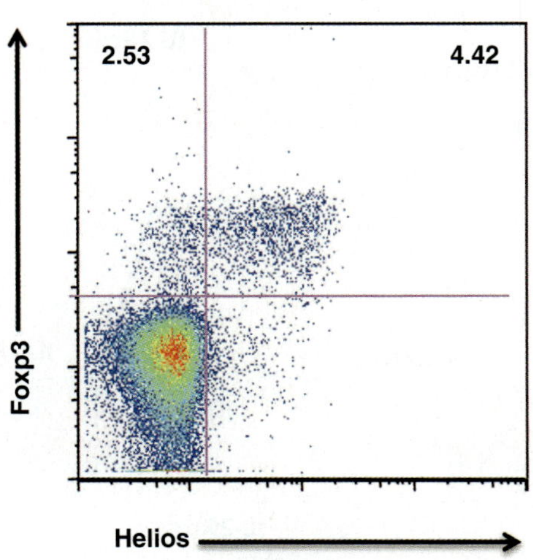

FIG. 29.10. ~70% of mouse Foxp3+ T cells express Helios. (Adapted from Thornton AM, Korty PE, Tran DQ, et al. Expression of Helios, and Ikaros transcription factor family member, differentiates thymic-derived from peripherally induced Foxp3+ T regulatory cells. *J Immunol.* 2010;184:3433-3441.)

regulatory populations induced by coculture with presumably Foxp3+ T cells actually were induced to express Foxp3. Other studies have directly demonstrated that activation of Foxp3− T cells in the presence of activated Foxp3+ T cells resulted in the induction of Foxp3 expression in the Foxp3− population.[93] The induction of Foxp3 expression was mediated by TGF-β produced by the Foxp3+ inducer population in a contact-dependent fashion. The newly induced Foxp3+ T cells mediated suppression both *in vivo* and *in vitro*. Foxp3+ T cells have also been shown to mediate infectious tolerance locally at the site of a transplanted skin allograft. The induced cells were suppressive and prevented potential effectors from rejecting grafts.[94]

Development of T_{reg} in the Thymus

The potential role of the thymus (see Chapter 28 for a complete discussion of T cell development in general) in the generation of regulatory T cells was first described several years ago.[95] Most studies strongly support the view that the tT_{reg} population is produced in the thymus as a functionally mature distinct T cell subpopulation. tT_{reg} are not derived from peripheral cells that have recirculated from the periphery to the thymus since T_{reg} developed *in vitro* in organ cultures of fetal thymus. Foxp3+ thymocytes are nonresponsive and suppress T cell activation *in vitro* in a manner similar to T_{reg} derived from the periphery.[24] The capacity of Foxp3+ T cells to migrate from the thymus to the periphery was documented by injection of FITC intrathymically. The percentage of Foxp3+ T cells within migrants and resident T cells was identical, suggesting that Foxp3+ T cells in the periphery can originate in the thymus.[96] Thymectomy at 4 to 5 weeks of age did not modify the number of Foxp3+ T cells in the periphery even when tested 19 months later.

The process of generating a diverse TCR repertoire leads to the development of many autoreactive T cells. It is widely accepted that conventional CD4+ and CD8+ T cells develop in the thymus by a process of positive and negative selection. Positive selection is mediated by interaction of developing T cells with MHC antigens on thymic cortical epithelium. Negative selection is mediated both by DC and thymic medullary epithelium. Since T_{reg} cells have an antigen-experienced phenotype in the absence of exposure to foreign antigens, it seems logical that their TCRs would be biased toward recognition of self-antigens. A number of studies have suggested that T_{reg} undergo a unique developmental process during their generation in the thymus. When TCR-transgenic mice bearing a TCR specific for a determinant (S1) derived from influenza hemagglutinin (HA) in association with I-E^d were crossed to mice expressing the HA transgene, the transgenic T cells were not deleted and a large proportion expressed CD25 and functioned as T_{reg}.[97] Radioresistant elements of the thymus were shown to be both necessary and sufficient for the selection of Foxp3+ T cells in these doubly transgenic mice. Similar results were obtained when TCR-transgenic mice on a RAG−/− background were mated to the HA-transgenic mice, clearly indicating that thymocytes that can only express a single transgenic TCR can undergo selection to become

T_{reg}. A second TCR-transgenic mouse was generated which expressed a variant determinant of HA but which recognized the S1 determinant with 100-fold less affinity. When these mice were bred to the HA-transgenic mouse (S1), they did not have an increased frequency of Foxp3+ T cells. Thus, thymocytes with a low-intrinsic affinity for the S1 peptide did not develop into Foxp3+ thymocytes in response to HA. These data are consistent with a model in which selection of T_{reg} that express a transgenic TCR depends on a high-affinity interaction of the TCR with its ligand. T_{reg} differentiation results from interactions that lie in between the signaling strength required on one side and clonal deletion on the other side. It could not be determined from these studies whether this selection process occurs on cortical or medullary epithelial cells (MEC).

A number of other models have been proposed[98] for T_{reg} differentiation in the thymus. Recent studies favor TCR specificity as playing a dominant role in thymic T_{reg} cell development.[99,100] Transgenic mice expressing TCRs derived from T_{reg} cells bore very few thymic Foxp3+ T cells in contrast to TCR-transgenic mice using foreign antigen-specific TCRs. The failure of these mice to develop significant numbers of Foxp3+ T cells suggested that developing T cells with the same antigenic specificity compete for a limited niche for T_{reg} cell development. In mixed bone marrow chimeras with varying proportions of polyclonal progenitors and T_{reg} cell-derived TCRs, a striking inverse correlation was observed between the percentage of TCR-transgenic cells and the propensity of these cells to give rise to Foxp3+ T cells. When TCR-transgenic cells accounted for 10% of total cells, 1% expressed Foxp3, but when TCR-transgenic cells were only 0.1% of the total cells, 30% expressed Foxp3. This result suggests the existence of a niche that limits T_{reg} cell development. Other studies have demonstrated efficient clonal deletion of CD4+Foxp3− T cells and the complete absence of Foxp3+ T cells in a transgenic mouse expressing a different T_{reg}-derived TCR transgene consistent with the view that T_{reg}-derived TCRs are self-reactive.[101] It is likely that both DCs and mTECS can facilitate T_{reg} development.[102]

The development of Foxp3+ thymocytes occurs in two stages (Fig. 29.11). A relatively small CD25+Foxp3− subset of CD4+ SP thymocytes is enriched in precursors of Foxp3+ T_{reg} cells. The first step is a TCR and CD28 driven instructive phase, while the second step is a cytokine driven consolidation phase.[103,104] Step one is driven by strong TCR stimulation in these CD4 SP thymocytes, which results in upregulation of the IL-2Rα, as well as multiple members of the TNFR superfamily including GITR, OX40, and TNFR2. In the second step, these T_{reg} progenitors are driven by cytokine-dependent conversion into mature T_{reg} via upregulation of Foxp3. These Foxp3− T_{reg} precursors require common γ-chain cytokines but no further TCR stimulation for acquisition of the mature Foxp3+ T_{reg} phenotype. The higher TCR self-reactivity of these T_{reg} precursors provides them with a selective advantage by allowing them to compete more effectively for IL-2, thereby biasing the T_{reg} cell repertoire toward self-reactivity. Analysis of reporter mice in which GFP is expressed coordinately with TCR signal strength demonstrated that T_{reg} cells were interacting more strongly with self-antigens.[105]

IL-2 and IL-15 are the major inducers of T_{reg} development, IL-7 has limited capacity, but IL-4 blocks T_{reg} cell development. Although multiple cytokines can induce Foxp3+ T_{reg} cell differentiation, they all do so via STAT5 activation. Activated STAT5 is likely to directly bind and regulate the Foxp3 gene at the promoter and at an intronic enhancer region.[104] STAT5−/− T cells are unable to differentiate into T_{reg} cells, but constitutive activation of STAT5 in STAT5b-CA transgenic mice results in a marked increase in T_{reg} cell development. An alternate pathway has defined a CD25−Foxp3^{lo} T_{reg} progenitor, but differentiation of these progenitors depends on the same two-step process.[106] It remains unclear

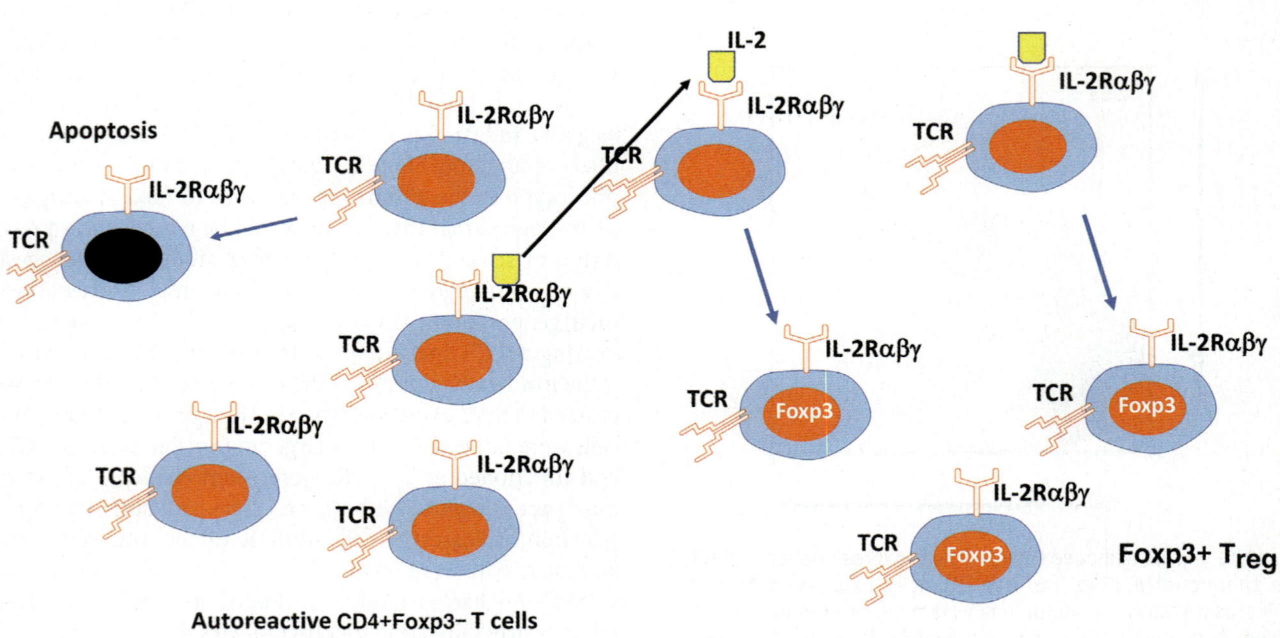

FIG. 29.11. Self-reactive T cells produce the IL-2 needed for T_{reg} differentiation.

how the CD25− Foxp3^{lo} T_{reg} progenitors are able to differentiate into mature T cells in response to low dose IL-2. In competitive intrathymic transfer studies, both T_{reg} precursor populations differentiated into mature T_{reg} cells with similar efficiency.[107] The TCR repertoires of these populations overlap with mature T_{reg} cells but much less so with each other. Surprisingly, T_{reg} cells derived from the CD25+ Foxp3^{lo} precursor population protected mice from EAE, while T_{reg} derived from CD25−Foxp3^{lo} precursors suppressed colitis.

The cellular source of cytokines needed for thymic T_{reg} development until recently has been incompletely understood, and both DCs and T cells have been implicated. The cellular sources of IL-7 and IL-15 are ill-defined. It has recently been shown that self-reactive CD4+CD25+Foxp3− SP thymocytes, the same subset that contains T_{reg} precursors, are the source of IL-2 that promotes T_{reg} cell generation.[108] T cells recirculating from the periphery are not the thymic IL-2 producers. The CD4+CD25+Foxp3− precursors had an active IL-2 locus as measured by Ataq-seq, and these may have received strong TCR stimulation during recognition of self-peptides that resulted in both IL-2 production and CD25 expression. Some of these cells become T_{reg} cells, some contribute to the autoreactive T cell pool, and others die by apoptosis (Fig. 29.11).

Events in addition to TCR and cytokine signaling are also required for the induction of Foxp3+ T cells. CD28 signaling also plays an important cell-intrinsic role in the development of Foxp3+ T cells in the thymus.[109] CD28 primarily improves the efficiency of T_{reg} cell development rather than enhancing the number of TCRs that can facilitate T_{reg} cells development. CD28 signaling is important for the generation of cytokine-responsive T_{reg} cell precursors, as the TCR repertoires of WT and CD28−/− mice were identical.[110] The ability of CD28 to support T_{reg} generation does not require an intact PI3K-binding motif as PI3K-signaling through Akt and mTOR will antagonize T_{reg} differentiation.[60,111]

How are thymic T_{reg} precursors exposed to tissue-specific antigens (TSAs) enabling the development of T_{reg} that control autoimmunity in all peripheral sites? Thymic MEC express TSAs. The presence of TSAs correlates with the expression of the transcriptional modulator regulator (AIRE), a gene linked to autoimmunity.[112] AIRE promotes immunological tolerance by inducing in thymic MEC, a large repertoire of mRNAs encoding proteins characteristic of differentiated cell types such as insulin. Peptide-MHCII complexes are displayed at the mTEC surface and negatively select thymocytes whose antigen receptors are avidly engaged. Mice that lacked expression of TSAs have increased number of autoreactive T cells in peripheral organs, which leads to multiorgan autoimmunity. The main role of AIRE in central tolerance is to mediate clonal deletion of tissue-specific effector cells. AIRE-mediated ectopic antigen expression in mTECs also plays an important role in tT_{reg} development. Analysis of the T cell repertoire in T_{conv} and T_{reg} cells in WT and AIRE−/− mice found that TCRs that are normally expressed in the T_{reg} cell lineage were found in the T_{conv} cell lineage in the AIRE−/− mice suggesting that some T_{reg} cells are selected via AIRE-driven antigens.[113] DCs also play a role in both clonal deletion and T_{reg} cell induction. The antigens presented by DCs are poorly defined because antigen can be transferred from AIRE expressing mTECs to medullary DCs.[114] CD8α+ DCs are the dominant cross-presenting DC subpopulation.

One population of tT_{reg}, which exits from the thymus exclusively in the perinatal period, is AIRE-dependent, persists in adult mice, has a unique transcriptome, and appears to protect a specific group of tissues from autoimmunity.[115] The group of tissues protected by these neonatal T_{reg} are different from tissues protected by adult T_{reg} and may be similar to the tissues impacted by 3dTx. This same population of AIRE-dependent T_{reg} may explain why AIRE-deficient mice only develop autoimmune disease in a subset of tissues. Older T_{reg} cells accumulate in the thymus as mice age and represent the majority of thymic T_{reg} by 8 weeks of age. These may be cells that never left the thymus or may be recirculating cells. Thymus transplantation studies demonstrate that T_{reg} cells migrate from the periphery to the thymus preferentially compared to T_{conv}.[116] The recirculating T_{reg} cells may compete with developing T_{reg} cells for IL-2 and limit their differentiation.[117]

IL-2 is also required for the maintenance and survival of Foxp3+ T cells in the periphery. The numbers of Foxp3+ T_{reg} cells in the periphery of IL-2−/−, IL-2Rα−/−, and IL-2Rβ−/−[118] mice are substantially decreased or almost absent.[119] Partial or more profound defects both in the number and function of Foxp3+ T cells have been reported in mice deficient for CD80/CD86, CD40,[120] CD40L,[121] and Stat5a.[122] The one common factor that characterizes these mice is that the products of all the deficient genes have important roles in the production or responsiveness to IL-2. Treatment of mice with anti-IL-2 or with CTLA-4Ig to inhibit costimulatory signals also leads to a rapid decline in the number of Foxp3+ T cells.[35] As Foxp3+ T cells do not produce IL-2, this deficiency may be secondary to the capacity of Foxp3− T cells to produce IL-2,[108,123] or to some intrinsic defect in the Foxp3+ T cells in their capacity to respond to IL-2. IL-2Rα−/− T cells in mixed bone marrow chimeras had a substantially decreased competitive fitness compared to WT Foxp3+ T_{reg}. The conclusion drawn from these studies was that IL-2 is an essential T_{reg} cell growth factor *in vivo* and that in the absence of IL-2, T_{reg} cells exhibit an impaired metabolic fitness. It appears that Foxp3 expression in T_{reg} establishes a gene expression program that renders T_{reg} critically dependent on paracrine IL-2 signaling resulting in repression of IL-2 production and the induction of IL-2Rα. IL-2 is not only required for the survival of Foxp3+ T_{reg} in the periphery but also appears to be critical for maintenance of the suppressive functions.[124-127] Deletion of the proapoptotic factor, Bim, results in the restoration of normal numbers of Foxp3+ T cells in IL-2−/− mice, but Bim-deleted mice still exhibited lethal autoimmunity.[126]

The potential role of TGF-β in the induction of Foxp3+ T cells during thymic development remains controversial. Liu et al[128] demonstrated that very few Foxp3+ cells could be detected in the thymus of TGF-βRI−/− mice between 3 and 5 days of age. Later in life, the number of Foxp3+ T cells in the thymus increased, and this expansion was driven by the production of IL-2. Very similar results were observed in TGF-βRII−/− mice with a reduced number of Foxp3+ thymocytes at 3 to 5 days of age with increased numbers at 2 to 3 weeks of age.[129] The few Foxp3+ T cells in the TGF-βRII−/− mice expressed high levels of caspase and enhanced apoptosis associated with high amounts of the proapoptotic Bcl-2 family members, Bim, Bak, and Bax. Bim deficiency enhanced the number of peripheral T_{regs}. Thus, TGF-β is probably not essential for the induction of Foxp3 expression during T_{reg} cell development, but the antiapoptotic effects of TGF-β are required for T_{reg} survival.

TCR Repertoire Analysis of Foxp3+ T_{reg}

As noted above, several lines of evidence suggest that T_{reg} development hinges upon a particular TCR specificity. Foxp3+ T_{reg}, in general, do not develop in TCR transgenic mice lacking RAG genes suggesting that only cells with certain TCR specificities can develop into T_{reg}. Several studies have addressed in detail the TCR repertoire of Foxp3+ T_{reg} and their antigen specificity.[130,131] Hsieh et al[130] have directly compared the sequences of the CDR3 region on T_{reg} and conventional T cells and found that they are only partially overlapping. When immunodeficient mice were reconstituted with T cells expressing diverse Vα2 TCR α-chains from T_{reg} or conventional T cells paired with a single Vβ chain, some of the T_{reg} TCR-reconstituted T cells were self-reactive, as they induced wasting disease in lymphopenic hosts but not in conventional recipients. Although 40% of the TCR were strongly self-reactive *in vivo*, others were only intermittently and moderately self-reactive or nonreactive as measured by T cell expansion in lymphopenic hosts. Other studies[131] have demonstrated a large degree of overlap in TCR usage between Foxp3+ and Foxp3− T cells. The experiments[99-101] demonstrating that TCR self-specificity is required for the intrathymic induction of Foxp3 expression strongly support the view that the TCRs of the Foxp3+ population are predominantly self-specific.

TCR repertoire analysis has also been used to determine the relationship between thymic Foxp3+ T_{reg} cells and Foxp3+ T_{reg} in the periphery. A comparison of the TCR repertoire of thymic and peripheral Foxp3− and Foxp3+ T cells revealed that the TCRs from Foxp3+CD4+ peripheral T cells resembled those of thymic precursors and differed from TCRs expressed by Foxp3−CD4+ T cells; in addition, the diversity of Foxp3+ TCRs was greater than TCRs on Foxp3−CD4+ naïve T cells.[132] These studies suggested that the great majority of peripheral Foxp3+ T_{reg} derive from thymic precursors and that peripheral conversion from Foxp3− to Foxp3+ is rare. TCR repertoire analysis has also been applied to autoreactive T cells in mice deficient in Foxp3.[133] These mice did not exhibit a defect in negative selection. Activated, but not naïve, T cells in Foxp3−/− mice often used TCRs found in the Foxp3+ T_{reg} TCR repertoire of normal mice, suggesting that T cells expressing these self-reactive TCR are not eliminated and might contribute to the pathology associated with Foxp3 deficiency.

If a significant percentage of Foxp3+ T cells are generated at peripheral sites, one should expect that the TCR repertoires of pT_{reg} and tT_{regs} to be different as peripheral conversion would allow for the generation of T_{reg} cells specific for antigens not presented in the thymus such as those derived from commensal bacteria. Lathrop et al[134] found a significant rate of conversion when CD4+Foxp3− T cells were transferred into lymphopenic host. There was considerable disparity in TCR usage between converted Foxp3+ and expanded Foxp3− T cell subsets within the same host. The converted TCRs were found in the normal non-T_{reg} cell subsets because the donor cells for these studies were normal Foxp3− T cells. However, some of these TCRs were also found in the normal T_{reg} subset and made up 5% of the sequences in the normal T_{reg} pool. It appears that some TCRs are able to facilitate both thymic and peripheral T_{reg} cell development. It is likely that TSAs influence the local T_{reg} TCR repertoires as several studies suggest that the presence of an organ facilitates T_{reg}-mediated tolerance to that organ.[135,136]

Most of the studies comparing TCR repertoires between Foxp3+ and Foxp3− T cells were done in mice with restricted TCR repertoires, and the results may not reflect the broader repertoires in normal mice. Wide variation (10%-60%) in overlap were also seen in studies attempting to address the issue of pT_{reg} generation by comparing thymic T_{reg} to peripheral T_{reg}. Since some studies have suggested that Helios is a marker of tT_{reg},[83] two studies have compared the repertoires of Helios+ and Helios− T_{reg}. One study[137] used high-throughput sequencing of T_{reg} from TCR minigene mice and concluded that most TCRs were expressed by both populations at similar frequencies. In contrast, a second study[138] examined the repertoires of Helios+ and Helios− T_{reg} from normal mice and concluded that the TCR repertoires of these two populations were different. However, this study was not directly able to demonstrate that the Helios− T_{reg} were generated from T_{conv} cells or that Helios− T_{reg} are the precursors of Helios+ T_{reg}. Further studies using deep-sequencing techniques of T_{reg} populations from different sites will be needed to address these questions.

T_{reg} Homeostasis and Activation

T_{conv} cells can be divided into activated (memory, effector) and resting (central, naïve) based on the expression of CD62L and CD44, and T_{reg} can be divided into similar subpopulations (Fig. 29.12).[139] $CD44^{low}CD62L^{high}$ resting T_{reg} cells actively recirculate through secondary lymphoid tissues, while $CD44^{high}CD62^{low}$ activated T_{reg} show limited recirculation and are predominantly found in nonlymphoid tissues. In addition, activated T_{reg} are highly enriched for the expression of adhesion and chemoattractant receptors that facilitate migration into nonlymphoid tissues. *In vivo*, the two populations markedly differ in their

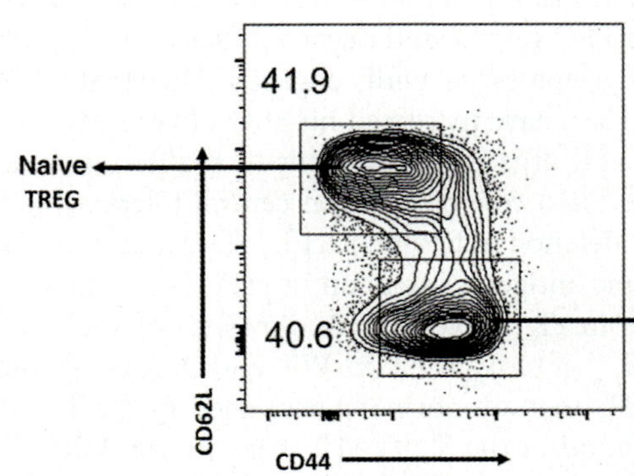

FIG. 29.12. Separation of T_{reg} subsets based on CD44 and CD62L expression.

capacity to proliferate as activated T_{reg} cells rapidly proliferate (~10%/day) as measured by Ki-67 staining, while resting T_{reg} are essentially nonproliferative.[140] Differential expression of the Ly-6C antigen has also been used to distinguish resting from activated T_{reg} (~30%, resting Ly-6C$^+$, ~70% activated Ly-6C−).[141] IL-2 is required to maintain the resting T_{reg} subpopulation in a quiescent state and the chemokine receptor CCR7 is critical for allowing resting T_{reg} access to IL-2 in secondary lymphoid organs. The activated T_{reg} subpopulation has reduced IL-2 signaling *in vivo*, and IL-2 plays no role in their maintenance, which may be controlled by DCs and ICOS.[139]

Single-cell gene expression analysis has shown that naïve/resting T_{reg} cells in lymphoid tissues express a unique signature including *Tcf7, Bcl2,* and *S1pr1.* In contrast, effector/activated T_{reg} cells in lymphoid tissues express a number of genes that are also expressed by T_{reg} in nonlymphoid tissues including *Tnfrsf9, Relb, Ikzf2,* and *Pdcd1.* A small proportion of activated T_{reg} cells in lymphoid tissues have a unique pattern of gene expression including a number of genes expressed by T_{reg} in nonlymphoid tissues (*Itgae, Rora, Fgl2, Klrg1*) and appear to have been primed to self-antigens resulting in their ability to migrate and adapt to different sites.[142] Effector T_{reg} can also be distinguished from resting T_{reg} cells by transcription factor expression. A distinct population of effector T_{reg} express the transcription factor Blimp-1 (encoded by *prdm1* a target of Foxp3), produce high levels of IL-10, express high levels of ICOS, and are localized in nonlymphoid tissues.[143] Blimp-1 is not required for differentiation of effector T_{reg} from resting T_{reg} but is absolutely required for IL-10 production. In contrast, IRF4-deficient T_{reg} cells fail to differentiate into effector T_{reg}, and IRF4 directly activates Blimp-1 expression.

One of the most critical factors in regulating the homeostasis of effector T_{reg} cells *in vivo* is that TCR engagement is required for their function.[144,145] Genetic ablation of TCR expression in T_{regs} was not required for Foxp3 expression or for high expression of most T_{reg} signature genes. T_{reg} with a loss of TCR expression remained naïve and did not enter nonlymphoid tissues. However, loss of TCR expression was required for T_{reg} cell activation, for maintenance of a limited set of genes normally expressed in activated T_{reg}, and T_{reg} suppressor function. Notably, one of the genes lost when the TCR was ablated was IRF4. Adult mice with inducible deletion of the T_{reg} TCR developed autoimmunity indicating that most of the suppressor function of T_{reg} *in vivo* is mediated by CD44^{high}CD62L^{low} effector T_{reg} cells.[144]

REGULATION OF FOXP3 GENE EXPRESSION

Transcriptional Expression

Both the initiation and maintenance of Foxp3 transcription are highly dependent on key conserved noncoding sequences (CNSs), which serve as binding sites for a number of transcription factors. These CNS elements encode information defining the size, composition, and stability of the T_{reg} population (Fig. 29.13). CNS3 is found between exon 1 and exon 2, and this region promotes the permissive histone modifications at the Foxp3 promoter, leaving it in an epigenetically poised state in both Foxp3+ T cells and T_{reg} precursors. Gene-targeting studies reveal an essential role for CNS3 in the induction of Foxp3 expression in the thymus and in the periphery. CNS3−/− mice remain healthy despite an approximately fivefold decrease in T_{reg} cells compared to WT mice. c-Rel binds to CNS3 and the Foxp3 promoter and is indispensable for initiating Foxp3 transcription and T_{reg} differentiation[146] but not for maintenance of Foxp3 expression. Genetic deletion of either c-Rel or RelA markedly reduces both Foxp3+ T_{reg} and CD25+Foxp3− T_{reg} precursors. In contrast to other CNS regions, CNS3 contains permissive chromatin in DP and CD4SP cells that can facilitate binding of c-Rel as a homodimer. This result suggests that c-Rel may be a pioneer transcription factor that opens the Foxp3 locus to other transcription factors. It is likely that that c-Rel binding to CNS3 in cooperation with other transcription factors, including NFAT, CREB, p65, and Smad, facilitates formation of a c-Rel containing enhanceosome at the Foxp3 promoter. Other members of the NF-κB pathway are important for T_{reg} differentiation, as mutations of PKCθ, CARMA1, Bcl-10, TAK1, and IKKβ greatly reduce thymic T_{reg} numbers.[147]

CNS2 (also known as T_{reg}-specific demethylated region, TSDR) is important for maintaining the expression of Foxp3 in tT_{reg} and contains a CpG island whose methylation status is correlated with the stability of Foxp3 expression.[148] The CNS2 region is located within the first intron of Foxp3, and CpG motifs within CNS2 become demethylated during tT_{reg}

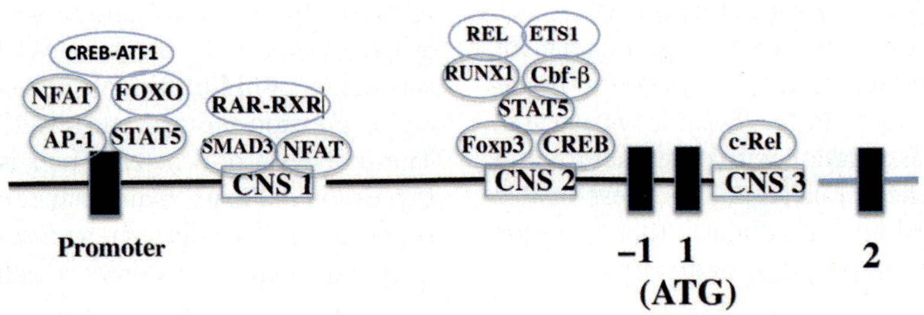

FIG. 29.13. Transcription factors controlling FoxP3 expression. (Adapted from Ruan Q, Kameswaran V, Tone Y, et al. Development of Foxp3+ regulatory T cells is driven by the c-Rel enhanceosome. *Immunity.* 2009;31:932-940, with permission from Elsevier and Adapted by permission from Springer Nature: Zheng Y, Josefowicz S, Chaudhry A, et al. Role of conserved non-coding DNA elements in the Foxp3 gene in regulatory T-cell fate. *Nature.* 2010;463:808-812.)

development but are fully methylated in Foxp3− T cells.[149] Demethylated CpGs within CNS2 enable the binding of several transcription factors including REL, CREB-ATF1, RUNX1-CBFb, ETS1, STAT5, as well as Foxp3 itself.[150] While Foxp3 induction in the thymus and periphery was unimpaired in CNS2−/− mice, T_{reg} cells lacking CNS2 progressively lost Foxp3 expression upon division, indicating that CNS2 is required for the maintenance of Foxp3 expression. Foxp3 enhances its own transcription by forming a Foxp3 autoregulatory circuit which involves the Foxp3-Runx1 interaction, and both Foxp3 and Runx1 bind CNS2. Foxp3 physically interacts with Runx1, and Runx1 is required for T_{reg} suppressive function. Foxp3 binding to CNS2 occurs after and is dependent on demethylation of CNS2. In the absence of Foxp3, Runx1 binds IL-2 enhancers in CD4+ T cells and is poised to activate IL-2 transcription.[151] Methyl-CpG binding domain protein 2 (MBD2) binds CNS2 and recruits ten-eleven translocation (TET) demethylases that maintain the hypomethylation of CNS2.[152] DNA methylase 1 (DNMT1) is known to promote methylation at CNS2. The proinflammatory cytokine IL-6 can trigger DNMT1 activity resulting in instability of Foxp3 expression, while IL-2 and vitamin C can mobilize TET enzymes to maintain the hypomethylation of CNS2.[153]

CNS1 is also located intronically and plays a critical role in the extrathymic induction of Foxp3 expression in response to TGFβ-induced SMAD signaling.[61] The binding of activated SMAD3 to CNS1 is a key event in the induction of pT_{reg} cells. CNS1 also contains NFAT, Smad3, and STAT5 binding sites. Analysis of mice deficient in CNS1 revealed that thymic differentiation was unaffected, while peripheral Foxp3 induction was severely impaired by CNS1 deficiency.[150] CNS1−/− mice have a defective pT_{reg} compartment that fails to maintain immune tolerance at barrier sites such as the gut and lung and defects at these sites do not develop until 1 year of age. The Foxp3 enhancer (CNS0), which is located 8.5 kb upstream of the transcription start site controls Foxp3 transcription by binding the histone methyltransferase MLLA.[154] CNS0 is also bound by the genome organizer Satb1, which promotes Foxp3 expression in the thymus. Foxp3 enhancers can interact with the Foxp3 promoter by forming chromatin loops.[155]

Foxp3 Transcription Factor Complex

Structurally, Foxp3 contains four major domains (Fig. 29.14): a proline-rich N-terminal domain (aa1-97) responsible for transcriptional activation and repression, a central zinc-finger domain, and leucine zipper domain (aa98-255),[68] implicated in oligomer formation and association with other factors, and a conserved forkhead domain (aa256-357)[74,84] responsible for DNA binding. Structural analyses suggest that a unique transcriptional function of Foxp3 is to mediate long-range interactions for Foxp3-targeted genes. Foxp3 is capable of binding to more than 2800 genomic sites, which correspond to approximately 700 to 1400 genes in developing and established T_{reg} cells. The number of genes bound by Foxp3 constitutes only a small proportion (~6%-10%) of those known to be under its control. Foxp3 can positively or negatively control the transcriptional activity of many target genes indirectly by interacting with a number of cofactors. Foxp3 and its binding partners form a large protein complex (400-800 kDa) in size involving more than 360 different factors including other transcription factors and chromatin-modifying factors.[156]

Studies of regulatory elements within the Foxp3 locus offer insights into the molecular mechanisms of Foxp3+ T-cell differentiation and maintenance.[150] The Foxp3 gene includes 11 exons, and a high degree of conservation is seen between the human and mouse genes, especially at the exon-intron boundaries. In response to signaling via the TCR and costimulation pathways, the Foxp3 promoter is bound and activated by transcription factors including NFAT, AP-1, and Nr4a.[157,158] Foxp3 binds NFAT and inhibits its transcriptional activities. When making a complex with NFATc2, the Foxp3 forkhead domain binds the DNA sequence that are occupied by Fos and Jun when they bind NFAT consistent with the model that Foxp3 competitively inhibits the NFATc2:Fos:Jun complex by binding NFATc2 and replacing Fos and Jun. Many of the genes regulated by Foxp3 are also target genes for NFAT. The IL-2 and IL-4 genes are activated by NFAT and repressed by Foxp3, while the CD25 and CTLA-4 genes are upregulated by both NFAT and Foxp3. It is likely that Foxp3 represses IL-2 expression by competing with NFAT for binding to DNA. Foxp3 may also repress NFAT-driven cytokine transcription by sequestering NFAT away from DNA. Examination of the crystal structure of an NFAT: FOXP2: DNA complex revealed an extensive protein-protein interaction interface between NFAT and Foxp2.[159] Structure-guided mutations of Foxp3, predicted to disrupt its interaction with NFAT, interfere in a graded manner with the ability of Foxp3 to repress expression of IL-2 and upregulate T_{reg} markers CTLA-4 and CD25. Thus, NFAT can switch its transcription partner from AP-1 that drives T cell activation to Foxp3 that programs T_{reg} function. Foxp3 also has been shown to interact with the transcription factor AML1/Runx1 that is required for normal hematopoiesis, as well as activation of IL-2 and IFN-γ gene expression.[151] The physical interaction of Foxp3 with RUNX1 suppresses IL-2 and IFN-γ production and induces T_{reg} cell-associated molecules such as CD25, CTLA-4, or GITR. T_{reg} cell deficiency of Cbf-β, a cofactor for all RUNX proteins, attenuated expression of Foxp3 and resulted in high expression of IL-4. The RUNX1-Cbf-β heterodimer is indispensable for optimal expression of Foxp3, maintenance of suppressive function, and suppression of T effector responses.[160] Foxp3 can form a transcriptional repressor complex with several posttranslational

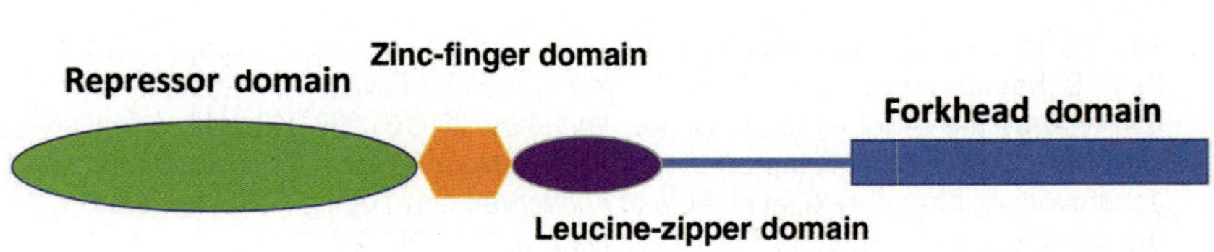

FIG. 29.14. Domains of Foxp3 gene.

modification enzymes such as histone deacetylase HDAC1, HDAC7, and HDAC9 or histone acetyltransferase TIP60.[161] In addition, the forkhead box O (FOXO) proteins FOXO1 AND FOXO3 bind to the promoter and to other regulatory elements of the Foxp3 gene. CREB-activating transcription factor 1 complexes drive activation of the promoter.[151]

IRF4 also interacts with Foxp3 and endows it with the ability to selectively regulate about 20% of the T_{reg} gene expression signature including genes that control Th2 responses as IRF4-deficient T_{reg} are selectively defective in their ability to suppress the production of the Th2 cytokines IL-4 and IL-5.[162] Foxp3 can also directly interact with c-JUN/AP-1 in T_{reg} cells sequestering activated AP-1 in the nucleus, altering the subnuclear distribution of activated AP-1, and disrupting chromatin binding of activated AP-1.[152] Foxp3 interacts with the TFs, GATA, REL, and RORγt and others. The association of Foxp3 with the genes that encode IL-2 and IFN-γ results in the deacetylation of histone H3, which is a modification that silences gene expression. In contrast, Foxp3 promotes the expression of the GITR, CD25, and CTLA-4 by inducing histone acetylation near the gene promoters. Foxp3 associates with molecules that mediate epigenetic modifications.[153] These interaction partners alter the methylation state of target loci with consequences for TF binding. They also execute histone modifications including TIP60, p300, and histone deacetylase 7 (HDAC7).

A group of Foxp3 cofactors (Eos, IRF4, SATB1, Lef1, and GATA1) help Foxp3 enforce the T_{reg} cell gene expression signature.[163] These cofactors have redundant roles in facilitating the regulatory action of Foxp3 as absence of one cofactor can be offset by the activity of another. Such functional overlap could make the capacity of Foxp3 to regulate gene expression quite resilient. Silencing of Eos in T_{reg} abrogates their ability to suppress *in vitro* immune responses and endows them with partial effector function, as they are capable of inducing IBD *in vivo*.[164] Knockdown of Eos reversed Foxp3-mediated suppression of IL-2 production. 90% of the genes suppressed by Foxp3 were no longer downregulated when Eos was knocked down in T_{regs}, while 95% of the Foxp3 upregulated genes were unaltered by Eos knockdown. When functioning as a repressor Foxp3 can recruit corepressors. Eos and HDAC3 contribute to IL-2 repression and both bind to the proline-rich N-terminal domain of Foxp3 protein.[152,164] While Foxp3 was initially characterized as a transcriptional repressor, Foxp3 can also activate transcription of multiple genes such as CD25.[151] It is still not clear how Foxp3 switches between activator and repressor. Foxp3 can function as a transcriptional activator by recruiting transcriptional coactivators including CBP/p300, which interact with Runx1 and thereby activate transcription of Foxp3 target genes.[165]

Another Foxp3 interaction partner and chromatin modifier is the histone methyltransferase enhancer of zeste homologue 2 (Ezh2), which is a component of the polycomb repressive complex 2. Foxp3 and Ezh2 form complexes in activated T_{reg} cells, and these complexes stabilize T_{reg} cell gene expression under conditions of inflammation and during activation. Activated T_{reg} cells that lack Ezh2 display unstable levels of Foxp3 expression and derepression of many genes.[166]

Foxp3 is a member of the Foxp transcription factor subfamily (Foxp1-4). All Foxp proteins contain a zinc-finger domain and a leucine zipper motif amino-(N)-terminal to the conserved DNA-binding forkhead domain. The Foxp transcription factors are capable of homo- and heterodimerization which is thought to enhance their binding to a relatively low-affinity consensus motif. Foxp1 is highly expressed in naïve T cells and undergoes downregulation on their activation. It also can heterodimerize with Foxp3 to form large protein complexes in T_{reg} cells that include Foxp1.[167] Foxp1 bound sites in naïve T cells were largely overlapping with those found in T_{reg} cells where the vast majority of Foxp1 bound sites were jointly occupied by Foxp3. T_{reg} cell cKO of Foxp1 resulted in a genome-wide reduction of Foxp3 binding and altered expression of a number of T_{reg} signature genes. Mice did not develop severe autoimmunity, but T_{reg} had diminished CD25 expression and impairment in IL-2 signaling. Foxp1 has a nonredundant role in T_{reg} cell fitness and suppressor function. Foxp1 binding was primarily at promoters in T_{reg} cells.

One other gene that appears to be important in the maintenance of Foxp3 suppressor function is the genome organizer, SATB1, which is one of the genes most repressed in human and mouse T_{reg} cells.[168] Foxp3 binds directly to the SATB1 locus in T_{reg} cells and prevents SATB1 transcription. Foxp3+ T cells overexpressing SATB1, despite expressing normal levels of Foxp3, lost their suppressive activity and produced T effector cytokines suggesting that downregulation of SATB1 in T_{reg} cells was necessary for a stable suppressive phenotype. Foxp1 plays a specific nonredundant role in repression of SATB1 in T_{reg} cells, and suppression of SATB1 is important in the functional competence of T_{reg} cells. Thus, T_{reg} cells function not only depends on the induction of Foxp3-dependent genes that mediate suppressor function but also on the specific repression of molecules such as SATB1 to prevent T effector function. Foxp1 probably acts synergistically and cooperatively with Foxp3 in T_{reg} cells and assists Foxp3 binding to DNA potentially through heterodimerization.

As mentioned previously, the molecular mechanisms for the differentiation of Th17 and T_{reg} are related. TCR stimulation of naïve CD4+ T cells in the presence of TGF-β induces the expression of both Foxp3 and RORγt. IL-2 promotes T_{reg} differentiation by enhancing Foxp3 expression but represses RORγt expression. In contrast, IL-6 promotes Th17 differentiation but inhibits T_{reg} differentiation.[68] Runx1 provides a platform for the competitive interaction between Foxp3 and RORγt. HIF1α plays an important role for the choice between Th17 and T_{reg} differentiation. Hypoxia stabilizes HIF1α, which enhances Th17 differentiation by making a complex between RORγt and p300. HIF1α binds and promotes degradation of Foxp3 protein.[169] On the other hand, the ubiquitin E3 ligase Deltex1 promotes the degradation of HIF1a and thereby stabilizes Foxp3 protein.[170]

Posttranslational Modification of Foxp3

The activity of Foxp3 is also regulated at the posttranslational level including acetylation, phosphorylation, and ubiquitylation. Foxp3 is degraded by the ubiquitin-proteasome system. HIF1α can interact with Foxp3 and trigger K48 polyubiquitylation.[169] In addition, the molecular chaperone heat shock 70 kDa protein (HSP70) interacts with Foxp3 and recruits the ubiquitin ligase Stub 1, which promotes polyubiquitination and proteosome-dependent degradation of Foxp3.[171] Conversely, the deubiquitinase USP7 can interact with Foxp3 protein removing ubiquitin and thereby increasing the stability of Foxp3 protein.[172] IL-6 upregulates Stub1 and decreases USP7, and thus induces Foxp3 degradation. Conditional deletion of USP7 in mouse T_{reg} cells leads to lethal autoimmunity, reduced T_{reg} suppressor function, and instability of Foxp3 expression.[173] The E3 ubiquitin ligase, Cbl-b, is responsible for the defective development of thymic T_{reg} cells in CD28−/− mice. Upon TCR stimulation, Cbl-b together with Stub1 targets Foxp3 for ubiquitination and degradation in the proteasome. The loss of Cbl-b rescues the defective development of T_{reg} in CD28−/− mice.[174] The E3 ligase, TRAF6, mediates Foxp3 K63 linked ubiquitination of the lysine 262 residue resulting in proper localization of Foxp3 and facilitation of Foxp3 transcriptional activity in T_{reg}.[175]

Acetylation modifications may prevent polyubiquitination that occurs on the same lysine residue through a competition-based mechanism. Acetylated Foxp3 molecules are more stable than those that are not acetylated. TIP60 is a histone acetyl transferase (HAT) that associates with a promoted Foxp3 acetylation. The N-terminal 106 to 190 aa of Foxp3 is required for TIP60-Foxp3 interaction.[161] Acetylation of Foxp3 by p300 prevents Foxp3 protein from proteasome-mediated degradation, and treatment with a histone deacetylase inhibitor increases numbers of Foxp3+ T_{reg}. TIP60 and p300 regulate Foxp3 acetylation and activity in a cooperative manner. While conditional deletion of p300 from T_{reg} cells demonstrates that p300 has only a modest effect on T_{reg} suppressive function, deletion of both TIP60 and p300 results in severe and fatal autoimmune disease early in life.[176] Treatment of mice with the pan-histone deactylase (HDAC) inhibitor, trichostatin A (TSA), increases the proportion and absolute numbers of T_{reg} and results in enhanced T_{reg} suppression by promoting the association of Foxp3 with target genes.[177] Butyrate, a short-chain fatty acid (SCFA), produced by commensal organisms by starch fermentation increases Foxp3 protein acetylation and facilitates pT_{reg} induction.[178] The sirtuins are NAD-dependent protein deacetylases that can regulate Foxp3. Ectopic expression of SIRT1 decreases Foxp3 acetylation and concomitantly increases K48-linked ubiquitylation reducing Foxp3 protein levels.[179]

Phosphorylation regulates multiple aspects of protein function including stability, localization, and trafficking. Foxp3 contains several conserved serine, threonine, and tyrosine residues which can serve as kinase interaction sites.[180] Pim-1 and Pim-2 kinases promote the rapamycin-resistant survival, growth, and proliferation of lymphocytes including T_{reg}.[181] Pim-1–induced phosphorylation of S422 interferes with Foxp3 activity,[182] and Pim-2 kinase negatively regulates T_{reg} suppressive function.[183] TCR stimulation downregulates Pim-1 expression but upregulates Foxp3 expression. Nemo-like-kinase (NLK) phosphorylates Foxp3 and prevents its interaction with Stub1 and its ubiquitination of Foxp3.[184] Cyclin-dependent kinase 2 (CDK2) which is activated by TCR signaling phosphorylates a number of CDK motifs in the N-terminal domain of Foxp3 and negatively affects Foxp3 expression and function.[185] Taken together, the Foxp3 protein can undergo different posttranslational modifications including ubiquitination, acetylation, phosphorylation, and methylation. These posttranslational modifications can affect Foxp3 protein conformation and interactions with other partners, which ultimately alters Foxp3 protein activity to modulate T_{reg} suppressive functions.

Analysis of Gene Expression by Foxp3+ T_{reg}

Both DNA microarray technology and serial analysis of gene expression (SAGE) technology have been used to compare patterns of gene expression between different cell types.[37] These technologies have also been applied to compare the patterns of gene expression in T_{reg} with Foxp3− T cells and other cell types with regulatory functions. One major goal of this approach is to determine whether T_{reg} simply represent a population of previously activated T cells or whether they display a unique pattern of gene expression that is correlated with their functional properties. Most of the results are consistent with the latter possibility. Some genes are differentially expressed between the resting Foxp3− T cells and T_{reg}, while others are differentially expressed following activation.

A comparison of a number of these studies has permitted the identification of groups of genes that are relatively T_{reg}-specific and allow one to begin to define a distinct T_{reg} molecular signature. The "signature" includes both genes encoding certain cell-surface receptors and genes encoding a wider array of transcription factors and other intracellular proteins.[186] Using a T_{reg} to T conventional fold ratio of 1.5, 603 genes were differentially expressed, with 407 overexpressed and 196 underexpressed in T_{reg}. The vast majority of the signature's differential elements were already present in the thymus. Thus, most of the T_{reg} cell signature is already specified at an early stage of lineage commitment, independent of peripheral influences. Many of the T_{reg} signature genes also identify activation antigens expressed on activated conventional T cells including CTLA-4, GITR, and other members of the TNF-receptor superfamily (TNFRSF) that are expressed by activated effector cells. Some including neuropilin, GITR, CD38, and CD5 appear to be shared with Foxp3− anergic T cells that do not exert suppressive functions.[187] On balance, the studies on global gene expression indicate that T_{reg} express a unique pattern of gene expression that may contribute to their survival and anergic state.

A Foxp3 chromatin immunoprecipitation method has been used to analyze the direct targets of Foxp3.[188,189] About 700 to 1000 genes contain Foxp3-binding regions, and many of these genes are up- or downregulated in Foxp3+ T cells confirming that Foxp3 can function as both a transcriptional activator and repressor. However, Foxp3 target genes comprise only a small portion of Foxp3-dependent gene expression suggesting that Foxp3 regulates a large part of its transcriptional program by acting on other transcription factors. It is likely that some of genes encoding molecules that could potentially mediate suppressor function (eg, IL-10, Granzyme B) are secondary Foxp3 targets, as they are not found in the group of genes that are direct Foxp3 targets.

Micro (mi)-RNAs act as posttranslational regulators of T_{reg} cell phenotype, as mice with a deletion of either DICER or DROSHA, RNAaseIII enzymes necessary for miRNA maturation, convert T_{reg} cells to Th1 or Th2 cells.[190] Conditional deletion of DICER in T_{reg} cells confirmed that miRNA are critical for the homeostatic potential of T_{reg} cells, as well as T_{reg} cell function in inflammation.[191] miRNA-155 is essential for the competitive fitness of T_{reg} cells by modulating SOCS-1 expression to regulate the expression of the transcription factor STAT5 in response to limiting amounts of IL-2.[192] In contrast, miRNA-146a is important to suppress pathogenic Th1 responses by downregulating STAT-1 expression in T_{reg} cells.[193]

One novel approach to understand the role of Foxp3 in directing T_{reg} function is to engineer a mouse expressing the coding sequence of GFP knocked into the *Foxp3* locus together with disruption of Foxp3 protein expression.[194,195] Surprisingly, the GFP+Foxp3− T cells in these mice shared many phenotypic features with their GFP+Foxp3+ counterparts. The former are what might be termed T_{reg} "wannabees," as they develop in normal mice and respond to signals required to induce Foxp3 expression. The GFP+Foxp3− cells were unable to produce IL-2, effector cytokines, expressed elevated levels of CD25, were relatively nonresponsive to TCR stimulation but lacked suppressor activity. T_{reg} "wannabees" expressed some (including transcriptional activity at the Foxp3 locus), but not all, of the genes of the T_{reg} signature. Thus, several of the characteristics of the T_{reg} cell lineage can be selected and persist in the absence of Foxp3. Thus, Foxp3 does not strictly function as the "master regulator" of the T_{reg} lineage but stabilizes or enhances certain T_{reg} characteristics, particularly its immunoregulatory activities including suppression. Foxp3 is also not sufficient to elicit the full T_{reg} signature after transfection or induction by TGF-β. T_{reg} "wannabees" do not become conventional effector cells, and transfer of these cells into RAG−/− recipients did not induce the scurfy phenotype.[133,196] It appears that WT mice have a population of self-reactive non-T_{reg} that express the T_{reg} self-reactive TCR repertoire that are not subject to negative selection but can potently induce autoimmune disease in the scurfy mouse.

Studies of the T_{reg} "wannabee" mice indicate that certain T_{reg} cell phenotypic properties including transcription of the Foxp3 locus and expression of some the T_{reg} signature genes can be independent of Foxp3. One candidate family of genes that can regulate these Foxp3-independent T_{reg} properties are the Foxo (Forkhead-box-O) family of transcription factors that play critical functions in the control of diverse cellular responses. Mice with a T cell-specific deletion of Foxo1 have defects in T cell survival and defects in expression of the IL-7 receptor α-chain and CD62L.[197] Foxo1 is required for the inhibition of T cell activation and Foxo1−/− develop a mild autoimmune disease. Foxo3 plays a role in DC function.[198] Mice with a deletion of both Foxo1 and Foxo3 develop a lethal inflammatory disease associated with a compromised T_{reg} cell differentiation. Foxo1 and Foxo3 function in a cell-intrinsic manner to suppress IFN-γ and IL-17 production. Foxo1/Foxo3−/− T_{reg} cells fail to protect lymphopenic mice from autoimmunity induced by scurfy T cells or from colitis induced by Foxp3− T cells.[197] Mixed bone marrow chimeras containing bone marrow from scurfy mice and Foxo1/Foxo3−/− developed a disease as severe as that observed with scurfy bone marrow alone and this was associated with fewer thymic and splenic T_{reg} cells. Thus, it appears that Foxo proteins control the expression of Foxp3 and a subset of T_{reg} cell-associated genes needed for suppressing inflammatory cytokine production in T_{reg} cells. In chromatin-immunoprecipitation assays using antibody to Foxo1 and Foxo3, T_{reg} cells were selectively enriched in genomic fragments containing the CNS1 and CNS3 regions, but not the CNS2 region of the Foxp3 promoter. Foxo transcription factors constitute one element of higher order control[197,198] of T_{reg} development and function.

Single Cell Gene Expression Analysis

Single-cell transcriptome analysis offers the potential in an unbiased manner that does not rely on assumptions of cell-type identities.[199] T_{reg} and T_{conv} have opposite functions suggesting sharp boundaries between their transcriptional programs. However, only a small core set of transcripts including FOXP3, LKZF2, TNFRSF1B, IL2RA, and IL2RB was uniformly expressed in all T_{reg} and consistently absent or underrepresented in T_{conv}. Some of these genes are involved in essential T_{reg} pathways such as IL-2R, other transcripts encode determining transcription factors (Foxp3 and Helios) and several members of the TNFRSF (OX-40 and GITR), others whose function is unknown (gpr83) or some that have not yet been explored. Beyond the small core of differentially expressed transcripts, gene expression in T_{reg} and T_{conv} are overlapping. About 25% of the T_{reg} population termed "furtive" were identified in the T_{conv} cell area. These "furtive" T_{reg} expressed the same level of Foxp3 protein as other T_{reg}, and other canonical T_{reg} signature genes were overexpressed in these cells. They expressed lower levels of IL-2Rα and helios but expressed some transcripts of genes normally associated with T_{conv}. These cells strongly expressed Satb1 and Bach2, two repressors of T effector functions. It remains unclear whether these cells represent pT_{reg} or resting T_{reg} that have poor suppressive activity. The conclusion drawn from these single-cell studies was that T_{reg} cell identity is modular. All T_{reg} express a very small core of T_{reg}-specific transcripts onto which additional programs are added as a function of differentiation and or location. Some activated T_{reg} and T_{conv} cells coexpressed similar transcriptional programs that overrode the canonical T_{reg}/T_{conv} cell difference.

SUPPRESSOR MECHANISMS UTILIZED BY FOXP3+ T_{REG}

General Characteristics of *In Vitro* Suppression Studies

An *in vitro* model system has been widely used for the analysis of T_{reg} cell function that allows rapid assays and which may offer some insights into the mechanism of action of T_{reg} function *in vivo*.[34,200] This approach also allows a comparison of the requirements for activation (costimulation, antigen concentration) of Foxp3+ T cells compared to Foxp3− T cells. Purified Foxp3+ T cells were completely unresponsive to high concentrations of IL-2 alone, to stimulation with plate-bound or soluble anti-CD3, or to the combination of anti-CD3 and anti-CD28. They could be induced to proliferate when stimulated with the combination of anti-CD3 and IL-2 but not by stimulation of endogenous IL-2 production with anti-CD28. The most striking property of the Foxp3+ T cells is their ability to suppress proliferative responses of both CD4+Foxp3− and CD8+ Foxp3− T cells (Fig. 29.15). The Foxp3+ T cells must be activated via their TCR to suppress. No suppression was seen when Foxp3+ T cells were separated by semipermeable membrane from the CD25− T cells. This demonstrates that cell contact between CD25+ and responder is required. Neutralization of the suppressor cytokines IL-4, IL-10, and TGF-β individually or in combination also had no effect on the T_{reg}-mediated suppression. Similarly, Foxp3+ T cells from mice deficient in IL-4, IL-10, or TGF-β were fully competent suppressors. Indo-1–loaded Foxp3+ T cells did not flux calcium in response to TCR stimulation,[201] suggesting that they have a block in proximal signaling similar to that seen in T cells rendered anergic *in vitro*.

A number of studies[202,203] have shown that T_{reg} exhibit diminished responses of both proximal and distal signaling pathways both in absolute amplitude and in duration when compared to the responses of CD25− T cells. T_{reg} have reduced phosphorylation of the TCR ζ-chain, reduced recruitment of ZAP-70, and reduced phosphorylation of SLP-76. Defects were seen in the activity of PLC-γ1 and in signals downstream of this enzyme including calcium mobilization, NFAT, NF-κB, and Ras-ERK-AP-1 activation. The diminished activity of PLC-γ1 in T_{reg} is likely to contribute not only to the delay and reduced appearance of NFAT in the nucleus but also to the attenuated activation of the PKC and Ras-ERK directed pathways as well. Other studies[202] have suggested that the proximal promoter of the

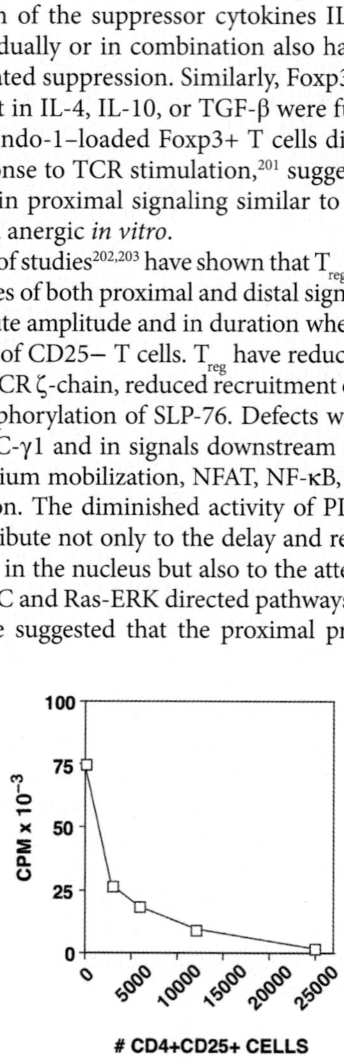

FIG. 29.15. CD4+ CD25+ T cells suppress the proliferative response of CD4+ CD25− T cells. Graded numbers of CD4+ CD25+ T cells were mixed with 5 × 10^4 CD4+ CD25− T cells, T-depleted spleen cells, and soluble anti-CD3. Proliferation was measured after 72 hours. (Adapted with permission of Rockefeller University Press, from Thornton AM, Shevach EM. CD4+CD25+ immunoregulatory T cells suppress polyclonal T cell activation *in vitro* by inhibiting interleukin 2 production. *J Exp Med.* 1998;188:287-296, permission conveyed through Copyright Clearance Center, Inc..)

IL-2 gene fails to undergo chromatin remodeling and remains in a closed chromatin configuration.

Despite the widely recognized importance of IL-2 in T_{reg} homeostasis, very little is known about the intracellular mechanisms that regulate IL-2R signaling in T_{reg}. Indeed, a defining characteristic of T_{reg} is their inability to expand *in vitro* upon stimulation with IL-2 alone despite expression of all three chains of the IL-2 receptor. It has been shown that JAK/STAT-dependent signaling is intact in T_{reg} stimulated with IL-2 but that downstream mediators of PI3K are not activated.[203,204] The nonresponsiveness of T_{reg} is associated with elevated levels of PTEN, a phosphoinositol 3,4,5-triphosphatase that catalyzes the reverse reaction of PI3K, negatively regulating the activation of downstream signaling pathways. In conventional T cells, the level of PTEN is downregulated after T cell activation, but PTEN remains highly expressed in T_{reg}. A critical role of PTEN, in the nonresponsiveness of T_{reg} to IL-2, was established by studying PTEN−/− mice.[205] PTEN−/− T_{reg} readily proliferated after stimulation with IL-2 alone *in vitro* and exhibited enhanced peripheral turnover *in vivo*. They retained their ability to suppress effector T cell responses both *in vitro* and *in vivo*. Forced expression of PTEN in recently activated non-T_{reg} inhibited their ability to expand in response to IL-2, confirming the ability of this lipid phosphatase to negatively regulate IL-2–dependent proliferation.

T_{regs} also have a defect in the activation of the PI3K-AKT pathway, as indicated by a reduction in AKT phosphorylation in T_{reg} cells.[60] Thymic-derived T_{regs} with established and stable Foxp3 levels were resistant to the effects of constitutively activated AKT, while activated AKT inhibits the induction of Foxp3 expression in Foxp3− T cells both *in vitro* and *in vivo*. IL-6 may block the differentiation of T_{regs} by activating an AKT-dependent pathway. The repression of Foxp3 expression by active AKT required its kinase activity and was partly counteracted by rapamycin treatment placing mTOR as a downstream target of AKT. mTORC1 is more sensitive to rapamycin treatment, so it is likely that AKT is exerting part of its effect via mTORC1. Conversely, inhibition of the PI3K-AKT-mTOR pathway results in de novo expression of Foxp3 in a TGF-β–independent manner.[206] Most studies have demonstrated that Foxp3+ T cells mediate suppression by inhibiting the induction of IL-2 mRNA and mRNA for other effector cytokines in the responder Foxp3− T cells. Suppression can be abrogated by the addition of IL-2, thereby circumventing the block. Although IL-2 gene transcription was inhibited in the presence or absence of exogenous IL-2, the addition of anti-CD28 overcomes suppression by potently stimulating the production of endogenous IL-2 and overriding the suppressive effects of the Foxp3+ cells. Surprisingly, transcription of IL-2 mRNA was also restored in the cocultures in the presence of anti–IL-2. These results suggest that T_{reg} do not suppress the initial activation of Foxp3− T cells but mediate their suppressive effects following production of IL-2 by the responder cells resulting in both the expansion of the T_{reg} and induction/enhancement of their suppressor functions.[125] Foxp3+ T cells do not directly mediate the death of the responders but induce a cell-cycle arrest at the G_1-S phase of the cell cycle. Such a cell-cycle arrest is often followed by cell death, and it is difficult to recover significant numbers of viable cells when the suppressors and responders are cocultured for periods longer than 48 hours. Suppression of T cell proliferation is the exclusive property of Foxp3+ T cells isolated from normal animals. The induction of CD25 expression by stimulating CD25− T cells via the TCR, in the absence of TGF-β, does not render the stimulated cells suppressive.

Foxp3+ T cells can be easily propagated *in vitro* for 3 to 14 days by stimulation initially with anti-CD3 and IL-2 and then expansion in IL-2 alone.[200] It has also been possible to clone murine Foxp3+ T cells by repeated stimulation with anti-CD3 and IL-2.[38] Following 7 to 14 days of activation and culture in IL-2, activated Foxp3+ T cells remain nonresponsive and cannot be induced to proliferate when restimulated via their TCR in the absence of IL-2. The activated Foxp3+ T cells have more potent suppressor activity on a per-cell basis (three- to fourfold) than freshly explanted Foxp3+ T cells. Antigen-specific Foxp3+ T cells can be identified in mice that express a transgenic TCR. In general, the percentage of Foxp3+ cells is reduced in TCR-transgenic mice (3%-5% compared to 10% in normal animals). The expression of the transgenic TCR α-chain is a convenient tool that allows activation of the Foxp3+ T cells with peptide/MHC rather than with anti-CD3.[200] When Foxp3+ T cells from TCR-transgenic mice are activated with their peptide/MHC ligand and then expanded *in vitro* in IL-2, the activated suppressors are subsequently capable of suppressing the proliferative responses of fresh Foxp3− T cells from mice that express a different transgenic TCR. There is no MHC restriction in the interaction of the activated suppressors and the Foxp3− responders. Therefore, the suppressor effector function of activated Foxp3+ T cells *in vitro* is completely nonspecific.

T_{reg}-mediated suppression is highly sensitive to antigenic stimulation. For example, when Foxp3+ and Foxp3− T cells are prepared from the same TCR-transgenic mouse and stimulated with a specific peptide, the antigen concentration required to stimulate the Foxp3+ T cells to suppress is 10- to 100-fold lower than that required for triggering the proliferation of the Foxp3− T cell population.[207] The partially activated phenotype of Foxp3+ T cells combined with their enhanced sensitivity to antigen stimulation suggests that they are highly differentiated in their function and are ready to mediate their suppressive functions immediately upon encounter with their target antigens. Their capacity to rapidly suppress responses *in vitro* suggests that they have been continuously stimulated by self-antigens in the normal physiologic state and continuously exert some degree of suppression *in vivo*.

Foxp3+ T cells suppress the proliferative responses of CD8+ T cells in a manner similar to that seen with CD4+ responders.[208] Marked suppression of the effector cytokine, IFN-γ, is also seen in the presence of Foxp3+ T cells. While suppression of the proliferation of CD4+ responders by Foxp3+ T cells can be completely reversed by the addition of IL-2 or anti-CD28, the suppression of CD8+ T cell responses is not reversed by the addition of IL-2 or anti-CD28. The failure of IL-2 to abrogate suppression is secondary to a failure of full upregulation of the expression of CD25 on the responder CD8+ T cells. Foxp3+ T cells thereby prevent responses mediated by CD8+ T cells both by inhibiting their ability to produce IL-2 and by inhibiting their ability to respond to IL-2, thus disrupting CD4+ T cell help for CD8+ T cells.

In general, murine T_{reg} failed to efficiently inhibit responses induced by plate-bound anti-CD3 but readily inhibited responses induced by soluble anti-CD3.[34] This finding raised the possibility that the cellular target of the T_{reg} was the antigen-presenting cells (APC) rather than the responder T cell, as responses to plate-bound anti-CD3 are relatively APC-independent, but other studies demonstrated that Foxp3+ T cells can mediate suppression *in vitro* via a T-T cell interaction and that the APC is not directly required for delivery of the suppressive signal.[209] Similar conclusions were drawn in other studies in which Foxp3+ T cells were shown to suppress responses of either mouse[208] or human Foxp3− T cells induced by anti-CD3 coupled to beads in the absence of APC and can efficiently directly target responder T cells. It remains unclear whether the mechanisms utilized by T_{reg} to suppress the proliferation of Foxp3− T cells in the presence or absence of APC are the same or whether assays of T_{reg} function under these distinct conditions measure distinct components of T_{reg}-mediated suppressor function. While multiple mechanisms of T_{reg}-mediated suppression have been proposed, suppression of APC function or suppression of the delivery of APC-derived costimulatory signals has achieved the greatest attention. If the APC is the primary target for T_{reg} suppression *in vivo*, it would be ideal to employ an *in vitro* assay that would mimic the *in vivo* action of T_{reg}.[210]

Role of IL-2 Consumption in T_{reg}-Mediated Suppression

The role of IL-2 consumption in the suppressive function of Foxp3+ T cells is controversial (Fig. 29.16). T_{reg} cells express high levels of the IL-2Rα but are unable to produce IL-2. IL-2 binds with low affinity to CD25 and with modest affinity to heterodimers of IL-2Rβ and IL-2Rγ, but receptor affinity increases 1000-fold when IL-2 binds to complexes of all three chains. The interaction of IL-2 with T_{regs} may result in a number of different effects including acting as an activation and maintenance signal due to triggering JAK-STAT and downstream pathways[152] or a directing signal to IL-2 producing self-reactive T cells.[211] Alternatively, Foxp3+ T cells by competing with Foxp3− T cells for IL-2, inhibit their proliferation resulting in the induction of apoptosis.[212] In a hybrid system in which human T_{reg} cells were shown to efficiently suppress mouse Foxp3− T cells, the addition of an antihuman CD25 mAb that blocks IL-2 binding had no effect on the function of the human T_{reg}.[213] Taken together with the data that Foxp3+ T_{reg} inhibit the production of IL-2, it is unlikely that T_{reg} function as IL-2 "sinks,"[214] at least *in vitro*.

The role of IL-2 consumption by T_{reg} in T_{reg} suppression *in vivo* has been partially addressed. WT T_{regs} suppress autoimmunity in IL-2Rβ−/− mice where all other cells in the recipient are nonresponsive to IL-2, directly suggesting that effective suppression is not simply due to depriving IL-2 growth-promoting signals to autoreactive T cells by T_{reg} consumption of this cytokine.[118] Mice with a T_{reg} conditional

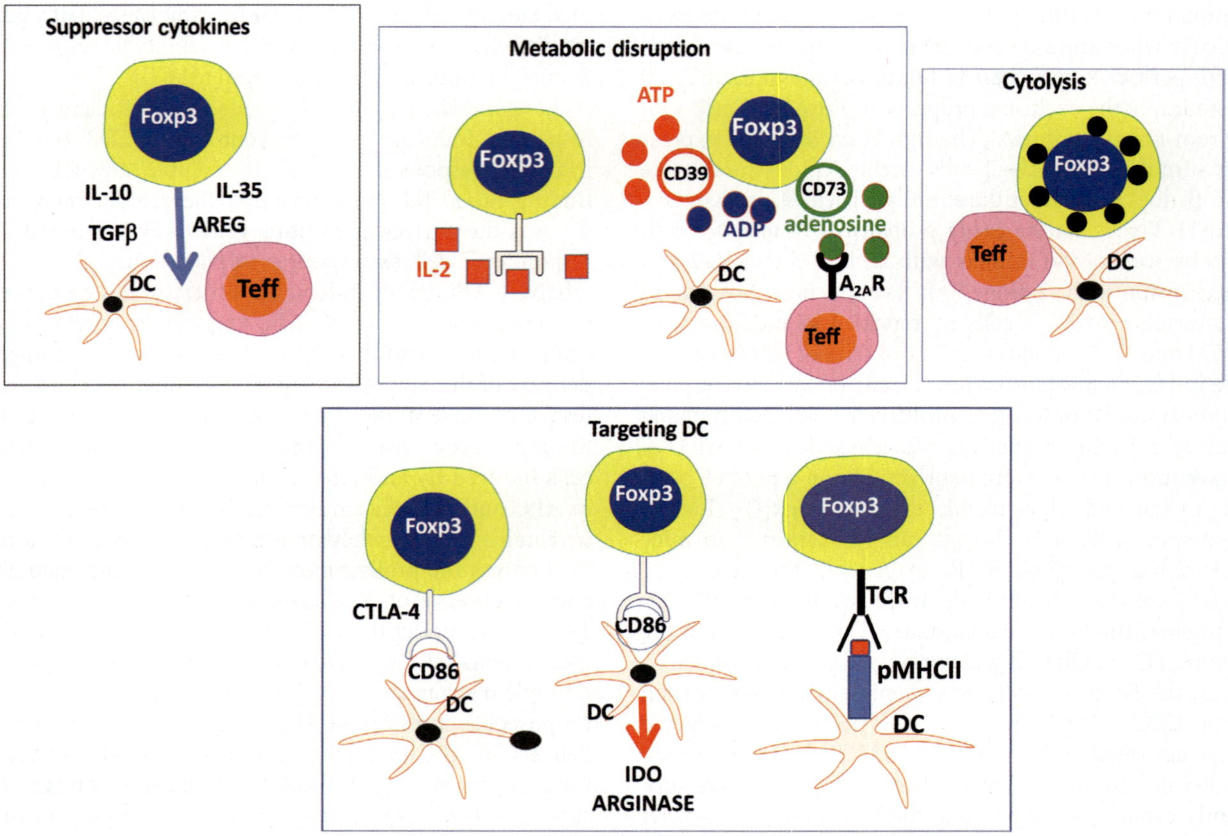

FIG. 29.16. Major T_{reg}-suppressor mechanisms.

deletion of either IL-2Rβ or IL-2Rα rapidly develop fatal autoimmunity. Introduction of a constitutively activated STAT5b gene to IL-2Rβ−/− T_{reg} cells prevented early fatal disease, but the T_{reg} cells remained unresponsive to IL-2. Similar results were observed when a constitutively active STAT5b gene was used to reconstitute IL-2Rα cKO mice. While the T_{reg} cells could not respond to IL-2, they could suppress the activity of autoreactive CD4+ T cells. However, very little suppression of autoreactive CD8+ T cells was observed, suggesting that IL-2 consumption may be involved in T_{reg} suppression of CD8+, but not CD4+ T cells.[127]

Very little is known about the molecular/biochemical signals generated in the responder cell following interaction with Foxp3+ T_{reg}. Detailed time course studies *in vitro* have shown that suppression is established after 6 to 12 hours of contact with T_{reg} and that responders are refractory to suppression after 12 hours of activation. As suggested by the requirements for IL-2 to activate suppressor function,[125] the IL-2 response of the responders is normal for the first 6 hours of the coculture and then abruptly terminates. Costimulation with anti-CD28 generated resistance to suppression by stabilizing IL-2 transcripts enabling the responder T cells to counterbalance transcriptional downregulation by T_{reg}.[118] T_{reg} do not inhibit the induction of CD69, and the viability of responder cells was excellent at 36 hours. Gene expression analysis in responder T cells using a microarray approach has shown that one set of genes was expressed after culture with T_{reg}, and this set of genes was distinct from genes induced in anergized T cells, T cells deprived of IL-2, or T cells treated with TGFβ.[215,216] Hopefully, further detailed studies of these genes will elucidate the specific pathways involved in T_{reg}-mediated inhibition.

Production of Suppressor Cytokines and Other Suppressive Factors by T_{reg}

TGF-β

Most of the studies with either human or mouse T_{reg} that have studied suppression of T cell activation *in vitro* have failed to identify a soluble suppressor cytokine. It is very difficult to rule out the involvement of a cytokine that acts over short distances or a cell-bound cytokine. Nakamura et al[217] have raised the possibility that TGF-β produced by T_{reg} and then bound to their cell surface by an as yet uncharacterized receptor might mediate suppression in a cell contact-dependent fashion. In their studies, TGF-β was detected on the surface of resting and activated CD25+ T cells, and suppression could be reversed by high concentrations of anti–TGF-β mAbs. They postulated that latent (inactive) TGF-β, bound to the cell surface of activated T_{reg}, is delivered directly to responder CD25− T cells and is then locally converted to its active form. In contrast to these studies, Piccirillo et al[218] were unable to show a requirement for either the production of TGF-β or responsiveness to TGF-β in T_{reg}-mediated suppression. CD25− T cells from *Smad3*−/− and from mice expressing a dominant negative form of the TGF-β receptor II (TGFβRII) that are completely resistant to the immunosuppressive effects of TGF-β were readily suppressed by T_{reg} cells from WT mice (Table 29.2). T_{reg} from TGF-β−/− mice were as efficient as T_{reg} from WT mice in mediating suppression of WT CD25− T cells. High concentrations of anti–TGF-β did not reverse suppression, nor did anti–TGF-β or a soluble form of the TGF-βRII inhibit suppression mediated by activated CD25+ T cells.

An alternative possibility is that TGF-β plays a role in the enhancement or maintenance of Foxp3+ T_{reg} suppressor activity. Indeed, Marie et al[219] have demonstrated that peripheral TGFβ1−/− T_{reg} expressed diminished levels of Foxp3. These studies suggested that TGF-β is required for maintenance of Foxp3 expression in peripheral T_{reg}, that the source of the TGF-β is not the T_{reg}, and that paracrine production of TGF-β by APC is needed for optimal T_{reg} survival and suppressor function. Further evidence for the role of TGF-β in the maintenance of T_{reg} was derived from studies of mice with T cell–specific deletion of TGF-βRII.[220] In these mice, TGF-β was not required for induction of Foxp3 expression in the thymus and for the thymic development of T_{reg} cells. In fact, TGF-βRII−/− mice have an increase in the percentage and number of Foxp3+ T cells in the thymus and increased BrdU uptake suggesting that TGF-β normally inhibits the proliferation of thymic T_{reg}. Increased T_{reg} proliferation was also seen in the periphery of TGF-βRII−/− mice, but these mice have a decrease in the percentage of Foxp3+ T_{reg} in the periphery, consistent with the possibility that TGF-β is essential for their survival in the periphery. When mixed bone marrow chimeras between TGF-βRII−/− and WT mice were generated, the knock-out Foxp3− T cells demonstrated complete resistance to suppression by WT T_{reg}. This result suggests that there is a defect in a cell-intrinsic mechanism of TGF-β–mediated control of T cell reactivity in the TGF-βRII−/− T cells.

The TGF-β gene has been deleted in activated T cells and T_{reg} cells or in T_{reg} cells alone to determine the exact source of TGF-β for regulating T cell tolerance and differentiation. Abrogation of TGF-β1 in activated T cells and T_{reg} cells, but not in T_{reg} cells alone, protected mice from EAE and was associated with compromised Th17 differentiation.[221] Th17 cells were the main producers of TGF-β1 *in vitro* and *in vivo*. TGF-β1 produced by T_{reg} cells is not essential for Th17 differentiation in the EAE model.[222] Mice with a specific deletion of TGF-β1 in T_{reg} showed no signs of inflammation even at the age of 9 months. These mice had increased frequencies of Foxp3+ T cells in peripheral and mesenteric nodes but not in the spleen, with no increase in conventional CD4+ or CD8+ T cells. A number of other experiments have demonstrated that autocrine TGF-β1 expression is not essential for T_{reg} suppression *in vivo*. Transfer of TGF-β1−/− splenocytes to RAG-2−/− mice induced disease with features similar to those of the TGF-β1−/− mice and disease transfer was accelerated by the depletion of TGF-β−/− CD25+ T cells. Importantly, cotransfer of TGF-β1−/− T_{reg} clearly attenuated disease in RAG2−/− recipients of CD25-depleted spleen and lymph nodes, but suppression was incomplete when compared to WT T_{reg}.[223] Although TGF-β appears to play a nonredundant role in control of intestinal inflammation, T_{reg} from TGF-β1−/− mice were capable of inhibiting IBD induced by CD45RB^{high} [223] or CD4+CD25− T cells[225] *in vivo*. Furthermore, T_{reg} from mice expressing a dominant negative form of the TGF-βRII are also capable of inhibiting colitis, suggesting that TGF-β is not required for the function of T_{reg}. Importantly, the suppressive function of TGF-β−/− T_{reg} cells was abrogated by anti–TGF-β indicating that TGF-β is absolutely required for protection from IBD in this model but can be provided by a non-T_{reg} cell source.

The potential role of TGF-β as a mediator of T_{reg} suppression must be reevaluated, as several studies[226-228] have identified GARP (LRRC32) as the cell surface receptor for latent TGF-β. Although the GARP/Latent TGF-β complex is also expressed by platelets, within the immune system, GARP expression is most often observed on activated Foxp3+ T_{regs} (Fig. 29.17). However, the function of the GARP/latent TGF-β complex on T_{regs} has remained elusive. T_{regs} that lacked the surface expression of LAP following treatment with TGF-β

TABLE 29.2 CD4+CD25+ Suppressor Function Occurs Independently of TGFβ

CD4+CD25+	CD4+CD25–	% Suppression
Wild type	Wild type	Yes
Wild type	Wild type + anti-TGF-β	Yes
Wild type	*SMAD3–/–*	Yes
SMAD3–/–	Wild type	Yes
Wild type	DNTGFβRII	Yes

CD4+CD25+ T cells were mixed with CD4+CD25– T cells in the presence of T-depleted spleen cells and soluble anti-CD3. Proliferation was measured at 72 hours.
Adapted with permission of Rockefeller University Press, from Piccirillo CA, Letterio JJ, Thornton AM, et al. CD4+CD25+ regulatory T cells can mediate suppressor function in the absence of transforming growth factor β1 production and responsiveness. *J Exp Med.* 2002;196:1-10, permission conveyed through Copyright Clearance Center, Inc.

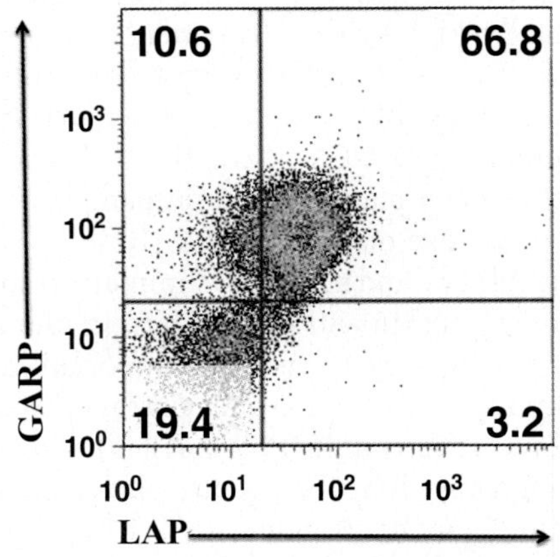

FIG. 29.17. A high percentage of activated human Foxp3+ T cells express GARP and LAP. (Adapted from Tran DQ, Andersson J, Wang R, et al. GARP(LRRC32) is essential for the surface expression of latent TGF-β on platelets and activated FOXP3+ regulatory T cells. *Proc Natl Acad Sci U S A.* 2009;106:13445-13450.)

siRNA or the expression of GARP and LAP following treatment with GARP siRNA were somewhat less suppressive than T_{regs} treated with the control siRNA in the *in vitro* suppression assay. A number of novel mAbs have been generated, which block the production of active TGF-β1 by human T_{regs}. These mAbs recognize a conformational epitope that requires amino acids GARP137-139 within GARP/TGF-β1 complexes. In a model of xenogeneic graft-versus-host disease (GVHD) in NSG mice, the blocking antibodies inhibited the immunosuppressive activity of human T_{regs}.[229] mAbs of this type may prove useful to reverse the suppressive effect of T_{reg} in infection or in tumor.[229]

TGF-β does appear to play a major role in the T_{reg}-NK cell interactions.[230] T_{reg} were shown to inhibit NKG2D-mediated cytolysis *in vitro* largely by a TGF-β–dependent, IL-10–independent mechanism. Similarly, T_{reg} directly inhibited NKG2D-mediated NK cell cytotoxicity *in vivo*, resulting in suppression of NK cell-mediated tumor rejection.[231] Other studies have shown that human T_{reg} could also directly inhibit NK cell effector functions by a TGF-β–dependent mechanism with a concomitant downregulation of NKG2D receptors on the NK cell surface.[232,233] NK cell cytotoxicity was restored in the presence of T_{reg} when the cocultures were performed in the presence of cytokines known to activate NK cells such as IL-2, IL-4, IL-7, and IL-12. Curiously, in both of these studies, T_{reg} cells did not induce a global inhibition of NK-activating receptors but only an inhibition of NKG2D. In other experiments, depletion of T_{reg} resulted in enhanced proliferation of NK cells as well as their cytotoxic potential. The role of T_{reg} in the regulation of NK cell function has also been studied in a bone marrow graft rejection model.[234] One explanation for the susceptibility of NK cells to T_{reg} control is that uninhibited NK cells might be a danger to the host in autoimmune diseases.

IL-10

The production of IL-10 by T_{reg} cells plays an important role in mediating the suppressive effects of T_{reg} in certain disease situations, particularly in models of inflammatory bowel disease (IBD), EAE, and in certain parasite infections.[235,236] IL-10 production by T_{reg} plays a particularly prominent role in autoimmune diseases characterized by a strong inflammatory component, such as IBD, but not for the much less inflammatory autoimmune (AIG) where IL-10–deficient T_{reg} are fully protective or to play a role in unmanipulated animals. Surprisingly, selective disruption of IL-10 in T_{reg} cells led to spontaneous colitis[237] and augmented immune response associated pathology in skin and lungs. Thus, in unmanipulated mice, IL-10 produced by T_{regs} plays an important role in suppressing immune inflammation at environmental interfaces.

One unique property of IL-10 is that it is not only a mediator of T_{reg} suppressive function but also plays a unique role by signaling into T_{reg}.[237] IL-10 induces STAT3 phosphorylation in T_{reg} to a greater extent than in T_{conv}. Mice with a cKO of IL-10Rα in T_{reg} developed severe colitis by 8 to 10 weeks of age but had a normal number of T_{reg} with normal expression of Foxp3 and normal *in vitro* suppressive function. However, T_{conv} cells in these mice produced higher amounts of IL-17, indicating that mice with a selective deletion of the IL-10Rα in T_{reg} were impaired in their ability to suppress inflammatory Th17 responses.[238] Importantly, IL-10Rα−/− T_{reg} were also defective in their ability to produce IL-10, indicating that IL-10 produced by T_{reg} and other cell types augments its own production by T_{reg}.

IL-35

The failure to observe suppression when T_{reg} are separated from responders does not rule out the possibility that T_{reg} secrete an as yet uncharacterized factor that functions in gradient fashion and requires proximity between suppressor and responder. IL-35, a heterodimeric member of the IL-12 family, composed of Epstein-Barr virus–induced gene 3 (Ebi3 which normally pairs with p28 to form IL-27) and p35 (which normally pairs with p40 to form IL-12) may subserve this function.[239] IL-35 is produced by activated T_{reg} and has been shown to act directly on responder T cells and suppress their proliferative responses. In addition, treatment of naïve mouse or human T cells with IL-35 induced a Foxp3− regulatory population (termed iTr35) that mediated suppression by secreting IL-35 but not IL-10 or TGF-β.[240] iTr35 were as effective as Foxp3+ T cells in preventing lethal autoimmunity in scurfy mice and completely protected mice from induction of EAE. Studies on the role of IL-35 have been hampered by the lack of a recombinant form of the cytokine, neutralizing antibodies, and limited characterization of its receptor.[241]

Other Secreted Mediators of Suppression

T_{reg} cells produce large quantities of exosomes which are 40 to 100 nM vesicles that are generated by the inward invagination of endosomal membranes forming multivesicular bodies. Multivesicular bodies then release exosomes which contain encapsulated cytosolic contents.

The release of exosomes from T_{reg} is hypoxia-sensitive and is mediated by the Rab27a and Rab27b GTPases. This pathway allows 1 cell to transfer bioactive materials including miRNAs from 1 cell to another, thus facilitating intercellular communication. Some of miRNA-containing exosomes released by T_{reg} can be taken up by T_{conv} cells or DCs to modulate gene expression. T_{reg} cells transferred miRNA, Let-7D, to Th1 cells, resulting in suppression of Th1 proliferation, IFN-γ secretion, and prevention of systemic disease.[242] The basis for the selective suppressive effect of T_{reg}-derived exosomes is that the miRNA content of T_{reg}-derived exosomes is distinct from the miRNA content of Th1 or Th2 cell-derived exosomes. Exosome miRNA-mediated transfer is a contact-independent T_{reg} cell-mediated suppression.

A number of novel suppressor mechanisms have recently been described. Human T_{reg} cells express high levels of 15-hydroxyprostaglandin dehydrogenase (HPGD). Expression requires IL-2 and Foxp3 but not TCR stimulation. The major function of HPGD is to catabolize the prostaglandin, PGE2, to 15-keto PGE2, which inhibits the proliferation of T_{conv} cells.[243] This pathway is also present in the mouse, most prominently in adipose tissue. 15-keto-PGE2 is a ligand for PPARγ, and PPARγ-deficient T_{reg} were less sensitive to inhibition by 15-keto-PGE2. Addition of PGE2 to *in vitro* suppression assays enhances suppression due to the metabolism of PGE2 to 15-keto PGE2, and in the presence of PGE2, HPGD−/− T_{reg} displayed diminished suppressor function. Mice with a selective deficiency of HPGD in T_{reg} did not develop any signs of autoimmunity, and no changes in the T_{reg} signature or to the core Foxp3-associated gene expression program were detected. The HPGD pathway appears to play a prominent role in adipose homeostasis as HPGD−/− mice developed a metabolic syndrome, had larger adipocytes, increased cellular infiltration with inflammatory macrophages but normal numbers of T_{reg}. The HPGD pathway may be a second example of a tissue-specific mechanism of T_{reg} suppression similar to the role of IL-10 for suppression of mucosal immunity.

T_{regs} can transfer cAMP to Teff via gap junctions, and elevating cAMP in T_{regs} renders them more suppressive.[244] Upon entrance to the target cell, cAMP inhibits IL-2 production and proliferation. T_{regs} also indirectly boost cAMP levels in target cells by augmenting adenosine-mediated signaling from extracellular nucleotides. One other metabolic pathway that can modulate T_{reg} function targets the P2X7 receptor that is expressed at high levels on Foxp3+ T cells.[245] Nicotinamide adenine dinucleotide (NAD+) released by cell lysis or nonlytic mechanisms can activate the P2X7 receptor and induce T_{reg} cell death. The systemic administration of NAD+ has resulted in selective depletion of T_{reg} cells and enhancement of an antitumor response.

Extracellular ATP functions as an indicator of tissue destruction and may exert its effects on DCs. The intracellular ATP concentration is so high that substantial amounts of the nucleotide are released upon cell damage. CD39 is the dominant ectoenzyme in the immune system that hydrolyzes ATP or ADP to AMP and is expressed by a number of different cell types including T_{regs}. ATP can activate DCs and lead to the upregulation of CD86. The catalytic inactivation of extracellular ATP by CD39 is a potential anti-inflammatory mechanism used by T_{regs}.[246] The immodulatory effects of removal of ATP by CD39 can be amplified by the generation of adenosine. Adenosine can be generated by CD39 in concert with the 5′-ectonucleoside CD73, which dephosphorylates the CD39 product, AMP.[247] Adenosine signals via the A2A receptor and can inhibit the function of DCs and activated T cells. The documented ability of T_{regs} to convert ATP to adenosine provides direct evidence of the functional capacity of these enzymes and supports the notion that adenosine is a mediator of T_{reg} function. Adenosine regulates T cell function indirectly by reducing the secretion of proinflammatory cytokines including IL-12 and TNFα from stimulated APC by 80% to 90%. The immunosuppressive function of adenosine may be mediated by increasing intracellular cyclic AMP.

Other secreted molecules may also play a role in T_{reg}-mediated suppression including galectin-1, a member of a highly conserved family of β-galactoside binding proteins.[248] Galectin-1 is secreted as a homodimer and binds to many glycoproteins including CD45, CD43, and CD7, resulting in cell cycle arrest, apoptosis, and the inhibition of proinflammatory cytokine production on target cells. Galectin-1 is preferentially expressed in T_{reg} cells and is upregulated upon TCR activation. Blocking galectin-1 markedly reduced T_{reg} suppressor function *in vitro*, and T_{reg} cells from galectin-1−/− mice had reduced suppressor function. Another member of the Galectin family, Galectin-10, is selectively expressed by T_{regs}.[249] However, Galectin-10 is primarily expressed intracellularly and is unlikely to be involved in suppression by T_{regs}.

Cell Contact-Dependent Mechanisms of Suppression

Cytolysis

Cytolysis of target cells must also be considered as another potential mechanism for T_{reg}-mediated suppression (Fig. 29.16). Human Foxp3+ T cells that are activated with a combination of antibodies to CD3 and CD46 express granzyme A and kill activated CD4+ and CD8+ T cells and other cell types in a perforin-dependent, Fas/Fas-L–independent manner.[250] Activation of mouse T_{reg} also results in upregulation of granzyme B expression, and one report[251] has claimed that activated T_{reg} kill responder cells by a granzyme B–dependent, perforin-independent mechanism. Other studies[252] demonstrated that activated mouse T_{reg} cells inhibited the proliferative responses of B cells to polyclonal B cell activators. Suppression of proliferation was due to increased cell death caused by the T_{reg} cells in a cell contact-dependent manner. The induction of B cell death was not mediated by Fas/Fas-L pathway but depended on the upregulation of perforin and granzymes in the T_{reg}. It remains to be determined whether T_{reg} can develop into cytotoxic cells *in vivo*. It has been difficult to demonstrate granzyme B expression by T_{reg} cells *in vivo*, but Cao et al[253] have demonstrated that 5% to 30% of T_{reg} cells in the tumor microenvironment express granzyme B and are lytic for NK cells and CTL in a granzyme B- and perforin-dependent manner. Foxp3+ T cells also reduce granule exocytosis in CTL. This inhibitory effect did not require long-lasting contacts between Foxp3+ T cells and CTL but was dependent on the secretion of TGF-β. It has also been shown that polyclonal Foxp3+ T cells limit DC survival in tumor-draining LN through an antigen-and perforin-dependent mechanism involving direct contacts between Foxp3+ T_{reg} and tumor antigen-presenting DCs.[254] Depletion of Foxp3+ T_{reg} resulted in an increase in the number of DCs in tumor-draining LN, a marked arrest in antitumor CD8+ T cell migration, increased T cell priming, and subsequent tumor rejection. This study concluded that the Foxp3+ T cells kill the DC directly but did not rule out indirect effects of T_{reg} on the survival of the DC. The studies demonstrating the cytotoxic activity of T_{regs} suggest that one should reevaluate the concept of "cytotoxic/suppressor cell" because it appears that T_{reg} can mediate both of these functions.

Inhibition of Costimulation

It was originally proposed that the T_{reg} target APC by inhibiting the induction of the expression of costimulatory molecules or perhaps by competition for costimulatory signals.[255] In short-term cultures, T_{reg} cells downregulated the expression of markers on immature bone marrow–derived DC and inhibited their ability to elicit antigen-specific CD8+ T cell responses *in vitro*.[256] Activation of immature DC with CpG DNA or LPS abrogated the ability of the T_{reg} to downregulate DC activation markers or to inhibit T cell activation. Similarly, systemic activation of DC with CpG DNA or an agonistic anti-CD40 antibody completely inhibited the antidiabetogenic activity of T_{reg} cells in nonobese diabetic (NOD) mice.[257] Although these results suggest that T_{reg} exert their suppressive effects by keeping DC in an immature state and that mature DC are resistant to the effects of T_{reg}, other studies have shown that T_{reg} are capable of inactivating CD40L-prestimulated human DC[256] and that T_{reg} can dampen immune responses induced by fully mature DC *in vivo*.[258] Some of the differences in these studies may be secondary to T_{reg} modulation of cytokine production by the DC. For example, although T_{reg} were unable to prevent DC maturation or induction of IL-6 transcription in the presence of strong inflammatory stimuli, they were still able to modulate mature DC responses by enhancing transcription of the anti-inflammatory cytokine IL-10.[259] T_{reg}-mediated enhancement of IL-10 production by fully mature DC may result in inhibition of T cell activation and is one mechanism by which T_{reg} can modulate the function of mature DC.

Considerable controversy exists regarding the significance of the expression of CTLA-4 on T_{reg} and its potential involvement in their suppressor function. Takahashi et al[36] have shown that the addition of the Fab fragment of anti-CTLA reverses suppression in cocultures of T_{reg} and CD4+CD25− T cells. Read et al[23] demonstrated that treatment of mice with anti-CTLA-4 abrogates suppression of IBD mediated by T_{reg} T cells. Reversal of suppression by anti-CTLA-4 or its Fab fragment *in vitro* has not been seen in all studies.[260] In studies[208] where T_{reg} suppress the activation of CD25− T cells in the absence of APC, the addition of anti–CTLA-4 or anti-CD80/CD86 did not reverse suppression, suggesting engagement of CTLA-4 is not required for suppressor function. T_{reg} from CTLA-4−/− mice are as efficient as WT T_{reg} in mediating suppressor function *in vitro*, but in contrast to WT T_{reg}, their suppressive effects were more dependent on TGF-β.[261]

The potential unique role of CTLA-4 as a mediator of T_{reg} suppression has been strengthened by the demonstration that animals with a selective deletion of the expression of CTLA-4 on T_{reg} develop systemic autoimmunity by 7 weeks of age.[262] Thus, CTLA-4 deficiency in T_{reg} cells alone is sufficient to cause fatal disease and maintenance of its expression in activated effector T cells is insufficient to prevent this outcome. Selective CTLA-4 deficiency does not alter the development or homeostasis of T_{reg} cells or render them pathogenic. Foxp3+ T cells with a selective loss of CTLA-4 expression remain anergic but are less suppressive *in vitro*. It has been proposed that the interaction of CTLA-4 on T_{regs} with its ligands, CD80 and CD86 on DCs blocks the subsequent increase of CD80 and CD86 expression or even downregulates CD80 and CD86 expression induced by CD40/CD40L interactions between T effector cells and DCs. T_{reg} cells from CTLA-4−/− mice are defective when compared T_{reg} cells from WT mice in preventing the upregulation of CD80/CD86 expression. The major conclusions from these studies are that inhibition of CD80/CD86 expression limits the delivery of costimulatory signals to naïve T cells via CD28, resulting in defective T effector cell activation. More recent studies have further complicated this interpretation as Paterson et al[263] have demonstrated that specific deletion of CTLA-4 from the adult mouse T_{reg} resulted in enhanced T_{reg} proliferation *in vivo* and was accompanied by increased T_{reg} suppressor function *in vivo*. It has also been shown that the homeostatic proliferation of T_{reg} *in vivo* can be markedly enhanced by treatment of mice with anti–CTLA-4.[140] The enhanced proliferation of T_{reg} in this model was accompanied by enhanced proliferation of memory phenotype CD4+ and CD8+ T cells consistent with a loss of T_{reg} suppressor function. Thus, the role of CTLA-4 as a mediator of the suppressor function of T_{reg} remains unclear.

Quereshi et al[264] have provided a molecular basis for the cell-extrinsic function of CTLA-4 (Fig. 29.16). They have postulated that because of its highly endocytic behavior, CTLA-4 can mediate the intercellular transfer of a ligand from 1 cell to its receptor on a different cell by a process termed transendocytosis. Following T_{reg} activation, CTLA-4 interacts with CD80/CD86, removes the costimulatory molecules from the cell surface, with subsequent degradation of these ligands inside the CTLA-4 expressing cells, thereby limiting CD28 driven costimulation. In contrast to CTLA-4, CD28 was not capable of transendocytosis. Although this model can potentially account for T_{reg}-mediated suppression and the phenotype of the T_{reg} conditional CTLA-4−/− mouse, it should also be emphasized that CTLA-4 expressed by activated T effector cells can also mediate transendocytosis of CD80/CD86 and transendocytosis of CD80/CD86 is not a T_{reg}-specific property.

One interesting disease which supports the transendocytosis model for CLTA-4 function and demonstrates the importance of cell surface expression of CTLA-4 on T_{reg} and/or activated T_{conv} cells is LPS-responsive beige-like anchor (LRBA) deficiency.[265] LRBA-deficient patients resemble patients with IPEX. LRBA plays a crucial role in the intracellular trafficking of CTLA-4 by shielding it from lysosomal degradation and allowing it to traffic back to the cell surface. Homozygous loss of function mutations in LRBA result in total loss of CTLA-4 expression and profound T_{reg} cell dysregulation, particularly Tfr cells. It remains unclear whether loss of LRBA on T_{reg} cells alone or both T_{reg} and activated T_{conv} cells is responsible for the clinical disease. In any case, patients rapidly respond to treatment with CTLA-4-Ig (Abetacept).

An additional pathway (Fig. 29.16) by which the interaction of CTLA-4 on T_{reg} with CD80/CD86 on DC can result in suppression is the activation of indoleamine 2, 3-dioxygenase (IDO). IDO is responsible for the metabolism of the essential amino acid tryptophan. Reduced amounts of tryptophan after induction of IDO are associated with decreased activation of T cells.[266] Other studies have demonstrated that in addition to IDO, T_{regs} may induce multiple other enzymes that consume essential amino acids (EAA).[267] The reduction in EAA levels resulted in reduced mTOR signaling that can potentially lead to the de novo induction of Foxp3+ T cells in a manner that is synergistic with TGF-β. This process may be primarily operative at sites of T_{reg}-mediated suppression, for example, protection against skin graft rejection.

Inhibition of Antigen Presentation

There are a number of major differences between the properties of T_{reg} cells *in vivo* compared to *in vitro*. Although T_{reg} cells are anergic or nonresponsive *in vitro* due to their failure to produce IL-2, T_{reg} cells expand *in vivo* after engagement of their TCR by its cognate antigen in a manner indistinguishable from conventional T cells,[268] and T_{reg} represent one of the most highly proliferative T lymphocyte population *in vivo*.[140] Taken together, it is very difficult to translate from *in vitro* studies of T_{reg} function to the *in vivo* situation. As a major goal of research in this area is the manipulation of T_{reg} function *in vivo* in disease, a clear understanding of T_{reg} biology *in vivo* is required. Specifically, which mechanisms from the long list of mechanisms proposed for T_{reg} suppressor function *in vitro* are used *in vivo*?

The potential pathways of T_{reg} suppression *in vivo* are likely to be more complex than those used *in vitro*. The T_{reg} must be able to localize to various parts of the body such that close contact with most effector T cells with a given antigen specificity either directly or through an APC intermediate becomes possible. Unlike the culture situation, there is ample space for T cells to evade suppression *in vivo* unless T_{reg} can localize with them in response to antigenic stimulation that begins in antigen-draining lymph nodes. A number of possible mechanisms for the suppressive function of T_{regs} *in vivo* have been proposed including whether they inhibit the priming and expansion of T effector cells, block the migration of T effector cells from the site of immunization (eg, the draining LN) to the target organ, or inhibit the differentiation of naïve precursors into pathogenic T effector (Th1, Th2, or Th17) cells.

The first studies to directly examine the effects of T_{reg} on priming immune responses *in vivo* used two-photon laser-scanning microscopy.[269,270] After entering the LN through high endothelial venules, T cells travel through a region near the paracortical T cell zone and B cell follicles where tissue emigrant DCs are found. In the absence of antigen, T cells move freely within the T cell zones of the LN. In contrast, in the presence of antigen, T cells interact with the antigen-loaded DC first by a process termed swarming that ultimately results in arrest of their locomotion for a period of hours before they again become motile. Taken together, these studies demonstrated that both T_{reg} cells and T effector cells formed similar clusters with DCs. In the absence of T_{reg}, cluster formation by T effector cells was more sustained. Cluster formation by T_{regs} preceded the inhibition of T effector activation. T_{reg} cells appeared to block T effector cell priming by decreasing the contact time between DCs and the naïve CD4+ T cells. Surprisingly, no direct interactions between T_{reg} and T effector cells were observed in these studies indicating that the major suppressive effects of the T_{reg} were directed to the DCs. The results of these studies are consistent with models that propose that T_{regs} cells can preferentially interact with DCs and either prevent the subsequent interaction of DCs with T effector cells or ultimately lead to inactivation of the DCs. It remains unclear why T_{regs} that express the same TCR as the T effector cells would preferentially interact with DCs. One explanation is the expression of higher levels of molecules mediating cell adhesion such as LFA-1[271] or Nrp-1.[272]

One approach to analyze T_{reg} suppressor function *in vivo* is to cotransfer antigen-specific T_{reg} from TCR-transgenic mice together with naïve antigen-specific T_{conv} of the same antigen specificity to normal immunocompetent mice. Studies of this type observed marked suppression of the expansion of the antigen-specific Teff cells, as well as marked inhibition of cytokine production (IL-2 and IFN-γ) 96 hours after cell transfer.[273-275] The failure of T effector cell expansion in the presence of antigen-specific T_{reg} suggested that the major target of the suppressive functions of the T_{reg} was the APC. Hanig and Lutz[275] cotransferred peptide-pulsed bone marrow DC that had been matured with LPS, with antigen-specific T effector cells with or without antigen-specific T_{reg} to normal recipients. Clustering of DC with both T effector and T_{reg} was observed in the spleen at 24 hours after cell transfer. However, when all populations were injected simultaneously, no effects of T_{reg} on total spleen cell size and T effector numbers were observed for the first 48 hours. In contrast, when T_{regs} were transferred 10 to 24 hours before the T effector cells, cluster formation was completely inhibited and marked suppression of T effector cell expansion was seen. Thus, in this model, T_{reg}-mediated suppression required that the T_{reg} be preactivated *in vivo*. When T_{reg} and DC were injected 24 hours before the T effector cells, marked downregulation of the expression of CD80/CD86, PD-L1, and PD-L2 was observed 48 hours later on DC. The conclusions drawn from these experiments were that the T_{reg}-mediated block in T effector cell expansion was mediated by downregulation of specific costimulatory molecules on the DC surface that, in turn, indirectly control the T effector cell expansion.

Similar results were obtained in cotransfer studies using preactivated antigen-specific iT_{regs} that had been induced *in vitro* by exposure to TGF-β and IL-2. Cotransfer of antigen-specific iT_{regs} resulted in a >90% inhibition of the expansion of the naïve T effector cells (Fig. 29.18) and complete inhibition of the induction of effector cytokine production.[276] While the expansion of the T effector cells was inhibited, the T_{regs} in these mice readily proliferated consistent with the earlier studies. Again, it is likely that the T_{reg} in this model were inhibiting antigen presentation by acting on DC. No evidence for T_{reg} killing of DC was observed, and the T_{reg} did not produce IL-10

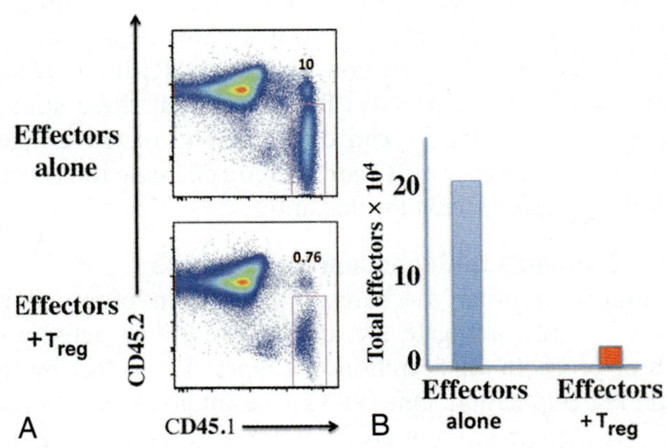

FIG. 29.18. Antigen-specific T_{reg} inhibits the expansion of antigen-specific T effector cells in response to immunization.

that could potentially suppress the capacity of DC to present antigen or deliver costimulatory signals. An alternative possibility is that the T_{reg} may inhibit chemokine production by the DCs resulting in less recruitment of the transferred DCs to the site of immunization and less T effector cell activation.[277] It is still difficult to rule out direct effects of the T_{regs} on the responder T cells *in vivo*. It is possible that T_{reg} and T effector cells could interact with the same antigen-presenting DC and complex interactions between all 3 cells would occur in this milieu.

These studies on T_{reg}-DC interactions were extended by Akkaya et al[278] who described a unique model by which T_{reg} may mediate suppression of DC function. Most studies indicate that TCR-signaling is required to induce the program of activities that mediate T_{reg} cell suppressor function in an antigen-nonspecific manner, resulting in suppression of DCs and/or T cells in their vicinity. In contrast, when antigen-specific T_{reg} cell function was analyzed *in vivo*, T_{reg} cells inhibited the interactions of antigen-specific T_{conv} cells with antigen-pulsed DCs, and the inhibition of this interaction *in vivo* was antigen-specific. Preincubation of T_{reg} cells *in vitro* with double-pulsed DCs inhibited the subsequent capacity of DCs to activate T_{conv} cells specific for their cognate antigen but not that of T_{conv} cells specific for the second antigen expressed on DCs. The defective stimulatory capacity of T_{reg} cell–treated DCs could be restored by repulsing DCs with the cognate peptide. This finding strongly suggests that mere competition for binding is not sufficient to completely explain antigen-specific T_{reg} cell–mediated suppression.

The strong binding of T_{reg} cells and their capacity to debilitate DC function in an antigen-specific manner raised the possibility that T_{reg} cells were removing their target antigen from the DC surface by a process similar to trogocytosis. Subsequent studies demonstrated that the uptake of peptide MHC-II (pMHC-II) complexes by T_{reg} *in vivo* and *in vitro* were antigen-specific. The noncognate antigen could not be detected on or in the antigen-specific T_{reg} cells. T_{reg} cells appeared to have a much greater capacity than that of T_{conv} cells or activated T cells to take up pMHC-II complexes. T_{reg} cells could take up several other DC-derived molecules, including CD86 and PD-L2. Therefore, it is likely that T_{reg} cell TCRs exist in a cluster as part of the immune synapse, interacting with their cognate pMHC-II together with surrounding DC molecules involved in costimulation and cell adhesion but excluding the noncognate pMHC-II complexes expressed on the same DC. Taken together, these studies strongly suggest that T_{reg} can deplete their cognate antigen from the surface of the DC and thereby decrease the activation of T_{conv} cells specific for that antigen.

Other Cell Contact–Dependent Mechanisms

A number of other mechanisms have been proposed by which T_{reg} cells abrogate the antigen-presenting activity of DCs or promote the secretion of suppressive factors by the target DC population. One cell surface antigen that may play in T_{reg} cell suppression of DC function is LAG-3 (CD223), a CD4 homolog that binds MHC class II molecules with very high affinity. Binding of LAG-3 to MHC class II molecules expressed by immature DCs induces an ITAM-mediated inhibitory signal that suppresses DC maturation and immunostimulatory capacity.[279] Because activated human, but not mouse, T cells can express MHC class II, T_{reg} cell-mediated ligation of LAG-3 on activated human T effector cells might also induce suppression of T cell function. A second molecule that has been proposed to play a role in the interaction of T_{reg} cells with DC is neuropilin-1 (Nrp-1). Nrp-1 is a receptor for class II semaphorins and a coreceptor for vascular endothelial growth factor. Nrp-1 is preferentially expressed on T_{regs} and can be induced by ectopic expression of Foxp3 in Foxp3– T cells.[280] Nrp-1 promotes long interactions between T_{reg} cells and immature DCs. Anti–Nrp-1 abrogates suppression of proliferation mediated by T_{reg} cells when the responder T cells are stimulated with low concentrations of antigen. These data suggest that the role of Nrp-1 is to give T_{reg} cells a head start over naïve responder T cells under conditions in which antigen is limiting. Nrp1 expression on T_{reg} has been proposed as a ligand for enhancing T_{reg} stability and suppressor function by interacting with semaphorin-4a[280] on T_{conv} cells. While Nrp1 is dispensable for T_{reg} suppressor function and homeostasis in the steady-state, it may play an important role in inflammatory sites such as tumor immunity or IBD. Sem4a ligation of Nrp1 prevented Akt phosphorylation in T_{reg} via the action of the phosphatase, PTEN, which resulted in nuclear localization of the transcription factor, Foxo3a. In addition to T_{conv} cells, Sem4a is expressed in the nervous system, the eye, and the intestine, which are all sites where potentiation of T_{reg} function may be desirable.

Reversal of and Resistance to Suppression

One approach to determining potential cell surface antigens involved in the process of T_{reg}-mediated suppression has been to reverse suppression with antibodies to candidate antigens. One member of the TNFRSF, the GITR (TNFRSF18), has been shown to play an important role in regulation of T cell suppressor activity. Both a polyclonal antiserum and a mAb to the GITR appeared to be able to reverse suppression mediated by freshly isolated CD25+ T cells.[37,38] However, the GITR is expressed on two populations of cells (resting T_{reg} and activated CD4+CD25– cells) that do not manifest suppressor activity. It is therefore very unlikely that the GITR is the molecule responsible for mediating suppressor effector function. Surprisingly, culturing CD25+ T cells with the anti-GITR in the presence of IL-2, but in the absence of anti-CD3, resulted in a vigorous proliferative response and breaking the nonresponsive state of the T_{reg} to stimulation with IL-2.

These studies were originally interpreted as indicating that anti-GITR functioned as an agonist for the GITR, resulting in a signal that instructs the T_{reg} not to mediate their suppressive functions. As both CD25– and T_{reg} can express the GITR, it was impossible to conclude that the anti-GITR mediated its effects by acting solely on the T_{reg}. When combinations of WT and GITR–/– CD4+CD25– and T_{reg} were used in coculture experiments, ligation of the GITR on the CD25– responders, not the T_{reg}, was required to abrogate suppression.[281] These results suggest that interactions of the GITR with its ligand, the GITR-L, provided a signal that renders effector T cells resistant to the inhibitory effects of T_{reg}. Similar conclusions were drawn from studies in which the costimulatory effects of anti-GITR *in vivo* were maintained following depletion of T_{reg}.[282] The GITR-L was shown to be expressed on a number of different resting APC populations including DC, macrophages, and B cells and to be downregulated on activation. Thus, engagement of the GITR on T effector cells by its ligand on APC early during the course of an immune response may render the responder cells resistant to suppression by T_{reg}.

Other studies have implicated additional members of the TNFRSF as potential targets for reversal of T_{reg}-mediated suppression. The capacity of T_{regs} to inhibit GVHD was reversed when T_{reg} were treated with an anti-OX40 mAb.[283] Injection of tumor-bearing mice with an agonistic anti-OX40 also appeared to inactivate intratumor T_{regs} and results in tumor rejection.[284] Stimulation of human T_{reg} *in vitro* in the presence of TNF or in the presence of an agonistic mAb to TNFRII reversed their *in vitro* suppressive activity and appeared to downregulate their expression of Foxp3.[285] One problem with the interpretation of these studies is that activated T_{conv} cells also express high levels of OX40 and TNFRII, and it is difficult to determine the precise site of action of the agonistic mAbs. As shown with the anti-GITR, the majority of the "reversal" of suppression may actually be the result of enhanced resistance of the activated T effector cells to the suppressive effects of the T_{reg}.[286] This explanation is also compatible with the observations of Pasare and Medzhitov[287] that soluble factors (eg, IL-6) released by activated DC can act directly on T effector cells, but not T_{reg}, to render them resistant to suppression. Other members of the TNFRSF appear to actually enhance T_{reg} function. Stimulation of T_{regs} with an agonistic mAb to TNFRSF25 (also known as DR3) resulted in marked expansion of T_{regs} *in vivo*, and these expanded T_{regs} were protected mice against the induction of allergic lung inflammation.[288]

In addition to the role of APC-derived signals in rendering responder cells resistant to suppression, a number of mutations in TCR signaling and costimulatory pathways also appear to render responder cells completely resistant to the inhibitory signals delivered by T_{reg}. One of the best examples of a defect in resistance to regulation is the cbl-b−/− mouse. Cbl-b is an E3 ubiquitin ligase that negatively regulates the CD28 costimulatory pathway and downregulates the PI3K/Akt pathway. In the absence of cbl-b, mice develop CD28-independent hyperactive responses and autoimmunity. T_{reg} from cbl-b−/− mice function normally *in vitro*, but CD25− T cells from cbl-b−/− mice are resistant to suppression *in vitro* and *in vivo* by wild type T_{reg}.[289] Effector T cells from a number of other mutant mouse strains exhibit a resistance to suppression similar to that seen with cbl-b−/− effectors (Table 29.3). Many of these strains have in common a hyperactivated PI3K/Akt pathway that may be a direct or indirect target for T_{reg}-mediated suppression.

T_{REG} METABOLISM

Metabolism plays an important role in the function of all immune cells including T_{reg} cells.[290,291] T_{reg} cells exhibit a unique metabolic signature in comparison to T helper subsets which sustains their proliferative and regulatory functions while constraining the development of effector programs normally associated with differentiated T helper (Th1, Th2, Th17) cells. While differentiated T helper cell subsets express high levels of the glucose transporter1 (Glut1) and display a heightened glycolytic rate, iT_{reg} cells were shown to preferentially use fatty acid oxidation (FAO) and pyruvate-dependent oxidative phosphorylation (OXPHOS) to satisfy their energetic demands.[292] Expression of Foxp3 results in upregulation of the electron transport chain, which drives an increase in the respiratory capacity and augments their ability to utilize fatty acids to fuel OXPHOS. Foxp3 confers a metabolic advantage on T_{reg} cells in low-glucose lactate-rich environments by suppressing Myc gene and T_{reg} glycolysis in order to maintain regulatory function. When T_{conv} cells are transduced with Foxp3, increased expression of genes involved in lipid and peptide hormone pathway metabolism pathways are increased, whereas those encoding products involved in glucose and nucleotide metabolism were downregulated.[293,294] In contrast, transgenic expression of Glut1 in T_{reg} cells reduced T cell suppressive capacity and resulted in downregulation of Foxp3.

To support the transition from quiescent naïve T cells to T effector cells, both T_{conv} cells and T_{reg} reprogram their cellular metabolism. In contrast to iT_{regs}, activated T_{reg} can engage aerobic glycolysis in a manner comparable to differentiated effector T cells. T_{reg} cells *in vivo* are more metabolically active than naive T_{conv} cell and undergo proliferation balanced by apoptosis. T_{reg} cells expressing high levels of Ki-67 have increased Glut1 expression, increased glycolysis, and increased mTORC1 activity. Upregulation of Glut1 expression dampens T_{reg} suppressor function and reduces the expression of Foxp3. To balance enhanced glycolysis, Foxp3 can program upregulation of multiple electron transport components, which results in an increase of spare respiratory capacity (SRC) and oxidative phosphorylation (OXPHOS) activity,[295] especially in tumors.[296] As inflammatory signals and T_{reg} activation decrease, Foxp3 levels increase and tilts the metabolic balance away from glycolysis to favor mitochondrial oxidative pathways, resulting in decreased proliferation and increased suppressive capacity.

The mechanistic target of rapamycin (mTOR) pathway integrates environmental signals and metabolic cues to direct T cell fate decisions. The activation of mTOR which is the catalytic subunit of the mTORC1 and mTORC2 complexes delivers an obligatory signal for the proper activation and differentiation of effector CD4+ T cells. T_{reg} cells have elevated steady-state mTORC1 activity compared to naïve

TABLE 29.3 Genetically Deficient Mice That Are Resistant to T_{reg} Suppression

Cbl-b−/−
TGF-β-RII−/−
TRAF6−/− T cell–specific
NFATc2−/−/NFATc3−/−

Data from Wohlfert EA, Gorelik L, Mittler R, et al. Cutting edge: deficiency in the E3 ubiquitin ligase Cbl-b results in a multifunctional defect in T cells TGF-β sensitivity in vitro and in vivo. *J Immunol.* 2006;176:1316-1320.

T_{conv} cells. Disruption of mTORC1 by deletion of Raptor (the protein required for mTORC1′ function) leads to loss of T_{reg} suppressive activity *in vivo* and the early onset of inflammatory disease.[297] Raptor-deficient T_{reg} have a deficiency in homeostatic proliferation upon transfer to normal mice, and this deficiency was associated with defective expression of ICOS and CTLA-4. Raptor-deficient T_{reg} had deficient ECAR and OCR, which mark glycolysis and mitochondrial respiration, respectively. Nutrient signals from the amino acids, arginine and leucine, are integrated by the small G proteins, Rag, and Rheb, which sustain mTORC1 activation in activated T_{reg} cells driving their accumulation, function, and suppressive gene signature.[298] However, inhibition of mTOR during T cell activation promotes the generation of long-lived T_{reg} cells with enhanced SRC and memory-like phenotype.[299]

The role of mTORC2 in T_{reg} function has also been investigated. Deletion of Rictor (the protein needed for mTORC2 function) does not result in any abnormalities in the steady-state.[300] It thus appears that mTORC2 has a less important function than mTORC1 in regulating T_{reg} metabolism in the steady-state. Notably, Foxp3-deficient cells of IPEX patients exhibited dysregulated mTORC2 signaling and augmented glycolysis and OXPHOS both of which were reversed by rapamycin.[301] Deletion of Rictor partially restored mutant T_{reg} cell suppressor function. Skewing of T_{reg} cells toward a Teff cell-like phenotype is also dependent on mTORC2 and glycolysis. Targeting these pathways may improve stability and function in autoimmune and inflammatory disease and may have a therapeutic effect in the TME.[302]

T$_{REG}$ SPECIALIZATION

Expression of T Effector Master Transcription Factors

The functional specialization of T effector cells is due to the expression of master transcription factors, namely T-bet, GATA-3, and RORγt, which control the differentiation of Th1, Th2, and Th17 subsets, respectively. In general, these subsets mediate proinflammatory and potential responses that are harmful. T_{reg} cells also use the Th cell–associated transcription factors to maintain or restore immune homeostasis (Fig. 29.19). T_{reg} cells expression of interferon regulatory factor 4 (IRF4), a transcription factor involved in the control of IL-4 production and Th2 cell differentiation, is required for T_{reg}-mediated control of Th2-type inflammatory responses. Mice in which IRF4 is specifically deleted in Foxp3+ T_{reg} develop a lymphoproliferative disease that is associated with a selective expansion of IL-4 and IL-5 producing Th2 cells, have elevated serum levels of IgG1 and IgE, and develop splenic germinal centers (GCs).[162] IRF4 is also crucial for the differentiation of T follicular helper (Tfh) cells. The increase in GCs in mice with a T_{reg} cell–specific deletion of IRF4 in the absence of an increase in IL-13 producing cells or eosinophil numbers suggested that T_{reg} cell–specific expression of IRF4 may be needed to control a specific component of Th2 cell–associated inflammation, Tfh cell activation, and high-affinity antibody production.

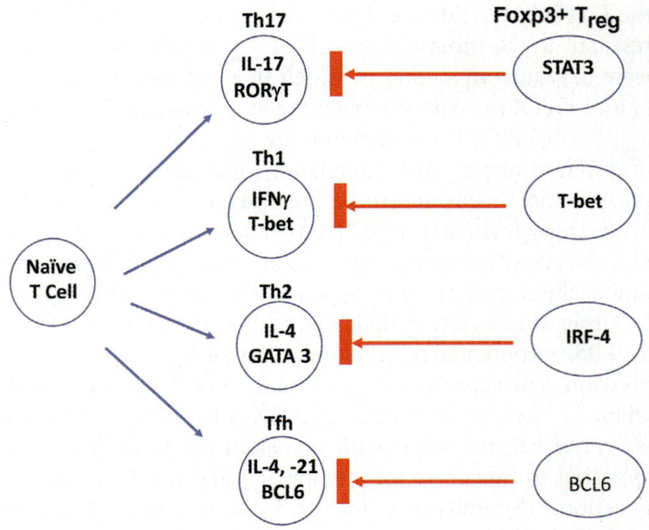

FIG. 29.19. T_{reg} specialization.

T-bet is the master transcription factor controlling Th1 responses and is also expressed by a subset of T_{reg} cells. T-bet expression is required for T_{reg} cells homeostasis and function during polarized Th1 responses. While IFN-γ can inhibit the peripheral generation of Foxp3+ T_{reg} cells from naïve CD4+ T cells,[69] IFN-γ signaling through STAT1 activation drives T-bet expression in T_{reg} cells resulting in control of Th1-type immune responses. IFN-γ may thus limit the magnitude and duration of Th1-type inflammatory responses. T-bet$^+$ T_{reg} accumulate at sites of Th1 inflammation, and T-bet−/− T_{reg} cells display impaired proliferation during Th1 type inflammation and do not control the expansion of IFN-γ–producing Th1 cells when transferred into scurfy mice.[303] The failure of T-bet−/− T_{reg} cells to control Th1-driven inflammatory responses may result from their inability to survive and proliferate in a highly polarized Th1-type environment.

Lastly, conditional deletion of STAT3 in T_{reg} cells results in the selective dysregulation of Th17 responses accompanied by development of a spontaneous fatal intestinal inflammation with elevated levels of IL-17 production.[304] The mechanisms by which T-bet, IRF4, and STAT3 control T_{reg} cell activity during Th1, Th2, and Th17 responses are still unclear but likely involve control of T_{reg} cell migration, function, and homeostasis. Selective expression of transcription factors associated with a given Th responses may induce molecular changes needed to restain a particular response. This hypothesis has important implications for the therapeutic uses of T_{reg} cells as it implies that specific subsets of T_{reg} cells may be needed in treating Th1-, Th2-, and Th17-mediated inflammatory diseases.

Tfr Cells

The discovery that a minority of CD4+ T cells, termed Tfh cells, can provide help to B cells, is one example of T effector cell specialization. Mice that lack Foxp3+ T cells develop spontaneous GCs and have high titers of autoantibodies. In many models of organ-specific autoimmune disease, T_{reg}

cells suppressed the activation of autoantigen-specific T cells and inhibited the production of autoantibodies to the target organ. As B cells and autoantibody production are prominent mediators of pathology in many autoimmune diseases, it is critical to understand the potential for T_{reg} to control autoantibody production. While it is clear that T_{reg} can suppress antibody production by blocking T helper functions, a number of other studies have suggested that T_{reg} can exert a direct effect on certain B cell functions. Bystry et al[305] were the first to suggest that T_{reg} cells could inhibit LPS-mediated B cell activation. Lim et al[306] demonstrated that human T_{reg} can suppress B cell–dependent immunoglobulin production and class switch recombination by a cell contact–dependent mechanism in the absence of T helper cells but the mechanism of suppression was not further characterized. Foxp3+ T_{reg} cells were identified in T-B area borders and within GCs of human lymphoid tissues.[307] T_{reg} cells were also able to suppress the expression of AID, a key enzyme involved in class switch recombination and affinity maturation. The biochemical mechanisms for the suppression of B cell function in these studies remain unclear.

The GC reaction is a multistep process during which naïve B cells become either plasma cells that produce high-affinity antibodies or memory B cells that can rapidly respond to antigen reexposure. B cells interact with Tfh cells which facilitates B cell activation, somatic hypermutation, affinity maturation, class switch recombination, and differentiation of B cells into memory cells and plasma cells.[308] Before the GC reaction can take place, CD4+ T cells differentiate into Tfh cells in the T cell zone when exposured to antigen presented by DCs and B cells primed by antigen presented on follicular DC. The Tfh cells then migrate to the B cell border via gradients of CXCL13 and interact with B cells. Tfh cells subsequently move deeper into the GC and interact with antigen-presenting B cells, resulting in B cell proliferation and GC formation (linked recognition). The bidirectional interaction between the Tfh cell and the B cells is characterized by the secretion of IL-21 and IL-4 by the Tfh cells and the delivery of costimulatory signals by CD40L on the Tfh cell to CD40 on the B cells. Similarly, B cells provide Tfh cells with antigenic stimulation via peptide-MHCII complexes, costimulatory signals through CD28 and ICOS on the Tfh cell. The antigen-specific B cells then undergo class switch recombination (CSR) and B cell receptor hypermutation. At the same time, Tfh cells facilitate affinity maturation of B cells by sending death signals through CD95/FAS on B cells so that only the highest affinity B cells overcome the death signal.

The GC needs to be highly regulated to prevent production of autoantibodies. Tfr cells control the GC reaction.[309-311] Tfr cells are a specialized subset of activated or effector T_{reg} that express high levels of PD-1, ICOS, and CXCR5, allowing them access to the GC to suppress B cell responses. Tfr cells express lower levels of the transcription factor, Bcl6, than Tfh cells and also express Blimp-1. Tfr cells originate from tT_{reg} and differentiate in response to antigen presented by DCs in lymph nodes, but complete differentiation of Tfr cells only occurs after interaction with cognate B cells within the follicle. Tfr cells constitute about 50% of all CD4+CXCR5+ cells, resulting in a 1:1 ratio of Tfr:Tfh cells prior to immunization. After immunization, both Tfh and Tfr cells differentiate, but at the peak of the response, Tfr cells are only ~20% of the CD4+CXCR5+ population. Both CD28 and ICOS are essential for Tfr differentiation as CD28−/− and ICOS−/− mice lack Tfr cells. Murine and human Tfr cells express low levels of CD25, and it is the CD25− Tfr cells that are preferentially localized to the GC.[312,313] CD25− Tfr cells have a gene expression profile more similar to Tfh cells, and it remains unclear whether the CD25− Tfr derive from CD25+ T_{reg}. One possibility is that Tfr cells lose CD25 expression throughout their differentiation pathway toward fully differentiated GC-resident Tfr cells.

PD-1 can inhibit Tfr suppressive function as well as Tfr-differentiation.[314] PD-1−/− Tfr cells are able to more potently suppress antibody production both *in vitro* and *in vivo*. CTLA-4 is expressed on Tfr cells at higher levels than non-Tfr T_{reg}. Deletion of CTLA-4 from Tfr cells markedly decreases their suppressive function. For example, in an *in vivo* adoptive transfer experiment, much higher levels of antigen-specific antibody were produced in the presence of CTLA-4−/− Tfr than in the presence of WT Tfr cells. These studies strongly imply that Tfr suppressive function depends on CTLA-4,[315,316] but the mechanism of CTLA-4–mediated regulation of Tfr function has not been characterized.

Tfr cells suppress both Tfh cells and B cells in the GC reaction, but it is not known whether Tfr cells can inhibit all stages of B cell activation. Tfr cells may inhibit initial T-B interactions at the T-B border and likely inhibit fully activated GC B cells. In addition, it has been shown that Tfr cells can inhibit plasma cells that have undergone class switch recombination.[317] Tfr cells might regulate the amount of antibody produced by plasma cells or the ability of plasma cells to migrate out of the lymph node. No methods have been developed to selectively modulate Tfr cells without affecting Tfh cells or non-Tfr T_{reg}. It would certainly be desirable to attenuate Tfr function during vaccine administration, as this might result in heightened prolonged antibody responses.

T_{REG} PLASTICITY

The studies on T_{reg} specialization have demonstrated the potential beneficial effects of Th cell–produced inflammatory cytokines on the ability of T_{regs} to control pathologic immune responses mediated by those cytokines. Other studies have challenged the notion of a committed Foxp3+ T_{reg} cell lineage and raised the possibility that T_{reg} cells can lose Foxp3 expression when present in an inflammatory milieu or following transfer to an immunodeficient environment and be "reprogrammed" to become pathogenic T effector cells. For example, during infection with *Toxoplasma gondii*, T_{reg} cells lose Foxp3 expression and become IL-12 responsive. They can then express high levels of T-bet, acquire Th1 effector characteristics, and produce IFN-γ contributing to lethality.[318] Furthermore, when T_{reg} cells were cultured under Th17- or Th1-polarizing conditions, a substantial fraction of IL-17+Foxp3+ and IFN-γ+Foxp3+ T cells were observed.[319] Thus, IL-6 may potentially convert T_{reg} cells into proinflammatory Th17 cells in a proinflammatory environment.[320,321]

No evidence of reciprocal conversion of Th17 cells to T_{reg} was observed.

The T_{reg} and Th17 programs of differentiation are clearly linked, and Foxp3 and RORγt transcription factors can interact with one another.[68] Whereas transduction of naïve CD4+ T cells with RORγt-induced IL-17 expression, cotransduction with Foxp3 abrogated induction when Foxp3 contains its exon 2 encoded domain, which is required for binding to RORγt. After *in vitro* induction of RORγt and Foxp3 by TGF-β, cells do not express IL-17 and have the potential to differentiate into either Th17 or T_{reg}, depending on the cytokine environment. In the presence of a proinflammatory cytokines (IL-6, IL-21, or IL-23) and low concentrations of TGF-β, RORγt expression is further elevated, and Foxp3 expression is reduced, resulting in Th17 differentiation. Conversely, in the absence of proinflammatory cytokines and high concentrations of TGF-β, T_{reg} induction is promoted. IL-2 suppresses RORγt expression and promotes the induction of T_{reg}.[322] Runx1 can also form a complex with RORγt to promote Th17 differentiation but can also interact with Foxp3 to suppress IL-2 and IFN-γ production, resulting in enhancement of T_{reg} suppressor function. Functional plasticity in cells that coexpress Runx1 and RORγt may be governed by the ability of Runx1 to cooperate with either transcription factor.[323] Foxp3+ T_{reg} cells in PP differentiate into Tfh cells that can then participate in GC formation and IgA synthesis in the gut.[324]

The potential plasticity of T_{reg} is supported by epigenetic studies of cytogene expression. Wei et al[325] have performed genome-wide examination of histone modifications and DNA methylation that accompany changes in gene expression in CD4+ T cell subsets. Trimethylation of histone H3 lysine 4 (H3K4me3) is a permissive mark found in the promoters and enhancers of active genes, whereas trimethylation of histone H3 lysine 27 (H3K427me3) is present in broad domains that encompass inactive genes. The epigenetic marks found at the IFN-γ, IL4, and IL17 genes correlate precisely with Th1 and Th2 lineages. However, these genes in T_{reg} were bivalently modified, and this result is consistent with their potential plasticity. These bivalent modifications may allow specific lineage regulator gene loci to be activated under different polarizing conditions, thus allowing reprogramming of T_{regs} into other lineages.

In addition to having permissive epigenetic modification of T effector cytokine genes, T_{regs} may express GATA3, the canonical Th2 transcription factor. GATA3 is highly expressed in T_{regs} at barrier sites such as the skin and the gut and is readily expressed in T_{regs} following TCR activation. T_{reg}-intrinsic expression of GATA3 is required for maintenance of high level of Foxp3 expression and promotes T_{reg} accumulation at inflamed sites.[326] Expression of GATA3 was not seen in T_{regs} expressing T-bet or RORγt. In one study, GATA3 expression by T_{regs} was required for T_{reg} accumulation in inflammatory conditions such as GI infection or EAE, and GATA3−/− T_{regs} failed to control colitis due to impaired accumulation. Another study demonstrated a more profound defect in T_{reg} function in T_{reg} conditional GATA−/− mice including the development of a generalized inflammatory disorder, decreased expression of Foxp3 and Foxp3 signature genes, and decreased suppressor function *in vitro*.[327]

Two groups generated mice that were engineered to track the fate of Foxp3+ cells *in vivo*. In these systems, mice expressing the Cre recombinase in T_{reg} cells under control of regulatory elements from the Foxp3 gene were crossed with mice in which the Cre-mediated recombination removes a stop codon from within a fluorescent protein encoding a reporter gene that was knocked into the ubiquitously expressed Rosa 26 locus. Thus, in these animals, cells even transiently expressing Foxp3 are permanently marked, and their phenotypic properties can be spatially and temporarily examined. Using this system, Zhou et al[328] reported that a portion of CD4+Foxp3+ T cells downregulate Foxp3 and acquire the ability to produce effector cytokines such as IFN-γ even in the absence of any experimental manipulation. The frequency of these ex-T_{reg} cells was increased in the pancreatic islets of NOD mice, indicating that they may contribute to immune pathology. These studies suggest that highly polarized inflammatory environments can subvert T_{reg} cell function by converting them to Foxp3− Teff cells. The other study demonstrated that Foxp3 expression by T_{reg} cells is remarkably stable even in highly inflammatory settings.[329] The discrepancies in these studies may be due to differences in the inflammatory systems used to examine T_{reg} stability *in vivo* or to differences in the way the reporter mice were constructed. Alternatively, T_{reg} that have lost Foxp3 may not be the product of reprogramming of committed T_{reg} cells but may represent the products of a minor population of uncommitted cells that transiently express Foxp3.[330] These cells can accumulate by selective expansion and under the appropriate environmental conditions differentiate into Th1, Th2, or Th17 T effector cells.[331]

Several studies have examined the potential molecular mechanisms that may be involved in the control of Foxp3 stability. The Wnt-signaling/β-catenin pathway appears to have a major role in controlling T_{reg} plasticity. Van Loosdregt et al[332] have shown that the activation of the Wnt pathway reduced T_{reg} suppressor function *in vivo* and *in vitro*. Similarly, Sumida et al[333] have demonstrated that T_{reg}-specific stabilization of β-catenin resulted in a loss of T_{reg} suppressive function and the induction of a disease resembling that seen in scurfy mice. Thus, the production of Wnt in the local inflammatory environment may result in impaired T_{reg} function.

T_{REG} FUNCTIONS IN TISSUES

In addition to subsets of T_{reg} specialized in the suppression of differentiated T helper populations, T_{reg} can be found in a number of nonlymphoid tissues where they play a role in suppressing inflammatory responses, but more importantly, play a role in modulating the physiologic functions of the tissue where they reside.[334]

Adipose Tissue

Adipose tissue is composed of adipocytes and stromal cells, vascular endothelial cells, and immune cells. The main function of white adipose tissue is to store lipid, while brown adipose tissue located around the neck and around large blood vessels of the thorax releases heat. Macrophages are the central players in obesity-associated inflammation. In the mouse, visceral adipose tissue (VAT) T_{regs} are found in the epididymal fat pad make up a high percentage of CD4+ T cells (up to 50%) at 20 to 25 weeks of age.[335] VAT T_{regs} arise in the thymus in the first week of life, seed VAT sparsely until 10 to 15 weeks of age, and then dominate the local CD4+ compartment. Foxp3+ T_{regs} are found in spaces between adipocytes often in structures containing macrophages. The TCRs expressed by T_{reg} in the VAT exhibit pauciclonal expansion suggesting that these T_{reg} may have been stimulated by VAT-specific antigens.[336] They express a unique transcriptome which differs from T_{reg} found in lymphoid organs. The transcription factor, PPARγ, is the master regulator of adipocyte differentiation and is also the major driver of the VAT T_{reg} phenotype.[337] Foxp3 and PPARγ interact either directly or as part of a complex. Conditional deletion of PPARγ in T_{regs} results in a reduction of VAT T_{regs}, but not lymphoid T_{regs}, and induction of inflammation and a metabolic syndrome. VAT T_{regs} function in a manner similar to lymphoid T_{regs} in suppression assays *in vitro* but have a distinct role *in vivo* in controlling VAT metabolism. The percentage of VAT T_{reg} in male mice is inversely correlated with insulin sensitivity in models of obesity. Depletion of T_{regs} provokes VAT tissue inflammation and a metabolic syndrome, while expansion of VAT T_{regs} improves metabolic health. The likely target cells of VAT T_{regs} are macrophages in the adipose tissue. VAT T_{reg}-derived IL-10 plays a prominent role in inhibiting proinflammatory cytokine production by adipose tissue macrophages.

The cytokine IL-33 also plays an important role in VAT T_{reg} biology.[338] IL-33 is a member of the IL-1 family and is primarily made by nonhematopoietic cells. The IL-33 receptor is termed, St2, and is encoded by *ILrl1*. St2 is expressed by various types of hematopoietic lineage cells. ILrl1associates with IL-1RAcP and signals through the Myd88 pathway. IL-33 is best known for promoting type 2 immunity by acting on ILC2s, Th2 cells, and anti-inflammatory macrophages. IL-33 can act on certain T_{reg} subpopulations, resulting in expansion in lymphoid organs and in colonic LP. More than 80% of VAT T_{regs} express St2. St2 or IL-33 deficient mice have a marked reduction in VAT T_{regs} but normal numbers of lymphoid T_{regs}. IL-33 injection can expand VAT T_{regs} with a reduction in inflammation and improvement in metabolic indices. Human omental fat has measurable levels of Foxp3 mRNA, which is inversely correlated with body mass index. The IL-33-St2 axis is also expressed in human omentum.

Skeletal Muscle

When skeletal muscle is damaged, regeneration is influenced by inflammatory events that accompany repair.[339] An initial population of proinflammatory or M1 macrophages is required for clearance of apoptotic, necrotic cells, and debris followed by M2-macrophages with proregenerative functions. A small population of Foxp3+ T_{regs} is normally present in muscle. Similar to VAT T_{regs}, skeletal muscle T_{regs} can be distinguished from their lymphoid organ counterparts by their TCR repertoire and transcriptome. Following mild cryoinjury or following the injection of cardiotoxin, this small population can expand and comprise 60% of CD4+ T cells. The rise in muscle T_{regs} coincides with the switch in macrophage phenotype from pro to anti-inflammatory. Depletion of T_{regs} compromises muscle regeneration including the pro to anti-inflammatory shift in macrophages. T_{regs} are located both within the inflammatory infiltrate of injured muscle and between remote myofibers. Clonal expansion of T_{reg} is also observed in response to both acute and chronic injury, and it is likely that the T_{regs} are responding to muscle antigens.[340] The transcriptome of muscle T_{regs} is very different from lymphoid T_{regs} but less distinct from VAT T_{regs}. Muscle T_{regs} overexpress amphiregulin, a member of the epidermal growth factor family, whose receptor EGFR is expressed on muscle satellite cells. Muscle T_{regs} also express high levels of IL-10. The IL-33-St2 axis plays an important role in the accumulation of T_{regs} in skeletal muscle as IL-33 rapidly increases within hours of muscle injury. Muscle of aged mice contains very few T_{regs}.[265] When muscle tissue in old mice is injured, injection of IL-33 expands T_{regs} and improves muscle regeneration. T_{regs} may also play a role in cardiac muscle.

Skin

The skin is the largest organ of the body and influences many physiological processes including thermoregulation, hydration, insulation, and sensation. Both mouse and human skin contain a large number of tissue-resident T_{regs} which are primarily found in the dermis, the layer below the epidermis.[341] One early observation that skin T_{regs} play an important role in skin physiology is that patients with IPEX frequently present with inflammatory skin disorders and eczema. T_{reg} cells in the skin are required for homeostasis with the microbiota and protection against pathogens. IL-7 rather than IL-2 is required for the maintenance of T_{reg} cells in the skin *in vivo*.[342] T_{regs} in skin express St2, GATA3, and amphiregulin and are poised to regulate Th2 immunity. GATA3 expression is important in skin T_{regs} and acts to reinforce Foxp3 expression in dividing T_{regs}. GATA3 mediates a transcriptional program in T_{regs} that acts to attenuate type 2 immunity responses in skin. Fibroblast activation is a key event in tissue remodeling after skin injury. IL-4 and IL-13 drive fibroblast activation, resulting in increased extracellular matrix (ECM) deposition leading to tissue fibrosis. T_{regs} in skin are uniquely differentiated to attenuate dermal fibroblast activation and skin fibrosis.[343] Depletion of GATA3 in T_{regs} resulted in an exacerbation of Th2 secretion preferentially in skin, resulting in enhanced fibroblast activation and dermal fibrosis.

The EGFR pathway plays a major role in skin wound healing by stimulating epidermal and dermal regeneration.[344] The EGFR is also expressed on T_{regs} where it plays a role in augmenting their suppressive capacity *in vitro*. Upon full thickness wounding, highly activated T_{regs} accumulate in skin, and T_{reg} play a major role in limiting IFN-γ production and proinflammatory macrophage accumulation. Specific ablation of T_{regs} early after wounding results in delayed wound closure. Similarly, mice with a specific deletion of the EGFR on T_{regs} had significantly delayed kinetics of wound closure with reduced percentages and numbers of T_{reg} in skin early after wounding and increased numbers of inflammatory macrophages.

Skin T_{reg} cells have been shown to stimulate regeneration of hair follicles by promoting the proliferation and differentiation of stem cells in the skin.[345] Hair follicles are a site for preferential accumulation of T_{reg} in the skin, and hair follicles house a major subset of skin stem cells. Skin-resident T_{regs} preferentially express high levels of the Notch ligand family member, Jagged 1. Jagged 1 expression by T_{reg} facilitated hair follicle epithelial stem cell function and efficient hair follicle regeneration. Signaling through the Notch pathway is a major mechanism by which T_{regs} promote hair follicle skin stem cell function.

Intestinal T_{reg} Cells

The major function of T_{reg} in the intestine is the control of proinflammatory responses to the microbiota and dietary antigens. Colonic T_{reg} can be divided into two distinct populations. One population is derived from CD4+Foxp3− T cells, represents a pT_{reg}, and expresses RORγt but lacks the expression of the TF Helios. The second develops in the thymus (tT_{reg}) and is characterized by expression of the TFs GATA3 and Helios. Tolerance toward the microbiota and dietary constituents is mediated mainly by pT_{reg}. Mice depleted of T_{reg} cells are more susceptible to the intestinal pathogen *Citrobacter rodentium*. The colonic pT_{reg} pool is enriched in a cell population that expresses RORγt and whose maintenance is dependent on the microbiota. The number of pT_{reg} cells in the colon is dramatically reduced in germ-free mice or mice treated with broad-spectrum antibiotics. IL-10 producing pT_{reg} cells are decreased in the colon of GF mice. TGF-β1 is regarded as a cytokine essential for the differentiation of pT_{reg} in the intestine. Mice deficient in CNS1, the Foxp3 intronic enhancer, which contains binding sites for Smad3, lack pT_{reg} cells and display a dysbiotic microbiota and elevated Th2 immunity in the colon.[346] pT_{reg} cells also prevent aberrant Th1 responses to food antigens in the gut.[74]

Vitamin A is found in high concentrations in the intestine and is metabolized to its bioactive form RA. RA facilitates the induction of pT_{reg}, their expression of CCR9, and the gut-homing integrin, α4β7. A CD103+ DC population modulates pT_{reg} induction in the gut by producing RA dehydrogenase, which mediates the generation of RA. This DC population also expresses the integrin, αVβ8, which activates latent TGF-β1. An RORγt− pT_{reg} subset responsive to food antigens is important in homeostasis of the small intestine.

A wide range of bacteria can promote pT_{reg} cells in the colon. Colonization of GF mice with a complex microbiota or a mixture of 17 human strains of Clostridia induces pT_{reg} cells in the colon.[84] Helicobacter species have been shown to drive antigen-specific T_{reg} cells in the colon.[347] It is likely that key bacterial drivers of antigen-specific pT_{reg} responses are mucosa-associated and thus constantly expose the immune system to their antigens. The microbiota also produce metabolites that influence intestinal T_{reg} responses. The best characterized are short-chain fatty acids (SCFA) such as butyrate, propionate, and acetate, which are fermented from dietary fiber by the gut microbiota such as *Clostridia* and *Bacteroides*.[348] SCFAs promote pT_{reg} differentiation in the colon and are recognized by the G-coupled receptors, GPR43 and GPR109 A. The generation of RORγt+ T_{reg} cells in the colon depends on IL-6 and intrinsic STAT3 signaling.[349,350] RORγt+ T_{reg} express high levels of c-Maf, and c-Maf is required for their development in the colon. Deletion of c-Maf in T_{reg} cells leads to intestinal inflammation, but specific deletion of RORγt does not.[351] RORγt+ T_{reg} cells are associated with high IL-10 production, and IL-10 can be regulated by c-Maf.

The colonic pool in GF mice is primarily comprised of Helios+ tT_{reg}. This population expresses GATA3 and has a phenotypic signature linked to tissue repair characterized by expression of St2 and production of amphiregulin.[352] Treatment with IL-33 increases both splenic and colonic numbers of RORγt− T_{reg}. GATA3 is important for Foxp3 expression and T_{reg} accumulation in the intestine. Intestinal St+ tT_{reg} have high expression of the gut-homing receptors, CCR9 and α4β7.[353] It is likely that GATA3+ tT_{reg} are adapted to home to sites of damage in the intestine and mediate repair. A role for T_{reg} cells in the renewal of epithelial stem cells in the small intestine has been documented. IL-2 is crucial for the maintenance of intestinal GATA3+ tT_{reg} cells as this subset is diminished in the small intestine of IL-2–deficient mice.

Placenta

During pregnancy, fetal antigens that are foreign to the mother's immune system and potentially represent a target for rejection of the fetus. Most animals lay eggs, so there is no need for the development of a mechanism of maternal-fetal tolerance. A generalized mechanism of immune suppression during pregnancy would not be optimal as it would expose the mother to the risk of infection. The immune system has therefore evolved a mechanism to prevent immune-mediated rejection of semiallogeneic fetuses. pT_{reg} cells play a critical role in the process of maternal-fetal tolerance.[354] CNS1 is present only in eutherian mammals, but not in marsupials or monotremes, and provides eutherian mammals with the advantage of pT_{reg} generation. As discussed previously, the CNS1 region of the Foxp3 enhancer is critical for the binding of Smad proteins, which are activated in response to TGF-β and are essential for the development of pT_{reg}. Interstrain breeding of CNS-1–deficient mice results in increased resorption of semiallogeneic fetuses, which is not seen with syngeneic fetuses. CNS1−/− mice also have impaired generation of paternal antigen-specific pT_{reg} at the fetal-maternal interface. Depletion of all T_{reg} did not produce greater fetal loss that depletion of pT_{reg} alone, demonstrating that tT_{reg} were not involved in maternal-fetal tolerance. It appears that tT_{reg} do not afford adequate protection of the fetus expressing allogeneic MHC alleles from immune-mediated attack, and pT_{reg} are absolutely required.

T_{reg} develop and proliferate in the decidua of pregnant females and specifically suppress the maternal immune system against the fetus. The proliferating cells are composed of both tT_{reg} and pT_{reg} specific for fetal antigens. However, the fraction of antifetal tT_{reg} cells (as determed by Helios expression) decline during gestation, and Helios− pT_{reg} were predominant at later stages. A pool of these pT_{reg} specific for fetal antigens persists, rapidly proliferates during subsequent pregnancies, and facilitates tolerance.[355] This reflects an antigen-specific response to a fetal antigen. The expanded cells are derived from preexisting T_{reg} cells retained from prior pregnancy. The nonresponsiveness of maternal T_{conv} cells that recognize fetal antigens is not cell-intrinsic but is maintained in the postpartum environment by T_{regs}.

T_{reg} cells are also involved in tolerance to noninherited maternal antigens (NIMAs). NIMAs are peptides derived from polymorphic genes such as MHC antigens that are expressed in the mother, but not in the offspring.[356] Maternal cells can be transferred to the fetus during both pregnancy and during breastfeeding, creating microchimerism. NIMAs represent targets for the fetal immune system resulting in the potential for fetus-versus-mother immune reactivity. Maternal cells transmitted to the fetus induce specific tolerance to NIMAs, which is mediated by NIMA antigen-specific Foxp3+ Helios− pT_{reg}. NIMA-specific tolerance is more robust in female than male offspring because greater maternal microchimerism is observed in the reproductive tract of females than males. The maintenance of tolerance and the NIMA-specific pT_{reg} cells requires maternal microchimerism that is sustained into adulthood.

FOXP3− T_{REG}

Foxp3+ tT_{reg} or pT_{reg} represent the major populations of T_{reg} that have been characterized in both normal physiological studies and in disease models. A number of other lymphoid populations have been described in both mouse and man that can exert potent regulatory functions in certain defined situations. One important distinction between these induced populations of T_{reg} and Foxp3+ tT_{reg} is that the TCR of the former are very frequently specific for known antigens, while the antigen specificity of the Foxp3+ T cells are likely preferentially specific for self-peptide MHC class II.

One of the first approaches used for the induction of T_{reg} was the administration of antigen via the oral route. Pretreatment with orally administered antigen induced suppressor populations that suppressed pathology in a number of different animal models of autoimmunity including experimental allergic EAE, collagen-induced arthritis (CIA), and uveitis.[357] The suppressor cells from orally tolerant mice have been termed Th3 cells and mediate their suppressive effects primarily by secreting TGF-β.[358] It is highly likely that Th3 cells do not represent a distinct population of T_{reg} but represent Foxp3+ pT_{reg}, which can also be induced by oral administration of antigen. A distinct population of CD4−CD8− (DN) regulatory cells have been described[359-361] and may play a role in transplantation immunity and GVHD[362] but have not been further characterized in depth.

T Regulatory 1 Cells

One important lesson that can be derived from the experiments on oral tolerance is that the milieu in which T cells are primed is critically important in determining whether regulatory rather than effector T cells will be generated. Decreased expression of costimulatory molecules on APC or the presence of suppressor cytokines such as IL-10 and TGF-β may generate suppressor T cells rather than effector T cells. The production of these suppressor cytokines by T_{reg} may lead to the generation or expansion of additional regulatory cells via a positive feedback loop.

One of the first studies demonstrating the potential importance of IL-10 in the generation of T_{reg} was derived from an analysis[363] of patients with severe combined immunodeficiency who received transplants of HLA-mismatched hematopoietic stem cells. Complete immunologic reconstitution was achieved in the absence of graft versus host disease. CD4+ T cell clones reactive with host MHC antigens from these patients produced IL-10, but not IL-2, after antigen-specific stimulation *in vitro*. It therefore seemed likely that endogenous IL-10 production in the transplanted patients was responsible for maintaining tolerance *in vivo*. The IL-10 may prevent the activation of host-reactive T cells or suppress APC function and cytokine production by host APC. The high IL-10 production *in vivo* may reflect a chronic activation of donor T cells and host monocytes.

Culture of mouse or human CD4+ T cells with antigen or alloantigen in the presence of IL-10 results in the generation of IL-10 producing T cell clones. Most of these T cell clones produce high levels of IL-10 and TGF-β, moderate amounts of IFN-γ and IL-5, but not IL-2 or IL-4. CD4+ T cells generated in this manner have been termed T regulatory 1 (Tr1) cells.[364] Tr1 cells proliferate poorly following polyclonal TCR-mediated or antigen-specific activation and do not expand significantly under standard T cell culture conditions. This low proliferative capacity is due in part to autocrine production of IL-10 since anti–IL-10 mAbs partially restore proliferative responses. The intrinsic low proliferative capacity of Tr1 cells has been a major limitation and has hindered their detailed characterization. The ability to generate human Tr1 cells is enhanced by the addition of IFN-α.[365] Tr1 cells can be generated from human cord blood with IFN-α alone, as cord blood T cells have the intrinsic ability to produce IL-10. IFN-α and IL-10 do not act as general antiproliferative agents but rather as factors that induce the differentiation of Tr1 cells and inhibit the growth of non-Tr1 cells in the culture.

Four criteria have been proposed to definitively identify Tr1 cells[366]: (1) they produce predominantly IL-10 and TGF-β but can produce variable amounts of other cytokines. (2) While multiple cells types can produce IL-10, only Tr1 cells produce IL-10 and have suppressor function. (3) They express LAG-3 and CD49b in both mouse and man. The expression of CD49b and LAG-3 facilitates the enrichment of both murine and human Tr1 cells.[367] (4) They do not constitutively express Foxp3, but activated human Tr1 cells can transiently express Foxp3.

Both human and mouse Tr1 clones suppress immune responses *in vitro*. Antigen-induced proliferation of naïve CD4+ T cells was dramatically reduced following coculture with activated Tr1 clones that were separated from the responding cells by a transwell insert. The capacity of either human or mouse Tr1 clones to suppress CD4+ T cells proliferation was reversed by the addition of anti–TGF-β and IL-10 mAbs.[368] Some IL-10 producing Tr1 cells suppress the proliferation of naïve CD4+ T cells by an IL-10–independent, cell contact–dependent mechanism. Human Tr1 cell clones also suppress the production of immunoglobulin by B cells, as well as the antigen-presenting capacity of monocytes and DC. Most importantly, it could be shown that mouse Tr1 clones have regulatory effects *in vivo* and suppress Th1-mediated colitis induced by transfer of CD45RB^{hi} cells into SCID mice. Suppression was only seen if the Tr1 clones were activated by antigen-specific stimulation via their TCR. Because the function of Tr1 cells is mediated by IL-10 and TGF-β, these studies imply that Tr1 clones can suppress active immune responses to unknown antigens in the microenvironment by an antigen-driven bystander suppression mechanism similar to the mechanism proposed for Th3 cells in oral tolerance. Although IL-10 was originally described as a product of mouse Th2 cells, mouse Tr1 clones are also capable of regulating Th2 responses including antigen-specific IgE production.[366] Anti–IL-10R antibodies reversed this inhibitory effect. One proposed mechanism of action of IL-10 is to inhibit the maturation and activation of DC and to generate what has been termed "tolerogenic DC."

The relationships between Tr1 cells that produce IL-10 and Foxp3+ T_{reg} that also produce IL-10 remain unclear. IL-10 reporter mice have been generated[369] in which expression of IL-10 was replaced with expression of a Thy-1.One reporter facilitating identification of cells producing IL-10 spontaneously *in vivo*. This mouse stably identifies all cells in which IL-10 alleles have been previously activated as well as those cells actively transcribing IL-10. Foxp3+IL-10− T_{reg} cells were more frequent in LN and spleen, while Foxp3+IL-10+ T_{reg} cells were more frequent in the lymphoid tissues of the large intestine. In the small intestine and PP, Foxp3−IL-10+ T_{reg} cells were the most predominant population and had a cytokine profile, proliferative response, and suppressive function typical of Tr1 cells. The Foxp3−IL-10+ population developed from Foxp3− precursors in all tissues, and its development was dependent on TGF-β but not IL-10. The presence of these cells in the gut may explain why IL-10−/− mice and mice treated with an anti–IL-10R succumb to intestinal inflammation despite having normal numbers of Foxp3+ cells in the intestines. Thus, IL-10 has a nonredundant function in maintaining immune homeostasis to the diverse intestinal microbiota. The relative contributions of the Foxp3+IL-10+ and Foxp3−IL-10+ producing subsets are still unclear. Tr1 cells are more abundant in small intestine, while Foxp3+ T cells are more prominent in the large intestine.[369] As the two populations are located in different parts of the intestinal tract, it remains possible that a distinct class of commensal organisms would preferentially induce both of the subsets.

IL-27, a member of the IL-12 cytokine family consisting of two subunits, p28 and EBI3, is a differentiation factor for the generation of IL-10 producing Tr1 cells.[370,371] IL-27 is secreted by tolerogenic DC that were conditioned *in vitro* or *in vivo* by Foxp3+ T_{reg} cells. These T_{reg}-modified DC express plasmacytoid-like markers. The addition of IL-27 to naïve T cells induced expansion and differentiation of CD4+ T cells that secreted high levels of IL-10.[372] IL-27–driven Tr1 cells proliferate poorly following TCR-mediated activation and suppress T effector responses through IL-10 production. Granzyme B is also induced by IL-27 during Tr1 differentiation and may mediate suppression of T effector responses by cytolysis. IL-27 also can induce IL-10 producing human Tr1 cells.[373] Signaling via the arylhydrocarbon receptor (AhR) may modulate Tr1 differentiation. Naïve T cells differentiated *in vitro* with IL-27 in the presence of the AhR ligand, TCDD, had enhanced secretion of both IL-10 and IL-21. While the role of IL-27 in the induction of human Tr1 cells is not clear, activation of human T cells via CD2, ICOS-L, or CD46 results in Tr1 induction in the absence of IL-10.[373]

While no master transcription factor has been identified in Tr1 cells, IL-27 promotes IL-10 production by CD4+ T cells via the activation of STAT1 and STAT3.[374] IL-27 also promotes EGR2, which in turn activates Blimp1 that transactivates the IL-10 gene. Overexpression of Blimp1 promotes the Tr1 phenotype in effector T cells.[374] IL-27 mediates the expression of c-Maf, which interacts with the Ahr, and together they promote the transactivation of IL-10 gene.[375] IRF1 and BATF have been shown to play a role in the differentiation of mouse Tr1 cells. BATF restructures the chromosomal landscape, while IRF1 is required for transactivation of the IL-10 gene.[375]

Tr1 cells regulate both antigen-specific and nonantigen-specific immune responses. Clinical trials are ongoing to evaluate the potential therapeutic effects of Tr1 cells in the prevention and treatment of GVHD post-BMT. The clinical protocol involves the transfer of *ex vivo* generated Tr1 cells to patients with hematological cancers treated with HSC transplantation. Immunotherapy with autoantigen-derived peptides in patients with autoimmune diseases results in the generation of antigen-specific Tr1 cells and amelioration of disease.[376] Specific immunotherapy in allergic patients promotes the generation of allergen-specific T cells that suppress pathogenic T cells responses via IL-10.[377] Treatment of mice with nanoparticles coated with disease-relevant peptides linked to MHCII may convert Th1 cells to Tr1 cells.[378]

CD8+ T_{reg}

Most of the early studies on T suppressor cells in the mouse demonstrated that they were confined to the CD8+ subset. Almost all of the recent studies on suppressor/regulatory T cells in mouse or man have focused on CD4+ T cells. Several studies have suggested that potent CD8+ T suppressor cells may also exist.[379] Mice genetically deficient or depleted of CD8+ T cells have demonstrated a clear role for CD8+ T cells in regulating EAE. CD8+ T cells from mice that had recovered from EAE downregulated or killed some CD4+ neuroantigen-specific T cell clones. Inhibition was blocked by antibodies to the MHC class Ib molecule Qa-1 but not by anti-MHC class Ia molecules.[380,381] A number of studies have demonstrated that a subset of CD8+ T cells that is restricted by class Ib molecules, Qa-1/HLA-E, have specialized regulatory functions.[79,382] CD8+ Qa-1–restricted T_{reg} cells represent one of the best characterized CD8+ regulatory populations. Approximately, 3% to 5% of CD8+ T cells mediate Qa-1–restricted regulatory activity, and they can be distinguished from nonregulatory CD8+ T cells by the cell surface expression of CD44, CD122, Ly49 and expression of the transcription factor, Helios.[383]

Qa-1 in mice is the functional homologue of HLA-E in man and represents a nonclassical MHC class Ib molecule with limited polymorphisms. Qa-1 is expressed on the cell surface with β2m and contains a peptide-binding pocket and can present a dominant peptide (derived from the leader sequences of H-2D and -L). Qa-1 is expressed on activated T cells, B cells, and DCs and is recognized by the TCR and NKG2A/CD94 receptor on NK cells. Activation, cellular stress, or transformation can promote presentation of characteristic peptides associated with Qa-1. Recognition of Qa-1 peptide complexes by CD8+ T_{reg} via the TCR can lead to suppression/deletion of these Qa-1^+ target cells. Qa-1–deficient mice developed exaggerated CD4+ T cell responses following herpes simplex infection or immunization with self-peptides but did not develop spontaneous autoimmune diseases because of the opposing influence of NK cell recognition of Qa-1.[384] Mice with a mutation in Qa-1 that disrupts Qa-1-CD8+ T cell interaction, but not Qa-1-NKG2A/CD94 interaction develop an SLE-like pathology late in life.[385] CD8+ T_{reg} express only inhibitory members of the Ly49 family and 90% of CD8+ T_{reg} express Ly49F which is not expressed by NK cells. Ly49 molecules may be required for CD8+ T_{reg} survival, expansion, and possibly function. A clonally expanded population of CD8+ T_{reg} cells has also been implicated in the control of EAE.[386] Although these CD8+ cells expressed CD44, CD122, and Ly49, their TCRs were not Qa-1b restricted. Surrogate peptides were identified that could trigger the CD8+ T cells to inhibit disease by suppressing the proliferation of myelin oligodendrocyte glycoprotein (MOG)-specific CD4+ T cells. CD8+ Foxp3− T_{reg} have also been implicated in the control of transplantation immunity and GVHD,[312] but many of these results remain controversial.[387]

T_{REG} AND DISEASE

Animal Models of Autoimmune Disease

Inflammatory Bowel Disease

One of the most widely used experimental models for the study of T_{reg} function is the animal model of IBD. Transfer of CD4+CD45RB^{high} cells to SCID mouse recipients resulted in the development of a wasting disease and colitis 6 to 8 weeks after T cell transfer. This disease was characterized pathologically by epithelial cell hyperplasia, goblet cell depletion, and transmural inflammation.[388] There was a 20- to 30-fold accumulation of Th1 cells in the intestine compared to normal mice. Treatment of recipients with anti–IFN-γ, anti–TNF-α, or anti–IL-12 inhibited the induction of disease. Transfer of CD45RB^{low} cells did not induce colitis, and cotransfer of RB^{high} and RB^{low} cells prevented the development of colitis. A ratio of 1:8 RB^{low} to RB^{high} was able to prevent disease. When CD45RB^{low} cells were fractionated into CD25+ and CD25− fractions, control of intestinal inflammation was primarily mediated by the CD25+ (Foxp3+) fraction.[23] CD45RB^{low} CD25− cells did exert some suppressive function when transferred at high cell concentrations.

Treatment of recipients of RB^{high} cells with IL-10 inhibited the development of colitis (Table 29.4). This treatment inhibited the accumulation of Th1 cells in the intestine, but did not

TABLE 29.4 Role of Cytokines in the Pathogenesis and Treatment of IBD

Treatment	Effect on Disease
Anti–IFN-γ (days 1 + 4)	Substantial protection
Anti–TNF-α (days 1 + 4)	No protection
Anti-TNF (weekly)	Protection during treatment only
IL-10	Protection during treatment only
IL-4	No protection
CD4+CD25+ T Cellls	Complete protection
CD4+CD25+ T Cells + anti–IL-4	Complete protection
CD4+CD25+ T Cells + anti–IL-10R	Protection with naïve CD45RB^{hi}
CD4+CD25+ T Cells + anti–IL-10R	No protection with naïve CD45RB^{low}
CD4+CD25+ T Cells + anti–TGF-β	No protection

Adapted from Powrie F, Leach MW, Mauze S, et al. Inhibition of TH1 responses prevents inflammatory bowel-disease in SCID mice reconstituted with CD45RBhigh T cells. *Immunity.* 1994;1:553-562 and Suri-Payer E, Cantor H. Differential cytokine requirements for regulation of autoimmune gastritis and colitis by CD4+CD25+ T cells. *J Autoimmun.* 2001;16:115-123.

induce T_{reg}, since colitis developed when IL-10 administration ceased. T_{reg} isolated from IL-10−/− mice are able to inhibit colitis induced by CD4+CD45RB^{hi} cells, indicating that IL-10 secretion by T_{reg} themselves is not required. However, control of colitis induced with CD25−CD45RB^{low} cells was found to be highly dependent on IL-10.[389] In contrast, T_{reg}-mediated suppression of AIG involves a cell contact–dependent mechanism of suppression, and IL-10 and TGF-β are not involved.[314] One reason for this difference is that the pathogenesis of AIG and IBD are quite distinct. Bacteria play a required role in IBD, while AIG can be induced by d3Tx of germ-free mice. It remains possible that suppression of IBD involves both the cell contact–dependent and the cytokine-mediated pathways of suppression. IL-10 is required to first dampen the inflammatory response induced by intestinal bacteria. Once this response is reduced, cell contact–mediated suppression can become operative.

One advantage of the IBD model is that CD4+CD25+ CD45RB^{low} (Foxp3+), but not CD4+CD25−CD45RB^{low}, T cells can cure ongoing colitis 4 weeks after transfer of CD4+CD45RB^{high} T cells.[390] T_{reg} proliferate and accumulate in the MLN and also in the colonic LP. At both sites, the progeny of the T_{reg} is in direct contact with CD11c+ DC as well as effector T cells. These findings suggest that regulation of an active immune response by T_{reg} occurs in the draining LN as well as at the site of inflammation. During the cure of intestinal inflammation, the majority of the Foxp3+ T cells are IL-10 secreting cells. Whereas Foxp3+ cells are present in similar frequencies in both the secondary lymphoid organs and LP of colitic animals, the IL-10 producing cells are selectively enriched in the colonic LP. The gut environment may condition the T_{reg} to differentiate into IL-10 producing T_{reg} in the colon. Alternatively, the accumulation of T_{reg} in the colon may be attributed to the preferential expansion of a minor subset of IL-10 producers contained in the transferred population.

Insulin-Dependent Diabetes Mellitus

The NOD mouse represents the best experimental model for autoimmune diabetes. It is widely accepted that Foxp3+ T_{reg} play a critical role in the regulation of disease in this animal model.[391] The islets of Langerhans become rapidly infiltrated with immune cells several weeks prior to the onset of diabetes. Diabetes progression may relate to a subsequent failure in the maintenance of T_{reg} function. Chatenoud et al[392] have examined whether CD4+CD25+ T_{reg} play a protective role in the NOD mouse. Young NOD/SCID recipients were injected i.v. with mixtures of splenocytes from diabetic NOD mice and T_{reg}. T_{reg} from young (6-week-old) prediabetic mice significantly protected against diabetes in this model. Protection could be reversed by treatment of the recipient mice with anti–TGF-β, but not anti-IL-4 or anti-IL-10. Interestingly, anti–TGF-β did not reverse protection mediated by T_{reg} derived from the thymus. Analysis of T_{reg} function in the NOD mouse is difficult because not all animals will develop disease, and it takes a long time for disease to develop.

Antigen-specific T_{reg} from the TCR-transgenic BDC2.5 mice that recognize an islet-derived antigen have been expanded with coimmobilized anti-CD3/anti-CD28 in the presence of IL-2.[393] These expanded T_{reg} prevented the development of diabetes in NOD mice, survived long-term *in vivo*, but required antigen exposure for functional activity. Islet autoantigen-specific BDC2.5 T_{reg} were significantly more effective than polyclonal T_{reg} in regulating autoimmune responses *in vivo*. Importantly, expanded BDC2.5 T_{reg} were able to reverse diabetes when transferred into NOD mice that had been diabetic for 2 weeks in conjunction with 500 syngeneic NOD islets. The expanded T_{reg} were also capable of inhibiting diabetes in newly diagnosed NOD mice.

These results raise the possibility that T_{reg} might be isolated from patients during remission or soon after disease onset, expanded, and used to treat patients at the time of maximal disease activity to moderate the inflammatory response. Indeed, Masteller et al[394] were able to expand T_{reg} with the BDC2.5 specificity by stimulating NOD T_{reg} with beads coated with a mimitope peptide complexed to MHC class II (I-A^{g7}) that is recognized by the BDC2.5 TCR. The expanded T_{reg} were more efficient than polyclonal T_{reg} in suppressing autoimmune diabetes. Thus, small numbers of autoantigen-specific T_{reg} may be clinically efficacious because they can suppress polyclonal pathogenic T cell responses either by bystander cytokine production or by recruitment of endogenous T_{reg}, while avoiding pan-immune suppression that might be induced by polyclonal T_{reg}.

The site of action and mechanisms of action of T_{reg} have been explored in autoimmune diabetes models.[395] T_{reg} had little or no effect on the expansion of effector cells but suppressed IFN-γ production by the effectors in the pancreatic LN and seemed to slow the migration of effector cells into the islets, perhaps through a reduction in CXCR3 in the pancreatic LN. One problem with studies of T_{reg}-mediated protection from autoimmune disease is that they frequently involve transfer into lymphopenic hosts where interpretation of the results may be difficult secondary to stimulation of the effector population that is seen in response to lymphopenia. Foxp3−/− NOD mice have been used to analyze where and how T_{reg} cells exert their control on diabetes under experimental conditions not influenced by homeostatic expansion of effector populations.[396] T_{reg} did not delay or hinder the activation and expansion of anti-islet T cells in the pancreatic LN nor the timing of their infiltration into the islets but primarily prevented locally damaging consequences of autoimmune infiltration in the islets. It appears the T_{reg} primarily impinge on autoimmune diabetes by suppressing destructive T cells inside the islets, more than during the initial activation in the draining LN. Further studies are clearly needed to resolve these conflicting findings as a detailed understanding of the site and mechanism of suppression of experimental autoimmune disease will facilitate translation of the therapeutic use of T_{reg} for treatment of autoimmune disease in man.

In studies of the relationships between T_{reg} cells and T effector cells in the pancreatic lymph nodes of NOD mice, it was noted that progression to diabetes was associated with an increase in T_{reg} cells in the pancreatic lymph node at the time of diabetes onset.[397] However, disease progression was associated with a decrease in the percentage of T_{reg} cells in the pancreatic islets with a concomitant loss of CD25 expression and expression of the antiapoptotic factor, Bcl-2, in the intraislet T_{regs}. CD4+ T cells from islets produced less IL-2 following stimulation *in vitro* than CD4+ T cells from spleen, suggesting that a deficiency in IL-2 production is responsible for the loss of T_{regs} in the islets. Administration of IL-2 promoted T_{reg} survival and protected the mice from diabetes. In the NOD mouse model of Type I diabetes, the idd3 region on mouse chromosome 3 is associated with a lower production of IL-2 and a decreased proportion of T_{reg} cells and increased susceptibility to disease. The agonistic activity of an IL-2/anti–IL-2 complex was exploited as a therapeutic agent for T1D in NOD mice. This complex extends the half-life of IL-2 from 30 m for free IL-2 to 48 to 72 hours for the complex. A low dose of complex increased expression of IL-2Rα and prosurvival factor, Bcl-2, by T_{reg} cells and prevented the onset of diabetes. A high dose of the complex accelerated the progression to T1D secondary to the enhancing autoreactive Teff cells that express the high-affinity IL-2Rα.[398] Similar results were observed in the animal model of multiple sclerosis, experimental autoimmune EAE. Pretreatment of mice with the complex modulated disease induction, while administration of the complex after disease initiation resulted in a more severe disease course.[399]

Encephalomyelitis

EAE is the major animal model for the study of multiple sclerosis, and T_{reg} have been shown to play a role in this model. When the T_{reg} population of naïve C57BL/6 mice was supplemented with CD4+CD25+ T_{reg} from a normal donor before the active induction of EAE by immunization with an encephalitogenic peptide from MOG, the T_{reg} conferred significant protection. Protection was associated with the presence of Th2 cytokines and a marked decreased CD4+ T cell, and APC infiltrates in the CNS.[400] These studies suggested that T_{reg} might inhibit the homing of pathogenic T cells to the CNS. McGeachy et al[236] demonstrated that that the accumulation of CD25+Foxp3+ cells in the CNS correlated with the recovery phase of EAE. As few as 10^4 CNS-derived CD25+ cells significantly reduced disease severity in recipients, whereas the same number of naïve CD25+ T cells had no effect. Although the mechanism of action of the T_{reg} in protection from EAE was not completely elucidated, CD25+ T cells were the major source of IL-10 in the CNS, and IL-10 production was crucial for full recovery from EAE. Using tetramers, it was shown that MOG-specific T_{regs} expanded alongside MOG-reactive T effector cells in the draining LN before migration to the spleen and accumulation in the CNS. MOG-reactive T_{regs} in peripheral sites showed suppressive activity against naïve MOG-reactive T cells, while T_{regs} derived from the CNS failed to suppress the *in vitro* proliferation of CNS-derived T effector cells due to the high levels of inflammatory cytokines coming from the T effector population. During recovery, there was a dramatic change in the T_{reg}:T effector ratio in the CNS in favor of T_{regs}.[401]

Allergy

T_{reg} suppress the differentiation and function of Th1 and Th2 cells *in vitro* and *in vivo*.[402,403] Other studies[404] have shown that freshly isolated T_{reg} inhibit the IL-4–induced development of Th2 cells but had no influence on established Th2 cells. In contrast, preactivated T_{reg} inhibited cytokine production and proliferation of established Th2 cells.

CD4+Foxp3+ T_{reg} modulate the development of experimentally induced asthma in mice genetically predisposed or comparatively resistant to the development of airway resistance. *In vivo* depletion of T_{reg} in resistant animals before initial antigen contact increased airway hypersensitivity reaction (AHR), airway eosinophilia, IgE synthesis, and was associated with Th2 cytokine production, while T_{reg} cell depletion had no effect on allergen-driven AHR in susceptible mice.[405] Enhancement of allergic phenotype in resistant mice was associated with stimulation of DC function, suggesting that the T_{reg} do not act on effector T cells but act indirectly via the DC. Th2 polarized CD4+ T cells depleted of CD25+ cells induce increased airway eosinophilia after adoptive transfer.[406] Allergen-specific CD4+CD25+ T cells from TCR-transgenic mice inhibited the classical pathology associated with allergic asthma including AHR, lung eosinophilia, and Th2 cytokine production in immunocompetent mice. Their effects were dependent on IL-10, but T_{reg} cells exerted their suppressive effects *in vivo* independently of their capacity to produce IL-10, by inducing IL-10 production from recipient CD4+ T cells.

Mice with an established helminth infection are less prone to allergic airway inflammation as measured by inflammatory infiltrate in bronchial lavage fluid or peribronchial and perivascular inflammation. Helminth infection with *H. polygyrus* protects from allergic airway responses by acting downstream from allergen sensitization and can be transferred from infected mice to uninfected presensitized animals by CD25+ T cells. Th2 effector cytokines (IL-5 and IL-13) were diminished, but allergen-specific IgE responses were not changed. Thus, chronic parasite infection may maintain a high level of activation in T_{reg} so that they primarily target the effector mechanisms resulting in decreased inflammation in a previously sensitized allergy-prone host.[407]

Allograft Rejection and GVHD

A considerable body of data exists that regulatory/suppressor T cells exist in patients and animals with long-term surviving allografts.[408] Analysis of the properties of these cells *in vitro* has been hampered because bulk populations of T cells from tolerant recipients proliferate normally when cultured *in vitro* with donor alloantigens and frequently secrete a proinflammatory Th1 pattern of cytokines. It appears that the T_{reg} are masked by the presence of naïve T cells responding to alloantigens. In animal models, a number of protocols have been used to induce tolerance to allografts (Fig. 29.20). Tolerance to cardiac allografts in mice can be induced by pretreatment with a donor-specific blood transfusion combined with depleting or nondepleting anti-CD4 antibody. It is likely that the protective effects of this protocol were mediated by the induction of CD4+ T_{reg} since complete depletion of CD4+ T cells fails to induce tolerance. Pretreated mice accepted cardiac grafts, and cells from these mice could transfer tolerance to naïve recipients. CD4+ T cells were responsible for both the induction and maintenance of tolerance.

In studies of allograft rejection across a full H-2 difference, the differential expression of CD45RB has facilitated separation of nontolerant naïve T cells from the induced regulatory T cells.[409] RB^{high} cells from tolerant mice responded normally to challenge with alloantigen *in vitro* and were able to reject allogeneic skin grafts when transferred alone to T cell–deficient mice. In contrast, $CD45RB^{low}$ cells failed to mount a

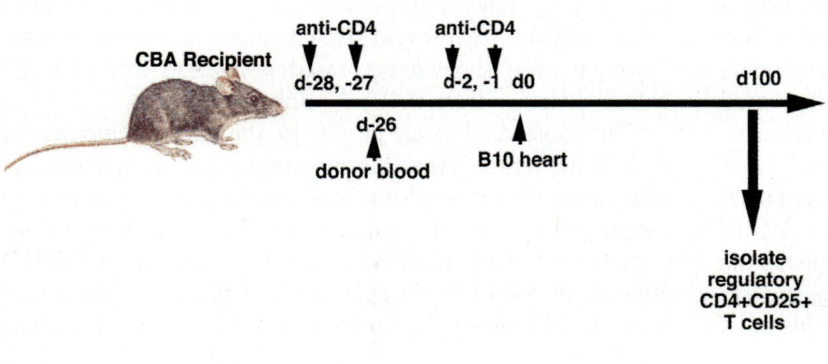

FIG. 29.20. Induction of regulatory T cells for tolerance to alloantigens. (Adapted from Waldmann H, Cobbold S. Regulating the immune response to transplants. A role for CD4+ regulatory cells? *Immunity*. 2001;14:399-406, with permission from Elsevier.)

proliferative response, to secrete cytokines in response to alloantigen *in vitro*, and to induce allograft rejection *in vivo*. The addition of $CD45RB^{low}$ cells to an MLR with $CD45RB^{high}$ cells as responders resulted in inhibition of proliferation. $CD45RB^{low}$ T cells from long-term tolerant mice were able to suppress responses to alloantigen *in vitro* only when the donor alloantigens were presented by the indirect pathway of allorecognition by the APC present in the culture. One explanation for this result is that it is potentially advantageous for the T_{reg} to respond to alloantigens via the indirect pathway because T_{reg} may require constant stimulation to maintain their function. Passenger leukocytes rapidly migrate out of the graft, and the graft tissues lack costimulatory molecules. Donor-derived allopeptides must be presented by recipient APC for stimulation of the T_{reg} *in vivo* and for the detection of functional activity of these cells *in vitro*. All of the suppressive activity of the $CD45RB^{low}$ cells was mediated by the CD4+CD25+ T cell subset contained in that pool. The suppressive capacity of the regulatory T cells in this model *in vitro* and *in vivo* was reversed by anti-IL-10R but not by anti–IL-4. T_{reg} can also be isolated from mice pretreated with donor alloantigen in combination with depleting anti-CD4 but not transplanted with an allograft.[409]

In studies examining induction of tolerance to minor histocompatibility antigens on allografts following protocols similar to those used above, both CD4+CD25+ and CD4+CD25− T cells from tolerant mice could mediate suppression, but suppression required 10 times more CD4+CD25− T cells.[410] Both populations may have a significant role in maintaining transplantation tolerance as the number of CD25− T cells is 10 times higher than CD25+ T cells in the tolerant mouse. It is possible that some of the CD25+ T cells have lost expression of CD25 during the tolerance induction protocol, that effector cell death in the CD25− T cell pool unmasked the presence of T_{reg}, or that some of the CD25− T cells express Foxp3. In contrast to the results with a major MHC difference, CD4+CD25+ T cells from naïve mice could also prevent naïve T cells from rejecting skin grafts although 5 times more cells were required than from tolerant donors. The enhanced potency of CD25+ T cells from tolerant mice may be secondary to expansion of an alloantigen-specific population. The capacity of T cells from tolerant donors or CD25+ T cells from normal mice to mediate their suppressive function could not be neutralized by anti–CTLA-4, by anti–IL-10, or anti–IL-4.

A short treatment of mice with saturating amounts of nondepleting anti-CD4 can induce tolerance to foreign protein antigens and enables long-term survival of minor antigen mismatched skin or MHC-mismatched cardiac grafts. The addition of anti-CD40L can extend this effect to full donor-specific tolerance such that second donor grafts are accepted at any later time, while third-party grafts are rapidly rejected. The maintenance of tolerance is dependent on CD4+ T_{reg} that can suppress both CD4+ and CD8+ naïve T cells after transfer into secondary recipients. T_{reg} cells are found not only in recipient lymphoid tissue after transplantation but also at the graft site. The localization of T_{reg} cells at more than one site *in vivo* is important if T_{reg} are to control aggressive immune reactivity to the graft effectively. In lymphoid tissues, T_{reg} cells might be effective at blocking the initiation of an aggressive response against the graft, whereas at the graft site, they can inhibit the effector activity of aggressive cells that have escaped regulation and migrated to the graft. Depletion of T_{reg} resulted in reversal of the tolerant state induced by anti-CD40L and allograft rejection.[411] Mast cells appear to play an important role in this mode of tolerance induction as tolerance induction by the combination of anti-CD40L and infusion of allogeneic cells could not be induced in mast cells deficient animals.[412]

Regulation observed in therapeutic tolerance probably is maintained both by preexisting T_{reg}, together with induced T_{reg}. In one model, antigen stimulation in the presence of a nondepleting anti-CD4 has been shown to convert CD4+Foxp3− T cells into Foxp3+ T_{reg} in a TGF-β–dependent manner *in vivo*, and this correlates with life-long acceptance of the graft.[413] T_{reg} cells that recognize donor MHC through the indirect pathway of allorecognition have been shown to be responsible for the phenomenon of linked unresponsiveness. Linked suppression *in vivo* seems to correlate with the *in vitro* finding that once T_{reg} are activated, they can suppress the activation of other T cells responding to antigen presented by the same APC in an antigen-nonspecific manner.

The inability of T cell populations containing large numbers of CD25+ T cells to mediate GVHD strongly raised the possibility that CD25+ T cells might actually be capable of inhibiting CD25− effectors in cotransfer studies. Freshly isolated CD25+ T cells only modestly inhibited GVHD, when mixed with CD25− T cells in equal numbers, while alloantigen-stimulated, cultured CD25+ T cells profoundly inhibited the capacity of CD25− T cells to inhibit rapidly lethal GVHD.[414] In murine models, transplantation of bone marrow into a lethally irradiated host does not cause GVHD, and cotransfer of naïve donor T cells (CD4+, CD8+, or both) is required to induce GVHD. Depletion of T_{reg} from the donor graft accelerated the course of GVHD and increased lethality. Transfer of T_{reg} with naïve T cells from the donor protected lethally irradiated recipients from GVHD morbidity and decreased GVHD-related mortality[415] across minor and major MHC class I and/or class II barriers in various mouse strain combinations. The effect of T_{reg} on GVHD is likely to be early and to affect the maturation and expansion of effector cells. $CD62L^{hi}$ T_{regs} are more potent suppressors of GVHD presumably due to their ability to migrate to secondary lymph nodes where they are primed and activated.[416] As host APC are necessary to trigger acute GVHD, it is possible that the activated T_{reg} function by downregulating costimulatory molecules and thereby prevent the activation of effector T cells. Hoffman et al showed decreased protection from GVHD with T_{reg} derived from IL-10−/− animals.[417]

The most analogous model to the murine models of GVHD induction is the administration of donor lymphocyte infusions to humans with relapsed leukemia after allogeneic hematopoietic cell transplantation (HCT) so as to obtain a graft-versus-tumor effect. The development of GVHD after donor lymphocyte infusion is associated with tumor clearance. However, T_{reg} controlling GVHD may also cause suppression of tumor immunity. Edinger and colleagues[418] found that adoptive transfer of donor T_{reg} effectively decreased the incidence and severity of GVHD without abrogating the graft versus tumor response. In this model, cotransfer of T_{reg} markedly decreased the proliferation of alloreactive effector cells in secondary lymphoid organs by >90% at day 7 but did not inhibit the activation of the effectors as judged by activation marker expression and cytokine production. One possibility is that the decrease in the proliferation and expansion of the alloreactive T effector cells was sufficient to control acute GVHD, but this decrease was not complete, and the remaining donor T effector cells were sufficient to effectively mount a graft versus tumor effect.

A model has been developed to demonstrate that human T_{reg} cells can inhibit GVHD and graft rejection *in vivo* in the mouse. When human PBMC are transferred to NOD/*SCID*/IL2Rγc[null] mice, the recipients develop a xeno-GVHD. When autologous expanded human T_{regs} are cotransferred with the PBMC, marked amelioration of the GVHD is observed.[419] One marker of chronic allograft dysfunction is transplant arteriosclerosis with neointimal thickening of arteries in the allograft. A chimeric human-mouse model that resembles transplant arteriosclerosis has been developed in which allogeneic human peripheral blood mononuclear cells elicit rejection of human arterial segments that have been transplanted into a mouse.[420] When human T_{reg} cells that had been expanded *in vitro* were cotransferred with the PBMC, abrogation of transplant arteriosclerosis was observed.

T_{reg} also play a facilitating role in bone marrow engraftment. Joffre et al[421] found that T_{reg} provided allospecific protection of transplanted bone marrow from host rejection. Enhanced engraftment with T_{reg} also increased long-term donor chimerism in animals that received T_{reg} compared to those that received bone marrow transplants only. Recipients of T_{reg} demonstrated tolerance to host and donor antigens but mounted responses to third party antigens.[422] T_{reg} therapy has been evaluated for the prevention of GVHD following hematopoietic stem cell transplantation in man.[423,424] In a safety trial of human *in vitro* expanded T_{regs} isolated from umbilical cord blood, no serious side effects were observed and some reduction of GVHD was observed. A more comprehensive trial to date was performed by Di Ianni et al.[424] This study involved 28 patients with high-risk hematologic malignancies who underwent HLA-haploidentical stem cell transplantation. No posttransplantation immunosuppression was given, and freshly isolated T_{regs} were administered without *in vitro* expansion. The incidence of GVHD was markedly decreased, and the T_{reg} treatment was not associated with an increased risk of leukemia relapse or with a decreased incidence of opportunistic pathogens.

Tumor Immunity

Many tumor-associated antigens recognized by autologous tumor-reactive lymphocytes are antigenically normal self-constituents. Tumor immunity is therefore autoimmunity, and the mechanisms for maintaining immunological self-tolerance may hinder effective tumor immunity. Studies by R. North and associates in the early 1980s[425] demonstrated the role of suppressor T cells in the prevention of tumor immunity. The suppressor T cells in the experiments of North and coworkers were shown to be CD4+CD8−, but were not characterized further.[426] Onizuki et al[427] were the first to suggest that Foxp3+ T cells played an important role in inhibiting tumor immunity. They first depleted CD4+CD25+ T cells by injecting a depleting anti-CD25 mAb and noted that this led to regression of a number of tumors that grew progressively in nondepleted mice. The depletion had to be performed not later than day 2 after injection of the tumor. Co-administration of anti-CD8 inhibited the tumor regression induced by anti-CD25 depletion, suggesting that CD8 T cells were responsible for tumor regression. Shimizu et al[231] drew similar conclusions when they transferred CD25-depleted or CD25-containing spleen cells to *nu/nu* recipients, and then challenged them with normally nonimmunogenic tumors. In the recipients of CD25-depleted cells, tumors grew but regressed; in contrast, all recipients of spleen cells containing CD25+ T cells died from rapidly growing tumors. Long-lasting tumor immunity to rechallenge could also be demonstrated in the recipients of the CD25-depleted spleen cells.

Some tumors are heavily infiltrated with immune cells, while other tumors exhibit a noninflammed TME with few immune cells. T_{regs} are prominently present among TILS in tumors that contain large infiltrations of immune cells but are rarely seen in the TME of tumors that are poorly infiltrated. Other models raise the possibility that T_{reg} may have a beneficial effect in preventing tumorigenesis. A widely used model for human colorectal carcinogenesis is the "multiple intestinal neoplasia" mouse, which has a germline mutation in the *Apc* tumor suppressor gene.[428] Transfer of T_{reg} cells reduced the multiplicity of epithelial adenomas in these mice by an IL-10–dependent mechanism. Recipients of T_{reg} showed increased apoptosis and downregulation of cyclooxygenase-2 within tumors coinciding with tumor regression. In this model, aspirin and cyclooxygenase inhibitors also decrease the risk for colon cancer by inhibiting cyclooxygenase-2. As T_{reg} can suppress bacterially triggered inflammatory responses in the bowel of mice, the ability of T_{reg} to traffic and suppress inflammation likely explains their therapeutic efficacy in this model. As inflammatory mediators drive tumor development, T_{reg} and anti-inflammatory drugs exert their effect by modulating the levels of these molecules.

The mechanisms responsible for the accumulation of T_{regs} in tumors remain poorly defined. The increased number of T_{regs} may be the result of increased trafficking to the TME, preferential expansion, or secondary to enhanced pT_{reg} differentiation in the TME or distally in tumor-draining LNs. Some of the factors present in the TME that may influence T_{reg} function are the metabolic nature of the TME including relative hypoxia, scarcity of nutrients, and heightened lactate production. Some tumor T_{reg} genes might be overexpressed because they are required by T_{regs} to survive in the TME. Woo et al[429] have shown that Foxp3+ T cells exist in high proportions (~33%) in the tumor-infiltrating lymphocytes of patients with non–small cell lung cancer. A high frequency of T_{reg} cells has been noted in many other human cancers. T_{reg} cell have been extensively studied in human ovarian cancer.[430] T_{reg} cells within the ovarian tumor microenvironment expressed Foxp3, inhibited tumor antigen-specific CD8+ T cell cytotoxicity, and contributed to tumor growth *in vivo* in a human-SCID mouse chimeric model. An accumulation of T_{reg} cells in the tumor predicted a striking reduction in survival.

A number of studies have described the transcriptional landscape of T_{regs} in the TME of different tumors.[431-433] Studies of CRC and NSCLC tumor-infiltrating T_{reg} demonstrated that their gene expression is different from circulating and tissue infiltrating T_{reg} consistent with the concept that the TME influences specific gene expression in T_{regs}. T_{reg} in the tumor TME express many of the checkpoint inhibitors including PD-1, PD-L1, PD-L2, IL1R2, GITR, OX40, TIGIT, LAG-3, and TIM-3. Notably, LAYN, MAGEH1, and CCR8 were three of the most enriched genes expressed by T_{reg} and expression of these genes in CRC and NSCLC patients correlated with a poor 5-year survival rate. CCL18 the ligand of CCR8 and was highly expressed in different tumors. Similar results were observed in the TME of breast cancer patients. In the breast cancer TME, tumor T_{reg} cells exhibited increased expression of genes involved in cell activation and cytokine and chemokine signaling including highly augmented expression of the chemokine receptor CCR8 and high expression of CCR8 was also associated with poor prognosis. Targeting the CCL18/CCR8 pathway may be an interesting therapeutic approach as Teff cells do not express CCR8. The frequency of T_{regs} was increased in the more aggressive triple-negative breast cancer.[434] While CCR8 was found to be the predominant chemokine receptor on T_{reg} in some tumors, CCR4 was highly expressed on T_{regs} in other tumors, and the tumors expressed the CCR4 ligands, CCL17 and CCL22. Overall, studies in different tumor models indicate that T_{reg} express a subset of transcripts which are tumor-specific. Tumor T_{reg} in breast cancer have distinct TCR repertoires compared to normal resting breast T_{regs}. Some of these genes are expressed in multiple patients with the same tumor and also exhibit enhanced expression in tumors in mice. Conceivably, it may be possible to define a common pathway which could be targeted in different types of tumors. Other studies have identified a very strong overlap between the transcriptomes of T_{regs} in the TME and T_{regs} in nonlymphoid tissues. Some of the signals that allow T_{regs} to accumulate in tissues may allow for their residence in tumors.

Taken together, the studies on depletion of CD4+CD25+ T cells and the transfer of CD4+CD25− T cells strongly suggest that the effectiveness of a tumor vaccine would be greatly enhanced by removal of T_{reg} suppressor activity. Indeed, Sutmuller et al[435] were able to demonstrate that antibody-mediated depletion of CD25+ T cells followed by vaccination with a GM-CSF transfected melanoma cell line resulted in enhanced tumor rejection. Tumor rejection was accompanied by skin depigmentation, suggesting that autoreactive immune responses are involved in this process. Depletion of CD25+ T cells followed by anti-CTLA-4 treatment and vaccination resulted in the most potent anti-tumor response (Table 29.5).

TABLE 29.5 Depletion of CD4+CD25+ T Cells Augments the Immune Response to a Tumor Vaccine

Treatment	% Surviving d90
None	0
Anti-CD25 (day 4)	20
GM-CSF Vaccine (day 0, 3, 6) +Anti-CD25 (day 4)	50
GM-CSF Vaccine (day 0, 3, 6) +Anti–CTLA-4 (day 0, 3, 6)	50
GM-CSF Vaccine (day 0, 3, 6) +Anti-CD25 (day 4) + anti–CTLA-4 (day 0, 3, 6)	100

Adapted with permission of Rockefeller University Press, from Sutmuller RP, van Duivenvoorde LM, van Elsas A, et al. Synergism of cytotoxic T lymphocyte-associated antigen 4 blockade and depletion of CD25+ regulatory T cells in antitumor therapy reveals alternative pathways for suppression of autoreactive cytotoxic T lymphocyte responses. *J Exp Med.* 2001;194:823-832, permission conveyed through Copyright Clearance Center, Inc.

Thus, CD25-depletion and anti–CTLA-4 treatment increase the immunogenicity of a tumor vaccine by distinct mechanisms involving nonredundant pathways.

Experiments with adoptively transferred mouse T_{reg} cells provided a direct link between T_{reg} cells and reduced tumor immunity. Tumor-specific CD8+ T cells were transferred with either T_{reg} cells or conventional T cells into mice-bearing B16 melanoma. In mice that received T_{reg} cells, but not in mice that received CD25− T cells, CD8+ T cell–mediated immunity was abolished. T_{reg} cells may mediate their suppressive effect by inhibiting T cell priming in lymphoid organs or by reducing the effector function of tumor antigen-specific cells. Antony et al[436] have shown that transfer of T_{reg} cells reduced the therapeutic efficiency of adoptively transferred tumor antigen-specific effector T cells in a mouse melanoma model. Depletion of intratumoral T_{reg} cells induced potent T-cell tumor immunity and resulted in regression of large established tumors. Administration of anti-GITR antibody protected mice from B16 tumor challenge and induced tumor regression in mice bearing carcinogen-induced sarcomas and colon carcinomas.[437] The anti-GITR likely costimulated T effector functions to make the effector cells resistant to T_{reg}-mediated suppression.

The spectrum of tumor-specific antigens recognized by T_{reg} has not yet been defined. In one case, cloned T_{reg} from a melanoma recognized LAGE-1, a cancer and testis-specific antigen, and suppressed LAGE-1-specific T cell activation.[438] It is important to identify the ligands for T_{reg} cells to compare them to the ligands for tumor-specific effector cells and to determine whether there are ligands that are exclusively recognized by T_{reg} cells and whether T_{reg} cells recognize mutated tumor antigens. T_{regs} are able to recognize some tumor antigens identified by serological identification of antigens by recombinant expression cloning (SEREX). Adoptive transfer of T_{regs} from mice immunized with SEREX antigens enhanced pulmonary metastasis and accelerated tumor development, indicating that acceleration of tumorigenesis and enhancement of metastasis may be secondary to the immunosuppressive effects of T_{regs} in the immunized hosts.[439] A major concern that remains to be addressed is whether vaccination with certain tumor antigens might promote T_{reg} cell clonal expansion.

The target cell of T_{reg} suppression in the TME and whether suppression is antigen-specific remain unknown. In addition, the functional redundancy of T_{reg} suppressor mechanisms complicates the selection of a specific mechanism as a therapeutic target. A number of mechanisms have been proposed for T_{reg}-mediated suppression of tumor immunity. As T_{reg} express CTLA-4, one possible model is that engagement of CD80/CD86 on tumor APC may induce production of the immunosuppressive molecule, IDO (Fig. 29.16). IDO expressing APC are found in human tumors and within draining lymph nodes. Alternatively, ovarian tumor-associated macrophages, but not normal macrophages, express B7-H4, a negative regulator of T cell responses.[440] T_{reg}, but not normal, T cells induced B7-H4 on monocytes, macrophages, and myeloid DC and rendered them immunosuppressive through B7-H4.

IL-10 and IL-35 are expressed by different T_{reg} subpopulations[441] in the TME. IL-35 expression was enriched on tumor-infiltrating T_{reg} and promotes the expression of inhibitory receptors (PD-1, TIM3, LAG3) on intratumoral T cells and limits T effector differentiation. IL-10 or L-35 targeted immunotherapy may be useful therapeutically. T_{reg} can shift between producing IL-10 or IL-35. The alarmin IL-33 also appears to have a unique promoting effect on T_{reg} function in the TME.[442,443] Direct signaling of IL-33 into T_{reg} was required for the development of both skin and colon cancer. Similarly, T_{reg}-specific deletion of the IL-33 receptor, St2, results in altered T_{reg} suppressor function, enhanced infiltration of CD8+ T cells into the TME, and decreases tumor burden. These studies suggest that targeting the IL-33/St2 complex may be an important area for tumor immunotherapy.

The concept of reversing immunosuppression in cancer has merit as a therapeutic approach. CTLA-4 blockade improves tumor immunity but also resulted in severe manageable autoimmune responses in patients.[444] The mechanistic link between the effects of treatment with anti-CTLA and T_{reg} function remain to be determined. CTLA-4 blockade by anti–CTLA-4 (Ipilimumab) releases effector cells from suppression and rejuvenates their antitumor activities, but T_{reg}-mediated suppression, phenotype, frequency, and stability were not altered. Ipilimumab is a fully human IgG1 and can engage FcγIIA (CD16) expressed on monocytes and NK cell and mediate ADCC. It is possible that Ipilimumab targets T_{reg} in tumor tissues that express high levels of CTLA-4, leaving T_{reg} with lower levels unperturbed. T_{reg} depletion may also be mediated by other mAbs that target antigens whose expression is selectively enhanced on T_{reg} in the tumor microenvironment[445] including anti-GITR, anti-4-1BB, and anti-OX40 whose depleting effects are also dependent on coengagement of activating FcγRs. Effects of anti–PD-1 (pembrolizumab), on T_{reg} which express levels of PD-1 as high as those on T effector cells remain unclear. Blocking anti–PD-1 does not reverse or decrease the suppressive functions of T_{reg} and the frequency and phenotype of T_{reg} is not altered.

Paradoxically, several of the mAbs to members of the TNFRSF that are preferentially expressed on T_{reg} also have agonistic effects on antitumor CD8+ T cells by engaging the inhibitory FcR, FcγRIIB. It has been possible to take advantage of the unique properties of different mAbs to the TNFRSF member, 4-1-BB, by first depleting T_{reg} cells using a mouse IgG2a mAb which engages activating FcRs and to later deliver an IgG1 anti-4-1-BB mAb which engages inhibitory FcRs. No effect was observed when the mAbs were given simultaneously.

While the mechanism of action of T_{reg} depleting and T effector cell enhancing mAbs have been well characterized in mice,[446] it remains controversial whether treatment of human tumors with mAbs such as anti–CTLA-4 drives depletion of T_{regs}. Clinical studies are limited to a single or to a couple of biopsies which due to intratumor heterogeneity provide only a partial representation of the whole tumor lesion. In man, depletion of T_{reg} is usually evaluated in biopsies performed more than 15 weeks after treatment with the mAb, while in the mouse depletion is evaluated after 3 to 10 days. After depletion, the remaining T_{regs} may proliferate to replenish their empty niche, and this rebound may be more prominent in patients with nonresponding or progressive lesions. Biopsies are also only usually performed in patients with residual disease. Overall, the mouse and human data are consistent with the T_{reg}-depleting capacity of anti–CTLA-4 mAbs, and the further development of anti–CTLA-4 mAbs with enhanced depleting properties is indicated.[447]

Immunity to Infectious Agents

T_{reg} cells dampen immune responses in a wide variety of infections caused by bacteria, fungi, parasites, and viruses especially those that persist in the host. One of the most important roles of T_{reg} may, in fact, involve modulation of the immune response to infectious agents to prevent the lethal consequences of an overwhelming inflammatory response during the course of a productive immune response to an invading microorganism. T_{reg} and effector cells must maintain equilibrium between no immunity at all and immunopathology. This critical role of T_{reg} is well illustrated by the immune response of mice to infection with *Pneumocystis carinii* (PC). When SCID mice chronically infected with PC are reconstituted with CD4+CD25− T cells, they develop a severe inflammatory response in their lungs that is ultimately fatal. Animals injected with CD25+ T cells alone did not become moribund and only manifested transient weight loss. Cotransfer of CD25+ T cells prevented not only the development of the PC-driven fatal pulmonary inflammation induced by CD25− T cells but also suppressed the elimination of PC mediated by the CD25− T cells. Protective CD25+ T cells are needed to inhibit the lethal immunopathologic response mediated by the PC-specific CD4+CD25− T cells, but they also inhibited complete clearing of the organism.[448] Although T_{reg} may target CD4+ and CD8+ T cells responding to an infectious challenge, T_{reg} have also been shown to act on the innate immune system in the response to *Helicobacter hepaticus*.[449]

The role of Foxp3+ T cells in the immune response to infection includes more than suppression of inflammation.[235,450] Foxp3+ T cells also maintain persistence of infection and promote chronicity. The persistence of pathogens following clinical cure is a hallmark of certain viral, bacterial, and parasitic infections. In clinical and experimental forms of leishmaniasis, small numbers of viable organisms persist within lymphoid tissue and within the site of former skin lesions following self-cure or successful chemotherapy. Since low numbers of parasites persisting in the dermis can be efficiently transmitted back to their vector sandflies, the expansion and/or recruitment of regulatory T cells to the site of *L. major* infection might reflect a parasite adaptive strategy to maintain its transmission cycle in nature. Despite the absence of sterilizing immunity, these individuals maintain strong life-long immunity to reinfection, a status similar to the concomitant immunity described in tumor models.[451]

In healed C57BL/6 mice, CD4+CD25+ regulatory T cells accumulate in sites of *L. major* infection in the skin.[452] These cells are exclusively derived from CD25+ T cells and not from activated CD25− T cells. They suppress the expansion of and killing mediated by *L. major*-specific effector cells. Although IL-10 produced by CD25+ T cells is essential to the establishment of persistent infection, early in the infectious process CD25+ T cells can promote parasite survival and growth in an IL-10–independent manner. Later in the course of infection, IL-10 is absolutely required for development of the chronic lesion since recipients of CD25+ T cells from IL-10−/− mice ultimately healed and completely cleared the parasite from the site. IL-10 produced by T_{reg} contributes directly to parasite persistence by either modulating APC function, inhibiting cytokine production by Th1 cells, or by rendering macrophages refractory to IFN-γ that is needed for intracellular killing.

When rechallenge studies were performed in IL-10−/− mice or in WT mice that were treated during the chronic stage of their primary infection with anti–IL-10R, conditions which result in complete clearance of parasites from the skin and draining lymph node, reinfection at a site distant from the initial infection resulted in parasite loads that were comparable to those following primary infection in naïve mice. Since healed mice treated with control antibody maintained strong immunity to reinfection, the maintenance of a residual source of infection, secondary to IL-10 production by CD25+ T cells at the lesion site, is required for preservation of acquired immunity to *L. major.* T_{reg} cell lines isolated from chronic *L. major* are able to respond to parasite-infected DC by proliferating and producing IL-10.[453] The cells that have undergone proliferation express and maintain Foxp3 expression. It appears that the parasite has specifically evolved to manipulate DC in a manner that favors and sustains T_{reg} proliferation. Most importantly, the *L. major*-specific cell lines maintain their suppressive functions *in vivo*, as transfer of the lines to chronically infected mice results in massive disease reactivation and dissemination.

Th1 effector cells are induced during the course of infection with *Bordetella pertussis* and ultimately play a critical role in the clearance of bacteria from the respiratory tract.[454] Antigen-specific Th1 responses in the lung and local lymph nodes are severely suppressed during the acute phase of infection. *B. pertussis* has evolved a number of strategies to circumvent protective immune responses. One bacterial component, filamentous hemagglutinin (FHA), is capable of inhibiting LPS-driven IL-12 production by macrophages and DC and stimulating IL-10 production. FHA may contribute to the suppressed Th1 responses during acute infection with *B. pertussis* by the induction of T cells with regulatory activity as a result of its interactions with cells of the innate immune system. Repeated stimulation of T cells from the lungs of mice acutely infected with *B. pertussis* resulted in the generation of Tr1 clones specific for FHA. Tr1 cells could only be generated from the lungs of infected animals and not from spleen. These Tr1 cells secreted high levels of IL-10 and inhibited protective immune responses against *B. pertussis in vivo* and *in vitro*. Suppression was substantially reversed by anti–IL-10 *in vivo*. The capacity to induce Tr1 cells is thereby exploited by a respiratory pathogen to evade protective immunity and suppress protective Th1 responses at local sites of infection.

Analysis of cytokine production by T cells from patients with chronic hepatitis C virus (HCV) infection identified antigen-specific regulatory T cells that secreted IL-10 in addition to IFN-γ circulating Th1 cells[455]; no IL-4 producing T cells were identified. IL-10 producing cells were detected in a higher proportion of patients with chronic infection than in those who had cleared the virus. Taken together with the studies on *L. major* and *B. pertussis*, these studies on HCV strongly support the general concept that many infectious agents have evolved mechanisms for selective activation of either T_{reg} cells or the generation of IL-10 producing Tr1 cells from CD25− T cells. The ultimate result is perpetuation of the chronic infectious state with incomplete clearing of the infection. Depending on the extent of suppression of effector T cells in the host, the consequences of the chronic state may be protective immunity (*L. major*) or continued pathogen-mediated organ destruction (HCV).

Many viruses such as herpesviruses, hepatitis viruses, and retroviruses evade immunological destruction during acute infection and establish chronic persistent infections that may culminate in life-threatening diseases. Friend virus (FV) infection in mice has been used as an experimental model to study retrovirus-induced immunosuppression and may offer insights into our understanding of immunosuppression associated with HIV. Mice chronically infected with FV are unable to reject both FV-induced and unrelated immunogenic tumors. CD8+ T cells from acutely infected mice produced perforin, granzyme A, and granzyme B and display recent evidence of degranulation and *in vivo* cytotoxicity. Activated T cells from chronically infected mice were deficient in cytolytic molecules and showed little evidence of recent degranulation and *in vivo* cytotoxicity. CD4+, but not CD8+, T cells from infected mice can transfer suppression to normal mice and can inhibit the generation of CTL in culture.[456] Suppression could be substantially reversed by the addition of anti-CTLA-4, but not anti–IL-10R, to the cultures. FV-induced T_{reg} suppress CD8+ T cells *in vitro* regardless of the TCR specificity of the CD8+ T cells.[457] It is unknown whether the T_{reg} in chronically infected mice are specific for any viral proteins. Depletion of Foxp3+ T cells early after infection resulted in an increase in virus-specific CD8+ T cells on day 10 postinfection, enhanced production of IFNγ and TNFα, and a 10-fold reduction in viral load.

As HIV infection in man in many respects mimics the animal models of chronic retroviral infections, several studies have examined T_{reg} function at different stages of HIV infection. Kinter et al[458] found that in the majority of HIV-infected, but still healthy, individuals, $CD25^{hi}$ T_{reg} cells significantly suppressed cellular proliferation and cytokine production by CD4+ and CD8+ T cells in response to HIV antigens *in vitro*. Suppression was cell contact-dependent and IL-10– and TGF-β–independent. Patients with strong HIV-specific suppression *in vitro* had lower levels of viremia and higher CD4/CD8 T cell ratios than patients who did not have T_{reg} activity. These data suggest that the suppression of CD4+ T cell activation by T_{reg} may make HIV replication less favorable as the virus must replicate within the CD4+ T cell itself. Thus, in this chronic infection model, the suppressive functions of T_{reg} may actually be beneficial to the patient.

The contribution of T_{reg} has been most clearly shown in responses to chronic infections. Much less is known about the role of T_{reg} in modulating acute viral infections. In the murine model of herpetic stromal keratitis, depletion of T_{reg} before infection resulted in lesions of greater severity and permitted the induction of disease with lower infecting doses of virus.[459] Co-transfer of T_{reg} with CD25− T cells reduced lesion severity and diminished the Ag-specific cytokine response of splenic CD4+ T cells. The mechanism of action of T_{reg} in this model is not fully elucidated. T_{reg} inhibit the induction of virus-specific CD4+ T cell induction but may also modify the expression of homing molecules involved in T cell migration to the ocular inflammatory site or the pathogenic T cells extralymphoid inflammatory sites. T_{reg} can also control the intensity of secondary responses to HSV and may also influence the magnitude of immunologic memory.

Pathogen-specific T_{reg} cells have been detected in a few studies using whole pathogen preparations as targets without the identification of specific target epitopes. The role of T_{reg} cells in facilitating the persistence of infections with *Mycobacterium tuberculosis* (Mtb) is unclear. Rigorous depletion of T_{reg} cells during early infection results in enhanced bacterial clearance.[460] Early arrival in the lung of Mtb-specific T cells is associated with enhanced immune control and lower bacterial burdens, and T_{reg} cells slow this recruitment process. This result differs from studies of infection with Herpes virus, which concluded that T_{reg} cells help facilitate effector T cell recruitment to tissue sites of infection.[461] Studies using TCR Tg cells specific for an epitope derived from Mtb attempted to identify T_{reg} specific for the same epitope recognized, but T_{reg} cells specific for this epitope were not found. Virus-specific T_{reg} cells have been detected in mice infected with a strain of mouse hepatitis virus. Polyclonal T_{reg} cells from normal mice will decrease weight loss, clinical scores, and demyelination when adoptively transferred into mice infected with this strain. T_{reg} cells for two viral epitopes were detected using peptide-MHC class II tetramers 6 to 7 days after infection, at the same time as CD4+ effector T cells. The virus-specific Foxp3+ cells expressed both IFN-γ and IL-10 in the CNS after infection and suppressed T cell proliferation of cognate epitope-specific CD4+ T cells.[462] IFN-γ expression by these cells was maintained during both acute and chronic phases of infection. The identification of T_{reg} specific for a pathogen-derived epitope is consistent with the view that exposure and immunization may expand not only protective T cell subsets but also T cell subsets that impede protection. Overall protection likely depends on the relative ratio of protective versus suppressive T cell subsets.

T_{REG} AND AUTOIMMUNE DISEASE IN MAN: APPROACHES TO THERAPY

In autoimmune disease in men, the major critical question which remains unresolved is whether T_{regs} play any role in the development of multifactorial polygenic common autoimmune diseases which afflict 10% of the population worldwide. It has been difficult to prove whether an anomaly or variation in T_{reg} function or number is a cause or consequence of an autoimmune disease. There is currently no clear evidence that a global deficiency in the number of T_{reg} cells is the source of failed regulation in the more common forms of autoimmunity.[463] There may be differences in T_{reg} phenotypes in disease including the numbers of CD4+CD45RA+Foxp3+ and CD4+CD45RA−Foxp3+ T_{regs} (Section III. B).

Increased numbers of T_{reg} cells are frequently observed in the affected tissues of patients with RA, IBD, and psoriasis, suggesting that T_{reg} function may be compromised by cell-extrinsic factors such as the cytokine milieu or overactive APC function, resulting in resistance to suppression. Several studies in patients with different autoimmune diseases have demonstrated a defect in the function *in vitro* of T_{reg} from peripheral blood raising the issue as to whether this may be a common denominator in the cause of human autoimmune disease (Table 29.6). In many of these studies, it was definitively shown that the decrease in T_{reg} function was due to defect in the T_{reg} subset rather than secondary to responder T cells that were refractory to suppression. One possibility is that T_{reg} may have migrated into the target tissue so that blood T_{reg} function is depressed, but T_{reg} function from the

TABLE 29.6 Human Diseases With Abnormal T_{reg} Function[466,471]

Disease
Multiple sclerosis
Myasthenia gravis
Rheumatoid arthritis
Autoimmune polyglandular II
Lupus erythematosis
Psoriasis
Type I diabetes
HIV/AIDS

target tissue may actually be enhanced. For example, the frequency of CD25^{hi} T cells was much greater in the synovial fluid as compared to the peripheral blood in adult patients with RA and synovial fluid, T_{reg} demonstrated normal suppressive activity *in vitro*.[464] Functional defects in the ability of T_{regs} from RA patients to suppress proliferation were not present, but the T_{reg} from RA patients appeared to have a selective defect in suppression of IFN-γ secretion. This selective suppression was associated with overexpression of TNF-α in the synovial tissues of RA subjects, decreased Foxp3 expression, and impaired CTLA-4 expression.[465-467] Treatment of patients with RA with anti-TNF-α restored the capacity of the T_{reg} to suppress cytokine production. It remains unclear whether the treatment acted directly on T_{reg} or whether it acted on the pathogenic T effector cells or modified the inflammatory milieu. Appropriate caution should be exercised in interpretation of these studies. As discussed above, identification of human Foxp3+ T_{reg} may be difficult, and many of the earlier published studies published have only utilized high levels of CD25 expression to identify human T_{reg} and did not validate the isolated populations for Foxp3 expression by intracellular staining.

As was observed with mouse T_{reg}, human T_{reg} cells possess some degree of plasticity and instability, particularly under inflammatory conditions. T_{reg} cells can be unstable and acquire effector phenotype after losing Foxp3 expression under inflammatory conditions.[468] Both T_{reg} cells with Th1- and Th17-like characteristics have been identified in a number of autoimmune and inflammatory diseases. Patients with different autoimmune diseases including type I diabetes, multiple sclerosis, autoimmune hepatitis, and Sjogren syndrome express an increased frequency of IFNγ+Foxp3+ T_{reg} cells in the periphery that display lower suppressive ability than that of control subjects. Patients with MS display an activated PI(3)-K-Akt-Foxo pathway *ex vivo*, and their suppressive activity is corrected by blockade of the PI-3-K pathway.[469] It is unknown whether Th-1–like T_{reg} cells have a role in disease pathogenesis. Patients with systemic sclerosis display an increased frequency of Th2-like T_{reg} cells in the skin, but not in the peripheral blood. These cells secrete IL-4 and IL-13 and upregulate GATA3 and IRF-4. Peripheral blood T_{reg} cells from these patients express high levels of St2, the receptor for IL-33.[470] Patients with food allergies may express Th2-like T_{reg} cells.[471] The significance of these effector-like T_{reg} populations in disease remains unclear. Does T_{reg} plasticity represent a general property of all T_{reg} cells, or does it reflect initial heterogeneity of the T_{reg} population so that only a subset is able to acquire effector properties? For example, pT_{reg} cells may be much more prone to potentially convert back to T effector cells than tT_{reg} cells. Although effector-like T_{reg} cells are present in disease their role in disease development and pathogenesis remains unclear.

Enhancement of either the numbers, function, or survival of T_{reg} represents a goal for the treatment of autoimmune and allergic diseases as well as for inhibition of allograft rejection.[423,424] One might expand either Foxp3+ T_{reg} or Tr1 T cells to generate sufficient numbers of cells for infusion back into patients.[472] If Foxp3+ T_{reg} recognize organ-specific antigens, and if these antigens can be identified, it may be possible to generate lines or clones of antigen-specific T_{reg} cells *in vitro* that could be used therapeutically. Organ-specific T_{reg} would home to their target, be activated by their target autoantigen, but mediate bystander suppression, since their effector function would be nonspecific. Alternatively, as a number of factors (eg, IL-6 or the GITR-L) can render effector cells resist to the suppressor function of T_{reg}, blocking or neutralizing these factors may enhance T_{reg} function. Ultimately, further studies of the molecular basis of T_{reg}-mediated suppression should allow the development of mAbs or small molecules[473] that could enhance their suppressor effector function.

Preclinical studies have demonstrated the capacity of T_{reg} to prevent or treat a number of diseases including GVHD,[474] solid organ transplantation, type I diabetes,[472] systemic lupus,[475] IBD, and MS.[400] Preclinical studies have shown that antigen-specific T_{reg} are more potent than polyclonal T_{reg} in T1D, autoimmune CNS disease, and transplantation. Antigen-specific T_{reg} predominantly localize at the site of antigen presentation, decreasing the risk of generalized immunosuppression. T_{reg} cells exert their suppressive function, both at the tissue site of inflammation and in local secondary lymphoid tissue. Adoptive T_{reg} cells may not have to persist indefinitely but just long enough to confer suppressive capacity to other immune cells located in the affected tissue (infectious tolerance).

Three different approaches are being developed for adoptive T_{reg} cell therapy. The simplest approach is to isolate polyclonal T_{reg} cells from the patient and expand them *in vitro* by stimulation with anti-CD3/anti-CD28 and high-dose IL-2.[476] One potential hazard of adoptive T_{reg} therapy is that normally a small subset of Foxp3+ T cells can lose Foxp3 expression (termed exFoxp3+ cells, Section VIII) especially in the presence of inflammatory cytokines leading to the acquisition of effector T cell phenotype. pT_{reg} are likely more susceptible to losing Foxp3 than tT_{reg}. If such conversion of T_{reg} cells to antigen-specific effector T cells occurs this may result in exacerbation of disease. A second approach primarily for use in transplantation is to expand alloantigen-specific T_{reg} from the transplant recipient *in vitro* using allogenic APC from the transplant donor. The third approach is to use designer T_{reg} which are specific for the autoantigen. One type of designer T_{regs} is generated by the transduction of a normal Foxp3+ T_{reg} from the patient with a TCR specific for the autoantigen derived from T effector cells from the same patient (Fig. 29.21).[477] A related approach is to transduce antigen-specific T effector cells with Foxp3, thereby converting pathogenic T cells to T_{reg}.

An extension of the third approach is to transduce normal Foxp3+ T_{reg} from the patient with chimeric antigen receptors (CARs) which are artificial receptors comprising an antigen-binding site of a mAb in their extracellular domains and T cell stimulatory and costimulatory intracellular domains (Fig. 29.21). CARs provide a unique opportunity to target T_{reg} cells to the site of tissue destruction or to transplanted tissue because they recognize whole proteins expressed in

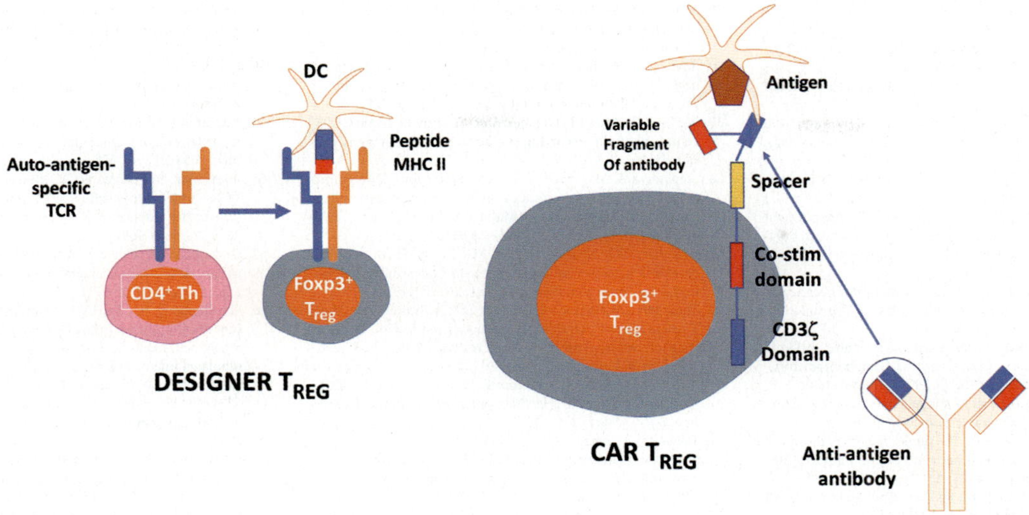

FIG. 29.21. Generation of designer T_{reg} and CAR T_{reg}.

tissues rather than peptide-MHC II complexes recognized by conventional TCRs.[478] T_{reg} cells engineered with a CAR targeting HLA-A2 can induce alloantigen-specific suppression in a humanized mouse transplantation model.[479] Activation of CAR T_{reg} may require higher levels of antigen expression than the low levels of antigen that are recognized by T_{reg} TCRs. It remains possible the direct recognition of an antigen on target cells by CARs could lead to T_{reg} cell–mediated killing via the perforin/granzyme B pathway.

An alternative approach to adoptive T_{reg} therapy would be to covert autoantigen-specific effector T_{reg} cells *in vitro* to iT_{regs} by culturing with antigen in the presence of TGF-β and IL-2. Although such an approach has shown effectiveness in animal models,[70] in most studies, it has been difficult to generate human iT_{reg}[49] and when generated human iT_{reg} lack T_{reg}-specific demethylation. Pharmacologic approaches to T_{reg} induction may prove to be more effective.

The use of IL-2 has also been proposed for the enhancement of T_{reg} cells numbers and function in autoimmunity. To avoid the severe side effects of high dose IL-2 and also to prevent the activation of T effector cells, it was thought that low-dose IL-2 would be beneficial[480] and that the high levels of expression of CD25 on T_{reg} cells could make them more sensitive than effector T cells to low dose IL-2. The administration of low-dose IL-2 has shown some promising results in treatment of autoimmune and autoinflammatory conditions.[481] Low-dose IL-2 induces robust expansion of T_{reg} cells, and the maximum expansion and duration of expansion are dose-dependent.[482] The expanded T_{reg} cells are activated as determined by expression of multiple markers, and they have enhanced suppressor activity.[483] Double-blind, placebo-controlled randomized trials of low-dose IL-2 therapy in multiple autoimmune diseases are now in progress.

An alternative approach to the use of low-dose IL-2 is the development of IL-2 variants (or muteins) that preferentially bind and activate cells expressing high levels of the IL-2Rαβγ receptor over those cells signaling primarily through the β and γ receptor chains. This can be achieved by either increasing IL-2 affinity to the α-chain, or decreasing IL-2 affinity to the β-chain.[484] A long-lived IL-2 mutein with reduced binding to the intermediate affinity IL-2Rβγ complex was shown to expand T_{reg} cells without expanding memory CD4+ or CD8+ cells in cynomolgus monkeys and humans.[485] A similar strategy was used to generate an IL-2/CD25 fusion protein which was shown to slow down the dissociation of IL-2 into an active monomer, thus improving its efficacy and decrease immunogenicity as compared to approaches utilizing introduction of mutations. A human anti–IL-2 mAb has been designed to stabilize IL-2 in a conformation that mediates selective binding to IL-2Rα and inhibits binding to IL-2Rβ. Treatment with this mAb in complex with IL-2 results in the preferential STAT5 phosphorylation of T_{reg} cells *in vitro* and expansion of T_{reg} cells in humanized NOD/SCID/IL2rγ^{null} (NSG) mice.[486] A more complex approach was used to preferentially expand CD8+ T cells engineered to express orthoIL-2Rβ by treatment with orthogonal IL-2 prior to adoptive transfer.[487] The efficacy of above-mentioned therapies needs to be evaluated further in man. The use of IL-2 muteins should be tested under inflammatory conditions where CD25 is upregulated on effector/memory cells to ensure that effector cell activation would not result in undesirable outcomes. It also remains unknown if the apparent T_{reg}-selectivity observed with low concentrations of IL-2 can be observed with high concentrations of IL-2 muteins.

ACKNOWLEDGMENT

This manuscript was supported by funds from the Division of Intramural Research, National Institute of Allergy and Infectious Diseases.

REFERENCES

1. Schwartz RH. A cell culture model for T lymphocyte clonal anergy. *Science*. 1990;248:1349-1356.
2. Miller JF, Heath WR. Self-ignorance in the peripheral T-cell pool. *Immunol Rev*. 1993;133:131-150.
3. Gershon RK, Kondo K. Cell interactions in the induction of tolerance: the role of thymic lymphocytes. *Immunology*. 1970;18:723-735.
4. Gershon RK, Kondo K. Infectious immunological tolerance. *Immunology*. 1971;21:903-914.
5. Baker PJ, Barth RF, Stashak PW, Amsbaugh DF. Enhancement of the antibody response to type III pneumococcal polysaccharide in mice treated with anti-lymphocyte serum. *J Immunol*. 1970;104:1313-1315.
6. Tada T, Taniguchi M, Okumura K. Regulation of homocytotropic antibody response in the rat. II. Effect of X-irradiation. *J Immunol*. 1971;106:1012-1018.
7. Okumura K, Tada T. Regulation of homocytotrophic antibody in the rat. VI. Inhibitory effect of thymocyte on homocytotrophic antibody response. *J Immunol*. 1971;107:1682-1689.
8. Tada T, Okumura O. The role of antigen-specific T cell factors in the immune response. *Adv Immunol*. 1979;28:1-87.
9. Germain RN, Benacerraf B. A single major pathway of T-lymphocyte interactions in antigen-specific immune suppression. *Scand J Immunol*. 1981;13:1-10.
10. Green DR, Flood PM, Gershon RK. Immunoregulatory T-cell pathways. *Annu Rev Immunol*. 1983;1:439-463.
11. Nishizuka Y, Sakakura T. Thymus and reproduction: sex-linked dysgenesia of the gonad after neonatal thymectomy in mice. *Science*. 1969;166:753-755.
12. Sakaguchi S, Sakaguchi N. *Thymus, T cells, and autoimmunity: various causes but a common mechanism of autoimmune disease*. In: *Autoimmunity: Physiology and Disease*. Wiley-Liss, Inc.; 1994:203-227.
13. Penhale WJ, Farmer A, Irvine WJ. Thyroiditis in T cell-depleted rats: influence of strain, radiation dose, adjuvants and antilymphocyte serum. *Clin Exp Immunol*. 1975;21:362-375.
14. Penhale WJ, Irvine WJ, Inglis JR, et al. Thyroiditis in T cell-depleted rats: suppression of the autoallergic response by reconstitution with normal lymphoid cells. *Clin Exp Immunol*. 1976;25:6-16.
15. Penhale WJ, Stumbles PA, Huxtable CR, et al. Induction of diabetes in PVG/c strain rats by manipulation of the immune system. *Autoimmunity*. 1990;7:169-179.
16. Sakaguchi S, Fukuma K, Kuribayashi K, et al. Organ-specific autoimmune diseases induced in mice by elimination of T cell subset. I. Evidence for the active participation of T cells in natural self-tolerance; deficit of a T cell subset as a possible cause of autoimmune disease. *J Exp Med*. 1985;161:72-87.
17. Sakaguchi S, Sakaguchi N, Asano M, et al. Immunologic self-tolerance maintained by activated T cells expressing IL-2 receptor α-chains (CD25). *J Immunol*. 1995;155:1151-1164.
18. Asano M, Toda M, Sakaguchi N, et al. Autoimmune disease as a consequence of developmental abnormality of a T cell subpopulation. *J Exp Med*. 1996;184:387-396.
19. Suri-Payer E, Amar AZ, Thornton AM, et al. CD4+CD25+ T cells inhibit both the induction and effector function of autoreactive T cells and represent a unique lineage of immunoregulatory cells. *J Immunol*. 1998;160:1212-1218.
20. Powrie F, Mason D. OX-22high CD4+ T cells induce wasting disease with multiple organ pathology: prevention by the OX-22low subset. *J Exp Med*. 1990;172:1701-1708.
21. Fowell D, Mason D. Evidence that the T cell repertoire of normal rats contains cells with the potential to cause diabetes. Characterization of the CD4+ T cell subset that inhibits this autoimmune potential. *J Exp Med*. 1993;177:627-636.
22. Powrie F, Leach MW, Mauze S, et al. Phenotypically distinct subsets of CD4+ T cells induce or protect from chronic intestinal inflammation in C. B-17 scid mice. *Int Immunol*. 1993;5:1461-1471.
23. Read S, Malmstrom V, Powrie F. Cytotoxic T lymphocyte-associated antigen 4 plays an essential role in the function of CD25+CD4+ regulatory cells that control intestinal inflammation. *J Exp Med*. 2000;192:295-302.
24. Itoh M, Takahashi T, Sakaguchi N, et al. Thymus and autoimmunity: production of CD25+CD4+ naturally anergic and suppressive T cells as a key function of the thymus in maintaining immunologic self-tolerance. *J Immunol*. 1999;162:5317-5326.
25. Baecher-Allan C, Brown JA, Freeman GJ, et al. CD4+CD25high regulatory cells in human peripheral blood. *J Immunol*. 2001;167:1245-1253.
26. Stephens LA, Mottet C, Mason D, et al. Human CD4+CD25+ thymocytes and peripheral T cells have immune suppression activity. *Eur J Immunol*. 2001;31:1247-1254.
27. Levings MK, Sangregorio R, Roncarolo MG. Human CD25+CD4+ T regulatory cells suppress naïve and memory T cell proliferation and can be expanded *in vitro* without loss of function. *J Exp Med*. 2001;193:1295-1301.
28. Jonuleit H, Schmitt E, Stassen M, et al. Identification and functional characterization of human CD4+CD25+ T cells with regulatory properties isolated from peripheral blood. *J Exp Med*. 2001;193:1285-1294.
29. Dieckmann D, Plottner H, Berchtold S, et al. *Ex vivo* isolation and characterization of CD4+CD25+ T cells with regulatory properties from human blood. *J Exp Med*. 2002;193:1303-1310.
30. Taams LA, Smith J, Rustin MH, et al. Human anergic/suppressive CD4+CD25+ T cells: a highly differentiated and apoptosis-prone population. *Eur J Immunol*. 2001;31:1122-1131.
31. Ng WF, Duggan PJ, Ponchel F, et al. Human CD4+CD25+ cells: a naturally occurring population of regulatory T cells. *Blood*. 2001;98:2736-2744.
32. Stephens LA, Mason D. CD25 is a marker for CD4+ thymocytes that prevent autoimmune diabetes in rats, but peripheral T cells with this function are found in both CD25+ and CD25- subpopulations. *J Immunol*. 2001;165:3105-3110.
33. Andersen KG, Nissen JK, Betz AG. Comparative genomics reveals key gain-of-function events in Foxp3 during regulatory T cell evolution. *Front Immunol*. 2012;3:113.
34. Thornton AM, Shevach EM. CD4+CD25+ immunoregulatory T cells suppress polyclonal T cell activation *in vitro* by inhibiting interleukin 2 production. *J Exp Med*. 1998;188:287-296.
35. Salomon B, Lenschow DJ, Rhee L, et al. B7/CD28 costimulation is essential for the homeostasis of the CD4+CD25+ immunoregulatory T cells that control autoimmune diabetes. *Immunity*. 2000;12:431-440.
36. Takahashi T, Tagami T, Yamazaki S, et al. Immunologic self-tolerance maintained by CD25+CD4+ regulatory T cells constitutively expressing cytotoxic T lymphocyte-associated antigen 4. *J Exp Med*. 2000;192:303-309.
37. McHugh RS, Whitters MJ, Piccirillo CA, et al. CD4+CD25+ immunoregulatory T cells: gene expression analysis reveals a functional role for the glucocorticoid-induced TNF receptor. *Immunity*. 2002;16:311-323.
38. Shimizu J, Yamazaki S, Takahashi T, Ishida Y, Sakaguchi S. Stimulation of CD25+CD4+ regulatory T cells through GITR breaks immunological self-tolerance. *Nat Immunol*. 2002;3:135-142.
39. Brunkow ME, Jeffrey EW, Hjerrild KA, et al. Disruption of a new forkhead/winged-helix protein, scurfin, results in the fatal lymphoproliferative disorder of the surfy mouse. *Nat Genet*. 2001;27:68-73.
40. Fontenot JD, Gavin MA, Rudensky AY. Foxp3 programs the development and function of CD4+CD25+ regulatory T cells. *Nat Immunol*. 2003;4:330-336.
41. Hori S, Nomura T, Sakaguchi S. Control of regulatory T cell development by the transcription factor *Foxp3*. *Science*. 2003;299:1057-1061.
42. Khattri R, Cox T, Yasayko SA, et al. An essential role for Scurfin in CD4+CD25+ T regulatory cells. *Nat Immunol*. 2003;4:337-342.
43. Fontenot JD, Rasmussen JP, Williams LM, et al. Regulatory T cell lineage specification by the forkhead transcription factor Foxp3. *Immunity*. 2005;22:329-341.
44. Fontenot JD, Dooley JL, Farr AG, Rudensky AY. Development regulation of Foxp3 expression during ontogeny. *J Exp Med*. 2005;202:901-906.
45. Kim JM, Rasmussen JP, Rudensky AY. Regulatory T cells prevent catastrophic autoimmunity throughout the lifespan of mice. *Nat Immunol*. 2007;8:191-197.
46. Wan YY, Flavell RA. Regulatory T-cell functions are subverted and converted owing to attenuated Foxp3 expression. *Nature*. 2007;445:766-770.
47. Allen SE, Crome SQ, Crellin NK, et al. Activation-induced FOXP3 in human T effector cells does not suppress proliferation or cytokine production. *Int Immunol*. 2007;190:345-354.
48. Wang J, Ioan-Facsinay A, van der Voort EI, Huizinga TW, Toes RE. Transient expression of FOXP3 in human activated nonregulatory CD4+ T cells. *Eur J Immunol*. 2007;37:129-138.
49. Tran DQ, Ramsey H, Shevach EM. Induction of FOXP3 expression in naïve human CD4+FOXP3-T cells by T cell receptor stimulation is TGF-β-dependent but does not confer a regulatory phenotype. *Blood*. 2007;110:2983-2990.
50. Liu WH, Putnam AL, Xu-Yu Z, et al. CD127 expression inversely correlates with FoxP3 and suppressive function of human CD4+ T_{reg} cells. *J Exp Med*. 2006;203:1701-1711.
51. Miyara M, Yoshioka Y, Kitoj A, et al. Functional delineation and differentiation dynamics of human CD4(+) T cells expressing the FoxP3 transcription factor. *Immunity*. 2009;30:899-911.
52. Golding A, Hasni A, Ilei G, et al. The percentage of FoxP3+Helios+ T_{reg} cells correlates positively with disease activity in systemic lupus erythematosus. *Arthritis Rheum*. 2013;65:2898-2906.
53. Ferreira RC, Simons HZ, Thompson WS, et al. Cells with T_{reg}-specific TSDR demethylation but low CD25 are prevalent in autoimmunity. *J Autoimmun*. 2017;84:75-86.
54. Shevach EM. Certified professionals: CD4+CD25+ suppressor T cells. *J Exp Med*. 2001;193:F41-F45.
55. Abbas AK, Benoist C, Bluestone JA, et al. Regulatory T cells: recommendations to simplify the nomenclature. *Nat Immunol*. 2013;14:307-308.
56. Apostolou I, Sarukhan A, Klein L, et al. Origin of regulatory T cells with known specificity for antigen. *Nat Immunol*. 2002;3:756-763.
57. Apostoulou I, von Boehmer H. *In vivo* instruction of suppressor commitment in naïve T cells. *J Exp Med*. 2004;199:1401-1408.
58. Kretschmer K, Apostolou I, Hawiger D, et al. Inducing and expanding regulatory T cell populations by foreign antigen. *Nat Immunol*. 2005;6:1219-1227.
59. Polansky JK, Kretschmer K, Freyer J, et al. DNA methylation controls Foxp3 gene expression. *Eur J Immunol*. 2008;38:1654-1663.
60. Haxhinasto S, Mathis D, Benoist C. The AKT-mTOR axis regulates de novo differentiation of CD4+Foxp3+ cells. *J Exp Med*. 2008;205:565-574.
61. Josefowicz SZ, Wilson CB, Rudensky AY. Cutting edge: TCR stimulation is sufficient for induction of Foxp3 expression in the absence of DNA methyltransferase I. *J Immunol*. 2009;182:6648-6652.
62. Gottschalk RA, Corse E, Allison JP. TCR ligand density and affinity determine peripheral induction of Foxp3 *in vivo*. *J Exp Med*. 2010;207:1701-1711.
63. Schallenberg S, Tsai PY, Riewaldt JR, Kretschmer K. Identification of an immediate Foxp3– precursor to Foxp3+ regulatory T cells in peripheral lymphoid organs on nonmanipulated mice. *J Exp Med*. 2010;207:1393-1407.
64. Chen WJ, Jin W, Hardegen N, et al. Conversion of peripheral CD4+CD25– naïve T cells to CD4+CD25+ regulatory T cells by TGF-β induction of transcription factor *Foxp3*. *J Exp Med*. 2003;198:1875-1886.
65. Fantini MC, Becker C, Monteleone G, Pallone F, Galle PR, Neurath MF. Cutting edge: TGF-β induces a regulatory phenotype in CD4+CD25– T cells through Foxp3 induction and down-regulation of Smad7. *J Immunol*. 2004;172:5149-5153.
66. Horwitz DA, Zheng SG, Gray JD, et al. Regulatory T cells generated *ex vivo* as an approach for the therapy of autoimmune disease. *Semin Immunol*. 2004;16:135-143.
67. Davidson T, DiPaolo RJ, Andersson J, et al. Cutting edge: IL-2 is essential for TGF-β-mediated induction of Foxp3+ regulatory cells. *J Immunol*. 2007;178:4027-4032.
68. Zhou L, Lopes JE, Chong MM, et al. TGF-β-induced Foxp3 inhibits Th17 cell differentiation by antagonizing RORγt function. *Nature*. 2008;453:236-240.
69. Caretto D, Katzman SD, Villarino AV, Gallo E, Abbas AK. Cutting edge: the Th1 response inhibits the generation of peripheral regulatory T cells. *J Immunol*. 2010;184:30-34.
70. DiPaolo RJ, Brinster C, Davidson TS, Andersson J, Glass D, Shevach EM. Autoantigen-specific TGF β-Induced Foxp3(+) regulatory T cells prevent autoimmunity by inhibiting dendritic cells from activating autoreactive T cells. *J Immunol*. 2007;179:4685-4693.
71. Hippen KL, Merkel SC, Schirm DK, et al. Generation and large-scale expansion of human inducible regulatory T cells that suppress graft-versus-host disease. *Am J Transplant*. 2011;11:1148-1157.
72. Lu L, Ma J, Li Z, et al. All-trans retinoic acid promotes TGF-β-induced T_{regs} via histone modification but not DNA demethylation on Foxp3 gene locus. *PLoS One*. 2011;6:e24590.
73. Schmidt A, Eriksson M, Shang MM, et al. Comparative analysis of protocols to induce human CD4+Foxp3+ regulatory T cells by combinations of IL-2, TGF-beta, retinoic acid, rapamycin and butyrate. *PLoS One*. 2016;11:e0148474.
74. Mucida D, Kutchukhidze N, Erazo A, Russo M, Lafaille JJ, Curotto de Lafaille MA. Oral tolerance in the absence of naturally occurring T_{regs}. *J Clin Invest*. 2005;115:1923-1933.
75. Mucida D, Park Y, Kim G, et al. Reciprocal Th17 and regulatory T cell differentiation mediated by retinoic acid. *Science*. 2007;317:256-260.
76. Coombes JL, Siddiqui KR, Arancibia-Carcamo CV, et al. A functionally specialized population of mucosal CD103+ DCs induces Foxp3+ regulatory T cell via a TGF-β-and retinoic acid-dependent mechanism. *J Exp Med*. 2007;204:1757-1764.
77. Francisco LM, Salinas VH, Brown KE, et al. PD-L1 regulates the development, maintenance, and function of induced regulatory T cells. *J Exp Med*. 2009;206:3015-3029.
78. Quintana FJ, Basso AS, Iglesias AH, et al. Control of T_{reg} and Th17 differentiation by the arylhydrocarbon receptor. *Nature*. 2008;453:65-71.
79. Akane K, Kojima S, Mak TW, Shiku H, Suzuki H. CD8(+) CD122(+)CD49d(low) regulatory T cells maintain T-cell homeostasis by killing activated T cells via Fas/FasL-mediated cytotoxicity. *Proc Nat Acad Sci U S A*. 2016;113:2460-2465.
80. Guo Z, Wang G, Lv Y, Wan YY, Zheng J. Inhibition of cdk8/cdk19 activity promotes T_{reg} cell differentiation and suppresses autoimmune diseases. *Front Immunol*. 2019;10:1988.
81. Haribhai D, Lin W, Edwards B, et al. A central role for induced regulatory T cells in tolerance induction in experimental colitis. *J Immunol*. 2009;182:3461-3468.
82. Haribhai D, Williams JB, Jia S, et al. A requisite role for induced regulatory T cells in tolerance based on expanding antigen receptor diversity. *Immunity*. 2011;35:109-122.
83. Thornton AM, Korty PE, Tran DQ, et al. Expression of Helios, and Ikaros transcription factor family member, differentiates thymic-derived from peripherally induced Foxp3+ T regulatory cells. *J Immunol*. 2010;184:3433-3441.
84. Atarashi K, Tanoue T, Shima T, et al. Induction of colonic regulatory T cells by indigenous *Clostridium* species. *Science*. 2011;331:337-341.

85. Geuking MB, Cahenzli J, Lawson MAE, et al. Intestinal colonization induces mutualistic regulatory T cell responses. *Immunity*. 2011;34:794-806.
86. Weiss JM, Bilate AM, Gobert M, et al. Neuropilin 1 is expressed on thymus-derived natural regulatory T cells, but not mucosa-generated induced Foxp3(+) T_{reg} cells. *J Exp Med*. 2012;209:1723-1742.
87. Yadav M, Louvet C, Davini D, et al. Neuropilin-1 distinguishes natural and inducible regulatory T cells among regulatory T cell subsets *in vivo*. *J Exp Med*. 2012;209:1713-1722.
88. Shevach EM, Thornton AM. tTregs, pTregs, and iTregs: similarities and differences. *Immunol Rev*. 2014;259:88-102.
89. Thorstenson KM, Khoruts A. Generation of anergic and potentially immunoregulatory CD25+CD4 T cells *in vivo* after induction of peripheral tolerance with intravenous or oral antigen. *J Immunol*. 2001;167:188-195.
90. Kalekar LA, Mueller DL. Relationship between CD4 regulatory T cells and anergy *in vivo*. *J Immunol*. 2017;198:2527-2533.
91. Dieckmann D, Bruett CH, Ploettner H, Lutz MB, Schuler G. Human CD4+CD25+ regulatory, contact-dependent T cells induce interleukin 10-producing, contact-independent type1-like regulatory T cells. *J Exp Med*. 2002;196:247-253.
92. Jonuleit H, Schmitt E, Kakirman H, Stassen M, Knop J, Enk AH. Infectious tolerance: human CD25+ regulatory T cells convey suppressor activity to conventional CD4+ T helper T cells. *J Exp Med*. 2002;196:255-260.
93. Andersson J, Tran DQ, Pesu M, et al. CD4+FoxP3+ regulatory T cells confer infectious tolerance in a TGF-β-dependent manner. *J Exp Med*. 2008;205:1975-1981.
94. Kendal AR, Chen Y, Regateiro FS, et al. Sustained suppression by Foxp3+ regulatory T cells is vital for infectious transplantation tolerance. *J Exp Med*. 2011;208:2043-2053.
95. Modigliani Y, Coutinho A, Pereira P, et al. Establishment of tissue-specific tolerance is driven by regulatory T cells selected by thymic epithelium. *Eur J Immunol*. 1996;26:1807-1815.
96. Papiernik M, de Moraes ML, Pontoux C, Vasseur F, Pénit C. Regulatory CD4 T cells: expression of IL-2Rα chain, resistance to clonal deletion and IL-2 dependency. *Int Immunol*. 1998;10:371-378.
97. Jordan MS, Boesteanu A, Reed AJ, et al. Thymic selection of CD4+CD25+ regulatory T cells induced by an agonist self-peptide. *Nat Immunol*. 2001;2:301-306.
98. Van Santen H-M, Benoist C, Mathis D. Number of T_{reg} that differentiate does not increase upon encounter of agonist ligand on thymic epithelial cells. *J Exp Med*. 2004;200:1221-1230.
99. Bautista JL, Lio CW, Lathrop SK, et al. Intraclonal competition limits the fate determination of regulatory T cells in the thymus. *Nat Immunol*. 2009;6:610-617.
100. Leung MWL, Shen SQ, Lafaille JJ. TCR-dependent differentiation of thymic Foxp3+ cells is limited to small clonal sizes. *J Exp Med*. 2009;206:2121-2130.
101. DiPaolo RA, Shevach EM. CD4+ T cell development in a mouse expressing a transgenic TCR derived from a regulatory T cell. *Eur J Immunol*. 2009;39:234-240.
102. Hinterberger M, Aichinger M, da Costa OP, Voehringer D, Hoffmann R, Klein L. Autonomous role of medullary thymic epithelial cells in central CD4+ T cell tolerance. *Nat Immunol*. 2010;11:512-519.
103. Burchill MA, Yang J, Vang KB, et al. Linked T cell receptor and cytokine signaling govern the development of the regulatory T cell repertoire. *Immunity*. 2008;28:112-121.
104. Lio CW, Hsieh CS. A two-step process for thymic regulatory T cell development. *Immunity*. 2008;28:100-110.
105. Moran AE, Holzapfel KL, Xing Y, et al. T cell receptor signal strength in T_{reg} and iNKT cell development demonstrated by a novel fluorescent reporter mouse. *J Exp Med*. 2011;208:1279-1289.
106. Tai X, Erman B, Alag A, et al. Foxp3 transcription factor is proapoptotic and lethal to developing regulatory T cells unless counterbalanced by cytokine survival signals. *Immunity*. 2013;38:1116-1128.
107. Owen DL, Mahmud SA, Sjaastad LE, et al. Thymic regulatory T cells arise via two distinct developmental programs. *Nat Immunol*. 2019;20:195-205.
108. Hemmers S, Schizas M, Azizi E, et al. IL-2 production by self-reactive CD4 thymocytes scales regulatory T cell generation in the thymus. *J Exp Med*. 2019;216:2466-2478.
109. Tai X, Cowan M, Feigenbaum L, Singer A. CD28 costimulation of developing thymocytes induces Foxp3 expression and regulatory T cell differentiation independently of interleukin 2. *Nat Immunol*. 2005;6:152-162.
110. Lio CW, Dodson LF, Deppong CM, Hsieh CS, Green JM. CD28 facilitates the generation of Foxp3− cytokine responsive regulatory T cell precursors. *J Immunol*. 2010;110:6007-6013.
111. Schmidt-Supprian M, Tian J, Grant EP, et al. Differential dependence of CD4+CD25+ regulatory and natural killer-like cells on signals leading to NF-κB activation. *Proc Natl Acad Sci U S A*. 2004;101:4566-4571.
112. Anderson MS, Venanzi ES, Klein L, et al. Projection of an immunological self-shadow within the thymus by the aire protein. *Science*. 2002;298:1395-1401.
113. Malchow S, Leventhal DS, Lee V, Nishi S, Socci ND, Savage PA. Aire enforces immune tolerance by directing autoreactive T cells into the regulatory T cell lineage. *Immunity*. 2016;44:1102-1113.
114. Hubert FX, Kinkel SA, Davey GM, et al. Aire regulates the transfer of antigen from mTECs to dendritic cells for induction of thymic tolerance. *Blood*. 2011;118:2462-2472.
115. Yang S, Fujikado N, Kolodin D, Benoist C, Mathis D. Regulatory T cells generated early in life play a distinct role in maintaining self-tolerance. *Science*. 2015;348:589-594.
116. Cowan JE, McCarthy NI, Anderson G. CCR7 controls thymus recirculation, but not production and emigration, of Foxp3(+) T cells. *Cell Rep*. 2016;14:1041-1048.
117. Weist BM, Kurd N, Bousser J, Chan SW, Robey EA. Thymic regulatory T cell niche size is dictated by limiting IL-2 from antigen-bearing dendritic cells and feedback competition. *Nat Immunol*. 2015;16:635-641.
118. Malek TR, Yu A, Vincek V, Scibelli P, Kong L. CD4 regulatory T cells prevent lethal autoimmunity in IL-2Rβ-deficient mice: implications for the nonredundant function of IL-2. *Immunity* 2002;17:167-178.
119. D'Cruz LM, Klein L. Development and function of agonist-induced CD25+Foxp3+ regulatory T cells in the absence of interleukin 2 signaling. *Nat Immunol*. 2005;6:1152-1159.
120. Kumanogoh A, Wang X, Lee I, et al. Increased T cell autoreactivity in the absence of CD40-CD40 ligand interactions: a role of CD40 in regulatory T cell development. *J Immunol*. 2001;166:353-360.
121. Singh B, Read S, Asseman C, Uhlig HH, Powrie F. Control of inflammation by regulatory T cells. *Immunol Rev*. 2001;182:190-200.
122. Kagami SI, Nakajima H, Suto A, et al. Stat5a regulates T helper cell differentiation by several distinct mechanisms. *Blood*. 2001;97:2358-2365.
123. Setoguchi R, Hori S, Takahashi T, Sakaguchi S. Homeostatic maintenance of natural Foxp3+CD25+CD4+ regulatory T cells by interleukin-2 and induction of autoimmune disease by IL-2 neutralization. *J Exp Med*. 2005;201:723-735.
124. Furtado GC, Curotto de Lafaille MA, Kutchukhidze N, Lafaille JJ. Interleukin 2 signaling is required for CD4+ regulatory T cell function. *J Exp Med*. 2002;196:851-857.
125. Thornton AM, Donovan EE, Piccirillo CA, Shevach EM. Cutting Edge: IL-2 is critically required for the in vitro activation of CD4+CD25+ T cell suppressor function. *J Immunol*. 2004;172:6519-6523.
126. Barron L, Dooms H, Hoyer KK, et al. Cutting edge:mechanisms of IL-2-dependent maintenance of functional regulatory T cells. *J Immunol*. 2010;185:6426-6430.
127. Chinen K, Kannan AK, Levine AG, et al. An essential role for the IL-2 receptor in T_{reg} cell function. *Nat Immunol*. 2016;17:1322-1333.
128. Liu Y, Zhang P, Li J, Kulkarni AB, Perruche S, Chen W. A critical function for TGF-beta signaling in the development of natural CD4+CD25+Foxp3+ regulatory T cells. *Nat Immunol*. 2008;9:632-640.
129. Ouyang W, Beckett O, Ma Q, Li MO. Transforming growth factor-β signaling curbs thymic negative selection promoting regulatory T cell development. *Immunity*. 2010;32:642-653.
130. Hsieh C-S, Liang Y, Tyznik AJ, Self SG, Liggitt D, Rudensky AY. Recognition of peripheral self by naturally arising CD25+CD4+ T cell receptors. *Immunity*. 2004;21:267-277.
131. Pacholczyk R, Ignatowicz H, Kraj P, Ignatowicz L. Origin and T cell receptor diversity of Foxp3+CD4+CD25+ T cells. *Immunity*. 2006;25:249-259.
132. Hsieh CS, Zheng Y, Liang Y, Fontenot JD, Rudensky AY. An intersection between self-reactive regulatory and nonregulatory T cell receptor repertoires. *Nat Immunol*. 2006;7:401-410.
133. Kuczma M, Podolsky R, Garge N, et al.. Foxp3-deficient regulatory T cells do not revert into conventional effector CD4+ T cells but constitute a unique cell subset. *J Immunol*. 2009;183:3731-3741.
134. Lathrop SK, Santacruz NA, Pham D, Luo J, Hsieh CS. Antigen-specific peripheral shaping of the natural regulatory T cell population. *J Exp Med*. 2008;205:3105-3117.
135. Taguchi O, Kontani K, Ikeda H, et al. Tissue-specific suppressor T cells involved in self-tolerance are activated extrathymically by self-antigens. *Immunology*. 1994;82:365-369.
136. Samy ET, Parker LA, Sharp CP, Tung KSK. Continuous control of autoimmune disease by antigen-dependent polyclonal CD4+CD25+ regulatory T cells in the regional lymph node. *J Exp Med*. 2005;202:771-782.
137. Szurek E, Cebula A, Wojciech L, et al. Differences in expression level of Helios and Neuropilin-1 do not distinguish thymus-derived from extrathymically-induced CD4(+) Foxp3(+) regulatory T cells. *PLoS One*. 2015;10:e0141161.
138. Thornton AM, Lu J, Korty PE, et al. Helios(+) and Helios(−) Treg subpopulations are phenotypically and functionally distinct and express dissimilar TCR repertoires. *Eur J Immunol*. 2019;49:398-412.
139. Smigiel KS, Richards E, Srivastava S, et al. CCR7 provides localized access to IL-2 and defines homeostatically distinct regulatory T cell subsets. *J Exp Med*. 2014;211:121-136.
140. Holt MP, Punkosdy GA, Glass DD, et al. TCR signaling and CD28/CTLA-4 signaling cooperatively modulate T regulatory cell homeostasis. *J Immunol*. 2017;198:1503-1511.
141. Delpoux A, Yakonowsky P, Durand A, et al. TCR signaling events are required for maintaining CD4 regulatory T cell numbers and suppressive capacities in the periphery. *J Immunol*. 2014;193:5914-5923.
142. Miragaia RJ, Gomes T, Chomka A, et al. Single-cell transcriptomics of regulatory T cells reveals trajectories of tissue adaptation. *Immunity*. 2019;50:493-504.
143. Cretney E, Xin A, Shi W, et al. The transcription factors Blimp-1 and IRF4 jointly control the differentiation and function of effector regulatory T cells. *Nat Immunol*. 2011;12:394-311.
144. Levine AG, Arvey A, Jin W, et al. Continuous requirement for the TCR in regulatory T cell function. *Nat Immunol*. 2014;15:1070-1078.
145. Vahl JC, Drees C, Heger K, et al. Continuous T cell receptor signals maintain a functional regulatory T cell pool. *Immunity*. 201;41:722-736.
146. Ruan Q, Kameswaran V, Tone Y, et al. Development of Foxp3+ regulatory T cells is driven by the c-Rel enhanceosome. *Immunity*. 2009;31:932-940.
147. Molinero LL, Yang J, Gajewski T, et al. CARMA1 controls an early checkpoint in the thymic development of FoxP3+ regulatory T cells. *J Immunol*. 2009;182:6736-6743.
148. Floess S, Freyer J, Siewert C, et al. Epigenetic control of the foxp3 locus in regulatory T cells. *PLoS Biol*. 2007;5:169-178.
149. Toker A, Engelbert D, Garg G, et al. Active demethylation of the Foxp3 locus leads to the generation of stable regulatory T cells within the thymus. *J Immunol*. 2013;190:3180-3188.
150. Zheng Y, Josefowicz S, Chaudhry A, et al. Role of conserved non-coding DNA elements in the Foxp3 gene in regulatory T-cell fate. *Nature*. 2010;463:808-812.
151. Ono M, Yaguchi H, Ohkura N, et al. Foxp3 controls regulatory T-cell function by interacting with AML1/Runx1. *Nature*. 2007;446:685-689.
152. Wang L, Liu Y, Han R, et al. Mbd2 promotes Foxp3 demethylation and T-regulatory-cell function. *Mol Cell Biol*. 2013;33:4106-4115.
153. Yue X, Trifari S, Aijo T, et al. Control of Foxp3 stability through modulation of TET activity. *J Exp Med*. 2016;213:377-397.
154. Placek K, Hu G, Cui KR, et al. MLL4 prepares the enhancer landscape for Foxp3 induction via chromatin looping. *Nat Immunol*. 2017;18:1035-1045.
155. Li X, Liang Y, LeBlanc M, et al. Function of a Foxp3 cis-element in protecting regulatory T cell identity. *Cell*. 2014;158:734-748.
156. Rudra D, deRoos P, Chaudhry A, et al. Transcription factor Foxp3 and its protein partners form a complex regulatory network. *Nat Immunol*. 2012;13:1010-1019.
157. Josefowicz SZ, Lu LF, Rudensky AY. Regulatory T cells: mechanisms of differentiation and function. *Annu Rev Immunol*. 2012;30:532-564.
158. Sekiya T, Hibino S, Saeki K, et al. Nr4a receptors regulate development and death of labile Treg precursors to prevent generation of pathogenic self-reactive cells. *Cell Rep*. 2018;24:1627-1638.
159. Wu Y, Borde M, Heissmeyer V, et al. FOXP3 controls regulatory T cell function through cooperation with NFAT. *Cell*. 2006;126:375-387.
160. Kitoh A, Ono M, Naoe Y, et al. Indispensable role of the Runx1-Cbfβ transcription complex for *in vivo*-suppressive function of FoxP3+ regulatory T cells. *Immunity*. 2009;31:609-620.
161. Li B, Samanta A, Song X, et al. FOXP3 interactions with histone acetyltransferase and class II histone deacetylases are required for repression. *Proc Nat Acad Sci U S A* 2007;104:4571-4576.
162. Zheng Y, Chaudhry A, Kas A, et al. Regulatory T-cell suppressor program co-opts transcription factor IRF4 to control T(H)2 responses. *Nature*. 2009;458:351-356.
163. Fu W, Ergun A, Lu T, et al. A multiply redundant genetic switch 'locks in' the transcriptional signature of regulatory T cells. *Nat Immunol*. 2012;13:972-980.
164. Pam F, Yu H, Dang EV, et al. Eos mediates Foxp3-dependent gene silencing in CD4+ regulatory T cells. *Science*. 2009;325:1142-1146.
165. Vo N, Goodman RH. CREB-binding protein and p300 in transcriptional regulation. *J Biol Chem*. 2001;276:13505-13508.
166. DuPage M, Chopra G, Quiros J, et al. The chromatin-modifying enzyme Ezh2 is critical for the maintenance of regulatory T cell identity after activation. *Immunity*. 2015;42:227-238.
167. Konopacki C, Pritykin Y, Rubtsov Y, et al. Transcription factor Foxp1 regulates Foxp3 chromatin binding and coordinates regulatory T cell function. *Nat Immunol*. 2019;20:232-242.
168. Beyer M, Thabet Y, Muller RU, et al. Repression of the genome organizer SATB1 in regulatory T cells is required for suppressive function and inhibition of cellular differentiation. *Nat Immunol*. 2011;12:898-908.
169. Dang EV, Barbi J, Yang HY, et al. Control of T(H)17/T_{reg} balance by hypoxia-inducible factor 1. *Cell*. 2011;146:772-784.
170. Hsiao HW, Hsu TS, Liu WH, et al. Deltex1 antagonizes HIF-1α and sustains the stability of regulatory T cells *in vivo*. *Nat Commun*. 2015;6:6353.
171. Chen CJ, Barbi J, Bu SR, et al. The ubiquitin ligase Stub1 negatively modulates regulatory T cell suppressive activity by promoting degradation of the transcription factor Foxp3. *Immunity*. 2013;39:272-285.
172. van Loosdregt J, Fleskens V, Fu J, et al. Stabilization of the transcription factor Foxp3 by the deubiquitinase USP7 increases treg-cell-suppressive capacity. *Immunity*. 2013;39:259-271.

173. Wang L, Kumar S, Dahiya S, et al. Ubiquitin-specific protease-7 inhibition impairs Tip60-dependent Foxp3+ T-regulatory cell function and promotes antitumor immunity. *EBioMedicine.* 2016;13:99-112.
174. Zhao Y, Guo H, Qiao GL, et al. E3 ubiquitin ligase Cbl-b regulates thymic-derived CD4(+)CD25(+) regulatory T cell development by targeting Foxp3 for ubiquitination. *J Immunol.* 2015;194:1639-1645.
175. Ni X, Kou W, Gu J, et al. TRAF6 directs FOXP3 localization and facilitates regulatory T-cell function through K63-linked ubiquitination. *EMBO J.* 2019;38:e99766.
176. Xiao Y, Nagai Y, Deng GP, et al. Dynamic interactions between TIP60 and p300 regulate FOXP3 function through a structural switch defined by a single lysine on TIP60. *Cell Rep.* 2014;7:1471-1480.
177. van Loosdregt J, Vercoulen Y, Guichelaar T, et al. Regulation of Treg functionality by acetylation-mediated Foxp3 protein stabilization. *Blood.* 2010;115:965-974.
178. Arpaia N, Campbell C, Fan XY, et al. Metabolites produced by commensal bacteria promote peripheral regulatory T-cell generation. *Nature.* 2013;504:451-455.
179. van Loosdregt J, Brunen D, Fleskens V, et al. Rapid temporal control of Foxp3 protein degradation by Sirtuin-1. *PLoS One.* 2012;6:e19047.
180. Deng G, Song X, Greene MI. FoxP3 in T-reg cell biology: a molecular and structural perspective. *Clin Exp Immunol.* 2020;199:255-262.
181. Basu S, Golovina T, Mikheeva T, et al. Cutting edge: Foxp3-mediated induction of Pim 2 allows human T regulatory cells to preferentially expand in rapamycin. *J Immunol.* 2008;180:5794-5798.
182. Li ZY, Lin F, Zhuo CH, et al. PIM1 kinase phosphorylates the human transcription factor FOXP3 at serine 422 to negatively regulate its activity under inflammation. *J Biol Chem.* 2014;289:26872-26881.
183. Deng G, Nagai Y, Xiao Y, et al. Pim-2 kinase influences regulatory T cell function and stability by mediating Foxp3 protein N-terminal phosphorylation. *J Biol Chem.* 2015;290:20211-201220.
184. Fleskens V, Minutti CM, Wu XM, et al. Nemo-like kinase drives Foxp3 stability and is critical for maintenance of immune tolerance by regulatory T cells. *Cell Rep.* 2019;26:3600-3612.
185. Morawski PA, Mehra P, Chen CX, et al. Foxp3 protein stability is regulated by cyclin-dependent kinase 2. *J Biol Chem.* 2013;288:24494-24502.
186. Hill JA, Feuerer M, Tash K, et al. Foxp3 transcription-factor–dependent and –independent regulation of the regulatory T cell signature. *Immunity.* 2007;27:786-800.
187. Knoechel B, Lohr J, Zhu S, et al. Functional and molecular comparison of anergic and regulatory T lymphocytes. *J Immunol.* 2006;176:6473-6483.
188. Zheng Y, Josefowicz SZ, Kas A, et al. Genome-wide analysis of Foxp3 target genes in developing and mature regulatory T cells. *Nature.* 2007;445:936-940.
189. Marson A, Kretchmer K, Frampton GM, et al. Foxp3 occupancy and regulation of key target genes during T-cell stimulation. *Nature.* 2007;445:931-935.
190. Zhou X, Jeker LT, Fife BT, et al. Selective miRNA disruption in T_{reg} cells leads to uncontrolled autoimmunity. *J Exp Med.* 2008;205:1983-1991.
191. Liston A, Lu LF, O'Caroll D, et al. Dicer-dependent micoRNA pathway safeguards regulatory T cell function. *J Exp Med.* 2008;205:1993-2004.
192. Lu LF, Thai TH, Calado DP, et al. Foxp3-dependent microRNA155 confers competitive fitness to regulatory T cells by targeting SOCS1 protein. *Immunity.* 2009;30:80-91.
193. Lu LF, Boldin MP, Chaudry A, et al. Function of miRNA-146a in controlling T_{reg} cell-mediated regulation of Th1 responses. *Cell.* 2010;142:914-929.
194. Gavin MA, Rasmussen JP, Fontenot JD, et al. Foxp3-dependent programme of regulatory T-cell differentiation. *Nature.* 2007;445:771-775.
195. Lin W, Haribhai D, Relland LM, et al. Regulatory T cell development in the absence of functional Foxp3. *Nat Immunol.* 2007;8:359-368.
196. Lahl K, Mayer CT, Bopp T, et al. Nonfunctional regulatory T cells and defective control of Th2 cytokine production in natural scurfy mutant mice. *J Immunol.* 2009;180:5662-5672.
197. Ouyang W, Beckett O, Ma Q, et al. Foxo proteins cooperatively control the differentiation of Foxp3+ regulatory T cells. *Nat Immunol.* 2010;11:618-627.
198. Kerdiles YM, Stone EL, Beisner DL, et al. Foxo transcription factors control regulatory T cell development and function. *Immunity.* 2010;33:890-894.
199. Zenmour D, Zilionis R, Kiner E, et al. Single-cell gene expression reveals a landscape of regulatory T cell phenotypes shaped by the TCR. *Nat Immunol.* 2018;19:291-301.
200. Thornton AM, Shevach EM. Suppressor effector function of CD4+CD25+ immunoregulatory T cells is antigen nonspecific. *J Immunol.* 2000;164:183-190.
201. Gavin MA, Clarke SR, Negrou E, et al. Homeostasis and anergy of CD4+CD25+ suppressor T cells *in vivo*. *Nat Immunol.* 2002;3:33-41.
202. Su L, Creusot RJ, Gallo EM, et al. Murine CD4+CD25+ regulatory T cells fail to undergo chromatin remodeling across the proximal promoter region of the IL-2 gene. *J Immunol.* 2004;173:4994-5001.
203. Hickman SP, Yang J, Thomas RM, et al. Defective activation of protein kinase C and Ras-ERK pathways limits IL-2 production and proliferation by CD4+CD25+ regulatory T cells. *J Immunol.* 2006;177:2186-2194.
204. Tsang JY, Camara NO, Eren E, et al. Altered proximal T cell receptor signaling in human CD4+CD25+ regulatory T cells. *J Leukoc Biol.* 2006;80:145-151.
205. Walsh PT, Buckler JL, Zhang J, et al. PTEN inhibits IL-2 receptor-mediated expansion of CD4+CD25+ T_{regs}. *J Clin Invest.* 2006;116:2521-2531.
206. Sauer S, Bruno L, Hertweck A, et al. T cell receptor signaling controls Foxp3 expression via PI3K, Akt, and mTOR. *Proc Nat Acad Sci U S A.* 2008;105:7797-7802.
207. Takahashi T, Kuniyasu Y, Toda M, et al. Immunologic self-tolerance maintained by CD25+CD4+ naturally anergic and autoimmune disease by breaking their anergic/suppressive state. *Int Immunol.* 1998;10:1969-1980.
208. Piccirillo CA, Shevach EM. Cutting Edge: control of CD8+ T cell activation by CD4+CD25+ immunoregulatory cells. *J Immunol.* 2000;167:1137-1140.
209. Ermann J, Szanya V, Ford GS, et al. CD4+CD25+ T cells facilitate the induction of T cell anergy. *J Immunol.* 2001;167:4271-4275.
210. Shevach EM. Foxp3(+) T regulatory cells: still many unanswered questions-A perspective after 20 years of study. *Front Immunol.* 2018;9:10.
211. Liu Z, Gerner MY, Van Panhuys N, et al. Immune homeostasis enforced by co-localized effector and regulatory T cells. *Nature.* 2015;528:225-230.
212. Padiyan P, Xheng L, Ishihara S, et al. CD4+CD25+Foxp3+ regulatory T cells induce cytokine deprivation-mediated apoptosis of effector CD4+ T cells. *Nat Immunol.* 2007;8:1353-1362.
213. Tran DQ, Glass DD, Uzel G, et al. Analysis of adhesion molecules, target cells, and role of IL-2 in human FOXP3+ regulatory T cell suppressor function. *J Immunol.* 2009;182:2929-2938.
214. Palacios R, Moller G. T-cell growth factor abrogates concanavalin A-induced suppressor cell-function. *J Exp Med.* 1981;153:1360-1365.
215. Sojka DK, Hughson A, Sukiennicki TL, et al. Early kinetic window of target T cell susceptibility to CD25+ regulatory T cell activity. *J Immunol.* 2005;175:7274-7280.
216. Sukiennicki TL, Fowell DJ. Distinct molecular program imposed on CD4+ T cell targets by CD4+CD25+ regulatory T cells. *J Immunol.* 2006;177:6952-6961.
217. Nakamura K, Kitani A, Strober W. Cell contact-dependent immunosuppression by CD4+CD25+ regulatory T cells is mediated by cell surface-bound transforming growth factor β. *J Exp Med.* 2001;194:629-644.
218. Piccirillo CA, Letterio JJ, Thornton AM, et al. CD4+CD25+ regulatory T cells can mediate suppressor function in the absence of transforming growth factor β1 production and responsiveness. *J Exp Med.* 2002;196:1-10.
219. Marie JC, Letterio JJ, Gavin M, et al. TGF-β1 maintains suppressor function and Foxp3 expression in CD4+CD25+ regulatory T cells. *J Exp Med.* 2005;201:1061-1067.
220. Marie JC, Liggitt D, Rudensky AY. Cellular mechanisms of fatal early-onset autoimmunity in mice with the T cell–specific targeting of transforming growth factor-β receptor. *Immunity.* 2006;25:441-454.
221. Li MO, Wan YY, Flavell RA. T cell–produced transforming growth factor-β1 controls T cell tolerance and regulates Th1- and Th17-cell differentiation. *Immunity.* 2007;26:579-591.
222. Gutcher I, Donkor MK, Ma Q, et al. Autocrine transforming growth factor-β1 promotes *in vivo* Th17 differentiation. *Immunity.* 2011;34:396-408.
223. Mamura M, Lee W, Sullivan TJ, et al. CD28 disruption exacerbates inflammation in TGF-β1−/− mice: in vivo suppression by CD4+CD25+ regulatory T cells independent of autocrine TGF-β1. *Blood.* 2002;103:4594-4601.
224. Fahlen L, Read S, Gorelik L, et al. T cells that cannot respond to TGF-β escape control by CD4+CD25+ regulatory T cells. *J Exp Med.* 2005;201:737-746.
225. Kullberg MC, Hay V, Cheever AW, et al. TGF-β1 production by CD4+CD25+ regulatory T cells is not essential for suppression of intestinal inflammation. *Eur J Immunol.* 2005;35:2886-2895.
226. Tran DQ, Andersson J, Wang R, et al. GARP(LRRC32) is essential for the surface expression of latent TGF-β on platelets and activated FOXP3+ regulatory T cells. *Proc Natl Acad Sci U S A.* 2009;106:13445-13450.
227. Wang R, Kozhaya L, Mercer F, et al. Expression of GARP selectively identifies activated human Foxp3+ regulatory T cells. *Proc Natl Acad Sci U S A.* 2009;106:13439-13444.
228. Stockis J, Colau D, Coulie PG, et al. Membrane protein GARP is a receptor for latent TGF-β on the surface of activated T_{reg}. *Eur J Immunol.* 2009;12:3315-3322.
229. Cuende J, Lienart S, Dedobbeleer O, et al. Monoclonal antibodies against GARP/TGF-β1 complexes inhibit the immunosuppressive activity of human regulatory T cells *in vivo*. *Sci Transl Med.* 2015;7(284):284ra56.
230. de Streel G, Bertrand C, Chalon N, et al. Selective inhibition of TGF-β1 produced by GARP-expressing T_{regs} overcomes resistance to PD-1/PD-L1 blockade in cancer. *Nat Commun.* 2020;11:4545.
231. Shimizu J, Yamazaki S, Sakaguchi S. Induction of tumor immunity by removing CD25+CD4+ T cells: a common basis between tumor immunity and autoimmunity. *J Immunol.* 1999;163:5211-5218.
232. Smyth MJ, Teng MWL, Swann J, et al. CD4+CD25+ T regulatory cells suppress NK cell-mediated immunotherapy of cancer. *J Immunol.* 2006;176:1582-1587.
233. Ghiringhelli F, Menard C, Terme M, et al. CD4+CD25+ regulatory T cells inhibit natural killer cell functions in a transforming growth-factor-β-dependent manner. *J Exp Med.* 2005;202:1075-1085.
234. Barao I, Hanash AM, Hallett W, et al. Suppression of natural killer cell-mediated bone marrow cell rejection by CD4+CD25+ regulatory T cells. *Proc Natl Acad Sci U S A.* 2006;103:5460-5465.
235. Belkaid Y, Rouse BT. Natural regulatory T cells in infectious disease. *Nat Immunol* 2005;6:353-360.
236. McGeachy MJ, Stephens LA, Anderton SM. Natural recovery and protection from autoimmune encephalomyelitis: contribution of CD4+CD25+ regulatory T cells within the central nervous system. *J Immunol* 2005;175:3025-3032.
237. Rubtsov YP, Rasmussen JP, Chi EY, et al. Regulatory T cell-derived Interleukin-10 limits inflammation at environmental interfaces. *Immunity.* 2008;28:546-558.
238. Chaudry A, Samstein RM, Treuting P, et al. Interleukin-10 signaling in regulatory T cells is required for suppression of Th17 cell-mediated inflammation. *Immunity.* 2011;34:566-578.
239. Collison LW, Workman CJ, Kuo TT, et al. The inhibitory cytokine IL-35 contributes to regulatory T-cell function. *Nature.* 2007;450:566-569.
240. Collison LW, Chaturvedi V, Henderson AL, et al. IL-35-mediated induction of a potent regulatory T cell population. *Nat Immunol.* 2010;11:1093-1101.
241. Collison LW, Delgoffe GM, Guy CS, et al. The composition and signaling of the IL-35 receptor are unconventional. *Nat Immunol.* 2012;13:290-299.
242. Okoye IS, Coomes SM, Pelly VS, et al. MicroRNA-containing T-regulatory-cell-derived exosomes suppress pathogenic T helper 1 cells. *Immunity.* 2014;41:89-103.
243. Schmidleithner L, Thabet Y, Schonfeld E. Enzymatic activity of HPGD in T_{reg} cells suppresses T_{conv} cells to maintain adipose tissue homeostasis and prevent metabolic dysfunction. *Immunity.* 2019;50:1232-1248.
244. Bopp T, Becker C, Klein M, et al. Cyclic adenosine monophosphate is a key component of regulatory T cell mediated suppression. *J Exp Med.* 2007;204:1303-1310.
245. Hubert S, Rissiek B, Klages K, et al. Extracellular NAD+ shapes Foxp3 regulatory T cell compartment through ART2-P2X7 pathway. *J Exp Med.* 2010;207:2561-2568.
246. Borsellino G, Kleinewietfeld M, Di Mitri D, et al. Expression of ecto nucleotidase CD39 by Foxp3+ Treg cells: hydrolysis of extracellular ATP and immune suppression. *Blood.* 2007;110:1225-1232.
247. Deaglio S, Dwyer KM, Gao W, et al. Adenosine generation catalyzed by CD39 and CD73 expressed on regulatory T cells mediates immune suppression. *J Exp Med.* 2007;204:1257-1265.
248. Garin MI, Chu CC, Golshayan D, et al. Galectin-1: a key effector of regulation mediated by CD4+CD25+ T cells. *Blood.* 2007;109:2058-2065.
249. Kubach J, Lutter P, Bopp T, et al. Human CD4+CD25+ regulatory T cells: proteome analysis identifies galectin-10 as a novel marker essential for their anergy and suppressive function. *Blood.* 2007;110:1550-1558.
250. Grossman WJ, Verbsky JW, Barchet W, et al. Human regulatory T cells can use the perforin pathway to cause autologous target cell death. *Immunity.* 2004;21:589-601.
251. Gondek DC, Lu LF, Quezada SA, et al. Cutting Edge: contact-mediated suppression by CD4+CD25+ regulatory cells involves a granzyme B-dependent, perforin-independent mechanism. *J Immunol.* 2005;174:1783-1786.
252. Zhao DM, Thornton AM, DiPaolo RJ, et al. Activated CD4+CD25+ regulatory T cells selectively kill B lymphocytes. *Blood.* 2006;107:3925-3932.
253. Cao X, Cai SF, Fehniger TA, et al. Granzyme B and perforin are important for regulatory T cell-mediated suppression of tumor clearance. *Immunity.* 2007;27:635-646.
254. Boissonnas A, Scholer-Dahirel A, Simon-Blancal V, et al. Foxp3+ T cells induce perforin-dependent dendritic cell death in tumor-draining lymph nodes. *Immunity.* 2010;32:266-278.
255. Cederbom L, Hall H, Ivars F. CD4+CD25+ regulatory T cells downregulate co-stimulatory molecules on antigen-presenting cells. *Eur J Immunol.* 2000;30:1538-1543.
256. Serra P, Amrani A, Yamanouchi J, et al. CD40 ligation releases immature dendritic cells from the control of regulatory CD4+CD25+ T cells. *Immunity.* 2003;19:877-889.
257. Misra N, Bayry J, Lacroix-Desmazes S, et al. Cutting edge: human CD4+CD25+ T cells restrain the maturation and antigen-presenting function of dendritic cells. *J Immunol.* 2004;172:4676-4680.

258. Oldenhove G, de Heusch M, Urbain-Vansanten G, et al. CD4+CD25+ regulatory T cells control T helper 1 responses to foreign antigens induced by mature dendritic cells *in vivo*. *J Exp Med.* 2003;198:259-266.
259. Veldhoen M, Moncrieffe H, Hocking RJ, et al. Modulation of dendritic cell function by naïve and regulatory CD4+ T cells. *J Immunol.* 2006;176:6202-6210.
260. Thornton AM, Piccirillo CA, Shevach EM. Activation requirements for the induction of CD4(+)CD25(+) T cell suppressor function. *Eur J Immunol.* 2004;34:366-376.
261. Tang Q, Boden EK, Henriksen KJ, et al. Distinct roles of CTLA-4 and TGF-β in CD4+CD25+ regulatory cell function. *Eur J Immunol.* 2004;34:2996-3005.
262. Wing K, Onishi Y, Prieto-Martin P, et al. CTLA-4 control over Foxp3+ regulatory cell function. *Science.* 2008;322:271-275.
263. Paterson AM, Lovitch SB, Sage PT, et al. Deletion of CTLA-4 on regulatory T cells during adulthood leads to resistance to autoimmunity. *J Exp Med.* 2015;212:1603-1621.
264. Qureshi OS, Zheng Y, Nakamura K, et al. Trans-endocytosis of CD80 and CD86: a molecular basis for the cell extrinsic function of CTLA-4. *Science.* 2011;332:600-603.
265. Lo B, Zhang KJ, Lu W, et al. Patients with LRBA deficiency show CTLA4 loss and immune dysregulation responsive to abatacept therapy. *Science.* 2015;349:436-440.
266. Grohmann U, Fallarino F, Puccetti P. Tolerance DCs and tryptophan: much ado about IDO. *Trends Immunol.* 2003;24:242-248.
267. Cobbold SP, Adams E, Farquhar CA, et al. Infectious tolerance via consumption of essential amino acids and mTOR signaling. *Proc Natl Acad Sci U S A.* 2009;106:12055-12060.
268. Walker LSK, Chodos A, Eggena M, et al. Antigen-dependent proliferation of CD4+CD25+ regulatory T cells *in vivo*. *J Exp Med.* 2003;198:249-258.
269. Tang Q, Adams JY, Tooley AJ, et al. Visualizing regulatory T cell control of autoimmune responses in nonobese diabetic mice. *Nat Immunol.* 2006;7:83-92.
270. Tadokoro CE, Shakhar G, Shen S, et al. Regulatory T cells inhibit stable contacts between CD4+ T cells and dendritic cells *in vivo*. *J Exp Med.* 2006;203:505-511.
271. Onishi Y, Fehervari Z, Yamaguchi T, et al. Foxp3+ natural regulatory T cells preferentially form aggregates on dendritic cells *in vitro* and actively inhibit their maturation. *Proc Natl Acad Sci U S A.* 2008;105:10113-10118.
272. Sarris M, Anderson KG, Randow F, et al. Neuropilin-1 expression on regulatory T cells enhances their interactions with dendritic cells during antigen recognition. *Immunity.* 2008;28:402-413.
273. Davidson TS, Shevach EM. Polyclonal T_{reg} cells modulate T effector cell trafficking. *Eur J Immunol.* 2011;41:2862-2870.
274. Billiard F, Litvinova E, Sadoun D, et al. Regulatory and effector T cell activation levels are prime determinants of *in vivo* immune regulation. *J Immunol.* 2006;177:2167-2174.
275. Hanig J, Lutz MB. Suppression of mature dendritic cell function by regulatory T cells *in vivo* is abrogated by CD40 licensing. *J Immunol.* 2008;180:1405-1413.
276. Shevach EM. Biological functions of regulatory T cells. *Adv Immunol.* 2011;112:137-176.
277. Morlacchi S, Dal Secco V, Soldani C, et al. Regulatory T cells target chemokine secretion by dendritic cells independently of their capacity to regulate T cell proliferation. *J Immunol.* 2011;186:6807-6814.
278. Akkaya B, Oya Y, Akkaya M, et al. Regulatory T cells mediate specific suppression by depleting peptide-MHC class II from dendritic cells. *Nat Immunol.* 2019;20:218-231.
279. Liang BT, Workman C, Lee J, et al. Regulatory T cells inhibit dendritic cells by lymphocyte activation gene-3 engagement of MHC class II. *J Immunol.* 2008;180:5916-5926.
280. Delgoffe GM, Woo SR, Turnis ME, et al. Stability and function of regulatory T cells is maintained by a neuropilin-1-semaphorin-4a axis. *Nature.* 2013;501:252-256.
281. Stephens GL, McHugh RS, Whitters MJ, et al. Engagement of glucocorticoid-induced TNFR family related receptor on effector T cells by its ligand mediates resistance to suppression by CD4+CD25+ T cells. *J Immunol.* 2004;173:5008-5020.
282. Kohm AP, Williams JS, Miller SD. Cutting edge: ligation of the glucocorticoid-induced TNF receptor enhances autoreactive CD4+ T cell activation and experimental autoimmune encephalomyelitis. *J Immunol.* 2004;172:4686-4690.
283. Valzasina B, Guiducci C, Dislich H, et al. Triggering of OX40 (CD134) on CD4+CD25+ T cells blocks their inhibitory activity: a novel regulatory role for OX40 and its comparison with GITR. *Blood.* 2005;105:2845-2851.
284. Piconese S, Valzasina B, Colombo MP. Ox40 triggering blocks suppression by regulatory T cells and facilitates tumor rejection. *J Exp Med.* 2008;205:825-839.
285. Valencia X, Stephens G, Goldsbach-Mansky R, et al. TNF downmodulates the function of human CD4+CD25hi T-regulatory cells. *Blood.* 2006;108:253-261.
286. Zaragoza B, Chen X, Oppenheim JJ, et al. Suppressive activity of human regulatory T cells is maintained in the presence of TNF. *Nat Med.* 2016;22:16-17.
287. Pasare C, Medzhitov R. Toll pathway-dependent blockade of CD4+CD25+ T cell-mediated suppression by dendritic cells. *Science.* 2003;299:1033-1036.
288. Schreiber TH, Wolf D, Tsai MS, et al. Therapeutic T_{reg} expansion in mice by TNFRSF25 prevents allergic inflammation. *J Clin Invest.* 2010;120:3629-3640.
289. Wohlfert EA, Gorelik L, Mittler R, et al. Cutting edge: deficiency in the E3 ubiquitin ligase Cbl-b results in a multifunctional defect in T cells TGF-β sensitivity in vitro and in vivo. *J Immunol.* 2006;176:1316-1320.
290. Galgani M, De Rosa V, La Cava A. Role of metabolism in the immunobiology of regulatory T cells. *J Immunol.* 2016;197:2567-2577.
291. Shi H, Chi H. Metabolic control of treg cell stability, plasticity, and tissue-specific heterogeneity. *Front Immunol.* 2019;10:27.
292. Michalek RD, Gerriets VA, Jacobs SR, et al. Cutting edge: distinct glycolytic and lipid oxidative metabolic programs are essential for effector and regulatory CD4(+) T cell subsets. *J Immunol.* 2011;186:3299-3303.
293. Gerriets VA, Kishton RJ, Johnson M, et al. Foxp3 and Toll-like receptor signaling balance T_{reg} cell anabolic metabolism for suppression. *Nat Immunol.* 2016;17:1459-1466.
294. Angelin A, Gil-de-Gomez L, Dahiya S, et al. Foxp3 reprograms T cell metabolism to function in low glucose high lactate environments. *Cell Metabol.* 2017;24:1282-1293.
295. Howie D, Cobbold SP, Adams E, et al. Foxp3 drives oxidative phosphorylation and protection from lipotoxicity. *JCI Insight.* 2017;2:e89160.
296. Pacella I, Procaccini C, Focaccetti C, et al. Fatty acid metabolism complements glycolysis in the selective regulatory T cell expansion during tumor growth. *Proc Nat Acad Sci U S A.* 2018;115:E6546-E6555.
297. Zeng H, Yang K, Cloer C, et al. mTORC1 couples immune signals and metabolic programming to establish T_{reg}-cell function. *Nature.* 2013;499:485-500.
298. Shi H, Chapman NM, Wen J, et al. Amino acids license kinase mTORC1 activity and treg cell function via small G proteins Rag and Rheb. *Immunity.* 2019;51:1012-1027.
299. Sun IH, Oh MH, Zhao L, et al. mTOR complex 1 signaling regulates the generation and function of central and effector Foxp3(+) regulatory T cells. *J Immunol.* 2018;201:481-492.
300. Delgoffe GM, Pollizzi KN, Waickman AT, et al. The kinase mTOR regulates the differentiation of helper T cells through the selective activation of signaling by mTORC1 and mTORC2. *Nat Immunol.* 2011;12:295-303.
301. Charbonnier LM, Cui Y, Stephen-Victor E, et al. Functional reprogramming of regulatory T cells in the absence of Foxp3. *Nat Immunol.* 2019;20:1208-1219.
302. Patel CH, Leone RD, Horton MR, et al. Targeting metabolism to regulate immune responses in autoimmunity and cancer. *Nat Rev Drug Discov.* 2019;18:669-698.
303. Koch MA, Tucker-Heard G, Perdue NR, et al. The transcription factor T-bet controls regulatory T cell homeostasis and function during type 1 inflammation. *Nat Immunol.* 2009;10:595-602.
304. Chaudhry A, Rudra D, Treuting P, et al. CD4+ regulatory T cells control Th17 responses in a Stat3-dependent manner. *Science.* 2009;326:986-991.
305. Brysty RS, Aluvihare V, Welch KA, et al. B cells and professional APC recruit regulatory T cells. *Nat Immunol.* 2001;2:1126-1132.
306. Lim HW, Hillsamer P, Banham AH, et al. Cutting Edge: direct suppression of B cells by CD4+CD25+ regulatory T cells. *J Immunol.* 2005;175:4180-4183.
307. Lim HW, Hillsamer P, Kim CH. Regulatory T cells can migrate to follicles upon T cell activation and suppress GC-Th and GC-Th cell driven B cell responses. *J Clin Invest.* 2002;114:1640-1649.
308. Crotty S. A brief history of T cell help to B cells. *Nat Rev Immunol.* 2015;15:185-189.
309. Chung Y, Tanaka S, Chu F, et al. Follicular regulatory T cells expressing Foxp3 and Bcl-6 suppress germinal center reactions. *Nat Med.* 2011;17:983-988.
310. Linterman MA, Pierson W, Lee SK, et al. Foxp3+ follicular regulatory T cells control the germinal center response. *Nat Med.* 2011;17:975-982.
311. Sage PT, Sharpe AH. T follicular regulatory cells. *Immunol Rev.* 2016;271:246-259.
312. Wing JB, Kitagawa Y, Locci M, et al. A distinct subpopulation of CD25(−) T-follicular regulatory cells localizes in the germinal centers. *Proc Nat Acad Sci U S A.* 2017;114:E6400-E6409.
313. Botta D, Fuller MJ, Marquez-Lago TT, et al. Dynamic regulation of T follicular regulatory cell responses by interleukin 2 during influenza infection. *Nat Immunol.* 2017;18:1249-1260.
314. Sage PT, Francisco LM, Carman CV, et al. The receptor PD-1 controls follicular regulatory T cells in the lymph nodes and blood. *Nat Immunol.* 2013;14:152-161.
315. Sage PT, Paterson AM, Lovitch SB, et al. The coinhibitory receptor CTLA-4 controls B cell responses by modulating T follicular helper, T follicular regulatory, and T regulatory cells. *Immunity.* 2014;41:1026-1039.
316. Wing JB, Ise W, Kurosaki T, et al. Regulatory T cells control antigen-specific expansion of Tfh cell number and humoral immune responses via the coreceptor CTLA-4. *Immunity.* 2014;41:1013-1025.
317. Sage PT, Alvarez D, Godec J, et al. Circulating T follicular regulatory and helper cells have memory-like properties. *J Clin Invest.* 2014;124:5191-5204.
318. Oldenhove G, Bouladoux N, Wohlfert EA. Decrease of Foxp3+ T_{reg} cell number and acquisition of effector cell phenotype during lethal infection. *Immunity.* 2009;31:772-786.
319. Xu L, Kitani A, Fuss I, et al. Cutting edge: regulatory T cells induce CD4+CD25-Foxp3-T cells are are self-induced to become Th17 cells in the absence of exogenous TGF-β. *J Immunol.* 2007;178:6725-6729.
320. Yang XO, Nurieva R, Martinez GJ, et al. Molecular antagonism and plasticity of regulatory and inflammatory T cell programs. *Immunity.* 2008;29:44-56.
321. Osorio F, Leibundgut-Landmann S, Lochner M, et al. DC activated via dectin-1 convert Treg into IL-17 producers. *Eur J Immunol.* 2008;38:3724-3281.
322. Laurence A, Tato CM, Davidson TS, et al. Interleukin-signaling via STAT5 constrains T-helper 17 generation. *Immunity.* 2007;26:371-381.
323. Bruno L, Mazzarella L, Hoogenkamp M, et al. Runx proteins regulate Foxp3 expression. *J Exp Med.* 2009;206:2329-2337.
324. Tsuji M, Komatsu N, Kawamoto S, et al. Preferential generation of follicular B helper T cells from Foxp3+ T cells in gut Peyer's patches. *Science.* 2009;323:1488-1492.
325. Wei G, Wei L, Zhu J, et al. Global mapping of H3K4me3 and H3K27me3 reveals specificity and plasticity in lineage fate determination of differentiating CD4+ T cells. *Immunity.* 2009;30:155-167.
326. Wohlfert EA, Grainger JR, Bouladoux N, et al. GATA3 controls Foxp3+ regulatory T cell fate during inflammation in mice. *J Clin Invest.* 2011;121:4503-4515.
327. Wang Y, Su MA, Wan YY. An essential role of the transcription factor GATA-3 for the function of regulatory T cells. *Immunity.* 2011;35:1-12.
328. Zhou X, Bailey-Bucktrout S, Jeker LT, et al. Instability of the transcription factor Foxp3 leads to the generation of pathogenic memory T cells *in vivo*. *Nat Immunol.* 2009;10:1000-1007.
329. Rubtsov YP, Niec RE, Josefowicz S, et al. Stability of the regulatory T cell lineage *in vivo*. *Science.* 2010;329:1667-1671.
330. Komatsue N, Mariotti-Ferandiz MA, Wang Y, et al. Heterogeneity of natural Foxp3+ T cells: a committed regulatory T-cell lineage and an uncommitted minor population retaining plasticity. *Proc Natl Acad Sci U S A.* 2009;106:1903-1908.
331. Hori S. Regulatory T cell plasticity: beyond the controversies. *Trends Immunol.* 2011;32:295-300.
332. van Loosdregt J, Fleskens V, Tiemessen MM, et al. Canonical Wnt signaling negatively modulates regulatory T cell function. *Immunity.* 2013;39:298-310.
333. Sumida T, Lincoln MR, Ukeje CM, et al. Activated β-catenin in Foxp3+ regulatory T cells links inflammatory environments to autoimmunity. *Nat Immunol.* 2018;19:1391-1402.
334. Panduro M, Benoist C, Mathis D, et al. Tissue T_{regs}. *Annu Rev Immunol.* 2016;34:609-633.
335. Feuerer M, Herrero L, Cipolletta D. Lean, but not obese, fat is enriched for a unique population of regulatory T cells that affect metabolic parameters. *Nat Med.* 2009;15:930-939.
336. Li C, DiSpirito JR, Zenmour D, et al. TCR transgenic mice reveal stepwise, multi-site acquisition of the distinctive fat-T_{reg} phenotype. *Cell.* 2018;174:285-299.
337. Cipolletta D, Feuerer M, Li A, et al. PPAR-gamma is a major driver of the accumulation and phenotype of adipose tissue T_{reg} cell. *Nature.* 2012;486:549-553.
338. Kolodin D, van Panhuys N, Li CR, et al. Antigen-and cytokine-driven accumulation of regulatory T cells in visceral adipose tissue of lean mice. *Cell Metabol.* 2015;21:543-557.
339. Burzyn D, Kuswanto W, Kolodin D, et al. A special population of regulatory T cells potentiates muscle repair. *Cell.* 2013;155:1282-1295.
340. Kuswanto W, Burzyn D, Panduro M, et al. Poor repair of skeletal muscle in aging mice reflects a defect in local, interleukin-33-dependent accumulation of regulatory T cells. *Immunity.* 2016;44:355-367.
341. Whibley N, Tucci A, Powrie F. Regulatory T cell adaptation in the intestine and skin. *Nat Immunol.* 2019;20:386-396.
342. Gratz IK, Truong HA, Yang SH, et al. Cutting edge: memory regulatory T cells require IL-7 and not IL-2 for their maintenance in peripheral tissues. *J Immunol.* 2013;190:4483-4487.
343. Kalekar LA, Cohen JN, Prevel N, et al. Regulatory T cells in skin are uniquely poised to suppress profibrotic immune responses. *Sci Immunol.* 2019;4:eaaw2910.
344. Nosbaum A, Prevel N, Truong HA, et al. Cutting edge: regulatory T cells facilitate cutaneous wound healing. *J Immunol.* 2016;196:2010-2014.
345. Ali N, Zirak B, Rodriguez RS, et al. Regulatory T cells in skin facilitate epithelial stem cell differentiation. *Cell.* 2017;169:1119-1128.
346. Josefowicz SZ, Niec RE, Kim HY, et al. Extrathymically generated regulatory T cells control mucosal T(H)2 inflammation. *Nature.* 2012;482:395-399.
347. Chai JN, Peng Y, Rengarajan S, et al. *Helicobacter* species are potent drivers of colonic T cell responses in homeostasis and inflammation. *Sci Immunol.* 2017;2:eaal5068.
348. Furusawa Y, Obata Y, Fukuda S, et al. Commensal microbe-derived butyrate induces the differentiation of colonic regulatory T cells. *Nature.* 2013;504:446-450.

349. Ohnmacht C, Park JH, Cording S, et al. The microbiota regulates type 2 immunity through RORγt(+) T cells. *Science*. 2015;349:989-993.
350. Sefik E, Geva-Zatorsky N, Oh S, et al. Individual intestinal symbionts induce a distinct population of RORγ(+) regulatory T cells. *Science*. 2015;349:993-997.
351. Xu M, Pokrovskii M, Ding Y, et al. c-MAF-dependent regulatory T cells mediate immunological tolerance to a gut pathobiont. *Nature*. 2018;554:373-377.
352. Yang BH, Hagemann S, Mamareli P, et al. Foxp3(+) T cells expressing RORγt represent a stable regulatory T-cell effector lineage with enhanced suppressive capacity during intestinal inflammation. *Mucosal Immunol*. 2016;9:444-457.
353. Siede J, Frohlich A, Datsi A, et al. IL-33 receptor-expressing regulatory T cells are highly activated, Th2 biased and suppress CD4 T cell proliferation through IL-10 and TGF beta release. *PLoS One*. 2016;11:e0161507.
354. Samstein RM, Josefowicz SZ, Arvey A, et al. Extrathymic generation of regulatory T cells in placental mammals mitigates maternal-fetal conflict. *Cell*. 2012;150:29-38.
355. Rowe JH, Ertelt JM, Xin LJ, et al. Pregnancy imprints regulatory memory that sustains anergy to fetal antigen. *Nature*. 2012;490:102-106.
356. Kinder JM, Jiang TT, Ertelt JM, et al. Cross-generational reproductive fitness enforced by microchimeric maternal cells. *Cell*. 2015;162:505-515.
357. Weiner HL, Friedman A, Miller A, et al. Oral tolerance: immunological mechanisms and treatment of animal and human organ-specific autoimmune-diseases by oral-administration of autoantigens. *Annu Rev Immunol*. 1994;12:809-837.
358. Chen Y, Kuchroo VK, Inobe JI, et al. Regulatory T cell clones induced by oral tolerance: suppression of autoimmune encephalomyelitis. *Science*. 1994;265:1237-1240.
359. Ford MS, Zhang ZX, Chen W, et al. Double-negative T regulatory cells can develop outside the thymus and do not mature from CD8+ T cell precursors. *J Immunol*. 2006;177:2803-2809.
360. Fischer K, Voelkl S, Heymann J, et al. Isolation and characterization of human antigen-specific TCR αβ+ CD4−CD8− double-negative regulatory T cells. *Blood*. 2005;105:2826-2835.
361. Zhang ZX, Yang L, Young KJ, et al. Identification of a previously unknown antigen-specific regulatory T cells and its mechanism of suppression. *Nat Med*. 2000;6:782-789.
362. McIver Z, Serio B, Dunbar A, et al. Double-negative regulatory T cells induce allotolerance when expanded after allogeneic haematopoietic stem cell transplantation. *Br J Haematol*. 2008;141:170-178.
363. Bacchetta R, Bigler M, Touraine JL, et al. High levels of interleukin 10 production *in vivo* are associated with tolerance in SCID patients transplanted with HLA mismatched hematopoietic stem cells. *J Exp Med*. 1994;179:493-502.
364. Groux H, O'Garra A, Bigler M, et al. A CD4+ T-cell subset inhibits antigen-specific T-cell responses and prevents colitis. *Nature*. 1997;389:737-742.
365. Levings MK, Sangregorio R, Galbiati F, et al. INF-α and IL-10 induce the differentiation of human type 1 T regulatory cells. *J Immunol*. 2001;166:5530-5539.
366. Roncarolo MG, Gregori S, Baccheta R, et al. The biology of T regulatory type 1 cells and their therapeutic application in immune-mediated diseases. *Immunity*. 2018;49:1004-1019.
367. Cottrez F, Hurst SD, Coffman RL, et al. T regulatory cells 1 inhibit a TH2-specific response *in vivo*. *J Immunol*. 2000;165:4848-4853.
368. Gagliani N, Magnani CF, Huber S, et al. Coexpression of CD49b and LAG-3 identifies human and mouse T regulatory type 1 cells. *Nat Med*. 2013;19:739-746.
369. Fitzgerald DC, Zhang GX, El-Behi M, et al. Suppression of autoimmune inflammation of the central nervous system by interleukin 10 secreted by interleukin 27-stimulated T cells. *Nat Immunol*. 2007;8:1372-1380.
370. Awasthi A, Carrier Y, Peron JPS, et al. A dominant function for interleukin 27 in generating interleukin10-producing anti-inflammatory cells. *Nat Immunol*. 2007;8:1380-1389.
371. Pot C, Jin H, Awasthi A, et al. Cutting edge: IL-27 induces the transcription factor c-Maf, cytokine IL-21, and the costimulatory receptor ICOS that coordinately act together to promote differentiation of IL-10-producing Tr1 cells. *J Immunol*. 2009;183:797-801.
372. Murugaiyan G, Mittal A, Lopez-Diego R, et al. IL-27 is a key regulator of IL-10 and IL-17 production by human CD4+ T cells. *J Immunol*. 2009;183:2435-2443.
373. Kemper C, Chan AC, Green JM, et al. Activation of human CD4(+) cells with CD3 and CD46 induces a T-regulatory cell 1 phenotype. *Nature*. 2003;421:388-392.
374. Heinemann C, Heink S, Petermann F, et al. IL-27 and IL-12 oppose pro-inflammatory IL-23 in CD4(+) T cells by inducing Blimp1. *Nat Commun*. 2014;5:3770.
375. Karwacz K, Miraldi ER, Pokrovskii M, et al. Critical role of IRF1 and BATF in forming chromatin landscape during type 1 regulatory cell differentiation. *Nat Immunol*. 2017;18:412-421.
376. Smith EL, Peakman M. Peptide immunotherapy for type 1 diabetes-clinical advances. *Front Immunol*. 2018;9:392.
377. Globinska A, Boonpiyathad T, Satitsuksanoa P, et al. Mechanisms of allergen-specific immunotherapy. *Ann Allergy Asthma Immunol* 2018;121:306-312.
378. Clemente-Casares X, Blanco J, Ambalavanan P, et al. Expanding antigen-specific regulatory networks to treat autoimmunity. *Nature*. 2016;530:434-440.
379. Nakagawa H, Wang L, Cantor H, et al. New insights into the biology of CD8 regulatory T cells. *Adv Immunol*. 2018;140:1-20.
380. Jiang H, Chess L. The specific regulation of immune responses by CD8+ T cells restricted by the MHC class IB molecule, Qa-1. *Annu Rev Immunol*. 2000;18:185-216.
381. Tang X, Maricic I, Purohit N, et al. Regulation of immunity by a novel population of Qa-1-restricted CD8αα+TCRαβ+ T cells. *J Immunol*. 2006;177:7645-7655.
382. Kim HJ, Cantor H. Regulation of self-tolerance by Qa-1-restricted CD8(+) regulatory T cells. *Semin Immunol*. 2011;23:446-452.
383. Kim HJ, Barnitz RA, Kreslavsky T, et al. Stable inhibitory activity of regulatory T cells requires the transcription factor Helios. *Science*. 2015;350:334-339.
384. Hu D, Ikizawa K, Lu LR, et al. Analysis of regulatory CD8 T cells in Qa-1-deficient mice. *Nat Immunol*. 2004;5:516-523.
385. Kim HJ, Verbinnen B, Tang XL, et al. Inhibition of follicular T-helper cells by CD8(+) regulatory T cells is essential for self-tolerance. *Nature*. 2010;467:328-332.
386. Saligrama N, Zhao F, Sikora MJ, et al. Opposing T cell responses in experimental autoimmune encephalomyelitis. *Nature*. 2019;572:481-487.
387. Bezie S, Anegon I, Guillonneau C, et al. Advances on CD8(+) T_{reg} cells and their potential in transplantation. *Transplantation*. 2018;10:1467-1478.
388. Powrie F, Leach MW, Mauze S, et al. Inhibition of TH1 responses prevents inflammatory bowel-disease in SCID mice reconstituted with CD45RBhigh T cells. *Immunity*. 1994;1:553-562.
389. Suri-Payer E, Cantor H. Differential cytokine requirements for regulation of autoimmune gastritis and colitis by CD4+CD25+ T cells. *J Autoimmun*. 2001;16:115-123.
390. Mottet C, Uhlig HH, Powrie F. Cutting edge: cure of colitis by CD4+CD25+ regulatory T cells. *J Immunol*. 2003;170:3939-3943.
391. Delovitch TL, Singh B. The non-obese diabetic mouse as a model of autoimmune diabetes: immune dysregulation gets the NOD. *Immunity*. 1997;7:727-738.
392. Chatenoud L, Salomon B, Bluestone JA. Suppressor T cells – they're back and critical for regulation of autoimmunity. *Immunol Rev*. 2001;182:149-163.
393. Tang Q, Henriksen KJ, Bi M, et al. *In vitro*-expanded antigen-specific regulatory T cells suppress autoimmune diabetes. *J Exp Med*. 2004;199:1455-1465.
394. Masteller EL, Warner MR, Tang Q, et al. Expansion of functional endogenous antigen-specific CD4+CD25+ regulatory T cells from nonobese diabetic mice. *J Immunol*. 2005;175:3053-3059.
395. Sarween N, Chodos A, Raykundalia C, et al. CD4+CD25+ cells controlling a pathogenic CD4 response inhibit cytokine differentiation, CXCR-3 expression, and tissue invasion. *J Immunol*. 2004;173:2942-2951.
396. Chen Z, Herman AE, Matos M, et al. Where CD4+CD25+ T_{reg} cells impinge on autoimmune diabetes. *J Exp Med*. 2005;202:1387-1397.
397. Tang Q, Adams JY, Penaranda C, et al. Central role of defective interleukin-2 production in the triggering of islet autoimmune destruction. *Immunity*. 2010;28:687-697.
398. Grinberg-Bleyer Y, Baeyens A, You S, et al. IL-2 reverses established type 1 diabetes in NOD mice by a local effect on pancreatic regulatory T cells. *J Exp Med*. 2010;207:1871-1878.
399. Webster KE, Walters S, Kohler RE, et al. In vivo expansion of Treg cells with IL-2 mAb complexes: induction of resistance to EAE and long-term acceptance of islet allografts without immunosuppression. *J Exp Med*. 2009;206:751-760.
400. Kohm AP, Carpentier PA, Anger HA, et al. Cutting edge: CD4+CD25+ regulatory T cells suppress antigen-specific autoreactive immune responses and central nervous system inflammation during active experimental autoimmune encephalomyelitis. *J Immunol*. 2002;169:4712-4716.
401. Korn T, Reddy J, Gao W, et al. Myelin-specific regulatory T cells accumulate in the CNS but fail to control autoimmune inflammation. *Nat Med*. 2007;13:423-437.
402. Asefa A, Gumy A, Launois P, et al. The early IL-4 response to *Leishmania major* and the resulting Th2 cell maturation steering progressive disease in BALB/c mice are subject to the control of regulatory CD4+CD25+ T cells. *J Immunol*. 2002;169:3232-3241.
403. Xu D, Lin H, Komai-Koma M, et al. CD4+CD25+ regulatory T cells suppress differentiation and functions of Th1 and Th2 cells, *Leishmania major* infection, and colitis in mice. *J Immunol*. 2003;170:394-399.
404. Stassen M, Jonuleit H, Muller C, et al. Differential regulatory capacity of CD25+ T regulatory cells and preactivated CD25+ T regulatory cells on development, functional activation, and proliferation of Th2 cells. *J Immunol*. 2004;173:267-274.
405. Lewkowich IP, Herman NS, Schliefer KW, et al. CD4+CD25+ T cells protect against experimentally induced asthma and alter pulmonary dendritic cell phenotype and function. *J Exp Med*. 2005;202:1549-1561.
406. Jaffar S, Sivakuru T, Roberts K. CD4+CD25+ T cells regulate airway eosinophilic inflammation by modulating Th2 cell phenotype. *J Immunol*. 2004;172:3842-3849.
407. Wilson MS, Taylor MD, Balic A, et al. Suppression of allergic airway inflammation by helminth-induced regulatory T cells. *J Exp Med*. 2005;202:1199-1212.
408. Waldmann H, Cobbold S. Regulating the immune response to transplants. A role for CD4+ regulatory cells? *Immunity*. 2001;14:399-406.
409. Hara M, Kingsley CI, Niimi M, et al. IL-10 is required for regulatory T cells to mediate tolerance to alloantigens *in vivo*. *J Immunol*. 2001;166:3789-3796.
410. Graca L, Thompson S, Lin CY, et al. Both CD4+CD25+ and CD4+CD25− regulatory cells mediate dominant transplantation tolerance. *J Immunol*. 2002;168:5558-5567.
411. Quezada SA, Bennett K, Blazar BR, et al. Analysis of the underlying cellular mechanisms of anti-CD154-induced graft tolerance: the interplay of clonal anergy and immune regulation. *J Immunol*. 2005;175:771-779.
412. Lu LF, Lind EF, Gondek DC, et al. Mast cells are essential intermediaries in regulatory T cell tolerance. *Nature*. 2006;442:997-1002.
413. Cobbold SP, Castejon R, Adams E, et al. Induction of *foxp3+* regulatory T cells in the periphery of T cell receptor transgenic mice tolerized to transplants. *J Immunol*. 2004;172:6003-6010.
414. Taylor PA, Lees CJ, Blazer BR. The infection of *ex vivo* activated and expanded CD4+CD25+ immune regulatory cells inhibits graft-vs-host disease (GVHD) lethality. *Blood*. 2002;99:4601-4609.
415. Cohen JL, Trenado A, Vasey D, et al. CD4+CD25+ immunoregulatory T cells: new therapeutics for graft-versus-host disease. *J Exp Med*. 2002;196:401-406.
416. Ermann J, Hoffman P, Edinger M, et al. Only the CD62L+ subpopulation of CD4+CD25+ regulatory T cells protects from lethal GVHD. *Blood*. 2005;105:2220-2226.
417. Hoffman P, Ermann J, Edinger M, et al. Donor-type CD4+CD25+ regulatory T cells suppress lethal acute graft-versus-host disease after allogeneic bone marrow transplantation. *J Exp Med*. 2002;196:389-399.
418. Edinger M, Hoffman P, Ermann J, et al. CD4+CD25+ regulatory T cells preserve graft-versus-tumor activity while inhibiting graft-versus-host disease after bone marrow transplantation. *Nat Med*. 2003;9:1144-1150.
419. Golovina TN, Mikheeva T, Suboski MM, et al. CD28 costimulation is essential for human T regulatory expansion and function. *J Immunol*. 2008;181:2855-2868.
420. Nadig SN, Wieckiewicz J, Wu DC, et al. *In vivo* prevention of transplant arteriosclerosis by *ex vivo*-expanded human regulatory T cells. *Nat Med*. 2010;16:809-813.
421. Joffre O, Gorsse N, Romagnoli P, et al. Induction of antigen-specific tolerance to bone marrow allografts with CD4+CD25+ T lymphocytes. *Blood*. 2004;103:4216-4221.
422. Hanash AM, Levy RB. Donor CD4+CD25+ T cells promote engraftment and tolerance following MHC-mismatched hematopoietic cell transplantation. *Blood*. 2005;105:1828-1836.
423. Brunstein CG, Miller JS, Cao Q, et al. Infusion of ex vivo expanded T regulatory cells in adults transplanted with umbilical cord blood: safety profile and detection kinetics. *Blood*. 2011;20:1061-1070.
424. Di Ianni M, Falzetti F, Carotti A, et al. T_{regs} prevent GVHD and promote immune reconstitution in HLA-haploidentical transplantation. *Blood*. 2011;117:3921-3928.
425. North RJ, Bursuker I. Generation and decay of the immune response to a progressive fibrosarcoma. I. Ly-1+2− suppressor T cells down-regulate the generation of Ly-1−2+ effector T cells. *J Exp Med*. 1984;159:1295-1311.
426. Awwad M, North RJ. Immunologically mediated regression of a murine lymphoma after treatment with anti-L3T4 antibody. A consequence of removing L3T4+ suppressor T cells from a host generating predominantly Lyt-2+ T cell-mediated immunity. *J Exp Med*. 1988;168:2193-2206.
427. Onizuka S, Tawara I, Shimizu J, et al. Tumor rejection by *in vivo* administration of anti-CD25 (interleukin-2 receptor α) monoclonal antibody. *Cancer Res*. 1999;59:3128-3133.
428. Erdman SE, Sohn JJ, Rao VP, et al. CD4+CD25+ regulatory lymphocytes induce regression of intestinal tumors in $Apc^{Min/+}$ mice. *Cancer Res*. 2005;65:3998-4004.
429. Woo EY, Yeh H, Chu CS, et al. Cutting edge: regulatory T cells from lung cancer patients directly inhibit autologous T cell proliferation. *J Immunol*. 2002;168:4272-4276.
430. Curiel TJ, Cooukos G, Zou LH, et al. Specific recruitment of regulatory T cells in ovarian carcinoma fosters immune privilege and predicts reduced survival. *Nat Med*. 2004;10:942-949.
431. De Simone M, Arrigoni A, Rossetti G, et al. Transcriptional landscape of human tissue lymphocytes unveils uniqueness of tumor-infiltrating T regulatory cells. *Immunity*. 2016;45:1135-1147.
432. Magnuson AM, Kiner E, Ergun A, et al. Identification and validation of a tumor-infiltrating T_{reg} transcriptional signature conserved across species and tumor types. *Proc Nat Acad Sci U S A*. 2018;115:E10672-E10681.
433. Plitas G, Konopacki C, Wu KM, et al. Regulatory T cells exhibit distinct features in human breast cancer. *Immunity*. 2016;45:1122-1134.

434. Wang L, Simons DL, Lug XY, et al. Connecting blood and intratumoral T_{reg} cell activity in predicting future relapse in breast cancer. *Nat Immunol.* 2019;20:1220-1230.
435. Sutmuller RP, van Duivenvoorde LM, van Elsas A, et al. Synergism of cytotoxic T lymphocyte-associated antigen 4 blockade and depletion of CD25+ regulatory T cells in antitumor therapy reveals alternative pathways for suppression of autoreactive cytotoxic T lymphocyte responses. *J Exp Med.* 2001;194:823-832.
436. Antony PA, Piccirillo CA, Akpinarli A, et al. CD8+ T cell immunity against a tumor/self-antigen is augmented by CD4+ T helper cells and hindered by naturally occurring T regulatory cells. *J Immunol.* 2005;174:2591-2601.
437. Turk MJ, Guevara-Patino JA, Rizzuto GA, et al. Concomitant tumor immunity to a poorly immunogenic melanoma is prevented by regulatory T cells. *J Exp Med.* 2004;200:771-782.
438. Wang HY, Lee DA, Peng GY, et al. Tumor-specific human CD4+ regulatory T cells and their ligands: implications for immunotherapy. *Immunity.* 2004;20:107-118.
439. Nishikawa H, Kato T, Tanida K, et al. CD4+CD25+ T cells responding to serologically defined autoantigens suppress antitumor immune responses. *Proc Natl Acad Sci U S A.* 2003;100:10902-10906.
440. Kryczek I, Wei S, Zou LH, et al. Cutting edge: induction of B7-H4 on APCs through IL-10. Novel suppressive mode for regulatory T cells. *J Immunol.* 2006;177:40-44.
441. Sawant DV, Yano H, Chikina M, et al. Adaptive plasticity of IL-10(+) and IL-35(+) T_{reg} cells cooperatively promotes tumor T cell exhaustion. *Nat Immunol.* 2019;20:724-735.
442. Ameri AH, Tuchayi SM, Zaalberg A, et al. IL-33/regulatory T cell axis triggers the development of a tumor-promoting immune environment in chronic inflammation. *Proc Nat Acad Sci U S A.* 2019;116:2646-2651.
443. Li A, Herbst RH, Canner D, et al. IL-33 signaling alters regulatory T cell diversity in support of tumor development. *Cell Rep.* 2019;29:2998-3008.
444. Phan GQ, Yang JC, Sherry RM, et al. Cancer regression and autoimmunity induced by cytotoxic T lymphocyte-associated antigen 4 blockade in patients with metastatic melanoma. *Proc Nat Acad Sci U S A.* 2003;100:8372-8377.
445. Buchan SL, Dou L, Remer M, et al. Antibodies to costimulatory receptor 4-1BB enhance anti-tumor immunity via T regulatory cell depletion and promotion of CD8 T cell effector function. *Immunity.* 2018;49:958-970.
446. Peggs KS, Quezada SA, Chambers CA, et al. Blockade of CTLA-4 on both effector and regulatory T cell compartments contributes to the antitumor activity of anti-CTLA-4 antibodies. *J Exp Med.* 2009;206:1717-1725.
447. Quezada SA, Peggs KS. Lost in translation: deciphering the mechanism of action of anti-human CTLA-4. *Clin Cancer Res.* 2019;25:1130-1132.
448. Hori S, Carvalho TL, Demengeot J. CD25+CD4+ regulatory T cells suppress CD4+ T cell-mediated pulmonary hyperinflammation driven by Pneumocystis carinii in immunodeficient mice. *Eur J Immunol.* 2002;32:1282-1291.
449. Maloy KJ, Salaun L, Cahill R, et al. CD4+CD25+ T(R) cells suppress innate immune pathology through cytokine-dependent mechanisms. *J Exp Med.* 2003;197:111-119.
450. Belkaid Y, Tarbell K. Regulatory T cells in the control of host-microorganism interactions. *Annu Rev Immunol.* 2009;27:551-589.
451. Belkaid Y, Piccirillo CA, Mendez S, et al. CD4+CD25+ immunoregulatory T lymphocytes control *Leishmania major* persistence and the development of concomitant immunity. *Nature.* 2002;420:502-507.
452. Suffia IJ, Reckling SK, Piccirillo CA, et al. Infected site-restricted Foxp3+ natural regulatory T cells are specific for microbial antigens. *J Exp Med.* 2006;203:777-788.
453. Belkaid Y, Hoffmann KF, Mendez S, et al. The role of interleukin (IL)-10 in the persistence of *Leishmania major* in the skin after healing and the therapeutic potential of anti-IL-10 receptor antibody for sterile cure. *J Exp Med.* 2001;194:1497-1506.
454. McGuirk P, McCann C, Mills KHG. Pathogen-specific T regulatory 1 cells induced in the respiratory tract by a bacterial molecule that stimulates interleukin 10 production by dendritic cells: a novel strategy for evasion of protective T helper type 1 responses by Bordetella pertussis. *J Exp Med.* 2002;195:221-231.
455. MacDonald AJ, Duffy M, Brady MT, et al. CD4 T helper type 1 and regulatory T cells induced against the same epitopes on the core protein in hepatitis C virus-infected persons. *J Infect Dis.* 2002;185:720-727.
456. Iwashiro M, Messer RJ, Peterson KE, et al. Immunosuppression by CD4+ regulatory T cells induced by chronic retroviral infection. *Proc Nat Acad Sci U S A.* 2001;98:9226-9230.
457. Robertson SJ, Messer RJ, Carmody AB, et al. *In vitro* suppression of CD8+ T cell function by Friend virus-induced regulatory T cells. *J Immunol.* 2006;176:3342-3349.
458. Kinter AL, Hennessey M, Bell A, et al. CD25+CD4+ regulatory T cells from the peripheral blood of asymptomatic HIV-infected individuals regulate CD4+ and CD8+ HIV-specific T cell immune responses *in vitro* and are associated with favorable clinical markers of disease status. *J Exp Med.* 2004;200:331-343.
459. Suvas S, Azkur AK, Kim BS, et al. CD4+CD25+ regulatory T cells control the severity of viral neuroinflammatory lesions. *J Exp Med.* 2004;172:4123-4132.
460. Shafiani S, Tucker-Heard G, Karoyone A, et al. Pathogen-specific regulatory T cells delay the arrival of effector T cells in the lung during early tuberculosis. *J Exp Med.* 2010;207:1409-1420.
461. Lund JM, Hsing L, Pham TT, et al. Coordination of early protective immunity to viral infection by regulatory T cells. *Science.* 2008;320:1220-1224.
462. Zhao J, Zhao J, Fett C, et al. IFN-γ- and IL-10-expressing virus epitope-specific Foxp3+ T_{reg} cells in the central nervous system during encephalomyelitis. *J Exp Med.* 2011;208:1571-1577.
463. Buckner JH. Mechanisms of impaired regulation by CD4+CD25+FOXP3+ regulatory T cells in human autoimmune diseases. *Nat Rev Immunol.* 2010;10:849-859.
464. Van Amelsfort JM, Jacobs, Billsma JW, et al. CD4+CD25+ regulatory T cells in rheumatoid arthritis: differences in the presence, phenotype, and function between peripheral blood and synovial fluid. *Arthritis Rheum.* 2004;50:2775-2785.
465. Notley CA, Ehrenstein MR. The yin and yang of regulatory T cells and inflammation in RA. *Nat Rev Rheumatol.* 2010;6:572-577.
466. Ehrenstein MR, Evans JG, Singh A, et al. Compromised function of regulatory T cells in rheumatoid arthritis and reversal by anti-TNF alpha therapy. *J Exp Med.* 2004;200:277-285.
467. Flores-Borja F, Jury EC, Mauri C, et al. Defects in CTLA-4 are associated with abnormal regulatory T cell function in rheumatoid arthritis. *Proc Natl Acad Sci U S A.* 2008;105:19396-19401.
468. Dominguez-Villar M, Hafler DA. Regulatory T cells in autoimmune disease. *Nat Immunol.* 2018;19:65-673.
469. Kitz A, de Marcken M, Gautron AS, et al. AKT isoforms modulate Th1-like T_{reg} generation and function in human autoimmune disease. *EMBO Rep.* 2016;17:1169-1183.
470. MacDonald KG, Dawson NAJ, Huang Q, et al. Regulatory T cells produce profibrotic cytokines in the skin of patients with systemic sclerosis. *J Allergy Clin Immunol.* 2015;135:946-955.
471. Rivas MN, Burton OT, Wise P, et al. Regulatory T cell reprogramming toward a Th2-cell-like lineage impairs oral tolerance and promotes food allergy. *Immunity.* 2015;42:512-523.
472. Tang QZ, Bluestone JA. Regulatory T-cell physiology and application to treat autoimmunity. *Immunol Rev.* 2006;212:217-237.
473. Zanin-Zhorov A, Ding Y, Kumari S, et al. Protein kinase C-θ mediates negative feedback on regulatory T cell function. *Science.* 2010;328:372-376.
474. Blazar BR, MacDonald KPA, Hill GR, et al. Immune regulatory cell infusion for graft-versus-host disease prevention and therapy. *Blood.* 2018;131:2651-2660.
475. Scalapino KJ, Tang Q, Bluestone JA, et al. Suppression of disease in New Zealand Black/New Zealand White lupus-prone mice by adoptive transfer of *ex vivo*, expanded regulatory T cells. *J Immunol.* 2006;177:1451-1459.
476. Bluestone JA, Buckner JH, Firtch M, et al. Type 1 diabetes immunotherapy using polyclonal regulatory T cells. *Sci Transl Med.* 2015;7:315ra189.
477. Kim YC, Zhang AH, Su Y, et al. Engineered antigen-specific human regulatory T cells: immunosuppression of FVIII-specific T-and B-cell responses. *Blood.* 2015;125:1107-1115.
478. Elinav E, Adam N, Waks T, et al. Amelioration of colitis by genetically engineered murine regulatory T cells redirected by antigen-specific chimeric receptor. *Gastroenterology.* 2009;136:1721-1731.
479. MacDonald KG, Hoeppli RE, Huang Q, et al. Alloantigen-specific regulatory T cells generated with a chimeric antigen receptor. *J Clin Invest.* 2016;126:1413-1424.
480. Klatzmann D, Abbas AK. The promise of low-dose interleukin-2 therapy for autoimmune and inflammatory diseases. *Nat Rev Immunol.* 2015;15:283-294.
481. Tahvildari M, Dana R. Low-dose IL-2 therapy in transplantation, autoimmunity, and inflammatory diseases. *J Immunol.* 2019;203:2749-2755.
482. Rosenzwajg M, Churlaud G. Low-dose interleukin-2 fosters a dose-dependent regulatory T cell tuned milieu in T1D patients. *J Autoimmun.* 2015;58:48-58.
483. Matsuoka K, Koreth J, Kim HT, et al. Low-dose interleukin-2 therapy restores regulatory T cell homeostasis in patients with chronic graft-versus-host disease. *Sci Transl Med.* 2013;5:179ra43.
484. Ghelani A, Bates D, Conner K, et al. Defining the threshold IL-2 signal required for induction of selective treg cell responses using engineered IL-2 muteins. *Front Immunol.* 2020;11:1106.
485. Peterson LB, Bell CJM, Howlett SK, et al. A long-lived IL-2 mutein that selectively activates and expands regulatory T cells as a therapy. *J Autoimmun.* 2018;95:1-14.
486. Trotta E, Bessette PH, Silveria SL, et al. A human anti-IL-2 antibody that potentiates regulatory T cells by a structure-based mechanism. *Nat Med.* 2018;24:1005-1014.
487. Sockolosky JT, Trotta E, Parisi G, et al. Selective targeting of engineered T cells using orthogonal IL-2 cytokine-receptor complexes. *Science.* 2018;359:1037-1042.

CHAPTER 30

CD8+ T Cell Memory

Francis R. Carbone, PhD • Thomas Gebhardt, MD, PhD

INTRODUCTION

The term "immunity" originates from the Latin immunitas, which meant the exemption from taxes or public service as granted by the Roman senate or emperor. Today, immunity is applied to the protection against infection that results from prior exposure to the pathogenic agent or a mimic, such as a vaccine. Immunity results from the body's adaptive response to antigen exposure and has the combined hallmarks of specificity and longevity. For example, exposure to the attenuated Oka varicella-zoster virus vaccine strain renders an individual protected against the childhood disease known as chickenpox but still leaves that same individual susceptible to the still too common infectious disease, measles, the latter having its own extremely effective vaccine. For both these pathogens, protection against renewed infection is maintained throughout life and this long-lived effect is viewed as "immunologic memory."

There are two key cellular components that form the foundation of the adaptive immune response, and these are B and T lymphocytes otherwise known as B and T cells. Here, we define lymphocytes that have never encountered antigen as naïve, whereas those that have been activated through their antigen-specific receptor and survive over the longer term as memory lymphocytes. In this chapter we deal exclusively with memory T cells and mainly focus on the CD8+ subset typically associated with killing of infected or transformed target cells. Cell survival and longevity is critical to T cell immunity, and survival is a key discriminator between two possible outcomes that follow antigen stimulation. One outcome of that stimulation is the massive expansion of populations of effector cells generated to immediately deal with the infectious agent followed by their subsequent elimination once infection is cleared and they are no longer needed for this purpose. The second outcome also involves specific T cell expansion, although on a smaller scale than seen with the effector response. More importantly, this more muted expansion results in a population of cells that persists almost indefinitely in the immune individual. These are the memory T cells that contribute to long-lasting immunity, and this selection between effector and memory cell fates is important for our understanding of T cell memory.

A comparison of B cells with T cells provides a useful insight into constraints that operate within T cell memory. Both T cells and B cells carry antigen-specific receptors formed as a consequence of somatic gene rearrangements. These give rise to a diverse population of lymphocytes that can recognize all manner of different infectious agents. Probably the most striking difference between the T cell and B cell receptors is that only the latter undergoes further changes during the immune response. These so-called somatic mutations mean that memory B cells have a vastly superior affinity than the antigen-specific precursors from which they are derived. This is not true for T cells, where memory progenitors are direct descendants of their naïve parent cells when it comes to antigen-specific receptors and affinity. However, like memory B cells, there are quantitatively more memory T cells for a given antigen than exist in the naïve T cell pool. Having more cells accelerates the response to infection, and it is an accelerated response that is central to immune memory. Another contributor to this acceleration is the ability of memory T cells to deliver effector molecules faster and in greater quantities than their naïve counterparts. Moreover, memory T cells are known to respond immediately on antigen recognition, whereas naïve T cells are far more constrained with a strict requirement for professional antigen-presenting cell and selective cytokine involvement. In addition, memory T cells take on new characteristics that further contribute to their enhanced effectiveness in the immune individual. One of the main examples relates to migration and tissue localization. Naïve T cells are only found in the blood, secondary lymphoid organs, or lymphatic vessels in what can be considered a closed circulatory loop. In contrast, memory T cells exist throughout the body in all manner of lymphoid and nonlymphoid compartments. Expanded levels of migration are necessary to protect against infections at portals of entry into the body, with the "diaspora" of memory cells analogous to that which exists at the peak of the effector T cell response.

Finally, all these enhanced characteristics persist long term in order to maintain the prolonged immunity that is evident after vaccination or recovery from infection. Increases in specific T cell frequencies require prolonged cell survival, and memory T cells have unique means to achieve this and avoid the programmed cell death that otherwise follows engagement of the T cell receptor. Enhanced effector function, altered migration patterns, and tissue residence also persist for the lifetime of the surviving memory T cells, and these are due to inheritable modification and epigenetic chromosomal alterations. In this chapter, we will explore what it takes to generate and maintain memory T cells, then describe the heterogeneity that exists within this broad population, and, finally, document the unique advantages and contributions made by the different memory T cell subsets.

GENERATION OF EFFECTOR AND MEMORY T CELLS

In their resting or naïve state, T cells specific for a given antigen are very rare, with frequencies estimated to be as low as 1 to 100 per million CD8+ T cells, the so-called precursor frequency.[1-3] Furthermore, their recirculation pattern is restricted to lymphoid tissues such as lymph nodes and spleen, with a notable exclusion from nonlymphoid barrier tissues such as skin and mucosa, the most common sites of pathogen invasion.[4] These traits of naïve T cells pose major hurdles for the timely deployment of enough numbers of armed effector T cells to sites of acute infection. Therefore, in order to contribute to infection clearance, naïve T cells must first undergo a proliferative burst to dramatically expand in numbers. In the case of systemic infections in mice, such as those with lymphocytic choriomeningitis virus (LCMV) or *Listeria monocytogenes*, activated T cells are estimated to undergo up to 15 to 20 rounds of rapid consecutive divisions, resulting in an overall expansion of the antigen-specific T cell pool by up to 10,000-fold compared with the preimmune status.[5] At the same time, T cells also need to acquire a set of new functional properties and they do so in a process referred to as "effector differentiation." This encompasses the *de novo* expression of a range of adhesion and chemokine receptors that ultimately guide the activated T cells to their target tissues.[4,6] Of importance, this process also arms activated T cells with effector molecules, including perforin, different types of granzymes, as well as a broad range of cytokines.[5,7,8] These effector molecules are released upon antigen reencounter in infected tissues and act to eliminate infected cells and to curb the spread of infection by sustaining tissue inflammation.

Upon resolution of infection, the overall population of effector cells rapidly declines in numbers, which involves various intrinsically and extrinsically triggered pathways of apoptotic cell death. However, 5% to 10% of activated T cells ultimately survive this contraction phase and form a heterogenous population of memory cells that persists for the lifetime of the host.[5,8,9] As discussed later, memory T cells possess several inheritably imprinted functions that differ from naïve cells and these traits explain their superior function in protection from reinfection with the same or antigenically related pathogens. In the following paragraphs we briefly summarize key steps in the T cell activation cascade that culminate in the generation of short-lived effector and long-lived memory T cells following infection or immunization.

T Cell Priming by Licensed Dendritic Cells in Lymphoid Tissues

Clonal expansion and effector differentiation of naïve CD8+ T cells occurs in the secondary lymphoid organs, such as the lymph nodes that drain sites of localized infection. Both processes are simultaneously initiated by dendritic cells that display peptide fragments of pathogen-derived antigens bound to MHC-I molecules for the activation of T cells via their highly specific T cell receptor (TCR). As such, for T cell activation to occur, relevant antigens first need to be directly introduced into the lymphoid tissues or shuttled from infected nonlymphoid tissues to lymph nodes. The latter may involve direct drainage of soluble antigens via lymph or, alternatively, transportation of antigens by dendritic cells from peripheral tissues into lymph nodes. These migrating dendritic cells may either directly activate naïve T cells or relay their antigenic cargo to distinct, lymph node–resident dendritic cells for subsequent T cell activation.[10] This type of antigen handover and subsequent cross-presentation of antigen by Batf3-dependent, CD11c+ CD8α+ CD103+/− classical type I dendritic cells in mice is often integral to efficient priming of CD8+ T cell responses to infection and tumor challenge.[11-14] The overall availability of antigens for processing and presentation by dendritic cells is linked to the magnitude of effector cell expansion.

Activation of naïve T cells via TCR recognition of cognate antigen displayed by dendritic cells initiates broad and rapid changes in chromatin accessibility of gene regulatory regions in responding T cells. These are the consequence of changes in DNA methylation patterns and the introduction or removal of diverse posttranslational modifications in DNA-packaging histones that together facilitate rapid induction or silencing of transcriptional activity.[15-17] Upstream transcription factors initiating these processes include TCR-induced members of the RHD (eg, NFAT) and bZIP families (eg, AP-1, BATF) that cooperate with co-induced transcription factors, such as IRF4, and constitutively expressed lineage-specific transcriptional regulators, such as Runx3.[15,16,18-20] In order to facilitate rapid induction of key transcriptional regulators of effector differentiation, including T-bet, Eomes, Blimp1, and IRF4,[21-26] the promotors of these genes are kept in a bivalent state in naïve T cells, whereby removal of suppressive histone marks upon activation promotes transcription in the continued presence of permissive histone marks.[27] The early epigenetic reconfiguration of naïve T cells ultimately sets up transcriptional networks encompassing members of diverse families of transcriptional regulators (Fig. 30.1), many of which continue to operate dynamically throughout subsequent effector differentiation and also after establishment of short-lived effector or long-lived memory fates.[15,16,19,28-31] Activated T cells also undergo dramatic metabolic reprogramming to meet the increased biosynthetic and bioenergetic demands inherent with proliferation and the acquisition of effector function.[32-34] This process involves networks of TCR-dependent and nutrient-sensing transcription factors, such as MYC, IRF4, mTOR, and HIF1α, whose coordinated activity promotes a shift of the catabolic state of naïve T cells toward an anabolic state in effector cells.[32,33,35-37] The latter is characterized by the upregulation of numerous metabolic pathways, including increased nutrient uptake, aerobic glycolysis (known as the Warburg effect), glutaminolysis, mitochondrial respiration, fatty acid synthesis, and the production of other biomolecules or cellular building blocks required for proliferative expansion.[32-34] Together, this early and intricately linked reprogramming of naïve T cells at the epigenetic, transcriptional, and metabolic level provides a molecular framework along which additional signals are integrated throughout the process of effector differentiation, resulting in reinforcement or suppression of respective transcriptional networks and metabolic pathways that ultimately determine the fate of the activated T cell.

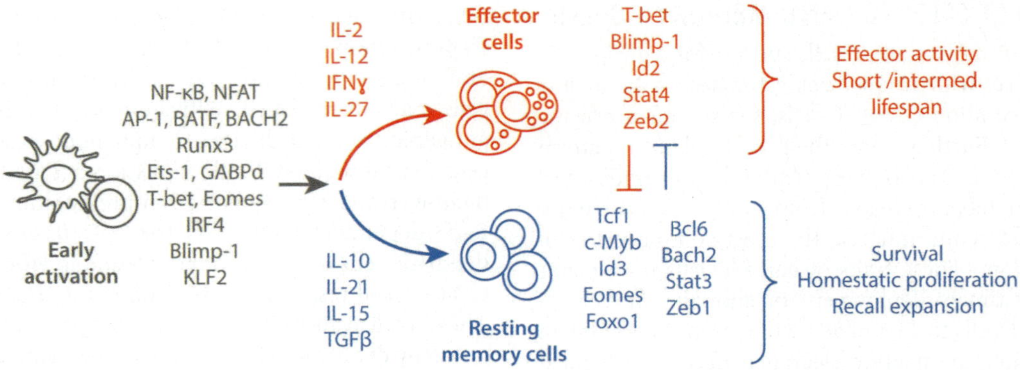

FIG. 30.1. Transcription factors and cytokines that program and preserve the fate of effector and memory CD8+ T cells. TCR-dependent activation results in rapid epigenetic remodeling and subsequent engagement and/or induction of a large number of transcriptional regulators (*gray*) that underpin effector differentiation and the generation and maintenance of memory T cells. Also shown are examples of cytokines that can direct T cell differentiation into either terminal effector (*red*) or memory T cells (*blue*), as well as transcription factors that preserve and reinforce the two opposing fates. Important, transcription factors operating in terminal effector or memory T cells are often expressed in a mutually antagonistic manner.

Integration of Multiple Signals During T Cell Priming

TCR engagement is the primary initiator of the activation cascade encompassing T cell proliferation, survival, and effector differentiation, and as such, it is often termed "signal 1." However, various other stimuli grouped together as a collective "signal 2" are also required for the complete progression down this pathway. These stimuli include provision of an array of costimulatory molecules that are expressed on the surface of dendritic cells and bind to their counterreceptors on CD8+ T cells. One such example is the B7 protein pair, CD80 and CD86, which bind to CD28 expressed by naïve T cells.[38] CD28 signaling introduces unique epigenetic, transcriptional, posttranslational, and metabolic adjustments that collectively support survival and differentiation of activated T cells.[38] As part of this, costimulation by CD28 increases glycolytic activity associated with effector differentiation and primes mitochondria for increased respiratory capacity,[39,40] the latter of which is thought to be required for optimal survival and recall functionality of the memory T cell progeny.[41-43] Later during the response, this stimulatory pathway is curtailed by T cell surface expression of CTLA4, a molecule that binds CD80 and CD86 with a different orientation and higher avidity than CD28 and thereby blocks further T cell stimulation through CD28.[44] Alongside CTLA4, other members of the CD28 subfamily of costimulatory molecules, such as PD-1, PD1H/VISTA, and TIGIT, also exert regulatory functions by restraining effector differentiation.[38,45] In addition to the CD28 pathway, there are several members of the tumor necrosis factor (TNF) receptor family and their respective ligands that mediate costimulation during T cell priming, including CD27/CD70, 41BB/41BB-L, OX40/OX40-L, HVEM-LIGHT, and CD30/CD30-L.[5,8,46] From this, it is evident that there exists a considerable level of redundancy such that genetic deletion of one or more activating pathways can usually be compensated by other receptor-ligand pairs in an infection or immunization-dependent manner.[8]

Dendritic cells constitutively process and present antigen for TCR engagement. However, they need to be activated in a manner that leads to a form of maturation or "licensing" in order to provide signal 2. Dendritic cell licensing can be mediated by primed CD4+ helper T cells expressing the TNF superfamily molecule, CD40L, which binds and activates its receptor, CD40, constitutively expressed by dendritic cells.[47-49] Functional maturation of dendritic cells is further promoted and shaped by the sensing of conserved pathogen-derived or tissue damage–associated molecules. This involves a broad range of pattern recognition receptors (PRRs) that include Toll-like receptors, C-type lectin receptors, NOD-like receptors, and RIG-I-like receptors.[50,51] These PPRs detect a range of extra- or intracellular pathogen-derived structural components and genetic materials released during the process of infection. In addition, dendritic cells respond to inflammatory cytokines such as type I interferons. Thus, dendritic cells integrate a multitude of signals from CD4+ T cells, pathogens, and the inflammatory tissue milieu to enhance their stimulatory capacity for CD8+ T cells.[51-53] Other cytokines that may be produced by licensed dendritic cells or by accessory immune cells including CD4+ helper T cells and innate lymphocytes, such as interleukin (IL)-1, IL-2, IL-12, IL-15, IL-27, and IFN-ɣ, variously support survival and functional differentiation of CD8+ T cells in a context-dependent manner.[15,16,30,54] Some of these cytokines, such as IL-2, IL-12, and IL-27, have known functions in directing activated T cells toward a short-lived effector cell fate by promoting expression of T-bet and Blimp-1 and repressing Tcf1, Bcl6, and the high-affinity IL-7 receptor α-chain, IL-7Rα (CD127).[23,30,55-58] Furthermore, type I IFN and IFNɣ can exert context-dependent roles in promoting the generation of either short-lived or long-lived progeny.[59-62] However, another set of cytokines, including TGFβ, IL-10, and IL-21, dampen effector T cell expansion and effector functionality and are therefore of particular importance in preventing overzealous T cell stimulation with the potential for autoimmune inflammation.[63-65] As part of this, IL-10 and IL-21 cooperate in signaling through the JAK-STAT-3 pathway to induce the suppressive cytokine signaling component, SOCS3.[65] This in turn shields activated T cells from excessive inflammatory signals such as IL-12 and thereby can support the generation of long-lived memory CD8+ T cells. Collectively, these cytokine-mediated signals are referred to as "signal 3." Of note, the extent to which all three signals are provided and the way they are dynamically integrated during T cell activation are intricately linked to the nature of the infectious pathogen, the extent of tissue inflammation, and the type of affected organ. This feedback loop serves to adjust the overall kinetics, magnitude, and quality of the effector T cell response relative to the current danger. Conversely, this adjustment also ensures the establishment of long-lived immunity in the form of memory T cells.

Primary Effector T Cell Contribution to Infection Resolution

Having received sufficient stimulation, activated T cells begin to reexpress the receptor for sphingosine-1-phosphate, S1P1, which is initially suppressed to facilitate transient retention in lymph nodes during T cell priming. Regaining chemoattraction to sphingosine-1-phosphate, a chemotactic lipid with elevated concentrations in blood and lymph compared with tissues, ultimately guides the egress of effector T cells from lymph nodes and the release into the blood circulation.[4,66] Two key mechanisms then ensure efficient targeting of effector T cells to infected tissues. First, inflammatory cytokines such as TNFα and IFNγ prompt important functional changes in the endothelial cells of blood vessels in proximity to infection foci. These endothelial cells strongly enhance expression of selectins and integrins and become decorated with tissue-specific chemokines, which results in a greatly enhanced ability to support the rolling, adhesion, and diapedesis cascade of T cell tissue infiltration.[67,68] Second, activated T cells are now equipped with a set of homing molecules that serve as binding partners for endothelial or tissue-derived ligands and receptors, including the chemokine receptors CCR2, CCR5, and CXCR3.[67,68]

Of interest, some organs express tissue-specific patterns of homing molecules and chemokines. Cutaneous venules preferentially display E-selectin and integrins, such as ICAM-1 and VCAM-1, and are coated with tissue-derived chemokines, such as CCL17 and CCL27.[67] Intestinal venules, on the other hand, preferentially express the addressin MAdCAM-1 and are coated with the chemokine CCL25.[67] A process known as tissue-specific conditioning can instruct the expression of T cell homing molecules that match those organ-specific sets of endothelial homing receptors and ligands.[67,69] For instance, effector T cells that arise in skin-draining lymph nodes express E-selectin ligands, LFA-1 and VLA-4 integrins, and the chemokine receptors CCR4 and CCR10 that result in their preferential recruitment into skin.[67,70] The latter is induced in T cells by exposure to the vitamin D metabolite, 1,25-dihydroxyvitamin D3 or calcitriol, which is converted from its precursor by exposure to sunlight.[71,72] By contrast, intestinal effector T cells primed under the control of the vitamin A metabolite, retinoic acid, are imprinted to express α4β7 and CCR9.[69,71,73,74]

Thus, although effector T cells released into the blood circulation can migrate to all manner of peripheral tissues, the aforementioned mechanisms ensure that they most efficiently target infected tissues. In some infections, this process is further facilitated by early infiltrating or "pioneering" CD4+ helper T cells that, upon antigen-dependent restimulation by dendritic cells in peripheral sites, sustain the expression of inflammatory chemokines, such as the CXCR3 ligands, CXCL9 and CXCL10.[75] These chemokines also guide CD8+ T cell migration within peripheral tissues toward infected cells.[76] In contrast to CD8+ T cell priming in lymph nodes, which strictly requires dendritic cells, the antigen and TCR-dependent elicitation of effector activities within infected has no such dependence on professional antigen-presenting cells and instead can be triggered by all types of infected cells.[77] Nevertheless, the considerable time it takes for T cells to undergo clonal expansion, effector differentiation, and migration to peripheral tissues means that, in the case of acute infections, their protective function is largely limited to the clearance or resolution phase from several days to a few weeks after initial infection (Fig. 30.2A).

Contraction of Primary Effectors and Emergence of Memory T Cells

Coinciding with infection control, the pool of effector cells rapidly decreases in numbers, affecting both activated T cells that have undergone extensive rounds of proliferation, as well as those with more muted clonal expansion (Fig. 30.2A).[78] The factors and molecular mechanisms underlying this phenomenon are incompletely understood. Depending on the type of infection and, by extension, the overall signals integrated during priming, this process may involve "programming" of cell death by early exposure to inflammatory cytokines such as INFγ,[60,61] as well as triggers such as signaling through death receptors by members of the TNF superfamily of cytokines, including TNFα, FAS-L, and TRAIL,[79-83] and competition for declining availability of survival-supporting cytokines, such as IL-2, IL-7, and IL-15.[84-87] Ligation of death receptors results in activation of the proapoptotic caspases 3 and 7, whereas cytokine withdrawal triggers the mitochondrial or intrinsic pathway of apoptosis involving mitochondrial membrane disruption, release of cytochrome c, activation of caspase 9, and, ultimately, downstream activation of caspases 3 and 7.[84,85] The latter death pathway is regulated by members of the Bcl-2 family of proteins, including some that serve as sensors for cell stress, such as Bim and PUMA, some that act as executioners of cell death on mitochondrial membranes, such as Bax and Bak, and others that can inhibit these processes to promote survival, such as Bcl-2, Bcl-X_L, and Mcl-1.[84,85] Accordingly, genetic deletion of cell stress sensors and executioner proteins can delay the onset and magnitude of contraction, whereas deletion of proapoptotic molecules can accelerate this process.[64,80,86-88] Emphasizing the central role of these pathways, activated T cells that survive the contraction phase to become long-lived memory cells are often characterized by elevated

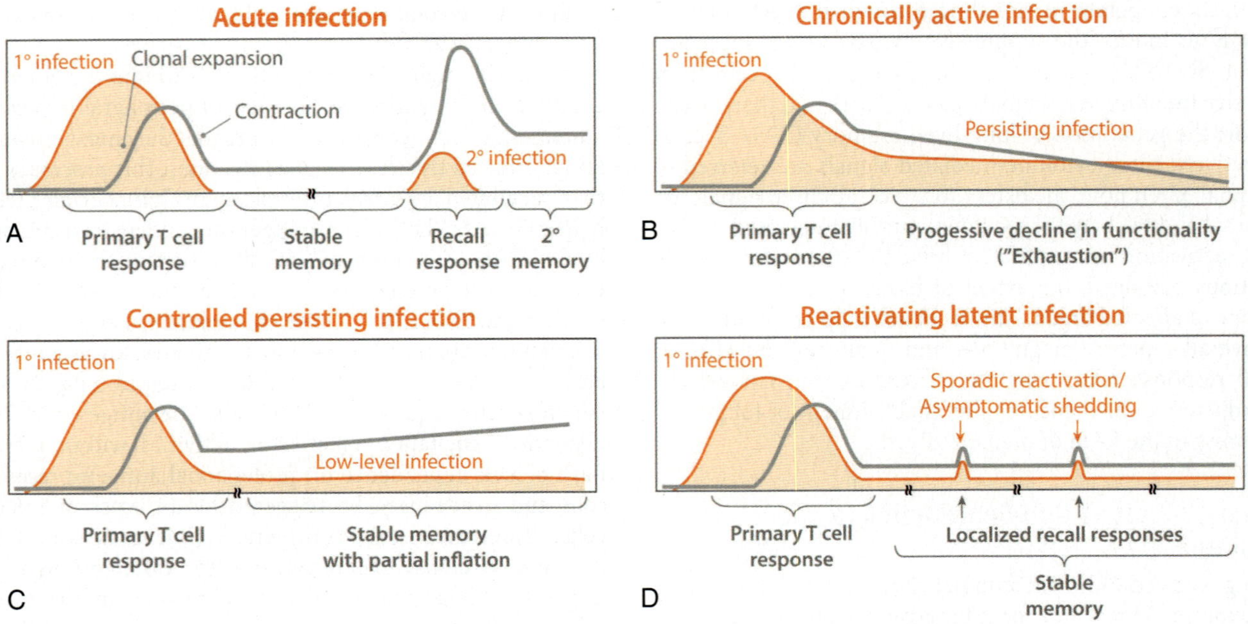

FIG. 30.2. T cell dynamics in acute and persisting infections. A: T cell responses (*gray line*) to acute infection (*orange area*) encompass distinct phases of clonal expansion, contraction, and stable memory. Upon reinfection, memory T cells are rapidly reactivated to suppress pathogen growth and spread, thereby ameliorating duration and severity of the secondary infection. As part of this, memory T cells undergo marked recall expansion, followed by a protracted phase of contraction and the establishment of a stable population of secondary memory T cells. **B-D:** T cell responses to different types of persisting infections also encompass stages of clonal expansion and contraction; however, ongoing or intermittent presence of replicating pathogens can have dramatic impact on T cell dynamics. **B:** During chronically active infections, ongoing pathogen replication drives constant production of effector cells from a pool of precursors with memory-like capacity. However, over time chronic T cell activation can result in loss of T cell clones and in a progressive and irreversible decline in functionality referred to as "exhaustion." Of importance, despite being functionally muted, such exhausted or functionally adjusted T cells contribute to pathogen control. **C:** In other persisting but more tightly controlled infections, the presence of latently infected tissue cells can drive accumulation or "inflation" of fully functional T_{EM}-like cells. However, this only applies to a fraction of the overall memory pool, with other T cell clones in the same individuals forming durable and quiescent T_{CM} cells. **D:** Other latent reactivating infections target specific anatomical niches such as neuronal tissues and, upon sporadic reactivation, can spread to peripheral tissues such as skin. T_{RM} cells established in these peripheral sites can rapidly extinguish reemerging pathogens, thereby effectively converting pathogen reactivation into short episodes of peripheral "asymptomatic shedding" without engagement of the systemic pool of memory cells.

expression of Bcl-2, as well as expression of receptors for the cytokines IL-7 and IL-15 that promote Bcl-2 expression and survival.[89,90]

In addition to expressing IL-7Rα and CD122 (IL-15Rβ), effector-like cells with potential for long-term survival also express high levels of CD27, CD62L, and CXCR3; low or intermediate levels of CX3CR1; and low levels of the IL-2R α-chain, CD25, and the killer cell lectin-like receptor G1, KLRG1. By contrast, more terminally differentiated short-term effectors are characterized by elevated expression of CD25, KLRG1, and an activated isoform of CD43.[23,89,91-94] This simple phenotypic dichotomy, however, may not apply to all infections or all conditions.[95] Some KLRG1^{hi} effector cells can reexpress IL-7Rα and assume an intermediate lifespan or extinguish KLRG1 expression during transition into long-lived cells, therefore posing caveats in predicting the ultimate fate of a cell based purely on surface marker expression.[96,97] Nevertheless, memory precursor effector cells can also be separated from their more short-lived counterparts by virtue of their heightened expression of memory fate-promoting transcription factors, such as Tcf1 and Id3.[98,99]

In summary, although downscaling the acute effector response is important to prevent excessive inflammation and damage to vital organs, it nevertheless leaves behind a heterogeneous pool of antigen-experienced T cells that persist in the host at frequencies that are several orders of magnitude higher than those of their naïve ancestors.[8] These longer-lived memory T cells possess functional traits that have inherently been imprinted during T cell priming as well as a programming central to their extended survival. Of importance, their contribution to T cell immunity is not limited merely to the dying stages of infection since the epigenetic reprogramming acquired during priming ensures a rapid reengagement in pathogen control in case of renewed infection. As we will discuss in more detail later in the chapter, this ability to more efficiently deal with subsequent infection is especially important for memory cells that permanently occupy peripheral tissues that may come under repeated attack by the same pathogen. In this case, memory cells can act as both "immune sentinels" and "first line defenders," reflecting what has aptly been coined their "sensing and alarm" function.[100] Thus, T cell responses serve two major purposes, that is, promoting infection resolution during a primary challenge and also establishing a more complex and potent line of defense against renewed infection.

MODELS FOR MEMORY DIFFERENTIATION

The bifurcation of postprimed T cell populations into either short-lived effector or long-lived memory cells, as well as the functional diversification within the latter, is critical for host protection from primary and renewed infections. How this robust heterogeneity is achieved through the dynamic interplay between extrinsic and T cell–intrinsic factors has been a matter of intensive debate.[101] There are numerous, often directly interrelated, factors that shape developmental trajectories of T cells during a primary infection or immunization. Their impact is directly related to the nature of the infection and inflammation and include, but by no means are limited to, the strength of TCR signaling (signal 1) and costimulation (signal 2) and the provision of cytokines (signal 3) and nutrients in the microanatomical niches where priming takes place. Thus, there is a myriad of diverse factors that can shape the fate of an activated T cell. Precisely when and the extent to which individual factors contribute to response diversification, however, awaits further clarification, although two combinations described below are favored in the literature at this point in time.

Early Selection Between Effector and Memory T Cell Fates

There is evidence for heterogeneity within the pool of naïve T cells such that, depending on their thymic development either early or late during life, some clones may be poised to preferentially give rise to either effector or memory progeny.[102] Separately, unequal distribution of transcriptional and metabolic molecular machinery between daughters of dividing T cells during their interaction with antigen-presenting dendritic cells may also shape fate diversification.[103] The concept of "asymmetric cell division" posits that a daughter cell arising in physical proximity to the activating dendritic cell inherits heightened expression of nutrient transporters, cytokine receptors (eg, IL-2Rα and IFNɣ-R) and transcription factors (eg, T-bet and c-Myc), resulting in enhanced activity of the PI3K and mTORC1 pathways associated with heightened metabolic activity and a more "effector"-biased differentiation toward cells with short (terminal effector) or intermediate (effector memory) life span. On the other hand, the distal daughter cell, which is furthest from the T cell–dendritic cell immune synapse, inherits less of these transcriptional and metabolic regulators and, therefore, may be more prone to give rise to long-lived memory T cells.[103-107] Thus, this biologically hard-wired mechanism may contribute to response diversification as early as the first division but possibly also during subsequent interactions with dendritic cells and upon antigen reencounter within infected tissues.

The Decreasing Potential Model

Experiments employing adoptive transfer of a single TCR-transgenic T cell clone or genetic barcoding of individual T cells have revealed that a single antigen inexperienced T cell can give rise to both effector and memory T cells (the "one cell–multiple fates" concept).[108-110] Of interest, individual naïve T cells can display considerable biases to one or the other developmental trajectory. However, when antigen-specific responses are integrated from a larger number of recruited naïve T cells, well within the estimated number of the 100 to 1000 naïve clones for a given antigen specificity,[1,3] they become highly reproducible in terms of heterogeneity and magnitude. Mathematic modeling described such results in a probabilistic framework with random integration of differentiation and division events. In this model, cell fates are established along a linear developmental pathway with progressive differentiation from slowly proliferating memory precursors to rapidly dividing short-lived effector cel ls.[101,109-111] Of note, there is evidence for both early and late "branching" of cells with memory potential from their counterparts developing along the linear effector differentiation pathway (Fig. 30.3). Thus, in some situations, memory precursor cells may undergo very little effector differentiation and branch off as early as during the first division.[31,103,105,112] However, memory cells can also develop from precursors that have undergone effector differentiation, as evidenced by fate-mapping experiments where T cells were irreversibly tagged upon transcription of effector-related genes such as INFɣ and KLRG1.[97,113] Likewise, memory T cells retain some epigenetic characteristics, that is, patterns of chromatin accessibility and DNA methylation. These are imprinted early during effector differentiation, whereas other parts of their epigenetic profile, including those affecting genes associated with longevity and lymph node homing, are further modified and even reversed to a more naïve-like state during development into long-lived cells.[114,115]

In Figure 30.3, we summarize some aspects of the early developmental models above, alongside the "decreasing potential" concept, to emphasize that different trajectories are not mutually exclusive but, instead, are likely to operate in parallel in an infection- and context-specific manner. Finally, some fate-permissive effector cells, after having infiltrated peripheral tissues, can respond to microenvironmental cues, including antigen and locally produced cytokines, to give rise to long-lived and sessile T_{RM} cells that will be described in more detail later in the chapter. This secondary maturation step resulting in molecular adaptation to a specific tissue microenvironment further adds to the heterogeneity within the overall pool of memory T cells. The existence and the nature of precursors for T_{RM} cells, and how they fit into the schematic shown in Figure 30.3, however, is currently not well understood.

DETECTION AND MARKERS OF MEMORY CD8+ T CELLS

Before we discuss in more detail the functional properties of memory T cells that determine their contribution to immune protection, we will first briefly summarize the main experimental strategies commonly employed for the detection and functional analysis of memory T cells. These include the

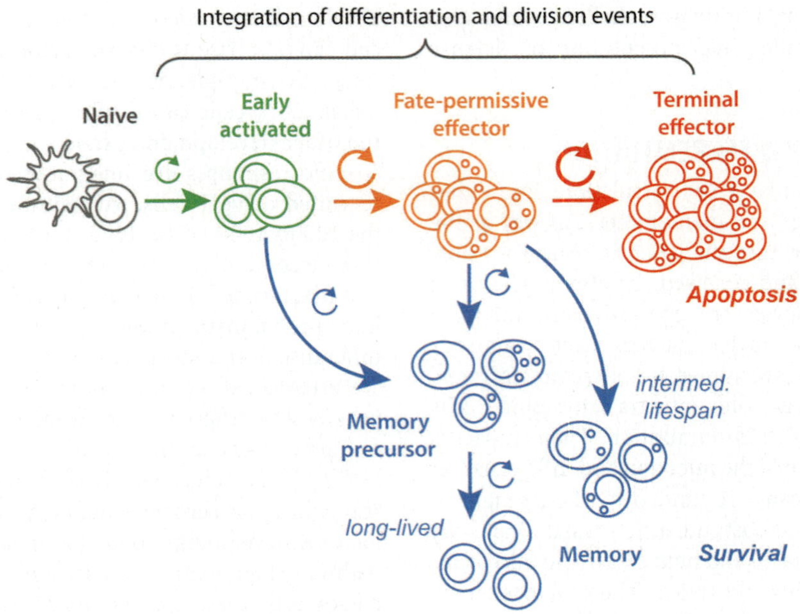

FIG. 30.3. The "decreasing potential" model of memory T cell differentiation. A single naïve T cell can give rise to distinct progeny with short, intermediate, and long lifespans, which is referred to as the "one cell–multiple fates" concept. Following activation by antigen-presenting dendritic cells, T cells undergo poorly defined stages of "early activated cells," "fate-permissive effector cells," and "terminally differentiated effector cells." Mathematic modeling has described this process along a linear developmental pathway with progressive differentiation and decline of developmental potential from slowly proliferating memory precursors (*blue cells*) to rapidly dividing short-lived effector cells (*red cells*). Proliferation speed is indicated by different sizes of the division symbol. In this model, various differentiation and division signals, including exposure to antigen, cytokines, costimulatory molecules, and nutrients, are integrated throughout the response and ultimately shape fate commitment and response diversity. Evidence suggests that precursors of long-lived memory T cells can branch off this linear pathway at various points during the response, including evidence for early emergence during the pre-effector stage, as well as late development from fully fledged but fate-permissive effector cells.

identification of memory T cells based on phenotypic markers, functional properties, or TCR-endowed specificity for a given antigen.

Phenotypic Characteristics of Memory T Cells

Phenotypic analyses take advantage of the observation that the transition from naïve to effector and memory CD8+ T cells is marked by transient or permanent changes in the expression of a broad variety of surface molecules. These include cytokine receptor subunits such as CD122 (IL-15Rβ), CD25 (IL-2Rα), and CD127 (IL-7Rα); chemokine receptors such as CCR7, CCR5, and CXCR3; adhesion molecules such as members of the β1 and β2 integrin families, CD49d, CD29, CD11a, CD11b and CD18, as well as CD44, CD62L, and E-selectin ligands. Combined with other molecules, such the GPI-anchored Ly6C, the killer lectin like receptor KLRG1, and isoforms of CD43 and CD45, they are commonly used to delineate effector and memory CD8+ T cells from their naïve counterparts. However, somewhat limiting their suitability as memory markers, expression of these molecules often reflects recent or intermittent stimulation and is subject to dynamic regulation over time. Thus, it should be kept in mind that, in certain cases and under certain circumstances, these markers may not always inform on memory potential or longevity.

The surface-expressed tyrosine phosphatase CD45 is often used as a marker for human antigen-experienced or memory CD8+ T cells. There exist different splice variants of CD45 that represent high- and low-molecular-weight isoforms and can be distinguished by antibody staining. Naïve cells express the high-molecular-weight form, CD45RA, whereas some memory T cells predominantly produce a low-molecular-weight variant, CD45RO. Accordingly, staining for the relative expression of CD45RA and CD45RO isoforms can identify a fraction of human CD8+ memory T cells, although this pattern can change over time such that some long-lived memory T cells become CD45RA-positive again. For instance, reexpression of CD45RA is seen in a large subset of human memory CD8+ T cells called CD45RA+ effector memory T cells. These are thought to represent a more terminal or senescent population of antigen-experienced cells that accumulate over time. Indeed, subsets of CD8+

memory T cells specific for human immunodeficiency virus (HIV), cytomegalovirus (CMV), Epstein-Barr virus (EBV), or parvovirus show gradual conversion to a CD45RA+ or CD45RA+/RO+ phenotype.[116-119] However, CD45RA+ memory CD8+ T cells have also been reported after vaccination against yellow fever virus, and these cells exhibit bona fide memory characteristics in that they persist for many years while retaining polyfunctionality without any signs of recent activation.[114,120]

In mice, high-level expression of the hyaluronic acid receptor, CD44, is the most commonly used marker for memory CD8+ T cells. Its elevated expression is induced early after priming and then maintained throughout the effector and memory phases. However, even specific pathogen-free mice that have not been immunized harbor CD44^{hi} CD8+ T cells and their proportion progressively increases with age. Such cells are thought to arise from cytokine-induced expansion in the absence of overt stimulation by foreign antigens, in either an IL-4-dependent manner within the thymus or an IL-15-dependent process within peripheral lymphoid tissues.[121,122] The latter phenomenon is prominent after lymphopenia-driven proliferation, although it can also occur in a more muted way in mice with a full repertoire of T cells. Although these memory phenotype cells receive only a weak signal 1, likely from MHC-I complexed with self-tissue-derived antigen, they nevertheless share functional traits with some conventional memory cells, including rapid IFNγ production in response to stimulation with the proinflammatory cytokines IL-12 and IL-18.[121,122] Accordingly, some of these CD44^{hi} CD8+ T cells are referred to as "natural" "virtual," or "innate" memory T cells (Table 30.1) and have been implicated in bystander immunity against systemic infection.[123] Finally, there also exists a population of so-called stem cell–like memory cells (T_{SCM}) that, like naïve cells, express only low levels of CD44 but are distinguishable from the latter by elevated expression of CD95 (Fas) and low expression of CD31.[124,125] T_{SCM} cells may develop from naïve cells without acquiring effector cell properties and, because of their increased proliferative capacity and developmental potential, are considered important components of adoptive cellular T cell therapies.[125] In summary, expression patterns of T cell surface molecules can inform on activation status or stimulation history and therefore at least in some cases, may serve as surrogates for memory potential. However, applied as stand-alone approach, phenotyping fails to reliably identify immunization-induced conventional memory T cells.

Detection of Memory T Cells Based on Functional Characteristics and Transgenic Tools

Instead of relying on phenotypic profiling, other commonly used assays for the detection of memory T cells measure functional properties or elevated frequencies of T cells with defined antigen specificity, that is, long-term persistence. As mentioned earlier, memory T cells remain poised for rapid proliferation and the production of effector molecules in response to reencounter with their cognate antigen. This distinct property can be read out after *ex vivo* antigen restimulation, and such assays, in particular those measuring cytokine production via intracellular staining or enzyme-linked immune absorbent spots (ELISPOT), have remained a gold standard for the assessment of antigen-specific T cell memory. In addition to these indirect functional assays, the advent of soluble peptide-MHC tetramer technology has enabled researchers to closely follow antigen-specific CD8+ T cell responses in mice and humans and has proven invaluable for the detection, isolation, and detailed phenotypic and functional analyses of resting memory T cells. Likewise, the generation of TCR-transgenic mice that produce T cells with defined specificities for pathogen-derived or model antigens has fueled fundamental discoveries regarding the biology of memory T cells and their protective function. TCR-transgenic T cells are typically used in low-number adoptive transfer settings where they are tracked in recipient mice using genetic or fluorescent markers. The use of TCR-transgenic cells has the added advantage that such cells have identical TCRs, meaning that variables such as differences in antigen affinity among the polyclonal repertoire of an endogenous antigen-specific T cell population can be controlled. Finally, many TCR-transgenic mouse strains are now available on backgrounds carrying defined genetic modifications, including gene deletions and gene expression reporter systems, and therefore allow for the *in vivo* analysis of T cell–intrinsic gene expression kinetics and gene functions, as well as lineage tracing.

Phenotype and Function Illustrate Subset Diversity Among Memory Cells

Combined with *in vivo* migration and protection assays, the aforementioned approaches have illuminated a vast heterogeneity of memory CD8+ T cells with regards to developmental, phenotypic, and functional traits. Thus, memory T cells are broadly categorized into various, sometimes overlapping, subsets (Table 30.1).[30,126-128] These include populations based on pattern or permissiveness of migration, such as the central (T_{CM}), effector (T_{EM}), and peripheral recirculating (T_{PM}) subsets, as well as the sessile tissue-resident (T_{RM}) subset, all of which are described in more detail later in the chapter. In addition, there are stem-like (T_{SCM}) and unconventional memory-phenotype (T_{VM}, T_{IM}) subsets, with respective key properties important for overall repertoire repopulation and functionality (Table 30.1; Fig. 30.4). Some of the aforementioned T cell surface markers can distinguish memory subsets within a given tissue, and this has proven valuable in defining the protective potential of the distinct memory subsets. Nevertheless, the extent to which current subset categories reflect the true diversity of the memory T cell pool has remained a matter of debate.[116,126] For instance, single cell genomics and mass cytometry analyses have emphasized a level of heterogeneity that is not reflected in the common classification.[19,105,129,130] Moreover, there is well-documented developmental plasticity such that cells from one "subset" can give rise to cells from another subset and *vice versa*, which may occur either

TABLE 30.1 Commonly Used Subset Classification of CD8+ Memory T Cells

Subset	Distribution	Key Properties	Phenotype
Central memory cells (T_{CM})	• Blood • Lymphoid tissues	• Recirculation • Self-renewal • Recall expansion • Pluripotency • Protection from systemic infection	$CD44^{hi}$ CD45RO+/− CD62L+ CCR7+ CX_3CR1^{lo} $CD27^{hi}$ $CD122^{hi}$ $CD127^{hi}$
Effector memory cells (T_{EM})	• Blood • Lymphoid tissues • Nonlymphoid tissues	• Recirculation • Recall expansion • Protection from systemic and peripheral infection	$CD44^{hi}$ CD45RO+/− CD62L− CCR7+/− CX_3CR1^{hi} $CD27^{lo}$ CD122+ $CD127^{int}$
Peripheral memory cells (T_{PM})	• Blood • Lymphoid tissues • Nonlymphoid tissues	• Recirculation • Self-renewal • Recall expansion • Pluripotency	$CD44^{hi}$ CD62L+/− CX_3CR1^{int} $CD127^{hi}$
Tissue-resident memory cells (T_{RM})	• Nonlymphoid tissues (including epithelial, stromal, and vascular compartments) • Lymphoid tissues	• No recirculation in steady state • Recall expansion • Protection from peripheral infection	CD69 CD103+/− CD62L− CCR7+/−
Stem cell–like memory cells (T_{SCM})	• Blood • Lymphoid tissues	• Self-renewal • Recall • Pluripotency	$CD44^{lo}$ CD45RO− CD45RA+ CD62L+ $CD122^{hi}$ $Sca\text{-}1^{hi}$ CD27+
Memory phenotype cells • Virtual (T_{VM}) • Innate (T_{IM})	• Blood • Lymphoid tissues	• Recirculation • Self-renewal • Bystander protection from infection	$CD44^{hi}$ CD62L+ $CD122^{hi}$ $CD49d^{low}$ CXCR3+

spontaneously or, more prominently, after renewed antigen stimulation.[92,131-134] Of note in this regard, intermediate- or long-lived memory cells generated by repeated antigen stimulation, for instance, in prime-boost settings with iterative infections, are phenotypically, transcriptionally, and functionally distinct from the memory subsets observed after a single infection.[135-139] Finally, immune responses in "microbially rich" or "dirty" environments differ dramatically from those observed in the "cleaner" environments commonly used in animal models,[140-142] and these differences extend to the kinetics and quality of CD8+ T cell responses.[143] Thus, although the widely applied categorization of memory cells into distinct subsets can provide a rough guide for functional specialization and protective potential, it does not capture the full complexity and diversity of memory T cell contribution to protective immunity.

CARDINAL FEATURES OF T CELL MEMORY

Memory T cells possess a number of key quantitative and qualitative properties that determine their superior protective function during renewed infection. These include (i) long-term persistence at elevated frequencies, (ii) rapid execution of effector functions, and (iii) broad distribution across the body (Table 30.2). However, reflecting the vast heterogeneity of memory T cell pool, the extent to which individual memory cells or memory subsets display these functional traits differs considerably. Nevertheless, within the overall population of memory cells, all these properties are usually evident. When combined, they underpin the rapid generation of secondary effector cells, as well as early engagement in pathogen control through immediate recall of effector activities. We will discuss three of the cardinal features of T cell memory in more detail in the following sections.

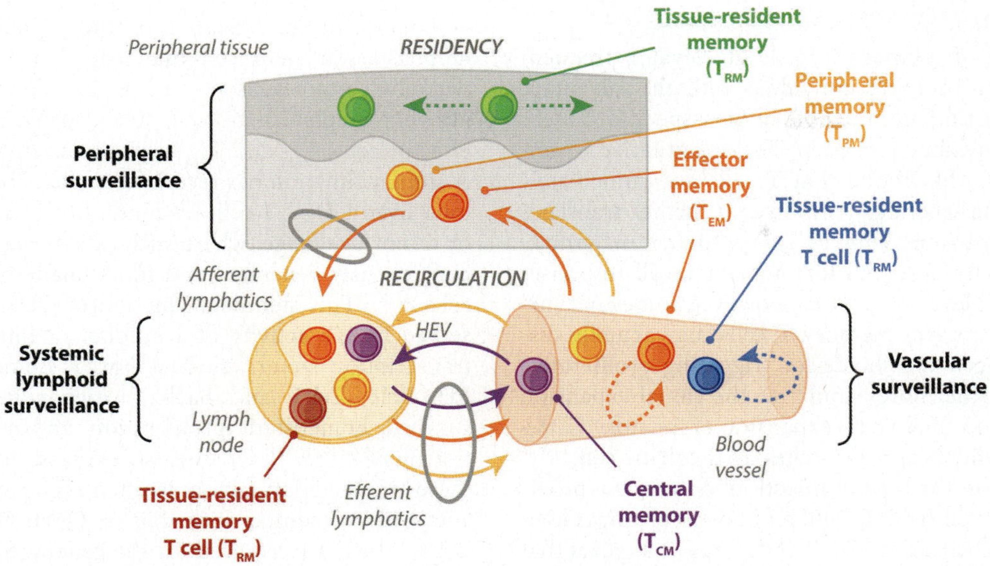

FIG. 30.4. **Anatomical distribution and migration patterns of major memory CD8+ T cell subsets.** Memory T cells are broadly categorized into subsets of recirculating and nonrecirculating cells that can populate all manners of tissue. Collectively, they provide peripheral, lymphoid, and vascular surveillance. Lymph nodes, for instance, contain memory T cells that arrive from peripheral tissues via afferent lymphatics as part of their peripheral migration program. These include classical T_{EM} cells, which overlap with a portion of the more recently defined T_{PM} cells. By contrast, classical T_{CM}, T_{SCM} (not shown), and a fraction of T_{PM} cells express the lymphoid targeting homing molecules, CD62L and CCR7, and enter lymph nodes from the blood via HEV. All recirculating memory cells eventually egress from lymph nodes and shuttle back to the blood via efferent lymphatics. Lymph nodes also contain nonrecirculating T_{RM} cells, some of which represent emigrants from peripheral tissues that may have been mobilized following local restimulation. T_{RM} cells populate all manners of other tissues, including barrier tissues such as skin and mucosa, where they form a first line of defense against invaders. Some T_{RM} cells take up residency in epithelial tissue compartments, whereas others reside in subepithelial and stromal compartments or the vasculature. Vascular surveillance (indicated by dotted lines with arrows) is a function of a subset of T_{RM} cells in the liver but has also been described for T_{EM} cells. Please refer to **Table 1** for phenotypic markers associated with the different types of recirculating and resident memory T cells. However, the definition and terminology to describe the memory subsets depicted herein remains in constant flux such that there exists considerable overlap between populations assigned to the T_{CM}, T_{EM}, and T_{PM} categories, as exemplified by their overlapping migratory patterns.

TABLE 30.2 Cardinal Features of CD8+ Memory T Cells That Determine Their Protective Function

Memory Features	Underpinning Molecular and Cellular Processes	Relevance for Protective Function
Persistence at elevated frequencies	• Prolonged survival • Homeostatic proliferation • Metabolic fitness	• Rapid and enhanced recall expansion providing secondary effector and memory cells
Rapid recall of effector functions	• Epigenetically, transcriptionally, and metabolically poised for rapid recall • TCR signaling modifications • Increased integrin expression for interaction with antigen-presenting cell (APC) • Responsive to stimulation by both professional and nonprofessional APC	• Immediate engagement in protective immunity
Broad tissue distribution	• Recirculation and residency in lymphoid and nonlymphoid tissues • Loss of egress receptors and molecular adaptation to tissue environments promotes long-term residency without recirculation	• Immediate engagement in protective immunity • Sensing and alarm function • Localized control of persisting or recurring infections

Memory Persistence at Elevated Frequencies

The persistence of memory T cells at elevated frequencies provides the immune individual with the advantage of a greater starting number of antigen-specific T cells against a given invading pathogen. This quantitative advantage increases the likelihood that T cells are immediately recruited into the secondary response. It further translates into faster reexpansion kinetics and a higher magnitude that are commonly observed for memory recall responses at the population level (Fig. 30.1). However, somewhat surprisingly in this regard, memory T cells may actually display higher antigen thresholds for triggering proliferative expansion.[144] Furthermore, similar to the clonal expansion of naïve T cells, optimal recall expansion of memory CD8+ T cells in lymphoid tissues also relies on dendritic cells[145,146] and, depending on the type of infection, also on the provision of CD4+ T cell help, CD40, and costimulatory molecules, such as CD28 and 4-1BB.[83,147-151] Thus, it appears that elevated starting frequencies are the main determinant for enhanced recall responses by memory T cells, regardless of whether these are recirculating memory cells or T_{RM} cells that remain permanently embedded in the tissues.

Following initial clonal expansion and survival throughout the "contraction" phase, the maintenance of memory T cells at the population level largely reflects a balance between cell loss and slow homeostatic proliferation. Unlike their naïve counterparts, CD8+ memory T cell maintenance is not dependent on MHC I expression and TCR signaling.[152,153] As such, it is independent of antigen or pathogen persistence. Instead, memory T cells survive through the action of the common γ-chain cytokine family members, IL-7 and IL-15, both of which suppress the mitochondrial pathway of apoptosis by inducing expression of antiapoptotic molecules, such as Bcl-2 and Mcl-1. Accordingly, subsets of memory T cells express high levels of receptors for both cytokines,[89,90] and genetic studies have demonstrated severe T cell developmental and memory defects in absence of IL-7 signaling, as well as rapid decay of antigen-specific memory T cells in the absence of IL-15 signaling.[153-158] Although these cytokines exert overlapping functions, IL-7 is thought to act predominantly as a prosurvival factor, whereas IL-15 additionally drives homeostatic proliferation.[153,156] The dependence on IL-15 as a maintenance factor, however, is not absolute, which is obvious when comparing cytokine requirements for nonrecirculating T_{RM} cells that take up residency in different types of tissues. T_{RM} cells in skin epidermis or salivary gland epithelium, for instance, strictly depend on local provision of IL-15.[138,159] The source of IL-15 in this case are radioresistant tissue cells such as keratinocytes[159] that, in analogy to what has been shown for dendritic and monocytic cells in lymphoid tissues, are likely to transpresent IL-15 in a cell-cell contact–dependent manner as a complex incorporating both the cytokine and the α-chain of its receptor.[160,161] By contrast, T_{RM} cells in other tissues, such as intestinal mucosa, kidney, and lymph nodes, survive normally in the absence of IL-15.[138,162] Finally, repeated infection or immunization in prime-boost settings yields memory T cells with an effector memory phenotype that become increasingly independent of IL-15.[138]

Long-term maintenance of the memory T cell pool emphasizes a stem cell–like behavior of memory cells, whereby certain subsets, such as T_{CM}, T_{SCM}, and T_{PM} cells, efficiently replenish lost cells, and also give rise to secondary effector cells, T_{EM} and T_{RM} cells, upon reinfection.[108,131,163,164] A similar function has recently been ascribed to a subset of long-lived CD8+ T cells in conditions of chronic infection or tumor challenge, where such cells serve as the source of continuously generated and functionally muted effector T cells.[165-171] This stemness function of CD8+ T cells is preserved by the activity of a number of interconnected key transcription factors, such as Foxo1, Eomes, Tcf1, c-Myb, Id3, Bcl6, Zeb1, and Bach2, which collectively support survival, homeostatic proliferation, and rapid recall functionality.[25,98,99,172-181] Of interest, expression and activity of these transcription factors is often regulated in a mutually antagonistic manner with that of T-bet, Blimp1, Id2, and Zeb2, which are required for the generation of T_{EM} cells or effector cells with intermediate or short lifespan but heightened effector activity[21,22,25,54,98,169,178,182,183] (Fig. 30.1). Indeed, conditional deletion of fate-defining transcription factors in experimental settings can reverse select phenotypic, transcriptional, and functional characteristics of short- or long-lived antigen-experienced T cells.[127,184] Such observations emphasize that T cell fate is actively and continually reinforced by key transcriptional networks and also offer an explanation for a level of developmental plasticity displayed by some memory CD8+ T cell subsets.

Metabolically, the maintenance of long-lived memory T cells has been linked to efficient mitochondrial respiration and increased fatty acid oxidation, both of which are facilitated by ultrastructural changes of mitochondria that involve fusion, elongation, and the formation of a tubular network with tight cristae. Mitochondrial remodeling is initiated early after activation in a manner dependent on IL-15 signaling and the fusion regulator, optic atrophy 1 (Opa1), and affords memory cells with a substantial spare respiratory capacity linked to enhanced survival and functionality.[39,41-43,185] In addition, IL-7 promotes memory T cell metabolism by facilitating glucose utilization and by controlling amino acid and glycerol uptake, processes that are regulated in part through induced expression of the glycerol channel, aquaporin 9, and enhanced expression and altered intracellular trafficking of the glucose transporter, Glut-1.[186,187] Some memory subsets utilize these substrates to support fatty acid metabolism and oxidative phosphorylation, whereas T_{RM} cells in peripheral tissues express tissue-dependent isoforms of fatty acid–binding proteins that allow for direct uptake of free fatty acids as fuel for mitochondrial respiration.[188,189] However, the metabolic switch from aerobic glycolysis in effector cells to predominant oxidative phosphorylation in memory cells is not always a strict requirement for the establishment of T cell memory. For instance, it has been shown that T_{EM} cells can fuel their energetic needs through sustained glycolysis and that CD8+ T cells rendered genetically deficient in fatty acid oxidation can give rise to memory T cells.[190] Metabolic fitness of both recirculating and tissue-resident memory cells is further dependent on their expression of the extracellular ATP sensor, P2RX7, thereby providing a metabolic link between tissue injury or inflammation, which result in ATP release, and the maintenance of T cell memory.[191]

Rapid Recall of Effector Functions

Another hallmark feature of memory T cells is their ability to rapidly exert effector function upon antigen reencounter without having to undergo renewed differentiation.[192-194] In addition, some memory cells can do this solely in response to proinflammatory cytokines, such as IL-12 and IL-18, without the need for antigen recognition.[195,196] At the molecular level, this is a consequence of a permissive or poised pattern of DNA demethylation and histone modifications at gene loci coding for major effector molecules, such as *Ifng*, *Gzmb*, and *Prf1*.[27,114,115,197,198] These epigenetic changes are induced at the time of T cell activation during initial priming and then stably passed on to their progeny, including short-lived effector and memory T cells. In other words, this epigenetic imprinting serves as a molecular basis for the ability of memory T cells to "remember" their previous "activated state" and lends credence to the notion that activated T cells that have undergone effector differentiation can give rise to long-lived memory T cells.[199] Although this permissive epigenetic state in resting memory T cells *per se* does not result in transcription of effector molecule genes,[114,115] some memory T cells actually contain preformed RNA transcripts encoding effector genes and these can be readily translated upon restimulation.[188,200-202] Finally, some memory CD8+ T cells, such as those residing in peripheral nonlymphoid organs, readily store preformed effector proteins, such as granzyme B, perforin, or the proinflammatory cytokine, CCL5.[201,203,204]

Memory T cells harbor, in addition to the molecular machinery enabling constitutive or rapid reinduction of effector molecule expression, higher-order TCR oligomers and alterations in components of the TCR signaling cascade that enhance their sensitivity to antigen.[205,206] Separately, memory T cells also express a diverse set of integrins that facilitate interactions with cells displaying MHC-antigen complexes on their surface. Importantly in this regard, the elicitation of effector function in memory T cells does not require professional antigen-presenting cells such as dendritic cells but instead can be readily triggered by infected tissue cells.[146,201] Finally, although T cells do not undergo affinity maturation of their TCRs through somatic mutation, high-affinity clones expand more vigorously during initial T cell priming than do low-affinity clones, which effectively translates into an overall increased affinity of the effector and memory T cell pools compared with the antigen-specific naïve T cell repertoire.[207] Thus, memory T cells display numerous molecular and cellular modifications compared with naïve T cells that collectively underpin their rapid recall of effector functions.

Broad Anatomical Distribution

Although immune surveillance by naïve T cell is largely restricted to lymphoid tissues, with their activation reliant upon dendritic cells to convey information about infection in peripheral tissues, memory T cells broadly distribute across all manners of lymphoid and, importantly, also nonlymphoid tissues (Fig. 30.4). In fact, the vast majority of conventional, or infection-induced, memory T cells localize to peripheral nonlymphoid tissues.[208-210] As we will discuss in more detail in the next section, many of these peripheral memory T cells belong to the T_{RM} category of locally migrating cells with restricted recirculation potential, at least in the steady state. Depending on the organ of residence, such T_{RM} cells patrol various microanatomical tissue compartments, including epithelial layers, subepithelial or stromal compartments, as well as the vasculature. Consequently, the peripheral distribution of memory T cells affords a considerably more efficient mode of immune surveillance than seen for their naïve counterparts. The combined features of strategic positioning within extralymphoid organs together with their ability to rapidly recall effector functions allow for immediate engagement in pathogen suppression and control in case of reinfection.[100,146,201,211-215] The broad anatomical distribution of memory T cells is therefore considered a major component of memory T cell–mediated immunity and will be discussed in more detail in the following section.

THE ROLE OF RECIRCULATION AND RESIDENCY IN T CELL MEMORY

Naïve cells enter lymph nodes through specialized structures, the high endothelial venules (HEVs), and once inside secondary lymphoid organs they interact with the antigen-bearing dendritic cells that promote initial priming of the immune response.[4] After activation, T cells begin to express a variety of homing molecules that completely change their strict blood-to-lymphoid pattern of migration. These include adhesions molecules such as the integrin α4β7 and chemokine receptors such as CCR9, which combine to allow selective entry into nonlymphoid tissues, which for the two mentioned molecules represents the small intestine. As such, the priming-induced alterations promote widespread distribution of these activated effector T cells throughout the body for the specific purpose of controlling an ensuing infection. From this, one can see that the different migration capabilities match the functional state of the distinct types of T cells. Naïve T cells are in a resting state involved in a tightly controlled central surveillance mechanism that efficiently detects the introduction of pathogen anywhere in the body. In contrast, effector T cells have been armed, numerically expanded, and then widely disseminated for the purpose of eliminating the specific pathogen wherever it has taken hold. As you will see in the following sections, when it comes to memory T cells as a group, they show recirculation elements of both these populations. They do so because they consist of a mixture of cells that differ in their function and migration potential.

The T_{EM} and T_{CM} Paradigm of Memory T Cell Demarcation

There are two ways that T cells enter the lymph node. One is directly from the blood after passing through the HEV, which is the means of entry for naïve T cells as mentioned above.

In addition, T cells can enter via the afferent lymphatic vessels, and this is the way cells return to the blood from the nonlymphoid tissues. Detailed early work examining the migration of T cells through the draining lymph nodes showed that distinct types of T cells were accessing these two modes of entry,[216,217] with one population involved in a blood-to-nodal pattern of migration via the HEV and another involved in a blood-to-tissue pattern entering the nodes via the afferent lymphatic vessels (Fig. 30.4). The two populations could be distinguished on the basis of CD44 and CD62L expression, molecules that discriminate naïve from and memory T cells.[218] From these studies it was deduced that the T cells emerging from the nonlymphoid tissues were likely memory T cells returning to the blood, arguing that at least some memory cells could adopt a pattern of recirculation that is effector-like in migration and differs from the strictly blood-to-nodal migration exhibited by their naïve counterparts.

Although some memory cells can recirculate through nonlymphoid compartments, this does not mean that all are involved in such a process. Pioneering studies by Federica Sallusto and Antonia Lanzavecchia established the fundamental concept that memory cells are heterogeneous and composed of cells with different functional and migratory capabilities.[219] Specifically, they identified two recirculating memory T cell populations in human blood and proposed that one was capable of migrating through nonlymphoid tissues, whereas the other was confined to the secondary lymphoid organs.[219] These populations differed in their expression of CCR7, which is the receptor that binds to the chemokines CCL19 and CCL21 and promotes T cell entry into lymph nodes via the HEV.[68,220] Since the CCR7-negative population had enhanced effector capabilities, these were termed effector memory T cells (T_{EM}), and they were proposed to be the memory cells that recirculate through nonlymphoid compartments. In contrast, the CCR7-positive memory T cells also expressed the lymph node homing molecule CD62L, mirroring the phenotype of the naïve T cell population. As a consequence, the latter were termed central memory T cells (T_{CM}) to denote a pattern of recirculation restricted to the secondary lymphoid organs.[219,221]

Memory T Cell Subsets Defined by Their Pattern of Recirculation

One problem with characterizing recirculating T cells based on CCR7 expression is that not only does CCR7 control lymph node entry via the high endothelial venule but it also promotes T cell entry into the lymphatics so they can exit nonlymphoid tissues,[222,223] something that was not known at the inception of the T_{EM}/T_{CM} paradigm. Moreover, it was subsequently shown that CCR7-expressing T cells could be found in peripheral tissues such as gut and skin.[208,224] Thus, it seemed likely that cells migrating through nonlymphoid tissue would express this chemokine receptor, calling into question the basis used for the original T_{EM} demarcation. As a consequence, other means of memory T cell categorization have been used to define different memory T cell subsets. One that appears to faithfully predict the true pattern of recirculation involves distinguishing memory T cells on the basis of varying expression levels of the fractalkine receptor, CX3CR1. Although the binding of this receptor to its ligand, CX3CL1, has functional importance for other cell types, for memory T cells it largely forms a convenient marker to identify cells with different migration and functional capabilities.[92,93] Using this marker, it was found that CX3CR1^{lo} memory T cells corresponded to the classically defined T_{CM} population in terms of their ability to recirculate between secondary lymphoid organs and the blood.[92] Surprising, when recirculation through nonlymphoid tissues was directly examined, the population that surveyed peripheral compartments was not the T_{EM}-like CX3CR1^{hi} subset as expected. Instead, recirculation through nonlymphoid tissues was dominated by CX3CR1^{int} cells that were termed peripheral memory (T_{PM}) cells, with the classic T_{EM}-like CX3CR1^{hi} subset involved in surveillance of the vasculature compartment.[92] Overall, the CX3CR1^{int} population has features that are T_{CM}-like, such as high levels of homeostatic proliferation and increasing prevalence in the memory pool over time. As a consequence, the original basis of T_{EM}/T_{CM} distinction based on CCR7 expression no longer appears to faithfully describe the migration characteristics for memory T cell populations. Nonetheless, the T_{EM} and T_{CM} nomenclature persists to this day and its core concept of functional heterogeneity within the broader memory T cell pool remains unchallenged (Table 30.1; Fig. 30.4).

Memory T Cell Subsets Defined by Functional and Activation Status

Not only are memory T cells heterogeneous in terms of recirculation but they also differ in other critical facets. These encompass some of the classical features that we associate with immune memory, such as the ability to mount a rapid recall response by enhanced proliferation and engagement of effector mechanisms. As a consequence, combinations of functionality and proliferative potential have also been used as a means of defining memory T cell subsets and their contribution toward protection against infection. Proliferation and numerical expansion on secondary antigen encounter are two of the major hallmarks of the recall immune response. However, not all circulating memory cells are equal in their ability to proliferate and this difference is a useful means of defining different memory T cell subpopulations. Hikono et al. separated memory T cells based on expression of the activation markers CD27 and CD43 to define memory cells with different proliferative capacity.[225] They showed that there was an inverse correlation between activation status and proliferative potential, with the CD27^{hi}CD43^{lo} cells having the strongest recall response. These cells accumulated within the classically defined T_{CM} pool over time, which is the subset that gradually dominates blood memory cells after infection has subsided.[131] In contrast, cells bearing the CD27^{lo}CD43^{lo} combination and expressing the senescent marker KLRG1 were gradually lost from the circulating memory T cell pool. It is the latter that are the ones that provide the greatest protection against systemic infection despite their limited proliferative potential.[96] Thus, memory T cells can be separated into two functional populations with T_{EM} and T_{CM} type qualities; one which is involved in immediate infection control without further expansion, whereas the second requires renewed stimulation and expansion in the lymphoid compartment in order to achieve its full potential during the secondary immune response.

Fixed Immune Surveillance and Tissue Resident Memory T Cells

Almost immediately after the T_{EM}/T_{CM} paradigm was first proposed, various groups described populations of memory cells in a variety of nonlymphoid tissues such as brain, gut, liver, lung, and skin.[226-228] Since these tissue T cells expressed low levels of the lymph node homing molecule CD62L and had enhanced effector capabilities it was assumed that they belonged to the T_{EM} class of recirculating memory T cells. However, although some were likely to be cells that passed through the tissues before returning to the blood, it was also clear that others were nonmigratory and instead appeared to remain fixed within these locations.[229-231] This was elegantly demonstrated by using the classical technique of parabiosis,[231] which involves the surgical linking of the vasculature of two animals so they share a single blood supply. By using this technique, it was shown that certain organs such as lung and liver appeared to predominantly contain recirculating T cells, whereas others, notably the small intestine and brain, showed minimal exchange via the blood.[231] It should be noted that it is now clear that the results from these early experiments were complicated by a failure to remove all the cells from the vasculature in organs such as the lung,[232] meaning that many cells thought at the time to be migrating through the parenchyma were in fact simply flowing through the blood vessels. Nonetheless, as a consequence of these early observations, it was originally proposed that some organs, but not others, restricted lymphocyte entry and exit in a way that resulted in the long-term residency of memory cells. Critically, it was assumed that both recirculating and nonrecirculating T cells belonged to the T_{EM} subset since these were thought to be the only T cells with access to the nonlymphoid tissues.

Further investigations revealed that this notion of tissue-dependent retention did not explain the behavior of memory T cells in nonlymphoid compartments. Nonmigratory T cells were found to coexist with migratory memory cells in the same tissue or organ,[211,233,234] arguing that it was not the location that restricted lymphocyte entry and exit but a feature intrinsic to the population. Since recirculating T cells were assumed to belong to the tissue-penetrating T_{EM} subset and the nonlymphoid T cells were not supposed to include T_{CM} cells, the nonrecirculating or fixed cells were called tissue-resident memory T cells or T_{RM} cells.[128,235-240] Such T_{RM} cells were found to have a unique phenotype, expressing the activation marker CD69 and one or more surface molecules absent from circulating memory cells, such as the adhesion molecules CD49a and CD103.[128,237-240] Combined, the results argued that the T_{RM} cells represented a unique memory subtype that was distinct from any found circulating in the blood.

T_{RM} Properties and Features of Fixed Immune Tissue Surveillance

T_{RM} cells have now been found in a variety of different tissues and organs, including brain, gut, kidney, salivary glands, lung, and liver, suggesting that they are likely to occupy almost every niche in the body.[128,237-241] They are commonly located at body surfaces such as the linings of the small intestine, female reproductive tract, and the skin. In such instances, they are often confined within or near the outermost epithelial layers[201,203,211,242-245] where they can act as a first line of defense against infection.[211-215,246] In the skin epithelia or epidermis these T cells are nearly sessile,[242] but with just enough lateral movement that combines with the projection of dendritic structures for efficient surveillance of the local environment.[245,247] T_{RM} cells in anatomical compartments such as the female reproductive tract, however, display more local motility than the ones in skin.[248] T_{RM} precursors can be widely distributed through the body immediately after infection, resulting in T_{RM} dispersion over large areas and to many organs.[214,243] Nonetheless, the presence of local inflammation tends to focus the greatest concentration at the site of original infection, resulting in biased or preferential protection in this vicinity.[211,246,249]

It should be noted that T_{RM} cells are not strictly confined to nonlymphoid tissues since parabiosis experiments have revealed the existence of nonmigratory populations of memory T cells in lymphoid organs such as the spleen and lymph nodes.[134,162] Peripheral T_{RM} cells have been shown capable of proliferating in response to homeostatic stimuli or antigenic challenge, while remaining within their anatomical confines.[233,248,250,251] Paradoxically, although T_{RM} cells are widely seen as resident, there is some level of migration out of the tissue and the emerging cells can be reprogrammed for residency in other organs.[132-134,203,252] Nonetheless, tissue exit and reprogramming appear relatively inefficient,[133] with the possible exception of the lung[252] where CD103+ T_{RM} cells show a relatively high rate of retrograde migration to the point where this depletes the memory cells from the tissue over time.[253] Overall, T_{RM} cells generally represent a stable, long-lived memory population involved in a highly localized and efficient form of immune surveillance and protection.

Mechanistic Control of T_{RM} Cell Development

T_{RM} precursor cells capable of seeding the various tissues are present in the circulation only for a short time after priming.[254] Once they reach the tissues, these precursors gradually adopt a mature T_{RM} cell phenotype over time. T_{RM} cells in a variety of tissues share a transcription profile that sets them apart from T_{EM} and T_{CM} cells.[130,134,159,188,255,256] This core transcriptional T_{RM} profile has common or overlapping elements, such as the upregulation of various genes involved in adhesion (*Itgae, Itga1, Cdh1*), immune regulation (*Ctla4, Icos, Tlr1*), as well as downregulation of others that would promote tissue egress such as *S1pr1, S1p5*, and *Ccr7*. The latter changes are critical, since expression of tissue egress molecules such as S1P1 or elimination of its functional inhibitor CD69 compromises T_{RM} development in some tissues.[257-261] Master transcription regulators control many of these changes, with Klf2, Runx3, and the T-box transcription factors T-bet and Eomes critical to the development process.[257,262,263] In addition, the transcription factor Blimp-1 and its homologue Hobit are particularly important.[264] Perhaps not surprising then, the latter also control long-term residency of other lymphocytes with similar cytokine dependence, specifically liver-resident NK cells and NKT cells, arguing for shared mechanistic control of residency that spans both the adaptive and innate immune systems.[264,265] Indeed, the concept of lymphocyte residency was accepted in the innate arm of the immune response before T_{RM} cells were shown to exist, with CD8αα intraepithelial T cells in the intestine and γδ-TCR dendritic epidermal T cells in the skin long-recognized as being nonmigratory cells that permanently reside in their respective tissues.

Within the tissues, T_{RM} precursors come under the influence of local factors that promote the change in transcription profile as they adapt their new environment. These include restimulation by local antigen, which in some tissues, such as brain, sensory ganglia, and lungs, is obligatory for T_{RM} differentiation.[261,266] In others, local antigen promotes T_{RM} formation but is not strictly required.[211,243,246,267,268] Other factors make a contribution to survival, such as local fatty acids that are taken up by fatty acid–binding proteins that are selectively upregulated on T_{RM} cells within the tissues,[188,189] while distinct molecules drive the combined development and survival processes. Two of the most prominent are the cytokines TGFβ and IL-15. TGFβ has widely been reported as being critical for T_{RM} development in many tissues and drives the expression of many prototypic T_{RM} markers such as CD103.[159,243,261] TGFβ signals through a relatively undefined Smad4-independent mechanism and often combines with the action of IL-15.[159,262,269] This cytokine combination is critical for the coordinated modulation of the T-bet and Eomes, the expression of Blimp1 and Hobit, as well as the downregulation of the Klf2-driven S1P1/CCR7 tissue egress mechanism.[262,264,265] In addition, IL-15 promotes the upregulation of the prosurvival molecule Bcl-2, critical for long-term T_{RM} cells persistence.[159] Finally, T_{RM} development in some tissues appears independent of these two cytokines, such as the TGFβ-independent T_{RM} formation in the intestinal mucosa.[270] It is likely that other cytokines, such as IL-33 and TNFα, which have variously been implicated in T_{RM} cell development,[243,257,271] play a dominant role in such situations. Overall, although T_{RM} cells share many common features associated with their property of long-term residence, their development and survival appear tuned to elements found in their localized environment.

THE CONTRIBUTION OF T CELL MEMORY TO PROTECTIVE IMMUNITY

The capacity of memory CD8+ T cells to afford durable protective immunity rests with their ability to kill infected target cells, to suppress pathogen replication, as well as to initiate and perturbate tissue inflammation to curb local spread of infection.[7] These effector functions entail the directed secretion of classical mediators of cell-dependent cytotoxicity toward target cells, such as perforin and granzymes; surface expression of death receptor ligands, such as Fas-L; as well as the secretion of a broad variety of cytokines and chemokines, including IFNγ, TNFα, and CCL5.[7] Furthermore, memory cells can undergo a marked proliferative burst upon reinfection, which generates secondary effector cells that can promote infection resolution in case pathogen eradication cannot be achieved immediately (Fig. 30.2A). Although the initial delineation between classical T_{EM} and T_{CM} cells emphasized distinctions in function and migration, both T_{EM} and T_{CM} cells can nevertheless undergo substantial recall expansion and exert rapid effector functions. Thus, both subsets can protect from blood-borne and/or systemic infections with viral and bacterial pathogens, although their relative contribution may depend on their preferential microanatomical distribution relative to the route of infection, target cell tropism of the attacking pathogen, and the overall kinetics of infection.[89,131,137,272,273] Separately, effector-like cells with a distinct $KLRG1^{hi}$ $CD27^{lo}$ $CD127^{int}$ $T\text{-}bet^{hi}$ $Eomes^{lo}$ phenotype can persist for several months until they eventually disappear. Such cells have been shown to be important for immunity against reinfection relatively early after primary infection but, interestingly, can become the predominant subset in certain prime-boost immunization settings.[96]

Although systemic infections can engage all manners of recirculating memory subsets, those pathogens that target peripheral or nonlymphoid tissues require the presence of T_{RM} cells for optimal protection (Fig. 30.4). Without the need for recruitment from the blood, T_{RM} cells can immediately recognize infected cells or pathogen-derived antigens presented by dendritic cells.[201,211,274] In addition to directly eliminating infected target cells, T_{RM} cells also release cytokines and chemokines that trigger a cascade of antimicrobial innate and adaptive effector mechanisms, including the induction of the interferon-induced transmembrane protein, IFITM3, in surrounding tissue cells.[213] As part of this response, T_{RM} cells also promote the recruitment of various types of innate immune cells and recirculating memory T cells. As such, T_{RM} cells act as important peripheral immune sentinels whose reactivation serves the dual purpose of target cell elimination and orchestration of inflammation,[100,212,213] the latter of which is referred to as the "sensing and alarm" function of T_{RM} cells.[100] As a final element in the local cascade, both locally proliferating T_{RM} cells and recruited recirculating memory cells can give rise to new T_{RM} cells, thereby bolstering the peripheral memory pool for future challenges.[233,248,251,275]

The aforementioned examples focused on acute infections where complete pathogen eradication is achieved either immediately or over the course of a few weeks. By contrast, other pathogens, such as HIV, herpesviruses, and hepatitis viruses, can establish persisting infections. Although persisting pathogens establish reservoirs that escape complete eradication, some of these are nevertheless constantly kept in check by antigen-specific T cells in what is referred to as a dynamic but metastable state of "host-pathogen equilibrium."[276] As shown in Figure 30.2B-D, pathogen persistence, and consequently chronic or intermittent antigenic stimulation, can have marked consequences for memory CD8+ T cell dynamics. Chronically active infections, including those with HIV, hepatitis virus, or the experimental model of LCMV clone 13 infection, result in a progressive decline in CD8+ T cell functionality, otherwise known as T cell "exhaustion" (Fig. 30.2B).[277] This involves progressive epigenetic, transcriptional, and metabolic alterations dependent on the transcriptional regulator Tox[278-280] as well as ligation of inhibitory checkpoint receptors, such as PD-1, Lag-3, and Tim3, whose expression is driven by chronic antigen stimulation and exposure to proinflammatory cytokines.[277] Despite this functional restraint, "exhausted" T cells nevertheless contribute to virus control and thereby promote the stalemate between the host and pathogen.[281,282] Accordingly, their irreversibly altered functional status can be interpreted as an adaptation to avoid potentially detrimental immunopathology.[282,283] As mentioned earlier, a distinct set of long-lived and functionally muted PD-1+ Tim3− Tox+ CD8+ T cells in chronic infection serves to replenish the pool of terminally exhausted effector cells. Echoing this partial stem cell–like and memory behavior, these precursors of exhausted T cells depend on transcriptional regulators that also instruct stemness and developmental plasticity in conventional memory cells, including Tcf1, Id3, and Bach-2.[165-171,181] Such cells have emerged as major targets for checkpoint blockade immunotherapies in chronic infection and cancer.[165,170,171]

Other persisting infections, such as those with herpes or some hepatitis viruses, are controlled more tightly by the immune system and, as such, are mostly clinically silent and cause overt disease mainly on immune suppression or in immunocompromised individuals. One such example is the latent infection with CMV, which is typically characterized by a state of viral latency with only sporadic low-level viral replication. As a consequence of intermittent antigen stimulation, a proportion of the CMV-specific pool of CD8+ memory T cells can accumulate to high numbers over time whereby these "inflated" memory cells retain a KLRG1+ CD127− CD62L− CD27− T_{EM}-like phenotype with robust functionality (Fig. 30.2C).[284,285] Some persisting pathogens target specific anatomical niches, including the central nervous system, the gastrointestinal tract, or the liver, and often require intermittent productive replication within epithelial tissues to facilitate transmission. Immune surveillance in such tissues is driven by a pool of T_{RM} cells that is generated during primary infection and then further shaped in size and specificity during the period of pathogen reemergence.[201,244,286-291] T_{RM} cells specific to herpes simplex virus (HSV), for instance, occupy sensory ganglia containing persistently HSV-infected neurons and also reside at the dermal-epidermal junction in skin, where they are in direct proximity to sensory nerve endings that connect the infected ganglia with the skin and genital mucosa.[201,289,292-294] Upon recurrent viral reactivation from latency, these T_{RM} cells can thus act to immediately extinguish the earliest rounds of virus replication in ganglia and barrier tissues, which at its most efficient, results in asymptomatic HSV shedding as opposed to full-blown lesion development.[201] Indeed, such episodes of HSV shedding are frequent and can be controlled within hours,[295] meaning that this type of memory response is likely restricted to temporal and highly localized T_{RM} reactivation without the involvement of recirculating memory T cells (Fig. 30.2D). Conversely, full-blown lesion development and recurrent herpetic disease upon reactivation in some patients have been linked with a reduction or temporally impaired functionality of HSV-specific CD8+ T_{RM} cells.[201] In a similar manner, T_{RM} cells contribute to control of another persistent herpesvirus, EBV. Although EBV establishes a latent infection in memory B cells that recirculate around the body, sporadic reactivation can result in productive infection of tonsillar epithelium as a means of transmission to other individuals. Much like in HSV infection, EBV-specific CD8+ CD69+ CD103+ T_{RM} cells reside within the tonsillar epithelium as the critical tissue for productive virus replication,[244,287,288] which may explain why intermittent low-level EBV shedding from the oropharyngeal epithelium usually does not result in overt tonsilitis.

Thus, in chronic and persisting infections, T_{RM} cells are focused on promoting virus containment, most likely because these pathogens have adapted to the host to such an extent that elimination becomes impossible. As such, T_{RM} cells promote the maintenance of sufficient and durable control while avoiding overt immune reactivation, immunopathology, or functional exhaustion.[240] Of note, CD8+ CD69+ CD103+ T_{RM} cells exert a related function in containing small cancer lesions that may never grow out but instead are controlled in a state of "cancer immune equilibrium."[296,297] Consistent with this, the accumulation of tumor-associated T_{RM} cells correlates with improved clinical outcomes in patients with cancer and has been linked with the success of checkpoint blockade immunotherapy.[297-299] Finally, T_{RM} cells are also generated in response to colonization with commensal microbiota. *Staphylococcus epidermidis*, for instance, is part of the normal microflora in humans but nevertheless can cause clinically relevant diseases when immune control at barrier tissues is compromised. Colonization of skin with *Staphylococcus epidermidis* results in the generation of epidermal IL-17-producing CD8+ CD69+ CD103+ T_{RM} cells, which can protect from invasion but, likely driven by intermittent antigen exposure, also establishes a regional state of alert that promotes cross-protection from unrelated microorganisms such as *Candida albicans*.[300] In summary, these examples illustrate the vastly different modes of immune surveillance and protection afforded by memory T cells. As such, they emphasize a level of specialization by subsets of recirculating and resident memory CD8+ T cells for the control of anatomically diverse acute and persisting threats.

CONCLUSIONS

Research over the past 3 decades has firmly established the concept of durable and protective T cell memory. This key aspect of T cell immunity is characterized by a vast diversity in epigenetic, transcriptional, metabolic, and functional states that appear to be tailored to specific needs in dealing with recurring and persisting infections and cancer. Although it is evident that the current, rather firm categorization of memory T cells into distinct subsets fails to portray this diversity, it nevertheless helped to define key functional properties required for immune protection against different types of threats. Such memory properties are harnessed clinically with vaccination and cancer immunotherapies, such as checkpoint blockade or adoptive T cell transfer. Our growing appreciation and understanding of the molecular pathways and regulatory factors underpinning the cardinal features of T cell memory will guide future refinement of such therapies and will open new avenues for the inception of novel and potentially more targeted approaches to the modulation of T cell immunity.

ACKNOWLEDGMENTS

T.G. has received fellowship support from the Sylvia and Charles Viertel Charitable and currently holds a Leadership Fellowship from the National Health and Medical Research Council (NHMRC; APP1194482).

REFERENCES

1. Blattman JN, Antia R, Sourdive DJ, et al. Estimating the precursor frequency of naive antigen-specific CD8 T cells. *J Exp Med.* 2002;195(5):657-664.
2. Jenkins MK, Chu HH, McLachlan JB, Moon JJ. On the composition of the preimmune repertoire of T cells specific for Peptide-major histocompatibility complex ligands. *Annu Rev Immunol.* 2010;28:275-294.
3. Obar JJ, Khanna KM, Lefrancois L. Endogenous naive CD8+ T cell precursor frequency regulates primary and memory responses to infection. *Immunity.* 2008;28(6):859-869.
4. Masopust D, Schenkel JM. The integration of T cell migration, differentiation and function. *Nat Rev Immunol.* 2013;13(5):309-320.
5. Williams MA, Bevan MJ. Effector and memory CTL differentiation. *Annu Rev Immunol.* 2007;25:171-192.
6. Woodland DL, Kohlmeier JE. Migration, maintenance and recall of memory T cells in peripheral tissues. *Nat Rev Immunol.* 2009;9(3):153-161.
7. Harty JT, Tvinnereim AR, White DW. CD8+ T cell effector mechanisms in resistance to infection. *Annu Rev Immunol.* 2000;18:275-308.
8. Kaech SM, Wherry EJ, Ahmed R. Effector and memory T-cell differentiation: implications for vaccine development. *Nat Rev Immunol.* 2002;2(4):251-262.
9. Harty JT, Badovinac VP. Shaping and reshaping CD8+ T-cell memory. *Nat Rev Immunol.* 2008;8(2):107-119.
10. Heath WR, Carbone FR. Dendritic cell subsets in primary and secondary T cell responses at body surfaces. *Nat Immunol.* 2009;10(12):1237-1244.
11. Hildner K, Edelson BT, Purtha WE, et al. Batf3 deficiency reveals a critical role for CD8alpha+ dendritic cells in cytotoxic T cell immunity. *Science.* 2008;322(5904):1097-1100.
12. Allan RS, Smith CM, Belz GT, et al. Epidermal viral immunity induced by CD8alpha+ dendritic cells but not by Langerhans cells. *Science.* 2003;301(5641):1925-1928.
13. Bedoui S, Whitney PG, Waithman J, et al. Cross-presentation of viral and self antigens by skin-derived CD103+ dendritic cells. *Nat Immunol.* 2009;10(5):488-495.
14. Allan RS, Waithman J, Bedoui S, et al. Migratory dendritic cells transfer antigen to a lymph node-resident dendritic cell population for efficient CTL priming. *Immunity.* 2006;25(1):153-162.
15. Diao H, Pipkin M. Stability and flexibility in chromatin structure and transcription underlies memory CD8 T-cell differentiation. *F1000Res.* 2019;8:F1000.
16. Chen Y, Zander R, Khatun A, Schauder DM, Cui W. Transcriptional and epigenetic regulation of effector and memory CD8 T cell differentiation. *Front Immunol.* 2018;9:2826.
17. Tough DF, Rioja I, Modis LK, Prinjha RK. Epigenetic regulation of T cell memory: recalling therapeutic implications. *Trends Immunol.* 2020;41(1):29-45.
18. Wang D, Diao H, Getzler AJ, et al. The transcription factor Runx3 establishes chromatin accessibility of cis-regulatory landscapes that drive memory cytotoxic T lymphocyte formation. *Immunity.* 2018;48(4):659.e6-674.e6.
19. Kakaradov B, Arsenio J, Widjaja CE, et al. Early transcriptional and epigenetic regulation of CD8(+) T cell differentiation revealed by single-cell RNA sequencing. *Nat Immunol.* 2017;18(4):422-432.
20. Kurachi M, Barnitz RA, Yosef N, et al. The transcription factor BATF operates as an essential differentiation checkpoint in early effector CD8+ T cells. *Nat Immunol.* 2014;15(4):373-383.
21. Kallies A, Xin A, Belz GT, Nutt SL. Blimp-1 transcription factor is required for the differentiation of effector CD8(+) T cells and memory responses. *Immunity.* 2009;31(2):283-295.
22. Rutishauser RL, Martins GA, Kalachikov S, et al. Transcriptional repressor Blimp-1 promotes CD8(+) T cell terminal differentiation and represses the acquisition of central memory T cell properties. *Immunity.* 2009;31(2):296-308.
23. Joshi NS, Cui W, Chandele A, et al. Inflammation directs memory precursor and short-lived effector CD8(+) T cell fates via the graded expression of T-bet transcription factor. *Immunity.* 2007;27(2):281-295.
24. Pearce EL, Mullen AC, Martins GA, et al. Control of effector CD8+ T cell function by the transcription factor Eomesodermin. *Science.* 2003;302(5647):1041-1043.
25. Intlekofer AM, Takemoto N, Wherry EJ, et al. Effector and memory CD8+ T cell fate coupled by T-bet and eomesodermin. *Nat Immunol.* 2005;6(12):1236-1244.
26. Yao S, Buzo BF, Pham D, et al. Interferon regulatory factor 4 sustains CD8(+) T cell expansion and effector differentiation. *Immunity.* 2013;39(5):833-845.
27. Russ BE, Olshanksy M, Smallwood HS, et al. Distinct epigenetic signatures delineate transcriptional programs during virus-specific CD8(+) T cell differentiation. *Immunity.* 2014;41(5):853-865.
28. Scott-Browne JP, Lopez-Moyado IF, Trifari S, et al. Dynamic changes in chromatin accessibility occur in CD8(+) T cells responding to viral infection. *Immunity.* 2016;45(6):1327-1340.
29. Yu B, Zhang K, Milner JJ, et al. Epigenetic landscapes reveal transcription factors that regulate CD8(+) T cell differentiation. *Nat Immunol.* 2017;18(5):573-582.
30. Kaech SM, Cui W. Transcriptional control of effector and memory CD8+ T cell differentiation. *Nat Rev Immunol.* 2012;12(11):749-761.
31. Chang JT, Wherry EJ, Goldrath AW. Molecular regulation of effector and memory T cell differentiation. *Nat Immunol.* 2014;15(12):1104-1115.
32. Geltink RIK, Kyle RL, Pearce EL. Unraveling the complex interplay between T cell metabolism and function. *Annu Rev Immunol.* 2018;36:461-488.
33. Man K, Kallies A. Synchronizing transcriptional control of T cell metabolism and function. *Nat Rev Immunol.* 2015;15(9):574-584.
34. Maciolek JA, Pasternak JA, Wilson HL. Metabolism of activated T lymphocytes. *Curr Opin Immunol.* 2014;27:60-74.
35. Man K, Miasari M, Shi W, et al. The transcription factor IRF4 is essential for TCR affinity-mediated metabolic programming and clonal expansion of T cells. *Nat Immunol.* 2013;14(11):1155-1165.
36. Finlay DK, Rosenzweig E, Sinclair LV, et al. PDK1 regulation of mTOR and hypoxia-inducible factor 1 integrate metabolism and migration of CD8+ T cells. *J Exp Med.* 2012;209(13):2441-2453.
37. Wang R, Dillon CP, Shi LZ, et al. The transcription factor Myc controls metabolic reprogramming upon T lymphocyte activation. *Immunity.* 2011;35(6):871-882.
38. Esensten JH, Helou YA, Chopra G, Weiss A, Bluestone JA. CD28 costimulation: from mechanism to therapy. *Immunity.* 2016;44(5):973-988.
39. Klein Geltink RI, O'Sullivan D, Corrado M, et al. Mitochondrial priming by CD28. *Cell.* 2017;171(2):385.e11-397.e11.
40. Frauwirth KA, Riley JL, Harris MH, et al. The CD28 signaling pathway regulates glucose metabolism. *Immunity.* 2002;16(6):769-777.
41. Buck MD, O'Sullivan D, Klein Geltink RI, et al. Mitochondrial dynamics controls T cell fate through metabolic programming. *Cell.* 2016;166(1):63-76.
42. van der Windt GJ, Everts B, Chang CH, et al. Mitochondrial respiratory capacity is a critical regulator of CD8+ T cell memory development. *Immunity.* 2012;36(1):68-78.
43. van der Windt GJ, O'Sullivan D, Everts B, et al. CD8 memory T cells have a bioenergetic advantage that underlies their rapid recall ability. *Proc Natl Acad Sci U S A.* 2013;110(35):14336-14341.
44. Engelhardt JJ, Sullivan TJ, Allison JP. CTLA-4 overexpression inhibits T cell responses through a CD28-B7-dependent mechanism. *J Immunol.* 2006;177(2):1052-1061.
45. Schildberg FA, Klein SR, Freeman GJ, Sharpe AH. Coinhibitory pathways in the B7-CD28 ligand-receptor family. *Immunity.* 2016;44(5):955-972.
46. Watts TH. TNF/TNFR family members in costimulation of T cell responses. *Annu Rev Immunol.* 2005;23:23-68.
47. Schoenberger SP, Toes RE, van der Voort EI, Offringa R, Melief CJ. T-cell help for cytotoxic T lymphocytes is mediated by CD40-CD40L interactions. *Nature.* 1998;393(6684):480-483.
48. Bennett SR, Carbone FR, Karamalis F, Flavell RA, Miller JF, Heath WR. Help for cytotoxic-T-cell responses is mediated by CD40 signalling. *Nature.* 1998;393(6684):478-480.
49. Ridge JP, Di Rosa F, Matzinger P. A conditioned dendritic cell can be a temporal bridge between a CD4+ T-helper and a T-killer cell. *Nature.* 1998;393(6684):474-478.
50. Reis e Sousa C. Dendritic cells as sensors of infection. *Immunity.* 2001;14(5):495-498.
51. Joffre O, Nolte MA, Sporri R, Reis e Sousa C. Inflammatory signals in dendritic cell activation and the induction of adaptive immunity. *Immunol Rev.* 2009;227(1):234-247.
52. Bedoui S, Heath WR, Mueller SN. CD4(+) T-cell help amplifies innate signals for primary CD8(+) T-cell immunity. *Immunol Rev.* 2016;272(1):52-64.
53. Reis e Sousa C. Dendritic cells in a mature age. *Nat Rev Immunol.* 2006;6(6):476-483.
54. Omilusik KD, Shaw LA, Goldrath AW. Remembering one's ID/E-ntity: E/ID protein regulation of T cell memory. *Curr Opin Immunol.* 2013;25(5):660-666.
55. Danilo M, Chennupati V, Silva JG, Siegert S, Held W. Suppression of Tcf1 by inflammatory cytokines facilitates effector CD8 T cell differentiation. *Cell Rep.* 2018;22(8):2107-2117.
56. Obar JJ, Jellison ER, Sheridan BS, et al. Pathogen-induced inflammatory environment controls effector and memory CD8+ T cell differentiation. *J Immunol.* 2011;187(10):4967-4978.
57. Kalia V, Sarkar S, Subramaniam S, Haining WN, Smith KA, Ahmed R. Prolonged interleukin-2Rα expression on virus-specific CD8+ T cells favors terminal-effector differentiation *in vivo*. *Immunity.* 2010;32(1):91-103.
58. Pipkin ME, Sacks JA, Cruz-Guilloty F, Lichtenheld MG, Bevan MJ, Rao A. Interleukin-2 and inflammation induce distinct transcriptional programs that promote the differentiation of effector cytolytic T cells. *Immunity.* 2010;32(1):79-90.
59. Kolumam GA, Thomas S, Thompson LJ, Sprent J, Murali-Krishna K. Type I interferons act directly on CD8 T cells to allow clonal expansion and memory formation in response to viral infection. *J Exp Med.* 2005;202(5):637-650.
60. Haring JS, Badovinac VP, Harty JT. Inflaming the CD8+ T cell response. *Immunity.* 2006;25(1):19-29.
61. Badovinac VP, Porter BB, Harty JT. CD8+ T cell contraction is controlled by early inflammation. *Nat Immunol.* 2004;5(8):809-817.
62. Badovinac VP, Tvinnereim AR, Harty JT. Regulation of antigen-specific CD8+ T cell homeostasis by perforin and interferon-gamma. *Science.* 2000;290(5495):1354-1358.
63. Sanjabi S, Mosaheb MM, Flavell RA. Opposing effects of TGF-beta and IL-15 cytokines control the number of short-lived effector CD8+ T cells. *Immunity.* 2009;31(1):131-144.
64. Weant AE, Michalek RD, Khan IU, Holbrook BC, Willingham MC, Grayson JM. Apoptosis regulators Bim and Fas function concurrently to control autoimmunity and CD8+ T cell contraction. *Immunity.* 2008;28(2):218-230.
65. Cui W, Liu Y, Weinstein JS, Craft J, Kaech SM. An interleukin-21-interleukin-10-STAT3 pathway is critical for functional maturation of memory CD8+ T cells. *Immunity.* 2011;35(5):792-805.
66. Cyster JG, Schwab SR. Sphingosine-1-phosphate and lymphocyte egress from lymphoid organs. *Annu Rev Immunol.* 2012;30:69-94.
67. Agace WW. Tissue-tropic effector T cells: generation and targeting opportunities. *Nat Rev Immunol.* 2006;6(9):682-692.

68. Griffith JW, Sokol CL, Luster AD. Chemokines and chemokine receptors: positioning cells for host defense and immunity. *Annu Rev Immunol.* 2014;32:659-702.
69. Mora JR, Cheng G, Picarella D, Briskin M, Buchanan N, von Andrian UH. Reciprocal and dynamic control of CD8 T cell homing by dendritic cells from skin- and gut-associated lymphoid tissues. *J Exp Med.* 2005;201(2):303-316.
70. Campbell JJ, Haraldsen G, Pan J, et al. The chemokine receptor CCR4 in vascular recognition by cutaneous but not intestinal memory T cells. *Nature.* 1999;400(6746):776-780.
71. Mora JR, Iwata M, von Andrian UH. Vitamin effects on the immune system: vitamins A and D take centre stage. *Nat Rev Immunol.* 2008;8(9):685-698.
72. Sigmundsdottir H, Pan J, Debes GF, et al. DCs metabolize sunlight-induced vitamin D3 to "program" T cell attraction to the epidermal chemokine CCL27. *Nat Immunol.* 2007;8(3):285-293.
73. Svensson M, Marsal J, Ericsson A, et al. CCL25 mediates the localization of recently activated CD8alphabeta(+) lymphocytes to the small-intestinal mucosa. *The Journal of clinical investigation.* 2002;110(8):1113-1121.
74. Iwata M, Hirakiyama A, Eshima Y, Kagechika H, Kato C, Song SY. Retinoic acid imprints gut-homing specificity on T cells. *Immunity.* 2004;21(4):527-538.
75. Nakanishi Y, Lu B, Gerard C, Iwasaki A. CD8(+) T lymphocyte mobilization to virus-infected tissue requires CD4(+) T-cell help. *Nature.* 2009;462(7272):510-513.
76. Ariotti S, Beltman JB, Borsje R, et al. Subtle CXCR3-dependent chemotaxis of CTLs within infected tissue allows efficient target localization. *J Immunol.* 2015;195(11):5285-5295.
77. Macleod BL, Bedoui S, Hor JL, et al. Distinct APC subtypes drive spatially segregated CD4+ and CD8+ T-cell effector activity during skin infection with HSV-1. *PLoS Pathogens.* 2014;10(8):e1004303.
78. Wherry EJ, Blattman JN, Murali-Krishna K, van der Most R, Ahmed R. Viral persistence alters CD8 T-cell immunodominance and tissue distribution and results in distinct stages of functional impairment. *J Virol.* 2003;77(8):4911-4927.
79. Singh A, Suresh M. A role for TNF in limiting the duration of CTL effector phase and magnitude of CD8 T cell memory. *J Leukoc Biol.* 2007;82(5):1201-1211.
80. Hughes PD, Belz GT, Fortner KA, Budd RC, Strasser A, Bouillet P. Apoptosis regulators Fas and Bim cooperate in shutdown of chronic immune responses and prevention of autoimmunity. *Immunity.* 2008;28(2):197-205.
81. Suresh M, Singh A, Fischer C. Role of tumor necrosis factor receptors in regulating CD8 T-cell responses during acute lymphocytic choriomeningitis virus infection. *J Virol.* 2005;79(1):202-213.
82. Badovinac VP, Messingham KA, Griffith TS, Harty JT. TRAIL deficiency delays, but does not prevent, erosion in the quality of "helpless" memory CD8 T cells. *J Immunol.* 2006;177(2):999-1006.
83. Janssen EM, Droin NM, Lemmens EE, et al. CD4+ T-cell help controls CD8+ T-cell memory via TRAIL-mediated activation-induced cell death. *Nature.* 2005;434(7029):88-93.
84. Marsden VS, Strasser A. Control of apoptosis in the immune system: bcl-2, BH3-only proteins and more. *Annu Rev Immunol.* 2003;21:71-105.
85. Strasser A. The role of BH3-only proteins in the immune system. *Nat Rev Immunol.* 2005;5(3):189-200.
86. Hildeman DA, Zhu Y, Mitchell TC, et al. Activated T cell death *in vivo* mediated by proapoptotic bcl-2 family member Bim. *Immunity.* 2002;16(6):759-767.
87. Strasser A, Pellegrini M. T-lymphocyte death during shutdown of an immune response. *Trends Immunol.* 2004;25(11):610-615.
88. Pellegrini M, Belz G, Bouillet P, Strasser A. Shutdown of an acute T cell immune response to viral infection is mediated by the proapoptotic Bcl-2 homology 3-only protein Bim. *Proc Natl Acad Sci U S A.* 2003;100(24):14175-14180.
89. Huster KM, Busch V, Schiemann M, et al. Selective expression of IL-7 receptor on memory T cells identifies early CD40L-dependent generation of distinct CD8+ memory T cell subsets. *Proc Natl Acad Sci U S A.* 2004;101(15):5610-5615.
90. Kaech SM, Tan JT, Wherry EJ, Konieczny BT, Surh CD, Ahmed R. Selective expression of the interleukin 7 receptor identifies effector CD8 T cells that give rise to long-lived memory cells. *Nat Immunol.* 2003;4(12):1191-1198.
91. Obar JJ, Lefrancois L. Early signals during CD8 T cell priming regulate the generation of central memory cells. *J Immunol.* 2010;185(1):263-272.
92. Gerlach C, Moseman EA, Loughhead SM, et al. The chemokine receptor CX3CR1 defines three antigen-experienced CD8 T cell subsets with distinct roles in immune surveillance and homeostasis. *Immunity.* 2016;45(6):1270-1284.
93. Bottcher JP, Beyer M, Meissner F, et al. Functional classification of memory CD8(+) T cells by CX3CR1 expression. *Nat Commun.* 2015;6:8306.
94. Harrington LE, Galvan M, Baum LG, Altman JD, Ahmed R. Differentiating between memory and effector CD8 T cells by altered expression of cell surface O-glycans. *J Exp Med.* 2000;191(7):1241-1246.
95. Croom HA, Denton AE, Valkenburg SA, et al. Memory precursor phenotype of CD8+ T cells reflects early antigenic experience rather than memory numbers in a model of localized acute influenza infection. *Eur J Immunol.* 2011;41(3):682-693.
96. Olson JA, McDonald-Hyman C, Jameson SC, Hamilton SE. Effector-like CD8(+) T cells in the memory population mediate potent protective immunity. *Immunity.* 2013;38(6):1250-1260.
97. Herndler-Brandstetter D, Ishigame H, Shinnakasu R, et al. KLRG1(+) effector CD8(+) T cells lose KLRG1, differentiate into all memory T cell lineages, and convey enhanced protective immunity. *Immunity.* 2018;48(4):716.e8-729.e8.
98. Yang CY, Best JA, Knell J, et al. The transcriptional regulators Id2 and Id3 control the formation of distinct memory CD8+ T cell subsets. *Nat Immunol.* 2011;12(12):1221-1229.
99. Jeannet G, Boudousquie C, Gardiol N, Kang J, Huelsken J, Held W. Essential role of the Wnt pathway effector Tcf-1 for the establishment of functional CD8 T cell memory. *Proc Natl Acad Sci U S A.* 2010;107(21):9777-9782.
100. Schenkel JM, Fraser KA, Vezys V, Masopust D. Sensing and alarm function of resident memory CD8(+) T cells. *Nat Immunol.* 2013;14(5):509-513.
101. Buchholz VR, Schumacher TN, Busch DH. T cell fate at the single-cell level. *Annu Rev Immunol.* 2016;34:65-92.
102. Smith NL, Patel RK, Reynaldi A, et al. Developmental origin governs CD8(+) T cell fate decisions during infection. *Cell.* 2018;174(1):117.e14-130.e14.
103. Chang JT, Palanivel VR, Kinjyo I, et al. Asymmetric T lymphocyte division in the initiation of adaptive immune responses. *Science.* 2007;315(5819):1687-1691.
104. Chang JT, Ciocca ML, Kinjyo I, et al. Asymmetric proteasome segregation as a mechanism for unequal partitioning of the transcription factor T-bet during T lymphocyte division. *Immunity.* 2011;34(4):492-504.
105. Arsenio J, Kakaradov B, Metz PJ, Kim SH, Yeo GW, Chang JT. Early specification of CD8+ T lymphocyte fates during adaptive immunity revealed by single-cell gene-expression analyses. *Nat Immunol.* 2014;15(4):365-372.
106. Verbist KC, Guy CS, Milasta S, et al. Metabolic maintenance of cell asymmetry following division in activated T lymphocytes. *Nature.* 2016;532(7599):389-393.
107. Pollizzi KN, Sun IH, Patel CH, et al. Asymmetric inheritance of mTORC1 kinase activity during division dictates CD8(+) T cell differentiation. *Nat Immunol.* 2016;17(6):704-711.
108. Stemberger C, Huster KM, Koffler M, et al. A single naive CD8+ T cell precursor can develop into diverse effector and memory subsets. *Immunity.* 2007;27(6):985-997.
109. Gerlach C, Rohr JC, Perie L, et al. Heterogeneous differentiation patterns of individual CD8+ T cells. *Science.* 2013;340(6132):635-639.
110. Buchholz VR, Flossdorf M, Hensel I, et al. Disparate individual fates compose robust CD8+ T cell immunity. *Science.* 2013;340(6132):630-635.
111. Kretschmer L, Flossdorf M, Mir J, et al. Differential expansion of T central memory precursor and effector subsets is regulated by division speed. *Nat Commun.* 2020;11(1):113.
112. Pais Ferreira D, Silva JG, Wyss T, et al. Central memory CD8(+) T cells derive from stem-like Tcf7(hi) effector cells in the absence of cytotoxic differentiation. *Immunity.* 2020;53(5):985-1000 e11.
113. Bannard O, Kraman M, Fearon DT. Secondary replicative function of CD8+ T cells that had developed an effector phenotype. *Science.* 2009;323(5913):505-509.
114. Akondy RS, Fitch M, Edupuganti S, et al. Origin and differentiation of human memory CD8 T cells after vaccination. *Nature.* 2017;552(7685):362-367.
115. Youngblood B, Hale JS, Kissick HT, et al. Effector CD8 T cells dedifferentiate into long-lived memory cells. *Nature.* 2017;552(7685):404-409.
116. Appay V, Dunbar PR, Callan M, et al. Memory CD8+ T cells vary in differentiation phenotype in different persistent virus infections. *Nat Med.* 2002;8(4):379-385.
117. Gillespie GM, Wills MR, Appay V, et al. Functional heterogeneity and high frequencies of cytomegalovirus-specific CD8(+) T lymphocytes in healthy seropositive donors. *J Virol.* 2000;74(17):8140-8150.
118. Hislop AD, Annels NE, Gudgeon NH, Leese AM, Rickinson AB. Epitope-specific evolution of human CD8(+) T cell responses from primary to persistent phases of Epstein-Barr virus infection. *J Exp Med.* 2002;195(7):893-905.
119. Isa A, Kasprowicz V, Norbeck O, et al. Prolonged activation of virus-specific CD8+T cells after acute B19 infection. *PLoS Med.* 2005;2(12):e343.
120. Akondy RS, Monson ND, Miller JD, et al. The yellow fever virus vaccine induces a broad and polyfunctional human memory CD8+ T cell response. *J Immunol.* 2009;183(12):7919-7930.
121. Jameson SC, Lee YJ, Hogquist KA. Innate memory T cells. *Adv Immunol.* 2015;126:173-213.
122. White JT, Cross EW, Kedl RM. Antigen-inexperienced memory CD8(+) T cells: where they come from and why we need them. *Nat Rev Immunol.* 2017;17(6):391-400.
123. White JT, Cross EW, Burchill MA, et al. Virtual memory T cells develop and mediate bystander protective immunity in an IL-15-dependent manner. *Nat Commun.* 2016;7:11291.
124. Gattinoni L, Lugli E, Ji Y, et al. A human memory T cell subset with stem cell-like properties. *Nat Med.* 2011;17(10):1290-1297.
125. Gattinoni L, Speiser DE, Lichterfeld M, Bonini C. T memory stem cells in health and disease. *Nat Med.* 2017;23(1):18-27.
126. Jameson SC, Masopust D. Understanding subset diversity in T cell memory. *Immunity.* 2018;48(2):214-226.
127. Omilusik KD, Goldrath AW. Remembering to remember: T cell memory maintenance and plasticity. *Curr Opin Immunol.* 2019;58:89-97.
128. Mueller SN, Gebhardt T, Carbone FR, Heath WR. Memory T cell subsets, migration patterns, and tissue residence. *Annu Rev Immunol.* 2013;31:137-161.
129. Wong MT, Ong DE, Lim FS, et al. A high-dimensional atlas of human T cell diversity reveals tissue-specific trafficking and cytokine signatures. *Immunity.* 2016;45(2):442-456.
130. Milner JJ, Toma C, He Z, et al. Heterogenous populations of tissue-resident CD8(+) T cells are generated in response to infection and malignancy. *Immunity.* 2020;52(5):808-824 e7.
131. Wherry EJ, Teichgraber V, Becker TC, et al. Lineage relationship and protective immunity of memory CD8 T cell subsets. *Nat Immunol.* 2003;4(3):225-234.
132. Behr FM, Parga-Vidal L, Kragten NAM, et al. Tissue-resident memory CD8(+) T cells shape local and systemic secondary T cell responses. *Nat Immunol.* 2020;21(9):1070-1081.
133. Fonseca R, Beura LK, Quarnstrom CF, et al. Developmental plasticity allows outside-in immune responses by resident memory T cells. *Nat Immunol.* 2020;21(4):412-421.
134. Beura LK, Wijeyesinghe S, Thompson EA, et al. T cells in nonlymphoid tissues give rise to lymph-node-resident memory T cells. *Immunity.* 2018;48(2):327.e5-338.e5.
135. Masopust D, Ha SJ, Vezys V, Ahmed R. Stimulation history dictates memory CD8 T cell phenotype: implications for prime-boost vaccination. *J Immunol.* 2006;177(2):831-839.
136. Wirth TC, Xue HH, Rai D, et al. Repetitive antigen stimulation induces stepwise transcriptome diversification but preserves a core signature of memory CD8(+) T cell differentiation. *Immunity.* 2010;33(1):128-140.
137. Nolz JC, Harty JT. Protective capacity of memory CD8+ T cells is dictated by antigen exposure history and nature of the infection. *Immunity.* 2011;34(5):781-793.
138. Schenkel JM, Fraser KA, Casey KA, et al. IL-15-Independent maintenance of tissue-resident and boosted effector memory CD8 T cells. *J Immunol.* 2016;196(9):3920-3926.
139. Jabbari A, Harty JT. Secondary memory CD8+ T cells are more protective but slower to acquire a central-memory phenotype. *J Exp Med.* 2006;203(4):919-932.
140. Rosshart SP, Herz J, Vassallo BG, et al. Laboratory mice born to wild mice have natural microbiota and model human immune responses. *Science.* 2019;365(6452):eaaw4361.
141. Rosshart SP, Vassallo BG, Angeletti D, et al. Wild mouse gut microbiota promotes host fitness and improves disease resistance. *Cell.* 2017;171(5):1015.e13-1028.e13.
142. Reese TA, Bi K, Kambal A, et al. Sequential infection with common pathogens promotes human-like immune gene expression and altered vaccine response. *Cell Host Microbe.* 2016;19(5):713-719.
143. Beura LK, Hamilton SE, Bi K, et al. Normalizing the environment recapitulates adult human immune traits in laboratory mice. *Nature.* 2016;532(7600):512-516.
144. Mehlhop-Williams ER, Bevan MJ. Memory CD8+ T cells exhibit increased antigen threshold requirements for recall proliferation. *J Exp Med.* 2014;211(2):345-356.
145. Zammit DJ, Cauley LS, Pham QM, Lefrancois L. Dendritic cells maximize the memory CD8 T cell response to infection. *Immunity.* 2005;22(5):561-570.
146. Low JS, Farsakoglu Y, Amezcua Vesely MC, et al. Tissue-resident memory T cell reactivation by diverse antigen-presenting cells imparts distinct functional responses. *J Exp Med.* 2020;217(8):e20192291.
147. Hernandez MG, Shen L, Rock KL. CD40 on APCs is needed for optimal programming, maintenance, and recall of CD8+ T cell memory even in the absence of CD4+ T cell help. *J Immunol.* 2008;180(7):4382-4390.
148. Bertram EM, Dawicki W, Sedgmen B, Bramson JL, Lynch DH, Watts TH. A switch in costimulation from CD28 to 4-1BB during primary versus secondary CD8 T cell response to influenza *in vivo*. *J Immunol.* 2004;172(2):981-988.
149. Borowski AB, Boesteanu AC, Mueller YM, et al. Memory CD8+ T cells require CD28 costimulation. *J Immunol.* 2007;179(10):6494-6503.
150. Shedlock DJ, Shen H. Requirement for CD4 T cell help in generating functional CD8 T cell memory. *Science.* 2003;300(5617):337-339.
151. Sun JC, Bevan MJ. Defective CD8 T cell memory following acute infection without CD4 T cell help. *Science.* 2003;300(5617):339-342.
152. Wiehagen KR, Corbo E, Schmidt M, Shin H, Wherry EJ, Maltzman JS. Loss of tonic T-cell receptor signals alters the generation but not the persistence of CD8+ memory T cells. *Blood.* 2010;116(25):5560-5570.
153. Surh CD, Sprent J. Homeostasis of naive and memory T cells. *Immunity.* 2008;29(6):848-862.

154. Judge AD, Zhang X, Fujii H, Surh CD, Sprent J. Interleukin 15 controls both proliferation and survival of a subset of memory-phenotype CD8(+) T cells. *J Exp Med.* 2002;196(7):935-946.
155. Fehniger TA, Suzuki K, Ponnappan A, et al. Fatal leukemia in interleukin 15 transgenic mice follows early expansions in natural killer and memory phenotype CD8+ T cells. *J Exp Med.* 2001;193(2):219-231.
156. Goldrath AW, Sivakumar PV, Glaccum M, et al. Cytokine requirements for acute and Basal homeostatic proliferation of naive and memory CD8+ T cells. *J Exp Med.* 2002;195(12):1515-1522.
157. Kennedy MK, Glaccum M, Brown SN, et al. Reversible defects in natural killer and memory CD8 T cell lineages in interleukin 15-deficient mice. *J Exp Med.* 2000;191(5):771-780.
158. Schluns KS, Kieper WC, Jameson SC, Lefrancois L. Interleukin-7 mediates the homeostasis of naive and memory CD8 T cells *in vivo*. *Nat Immunol.* 2000;1(5):426-432.
159. Mackay LK, Rahimpour A, Ma JZ, et al. The developmental pathway for CD103(+)CD8+ tissue-resident memory T cells of skin. *Nat Immunol.* 2013;14(12):1294-1301.
160. Dubois S, Mariner J, Waldmann TA, Tagaya Y. IL-15Ralpha recycles and presents IL-15 in trans to neighboring cells. *Immunity.* 2002;17(5):537-547.
161. Schluns KS, Klonowski KD, Lefrancois L. Transregulation of memory CD8 T-cell proliferation by IL-15Ralpha+ bone marrow-derived cells. *Blood.* 2004;103(3):988-994.
162. Schenkel JM, Fraser KA, Masopust D. Cutting edge: resident memory CD8 T cells occupy frontline niches in secondary lymphoid organs. *J Immunol.* 2014;192(7):2961-2964.
163. Graef P, Buchholz VR, Stemberger C, et al. Serial transfer of single-cell-derived immunocompetence reveals stemness of CD8(+) central memory T cells. *Immunity.* 2014;41(1):116-126.
164. Bachmann MF, Wolint P, Schwarz K, Jager P, Oxenius A. Functional properties and lineage relationship of CD8+ T cell subsets identified by expression of IL-7 receptor alpha and CD62L. *J Immunol.* 2005;175(7):4686-4696.
165. Im SJ, Hashimoto M, Gerner MY, et al. Defining CD8+ T cells that provide the proliferative burst after PD-1 therapy. *Nature.* 2016;537(7620):417-421.
166. Utzschneider DT, Charmoy M, Chennupati V, et al. T cell factor 1-expressing memory-like CD8(+) T cells sustain the immune response to chronic viral infections. *Immunity.* 2016;45(2):415-427.
167. He R, Hou S, Liu C, et al. Follicular CXCR5− expressing CD8(+) T cells curtail chronic viral infection. *Nature.* 2016;537(7620):412-428.
168. Leong YA, Chen Y, Ong HS, et al. CXCR5(+) follicular cytotoxic T cells control viral infection in B cell follicles. *Nat Immunol.* 2016;17(10):1187-1196.
169. Wu T, Ji Y, Moseman EA, et al. The TCF1-Bcl6 axis counteracts type I interferon to repress exhaustion and maintain T cell stemness. *Sci Immunol.* 2016;1(6):eaai8593.
170. Miller BC, Sen DR, Al Abosy R, et al. Subsets of exhausted CD8(+) T cells differentially mediate tumor control and respond to checkpoint blockade. *Nat Immunol.* 2019;20(3):326-336.
171. Siddiqui I, Schaeuble K, Chennupati V, et al. Intratumoral Tcf1(+) PD-1(+) CD8(+) T cells with stem-like properties promote tumor control in response to vaccination and checkpoint blockade immunotherapy. *Immunity.* 2019;50(1):195. e10-211.e10.
172. Rao RR, Li Q, Gubbels Bupp MR, Shrikant PA. Transcription factor Foxo1 represses T-bet-mediated effector functions and promotes memory CD8(+) T cell differentiation. *Immunity.* 2012;36(3):374-387.
173. Kim MV, Ouyang W, Liao W, Zhang MQ, Li MO. The transcription factor Foxo1 controls central-memory CD8+ T cell responses to infection. *Immunity.* 2013;39(2):286-297.
174. Utzschneider DT, Delpoux A, Wieland D, et al. Active maintenance of T cell memory in acute and chronic viral infection depends on continuous expression of FOXO1. *Cell Reports.* 2018;22(13):3454-3467.
175. Zhou X, Yu S, Zhao DM, Harty JT, Badovinac VP, Xue HH. Differentiation and persistence of memory CD8(+) T cells depend on T cell factor 1. *Immunity.* 2010;33(2):229-240.
176. Gautam S, Fioravanti J, Zhu W, et al. The transcription factor c-Myb regulates CD8(+) T cell stemness and antitumor immunity. *Nat Immunol.* 2019;20(3):337-349.
177. Ichii H, Sakamoto A, Hatano M, et al. Role for Bcl-6 in the generation and maintenance of memory CD8+ T cells. *Nat Immunol.* 2002;3(6):558-563.
178. Guan T, Dominguez CX, Amezquita RA, et al. ZEB1, ZEB2, and the miR-200 family form a counterregulatory network to regulate CD8(+) T cell fates. *J Exp Med.* 2018;215(4):1153-1168.
179. Roychoudhuri R, Clever D, Li P, et al. BACH2 regulates CD8(+) T cell differentiation by controlling access of AP-1 factors to enhancers. *Nat Immunol.* 2016;17(7):851-860.
180. Utzschneider DT, Gabriel SS, Chisanga D, et al. Early precursor T cells establish and propagate T cell exhaustion in chronic infection. *Nat Immunol.* 2020;21(10):1256-1266.
181. Yao C, Lou G, Sun HW, et al. BACH2 enforces the transcriptional and epigenetic programs of stem-like CD8(+) T cells. *Nat Immunol.* 2021;22(3):370-380.
182. Ji Y, Pos Z, Rao M, et al. Repression of the DNA-binding inhibitor Id3 by Blimp-1 limits the formation of memory CD8+ T cells. *Nat Immunol.* 2011;12(12):1230-1237.
183. Dominguez CX, Amezquita RA, Guan T, et al. The transcription factors ZEB2 and T-bet cooperate to program cytotoxic T cell terminal differentiation in response to LCMV viral infection. *J Exp Med.* 2015;212(12):2041-2056.
184. Omilusik KD, Nadjsombati MS, Shaw LA, Yu B, Milner JJ, Goldrath AW. Sustained Id2 regulation of E proteins is required for terminal differentiation of effector CD8(+) T cells. *J Exp Med.* 2018;215(3):773-783.
185. O'Sullivan D, van der Windt GJ, Huang SC, et al. Memory CD8(+) T cells use cell-intrinsic lipolysis to support the metabolic programming necessary for development. *Immunity.* 2014;41(1):75-88.
186. Cui G, Staron MM, Gray SM, et al. IL-7-Induced glycerol transport and TAG synthesis promotes memory CD8+ T cell longevity. *Cell.* 2015;161(4):750-761.
187. Wofford JA, Wieman HL, Jacobs SR, Zhao Y, Rathmell JC. IL-7 promotes Glut1 trafficking and glucose uptake via STAT5-mediated activation of Akt to support T-cell survival. *Blood.* 2008;111(4):2101-2111.
188. Pan Y, Tian T, Park CO, et al. Survival of tissue-resident memory T cells requires exogenous lipid uptake and metabolism. *Nature.* 2017;543(7644):252-256.
189. Frizzell H, Fonseca R, Christo SN, et al. Organ-specific isoform selection of fatty acid-binding proteins in tissue-resident lymphocytes. *Sci Immunol.* 2020;5(46):eaay9283.
190. Phan AT, Doedens AL, Palazon A, et al. Constitutive glycolytic metabolism supports CD8(+) T cell effector memory differentiation during viral infection. *Immunity.* 2016;45(5):1024-1037.
191. Borges da Silva H, Beura LK, Wang H, et al. The purinergic receptor P2RX7 directs metabolic fitness of long-lived memory CD8(+) T cells. *Nature.* 2018;559(7713):264-268.
192. Zimmermann C, Prevost-Blondel A, Blaser C, Pircher H. Kinetics of the response of naive and memory CD8 T cells to antigen: similarities and differences. *Eur J Immunol.* 1999;29(1):284-290.
193. Veiga-Fernandes H, Walter U, Bourgeois C, McLean A, Rocha B. Response of naive and memory CD8+ T cells to antigen stimulation *in vivo*. *Nat Immunol.* 2000;1(1):47-53.
194. Slifka MK, Whitton JL. Functional avidity maturation of CD8(+) T cells without selection of higher affinity TCR. *Nat Immunol.* 2001;2(8):711-717.
195. Berg RE, Cordes CJ, Forman J. Contribution of CD8+ T cells to innate immunity: IFN-gamma secretion induced by IL-12 and IL-18. *Eur J Immunol.* 2002;32(10):2807-2816.
196. Berg RE, Crossley E, Murray S, Forman J. Memory CD8+ T cells provide innate immune protection against *Listeria monocytogenes* in the absence of cognate antigen. *J Exp Med.* 2003;198(10):1583-1593.
197. Kersh EN, Fitzpatrick DR, Murali-Krishna K, et al. Rapid demethylation of the IFN-gamma gene occurs in memory but not naive CD8 T cells. *J Immunol.* 2006;176(7): 4083-4093.
198. Northrop JK, Thomas RM, Wells AD, Shen H. Epigenetic remodeling of the IL-2 and IFN-gamma loci in memory CD8 T cells is influenced by CD4 T cells. *J Immunol.* 2006;177(2):1062-1069.
199. Omilusik KD, Goldrath AW. The origins of memory T cells. *Nature.* 2017;552(7685):337-339.
200. Hombrink P, Helbig C, Backer RA, et al. Programs for the persistence, vigilance and control of human CD8+ lung-resident memory T cells. *Nat Immunol.* 2016;17(12):1467-1478.
201. Zhu J, Peng T, Johnston C, et al. Immune surveillance by CD8αα+ skin-resident T cells in human herpes virus infection. *Nature.* 2013;497(7450):494-497.
202. Kaech SM, Hemby S, Kersh E, Ahmed R. Molecular and functional profiling of memory CD8 T cell differentiation. *Cell.* 2002;111(6):837-851.
203. Masopust D, Vezys V, Wherry EJ, Barber DL, Ahmed R. Cutting edge: gut microenvironment promotes differentiation of a unique memory CD8 T cell population. *J Immunol.* 2006;176(4):2079-2083.
204. Collins N, Jiang X, Zaid A, et al. Skin CD4(+) memory T cells exhibit combined cluster-mediated retention and equilibration with the circulation. *Nat Commun.* 2016;7:11514.
205. Kumar R, Ferez M, Swamy M, et al. Increased sensitivity of antigen-experienced T cells through the enrichment of oligomeric T cell receptor complexes. *Immunity.* 2011;35(3):375-387.
206. Kersh EN, Kaech SM, Onami TM, et al. TCR signal transduction in antigen-specific memory CD8 T cells. *J Immunol.* 2003;170(11):5455-5463.
207. Zehn D, Lee SY, Bevan MJ. Complete but curtailed T-cell response to very low-affinity antigen. *Nature.* 2009;458(7235):211-214.
208. Clark RA, Chong B, Mirchandani N, et al. The vast majority of CLA+ T cells are resident in normal skin. *J Immunol.* 2006;176(7):4431-4439.
209. Steinert EM, Schenkel JM, Fraser KA, et al. Quantifying memory CD8 T cells reveals regionalization of immunosurveillance. *Cell.* 2015;161(4):737-749.
210. Sathaliyawala T, Kubota M, Yudanin N, et al. Distribution and compartmentalization of human circulating and tissue-resident memory T cell subsets. *Immunity.* 2012;38(1):187-197.
211. Gebhardt T, Wakim LM, Eidsmo L, Reading PC, Heath WR, Carbone FR. Memory T cells in nonlymphoid tissue that provide enhanced local immunity during infection with herpes simplex virus. *Nat Immunol.* 2009;10(5):524-530.
212. Schenkel JM, Fraser KA, Beura LK, Pauken KE, Vezys V, Masopust D. T cell memory. Resident memory CD8 T cells trigger protective innate and adaptive immune responses. *Science.* 2014;346(6205):98-101.
213. Ariotti S, Hogenbirk MA, Dijkgraaf FE, et al. T cell memory. Skin-resident memory CD8(+) T cells trigger a state of tissue-wide pathogen alert. *Science.* 2014;346(6205):101-105.
214. Jiang X, Clark RA, Liu L, Wagers AJ, Fuhlbrigge RC, Kupper TS. Skin infection generates non-migratory memory CD8+ T(RM) cells providing global skin immunity. *Nature.* 2012;483(7388):227-231.
215. Hofmann M, Pircher H. E-cadherin promotes accumulation of a unique memory CD8 T-cell population in murine salivary glands. *Proc Natl Acad Sci U S A.* 2011;108(40):16741-16746.
216. Cahill RN, Poskitt DC, Frost DC, Trnka Z. Two distinct pools of recirculating T lymphocytes: migratory characteristics of nodal and intestinal T lymphocytes. *J Exp Med.* 1977;145(2):420-428.
217. Mackay CR, Kimpton WG, Brandon MR, Cahill RN. Lymphocyte subsets show marked differences in their distribution between blood and the afferent and efferent lymph of peripheral lymph nodes. *J Exp Med.* 1988;167(6):1755-1765.
218. Mackay CR, Marston WL, Dudler L. Naive and memory T cells show distinct pathways of lymphocyte recirculation. *J Exp Med.* 1990;171(3):801-817.
219. Sallusto F, Lenig D, Forster R, Lipp M, Lanzavecchia A. Two subsets of memory T lymphocytes with distinct homing potentials and effector functions. *Nature.* 1999;401(6754):708-712.
220. Forster R, Schubel A, Breitfeld D, et al. CCR7 coordinates the primary immune response by establishing functional microenvironments in secondary lymphoid organs. *Cell.* 1999;99(1):23-33.
221. Sallusto F, Geginat J, Lanzavecchia A. Central memory and effector memory T cell subsets: function, generation, and maintenance. *Annu Rev Immunol.* 2004;22:745-763.
222. Bromley SK, Thomas SY, Luster AD. Chemokine receptor CCR7 guides T cell exit from peripheral tissues and entry into afferent lymphatics. *Nat Immunol.* 2005;6(9):895-901.
223. Debes GF, Arnold CN, Young AJ, et al. Chemokine receptor CCR7 required for T lymphocyte exit from peripheral tissues. *Nat Immunol.* 2005;6(9):889-894.
224. Campbell JJ, Murphy KE, Kunkel EJ, et al. CCR7 expression and memory T cell diversity in humans. *J Immunol.* 2001;166(2):877-884.
225. Hikono H, Kohlmeier JE, Takamura S, Wittmer ST, Roberts AD, Woodland DL. Activation phenotype, rather than central- or effector-memory phenotype, predicts the recall efficacy of memory CD8+ T cells. *J Exp Med.* 2007;204(7):1625-1636.
226. Masopust D, Vezys V, Marzo AL, Lefrancois L. Preferential localization of effector memory cells in nonlymphoid tissue. *Science.* 2001;291(5512):2413-2417.
227. Reinhardt RL, Khoruts A, Merica R, Zell T, Jenkins MK. Visualizing the generation of memory CD4 T cells in the whole body. *Nature.* 2001;410(6824):101-105.
228. Marshall DR, Turner SJ, Belz GT, et al. Measuring the diaspora for virus-specific CD8+ T cells. *Proc Natl Acad Sci U S A.* 2001;98(11):6313-6318.
229. Hawke S, Stevenson PG, Freeman S, Bangham CR. Long-term persistence of activated cytotoxic T lymphocytes after viral infection of the central nervous system. *J Exp Med.* 1998;187(10):1575-1582.
230. Harris NL, Watt V, Ronchese F, Le Gros G. Differential T cell function and fate in lymph node and nonlymphoid tissues. *J Exp Med.* 2002;195(3):317-326.
231. Klonowski KD, Williams KJ, Marzo AL, Blair DA, Lingenheld EG, Lefrancois L. Dynamics of blood-borne CD8 memory T cell migration *in vivo*. *Immunity.* 2004;20(5):551-562.
232. Anderson KG, Sung H, Skon CN, et al. Cutting edge: intravascular staining redefines lung CD8 T cell responses. *J Immunol.* 2012;189(6):2702-2706.
233. Wakim LM, Waithman J, van Rooijen N, Heath WR, Carbone FR. Dendritic cell-induced memory T cell activation in nonlymphoid tissues. *Science.* 2008;319(5860):198-202.
234. Gebhardt T, Carbone FR. Immunology: a helpers' guide to infection. *Nature.* 2009;462(7272):418-419.
235. Gebhardt T, Mackay LK. Local immunity by tissue-resident CD8(+) memory T cells. *Front Immunol.* 2012;3:340.
236. Ariotti S, Haanen JB, Schumacher TN. Behavior and function of tissue-resident memory T cells. *Adv Immunol.* 2012;114:203-216.
237. Schenkel JM, Masopust D. Tissue-resident memory T cells. *Immunity.* 2014;41(6):886-897.
238. Turner DL, Gordon CL, Farber DL. Tissue-resident T cells, *in situ* immunity and transplantation. *Immunol Rev.* 2014;258(1):150-166.
239. Park CO, Kupper TS. The emerging role of resident memory T cells in protective immunity and inflammatory disease. *Nat Med.* 2015;21(7):688-697.

240. Gebhardt T, Palendira U, Tscharke DC, Bedoui S. Tissue-resident memory T cells in tissue homeostasis, persistent infection, and cancer surveillance. *Immunol Rev*. 2018;283(1):54-76.
241. Szabo PA, Miron M, Farber DL. Location, location, location: tissue resident memory T cells in mice and humans. *Sci Immunol*. 2019;4(34):eaas9673.
242. Gebhardt T, Whitney PG, Zaid A, et al. Different patterns of peripheral migration by memory CD4+ and CD8+ T cells. *Nature*. 2011;477(7363):216-219.
243. Casey KA, Fraser KA, Schenkel JM, et al. Antigen-independent differentiation and maintenance of effector-like resident memory T cells in tissues. *J Immunol*. 2012;188(10):4866-4875.
244. Woon HG, Braun A, Li J, et al. Compartmentalization of total and virus-specific tissue-resident memory CD8+ T cells in human lymphoid organs. *PLoS Pathogens*. 2016;12(8):e1005799.
245. Ariotti S, Beltman JB, Chodaczek G, et al. Tissue-resident memory CD8+ T cells continuously patrol skin epithelia to quickly recognize local antigen. *Proc Natl Acad Sci U S A*. 2012;109(48):19739-19744.
246. Mackay LK, Stock AT, Ma JZ, et al. Long-lived epithelial immunity by tissue-resident memory T (TRM) cells in the absence of persisting local antigen presentation. *Proc Natl Acad Sci U S A*. 2012;109(18):7037-7042.
247. Zaid A, Mackay LK, Rahimpour A, et al. Persistence of skin-resident memory T cells within an epidermal niche. *Proc Natl Acad Sci U S A*. 2014;111(14):5307-5312.
248. Beura LK, Mitchell JS, Thompson EA, et al. Intravital mucosal imaging of CD8(+) resident memory T cells shows tissue-autonomous recall responses that amplify secondary memory. *Nat Immunol*. 2018;19(2):173-182.
249. Davies B, Prier JE, Jones CM, Gebhardt T, Carbone FR, Mackay LK. Cutting edge: tissue-resident memory T cells generated by multiple immunizations or localized deposition provide enhanced immunity. *J Immunol*. 2017;198(6):2233-2237.
250. Cuburu N, Graham BS, Buck CB, et al. Intravaginal immunization with HPV vectors induces tissue-resident CD8+ T cell responses. *The Journal of clinical investigation*. 2012;122(12):4606-4620.
251. Park SL, Zaid A, Hor JL, et al. Local proliferation maintains a stable pool of tissue-resident memory T cells after antiviral recall responses. *Nat Immunol*. 2018;19(2):183-191.
252. Stolley JM, Johnston TS, Soerens AG, et al. Retrograde migration supplies resident memory T cells to lung-draining LN after influenza infection. *J Exp Med*. 2020;217(8):e20192197.
253. Slutter B, Van Braeckel-Budimir N, Abboud G, Varga SM, Salek-Ardakani S, Harty JT. Dynamics of influenza-induced lung-resident memory T cells underlie waning heterosubtypic immunity. *Sci Immunol*. 2017;2(7):eaag2031.
254. Masopust D, Choo D, Vezys V, et al. Dynamic T cell migration program provides resident memory within intestinal epithelium. *J Exp Med*. 2010;207(3):553-564.
255. Wakim LM, Woodward-Davis A, Liu R, et al. The molecular signature of tissue resident memory CD8 T cells isolated from the brain. *J Immunol*. 2012;189(7):3462-3471.
256. Kumar BV, Ma W, Miron M, et al. Human tissue-resident memory T cells are defined by core transcriptional and functional signatures in lymphoid and mucosal sites. *Cell Reports*. 2017;20(12):2921-2934.
257. Skon CN, Lee JY, Anderson KG, Masopust D, Hogquist KA, Jameson SC. Transcriptional downregulation of S1pr1 is required for the establishment of resident memory CD8+ T cells. *Nat Immunol*. 2013;14(12):1285-1293.
258. Mackay LK, Braun A, Macleod BL, et al. Cutting edge: CD69 interference with sphingosine-1-phosphate receptor function regulates peripheral T cell retention. *J Immunol*. 2015;195(5):2059-2063.
259. Walsh DA, Borges da Silva H, Beura LK, et al. The functional requirement for CD69 in establishment of resident memory CD8(+) T cells varies with tissue location. *J Immunol*. 2019;203(4):946-955.
260. Takamura S, Yagi H, Hakata Y, et al. Specific niches for lung-resident memory CD8+ T cells at the site of tissue regeneration enable CD69-independent maintenance. *J Exp Med*. 2016;213(13):3057-3073.
261. Lee YT, Suarez-Ramirez JE, Wu T, et al. Environmental and antigen receptor-derived signals support sustained surveillance of the lungs by pathogen-specific cytotoxic T lymphocytes. *J Virol*. 2011;85(9):4085-4094.
262. Mackay LK, Wynne-Jones E, Freestone D, et al. T-box transcription factors combine with the cytokines TGF-beta and IL-15 to control tissue-resident memory T cell fate. *Immunity*. 2015;43(6):1101-1111.
263. Milner JJ, Toma C, Yu B, et al. Runx3 programs CD8(+) T cell residency in non-lymphoid tissues and tumours. *Nature*. 2017;552:253-257.
264. Mackay LK, Minnich M, Kragten NA, et al. Hobit and Blimp1 instruct a universal transcriptional program of tissue residency in lymphocytes. *Science*. 2016;352(6284):459-463.
265. Mackay LK, Kallies A. Transcriptional regulation of tissue-resident lymphocytes. *Trends Immunol*. 2017;38(2):94-103.
266. Wakim LM, Woodward-Davis A, Bevan MJ. Memory T cells persisting within the brain after local infection show functional adaptations to their tissue of residence. *Proc Natl Acad Sci U S A*. 2010;107(42):17872-17879.
267. Muschaweckh A, Buchholz VR, Fellenzer A, et al. Antigen-dependent competition shapes the local repertoire of tissue-resident memory CD8+ T cells. *J Exp Med*. 2016;213(13):3075-3086.
268. Khan TN, Mooster JL, Kilgore AM, Osborn JF, Nolz JC. Local antigen in nonlymphoid tissue promotes resident memory CD8+ T cell formation during viral infection. *J Exp Med*. 2016;213(6):951-966.
269. Hu Y, Lee YT, Kaech SM, Garvy B, Cauley LS. Smad4 promotes differentiation of effector and circulating memory CD8 T cells but is dispensable for tissue-resident memory CD8 T cells. *J Immunol*. 2015;194(5):2407-2414.
270. Bergsbaken T, Bevan MJ. Proinflammatory microenvironments within the intestine regulate the differentiation of tissue-resident CD8 T cells responding to infection. *Nat Immunol*. 2015;16(4):406-414.
271. Bergsbaken T, Bevan MJ, Fink PJ. Local inflammatory cues regulate differentiation and persistence of CD8(+) tissue-resident memory T cells. *Cell Reports*. 2017;19(1):114-124.
272. Bachmann MF, Wolint P, Schwarz K, Oxenius A. Recall proliferation potential of memory CD8+ T cells and antiviral protection. *J Immunol*. 2005;175(7):4677-4685.
273. Badovinac VP, Messingham KA, Jabbari A, Haring JS, Harty JT. Accelerated CD8+ T-cell memory and prime-boost response after dendritic-cell vaccination. *Nat Med*. 2005;11(7):748-756.
274. Shin H, Kumamoto Y, Gopinath S, Iwasaki A. CD301b+ dendritic cells stimulate tissue-resident memory CD8+ T cells to protect against genital HSV-2. *Nat Commun*. 2016;7:13346.
275. Enamorado M, Iborra S, Priego E, et al. Enhanced anti-tumour immunity requires the interplay between resident and circulating memory CD8(+) T cells. *Nat Commun*. 2017;8:16073.
276. Virgin HW, Wherry EJ, Ahmed R. Redefining chronic viral infection. *Cell*. 2009;138(1):30-50.
277. Wherry EJ. T cell exhaustion. *Nat Immunol*. 2011;12(6):492-499.
278. Khan O, Giles JR, McDonald S, et al. TOX transcriptionally and epigenetically programs CD8(+) T cell exhaustion. *Nature*. 2019;571(7764):211-218.
279. Scott AC, Dundar F, Zumbo P, et al. TOX is a critical regulator of tumour-specific T cell differentiation. *Nature*. 2019;571(7764):270-274.
280. Alfei F, Kanev K, Hofmann M, et al. TOX reinforces the phenotype and longevity of exhausted T cells in chronic viral infection. *Nature*. 2019;571(7764):265-269.
281. Paley MA, Kroy DC, Odorizzi PM, et al. Progenitor and terminal subsets of CD8+ T cells cooperate to contain chronic viral infection. *Science*. 2012;338(6111):1220-1225.
282. Utzschneider DT, Legat A, Fuertes Marraco SA, et al. T cells maintain an exhausted phenotype after antigen withdrawal and population reexpansion. *Nat Immunol*. 2013;14(6):603-610.
283. Speiser DE, Utzschneider DT, Oberle SG, Munz C, Romero P, Zehn D. T cell differentiation in chronic infection and cancer: functional adaptation or exhaustion?. *Nat Rev Immunol*. 2014;14(11):768-774.
284. O'Hara GA, Welten SP, Klenerman P, Arens R. Memory T cell inflation: understanding cause and effect. *Trends Immunol*. 2012;33(2):84-90.
285. Klenerman P, Oxenius A. T cell responses to cytomegalovirus. *Nat Rev Immunol*. 2016;16(6):367-377.
286. Pallett LJ, Davies J, Colbeck EJ, et al. IL-2(high) tissue-resident T cells in the human liver: sentinels for hepatotropic infection. *J Exp Med*. 2017;214(6):1567-1580.
287. Hislop AD, Kuo M, Drake-Lee AB, et al. Tonsillar homing of Epstein-Barr virus-specific CD8+ T cells and the virus-host balance. *The Journal of clinical investigation*. 2005;115(9):2546-2555.
288. Woodberry T, Suscovich TJ, Henry LM, et al. Alpha E beta 7 (CD103) expression identifies a highly active, tonsil-resident effector-memory CTL population. *J Immunol*. 2005;175(7):4355-4362.
289. Zhu J, Koelle DM, Cao J, et al. Virus-specific CD8+ T cells accumulate near sensory nerve endings in genital skin during subclinical HSV-2 reactivation. *J Exp Med*. 2007;204(3):595-603.
290. Piet B, de Bree GJ, Smids-Dierdorp BS, et al. CD8(+) T cells with an intraepithelial phenotype upregulate cytotoxic function upon influenza infection in human lung. *The Journal of clinical investigation*. 2011;121(6):2254-2263.
291. Jozwik A, Habibi MS, Paras A, et al. RSV-specific airway resident memory CD8+ T cells and differential disease severity after experimental human infection. *Nat Commun*. 2015;6:10224.
292. Verjans GM, Hintzen RQ, van Dun JM, et al. Selective retention of herpes simplex virus-specific T cells in latently infected human trigeminal ganglia. *Proc Natl Acad Sci U S A*. 2007;104(9):3496-3501.
293. Khanna KM, Bonneau RH, Kinchington PR, Hendricks RL. Herpes simplex virus-specific memory CD8+ T cells are selectively activated and retained in latently infected sensory ganglia. *Immunity*. 2003;18(5):593-603.
294. Theil D, Derfuss T, Paripovic I, et al. Latent herpesvirus infection in human trigeminal ganglia causes chronic immune response. *Am J Pathol*. 2003;163(6):2179-2184.
295. Mark KE, Wald A, Magaret AS, et al. Rapidly cleared episodes of herpes simplex virus reactivation in immunocompetent adults. *J Infect Dis*. 2008;198(8):1141-1149.
296. Park SL, Buzzai A, Rautela J, et al. Tissue-resident memory CD8(+) T cells promote melanoma-immune equilibrium in skin. *Nature*. 2019;565(7739):366-371.
297. Park SL, Gebhardt T, Mackay LK. Tissue-resident memory T cells in cancer immunosurveillance. *Trends Immunol*. 2019;40(8):735-747.
298. Amsen D, van Gisbergen K, Hombrink P, van Lier RAW. Tissue-resident memory T cells at the center of immunity to solid tumors. *Nat Immunol*. 2018;19(6):538-546.
299. Edwards J, Wilmott JS, Madore J, et al. CD103+ tumor-resident CD8+ T cells are associated with improved survival in immunotherapy naive melanoma patients and expand significantly during anti-PD1 treatment. *Clin Cancer Res*. 2018;24(13):3036-3045.
300. Naik S, Bouladoux N, Linehan JL, et al. Commensal-dendritic-cell interaction specifies a unique protective skin immune signature. *Nature*. 2015;520(7545):104-108.

CHAPTER 31

Cytotoxic Effector Functions

Morgan Huse, PhD

INTRODUCTION

Among the more unique qualities of the immune system is its capacity to destroy specific groups of target cells in a selective manner. A substantial fraction of immune-mediated killing is carried out by cytotoxic lymphocytes, comprising CD8+ cytotoxic T lymphocytes (CTLs) and natural killer (NK) cells, which are critical for immune responses against intracellular pathogens and cancer (Fig. 31.1). They are robust serial killers, capable of destroying approximately 20 infected target cells per lymphocyte per day in vivo.[1,2] Cytotoxic lymphocytes identify target cells based on the expression of cell surface molecules indicative of stress, infection, or nonself. The recognition of these ligands by activating immunoreceptors on the lymphocyte induces the formation of a stereotyped cell–cell interaction known as the cytolytic immune synapse.[3] The lymphocyte then releases a cellular venom into this interface containing the pore-forming protein perforin and several granzyme proteases. Perforin damages the target cell membrane, enabling granzymes to enter the cytoplasm and induce programmed cell death.[4] Activated CTLs and NK cells also release the transmembrane protein Fas-ligand (FasL) into the synaptic space, which engages the death receptor Fas on the target cell surface to induce canonical apoptosis.[5] The perforin/granzyme and Fas pathways account for essentially all of the killing activity of cytotoxic lymphocytes,[6] and their importance for immune function has been substantiated by loss-of-function studies in animal models and by human diseases in which key cytotoxic components are lost. This chapter will provide a mechanistic overview of cellular cytotoxicity, including description of the molecular components, an account of the dynamic cell biological process, and a discussion of how cellular cytotoxicity contributes to immune responses against intracellular pathogens, antitumor immunosurveillance, and the maintenance of immune homeostasis.

CYTOTOXIC LYMPHOCYTES AND TRANSCRIPTIONAL CONTROL OF THE CYTOLYTIC PROGRAM

Both CTLs and NK cells acquire cytotoxic potential as part of their effector differentiation program. Naïve CD8+ T cells lack killing activity and acquire it only after T cell receptor (TCR)–mediated recognition of cognate peptide-major histocompatibility complex (pMHC) in the presence of suitable levels of costimulation.[7] This priming stimulus, which is typically provided by professional antigen-presenting cells in secondary lymphoid organs, induces dramatic clonal expansion accompanied by profound changes in gene expression and metabolism. The fully differentiated CTLs that emerge 5 to 7 days later migrate out into the periphery to engage the source of antigen, typically a viral or bacterial infection. In these peripheral locations, TCR engagement induces the killing of infected target cells and the secretion of inflammatory cytokines such as interferon-γ (IFN-γ) and tumor necrosis factor (TNF).

CTL differentiation depends on the cytokine interleukin 2 (IL-2), which is generated by activated T cells within hours of primary stimulation.[8] Secreted IL-2 binds, in an autocrine/paracrine fashion, to receptors on the T cell surface, driving phosphorylation and nuclear import of the transcription factor STAT5, which regulates a multiplicity of genes to sustain expansion and differentiation. Several other transcription factors play essential roles in the acquisition of cytotoxic effector function. Runx3 serves as a master regulator of the CD8+ lineage by repressing components required for CD4+ T cell differentiation, most conspicuously *CD4* itself.[9,10] During CTL differentiation, Runx3 drives the expression of eomesodermin (Eomes), a member of the T-box family of transcription factors.[11] Both Runx3 and Eomes have been implicated in the direct regulation of effector genes that define the CTL subset, including *PRF1* (perforin) and *IFNG* (IFN-γ).[11,12] CTL differentiation also requires a second T-box transcription factor, T-bet, which is upregulated before Eomes in a Runx3-independent manner.[11,13,14] Although T-bet shares certain direct targets with Eomes, such as *IFNG*, it appears to have a nonredundant role in the induction of at least some of the granzymes.[15]

Effector differentiation is a huge metabolic undertaking, involving not only rapid cell division but also a marked increase in cell size.[16] CTLs must also synthesize, within days, copious amounts of cytotoxic proteins, including perforin, granzymes, and the small pore-forming molecule granulysin, which are trafficked and stored in newly assembled secretory lysosomes called lytic granules. Proteomic studies have shown that fully armed CTLs contain just as many molecules of granzyme B as actin,[17] starkly illustrating the metabolic cost of this process. To fuel anabolic expansion, differentiating T cells switch from oxidative phosphorylation to glycolysis, increase their uptake of glucose and amino acids, and upregulate protein translation and lipid biosynthesis. The mammalian Tor complex 1 (MTORC1) guides this metabolic transition by stimulating translation and activating transcription factors that drive the expression of nutrient transporters and lipid biosynthesis machinery.[18] CTLs treated with MTORC1

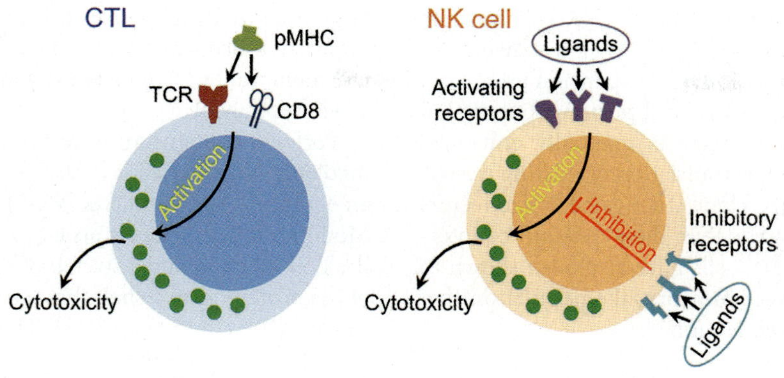

FIG. 31.1. Cytotoxic T lymphocytes (CTLs) and natural killer (NK) cells. CTL activity is controlled by the TCR, which binds to cognate pMHC in conjunction with the coreceptor CD8. NK cells are controlled by a diverse set of activating and inhibitory receptors, and whether or not they mount cytotoxic responses depends on the balance of activating and inhibitory signals they receive.

inhibitors dramatically decrease their expression of perforin, granzyme B, and IFN-γ, demonstrating the dynamic coupling between metabolic state and cytotoxic effector function.[17]

The vast majority of effector CTLs die during the resolution phase of the immune response, when they are no longer necessary to engage pathogen.[7] The small numbers of antigen-specific CD8+ T cells that remain adopt a memory phenotype characterized by long-term persistence. A number of distinct memory subsets exist, which confer protection from pathogen reinfection in different ways.[19] Effector memory T (T_{em}) cells are the most effector like. They retain expression of perforin and granzyme protein and circulate through the body poised to mount cytotoxic responses against antigen positive target cells. By contrast, central memory (T_{cm}) and resident memory (T_{rm}) cells, which reside in secondary lymphoid organs and specific peripheral tissues, respectively, downregulate cytotoxic effector proteins. Antigen reencounter drives the clonal expansion of T_{cm} and T_{rm} cells, which quickly reacquire cytolytic capacity in a matter of hours. This process is facilitated by the presence of preexisting perforin and granzyme mRNA, which is retained by memory T cells at rest.[20]

Although CD4+ T cells are, in general, not cytotoxic, certain subsets express cytotoxic effector proteins and kill target cells via both FasL and perforin-granzyme–mediated mechanisms.[21] Invariant NK T cells, which recognize specific glycolipid ligands bound to the MHC-related molecule CD1d, kill multiple CD1d+ tumor cell lines in vitro, although the importance of this activity in vivo is less clear.[22] Conventional CD4+ T cells with cytotoxic capacity have also been observed in the context of viral infection and may play a role in antitumor immunity. Finally, a fraction of activated Foxp3+ regulatory T cells (T_{reg}) expresses perforin and granzymes.[21] These T_{reg} cells are thought to employ cytotoxicity to suppress conventional CD4+ T cell responses, either by killing antigen-presenting cells or the conventional CD4+ T cells themselves.

In contrast to naïve CD8+ T cells, NK cells express cytolytic machinery constitutively and are classically conceptualized as early effectors, capable of mounting rapid cytotoxic and inflammatory responses without a priming step.[23] Many of the key regulatory factors required for the acquisition and maintenance of NK cell cytotoxicity have been identified.[24] Unsurprisingly, certain regulators that are important for CTL differentiation also control NK cell fate. Eomes, for instance, is absolutely required for the final maturation of NK cells into fully armed effectors with high expression of perforin and granzyme.[25] Other transcription factors appear to play an outsize role in NK cells. Among these is the Ets family protein MEF, which directly binds the *PRF1* promoter and maintains high expression of the gene.[26] As with T cells, NK cell maturation is tightly controlled by cytokines. IL-15, a close homolog of IL-2 that also activates STAT5, is critical not only for the maintenance of the cytotoxic program but also for NK cell survival.[27,28]

NK cell–mediated killing is controlled by a diverse set of cell surface receptors that transduce either activating or inhibitory signals.[29] NK cells integrate competing stimuli from theses receptors and respond based on the balance of signals they receive. In general, inhibitory NK receptors contain cytoplasmic immunotyrosine-based inhibitory motifs (ITIMs) that recruit protein and lipid phosphatases. The mechanism(s) by which they regulate cytotoxicity are discussed in more detail later in the chapter. Activating NK receptors comprise a molecularly diverse set of transmembrane proteins that includes integrins, immunoglobulin family receptors, and C-type lectins. Unlike the TCR, which recognizes features specific to each infectious pathogen, most activating NK receptors bind to germline-encoded protein ligands indicative of infection, stress, or immune targeting. The prototypical NK receptor NKG2D, for example, binds to MHC class I–related (MIC) and UL16 binding protein (ULBP) family ligands, which are upregulated by stressed, damaged, or infected cells. The low-affinity Fc receptor CD16 is also expressed by many NK cells and enables them to recognize and destroy opsonized target cells by antibody-induced cellular cytotoxicity.

Although mature NK cells are typically thought of as terminally differentiated, in recent years it has become clear that they undergo a differentiation cycle in response to pathogens that is not unlike what occurs in T cells.[30] NK cells specific for the mouse cytomegalovirus (MCMV), for instance, have been shown to undergo dramatic clonal expansion upon viral infection, somewhat analogous to T cell priming. Expression of perforin and granzyme proteins is markedly enhanced during activation, supported by translation of existing mRNA and de novo transcription.[31] Hence, activated NK effectors are substantially more cytotoxic than their "resting" counterparts. Subsets of activated NK cells have also been shown to differentiate into a long-lived, memory-like population that confers protection against reinfection.[32]

CYTOTOXIC EFFECTORS

To mediate effective immunity, cytotoxic lymphocytes must be able to destroy a variety of cell types in a wide range of conditions. Accordingly, the killing mechanisms that have evolved are notable both for their biochemical complementarity and their functional redundancy. Cytotoxic lymphocytes routinely express at least five granzyme isoforms.[4] Although most granzymes can drive programmed cell death on their own, they do so by targeting distinct pools of cytoplasmic substrates. Hence, it is difficult to evolve resistance mechanisms that inhibit them all. Furthermore, perforin pore formation is a robust and rapid process that affects all eukaryotic membranes. Finally, Fas signaling provides a granzyme-independent mechanism for target cell killing, which, as discussed later in the chapter, is particularly useful for homeostatic regulation of other immune cells, including the cytotoxic lymphocytes themselves. This section describes the molecular components of cytotoxicity, focusing on their structures, enzymatic activities, and how they elicit target cell death.

Perforin

Perforin is a ~65 kD member of the membrane attack complex and perforin (MACPF) family of pore-forming toxins, which also includes cholesterol-dependent cytolysins (CDCs) and complement protein C9.[33,34] MACPF pore formation is highly toxic to cells and, as such, the activity of family members tends to be tightly regulated at a posttranslational level. Bacterial cytolysins must undergo specific proteolytic processing steps both before and after secretion to remove inhibitory pre- and propeptides.[35] Complement C9 activity is also regulated by proteolysis, but in this case it is the proteolysis of upstream components of the complement cascade, leading eventually to the deposition of C5b, C6, C7, and C8 onto the target membrane.[36] C9 then associates with these cofactors and oligomerizes to form the membrane attack complex. By contrast, perforin activity is controlled by pH and access to divalent cations. At low pH and in the absence of calcium (Ca^{2+}), the conditions within lytic granules, perforin exists as a metastable monomer. Secretion into the IS neutralizes the pH around perforin and exposes it to extracellular Ca^{2+}, thereby inducing dramatic conformational changes that lead to the formation of oligomeric perforin pores in the target cell membrane. Although the presence of cholesterol in the target membrane enhances this process,[37] it is not absolutely required. Indeed, perforin can attack a wide variety of biological membranes,[38,39] consistent with its broad applicability as a cytotoxic agent.

Perforin contains an N-terminal MACPF domain, which mediates oligomerization and pore formation, followed by an epidermal growth factor (EGF) domain and a C2 domain. Monomeric perforin adopts a key-shaped structure in which the EGF domain and a stretch of sequence at the C-terminus of the protein form a shelf that separates the C2 domain from the MACPF core[40] (Fig. 31.5). The perforin C2 domain, like the C2 domains of other proteins, binds membranes in a Ca^{2+}-dependent manner, and it is generally thought that this recognition event mediates the initial accumulation of perforin on the target membrane.[33,34,41] This is followed by perforin oligomerization and membrane penetration. Although the structure of fully activated, oligomerized perforin has not been determined at atomic resolution, a structural mechanism for pore formation has been inferred based on electron microscopy (EM) of assembled perforin pores and crystal structures of MACPF and related CDC family proteins.[40,42-45] The critical step in this process is a dramatic conformational change in which two helical clusters within the MACPF domain reorganize into two long, hydrophobic β-hairpins that project into the membrane beneath the C2 domain. These β-hairpins line the pore in the fully assembled, oligomeric structure (Fig. 31.5), forming a cylindrical antiparallel β-barrel that is the hallmark feature of MACPF family.

Time-resolved analyses of the MACPF/CDC pore-forming process suggest that oligomerization on the membrane precedes conformational change and membrane insertion. "Pre-pore" structures containing 20 to 50 monomers have been observed by EM,[45-48] and recent real-time atomic force microscopy (AFM) experiments have documented even smaller structures containing approximately eight loosely bound subunits.[49] These results are consistent with the idea that oligomerization promotes perforin conformational change. Perforin pores contain a variable number of subunits and range in diameter from 5 to 30 nm (implying a lumen diameter of 1.3-20 nm), based on EM and AFM work. In addition, multiple labs have observed the formation of incomplete, arc-shaped pores that, despite being open on one side, still achieve membrane perforation.[49-51] This variation in structure and size is consistent with electrophysiological experiments, which have documented dramatic heterogeneity in the conductance properties of individual perforin pores.[52] Importantly, a 30-nm diameter pore is large enough to accommodate granzyme monomers or, in the case of granzyme A (see below), dimers. Hence, the existing structural and biochemical data are consistent with the idea that oligomeric perforin pores enable granzymes and other cytotoxic effectors to enter the target cell.

The simplest way for this to occur would be for granzymes to move directly through pores in the plasma membrane. Pores of sufficient size to accommodate granzyme have been observed in plasma membranes from perforin exposed cells.[53,54] In addition, imaging experiments have revealed that target cells become transiently permeable to ions and fluorescent dyes shortly after IS formation.[55-57] This brief permeabilization depends on perforin secretion, and it has been argued that it is sufficient to enable transfer of toxic levels of granzyme to the target cytoplasm. Other studies, however, have revealed an alternative mechanism in which granzymes are trafficked through target cell endosomes[58,59] (Fig. 31.2). In this indirect pathway, the variably sized pores formed by perforin in the plasma membrane function less to provide avenues for granzyme traffic than to induce damage. This leads, within minutes, to a stereotyped, Ca^{2+}-dependent damage response in which endosomal and lysosomal membranes are used to reestablish plasma membrane integrity. In the process, perforin and granzymes are taken up into endosomes, which then fuse into atypically large vesicles that have been termed "gigantosomes." Imaging experiments indicate that perforin forms oligomers in these compartments, inhibits their acidification, and promotes granzyme escape into the cytoplasm.[59] Lending further support to the endosomal model, it has been shown that disrupting Ca^{2+} signaling or clathrin-mediated endocytosis in target cells inhibits the apoptotic cell death typically induced by perforin and granzyme;[58,60] instead, the target cells die from necrosis, most likely due to unrepaired perforin-mediated damage to the membrane. Notably, in certain melanoma cell lines, perforin-induced membrane damage actually inhibits cytotoxicity, presumably by delivering lysosomal enzymes to the intercellular space that degrade cytotoxic molecules.[61] It remains unclear precisely why the endolysosomal fusion response that leads to perforin and granzyme uptake in most cells actually enables these melanoma lines to avoid killing.

Although the relative importance of direct access versus endosomal trafficking of granzymes remains a matter of debate, it is not hard to imagine that both pathways contribute to target cell killing in a physiological setting (Fig. 31.2). Targets that are particularly sensitive to perforin and granzyme may succumb quickly to the small amount of granzyme that enters through plasma membrane pores. By contrast, targets with robust membrane repair and a degree of granzyme resistance would only undergo apoptosis after granzyme influx from endosomes.

Perforin is the only MACPF protein expressed by cytotoxic lymphocytes, making its expression and activity essential for proper immune function. Mice lacking perforin are vulnerable to intracellular pathogens and exhibit a high incidence of spontaneous lymphoma (see below). Furthermore, loss-of-function mutations in human *PRF1* cause type I familial hemophagocytic lymphohistiocytosis (FHL), a combined immune deficiency/autoimmune syndrome characterized by massive infiltration of activated leukocytes into peripheral organs.[62] FHL patients are also highly susceptible to viral and bacterial infections. Indeed, the inability to control these infections is widely thought to be the proximal cause of the pathogenic leukocyte activation.[63] Hence, perforin-mediated killing protects the host not only from foreign pathogens but also from destructive downstream immune activation resulting from the growth and expansion of infection.

The transcriptional regulation of the *PRF1* gene is well studied and represents a model for lineage-restricted gene expression in the immune system.[64] *PRF1* contains three closely spaced exons and is located on chromosome 10 (also 10 in mouse) between two functionally unrelated genes (*ADAMTS14* and *PALD*). Mapping studies have identified four cis-acting regions that control *PRF1* expression, a promoter, two enhancers located 1 and 15 kb upstream of the transcriptional start site, respectively, and a locus control region (LCR) in the distal 5′ flanking region of the gene (>50

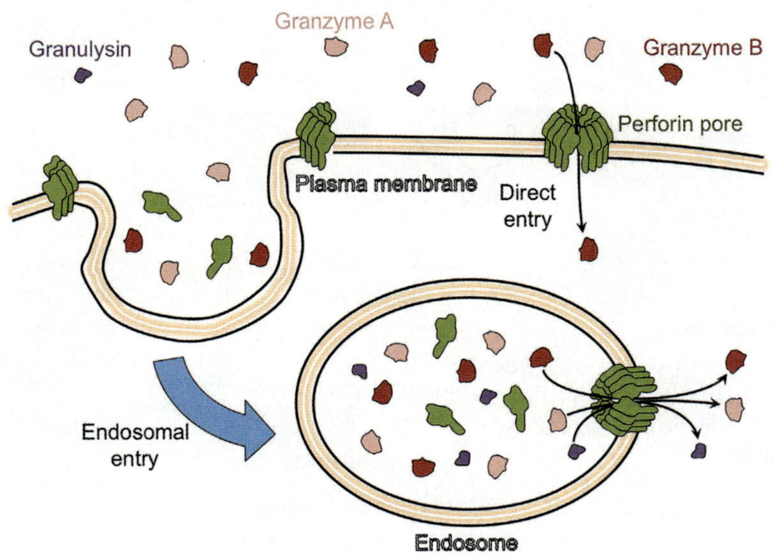

FIG. 31.2. Granzymes and granulysin access the target cell cytoplasm through perforin pores. Perforin pores in the plasma membrane enable direct entry of granzymes into the cytoplasm (direct entry). Perforin also induces a membrane repair response during which perforin, granzymes, and granulysin are taken up into endosomes. Subsequent perforin pore formation in the endosomal membrane provides an additional avenue for cytoplasmic access (endosomal entry).

kb upstream). The promoter contains multiple binding sites for Ets family proteins, which makes sense given the critical role for MEF in driving *PRF1* expression in NK cells.[65] Runx3 also binds to the *PRF1* promoter region,[11] consistent with the capacity of this transcription factor to drive perforin expression in T cells. The *PRF1* enhancers are bound by a number transcription factors with well-established roles in lymphocyte activation. These include activator protein 1 (AP1), nuclear factor κB (NF-κB), and nuclear factor of activated T cells (NFAT), which transduce TCR-induced activating signals to the nucleus.[66,67] Both enhancers also contain binding sites for STAT5 and are likely the control regions that are responsive to IL-2 and IL-15.[67] The LCR was established by chromosome transfer experiments and comprises a cluster of DNAse hypersensitive sites that appear under conditions of high *PRF1* expression.[68] Runx3 has been observed to associate with multiple DNAse hypersensitive domains within the *PRF1* locus, including the LCR, and it has been proposed that it may bring activating cis-elements and the promoter into close physical apposition.[11,64] Interestingly, while Eomes had been shown to associate with both the promoter and the proximal enhancer, T-bet does not appear to regulate *PRF1* transcription directly.[11,64] Rather, T-bet is thought to enhance perforin expression indirectly by inducing expression of the IL-2 receptor β subunit.

Granzymes

Granzymes are organized into three gene clusters in both humans and mice. *GZMA* and *GZMK* are located between two functionally unrelated genes (*ESM1* and *CDC20B*) on chromosome 5 (mouse chromosome 13). *GZMB* and *GZMH* form part of the chymase locus (chromosome 14), which includes proteases (such as *CMA1*) that are expressed in myeloid cells. Due to gene duplications, the mouse chymase locus (chromosome 14) contains seven granzyme genes, *Gzmb*, *Gzmc* (the proposed ortholog of human *GZMH*), *Gzme*, *Gzmd*, *Gzmg*, *Gzmn*, and *Gzmf*. Finally, *GZMM* is located by itself on chromosome 19 (mouse chromosome 10). Granzyme proteins are all trypsin-like serine proteases, containing canonical histidine–serine–aspartic acid catalytic triads for peptide cleavage.[4] All family members that have been analyzed cut in a sequence-specific manner, with the most important determinant being the P1 residue located just N-terminal to the scissile peptide bond. In general, granzymes are initially translated as inactive zymogens (pro-forms) containing inhibitory N-terminal dipeptide extensions. These extensions are subsequently removed by the lytic granule resident protease cathepsin C (also known as dipeptidyl peptidase I, or DPPI).[69] This mechanism ensures that granzymes reach their functional configuration only within the confines of the lytic granule, where their activities are constrained by low pH and other mechanisms (see below).

Granzymes A and B are the most highly expressed members of the family. Indeed, a recent proteomics study found that these two isoforms account for >99% of the granzyme in murine CTLs,[17] implying an outsize role in target cell killing. Granzyme B is the more extensively studied of the two and is understood to be a powerful inducer of canonical apoptosis (Fig. 31.3). It is the only known granzyme to cleave after acidic residues, typically aspartate.[4] This specificity is shared by most caspase family proteases, the principle drivers of apoptosis. Indeed, the consensus recognition motifs for human and mouse granzyme B (IEPD and IEFD, respectively)[70,71] closely

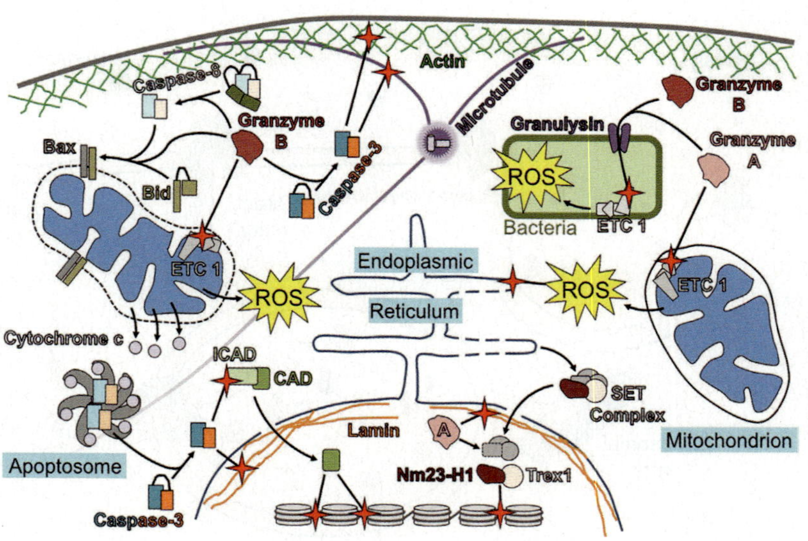

FIG. 31.3. Target cell and microbe death pathways induced by granzyme A, granzyme B, and granulysin. Granzymes A and B both target ETC1, disrupting mitochondrial respiration. Granzyme A also activates the SET complex, leading to DNA degradation. Granzyme B has evolved to induce canonical apoptosis through caspase-3 and Bid, leading to mitochondrial permeabilization and CAD-dependent DNA fragmentation. Granulysin forms pores in bacterial, fungal, and parasite membranes, allowing granzymes to induce microbe death. Red crosses indicate cleavage events.

resemble canonical cleavage sites for caspase-3 (DEVD) and caspase-8 (IETD). These similarities in target specificity allow granzyme B to target a subset of caspase substrates. In some cases, this can circumvent the need for caspase activation in granzyme B–induced cell death.[72-75] That being said, a predominant mechanism by which granzyme B kills target cells is by activating caspases, including caspase-7, caspase-8, caspase-10, and most importantly caspase-3, which is the principle effector caspase in most cells.[76-78] Like granzymes, caspases must undergo proteolytic processing in their catalytic domains to acquire full activity.[79,80] Other than the caspases themselves, granzyme B is the only known enzyme capable of this processing step. Once activated, caspase-3 targets a number of key structural proteins, including actin, tubulin, and the nuclear lamins, thereby undermining cellular architecture. Caspase-3 also cleaves the cytoplasmic inhibitor of the caspase-activated DNAse (ICAD). This liberates CAD to enter the nucleus, where it begins to reduce chromosomal DNA into nucleosomal fragments, a characteristic feature of canonical apoptosis.

Another critical substrate in granzyme B–induced apoptosis is Bid, a "BH3-only" member of the Bcl2 family. Bid is cleaved directly by human granzyme B and can be targeted by mouse granzyme B via activation of caspase-8.[72,73,81] Full-length Bid adopts an autoinhibited conformation in the cytoplasm. Proteolytic processing generates an active, truncated form (t-Bid) that incorporates into the mitochondrial membrane, where it forms complexes with another proapoptotic member of its family, Bax.[82] This triggers mitochondrial permeabilization and release of the small respiratory protein cytochrome c into the cytoplasm, where it assembles with the adapter molecule Apaf-1 to form a ring-shaped complex called the apoptosome.[83] The apoptosome then recruits and activates caspase-9, leading in turn to the activation of caspase-3 and the downstream effects described above. Although the Bid–cytochrome c–apoptosome pathway is important for the induction of canonical apoptosis, it is not the only way in which granzyme B damages mitochondria. Recent work has shown that granzyme B and active caspase-3 can directly access the mitochondrial matrix and intermembrane space by translocating through channels in the inner and outer mitochondrial membranes.[84] Once localized in this manner, both proteases attack components of the electron transport chain complex I (ETC1). Granzyme B targets NDUFS3, NDUFV1, and NDUFS2, while caspase-3 cleaves NDUFS1.[85] These events disable ETC1, leading to depolarization of the mitochondrial membrane potential ($\Delta\Psi_m$) and the release of reactive oxygen species (ROS), which go on to damage surrounding organelles. ROS scavengers significantly inhibit granzyme B and caspase-3–induced cytotoxicity,[85,86] highlighting the importance of ETC1 disruption and mitochondrial damage for this killing pathway.

Granzyme A is the most evolutionarily ancient member of the granzyme family and the only isoform that exists as a disulfide-linked dimer.[4] Unlike granzyme B, granzyme A cleaves peptides containing positively charged amino acids in the P1 position.[87] This specificity targets granzyme A to a different pool of cytoplasmic substrates, leading ultimately to a form of programmed cell death, called athetosis,[88] that is morphologically similar to apoptosis but molecularly distinct. The mitochondria play an essential role in this process, but not as targets for t-Bid. Instead, granzyme A, like granzyme B, enters mitochondria and cleaves ETC1 components, inducing respiratory dysfunction, $\Delta\Psi_m$ depolarization, and ROS generation without mitochondrial permeabilization and the release of cytochrome c[89,90] (Fig. 31.3). A critical subsequent step in granzyme A–induced killing is the activation of the SET complex, a heteromeric assembly that contains three DNA nucleases, Ape1, Nm23-H1, and Trex1.[90-95] In resting cells, the SET complex is associated with the endoplasmic reticulum (ER). Oxidative damage induced by ROS promotes release of the SET complex from the ER and its translocation to the nucleus. Granzyme A then cleaves multiple SET complex components, including Ape1 and the regulatory subunits SET and HMGB2. This unleashes NM23-H1 and Trex1 to digest nuclear DNA; NM23-H1 induces single-stranded nicks, which are then extended by Trex1. This process generates rather large fragments of genomic DNA, which are easily distinguished from the oligonucleosomal ladder typically observed in apoptotic cells. Granzyme A also targets DNA repair factors, such as Ku70,[96] further compromising genomic integrity, and attacks nuclear lamins,[97] thereby destabilizing nuclear architecture.

The mechanistic distinctions between granzyme A– and granzyme B–mediated cell death enable the pathways to function in a complementary manner, thereby enhancing the overall robustness of the killing response. Hence, it is perhaps unsurprising that loss of either granzyme A or granzyme B has only subtle effects on in vitro cytotoxicity and in vivo immune responses requiring cytotoxic lymphocytes.[98-105] Indeed, even mice lacking both granzyme A and granzyme B are largely resistant to many model pathogens, likely due to the compensatory function of other granzymes. One notable exception is the poxvirus ectromelia. Viral clearance is markedly impaired in mice lacking either granzyme A or granzyme B, and loss of both isoforms compromises immunity to the same extent as perforin deficiency.[106-108] It remains unclear precisely why high granzyme loads are so critical for immune responses against ectromelia, while being less important for other perforin-dependent processes, such as resistance to lymphocytic choriomeningitis virus (LCMV).[102,109]

Granzyme isoforms other than granzyme A and granzyme B are generally expressed at much lower levels, and considerably less is known about their functionality.[110] Indeed, the absence of clearly defined physiological roles has earned these enzymes the collective moniker "orphan granzymes." Humans express three orphan granzymes, granzyme K, granzyme H, and granzyme M. Granzyme K is quite similar to granzyme A, although, unlike granzyme A, it is a monomer. Granzyme K cleaves peptides with basic residues in the P1 position and has been shown to target canonical granzyme A substrates, including the SET complex.[111-113] Granzyme H requires aromatic residues in the P1 position, a substrate specificity distinct from that of both granzyme A and granzyme B.[114] Nevertheless, it has been reported to target two critical substrates in the granzyme B–caspase cascade, Bid and ICAD, although these results are controversial.[115,116] Granzyme M, which cleaves after leucine or methionine residues, processes a largely distinct group of substrates.[117,118] To the extent they have been studied, the mouse orthologs of granzymes K, H (C in mice), and M exhibit analogous biochemical properties. Less is known about the other orphan granzymes in mice (E, D, G, N, and F), and they will not be discussed further here.

Although the physiological roles of orphan granzymes are poorly characterized, loss-of-function studies have hinted at their functional importance. The targeting strategy initially used to knock out *Gzmb* in mice markedly reduced the expression of *Gzmc* and *Gzmf*, as well. CTLs and NK cells from these "granzyme B cluster" deficient animals are significantly less cytotoxic than analogous cells from mice lacking granzyme B alone, implying a role for these other isoforms in compensating for reduced granzyme B activity.[119] Likewise, the biochemical similarities between granzymes A and K are consistent with at least some functional overlap. Taken together, these data imply that orphan granzymes serve as backup cytolytic effectors in the event that the predominant isoforms are missing or incapacitated. That being said, certain studies have revealed highly specialized functions for orphan granzymes, suggesting that they evolved to counter specific pathogens. Granzyme H, for instance, plays a critical role in immune responses against adenovirus by cleaving 100k, an adenoviral protein that inhibits granzyme B.[120] This has the effect of derepressing granzyme B–mediated apoptosis in adenovirally infected cells. Similarly, granzyme M appears to be particularly important for fighting CMV infections.[121] Interestingly, granzyme M is predominantly expressed by NK cells,[122] consistent with the established importance of this cell type for CMV immunity.

Granulysin

Granulysin is a small, pore-forming molecule that resembles saposin.[123,124] It is expressed in primates and ruminants, but interestingly, not rodents. Granulysin is initially synthesized as a 15-kD protein that is then processed into a 9-kD form, which is stored with perforin and granzymes in lytic granules. Both forms bind to membranes and generate pores. Granulysin has particularly high affinity for the membranes of bacteria, fungi, and parasites, and it has long been thought that it plays an important role in the destruction of these pathogens.[125] Recent work has provided critical mechanistic insight by demonstrating that granulysin allows granzymes A and B to access the bacterial cytoplasm[126] (Fig. 31.3). The granzymes then cleave components of the bacterial ETC1, inducing copious ROS production and ultimately lethal oxidative damage. This killing mechanism is broadly effective against not only gram-positive (eg, *Listeria monocytogenes*) and gram-negative (eg, *Escherichia coli*) bacteria but also intracellular protozoa such as *Trypanosoma cruzi* and *Leishmania major*.[127] Importantly, granzyme and granulysin on their own are insufficient to destroy intracellular pathogens living within infected target cells.[126,127] In these contexts, perforin is also required, consistent with a model in which granulysin and granzymes first move into the target cell cytoplasm via perforin pores, and then granulysin forms pores in the bacterial or parasite membrane to enable granzyme entry. Interestingly, when bacterially infected target cells are treated with mixtures of perforin, granulysin, and granzyme, the bacterial death occurs before target apoptosis,[126] presumably to minimize the release of live pathogens by dying cells.

Evaluating the importance of granulysin-mediated killing in vivo was initially hampered by the fact that granulysin is not expressed in mice. To circumvent this problem, mice were generated that expressed granulysin from a bacterial artificial chromosome transgene.[128] These mice exhibit enhanced resistance to both intracellular bacterial and protozoa.[126,127] Importantly, this phenotype depends on *Prf1* expression, indicating that the two-step mechanism of perforin pore formation in the plasma membrane followed by granulysin pore formation in the pathogen membrane also applies in vivo.

The targeting of bacterial ETC1 by granulysin and granzymes strikingly parallels the granzyme and caspase-dependent targeting of mitochondrial ETC1. Considering that mitochondria are prokaryotically derived, these similarities should perhaps be unsurprising. When considered together, these two pathways suggest a common evolutionary basis for cytotoxicity that is built on granzyme-induced oxidative damage.[124] Over the course of evolution, the proteolytic activity of these enzymes was coupled to programmed cell death pathways and pore-forming proteins, generating the robust and redundant killing program that exists today.

FasL

In addition to perforin-granzyme-granulysin–mediated destruction, cytotoxic lymphocytes can also induce apoptosis via the FasL-Fas death receptor pathway.[5] Even at rest, NK cells and armed CTLs express substantial amounts of FasL.[129-131] The vast majority of this protein, however, is not found on the cell surface, but rather in intracellular compartments that fuse with the plasma membrane upon lymphocyte activation and immune synapse formation.[129,132] FasL is a TNF-related molecule and, like other members of this superfamily, it is expressed as a homotrimeric type II transmembrane protein. Signal transduction occurs when FasL binds to three Fas chains leading to assembly of a heterohexameric complex. This typically occurs within the synaptic space, with membrane-anchored FasL on one side and Fas on the other. FasL can also be cleaved from its transmembrane domain, forming a soluble ligand that can, at least in principle, induce target cell death at a distance. Comparative studies have shown, however, that membrane-anchored FasL transduces significantly stronger signals than its soluble counterpart,[133-135] implying an important, signal potentiating role either for the clustering of FasL trimers within the plasma membrane or for mechanical/architectural advantages conferred by the synapse.

Fas (also called CD95) is widely expressed, particularly among cells of lymphoid lineage.[136] It is a type I transmembrane protein composed of an extracellular ligand binding domain, a membrane spanning helix, and a signal

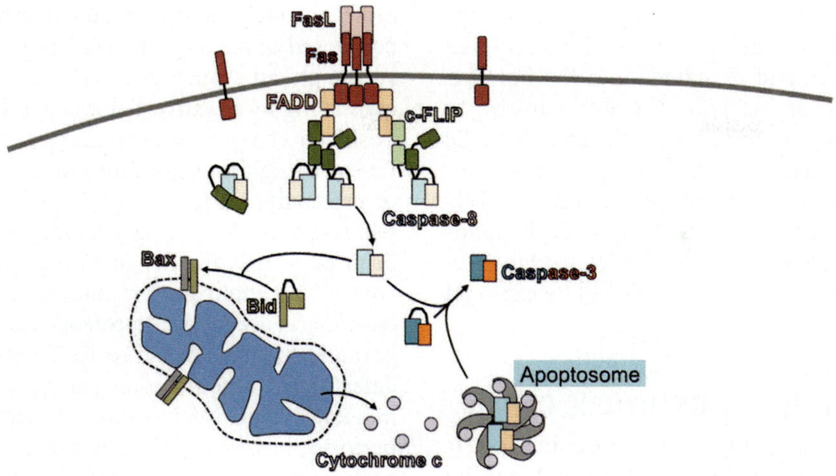

FIG. 31.4. FasL-Fas–mediated apoptosis. Fas activation induces the assembly of a DISC containing FADD, c-Flip, and caspase-8, leading to caspase-8 processing and the activation of canonical apoptosis via caspase-3 and the mitochondrial pathway.

transducing cytoplasmic tail that contains a conserved death domain.[5,137] Upon FasL binding, the death domains within the Fas trimer recruit multiple copies of the adapter molecule FADD, which interacts with Fas via its own death domain in a homotypic-style contact. FADD also contains a module called a death effector domain (DED), which functions by recruiting caspase-8 (and in humans, caspase-10) into the receptor complex (Fig. 31.4). Caspase-8 and caspase-10 are canonical members of the receptor-regulated caspase subfamily.[79,80] At rest, they adopt an autoregulated conformation in which the catalytic activity of their C-terminal protease domain is inhibited by intramolecular interactions with the N-terminal regulatory region. This regulatory region contains two DEDs; one binds to the DED of FADD, while the other mediates homodimerization with other caspase-8 and caspase-10 protomers. This oligomeric assembly of Fas, FADD, and caspase-8/10 has been termed the "death-inducing signaling complex" (DISC). It is important to note that the precise stoichiometry of the DISC is not fixed. The multiplicity of DED domains within the receptor-regulated caspases and their capacity to interact with each other enables the DISC to propagate outward in chains,[138,139] incorporating additional caspase monomers as it grows. Once they are recruited into the DISC, caspase-8 and caspase-10 undergo conformational changes that derepress their catalytic activity, leading to the cleavage of their catalytic domains from their respective regulatory regions and additional processing to generate fully active caspase enzymes. The activated proteases then diffuse into the cytoplasm, where they target their apoptotic effectors, caspase-3 and Bid (Fig. 31.4). This leads, in turn, to cytoskeletal collapse, nuclear DNA reduction, and mitochondrial permeabilization by the pathways described above. FasL-induced DISC assembly does not necessarily translate into apoptosis. The DED containing adapter molecule c-FLIP controls DISC lethality by infiltrating into the complex.[140] The N-terminal region of c-FLIP resembles that of a receptor-regulated caspase, and it binds to caspase-8 and caspase-10 with high affinity. At low levels, these interactions enhance the strength of DISC assembly, leading to increased caspase activation and downstream signaling. At higher levels, however, c-FLIP effectively limits caspase incorporation, reducing the capacity of the complex to generate active proteases.

Cytotoxic lymphocytes can also induce target cell apoptosis through the FasL homolog TRAIL.[141,142] Like FasL, TRAIL is expressed as a transmembrane homotrimer that can undergo metalloprotease-mediated cleavage to generate a soluble form. It signals through the TNF receptor superfamily members DR4 and DR5, which, like Fas, stimulate apoptosis via assembly of the FADD-caspase-8/10 DISC. TRAIL differs from FasL in that it is expressed not only by T cells and NK cells but also by certain subsets of monocytes, dendritic cells, and macrophages. Another point of difference between the two systems is the inducing stimulus. While FasL is trafficked to the synapse in response to the engagement of activating receptors, the cell surface expression and secretion of TRAIL is triggered by inflammatory cytokines. The release of TRAIL outside the confines of the synapse implies that a broader swath of potential target cells, including some innocent, uninfected bystanders, could be exposed to this apoptotic ligand. Several mechanisms have evolved that address this potential specificity issue. During viral infections, for instance, it has been shown that DR4 and DR5 are upregulated in infected cells and downregulated in uninfected cells.[143-146] Increased expression of endogenous caspase inhibitors and c-FLIP has also been shown to reduce susceptibility to TRAIL-mediated killing.[147-149]

In general, death receptor–induced apoptosis occurs more slowly than perforin and granzyme-mediated lysis. Recent work suggests that these kinetic differences effectively produce a division of labor during episodes of serial killing.[150] Unsurprisingly, the perforin-granzyme pathway dominates the early killing events, but as lytic granule stores are depleted, FasL-dependent killing becomes increasingly prevalent. The two killing mechanisms can also be distinguished by their molecular requirements and sensitivities. Death receptor pathways are highly dependent on caspases and are therefore sensitive to protein and small molecule caspase inhibitors.[5,141,142] By contrast, Ca^{2+} influx is required for the release of perforin and granzyme into the immune synapse.[151,152] Perforin/granzyme killing is also sensitive to concanamycin A,[153] which blocks the acidification of lytic granules, and ROS scavengers,[85,90] which mitigate mitochondrial damage. Compensation between these pathways likely contributes to the overall robustness of cytotoxic lymphocyte–dependent killing. That being said, in vivo studies indicate some degree of nonredundancy, which will be explored later in this chapter.

COMPARTMENTALIZATION OF CYTOTOXIC CARGO

Lytic granules comprise a structurally and molecularly heterogeneous group of secretory lysosomes responsible for the storage and release of perforin, granzymes, and other cytotoxic factors. It is generally thought that they are derived from the endolysosomal network and also that they serve as hybrid organelles that carry out canonical lysosomal functions in addition to their roles in target killing[154-156] (Fig. 31.5). Activating signals drive a stereotyped process by which lytic granules dock and fuse with the synaptic membrane, releasing their contents into the intercellular space. Over the past three decades, cell biological methods, high-resolution imaging, and the study of heritable cytotoxicity defects have provided critical insights into the ontogeny and composition of lytic granules. Progress in this area, however, has been and continues to be hampered by the small size and structural plasticity of lymphocytes, features that complicate the analysis of intracellular compartments.

The defining property of a lytic granule is its electron dense core, which is readily apparent in transmission electron micrographs[157,158] (Fig. 31.5). The framework for the core is established by chondroitin sulfate proteoglycans,[159] which are composed of anchor proteins linked to multiple chondroitin sulfate glycan chains. The matrix formed by these proteoglycans is highly negatively charged and consequently binds to positively charged perforin and granzymes,[160,161] which accumulate at high concentrations within the lattice. The majority of proteoglycans in lytic granules contain the lumenal protein serglycin. Accordingly, targeting of the *Serglycin* gene has been used to probe the functional importance of the dense core.[162,163] Cytotoxic lymphocytes from *Serglycin*$^{-/-}$ mice have markedly reduced proteoglycan content and form lytic granules that lack dense cores. They also exhibit a profound defect in the accumulation granzyme B protein, despite normal levels of mRNA message. The storage of granzyme A and perforin, however, is largely normal, indicating either that proteoglycans are dispensable for the compartmentalization of these factors or that another proteoglycan species compensates for the absence of serglycin. Indeed, certain glypicans and syndecans are expressed by CTLs, but their roles in lytic granule biogenesis have not been investigated. *Serglycin*$^{-/-}$ CTLs and NK cells exhibit modest reductions in cytotoxic efficiency that are consistent with their reduced granzyme B expression and the fact that they retain both granzyme A and perforin. Hence, serglycin scaffolding appears to boost granzyme B storage capacity specifically and thereby enhance the ability of patrolling cytotoxic lymphocytes to function as effective serial killers.

Lytic granules are a subclass of lysosome and therefore contain canonical lysosomal enzymes, such as cathepsins and acid hydrolases, in addition to the cytotoxic factors described above.[164,165] They can also be identified by expression of integral membrane glycoproteins Lamp1 (lysosomal-associated membrane protein 1, also called CD107a) and Lamp2. Lamp1 and Lamp2 are particularly abundant in lysosomes,

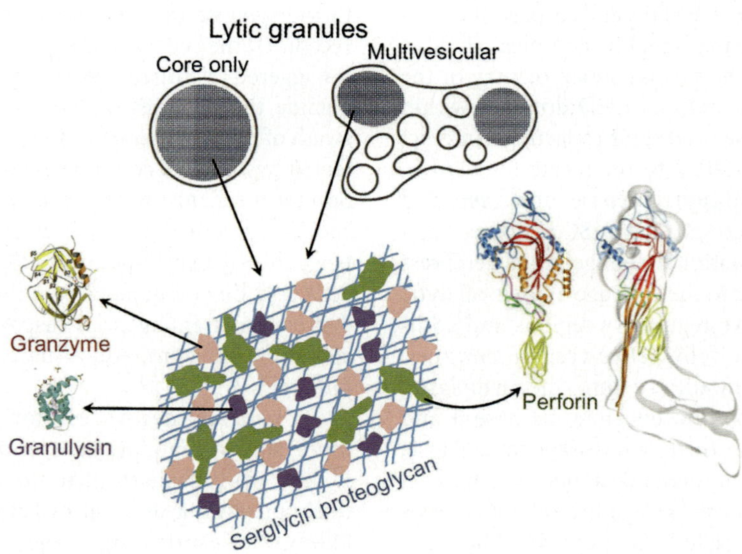

FIG. 31.5. Lytic granule composition. Lytic granules contain dense proteoglycan cores and exist either as simple, core-only granules or multivesicular compartments containing dense cores in addition to intralumenal vesicles. Cytotoxic proteins, in particular granzyme B, are concentrated within lytic granules by binding to the negatively charged proteoglycan matrix.

accounting for ~50% of the membrane protein,[166] and they are used as standard markers for this organelle. Lysosomes are the end point of the endosomal network, which is responsible for the uptake and degradation of extracellular material.[156,167] The first stage in this pathway is the early endosome, which sorts internalized cargo and sends material not destined for destruction either to the Golgi or back to the cell surface. Late endosomes, the next step in the maturation process, are characterized by a multivesicular structure, the product of ubiquitin-induced membrane involution catalyzed by the ESCRT (endosomal sorting complexes required for transport) machinery. Late endosomes also exhibit markedly acidic pH (4.5-5.0, compared to 6.5 for early endosomes), which is driven by the vacuolar-type H^+ ATPase (V-ATPase) in the endosomal membrane. V-ATPase–mediated acidification continues as late endosomes transition into lysosomes, reaching a final pH of 4.0 to 4.5. Concomitantly, resident lysosomal proteins, including cathepsins, acid hydrolases, and Lamps, are delivered from the Golgi network. Perforin and granzymes also reach the lysosome at this stage. Granzymes utilize the same trafficking pathway as other lysosomal proteases, involving N-linked glycosylation with mannose 6-phosphate.[168,169] Proteins glycosylated in this manner are bound in the Golgi by the mannose 6-phosphate receptor (MPR) and then delivered to the lysosome. Perforin trafficking is controlled by a short C-terminal extension, which drives rapid export from the ER to the Golgi.[170] This motif is then glycosylated, enabling transport to lysosomes, where it is removed by proteolytic cleavage. This entire compartmentalization process has evolved so as to limit the potential of perforin and granzymes to damage the cytotoxic lymphocyte. The rapid removal of perforin from the ER minimizes its exposure to activating Ca^{2+} and neutral pH. Indeed, expression of a perforin mutant lacking the C-terminal export motif leads to ER retention and lymphocyte death.[170] Similarly, granzymes are only processed into their fully active states upon reaching the lysosome,[69] where their activity is inhibited both by acidic pH and by binding to the proteoglycan core.

The importance of proper lysosomal trafficking and maturation for cytotoxic lymphocyte function is highlighted by foundational genetic and pharmacological observations. The V-ATPase inhibitor concanamycin A, for instance, which deacidifies lysosomes, is well known to block perforin and granzyme-mediated killing.[153] Similarly, loss-of-function mutations in LYST and AP3B1, two proteins involved in trafficking between lysosomes and other intracellular compartments, are causative for Chediak-Higashi and Hermansky-Pudlak syndromes, respectively.[171,172] Both diseases are immune deficiencies characterized by profoundly impaired lymphocyte cytotoxicity and susceptibility to microbial infection.

Electron microscopy studies have identified at least two different types of lytic granule.[157,158] The first type contains only the proteoglycan core surrounded by a vesicle membrane, while the second is a more complex compartment that exhibits both core regions and intraluminal vesicles (Fig. 31.5). It is tempting to speculate that these two compartments correspond to different stages along the granule maturation pathway. The multivesicular structure of the complex granules, for instance, is consistent with the notion that they are responsible both for routine lysosomal functions and also for producing specialized, core-only granules designed for the storage and secretion of cytotoxic molecules. Alternatively, one could imagine a scenario in which core-only compartments fuse with multivesicular endosomes to generate complex granules, which are then secreted. In that regard, it is interesting to note that most granules containing perforin and granzyme lack the small GTPase Rab27a and the adapter protein Munc13-4,[173,174] which are required for docking and priming, respectively, on the plasma membrane (see below). Indeed, Munc13-4, which accumulates on recycling endosomes, actually drives fusion between recycling endosomes and Rab27a$^+$ late endosomes.[173] The resulting hybrid compartment then fuses with perforin$^+$granzyme$^+$ vesicles to produce fusion competent lytic granules (Fig. 31.9). The merging of granules containing cytotoxic cargo with fusion factors derived from late endosomes is consistent with the idea that core-only granules might mature to more complex compartments prior to secretion. Total internal reflection fluorescence (TIRF) imaging of individual degranulation events, however, has indicated that the granules that actually fuse with the plasma membrane tend to be small and uniform in size.[175] Similarly, recent correlative light and electron microscopy studies have found that Vamp2, a v-SNARE protein required for lytic granule fusion with the plasma membrane (see below), accumulates exclusively in a population of small (~350 nm diameter) dense core granules.[176] Hence, the core-only subset may be the pool that is secreted, after all. Additional high-resolution imaging, coupled with precise loss-of-function methods, will be required to resolve the functional heterogeneity of lytic granules and decipher the ontogeny of the various compartmental subgroups.

With respect to ontogeny, the transcription factors TFEB and TFE have recently emerged as master regulators of lysosome biogenesis.[177] As such, it will be interesting to explore their roles in the generation and maintenance of lytic granules. Also poorly understood is the importance of the cytoskeleton for granule biogenesis. Lysosomes traffic along microtubules, and their specific location within the microtubule network has been linked to their maturation and function in nonimmune cell types.[178,179] Interestingly, CTLs lacking centrioles, the core structural components of the centrosome, exhibit a marked defect in granule formation and persistence,[180] implying a role for the cytoskeleton in critical maturation and/or maintenance programs.

It is now generally accepted that cytotoxic lymphocytes bear preexisting pools of FasL, which they store in specialized intracellular compartments.[129,132] In response to activating signals, these compartments fuse with the synaptic membrane, releasing their cargo into the immune synapse. Although FasL was initially thought to occupy perforin$^+$granzyme$^+$ lytic granules, more recent proteomic and imaging-based studies demonstrated that FasL occupies a distinct subset of compartments.[132,165,181] These results are consistent with functional experiments indicating that, unlike perforin and granzyme, FasL does not require Ca^{2+} influx for secretion.[182] Beyond that, the architectural and molecular features of FasL^{+} compartments are largely unknown.

CHOREOGRAPHY OF KILLING

Cytotoxic killing is a contact-dependent process that moves through a stereotyped sequence of steps (Fig. 31.6). Activating ligands on the target cell surface induce strong polarized signaling within the lymphocyte, leading to formation of the immune synapse the secretion of cytotoxic proteins into the intercellular space. This section will describe the molecular and cellular events that characterize each of the steps in the killing process, with particular emphasis on features that contribute to the potency and the specificity of the response.

Cytotoxic Lymphocyte Activation and Signaling

Canonical cytotoxic killing requires target cell adhesion, immune synapse formation, cytoskeletal polarization, and the elevation of intracellular Ca^{2+}.[3] The activating signals that drive these processes are best understood in the case of CTLs, which are activated by TCR-mediated recognition of cognate pMHC[183,184] (Fig. 31.7). CD8 enhances the avidity of this interaction by binding to the side of the pMHC complex. TCR engagement induces rapid phosphorylation of immunotyrosine-based activation motifs (ITAMs, consensus sequence YxxL/Ix$_{(6\text{-}8)}$YxxL/I) within the signaling chains of the associated CD3 complex by the Src family kinase Lck. Phosphorylated ITAMs recruit the Syk family kinase Zap70, which, together with Lck, catalyzes the phosphorylation of multiple residues within the transmembrane adapter protein LAT. This induces the formation of a complex between LAT and a second large adapter, Slp76, which together serve as a platform for the recruitment and activation of other proteins that propagate activating signals downstream. One such protein is the guanine nucleotide exchange factor SOS, which catalyzes the formation of the active, GTP bound form of the small GTPase Ras. Ras-GTP then stimulates signaling through both the MAP kinase (MAPK) and phosphoinositide 3-kinase (PI3K) pathways. Another exchange factor, Vav, activates Rho family GTPases and thereby transmits signals to the filamentous actin (F-actin) cytoskeleton.

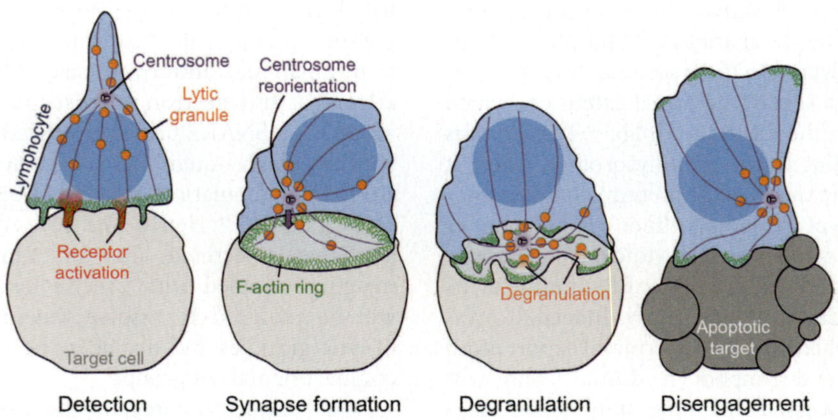

FIG. 31.6. The choreography of cytotoxic killing. Lymphocytes detect target cells with microvillar-like structures that are enriched in activating receptors. Cognate ligand binding induces synapse formation on the target cell surface, which is driven by a radially symmetric lamellipodium. Concomitantly, the centrosome reorients to the interface, carrying with it most of the lytic granules in the lymphocyte. Degranulation then occurs at the base of F-actin–rich protrusions that exert mechanical force against the target cell surface. Target cell death induces synapse dissolution via poorly understood mechanisms, enabling the lymphocyte to move on to other targets.

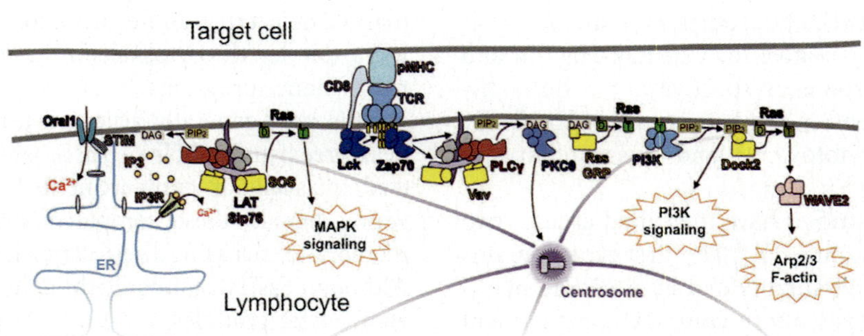

FIG. 31.7. Schematic diagram showing membrane proximal TCR signaling events and how they couple to downstream pathways.

One of the most critical effector proteins recruited to the LAT-Slp76 complex is phospholipase C gamma (PLCγ).[183,184] This lipid-processing enzyme cleaves phosphatidylinositol-4,5-bisphosphate (PIP_2) in the plasma membrane to generate inositol trisphosphate (IP_3) and diacylglycerol (DAG). IP_3, the soluble headgroup of PIP_2, diffuses into the cytoplasm, where it binds to IP_3-sensitive Ca^{2+} channels (called IP_3 receptors) that are embedded in the ER membrane (Fig. 31.7). Ca^{2+} binding induces IP_3 receptor opening, leading to the depletion of ER Ca^{2+} stores, which in turn drives the clustering of the ER resident protein STIM. Clusters of STIM then accumulate at ER–plasma membrane junctions, where they recruit and activate the ion channel Orai1, whose opening leads to Ca^{2+} influx from the extracellular space and elevation of intracellular Ca^{2+} levels. DAG, the lipid remnant left after PIP_2 cleavage, triggers a complementary set of signaling events at the membrane by recruiting proteins containing DAG-binding C1 domains. In T cells, these include protein kinase C (PKC) θ, which activates the NF-κB pathway, and RasGRP, a Ras-specific exchange factor.

The stereotyped set of TCR-induced signaling events described above represents the most extensively studied example of ITAM signaling, and the general features of this network, including the initial choreography of Src and Syk kinases, the assembly of a scaffolding complex like LAT-Slp76, and activation of a PLC isoform are shared by other ITAM-coupled receptors. NK cells express multiple such receptors, including CD16 and members of the natural cytotoxicity receptor (NCR) family.[29] Engagement of these receptors induces robust membrane proximal signaling, leading to strong Ca^{2+} flux and degranulation. NK cells also express other types of activating receptor, however, which transduce signals via distinct mechanisms. NKG2D, for instance, signals though YINM motifs in its associated DAP10 signaling chain, leading to the direct recruitment of PI3K and Vav.[185] Another common activating NK receptor, 2B4, contains immunotyrosine-based switch motifs, which, upon phosphorylation, bind to the Src kinase Fyn through the adapter protein SAP.[186] In general, non-ITAM classes of NK receptor induce weaker activating signals than their ITAM containing counterparts. Simultaneous engagement of multiple non-ITAM receptors, however, has been shown to boost NK cell activation and cytotoxicity to levels commensurate with the effects of ITAM signaling.[187] This makes sense given the multifactorial nature of NK cell target selection, which requires the quasi-proportional enumeration of countervailing inhibitory and activating stimuli.

Although synapse formation is initiated by activating immunoreceptors like the TCR, it is the integrins that establish strong binding to the target cell. Integrins constitute a family of large, heterodimeric proteins that couple recognition of extracellular ligands, typically extracellular matrix molecules or adhesion proteins, to the cortical F-actin cytoskeleton.[188] The $\alpha_L\beta_2$ integrin LFA1, which binds to intercellular adhesion molecule 1 (ICAM1) or ICAM2 on the target cell, plays a particularly important role in target cell engagement and is also known to maintain synapse organization and boost the strength of activating signals.[187,189-195] Indeed, LFA1 is often likened to classical costimulatory receptors for its capacity to enhance the effects of TCR signaling and to synergize with activating NK receptors. At rest, integrins occupy a bent, inactive conformation in which their ligand binding head domain is occluded. Signaling through the TCR and other activating receptors induces the jackknife-like opening of integrins on the cell surface, exposing the headgroup for ligand engagement.[196,197] A critical component of this "inside-out" integrin activation pathway is the small GTPase Rap1, which is activated by two exchange factors, C3G and RAPGEF2, within the TCR signaling network.[198,199] Activated, GTP-bound Rap1 modulates integrin activity through its effector proteins RapL and RIAM.[200,201] RIAM plays a particularly important role by recruiting the large cytoskeletal adapter protein talin, which binds to the cytoplasmic tail of the integrin β subunit.[202,203] This interaction separates the β tail from the corresponding α tail, inducing a conformational change that propagates through the plasma membrane to elicit integrin opening. Subsequent ligand binding induces the assembly of protein complexes around talin and the integrin tails. These complexes, which contain the cytoskeletal adapters paxillin and vinculin, transmit "outside-in" signals that further boost lymphocyte activation.[204] The functional relevance of both inside-out and the outside-in integrin signaling is well established. Of particular relevance to this chapter, blocking integrin engagement reduces cytotoxicity by a factor of three.[205] It remains unclear, however, whether this loss-of-function phenotype reflects the importance of integrins for general T cell activation or a more specific role in target cell adhesion or degranulation. Unfortunately, the pleiotropy of integrin signaling and the challenge of experimentally separating the effects of the inside-out and the outside-in pathways make focused mechanistic analyses more challenging.

While costimulatory receptors, most obviously CD28, are critical for the priming and differentiation of naïve T cells,[206] their role in the activation of cytotoxic effectors is less well understood. The restricted expression pattern of CD28 ligands, which are found predominantly on B cells and antigen-presenting myeloid cells, does not fit with the broad applicability of CTL-mediated killing, implying that CD28 signaling is dispensable for cytotoxicity. Consistent with this interpretation, CTLs kill target cells robustly in vitro in the absence of CD28 engagement. In vivo studies, however, suggest a potential role for CD28 in the maintenance of cytotoxic potential in peripheral locations. Indeed, dendritic cells recruited to sites of infection and tumor growth have been shown to interact physically with infiltrating $CD8^+$ T cells, thereby promoting their effector function.[207-209] It seems likely that the supportive activity of these T cell–dendritic cell synapses depends, at least in part, on CD28. Consistent with this conjecture, therapeutic rejuvenation of $CD8^+$ T cells by blockade of the inhibitory receptor PD1 (see below) requires both interactions with peripheral dendritic cells[210] and CD28.[211] Hence, although classical costimulation may not be involved in cytotoxicity induction by target cells, it does appear to play a role in maintaining CTL responsiveness in critical immune environments.

Within the CTL–target cell synapse itself, the role of "costimulation" appears to have been subsumed by activating receptors canonically found on NK cells. CTLs express a variety of these proteins, as well, including members of the SLAM family, the NCRs, and NKG2D. The costimulatory function of activating NK receptors is best understood for NKG2D, which is widely expressed among both human and mouse CTL subsets.[212] NKG2D engagement does not induce T cell activation and cytotoxicity on its own; rather, it enhances TCR-induced responses in a manner analogous to the role of CD28 during naïve T cell priming.[213] Given this functional parallel, it is interesting to note that both CD28 and NKG2D signal via cytoplasmic YINM motifs. CD28 contains one such motif in its cytoplasmic tail, while NKG2D is noncovalently coupled to the YINM in DAP10. This confers a certain degree of similarity to CD28 and NKG2D-induced signaling, particularly with regard to activation of the PI3K pathway. Other mechanistic aspects differ, however. NKG2D recruits and activates PLCγ,[214] for instance, which is not known to function downstream of CD28. Ligands for NKG2D are often upregulated in virally infected or transformed tissue,[215] implying that NKG2D engagement is particularly critical for CTL function in these contexts. Consistent with this notion, genetic deletion or blockade of NKG2D promotes tumor growth in multiple experimental systems.[216,217] Similarly, NKG2D$^+$ CD8 T cells have been associated with antiviral responses in both mice and humans.[218,219]

The identification of target cells by cytotoxic lymphocytes is complicated by the fact that activating ligands are embedded within the glycocalyx, a dense mixture of glycoproteins and glycolipids that covers the surfaces of most cells.[220] In order to detect these ligands efficiently, lymphocytes have developed a physical mechanism for infiltrating this matrix. It has been known for some time that lymphocytes are covered with microvilli, small, F-actin–based structures that project ~1 μm from the surface. In recent years, it has become apparent that the distribution of cell surface proteins along these structures is nonuniform.[221,222] The TCR, in particular, is concentrated at the microvilli tips, implying that they might be designed to penetrate the glycocalyx in search of cognate pMHC. Consistent with this model, it has been shown that limited contact between stimulatory surfaces and microvilli tips is sufficient to induce strong TCR signaling.[223] Furthermore, high-resolution analysis of T cell microvilli has revealed that they are highly dynamic and that they move in a diffusive manner, seemingly scanning the space in front of them.[224] Hence, T cells facilitate antigen detection by concentrating activating receptors at the tips of motile protrusions that they use to probe their environment (Fig. 31.6). Whether NK receptors exhibit an analogous organization pattern remains to be seen.

Synapse Formation and F-Actin Architecture

Within the context of the cytotoxic cascade, the purpose of lymphocyte activation is to induce immune synapse assembly and subsequent degranulation. Synapse formation is initiated by the spreading of a radially symmetric lamellipodium over the target cell surface[225,226] (Fig. 31.6). This process depends on the Arp2/3 nucleation complex, which catalyzes actin polymerization from the sides of existing F-actin fibers.[227] Arp2/3 recruitment to the incipient synapse is mediated by WAVE2, an actin nucleation promoting factor that itself localizes to the periphery of the expanding contact.[195,228,229] The activation of Arp2/3 in this zone induces the formation of a dense, branched F-actin array that expands the interface to a size of ~100 to 200 μm^2 within a minute or so. The mature synapse exhibits stereotyped concentric organization.[225] At the periphery of the contact lies the abovementioned band of lamellipodial F-actin. As with most lamellipodia, this peripheral F-actin sheet undergoes continuous retrograde flow (treadmilling) toward the center of the contact. Within the immune synapse, this centripetal flow is further reinforced by nonmuscle myosin II (NMII), which accumulates just within the inner aspect of the peripheral F-actin ring.[230,231] Here, it exerts contractile forces that collapse the dendritic lamellipodial array into arcs. Collectively, these dynamics sweep clusters of ligand-bound activating receptors and integrins toward more centrally located synaptic domains. Ligand-bound integrins, most conspicuously LFA1, accumulate in an annular cluster just inside the F-actin ring,[231-233] while smaller activating receptors, such as the TCR, traffic to the very center of the synapse, where they are downregulated and endocytosed either for degradation or for recycling to the surface.[234,235] The center of the synapse tends to be devoid of lamellipodial F-actin. Instead, punctate, protrusive F-actin foci appear in this domain.[236] These structures, which depend on the NPF family member WASP (Wiskott-Aldrich syndrome protein), have been implicated in the maintenance of contact symmetry and, as described below, play an important role in the mechanical potentiation of killing responses.[237,238]

The patterning of synaptic domains is dictated by lipid second messengers in the synaptic membrane[239] (Fig. 31.8). The periphery of the contact is dominated by an annular accumulation of phosphatidylinositol-3,4,5-trisphosphate

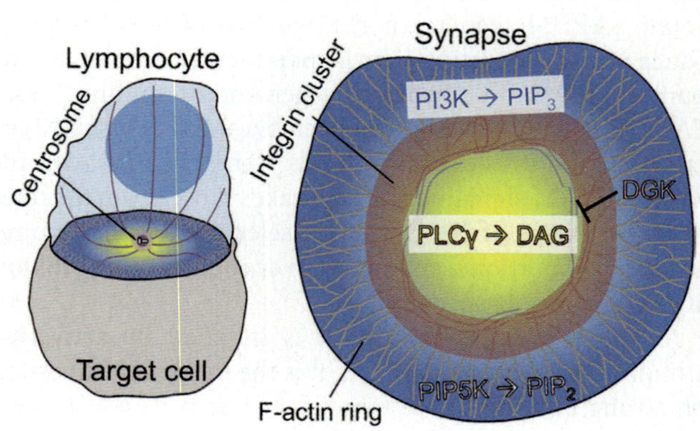

FIG. 31.8. Lipid second messengers pattern cytoskeletal architecture at the immune synapse. A lymphocyte–target cell conjugate is depicted on the left and an end face view of the immune synapse on the right. DAG accumulation at the center of synapse, which is established by the combined activities of PLCγ and DGK, functions to recruit the centrosome. In the periphery, annular gradients of PIP_3 and PIP_2 promote the production and stabilization of lamellipodial F-actin.

(PIP_3),[195,240,241] which is generated from PIP_2 by class IA PI3K isoforms downstream of activated Ras-GTP.[195] This PIP_3 gradient stimulates F-actin polymerization in the periphery by recruiting Dock2, an exchange factor for the Rho family GTPase Rac. Subsequent GTP loading of Rac leads to the recruitment of WAVE2 in this zone and the stimulation of Arp2/3-dependent F-actin polymerization. The synapse periphery is also enriched in PIP_2,[242] which stabilizes the branched F-actin array by interacting with a number of F-actin binding proteins and also by serving as the substrate for more PI3K-mediated PIP_3 production. High PIP_2 levels are maintained in this region by phosphatidylinositol-4-phosphate 5-kinases (PIP5Ks), which accumulate on the negatively charged inner leaflet of the peripheral synaptic membrane.[242] By contrast, the center of the synapse contains lower levels of PIP_2 and PIP_3 and is instead enriched in PLCγ-derived DAG.[243] This centralized DAG gradient results not only from polarized PLCγ activation at the synapse but also from rapid DAG conversion to phosphatidic acid by DAG kinases at the periphery of the contact.[244] As detailed below, PLCγ-dependent DAG production is particularly important for inducing polarization of the microtubule cytoskeleton.[245] It also contributes to the marked hypodensity of F-actin at the center of the synapse by depleting negative charge from the plasma membrane. This induces dissociation of PIP5K, leading in turn to reduced PIP_2 levels and F-actin accumulation.[242]

It must be emphasized that most high-resolution studies of synaptic architecture have employed not lymphocyte–target cell conjugates but rather model systems in which the target cell is replaced by a stimulatory glass surface or supported lipid bilayer coated with cognate pMHC and/or antibodies against the TCR, occasionally accompanied by purified adhesion proteins.[246,247] These systems are useful because they facilitate the application of imaging modalities such as TIRF, which can effectively image features that are difficult, if not impossible, to resolve by confocal microscopy of more complex, three-dimensional samples. The synapses that form on flat, stimulatory surfaces exhibit clear concentric organization and a well-defined peripheral F-actin ring. Retrograde flow within this ring is consistent and uniform across the entire circumference of the contact. These experimental conditions, however, do not fully replicate the chemical and mechanical properties of a target cell. As such, the contacts that they induce only partially capture the architectural complexity of bona fide cytotoxic synapses. Peripheral F-actin rings are much less apparent in lymphocyte–target cell interactions, and when they do appear they are much more transient. Furthermore, bone fide synapses exhibit a wider variety of F-actin–based structures, including dynamic protrusions and explosive anterograde waves that emerge from the center of the contact.[224,238,248,249] The molecular mechanisms governing the interplay between these distinct cytoskeletal processes are largely unknown.

Lytic Granule Polarization and Cytotoxic Specificity

Cytotoxic lymphocytes efficiently eliminate target cells while leaving healthy bystander cells untouched. This specificity of killing is remarkable given that lymphocytes operate in densely packed intercellular environments in which essentially all cells are sensitive to perforin and granzyme. To prevent bystander killing under these conditions, lymphocytes must deliver cytotoxic factors such that only the intended target cell is exposed. The immune synapse ensures this specificity by channeling perforin and granzymes into the lymphocyte–target cell interface.[154,250] This directional secretion process is supported both by the polarized trafficking of lytic granules within the lymphocyte and the creation of dedicated domains within the synaptic membrane for docking and fusion.

Perhaps the most characteristic feature of an activated, synaptically engaged lymphocyte is the polarization of the microtubule cytoskeleton toward the target cell[154,250] (Figs. 31.6 and 31.8). Lymphocyte microtubules radiate from the centrosome (also called the microtubule organizing center), a membraneless organelle containing a central pair of centrioles surrounded by a cloud of proteinaceous pericentriolar material. Within minutes of initial lymphocyte activation, the centrosome reorients to a position just beneath the center of the synapse. Concomitantly, lytic granules converge on the centrosome by trafficking along microtubules. The end result is the close apposition of the granule pool with docking and fusion sites in the synaptic membrane. The presumed importance of this polarization process for cytotoxic specificity has made it the subject of intense research for a number of years.

In T cells, centrosome polarization depends on TCR signaling and is spatially coupled to regions of TCR, but not LFA1 activation.[251] By contrast, LFA1 engagement is sufficient to induce centrosome polarization in NK cells.[252] The basis for this discrepancy is not well understood. Studies in T cells have revealed several key steps in the polarization process. PLCγ plays a critical role by generating a gradient of DAG centered at the synapse[245] (Figs. 31.7 and 31.8). This DAG accumulation induces the ordered recruitment of multiple PKC isoforms,[243] which in turn drive the accumulation of the microtubule motor dynein in the region of TCR stimulation.[245,253] Dynein is a minus end–directed motor, and microtubules emanate from the centrosome with their plus ends outward and minus ends inward. Hence, the movement of membrane-anchored dynein molecules along microtubules would have the effect of dragging the centrosome toward the synapse. Dynein is also capable of inducing microtubule shrinkage from the plus end while maintaining contact with the residual filament. Close analysis of microtubule dynamics in polarizing T cells has indicated that this capture-shrinkage mechanism is critical for centrosome reorientation to the synapse.[254] As dynein pulls on the centrosome from the front, formin proteins promote the generation of stabilized microtubules on the opposite side of the T cell, which "push" the centrosome forward.[255] Contractile clusters of cortical NMII on the sides and back of the cell also appear to contribute to the process by shifting the entire microtubule cytoskeleton in the direction of the target cell.[253] Transmission electron micrographs of CTL–target cell conjugates indicate that the polarized centrosome actually comes into physical contact with the synaptic membrane.[256,257] This docking interaction is reminiscent of the primary cilium, a microtubule-based sensory organelle found in most cell types, although, interestingly, not lymphocytes.[258] Recent work suggests that centriolar distal appendages, which mediate centrosome anchoring at the base of the primary cilium, also control centrosome docking at the immune synapse.[259] Hence, it appears that lymphocytes have coopted at least some of the machinery responsible for primary cilium formation in order to build a polarized secretory contact.

Lytic granules are closely associated with microtubules and use them to access different parts of the cell.[154] Upon lymphocyte activation, granules move in a minus end–directed manner so as to cluster around the centrosome. This trafficking is mediated by dynein,[260] which is physically coupled to lytic granules via Rab7 and the adapter protein RILP.[261] In CTLs, granule convergence also requires Ca^{2+} influx,[151] paralleling the requirement of DAG for centrosome polarization and further emphasizing the importance of PLCγ signaling for the delivery of granules to the synapse. Interestingly, Ca^{2+} signaling is not required for granule clustering in NK cells.[262] While the basis for this discrepancy is not understood, it is consistent with the fact that ITAM signaling is a strict requirement for synapse assembly in T cells, but not NK cells. Once they reach the synapse, lytic granules in the centrosomal cluster must find suitable plasma membrane sites for fusion. Precisely how they access these sites is poorly understood. It has been argued that the centrosome itself delivers granules to their final destinations.[154] This model is consistent with the close apposition between the centrosome and the plasma membrane and experiments indicating that plus end–directed microtubule motion is dispensable for killing.[257] Other lines of investigation, however, suggest that granules do in fact move along synaptic microtubules in the minus to plus direction to reach their ultimate sites of fusion.[263,264] These mechanisms are, of course, not mutually exclusive; each might apply to separate groups of cytotoxic lymphocytes or to separate granule pools within the same cell.

Although centrosome polarization is clearly important for directional cytotoxic secretion, it should be noted that it is not an absolute requirement. Indeed, studies from multiple labs have demonstrated that cytotoxic lymphocytes are capable of specific killing independent of the centrosome. Videomicroscopy-based analyses of CTL–target cell interactions, for example, have documented numerous cases of CTLs secreting granule contents and killing target cells prior to centrosome reorientation.[265] Furthermore, NK cells can secrete cytotoxic cargo via thin plasma membrane nanotubes in the complete absence of an immune synapse.[266] Recently, a genetic strategy was used to delete centrioles from interphase CTLs.[180] Although these cells lacked functional centrosomes and exhibited perturbed microtubule architecture, they were nevertheless capable of specific killing and, importantly, directional secretion. In fact, even pharmacologic depolymerization of the entire microtubule cytoskeleton failed to alter the directionality of granule release, although it did substantially reduce the magnitude of the secretion.[180] Hence, while centrosome polarization enhances the efficiency of cytotoxic exocytosis by delivering granules to the synapse, other control mechanisms are also critical for establishing the directionality of the response.

It seems likely that these additional mechanisms operate at the synaptic membrane itself by establishing permissive sites for granule docking and fusion. Little is known about these putative docking sites other than the fact that they must be devoid of cortical F-actin, which would otherwise impede granule approach. Based on this criterion, it was initially proposed that granule fusion occurs in the central region within the adhesive integrin ring, which was thought to be relatively clear of F-actin. Higher resolution imaging studies, however, revealed that this central domain is actually crisscrossed by numerous individual F-actin fibers.[267,268] Granule fusion was shown to occur in small, transient gaps that appear within the F-actin mesh. Although the mechanisms controlling these transient clearing events are unknown, it is tempting to speculate that lipid second messenger signaling might be involved, particularly because PIP_2, PIP_3, and DAG are known to pattern synaptic F-actin at a larger scale.[239,242] In that regard, it is interesting to note that key components of the granule docking and fusion machinery, such as Munc13-4 and the synaptotagmin-like proteins, contain membrane-binding domains that could engage specific lipids. Studies have also linked LFA1 to the spatial organization of granule exocytosis. In NK cells, LFA1 engagement was found to be critical for establishing a degranulation domain at the center of the synapse.[269] More recently, super-resolution imaging experiments documented an intriguing link between granule docking and regions of LFA1 activation.[270] An important role for integrin-mediated contacts in cytotoxic secretion makes sense because they would provide a straightforward way for the lymphocyte to ensure that toxic cargo is delivered into a region of close contact with the target, thereby minimizing damage to surrounding bystander cells.

Lytic Granule Docking and Exocytosis

Vesicle fusion is an energetically demanding process that requires the close apposition and subsequent distortion of vesicle and target membranes. In eukaryotic cells, a standard fusion mechanism has evolved that is based on heterotypic interactions between members of the SNARE family.[271-273] Each class of intracellular membrane contains a distinct cohort of vesicle and target SNAREs (v-SNAREs and t-SNAREs, respectively) that essentially dictates its fusion potential. If v-SNAREs and t-SNAREs on opposing membranes are compatible, they anneal into a four-helix bundle, called a SNAREpin, which pulls the membranes close together and drives their fusion. In vivo, SNARE activity is controlled by accessory factors that regulate both the initiation and the assembly of SNAREpins. Often, the fusion process moves through an obligate "docking" step mediated by cognate protein–protein and protein–lipid interactions between the opposing membranes. Rab family GTPases play an outsize role in docking by recruiting proteins that specify both the identity of a compartment and its fusion competence.[274] In certain situations, docked vesicles must also undergo "priming" prior to secretion. Priming is defined somewhat nebulously, but is generally thought to involve the partial assembly of SNAREpins. Adaptor proteins in the Munc18 and Munc13 families regulate priming at both neuronal and immune synapses.[155,156,275]

Many of the specific mediators of lytic granule fusion were initially identified by studying human immunodeficiencies. Loss-of-function mutations in the t-SNARE syntaxin-11 and the priming adapters Munc18-2 and Munc13-4 lead to FHL, a family of disorders characterized by susceptibility to viral infections and multicompartment immune activation.[276-279] Cytotoxic lymphocytes derived from FHL patients exhibit profound defects in degranulation and killing. Similarly, loss of the critical docking factor Rab27a causes Griscelli syndrome, a disease characterized by both immunodeficiency and albinism.[280] This distinctive confluence of symptoms results from the fact that Rab27a is necessary not only for lymphocyte cytotoxicity but also for pigment release by melanocytes. Naturally occurring null mutations of murine Rab27a and Munc13-4 have been characterized (*ashen* and *jinx* mice, respectively),[281,282] and it has been shown that they also abrogate granule release.

Cytotoxic lymphocytes lacking Rab27a display a striking killing phenotype in which lytic granules polarize toward the target but fail to make contact with the synaptic membrane.[283] This observation, taken together with the fact that Rab27a localizes to a subset of lytic granules,[173] indicates that the protein is required for granule docking. Precisely how Rab27a operates in this context is unclear, but likely involves synaptotagmin-like protein 1 (Slp1) and Slp2, which bind selectively to active, GTP-loaded Rab27a and are therefore thought to be Rab27a effectors.[284,285] In the neuronal synapse, synaptotagmin mediates Ca^{2+}-dependent docking to the synaptic membrane through its C2 domains.[275] Slp2 has been shown to mediate phosphatidylserine-dependent membrane docking in melanosomes.[286] Hence, it is reasonable to expect that Rab27a-associated Slp1 and Slp2 operate as docking factors in lymphocytes (Fig. 31.9), although this hypothesis has not been rigorously tested.

Lytic granule fusion with the plasma membrane is mediated by a SNAREpin containing the v-SNARE Vamp2 (Vamp7 in humans) and the t-SNARE syntaxin-11.[176,287,288] SNAP-23, a ubiquitously expressed member of the SNAP-25 subclass of t-SNAREs, likely serves as the third obligate member of this complex[288] (Fig. 31.9). Appreciable fusion still occurs in syntaxin-11–deficient cells, implying that other members of the syntaxin family also operate in this context. Functional assembly of the Vamp2/syntaxin-11/SNAP-23 SNAREpin also requires Munc13-4 and Munc18-2. In cytotoxic lymphocytes lacking either protein, lytic granules dock at the immune synapse but fail to fuse.[278,279] Munc13-4–dependent priming has been specifically linked to the immobilization of granules on the synaptic membrane.[289] The mechanistic basis for this effect, however, is not well understood. Munc13-4 interacts physically with Rab27a and SNARE complexes, and in cells, it localizes to both lytic granules and the plasma membrane.[290,291] It contains two C2 domains, which are thought to mediate membrane binding, and two MHD domains, which engage in protein–protein interactions. Both C2 domains are required for priming, and it is likely that the MHD domains also function in this context.[279,289,290] These results are consistent with a model in which Munc13-4 engages in multiple distinct interactions either sequentially or simultaneously to couple Rab27a-dependent docking to productive SNAREpin assembly. Munc18-2, for its part, appears to function as a regulator of syntaxin-11. The two proteins interact directly,[277,278] and Munc18-2 has been shown to drive syntaxin-11–dependent granule fusion to completion in cell-free assays.[292]

Given the toxic capacity of granule contents, it is not surprising that granule release is subject to additional control mechanisms above and beyond docking and priming. Syntaxin-11 availability appears to be a particularly important nexus of regulation.[156] In resting cytotoxic lymphocytes, most syntaxin-11 localizes to recycling endosomes.[288] Upon activation, these endosomes fuse with the synaptic membrane in a manner that depends on the v-SNARE Vamp8[293]

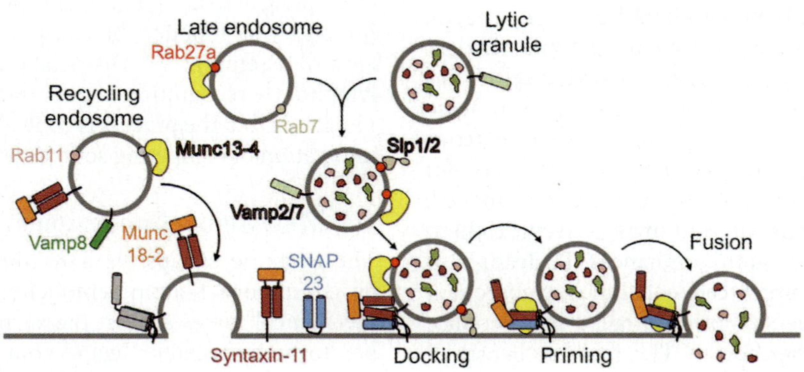

FIG. 31.9. Molecular mechanisms of cytotoxic secretion. Lytic granule fusion is mediated by a SNAREpin composed of syntaxin-11, SNAP-23, and Vamp2/7. Rab27a controls granule docking via Slp1/2, while Munc13-4 and Munc18-2 regulate the priming step. Late endosomes provide key components to lytic granules to confer fusion competence. Finally, recycling endosomes deliver syntaxin-11 to fusion competent regions of the plasma membrane, establishing zones for subsequent lytic granule docking.

(Fig. 31.9). This creates domains of syntaxin-11 enrichment, which are then targeted by lytic granules. The direct fusion of recycling endosomes with the plasma membrane may also be important for delivering additional Munc13-4 for the priming step.[156,173] Earmarking granule fusion sites in this manner provides an additional layer of spatiotemporal control and likely contributes to the maintenance of cytotoxic specificity. Granule exocytosis is also highly dependent on Ca^{2+} influx. Ca^{2+} chelators profoundly suppress cytotoxic secretion,[294] and combined stimulation with phorbol esters and calcium ionophores can induce degranulation in the absence of activating receptor engagement.[295] Although this requirement for Ca^{2+} has been known for some time, identifying the proteins responsible for it has been challenging. Based on their homology with synaptotagmin, one might expect Slp1 and Slp2 to be Ca^{2+} responsive. Neither protein, however, binds membranes in a Ca^{2+}-dependent manner,[296] implying that other components confer Ca^{2+} sensitivity. Munc13-4 has recently emerged as a candidate for this role. Mutation of putative Ca^{2+}-binding sites within the Munc13-4 C2 domains inhibits both Ca^{2+}-dependent degranulation and target cell killing by NK cells.[290] Hence, cellular cytotoxicity is regulated by Ca^{2+} at the priming step. Pharmacological studies also indicate that the Ca^{2+}/calmodulin-dependent phosphatase calcineurin is required for cytotoxic secretion.[297-299] The relevant calcineurin substrates in this context, however, remain to be defined.

As they kill, cytotoxic lymphocytes largely escape the toxic effects of their own secretions, a property that enables them to excel as serial killers. Despite years of study, precisely how CTLs and NK cells protect themselves from perforin and granzyme-mediated killing remains poorly understood. CTLs express high levels of the granzyme B inhibitor serpin B9,[300] suggesting that they might be more resistant to granzyme-mediated killing than other cells. However, antigen-loaded CTLs are still killed robustly by otherwise identical, antigen-free CTLs, indicating that intrinsic resistance is not sufficient to explain CTL protection.[57] Rather, the synapse itself appears to confer directionality to the cytotoxic response, with the degranulating lymphocyte remaining unscathed and the nondegranulating lymphocyte undergoing cell death. Accordingly, much work has focused on granule resident factors that could selectively protect the attacking cell. It has been suggested that cathepsin B, a lysosomal cysteine protease that moves to the cell surface during degranulation, serves as an enzymatic line of defense by cleaving cytotoxic molecules that diffuse back toward the synaptic membrane.[301] Indeed, cathepsin B can degrade perforin, and small molecule cathepsin inhibitors were initially found to enhance CTL death (also called suicide) in CTL-target cell cocultures. Subsequent pharmacological experiments, however, failed to reproduce these findings, and furthermore, CTLs from cathepsin B knockout mice were found to kill target cells normally without enhanced suicide.[302] More recent work has focused on the highly expressed glycoprotein Lamp1, which also becomes surface exposed during degranulation. Overexpression of Lamp1 in heterologous cells protects them from NK cell–mediated killing, and NK cells lacking Lamp1 exhibit higher levels of suicide during cytotoxic interactions.[303] These results are consistent with the idea that cell surface Lamp1 functions as a physical barrier to perforin binding. The Lamp1 loss-of-function phenotypes are quite subtle, however, implying the existence of other mechanisms for cytotoxic lymphocyte protection. Efforts in this area have thus far focused almost exclusively on identifying putative protective proteins that are asymmetrically distributed between killer cell and target. It is also conceivable, however, that the dynamic physical architecture of the synapse is itself protective. Indeed, it has been shown that perforin binding is modulated by the spacing between phospholipid headgroups in the target membrane.[38] Hence, one can imagine a situation in which the directionality of killing is dictated by asymmetric membrane stretch and/or curvature across the synapse.

Lymphocyte Dissociation

Potent serial killing demands not only rapid and specific destruction of synaptically engaged target cells but also efficient dissociation so that the lymphocyte can move on to find the next target (Fig. 31.6). As with lymphocyte protection from cytotoxic secretions (see above), we know little about the molecular mechanisms controlling synapse termination and lymphocyte dissociation. What is clear, however, is that the process is linked to target cell death. Genetic or pharmacologic blockade of perforin/granzyme-mediated killing extends synapse lifetime, promoting sustained lymphocyte activation and enhanced secretion of inflammatory cytokines.[304-306] The apoptotic cascade appears to be particularly important for this process, as caspase inhibitors also prolong contact time. It is possible that a molecular feature associated with apoptosis, such as the surface exposure of phosphatidylserine, could function as a signal to drive synapse disassembly. In that regard, it is interesting that mature CTLs and NK cells both express the phosphatidylserine receptor Tim-3, although a role for this protein in lymphocyte dissociation has not been reported. It is also possible that dissociation results not from a specific inhibitory signal, but rather from the cessation of activating signals. Indeed, acute interdiction of ongoing activating receptor signaling with small molecule inhibitors or function blocking antibodies accelerates synapse disassembly.[234,240] Dissociation can also be induced by lymphocyte recognition of a second, more stimulatory target, indicating that the process is controlled, at least in part, by the integration of competing sources of activating stimulation.

The Mechanical Dimension of Killing

The immune synapse is a roiling interface characterized by continuous F-actin remodeling. These dynamics exert mechanical forces against the surface of the target cell that are borne by receptor ligand contacts within the interface. In recent years, it has become clear that force regulates multiple aspects of the cytotoxic cascade, including initial lymphocyte activation, adhesion, and the activity of cytotoxic factors.[307] These findings have highlighted the importance of treating the synapse as a biophysical entity with not only chemical but also physical inputs and outputs that dictate its function.

Transmembrane receptors within the synapse, particularly those with membrane-associated ligands, are subject to cytoskeletally derived pushing, pulling, and shear forces. For at least some of these receptors, interfacial forces play a critical role in ligand binding and subsequent signal transduction. Most intermolecular interactions are "slip bonds," in that they become weaker with applied force. Some proteins, however, form "catch bonds" with their ligands that actually increase in affinity up to an optimal tension.[308,309] Integrins are perhaps the best-characterized catch bond receptors, achieving maximal affinity only when they are subjected to pulling forces.[310-312] A number of proteins within integrin adhesion complexes, including talin and CasL, are also mechanosensitive;[313-315] applied force physically extends these molecules, revealing cryptic sites that are then phosphorylated by cytoplasmic kinases and/or function as binding sites for downstream factors. Within the immune synapse, the driving force behind integrin mechanotransduction derives from retrograde F-actin flow within the peripheral lamellipodium. This flow takes LFA1 molecules, which are bound to ICAM1 proteins on the target cell, and drags them centripetally, imparting tension that fully extends the integrin.[231,316,317] Pharmacological and genetic perturbations that reduce this tension preclude integrin activation and also blunt T cell activation by reducing outside-in integrin signaling. Synaptic force therefore functions as a critical promoter of integrin function.

Over the past decade, this mechanotransduction paradigm has been expanded to include both the TCR and CD16, both of which form catch bonds with their cognate ligands.[318-320] In the case of the TCR, catch bond formation is strongly associated with the capacity of the pMHC ligand to drive biological responses. Highly stimulatory agonist pMHC complexes exhibit catch bond behavior, while weak agonist peptides form slip bonds with the TCR.[318] Hence, mechanical force not only prolongs TCR–pMHC interactions but also enhances their discriminatory power. Recent in silico molecular dynamics studies have suggested a basis for TCR catch bonds. When placed under simulated tension, catch bond pMHC ligands, but not slip bond pMHCs, form additional stabilizing contacts with the TCR.[321,322] This mechanically induced shift appears to be relevant to actual TCR–pMHC interactions, as mutation of amino acids that mediate the induced contacts was found to abrogate catch bond behavior.[322] The centrality of applied force during TCR ligand binding raises the possibility that it might also be involved in subsequent signal transduction steps. It is known that pMHC recognition by the TCR induces a conformational change that exposes the cytoplasmic CD3 ITAMs for phosphorylation by Lck.[323-326] The close temporal correlation between this conformational change and TCR catch bond formation implies that the two events might be causally related, and future studies will no doubt address this hypothesis.

Although they are abundantly produced by cytotoxic lymphocytes, effector proteins like perforin and granzyme can still be depleted during sustained bouts of serial killing.[150] The efficient use of these molecules is therefore of paramount importance. Recent studies have revealed an important link between synaptic force exertion and the cytolytic power of degranulation.[238,327] Using single cell assays of biophysical output, it was found that perturbations of the lymphocyte cytoskeleton that increase synaptic force exertion enhance cytotoxicity, whereas perturbations that weaken physical output have the opposite effect. Force functions in this context by accelerating the formation of perforin pores on the target cell surface. As with other pore-forming peptides, perforin activity depends on the tension of the target membrane. Increasing this parameter by pharmacological or biophysical means was found to sensitize target cells to perforin-mediated lysis, and vice versa. These results suggested that cytotoxic lymphocytes use synaptic force to strain the target cell surface, increasing membrane tension and thereby potentiating the activity of secreted perforin. Consistent with this model, it was found that CTLs degranulate close to regions of ongoing force exertion. Subsequent high-resolution imaging of these domains has revealed that degranulation occurs at the base of F-actin–rich protrusions that depend on WASP for their growth (Fig. 31.6). Importantly, WASP-deficient CTLs, which lack these protrusions, exhibit reduced synaptic force exertion and killing efficiency. Collectively, these results indicate that cytotoxic lymphocytes employ mechanically active synaptic protrusions to maximize synergy between target cell deformation and perforin activity.

It is important to emphasize that the consequences of mechanical force within the synapse depend on not only the mechanical activity of the lymphocyte but also the physical properties of the target cell. Indeed, effective mechanotransduction through integrins and mechanosensitive activating receptors can only occur if ligands on the opposing surface provide sufficient mechanical resistance to induce catch bond formation and conformational change. Thus, the degree of lymphocyte activation induced by activating ligands depends on the rigidity of the surface on which ligands are presented, with stiffer surfaces being more stimulatory. This principle, which was first demonstrated using deformable plastic or polyacrylamide substrates,[328-331] has more recently been extended to actual antigen-presenting cells.[317,332] As dendritic cells mature, their cortical stiffness increases by a factor of two. Concomitantly, cell surface ICAM1 binds to sites within the cortical F-actin cytoskeleton and becomes immobilized. Collectively, these biophysical changes allow dendritic cells to engage mechanosensitive receptors, in particular LFA1, across the immune synapse and thereby drive stronger lymphocyte activation. Based on this work, it is reasonable to assume that stiffer target cells would elicit more robust responses from cytotoxic lymphocytes. Similarly, in line with the studies described above on the potentiation of perforin, one would expect that targets cells with higher membrane tension would be more vulnerable to perforin-dependent killing. Accordingly, it is tempting to speculate that infected or transformed cells might elude immune attack by modulating their biophysical properties or, conversely, that immune cells might use biophysical features to detect and eliminate dysregulated targets.

The study of immune synapse mechanics is in its infancy, and consequently there is much about how synaptic force regulates cellular cytotoxicity that remains to be investigated. One can easily imagine, for instance, force exertion influencing lymphocyte protection from cytotoxic secretions and dissociation from target cells. It also seems plausible that synapse mechanics might regulate FasL-mediated killing, given that both FasL and Fas are transmembrane proteins that are physically coupled to the surface of the lymphocyte and target, respectively. In that regard, it is interesting to note that full-length FasL stimulates much stronger Fas signaling than does the cleaved, soluble version of the protein.[133-135] This disparity has been attributed to potential cell surface clustering by full-length FasL, but the available data do not exclude a role for mechanotransduction.

INHIBITION OF CELLULAR CYTOTOXICITY

Cellular cytotoxicity is a complex process with many potential nodes of regulation. Various inhibitory mechanisms have evolved that target these nodes to suppress or redirect killing responses. Pathways emanating from inhibitory cell surface receptors play particularly important roles in preventing nonspecific or otherwise inappropriate attack of healthy cells. Disabling cytotoxic lymphocyte function is also a mainstay of immune evasion by viruses. Several well-established inhibitory mechanisms are described below.

Inhibitory Cell Surface Receptors

One of the most prevalent and important mechanisms for suppressing cellular cytotoxicity involves the engagement of cell surface receptors containing cytoplasmic ITIMs (consensus sequence S/I/V/LxYxxI/V/L).[333] Upon ligand binding, ITIMs become phosphorylated in a Src family kinase–dependent manner, whereupon they recruit downstream effectors. The best characterized ITIM effectors are the lipid phosphatase SHIP and protein tyrosine phosphatases SHP1 and SHP2, which function by dephosphorylating lipid second messengers and signaling proteins, respectively. ITIM receptors are particularly important for the regulation of NK cells, which are activated by generic molecular features of dysregulation and are therefore inherently less specific than CTLs. There are multiple distinct classes of ITIM-coupled NK receptor, including inhibitory members of the killer immunoglobulin receptor (KIR) family and C-type lectins such as NKG2A and inhibitory Ly49 proteins.[29] Most of these proteins recognize either classical or nonclassical class I MHC molecules, which are often downregulated in infected or transformed cells, and therefore serve as an index for good cellular health.

Inhibitory KIRs, in particular KIR2DL1 and KIR2DL2, have been extensively studied, and they represent the archetype of how ITIM receptors suppress killing responses. The phosphorylated ITIM motifs within these proteins recruit SHP-1,[334,335] a rather broad specificity enzyme that can dephosphorylate a variety of substrates in vitro. In NK cells, SHP-1 primarily targets Vav,[336] a large scaffolding protein that also functions as a critical regulator of the F-actin cytoskeleton by activating the Rho family GTPases Rac and Cdc42.[337,338] Tyrosine dephosphorylation downregulates Vav and would therefore be expected to blunt actin polymerization within the immune synapse. Consistent with the idea that KIRs specifically target the cytoskeleton, studies using micropatterned stimulatory surfaces have documented a remarkable anticorrelation between regions of KIR engagement and regions of F-actin accumulation.[339] Furthermore, acute stimulation of KIR signaling within an established synapse has been shown to drive precipitous collapse of the peripheral F-actin ring.[340] Importantly, the effects of inhibitory KIR signaling remain localized to the interface in which KIRs are engaged. Localized cytoskeletal disruption in the absence of global inhibition is a particularly appropriate mechanism for the regulation of NK cell function. It rapidly undermines the structural support required for target cell adhesion, force exertion, and the maintenance of activating signals, and it does so in a way that is independent of specific activating receptors or signaling pathways. Because the response is spatially constrained, however, it does not limit activating responses that may be developing or ongoing at other cell–cell interfaces. Hence, inhibitory KIRs not only prevent the killing of normal healthy cells but also direct cytotoxic responses toward the correct targets within a population.

Inhibitory receptors play prominent roles in the regulation of T cell immunity, as well. Much recent attention has focused on the so-called checkpoint receptors PD1 and CTLA4, which recognize cell surface immunoglobulin domain ligands of the B7 family.[341] Both PD1 and CTLA4 attenuate T cell responses and thereby act as safeguards against autoimmunity. Their activity, however, also contributes to a prolonged state of T cell hyporesponsiveness, known as T cell exhaustion, which is associated with pathological conditions like chronic infection and cancer.[342] Indeed, antibody blockade of CTLA4 and/or PD1 "rejuvenates" T cell immunity against both viruses and tumors [343-345] and is currently used to treat both melanoma and lung cancer.[346] PD1 transduces signals via its cytoplasmic ITIM, which becomes phosphorylated upon recognition of the cognate ligands PD-L1 or PD-L2. The signaling cascade induced by PD1 differs markedly from that of the inhibitory KIRs. Phosphorylated PD1 recruits SHP-2,[347] rather than SHP-1, and it does not appear to target the cytoskeleton directly. Indeed, synapses containing ligand-bound PD1 exhibit canonical radially symmetric architecture.[348] Instead, the inhibitory effects of PD1 appear to manifest further downstream, as reduced cytokine secretion and antigen-induced proliferation. These differences likely reflect the distinct pool of substrate proteins targeted by SHP-2 at the plasma membrane. Recent studies indicate that SHP-2 preferentially dephosphorylates the costimulatory receptor CD28,[211,349] which is particularly important for supporting activation-induced transcriptional responses in T cells. Hence, the consequences of ITIM engagement depend on both the inhibitory effector recruited and the constellation of downstream substrates.

CTLA4 also targets CD28 signaling, but in a manner that is mechanistically distinct from PD1. In resting T cells, the vast majority of CTLA4 is compartmentalized intracellularly in endosomes.[350] Upon TCR activation, CTLA4 moves to the plasma membrane, where it dampens activating signals by outcompeting CD28 for its ligands, B7-1 and B7-2.[350,351]

This phosphatase-independent mechanism blocks signaling at the level of extracellular ligand binding, rather than intracellular receptor phosphorylation. CTLA4 and PD1 are also thought to function in different physiological contexts.[341] CTLA4 operates during naïve T cell priming, thereby restricting the initial expansion and differentiation of effector CD4$^+$ and CD8$^+$ T cells. Conversely, because PD-L1 and PD-L2 are widely expressed in nonlymphoid tissues, PD1 primarily regulates effector T cell activation in pertinent peripheral locations, such as infected organs and the tumor microenvironment. This discrepancy explains, at least in part, the differences in therapeutic efficacy and adverse effects associated with anti-CTLA4 and anti-PD1 therapy in the clinic.

Viral Inhibition of Cellular Cytotoxicity

The importance of cellular cytotoxicity for antiviral immunity is highlighted by the number and variety of mechanisms used by viruses to counteract this process. Many viruses (and some intracellular bacteria) induce the downregulation of class I MHC, thereby rendering infected cells undetectable by CTLs.[352] The absence of class I MHC, however, exposes infected cells to NK cell–mediated killing, which limits the overall effectiveness of this strategy. Accordingly, other mechanisms have evolved that specifically target the cytotoxic cascade, in particular the induction of apoptosis.[353]

As aforementioned, the adenoviral 100K protein is a specific inhibitor of human granzyme B.[120] 100K binds with high affinity to granzyme B, whereupon it is slowly cleaved at a near consensus recognition motif (IEQ**D**) close to the N-terminus of the protein. Because 100K is typically in vast excess of granzyme B in infected cells, it has the effect of sopping up granzyme B activity, thereby preventing the cleavage of canonical substrates such as Bid. The cleavage of 100K by granzyme H relieves this inhibitory mechanism [354] and represents a textbook example of the evolutionary arms race between the immune system and the pathogens that seek to circumvent it.

Viruses use a wide variety of mechanisms to inhibit the apoptotic cascade.[353] Certain poxviruses encode viral inhibitors of apoptosis (vIAPs) that contain Cys/His-rich motifs called BIR domains (for baculoviral IAP repeat), which bind directly to caspases and inhibit their activity.[355] Poxviruses also encode homologs of human serpins. The archetypal viral serpin is CrmA, which can inhibit both caspase-8 and, to a lesser extent, granzyme B.[356,357] Other antiapoptotic mechanisms target mitochondrial apoptosis. The adenoviral 19K protein, for example, inhibits mitochondrial permeabilization and cytochrome c release by sequestering the proapoptotic Bcl2 family members Bak and Bax.[358,359] Gamma herpes viruses and CMV encode Bcl2-like proteins that function in a similar manner.[360,361] CMV also expresses a noncoding RNA called β2.7, which binds to the mitochondrial ETC1 in order to maintain ATP production and guard against respiratory failure.[362] The convergence of cytotoxic cascades and viral resistance on ETC1 highlights the emergence of this complex as a critical node for cell death. Finally, certain herpes viruses express homologs of c-FLIP,[363] which block receptor-induced apoptosis (downstream of Fas, for example) by incorporating into the caspase-8 DISC and attenuating its activity.

PHYSIOLOGICAL ROLE OF CELLULAR CYTOTOXICITY

Cytotoxic lymphocytes are clearly critical for immunity. To identify the specific contribution of cellular cytotoxicity, however, perforin-granzyme and FasL-Fas–mediated killing must be delineated from the production of inflammatory cytokines, the other primary function of cytotoxic lymphocytes. Loss-of-function approaches that target cytotoxic effector pathways, rather than the lymphocytes themselves, have been instrumental in addressing this issue. Mice lacking perforin (*Prf1*$^{-/-}$), FasL (*Gld*), and Fas (*Lpr*) have been particularly important. This area has also benefited substantially from the analysis of human immunodeficiencies that cripple the lytic granule secretion pathway. Collectively, loss-of-function studies in mice and humans have identified two broadly defined functions for cellular cytotoxicity: (1) the eradication of intracellular pathogens and tumors, and (2) the homeostatic regulation of immune cells. Although perforin-granzyme and FasL-Fas contribute to both of these functions, on balance, perforin-granzyme appears to be more important for the former, while FasL-Fas plays the predominant role in the latter.

Studies in mouse models have highlighted the importance of perforin-granzyme killing for antiviral immunity while revealing some interesting complexities. Foundational experiments demonstrated that *Prf1*$^{-/-}$ animals were more vulnerable to the noncytopathic viruses LCMV and MCMV,[364-366] but mounted normal or even enhanced responses against cytopathic species like vaccinia and vesicular stomatitis virus.[366] This pattern of sensitivity appeared to match the predicted effects of cellular cytotoxicity on the viral life cycle. While cytopathic viruses use infected cell lysis for escape and transmission, noncytopathic viruses replicate without lysis, turning infected cells into factories that persistently secrete viral particles. Accordingly, perforin-granzyme–mediated killing was interpreted to be much more detrimental to noncytopathic viruses because keeping infected cells alive is so central to their lifestyle. Conversely, cellular cytotoxicity would be expected to have less of an adverse effect on cytopathic viruses, possibly even accelerating their transmission by inducing lysis. Although this model made a great deal of sense, subsequent studies revealed that perforin actually does contribute to the control of some cytopathic viruses, like ectromelia and Ebola.[108,367] It has been proposed that, in these cases, cytotoxic lymphocytes are able to kill infected cells before they can generate high levels of virus. Hence, the effectiveness of cellular cytotoxicity in combatting cytopathic viruses depends on the relative rates of cell death versus viral reproduction. Host–pathogen interaction studies also support a role for perforin-granzyme–mediated killing in resistance to intracellular bacteria. Loss of perforin has been found to sensitize mice to *L. monocytogenes*.[368] This defect is particularly apparent in previously infected mice that are rechallenged with a high dose of the pathogen, implying a predominant role for cellular cytotoxicity during recall responses. Perforin is also required for the control *Mycobacterium tuberculosis*, a killing response that requires granulysin, as well.[125] Finally, *Prf1*$^{-/-}$ mice are vulnerable to both *Histoplasma capsulatum* and *T. cruzi*, indicating that the perforin-granzyme pathway is critical for immune responses against intracellular fungi and eukaryotic parasites, respectively.[369,370]

In humans, defective perforin-granzyme–mediated killing leads to FHL, a combined immunodeficiency/autoimmune syndrome.[63] Loss-of-function mutations in perforin, Munc13-4, syntaxin-11, and Munc18-2 cause canonical FHL, while disruption of Rab27a expression or localization cause Griscelli syndrome and Hermansky-Pudlak syndrome, which are FHL variants that combine immune defects with depigmentation. FHL patients exhibit massive infiltration of inflammatory lymphocytes and myeloid cells into the liver and other peripheral organs. This uncontrolled immune activation is thought to result secondarily from an underlying cytotoxicity defect. FHL patients are susceptible to infection by herpes viruses, in particular Epstein-Barr virus, because they lack the capacity to destroy infected cells. Consequently, the infections progress, inducing sustained lymphocyte activation and the release of inflammatory cytokines, which in turn promotes generalized myeloid cell activation and tissue damage. Although the failure to kill infected cells lies at the heart of FHL pathogenesis, it is possible that perforin-granzyme–mediated killing is also required in this context for homeostatically regulating the immune response. Indeed, studies in mouse models have implicated perforin in the culling of expanded effector lymphocytes [364] and the destruction of antigen-presenting cells.[371] As described above, perforin is also expressed by T_{reg}, which employ perforin-granzyme–mediated killing as one mechanism for controlling spurious conventional T cell activation.[21] That being said, the fact that *Prf1*$^{-/-}$ mice exhibit largely normal immune homeostasis (this is not the case for *Lpr* and *Gld* mice) implies that perforin-dependent regulation of immune subsets is, in general, of lesser importance than perforin-dependent pathogen clearance.

Perforin deficiencies in mouse and man have also revealed an important role for granule-mediated lysis in antitumor responses. Aged *Prf1*$^{-/-}$ mice are highly susceptible to spontaneous lymphomas, which tend to be of B cell origin.[372] They also respond poorly to transplanted lymphoma cells, which are targeted by $CD8^+$ T cells in wild-type recipients. Interestingly, human patients carrying a partial loss-of-function mutation (A91V) in perforin also exhibit enhanced incidence of hematological malignancy,[373] confirming the importance of perforin-granzyme–dependent killing for combatting this type of disease. Perforin has also been implicated in the control of epithelially derived cancers, in particular the metastasis of tumor cells to the lung.[374] In addition, *Prf1*$^{-/-}$ mice are more susceptible to sarcoma induction by the carcinogen methylcholanthrene.[375] Perforin does not appear to be required for robust tumor immunity in all contexts, however.[376-378] Indeed, IFN-γ production by cytotoxic lymphocytes is, on balance, more critical for antitumor function.[379] The absence of a strict requirement for perforin likely reflects the fact that immune cells can kill tumor cells by a variety of mechanisms, such as cytokine-induced apoptosis, direct phagocytosis, nutrient withdrawal, and ROS release.

Lpr and *Gld* mice have enabled the study of FasL-Fas–dependent killing in multiple disease models. Although the pathway is dispensable for responses to acute LCMV infection,[366] it does appear to contribute to the clearance of other viruses, including ectromelia, West Nile virus, and influenza.[380-382] The *Lpr* and *Gld* loss-of-function phenotypes in some of these model systems are quite subtle, but they become more obvious upon concomitant deletion of *Prf1*, implying compensation between the two cytotoxicity pathways. The FasL-Fas pathway also contributes to responses against *L. monocytogenes*,[368] although its primary role in this context may be to promote inflammatory cytokine secretion by macrophages rather than the direct killing of bacterially infected cells.[383] With respect to cancer, both humans and mice lacking FASL or FAS are prone to lymphoma, suggesting a requirement for the pathway in antitumor immunity.[384,385] This role appears to be specific for hematological malignancies, however, as a number of other tumors appear to be resistant to FasL-Fas–induced killing. Indeed, Fas signaling has been found to promote survival in some cancer cell lines.[386] Hence, whether FasL-dependent killing of transformed cells represents a substantial part of its physiological function remains somewhat ambiguous.

By contrast, the importance of FasL and Fas for immune homeostasis in vivo is clear and well established.[5] In humans, loss-of-function mutations in *FAS* and *FASL* cause ALPS (autoimmune lymphoproliferative syndrome), a lymphoproliferative disease associated with swollen secondary lymphoid organs, anemia, thrombocytopenia, autoantibodies, and susceptibility to lymphoma.[387] Importantly, ALPS patients do not exhibit the same degree of susceptibility to infectious agents seen in FHL patients, consistent with the interpretation that Fas signaling is primarily required for homeostasis, rather than pathogen control. *Lpr* and *Gld* mice exhibit similar phenotypes and are routinely used to model both ALPS and lupus experimentally. The lymphoproliferation observed in these mice and in human patients appears to be caused by autoreactive T cells that escape thymic selection and become activated in the periphery.[388] Under normal circumstances, repeated T cell stimulation triggers activation-induced cell death (AICD), a stereotyped culling process that is highly dependent on fratricidal FasL-Fas–dependent killing.[389-391] In the absence of Fas or FasL, these populations of T cells expand and induce autoimmunity.[392-394] Interestingly, the Fas pathway appears to be quite specific for AICD. It is not required for the negative selection of autoreactive thymocytes during T cell development.[395] Furthermore, it appears to be dispensable for the contraction of effector T cell populations during the resolution of acute immune responses.[396,397] In this context, T cell death results primarily from the withdrawal of the survival promoting cytokines, such as IL-7, rather than cellular cytotoxicity. FasL-Fas–dependent killing, however, does appear to play a role in reversing T cell expansion in the context of persistent infection. For instance, *Lpr* mice chronically infected with herpes virus exhibit ~10-fold more virus-specific T cells in their secondary lymphoid organs compared to control animals.[398] By attenuating T cell responses to chronic infection in this manner, the Fas pathway limits the tissue damage that would result from a sustained inflammatory response. FasL-Fas–dependent killing also mediates the deletion of autoreactive B cells in germinal centers,[399] thereby blocking the generation of autoantibodies resulting from somatic hypermutation.

Finally, both perforin-granzyme and FasL-Fas–induced cytotoxicity have been examined in the context of graft-versus-host disease (GvHD), a life-threatening complication of allogeneic bone marrow transplantation (BMT) in which host tissue is attacked by MHC mismatched T cells derived from the graft.[400,401] Interestingly, when patients with hematologic malignancies are treated with allogeneic BMT, MHC mismatch can also drive the eradication of residual tumor cells, which is called the graft versus tumor (GvT) effect. Studies in mouse models have implicated both the perforin and the Fas pathways in GvHD progression.[402-404] Notably, the perforin pathway appears to be more important for the GvT response.[405,406] Hence, pharmacologic blockade of FasL would be expected to substantially reduce the harmful autoimmune effects of allogeneic BMT without altering its antitumor potential to the same extent.

REFERENCES

1. Ganusov VV, De Boer RJ. Estimating in vivo death rates of targets due to CD8 T-cell-mediated killing. *J Virol.* 2008;82:11749-11757. doi:10.1128/JVI.01128-08.
2. Halle S, et al. In vivo killing capacity of cytotoxic T cells is limited and involves dynamic interactions and T cell cooperativity. *Immunity.* 2016;44:233-245. doi:10.1016/j.immuni.2016.01.010.
3. Dustin ML, Long EO. Cytotoxic immunological synapses. *Immunol Rev.* 2010;235:24-34. doi:IMR904 [pii] 10.1111/j.0105-2896.2010.00904.x.
4. Chowdhury D, Lieberman J. Death by a thousand cuts: granzyme pathways of programmed cell death. *Annu Rev Immunol.* 2008;26:389-420. doi:10.1146/annurev.immunol.26.021607.090404.
5. Strasser A, Jost PJ, Nagata S. The many roles of FAS receptor signaling in the immune system. *Immunity.* 2009;30:180-192. doi:10.1016/j.immuni.2009.01.001.
6. Kagi D, et al. Fas and perforin pathways as major mechanisms of T cell-mediated cytotoxicity. *Science.* 1994;265:528-530. doi:10.1126/science.7518614.
7. Zhang N, Bevan MJ. CD8(+) T cells: foot soldiers of the immune system. *Immunity.* 2011;35:161-168. doi:10.1016/j.immuni.2011.07.010.
8. Ross SH, Cantrell DA. Signaling and function of interleukin-2 in T lymphocytes. *Annu Rev Immunol.* 2018;36:411-433. doi:10.1146/annurev-immunol-042617-053352.
9. Taniuchi I. CD4 helper and CD8 cytotoxic T cell differentiation. *Annu Rev Immunol.* 2018;36:579-601. doi:10.1146/annurev-immunol-042617-053411.
10. Taniuchi I, et al. Differential requirements for Runx proteins in CD4 repression and epigenetic silencing during T lymphocyte development. *Cell.* 2002;111:621-633. doi:10.1016/s0092-8674(02)01111-x.
11. Cruz-Guilloty F, et al. Runx3 and T-box proteins cooperate to establish the transcriptional program of effector CTLs. *J Exp Med.* 2009;206:51-59. doi:10.1084/jem.20081242.
12. Pearce EL, et al. Control of effector CD8+ T cell function by the transcription factor eomesodermin. *Science.* 2003;302:1041-1043. doi:10.1126/science.1090148.
13. Intlekofer AM, et al. Effector and memory CD8+ T cell fate coupled by T-bet and eomesodermin. *Nat Immunol.* 2005;6:1236-1244. doi:10.1038/ni1268.
14. Joshi NS, et al. Inflammation directs memory precursor and short-lived effector CD8(+) T cell fates via the graded expression of T-bet transcription factor. *Immunity.* 2007;27:281-295. doi:10.1016/j.immuni.2007.07.010.
15. Kallies A, Good-Jacobson KL. Transcription factor T-bet orchestrates lineage development and function in the immune system. *Trends Immunol.* 2017;38:287-297. doi:10.1016/j.it.2017.02.003.
16. Buck MD, O'Sullivan D, Pearce EL. T cell metabolism drives immunity. *J Exp Med.* 2015;212:1345-1360. doi:10.1084/jem.20151159.
17. Hukelmann JL, et al. The cytotoxic T cell proteome and its shaping by the kinase mTOR. *Nat Immunol.* 2016;17:104-112. doi:10.1038/ni.3314.
18. Navarro MN, Cantrell DA. Serine-threonine kinases in TCR signaling. *Nat Immunol.* 2014;15:808-814. doi:10.1038/ni.2941.
19. Jameson SC, Masopust D. Understanding subset diversity in T cell memory. *Immunity.* 2018;48:214-226. doi:10.1016/j.immuni.2018.02.010.
20. Kaech SM, Hemby S, Kersh E, Ahmed R. Molecular and functional profiling of memory CD8 T cell differentiation. *Cell.* 2002;111:837-851. doi:10.1016/s0092-8674(02)01139-x.
21. Juno JA, et al. Cytotoxic CD4 T cells-friend or foe during viral infection? *Front Immunol.* 2017;8:19. doi:10.3389/fimmu.2017.00019.
22. Bedard M, Salio M, Cerundolo V. Harnessing the power of invariant natural killer T cells in cancer immunotherapy. *Front Immunol.* 2017;8:1829. doi:10.3389/fimmu.2017.01829.
23. Geiger TL, Sun JC. Development and maturation of natural killer cells. *Curr Opin Immunol.* 2016;39:82-89. doi:10.1016/j.coi.2016.01.007.
24. Male V, Brady HJ. Transcriptional control of NK cell differentiation and function. *Curr Top Microbiol Immunol.* 2014;381:173-187. doi:10.1007/82_2014_376.
25. Gordon SM, et al. The transcription factors T-bet and eomes control key checkpoints of natural killer cell maturation. *Immunity.* 2012;36:55-67. doi:10.1016/j.immuni.2011.11.016.
26. Lacorazza HD, et al. The ETS protein MEF plays a critical role in perforin gene expression and the development of natural killer and NK-T cells. *Immunity.* 2002;17:437-449. doi:10.1016/s1074-7613(02)00422-3.
27. Gilmour KC, et al. Defective expression of the interleukin-2/interleukin-15 receptor beta subunit leads to a natural killer cell-deficient form of severe combined immunodeficiency. *Blood.* 2001;98:877-879. doi:10.1182/blood.v98.3.877.
28. Kennedy MK, et al. Reversible defects in natural killer and memory CD8 T cell lineages in interleukin 15-deficient mice. *J Exp Med.* 2000;191:771-780. doi:10.1084/jem.191.5.771.
29. Lanier LL. NK cell recognition. *Annu Rev Immunol.* 2005;23:225-274.
30. O'Sullivan TE, Sun JC, Lanier LL. Natural killer cell memory. *Immunity.* 2015;43:634-645. doi:10.1016/j.immuni.2015.09.013.
31. Fehniger TA, et al. Acquisition of murine NK cell cytotoxicity requires the translation of a pre-existing pool of granzyme B and perforin mRNAs. *Immunity.* 2007;26:798-811. doi:10.1016/j.immuni.2007.04.010.
32. Sun JC, Beilke JN, Lanier LL. Adaptive immune features of natural killer cells. *Nature.* 2009;457:557-561. doi:10.1038/nature07665.
33. Thiery J, Lieberman J. Perforin: a key pore-forming protein for immune control of viruses and cancer. *Subcell Biochem.* 2014;80:197-220. doi:10.1007/978-94-017-8881-6_10.
34. Voskoboinik I, Whisstock JC, Trapani JA. Perforin and granzymes: function, dysfunction and human pathology. *Nat Rev Immunol.* 2015;15:388-400. doi:10.1038/nri3839.
35. Yamamoto K, et al. 2-Step processing for activation of the cytolysin hemolysin of vibrio-cholerae O1 biotype El-Tor - nucleotide-sequence of the structural gene (hlyA) and characterization of the processed products. *Infect Immun.* 1990;58:4106-4116. doi:10.1128/Iai.58.12.4106-4116.1990.
36. Bayly-Jones C, Bubeck D, Dunstone MA. The mystery behind membrane insertion: a review of the complement membrane attack complex. *Philos Trans R Soc Lond B Biol Sci.* 2017;372:20160221. doi:ARTN 2016022110.1098/rstb.2016.0221.
37. Stewart SE, et al. Analysis of perforin assembly by quartz crystal microbalance reveals a role for cholesterol and calcium-independent membrane binding. *J Biol Chem.* 2015;290:31101-31112. doi:10.1074/jbc.M115.683078.
38. Antia R, Schlegel RA, Williamson P. Binding of perforin to membranes is sensitive to lipid spacing and not headgroup. *Immunol Lett.* 1992;32:153-157. doi:10.1016/0165-2478(92)90108-z.
39. Rochel N, Cowan JA. Dependence of the lytic activity of the N-terminal domain of human perforin on membrane lipid composition—implications for T-cell self-preservation. *Eur J Biochem.* 1997;249:223-231. doi:10.1111/j.1432-1033.1997.00223.x.
40. Law RH, et al. The structural basis for membrane binding and pore formation by lymphocyte perforin. *Nature.* 2010;468:447-451. doi:10.1038/nature09518.
41. Voskoboinik I, et al. Calcium-dependent plasma membrane binding and cell lysis by perforin are mediated through its C2 domain: a critical role for aspartate residues 429, 435, 483, and 485 but not 491. *J Biol Chem.* 2005;280:8426-8434. doi:10.1074/jbc.M413303200.
42. Hadders MA, Beringer DX, Gros P. Structure of C8alpha-MACPF reveals mechanism of membrane attack in complement immune defense. *Science.* 2007;317:1552-1554. doi:10.1126/science.1147103.
43. Rosado CJ, et al. A common fold mediates vertebrate defense and bacterial attack. *Science.* 2007;317:1548-1551. doi:10.1126/science.1144706.
44. Shatursky O, et al. The mechanism of membrane insertion for a cholesterol-dependent cytolysin: a novel paradigm for pore-forming toxins. *Cell.* 1999;99:293-299. doi:10.1016/s0092-8674(00)81660-8.
45. Tilley SJ, Orlova EV, Gilbert RJ, Andrew PW, Saibil HR. Structural basis of pore formation by the bacterial toxin pneumolysin. *Cell.* 2005;121:247-256. doi:10.1016/j.cell.2005.02.033.
46. Shepard LA, Shatursky O, Johnson AE, Tweten RK. The mechanism of pore assembly for a cholesterol-dependent cytolysin: formation of a large prepore complex precedes the insertion of the transmembrane beta-hairpins. *Biochemistry.* 2000;39:10284-10293. doi:10.1021/bi000436r.
47. Hotze EM, et al. Arresting pore formation of a cholesterol-dependent cytolysin by disulfide trapping synchronizes the insertion of the transmembrane beta-sheet from a prepore intermediate. *J Biol Chem.* 2001;276:8261-8268. doi:10.1074/jbc.M009865200.
48. Leung C, et al. Stepwise visualization of membrane pore formation by suilysin, a bacterial cholesterol-dependent cytolysin. *Elife.* 2014;3:e04247. doi:10.7554/eLife.04247.
49. Leung C, et al. Real-time visualization of perforin nanopore assembly. *Nat Nanotechnol.* 2017;12:467-473. doi:10.1038/nnano.2016.303.
50. Sonnen AF, Plitzko JM, Gilbert RJ. Incomplete pneumolysin oligomers form membrane pores. *Open Biol.* 2014;4:140044. doi:10.1098/rsob.140044.
51. Ruan Y, Rezelj S, Bedina Zavec A, Anderluh G, Scheuring S. Listeriolysin O membrane damaging activity involves arc formation and lineaction—implication for Listeria monocytogenes escape from phagocytic vacuole. *PLoS Pathog.* 2016;12:e1005597. doi:10.1371/journal.ppat.1005597.
52. Praper T, et al. Human perforin employs different avenues to damage membranes. *J Biol Chem.* 2011;286:2946-2955. doi:10.1074/jbc.M110.169417.
53. Dourmashkin RR, Deteix P, Simone CB, Henkart P. Electron microscopic demonstration of lesions in target cell membranes associated with antibody-dependent cellular cytotoxicity. *Clin Exp Immunol.* 1980;42:554-560.
54. Liu Y, et al. Visualization of perforin/gasdermin/complement-formed pores in real cell membranes using atomic force microscopy. *Cell Mol Immunol.* 2019;16:611-620. doi:10.1038/s41423-018-0165-1.
55. Poenie M, Tsien RY, Schmitt-Verhulst AM. Sequential activation and lethal hit measured by [Ca2+]i in individual cytolytic T cells and targets. *EMBO J.* 1987;6:2223-2232.
56. Lopez JA, et al. Perforin forms transient pores on the target cell plasma membrane to facilitate rapid access of granzymes during killer cell attack. *Blood.* 2013;121:2659-2668. doi:10.1182/blood-2012-07-446146.
57. Lopez JA, et al. Rapid and unidirectional perforin pore delivery at the cytotoxic immune synapse. *J Immunol.* 2013;191:2328-2334. doi:10.4049/jimmunol.1301205.
58. Keefe D, et al. Perforin triggers a plasma membrane-repair response that facilitates CTL induction of apoptosis. *Immunity.* 2005;23:249-262. doi:10.1016/j.immuni.2005.08.001.
59. Thiery J, et al. Perforin pores in the endosomal membrane trigger the release of endocytosed granzyme B into the cytosol of target cells. *Nat Immunol.* 2011;12:770-777. doi:10.1038/ni.2050.
60. Thiery J, et al. Perforin activates clathrin-and dynamin-dependent endocytosis, which is required for plasma membrane repair and delivery of granzyme B for granzyme-mediated apoptosis. *Blood.* 2010;115:1582-1593. doi:10.1182/blood-2009-10-246116.
61. Khazen R, et al. Melanoma cell lysosome secretory burst neutralizes the CTL-mediated cytotoxicity at the lytic synapse. *Nat Commun.* 2016;7:10823. doi:10.1038/ncomms10823.
62. Stepp SE, et al. Perforin gene defects in familial hemophagocytic lymphohistiocytosis. *Science.* 1999;286:1957-1959. doi:10.1126/science.286.5446.1957.

63. Sieni E, et al. Familial hemophagocytic lymphohistiocytosis: when rare diseases shed light on immune system functioning. *Front Immunol.* 2014;5:167. doi:10.3389/fimmu.2014.00167.
64. Pipkin ME, Rao A, Lichtenheld MG. The transcriptional control of the perforin locus. *Immunol Rev.* 2010;235:55-72. doi:10.1111/j.0105-2896.2010.00905.x.
65. Koizumi H, et al. Identification of a killer cell-specific regulatory element of the mouse perforin gene: an Ets-binding site-homologous motif that interacts with Ets-related proteins. *Mol Cell Biol.* 1993;13:6690-6701. doi:10.1128/mcb.13.11.6690.
66. Zhou J, Zhang J, Lichtenheld MG, Meadows GG. A role for NF-kappa B activation in perforin expression of NK cells upon IL-2 receptor signaling. *J Immunol.* 2002;169:1319-1325. doi:10.4049/jimmunol.169.3.1319.
67. Zhang J, Scordi I, Smyth MJ, Lichtenheld MG. Interleukin 2 receptor signaling regulates the perforin gene through signal transducer and activator of transcription (Stat)5 activation of two enhancers. *J Exp Med.* 1999;190:1297-1308. doi:10.1084/jem.190.9.1297.
68. Pipkin ME, et al. Chromosome transfer activates and delineates a locus control region for perforin. *Immunity.* 2007;26:29-41. doi:10.1016/j.immuni.2006.11.009.
69. Kummer JA, Kamp AM, Citarella F, Horrevoets AJ, Hack CE. Expression of human recombinant granzyme A zymogen and its activation by the cysteine proteinase cathepsin C. *J Biol Chem.* 1996;271:9281-9286. doi:10.1074/jbc.271.16.9281.
70. Casciola-Rosen L, et al. Mouse and human granzyme B have distinct tetrapeptide specificities and abilities to recruit the bid pathway. *J Biol Chem.* 2007;282:4545-4552. doi:10.1074/jbc.M606564200.
71. Cullen SP, Adrain C, Luthi AU, Duriez PJ, Martin SJ. Human and murine granzyme B exhibit divergent substrate preferences. *J Cell Biol.* 2007;176:435-444. doi:10.1083/jcb.200612025.
72. Heibein JA, et al. Granzyme B-mediated cytochrome c release is regulated by the Bcl-2 family members bid and Bax. *J Exp Med.* 2000;192:1391-1402. doi:10.1084/jem.192.10.1391.
73. Sutton VR, et al. Initiation of apoptosis by granzyme B requires direct cleavage of bid, but not direct granzyme B-mediated caspase activation. *J Exp Med.* 2000;192:1403-1414. doi:10.1084/jem.192.10.1403.
74. Thomas DA, Du C, Xu M, Wang X, Ley TJ. DFF45/ICAD can be directly processed by granzyme B during the induction of apoptosis. *Immunity.* 2000;12:621-632. doi:10.1016/s1074-7613(00)80213-7.
75. Sharif-Askari E, et al. Direct cleavage of the human DNA fragmentation factor-45 by granzyme B induces caspase-activated DNase release and DNA fragmentation. *EMBO J.* 2001;20:3101-3113. doi:10.1093/emboj/20.12.3101.
76. Darmon AJ, Nicholson DW, Bleackley RC. Activation of the apoptotic protease CPP32 by cytotoxic T-cell-derived granzyme B. *Nature.* 1995;377:446-448. doi:10.1038/377446a0.
77. Adrain C, Murphy BM, Martin SJ. Molecular ordering of the caspase activation cascade initiated by the cytotoxic T lymphocyte/natural killer (CTL/NK) protease granzyme B. *J Biol Chem.* 2005;280:4663-4673. doi:10.1074/jbc.M410915200.
78. Medema JP, et al. Cleavage of FLICE (caspase-8) by granzyme B during cytotoxic T lymphocyte-induced apoptosis. *Eur J Immunol.* 1997;27:3492-3498. doi:10.1002/eji.1830271250.
79. Boatright KM, Salvesen GS. Mechanisms of caspase activation. *Curr Opin Cell Biol.* 2003;15:725-731. doi:10.1016/j.ceb.2003.10.009.
80. Julien O, Wells JA. Caspases and their substrates. *Cell Death Differ.* 2017;24:1380-1389. doi:10.1038/cdd.2017.44.
81. Li H, Zhu H, Xu CJ, Yuan J. Cleavage of BID by caspase 8 mediates the mitochondrial damage in the Fas pathway of apoptosis. *Cell.* 1998;94:491-501. doi:10.1016/s0092-8674(00)81590-1.
82. Pena-Blanco A, Garcia-Saez AJ. Bax, Bak and beyond—mitochondrial performance in apoptosis. *FEBS J.* 2018;285:416-431. doi:10.1111/febs.14186.
83. Riedl SJ, Salvesen GS. The apoptosome: signalling platform of cell death. *Nat Rev Mol Cell Biol.* 2007;8:405-413. doi:10.1038/nrm2153.
84. Chiusolo V, et al. Granzyme B enters the mitochondria in a Sam50-, Tim22- and mtHsp70-dependent manner to induce apoptosis. *Cell Death Differ.* 2017;24:747-758. doi:10.1038/cdd.2017.3.
85. Jacquemin G, et al. Granzyme B-induced mitochondrial ROS are required for apoptosis. *Cell Death Differ.* 2015;22:862-874. doi:10.1038/cdd.2014.180.
86. Ricci JE, et al. Disruption of mitochondrial function during apoptosis is mediated by caspase cleavage of the p75 subunit of complex I of the electron transport chain. *Cell.* 2004;117:773-786. doi:10.1016/j.cell.2004.05.008.
87. Van Damme P, et al. The substrate specificity profile of human granzyme A. *Biol Chem.* 2010;391:983-997. doi:10.1515/BC.2010.096.
88. Susanto O, et al. Mouse granzyme A induces a novel death with writhing morphology that is mechanistically distinct from granzyme B-induced apoptosis. *Cell Death Differ.* 2013;20:1183-1193. doi:10.1038/cdd.2013.59.
89. Martinvalet D, Dykxhoorn DM, Ferrini R, Lieberman J. Granzyme A cleaves a mitochondrial complex I protein to initiate caspase-independent cell death. *Cell.* 2008;133:681-692. doi:10.1016/j.cell.2008.03.032.
90. Martinvalet D, Zhu P, Lieberman J. Granzyme A induces caspase-independent mitochondrial damage, a required first step for apoptosis. *Immunity.* 2005;22:355-370. doi:10.1016/j.immuni.2005.02.004.
91. Beresford PJ, Xia Z, Greenberg AH, Lieberman J. Granzyme A loading induces rapid cytolysis and a novel form of DNA damage independently of caspase activation. *Immunity.* 1999;10:585-594. doi:10.1016/s1074-7613(00)80058-8.
92. Chowdhury D, et al. The exonuclease TREX1 is in the SET complex and acts in concert with NM23-H1 to degrade DNA during granzyme A-mediated cell death. *Mol Cell.* 2006;23:133-142. doi:10.1016/j.molcel.2006.06.005.
93. Fan Z, Beresford PJ, Oh DY, Zhang D, Lieberman J. Tumor suppressor NM23-H1 is a granzyme A-activated DNase during CTL-mediated apoptosis, and the nucleosome assembly protein SET is its inhibitor. *Cell.* 2003;112:659-672. doi:10.1016/s0092-8674(03)00150-8.
94. Fan Z, Beresford PJ, Zhang D, Lieberman J. HMG2 interacts with the nucleosome assembly protein SET and is a target of the cytotoxic T-lymphocyte protease granzyme A. *Mol Cell Biol.* 2002;22:2810-2820. doi:10.1128/mcb.22.8.2810-2820.2002.
95. Fan Z, et al. Cleaving the oxidative repair protein Ape1 enhances cell death mediated by granzyme A. *Nat Immunol.* 2003;4:145-153. doi:10.1038/ni885.
96. Zhu P, et al. Granzyme A, which causes single-stranded DNA damage, targets the double-strand break repair protein Ku70. *EMBO Rep.* 2006;7:431-437. doi:10.1038/sj.embor.7400622.
97. Zhang D, Beresford PJ, Greenberg AH, Lieberman J. Granzymes A and B directly cleave lamins and disrupt the nuclear lamina during granule-mediated cytolysis. *Proc Natl Acad Sci U S A.* 2001;98:5746-5751. doi:10.1073/pnas.101329598.
98. Simon MM, et al. In vitro-and ex vivo-derived cytolytic leukocytes from granzyme A × B double knockout mice are defective in granule-mediated apoptosis but not lysis of target cells. *J Exp Med.* 1997;186:1781-1786. doi:10.1084/jem.186.10.1781.
99. Davis JE, Smyth MJ, Trapani JA. Granzyme A and B-deficient killer lymphocytes are defective in eliciting DNA fragmentation but retain potent in vivo anti-tumor capacity. *Eur J Immunol.* 2001;31:39-47. doi:10.1002/1521-4141(200101)31:1<39::aid-immu39>3.0.co;2-1.
100. Pardo J, Balkow S, Anel A, Simon MM. The differential contribution of granzyme A and granzyme B in cytotoxic T lymphocyte-mediated apoptosis is determined by the quality of target cells. *Eur J Immunol.* 2002;32:1980-1985. doi:10.1002/1521-4141(200207)32:7<1980::AID-IMMU1980>3.0.CO;2-Z.
101. Zelinskyy G, et al. Independent roles of perforin, granzymes, and Fas in the control of Friend retrovirus infection. *Virology.* 2004;330:365-374. doi:10.1016/j.virol.2004.08.040.
102. Zajac AJ, Dye JM, Quinn DG. Control of lymphocytic choriomeningitis virus infection in granzyme B deficient mice. *Virology.* 2003;305:1-9. doi:10.1006/viro.2002.1754.
103. Loh J, Thomas DA, Revell PA, Ley TJ, Virgin HW. Granzymes and caspase 3 play important roles in control of gammaherpesvirus latency. *J Virol.* 2004;78:12519-12528. doi:10.1128/JVI.78.22.12519-12528.2004.
104. van Dommelen SL, et al. Perforin and granzymes have distinct roles in defensive immunity and immunopathology. *Immunity.* 2006;25:835-848. doi:10.1016/j.immuni.2006.09.010.
105. Salti SM, et al. Granzyme B regulates antiviral CD8+ T cell responses. *J Immunol.* 2011;187:6301-6309. doi:10.4049/jimmunol.1100891.
106. Mullbacher A, et al. Granzyme A is critical for recovery of mice from infection with the natural cytopathic viral pathogen, ectromelia. *Proc Natl Acad Sci U S A.* 1996;93:5783-5787. doi:10.1073/pnas.93.12.5783.
107. Mullbacher A, et al. Granzymes are the essential downstream effector molecules for the control of primary virus infections by cytolytic leukocytes. *Proc Natl Acad Sci U S A.* 1999;96:13950-13955. doi:10.1073/pnas.96.24.13950.
108. Mullbacher A, Hla RT, Museteanu C, Simon MM. Perforin is essential for control of ectromelia virus but not related poxviruses in mice. *J Virol.* 1999;73:1665-1667.
109. Kagi D, et al. Cytotoxicity mediated by T cells and natural killer cells is greatly impaired in perforin-deficient mice. *Nature.* 1994;369:31-37. doi:10.1038/369031a0.
110. Bovenschen N, Kummer JA. Orphan granzymes find a home. *Immunol Rev.* 2010;235:117-127. doi:10.1111/j.0105-2896.2010.00889.x.
111. Zhao T, Zhang H, Guo Y, Fan Z. Granzyme K directly processes bid to release cytochrome c and endonuclease G leading to mitochondria-dependent cell death. *J Biol Chem.* 2007;282:12104-12111. doi:10.1074/jbc.M611006200.
112. Bovenschen N, et al. Granzyme K displays highly restricted substrate specificity that only partially overlaps with granzyme A. *J Biol Chem.* 2009;284:3504-3512. doi:10.1074/jbc.M806716200.
113. Mahrus S, Craik CS. Selective chemical functional probes of granzymes A and B reveal granzyme B is a major effector of natural killer cell-mediated lysis of target cells. *Chem Biol.* 2005;12:567-577. doi:10.1016/j.chembiol.2005.03.006.
114. Edwards KM, Kam CM, Powers JC, Trapani JA. The human cytotoxic T cell granule serine protease granzyme H has chymotrypsin-like (chymase) activity and is taken up into cytoplasmic vesicles reminiscent of granzyme B-containing endosomes. *J Biol Chem.* 1999;274:30468-30473. doi:10.1074/jbc.274.43.30468.
115. Hou Q, et al. Granzyme H induces apoptosis of target tumor cells characterized by DNA fragmentation and Bid-dependent mitochondrial damage. *Mol Immunol.* 2008;45:1044-1055. doi:10.1016/j.molimm.2007.07.032.
116. Fellows E, Gil-Parrado S, Jenne DE, Kurschus FC. Natural killer cell-derived human granzyme H induces an alternative, caspase-independent cell-death program. *Blood.* 2007;110:544-552. doi:10.1182/blood-2006-10-051649.
117. Smyth MJ, O'Connor MD, Trapani JA, Kershaw MH, Brinkworth RI. A novel substrate-binding pocket interaction restricts the specificity of the human NK cell-specific serine protease, Met-ase-1. *J Immunol.* 1996;156:4174-4181.
118. Kelly JM, et al. Granzyme M mediates a novel form of perforin-dependent cell death. *J Biol Chem.* 2004;279:22236-22242. doi:10.1074/jbc.M401670200.
119. Revell PA, et al. Granzyme B and the downstream granzymes C and/or F are important for cytotoxic lymphocyte functions. *J Immunol.* 2005;174:2124-2131. doi:10.4049/jimmunol.174.4.2124.
120. Andrade F, et al. Adenovirus L4-100K assembly protein is a granzyme B substrate that potently inhibits granzyme B-mediated cell death. *Immunity.* 2001;14:751-761. doi:10.1016/s1074-7613(01)00149-2.
121. Pao LI, et al. Functional analysis of granzyme M and its role in immunity to infection. *J Immunol.* 2005;175:3235-3243. doi:10.4049/jimmunol.175.5.3235.
122. de Koning PJ, et al. The cytotoxic protease granzyme M is expressed by lymphocytes of both the innate and adaptive immune system. *Mol Immunol.* 2010;47:903-911. doi:10.1016/j.molimm.2009.10.001.
123. Krensky AM, Clayberger C. Biology and clinical relevance of granulysin. *Tissue Antigens.* 2009;73:193-198. doi:10.1111/j.1399-0039.2008.01218.x.
124. Dotiwala F, Lieberman J. Granulysin: killer lymphocyte safeguard against microbes. *Curr Opin Immunol.* 2019;60:19-29. doi:10.1016/j.coi.2019.04.013.
125. Stenger S, et al. An antimicrobial activity of cytolytic T cells mediated by granulysin. *Science.* 1998;282:121-125. doi:10.1126/science.282.5386.121.
126. Walch M, et al. Cytotoxic cells kill intracellular bacteria through granulysin-mediated delivery of granzymes. *Cell.* 2014;157:1309-1323. doi:10.1016/j.cell.2014.03.062.
127. Dotiwala F, et al. Killer lymphocytes use granulysin, perforin and granzymes to kill intracellular parasites. *Nat Med.* 2016;22:210-216. doi:10.1038/nm.4023.
128. Huang LP, Lyu SC, Clayberger C, Krensky AM. Granulysin-mediated tumor rejection in transgenic mice. *J Immunol.* 2007;178:77-84. doi:10.4049/jimmunol.178.1.77.
129. Bossi G, Griffiths GM. Degranulation plays an essential part in regulating cell surface expression of Fas ligand in T cells and natural killer cells. *Nat Med.* 1999;5:90-96. doi:10.1038/4779.
130. Kojima Y, et al. Localization of Fas ligand in cytoplasmic granules of CD8+ cytotoxic T lymphocytes and natural killer cells: participation of Fas ligand in granule exocytosis model of cytotoxicity. *Biochem Biophys Res Commun.* 2002;296:328-336. doi:10.1016/s0006-291x(02)00841-0.
131. Li JH, et al. The regulation of CD95 ligand expression and function in CTL. *J Immunol.* 1998;161:3943-3949.
132. He JS, Ostergaard HL. CTLs contain and use intracellular stores of FasL distinct from cytolytic granules. *J Immunol.* 2007;179:2339-2348. doi:10.4049/jimmunol.179.4.2339.
133. Schneider P, et al. Conversion of membrane-bound Fas(CD95) ligand to its soluble form is associated with downregulation of its proapoptotic activity and loss of liver toxicity. *J Exp Med.* 1998;187:1205-1213. doi:10.1084/jem.187.8.1205.
134. Suda T, Hashimoto H, Tanaka M, Ochi T, Nagata S. Membrane Fas ligand kills human peripheral blood T lymphocytes, and soluble Fas ligand blocks the killing. *J Exp Med.* 1997;186:2045-2050. doi:10.1084/jem.186.12.2045.
135. Tanaka M, Itai T, Adachi M, Nagata S. Downregulation of Fas ligand by shedding. *Nat Med.* 1998;4:31-36. doi:10.1038/nm0198-031.
136. Miyawaki T, et al. Differential expression of apoptosis-related Fas antigen on lymphocyte subpopulations in human peripheral blood. *J Immunol.* 1992;149:3753-3758.
137. Yamada A, Arakaki R, Saito M, Kudo Y, Ishimaru N. Dual role of fas/FasL-mediated signal in peripheral immune tolerance. *Front Immunol.* 2017;8:403. doi:10.3389/fimmu.2017.00403.
138. Dickens LS, et al. A death effector domain chain DISC model reveals a crucial role for caspase-8 chain assembly in mediating apoptotic cell death. *Mol Cell.* 2012;47:291-305. doi:10.1016/j.molcel.2012.05.004.
139. Schleich K, et al. Stoichiometry of the CD95 death-inducing signaling complex: experimental and modeling evidence for a death effector domain chain model. *Mol Cell.* 2012;47:306-319. doi:10.1016/j.molcel.2012.05.006.
140. Tsuchiya Y, Nakabayashi O, Nakano H. FLIP the switch: regulation of apoptosis and necroptosis by cFLIP. *Int J Mol Sci.* 2015;16:30321-30341. doi:10.3390/ijms161226232.

141. Shepard BD, Badley AD. The biology of TRAIL and the role of TRAIL-based therapeutics in infectious diseases. *Antiinfect Agents Med Chem.* 2009;8:87-101. doi:10.2174/187152109787846060.
142. Trivedi R, Mishra DP. Trailing TRAIL resistance: novel targets for TRAIL sensitization in cancer cells. *Front Oncol.* 2015;5:69. doi:10.3389/fonc.2015.00069.
143. Lum JJ, Schnepple DJ, Badley AD. Acquired T-cell sensitivity to TRAIL mediated killing during HIV infection is regulated by CXCR4-gp120 interactions. *AIDS.* 2005;19:1125-1133. doi:10.1097/01.aids.0000176212.16205.23.
144. Sedger LM, et al. IFN-gamma mediates a novel antiviral activity through dynamic modulation of TRAIL and TRAIL receptor expression. *J Immunol.* 1999;163:920-926.
145. Saitou Y, et al. Involvement of tumor necrosis factor-related apoptosis-inducing ligand and tumor necrosis factor-related apoptosis-inducing ligand receptors in viral hepatic diseases. *Hum Pathol.* 2005;36:1066-1073. doi:10.1016/j.humpath.2005.07.019.
146. Mundt B, et al. Involvement of TRAIL and its receptors in viral hepatitis. *FASEB J.* 2003;17:94-96. doi:10.1096/fj.02-0537fje.
147. Golks A, Brenner D, Fritsch C, Krammer PH, Lavrik IN. c-FLIPR, a new regulator of death receptor-induced apoptosis. *J Biol Chem.* 2005;280:14507-14513. doi:10.1074/jbc.M414425200.
148. Krueger A, Schmitz I, Baumann S, Krammer PH, Kirchhoff S. Cellular FLICE-inhibitory protein splice variants inhibit different steps of caspase-8 activation at the CD95 death-inducing signaling complex. *J Biol Chem.* 2001;276:20633-20640. doi:10.1074/jbc.M101780200.
149. Falschlehner C, Emmerich CH, Gerlach B, Walczak H. TRAIL signalling: decisions between life and death. *Int J Biochem Cell Biol.* 2007;39:1462-1475. doi:10.1016/j.biocel.2007.02.007.
150. Prager I, et al. NK cells switch from granzyme B to death receptor-mediated cytotoxicity during serial killing. *J Exp Med.* 2019;216:2113-2127. doi:10.1084/jem.20181454.
151. Beal AM, et al. Kinetics of early T cell receptor signaling regulate the pathway of lytic granule delivery to the secretory domain. *Immunity.* 2009;31:632-642. doi:S1074-7613(09)00416-6 [pii] 10.1016/j.immuni.2009.09.004.
152. Ostergaard HL, Kane KP, Mescher MF, Clark WR. Cytotoxic T lymphocyte mediated lysis without release of serine esterase. *Nature.* 1987;330:71-72.
153. Kataoka T, et al. Concanamycin A, a powerful tool for characterization and estimation of contribution of perforin- and Fas-based lytic pathways in cell-mediated cytotoxicity. *J Immunol.* 1996;156:3678-3686.
154. Stinchcombe JC, Griffiths GM. Secretory mechanisms in cell-mediated cytotoxicity. *Annu Rev Cell Dev Biol.* 2007;23:495-517.
155. de Saint Basile G, Menasche G, Fischer A. Molecular mechanisms of biogenesis and exocytosis of cytotoxic granules. *Nat Rev Immunol.* 2010;10:568-579. doi:10.1038/nri2803.
156. Chang HF, et al. Preparing the lethal hit: interplay between exo- and endocytic pathways in cytotoxic T lymphocytes. *Cell Mol Life Sci.* 2017;74:399-408. doi:10.1007/s00018-016-2350-7.
157. Burkhardt JK, Hester S, Lapham CK, Argon Y. The lytic granules of natural killer cells are dual-function organelles combining secretory and pre-lysosomal compartments. *J Cell Biol.* 1990;111:2327-2340.
158. Peters PJ, et al. Cytotoxic T lymphocyte granules are secretory lysosomes, containing both perforin and granzymes. *J Exp Med.* 1991;173:1099-1109. doi:10.1084/jem.173.5.1099.
159. MacDermott RP, et al. Proteoglycans in cell-mediated cytotoxicity. Identification, localization, and exocytosis of a chondroitin sulfate proteoglycan from human cloned natural killer cells during target cell lysis. *J Exp Med.* 1985;162:1771-1787. doi:10.1084/jem.162.6.1771.
160. Masson D, Peters PJ, Geuze HJ, Borst J, Tschopp J. Interaction of chondroitin sulfate with perforin and granzymes of cytolytic T-cells is dependent on pH. *Biochemistry.* 1990;29:11229-11235. doi:10.1021/bi00503a011.
161. Metkar SS, et al. Cytotoxic cell granule-mediated apoptosis: perforin delivers granzyme B-serglycin complexes into target cells without plasma membrane pore formation. *Immunity.* 2002;16:417-428. doi:10.1016/s1074-7613(02)00286-8.
162. Grujic M, et al. Serglycin-deficient cytotoxic T lymphocytes display defective secretory granule maturation and granzyme B storage. *J Biol Chem.* 2005;280:33411-33418. doi:10.1074/jbc.M501708200.
163. Sutton VR, et al. Serglycin determines secretory granule repertoire and regulates natural killer cell and cytotoxic T lymphocyte cytotoxicity. *FEBS J.* 2016;283:947-961. doi:10.1111/febs.13649.
164. Peters PJ, et al. Molecules relevant for T cell-target cell interaction are present in cytolytic granules of human T lymphocytes. *Eur J Immunol.* 1989;19:1469-1475. doi:10.1002/eji.1830190819.
165. Schmidt H, et al. Effector granules in human T lymphocytes: the luminal proteome of secretory lysosomes from human T cells. *Cell Commun Signal.* 2011;9:4. doi:10.1186/1478-811X-9-4.
166. Hunziker W, Simmen T, Honing S. Trafficking of lysosomal membrane proteins in polarized kidney cells. *Nephrologie.* 1996;17:347-350.
167. Scott CC, Vacca F, Gruenberg J. Endosome maturation, transport and functions. *Semin Cell Dev Biol.* 2014;31:2-10. doi:10.1016/j.semcdb.2014.03.034.
168. Burkhardt JK, Hester S, Argon Y. Two proteins targeted to the same lytic granule compartment undergo very different posttranslational processing. *Proc Natl Acad Sci U S A.* 1989;86:7128-7132. doi:10.1073/pnas.86.18.7128.
169. Griffiths GM, Isaaz S. Granzymes A and B are targeted to the lytic granules of lymphocytes by the mannose-6-phosphate receptor. *J Cell Biol.* 1993;120:885-896. doi:10.1083/jcb.120.4.885.
170. Brennan AJ, et al. Protection from endogenous perforin: glycans and the C terminus regulate exocytic trafficking in cytotoxic lymphocytes. *Immunity.* 2011;34:879-892. doi:10.1016/j.immuni.2011.04.007.
171. Nagle DL, et al. Identification and mutation analysis of the complete gene for Chediak-Higashi syndrome. *Nat Genet.* 1996;14:307-311. doi:10.1038/ng1196-307.
172. Barbosa MD, et al. Identification of the homologous beige and Chediak-Higashi syndrome genes. *Nature.* 1996;382:262-265. doi:10.1038/382262a0.
173. Menager MM, et al. Secretory cytotoxic granule maturation and exocytosis require the effector protein hMunc13-4. *Nat Immunol.* 2007;8:257-267. doi:10.1038/ni1431.
174. Wood SM, et al. Different NK cell-activating receptors preferentially recruit Rab27a or Munc13-4 to perforin-containing granules for cytotoxicity. *Blood.* 2009;114:4117-4127. doi:10.1182/blood-2009-06-225359.
175. Ming M, Schirra C, Becherer U, Stevens DR, Rettig J. Behavior and properties of mature lytic granules at the immunological synapse of human cytotoxic T lymphocytes. *PloS One.* 2015;10:e0135994. doi:10.1371/journal.pone.0135994.
176. Matti U, et al. Synaptobrevin2 is the v-SNARE required for cytotoxic T-lymphocyte lytic granule fusion. *Nat Commun.* 2013;4:1439. doi:10.1038/ncomms2467.
177. Raben N, Puertollano R. TFEB and TFE3: linking lysosomes to cellular adaptation to stress. *Annu Rev Cell Dev Biol.* 2016;32:255-278. doi:10.1146/annurev-cellbio-111315-125407.
178. Pu J, Guardia CM, Keren-Kaplan T, Bonifacino JS. Mechanisms and functions of lysosome positioning. *J Cell Sci.* 2016;129:4329-4339. doi:10.1242/jcs.196287.
179. Korolchuk VI, et al. Lysosomal positioning coordinates cellular nutrient responses. *Nat Cell Biol.* 2011;13:453-460. doi:10.1038/ncb2204.
180. Tamzalit F, et al. Centrioles control the capacity, but not the specificity, of cytotoxic T cell killing. *Proc Natl Acad Sci U S A.* 2020;117:4310-4319. doi:10.1073/pnas.1913220117.
181. Schmidt H, et al. Effector granules in human T lymphocytes: proteomic evidence for two distinct species of cytotoxic effector vesicles. *J Proteome Res.* 2011;10:1603-1620. doi:10.1021/pr100967v.
182. Rouvier E, Luciani MF, Golstein P. Fas involvement in Ca(2+)-independent T cell-mediated cytotoxicity. *J Exp Med.* 1993;177:195-200. doi:10.1084/jem.177.1.195.
183. Huse M. The T-cell-receptor signaling network. *J Cell Sci.* 2009;122:1269-1273.
184. Courtney AH, Lo WL, Weiss A. TCR signaling: mechanisms of initiation and propagation. *Trends Biochem Sci.* 2018;43:108-123. doi:10.1016/j.tibs.2017.11.008.
185. Upshaw JL, Leibson PJ. NKG2D-mediated activation of cytotoxic lymphocytes: unique signaling pathways and distinct functional outcomes. *Semin Immunol.* 2006;18:167-175. doi:S1044-5323(06)00037-6 [pii] 10.1016/j.smim.2006.03.001.
186. Wu N, Veillette A. SLAM family receptors in normal immunity and immune pathologies. *Curr Opin Immunol.* 2016;38:45-51. doi:10.1016/j.coi.2015.11.003.
187. Bryceson YT, Ljunggren HG, Long EO. Minimal requirement for induction of natural cytotoxicity and intersection of activation signals by inhibitory receptors. *Blood.* 2009;114:2657-2666.
188. Luo BH, Carman CV, Springer TA. Structural basis of integrin regulation and signaling. *Annu Rev Immunol.* 2007;25:619-647. doi:10.1146/annurev.immunol.25.022106.141618.
189. Anikeeva N, et al. Distinct role of lymphocyte function-associated antigen-1 in mediating effective cytolytic activity by cytotoxic T lymphocytes. *Proc Natl Acad Sci U S A.* 2005;102:6437-6442. doi:10.1073/pnas.0502467102.
190. Liu D, et al. Integrin-dependent organization and bidirectional vesicular traffic at cytotoxic immune synapses. *Immunity.* 2009;31:99-109.
191. Morgan MM, et al. Superantigen-induced T cell: B cell conjugation is mediated by LFA-1 and requires signaling through Lck, but not ZAP-70. *J Immunol.* 2001;167:5708-5718.
192. van Triest M, Bos JL. Pull-down assays for guanoside 5′-triphosphate-bound Ras-like guanosine 5′-triphosphatases. *Methods Mol Biol.* 2004;250:97-102. doi:1-59259-671-1:97 [pii] 10.1385/1-59259-671-1:97.
193. Dubey C, Croft M, Swain SL. Costimulatory requirements of naive CD4+ T cells. ICAM-1 or B7-1 can costimulate naive CD4 T cell activation but both are required for optimum response. *J Immunol.* 1995;155:45-57.
194. Bachmann MF, et al. Distinct roles for LFA-1 and CD28 during activation of naive T cells: adhesion versus costimulation. *Immunity.* 1997;7:549-557. doi:10.1016/s1074-7613(00)80376-3.
195. Le Floc'h A, et al. Annular PIP3 accumulation controls actin architecture and modulates cytotoxicity at the immunological synapse. *J Exp Med.* 2013;210:2721-2737. doi:10.1084/jem.20131324.
196. Springer TA, Dustin ML. Integrin inside-out signaling and the immunological synapse. *Curr Opin Cell Biol.* 2012;24:107-115. doi:10.1016/j.ceb.2011.10.004.
197. Kinashi T. Intracellular signalling controlling integrin activation in lymphocytes. *Nat Rev Immunol.* 2005;5:546-559. doi:nri1646 [pii] 10.1038/nri1646.
198. Nolz JC, et al. The WAVE2 complex regulates T cell receptor signaling to integrins via Abl-and CrkL-C3G-mediated activation of Rap1. *J Cell Biol.* 2008;182:1231-1244.
199. Letschka T, et al. PKC-theta selectively controls the adhesion-stimulating molecule Rap1. *Blood.* 2008;112:4617-4627.
200. Katagiri K, Maeda A, Shimonaka M, Kinashi T. RAPL, a Rap1-binding molecule that mediates Rap1-induced adhesion through spatial regulation of LFA-1. *Nat Immunol.* 2003;4:741-748. doi:10.1038/ni950.
201. Lafuente EM, et al. RIAM, an Ena/VASP and Profilin ligand, interacts with Rap1-GTP and mediates Rap1-induced adhesion. *Dev Cell.* 2004;7:585-595. doi:10.1016/j.devcel.2004.07.021.
202. Han J, et al. Reconstructing and deconstructing agonist-induced activation of integrin alphaIIbbeta3. *Curr Biol.* 2006;16:1796-1806. doi:10.1016/j.cub.2006.08.035.
203. Lee HS, Lim CJ, Puzon-McLaughlin W, Shattil SJ, Ginsberg MH. RIAM activates integrins by linking talin to ras GTPase membrane-targeting sequences. *J Biol Chem.* 2009;284:5119-5127. doi:10.1074/jbc.M807117200.
204. Legate KR, Wickstrom SA, Fassler R. Genetic and cell biological analysis of integrin outside-in signaling. *Genes Dev.* 2009;23:397-418. doi:23/4/397 [pii]10.1101/gad.1758709.
205. Beal AM, et al. Protein kinase C theta regulates stability of the peripheral adhesion ring junction and contributes to the sensitivity of target cell lysis by CTL. *J Immunol.* 2008;181:4815-4824.
206. Esensten JH, Helou YA, Chopra G, Weiss A, Bluestone JA. CD28 costimulation: from mechanism to therapy. *Immunity.* 2016;44:973-988. doi:10.1016/j.immuni.2016.04.020.
207. Broz ML, et al. Dissecting the tumor myeloid compartment reveals rare activating antigen-presenting cells critical for T cell immunity. *Canc Cell.* 2014;26:638-652. doi:10.1016/j.ccell.2014.09.007.
208. GeurtsvanKessel CH, et al. Clearance of influenza virus from the lung depends on migratory langerin(+)CD11b(−) but not plasmacytoid dendritic cells. *J Exp Med.* 2008;205:1621-1634. doi:10.1084/jem.20071365.
209. McGill J, Van Rooijen N, Legge KL. Protective influenza-specific CD8 T cell responses require interactions with dendritic cells in the lungs. *J Exp Med.* 2008;205:1635-1646. doi:10.1084/jem.20080314.
210. Garris CS, et al. Successful anti-PD-1 cancer immunotherapy requires T cell-dendritic cell crosstalk involving the cytokines IFN-gamma and IL-12. *Immunity.* 2018;49:1148-1161. doi:10.1016/j.immuni.2018.09.024.
211. Kamphorst AO, et al. Rescue of exhausted CD8 T cells by PD-1-targeted therapies is CD28-dependent. *Science.* 2017;355:1423-1427. doi:10.1126/science.aaf0683.
212. Prajapati K, Perez C, Rojas LBP, Burke B, Guevara-Patino JA. Functions of NKG2D in CD8(+) T cells: an opportunity for immunotherapy. *Cell Mol Immunol.* 2018;15:470-479. doi:10.1038/cmi.2017.161.
213. Jamieson AM, et al. The role of the NKG2D immunoreceptor in immune cell activation and natural killing (vol 17, pg 19, 2002). *Immunity.* 2004;20:799. doi:10.1016/j.immuni.2004.05.003.
214. Upshaw JL, Schoon RA, Dick CJ, Billadeau DD, Leibson PJ. The isoforms of phospholipase C-gamma are differentially used by distinct human NK activating receptors. *J Immunol.* 2005;175:213-218. doi:10.4049/jimmunol.175.1.213.
215. Raulet DH, Gasser S, Gowen BG, Deng WW, Jung HY. Regulation of ligands for the NKG2D activating receptor. *Annu Rev Immunol.* 2013;31:413-441. doi:10.1146/annurev-immunol-032712-095951.
216. Guerra N, et al. NKG2D-deficient mice are defective in tumor surveillance in models of spontaneous malignancy. *Immunity.* 2008;28:571-580. doi:S1074-7613(08)00112-X [pii] 10.1016/j.immuni.2008.02.016.
217. Smyth MJ, et al. NKG2D function protects the host from tumor initiation. *J Exp Med.* 2005;202:583-588. doi:10.1084/jem.20050994.
218. Groh V, et al. Costimulation of CD8 alpha beta T cells by NKG2D via engagement by MIC induced on virus-infected cells. *Nat Immunol.* 2001;2:255-260. doi:10.1038/85321.
219. Kennedy PTF, et al. The expression and function of NKG2D molecule on intrahepatic CD8+T cells in chronic viral hepatitis. *J Viral Hepatitis.* 2008;15:901-909. doi:10.1111/j.1365-2893.2008.01049.x.
220. Mockl L. The emerging role of the mammalian glycocalyx in functional membrane organization and immune system regulation. *Front Cell Dev Biol.* 2020;8:253. doi:10.3389/fcell.2020.00253.

221. Jung Y, et al. Three-dimensional localization of T-cell receptors in relation to microvilli using a combination of superresolution microscopies. *Proc Natl Acad Sci U S A.* 2016;113:E5916-E5924. doi:10.1073/pnas.1605399113.
222. Ghosh S, et al. ERM-dependent assembly of T cell receptor signaling and co-stimulatory molecules on microvilli prior to activation. *Cell Rep.* 2020;30:3434-3447.e6. doi:10.1016/j.celrep.2020.02.069.
223. Bunnell SC, et al. T cell receptor ligation induces the formation of dynamically regulated signaling assemblies. *J Cell Biol.* 2002;158:1263-1275.
224. Cai E, et al. Visualizing dynamic microvillar search and stabilization during ligand detection by T cells. *Science.* 2017;356:eaal3118. doi:10.1126/science.aal3118.
225. Fooksman DR, et al. Functional anatomy of T cell activation and synapse formation. *Annu Rev Immunol.* 2010;28:79-105. doi:10.1146/annurev-immunol-030409-101308.
226. Le Floc'h A, Huse M. Molecular mechanisms and functional implications of polarized actin remodeling at the T cell immunological synapse. *Cell Mol Life Sci.* 2015;72:537-556. doi:10.1007/s00018-014-1760-7.
227. Goley ED, Welch MD. The ARP2/3 complex: an actin nucleator comes of age. *Nat Rev Mol Cell Biol.* 2006;7:713-726. doi:10.1038/nrm2026.
228. Nolz JC, et al. The WAVE2 complex regulates actin cytoskeletal reorganization and CRAC-mediated calcium entry during T cell activation. *Curr Biol.* 2006;16:24-34.
229. Zipfel PA, et al. Role for the Abi/wave protein complex in T cell receptor-mediated proliferation and cytoskeletal remodeling. *Curr Biol.* 2006;16:35-46. doi:10.1016/j.cub.2005.12.024.
230. Murugesan S, et al. Formin-generated actomyosin arcs propel T cell receptor microcluster movement at the immune synapse. *J Cell Biol.* 2016;215:383-399. doi:10.1083/jcb.201603080.
231. Yi J, Wu XS, Crites T, Hammer JA 3rd. Actin retrograde flow and actomyosin II arc contraction drive receptor cluster dynamics at the immunological synapse in Jurkat T cells. *Mol Biol Cell.* 2012;23:834-852. doi:mbc.E11-08-0731 [pii] 10.1091/mbc.E11-08-0731.
232. Grakoui A, et al. The immunological synapse: a molecular machine controlling T cell activation. *Science.* 1999;285:221-227.
233. Sims TN, et al. Opposing effects of PKCtheta and WASp on symmetry breaking and relocation of the immunological synapse. *Cell.* 2007;129:773-785.
234. Varma R, Campi G, Yokosuka T, Saito T, Dustin ML. T cell receptor-proximal signals are sustained in peripheral microclusters and terminated in the central supramolecular activation cluster. *Immunity.* 2006;25:117-127.
235. Vardhana S, Choudhuri K, Varma R, Dustin ML. Essential role of ubiquitin and TSG101 protein in formation and function of the central supramolecular activation cluster. *Immunity.* 2010;32:531-540. doi:S1074-7613(10)00130-5 [pii] 10.1016/j.immuni.2010.04.005.
236. Kumari S, et al. Actin foci facilitate activation of the phospholipase C-gamma in primary T lymphocytes via the WASP pathway. *Elife.* 2015;4:e04953. doi:10.7554/eLife.04953.
237. Kumari S, et al. Cytoskeletal tension actively sustains the migratory T-cell synaptic contact. *EMBO J.* 2020;39:e102783. doi:10.15252/embj.2019102783.
238. Tamzalit F, et al. Interfacial actin protrusions mechanically enhance killing by cytotoxic T cells. *Sci Immunol.* 2019;4:eaav5445. doi:10.1126/sciimmunol.aav5445.
239. Huse M. Lipid-based patterning of the immunological synapse. *Biochem Soc Trans.* 2014;42:1506-1511. doi:10.1042/BST20140191.
240. Huppa JB, Gleimer M, Sumen C, Davis MM. Continuous T cell receptor signaling required for synapse maintenance and full effector potential. *Nat Immunol.* 2003;4:749-755.
241. Harriague J, Bismuth G. Imaging antigen-induced PI3K activation in T cells. *Nat Immunol.* 2002;3:1090-1096.
242. Gawden-Bone CM, et al. PIP5 kinases regulate membrane phosphoinositide and actin composition for targeted granule secretion by cytotoxic lymphocytes. *Immunity.* 2018;49:427-437.e4. doi:10.1016/j.immuni.2018.08.017.
243. Quann EJ, Liu X, Altan-Bonnet G, Huse M. A cascade of protein kinase C isozymes promotes cytoskeletal polarization in T cells. *Nat Immunol.* 2011;12:647-654. doi:ni.2033 [pii] 10.1038/ni.2033.
244. Chauveau A, Le Floc'h A, Bantilan NS, Koretzky GA, Huse M. Diacylglycerol kinase alpha establishes T cell polarity by shaping diacylglycerol accumulation at the immunological synapse. *Sci Signal.* 2014;7:ra82. doi:10.1126/scisignal.2005287.
245. Quann EJ, Merino E, Furuta T, Huse M. Localized diacylglycerol drives the polarization of the microtubule-organizing center in T cells. *Nat Immunol.* 2009;10:627-635.
246. Dustin ML. Insights into function of the immunological synapse from studies with supported planar bilayers. *Curr Top Microbiol Immunol.* 2010;340:1-24. doi:10.1007/978-3-642-03858-7_1.
247. Balagopalan L, Sherman E, Barr VA, Samelson LE. Imaging techniques for assaying lymphocyte activation in action. *Nat Rev Immunol.* 2011;11:21-33. doi:10.1038/nri2903.
248. Sage PT, et al. Antigen recognition is facilitated by invadosome-like protrusions formed by memory/effector T cells. *J Immunol.* 2012;188:3686-3699. doi:10.4049/jimmunol.1102594.
249. Ritter AT, et al. Actin depletion initiates events leading to granule secretion at the immunological synapse. *Immunity.* 2015;42:864-876. doi:10.1016/j.immuni.2015.04.013.
250. Huse M, Le Floc'h A, Liu X. From lipid second messengers to molecular motors: microtubule-organizing center reorientation in T cells. *Immunol Rev.* 2013;256:95-106. doi:10.1111/imr.12116.
251. Sedwick CE, et al. TCR, LFA-1, and CD28 play unique and complementary roles in signaling T cell cytoskeletal reorganization. *J Immunol.* 1999;162:1367-1375.
252. Bryceson YT, March ME, Barber DF, Ljunggren HG, Long EO. Cytolytic granule polarization and degranulation controlled by different receptors in resting NK cells. *J Exp Med.* 2005;202:1001-1012.
253. Liu X, Kapoor TM, Chen JK, Huse M. Diacylglycerol promotes centrosome polarization in T cells via reciprocal localization of dynein and myosin II. *Proc Natl Acad Sci U S A.* 2013;110:11976-11981. doi:10.1073/pnas.1306180110.
254. Yi J, et al. Centrosome repositioning in T cells is biphasic and driven by microtubule end-on capture-shrinkage. *J Cell Biol.* 2013;202:779-792. doi:10.1083/jcb.201301004.
255. Andres-Delgado L, et al. INF2 promotes the formation of detyrosinated microtubules necessary for centrosome reorientation in T cells. *J Cell Biol.* 2012;198:1025-1037. doi:10.1083/jcb.201202137.
256. Stinchcombe JC, Bossi G, Booth S, Griffiths GM. The immunological synapse of CTL contains a secretory domain and membrane bridges. *Immunity.* 2001;15:751-761.
257. Stinchcombe JC, Majorovits E, Bossi G, Fuller S, Griffiths GM. Centrosome polarization delivers secretory granules to the immunological synapse. *Nature.* 2006;443:462-465.
258. de la Roche M, Asano Y, Griffiths GM. Origins of the cytolytic synapse. *Nat Rev Immunol.* 2016;16:421-432. doi:10.1038/nri.2016.54.
259. de la Roche M, et al. Hedgehog signaling controls T cell killing at the immunological synapse. *Science.* 2013;342:1247-1250. doi:10.1126/science.1244689.
260. Mentlik AN, Sanborn KB, Holzbaur EL, Orange JS. Rapid lytic granule convergence to the MTOC in natural killer cells is dependent on dynein but not cytolytic commitment. *Mol Biol Cell.* 2010;21:2241-2256. doi:E09-11-0930 [pii] 10.1091/mbc.E09-11-0930.
261. Daniele T, et al. A role for Rab7 in the movement of secretory granules in cytotoxic T lymphocytes. *Traffic.* 2011;12:902-911. doi:10.1111/j.1600-0854.2011.01194.x.
262. James AM, et al. Rapid activation receptor-or IL-2-induced lytic granule convergence in human natural killer cells requires Src, but not downstream signaling. *Blood.* 2013;121:2627-2637. doi:10.1182/blood-2012-06-437012.
263. Butler B, Cooper JA. Distinct roles for the actin nucleators Arp2/3 and hDia1 during NK-mediated cytotoxicity. *Curr Biol.* 2009;19:1886-1896. doi:10.1016/j.cub.2009.10.029.
264. Kurowska M, et al. Terminal transport of lytic granules to the immune synapse is mediated by the kinesin-1/Slp3/Rab27a complex. *Blood.* 2012;119:3879-3889. doi:10.1182/blood-2011-09-382556.
265. Bertrand F, et al. An initial and rapid step of lytic granule secretion precedes microtubule organizing center polarization at the cytotoxic T lymphocyte/target cell synapse. *Proc Natl Acad Sci U S A.* 2013;110:6073-6078. doi:10.1073/pnas.1218640110.
266. Chauveau A, Aucher A, Eissmann P, Vivier E, Davis DM. Membrane nanotubes facilitate long-distance interactions between natural killer cells and target cells. *Proc Natl Acad Sci U S A.* 2010;107:5545-5550. doi:10.1073/pnas.0910074107.
267. Rak GD, Mace EM, Banerjee PP, Svitkina T, Orange JS. Natural killer cell lytic granule secretion occurs through a pervasive actin network at the immune synapse. *PLoS Biol.* 2011;9:e1001151. doi:10.1371/journal.pbio.1001151 PBIOLOGY-D-11-00213 [pii].
268. Brown AC, et al. Remodelling of cortical actin where lytic granules dock at natural killer cell immune synapses revealed by super-resolution microscopy. *PLoS Biol.* 2011;9:e1001152. doi:10.1371/journal.pbio.1001152 PBIOLOGY-D-11-00271 [pii].
269. Liu D, Martina JA, Wu XS, Hammer JA 3rd, Long EO. Two modes of lytic granule fusion during degranulation by natural killer cells. *Immunol Cell Biol.* 2011;89:728-738. doi:10.1038/icb.2010.167.
270. Houmadi R, et al. The Wiskott-Aldrich syndrome protein contributes to the assembly of the LFA-1 nanocluster belt at the lytic synapse. *Cell Rep.* 2018;22:979-991. doi:10.1016/j.celrep.2017.12.088.
271. Weber T, et al. SNAREpins: minimal machinery for membrane fusion. *Cell.* 1998;92:759-772.
272. Sudhof TC, Rothman JE. Membrane fusion: grappling with SNARE and SM proteins. *Science.* 2009;323:474-477. doi:10.1126/science.1161748.
273. Han J, Pluhackova K, Bockmann RA. The multifaceted role of SNARE proteins in membrane fusion. *Front Physiol.* 2017;8:5. doi:10.3389/fphys.2017.00005.
274. Stenmark H. Rab GTPases as coordinators of vesicle traffic. *Nat Rev Mol Cell Biol.* 2009;10:513-525. doi:10.1038/nrm2728.
275. Sudhof TC. Neurotransmitter release: the last millisecond in the life of a synaptic vesicle. *Neuron.* 2013;80:675-690. doi:10.1016/j.neuron.2013.10.022.
276. zur Stadt U, et al. Linkage of familial hemophagocytic lymphohistiocytosis (FHL) type-4 to chromosome 6q24 and identification of mutations in syntaxin 11. *Hum Mol Genet.* 2005;14:827-834. doi:10.1093/hmg/ddi076.
277. zur Stadt U, et al. Familial hemophagocytic lymphohistiocytosis type 5 (FHL-5) is caused by mutations in Munc18-2 and impaired binding to syntaxin 11. *Am J Hum Genet.* 2009;85:482-492. doi:10.1016/j.ajhg.2009.09.005.
278. Cote M, et al. Munc18-2 deficiency causes familial hemophagocytic lymphohistiocytosis type 5 and impairs cytotoxic granule exocytosis in patient NK cells. *J Clin Invest.* 2009;119:3765-3773. doi:10.1172/JCI40732.
279. Feldmann J, et al. Munc13-4 is essential for cytolytic granules fusion and is mutated in a form of familial hemophagocytic lymphohistiocytosis (FHL3). *Cell.* 2003;115:461-473.
280. Menasche G, et al. Mutations in RAB27A cause Griscelli syndrome associated with haemophagocytic syndrome. *Nat Genet.* 2000;25:173-176.
281. Wilson SM, et al. A mutation in Rab27a causes the vesicle transport defects observed in ashen mice. *Proc Natl Acad Sci U S A.* 2000;97:7933-7938. doi:10.1073/pnas.140212797.
282. Crozat K, et al. Jinx, an MCMV susceptibility phenotype caused by disruption of Unc13d: a mouse model of type 3 familial hemophagocytic lymphohistiocytosis. *J Exp Med.* 2007;204:853-863. doi:10.1084/jem.20062447.
283. Stinchcombe JC, et al. Rab27a is required for regulated secretion in cytotoxic T lymphocytes. *J Cell Biol.* 2001;152:825-834.
284. Holt O, et al. Slp1 and Slp2-a localize to the plasma membrane of CTL and contribute to secretion from the immunological synapse. *Traffic.* 2008;9:446-457.
285. Kuroda TS, Fukuda M, Ariga H, Mikoshiba K. The Slp homology domain of synaptotagmin-like proteins 1-4 and Slac2 functions as a novel Rab27A binding domain. *J Biol Chem.* 2002;277:9212-9218. doi:10.1074/jbc.M112414200.
286. Kuroda TS, Fukuda M. Rab27A-binding protein Slp2-a is required for peripheral melanosome distribution and elongated cell shape in melanocytes. *Nat Cell Biol.* 2004;6:1195-1203. doi:10.1038/ncb1197.
287. Chitirala P, et al. Cytotoxic granule exocytosis from human cytotoxic T lymphocytes is mediated by VAMP7. *Front Immunol.* 2019;10:1855. doi:10.3389/fimmu.2019.01855.
288. Halimani M, et al. Syntaxin11 serves as a t-SNARE for the fusion of lytic granules in human cytotoxic T lymphocytes. *Eur J Immunol.* 2014;44:573-584. doi:10.1002/eji.201344011.
289. Dudenhoffer-Pfeifer M, et al. Different Munc13 isoforms function as priming factors in lytic granule release from murine cytotoxic T lymphocytes. *Traffic.* 2013;14:798-809. doi:10.1111/tra.12074.
290. Bin NR, et al. C2 domains of munc13-4 are crucial for Ca(2+)-dependent degranulation and cytotoxicity in NK cells. *J Immunol.* 2018;201:700-713. doi:10.4049/jimmunol.1800426.
291. Krzewski K, Coligan JE. Human NK cell lytic granules and regulation of their exocytosis. *Front Immunol.* 2012;3:335. doi:10.3389/fimmu.2012.00335.
292. Spessott WA, Sanmillan ML, McCormick ME, Kulkarni VV, Giraudo CG. SM protein Munc18-2 facilitates transition of Syntaxin 11-mediated lipid mixing to complete fusion for T-lymphocyte cytotoxicity. *Proc Natl Acad Sci U S A.* 2017;114:E2176-E2185. doi:10.1073/pnas.1617981114.
293. Marshall MR, et al. VAMP8-dependent fusion of recycling endosomes with the plasma membrane facilitates T lymphocyte cytotoxicity. *J Cell Biol.* 2015;210:135-151. doi:10.1083/jcb.201411093.
294. Ostergaard HL, Clark WR. Evidence for multiple lytic pathways used by cytotoxic T lymphocytes. *J Immunol.* 1989;143:2120-2126.
295. Takayama H, Sitkovsky MV. Antigen receptor-regulated exocytosis in cytotoxic T lymphocytes. *J Exp Med.* 1987;166:725-743.
296. Fukuda M, Mikoshiba K. Synaptotagmin-like protein 1-3: a novel family of C-terminal-type tandem C2 proteins. *Biochem Biophys Res Commun.* 2001;281:1226-1233. doi:10.1006/bbrc.2001.4512.
297. Dutz JP, Fruman DA, Burakoff SJ, Bierer BE. A role for calcineurin in degranulation of murine cytotoxic T lymphocytes. *J Immunol.* 1993;150:2591-2598.
298. Grybko MJ, Bartnik JP, Wurth GA, Pores-Fernando AT, Zweifach A. Calcineurin activation is only one calcium-dependent step in cytotoxic T lymphocyte granule exocytosis. *J Biol Chem.* 2007;282:18009-18017. doi:10.1074/jbc.M702222200.
299. Trenn G, et al. Biochemical characterization of the inhibitory effect of CsA on cytolytic T lymphocyte effector functions. *J Immunol.* 1989;142:3796-3802.
300. Bird CH, et al. Selective regulation of apoptosis: the cytotoxic lymphocyte serpin proteinase inhibitor 9 protects against granzyme B-mediated apoptosis without perturbing the Fas cell death pathway. *Mol Cell Biol.* 1998;18:6387-6398. doi:10.1128/mcb.18.11.6387.

301. Balaji KN, Schaschke N, Machleidt W, Catalfamo M, Henkart PA. Surface cathepsin B protects cytotoxic lymphocytes from self-destruction after degranulation. *J Exp Med.* 2002;196:493-503.
302. Baran K, et al. Cytotoxic T lymphocytes from cathepsin B-deficient mice survive normally in vitro and in vivo after encountering and killing target cells. *J Biol Chem.* 2006;281:30485-30491. doi:10.1074/jbc.M602007200.
303. Cohnen A, et al. Surface CD107a/LAMP-1 protects natural killer cells from degranulation-associated damage. *Blood.* 2013;122:1411-1418. doi:10.1182/blood-2012-07-441832.
304. Jenkins MR, et al. Failed CTL/NK cell killing and cytokine hypersecretion are directly linked through prolonged synapse time. *J Exp Med.* 2015;212:307-317. doi:10.1084/jem.20140964.
305. Netter P, Anft M, Watzl C. Termination of the activating NK cell immunological synapse is an active and regulated process. *J Immunol.* 2017;199:2528-2535. doi:10.4049/jimmunol.1700394.
306. Anft M, et al. NK cell detachment from target cells is regulated by successful cytotoxicity and influences cytokine production. *Cell Mol Immunol.* 2020;17:347-355. doi:10.1038/s41423-019-0277-2.
307. Huse M. Mechanical forces in the immune system. *Nat Rev Immunol.* 2017;17:679-690. doi:10.1038/nri.2017.74.
308. Dembo M, Torney DC, Saxman K, Hammer D. The reaction-limited kinetics of membrane-to-surface adhesion and detachment. *Proc R Soc Lond B Biol Sci.* 1988;234:55-83.
309. Thomas WE, Vogel V, Sokurenko E. Biophysics of catch bonds. *Annu Rev Biophys.* 2008;37:399-416. doi:10.1146/annurev.biophys.37.032807.125804.
310. Kong F, Garcia AJ, Mould AP, Humphries MJ, Zhu C. Demonstration of catch bonds between an integrin and its ligand. *J Cell Biol.* 2009;185:1275-1284. doi:10.1083/jcb.200810002.
311. Chen W, Lou J, Zhu C. Forcing switch from short-to intermediate- and long-lived states of the alphaA domain generates LFA-1/ICAM-1 catch bonds. *J Biol Chem.* 2010;285:35967-35978. doi:10.1074/jbc.M110.155770.
312. Rosetti F, et al. A lupus-associated mac-1 variant has defects in integrin allostery and interaction with ligands under force. *Cell Rep.* 2015;10:1655-1664. doi:10.1016/j.celrep.2015.02.037.
313. Sawada Y, et al. Force sensing by mechanical extension of the Src family kinase substrate p130Cas. *Cell.* 2006;127:1015-1026. doi:10.1016/j.cell.2006.09.044.
314. del Rio A, et al. Stretching single talin rod molecules activates vinculin binding. *Science.* 2009;323:638-641. doi:10.1126/science.1162912.
315. Janostiak R, Pataki AC, Brabek J, Rosel D. Mechanosensors in integrin signaling: the emerging role of p130Cas. *Eur J Cell Biol.* 2014;93:445-454. doi:10.1016/j.ejcb.2014.07.002.
316. Comrie WA, Babich A, Burkhardt JK. F-actin flow drives affinity maturation and spatial organization of LFA-1 at the immunological synapse. *J Cell Biol.* 2015;208:475-491. doi:10.1083/jcb.201406121.
317. Comrie WA, Li S, Boyle S, Burkhardt JK. The dendritic cell cytoskeleton promotes T cell adhesion and activation by constraining ICAM-1 mobility. *J Cell Biol.* 2015;208:457-473. doi:10.1083/jcb.201406120.
318. Liu B, Chen W, Evavold BD, Zhu C. Accumulation of dynamic catch bonds between TCR and agonist peptide-MHC triggers T cell signaling. *Cell.* 2014;157:357-368. doi:10.1016/j.cell.2014.02.053.
319. Kim ST, et al. The alphabeta T cell receptor is an anisotropic mechanosensor. *J Biol Chem.* 2009;284:31028-31037. doi:10.1074/jbc.M109.052712.
320. Gonzalez C, et al. Nanobody-CD16 catch bond reveals NK cell mechanosensitivity. *Biophys J.* 2019;116:1516-1526. doi:10.1016/j.bpj.2019.03.012.
321. Sibener LV, et al. Isolation of a structural mechanism for uncoupling T cell receptor signaling from peptide-MHC binding. *Cell.* 2018;174:672-687.e27. doi:10.1016/j.cell.2018.06.017.
322. Wu P, et al. Mechano-regulation of peptide-MHC class I conformations determines TCR antigen recognition. *Mol Cell.* 2019;73:1015-1027.e17. doi:10.1016/j.molcel.2018.12.018.
323. Xu C, et al. Regulation of T cell receptor activation by dynamic membrane binding of the CD3 epsilon cytoplasmic tyrosine-based motif. *Cell.* 2008;135:702-713. doi:10.1016/j.cell.2008.09.044.
324. Aivazian D, Stern LJ. Phosphorylation of T cell receptor zeta is regulated by a lipid dependent folding transition. *Nat Struct Biol.* 2000;7:1023-1026.
325. Gil D, Schamel WW, Montoya M, Sanchez-Madrid F, Alarcon B. Recruitment of Nck by CD3 epsilon reveals a ligand-induced conformational change essential for T cell receptor signaling and synapse formation. *Cell.* 2002;109:901-912.
326. Lee MS, et al. A mechanical switch couples T cell receptor triggering to the cytoplasmic juxtamembrane regions of CD3zetazeta. *Immunity.* 2015;43:227-239. doi:10.1016/j.immuni.2015.06.018.
327. Basu R, et al. Cytotoxic T cells use mechanical force to potentiate target cell killing. *Cell.* 2016;165:100-110. doi:10.1016/j.cell.2016.01.021.
328. Saitakis M, et al. Different TCR-induced T lymphocyte responses are potentiated by stiffness with variable sensitivity. *Elife.* 2017;6:e23190. doi:10.7554/eLife.23190.
329. Wan Z, et al. B cell activation is regulated by the stiffness properties of the substrate presenting the antigens. *J Immunol.* 2013;190:4661-4675. doi:10.4049/jimmunol.1202976.
330. Judokusumo E, Tabdanov E, Kumari S, Dustin ML, Kam LC. Mechanosensing in T lymphocyte activation. *Biophys J.* 2012;102:L5-L7. doi:10.1016/j.bpj.2011.12.011.
331. O'Connor RS, et al. Substrate rigidity regulates human T cell activation and proliferation. *J Immunol.* 2012;189:1330-1339. doi:10.4049/jimmunol.1102757.
332. Blumenthal D, Chandra V, Burkhardt JK. Mouse T-cell priming is enhanced by maturation-dependent stiffening of the dendritic cell cortex. *Elife.* 2020;9:e55995. doi:10.7554/eLife.55995.
333. Long EO. Negative signaling by inhibitory receptors: the NK cell paradigm. *Immunol Rev.* 2008;224:70-84.
334. Burshtyn DN, et al. Recruitment of tyrosine phosphatase HCP by the killer cell inhibitor receptor. *Immunity.* 1996;4:77-85.
335. Olcese L, et al. Human and mouse killer-cell inhibitory receptors recruit PTP1C and PTP1D protein tyrosine phosphatases. *J Immunol.* 1996;156:4531-4534.
336. Stebbins CC, et al. Vav1 dephosphorylation by the tyrosine phosphatase SHP-1 as a mechanism for inhibition of cellular cytotoxicity. *Mol Cell Biol.* 2003;23:6291-6299.
337. Bustelo XR. Regulatory and signaling properties of the Vav family. *Mol Cell Biol.* 2000;20:1461-1477.
338. Riteau B, Barber DF, Long EO. Vav1 phosphorylation is induced by beta2 integrin engagement on natural killer cells upstream of actin cytoskeleton and lipid raft reorganization. *J Exp Med.* 2003;198:469-474.
339. Culley FJ, et al. Natural killer cell signal integration balances synapse symmetry and migration. *PLoS Biol.* 2009;7:e1000159.
340. Abeyweera TP, Merino E, Huse M. Inhibitory signaling blocks activating receptor clustering and induces cytoskeletal retraction in natural killer cells. *J Cell Biol.* 2011;192:675-690. doi:-jcb.201009135 [pii] 10.1083/jcb.201009135.
341. Wei SC, Duffy CR, Allison JP. Fundamental mechanisms of immune checkpoint blockade therapy. *Cancer Discov.* 2018;8:1069-1086. doi:10.1158/2159-8290.CD-18-0367.
342. McLane LM, Abdel-Hakeem MS, Wherry EJ. CD8 T cell exhaustion during chronic viral infection and cancer. *Annu Rev Immunol.* 2019;37:457-495. doi:10.1146/annurev-immunol-041015-055318.
343. Barber DL, et al. Restoring function in exhausted CD8 T cells during chronic viral infection. *Nature.* 2006;439:682-687. doi:10.1038/nature04444.
344. Leach DR, Krummel MF, Allison JP. Enhancement of antitumor immunity by CTLA-4 blockade. *Science.* 1996;271:1734-1736. doi:10.1126/science.271.5256.1734.
345. Strome SE, et al. B7-H1 blockade augments adoptive T-cell immunotherapy for squamous cell carcinoma. *Canc Res.* 2003;63:6501-6505.
346. Postow MA, Callahan MK, Wolchok JD. Immune checkpoint blockade in cancer therapy. *J Clin Oncol.* 2015;33:1974-1982. doi:10.1200/JCO.2014.59.4358.
347. Okazaki T, Maeda A, Nishimura H, Kurosaki T, Honjo T. PD-1 immunoreceptor inhibits B cell receptor-mediated signaling by recruiting src homology 2-domain-containing tyrosine phosphatase 2 to phosphotyrosine. *Proc Natl Acad Sci U S A.* 2001;98:13866-13871. doi:10.1073/pnas.231486598 231486598 [pii].
348. Yokosuka T, et al. Programmed cell death 1 forms negative costimulatory microclusters that directly inhibit T cell receptor signaling by recruiting phosphatase SHP2. *J Exp Med.* 2012;209:1201-1217. doi:jem.20112741 [pii] 10.1084/jem.20112741.
349. Hui E, et al. T cell costimulatory receptor CD28 is a primary target for PD-1-mediated inhibition. *Science.* 2017;355:1428-1433. doi:10.1126/science.aaf1292.
350. Egen JG, Allison JP. Cytotoxic T lymphocyte antigen-4 accumulation in the immunological synapse is regulated by TCR signal strength. *Immunity.* 2002;16:23-35.
351. Pentcheva-Hoang T, Egen JG, Wojnoonski K, Allison JP. B7-1 and B7-2 selectively recruit CTLA-4 and CD28 to the immunological synapse. *Immunity.* 2004;21:401-413.
352. Yewdell JW, Bennink JR. Mechanisms of viral interference with MHC class I antigen processing and presentation. *Annu Rev Cell Dev Biol.* 1999;15:579-606. doi:10.1146/annurev.cellbio.15.1.579.
353. Kaminskyy V, Zhivotovsky B. To kill or be killed: how viruses interact with the cell death machinery. *J Intern Med.* 2010;267:473-482. doi:10.1111/j.1365-2796.2010.02222.x.
354. Andrade F, Fellows E, Jenne DE, Rosen A, Young CS. Granzyme H destroys the function of critical adenoviral proteins required for viral DNA replication and granzyme B inhibition. *EMBO J.* 2007;26:2148-2157. doi:10.1038/sj.emboj.7601650.
355. Eckelman BP, Salvesen GS, Scott FL. Human inhibitor of apoptosis proteins: why XIAP is the black sheep of the family. *EMBO Rep.* 2006;7:988-994. doi:10.1038/sj.embor.7400795.
356. Quan LT, Caputo A, Bleackley RC, Pickup DJ, Salvesen GS. Granzyme B is inhibited by the cowpox virus serpin cytokine response modifier A. *J Biol Chem.* 1995;270:10377-10379. doi:10.1074/jbc.270.18.10377.
357. Ray CA, et al. Viral inhibition of inflammation - cowpox virus encodes an inhibitor of the interleukin-1-beta converting enzyme. *Cell.* 1992;69:597-604. doi:10.1016/0092-8674(92)90223-Y.
358. Rao L, et al. The adenovirus E1A proteins induce apoptosis, which is inhibited by the E1B 19-kDa and Bcl-2 proteins. *Proc Natl Acad Sci U S A.* 1992;89:7742-7746. doi:10.1073/pnas.89.16.7742.
359. Cuconati A, Degenhardt K, Sundararajan R, Anschel A, White E. Bak and Bax function to limit adenovirus replication through apoptosis induction. *J Virol.* 2002;76:4547-4558. doi:10.1128/jvi.76.9.4547-4558.2002.
360. Arnoult D, et al. Cytomegalovirus cell death suppressor vMIA blocks Bax- but not Bak-mediated apoptosis by binding and sequestering Bax at mitochondria. *Proc Natl Acad Sci U S A.* 2004;101:7988-7993. doi:10.1073/pnas.0401897101.
361. Henderson S, et al. Epstein-Barr virus-coded BHRF1 protein, a viral homologue of Bcl-2, protects human B cells from programmed cell death. *Proc Natl Acad Sci U S A.* 1993;90:8479-8483. doi:10.1073/pnas.90.18.8479.
362. Reeves MB, Davies AA, McSharry BP, Wilkinson GW, Sinclair JH. Complex I binding by a virally encoded RNA regulates mitochondria-induced cell death. *Science.* 2007;316:1345-1348. doi:10.1126/science.1142984.
363. Thome M, et al. Viral FLICE-inhibitory proteins (FLIPs) prevent apoptosis induced by death receptors. *Nature.* 1997;386:517-521. doi:10.1038/386517a0.
364. Matloubian M, et al. A role for perforin in downregulating T-cell responses during chronic viral infection. *J Virol.* 1999;73:2527-2536.
365. Riera L, et al. Murine cytomegalovirus replication in salivary glands is controlled by both perforin and granzymes during acute infection. *Eur J Immunol.* 2000;30:1350-1355. doi:10.1002/(SICI)1521-4141(200005)30:5<1350::AID-IMMU1350>3.0.CO;2-J.
366. Kagi D, et al. The roles of perforin-and Fas-dependent cytotoxicity in protection against cytopathic and noncytopathic viruses. *Eur J Immunol.* 1995;25:3256-3262. doi:10.1002/eji.1830251209.
367. Gupta M, et al. CD8-mediated protection against Ebola virus infection is perforin dependent. *J Immunol.* 2005;174:4198-4202. doi:10.4049/jimmunol.174.7.4198.
368. Jensen ER, et al. Fas (CD95)-dependent cell-mediated immunity to Listeria monocytogenes. *Infect Immun.* 1998;66:4143-4150.
369. Zhou P, Freidag BL, Caldwell CC, Seder RA. Perforin is required for primary immunity to Histoplasma capsulatum. *J Immunol.* 2001;166:1968-1974. doi:10.4049/jimmunol.166.3.1968.
370. Nickell SP, Sharma D. Trypanosoma cruzi: roles for perforin-dependent and perforin-independent immune mechanisms in acute resistance. *Exp Parasitol.* 2000;94:207-216. doi:10.1006/expr.2000.4498.
371. Yang J, Huck SP, McHugh RS, Hermans IF, Ronchese F. Perforin-dependent elimination of dendritic cells regulates the expansion of antigen-specific CD8+ T cells in vivo. *Proc Natl Acad Sci U S A.* 2006;103:147-152. doi:10.1073/pnas.0509054103.
372. Smyth MJ, et al. Perforin-mediated cytotoxicity is critical for surveillance of spontaneous lymphoma. *J Exp Med.* 2000;192:755-760. doi:10.1084/jem.192.5.755.
373. Mehta PA, et al. Perforin polymorphism A91V and susceptibility to B-precursor childhood acute lymphoblastic leukemia: a report from the Children's Oncology Group. *Leukemia.* 2006;20:1539-1541. doi:10.1038/sj.leu.2404299.
374. Smyth MJ, et al. Perforin is a major contributor to NK cell control of tumor metastasis. *J Immunol.* 1999;162:6658-6662.
375. van den Broek ME, et al. Decreased tumor surveillance in perforin-deficient mice. *J Exp Med.* 1996;184:1781-1790.
376. Seki N, et al. Tumor-specific CTL kill murine renal cancer cells using both perforin and Fas ligand-mediated lysis in vitro, but cause tumor regression in vivo in the absence of perforin. *J Immunol.* 2002;168:3484-3492. doi:10.4049/jimmunol.168.7.3484.
377. Peng L, et al. T cell-mediated tumor rejection displays diverse dependence upon perforin and IFN-gamma mechanisms that cannot be predicted from in vitro T cell characteristics. *J Immunol.* 2000;165:7116-7124. doi:10.4049/jimmunol.165.12.7116.
378. Dobrzanski MJ, Reome JB, Dutton RW. Role of effector cell-derived IL-4, IL-5, and perforin in early and late stages of type 2 CD8 effector cell-mediated tumor rejection. *J Immunol.* 2001;167:424-434. doi:10.4049/jimmunol.167.1.424.
379. Dunn GP, Koebel CM, Schreiber RD. Interferons, immunity and cancer immunoediting. *Nat Rev Immunol.* 2006;6:836-848. doi:nri1961 [pii]10.1038/nri1961.
380. Topham DJ, Tripp RA, Doherty PC. CD8+ T cells clear influenza virus by perforin or Fas-dependent processes. *J Immunol.* 1997;159:5197-5200.
381. Bien K, et al. Fas/FasL pathway participates in regulation of antiviral and inflammatory response during mousepox

infection of lungs. *Mediat Inflamm.* 2015;2015:281613. doi:10.1155/2015/281613.
382. Wang Y, Lobigs M, Lee E, Mullbacher A. Exocytosis and Fas-mediated cytolytic mechanisms exert protection from West Nile virus induced encephalitis in mice. *Immunol Cell Biol.* 2004;82:170-173. doi:10.1046/j.0818-9641.2004.01227.x.
383. Uchiyama R, Yonehara S, Tsutsui H. Fas-mediated inflammatory response in Listeria monocytogenes infection. *J Immunol.* 2013;190:4245-4254. doi:10.4049/jimmunol.1203059.
384. Straus SE, et al. The development of lymphomas in families with autoimmune lymphoproliferative syndrome with germline Fas mutations and defective lymphocyte apoptosis. *Blood.* 2001;98:194-200. doi:10.1182/blood.v98.1.194.
385. Davidson WF, Giese T, Fredrickson TN. Spontaneous development of plasmacytoid tumors in mice with defective Fas-Fas ligand interactions. *J Exp Med.* 1998;187:1825-1838. doi:10.1084/jem.187.11.1825.
386. Peter ME, et al. The role of CD95 and CD95 ligand in cancer. *Cell Death Differ.* 2015;22:549-559. doi:10.1038/cdd.2015.3.
387. Teachey DT, Seif AE, Grupp SA. Advances in the management and understanding of autoimmune lymphoproliferative syndrome (ALPS). *Br J Haematol.* 2010;148:205-216. doi:10.1111/j.1365-2141.2009.07991.x.
388. Krammer PH. CD95's deadly mission in the immune system. *Nature.* 2000;407:789-795. doi:10.1038/35037728.
389. Roberts AI, et al. The role of activation-induced cell death in the differentiation of T-helper-cell subsets. *Immunol Res.* 2003;28:285-293. doi:10.1385/IR:28:3:285.
390. Alderson MR, et al. Fas ligand mediates activation-induced cell death in human T lymphocytes. *J Exp Med.* 1995;181:71-77. doi:10.1084/jem.181.1.71.
391. Varadhachary AS, Perdow SN, Hu C, Ramanarayanan M, Salgame P. Differential ability of T cell subsets to undergo activation-induced cell death. *Proc Natl Acad Sci U S A.* 1997;94:5778-5783. doi:10.1073/pnas.94.11.5778.
392. Steinberg AD. MRL-lpr/lpr disease: theories meet Fas. *Semin Immunol.* 1994;6:55-69. doi:10.1006/smim.1994.1009.
393. Roths JB, Murphy ED, Eicher EM. A new mutation, gld, that produces lymphoproliferation and autoimmunity in C3H/HeJ mice. *J Exp Med.* 1984;159:1-20. doi:10.1084/jem.159.1.1.
394. Ito MR, et al. Rheumatic diseases in an MRL strain of mice with a deficit in the functional Fas ligand. *Arthritis Rheum.* 1997;40:1054-1063. doi:10.1002/art.1780400610.
395. Sidman CL, Marshall JD, Von Boehmer H. Transgenic T cell receptor interactions in the lymphoproliferative and autoimmune syndromes of lpr and gld mutant mice. *Eur J Immunol.* 1992;22:499-504. doi:10.1002/eji.1830220231.
396. Hildeman DA, et al. Activated T cell death in vivo mediated by proapoptotic bcl-2 family member bim. *Immunity.* 2002;16:759-767. doi:10.1016/s1074-7613(02)00322-9.
397. Pellegrini M, et al. FADD and caspase-8 are required for cytokine-induced proliferation of hemopoietic progenitor cells. *Blood.* 2005;106:1581-1589. doi:10.1182/blood-2005-01-0284.
398. Hughes PD, et al. Apoptosis regulators Fas and Bim cooperate in shutdown of chronic immune responses and prevention of autoimmunity. *Immunity.* 2008;28:197-205. doi:10.1016/j.immuni.2007.12.017.
399. Rathmell JC, et al. CD95 (Fas)-dependent elimination of self-reactive B cells upon interaction with CD4+ T cells. *Nature.* 1995;376:181-184. doi:10.1038/376181a0.
400. Shlomchik WD. Graft-versus-host disease. *Nat Rev Immunol.* 2007;7:340-352. doi:10.1038/nri2000.
401. Du W, Cao X. Cytotoxic pathways in allogeneic hematopoietic cell transplantation. *Front Immunol.* 2018;9:2979. doi:10.3389/fimmu.2018.02979.
402. Via CS, Nguyen P, Shustov A, Drappa J, Elkon KB. A major role for the Fas pathway in acute graft-versus-host disease. *J Immunol.* 1996;157:5387-5393.
403. Braun MY, Lowin B, French L, Acha-Orbea H, Tschopp J. Cytotoxic T cells deficient in both functional fas ligand and perforin show residual cytolytic activity yet lose their capacity to induce lethal acute graft-versus-host disease. *J Exp Med.* 1996;183:657-661. doi:10.1084/jem.183.2.657.
404. Baker MB, Altman NH, Podack ER, Levy RB. The role of cell-mediated cytotoxicity in acute GVHD after MHC-matched allogeneic bone marrow transplantation in mice. *J Exp Med.* 1996;183:2645-2656. doi:10.1084/jem.183.6.2645.
405. Tsukada N, Kobata T, Aizawa Y, Yagita H, Okumura K. Graft-versus-leukemia effect and graft-versus-host disease can be differentiated by cytotoxic mechanisms in a murine model of allogeneic bone marrow transplantation. *Blood.* 1999;93:2738-2747.
406. Schmaltz C, et al. Differential use of Fas ligand and perforin cytotoxic pathways by donor T cells in graft-versus-host disease and graft-versus-leukemia effect. *Blood.* 2001;97:2886-2895. doi:10.1182/blood.v97.9.2886.

SECTION VI Integrative Immunology

CHAPTER 32 Immunologic Tolerance

Colin C. Anderson, PhD • Troy A. Baldwin, PhD • Peter A. Bretscher, PhD

INTRODUCTION

Paul Ehrlich reported in 1901 that goats, immunized with red blood cells from another goat, produced antibodies against these red cells, but the antibody did not react with the immunized goat's own red cells. Ehrlich surmised that such antibody, if produced, would be much more damaging than most external threats. Ehrlich's thought is the first known to recognize that the immune system responds against foreign invaders and other foreign entities much better than to components of the body to which it belongs. The discovery of autoimmunity in the 1940s made immunologists realize how damaging immune reactivity against self can be. It was this recognition that led all theorists, attempting to describe the general nature of immune responses, to speculate on how the attribute of self-nonself (SNS) recognition is realized.

However, studies in immune tolerance have become increasingly more descriptive with time. This has occurred for two different reasons. We have become aware with time that there are a number of distinct mechanisms contributing to how the attribute of SNS discrimination is realized. Thus, a proposal as to one of the contributing mechanisms cannot be expected to explain all the diverse observations. This recognition has naturally made immunologists open to new observations inexplicable by current conceptual proposals. Secondly, advances in technology have resulted in an exponentially growing amount of valuable information about the immune system. This explosion of information at the same time creates a challenge. How is one to put order in or make sense of all of the descriptive findings? Sorting the information in an effort to move beyond description toward an evolutionary understanding of the subject is benefited by considering the pertinence of various theoretical frameworks; these attempt to integrate the observations in a manner that makes physiological sense. Nevertheless, because there are multiple mechanisms contributing to tolerance, it is unrealistic to demand any particular framework will be consistent with or account for all of the observations. On the positive side, a plausible framework may raise doubts about the interpretation of data that do not appear consistent with it, stimulating a critical reanalysis of the data or of the experimental systems employed. We therefore begin with a history of some of the useful theories of how tolerance is achieved and several contemporary frameworks that were initially proposed as competing models. Rather than competing, the reader may instead come to see how certain aspects of each of these frameworks could be fused to create our most accurate current understanding of immunologic tolerance. The reader may wish to avoid developing a bias toward one or other of the preexisting frameworks and skip directly to the later sections beginning with that titled "Central Tolerance." This would allow the reader to face all the experimental findings/conclusions without bias and develop their own framework for understanding immunologic tolerance. This understanding can then be compared with the models described at the beginning of this chapter.

In immunology, the word "tolerance" has been used in many different contexts, often with subtle but important differences in meaning (see Box 32.1 for a list of definitions of frequently used terms). In this chapter, we will focus primarily on tolerance in the adaptive immune system. While the innate immune system is also tolerant of self, this is mostly due to quite distinct mechanisms when compared to adaptive immunity. Innate cells are largely self-tolerant due to germline selected mechanisms. The receptors used to recognize what to respond to (eg, toll-like receptors [TLRs]) have been inherited in a conserved manner through the germline and selected during evolution to respond to molecules not present on healthy tissues of the species. This makes innate cells tolerant of not only an individual's own cells but also tolerant of cells from other individuals of the

BOX 32.1 Definition of tolerance related terms

- Tolerance: "An absence of immunity toward what would otherwise be an immunogenic antigen challenge and resulting from prior exposure to the specific antigen(s)."[1]
- Operational tolerance: persistence of antigen despite production of an immune response to the antigen.
- Ignorance: "A state where lymphocytes in an animal appear neither to be tolerant of, nor immunized (primed) to, an antigen present in the animal; the animal's immune system appears 'naïve' despite the presence of antigen."[1]
- Deletion: Physical elimination (death) of an antigen-specific lymphocyte upon recognition of antigen. Deletion is a recessive form of tolerance.
- Anergy: A state of T or B cell unresponsiveness induced by prior antigen exposure without death of the cell (death may occur later). Depending on the specific circumstances, the anergic state may involve unresponsiveness to antigen in terms of all known response types (eg, production of particular cytokines) or only select types of responses. Related terms used in specific settings include exhaustion, adaptive tolerance, and tuning (see text). The anergic state can be associated with either recessive or dominant tolerance.
- Suppressive (regulatory) tolerance: A reduced response by specific lymphocytes toward an antigen that is due to suppression by other lymphocytes (typically T_{reg} cells). This is a dominant form of tolerance.
- Linked recognition/linked suppression: An immune response (linked recognition) or lack of response (linked suppression) that occurs as a result of two or more different lymphocytes recognizing epitopes that are physically linked to one another or are presented by the same antigen-presenting cell.

same species. There are exceptions to this rule, however, as certain innate cells (eg, NK cells and macrophages) possess receptor systems involved in a somatic learning process (eg, MHC/KIR-Ly49 and CD47/SIRPα) that allows these cells to distinguish self-cells from cells of another individual, as detailed in Chapters 15 and 18). Unless accompanied by additional explanations, the definition of immunologic tolerance in this chapter will relate to adaptive immunity and be as described in[1] and shown in Box 32.1. "An absence of immunity toward what would otherwise be an immunogenic antigen challenge and resulting from prior exposure to the specific antigen(s). Establishment of tolerance can therefore be considered a negative response to antigen. The existence of a state of tolerance may either be directly tested by experimental challenge with an immunogen or inferred by survival or persistence of otherwise immunogenic antigen(s) in the putatively tolerant host." However, persistence of the antigen alone is insufficient to conclude there is a state of tolerance. Persistence of antigen together with a lack of all classes of effector responses (both cellular and humoral) would suggest there is a state of tolerance. This is to be distinguished from states known as "operational tolerance," where there can be persistence of antigen accompanied by production of an immune response to the antigen. This operational tolerance is typically a result of generating an ineffective class of immune response; for example, the generation of humoral immunity when instead a delayed-type hypersensitivity (DTH) response is the class of response capable of eliminating/containing the antigen/pathogen. Operational tolerance can occur in a number of settings. Animals can become chronically infected under conditions where an ineffective immune response was made to the infectious agent. Similarly, allogeneic tissue transplants can sometimes be accepted despite the presence of an immune response, and this is referred to as graft accommodation.[2] Tolerance can be distinguished from a state known as ignorance, which is defined as: "A state where lymphocytes in an animal appear neither to be tolerant of, nor immunized (primed) to, an antigen present in the animal; the animal's immune system appears 'naïve' despite the presence of antigen. Ignorance could truly reflect a naïve state of all relevant specific lymphocytes or instead the sum total of positive and negative interactions of individual lymphocytes with antigen, giving the overall appearance of a naïve state (a zero sum)." Both tolerance and ignorance are considered to contribute, in varying degrees, to a lack of immune system attack on self-antigens, as will be discussed in more detail below.

When an animal "reads" out as tolerant of a given antigen, the tolerance can be a result of several different states or the combination of these states. Tolerance can be a result of physical elimination of the relevant specificities from the lymphocyte repertoires, resulting from a process known as clonal deletion. Alternatively, the relevant lymphocyte specificities may be present in normal or near-normal numbers and yet those lymphocytes have an intrinsic reduced ability to respond. This state or states, is referred to by various monitors, including anergy, exhaustion, adaptive tolerance, or tuning. Though the named states may have some differences, all have in common the presence of cells that are intrinsically reduced in the ability to respond to antigen compared to naive cells. Lastly, tolerance may result from the ability of a lymphocyte to suppress the response of another lymphocyte. This suppressed lymphocyte is not intrinsically unresponsive, as, in the absence of the suppressive lymphocyte (known as a regulatory cell), it is fully activatable. In some experimental settings, the suppressed lymphocyte may itself become suppressive, a process known as infectious tolerance. However, it seems unlikely that infectious tolerance is a mechanism involved in self-tolerance (discussed in more detail below).

A HISTORY OF MODELS

It was known in the early 1900s that the immune system could respond to diverse molecules, including molecules that presumably would not have been a selection pressure on the evolution of the immune system to respond to. In other words, immune systems have evolved to deal with everything foreign, including the unexpected, an attribute called universality. Universality is characteristic of a number of biologic functions, another example being the ability of our brains to learn any language, in contrast to the more limited ability to learn a particular language. Ehrlich recognized a corollary of this attribute of universality, namely that the generation of seemingly random specificities in order to be able to respond to everything, was the generation of potentially lethal autoreactive specificities, what he referred to as "horror autotoxicus."[3,4] Thus, it was clear that the immune system must have one or more mechanisms that ablate the potential to respond to self-antigens, resulting in the attribute of SNS discrimination, in order to prevent horror autotoxicus. Ehrlich envisaged mechanisms that would either prevent the generation of autoreactive specificities or block effector functions of antiself antibodies.[3] Figure 32.1 provides a timeline of some of the major concepts and models of tolerance versus immunity, beginning with Ehrlich. Moving beyond Ehrlich's identification of the problem of SNS discrimination, an understanding was required as to how, from among the diverse repertoire of antibodies, the correct antibody could be selected and mass produced. All major and contemporary models of how SNS discrimination is achieved are based upon the clonal selection theory (see the basics of select models in Box 32.2), the solution as to how the mass production of the correct specificity is achieved, and as recognized later, the correct activation of appropriately specific T cells. Macfarlane Burnet and David Talmage, in 1957, proposed the central concept of the clonal selection theory that each lymphocyte harbors a single specificity of receptor and that it is the cell with the appropriate and specific receptor that is selected to replicate.[5,6] The problem then became, how does the immune system select for antiforeign cells and against potentially lethal antiself cells, that is, what mechanism(s) generate an appropriate somatic selection of lymphocytes?

Timing of Antigen Exposure as the Determining Factor in SNS Discrimination

Even before the need for a somatic selection was realized, Burnet and Fenner had proposed a partial solution to SNS discrimination in 1949, when they suggested that antigens present early in life (expressed during the fetal period) would be treated as self (ie, induce tolerance) while antigens appearing later would be treated

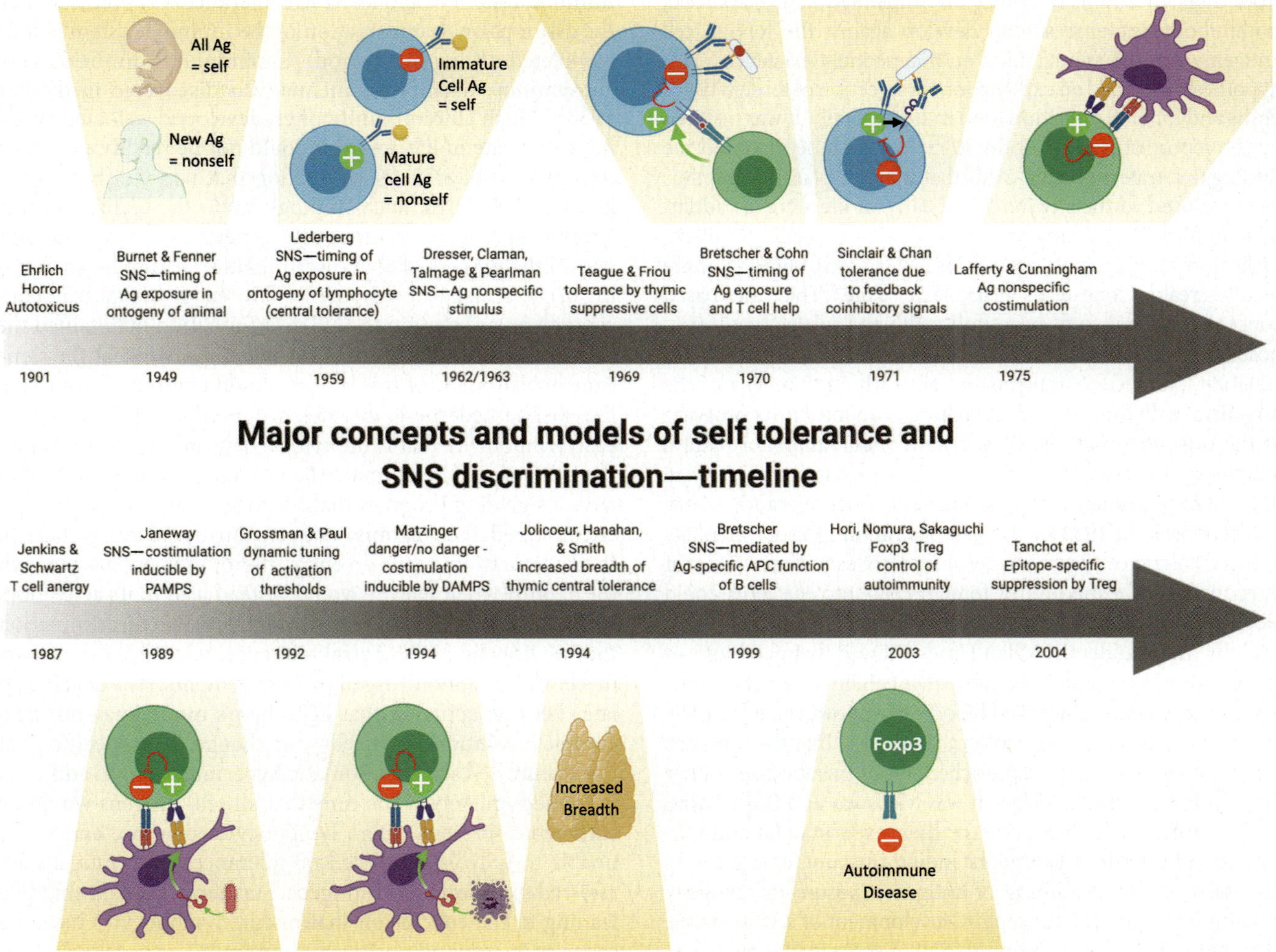

FIG. 32.1. A timeline of the origins of selected major concepts and models of tolerance and self-nonself discrimination. See the text for a more detailed discussion of each concept or model.

BOX 32.2 Models of Self-Nonself Discrimination and Selected Key Postulates (see also Fig. 32.1)

- [a]Timing of antigen exposure/central tolerance: In the immature stage, lymphocytes are programmed to become tolerant (deleted) upon antigen encounter. Mature lymphocytes instead are programmed to become activated by antigen encounter.
- Antigen specific help (two signals, quorum): Peripheral tolerance is due to the receipt of antigen receptor signals (signal one) without a second signal arising from "helper" cells. Immunity is induced when a lymphocyte receives both signal one as well as signal two, the latter generated by another antigen-specific lymphocyte recognizing a linked antigen/epitope.
- Pathogen-associated molecular patterns (PAMPs)/damage-associated molecular patterns (DAMPs) (two signals, costimulation): Peripheral tolerance is due to the receipt of antigen receptor signals (signal one) without the second signal delivered by costimulatory ligands on antigen-presenting cells (APCs). Immunity is induced when a lymphocyte receives both signal one as well as signal two, the latter generated by costimulatory ligands induced on APCs by PAMPs or DAMPs.
- Coinhibition/costimulation: Peripheral tolerance is due to the coengagement of antigen receptors and coinhibitory receptors, with coinhibitory receptor expression being enhanced by chronic antigen exposure. Immunity is induced when a lymphocyte receives both signal one and sufficient signal two (costimulation and/or help) to overcome coinhibition.
- Tuning: Lymphocytes can adjust their signaling thresholds. Tolerance is the result of raising the lymphocyte's threshold level of signals needed for a response to a level that is higher than the ambient level of signals in homeostatic conditions. Immunity is induced by an increase in the levels of signals at a rate that is too rapid for the lymphocyte to readjust their signaling threshold.

[a]All of the other models shown also incorporate the central tolerance concept; however, they propose that mature lymphocytes have the potential to be activated or made tolerant.

as foreign.[7] They stated, "If in embryonic life expendable cells from a genetically distinct race are implanted and established, no antibody response should develop against the foreign cell antigen when the animal takes on independent existence." This hypothesis provided an explanation for rejection of foreign tissue grafts and for SNS discrimination more generally. It was inspired by the report of Ray Owen in 1945, in which he recorded the finding that fraternal cattle twins that shared a placenta and thus were exposed to their twin's blood early in life were, as adults, unable to make antibody to each other's blood cells.[8] Similarly, earlier (in utero) exposure to LCMV virus was found to more readily create a long-term virus carrier state.[9] This time-based concept of tolerance fit beautifully with the subsequently proposed clonal selection theory. A developmental program would generate tolerance to antigens present early in ontogeny of the organism, with this program switching to an immunity program as the organism matured. Ray Owen's observations of natural tolerance associated with early exposure to foreign antigens were brought to prominence and the Burnet/Fenner tolerance theory was supported in 1953 by the demonstration of experimentally induced tolerance. Peter Medawar and colleagues showed that introduction of a mixture of foreign cells into the fetus could block rejection of donor skin grafts when the recipient was an adult.[10,11] Independently, Milan Hasek showed that parabiosis of chick embryos prevented the subsequent ability to generate antibodies against each other's red blood cells[12]; however, he initially misinterpreted these observations, suggesting that the outcome was a consequence of a "rapprochement of metabolism" rather than immune tolerance. Thus, it was Medawar and Burnet who were awarded the Nobel prize for their experimental and theoretical contributions to understanding immune tolerance. As discussed below, the timing of antigen exposure in ontogeny remains an important concept in development of self-tolerance although not the sole solution to SNS discrimination as envisaged by Burnet.

Burnet's window of time (fetal/neonatal period) where the immune system would learn self-tolerance was predicated on the assumption that formation of the immune system is fully completed during this period. Autoimmunity, in the form of autoimmune haemolytic anemia, was discovered in the late 1930s.[13] Such autoimmunity often developed well after birth, not to say late in life, and this could not be readily explained by a process of SNS discrimination that was uniquely established before birth. Such findings led to the realization that lymphocytes were continuously generated throughout life, and so the process of SNS discrimination had to be an ongoing process, also throughout life. Two major advances in conceptualizing a solution to SNS discrimination maintained the concept of a window in time for tolerance, while at the same time explaining how this process could continue throughout life. Joshua Lederberg, in 1959, proposed that it is the age of each lymphocyte that is the critical determining factor rather than the age of the animal.[14] Lymphocytes were proposed to have a signaling program that led to their elimination if they encountered their agonist antigen during an early stage in their maturation, while an encounter of the lymphocyte with the agonist when mature would instead lead to its activation. Thus, Lederberg proposed the solution to SNS discrimination that we now know as central tolerance, tolerance that occurs in central lymphoid organs where lymphocytes develop. It later became apparent that Lederberg's model may not be a complete solution, as B cells can change their specificity at the mature cell stage via somatic hypermutation. In addition, it seemed unlikely at the time that all self-antigens would be sufficiently present where lymphocytes develop, and so be unable to fully establish central tolerance. Such antigens are referred to as peripheral antigens. Furthermore, data emerged leading to the conclusion that mature lymphocytes have the capacity to respond positively (immunity) or negatively (tolerance) to antigens.[15] Thus, in addition to the concept of central tolerance, a model for how mature peripheral lymphocytes could be activated or inactivated was needed.

The problem of what could constitute a mechanism of peripheral SNS discrimination that would operate throughout life was tackled by Peter Bretscher and Melvin Cohn. They proposed a model explaining both peripheral tolerance and activation. In their model, for a lymphocyte to respond, it would need to collaborate with another lymphocyte specific for the same antigen. At the time (1970), there was evidence that cells that were generated in the bone marrow and thymus collaborate in generating an antibody response.[16] This two-signal model postulated that a single lymphocyte receiving only an antigen receptor signal (B cell receptor [BCR] or T cell receptor [TCR]), termed signal 1, would lead to tolerance, while the second signal obtained by collaboration of lymphocytes leads to activation.[17] The activation of the responding lymphocyte was envisaged to be dependent on its receiving signal 2; the initiation of the generation of signal 2 was envisaged to follow the binding of a "helper lymphocyte" to antigen. Thus, the activation of the responding lymphocyte would be helper lymphocyte dependent according to this mechanism. It was also proposed that signal 2 might be mediated by short-range molecules and/or membrane/membrane interactions. Collaboration, commonly referred to as T cell help, was proposed to require *the linked recognition* of antigenic determinants by the collaborating lymphocytes (see Fig. 32.2); and substantial evidence supports this proposition for the interaction between a B cell and T helper cells[18-21] and for the interaction between CD8 T cells and CD4 T cells[22-24] and for CD4 T cell cooperation.[24-26] Linked recognition means that the collaborating cells (eg, a CD4 T cell helping a B cell or CD8 T cell or a CD4 T cell) must see the antigen in a linked fashion (Fig. 32.2). The epitopes seen by each cell must be sufficiently linked (eg, by being part of the same antigenic complex) that they are presented to the collaborating cells by the same APC.

It was more recently shown that antigen can activate a sufficient number of CD8 T cells but that the activation of a lesser number of CD8 T cells requires CD4 T cells.[27] These observations were interpreted as showing that a quorum, that is, a minimal number, of T cells is required for the activation of CD8 T cells. The quorum concept and the proposal that lymphocyte cooperation is required for lymphocyte activation are different ways of expressing the same basic idea. This possibility, however, raises a further question. Naive T cells must be activated before they can optimally facilitate the activation of B cells. Evidence similarly suggests that activated, radiation-resistant CD4 T cells are required to optimally facilitate the activation of naïve CD4 T cells, whose activation is radiation-sensitive.[25] Although the antigen-dependent generation of these activated CD4 T cells appears to also be CD4 T cell cooperation dependent,[24-26] this conclusion raises the obvious question of the origin of the first activated CD4 T cells. This question, referred to as the "priming problem," is not resolved though suggestions as to its solution have been made.[28] We outline the situation in such detail to be explicit about the open questions associated with this framework.

The importance of linked recognition was first demonstrated in experiments by Rajwesky and Rottlander for the primary antibody response in rabbits[19] and for the secondary antibody response of mice by Mitchison.[21] The scenario for demonstrating this requirement can be readily illustrated by the rabbit experiments. This involved examining antibody responses to isozymes of a porcine enzyme that existed in various tetrameric forms, including A_4, B_4, and A_2B_2. Some rabbits are unresponsive to B_4. Immunization of these rabbits with A_4 led to the production of antibody that bound to A_4 but not to B_4. Immunization of all rabbits with A_2B_2 led to some antibodies that bound to A_4 and other antibodies that bound to B_4. Thus, in B_4 unresponsive rabbits, the B subunit appeared to be a macromolecular hapten and the A subunit acts as a carrier. It was found that the simultaneous immunization of these rabbits with A_4 and B_4 resulted only in antibody able to bind A_4. Thus, an antibody response to the hapten B could only be facilitated by helper T cells specific for the carrier A in the presence of a conjugate, A_2B_2 in this case, but not when both A and B are present, but unlinked, in the form of A_4 and B_4.

The two-signal model explained peripheral tolerance in the following manner. The first waves of T and B cells exported from central lymphoid organs early in life would contain too low a number of cells specific for a peripheral self-antigen for the antigen to activate these cells; these cells would receive signal 1 without help and so be inactivated. Further cells specific for this peripheral self-antigen would be inactivated as generated, one or a few at a time, unless a sufficiently foreign antigen were introduced that also contained a linked self epitope(s). T cells specific for the foreign epitopes might allow the foreign antigen to break tolerance to the linked self-epitopes, particularly when the expression level of self-antigen is low and consequently the self-specific cells have not been fully eliminated or tolerized. With a highly expressed self-antigen, only the newly generated lymphocytes would be in a nontolerant state, not having undergone peripheral tolerance; and thus multiple exposures to this mixed foreign/self-antigen would likely be needed to sufficiently boost the number of antiself T cells such that antiself T cells alone provide sufficient help to each other and cause autoimmune disease. It is interesting in this context that infection by group A streptococci, associated with rheumatic heart disease, not only gives rise to cross-reactive antibody,[29] that binds to both heart and bacterial components, but CD4 T cells that also cross-react.[30] Infectious agents engineered to contain a model tissue-specific self-antigen can recruit T cells from a normal polyclonal repertoire that escaped central tolerance due to their low affinity to the self-antigen. Upon two immunizations with the mixed foreign/self-antigen expressing infection, these low-affinity T cells were eventually sufficiently activated and expanded to cause autoimmune disease.[31] The ability of T cell help to break tolerance also predicts that an initial response to one epitope can facilitate/help generate responses to other epitopes, a phenomena known as epitope spreading, which is known to occur in many autoimmune disease states.[32,33]

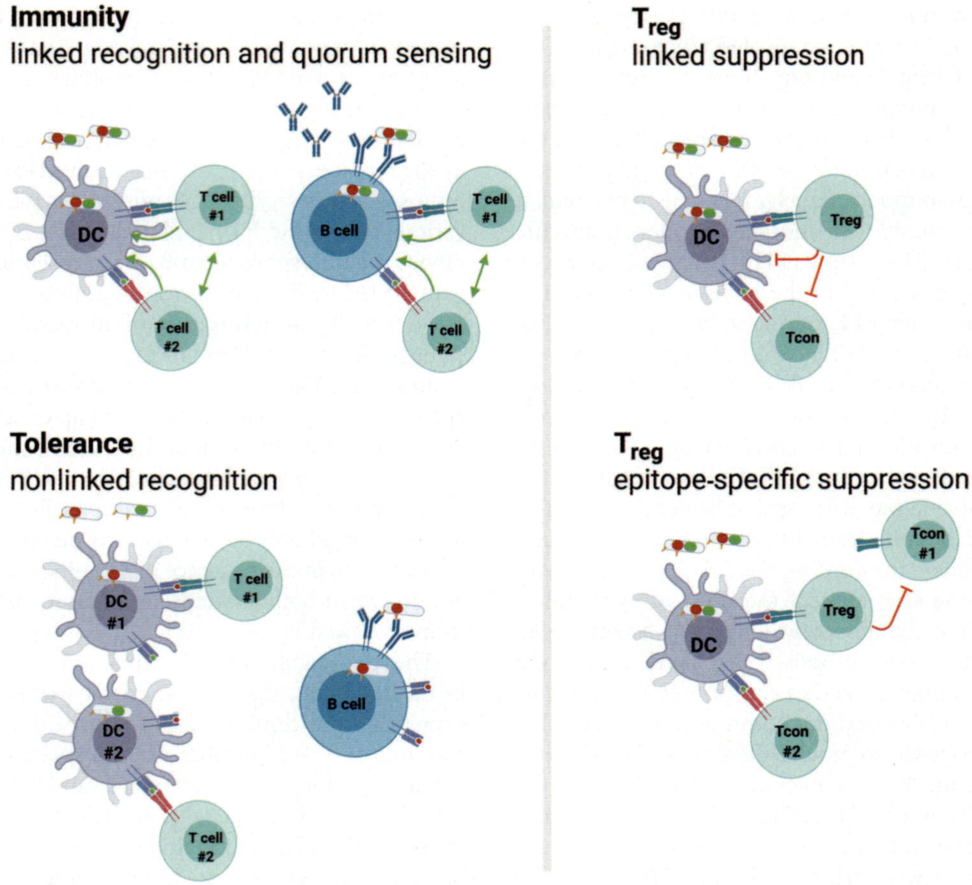

FIG. 32.2. Linked recognition mediating quorum sensing's role as a determining factor in self-nonself discrimination and a comparison of linked suppression to epitope-specific suppression. *Left panel:* If a single dendritic cell takes in and presents two different epitopes to two different T cells or a B cell specific to an epitope takes in the antigenic structure via its specific B cell receptor (BCR) and presents two different epitopes to two different T cells (*upper panel*), an immune response is generated. In contrast, if a single T cell recognizes its epitope (*lower panel*), tolerance is induced in the T cell, and similarly, if the B cell recognizes its epitope without presenting an epitope to a helper T cell, the B cell becomes tolerant. While quorum is shown as two T cells recognizing their epitopes in a linked fashion, the number of T cells needed to achieve quorum has not been fully determined and may be more than two. The T cell collaboration is depicted as simultaneous epitope recognition by two different T cells; however, the two recognition events can be separated in time; for example, when T cell #1 is a CD4 effector T cell recognizing its epitope, it can send a signal to the antigen-presenting cell (APC), allowing the APC to go on an activate a CD8 T cell recognizing another epitope on that APC even though the CD4 cell is no longer present. Under physiologic conditions, whether the APC mediating linked presentation of epitopes to *initiate* an immune response can be a cell that lacks antigen specificity (eg, a DC) or if it must be an antigen-specific B cell has not been fully resolved. *Right panel:* Linked suppression (*upper panel*) is depicted and requires that the same APC presents an epitope to both a suppressive T cell (T_{reg}) and a potential responding cell (ie, a conventional T cell; Tcon); the two recognition events need not be simultaneous if the T_{reg} "educates" the APC to be suppressive. The T cell that has been suppressed, under some conditions, may itself become suppressive; a process known as infectious tolerance. In epitope-specific suppression (*lower panel*), a T cell (Tcon#2) recognizing a different epitope than the T_{reg} is not suppressed and can respond, only a Tcon (Tcon#1) recognizing the same epitope as the T_{reg} is blocked from responding.

Models Proposing the Context of Antigen Presentation Determines Peripheral Tolerance Versus Immunity

Several common findings in immunology appeared difficult to reconcile with the concept that the timing of antigen exposure, either in the life of the animal or the life of the lymphocyte, was the factor determining SNS discrimination. The use of adjuvants to help stimulate responses to foreign antigens would not be required if all that were needed was foreign epitopes on the antigen. In addition, it was found that grafts bearing foreign histocompatibility antigens were sometimes spontaneously accepted and induced tolerance, when given to immunologically mature animals.[34,35] Mitogens, including portions of microbes, such as endotoxin (lipopolysaccharide;

LPS), could promote antigen-specific responses and block humoral tolerance.[36] Foreign tissue transplanted into a fetus could, in some cases, be rejected despite being present early in ontogeny and the semiallogeneic fetus is not rejected despite only appearing later in life in the mother. These and other findings suggested to some researchers that antigen nonspecific signals may be key in stimulating adaptive immune responses and that an antigen-specific signal without a nonspecific second signal leads to tolerance.[36-38] Coutinho and Moller further emphasized the importance of a nonspecific second signal when they suggested that the BCR simply focuses antigen to the correct cell and that a mitogen linked to the antigen was responsible for activating the B cell via mitogen receptors (eg, LPS receptors)[39]; note, later findings showed that the BCR does more than focus the antigen to the correct cell[40] (see also Chapter 26). Similar to these models of B cell responses, Lafferty and Cunningham proposed that for T cell responses, in addition to the antigen receptor signal 1, a second antigen nonspecific costimulatory signal must be received by lymphocytes for activation.[41] Charles Janeway much later linked these two concepts, proposing how costimulatory signals would occur for infectious nonself entities, but not for self[42]; in the absence of infection, self-antigens would provide signal 1 without costimulation, leading to tolerance. During infection, the recognition of conserved PAMPs, such as LPS via TLRs on APCs, would upregulate the expression of these costimulatory signals leading to immune responses to pathogenic microbes. However, this proposal did not seem to fully explain how the immune system determines what to attack and what to become tolerant of. To explain why sterile transplants are rejected but the semiallogeneic fetus is not, among other paradoxical observations, Polly Matzinger proposed that healthy tissues induce tolerance (signal 1 without costimulation) while tissues that have been damaged release endogenous DAMPs that trigger costimulatory signals by APCs. Thus, transplants were postulated to trigger immunity because of surgical tissue damage and lack of the rejection of the semiallogeneic fetus was explained by the healthy nature of fetal tissue. Peripheral tolerance induced by healthy tissues was again postulated to be the result of signal 1 (TCR/BCR) without signal 2 (costimulation).[476]

One aspect similar in both the time- and context-based models discussed to this point is the postulate that antigen-specific signal 1 without a second signal is responsible for induction of peripheral tolerance. However, it was known that immune responses could be shut down by feedback mechanisms, such as antigen-specific antibody-mediated feedback of B cell responses. If antibody feedback were to work by masking antigen (preventing signal 1), such feedback should, based on the above models, prevent both immunity and tolerance to the antigen. However, Sinclair and colleagues found that feedback suppression by specific antibody to antigen could depend on the Fc portion of IgG antibody and that Fc-dependent suppression occurred at certain ratios of antibody to antigen.[43-45] These findings led them to propose that antibody feedback shut down of the B cell response was not the result of signal 1 alone or antigen masking but instead through the coligation of the BCR and a receptor for the Fc portion of IgG on the B cell.[45] Sinclair termed this process coinhibition, the opposite of costimulation, and proposed that both T and B cell tolerance were caused by a series of coinhibitory receptors.[46,47]

The last model to be discussed, known as tuning, has some similarities to the costimulation/coinhibition model. Not all antiself lymphocytes are eliminated by central tolerance, and for T cells, every T cell exported from the thymus has at least some low-level ability to engage self-peptide/MHC (a consequence of thymic positive selection). What prevents this engagement from generating an autoimmune response? Grossman and Paul proposed that the decision between peripheral tolerance and immunity is dependent on an ongoing tuning of the threshold level of signals needed for activation (tunable activation thresholds model). Lymphocytes are proposed to continuously set their signaling threshold based on the ambient level of signaling over a period of time, and this is postulated to occur during lymphocyte development and in the periphery.[48,49] Thus, a relatively constant level of ambient signals (TCR/BCR, costimulation, cytokines, etc.) sets the signaling threshold, and when these signals increase beyond the threshold in a rapid fashion, too fast for the T cells to retune, the specific lymphocytes are activated. Thus, the constant signaling from self-antigens sets a high threshold level of signals needed to activate these cells. In contrast, antiforeign lymphocytes are set at a lower threshold due to the absence of their agonist under homeostatic conditions.

An Introduction to Data Supporting or Conflicting With the Above Minimal Models of SNS Discrimination

Two Antigen-Specific Signals (Quorum)

Lederberg's concept that early in the life of a lymphocyte it is programmed to become tolerant if it encounters antigen at that stage is now well accepted and is now known as the mechanism underlying central tolerance. However, unlike Lederberg's additional proposal (mature cells are activated by antigen), we now know that in the next stage the mature lymphocytes have the potential to become tolerant or activated by encounter with antigen. The two-signal model proposing a quorum of lymphocytes must collaborate for activation, and tolerance ensues in the absence of quorum (eg, a single lymphocyte encountering antigen) is supported by many findings over the past 50 years.[24,50] A very early study showed that B cells, exposed to antigen in the absence of helper T cells for about 2 days, could no longer be activated.[51] Goodnow and colleagues showed that BCR transgenic B cells specific for hen egg lysozyme (HEL) were inactivated by HEL in the absence of T cell help.[52] Guerder and Matzinger showed, in a system where the activation of CD8 T cells was known to be CD4 T cell dependent,[22] that exposure to antigen in the absence of CD4 T cells led to the inactivation of the CD8 T cells.[23] Although evidence shows that CD4 T cell cooperation facilitates the activation of CD4 T cells,[24-26] there is no reported evidence that CD4 T cells are inactivated by antigen in the absence of CD4 T cell cooperation.

The most direct challenge to the two-signal model has been related to its postulate that there is a period early in life where encountering antigen can only induce tolerance (fetal/neonatal tolerance period) because of the paucity of help. While the studies of Medawar and colleagues demonstrating the induction of tolerance by injection of allogeneic cells early in life are well known, when one considers the many additional published studies, the outcome of introducing foreign antigen/cells early in life is frequently immunity rather than tolerance. For example, a skin graft from an allogeneic fetus to a sheep fetus stimulates immunity/rejection rather than tolerance.[53] Similarly, an allogeneic or xenogeneic fetal wing limb bud transplanted to a fetal chick triggers an immune response posthatching.[54] The paradox that allogeneic tissue transplanted before the recipient's immune system develops does not always induce tolerance appears to be explained primarily by the location of donor antigens.[1] Antigens that reach central lymphoid organs generate tolerance (eg, Medawar studies), whereas antigens that are present only outside the central lymphoid organs usually trigger immunity (eg, skin graft or limb bud). The data appear to support the concept of central tolerance and deny the capacity of peripheral tolerance, contradicting the concept of a period when lymphocytes first seed the periphery and are inevitably inactivated on interacting with peripheral self-antigens. However, a deeper look at the concepts and data suggest another view. It is anticipated, according to the two-signal (quorum) model outlined above, that antigen will inactivate lymphocytes if they are insufficient in number to result in lymphocyte cooperation and lymphocyte activation. Thus, a mature lymphocyte, early in its life, is not inevitably inactivated on interacting with antigen, but its fate is instead defined by whether a quorum of lymphocytes is present upon antigenic challenge. Whether quorum is reached is in turn determined by the number of lymphocytes with appropriate specificity in the repertoire. The higher the number of foreign epitopes on an antigen the greater the number of potential responding cells and the more likely quorum will be achieved. This has been observed, for example, in studying responses to xenogeneic serum albumins.[24,55] Experiments testing whether grafts given before the recipient's immune system develops induce peripheral tolerance or immunity, also show this depends on the number of foreign antigens present on the graft, and appear consistent with this hypothesis[1]; see[56] for a more detailed discussion. When there are a large number of foreign antigens on a donor cell, the donor cells must reach central lymphoid organs to establish tolerance, while donor tissue with a small number of foreign antigens can induce tolerance in the periphery, that is, without donor cells reaching central lymphoid organs. Another potential challenge to the two-signal model was the contention that antibody responses to carbohydrate antigens, such as ABO blood group antigens, occurs without T cell help (ie, without the second signal).[57] If self ABO antigens are indeed T cell independent, then the two-signal model could not explain SNS discrimination of blood group antigens. However, more recent data has indicated that in fact ABO antigens are T cell–dependent antigens.[58] Although antigens derived from xenogeneic sources (eg, microbes) may have cross-reactive antigens[59] that stimulate ABO antibody production in a T independent fashion, there have not been any mammalian self-antigens identified that stimulate antibody production in a T independent fashion when introduced into an allogeneic recipient. Thus, all responses to "within species" antigens are governed by a two-signal mechanism, supporting the two-signal model as the means of achieving peripheral SNS discrimination. The need for cellular collaboration in B cell and CD8 T cell responses to nominal protein antigens[22,60] and the tolerance that ensues in the absence of T cell help[23,61] are all consistent with the two antigen-specific signals model. Thus, a generalization that appears valid is that destructive adaptive humoral and cellular responses to our own antigens, in their normal context, are T cell help dependent (the absence of help prevents these responses). In contrast, these same types of responses, when directed against antigens of particular infectious agents, can occur in a T cell help independent manner. Additional factors unique to microbes, such as the structure of the antigen,[62,63] or their interaction with additional receptors may allow T cell help independent immune responses.

Antigen Nonspecific Second Signals Regulated by PAMPs or DAMPs

The concept that the presence of PAMPs or DAMPs, and consequently, the presence of antigen nonspecific costimulatory signals, determines the decision between tolerance and immunity is also supported by extensive data. It is clear that triggering TLRs, whether by PAMPs or by DAMPs, can in many cases increase immune responses substantially.[64] In terms of explaining graft rejection, there is evidence that DAMPS promote rejection,[65] consistent with the fact that the highest doses of immunosuppressive drugs are needed early after transplantation. Based on these data, it is widely believed that the presence of PAMPs/DAMPs is required to induce immune responses and the absence of PAMPs/DAMPs leads to tolerogenic antigen exposure. However, direct tests of whether PAMPs/DAMPs are absolutely required to trigger immune responses do not support such a requirement. For example, allogeneic tissue, even when MHC-matched and allowed to heal-in, for many months in an immunodeficient recipient (ie, a PAMP- and DAMP-free graft), is able to trigger rejection when a recipient genotype immune system develops in the presence of a sufficiently foreign graft.[1,66,67] The danger model has proposed that long-healed in grafts can be rejected because the immunosuppressive drugs that patients take block tolerogenic signals (eg, rejection of kidney transplants many years after transplantation when the patient stops taking tacrolimus). However, this explanation does not hold for the above healed in graft studies in mice, where no immunosuppressive drugs were employed. Development of autoimmunity in AutoImmune REgulator (AIRE)–deficient mice, including those housed as germ free, also seems to defy the hypothesis that PAMPs/DAMPs are required to trigger an immune response.[68] A further difficulty for both Janeway's and Matzinger's proposals was that injection with a sharp needle of sterile and sufficiently foreign vertebrate antigens, such as sheep red blood cells into mice, could generate immune responses. Such an immunogenic challenge should not generate danger signals nor involve the introduction of PAMPs. However, recent data has shown the innate immune system, via the CD47/SIRPα receptor/ligand system, can recognize and respond to foreign cells, including allogeneic cells.[69] This opens the possibility that the innate system controls these responses even without stimulation by PAMPs or DAMPs. Further evidence against a need for PAMPs/DAMPs, the CD4 T cell–dependent antibody response to blood group antigens does not require MyD88 or TRIF, essential components of TLR-signaling pathways.[58] The danger model also proposes that the semiallogeneic fetus is not rejected because it is a healthy tissue (no danger signals) and that there is no need to hide fetal antigens from the maternal immune system. In contrast to this view, there is a reduction of classical MHC/HLA expression on fetal placental tissue in humans and in mice. In each species, only one of the three classical MHC class I molecules is expressed on fetal placental cells.[70,71] In humans, it is precisely those highly polymorphic HLA that are critical in tissue matching for transplantation (HLA-A, HLA-B, and HLA-DR) that are not expressed by fetal placental trophoblast cells. Mammalian reproduction does however eventually lead to a maternal immune response against paternal antigens of the fetus (eg, antipaternal HLA-specific antibodies). Whether it is the DAMPs released during parturition that leads to this immune response, as proposed in the danger model, has not been resolved.

Distinct Roles for PAMPs/DAMPs and Quorum

Do both the antigen-specific two-signal mechanism and the antigen nonspecific PAMPs/DAMPs/costimulation mechanism act as mechanisms that decide between tolerance versus immunity? From an evolutionary perspective, it seems unlikely such distinct mechanisms serve precisely the same purpose. Thus, it could be that the decision between immunity and tolerance is made by one of the two mechanisms and the other mechanism serves only as an amplifying mechanism, solidifying the decision that has already been made. For example, quorum could be the decision point, with a sufficiently foreign antigen triggering a weak immune response, while the addition of PAMP/DAMP triggered costimulation (ie, addition of an adjuvant) amplifies the response. We summarize three sets of experiments that appear to favor the idea that lymphocyte collaboration is required to induce helper function and that the presence of PAMPs/DAMPs is not the determining factor as to whether antigen induces helper function. The first set of experiments was carried out by Weigle and reported in 1961.[55] Weigle did not interpret his observations in the way discussed here. His experiments were carried out even before it was recognized that there were different types of lymphocyte. A reader might well wonder why we discuss such old experiments rather than more recent and incisive ones. The fact is that Weigle's experiments were minimally invasive, carried out by giving antigen in a manner that prevented subsequent responses or resulted in immunity. Thus, he did not do any complex manipulations that might lead one to question the physiological relevance of his findings. This is why such old-fashioned experiments are worthy of examination for their implications in a contemporary context. Such an examination is conducted in the context of modern findings and knowledge.

Weigle exposed neonatal rabbits to a series of injections of bovine serum albumin (BSA) that rendered them unresponsive to a subsequent BSA challenge, given at either 3 or 6 months of age, which produced a robust antibody response in unexposed rabbits. He found that immunization with a series of challenges of human serum albumin (HSA) at 3 months of age resulted in the production of antibody in almost all the rabbits, all of which bound to HSA, and a minority of which also bound to BSA. This means, in the context of modern knowledge that B cells, with antibody receptors able to bind both BSA and HSA, must exist in 3-month-old, unresponsive rabbits. The lack of a response on challenge with BSA is presumably due to the insufficient presence or activation of BSA-specific helper T cells. Weigle later showed that immunization of these rabbits with HSA would allow the unresponsive rabbits to now respond to a BSA challenge to produce BSA-specific antibody. Rabbits unexposed to HSA were unresponsive. Thus, exposure to HSA endowed the "unresponsive rabbits" with the ability to produce antibody on a subsequent challenge with BSA. We suppose this BSA-specific antibody response to be helper T cell dependent. In this case, there must be BSA/HSA-specific precursor helper T cells in 3-month-old, unresponsive rabbits. These helper T cells can be induced by HSA but not by BSA. Further studies/observations made by Weigle indicate why this might be so.[24,72]

Weigle examined whether a similar immunization regimen, as that used with HSA but with other foreign serum albumins, could break the unresponsive state. No unresponsive rabbits challenged with BSA produced antibody (0/9). Another albumin employed was sheep serum albumin (SSA). Of the antibodies raised against SSA in normal rabbits, 75% also bound to BSA. This presumably means that the neonatal ablation of B cells specific for BSA affects approximately 75% of the B cells specific for SSA. Immunization with SSA did not produce antibody in any (0/7) of the BSA-unresponsive rabbits. Weigle also employed pig serum albumin (PSA). Of the antibodies raised in normal rabbits to PSA, 35% bound to BSA. Challenge with PSA resulted in 2 of 5 unresponsive rabbits producing antibody to PSA, some of which reacted with BSA. Of the antibodies raised in normal rabbits to HSA, 15% bound to BSA. A similar challenge with HSA resulted in most rabbits (12/14) responding to produce antibody to HSA, some of which bound to BSA.

It seems inevitable that the neonatal administration of BSA affects the number and/or state of the B cells specific for BSA more than those specific for any other serum albumin and affects about 75% of B cells specific for SSA, about 35% of the B cells specific for PSA and about 15% of the B cells specific for HSA. These differential effects on the B cells specific for different albumins reflects the degree of similarity of the different albumins with BSA. It seems natural to suppose that the helper lymphocyte compartment specific for these different albumins are also negatively affected differently, increasing in the order HSA, PSA, SSA, and BSA. All the albumins were administered in the same way, without deliberate adjuvant, and so most likely involved the rabbits being exposed to similar PAMPs/DAMPs upon immunization. Given that we have concluded there are some helper lymphocytes specific for BSA, as well as for the other foreign albumins, one would expect, if there was sufficient PAMPs/DAMPs to activate helper cells, that all challenges would have resulted in antibody production, and, if there was insufficient PAMPs/DAMPs, none would have. In fact, the albumins most different from BSA are the most efficient at breaking the unresponsive state, a result anticipated on the proposition that the activation of helper lymphocytes requires lymphocyte cooperation. It might be felt that this conclusion is a bit nebulous, without a definition of the lymphocyte subsets involved. We next outline later experiments that allow one to better define the types of lymphocytes involved.

The second example came from Janeway's laboratory.[73] The investigators found conditions under which they could activate in mice CD4 T cells specific for the self-antigen, mouse cytochrome C (MCC). Critical to this study is the finding that immunization of mice with human cytochrome C (HCC) can result in the production of antibody cross-reactive with MCC, demonstrating the existence in normal mice of B cells specific for both MCC and HCC. Immunization of mice with MCC in CFA was ineffective in raising MCC-specific CD4 T cells. However, mice similarly immunized, but given in addition activated B cells specific for MCC, generated activated, MCC-specific CD4 T cells. The MCC-specific activated B cells were raised by immunizing syngeneic mice with HCC. The activation of these cross-reacting B cells is presumably helper T cell dependent. These observations are reminiscent of those showing that CD4 T cells can license APC, presenting the antigen, to activate CD8 T cells.[74] Thus, these observations support the idea that the activation of CD4 T cells requires CD4 T cell cooperation mediated by an antigen-specific B cell. This is exactly what is proposed in the two-step, two-signal model of CD4 T cell activation.[50] These observations collectively show that MCC-specific CD4 T cells exist in mice. The inability to activate them by immunizing with MCC in CFA is paradoxical within the context of the PAMP/DAMP model. These findings naturally lead to the question of why B cells, rather than other APC, appear to be required to mediate the cooperation between CD4 T cells.

We have already noted above, in the context of the quorum hypothesis for lymphocyte activation, that the antigen-mediated interaction between a responder lymphocyte and helper lymphocyte has been shown in many cases to require the recognition of linked antigenic epitopes. This feature of the interaction between lymphocytes seems physiologically important, as it ensures that the activation of a responder lymphocyte specific for one antigen cannot be helped by lymphocytes specific for an unrelated antigen in the presence of this unrelated antigen. In the context of the inactivation of CD4 T cells, a requirement for linked recognition means that the inactivation of a CD4 T cell by a peripheral self-antigen cannot be subverted by CD4 T cells specific for an unrelated, foreign antigen in the presence of this antigen. Thus, linked recognition (Fig. 32.2) ensures this principle of noninterference in the inactivation of lymphocytes.[75] B cells primarily present peptides derived from internal self-proteins and peptides derived from nominal, exogenous antigens that can bind to the BCR and so be endocytosed. In this sense, B cells can act as antigen-specific APC specific for exogenous antigens. It was on this basis that it was proposed that the antigen-mediated CD4 T cell cooperation, envisaged to be required to activate a responder CD4 T cell, must be mediated by a B cell acting as an antigen-specific APC.[50] This would ensure that the CD4 T cell cooperation occurred by a mechanism involving the linked recognition of epitopes. The observations of Janeway and colleagues,[73] already described, as well as other observations,[76] show a central role of B cells in mediating CD4 T cell cooperation. Studies in mice also show an antibody-independent role of B cells in murine lupus.[77] It seems likely that this role of B cells is also pertinent to understanding the efficacy of rituximab therapy, resulting in B cell depletion, in treating cell-mediated autoimmunity.[75,78–80] Such observations collectively support a central role for B cells in mediating CD4 T cell cooperation. Further support for the role of B cells as an APC for naive T cell activation comes from examining the response to B cell idiotypes. Bogen and colleagues showed that naive CD4 T cells specific to a B cell idiotype were stimulated and provided T cell help to the corresponding B cell only when its BCR (that possesses the idiotype sequence) had been ligated.[81] In an earlier study employing a similar system, they showed that T cell activation occurred by B cell and not dendritic cell (DC)-mediated antigen presentation.

It is possible to define the peptide specificity of all the CD4 T cells, generated upon immunization with a protein antigen, employing an ELISPOT assay to enumerate the number of antigen- or peptide-specific cytokine-producing CD4 T cells. Such a definition in BALB/c mice of the peptide specificity of all the CD4 T cells generated upon immunization with HEL demonstrated that roughly half the HEL-specific CD4 T cells were specific for a dominant peptide, and the other CD4 T cells were specific for minor peptides.[82] HEL is not highly immunogenic. It was anticipated that CD4 T cell cooperation might be limiting the initiation of the anti-HEL response, and so a strategy was developed to test this proposition. The CD4 T cells specific for the major peptide were eliminated by giving this peptide to mice in a manner known to ablate peptide-specific CD4 T cells. Marc Jenkins and colleagues had shown that TcR transgenic CD4 T cells, transferred to a syngeneic, normal recipient, could be stimulated to divide and then go into apoptosis upon sufficient exposure to the pertinent peptide.[83] The substantial amount of peptide administered most likely decorates very many APC "from the outside," precluding the possibility of APC-mediated CD4 T cell cooperation, in the sense that the presenting APC would be in great excess. The major peptide of HEL was administered to mice in the same manner and exposed mice could no longer respond to the peptide given in CFA. Such extensive ablation naturally greatly reduced the CD4 T cells generated specific for the major peptide on immunization with HEL, but it also drastically reduced the generation of cytokine-producing cells specific for the minor HEL peptides.[82] Thus, it appears that CD4 T cells specific for the major peptide facilitate the activation of CD4 T cells specific for the minor peptides. Normal mice and mice exposed to the major peptide in a manner that made them unresponsive against it were immunized with HEL conjugated to ovalbumin (OVA). Identical CD4 T cell responses to minor

HEL peptides were found in both types of mice, but there was a greatly diminished CD4 T cell response to the major peptide in peptide-treated mice, consistent with a hole in the CD4 T cell repertoire. CD4 T cells specific for OVA appeared to have helped the activation of CD4 T cells specific for minor HEL peptides in mice exposed to the major peptide. These observations were interpreted as showing that the administration of the major peptide resulted in the ablation of the pertinent CD4 T cells rather than the generation of peptide-specific inhibitory cells. We suggest these observations are collectively difficult to reconcile with the PAMP/DAMP models. As CD4 T cells were activated in normal mice on challenge with HEL, the immunization must, on these models, involve PAMP/DAMP signals. Yet the very same immunization, in mice exposed to the major HEL peptide, did not result in robust generation of CD4 T cells specific for minor peptides, as would be expected on PAMP/DAMP models.

Tolerance Via Coinhibitory Signals and Tuning

In terms of the origin of tolerogenic signals, the answer is also not fully settled. While there is agreement that coinhibitory receptors generally function to reduce immune responses and allow tolerance, whether peripheral tolerance can occur via signal 1 (TCR/BCR) alone, that is, without involvement of a coinhibitory signal, is not known. However, coinhibitory signals were postulated to be involved in both B[84] and T cells[47] when antigen exposure becomes chronic. This postulate has been supported by many recent observations, where chronic antigen exposure, particularly in the setting of chronic viral infection leads to upregulation of coinhibitory receptors. T cells exposed chronically to these viruses progressively loose function, a state often referred to as exhaustion, and blocking the coinhibitors can in some settings reverse this unresponsive state.[85,86] In other settings, coinhibitors have not been identified as contributors to exhaustion.[87] However, such studies only examine one or a few of the potential coinhibitors that may be involved. There are data that support the concept of tunable activation thresholds both in thymocytes[88,89] and in the periphery. Preselection thymocytes have greater TCR sensitivity to low-affinity ligands (potential self-mimics) than the corresponding mature single positive T cell. Despite having reduced sensitivity to low-affinity ligands, the mature T cell maintains high sensitivity to the agonist peptide ligand (ie, tuning of the thymocyte reduces the ability of the mature T cell to respond to low-affinity ligands but not agonist).[90] Tuning can also be achieved in the thymus by reducing the level of coreceptor (eg, CD8) expression[91,92] or by the upregulation of coinhibitory receptors such as CD5.[93-98] CD5 and CD8 levels have been implicated in reversible peripheral tuning of CD8 responses.[99,100] Substantial evidence for tuning of mature T cells is found in models of adoptive cell transfer. Although it was proposed that tuning of the TCR sets thresholds to allow responses to qualitatively different antigens (eg, a different affinity ligand),[49] there are also data supporting a tuning that allows responses to a change in the available quantity of the same antigen. As discussed in more detail in a subsequent section, increasing the exposure to a chronic antigen leads to increasingly deeper states of tolerance in a monoclonal population of T cells responding to the same peptide.[101,102] Similarly, reduced IgM BCR expression tunes B cells,[103,104] preventing responses to self while potentially allowing responses to the same epitopes presented at higher valency. At this point, it is unclear whether tuning could be readily achieved for a tissue-specific peripheral self-antigen, and data addressing this question is lacking.

In summary, the top contending models of SNS discrimination incorporate a central tolerance mechanism and a peripheral tolerance/immunity decision controlled by either two antigen-specific signals (quorum sensing), the presence or absence of PAMPs/DAMPs/costimulation or the balance between second-signals/costimulation and coinhibition. The primary question that needs to be answered to understand which model is correct centers around the question of whether a SNS discrimination can be achieved via mechanisms that are not fully antigen specific (PAMPs/DAMPs/costimulation) or if SNS discrimination requires an antigen-specific mechanism (ie, two antigen-specific signals/quorum sensing). Is the association of PAMPs and DAMPs with entities that we would want our immune system to attack (pathogenic microbes etc.) and the reduced association with entities we do not want attacked (healthy self-tissue) sufficient to provide an appropriate SNS discrimination? Consensus has not been reached on these important questions.

A Role for T_{reg} Cells in SNS Discrimination?

At this point the reader may well be wondering, what about the balance between regulatory T cells (T_{reg}) and conventional T cells as the determining factor in SNS discrimination? T_{reg} cells were previously not considered the determining factor in any clearly enunciated models but instead were known to have a number of roles that will be discussed in more detail in subsequent sections. T_{reg} are critical for tolerance and homeostasis in general in the immune system. However, from a theoretical standpoint, T_{reg} were not considered the major determining factor of SNS discrimination because they were thought to function solely through mechanisms that involve bystander suppression (including linked suppression or infectious tolerance; see Fig. 32.2), and SNS discrimination cannot, in principle, be governed by bystander suppression. Bystander suppression usually means that the T_{reg} recognizing its specific epitope on an APC suppresses other T cells seeing different epitopes presented by the same APC. This is also known as linked suppression, as the epitopes are linked by being presented on the same APC. However, bystander suppression is defined by some authors as being different from linked suppression, with bystander suppression being considered suppression over greater "distances;" the T_{reg} suppresses T cells seeing other antigens presented by a different, usually adjacent, APC. Since the self-tolerance mechanism(s) that allows a SNS discrimination must be on an epitope by epitope basis, it cannot be governed by linked or bystander suppression as the primary mechanism. How do we know this? As a simple example, we know that a small number of minor histocompatibility mismatches is sufficient to trigger an immune response and rejection of a skin graft. Were tolerance to self to be accomplished in a major way via linked suppression, self-specific T_{reg} would prevent rejection of such skin grafts by preventing attack on these linked transplant antigens (the skin graft cells present both foreign and self-epitopes in a linked fashion). Thus, linked suppression from T_{reg} may even limit our ability to discriminate self from nonself. There are in fact situations where an otherwise normally functioning immune system seems unable to successfully discriminate self from nonself, and this can be associated with induction of T_{reg} cell activity. Spontaneous acceptance of a foreign graft is one such example. While the grafts with a small number of minor histocompatibility mismatches were rejected when the donor tissue was skin, they were spontaneously accepted when the graft was from a less immunogenic tissue, such as islet cells. This spontaneous acceptance of foreign tissue is associated with a dominant T_{reg}-dependent suppression[105] and may reflect an ineffective response rather than an inability to distinguish self from nonself. Observations from studies of allogeneic chimerism or bone marrow transplantation provide further evidence that the mechanism of SNS discrimination is highly specific and does not result in unresponsiveness to linked antigens. Different tissues express not only antigens common to all tissues but also different tissue-specific antigens that may be polymorphic (differ between donor and recipient). Tolerance to the donor antigens of one tissue does not guarantee tolerance to other allogeneic tissues of the same donor. Allogeneic hematopoietic cell chimerism, as can occur spontaneously in the case of fraternal cattle twins (as discussed earlier), or in the setting of allogeneic bone marrow transplantation, generates a state of tolerance to allogeneic hematopoietic cells but other donor allogeneic tissues, such as skin, can be rejected.[106,107] This is known as a state of split tolerance, a state that would not occur if SNS discrimination were mediated by linked suppression. In addition, a mechanism of linked suppression underlying peripheral tolerance would inhibit immune responses to pathogenic microbes.

This is not to say that T_{reg} cells have no role in achieving a SNS discrimination. The ability of T cells to suppress was suggested as early as 1969.[108] A number of other studies provided evidence that tolerance to self can be dominant rather than recessive.[109-112] For example, fetal rats given radioiodine to eliminate their thyroid were autoimmune against a subsequent syngeneic thyroid graft, and this autoimmunity was prevented by parabiosis to a normal rat. Such restoration of tolerance was abrogated by sublethal irradiation of the normal rat, strongly suggesting that irradiation sensitive lymphocytes could confer tolerance.[109] By recessive tolerance, we mean a tolerance that remains intrinsic to the tolerant population such that when one mixes tolerant cells with naive cells, tolerance is not imposed on the whole population. Conversely, the dominant tolerance that T_{reg} cells are considered to generate would cause the whole mixture of cells to be tolerant of the same antigens. While linked suppression has been considered a form of dominant tolerance, as mentioned, this mechanism cannot be a primary means to achieve a SNS discrimination. However, more recent data has identified a mechanism of T_{reg}-mediated dominant suppression that can contribute to achieving a SNS discrimination; namely, epitope-specific suppression[113,114]; shown in Figures 32.1 and 32.2. In epitope-specific suppression, T_{reg} suppress conventional T cells seeing the same epitope but do not suppress T cells that see different epitopes even if they are presented on the same APC. This type of epitope-specific suppression would reduce the need for a complete deletion of self-specific cells and in this way contribute to SNS discrimination. It will be critical in future research to determine the conditions under which T_{reg} cells act via epitope-specific suppression versus linked or bystander suppression, particularly if we are to utilize these cells with the goal of inducing antigen-specific tolerance (eg, to shut down autoimmune disease or induce donor-specific tolerance for transplants). Additional raison d'être for T_{reg} cells have been postulated, such as the promotion of survival of appropriate commensal flora,[115-117] preventing the complete elimination of antigen so that memory is maintained,[118,119] tuning down the stimulatory capacity of APCs,[49] and the hypothesis that T_{reg} are simply another "flavor" of helper cell that promotes an appropriate class of response (eg, IgA production) while suppressing other classes of response (eg, DTH responses). This type of class-switching/suppression also involves linked-recognition of antigenic determinants.[120-122]

Self-tolerance mechanisms may be encountered throughout the life of a lymphocyte and occur primarily at two distinct stages, first in the central lymphoid organs and then in peripheral lymphoid organs (Fig. 32.3). The following sections detail current knowledge regarding the mechanisms of central and peripheral tolerance and their relevance to human health.

CENTRAL TOLERANCE

When, in 1959, Lederberg proposed that antigen encounter during an early stage in lymphocyte maturation would generate tolerance, he did not know where lymphocytes are generated. B and T cell development were later found to occur in the bursa (in birds)[123] or bone marrow (in mammals, preceded by the fetal liver)[124,125] and thymus,[126,127] respectively. An allogeneic thymus graft given to neonatally thymectomized recipients was shown to generate specific tolerance of the donor (acceptance of donor skin grafts).[128] Thus, the thymus was a tissue needed not only for T cell generation but also was a site of tolerance and SNS discrimination. Clear evidence for a tolerance prone maturation stage in thymocytes would not arrive for several decades until appropriate tools were developed. It was assumed that this "central" tolerance process for developing thymocytes would be limited in scope, covering only those antigens present in the thymus (see Chapter 27 for developmental steps occurring in the thymus). While this reasoning was correct, as will be described in more detail below, the thymus was found to express many more antigens than expected, including many antigens that were initially thought to be unique to peripheral tissues.[129-133] Expression by thymic cells could even explain how robust tolerance could be achieved to serum proteins whose levels vary drastically in the circulation (eg, C-reactive protein). Such "promiscuous" expression of tissue-specific antigens substantially reduces susceptibility to autoimmune disease.[134] Thus, the scope of Lederberg's solution, the central tolerance mechanism (Fig. 32.4), and its overall importance in self-tolerance was not sufficiently appreciated prior to these discoveries. However, while broader than initially anticipated, central tolerance is incomplete not only due to a lack of certain antigens in the thymus but also because there is an affinity cutoff. Developing T cells with low affinity to an antigen

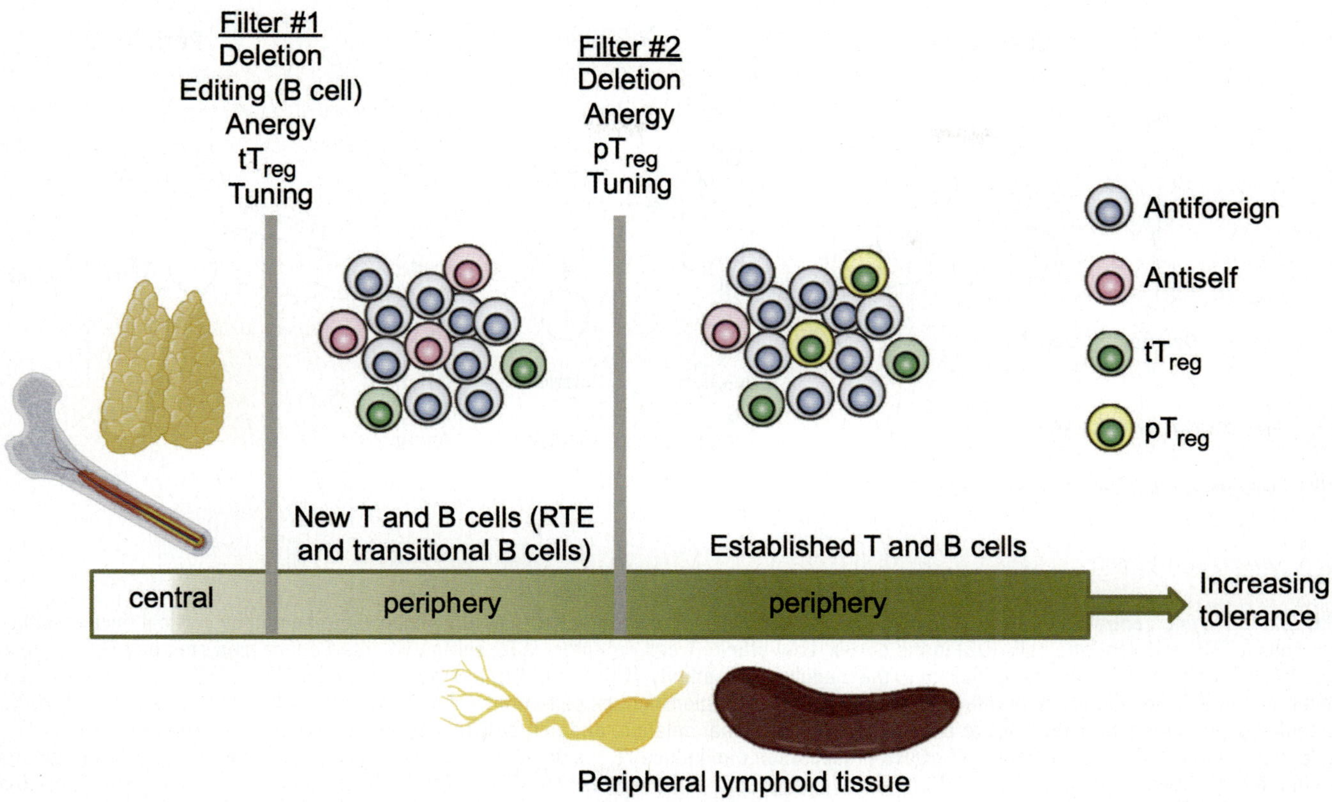

FIG. 32.3. The major outcomes for cells receiving tolerance-inducing signals occurring over cellular developmental time in the central and peripheral lymphoid tissue. There are two major stages, central and peripheral, at which self-reactivity is "filtered" out of lymphocyte repertoire. This filtering can involve physical elimination of cells with self-specific receptors (*deletion*), editing of the receptor in the case of the B cell receptor (BCR) in immature B cells, induction of an intrinsic unresponsive state (*anergy*), conversion to a T_{reg} phenotype, or an increase in the threshold signals needed for activation (*tuning*). In addition, for some self-specific lymphocytes, they do not receive tolerance-inducing signals and remain in a state of ignorance. The overall degree of tolerance to self within the lymphocyte repertoire increases, becoming maximal once T and B cells become fully mature (*established*) in the periphery, a process that can take up to a few weeks following export from the central lymphoid organs. The repertoire of new T cells (the recent thymic emigrants; RTE) and new B cells (transitional B cells) that seeds the periphery has not yet achieved a maximal state of self-tolerance. Tolerance is never complete, as new lymphocytes are continuously generated. pT_{reg}, perhipheral T_{reg}; tT_{reg}, thymic T_{reg}.

can escape central deletion and become effector cells in the periphery, as was recently demonstrated for T cells specific to the hematopoietic cell expressed minor histocompatibility antigen H60.[135] The incomplete nature of central tolerance may be of clinical utility, allowing newly generated donor T cells to generate a graft versus leukemia response following bone marrow transplantation.

Evidence for Central Deletion and Receptor Editing of Self-Specific Lymphocytes

The study of how self-specific lymphocytes become tolerant is made challenging by the low frequency of these cells and the need for tools to track them. The first such tool that became available was antibodies that recognize particular families of TCR Vβ chains. In the late 1980s, it was found that particular Vβ families were absent in certain strains of mice, including in SP thymocytes, and this was due to the ability of an endogenously produced retroviral gene product of mouse mammary tumor virus to bind both a particular MHC molecule and a particular Vβ family, thus linking the two.[136,137] The proviral protein was able to stimulate all T cells that use a particular Vβ family (eg, a viral antigen that binds MHC class II I-E and all T cells using Vβ17a). This had the tremendous advantage that the frequency of T cells able to bind the viral antigen was much higher than for normal antigens, on the order of 1/10 T cells, rather than the typical frequency of less than 1/100,000. Thus, the antigen binding T cells could easily be tracked by flow cytometry using a Vβ family specific monoclonal antibody. Given the ability of these provirus antigens to stimulate a large fraction of T cells, they have been termed superantigens. These superantigens induced a physical loss of the specific thymocytes, a process known as thymic-negative selection via deletion. Deletion of these immature cells was found to be caused by an apoptotic mechanism.[138,139] Important corroborating evidence for clonal deletion was provided in the context of a TCR recognizing a conventional peptide/MHC complex. This required the development of TCR transgenic mice to increase the frequency of T cells to a particular peptide/MHC complex. Compared to female mice, male mice with a transgenic TCR specific to the Y chromosome derived histocompatibility

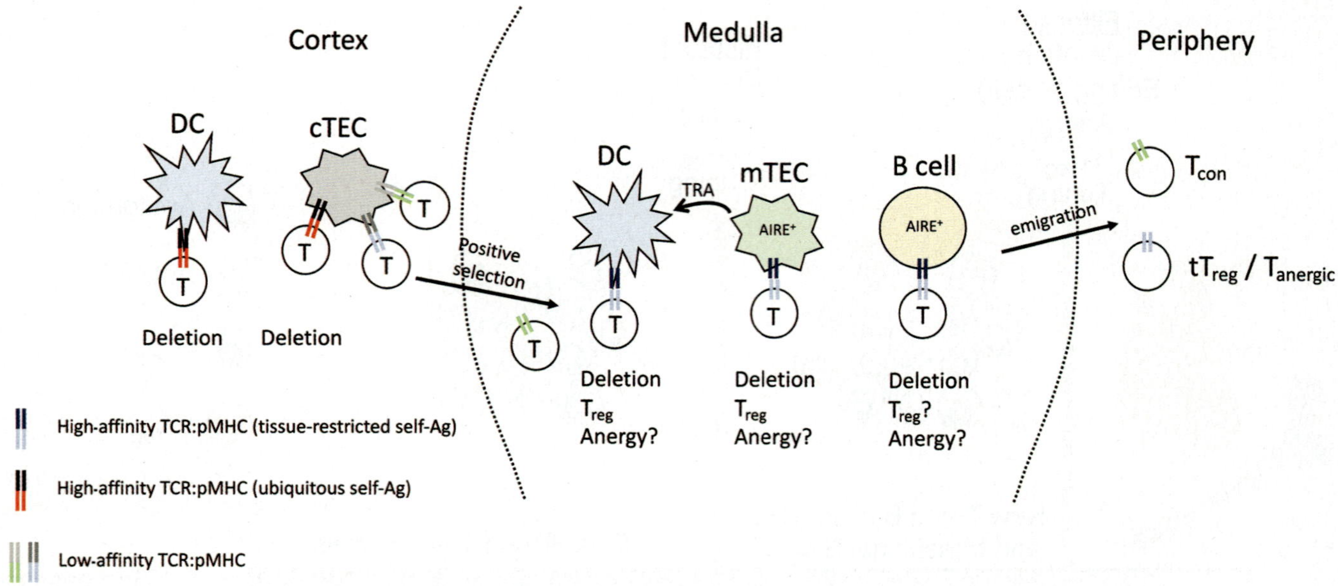

FIG. 32.4. Thymic central tolerance. Developing DP thymocytes first encounter self-pMHC complexes presented by cortical thymic epithelial cells (cTEC) and dendritic cells (DC) in the cortex. Low-affinity T cell receptor (TCR): self-pMHC interactions mediated by cTEC result in positive selection and thymocyte trafficking to the medulla. High-affinity TCR: self-pMHC interactions mediated by cTEC or DC induce clonal deletion. Once in the medulla, high-affinity TCR: self-pMHC interactions induce cell-intrinsic and cell-extrinsic forms of tolerance. Self-pMHC complexes presented by mTEC, DC, or B cells can induce clonal deletion, and self-pMHC presentation by mTEC or DC can result in tT_{reg} selection. Whether B cell–mediated self-pMHC presentation can induce tT_{reg} is unclear, as is the ability of different thymic APCs to induce thymic anergy. AIRE expression by mTEC and B cells can result in the expression of self-pMHC complexes derived from tissue-restricted antigens (TRA). TRA from mTEC can be presented indirectly via DC for central tolerance induction. Whether TRA from B cells can be indirectly presented by DC remains to be elucidated. Following thymocyte selection, conventional T cells (Tcon), tT_{reg}, and anergic T cells (Tanergic) emigrate from the thymus to populate the periphery.

antigen known as H-Y (in the context of H-2D^b) had greatly reduced CD8 SP T cells in the periphery and the thymus, as well as reduced DP T cells in the thymus.[140] Central deletion of class II restricted CD4 T cell precursors against a transgenic antigen was similarly demonstrated.[141] Evidence from the above models suggested that clonal deletion could occur late in the DP stage, at the transition to SP cells in the corticomedullary junction, or at an early DP stage, depending on the system. However, each system had its caveats, such as the unusual nature of the superantigen-TCR interaction and, in the case of TCR transgenic mice, the precocious expression of the TCRα chain at the DN stage of development (normally it is not expressed until the DP stage). A more sophisticated model has been developed that allows appropriate timing of transgenic TCR expression. In this setting, negative selection of class I restricted T cells specific to ubiquitously expressed H-Y antigen occurred at the DP to SP transition stage.[142] While the physical elimination of a self-specific clone may not take place until migration from the cortex to the medulla, antigen encounter in the cortex alone can induce tolerance.[143] In fact, with the above model, that has appropriate timing of anti–H-Y transgenic TCR expression, the deletion is triggered in the cortex in association with cortical DCs.[144] Central T cell deletion does not appear to be highly dependent on the location the antigen is normally expressed in the "deleting" cell as deletion occurs to both intracellular antigens (eg, H-Y), plasma membrane bound antigens such as MHC,[145] and soluble antigens present in serum.[146]

Similar to the above studies in T cells, central deletion of self-specific B cells was demonstrated using BCR transgenic mice crossed to mice that expressed the cognate self-antigen, in this case a self MHC class I molecule[147,148] (see Chapter 25 for stages of B cell development). This *in vivo* finding was presaged by earlier *in vitro* experiments where antibody cross-linking of the BCR on immature B cells led to a block in their development[149] and by studies of *in vivo* administration to chickens of anti-IgM/BCR antibody that blocked the later generation of plasma cells (likely due to developmental arrest).[150] Another BCR transgenic system employed a BCR specific for the antigen HEL. Interestingly, for B cells, this central deletion (ie, deletion of bone marrow precursors) required the protein antigen to be present in a plasma membrane bound form.[148] In contrast, when present in a soluble form, the B cells were not eliminated in the bone marrow but became anergic and expressed IgD but at least 10-fold less surface IgM BCR.[151] Anergic B cells would not respond to either stimulation by HEL antigen or to mitogenic stimulation (CpG-DNA/LPS).[152] While soluble antigen caused an anergic state, the anergic cells had a substantially shortened lifespan, indicating that anergy was followed by deletion.[153] Although protein expressed on the plasma membrane readily induced deletion of the corresponding specific B cells, when the same protein was engineered to be retained intracellularly, B cell deletion did not occur. This is consistent with the fact that naturally occurring autoantibodies, present even in healthy individuals, are frequently specific to intracellular components.[154]

However, this does not mean that antigens that are normally present intracellularly are incapable of inducing deletion.[155] For example, B cells specific for dsDNA, like those to membrane bound protein, were centrally deleted.[156] It seems likely that there is sufficient extracellular dsDNA to engage immature B cells and the multimeric nature (repeating epitopes) makes this soluble antigen capable of causing deletion.

There is not yet a consensus on the nature of the BCR signal, signal 1 (a detailed discussion of signaling is presented in Chapter 24). Classical observations show that antibody to surface immunoglobulin (sIg) is mitogenic for B cells and gave rise to the sIg cross-linking model for the sIg-mediated signal,[157] implying monovalent antigens cannot generate this signal. In contrast, Reth and colleagues have argued that sIg are naturally associated and that monovalent antigens binding to sIg can cause sIg disassociation and signaling.[158] Evidence supports the idea that monovalent binding of antigen to sIg can cause signaling,[159,160] explaining the potency of monomeric antigens in causing B cell inactivation.[161] Tolerance at the B cell level will be complete only if monovalent binding to sIg, and not its cross-linking, is sufficient to generate an inactivating signal.[162]

Central B cell deletion has at least two stages, a transiently reversible arrest in development, followed by death of arrested B cells.[163] While the reversible arrest might allow a self-specific B cells to escape deletion, it appears the risk of this potential danger was worth taking, as it allows an immature B cell that might otherwise be wasted to instead be rescued in a process known as receptor editing. The potential level of waste is quite high, as it has been estimated that for more than half of immature B cells, their first attempt to generate a functional BCR results in an autoreactive BCR.[164] To counter this, immature B cells that have generated a self-specific BCR, upon engagement of the self-ligand, can upregulate RAG gene expression leading to secondary rearrangements of Ig light chains and thus development of a new specificity. This process was demonstrated in BCR transgenic mice specific for self MHC class I or chromatin-DNA complexes.[165-167] Secondary rearrangement involves deletion or inversion of intervening gene segments, allowing upstream Vκ to be inserted beside downstream Jκ, eliminating expression of the original light chain.[168] Secondary rearrangements are sequential, beginning with Igκ and then Igλ. IgH chain editing may also occur, as seen in mice with an anti-dsDNA VDJ inserted in the J_H locus[169] and in analysis of V_H genes of human immature B cells.[170] Receptor editing is a process that occurs in immature B cells in the bone marrow and not at the mature cell stage, as demonstrated by the presence of receptor editing when the target self-antigen was in the bone marrow but not when its expression was restricted to the liver, deletion occurring without editing in the latter case.[165] Since the IgH chain plays a greater role in determining specificity (the same IgH can sometimes retain specificity with different IgL chains), multiple secondary light chain rearrangements may be needed before self-specificity of the BCR is lost; multiple secondary rearrangements would also be expected based on the two-third chance of out of frame rearrangement. Based on BCR knockin models of antiself BCR, receptor editing appeared to be the primary means of removing self-specificity from the immature B cell pool.[168,171,172] However, efficient receptor editing may be dependent on the form of antigen, as it seems to occur less frequently in B cells specific for soluble proteins than for ssDNA, where the latter B cells emerge from the bone marrow in an anergic state.[173,174]

Receptor editing has an important role in giving a developing T cell a second chance at positive selection if the receptor it expresses does not sufficiently engage self-peptide/MHC. Similar to the process in B cells, receptor editing is also able to rescue self-specific T cells in transgenic TCR/antigen models through additional rearrangements at the TCRα locus.[175,176] However, other experimental models have led to different inferences. For example, if receptor editing were a significant means of avoiding deletional T cell central tolerance, then there should be an increase in in-frame TCR excision circles (TRECs) in thymocytes undergoing central tolerance (if instead cells undergo deletion the TRECs would be lost). Hogquist and colleagues used TCRβ transgenic mice specific to the male H-Y antigen and found there were lower in-frame TCRα chain TRECs in male compared to female mice (ie, controls that lack H-Y self-antigen), providing evidence against receptor editing as a means of avoiding deletional T cell central tolerance.[177] Thus, unlike the situation for central tolerance of B cells, the relative contribution of receptor editing to self-tolerance in developing T cells, when compared to deletion, is likely to be quite low.

TCR affinity for pMHC is important for determining positive versus negative selection but is not the only determining factor. Developing T cells appear to be able to adopt different fates based in part on different levels of presentation of the same peptide, where low levels of peptide can induce positive selection and high levels induce deletion.[178,179] Although these data may only apply to rare T cells near the boundary of positive/negative selection, they have been interpreted to support a differential affinity-avidity model of positive versus negative selection. It would be dangerous if mature lymphocytes had greater sensitivity to antigen than immature cells, as the mature cell could respond to a particular level of self-antigen to which immature cells would be blind (ie, could not become centrally tolerant of). It therefore may not be surprising to learn that in fact immature cells have greater sensitivity than mature cells, providing a margin for safety. The BCR of immature B cells has greater sensitivity to antigen than that of mature B cells, as assessed by increased tyrosine phosphorylation (including, among others, the BCR, Lyn, and PLCγ2) and calcium entry.[180] Early *in vitro* studies using a bacterial superantigen, staphylococcal enterotoxin B (SEB), showed that developing TCR transgenic T cells may be more than 10-fold more sensitive to deletion by SEB than peripheral T cells are to activation by the same antigen.[181] Importantly, a heightened sensitivity to antigen dose for tolerance in developing thymocytes, when compared to the dose for activation of mature cells, was also observed for a conventional self-antigen found in serum, the F liver protein.[182] Thymic deletion to a high-affinity ligand could occur with as few as several specific peptide-MHC complexes per APC.[183] *In vivo* studies also supported the conclusion of greater sensitivity to antigen by developing T cells.[184]

A significant portion of T cell precursors undergoing negative selection in the thymus are those that have generated promiscuous TCRs capable of recognizing MHC more generally, independent of specificity for a particular peptide or even MHC allele.[185,186] A substantial fraction of peripheral T cells expresses a second TCRα chain due to less stringent allelic exclusion compared to the β chain.[187] Such dual TCR expression can compromise thymic negative selection and participate in autoimmune disease development.[188]

Nondeletional T Cell Central Tolerance

While substantial evidence exists for the physical elimination of T and B cells that express a strongly self-reactive antigen receptor and strongly self-reactive B cells can alter their antigen receptor through receptor editing, these central tolerance mechanisms do not eliminate all cells that express a highly self-reactive antigen receptor. In fact, autoreactive B and T cells are a part of the peripheral repertoire[30,154] but do not normally cause, and in many cases are thought to limit, autoimmune diseases at homeostasis in most individuals. Two additional nondeletional outcomes for thymocytes that receive strong TCR signals during development are clonal anergy and clonal diversion, and these fates likely account for the presence of substantial numbers of autoreactive peripheral T cells. While clonal anergy is certainly induced in B cells, it has been examined primarily in the context of peripheral tolerance and will be discussed later. In the late 1980s, around the same time as thymic clonal deletion was being studied, Ramsdell and colleagues used radiation chimeras to understand the contribution of different APCs to central tolerance.[189] Using mAbs that recognized TCRVβ families that specifically bound to superantigens expressed by particular mouse strains, it was discovered that rather than being deleted during development, some superantigen-reactive thymocytes were present but rendered functionally inactive. Both CD4SP and CD8SP thymocyte subsets were found in the thymus and corresponding mature T cells were found in lymph nodes. The mice harboring the superantigen-reactive thymocytes and T cells appeared healthy without any spontaneous deaths suggesting a lack of graft-versus-host disease and stable tolerance. This was corroborated with adoptive transfer approaches using lymph node cells obtained from the chimeras. To confirm the surviving thymocytes and mature T cells were indeed anergic, the authors stimulated thymocytes and lymph node cells with host-derived stimulators and found a lack of proliferation. A follow-up study from Ramsdell and Fowlkes determined that the maintenance of tolerance required continued access to self-Ag.[190] What was unclear from these studies was whether the lack of a response by the "tolerized" thymocytes or T cells was cell-intrinsic in nature (anergy) or whether it was wholly, or in part, mediated by cell-extrinsic mechanisms such as the generation of regulatory T cells.

In addition to thymic anergy induction, another form of nondeletional central tolerance is clonal diversion of thymocytes to regulatory lineages. The presence of thymically derived "regulatory T cells" (T_{reg}) was supported by neonatal thymectomy experiments where thymectomy of neonatal rats or mice resulted in the development of autoimmune diseases targeting the thyroid and/or ovaries.[191-193] Subsequently, Mason's group determined that coadoptive transfer of a subset of CD4+ T cells that expressed markers of T cell activation obtained from unimmunized rats supressed the development of autoimmune diseases elicited by adoptive transfer of naïve CD4+ T cells into immunodeficient recipients.[194,195] These findings were recapitulated in mice by Powrie.[196] Sakaguchi refined the phenotypic definition of T_{reg} by demonstrating that the CD4+CD25+ subset of mature T cells contained the suppressive T_{reg} population.[197]

A major breakthrough in the understanding of self-tolerance and in the identification of CD4+CD25+ T_{reg} cells was achieved with the cloning of the transcription factor Foxp3. Using the spontaneous mutant mouse model called scurfy that was autoimmune disease prone, Foxp3 was cloned by genetic mapping and sequenced.[198] At the same time, it was determined that human patients with an autoimmune/autoinflammatory disease called IPEX (immune dysregulation, polyendocrinopathy, enteropathy, X-linked syndrome) had mutations in the *FOXP3* gene indicating that the scurfy mouse was a model for IPEX patients.[199-201] Cementing the importance of Foxp3 in the development and function of CD4+CD25+ T_{reg}, Foxp3-deficient mice were generated. While a few CD4+CD25+ cells were present in these mice, they lacked suppressive ability.[202-204] This defect in the generation and function of CD4+CD25+ T_{reg} resulted in the development of autoimmune disease, allergy, and inflammation.[205,206] Furthermore, it was demonstrated that ectopic expression of Foxp3 in naïve CD4+ T cells by retroviral transduction "converted" conventional CD4+ T cells into CD4+ T cells with suppressive capabilities highlighting the central role of Foxp3 in T_{reg} physiology.[202,203]

A current outstanding question surrounds the factors that regulate the different central tolerance fates. TCR affinity for self-pMHC is one potential factor. Early experiments with TCR transgenic mice demonstrated that TCR agonist peptide expressed in the thymus can increase the proportion of CD4 cells that are of the T_{reg} lineage.[207] Additional experiments by Lee et al. examined thymocytes with a fixed TCRβ chain from the OVA-specific DO11.10 TCR paired with different endogenous TCRα chains cloned from Foxp3+ CD4SP thymocytes specific for OVA. It was observed that selection into the Foxp3+ T_{reg} lineage occurred over a broad TCR affinity window.[208] Importantly, thymocytes expressing TCRs with an affinity below that required for clonal deletion could be selected into the T_{reg} lineage; however, the efficiency of T_{reg} selection increased with increased TCR affinity. A TCR was also identified that could induce both T_{reg} selection and clonal deletion further supporting an overlap in TCR affinities that can drive clonal deletion and T_{reg} selection. While TCR affinity in part may regulate clonal deletion versus T_{reg} selection, discussed in more detail in Chapter 29, many other factors contribute.[209] Additionally, the factors that influence the induction of thymic anergy and how those relate to clonal deletion and T_{reg} generation is currently unclear, but since thymic anergy was originally described in a superantigen system that is known to be a model of high-affinity antigen-driven tolerance, it is likely there is overlap in the TCR affinity of thymocytes that undergo all forms of thymic central tolerance.

Antigens and APCs in T Cell Central Tolerance

The thymus is primarily composed of epithelial cells (cortical and medullary), T cell precursors, and a small number of DCs, macrophages, and B cells. The ability of antigen presentation to induce central tolerance is not due to a particular capacity of APCs in the thymus, as peripheral APCs that stimulate responses in the periphery instead induce tolerance when injected into the thymus.[210] This is consistent with Lederberg's model that postulated lymphocytes are programmed to be tolerizable only (not inducible) when at this early developmental stage (ie, when they are precursors to mature lymphocytes).[14] For a long-time, it was thought that positive and negative selection were spatially separated events, with positive selection being driven by cortical thymic epithelial cells (cTECs) and negative selection being driven by subsequent antigen encounter in the DC rich medulla or at the corticomedullary junction.[211] While it has long been clear that hematopoietically derived cells are efficient at inducing thymocyte clonal deletion, the question of how tolerance could be induced to self-antigens specific to the thymic cortex remained. Subsequent studies, including restricting MHC expression to cTECs, showed that T cells do indeed become centrally tolerant of antigens presented on cTECs,[143] and that the majority of deletion takes place in the cortex.[144,212] Monitoring cleaved caspase-3, CD5 and TCR expression (both upregulated by TCR signals) as well as CCR7 (needed for migration to the medulla[213,214]) allowed quantitative assessment of clonal deletion and its anatomic location within a polyclonal thymocyte repertoire. Employing this method in bone marrow chimeras that lacked MHC II on bone marrow cells, it appeared that at least half of clonal deletion occurred independent of antigen presentation by bone marrow–derived cells.[212] While previously contentious, these studies also supported a prominent role for CD28-CD80/86 costimulatory interactions in clonal deletion.

While clonal deletion of thymocytes specific for common or ubiquitous self-proteins was relatively easy to understand, whether and how central tolerance was enforced against proteins that are expressed in a tissue-restricted manner was more difficult to envision. Migration of APCs carrying self-antigens from the periphery to the thymus represented one possible solution. Although many thymic DCs are generated in the thymus, about half of the DCs in the thymus are from several subsets of DCs that migrate to the thymus.[215] A number of approaches provided evidence that DCs could carry antigen to the thymus and induce tolerance.[216] pDCs, via CCR9, have been shown to take up antigen from the periphery to the thymus, resulting in central tolerance; interestingly, this capacity is prevented by TLR activation.[217] While there is some evidence that such pDC trafficking from the colon to the thymus alters innate lymphocyte cellularity in the thymus and affects susceptibility to experimental colitis,[218] evidence that self-tolerance depends on APCs taking antigen from the periphery to the thymus is limited.[219] Instead, early experiments by the Hanahan and Smith group provided the critical new concept answering the above questions. They demonstrated that neo-self-antigens expressed under control of the rat insulin promoter resulted in T and B cell tolerance to the neo-self-antigen and that expression of the neo-self-antigen could be detected in thymic tissue by cells he called "peripheral antigen-expressing" cells.[220,221] They were also able to detect the expression of other tissue-restricted antigens including insulin and somatostatin in the medullary region of the thymus in wildtype mice.[220] Gery and colleagues showed that the resistance and susceptibility to experimentally induced autoimmunity in the eye, in different species and strains, was associated with thymic expression and lack of expression, respectively, of a natural ocular autoantigen.[134] Similarly, a genetic variant with increased insulin expression in the thymus is associated with dominant protection from type-1 diabetes.[222] The ability of medullary thymic epithelial cells (mTECs) to express a variety of tissue-restricted antigens, both at the mRNA and protein level, was later described by Derbinski et al.[133] What was unclear from these studies was the mechanism that underpinned this promiscuous peripheral antigen expression in the thymus.

A major breakthrough in our understanding of the molecular mechanisms in central tolerance occurred when the AIRE gene was identified. Coincident with the findings that mTECs could express tissue-restricted antigens, two groups cloned a novel gene from humans that, when found to be mutated, resulted in a devastating, autosomal recessive multiorgan autoimmune disease called autoimmune polyendocrinopathy-candidiasis ectodermal dystrophy (APECED) or autoimmune polyendocrinopathy syndrome type 1 (APS-1).[223,224] This gene, *AIRE*, contained features that suggested it may play a role in central tolerance induction. AIRE has zinc finger binding motifs commonly found in transcription factors,[223] it's expression is largely restricted to the thymus and lymph node,[223] and it is localized to the nucleus in mTECs.[133,225] While all these data circumstantially suggested that AIRE was the protein that regulated the expression of tissue-restricted antigen in mTECs leading to central tolerance, this was not directly demonstrated until Anderson et al. generated and studied AIRE-deficient mice.[226] The immune system of AIRE-deficient mice was grossly normal with no overt changes in thymocyte and B cell numbers or composition or in DC and cytokine profiles. Elevated frequencies of CD44^{hi} memory phenotype CD4+ T cells were detected in unmanipulated *AIRE*$^{-/-}$ mice suggesting the possibility of an autoimmune response. Consistent with this, the mice displayed overt signs of autoimmunity including the production of autoantibodies that targeted discrete substructures within various tissues including the retina, salivary gland, pancreas, stomach, ovary, and thyroid. This was accompanied by lymphocytic infiltrates in various tissues. Reciprocal bone marrow chimeras and thymic transplant experiments elegantly determined that AIRE-deficient thymic stroma was required to precipitate autoimmunity. To directly link AIRE with altered gene expression, mRNA analysis from control and AIRE-deficient mTECs revealed dramatic reductions in the expression of a substantial number of genes, many of which display a tissue-restricted expression pattern, suggesting AIRE indeed regulates the expression of tissue-restricted antigens. However, not all mTECs-expressing tissue-restricted antigens appear to be dependent upon AIRE for this expression suggesting the existence of another gene(s) that may function similarly to AIRE.[226] Subsequent work from Liston et al. directly determined that AIRE-regulated tissue-restricted antigen expression was required for clonal deletion of thymocytes specific for an ectopically expressed tissue-restricted antigen.[227] Elegant studies by DeVoss et al demonstrated that AIRE-dependent thymic expression of a natural retinal autoantigen, interphotoreceptor retinoid-binding protein (IRBP), was necessary to prevent the retinal autoimmune disease, uveitis. Furthermore, by transplanting IRBP-deficient fetal thymus into nude mice, they showed that uveitis developed when both the thymus and recipient were AIRE-sufficient.[228] Collectively, these data demonstrated the profound impact of AIRE-dependent expression of self-antigens by mTECs on self-tolerance. This raised the question as to whether immature thymocytes needed to interact with mTECs for tolerization of the T cell repertoire. Gallegos et al. showed that AIRE-dependent tissue-restricted antigens could be both directly presented by mTECs and indirectly presented by hematopoietically derived APCs to induce clonal deletion.[229] This finding is important when one considers that individual mTEC express a variable fraction of AIRE-dependent tissue-restricted antigens.[133] By spreading self-antigens to other medullary APCs, immature thymocytes are more likely to be tolerized than if they had to encounter the rare mTEC presenting the tissue-restricted antigen for which their TCR is specific. In addition to mTEC, Yamano et al. used AIRE reporter mice to discover that thymic B cells also expressed AIRE and contributed to thymic clonal deletion.[230] Through a series of elegant experiments, it was determined that peripheral B cells trafficked to the thymus where they were "licensed" to express AIRE through cognate interactions with CD40L-expressing CD4SP thymocytes. Importantly, BCR signaling prevented AIRE expression in B cells and exclusively antiforeign CD4SP thymocyte specificity (OT-II CD4SP) was not sufficient to license B cells. Finally, direct B cell pMHC II presentation of an AIRE-regulated neo-self-antigen resulted in clonal deletion of neo-self-antigen-specific CD4SP thymocytes. Perera et al also determined that thymic polyclonal B cells could induce deletion of thymocytes specific for a ubiquitous self-protein glucose-6-phosphate isomerase (GPI), and mice with autoantigen binding BCR transgenic B cells were more efficient inducers of deletion.[231]

The finding that AIRE regulated the expression of tissue-restricted antigens, leading to thymocyte clonal deletion, was used as strong evidence for clonal deletion in establishing T cell central tolerance. However, it was also possible that the absence of AIRE and subsequent decrease in tissue-restricted antigen expression could impair the generation of tissue-restricted antigen-specific T_{reg} cells, and this impaired antigen-specific T_{reg} cell generation contributed to the autoimmune disease seen in AIRE-deficient mice. Grossly, AIRE deficiency did not impact the frequency or number of T_{reg}[232,233]. However, to address this issue more definitively, wildtype and AIRE-deficient fetal thymic lobes were cotransplanted under the kidney capsule of nude mice.[232,233] If autoimmunity in AIRE-deficient mice was a result of impaired T_{reg} generation, the presence of a wildtype thymus would provide T_{reg} that would then prevent autoimmunity. If, however, autoimmunity was still observed, it would suggest that a lack of clonal deletion in the AIRE-deficient thymus allowed the export of autoreactive T cells that caused the multiorgan autoimmunity. They found that autoimmunity occurred similarly when either a single AIRE-deficient thymus was transplanted into nude mice or when an AIRE-deficient and a wildtype thymus were cotransplanted into nude mice suggesting that a defect in T_{reg} cell generation did not underpin the autoimmunity observed when AIRE was absent. However, when they cotransplanted four wildtype thymi with a single AIRE-deficient thymus, tolerance was largely restored.[232] These findings could suggest that the "extra" T_{reg} provided by the four wildtype thymi suppressed the autoimmunity caused by T cells that were generated in the AIRE-deficient thymus. However, at the time, the authors concluded it was more likely a nonspecific crowding effect. In contrast, if we assume that AIRE-dependent expression of tissue-restricted antigens controls both clonal deletion of autoreactive thymocytes and selection of T_{reg}, there is likely a greater number of autoreactive T cells released into the periphery of AIRE-deficient mice (sum of cells that would be deleted plus those that would become T_{reg}) than the number of T_{reg} generated from a wildtype mouse. Therefore, the findings from Anderson et al. could be reinterpreted as support for AIRE-dependent generation of tissue-restricted antigen specific T_{reg}. In fact, support for this possibility was recently provided by Malchow et al.[234] More direct support for the involvement of AIRE in T_{reg} selection was provided by Perry et al when they sequenced TCRs from T_{reg} generated in AIRE-sufficient or -deficient thymus.[235] To restrict TCR diversity, they used a transgenic mouse that expressed the TCRβ chain from a TCR specific for a peptide from human CLIP presented by mouse I-A^{b}[236] and was also heterozygous for TCRα. They sequenced TRAV14 (Vα2), which is known to pair well with TCli TCRβ, from Foxp3− or Foxp3+ CD4SP thymocytes from this mouse. While some TCRs were selected in an AIRE-independent fashion, there was a loss of some T_{reg} TCR sequences in AIRE-deficient mice. In particular, AIRE deficiency appeared to disproportionately impact the selection of less abundant T_{reg} clones. The issue of whether or not AIRE regulated the expression of proteins that induced the selection of T_{reg} was more fully resolved when Malchow identified a self-protein regulated by AIRE that instructed the selection of antigen-specific T_{reg}.[237] The experiments from Perry et al also, not surprisingly, found an imprint of AIRE on CD4SP thymocyte clonal deletion and through the use of bone marrow chimeras they determined that thymic DCs and mTECs make nonredundant contributions to both thymocyte clonal deletion and T_{reg} selection.[235]

While AIRE is clearly required for the expression of a variety of tissue-restricted antigens, it has also been realized that some tissue-restricted antigens are expressed in an AIRE-independent fashion. As such, the existence of an additional gene that regulates the expression of tissue-restricted antigens in mTECs has been hypothesized. Recently, Fezf2 was identified as a gene whose expression is restricted to mTEC in the thymus and is a critical regulator of tissue-restricted antigen expression required for tolerance induction.[238] As the role of

Fezf2 in thymic tolerance was only recently determined, there is still much to be learnt regarding the biology of Fezf2 in central tolerance.

While we have learned much regarding the transcriptional regulation of promiscuous peripheral gene expression in the thymus, it is ultimately the peptides presented in the context of self-MHC molecules that regulate thymic central tolerance. Therefore, it is important that we understand the identity and source of the peptides presented by self-MHC in the thymus. In terms of self-peptide–MHC presentation in the thymus, of course, hematopoietically derived APC present self-peptide-MHC I and -MHC II complexes to developing thymocytes. Additionally, as with all cells, TECs present self-peptide-MHC I complexes. Importantly, and uniquely, TECs also present self-peptide-MHC II complexes under steady state conditions endowing them with the ability to make nonredundant contributions to both positive selection and central tolerance. Direct peptide-MHC I/II presentation by APC (both hematopoeitically derived and TEC) as well as indirect presentation of TEC-derived proteins by hematopoietically derived APC contribute to deletional and nondeletional forms of central tolerance.[229,239,240] The identity of and the physiological processes that produce the peptides loaded into MHC molecules are being investigated but much is left to discover in this area. It is currently thought that ubiquitin-mediated degradation of newly synthesized proteins rather than "old" proteins is the main contributor to the pool of peptides presented by MHC I. A large proportion of these newly synthesized, rapidly degraded proteins are in fact defective ribosomal products (DRiPs).[241] The use of mass spectrometry to identify MHC I–associated peptides has relied on peptide-identification based upon information from reference databases that use "canonical" protein sequences derived from conventional reading frames. Importantly, these databases will, by definition, miss many potential MHC I bound peptides due to the presences of single nucleotide polymorphisms (SNPs) (minor histocompatibility antigens) and translation in alternate reading frames or from cryptic translational start sites.[242,243] Therefore, the MHC I–associated peptidome only partially overlaps with the proteome. For peptide-MHC II presentation, thymic APC of hematopoietic origin can present peptides derived from exogenous proteins through the typical MHC II presentation pathway. However, for TECs, which do not efficiently endocytose extracellular material, MHC II peptides must be sourced from endogenous proteins. Macroautophagy, a normally inducible process used to degrade proteins during times of nutrient depravation to maintain metabolism, is constitutively active in a portion of TECs (both cortical and medullary) and has been shown to be critical for thymic central tolerance.[244,245] In addition to the peptide sequences involved in central tolerance, posttranslational modifications (eg, citrullination, glycosylation) may alter the level of tolerance achieved to a given peptide. These modified peptides can be a target in autoimmune diseases, and a deficiency in central tolerance to a modified peptide has been observed.[246,247] While our understanding of the nature of the peptides involved in central tolerance is expanding, much is left to uncover including improving our understanding of the relationship between the peptides presented in the thymus for central tolerance induction and those involved in driving autoimmune diseases.

Collectively, a consensus is emerging whereby many different cell types (cTECs, mTECs, DCs) in the thymus are able to influence central tolerance. Furthermore, these different cell types make nonredundant contributions to both cell-intrinsic and cell-extrinsic forms of central tolerance and support the concept that impairments in any of these different cell types can negatively impact T cell central tolerance.

PERIPHERAL TOLERANCE

Given the wide coverage of tolerance to self-antigens provided by central tolerance, including promiscuous tissue-specific gene expression driven by AIRE and Fezf2, one might ask whether a peripheral tolerance mechanism is needed at all. Since central tolerance has an affinity/avidity cutoff, even if all relevant self-antigens were present in the central lymphoid organs, lymphocytes with below threshold affinity for tolerance induction to these antigens would escape central tolerance. The ability of infections to trigger autoimmunity provides clear evidence of self-specific cells escaping central tolerance; for example, group A Streptococci infection gives rise to T and B cells cross-reactive to cardiac antigens and the development of Rheumatic fever.[248] Direct evidence that self-antigens exist in the periphery that are not present in central lymphoid organs is quite limited; instead, transgenic "neo" self-antigens have been used frequently to model the existence of tissue-restricted peripheral self-antigens. There is substantial evidence that mature peripheral lymphocytes can be made tolerant experimentally.[249,250] However, these data are also indirect in terms of whether lymphocytes naturally become tolerant of self-antigens that are found only in the periphery. To more directly address whether there are natural organ/tissue-specific antigens whose expression in these tissues is needed for peripheral tolerance of the same tissue, a number of studies have examined the effect of removing a tissue on tolerance to the same tissue. Peripheral tissues such as ovary, thyroid, and hypophysis were removed, and at later time points, the same tissue was returned as a syngeneic graft or immunization to tissue antigen in adjuvant was tested. Since lymphocytes are continuously generated, new lymphocytes would not have a chance to become peripherally tolerant once the tissue is removed, and rejection of a syngeneic graft would then be expected based on a need for peripheral tolerance. Data in these tissue removal models have been mixed, with some but not all providing evidence for an immune response to the tissue antigens and the syngeneic graft.[251-256] Together, the data suggest that there is some need for peripheral tolerance that is maintained by the peripheral tissue itself. Furthermore, in addition to a tolerance sensitive programming in immature lymphocytes within central lymphoid organs, lymphocytes newly exported from central lymphoid organs, the recent thymic emigrant and transitional B cell stages, respectively, maintain a heightened capacity for tolerance (Fig. 32.3). Thus, B cells also have a second tolerance checkpoint that occurs in the periphery.[164] Recently, Watanabe et al used a small panel of foreign- and self-antigens and compared IgGs derived from individual transitional (CD19+CD27–CD38^{hi}CD10+IgD+) versus mature (CD19+CD27–CD38^{lo}CD10–IgD+) human B cells. Of the foreign reactive transitional B cells, approximately 70% were found to have cross-reactivity to self-antigens and half of this cross-reactivity was absent from the mature B cell stage.[257] As a result of this cross-reactivity, self-tolerance, including peripheral self-tolerance, ends up creating "holes" in the repertoire of foreign reactivity. Holes in the repertoire of high-affinity cross-reactive CD4 T cells specific to defined foreign-peptides caused by central tolerance to systemic self-peptides have been demonstrated.[258] Peripheral T cell tolerance may similarly create such holes, as mature naive T cells demonstrate shorter average CDR3 length compared to recent thymic emigrants.[259] It has long been known that the form and route of delivery of antigen can determine whether productive immunity or a state that appears like tolerance ensues. Deaggregation of antigen (ie, monomeric antigen) along with systemic distribution of antigen, as is achieved by intravenous or intraperitoneal injection, favors a tolerant state.[37,83,260] Conversely, local delivery of aggregated or cell bound antigen favors immunity (eg, subcutaneous injection). "Tolerance" induction based on particular forms of antigen and routes of injection are typically complex states, in some cases involving the generation of suppressive T cells. It is likely that such unresponsive states involve a mixture of responses, including real tolerance in a subset of antigen-specific cells and a switch in the class of response in the remaining cells. We therefore focus below on studies of tolerance to antigens generated by the animal itself to gain an understanding of natural mechanisms of peripheral self-tolerance.

Tolerance Versus Ignorance, Trafficking, and the Role of the Fetal/Neonatal Period

While it intuitively seems important that our lymphocytes become tolerant of our own antigens to prevent autoimmune disease, it should be kept in mind that each beneficial evolutionary "advance" in our biology may also come at a cost. In the context of lymphocytes, complete elimination of self-specific lymphocytes could compromise our ability to respond to a particular pathogenic microbe and that microbe may be a greater threat to our survival than the potential autoimmune reaction. Thus, it should not be surprising that healthy individuals harbor a significant number of lymphocytes whose antigen receptors can bind self-molecules with substantial affinity. A number of experimental models have indicated that B cells and CD8 T cells can be ignorant of the presence of a particular natural or model self-antigen.[261-263] Figure 32.5 shows some of the conditions and signals leading to ignorance versus tolerance of a peripheral self-antigen. Since CD4 helper T cells are critical in generating many (perhaps most) immune responses, including humoral and CTL responses, it has been argued that tolerance in CD4 T cells is essential, while ignorance rather than tolerance may occur more frequently for B cells and CD8 T cells.[264] For example, B cells specific for the self-complement component C5 appear to be ignorant, as they can generate autoantibodies that functionally inhibit C5 activity if they are provided T cell help (ie, tolerance is at the level of the T cell and not the B cell).[261] Similarly, B cells appear ignorant of F liver protein.[182] However, there is evidence that B cells and CD8 T cells can be intrinsically tolerant of self-antigens.[265,266] In contrast to the findings with C5, B cells specific to the transgenic model soluble self-antigen HEL were anergic rather than ignorant. It is unclear why the outcome was different for these two soluble peripheral self-antigens; however, it may have to do with whether the antigen is presented to B cells in a multimeric form or the association of the antigen with inflammatory processes (ie, C5 components are part of an inflammatory pathway and would be associated with antibody, foreign antigen, and T cell help). Consistent with the latter possibility, B cells specific to the Fc portion of IgG also do not appear to be fully tolerant. These B cells are stimulated by immune complexes to produce antibodies known as rheumatoid factor (RF),[267] and this response can be dependent on the immune complex containing a TLR ligand (CpG in chromatin) and can be augmented by specific T cell help to antigens contained in the complex.[268-270] This T cell help or the presence of a TLR ligand in the immune complex may block tolerance in RF generating B cells. In addition, whether or not lymphocytes are tolerant or ignorant of a given self-antigen is likely to depend on the level of expression of the antigen and consequently the level of presentation of the antigen as well as the location of the antigen and the relevant lymphocytes. As an example of the latter, mice expressing a transgenic BCR specific to erythrocytes largely deleted these cells in the bone marrow. However, some B-1 cells could escape tolerance and survive in the peritoneal cavity, a site free of the antigen, resulting in autoantibody production and hemolytic anemia.[271] One must keep in mind, however, that under physiologic conditions of a polyclonal repertoire, the frequency of antierythrocyte B cells is much lower, and hence autoimmunity as a result of escape of these B cells to the antigen-free peritoneum does not usually occur. Instead, these studies highlight the possibility that, in very rare cases, B cells that escape central tolerance may "hide" in a location where the relevant autoantigen is absent and thus also escape peripheral tolerance as well.

Is peripheral tolerance induced in parenchymal tissue of organs or in the lymphoid tissue that drains these organs? It appears that peripheral tolerance largely depends on presentation of self-antigens in lymphoid tissue. This fits with the trafficking of naive lymphocytes as they circulate from blood to lymph and do not migrate into parenchymal tissue. A number of pathways can lead to presentation of self-antigens in lymph nodes. Antigens may be shed or released during normal cellular turnover within tissues, and these antigens can go directly to the lymph node or be picked-up by tissue DCs that then migrate to the draining lymph node. This would be the conventional presentation process for antigens presented on MHC class II. In addition, these antigens can be cross-presented in MHC class I in specific subsets of DCs.[250] Thus, peripheral self-antigens can be presented to both CD4 and CD8 T cells in draining lymph nodes. In addition, like in the thymus, there is evidence that tissue-specific peripheral self-antigens can be expressed constitutively in a specific subpopulation of bone marrow–derived lymph node APCs driven by

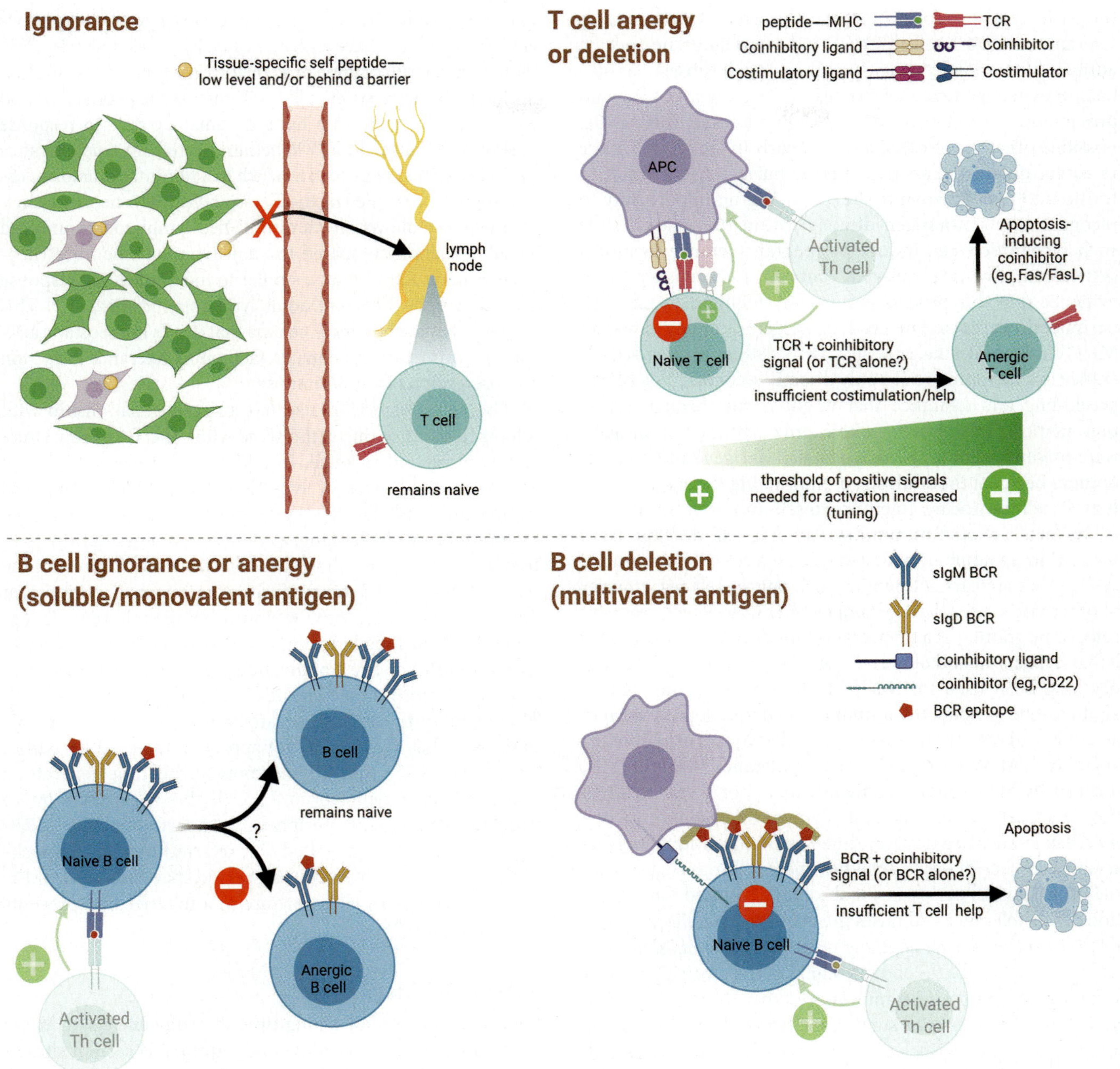

FIG. 32.5. Ignorance and signals leading to the peripheral tolerance outcomes of anergy and deletion. The main context and signals that lead to peripheral tolerance or ignorance are depicted. Whether the tolerant states (anergy or apoptosis/deletion) are always due to the antigen receptor signal (T cell receptor [TCR] or B cell receptor [BCR]) signal together with a coinhibitory receptor signal or can be induced by the TCR/BCR without any coinhibitory signal, remains controversial; therefore, their combined signal is shown as a negative signal for simplicity. Shown in a faded manner are the cells/signals that need to be absent for tolerance (ie, they would prevent tolerance and lead to activation were they to be present in sufficient quantity). Anergic cells are hyporesponsive to stimulation by antigen together with the activating signals (costimulation and/or T cell help). *Upper left:* Lymphocytes can remain naive due to insufficient encounter with a peripheral self-antigen due to very low expression of the antigen or due to barriers that prevent access of the antigen to lymphoid tissue. *Upper right:* Mature naive T cells encountering antigen in the periphery can enter an anergic state when they do not receive activating signals from other T cells (see quorum sensing Fig. 32.2) and/or costimulation from the APC. The signaling threshold for activation is "tuned" higher in anergic cells. If coinhibitory receptors that induce apoptosis are engaged (eg, Fas), the T cell may become apoptotic. *Lower left:* Mature naive B cells encountering soluble monomeric antigens can induce anergy that is associated with a downregulation of the surface IgM (sIgM) BCR while maintaining surface IgD (sIgD) BCR. However, some soluble monomeric antigens appear to leave the B cells in a naive state and the factors that determine this outcome have not been elucidated. *Lower right:* Engagement of a multivalent antigen (eg, membrane-bound antigen) leads to deletion of mature B cells in the absence of T cell help.

the AIRE transcription factor, and this presentation induces tolerance in CD4 T cells.[272] AIRE-independent tissue-specific antigen expression by other lymph node cell subsets has also been shown to generate tolerance in CD8 T cells.[273,274] Beyond presentation of self-antigens in lymphoid tissue, there is the possibility that the rules are different early in ontogeny. Unlike in adults, during the neonatal period naive T cells are able to traffic into parenchymal tissue, providing an opportunity to recognize the self-antigen directly within the tissue. While such recognition can induce peripheral tolerance in model systems employing a transgenic antigen,[275] there is as yet no evidence that this process is normally required to establish peripheral tolerance. The model transgenic employed was an MHC class I molecule, and the CD8 T cells examined for tolerance in this setting could only directly recognize this MHC, precluding any tolerance that might occur through cross-presentation. In a system where only natural self-antigens were present (ie, no transgenic neo-self-antigen) and T development began at the adult stage, eliminating the stage of neonatal T cell trafficking, tolerance to self-appeared intact.[67]

When an adaptive immune system or T cells develop "from scratch" in an adult animal, it results in a functional immune system that appears substantially tolerant of self (eg, transfer of otherwise syngeneic *Rag* competent HSC to *Rag*$^{-/-}$ recipient mice or by grafting of a thymus into a nude mouse).[67] Similarly, lethal irradiation and allogeneic bone marrow transplantation has long been used to treat patients with leukemia. This is another setting where the immune system develops *de novo* in an adult, and the immune system that develops is usually tolerant of the host. When there is a lack of tolerance, that is GVHD, it is usually the result of already mature donor T cells that had not developed within the host thymus. Interestingly, in one model, tolerance to a transgenic neo-self-antigen could only be established later in life.[276] Together, these findings suggest that the fetal/neonatal period is not absolutely required for generation of self-tolerance and raises the question, *does the fetal/neonatal period contribute anything unique to the establishment of self-tolerance?* While the above findings of tolerance in immune systems that develop in an adult suggest there is no gross deficit in tolerance, there may be subtle deficits in tolerance due to the absence of characteristics unique to the fetal/neonatal period. The hematopoietic precursors that generate T and B cells are known to change with age,[277] from fetal/neonatal to adult precursors,[278-280] resulting in different types of lymphocytes being produced.[279,281-283] There is an increased frequency of T_{reg} CD4 T cells during the fetal period in humans, and these cells may help establish tolerance to self and to noninherited maternal antigens.[279,284-286] In fact, lack of T_{reg} cells during the fetal period leads to multiorgan autoimmunity that develops before birth.[287] There are many other differences in the fetal/neonatal period that may help promote the establishment of self-tolerance. Fetal/neonatal lymphocytes have less diversity in their TCRs and BCRs due to lower expression of terminal deoxynucleotidyl transferase and hence a lack of nontemplated nucleotide additions in CDR3 regions. Neonatal innate cells respond differently to TLR ligands compared to corresponding adult cells, in a manner that may diminish Th1 responses.[288,289] The neonatal period is dominated by recent thymic emigrants, cells that more readily become tolerant of tissue-restricted peripheral antigen[290] and have a greater capacity to become pT_{reg}.[291,292] Perhaps as a result of these propensities, the greater colonization of skin by bacteria-specific T_{reg} cells during the neonatal period allows establishment of a noninflammatory state in response to skin commensal flora.[293] Whether this reflects true tolerance (ie, in all classes) or instead a switch to a noninflammatory class of response, controlled by the T_{reg}, is unclear. The neonatal environment also allows only a diminished lymphopenia-induced T cell activation compared to adult lymphopenic mice (neonatal vs adult *Rag*$^{-/-}$ mice), similar to the diminished response in lymph node–deficient adult lymphopenic mice.[294,295] This suggests that the lymphoid stroma that drives homeostatic activation is immature in neonates, facilitating tolerance induction and prevention of autoimmunity.

The neonatal period also has a special relationship with tolerance because this is the stage when the T cell repertoire is first generated in mice. In addition, the neonatal period has a transient state of lymphopenia until the lymphoid compartment is filled up. Lymphopenia makes induction of tolerance more challenging, as it induces homeostatic activation of lymphocytes. Thus, there is a potential to prevent the establishment of tolerance during the neonatal period that does not exist in the same way in normal adult animals. For example, it has long been known that thymectomy of certain strains of mice at day 3 of life, but not thymectomy of adults, leads to various tissue-specific autoimmune diseases.[191,296] This inability to establish self-tolerance may be attributable to the fact that T_{reg} cells do not appear in the periphery until just after day 3[297]; thus, thymectomy at this time generates a T cell repertoire deficient in T_{reg} cells that is exported into a neonatal lymphopenic peripheral compartment. The AIRE-dependent thymus generated T_{reg}, referred to as tT_{reg}, specifically generated in the neonatal period may thus control the newly generated T cell repertoire in a manner that prevents autoimmunity.[298]

Immune-Privileged Sites

It has been known for a long time that allogeneic skin is not rejected in the usual fashion when grafted to certain sites of the body, such as the brain or anterior chamber of the eye[299] or the testes,[300] so-called "privileged sites." Subsequent studies have implicated many different mechanisms as contributing to the lack of rejection. An early finding was that allogeneic grafts in the anterior chamber of the eye that were not vascularized were not rejected, but vascularized grafts were.[299] This observation led to the idea that sensitization for rejection requires vascularization. However, a state of "tolerance" to allogeneic grafts in the anterior chamber of the eye was found to be associated with a systemic inability to generate DTH to the alloantigens of the graft and the production of noncomplement fixing antibodies. The implication drawn was that the alloantigens of the graft were immunogenic. Indeed, the phenomenon was called anterior chamber-associated immune deviation,[301] with the implication that the phenomenon might be related to the phenomenon uncovered by Asherson and Stone upon immunizing guinea pigs at nonprivileged sites: immunization leading to the production of antibody to an antigen undermined subsequent attempts to induce DTH against the antigen.[302] Such studies led to the suggestion that allografts in privileged sites may lead a deviation of the response into a noninflammatory mode.

Other studies, though not countering such speculation, led to an appreciation of further mechanisms. The expression of FasL by cells can protect them from attack by effector T cells expressing Fas. The interaction between the Fas-bearing effector T cells and target cells bearing FasL results in the death of the effector T cells. The transfer of FasL-bearing Sertoli cells of the testes to nonprivileged sites of allogenic mice were not rejected but similar transplants of Sertoli cells from FasL deficient mice were.[303] FasL is expressed by corneal epithelial and endothelial cells,[304] presumably providing them with protection against effector T cells.

The presence of TGFβ and retinoic acid in ocular fluid has been thought for some time to favor the induction of peripheral T_{reg} (pT_{reg}) cells, based upon the influence of such molecules on the *in vitro* generation of T_{reg} (iT_{reg}) cells.[305] This subset of CD4 T cells is thought to inhibit other CD4 T cell subsets, and so implement a noninflammatory agenda. *In vivo* studies support such an idea.[306] In addition, neuropeptides, in particular vasoactive intestinal peptide, facilitate the establishment of anterior chamber-associated immune deviation.[307] The literature on sites of immune privilege has broadened with time, in the sense that, though the mechanisms of protecting antigens in privileged sites, such as the anterior chamber of the eye, are multifaceted, there is an increasing awareness that these mechanisms may be of more general significance than initially supposed. Some even relate work on immune privilege to how immunity to infections and to tumors is regulated. We make some speculative suggestions in our attempt to integrate the findings described above into a broader context.

The noninflammatory properties of certain antibodies are striking. IgA antibodies are preferentially generated following prolonged mucosal stimulation but, remarkably, are unable to activate diverse effector mechanisms. Human IgG4 antibody molecules not only do not activate effector mechanisms but they also have a remarkable lability in disassociating into two halves and recombining, thus giving rise to molecules that are monovalent for two different antigens. As such, they can block the ability of antibodies from other classes from activating effector functions.[308] *Prolonged* immunization by mucosal and parenteral (nonmucosal) routes can give rise in humans to IgG4 antibody production.[309]

It has been known since the 1950s that immune responses evolve and change their nature with time. Thus, most often, if immunization is chronic, cell-mediated responses evolve from a cell-mediated to humoral mode.[310] Successful desensitization of allergies[311] or the natural out-growing of an allergic state that sometimes occurs[312] involves an evolution of the immune response from an allergic, IgE, to nonallergic, IgA/IgG4 mode. Chronic immunization via the oral route involves sensitization by retinoic acid-producing DCs, giving rise to pT_{reg} cells[313] that, through their production of IL-10 and TGFβ, respectively, facilitate the production of IgG4 and IgA antibodies.[311]

We suggest that many mechanisms contribute to immune privilege, of which we would like to highlight three in our attempt to initiate an integrated picture. Consider an allogeneic graft into the anterior chamber of the eye. From the evidence outlined above, it appears that the ambient environment at the anterior chamber of the eye will tend to skew the response away from an inflammatory, T cell response that initially occurs. Even if such a response occurs to some extent, the presence of FasL on local tissues protects these tissues from such an attack. The local milieu, through the production of such factors as retinoic acid and TGFβ, favors the production of noninflammatory responses. These likely include the generation of pT_{reg} cells and the production of IgA and IgG4 antibodies. The associated T cells and antibodies are not only noninflammatory but also counteract inflammatory T cells and antibodies. It is obvious that it will be important, in testing this proposal, to more carefully assess the class/subclass of any antibodies associated with states of immune privilege. We draw attention to a recent perspective[314] to readers who wish a broader view of immune privilege than that possible here.

Deletion, Intrinsic Unresponsiveness, and Receptor Editing in the Periphery

It is clear that T cells encountering antigen in the periphery can become tolerant of the antigen. What is less clear is which of the potential fates of these cells is the most common and what determines their fate, be it deletion, various forms of unresponsiveness (anergy etc), or receptor editing. While the physical elimination (deletion) of self-specific cells is a major mechanism of tolerance in central lymphoid organs, this is less clear for peripheral tolerance. Very few model systems have provided a clear demonstration of deletion of self-specific lymphocytes in the periphery without substantial caveats. In most model systems where deletion has been observed, it is preceded by activation and proliferation, much like a conventional immune response. Similar to the examination of central tolerance mechanisms, tracking the fate of lymphocytes specific for a peripheral self-antigen was difficult due to the low frequency of cells specific for a given antigen. This difficulty was tackled by employing systems that artificially upped the frequency of these cells, beginning with the use of superantigens and followed by TCR transgenic mice crossed with mice that express the cognate antigen in the periphery or by transfer of the transgenic T cells into an antigen-expressing host. We will cover key insights from several of these model systems. A more detailed listing of the many TCR, BCR, and antigen transgenic models that have been created in the past several decades for the purpose of examining tolerance mechanisms can be found in earlier editions of this textbook.

Using a superantigen model and by blocking I-E dependent deletion in the thymus, Jones et al showed that T cells reactive to this widely distributed antigen could be deleted in the periphery.[315] Similarly, when TCR transgenic female CD8 T cells specific to the male antigen HY were transferred to male recipients, they underwent proliferation and then death. Although the rapid decline in specific T cell numbers might be considered to reflect deletional tolerance, substantial numbers of HY-specific cells remained and were unresponsive to HY.[316] Singh and colleagues found that transfer of monoclonal CD4 T cells into lymphoreplete syngeneic recipients expressing cognate antigen lead to proliferation followed by anergy and a slow deletion that was eventually nearly complete.[317] However, tolerance via an anergic state, without substantial deletion, could occur if a lower number of monoclonal T cells was injected.[318] Thus, deletion in this setting depends on intercellular competition and the findings argue that at normally low physiologic specific T cell frequencies, an anergic-like state is the dominant mechanism. Deletion becomes visible when the self-specific T cell frequency is increased and involves a competition for survival signals,[319] as discussed further below. The above models examined peripheral tolerance to a widely distributed cell-associated antigen. In addition to deletion, peripheral tolerance in CD8 T cells specific to a widely distributed peripheral transgenic self-antigen could be maintained with reduced expression of the TCR and the coreceptor CD8[316,320]; reduced BCR expression in B cells is also associated with their tolerance (discussed below). Downregulation of coreceptors as a potential mechanism involved in peripheral tolerance to widely distributed antigen has also been observed for CD4 T cells.[321]

To address the mechanisms of tolerance to tissue-restricted peripheral self-antigen, Kurts and colleagues employed TCR transgenic CD8 T cells specific to OVA and OVA expression driven by the rat insulin promoter and demonstrated that CD8 T cells encountering OVA as a peripheral self-antigen could be deleted from the repertoire over time.[250] Deletional tolerance was due to cross-presentation of the antigen in the draining lymph node and required Fas expression on the CD8 T cells.[322] However, they also showed that addition of transgenic OVA-specific CD4 T cells could block deletion and result in autoimmunity.[323] Thus, it was not clear whether there would be deletion of CD8 T cells under conditions where there is a physiologic number of peripheral self-antigen specific CD4 and CD8 T cells. Supporting a role for peripheral deletion, monoclonal CD4 T cells specific to a peripheral tissue antigen (HY expressed by an islet graft), that developed within a polyclonal repertoire, demonstrated an antigen-dependent reduction in frequency over time.[324] While studies with monoclonal T cells have the advantage of identical TCR affinity when comparing antigen-expressing hosts versus control hosts, even when mixed with polyclonal cells, the frequency of the antigen-specific cells is abnormally high. Recent elegant studies by Jenkins and colleagues addressed this issue by examining, within a completely normal repertoire, the frequency of CD4 T cells specific to a neo-self-antigen (eGFP) expressed in various locations. Under these conditions the presence of the neo-self-antigen did not appear to cause a substantial level of peripheral deletion or peripheral tolerance; instead, the CD4 T cells appeared ignorant, as assessed by their ability to respond to antigen in CFA.[325] However, the ability to cause deletion is affected by the level of expression of the antigen,[326] deletion can be masked by rapid replacement of specific T cells by recent thymic emigrants,[250] or by antigen-dependent selection within the polyclonal self-specific repertoire favoring survival/expansion of T cells with lower affinity TCR. In addition, whether ignorance or tolerance ensues may depend on the location within the cell that the antigen is expressed; cytosolic expression, as in the studies above using eGFP, may not be sufficiently presented in MHC II to generate deletion of CD4 T cells.[325] Thus, for tissue-restricted peripheral antigens the predominant fate of their cognate T cells remains to be fully resolved. For B cells specific to a peripheral membrane-bound antigen, Nemazee and colleagues showed the outcome for these cells can be deletional tolerance.[327]

When lymphocytes made tolerant by antigen encounter are not physically eliminated from the repertoire, they can become intrinsically unresponsive, a state frequently referred to as anergy. This anergic state was first proposed for B cells although the data supporting it were quite limited at the time.[265,328] However, double-transgenic mice expressing a transgenic BCR specific to a transgenic soluble neoantigen (HEL) demonstrated that B cells could indeed become anergic,[151] and this anergy could be induced not only in immature bone marrow B cells but also in mature B cells in the periphery with an accompanying downregulation of the IgM (but not IgD) BCR.[52] Provision of T cell help did not overcome the anergy. A similar IgM^{low} IgD+ anergic B cell population exists in mice and humans, making up a substantial fraction of the normal B cell repertoire.[329] Despite lowered BCR expression, continued BCR signaling by antigen is required to maintain anergy.[330,331] Anergic B cells may primarily be composed of B cells with polyreactive BCR, cells typically involved in generating an early innate-like response to infection.[332] Peripheral B cell tolerance appears to depend on T_{reg} cells, as patients with mutations in Foxp3 or deficient in T cells generally have increased autoreactive B cells despite intact central B cell tolerance.[333,334] Interestingly, the same self-antigen can generate different fates for different B cell subsets. For example, a highly expressed glycosyl epitope of the T cell antigen Thy1 can promote generation of peritoneal and lamina propria B1 cells and IgM autoantibody specific to the antigen while at the same time causing maturational arrest (deletion) of immature conventional (B2) B cells of the same specificity in the spleen.[104,335] This is consistent with the idea that self-reactivity helps positively select innate like B cells (B1 cells) that are known to have BCRs cross-reactive with epitopes on pathogens, while at the same time preventing development of more dangerous autoreactive conventional B2 cells. B2 cells with low-affinity self-reactivity can also provide a source of foreign reactive cells. Immunization with the foreign antigen efficiently recruits these cells into the germinal center where there is selection for BCR mutations that have greater affinity to the foreign antigen and reduced affinity for self.[336] Foreign reactive B cells can gain antiself specificity during somatic hypermutation in the germinal center. Tolerance in these new antiself B cells depended on the antigen being present in or near the germinal center. The new antiself B cells escaped tolerance when the model self-antigen was localized to a distant tissue,[337] supporting the possibility that autoimmunity may be induced by foreign antigens that stimulate B cells cross-reactive to tissue-specific self-antigens.

Similar to B cell anergy, T cell anergy represents a state of unresponsiveness intrinsic to the T cell. T cell anergy was first described by Jenkins and Schwartz, when they showed that antigen presentation by chemically treated APC did not stimulate proliferation or IL-2 production in CD4 T cell clones and instead induced an unresponsive state.[338] This anergic state could be triggered by peptide/MHC complexes alone in a planar membrane without an APC,[339] and it could be blocked by addition of APC that were unable themselves to present the antigen but retained the capacity to generate the nonspecific second signal.[340] While certain cytokines, such as IL-1, can strongly promote T cell activation,[341] the costimulatory second signal needed for IL-2 production was shown to be mediated by engagement of cell surface receptors and ligands between cells; namely, CD28 engagement on the T cell with CD80/CD86, also known as B7.1 and B7.2, on the APC.[342] While many other costimulatory receptors have been identified and are discussed in other chapters, the CD28/CD80/CD86 and CD40/CD40L receptor-ligand pairs appear the most critical in promoting immunity over tolerance. Anergy generated *in vitro* as a result of the lack of costimulation, however, is not simply an unresponsive state. It was later shown that such anergized cells are in a more complex state, retaining the ability to produce certain cytokines and chemokines[343]; thus, anergy *in vitro* did not grossly compromise effector functions. Anergy *in vitro* was readily achieved with CD4 T cells that had been previously activated, such as T cells clones, but not with primary naive T cells.[344] This contrasts with the unresponsive state that could be generated in primary naive T cells *in vivo*, as discussed in the next section. Anergic CD4 T cells that are hyporesponsive to stimulation, in terms of proliferation and cytokine production (including IL-2, IFNγ, IL-17a, IL-10, and IL-21), were phenotypically defined as Foxp3 negative cells having high expression of CD44, CD73, Fr4, CD69, PD-1, CTLA-4, CD5, and Nrp1. Many of these markers suggested the possibility these cells are continually encountering self-antigen and high expression of Nur77 (a transcription factor induced by TCR signals) in these cells, when using Nur77 reporter mice, further supported this idea. These *in vivo* anergic cells appear to be 2% to 5% of the normal CD4 T cell repertoire *in vivo*. As would be expected for self-reactive cells, the frequency of the anergic cells was found to be increased in AIRE$^{-/-}$ mice where there is a deficiency in thymic-negative selection.[345] The anergic CD4 T cells in these studies were found to have a propensity to become pT_{reg} (Foxp3+). Since recent thymic emigrants also have a propensity to become pT_{reg},[291,292] it will be of interest to determine if the anergic cells are part of the recent thymic emigrant pool.

While the role of receptor editing in thymic central tolerance is controversial, there is also some evidence that under special conditions, peripheral tolerance in CD4 T cells can involve receptor editing. In mice with an endogenous superantigen that binds Vβ5 as well as a Vβ5 containing TCR transgene, the chronic exposure to the superantigen leads to reexpression of *Rag* genes and expression of an endogenous TCRβ chain.[346] This peripheral receptor editing is associated with cells that take on a Th17 phenotype.[347]

Antigen Dose, T Cell Tuning, and Competition

Early studies examined tolerance in different "zones" of antigen, referred to as high- and low-zone tolerance. Mitchison examined in the mid-1960s the effect of administering to immunologically mature mice a weak antigen, BSA, several times a week for several weeks, on a subsequent challenge of BSA that induced the production of antibody in naïve mice. He preexposed different mice to different amounts of antigen and covered a wide range of doses. He demonstrated that mice preexposed to medium doses made a heightened antibody response; in contrast, mice preexposed to lower and higher doses made diminished responses on challenge.[348] Dresser and Mitchison interpreted their observations as reflecting states of (peripheral) tolerance.[15] However, in similar experiments, Parish measured the state of DTH to the antigen and found that the rats preexposed to low and high doses of the antigen, but not to medium doses, expressed DTH to the antigen in question.[349] Parish interpreted these data as demonstrating the existence of states of low-zone and high-zone cell-mediated immune deviation. Mitchison's and Parish's observations are very important, in part because they are quantitative and because, taken together, they identify a major difficulty in assessing tolerance. The findings highlight the importance of examining the unresponsiveness for different classes of immunity before identifying an experimentally established unresponsive state as an appropriate model of peripheral tolerance.

A specific clone of T cells can be made tolerant of a self-antigen and remain susceptible to activation by other antigens. For example, CD8 TCR transgenic cells capable of being stimulated by a superantigen *Mls^a* become tolerant of *Mls^a* if they grow up in an *Mls^a* mouse. Nevertheless, these T cells remain fully responsive to both an LCMV peptide or a bacterial superantigen.[88] Similarly, tolerance can be induced to a partial agonist peptide without compromising the ability to respond to the full agonist.[89] Deprivation of self-peptide/MHC signals can lead to a reduced response to agonist ligands.[350] These findings and others suggested that tuning generates increased sensitivity to changes in peptide sequence rather than raising the threshold for responsiveness to precisely the same peptide that mediated the tuning[49]; that is, once tuned to a particular pMHC, the cell could remain tolerant of it even if the pMHC presentation were increased. However, deprivation of self-peptide/MHC signals can also lead to a loss of self-tolerance.[351,352] In addition, as mentioned previously, there are data indicating that tuning sets a threshold for a particular peptide-MHC and that threshold can be breached leading to activation. For example, transfer of monoclonal CD4 transgenic T cells into a lymphopenic (T cell deficient) host that expresses systemically the cognate antigen for those T cells leads to T cell activation/proliferation followed by a tuning that raises the threshold for activation.[101,353] The tolerance that developed was incomplete in the lymphopenic recipient as the T cells retained some capacity to provide help to B cells that generate autoantibodies, causing a mild arthritis. The tolerance was more profound when the monoclonal T cells were transferred to a T cell replete, antigen-expressing recipient,[317] with both an intrinsic decrease in response to TCR signals and a slow deletion of the antigen-specific cells that occurred due to competition for a nonagonist self-peptide/MHC ligand.[318] In the lymphopenic recipient, as the T cells increased in numbers, there was increased competition for antigen recognition and hence the equivalent of a decrease in antigen dose. Despite the continued presence of the agonist antigen, T cell proliferation ceased with high T cell numbers and so plateaued. Retransfer of the T cells to a new lymphopenic antigen-expressing host exposed the T cells to a higher level of antigen, and they again proliferated but to a lower degree, and thus reached a lower T cell plateau. Thus, the T cell plateau in the secondary recipients was lower than the plateau in primary recipients, and the T cells in secondary recipients were therefore exposed to a higher level of antigen, leading to tuning to an even higher threshold for activation (ie, a deeper state of tolerance).[102] This tuning-induced increase in activation threshold that can be breached by increased antigenic signals was directly demonstrated for transgenic B cells specific to HEL. Anergic anti-HEL B cells from soluble HEL transgenic mice, unlike naive B cells, were unresponsive to HEL together with T cell help except when given HEL as a membrane bound multimeric form.[354] These results can be interpreted as chronic-soluble antigen tuning the BCR to have a signaling threshold for activation that can be reached by multimeric but not soluble antigen. Thus, competition and its effect on tuning are also likely to be important in B cell tolerance. The shortened life-span of anergic B cells involves competition in the periphery with other B cells for survival factors (eg, BAFF/BLyS) and exclusion from follicles.[104] The tuning of activation thresholds can be affected by the local environment. For example, gut commensal flora can alter the threshold of CD4 T cells tuned to a model systemic self-antigen, promoting an autoimmune response.[355]

T_{reg} Cells and APCs that Generate Peripheral Tolerance

Next, we provide a brief description of APCs and regulatory T cells that suppress immune responses. More detailed coverage of such regulatory mechanisms can be found in Chapter 29.

Although it had been known for a long time that some T cells were able to suppress the response of other lymphocytes, the inability to identify markers specific to such suppressive/regulatory T cells hampered progress. Only with the discovery that the high-affinity receptor for IL-2, CD25, was expressed constitutively on CD4 cells with regulatory capacity,[197,356] did progress become more rapid. However, CD25 is also transiently expressed on most T cells following activation, and therefore it did not represent a perfect marker to define T_{reg} cells. Nevertheless, selection for CD4 T cells expressing CD25 provides a means of obtaining a population that in most cases has overall suppressive capacity, while conversely depletion of cells expressing CD25 reveals the autoimmune potential of a polyclonal T cell population from otherwise healthy mice. The most well-characterized CD4 T_{reg} cells express the transcription factor Foxp3,[202] and, at least in mice, the expression of Foxp3 is relatively limited to this subset although not exclusively. In humans, only cells with very high levels of Foxp3 provide some confidence that the cell is a T_{reg} cell, as activated human T cells can transiently express Foxp3.[357] Although a better marker of T_{reg} cells than CD25, Foxp3 is intracellular and therefore could not be used to sort live T_{reg} cells until fluorescent Foxp3 reporter mice were developed (ie, expressing GFP driven by the Foxp3 promoter).

CD4 T_{reg} cells come in a number of varieties. Some of the Foxp3 expressing CD4 T_{reg} develop as T_{reg} in the *thymus*, the tT_{reg} (note: initially they were referred to as natural T_{reg}, nT_{reg}), while conventional CD4 T cells that become Foxp3 expressing T_{reg} after export from the thymus into the *periphery* are known as pT_{reg}. Both tT_{reg} and pT_{reg} are needed to prevent autoimmune disease.[358] When Foxp3 T_{reg} are induced *in vitro* from conventional T cells, they are frequently referred to as iT_{reg}. In addition to these Foxp3 expressing CD4 T_{reg}, there are other CD4 T cells that have regulatory/suppressive capacity and do not constitutively express Foxp3. Tr1 cells are a CD4 T cell subset that expresses IL-10, a cytokine that in many (but not all) settings suppresses the response of other T cells.[359] For example, IL-10–producing Tr1 cells are involved in dampening DTH, Th1, and Th2 responses to bee venom in nonallergic beekeepers continually exposed to high doses of venom and associated with elevated IgG4 antibody specific to the venom.[360] IgG4 is a subclass of antibody that is able to neutralize an antigen without triggering other inflammatory effector mechanisms (eg, complement activation or FcR-driven mechanisms). Another subset of CD4 T cells, sometimes referred to as Th3 cells,[361] express TGFβ but not Foxp3 and are involved in a process known as oral tolerance, where exposure to antigens via the intestinal mucosa generates a Th3 response that suppresses T cells that mediate delayed-type hypersensitivity Th1 responses. While TGFβ, like IL-10, is suppressive of a number of types of immune responses, it is a cytokine that promotes IgA production. The generation of Foxp3+ pT_{reg} is needed to maintain tolerance/homeostasis to certain bacterial flora, and this depends on the appropriate timing of exposure to these flora antigens early in life.[362] Although little is known about the specificity of T_{reg} cells, these pT_{reg} cells that develop primarily in mesenteric lymph nodes include within their repertoire TCR specific to commensal flora.[363] Foxp3 T_{reg}-mediated tolerance of our own flora and food antigens does not appear to be true tolerance but instead promotion of secretory IgA in the gut lumen, as well as IgG4 responses to these antigens, while dampening other classes of response and limiting systemic responses. The IgA responses are in part dependent on Foxp3+ pT_{reg} cells that transition into follicular helper T cells.[364] The T_{reg}-dependent IgA response not only keeps flora from penetrating the mucus layer, it is also needed to maintain a diversity of flora. Immunodeficient mice have reduced diversity of flora that can be restored by transfer of T_{reg} cells.[365] There is a complex network of cells controlling these responses, including the Th3 cells and specific DC populations that promote pT_{reg} (Foxp3+) generation.[366,367] Distinct populations of DCs in the gut-associated lymphoid tissue express or generate factors that promote T_{reg} cell generation, including IL-10, TGFβ, indoleamine 2,3-dioxygenase (catabolizes tryptophan), and retinoic acid. B cells producing IgG4 are also able to produce regulatory cytokines, and evidence for their role in oral tolerance is being uncovered.[367] This complex network of interactions is also critical to prevent allergic responses (eg, IgE) to dietary antigens and instead promote a healthy IgA response to these antigens. Mice fed a chemically defined (antigen free) diet have substantially reduced IgA in serum and the small intestine.[368]

Given the ability of costimulatory signals provided by DC APCs to promote immune responses, it was initially thought that DCs were incapable of tolerogenic antigen presentation. A landmark study by Finkelman and colleagues demonstrated that antigen targeted specifically to DCs *in vivo* could in fact induce tolerance.[369] They injected into mice a rat IgG specific to a surface receptor on some conventional DCs and showed the mice became tolerant to the rat IgG at both the T and B cell level. This tolerance could be converted to immunity by various means, such as provision of an inflammatory stimulus in the form of IL-1 or by aggregation of the anti-DC antibody. Stimulation of CD40 was also later shown to prevent this tolerance and instead triggered immunity. For T cells responding to tolerogenic DC antigen presentation, all anticipated fates have been observed, including deletion, anergy, generation of pT_{reg}, and expansion of tT_{reg}.[370] Given the prevailing concept that peripheral tolerance would be induced by cells that lacked costimulatory activity, tolerance to self-antigens expressed by tissue parenchymal cells was expected to be induced by antigen presentation on the parenchymal cells themselves. However, studies employing transgenic tissue-specific neo-self-antigens showed peripheral tolerance was instead induced in CD4 and CD8 T cells by antigen processed by bone marrow APCs.[250,371] Later studies have shown that approaches that ablate DC function can block tolerance in various settings and can lead to autoimmune disease.[372] Tolerogenic presentation by DC can involve engagement of coinhibitory receptors and ligands (eg, DC/T cell: PD-L1/PD-1, BTLA/HVEM and B7/CTLA-4). While DC2s and plasmacytoid DCs can also be tolerogenic in certain situations, of the various DC subsets, DC1 (CD8α+ cDC) that reside in or migrate to lymphoid organs and take up antigens from apoptotic cells, via CD36, appear to play the greatest role in peripheral tolerance under homeostatic conditions.[373] These are the same DCs that are known to be potent in cross-priming of CD8 T cell responses. Thus, DC1 are important both in cross-priming and in cross-tolerance, that is, the presentation of exogenous antigen in MHC class I for immunity and tolerance, respectively. *Eventually, a simple concept emerged suggesting that immature DCs induce tolerance while mature* (eg, *TLR activated) DCs induce immunity.* Immature DC could migrate from parenchymal tissues carrying tissue antigens to the draining lymph node to induce tolerance under steady stage conditions.[370] However, many diverse factors in the local tissue environment can promote the tolerogenic function of DC1 cells, including stimulating IL-10 production, perhaps most surprisingly certain TLR ligands have this effect.[372] Thus, while a number of factors often described as "inflammatory" can block tolerance induced by DC antigen presentation, the stimuli that will create tolerogenic versus immunogenic DCs cannot be easily categorized. This raises questions about the validity of the immature versus mature DC concept in determining tolerance versus immunity. The malleability of DC presentation also raises the question of whether tolerogenic self-antigen presentation in the periphery always has the potential to be in the dangerous situation of being blocked in the context of inflammatory stimulation. This danger might be reduced by a bone marrow–derived population of APC in lymph nodes that expresses AIRE-driven antigens as it appears resistant to inflammatory stimuli. These APC express MHC class II but have low B7 expression that remains low even in the presence of TLR3 ligand poly I:C or anti-CD40 stimulation.[272]

Coinhibitory Signals for Peripheral Tolerance

The role of coinhibitory receptors as "breaks" on the immune system has recently become more highly appreciated because these receptors and their ligands are the targets of checkpoint blockade antibodies employed in cancer immunotherapy. The first coinhibitory receptor identified was FcγRIIb on B cells, and this receptor was shown to inhibit B cell activation when the BCR was coligated with the FcγRIIb via an IgG containing immune complex.[44,45,374-378] Thus, FcγRIIb is involved in the process known as antibody-mediated feedback suppression and depends on a ratio of IgG to antigen that leaves some epitopes of the immune complex free to engage the BCR; excess IgG simply leads to antigen masking and not coligation of the BCR and FcγRIIb. Coinhibition via FcγRIIb provides a quorum sensing mechanism for B cells, controlling the number of responding B cells[379] as well as negatively regulating later stages in the generation of IgG autoantibodies.[380] B cell tolerance can involve coinhibitory signals from a number of different coinhibitory receptors (many containing ITIM sequences) including, but not limited to, FcγRIIb, CD22 (Siglec-2),[381,382] Siglec-G (Siglec-10 in humans),[383,384] CD72,[385,386] PIR-B (LILRB1 in humans),[387,388] BTLA (CD272),[389] CD31 (PECAM-1), CD66a (CEACAM-1), LAG-3, and Fas (CD95). Genetic knockouts of many of these coinhibitors lead either to spontaneous autoimmune disease, systemic lupus erythematosus (SLE), or enhanced SLE in the context of other genetic factors (eg, $Fas^{lpr/lpr}$). CD72 is of particular importance, as it binds an RNA-containing self-antigen, SmRNP, a target autoantigen of antibodies in SLE and an agonist of a critical TLR in lupus, TLR7. Substantial spontaneous SLE occurs in CD72 deficient mice and a hypomorphic allele of CD72 contributes to SLE in MRL-lpr mice.[385,390] Thus, there appears to be a highly specific role for CD72 in coinhibiting the response to SmRNP but not to other antigens. Coinhibitory ligands can be expressed by many different cells (some predominately on hematopoietic cells, others also expressed on tissue "parenchymal" cells) and can be cell tethered sugars (eg, α2,6 sialic acid for CD22) and glycoproteins (eg, HVEM for BTLA and MHC class I for PIR-B), or soluble, such as cytokines (eg, TGFβ). B-1 cells express high levels of α2,3 sialic acids, including on their BCR, and Siglec-G in turn binds these sialic acids and regulates BCR signals in these cells.[391] The diverse regulation of coinhibitory receptors and their ligands is complex, and specific roles for each coinhibitory pathway are still being elucidated. The complexity is increased by the fact that, unlike the first coinhibitor pathway identified, IgG binding to FcγRIIb, many coinhibitors appear to have multiple ligands that are quite distinct.[392] The biology of coinhibitors is also complex because some coinhibitors appear to play a positive role in immune responses under some conditions. Thus, some coinhibitors may function more in the regulation of the type of response (ie, immune class regulation) than in generation of immune tolerance. Given the wide distribution of expression of many coinhibitors, cell subset specific knockout studies are needed to more fully examine the role of each coinhibitor in T cells versus B cells versus innate cells. In addition, many coinhibitors are expressed by both conventional T cells and T_{reg} cells, making it important to distinguish whether functional outcomes are due to coinhibitor function in suppressive cells or effector cells.

Coinhibitors in the Establishment of Tolerance

The B cell coinhibitors discussed above are constitutively and widely expressed on B cells. CD5 is also constitutively expressed by the B1a B cell subset, while it is inducible on conventional B2 cells and is needed for the process of anergy to soluble self-antigen.[393] In contrast to B cells, many of the coinhibitors on T cells are not expressed on naive T cells but are induced upon activation. CTLA-4 was the first coinhibitory receptor to be identified on T cells and may have a role at several stages of the immune response. T cell recognition of cognate antigen presented in lymphoid tissue involves at least three phases, a first phase, over about 8 hours, of short-lived interactions with DCs, a second phase over the next 12 hours of much longer-lived interactions, and finally a return to short-lived interactions beginning 1 day after initial antigen encounter.[394] Tolerogenic presentation, such as when CTLA-4 coinhibition occurs, prevents activated T cells from exiting the first phase of stimulation.[395] There is also an early role for CTLA-4 in terms of the ontogeny of tolerance. A genetic deficiency in CTLA-4 results in multiorgan autoimmunity that is lethal within a few weeks after birth[396] and the disease is CD28 dependent[397]; whether disease is quorum dependent has not been examined. The ensuing autoimmunity involves T cells reactive to tissue-specific self-antigens and CTLA-4 function is needed in conventional T cells to prevent this autoimmunity.[398] CTLA-4's function in T_{reg} may be more complex, promoting or inhibiting their function in different settings.[399] Thus, although not constitutively present, CTLA-4's induced expression from storage in intracellular vesicles early after antigen engagement allows it to control responses quite rapidly. The mechanism of inhibition of T cell responses by CTLA-4 is complex, with evidence for altered TCR mediated signals, competition for ligands with CD28, including stripping of its ligands, CD80/CD86, from APCs; the latter mechanism reduces the ability of APCs to activate T cells.[400,401]

While there has been substantial investigation of the role of various other coinhibitors controlling the CD8 T cell response to antigens that are presented chronically, there are a number of studies that suggest that CTLA-4 is not the only coinhibitor important early on in the establishment of tolerance. For example, the inducible expression of PD-1 (CD279) can be important in establishing tolerance in naive CD4[294,324] and CD8 T cells[402,403]; PD-1 is also transiently upregulated on activated B cells.[404,405] B6 mice deficient in PD-1 develop a late life lupus-like autoimmune disease.[406] However, the most severe autoimmune phenotype caused by PD-1 deficiency is in the setting of newly generated T cells lacking PD-1 migrating from the thymus into a lymphopenic periphery, where they cause a rapidly lethal MHC class II dependent multiorgan autoimmunity.[294,324] Together with the observation that PD-1 is highly expressed on normal CD4 cells naturally undergoing homeostatic proliferation during the neonatal period,[324] the data suggest that PD-1 is important during the establishment of tolerance in recent thymic emigrants when the T cell repertoire is first generated.[295] Most of the tolerance promoting effects of PD-1 identified involve its interaction with PD-L1 (CD274) rather than PD-L2 (CD273), and PD-L1 on parenchymal cells in some tissues (eg, islets) can contribute to tolerance.[407] Furthermore, both T cells and APCs can express PD-L1 and CD80, and PD-L1 binds CD80 in trans and in cis. Cis interaction on APCs reduces interaction of PD-L1 with PD-1 on the T cell although the magnitude of the impact of this interaction is not yet clear.[408] While trans interaction, can, in some settings, promote T cell proliferation and PD-1 expression, it leads to enhanced T cell apoptosis that is PD-1 dependent.[409,410] Overall PD-L1's effects are primarily tolerance promoting. The effect of PD-1 across the repertoire of responding cells is not equal. Instead, PD-1 has a prominent role in limiting epitope spreading during CD8 T cells responses by controlling subdominant clones.[411] While PD-1 is upregulated after activation, several other receptors with coinhibitory function are already expressed on naive T cells and therefore may also play a role very early in the T cell response, including BTLA (CD272), CD5, TGFβR, and VISTA. TGFβR expression on T cells, like CTLA-4, appears critical in the early establishment of tolerance. Mice deficient in TGFβ1 or specifically deficient in TGFβRII in T cells also die within several weeks of birth from a generalized lymphoproliferative autoimmune disease.[412,413] BTLA coinhibition is complex, as its ligand, HVEM, which itself is an activating receptor with several other ligands (LIGHT, LTα, CD160), is coexpressed on T cells with BTLA. This allows interaction between BTLA and HVEM in cis (ie, on the same cell), blocking activation of HVEM via its other ligands in trans (ie, on other cells)[389,414,415] explaining why T cells are not constitutively activated via their HVEM expression. BTLA was found to be involved in T cell tolerance to high-dose soluble antigen injected i.v., and tolerance of transgenic neo-self-antigen in pancreatic islets.[416] Conversely, BTLA expression by DCs can have a positive role, causing trans signaling through HVEM on T cells, promoting their survival in a number of settings.[417,418] Thus, BTLA provides an example of evolutionary forces driving multiple roles for a single gene product. Despite its high expression on B cells, BTLA's role in B cells has not yet been examined in detail.

Coinhibitors in Chronic Antigen Stimulation

Many coinhibitors are transiently upregulated upon acute antigen-mediated stimulation. In contrast, some of these normally transient coinhibitors can also become constitutively upregulated on T cells in states of chronic TCR signaling, such as with chronic viral infection (eg, HIV, LCMV) and some tumor antigens. These "exhaustion" related coinhibitors include PD-1, TIGIT, LAG-3, TIM-3, and others.[85,392,419,420]

Terminology in this area can be a source of confusion. Although the term exhaustion is commonly used to refer to hyporesponsive T cells chronically stimulated by antigen, this is often a misnomer, as in many cases these cells can have responsiveness restored by coinhibitor blockade. Furthermore, addition of other activating signals (eg, T cell help) does not promote further exhaustion, contrary to what one would expect based on the term exhaustion but instead can help reverse exhaustion.[421] However, chronic antigen signaling can eventually lead to epigenetic changes and an hyporesponsiveness that is not reversed by coinhibitor blockade.[422] Thus, it is increasingly being uncovered that "exhaustion" involves a heterogenous state of T cells encountering chronic antigen. This may be a continuum of states, involving altered transcription factors and epigenetic changes, from a selective and reversible loss of some effector functions (particular cytokines) in self renewing cells to a terminal differentiation involving an irreversible loss of nearly all effector function and proliferation.[423] The inability of PD-L1 blockade to reverse the epigenetic genetic changes further suggests that its capacity to temporarily "reverse" exhaustion relates only to a subset of the heterogenous population of "exhausted" cells.[424] These states are likely to be given distinct names in the future. Exhaustion arises after initiation of a conventional immune response (ie, the signals needed for a response are present) distinguishing it from anergy, where needed signals are absent at the beginning of the antigen challenge. Many coinhibitors may be coregulated on CD4 and CD8 T cells by IL-27 in these chronic antigen states and can depend on T_{reg} generated IL-35, together with IL-10 production; IL-27R–deficient T cells did not have upregulated coinhibitors.[425,426] Interestingly, the IL-35 that upregulates coinhibitors on conventional T cells can be transmitted to them through exosomes from a small subset of T_{reg} cells.[427] PD-1 ligands, PD-L1 and PD-L2, are respectively expressed widely and restricted to hematopoietic cells and can be upregulated by inflammatory signals.[392,428] TIM-3 expressed on Th1, Tc1, and T_{reg} cells may have several ligands, with the soluble lectin galectin-9 playing a prominent role.[429,430] LAG-3 (CD223) coinhibits T cell activation.[431] It is upregulated on T cells postactivation; it is present on a subset of T_{reg} cells and, like CD4, can bind to MHC class II.[432] However, LAG-3 is a negative regulator of both CD4 and CD8 T cells, suggesting another ligand may play a more prominent role in LAG-3 mediated coinhibition.[433,434] Fibrinogen-like protein 1 (FGL1), a soluble protein in circulation made primarily in the liver and overexpressed in some human cancers was identified as a ligand for LAG-3 capable of mediating coinhibition.[435] Metalloproteases cleave LAG-3, releasing T cells from inhibition.[436] TIGIT is expressed upon activation and has similarities to CTLA-4, in that, it competes with a costimulator, in this case CD266, for binding of their ligands, PVR (CD155) and PVRL2 (CD112). While TIGIT deficiency by itself does not cause spontaneous autoimmunity, blocking many of the above coinhibitors, or their ligands, can have substantial effects in restoring T responses, and polymorphisms or engineered mutations in these signaling pathways are associated with the development or increased severity of many autoimmune diseases.[399,430] A lack of overt spontaneous autoimmune disease in a given coinhibitor knockout animal may be due to compensatory upregulation of other coinhibitors.[437] Deficiencies in multiple coinhibitors can synergize and generate severe autoimmune disease; for example, LAG-3 and PD-1 deficiency.[438] Mutations causing Tim-3 deficiency in humans are associated with development of an autoinflammatory condition (including autoantibodies), subcutaneous panniculitis-like T cell lymphoma, and haemophagocytic lymphohistiocytosis.[439]

CD95 (Fas) can cause apoptosis in various cell types; within T cells, it only induces apoptosis under a limited set of conditions. These conditions, including simultaneous or recent TCR engagement, particularly chronic engagement, allow CD95 to act as a coinhibitory receptor. Although low level CD95 (Fas)/CD178 (Fas-L) signaling can promote initial T cell activation/proliferation (acting as a costimulator), it is thought to have a prominent role in termination of immune responses, in a process named activation-induced cell death (AICD). CD95 is relatively ubiquitously expressed while its ligand is present on activated T cells and NK cells and on some parenchymal tissue cells within "immune-privileged" sites as well as some tumor cells.[440,441] The membrane but not soluble form of the ligand induces apoptosis in T cells,[442] and a lack of CD95-mediated apoptosis results in increased CD4/CD8 double-negative T cells. This CD95 mediated termination of T cell survival involves IL-2 and depends on recent, if not simultaneous, TCR signaling. The importance of a CD95 mediated coinhibition is strongly suggested by the very similar lymphoproliferative and autoimmune diseases (eg, lupus, arthritis, and autoimmune lymphoproliferation syndrome) seen with mutations of CD95 or its ligand.[441,443,444] Consistent with a role for IL-2 in AICD, IL-2–deficient mice display some autoimmune-like manifestations (primarily colitis); however, the disease is quite different from the autoimmune diseases seen with mutations in CD95/CD178. Thus, autoimmunity due to IL-2 deficiency may relate more to its role in control of T_{reg} stability (see Chapter 29). CD95 is also expressed on germinal center B cells, and B cell specific loss of CD95 leads to increased autoantibodies and IgE.[445,446] B cells receiving acute BCR signals, as occurs initially upon infection, are resistant to CD95-mediated apoptosis. In contrast, chronic BCR signaling makes B cells susceptible to CD95-dependent killing by the CD4 T cells recognizing the antigen the B cell takes in via the BCR and presents on MHC class II.[447] Deletion of CD8 T cells encountering a peripheral "neo-self-antigen" was also found to be CD95 dependent.[322]

Coinhibition and T_{reg} Cells

As mentioned above, many coinhibitors, including CTLA-4, LAG-3, TIGIT, TIM-3, and PD-1, are expressed in various regulatory T cells populations and in some cases have been shown to be involved in their function or generation. In addition, some coinhibitory ligands are expressed by T_{reg}; for example, HVEM and PD-L1. Although CTLA-4 limits T_{reg} numbers, it has an important function in T_{reg} cells,[206] and LAG-3 expression is associated with IL-10 producing Tr1 cells and is expressed on tT_{reg} cells.[432] TIGIT is involved in generation of iT_{reg} cells. T_{reg} cells increase their HVEM expression postactivation, and their suppressive capacity in part depends on BTLA expression on the T cell that is being suppressed.[448] pT_{reg} generation is facilitated by antigen presented by the DEC205+CD8+BTLA+ DC subset in a manner that was partially dependent on BTLA/HVEM.[449] However, so far only CTLA-4 deficiency in T_{reg} causes a severe spontaneous autoimmune phenotype, suggesting that the functional requirements for these other coinhibitors (or their ligands) in T_{reg} cells is much more limited. Furthermore, some coinhibitors limit rather than promote the generation of T_{reg} cells, as is the case for PD-1 in pT_{reg} generation.[292]

Summary and Open Questions on the Role of Coinhibitors

Overall, coinhibitory signals have an important role in negatively regulating immune responses at almost all stages, if not all stages, of peripheral lymphocyte encounter with antigen. The growing list of coinhibitory receptors raises the question of the distinct functions of each coinhibitor. Differential tissue expression of coinhibitory ligands and differential timing and duration of expression of coinhibitors suggest unique selection pressures for specific coinhibitors to act in unique sites and at unique times during antigen encounter (eg, the role of Fas/FasL in immune privilege of certain tissues). Many critical questions remain unanswered; for example, does blockade of coinhibitory signals reduce the requirement for quorum sensing (T cell help) in immune responses[24,374,450]? Are the actions of a particular coinhibitory pathway on immune system cells independent of or instead linked to their role in the nervous system[451,452]? Is the purpose of some coinhibitors to regulate the class of immune response rather than a general inhibition of the immune response? Do specific coinhibitors establish the tuning of the activation threshold through TCR/BCR and can a change in coinhibitor expression reset the threshold over time? In addition, while some coinhibitors are known to modulate thymic-positive selection, they do not appear to have a major effect on thymic-negative selection to ubiquitously expressed antigens. Whether coinhibitors affect central tolerance to tissue-specific antigens has not been investigated.

Intracellular Signals in Tolerance

We focus here on intracellular signals controlling the development of tolerance; signaling is covered in greater detail in Chapter 24. Strong TCR/BCR signaling is a critical regulator of tolerance both in developing lymphocytes in central lymphoid organs as well as in mature peripheral lymphocytes. In addition, ligation of other cell surface receptors can also modulate the tolerance versus immunity decision by altering the nature of the intracellular signals received by the lymphocyte. Protein tyrosine kinases (PTKs) and protein tyrosine phosphatases (PTPs) are key enzymes that translate the signals that emanate from cell surface receptors into the physiological response of the lymphocyte. As such, it is not surprising that mutations in PTKs and PTPs influence tolerance. Mutations in enzymes involved in the most membrane proximal signal transduction events result in a loss of tolerance. Global deficiency of Lyn, an important Src-family kinase in B cells, leads to a lupus-like autoimmunity despite increasing developmental arrest of HEL-specific B cells in the bone marrow of mice expressing sHEL as a self-antigen.[381] Deficiency in the PTP SHP-1 results in a similar outcome. B cell–specific deficiency of Lyn leads to autoantibody production and a lupus-like autoimmunity pointing to a cell-intrinsic role for Lyn in B cell tolerance.[453] Mechanistically, Lyn-deficiency is linked to reduced phosphorylation of the ITIM-containing and sialic acid–binding protein CD22 and subsequent impaired recruitment of SHP-1 to CD22.[381] Similarly, in T cells, phosphorylation of the coinhibitory receptor PD-1 on a tyrosine residue with an immunotyrosine-based switch motif (ITSM), likely by the Src-family kinase Lck, recruits SHP-2 to dampen TCR and CD28 signaling.[454-456] In addition to the PTPs mentioned above, there are many other PTPs that play a critical role in T and B cell tolerance. SNPs in PTPN22 (also known as "Lyp") are correlated with the development of many different autoimmune diseases in humans (for review see[457]). PTPN22 deficiency results in increased proximal TCR/BCR signals. Mechanistically, PTPN22 is reported to bind to the Src-family kinase Csk and the Csk-PTPN22 complex negatively regulates proximal TCR/BCR signaling.[458,459] Precisely, how the different PTPN22 SNPs predispose individuals to autoimmunity is currently unclear and further investigation is needed.[457] Collectively, these findings suggest that proximal signaling molecules that regulate signal strength downstream of antigen receptors are important for the establishment of tolerance.

Phosphoinositides are lipid molecules found in membranes that are also involved in signaling and impact tolerance. The phosphoinositide-3-kinase family (PI3Ks) and the inositol phosphatases, PTEN and SHIP-1, control the phosphorylation status of phosphoinositides. PI3Ks phosphorylate PI(4,5)P2 to create PI(3,4,5)P3 which is an important docking site for many signaling proteins that contain PH domains, including AKT, leading to their activation. PTEN directly counters the activity of PI3K while SHIP-1 dephosphorylates PI(3,4,5)P3 to create PI(3,4)P2. When unligated, basal BCR signaling results in activation of PI3K, subsequent activation of AKT, phosphorylation and nuclear exclusion of Foxo1, and decreased Rag expression.[460,461] However, when ligated by self-Ag, the BCR is internalized, PI3K activity is reduced leading to increased Rag expression and receptor editing in immature B cells.[461,462] If receptor editing does not result in the production of a non–self-reactive BCR, clonal deletion through the Bim pathway results.[463,464] In thymocytes, Bim induction in response to strong TCR signaling similarly induces clonal deletion[465] although clonal deletion can occur independently of Bim.[466] Additionally, Bim induction in thymocytes has been reported to require CD28 costimulatory signals.[467] In addition to regulating receptor editing, control of the PI3K pathway is also important in maintaining anergy in mature B cells. Using B cell specific, conditional deletion approaches, Getahun et al. showed that deletion of SHIP-1 or PTEN abrogated established B cell anergy.[468] Consistent with these findings, conditional expression of a constitutively active PI3K p110 subunit in anergic B cells broke anergy. Finally, Getahun also demonstrated that the SHIP-1 and SHP-1 operate in parallel pathways to maintain B cell anergy.[468]

In addition to membrane proximal signal transduction pathways being regulated by PTK and PTP, ubiquitylation of signaling proteins is another means by which intracellular signals essential for tolerance can be controlled. In the context of T cell and B cell tolerance, the Casitas B Lineage lymphoma (Cbl) family of ubiquitin ligases, and more specifically c-Cbl and Cbl-b, have been shown to be important. Cbl-b–deficient mice show evidence of autoimmunity including elevated anti-dsDNA IgG antibodies, and T and B cell infiltration of various organs including the salivary gland and pancreas.[469] Lymph node T cells from Cbl-b–deficient mice could proliferate in a syngeneic mixed lymphocyte reaction and B cells showed increased proliferation upon BCR stimulation. From a mechanistic perspective, deficiency of Cbl-b resulted in mature T cells proliferating and producing IL-2 independent of CD28 costimulation.[469] Subsequent work suggested that a lack of Cbl-b resulted in enhanced PI3K association with CD28 and CD3z.[470] In addition to the nonproteolytic roles of Cbl-b, it has been reported that c-Cbl and Cbl-b can bind tyrosine phosphorylated, membrane proximal signaling components; however, the significance of this binding is currently unclear.[354,471]

In addition to the pathways discussed above, calcium/NF-AT signaling has also been shown to regulate tolerance. In fact, one of the common features of T and B cell anergy is a defect in Ca2+ flux following antigen stimulation,[354,471] and in T cells, anergy can be induced experimentally by treating the cells with a calcium ionophore alone.[471] Furthermore, anergic B cells display a small elevation in their basal level of Ca2+.[472] When conventional (B2) B cells recognize antigen, they upregulate their metabolism. However, if they do not receive signals from helper T cells within approximately 9 hours, the accumulation of intracellular Ca2+ leads to mitochondrial dysfunction and apoptosis.[473] NF-AT is a transcription factor whose nuclear localization is regulated by dephosphorylation by the Ca2+/calmodulin-dependent phosphatase calcineurin. Therefore, elevated intracellular Ca2+ causes nuclear translocation of NF-AT where it can regulate gene transcription. Induction of T cell anergy results in increased expression of various genes, and these gene expression changes and the ability of T cells to be anergized was impaired in the absence of NF-AT1.[474] Of note, NF-AT mediated anergy induction appears to occur only when AP-1 transcription factor family activity is absent further supporting a role for differential signal transduction in anergy induction.[474] Finally, some evidence suggests a role for NF-AT1 in regulating B cell anergy,[475] but this issue is not fully resolved.

Collectively, dysregulation of signal transduction in various pathway contributes to a breakdown in tolerance. Above are only a few examples of different pathways that have been shown to be important in tolerance. This diversity in signal transduction pathways that regulate tolerance is reflected in the human population by the identification of a number of SNPs in various genes that predispose individuals to variety of autoimmune diseases. Understanding the role of these various genes and the effect of the SNP on the function and/or regulation of the gene in immune tolerance will be critical as we look for ways to regulate tolerance.

SUMMARY

In the last 60 years, since the clonal selection theory was proposed, the primary mechanism accounting for the ability of the immune system to successfully make a SNS discrimination has been debated. We now recognize there are multiple levels, each of which, if not functioning, can lead to loss of SNS discrimination. The coordinated activity of these mechanisms can be likened to the different layers of an onion. All layers, be they central or peripheral, act cooperatively and are interdependent. They have their physiological thresholds of applicability. For example, AIRE deficiency results in partial ablation of central tolerance, thereby likely undermining the protection normally provided by the need to achieve quorum, as the rate of emigration of peripheral antigen-specific T cells goes above the "quorum" threshold that normally ensures tolerance. While reducing the frequency of self-specific lymphocytes by deleting them is a means to prevent these cells from achieving quorum and causing autoimmunity, deletion is far from complete. There may even be a selection pressure to maintain some self-specific lymphocytes, as their ability to be recruited to combat some infections may outweigh their danger in terms of autoreactivity. This may be the reason the process of immunologic tolerance evolved such that it does not reach its maximum until lymphocytes have fully matured after residence in the periphery for a substantial period of time (Fig. 32.3). Deletion of self-specific cells need not be complete to maintain tolerance, as a number of mechanisms effectively work together to deal with the remaining self-specific cells. The recognition of antigen by antigen receptors (signal 1) is not completely determined by the affinity of the antigen receptor; their signaling can be tuned down (increased signaling threshold), effectively making them nonresponsive to the level of self-antigen and other signals present under homeostatic conditions. A complementary mechanism, epitope-specific suppression by T_{reg} cells effectively outcompetes the remaining self-specific cells. The presence of antigen nonspecific signals from PAMPs and DAMPs, ultimately culminating in costimulatory signals from APCs, may be perceived as a sign to the immune system that the current conditions (eg, infection) warrant a ramping up of immunity even if it is by mechanisms that may temporarily compromise SNS discrimination. Feedback coinhibitory signals provide a means to limit both the antigen-specific signal 1 (BCR and TCR signals), and antigen-nonspecific costimulatory signals. This feedback is important early to establish tolerance in newly generated lymphocytes and as a means to control lymphocytes undergoing chronic stimulation that might otherwise cause substantial collateral damage. Chronic stimulation, particularly at mucosal sites (eg, during oral tolerance induction), can also create a state that appears to be tolerance, in that cell mediated immunity and most classes of humoral immunity are suppressed by T_{reg} cells that function via linked suppression. However, this is not "true" tolerance as other humoral classes of immunity are generated (particularly IgA). These responses may even help foster beneficial flora. Uncovering the mechanisms involved in our flora's ability to alter our immune system's interaction with self is one of the key challenges for research in immunologic tolerance and autoimmune disease.

ACKNOWLEDGMENTS

The majority of the figures were created with BioRender.com.

REFERENCES

1. Chan WFN, Perez-Diez A, Razavy H, Anderson CC. The ability of natural tolerance to be applied to allogeneic tissue: determinants and limits. *Biol Direct.* 2007;2:10.
2. Koch CA, Khalpey ZI, Platt JL. Accommodation: preventing injury in transplantation and disease. *J Immunol.* 2004;172:5143-5148.
3. Ehrlich P, Morgenroth J. On haemolysins: fifth communication. *Berl Klin Wochenschr.* 1901;38:251-257.
4. Ehrlich P. *Collected Studies on Immunity.* John Wiley & Sons; 1906.
5. Burnet FM. A modification of Jerne's theory of antibody production using the concept of clonal selection. *Aust J Sci.* 1957;20:67-69.
6. Talmage DW. Allergy and immunology. *Annu Rev Med.* 1957;8:239-256.
7. Burnet FM, Fenner F. *The Production of Antibodies.* Macmillan and Company Limited; 1949.
8. Owen RD. Immunogenetic consequences of vascular anastomoses between bovine twins. *Science.* 1945;102:400-401.
9. Traub E. Factors influencing the persistence of choriomeningitis virus in the blood of mice after clinical recovery. *J Exp Med.* 1938;68:229-250.
10. Billingham RE, Brent L, Medawar PB. Actively acquired tolerance of foreign cells. *Nature.* 1953;172:603-606.
11. Billingham RE, Brent L, Medawar PB. Quantitative studies on tissue transplantation immunity. III. Actively acquired tolerance. Philos Trans R Soc B. 1956;239:357-414.
12. Hasek M. Vegetative hybridization of animals by joint blood circulation during embryonal development. *Cs Biol.* 1953;2:265-277.
13. Dameshek W, Schwartz SO. The presence of hemolysins in acute hemolytic anemia. *N Engl J Med.* 1938;218:75-80.
14. Lederberg J. Genes and antibodies. *Science.* 1959;129:1649-1653.
15. Dresser DW, Mitchison NA. The mechanism of immunological paralysis. *Adv Immunol.* 1968;8:129-181.
16. Claman HN, Chaperon EA, Triplett RF. Thymus-marrow cell combinations. Synergism in antibody production. *Proc Soc Exp Biol Med.* 1966;122:1167-1178.
17. Bretscher P, Cohn M. A theory of self-nonself discrimination. *Science.* 1970;169:1042-1049.
18. Mitchison NA. Antigen recognition responsible for the induction in vitro of the secondary response. *Cold Spring Harb Symp Quant Biol.* 1967;32:431-439.
19. Rajewsky K, Rottländer E. Tolerance specificity and the immune response to lactic dehydrogenase isoenzymes. *Cold Spring Harb Symp Quant Biol.* 1967;32:547-554.
20. Rajewsky K, Schirrmacher V, Nase S, Jerne NK. The requirement of more than one antigenic determinant for immunogenicity. *J Exp Med.* 1969;129:1131-1143.
21. Mitchison NA. The carrier effect in the secondary response to hapten-protein conjugates. II. Cellular cooperation. *Eur J Immunol.* 1971;1:18-27.
22. Keene JA, Forman J. Helper activity is required for the in vivo generation of cytotoxic T lymphocytes. *J Exp Med.* 1982;155:768-782.
23. Guerder S, Matzinger P. A fail-safe mechanism for maintaining self-tolerance. *J Exp Med.* 1992;176:553-564.
24. Al-Yassin GA, Bretscher PA. Does T cell activation require a quorum of lymphocytes? *J Immunol.* 2018;201:2855-2861.
25. Tucker MJ, Bretscher PA. T cells cooperating in the induction of delayed-type hypersensitivity act via the linked recognition of antigenic determinants. *J Exp Med.* 1982;155:1037-1049.
26. Bretscher PA. A cascade of T-T interactions, mediated by the linked recognition of antigen, in the induction of T cells able to help delayed-type hypersensitivity responses. *J Immunol.* 1986;137:3726-3733.
27. Bosch AJT, Bolinger B, Keck S, et al. A minimum number of autoimmune T cells to induce autoimmunity? *Cell Immunol.* 2017;316:21-31.
28. Bretscher P. The activation and inactivation of mature CD4 T cells: a case for peripheral self-nonself discrimination. *Scand J Immunol.* 2014;79(6):348-360. doi:10.1111/sji.12173.
29. Zabriskie JB, Hsu KC, Seegal BC. Heart-reactive antibody associated with rheumatic fever: characterization and diagnostic significance. *Clin Exp Immunol.* 1970;7:147-159.
30. Guilherme L, Cunha-Neto E, Coelho V, et al. Human heart-infiltrating T-cell clones from rheumatic heart disease patients recognize both streptococcal and cardiac proteins. *Circulation.* 1995;92:415-420.
31. Enouz S, Carrié L, Merkler D, Bevan MJ, Zehn D. Autoreactive T cells bypass negative selection and respond to self-antigen stimulation during infection. *J Exp Med.* 2012;209:1769-1779.
32. Lehmann PV, Forsthuber T, Miller A, Sercarz EE. Spreading of T-cell autoimmunity to cryptic determinants of an autoantigen. *Nature.* 1992;358:155-157.
33. Venkatesha SH, Durai M, Moudgil KD. Chapter 4-epitope spreading in autoimmune diseases. In: Shoenfeld Y, Agmon-Levin N, Rose NR, eds. *Infection and Autoimmunity.* 2nd ed. Academic Press; 2015:45-68.
34. Wang C, Cordoba S, Hu M, et al. Spontaneous acceptance of mouse kidney allografts is associated with increased Foxp3 expression and differences in the B and T cell compartments. *Transpl Immunol.* 2011;24:149-156.
35. Thangavelu G, Murphy KM, Yagita H, Boon L, Anderson CC. The role of co-inhibitory signals in spontaneous tolerance of weakly mismatched transplants. *Immunobiology.* 2011;216:918-924.
36. Claman HN. Tolerance to a protein antigen in adult mice and the effect of nonspecific factors. *J Immunol.* 1963;91:833-839.
37. Dresser DW. Specific inhibition of antibody production. II. Paralysis induced in adult mice by small quantities of protein antigen. *Immunology.* 1962;5:378-388.
38. Talmage DW, Pearlman DS. The antibody response: a model based on antagonistic actions of antigen. *J Theor Biol.* 1963;5:321-339.
39. Coutinho A, Möller G. Editorial: immune activation of B cells – evidence for "one nonspecific triggering signal" not delivered by the Ig receptors. *Scand J Immunol.* 1974;3:133-146.
40. Bretscher PA. Requirement for antigen in lipopolysaccharide-dependent induction of B cells. *Eur J Immunol.* 1978;8:534-537.
41. Lafferty KJ, Cunningham AJ. A new analysis of allogeneic interactions. *Aust J Exp Biol Med Sci.* 1975;53:27-42.
42. Janeway CA. Approaching the asymptote? Evolution and revolution in immunology. *Cold Spring Harb Symp Quant Biol.* 1989;54 Pt 1:1-13.
43. Sinclair NR, Lees RK, Elliott EV. Role of the Fc fragment in the regulation of the primary immune response. *Nature.* 1968;220:1048-1049.
44. Sinclair NRS. Regulation of the immune response. I. Reduction in ability of specific antibody to inhibit long-lasting IgG immunological priming after removal of the Fc fragment. *J Exp Med.* 1969;129:1183-1201.
45. Sinclair NRS, Chan PL. Regulation of the immune response. IV. The role of the Fc-fragment in feedback inhibition by antibody. *Adv Exp Med Biol.* 1971;12:609-615.
46. Sinclair NR. How many signals are enough? *Cell Immunol.* 1990;130:204-212, discussion 213-235.
47. Sinclair NR, Anderson CC. Co-stimulation and co-inhibition: equal partners in regulation. *Scand J Immunol.* 1996;43:597-603.
48. Grossman Z, Paul WE. Adaptive cellular interactions in the immune system: the tunable activation threshold and the significance of subthreshold responses. *Proc Natl Acad Sci U S A.* 1992;89:10365-10369.
49. Grossman Z, Paul WE. Dynamic tuning of lymphocytes: physiological basis, mechanisms, and function. *Annu Rev Immunol.* 2015;33:677-713.
50. Bretscher PA. A two-step, two-signal model for the primary activation of precursor helper T cells. *Proc Natl Acad Sci U S A.* 1999;96:185-190.
51. Metcalf ES, Klinman NR. In vitro tolerance induction of neonatal murine B cells. *J Exp Med.* 1976;143:1327-1340.
52. Goodnow CC, Crosbie J, Jorgensen H, Brink RA, Basten A. Induction of self-tolerance in mature peripheral B lymphocytes. *Nature.* 1989;342:385-391.
53. McCullagh P. Inability of fetal skin to induce allograft tolerance in fetal lambs. *Immunology.* 1989;67:489-495.
54. Ohki H, Martin C, Corbel C, Coltey M, Douarin NML. Tolerance induced by thymic epithelial grafts in birds. *Science.* 1987;237:1032-1035.
55. Weigle WO. The immune response of rabbits tolerant to bovine serum albumin to the injection of other heterologous serum albumins. *J Exp Med.* 1961;114:111-125.
56. Anderson CC, Bretscher P, Corthay A, et al. Immunological tolerance. Part I of a report of a Workshop on Foundational Concepts of Immune Regulation. *Scand J Immunol.* 2017;85:84-94.
57. Néron S, Lemieux R. Type 2 T-cell-independent murine immune response to the human ABO blood group antigens. *Vox Sang.* 1994;67:68-74.
58. Christiansen D, Vaughan HA, Milland J, et al. Antibody responses to glycolipid-borne carbohydrates require CD4+ T cells but not CD1 or NKT cells. *Immunol Cell Biol.* 2011;89:502-510.
59. Springer GF, Horton RE. Blood group isoantibody stimulation in man by feeding blood group-active bacteria. *J Clin Invest.* 1969;48:1280-1291.
60. Boak JL, Mitchison NA, Pattisson PH. The carrier effect in the secondary response to hapten-protein conjugates. 3. The anatomical distribution of helper cells and antibody-forming-cell-precursors. *Eur J Immunol.* 1971;1:63-65.
61. Rees MA, Rosenberg AS, Munitz TI, Singer A. In vivo induction of antigen-specific transplantation tolerance to Qa1a by exposure to alloantigen in the absence of T-cell help. *Proc Natl Acad Sci U S A.* 1990;87:2765-2769.
62. Fehr T, Bachmann MF, Bluethmann H, Kikutani H, Hengartner H, Zinkernagel RM. T-independent activation of B cells by vesicular stomatitis virus: no evidence for the need of a second signal. *Cell Immunol.* 1996;168:184-192.
63. Zinkernagel RM. Immunology taught by viruses. *Science.* 1996;271:173-178.
64. Maisonneuve C, Bertholet S, Philpott DJ, De Gregorio E. Unleashing the potential of NOD- and Toll-like agonists as vaccine adjuvants. *Proc Natl Acad Sci U S A.* 2014;111:12294-12299.
65. Land WG, Agostinis P, Gasser S, Garg AD, Linkermann A. Transplantation and damage-associated molecular patterns (DAMPs). *Am J Transplant.* 2016;16:3338-3361.
66. Bingaman AW, Ha J, Waitze SY, et al. Vigorous allograft rejection in the absence of danger. *J Immunol.* 2000;164:3065-3071.
67. Anderson CC, Carroll JM, Gallucci S, Ridge JP, Cheever AW, Matzinger P. Testing time-, ignorance-, and danger-based models of tolerance. *J Immunol.* 2001;166:3663-3671.
68. Gray DHD, Gavanescu I, Benoist C, Mathis D. Danger-free autoimmune disease in Aire-deficient mice. *Proc Natl Acad Sci U S A.* 2007;104:18193-18198.
69. Dai H, Friday AJ, Abou-Daya KI, et al. Donor SIRPα polymorphism modulates the innate immune response to allogeneic grafts. *Sci Immunol.* 2017;2:eaam6202.
70. Apps R, Murphy SP, Fernando R, Gardner L, Ahad T, Moffett A. Human leucocyte antigen (HLA) expression of primary trophoblast cells and placental cell lines, determined using single antigen beads to characterize allotype specificities of anti-HLA antibodies. *Immunology.* 2008;127:26-39.
71. Madeja Z, Yadi H, Apps R, et al. Paternal MHC expression on mouse trophoblast affects uterine vascularization and fetal growth. *Proc Natl Acad Sci U S A.* 2011;108:4012-4017.
72. Weigle WO. The immune response of BSA tolerant rabbits to injections of BSA following the termination of the tolerant state. *J Immunol.* 1964;92:791-797.
73. Lin RH, Mamula MJ, Hardin JA, Janeway CA. Induction of autoreactive B cells allows priming of autoreactive T cells. *J Exp Med.* 1991;173:1433-1439.
74. Ridge JP, Di Rosa F, Matzinger P. A conditioned dendritic cell can be a temporal bridge between a CD4+ T-helper and a T-killer cell. *Nature.* 1998;393:474-478.
75. Bretscher P. *The Foundations of Immunology and their Pertinence to Medicine.* FriesenPress; 2016.
76. Kroeger DR, Rudulier CD, Bretscher PA. Antigen presenting B cells facilitate CD4 T cell cooperation resulting in enhanced generation of effector and memory CD4 T cells. *PloS One.* 2013;8:e77346.
77. Chan OT, Hannum LG, Haberman AM, Madaio MP, Shlomchik MJ. A novel mouse with B cells but lacking serum antibody reveals an antibody-independent role for B cells in murine lupus. *J Exp Med.* 1999;189:1639-1648.
78. Silverman GJ, Weisman S. Rituximab therapy and autoimmune disorders: prospects for anti-B cell therapy. *Arthritis Rheum.* 2003;48:1484-1492.
79. Townsend MJ, Monroe JG, Chan AC. B-cell targeted therapies in human autoimmune diseases: an updated perspective. *Immunol Rev.* 2010;237:264-283.
80. Pillai S, Mattoo H, Cariappa A. B cells and autoimmunity. *Curr Opin Immunol.* 2011;23:721-731.
81. Huszthy PC, Gopalakrishnan RP, Jacobsen JT, et al. B cell receptor ligation induces display of V-region peptides on MHC class II molecules to T cells. *Proc Natl Acad Sci U S A.* 2019;116:25850-25859.
82. Peters NC, Hamilton DH, Bretscher PA. Analysis of cytokine-producing Th cells from hen egg lysozyme-immunized mice reveals large numbers specific for "cryptic" peptides and different repertoires among different Th populations. *Eur J Immunol.* 2005;35:56-65.
83. Kearney ER, Pape KA, Loh DY, Jenkins MK. Visualization of peptide-specific T cell immunity and peripheral tolerance induction in vivo. *Immunity.* 1994;1:327-339.
84. Sinclair NR, Panoskaltsis A. The immunoregulatory apparatus and autoimmunity. *Immunol Today.* 1988;9:260-265.
85. Blackburn SD, Shin H, Haining WN, et al. Coregulation of CD8+ T cell exhaustion by multiple inhibitory receptors during chronic viral infection. *Nat Immunol.* 2009;10:29-37.
86. Marraco SAF, Neubert NJ, Verdeil G, Speiser DE. Inhibitory receptors beyond T cell exhaustion. *Front Immunol.* 2015;6:310.
87. Han S, Asoyan A, Rabenstein H, Nakano N, Obst R. Role of antigen persistence and dose for CD4+ T-cell exhaustion and recovery. *Proc Natl Acad Sci U S A.* 2010;107:20453-20458.
88. Kawai K, Ohashi PS. Immunological function of a defined T-cell population tolerized to low-affinity self antigens. *Nature.* 1995;374:68-69.
89. Sebzda E, Kündig TM, Thomson CT, et al. Mature T cell reactivity altered by peptide agonist that induces positive selection. *J Exp Med.* 1996;183:1093-1104.
90. Davey GM, Schober SL, Endrizzi BT, Dutcher AK, Jameson SC, Hogquist KA. Preselection thymocytes are more sensitive to T cell receptor stimulation than mature T cells. *J Exp Med.* 1998;188:1867-1874.
91. Park J-H, Adoro S, Lucas PJ, et al. "Coreceptor tuning": cytokine signals transcriptionally tailor CD8 coreceptor expression to the self-specificity of the TCR. *Nat Immunol.* 2007;8:1049-1059.
92. Jameson SC, Hogquist KA, Bevan MJ. Specificity and flexibility in thymic selection. *Nature.* 1994;369:750-752.
93. Azzam HS, Grinberg A, Lui K, Shen H, Shores EW, Love PE. CD5 expression is developmentally regulated by T cell receptor (TCR) signals and TCR avidity. *J Exp Med.* 1998;188:2301-2311.
94. Peña-Rossi C, Zuckerman LA, Strong J, et al. Negative regulation of CD4 lineage development and responses by CD5. *J Immunol.* 1999;163:6494-6501.
95. Mandl JN, Monteiro JP, Vrisekoop N, Germain RN. T cell-positive selection uses self-ligand binding strength to optimize repertoire recognition of foreign antigens. *Immunity.* 2013;38:263-274.

96. Persaud SP, Parker CR, Lo W-L, Weber KS, Allen PM. Intrinsic CD4+ T cell sensitivity and response to a pathogen are set and sustained by avidity for thymic and peripheral complexes of self peptide and MHC. *Nat Immunol.* 2014;15:266-274.
97. Tarakhovsky A, Kanner SB, Hombach J, et al. A role for CD5 in TCR-mediated signal transduction and thymocyte selection. *Science.* 1995;269:535-537.
98. Voisinne G, de Peredo AG, Roncagalli R. CD5, an undercover regulator of TCR signaling. *Front Immunol.* 2018;9:2900.
99. Stamou P, de Jersey J, Carmignac D, Mamalaki C, Kioussis D, Stockinger B. Chronic exposure to low levels of antigen in the periphery causes reversible functional impairment correlating with changes in CD5 levels in monoclonal CD8 T cells. *J Immunol.* 2003;171:1278-1284.
100. Takada K, Jameson SC. Self-class I MHC molecules support survival of naive CD8 T cells, but depress their functional sensitivity through regulation of CD8 expression levels. *J Exp Med.* 2009;206:2253-2269.
101. Singh NJ, Schwartz RH. The strength of persistent antigenic stimulation modulates adaptive tolerance in peripheral CD4+ T cells. *J Exp Med.* 2003;198:1107-1117.
102. Chuang E, Augustine M, Jung M, Schwartz RH, Singh NJ. Density dependent re-tuning of autoreactive T cells alleviates their pathogenicity in a lymphopenic environment. *Immunol Lett.* 2017;192:61-71.
103. DeFranco AL. Multilayer control of B cell activation by the B cell antigen receptor: following themes initiated with Bill Paul. *Front Immunol.* 2018;9:739.
104. Tan C, Noviski M, Huizar J, Zikherman J. Self-reactivity on a spectrum: a sliding scale of peripheral B cell tolerance. *Immunol Rev.* 2019;292:37-60.
105. Yoon I-H, Choi S-E, Kim Y-H, et al. Pancreatic islets induce CD4+ CD25-Foxp3+ T-cell regulated tolerance to HY-mismatched skin grafts. *Transplantation.* 2008;86:1352-1360.
106. Emery D, McCullagh P. Immunological reactivity between chimeric cattle twins. I. Homograft reaction. *Transplantation.* 1980;29:4-9.
107. Luo B, Chan WFN, Shapiro AMJ, Anderson CC. Non-myeloablative mixed chimerism approaches and tolerance, a split decision. *Eur J Immunol.* 2007;37:1233-1242.
108. Teague PO, Friou GJ. Antinuclear antibodies in mice. II. Transmission with spleen cells; inhibition or prevention with thymus or spleen cells. *Immunology.* 1969;17:665-675.
109. McCullagh P. Curtailment of autoimmunity following parabiosis with a normal partner. *Immunology.* 1990;71:595-597.
110. Chen X, Shelton J, McCullagh P. Suppression of anti-thyrocyte autoreactivity by the lymphocytes of normal fetal lambs. *J Autoimmun.* 1995;8:539-559.
111. Zamoyska R, Waldmann H, Matzinger P. Peripheral tolerance mechanisms prevent the development of autoreactive T cells in chimeras grafted with two minor incompatible thymuses. *Eur J Immunol.* 1989;19:111-117.
112. Douarin NL, Corbel C, Bandeira A, et al. Evidence for a thymus-dependent form of tolerance that is not based on elimination or anergy of reactive T cells. *Immunol Rev.* 1996;149:35-53.
113. Tanchot C, Vasseur F, Pontoux C, Garcia C, Sarukhan A. Immune regulation by self-reactive T cells is antigen specific. *J Immunol.* 2004;172:4285-4291.
114. Akkaya B, Oya Y, Akkaya M, et al. Regulatory T cells mediate specific suppression by depleting peptide-MHC class II from dendritic cells. *Nat Immunol.* 2019;20:218-231.
115. Dembic Z. Response to Cohn: the immune system rejects the harmful, protects the useful and neglects the rest of microorganisms. *Scand J Immunol.* 2004;60:3-5, discussion 6-8.
116. Wang S, Charbonnier L-M, Rivas MN, et al. MyD88 adaptor-dependent microbial sensing by regulatory T cells promotes mucosal tolerance and enforces commensalism. *Immunity.* 2015;43:289-303.
117. Russler-Germain EV, Rengarajan S, Hsieh CS. Antigen-specific regulatory T-cell responses to intestinal microbiota. *Mucosal Immunol.* 2017;10:1375-1386.
118. Belkaid Y, Piccirillo CA, Mendez S, Shevach EM, Sacks DL. CD4+CD25+ regulatory T cells control Leishmania major persistence and immunity. *Nature.* 2002;420:502-507.
119. Anderson CC. Time, space and contextual models of the immunity tolerance decision: bridging the geographical divide of Zinkernagel and Hengartner's "Credo 2004". *Scand J Immunol.* 2006;63:249-256.
120. Akdis M, Verhagen J, Taylor A, et al. Immune responses in healthy and allergic individuals are characterized by a fine balance between allergen-specific T regulatory 1 and T helper 2 cells. *J Exp Med.* 2004;199:1567-1575.
121. Alpan O, Bachelder E, Isil E, Arnheiter H, Matzinger P. "Educated" dendritic cells act as messengers from memory to naive T helper cells. *Nat Immunol.* 2004;5:615-622.
122. Ramshaw IA, Bretscher PA, Parish CR. Regulation of the immune response. I. Suppression of delayed-type hypersensitivity by T cells from mice expressing humoral immunity. *Eur J Immunol.* 1976;6:674-679.
123. Glick B, Chang TS, Jaap R. The bursa of Fabricius and antibody production. *Poult Sci.* 1956;35:224-225.
124. Owen JJ, Cooper MD, Raff MC. In vitro generation of B lymphocytes in mouse foetal liver, a mammalian "bursa equivalent". *Nature.* 1974;249:361-363.
125. Osmond DG, Nossal GJ. Differentiation of lymphocytes in mouse bone marrow. II. Kinetics of maturation and renewal of antiglobulin-binding cells studied by double labeling. *Cell Immunol.* 1974;13:132-145.
126. Miller JF. Immunological function of the thymus. *Lancet.* 1961;2:748-749.
127. Miller JF. The discovery of the immunological function of the thymus. *Immunol Today.* 1991;12:42-45.
128. Miller JFAP. Effect of neonatal thymectomy on the immunological responsiveness of the mouse. *Proc R Soc Lond B Biol Sci.* 1962;156:415-428.
129. Hanahan D. Peripheral-antigen-expressing cells in thymic medulla: factors in self-tolerance and autoimmunity. *Curr Opin Immunol.* 1998;10:656-662.
130. Werdelin O, Cordes U, Jensen T. Aberrant expression of tissue-specific proteins in the thymus: a hypothesis for the development of central tolerance. *Scand J Immunol.* 1998;47:95-100.
131. Heath VL, Moore NC, Parnell SM, Mason DW. Intrathymic expression of genes involved in organ specific autoimmune disease. *J Autoimmun.* 1998;11:309-318.
132. Klein L, Klein T, Rüther U, Kyewski B. CD4 T cell tolerance to human C-reactive protein, an inducible serum protein, is mediated by medullary thymic epithelium. *J Exp Med.* 1998;188:5-16.
133. Derbinski J, Schulte A, Kyewski B, Klein L. Promiscuous gene expression in medullary thymic epithelial cells mirrors the peripheral self. *Nat Immunol.* 2001;2:1032-1039.
134. Egwuagu CE, Charukamnoetkanok P, Gery I. Thymic expression of autoantigens correlates with resistance to autoimmune disease. *J Immunol.* 1997;159:3109-3112.
135. Ju J-M, Jung MH, Nam G, et al. Escape from thymic deletion and anti-leukemic effects of T cells specific for hematopoietic cell-restricted antigen. *Nat Commun.* 2018;9:225.
136. Kappler JW, Roehm N, Marrack P. T cell tolerance by clonal elimination in the thymus. *Cell.* 1987;49:273-280.
137. MacDonald HR, Schneider R, Lees RK, et al. T-cell receptor V beta use predicts reactivity and tolerance to Mlsa-encoded antigens. *Nature.* 1988;332:40-45.
138. Smith CA, Williams GT, Kingston R, Jenkinson EJ, Owen JJ. Antibodies to CD3/T-cell receptor complex induce death by apoptosis in immature T cells in thymic cultures. *Nature.* 1989;337:181-184.
139. Surh CD, Sprent J. T-cell apoptosis detected in situ during positive and negative selection in the thymus. *Nature.* 1994;372:100-103.
140. Kisielow P, Blüthmann H, Staerz UD, Steinmetz M, Boehmer H von. Tolerance in T-cell-receptor transgenic mice involves deletion of nonmature CD4+8+ thymocytes. *Nature.* 1988;333:742-746.
141. Vasquez NJ, Kaye J, Hedrick SM. In vivo and in vitro clonal deletion of double-positive thymocytes. *J Exp Med.* 1992;175:1307-1316.
142. Baldwin TA, Sandau MM, Jameson SC, Hogquist KA. The timing of TCR alpha expression critically influences T cell development and selection. *J Exp Med.* 2005;202:111-121.
143. Goldman KP, Park C-S, Kim M, Matzinger P, Anderson CC. Thymic cortical epithelium induces self tolerance. *Eur J Immunol.* 2005;35:709-717.
144. McCaughtry TM, Baldwin TA, Wilken MS, Hogquist KA. Clonal deletion of thymocytes can occur in the cortex with no involvement of the medulla. *J Exp Med.* 2008;205:2575-2584.
145. Sha WC, Nelson CA, Newberry RD, Kranz DM, Russell JH, Loh DY. Positive and negative selection of an antigen receptor on T cells in transgenic mice. *Nature.* 1988;336:73-76.
146. Bogen B, Dembic Z, Weiss S. Clonal deletion of specific thymocytes by an immunoglobulin idiotype. *EMBO J.* 1993;12:357-363.
147. Nemazee D, Buerki K. Clonal deletion of autoreactive B lymphocytes in bone marrow chimeras. *Proc Natl Acad Sci U S A.* 1989;86:8039-8043.
148. Hartley SB, Crosbie J, Brink R, Kantor AB, Basten A, Goodnow CC. Elimination from peripheral lymphoid tissues of self-reactive B lymphocytes recognizing membrane-bound antigens. *Nature.* 1991;353:765-769.
149. Raff MC, Owen JJ, Cooper MD, Lawton AR, Megson M, Gathings WE. Differences in susceptibility of mature and immature mouse B lymphocytes to anti-immunoglobulin-induced immunoglobulin suppression in vitro. Possible implications for B-cell tolerance to self. *J Exp Med.* 1975;142:1052-1064.
150. Kincade PW, Lawton AR, Bockman DE, Cooper MD. Suppression of immunoglobulin G synthesis as a result of antibody-mediated suppression of immunoglobulin M synthesis in chickens. *Proc Natl Acad Sci U S A.* 1970;67:1918-1925.
151. Goodnow CC, Crosbie J, Adelstein S, et al. Altered immunoglobulin expression and functional silencing of self-reactive B lymphocytes in transgenic mice. *Nature.* 1988;334:676-682.
152. Rui L, Vinuesa CG, Blasioli J, Goodnow CC. Resistance to CpG DNA-induced autoimmunity through tolerogenic B cell antigen receptor ERK signaling. *Nat Immunol.* 2003;4:594-600.
153. Fulcher DA, Basten A. Reduced life span of anergic self-reactive B cells in a double-transgenic model. *J Exp Med.* 1994;179:125-134.
154. Pfueller SL, Logan D, Tran TT, Bilston RA. Naturally occurring human IgG antibodies to intracellular and cytoskeletal components of human platelets. *Clin Exp Immunol.* 1990;79:367-373.
155. Ferry H, Jones M, Vaux DJ, Roberts ISD, Cornall RJ. The cellular location of self-antigen determines the positive and negative selection of autoreactive B cells. *J Exp Med.* 2003;198:1415-1425.
156. Chen C, Nagy Z, Radic MZ, et al. The site and stage of anti-DNA B-cell deletion. *Nature.* 1995;373:252-255.
157. Woodruff MFA, Reid B, James K. Effect of antilymphocytic antibody and antibody fragments on human lymphocytes in vitro. *Nature.* 1967;215:591-594.
158. Yang J, Reth M. Receptor dissociation and B-cell activation. *Curr Top Microbiol Immunol.* 2015;393:27-43.
159. Kim Y-M, Pan JY-J, Korbel GA, Peperzak V, Boes M, Ploegh HL. Monovalent ligation of the B cell receptor induces receptor activation but fails to promote antigen presentation. *Proc Natl Acad Sci U S A.* 2006;103:3327-3332.
160. Avalos AM, Bilate AM, Witte MD, et al. Monovalent engagement of the BCR activates ovalbumin-specific transnuclear B cells. *J Exp Med.* 2014;211:365-379.
161. Eynon EE, Parker DC. Small B cells as antigen-presenting cells in the induction of tolerance to soluble protein antigens. *J Exp Med.* 1992;175:131-138.
162. Bretscher PA, Al-Yassin GA. On how the immune system preferentially interacts with antigen-specific molecules bound to antigen over unbound molecules, with emphasis on B cell receptor signalling. *Scand J Immunol.* 2019;90:e12795.
163. Hartley SB, Cooke MP, Fulcher DA, et al. Elimination of self-reactive B lymphocytes proceeds in two stages: arrested development and cell death. *Cell.* 1993;72:325-335.
164. Wardemann H, Yurasov S, Schaefer A, Young JW, Meffre E, Nussenzweig MC. Predominant autoantibody production by early human B cell precursors. *Science.* 2003;301:1374-1377.
165. Tiegs SL, Russell DM, Nemazee D. Receptor editing in self-reactive bone marrow B cells. *J Exp Med.* 1993;177:1009-1020.
166. Radic MZ, Erikson J, Litwin S, Weigert M. B lymphocytes may escape tolerance by revising their antigen receptors. *J Exp Med.* 1993;177:1165-1173.
167. Gay D, Saunders T, Camper S, Weigert M. Receptor editing: an approach by autoreactive B cells to escape tolerance. *J Exp Med.* 1993;177:999-1008.
168. Nemazee D. Receptor editing in lymphocyte development and central tolerance. *Nat Rev Immunol.* 2006;6:728-740.
169. Chen C, Nagy Z, Prak EL, Weigert M. Immunoglobulin heavy chain gene replacement: a mechanism of receptor editing. *Immunity.* 1995;3:747-755.
170. Zhang Z, Zemlin M, Wang Y-H, et al. Contribution of Vh gene replacement to the primary B cell repertoire. *Immunity.* 2003;19:21-31.
171. Pelanda R, Schwers S, Sonoda E, Torres RM, Nemazee D, Rajewsky K. Receptor editing in a transgenic mouse model: site, efficiency, and role in B cell tolerance and antibody diversification. *Immunity.* 1997;7:765-775.
172. Novobrantseva T, Xu S, Tan JE-L, et al. Stochastic pairing of Ig heavy and light chains frequently generates B cell antigen receptors that are subject to editing in vivo. *Int Immunol.* 2005;17:343-350.
173. Hippen KL, Schram BR, Tze LE, Pape KA, Jenkins MK, Behrens TW. In vivo assessment of the relative contributions of deletion, anergy, and editing to B cell self-tolerance. *J Immunol.* 2005;175:909-916.
174. Erikson J, Radic MZ, Camper SA, Hardy RR, Carmack C, Weigert M. Expression of anti-DNA immunoglobulin transgenes in non-autoimmune mice. *Nature.* 1991;349:331-334.
175. Wang F, Huang CY, Kanagawa O. Rapid deletion of rearranged T cell antigen receptor (TCR) Valpha-Jalpha segment by secondary rearrangement in the thymus: role of continuous rearrangement of TCR alpha chain gene and positive selection in the T cell repertoire formation. *Proc Natl Acad Sci U S A.* 1998;95:11834-11839.
176. McGargill MA, Derbinski JM, Hogquist KA. Receptor editing in developing T cells. *Nat Immunol.* 2000;1:336-341.
177. Holman PO, Walsh ER, Hogquist KA. The central tolerance response to male antigen in normal mice is deletion and not receptor editing. *J Immunol.* 2003;171:4048-4053.
178. Ashton-Rickardt PG, Bandeira A, Delaney JR, et al. Evidence for a differential avidity model of T cell selection in the thymus. *Cell.* 1994;76:651-663.
179. Sebzda E, Wallace VA, Mayer J, Yeung RS, Mak TW, Ohashi PS. Positive and negative thymocyte selection induced by different concentrations of a single peptide. *Science.* 1994;263:1615-1618.
180. Benschop RJ, Brandl E, Chan AC, Cambier JC. Unique signaling properties of B cell antigen receptor in mature and immature B cells: implications for tolerance and activation. *J Immunol.* 2001;167:4172-4179.
181. Yagi J, Janeway CA. Ligand thresholds at different stages of T cell development. *Int Immunol.* 1990;2:83-89.
182. Robertson K, Simon K, Schneider S, Timms E, Mitchison A. Tolerance of self induced in thymus organ culture. *Eur J Immunol.* 1992;22:207-211.
183. Peterson DA, DiPaolo RJ, Kanagawa O, Unanue ER. Cutting edge: negative selection of immature thymocytes by a few peptide-MHC complexes – differential sensitivity of immature and mature T cells. *J Immunol.* 1999;162:3117-3120.
184. Pircher H, Rohrer UH, Moskophidis D, Zinkernagel RM, Hengartner H. Lower receptor avidity required for thymic clonal deletion than for effector T-cell function. *Nature.* 1991;351:482-485.

185. Huseby ES, Crawford F, White J, Kappler J, Marrack P. Negative selection imparts peptide specificity to the mature T cell repertoire. *Proc Natl Acad Sci U S A*. 2003;100:11565-11570.
186. McDonald BD, Bunker JJ, Erickson SA, Oh-Hora M, Bendelac A. Crossreactive αβ T cell receptors are the predominant targets of thymocyte negative selection. *Immunity*. 2015;43:859-869.
187. Corthay A, Nandakumar KS, Holmdahl R. Evaluation of the percentage of peripheral T cells with two different T cell receptor alpha-chains and of their potential role in autoimmunity. *J Autoimmun*. 2001;16:423-429.
188. Schuldt NJ, Auger JL, Spanier JA, et al. Dual TCRα expression poses an autoimmune hazard by limiting Treg cell generation. *J Immunol*. 2017;199:33-38.
189. Ramsdell F, Lantz T, Fowlkes B. A nondeletional mechanism of thymic self tolerance. *Science*. 1989;246:1038-1041.
190. Ramsdell F, Fowlkes BJ. Maintenance of in vivo tolerance by persistence of antigen. *Science*. 1992;257:1130-1134.
191. Nishizuka Y, Sakakura T. Thymus and reproduction: sex-linked dysgenesia of the gonad after neonatal thymectomy in mice. *Science*. 1969;166:753-755.
192. Sakaguchi S, Fukuma K, Kuribayashi K, Masuda T. Organ-specific autoimmune diseases induced in mice by elimination of T cell subset. I. Evidence for the active participation of T cells in natural self-tolerance; deficit of a T cell subset as a possible cause of autoimmune disease. *J Exp Med*. 1985;161:72-87.
193. Penhale WJ, Farmer A, McKenna RP, Irvine WJ. Spontaneous thyroiditis in thymectomized and irradiated Wistar rats. *Clin Exp Immunol*. 1973;15:225-236.
194. Fowell D, Mason D. Evidence that the T cell repertoire of normal rats contains cells with the potential to cause diabetes. Characterization of the CD4+ T cell subset that inhibits this autoimmune potential. *J Exp Med*. 1993;177:627-636.
195. Powrie F, Mason D. OX-22high CD4+ T cells induce wasting disease with multiple organ pathology: prevention by the OX-22low subset [published erratum appears in J Exp Med 1991 Apr 1;173(4):1037]. *J Exp Med*. 1990;172:1701-1708.
196. Powrie F, Leach MW, Mauze S, Menon S, Caddle LB, Coffman RL. Inhibition of Thl responses prevents inflammatory bowel disease in scid mice reconstituted with CD45RBhi CD4+ T cells. *Immunity*. 1994;1:553-562.
197. Sakaguchi S, Sakaguchi N, Asano M, Itoh M, Toda M. Immunologic self-tolerance maintained by activated T cells expressing IL-2 receptor alpha-chains (CD25). Breakdown of a single mechanism of self-tolerance causes various autoimmune diseases. *J Immunol*. 1995;155:1151-1164.
198. Brunkow ME, Jeffery EW, Hjerrild KA, et al. Disruption of a new forkhead/winged-helix protein, scurfin, results in the fatal lymphoproliferative disorder of the scurfy mouse. *Nat Genet*. 2001;27:68-73.
199. Wildin RS, Ramsdell F, Peake J, et al. X-linked neonatal diabetes mellitus, enteropathy and endocrinopathy syndrome is the human equivalent of mouse scurfy. *Nat Genet*. 2001;27:18-20.
200. Bennett CL, Christie J, Ramsdell F, et al. The immune dysregulation, polyendocrinopathy, enteropathy, X-linked syndrome (IPEX) is caused by mutations of FOXP3. *Nat Genet*. 2001;27:20-21.
201. Chatila TA, Blaeser F, Ho N, et al. JM2, encoding a fork head-related protein, is mutated in X-linked autoimmunity-allergic disregulation syndrome. *J Clin Invest*. 2000;106:R75-R81.
202. Hori S, Nomura T, Sakaguchi S. Control of regulatory T cell development by the transcription factor Foxp3. *Science*. 2003;299:1057-1061.
203. Fontenot JD, Gavin MA, Rudensky AY. Foxp3 programs the development and function of CD4+CD25+ regulatory T cells. *Nat Immunol*. 2003;4:330-336.
204. Khattri R, Cox T, Yasayko S-A, Ramsdell F. An essential role for Scurfin in CD4+CD25+ T regulatory cells. *Nat Immunol*. 2003;4:337-342.
205. Kim JM, Rasmussen JP, Rudensky AY. Regulatory T cells prevent catastrophic autoimmunity throughout the lifespan of mice. *Nat Immunol*. 2007;8:191-197.
206. Wing K, Onishi Y, Prieto-Martin P, et al. CTLA-4 control over Foxp3+ regulatory T cell function. *Science*. 2008;322:271-275.
207. Jordan MS, Boesteanu A, Reed AJ, et al. Thymic selection of CD4+CD25+ regulatory T cells induced by an agonist self-peptide. *Nat Immunol*. 2001;2:301-306.
208. Lee H-M, Bautista JL, Scott-Browne J, Mohan JF, Hsieh C-S. A broad range of self-reactivity drives thymic regulatory T cell selection to limit responses to self. *Immunity*. 2012;37:475-486.
209. Klein L, Robey EA, Hsieh C-S. Central CD4+ T cell tolerance: deletion versus regulatory T cell differentiation. *Nat Rev Immunol*. 2019;19:7-18.
210. Matzinger P, Guerder S. Does T-cell tolerance require a dedicated antigen-presenting cell? *Nature*. 1989;338:74-76.
211. Palmer E. Negative selection-clearing out the bad apples from the T-cell repertoire. *Nat Rev Immunol*. 2003;3:383-391.
212. Breed ER, Watanabe M, Hogquist KA. Measuring thymic clonal deletion at the population level. *J Immunol*. 2019;202:3226-3233.
213. Kwan J, Killeen N. CCR7 directs the migration of thymocytes into the thymic medulla. *J Immunol*. 2004;172:3999-4007.
214. Nitta T, Nitta S, Lei Y, Lipp M, Takahama Y. CCR7-mediated migration of developing thymocytes to the medulla is essential for negative selection to tissue-restricted antigens. *Proc Natl Acad Sci U S A*. 2009;106:17129-17133.
215. Li J, Park J, Foss D, Goldschneider I. Thymus-homing peripheral dendritic cells constitute two of the three major subsets of dendritic cells in the steady-state thymus. *J Exp Med*. 2009;206:607-622.
216. Bonasio R, Scimone M, Schaerli P, Grabie N, Lichtman A, von Andrian U. Clonal deletion of thymocytes by circulating dendritic cells homing to the thymus. *Nat Immunol*. 2006;7:1092-1100.
217. Hadeiba H, Lahl K, Edalati A, et al. Plasmacytoid dendritic cells transport peripheral antigens to the thymus to promote central tolerance. *Immunity*. 2012;36:438-450.
218. Ennamorati M, Vasudevan C, Clerkin K, et al. Intestinal microbes influence development of thymic lymphocytes in early life. *Proc Natl Acad Sci U S A*. 2020;117:2570-2578.
219. Santamaria J, Darrigues J, van Meerwijk JPM, Romagnoli P. Antigen-presenting cells and T-lymphocytes homing to the thymus shape T cell development. *Immunol Lett*. 2018;204:9-15.
220. Smith KM, Olson DC, Hirose R, Hanahan D. Pancreatic gene expression in rare cells of thymic medulla: evidence for functional contribution to T cell tolerance. *Int Immunol*. 1997;9:1355-1365.
221. Jolicoeur C, Hanahan D, Smith KM. T-cell tolerance toward a transgenic beta-cell antigen and transcription of endogenous pancreatic genes in thymus. *Proc Natl Acad Sci U S A*. 1994;91:6707-6711.
222. Vafiadis P, Bennett ST, Todd JA, et al. Insulin expression in human thymus is modulated by INS VNTR alleles at the IDDM2 locus. *Nat Genet*. 1997;15:289-292.
223. Nagamine K, Peterson P, Scott HS, et al. Positional cloning of the APECED gene. *Nat Genet*. 1997;17:393-398.
224. Aaltonen J, Björses P, Perheentupa J, et al. An autoimmune disease, APECED, caused by mutations in a novel gene featuring two PHD-type zinc-finger domains. *Nat Genet*. 1997;17:399-403.
225. Heino M, Peterson P, Kudoh J, et al. Autoimmune regulator is expressed in the cells regulating immune tolerance in thymus medulla. *Biochem Biophys Res Commun*. 1999;257:821-825.
226. Anderson MS, Venanzi ES, Klein L, et al. Projection of an immunological self shadow within the thymus by the aire protein. *Science*. 2002;298:1395-1401.
227. Liston A, Lesage S, Wilson J, Peltonen L, Goodnow CC. Aire regulates negative selection of organ-specific T cells. *Nat Immunol*. 2003;4:350-354.
228. DeVoss J, Hou Y, Johannes K, et al. Spontaneous autoimmunity prevented by thymic expression of a single self-antigen. *J Exp Med*. 2006;203:2727-2735.
229. Gallegos AM, Bevan MJ. Central tolerance to tissue-specific antigens mediated by direct and indirect antigen presentation. *J Exp Med*. 2004;200:1039-1049.
230. Yamano T, Nedjic J, Hinterberger M, et al. Thymic B cells are licensed to present self antigens for central T cell tolerance induction. *Immunity*. 2015;42:1048-1061.
231. Perera J, Meng L, Meng F, Huang H. Autoreactive thymic B cells are efficient antigen-presenting cells of cognate self-antigens for T cell negative selection. *Proc Natl Acad Sci U S A*. 2013;110:17011-17016.
232. Anderson MS, Venanzi ES, Chen Z, Berzins SP, Benoist C, Mathis D. The cellular mechanism of aire control of T cell tolerance. *Immunity*. 2005;23:227-239.
233. Kuroda N, Mitani T, Takeda N, et al. Development of autoimmunity against transcriptionally unrepressed target antigen in the thymus of aire-deficient mice. *J Immunol*. 2005;174:1862-1870.
234. Malchow S, Leventhal DS, Lee V, Nishi S, Socci ND, Savage PA. Aire enforces immune tolerance by directing autoreactive T cells into the regulatory T cell lineage. *Immunity*. 2016;44:1102-1113.
235. Perry JSA, Lio C-WJ, Kau AL, et al. Distinct contributions of Aire and antigen-presenting-cell subsets to the generation of self-tolerance in the thymus. *Immunity*. 2014;41:414-426.
236. Wong P, Goldrath AW, Rudensky AY. Competition for specific intrathymic ligands limits positive selection in a TCR transgenic model of CD4+ T cell development. *J Immunol*. 2000;164:6252-6259.
237. Malchow S, Leventhal DS, Nishi S, et al. Aire-dependent thymic development of tumor-associated regulatory T cells. *Science*. 2013;339:1219-1224.
238. Takaba H, Morishita Y, Tomofuji Y, et al. Fezf2 orchestrates a thymic program of self-antigen expression for immune tolerance. *Cell*. 2015;163:975-987.
239. Humblet C, Rudensky AY, Kyewski B. Presentation and intercellular transfer of self antigen within the thymic microenvironment: expression of the E alpha peptide-I-Ab complex by isolated thymic stromal cells. *Int Immunol*. 1994;6:1949-1958.
240. Aschenbrenner K, D'Cruz LM, Vollmann EH, et al. Selection of Foxp3+ regulatory T cells specific for self antigen expressed and presented by Aire+ medullary thymic epithelial cells. *Nat Immunol*. 2007;8:351-358.
241. Antón LC, Yewdell JW. Translating DRiPs: MHC class I immunosurveillance of pathogens and tumors. *J Leukoc Biol*. 2014;95:551-562.
242. Granados DP, Laumont CM, Thibault P, Perreault C. The nature of self for T cells-a systems-level perspective. *Curr Opin Immunol*. 2015;34:1-8.
243. Laumont CM, Daouda T, Laverdure J-P, et al. Global proteogenomic analysis of human MHC class I-associated peptides derived from non-canonical reading frames. *Nat Commun*. 2016;7:10238.
244. Aichinger M, Wu C, Nedjic J, Klein L. Macroautophagy substrates are loaded onto MHC class II of medullary thymic epithelial cells for central tolerance. *J Exp Med*. 2013;210:287-300.
245. Nedjic J, Aichinger M, Emmerich J, Mizushima N, Klein L. Autophagy in thymic epithelium shapes the T-cell repertoire and is essential for tolerance. *Nature*. 2008;455:396-400.
246. Hill JA, Bell DA, Brintnell W, et al. Arthritis induced by posttranslationally modified (citrullinated) fibrinogen in DR4-IE transgenic mice. *J Exp Med*. 2008;205:967-979.
247. Raposo B, Merky P, Lundqvist C, et al. T cells specific for post-translational modifications escape intrathymic tolerance induction. *Nat Commun*. 2018;9:353.
248. Cunningham MW. Rheumatic fever, autoimmunity, and molecular mimicry: the streptococcal connection. *Int Rev Immunol*. 2014;33:314-329.
249. Lanoue A, Bona C, von Boehmer H, Sarukhan A. Conditions that induce tolerance in mature CD4+ T cells. *J Exp Med*. 1997;185:405-414.
250. Kurts C, Kosaka H, Carbone FR, Miller JF, Heath WR. Class I-restricted cross-presentation of exogenous self-antigens leads to deletion of autoreactive CD8(+) T cells. *J Exp Med*. 1997;186:239-245.
251. Triplett EL. On the mechanism of immunologic self recognition. *J Immunol*. 1962;89:505-510.
252. Rollins-Smith LA, Cohen N. Self-pituitary grafts are not rejected by frogs deprived of their pituitary anlagen as embryos. *Nature*. 1982;299:820-821.
253. McCullagh P. Interception of the development of self tolerance in fetal lambs. *Eur J Immunol*. 1989;19:1387-1392.
254. McCullagh P. The significance of immune suppression in normal self tolerance. *Immunol Rev*. 1996;149:127-153.
255. Vidal-Puig A, Faustman DL. Tolerance to peripheral tissue is transient and maintained by tissue-specific class I expression. *Transplant Proc*. 1994;26:3314-3316.
256. Anderson CC. Conceptual models in immunity/tolerance: application to transplantation. In: Wilkes D, Burlingham W, eds. *Immunobiology of Organ Transplant*. Springer; 2004:171-190.
257. Watanabe A, Su K-Y, Kuraoka M, et al. Self-tolerance curtails the B cell repertoire to microbial epitopes. *JCI Insight*. 2019;4:e122551.
258. Nelson RW, Beisang D, Tubo NJ, et al. T cell receptor cross-reactivity between similar foreign and self peptides influences naive cell population size and autoimmunity. *Immunity*. 2015;42:95-107.
259. Houston EG, Fink PJ. MHC drives TCR repertoire shaping, but not maturation, in recent thymic emigrants. *J Immunol*. 2009;183:7244-7249.
260. Miller SD, Wetzig RP, Claman HN. The induction of cell-mediated immunity and tolerance with protein antigens coupled to syngeneic lymphoid cells. *J Exp Med*. 1979;149:758-773.
261. Harris DE, Cairns L, Rosen FS, Borel Y. A natural model of immunologic tolerance. Tolerance to murine C5 is mediated by T cells, and antigen is required to maintain unresponsiveness. *J Exp Med*. 1982;156:567-584.
262. Ohashi PS, Oehen S, Buerki K, et al. Ablation of "tolerance" and induction of diabetes by virus infection in viral antigen transgenic mice. *Cell*. 1991;65:305-317.
263. Steinaa L, Rasmussen PB, Gautam A, Mouritsen S. Breaking B-cell tolerance and CTL tolerance in three OVA-transgenic mouse strains expressing different levels of OVA. *Scand J Immunol*. 2008;67:113-120.
264. Cowing C. Does T-cell restriction to Ia limit the need for self-tolerance? *Immunol Today*. 1985;6:72-74.
265. Yarkoni Y, Getahun A, Cambier JC. Molecular underpinning of B-cell anergy. *Immunol Rev*. 2010;237:249-263.
266. Parish IA, Heath WR. Too dangerous to ignore: self-tolerance and the control of ignorant autoreactive T cells. *Immunol Cell Biol*. 2008;86:146-152.
267. Rifkin IR, Leadbetter EA, Beaudette BC, et al. Immune complexes present in the sera of autoimmune mice activate rheumatoid factor B cells. *J Immunol*. 2000;165:1626-1633.
268. Viglianti GA, Lau CM, Hanley TM, Miko BA, Shlomchik MJ, Marshak-Rothstein A. Activation of autoreactive B cells by CpG dsDNA. *Immunity*. 2003;19:837-847.
269. Sweet RA, Ols ML, Cullen JL, Milam AV, Yagita H, Shlomchik MJ. Facultative role for T cells in extrafollicular Toll-like receptor-dependent autoreactive B-cell responses in vivo. *Proc Natl Acad Sci U S A*. 2011;108:7932-7937.
270. Nemazee DA. Immune complexes can trigger specific, T cell-dependent, autoanti-IgG antibody production in mice. *J Exp Med*. 1985;161:242-256.
271. Murakami M, Tsubata T, Okamoto M, et al. Antigen-induced apoptotic death of Ly-1 B cells responsible for autoimmune disease in transgenic mice. *Nature*. 1992;357:77-80.
272. Gardner JM, Metzger TC, McMahon EJ, et al. Extrathymic Aire-expressing cells are a distinct bone marrow-derived population that induce functional inactivation of CD4+ T cells. *Immunity*. 2013;39:560-572.

273. Cohen JN, Guidi CJ, Tewalt EF, et al. Lymph node-resident lymphatic endothelial cells mediate peripheral tolerance via Aire-independent direct antigen presentation. *J Exp Med.* 2010;207:681-688.
274. Fletcher AL, Lukacs-Kornek V, Reynoso ED, et al. Lymph node fibroblastic reticular cells directly present peripheral tissue antigen under steady-state and inflammatory conditions. *J Exp Med.* 2010;207:689-697.
275. Alferink J, Tafuri A, Vestweber D, Hallmann R, Hammerling GJ, Arnold B. Control of neonatal tolerance to tissue antigens by peripheral T cell trafficking. *Science.* 1998;282:1338-1341.
276. Morgan DJ, Kurts C, Kreuwel HT, Holst KL, Heath WR, Sherman LA. Ontogeny of T cell tolerance to peripherally expressed antigens. *Proc Natl Acad Sci U S A.* 1999;96:3854-3858.
277. Herzenberg LA. Toward a layered immune system. *Cell.* 1989;59:953-954.
278. Kincade PW, Owen JJT, Igarashi H, Kouro T, Yokota T, Rossi MID. Nature or nurture? Steady-state lymphocyte formation in adults does not recapitulate ontogeny. *Immunol Rev.* 2002;187:116-125.
279. Mold JE, Venkatasubrahmanyam S, Burt TD, et al. Fetal and adult hematopoietic stem cells give rise to distinct T cell lineages in humans. *Science.* 2010;330:1695-1699.
280. Mold JE, McCune JM. At the crossroads between tolerance and aggression: revisiting the "layered immune system" hypothesis. *Chimerism.* 2011;2:35-41.
281. Havran WL, Allison JP. Developmentally ordered appearance of thymocytes expressing different T-cell antigen receptors. *Nature.* 1988;335:443-445.
282. Ikuta K, Kina T, MacNeil I, et al. A developmental switch in thymic lymphocyte maturation potential occurs at the level of hematopoietic stem cells. *Cell.* 1990;62:863-874.
283. Zhang X, Zhivaki D, Lo-Man R. Unique aspects of the perinatal immune system. *Nat Rev Immunol.* 2017;17:495-507.
284. Kinder JM, Jiang TT, Ertelt JM, et al. Cross-generational reproductive fitness enforced by microchimeric maternal cells. *Cell.* 2015;162:505-515.
285. Thome JJC, Bickham KL, Ohmura Y, et al. Early-life compartmentalization of human T cell differentiation and regulatory function in mucosal and lymphoid tissues. *Nat Med.* 2016;22:72-77.
286. Burlingham WJ, Grailer AP, Heisey DM, et al. The effect of tolerance to noninherited maternal HLA antigens on the survival of renal transplants from sibling donors. *N Engl J Med.* 1998;339:1657-1664.
287. Allenspach EJ, Finn LS, Rendi MH, et al. Absence of functional fetal regulatory T cells in humans causes in utero organ-specific autoimmunity. *J Allergy Clin Immunol.* 2017;140:616-619.e7.
288. Upham JW, Lee PT, Holt BJ, et al. Development of interleukin-12-producing capacity throughout childhood. *Infect Immun.* 2002;70:6583-6588.
289. Kollmann TR, Crabtree J, Rein-Weston A, et al. Neonatal innate TLR-mediated responses are distinct from those of adults. *J Immunol.* 2009;183:7150-7160.
290. Friesen TJ, Ji Q, Fink PJ. Recent thymic emigrants are tolerized in the absence of inflammation. *J Exp Med.* 2016;213:913-920.
291. Paiva RS, Lino AC, Bergman M-L, et al. Recent thymic emigrants are the preferential precursors of regulatory T cells differentiated in the periphery. *Proc Natl Acad Sci U S A.* 2013;110:6494-6499.
292. Ellestad KK, Thangavelu G, Ewen CL, Boon L, Anderson CC. PD-1 is not required for natural or peripherally induced regulatory T cells: severe autoimmunity despite normal production of regulatory T cells. *Eur J Immunol.* 2014;44:3560-3572.
293. Scharschmidt TC, Vasquez KS, Truong H-A, et al. A wave of regulatory T cells into neonatal skin mediates tolerance to commensal microbes. *Immunity.* 2015;43:1011-1021.
294. Thangavelu G, Parkman JC, Ewen CL, Uwiera RR, Baldwin TA, Anderson CC. Programmed death-1 is required for systemic self-tolerance in newly generated T cells during the establishment of immune homeostasis. *J Autoimmun.* 2011;36:301-312.
295. Ellestad KK, Anderson CC. Two strikes and you're out? The pathogenic interplay of coinhibitor deficiency and lymphopenia-induced proliferation. *J Immunol.* 2017;198:2534-2541.
296. del Rio R, Sun Y, Alard P, Tung KSK, Teuscher C. H2 control of natural T regulatory cell frequency in the lymph node correlates with susceptibility to day 3 thymectomy-induced autoimmune disease. *J Immunol.* 2011;186:382-389.
297. Asano M, Toda M, Sakaguchi N, Sakaguchi S. Autoimmune disease as a consequence of developmental abnormality of a T cell subpopulation. *J Exp Med.* 1996;184:387-396.
298. Yang S, Fujikado N, Kolodin D, Benoist C, Mathis D. Regulatory T cells generated early in life play a distinct role in maintaining self-tolerance. *Science.* 2015;348:589-594.
299. Medawar PB. Immunity to homologous grafted skin; the fate of skin homografts transplanted to the brain, to subcutaneous tissue, and to the anterior chamber of the eye. *Br J Exp Pathol.* 1948;29:58-69.
300. Head JR, Neaves WB, Billingham RE. Immune privilege in the testis. *Transplantation.* 1983;36:423-430.
301. Niederkorn JY, Stein-Streilein J. History and physiology of immune privilege. *Ocul Immunol Inflamm.* 2010;18:19-23.
302. Asherson GL, Stone SH. Selective and specific inhibition of 24 hour skin reactions in the Guinea-pig. I. Immune deviation: description of the phenomenon and the effect of splenectomy. *Immunology.* 1965;9:205-217.
303. Bellgrau D, Gold D, Selawry H, Moore J, Franzusoff A, Duke RC. A role for CD95 ligand in preventing graft rejection. *Nature.* 1995;377:630-632.
304. Wilson SE, Li Q, Weng J, et al. The Fas-Fas ligand system and other modulators of apoptosis in the cornea. *Invest Ophthalmol Vis Sci.* 1996;37:1582-1592.
305. Zhou R, Horai R, Mattapallil MJ, Caspi RR. A new look at immune privilege of the eye: dual role for the vision-related molecule retinoic acid. *J Immunol.* 2011;187:4170-4177.
306. Zhou R, Horai R, Silver PB, et al. The living eye "disarms" uncommitted autoreactive T cells by converting them to Foxp3(+) regulatory cells following local antigen recognition. *J Immunol.* 2012;188:1742-1750.
307. Ferguson TA, Fletcher S, Herndon J, Griffith TS. Neuropeptides modulate immune deviation induced via the anterior chamber of the eye. *J Immunol.* 1995;155:1746-1756.
308. Kolfschoten Mvan der N, Schuurman J, Losen M, et al. Anti-inflammatory activity of human IgG4 antibodies by dynamic Fab arm exchange. *Science.* 2007;317:1554-1557.
309. James LK, Shamji MH, Walker SM, et al. Long-term tolerance after allergen immunotherapy is accompanied by selective persistence of blocking antibodies. *J Allergy Clin Immunol.* 2011;127:509-516.e5.
310. Salvin SB. Occurrence of delayed hypersensitivity during the development of arthus type hypersensitivity. *J Exp Med.* 1958;107:109-124.
311. Jutel M, Akdis M, Budak F, et al. IL-10 and TGF-β cooperate in the regulatory T cell response to mucosal allergens in normal immunity and specific immunotherapy. *Eur J Immunol.* 2003;33:1205-1214.
312. Karlsson MR, Rugtveit J, Brandtzaeg P. Allergen-responsive CD4+CD25+ regulatory T cells in children who have outgrown cow's milk allergy. *J Exp Med.* 2004;199:1679-1688.
313. Sun C-M, Hall JA, Blank RB, et al. Small intestine lamina propria dendritic cells promote de novo generation of Foxp3 T reg cells via retinoic acid. *J Exp Med.* 2007;204:1775-1785.
314. Niederkorn JY. The eye sees eye to eye with the immune system: the 2019 proctor lecture. *Invest Ophthalmol Vis Sci.* 2019;60:4489-4495.
315. Jones L, Chin L, Longo D, Kruisbeek A. Peripheral clonal elimination of functional T cells. *Science.* 1990;250:1726-1729.
316. Rocha B, von Boehmer H. Peripheral selection of the T cell repertoire. *Science.* 1991;251:1225-1228.
317. Singh NJ, Chen C, Schwartz RH. The impact of T cell intrinsic antigen adaptation on peripheral immune tolerance. *PLoS Biol.* 2006;4:e340.
318. Steinert EM, Schwartz RH, Singh NJ. At low precursor frequencies, the T-cell response to chronic self-antigen results in anergy without deletion. *Eur J Immunol.* 2012;42:2875-2880.
319. Singh NJ, Bando JK, Schwartz RH. Subsets of nonclonal neighboring CD4+ T cells specifically regulate the frequency of individual antigen-reactive T cells. *Immunity.* 2012;37:735-746.
320. Schonrich G, Kalinke U, Momburg F, et al. Down-regulation of T cell receptors on self-reactive T cells as a novel mechanism for extrathymic tolerance induction. *Cell.* 1991;65:293-304.
321. Chan WFN, Razavy H, Anderson CC. Differential susceptibility of allogeneic targets to indirect CD4 immunity generates split tolerance. *J Immunol.* 2008;181:4603-4612.
322. Kurts C, Heath WR, Kosaka H, Miller JF, Carbone FR. The peripheral deletion of autoreactive CD8+ T cells induced by cross-presentation of self-antigens involves signaling through CD95 (Fas, Apo-1). *J Exp Med.* 1998;188:415-420.
323. Kurts C, Carbone FR, Barnden M, et al. CD4+ T cell help impairs CD8+ T cell deletion induced by cross-presentation of self-antigens and favors autoimmunity. *J Exp Med.* 1997;186:2057-2062.
324. Ellestad KK, Thangavelu G, Haile Y, Lin J, Boon L, Anderson CC. Prior to peripheral tolerance, newly generated CD4 T cells maintain dangerous autoimmune potential: Fas- and perforin-independent autoimmunity controlled by programmed death-1. *Front Immunol.* 2018;9:12.
325. Malhotra D, Linehan JL, Dileepan T, et al. Tolerance is established in polyclonal CD4(+) T cells by distinct mechanisms, according to self-peptide expression patterns. *Nat Immunol.* 2016;17:187-195.
326. Morgan DJ, Kreuwel HT, Sherman LA. Antigen concentration and precursor frequency determine the rate of CD8+ T cell tolerance to peripherally expressed antigens. *J Immunol.* 1999;163:723-727.
327. Russell DM, Dembić Z, Morahan G, Miller JF, Bürki K, Nemazee D. Peripheral deletion of self-reactive B cells. *Nature.* 1991;354:308-311.
328. Nossal GJ, Pike BL. Clonal anergy: persistence in tolerant mice of antigen-binding B lymphocytes incapable of responding to antigen or mitogen. *Proc Natl Acad Sci U S A.* 1980;77:1602-1606.
329. Burnett DL, Reed JH, Christ D, Goodnow CC. Clonal redemption and clonal anergy as mechanisms to balance B cell tolerance and immunity. *Immunol Rev.* 2019;292:61-75.
330. Goodnow CC, Brink R, Adams E. Breakdown of self-tolerance in anergic B lymphocytes. *Nature.* 1991;352:532-536.
331. Gauld SB, Benschop RJ, Merrell KT, Cambier JC. Maintenance of B cell anergy requires constant antigen receptor occupancy and signaling. *Nat Immunol.* 2005;6:1160-1167.
332. Ochsenbein AF, Fehr T, Lutz C, et al. Control of early viral and bacterial distribution and disease by natural antibodies. *Science.* 1999;286:2156-2159.
333. Kinnunen T, Chamberlain N, Morbach H, et al. Accumulation of peripheral autoreactive B cells in the absence of functional human regulatory T cells. *Blood.* 2012;121:1595-1603.
334. Sng J, Ayoglu B, Chen JW, et al. AIRE expression controls the peripheral selection of autoreactive B cells. *Sci Immunol.* 2019;4:eaav6778.
335. Hayakawa K, Asano M, Shinton SA, et al. Positive selection of anti-thy-1 autoreactive B-1 cells and natural serum autoantibody production independent from bone marrow B cell development. *J Exp Med.* 2003;197:87-99.
336. Burnett DL, Langley DB, Schofield P, et al. Germinal center antibody mutation trajectories are determined by rapid self/foreign discrimination. *Science.* 2018;360:223-226.
337. Chan TD, Wood K, Hermes JR, et al. Elimination of germinal-center-derived self-reactive B cells is governed by the location and concentration of self-antigen. *Immunity.* 2012;37:893-904.
338. Jenkins MK, Schwartz RH. Antigen presentation by chemically modified splenocytes induces antigen-specific T cell unresponsiveness in vitro and in vivo. *J Exp Med.* 1987;165:302-319.
339. Quill H, Schwartz RH. Stimulation of normal inducer T cell clones with antigen presented by purified Ia molecules in planar lipid membranes: specific induction of a long-lived state of proliferative nonresponsiveness. *J Immunol.* 1987;138:3704-3712.
340. Jenkins MK, Ashwell JD, Schwartz RH. Allogeneic non-T spleen cells restore the responsiveness of normal T cell clones stimulated with antigen and chemically modified antigen-presenting cells. *J Immunol.* 1988;140:3324-3330.
341. Ben-Sasson SZ, Hu-Li J, Quiel J, et al. IL-1 acts directly on CD4 T cells to enhance their antigen-driven expansion and differentiation. *Proc Natl Acad Sci U S A.* 2009;106:7119-7124.
342. Schwartz RH. T cell anergy*. *Annu Rev Immunol.* 2003;21:305-334.
343. Lerner CG, Horton MR, Schwartz RH, Powell JD. Distinct requirements for C-C chemokine and IL-2 production by naive, previously activated, and anergic T cells. *J Immunol.* 2000;164:3996-4002.
344. Hayashi RJ, Loh DY, Kanagawa O, Wang F. Differences between responses of naive and activated T cells to anergy induction. *J Immunol.* 1998;160:33-38.
345. Kalekar LA, Schmiel SE, Nandiwada SL, et al. CD4+ T cell anergy prevents autoimmunity and generates regulatory T cell precursors. *Nat Immunol.* 2016;17:304-314.
346. McMahan CJ, Fink PJ. RAG reexpression and DNA recombination at T cell receptor loci in peripheral CD4+ T cells. *Immunity.* 1998;9:637-647.
347. Zehn D, Bevan MJ, Fink PJ. Cutting edge: TCR revision affects predominantly Foxp3 cells and skews them toward the Th17 lineage. *J Immunol.* 2007;179:5653-5657.
348. Mitchison NA. Induction of immunological paralysis in two zones of dosage. *Proc R Soc Lond B Biol Sci.* 1964;161:275-292.
349. Parish CR. The relationship between humoral and cell-mediated immunity. *Immunol Rev.* 1972;13:35-66.
350. Stefanová I, Dorfman JR, Germain RN. Self-recognition promotes the foreign antigen sensitivity of naive T lymphocytes. *Nature.* 2002;420:429-434.
351. Bhandoola A, Tai X, Eckhaus M, et al. Peripheral expression of self-MHC-II influences the reactivity and self-tolerance of mature CD4(+) T cells: evidence from a lymphopenic T cell model. *Immunity.* 2002;17:425-436.
352. Jubala CM, Lamerato-Kozicki AR, Borakove M, et al. MHC-dependent desensitization of intrinsic anti-self reactivity. *Cancer Immunol Immunother.* 2009;58:171-185.
353. Tanchot C, Barber DL, Chiodetti L, Schwartz RH. Adaptive tolerance of CD4+ T cells in vivo: multiple thresholds in response to a constant level of antigen presentation. *J Immunol.* 2001;167:2030-2039.
354. Cooke MP, Heath AW, Shokat KM, et al. Immunoglobulin signal transduction guides the specificity of B cell-T cell interactions and is blocked in tolerant self-reactive B cells. *J Exp Med.* 1994;179:425-438.
355. Chappert P, Bouladoux N, Naik S, Schwartz RH. Specific gut commensal flora locally alters T cell tuning to endogenous ligands. *Immunity.* 2013;38:1198-1210.
356. Hall BM, Pearce NW, Gurley KE, Dorsch SE. Specific unresponsiveness in rats with prolonged cardiac allograft survival after treatment with cyclosporine. III. Further characterization of the CD4+ suppressor cell and its mechanisms of action. *J Exp Med.* 1990;171:141-157.
357. Allan SE, Crome SQ, Crellin NK, et al. Activation-induced FOXP3 in human T effector cells does not suppress proliferation or cytokine production. *Int Immunol.* 2007;19:345-354.

358. Haribhai D, Williams JB, Jia S, et al. A requisite role for induced regulatory T cells in tolerance based on expanding antigen receptor diversity. *Immunity*. 2011;35:109-122.
359. Gregori S, Roncarolo MG. Engineered T regulatory type 1 cells for clinical application. *Front Immunol*. 2018;9:233. doi:10.3389/fimmu.2018.00233.
360. Meiler F, Zumkehr J, Klunker S, Rückert B, Akdis CA, Akdis M. In vivo switch to IL-10-secreting T regulatory cells in high dose allergen exposure. *J Exp Med*. 2008;205:2887-2898.
361. Rezende RM, Weiner HL. History and mechanisms of oral tolerance. *Semin Immunol*. 2017;30:3-11.
362. Knoop KA, Gustafsson JK, McDonald KG, et al. Microbial antigen encounter during a preweaning interval is critical for tolerance to gut bacteria. *Sci Immunol*. 2017;2:eaao1314. doi:10.1126/sciimmunol.aao1314.
363. Whibley N, Tucci A, Powrie F. Regulatory T cell adaptation in the intestine and skin. *Nat Immunol*. 2019;20:386-396.
364. Tsuji M, Komatsu N, Kawamoto S, et al. Preferential generation of follicular B helper T cells from Foxp3+ T cells in gut Peyer's patches. *Science*. 2009;323:1488-1492.
365. Kawamoto S, Maruya M, Kato LM, et al. Foxp3(+) T cells regulate immunoglobulin a selection and facilitate diversification of bacterial species responsible for immune homeostasis. *Immunity*. 2014;41:152-165.
366. Cong Y, Feng T, Fujihashi K, Schoeb TR, Elson CO. A dominant, coordinated T regulatory cell-IgA response to the intestinal microbiota. *Proc Natl Acad Sci U S A*. 2009;106:19256-19261.
367. Wawrzyniak M, O'Mahony L, Akdis M. Role of regulatory cells in oral tolerance. *Allergy Asthma Immunol Res*. 2017;9:107-115.
368. Hara S, Sasaki T, Satoh-Takayama N, et al. Dietary antigens induce germinal center responses in Peyer's patches and antigen-specific IgA production. *Front Immunol*. 2019;10:2432.
369. Finkelman FD, Lees A, Birnbaum R, Gause WC, Morris SC. Dendritic cells can present antigen in vivo in a tolerogenic or immunogenic fashion. *J Immunol*. 1996;157:1406-1414.
370. Iberg CA, Jones A, Hawiger D. Dendritic cells as inducers of peripheral tolerance. *Trends Immunol*. 2017;38:793-804.
371. Adler AJ, Marsh DW, Yochum GS, et al. CD4+ T cell tolerance to parenchymal self-antigens requires presentation by bone marrow-derived antigen-presenting cells. *J Exp Med*. 1998;187:1555-1564.
372. Iberg CA, Hawiger D. Natural and induced tolerogenic dendritic cells. *J Immunol*. 2020;204:733-744.
373. Belz GT, Behrens GMN, Smith CM, et al. The CD8alpha(+) dendritic cell is responsible for inducing peripheral self-tolerance to tissue-associated antigens. *J Exp Med*. 2002;196:1099-1104.
374. Sinclair NR, Panoskaltsis A. Immunoregulation by Fc signals A mechanism for self-nonself discrimination. *Immunol Today*. 1987;8:76-79.
375. Phillips NE, Parker DC. Cross-linking of B lymphocyte Fc gamma receptors and membrane immunoglobulin inhibits anti-immunoglobulin-induced blastogenesis. *J Immunol*. 1984;132:627-632.
376. Phillips NE, Parker DC. Fc-dependent inhibition of mouse B cell activation by whole anti-mu antibodies. *J Immunol*. 1983;130:602-606.
377. Phillips NE, Parker DC. Subclass specificity of Fc gamma receptor-mediated inhibition of mouse B cell activation. *J Immunol*. 1985;134:2835-2838.
378. Klaus GG, Bijsterbosch MK, O'Garra A, Harnett MM, Rigley KP. Receptor signalling and crosstalk in B lymphocytes. *Immunol Rev*. 1987;99:19-38.
379. Montaudouin C, Anson M, Hao Y, et al. Quorum sensing contributes to activated IgM-secreting B cell homeostasis. *J Immunol*. 2012;190:106-114.
380. Verbeek JS, Hirose S, Nishimura H. The complex association of FcγRIIb with autoimmune susceptibility. *Front Immunol*. 2019;10:2061.
381. Cornall RJ, Cyster JG, Hibbs ML, et al. Polygenic autoimmune traits: Lyn, CD22, and SHP-1 are limiting elements of a biochemical pathway regulating BCR signaling and selection. *Immunity*. 1998;8:497-508.
382. Cyster JG, Goodnow CC. Tuning antigen receptor signaling by CD22: integrating cues from antigens and the microenvironment. *Immunity*. 1997;6:509-517.
383. Hoffmann A, Kerr S, Jellusova J, et al. Siglec-G is a B1 cell-inhibitory receptor that controls expansion and calcium signaling of the B1 cell population. *Nat Immunol*. 2007;8:695-704.
384. Müller J, Lunz B, Schwab I, et al. Siglec-G deficiency leads to autoimmunity in aging C57BL/6 mice. *J Immunol*. 2015;195:51-60.
385. Tsubata T. Inhibitory B cell co-receptors and autoimmune diseases. *Immunol Med*. 2019;42:108-116. doi:10.1080/25785826.2019.1660038:1-9.
386. Akatsu C, Shinagawa K, Numoto N, et al. CD72 negatively regulates B lymphocyte responses to the lupus-related endogenous toll-like receptor 7 ligand Sm/RNP. *J Exp Med*. 2016;213:2691-2706.
387. Burshtyn DN, Morcos C. The expanding spectrum of ligands for leukocyte Ig-like receptors. *J Immunol*. 2016;196:947-955.
388. Blery M, Kubagawa H, Chen C-C, Vely F, Cooper MD, Vivier E. The paired Ig-like receptor PIR-B is an inhibitory receptor that recruits the protein-tyrosine phosphatase SHP-1. *Proc Natl Acad Sci U S A*. 1998;95:2446-2451.
389. Ward-Kavanagh LK, Lin WW, Šedý JR, Ware CF. The TNF receptor superfamily in co-stimulating and co-inhibitory responses. *Immunity*. 2016;44:1005-1019.
390. Qu W-M, Miyazaki T, Terada M, et al. Genetic dissection of vasculitis in MRL/lpr lupus mice: a novel susceptibility locus involving theCD72c allele. *Eur J Immunol*. 2000;30:2027-2037.
391. Müller J, Nitschke L. The role of CD22 and Siglec-G in B-cell tolerance and autoimmune disease. *Nat Rev Rheumatol*. 2014;10:422-428.
392. Linhares ADS, Leitner J, Grabmeier-Pfistershammer K, Steinberger P. Not all immune checkpoints are created equal. *Front Immunol*. 2018;9:1909.
393. Hippen KL, Tze LE, Behrens TW. Cd5 maintains tolerance in anergic B cells. *J Exp Med*. 2000;191:883-890.
394. Mempel TR, Henrickson SE, von Andrian UH. T-cell priming by dendritic cells in lymph nodes occurs in three distinct phases. *Nature*. 2004;427:154-159.
395. Henrickson SE, Mempel TR, Mazo IB, et al. T cell sensing of antigen dose governs interactive behavior with dendritic cells and sets a threshold for T cell activation. *Nat Immunol*. 2008;9:282-291.
396. Waterhouse P, Penninger JM, Timms E, et al. Lymphoproliferative disorders with early lethality in mice deficient in Ctla-4. *Science*. 1995;270:985-988.
397. Tai X, Laethem FV, Sharpe AH, Singer A. Induction of autoimmune disease in CTLA-4/mice depends on a specific CD28 motif that is required for in vivo costimulation. *Proc Natl Acad Sci U S A*. 2007;104:13756-13761.
398. Ise W, Kohyama M, Nutsch KM, et al. CTLA-4 suppresses the pathogenicity of self antigen-specific T cells by cell-intrinsic and cell-extrinsic mechanisms. *Nat Immunol*. 2010;11:129-135.
399. Huang C, Zhu H-X, Yao Y, et al. Immune checkpoint molecules. Possible future therapeutic implications in autoimmune diseases. *J Autoimmun*. 2019;104:102333.
400. Masteller EL, Chuang E, Mullen AC, Reiner SL, Thompson CB. Structural analysis of CTLA-4 function in vivo. *J Immunol*. 2000;164:5319-5327.
401. Qureshi OS, Zheng Y, Nakamura K, et al. Trans-endocytosis of CD80 and CD86: a molecular basis for the cell-extrinsic function of CTLA-4. *Science*. 2011;332:600-603.
402. Martin-Orozco N, Wang Y-H, Yagita H, Dong C. Cutting edge: programed death (PD) ligand-1/PD-1 interaction is required for CD8+ T cell tolerance to tissue antigens. *J Immunol*. 2006;177:8291-8295.
403. Pauken KE, Nelson CE, Martinov T, et al. Cutting edge: identification of autoreactive CD4+ and CD8+ T cell subsets resistant to PD-1 pathway blockade. *J Immunol*. 2015;194:3551-3555.
404. Zhao P, Wang P, Dong S, et al. Depletion of PD-1-positive cells ameliorates autoimmune disease. *Nat Biomed Eng*. 2019;3:292-305.
405. Agata Y, Kawasaki A, Nishimura H, et al. Expression of the PD-1 antigen on the surface of stimulated mouse T and B lymphocytes. *Int Immunol*. 1996;8:765-772.
406. Nishimura H, Nose M, Hiai H, Minato N, Honjo T. Development of lupus-like autoimmune diseases by disruption of the PD-1 gene encoding an ITIM motif-carrying immunoreceptor. *Immunity*. 1999;11:141-151.
407. Keir ME, Liang SC, Guleria I, et al. Tissue expression of PD-L1 mediates peripheral T cell tolerance. *J Exp Med*. 2006;203:883-895.
408. Sugiura D, Maruhashi T, Okazaki I, et al. Restriction of PD-1 function by cis -PD-L1/CD80 interactions is required for optimal T cell responses. *Science*. 2019;364:558-566.
409. Dong H, Zhu G, Tamada K, Chen L. B7-H1, a third member of the B7 family, co-stimulates T-cell proliferation and interleukin-10 secretion. *Nat Med*. 1999;5:1365-1369.
410. Deng R, Cassady K, Li X, et al. B7H1/CD80 interaction augments PD-1–dependent T cell apoptosis and ameliorates graft-versus-host disease. *J Immunol*. 2014;194:560-574.
411. Memarnejadian A, Meilleur CE, Shaler CR, et al. PD-1 blockade promotes epitope spreading in anticancer CD8 + T cell responses by preventing fratricidal death of subdominant clones to relieve immunodomination. *J Immunol*. 2017;199:3348-3359.
412. Shull MM, Ormsby I, Kier AB, et al. Targeted disruption of the mouse transforming growth factor-β1 gene results in multifocal inflammatory disease. *Nature*. 1992;359:693-699.
413. Zhang N, Bevan MJ. TGF-β signaling to T cells inhibits autoimmunity during lymphopenia-driven proliferation. *Nat Immunol*. 2012;13:667-673.
414. Cheung TC, Oborne LM, Steinberg MW, et al. T cell intrinsic heterodimeric complexes between HVEM and BTLA determine receptivity to the surrounding microenvironment. *J Immunol*. 2009;183:7286-7296.
415. Yu X, Zheng Y, Mao R, Su Z, Zhang J. BTLA/HVEM signaling: milestones in research and role in chronic hepatitis B virus infection. *Front Immunol*. 2019;10:617.
416. Liu X, Alexiou M, Martin-Orozco N, et al. Cutting edge: a critical role of B and T lymphocyte attenuator in peripheral T cell tolerance induction. *J Immunol*. 2009;182:4516-4520.
417. Flynn R, Hutchinson T, Murphy KM, Ware CF, Croft M, Salek-Ardakani S. CD8 T cell memory to a viral pathogen requires trans cosignaling between HVEM and BTLA. *PloS One*. 2013;8:e77991.
418. Hurchla MA, Sedy JR, Murphy KM. Unexpected role of B and T lymphocyte attenuator in sustaining cell survival during chronic allostimulation. *J Immunol*. 2007;178:6073-6082.
419. Monney L, Sabatos CA, Gaglia JL, et al. Th1-specific cell surface protein Tim-3 regulates macrophage activation and severity of an autoimmune disease. *Nature*. 2002;415:536-541.
420. Triebel F, Jitsukawa S, Baixeras E, et al. LAG-3, a novel lymphocyte activation gene closely related to CD4. *J Exp Med*. 1990;171:1393-1405.
421. Aubert RD, Kamphorst AO, Sarkar S, et al. Antigen-specific CD4 T-cell help rescues exhausted CD8 T cells during chronic viral infection. *Proc Natl Acad Sci U S A*. 2011;108:21182-21187.
422. Ghoneim HE, Fan Y, Moustaki A, et al. De novo epigenetic programs inhibit PD-1 blockade-mediated T cell rejuvenation. *Cell*. 2017;170:142-157.e19.
423. Blank CU, Haining WN, Held W, et al. Defining "T cell exhaustion." *Nat Rev Immunol*. 2019;19:665-674.
424. Pauken KE, Sammons MA, Odorizzi PM, et al. Epigenetic stability of exhausted T cells limits durability of reinvigoration by PD-1 blockade. *Science*. 2016;354:1160-1165.
425. Chihara N, Madi A, Kondo T, et al. Induction and transcriptional regulation of the co-inhibitory gene module in T cells. *Nature*. 2018;558:454-459.
426. Turnis ME, Sawant DV, Szymczak-Workman AL, et al. Interleukin-35 limits anti-tumor immunity. *Immunity*. 2016;44:316-329.
427. Sullivan JA, Tomita Y, Jankowska-Gan E, et al. Treg-cell-derived IL-35-coated extracellular vesicles promote infectious tolerance. *Cell Rep*. 2020;30:1039-1051.e5.
428. Rodig N, Ryan T, Allen JA, et al. Endothelial expression of PD-L1 and PD-L2 down-regulates CD8+ T cell activation and cytolysis. *Eur J Immunol*. 2003;33:3117-3126.
429. Zhu C, Anderson AC, Schubart A, et al. The Tim-3 ligand galectin-9 negatively regulates T helper type 1 immunity. *Nat Immunol*. 2005;6:1245-1252.
430. Mercier IL, Lines JL, Noelle RJ. Beyond CTLA-4 and PD-1, the generation Z of negative checkpoint regulators. *Front Immunol*. 2015;6:418.
431. Huard B, Prigent P, Pagès F, Bruniquel D, Triebel F. T cell major histocompatibility complex class II molecules down-regulate CD4+ T cell clone responses following LAG-3 binding. *Eur J Immunol*. 1996;26:1180-1186.
432. Huang C-T, Workman CJ, Flies D, et al. Role of LAG-3 in regulatory T cells. *Immunity*. 2004;21:503-513.
433. Kouo T, Huang L, Pucsek AB, et al. Galectin-3 shapes antitumor immune responses by suppressing CD8+ T cells via LAG-3 and inhibiting expansion of plasmacytoid dendritic cells. *Cancer Immunol Res*. 2015;3:412-423.
434. Grosso JF, Kelleher CC, Harris TJ, et al. LAG-3 regulates CD8+ T cell accumulation and effector function in murine self- and tumor-tolerance systems. *J Clin Invest*. 2007;117:3383-3392.
435. Wang J, Sanmamed MF, Datar I, et al. Fibrinogen-like protein 1 is a major immune inhibitory ligand of LAG-3. *Cell*. 2019;176:334-347.e12.
436. Li N, Wang Y, Forbes K, et al. Metalloproteases regulate T-cell proliferation and effector function via LAG-3. *EMBO J*. 2007;26:494-504.
437. Huang R-Y, Francois A, McGray AR, Miliotto A, Odunsi K. Compensatory upregulation of PD-1, LAG-3, and CTLA-4 limits the efficacy of single-agent checkpoint blockade in metastatic ovarian cancer. *OncoImmunology*. 2016;6:e1249561.
438. Okazaki T, Okazaki I, Wang J, et al. PD-1 and LAG-3 inhibitory co-receptors act synergistically to prevent autoimmunity in mice. *J Exp Med*. 2011;208:395-407.
439. Gayden T, Sepulveda FE, Khuong-Quang D-A, et al. Germline HAVCR2 mutations altering TIM-3 characterize subcutaneous panniculitis-like T cell lymphomas with hemophagocytic lymphohistiocytic syndrome. *Nat Genet*. 2018;50:1650-1657.
440. Griffith TS, Brunner T, Fletcher SM, Green DR, Ferguson TA. Fas ligand-induced apoptosis as a mechanism of immune privilege. *Science*. 1995;270:1189-1192.
441. Yamada A, Arakaki R, Saito M, Kudo Y, Ishimaru N. Dual role of Fas/FasL-mediated signal in peripheral immune tolerance. *Front Immunol*. 2017;8:403.
442. Suda T, Hashimoto H, Tanaka M, Ochi T, Nagata S. Membrane Fas ligand kills human peripheral blood T lymphocytes, and soluble Fas ligand blocks the killing. *J Exp Med*. 1997;186:2045-2050.
443. Krueger A, Fas SC, Baumann S, Krammer PH. The role of CD95 in the regulation of peripheral T-cell apoptosis. *Immunol Rev*. 2003;193:58-69.
444. Brint E, O'Callaghan G, Houston A. Life in the Fas lane: differential outcomes of Fas signaling. *Cell Mol Life Sci*. 2013;70:4085-4099.

445. Hao Z, Duncan GS, Seagal J, et al. Fas receptor expression in germinal-center B cells is essential for T and B lymphocyte homeostasis. *Immunity*. 2008;29:615-627.
446. Butt D, Chan TD, Bourne K, et al. FAS inactivation releases unconventional germinal center B cells that escape antigen control and drive IgE and autoantibody production. *Immunity*. 2015;42:890-902.
447. Rathmell JC, Townsend SE, Xu JC, Flavell RA, Goodnow CC. Expansion or elimination of B cells in vivo: dual roles for CD40- and Fas (CD95)-ligands modulated by the B cell antigen receptor. *Cell*. 1996;87:319-329.
448. Tao R, Wang L, Murphy KM, Fraser CC, Hancock WW. Regulatory T cell expression of herpesvirus entry mediator suppresses the function of B and T lymphocyte attenuator-positive effector T cells. *J Immunol*. 2008;180:6649-6655.
449. Jones A, Bourque J, Kuehm L, et al. Immunomodulatory functions of BTLA and HVEM govern induction of extrathymic regulatory T cells and tolerance by dendritic cells. *Immunity*. 2016;45:1066-1077.
450. Snapper CM, Hooley JJ, Atasoy U, Finkelman FD, Paul WE. Differential regulation of murine B cell Fc gamma RII expression by CD4+ T helper subsets. *J Immunol*. 1989;143:2133-2141.
451. Chen G, Kim YH, Li H, et al. PD-L1 inhibits acute and chronic pain by suppressing nociceptive neuron activity via PD-1. *Nat Neurosci*. 2017;20:917-926.
452. Zhu Y, Yao S, Augustine MM, et al. Neuron-specific SALM5 limits inflammation in the CNS via its interaction with HVEM. *Sci Adv*. 2016;2:e1500637.
453. Lamagna C, Hu Y, DeFranco AL, Lowell CA. B cell-specific loss of Lyn kinase leads to autoimmunity. *J Immunol*. 2013;192:919-928.
454. Yokosuka T, Takamatsu M, Kobayashi-Imanishi W, Hashimoto-Tane A, Azuma M, Saito T. Programmed cell death 1 forms negative costimulatory microclusters that directly inhibit T cell receptor signaling by recruiting phosphatase SHP2. *J Exp Med*. 2012;209:1201-1217.
455. Hui E, Cheung J, Zhu J, et al. T cell costimulatory receptor CD28 is a primary target for PD-1–mediated inhibition. *Science*. 2017;355:1428-1433.
456. Kamphorst AO, Wieland A, Nasti T, et al. Rescue of exhausted CD8 T cells by PD-1–targeted therapies is CD28-dependent. *Science*. 2017;355:1423-1427.
457. Stanford SM, Bottini N. PTPN22: the archetypal non-HLA autoimmunity gene. *Nat Rev Rheumatol*. 2014;10:602-611.
458. Cloutier J-F, Veillette A. Cooperative inhibition of T-cell antigen receptor signaling by a complex between a kinase and a phosphatase. *J Exp Med*. 1999;189:111-121.
459. Arechiga AF, Habib T, He Y, et al. Cutting edge: the PTPN22 allelic variant associated with autoimmunity impairs B cell signaling. *J Immunol*. 2009;182:3343-3347.
460. Amin RH, Schlissel MS. Foxo1 directly regulates the transcription of recombination-activating genes during B cell development. *Nat Immunol*. 2008;9:613-622.
461. Herzog S, Hug E, Meixlsperger S, et al. SLP-65 regulates immunoglobulin light chain gene recombination through the PI(3)K-PKB-Foxo pathway. *Nat Immunol*. 2008;9:623-631.
462. Hayashi K, Nojima T, Goitsuka R, Kitamura D. Impaired receptor editing in the primary B cell repertoire of BASH-deficient mice. *J Immunol*. 2004;173:5980-5988.
463. Lang J, Arnold B, Hammerling G, et al. Enforced Bcl-2 expression inhibits antigen-mediated clonal elimination of peripheral B cells in an antigen dose–dependent manner and promotes receptor editing in autoreactive, immature B cells. *J Exp Med*. 1997;186:1513-1522.
464. Enders A, Bouillet P, Puthalakath H, Xu Y, Tarlinton DM, Strasser A. Loss of the Pro-apoptotic BH3-only Bcl-2 family member Bim inhibits BCR stimulation–induced apoptosis and deletion of autoreactive B cells. *J Exp Med*. 2003;198:1119-1126.
465. Bouillet P, Purton JF, Godfrey DI, et al. BH3-only Bcl-2 family member Bim is required for apoptosis of autoreactive thymocytes. *Nature*. 2002;415:922-926.
466. Hu Q, Sader A, Parkman JC, Baldwin TA. Bim-mediated apoptosis is not necessary for thymic negative selection to ubiquitous self-antigens. *J Immunol*. 2009;183:7761-7767.
467. Graham DB, Bell MP, Huntoon CJ, et al. CD28 ligation costimulates cell death but not maturation of double-positive thymocytes due to defective ERK MAPK signaling. *J Immunol*. 2006;177:6098-6107.
468. Getahun A, Beavers NA, Larson SR, Shlomchik MJ, Cambier JC. Continuous inhibitory signaling by both SHP-1 and SHIP-1 pathways is required to maintain unresponsiveness of anergic B cells. *J Exp Med*. 2016;213:751-769.
469. Bachmaier K, Krawczyk C, Kozieradzki I, et al. Negative regulation of lymphocyte activation and autoimmunity by the molecular adaptor Cbl-b. *Nature*. 2000;403:211-216.
470. Fang D, Liu Y-C. Proteolysis-independent regulation of PI3K by Cbl-b–mediated ubiquitination in T cells. *Nat Immunol*. 2001;2:870-875.
471. Jenkins MK, Pardoll DM, Mizuguchi J, Chused TM, Schwartz RH. Molecular events in the induction of a nonresponsive state in interleukin 2-producing helper T-lymphocyte clones. *Proc Natl Acad Sci U S A*. 1987;84:5409-5413.
472. Healy JI, Dolmetsch RE, Timmerman LA, et al. Different nuclear signals are activated by the B cell receptor during positive versus negative signaling. *Immunity*. 1997;6:419-428.
473. Akkaya M, Traba J, Roesler AS, et al. Second signals rescue B cells from activation-induced mitochondrial dysfunction and death. *Nat Immunol*. 2018;19:871-884.
474. Macián F, García-Cózar F, Im S-H, Horton HF, Byrne MC, Rao A. Transcriptional mechanisms underlying lymphocyte tolerance. *Cell*. 2002;109:719-731.
475. Barrington RA, Borde M, Rao A, Carroll MC. Involvement of NFAT1 in B cell self-tolerance. *J Immunol*. 2006;177:1510-1515.
476. Matzinger P. Tolerance, danger, and the extended family. *Annu Rev Immunol*. 1994;12:991-1045.

CHAPTER 33

Fc Receptors and Their Role in Immune Regulation and Inflammation

Jeffrey V. Ravetch, MD, PhD • Aaron Gupta, BS, PhD • Falk Nimmerjahn, PhD

HISTORICAL BACKGROUND

Cellular receptors for immunoglobulins (Igs) were anticipated by the description of cytophilic antibodies of the Ig G class, identified by Boyden and Sorkin in 1960.[1] These antibodies conferred upon normal cells, like macrophages, the capacity to specifically absorb antigen. Using sheep red blood cells (RBCs) as the antigen resulted in rosette formation between the cytophilic anti-sheep RBC antibodies and macrophages and provided a convenient means of visualization of the binding of cytophilic antibodies with normal cells. Subsequent studies by Berken and Benacerraf[2] suggested that the crystallized fragment (Fc) of the cytophilic antibody interacted with a cell surface receptor on macrophages. Similar studies on B lymphocytes extended the generality of these receptors and led to the term Fc receptor (FcR) to denote the surface molecules on lymphoid and myeloid cells that are capable of interacting with the Fc of Ig molecules.[3] Studies on IgE, IgM, and IgA demonstrated the existence of distinct receptors for those isotypes as well on various immune cell types. Detailed biochemical characterization of FcRs was inaugurated by the studies of Kulczycki et al.[4] on the high-affinity IgE FcR of mast cells, revealing a heterooligomeric $\alpha\beta\gamma 2$ subunit structure. A distinction between FcRs for the IgE and IgG isotypes emerged with the observation of the very high (10^{10} M^{-1}) binding affinity of IgE for its receptor in comparison with the low binding (10^{6} M^{-1}) of mouse IgG1 to its receptor. This distinction led to the realization that the functional IgG1 ligand was exclusively in the form of an immune complex (IC), whereas IgE binding occurred through monomer interaction with its receptor. This difference in binding affinity had significant functional implications for the structures of these receptors and mechanisms by which each isotype activated its target cell. Determination of the structure of these receptors was facilitated by their molecular cloning, beginning with the IgG FcRs,[5,6] followed by the IgE FcRs.[7] Two distinct types of IgG receptors, differing in their transmembrane and cytoplasmic sequences, were identified, which thus offered a molecular explanation for the apparent contradictory activation and inhibitory activities attributed to IgG FcRs. The primary structure of the subunits of the high-affinity IgE FcR revealed homology in the ligand binding α subunit to its IgG counterparts. However, the extent of similarity between these receptors became apparent with the observation that the γ chain subunit was common to both IgG and IgE FcRs, providing both assembly and signaling functions to these activation receptors.[8,9] This common structure suggested a functional link between IC diseases and allergic reactions, a prediction that was confirmed through mouse knockout studies of IgG FcRs.[10,11] The FcRs, through their dependence on the immunoreceptor tyrosine-based activation motif (ITAM) pathway of cellular activation, belonged to the family of immunoreceptors that included the antigen receptors on B cells and T cells, and NK cell receptors on natural killer (NK) cells. Three-dimensional crystal structures have been solved for the low-affinity IgG FcRs[12,13] and the high-affinity IgE FcRs,[14] alone and in complex with their Ig ligands,[15,16] further establishing the close structural link between these Ig receptors.

The functional roles of IgG FcRs were suggested by the distribution of these receptors on both lymphoid and myeloid cells.[17] On myeloid cells, they were presumed to mediate effector cell activation, resulting in phagocytosis, antibody-dependent cellular cytotoxicity (ADCC), and release of inflammatory mediators. However, the well-known ability of the classical pathway of complement to generate activated fragments in response to ICs capable of inducing inflammatory responses by myeloid cells complicated the interpretation of the physiological role of IgG FcRs. Thus, the contribution of IgG FcRs to the mechanism of IC-mediated inflammation, as distinct from the role of complement, remained uncertain. Insight into this distinction was gained through the generation of mouse strains specifically deficient or blocked in either FcRs[10,18-20] or components of the classical complement pathway.[21] Studies on IC-mediated inflammatory responses in these animals, such as the Arthus reaction, led to the realization that IgG FcRs and not the classical pathway of complement activation were the functional mediators of inflammatory responses triggered by ICs.[20,22,23] The situation for IgE was less confounding, and the identification and characterization of a high-affinity receptor for this isotype on mast cells offered a plausible explanation for many of the inflammatory features of allergic reactions,[24] validated later by mouse knockouts of this receptor. IgG ICs had also been observed to mediate suppression of B cell responses; thus, the presence of an IgG FcR activity on B cells provided a possible, but uncharacterized, mechanism for this inhibitory activity. Molecular characterization of this inhibitory activity for the B cell FcR, FcγRIIB, resulted in the first detailed description of an inhibitory motif, now termed the immunoreceptor tyrosine-based inhibitory motif (ITIM),[25,26] and the signaling pathway by which it abrogates ITAM-triggered activation. The ITIM mechanism is now recognized as ubiquitous and has resulted in the recognition of a large family of inhibitory receptors on immune cells that function to maintain proper thresholds for activation and abrogate activation responses to terminate an immune reaction.[27]

FcRs are now recognized as central mediators of antibody-triggered responses, coupling the innate and adaptive immune responses in effector cell activation.[28] In addition to these specialized roles, the IgG FcRs have served as an example of the emerging class of balanced immunoreceptors, in which activation and inhibition are tightly coupled in response to ligand binding. Perturbations in either arm of the response have been shown to lead to pathological consequences and have been taken as a paradigm of how these systems are likely to work for those paired immunoreceptors with unknown ligand-binding functions. The newly described roles for FcRs in maintaining peripheral tolerance, shaping the antibody repertoire, regulating antigen-presenting cell (APC) maturation, and promoting mast cell survival indicate the diversity of functions that these receptors possess and their central role in modulating both afferent and efferent responses in the immune response.

This chapter focuses primarily on the IgG and IgE FcRs, for which substantial data on their structure, function, regulation, and role in a variety of physiological and pathological conditions are now available. The similarity in structure and signaling between those receptors and other members of this family, such as the IgA FcR and FcαR/CD89, is also discussed. The canonical Fcγ receptors will be referred to as "type I Fcγ receptors," reflecting their shared structural and functional properties. In contrast, C-type lectins such as DC-SIGN (CD209) and CD23 (FcεRII), which bind a sialylated form of IgG Fc and, in the case of CD23, IgE with low affinity, represent the type II Fcγ receptors and will be discussed in detail below. Other Ig receptors with specialized functions, such as the FcRn[29] and the polymeric Ig FcR,[30] will be discussed in Chapter 19 in greater detail. In addition, Fc receptors for both mouse and human IgM (FcμR) and IgM and IgA (Fcα/μR) have been identified. While all of the previous Fc receptors are cell surface proteins, a cytosolic IgG Fc receptor called TRIM21 has been identified more recently. TRIM21 belongs to the large family of tripartite motif (TRIM) containing proteins, which are characterized by an N-terminal RING domain and by their E3 ubiquitin ligase activity. With respect to its function, it has been suggested that TRIM21 acts as an intracellular pattern recognition receptor detecting cytosolic antibodies (IgM, IgA, IgG) present on opsonized viruses or bacteria, which have escaped endosomal/lysosomal degradation. Upon TRIM21 binding, the opsonized pathogen is targeted for proteasomal degradation, preventing pathogen replication.[31]

STRUCTURE AND EXPRESSION

Molecular Genetics

Two general classes of canonical FcRs are now recognized: the activation receptors, characterized by the presence of a cytoplasmic ITAM sequence, either encoded within the receptor polypeptide chain itself or via association of the activating FcRs with ITAM-containing adaptor molecules, and the inhibitory receptor, characterized by the presence of an ITIM sequence present in the cytosolic tail of the receptor. These two classes of receptors function in concert and are usually found coexpressed on the cell surface. Thus coengagement of both signaling pathways is the rule, setting thresholds for and ultimately determining the physiological outcome of effector cell responses. Among the factors that determine this threshold level are the actual affinities of the individual activating and inhibitory receptors for a specific IgG ligand and the expression level of the receptor pairs on immune effector cells. Importantly, the affinity of different antibody isotypes and subclasses for their respective activating and inhibitory FcRs varies significantly, thus explaining the differential activity of antibody isotypes *in vivo*.[20,32,33] In addition, alleles of specific FcRs have been described, which alter their affinity for individual subclasses, thus accounting for the variable responses seen in a population to an antibody (see below). Noncanonical FcRs such as the IgM receptors FcμR, Fcα/μR, the poly-Ig receptor, and the type II FcγRs DC-SIGN and CD23 contain neither ITAM nor ITIM motifs, but balance activating and inhibitory signaling indirectly as discussed below.

Subunit Composition

Canonical FcRs are typically type I integral membrane glycoproteins consisting of, at the least, a ligand recognition α subunit that confers isotype specificity for the receptor. α subunits for IgG, IgE, IgM, and IgA have been described.[17,34-36] These subunits typically consist of two extracellular domains of the Ig V type superfamily: a single transmembrane domain and a relatively short intracytoplasmic domain. In activation FcRs, a signaling subunit of the γ family, also termed FcR common γ chain or FcRγ, is often found to be noncovalently associated with the FcR α chain, resulting in an αγ2 complex. It is important not to confuse the FcR common γ chain (FcRγ) with the common cytokine γ chain (γc), which is associated with cytokine receptors such as the receptors for IL-2, 4, 9, 15, or 21.[37] The ITIM containing inhibitory FcγRIIB molecule, in contrast, is expressed as a single-chain receptor. The α subunits have apparent molecular weights of between 40 and 75 kDa and share significant amino acid sequence homology in their extracellular domains. There is evidence for a rapid evolution within the human FcR locus, which is characterized by allelic receptor variants, pseudogenes, and copy number polymorphisms. Alternatively spliced forms of FcγRIIB modify the intracytoplasmic domain of this molecule. For example, the

B2 form lacks sequences that inhibit internalization and thus demonstrates enhanced internalization of ICs, in comparison with FcγRIIB1. However, all the splice variants contain the ITIM motif, a necessary and sufficient domain for mediating inhibitory signaling.[34] The conservation of this inhibitory sequence motif in mice and humans, its presence in all splice variants, and the hyperresponsive phenotypes generated in mice deficient in this receptor all support inhibition as the central function of FcγRIIB. The specific structures of the α subunits of the canonical FcRs are shown in Figure 33.1. The notable exceptions to the general structure just outlined are seen for the high-affinity FcγRIα subunit, which has three extracellular domains; the activation FcγRIIA/FcγRIIC α subunits, which does not require additional subunits for assembly or signaling; and the GPI-linked FcγRIIIB, which attaches to the cell surface through a glycosyl-phosphatidylinositol linkage, rather than through a transmembrane domain. Secreted forms of FcRs, most likely derived from neutrophils generated via protease cleavage from the cell surface, have been detected in the serum, although their function remains to be established.

The γ subunit is found associated with activation of IgG, IgE, and IgA FcRs, as well as with non-FcR molecules, such as paired Ig-like receptor A (PIR-A), macrophage-inducible C-type lectin (Mincle), Dectin-2, NK cell cytotoxicity receptors, the T cell receptor, but not with IgM receptors.[38-40] It is required for assembly of the α subunits of these receptors by protecting these subunits from degradation in the endoplasmic reticulum. The FcRγ chain is found as a disulfide-linked homodimer, with a short extracellular domain containing the cysteine involved in dimerization, a transmembrane domain, and an intracytoplasmic domain containing the ITAM. An

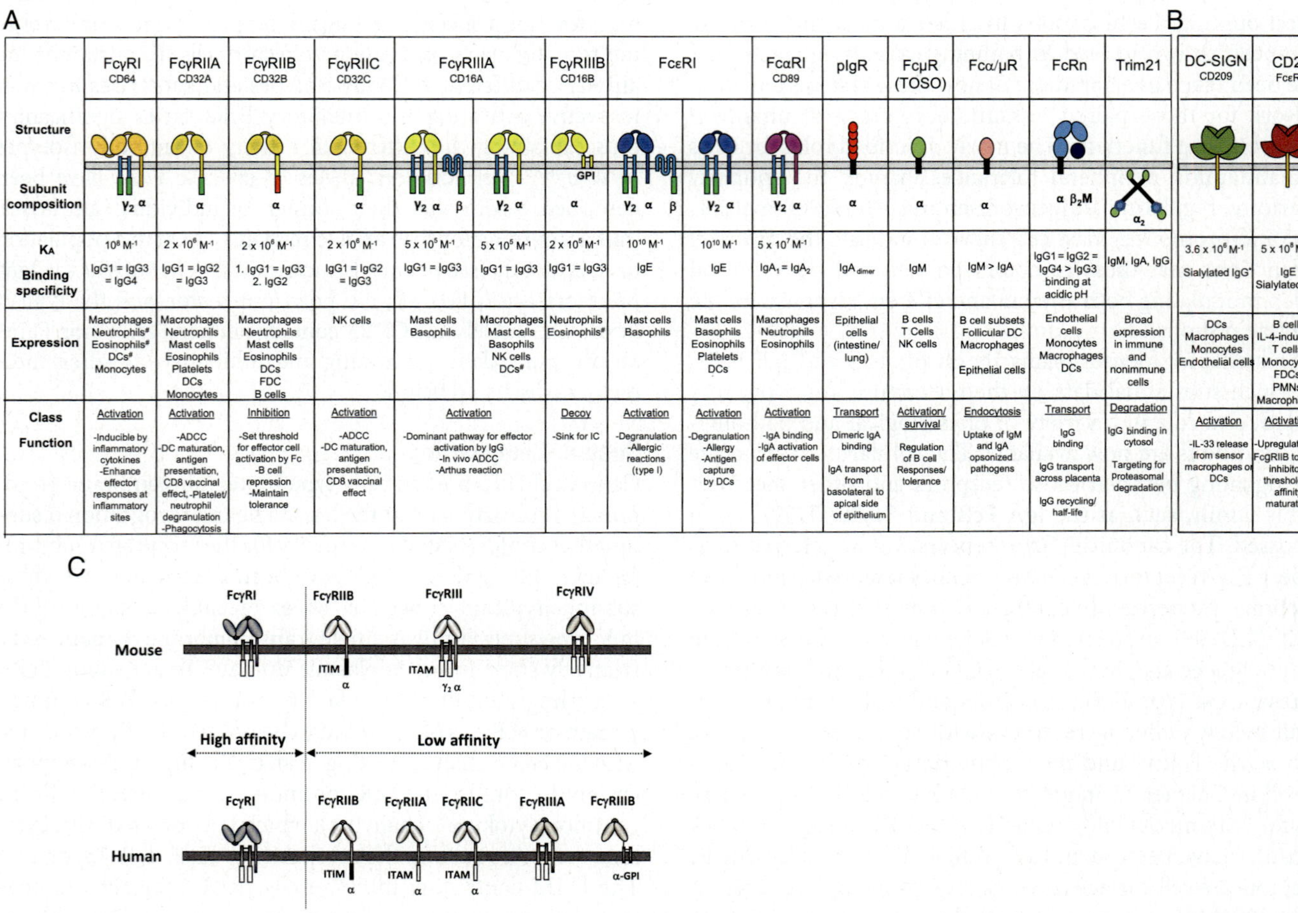

FIG. 33.1. A: Summary of canonical and noncanonical Fc receptor structures, expression patterns, and *in vivo* functions. The immunoreceptor tyrosine-based activation motif (ITAM) is indicated by the *green rectangle*; the immunoreceptor tyrosine-based inhibitor motif (ITIM) is indicated as a *red rectangle*. Alleles of FcγRIIA and FcγRIIIB and their binding properties are discussed in the text. # Expression is inducible on this cell subset. **B:** Summary of type II FcγR structures, expression patterns, and *in vivo* functions. *DC-SIGN has a number of high-mannose glycoprotein ligands including HIV gp120 and dengue virus E protein. **C:** Comparison of the human and mouse canonical Fcγ receptor protein family. In both species, canonical FcγRs can be distinguished by their affinity for the antibody Fc portion (high or low affinity) and by the signaling pathways they trigger (activating vs inhibitory). ADCC, antibody-dependent cellular cytotoxicity; DC, dendritic cell; FDC, follicular dendritic cell; IC, immune complex; NK, natural killer; PMNs, polymorphonuclear leukocytes.

aspartic acid residue found in the transmembrane domain of the FcRγ chain is often associated with a basic amino acid residue in the transmembrane domain of the α subunit. The FcRγ subunit belongs to a gene family that includes the T cell receptor–associated CD3 ζ chain and the NK receptor DAP-10– and DAP-12–associated molecules.[41] FcγRIIIA can associate with the ζ chain, resulting in the αζ2 complex found in human NK cells.

A third β subunit is found associated with the activation FcRs FcεRI and FcγRIII. This 33-kDa subunit has four transmembrane-spanning domains and amino and carboxy intracytoplasmic domains, belonging to the CD20 family of tetraspan molecules (Fig. 33.1).[24] An ITAM sequence is found in the intracytoplasmic carboxy domain. In mast cells and basophils, the β chain assembles into an αβγ2 complex with the α chain belonging to either FcγRIII or FcεRI. Its presence is required for assembly of FcεRI in rodents. In humans, however, αγ2 complexes of FcεRI are found in monocytes, Langerhans cells, and dendritic cells (DCs), in addition to the αβγ2 complexes found in mast cells and basophils. The ITAM motif found in the β subunit is not an autonomous activation sequence but functions as a signaling amplifier of the ITAM found in the γ subunits.[42]

The type II FcRs (Fig. 33.1B) do not belong to the Ig superfamily and therefore have a distinct subunit composition from canonical FcRs. CD23, a low-affinity receptor for IgE and sialylated IgG, is a type II membrane glycoprotein in the C-type lectin family. It comprises three C-type lectin head domains connected to the membrane by a trimeric α-helical coiled-coil stalk, a transmembrane domain, and a short N-terminal cytoplasmic domain. Membrane-bound CD23 can be cleaved from the surface by endogenous proteases to yield soluble CD23 that helps regulate IgE synthesis by B cells. Like CD23, DC-SIGN is a type II membrane glycoprotein belonging to the C-type lectin family. Its long extracellular domain is divided into two structurally and functionally distinct portions: the neck domain responsible for receptor tetramerization and a calcium-dependent carbohydrate recognition domain responsible for binding heavily glycosylated pathogens such as HIV-1 and Ebola virus.

Gene Organization, Linkage, and Polymorphisms

All α subunits of canonical FcRs share a common gene organization, which indicates that the evolution of this family of receptors resulted from gene duplication from a common ancestor.[43] Sequence divergence then resulted in the acquisition of distinctive specificities for these related sequences. Most of the genes belonging to the expanded FcR family including the recently identified Fcμ receptor; the Fcα/μR; the poly-Ig receptor; the group of FcR homologous or FcR-like (FCRL) proteins; and the α chains of FcγRI, FcγRII, FcγRIII, and FcεRI including the common FcRγ chain (Fig. 33.2) are found on the long arm of chromosome 1.[35,36,44-46] This region is syntenic with a comparable region on mouse chromosome 1; however, FcγRIα and several FCRLs are found on mouse chromosome 3. In humans, the α subunit of the IgA receptor is found on chromosome 19, while the β subunit is on chromosome 11. The type II FcRs CD23 and DC-SIGN are part of a cluster of C-type lectins on chromosome 19 including Clec4m and Clec4g. The FcγRII-FcγRIII locus on chromosome 1 is further linked to a variety of lupus susceptibility genes found in that region, including the Sle1 cluster.[47] A locus linked to atopy has been identified at 11q12-13 and further delineated polymorphisms of the β chain (I181V and V183L) that are associated with a heightened risk of atopy. However, a direct functional association of these polymorphisms with the known biological activities of the β chain has not been found.[48]

Polymorphisms and copy number variations (CNVs) in the α chains of FcγRs have been described, most notably in FcγRIIA and FcγRIIIA; these polymorphisms result in differences in binding affinity to specific IgG subclasses, whereas CNV may result in increased levels of FcγR expression.[49] With respect to polymorphisms, for example, a histidine at position 131 in FcγRIIA results in higher affinity binding to IgG2 and IgG3 than does an arginine at that position. Similarly, FcγRIIIA with valine at position 158 of the α chain has a higher binding affinity for IgG1 and IgG3 than does the polymorphic form with phenylalanine at that position. These polymorphisms translate into a more robust ADCC response for the FcγRIIIA val/val and the FcγRIIA his/his haplotype *in vitro* and have been positively correlated with a better clinical response to antitumor antibodies including the CD20-specific antibody rituximab, the Her2/neu-specific antibody trastuzumab, and the epidermal growth factor receptor–specific antibody cetuximab in human lymphoma, metastatic breast cancer, and colorectal cancer patient cohorts, respectively[50-55] These data are consistent with previous results obtained with rituximab and trastuzumab in mouse models, demonstrating an important role of cellular FcγRs for the activity of these therapeutic antibodies *in vivo*.[56,57] Four amino acids are polymorphic for FcγRIIIB at positions 18, 47, 64, and 88, which contribute to the neutrophil antigen polymorphisms for this receptor. Several studies have attempted to link specific FcR polymorphisms or CNVs to autoimmune diseases, specifically to systemic lupus erythematosus (SLE).[58] Thus, studies have reported associations in susceptibility to SLE in both murine and human populations with levels of FcγRIIB expression or alleles of FcγRIIB.[59] Reduced expression of FcγRIIB on activated B cells, such as memory cells, has been seen in patients with SLE and chronic inflammatory demyelinating polyneuropathy (CIDP) and is associated with a promoter polymorphism.[60,61] In mouse strains that develop a spontaneous, lupus-like disease, restoring the level of FcγRIIB expression on their B cells to a wild-type level reverses disease.[62,63] A higher incidence of an allele of FcγRIIB has been reported in several populations with autoimmune disease. This allele, found in the transmembrane domain of the receptor, is suggested to result in a hypomorphic phenotype, similar to the reduced expression observed.[59] Of note, despite increasing the risk for development of SLE, this FcγRIIB allele seems to decrease the likelihood of malaria infections and the severity of disease and can be found at increased frequencies in areas of the world where malaria is endemic.[64,65] Two primary alleles for FcγRIIC exist as a result of a SNP in exon 3: the first codes for an intact gene, named FCGR2C-Orf, and is expressed in 15% to 20%

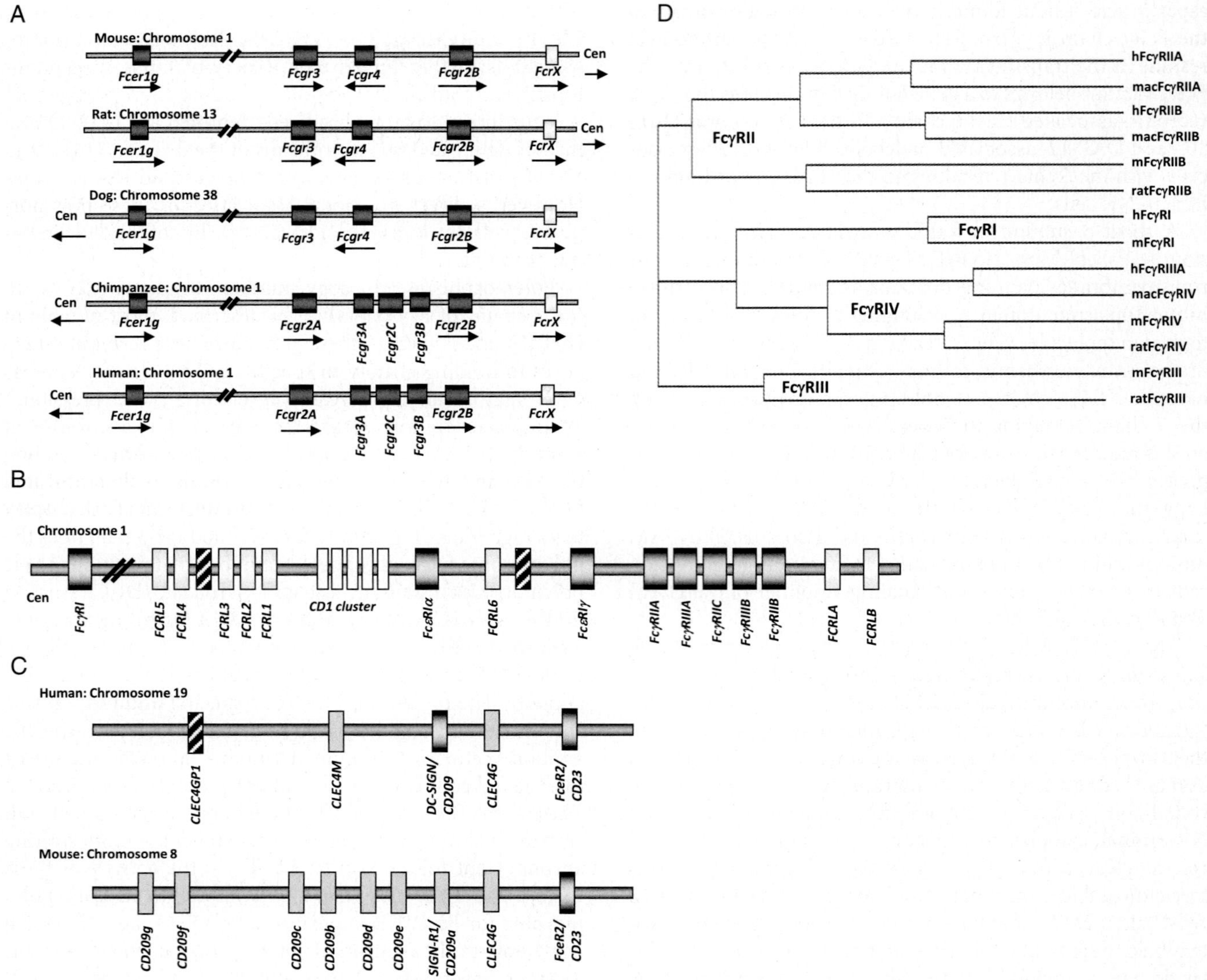

FIG. 33.2. Chromosomal organization of Fc receptor (FcR) and FCRL genes. A: Localization of FcR genes in different species. **B:** The human FcR locus on chromosome 1. Classical FcR genes are shown in *dark gray*, FCRL genes as *light gray*, and the CD1 gene cluster on chromosome 1 as *white boxes*. Pseudogenes are indicated as hatched boxes. (adapted from[44]). **C:** Chromosomal organization of type II FcR genes. Other nearby C-type lectins shown in *light gray* and pseudogenes shown as hatched boxes. The mouse has a number of SIGN family orthologues with unknown function. **D:** The cladogram shows the alignment of selective classical FcRs of humans (h), macaques (mac), mice (m), and rats.

of healthy individuals; the second has a premature stop codon and is referred to as FCGR2C-Stop. These alleles have been associated with some autoimmune conditions.[66] FcRs are also subject to CNV, particularly in the γ chain locus, affecting FcγRIIC, FcγRIIIA, and FcγRIIIB.[58] For example, FcγRIIIB expression on neutrophils has been shown to correlate with copy number and may have implications for SLE pathogenesis.[67] Similarly, FcγRIIC copy number may cause imbalance of activation and inhibitory signaling in patients with idiopathic thrombocytopenic purpura.[66]

Species Comparisons

Detailed comparisons between the canonical FcRs in mice and humans have revealed several notable differences in both structure and expression of these molecules (Fig. 33.1C). Whereas IgG, IgM, and IgE FcRs are conserved in these species, IgA FcRs are not. To date, a murine homolog for the IgA FcR has not been identified. In general, murine and human IgG FcRs display comparable degrees of heterogeneity and complexity.[43,68] Specific differences, however, have been noted. For example, FcγRI is encoded by a single gene in the mouse, in comparison with three genes in the human, though two of them do not code for a full-length transmembrane receptor.[46] Two genes for activation FcRs, FcγRIIA and FcγRIIC, are found in human and not rodents, which is notable because of their unusual single-chain activation structure.[69] Neutrophils and DCs express FcγRIIA as their activation FcR, while myeloid cells such as macrophages, mast cells, and NK cells express FcγRIIIA for activation by ICs. Both mouse and human encode a gene referred to as FcγRIIIA, although studies have identified a novel mouse FcR with higher homology to the human FcγRIIIA called FcγRIV.[32,68,70] As mentioned previously, FcγRIIIB is unique among FcRs in being expressed as a GPI-anchored protein (Fig. 33.1A). Its expression is limited to human neutrophils, in comparison to FcγRIIIA, which is expressed widely on cells of the myeloid lineage, such as macrophages, NK cells, and mast cells. Finally, both mice and humans have only a single gene encoding the inhibitory FcγRIIB molecule. Among the type II FcγRs, the lectin DC-SIGN is the functional homologue of the murine SIGNR1 molecule with respect to its ability to bind sialylated IgG Fc and mediate an anti-inflammatory response.[71,72] While the overall structure of these proteins are similar, their patterns of expression are quite distinct. SIGNR1 expression is restricted to marginal zone macrophages of the spleen and lymph node, while DC-SIGN expression is seen on myeloid DCs and macrophages in a variety of tissues, including spleen, bone marrow, and lymph node.[73] The expression pattern of CD23 in mice is similar to that in humans. The genes for the IgE FcR are also conserved in mice and humans. The difference that is observed relates to the requirement for the β chain to achieve surface expression in mice, precluding the expression of the αγ2 complex.[24] In humans, this form of the receptor is widely expressed on monocytes, Langerhans cells, and DCs and is likely to be found on mast cells and basophils as well. This difference in FcεRI subunit composition is likely to result in functional differences as well.

Although these specific interspecies differences are important, the fundamental organization of the canonical FcR system, with activation and inhibitory signaling through a shared ligand specificity coupled to opposing signaling pathways, is well conserved, as is the role of the type II FcγRs in mediating the immunomodulatory functions of sialylated IgG. Thus, conclusions regarding the function of this expanded FcR system in immunity by the analysis of murine models are relevant to an understanding of the role of these receptors to human immunity. However, the fine specificity of IgG subclasses from different species binding to cognate and noncognate FcRs has revealed the inherent difficulties in extrapolating from one species to another (Table 33.2). Thus, despite the proven utility of murine models, there remained interest in the direct study of human antibodies in the context of their native receptors. To close this gap, mouse models that generally recapitulate human FcγR structure, expression, and effector function either through the deletion of native mouse FcγRs and transgenic expression of human orthologues or via reconstitution of immunodeficient mice lacking the mouse FcRγ with a human immune system have become key platforms for mechanistic and therapeutic discovery.[74,75] These models have been used widely to study the therapeutic potential of human antibodies in a variety of contexts including autoimmunity, cancer, and infectious diseases.[76-78]

Expression

Canonical FcRs are expressed widely, including on monocytes, macrophages, DCs, mast cells, basophils, neutrophils, eosinophils, platelets, and NK cells.[17,34] In addition, B cells, follicular dendritic cells (FDCs), and liver sinusoidal endothelial cells express the inhibitory FcγRIIB receptor. With the possible exception of FcγRIIB expression on subsets of CD8 and CD4 memory T cells, the majority of T cells are negative for FcγR expression, but may express the Fc receptor for IgM.[79,80] The specific expression pattern for each Fc receptor varies, and these patterns are summarized in Figure 33.1A. Because canonical FcRs represent a balanced system of activation and inhibition, the general rule of coexpression of FcRs of these classes is maintained. B cells use the B cell antigen receptor as the activation coreceptor for FcγRIIB, whereas NK cells appear to utilize NK inhibitory receptors to modulate FcγRIIIA activation. The decoy FcγR, FcγRIIIB, is expressed exclusively on human neutrophils, on which it functions to concentrate ICs without directly triggering cell activation, perhaps also playing role in neutrophil recruitment.[81] IC binding to FcγRIIIB may activate neutrophils via concomitant binding to FcγRIIA, however. The FcγRIIA-FcγRIIB pair functions on neutrophils and DCs to modulate IC activation. FcεRI can be modulated by FcγRIIB, as demonstrated both *in vitro* and *in vivo*; mice deficient in FcγRIIB display enhanced IgE-triggered anaphylaxis[82] by virtue of the ability of IgE to bind with high affinity to FcεRI and with low affinity for FcγRIIB. Other mast cell inhibitory receptors, such as glycoprotein 49B1, modulate mast cell sensitivity to IgE: Mice deficient in this molecule display enhanced anaphylactic responses to IgE stimulation.[83] Expression of FcRγ chain is broad: it has been found on all myeloid and lymphoid cells examined to date. In contrast, the β chain appears to be quite restricted in its expression: it has been found only on mast cells and basophils. CD23, the aforementioned type II FcγR capable of engaging sialylated IgG Fc and IgE, is expressed as two isoforms, CD23a and CD23b. CD23a is exclusively expressed by B cells, while CD23b expression is induced by IL-4 and other stimuli in a variety of hematopoietic cell types, including B and T cells, monocytes, and polymorphonuclear leukocytes. DC-SIGN is expressed on DCs and macrophages in a variety of tissues including spleen, bone marrow, and lymph node, where it participates in the innate immune response (Fig. 33.1B).

Regulation of canonical FcR expression can occur at several levels. In general, cytokines involved in activation of inflammatory responses induce expression of activation FcγRs, whereas inhibitory cytokines downregulate these activation receptors. Transcriptional regulation of α chain levels has been documented for a variety of cytokines, including interferon-γ, IL-4, and transforming growth factor β.[84,85] In addition, complement component C5a binding to its receptor, C5aR, results in the induction of expression of activation FcγRs.[86] Induction of FcγRI, FcγRIIA, and FcγRIIIA α and γ chains in myeloid cells occur upon interferon-γ treatment: IL-4 generally inhibits expression of these activation receptors but induces expression of the inhibitory FcγRIIB.[87] Administration of intravenous gamma globulin (IVIG), a widely used treatment for inflammatory diseases, has been shown to induce expression of the inhibitory FcγRIIB on effector macrophages and B cells in mice and humans.[61,88-90] This induction is not direct but mediated through type II FcγR pathways (Fig. 33.7A). Involvement of DC-SIGN (or its mouse orthologue SIGNR1) by sialylated

IgG results in an intrinsic Th2 pathway signaling, ultimately leading to the expression of IL-4 that induces FcγRIIB expression on inflammatory macrophages and thus immunosuppression.[71,72,91,92] The situation in B cells is likely to be more complex, whereby regulation of FcγRIIB is critical for the maintenance of peripheral tolerance and for setting the inhibitory threshold for B cell receptor (BCR) signaling during activation of naïve B cells and affinity maturation in the germinal center. Recent studies have shown that engagement of the type II FcγR CD23 on germinal center B cells by sialylated IgG Fc in ICs can upregulate FcγRIIB on those cells, resulting in selection of higher affinity IgG clones. The contribution of CD23 to the afferent immune response is summarized in a subsequent section of this chapter.

Regulation of FcR expression has also been documented to occur upon binding of ligand. IgE regulates the expression of FcεRI by stabilizing the intracellular pool of receptor upon receptor engagement.[93] Thus, high IgE levels result in the induction of surface expression of FcεRI. However, this same mechanism of regulation is not seen for FcγRs: mice deficient in IgG have FcγR levels comparable with those of wild-type animals. Competition for limiting subunits also contributes to regulation of receptor expression. In mast cells, it appears that the level of γ chain is limiting. Competition between α chains for the limiting concentration of γ chain has been documented in mouse knockouts, whereby levels of one receptor increase if the α chain of the other receptor is reduced.[94] This type of reciprocal regulation has recently also been observed for FcγRs in neutrophils and is likely to be significant in the cross-regulation of FcRs by different isotypes of Ig.[95]

Three-Dimensional Structure

The crystal structures of human FcγRI, FcγRIIA, FcγRIIB, FcγRIIIA, and FcεRI have been solved, as have the cocrystals of FcγRI-IgG1 Fc, FcγRIIA-IgG1 Fc, FcγRIIIA-IgG1 Fc, FcεRI-IgE Fc, and CD23-IgE Fc[96-98] (Fig. 33.3). These studies demonstrate that the canonical low-affinity receptors have a common structure in which the two extracellular Ig domains fold in a strongly bent overall structure, arranged into a heart-shaped domain structure. A 1:1 stoichiometry between the receptor and ligand is observed, with the receptor inserted into the cleft formed by the two chains of the Fc fragment (Cγ2 or Cε3). The binding region of the FcR to Fc fragments consists mainly of rather flexible loops that rearrange upon complex formation. Only domain 2 and the linker region connecting domains 1 and 2 interact in the complex with different regions of both chains of the Fc. Conserved tryptophans located on the FcRs interact with proline to form a "proline sandwich." A solvent-exposed hydrophobic residue at position 155 is conserved among all FcRs and represents a binding site for the important IgG1 residue Leu 235 (not found in IgE). Specificity is generated among the receptor-ligand pairs in a variable region connecting the two extracellular domains that are in contact with the lower hinge region of the Fc fragment (residues 234-238), a region not conserved among the IgGs and IgE. While the binding region of FcRs to their Ig ligands partially overlaps with the binding site of C1q, it does not overlap with other Fc binding molecules such as protein A, protein G, and FcRn, which bind to the IgG CH3 domain.

The structure of the FcR bound to its ligand reveals that the antigen-binding fragment (Fab) arms are quite sharply bent and may adopt a perpendicular orientation toward the Fc. This arrangement would give the Fab arms maximal flexibility to bind antigen when the Fc fragment is oriented parallel to the membrane of the FcR-expressing cell. The asymmetrical interaction of the two Fc chains with a single FcR prevents a single antibody molecule from triggering dimerization of

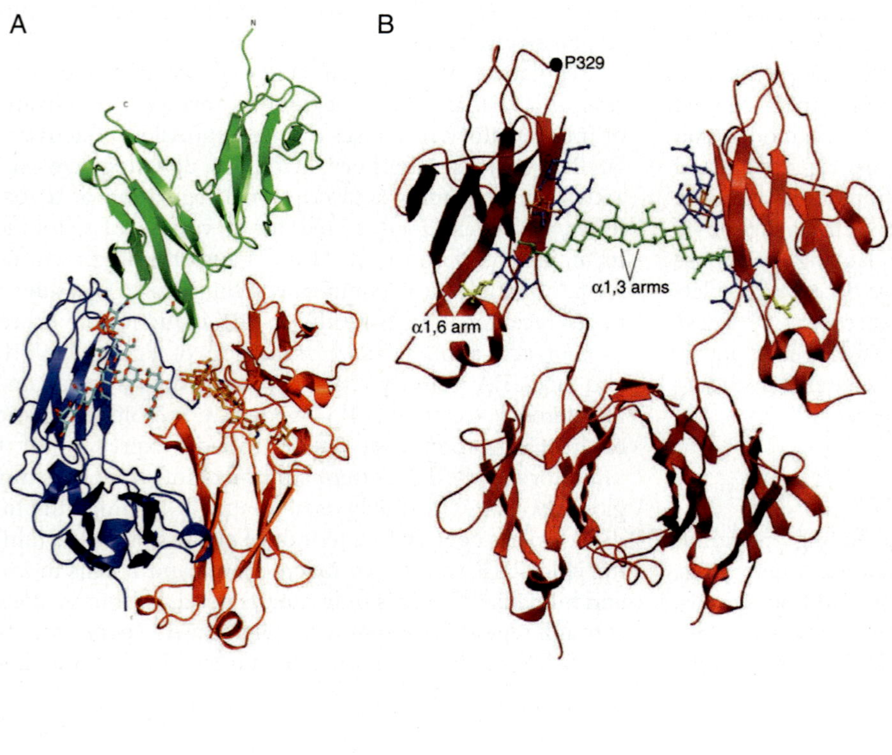

FIG. 33.3. A: Ribbon diagram of crystallizable fragment (Fc) receptor III–immunoglobulin G1 (IgG1) Fc structure. The carbohydrate moiety attached to both C_H2 domains of the IgG Fc is shown as a stick and ball model. The extracellular domains of Fc receptor III are shown in green, together with the Fc fragment of IgG1, depicted in blue and red. See text for details. (Adapted from Sondermann et al,[15] with permission). **B:** Crystal structure of IgG1 Fc (PDB 1H3Y) showing the α1,3 arms of the glycan, which, in the absence of sialylation, keeps the molecule in its "open" conformation.

receptors and initiating signaling. Instead, dimerization is initiated by the interaction of antigen with the Fab arms, thus linking adaptive responses to effector cell triggering.

Fc Glycosylation

It is important to note that the majority of the IgG-FcγR cocrystals discussed above have been generated with protein produced in bacteria and therefore in an aglycosylated form. However, the past few decades of research have shed light on the structural diversity of IgG introduced by Fc glycosylation and the significant impact it has on FcγR binding and thus FcγR effector function.[99] In general, IgG's complex, biantennary N-linked glycan at Asn297, is critical for FcγR binding. Removing the sugar moiety by enzymatic deglycosylation or via mutation of the Asn297 residue shifts the open horse shoe shape of the IgG C_H2 domains toward a more closed conformation preventing an efficient interaction of FcγRs with the lower hinge region. Moreover, modulation of FcγR binding by glycan modification has been observed for several classes of glycan modifications, including sialylation, fucosylation, and branching N-acetyl glucosamine (Fig. 33.4A).[100,101] Similarly, IgE harbors six N-linked glycosylation sites, and the oligomannose glycan at Asn394 has been demonstrated to be essential to FcεRI binding, mast cell degranulation, and the ensuing anaphylactic response.[102] Furthermore, sialylation of IgE was found to be a key regulator of allergic pathogenicity, where allergic individuals had increased sialic acid content on allergen-specific IgE compared to nonatopic individuals.[103] Interestingly, administration of proteins with asialylated complex, biantennary N-linked glycans, attenuated anaphylaxis in multiple models of allergic disease. These modifications in Fc glycan composition are regulated in homeostasis and disease, though the precise mechanisms require further study.

The first IgG-FcγR cocrystal structure with differentially glycosylated IgG variants demonstrated that carbohydrate-carbohydrate interactions between IgG and FcγR are

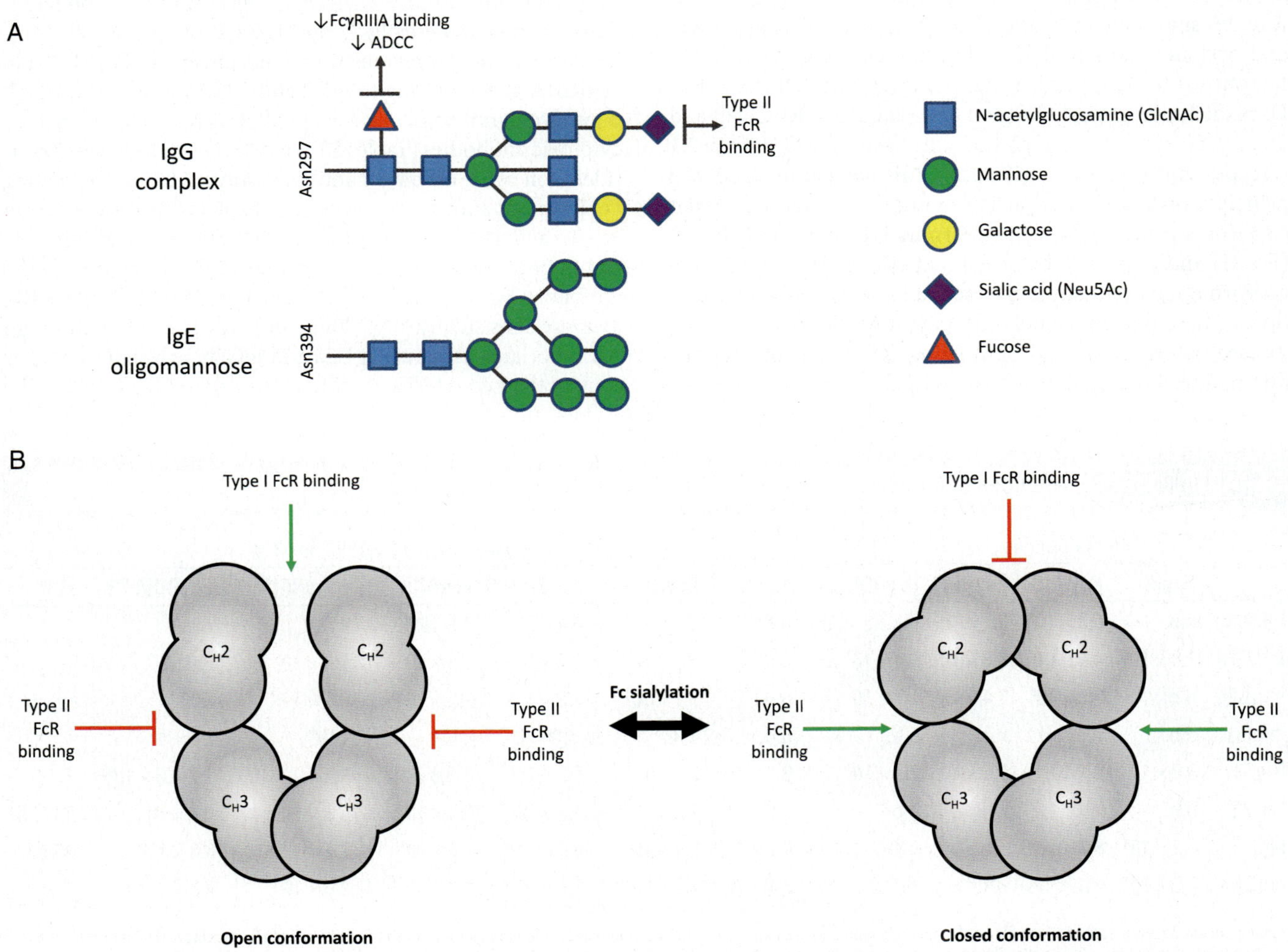

FIG. 33.4. **A:** The highly conserved complex, biantennary glycan on IgG Fc Asn297 and oligomannose glycan of IgE Asn394. Modifications of the glycan structure significantly influence Fc-FcR interactions and associated effector functions. On IgG Fc, the presence of core fucose decreases FcγRIIIA binding and its associated effector function, antibody-dependent cellular cytotoxicity. Terminal sialylation enhances type II FcR binding and reduces type I FcγR binding. Glycosylation of IgE at Asn394 is essential for FcεRI engagement and anaphylaxis. **B:** Sialylation of IgG Fc forces the molecule into the "closed" conformation, occluding the type I FcγR binding site and revealing the type II FcR binding site. FcR, Fc receptor.

essential for high-affinity recognition of afucosylated IgG glycovariants by FcγRIIIA.[104] In contrast, low-affinity binding of FcγRIIIA to fucosylated IgG is independent of these carbohydrate interactions. The pathological relevance of the differing strength of these interactions is borne out in secondary dengue infection, where titers of afucosylated IgG1—and therefore degree of FcγRIIIA engagement—correlate with disease severity.[105]

Similarly, the structural impact of IgG Fc sialylation was found to be the primary determinant of type I FcγR versus type II FcγR binding. Due to the biantennary nature of the Fc glycan, the two arms have distinct roles in determining IgG conformation. The α1,6 arm follows the protein backbone of the Cγ2 domain, while the α1,3 arm occupies the space between Cγ2 domains (Fig. 33.4B). Occupying this cavity is what restrains the interdomain flexibility of the Cγ2 domains and stabilizes the "open" conformation of IgG Fc. However, sialylation triggers a destabilization of the Cγ2 domain, converting IgG Fc from its "open" to "closed" conformation. This conformational change, by analogy to the CD23-IgE structure, 96 simultaneously allows for type II FcγR engagement and occludes the type I FcγR binding site (Fig. 33.4B).[106,107] In contrast to type I FcγRs that bind the "open" form of IgG Fc at the hinge proximal region, the type II FcγRs CD23 and DC-SIGN bind at the Cγ2-Cγ3 interface of IgG with a 2:1 (receptor:antibody) stoichiometry. This newfound structural principle of IgG sialylation is reminiscent of IgE's inherent Cε3 domain flexibility, which permits it to bind both type I (FcεRI) and type II (CD23) FcRs in two mutually exclusive conformations and similarly mediate diverse effector functions. These novel insights offer new ways of understanding disease pathogenesis and optimizing IgG-FcγR interactions for enhanced therapeutic activity of IgG.

IN VITRO ACTIVITY

Binding Properties

As outlined in Fig. 33.1, Ig binding to canonical FcRs falls into either high- or low-affinity binding classes. The high-affinity binding class is typified by FcεRI, with a binding affinity of 1010 M^{-1} for IgE, which ensures a monomeric interaction between IgE and its receptor. Human and mouse FcγRI bind with relatively high affinity for IgG1 and IgG3 (in humans) or IgG2a (in mouse) with an affinity constant of 108 M^{-1}. In contrast to this high-affinity FcRs, the low- and intermediate-affinity receptors, such as the human FcγRIIA, FcγRIIB, FcγRIIIA, FcγRIIIB, and FcαRI and the mouse FcγRIIB, FcγRIII, and FcγRIV, bind with affinities ranging from 5 × 105 M^{-1} (FcγRIII) to 5 × 107 M^{-1}. This low-affinity binding ensures that these receptors interact with ICs and not monomeric ligands. As described later, this dependence on high-avidity and low-affinity interactions ensures that these receptors are activated only by physiologically relevant ICs and not by circulating monomeric Ig, thus avoiding inappropriate activation of effector responses. In general, low-affinity FcγRs bind IgG1 and IgG3 preferentially; their binding to IgG2 and IgG4 is of even lower affinity. As mentioned previously, polymorphisms in FcγRIIA and FcγRIIIA affect binding to IgG2 and IgG1, respectively, which may have significance *in vivo* in predicting responses to specific cytotoxic antibodies (Tables 33.1 and 33.2).[115] The isolated consideration of the affinity of antibody isotypes to their activating FcγRs is not sufficient, however, to explain the differences in *in vivo* activity. More importantly, the ratio of the affinities of an antibody isotype to the activating FcγRs compared to the inhibitory FcγRIIB (termed A/I ratio) predicts antibody activity more consistently.[20,68,116] Thus, antibody isotypes with a high A/I ratio, such as mouse IgG2a and IgG2b, will show a greater activity than counterparts with a lower ratio (Table 33.1). The

TABLE 33.1 Affinities of Mouse and Human FcRs to Different Antibody Isotypes [32,108-110]

	Mouse Soluble FcγR (K_A in M^{-1})					Human Soluble FcγR (K_A in M^{-1})						
	FcγRI	**FcγRIIB**	**FcγRIII**	**FcγRIV**	**A/I**	**FcγRI**	**FcγRIIA^{131R}**	**FcγRIIA^{131H}**	**FcγRIIB**	**FcγγRIIIA^{158F}**	**FcγRIIIA^{158V}**	**A/I**[a]
mIgG1	n.b.	3.3×10^{6}	3.1×10^{5}	n.b.	0.1	n.d.	2.5×10^{5}	0.4×10^{5}	1×10^{5}	$<10^{4}$[b]	n.d.	
mIgG2a	1.8×10^{8}	0.42×10^{6}	6.8×10^{5}	2.9×10^{7}	69	n.d.	3.2×10^{5}	1.7×10^{5}	1.6×10^{5}	1.0×10^{5}	n.d.	
mIgG2b	n.b.	2.2×10^{6}	6.4×10^{5}	1.7×10^{7}	7	n.d.	9.1×10^{4}	1.2×10^{5}	1.2×10^{5}	0.1×10^{5}	n.d.	
mIgG3	n.b.	n.b.	n.b.	n.b.	—	n.d.	$<10^{4}$[b]	$<10^{4}$[b]	$<10^{4}$[b]	$<10^{4}$[b]	n.d.	
hIgG1	3.8×10^{6}	2×10^{5}	3.5×10^{4}	2.2×10^{6}		9.1×10^{8}[d]	3.5×10^{5}[e]	5.2×10^{5}[e]	$1\text{-}3.8 \times 10^{5}$[c,e]	$0.4\text{-}1.1 \times 10^{6}$[e]	$1.9\text{-}4.8 \times 10^{6}$[c,e]	1/13
hIgG2	n.b.	$<10^{4}$[b]	$<10^{4}$[b]	n.b		n.b.	1.0×10^{5}[e]	4.5×10^{5}[e]	0.2×10^{5}[e]	0.3×10^{5}[e]	0.7×10^{5}[e]	1.5/3.5
hIgG3	1.2×10^{6}	8.3×10^{4}	n.b.	$<10^{4}$[b]		6.1×10^{7}[d]	9.1×10^{5}[e]	8.9×10^{5}[e]	1.7×10^{5}[e]	7.7×10^{6}[e]	9.8×10^{6}[e]	45/57
hIgG4	7.2×10^{4}	$<10^{4}$[b]	$<10^{4}$[b]	$<10^{4}$[b]		3.4×10^{7}[e]	2.1×10^{5}[e]	1.7×10^{5}[e]	2×10^{5}[e]	2.0×10^{5}[e]	2.5×10^{5}[e]	1

Binding constants were obtained by surface plasmon resonance analysis with immobilized antibodies (FITC-isotype switch variants) and soluble Fc receptors (FcRs) produced by transient transfection in 293T cells. Shown are the association constants (K_A in M^{-1}) of soluble FcR binding to the indicated antibody isotypes as determined by surface plasmon resonance analysis.[31,111-113] Ig, immunoglobulin; n.b., no detectable binding or binding that is too low to be evaluated; n.d., no surface plasmon resonance or other quantitative data are available.

[a]The two numbers indicate the A/I ratios for the low- and high-affinity allele of FcγRIII.

[b]Indicates detectable but very low binding that did not allow to determine exact binding constants by surface plasmon resonance.

[c]Maenaka et al[111]; Okazaki et al.[112]

[d]Paetzetal.[113]

[e]Bruhns et al.[114]

TABLE 33.2 Relative Binding of Mouse and Human IgG Subclasses to Fc Receptors

	Mouse					Human			
	IgG1	IgG2a	IgG2b	IgG3		IgG1	IgG2	IgG3	IgG4
FcγRI	–	++++	–	–	FcγRI	++++	-	++++	+++
FcγRIIB	++	+	+	–	FcγRIIB	+	+	+	+
FcγRIII	+	+	+	–	FcγRIIA	++	+(+)[a]	++	+
FcγRIV	–	+++(+)[b]	++(+)[b]	–	FcγRIIIA	++(+)[a,b]	+(+)[b]	++(+)[a,b]	+(+)[b]

[a]Allelic variants of FcγRIIA and FcγRIIIA with enhanced binding to the respective IgG subclass exist.
[b]Afucosylated IgG subclasses show enhanced binding to mouse FcγRIV and human FcγRIIIA.

sugar moiety of the antibody Fc portion is essential for FcR binding, and the presence or absence of certain sugar residues can significantly impact on antibody-FcR binding.[20,92,117-120] The absence of fucose, for example, will selectively increase the affinity of human IgG1 or mouse IgG2b for human FcγRIIIA or mouse FcγRIV, respectively. In contrast, the presence of terminal sialic acid on the N-linked Fc glycan may reduce the affinity of IgG for type I FcγRs with a concomitant reduction in *in vivo* activity.[92] However, as noted in the previous section, sialylation converts IgG to its "closed" conformation, enhancing type II FcγR binding and its associated effector functions. Besides the influence of glycosylation on binding, subunit interactions have also been reported to influence affinity for ligand, as demonstrated for the common γ chain associating with FcγRIIIA.[121] Its affinity for IgG1 is higher than the GPI-anchored form of this receptor, FcγRIIIB.

The crystal structures of IgG1-FcγRIIIA and IgE-FcεRI reveal similarities in the binding properties of these two complexes. Of significance is the 1:1 stoichiometry of the complexes, which ensures that a single receptor binds to a single Ig molecule.[15,16] This property in turn ensures that activation occurs upon cross-linking of receptor complexes by multivalent ligands. Two binding sites on the receptor interact asymmetrically with two sites on the Fc molecule. The FcR inserts into the cleft formed by the two chains of the Fc molecule, burying a binding surface of 895 Å for each binding site. Alterations in the Fc structure that reduce the cleft, such as complete deglycosylation of IgG, inhibit FcR binding. Four distinct regions have been defined in the Fc domains involved in FcR interactions. For IgE, this includes residues 334 to 336, 362 to 365, 393 to 396, and 424. The homologous regions for IgG are residues 234 to 239, 265 to 269, 297 to 299, and 327 to 332. Interactions of these residues occur with the carboxy-terminal domain 2 of the respective FcRs. In view of the similarities of these complexes and the homologies among the receptors and their ligands, an obvious question that arises concerns the molecular basis for specificity. Attempts to resolve that question have relied on mutagenesis studies of the ligands and domain exchanges between receptors. For example, exchange of the FG loop in domain 2 of Fcε to Fcγ receptors confers detectable IgE binding; similarly, variation in this loop in FcγRs may provide interactions that determine IgG specificity for these receptors. Mutagenesis of IgG1 revealed that a common set of residues is involved in binding to all FcγRs, but FcγRII and FcγRIII also utilize distinct residues.[122] Several IgG1 residues not found at the IgG-FcR interface by crystallographic determination had a profound effect on binding, which indicates the greater complexity of these interactions in solution.

Recent studies have uncovered the precise structural underpinnings of type II FcγR binding. Specifically, the biantennary N-linked glycan on IgG Fc–Cγ2 plays a crucial role in regulating the conformational state of the IgG molecule. In the absence of sialylation, the α1,3 arm protrudes into the Fc cleft and contacts the α1,3 arm of the second Cγ2 domain, forcing an "open" conformation that is preferential for type I FcγR binding. Upon sialylation, the Cγ2 domain is destabilized and the sialic acid is positioned in a pocket formed by Lys 340 and Glu 318, transitioning the Fc fragment to the "closed" conformation. This shift introduces stabilizing interactions between the Fc glycan and Asn370 of DC-SIGN, additional salt bridge contacts in the Fc-Cγ3 domain, and hydrogen bonds in the Cγ2 domain.

The implications of these structural studies are that the Fc domain of IgG may be selectively mutated to direct its binding to specific FcγRs. Fc mutants that selectively engage activation FcγRs (IIIA and IIA) while minimally interacting with the inhibitory FcγRIIB would confer optimal cytotoxic potential for tumoricidal applications, as well as promoting DC maturation with subsequent T cell activation. Indications that such Fc engineering is possible are suggested by IgG mutants with selective binding to FcγRIII or FcγRII.[123-125] Selective binding to the inhibitory receptor FcγRIIB has also been demonstrated and provides the basis for *in vivo* cross-linking of agonistic antibodies, such as CD40.[126,127] In addition, IgG Fc can be tuned to distinguish between type I and type II FcγRs, resulting in diverse downstream proinflammatory, anti-inflammatory, or immunomodulatory consequences. This principle is captured by the IgG Fc F241A mutant, which has increased mobility of its α1,3 arm, recapitulating the type II FcγR binding pattern and anti-inflammatory activity of sialylated Fc.[128]

Effector Cell Activation

The critical step in triggering effector cell response by canonical FcRs is mediated by the cross-linking of these receptors by Ig. This can occur either by interactions of low-affinity, high-avidity IgG ICs or of IgG-opsonized cells with activation FcγRs or by the cross-linking of monomeric IgG or IgE bound to FcγRI or FcεR, respectively, by multivalent antigens binding to the Fab of the antibody. Cross-linking of ITAM-bearing FcRs results in common cellular responses, determined by the cell type, rather than the FcR. Thus, for example, FcεRI or FcγRIII cross-linking of mast cells results in degranulation of these cells, whereas cross-linking of macrophage-expressed FcαRI or FcγRIII by opsonized cells triggers phagocytosis. These functions underlie the functional similarity of activation FcRs in which cross-linking mediates cellular responses by ITAM-mediated tyrosine kinase cascades. In addition to degranulation and phagocytosis, activation FcR cross-linking has been demonstrated to induce ADCC, the oxidative burst, and the release of cytokines and other inflammatory cell mediators. A sustained calcium influx is associated with these functions, as is transcription of genes associated with the activated state.

Cellular activation initiated by ITAM-bearing activation FcRs can be enhanced by coengagement with integrin and complement receptors. Although the ability of these receptors to mediate phagocytosis, for example, is modest, synergistic interactions with FcRs result in sustained activation and enhancement. Synergistic interactions between activation FcRs and toll receptors, mannose receptors, and other pattern recognition molecules have also been reported *in vitro* and suggest that interplay between the innate and adaptive effector mechanisms of an immune response are involved in mediating efficient protection from microbial pathogens.

In contrast to the activation of effector cell responses triggered by cross-linking of ITAM-bearing FcRs *in vitro*, cross-linking of an ITIM-bearing inhibitory receptor to an ITAM-bearing receptor results in the arrest of these effector responses. Homoaggregation of FcγRIIB by its cross-linking on effector cells by ICs does not result in cellular responses; rather, it is the coengagement of ITAM- and ITIM-bearing receptors that results in the functional generation of an inhibitory signal. *In vitro*, it is possible to ligate any ITAM-bearing receptor to any ITIM-bearing receptor with a resulting inhibitory response. This activity is used functionally to define putative ITIMs and has proved to be a useful device in dissecting the signaling pathways induced by ITAM-ITIM colligation.[18,129]

B Lymphocyte Suppression

B cell stimulation through the B cell antigen receptor can be arrested by the colligation of FcγRIIB to the BCR. This occurs naturally when ICs, retained on FDCs in the germinal center, interact with both the BCR and FcγRIIB during the affinity maturation of an antibody response. *In vitro* suppression of B cell activation has been demonstrated by colligation of BCR and FcγRIIB, resulting in arrest of calcium influx and proliferative responses triggered by the BCR,[25,130] the result of recruitment of the SH2-containing inositol 5′-phosphatase (SHIP)−1.[131] Calcium release from the endoplasmic reticulum is not affected, and there is thus an initial rise in intracellular calcium; however, this calcium flux is not sustained because SHIP recruitment blocks calcium influx by uncoupling of the capacitance channel. Homoaggregation of FcγRIIB by ICs can trigger apoptosis in B cells, as demonstrated in the DT40 B cell line and in murine splenocyte preparations.[132] This activity is retained in ITIM mutants and is dependent on the transmembrane sequence of FcγRIIB. Consistent with this, other studies showed that a SHIP-independent but BTK-, JNK1-, and cABL-dependent pathway is involved in induction of apoptosis upon FcγRIIB homoaggregation.[133] The potent inhibitory effect of FcγRIIB on B cell activation has recently been used therapeutically to suppress autoantibody production by B cells. By engineering the Fc fragment of CD19-specific antibodies to have increased affinity to FcγRIIB, resulting in an enhanced coengagement of the inhibitory FcγRIIB with the BCR signaling complex, it was possible to suppress humoral immunity in mice and to shut down autoantibody production by B cells.[125] In contrast to mature B cells that coexpress the BCR and FcγRIIB, plasma cells responsible for high level antibody production downregulate BCR expression but maintain FcγRIIB expression. It was suggested that cross-linking of FcγRIIB by ICs on bone marrow plasma cells might be involved in inducing apoptosis in a fraction of plasma cells, thereby creating niches for newly generated plasma cells during exposure to a new antigenic stimulus.[134]

SIGNALING

ITAM Pathways

The general features of signal transduction through ITAM receptors are conserved among all members of this family, including T cell receptors, BCRs, NK cell receptors, and various FcRs. The 19-amino acid–conserved ITAM is necessary and sufficient to generate an activation response, as demonstrated by the analysis of chimeric receptors. With a single exception, FcRs associate with accessory subunits that contain these signaling motifs. As described previously, the FcRγ chain contains an ITAM and is associated with FcεRI, FcγRI, FcγRIII, FcγRIV, and FcαRI. In addition, both FcεRI and FcγRIII may associate with the β subunit in mast cells. The ITAM found in the β chain does not function as an autonomous activation cassette, as has been found for most other ITAMs. Rather, it functions to amplify the activation response generated by the FcRγ chain ITAM by increasing the local concentration of Src family kinase Lyn available for activation upon aggregation of the receptor.[42] FcγRIIA and FcγRIIC contain an ITAM in the cytoplasmic domain of the ligand recognition α subunit and are thus able to activate in the absence of any associated subunit. Their α subunits also contain a unique ITAM sequence with three additional aspartic acid residues compared to the common gamma chain, though the significance of this difference is unknown.[26,135]

Upon sustained receptor aggregation, Src family kinases that may be associated with the receptor in an inactive form become activated and rapidly tyrosine-phosphorylate the ITAM sequences, creating SH2 sites for the docking and subsequent activation of Syk kinases. Ligands that rapidly dissociate from their receptors result in nonproductive signaling complexes that fail to couple to downstream events and behave as antagonists.[136] The specific Src kinase involved for each FcR depends on the receptor and cell type in which it is studied. Thus, Lyn is associated with the FcεRI pathway in mast cells, Lck is associated with FcγRIIIA in NK cells, and both of these kinases as well as Hck are associated with FcγRI and FcγRIIA in macrophages. After activation of the Src kinase, tyrosine phosphorylation of the ITAM motif rapidly ensues, leading to the recruitment and activation of Syk kinases. This two-step process is absolutely necessary to transduce the aggregation signal to a sustainable intracellular response. Once activated, Syk kinases lead to the phosphorylation or recruitment of a variety of intracellular substrates, including PI3K, Btk, and other Tec family kinases; phospholipase C-γ (PLCγ); and adaptor proteins such as SLP-76 and BLNK. The Ras pathway is also activated through Sos bound to Grb2 that is recruited upon phosphorylation of Shc. Ras phosphorylates Raf, which in turn leads to MEK kinase and MAP kinase activation. A summary of these intracellular pathways is shown in Figure 33.5A. A crucial step in this sequential activation cascade occurs with the activation of PI3K by Syk. By generating phosphatidylinositol polyphosphates, such as PIP3, PI3K leads to the recruitment of pleckstrin homology (PH) domain–expressing proteins such as Btk and PLCγ, which in turn leads to the generation of inositol triphosphate (IP3) and diacylglycerol, intermediates crucial to the mobilization of intracellular calcium and activation of protein kinase C, respectively.

ITIM Pathways

The inhibitory motif, embedded in the cytoplasmic domain of the single-chain FcγRIIB molecule, was defined as a 13-amino acid sequence AENTITYSLLKHP, shown to be both necessary and sufficient to mediate the inhibition of BCR-generated calcium mobilization and cellular proliferation.[25,26] Significantly, phosphorylation of the tyrosine of this motif was shown to occur upon BCR colligation and was required for its inhibitory activity. This modification generated an SH2 recognition domain that is the binding site for the inhibitory signaling molecule SHIP.[131,137] In addition to its expression on B cells, where it is the only IgG FcR, FcγRIIB is widely expressed on macrophages, neutrophils, mast cells, DCs, and FDCs, absent only from NK cells. Studies on FcγRIIB provided the impetus to identify similar sequences in other surface molecules that mediated cellular inhibition and resulted in the description of the ITIM, a general feature of inhibitory receptors.

FcγRIIB displays multiple inhibitory activities. Coengagement of FcγRIIB to an ITAM-containing receptor leads to tyrosine phosphorylation of the ITIM by the Lyn kinase, recruitment of SHIP, and the inhibition of ITAM-triggered calcium mobilization and cellular proliferation (Fig. 33.5B).[131,138,139] These two activities result from different signaling pathways; calcium inhibition requires the phosphatase activity of SHIP to hydrolyze PIP3 and the ensuing dissociation of PH domain—containing proteins such as Btk and PLCγ.[140] The net effect is to block calcium influx and prevent sustained calcium signaling. Calcium-dependent processes such as degranulation, phagocytosis, ADCC, cytokine release, and

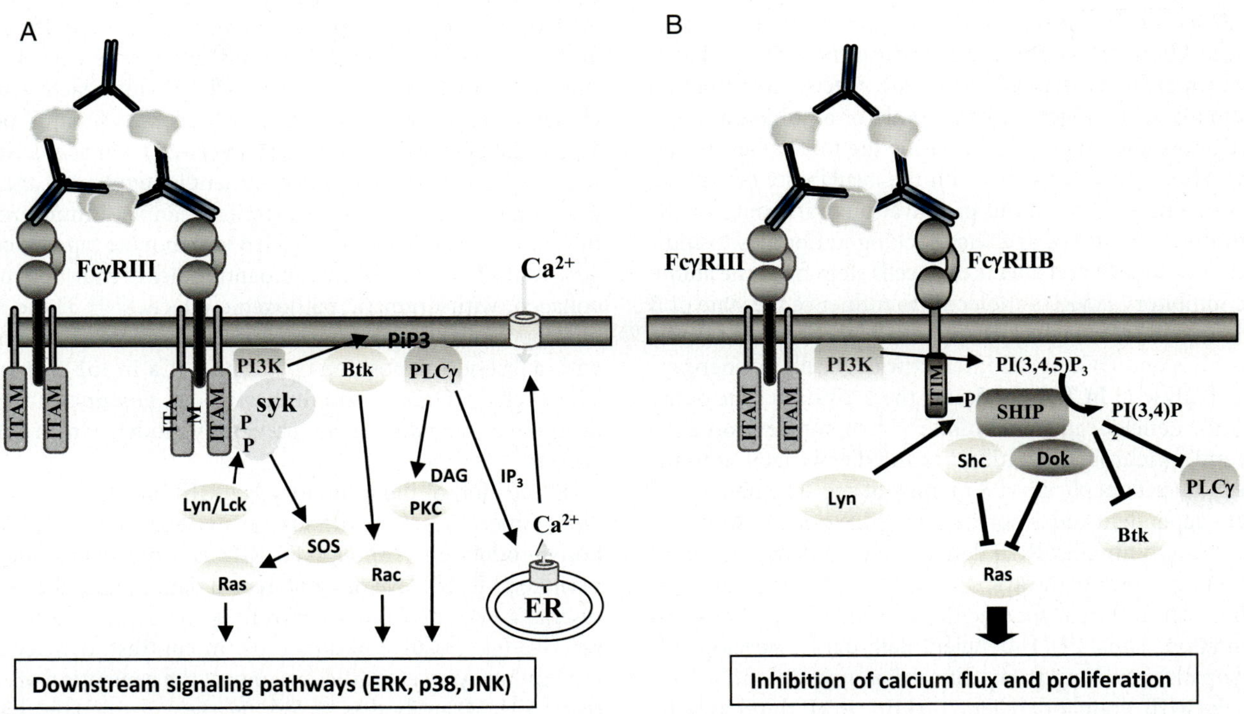

FIG. 33.5. Signaling by activation and inhibitory Fc receptors (FcRs). A: FcγRIII signaling in natural killer cells is shown as an example of the activation class of FcRs. Cross-linking by an immune complex initiates the signaling cascade. The specific Src family kinase varies, depending on the cell type. **B:** FcR signaling in macrophages is shown as an example of simultaneous triggering of activation and inhibitory FcRs.

proinflammatory activation are all blocked. Arrest of proliferation in B cells is also dependent on the ITIM pathway, through the activation of the adaptor protein Dok and subsequent inactivation of MAP kinases.[141,142] The role of SHIP in this process has not been fully defined, inasmuch as it can affect proliferation in several ways. SHIP, through its catalytic phosphatase domain, can prevent activation of the PH domain survival factor Akt by hydrolysis of PIP3.[143,144] SHIP also contains PTB domains that could act to recruit Dok to the membrane and provide access to the Lyn kinase that is involved in its activation. Dok-deficient B cells are unable to mediate FcγRIIB-triggered arrest of BCR-induced proliferation, while retaining their ability to inhibit a calcium influx, which demonstrates the dissociation of these two ITIM-dependent pathways.

Another inhibitory activity displayed by FcγRIIB is independent of the ITIM sequence and is displayed upon homoaggregation of the receptor. Under these conditions of FcγRIIB clustering, a proapoptotic signal is generated through the transmembrane sequence. This proapoptotic signal is blocked by recruitment of SHIP, which occurs upon colligation of FcγRIIB to the BCR, because of the Btk requirement for this apoptotic pathway.[132] This activity has been reported only in B cells and has been proposed to act as a means of maintaining peripheral tolerance for B cells that have undergone somatic hypermutation or to delete a fraction of established bone marrow resident plasma cells during subsequent immune responses, allowing newly generated plasma cells to access a limited number of bone marrow niches to become long-lived.

IN VIVO FUNCTIONS

Fcγ Receptors in the Afferent Response

The ability of IgG ICs to influence the afferent response has been known since the 1950s and can be either enhancing or suppressive, depending on the precise combination of antibody and antigen and the mode of administration.[145] Investigators have attempted to define the molecular mechanisms behind these activities with the availability of defined mouse strains with mutations in activation or inhibitory FcRs and through the use of specific blocking antibodies to individual receptors. Direct effects on B cells stem from the ability of the inhibitory FcγRIIB molecule to influence the state of B cell activation and survival. Because antigen is retained in the form of ICs on FDCs, it can interact with B cells by coengaging FcγRIIB with BCR, modulating the activation state of the cell. Mice deficient in this inhibitory receptor develop anti-DNA and antichromatin antibodies and die of a fatal, autoimmune glomerulonephritis at 8 months of age. The phenotype is strain dependent and is not seen in BALB/c or 129 strains of mice.[146] Combining FcγRIIB deficiency with defects in other inhibitory receptor pathways, such as PD-1, results in autoantibodies with different specificities and distinct pathological presentation. Thus, PD-1–deficient Balb/c mice develop cardiomyopathy resulting from anticardiac myosin antibodies. PD-1/FcγRIIB double-deficient BALB/c mice develop anti-uroepithelial antibodies and hydronephrosis.[147] FcγRIIB thus acts as a modifier of autoimmune disease, a conclusion further supported by studies that determined the contribution of the C57BL/6 background to the spontaneous lupus-like disease observed in those mice. The B6 background, by virtue of an incomplete light chain–editing pathway, provides a source of autoreactive B cells in the periphery, which, when combined with defects in the inhibitory FcγRIIB pathway, lead to the accumulation of autoantibodies, pathogenic ICs, and disease.[148] Further support for this conclusion is provided by the observations that autoimmune disease–prone strains of mice, such as New Zealand black (NZB), BXSB, SB/Le, MRL, and nonobese diabetic (NOD), have reduced surface expression of FcγRIIB on activated B cells, attributed to DNA polymorphisms in the promoter region of the gene encoding this receptor.[149,150] This reduced expression of FcγRIIB is thus suggested to contribute to the increased susceptibility of these animals to the development of autoantibodies and autoimmune disease. Direct evidence that this is indeed the case comes from studies where FcγRIIB expression levels have been restored by retrovirus-mediated gene transfer in BXSB, NZM, and FcγRIIB knockout animals. These mice had dramatically reduced levels of autoreactive antibodies and were protected from the development of fatal autoimmune disease.[62] Consistent with these results obtained in mouse model systems, patients with SLE and CIDP were demonstrated to have a reduced FcγRIIB expression level on B cells.[60,61]

Moreover, if FcγRIIB indeed functions *in vivo* to maintain peripheral tolerance, then its loss should allow for the emergence of autoantibodies when otherwise resistant animals are challenged with potentially cross-reactive antigens. This hypothesis has been validated in models of collagen-induced arthritis and Goodpasture syndrome. FcγRIIB-deficient mice, with the nonpermissive H-2b haplotype, develop arthritis when immunized with bovine type II collagen.[151] The loss of FcγRIIB thus bypasses the requirement for the specific H-2q and H-2r alleles previously demonstrated to be necessary in this model by allowing FcγRIIB-deficient autoreactive B cell clones to expand and produce pathogenic autoantibodies. When the permissive DBA/1 strain (H-2q) is made deficient in FcγRIIB, autoantibody development is augmented and disease is greatly enhanced. In a similar manner, immunization of H-2b mice deficient in FcγRIIB with bovine type IV collagen results in cross-reactive autoantibodies to murine type IV collagen, with dramatic pathogenic effects.[151,152] These mice develop hemorrhagic lung disease and glomerulonephritis with a "ribbon deposition" pattern of ICs in the glomeruli. These characteristics are indicative of Goodpasture syndrome, a human disease not previously modeled in an animal species.

Expression of the inhibitory FcγRIIB on B cells thus provides a mechanism for the suppressive effects of ICs on antibody production. Although FcγRIIB is expressed throughout peripheral B cell development, recent data suggest that it represents a late checkpoint controlling the expansion of autoreactive IgG positive plasma cells. In contrast, deficiency of the inhibitory receptor did not impact the generation of autoreactive IgM antibodies.[148] Taking the considerably higher pathogenic potential of IgG compared to IgM antibodies into account, this late stage of FcγRIIB-mediated regulation seems to be sufficient to prevent severe autoreactive processes.

FcγRIIB also exerts its inhibitory effects to modulate B cell affinity maturation in the germinal center (Fig. 33.7B). Recent studies have shown that seasonal influenza vaccination elicits anti-hemagglutinin (HA) antibodies enriched for sialylated Fc glycans. The resulting sialylated ICs feed back into the afferent response by engaging the type II FcR, CD23, on the surface of B cells in the germinal center, triggering FcγRIIB upregulation. This enhanced inhibitory signaling raises the threshold for activating BCR signaling, leading to selection of a higher affinity BCR. In this vein, the titer of sialylated IgG Fc following vaccination was shown to predict the affinity of anti-HA antibodies, and thus the quality of vaccine response. Exploiting this novel noncanonical FcR pathway by immunizing with sialylated ICs leads to broadly protective CD23-dependent responses in models of homologous and heterologous challenge with the H1 family influenza viruses.[153,154] However, the precise signaling mechanisms connecting CD23 engagement to FcγRIIB upregulation remain unsolved, as does the complex regulation of IgG Fc glycosylation during humoral immune responses.

The enhancing property of ICs on the afferent response may arise from the expression of FcRs on APCs, such as DCs.[155-157] DCs express the type I activation FcR, FcRIIA, and the inhibitory receptor FcRIIB, as well as the type II receptor DC-SIGN. Consistent with *in vitro* studies that have suggested that triggering of activation FcRs can induce DC maturation, *in vivo* studies of this pathway has demonstrated the ability of FcRIIA engagement to result in a vaccinal effect, with induction of CD8 memory responses.[158,159] The ability of FcRs, particularly of activating FcγRs to internalize ICs, could provide a mechanism for enhanced presentation and augmented antibody responses, whereas the presence of the inhibitory FcγRIIB molecule appears to reduce the enhancing effect. Indeed, targeting antigens to activating FcγRs expressed on CD11c/CD8+ and CD11c/CD8 DC subsets in mice *in vivo* triggered strong T cell responses. Of note, despite expression of activating FcγRs on both CD11c/CD8+ and CD11c/CD8 DC subsets, T helper cell responses were selectively induced via CD11c/CD8 DCs. In contrast, CD11c/CD8+ DCs triggered expansion of cytotoxic T cells, suggesting that DC subset–specific pathways determine the outcome of FcγR-dependent antigen uptake.[160] Consistent with these results, *in vitro* studies suggest that internalization through specific FcRs on APCs may influence the epitopes presented and T cell response generated as a result.

Mice deficient in FcγRIIB display enhanced antibody responses to soluble antibody-antigen complexes, in some cases dramatically so, which is likely to result from enhanced presentation.[161,162] At present, a growing body of data suggests that FcRs are indeed involved in enhancement of the afferent response, by influencing antigen presentation and cognate T cell interactions. FcγRIIB-deficient DCs pulsed *ex vivo* with antigen in the form of ICs induce a strong and protective cytotoxic immune response after transfer into naïve mice.[163] In contrast, wild-type DCs induce a much smaller and nonprotective response, indicating that the threshold set by co-cross-linking of the inhibitory and activating Fc receptors prevents complete DC maturation. Moreover, blocking of IC binding to FcγRIIB on human DCs with an FcγRIIB-specific antibody resulted in spontaneous maturation of the cells by ICs present in low amounts in human serum.[87,164] Taken together, these data indicate that the inhibitory receptor is an important regulator of DC activation. As the DC maturation state will determine whether an activating or a tolerogenic signal will be delivered to T cells, FcRs might be important factors for the maintenance of peripheral tolerance in the cellular immune system. Transiently blocking FcγRIIB activity with monoclonal antibodies (mAbs) *in vivo* might thus be an interesting strategy to optimize immunotherapeutic and vaccination approaches. Alternatively, Fc engineering of human IgG1 to selectively enhance FcRIIA binding, while reducing FcRIIB engagement, results in enhanced CD8 responses in mice humanized for FcRs and in humans.[158]

Fcγ Receptors in the Efferent Response

The first canonical FcR knockout to be described was for the common activation subunit, the γ chain, which resulted in the loss of surface assembly and signaling of FcγRI, FcγRIII, and FcγRIV as well as FcεRI.[10] Mice deficient in the common γ chain were systematically studied in diverse models of inflammation and found to be unable to mediate IgG-triggered inflammatory responses.[22,165-167] The results were further confirmed by comparisons of mice deficient or blocked for either FcγRI, FcγRIII, or FcγRIV.[19,20,33,168] The loss of FcεRI ablated IgE-mediated anaphylaxis; this was demonstrated independently by gene disruption in the α subunit of that receptor.[11] Subsequent studies on mice deficient in the inhibitory FcγRIIB molecule established the opposing action of this receptor, in which mice deficient in that receptor displayed enhanced B cell responses, autoimmunity, and augmented IgG-mediated inflammation in a subclass and effector cell–dependent manner.[18,20,33,57,167] The general finding, which is discussed in detail later, illustrates that IgGs initiate their effector responses *in vivo* through coengagement of activating and inhibitory FcRs. The physiological response is thus the net of the opposing activation and inhibitory signaling pathways that each receptor triggers and is determined by the level of expression of each receptor and the selective avidity of the IgG ligand (Table 33.1). This also explains the long-standing observation that different IgG isotypes have a differential activity *in vivo* (Fig. 33.6A, B). The absence of a murine homolog for FcαRI has precluded similar studies for that receptor. Studies on mice bearing a human transgene of FcαRI suggest that this receptor is involved in IgA nephropathy (Berger disease).[169]

Type I Hypersensitivity

Both cutaneous and systemic models of passive anaphylaxis, induced by IgE, were studied in FcRγ chain–deficient mice and were found to be absent, a finding fully consistent with the observations obtained in FcεRI-deficient mice and confirming the role of the high-affinity IgE receptor in mediating IgE-induced anaphylactic responses.[10,11,42,170] FcγRIIB-deficient mice, challenged in this model, displayed an unexpected enhancement of IgE-mediated anaphylaxis,

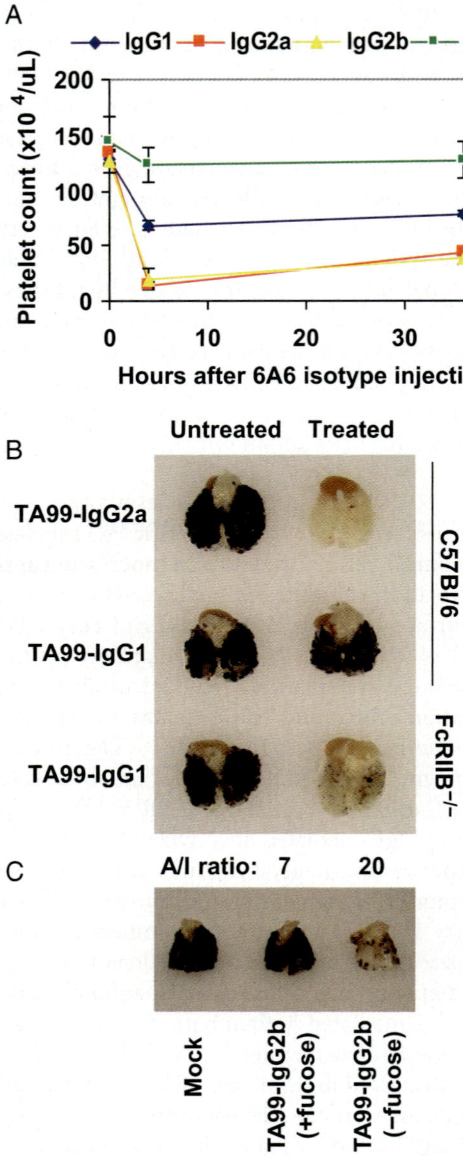

FIG. 33.6. Antibody activity is determined by activating and inhibitory Fc receptors (FcRs). A: Mice were injected with the same amount of different 6A6 antibody isotypes, which recognize an integrin on mouse platelets and lead to FcR-dependent clearance of platelets from the blood as observed by the drop in platelet count after antibody injection. **B:** Wild-type or FcγRIIB knockout mice were injected intravenously with B16 melanoma cells on day 0 and were either left untreated or were treated with IgG1 and IgG2a isotype switch variants of the melanoma-specific monoclonal antibody TA99 on alternate days. **C:** Mice were injected with antitumor antibodies (TA99) containing or lacking branching fucose residues, thereby enhancing binding to activating FcRs and changing the A/I ratio (see text for details). Lungs were harvested on day 14. (From Nimmerjahn F, Ravetch JV. Divergent immunoglobulin g subclass activity through selective Fc receptor binding. *Science*. 2005;310(5753):1510-1512. Reprinted with permission from AAAS.)

which suggests a physiological interaction between this inhibitory receptor and FcεRI.[82] The molecular basis for this modulation of FcεRI signaling by FcγRIIB has not been determined, although previous studies have indicated that IgE can bind with low affinity to FcγRII/FcγRIII, which suggests that there exists a mechanism for coengagement of these receptors. Deletion of the mast cell inhibitory receptor glycoprotein 49B1 also results in enhanced IgE-induced anaphylaxis.[83] In addition to FcεRI, mast cells also express the IgG FcγRIIB and FcγRIII, but not FcγRIV. Passive systemic anaphylaxis induced by IgG was attenuated in FcRγ chain–deficient and FcγRIII-deficient mice, which indicates the capacity of IgG and FcγRIII to mediate mast cell activation *in vivo*. FcγRIIB-deficient mice displayed enhanced IgG-induced anaphylaxis. Active anaphylaxis, induced by immunization with antigen in alum, was enhanced in FcεRI-, FcγRIIB-, and glycoprotein 49B1–deficient mice and attenuated in FcRγ- and FcγRIII-deficient mice. All these animals displayed antigen-specific antibodies for IgE and IgGs, which indicates that the active anaphylaxis seen was attributed primarily to IgG antibodies. The reason for the enhancement of anaphylactic responses in FcεRI-deficient animals resulted from the increased expression of FcγRIII on mast cells in these mice, normally limited by competition of α chains for the available pool of the common γ chain.[94] In the absence of FcεRI α chain, FcRγ chain is available to associate with FcγRIII α chain and assemble on the cell surface as a functional signaling receptor. A similar type of regulation was also described for expression of FcγRIII and FcγRIV on mouse neutrophils, where deletion of one receptor resulted in enhanced expression of the other γ chain–dependent molecule.[95] These studies indicated the importance of the γ chain in regulating the level of surface expression of FcεRI and FcγRIII. Because γ chain is also associated with other members of the activation/inhibition paired receptors expressed on mast cells, such as PIR-A/PIR-B, the intracellular competition between these diverse α subunits and the common γ chain determines the level of surface expression of individual receptors and thus their ability to respond to specific biological stimuli. The absolute level of surface expression of FcRs on mast cells is clearly of therapeutic significance in both IgE- and IgG-mediated inflammatory responses; modulation of γ chain expression could thus represent a new therapeutic avenue for intervention in diseases such as anaphylaxis and asthma.

Type II Hypersensitivity: Cytotoxic Immunoglobulin G

Cytotoxic IgGs are found in a variety of autoimmune disorders and have been developed for therapeutic indications in the treatment of infectious and neoplastic diseases. The mechanisms by which these antibodies trigger cytotoxicity *in vivo* have been investigated in FcR knockout mice. Anti-RBC antibodies trigger erythrophagocytosis of IgG-opsonized RBCs in an FcR-dependent manner; FcRγ chain–deficient mice were protected from the pathogenic effect of these antibodies, whereas complement C3–deficient mice were indistinguishable from wild-type animals in their ability to clear the targeted RBCs.[171,172] FcγRIII plays the exclusive role in this process for the mouse IgG1 isotype. Murine IgG2a anti-RBC antibodies utilize primarily the FcγRIV receptor pathway despite the singular ability of murine IgG2a antibodies to bind as monomers to FcγRI. These and other studies suggest that the role of the high-affinity FcγRI in IgG-mediated inflammation is likely to be restricted to augmenting the effector response (determined by FcγRIII and IV) in situations that involve high concentrations of murine IgG2a antibodies that are found at localized inflammatory sites where FcγRI expression is on recruited macrophages.

Experimental models of immune thrombocytopenic purpura (ITP), in which murine IgG1 antiplatelet antibodies trigger thrombocytopenia, yielded results similar to those of the anti-RBC studies cited previously. The specific FcγR involved depended on the subclass of antibody used. IgG1 antibodies mediated their activity exclusively through FcγRIII, while IgG2a and 2b were FcγRIV dependent. In contrast, FcγRI- or C3-deficient mice were fully susceptible to antibody-induced thrombocytopenia.[20,32] FcγRIIB-deficient mice showed an isotype-specific enhancement of antibody-mediated platelet depletion, with the strongest impact on IgG1 and much smaller increases for IgG2a and IgG2b isotypes. This is consistent with the affinities of these isotypes for their specific activating and the inhibitory receptor, which will determine antibody activity *in vivo* (Table 33.1). In a passive protection model of *Cryptococcus neoformans*–induced disease, passive immunization with mouse IgG1, IgG2a, and IgG2b antibodies resulted in protection in wild-type animals but not in FcRγ chain–deficient animals.[173]

IgG antibodies raised to murine glomerular basement membrane preparations induce acute glomerulonephritis in a model of Goodpasture disease in wild-type but not FcRγ- or FcγRIV-deficient animals.[89,174,175] FcγRIIB-deficient animals displayed enhanced disease in this model, which indicates that the effector cells involved were constitutively expressing significant levels of FcγRIIB. Similar results were obtained when DBA/1 animals were immunized with bovine type II collagen to induce arthritis. Deficiency of FcRγ chain protected these mice from the pathogenic effects of the anticollagen antibodies that were generated.[176] As mentioned previously, deficiency of FcγRIIB in the DBA/1 collagen-induced arthritis model resulted in enhanced disease, through increased autoantibody production and elevated effector responses.

A dramatic example of the importance of these pathways in determining the *in vivo* activity of cytotoxic antibodies was obtained in models of antitumor antibody response. In a syngeneic murine model of metastatic melanoma, a murine IgG2a antimelanocyte antibody was able to reduce tumor metastasis in wild-type animals but was ineffective in FcRγ- or FcγRIV-deficient mice.[20,95,177] In the absence of FcγRIIB, the activity of an IgG1 antibody, matched in its antigen-binding domain, was enhanced, which indicates that the *in vivo* cytotoxic activity of the antibody was the net of activation and inhibitory receptor engagement[20,57] (Fig. 33.6B). These studies, together with similar studies performed with antiplatelet antibodies or defucosylated antibodies, demonstrated that the *in vivo* activity of a cytotoxic antibody could be predicted by a simple equilibrium binding model in which the ratio of the monomeric affinity constants for the activation and inhibitory FcRs (A/I ratio) are the dominant parameters, as demonstrated for TCR-MHC interactions.[178] An example for the enhanced cytotoxic activity of an antitumor antibody due to increased affinity for activating FcRs is shown in Figure 33.6C. The generation of an afucosylated antitumor antibody resulted in an increase of the A/I ratio of 7 to 20 for this IgG subclass, resulting in enhanced antitumor activity *in vivo*. Xenograft models of human tumors transplanted into nude mice demonstrated, for a variety of tumors and cytotoxic antibodies, the requirement for FcR effector activity. For example, human breast carcinoma or lymphoma lines transplanted into nude mice and treated with either the humanized IgG1 or chimerized IgG1 antibodies (trastuzumab [Herceptin] and rituximab [Rituxan]), respectively, revealed that the ability of these antibodies to modulate tumor growth was abrogated in FcRγ chain–deficient mice. A point mutation that eliminated FcR binding of the anti-Her2/neu murine IgG1 antibody 4D5 abolished the *in vivo* cytotoxic activity of the antibody against a human xenograft but did not affect the *in vitro* growth inhibitory activity; this again illustrates the difference between *in vivo* and *in vitro* mechanisms. Similar results were obtained for depletion of B cells with CD20-specific IgG subclass variants and for T cell lymphoma xenograft models and anti-CD2 antibodies, among others.[179,180] More recently, it has become clear that some of the checkpoint blockade antibodies, which have shown significant therapeutic activity in cancer therapy for some types of tumors may work, at least in part, via cytotoxic mechanisms. For example, a CTLA4-specific antibody was demonstrated to deplete intratumoral T_{regs} in mice in a strictly FcγR-dependent fashion. However, the mechanism of enhancing CD8 activation in human therapy is still under investigation.

With respect to the effector cell populations responsible for cytotoxic IgG activity, NK cells were thought to be of major importance. However, several more recent independent lines of evidence point toward myeloid cells as an important effector cell population. Firstly, CD20-specific antibodies were shown to depend on liver resident macrophages for depleting B cells from the circulation.[111] In a similar manner, depletion of macrophages or monocytes, but not of NK cells, inhibited the cytotoxic activity of CD20-specific, CTLA4-specific as well as of antibodies directed against the melanoma antigen gp75, which was discussed earlier. Secondly, the modulation of cytotoxic IgG activity via the inhibitory FcγRIIB, which was observed in several *in vivo* models, could not be explained if NK cells were the major effector cell population, as NK cells do not express the inhibitory FcγRIIB. Thirdly, the critical importance of FcγRIV, which is expressed on myeloid cells, but not NK cells, is another argument for myeloid cells as effector cells. Finally, it was demonstrated that the CD47-Sirpα axis modulates cytotoxic IgG activity. CD47 is a ubiquitously expressed molecule, which interacts with Sirpa, which is expressed on myeloid cells and macrophages, but not on NK cells. Sirpα is an inhibitory receptor that prevents macrophages from phagocytosing cells that express the "do not eat me" signal CD47, and antibodies blocking this interaction were demonstrated to enhance cytotoxic IgG activity. 112 Thus, myeloid cells seem to be a major effector cell population for cytotoxic IgG activity *in vivo*.[113]

The general conclusions that can be drawn from these studies support a dominant role for the low-affinity activating FcγRs in mediating cytotoxicity by IgG antibodies.

FcγRIIB restricts the effector response for those antibodies with low A/I ratios in situations in which the effector cell, such as monocytes and macrophages, expresses this inhibitory molecule.

The relevance of these murine *in vivo* studies to the treatment of human populations with antitumor cytotoxic antibodies has been demonstrated in two studies that investigated the differential responses of patients treated with anti-CD20 (rituximab) for lymphoma.[50,54] Both studies demonstrated a highly significant correlation between patient response, as measured by the time to relapse, and alleles of FcγRIIIA. In patients with an allele of this low-affinity activation receptor (158V) that confers higher binding affinity for human IgG1 Fc, improved outcome was observed, as compared to those with a lower binding allele of this receptor (158F). These observations have been translated into clinical applications. For example, a glycoengineered, afucosylated anti-CD20 antibody with higher affinity to FcγRIIIA demonstrated improved survival in CLL as compared to patients receiving anti-CD20 antibodies with fucosylated Fc.[114,181] A similar strategy is now in wide use for the improvement of cytotoxic antitumor antibodies, including mogamulizumab, benralizumab, and margetuximab.[182-184] In addition, lymphoma patients with the high-affinity allele showed a significantly better clinical response after receiving anti-idiotype vaccination.[53] FcγRIIIA, which is the orthologue of mouse FcγRIV, is expressed on NK cells, monocyte subsets, and macrophages. Which of these potential effector cell subsets are critical for cytotoxic IgG activity in humans remains to be established. However, it has also become clear that alleles of FcγRIIA, which is expressed on myeloid cells, DCs, and neutrophils, associate with the activity of some cytotoxic antibodies in humans, suggesting that similar to mice, myeloid cells may participate in removing opsonized target cells in humans and in augmenting CD8 antitumor responses.

Type III Hypersensitivity: IC-Mediated Inflammation

The classic example of this reaction, the Arthus reaction, has been studied in a variety of FcR- and complement-deficient animals. The initial studies were performed by using the cutaneous reverse passive Arthus reaction, in which antibody was injected intradermally and antigen was given intravenously. An inflammatory response, characterized by edema, hemorrhage, and neutrophil infiltration, developed within 2 hours. This reaction was elicited in a variety of complement- and FcR-deficient animals. The results from several independent studies confirmed the initial observations: that IgG ICs triggered cutaneous inflammatory reactions even in the absence of complement but displayed an absolute requirement for FcγR activation.[22] FcγRIIB modulated the magnitude of the response, with enhanced Arthus reactions observed in FcγRIIB-deficient strains.[18] The effector cell in the cutaneous reaction was determined to be the mast cell, as demonstrated by the use of mast cell–deficient strains and by mast cell reconstitution studies.[170] The generality of this result was demonstrated in similar reactions performed in the lung, illustrating the FcR dependence and relative complement independence of this response.[152] Thus, all studies have demonstrated an absolute dependence on FcR expression in the Arthus reaction. One model for IC-induced arthritis, the KRN/NOD model, in which IgG1 anti-GPI antibodies are responsible for IC deposition in the synovium[185] has been shown to depend on both FcγRIII and C3 but not on components of the classical complement pathway, such as C1q and C4; transfer of serum to animals deleted for FcγRIII or C3 prevented the development of disease.[186,187] Deficiency in the late components of complement, such as C5a or its receptor, has also been reported to result in a partial reduction in the magnitude of the response in IC-induced lung inflammation[188] and to result in a complete block in the KRN/NOD arthritis model. The likely mechanism by which C5a exerts its effects is through upregulation of activating FcγRs, resulting in an amplification loop.[86] C5a may be generated in this system locally by the effector cell itself as a result of FcR activation and is independent of the classical, alternative, and MBP pathways.[189] Binding of C5a to the C5aR results in upregulation of activation receptors on these effector macrophages, thus augmenting the inflammatory response triggered by FcRs. These studies have led to a revision of the hypotheses about the mechanism of IC-mediated inflammation, typified by the Arthus reaction, in which there is an absolute requirement for activating FcγRs in initiating mast cell activation by ICs. FcγR activation is, in turn, modulated by the inhibitory receptor FcγRIIB. The A/I value of a specific IgG antibody and the densities of these opposing signaling receptors determines the concentration threshold for IC activation and the magnitude of the effector response that can be obtained. The classical pathway of complement activation is not required; however, C5a activation, through the FcγR pathway, may enhance the response under some circumstances through an amplifying loop. The release of inflammatory mediators such as vasoactive amines, chemokines, and cytokines leads to the hallmarks of this reaction: edema, hemorrhage, and neutrophil infiltration at the site of IC deposition.

The significance of the FcR pathway in initiating IC inflammation in autoimmune disease was further established by investigating a spontaneous murine model of lupus, the B/W F1 mouse. The Arthus reaction results predicted the absolute requirement of activation FcγR in initiating inflammation and tissue damage in IC diseases such as lupus. The FcRγ chain deletion was backcrossed onto the NZB and New Zealand white strains for eight generations, and the intercrossed progeny was segregated into B/W FcRγ−/− and FcRγ+/−. Anti-DNA antibodies and circulating ICs developed in all animals; IC and complement C3 deposition was similarly observed in all animals. However, mice deficient in the common γ chain showed no evidence of glomerulonephritis and had normal life expectancy, despite comparable levels of circulating ICs and glomerular deposition of these complexes along with complement C3. Mice heterozygous for the γ chain mutation were indistinguishable from B/W F1 animals with wild-type γ chains in developing glomerulonephritis and displaying reduced viability.[166] This spontaneous model supports the conclusions stated previously about the absolute requirement for FcγRIII in the activation of inflammatory disease by ICs: in the absence of this receptor, deposited ICs and C3 are not sufficient to trigger effector cell activation, which indicates that it is possible to uncouple pathogenic ICs from inflammatory disease by removing activating FcR engagement. Similar conclusions were reached in a murine model of Goodpasture disease where IC deposition, composed of mouse IgG2b antibodies, resulted in a fulminant glomerulonephritis.[89] Blocking the relevant activation FcR, FcγRIV, with a mAb, protected the animals from fatal disease. These results further indicate that intervention in the effector stage of IC diseases, such as lupus and rheumatoid arthritis, would be accomplished by blocking activation FcγRs to prevent initiation of effector cell responses.

Type IV Hypersensitivity: Cell-Mediated Memory Response

The past several decades of research have uncovered an essential role of Fc-FcR interactions in mediating antitumor responses to mAbs. As outlined in the discussion of type II hypersensitivity, passive administration of these therapeutics results in FcγRIIIA engagement and subsequent acute ADCC or antibody-dependent cellular phagocytosis, likely facilitated by NK and/or myeloid cells, respectively. This effector function is required for the *in vivo* efficacy of these antibodies. However, some patients see a long-term response to therapeutic mAbs, whereby a single course of treatment may result in a durable, long-lasting memory response.[190] Recent lines of evidence have pointed to FcγRIIA, the activating receptor found on a number of myeloid lineages, most significantly on DCs, as the mediator of this response. When mice bearing human FcγRs were challenged with a human CD20-expressing lymphoma cell line, only treatment with rituximab (anti-CD20) engineered for FcγRIIIA binding provided adequate tumor clearance. Upon rechallenge 60 days later with a 10-fold increase in inoculant, however, only antibodies engineered for FcγRIIIA binding and FcγRIIA binding were largely protective.[158] DCs express FcγRIIA as their sole activating FcR and are the APCs that drive T cell–mediated memory responses, thereby implicating this cell type and its FcR-dependent effector functions in long-lasting antitumor immunity. While therapeutic antibodies both approved and in ongoing trials are often optimized for ADCC, recent results have highlighted the enhanced clinical responses that are elicited by engineering mAbs to facilitate a long-term cellular response.[191]

IgG-Dependent Neutralization and Induction of Agonistic Signaling

In contrast to these FcR-dependent activities, IgG molecules, by virtue of their exquisite specificity for discrete antigens, are expected to mediate FcγR-independent effects, such as neutralization of bacterial toxins or viruses. Here, IgG molecules would prevent the toxin or microorganism from binding to its cellular receptor, thereby protecting the host from the pathogenic effects of the infecting microbe. This picture changed, however, as it was demonstrated that both IgG-mediated toxin and virus neutralization can be dependent on cellular FcRs *in vivo*. Thus, neutralization of the anthrax toxin (called protective antigen [PA] component) by anti-PA antibodies required FcR expression *in vivo*, and anti-PA antibodies that differed only by their Fc subclasses showed a hierarchy of protection with IgG2a and IgG2b antibodies affording a better protection than IgG1 antibodies, consistent with previous observation for cytotoxic IgG.[192,193] Similar results were obtained for the mechanism of activity of neutralizing antibodies protecting from influenza or HIV infection.[194,195] These results suggest that current *in vitro* assays using only the pathogen-specific antibody and a target cell line might miss an important component required for antibody activity *in vivo* and that novel assays including the FcR-dependent component might provide a better predictability of IgG activity.[196]

Beyond the central function of IgG in host defense, agonistic antibodies triggering signaling pathways in target cells are being evaluated for their therapeutic potential. For example, agonistic antibodies for TNFR family members such as anti-CD40 or anti-DR5 (drozitumab) are under evaluation for their ability to enhance tumor cell clearance by both direct and indirect mechanisms. Interestingly, the apoptosis-inducing activity of anti-DR5 agonistic antibodies was found to be dependent on FcRs *in vivo*.[197] A similar dependence on FcRs for *in vivo* activity was demonstrated for agonistic anti-Fas antibodies.[198] A variety of CD40-specific agonistic antibodies have been used to deliver costimulatory signals to augment and sustain T cell responses to result in tumor clearance.[197] Recent studies showed that the inhibitory FcγRIIB was essential for this agonistic activity and could not be replaced by activation FcγRs.[199,200] Mice with immune cell subset–specific deletions of FcγRIIB revealed that B cells played a major role for the CD40-specific immunomodulatory antibody. In contrast, DR5-specific agonistic antibodies rather required FcγRIIB expression on myeloid cells, suggesting that the immunological environment in which an agonistic antibody mediates its activity determines which immune cell subset contributes to its activity via FcγRIIB engagement. Of note, FcγRIIB-dependent induction of inhibitory signaling pathways was not critical for agonistic antibody activity, suggesting that FcγRIIB expressed on neighboring immune cells seems to act as a platform to cross-link agonistic antibodies bound to their target antigen.[126] Although the exact mechanism of these FcR-dependent pathways remains to be established, these findings open the path toward generating improved agonistic anti-TNFR antibodies by Fc engineering. Consistent with this notion, a recent study in mice showed that generating anti-CD40 antibodies with enhanced FcγRIIB binding considerably augmented their *in vivo* adjuvant activity[200] Similarly, in a mouse model humanized for FcγRs and CD40, FcγRIIB was essential for anti-CD40 mAb activity, and variants enhanced for FcγRIIB binding showed improved antitumor immunity.[76]

Antibody-Dependent Enhancement

Antibody-dependent enhancement (ADE) represents a peculiar effector function of the activating FcγRs, whereby non-neutralizing antibodies existing in circulation can exacerbate clinical disease. While antibodies are generally thought to opsonize pathogens and assist with host clearance, they occasionally facilitate enhanced infection of host cells. The best studied example of this phenomenon is secondary dengue virus infection. Primary infection with one serotype of the virus produces nonneutralizing serum titers of antidengue antibodies. Upon reexposure with a different serotype, the existing antibodies are unable to neutralize the infecting virus, but form ICs with viral particles and bind activating FcRs on monocytes and macrophages to initiate internalization of the complexes through FcRs, rather than canonical viral receptors. This enhances infection in FcR-bearing cells, such as myeloid cells, and can trigger a cascade of massive cytokine release, vascular leakage, and gastrointestinal hemorrhage, which characterize a severe, but rare clinical response to secondary dengue virus infection. Both *in vitro* and *in vivo* experiments have demonstrated that ADE only occurs within a middle range of nonneutralizing antibody titers: if titers are too high, neutralization occurs with few clinical symptoms; if they are too low, the pathology mimics a primary infection where existing antibodies are scarce or none.[201] In dengue virus infection, nonneutralizing antidengue antibodies with high affinity for FcγRIIIA have been particularly associated with ADE.[105] This affinity is determined both by subclass and Fc glycoform. A second subtype of enhancement, vaccine-associated enhanced respiratory disease, is independent of FcR-mediated pathways that do not facilitate host cell infection or increased viral replication has been demonstrated in respiratory syncytial virus, measles, and H1N1 influenza. The pathology in this case is caused by nonneutralizing IgG IC deposition, complement deposition, and subsequent inflammation[202] with skewing to a Th2 response.[203]

DISEASE ASSOCIATIONS

Autoimmunity and Tolerance

In view of their functional capacity to link autoantibodies to effector cells, FcRs have naturally been considered to have a pathogenic role in the development of autoimmune diseases. Several studies have attempted to correlate specific polymorphisms in FcγRIIA, FcγRIIIA, or FcγRIIB with incidence or severity of lupus or rheumatoid arthritis.[204] In view of the heterogeneity of these diseases, it is perhaps not surprising that inconsistent results have been obtained. Alleles that increase the ability of FcγRIIA to bind IgG2 or FcγRIIIA to bind IgG1 might be expected to correlate with disease severity in some populations. Indeed, these types of associations have been reported in some studies but not in others.[205] These variable results have often been explained as an indication that other genes may be in linkage disequilibrium with the FcR alleles under investigation. This is a plausible explanation when viewed in light of the autoimmunity susceptibility genes mapping in or near the region of the FcR genes, chromosome 1q21 to 1q24.[206] This region of chromosome 1 has been implicated in a variety of human and murine linkage studies. For example, the Sle1 alleles derived from NZB flank the FcγRIIB gene and form a linkage group with the ability to break tolerance to nuclear antigens, resulting in production of antichromatin antibodies. Epistatic interactions between FcγRIIB and lupus susceptibility genes have been demonstrated in the murine lupus model of B6.RIIB. Crossing the Yaa gene to this strain accelerates the development of disease; 50% survival is decreased from 8 to 4 months, with 100% fatality by 8 months. This increase in severity correlates with a change in the specificity of the autoantibodies, from diffuse antinuclear antibodies to antibodies that stain with a punctate, nucleolar pattern on antinuclear antibody staining. Recently, it has been shown that the Yaa susceptibility locus contains a duplication of the Toll-like receptor 7 (TLR7) gene in the pseudoautosomal region of the Y chromosome.[207] Moreover, TLR7 and TLR9 with their ability to recognize potential self-antigens such as RNA or DNA, respectively, were shown to be important components in the generation of pathogenic autoreactive antibodies.[208,209] Compared to other Yaa-independent SLE susceptibility loci, FcγRIIB may be a dominant factor for the development of autoantibodies and the systemic activation of the immune system as demonstrated by a study using NZB/BXSB F1 animals congenic for the wild-type Fcgr2b gene. In these mice, a dramatic reduction in the production of autoantibodies and immune cell activation was noted, compared to mice with impaired FcγRIIB expression.[210] These data are consistent with studies that overexpressed FcγRIIB in autoimmune-prone mouse strains resulting in suppression of autoantibody production.[62,63]

Two types of polymorphisms in the FcγRIIB gene have been associated with SLE. Promoter polymorphisms have been described resulting in reduced expression of FcγRIIB on activated B cells in both mouse and human SLE and CIDP populations.[61,149,211] In addition, a polymorphism in the transmembrane domain of FcγRIIB that is associated with susceptibility to SLE in Japanese populations has been identified. This polymorphism results in an FcγRIIB protein with reduced ability to enter lipid rafts and thus behaves as a hypomorphic allele for inhibitory function.[212]

Together, these studies point to FcγRIIB as a susceptibility factor in the development of autoimmunity with the ability to interact with other susceptibility factors to modify both the afferent and efferent limbs of the autoimmune response.

Inflammation

Antibody-mediated inflammatory diseases have been clearly demonstrated to involve the coupling of pathogenic autoantibodies or ICs to cellular FcRs. Therapeutics targeted to disrupt these interactions are in development, beginning with an mAb to human IgE that functions to reduce IgE binding to its high-affinity receptor and thereby prevent allergic and anaphylactic reactions.[213] Because IgE is required for the survival of mast cells as well as in the regulation of FcεRI expression, reduction in IgE has synergistic effects on the ligand, receptor, and effector cell. The success of this approach will undoubtedly lead to other approaches that target the receptor or its signaling pathway. Blocking FcγRIIIA or IIA is expected to mimic the phenotype of FcRγc chain–deficient animals in models of IgG-induced disease. Early attempts to use this approach in ITP were promising but limited by the cross-reactivity to receptors on neutrophils, which led to neutropenia and the development of immune response to the murine antibody.[214] Development of second-generation anti-FcγRIIIA antibodies with greater specificity and reduced toxicity now appears to be a viable approach for the treatment of autoimmune diseases. Moreover, using small nonactivating

ICs to interfere with autoantibody binding to activating FcγRs or small molecules to inhibit signaling molecules such as Syk seems to have promising effects in preventing autoantibody-dependent platelet depletion in mice and humans.[215]

An alternative approach to limiting the activation of FcRs is to utilize the endogenous inhibitory pathway to abrogate IgE or IgG activation of their cognate receptors through colligation to FcγRIIB. This mechanism has been proposed to explain the ability to induce desensitization for the treatment of allergic diseases.[34,216] Inducing production of IgG antibodies to an allergen may facilitate cross-linking of FcγRIIB to FcεRI. A similar approach involving anti-CD19 antibodies containing an engineered Fc fragment with enhanced binding to FcγRIIB has been successfully used to shut down autoantibody production by human B cells in a xenograft mouse model system.[217] The ability to exploit the inhibitory pathway to reduce the activity of activation FcγRs has been demonstrated to account for some of the anti-inflammatory activity associated with high-dose IVIG, which consists of the pooled IgG fraction of serum from thousands of donors.[88] The use of IVIG for the treatment of ITP and other autoimmune diseases is well established, although the mechanism of action has been elusive. In murine models of ITP, arthritis, and nephritis, it has been demonstrated that protection by IVIG is dependent on the presence of FcγRIIB; deletion of FcγRIIB or blocking FcγRIIB by a mAb eliminates the ability of IVIG to protect the animal against an inflammatory response.[88,90,92] IVIG was demonstrated to lead to the *in vivo* induction of FcγRIIB on splenic effector macrophages, which would raise the threshold required for platelet clearance by activating FcγRs on these cells.[20,32,90,95] A similar upregulation of the human inhibitory FcγRIIB was demonstrated in patients with CIDP receiving IVIG treatment as a first-line therapy.[61]

In the past decade, it has become clear that the sialic acid–rich fraction of IgG antibodies in IVIG preparations is responsible for anti-inflammatory activity and FcγRIIB upregulation. For the first time, this has enabled the generation of a recombinant IVIG replacement consisting of a sialic acid–rich IgG Fc fragment, which has now entered clinical trials.[91,92] As mentioned earlier, sialylated IgG Fc has a reduced affinity for type I FcγRs and acquires the capacity to bind type II FcγRs, of which mouse SIGNR1 is strongly expressed on splenic marginal zone macrophages.[72,92] Without SIGNR1, mice challenged with arthritogenic K/BxN serum were shown to be unresponsive to sialylated IgG Fc. By using SIGNR1-deficient mice expressing human DC-SIGN, it was demonstrated that human DC-SIGN expressed on macrophages or DCs can substitute for mouse SIGNR1, affording a candidate molecule essential for the anti-inflammatory activity of IVIG in humans.[71] Both molecules belong to the C-type lectin family and have been implicated in recognition of pathogenic microorganisms including HIV and *Mycobacterium tuberculosis*.[218] These studies provide evidence that human DC-SIGN and mouse SIGNR1 are type II FcγRs capable of recognizing specific IgG glycovariants and, together with the family of TLRs, are a prime example for the dual usage of one molecule in the recognition of self-ligands and non–self-ligands. Moreover, infusion of sialic acid–rich IgG resulted in an IL-33–dependent expansion of basophils and IL-4 production. IL-4 is one of the interleukins known to upregulate FcγRIIB on innate immune effector cells, consistent with the upregulation of FcγRIIB noted in mouse model systems and humans. Together with previous data on the dependence of FcγRIIB upregulation on CSF-1–dependent macrophages, so-called sensor macrophages, this suggests the following model for the anti-inflammatory activity of IVIG (Fig. 33.7A). In this model, sialic acid–rich IVIG binds to mouse SIGNR1 or human DC-SIGN on sensor macrophages or DCs, which results in the production of IL-33 that in turn leads to IL-4 production by basophils, inducing the upregulation of FcγRIIB on effector macrophages, thereby raising the threshold for activation.[71] There is evidence that this anti-inflammatory pathway triggered by sialylated IgG is of general importance for maintaining immune homeostasis. In this vein, patients with rheumatoid arthritis and autoimmune-prone mouse strains have reduced levels of serum IgG–rich containing terminal sialic acid and galactose residues during active phases of the disease.[219-221] A similar reduction of this anti-inflammatory IgG variant is observed during aging, consistent with an increased probability of developing autoimmune disease.[222] This downmodulation of sialic acid–rich IgG was also observed during normal immunization conditions in mice, suggesting the general importance of this phenomenon during proinflammatory conditions.[92] The opposite type of regulation is seen in women with rheumatoid arthritis during pregnancy.[223] Paralleling the decreased frequency of arthritic flares, an increased level of IgG glycoforms rich in terminal galactose and sialic acid residues was observed. More recent studies demonstrated that the individual IgG subclasses show distinct glycosylation patterns during the steady state. Thus, the majority of sialic acid residues present in mouse serum IgG was present on IgG2a antibodies, whereas IgG1 antibodies had only very low amounts of terminal sialic acid residues.[224] This may indicate that besides the differential binding of IgG subclasses to activating and inhibitory FcRs (the A/I ratio concept), a second level of regulation via sialylation may exist. Thus, IgG1 antibodies, which bind strongly to the inhibitory FcγRIIB, are not further inhibited via sialylation. In contrast, IgG2a antibodies, which weakly bind to FcγRIIB, may be prevented from an unwanted activation of activating FcγRs during the steady state via sialylation. During inflammation, both sialylation and FcγRIIB expression are downmodulated, allowing a full blown activation of innate immune effector cells via IgG binding to activating FcγRs. With respect to autoimmunity, inducing expression of FcγRIIB or triggering high levels of autoantibody sialylation might be a clinically feasible approach and could be effective at modulating the activating potential of pathogenic autoantibodies. Indeed, when soluble galacto- and sialyltransferases were administered in mouse models of autoimmune disease, they selectively sialylated autoantibodies and attenuated inflammation, using platelets as nucleotide-sugar donors.[225]

Studies on the FcαRI receptor have demonstrated a role for this molecule in the pathogenesis of IgA nephropathy, in which circulating macromolecular complexes are deposited in the mesangium, resulting in hematuria and eventually

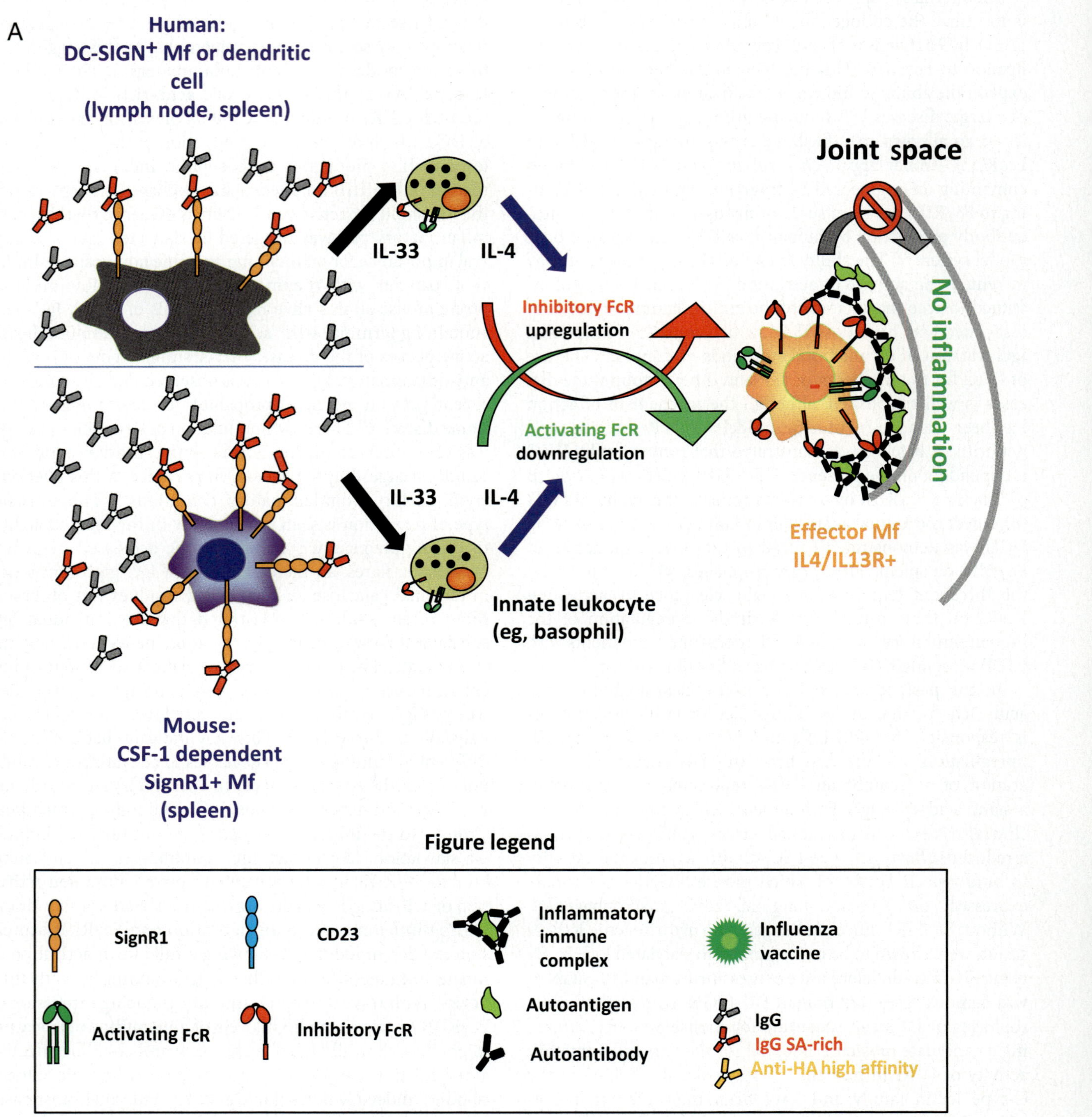
A
Human:
DC-SIGN+ Mf or dendritic cell
(lymph node, spleen)
IL-33
IL-4
Inhibitory FcR
upregulation
Activating FcR
downregulation
IL-33
IL-4
Innate leukocyte
(eg, basophil)
Joint space
No inflammation
Effector Mf
IL4/IL13R+
Mouse:
CSF-1 dependent
SignR1+ Mf
(spleen)
Figure legend
SignR1
CD23
Inflammatory
immune
complex
Influenza
vaccine
Activating FcR
Inhibitory FcR
Autoantigen
Autoantibody
IgG
IgG SA-rich
Anti-HA high affinity

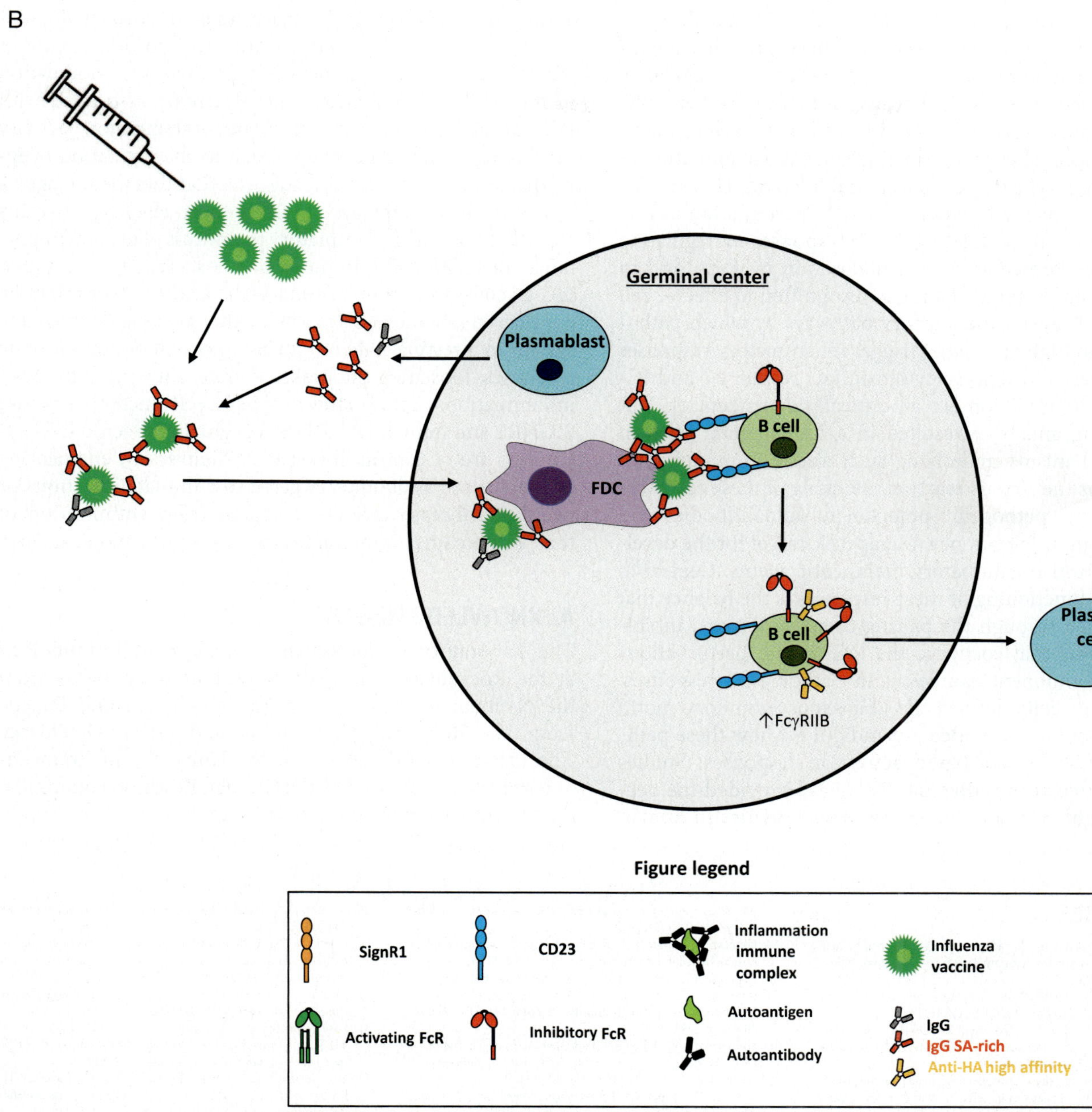

FIG. 33.7. A: Model for the anti-inflammatory activity of intravenous gamma globulin (IVIG) in mice and humans. Sialic acid–rich antibodies in the IVIG preparation lose affinity for classical FcγRs and acquire the capacity to bind to molecules such as mouse SIGNR1 or human DC-SIGN on the surface of CSF-1–dependent sensor macrophages or dendritic cells (DCs). This results in the production of IL-33 and the expansion of innate leukocytes, such as basophils, followed by IL-4 release resulting in the upregulation of the inhibitory FcγRIIB and downmodulation of activating FcγRs on effector macrophages inhibiting the release of inflammatory mediators and tissue destruction. **B:** Model for the role of sialylated IgG Fc in the development of high-affinity antibody responses. Shortly after vaccination, plasmablasts create sialylated IgG that binds antigen, is presented by DCs, and engages CD23 on the surface of germinal center B cells. This in turn upregulates the expression of FcγRIIB, increasing the threshold for affinity maturation, and results in higher affinity IgG responses. FDC, follicular dendritic cell.

leading to renal failure. Soluble FcαRI is found in the circulating IgA complexes, which suggests a role for the receptor in the formation of these pathogenic complexes. A transgenic mouse expressing FcαRI spontaneously develops IgA nephropathy resulting from the interaction of polymeric mouse IgA and the human FcαRI receptor to release soluble receptor-IgA complexes, which leads to deposition in the mesangium and the sequelae of IgA neuropathy.

SUMMARY AND CONCLUSIONS

Receptors for the Fc of Igs provide an essential link between the humoral and adaptive response, translating the specificity of antibody diversity into cellular responses. The canonical Fc receptors mediate their biological responses through the coupling of Fc recognition to ITAM/ITIM-based signaling motifs. A diverse array of biological responses depends on the FcR system, influencing both the afferent and efferent limbs of the immune response. Detailed biochemical, structural, and molecular biological data have provided an in-depth understanding of how these receptors are regulated, are assembled, bind their ligands, and transduce specific cellular signals. FcRs play a significant role *in vivo* in maintaining peripheral tolerance by limiting the accumulation of autoreactive B cells that escape central tolerance checkpoints such as light chain editing or somatic hypermutation in germinal centers, in modulating T cell responses by regulating both antigen presentation and maturation by DCs, and in mediating the coupling of antigen recognition to effector cell activation. They are the primary pathways by which pathogenic IgG and IgE antibodies trigger inflammatory responses *in vivo.* Allergic reactions, cytotoxic IgG responses, and IC-mediated inflammation are all critically dependent on FcR cross-linking and have resulted in a fundamental revision of the mechanisms underlying such classic immunological responses as the Arthus reaction. Blocking of these receptors uncouples the pathogenic potential of autoantibodies and represents an important new therapeutic target for the development of anti-inflammatory therapeutic agents. Central to the correct functioning of these responses is the balance that is maintained through the pairing of activation and inhibitory receptors that coengage the IgG ligand; perturbations in either component can result in pathological responses. The study of FcRs defined the ubiquitous inhibitory motif, the ITIM, and has provided a paradigm for how these pathways modulate ITAM-based activation responses. Studies in mice deficient in individual FcRs have provided the necessary insights for defining comparable activities in human autoimmune diseases and suggest ways in which manipulation of the IgG-FcR interaction may lead to new classes of therapeutics for the treatment of these diseases. Modulation of the inhibitory response, a novel activity associated with IVIG to account for some of its anti-inflammatory activity *in vivo,* represents a novel approach to the regulation of Ig-mediated inflammation and suggests that therapeutic agents based on those pathways are likely to be effective. It seems clear that this anti-inflammatory pathway is of general importance for maintaining immune homeostasis, suggesting that IgG glycoforms rich in terminal sialic acid residues might be considered as a molecular switch either keeping the immune system in a resting state or, during infection or autoimmune responses, loosening the brakes thereby allowing full-blown inflammation. In this pathway, type II FcRs including mouse SIGNR1 and human DC-SIGN recognizing specific IgG glycoforms are of central importance. Conversely, engineering of therapeutic antibodies targeted to eliminate infectious or neoplastic disease will likely benefit from optimization of their Fc domains for interaction with specific canonical FcRs.

ACKNOWLEDGMENTS

The Laboratory of Molecular Genetics and Immunology at the Rockefeller University is supported by grants from the National Institutes of Health, by Theresa and Eugene Lang, the Robertson Foundation, and Eileen Greenland. The Laboratory of Genetics at the University of Erlangen-Nuremberg is supported by the German Research Foundation and the Elite Network of Bavaria.

REFERENCES

1. Boyden SV, Sorkin E. The adsorption of antigen by spleen cells previously treated with antiserum *in vitro*. *Immunology.* 1960;3:272-283.
2. Berken A, Benacerraf B. Properties of antibodies cytophilic for macrophages. *J Exp Med.* 1966;123(1):119-144.
3. Paraskevas F, Lee ST, Orr KB, Israels LG. A receptor for Fc on mouse B-lymphocytes. *J Immunol.* 1972;108(5):1319-1327.
4. Kulczycki A Jr, Isersky C, Metzger H. The interaction of IgE with rat basophilic leukemia cells. I. Evidence for specific binding of IgE. *J Exp Med.* 1974;139(3):600-616.
5. Lewis VA, Koch T, Plutner H, Mellman I. A complementary DNA clone for a macrophage-lymphocyte Fc receptor. *Nature.* 1986;324(6095):372-375.
6. Ravetch JV, Luster AD, Weinshank R, et al. Structural heterogeneity and functional domains of murine immunoglobulin G Fc receptors. *Science.* 1986;234(4777):718-725.
7. Kinet JP, Metzger H, Hakimi J, Kochan J. A cDNA presumptively coding for the alpha subunit of the receptor with high affinity for immunoglobulin E. *Biochemistry.* 1987;26(15):4605-4610.
8. Ra C, Jouvin MH, Blank U, Kinet JP. A macrophage Fcγ receptor and the mast cell receptor for IgE share an identical subunit. *Nature.* 1989;341(6244):752-754.
9. Kurosaki T, Ravetch JV. A single amino acid in the glycosyl phosphatidylinositol attachment domain determines the membrane topology of Fcγ RIII. *Nature.* 1989;342(6251):805-807.
10. Takai T, Li M, Sylvestre D, Clynes R, Ravetch JV. FcRγ chain deletion results in pleiotrophic effector cell defects. *Cell.* 1994;76(3):519-529.
11. Dombrowicz D, Flamand V, Brigman KK, Koller BH, Kinet JP. Abolition of anaphylaxis by targeted disruption of the high affinity immunoglobulin E receptor alpha chain gene. *Cell.* 1993;75(5):969-976.
12. Maxwell KF, Powell MS, Hulett MD, et al. Crystal structure of the human leukocyte Fc receptor, FcγRIIa. *Nat Struct Biol.* 1999;6(5):437-442.
13. Sondermann P, Huber R, Jacob U. Crystal structure of the soluble form of the human fcγ-receptor IIb: a new member of the immunoglobulin superfamily at 1.7 A resolution. *EMBO J.* 1999;18(5):1095-1103.
14. Garman SC, Kinet JP, Jardetzky TS. Crystal structure of the human high-affinity IgE receptor. *Cell.* 1998;95(7):951-961.
15. Garman SC, Wurzburg BA, Tarchevskaya SS, Kinet JP, Jardetzky TS. Structure of the Fc fragment of human IgE bound to its high-affinity receptor FcεRIα. *Nature.* 2000;406(6793):259-266.
16. Sondermann P, Huber R, Oosthuizen V, Jacob U. The 3.2-A crystal structure of the human IgG1 Fc fragment-FcγRIII complex. *Nature.* 2000;406(6793):267-273.
17. Ravetch JV, Kinet JP. Fc receptors. *Annu Rev Immunol.* 1991;9:457-492.
18. Takai T, Ono M, Hikida M, Ohmori H, Ravetch JV. Augmented humoral and anaphylactic responses in FcγRII-deficient mice. *Nature.* 1996;379(6563):346-349.
19. Hazenbos WL, Gessner JE, Hofhuis FM, et al. Impaired IgG-dependent anaphylaxis and Arthus reaction in FcγRIII (CD16) deficient mice. *Immunity.* 1996;5(2):181-188.
20. Nimmerjahn F, Ravetch JV. Divergent immunoglobulin g subclass activity through selective Fc receptor binding. *Science.* 2005;310(5753):1510-1512.
21. Carroll MC. The role of complement and complement receptors in induction and regulation of immunity. *Annu Rev Immunol.* 1998;16:545-568.
22. Sylvestre DL, Ravetch JV. Fc receptors initiate the Arthus reaction: redefining the inflammatory cascade. *Science.* 1994;265(5175):1095-1098.
23. Ravetch JV, Clynes RA. Divergent roles for Fc receptors and complement *in vivo*. *Annu Rev Immunol.* 1998;16:421-432.
24. Kinet JP. The high-affinity IgE receptor (Fc epsilon RI): from physiology to pathology. *Annu Rev Immunol.* 1999;17:931-972.
25. Muta T, Kurosaki T, Misulovin Z, Sanchez M, Nussenzweig MC, Ravetch JV. A 13-amino-acid motif in the cytoplasmic domain of FcγRIIB modulates B-cell receptor signalling. *Nature.* 1994;368(6466):70-73.
26. Amigorena S, Bonnerot C, Drake JR, et al. Cytoplasmic domain heterogeneity and functions of IgG Fc receptors in B lymphocytes. *Science.* 1992;256(5065):1808-1812.
27. Ravetch JV, Lanier LL. Immune inhibitory receptors. *Science.* 2000;290(5489):84-89.
28. Ravetch JV, Bolland S. IgG Fc receptors. *Annu Rev Immunol.* 2001;19:275-290.
29. Ghetie V, Ward ES. Multiple roles for the major histocompatibility complex class I-related receptor FcRn. *Annu Rev Immunol.* 2000;18:739-766.
30. Mostov KE. Transepithelial transport of immunoglobulins. *Annu Rev Immunol.* 1994;12:63-84.
31. Foss S, Watkinson R, Sandlie I, James LC, Andersen JT. TRIM21: a cytosolic Fc receptor with broad antibody isotype specificity. *Immunol Rev.* 2015;268(1):328-339.
32. Nimmerjahn F, Bruhns P, Horiuchi K, Ravetch JV. FcgammaRIV: a novel FcR with distinct IgG subclass specificity. *Immunity.* 2005;23(1):41-51.
33. Hamaguchi Y, Xiu Y, Komura K, Nimmerjahn F, Tedder TF. Antibody isotype-specific engagement of Fcγ receptors regulates B lymphocyte depletion during CD20 immunotherapy. *J Exp Med.* 2006;203(3):743-753.
34. Daeron M. Fc receptor biology. *Annu Rev Immunol.* 1997;15:203-234.
35. Kubagawa H, Oka S, Kubagawa Y, et al. Identity of the elusive IgM Fc receptor (FcμR) in humans. *J Exp Med.* 2009;206(12):2779-2793.
36. Shima H, Takatsu H, Fukuda S, et al. Identification of TOSO/FAIM3 as an Fc receptor for IgM. *Int Immunol.* 2010;22(3):149-156.
37. Brandsma AM, Hogarth PM, Nimmerjahn F, Leusen JH. Clarifying the confusion between cytokine and Fc receptor "common gamma chain." *Immunity.* 2016;45(2):225-226.
38. Kubagawa H, Chen CC, Ho LH, et al. Biochemical nature and cellular distribution of the paired immunoglobulin-like receptors, PIR-A and PIR-B. *J Exp Med.* 1999;189(2):309-318.
39. Maeda A, Kurosaki M, Kurosaki T. Paired immunoglobulin-like receptor (PIR)-A is involved in activating mast cells through its association with Fc receptor gamma chain. *J Exp Med.* 1998;188(5):991-995.
40. Moretta A, Bottino C, Vitale M, et al. Activating receptors and coreceptors involved in human natural killer cell-mediated cytolysis. *Annu Rev Immunol.* 2001;19:197-223.
41. Lanier LL. Face off—The interplay between activating and inhibitory immune receptors. *Curr Opin Immunol.* 2001;13(3):326-331.

42. Dombrowicz D, Lin S, Flamand V, Brini AT, Koller BH, Kinet JP. Allergy-associated FcRβ is a molecular amplifier of IgE- and IgG-mediated *in vivo* responses. *Immunity*. 1998;8(4):517-529.
43. Qiu WQ, de Bruin D, Brownstein BH, Pearse R, Ravetch JV. Organization of the human and mouse low-affinity FcγR genes: duplication and recombination. *Science*. 1990;248(4956):732-735.
44. Davis RS, Dennis G Jr, Odom MR, et al. Fc receptor homologs: newest members of a remarkably diverse Fc receptor gene family. *Immunol Rev*. 2002;190:123-136.
45. Su Y, Brooks DG, Li L, et al. Myelin protein zero gene mutated in Charcot-Marie-tooth type 1B patients. *Proc Natl Acad Sci U S A*. 1993;90(22):10856-10860.
46. Maresco DL, Chang E, Theil KS, Francke U, Anderson CL. The three genes of the human FCGR1 gene family encoding FcγRI flank the centromere of chromosome 1 at 1p12 and 1q21. *Cytogenet Cell Genet*. 1996;73(3):157-163.
47. Morel L, Blenman KR, Croker BP, Wakeland EK. The major murine systemic lupus erythematosus susceptibility locus, Sle1, is a cluster of functionally related genes. *Proc Natl Acad Sci U S A*. 2001;98(4):1787-1792.
48. Sandford AJ, Moffatt MF, Daniels SE, et al. A genetic map of chromosome 11q, including the atopy locus. *Eur J Hum Genet*. 1995;3(3):188-194.
49. van der Pol W, van de Winkel JG. IgG receptor polymorphisms: risk factors for disease. *Immunogenetics*. 1998;48(3):222-232.
50. Cartron G, Dacheux L, Salles G, et al. Therapeutic activity of humanized anti-CD20 monoclonal antibody and polymorphism in IgG Fc receptor FcγRIIIa gene. *Blood*. 2002;99(3):754-758.
51. Musolino A, Naldi N, Bortesi B, et al. Immunoglobulin G fragment C receptor polymorphisms and clinical efficacy of trastuzumab-based therapy in patients with HER-2/neu-positive metastatic breast cancer. *J Clin Oncol*. 2008;26(11):1789-1796.
52. Paiva M, Marques H, Martins A, Ferreira P, Catarino R, Medeiros R. FcγRIIa polymorphism and clinical response to rituximab in non-Hodgkin lymphoma patients. *Cancer Genet Cytogenet*. 2008;183(1):35-40.
53. Weng WK, Czerwinski D, Timmerman J, Hsu FJ, Levy R. Clinical outcome of lymphoma patients after idiotype vaccination is correlated with humoral immune response and immunoglobulin G Fc receptor genotype. *J Clin Oncol*. 2004;22(23):4717-4724.
54. Weng WK, Levy R. Two immunoglobulin G fragment C receptor polymorphisms independently predict response to rituximab in patients with follicular lymphoma. *J Clin Oncol*. 2003;21(21):3940-3947.
55. Bibeau F, Lopez-Crapez E, Di Fiore F, et al. Impact of FcγRIIa-FcγRIIIa polymorphisms and KRAS mutations on the clinical outcome of patients with metastatic colorectal cancer treated with cetuximab plus irinotecan. *J Clin Oncol*. 2009;27(7):1122-1129.
56. Houghton AN, Scheinberg DA. Monoclonal antibody therapies-a "constant" threat to cancer. *Nat Med*. 2000;6(4):373-374.
57. Clynes RA, Towers TL, Presta LG, Ravetch JV. Inhibitory Fc receptors modulate *in vivo* cytoxicity against tumor targets. *Nat Med*. 2000;6(4):443-446.
58. Bournazos S, Woof JM, Hart SP, Dransfield I. Functional and clinical consequences of Fc receptor polymorphic and copy number variants. *Clin Exp Immunol*. 2009;157(2):244-254.
59. Smith KG, Clatworthy MR. FcγRIIB in autoimmunity and infection: evolutionary and therapeutic implications. *Nat Rev Immunol*. 2010;10(5):328-343.
60. Mackay M, Stanevsky A, Wang T, et al. Selective dysregulation of the FcγIIB receptor on memory B cells in SLE. *J Exp Med*. 2006;203(9):2157-2164.
61. Tackenberg B, Jelcic I, Baerenwaldt A, et al. Impaired inhibitory Fcγ receptor IIB expression on B cells in chronic inflammatory demyelinating polyneuropathy. *Proc Natl Acad Sci U S A*. 2009;106(12):4788-4792.
62. McGaha TL, Sorrentino B, Ravetch JV. Restoration of tolerance in lupus by targeted inhibitory receptor expression. *Science*. 2005;307(5709):590-593.
63. Brownlie RJ, Lawlor KE, Niederer HA, et al. Distinct cell-specific control of autoimmunity and infection by FcγRIIb. *J Exp Med*. 2008;205(4):883-895.
64. Clatworthy MR, Willcocks L, Urban B, et al. Systemic lupus erythematosus-associated defects in the inhibitory receptor FcγRIIb reduce susceptibility to malaria. *Proc Natl Acad Sci U S A*. 2007;104(17):7169-7174.
65. Waisberg M, Tarasenko T, Vickers BK, et al. Genetic susceptibility to systemic lupus erythematosus protects against cerebral malaria in mice. *Proc Natl Acad Sci U S A*. 2011;108(3):1122-1127.
66. Breunis WB, van Mirre E, Bruin M, et al. Copy number variation of the activating FCGR2C gene predisposes to idiopathic thrombocytopenic purpura. *Blood*. 2008;111(3):1029-1038.
67. Willcocks LC, Lyons PA, Clatworthy MR, et al. Copy number of FCGR3B, which is associated with systemic lupus erythematosus, correlates with protein expression and immune complex uptake. *J Exp Med*. 2008;205(7):1573-1582.
68. Nimmerjahn F, Ravetch JV. Fcγ receptors: old friends and new family members. *Immunity*. 2006;24(1):19-28.
69. Brooks DG, Qiu WQ, Luster AD, Ravetch JV. Structure and expression of human IgG FcRII(CD32). Functional heterogeneity is encoded by the alternatively spliced products of multiple genes. *J Exp Med*. 1989;170(4):1369-1385.
70. Ravetch JV, Perussia B. Alternative membrane forms of FcγRIII(CD16) on human natural killer cells and neutrophils. Cell type-specific expression of two genes that differ in single nucleotide substitutions. *J Exp Med*. 1989;170(2):481-497.
71. Anthony RM, Kobayashi T, Wermeling F, Ravetch JV. Intravenous gammaglobulin suppresses inflammation through a novel T(H)2 pathway. *Nature*. 2011;475(7354):110-113.
72. Anthony RM, Wermeling F, Karlsson MC, Ravetch JV. Identification of a receptor required for the anti-inflammatory activity of IVIG. *Proc Natl Acad Sci U S A*. 2008;105(50):19571-19578.
73. Geijtenbeek TB, Gringhuis SI. Signalling through C-type lectin receptors: shaping immune responses. *Nat Rev Immunol*. 2009;9(7):465-479.
74. Smith P, DiLillo DJ, Bournazos S, Li F, Ravetch JV. Mouse model recapitulating human Fcγ receptor structural and functional diversity. *Proc Natl Acad Sci U S A*. 2012;109(16):6181-6186.
75. Lux A, Seeling M, Baerenwaldt A, et al. A humanized mouse identifies the bone marrow as a niche with low therapeutic IgG activity. *Cell Rep*. 2014;7(1):236-248.
76. Dahan R, Barnhart BC, Li F, Yamniuk AP, Korman AJ, Ravetch JV. Therapeutic activity of agonistic, human anti-CD40 monoclonal antibodies requires selective FcγR engagement. *Cancer Cell*. 2016;29(6):820-831.
77. Bournazos S, Gazumyan A, Seaman MS, Nussenzweig MC, Ravetch JV. Bispecific anti-HIV-1 antibodies with enhanced breadth and potency. *Cell*. 2016;165(7):1609-1620.
78. Schwab I, Lux A, Nimmerjahn F. Pathways responsible for human autoantibody and therapeutic intravenous IgG activity in humanized mice. *Cell Rep*. 2015;13(3):610-620.
79. Starbeck-Miller GR, Badovinac VP, Barber DL, Harty JT. Cutting edge: expression of FcγRIIB tempers memory CD8 T cell function in vivo. *J Immunol*. 2014;192(1):35-39.
80. Nakamura T, Kubagawa H, Ohno T, Cooper MD. Characterization of an IgM Fc-binding receptor on human T cells. *J Immunol*. 1993;151(12):6933-6941.
81. Coxon A, Cullere X, Knight S, et al. FcγRIII mediates neutrophil recruitment to immune complexes. a mechanism for neutrophil accumulation in immune-mediated inflammation. *Immunity*. 2001;14(6):693-704.
82. Ujike A, Ishikawa Y, Ono M, et al. Modulation of immunoglobulin (Ig)E-mediated systemic anaphylaxis by low-affinity Fc receptors for IgG. *J Exp Med*. 1999;189(10):1573-1579.
83. Daheshia M, Friend DS, Grusby MJ, Austen KF, Katz HR. Increased severity of local and systemic anaphylactic reactions in gp49B1-deficient mice. *J Exp Med*. 2001;194(2):227-234.
84. de Andres B, Mueller AL, Verbeek S, Sandor M, Lynch RG. A regulatory role for Fcγ receptors CD16 and CD32 in the development of murine B cells. *Blood*. 1998;92(8):2823-2829.
85. Pricop L, Redecha P, Teillaud JL, et al. Differential modulation of stimulatory and inhibitory Fcγ receptors on human monocytes by Th1 and Th2 cytokines. *J Immunol*. 2001;166(1):531-537.
86. Schmidt RE, Gessner JE. Fc receptors and their interaction with complement in autoimmunity. *Immunol Lett*. 2005;100(1):56-67.
87. Boruchov AM, Heller G, Veri MC, Bonvini E, Ravetch JV, Young JW. Activating and inhibitory IgG Fc receptors on human DCs mediate opposing functions. *J Clin Invest*. 2005;115(10):2914-2923.
88. Samuelsson A, Towers TL, Ravetch JV. Anti-inflammatory activity of IVIG mediated through the inhibitory Fc receptor. *Science*. 2001;291(5503):484-486.
89. Kaneko Y, Nimmerjahn F, Madaio MP, Ravetch JV. Pathology and protection in nephrotoxic nephritis is determined by selective engagement of specific Fc receptors. *J Exp Med*. 2006;203(3):789-797.
90. Bruhns P, Samuelsson A, Pollard JW, Ravetch JV. Colony-stimulating factor-1-dependent macrophages are responsible for IVIG protection in antibody-induced autoimmune disease. *Immunity*. 2003;18(4):573-581.
91. Anthony RM, Nimmerjahn F, Ashline DJ, Reinhold VN, Paulson JC, Ravetch JV. Recapitulation of IVIG anti-inflammatory activity with a recombinant IgG Fc. *Science*. 2008;320(5874):373-376.
92. Kaneko Y, Nimmerjahn F, Ravetch JV. Anti-inflammatory activity of immunoglobulin G resulting from Fc sialylation. *Science*. 2006;313(5787):670-673.
93. Yamaguchi M, Lantz CS, Oettgen HC, et al. IgE enhances mouse mast cell FcεRI expression in vitro and in vivo: evidence for a novel amplification mechanism in IgE-dependent reactions. *J Exp Med*. 1997;185(4):663-672.
94. Dombrowicz D, Flamand V, Miyajima I, Ravetch JV, Galli SJ, Kinet JP. Absence of FcεRIα chain results in upregulation of FcγRIII-dependent mast cell degranulation and anaphylaxis. Evidence of competition between FcεRI and FcγRIII for limiting amounts of FcR beta and gamma chains. *J Clin Invest*. 1997;99(5):915-925.
95. Nimmerjahn F, Lux A, Albert H, et al. FcγRIV deletion reveals its central role for IgG2a and IgG2b activity *in vivo*. *Proc Natl Acad Sci U S A*. 2010;107(45):19396-19401.
96. Dhaliwal B, Yuan D, Pang MO, et al. Crystal structure of IgE bound to its B-cell receptor CD23 reveals a mechanism of reciprocal allosteric inhibition with high affinity receptor FcεRI. *Proc Natl Acad Sci U S A*. 2012;109(31):12686-12691.
97. Sondermann P, Kaiser J, Jacob U. Molecular basis for immune complex recognition: a comparison of Fc-receptor structures. *J Mol Biol*. 2001;309(3):737-749.
98. Ramsland PA, Farrugia W, Bradford TM, et al. Structural basis for FcγRIIa recognition of human IgG and formation of inflammatory signaling complexes. *J Immunol*. 2011;187(6):3208-3217.
99. Wang TT, Ravetch JV. Functional diversification of IgGs through Fc glycosylation. *J Clin Invest*. 2019;129(9):3492-3498.
100. Anthony RM, Nimmerjahn F. The role of differential IgG glycosylation in the interaction of antibodies with FcγRs *in vivo*. *Curr Opin Organ Transplant*. 2011;16(1):7-14.
101. Ferrara C, Stuart F, Sondermann P, Brunker P, Umana P. The carbohydrate at FcγRIIIa Asn-162. An element required for high affinity binding to non-fucosylated IgG glycoforms. *J Biol Chem*. 2006;281(8):5032-5036.
102. Shade KT, Platzer B, Washburn N, et al. A single glycan on IgE is indispensable for initiation of anaphylaxis. *J Exp Med*. 2015;212(4):457-467.
103. Shade K-TC, Conroy ME, Washburn N, et al. Sialylation of immunoglobulin E is a determinant of allergic pathogenicity. *Nature*. 2020;582:265-270.
104. Ferrara C, Grau S, Jager C, et al. Unique carbohydrate-carbohydrate interactions are required for high affinity binding between FcγRIII and antibodies lacking core fucose. *Proc Natl Acad Sci U S A*. 2011;108(31):12669-12674.
105. Wang TT, Sewatanon J, Memoli MJ, et al. IgG antibodies to dengue enhanced for FcγaRIIIA binding determine disease severity. *Science*. 2017;355(6323):395-398.
106. Pincetic A, Bournazos S, DiLillo DJ, et al. Type I and type II Fc receptors regulate innate and adaptive immunity. *Nat Immunol*. 2014;15(8):707-716.
107. Sondermann P, Pincetic A, Maamary J, Lammens K, Ravetch JV. General mechanism for modulating immunoglobulin effector function. *Proc Natl Acad Sci U S A*. 2013;110(24):9868-9872.
108. Paetz A, Sack M, Thepen T, et al. Recombinant soluble human Fcγ receptor I with picomolar affinity for immunoglobulin G. *Biochem Biophys Res Commun*. 2005;338(4):1811-1817.
109. Okazaki A, Shoji-Hosaka E, Nakamura K, et al. Fucose depletion from human IgG1 oligosaccharide enhances binding enthalpy and association rate between IgG1 and FcγRIIIa. *J Mol Biol*. 2004;336(5):1239-1249.
110. Maenaka K, van der Merwe PA, Stuart DI, Jones EY, Sondermann P. The human low affinity Fcγ receptors IIa, IIb, and III bind IgG with fast kinetics and distinct thermodynamic properties. *J Biol Chem*. 2001;276(48):44898-44904.
111. Biburger M, Aschermann S, Schwab I, et al. Monocyte subsets responsible for immunoglobulin G-dependent effector functions *in vivo*. *Immunity*. 2011;35(6):932-944.
112. Oldenborg PA, Gresham HD, Lindberg FP. CD47-signal regulatory protein alpha (SIRPα) regulates Fcγ and complement receptor-mediated phagocytosis. *J Exp Med*. 2001;193(7):855-862.
113. Biburger M, Lux A, Nimmerjahn F. How immunoglobulin G antibodies kill target cells: revisiting an old paradigm. *Adv Immunol*. 2014;124:67-94.
114. Goede V, Fischer K, Busch R, et al. Obinutuzumab plus chlorambucil in patients with CLL and coexisting conditions. *N Engl J Med*. 2014;370(12):1101-1110.
115. Lux A, Yu X, Scanlan CN, Nimmerjahn F. Impact of immune complex size and glycosylation on IgG binding to human FcγRs. *J Immunol*. 2013;190(8):4315-4323.
116. Robinett RA, Guan N, Lux A, Biburger M, Nimmerjahn F, Meyer AS. Dissecting FcgammaR regulation through a multivalent binding model. *Cell Syst*. 2018;7(1):41-48 e5.
117. Shinkawa T, Nakamura K, Yamane N, et al. The absence of fucose but not the presence of galactose or bisecting N-acetylglucosamine of human IgG1 complex-type oligosaccharides shows the critical role of enhancing antibody-dependent cellular cytotoxicity. *J Biol Chem*. 2003;278(5):3466-3473.
118. Shields RL, Lai J, Keck R, et al. Lack of fucose on human IgG1 N-linked oligosaccharide improves binding to human Fcγ RIII and antibody-dependent cellular toxicity. *J Biol Chem*. 2002;277(30):26733-26740.
119. Li T, DiLillo DJ, Bournazos S, Giddens JP, Ravetch JV, Wang LX. Modulating IgG effector function by Fc glycan engineering. *Proc Natl Acad Sci U S A*. 2017;114(13):3485-3490.
120. Kao D, Danzer H, Collin M, et al. A monosaccharide residue is sufficient to maintain mouse and human IgG subclass Activity and directs IgG effector functions to cellular Fc receptors. *Cell Rep*. 2015;13(11):2376-2385.
121. Miller KL, Duchemin AM, Anderson CL. A novel role for the Fc receptor gamma subunit: enhancement of FcγR ligand affinity. *J Exp Med*. 1996;183(5):2227-2233.
122. Shields RL, Namenuk AK, Hong K, et al. High resolution mapping of the binding site on human IgG1 for FcγRI, FcγRII, FcγRIII, and FcRn and design of IgG1

variants with improved binding to the FcγR. *J Biol Chem.* 2001;276(9):6591-6604.
123. Zalevsky J, Leung IW, Karki S, et al. The impact of Fc engineering on an anti-CD19 antibody: increased Fcγ receptor affinity enhances B-cell clearing in nonhuman primates. *Blood.* 2009;113(16):3735-3743.
124. Lazar GA, Dang W, Karki S, et al.. Engineered antibody Fc variants with enhanced effector function. *Proc Natl Acad Sci U S A.* 2006;103(11):4005-4010.
125. Horton HM, Bernett MJ, Peipp M, et al. Fc-engineered anti-CD40 antibody enhances multiple effector functions and exhibits potent *in vitro* and *in vivo* antitumor activity against hematologic malignancies. *Blood.* 2010;116(16):3004-3012.
126. Li F, Ravetch JV. Antitumor activities of agonistic anti-TNFR antibodies require differential FcγRIIB coengagement *in vivo. Proc Natl Acad Sci U S A.* 2013;110(48):19501-19506.
127. Woyach JA, Johnson AJ, Byrd JC. The B-cell receptor signaling pathway as a therapeutic target in CLL. *Blood.* 2012;120(6):1175-1184.
128. Fiebiger BM, Maamary J, Pincetic A, Ravetch JV. Protection in antibody- and T cell-mediated autoimmune diseases by anti-inflammatory IgG Fcs requires type II FcRs. *Proc Natl Acad Sci U S A.* 2015;112(18):E2385-E2394.
129. Daeron M, Lesourne R. Negative signaling in Fc receptor complexes. *Adv Immunol.* 2006;89:39-86.
130. Choquet D, Partiseti M, Amigorena S, Bonnerot C, Fridman WH, Korn H. Cross-linking of IgG receptors inhibits membrane immunoglobulin-stimulated calcium influx in B lymphocytes. *J Cell Biol.* 1993;121(2):355-363.
131. Ono M, Bolland S, Tempst P, Ravetch JV. Role of the inositol phosphatase SHIP in negative regulation of the immune system by the receptor FcγRIIB. *Nature.* 1996;383(6597):263-266.
132. Pearse RN, Kawabe T, Bolland S, Guinamard R, Kurosaki T, Ravetch JV. SHIP recruitment attenuates FcγRIIB-induced B cell apoptosis. *Immunity.* 1999;10(6):753-760.
133. Tzeng SJ, Bolland S, Inabe K, Kurosaki T, Pierce SK. The B cell inhibitory Fc receptor triggers apoptosis by a novel c-Abl family kinase-dependent pathway. *J Biol Chem.* 2005;280(42):35247-35254.
134. Xiang Z, Cutler AJ, Brownlie RJ, et al. FcγRIIb controls bone marrow plasma cell persistence and apoptosis. *Nat Immunol.* 2007;8(4):419-429.
135. Anania JC, Chenoweth AM, Wines BD, Hogarth PM. The human FcγRII (CD32) family of leukocyte FcR in Health and disease. *Front Immunol.* 2019;10:464.
136. Torigoe C, Inman JK, Metzger H. An unusual mechanism for ligand antagonism. *Science.* 1998;281(5376):568-572.
137. Tridandapani S, Pradhan M, LaDine JR, Garber S, Anderson CL, Coggeshall KM. Protein interactions of Src homology 2 (SH2) domain-containing inositol phosphatase (SHIP): association with Shc displaces SHIP from FcγRIIb in B cells. *J Immunol.* 1999;162(3):1408-1414.
138. Daeron M, Latour S, Malbec O, et al. The same tyrosine-based inhibition motif, in the intracytoplasmic domain of FcγRIIB, regulates negatively BCR-, TCR-, and FcR-dependent cell activation. *Immunity.* 1995;3(5):635-646.
139. Malbec O, Fong DC, Turner M, et al. Fc epsilon receptor I-associated lyn-dependent phosphorylation of Fcγreceptor IIB during negative regulation of mast cell activation. *J Immunol.* 1998;160(4):1647-1658.
140. Bolland S, Pearse RN, Kurosaki T, Ravetch JV. SHIP modulates immune receptor responses by regulating membrane association of Btk. *Immunity.* 1998;8(4):509-516.
141. Tamir I, Stolpa JC, Helgason CD, et al. The RasGAP-binding protein p62dok is a mediator of inhibitory FcγRIIB signals in B cells. *Immunity.* 2000;12(3):347-358.
142. Yamanashi Y, Tamura T, Kanamori T, et al. Role of the rasGAP-associated docking protein p62(dok) in negative regulation of B cell receptor-mediated signaling. *Genes Dev.* 2000;14(1):11-16.
143. Aman MJ, Lamkin TD, Okada H, Kurosaki T, Ravichandran KS. The inositol phosphatase SHIP inhibits Akt/PKB activation in B cells. *J Biol Chem.* 1998;273(51):33922-33928.
144. Liu Q, Sasaki T, Kozieradzki I, et al. SHIP is a negative regulator of growth factor receptor-mediated PKB/Akt activation and myeloid cell survival. *Genes Dev.* 1999;13(7):786-791.
145. Heyman B. Regulation of antibody responses via antibodies, complement, and Fc receptors. *Annu Rev Immunol.* 2000;18:709-737.
146. Bolland S, Ravetch JV. Spontaneous autoimmune disease in FcγRIIB-deficient mice results from strain-specific epistasis. *Immunity.* 2000;13(2):277-285.
147. Okazaki T, Otaka Y, Wang J, et al. Hydronephrosis associated with antiurothelial and antinuclear autoantibodies in BALB/c–Fcgr2b–/–Pdcd1–/–mice. *J Exp Med.* 2005;202(12):1643-1648.
148. Fukuyama H, Nimmerjahn F, Ravetch JV. The inhibitory Fcγ receptor modulates autoimmunity by limiting the accumulation of immunoglobulin G+ anti-DNA plasma cells. *Nat Immunol.* 2005;6(1):99-106.
149. Jiang Y, Hirose S, Abe M, et al. Polymorphisms in IgG Fc receptor IIB regulatory regions associated with autoimmune susceptibility. *Immunogenetics.* 2000;51(6):429-435.
150. Pritchard NR, Cutler AJ, Uribe S, Chadban SJ, Morley BJ, Smith KG. Autoimmune-prone mice share a promoter haplotype associated with reduced expression and function of the Fc receptor FcγRII. *Curr Biol.* 2000;10(4):227-230.
151. Yuasa T, Kubo S, Yoshino T, et al. Deletion of Fcγ receptor IIB renders H-2(b) mice susceptible to collagen-induced arthritis. *J Exp Med.* 1999;189(1):187-194.
152. Nakamura A, Yuasa T, Ujike A, et al. Fcγ receptor IIB-deficient mice develop Goodpasture's syndrome upon immunization with type IV collagen: a novel murine model for autoimmune glomerular basement membrane disease. *J Exp Med.* 2000;191(5):899-906.
153. Maamary J, Wang TT, Tan GS, Palese P, Ravetch JV. Increasing the breadth and potency of response to the seasonal influenza virus vaccine by immune complex immunization. *Proc Natl Acad Sci U S A.* 2017;114(38):10172-10177.
154. Wang TT, Maamary J, Tan GS, et al. Anti-HA glycoforms drive B cell affinity selection and determine influenza vaccine efficacy. *Cell.* 2015;162(1):160-169.
155. Banchereau J, Steinman RM. Dendritic cells and the control of immunity. *Nature.* 1998;392(6673):245-252.
156. Amigorena S, Bonnerot C. Fc receptors for IgG and antigen presentation on MHC class I and class II molecules. *Semin Immunol.* 1999;11(6):385-390.
157. Hamano Y, Arase H, Saisho H, Saito T. Immune complex and Fc receptor-mediated augmentation of antigen presentation for *in vivo* Th cell responses. *J Immunol.* 2000;164(12):6113-6119.
158. DiLillo DJ, Ravetch JV. Differential Fc-receptor engagement drives an anti-tumor vaccinal effect. *Cell.* 2015;161(5):1035-1045.
159. Regnault A, Lankar D, Lacabanne V, et al. Fcγ receptor-mediated induction of dendritic cell maturation and major histocompatibility complex class I-restricted antigen presentation after immune complex internalization. *J Exp Med.* 1999;189(2):371-380.
160. Lehmann CHK, Baranska A, Heidkamp GF, et al. DC subset-specific induction of T cell responses upon antigen uptake via Fcγ receptors *in vivo. J Exp Med.* 2017;214(5):1509-1528.
161. Wernersson S, Karlsson MC, Dahlstrom J, Mattsson R, Verbeek JS, Heyman B. IgG-mediated enhancement of antibody responses is low in Fc receptor gamma chain-deficient mice and increased in FcγRII-deficient mice. *J Immunol.* 1999;163(2):618-622.
162. Baiu DC, Prechl J, Tchorbanov A, et al. Modulation of the humoral immune response by antibody-mediated antigen targeting to complement receptors and Fc receptors. *J Immunol.* 1999;162(6):3125-3130.
163. Kalergis AM, Ravetch JV. Inducing tumor immunity through the selective engagement of activating Fcγ receptors on dendritic cells. *J Exp Med.* 2002;195(12):1653-1659.
164. Dhodapkar KM, Kaufman JL, Ehlers M, et al. Selective blockade of inhibitory Fcγ receptor enables human dendritic cell maturation with IL-12p70 production and immunity to antibody-coated tumor cells. *Proc Natl Acad Sci U S A.* 2005;102(8):2910-2915.
165. Clynes R, Ravetch JV. Cytotoxic antibodies trigger inflammation through Fc receptors. *Immunity.* 1995;3(1):21-26.
166. Clynes R, Dumitru C, Ravetch JV. Uncoupling of immune complex formation and kidney damage in autoimmune glomerulonephritis. *Science.* 1998;279(5353):1052-1054.
167. Clynes R, Maizes JS, Guinamard R, Ono M, Takai T, Ravetch JV. Modulation of immune complex-induced inflammation *in vivo* by the coordinate expression of activation and inhibitory Fc receptors. *J Exp Med.* 1999;189(1):179-185.
168. Hazenbos WL, Heijnen IA, Meyer D, et al. Murine IgG1 complexes trigger immune effector functions predominantly via FcγRIII (CD16). *J Immunol.* 1998;161(6):3026-3032.
169. Launay P, Grossetete B, Arcos-Fajardo M, et al. Fcα receptor (CD89) mediates the development of immunoglobulin A (IgA) nephropathy (Berger's disease). Evidence for pathogenic soluble receptor-Iga complexes in patients and CD89 transgenic mice. *J Exp Med.* 2000;191(11):1999-2009.
170. Miyajima I, Dombrowicz D, Martin TR, Ravetch JV, Kinet JP, Galli SJ. Systemic anaphylaxis in the mouse can be mediated largely through IgG1 and FcγRIII. Assessment of the cardiopulmonary changes, mast cell degranulation, and death associated with active or IgE-or IgG1-dependent passive anaphylaxis. *J Clin Invest.* 1997;99(5):901-914.
171. Sylvestre D, Clynes R, Ma M, Warren H, Carroll MC, Ravetch JV. Immunoglobulin G-mediated inflammatory responses develop normally in complement-deficient mice. *J Exp Med.* 1996;184(6):2385-2392.
172. Fossati-Jimack L, Ioan-Facsinay A, Reininger L, et al. Markedly different pathogenicity of four immunoglobulin G isotype-switch variants of an antierythrocyte autoantibody is based on their capacity to interact *in vivo* with the low-affinity Fcγ receptor III. *J Exp Med.* 2000;191(8):1293-1302.
173. Yuan R, Clynes R, Oh J, Ravetch JV, Scharff MD. Antibody-mediated modulation of *Cryptococcus neoformans* infection is dependent on distinct Fc receptor functions and IgG subclasses. *J Exp Med.* 1998;187(4):641-648.
174. Suzuki Y, Shirato I, Okumura K, et al. Distinct contribution of Fc receptors and angiotensin II-dependent pathways in anti-GBM glomerulonephritis. *Kidney Int.* 1998;54(4):1166-1174.
175. Park SY, Ueda S, Ohno H, et al.. Resistance of Fc receptor-deficient mice to fatal glomerulonephritis. *J Clin Invest.* 1998;102(6):1229-1238.
176. Kleinau S, Martinsson P, Heyman B. Induction and suppression of collagen-induced arthritis is dependent on distinct Fcγ receptors. *J Exp Med.* 2000;191(9):1611-1616.
177. Clynes R, Takechi Y, Moroi Y, Houghton A, Ravetch JV. Fc receptors are required in passive and active immunity to melanoma. *Proc Natl Acad Sci U S A.* 1998;95(2):652-656.
178. Stone JD, Cochran JR, Stern LJ. T-cell activation by soluble MHC oligomers can be described by a two-parameter binding model. *Biophys J.* 2001;81(5):2547-2557.
179. Zhang Z, Zhang M, Ravetch JV, Goldman C, Waldmann TA. Effective therapy for a murine model of adult T-cell leukemia with the humanized anti-CD2 monoclonal antibody, MEDI-507. *Blood.* 2003;102(1):284-288.
180. Zhang M, Zhang Z, Garmestani K, et al. Activating Fc receptors are required for antitumor efficacy of the antibodies directed toward CD25 in a murine model of adult t-cell leukemia. *Cancer Res.* 2004;64(16):5825-5829.
181. Mössner E, Brünker P, Moser S, et al. Increasing the efficacy of CD20 antibody therapy through the engineering of a new type II anti-CD20 antibody with enhanced direct and immune effector cell-mediated B-cell cytotoxicity. *Blood.* 2010;115(22):4393-4402.
182. Kim YH, Bagot M, Pinter-Brown L, et al. Mogamulizumab versus vorinostat in previously treated cutaneous T-cell lymphoma (MAVORIC): an international, open-label, randomised, controlled phase 3 trial. *Lancet Oncol.* 2018;19(9):1192-1204.
183. Ghazi A, Trikha A, Calhoun WJ. Benralizumab—A humanized mAb to IL-5Rα with enhanced antibody-dependent cell-mediated cytotoxicity—A novel approach for the treatment of asthma. *Expet Opin Biol Ther.* 2012;12(1):113-118.
184. Bang YJ, Giaccone G, Im SA, et al. First-in-human phase 1 study of margetuximab (MGAH22), an Fc-modified chimeric monoclonal antibody, in patients with HER2-positive advanced solid tumors. *Ann Oncol.* 2017;28(4):855-861.
185. Ji H, Ohmura K, Mahmood U, et al. Arthritis critically dependent on innate immune system players. *Immunity.* 2002;16(2):157-168.
186. Korganow AS, Ji H, Mangialaio S, et al. From systemic T cell self-reactivity to organ-specific autoimmune disease via immunoglobulins. *Immunity.* 1999;10(4):451-461.
187. Matsumoto I, Staub A, Benoist C, Mathis D. Arthritis provoked by linked T and B cell recognition of a glycolytic enzyme. *Science.* 1999;286(5445):1732-1735.
188. Baumann U, Kohl J, Tschernig T, et al. A codominant role of FcγRI/III and C5aR in the reverse Arthus reaction. *J Immunol.* 2000;164(2):1065-1070.
189. Hawlisch H, Wills-Karp M, Karp CL, Kohl J. The anaphylatoxins bridge innate and adaptive immune responses in allergic asthma. *Mol Immunol.* 2004;41(2-3):123-131.
190. Cartron G, Watier H, Golay J, Solal-Celigny P. From the bench to the bedside: ways to improve rituximab efficacy. *Blood.* 2004;104(9):2635-2642.
191. Arce-Vargas F, Furness AJS, Litchfield K, et al. Fc effector function contributes to the activity of human anti-CTLA-4 antibodies. *Cancer Cell.* 2018;33(4):649-663.
192. Nimmerjahn F, Ravetch JV. Antibody-mediated modulation of immune responses. *Immunol Rev.* 2010;236:265-275.
193. Abboud N, Chow SK, Saylor C, et al. A requirement for FcγR in antibody-mediated bacterial toxin neutralization. *J Exp Med.* 2010;207(11):2395-2405.
194. Huber VC, Lynch JM, Bucher DJ, Le J, Metzger DW. Fc receptor-mediated phagocytosis makes a significant contribution to clearance of influenza virus infections. *J Immunol.* 2001;166(12):7381-7388.
195. Hessell AJ, Hangartner L, Hunter M, et al. Fc receptor but not complement binding is important in antibody protection against HIV. *Nature.* 2007;449(7158):101-104.
196. Verma A, Ngundi MM, Meade BD, De Pascalis R, Elkins KL, Burns DL. Analysis of the Fcγ receptor-dependent component of neutralization measured by anthrax toxin neutralization assays. *Clin Vaccine Immunol.* 2009;16(10):1405-1412.
197. Wilson NS, Yang B, Yang A, et al. An Fcγ receptor-dependent mechanism drives antibody-mediated target-receptor signaling in cancer cells. *Cancer Cell.* 2011;19(1):101-113.
198. Xu Y, Szalai AJ, Zhou T, et al. FcγRs modulate cytotoxicity of anti-Fas antibodies: implications for agonistic antibody-based therapeutics. *J Immunol.* 2003;171(2):562-568.
199. White AL, Chan HT, Roghanian A, et al. Interaction with FcγRIIB is critical for the agonistic activity of anti-CD40 monoclonal antibody. *J Immunol.* 2011;187(4):1754-1763.
200. Li F, Ravetch JV. Inhibitory Fcγ receptor engagement drives adjuvant and anti-tumor activities of agonistic CD40 antibodies. *Science.* 2011;333(6045):1030-1034.
201. Katzelnick LC, Gresh L, Halloran ME, et al. Antibody-dependent enhancement of severe dengue disease in humans. *Science.* 2017;358(6365):929-932.
202. Monsalvo AC, Batalle JP, Lopez MF, et al. Severe pandemic 2009 H1N1 influenza disease due to pathogenic immune complexes. *Nat Med.* 2011;17(2):195-199.
203. Rajão DS, Loving CL, Gauger PC, Kitikoon P, Vincent AL. Influenza A virus hemagglutinin protein subunit vaccine elicits vaccine-associated enhanced respiratory disease in pigs. *Vaccine.* 2014;32(40):5170-5176.
204. Dijstelbloem HM, van de Winkel JG, Kallenberg CG. Inflammation in autoimmunity: receptors for IgG revisited. *Trends Immunol.* 2001;22(9):510-516.

205. Nimmerjahn F, Ravetch JV. Fcγ receptors as regulators of immune responses. *Nat Rev Immunol.* 2008;8(1):34-47.
206. Wakeland EK, Liu K, Graham RR, Behrens TW. Delineating the genetic basis of systemic lupus erythematosus. *Immunity.* 2001;15(3):397-408.
207. Pisitkun P, Deane JA, Difilippantonio MJ, Tarasenko T, Satterthwaite AB, Bolland S. Autoreactive B cell responses to RNA-related antigens due to TLR7 gene duplication. *Science.* 2006;312(5780):1669-1672.
208. Pasare C, Medzhitov R. Control of B-cell responses by Toll-like receptors. *Nature.* 2005;438(7066):364-368.
209. Ehlers M, Fukuyama H, McGaha TL, Aderem A, Ravetch JV. TLR9/MyD88 signaling is required for class switching to pathogenic IgG2a and 2b autoantibodies in SLE. *J Exp Med.* 2006;203(3):553-561.
210. Lin Q, Hou R, Sato A, et al. Inhibitory IgG Fc receptor promoter region polymorphism is a key genetic element for murine systemic lupus erythematosus. *J Autoimmun.* 2010;34(4):356-363.
211. Blank MC, Stefanescu RN, Masuda E, et al. Decreased transcription of the human FCGR2B gene mediated by the-343 G/C promoter polymorphism and association with systemic lupus erythematosus. *Hum Genet.* 2005;117(2-3): 220-227.
212. Floto RA, Clatworthy MR, Heilbronn KR, et al. Loss of function of a lupus-associated FcγRIIb polymorphism through exclusion from lipid rafts. *Nat Med.* 2005;11(10):1056-1058.
213. Heusser C, Jardieu P. Therapeutic potential of anti-IgE antibodies. *Curr Opin Immunol.* 1997;9(6):805-813.
214. Bussel JB. Fc receptor blockade and immune thrombocytopenic purpura. *Semin Hematol.* 2000;37(3):261-266.
215. Podolanczuk A, Lazarus AH, Crow AR, Grossbard E, Bussel JB. Of mice and men: an open-label pilot study for treatment of immune thrombocytopenic purpura by an inhibitor of Syk. *Blood.* 2009;113(14):3154-3160.
216. Strait RT, Morris SC, Finkelman FD. IgG-blocking antibodies inhibit IgE-mediated anaphylaxis *in vivo* through both antigen interception and FcγRIIb cross-linking. *J Clin Invest.* 2006;116(3):833-841.
217. Horton HM, Chu SY, Ortiz EC, et al. Antibody-mediated coengagement of FcγRIIb and B cell receptor complex suppresses humoral immunity in systemic lupus erythematosus. *J Immunol.* 2011;186(7):4223-4233.
218. Geijtenbeek TB, den Dunnen J, Gringhuis SI. Pathogen recognition by DC-SIGN shapes adaptive immunity. *Future Microbiol.* 2009;4(7):879-890.
219. Parekh RB, Dwek RA, Sutton BJ, et al. Association of rheumatoid arthritis and primary osteoarthritis with changes in the glycosylation pattern of total serum IgG. *Nature.* 1985;316(6027):452-457.
220. Bond A, Cooke A, Hay FC. Glycosylation of IgG, immune complexes and IgG subclasses in the MRL-lpr/lpr mouse model of rheumatoid arthritis. *Eur J Immunol.* 1990;20(10):2229-2233.
221. Pfeifle R, Rothe T, Ipseiz N, et al. Regulation of autoantibody activity by the IL-23-TH17 axis determines the onset of autoimmune disease. *Nat Immunol.* 2017;18(1):104-113.
222. Parekh R, Roitt I, Isenberg D, Dwek R, Rademacher T. Age-related galactosylation of the N-linked oligosaccharides of human serum IgG. *J Exp Med.* 1988;167(5):1731-1736.
223. van de Geijn FE, Wuhrer M, Selman MH, et al. Immunoglobulin G galactosylation and sialylation are associated with pregnancy-induced improvement of rheumatoid arthritis and the postpartum flare: results from a large prospective cohort study. *Arthritis Res Ther.* 2009;11(6):R193.
224. Kao D, Lux A, Schaffert A, Lang R, Altmann F, Nimmerjahn F. IgG subclass and vaccination stimulus determine changes in antigen specific antibody glycosylation in mice. *Eur J Immunol.* 2017;47(12):2070-2079.
225. Pagan JD, Kitaoka M, Anthony RM. Engineered sialylation of pathogenic antibodies *in vivo* attenuates autoimmune disease. *Cell.* 2018;172(3):564.e13-577.e13.

CHAPTER 34

Dendritic cells

Roxane Tussiwand, PhD • Pierre Guermonprez, PhD • Simon Yona, PhD

INTRODUCTION

Dendritic cells (DC) form an order of bone marrow–derived leukocytes that act as the critical bridge between the innate sensing of pathogens and the induction of adaptive immunity.

THE DISCOVERY OF DENDRITIC CELLS

Innate immunity provides the early defense mechanism against pathogen invasion. Epithelial barriers prevent direct exposure to the external environment and pathogens; however, when a lesion occurs during an infection, tissues become vulnerable, and an immediate response is required. The main feature of innate immunity is its ability to act promptly and recognize common danger signals. A large set of molecules, in the case of humoral immunity, and receptors, in the case of cellular immunity, define the spectrum of innate responses that can be induced. These molecules and receptors are germline-encoded and conserved across individuals and species (Chapter 12). Among those are receptors for so-called danger signals, which detect molecules found in damaged tissues, and "pathogen-associated receptors," which have a broad recognition potential for foreign molecules.[1,2] Following the early response to contain the injury and/or infection, a highly pathogen-specific adaptive immune response is initiated. A key process for the induction of adaptive immunity is the ability to discriminate between self and nonself. This requires the simultaneous recognition of foreign antigens presented by specialized cells, in the context of self-molecules that are collectively referred as the major histocompatibility complex (MHC) class I and class II. Peter Medawar's seminal studies on graft rejection highlighted that an individual could distinguish between self and nonself, by accepting autologous skin grafts while rejecting foreign skin grafts.[3-5] Furthermore, the rate of rejection following a second graft from the same foreign donor was remarkably rapid compared to the rejection rate of a second graft from a different individual, demonstrating immunological memory. Experiments performed by Doherty and Zinkernagel revealed how T cells recognize viral antigens only in class I haploidentical recipients, suggesting that T cell receptors required simultaneous self-recognition of the MHC along with foreign antigens, now defined as MHC restriction.[6] The mechanism of MHC restriction, or presentation of antigen in the context of self MHC molecules, became clear when the crystal structure of several human MHC class I molecules complexed with peptides was obtained in 1987 by Bjorkman, a postdoctoral research fellow working with Strominger and Wiley.[7] Several landmark discoveries by Emil Unanue highlighted that the immunogenicity of proteins was enhanced by macrophage phagocytosis and lysosomal proteolysis. In collaboration with Allen and Babbitt, they demonstrated that following phagocytosis, macrophages were capable of loading peptides into MHC class II molecules.[8] These MHC-peptide complexes were then transported to the cell surface and could induce T cell activation. A few years later, the time was ripe for the discovery of DC. Several studies had indicated the existence of so-called accessory cells that could induce T cell proliferation in mixed lymphocyte reactions.[9,10] The nature of these cells remained obscure until their formal identification in 1973 by contrast microscopy in the mouse spleen, as stellate-shaped cells by Steinman and Cohn who named them dendritic cells (DC). Their morphology and capacity to stimulate T cells aligned with the description of similar cells, including Langerhans cells (LC)[11] in the skin, interdigitating cells in the thymus,[12] and veiled cells in draining lymphatics.[13-15] Following their discovery, the DC gained center stage and studies continue to characterize this intriguing cell type. Currently, two principal families of DC have been identified: plasmacytoid DC (pDC) and conventional DC (cDC). The latter is further partitioned into type 1 DC (or cDC1) and the heterogeneous family of type 2 DC (cDC2).

TYPE 1 DENDRITIC CELLS

DC heterogeneity had been evident from the outset. Shortly after their discovery in 1973,[16] it was noted that several subsets of cells similar to DC existed and shared the ability to induce a potent T cell response. In 1982, Tew, Thorbecke, and Steinman attempted an initial classification of DC heterogeneity referred to as[17] D-cells and included follicular dendritic cells (FDCs), lymphoid dendritic cells, interdigitating cells, and LCs. Currently, only lymphoid dendritic cells and interdigitating cells remain members of the exclusive "DC club," while FDCs are recognized as a nonhematopoietic population and LC are a tissue-specific macrophage subset. However, two populations that were not included in the original DC classification are now ascribed to as *bona fide* DC: nurse cells in the thymus[18] and veiled cells described in the lymphatics by S. Knight and B. Balfour.[19]

The lack of precise tools to specifically isolate and examine DC during the 1970s hampered a more detailed examination. At the time their defining features were the expression of McDevitt's Immune response gene-associated antigen (Ia),[4] later recognized as MHC class II; the ability to induce strong T cell response; the ruffled membrane appearance; and stellate morphology, which was the reason for their designation as *Dendritic* cells from the Greek δένδρον dendron or branching protrusions.

The need to better understand DC functional properties and examine them across different tissues led to the development of specific monoclonal antibodies including 33D1[20] and DEC205.[21] Nevertheless, it was the introduction of the pan-DC monoclonal antibody N418 that recognizes the integrin CD11c,[22] when the realization of DC subset heterogeneity was confirmed by the Shortman group.[23] Vremec and colleagues observed that monoclonal antibodies directed against CD4 and CD8 enabled the discrimination of three DC subsets: CD8α+, CD4+, and double-negative DC. CD8+ DC, now referred as conventional type 1 DC or cDC1, express CD8α in secondary lymphoid tissues and are abundant in T cell areas of spleen and lymph nodes. In peripheral tissues cDC1 do not express CD8α, yet can be identified using the epithelial homing receptor CD103 and, depending on the tissue, express DEC205 and langerin. Their discrimination from type 2 DC (cDC2) requires the use of Sirp-α or CD11b, with cDC1 being negative or low for these markers.[24,25] Following their transcriptional characterization, initially by microarrays and later by RNA sequencing, XCR1 became recognized as the most reliable marker to identify cDC1 across all tissues and species.[26,27] Within human tissues, besides XCR1, expression of CD141 or BDCA3 is conventionally used to identify them.[28,29]

The functional characteristics of cDC1 are intimately linked with their selective expression of specific danger-associated and pathogen recognition receptors (DAMPs and PRRs). Specifically, cDC1 express Toll-like receptor 3 (TLR3), TLR7, and the C-type lectin receptor CLEC9A, all crucial for the recognition of virally infected, apoptotic or necrotic cells. Phagocytosis by cDC1 of infected cells expressing foreign antigens, or dying cells expressing transformed self-antigens in the context of tumors, results in "cross-presentation" (see below) on MHC class I. cDC1 are therefore potent activators of cytotoxic CD8 T cells.[30,31]

In addition to their role in promoting adaptive immunity, cDC1 orchestrate the innate immune response against intracellular pathogens (see below). Mouse cDC1 are characterized by the expression of TLR11 and TLR12, enabling them to recognize *Toxoplasma gondii* profilin.[32] Several other intracellular pathogens are targeted by cDC1 and limit the spread of infection through the secretion of IL-12, which in turn activates NK cells, promoting the induction of IFN-γ and interferon-responsive genes (IRG).

Taken together, cDC1 are essential for the recognition of intracellular pathogens or altered self. They induce a potent cytotoxic T lymphocyte (CTL) response through cross-presentation, promote NK cell activation through the secretion of IL-12, and collectively prime IFN-γ-mediated immunity.

TYPE 2 DENDRITIC CELLS

cDC2 represent a heterogeneous DC population. Historically, it has proven difficult to reach a consensus to the boundaries of the DC lineage spectrum, particularly for cDC2. Membrane expression of CD172a/SIRP-α and CD11b combined with high expression of CD11c and MHC class II help to identify cDC2. However, all these markers are not exclusive but shared with several other mononuclear phagocytes. Another criterion used to define members of the cDC2 family is the expression of the transcription factor IRF4,[33-35] essential for the coordination of MHC class II expression, antigen presentation,[36] and cDC2 migration.[37] Under inflammatory conditions, the edges between individual cDC2 subsets, inflammatory DC, and monocyte-derived DC are especially thin, as profound changes in surface marker expression occurs. However, their developmental origin turned out to be key in defining cDC subsets. In particular the expression of FLT3[38] and their dependency on FLT3L are clearly DC-defining criteria,[39-41] while monocytes and monocyte-derived DC depend on CSF1 and the chemokine receptor CCR2.[39,42] As described in more detail below, cDC2 development proceeds from common DC progenitors (CDP)[43-45] present in the bone marrow and in peripheral tissues, together with a circulating immediate pre-cDC2-committed precursor.[46-50]

Examination of several transcription factor expression profiles was instrumental in highlighting cDC2 heterogeneity. NOTCH2 and RBPJ are comparably expressed across cDC subsets.[51] However, mice carrying a NOTCH2 or RBPJ deletion in CD11c-expressing cells—which includes all cDC—are selectively deficient in splenic ESAM1+ cDC2, a fraction of lung cDC2 and gut-associated cDC2, that co-express CD11b+ and CD103+.[51-53] The mouse intestinal tissue NOTCH2-dependent DC subset helps to control bacterial infections, such as *Citrobacter rodentium* and *Salmonella typhimurium;* cDC2 transport most of the bacterially derived antigens to the draining lymph nodes in a CCR7-dependent manner.[54] The human NOTCH2-dependent cDC2 counterpart[55] has been identified with similar features.

Conditional deficiency of IRF4 results in fewer cDC2 in several tissues with profound functional defects. In these mice, Th2 responses against papain or *Nippostrongylus brasiliensis* are impaired.[56] This observation resulted in identification of a NOTCH2-independent cDC2 subset in skin-draining lymph nodes that requires the transcription factor KLF4 for its development and lacks the expression of CD11b.[57] This subset has been characterized across multiple tissues showing expression of heterogenous markers depending on the tissue of origin.[57,58]

A reevaluation of the diversity of murine and human cDC2 has identified the expression of T-bet and RORγt within the cDC2 compartment.[59] T-bet+ cDC2 express reduced levels of CD86 while RORγt+ cDC2 express Clec10A and Clec12A and have a more proinflammatory phenotype. It remains unclear how T-bet+ and RORγt+ cDC2 align to mouse NOTCH2- and KLF4-dependent subsets.

In human peripheral blood, cDC2 are characterized by the high expression of CD1c, FceRIA, and Clec10A.[60,61] Similar to that in mice, this subset is heterogenous and shows a wide range of expression levels for the surface markers CD5, CD163, and CD14, among others.[60-62] Furthermore, a subset of these cells has been identified and referred to as "DC3," which shares part of their transcriptome with monocytes (eg, CD14,

S100A8, and S100A9), lacks the expression of the monocyte-identifying marker C5AR1/CD88,[60-62] and exhibits functional DC-specific features including the upregulation of CCR7 upon activation and the capacity to efficiently prime naïve T cells.[60] Differences between human and mouse subsets make attempts to identify cross-species alignment challenging.

Collectively the complexity and heterogeneity of the cDC2 subset is far from being resolved developmentally and functionally, especially during inflammation.

INFLAMMATORY DENDRITIC CELLS

The DC network evolved to be responsive to changes in the environment. Acute and chronic inflammatory reactions lead to the mobilization of new a category of DC, termed with a subjective prefix "inflammatory DC" (iDC). The common defining feature of these cells is their generation upon inflammatory challenge. Human iDC coexpress cDC2 (CD1c, CD172a/SIRP-α) and monocytic markers such as CD14, CD11b, and FcγRI/CD64 (Fig. 34.1). This subset is abundant in inflamed skin, tonsils, synovial fluids, and ovarian cancer ascites and infiltrates solid tumors and their draining lymph nodes[63-70] and therefore have been investigated as a potential therapeutic target.

Mouse iDC express the membrane markers Ly6C, CD209, FceRI, CD206[71-73] and have been reported in colitis,[74] inflamed lungs,[73] adjuvant-[72] and TLR-induced inflammation.[71] Adoptive transfer of CCR2-dependent classical monocytes gives rise to CCR2-dependent colitis-associated iDC.[74,75] In this setting, monocytes acquire MHC class II upon their differentiation into iDC within the inflamed tissues.[74,75] Despite evidence that monocytes can generate iDC *in vitro* and *in vivo*, progenitors distinct from both "classical" inflammatory monocytes or DC have been reported to give rise to iDC.[39,60] Human iDC were traditionally generated from monocytes upon exposure to GM-CSF and IL-4,[67] where IL-4 induces the transcriptional regulator NCOR2, which supports iDC specification[76] and GM-CSF signaling supports the "inflammatory program." Other inflammogens promote iDC development from monocytes such as agonists of the aryl hydrocarbon receptor (AHR)[77] and viral infections, such as *Pneumovirus* through type I IFN signaling. The "inflammatory" phenotype characterized by the intermediate expression of CD64/FcγRI and FceRI[78] will be driven by different transcriptional programs to accommodate a wide range of inflammatory triggers and immune functions required to respond to them.

PLASMACYTOID DENDRITIC CELLS

Plasmacytoid dendritic cells (pDC) were first described in 1958 in lymph nodes as a cluster of lymphoid cells, characterized by a well-structured rough endoplasmic reticulum (ER).[79] The lack of lineage markers and the presence

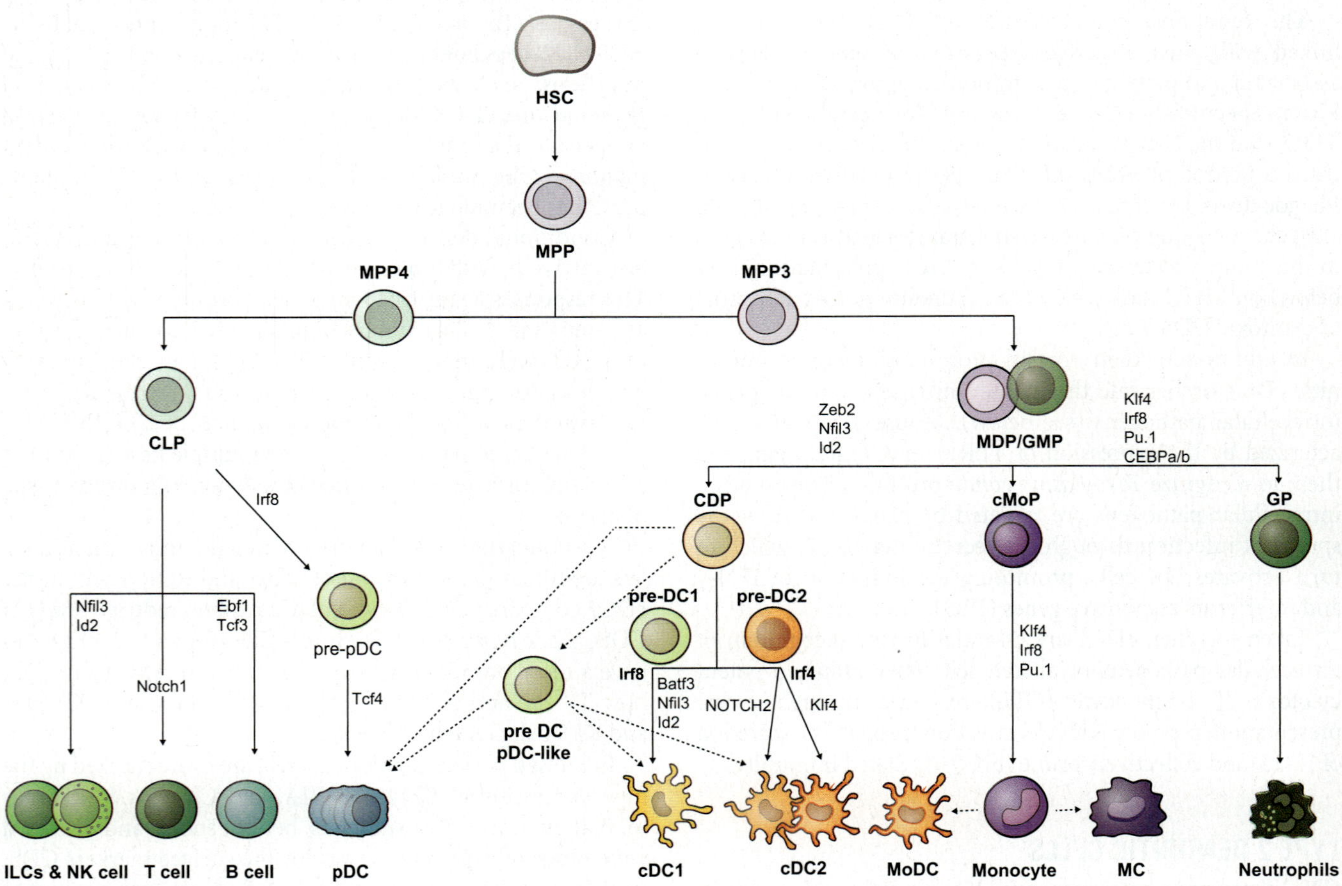

FIG. 34.1. Dendritic cell development. Dendritic cells develop from a hematopoietic progenitor (HSC) in the bone marrow. Progressive lineage commitment is obtained through the expression of selective transcription factors during each stage of development.

of extended ER membranes were suggestive of a new variety of plasma cells.[80] In 1983, the close localization of these cells with T cell areas in secondary lymphoid tissues and their expression of T cell markers, including the glycoprotein CD4 led to their designation as plasmacytoid T cells.[81] A few years later in 1988 a comprehensive characterization was performed, showing the expression of granulocyte-associated antigens and myelomonocytic markers including CD36, CD68, the IL-3R, HLA-DR, and the invariant chain CD74.[82] Importantly these experiments conducted by Facchetti et al highlighted the close association of these cells with high endothelial venules (HEV) in human lymph nodes, suggesting that a migration pattern similar to monocytes was taking place from the peripheral blood, hence their relabeling as plasmacytoid monocytes.[82] Around the same time several groups observed that within peripheral blood, a subset of cells was capable of producing high quantities of type I interferons (IFNs including IFN-α and IFN-β) in response to enveloped viruses, bacteria, and tumor cells and therefore referred to as "natural IFN-producing cells" (NIPC).[83-87] Through depletion experiments, Perussia and colleagues demonstrated that this small subset of cells was not contained within T, B, or granulocyte fractions and thus represented a new subpopulation of leukocytes.[88] Only at the turn of this century the work of Cella et al and Siegal et al associated the mysterious plasmacytoid cells present in human lymph nodes to these NIPC.[89,90] Their assignment to the dendritic cell family was predominately motivated by their ability to differentiate into antigen-presenting cells in the presence of interleukin-3 (IL-3) and CD40-L, resulting in the current nomenclature "plasmacytoid dendritic cells."[91-93] A breakthrough in pDC biology was their identification within murine tissues, which enabled experiments to move beyond descriptive studies of human samples.[91,94,95] However, despite the realization of their key ability to produce high quantities of IFN-a and b, a more comprehensive characterization of their functional properties was nevertheless only conceivable following the identification of specific membrane markers. To date, murine pDC are defined as CD11c+, B220+, MHC-II+, and Ly6C+ while negative for most other lineage markers. This combination of markers was, however, insufficient to obtain the required purity for rigorous analyses. Only the identification of Siglec-H[96] and BST2[97] in murine pDC and of BDCA-2 (CD303/CLEC4C)[98] and the immunoglobulin-like transcript 7 (ILT7)[99-101] in human pDC enabled functional studies based on depletion models in mice and improved isolation in human samples.[102] Currently, pDC remain an intriguing leukocyte subset that expresses lymphoid as well as myeloid-associated antigens, for which the functional properties are still a matter of investigation. Their ability to produce high amounts of type I and type III IFNs (ie, IFN-l) makes them early responders during viral infections. This in combination with their antigen-presenting properties has suggested an involvement of pDC in the pathogenesis of autoimmune diseases.[103] While pDC were long recognized to develop from both myeloid and lymphoid progenitors, recent evidence has challenged their affiliation to the DC family, suggesting that they are more related to innate lymphocytes (ILC).[104-106]

DENDRITIC CELLS DEVELOPMENT

DC develop in the bone marrow from hematogenic progenitors that are constantly being replenished throughout life.[107] The DC pool is replaced every few weeks in contrast to long-lived macrophages.[108-111] The commitment and development of dendritic cells predominantly relies on a single cytokine: Fms-like tyrosine kinase 3 ligand (Flt3L). Mice lacking this cytokine or its receptor FLT3 have a profound reduction of cDC and pDC.[112-114] Likewise, injection of FLT3L expands the DC pool.[41,112] This Flt3-DC axis is conserved across species, and in humans, FLT3L promotes the development of DC from bone marrow or umbilical cord blood progenitor cells.[115] The fact that few cDC and pDC remain in FLT3- and FLT3R-deficient mice supports speculations on the existence of another "DC" cytokine. Initial studies pointed toward granulocyte macrophage colony stimulating factor (GM-CSF/CSF2), based on its ability to induce differentiation of peripheral blood monocytes into dendritic cells.[116] GM-CSF-deficient mice have typical levels of lymphoid tissue DC; however, deficiency of this cytokine prevents the induction of CD103, which is required for DC homing to epithelial tissues, resulting in reduced CD103 expression or reduced peripheral tissue DC in the lungs, skin, and intestine.[117,118] Collectively, while FLT3L is necessary and sufficient for DC homeostasis, GM-CSF appears to promote the survival of peripheral tissue DC and the development of inflammatory monocyte-derived DC *in vivo*.[39] FLT3 is expressed on early uncommitted progenitors and is maintained during all stages of DC differentiation including mature cells, where it supports their survival. FLT3L is highly expressed on BM stromal cells as membrane bound or cleaved. Both forms actively engage the receptor leading to the phosphorylation of phosphoinositide 3-kinase (PI3K) and signal transducer and activator of transcription 3 (STAT3) and thereby mediating the activation of mammalian target of rapamycin (mTOR). Development of DC occurs via defined stages of precursor populations. FLT3-expressing multipotent progenitors branch into two separate pathways, one characterized by its lymphoid bias and the other by its myeloid bias, where common lymphoid or multipotent progenitors 4 and common myeloid or multipotent progenitors 3 (CLP/MPP4 and CMP/MPP3) are respectively the earliest identified subsets. The majority of cDC arise from CMP or MPP3, which further differentiate—in the presence of FLT3L—into a common monocyte/DC precursor (MDP) that has lost the potential to develop into granulocytes (Fig. 34.2). MDPs make a binary decision to either divide into the dedicated common DC precursors (CDP) or into the unipotent common monocyte progenitors (cMoP). While cMoP give rise to classical circulating monocytes that convert into nonclassical monocytes,[119,120] CDP were shown to develop into cDC directly or through a pre-DC stage and also partially contribute to the mature pDC pool. pDC-committed precursors or pre-pDC are characterized by the expression of FLT3 and by the induction of pDC specific markers, such as SiglecH and Ly6D. While pDC complete their maturation in the BM, where they can be identified as mature cells, pre-DC leave the bone marrow and complete their maturation within secondary lymphoid tissues. Circulating pre-DC, the immediate

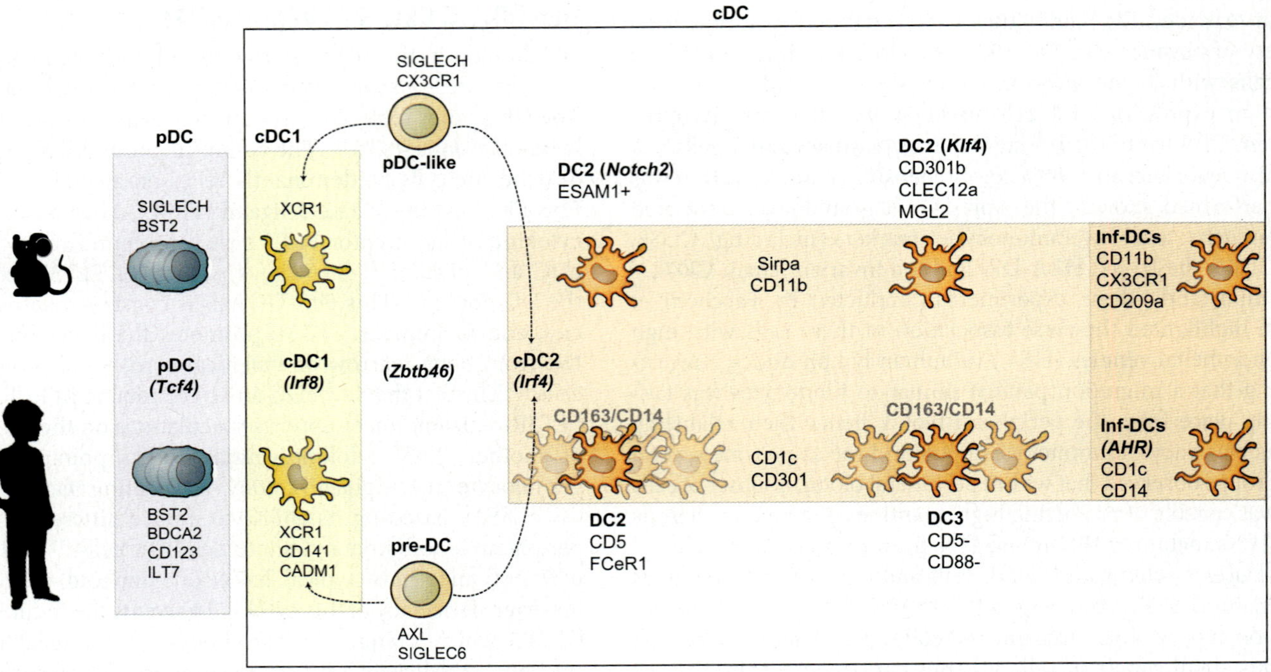

FIG. 34.2. Dendritic cell subsets. Dendritic cells in human and mouse comprise plasmacytoid dendritic cells (pDCs) and conventional dendritic cells (cDCs). The latter can be further subdivided based on the expression of specific surface markers and transcription factors.

cDC precursors, are characterized[43,44,46,47,121] by their lack of expression of lineage markers including MHC class II, and the expression of high of levels of FLT3, the DC-transcription factor Zbtb46, the immature DC marker CD226, and the pan DC marker CD26. Several other DC markers are present at low levels including the integrin CD11c. Pre-DC were also identified in human peripheral blood samples[48] and characterized as a subset expressing the IL-3R (CD123), Siglec6, AXL, and CX_3CR1.

The final stage of cDC development is their maturation, which occurs in response to sterile inflammation or infection. Immature cDC characterized by low expression of costimulatory molecules, high phagocytic capacity, and low T cell activation potential will mature and migrate to the T cell areas of draining lymph nodes to become efficient antigen-presenting cells as described later.

TRANSCRIPTIONAL REGULATION OF DENDRITIC CELLS COMMITMENT AND IDENTITY

The commitment to the DC lineage is influenced and maintained by various hematopoietic growth and transcription factors (TFs). Despite the common expression of several cytokine receptors and TFs, accumulating evidence suggests lineage development is regulated through the action of TF at specific enhancer elements. As mentioned earlier, FLT3L is the central growth factor regulating DC development from early uncommitted progenitors in the bone marrow. Induction of FLT3 expression is controlled by the transcription factors Bcl11a and Gfi-1.[122,123] Ligation of FLT3L and GM-CSF leads to the phosphorylation of the corresponding tyrosine kinase receptors, STAT3 and STAT5, respectively, which activate AKT and mTOR.[124,125] While the absence of STAT3 results in a drastic reduction of all DC subsets, the function of STAT5 in DC development remains unclear. Only Th2 priming is compromised in the STAT5 DC-specific knockout mouse.[126] The principal targets of STAT3 and STAT5 helping define DC lineage commitment are the CCAAT/enhancer binding protein α (C/EBPα), the ETS factor PU.1, and the interferon response factor IRF8. The levels of these downstream TF are essential in promoting lineage fate.

IRF8 was the first transcription factor shown to impair DC development, specifically cDC1 and pDC.[127,128] For this reason, IRF8 is regarded as a "terminal selector" for cDC1 in mice.[129] However, other leukocyte subsets are defective in IRF8-deficient mice, such as monocytes, germinal center B cells, and eosinophils.[130-133] In mice, a spontaneous mutation of IRF8 alters its binding properties to other TF, such as PU.1, resulting in complete deficiency of the cDC1 compartment and the expansion of a myelomonocytic progenitor.[134] Patients harboring a mutation in IRF8 (K108E) have an autosomal recessive immunodeficiency also characterized by the loss of circulating monocytes, pDC and cDC.[135] Collectively, clinical data have shown that a graded reduction of IRF8 can selectively impair cDC1 development, while patients with haploinsufficiency or carrying a dominant negative mutation are compromised across multiple DC and monocyte subsets. Hence, IRF8 is essential in mice and humans for cDC1, pDC, monocytes, and granulocytes development, yet only within the human hematopoietic system; it also controls cDC2 commitment.[136] Collectively, across multiple species IRF8 regulates myeloid development, and depending on the nature of the mutation, different outcomes are expected, most likely reflecting the tight regulation of this TF in the context of alterative interacting partners.

Additional transcription factors impair cDC1 development including NFIL3,[137] ID2,[138] and BATF3.[139] These TFs also display a broad expression pattern across several hematopoietic compartments, and therefore developmental defects are not exclusive to cDC1; that is, ID2 as well as NFIL3 deficiency leads to defects in ILC and NK cells.[140-144] Defects in BATF3 also impair the formation of memory CD8 T cells.[145] Taken together these studies highlight how transcriptional pathways are shared across multiple hematopoietic lineages and subsets. What appears to be lineage-specific is their "*in cis*" regulation, such that specific enhancer elements have been identified and are uniquely accessible to each subset.[146] Batf3 interacts with the AP-1 transcription factor Jun forming heterodimers at specific IRF8 enhancer elements, thereby stabilizing its expression during cDC1 development.[121,147-149] While cDC1 uses a +32Kb enhancer, the −50 enhancer acts on monocytes.[146] Collectively, the interplay of PU.1, IRF8, C/EBPα, BATF3, NFIL3, ID2, several E-proteins, such as E2-2 and ZEB2, at specific stages during development, acts in a concerted manner ensuring the development of MDP toward DC, pDC, macrophages, or monocytes.[129,146,150-156] Beyond the developmental aspect, several TFs regulate the survival or function of the mature cDC1 compartment: L-Myc,[157] Mxd1, and XBP1.[154,158]

While the pathway toward cDC1 commitment has been traced from the MDP progenitor to the cDC1, cDC2 development remains unclear. This is mainly due to the heterogeneity of this compartment and partially due to the lack of selective TFs that impair their development. Several TFs were nevertheless identified that result in a partial developmental and/or functional defect of the cDC2 compartment including IRF4, KLF4, RELB, NOTCH2, and TBX21. IRF4-deficient mice have a profound reduction of cDC2 within the lungs,[37,46] yet across other tissues their development remains normal but their ability to migrate to draining lymph nodes and express MHC class II-complexed antigen is highly compromised.[152,159] As a consequence, the lack of IRF4 leads to defects in the modulation of Th2 and Th17 responses.[33-35,37,56,160-163] Similarly, selective cDC deletion of KLF4 and NOTCH2 abrogates Th2 and Th17 immunity, respectively, suggesting that each response is regulated by a specific cDC2 subset or by the expression of each TF within multiple subsets. KLF4-dependent DC, required for Th2 priming, are characterized by Mgl2, reduced CD11b, and CD24 expression within skin-draining lymph nodes.[57] NOTCH-dependent DC that prime Th17 responses are characterized by ESAM expression in the spleen and by the dual expression of CD11b and CD103 in the lamina propria.[34,51] Similarly, *Tbx21* expression has been linked to a small fraction of cDC2, which appears to be proinflammatory given their ability to induce IFN-γ production.[59]

The development of pDC requires the expression of the E protein, E2-2 (encoded by *Tcf4*),[153] and NFIL3. Similar to other TFs, E2-2 is broadly expressed and is essential for early hematopoiesis, resulting in multiple defects across all lineages.[153] In humans, patients with Pitt-Hopkins syndrome, who have haploinsufficiency in the E2-2 gene, are characterized by severe intellectual impairment, several craniofacial deformations, and the reduction of pDC that have lost the ability to produce IFN.[164,165] Within the CDP compartment, the expression of ID2 or E2-2, which antagonize each other, will determine the developmental trajectory toward either cDC1 or pDC.[166] ZEB2 is an additional TF that appears to regulate several lineages, including pDC, DC, and monocytes. Within the DC compartment, it has been proposed to modulate ID2 expression and thereby defining the commitment of pDC within the pDC/cDC1 axes and cDC2 within the cDC1/cDC2 axes. Lack of *Zeb2* results in reduced cDC2 and pDC.[167,168] At least two modes of action have been proposed: repression of ZEB2 by NFIL3 will enable sustained ID2 expression essential for cDC1 development,[152,168] alternatively the induction of ZEB2 inhibits ID2 leading to pDC development and repression of cDC1 fate.[167] Independent of the way these three TFs modulate DC lineage commitment, their expression levels and developmental stage at which they are expressed are crucial to determine the developmental path toward cDC1, cDC2, or pDC. A complex interplay of multiple TF at specific developmental stages is involved in deciding their fate and function throughout a cell's life.[169]

SINGLE CELL ANALYSIS OF DENDRITIC CELLS

Over the past few years single cell sequencing has become the gold standard in dissecting the heterogeneity of leukocyte populations. A pioneering study isolated and performed RNA sequencing on murine splenocytes including approximately 1000 DC.[170] Within the DC pool, 15% could clearly be classified as cDC1 and 4% as pDC; the remaining cells comprised a heterogenous cDC2 population, postulating the existence of at least three distinct cDC2 subsets based on clustering. A few years later, two independent laboratories characterized at a single cell level DC from human peripheral blood samples.[50,62] The first study extended the circulating DC pool beyond the cDC1 (CD141+, BDCA-3+ DC), the cDC2 subsets (CD1C+ BDCA-1+)/(CD11c+ CD1−CD141−), and pDC (BDCA-4+BDCA-2+) adding more complexity to the DC compartment.[62] A newly identified DC5 subset referred as AS-DC based on the expression of AXL and SIGLEC6 shared several markers with mature pDC, such as the IL-3R. It was later suggested to be an immature population and therefore renamed pre-DC. A similar subset has been identified in mice; the contribution of this progenitor and CDP to the different mature cells remains to be addressed.[171]

Collectively, single cell sequencing approaches have further highlighted the heterogeneity of the myeloid and specifically the DC compartment. While cDC1 and pDC remain well-defined subsets across tissues and species, the characterization of the cDC2 compartment is ambiguous, relying on multiple surfaces as well as intracellular markers which may vary according to their tissue of origin and their activation status.

PERIPHERAL TISSUE DENDRITIC CELLS

cDC reside within peripheral tissues throughout an organism. The relative frequencies of each subset vary according to tissue-specific properties and reflect, on one hand, the degree of exposure to environmental cues and, on the other hand, the cytokine or chemokine levels of the tissue of residence.[107] Epithelial tissues such as the small intestine contain cDC with high expression of MHC class II, due to the constant exposure of exogenous antigens. As mentioned previously the seeding of peripheral tissues occurs from a short-lived circulating precursor, whether pre-cDC or CDP.[172] In contrast to humans, in mice, mature circulating DC are extremely rare, suggesting that the DC pool is predominantly regulated by the frequency of circulating immature DC capable of responding to microenvironmental cues, which will guide them to the periphery. Pre-cDC express the chemokine receptor CCR5 and can be further subdivided into CXCR3 expressing pre-cDC1 and CX_3CR1 expressing pre-DC2.[173] As a consequence, pre-cDC1 respond to the IFN-γ-induced chemokines CXCL9 and CXCL10. Once recruited into the tissue, pre-cDC have the ability to proliferate[47,172,174,175] and terminally differentiate into peripheral nonlymphoid cDC, thereby forming a network of clones of sister cells derived from individual pre-cDC clones of cDC1 or cDC2.[26,28,175-180] The induction of XCR1 marks the maturation of pre-DC1 into cDC1 and makes them highly responsive to XCL1, CCL5, and XCL2, secreted by NK and effector CD8+ T cells.[181-183] This XCR1-XCL1/CCL5 loop was recently shown to be an essential mechanism by which cDC1 are recruited within tumor tissues, enhancing cytotoxic response of NK cells and CD8 T cells.[183]

Mature DC constantly migrate from peripheral tissues to secondary lymphoid organs maintaining the pool of peripheral cDC. Their seeding within secondary lymphoid organs is mediated by the adhesion molecule L-Selectin (CD62L), which enables their binding to specialized high endothelial venules defining the frequency of resident cDC in spleen and lymph nodes.[47]

pDC, in contrast to cDC but similar to monocytes, enter secondary lymphoid organs via HEV residing in their proximity.[82] Under inflammatory conditions pDC are recruited via circulation to sites of inflammation. This different behavior of cDC and pDC reflects their unique functional properties that will be discussed later.

PATHOGEN SENSING

An invading pathogen must withstand various protective hurdles set in place by the host, including the skin and internal mucosal surfaces to gain access. In 1989, Charles Janeway proposed a group of receptors ought to exist, that would act as invariant innate PRRs. He predicted that conserved pathogenic structures, known as pathogen-associated molecular patterns (PAMPs) could be sensed by the host.[2] These PAMPs would help discriminate self from nonself, allowing the host to recognize and eliminate the invading pathogen. Recognition of pathogens ranging from viruses, bacteria, fungi, and multicellular helminths is supported by the expression of these PRR. PRR sensors can also detect tissue damage or damage-associated molecular patterns (DAMPs), induced by pathogens or inflammation (Chapter 12). DC are equipped with an array of PRR including Toll-like receptors (TLR), C-type lectins, NOD-like receptors, and RIG-I-like receptors. These receptors have a recognition domain (the sensor) that is either extracellular, cytoplasmic, or facing intracellular vesicles and can possess a dimerization domain and/or a signaling domain. Each PRR discriminates between distinctive categories of infectious agents; for example, muramyl dipeptide a component of *Streptococcus pneumoniae* can be recognized by NOD2. TLR4 binds lipopolysaccharide (LPS), a component of the outer membrane of gram-negative bacteria. Most PRRs signal via intracellular adaptor proteins, which are recruited by the adaptor domain. In the case of TLR signaling, there are two major pathways: the Myd88-dependent and Myd88-independent TRIF-dependent pathway, which result in the activation of NF-κB or IRF3. NOD-like receptors or C-type lectins will activate the downstream cascade via Caspase-1 or Syk.

The activation of PRR can promote phagocytosis of the infectious agent. In most cases, triggering of PRR initiates innate immune defense mechanisms such as the secretion of proinflammatory cytokines.[184] The extent of the response is determined by additional receptors that set effector thresholds, to ensure a controlled response.

Under healthy homeostasis, cDC are poised to capture and present antigens to T cells, and the consequence of antigen presentation is tolerance (Fig. 34.3). During an infection, detection of alterations in the microenvironment, for example, pathogens or the presence of proinflammatory cytokines, secreted by damaged epithelial cells including TNF-α or IL-1β, result in the engagement of multiple PRR. Collectively these signals provoke widespread DC differentiation from an immature to a mature/activated state characterized by increased membrane expression of MHC class II, costimulatory molecules such as B7-1 (CD80) and B7-2 (CD86), upregulation of certain chemokine receptors, cytokine secretion, and increased migration to draining lymph nodes (Fig. 34.3).

This activated status generates the necessary signals that initiate effective naïve T cell differentiation: presentation of peptide MHC (signal 1); high expression of costimulatory molecules (signal 2); and production of polarizing cytokines (signal 3). Taken together, pathogen recognition by DC ties innate pathogen sensing to adaptive immunity.

cDC1 and cDC2 each express their own unique repertoire of PRR to detect pathogen and danger signals.[185] The functional properties of cDC1 arise from their ability to recognize necrotic cells via Clec9a that binds filamentous actin (F-actin)[186] and viral RNA via TLR3.[187] cDC2 express the C-type lectin Dectin-1 (Clec7a) that recognizes β-glucans on fungi[188,189] and TLR5 that recognizes flagellin. Proteomic analysis revealed that cDC2 express high levels of cytoplasmic viral sensors including RIG-I and MDA5.[190] pDC are equipped with TLR7 and 9 and are therefore efficient in sensing single stranded RNA and double stranded DNA, present during viral infections, practically explaining their high production of type I IFN.[191]

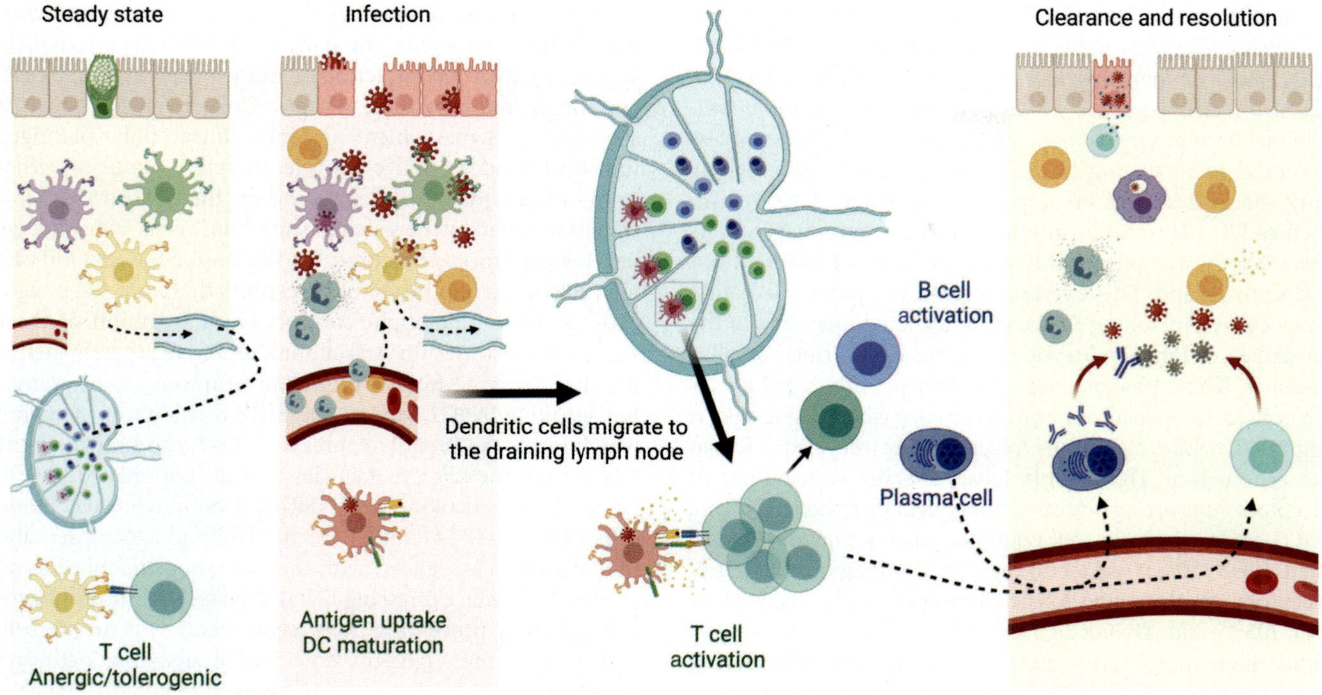

FIG. 34.3. Immune response at epithelial Barriers. Antigen capture, processing, and presentation by dendritic cells result in the induction of tolerance in the absence of inflammation. Upon infection, DC will sense pathogens and tissue damage and acquire an activated phenotype. Antigen presentation under inflammatory conditions will lead to the induction of innate and adaptive immunity. Resolution of inflammation restores the tolerogenic environment, once the pathogen is cleared and the tissue healed. (Created with BioRender.com.)

Furthermore, DC bone marrow progenitors have characteristic PRR expression profiles; in particular CDP express TLR2, TLR4, and TLR9 in the bone marrow.[192] The reason for the expression of functional PRR receptors on BM progenitors remains to be fully determined, yet ligation of LPS recognition by TLR4 upregulates CCR7 and downregulates CXCR4, promoting CDP exit from the bone marrow and homing to inflamed lymph nodes. During viral respiratory infection, the local increase of cDC is a consequence of pre-DC efflux from the bone marrow and influx into the lung parenchyma, implying a communication from the inflamed tissue to mobilize an emergency "squad" of cDC.[175,193]

Collectively the selective expression of PRR on each DC subset defines the sensing capacity to mediate the appropriate response. In the context of infection or inflammation, multiple pathways will be triggered and lead to a concerted action, where DC will capture pathogens and pathogen-derived antigens, home to the draining lymph nodes, and acquiring an activated phenotype necessary to induce the desired T cell response.

ANTIGEN CAPTURE AND PRESENTATION BY MHC MOLECULES

Dendritic cells are occasionally referred to as "professional antigen-presenting cells." One feature that is linked to this superior capacity to present antigen is their constitutive expression of a distinct proteasome, the "*immunoproteasome,*" which on other cells is only induced under inflammatory conditions. The catalytic core of the proteasome is composed of two outer α rings and two inner catalytic β rings (β1 and β5). IFN-γ or TNF-α exposure induce the synthesis of three alternatives β subunits: β1i (LMP2), β5i (LMP7), and β2i (LMP10) (Chapter 20). This distinctive immuneproteasome is expressed at steady state by DC and appears to improve peptide generation for antigen presentation.[194,195]

Antigen Uptake in the Endocytic Pathway

DC capture exogenous antigens using distinctive endocytic mechanisms which can be broadly divided into phagocytosis, the uptake of large particles (>0.5 μm), and pinocytosis, the uptake of fluids or solutes.[196] Phagocytosis is a receptor-dependent endocytosis exclusive of certain subsets of cells. Pinocytosis includes clathrin-, caveolae-mediated endocytosis and actin-dependent macropinocytosis.

DC and in particular antigen presentation by DC occurs mostly via phagocytosis, clathrin-mediated endocytosis, and macropinocytosis.[197-199] In the case of micropinocytosis, actin-based processes are responsible for the formation of large vesicles containing extracellular fluids together with soluble antigens. In immature patrolling DC, actomyosin is recruited to the cell front to build macropinosomes.[200,201] Upon innate activation, micropinocytosis is shut down to enable DC migration toward lymph nodes.[200,201]

Ilya Metchnikoff received the Nobel prize in Physiology or Medicine for his discovery of phagocytosis in 1908, as process by which cells capture foreign particles. "*Whenever the organism enjoys immunity, the introduction of infectious microbes is followed by the accumulation of mobile cells, of* *white corpuscles of the blood in particular which absorb the microbes and destroy them. The white corpuscles and the other cells capable of doing this have been designated 'phagocytes,' (*ie, *devouring cells) and the whole function that ensures immunity has been given the name of 'phagocytosis'.*" This process is central in mediating host defense and tissue remodeling; only specialized cells are capable of efficient phagocytosis, such as DC neutrophils, monocytes, and macrophages.[202] As described earlier, phagocytosis occurs via receptor-mediated endocytosis and DC express several receptors mediating phagocytosis including PRRs. Phagocytic uptake leads to the formation of large endocytic compartments termed as phagosomes. These phagosomes as well as all endocytic vesicles are routed to specialized compartments where the cargo is degraded, while membrane receptors are recycled back to the cell surface. DC exhibit subset-specific distribution of receptors supporting selective phagocytic functions. Murine cDC1 are particularly well equipped for the phagocytosis of dead cells.[203-206] Several lectin receptors, among other types of receptors, are dedicated to the sensing of dead cells, such as DEC205[207] and DNGR1/Clec9a.[30,31,180,208,209] DNGR1/Clec9a binds filamentous actin released from necrotic cells during immunogenic cell death.[186,210,211] While the uptake of apoptotic or necrotic cells by macrophages leads to the endocytic release of antigens, cDC1 appear to preserve their cargo, enabling cross-presentation of exogenous antigens on MHC class I. This pathway is required for the presentation of intracellular pathogens[212,213] as well as transformed self or tumors,[30,138,214] for which cDC1 priming of cytotoxic response is critical.

cDC2 and iDC efficiently promote the activation of Th cells and are specialized at phagocytosing bacteria, fungi, viruses, and antibody-opsonized particles.[212,215,216] cDC2 therefore express a wide array of C-type lectins dedicated to the sensing of glycans, such as Clec10A, Clec4A4/DC immunoreceptor 2 (DCIR2), and mannose receptors (MRs).[59,189,213,217] C-type lectins have a characteristic carbohydrate recognition domain (CRD) allowing the binding of sugars in a calcium-dependent fashion; that is, MR recognizes pathogen-associated high-mannose structures of bacteria, yeast, and viruses. Upon recognition and internalization, the cargo will be delivered to specialized vesicles for its degradation and for the loading of the generated antigens to the MHC class II.

DC are constantly sampling tissues and, when activated, migrate to draining lymph nodes. The fine balance in discriminating between self and nonself enables the appropriate induction of immunity while preserving integrity of tissues. This task is extremely challenging in the context of transformed self or infected self, where both endogenous and exogenous antigens are present. Similarly, recognition of dying self is essential for preservation of tolerance as demonstrated by the vast redundancy in dedicated receptors supporting efferocytosis such as DNGR1/Clec9a, DEC205, *CD36*, and *B3B5*.[30,180,204,208,218-220] Furthermore, several receptors that recognize ligands released by dying cells decorate the surface of DC and macrophages promoting phagocytosis, such as TIM3 which will recognize galectin-9 and HMGB1.[221] Increased membrane exposure of phosphatidylserine (PS) is common for altered self or during apoptosis. While inhibition of PS recognition *in vivo* impairs macrophage phagocytosis, it promotes efferocytosis by spleen cDC1.[222] This mechanism appears to favor the induction of antigen presentation by DC over degradation by macrophages. Collectively, apoptosis is a phagocytic signal which can hint to intracellular pathogen infection besides cellular damage, therefore requiring either immunity or tolerance depending on the context. Multiple signals initiated by the sensing of apoptotic cells will converge into the appropriate outcome. For this reason, DC exhibit various activation and inhibitory receptors that keep the phagocytic process under tight control. The integration of these signals orchestrates the activation level of DC, thereby defining the threshold for the induction of immunity or anergy. For example, cDC2 express the SIRP-α inhibitory receptor, which binds the broadly expressed CD47 also known as the "do not eat me" signal. Particles that do not engage SIRP-α (such sheep red blood cells [SRBCs] or murine red blood cells from CD47−/− mice) are efficiently phagocytosed and this process is associated with the immunogenic maturation of cDC2.[223] Cells expressing CD47 engage SIRP-α on phagocytes, thereby preventing their engulfment. This mechanism and many more highlight how innate signaling pathways coordinate phagocytic uptake in order to regulate T cell activation versus anergy by DC.

Antigen Retention and Degradation in the Endocytic Pathway of Dendritic Cells

From a general standpoint, endocytic compartments formed after the engulfment of antigens will fuse with other vesicular compartments including degradative lysosome. A result of lysosomal fusion with endocytic vesicles leads to progressive acidification of the lumen, enabling proteolysis and degradation of their cargo.[224] Immature DC have a rather low degradative activity following phagocytosis, as compared to other phagocytes such as macrophages or neutrophils, ensuring preservation of antigens.[221,225-228] Various mechanisms have been proposed to explain the overall low proteolytic activity of immature DC, such as low expression of cathepsin proteases,[226] inefficient luminal acidification,[225,227,229] or the formation of antigen retention compartments.[228,230,231] Collectively, the mild degradative activity promotes the preservation, delivery, and efficient presentation of antigens on MHC in draining lymph nodes.[221,225]

Antigen Presentation by MHC Class II

The loading of peptides produced in phagolysosomal compartments also occurs in specialized vesicles. Newly synthesized MHC class II is targeted to the endocytic compartments by the virtue of its interaction with the invariant chain chaperone (Ii). After fusion, the invariant chain is cleaved, progressively releasing the MHC class II, and enabling chaperone-assisted loading with the antigenic peptides generated in the phagolysosome.

DC maturation is linked to increased MHC class II surface expression and half-life, achieved through neo-synthesis and cell surface stabilization.[232,233] Under steady-state conditions, the cytosolic tail of the MHC class II β-chain undergoes constitutive polyubiquitination and degradation. Upon DC activation, polyubiquitination and degradation of the neosynthesized MHC-II is prevented resulting in its accumulation at cell surface.[234-238]

Antigen Presentation and Cross-Presentation by MHC

MHC class I is expressed by all cells. Peptides presented on MHC class I reflect the protein composition of a cell. Ubiquitinylated proteins are degraded by the proteasomes and the resulting peptides are transported into the endoplasmic reticulum (ER) by TAP. In the ER, peptides are loaded onto MHC class I molecules by molecular chaperones (Chapter 20). The MHC class I peptide complex is subsequently transported to the cell surface for presentation to CD8+ T cells.[239] This pathway is essential to provide a means of recognition of microbe-infected cells. CD8+ T cells recognize foreign peptides and activate their cytotoxic potential, promoting the lysis of target cells. The downregulation of MHC class I expression is a typical mechanism by which intracellular pathogens prevent recognition by immune cells. Therefore, in the absence of direct presentation, CD8 T cells rely on a specialized pathway referred as cross-presentation. Infected cells are phagocytosed by DC but exogenous, foreign antigens are then loaded on MHC class I. Cross-presentation is not only important to mount CD8+ T cell responses against viral[240,241] and intracellular parasites,[242] but also is essential to promote immunity against tumor-associated antigens (transformed self)[139] and contributes to the maintenance of peripheral tolerance to tissue-specific self-antigens.[243,244] Murine and human cDC1 are the most efficient DC subset for cross-presentation,[139,206] showing not only enhanced expression of the MHC class I peptide-loading complex[36,189] but their endocytic pathway is well suited, such that acidification is limited and antigen persistence increased.[36,189,245-248] The endocytic pathway that leads to efficient cross-presentation on cDC1 is regulated by IRF8 and BATF3[214] and is defined by the selective expression of a set of vesicular regulators, such as Plin2,[249] Rab43,[250] Rab39a,[251] and WDFY4.[252]

Although a division of labor exists between DC subsets where cDC2 are potent mediators of antigen presentation on MHC class II[189] and cDC1 are specialized in cross-presentation, both subsets have the capacity of performing either function under the appropriate circumstances. Cross-presentation by cDC2 occurs for soluble antigens or antigens targeted via specific endocytic receptors,[247,248,253-255] that is, following internalization of immune complexes via FcγR.[212,215] Presentation on MHC class II by cDC1 is particularly important to generate the necessary CD4 T help (IL-2 secretion) required by naïve CD8 T cells in the context of tumor rejection[256] and *Toxoplasma*-induced immunity.[257]

DENDRITIC CELL MIGRATION

Naïve T cell activation requires the migration of mature cDC from nonlymphoid tissue to lymphoid organs (Fig. 34.3). These mature cDC will have processed antigen from the periphery before transporting and concentrating these antigens in lymph nodes, enhancing the process of antigen presentation to T cells. DC are strategically positioned at barrier sites such as the lung and intestine where they interact with the epithelium forming a functional network. Upon activation sentinel cDC undergo a transformation from immature cells continuously sampling foreign and self-antigens in their confined extracellular milieu by pinocytosis, endocytosis, and phagocytosis to a highly mobile mature cDC that has captured antigen (Fig. 34.3). In the presence of pathogens or during inflammation, peripheral cDC upregulate MHC class II, costimulatory molecules, and the chemokine receptor CCR7, which guides their migration to the T cell zone through the afferent lymphatics.[258] CCL21 and CCL19 act as the endogenous functional chemokine ligands for CCR7.[259,260] These same chemokines also guide T cell migration resulting in the co-localization of DC presenting and T cells sampling the antigens. CCL21 is predominantly expressed in lymphoid organs and lymphatic endothelial cells. DC were observed and isolated from the lymphatics since their discovery,[14,15] and skilled surgical techniques honed by members of the McPherson Group in Oxford progressively improved their collection, before better isolation protocols based on antibodies were available.[261,262] More recently, state-of-the-art microscopy studies on lymphatic trafficking showed how cDC crawl across endothelial cell barriers gaining access to the lymphatics through portals or crevice-like gaps of the basement membrane.[263-265] Their migration path to the draining lymph nodes starts with their passive transport by laminar flow within the lymphatics where they reach the subcapsular sinus. Once inside the lymph node, DC will migrate over a network of reticular stromal cells and extracellular matrix toward the T cell zone.[266-268] Migration of DC requires not only the expression of CCR7 and its ligand CCL21[269] but also a highly regulated migratory behavior dependent of other G-protein-coupled receptors and proteases[270] yet independent of integrins.[265]

THE IMMUNOGENIC FUNCTION OF DC1

The development of genetic tools has enabled immunologists to target and dissect the function of cDC1. Several models based on the transgenic expression of diphtheria toxin receptor (DTR) allowed for selective DC depletion of these cells.[271] Unlike humans, rodents are naturally resistant to bacterial diphtheria toxin because they lack a high-affinity DTR. Therefore, genetic insertion of the DTR gene under a DC-specific promoter gene will render DC susceptible to the administered toxin. The first engineered mouse model was the pan-DC deleter mouse created by Jung et al, in 2002, where the DTR was induced under the control of the CD11c promoter.[272] Later studies, aimed at subset specificity, such that cDC1 depletion was achieved by introducing the DTR under the control of the langerin gene ($Lang^{\mathrm{DTR}}$)[273] or Clec9a ($Clec9a^{\mathrm{DTR}}$).[274] However, in both cases depletion was not exclusive impairing also LCs or pDC, respectively. To date the best genetic model, reflecting the specificity of its expression on cDC1, is the $Xcr1^{\mathrm{DTR}}$ mouse.[26,275,276] Beside the DTR-based depletion systems, cDC1-deficient *Batf3−/−* mice have been widely used to examine cDC1 function.[139] Independently of which mouse model is employed, mice lacking this compartment have impaired CD8 T cell priming, resulting in defective immunity against West Nile virus,[139] *cytomegalovirus* (CMV),[277] influenza, and *herpes simplex virus* (HSV) infections.[240,277,278] Similar to viral infections, CD8 T cell responses to intracellular parasites are also impaired such as in the case of cerebral malaria in *Lang*^{DTR}[279] and *Clec9a*^{DTR} mice.[274] Collectively all the immune defects related to the absence of cDC1 are related to their unique capacity to uptake, preserve, and cross-present antigen to CD8+ T cells.[26,28,31,176,180,240,246,250,252,280]

As mentioned earlier, cross-presentation is also important to respond to transformed self, and mice deficient in cDC1 have defective antitumor immunity.[139] Their capacity to mount an efficient antitumoral response is linked to their ability to cross-present cell-associated antigens (often neoantigens) acquired via the uptake of apoptotic tumor cells and cell debris.[203,205,219,240]

Besides their ability to mount a robust CD8 T cell immune response, cDC1 also play a pivotal role in initiating innate immunity against a series of intracellular pathogens. In mice, cDC1 secrete high amounts of IL-12 in response to triggering by TLR11 and 12, which recognize the *T. gondii* protein profilin.[281] IL-12 activates NK and ILC1 cells, resulting in the induction of IFN-γ. In general pathogen that trigger cDC1 will promote the secretion of high levels of IL-12 and type III IFNs, which on one hand initiate innate IFN-γ-mediated immunity and on the other hand represent "signal 3" for the differentiation of naïve CD8 T cells.[32,282-284] As mentioned earlier, cDC1 can also present antigen on MHC class II to CD4 T cells, enabling the engagement of the CD40-CD40L axes, the production of IL-12[285] and the delivery of appropriate help for naïve CD8 T cell activation.[240,285-288] The efficiency of CD8 T cell activation is also reflected by the correct positioning of these cells within secondary lymphoid tissues mediated by XCR1/XCL1.[181,289]

A further function of cDC1 is the trans-presentation of IL-15 to NK and memory T cells, which is achieved by presenting IL-15, pre-bound to its receptor, on their membrane.[257,290,291]

Upon secondary exposure to the same antigen, cDC1 secret high levels of CXCL9, which engages CXCR3 on effector and memory CD8 T cells ensuring the recruitment of antigen-experienced memory cells.[287,292,293] Collectively, cDC1 orchestrate "type I" immune responses, that culminate in the secretion of IFN-γ by CD8 and CD4 T cells (Th1), NK cells, and ILC1. Type I innate and adaptive immunity is a prerequisite for combating intracellular pathogens, whether viral, bacterial, or parasites, and is critical to maintain tumor immune surveillance. The immunological context or the type of insult determines the extent of innate immunity; priming and help that is induced and achieved by cDC1 reflect the interplay of multiple immune and nonimmune subsets.

PRIMING BY DC2

cDC2 have a superior aptitude in presenting MHC class II-loaded antigen to CD4 T cells, compared to cDC1. Mounting evidence suggests cDC2 are endowed with the specific machinery required for the priming and polarizing of T-helper subsets (signals 2 and 3). Specifically, IRF4- and NOTCH2-dependent cDC2 crucially regulate Th17/Th22 T cell responses in mucosal tissue, while also activating the corresponding innate immune subset ILC3. DC-specific inactivation of IRF4 or NOTCH2 restricts the survival and the function of cDC2 or depletes mucosal CD103+CD11b+ migratory cDC2, respectively, thereby limiting Th17 priming[34,35],[53]. This subset, if activated by the triggering of specific PRRs will produce high levels of IL-23 and IL-6 essential for the production of both IL-17 and IL-22 by T cells and ILC.[35,51,294] This IL-23/IL-22-17 axis mediates the resistance to several pathogens including the enterobacterium *C. rodentium* in the intestine[295] and the fungus *Aspergillus fumigatus* in the lungs.[34] The mechanism of action of this subset is conserved across species, with CD103+ SIRPα+ cDC2 patrolling the human intestinal mucosal tissue.[55]

The activation of CD4 T cell responses and the capacity of cDC2 to produce IL-6 are also important in regulating follicular helper T cells (Tfh) priming, essential for the formation of germinal center (GC) reactions and high-affinity antibody (Ab) production. In the absence of NOTCH2, splenic ESAM+cDC2 are absent and immunization against SRBC shows profoundly reduced Ab titers.[296] Human tonsils have abundant cDC2, which appear to similarly control GC reactions.[297]

By contrast, IRF4-dependent NOTCH2-independent cDC2 control Th2 responses to skin-derived antigens.[56] Depletion of this cDC2 subset in CD301b^{-DTR} mice prevents the development of Th2 responses to papain immunization or *N. brasiliensis* infection.[56,163,298] KLF4 controls the differentiation of this cDC2 subset essential for the induction of Th2 responses. Indeed, mice defective for KLF4 have impaired Th2 responses and are therefore susceptible to the infection with *Schistosoma mansoni* but resistant to house dust mite–induced allergic reactions.[57]

cDC2 appear to also regulate the restimulation of tissue-resident T_{RM} (CD8+CD103+CD69+) cells upon reinfection as shown in HSV type 2 and influenza infections.[299] These findings might apply to human cDC, where CD1c+ DC activate CD103+CD8+ T cells upon influenza vaccination.[300] Recent studies have demonstrated that within the CD1c+ DC2 compartment, expression of CD163 identifies the subset with the higher ability in differentiating naïve CD8+ T cells to T_{RM} cells.[60]

Taken together, cDC2 are important in the regulation of Th cell immunity. Similar to cDC1 the context of infection will activate the appropriate T cell subset hopefully, resulting in the most appropriate immune response to the invading pathogen.

pDC IN IMMUNE RESPONSES AND DISEASE

A longstanding question that remains unanswered in DC biology is the contribution of pDC to antigen presentation and to the pathogenesis of autoimmune diseases. While cDC are indisputably the most efficient antigen-presenting cell capable of inducing potent T cell activation, pDC's presenting role has been controversial. The identification of a cDC precursor within the conventional pDC pool has further nurtured doubts concerning pDC's ability to prime T cells and raised questions on the historical data.[301] Nevertheless, studies on the pDC capacity to present antigen has relied on antibody-mediated delivery that targeted different pDC receptors, such as BST2, CLEC4A, or SiglecH.[301,302] Collectively, the outcome of the responses was dependent on which receptor was targeted and the adjuvant applied, showing either induction of tolerance, delayed onset of autoimmunity, or protection against subsequent viral infection or tumor growth.[103] BDCA2, CLEC4A, and SiglecH delivery promoted tolerance by suppressing CD4 T cell responses and inducing T_{reg} expansion.[303-306] In contrast, delivery via BST2 promoted humoral and cellular immunity with increased protection against subsequent viral challenges.[303,307] Importantly the ability of pDC to upregulate costimulatory molecules and to produce type-I IFNs in response to viral infections also suggested that pDC may play a role in directly presenting or cross-presenting viral antigens to CD8 T cells.[308,309]

Attempts to understand the function of pDC *in vivo* has also relied on mouse models using DTR or antibody-mediated depletion.[310] The major function of pDC is undoubtedly the production of type I and type III-IFNs (IFN-α, IFN-β, and IFN-λ) in response to viral infections or sensing of endogenous RNA and DNA.[302,310] The signaling pathway following TLR7 and 9 triggering occurs mostly via IRF7.[311] Most cells are capable of producing type I IFNs and thereby limit the spread of viral infections. The requirement for pDC *in vivo* has been investigated for different pathogens. Only in the case of mouse hepatitis virus (MHV) infection, pDC appear to be indispensable.[312] In the context of other infections, including dengue (DENV) and chikungunya (CHIKV), IFN production by pDC is sufficient to contain the viral burden, but not essential.[313,314] Similar results were obtained in cases of murine cytomegalovirus (MCMV) and vesicular stomatitis virus (VSV) infection.[312] pDC depletion reduced early IFN-I production leading to an expansion of NK cells and a reduction of virus-specific CTL, but ultimately there was no difference in the survival of mice.[313] It is possible that the reduction of pDC may have boosted the innate response by NK cells but reduced their presentation of viral antigens.[313] A similar reduction of T cell responses and increased viral titers were shown also upon chronic LCMV infection.[312] It is likely that pDC are important to control early events during viral infections, limiting the spread. Unfortunately, most pDC markers are nonspecific and therefore depletion-based models result in the depletion of other subsets making it difficult to ascribe specific function to pDC during immune responses.[315]

pDCs are also capable of secreting other proinflammatory cytokines and chemokines, including IL-6, IL-12, TNF-α, CXC-chemokine ligand 8 (CXCL8), CXCL10, CC-chemokine ligand 3 (CCL3), and CCL4,[302,316,317] suggesting that they may help to sustain the inflammatory milieu where they are recruited via their expression CXCR3 and CCR5.[308,318] Indeed, one of the major differences between cDC and pDC is in their migratory properties: while cDC reach draining lymph nodes via afferent lymphatics, mature pDC leave the bone marrow, enter the peripheral circulation, and reach either T cell areas of lymph nodes through high endothelial venules, or sites of inflammation.[319,320] The abundance of pDC in inflamed tissues and the elevated levels of type-I IFN in autoimmune diseases have suggested a role of pDC in the pathogenesis of several autoimmune diseases where their ability to present autoantigens could be harmful.[321-323] Early depletion of pDC in lupus models delayed the onset of disease, suggesting that type I IFN production by pDC may contribute detrimentally to the overall effect without being necessary for the pathogenesis.[324,325] It was shown that internalization via Fc-receptors of DNA/RNA-anti nuclear antibody complexes is capable of triggering TLR7 and 9 and lead to the production of IFNs. This mechanism will promote the activation of both B cells as well as pDC.[326] While B cells are certainly responsible for the production of autoantibodies, the role of pDC in the initiation or in sustaining the pathogenic effect is under debate.[327]

Taken together, pDC appear capable of presenting antigen in the appropriate situation but their classical function is clearly the production of type I and type III IFNs in response to viral infections thereby containing viral spread at early timepoints.

DENDRITIC CELLS SUPPORT CENTRAL TOLERANCE IN THE THYMUS

T cell development in the thymus requires the removal of autoreactive thymocytes in a process referred to as negative selection to avoid autoimmunity. This process relies on the deletion of thymocytes that express a TCR with high affinity for self-peptide-MHC complexes (Chapter 27). The thymic medulla is home to two distinct antigen-presenting cells expressing MHC class I and class II: the medullary thymic epithelial cells (mTECs) and the DC, both of which are essential for the establishment of central tolerance supported by the occurrence of autoimmunity in their absence.[328] mTEC express an enormous variety of tissue-specific antigens (TSA) that are typically restricted to peripheral tissues; this process is regulated by the auto-immune-regulator (AIRE). DC are predominantly located in the medulla of the thymus and are important in the presentation of a wide range of self-antigens.[329] Thymic DC encompass resident DC (CD8α+ SIRPα−), migratory DC (CD8α− CD11b+ SIRPα+), and pDC ($CD11c^{int}$ $CD45RA^{int}$).[330] The origin of antigens presented by thymic DC is diverse and includes the acquisition of antigens from mTECs, from the circulation, and from peripheral tissues. Thymic DC contribute to negative selection by presenting and cross-presenting tissue-specific antigens (TSA) and tissue nonspecific antigens captured from mTEC.[331] DC-targeted transgenic expression of self-antigens is sufficient to induce the deletion of autoreactive CD4 T cells.[332] Thymic DC also acquire circulating antigen. Intravenous injection of soluble antigens will label all subsets: cDC1, cDC2, and pDCs. However, the proximity of CD172+ cDC2 to the highly vascularized cortico-medullary junctions enables them to capture most of the permeating antigen.[333,334] Finally, peripherally collected antigens can also be presented by all subsets after homing to the thymus, promoting negative selection of CD4+ T cells and the generation of T_{regs}.[329,335,336] The outcome of thymic DC antigen presentation is typically regulatory. As alluded to earlier, in addition to negative selection, antigen presentation by thymic cDC supports the development of thymic-derived Foxp3+ T_{reg} that suppress autoreactive peripheral T cells escaping negative selection,[337-339] collectively shaping central tolerance.

PERIPHERAL TOLERANCE

A major challenge for the immune system is the ability to detect danger while ignoring innocuous antigens. This task is particularly complex at epithelial barrier sites in direct contact with the external environment where food- and microbiome-derived antigens are constantly present. The mechanisms DC employ to initiate adaptive immunity can become a threat and lead to the development of debilitating autoimmunity, allergy toward harmless agents, or chronic inflammation. DC regulate the balance between immunity and tolerance. As mentioned above, DC crucially participate in the establishment of central tolerance via negative selection of autoreactive T cells and the differentiation of Foxp3+ regulatory T cells.[329] Furthermore, DC continuously ensure the functional inactivation or induction of anergy of peripheral T cells that escape negative selection or that recognize innocuous environmental antigens.

This control is achieved by ensuring an appropriate expansion of effector T cells, thereby limiting potential tissue damage and immunopathology.[340] DC continuously phagocytose cellular debris which leads to presentation of self-antigens on MHC class I and II.[341] The absence of inflammation or infection prevents DC activation, limiting their capacity to express signal 2 and 3, resulting in suboptimal T cell activation. Seminal studies performed by Kurts et al illustrated how tolerance is preserved by continuous cross-presentation by cDC1 of β-islet-derived antigens to cognate CD8 T cells in draining lymph nodes.[243,342-345] This process termed "cross-tolerance" leads to transient activation and subsequent deletion of self-reactive mature T cells.[344] These observations have been extended to antigens expressed in other tissues such as the skin and gut where migratory CD103+ cDC1 are the most efficient cross-presenting cells.[346,347] CD103+ DC migrating from the gut to mesenteric lymph nodes are also efficient inducers of adaptive Foxp3+ T_{reg} specific for food- or microbiota-derived antigens.[348,349] This process is dependent on TGF-β and retinoic acid (RA).[348,349] DC express retinal dehydrogenase (RALDH2) which produces RA from dietary vitamin A.[348-350] RA not only promotes T_{reg} induction but also imprints a program of gut residency by driving the expression of CCR9 and α4β7. These receptors attract and confine T_{regs} at the intestinal barrier via the gut-expressed chemokine CCL25 and the mucosal vascular addressin MAdCAM1.[351-353]

The contribution of DC to the maintenance of skin homeostasis obeys similar principles of cross-tolerance.[354] Migratory DC from the skin induce tolerance, taking advantage of the sunlight-induced vitamin D3 (vitD3) metabolite. Both vitD3-primed LCs and dermal DC induce Foxp3+ and Foxp3− T_{regs} via TGF-β or IL-10, respectively. The expression of CCR10 by the generated T_{regs} ensures skin homing.[355,356]

Thus, DC preserve the integrity of the tissues by shaping the peripheral T_{reg} pool on one hand through the induction of adaptive regulatory T cells from naïve T cells and on the other hand through the expansion of natural T_{reg} generated in the thymus.

Constitutive ablation of DC reduces T_{reg} numbers and in turn induces inflammation by both Th1 and Th17 T cells.[357] Conversely, expansion of the DC pool by administration of FLT3L triggers the expansion of T_{regs} in the periphery via a mechanism dependent on MHC class II expression in DC.[358,359] DC also support the proliferative expansion of natural T_{regs} in re-population settings following T_{reg} depletion.[360] Importantly, interactions between DC and T_{reg} are reciprocal. Antigen presentation on MHC class II is essential to maintain the tolerogenic phenotype of DC. Indeed, T_{regs} restrict DC activation, for instance, T_{regs} downregulate CD80 and CD86 costimulatory molecules via CTLA4-dependent trans-endocytosis thereby lowering their ability to activate naïve T cells.[361,362] Furthermore, T_{reg}-educated DC produce IL-27, further expanding their ability to induce adaptive Foxp3⁻ IL-10+ regulatory T cells.[363] MHC class II inactivation in DC prevents this interplay and breaks CD8+ T cell tolerance, resulting in overt gut inflammation.[364,365] Similarly, deletion of MHC class II on cross-presenting cDC1 swings the balance from cross-tolerance to cross-priming of self-reactive CD8+ T cells, resulting in a lethal skin and intestinal autoimmune pathology.[366] These findings highlight the fine-tuning that constantly shapes the pool of peripheral T_{regs} and that depends on MHC class II presentation by peripheral tolerogenic DC. Certain fundamental molecules regulate this balance, such as the cytokines IL-10 and TGF-β, IL-27, IL-35, and the vitamin A metabolite RA. Inactivation of β-catenin in DC abrogates the production of the anti-inflammatory cytokines like IL-10 and TGF-β in the gut, decreases T_{regs} numbers, and precipitates inflammation.[367] Genetic removal of the TGF-β activating integrin $\alpha_v\beta_8$ on DC results in the failure to induce T_{reg} leading to inflammatory bowel disease and autoimmunity.[368]

The tight balance that ensures an appropriate immune response while preserving tissue integrity and self-tolerance is the result of the complex crosstalk and interplay between DC, T_{regs}, effector T cells, and their surrounding microenvironment.

HARNESSING DENDRITIC CELLS TO INDUCE ADAPTIVE IMMUNITY

The concept of acquired immunity was acknowledged over 2000 years ago by Thucydides during the *Peloponnesian* wars. He described how plague survivors in Athens ~430 BCE were resistant to secondary challenges of plague but no other diseases. Similarly, the introduction of variolation in 11th-century China were all early attempts at vaccination. Healthy individuals would inhale the powder from ground smallpox

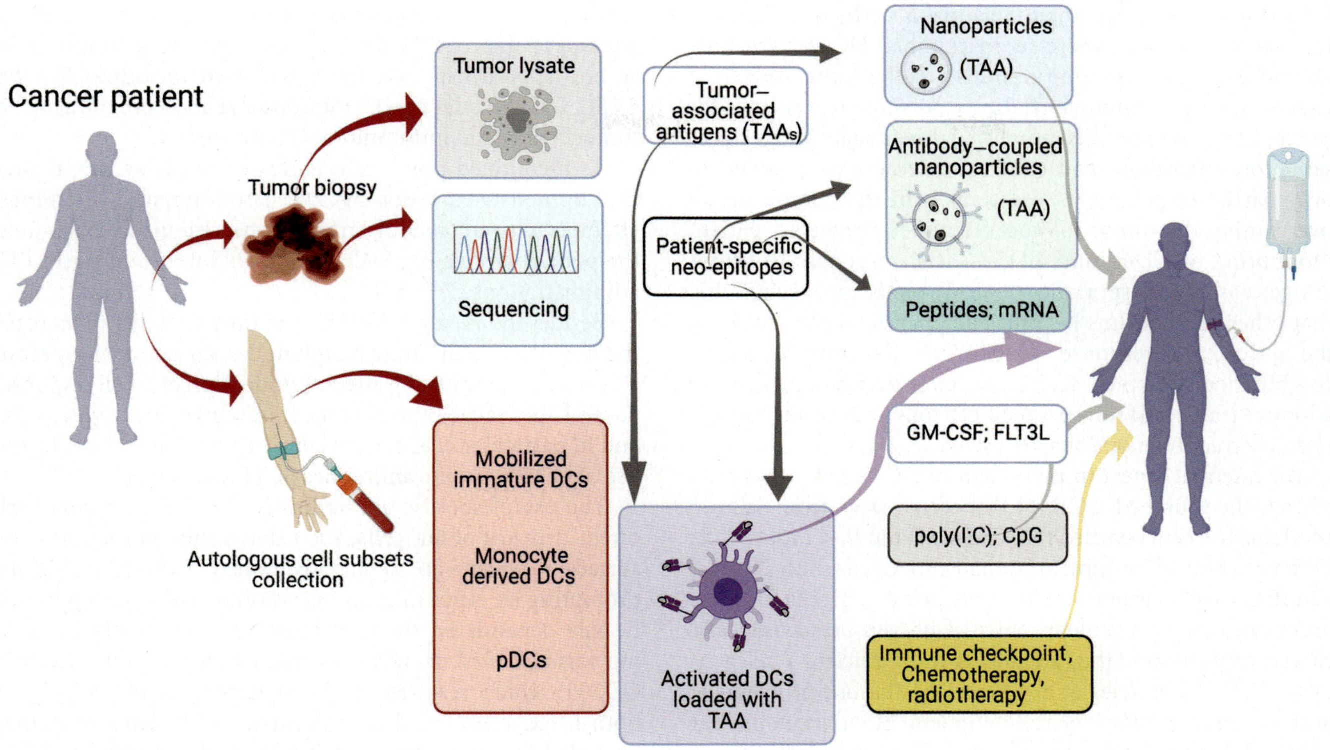

FIG. 34.4. Dendritic cell-based Vaccines. Several approaches are currently used to exploit DC in cancer treatments. DC isolated from patients are subsequently loaded with tumor antigens and activated prior reinjection into patients. DC can also be directly activated *in vivo* using different approaches such as growth factors (FLT3L/GM-CSF, STAT3 inhibitors), adjuvants (Poly(I:C), CpG, other TLR agonists, etc) alone or in combination with tumor lysates, tumor-associated antigens (TAA), patient-specific tumor antigens in forms of peptides or RNA. Alternatively, TAA or patient-specific tumor antigens are delivered via nanoparticles targeted to DC via direct injection into lymph nodes or using antibody-coated nanoparticles directed to recognize DC-surface receptors. All these approaches are differently combined with each other and integrated into conventional standard care therapies such as immunotherapies (immune checkpoint inhibitors), radiation therapy, chemotherapy, etc. (Created with BioRender.com.)

scabs and be protected from severe disease exhibiting only mild symptoms (Chapter 41).

Classical vaccines were developed without knowledge of DC biology, yet it is clear that DC are an essential component of their mechanism of action. They are indispensable for initiation of the adaptive immune response, providing the necessary signals to ensure a strong T cell response.[369] Indeed, the selection of adjuvants that trigger PRR on DC and induce innate signals is key in determining an appropriate response and therefore in vaccine design.

More recently, the interference of DC-T cells interactions has been exploited for cancer treatment. Immune checkpoint blockade (ICB) immunotherapies have significantly improved the survival of certain cancer patients. Anti-PD1 antibodies trigger a rapid IFN-γ-dependent activation of intratumoral DC enhancing their IL-12 production.[370] In addition, depletion of DC1 abrogates the antitumoral efficiency of ICB in preclinical models.[371,372] The concept to therapeutically harness the attributes of DC to enhance adaptive immunity against infectious diseases or cancer emerged in the 1990s. This idea was explored following three mechanistic lines: the injection of antigen-loaded DC, the delivery of patient-specific antigens, and via DC-targeting of selected antigens (Fig. 34.4).

Injection of Antigen-Loaded Autologous DC Cellular Vaccines

DC-based cellular vaccines involve autologous DC that are cultured or generated *ex vivo*. These DC are subsequently exposed to cancer cells, tumor-derived lysates, or established tumor-specific antigens. Following antigen loading and activation, these autologous DC are reinjected back into patients as "reprogrammed" DC. The rationale behind this approach lies both in the rarity of DC-infiltrating tumors[373,374] and in the impairment of DC function associated with chronic infections or cancer. Cancer-derived cytokines including VEGF, IL-6, TGF-β, and IL-10 impair DC activation and lead to tumor progression through several mechanisms. VEGF inhibits FLT3L and thereby blocks DC differentiation.[375,376] As described above, TGF-β and IL-10 have tolerogenic properties during DC-T cell interactions. Tumor cells also express on their membrane high levels of the "do not eat me" signal CD47 to circumvent their phagocytosis by macrophage and DC.[375] The tumor microenvironment comprises immune subsets, stromal and vascular components characterized by their immune suppressive nature, and attempts to override this suppression are exploited in tumor-based vaccines. On the one hand, attempts are based on the intratumoral recruitment of DC and on the other hand based on *in vitro* manipulation following delivery. Steinman and Inaba performed the first DC-vaccine in mice where reprogrammed DC *in vitro* were able to prime an immune response to the same antigen.[377] However, despite multiple trials, harnessing DC vaccines for cancer has so far been disappointing. Inadequate antigen presentation, migration, and cytokine release have resulted in only partial responses,[378] consistent with the complexity of overcoming the tumor tolerogenic milieu. However, patient monitoring of CD4+ and CD8+ T cell responses to tumor antigens showed sporadic clinical remission, suggesting that other mechanisms beyond DC priming were impairing the antitumoral response.[379] Therefore, the introduction of combination therapies, such as radio/chemotherapies or the adoptive transfer of tumor-specific T cells, may result in more effective treatment options.[378,380,381]

An essential aspect in the design of DC-based cancer vaccines is the source of DC.[382,383] Pioneering work from Sallusto and Lanzavecchia revealed that GM-CSF and IL-4 induced the differentiation of peripheral human monocytes into DC-like ("moDC") cells, capable of efficiently priming T cells.[116] While monocytes are a convenient source of antigen-presenting cells, clinical trials showed they are not the most efficient T cell activators.[379,381,383] De Vries et al introduced the use of magnetic beads to enrich donor-derived peripheral blood mononuclear cells in different circulating DC subsets for immunotherapy.[383] Their study suggested that patients receiving antigen-loaded CD1c+ cDC2 had higher tumor-specific CD8+ T cells, which correlated well with prolonged progression-free survival.[384] This study highlights the potential of harnessing DC-based vaccines in immunotherapies underlying the necessity to improve its design and use in combination therapies.[385-390]

Personalized Vaccination

In the process of cancer initiation and progression, tumors acquire multiple genetic aberrations, including mutations and translocations.[391,392] These mutations, referred as "neoantigens" or "neo-epitopes," are often the cause of oncogenic transformation and represent an ideal tumor-specific target for immunotherapies (Chapter 49).[393-395] High-throughput analysis of patient's mutated genome or "mutanome" is cost effective and achievable within a short time frame. Furthermore, bioinformatic methodologies of epitope prediction have been established, opening new avenues in the guided design of personalized immunotherapies.[394,395] Mutanome identification coupled to vaccination has been pioneered in mice model of cancer and has been recently extended to human trials.[396] What remains to be defined is the optimal strategy to deliver these antigens to DC to initiate the desired response.

Targeted DC Delivery

The presence of cDC1 within tumor tissues is associated with increased T cell infiltration and improved prognosis.[397] For this reason, exploring the potential of manipulating DC migration and local recruitment into the tumor tissue represents an important avenue of research. DC migration to lymph nodes is critical for the delivery of tumor-associated antigens to T cells and reliant on CCR7 expression. In an attempt to improve Ag delivery to the draining lymph node, several approaches have been evaluated including *in vitro* CCR7 DC transduction,[398] intranodal delivery of antigens or tumor lysate, and intratumoral DC injections.[399]

As mentioned previously, increased levels of VEGF alter DC homeostasis reducing DC development. Therefore, attempts to overcome this inhibition were done by expanding the pool of circulating DC and pre-DC *in vivo* using FLT3L administrations.[279,400-402]

Besides the expansion of DC and the possibility to increase the migration to draining lymph nodes, an essential aspect of DC-based vaccine is the effective induction of T cell response. Several approaches aimed at the immunogenic activity of DC and in particular cDC1, have been explored, for example, the use of CD40-specific antibodies or TLR3 agonist.[372,403,404]

The use of specific antibodies is also being explored for direct delivery of antigens, such that antibodies specific for surface receptors are covalently linked to tumor antigens, prompting an efficient internalization by DC. This approach heavily depends on the type of adjuvant used. cDC targeting was achieved using the C-type lectin receptor DEC205/CD205, which resulted in the presentation of antigen on both MHC class I and II molecules.[189,405-409] Further studies extended the potential of antibody-mediated antigen delivery to DC via DC-SIGN, DNGR1/CLEC9A, CD207, CD206, and CD40 and optimized adjuvants regimen for tailored T cell responses.[30,247,248,410-413] Protective immunity against intracellular pathogens such as *T. gondii* or *Leishmania major*[414,415] and cancer was obtained by DC-targeted vaccine delivery.[30,412] However, when DC-targeting is performed in the absence of adjuvants, antibody-mediated delivery can result in the induction of tolerogenic T cells and deletion of CD8+ T cells.[405,408] Therefore, selecting the receptor to be targeted has to be linked to the use of the appropriate adjuvant. While in one case the goal may be to generate CTL responses and NK cell activation by targeting cDC1-driven immunity, alternative strategies, targeting cDC2-mediated tolerance, could be exploited for the induction of T_{reg} in autoimmune diseases.[416]

Collectively, several combinations of tailored DC-based vaccines are being explored. The intrinsic T cell activation capacity of DC makes them theoretically the ideal target. However, an underlying theme throughout this chapter has been that DC contribute to the fine-tuning of immunity versus tolerance. Therefore, combining all these different aspects of DC biology will be key in unraveling their therapeutic potential.[375,413,417]

ACKNOWLEDGMENTS

All authors contributed equally to this work and can be cited in any order.

We would like to thank Prof. Martin Flajnik and Prof. Siamon Gordon for critically evaluating the manuscript, the useful discussions, and precious inputs. We apologize for any references omitted due to space or ignorance. SY is supported by an Israel Science Foundation Grant. RT is supported by the Intramural Research Program of the NIH, NIDCR. PG is a CNRS investigator.

REFERENCES

1. Paul WE. Bridging innate and adaptive immunity. *Cell.* 2011;147:1212-1215.
2. Janeway CA Jr. Approaching the asymptote? Evolution and revolution in immunology. *Cold Spring Harb Symp Quant Biol.* 1989;54(pt 1):1-13.
3. Billington WD. The immunological problem of pregnancy: 50 years with the hope of progress. A tribute to Peter Medawar. *J Reprod Immunol.* 2003;60:1-11.
4. McDevitt HO. Discovering the role of the major histocompatibility complex in the immune response. *Annu Rev Immunol.* 2000;18:1-17.
5. Billingham RE, Krohn PL, Medawar PB. Effect of locally applied cortisone acetate on survival of skin homografts in rabbits. *Br Med J.* 1951;2:1049-1053.
6. Zinkernagel RM, Doherty PC. Restriction of *in vitro* T cell-mediated cytotoxicity in lymphocytic choriomeningitis within a syngeneic or semiallogeneic system. *Nature.* 1974;248:701-702.
7. Bjorkman PJ, Saper MA, Samraoui B, Bennett WS, Strominger JL, Wiley DC. Structure of the human class I histocompatibility antigen, HLA-A2. *Nature.* 1987;329:506-512.
8. Babbitt BP, Allen PM, Matsueda G, Haber E, Unanue ER. Binding of immunogenic peptides to Ia histocompatibility molecules. *Nature.* 1985;317:359-361.
9. Bach F, Hirschhorn K. Lymphocyte interaction: a potential histocompatibility test in vitro. *Science.* 1964;143:813-814.
10. Bain B, Vas MR, Lowenstein L. The development of large immature mononuclear cells in mixed leukocyte cultures. *Blood.* 1964;23:108-116.
11. Langerhans P. Uber die nerven der menschlichen haut. *Archives of Pathological Anatomy.* 1868;44:325-337.
12. Veldman JE, Keuning FJ, Lennert K, vd Meer JB, Feltkamp-Vroom TM. Interdigitating cells. A guiding cell line in T-cell reactivity. *Adv Exp Med Biol.* 1982;149:407-414.
13. Hoefsmit EC, Duijvestijn AM, Kamperdijk EW. Relation between langerhans cells, veiled cells, and interdigitating cells. *Immunobiology.* 1982;161:255-265.
14. Kelly RH, Wolstencroft RA, Dumonde DC, Balfour BM. Role of lymphocyte activation products (LAP) in cell-mediated immunity. II. Effects of lymphocyte activation products on lymph node architecture and evidence for peripheral release of LAP following antigenic stimulation. *Clin Exp Immunol.* 1972;10:49-65.
15. McFarlin DE, Balfour B. Contact sensitivity in the pig. *Immunology.* 1973;25:995-1009.
16. Steinman RM, Cohn ZA. Identification of a novel cell type in peripheral lymphoid organs of mice. I. Morphology, quantitation, tissue distribution. *J Exp Med.* 1973;137:1142-1162.
17. Tew JG, Thorbecke GJ, Steinman RM. Dendritic cells in the immune response: characteristics and recommended nomenclature (A report from the Reticuloendothelial Society Committee on Nomenclature). *J Reticuloendothel Soc.* 1982;31:371-380.
18. Wekerle H, Ketelsen UP. Thymic nurse cells – Ia-bearing epithelium involved in T-lymphocyte differentiation? *Nature.* 1980;283:402-404.
19. Knight SC, Balfour BM, O'brien J, Buttifant L, Sumerska T, Clarke J. Role of veiled cells in lymphocyte activation. *Eur J Immunol.* 1982;12:1057-1060.
20. Steinman RM, Gutchinov B, Witmer MD, Nussenzweig MC. Dendritic cells are the principal stimulators of the primary mixed leukocyte reaction in mice. *J Exp Med.* 1983;157:613-627.
21. Inaba K, Swiggard WJ, Inaba M, et al. Tissue distribution of the DEC-205 protein that is detected by the monoclonal antibody NLDC-145. I. Expression on dendritic cells and other subsets of mouse leukocytes. *Cell Immunol.* 1995;163:148-156.
22. Metlay JP, Witmer-Pack MD, Agger R, Crowley MT, Lawless D, Steinman RM. The distinct leukocyte integrins of mouse spleen dendritic cells as identified with new hamster monoclonal antibodies. *J Exp Med.* 1990;171:1753-1771.
23. Vremec D, Zorbas M, Scollay R, et al. The surface phenotype of dendritic cells purified from mouse thymus and spleen: investigation of the CD8 expression by a subpopulation of dendritic cells. *J Exp Med.* 1992;176:47-58.
24. Edelson BT, Bradstreet TR, Hildner K, et al. CD8α(+) dendritic cells are an obligate cellular entry point for productive infection by *Listeria monocytogenes. Immunity.* 2011;35:236-248.
25. Ginhoux F, Liu K, Helft J, et al. The origin and development of nonlymphoid tissue CD103+ DCs. *J Exp Med.* 2009;206:3115-3130.
26. Crozat K, Guiton R, Contreras V, et al. The XC chemokine receptor 1 is a conserved selective marker of mammalian cells homologous to mouse CD8α+ dendritic cells. *J Exp Med.* 2010;207:1283-1292.
27. Guilliams M, Dutertre CA, Scott CL, et al. Unsupervised high-dimensional analysis aligns dendritic cells across tissues and species. *Immunity.* 2016;45:669-684.
28. Jongbloed SL, Kassianos AJ, McDonald KJ, et al. Human CD141+ (BDCA-3)+ dendritic cells (DCs) represent a unique myeloid DC subset that cross-presents necrotic cell antigens. *J Exp Med.* 2010;207:1247-1260.
29. MacDonald KP, Munster DJ, Clark GJ, Dzionek A, Schmitz J, Hart DNJ. Characterization of human blood dendritic cell subsets. *Blood.* 2002;100:4512-4520.
30. Sancho D, Mourão-Sá D, Joffre OP, et al. Tumor therapy in mice via antigen targeting to a novel, DC-restricted C-type lectin. *J Clin Invest.* 2008;118:2098-2110.
31. Sancho D, Joffre OP, Keller AM, et al. Identification of a dendritic cell receptor that couples sensing of necrosis to immunity. *Nature.* 2009;458:899-903.
32. Yarovinsky F. TLR11 activation of dendritic cells by a protozoan profilin-like protein. *Science.* 2005;308:1626-1629.
33. Suzuki S, Honma K, Matsuyama T, et al. Critical roles of interferon regulatory factor 4 in CD11b^{high}CD8α- dendritic cell development. *Proc Natl Acad Sci U S A.* 2004;101:8981-8986.
34. Schlitzer A, McGovern N, Teo P, et al. IRF4 transcription factor-dependent CD11b+ dendritic cells in human and mouse control mucosal IL-17 cytokine responses. *Immunity.* 2013;38:970-983.
35. Persson EK, Uronen-Hansson H, Semmrich M, et al. IRF4 transcription-factor-dependent CD103(+)CD11b(+) dendritic cells drive mucosal T helper 17 cell differentiation. *Immunity.* 2013;38:958-969.
36. Vander Lugt B, Khan AA, Hackney JA, et al. Transcriptional programming of dendritic cells for enhanced MHC class II antigen presentation. *Nat Immunol.* 2014;15:161-167.
37. Bajana S, Roach K, Turner S, Paul J, Kovats S. IRF4 promotes cutaneous dendritic cell migration to lymph nodes during homeostasis and inflammation. *J Immunol.* 2012;189:3368-3377.
38. Guilliams M, Ginhoux F, Jakubzick C, et al. Dendritic cells, monocytes and macrophages: a unified nomenclature based on ontogeny. *Nat Rev Immunol.* 2014;14:571-578.
39. Menezes S, Melandri D, Anselmi G, et al. The heterogeneity of Ly6Chi monocytes controls their differentiation into iNOS+ macrophages or monocyte-derived dendritic cells. *Immunity.* 2016;45:1205-1218.
40. Schraml BU, van Blijswijk J, Zelenay S, et al. Genetic tracing via DNGR-1 expression history defines dendritic cells as a hematopoietic lineage. *Cell.* 2013;154:843-858.
41. McKenna HJ, Stocking KL, Miller RE, et al. Mice lacking flt3 ligand have deficient hematopoiesis affecting hematopoietic progenitor cells, dendritic cells, and natural killer cells. *Blood.* 2000;95:3489-3497.
42. Serbina NV, Pamer EG. Monocyte emigration from bone marrow during bacterial infection requires signals mediated by chemokine receptor CCR2. *Nat Immunol.* 2006;7:311-317.
43. Onai N, Obata-Onai A, Schmid MA, Manz MG. Flt3 in regulation of type I interferon-producing cell and dendritic cell development. *Ann N Y Acad Sci.* 2007;1106:253-261.
44. Naik SH, Sathe P, Park HY, et al. Development of plasmacytoid and conventional dendritic cell subtypes from single precursor cells derived *in vitro* and *in vivo. Nat Immunol.* 2007;8:1217-1226.
45. Lee J, Breton G, Oliveira TY, et al. Restricted dendritic cell and monocyte progenitors in human cord blood and bone marrow. *J Exp Med.* 2015;212(3):385-399.
46. Schlitzer A, Sivakamasundari V, Chen J, et al. Identification of cDC1- and cDC2-committed DC progenitors reveals early lineage priming at the common DC progenitor stage in the bone marrow. *Nat Immunol.* 2015;16:718-728.
47. Liu K, Victora GD, Schwickert TA, et al. *In vivo* analysis of dendritic cell development and homeostasis. *Science.* 2009;324:392-397.
48. Breton G, Zheng S, Valieris R, et al. Human dendritic cells (DCs) are derived from distinct circulating precursors that are precommitted to become CD1c+ or CD141+ DCs. *J Exp Med.* 2016;213:2861-2870.
49. Breton G, Lee J, Zhou YJ, et al. Circulating precursors of human CD1c(+) and CD141(+) dendritic cells. *J Exp Med.* 2015;212:401-413.
50. See P, Dutertre CA; Chen J, et al. Mapping the human DC lineage through the integration of high-dimensional techniques. *Science.* 2017;356.
51. Satpathy AT, Briseño CG, Lee JS, et al. Notch2-dependent classical dendritic cells orchestrate intestinal immunity to attaching-and-effacing bacterial pathogens. *Nat Immunol.* 2013;14:937-948.
52. Caton ML, Smith-Raska MR, Reizis B. Notch-RBP-J signaling controls the homeostasis of CD8- dendritic cells in the spleen. *J Exp Med.* 2007;204:1653-1664.
53. Lewis KL, Caton ML, Bogunovic M, et al. Notch2 receptor signaling controls functional differentiation of dendritic cells in the spleen and intestine. *Immunity.* 2011;35:780-791.
54. Bogunovic M, Ginhoux F, Helft J, et al. Origin of the lamina propria dendritic cell network. *Immunity.* 2009;31:513-525.
55. Watchmaker PB, Lahl K, Lee M, et al. Comparative transcriptional and functional profiling defines conserved programs of intestinal DC differentiation in humans and mice. *Nat Immunol.* 2014;15:98-108.
56. Gao Y, Nish SA, Jiang R, et al. Control of T helper 2 responses by transcription factor IRF4-dependent dendritic cells. *Immunity.* 2013;39:722-732.
57. Tussiwand R, Everts B, Grajales-Reyes GE, et al. Klf4 expression in conventional dendritic cells is required for T helper 2 cell responses. *Immunity.* 2015;42:916-928.
58. Connor LM, Tang SC, Cognard E, et al. Th2 responses are primed by skin dendritic cells with distinct transcriptional profiles. *J Exp Med.* 2017;214:125-142.
59. Brown CC, Gudjonson H, Pritykin Y, et al. Transcriptional basis of mouse and human dendritic cell heterogeneity. *Cell.* 2019;179:846.e824-863.e824.
60. Bourdely P, Anselmi G, Vaivode K, et al. Transcriptional and functional analysis of CD1c(+) human dendritic cells identifies a CD163(+) subset priming CD8(+)CD103(+) T cells. *Immunity.* 2020;53:335.e338-352.e338.
61. Dutertre CA, Becht E, Irac SE, et al. Single-cell analysis of human mononuclear phagocytes reveals subset-defining markers and identifies circulating inflammatory dendritic cells. *Immunity.* 2019;51:573.e578-589.e578.
62. Villani AC, Satija R, Reynolds G, et al. Single-cell RNA-seq reveals new types of human blood dendritic cells, monocytes, and progenitors. *Science.* 2017;356.
63. Segura E, Touzot M, Bohineust A, et al. Human inflammatory dendritic cells induce Th17 cell differentiation. *Immunity.* 2013;38:336-348.
64. Bakdash G, Buschow SI, Gorris MAJ, et al. Expansion of a BDCA1+CD14+ myeloid cell population in melanoma patients may attenuate the efficacy of dendritic cell vaccines. *Cancer Res.* 2016;76:4332-4346.
65. Lavin Y, Kobayashi S, Leader A, et al. Innate immune landscape in early lung adenocarcinoma by paired single-cell analyses. *Cell.* 2017;169:750.e717-765.e717.
66. Segura E, Valladeau-Guilemond J, Donnadieu MH, Sastre-Garau X, Soumelis V, Amigorena S. Characterization of resident and migratory dendritic cells in human lymph nodes. *J Exp Med.* 2012;209:653-660.
67. Granot T, Senda T, Carpenter DJ, et al. Dendritic cells display subset and tissue-specific maturation dynamics over human life. *Immunity.* 2017;46:504-515.
68. Wollenberg A, Kraft S, Hanau D, Bieber T. Immunomorphological and ultrastructural characterization of Langerhans cells and a novel, inflammatory dendritic epidermal cell (IDEC) population in lesional skin of atopic eczema. *J Invest Dermatol.* 1996;106:446-453.
69. Zaba LC, Fuentes-Duculan J, Eungdamrong NJ, et al. Psoriasis is characterized by accumulation of immunostimulatory and Th1/Th17 cell-polarizing myeloid dendritic cells. *J Invest Dermatol.* 2009;129:79-88.
70. Binnewies M, Mujal AM, Pollack JL, et al. Unleashing type-2 dendritic cells to drive protective antitumor CD4(+) T cell immunity. *Cell.* 2019;177:556-571 e516.
71. Cheong C, Matos I, Choi JH, et al. Microbial stimulation fully differentiates monocytes to DC-SIGN/CD209(+) dendritic cells for immune T cell areas. *Cell.* 2010;143:416-429.
72. Langlet C, Tamoutounour S, Henri S, et al. CD64 expression distinguishes monocyte-derived and conventional dendritic cells and reveals their distinct role during intramuscular immunization. *J Immunol.* 2012;188:1751-1760.
73. Plantinga M, Guilliams M, Vanheerswynghels M, et al. Conventional and monocyte-derived CD11b(+) dendritic cells initiate and maintain T helper 2 cell-mediated immunity to house dust mite allergen. *Immunity.* 2013;38:322-335.
74. Zigmond E, Varol C, Farache J, et al. Ly6C hi monocytes in the inflamed colon give rise to proinflammatory effector cells and migratory antigen-presenting cells. *Immunity.* 2012;37:1076-1090.
75. Bain CC, Scott CL, Uronen-Hansson H, et al. Resident and pro-inflammatory macrophages in the colon represent alternative context-dependent fates of the same Ly6Chi monocyte precursors. *Mucosal Immunol.* 2013;6:498-510.
76. Sander J, Schmidt SV, Cirovic B, et al. Cellular differentiation of human monocytes is regulated by time-dependent interleukin-4 signaling and the transcriptional regulator NCOR2. *Immunity.* 2017;47:1051.e1012-1066.e1012.
77. Goudot C, Coillard A, Villani AC, et al. Aryl hydrocarbon receptor controls monocyte differentiation into dendritic cells versus macrophages. *Immunity.* 2017;47:582.e586-596.e586.
78. Bosteels C, Neyt K, Vanheerswynghels M, et al. Inflammatory type 2 cDCs acquire features of cDC1s and macrophages to orchestrate immunity to respiratory virus infection. *Immunity.* 2020;52:1039.e1039-1056.e1039.
79. Lennert K, Remmele W. Karyometric research on lymph node cells in man. I. Germinoblasts, lymphoblasts & lymphocytes. Article in German. *Acta Haematol.* 1958;19, 99-113.
80. Muller-Hermelink HK, Kaiserling E, Lennert K. Pseudofollicular nests of plasmacells (of a special type?) in paracortical pulp of human lymph nodes (author' s transl). Article in German. *Virchows Arch B Cell Pathol.* 1973;14:47-56.
81. Papadimitriou CS, Stephanaki-Nikou SN, Malamou-Mitsi VD. Comparative immunostaining of T-associated plasma cells and other lymph-node cells in paraffin sections. *Virchows Arch B Cell Pathol Incl Mol Pathol.* 1983;43:31-36.
82. Facchetti F, de Wolf-Peeters C, Mason DY, Pulford K, van den Oord JJ, Desmet VJ. Plasmacytoid T cells. Immunohistochemical evidence for their monocyte/macrophage origin. *Am J Pathol.* 1988;133:15-21.
83. Ronnblom L, Ramstedt U, Alm GV. Properties of human natural interferon-producing cells stimulated by tumor cell lines. *Eur J Immunol.* 1983;13:471-476.

84. Chehimi J, Starr SE, Kawashima H, et al. Dendritic cells and IFN-alpha-producing cells are two functionally distinct non-B, non-monocytic HLA-DR+ cell subsets in human peripheral blood. *Immunology*. 1989;68:486-490.
85. Feldman M, Fitzgerald-Bocarsly P. Sequential enrichment and immunocytochemical visualization of human interferon-alpha-producing cells. *J Interferon Res*. 1990;10:435-446.
86. Sandberg K, Eloranta ML, Johannisson A, Alm GV. Flow cytometric analysis of natural interferon-alpha producing cells. *Scand J Immunol*. 1991;34:565-576.
87. Fitzgerald-Bocarsly P. Human natural interferon-alpha producing cells. *Pharmacol Ther*. 1993;60:39-62.
88. Perussia B, Fanning V, Trinchieri G. A leukocyte subset bearing HLA-DR antigens is responsible for *in vitro* alpha interferon production in response to viruses. *Nat Immun Cell Growth Regul*. 1985;4:120-137.
89. Cella M, Jarrossay D, Facchetti F, et al. Plasmacytoid monocytes migrate to inflamed lymph nodes and produce large amounts of type I interferon. *Nat Med*. 1999;5:919-923.
90. Siegal FP. The nature of the principal type 1 interferon-producing cells in human blood. *Science*. 1999;284:1835-1837.
91. Grouard G, Rissoan MC, Filgueira L, Durand I, Banchereau J, Liu YJ. The enigmatic plasmacytoid T cells develop into dendritic cells with interleukin (IL)-3 and CD40-ligand. *J Exp Med*. 1997;185:1101-1111.
92. Cella M, Facchetti F, Lanzavecchia A, Colonna M. Plasmacytoid dendritic cells activated by influenza virus and CD40L drive a potent TH1 polarization. *Nat Immunol*. 2000;1:305-310.
93. Asselin-Paturel C, Boonstra A, Dalod M, et al. Mouse type I IFN-producing cells are immature APCs with plasmacytoid morphology. *Nat Immunol*. 2001;2:1144-1150.
94. Nakano H, Yanagita M, Gunn MD. CD11c(+)B220(+) Gr-1(+) cells in mouse lymph nodes and spleen display characteristics of plasmacytoid dendritic cells. *J Exp Med*. 2001;194:1171-1178.
95. Asselin-Paturel C, Brizard G, Pin JJ, Briere F, Trinchieri G. Mouse strain differences in plasmacytoid dendritic cell frequency and function revealed by a novel monoclonal antibody. *J Immunol*. 2003;171:6466-6477.
96. Blasius AL, Cella M, Maldonado J, Takai T, Colonna M. Siglec-H is an IPC-specific receptor that modulates type I IFN secretion through DAP12. *Blood*. 2006;107:2474-2476.
97. Blasius AL, Giurisato E, Cella M, Schreiber RD, Shaw AS, Colonna M. Bone marrow stromal cell antigen 2 is a specific marker of type I IFN-producing cells in the naive mouse, but a promiscuous cell surface antigen following IFN stimulation. *J Immunol*. 2006;177:3260-3265.
98. Dzionek A, Fuchs A, Schmidt P, et al. BDCA-2, BDCA-3, and BDCA-4: three markers for distinct subsets of dendritic cells in human peripheral blood. *J Immunol*. 2000;165:6037-6046.
99. Rissoan MC, Duhen T, Bridon JM, et al. Subtractive hybridization reveals the expression of immunoglobulin-like transcript 7, Eph-B1, granzyme B, and 3 novel transcripts in human plasmacytoid dendritic cells. *Blood*. 2002;100:3295-3303.
100. Ju XS, Hacker C, Scherer B, et al. Immunoglobulin-like transcripts ILT2, ILT3 and ILT7 are expressed by human dendritic cells and down-regulated following activation. *Gene*. 2004;331:159-164.
101. Cao W, Rosen DB, Ito T, et al. Plasmacytoid dendritic cell-specific receptor ILT7-Fc epsilonRI γ inhibits Toll-like receptor-induced interferon production. *J Exp Med*. 2006;203:1399-1405.
102. Facchetti F, Vermi W, Mason D, Colonna M. The plasmacytoid monocyte/interferon producing cells. *Virchows Arch*. 2003;443:703-717.
103. Reizis B, Bunin A, Ghosh HS, Lewis KL, Sisirak V. Plasmacytoid dendritic cells: recent progress and open questions. *Annu Rev Immunol*. 2011;29:163-183.
104. Dress RJ, Dutertre CA, Giladi A, et al. Plasmacytoid dendritic cells develop from Ly6D(+) lymphoid progenitors distinct from the myeloid lineage. *Nat Immunol*. 2019;20:852-864.
105. Rodrigues PF, Alberti-Servera L, Eremin A, Grajales-Reyes GE, Ivanek R, Tussiwand R. Distinct progenitor lineages contribute to the heterogeneity of plasmacytoid dendritic cells. *Nat Immunol*. 2018;19:711-722.
106. Rodrigues PF, Tussiwand R. Novel concepts in plasmacytoid dendritic cell (pDC) development and differentiation. *Mol Immunol*. 2020;126:25-30.
107. Merad M, Manz MG. Dendritic cell homeostasis. *Blood*. 2009;113:3418-3427.
108. Patel AA, Yona S. *Phagocyte Development: eLS*. John Wiley & Sons, Ltd; 2018:1-13.
109. Varol C, Mildner A, Jung S. Macrophages: development and tissue specialization. *Annu Rev Immunol*. 2015;33:643-675.
110. Geissmann F, Manz MG, Jung S, Sieweke MH, Merad M, Ley K. Development of monocytes, macrophages, and dendritic cells. *Science*. 2010;327:656-661.
111. Kamath AT, Henri S, Battye F, Tough DF, Shortman K. Developmental kinetics and lifespan of dendritic cells in mouse lymphoid organs. *Blood*. 2002;100:1734-1741.
112. Waskow C, Liu K, Darrasse-Jèze G, et al. The receptor tyrosine kinase Flt3 is required for dendritic cell development in peripheral lymphoid tissues. *Nat Immunol*. 2008;9:676-683.
113. Durai V, Bagadia P, Briseño CG, et al. Altered compensatory cytokine signaling underlies the discrepancy between Flt3(−/−) and Flt3l(−/−) mice. *J Exp Med*. 2018;215:1417-1435.
114. Kingston D, Schmid MA, Onai N, Obata-Onai A, Baumjohann D, Manz MG. The concerted action of GM-CSF and Flt3-ligand on *in vivo* dendritic cell homeostasis. *Blood*. 2009;114:835-843.
115. Chicha L, Jarrossay D, Manz MG. Clonal type I interferon-producing and dendritic cell precursors are contained in both human lymphoid and myeloid progenitor populations. *J Exp Med*. 2004;200:1519-1524.
116. Sallusto F, Lanzavecchia A. Efficient presentation of soluble antigen by cultured human dendritic cells is maintained by granulocyte/macrophage colony-stimulating factor plus interleukin 4 and downregulated by tumor necrosis factor alpha. *J Exp Med*. 1994;179:1109-1118.
117. Edelson BT, Bradstreet TR, Kc W, et al. Batf3-dependent CD11b(low/−) peripheral dendritic cells are GM-CSF-independent and are not required for Th cell priming after subcutaneous immunization. *PLoS One*. 2011;6:e25660.
118. Greter M, Helft J, Chow A, et al. GM-CSF controls nonlymphoid tissue dendritic cell homeostasis but is dispensable for the differentiation of inflammatory dendritic cells. *Immunity*. 2012;36:1031-1046.
119. Patel AA, Zhang Y, Fullerton JN, et al. The fate and lifespan of human monocyte subsets in steady state and systemic inflammation. *J Exp Med*. 2017;214:1913-1923.
120. Yona S, Kim KW, Wolf Y, et al. Fate mapping reveals origins and dynamics of monocytes and tissue macrophages under homeostasis. *Immunity*. 2013;38:79-91.
121. Grajales-Reyes GE, Iwata A, Albring J, et al. Batf3 maintains autoactivation of Irf8 for commitment of a CD8 α(+) conventional DC clonogenic progenitor. *Nat Immunol*. 2015;16:708-717.
122. Ippolito GC, Dekker JD, Wang YH, et al. Dendritic cell fate is determined by BCL11A. *Proc Natl Acad Sci U S A*. 2014;111:E998-E1006.
123. Wu XL, Yu Zh, Qiu J, Yang Yh, Shen Xl, Su P. Postcoital administration of asoprisnil inhibited embryo implantation and disturbed ultrastructure of endometrium in implantation window in mice. *J Huazhong Univ Sci Technolog Med Sci*. 2013;33:277-283.
124. Laouar Y, Welte T, Fu XY, Flavell RA. STAT3 is required for Flt3L-dependent dendritic cell differentiation. *Immunity*. 2003;19:903-912.
125. Li HS, Yang CY, Nallaparaju KC, et al. The signal transducers STAT5 and STAT3 control expression of Id2 and E2-2 during dendritic cell development. *Blood*. 2012;120:4363-4373.
126. Bell BD, Kitajima M, Larson RP, et al. The transcription factor STAT5 is critical in dendritic cells for the development of TH2 but not TH1 responses. *Nat Immunol*. 2013;14:364-371.
127. Aliberti J, Schulz O, Pennington DJ, et al. Essential role for ICSBP in the *in vivo* development of murine CD8α+ dendritic cells. *Blood*. 2003;101:305-310.
128. Schiavoni G, Mattei F, Sestili P, et al. ICSBP is essential for the development of mouse type I interferon-producing cells and for the generation and activation of CD8α(+) dendritic cells. *J Exp Med*. 2002;196:1415-1425.
129. Sichien D, Scott CL, Martens L, et al. IRF8 transcription factor controls survival and function of terminally differentiated conventional and plasmacytoid dendritic cells, respectively. *Immunity*. 2016;45:626-640.
130. Kurotaki D, Osato N, Nishiyama A, et al. Essential role of the IRF8-KLF4 transcription factor cascade in murine monocyte differentiation. *Blood*. 2013;121:1839-1849.
131. Wang H, Lee CH, Qi C, et al. IRF8 regulates B-cell lineage specification, commitment, and differentiation. *Blood*. 2008;112:4028-4038.
132. Sasaki H, Kurotaki D, Osato N, et al. Transcription factor IRF8 plays a critical role in the development of murine basophils and mast cells. *Blood*. 2015;125:358-369.
133. Milanovic M, Terszowski G, Struck D, Liesenfeld O, Carstanjen D. IFN consensus sequence binding protein (Icsbp) is critical for eosinophil development. *J Immunol*. 2008;181:5045-5053.
134. Tailor P, Tamura T, Morse HC III, Ozato K. The BXH2 mutation in IRF8 differentially impairs dendritic cell subset development in the mouse. *Blood*. 2008;111:1942-1945.
135. Hambleton S, Salem S, Bustamante J, et al. IRF8 mutations and human dendritic-cell immunodeficiency. *N Engl J Med*. 2011;365:127-138.
136. Cytlak U, Resteu A, Pagan S, et al. Differential IRF8 transcription factor requirement defines two pathways of dendritic cell development in humans. *Immunity*. 2020;53:353.e358-370.e358.
137. Kashiwada M, Pham NL, Pewe LL, Harty JT, Rothman PB. NFIL3/E4BP4 is a key transcription factor for CD8α(+) dendritic cell development. *Blood*. 2011;117:6193-6197.
138. Hacker C, Kirsch RD, Ju XS, et al. Transcriptional profiling identifies Id2 function in dendritic cell development. *Nat Immunol*. 2003;4:380-386.
139. Hildner K, Edelson BT, Purtha WE, et al. Batf3 deficiency reveals a critical role for CD8α+ dendritic cells in cytotoxic T cell immunity. *Science*. 2008;322:1097-1100.
140. Serafini M, Sacchetti B, Pievani A, et al. Establishment of bone marrow and hematopoietic niches *in vivo* by reversion of chondrocyte differentiation of human bone marrow stromal cells. *Stem Cell Res*. 2014;12:659-672.
141. Yokota Y, Mansouri A, Mori S, et al. Development of peripheral lymphoid organs and natural killer cells depends on the helix-loop-helix inhibitor Id2. *Nature*. 1999;397:702-706.
142. Seillet C, Huntington ND, Gangatirkar P, et al. Differential requirement for Nfil3 during NK cell development. *J Immunol*. 2014;192:2667-2676.
143. Geiger TL, Abt MC, Gasteiger G, et al. Nfil3 is crucial for development of innate lymphoid cells and host protection against intestinal pathogens. *J Exp Med*. 2014;211:1723-1731.
144. Gascoyne DM, Long E, Veiga-Fernandes H, et al. The basic leucine zipper transcription factor E4BP4 is essential for natural killer cell development. *Nat Immunol*. 2009;10:1118-1124.
145. Ataide MA, Komander K, Knöpper K, et al. BATF3 programs CD8(+) T cell memory. *Nat Immunol*. 2020;21:1397-1407.
146. Durai V, Bagadia P, Granja JM, et al. Cryptic activation of an Irf8 enhancer governs cDC1 fate specification. *Nat Immunol*. 2019;20:1161-1173.
147. Tussiwand R, Lee WL, Murphy TL, et al. Compensatory dendritic cell development mediated by BATF-IRF interactions. *Nature*. 2012;490:502-507.
148. Tussiwand R, Rauch M, Fluck LA, Rolink AG. BAFF-R expression correlates with positive selection of immature B cells. *Eur J Immunol*. 2012;42:206-216.
149. Glasmacher E, Agrawal S, Chang AB, et al. A genomic regulatory element that directs assembly and function of immune-specific AP-1-IRF complexes. *Science*. 2012;338:975-980.
150. Schonheit J, Kuhl C, Gebhardt ML, et al. PU.1 level-directed chromatin structure remodeling at the Irf8 gene drives dendritic cell commitment. *Cell Rep*. 2013;3:1617-1628.
151. Seillet C, Jackson JT, Markey KA, et al. CD8α+ DCs can be induced in the absence of transcription factors Id2, Nfil3, and Batf3. *Blood*. 2013;121:1574-1583.
152. Bagadia P, Huang X, Liu TT, et al. An Nfil3-Zeb2-Id2 pathway imposes Irf8 enhancer switching during cDC1 development. *Nat Immunol*. 2019;20:1174-1185.
153. Cisse B, Caton ML, Lehner M, et al. Transcription factor E2-2 is an essential and specific regulator of plasmacytoid dendritic cell development. *Cell*. 2008;135:37-48.
154. Anderson DA III, Dutertre CA, Ginhoux F, Murphy KM. Genetic models of human and mouse dendritic cell development and function. *Nat Rev Immunol*. 2021;21(2):101-115.
155. Murphy TL, Grajales-Reyes GE, Wu X, et al. Transcriptional control of dendritic cell development. *Annu Rev Immunol*. 2016;34:93-119.
156. Belz GT, Nutt SL. Transcriptional programming of the dendritic cell network. *Nat Rev Immunol*. 2012;12:101-113.
157. Kc W, Satpathy AT, Rapaport AS, et al. L-Myc expression by dendritic cells is required for optimal T-cell priming. *Nature*. 2014;507:243-247.
158. Osorio F, Tavernier SJ, Hoffmann E, et al. The unfolded-protein-response sensor IRE-1α regulates the function of CD8α+ dendritic cells. *Nat Immunol*. 2014;15:248-257.
159. Vander Lugt B, Riddell J, Khan AA, et al. Transcriptional determinants of tolerogenic and immunogenic states during dendritic cell maturation. *J Cell Biol*. 2017;216:779-792.
160. Tamura T, Tailor P, Yamaoka K, et al. IFN regulatory factor-4 and -8 govern dendritic cell subset development and their functional diversity. *J Immunol*. 2005;174:2573-2581.
161. Bajana S, Turner S, Paul J, Ainsua-Enrich E, Kovats S. IRF4 and IRF8 act in CD11c+ cells to regulate terminal differentiation of lung tissue dendritic cells. *J Immunol*. 2016;196:1666-1677.
162. Krishnaswamy JK, Gowthaman U, Zhang B, et al. Migratory CD11b(+) conventional dendritic cells induce T follicular helper cell-dependent antibody responses. *Sci Immunol*. 2017;2:eaam9169.
163. Kumamoto Y, Linehan M, Weinstein JS, Laidlaw BJ, Craft JE, Iwasaki A. CD301b(+) dermal dendritic cells drive T helper 2 cell-mediated immunity. *Immunity*. 2013;39:733-743.
164. Pitt D, Hopkins I. A syndrome of mental retardation, wide mouth and intermittent overbreathing. *Aust Paediatr J*. 1978;14:182-184.
165. Nagasawa M, Schmidlin H, Hazekamp MG, Schotte R, Blom B. Development of human plasmacytoid dendritic cells depends on the combined action of the basic helix-loop-helix factor E2-2 and the Ets factor Spi-B. *Eur J Immunol*. 2008;38:2389-2400.
166. Ghosh HS, Cisse B, Bunin A, Lewis KL, Reizis B. Continuous expression of the transcription factor e2-2 maintains the cell fate of mature plasmacytoid dendritic cells. *Immunity*. 2010;33:905-916.
167. Scott CL, Soen B, Martens L, et al. The transcription factor Zeb2 regulates development of conventional and plasmacytoid DCs by repressing Id2. *J Exp Med*. 2016;213:897-911.
168. Wu X, Briseño CG, Grajales-Reyes GE, et al. Transcription factor Zeb2 regulates commitment to plasmacytoid dendritic cell and monocyte fate. *Proc Natl Acad Sci U S A*. 2016;113:14775-14780.
169. Patel AA, Yona S. *Myeloid cell development*. In: *Encyclopedia of Life Sciences*. Wiley Online library; 2018.

170. Jaitin DA, Kenigsberg E, Keren-Shaul H, et al. Massively parallel single-cell RNA-seq for marker-free decomposition of tissues into cell types. *Science*. 2014;343:776-779.
171. Leylek R, Alcántara-Hernández M, Lanzar Z, et al. Integrated cross-species analysis identifies a conserved transitional dendritic cell population. *Cell Rep*. 2019;29: 3736-3750 e3738.
172. Liu K, Waskow C, Liu X, Yao K, Hoh J, Nussenzweig M. Origin of dendritic cells in peripheral lymphoid organs of mice. *Nat Immunol*. 2007;8:578-583.
173. Cook SJ, Lee Q, Wong AC, et al. Differential chemokine receptor expression and usage by pre-cDC1 and pre-cDC2. *Immunol Cell Biol*. 2018;96:1131-1139.
174. Kabashima K, Banks TA, Ansel KM, Lu TT, Ware CF, Cyster JG. Intrinsic lymphotoxin-beta receptor requirement for homeostasis of lymphoid tissue dendritic cells. *Immunity*. 2005;22:439-450.
175. Cabeza-Cabrerizo M, van Blijswijk J, Wienert S, et al. Tissue clonality of dendritic cell subsets and emergency DCpoiesis revealed by multicolor fate mapping of DC progenitors. *Sci Immunol*. 2019;4:eaaw1941.
176. Bachem A, Güttler S, Hartung E, et al. Superior antigen cross-presentation and XCR1 expression define human CD11c+CD141+ cells as homologues of mouse CD8+ dendritic cells. *J Exp Med*. 2010;207:1273-1281.
177. Balan S, Arnold-Schrauf C, Abbas A, et al. Large-scale human dendritic cell differentiation revealing notch-dependent lineage bifurcation and heterogeneity. *Cell Rep*. 2018;24:1902. e1906-1915.e1906.
178. Balan S, Ollion V, Colletti N, et al. Human XCR1+ dendritic cells derived *in vitro* from CD34+ progenitors closely resemble blood dendritic cells, including their adjuvant responsiveness, contrary to monocyte-derived dendritic cells. *J Immunol*. 2014;193:1622-1635.
179. Becker M, Güttler S, Bachem A, et al. Ontogenic, phenotypic, and functional characterization of XCR1(+) dendritic cells leads to a consistent classification of intestinal dendritic cells based on the expression of XCR1 and SIRPα. *Front Immunol*. 2014;5:326.
180. Poulin LF, Salio M, Griessinger E, et al. Characterization of human DNGR-1+ BDCA3+ leukocytes as putative equivalents of mouse CD8α+ dendritic cells. *J Exp Med*. 2010;207:1261-1271.
181. Dorner BG, Dorner MB, Zhou X, et al. Selective expression of the chemokine receptor XCR1 on cross-presenting dendritic cells determines cooperation with CD8+ T cells. *Immunity*. 2009;31:823-833.
182. Brewitz A, Eickhoff S, Dähling S, et al. CD8(+) T cells orchestrate pDC-XCR1(+) dendritic cell spatial and functional cooperativity to optimize priming. *Immunity*. 2017;46:205-219.
183. Bottcher JP, Bonavita E, Chakravarty P, et al. NK cells stimulate recruitment of cDC1 into the tumor microenvironment promoting cancer immune control. *Cell*. 2018;172:1022. e1014-1037.e1014.
184. Trinchieri G, Sher A. Cooperation of Toll-like receptor signals in innate immune defence. *Nat Rev Immunol*. 2007;7:179-190.
185. Merad M, Sathe P, Helft J, Miller J, Mortha A. The dendritic cell lineage: ontogeny and function of dendritic cells and their subsets in the steady state and the inflamed setting. *Annu Rev Immunol*. 2013;31:563-604.
186. Ahrens S, Zelenay S, Sancho D, et al. F-actin is an evolutionarily conserved damage-associated molecular pattern recognized by DNGR-1, a receptor for dead cells. *Immunity*. 2012;36:635-645.
187. Edwards AD, Diebold SS, Slack EMC, et al. Toll-like receptor expression in murine DC subsets: lack of TLR7 expression by CD8α+ DC correlates with unresponsiveness to imidazoquinolines. *Eur J Immunol*. 2003;33:827-833.
188. Brown GD, Gordon S. Immune recognition. A new receptor for beta-glucans. *Nature*. 2001;413:36-37.
189. Dudziak D, Kamphorst AO, Heidkamp GF, et al. Differential antigen processing by dendritic cell subsets *in vivo*. *Science*. 2007;315:107-111.
190. Luber CA, Cox J, Lauterbach H, et al. Quantitative proteomics reveals subset-specific viral recognition in dendritic cells. *Immunity*. 2010;32:279-289.
191. Megjugorac NJ, Young HA, Amrute SB, Olshalsky SL, Fitzgerald-Bocarsly P. Virally stimulated plasmacytoid dendritic cells produce chemokines and induce migration of T and NK cells. *J Leukoc Biol*. 2004;75:504-514.
192. Schmid MA, Takizawa H, Baumjohann DR, Saito Y, Manz MG. Bone marrow dendritic cell progenitors sense pathogens via Toll-like receptors and subsequently migrate to inflamed lymph nodes. *Blood*. 2011;118:4829-4840.
193. Wei Q, Frenette PS. Niches for hematopoietic stem cells and their progeny. *Immunity*. 2018;48:632-648.
194. Murata S, Takahama Y, Kasahara M, Tanaka K. The immunoproteasome and thymoproteasome: functions, evolution and human disease. *Nat Immunol*. 2018;19:923-931.
195. York IA, Rock KL. Antigen processing and presentation by the class I major histocompatibility complex. *Annu Rev Immunol*. 1996;14:369-396.
196. Conner SD, Schmid SL. Regulated portals of entry into the cell. *Nature*. 2003;422:37-44.
197. Sallusto F, Cella M, Danieli C, Lanzavecchia A. Dendritic cells use macropinocytosis and the mannose receptor to concentrate macromolecules in the major histocompatibility complex class II compartment: downregulation by cytokines and bacterial products. *J Exp Med*. 1995;182:389-400.
198. Savina A, Amigorena S. Phagocytosis and antigen presentation in dendritic cells. *Immunol Rev*. 2007;219:143-156.
199. Garrett WS, Chen LM, Kroschewski R, et al. Developmental control of endocytosis in dendritic cells by Cdc42. *Cell*. 2000;102:325-334.
200. Moreau HD, Blanch-Mercader C, Attia R, et al. Macropinocytosis overcomes directional bias in dendritic cells due to hydraulic resistance and facilitates space exploration. *Dev Cell*. 2019;49:171-188 e175.
201. Bretou M, Sáez PJ, Sanséau D, et al. Lysosome signaling controls the migration of dendritic cells. *Sci Immunol*. 2017;2.
202. Yona S, Gordon S. From the reticuloendothelial to mononuclear phagocyte system - the unaccounted years. *Front Immunol*. 2015;6:328.
203. den Haan JM, Lehar SM, Bevan MJ. CD8(+) but not CD8(−) dendritic cells cross-prime cytotoxic T cells *in vivo*. *J Exp Med*. 2000;192:1685-1696.
204. Iyoda T, Shimoyama S, Liu K, et al. The CD8+ dendritic cell subset selectively endocytoses dying cells in culture and *in vivo*. *J Exp Med*. 2002;195:1289-1302.
205. Desch AN, Randolph GJ, Murphy K, et al. CD103+ pulmonary dendritic cells preferentially acquire and present apoptotic cell-associated antigen. *J Exp Med*. 2011;208:1789-1797.
206. Sancho D, Reis e Sousa C. Signaling by myeloid C-type lectin receptors in immunity and homeostasis. *Annu Rev Immunol*. 2012;30:491-529.
207. Shrimpton RE, Butler M, Morel AS, Eren E, Hue SS, Ritter MA. CD205 (DEC-205): a recognition receptor for apoptotic and necrotic self. *Mol Immunol*. 2009;46:1229-1239.
208. Huysamen C, Willment JA, Dennehy KM, Brown GD. CLEC9A is a novel activation C-type lectin-like receptor expressed on BDCA3+ dendritic cells and a subset of monocytes. *J Biol Chem*. 2008;283:16693-16701.
209. Poulin LF, Reyal Y, Uronen-Hansson H, et al. DNGR-1 is a specific and universal marker of mouse and human Batf3-dependent dendritic cells in lymphoid and nonlymphoid tissues. *Blood*. 2012;119:6052-6062.
210. Zhang JG, Czabotar PE, Policheni AN, et al. The dendritic cell receptor Clec9A binds damaged cells via exposed actin filaments. *Immunity*. 2012;36:646-657.
211. Hanc P. Structure of the complex of F-actin and DNGR-1, a C-type lectin receptor involved in dendritic cell cross-presentation of dead cell-associated antigens. *Immunity*. 2015;42:839-849.
212. den Haan JM, Bevan MJ. Constitutive versus activation-dependent cross-presentation of immune complexes by CD8(+) and CD8(−) dendritic cells *in vivo*. *J Exp Med*. 2002;196:817-827.
213. Lee SJ. Mannose receptor-mediated regulation of serum glycoprotein homeostasis. *Science*. 2002;295:1898-1901.
214. Theisen DJ, Ferris ST, Briseño CG, et al. Batf3-Dependent genes control tumor rejection induced by dendritic cells independently of cross-presentation. *Cancer Immunol Res*. 2019;7:29-39.
215. Regnault A, Lankar D, Lacabanne V, et al. Fcγ receptor-mediated induction of dendritic cell maturation and major histocompatibility complex class I-restricted antigen presentation after immune complex internalization. *J Exp Med*. 1999;189:371-380.
216. Kalergis AM, Ravetch JV. Inducing tumor immunity through the selective engagement of activating Fcγ receptors on dendritic cells. *J Exp Med*. 2002;195:1653-1659.
217. Heger L, Balk S, Lühr JJ, et al. CLEC10A is a specific marker for human CD1c(+) dendritic cells and enhances their Toll-like receptor 7/8-induced cytokine secretion. *Front Immunol*. 2018;9:744.
218. Luckashenak N, chroeder S, Endt K, et al. Constitutive crosspresentation of tissue antigens by dendritic cells controls CD8+ T cell tolerance *in vivo*. *Immunity*. 2008;28:521-532.
219. Belz GT, Vremec D, Febbraio M, et al. CD36 is differentially expressed by CD8+ splenic dendritic cells but is not required for cross-presentation *in vivo*. *J Immunol*. 2002;168:6066-6070.
220. Schulz O, Pennington DJ, Hodivala-Dilke K, Febbraio M, Reis e Sousa C. CD36 or $\alpha_v\beta_3$ and $\alpha_v\beta_5$ integrins are not essential for MHC class I cross-presentation of cell-associated antigen by CD8 α+ murine dendritic cells. *J Immunol*. 2002;168:6057-6065.
221. Delamarre L, Couture R, Mellman I, Trombetta ES. Enhancing immunogenicity by limiting susceptibility to lysosomal proteolysis. *J Exp Med*. 2006;203:2049-2055.
222. Bondanza A, Zimmermann VS, Rovere-Querini P, et al. Inhibition of phosphatidylserine recognition heightens the immunogenicity of irradiated lymphoma cells *in vivo*. *J Exp Med*. 2004;200:1157-1165.
223. Yi T, Li J, Chen H, et al. Splenic dendritic cells survey red blood cells for missing self-CD47 to trigger adaptive immune responses. *Immunity*. 2015;43:764-775.
224. Maric M. Defective antigen processing in GILT-free mice. *Science*. 2001;294:1361-1365.
225. Trombetta ES, Ebersold M, Garrett W, Pypaert M, Mellman I. Activation of lysosomal function during dendritic cell maturation. *Science*. 2003;299:1400-1403.
226. Delamarre L, Pack M, Chang H, Mellman I, Trombetta ES. Differential lysosomal proteolysis in antigen-presenting cells determines antigen fate. *Science*. 2005;307:1630-1634.
227. Savina A, Jancic C, Hugues S, et al. NOX2 controls phagosomal pH to regulate antigen processing during crosspresentation by dendritic cells. *Cell*. 2006;126:205-218.
228. Thacker RI, Janssen EM. Cross-presentation of cell-associated antigens by mouse splenic dendritic cell populations. *Front Immunol*. 2012;3:41.
229. Foote JR, Levine AP, Behe P, Duchen MR, Segal AW. Imaging the neutrophil phagosome and cytoplasm using a ratiometric pH indicator. *J Vis Exp*. 2017;(122):55107.
230. van Montfoort N, Camps MG, Khan S, et al. Antigen storage compartments in mature dendritic cells facilitate prolonged cytotoxic T lymphocyte cross-priming capacity. *Proc Natl Acad Sci U S A*. 2009;106:6730-6735.
231. Lutz MB, Rovere P, Kleijmeer MJ, et al. Intracellular routes and selective retention of antigens in mildly acidic cathepsin D/lysosome-associated membrane protein-1/MHC class II-positive vesicles in immature dendritic cells. *J Immunol*. 1997;159:3707-3716.
232. ten Broeke T, Wubbolts R, Stoorvogel W. MHC class II antigen presentation by dendritic cells regulated through endosomal sorting. *Cold Spring Harb Perspect Biol*. 2013;5:a016873.
233. Wilson NS, El-Sukkari D, Villadangos JA. Dendritic cells constitutively present self antigens in their immature state *in vivo* and regulate antigen presentation by controlling the rates of MHC class II synthesis and endocytosis. *Blood*. 2004;103:2187-2195.
234. De Gassart A, Camosseto V, Thibodeau J, et al. MHC class II stabilization at the surface of human dendritic cells is the result of maturation-dependent MARCH I down-regulation. *Proc Natl Acad Sci U S A*. 2008;105:3491-3496.
235. Shin JS, Ebersold M, Pypaert M, Delamarre L, Hartley A, Mellman I. Surface expression of MHC class II in dendritic cells is controlled by regulated ubiquitination. *Nature*. 2006;444:115-118.
236. Young LJ, Wilson NS, Schnorrer P, et al. Differential MHC class II synthesis and ubiquitination confers distinct antigen-presenting properties on conventional and plasmacytoid dendritic cells. *Nat Immunol*. 2008;9:1244-1252.
237. Cho KJ, Walseng E, Ishido S, Roche PA. Ubiquitination by March-I prevents MHC class II recycling and promotes MHC class II turnover in antigen-presenting cells. *Proc Natl Acad Sci U S A*. 2015;112:10449-10454.
238. Kleijmeer M, Ramm G, Schuurhuis D, et al. Reorganization of multivesicular bodies regulates MHC class II antigen presentation by dendritic cells. *J Cell Biol*. 2001;155:53-63.
239. Groettrup M, Kirk CJ, Basler M. Proteasomes in immune cells: more than peptide producers? *Nat Rev Immunol*. 2010;10:73-78.
240. Helft J, Manicassamy B, Guermonprez P, et al. Cross-presenting CD103+ dendritic cells are protected from influenza virus infection. *J Clin Invest*. 2012;122:4037-4047.
241. Carbone FR, Kurts C, Bennett SR, Miller JF, Heath WR. Cross-presentation: a general mechanism for CTL immunity and tolerance. *Immunol Today*. 1998;19:368-373.
242. Lundie RJ, de Koning-Ward TF, Davey GM, et al. Blood-stage Plasmodium infection induces CD8+ T lymphocytes to parasite-expressed antigens, largely regulated by CD8α+ dendritic cells. *Proc Natl Acad Sci U S A*. 2008;105:14509-14514.
243. Ferris ST, Carrero JA, Mohan JF, Calderon B, Murphy KM, Unanue ER. A minor subset of Batf3-dependent antigen-presenting cells in islets of Langerhans is essential for the development of autoimmune diabetes. *Immunity*. 2014;41:657-669.
244. Kurts C, Heath WR, Kosaka H, Miller JF, Carbone FR. The peripheral deletion of autoreactive CD8+ T cells induced by cross-presentation of self-antigens involves signaling through CD95 (Fas, Apo-1). *J Exp Med*. 1998;188:415-420.
245. Schnorrer P, Behrens GMN, Wilson NS, et al. The dominant role of CD8+ dendritic cells in cross-presentation is not dictated by antigen capture. *Proc Natl Acad Sci U S A*. 2006;103:10729-10734.
246. Savina A, Peres A, Cebrian I, et al. The small GTPase Rac2 controls phagosomal alkalinization and antigen crosspresentation selectively in CD8(+) dendritic cells. *Immunity*. 2009;30:544-555.
247. Cohn L, Chatterjee B, Esselborn F, et al. Antigen delivery to early endosomes eliminates the superiority of human blood BDCA3+ dendritic cells at cross presentation. *J Exp Med*. 2013;210:1049-1063.
248. Chatterjee B, Smed-Sörensen A, Cohn L, et al. Internalization and endosomal degradation of receptor-bound antigens regulate the efficiency of cross presentation by human dendritic cells. *Blood*. 2012;120:2011-2020.
249. Wachler BS, Krueger RR. Normalized contrast sensitivity values. *J Refract Surg*. 1998;14:463-466.
250. Kretzer NM, Theisen DJ, Tussiwand R, et al. RAB43 facilitates cross-presentation of cell-associated antigens by CD8α+ dendritic cells. *J Exp Med*. 2016;213:2871-2883.

251. Cruz FM, Colbert JD, Rock KL. The GTPase Rab39a promotes phagosome maturation into MHC-I antigen-presenting compartments. *EMBO J*. 2020;39:e102020.
252. Theisen DJ, Davidson JT, Briseño CG, et al. WDFY4 is required for cross-presentation in response to viral and tumor antigens. *Science*. 2018;362, 694-699.
253. Kamphorst AO, Guermonprez P, Dudziak D, Nussenzweig MC. Route of antigen uptake differentially impacts presentation by dendritic cells and activated monocytes. *J Immunol*. 2010;185:3426-3435.
254. Trombetta ES, Mellman I. Cell biology of antigen processing *in vitro* and *in vivo*. *Annu Rev Immunol*. 2005;23:975-1028.
255. Guermonprez P, Valladeau J, Zitvogel L, Thery C, Amigorena S. Antigen presentation and T cell stimulation by dendritic cells. *Annu Rev Immunol*. 2002;20:621-667.
256. Ferris ST, Durai V, Wu R, et al. cDC1 prime and are licensed by CD4(+) T cells to induce anti-tumour immunity. *Nature*. 2020;584:624-629.
257. Sosinowski T, White JT, Cross EW, et al. CD8α+ dendritic cell trans presentation of IL-15 to naive CD8+ T cells produces antigen-inexperienced T cells in the periphery with memory phenotype and function. *J Immunol*. 2013;190:1936-1947.
258. Krappmann D, Wegener E, Sunami Y, et al. The IκB kinase complex and NF-κB act as master regulators of lipopolysaccharide-induced gene expression and control subordinate activation of AP-1. *Mol Cell Biol*. 2004;24:6488-6500.
259. Yoshida R, Nagira M, Kitaura M, Imagawa N, Imai T, Yoshie O. Secondary lymphoid-tissue chemokine is a functional ligand for the CC chemokine receptor CCR7. *J Biol Chem*. 1998;273:7118-7122.
260. Forster R, Schubel A, Breitfeld D, et al. CCR7 coordinates the primary immune response by establishing functional microenvironments in secondary lymphoid organs. *Cell*. 1999;99:23-33.
261. Milling S, Yrlid U, Cerovic V, MacPherson G. Subsets of migrating intestinal dendritic cells. *Immunol Rev*. 2010;234:259-267.
262. Liu LM, MacPherson GG. Antigen acquisition by dendritic cells: intestinal dendritic cells acquire antigen administered orally and can prime naive T cells in vivo. *J Exp Med*. 1993;177:1299-1307.
263. Worbs T, Hammerschmidt SI, Forster R. Dendritic cell migration in health and disease. *Nat Rev Immunol*. 2017;17:30-48.
264. Pflicke H, Sixt M. Preformed portals facilitate dendritic cell entry into afferent lymphatic vessels. *J Exp Med*. 2009;206:2925-2935.
265. Lammermann T, Bader BL, Monkley SJ, et al. Rapid leukocyte migration by integrin-independent flowing and squeezing. *Nature*. 2008;453:51-55.
266. Junt T, Scandella E, Ludewig B. Form follows function: lymphoid tissue microarchitecture in antimicrobial immune defence. *Nat Rev Immunol*. 2008;8:764-775.
267. Yamada KM, Sixt M. Mechanisms of 3D cell migration. *Nat Rev Mol Cell Biol*. 2019;20:738-752.
268. Thierry GR, Gentek R, Bajenoff M. Remodeling of reactive lymph nodes: dynamics of stromal cells and underlying chemokine signaling. *Immunol Rev*. 2019;289:42-61.
269. Braun A, Worbs T, Moschovakis GL, et al. Afferent lymph-derived T cells and DCs use different chemokine receptor CCR7-dependent routes for entry into the lymph node and intranodal migration. *Nat Immunol*. 2011;12:879-887.
270. Kopf A. Microtubules control cellular shape and coherence in amoeboid migrating cells. *J Cell Biol*. 2020;219.
271. Durai V, Murphy KM. Functions of murine dendritic cells. *Immunity*. 2016;45:719-736.
272. Jung S, Unutmaz D, Wong P, et al. *In vivo* depletion of CD11c+ dendritic cells abrogates priming of CD8+ T cells by exogenous cell-associated antigens. *Immunity*. 2002;17:211-220.
273. Kissenpfennig A, Aït-Yahia S, Clair-Moninot V, et al. Disruption of the langerin/CD207 gene abolishes Birbeck granules without a marked loss of Langerhans cell function. *Mol Cell Biol*. 2005;25:88-99.
274. Piva L, Tetlak P, Claser C, Karjalainen K, Renia L, Ruedl C. Cutting edge: Clec9A+ dendritic cells mediate the development of experimental cerebral malaria. *J Immunol*. 2012;189:1128-1132.
275. Crozat K, Tamoutounour S, Vu Manh TP, et al. Cutting edge: expression of XCR1 defines mouse lymphoid-tissue resident and migratory dendritic cells of the CD8α+ type. *J Immunol*. 2011;187:4411-4415.
276. Yamazaki C, Sugiyama M, Ohta T, et al. Critical roles of a dendritic cell subset expressing a chemokine receptor, XCR1. *J Immunol*. 2013;190:6071-6082.
277. Torti N, Walton SM, Murphy KM, Oxenius A. Batf3 transcription factor-dependent DC subsets in murine CMV infection: differential impact on T-cell priming and memory inflation. *Eur J Immunol*. 2011;41:2612-2618.
278. Waithman J, Zanker D, Xiao K, et al. Resident CD8(+) and migratory CD103(+) dendritic cells control CD8 T cell immunity during acute influenza infection. *PLoS One*. 2013;8:e66136.
279. Guermonprez P, Helft J, Claser C, et al. Inflammatory Flt3l is essential to mobilize dendritic cells and for T cell responses during *Plasmodium* infection. *Nat Med*. 2013;19:730-738.
280. Bougneres L, Helft J, Tiwari S, et al. A role for lipid bodies in the cross-presentation of phagocytosed antigens by MHC class I in dendritic cells. *Immunity*. 2009;31:232-244.
281. Koblansky AA, Jankovic D, Oh H, et al. Recognition of profilin by Toll-like receptor 12 is critical for host resistance to *Toxoplasma gondii*. *Immunity*. 2013;38:119-130.
282. Lauterbach H, Bathke B, Gilles S, et al. Mouse CD8α+ DCs and human BDCA3+ DCs are major producers of IFN-lambda in response to poly IC. *J Exp Med*. 2010;207:2703-2717.
283. Yoshio S, Kanto T, Kuroda S, et al. Human blood dendritic cell antigen 3 (BDCA3)(+) dendritic cells are a potent producer of interferon-lambda in response to hepatitis C virus. *Hepatology*. 2013;57:1705-1715.
284. Nizzoli G, Krietsch J, Weick A, et al. Human CD1c+ dendritic cells secrete high levels of IL-12 and potently prime cytotoxic T-cell responses. *Blood*. 2013;122:932-942.
285. Schulz O, Edwards AD, Schito M, et al. CD40 triggering of heterodimeric IL-12 p70 production by dendritic cells *in vivo* requires a microbial priming signal. *Immunity*. 2000;13:453-462.
286. Eickhoff S, Brewitz A, Gerner MY, et al. Robust anti-viral immunity requires multiple distinct T cell-dendritic cell interactions. *Cell*. 2015;162:1322-1337.
287. Kastenmuller W, Brandes M, Wang Z, Herz J, Egen JG, Germain RN. Peripheral prepositioning and local CXCL9 chemokine-mediated guidance orchestrate rapid memory CD8+ T cell responses in the lymph node. *Immunity*. 2013;38:502-513.
288. Tussiwand R, Behnke MS, Kretzer NM, et al. An important role for CD4(+) T cells in adaptive immunity to Toxoplasma gondii in mice lacking the transcription factor Batf3. *mSphere*. 2020;5:e00634-e00720.
289. Calabro S, Liu D, Gallman A, et al. Differential intrasplenic migration of dendritic cell subsets tailors adaptive immunity. *Cell Rep*. 2016;16:2472-2485.
290. Ng SL, Teo YJ, Setiagani YA, Karjalainen K, Ruedl C. Type 1 conventional CD103(+) dendritic cells control effector CD8(+) T cell migration, survival, and memory responses during influenza infection. *Front Immunol*. 2018;9:3043.
291. Lucas M, Schachterle W, Oberle K, Aichele P, Diefenbach A. Dendritic cells prime natural killer cells by trans-presenting interleukin 15. *Immunity*. 2007;26:503-517.
292. Alexandre YO, Ghilas S, Sanchez C, Le Bon A, Crozat K, Dalod M. XCR1+ dendritic cells promote memory CD8+ T cell recall upon secondary infections with *Listeria monocytogenes* or certain viruses. *J Exp Med*. 2016;213:75-92.
293. Spranger S, Dai D, Horton B, Gajewski TF. Tumor-residing Batf3 dendritic cells are required for effector T cell trafficking and adoptive T cell therapy. *Canc Cell*. 2017;31:711.e714-723.e714.
294. Kinnebrew MA. Interleukin 23 production by intestinal CD103(+)CD11b(+) dendritic cells in response to bacterial flagellin enhances mucosal innate immune defense. *Immunity*. 2012;36(2):276-287.
295. Ivanov II, Atarashi K, Manel N, et al. Induction of intestinal Th17 cells by segmented filamentous bacteria. *Cell*. 2009;139:485-498.
296. Briseno CG, Satpathy AT, Davidson JT, et al. Notch2-dependent DC2s mediate splenic germinal center responses. *Proc Natl Acad Sci U S A*. 2018;115:10726-10731.
297. Durand M, Walter T, Pirnay T, et al. Human lymphoid organ cDC2 and macrophages play complementary roles in T follicular helper responses. *J Exp Med*. 2019;216:1561-1581.
298. Hilligan KL, Tang SC, Hyde EJ, et al. Dermal IRF4+ dendritic cells and monocytes license CD4+ T helper cells to distinct cytokine profiles. *Nat Commun*. 2020;11:5637.
299. Ainsua-Enrich E, Hatipoglu I, Kadel S, et al. IRF4-dependent dendritic cells regulate CD8(+) T-cell differentiation and memory responses in influenza infection. *Mucosal Immunol*. 2019;12:1025-1037.
300. Yu CI, Becker C, Wang Y, et al. Human CD1c+ dendritic cells drive the differentiation of CD103+ CD8+ mucosal effector T cells via the cytokine TGF-β. *Immunity*. 2013;38:818-830.
301. Villadangos JA, Young L. Antigen-presentation properties of plasmacytoid dendritic cells. *Immunity*. 2008;29:352-361.
302. Swiecki M, Colonna M. The multifaceted biology of plasmacytoid dendritic cells. *Nat Rev Immunol*. 2015;15:471-485.
303. Loschko J, Schlitzer A, Dudziak D, et al. Antigen delivery to plasmacytoid dendritic cells via BST2 induces protective T cell-mediated immunity. *J Immunol*. 2011;186:6718-6725.
304. Loschko J, Heink S, Hackl D, et al. Antigen targeting to plasmacytoid dendritic cells via Siglec-H inhibits Th cell-dependent autoimmunity. *J Immunol*. 2011;187:6346-6356.
305. Meyer-Wentrup F, Benitez-Ribas D, Tacken PJ, et al. Targeting DCIR on human plasmacytoid dendritic cells results in antigen presentation and inhibits IFN-alpha production. *Blood*. 2008;111:4245-4253.
306. Chappell CP, Giltiay NV, Draves KE, et al. Targeting antigens through blood dendritic cell antigen 2 on plasmacytoid dendritic cells promotes immunologic tolerance. *J Immunol*. 2014;192:5789-5801.
307. Mouries J, Moron G, Schlecht G, Escriou N, Dadaglio G, Leclerc C. Plasmacytoid dendritic cells efficiently cross-prime naive T cells *in vivo* after TLR activation. *Blood*. 2008;112:3713-3722.
308. Krug A, Veeraswamy R, Pekosz A, et al. Interferon-producing cells fail to induce proliferation of naive T cells but can promote expansion and T helper 1 differentiation of antigen-experienced unpolarized T cells. *J Exp Med*. 2003;197:899-906.
309. Hoeffel G, Ripoche AC, Matheoud D, et al. Antigen crosspresentation by human plasmacytoid dendritic cells. *Immunity*. 2007;27:481-492.
310. Reizis B. Plasmacytoid dendritic cells: development, regulation, and function. *Immunity*. 2019;50:37-50.
311. Honda K, Ohba Y, Yanai H, et al. Spatiotemporal regulation of MyD88-IRF-7 signalling for robust type-I interferon induction. *Nature*. 2005;434:1035-1040.
312. Cervantes-Barragan L, Lewis KL, Firner S, et al. Plasmacytoid dendritic cells control T-cell response to chronic viral infection. *Proc Natl Acad Sci U S A*. 2012;109:3012-3017.
313. Swiecki M, Gilfillan S, Vermi W, Wang Y, Colonna M. Plasmacytoid dendritic cell ablation impacts early interferon responses and antiviral NK and CD8(+) T cell accrual. *Immunity*. 2010;33:955-966.
314. Webster B, Werneke SW, Zafirova B, et al. Plasmacytoid dendritic cells control dengue and Chikungunya virus infections via IRF7-regulated interferon responses. *Elife*. 2018;7:e34273.
315. Swiecki M, Wang Y, Riboldi E, et al. Cell depletion in mice that express diphtheria toxin receptor under the control of SiglecH encompasses more than plasmacytoid dendritic cells. *J Immunol*. 2014;192:4409-4416.
316. Gilliet M, Cao W, Liu YJ. Plasmacytoid dendritic cells: sensing nucleic acids in viral infection and autoimmune diseases. *Nat Rev Immunol*. 2008;8:594-606.
317. Tomasello E, Naciri K, Chelbi R, et al. Molecular dissection of plasmacytoid dendritic cell activation *in vivo* during a viral infection. *EMBO J*. 2018;37.
318. Diacovo TG, Blasius AL, Mak TW, Cella M, Colonna M. Adhesive mechanisms governing interferon-producing cell recruitment into lymph nodes. *J Exp Med*. 2005;202:687-696.
319. Sozzani S, Vermi W, Del Prete A, Facchetti F. Trafficking properties of plasmacytoid dendritic cells in health and disease. *Trends Immunol*. 2010;31:270-277.
320. Penna G, Sozzani S, Adorini L. Cutting edge: selective usage of chemokine receptors by plasmacytoid dendritic cells. *J Immunol*. 2001;167:1862-1866.
321. Obermoser G, Pascual V. The interferon-alpha signature of systemic lupus erythematosus. *Lupus*. 2010;19:1012-1019.
322. Banchereau R, Hong S, Cantarel B, et al. Personalized immunomonitoring uncovers molecular networks that stratify lupus patients. *Cell*. 2016;165:1548-1550.
323. Palucka AK, Ueno H, Connolly J, et al. Dendritic cells loaded with killed allogeneic melanoma cells can induce objective clinical responses and MART-1 specific CD8+ T-cell immunity. *J Immunother*. 2006;29:545-557.
324. Sisirak V, Ganguly D, Lewis KL, et al. Genetic evidence for the role of plasmacytoid dendritic cells in systemic lupus erythematosus. *J Exp Med*. 2014;211:1969-1976.
325. Rowland SL, Riggs JM, Gilfillan S, et al. Early, transient depletion of plasmacytoid dendritic cells ameliorates autoimmunity in a lupus model. *J Exp Med*. 2014;211:1977-1991.
326. Moody KL, Uccellini MB, Avalos AM, Marshak-Rothstein A, Viglianti GA. Toll-like receptor-dependent immune complex activation of B cells and dendritic cells. *Methods Mol Biol*. 2016;1390:249-272.
327. Marshak-Rothstein A. Toll-like receptors in systemic autoimmune disease. *Nat Rev Immunol*. 2006;6:823-835.
328. Herbin O, Bonito AJ, Jeong S, et al. Medullary thymic epithelial cells and CD8α(+) dendritic cells coordinately regulate central tolerance but CD8α(+) cells are dispensable for thymic regulatory T cell production. *J Autoimmun*. 2016;75:141-149.
329. Oh J, Shin JS. The role of dendritic cells in central tolerance. *Immune Netw*. 2015;15:111-120.
330. Hasegawa H, Matsumoto T. Mechanisms of tolerance induction by dendritic cells *in vivo*. *Front Immunol*. 2018;9:350.
331. Gallegos AM, Bevan MJ. Central tolerance to tissue-specific antigens mediated by direct and indirect antigen presentation. *J Exp Med*. 2004;200:1039-1049.
332. Brocker T, Riedinger M, Karjalainen K. Targeted expression of major histocompatibility complex (MHC) class II molecules demonstrates that dendritic cells can induce negative but not positive selection of thymocytes *in vivo*. *J Exp Med*. 1997;185:541-550.
333. Baba T, Nakamoto Y, Mukaida N. Crucial contribution of thymic Sirpα+ conventional dendritic cells to central tolerance against blood-borne antigens in a CCR2-dependent manner. *J Immunol*. 2009;183:3053-3063.
334. Bonasio R, Scimone ML, Schaerli P, Grabie N, Lichtman AH, von Andrian UH. Clonal deletion of thymocytes by circulating dendritic cells homing to the thymus. *Nat Immunol*. 2006;7:1092-1100.
335. Hadeiba H, Lahl K, Edalati A, et al. Plasmacytoid dendritic cells transport peripheral antigens to the thymus to promote central tolerance. *Immunity*. 2012;36:438-450.
336. Atibalentja DF, Murphy KM, Unanue ER. Functional redundancy between thymic CD8α+ and Sirpα+ conventional dendritic cells in presentation of blood-derived lysozyme by MHC class II proteins. *J Immunol*. 2011;186:1421-1431.

337. Jordan MS, Boesteanu A, Reed AJ, et al. Thymic selection of CD4+CD25+ regulatory T cells induced by an agonist self-peptide. *Nat Immunol.* 2001;2:301-306.
338. Proietto AI, Lahoud MH, Wu L. Distinct functional capacities of mouse thymic and splenic dendritic cell populations. *Immunol Cell Biol.* 2008;86:700-708.
339. Coquet JM, Ribot JC, Bąbała N, et al. Epithelial and dendritic cells in the thymic medulla promote CD4+Foxp3+ regulatory T cell development via the CD27-CD70 pathway. *J Exp Med.* 2013;210:715-728.
340. Steinman RM, Hawiger D, Liu K, et al. Dendritic cell function in vivo during the steady state: a role in peripheral tolerance. *Ann N Y Acad Sci.* 2003;987:15-25.
341. Inaba K, Turley S, Yamaide F, et al. Efficient presentation of phagocytosed cellular fragments on the major histocompatibility complex class II products of dendritic cells. *J Exp Med.* 1998;188:2163-2173.
342. Kurts C, Heath WR, Carbone FR, Allison J, Miller JF, Kosaka H. Constitutive class I-restricted exogenous presentation of self antigens *in vivo*. *J Exp Med.* 1996;184:923-930.
343. Kurts C, Carbone FR, Barnden M, et al. CD4+ T cell help impairs CD8+ T cell deletion induced by cross-presentation of self-antigens and favors autoimmunity. *J Exp Med.* 1997;186:2057-2062.
344. Heath WR, Kurts C, Miller JF, Carbone FR. Cross-tolerance: a pathway for inducing tolerance to peripheral tissue antigens. *J Exp Med.* 1998;187:1549-1553.
345. Belz GT, Behrens GMN, Smith CM, et al. The CD8α(+) dendritic cell is responsible for inducing peripheral self-tolerance to tissue-associated antigens. *J Exp Med.* 2002;196:1099-1104.
346. Henri S, Guilliams M, Poulin LF, et al. Disentangling the complexity of the skin dendritic cell network. *Immunol Cell Biol.* 2010;88:366-375.
347. Bedoui S, Whitney PG, Waithman J, et al. Cross-presentation of viral and self antigens by skin-derived CD103+ dendritic cells. *Nat Immunol.* 2009;10:488-495.
348. Coombes JL, Siddiqui KRR, Arancibia-Cárcamo CV, et al. A functionally specialized population of mucosal CD103+ DCs induces Foxp3+ regulatory T cells via a TGF-β and retinoic acid-dependent mechanism. *J Exp Med.* 2007;204:1757-1764.
349. Sun CM, Hall JA, Blank RB, et al. Small intestine lamina propria dendritic cells promote de novo generation of Foxp3 T_{reg} cells via retinoic acid. *J Exp Med.* 2007;204:1775-1785.
350. Hall JA, Grainger JR, Spencer SP, Belkaid Y. The role of retinoic acid in tolerance and immunity. *Immunity.* 2011;35:13-22.
351. Papadakis KA, Prehn J, Nelson V, et al. The role of thymus-expressed chemokine and its receptor CCR9 on lymphocytes in the regional specialization of the mucosal immune system. *J Immunol.* 2000;165:5069-5076.
352. Kunkel EJ, Campbell JJ, Haraldsen G, et al. Lymphocyte CC chemokine receptor 9 and epithelial thymus-expressed chemokine (TECK) expression distinguish the small intestinal immune compartment: epithelial expression of tissue-specific chemokines as an organizing principle in regional immunity. *J Exp Med.* 2000;192:761-768.
353. Wurbel MA, Philippe JM, Nguyen C, et al. The chemokine TECK is expressed by thymic and intestinal epithelial cells and attracts double- and single-positive thymocytes expressing the TECK receptor CCR9. *Eur J Immunol.* 2000;30:262-271.
354. Guilliams M, Crozat K, Henri S, et al. Skin-draining lymph nodes contain dermis-derived CD103(−) dendritic cells that constitutively produce retinoic acid and induce Foxp3(+) regulatory T cells. *Blood.* 2010;115:1958-1968.
355. Sigmundsdottir H, Pan J, Debes GF, et al. DCs metabolize sunlight-induced vitamin D3 to "program" T cell attraction to the epidermal chemokine CCL27. *Nat Immunol.* 2007;8:285-293.
356. van der Aar AM, Sibiryak DS, Bakdash G, et al. Vitamin D3 targets epidermal and dermal dendritic cells for induction of distinct regulatory T cells. *J Allergy Clin Immunol.* 2011;127:1532.e1537-1540.e1537.
357. Ohnmacht C, Pullner A, King SBS, et al. Constitutive ablation of dendritic cells breaks self-tolerance of CD4 T cells and results in spontaneous fatal autoimmunity. *J Exp Med.* 2009;206:549-559.
358. Darrasse-Jeze G, Deroubaix S, Mouquet H, et al. Feedback control of regulatory T cell homeostasis by dendritic cells *in vivo*. *J Exp Med.* 2009;206:1853-1862.
359. Swee LK, Bosco N, Malissen B, Ceredig R, Rolink A. Expansion of peripheral naturally occurring T regulatory cells by Fms-like tyrosine kinase 3 ligand treatment. *Blood.* 2009;113:6277-6287.
360. Suffner J, Hochweller K, Kühnle MC, et al. Dendritic cells support homeostatic expansion of Foxp3+ regulatory T cells in Foxp3.LuciDTR mice. *J Immunol.* 2010;184:1810-1820.
361. Yi J, Jung J, Hong SW, et al. Unregulated antigen-presenting cell activation by T cells breaks self tolerance. *Proc Natl Acad Sci U S A.* 2019;116:1007-1016.
362. Qureshi OS, Zheng Y, Nakamura K, et al. Trans-endocytosis of CD80 and CD86: a molecular basis for the cell-extrinsic function of CTLA-4. *Science.* 2011;332:600-603.
363. Awasthi A, Carrier Y, Peron JPS, et al. A dominant function for interleukin 27 in generating interleukin 10-producing anti-inflammatory T cells. *Nat Immunol.* 2007;8:1380-1389.
364. Muth S, Schutze K, Schild H, Probst HC. Release of dendritic cells from cognate CD4+ T-cell recognition results in impaired peripheral tolerance and fatal cytotoxic T-cell mediated autoimmunity. *Proc Natl Acad Sci U S A.* 2012;109:9059-9064.
365. Loschko J, Schreiber HA, Rieke GJ, et al. Absence of MHC class II on cDCs results in microbial-dependent intestinal inflammation. *J Exp Med.* 2016;213:517-534.
366. Wohn C, Le Guen V, Voluzan O, Fiore F, Henri S, Malissen B. Absence of MHC class II on cDC1 dendritic cells triggers fatal autoimmunity to a cross-presented self-antigen. *Sci Immunol.* 2020;5:eaba1896.
367. Manicassamy S, Reizis B, Ravindran R, et al. Activation of β-catenin in dendritic cells regulates immunity versus tolerance in the intestine. *Science.* 2010;329:849-853.
368. Travis MA, Reizis B, Melton AC, et al. Loss of integrin $\alpha_v\beta_8$ on dendritic cells causes autoimmunity and colitis in mice. *Nature.* 2007;449:361-365.
369. Kool M, Soullié T, van Nimwegen M, et al. Alum adjuvant boosts adaptive immunity by inducing uric acid and activating inflammatory dendritic cells. *J Exp Med.* 2008;205:869-882.
370. Garris CS, Arlauckas SP, Kohler RH, et al. Successful anti-PD-1 cancer immunotherapy requires T cell-dendritic cell crosstalk involving the cytokines IFN-γ and IL-12. *Immunity.* 2018;49:1148.e1147-1161.e1147.
371. Sanchez-Paulete AR, Cueto FJ, Martínez-López M, et al. Cancer immunotherapy with immunomodulatory anti-CD137 and anti-PD-1 monoclonal antibodies requires BATF3-dependent dendritic cells. *Cancer Discov.* 2016;6:71-79.
372. Salmon H, Idoyaga J, Rahman A, et al. Expansion and activation of CD103(+) dendritic cell progenitors at the tumor site enhances tumor responses to therapeutic PD-L1 and BRAF inhibition. *Immunity.* 2016;44:924-938.
373. Broz ML, Binnewies M, Boldajipour B, et al. Dissecting the tumor myeloid compartment reveals rare activating antigen-presenting cells critical for T cell immunity. *Cancer Cell.* 2014;26:638-652.
374. Barry KC, Hsu J, Broz ML, et al. A natural killer-dendritic cell axis defines checkpoint therapy-responsive tumor microenvironments. *Nat Med.* 2018;24:1178-1191.
375. Wculek SK, Cueto FJ, Mujal AM, Melero I, Krummel MF, Sancho D. Dendritic cells in cancer immunology and immunotherapy. *Nat Rev Immunol.* 2020;20:7-24.
376. Johnson JS, Lucas SY, Amon LM, et al. Reshaping of the dendritic cell chromatin landscape and interferon pathways during HIV infection. *Cell Host Microbe.* 2018;23:366.e369-381.e369.
377. Inaba K, Metlay JP, Crowley MT, Steinman RM. Dendritic cells pulsed with protein antigens *in vitro* can prime antigen-specific, MHC-restricted T cells *in situ*. *J Exp Med.* 1990;172:631-640.
378. Saxena M, Bhardwaj N. Re-emergence of dendritic cell vaccines for cancer treatment. *Trends Cancer.* 2018;4:119-137.
379. Palucka K, Banchereau J. Human dendritic cell subsets in vaccination. *Curr Opin Immunol.* 2013;25:396-402.
380. Palucka K. Q&A: evidence presenter. Interview by Marian Turner. *Nature.* 2013;504:S9.
381. Bol KF, Schreibelt G, Gerritsen WR, de Vries IJ, Figdor CG. Dendritic cell-based immunotherapy: state of the art and beyond. *Clin Cancer Res.* 2016;22:1897-1906.
382. Palucka K, Banchereau J. Dendritic-cell-based therapeutic cancer vaccines. *Immunity.* 2013;39:38-48.
383. Bol KF, van den Bosch T, Schreibelt G, et al. Adjuvant dendritic cell vaccination in high-risk uveal melanoma. *Ophthalmology.* 2016;123:2265-2267.
384. Schreibelt G, Bol KF, Westdorp H, et al. Effective clinical responses in metastatic melanoma patients after vaccination with primary myeloid dendritic cells. *Clin Cancer Res.* 2016;22:2155-2166.
385. Prue RL, Vari F, Radford KJ, et al. A phase I clinical trial of CD1c (BDCA-1)+ dendritic cells pulsed with HLA-A*0201 peptides for immunotherapy of metastatic hormone refractory prostate cancer. *J Immunother.* 2015;38:71-76.
386. Senju S, Haruta M, Matsumura K, et al. Generation of dendritic cells and macrophages from human induced pluripotent stem cells aiming at cell therapy. *Gene Ther.* 2011;18:874-883.
387. Motomura Y, Senju S, Nakatsura T, et al. Embryonic stem cell-derived dendritic cells expressing glypican-3, a recently identified oncofetal antigen, induce protective immunity against highly metastatic mouse melanoma, B16-F10. *Cancer Res.* 2006;66:2414-2422.
388. Matsuyoshi H, Senju S, Hirata S, Yoshitake Y, Uemura Y, Nishimura Y. Enhanced priming of antigen-specific CTLs in vivo by embryonic stem cell-derived dendritic cells expressing chemokine along with antigenic protein: application to antitumor vaccination. *J Immunol.* 2004;172:776-786.
389. Yanagimachi MD, Niwa A, Tanaka T, et al. Robust and highly-efficient differentiation of functional monocytic cells from human pluripotent stem cells under serum- and feeder cell-free conditions. *PLoS One.* 2013;8:e59243.
390. Silk KM, Leishman AJ, Nishimoto KP, Reddy A, Fairchild PJ. Rapamycin conditioning of dendritic cells differentiated from human ES cells promotes a tolerogenic phenotype. *J Biomed Biotechnol.* 2012;2012:172420.
391. Schumacher FR, Schmit SL, Jiao S, et al. Genome-wide association study of colorectal cancer identifies six new susceptibility loci. *Nat Commun.* 2015;6:7138.
392. Schumacher TN, Scheper W, Kvistborg P. Cancer neoantigens. *Annu Rev Immunol.* 2019;37:173-200.
393. Schumacher TN, Schreiber RD. Neoantigens in cancer immunotherapy. *Science.* 2015;348:69-74.
394. Sahin U, Schuler M, Richly H, et al. A phase I dose-escalation study of IMAB362 (Zolbetuximab) in patients with advanced gastric and gastro-oesophageal junction cancer. *Eur J Cancer.* 2018;100:17-26.
395. Gubin MM, Artyomov MN, Mardis ER, Schreiber RD. Tumor neoantigens: building a framework for personalized cancer immunotherapy. *J Clin Invest.* 2015;125:3413-3421.
396. Sahin U, Derhovanessian E, Miller M, et al. Personalized RNA mutanome vaccines mobilize poly-specific therapeutic immunity against cancer. *Nature.* 2017;547:222-226.
397. Spranger S, Gajewski TF. A new paradigm for tumor immune escape: β-catenin-driven immune exclusion. *J Immunother Cancer.* 2015;3:43.
398. Okada N, Mori N, Koretomo R, et al. Augmentation of the migratory ability of DC-based vaccine into regional lymph nodes by efficient CCR7 gene transduction. *Gene Ther.* 2005;12:129-139.
399. Gardner A, Ruffell B. Dendritic cells and cancer immunity. *Trends Immunol.* 2016;37:855-865.
400. Pulendran B, Smith JL, Jenkins M, Schoenborn M, Maraskovsky E, Maliszewski CR. Prevention of peripheral tolerance by a dendritic cell growth factor: flt3 ligand as an adjuvant. *J Exp Med.* 1998;188:2075-2082.
401. Anandasabapathy N, Breton G, Hurley A, et al. Efficacy and safety of CDX-301, recombinant human Flt3L, at expanding dendritic cells and hematopoietic stem cells in healthy human volunteers. *Bone Marrow Transplant.* 2015;50:924-930.
402. Breton G, Lee J, Liu K, Nussenzweig MC. Defining human dendritic cell progenitors by multiparametric flow cytometry. *Nat Protoc.* 2015;10:1407-1422.
403. Hegde S, Krisnawan VE, Herzog BH, et al. Dendritic cell paucity leads to dysfunctional immune surveillance in pancreatic cancer. *Cancer Cell.* 2020;37:289.e289-307.e289.
404. Oba T, Long MD, Keler T, et al. Overcoming primary and acquired resistance to anti-PD-L1 therapy by induction and activation of tumor-residing cDC1s. *Nat Commun.* 2020;11:5415.
405. Hawiger D, Inaba K, Dorsett Y, et al. Dendritic cells induce peripheral T cell unresponsiveness under steady state conditions *in vivo*. *J Exp Med.* 2001;194:769-779.
406. Boscardin SB, Hafalla JCR, Masilamani RF, et al. Antigen targeting to dendritic cells elicits long-lived T cell help for antibody responses. *J Exp Med.* 2006;203:599-606.
407. Bozzacco L, Trumpfheller C, Siegal FP, et al. DEC-205 receptor on dendritic cells mediates presentation of HIV gag protein to CD8+ T cells in a spectrum of human MHC I haplotypes. *Proc Natl Acad Sci U S A.* 2007;104:1289-1294.
408. Bonifaz L, Bonnyay D, Mahnke K, Rivera M, Nussenzweig MC, Steinman RM. Efficient targeting of protein antigen to the dendritic cell receptor DEC-205 in the steady state leads to antigen presentation on major histocompatibility complex class I products and peripheral CD8+ T cell tolerance. *J Exp Med.* 2002;196:1627-1638.
409. Park JJ, Anand S, Zhao Y, et al. Expression of anti-HVEM single-chain antibody on tumor cells induces tumor-specific immunity with long-term memory. *Cancer Immunol Immunother.* 2012;61:203-214.
410. Tullett KM. Targeting CLEC9A delivers antigen to human CD141(+) DC for CD4(+) and CD8(+)T cell recognition. *JCI Insight.* 2016;1:e87102.
411. Macri C, Dumont C, Panozza S, et al. Antibody-mediated targeting of antigen to C-type lectin-like receptors Clec9A and Clec12A elicits different vaccination outcomes. *Mol Immunol.* 2017;81:143-150.
412. Neubert K, Lehmann CHK, Heger L, et al. Antigen delivery to CD11c+CD8− dendritic cells induces protective immune responses against experimental melanoma in mice *in vivo*. *J Immunol.* 2014;192:5830-5838.
413. Lehmann CH, Heger L, Heidkamp GF, et al. Direct delivery of antigens to dendritic cells via antibodies specific for endocytic receptors as a promising strategy for future therapies. *Vaccines (Basel).* 2016;4:8.
414. Lakhrif Z, Moreau A, Hérault B, et al. Targeted delivery of *Toxoplasma gondii* antigens to dendritic cells promote immunogenicity and protective efficiency against toxoplasmosis. *Front Immunol.* 2018;9:317.
415. Soares H, Waechter H, Glaichenhaus N, et al. A subset of dendritic cells induces CD4+ T cells to produce IFN-γ by an IL-12-independent but CD70-dependent mechanism *in vivo*. *J Exp Med.* 2007;204:1095-1106.
416. Cintolo JA, Datta J, Mathew SJ, Czerniecki BJ. Dendritic cell-based vaccines: barriers and opportunities. *Future Oncol.* 2012;8:1273-1299.
417. Perez CR, De Palma M. Engineering dendritic cell vaccines to improve cancer immunotherapy. *Nat Commun.* 2019;10:5408.

CHAPTER 35

Gamma Delta T Cells

Adrian C. Hayday, PhD

INTRODUCTION

Gamma delta (γδ) T cells are named with respect to their defining cell surface receptor comprising a single γ chain paired to a single δ chain, and to their (mostly) thymic origin. As is true for T-cell receptor (TCR) alpha (α) and beta (β) chains, TCRγ and TCRδ chains each comprise a single Ig-like variable (V) domain located N terminal to an Ig-like constant (C) domain that connects to a single-pass transmembrane domain that sits N terminal to a very short cytoplasmic tail. Also akin to αβ TCRs, each γδ heterodimer is associated with membrane-spanning CD3 chains that facilitate signal transduction upon TCR engagement (Fig. 35.1A). Hence, all γδ T cells are CD3+,[1] which has become an operational definition of a T cell.

Prior to their discovery, there was no scientific report either predicting the existence of γδ T cells or anticipating a need for them. Rather, the initiation of their discovery was possibly the first example of the power of molecular biology to identify unanticipated cell lineage-defining genes, in this case the TCRγ-chain genes, described in 1985.[2] As is the case for TCRα and for Ig light (L) chains, TCRγ chains are encoded by discontiguous variable (V) and junctional (J) gene segments[2] that are somatically recombined in lymphocyte progenitors via Recombinase Activating Gene (RAG) 1 and RAG2-dependent DNA rearrangements (Fig. 35.1B).

The first TCRγ cDNA to be described was isolated from a murine αβ T cell line (2C) and was initially deduced to be a TCRα cDNA.[3] Shortly after this, it was reported that many cultured αβ T cell lines expressed TCRγ chain RNAs,[4] mostly transcribed from nonproductively rearranged gene segments. The evident dispensability of TCRγ in αβ T cells fueled wide-ranging speculation about its importance, ranging from it being an evolutionary artifact to its playing an intermediate role in αβ T cell development.[5-7] However, within 1 year, an unexpected third set of T cells (in addition to CD4+ and CD8+ T cells) had been reported in sheep,[8] and within a further 6 months, productive TCRγ chain gene rearrangements had been reported for primary murine splenocytes and thymocytes,[9-11] while a third (and small) subset of unique TCRαβ–CD3+ lymphocytes expressing TCRγ chains was identified in human peripheral blood,[1] with further descriptions made soon afterward of human lymphocytes and thymocytes bearing TCRγ chain–based TCRs.[12-15] Those lymphocytes were subsequently validated as γδ T cells following the discovery of the TCRδ chain genes.[16-21] Akin to TCRβ and Ig heavy (H) chains, TCRδ chains are encoded by genes from somatic recombination of V, diversity (D), and J gene segments (Fig. 35.1B). Like TCRαβ but unlike Ig, rearranged TCRγ/δ genes in most species show no evidence of further, AID-dependent somatic mutation, with the evident exceptions to date of sharks and dromedary camels.[22-24]

Given that their discovery was driven mostly by molecular immunology rather than by function, the biology of γδ T cells has required elucidation from scratch, an endeavor that has exposed some overt paradoxes. For example, because of the combinations of gene segments available, the quasi-random flexibility and imprecision of their rearrangement, and the lack of evident restriction of CDR3 lengths (particularly those generated by TCRδ gene segment recombinations) there seems to be at least 1000-fold greater potential for diversity in TCRγδ than in either Ig or TCRαβ.[25] And yet, γδ T cells set the historic precedent for T cell compartments expressing canonical, monoclonal TCRs.[25,26] In approaching such paradoxes, one should consider that γδ T cells are not a single entity but that they comprise qualitatively diverse lymphocytes that may be deployed to different degrees in different species. This notwithstanding γδ T cells are broadly conserved across all jawed vertebrates (gnathostoma), and our consideration of γδ T cell biology will therefore begin by considering the possible basis of that.

γδ T CELL CONSERVATION

The conservation of γδ T cells, B cells, and αβ T cells in most jawed vertebrates is seemingly mirrored by an analogous tripartite conservation of lymphocyte-like cells in jawless vertebrates (agnatha), for example, the lamprey.[27] Agnatha employ an entirely unrelated molecular mechanism to generate receptor diversity in one lineage (VLR-B), which secretes its antigen receptor, and in two other lineages (VLR-A and VLR-C), which retain their antigen receptors on the cell surface.[27] Thus, parallels have been drawn between VLR-B and B cells, between VLR-A and TCRαβ+ cells, and between VLR-C and TCRγδ+ cells, with which they share some conspicuous traits, considered further below.

It is commonly considered that B-cell conservation reflects the capacity of Ig to recognize highly diverse, extracellular native antigens, particularly those released from or displayed by microbial pathogens. Such recognition can block the biological activity of the antigen (eg, cell entry impairment, or toxin neutralization) and/or initiate the recruitment of additional effector mechanisms (eg, complement fixation or NK cell–mediated, antibody-dependent cell cytotoxicity [ADCC]) that target antigen-expressing microbes or infected

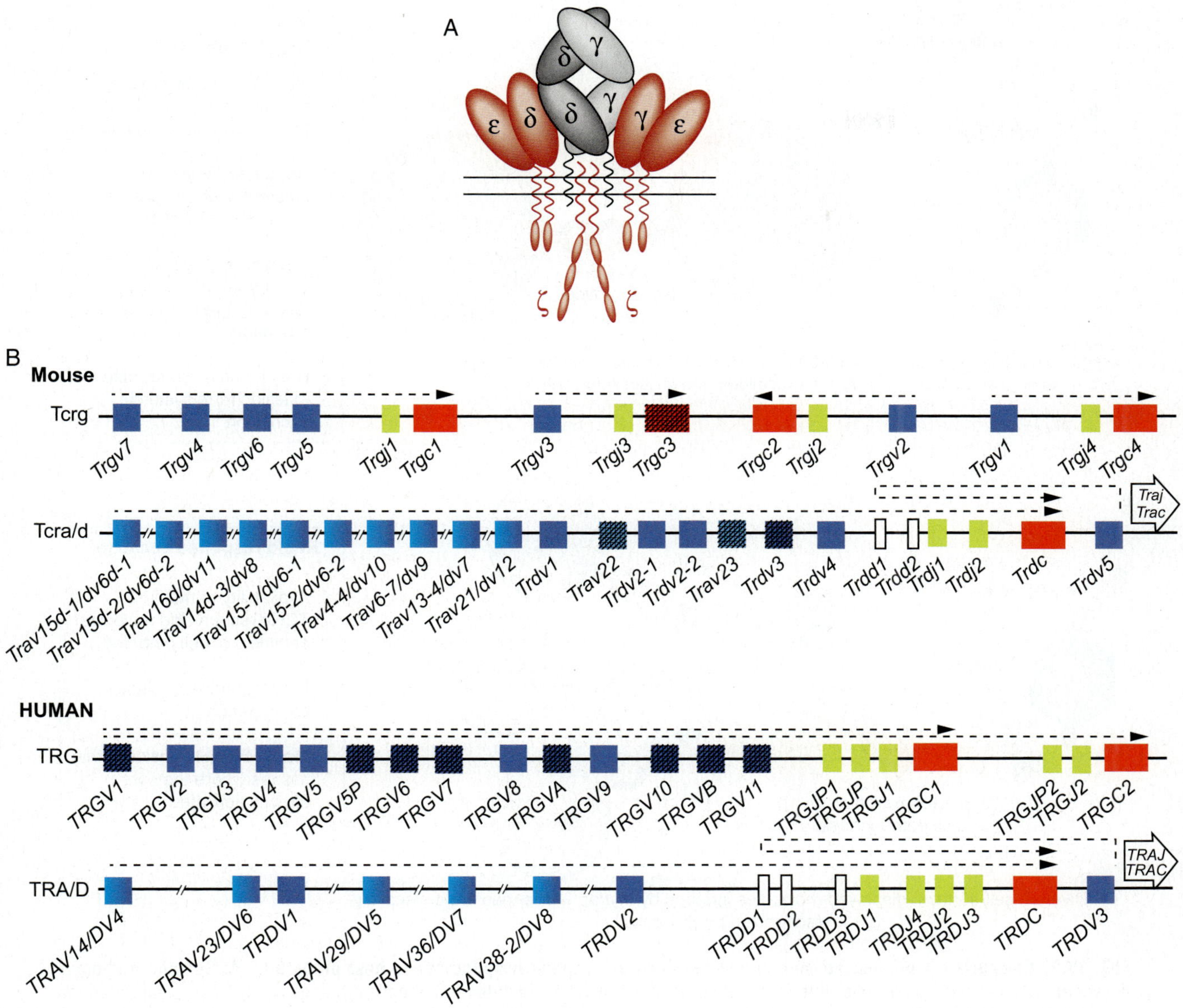

FIG. 35.1. The γδ T-cell receptor and the genes encoding it. **A:** Schematic of the human γδ TCR chains (black) complexed with CD3 chains (red). **B:** Mouse and human γ and δ gene segment families: blue, V gene segments; light and dark blue indicates Vδ segments that can evidently be productively rearranged to Jα; white, D gene segments; yellow, J gene segments; red, C gene segments; hatched denotes pseudogene segments. Arrows denote direction of transcription and orientation of rearrangement. Gene nomenclature: human, HGNC; mouse, MGI based on segment nomenclature of Heilig J. and Tonegawa[43]. (**A:** Adapted from Marin AV, et al. Human congenital T-cell receptor disorders. *LymphoSign J.* 2015;2:3-19.)

cells (Fig. 35.2A). A capacity for Ig to improve its engagement of antigen is provided by somatic mutation that underpins affinity maturation. Conversely, αβ T cell conservation primarily reflects a capacity of TCRαβ to survey the intracellular space for infections or genomic damage following the processing of antigens and their display on the cell surface by autologous antigen presenting proteins; thus, major histocompatibility complex (MHC), CD1, and MR1 proteins present peptides, glycolipids, and microbial metabolites, respectively (Fig. 35.2A). There is no evident means by which TCRαβ can improve its engagement of presented antigen, and the architecture of presented antigen complexes de facto limits TCRαβ diversity.

With the extracellular and intracellular antigenic universes jointly surveyed by B cells and αβ T cells, the unresolved question is "what do γδ T cells uniquely provide?" The solution presumably lies with the TCRγδ heterodimer since it is the only molecule uniquely expressed by γδ T cells. Indeed, γ chains can be paired in marsupials with so-called μ-chains[28] and in sharks with so-called NAR-TCR chains,[29] both of which are TCRδ-related, 3 Ig-domain chains, but neither of which is conserved across gnathostome evolution. Likewise, because the TCRδ gene segments sit within a combined TCRα/δ locus, that in humans comprises ~1mb, V-delta gene segments can be productively rearranged to Jα-Cα (Fig. 35.1B) and expressed in αβ T cells, thereby contributing to

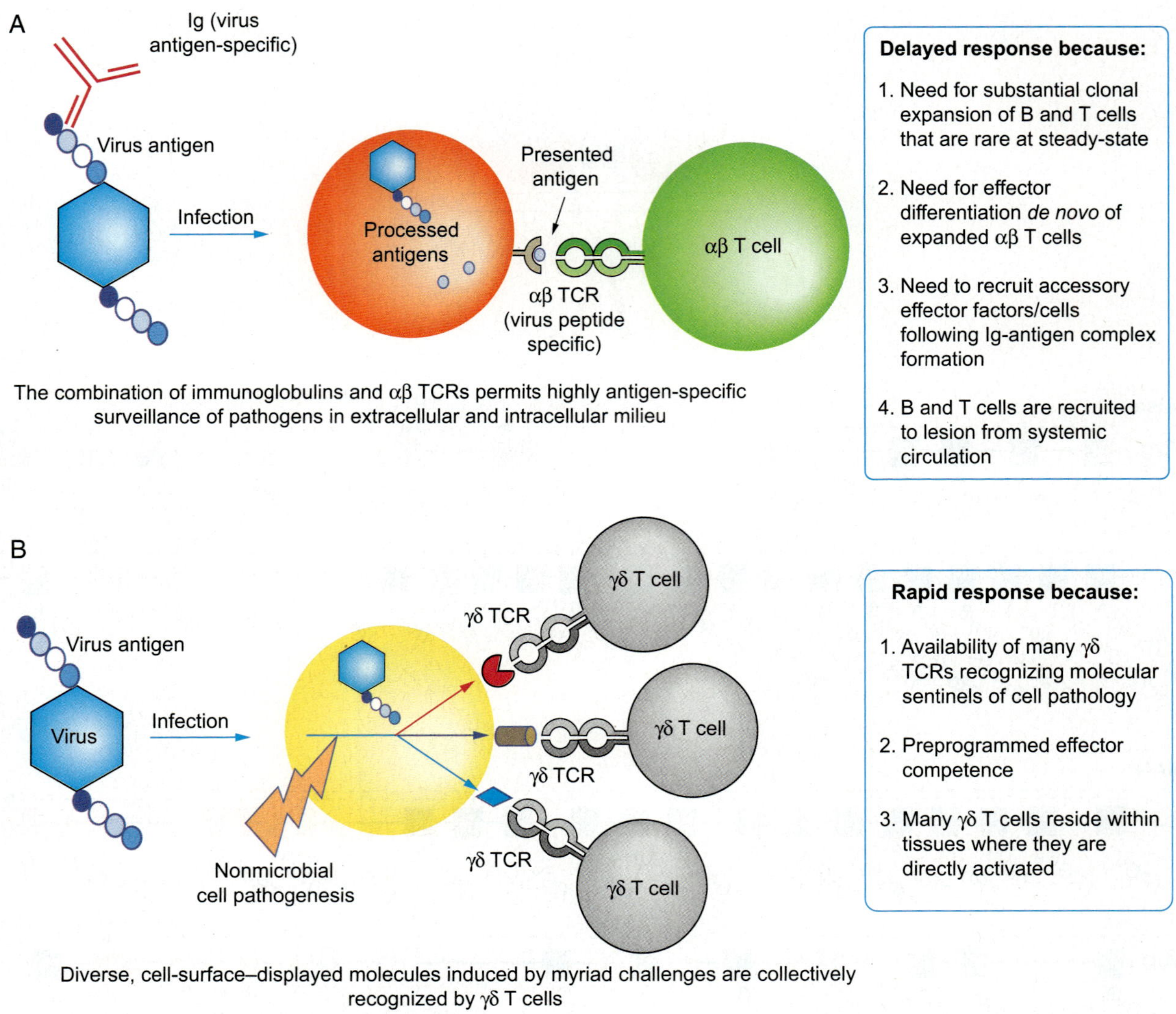

FIG. 35.2. Comparing hypothesized unique contributions of adaptive lymphocytes to host protection. A: Specific pathogen detection by B cells and αβ T cells, illustrated for virus infection. **B:** Rapid detection of cell pathology by γδ T cells, illustrated for virus infection or nonmicrobial dysregulation.

TCRα chain diversity. Thus, the sole unique property of γδ T cells is the TCRγδ heterodimer.

Notwithstanding some examples of MHC-restricted γδ TCRs,[30] TCRγδ is neither limited by MHC restriction nor primarily focused on processed antigens. Therefore, its unique contribution may be to facilitate recognition of highly diverse, native cell surface antigens of diverse composition displayed by cells recently infected or dysregulated by other means, for example, genotoxic damage, and, as a consequence, to trigger rapid cell-intrinsic effector responses (Fig. 35.2B). Because the first trait requires neither processing nor professional antigen-presenting cells, essentially any cell type might be recognized by γδ T cells, soon after infection or dysregulation. Moreover, meeting the requirement for rapid responsiveness to infected or dysregulated cells would not be delayed by time-consuming peripheral differentiation, as occurs for CD4+ αβ T cells. Indeed, discrete γδ T effector modalities are commonly preprogrammed during the cells' development,[31] as will be considered further below. Rapid effector responsiveness likewise distinguishes γδ T cells from most Ig-mediated recognition of infected cells that requires time-dependent B-cell maturation followed by the recruitment of accessory effectors (eg, ADCC), as considered above.

For adaptive lymphocytes, the delivery of effector responses is also delayed by the time required for sufficient clonal expansion of αβ T cells and B cells with appropriate specificities. γδ T cells display two traits that seemingly limit this delay. One trait is to favor the generation of particular TCRs, thereby increasing the cells' precursor frequency at steady state.[32-34] A second trait is for very many specificities to be relevant to any one challenge (Fig. 35.2B). Thus, by comparison with prototypic adaptive responses, there will be concurrent responses of several clones, any single one of which is not essential. This might reflect the fact that, in most species, there is no evident means by which TCRγδ can improve its engagement of antigen, for example, by affinity maturation. These issues will be considered further when TCR specificities are reviewed, below. Their importance may be considerable, given that the evolution of γμ receptors in marsupials and of NAR-TCRs in sharks seems to be based on increasing diversity.

Of note, the recognition of native surface antigens, of diverse composition, displayed by infected cells or cells dysregulated by other means would be consistent with the broad anatomical distribution of γδ T cells that is also commonly established during the development of different tissues (eg, skin, lung, gut, uterus, liver) and that contrasts with αβ T cells and B cells, which prior to antigen exposure are primarily localized within lymphoid tissues. Moreover, it would place γδ T cells into the early, afferent phase of immune responses, commonly considered to be the domain of innate immune myeloid cells. Like those cells, the rapid responses of γδ T cells, including some potential to present antigen, can shape immunogenicity, orchestrating downstream adaptive immunity. This has been termed "lymphoid stress surveillance."[35] Other afferent phase lymphocytes include natural killer (NK) cells and innate lymphoid cells (ILCs), but whereas NK cells can recognize surface molecular sentinels of infected or transformed cells, the breadth of γδ TCR diversity has the potential to greatly expand the types of cell pathology recognized by γδ T cells.

If the lineage-defining role for γδ T cells is the rapid recognition of cells dysregulated by infection or other means, then the implication is that the cells can discriminate dysregulated self from normal self. NK cells seemingly achieve this by being suppressed by normal cell expression of MHC,[36] whereas we shall consider below the prospect that γδ T cells can positively engage normal cells, thereby contributing to tissue-specific physiology and homeostasis as well as to immunoprotection. Indeed, γδ T cells make critical nonredundant contributions to adaptive physiologic functions as wide ranging as thermal regulation, dietary adaptation, water retention, wound healing, orthodontic tooth movement, and short-term memory.[37-42] Such contributions will have been underestimated by traditional immunologic approaches that commonly paid little attention to physiology, for example, failing to use natural routes of infection and failing to assess physiological outcome measures.

Going forward, it will be interesting to determine which contributions reflect intrinsic γδ T cell competences that do not rely on collaboration with other lymphocytes versus those that require collaboration with αβ T cells and/or B cells. Clearly, the latter category has implications well beyond those strands of immunology focused only on γδ T cells. Moreover, some nonredundant contributions of γδ T cells may only become apparent contingent on the loss of another function, for example, αβ T cell impotence in the setting of MHC-deficient tumors or immunosuppressed transplant recipients. Any hypothesized functions for γδ T cells are necessarily constrained by the cells' anatomical distribution(s), and for this reason we shall next consider where γδ T cells are found.

DISCRETE γδ T CELL COMPARTMENTS

A signature feature of γδ T cells, by contrast to αβ T cells and B cells, is the conspicuous association of Vγ and Vδ gene segment usages with particular anatomical subcompartments, particularly in the mouse, but not exclusively so (Fig. 35.3). To aid the consideration of γδ T cell anatomy, please refer to the gene families encoding murine and human TCRγ and TCRδ chains that are shown in Figure 35.1B (note that the mouse nomenclature adopted is that of Heilig and Tonegawa[43] as was employed in Chapter 23).

Lymphoid

Like most B cells and αβ T cells, γδ T cells can be found in lymphoid organs, including peripheral lymph nodes (pLNs), spleen, and peripheral blood. In mice they account for approximately 1% to 5% of splenic or pLN T cells (Table 35.1) and primarily comprise Vγ1+ and Vγ4+ T cells, each paired with a spectrum of Vδ chains.[44] In humans, γδ T cells compose from ~0.1% to ~15% of peripheral blood T cells and most often the great majority express Vγ9Vδ2 TCRs.[44] However, in chickens, in fish, and in very many animals from order *Artiodactyla* that collectively compose most of the world's species of large land mammals (eg, cattle, pigs, goats, sheep, deer, antelopes, and camels), the representation of lymphoid γδ T cells may be higher, particularly in younger animals (Table 35.1).[44-56] Such species are commonly termed "γδ-high."[46] Nonetheless, even in those species, many lymphoid γδ T cells display noticeably atypical distributions, outside of the conventional T- and B-cell zones, for example, throughout the splenic red pulp.

This highlights the fact that much of γδ T cell biology is "unconventional," for example, not accommodated by the typical routes of αβ T cell and B cell recirculation that underpin adaptive immunity. Those typical routes allow highly polyclonal, antigen-naïve αβ T cells and B cells to accumulate in sites (eg, LNs, spleen) where they can be exposed to vast numbers of antigens and APC draining from peripheral tissues. Thereupon cognate, antigen-specific lymphocytes are triggered to undergo massive clonal expansion, which can be accommodated by LN and splenic anatomies, followed by differentiation and migration to the peripheral sites of antigenic challenge. Beyond delivering local effector functions, some such cells may remain in the tissues for long periods of time, so-called tissue-resident memory (T_{RM}) cells.

While some lymphoid γδ T cells present in the white pulp may adopt much of this adaptive immune modality, many γδ T cells do not.[45,57] Thus, the dominant lymphoid γδ T cell subtype in humans expresses polyclonal TCRs, but they are conspicuously limited to Vγ9Vδ2 pairings,[58-61] and most share a single reactivity: a signature response to microbial or endogenous low-molecular-weight lipidic pyrophosphates, known collectively as phosphoantigens (PAgs)[62-68]: such cells are frequently considered "innate-like" for reasons considered further below. The other lymphoid γδ subtype is highly polyclonal, albeit mostly lacking Vδ2+ cells, and is commonly classified as adaptive by the same criteria as apply to

	Mouse	**Human**
Epidermis	**+(Vγ5Vδ1)**	**-**
Dermis	**+(Vγ6Vδ1;Vγ4)**	**+(Vδ1)**
Intestinal epithelium	**+ (Vγ7)**	**+(Vγ4,Vδ1)**
Lung	+(Vγ6Vδ1)	+(Vδ1)
Uterus/decidua	+(Vγ6Vδ1)	+(Vδ1)
Liver	+(Vγ6Vδ1;Vγ4;Vγ1Vδ6.3)	+(Vδ1,Vδ3,Vδ5;Vδ2)
Breast	+(Vγ6;Vδ4)	+(Vδ1)

Other mouse γδ tissue associations, particularly for Vγ6 cells, that have been reported include gingiva, meninges, adipose tissue, and peritoneal cavity

IV

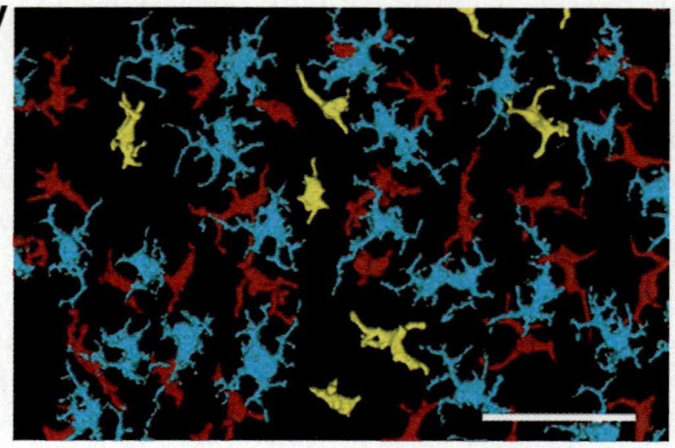

I II III

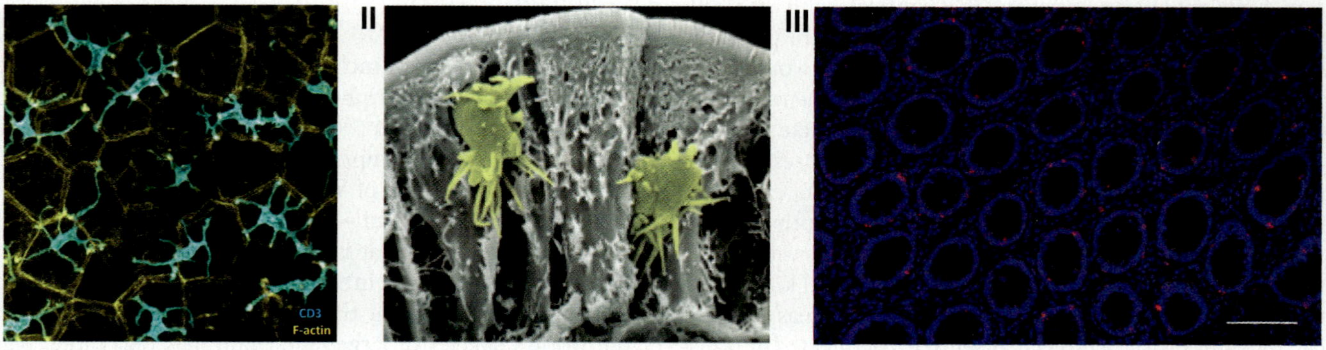

FIG. 35.3. Signature associations of γδ T cells with discrete extralymphoid tissues. Table lists examples of mouse and human TCR Vγ and Vδ region expression documented for indicated organs. Bold type denotes the most widely studied examples. Additional notes highlight widespread γδ T cell distribution in mice, particularly for Vγ6+ cells. Panels illustrate γδ T cells in situ: [I] Mouse Dendritic Epidermal T cells (DETC) visualized in confocal microscopy of epidermal whole mounts—anti-CD3 stain, green; F-actin stain, yellow; colocalization, white. [II] Scanning electron micrograph reveals two mouse small intestinal TCRγδ+ intraepithelial lymphocytes (yellow) [from Ishikawa, H et al. (2007) *Curriculum vitae* of intestinal intraepithelial T cells: their developmental and behavioral characteristics, *Immunol Rev.* 215: p154 to 165]. [III] Human intraepithelial lymphocytes, denoted by TCRδ transcripts detected by RNAScope (pink, ACD probe Hs- TRDC, 433671) and DAPI counterstain in paraffin-embedded normal colonic tissue (US Biomax, BC05002) captured using a Nikon Eclipse Ti-2 inverted microscope and a Nikon DS-Qi2 sCMOS camera (scale bar denotes 100 μm). [IV] Most DETCs visualized by confocal microscopy (detected with anti-TCRδ) (red) directly contact epidermal Langerhans cells (detected with anti-MHC Class II) (cyan), at any one time, while a minority (yellow) are not in contact (scale bar denotes 50 μm; image processed by *Definiens* software).

conventional αβ T cells.[69-71] In mice, lymphoid Vγ1+ T cells and Vγ4+ T cells also display different biologies, as is considered below. Lymphoid γδ T cells in γδ-high cattle likewise fall into discrete categories with different properties.[72]

Extralymphoid

The prototypic example of atypical lymphocyte localization was presented by murine intraepithelial γδ T cells that compose ~90% of T cells in the mouse epidermis and that are morphologically dendritic: hence the term dendritic epidermal T cell (DETC)[26,73-75] (Fig. 35.3). The discovery and study of DETC introduced some key concepts that have since dominated perspectives about γδ T cell biology. First, DETCs do not recirculate to the lymphoid organs, including the tissue-draining LNs. This poses a challenge for any requirement to sample diverse antigens, seemingly consistent with which is the second concept, namely, that there exist T cell repertoires with minimal diversity. Indeed, most DETCs express a "canonical" γδ TCR comprising monoclonal Vγ5 and Vδ1 chains[26,76] (Fig. 35.4). The third concept is that γδ T cells in general are disproportionately represented in intraepithelial lymphocyte (IEL) compartments, where they intimately engage epithelial cells (Fig. 35.3, panels **I-III**), by contrast to the common association of αβ T cells with dendritic cells (DCs). Interesting, agnathan VLR-C+ lymphocyte-like cells (putative γδ T cell counterparts) can also display relatively little diversity and be enriched among intestinal and cutaneous IELs.[77]

DETCs derive exclusively from fetal thymic progenitors.[78,79] Thus, because they cannot be replaced if depleted, they must combine differentiated function with a potential for lifelong renewal. This introduced a fourth concept, namely, of embryo-derived self-renewing tissue-associated immune cell compartments, a concept that later proved applicable to several myeloid compartments including Langerhans cells (LCs) that share the epidermis with DETC. Indeed, at steady state, ~70% of DETCs physically contact LCs (Fig. 35.3 panel **IV**), although neither the basis of this nor its biological implications for either DETC or LC are known.

These four immunological concepts apply variably to γδ T cells found in other murine extralymphoid tissues. For

TABLE 35.1 Representation of γδ T Cells in the Lymphoid Organs/Peripheral Blood in Different Species

Humans	0.1%–15%[46]
Mouse	1%–5%[46]
Rabbit	~20%[46]
Sheep	2%–15%,[379] 30%[46]
Cattle	
Adults	2%–15%,[379] 20%–40%,[45] 30%[46]
Calves	50%–60%
Chicken	20%[46]
Pigs	15%–35%,[47] ~45%[48]
Goats	8%–16%[49]; ~20%[50]
Deer	15%–28%[51]
Camels	~7% (of lymphocytes; WC1+)[52]
Zebrafish	8%–20%[53]
Nurse fish	"high" percentage[54]

Note that shark classification is inferred from molecular data rather than direct cell counting.

example, although γδ T cells compose ~25% to 45% of mouse small intestinal IEL (a far greater γδ:αβ T cell ratio than occurs in lymphoid tissues), and although ~60% to 70% of them express Vγ7 (Fig. 35.3), the limitations on their TCR diversity are very different from DETC. Thus, intestinal Vγ7+ IELs express myriad Vδ genes, and the Vγ7 chains are themselves polyclonal, collectively displaying moderately diverse V-J junctional sequences.[80-82]

Variation in γδ IEL compartments is more evident cross-species. The total numbers of T cells in human skin and gut exceed those of circulating T cells by approximately 2-fold and 3-fold, respectively, and those sites are often relatively enriched in γδ T cells: thus, akin to murine Vγ7+ IEL, the human colonic epithelium is enriched in Vγ4+ cells[83,84] (Fig. 35.3). However, Vγ4+ cells account for only ~25% to 30% of γδ IEL, which in turn comprise only ~10% to 15% of colonic IEL versus 85% to 90% αβ T cells. Second, human tissue-associated γδ T cells irrespective of anatomical site are strikingly enriched in expression of the Vδ1 chain, albeit with some expression of Vδ3 (particularly in the liver) or Vδ5[85,86] (Fig. 35.3), whereas there is no equivalent pan tissue-association of a TCRδ chain in mice. Cattle (as representative of *Artiodactyla*) and chickens also harbor substantial numbers of TCRγδ+ intestinal IEL.[45,87]

Cross-species divergence is extreme in the epidermis, with primates lacking an overt equivalent of a DETC compartment.

Mouse

DETC

```
Trgv5 (Vγ5)
nt:  ...ACGTACTACTGTGCCTGCTGGGAct
AA:  ... T  Y  Y  C  A  C  W  D

Trgj1 (Jγ1)
nt:  atAGCTCAGGTTTTCACAAGGTATTTGCA...
AA:     S  S  G  F  H  K  V  F  A ...

Trdv4 (Vδ1)
nt:  ...ACGTACTACTGTGGGTCAGATATC
AA:  ... T  Y  Y  C  G  S  D  I

Trdd2 (Dδ2)
nt:  atcGGAGGGATAcgag
AA:      G  G  S

Trdj2 (Jδ2)
nt:  cTCCTGGGACACCCGACAGATGTTTTTT...
AA:   S  W  D  T  R  Q  M  F  F ...
```

Uterine/lung/dermal

```
Trgv6 (Vγ6)
nt:  ...ACTTATTACTGTGCATGCTGGGATa
AA:  ... T  Y  Y  C  A  C  W  D

Trgj1 (Jγ1)
nt:  atAGCTCAGGTTTTCACAAGGTATTTGCA...
AA:     S  S  G  F  H  K  V  F  A ...

Trdv4 (Vδ1)
nt:  ...ACGTACTACTGTGGGTCAGATATC
AA:  ... T  Y  Y  C  G  S  D  I

Trdd2 (Dδ2)
nt:  atcGGAGGGATAcgag
AA:      G  G  S

Trdj2 (Jδ2)
nt:  cTCCTGGGACACCCGACAGATGTTTTTT...
AA:   S  W  D  T  R  Q  M  F  F ...
```

Human

Peripheral blood (fetal)

```
TRGV9 (Vγ9)
nt:  ...ACCTACTACTGTGCCTTGTGGGACGTG
AA:  ... T  Y  Y  C  A  L  W  E  V

TRGJP (JγP)
nt:  tgggCAAGAGTTGGGCAAAAAAATCAAGGTATTT...
AA:      Q  E  L  G  K  K  I  K  V  F ...

TRDV2 (Vδ2)
nt:  ...TCTTACTACTGTGCCTGTGACACC
AA:  ... S  Y  Y  C  A  C  D  T

TRDD3 (Dδ3)
nt:  GTACTGGGGGATACg
AA:  V  L  G  D  T

TRDJ3 (Jδ3)
nt:  ctcCTGGGACACCCGACAGATGTTTTTCGGAACT...
AA:      W  D  T  R  Q  M  F  F  G  T ...
```

Peripheral blood (adult)

```
TRDV2 (Vδ2 )
nt:  ...TCTTACTACTGTGCCTGTGACACC
AA:  ... S  Y  Y  C  A  C  D  T

TRDD3 (Dδ3)
nt:  aCTGGGGGATacg
AA:   L  G  D  T

TRDJ1 (Jδ1)
nt:  acACCGATAAACTCATCTTTGGAAAAGGAACCCGT...
AA:    T  D  K  L  I  F  G  K  G  T  R ...
```

FIG. 35.4. **Canonical TCR sequences of innate-like γδ T cell compartments.** Nucleotide and protein sequences shown for invariant Vγ5Vδ1+ DETCs, mouse Vγ6Vδ1+ cells, and commonly identified fetal and adult Vγ9Vδ2+ cells. Blue type denotes CDR3 sequences; lower case denotes gene segment residues not included in the productive rearrangements; underlined denotes P-nucleotide-derived; red arrows (in Vδ2) denote amino acid 97, which is almost always L, V, or I in PAg-reactive TCRs. Note that gene nomenclature does not always align with encoded chain, for example, mouse Vδ1 is encoded by the Trgv4 gene segment.

However, the dermis of primates and of rodents contains sizable γδ T cell compartments[88-93] (Fig. 35.3), wherein the cells predominantly interact with fibroblasts and/or immune cells rather than with epithelial cells. In cattle, γδ T cells are found in dermis and to a lesser extent epidermis,[56,94] but they are rare in the skin of sheep or chickens. In mice, the dermal γδ cell compartment includes a subset of cells expressing an invariant Vγ6Vδ1 TCR that features an identical Vδ1 chain to that paired with Vγ5 in DETC (Figs. 35.3 and 35.4) and that, like DETC, also derive exclusively from fetal progenitors.[88,95] Of note, cells expressing the invariant Vγ6Vδ1 TCR share the dermis with a small number of semi-invariant Vγ6Vδ1+ clones and an oligoclonal population expressing Vγ4 paired with different TCRδ chains (Fig. 35.3),[95] considered further below.

Unlike the canonical Vγ5Vδ1 DETC TCR and intestinal IEL Vγ7+ TCRs, the invariant Vγ6Vδ1 TCR is widely distributed, being reported in (at least) lung, tongue, reproductive tract, liver, peritoneum, entheses, gingiva (mucosal tissue surrounding the teeth), adipose tissue, meninges, and mammary fat pad (Fig. 35.3).[37,96-106] Note that some of these anatomical distributions seem conspicuously conserved across broad evolutionary distances, as illustrated by the predominance of γδ T cells in the lactating mammary tissue of opossums.[107] Also of note, uterine Vγ6Vδ1+ cells do not appear to be primarily intraepithelial (as was originally claimed),[97] and this may be true for Vγ6Vδ1+ cells in other sites, with the possible exception of the gingiva (below): indeed, most properties of Vγ6Vδ1+ cells are distinct from those of IEL, as will be considered below. This highlights an important point, namely, that extralymphoid γδ T cell compartments are not limited to IEL but include those in the subepithelial stroma. Thus, the intestinal lamina propria may harbor γδ T cells lying just microns away from Vγ7+ IEL from which they are distinct by their TCR usage, functional potentials, and developmental origins, as will be considered below.[108,109]

In addition, the phenotypes of tissue-associated γδ T cells can be strongly influenced by their anatomical site, as is true for other immune cells. For example, mouse uterine Vγ6Vδ1+ cells express progesterone receptor to a much higher degree than counterpart lung γδ T cells, whereas the latter express genes encoding surfactant proteins (eg, Sftpb, Sftpc) and regulators of surfactant protein maturation (eg, Lamp3).[97] Likewise, the transcriptome of dermal Vγ6Vδ1+ γδ T cells depicts strikingly homogeneous cells that can be distinguished from thymic or peripheral lymph node counterparts, for example, by their steady-state expression of RNAs for effector molecules, IL-17 and amphiregulin.[93] However, the tissue of residence does not override cell-intrinsic programs: thus, dermal Vγ6Vδ1+ T cells invariably express Scart1 (a scavenger receptor), by contrast to colocalized dermal Vγ4+ cells, which express Scart2.[93] Another example of this is seen in the gingiva, in which the local γδ T cell compartment is conspicuously diverse, including Vγ1+, Vγ4+, Vγ5+, and Vγ6+ T cells with distinct functional phenotypes.[100]

Not all organs, for example, human epidermis, contain overt γδ T cell compartments: indeed, it is unclear what limits the cells' association with certain tissues. Nonetheless, the striking breadth of organs with which γδ T cells can be associated (Fig. 35.3) presumably reflects the cells' collective capacity to directly interact with many cell types, for example, epithelial, fibroblastic, myeloid, neuronal, and lymphoid cells. It is reasonable to assume that these different interactions will provoke effector functions suited to the cells' locations. Given the association of defined TCRs with discrete anatomical sites, we can now consider the evidence that defined γδ TCRs are associated with discrete effector functions.

γδ TCR-ASSOCIATED EFFECTOR POTENTIALS

Experimental Preface

Studies of γδ T cell biology have consistently employed state-of-the-art experimental systems but nonetheless have been confounded by several factors that have limited many studies to being descriptive and that have qualified some conclusions. For example, unlike the restriction of αβ T cells by MHC, there is to date no natural genetic system for validating and for better understanding γδ T cell responses. Moreover, there has been no γδ T cell–specific primary immunodeficiency to offer insights comparable with those provided for B cells by *BTK* and other haplotypes. Third, there has been relatively little impact of adoptive γδ T cell transfer experiments, in part because the identification of biologically relevant antigen specificities has been rare (see below), and in part because γδ T cell compartments under study are often tissue intrinsic rather than systemic and not readily reconstituted from adult hematopoietic cells. Because of this, conclusions drawn from γδ T cell–deficient mice are seldom validated by rescue studies and hence are open to different interpretations; for example, an observed phenotype might reflect an absence of γδ T cells or poorly understood activities of poorly characterized "replacement" lymphocytes that occupy niches that would ordinarily be filled by γδ T cells. This situation is changing with the introduction of newer models, for example, inducible depletion of γδ T cells or their ligands from adult animals, from which new insights are already being gained. Finally, whereas much has been learned about other immune cells by virtue of immunoevasion strategies employed by microbes and viruses, for example, downregulation of MHC and NKG2D ligands, it has not yet been possible to link a molecular means of immunoevasion specifically and uniquely to γδ T cell biology. These important caveats notwithstanding, sufficient has been learned about γδ T cells to justify their classification as a unique and essential lymphocytic lineage.

TCR Vγ Regions Associate With IL-4, IL-17, and IFNγ

Historically, a small number of independently conducted studies of systemic γδ T cells found associations, albeit inexact, of Vγ gene segment usage with their effector functions.[110-113] For example, Vγ4+ cells promoted coxsackie B3-virus-induced myocarditis in mice, whereas Vγ1+ cells suppressed it.[113] Conversely, Vγ1+ cells were observed to promote airway hyperresponsiveness and Th2 cytokine release, whereas these phenotypes were suppressed by Vγ4+ cells.[110] In parallel studies, IL-4 production by mouse splenic and liver γδ T cells was strikingly associated with the use of Vγ1Vδ6.3 or Vγ1Vδ6.4 TCRs, frequently of limited junctional diversity.[114-116] Those IL-4-producing cells were later found to express PLZF, a BTB zinc-finger transcription factor (Zbtb16) that in part determines the cells' rapid and robust "innate-like" responsiveness to stimulation and that regulates expression of phenotypic markers ordinarily associated with antigen-experienced cells and with IL-15-stimulated NK cells.[117] Those markers include receptors for cytokines to which the cells thereby become highly responsive. Because this "innate" phenotype is displayed by TCRαβ+ NKT cells, and because they express the marker NK1.1, IL-4-producing Vγ1+ T cells have been termed γδ NKT cells.[114]

In another illustrative setting, immunopathology in a mouse model of collagen-induced arthritis was exacerbated by IL-17 that was produced overwhelmingly by Vγ4+ γδ T cells, rather than by Vγ1+ T cells.[118] Such observations were strikingly validated when an activation marker, CD27, was shown to functionally segregate peripheral γδ T cells; thus, CD27+ cells (the majority) were mostly CD62L+CD44−, were disproportionately enriched in Vγ1+ T cells, and mostly produced IFNγ and TNFα upon polyclonal stimulation. Conversely, CD27− cells were mostly CD44+ CD62L^{lo}, often expressed Scart2, were ∼5-fold enriched in Vγ4+ T cells, and (within only 4 hours of stimulation) mostly produced IL-17.[31] IFNγ production by CD27+ cells was promoted by Tcf1 (encoded by the *Tcf7* gene), but this was suppressed in IL-17-producing γδ T cells by the action of the transcription factor *c-maf* that concurrently promoted upregulation of Rorγt, a nuclear receptor and master regulator of IL-17 production by T cells.[119-121] This IL-17-skewed phenotype is also associated with the expression of sox13, which was once considered a marker for γδ T cells,[120-122] and upon which IL-17-producing Vγ4+ T cells completely depend.[120,121] Clearly, Vγ4+ cell skewing toward IL-17 could explain their promotion of myocarditis, as considered above.

Because of these effector polarizations, it has become commonplace to denote γδ T cells as γδ17, γδTc1, and γδTc2 cells, respectively. However, this terminology is not used here because it can obscure the pleiotropic functional potentials of γδ T cells that may also segregate with TCR usage. For example, mucosal γδ T cells with the potential to produce IL-17 may, in particular circumstances, be skewed toward the production of IL-22 or amphiregulin that are implicated in tissue repair, rather than inflammation.[100] In the same vein, parasite recrudescence in mice infected with *Plasmodium chabaudi* was limited by M-CSF that was almost exclusively attributable to expanded Vγ1Vδ6.3+ cells, ordinarily associated with IL-4 production.[123]

CD27+ and CD27− γδ T cells are also regulated very differently. Thus, CD27+ γδ T cells, which are primarily dependent on glycolysis, proliferate strongly in response to TCR-mediated stimulation.[124,125] By contrast, CD27− IL-17-producing cells mostly display oxidative metabolism and increased mitochondrial activity, and they respond less well to TCR-mediated stimulation.[125] Indeed, following *Plasmodium berghei* infection, the response of IL-17-producing CD27− cells was attenuated by MyD88 deficiency, which would ordinarily transduce IL1β signals that are associated with innate immunity.[124] Conversely, responding IFNγ-producing cells were mostly unaffected by MyD88 deficiency but were impaired by deficiency in CD27, which ordinarily limits apoptosis of cells receiving strong TCR-mediated stimulation.[124] For these and additional reasons considered below, CD27−CD44+ IL-17-skewed cells are often considered more innate-like, compared with an adaptive biology of CD27+ IFNγ-committed cells. Of note, the phenotypic segregation of IFNγ with CD27+ cells versus IL-17 with CD27− γδ T cells was de facto magnified by malaria infection, consistent with its hypothesized value in rapidly facilitating appropriate, preprogrammed responses to challenge.[124] Underlining this, CD27+ γδ T cells showed epigenetic suppression of the gene regulatory network associated with IL-17 production.[126] Nonetheless, this is not universal since the CD27− compartment can display IL-17+IFNγ+ double producing cells,[108] particularly at the outset of a response, as is considered toward the end of this chapter.

Of importance, functional segregation according to CD27 (and CD44 and CD62 L) was already apparent in thymic γδ T cells, including fetal thymic progenitors that can constitutively express transcripts for either IL-17 and Rorγt (CD27− cells) or IFNγ and T-bet, a master regulator of IFNγ production (CD27+ cells).[31] The subsets were approximately equally represented at embryonic day (E) 14, whereas a numerical dominance of CD27+ cells became apparent between E15 and E18. Given their origins in the fetus, these phenotypically skewed cells evidently reflect developmental precommitment, rather than mature polarized effector T cells recirculating through the thymus.[31]

Polarized Functions of Epithelial and Nonepithelial Invariant γδ T Cells

Notwithstanding the common association of IL-17 production with Vγ4 usage, many IL-17-committed γδ T cells express canonical Vγ6Vδ1 rearrangements (Fig. 35.4). Such cells, which can show more variability in CD27 expression, are derived from the fetal thymus (as considered above),[127,128] whereafter they primarily populate subepithelial stroma of mucosal tissues and the dermis (Fig. 35.3). Of interest, they also phenocopy γδ NKT cells (above) in expressing PLZF, which was found to be required for the proliferation of their progenitors and the acquisition of a mature, innate-like IL-17-producing phenotype.[129]

PLZF was also expressed by Vγ5+ DETC progenitors that develop slightly in advance of canonical Vγ6Vδ1 thymocytes. However, PLZF deletion had no overt impact on the phenotypic or functional differentiation of DETCs, which do not make IL-17.[127,129] Rather, DETCs have been observed under different circumstances to produce IL-13, chemokine Xcl1, keratinocyte growth factor (KGF, a.k.a. FGFVII), vascular endothelial growth factor (VEGF), and insulin-like growth factor (IGF) and to have cytolytic potentials.[130-135] This skew of DETC away from IL-17 production is phenocopied by murine Vγ7+ intestinal IEL, human colonic Vγ4+ IEL, and possibly all IELs.[84,136] Murine and human IELs have been associated with an IFNγ-producing cytolytic effector phenotype, and the production of chemokine RANTES (CCL5),[137,138] while murine TCRγδ+ IELs have also been reported to produce KGF.[130] At a distance of only a few microns away from IEL, subepithelial dermal and intestinal lamina propria γδ T cells, respectively, commonly express Vγ4 and primarily produce IL-17[95,139] (Fig. 35.5).

In humans, the most obviously tissue-associated TCR chain is Vδ1 (Fig. 35.3). For the most part, such cells fail to make IL-17 and can be defined by a rapidly deployed, cytolytic, IFNγ-producing phenotype. Not only does this hold for intestinal Vγ4+ IEL but also seems true for human skin and breast Vδ1+ cells that are not overtly intraepithelial.[140-142] This contrasts with the skewing of murine subepithelial γδ T cells toward IL-17. Likewise, the majority of human lymphoid γδ T cells seem skewed against IL-17 production, with IL-17-producing human γδ T cells having proved challenging to identify, even by use of relatively elaborate, IL-17-skewed protocols.[143] Similarly, human fetal Vγ9Vδ2+ T cells were reportedly preprogrammed for IFNγ production.[144] Nonetheless, by using transcriptomics a cluster of Vγ9Vδ2+ T cells was identified in neonatal and adult blood, that coexpressed CCR6, RORC, and IL23R, which are markers of IL-17 producing potentials in mice, although it has not proven straightforward to elicit IL-17 production from those cells.[145] These findings have important implications because the rapid and considerable contributions of mouse IL-17-producing γδ T cells to protective and immunopathologic inflammatory responses (to be considered below) may not be echoed to the same degree in human pathophysiology. Conversely, some data suggest that expansion of human γδ T cells atypically producing IL-17 may be a consequence of inflammatory environments. Such human IL-17 producers may further exacerbate inflammation, for which they act as a sentinel marker.[146,147] Moreover, there may be a higher likelihood of IL17 skewing in newborns, which would have some parallels with mice.[147,148] Likewise, transcriptomics identified human γδ T cells with Th2 signatures that were enriched in neonates,

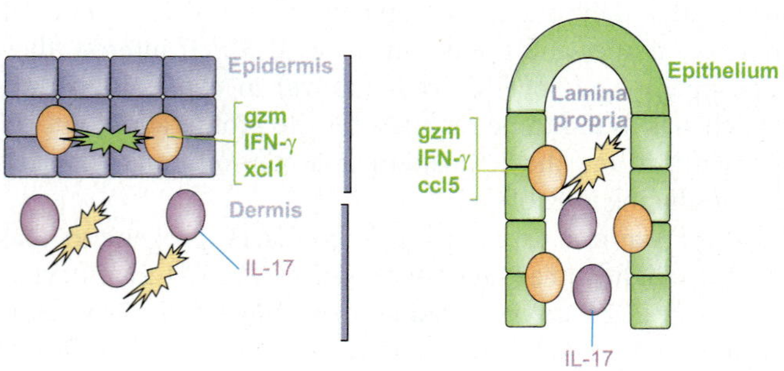

FIG. 35.5. Mouse extralymphoid γδ T cell compartments are distinguished by discrete preprogrammed effector potentials. A schematic of the mouse skin and small intestine indicates the contrasting effector potentials of subepithelial γδ T cells (dermis; lamina propria) (purple), which are skewed toward IL-17, and intraepithelial γδ T cells (orange), which in the same organs are skewed away from IL-17 and toward a cytolytic/IFNγ phenotype that includes elaboration of signature chemokines, as indicated.

as is the case for αβ T cells,[145] consistent with earlier studies of fetal-derived versus postnatal-derived γδ T cell clones.[149]

Thus, there is an emerging picture of γδ T cells showing discrete γδ TCR-associated competences to rapidly deliver site-specific effector functions in highly diverse settings. As in all biological arenas, the granularity of these definitions is being increased by single cell analyses.[93,121,145] Thus, it is now appropriate to consider the forms of TCRγδ ligand recognition that will contribute to the collective pleiotropic effector potentials of γδ T cells. Given the coverage devoted to TCRs (αβ and γδ) in Chapter 23, this consideration of TCRγδ will focus on how ligands relate to γδ T cell biology rather than on the biochemical and biophysical parameters of ligand engagement.

T CELL RECEPTOR γδ

Unrestricted Adaptive Diversity

As noted, murine and human γδ TCRs have very high potential for diversity, primarily attributable to TCRδ, whereas murine and human TCRγ chain composition can draw upon only six functional Vγ gene segments, that each recombine to one of relatively few Jγ gene segments (Fig. 35.1B, above). As mentioned, γδ cells in some species compose a greater percentage of circulating T cells, correlating with which are expansions of the TCRγ locus, frequently as duplicated V-J-C cassettes: three in dromedary camels, six in sheep, seven in cattle and goats.[150] These can be sited in two distinct loci on the same chromosome. Nonetheless, no functional significance has yet been assigned to the use of different Cγ-regions.

The murine and human TCRδ genes also comprise relatively few V-gene segments, although they too can be expanded in some animals: 65 in sheep, 50 in cattle, 31 in chickens.[151] This notwithstanding, the uniquely high potential for TCRδ diversity resides within the V-D-J junctions, which in humans can additionally include V-D-D-J and V-D-D-D-J rearrangements (for which the only precedent is in shark IgNAR receptor genes) and which can be translated in multiple reading frames. Added to this, some ruminant genomes contain nine Dδ and four Jδ gene segments.

Thereby, the potential for recognizing immensely diverse antigens is realized without devoting much genomic space to multiple V-gene segments (as is the case for IgH) or multiple V- and J-gene segments (as is the case for TCRα). This constraint on Vδ expansion may in part reflect the location of the TCRδ locus within the TCRα locus (Fig. 35.1B, above): thus, Vδ gene duplication could push Vα gene segments to very long distances 5′ of Jα, reducing the efficiency of rearrangement. Indeed, the TCRα/δ locus of sheep comprises 2882 mb.[151] This notwithstanding, there is some interchangeable usage of Vα and Vδ (Fig. 35.1B, above) and even of VH gene segments, in recombinations to Dδ and Jα, respectively. This interchangeability may imply a core structural similarity of Vδ and Vα that may underpin the propensity to identify MHC and MHC-like molecules, including CD1 and MR1, among the few candidate ligands proven to bind TCRγδ. Because γδ T cells were unanticipated, many initial and subsequent investigations sought parallels with αβ T cells, foremost among which was whether TCRγδ recognized processed antigens presented by MHC, CD1, and/or some other element(s).

γδ T Cells and MHC

Given early indications that γδ T cells were not restricted by polymorphic MHCI or MHCII, there was much interest in nonpolymorphic MHCI-related proteins, sometimes termed MHC class 1b. Those include Qa1, which was reported to present a synthetic copolymer poly(Glu50Tyr50) to a γδ T cell hybridoma.[152] Shortly thereafter, murine γδ T cells were reported to recognize peptides derived from mycobacterial heat shock protein, Hsp65, that shared properties with an Hsp65-derived peptide recognized by αβ T cells, but that were presented by an undetermined, nonpolymorphic molecule(s).[153] Moreover, cross-reactivity was reported to peptides derived from endogenous Hsp63.[153] Likewise, it was reported that autologous APCs were optimal for the recognition by human Vγ9Vδ2+ T cells of *Mycobacteria leprae* cell wall antigens or tuberculin-derived purified protein derivative (PPD), although it was again not possible to implicate MHCI or MHCII as the basis for this recognition.[60] Subsequently, there have been several instances of γδ TCR recognition of lipid-CD1 complexes[154-156]; examples of γδ recognition of riboflavin-derived metabolites presented by MR-1,[157] a murine γδ TCR derived from a nonobese diabetic (NOD) mouse that recognizes an insulin-derived peptide,[158] and more recently, human γδ TCRs with specificity for melanoma-derived peptides presented by HLA-A.[30] Conspicuously, most "cargo specificities" described to date are for self-antigens.

Notwithstanding these reports, processed antigens are neither the primary nor obligatory targets of TCRγδ recognition. Moreover, there is no evidence for the cells' obligate restriction by MHCI or MHCII or by any other polymorphic antigen-presenting molecule in either mice or humans.[153,159] Indeed, in relation to Hsp peptide recognition, contemporaneous datasets indicated that human γδ T cells could recognize human hsp homologs (eg, Hsp58) as native proteins presented on the surface of tumor cells (below; see also Fig. 35.2B).[160,161] Furthermore, murine γδ cell compartments develop seemingly normally in mice lacking β2-microglobulin (β2m) (an obligate cofactor for most MHCI-related molecules, including CD1 and many class Ib molecules) and in MHC II-deficient mice.[162,163] In addition, human peripheral blood γδ T cell repertoires showed no obvious association with MHC haplotype and moreover could be very different in identical twins.[61] These data do not preclude roles for MHCII in the regulation of lymphoid and extralymphoid γδ T cell compartments, as has been reported,[164,165] but they are consistent with the fact that most γδ T cells express neither CD4, which facilitates TCRαβ responses to MHCII-bound ligands, nor CD8αβ, which facilitates TCRαβ responses to MHCI-bound ligands. It can also explain the striking contrast in CDR3 length diversity, which is conspicuously limited in αβ TCRs, reflecting the constraints of binding to MHC proteins, and highly variable in γδ TCRs, particularly TCRδ, which is most similar to IgVH (above), signifying no evident structural constraints imposed by an antigen-presenting element.[166]

Thus, TCRγδ shares with Ig an adaptive potential to recognize qualitatively heterogeneous antigens that may be extremely diverse in size and composition, including native proteins, presented peptides, and haptens, at minimum, and that may be expressed by myriad target cell types. Important evidence that this adaptive potential is realized has been provided by next-generation sequencing of human thymic and peripheral blood γδ T cells. The thymic TCRγ/δ sequences were extremely diverse, and although a small fraction of TCRγ clonotypes was shared across donors, the TCRδ repertoires were almost exclusively private, that is, specific to single individuals.[167] One peripheral blood analysis excluded Vδ2+ cells because of their innate-like responsiveness to PAgs (above, and to be considered further below), and the residual cells (primarily Vδ1+) showed profound clonal expansions that strikingly reduced D50 values (the number of clones accounting for 50% of the repertoire sequences), and that also were mostly private.[69] Such expansions were enriched in cells displaying an "antigen-experienced" $CD27^{lo}$CXCR3+ phenotype, with enhanced expression of granzymes, consistent with their effector differentiation following clonal expansion,[69] a defining hallmark of adaptive immunity. Moreover, the expansions could be strikingly durable, consistent with adaptive immunological memory. Similar clonal expansions of effector-phenotype were observed to dominate the liver Vδ2− T cell compartment,[168] and there were likewise striking clonal expansions among peripheral blood γδ T cells that expressed Vδ2, but not paired with Vγ9.[169] Predictably, considerable effort is being devoted to elucidating the targets of such clonally expanded TCRs. Based on the nature of validated TCRγδ ligands that will be considered next, those TCRγδ targets are likely to include native cell surface proteins (see Fig. 35.2B).

Cell Surface Ligands

The relatively few examples of activating ligands for which there is biochemical evidence of γδ TCR engagement are conspicuously enriched in native membrane-associated polypeptides. Lists of these have been presented in various reviews and detailed considerations offered in Chapter 23. Those proteins include mouse MHCII[170]; the closely related MHCIb molecules, T10 and T22[171-173]; CD1[154]; endothelial protein C receptor (EPCR), which is CD1-related[71]; MR-1[157]; and MICA.[174] Moreover, in some cases lipid and peptide "antigens" appear primarily to stabilize CD1 or MHC, which are the primary targets of TCR recognition, rather than being recognized as presented cargoes.[154,155]

These protein antigens share three key features: they are seemingly recognized by relatively few γδ TCRs, consistent with their being adaptive, clonotypic ligands; they are self-encoded; and they can serve as sentinels of target cell status, being altered in expression by a variety of "stressors" including infection and cell transformation, although the basis for such regulation is largely unknown. Likewise, a human

Vγ8Vδ3 TCR isolated from a CMV-infected transplant recipient bound native annexin A2 that was increased on the surface of tumor cells by reactive oxygen stress and that could seemingly stimulate polyclonal Vδ2$^-$ peripheral blood γδ T cells.[175] In addition, although direct TCR binding was not evident, a Vγ9Vδ1 TCR isolated from a CMV-infected kidney transplant patient was specific for native Ephrin A2 (EphA2), a GPI-liked membrane protein upregulated by metabolic reprogramming of cancer cells and of CMV-infected cells.[176] A triangular complex was proposed whereby EphA2 bound to its ligand Ephrin A expressed by γδ T cells, while an additional loop on EphA2 bound to the TCR: hence Ephrin A functioned akin to CD8β as a coreceptor for the TCR.[176]

Binding was reported of a Vγ9Vδ2 TCR to a form of F1-ATPase that was surface expressed by tumor cells,[177] and likewise binding of Vγ9Vδ2 TCRs to Butyrophilin (BTN) 2A1 has been clearly shown.[178,179] Neither of these Vγ9Vδ2 antigens can be classified as an adaptive, clonotypic ligand. Rather, they seemingly facilitate polyclonal Vγ9Vδ2+ T cell recognition of infected cells harboring PAgs, as is considered further below.

In short, documented TCR ligands would seem to provide γδ T cells with various means to recognize and target dysregulated cells without requisite intermediaries of antigen processing and antigen-presenting cells (see Fig. 35.2B). This is consistent with the implications of γδ T cells in lymphoid stress surveillance and cancer detection, considered below, but the circle is not yet closed in that no clear biological consequence has been shown for either preventing or ablating any of the γδTCR-ligand interactions described. Moreover, the cited antigens are not obviously exclusive to dysregulated cells, and whether or how their expression on normal cells is ignored by γδ T cells remains to be clarified. In sum, the ligands cited require much clearer biological contextualization.

A possible solution to this conundrum is that the fine clonotypic specificity of any single γδ T cell is not essential for the response to infection or toxic dysregulation of cells, since repertoire diversity ensures that another γδ T cell could mediate the response by engaging another molecular moiety expressed on the target cell surface as a result of cellular dysregulation (see Fig. 35.2B). In sum, the primary requirement may be for γδ T cells to recognize changes in the cell surface expression profile that are rapidly induced by infection or dysregulation, rather than to recognize changes in any one specific antigen. In other words, γδ T cells commonly recognize atypical cellular states (cell pathology) rather than recognizing the perpetrators of change that are instead recognized by αβ T cells and B cells (Figs. 35.2A, B).

γδ TCRs can also be specific for low-molecular-mass moieties, including the haptens, 4-hydroxy-3-nitrophenylacetyl (NP) and cyanine 3.[180] For a cell-surface TCR to recognize free low-molecular-mass moieties would seem of questionable value; thus, it seems likely that such reactivities reflect recognition of modifications to cell-associated proteins and possibly other macromolecules such as glycans. Hence, these reactivities broaden the case for γδ TCRs being sensitive to altered states of endogenous cell surface molecules.

Self Versus Foreign Ligands

In the 1990s, the TCR of a γδ T cell clone derived from mice infected with herpes simplex virus (HSV) 1 was shown to recognize backbone conformational determinants of the HSV glycoprotein, gI.[181,182] Moreover, a broader capacity of TCRγδ to recognize microbial antigens was established by the finding that human, bovine, and mouse γδ TCRs can bind to phycoerythrin (PE), a protein encoded by alga and cyanobacteria.[183] Indeed, by use of multimerized antigen, PE reactivities were deduced to be shared by between 1/1000 and 1/10,000 peripheral blood T cells, consistent with the adaptive immunity paradigm, albeit that PE stimulation drove effector differentiation more obviously than clonal expansion.[183] However, as was considered to be the case for self-encoded cell surface ligands, PE reactivity remains of unresolved biological significance.

To date, these two examples of microbial γδ TCR ligands seem highly exceptional. Despite irrefutable evidence of γδ T cell responses in myriad virus infections of mice and humans, for example, CMV,[184-186] HIV,[187,188] HSV,[189] West Nile virus (WNV),[190] Vaccinia,[191] and SARS-CoV-2[192,193]; bacterial infections, for example, *Listeria*,[102,108,194-196] *Borrelia*,[197] *Salmonella*,[103,194] *Brucella*,[198] *Erlichia*,[199] tularaemia,[200,201] and mycobacteria[60,202-206]; parasite infections, for example, malaria,[123,124,207-210] *Cryptosporidium*,[211] *Toxoplasma*,[212,213] and coccidiosis[214,215]; and fungal infections such as *Candidiasis*,[97] pathogen-specific γδ TCR reactivities have only very rarely been convincingly demonstrated. Consistent with this, there have been very few examples of protection of naïve animals achieved by adoptive transfer of γδ T cells from infected or vaccinated mice, and in such cases, no clear evidence for pathogen specificity. Clearly, absence of evidence is not evidence of absence, and microbial specificities may emerge as ligand searches are more intensively pursued, particularly in animals with expanded γδ T cell compartments. Nonetheless, to highlight the paradox we shall consider three illustrative examples of infections by a virus, bacterium, and protozoan, respectively.

Ligands in Context: Lessons From CMV, *Borrelia*, and *Plasmodium*

In patients receiving hematopoietic stem cell transplants, peripheral blood γδ cells were rapidly reconstituted and showed striking, patient-specific, clonal expansions that frequently coincided with CMV reactivation.[70] Major clonal expansions of Vγ8Vδ1+ T cells were likewise associated with vertical CMV transmission in utero,[186] and clonal expansions of Vδ1+ and Vδ3+ T cells were associated with cell activation markers in a CMV-exposed cohort of Ugandan infants.[216] Furthermore, major clonal expansions were observed in heavily immunosuppressed recipients of solid organ grafts displaying CMV infection, and among which several expanded clones (including ones known as "LES" and "MAU") lysed CMV-infected cells.[71,217] Paralleling these studies, mouse γδ and αβ T cells responded to infection with murine CMV (that is ordinarily controlled by NK cells), and genetic ablation of both T cell types compromised virus control, whereas loss of either alone did not.[185,218] *Prima facie* these data collectively offer a strong case for an adaptive, CMV-specific, immunoprotective γδ T cell response.

However, when the TCR specificity was deduced of the human LES clone, it proved to be for a self-encoded CD1-related protein, EPCR, that was recognized in native conformation, independent of lipid, and with a Kd of ~90 μm.[71] Essentially the same circumstances led to the identification of human γδ T cell specificities for EphaA2 (for the clone MAU) and Annexin-2 (above). Each antigenic target can be upregulated by CMV infection. Moreover, LES, MAU, and other γδ T cell clones also showed CMV-independent recognition of tumor cells overexpressing their target ligands.[71,175] Indeed, there are data that EPCR, Annexin-2, and EphA2 can be naturally overexpressed in a variety of human tumor samples. Thus, the human γδ T cell response to CMV is characterized by rare, clonotypic expansions, but those so far characterized are not pathogen specific and instead show immunosurveillance responses against transformed cells as well as infected cells.

In a second setting, expansions of diverse Vδ1+ T cells can be observed in the synovial fluids of persons with Lyme arthritis, a chronic rheumatic disease induced by *Borrelia burgdorferi*.[219] Moreover, γδ T cell–deficient mice show increased bacterial burden and heightened cardiac inflammation following *B. burgdorferi* infection.[197] Again, these data collectively portend the existence of adaptive *B. burgdorferi*–specific γδ T cells in humans and mice, but again this seems not to be the case. Indeed, a Vγ9Vδ1 TCR derived from one such clone was reactive to uninfected human epithelial cells and fibroblasts and to IL1β-treated monocytes and T cells.[220]

In the third case, γδ T cells were, as already mentioned, expanded in the livers, lungs, and spleens of *Plasmodium chabaudi*–infected mice and contributed substantially to limiting malarial recrudescence by their production of M-CSF.[123] However, this was one of several settings (considered further below) that have been termed "quasi-adaptive" in which expanded γδ T cells were not diverse, in this case mostly expressing TCR Vγ1Vδ6.3, with no evident pathogen-specific reactivities.[123] In humans, some subjects repeatedly vaccinated with malarial sporozoites showed efficacious immunoprotective responses, and this was conspicuously associated with γδ cell activation.[221] However, neither this nor any other vaccination setting in mice or humans has evidently evoked vaccine antigen-specific γδ T cells.

Note that vaccine-specific γδ T cell responses have been reported for ruminants that have expanded γδ T cell compartments. In cattle, γδ T cells are the first to respond to vaccines against the important zoonotic pathogen, *Leptospira* (see below).[222,223] Similarly, the higher production of IFNγ and the higher percentage of γδ T cells in neonatal cattle were both correlated with an increased efficacy of *Mycobacterium bovis* BCG vaccination when compared with its relatively poor efficacy in older animals.[222,224] In addition, goats administered a *Salmonella*-vectored *Mycobacterium paratuberculosis* vaccine displayed γδ T cells that responded to *M. paratuberculosis* antigens in recall responses,[225] while γδ T cells in sheep responded to tuberculosis antigens in both primary and secondary responses.[226] Bovine γδ T cells have also been reported to respond to cells infected with the protozoan parasite *Theileria parva*.[227] However, assessing foreign antigen specificities of γδ T cells in ruminants is complicated by the coexpression of the TCR with a family of molecules known as WC1, a scavenger receptor that has its own capacities to recognize microbial determinants,[72,222] as is discussed further below.

Collectively, these observations raise fundamental questions, including how potentially diverse γδ TCRs are seemingly focused on self-antigenic specificities in multiple settings of antigen-rich pathogen exposures. It seems unlikely that TCRγδ repertoires, which can be so diverse, have blind spots for pathogen-specific antigens. Instead, it may be the case that such responses were outcompeted by those to self-encoded antigens, because early after cell infection or dysregulation, the altered expression of self-encoded cell surface molecules may be more overt than the surface display of any microbe-encoded or virus-encoded proteins. Thus, high TCR avidity may drive the cells' rapid responses to high quantities of self-antigens (not limited to proteins) that are quickly upregulated on infected cells and the fine specificity of which is not of paramount importance. This rapid recognition might be aided by ligand-receptor engagements that do not involve the TCR, for example, costimulators, scavenger receptors, or cytokine receptors. But if γδ T cells were not to be tolerized to self-antigens (see below), then what checkpoints limit their precocious responses against innocuous targets? To address these questions will require more detailed study of how and where γδ T cell responses to infection and/or cell dysregulation are naturally initiated.

γδ TCR: Lessons From Mycobacteria

γδ T cells have been very strongly associated with mycobacterial responses in humans and other animals. Indeed, human Vδ1+ T cells were first associated with human disease by the finding that the cells' representation among total T cells increased 5- to 8-fold in early active granulomatous responses to *Mycobacteria leprae*.[60] Those expanded γδ T cells responded to mycobacterial extracts that seemingly required some form of presentation, although there was no evidence for MHC restriction.[60] In parallel, mouse γδ T cells were reported to respond to mycobacterial hsp65, although such responses were difficult to distinguish from self-reactivities to hsp63, as was considered above. Moreover, added to the several implications of γδ T cells in mycobacterial responses in ruminants (above), γδ T cells were reported to be the first cells to arrive at sites of mycobacterial purified protein derivative (PPD) injection into cattle, correlating with increased IFNγ production and preceding the $CD4^+$ αβ T cell response.[228] Thus, with growing associations of γδ T cells in mycobacterial infections including tuberculosis, an intensive search for TCR ligands commenced that focused on human Vγ9Vδ2+ cells, since those are readily accessible from human peripheral blood and are frequently found in mycobacterial lesions.[62,63,229]

Somewhat surprisingly, that search converged on low-molecular-mass, nonpeptidic pyrophosphates,[62,67,68,229] specifically hydroxymethylbut-2-enyl pyrophosphate (HMBPP),[66] which is an intermediate in the methylerythritol phosphate (MEP) pathway that most (but not all) bacteria, including *M. Tb* and *M. leprae*, and some protozoa, including *Plasmodium*, use to synthesize isoprenoids (a.k.a. terpenoids), starting from glyceraldehyde 3-phosphate and pyruvate (Fig. 35.6). Isoprenoids are essential sources of steroids and sterols and facilitate the critical isoprenylation of proteins such as ras, thereby permitting their membrane association. However, in vertebrates, and in some microbes, isoprenoids are generated by the mevalonate pathway that starts with acetyl-CoA and does not feature HMBPP as an intermediate (Fig. 35.6). Thus, HMBPP specificity is a clear example of microbe-associated molecular pattern (MAMP) recognition that is ordinarily assigned to innate immunity rather than to adaptive immunity.

HMBPP reactivity requires the pairing of Vγ9 with Vδ2 cells and a CDR3δ that in most cases includes a leucine, valine, or isoleucine residue at position 97 ("d97LVI"). As mentioned, Vγ9Vδ2 cells typically compose the dominant human peripheral blood γδ T cell subset and ~80% of them include d97LVI (see Fig. 35.4). Therefore, the HMBPP response is nonclonal, which is a defining hallmark of innate immunity that contrasts with the clonotypic nature of adaptive responses. Furthermore, whereas Vγ9Vδ2+ cells respond to picomolar HMBPP, they will also respond to isopentenyl pyrophosphate (IPP) (albeit in higher concentrations),[62,68] which is an intermediate common to the MEP and mevalonate pathways (Fig. 35.6). Hence, the application of the term phosphoantigen (PAg) to the multiplicity of organic pyrophosphates that can provoke TCR-dependent Vγ9Vδ2 T cell activation and that can be of microbial, endogenous, and synthetic origins.[62,67]

The mechanism of PAg recognition is not fully understood, but it depends on BTN3A1 and BTN2A1, with the latter shown to directly bind Vγ9Vδ2 TCRs via Vγ9 (also referred to as Vγ2 in some papers)[178,179,230-234](Fig. 35.7A). BTN3A1 and BTN2A1 heterodimers are expressed essentially ubiquitously, and many cell types, including Vγ9Vδ2+ T cells, can support Vγ9Vδ2 responses to PAgs. However, this is unlikely to reflect antigen presentation akin to MHC presentation of peptides. Rather, the cytoplasmic tail of BTN3A1 includes a B30.2 domain that can seemingly bind PAg, acting de facto as a sensor for the levels of MAMPs or DAMPs (danger-associated molecular patterns),[230,235-237] for example, high concentrations of IPP, which have been associated with virus infection of cells and cell transformation.[238] It is hypothesized that BTN2A1 can engage Vγ9 chains at steady state but that this binding is altered upon PAg binding to the B30.2 domain of BTN3A1, with which BTN2A1 dimerizes, thereby eliciting effector responses from Vγ9Vδ2+ T cells. Nonetheless,

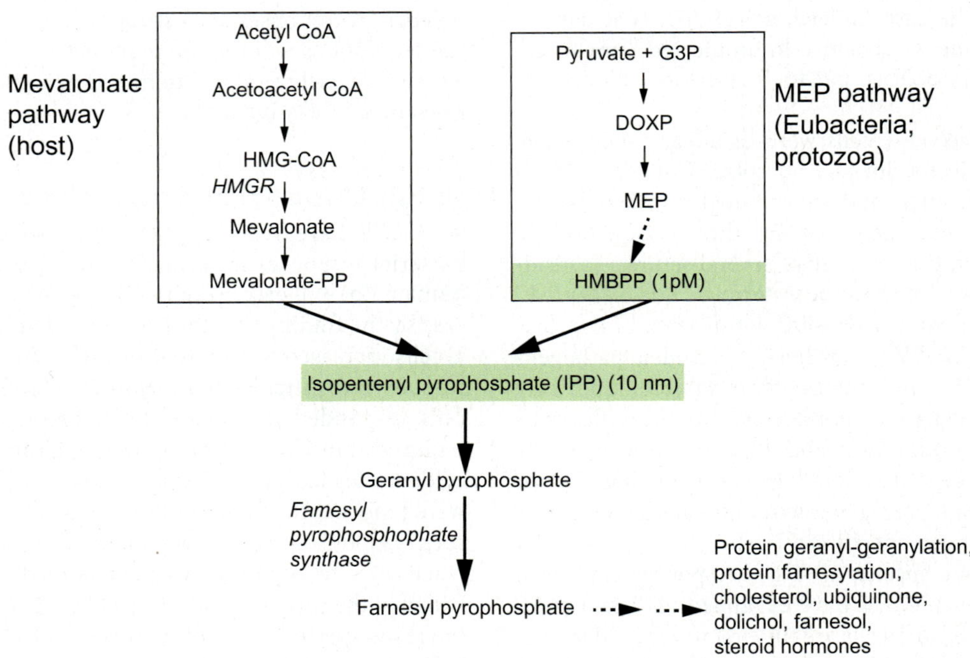

FIG. 35.6. The generation of microbial and endogenous phosphoantigen. The host "mevalonate pathway," which is essential for the biosynthesis of isoprenoids from which sterols and steroid hormones derive, is compared with an analogous "MEP pathway," which is utilized by many microbes, albeit not all. The two pathways converge at isopentenyl pyrophosphate (IPP), which at 10 nm concentrations (indicated) provokes BTN3A1-dependent, BTN2A1-mediated engagement of human Vγ9Vδ2+ TCRs. This "phosphoantigen" property is shared with HMBPP, which is active in the picomolar range (indicated). The enzyme farnesyl pyrophosphate synthase (indicated) is inhibited by amino bisphosphonates (ABP), including those in clinical use, for example, zoledronate. By provoking upstream IPP accumulation, ABP-treated cells become activators of Vγ9Vδ2+ T cell responses.

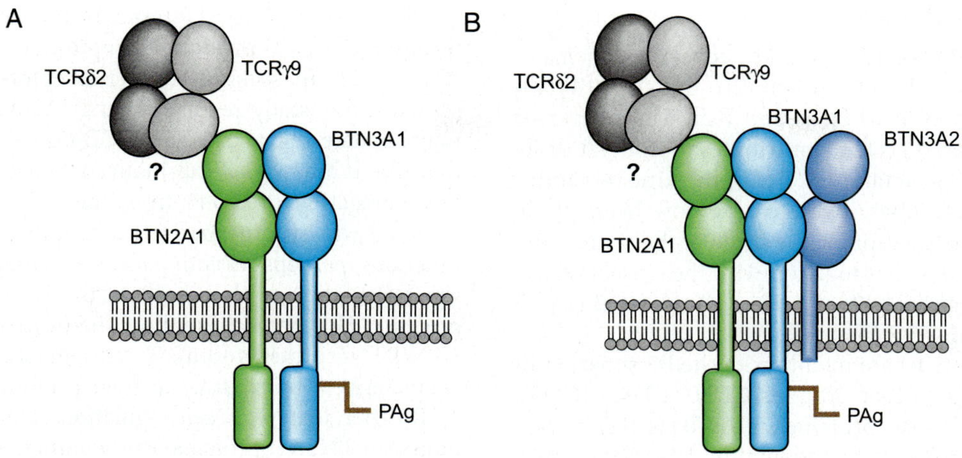

FIG. 35.7. Models for BTN2A1 engagement of TCRVγ9. A: Cartoon depicting the documented engagement of the TCR Vγ9 chain by BTN2A1, which associates with BTN3A1, which in turn is regulated by PAg binding to its cytoplasmic B30.2 domain. The question mark (?) denotes an unidentified ligand for CDR3 sequences in PAg-reactive Vγ9Vδ2 TCRs. **B:** Cartoon as in [a], but additionally speculating the involvement of BTN3A2, since genetic evidence indicates that PAg-induced activation of Vγ9Vδ2+ T cells can also be regulated by BTN3A2 and/or BTN3A3, which are documented to form complexes with BTN3A1. It is not excluded that Vγ9Vδ2 CDR3 is bound by a BTN-family moiety.

this process may be complicated, in that the BTN3A1 locus encodes BTN3A1, 3A2, and 3A3, each of which has also been implicated in PAg-dependent responses of Vγ9Vδ2+ T cells in different settings (Fig. 35.7B).[231] Moreover, whether one or more component(s) of the BTN2A1-dependent antigenic complex also binds to the Vγ9Vδ2 TCR via TCRVδ2 and/or CDR3 remains to be determined (Figs. 35.7A, B), since there are stringent constraints on CDR3 motifs that permit PAg recognition, as considered above. Of note, Vγ9Vδ2+ T cells respond to PAg exposure with very fast kinetics, seemingly illustrating the rapidity by which responses may be mounted to self-encoded cell surface proteins affected by infection.

This dominant recognition pattern of human peripheral blood γδ T cells may pertain to other γδ TCR specificities. Thus, F1-ATPase, which can reportedly bind to a Vγ9Vδ2 TCR (above), can interact with ApppI, an adenosine derivative of IPP, and thus might be induced to engage Vγ9Vδ2 by changes in IPP concentrations.[239] Also, IPP-ApppI interconversion may be catalyzed by an aminoacyl-tRNA synthetase, a fact that may pertain to the reactivity of a human Vγ3Vδ2 TCR for a conformational epitope on histidyl-tRNA synthetase.[240] Collectively, these observations may reflect human γδ T cell–mediated surveillance of alterations in isoprenoid metabolism that serve as useful sentinels of cells dysregulated by infection or other stressors. Thus, it might make important contributions to the prompt development of granulomas that facilitate our protection against *Mycobacterium tuberculosis* by walling it off. Likewise, this responsiveness evidently confers on Vγ9Vδ2+ T cells the capacity to make rapid polyclonal, highly sensitive responses to myriad cell types infected with highly diverse bacteria, viruses, and protozoa, potentially offering a generic first line of defense.

Nevertheless, several questions are unanswered, including how Vγ9Vδ2+ T cells remain tolerant of commensals: is it because Vγ9Vδ2+ T cells are confined to sites wherein microbial exposure would reflect pathogenic penetration? Likewise, what is the biological significance of the Vγ9Vδ2 PAg response? Studies in humanized SCID mice demonstrated that transferred Vγ9Vδ2+ cells contributed very rapid immunoprotection against gram-negative (*Escherichia coli*, *Morganella morgani*) and gram-positive (*Staphylococcus aureus*) bacterial infections,[241] but for immunocompetent humans the relative importance of these cells versus other lines of defense is unresolved.

Conceivably this mechanism is important in cattle, where IPP responses have been reported for cells from vaccinated animals, although the abundance of those cells is rare and variable. Other placental mammals with conserved counterparts both of BTN3A1 and of the Vγ9Vδ2 TCR include the sloth, the armadillo, the gray mouse lemur and its relative the aye-aye, the alpaca, the bottlenose dolphin, and the killer whale.[234] Of importance, the likelihood that the primary role of BTN3 gene products is to facilitate PAg-dependent activation of Vγ9Vδ2+ T cells is suggested by the fact that nearly all species with a functional open reading frame for *BTN3* also possess functional open reading frames for genes classifiable as Vγ9 and Vδ2.

Nonetheless, none of BTN3A1, TCR Vγ9Vδ2, or PAg responses is seemingly conserved in rodents. Whether or not it reflects the loss of the PAg response from rodents and some other orders, this fact has fueled the view that γδ T cell biology, like many aspects of innate immunity, is not strongly conserved cross-species. Ironically, however, the PAg response illustrates one of the most conspicuously conserved cores of γδ T cell biology, at least among mammals, namely, the cells' TCR-dependent regulation by BTN and Butyrophilin-like (BTNL) proteins that first came to light when DETC development was shown to depend on the product of an hitherto unannotated murine *Btnl* gene termed *Skint1*.[242,243]

TCRγδ and BTNL Proteins

Genes encoding BTN and BTNL are conserved in mammals (Fig. 35.8A), with the possibility that functionally related genes also exist in birds and fish. The conserved prototypic member, BTN1a1, contributes to milk fat droplet formation during lactation,[244] but of the nine remaining human *BTN/BTNL* genes, five have by now been implicated in γδ T cell regulation, while three of the six remaining murine *Btn/Btnl* genes together with the closely related *Skint1* and *Skint2* genes have so far been implicated in γδ T cell regulation (Fig. 35.8A).

BTN/BTNL proteins are members of the B7 superfamily that includes among other proteins CD80, CD86, ICOSL, and PD-L1. Like those proteins, most BTN/BTNL proteins comprise two Ig-like ectodomains (V + C), a transmembrane pass, and variable intracytoplasmic tails that in several BTN/BTNL proteins includes a functionally important B30.2 domain (Fig. 35.8B). Of note, by use of its B30.2 motif to sense intracellular PAg levels, as considered above, BTN3A1 can communicate the dysregulated status of a cell, just as other B7 proteins (eg, CD80, CD86, ICOSL, and PD-L1) communicate various states of dysregulation (eg, infection, hyperimmune activation) to αβ T cells, thereby influencing the nature and strength of the T cells' TCR-dependent responses to antigen. The striking similarity of the N-terminal ectodomain of Skint1 and PD-L1 was revealed by NMR spectroscopy[245] (Fig. 35.8C).

Well-characterized B7 proteins mediate such communication via "costimulatory" receptors, for example, CD28, ICOS, PD1. In seeking counter-receptors for BTN/BTNL proteins, one study reported that BTN3A1 could coimmunoprecipitate the phosphatase, CD45, from αβ T cells and that this inhibited TCRαβ-induced signaling by preventing the segregation of CD45 from the TCR synapse.[246] Thus, BTN3A1 was hypothesized to have multiple T cell regulatory functions, perhaps reflecting different complexes in which it sits, for example, with or without BTN2A1 or other BTN3A proteins. This notwithstanding, the overall contributions of BTN/BTNL proteins to αβ T cell regulation remain poorly characterized, by contrast to their profound importance in TCR-dependent γδ T cell regulation. Thus, it is clear that human BTN2A1 proteins can communicate directly via Vγ9Vδ2 TCR engagement, as considered above, with possible parallels to the recognition of B7H6 (another B7 family member) by the activating NK receptor, NKp30,[247,248] and to the recognition of B7H7 by killer inhibitory receptors (KIRs) of NK cells.[249]

Furthermore, human BTNL3 is expressed specifically by intestinal epithelial cells and binds to human TCR Vγ4 chains that are common among human colonic γδ IELs.[83,156,250] Most interesting, the Vγ-chain engagement site for BTNL3 (which exists as a heterodimer with BTNL8) has been mapped to an exposed loop displaced to the side of CDR3 (Fig. 35.9A), comprising amino acids from within and around CDR2 and

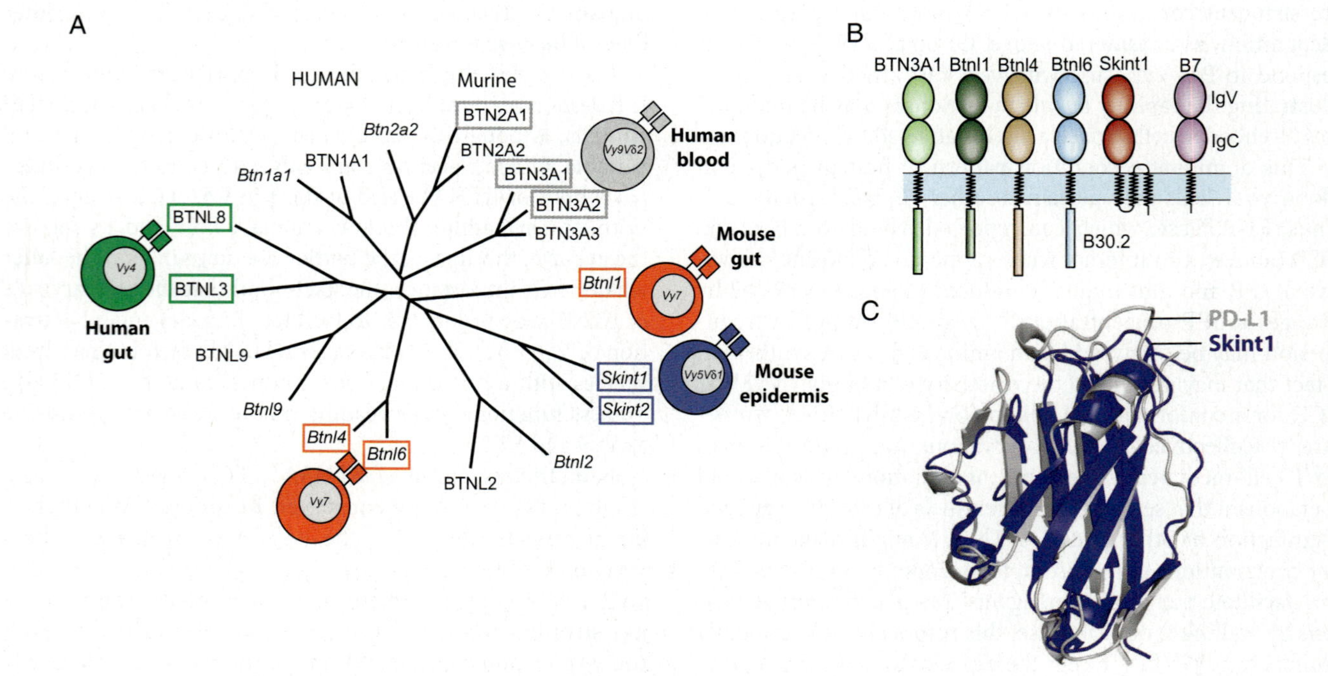

FIG. 35.8. Mouse and HUMAN Btn/BTN and Btnl/BTNL genes and proteins. A: Unrooted phylogenetic tree depicting the relationship between the known mouse (lower case) and human (upper case) genes of the butyrophilin (BTN) and butyrophilin-like (BTNL) families. Those boxed have been implicated in the regulation of γδ T cell subsets denoted by the corresponding color. **B:** Cartoon of plasma membrane–spanning BTN and Btnl protein structures, by comparison with a prototypic B7 protein. Ovals denote Ig superfamily domains; rectangles denote B30.2 domains. **C:** Overlay of Skint1 N-terminal Ig-variable domain (blue), derived from NMR-based structural analysis,[245] with equivalent domain of human PD-L1 (grey; PDB ID 3FN3).

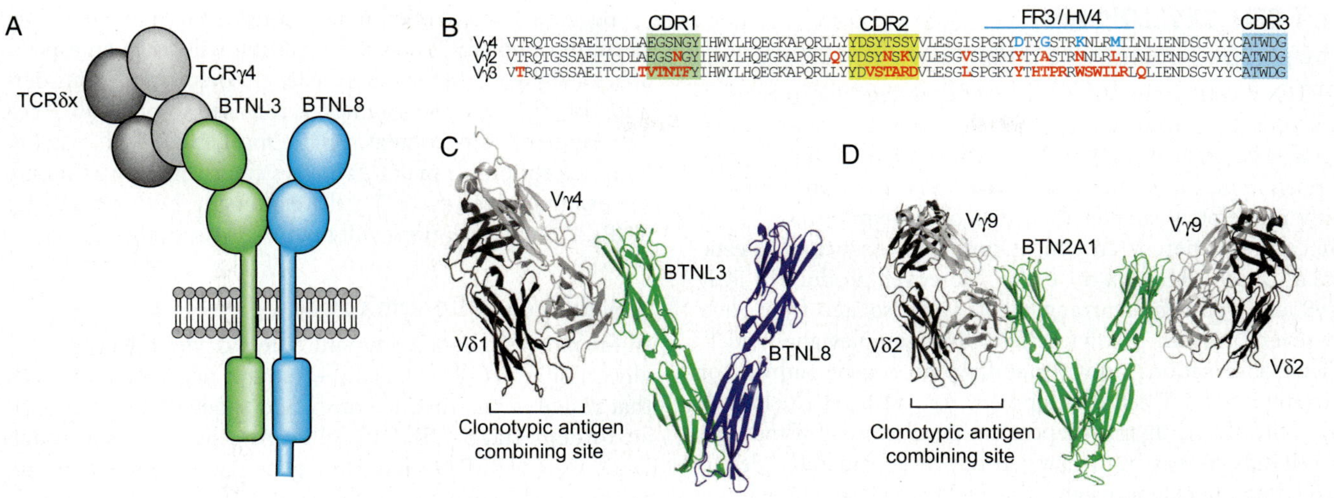

FIG. 35.9. Nonclonotypic TCR ligands: Human BTNL3 binds Vγ4 as part of a BTNL3-BTNL8 complex. A: Cartoon depicting the documented engagement of the human TCR Vγ4 chain by BTNL3, which associates with BTNL8. **B:** Sequence alignment of human TCR Vγ4 with closely related Vγ gene segments (Vγ2, Vγ3): mismatches relative to Vγ4 shown in red; CDR1, CDR2 and CDR3 highlighted; Hypervariable region 4 (HV4) within Framework 3 (FR3) is denoted, although its boundaries are approximate. Vγ4 HV4 residues in blue are critical for BTNL3-mediated stimulation of human Vγ4+ T cells. **C:** Biomolecular modeling of BTNL3 engagement of Vγ4, as shown by Melandri et al,[156] TCRγ (silver), TCRδ (black), BTNL3 (green), BTNL8 (blue). For comparison, the TCR face typically utilized for clonotypic antigen binding is shown. **D.** For comparison with (C), biomolecular modeling of BTN2A1 engagement of Vγ9, as shown by Karunakaran et al,[179] TCRγ (silver), TCRδ (black), BTN2A1 (green).

HV4 (hypervariable region 4),[156,250] which is an additional region of inherited hypervariability across TCR and Ig V gene segments (Fig. 35.9B). Over its whole length (~100 amino acids) human Vγ2 differs from Vγ4 by only nine amino acids, two of which are in CDR2 and four of which are in HV4 (Fig. 35.9B), but these differences are sufficient to abrogate binding to BTNL3.[156,250] In binding germline-encoded motifs of essentially all Vγ4 chains, largely irrespective of either CDR3 sequence or the nature of paired Vδ chain (Fig. 35.9C), BTNL3 interactions are nonclonotypic and by that definition should be classified as examples of innate immunity, as considered above for Vγ9 binding.[251] Indeed, Vγ9 engagement by BTN2A1 is likely mediated by amino acids from within and around CDR2 and HV4[179](Fig. 35.9D).

By contrast to all other TCRγδ ligands considered above, the biological significance of BTNL3-binding seems clear, namely, it probably determines the characteristic prevalence of human Vγ4+ cells in the colonic IEL compartment.[84] This developmental role in T cell selection, rather than T cell activation, is suggested by analogous roles established for mouse Btnl1, Btnl4, and Btnl6 in mouse Vγ7+ IEL development,[84,156] as will be considered below. Although their direct TCR engagement has not yet been demonstrated, site-directed mutagenesis studies indicate that Btnl6 and Btnl4 can each make nonclonal, innate interactions with the γδ TCR via mouse Vγ7, analogous to BTNL3 engagement of human Vγ4.[156,252]

Precedent for TCRs acting as innate receptors exists in the nonclonotypic engagement of αβ TCRs by superantigens (SAg) that bind germline-encoded Vβ chain motifs that also commonly comprise CDR2 and HV4 residues.[253] However, prototypic SAgs connect a T cell and an APC by simultaneously engaging TCR Vβ and MHCII,[253] whereas no additional binding partner has yet been identified for BTNL3+BTNL8 heterodimers beyond Vγ4. Again, however, absence of evidence is not evidence of absence. Interestingly, SAgs are either microbial or encoded by infectious or endogenous viruses, and their biological significance has mostly been viewed as being beneficial to microbes via the activation of lymphocytes that may aid microbial transmission and/or the dysregulation of the T cell repertoire that thereby impairs immunoprotection. By contrast, BTNL3-Vγ4 binding demonstrates that SAgs may be self-encoded and that their interactions with γδ TCRs are of profound physiologic significance to the host.[253]

By emphasizing the contribution of TCRγ chains to TCRγδ reactivities, this perspective counterbalances an orthodox, CDR3-centric view that the greater share of TCRγδ specificities is attributable to TCRδ. Indeed, TCR Vγ-mediated engagement of discrete self-encoded ligands, including but not necessarily limited to BTN/BTNL proteins, might explain the common association of γδ T cells expressing specific Vγ chains with discrete anatomical sites, functions, or responsiveness, considered above. Moreover, given that the localization and responsiveness of human Vδ1+ and Vδ2+ γδ T cells can be segregated by several criteria, it remains formally possible that there exist nonclonal ligands for germline-encoded Vδ motifs. Likewise, there may exist microbe-encoded SAgs for TCRγδ, which might effectively disrupt BTN/BTNL-mediated regulation of γδ T cell activation and/or development. Beyond such speculation, BTNL3 binding to TCR Vγ4 offers an excellent example of the fact that TCRγδ signaling plays profound roles in γδ T cell development, which we shall now consider.

γδ T CELL DEVELOPMENT

Thymus-Dependent γδ T Cells

DETCs illustrate the strict thymus dependence of most γδ T cell subsets.[79] However, it is not clear that this applies universally because *nu/nu* FoxN1-deficient athymic mice harbor a phenotypically mature, intestinal Vγ7+ IEL compartment only moderately smaller than that of euthymic mice.[84,254,255] By contrast, there is an almost complete absence of intestinal αβ T cells in *nu/nu* mice.[256,257] Moreover, productive TCR Vγ9 and Vδ2 gene rearrangements were isolated from each of several human prethymic fetal liver samples analyzed.[258] These observations notwithstanding, no cogent pathway of extrathymic γδ T cell development has yet been elucidated. By contrast, the thymus-dependent development of most γδ T cell subsets has been mapped out in considerable detail, providing important insights that will be reviewed here.[259]

A series of independent studies in mice established early on that in most species γδ T cell development precedes αβ T cell development, originating in the fetal thymus.[260-263] Lymphocyte progenitors originating in the fetal liver first develop into DETC progenitors, but even in species lacking overt DETC counterparts, γδ T cells are commonly the first T cells to mature,[55,149] the single exception to date reportedly being the gray short-tailed opossum. Some γδ T cell subsets, including Vγ5Vγ1+ DETC, and mucosal T cells expressing the canonical Vγ6Vδ1 TCR derive purely from the fetal or perinatal thymus,[264,265] whereas others continue to be produced postnatally, together with αβ T cells, which numerically become the predominant output in mice and humans, as is considered in greater detail below. However, in neither the fetal nor the postnatal thymus has an undifferentiated progenitor thymocyte yet been clearly identified that is irrevocably committed solely to a γδ T cell fate. While bourgeoning single cell transcriptomic studies may yet identify precommited progenitors, their existence is not mutually exclusive of there being one or more common progenitors to γδ and αβ T cells, which become skewed toward either an αβ or γδ T cell fate by a combination of interactions with temporally dynamic stromal determinants. Such interactions include signals of various strengths transmitted via nascent TCR-CD3 complexes.

Because T cell development is considered in much greater detail in Chapter 27, our consideration will focus on aspects specifically germane to γδ T cells, although some consideration of αβ T cell development is required to place γδ T cell development into context. As in most biological systems, some exceptions to major pathways can seemingly be found, for example, reports of T cells expressing both γδ and αβ TCRs.[266] Such exceptions will not be a focus of this chapter.

γδ-αβ Lineage Determination

In the mouse, the most immature thymocytes are operationally classified as CD4−CD8αβ−, "double negative" (DN) cells that reside in the thymic cortex and sequentially differentiate through stages I, II, III, and IV. As cells progress through DNII to DNIII, RAG-mediated gene rearrangements occur, seemingly concurrently, at the TCRβ, TCRγ, and TCRδ loci. In instances of productive V-D-Jβ rearrangements, the resultant TCRβ chains can form a preTCR by pairing with pTα, a thymocyte-specific surrogate TCRα chain.[267] The preTCR provides a nonclonal checkpoint for productive TCRβ chain expression, termed "β-selection," which acts on small-sized cells in the DNIII pool (Fig. 35.10A).[268-270]

CD3-mediated preTCR-initiated signaling is most likely induced independently of any specific ligand engagement.[271] Its consequences are to temporarily arrest RAG-mediated gene rearrangements, enforcing allelic exclusion at the TCRβ, TCRγ, and TCRδ loci and promoting DNIII cells to enlarge and to transition into DNIV cells, which enter into a short period of very rapid proliferation, with generation times of ~6 h.[272,273] This greatly expanded thymocyte pool acquires CD4 and CD8αβ expression as so-called double positive (DP) cells and resumes gene rearrangements now targeted at the TCRα locus at which a key transcriptional enhancer element becomes active. These outcomes require that basic-helix-loop-helix transcription factors known as E proteins are suppressed, which is achieved by a combination of preTCR and Notch signaling, since preTCR signaling alone is insufficient to achieve this.[274,275] Note that neither productive nor nonproductive TCRγ gene segment rearrangements preclude DNIII thymocytes from passing the β-selection checkpoint, subsequently emerging as αβ T cells that may continue to express

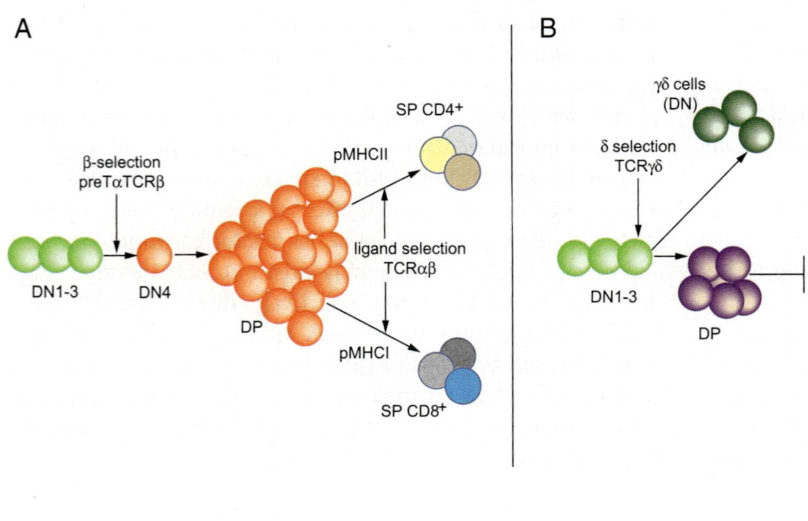

FIG. 35.10. Critical selection points in T cell development. A: Thymic αβ T cell development is shown to pass through two checkpoints mediated by the preTCR and TCRαβ, respectively. **B:** Thymic γδ T cell development is shown to pass through one checkpoint mediated by TCRγδ (DN, CD4−CD8− "double negative"; DP, CD4+CD8+ "double positive"; SP, single positive for CD4 or CD8).

rearranged TCRγ genes, as illustrated by the cell line 2C, the source of the first described TCRγ cDNA.[3,4]

Given the architecture of the TCRα locus (see Fig. 35.1), Vα-Jα rearrangements necessarily delete the intervening TCRδ chain gene segments and, given that the likelihood of productive TCRα rearrangement on a single allele is <33%, most DP thymocytes will proceed to rearranging the second allele, thereby deleting D-J-Cδ from both chromosomes: a point of no return for αβ versus γδ commitment. Indeed, given that some cells that traverse β-selection can harbor TCRγ and TCRδ genes that were productively rearranged concurrently with preTCR signaling, TCRδ deletion rather than β-selection may be the fundamental αβ/γδ-lineage determining factor for progenitors that were not precommitted. Other factors that may promote αβ versus γδ T cell differentiation from DNIV and DP thymocytes include the developmentally programmed suppression of an intronic transcriptional enhancer element in the TCRδ locus coincident with the activation of an enhancer element downstream of TCR-Cα.[276] Likewise, there is an element that can seemingly silence TCRγ gene expression,[277,278] although this may in part be to prevent concurrent TCRγ gene rearrangements when RAG1/RAG2-catalyzed recombination events recommence at the TCRα locus. In addition, TCRδ gene segments can be lost via deletional recombination between an element, "TEA" (T early alpha) (located 5′ to D-J-Cδ), and a pseudo-Jα element, 3′ to D-J-Cδ.[279] More importantly, however, this deletion may serve to enhance the breadth of the TCR Vα chain repertoire available to developing αβ T cells.[280] Nonetheless, several other commonly occurring gene segment rearrangements de facto limit γδ differentiation potential by deleting the TCRδ locus.[34]

The primary role of TCRαβ in contributing to T cell development is played in the thymic medulla where DP thymocytes with newly assembled αβ TCRs accumulate and are "selected" according to the strength of signaling by which they engage self pMHC complexes.[281] Moderate affinity interactions promote further maturation, including the transition to being either CD4+ "single positive" (SP) cells, if selected on MHCII, or CD8+ SP cells if selected on MHCI, followed by thymic egress. High-affinity interactions promote thymocyte apoptosis, anergy, or acquisition of a regulatory T cell fate. Very-low-affinity interactions fail to rescue from apoptosis (death-by-neglect), which is considered the most common outcome for DP thymocytes. Of note, the broad spectrum of peptides presented by MHC molecules in the medulla depends largely on AIRE (the AutoImmuneREgulator), a chromatin regulatory factor that is expressed primarily by medullary thymic epithelial cells (mTECs) and that facilitates transcription of genes ordinarily expressed in adults by differentiated, tissue-specific cell types. As we shall consider later, some activities of TCRγδ+ thymocytes may be integrated with AIRE-dependent T cell regulation. Finally, while strength of signaling from the preTCR/TCRs is evidently a major instructive influence on the outcomes of thymocyte development, it is important to note that those outcomes may be substantially influenced via one or more of many other receptors including Notch (as above), CD28, cytokine and chemokine receptors, and homotypic SLAM receptors that signal via SAP. Those outcomes may also be regulated by variation in microenvironments of different anatomical niches in which thymocytes localize, commonly under the influence of chemokines, although this is a relatively underexplored topic.

δ-Selection

In striking contrast to αβ T cell development and B cell development, TCRγδ development does not include a preTCR/preBCR checkpoint.[282] Rather, DNIII thymocytes with productive TCRδ gene rearrangements traverse a "δ-selection" checkpoint, which is facilitated by TCRδ pairing to TCRγ, that is, TCRγδ (Fig. 35.10B).[283,284] However, different Vγ chains contribute to δ-selection to different degrees with consequent variation in outcomes,[283,284] as is considered below. Conversely, there is no evidence for γ-selection. Hence, productive γ-chain gene rearrangements do not preclude further γ-chain gene rearrangements until either β-selection or δ-selection arrests RAG1/RAG2-catalyzed recombination events. As a consequence, γδ T cells can express >1 TCRγ chain, just as shark B cells and mouse B1 B cells can commonly express two Ig light chains.

Most cells traversing δ-selection are signaled to proliferate, albeit less than cells signaled by the preTCR, and to further differentiate toward a DN TCRγδ+ fate and away from an αβ T cell fate.[283,285] This in large part reflects greater signal strength from TCRγδ, which is sufficiently strong to suppress E proteins via ID3 upregulation, and thereby promotes DNIII cell differentiation without any additional dependence on Notch (not discounting the epistatic requirement for Notch signaling earlier in thymocyte differentiation).[274,286] In adult mice, most productive rearrangements facilitating this transition are of Vγ1-Jγ4 gene segments.[283] An important interim conclusion is that emergence as a γδ T cell from any bipotent γδ/αβ DNII-DNIII thymocyte progenitor requires that the progenitor be tolerant of γδ TCR-mediated signaling that is of sufficient strength to suppress an αβ T cell fate.[287-289] Commitment of murine progenitors to the γδ lineage is commonly monitored by assaying for CD73 upregulation, following which CD24 expression declines, marking γδ lineage maturation.[289,290]

Signaling from a precociously expressed TCRαβ can provoke the development of CD4−CD8− TCRαβ+ cells with γδ-lineage phenotypes,[291] indicating that the qualitative nature of the TCR isoform (TCRαβ/preTCR/TCRγδ) is not per se a determinant of lineage. Indeed, δ-selection can also phenocopy β-selection in promoting very active proliferation followed by DP differentiation.[283,284] Hence, DP thymocytes are essentially absent from TCRβδ$^{-/-}$ mice, whereas small numbers of DP cells are variably present in TCRβ$^{-/-}$ mice (that cannot pass β-selection) and they are enriched for in-frame TCRδ gene rearrangements.[283,284] Nonetheless, γδ T cells will rarely differentiate from a DP pool, in part because TCRγ gene silencers may be activated in this pathway, in part because the TCRδ enhancer is suppressed, and in part because TCRδ locus may be subsequently deleted. Reflecting this, γδ T cells rarely express CD4 or CD8αβ and are not selected on MHC (above), even in the case of a poorly studied small subset of CD8αβ+ γδ T cells that can be found in perinatal mice. Nonetheless, it was striking that most productive rearrangements facilitating the DNIII-DNIV-DP transition in adult mice are of Vγ4-Jγ1 gene segments. This suggests that signaling from Vγ4+ γδ TCRs is often weaker than from Vγ1+ γδ TCRs,[283] a point with important implications for effector function precommitment that will be considered below. Of note, productive Vγ5-Jγ1 rearrangements can also contribute to δ-selection in neonatal mice, but those contributions cease within 2 weeks, despite some continued generation of productive Vγ5-Jγ1 rearrangements.[283] While this is consistent with the limitation of Vγ5+ DETC development to the fetal and perinatal periods, it suggests that there are additional restrictions on the postnatal use of TCR Vγ5 chains.

Because δ-selection suppresses further RAG1/RAG2-catalyzed recombination events, it extinguishes any potential for further TCRα/TCRβ gene segment rearrangements. That this is a major factor in γδ/αβ lineage determination is evident from the observation that residual V-D-Jδ gene rearrangements found in αβ T cells and their DNIV and DP progenitors are mostly out of the frame, whereas this is not the case in cells from TCRδ−/− mice in which productively rearranged TCRδ chain genes cannot encode a functional TCRδ chain.[292] Hence, the relatively low ratio of γδ T:αβ T cells emerging from the thymus, at least in mice and humans, seems mostly to reflect the reduced likelihood of productive rearrangements, given the small size of the TCRγ and TCRδ loci, coupled with the failure of γδ thymocytes to accumulate in a large DP pool. Indeed, γδ T cell numbers are rarely substantially increased in TCRα−/− or TCRβ−/− mice, and γδ T cell numbers in TCRδ+/− heterozygotes are commonly reduced by ~50% when compared with TCRδ+/+ mice, consistent with γδ T cell numbers being dictated by the probability of productive gene rearrangements. Indeed, there is an increased representation of γδ T cells in species with considerably more expansive TCRγ and/or TCRδ chain loci that may offer greater probability of successful TCRγ/δ gene rearrangements. This notwithstanding, the relative representations of γδ and αβ T cells might also be influenced by the frequency of lineage-precommitted progenitors that may exist.

Developmental Waves of γδ T Cells

A signature facet of murine γδ T cell biology is the appearance in sequential waves of different γδ T cell types, defined by their usage of discrete V-gene segments, commonly rearranged to specific D and J segments[78,264,265,293] (Fig. 35.11). Two mechanisms underlying this are, respectively, the temporal regulation of chromatin accessibility[294-298] and the enhancement of productive rearrangements by specific sequences (eg, direct repeats) at sites of V-(D)-J DNA breakage and reunion events. The latter mechanism is particularly important in fetal development, prior to the onset of expression of terminal deoxynucleotidyl transferase (TdT) that introduces greater complexity into V-(D)-J junctions. Thus, fetal DNA breakage and reunion events are biased toward producing canonical Vγ5-Jγ1 and Vγ6-Jγ1, both paired to identical Vδ1-Dδ2-Jδ2 rearrangements in DETC and invariant Vγ6+ cells,

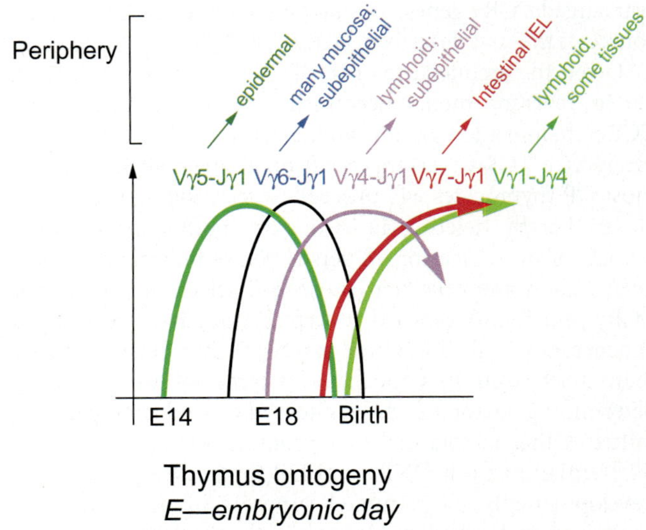

FIGURE 35.11. Developmentally ordered rearrangements of mouse γ gene segments. Cartoon depicts the TCRγ chain gene segment rearrangements that are most prevalent as mouse thymic ontogeny proceeds. The peripheral sites primarily seeded by the emerging γδ T cells are indicated.

respectively[32,33] (see Fig. 35.4, above). In addition, qualitative bias of gene segment rearrangements shows temporal variation in T cell progenitors utilized sequentially during mouse and human ontogeny.[259,265,299,300]

Indeed, human fetal thymic γδ T cell development displays many of the same temporally ordered features as murine γδ T cell development. Temporally ordered utilization of Vδ2, Dδ3, and Jδ and of Vγ9 gene segments favors the early development of Vγ9Vδ2 T cells, although Vγ8 and Vδ1, respectively, can also be rearranged early.[149,301] V-(D)-J rearrangements are initially simple, reflecting low TdT activity and the strong influence of direct repeats[149,301] (Fig. 35.12). This results in several examples of public (ie, shared across individuals), "invariant" TCRs, specifically including Vδ2, Vδ1, and Vγ8 gene segments (see Fig. 35.4).[299] Provocatively, resultant invariant Vγ8 and Vδ2 chain sequences have been implicated in fetal and newborn responses to infections in utero with CMV and *Toxoplasma*.[186,216] Hence, it is hypothesized that DNA sequences and other factors promoting particular rearrangements of specific gene segments have been selected under the pressure of viral, protozoan, and other pathogen infections. By week 17 of ontogeny, rearrangements of Vδ1, Vδ3, Vδ5, Dδ1, and Dδ3 are common and increasingly complex, as are rearrangements of various Vγ gene segments,[149,301] whereas postnatally there is a relative decline of Vγ9Vδ2 cell development, although it does not cease entirely.[61,144,145,300,301] Hence, substantial environmentally induced Vγ9Vδ2+ T cell expansions that are a common feature of early life may derive from fetal and postnatal developmental programs.[61,300]

As has been introduced above, Vγ5 and Vδ1 rearrangements are the first to occur in mouse fetal ontogeny, at E14-E16, closely followed by Vγ6 together with ongoing Vδ1 rearrangements, which are temporally followed by rearrangements pairing various V-D-Jδ chains with Vγ7-Jγ1 rearrangements

```
  V1     P    N1      P  D1    P  N2   P     D2     P       N3          P    D3          P   N4  P     J
  GG GAA              GAAATAGT            CCTTCCTAC                        ACTGGGGGATACG             AC ACC 1
                                                                                                     CT TTG 2
                                                                                                      C TCC 3

  11 wk.
- GG GAA                                       CTA  G                       CTGGGGGATA       T            C 3
+ GG GA                                                                    ACTGGGGGATA               CT TTG 2
+ GG GA                                                                    ACTGGGGGATACG C           AC ACC 1
+ GG GA                                                                    ACTGGGGGATACG           T AC ACC 1
+ GG GAA                                                                GT ACTGGGGGATAC                   C 3
+ GG GA       G                                                             CTGGGGGAT        TT       T TTG 2
- GG GA                                                                    ACTGGGG      AGCTCCTA       ... 3
  17 wk.
+ GG GA                                                                    ACTGGGGG          AG  AG   C TCC 3
+ GG GA                                                                     CTGGGGGATACG              C TCC 3
- GG GA                                                                    ACTGGGGG      C   TAGT     T TTG 2
+ GG GA       G                                                             CTGGG                     C TCC 3
- GG G                                                                          GGGATACG              C TCC 3
- GG GA                                                                    ACTGGGGGAT             GT AC ACC 1
+ GG GA       GTATT      ATA     TGCAC                                     ACTGGGGGATA                C TCC 3
  3 yr.
+ GG GAA                                   CTTCC       CC                      GG            CT   GT AC ACC 1
- GG GAA      CCCA     GA          C        TCC    GT CCGGGAT          T ACTGGGGGAT            GT       ... 1
- GG GAA                                       CTA  G GAAAGGCCCCC      T ACTGGGGGATA                   CC 1
- GG GA                GA                  CTT      G CCATTATCGTCCGTTTCCCAC    GG        TTATCTGTGG     ... 1
+ GG GA                GA         CCCTCC      CCTAC                            GGGG     C CCTGAGC       ... 1
+ GG GA                                        CC                          CTGGGGGATACG                CC 1
+ GG GAA                                       CTA      TTTCTA               GGGGGA         GAA         ACC 1
+ GG      TT  TC     C                 GG CCTTCCTA    AGGCCCTTCGG            TGGGGGATACGC CGGAAA GT AC ACC 1
- GG G               C                    CCTTCC                                 GGA     C AT-15-ATC     ... 1
```

FIG. 35.12. **Developmentally ordered rearrangements and complexity of human δ gene segments.** DNA sequences of human Vδ1 gene segment rearrangements from thymi at 11 and 17 weeks post conception and at 3 years postpartum. Early fetal rearrangements are dominated by use of Dδ3 and sparse N nucleotide addition (underlined). Postnatal joins show frequent inclusion of multiple D gene segments. The J segments used are indicated by the numbers to the right. (−) and (+) refer to abortive and productive gene rearrangements, respectively.[301]

that continue postnatally (Fig. 35.11). Following their development and thymic egress, these fetal/neonatal γδ T cell subsets can establish lifelong and largely exclusive associations with tissues, as considered above, for example, epidermal Vγ5Vδ1+ DETC, uterine Vγ6Vδ1+ T cells, and intestinal Vγ7+ IEL[82,84,96,97,264,265] (Fig. 35.11). However, in some tissues, they are subsequently joined by relatively large numbers of cells carrying Vγ4-Jγ1 and/or Vγ1-Jγ4 rearrangements, each paired with diverse V-D-Jδ chains, that begin to emerge in late fetal ontogeny and continue postnatally, by which time the thymus predominantly produces αβ T cells. Thus, the dermis harbors Vγ6Vδ1+ cells and Vγ4+ T cells,[95] while in the gingiva, small numbers of Vγ5Vδ1+ and Vγ6Vδ1+ T cells that enter the presumptive tissue prenatally and perinatally cede space to Vγ4+ and Vγ1+ cells within 7 days of birth.[100]

The developmental waves were elegantly recapitulated in so-called RBPJ^{ind} mice in which regulated Notch-pathway activation allowed T cell development to be precisely limited to the fetus, or to the first 3 weeks of life, or to 6 to 9 weeks post partum.[259] That system clearly showed how Vγ4-Jγ1 gene rearrangements become overshadowed shortly after birth by Vγ1-Jγ4 cells. This has functional implications because of the respective associations of Vγ4 and Vγ1 with IL-17 and IFNγ production (above), immediately foretelling that the activities of IL-17-producing γδ T cells may be enriched in the fetal and neonatal periods. Indeed, as was considered above, thymic γδ T cells, including those in the fetus, already show segregation of cells with IFNγ-producing and IL-17-producing potentials, respectively. Developmental imprinting of effector functions may not apply to all γδ T cells, but where it does apply it permits mature γδ T cells to deliver effector functions immediately, in stark contrast to CD4$^+$ TCRαβ$^+$ MHCII-restricted helper T cells that mostly delay commitment to defined cytokine profiles (eg, IFNγ [Th1]; IL-4 [Th2]; IL-17 [Th17]) until antigen-driven expansion and differentiation in the periphery.

In relation to this the PLZF-regulated "innate-like" phenotype, considered above for IL-4-biased Vγ1Vδ6.3+ cells and for IL-17-biased Vγ6+ cells, is another illustration of the waves by which γδ T cells develop (Fig. 35.13).[127,128] The molecular determinants of this innate state include lin28b, an RNA-binding protein, whose effects can be to suppress TdT, thereby maintaining progenitor cell skewing toward low junctional diversity.[259,302] Moreover, lin28b suppresses the Let7 family of miRNAs, which in turn regulate PLZF (Zbtb16), and Igf2bp3, which is an RNA-binding protein that is also implicated in the innate-like phenotype.

The lin28b innate-like phenotype appears to be an intrinsic property of fetal thymocyte progenitors. Thus, when T cell development was limited to the fetus in RBPJ^{ind} mice, ~40% of Vγ1+ splenocytes that emerged were innate-like, expressing PLZF, whereas only ~5% of Vγ1+ cells were PLZF+ when

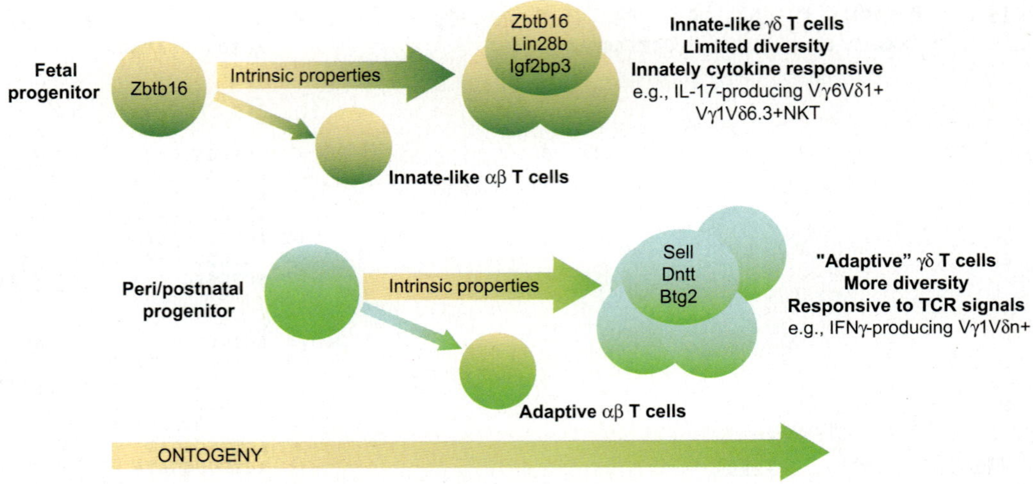

FIG. 35.13. Distinct properties of fetal and postnatal progenitors imprint innate-like and adaptive phenotypes. The upper part of the cartoon depicts mouse γδ T cells maturing from a fetal thymic progenitor, the intrinsic properties of which drive γδ T cell differentiation toward innate-like phenotypes described by signature genes and properties, indicated. The bipotential progenitor may likewise give rise to αβ T cells with innate-like properties. The lower part of the cartoon depicts a thymic progenitor arising later in development, the intrinsic properties of which drive γδ T cell differentiation toward a more conventional, "adaptive" phenotype, described by signature genes and properties, indicated, including the use of TCR signaling as the primary checkpoint for cell activation in the periphery. The bipotential progenitor may likewise give rise to αβ T cells with adaptive properties.

T cell development was restricted to neonates or adults (Fig. 35.13).[259] Of interest, the innate-like Vγ1+ cells were enriched in *lin28b*, *Blk*, *Il17a*, and *Il17f* transcripts, suggesting that whereas Vγ1+ γδ cells are generally skewed away from IL-17 production, innate-like IL-17-producing Vγ1+ γδ cells can be found that derive from fetal progenitors.[259]

Likewise in RBPJ^{ind} mice, lung IL-17-producing γδ T cells could be produced when T cell development was restricted to either the fetal or postnatal periods,[259] countering a commonly held impression that IL-17-producing γδ T cells derive only from fetal progenitors,[127,128] but consistent with other data that Vγ4+ IL-17-producing γδ T cells could be produced in adult mice.[303] However, only cells generated in the fetus were innate-like as judged by their capacity to produce IL-17 rapidly in response to innate signals alone (IL1β and IL-23)[259] (Fig. 35.13). Indeed, fetal sox13^{+} Rorγt^{+} γδ T cell progenitors express RNAs for IL-17a and IL-17f, seemingly building the competence for the rapid and innate release of IL-17 by mature γδ T cells in the periphery. Conversely, cytokine RNAs are not expressed by adult sox13+Rorγt+ γδ T cell progenitors, consistent with additional checkpoints existing for IL-17 release from adult-derived γδ T cells. Collectively, these data seem to offer strong endorsement of the findings that, prior to their commitment to either αβ or γδ TCR expression, fetal bipotent thymocyte progenitors predispose TCRαβ+ and TCRγδ+ progeny to innate behavior, whereas postnatal bipotent progenitors predispose their progeny toward differentiation as adaptive immune cells (Fig. 35.13).[304]

Strikingly, fetal innate-like imprinting appears evolutionarily conserved, with γδ T cells arising from human fetal thymic development expressing an innate-like, Lin28-associated phenotype and displaying invariant TCR gene rearrangements.[299] Note, however, that there are additional innate traits of γδ T cell subsets that are not obviously PLZF dependent and that are regulated by TCR signaling.

Developmental Functional Imprinting by γδ TCRs

Several key aspects of γδ TCR biology are differentially regulated by γδ TCRs, during the cells' development (Fig. 35.14A). Much evidence has been obtained from independent studies tracking γδ T cells specific for T10 and T22 (above), both of which largely depend on β2m for their expression, and from studies of TCRγδ transgenic mice, including those expressing either of two Vγ4Vδ5 TCRs, "G8" or "KN6," which each show high-affinity reactivity toward T22 and lower-affinity reactivity toward T10. From some seemingly comparable experimental systems, contrasting results have been obtained, and as such this summary will focus on consensus areas of agreement.

First, multiple sets of evidence argue that, akin to negative selection of αβ T cells, ligand exposure in the thymus can delete or anergise high-affinity T22-reactive T cells, specifically those developing from adult versus fetal thymocytes.[305,306] Hence, the self-reactivities of mature γδ T cells, considered above, are presumably in the moderate affinity range, which could bias functional reactivities in vivo toward avidity-based recognition of ligands overexpressed on dysregulated cells, or else posttranslationally modified to forms that engage that TCR with higher affinity. Alternatively, self-reactivities of fetal-derived cells may not require such stringent tolerance because their innate-like properties may make them unlikely sources of durable autoimmune pathology. In

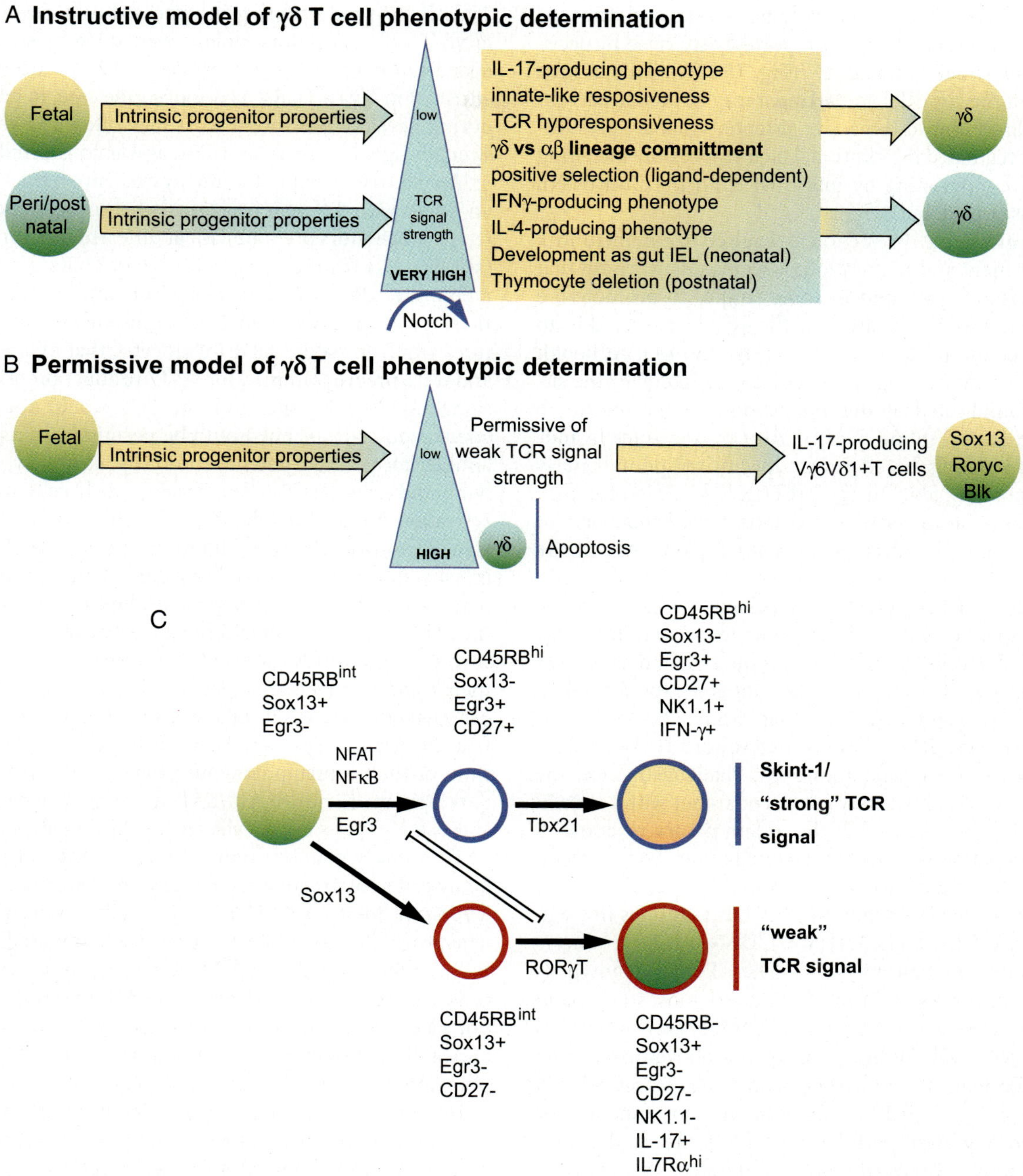

FIG. 35.14. Roles of the TCR in shaping γδ T cell biology. A: According to the instructive model, contingent on the strength of signal received from the newly expressed TCR (low at the top; highest at the bottom), developing γδ+ thymocytes will show a number of responses and differentiation phenotypes. As indicated, the penetrance of these effects varies according to whether the thymocytes are developing from early fetal progenitors, or are developing later, and is further influenced by the level of signals received via Notch. Of the properties shown, γδ/αβ lineage commitment is evidently essential for all γδ T cells and is shown in bold face. Note that this does not formally preclude the existence of some progenitors precommited to TCRαβ and TCRγδ fates. **B:** According to the permissive model, the phenotypes of discrete subsets of γδ T cells (illustrated here by mouse fetal thymus-derived IL-17-expressing cells) are predetermined by their progenitors, which are merely permissive of a specified level of TCR signaling that may be required for γδ/αβ lineage commitment and to distinguish the progeny T cells from ILC. By this model, the development of different innate-like γδ T cell subtypes would seemingly require different progenitors with distinct properties. **C:** Instructive model by which fetal thymic DETC progenitor selection is mediated by strong signaling from newly expressed Vγ5Vδ1 TCRs in response to thymic epithelial Skint1. In the presence of Skint1, this process imposes a new gene regulatory network and surface phenotype selectively on Vγ5Vδ1+ thymocytes, as indicated. In the absence of Skint1, a default state retains a Sox13+Rorγt+ phenotype associated with IL-17 production. (Adapted from Turchinovich G, Hayday AC. Skint-1 identifies a common molecular mechanism for the development of interferon-gamma-secreting versus interleukin-17-secreting gammadelta T cells. *Immunity.* 2011;35(1):59-68.)

this regard, "high-affinity" G8 cells emerging from the neonatal (as opposed to adult) thymus reportedly survived by occupying the intestinal epithelium[307] (Fig. 35.14A). This in turn suggests that the gut IEL compartment may be seeded early in life with cells of high autoreactive potentials that are presumably tightly regulated by additional factors, possibly including phenotypic conditioning by intestinal epithelial Btnl/BTNL selection, as considered below.

Second, given that δ-selection is required to complete γδ T cell commitment and maturation, it seems unsurprising that ligand exposure was found in some studies to promote the development and composition of the γδ repertoire, akin to positive selection of αβ T cells[308,309] (Fig. 35.14A). Although some of those conclusions have been challenged,[136,310] the situation is complicated by the interesting fact that mouse γδ TCRs, with the exception of Vγ5Vδ1, can transduce ligand-independent signals based on a capacity to spontaneously dimerize, possibly akin to the preTCR.[136,271] This of course raises the issue of how such precocious, ligand-independent signaling would be prevented in mature γδ T cells in the periphery.

By contrast, where greatest agreement exists is on the reported capacity of ligand exposure to skew some large fraction of γδ T cells toward IFNγ production and a $CD122^{hi}$ phenotype typically displayed by antigen-experienced T cells, whereas ligand-naïve cells (eg, from β2m−/− mice) mostly remained $CD122^{lo}$ and skewed toward IL-17 production.[136] Moreover, in hapten-adjuvant challenged mice, the IL-17 response was extremely rapid, consistent with the cells' placement in the vanguard of the inflammatory response.[183] Nonetheless, some issues need reconciliation. For example, two studies associated the skew toward $CD122^{lo}$ IL-17 producers with a $CD44^{lo}$ phenotype,[136,183] but that does not reconcile with the $CD44^{hi}$ phenotype of CD27−, IL-17 producers in several other studies.[31,124,129,189,311] The $CD44^{lo}$ phenotype is displayed by most γδ T cells in adult mice of most strains, and the current consensus is that this represents two subsets: [i] $CD45RB^{hi}$ cells that are likely (albeit not obligatorily) to be antigen experienced, rapidly responsive cells biased toward IFNγ; and [ii] $CD45RB^{lo/-}$ cells that are not functionally mature, and that await peripheral priming, that is, that may be truly adaptive. A second point to reiterate is that irreversible effector function imprinting may be more rigid in fetal progenitors than in cells derived from postnatal progenitors. For example, some CD27− cells biased toward IL-17 were shown to "reactivate" IFNγ production, albeit transiently, as is considered later in this chapter in the context of *Listeria* infection.[108]

Notwithstanding these issues, the case for developmental dose-dependent impacts of TCR ligands on effector skewing was subsequently confirmed and extended by other studies of KN6 transgenic mice. Specifically, developmental TCR exposure to high-affinity ligand, T22, either skewed toward IL-4, contingent on Notch signaling (perhaps to suppress activation induced cell death), or toward IFNγ, independently of Notch[259] (Fig. 35.14A). Cells exposed only to lower-affinity ligand, T10, showed either IFNγ skewing, which was inhibited by Notch, or Notch-independent IL-17 production.[259] Thus IL-17-production again appeared as a default capacity of γδ T cell progenitors, unless overridden by stronger TCR ligand exposure.[136,311] By this model, IL-17-producers would be enriched in Vγ4+ TCRs, as is indeed the case, in part because during development those TCRs may often transduce lower-strength signals than Vγ1+ TCRs, as was considered above.[283]

An alternative but not mutually exclusive model posits that γδ T cells with different functional phenotypes arise from different progenitors predominant at different times in ontogeny (see Fig. 35.14B) that are permissive to TCRs of varying signaling strength, rather than being instructed by them.[312] Thus, IL-17-producing Vγ4+ γδ T cells arose in thymic organ cultures from immature fetal thymocytes that expressed sox13 and were preprogrammed for IL-17 production prior to TCR signaling.[312] In this situation, the TCR would not determine effector outcome,[120] but would be permissive, simply adding antigen recognition potentials to cells akin to type 3 innate lymphoid cells (ILC3). Conceivably, such cells might enter activation-induced cell death (AICD) in response to stronger signals from a qualitatively different type of γδ TCR.[311] Origins in separate progenitors could be consistent with the ontogenetic waves of γδ cell development (above) and might explain the substantial qualitative differences between IL-17-skewed thymocytes and IFNγ-skewed thymocytes (above), including their respective metabolic profiles that are already evident in early fetal thymocytes.[125] Likewise, it might be considered that fetal invariant Vγ5+ and Vγ6+ cells, respectively, originate from distinct contemporaneous progenitors in that Vγ6+ cell development is not SLAM/SAP dependent, whereas Vγ4+, Vγ5+, and Vγ1+Vδ6.3+ cells are SLAM/SAP dependent.[313]

Conversely, counter arguments oppose the concept of there being different progenitors. For example, invariant Vγ5+ and Vγ6+ cells jointly arise from mouse yolk sac–derived thymocyte progenitors that likewise give rise to thymic ILC but not to any other γδ cells.[314,315] Moreover, many invariant Vγ6+ cells harbor out-of-the-frame Vγ5-Jγ1 rearrangements but not vice versa, consistent with there being a common precursor that attempts first to differentiate as a Vγ5+ DETC progenitor, but which may subsequently mature as a Vγ6+ cell if both Vγ5 alleles are nonproductively rearranged.[316] In another example, Vγ1+ progenitors and Vγ4+ progenitors coexpress SLAM1 and SLAM6, while their progeny mostly express either SLAM6 (IFNγ-skewed cells) or SLAM1 (IL-17 skewed cells).[313] In fact, the phenotypic divergence of progeny of a common progenitor can develop rapidly under the influence of secondary factors. For example, murine IL-17-producing thymocytes and mature IL-17-skewed γδ T cells are disproportionately expanded by IL-7,[317] in part because sox13 upregulates IL-7R expression[318] that in these cells atypically signals via STAT3.[317] IL-7 increased the cells' response to TCR agonists and also promoted the expansion of IL17-producing human γδ T cells that have otherwise been challenging to identify.[317]

In sum, the capacity of a large fraction of γδ T cells to make uniquely rapid and appropriate effector responses to a broad range of dysregulated target cells is imprinted developmentally. Cell-intrinsic factors, for example, those of fetal versus postnatal thymocytes, skew the response modality toward innate versus adaptive, while the effector modality is skewed at least in part by γδ TCR signal strength, with or without Notch and other factors such as cytokines.[259] Developmental skewing may apply to other functions, such as IL-22 and amphiregulin that may be produced particularly by IL-17 skewed cells and m-CSF that may be enriched in IL-4-skewed Vγ1+ cells, as considered above. Nonetheless, there is no reported imprinting of an obviously suppressive modality, akin to self-reactive TCRαβ+ thymocytes being skewed toward T_{reg} cells. Also, clarification is required of the degree to which developmental skewing may determine the potent granzyme-dependent cytotoxic activities of many different types of γδ T cell.

As we have considered, several traits of mouse fetal thymic γδ T cell development are conserved in humans: the formation of simple V-J and V-D-J junctions, the emergence of invariant TCRs, the development of innate-like cells defined by a Lin28b gene regulatory network,[145,299] and temporally ordered gene segment usage as junctional complexity incre ases.[149,300,301] However, these traits appear less prominently in the peripheral γδ T cell compartments of humans. Likewise, it is not yet clear whether the relative scarcity of human IL-17-producing γδ T cells reflects an altered balance of developmental potentials. In that regard, analysis of single cell transcriptomic data from human fetal thymus identified the coclustering on UMAP of a very small number of cells with transcripts for Maf, Blk, Sox13, PLZF (ZBTB16), and TCRγ/δ that might portend development as IL-17-producing cells.[145] No *IL17* transcripts were detected, but this might reflect relatively low sequence depth. A similar conclusion was reached from another analysis in which the Sox13$^+$ cluster was limited to samples obtained prior to 10 weeks post conception.[319] Although this does not preclude those thymocytes emerging as mature cells with lifelong durability, it seems to contrast with the mouse in which there are multiple origins of IL-17-producing cells,[91,95,120] as considered above. Possibly, a role for innate-like proinflammatory γδ T cells in human neonates may be less critical because of the existence of polyclonal αβ T cells that produce IL-8,[320] which (like IL-17) is a neutrophil activator and attractant but which is not conserved in rodents.

SELECTION OF γδ T CELL COMPARTMENTS

Critical effector functions clearly need to be appropriate to discrete anatomical sites. Because those functions appear to be developmentally imprinted, at least in part by quantitatively and qualitatively distinct TCR signals reflected in different TCR Vγ gene usage, and because different TCR Vγ gene usages also show strong associations with discrete anatomical sites, it is reasonable to ask whether those anatomical associations might also be determined, at least in part, by qualitatively distinct TCR signals.

DETC Selection

Several independently obtained datasets suggesting that the canonical TCR Vγ5Vδ1+ DETC repertoire was shaped by a thymic selecting element[321-323] were validated by the finding that DETC development relies upon thymic epithelial cell expression of Skint1, the prototypic member of an hitherto unannotated Skint gene family, and upon Skint2.[243,252,318,324] DETCs are >90% ablated in Skint1 or Skint 2 mutant mice, but no other T cells are affected. In humans, which lack DETC, *SKINT1L* exists as a pseudogene. As considered briefly above, Skint gene products are Btnl family members (Fig. 35.8, above) that are themselves members of the B7-superfamily (B7SF) (Fig. 35.8, above). Nonetheless, Skint proteins are exceptional among B7SF members in linking B7-like ectodomains to multiple transmembrane (TM) passes (usually 3 or 5) and a short cytoplasmic tail. The significance of those TM passes is unknown, although they evoke the five TM passes of CD47, another Ig-superfamily member gene product that communicates the healthy state of a cell to macrophages, that is, "don't eat me."

Most probably, Skint1 and Skint2 function as heteromers, akin to other Btnl/BTN proteins,[231,252,325] and by engaging E14-E15 Vγ5Vδ1+ thymocytes, they induce overt phenotypic changes, including a signature cell surface (CD45RB^{hi}CD122^{hi}CD27+TIGIT+CD5^{lo}) that is underpinned by induction of Egr3, which induces Id3, thereby suppressing E-box proteins, as was considered above[242,318] (Fig. 35.14C). Concurrently, selection induces RNAs for Tnfsfr9 (4-1BB) and Tbx21 (Tbet) and skewing toward IFNγ and the chemokine, Xcl1, while it downregulates sox13, Blk, Bcl11b, Rorγt, and an IL-17-producing capacity that are collectively present in unselected DETC progenitors[318] (Fig. 35.14C). Indeed, some Vγ5Vδ1+ progenitors on a nonselectable Skint1-mutant background showed PLZF-dependent maturation as IL-17 producers that homed to mucosa rather than to the epidermis[318,326](Fig. 35.14C), offering further evidence for there being a common fetal progenitor of DETC and IL-17-skewed, invariant Vγ6Vδ1+ T cells. The impacts of Skint1 on DETC progenitors seem in some part to be generalizable, since thymic TCRγδ+ cells of adult mice showed enrichment of Egr3 in CD27, NK1.1, CD45RB^{hi} cells with potential to make IFNγ, versus Sox13 enrichment in CD27$^-$, NK1.1− cells with potential to make IL-17.[318] Furthermore, overnight TCRγδ-mediated stimulation of thymocytes upregulated egr3 and downregulated sox13 and Rorγt.[318]

In relation to signature tissue associations, it should be noted that there is still an incomplete understanding of how γδ T cells with specific TCRs home directly to anatomical sites, without requiring prior antigen exposure in secondary lymphoid tissues, as is the case for most αβ T cells.[326-328] Moreover, we do not know how most newly developed human γδ T cells expressing Vδ1 home to extralymphoid tissues, while those cells expressing Vδ2 are primarily lymphoid. Of importance, this behavior is unlikely to relate solely to the expression of chemokine receptors and addressins, and the prospect exists that γδ TCR chains play critical and direct roles. Indeed, there are data linking Skint-mediated selection of Vγ5Vδ1 TCRs to homing via its upregulation of CCR10, a ligand for which, CCL27, is primarily expressed by keratinocytes.[327,328] Of interest, this is another ontogeny-dependent potential, because the CCR10 gene is in open chromatin only in fetal thymocyte progenitors. Because CCR10 can also bind CCL28 expressed by epithelial cells in other sites, its TCR-dependent induction in other fetal thymocytes may permit them to seed other IEL compartments.[327,329] Conversely, Skint selection downregulates CCR6, which is characteristic of IL-17-producing γδ T cells, including fetal invariant Vγ6Vδ1 cells, and which facilitates homing to the dermis and mucosa, which are rich in CCL20, the CCR6 ligand.[95,330] This downregulation is another key component of TCR signaling overriding the default state, since enforced transgenic CCR6 expression in maturing DETC promoted their relocalization to the dermis, albeit it did not affect their effector potentials.[331] Likewise, as considered above, Vγ5Vδ1+ progenitors maturing on a nonselectable Skint1-mutant background homed to mucosa and not to the epidermis.

Skint-mediated selection also imposes on maturing DETC progenitors an attenuated response to TCR stimulation (Fig. 35.14A), evident in impaired calcium mobilization and erk and src phosphorylation[332]. Such attenuation may reflect the cells' skewing to nonclonal innate responses, although these are PLZF-independent traits. Indeed, DETC could respond in situ solely to the local upregulation on keratinocytes of Rae1ε, a stress-induced ligand for the activating receptor NKG2D,[333] which for conventional T cells acts only as a TCR costimulator.

Interestingly, adult Vγ4+ and neonatal Vγ6+ IL-17-skewed cells also showed developmentally acquired TCR hyporesponsiveness[332] (Figure 35.14A). Indeed, >80% of CD27−CD44+ Vγ4+/Vγ6+ cells produced IL-17 in response solely to IL-1β and IL-23,[332] an innate response that was likewise seen in lung cells derived from fetal but not from adult progenitors in RBPJ^{ind} mice,[259] as introduced above. This shared trait of innate responsiveness seems profound given that Skint-selected Vγ5Vδ1+ DETC and Vγ6Vδ1+ IL-17-producing cells are very different by many other criteria. Possibly it reflects a gene regulatory network commonly induced by TCR signaling in fetal progenitors, permitting them to be "safely autoreactive." In seeming to challenge the view that IL-17-producing γδ cells are purely a default state, these data are joined by the observation that the cells' signature phenotype, CD44^{hi}CD62L^{lo}, is associated with TCR activation and that CD27− IL-17-producing T cells were severely depleted in mice expressing a hypomorphic Zap70 allele that would be unable to sustain TCR signaling.[332] Indeed, more detailed signaling studies assigned a key role for Zap70 in the thymic maturation of Vγ6+ IL-17-skewed cells and in the peripheral maintenance of Vγ4+ IL-17-skewed cells.[334]

The need for Skint1+Skint2-mediated DETC progenitor selection might reflect the fact that Vγ5Vδ1 TCRs do not obviously activate by ligand-independent dimerization (see above). Skint1 deficiency can be effectively complemented by TCR signaling,[242] and although there is as yet no biochemical proof, it seems likely that either Skint1 or Skint2 is a Vγ5Vδ1 TCR ligand, given evidence that the seemingly analogous process of BTNL-dependent intestinal IEL regulation involves direct TCR engagement.

Developmental γδ IEL Selection

The Vγ7+ IEL compartment is severely depleted, albeit not absent, in mice lacking Btnl1, whereas no other T cells, including other intestinal γδ IEL or αβ IEL, are affected.[84] Btnl1 can heteromerize with Btnl4 or Btnl6.[252,325] Btnl4-deficient mice harbor a normal Vγ7+ IEL compartment, presumably because Btnl6 can compensate for its loss. However, Btnl6-deficient mice reproducibly show a loss of ~50% of Vγ7+ IEL and the residual TCR repertoire is slightly but significantly different from WT mice.[252] Thus, Vγ7+ IEL progenitors expressing different TCRδ chains may vary in their capacity to interact with different Btnl1-based heterodimers, although neither the mechanism of how TCRVδ might influence Vγ7-Btnl interactions nor its functional significance is known. One set of findings that may be germane to this is the extrathymic influence on IEL TCRVδ usage of MHCII-linked genes, although no connection between the two modes of selection is as yet apparent.[165]

Btnl1-based heteromers act specifically on immature Vγ7+ IEL progenitors, promoting a unique phenotype that is nonetheless strikingly similar to that induced in DETC progenitors by Skint1+Skint2, that is, CD122+TIGIT+CD5^{lo}, *Xcl1*+, *Tnfrsf9*+, with concurrent downregulation of *sox13* and *Bcl11b*.[84] Mature IELs also show profound innate-like hyporesponsiveness to TCR-mediated activation, to which contributory factors may include the cells' origins in early life progenitors, (since in mice the compartment largely builds between days 20 and days 35 post partum), and/or suppressive effects of sustained contact with epithelium.[84,332,335] A question that arises is whether the suppression of *sox13*, *Bcl11b*, and *Il17* RNAs, that is common to Skint-mediated and Btnl-mediated TCR-dependent selection (Fig. 35.14C), asserts that invariant Vγ6+Vδ1+ cells cannot be a product of ligand-dependent selection: a question that will surely be answered by the identification of the Vγ6Vδ1 ligand.

Of note, Btnl1, Btnl4, and Btnl6 are not expressed in thymic epithelium but are largely limited to intestinal epithelial cells.[252,325,336] Thus, Vγ7+ thymocytes might use ligand-independent signaling to partly mature and exit the thymus, whereafter they are phenotypically matured by ligand-dependent TCR signaling in the gut. Alternatively, high-affinity autoreactive interactions of Vγ7+ thymocytes in the neonate might induce a transcriptional program that orchestrates their escape from negative selection and their homing to the gut where they would mature under the influence of Btnl1-based heteromers.[307,325,337] Yet again, some Vγ7+ IEL development might be wholly extrathymic, in support of which there is evidence for phenotypically mature Vγ7+ IEL in athymic *nu/nu* mice (above), and for conspicuous expression of rearranged Vγ7 genes in the fetal liver of *nu/nu* mice.[338] Likewise, in-frame TCRδ gene rearrangements could be detected in human gut as early as 6 weeks post conception.[258] Intestinal Vγ7+ IEL-specific regulation by Btnl1+Btnl6 or Btnl1+Btnl4 is seemingly phenocopied by the regulation of human Vγ4+ colonic IEL by human BTNL3+BTNL8, which are also largely limited in expression to gut epithelial cells.[84] By investigating BTNL3 and/or BTNL8 deficiency haplotypes,[339] it should be possible to determine whether BTNL3+BTNL8 are responsible for the local expansion and differentiation of human Vγ4+ IEL, akin to *Btnl1*-dependent selection of mouse Vγ7+ IEL.

Using site-directed mutagenesis, Btnl1+6-mediated and BTNL3+BTNL8-mediated IEL regulation were found to

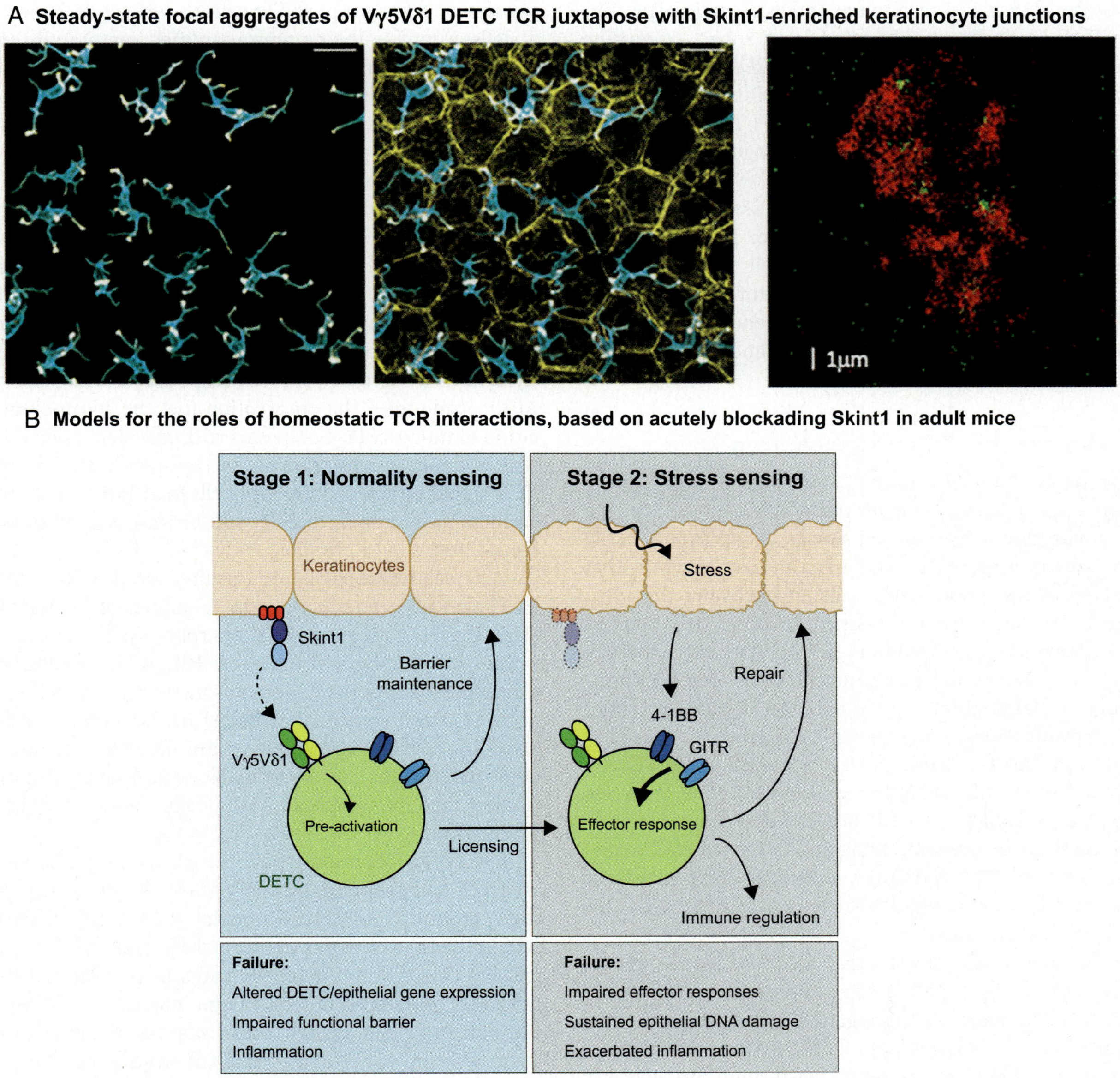

FIG. 35.16. Contributions of the TCR to homeostatic interactions with tissues. A: Left and center panels show confocal imaging of mouse epidermal whole mounts in which steady-state Vγ5Vδ1 TCR focal aggregates are shown as white areas of colocalization of CD3 (cyan) and F-actin (yellow); the epithelial cells are masked in the left panel to permit clearer visualization of the focal aggregates, whereas in the middle panel, the focal aggregates are observed to align with juxtaposed epithelial cell membranes. Right panel. dSTORM with Abbelight, super-resolution microscopy applied to epidermal sheets; red denotes anti-CD3; green denotes anti-Skint1 (size marker indicated). **B:** Cartoon depicting the biological significance of "normality sensing" via the interactions shown in [A], as assessed by acute intradermal perturbation of the interaction using a blocking anti-Skint1 antibody (see McKenzie, et al[135]).

to upregulated NKG2D ligands (Fig. 35.16B). Resolution of these issues can provide considerable insight into how γδ T cells maintain body-surface immunosurveillance.

Because Skint and Btnl/BTNL gene products are expressed by differentiated, postmitotic epithelial cells, their engagement would de facto permit TCRγδ$^+$ IEL to "sense normality"[135] (Fig. 35.16A, B). This would seem distinct from normality sensing by NK cells, since the latter is mediated by inhibitory signals transduced following MHC engagement. Likewise, SIGLEC-dependent sensing of glycans by innate immune cells is also inhibitory, whereas there is no evidence that Skint/Btnl/BTNL engagement initiates inhibitory signaling. If normality sensing is a positive Skint-/Btnl-/BTNL-dependent response of IEL to the healthy status of adjoining epithelial cells, one can ask whether this is critically contributed to by discrete domains of the proteins, for example, the multiple TM passes of Skint1, or the Btnl/BTNL B30.2 domains that might act as sensors of metabolites, akin to PAg sensing by BTN3A1.

For γδ T cells that are seemingly independent of Skint/Btnl/BTN, some as yet unelucidated molecular interactions presumably maintain the competence to make rapid yet safe responses to a dynamic environment. Indeed, sustained peripheral TCR dependence was reported for a subset of murine lymphoid γδ T cells defined by being CD27+CD4RB^{hi} [350] and that were previously shown to be innate-like, akin to other, TCR-selected CD45RB^{hi} γδ T cells.[332] Thus, other self-encoded molecules may emerge with analogous functions to Btnl proteins. In addition, cytokines play key roles, manifest in the strong, selective responses of IL17-skewed γδ T cells to IL-7 (above), and in the requirement for IL-15 to sustain TCRγδ+ IEL that is consistent with their selection-based upregulation of CD122, the IL-15 receptor. Indeed, IL-15 may be important for maintaining the phenotypes of IFNγ-skewed γδ T cells that upregulate CD122 upon initial ligand exposure.

Xenobiotics and γδ T Cell Maintenance

It seems likely that environmental sensing also contributes to the maintenance of γδ T cells, particularly those localized within environmentally-exposed tissues. As an example, the aryl hydrocarbon receptor (AHR) is a nuclear receptor that is activated by xenobiotic chemicals and activates the cytochrome P450 detoxification system. In αβ T cells and in ILC3, AHR signaling can promote IL-17 responses,[351] thereby acting as a cofactor for immunoprotective and immunopathologic inflammation. Likewise, AHR appears essential for IL-22 production by murine IL-17-skewed γδ T cells.[352] Provocatively, however, AHR is also expressed by DETC and TCRγδ$^+$ intestinal IEL, and although those cell compartments can develop seemingly normally in AHR-deficient mice, they do not sustain over time, implying an AHR-mediated maintenance function.[353,354] Moreover, chemical components of dietary vegetables could regulate intestinal TCRγδ$^+$ IEL in an AHR-dependent capacity.[353]

The possibility also exists that host-produced ligands for AHR and/or other nuclear receptors might contribute to DETC/IEL maintenance. In this regard, the neonatal influx and local expansion of Vγ4+ and Vγ1+ T cells in the mouse gingiva seemed attributable to the onset of mastication, possibly via responses to endogenous molecules upregulated by mechanical stresses.[100] In addition, a role for microbial colonization was suggested by the reduced numbers of gingival Vγ6+ T cells in germfree mice.[355] However, this reflects a relatively rare dependence of mouse γδ T cells on microbes, as is considered next.

Microbes and γδ T Cell Maintenance

Only in rare instances does bacterial colonization seem to be required for typical γδ T cell compartments to form in mice.[356] Thus, for IL-17-skewed Vγ6+ T cells, microbes are a driver of postnatal expansion in the gingiva[355] and testis,[357] although they are not required for the cells' initial development. Moreover, seemingly normal IL-17-skewed Vγ6$^+$ cell compartments form in the female reproductive tract,[97] meninges,[41] and lung[358] without any obvious requirements for microbes. Commensals were required to maintain IL-17-skewed γδ T cell numbers in the liver and in the peritoneum, although it was unclear whether this pertained to Vγ6+ and/or to Vγ4+ T cells.[358,359] Considering the selective impacts of microbes, it was proposed that fetal-derived, invariant Vγ6+ T cells complete their preprogrammed maturation and tissue homing as innate IL-17-producing cells independently of microbial influences, whereas in the gingiva and testis (and possibly liver), microbes may compose one of several critical checkpoints for the continuing postnatal maturation and expansion of Vγ6+ IL-17-skewed cells. The selective impacts of microbes were also evident in the finding that *Corynebacterium* spp. drove expansions of Vγ4+ cells in the ocular mucosa and in dermis, but did not substantively impact upon the innate Vγ6+ dermal subset.[360,361] The underlying mechanisms are unresolved, although bacterial mycolic acid, dendritic cells, and IL-23 are required, collectively indicating that γδ T cell responses may be indirect.[356] At the same time, the observation that the commensal regulation of mucosal IL-17-skewed γδ T cells depended on Vav1, which signals downstream of the TCR, might indicate that γδ TCR signaling was required for cells to acquire innate responsiveness to microbes, directly and/or indirectly,[358] as we have considered.

Whereas intestinal lamina propria γδ T cells, which are IL-17 skewed, are reduced in frequency in germ-free mice or those treated with antibiotics, microbes did not substantively impact the numbers of intestinal IEL, which might be considered to be closer to the environmental "front line." This point is considered further below in the context of *Listeria* infection. DETCs are likewise uninfluenced by microbes, while insufficient data are available regarding the importance of microbes for lymphoid Vγ1+ cells skewed toward IFNγ production.

The typical dominance of the human peripheral blood γδ T cell compartment by Vγ9Vδ2+ cells completely reflects their exposure to environmental antigens.[61,346] Exposure within the first 10 weeks of life drove polyclonal proliferation and effector differentiation that was largely limited to public clones derived from fetal thymic progenitors.[346] Although the nature of the stimuli could not be precisely resolved, there was an enrichment for TCR sequences that predict PAg reactivity. Nonetheless, there seemed to be little influence of PAg-containing BCG vaccines.[346] Moreover, despite the cells' cross-reactivity to endogenous PAgs, such as IPP, it seems clear that endogenous PAgs were insufficient to expand the repertoire. Conversely, expansion is likely dependent on BTN3A1 and BTN2A1 (and possibly on BTN3A2 and BTN3A3): hence, this example of repertoire maturation may establish another parallel with BTNL3+BTNL8-mediated IEL repertoire sculpting. In addition, as the authors of these studies noted, environmentally driven neonatal Vγ9Vδ2+ T cell expansions may reflect unique functional roles for γδ T cells in neonatal immunoprotection, and possibly in establishing a commensal microbiome to the exclusion of pathogens.[362] Certainly, human neonatal γδ T cells are precocious IFNγ producers by comparison with αβ T cells.[363] Furthermore, comparison of European and sub-Saharan neonates revealed some differential expansions of blood Vδ1+ and Vδ3+ cells that may reflect geographically distinct antigenic drivers.[364] Again, how TCR subtype might relate to discrete microbial species is unresolved.

The capacity of some γδ T cell compartments to develop phenotypically normally in the absence of microbes does not preclude those cells being reactive to commensals at steady state. For example, dermal Vγ6Vδ1+ T cells expand in response to *Staphylococcus epididymis,* which is a notable skin commensal.[365] Likewise, commensals drove the expansion of lung Vγ6Vδ1+ T cells, promoting an inflammatory response that contributed to lung cancer progression.[366] Again, however, such responses are not universal, since environmentally enriched housing that included farmyard fecal microbiota had no significant impact on uterine Vγ6Vδ1+ T cells.[97]

Influences of microbes on mouse γδ T cells are not limited to IL-17-producing cells. Thus, exposure of intestinal TCRγδ+ IELs to microbes in situ induced transcriptional changes and a glycolytic program that facilitated the IELs' increased local motility within the villi.[367] This capacity to seemingly enhance the cells' immunosurveillance potential might suggest that microbial exposure could influence the efficacy with which IELs contribute to maintaining tissue integrity, with possible implications for cancer control (below). In this light, it is noteworthy that the IEL recognition of infection was indirect, relying on enterocytes as the primary sensors,[367] and independent studies showed migration of IEL toward enterocytes that were in contact with bacteria.[368] The molecules mediating IEL sensing of microbe-responsive enterocytes were not elucidated, but the implication is that deficiencies in microbe sensing by enterocytes could also impact on the tissue surveillance potential of IELs that are intrinsically perfectly functional.

The mechanisms underlying the variable developmental and effector responses of γδ T cells to microbes are for the most part unresolved and may include indirect impacts of cytokines released by myeloid cells responding to MAMPs. They may also reflect roles for short-chain fatty acids or secondary metabolites that can be translocated from the gut. By these means, microbes and/or their derivatives can clearly act as important cofactors for γδ T cell activation, expansion, and effector function, particularly, but not exclusively, IL-17 production. These considerations need to be placed into the context of the activation requirements for γδ T cells.

γδ T CELL ACTIVATION

Having considered that γδ TCR diversity (particularly in TCRδ) is a major driver of the cells' evolutionary conservation, it is reasonable to assume that the γδ TCR, which is a cell-activating signal transduction unit, is a central conduit for γδ T cell activation. There are some differences, in the stoichiometry with which CD3 associates with mouse TCRγδ and TCRαβ, respectively, particularly the absence from the former of CD3δ: hence, TCRαβCD3ε2γδζ2 versus TCRγδCD3ε2γ2ζ2.[369] This distinction is not evident in humans, where both receptor complexes phenocopy the mouse TCRαβ complex, but nonetheless, a leaky mutation in CD3 delta had differential and complex impacts on αβ T cells and γδ T cells, respectively.[370] Moreover, human TCRγδ and TCRαβ complexes are glycosylated differently and may be positioned differently in the membrane with potential to physically alter the accessibility of CD3-ITAM motifs for phosphorylation events that initiate signal transduction.[371]

Mechanoreception

With regard to positioning CD3-ITAM motifs to initiate signal transduction, a significant distinction between TCRαβ and TCRγδ may be the former's capacity to act as a mechanoreceptor.[372,373] Specifically, the engagement of membrane-bound pMHC by a membrane-bound TCR will be subject to forces exerted transversely by lateral membrane flow and longitudinally by underlying cytoskeletal dynamics. As TCRαβ undergoes conformational change in response to pMHC engagement, these forces are captured by catch bonds that enhance signaling by decreasing the off-rate.[373,374] The aggregate result is an enhanced capacity of the T cell to discriminate between low numbers of antigenic peptides (eg, microbe derived) from larger numbers of partial agonists (eg, self-peptides), thereby facilitating digital recognition of very small numbers of pMHC complexes.[373] Moreover, CD3ζ extrusion from the cytoplasmic side of the plasma membrane is enhanced, increasing ITAM accessibility. The region responsible for mechanosensing is an FG loop in the Cβ region, which has a major allosteric impact on how Vα and Vβ align. Of note, TCRγδ contains no counterpart of FG, limiting its capacity to act as a mechanoreceptor.[372] Therefore, TCRγδ may be more influenced by ligand abundance, that is, avidity-based discrimination of bona fide target cells from their healthy counterparts.[372] This is an attractive prospect for a system required to detect changes in the status of self-encoded cell surface molecules (see Fig. 35.2). Nonetheless, there is little doubt that force will contribute in some ways to γδ TCR-ligand interactions, particularly for T cells constrained within epithelia. This merits further investigation.

Beyond the TCR-CD3 complex, signaling transduced by TCRγδ is considered to be largely comparable with that transduced by TCRαβ, with involvement of well-studied molecules, for example, Lck, Zap70, Lat, and NFAT, that initiate downstream activation of well-documented pathways centered on Erk, NFκB, and mTOR.[369] Indeed, rapid transcriptional upregulation of nur77 (a.k.a. the nuclear receptor Nr4a1) is used as a marker of both TCRαβ and TCRγδ stimulation. Nonetheless, one may consider that γδ TCR-proximal signaling may be affected by the absence from most γδ T cells of the CD4 and CD8αβ coreceptors that bind and effectively sequester Lck. When this was enforced on TCRαβ cells by the creation of "quad-deficient" mice, lacking CD4, CD8, and MHCI and MHCII, the resulting αβ T cells were seemingly reactive to a broad range of native, self-encoded cell surface proteins, which is de facto a trait of γδ T cells.[375,376]

Costimulation and Context

TCR-driven activation seems manifest in the adaptive behavior of some fraction of lymphoid γδ T cells, specifically their clonal expansions from low-frequency precursors, their associated differentiation, and their long-term durability. Presumably such TCR-dependent activation is regulated by costimulators of the B7SF and/or TNFRSF, for example, CD28 and 4-1BB, respectively, which are variably expressed by γδ T cells. Nonetheless, the expression of CD28 by peripheral γδ T cells does not match its near-universal expression by naïve lymphoid αβ T cells. This may reflect the prospect that γδ T cell costimulation is mediated by diverse types of target cells that collectively express relevant costimulators concurrently with appropriate ligands for the TCR. In this regard, neither the human EPCR-specific Vγ4Vδ5+ LES T cell clone (above) nor Jurkat T cells transduced with its TCR responded to U251 cells that either overexpressed EPCR or were CMV infected, but they did respond to CMV-infected cells overexpressing EPCR.[71] Hence, a biological context was required for signals to be transduced from ligand-engaged TCRs. Likewise, it was considered above that critical costimulation may be required to overcome attenuated IEL responses to clonotypic antigens.[335,344] To resolve this issue requires that we elucidate which cells initiate functional contacts with γδ T cells and by which molecular axes. To this end, functional γδ T cell response studies should be combined with the dynamic application of ever-improving spatially restricted imaging and -omics technologies.

Costimulation by WC1

A striking illustration of the likely importance of costimulation in γδ T cell activation is provided by WC1 molecules, which are cell surface expressed pattern recognition receptors of the group B "scavenger receptor cysteine-rich" (SRCR) glycoprotein superfamily, related to CD163.[72,377] They are large molecules, frequently >200 kD, composed of multiple SRCR domains of ~100 amino acids, in which 6 to 8 highly conserved cysteine residues usually form 3 to 4 disulfide bonds. The intracellular domains of WC1 proteins are variable, but each type includes tyrosine residues that mediate signal transduction upon phosphorylation. Highly similar genes encoding multiple WC1 isoforms have seemingly been conserved across many species for over millions of years of evolution, including in agnatha, evidently pre-dating γδ T cells. Within any species the genes are mostly nonpolymorphic, and across species homologous WC1 SRCR domains show very high similarity.[378]

Although a description of WC1 ligands is currently incomplete, there is potential for them to include MAMPs encoded by gram-positive and gram-negative bacteria and yeast. To date, binding has been reported of cattle SRCR domains to molecules expressed by *Borrelia* and *Mycobacteria*, and to two outer membrane proteins encoded by *Leptospira*, to which bovine γδ T cells strongly respond, as was considered above. In short, WC1 proteins and their relatives are likely to compose a major innate sensing system for microbes. Because WC1 proteins are uniquely expressed by γδ T cells in pigs, cattle, goats, and sheep (wherein it is also known as T19), camels, antelopes, and probably many other ruminants, for which they were initially believed to be a defining γδ T cell marker,[45,56,72] they may provide contextual information that, at least in part facilitates, the responses of particular γδ T cell subtypes to discrete subtypes of microbes. In essence, WC1 on any single γδ T cell may combine the roles of pattern recognition receptor and costimulatory signal transducer.

WC1 molecules are not expressed by human or rodent γδ T cells, although WC1 genes are conserved in some rodent species.[72] Many γδ T cells in goats are WC1−, as are most bovine intestinal TCRγδ+ IELs and splenic and uterine γδ T cells.[50,379] WC1+ and WC1− γδ T cells are commonly CD8− and CD8+, respectively, and display distinct patterns of responsiveness, considered briefly below, but the strict association of WC1 with γδ T cells in *Artiodactyla* implies the functional integration of WC1 with γδ TCR responsiveness. Indeed, WC1+ γδ T cell activation is achieved experimentally by co-cross-linking WC1 and TCR molecules that are physically colocalized, reportedly within lipid rafts, whereas engaging either molecule alone fails to drive full activation, and solely ligating WC1 may be inhibitory.[380,381]

Under normal circumstances, co-cross-linking of WC1 and TCRγδ would presumably be achieved by corecognition of their respective ligands. Hence, WC1 might focus γδ T cells toward classes of microbial targets. Moreover, a single γδ T cell can express multiple isoforms of WC1, thereby diversifying their pattern sensing potentials. Indeed, γδ T cells in neonatal cattle together transcribe all 13 WC1 genes, collectively expressing 137 SRCR domains, but as animals age, the repertoires decrease. This suggests that γδ T cells might be very broadly immunoprotective in young animals, with subsequent focusing based on selection and/or immunological memory. This perspective has evident parallels to several key findings that have been considered, above, in relation to human and mouse γδ T cells.

Of interest, bovine WC1+ cells can express a spectrum of Vδ gene segments but use only 4 Vγ-gene segments of 16 available, and these are all drawn from a single V-J-Cγ cluster, which encodes TCR Cγ5. This association evokes several other associations of Vγ gene segment usage with γδ T cell biology, considered throughout this chapter. Possibly, only a limited set of Vγ regions can form stable protein complexes with WC1 that preserve the ligand-binding potentials of each receptor. Alternatively, the specific Vγ regions might engage a maintenance ligand(s) that at steady state confers cellular competence to respond to WC1 ligands: a form of TCR-mediated licensing that is considered further below. Clearly, several issues remain to be resolved: is the primary target of the activated γδ T cell microbial or is it an infected cell; if it is the former, does the γδ T cell phagocytose the target, and does it then present antigen (see below); is there a major role for WC1 in facilitating pathogen recognition; and can WC1 act as a DAMP sensor as well as an MAMP sensor? Whatever the answers, WC1 provides a powerful illustration of how γδ TCR-mediated cell activation is shaped and constrained by context.

Costimulation Beyond WC1

Important insights into γδ T cell regulation have been gained from the elucidation of costimulatory axes that had not emerged from studies of αβ T cells. For example, DETC activation is greatly enhanced by the engagement of a cell surface molecule, JAML (junctional adhesion molecule like), by the coxsackie and adenovirus receptor (CAR) that is expressed by keratinocytes.[382] Such engagement leads to the recruitment of PI3-kinase to a motif, YMxM, that provocatively is shared by JAML and CD28, the prototypic costimulatory receptor.[382,383] This then promotes Akt activation, in concert with jnk activation.[382] The context for costimulation in vivo is suggested by the fact that CAR expression is upregulated by epidermal wounding and that blockade of JAML interactions impairs local wound healing.[382] In addition, JAML can costimulate TCRγδ+ intestinal IEL but has negligible impacts on lymphoid γδ T cells. By contrast, potent costimulatory impacts of JAML costimulation have been shown for activated CD8+ αβ T cells in the context of a B16F10 melanoma model.[384]

A similar costimulatory axis is mediated via engagement of the semaphorin, CD100, expressed by DETC, by plexin B2 expressed by keratinocytes.[385] The consequences of this interaction in the context of TCR stimulation include Erk activation and the morphologic rounding-up that is a signature of activated DETC.[385] As was the case for JAML-CAR, blockade of this axis impaired the epidermal wound-healing response.[385] CD100 is widely expressed among T cells and, albeit to a lesser extent, B cells and myeloid cells, and is expressed as transmembrane and secreted isoforms. Plexin B2 is a member of a family of plasma-membrane molecules with conserved intracytoplasmic domains that, at least in the nervous system, can regulate cytoskeletal architecture via interactions with small GTP-binding proteins. Hence, it is reasonable to ask whether this axis, and/or other costimulatory axes, reciprocally signals back to the keratinocyte or other cell type that initiates costimulation, and what the consequences of this might be. The elaboration of additional costimulatory axes is ongoing with the recent highlighting of Hsp8a and ICAM expressed by stressed keratinocytes, which engage CD316 and LFA-1, respectively, that can be expressed by DETC.[386]

Activation of Quasi-Adaptive and Innate γδ T Cell Responses

Because of the striking oligoclonality of murine γδ T cells expanding and undergoing effector maturation in the context of *Plasmodium* infection, we define the response as quasi-adaptive,[343] as was considered above. This argues that infected target cells probably express a ligand(s) for a discrete subset of γδ T cells defined by their TCRs, which may or may not require additional costimulators to achieve T cell activation. Likewise, despite existing at very high precursor frequencies compared with adaptive αβ T cells, PAg-reactive Vγ9Vδ2+ cells also evidently respond primarily via the TCR to a ligand that includes BTN2A1 (see above). Although no key costimulator is evidently necessary, Vγ9Vδ2+ cell responses are stronger in the presence of myeloid APC. A practical outcome of characterizing this form of γδ T cell activation has been to stimulate human Vγ9Vδ2 cells in vivo by use of zoledronate (zol) and other aminobisphosphonates that are clinically approved for treating osteoporosis and bone metastases, and which increase intracellular IPP concentrations via their inhibition of farnesyl pyrophosphate synthase[387,388] (see Fig. 35.6).

Likewise, TCR signaling may contribute to the very rapid, innate-like responses of γδ T cells generated during fetal thymic development, including dermal Vγ6+ IL-17-producing cells and Vγ1+ IL-4-producing cells. As has been considered,[389] those responses are regulated by core transcription factors, Rorγc, c-Maf, and IRF4 operative in seemingly all IL-17-producing γδ T cells, and factors, including sox13, sox4, PLZF, Stat3, and HEB, which are differentially important in different IL-17-producing γδ T cell subsets. Of note, c-Maf regulation of Rorγt is dependent on TCR signaling, at least during the cells' maturation.[119] Signals immediately downstream of the TCR are transduced via syk, rather than by Zap70, which is required for the cells' developmental maturation (Vγ6+ cells) or peripheral maintenance (Vγ4+ cells).[334] Syk signaling initiates Lat-dependent activation of the MAPK cascade and LAT-independent activation of PI3-kinase and AKT.[334] Downstream consequences will include mTOR activation,[390] which will promote mitochondrial activation and energy utilization, which is skewed toward oxidative phosphorylation,[125] and protein translation, including that of mRNAs accumulated at steady state.

These data notwithstanding, it was considered above that IL-17-producing γδ T cells commonly show attenuated TCR signaling responses, by contrast to which very high percentages of such cells can be rapidly activated to full effector function by concurrent provision of IL-1β and IL-23[332.] Similarly, $CD44^{lo}CD45RB^{hi}$ TCRγδ+ T cells and thymocytes can produce IFNγ in response to IL-12+IL-18 alone.[332] In addition to these innate activating cytokines, maintenance cytokines may also promote subset-specific γδ T cell activation, as reflected in IL-7-mediated activation of IL-17-skewed cells[317] and in the activation of murine $TCR\gamma\delta^{+}$ IEL by high levels of IL-15 copresented with its receptor (IL-15Rα), which is a common form of IL-15 release by stressed epithelial cells.[344,391,392]

Adaptate γδ T Cell Activation Potentials

As considered above, distinct antigen specificities have been identified for the γδ TCRs of human (CD1+lipids) and mouse (T10/T22) IELs.[156,172] This raises the question as to whether TCRγδ+ IEL, and possibly other γδ T cells, have adaptate capacity to alternately transduce innate and adaptive signals via the TCR[343] (Fig. 35.15). There may likewise be adaptate potential for the invariant Vγ5Vδ1 DETC TCR, albeit that no direct binding ligand has yet been demonstrated. Thus, it may mediate selection and phenotypic maintenance via interactions with Skint1+Skint2, and it might mediate full activation by engaging another type of ligand(s), as indicated by the capacity of Vγ5Vδ1 multimers to stain stressed keratinocytes.[393] Of interest, any adaptive ligand-binding potential of CDR3 and/or Vδ1 residues would presumably be shared with canonical Vγ6Vδ1 TCRs, which carry the same CDR3, only differing from the DETC TCR in the germline-encoded residues of their Vγ chains.

This notwithstanding, it was demonstrated in an inducible transgenic system that DETCs could make innate responses solely to upregulation on suprabasal keratinocytes of Rae1ε, an activating NKG2D ligand.[333] Although no overt TCR manipulation was required to achieve this, it cannot be precluded that constitutive Skint1-dependent TCR signaling (above) licensed the response.[135,348] Of note, as TCRαβ+ CD8 T cells become chronically exposed to antigens in the celiac gut, they acquire innate competence to respond to MICA (an NKG2D ligand) and IL-15 alone.[391,392]

This revisits the hypothesis (above) that TCR-mediated maintenance functions, such as those mediated by Vγ-engagement, sustain an "activated-yet-resting" state that licenses T cells for very rapid responses via innate receptors, including but not limited to NKG2D and IgSF and TnfrSF receptors that are more commonly regarded as T cell costimulators.[135] By supplementing TCR expression with the developmental acquisition of responsiveness to a broad range of receptors that would costimulate conventional T cells, γδ T cells could be ideally placed to rapidly integrate multiple information inputs that collectively define the dysregulated state of a target cell (Fig. 35.15). As considered above, this may also be the case in ruminants where prior TCR signaling might license γδ T cells to respond to a broad range of WC1 antigens. It might even be true in marsupials, where the γ chain might license responses of γμ+ lymphocytes.

Pleiotropic γδ T Cell Functions

Once activated, γδ T cells can collectively secrete a very broad spectrum of effector molecules, including growth factors, cytolytic mediators, cytokines, and chemokines that are referred to throughout this chapter. These can target diverse cell types in pleiotropic ways, including but not limited to perforin-mediated target cell killing; gzm-mediated modification of the extracellular matrix; chemokine-mediated immune cell recruitment; IL-17/TNF-mediated proinflammatory innate immune activation; and promotion of epithelial growth, repair, and differentiation, mediated, for example, by KGF, IL-22, or amphiregulin. RNAs for some effectors can seemingly be accumulated at steady state and translated immediately upon cell activation, thereby accelerating the delivery of effector potential. Those steady-state RNAs can include extremely high quantities of transcripts for different granzymes,[137,138] although the different roles played by different granzymes are unresolved. Because epithelia generally lack plasmacytoid DC, TCRγδ+ IELs may also be important sources of type I IFN critical to antiviral responses and possibly to other forms of immunosurveillance.[394]

In the context of this pleiotropy, it seems striking that no overtly immunosuppressive cytokines, such as IL10, are the principal products of any functionally identified human or mouse γδ T cell subset. Steady-state TCRγδ+ IELs express a relatively high level of mRNA for IL-16, which has been associated with T cell suppression[395] and which is downregulated by TCR-mediated activation, but this is not a trait unique to either γδ T cells or IEL. Thus, γδ T cells appear primarily to be positively acting immune effector cells, even in cases where their aggregate outcomes may be anti-inflammatory, for example, wound healing or attenuation of αβ T cell responses.[396,397] Nonetheless, this is not to preclude a capacity of γδ T cells to mediate immune suppression by cell-cell contact, for example, via expression of PD-L1. Moreover, there are subpopulations of bovine WC1+ and WC− γδ T cells that seemingly produce IL-10 and TGFβ and that have therefore been considered regulatory, albeit independent of FoxP3.[72,398]

One striking trait of some bovine γδ T cells and human Vγ9Vδ2+ T cells is the capacity to present peptide antigens to αβ T cells.[399-405] Although this trait might a priori co-classify Vγ9Vδ2 T cells together with innate myelomonocytic cells (eg, DC) that respond during the early stages of infection, it is important to recall that other human lymphocytes including B cells, NK cells, and some activated αβ T cells can also present antigen.[406,407] The APC competence of γδ T cells relates in part to their striking capacity to phagocytose and traffic particulate antigens,[402,404] whereupon they are processed and presented on MHCI or MHCII, the latter being expressed by activated human γδ T cells, although not by activated mouse γδ T cells.

With no mouse model, the APC functions of γδ T cells remain poorly understood but may be particularly relevant to particular outcomes. For example, blood and intestinal Vγ9Vδ2 T cells that acquired APC functions following microbial activation in the presence of IL-15 could skew responding TCRαβ CD4+ T cells toward IL-22 without promoting IL-17 responses.[408] Likewise, the capacity of human γδ T cells to cross-present antigen to CD8+ T cells might accelerate the adaptive response to recently infected or transformed cells harboring neoantigens.[400]

γδ T CELL REGULATION

Inevitably, the question arises as to how γδ T cells are regulated, thereby preventing their potentially dangerous chronic activation. Currently, little is known in this regard. Attempts to promote human Vγ9Vδ2+ T cell activation using zol (above) frequently led to the cells' exhaustion. Nonetheless, as perspectives on αβ T cell exhaustion are evolving, it is reasonable to ask whether such cells are permanently inactivated or enter a temporarily refractory state from which they may to some extent recover.[409,410] This could be particularly important for γδ T cell compartments that require life-long peripheral self-renewal. In this regard, it would seem appropriate to track superenhancer activities and chromatin accessibilities that collectively define γδ T cell states, from steady state through antigen activation to recovery.

AICD may be a prominent means of γδ T cell regulation in vivo. For example, individuals infected with HIV show conspicuous losses from the peripheral blood of Vγ9Vδ2+ T cells, especially those with particular CDR3 sequences, leading to speculation that such cells were chronically activated by infection-associated antigens.[187] This was set against a backdrop of expanded peripheral blood Vδ1+ cells, possibly manifesting adaptive responses to infection-associated antigens, but without clear evidence for AICD. Provocatively, a very similar situation appears to occur in SARS-CoV-2-driven COVID-19, in which profound depletions of Vγ9Vδ2+ T cells have been set against clonal Vδ1+ T cell expansions that composed one of very few immunological correlates of virus load.[192,193]

γδ T cells can express Fas, which permits their regulation by FasL-expressing cells,[411-413] and in Fas-deficient *lpr/lpr* mice, lymphoid γδ T cell numbers were substantially increased from 0.6×10^6 cells (in WT mice) to 32×10^6 cells.[411] Moreover, when Fas deficiency was crossed onto $TCR\alpha^{-/-}$ mice, the LN yields exceeded 44×10^6 cells.[411] Intriguingly, Vγ7+ T cells, which are ordinarily confined to the TCRγδ+ IEL compartment, were readily detected among LN cells, suggesting that FasL-expressing cells may somehow limit IEL access to systemic compartments, whereas there was no analogous relocation of DETC.[411]

In addition, when HMBPP was used to stimulate Vγ9Vδ2+ T cells in the presence of IL-2, IFNγ release was significantly diminished by coengagement of the complement receptor CD46, which can be bound by complement opsonins C3b and C4b that are generated when αβ T cells interact with DCs.[414] Hence, this might offer a means to suppress Vγ9Vδ2 T cell activity as αβ T cell–mediated immune responses gain momentum following their initiation by responding γδ T cells.

An additional issue is whether γδ T cell responses can be inhibited by TCRαβ+ T_{reg}. Given that at least some TCRαβ+ T_{reg} cells are considered to function by removing CD28 ligands from APC, thereby limiting CD28-mediated αβ T cell costimulation, they would be predicted to have marginal impact on the bulk of γδ T cells that are naturally $CD28^-$. Likewise, the capacity of TCRαβ+ T_{reg} cells to act as a sink for IL-2 is unlikely to have much impact on γδ T cells that depend on IL-15. However, γδ T cells are reportedly susceptible to other forms of T_{reg} suppression, such as that mediated by IL-35 and IL-10.[415,416] Interesting, intestinal CD8α+ TCRγδ+ IELs derepressed by the absence of competent T_{reg} cells were atypically enriched in Vγ4+ IL-17-producing T cells that were implicated in consequent colitis.[416] This emphasizes the complex regulatory ecology required to prevent the delivery of discrete effectors to inappropriate anatomical sites. In a similar vein, keratinocyte expression of a GPI-linked surface protein, CD109, is required to limit pathologic epidermal infiltration by dermal, IL-17-producing γδ T cells.[417]

Activated γδ T cells can also express PD1, as can some steady-state intestinal IELs. Hence, γδ T cell activation may be limited by one or more cell types that upregulate PD-L1, particularly in the setting of interferons that are known upregulators of PD-L1 expression. In this regard, it is interesting to consider that one consequence of PD-L1 engagement by PD1+ αβ T cells is to increase their motility, limiting strong-stop interactions with APC in all but instances of very high affinity. Thereby, regulated motility of αβ and of γδ T cells may contribute to tissue tolerance.[418] Indeed, acute myc activation in vivo in lung epithelial cells expressing activated ras was rapidly followed by the PD1-dependent relocation *pro tem* of local T cells.[419]

In this light, it is striking that the rapid response of Vγ5Vδ1+ DETC to local tissue dysregulation, for example, induced by tape-stripping or ultraviolet irradiation or inducible Rae1 upregulation (above), invariably includes the rounding-up of cells and their migration, to such an extent that DETC-free areas of the epidermis become apparent.[333,385] Neither the destination nor the fate of those DETCs is clearly resolved, but the possibility exists that they relocate to a site of self-renewal, for example, the hair follicles, in order to downregulate their effector status and to preserve the compartment for which there are no postnatal progenitors. DETCs likewise cede areas of the epidermis to incoming TCRαβ+CD8+ T_{RM} cells following epicutaneous virus infection,[420] conceivably for similar reasons. Furthermore, at another anatomical site, there was a marked reduction of TCRγδ+ IEL recoverable from the intestinal epithelium by 3 days following gavage infection with the small intestinal-tropic coccidian parasite, *Eimeria vermiformis*, although the cell numbers recovered within 1 week.[214] Thus, one wonders whether γδ T cells comply with a general rule by which physical relocation of T cells *vis-à-vis* tissue lesions tissue serves both to limit chronic T cell activation and to protect local T cell pools. Of note, studies of DETC recovery in epidermal areas subject to slight or to gross DETC ablation showed that "replacement" cells adopt a highly ordered tiling pattern and are strictly influenced by the density of the epithelial cells.[421] This can be assumed to be critical because, unlike their intestinal IEL equivalents, which are motile, DETCs are largely sessile, performing surveillance via the dynamic activities of their dendrite tips and associated TCR foci (Fig. 35.15). Nonetheless, there are clearly factors that likewise maintain a relatively uniform IEL:IEC ratio in the intestine.

PROTECTION AGAINST INFECTION BY γδ T CELLS

As mentioned, myriad settings of bacterial, viral, parasitic, and fungal infections are associated with γδ T cell responses. Moreover, important immunoprotective roles have been assigned to human γδ T cells in the settings of TB, CMV, malaria, and HIV1, although in every case there is imprecision over the relevant antigens and the unique host-beneficial contributions of γδ T cells. Likewise, γδ T cells have been assigned major contributions to infection-associated immunopathologies, for example, coxsackie virus myocarditis (above), although these may be exaggerated in the mouse relative to humans by the greater representation of IL-17-secreting cells. Indeed, much data on the roles of γδ T cells in protecting against infection remain phenomenological because for any one single infection there is very little understanding of how the γδ T cell response unfolds in relation to well-established temporal landmarks, starting with innate immune recognition and antigen presentation and culminating in pathogen clearance and immune downregulation. Because of this, we shall limit our consideration to illustrative examples of key concepts.

Listeria Monocytogenes

Recent studies have emphasized that the antigen specificities of responding γδ T cells may often not be pathogen specific, and that in place of this, "context" is likely to be important in determining whether a γδ T cell response will be provoked. This would be consistent with the plethora of innate receptors, considered above, which are used either as costimulators (eg, WC1) or as direct agonists (eg, NKG2D-ligands) for appropriately licensed cells. Moreover, responding γδ T cells include those that are resident at the site of infection and those that infiltrate a site de novo. Illustrating this, from 5 to 10 days following oral infection with an epithelial-tropic *Listeria monocytogenes* (*L.m.*) strain, there were striking intestinal accumulations of proliferative CD27−CD44^{hi} Vγ6Vδ1+ T cells in the lamina propria and in the mesenteric LNs (MLNs), with the latter being a site in which Vγ6Vδ1+ T cells are seldom found at steady state[108] (Fig. 35.17). Such an expansion and migration of T cells into a lesional tissue is a criterion of an adaptive response. There was no obvious concurrent response of TCRγδ+ IEL, despite the *L.m.* strain being gut homing and epithelial tropic.[108] Conversely, it was reported 20 years earlier that mice infected orally with virulent *L.m.* at doses that caused bacterial invasion through the intestinal epithelium provoked increases in the number of IFNy-secreting TCRγδ+ IELs, which could secrete cytokines in response to target cells pulsed with listerial antigens.[195] Thus, the full nature of the gut γδ T cell response to bacterial infection via the epithelium in vivo remains unresolved.

Like adaptive T_{RM} cells, the expanded MLN CD27−CD44^{hi} Vγ6Vδ1+ T cells did not recirculate, as judged by parabiosis experiments, and they remained as expanded subcompartments for over 5 months.[108,422] They were rapidly reactivated upon oral *L.m.* reinfection (Fig. 35.17), whereupon they reorganized into clusters with myeloid cells surrounding sites of *L.m.* replication.[108,422] Because the expanded cells were unresponsive to oral infection by live or attenuated *Salmonella*, antigen specificity was implied,[108] with pathogen-specific memory responses being another defining trait of adaptive immunity. However, there was no response of gut, liver, or spleen Vγ6Vδ1+ T cells when mice were intravenously

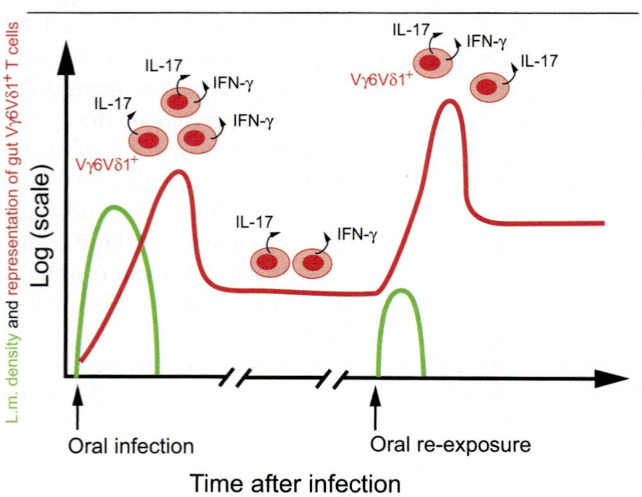

FIG. 17. Quasi-adaptive response of Vγ6Vδ1+ cells to oral *Listeria monocytogenes*. Cartoon depicting recruitment, expansion, retention, and re-expansion of murine Vγ6Vδ1+ T cells in the MLNs following oral gavage infection, and subsequent reinfection, using mouse-adapted, epithelial-tropic *L.m.* (Adapted from Sheridan BS, Romagnoli PA, Pham QM, et al. γδ T cells exhibit multifunctional and protective memory in intestinal tissues. *Immunity.* 2013;39(1):184-195.)

infected with *L.m.*, despite there being substantial bacterial growth in the liver[108] and despite the liver previously being reported as a site for γδ T cell expansion (including invariant Vγ6Vδ1+ T cells) in response to *L.m.* infection.[102,196] In no study were *Listeria*-specific γδ T cell antigen reactivities identified.

Although the expanded Vγ6Vδ1+ T cells had the stereotypical phenotype (CD27−CD44^{hi}) of IL-17-skewed cells, they unexpectedly produced IFNγ as well as IL-17 in response to TCR-mediated stimulation (Fig. 35.17), and included a distinct set of double producers, although those disappeared over time.[108] This reflects some level of effector plasticity of the cells, which, based on the diverse Vδ chain sequences, may have been produced after the main fetal Vγ6 wave in which IL-17 production seems to be more firmly imprinted. Nonetheless, the expanded Vγ6Vδ1+ T cells were the predominant source of IL-17, which was critical to limiting bacterial growth in primary and secondary infection, in combination with the CD4+ and CD8+ αβ T cell responses; that is, the functional response was a collaboration of γδ T cells and αβ T cells.

In short, the γδ T cell response contributed to the efficacy of the primary and memory phases, but its basis remains enigmatic, seemingly straddling standard descriptions of innate and adaptive immunity, and with uncertain dependence on pathogen recognition.[343] Related to this, intratesticular *L.m.* infection evoked and expanded in the infected testis and in the uninfected contralateral testis αβ T cells and Vγ6Vδ1+ cells, at least some of which are tissue-intrinsic testicular cells.[357,423,424] γδ T cell deficiency impaired control of bacterial growth,[357,424] but an equally conspicuous phenotype of γδ-deficiency was an accelerated orchitis.[424] Moreover, the intratesticular Vγ6Vδ1+ cell response could be induced in the absence of either *L.m* or microbial adjuvants by using testicular cell inocula to induce autoimmune orchitis.[425] Similarly, infection of γδ T cell–deficient mice with *Nocardia asteroides*, an intracellular gram-positive bacterium that targets nonciliated bronchial and bronchiolar epithelial cells, resulted in lethally exacerbated local inflammation, which also could be phenocopied in the nonmicrobial setting of ozone stress.[426] In both cases, γδ T cell–deficient mice failed to mobilize neutrophils required to clear bacteria and damaged epithelium, respectively. Such aggregate anti-inflammatory impacts of γδ T cells are noteworthy, given a general tendency to view their rapid production of IL-17 as wholly proinflammatory. Indeed, heightened inflammatory immunopathology is a common aggregate phenotype of γδ T cell deficiency that likewise extends to aerosolized mycobacterial infection and to parasites (below).[427,428]

Thus, host-beneficial, tissue-specific γδ T cell responses to bacterial infection appear to depend on "context" as much as on TCR antigen specificity and have aggregate immunoregulatory as well as immunoprotective impacts. This assessment may apply more generally, given that an appropriate context of virus infection was required for an EPCR-specific human γδ T cell clone to respond to cells displaying its target antigen, as considered above. The requirement for appropriate context might likewise explain why repeated inhalations of *Bacillus subtilis* rapidly induced lung Vγ6Vδ1+ cells to expand and produce IL-17 that promoted bacterial clearance and IL-22 that helped resolve fibrosis,[429] whereas *Saccharopolyspora rectivirgula*–induced pulmonary fibrosis did not induce lung Vγ6Vδ1+ cell activation, with IL-17A production in this setting being attributable to expanded TCRαβ+ Th17 cells.[430] We therefore consider that context is an aggregate of the anatomical sites of antigen exposure, innate immune stimuli comprising DAMPs as well as MAMPs, pathogenicity, and the level, quality, and kinetics of responsiveness of other immune cell types within the "immunological ecosystem."

West Nile Virus and Other Viruses

Studies with WNV (a mosquito-borne encephelopathogenic virus of humans) illustrate how IFNγ-producing γδ T cells can profoundly limit virus growth and improve the quality of host immunity. In WT mice, γδ T cell expansions were observed within 2 days of infection, and at least some fraction of those cells made IFNγ, which was essential for limiting virus growth and neurological dissemination.[431] Thus, TCRδ−/− mice were highly susceptible to lethal meningioencephalitis because of increased viral load and because of increased virus dissemination to the brain.[431] Strikingly, reconstitution of TCRδ−/− mice 24 h prior to WNV infection with splenocytes from naïve TCRβ−/− mice partially restored protection, unless those splenocytes were drawn from IFNγ−/− mice.[431] Like the epithelial-tropic *L.m.* study (above), this study failed to find any evidence of pathogen-specific γδ T cells.

Subsequently it was found that TCRδ−/− mice that survived primary infection were more susceptible to reinfection than counterpart WT mice, which in turn was because of impaired quantity and quality of WNV-specific CD8+ TCRαβ+ memory T cells that developed in TCRδ−/− mice.[190] γδ T cells exert a seemingly analogous influence over CD4+ T cell memory generated during intravaginal infection by HSV2.[432] These findings legitimize the concept of γδ T cells as rapidly responding orchestrators of downstream adaptive responses. Nonetheless, contexts vary, and although splenic, IFNγ-producing γδ T cells also expanded within 2 days following intraperitoneal vaccinia virus infection, limiting virus titers and virus-associated mortality, there was no obvious impact on recall responses.[191]

Protozoan Parasites

Ruminants carry high parasitic burdens, with coccidiosis being an important disease of farmed sheep, cattle, and goats. It likewise is a major economically impactful disease of chickens. Thus, it has been considered that the expansion of γδ T cells in γδ-high species, particularly in young animals, might be driven by parasites, and there are ongoing efforts to develop γδ T cell adjuvants to promote immunoprotection against parasitic and other infectious diseases of livestock.[433] This perspective has been supported by two independent but related studies of parasite infections in mice.

In the first study, TCRδ−/− and TCRα−/− mice aged 1 week and 8 weeks, respectively, were infected orally with *Cryptosporidium parvum*, an intracellular protozoan parasite that causes diarrheal disease in immunosuppressed humans and other mammals. One-week-old TCRδ−/− mice showed the lowest capacity to control parasite growth over the next 2 weeks following infection, although they did clear the infection by 7 weeks, in contrast to persistent oocyst shedding by TCRα−/− mice.[211] In adult mice, the absence of γδ T cells had no significant impact on parasite load, whereas αβ T cells were again required for parasite clearance.[211] Although not contributing to adult immunoprotection, γδ T cells were involved in the adult response, since the pathology of infected TCRβδ−/− mice (lacking all T cells) was qualitatively different from the pathology of αβ T cell–deficient mice.[211]

In the second study, γδ T cells were nonredundantly required for the protection of young mice following gavage infection with the coccidian *E. vermiformis*, but not for the protection of adult mice.[215,434] Nonetheless, the existence of an immunoprotective, IFNγ-mediated γδ T cell potential in adults was evident from the increased susceptibility of TCRβδ−/− mice compared with TCRβ−/− mice.[435,436] This phenocopies the data for mouse CMV infection in which mice lacking all T cells were more severely affected than mice lacking either αβ T cells or γδ T cells (above). Nonetheless, the most striking impact of γδ T cell deficiency in adult *Eimieria*-infected mice was an exaggerated immunopathologic αβ T cell response,[215] seemingly akin to the models of *L.m.*-induced orchitis and of infection-induced and noninfectious pneumonitis models, considered above.

As mentioned above, γδ T cells showed nonredundant importance in suppressing parasite recrudescence in *Plasmodium*-infected mice.[123] With regard to humans, a 6-month longitudinal cohort study involving 206 malaria-exposed school-aged children in Papua and New Guinea showed that IFNγ production by CD56+ γδ T cells in response to *P. falciparum*–infected erythrocytes was a prognostic marker for protection from symptomatic infection.[209] This reinforced the potential importance of developing γδ T cell adjuvants that could enhance vaccine efficacy in humans and agricultural animals.

Fungal Infections

Fungal diseases are of increasing prevalence, and there is much need for new therapeutic approaches. *Candidiasis* susceptibility in humans is associated with IL-17 deficiency; hence, given the propensity of murine γδ T cells to secrete IL-17, it would seem surprising if γδ T cells were not immunoprotective against murine fungal infections. Nonetheless, their study has highlighted the challenges of finding agreement over the cells' contributions. Thus, in one study, the combined intravaginal and intrauterine fungal load was significantly increased in TCRδ−/− mice versus WT mice following intravaginal infection with *Candida albicans*, with female reproductive tract Vγ6Vδ1+ cells implicated in immunoprotection.[97] Likewise, γδ T cell deficiency increased the growth of *Candida* in the lungs, shortly after systemic infection, in which setting γδ T cells were shown to promote antifungal neutrophil activity.[437] Added to this, γδ T cells were required to control *C. albicans* after i.p. infection by enhancing macrophage nitric oxide production and anticandida activity in vitro.[438] Nonetheless, no profound impacts of γδ deficiencies were observed in other studies of intravaginal *Candida* infection,[439] possibly because even closely related systems differed with regard to the enigmatic issues of "context," introduced above.

IFNγ can also contribute to antifungal protection, and human Vδ1+ peripheral blood T cells that were expanded in HIV1-infected patients, particularly those with *Candidiasis*, collectively produced IFNγ and IL-17 ex vivo in response to *C. albicans*, as judged by intracellular flow cytometry.[440] Indeed, there again seemed to be considerable numbers of "double-producer" cells. These assays were undertaken in peripheral blood and so the γδ T cell responses to *Candida* may have been indirect. Likewise, there are reports of human γδ T cell responses to *Aspergillus fumigatus* and *Paracoccidioides brasiliensis*, but again the γδ T cell responses may be indirect with no case reported to date of fungus-specific antigen reactivities.[441]

γδ T Cells, the Microbiome, and Models of Inflammation

Inflammation can commonly result from innate immune responses to dysbiosis. We have already considered that γδ T cells from essentially all species seem able to respond, albeit by different means, to many types of bacteria, providing them the potential to greatly influence microbiome composition, although not obviously via pathogen-specific antigen recognition. At least in mice, this potential is compounded by the proinflammatory capacity of γδ T cells to produce IL-17, which, albeit indirectly, activates neutrophils, increasing their survival and promoting their production of IL-1β and TNFα, which, in a seemingly vicious circle, can enhance the activity of IL-17-producing γδ T cells. It is therefore not surprising that murine γδ T cells that contribute greatly to regulating commensals and to limiting pathogenic microbes are also causally implicated in autoinflammatory diseases, including collagen-induced arthritis (above) and three other widely used mouse models, as follows.

Mouse dermal IL-17-producing γδ T cells promote a psoriasiform skin disease induced by cutaneous administration of imiquimod, a TLR7 agonist.[89,442] This response was exacerbated by so-called trained immunity, in that the responding γδ T cells formed expanded compartments of IL-1R^{hi} cells in nonlesional skin that facilitated hyperresponsiveness to imiquimod re-exposure.[443] In addition, activation of dermal γδ T cells can facilitate their sensing of and contributions to other inflammatory sites, since they downregulate CCR6 that determines their homing to healthy tissues, leaving them to use CCR2 to home into inflammatory settings.[330]

In the second example, γδ T cells collaborated with TCRαβ+ Th17 cells to promote experimental allergic encephalomyelitis (EAE), a progressively debilitating mouse model of multiple sclerosis induced by immunization with Myelin Oligodendrocyte Glycoprotein peptides in microbial adjuvants. Large numbers of IL-17-producing γδ T cells were found in the brains of diseased mice, and they coproduced IL-21, which promoted the activity of Th17 cells.[444]

In the third case, γδ T cells were identified as causal effectors in the NOD mouse model of type I diabetes (T1D).[445] Specifically, IFNγ-producing CD27+CD44^{lo} γδ T cells and IL-17-producing CD27−CD44^{hi} γδ T cells accumulated in pancreatic islet immune infiltrates. Furthermore, CD27− γδ T cells (sources of IL-17) from NOD mice significantly enhanced the capacity of NOD αβ T cells to induce T1D following adoptive transfer to recipient NOD mice deficient in both B and αβ T cells. Added to this, introgression of the TCRδ knockout allele onto the NOD background protected mice from disease, showing the causal pathogenic role(s) played by γδ T cells.[445]

At minimum, these three examples prove the immunopathogenic potency of γδ T cells with rapid propensity to produce IL-17. Since those γδ cells are strongly responsive to IL-1β and IL-23 (above), there is a high likelihood that their pathogenic potentials could be indirectly initiated by such cytokines released from innate immune cells responding to microbial infections and/or dysbiosis. In this regard, it should be noted that there can also be aggregate anti-inflammatory effects of IL-17-producing γδ T cells. Thus, a deficiency in fetal-derived dermal, IL-17-producing γδ T cells led to spontaneous atopic dermatitis (AD), attributable to inappropriate regulation of skin commensals, dysbiosis, and consequent aggressive αβ T cell infiltration.[396] Of interest, αβ T cell–dependent AD developed spontaneously in FVB.TCRδ−/− mice and could be rescued by transfer specifically of DETC progenitors.[397] In this case, the role of the DETC was not to limit αβ T cell priming but the manifestation of their effector functions in the tissues.[397] Based on this, a collaborative ecological model might be proposed by which dermal γδ T cells limit inappropriate priming of αβ T cells by appropriately regulating the skin microbiome, whereas epidermal γδ T cells limit excessive local αβ T cell activity.

Thus, IL-17-producing γδ T cells do not invariably promote inflammatory diseases, a point considered above in relation to the gingiva.[100] Indeed, spontaneous AD also developed spontaneously in transgenic mice overexpressing the Vγ1Vδ6.4 TCR from PLZF-dependent IL-4-producing T cells.[117] Hence, irrespective of the relative underrepresentation of IL-17-producing γδ T cells in humans, the rapid direct and indirect responses of human γδ T cells to microbial and sterile inflammatory factors makes them logical candidates for promoting inflammatory disease, and consequently potential therapeutic targets.

γδ T Cells and Human Gut Inflammation

The immunopathologic potential of human IFNγ-producing γδ T cells was recently emphasized by studies of celiac disease, a commonly occurring, αβ T cell–dependent, allergic response to gliadins in dietary gluten, that can transform into an autoinflammatory disease targeting tissue transglutaminase. For decades, celiac disease has been a poster-child for an association of γδ T cells with human disease since one of its pathognomonic features is a substantial increase in the numbers TCRγδ+ IELs in the small intestinal villi, which are morphologically altered in the celiac gut.[446,447] However, during epithelial reconfiguration, enterocyte expression of BTNL8 was reduced with concomitant diminution of Vγ4Vδ1+ IELs and their replacement by highly proinflammatory IFNγ-secreting Vγ4−Vδ1+ T cells that were associated with disease relapses.[448]

This introduces into human disease the concept considered above in the context of murine AD, namely, that IEL and other homeostatic γδ T cell compartments may suppress inflammation, which is thereby promoted by the suppression or loss of such cells. This concept also applies to a mouse model of lupus, considered below. There are at least three interrelated means by which the cells can be anti-inflammatory: by direct immune suppression, for which there is currently little clear evidence; by clearance or regulation of microbial drivers of inflammation, for which there exits ample evidence (reviewed above); and by promotion of wound resolution, for which tissue-localized T cells might seem well placed.

γδ T CELLS, WOUND HEALING, AND TISSUE HOMEOSTASIS

When DETCs are absent or when there are deficiencies in DETC regulators such as JAML and CD100 (above), there is delayed resolution of full-thickness cutaneous wounds.[40,382,385] In WT mice, activated DETCs accumulate adjacent to wound edges and reportedly produce KGF and possibly other factors, for example, IGF1 and EGF, relevant to local epithelial repair.[40,132] Implying TCR-mediated recognition, multimerized DETC Vγ5Vδ1+ TCRs stained juxtalesional epithelium.[393] This trait has been cited to explain, in part, the delayed resolution in TCRδ−/− mice of intestinal epithelial damage induced by either dextran sodium sulfate (DSS) or TNBS, which are both commonly used models of colitis, or by *Eimeria* infection (above).[215,449,450] Hence, a wound-repair function has been associated with IEL. Nonetheless, this may not be exclusively the case, since gingival IL-17-skewed subepithelial γδ T cells could also promote wound healing, as was clearly shown in relation to periodontal disease that either developed gradually with age or that was induced acutely. The wound repair was attributable to the production by IL-17-skewed γδ T cells of amphiregulin, which is both an epithelial growth factor and an immune regulator.[100] IL-17-skewed γδ T cells could also produce IL-22, which promotes barrier-surface antimicrobial peptide production and epithelial repair. Thus, inflammation induced either by *Citrobacter* infection or by DSS was reduced by retinoic acid in part by skewing γδ T cells toward IL-22 production.[451]

Sometimes the wound-healing impacts of γδ T cells are indirect. Thus, neonatal influenza infection provoked IL-17 production by γδ T cells that in turn promoted IL-33 production by lung epithelial cells. IL-33 promoted lung infiltration by ILC2 and T_{reg} cells that produced amphiregulin, which again mediated critical contributions to tissue repair. Interestingly, IL-17A, IL-33, and amphiregulin expression levels were correlated in influenza-infected children.[452]

γδ T Cells, Adipogenesis, and Dietary Adaptation

The contributions of γδ T cells to wound resolution in the absence of overt microbial drivers suggests their greater potential to regulate physiological processes that maintain homeostasis.[343] In this regard, murine PLZF+ Vγ6+ cells resident in adipose tissue produce TNF and IL-17A that, also by regulating local IL-33 production, maintain local ILC2 and T_{reg} cells that jointly regulate adipose tissue homeostasis.[37] Furthermore, γδ T cell–derived IL-17A and TNF promoted adipocyte expression of uncoupling protein 1, favoring thermogenesis: thus, γδ T cell–deficient mice displayed low body temperature and increased breathing at thermoneutrality and particularly after cold challenge.[37]

More recently, intestinal γδ T cells were shown to be essential for adaptation to an acute shift from a high-protein to a high-fat or high-carbohydrate diet. This adaptation occurs within 5 days via a switch from Fabp6+ enterocytes to Fabp1+ enterocytes that express a transcriptional program appropriate to carbohydrate metabolism.[38] This program is suppressed by IL-22, which has profound diet-related effects on the gut epithelium, including lipid uptake, and which is locally produced by ILC. γδ T cells exert their critical role via suppression of IL-22 production by ILC.[38] Whether this is attributable to TCRγδ+ LP or IEL is unknown, but during the first 5 days of high-carbohydrate intake the transcriptomes of both cell types were changed; the numbers of LP γδ T cells increased; and the mobility of TCRγδ+ IEL was enhanced, with increasing concentrations of γδ T cells close to the crypts, where ILC might conceivably impact upon enterocyte differentiation into Fabp1+ or Fabp6+ cells.[38] Although the afferent signals for responding γδ T cells are unresolved, including the role of the TCR, one candidate is prostaglandin made by intestinal tuft cells.[38,453]

Such codependent association of gut γδ T cells and intestinal epithelial cell biology was first suggested by the report that γδ T cell–deficient mice have decreased intestinal epithelial cell turnover and impaired epithelial cell differentiation.[454] Likewise, $TCR\gamma\delta^+$ IELs have been implicated in cytokine-dependent extrusion of enterocytes from villus apical tips.[455]

Anatomically upstream of enterocyte regulation, IL-17-biased Vγ6+ T cells are enriched in the periodontal ligament, which is the conduit by which information about the status of orthodontic tooth movement (OTM) is relayed to the bone. Thereupon, osteoclast mobilization permits the tooth to adjust position in the alveolar bone, in a tissue repair process reflecting adjustment to the qualitative nature of food stuffs. Strikingly, in inducible γδ T cell–deficient mice, this information relay was impaired, resulting in reduced numbers of osteoclasts in the pressure site during OTM.[42]

A Neurophysiologic γδ Loop: Short-Term Memory and Anxiety

Physiologically downstream of dietary adaptation are neurological impacts, in relation to which Vγ6+ T cells are enriched in the meninges.[41] There they account for most IL-17 produced at steady state, which promotes microglia to produce brain-derived neurotrophic factor that is a critical regulator of glutamatergic synapse plasticity and short-term memory: hence, both are impaired in TCRδ−/− mice.[41] In addition, IL-17 signaling of glutamatergic cortical neurons promoted anxiety responses,[456] possibly in relation to microbe-driven fluctuations of IL-17. Thus a "γδ loop" may arise, by which gut-resident γδ T cells may make immediate responses to environmental change, including dysbiosis or altered diet, which are transduced to distal anatomical sites, for example, the brain, where they are received by resident γδ T cells that promote site-specific neurophysiologic adaptation.

Their overt physiologic roles reemphasize the likelihood that γδ T cells engage in normality sensing from which they build a frame of reference for detecting dysregulation, whereupon they effect changes that can preserve or restore normality.[135] This classifies defense against pathogens as only one of the causes of change to which γδ T cells will respond, and in restoring normality, for example, by wound resolution, it emphasizes the myriad cell types and physiologic axes with which γδ T cells can interact at the effector level. Among those very many target cell types are αβ T cells.

LYMPHOCYTE REGULATION BY γδ T CELLS AND OF γδ T CELLS

Given that some effector functions are imprinted and precociously initiated in developing γδ T cells, particularly in the fetal thymus, it seems unsurprising that thymic γδ T cells may influence αβ T cell development. The relevant studies will be reviewed briefly, because in no case has the mechanistic basis been thoroughly explored. Nonetheless, an intriguing early observation was that introduction of mature γδ T cells (but not αβ T cells) into a SCID thymus could act in *trans* to drive recombinase-deficient thymocytes through the β-selection block, permitting their maturation from the DN into DP stage.[457,458] Note that this was distinct from the progression of TCRγδ+ thymocytes into the DP stage that depends on δ-selection (above).

Thymocyte Cross Talk and Tolerance

Reciprocally, β-selected DP cells seemingly use lymphotoxin beta (LTβ) to promote the functional maturation of γδ T cells, a process termed *trans*-conditioning.[459] As a result, γδ T cells are functionally abnormal in TCRβ−/− mice, as reflected in reduced IFNγ production and a corresponding increase in $CD44^{hi}$ and $IL\text{-}7R^{hi}$ cells, associated with IL-17 production (although this was not known at the time of the studies).[459] *Trans*-conditioning was independently validated, although its impact was claimed to be greater on IL-17 production by Vγ4+ thymocytes.[460] In considering the potential for conservation of this pathway, the expression of LTβ-receptor transcripts among human fetal thymic hematopoietic cells appears to be limited to a small number of γδ T cells and their DN progenitors.[461] Operationally, this finding adds some qualification to the use of γδ T cells from TCRβ−/− (from which they are simpler to isolate) as a perfect phenocopy of WT γδ T cells.

AIRE plays a fundamental role in establishing αβ T cell tolerance, as was considered briefly above. By E16, its expression is induced in medullary thymic epithelial cells (mTECs) by RANK-ligand, prior to the emergence of the first TCRαβ+ thymocytes. This induction had been ascribed to RANK-L+ Lymphoid Tissue Inducer (LTI) cells but was instead found to be orchestrated, at least in part, by differentiating DETC progenitors that express RANK-L.[462] Thereupon, the maturing mTECs upregulate Skint1+Skint2 that facilitate DETC selection, in a striking example of apparent thymic symbiosis.[462] Humans do not develop DETCs, but this may be because of the lack of Skint1 function; hence, they too may develop an initial $TCR\gamma\delta^{+}$ thymocyte pool that critically promotes AIRE upregulation, but does not mature further.

Although AIRE tolerizes αβ T cells to myriad tissue-restricted antigens, it is unclear how αβ T cells become tolerized to cytokines, which are inevitably abundant in settings of T cell activation. Despite not being AIRE target genes, cytokines are common targets of autoimmune B and T cells in humans with AIRE deficiency.[463,464] In considering this, it seems conceivable that the precocious production of cytokines by developing TCRγδ+ fetal thymocytes provides antigens against which cells could be tolerized. Moreover, AIRE deficiency is associated with an imbalance in the γδ T cell compartment that skews toward proinflammatory IL-17-associated functions that may further exacerbate autoinflammatory diseases associated with defects in tolerance.[465]

In short, γδ T cells seem from the outset to contribute to αβ T cell regulation. An overt manifestation of this in adult mice is the heightened αβ T cell–dependent inflammatory response of TCRδ−/− FVB mice, that may be spontaneous or that is exacerbated by epicutaneous application of chemical irritants such as dinitrofluorobenzene.[397] Those immunopathologies reflect exaggerated local effector activities of responding αβ T cells that would ordinarily be suppressed by local γδ T cells. Likewise, the autoaggressive behavior of Fas-deficient αβ T cells in *MRL/lpr* mice was greatly enhanced on the TCRδ-deficient background, with augmented hypergammaglobulinemia and autoantibody production, exacerbated renal disease, and increased mortality, associated with a polyclonal expansion of CD4+ αβ T cells.[413] Nonetheless, this same study identified the fine balance between the anti-inflammatory and proinflammatory roles of γδ T cells (considered above in the context of celiac disease), since αβ T cell–deficient *Mrl/lpr* mice developed isotype-specific hypergammaglobulinemia, incompletely penetrant autoantibodies, and mild immune complex renal disease, all of which were driven by γδ T cell help for B cells.[412,413]

γδ T Cells and B Cells

The studies of lupus models clearly highlighted a capacity of γδ T cells (primarily IL-4-producing Vγ1+ T cells) to provide B cell help in αβ T cell–deficient mice. In conventional animal facilities, this help is evident spontaneously and includes the formation of germinal centers, albeit small, and the promotion of high levels of serum IgG1 and IgE.[466-468] Observationally, WT inbred mice in SPF conditions frequently express very varied levels of IgE at steady state, but those levels are uniformly reduced to baseline by crossing onto a TCRδ−/− background. Moreover, alterations in γδ T cell repertoires were accompanied by significant changes in the peripheral B cell compartment, associated with altered levels of IL-4.[469] Likewise, human activated γδ T cells express CD40L and can help IgE synthesis.[470] Indeed, a Vγ9Vδ2 subset displays many phenotypic features of T-follicular helper (T_{FH}) cells, including ICOS, and CD40L expression, and production of CXCL13, CXCR5, and IL-4.[471] Moreover, the capacity to provide B cell help is promoted by IL-21.[472] Reciprocal interactions between B cells and $V\delta3^+$ cells have also been reported.[473] Indirectly, γδ T cells can help B cells by promoting T_{FH} differentiation.[474] However, although T_{FH}-B cell interactions are driven by the linked specificities of the TCR and BCRs, the mechanisms governing the specificities of γδ T-B cell interactions are entirely unresolved. Evidently, the resulting B cell specificities were very commonly autoreactive, even when *E. vermiformis,* a live polyantigenic parasite was used to initiate the response.[468] Hence, γδ-B interactions are hypothesized to promote high levels of autoantibodies that are commonly associated with αβ T cell deficiencies in mice and humans.[466] Indeed, a patient with a hypomorphic RAG1 mutation that largely ablated αβ T cells displayed approximately normal numbers of clonally diversified γδ T cells and sustained antibody synthesis despite having very few B cells.[475] Those antibodies included ones with hallmarks of T cell help, for example, class-switched to all isotypes, and included autoantibodies, although there were also ones targeting vaccines (eg, diphtheria) and infections (eg, CMV-reactive IgG).[475] These data seemingly attest to a clear capacity of γδ T cells to regulate B cells, posing the question of under which circumstances this may occur in immunocompetent individuals.

More recently, DETCs responding to local administration of DMBA, a carcinogenic xenobiotic, were shown to promote a predominantly autoreactive, local IgE response. Of interest, this effect of DMBA was not mediated by the AHR but by a response to DMBA-induced DNA damage.[476] Moreover, the presence of local cells expressing FcεR1 (most likely basophils) that could be activated by local IgE was associated with protection from squamous cell carcinoma.[476] This capacity of DETCs may relate to their production of IL-13, which was induced following recognition of local upregulation of Rae-1ε, an NKG2D ligand, that is, as part of the lymphoid stress surveillance response (LSS).[35,477] When epicutaneous exposure to antigen (ovalbumin) occurred in the context of LSS, DETC-derived IL-13 was associated with a striking increase in ovalbumin-specific IgE and in total IgE.[477] Hence, the local γδ T cell response to epithelial dysregulation was an adjuvant for local B cell activation.

Possibly this humoral component of LSS might limit tissue damage by targeting toxins that are common causes of local tissue dysregulation. Perhaps too the IgE effector response could contribute to toxin expulsion and additionally limit toxin dissemination by reducing blood pressure, as originally hypothesized by Profet.[478] The counterbalancing cost might be the development of potent IgE responses to benign antigens encountered during episodes of tissue disruption and LSS. This may be particularly relevant to early life, since B cells helped by γδ T cells seem often to have switched directly to IgE production without transiting through an IgG1 intermediate, a process more common in very young mice.[479] Thus, the highest dependence of IgE-based immunity on γδ T cells may occur in newborns, which has inescapable implications for the genesis of childhood atopy.

γδ T CELLS AND CANCER

Interactions with B cells opened a very substantial chapter in γδ T cell biology, namely, the frequent proliferative and cytolytic responses of human γδ T cells to B cell lymphomas without prior intentional priming in vitro or in vivo.[12,160,480-483] As a prototypic example, the human MHCI-deficient Burkitt lymphoma cell line, Daudi, is a common target of Vγ9Vδ2+ T cells but is only rarely targeted by Vγ9Vδ1+ cells, which by contrast can target other human B cell lymphomas.[482] Such findings, together with cold target inhibition assays, led to five conclusions: that human γδ T cell recognition was frequently TCR-dependent, that for Vγ9Vδ2 TCRs this may involve GroEL/hsp recognition, that there was no MHC restriction, that targeting was also boosted by innate receptor-ligand axes shared by different tumor lines, and that killing could be inhibited by MHC (via inhibitory NK receptors) and by PD-L1. The broader relevance of γδ T cell surveillance of lymphomas in vivo was suggested by the finding that mice transgenic for a Vγ1-Jγ4-Cγ4 TCR chain were resistant to T cell lymphomas but not to solid tumor formation.[484] The resistance reflected TCR-dependent recognition, independent of Hsp65, MHCI, and MHCII. Nonetheless, it was another 6 years before the key observation was made that γδ T cell–deficient mice are overtly more susceptible to solid tumors, at least in conventional housing conditions.[133]

Cancer Immunosurveillance by γδ T Cells

Relative to WT controls, TCRδ−/− mice were shown to be highly susceptible to skin cancers induced by each of three regimens: squamous cell carcinomas (SCCs) induced by DMBA + TPA, methylcholanthrene (MCA)-induced fibrosarcomas, and engraftment of an SCC cell line, PDV.[133] Some evidence was provided for direct, TCR-dependent, and NKG2D-dependent PDV targeting by activated DETC. Of note, the increased susceptibilities of TCRδ−/− mice were more robust and more routinely observed than equivalent susceptibilities of TCRβ−/− mice.[133,485]

These increased incidences of cancer in TCRδ−/− mice were reported at a time when tumor immunology was subject to broad-brush skepticism, based on the failure to note increased cancers in athymic *nu/nu* mice or in immunosuppressed patients with AIDS, and based on the frequent restriction of cancers observed in iatrogenically immunosuppressed persons to those with overt viral etiologies. Nonetheless, a contemporaneous study showed increased tumor susceptibility of mice deficient in RAG-dependent lymphocytes and/or IFNγ signaling.[486] Since then, an overwhelmingly strong case for tumor immunology has been made by many seminal findings in many different settings, including evidence that increased numbers of tumor-infiltrating lymphocytes, defining "hot tumors," are a robust correlate of positive patient outcomes,[487] and that the activity of such cells can be unleashed clinically by checkpoint (PD1/CTLA4/TIM3/TIGIT/LAG3) blockade. During this time, several additional examples of γδ T cell contributions to cancer surveillance have been reported, including the increased aggressiveness in γδ T cell–deficient mice of prostate cancers in the SV40 T-driven TRAMP mouse model.[488]

In all cases, the core mechanism(s) underlying γδ T cell–mediated tumor surveillance in vivo have remained largely unelucidated, although IFNγ production has been strongly implicated.[489] Given the pleiotropic effector potentials of γδ T cells and the promiscuity of their cell-cell interactions (above), many factors may contribute, including direct tumor cytolysis, tumor antigen-presentation, and promotion of efficacious effector/memory states of αβ T cells. While not excluding any of these, a very probable modus operandi is the limitation of inflammation coupled to growth factor–mediated wound healing, since these integrated activities sit at the core of maintaining and restoring tissue homeostasis, which can be considered a fundamental biological property of tissue-resident γδ T cells. Indeed, cancers are commonly referred to as wounds that never heal.[490] As an example of how these two processes may be integrated at the effector level, IL-13 produced by tissue-resident γδ T cells (above) is de facto a "crossover" effector since it can contribute to B cell immunoregulation and can regulate epithelial cell growth and differentiation. Given that malignancies, by definition, expand beyond their typical anatomical compartment, for example, the epithelium, it is also probable that different subtypes of γδ T cells, specifically IEL and subepithelial cells, respectively, engage precancerous and cancerous lesions in a temporal order, thereby regulating cancer via a spectrum of biological activities. This inevitably adds complexity to attempts to predict cancer progression based on the detection of tumor-associated γδ T cells. Indeed, overactivity of some γδ T cell subsets might be predicted to promote tumor progression, for example, via proinflammatory effects that ordinarily clear a tissue of debris, prior to wound resolution.

γδ T Cells and Human Cancer

The potential for IL-17-producing γδ T cells to promote tumor promotion has been borne out in several murine models.[491] Although some parallel settings have been reported in humans, particularly but not exclusively in colorectal cancers,[492-494] the uncertain importance of IL-17-skewed γδ T cells in humans (above) leaves open the question of whether or not the cells commonly contribute to clinical cancer progression. Furthermore, human IL-17-producing γδ T cells have been reported to be toxic toward ovarian cancer stem cells.[495] Conversely, there are studies, albeit small, in which increased numbers of IFNγ-producing, cytolytic γδ T cells within human tumors, as well as in associated extralesional healthy tissue, have been associated with improved patient outcome over several years.[142,496] On a much larger scale, some very strong associations of intratumoral or juxtatumoral γδ T cell signatures with positive patient outcomes have been inferred from -omics analyses,[497] but these are somewhat confounded by the lack of γδ T cell–unique genes,[498] as was considered at the beginning of this chapter.

Indeed, for γδ T cells to make unique patient-beneficial contributions to cancer presumably relies on their unique TCR and their rapidly deployed intrinsic effector functions that this chapter has reviewed. Possibly γδ T cells are the body's natural mechanism for discriminating between normal and generically dysregulated cells and for targeting the latter, both directly and indirectly. Consistent with this, putative normality signals, for example, BTNL genes, are broadly downregulated in mouse and human cancers, but this notwithstanding, the community lacks a detailed understanding of how the γδ TCR orchestrates cancer immune surveillance.

Clinical Applications of γδ T Cells

From a practical standpoint, here are several attractive aspects of γδ T cells as cancer-targeting agents, including their capacity to straddle innate and adaptive immunity; their capacity to home to and mediate function within tissues that are the sites of solid cancer formation or metastasis; their diverse TCR specificities that may enable them to function more broadly than NK cells; and their lack of MHC restriction, which may enable them to function in settings where αβ T cells are impotent. Indeed, the frequent identification of activated γδ T cells in immunosuppressed transplant recipients emphasizes the capacity of γδ T cells to offer immunoprotective functions in the functional absence de facto of αβ T cells. A further illustration of this may be the critical roles played by γδ T cells in neonates, prior to the maturation of αβ T cell responses. Thus, γδ T cells may be able to provide patient benefit in cancers refractory to the impacts of other lymphocytes. In addition, the lack of MHC restriction theoretically permits γδ T cells to function efficaciously in unmatched recipients, opening a door on the potential for off-the-shelf, on-demand treatments.

For these and many other reasons, the clinical opportunities of γδ T cells are being intensively pursued from the perspectives of pharmacological cell activation and autologous and allogeneic adoptive γδ cell therapy.[499] Vγ9Vδ2+ cell activation by administering zol (above) ± IL-2 showed marginal efficacy at best, possibly because the cells apparently entered an exhausted state, as considered above. Moreover, the use of PAgs as adjuvants was confounded by their very short half-lives in vivo. Hence, alternative approaches have been developed. Those include the use of other cell types, specifically αβ T cells, to act as vectors for transduced Vγ9Vδ2 TCRs, and a greater focus on Vδ1+ T cells, which naturally operate within the tissues in which solid tumors form. Such cells may show clinical efficacy based on their intrinsic potentials to recognize and respond to cancer cells and/or by virtue of chimeric antigen receptor constructs with which they are transduced.

Carefully designed human trials should offer valuable opportunities to study human γδ T cell biology in vivo, possibly offering unprecedented insight into the mechanisms of immunoevasion against γδ T cell surveillance. PD-L1 upregulation and the downregulation of MHC and NKG2D ligand expression by tumor cells provide evidence for the importance of those pathways in tumor immunosurveillance by αβ T cells and NK cells, whereas the hierarchy of molecular pathways involved in cancer surveillance by γδ T cells is less clear. Possibly, tumors and pathogens might evade γδ T cells by upregulating normality signals that would thereby perturb the cells' capacity to discriminate target cells as dysregulated.

CONCLUSIONS

The burgeoning clinical interest in γδ T cells reflects the growing realization that γδ T cells make unique contributions to immune responses that reflect at least two of their core traits: the nature of their unique TCR and their rapid deployment of effector functions at body surfaces, in the viscera, and even in the brain. These capabilities go well beyond the conventional immunological perspectives of protection from infection, and responsiveness to foreign antigens, including mutated self. As a result, many improvements may need to be made in experimental systems, particularly the refinement of physiologic read-outs, before the full spectrum of γδ T cell biology is appreciated. In most of the world's species of large land mammals, γδ T cells are abundant, particularly in the young, but this is rarely the case in rodents or primates in which γδ T cells are most commonly studied. Nonetheless, many signatures of γδ T cell biology are evidently highly conserved across γδ-high and γδ-low animals. Although γδ T cells have myriad, pleiotropic effector potentials, their most profound roles may be in orchestrating downstream adaptive response by, for example, orchestrating a balance of immunogenicity of dysregulated cells and tissues versus tolerance toward normality. By telescoping that discrimination into the activation or tolerance of αβ T cell memory, the impacts of γδ T cell recognition and activation may be long-lasting. For this reason, there will be much interest in γδ T cell adjuvants for vaccines in humans and animals, particularly given the increased concerns over zoonoses and the renaissance of interest in cancer vaccines.

REFERENCES

1. Brenner MB, McLean J, Dialynas DP, et al. Identification of a putative second T-cell receptor. *Nature*. 1986;322(6075):145-149.
2. Hayday AC, Saito H, Gillies SD, et al. Structure, organization, and somatic rearrangement of T cell gamma genes. *Cell*. 1985;40(2):259-269.
3. Saito H, Kranz DM, Takagaki Y, Hayday AC, Eisen HN, Tonegawa S. Complete primary structure of a heterodimeric T-cell receptor deduced from cDNA sequences. *Nature*. 1984;309(5971):757-762.
4. Kranz DM, Saito H, Heller M, et al. Limited diversity of the rearranged T-cell gamma gene. *Nature*. 1985;313(6005):752-755.
5. Pernis B, Axel R. A one and a half receptor model for MHC-restricted antigen recognition by T lymphocytes. *Cell*. 1985;41(1):13-16.
6. Born W, Rathbun G, Tucker P, et al. Synchronized rearrangement of T-cell gamma and beta chain genes in fetal thymocyte development. *Science*. 1986;234(4775):479-482.
7. Raulet DH, Garman RD, Saito H, et al. Developmental regulation of T-cell receptor gene expression. *Nature*. 1985;314(6006):103-107.
8. Mackay CR, Maddox JF, Brandon MR. Three distinct subpopulations of sheep T lymphocytes. *Eur J Immunol*. 1986;16(1):19-25.
9. Jones B, Mjolsness S, Janeway C, et al. Transcripts of functionally rearranged gamma genes in primary T cells of adult immunocompetent mice. *Nature*. 1986;323(6089):635-638.
10. Yoshikai Y, Reis MD, Mak TW. Athymic mice express a high level of functional gamma-chain but greatly reduced levels of alpha- and beta-chain T-cell receptor messages. *Nature*. 1986;324(6096):482-485.
11. Garman RD, Doherty PJ, Raulet DH. Diversity, rearrangement, and expression of murine T cell gamma genes. *Cell*. 1986;45(5):733-742.
12. Borst J, van de Griend RJ, van Oostveen JW, et al. A T-cell receptor gamma/CD3 complex found on cloned functional lymphocytes. *Nature*. 1987;325(6106):683-688.
13. Moingeon P, Jitsukawa S, Faure F, et al. A gamma-chain complex forms a functional receptor on cloned human lymphocytes with natural killer-like activity. *Nature*. 1987;325(6106):723-726.
14. Bank I, DePinho RA, Brenner MB, Cassimeris J, Alt FW, Chess L. A functional T3 molecule associated with a novel heterodimer on the surface of immature human thymocytes. *Nature*. 1986;322(6075):179-181.
15. Van Dongen JJ, Wolvers-Tettero IL, Seidman JG, et al. Two types of gamma T cell receptors expressed by T cell acute lymphoblastic leukemias. *Eur J Immunol*. 1987;17(12):1719-1728.
16. Chien YH, Iwashima M, Kaplan KB. et al. A new T-cell receptor gene located within the alpha locus and expressed early in T-cell differentiation. *Nature*. 1987;327(6124):677-682.
17. Loh EY, Lanier LL, Turck CW, et al. Identification and sequence of a fourth human T cell antigen receptor chain. *Nature*. 1987;330(6148):569-572.
18. Hata S, Brenner MB, Krangel MS. Identification of putative human T cell receptor delta complementary DNA clones. *Science*. 1987;238(4827):678-682.
19. Band H, Hochstenbach F, McLean J, Hata S, Krangel MS, Brenner MB. Immunochemical proof that a novel rearranging gene encodes the T cell receptor delta subunit. *Science*. 1987;238(4827):682-684.
20. Born W, Miles C, White J, et al. Peptide sequences of T-cell receptor delta and gamma chains are identical to predicted X and gamma proteins. *Nature*. 1987;330(6148):572-574.
21. Bonyhadi M, Weiss A, Tucker PW, et al. Delta is the Cx-gene product in the gamma/delta antigen receptor of dendritic epidermal cells. *Nature*. 1987;330(6148):574-576.
22. Chen H, Kshirsagar S, Jensen I, et al. Characterization of arrangement and expression of the T cell receptor gamma locus in the sandbar shark. *Proc Natl Acad Sci U S A*. 2009;106(21):8591-8596.
23. Chen H, Bernstein H, Ranganathan P, et al. Somatic hypermutation of TCR gamma V genes in the sandbar shark. *Dev Comp Immunol*. 2012;37(1):176-183.
24. Ciccarese S, Vaccarelli G, Lefranc MP, et al. Characteristics of the somatic hypermutation in the Camelus dromedarius T cell receptor gamma (TRG) and delta (TRD) variable domains. *Dev Comp Immunol*. 2014;46(2):300-313.
25. Davis MM, Bjorkman PJ. T-cell antigen receptor genes and T-cell recognition. *Nature*. 1988;334(6181):395-402.
26. Asarnow DM, Kuziel WA, Bonyhadi M, Tigelaar RE, Tucker PW, Allison JP. Limited diversity of gamma delta antigen receptor genes of Thy-1+ dendritic epidermal cells. *Cell*. 1988;55(5):837-847.
27. Boehm T, Hirano M, Holland SJ, Das S, Schorpp M, Cooper MD. Evolution of alternative adaptive immune systems in vertebrates. *Annu Rev Immunol*. 2018;36:19-42.
28. Morrissey KA, Wegrecki M, Praveena T, et al. The molecular assembly of the marsupial gammamu T cell receptor defines a third T cell lineage. *Science*. 2021;371(6536):1383-1388.
29. Criscitiello MF, Saltis M, Flajnik MF. An evolutionarily mobile antigen receptor variable region gene: doubly rearranging NAR-TcR genes in sharks. *Proc Natl Acad Sci U S A*. 2006;103(13):5036-5041.
30. Benveniste PM, Roy S, Nakatsugawa M, et al. Generation and molecular recognition of melanoma-associated antigen-specific human gammadelta T cells. *Sci Immunol*. 2018;3(30):eaav4036.
31. Ribot JC, deBarros A, Pang DJ, et al. CD27 is a thymic determinant of the balance between interferon-gamma- and interleukin 17-producing gammadelta T cell subsets. *Nat Immunol*. 2009;10(4):427-436.
32. Asarnow DM, Cado D, Raulet DH. Selection is not required to produce invariant T-cell receptor gamma-gene junctional sequences. *Nature*. 1993;362(6416):158-160.
33. Zhang Y, Cado D, Asarnow DM, et al. The role of short homology repeats and TdT in generation of the invariant gamma delta antigen receptor repertoire in the fetal thymus. *Immunity*. 1995;3(4):439-447.
34. Zhao L, Frock RL, Du Z, et al. Orientation-specific RAG activity in chromosomal loop domains contributes to Tcrd V(D)J recombination during T cell development. *J Exp Med*. 2016;213(9):1921-1936.
35. Hayday AC. Gammadelta T cells and the lymphoid stress-surveillance response. *Immunity*. 2009;31(2):184-196.
36. Ljunggren HG, Karre K. In search of the 'missing self': MHC molecules and NK cell recognition. *Immunol Today*. 1990;11(7):237-244.
37. Kohlgruber AC, Gal-Oz ST, LaMarche NM, et al. Gammadelta T cells producing interleukin-17A regulate adipose regulatory T cell homeostasis and thermogenesis. *Nat Immunol*. 2018;19(5):464-474.
38. Sullivan ZA, Sullivan ZA, Sullivan ZA, et al. Gammadelta T cells regulate the intestinal response to nutrient sensing. *Science*. 2021;371(6535):eaba8310.
39. Girardi M, Lewis JM, Filler RB, et al. Environmentally responsive and reversible regulation of epidermal barrier function by gammadelta T cells. *J Invest Dermatol*. 2006;126(4):808-814.
40. Jameson J, Ugarte K, Chen N, et al. A role for skin gammadelta T cells in wound repair. *Science*. 2002;296(5568):747-749.
41. Ribeiro M, Brigas HC, Temido-Ferreira M, et al. Meningeal gammadelta T cell-derived IL-17 controls synaptic plasticity and short-term memory. *Sci Immunol*. 2019;4(40):eaay5199.
42. Wald S, Leibowitz A, Aizenbud Y, et al. gammadeltaT cells are essential for orthodontic tooth movement. *J Dent Res*. 2021;100(7):731-738.
43. Heilig JS, Tonegawa S. Diversity of murine gamma genes and expression in fetal and adult T lymphocytes. *Nature*. 1986;322(6082):836-840.
44. Haas W, Pereira P, Tonegawa S. Gamma/delta cells. *Annu Rev Immunol*. 1993;11:637-685.
45. Mackay CR, Hein WR. A large proportion of bovine T cells express the gamma delta T cell receptor and show a distinct tissue distribution and surface phenotype. *Int Immunol*. 1989;1(5):540-545.
46. Su C, Jakobsen I, Gu X, et al. Diversity and evolution of T-cell receptor variable region genes in mammals and birds. *Immunogenetics*. 1999;50(5-6):301-308.
47. Takamatsu HH, Denyer MS, Stirling C, et al. Porcine gammadelta T cells: possible roles on the innate and adaptive immune responses following virus infection. *Vet Immunol Immunopathol*. 2006;112(1-2):49-61.
48. Wu M, Jiang Q, Nazmi A, Yin J, Yang G. Swine unconventional T cells. *Dev Comp Immunol*. 2021;128:104330.
49. Yirsaw A, Baldwin CL. Goat gammadelta T cells. *Dev Comp Immunol*. 2021;114:103809.
50. Yirsaw AW, Gillespie A, Britton E, et al. Goat gammadelta T cell subpopulations defined by WC1 expression, responses to pathogens and cytokine production. *Dev Comp Immunol*. 2021;118:103984.
51. Cross ML, Chinn ND, Griffin JF, et al. T cell responses to Mycobacterium bovis in red deer, a large animal model for tuberculosis. *Immunol Cell Biol*. 1996;74(1):32-37.
52. Hussen J, Schuberth HJ. Recent advances in camel immunology. *Front Immunol*. 2020;11:614150.

53. Wan F, Hu CB, Ma JX, Gao K, Xiang LX, Shao JZ. Characterization of gammadelta T cells from zebrafish provides insights into their important role in adaptive humoral immunity. *Front Immunol.* 2016;7:675.
54. Criscitiello MF, Ohta Y, Saltis M, et al. Evolutionarily conserved TCR binding sites, identification of T cells in primary lymphoid tissues, and surprising trans-rearrangements in nurse shark. *J Immunol.* 2010;184(12):6950-6960.
55. Sowder JT, Chen CL, Ager LL, et al. A large subpopulation of avian T cells express a homologue of the mammalian T gamma/delta receptor. *J Exp Med.* 1988;167(2):315-322.
56. Clevers H, MacHugh ND, Bensaid A, et al. Identification of a bovine surface antigen uniquely expressed on CD4-CD8- T cell receptor gamma/delta+ T lymphocytes. *Eur J Immunol.* 1990;20(4):809-817.
57. Bucy RP, Chen CL, Cooper MD. Tissue localization and CD8 accessory molecule expression of T gamma delta cells in humans. *J Immunol.* 1989;142(9):3045-3049.
58. Borst J, Wicherink A, Van Dongen JJ, et al. Non-random expression of T cell receptor gamma and delta variable gene segments in functional T lymphocyte clones from human peripheral blood. *Eur J Immunol.* 1989;19(9):1559-1568.
59. Tribel F, Lefranc MP, Hercend T. Further evidence for a sequentially ordered activation of T cell rearranging gamma genes during T lymphocyte differentiation. *Eur J Immunol.* 1988;18(5):789-794.
60. Modlin RL, Pirmez C, Hofman FM, et al. Lymphocytes bearing antigen-specific gamma delta T-cell receptors accumulate in human infectious disease lesions. *Nature.* 1989;339(6225):544-548.
61. Parker CM, Groh V, Band H, et al. Evidence for extrathymic changes in the T cell receptor gamma/delta repertoire. *J Exp Med.* 1990;171(5):1597-1612.
62. Tanaka Y, Morita CT, Nieves E, Brenner MB, Bloom BR. Natural and synthetic non-peptide antigens recognized by human gamma delta T cells. *Nature.* 1995;375(6527):155-158.
63. Constant P, Davodeau F, Peyrat MA, et al. Stimulation of human gamma delta T cells by nonpeptidic mycobacterial ligands. *Science.* 1994;264(5156):267-270.
64. Jomaa H, Feurle J, Luhs K, et al. Vgamma9/Vdelta2 T cell activation induced by bacterial low molecular mass compounds depends on the 1-deoxy-D-xylulose 5-phosphate pathway of isoprenoid biosynthesis. *FEMS Immunol Med Microbiol.* 1999;25(4):371-378.
65. Altincicek B, Moll J, Campos N, et al. Cutting edge: human gamma delta T cells are activated by intermediates of the 2-C-methyl-D-erythritol 4-phosphate pathway of isoprenoid biosynthesis. *J Immunol.* 2001;166(6):3655-3658.
66. Hintz M, Reichenberg A, Altincicek B, et al. Identification of (E)-4-hydroxy-3-methyl-but-2-enyl pyrophosphate as a major activator for human gammadelta T cells in *Escherichia coli. FEBS Lett.* 2001;509(2):317-322.
67. Hayday AC. [gamma][delta] cells: a right time and a right place for a conserved third way of protection. *Annu Rev Immunol.* 2000;18:975-1026.
68. Puan KJ, Jin C, Wang H, et al. Preferential recognition of a microbial metabolite by human Vgamma2Vdelta2 T cells. *Int Immunol.* 2007;19(5):657-673.
69. Davey MS, Willcox CR, Joyce SP, et al. Clonal selection in the human Vdelta1 T cell repertoire indicates gammadelta TCR-dependent adaptive immune surveillance. *Nat Commun.* 2017;8:14760.
70. Ravens S, Schultze-Florey C, Raha S, et al. Human gammadelta T cells are quickly reconstituted after stem-cell transplantation and show adaptive clonal expansion in response to viral infection. *Nat Immunol.* 2017;18(4):393-401.
71. Willcox CR, Pitard V, Netzer S, et al. Cytomegalovirus and tumor stress surveillance by binding of a human gammadelta T cell antigen receptor to endothelial protein C receptor. *Nat Immunol.* 2012;13(9):872-879.
72. Telfer JC, Baldwin CL. Bovine gamma delta T cells and the function of gamma delta T cell specific WC1 co-receptors. *Cell Immunol.* 2015;296(1):76-86.
73. Stingl G, Koning F, Yamada H, et al. Thy-1+ dendritic epidermal cells express T3 antigen and the T-cell receptor gamma chain. *Proc Natl Acad Sci U S A.* 1987;84(13):4586-4590.
74. Kuziel WA, Takashima A, Bonyhadi M, et al. Regulation of T-cell receptor gamma-chain RNA expression in murine Thy-1+ dendritic epidermal cells. *Nature.* 1987;328(6127):263-266.
75. Shimada S, Shibagaki N, Tanaka T, Kubota Y, Hashimoto Y, Tamaki K. Identification of an equivalent to murine Thy-1+ dendritic epidermal cells in the rat epidermis. *J Dermatol Sci.* 1992;3(1):68-71.
76. Havran WL, Grell S, Duwe G, et al. Limited diversity of T-cell receptor gamma-chain expression of murine Thy-1+ dendritic epidermal cells revealed by V gamma 3-specific monoclonal antibody. *Proc Natl Acad Sci U S A.* 1989;86(11):4185-4189.
77. Hirano M, Guo P, McCurley N, et al. Evolutionary implications of a third lymphocyte lineage in lampreys. *Nature.* 2013;501(7467):435-438.
78. Havran WL, Allison JP. Developmentally ordered appearance of thymocytes expressing different T-cell antigen receptors. *Nature.* 1988;335(6189):443-445.
79. Havran WL, Allison JP. Origin of Thy-1+ dendritic epidermal cells of adult mice from fetal thymic precursors. *Nature.* 1990;344(6261):68-70.
80. Bonneville M, Janeway CA, Ito K, et al. Intestinal intraepithelial lymphocytes are a distinct set of gamma delta T cells. *Nature.* 1988;336(6198):479-481.
81. Kyes S, Carew E, Carding SR, et al. Diversity in T-cell receptor gamma gene usage in intestinal epithelium. *Proc Natl Acad Sci U S A.* 1989;86(14):5527-5531.
82. Goodman T, Lefrancois L. Expression of the gamma-delta T-cell receptor on intestinal CD8+ intraepithelial lymphocytes. *Nature.* 1988;333(6176):855-858.
83. Landau SB, Aziz WI, Woodcock-Mitchell J, et al. V gamma (I) expression in human intestinal lymphocytes is restricted. *Immunol Invest.* 1995;24(6):947-955.
84. Di Marco Barros R, Roberts NA, Dart RJ, et al. Epithelia use butyrophilin-like molecules to shape organ-specific gammadelta T cell compartments. *Cell.* 2016;167(1):203-218 e17.
85. Deusch K, Luling F, Reich K, Classen M, Wagner H, Pfeffer K. A major fraction of human intraepithelial lymphocytes simultaneously expresses the gamma/delta T cell receptor, the CD8 accessory molecule and preferentially uses the V delta 1 gene segment. *Eur J Immunol.* 1991;21(4):1053-1059.
86. Christmas SE. T-cell receptor gene expression by human gamma delta T-cell clones from peripheral blood and reproductive tissues in relation to non-MHC-restricted cytotoxic function. *Scand J Immunol.* 1991;33(6):627-637.
87. Bucy RP, Chen CL, Cihak J, et al. Avian T cells expressing gamma delta receptors localize in the splenic sinusoids and the intestinal epithelium. *J Immunol.* 1988;141(7):2200-2205.
88. Sumaria N, Roediger B, Ng LG, et al. Cutaneous immunosurveillance by self-renewing dermal gammadelta T cells. *J Exp Med.* 2011;208(3):505-518.
89. Cai Y, Shen X, Ding C, et al. Pivotal role of dermal IL-17-producing gammadelta T cells in skin inflammation. *Immunity.* 2011;35(4):596-610.
90. Gray EE, Suzuki K, Cyster JG. Cutting edge: identification of a motile IL-17-producing gammadelta T cell population in the dermis. *J Immunol.* 2011;186(11):6091-6095.
91. O'Brien RL, Born WK. Dermal gammadelta T cells—What have we learned? *Cell Immunol.* 2015;296(1):62-69.
92. Kisielow J, Kopf M, Karjalainen K. SCART scavenger receptors identify a novel subset of adult gammadelta T cells. *J Immunol.* 2008;181(3):1710-1716.
93. Tan L, Sandrock I, Odak I, et al. Single-cell transcriptomics identifies the adaptation of Scart1(+) Vgamma6(+) T cells to skin residency as activated effector cells. *Cell Rep.* 2019;27(12):3657-3671 e4.
94. Hein WR, Dudler L. TCR gamma delta+ cells are prominent in normal bovine skin and express a diverse repertoire of antigen receptors. *Immunology.* 1997;91(1):58-64.
95. Cai Y, Xue F, Fleming C, et al. Differential developmental requirement and peripheral regulation for dermal Vgamma4 and Vgamma6T17 cells in health and inflammation. *Nat Commun.* 2014;5:3986.
96. Itohara S, Farr AG, Lafaille JJ, et al. Homing of a gamma delta thymocyte subset with homogeneous T-cell receptors to mucosal epithelia. *Nature.* 1990;343(6260):754-757.
97. Monin L, Ushakov DS, Arnesen H, et al. Gammadelta T cells compose a developmentally regulated intrauterine population and protect against vaginal candidiasis. *Mucosal Immunol.* 2020;13(6):969-981.
98. Koren N, Zubeidat K, Saba Y, et al. Maturation of the neonatal oral mucosa involves unique epithelium-microbiota interactions. *Cell Host Microbe.* 2021;29(2):197-209 e5.
99. Reinhardt A, Yevsa T, Worbs T, et al. Interleukin-23-Dependent gamma/delta T cells produce interleukin-17 and accumulate in the enthesis, aortic valve, and ciliary body in mice. *Arthritis Rheumatol.* 2016;68(10):2476-2486.
100. Krishnan S, Prise IE, Wemyss K, et al. Amphiregulin-producing gammadelta T cells are vital for safeguarding oral barrier immune homeostasis. *Proc Natl Acad Sci U S A.* 2018;115(42):10738-10743.
101. Augustin A, Kubo RT, Sim GK. Resident pulmonary lymphocytes expressing the gamma/delta T-cell receptor. *Nature.* 1989;340(6230):239-241.
102. Roark CE, Vollmer MK, Campbell PA, et al. Response of a gamma delta+ T cell receptor invariant subset during bacterial infection. *J Immunol.* 1996;156(6):2214-2220.
103. Emoto M, Danbara H, Yoshikai Y. Induction of gamma/delta T cells in murine salmonellosis by an avirulent but not by a virulent strain of *Salmonella choleraesuis. J Exp Med.* 1992;176(2):363-372.
104. Ribot JC, Lopes N, Silva-Santos B. γδ T cells in tissue physiology and surveillance. *Nat Rev Immunol.* 2021;21(4):221-232.
105. Reardon C, Lefrancois L, Farr A, Kubo R, O'Brien R, Born W. Expression of gamma/delta T cell receptors on lymphocytes from the lactating mammary gland. *J Exp Med.* 1990;172(4):1263-1266.
106. Shibata K, Yamada H, Hara H, et al.Resident Vdelta1+ gammadelta T cells control early infiltration of neutrophils after *Escherichia coli* infection via IL-17 production. *J Immunol.* 2007;178(7):4466-4472.
107. Fehrenkamp BD, Miller RD. γδ T cells are the predominant T cell type in opossum mammaries during lactation. *Dev Comp Immunol.* 2019;95:96-100.
108. Sheridan BS, Romagnoli PA, Pham QM, et al. γδ T cells exhibit multifunctional and protective memory in intestinal tissues. *Immunity.* 2013;39(1):184-195.
109. Asigbetse KE, Eigenmann PA, Frossard CP. Intestinal lamina propria TcRgammadelta+ lymphocytes selectively express IL-10 and IL-17. *J Investig Allergol Clin Immunol.* 2010;20(5):391-401.
110. Hahn YS, Taube C, Jin N, et al. Different potentials of gamma delta T cell subsets in regulating airway responsiveness: V gamma 1+ cells, but not V gamma 4+ cells, promote airway hyperreactivity, Th2 cytokines, and airway inflammation. *J Immunol.* 2004;172(5):2894-2902.
111. Bonneville M, O'Brien RL, Born WK. Gammadelta T cell effector functions: a blend of innate programming and acquired plasticity. *Nat Rev Immunol.* 2010;10(7):467-478.
112. Fahl SP, Coffey F, Wiest DL. Origins of γδ T cell effector subsets: a riddle wrapped in an enigma. *J Immunol.* 2014;193(9):4289-4294.
113. Huber SA, Graveline D, Newell MK. V gamma 1+ T cells suppress and V gamma 4+ T cells promote susceptibility to coxsackievirus B3-induced myocarditis in mice. *J Immunol.* 2000;165(8):4174-4181.
114. Azuara V, Levraud JP, Lembezat MP, et al. A novel subset of adult gamma delta thymocytes that secretes a distinct pattern of cytokines and expresses a very restricted T cell receptor repertoire. *Eur J Immunol.* 1997;27(2):544-553.
115. Gerber DJ, Azuara V, Levraud JP, Huang SY, Lembezat MP, Pereira P. IL-4-producing gamma delta T cells that express a very restricted TCR repertoire are preferentially localized in liver and spleen. *J Immunol.* 1999;163(6):3076-3082.
116. Wen L, Barber DF, Pao W, Wong FS, Owen MJ, Hayday A. Primary gamma delta cell clones can be defined phenotypically and functionally as Th1/Th2 cells and illustrate the association of CD4 with Th2 differentiation. *J Immunol.* 1998;160(4):1965-1974.
117. Kreslavsky T, Savage AK, Hobbs R, et al. TCR-inducible PLZF transcription factor required for innate phenotype of a subset of gammadelta T cells with restricted TCR diversity. *Proc Natl Acad Sci U S A.* 2009;106(30):12453-12458.
118. Roark CL, French JD, Taylor MA, Bendele AM, Born WK, O'Brien RL. Exacerbation of collagen-induced arthritis by oligoclonal, IL-17-producing gamma delta T cells. *J Immunol.* 2007;179(8):5576-5583.
119. Zuberbuehler MK, Parker ME, Wheaton JD, et al. The transcription factor c-Maf is essential for the commitment of IL-17-producing gammadelta T cells. *Nat Immunol.* 2019;20(1):73-85.
120. Malhotra N, Narayan K, Cho OH, et al. A network of high-mobility group box transcription factors programs innate interleukin-17 production. *Immunity.* 2013;38(4):681-693.
121. Sagar, Pokrovskii M, Herman JS. et al. Deciphering the regulatory landscape of fetal and adult gammadelta T-cell development at single-cell resolution. *EMBO J.* 2020;39(13):e104159.
122. Melichar HJ, Narayan K, Der SD, et al. Regulation of gammadelta versus alphabeta T lymphocyte differentiation by the transcription factor SOX13. *Science.* 2007;315(5809):230-233.
123. Mamedov MR, Scholzen A, Nair RV, et al. A macrophage colony-stimulating-factor-producing gammadelta T cell subset prevents malarial parasitemic recurrence. *Immunity.* 2018;48(2):350-363.e7.
124. Ribot JC, Chaves-Ferreira M, d'Orey F, et al. Cutting edge: adaptive versus innate receptor signals selectively control the pool sizes of murine IFN-gamma- or IL-17-producing gammadelta T cells upon infection. *J Immunol.* 2010;185(11):6421-6425.
125. Lopes N, McIntyre C, Martin S, et al. Distinct metabolic programs established in the thymus control effector functions of gammadelta T cell subsets in tumor microenvironments. *Nat Immunol.* 2021;22(2):179-192.
126. Schmolka N, Serre K, Grosso AR, et al. Epigenetic and transcriptional signatures of stable versus plastic differentiation of proinflammatory gammadelta T cell subsets. *Nat Immunol.* 2013;14(10):1093-1100.
127. Shibata K, Yamada H, Nakamura R, Sun X, Itsumi M, Yoshikai Y. Identification of CD25+ gamma delta T cells as fetal thymus-derived naturally occurring IL-17 producers. *J Immunol.* 2008;181(9):5940-5947.
128. Haas JD, Ravens S, Duber S, et al. Development of interleukin-17-producing gammadelta T cells is restricted to a functional embryonic wave. *Immunity.* 2012;37(1):48-59.
129. Lu Y, Cao X, Zhang X, et al. PLZF controls the development of fetal-derived IL-17+Vgamma6+ gammadelta T cells. *J Immunol.* 2015;195(9):4273-4281.
130. Boismenu R, Havran WL. Modulation of epithelial cell growth by intraepithelial gamma delta T cells. *Science.* 1994;266(5188):1253-1255.
131. Boismenu R, Feng L, Xia YY, et al. Chemokine expression by intraepithelial gamma delta T cells. Implications for the recruitment of inflammatory cells to damaged epithelia. *J Immunol.* 1996;157(3):985-992.

132. Sharp LL, Jameson JM, Cauvi G, et al. Dendritic epidermal T cells regulate skin homeostasis through local production of insulin-like growth factor 1. *Nat Immunol.* 2005;6(1):73-79.
133. Girardi M, Oppenheim DE, Steele CR, et al. Regulation of cutaneous malignancy by gammadelta T cells. *Science.* 2001;294(5542):605-609.
134. Dalessandri T, Crawford G, Hayes M, et al. IL-13 from intraepithelial lymphocytes regulates tissue homeostasis and protects against carcinogenesis in the skin. *Nat Commun.* 2016;7:12080.
135. McKenzie DHR, Bah N, Ushakov D, Munoz-Ruiz M, Feederle R, Hayday AC. Normality-sensing licenses local T cells for innate-like tissue-surveillance. *Nat Immunol,* 2022;23(3):411-422.
136. Jensen KD, Su X, Shin S, et al. Thymic selection determines gammadelta T cell effector fate: antigen-naive cells make interleukin-17 and antigen-experienced cells make interferon gamma. *Immunity.* 2008;29(1):90-100.
137. Shires J, Theodoridis E, Hayday AC. Biological insights into TCRgammadelta+ and TCRalphabeta+ intraepithelial lymphocytes provided by serial analysis of gene expression (SAGE). *Immunity.* 2001;15(3):419-434.
138. Fahrer AM, Konigshofer Y, Kerr EM, et al. Attributes of gammadelta intraepithelial lymphocytes as suggested by their transcriptional profile. *Proc Natl Acad Sci U S A.* 2001;98(18):10261-10266.
139. Muzaki A, Soncin I, Setiagani YA, et al. Long-lived innate IL-17-producing gamma/delta T cells modulate antimicrobial epithelial host defense in the colon. *J Immunol.* 2017;199(10):3691-3699.
140. Mikulak J, Oriolo F, Bruni E, et al. NKp46-expressing human gut-resident intraepithelial Vdelta1 T cell subpopulation exhibits high antitumor activity against colorectal cancer. *JCI Insight.* 2019;4(24):e125884.
141. Ebert LM, Meuter S, Moser B. Homing and function of human skin gammadelta T cells and NK cells: relevance for tumor surveillance. *J Immunol.* 2006;176(7):4331-4336.
142. Wu Y, Kyle-Cezar F, Woolf RT, et al. An innate-like Vdelta1(+) gammadelta T cell compartment in the human breast is associated with remission in triple-negative breast cancer. *Sci Transl Med.* 2019;11(513):eaax9364.
143. Caccamo N, La Mendola C, Orlando V, et al. Differentiation, phenotype, and function of interleukin-17-producing human Vgamma9Vdelta2 T cells. *Blood.* 2011;118(1):129-138.
144. Dimova T, Brouwer M, Gosselin F, et al. Effector Vgamma9Vdelta2 T cells dominate the human fetal gammadelta T-cell repertoire. *Proc Natl Acad Sci U S A.* 2015;112(6):E556-E565.
145. Tan L, Fichtner AS, Bruni E, et al. A fetal wave of human type 3 effector gammadelta cells with restricted TCR diversity persists into adulthood. *Sci Immunol.* 2021;6(58):eabf0125.
146. Kessel C, Lippitz K, Weinhage T, et al. Proinflammatory cytokine environments can drive interleukin-17 overexpression by gamma/delta T cells in systemic Juvenile idiopathic arthritis. *Arthritis Rheumatol.* 2017;69(7):1480-1494.
147. Moens E, Brouwer M, Dimova T, Goldman M, Willems F, Vermijlen D. IL-23R and TCR signaling drives the generation of neonatal Vgamma9Vdelta2 T cells expressing high levels of cytotoxic mediators and producing IFN-gamma and IL-17. *J Leukoc Biol.* 2011;89(5):743-752.
148. Ness-Schwickerath KJ, Jin C, Morita CT. Cytokine requirements for the differentiation and expansion of IL-17A- and IL-22-producing human Vgamma2Vdelta2 T cells. *J Immunol.* 2010;184(12):7268-7280.
149. Krangel MS, Yssel H, Brocklehurst C, et al. A distinct wave of human T cell receptor gamma/delta lymphocytes in the early fetal thymus: evidence for controlled gene rearrangement and cytokine production. *J Exp Med.* 1990;172(3):847-859.
150. Antonacci R, Massari S, Linguiti G, et al. Evolution of the T-cell receptor (TR) loci in the adaptive immune response: the tale of the TRG locus in mammals. *Genes.* 2020;11(6):624.
151. Massari S, Linguiti G, Giannico F, D'Addabbo P, Ciccarese S, Antonacci R. The genomic organisation of the TRA/TRD locus validates the peculiar characteristics of dromedary delta-chain expression. *Genes.* 2021;12(4):544.
152. Vidovic D, Roglic M, McKune K, Guerder S, MacKay C, Dembic Z. Qa-1 restricted recognition of foreign antigen by a gamma delta T-cell hybridoma. *Nature.* 1989;340(6235):646-650.
153. Born W, Hall L, Dallas A, et al. Recognition of a peptide antigen by heat shock--reactive gamma delta T lymphocytes. *Science.* 1990;249(4964):67-69.
154. Hayday A, Vantourout P. A long-playing CD about the gammadelta TCR repertoire. *Immunity.* 2013;39(6):994-996.
155. Reijneveld JF, Ocampo TA, Shahine A, et al. Human gammadelta T cells recognize CD1b by two distinct mechanisms. *Proc Natl Acad Sci U S A.* 2020;117(37):22944-22952.
156. Melandri D, Zlatareva I, Chaleil RAG, et al. The gammadeltaTCR combines innate immunity with adaptive immunity by utilizing spatially distinct regions for agonist selection and antigen responsiveness. *Nat Immunol.* 2018;19(12):1352-1365.
157. Le Nours J, Gherardin NA, Ramarathinam SH, et al. A class of gammadelta T cell receptors recognize the underside of the antigen-presenting molecule MR1. *Science.* 2019;366(6472):1522-1527.
158. Aydintug MK, Zhang L, Wang C, et al. Gammadelta T cells recognize the insulin B:9-23 peptide antigen when it is dimerized through thiol oxidation. *Mol Immunol.* 2014;60(2):116-128.
159. Kabelitz D, Bender A, Schondelmaier S, et al. A large fraction of human peripheral blood gamma/delta + T cells is activated by Mycobacterium tuberculosis but not by its 65-kD heat shock protein. *J Exp Med.* 1990;171(3):667-679.
160. Fisch P, Malkovsky M, Kovats S, et al. Recognition by human V gamma 9/V delta 2 T cells of a GroEL homolog on Daudi Burkitt's lymphoma cells. *Science.* 1990;250(4985):1269-1273.
161. O'Brien RL, Born W. Heat shock proteins as antigens for gamma delta T cells. *Semin Immunol.* 1991;3(2):81-87.
162. Correa I, Bix M, Liao NS, Zijlstra M, Jaenisch R, Raulet D. Most gamma delta T cells develop normally in beta 2-microglobulin-deficient mice. *Proc Natl Acad Sci U S A.* 1992;89(2):653-657.
163. Bigby M, Markowitz JS, Bleicher PA, et al. Most gamma delta T cells develop normally in the absence of MHC class II molecules. *J Immunol.* 1993;151(9):4465-4475.
164. Nishimura H, Washizu J, Naiki Y, et al. MHC class II-dependent NK1.1+ gammadelta T cells are induced in mice by Salmonella infection. *J Immunol.* 1999;162(3):1573-1581.
165. Lefrancois L, LeCorre R, Mayo J. Extrathymic selection of TCR gamma delta + T cells by class II major histocompatibility complex molecules. *Cell.* 1990;63(2):333-340.
166. Rock EP, Sibbald PR, Davis MM, et al. CDR3 length in antigen-specific immune receptors. *J Exp Med.* 1994;179(1):323-328.
167. Di Lorenzo B, Ravens S, Silva-Santos B. High-throughput analysis of the human thymic Vδ1+ T cell receptor repertoire. *Sci Data.* 2019;6(1):115.
168. Hunter S, Willcox CR, Davey MS, et al. Human liver infiltrating gammadelta T cells are composed of clonally expanded circulating and tissue-resident populations. *J Hepatol.* 2018;69(3):654-665.
169. Davey MS, Willcox CR, Hunter S, et al. The human Vdelta2(+) T-cell compartment comprises distinct innate-like Vgamma9(+) and adaptive Vgamma9(-) subsets. *Nat Commun.* 2018;9(1):1760.
170. Schild H, Mavaddat N, Litzenberger C, et al. The nature of major histocompatibility complex recognition by gamma delta T cells. *Cell.* 1994;76(1):29-37.
171. Ito K, Van Kaer L, Bonneville M, Hsu S, Murphy DB, Tonegawa S. Recognition of the product of a novel MHC TL region gene (27b) by a mouse gamma delta T cell receptor. *Cell.* 1990;62(3):549-561.
172. Crowley MP, Fahrer AM, Baumgarth N, et al. A population of murine gammadelta T cells that recognize an inducible MHC class Ib molecule. *Science.* 2000;287(5451):314-316.
173. Adams EJ, Chien YH, Garcia KC. Structure of a gammadelta T cell receptor in complex with the nonclassical MHC T22. *Science.* 2005;308(5719):227-231.
174. Xu B, Pizarro JC, Holmes MA, et al. Crystal structure of a gammadelta T-cell receptor specific for the human MHC class I homolog MICA. *Proc Natl Acad Sci U S A.* 2011;108(6):2414-2419.
175. Marlin R, Pappalardo A, Kaminski H, et al. Sensing of cell stress by human γδ TCR-dependent recognition of annexin A2. Sensing of cell stress by human gammadelta TCR-dependent recognition of annexin A2. *Proc Natl Acad Sci U S A.* 2017;114(12):3163-3168.
176. Harly C, Joyce SP, Domblides C, et al. Human gammadelta T cell sensing of AMPK-dependent metabolic tumor reprogramming through TCR recognition of EphA2. *Sci Immunol.* 2021;6(61):eaba9010.
177. Scotet E, Martinez LO, Grant E, et al. Tumor recognition following Vgamma9Vdelta2 T cell receptor interactions with a surface F1-ATPase-related structure and apolipoprotein A-I. *Immunity.* 2005;22(1):71-80.
178. Rigau M, Ostrouska S, Fulford TS, et al. Butyrophilin 2A1 is essential for phosphoantigen reactivity by gammadelta T cells. *Science.* 2020;367(6478):eaay5516
179. Karunakaran MM, Willcox CR, Salim M, et al. Butyrophilin-2A1 directly binds germline-encoded regions of the Vγ9Vδ2 TCR and is essential for phosphoantigen sensing. *Immunity.* 2020;52(3):487-498 e6.
180. Zeng X, Meyer C, Huang J, et al. Gamma delta T cells recognize haptens and mount a hapten-specific response. *Elife.* 2014;3:e03609.
181. Johnson RM, Lancki DW, Sperling AI, et al. A murine CD4-, CD8- T cell receptor-gamma delta T lymphocyte clone specific for herpes simplex virus glycoprotein I. *J Immunol.* 1992;148(4):983-988.
182. Sciammas R, Bluestone JA. HSV-1 glycoprotein I-reactive TCR gamma delta cells directly recognize the peptide backbone in a conformationally dependent manner. *J Immunol.* 1998;161(10):5187-5192.
183. Zeng X, Wei YL, Huang J, et al. γδ T cells recognize a microbial encoded B cell antigen to initiate a rapid antigen-specific interleukin-17 response. *Immunity.* 2012;37(3):524-534.
184. Kaminski H, Marseres G, Cosentino A, et al. Understanding human γδ T cell biology toward a better management of cytomegalovirus infection. *Immunol Rev.* 2020;298(1):264-288.
185. Sell S, Dietz M, Schneider A, Holtappels R, Mach M, Winkler TH. Control of murine cytomegalovirus infection by γδ T cells. *PLoS Pathog.* 2015;11(2):e1004481.
186. Vermijlen D, Brouwer M, Donner C, et al. Human cytomegalovirus elicits fetal gammadelta T cell responses in utero. *J Exp Med.* 2010;207(4):807-821.
187. Pauza CD, Poonia B, Li H, et al. γδ T cells in HIV disease: past, present, and future. *Front Immunol.* 2014;5:687.
188. Poccia F, Gougeon ML, Agrati C, et al. Innate T-cell immunity in HIV infection: the role of Vgamma9Vdelta2 T lymphocytes. *Curr Mol Med.* 2002;2(8):769-781.
189. Fitzpatrick S, Lausch R, Barrington RA. CCR6-positive γδ T cells provide protection against intracorneal HSV-1 infection. *Invest Ophthalmol Vis Sci.* 2019;60(12):3952-3962.
190. Wang T, Gao Y, Scully E, et al. Gamma delta T cells facilitate adaptive immunity against West Nile virus infection in mice. *J Immunol.* 2006;177(3):1825-1832.
191. Selin LK, Santolucito PA, Pinto AK, et al. Innate immunity to viruses: control of vaccinia virus infection by gamma delta T cells. *J Immunol.* 2001;166(11):6784-6794.
192. von Massow G, Oh S, Lam A, et al. Gamma delta T cells and their involvement in COVID-19 virus infections. *Front Immunol.* 2021;12:741218.
193. Laing AG, Lorenc A, Del Molino Del Barrio I, et al. A dynamic COVID-19 immune signature includes associations with poor prognosis. *Nat Med.* 2020;26(10):1623-1635.
194. Skeen MJ, Ziegler HK. Induction of murine peritoneal gamma/delta T cells and their role in resistance to bacterial infection. *J Exp Med.* 1993;178(3):971-984.
195. Yamamoto S, Russ F, Teixeira HC, et al. Listeria monocytogenes-induced gamma interferon secretion by intestinal intraepithelial gamma/delta T lymphocytes. *Infect Immun.* 1993;61(5):2154-2161.
196. Hiromatsu K, Matsuzaki G, Tauchi Y, et al. Sequential analysis of T cells in the liver during murine listerial infection. *J Immunol.* 1992;149(2):568-573.
197. Shi C, Sahay B, Russell JQ, et al. Reduced immune response to Borrelia burgdorferi in the absence of gammadelta T cells. *Infect Immun.* 2011;79(10):3940-3946.
198. Bertotto A, Gerli R, Spinozzi F, et al. Lymphocytes bearing the gamma delta T cell receptor in acute Brucella melitensis infection. *Eur J Immunol.* 1993;23(5):1177-1180.
199. Caldwell CW, Everett ED, McDonald G, et al. Lymphocytosis of gamma/delta T cells in human ehrlichiosis. *Am J Clin Pathol.* 1995;103(6):761-766.
200. Poquet Y, Kroca M, Halary F, et al. Expansion of Vgamma9 Vdelta2 T cells is triggered by Francisella tularensis-derived phosphoantigens in tularemia but not after tularemia vaccination. *Infect Immun.* 1998;66(5):2107-2114.
201. Kroca M, Tarnvik A, Sjostedt A. The proportion of circulating gammadelta T cells increases after the first week of onset of tularaemia and remains elevated for more than a year. *Clin Exp Immunol.* 2000;120(2):280-284.
202. Munk ME, Gatrill AJ, Kaufmann SH. Target cell lysis and IL-2 secretion by gamma/delta T lymphocytes after activation with bacteria. *J Immunol.* 1990;145(8):2434-2439.
203. Shen Y, Zhou D, Qiu L, et al. Adaptive immune response of Vgamma2Vdelta2+ T cells during mycobacterial infections. *Science.* 2002;295(5563):2255-2258.
204. Barnes PF, Grisso CL, Abrams JS, Band H, Rea TH, Modlin RL. Gamma delta T lymphocytes in human tuberculosis. *J Infect Dis.* 1992;165(3):506-512.
205. Balbi B, Valle MT, Oddera S, et al. T-lymphocytes with gamma delta+ V delta 2+ antigen receptors are present in increased proportions in a fraction of patients with tuberculosis or with sarcoidosis. *Am Rev Respir Dis.* 1993;148(6 pt 1):1685-1690.
206. Bertotto A, Gerli R, Castellucci G, et al. Mycobacteria-reactive gamma/delta T cells are present in human colostrum from tuberculin-positive, but not tuberculin-negative nursing mothers. *Am J Reprod Immunol.* 1993;29(2):131-134.
207. Langhorne J, Mombaerts P, Tonegawa S. Alpha beta and gamma delta T cells in the immune response to the erythrocytic stages of malaria in mice. *Int Immunol.* 1995;7(6):1005-1011.
208. D'Ombrain MC, Hansen DS, Simpson KM, et al. gammadelta-T cells expressing NK receptors predominate over NK cells and conventional T cells in the innate IFN-gamma response to Plasmodium falciparum malaria. *Eur J Immunol.* 2007;37(7):1864-1873.
209. D'Ombrain MC, Robinson LJ, Stanisic DI, et al. Association of early interferon-gamma production with immunity to clinical malaria: a longitudinal study among Papua New Guinean children. *Clin Infect Dis.* 2008;47(11):1380-1387.
210. Ho M, Webster HK, Tongtawe P, et al. Increased gamma delta T cells in acute Plasmodium falciparum malaria. *Immunol Lett.* 1990;25(1-3):139-141.
211. Waters WR, Harp JA. Cryptosporidium parvum infection in T-cell receptor (TCR)-alpha- and TCR-delta-deficient mice. *Infect Immun.* 1996;64(5):1854-1857.
212. Scalise F, Gerli R, Castellucci G, et al. Lymphocytes bearing the gamma delta T-cell receptor in acute toxoplasmosis. *Immunology.* 1992;76(4):668-670.
213. Ma L, Papadopoulou M, Taton M, et al. Effector Vgamma9Vdelta2 T cell response to congenital *Toxoplasma gondii* infection. *JCI Insight.* 2021;6(16):e138066.
214. Findly RC, Roberts SJ, Hayday AC. Dynamic response of murine gut intraepithelial T cells after infection by the coccidian parasite Eimeria. *Eur J Immunol.* 1993;23(10):2557-2564.

215. Roberts SJ, Smith AL, West AB, et al. T-cell alpha beta + and gamma delta + deficient mice display abnormal but distinct phenotypes toward a natural, widespread infection of the intestinal epithelium. *Proc Natl Acad Sci U S A.* 1996;93(21):11774-11779.
216. Tuengel J, Ranchal S, Maslova A, et al. Characterization of adaptive-like gammadelta T cells in Ugandan infants during primary cytomegalovirus infection. *Viruses.* 2021;13(10).
217. Halary F, Pitard V, Dlubek D, et al. Shared reactivity of V{delta}2(neg) {gamma}{delta} T cells against cytomegalovirus-infected cells and tumor intestinal epithelial cells. *J Exp Med.* 2005;201(10):1567-1578.
218. Khairallah C, Netzer S, Villacreces A, et al. Gammadelta T cells confer protection against murine cytomegalovirus (MCMV). *PLoS Pathog.* 2015;11(3):e1004702.
219. Vincent MS, Roessner K, Sellati T, et al. Lyme arthritis synovial gamma delta T cells respond to Borrelia burgdorferi lipoproteins and lipidated hexapeptides. *J Immunol.* 1998;161(10):5762-5771.
220. Collins C, Lui Y, Santos AM, et al. Detection of cell surface ligands for human synovial gammadelta T cells. *J Immunol.* 2019;203(9):2369-2376.
221. Seder RA, Chang LJ, Enama ME, et al. Protection against malaria by intravenous immunization with a nonreplicating sporozoite vaccine. *Science.* 2013;341(6152):1359-1365.
222. Baldwin CL, Yirsaw A, Gillespie A, et al. Gammadelta T cells in livestock: responses to pathogens and vaccine potential. *Transbound Emerg Dis.* 2020;67(suppl 2):119-128.
223. Blumerman SL, Herzig CT, Baldwin CL. WC1+ gammadelta T cell memory population is induced by killed bacterial vaccine. *Eur J Immunol.* 2007;37(5):1204-1216.
224. Hope JC, Thom ML, Villarreal-Ramos B, Vordermeier HM, Hewinson RG, Howard CJ. Vaccination of neonatal calves with Mycobacterium bovis BCG induces protection against intranasal challenge with virulent M. bovis. *Clin Exp Immunol.* 2005;139(1):48-56.
225. Faisal SM, Yan F, Chen TT, et al. Evaluation of a Salmonella vectored vaccine expressing Mycobacterium avium subsp. paratuberculosis antigens against challenge in a goat model. *PLoS One.* 2013;8(8):e70171.
226. Evans CW, Lund BT, McConnell I, et al. Antigen recognition and activation of ovine gamma delta T cells. *Immunology.* 1994;82(2):229-237.
227. Daubenberger CA, Taracha EL, Gaidulis L, et al. Bovine gammadelta T-cell responses to the intracellular protozoan parasite Theileria parva. *Infect Immun.* 1999;67(5):2241-2249.
228. Buza J, Kiros T, Zerihun A. et al. Vaccination of calves with Mycobacteria bovis Bacilli Calmete Guerin (BCG) induced rapid increase in the proportion of peripheral blood gammadelta T cells. *Vet Immunol Immunopathol.* 2009;130(3-4):251-255.
229. Tanaka Y, Sano S, Nieves E, et al. Nonpeptide ligands for human gamma delta T cells. *Proc Natl Acad Sci U S A.* 1994;91(17):8175-8179.
230. Sandstrom A, Peigne CM, Leger A, et al. The intracellular B30.2 domain of butyrophilin 3A1 binds phosphoantigens to mediate activation of human Vgamma9Vdelta2 T cells. *Immunity.* 2014;40(4):490-500.
231. Vantourout P, Laing A, Woodward MJ, et al. Heteromeric interactions regulate butyrophilin (BTN) and BTN-like molecules governing gammadelta T cell biology. *Proc Natl Acad Sci U S A.* 2018;115(5):1039-1044.
232. Wang H, Henry O, Distefano MD, et al. Butyrophilin 3A1 plays an essential role in prenyl pyrophosphate stimulation of human Vgamma2Vdelta2 T cells. *J Immunol.* 2013;191(3):1029-1042.
233. Harly C, Guillaume Y, Nedellec S, et al. Key implication of CD277/butyrophilin-3 (BTN3A) in cellular stress sensing by a major human gammadelta T-cell subset. *Blood.* 2012;120(11):2269-2279.
234. Herrmann T, Fichtner AS, Karunakaran MM. An update on the molecular basis of phosphoantigen recognition by Vgamma9Vdelta2 T cells. *Cells.* 2020;9(6):1433.
235. Wang H, Morita CT. Sensor function for butyrophilin 3A1 in prenyl pyrophosphate stimulation of human Vgamma2Vdelta2 T cells. *J Immunol.* 2015;195(10):4583-4594.
236. Yang Y, Li L, Yuan L, et al. A structural change in butyrophilin upon phosphoantigen binding underlies phosphoantigen-mediated Vgamma9Vdelta2 T cell activation. *Immunity.* 2019;50(4):1043-1053 e5.
237. Salim M, Knowles TJ, Baker AT, et al. BTN3A1 discriminates gammadelta T cell phosphoantigens from nonantigenic small molecules via a conformational sensor in its B30.2 domain. *ACS Chem Biol.* 2017;12(10):2631-2643.
238. Gober HJ, Kistowska M, Angman L, Jeno P, Mori L, De Libero G. Human T cell receptor gammadelta cells recognize endogenous mevalonate metabolites in tumor cells. *J Exp Med.* 2003;197(2):163-168.
239. Vantourout P, Radojkovic C, Lichtenstein L, Pons V, Champagne E, Martinez LO. Ecto-F(1)-ATPase: a moonlighting protein complex and an unexpected apoA-I receptor. *World J Gastroenterol.* 2010;16(47):5925-5935.
240. Bruder J, Siewert K, Obermeier B, et al. Target specificity of an autoreactive pathogenic human gammadelta-T cell receptor in myositis. *J Biol Chem.* 2012;287(25):20986-20995.
241. Wang L, Kamath A, Das H, et al. Antibacterial effect of human V gamma 2V delta 2 T cells in vivo. *J Clin Invest.* 2001;108(9):1349-1357.
242. Lewis JM, Girardi M, Roberts SJ, Barbee SD, Hayday AC, Tigelaar RE. Selection of the cutaneous intraepithelial gammadelta+ T cell repertoire by a thymic stromal determinant. *Nat Immunol.* 2006;7(8):843-850.
243. Boyden LM, Lewis JM, Barbee SD, et al. Skint1, the prototype of a newly identified immunoglobulin superfamily gene cluster, positively selects epidermal gammadelta T cells. *Nat Genet.* 2008;40(5):656-662.
244. Ogg SL, Weldon AK, Dobbie L, et al. Expression of butyrophilin (Btn1a1) in lactating mammary gland is essential for the regulated secretion of milk-lipid droplets. *Proc Natl Acad Sci U S A.* 2004;101(27):10084-10089.
245. Salim M, Knowles TJ, Hart R, et al. Characterization of a putative receptor binding surface on Skint-1, a critical determinant of dendritic epidermal T cell selection. *J Biol Chem.* 2016;291(17):9310-9321.
246. Payne KK, Mine JA, Biswas S, et al. BTN3A1 governs antitumor responses by coordinating alphabeta and gammadelta T cells. *Science.* 2020;369(6506):942-949.
247. Brandt CS, Baratin M, Yi EC, et al. The B7 family member B7-H6 is a tumor cell ligand for the activating natural killer cell receptor NKp30 in humans. *J Exp Med.* 2009;206(7):1495-1503.
248. Flajnik MF, Tlapakova T, Criscitiello MF, et al. Evolution of the B7 family: co-evolution of B7H6 and NKp30, identification of a new B7 family member, B7H7, and of B7's historical relationship with the MHC. *Immunogenetics.* 2012;64(8):571-590.
249. Bhatt RS, Berjis A, Konge JC, et al. KIR3DL3 is an inhibitory receptor for HHLA2 that mediates an alternative Immunoinhibitory pathway to PD1. *Cancer Immunol Res.* 2021;9(2):156-169.
250. Willcox CR, Vantourout P, Salim M, et al. Butyrophilin-like 3 directly binds a human Vgamma4(+) T cell receptor using a modality distinct from clonally-restricted antigen. *Immunity.* 2019;51(5):813-825 e4.
251. Medzhitov R, Janeway CA Jr. Innate immunity: the virtues of a nonclonal system of recognition. *Cell.* 1997;91(3):295-298.
252. Jandke A, Melandri D, Monin L, et al. Butyrophilin-like proteins display combinatorial diversity in selecting and maintaining signature intraepithelial gammadelta T cell compartments. *Nat Commun.* 2020;11(1):3769.
253. Hayday AC, Vantourout P. The innate biologies of adaptive antigen receptors. *Annu Rev Immunol.* 2020;38:487-510.
254. Guy-Grand D, Azogui O, Celli S, et al. Extrathymic T cell lymphopoiesis: ontogeny and contribution to gut intraepithelial lymphocytes in athymic and euthymic mice. *J Exp Med.* 2003;197(3):333-341.
255. Nonaka S, Naito T, Chen H, et al. Intestinal gamma delta T cells develop in mice lacking thymus, all lymph nodes, Peyer's patches, and isolated lymphoid follicles. *J Immunol.* 2005;174(4):1906-1912.
256. Eberl G, Littman DR. Thymic origin of intestinal alphabeta T cells revealed by fate mapping of RORgammat+ cells. *Science.* 2004;305(5681):248-251.
257. Hayday A, Gibbons D. Brokering the peace: the origin of intestinal T cells. *Mucosal Immunol.* 2008;1(3):172-174.
258. McVay LD, Jaswal SS, Kennedy C, et al. The generation of human gammadelta T cell repertoires during fetal development. *J Immunol.* 1998;160(12):5851-5860.
259. Chen ELY, Lee CR, Thompson PK, Wiest DL, Anderson MK, Zuniga-Pflucker JC. Ontogenic timing, T cell receptor signal strength, and Notch signaling direct gammadelta T cell functional differentiation in vivo. *Cell Rep.* 2021;35(10):109227.
260. Pardoll DM, Fowlkes BJ, Bluestone JA, et al. Differential expression of two distinct T-cell receptors during thymocyte development. *Nature.* 1987;326(6108):79-81.
261. Nakanishi N, Maeda K, Ito K, et al. T gamma protein is expressed on murine fetal thymocytes as a disulphide-linked heterodimer. *Nature.* 1987;325(6106):720-723.
262. Bluestone JA, Pardoll D, Sharrow SO, et al. Characterization of murine thymocytes with CD3-associated T-cell receptor structures. *Nature.* 1987;326(6108):82-84.
263. Haars R, Kronenberg M, Gallatin WM, Weissman IL, Owen FL, Hood L. Rearrangement and expression of T cell antigen receptor and gamma genes during thymic development. *J Exp Med.* 1986;164(1):1-24.
264. Allison JP, Havran WL. The immunobiology of T cells with invariant gamma delta antigen receptors. *Annu Rev Immunol.* 1991;9:679-705.
265. Ikuta K, Kina T, MacNeil I, et al. A developmental switch in thymic lymphocyte maturation potential occurs at the level of hematopoietic stem cells. *Cell.* 1990;62(5):863-874.
266. Edwards SC, Sutton CE, Ladell K, et al. A population of proinflammatory T cells coexpresses alphabeta and gammadelta T cell receptors in mice and humans. *J Exp Med.* 2020;217(5):e20190834.
267. Groettrup M, Ungewiss K, Azogui O, et al. A novel disulfide-linked heterodimer on pre-T cells consists of the T cell receptor beta chain and a 33 kd glycoprotein. *Cell.* 1993;75(2):283-294.
268. Shinkai Y, Koyasu S, Nakayama K, et al. Restoration of T cell development in RAG-2-deficient mice by functional TCR transgenes. *Science.* 1993;259(5096):822-825.
269. Mallick CA, Dudley EC, Viney JL, et al. Rearrangement and diversity of T cell receptor beta chain genes in thymocytes: a critical role for the beta chain in development. *Cell.* 1993;73(3):513-519.
270. Dudley EC, Petrie HT, Shah LM, et al. T cell receptor beta chain gene rearrangement and selection during thymocyte development in adult mice. *Immunity.* 1994;1(2):83-93.
271. Yamasaki S, Ishikawa E, Sakuma M, et al. Mechanistic basis of pre-T cell receptor-mediated autonomous signaling critical for thymocyte development. *Nat Immunol.* 2006;7(1):67-75.
272. Hoffman ES, Passoni L, Crompton T, et al. Productive T-cell receptor beta-chain gene rearrangement: coincident regulation of cell cycle and clonality during development in vivo. *Genes Dev.* 1996;10(8):948-962.
273. Kreslavsky T, Gleimer M, Miyazaki M, et al. beta-Selection-induced proliferation is required for alphabeta T cell differentiation. *Immunity.* 2012;37(5):840-853.
274. Lauritsen JP, Wong GW, Lee SY, et al. Marked induction of the helix-loop-helix protein Id3 promotes the gammadelta T cell fate and renders their functional maturation Notch independent. *Immunity.* 2009;31(4):565-575.
275. Engel I, Johns C, Bain G, et al. Early thymocyte development is regulated by modulation of E2A protein activity. *J Exp Med.* 2001;194(6):733-745.
276. Hernandez-Munain C, Sleckman BP, Krangel MS. A developmental switch from TCR delta enhancer to TCR alpha enhancer function during thymocyte maturation. *Immunity.* 1999;10(6):723-733.
277. Ishida I, Verbeek S, Bonneville M, Itohara S, Berns A, Tonegawa S. T-cell receptor gamma delta and gamma transgenic mice suggest a role of a gamma gene silencer in the generation of alpha beta T cells. *Proc Natl Acad Sci U S A.* 1990;87(8):3067-3071.
278. Bonneville M, Ishida I, Mombaerts P, et al. Blockage of alpha beta T-cell development by TCR gamma delta transgenes. *Nature.* 1989;342(6252):931-934.
279. de Villartay JP, Hockett RD, Coran D, et al. Deletion of the human T-cell receptor delta-gene by a site-specific recombination. *Nature.* 1988;335(6186):170-174.
280. Villey I, Caillol D, Selz F, et al. Defect in rearrangement of the most 5' TCR-J alpha following targeted deletion of T early alpha (TEA): implications for TCR alpha locus accessibility. *Immunity.* 1996;5(4):331-342.
281. Daniels MA, Teixeiro E, Gill J, et al. Thymic selection threshold defined by compartmentalization of Ras/MAPK signalling. *Nature.* 2006;444(7120):724-729.
282. Fehling HJ, Krotkova A, Saint-Ruf C, et al. Crucial role of the pre-T-cell receptor alpha gene in development of alpha beta but not gamma delta T cells. *Nature.* 1995;375(6534):795-798.
283. Passoni L, Hoffman ES, Kim S, et al. Intrathymic delta selection events in gammadelta cell development. *Immunity.* 1997;7(1):83-95.
284. Kang J, Coles M, Cado D, et al. The developmental fate of T cells is critically influenced by TCRgammadelta expression. *Immunity.* 1998;8(4):427-438.
285. Taghon T, Yui MA, Pant R, et al. Developmental and molecular characterization of emerging beta- and gammadelta-selected pre-T cells in the adult mouse thymus. *Immunity.* 2006;24(1):53-64.
286. Garbe AI, Krueger A, Gounari F, et al. Differential synergy of Notch and T cell receptor signaling determines alphabeta versus gammadelta lineage fate. *J Exp Med.* 2006;203(6):1579-1590.
287. Haks MC, Lefebvre JM, Lauritsen JPH, et al. Attenuation of gammadeltaTCR signaling efficiently diverts thymocytes to the alphabeta lineage. *Immunity.* 2005;22(5):595-606.
288. Hayes SM, Li L, Love PE. TCR signal strength influences alphabeta/gammadelta lineage fate. *Immunity.* 2005;22(5):583-593.
289. Zarin P, Wong GW, Mohtashami M, et al. Enforcement of gammadelta-lineage commitment by the pre-T-cell receptor in precursors with weak gammadelta-TCR signals. *Proc Natl Acad Sci U S A.* 2014;111(15):5658-5663.
290. Coffey F, Lee SY, Buus TB, et al. The TCR ligand-inducible expression of CD73 marks gammadelta lineage commitment and a metastable intermediate in effector specification. *J Exp Med.* 2014;211(2):329-343.
291. Bruno L, Fehling HJ, von Boehmer H. The alpha beta T cell receptor can replace the gamma delta receptor in the development of gamma delta lineage cells. *Immunity.* 1996;5(4):343-352.
292. Dudley EC, Girardi M, Owen MJ. Alpha beta and gamma delta T cells can share a late common precursor. *Curr Biol.* 1995;5(6):659-669.
293. Ito K, Bonneville M, Takagaki Y, et al. Different gamma delta T-cell receptors are expressed on thymocytes at different stages of development. *Proc Natl Acad Sci U S A.* 1989;86(2):631-635.
294. McMurry MT, Krangel MS. A role for histone acetylation in the developmental regulation of VDJ recombination. *Science.* 2000;287(5452):495-498.
295. Wagatsuma K, Tani-ichi S, Liang B, et al. STAT5 orchestrates local epigenetic changes for chromatin accessibility and rearrangements by direct binding to the TCRgamma locus. *J Immunol.* 2015;195(4):1804-1814.
296. Schlissel MS, Durum SD, Muegge K. The interleukin 7 receptor is required for T cell receptor gamma locus accessibility to the V(D)J recombinase. *J Exp Med.* 2000;191(6):1045-1050.

297. Agata Y, Katakai T, Ye SK, et al. Histone acetylation determines the developmentally regulated accessibility for T cell receptor gamma gene recombination. *J Exp Med*. 2001;193(7):873-880.
298. Ye SK, Maki K, Kitamura T, et al. Induction of germline transcription in the TCRgamma locus by Stat5: implications for accessibility control by the IL-7 receptor. *Immunity*. 1999;11(2):213-223.
299. Tieppo P, Papadopoulou M, Gatti D, et al. The human fetal thymus generates invariant effector gammadelta T cells. *J Exp Med*. 2020;217(3):jem.20190580.
300. Papadopoulou M, Tieppo P, McGovern N, et al. TCR sequencing reveals the distinct development of fetal and adult human Vgamma9Vdelta2 T cells. *J Immunol*. 2019;203(6):1468-1479.
301. McVay LD, Carding SR, Bottomly K, et al. Regulated expression and structure of T cell receptor gamma/delta transcripts in human thymic ontogeny. *EMBO J*. 1991;10(1):83-91.
302. Yuan J, Nguyen CK, Liu X, et al. Lin28b reprograms adult bone marrow hematopoietic progenitors to mediate fetal-like lymphopoiesis. *Science*. 2012;335(6073):1195-1200.
303. Shibata K, Yamada H, Sato T, et al. Notch-Hes1 pathway is required for the development of IL-17-producing gammadelta T cells. *Blood*. 2011;118(3):586-593.
304. Kisielow J, Tortola L, Weber J, et al. Evidence for the divergence of innate and adaptive T-cell precursors before commitment to the alphabeta and gammadelta lineages. *Blood*. 2011;118(25):6591-6600.
305. Bonneville M, Ishida I, Itohara S, et al. Self-tolerance to transgenic gamma delta T cells by intrathymic inactivation. *Nature*. 1990;344(6262):163-165.
306. Dent AL, Matis LA, Bluestone JA, et al. Evidence for programmed cell death of self-reactive gamma delta T cell receptor-positive thymocytes. *Eur J Immunol*. 1993;23(10):2482-2487.
307. Lin T, Yoshida H, Matsuzaki G, et al. Autospecific gammadelta thymocytes that escape negative selection find sanctuary in the intestine. *J Clin Invest*. 1999;104(9):1297-1305.
308. Wells FB, Gahm SJ, Hedrick SM, Bluestone JA, Dent A, Matis LA. Requirement for positive selection of gamma delta receptor-bearing T cells. *Science*. 1991;253(5022):903-905.
309. Pereira P, Zijlstra M, McMaster J, Loring JM, Jaenisch R, Tonegawa S. Blockade of transgenic gamma delta T cell development in beta 2-microglobulin deficient mice. *EMBO J*. 1992;11(1):25-31.
310. Schweighoffer E, Fowlkes BJ. Positive selection is not required for thymic maturation of transgenic gamma delta T cells. *J Exp Med*. 1996;183(5):2033-2041.
311. Sumaria N, Grandjean CL, Silva-Santos B, et al. Strong TCRgammadelta signaling prohibits thymic development of IL-17a-secreting gammadelta T cells. *Cell Rep*. 2017;19(12):2469-2476.
312. Spidale NA, Sylvia K, Narayan K, et al. Interleukin-17-Producing gammadelta T Cells originate from SOX13(+) progenitors that are independent of gammadeltaTCR signaling. *Immunity*. 2018;49(5):857-872 e5.
313. Dienz O, DeVault VL, Musial SC, et al. Critical role for SLAM/SAP signaling in the thymic developmental programming of IL-17- and IFN-gamma-producing gammadelta T cells. *J Immunol*. 2020;204(6):1521-1534.
314. Elsaid R, Meunier S, Burlen-Defranoux O, et al. A wave of bipotent T/ILC-restricted progenitors shapes the embryonic thymus microenvironment in a time-dependent manner. *Blood*. 2021;137(8):1024-1036.
315. Gentek R, Ghigo C, Hoeffel G, et al. Epidermal gammadelta T cells originate from yolk sac hematopoiesis and clonally self-renew in the adult. *J Exp Med*. 2018;215(12):2994-3005.
316. Heyborne K, Fu YX, Kalataradi H, et al. Evidence that murine V gamma 5 and V gamma 6 gamma delta-TCR+ lymphocytes are derived from a common distinct lineage. *J Immunol*. 1993;151(9):4523-4527.
317. Michel ML, Pang DJ, Haque SFY, Potocnik AJ, Pennington DJ, Hayday AC. Interleukin 7 (IL-7) selectively promotes mouse and human IL-17-producing gammadelta cells. *Proc Natl Acad Sci U S A*. 2012;109(43):17549-17554.
318. Turchinovich G, Hayday AC. Skint-1 identifies a common molecular mechanism for the development of interferon-gamma-secreting versus interleukin-17-secreting gammadelta T cells. *Immunity*. 2011;35(1):59-68.
319. Zeng Y, Liu C, Gong Y, et al. Single-cell RNA sequencing resolves spatiotemporal development of pre-thymic lymphoid progenitors and thymus organogenesis in human embryos. *Immunity*. 2019;51(5):930-948 e6.
320. Gibbons D, Fleming P, Virasami A, et al. Interleukin-8 (CXCL8) production is a signatory T cell effector function of human newborn infants. *Nat Med*. 2014;20(10):1206-1210.
321. Mallick-Wood CA, Lewis JM, Richie LI, Owen MJ, Tigelaar RE, Hayday AC. Conservation of T cell receptor conformation in epidermal gammadelta cells with disrupted primary Vgamma gene usage. *Science*. 1998;279(5357):1729-1733.
322. Kyes S, Pao W, Hayday A. Influence of site of expression on the fetal gamma delta T-cell receptor repertoire. *Proc Natl Acad Sci U S A*. 1991;88(17):7830-7833.
323. Bonneville M, Itohara S, Krecko EG, et al. Transgenic mice demonstrate that epithelial homing of gamma/delta T cells is determined by cell lineages independent of T cell receptor specificity. *J Exp Med*. 1990;171(4):1015-1026.
324. Barbee SD, Woodward MJ, Turchinovich G, et al. Skint-1 is a highly specific, unique selecting component for epidermal T cells. *Proc Natl Acad Sci U S A*. 2011;108(8):3330-3335.
325. Lebrero-Fernandez C, Bergstrom JH, Pelaseyed T, et al. Murine butyrophilin-like 1 and Btnl6 form heteromeric complexes in small intestinal epithelial cells and promote proliferation of local T lymphocytes. *Front Immunol*. 2016;7:1.
326. Jin Y, Xia M, Saylor CM, et al. Cutting edge: intrinsic programming of thymic gammadeltaT cells for specific peripheral tissue localization. *J Immunol*. 2010;185(12):7156-7160.
327. Jin Y, Xia M, Sun A, et al. CCR10 is important for the development of skin-specific gammadeltaT cells by regulating their migration and location. *J Immunol*. 2010;185(10):5723-5731.
328. Xia M, Qi Q, Jin Y, Wiest DL, August A, Xiong N. Differential roles of IL-2-inducible T cell kinase-mediated TCR signals in tissue-specific localization and maintenance of skin intraepithelial T cells. *J Immunol*. 2010;184(12):6807-6814.
329. Xiong N, Fu Y, Hu S, et al. CCR10 and its ligands in regulation of epithelial immunity and diseases. *Protein Cell*. 2012;3(8):571-580.
330. McKenzie DR, Kara EE, Bastow CR, et al. IL-17-producing gammadelta T cells switch migratory patterns between resting and activated states. *Nat Commun*. 2017;8:15632.
331. Hu S, Xiong N. Programmed downregulation of CCR6 is important for establishment of epidermal gammadeltaT cells by regulating their thymic egress and epidermal location. *J Immunol*. 2013;190(7):3267-3275.
332. Wencker M, Turchinovich G, Di Marco Barros R, et al. Innate-like T cells straddle innate and adaptive immunity by altering antigen-receptor responsiveness. *Nat Immunol*. 2014;15(1):80-87.
333. Strid J, Roberts SJ, Filler RB, et al. Acute upregulation of an NKG2D ligand promotes rapid reorganization of a local immune compartment with pleiotropic effects on carcinogenesis. *Nat Immunol*. 2008;9(2):146-154.
334. Muro R, Nitta T, Nakano K, Okamura T, Takayanagi H, Suzuki H. gammadeltaTCR recruits the Syk/PI3K axis to drive proinflammatory differentiation program. *J Clin Invest*. 2018;128(1):415-426.
335. Brenes AJ, Vandereyken M, James OJ, et al. Tissue environment, not ontogeny, defines murine intestinal intraepithelial T lymphocytes. *Elife*. 2021;10:e70055.
336. Bas A, Swamy M, Abeler-Dorner L, et al. Butyrophilin-like 1 encodes an enterocyte protein that selectively regulates functional interactions with T lymphocytes. *Proc Natl Acad Sci U S A*. 2011;108(11):4376-4381.
337. Lebrero-Fernandez C, Bas-Forsberg A. The ontogeny of Butyrophilin-like (Btnl) 1 and Btnl6 in murine small intestine. *Sci Rep*. 2016;6:31524.
338. Carding SR, Kyes S, Jenkinson EJ, et al. Developmentally regulated fetal thymic and extrathymic T-cell receptor gamma delta gene expression. *Genes Dev*. 1990;4(8):1304-1315.
339. Aigner J, Villatoro S, Rabionet R, et al. A common 56-kilobase deletion in a primate-specific segmental duplication creates a novel butyrophilin-like protein. *BMC Genet*. 2013;14:61.
340. Mallick-Wood CA, Pao W, Cheng AM, et al. Disruption of epithelial gamma delta T cell repertoires by mutation of the Syk tyrosine kinase. *Proc Natl Acad Sci U S A*. 1996;93(18):9704-9709.
341. Liu CP, Ueda R, She J, et al. Abnormal T cell development in CD3-zeta-/- mutant mice and identification of a novel T cell population in the intestine. *EMBO J*. 1993;12(12):4863-4875.
342. Rodewald HR, Arulanandam AR, Koyasu S, et al. The high affinity Fc epsilon receptor gamma subunit (Fc epsilon RI gamma) facilitates T cell receptor expression and antigen/major histocompatibility complex-driven signaling in the absence of CD3 zeta and CD3 eta. *J Biol Chem*. 1991;266(24):15974-15978.
343. Hayday AC. Gammadelta T cell update: adaptate orchestrators of immune surveillance. *J Immunol*. 2019;203(2):311-320.
344. James OJ, Vandereyken M, Marchingo JM, et al. IL-15 and PIM kinases direct the metabolic programming of intestinal intraepithelial lymphocytes. *Nat Commun*. 2021;12(1):4290.
345. Jensen KD, Shin S, Chien YH. Cutting edge: gammadelta intraepithelial lymphocytes of the small intestine are not biased toward thymic antigens. *J Immunol*. 2009;182(12):7348-7351.
346. Papadopoulou M, Dimova T, Shey M, et al. Fetal public Vgamma9Vdelta2 T cells expand and gain potent cytotoxic functions early after birth. *Proc Natl Acad Sci U S A*. 2020;117(31):18638-18648.
347. Fenzl L, Gobel TW, Neulen ML, et al. Gammadelta T cells represent a major spontaneously cytotoxic cell population in the chicken. *Dev Comp Immunol*. 2017;73:175-183.
348. Chodaczek G, Papanna V, Zal MA, et al. Body-barrier surveillance by epidermal gammadelta TCRs. *Nat Immunol*. 2012;13(3):272-282.
349. Montufar-Solis D, Garza T, Klein JR, et al. T-cell activation in the intestinal mucosa. *Immunol Rev*. 2007;215:189-201.
350. Tani-Ichi S, Wagatsuma K, Hara T, et al. Innate-like CD27(+) CD45RB(high) gammadelta T cells require TCR signaling for homeostasis in peripheral lymphoid organs. *J Immunol*. 2020;204(10):2671-2684.
351. Veldhoen M, Hirota K, Westendorf AM, et al. The aryl hydrocarbon receptor links TH17-cell-mediated autoimmunity to environmental toxins. *Nature*. 2008;453(7191):106-109.
352. Martin B, Hirota K, Cua DJ, et al. Interleukin-17-producing gammadelta T cells selectively expand in response to pathogen products and environmental signals. *Immunity*. 2009;31(2):321-330.
353. Li Y, Innocentin S, Withers DR, et al. Exogenous stimuli maintain intraepithelial lymphocytes via aryl hydrocarbon receptor activation. *Cell*. 2011;147(3):629-640.
354. Kadow S, Jux B, Zahner SP, et al. Aryl hydrocarbon receptor is critical for homeostasis of invariant gammadelta T cells in the murine epidermis. *J Immunol*. 2011;187(6):3104-3110.
355. Wilharm A, Tabib Y, Nassar M, et al. Mutual interplay between IL-17-producing gammadeltaT cells and microbiota orchestrates oral mucosal homeostasis. *Proc Natl Acad Sci U S A*. 2019;116(7):2652-2661.
356. Papotto PH, Yilmaz B, Silva-Santos B. Crosstalk between gammadelta T cells and the microbiota. *Nat Microbiol*. 2021;6(9):1110-1117.
357. Wilharm A, Brigas HC, Sandrock I, et al. Microbiota-dependent expansion of testicular IL-17-producing Vgamma6(+) gammadelta T cells upon puberty promotes local tissue immune surveillance. *Mucosal Immunol*. 2021;14(1):242-252.
358. Duan J, Chung H, Troy E, et al. Microbial colonization drives expansion of IL-1 receptor 1-expressing and IL-17-producing gamma/delta T cells. *Cell Host Microbe*. 2010;7(2):140-150.
359. Li F, Hao X, Chen Y, et al. The microbiota maintain homeostasis of liver-resident gammadeltaT-17 cells in a lipid antigen/CD1d-dependent manner. *Nat Commun*. 2017;7:13839.
360. Ridaura VK, Bouladoux N, Claesen J, et al. Contextual control of skin immunity and inflammation by Corynebacterium. *J Exp Med*. 2018;215(3):785-799.
361. St Leger AJ, Desai JV, Drummond RA, et al. An ocular commensal protects against corneal infection by driving an interleukin-17 response from mucosal gammadelta T cells. *Immunity*. 2017;47(1):148-158 e5.
362. Papadopoulou M, Sanchez Sanchez G, Vermijlen D. Innate and adaptive gammadelta T cells: how, when, and why. *Immunol Rev*. 2020;298(1):99-116.
363. Gibbons DL, Haque SFY, Silberzahn T, et al. Neonates harbour highly active gammadelta T cells with selective impairments in preterm infants. *Eur J Immunol*. 2009;39(7):1794-1806.
364. Ravens S, Fichtner AS, Willers M, et al. Microbial exposure drives polyclonal expansion of innate gammadelta T cells immediately after birth. *Proc Natl Acad Sci U S A*. 2020;117(31):18649-18660.
365. Naik S, Bouladoux N, Wilhelm C, et al. Compartmentalized control of skin immunity by resident commensals. *Science*. 2012;337(6098):1115-1119.
366. Jin C, Lagoudas GK, Zhao C, et al. Commensal microbiota promote lung cancer development via gammadelta T cells. *Cell*. 2019;176(5):998-1013 e16.
367. Hoytema van Konijnenburg DP, Reis BS, Pedicord VA, Farache J, Victora GD, Mucida D. Intestinal epithelial and intraepithelial T cell crosstalk mediates a dynamic response to infection. *Cell*. 2017;171(4):783-794 e13.
368. Edelblum KL, Singh G, Odenwald MA, et al. Gammadelta intraepithelial lymphocyte migration limits transepithelial pathogen invasion and systemic disease in mice. *Gastroenterology*. 2015;148(7):1417-1426.
369. Hayes SM, Love PE. Distinct structure and signaling potential of the gamma delta TCR complex. *Immunity*. 2002;16(6):827-838.
370. Gil J, Busto EM, Garcillan B, et al. A leaky mutation in CD3D differentially affects alphabeta and gammadelta T cells and leads to a Talphabeta-Tgammadelta+B+NK+ human SCID. *J Clin Invest*. 2011;121(10):3872-3876.
371. Morath A, Schamel WW. Alphabeta and gammadelta T cell receptors: similar but different. *J Leukoc Biol*. 2020;107(6):1045-1055.
372. Mallis RJ, Duke-Cohan JS, Das DK, et al. Molecular design of the gammadeltaT cell receptor ectodomain encodes biologically fit ligand recognition in the absence of mechanosensing. *Proc Natl Acad Sci U S A*. 2021;118(26):e2023050118.
373. Feng Y, Brazin KN, Kobayashi E, Mallis RJ, Reinherz EL, Lang MJ. Mechanosensing drives acuity of alphabeta T-cell recognition. *Proc Natl Acad Sci U S A*. 2017;114(39):E8204-E8213.
374. Malissen B, Bongrand P. Early T cell activation: integrating biochemical, structural, and biophysical cues. *Annu Rev Immunol*. 2015;33:539-561.
375. Van Laethem F, Sarafova SD, Park JH, et al. Deletion of CD4 and CD8 coreceptors permits generation of alphabetaT cells that recognize antigens independently of the MHC. *Immunity*. 2007;27(5):735-750.
376. Tikhonova AN, Van Laethem F, Hanada KI, et al. Alphabeta T cell receptors that do not undergo major histocompatibility complex-specific thymic selection possess antibody-like recognition specificities. *Immunity*. 2012;36(1):79-91.
377. Hsu H, Chen C, Nenninger A, Holz L, Baldwin CL, Telfer JC. WC1 is a hybrid gammadelta TCR coreceptor and pattern recognition receptor for pathogenic bacteria. *J Immunol*. 2015;194(5):2280-2288.
378. Chen C, Herzig CTA, Alexander LJ, et al. Gene number determination and genetic polymorphism of the gamma delta T cell co-receptor WC1 genes. *BMC Genet*. 2012;13:86.
379. Baldwin CL, Damani-Yokota P, Yirsaw A, Loonie K, Teixeira AF, Gillespie A. Special features of gammadelta T cells in ruminants. *Mol Immunol*. 2021;134:161-169.

380. Wang F, Herzig C, Ozer D, et al. Tyrosine phosphorylation of scavenger receptor cysteine-rich WC1 is required for the WC1-mediated potentiation of TCR-induced T-cell proliferation. *Eur J Immunol.* 2009;39(1):254-266.
381. Hanby-Flarida MD, Trask OJ, Yang TJ, et al. Modulation of WC1, a lineage-specific cell surface molecule of gamma/delta T cells augments cellular proliferation. *Immunology.* 1996;88(1):116-123.
382. Witherden DA, Verdino P, Rieder SE, et al. The junctional adhesion molecule JAML is a costimulatory receptor for epithelial gammadelta T cell activation. *Science.* 2010;329(5996):1205-1210.
383. Verdino P, Witherden DA, Havran WL, et al. The molecular interaction of CAR and JAML recruits the central cell signal transducer PI3K. *Science.* 2010;329(5996):1210-1214.
384. McGraw JM, Thelen F, Hampton EN, et al. JAML promotes CD8 and gammadelta T cell antitumor immunity and is a novel target for cancer immunotherapy. *J Exp Med.* 2021;218(10):e20202644.
385. Witherden DA, Watanabe M, Garijo O, et al. The CD100 receptor interacts with its plexin B2 ligand to regulate epidermal gammadelta T cell function. *Immunity.* 2012;37(2):314-325.
386. Johnson MD, Otuki MF, Cabrini DA, Rudolph R, Witherden DA, Havran WL. Hspa8 and ICAM-1 as damage-induced mediators of gammadelta T cell activation. *J Leukoc Biol.* 2022;111(1):135-145.
387. Kunzmann V, Bauer E, Wilhelm M. Gamma/delta T-cell stimulation by pamidronate. *N Engl J Med.* 1999;340(9):737-738.
388. Kunzmann V, Bauer E, Feurle J, Weissinger F, Tony HP, Wilhelm M. Stimulation of gammadelta T cells by aminobisphosphonates and induction of antiplasma cell activity in multiple myeloma. *Blood.* 2000;96(2):384-392.
389. Chen X, Morrissey S, Chen F. Novel insight into the molecular and metabolic mechanisms orchestrating IL-17 production in gammadelta T cells. *Front Immunol.* 2019;10:2828.
390. Cai Y, Xue F, Qin H, et al. Differential roles of the mTOR-STAT3 signaling in dermal gammadelta T cell effector function in skin inflammation. *Cell Rep.* 2019;27(10):3034-3048 e5.
391. Meresse B, Chen Z, Ciszewski C, et al. Coordinated induction by IL15 of a TCR-independent NKG2D signaling pathway converts CTL into lymphokine-activated killer cells in celiac disease. *Immunity.* 2004;21(3):357-366.
392. Hue S, Mention JJ, Monteiro RC, et al. A direct role for NKG2D/MICA interaction in villous atrophy during celiac disease. *Immunity.* 2004;21(3):367-377.
393. Komori HK, Witherden DA, Kelly R, et al. Cutting edge: dendritic epidermal gammadelta T cell ligands are rapidly and locally expressed by keratinocytes following cutaneous wounding. *J Immunol.* 2012;188(7):2972-2976.
394. Swamy M, Abeler-Dorner L, Chettle J, et al. Intestinal intraepithelial lymphocyte activation promotes innate antiviral resistance. *Nat Commun.* 2015;6:7090.
395. Ogasawara H, Takeda-Hirokawa N, Sekigawa I, Hashimoto H, Kaneko Y, Hirose S. Inhibitory effect of interleukin-16 on interleukin-2 production by CD4+ T cells. *Immunology.* 1999;96(2):215-219.
396. Spidale NA, Malhotra N, Spidale NA, Frascoli M, et al. Neonatal-derived IL-17 producing dermal gammadelta T cells are required to prevent spontaneous atopic dermatitis. *Elife.* 2020;9:e51188.
397. Girardi M, Lewis J, Glusac E, et al. Resident skin-specific gammadelta T cells provide local, nonredundant regulation of cutaneous inflammation. *J Exp Med.* 2002;195(7):855-867.
398. Guzman E, Hope J, Taylor G, Smith AL, Cubillos-Zapata C, Charleston B. Bovine gammadelta T cells are a major regulatory T cell subset. *J Immunol.* 2014;193(1):208-222.
399. Collins RA, Werling D, Duggan SE, Bland AP, Parsons KR, Howard CJ. Gammadelta T cells present antigen to CD4+ alphabeta T cells. *J Leukoc Biol.* 1998;63(6):707-714.
400. Brandes M, Willimann K, Bioley G, et al. Cross-presenting human gammadelta T cells induce robust CD8+ alphabeta T cell responses. *Proc Natl Acad Sci U S A.* 2009;106(7):2307-2312.
401. Brandes M, Willimann K, Moser B. Professional antigen-presentation function by human gammadelta T Cells. *Science.* 2005;309(5732):264-268.
402. Wu Y, Wu W, Wong WM, et al. Human gamma delta T cells: a lymphoid lineage cell capable of professional phagocytosis. *J Immunol.* 2009;183(9):5622-5629.
403. Himoudi N, Morgenstern DA, Yan M, et al. Human gammadelta T lymphocytes are licensed for professional antigen presentation by interaction with opsonized target cells. *J Immunol.* 2012;188(4):1708-1716.
404. Meuter S, Eberl M, Moser B. Prolonged antigen survival and cytosolic export in cross-presenting human gammadelta T cells. *Proc Natl Acad Sci U S A.* 2010;107(19):8730-8735.
405. Moser B, Eberl M. Gammadelta T cells: novel initiators of adaptive immunity. *Immunol Rev.* 2007;215:89-102.
406. Lanzavecchia A. Antigen-specific interaction between T and B cells. *Nature.* 1985;314(6011):537-539.
407. Costa-Garcia M, Ataya M, Moraru M, Vilches C, Lopez-Botet M, Muntasell A. Human cytomegalovirus antigen presentation by HLA-DR+ NKG2C+ adaptive NK cells specifically activates polyfunctional effector memory CD4+ T lymphocytes. *Front Immunol.* 2019;10:687.
408. Tyler CJ, McCarthy NE, Lindsay JO, Stagg AJ, Moser B, Eberl M. Antigen-presenting human gammadelta T cells promote intestinal CD4(+) T cell expression of IL-22 and mucosal release of calprotectin. *J Immunol.* 2017;198(9):3417-3425.
409. Utzschneider DT, Legat A, Fuertes Marraco SA, et al. T cells maintain an exhausted phenotype after antigen withdrawal and population reexpansion. *Nat Immunol.* 2013;14(6):603-610.
410. Blank CU, Haining WN, Held W, et al. Defining 'T cell exhaustion'. *Nat Rev Immunol.* 2019;19(11):665-674.
411. Hughes DP, Hayday A, Craft JE, et al. T cells with gamma/delta T cell receptors (TCR) of intestinal type are preferentially expanded in TCR-alpha-deficient lpr mice. *J Exp Med.* 1995;182(1):233-241.
412. Peng SL, Madaio MP, Hughes DP, et al. Murine lupus in the absence of alpha beta T cells. *J Immunol.* 1996;156(10):4041-4049.
413. Peng SL, Madaio MP, Hayday AC, et al. Propagation and regulation of systemic autoimmunity by gammadelta T cells. *J Immunol.* 1996;157(12):5689-5698.
414. Cardone J, Le Friec G, Vantourout P, et al. Complement regulator CD46 temporally regulates cytokine production by conventional and unconventional T cells. *Nat Immunol.* 2010;11(9):862-871.
415. Faustino LD, Griffith JW, Rahimi RA, et al. Interleukin-33 activates regulatory T cells to suppress innate gammadelta T cell responses in the lung. *Nat Immunol.* 2020;21(11):1371-1383.
416. Park SG, Mathur R, Long M, et al. T regulatory cells maintain intestinal homeostasis by suppressing gammadelta T cells. *Immunity.* 2010;33(5):791-803.
417. Zhang H, Carnevale G, Polese B, et al. CD109 Restrains activation of cutaneous IL-17-producing gammadelta T cells by commensal microbiota. *Cell Rep.* 2019;29(2):391-405 e5.
418. Fife BT, Pauken KE, Eagar TN, et al. Interactions between PD-1 and PD-L1 promote tolerance by blocking the TCR-induced stop signal. *Nat Immunol.* 2009;10(11):1185-1192.
419. Kortlever RM, Sodir NM, Wilson CH, et al. Myc Cooperates with ras by programming inflammation and immune suppression. *Cell.* 2017;171(6):1301-1315 e14.
420. Zaid A, Mackay LK, Rahimpour A, et al. Persistence of skin-resident memory T cells within an epidermal niche. *Proc Natl Acad Sci U S A.* 2014;111(14):5307-5312.
421. Park S, Matte-Martone C, Gonzalez DG, et al. Skin-resident immune cells actively coordinate their distribution with epidermal cells during homeostasis. *Nat Cell Biol.* 2021;23(5):476-484.
422. Romagnoli PA, Sheridan BS, Pham QM, et al. IL-17A-producing resident memory gammadelta T cells orchestrate the innate immune response to secondary oral Listeria monocytogenes infection. *Proc Natl Acad Sci U S A.* 2016;113(30):8502-8507.
423. Mukasa A, Lahn M, Pflum EK, et al. Evidence that the same gamma delta T cells respond during infection-induced and autoimmune inflammation. *J Immunol.* 1997;159(12):5787-5794.
424. Mukasa A, Hiromatsu K, Matsuzaki G, O'Brien R, Born W, Nomoto K. Bacterial infection of the testis leading to autoaggressive immunity triggers apparently opposed responses of alpha beta and gamma delta T cells. *J Immunol.* 1995;155(4):2047-2056.
425. Mukasa A, Born WK, O'Brien RL. Inflammation alone evokes the response of a TCR-invariant mouse gamma delta T cell subset. *J Immunol.* 1999;162(8):4910-4913.
426. King DP, Hyde DM, Jackson KA, et al. Cutting edge: protective response to pulmonary injury requires gamma delta T lymphocytes. *J Immunol.* 1999;162(9):5033-5036.
427. Hayday A, Tigelaar R. Immunoregulation in the tissues by gammadelta T cells. *Nat Rev Immunol.* 2003;3(3):233-242.
428. Born W, Cady C, Jones-Carson J, Mukasa A, Lahn M, O'Brien R. Immunoregulatory functions of gamma delta T cells. *Adv Immunol.* 1999;71:77-144.
429. Simonian PL, Wehrmann F, Roark CL, Born WK, O'Brien RL, Fontenot AP. Gammadelta T cells protect against lung fibrosis via IL-22. *J Exp Med.* 2010;207(10):2239-2253.
430. Simonian PL, Roark CL, Wehrmann F, et al. Th17-polarized immune response in a murine model of hypersensitivity pneumonitis and lung fibrosis. *J Immunol.* 2009;182(1):657-665.
431. Wang T, Scully E, Yin Z, et al. IFN-gamma-producing gamma delta T cells help control murine West Nile virus infection. *J Immunol.* 2003;171(5):2524-2531.
432. Nishimura H, Yajima T, Kagimoto Y, et al. Intraepithelial gammadelta T cells may bridge a gap between innate immunity and acquired immunity to herpes simplex virus type 2. *J Virol.* 2004;78(9):4927-4930.
433. Hedges JF, Holderness J, Jutila MA. Adjuvant materials that enhance bovine gammadelta T cell responses. *Vet Immunol Immunopathol.* 2016;181:30-38.
434. Ramsburg E, Tigelaar R, Craft J, et al. Age-dependent requirement for gammadelta T cells in the primary but not secondary protective immune response against an intestinal parasite. *J Exp Med.* 2003;198(9):1403-1414.
435. Smith AL, Hayday AC. An alphabeta T-cell-independent immunoprotective response towards gut coccidia is supported by gammadelta cells. *Immunology.* 2000;101(3):325-332.
436. Smith AL, Hayday AC. Genetic dissection of primary and secondary responses to a widespread natural pathogen of the gut, Eimeria vermiformis. *Infect Immun.* 2000;68(11):6273-6280.
437. Dejima T, Shibata K, Yamada H, et al. Protective role of naturally occurring interleukin-17A-producing gammadelta T cells in the lung at the early stage of systemic candidiasis in mice. *Infect Immun.* 2011;79(11):4503-4510.
438. Jones-Carson J, Vazquez-Torres A, van der Heyde HC, Warner T, Wagner RD, Balish E. Gamma delta T cell-induced nitric oxide production enhances resistance to mucosal candidiasis. *Nat Med.* 1995;1(6):552-557.
439. Wormley FL Jr, Steele C, Wozniak K, Fujihashi K, McGhee JR, Fidel PL. Resistance of T-cell receptor delta-chain-deficient mice to experimental Candida albicans vaginitis. *Infect Immun.* 2001;69(11):7162-7164.
440. Fenoglio D, Poggi A, Catellani S, et al. Vdelta1 T lymphocytes producing IFN-gamma and IL-17 are expanded in HIV-1-infected patients and respond to Candida albicans. *Blood.* 2009;113(26):6611-6618.
441. Dunne MR, Wagener J, Loeffler J, et al. Unconventional T cells - new players in antifungal immunity. *Clin Immunol.* 2021;227:108734.
442. Pantelyushin S, Haak S, Ingold B, et al. Rorgammat+ innate lymphocytes and gammadelta T cells initiate psoriasiform plaque formation in mice. *J Clin Invest.* 2012;122(6):2252-2256.
443. Ramirez-Valle F, Gray EE, Cyster JG. Inflammation induces dermal Vgamma4+ gammadeltaT17 memory-like cells that travel to distant skin and accelerate secondary IL-17-driven responses. *Proc Natl Acad Sci U S A.* 2015;112(26):8046-8051.
444. Sutton CE, Lalor SJ, Sweeney CM, Brereton CF, Lavelle EC, Mills KHG. Interleukin-1 and IL-23 induce innate IL-17 production from gammadelta T cells, amplifying Th17 responses and autoimmunity. *Immunity.* 2009;31(2):331-341.
445. Markle JG, Mortin-Toth S, Wong ASL, Geng L, Hayday A, Danska JS. Gammadelta T cells are essential effectors of type 1 diabetes in the nonobese diabetic mouse model. *J Immunol.* 2013;190(11):5392-5401.
446. Halstensen TS, Scott H, Brandtzaeg P. Intraepithelial T cells of the TcR gamma/delta+ CD8- and V delta 1/J delta 1+ phenotypes are increased in coeliac disease. *Scand J Immunol.* 1989;30(6):665-672.
447. Halstensen TS, Scott H, Farstad IN, et al. In situ two- and three-color immunofluorescence staining of mucosal T-cells in celiac disease. Increase of TCR gamma/delta+CD8- and TCR alpha/beta+CD45R0+ intraepithelial lymphocytes and IL-2R+ TCR alpha/beta+CD4+CD45R0+ lamina propria lymphocytes. *Prog Histochem Cytochem.* 1992;26(1–4):201-210.
448. Mayassi T, Ladell K, Gudjonson H, et al. Chronic inflammation permanently reshapes tissue-resident immunity in celiac disease. *Cell.* 2019;176(5):967-981 e19.
449. Inagaki-Ohara K, Chinen T, Matsuzaki G, et al. Mucosal T cells bearing TCRgammadelta play a protective role in intestinal inflammation. *J Immunol.* 2004;173(2):1390-1398.
450. Kober OI, Ahl D, Pin C, Holm L, Carding SR, Juge N. Gammadelta T-cell-deficient mice show alterations in mucin expression, glycosylation, and goblet cells but maintain an intact mucus layer. *Am J Physiol Gastrointest Liver Physiol.* 2014;306(7):G582-G593.
451. Mielke LA, Jones SA, Raverdeau M, et al. Retinoic acid expression associates with enhanced IL-22 production by gammadelta T cells and innate lymphoid cells and attenuation of intestinal inflammation. *J Exp Med.* 2013;210(6):1117-1124.
452. Guo XZJ, Dash P, Crawford JC, et al. Lung gammadelta T Cells mediate protective responses during neonatal influenza infection that are associated with type 2 immunity. *Immunity.* 2018;49(3):531-544 e6.
453. Schneider C, O'Leary CE, Locksley RM. Regulation of immune responses by tuft cells. *Nat Rev Immunol.* 2019;19(9):584-593.
454. Komano H, Fujiura Y, Kawaguchi M, et al. Homeostatic regulation of intestinal epithelia by intraepithelial gamma delta T cells. *Proc Natl Acad Sci U S A.* 1995;92(13):6147-6151.
455. Hu MD, Golovchenko NB, Burns GL, et al. Gammadelta intraepithelial lymphocytes facilitate pathological epithelial cell shedding via CD103-mediated granzyme release. *Gastroenterology.* 2021;162:877-889.e7.
456. Alves de Lima K, Rustenhoven J, Da Mesquita S, et al. Meningeal gammadelta T cells regulate anxiety-like behavior via IL-17a signaling in neurons. *Nat Immunol.* 2020;21(11):1421-1429.
457. Lynch F, Shevach EM. Gamma delta T cells promote CD4 and CD8 expression by SCID thymocytes. *Int Immunol.* 1993;5(8):991-995.
458. Shores EW, Sharrow SO, Uppenkamp I, et al. T cell receptor-negative thymocytes from SCID mice can be induced to enter the CD4/CD8 differentiation pathway. *Eur J Immunol.* 1990;20(1):69-77.
459. Silva-Santos B, Pennington DJ, Hayday AC. Lymphotoxin-mediated regulation of gammadelta cell differentiation by alphabeta T cell progenitors. *Science.* 2005;307(5711):925-928.
460. Powolny-Budnicka I, Riemann M, Tanzer S, Schmid RM, Hehlgans T, Weih F. RelA and RelB transcription factors in distinct thymocyte populations control lymphotoxin-dependent interleukin-17 production in gammadelta T cells. *Immunity.* 2011;34(3):364-374.
461. Park JE, Botting RA, Dominguez Conde C, et al. A cell atlas of human thymic development defines T cell repertoire formation. *Science.* 2020;367(6480):eaay3224.

462. Roberts NA, White AJ, Jenkinson WE, et al. Rank signaling links the development of invariant gammadelta T cell progenitors and Aire(+) medullary epithelium. *Immunity*. 2012;36(3):427-437.
463. Meyer S, Woodward M, Hertel C, et al. AIRE-deficient patients harbor unique high-affinity disease-ameliorating autoantibodies. *Cell*. 2016;166(3):582-595.
464. Kisand K, Lilic D, Casanova JL, Peterson P, Meager A, Willcox N. Mucocutaneous candidiasis and autoimmunity against cytokines in APECED and thymoma patients: clinical and pathogenetic implications. *Eur J Immunol*. 2011;41(6):1517-1527.
465. Fujikado N, Mann AO, Bansal K, et al. Aire Inhibits the generation of a perinatal population of interleukin-17a-producing gammadelta T cells to promote immunologic tolerance. *Immunity*. 2016;45(5):999-1012.
466. Wen L, Roberts SJ, Viney JL, et al. Immunoglobulin synthesis and generalized autoimmunity in mice congenitally deficient in alpha beta(+) T cells. *Nature*. 1994;369(6482):654-658.
467. Wen L, Pao W, Wong FS, et al. Germinal center formation, immunoglobulin class switching, and autoantibody production driven by "non alpha/beta" T cells. *J Exp Med*. 1996;183(5):2271-2282.
468. Pao W, Wen L, Smith AL, et al. Gamma delta T cell help of B cells is induced by repeated parasitic infection, in the absence of other T cells. *Curr Biol*. 1996;6(10):1317-1325.
469. Huang Y, Getahun A, Heiser RA, et al. Gammadelta T cells shape preimmune peripheral B cell populations. *J Immunol*. 2016;196(1):217-231.
470. Horner AA, Jabara H, Ramesh N, et al. gamma/delta T lymphocytes express CD40 ligand and induce isotype switching in B lymphocytes. *J Exp Med*. 1995;181(3):1239-1244.
471. Caccamo N, Battistini L, Bonneville M, et al. CXCR5 identifies a subset of Vgamma9Vdelta2 T cells which secrete IL-4 and IL-10 and help B cells for antibody production. *J Immunol*. 2006;177(8):5290-5295.
472. Bansal RR, Mackay CR, Moser B, et al. IL-21 enhances the potential of human gammadelta T cells to provide B-cell help. *Eur J Immunol*. 2012;42(1):110-119.
473. Petrasca A, Melo AM, Breen EP, et al. Human Vdelta3(+) gammadelta T cells induce maturation and IgM secretion by B cells. *Immunol Lett*. 2018;196:126-134.
474. Rezende RM, Lanser AJ, Rubino S, et al. Gammadelta T cells control humoral immune response by inducing T follicular helper cell differentiation. *Nat Commun*. 2018;9(1):3151.
475. Ehl S, Schwarz K, Enders A, et al. A variant of SCID with specific immune responses and predominance of gamma delta T cells. *J Clin Invest*. 2005;115(11):3140-3148.
476. Crawford G, Hayes MD, Seoane RC, et al. Epithelial damage and tissue gammadelta T cells promote a unique tumor-protective IgE response. *Nat Immunol*. 2018;19(8):859-870.
477. Strid J, Sobolev O, Zafirova B, et al. The intraepithelial T cell response to NKG2D-ligands links lymphoid stress surveillance to atopy. *Science*. 2011;334(6060):1293-1297.
478. Profet M. The function of allergy: immunological defense against toxins. *Q Rev Biol*. 1991;66(1):23-62.
479. Wesemann DR, Magee JM, Boboila C, et al. Immature B cells preferentially switch to IgE with increased direct Smu to Sepsilon recombination. *J Exp Med*. 2011;208(13):2733-2746.
480. Sturm E, Braakman E, Fisch P, Vreugdenhil RJ, Sondel P, Bolhuis RL. Human V gamma 9-V delta 2 T cell receptor-gamma delta lymphocytes show specificity to Daudi Burkitt's lymphoma cells. *J Immunol*. 1990;145(10):3202-3208.
481. Bosnes V, Halvorsen R, Skaftadottir I, et al. Specificity of gamma delta receptor-bearing cytotoxic T lymphocytes isolated from human peripheral blood. *Scand J Immunol*. 1989;29(6):723-731.
482. Fisch P, Oettel K, Fudim N, Surfus JE, Malkovsky M, Sondel PM. MHC-unrestricted cytotoxic and proliferative responses of two distinct human gamma/delta T cell subsets to Daudi cells. *J Immunol*. 1992;148(8):2315-2323.
483. Fisch P, Malkovsky M, Braakman E, et al. Gamma/delta T cell clones and natural killer cell clones mediate distinct patterns of non-major histocompatibility complex-restricted cytolysis. *J Exp Med*. 1990;171(5):1567-1579.
484. Penninger JM, Wen T, Timms E, et al. Spontaneous resistance to acute T-cell leukaemias in TCRV gamma 1.1J gamma 4C gamma 4 transgenic mice. *Nature*. 1995;375(6528):241-244.
485. Girardi M, Glusac E, Filler RB, et al. The distinct contributions of murine T cell receptor (TCR)gammadelta+ and TCRalphabeta+ T cells to different stages of chemically induced skin cancer. *J Exp Med*. 2003;198(5):747-755.
486. Shankaran V, Ikeda H, Bruce AT, et al. IFNgamma and lymphocytes prevent primary tumour development and shape tumour immunogenicity. *Nature*. 2001;410(6832):1107-1111.
487. Galon J, Costes A, Sanchez-Cabo F, et al. Type, density, and location of immune cells within human colorectal tumors predict clinical outcome. *Science*. 2006;313(5795):1960-1964.
488. Liu Z, Eltoum IEA, Guo B, Beck BH, Cloud GA, Lopez RD. Protective immunosurveillance and therapeutic antitumor activity of gammadelta T cells demonstrated in a mouse model of prostate cancer. *J Immunol*. 2008;180(9):6044-6053.
489. Gao Y, Yang W, Pan M, et al. Gamma delta T cells provide an early source of interferon gamma in tumor immunity. *J Exp Med*. 2003;198(3):433-442.
490. Dvorak HF. Tumors: wounds that do not heal. Similarities between tumor stroma generation and wound healing. *N Engl J Med*. 1986;315(26):1650-1659.
491. Lopes N, Silva-Santos B. Functional and metabolic dichotomy of murine gammadelta T cell subsets in cancer immunity. *Eur J Immunol*. 2021;51(1):17-26.
492. Wu P, Wu D, Ni C, et al. gammadeltaT17 cells promote the accumulation and expansion of myeloid-derived suppressor cells in human colorectal cancer. *Immunity*. 2014;40(5):785-800.
493. Meraviglia S, Lo Presti E, Tosolini M, et al. Distinctive features of tumor-infiltrating gammadelta T lymphocytes in human colorectal cancer. *OncoImmunology*. 2017;6(10):e1347742.
494. Patil RS, Shah SU, Shrikhande SV, Goel M, Dikshit RP, Chiplunkar SV. IL17 producing gammadeltaT cells induce angiogenesis and are associated with poor survival in gallbladder cancer patients. *Int J Cancer*. 2016;139(4):869-881.
495. Lai D, Wang F, Chen Y, et al. Human ovarian cancer stem-like cells can be efficiently killed by gammadelta T lymphocytes. *Cancer Immunol Immunother*. 2012;61(7):979-989.
496. Foord E, Arruda LCM, Gaballa A, et al. Characterization of ascites- and tumor-infiltrating gammadelta T cells reveals distinct repertoires and a beneficial role in ovarian cancer. *Sci Transl Med*. 2021;13(577):eabb0192.
497. Gentles AJ, Newman AM, Liu CL, et al. The prognostic landscape of genes and infiltrating immune cells across human cancers. *Nat Med*. 2015;21(8):938-945.
498. Pizzolato G, Kaminski H, Tosolini M, et al. Single-cell RNA sequencing unveils the shared and the distinct cytotoxic hallmarks of human TCRVdelta1 and TCRVdelta2 gammadelta T lymphocytes. *Proc Natl Acad Sci U S A*. 2019;116(24):11906-11915.
499. Kabelitz D, Serrano R, Kouakanou L, et al. Cancer immunotherapy with gammadelta T cells: many paths ahead of us. *Cell Mol Immunol*. 2020;17(9):925-939.

CHAPTER 36

CD1-Restricted T Cells

Pankaj Sharma, PhD • Florian Winau, MD

INTRODUCTION

In the 1970s, the development of monoclonal antibody technology led to the identification of CD1 proteins as the first members of the cluster of differentiation system. This prominent numbering was due to the fact that human thymocytes, which abundantly express CD1, were used for immunization of mice. The structural similarities with MHC-I proteins suggested that CD1 molecules might play a role in antigen presentation. A decade later, CD1a and CD1c were shown to specifically activate T lymphocytes, followed by the discovery that CD1b presented a bacterial lipid antigen. Subsequent work in the murine system identified invariant T cells expressing natural killer (NK) cell markers that were restricted by CD1d, prior to the description of their antigenic ligands as glycosylceramides.

Thus, in contrast to peptide presentation by MHC molecules, the CD1 system evolved to present lipid antigens to T lymphocytes. Based on sequence homology, group 1 CD1 molecules comprise CD1a, b, and c, whereas CD1d is the only member of group 2. Although CD1 proteins share common principles in lipid presentation, the responding T cells are biologically quite distinct. Accordingly, CD1d selects and activates natural killer T (NKT) cells, which represent innate-like T cells with a limited T cell receptor (TCR) repertoire. Upon activation, they rapidly produce substantial amounts of a broad range of cytokines. This versatility endows NKT cells with important functions in infection, inflammation, autoimmunity, and cancer. The biology of NKT cells will be discussed in the first part of this chapter.

By contrast, T lymphocytes that recognize CD1a-c express a diverse TCR repertoire and rather behave like adaptive T cells. Notably, CD1a-c are mainly expressed in humans and lacking in mice. Therefore, our knowledge on group 1 CD1-restricted T cells is principally based on human studies, as exemplified by the discovery of lipid antigens from mycobacteria that stimulate human T cell clones. Thus, bacterial lipid display via CD1a-c suggests a role of this antigen presentation system in infectious disease, such as tuberculosis. Moreover, recent and emerging evidence indicates that CD1a-c present self-lipid antigens to autoreactive T lymphocytes as part of physiological homeostasis. However, depending on the biological context, these autoreactive CD1-restricted T cell responses are able to cause pathology, as observed in inflammatory skin disease. The second part of this chapter is dedicated to the biology of group 1 CD1.

BASIC FEATURES OF NKT CELLS

Most NKT cells in mice express an invariant TCRα chain consisting of Vα14-Jα18 and are hence called invariant NKT (iNKT) cells. The murine invariant α chain is paired with Vβ8, Vβ7, or Vβ2, while in humans, the homologous Vα24-Jα18 chain is coupled with Vβ11.[1] Other CD1d-restricted NKT cells, termed type II NKT cells, exhibit a more diverse TCR repertoire than iNKT cells (also called type I) and are rare in mice but more abundant in the human system. The canonical TCR of iNKT cells rearranges during thymic development in a random and infrequent fashion. However, upon TCR expression and positive selection, iNKT cells drastically expand and differentiate into long-lived effector cells.[2] Exiting the thymus, iNKT cells colonize lymphoid tissues, such as spleen and lymph nodes, as well as peripheral organs, including liver, lung, and the gut. In mice, iNKT cells are especially abundant in hepatic tissue. Due to their innate-like responsiveness and tissue residency, iNKT cells are poised to quickly deploy their effector functions after recognition of microbial or self-lipid antigens presented by CD1d. In a larger context, the iNKT cell lineage belongs to a family of innate-like lymphocytes, including γδ T cells and MR1-restricted mucosal-associated invariant T (MAIT) cells, which all express antigen-specific TCRs like adaptive T lymphocytes, while showing immediate effector functions and tissue localization comparable to innate lymphoid cells (ILCs). Thus, iNKT cells might represent an evolutionary link between innate and adaptive immunity.

ANTIGEN PRESENTATION

CD1d Expression and Presentation of Lipids

CD1d is an MHC-I-like protein that is noncovalently associated with β_2-microglobulin and the only member of the CD1 family conserved in mice.[3] The murine genome contains two genes: *cd1d1* is productively expressed as a surface glycoprotein, whereas *cd1d2* is either weakly expressed or inactivated by mutation in the most widely used experimental strain C57BL/6.[4] CD1d is expressed on cortical thymocytes,[5] which is critical for positive selection of iNKT cells in the thymus. With regard to peripheral activation of iNKT cells, CD1d is expressed on most types of antigen-presenting cells, such as dendritic cells (DCs) and B cells.[6,7] Moreover, epithelial cells are described to express CD1d, including hepatocytes,[8] cholangiocytes,[9] and intestinal epithelial cells (IECs).[10]

Nascent CD1d proteins are assembled in the endoplasmic reticulum (ER), coupled to β_2-microglobulin, and loaded with self-lipid ligands, prior to trafficking to the cell surface[11,12]

(Fig. 36.1). In the ER, microsomal triglyceride transfer protein (MTP) functions as chaperone for lipid loading.[13] Although it normally serves in triglyceride transfer for apolipoprotein B biosynthesis in the liver, in antigen presentation, MTP transfers phospholipids onto CD1d.[14] Following the secretory route, CD1d mainly harbors phospholipids, such as phosphatidylcholine.[15] Subsequent to expression on the surface, CD1d gets effectively endocytosed through interaction of a tyrosine motif in its cytoplasmic tail with adaptor protein (AP)-2 and AP-3 to facilitate trafficking to late endosomes and lysosomes.[16-19] This acidic compartment is well equipped to mediate processing of exogenous lipid antigens, using enzymes like α-galactosidase A.[20,21] Subsequently, potential lipid antigens are loaded onto CD1d with the help of lipid transfer proteins, including saposins,[22-24] Gm_2 activator,[22] and Niemann-Pick type C2 protein.[25] Especially saposins have prominent amphiphilic features, which allows them to closely interact with lipid bilayers to mobilize lipid antigens from the hydrophobic phase while simultaneously binding to CD1d. The lysosomal compartment is also a reservoir for exogenous lipids following their uptake by antigen-presenting cells. Accordingly, very-low-density lipoproteins containing apolipoprotein E are able to shuttle lipid antigens, prior to engulfment via the LDL receptor.[26] Furthermore, scavenger receptors show a broad aptitude for uptake of diverse lipid classes.[27] Following successful loading of lipid antigens in late endosome and lysosomes, the CD1d/lipid complexes are transported to the cell surface for subsequent recognition by iNKT cells (Fig. 36.1).

CD1d Signaling and Regulation

In addition to antigen display to T lymphocytes, CD1d also has the ability for reverse outside-in signaling. Engagement of CD1d molecules on the surface of IECs triggered a STAT3-dependent downstream signaling cascade that led to induced transcription of IL-10 and CD1d itself.[10] Accordingly, IEC-specific ablation of CD1d caused fulminant iNKT cell-mediated colitis. Taken together, reverse signaling through CD1d on intestinal epithelium proved to be vital for mucosal IL-10 induction in order to suppress inflammatory bowel disease.

The strength of iNKT cell activation depends on the abundance of CD1d/lipid complexes on the surface of antigen-presenting cells. In this context, the actin cytoskeleton grouped lipid-loaded CD1d proteins in defined nanoclusters, preventing their uncontrolled coalescence to ensure fine-tuned iNKT cell activation.[28] Under inflammatory conditions, interfering with the interaction between the CD1d cytosolic tail and the actin cytoskeleton could lead to larger nanoclusters and increased stimulation of iNKT cells. Thus, the actin cytoskeleton is an important mechanistic regulator

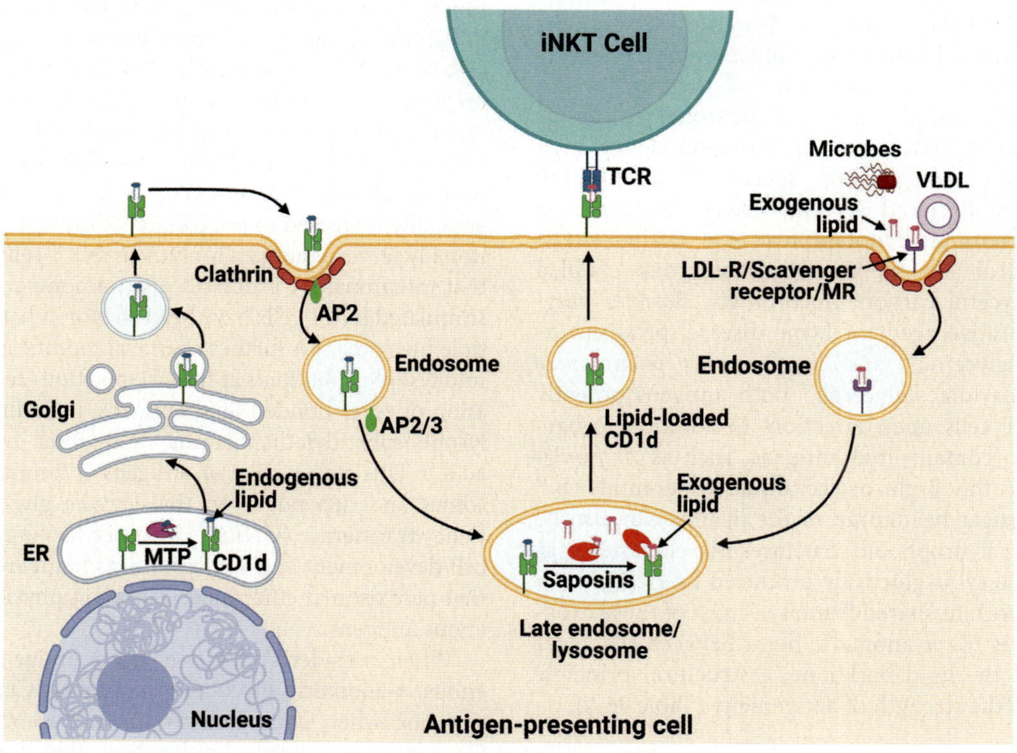

FIG. 36.1. CD1d-mediated presentation of lipid antigens. Nascent CD1d proteins are loaded with lipid ligands in the endoplasmic reticulum (ER) with the help of microsomal triglyceride transfer protein (MTP). Subsequently, CD1d/lipid complexes follow the secretory route to the surface of an antigen-presenting cell. After endocytosis and trafficking to late endosomes/lysosomes, lipid transfer proteins like saposins facilitate the loading of CD1d molecules with lipid antigens that were derived from the depicted extracellular sources. Eventually, complexes of CD1d and lipid antigen are transported to the cell surface for antigen presentation and activation of iNKT cells. AP2/3, adaptor protein 2/3; LDL-R, low-density lipoprotein receptor; MR, mannose receptor; TCR, T cell receptor.

of the microanatomical organization of antigen-presenting CD1d molecules.

Lastly, regulation of CD1d endocytosis can affect lipid presentation to T cells. Accordingly, lack of the central autophagy gene Atg5 specifically in dendritic cells led to failed recruitment of AP-2 to CD1d, causing increased CD1d surface expression and iNKT cell activation.[29] Therefore, the autophagy pathway in antigen-presenting cells balances the surface availability of CD1d by influencing its internalization.

Exogenous Lipid Antigens

The first and most potent antigen for iNKT cells is α-galactosylceramide (α-GalCer),[30] which was initially described in the marine sponge *Agelas mauritianus* when testing its extracts for antitumor activity. Chemically, α-GalCer belongs to the family of glycosphingolipids, which consist of a ceramide backbone, including the amino alcohol sphingosine esterified with a fatty acid, coupled to a carbohydrate headgroup. The α-anomeric configuration of the sugar is critical to render the glycolipid antigenic for iNKT cells. α-GalCer can be readily loaded onto CD1d-tetramers to facilitate the staining of the majority of murine and human iNKT cells.[31,32]

Since the importance of a sponge-derived antigen for murine and human physiology was not obvious, subsequent studies discovered bacterial glycolipid antigens as powerful stimulators of iNKT cell responses. The gram-negative α-proteobacteria *Sphingomonas* are devoid of lipopolysaccharide (LPS) and therefore use glycuronosylceramides instead to build the outer membrane of their cell wall.[33] α-Galacturonosylceramide and α-glucuronosylceramide represent powerful iNKT cell antigens upon infection of the host with *Sphingomonas* species.[34,35] In addition to sphingolipids, bacterial glycerol esters also serve as iNKT cell antigens and contain two acyl chains esterified with glycerol and the carbohydrate group coupled to the third glycerol carbon. Accordingly, *Borrelia burgdorferi*, the causative agent of Lyme disease, produces α-galactosyldiacylglycerol,[36] and *Streptococcus pneumoniae* harbors α-glucosyldiacylglycerol,[37] both antigens potently activating iNKT cells upon infection. In addition to bacteria, fungi also contain lipid antigens, such as *A. fumigatus* producing the β-glucosylceramide asperamide B.[38] Moreover, the sugar headgroup of the ligands can also be coupled to other hydrophobic structures like cholesterol, as found in cholesteryl α-glucoside produced by *Helicobacter pylori*.[39] An important shared feature of most of these exogenous antigens is the α-anomeric bond between the sugar headgroup and the lipid backbone, a structural principle that determines the strength of antigenicity (Table 36.1).

Endogenous Lipid Antigens

iNKT cells display a certain degree of autoreactivity, because they are able to respond to CD1d-expressing cells in the absence of extrinsic antigen.[40] This autoreactivity is essential for iNKT cell development in the thymus, where self-lipid ligands allow for positive selection.[41] Furthermore, autoreactive responses are also vital for peripheral iNKT cell activation, especially in the context of sterile inflammation or in infections with pathogens that lack lipid antigens.[42] The identification of the endogenous lipid antigens is therefore crucial to understand iNKT cell development and activation, and the quest for their entity is still ongoing.

Since the endosomal trafficking of CD1d is critically required for the activation of self-reactive iNKT cells,[41] it was hypothesized that the endogenous lipid antigens are localized in the endosomal and lysosomal compartment. Indeed, this approach led to the discovery of the glycosphingolipid isoglobotriaosylceramide (iGb3) as a prominent candidate for the elusive endogenous antigen for iNKT cells.[43] iGb3 is produced in the Golgi by addition of a terminal galactose to lactosylceramide, catalyzed by iGb3-synthase, and is also generated upon degradation of iGb4 by β-hexosaminidase A/B in the lysosome. Mice deficient for this enzyme required for lysosomal iGb3 production lacked iNKT cells, and purified iGb3 proved to be stimulatory for iNKT cells.[43] However, the subsequent analysis of iGb3-synthase knock-out animals revealed normal iNKT cell frequencies,[44] indicating that iGb3 is not the only intrinsic antigen. Moreover, it proved to be difficult to detect iGb3 in tissues using sensitive HPLC methods,[45,46] although minute amounts of the lipid have been described in mouse and human thymus based on ion trap mass spectrometry.[47] Complicating the situation were findings in the human system, in which the gene for iGb3-synthase turned out to be nonfunctional,[48] and a single amino acid substitution in human CD1d prevented the binding and presentation of iGb3.[49] Overall, iGb3 is clearly an antigenic lipid for iNKT cells but is certainly not the exclusive endogenous antigen in mice and unlikely plays a major role in humans.

In addition to sphingolipids, phospholipids represent another group of candidates for self-antigens. Accordingly, lysophosphatidylcholine (LPC) is a lipid messenger that is generally increased in the context of inflammation and additionally serves as antigen for human iNKT cells.[50] Considering that inflammatory processes often accompany cancer, LPC-stimulated iNKT cells have been shown in patients with multiple myeloma.[51] A further chemical modification introduced into lysophospholipids at the sn1-position leads to the generation of ether-bonded lipid antigens, including plasmalogen lysophosphatidylethanolamine and ether lysophosphatidic acid.[52] This novel class of antigens is produced in peroxisomes and dependent on the enzyme glyceronephosphate O-acyltransferase (GNPAT). In mice lacking GNPAT, iNKT cell development and activation was hampered,[52] suggesting that peroxisomal ether lipids represent physiological endogenous antigens.

When considering the strength of antigenicity, the exogenous α-anomeric glycolipids like α-GalCer appear to be superior when compared to the endogenous β-anomers. Thus, a general conclusion has been that the self-lipid antigens probably represent weaker agonists overall. For many years, the scientific community was satisfied with the search for β-anomeric glycosphingolipids, because it was thought that the α-anomers simply do not exist in mammals. In the synthetic pathway of glycosylceramides, there are two genes known to code for the respective β-glycosyltransferases, with Ugcg facilitating β-glucosylceramide and Ugt8a allowing for

TABLE 36.1 Lipid Antigens for iNKT Cells

Antigen	Structure	Source
α-Galactosylceramides	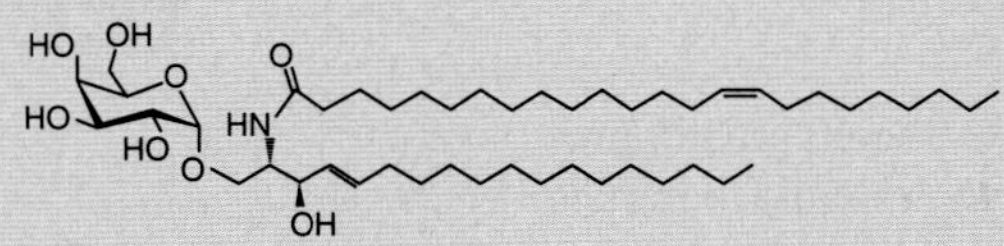	*Agelas mauritianus, Bacteroides fragilis, Self*
α-Glucosylceramide	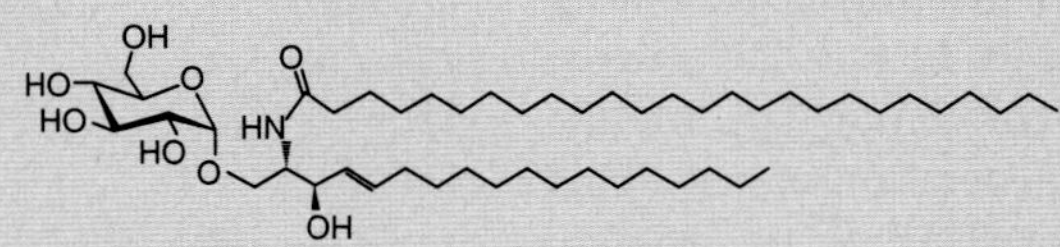	*Self*
α-Glucuronosylceramide	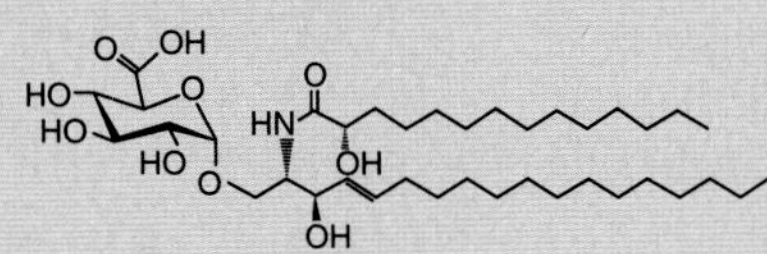	*Sphingomonas* spp.
α-Galactosyldiacylglycerol	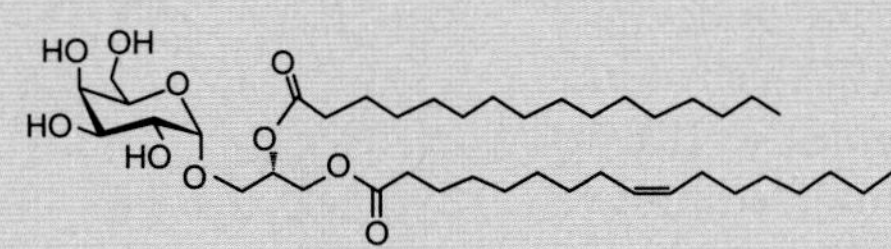	*Borrelia burgdorferi*
α-Glucosyldiacylglycerol	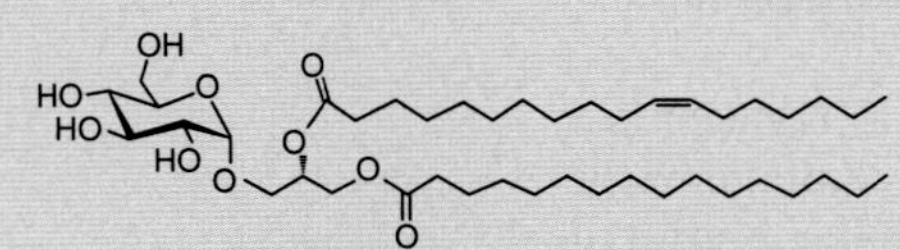	*Streptococcus pneumoniae*
Lyso-phosphatidycholine	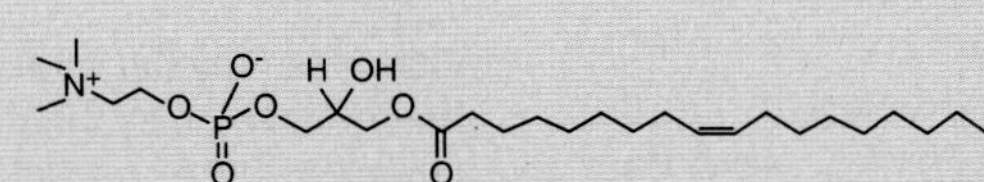	*Self*
Plasmalogen lysophosphatidylethanol-amine (pLPE)	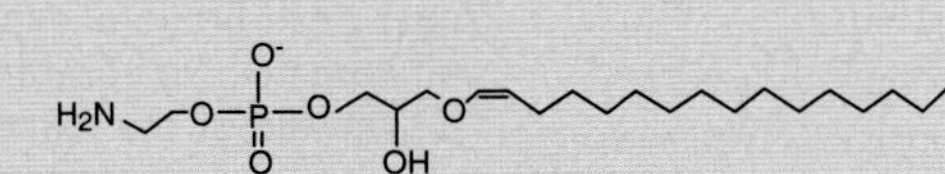	*Self*
Asperamide B	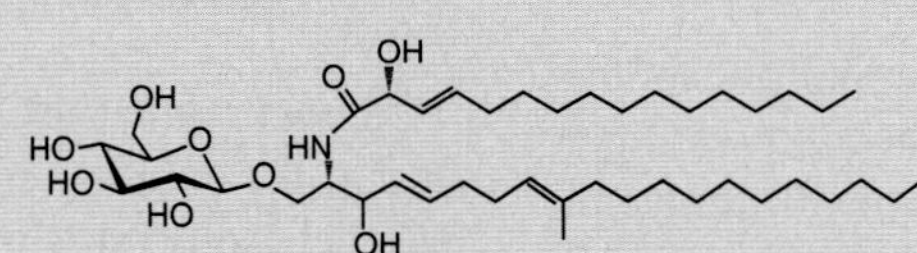	*Aspergillus fumigatus*
Cholesteryl α-glucoside	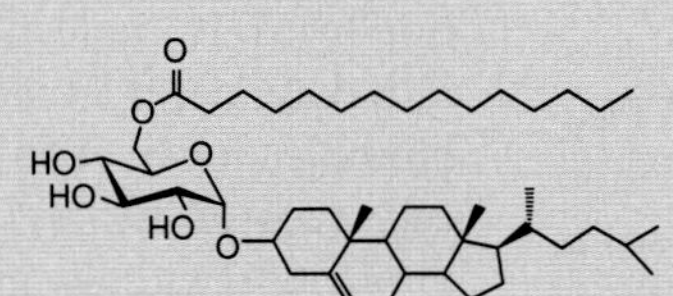	*Helicobacter pylori*

β-galactosylceramide production. However, enzymes making the corresponding α-anomers are not known to date. Could it be possible that α-anomeric glycolipids are produced by unknown enzymes or by the known enzymes that synthesize them in an unfaithful manner?

Indeed, recent findings demonstrated the presence of mammalian α-glycosylceramides.[53] Using glycolipid-specific antibodies for immunoprecipitation of lipid extracts from mammalian cell lines, α-GalCer proved to exist in minute amounts. Moreover, degradative enzymes were involved in regulating the abundance of endogenous lipid antigen, with acid ceramidase cleaving first the fatty acid, and α-galactosidase A subsequently removing the sugar head group from the sphingosine base of endogenous α-GalCer.[53] Furthermore, subsequent studies revealed the presence of α-glucosylceramide (α-GlcCer) in cow's milk.[54,55] Recent progress in analytical methods introduced cryogenic infrared spectroscopy for differentiating anomeric glycolipids as well as identifying different sphingosine bases and lipid moiety backbones. Notably, this technique was able to detect endogenous α-GlcCer that accumulated in spleens from mice lacking its degrading enzyme, acid α-glucosidase.[56] Taken together, our current understanding of the realm of intrinsic iNKT cell antigens is that α-anomeric GalCer and GlcCer play vital roles. Thus, the initial marine sponge-derived ligands represent also our endogenous lipid antigens after all (so, we are probably more similar to sponges than expected) (Table 36.1). However, the enzymatic pathways that produce endogenous α-GalCer and α-GlcCer still need to be elucidated and are the subject of current investigation.

Lipid Ligands and Regulation of Antigen Presentation

Even under conditions in which the lipid antigens are not precisely defined, external stimuli are nevertheless able to induce the production of CD1d self-ligands in antigen-presenting cells to subsequently activate iNKT cells. In this line, microbial products, such as LPS and DNA-containing CpG motifs, engaged TLR receptors to elicit the generation of self-lipids and IL-12 in DCs, which both were required to facilitate iNKT cell activation.[21,42,57] Moreover, TLR-9 triggering in DCs using CpG DNA induced stimulation of iNKT cells that was dependent on type I interferon and generation of ganglioside species.[58,59]

Furthermore, when antigen-presenting cells underwent ER stress, their lipidomic changes promoted iNKT cell induction.[60] Specifically, the inositol-requiring enzyme-1α (IRE1α) and protein kinase R-like ER kinase of the unfolded protein response were both required to drive the accumulation of neutral lipids.[61]

CD1d is also capable of binding various nonagonistic lipid ligands that fail to activate iNKT cells. Sphingomyelin has been shown to accumulate in acid sphingomyelinase deficiency and to consequently block antigen presentation by CD1d, thus suppressing iNKT cell-mediated inflammation.[62] Similarly, phospholipids like phosphatidic acid competed with agonistic lipid antigens for CD1d binding, which resulted in inhibition of iNKT cell functions.[63] Therefore, nonagonistic lipid ligands for CD1d can control iNKT cell-dependent immune responses.

With regard to antigen processing, lysosomal phospholipase A2 (PLA2) was required for efficient endogenous antigen presentation through CD1d,[64] suggesting a role for antigenic lysophospholipids whose generation needs cleavage by PLA2. Furthermore, proper lysosomal trafficking of lipid antigens for effective CD1d loading depended on the Arf-like GTPase Arl8b. Mechanistically, Arl8b recruited the homotypic fusion and vacuole protein sorting (HOPS) complex to the lysosome and thus facilitated cargo delivery and subsequent CD1d-mediated antigen presentation.[65]

Structural Biology of Lipid Presentation

Abundant work resolving crystal structures of glycolipid bound to CD1d and the ternary complexes involving the NKT cell receptors revealed the atomic anatomy of lipid presentation. Glycosphingolipids like α-GalCer seem predisposed to fit well into the antigen-binding pockets of CD1d, with the acyl chain buried in the A′-pocket and the sphingosine base binding along the F′-pocket.[66] This constellation exposes the sugar headgroup beyond the CD1d surface and in parallel to the plane of its α-helices for recognition by the iNKT cell receptor.

Elucidation of the ternary complex between human iNKT cell receptor, CD1d, and the lipid α-GalCer showed that the TCR docks onto the CD1d binding groove in a parallel fashion and at the extreme end over the F′-pocket[67] (Fig. 36.2). This parallel footprint is in contrast to classical peptide presentation by MHC molecules, in which the TCRs dock onto MHC in a diagonal manner and more centralized on the peptide-binding groove. The characteristic orientation of the iNKT cell receptor on the CD1d surface facilitates a lock-and-key interaction with the lipid antigen.[67] Specifically, the conserved Jα18 segment coding for CDR3α mediates most of the contacts, including the antigen-binding cleft, α1 and α2 helices of CD1d, and α-GalCer itself. CDR1α interacts exclusively with α-GalCer, while CDR2β binds mainly to the CD1d α1 helix.

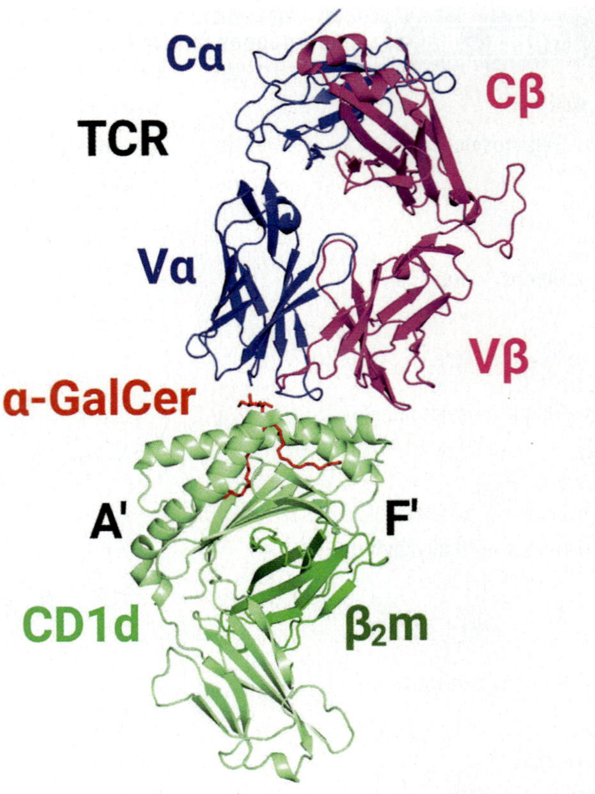

FIG. 36.2. Crystal structure of the iNKT cell receptor recognizing lipid presented by CD1d. CD1d displays α-GalCer with its acyl chain buried in the A′ pocket and the sphingosine in the F′ pocket. The sugar head group is exposed for engagement by the human iNKT cell receptor. Notably, the T cell receptor (TCR) binds in a parallel fashion and at the extreme end over the F′-pocket, facilitating a lock-and-key type interaction with α-GalCer. β_2m, β_2-microglobulin.

Structural studies on further microbial antigens and self-ligands explored a conserved binding mode of the iNKT cell receptor to CD1d/lipid complexes. In principle, diverse structural orientations of lipid ligands bound to CD1d undergo three-dimensional changes when the iTCR docks onto the CD1/lipid complex, in a way that the overall configuration resembles the binding mode to α-GalCer. In this line, the α-anomeric glycosyldiacylglycerol antigens from *Borrelia* and *Streptococcus* exhibited a configuration comparable to α-GalCer, with an additional induced fit upon TCR binding.[68] Likewise, β-anomeric self-glycolipids, such as β-GalCer, were recognized by iNKT cells when the TCR bended the β-linked head groups, basically approximating the configuration of α-GalCer in a mechanism of induced-fit molecular mimicry.[69]

Beyond glycolipids, the presentation of other entities, such as phospholipids, has been structurally resolved. In this context, the ternary crystal structure of a human iNKT cell receptor in complex with CD1d presenting LPC revealed that the α1 helix of CD1d was required to shift in order to accommodate LPC in an open A′-pocket.[70] The binding of the TCR then mediated a displacement of the LPC headgroup to allow for closed conformation of the whole complex, yet another example of induced fit in antigen recognition.

Beyond iNKT cells, other semi-invariant NKT cells are able to recognize α-GalCer presented by CD1d. T cells expressing Vα10-Jα50 responded to α-GalCer but showed a stronger reactivity to α-GlcCer.[71] Although the sequence similarity of the TCR α-chain was low when compared to iNKT cells, the crystal structure revealed a similar binding mode like the iTCR to α-GalCer. Moreover, in the human system, NKT cells with noninvariant TCRs and T lymphocytes harboring TCR hybrids (containing Vδ1 fused to Jα and Cα coupled with TCR β) have been structurally described to recognize α-GalCer/CD1d complexes in an atypical fashion, namely docking onto CD1d orthogonally and over the A′ hydrophobic pocket.[72,73] This indicates that the universe of lipid recognition modes is more diverse than the conserved binding mechanism of iNKT cells. On the other hand, considering that microbial lipid antigens assume the conformation of α-GalCer, semi-invariant NKT cells recognize lipids in similar structural orientation like α-GalCer, and β-linked lipids are shaped by the docking iNKT cell receptor to render them structurally similar to α-anomeric ligands; this might indicate that the archetypical ligand α-GalCer also represents the physiological binding mechanism of iNKT cell antigens.

CD1 AND EVOLUTION

The family of CD1 proteins is widely distributed in mammalian species although the size and composition of the CD1 gene clusters can vary substantially.[74] Accordingly, one can find CD1 genes in a broad spectrum of mammals, such as microbats, horses, and elephants. Interestingly, while not necessarily all five CD1 isoforms are present in each species, certain animals express multiple copies of specific CD1 molecules.[74] The discovery of two CD1 genes in chicken demonstrated that CD1 is evolutionarily old and must have been present in the last common ancestor of birds and mammals over 300 million years ago.[75,76] Notably, chicken CD1.1 and CD1.2 mapped to the MHC, suggesting that CD1 molecules diverged from classical MHC genes. The crystal structure of chicken CD1.2 revealed a narrow hydrophobic binding groove that could only fit a single lipid tail,[77] which is in sharp contrast to the deeper A′ and F′ pockets in mammalian CD1. Therefore, the shape of chicken CD1.2 has been considered as primordial.[77] However, the structural biology of the other chicken CD1 molecule (CD1.1) showed two deep binding pockets that were separated from each other but connected to a hydrophobic surface cleft.[78] Moreover, chicken CD1.1 was able to harbor long lipids, such as mycolic acid, as well as glycolipids equipped with two hydrophobic tails. Analysis of CD1.1 protein expression in chicken demonstrated a high abundance in spleen and the bursa of Fabricius.[79] CD1.1-expressing cells included splenic B cells, Langerhans cell-like cells in the skin, and medullary thymus. Furthermore, CD1.1 molecules colocalized with lysosomal markers, indicating an endosomal trafficking pattern.[79] Overall, the tissue localization in chicken proved to be strikingly similar to the broad distribution of CD1 molecules in mammals. Beyond mammals and birds, CD1 genes were also found in reptiles, such as the green anole lizard and crocodiles.[80] CD1 was located on the same chromosome like the MHC in the lizard, whereas in crocodiles, CD1 genes mapped to two loci, linked to the MHC and an MHC paralogous region on a different chromosome. Based on their existence in mammals, birds, and reptiles, several models have been proposed to explain the origin and evolution of CD1 molecules.[81] One possibility describes the presence of CD1 in the primordial MHC, prior to the two rounds of whole-genome duplications (2R) early in vertebrate evolution. Following 2R, four MHC paralogous regions are generated corresponding to human chromosomes 1, 6, 9, and 19. Subsequently, differential silencing of MHC and CD1 genes located on separate chromosomes could explain their different locations in various species. Alternatively, CD1 might have developed after 2R by tandem duplication of MHC class I genes and subsequent neofunctionalization, followed by chromosomal translocation and differential gene silencing. A recent hypothesis integrates both models into a unifying scenario,[82] in which CD1 is initially generated by tandem duplication and neofunctionalization of MHC class I/II precursor genes in the MHC, prior to 2R. Subsequently, a part of the MHC region is translocated on the same chromosome (MHC *trans*), followed by differential gene silencing that keeps MHC class I/II genes inside the MHC and CD1 in the translocated MHC region. Eventually, this MHC *trans* is further translocated to a different chromosome, which is by coincidence an MHC paralogous region (like on human chromosome 1).

Sharks are cartilaginous fish and represent the most ancient group of vertebrates that harbors MHC and other important immune genes pertaining to adaptive immunity. Although bona fide CD1 genes have not been described in cartilaginous fish, sharks do express nonclassical MHC-I genes, whose degree of divergence to MHC-I molecules is reminiscent of mammalian CD1.[83] In fact, sharks possess very large gene families of nonclassical MHC-I molecules, which are predicted to express rather hydrophobic ligand-binding regions,[84] suggesting a possible role in lipid presentation comparable to the function of CD1 proteins. Furthermore, such complex nonclassical MHC-I gene organization can be found in the amphibian *Xenopus* (XNC genes).[85]

Notably, tetramers built of XNC-10 molecules were able to detect a Vα6⁺ invariant T cell population in the frog *Xenopus laevis*.[86] These Vα6⁺ T lymphocytes appeared already early during ontogeny as they were present at the tadpole stage. Deficiency in XNC-10 led to reduced Vα6⁺ T cell numbers, which caused increased susceptibility of tadpoles to frog virus (FV) 3 infection.[86] These findings suggested that the system comprising nonclassical MHC-I molecules and invariant T cells is critical to protect the host against infection when adaptive immunity is not yet fully developed. Moreover, infection of adult frogs with FV3 triggered rapid activation of Vα6$^+$ T lymphocytes that responded in a similar fashion like mammalian iNKT cells, quickly producing cytokines, activating myeloid cells, and controlling viral burden and associated tissue pathology.[87] Interestingly, in tadpoles, the TCRα repertoire proved to be limited and showed some dominantly used invariant TCRα chains, indicating the presence of further innate-like T cell populations. In this context, invariant Vα45$^+$ T cells were shown to be restricted by the nonclassical MHC molecule XNC-4 and exhibited important functions in mycobacterial infection.[88] Accordingly, tadpoles lacking XNC-4 and/or Vα45$^+$ T lymphocytes demonstrated less survival when infected with *Mycobacterium marinum*, whereas their protection against FV3 infection was not affected, highlighting specific functions of diverse invariant T cell subsets. Whereas adult frogs responded to mycobacterial infection based on inflammatory CD8 T cell responses and granuloma formation, in tadpoles, the immune response was mainly driven by invariant T cells, rather lacked granulomatous inflammation, and exhibited increased tolerance toward pathogen burden.[89] Lastly, using tumor models in the amphibian *Xenopus*, XNC-10 and invariant T cells also showed a crucial impact on regulating cancer growth.[90] Taken together, considering that a common ancestor of amphibians and mammals dates back more than 350 million years, frogs developed a surprisingly diverse system of nonclassical MHC-I molecules restricting the corresponding invariant T cells, which behave in a comparable manner as their mammalian iNKT cell counterparts. This emphasizes the prominent position of innate-like T lymphocytes in the genesis and evolution of the immune system.

DEVELOPMENT OF iNKT CELLS IN THE THYMUS

Critical for the induction of the iNKT cell developmental program is the expression of the invariant TCR. Its engagement of self-lipid antigens presented by CD1d expressed on cortical thymocytes triggers a gene expression program that facilitates subsequent differentiation of iNKT cells. Central to this trajectory is the master transcription factor PLZF (Promyelocytic Leukemia Zinc Finger) that allows for lineage specification. Following their positive selection by thymocytes in the cortex, iNKT cell precursors migrate into the thymic medulla for further expansion and generation of different functional subsets, prior to their exit and seeding of peripheral tissues.

Positive and Negative Selection

The advent of α-GalCer–loaded CD1d tetramers allowed the specific tracking of iNKT cells and revealed the intricacies of their thymic development. Initially, like conventional T lymphocytes, mainstream thymocyte precursors transition through the pre-TCR$\alpha\beta^+$ stage to become CD4$^+$CD8$^+$ double-positive (DP) thymocytes. At the DP stage occur random Vα-Jα rearrangements that form the canonical iNKT cell receptor, consisting of Vα14-Jα18 coupled to Vβ8, Vβ7, or Vβ2.[91-93] Since these rearrangements combine distant gene segments, the recombination process is supported by the transcription factor HEB, which induces Bcl-xL to allow for prolonged thymocyte survival.[94] Thus, the iNKT cell lineage is selected from DP thymocyte precursors that recognize self-lipid/CD1d complexes expressed on DP thymocytes[91-93] (Fig. 36.3). Although in principle, all Vβ families can pair with Vα14-Jα18, only the limited set of Vβs supports the recognition of the selecting endogenous ligands. While Vβ8 is most abundant, Vβ7 is most strongly enriched upon selection and shows the highest affinity for endogenous antigens.[95]

In addition to positive selection, there is also some evidence for the existence of negative selection of iNKT cells. Accordingly, transgenic overexpression of CD1d on dendritic cells[96] or expression of a TCR-β chain that confers high affinity for self-lipid/CD1d complexes[97] led to reduced iNKT cell frequencies. In the latter scenario, iNKT cells that avoided negative selection exhibited natural sequence variants in the canonical TCR-α chain to reduce overall affinity to self.[97]

Although positive and negative selection are at play in iNKT cell development, it remains unclear which self-lipids are required for each process and how these two mechanisms are physically segregated in the microenvironment of the thymus, with possible different roles of cortex and medulla. The diverse functions of endogenous lipid antigens in positive versus negative selection are currently the subject of investigation.

Linear Maturation and Lineage Differentiation

There are currently two models available that aim at describing the developmental pathway of iNKT cells in the thymus. One focusses on the linear maturation of iNKT cells and describes sequential developmental stages.[98] The other centers around a common iNKT cell precursor that gives rise to three different lineages whose cytokine effector functions are determined by specific transcription factors.[99]

Following the maturation model,[98] recently selected DP cells that bear the iNKT cell receptor become CD4 single-positive, express CD24 (heat-stable antigen, HSA) characteristic of immature cortical thymocytes, and produce CD69, an activation marker and indicator of recent positive selection (stage 0). Subsequently, iNKT cells undergo a massive proliferation wave dependent on the transcription factor c-Myc[100] and downregulate CD24 (stage 1). Developing iNKT cells first induce expression of interleukin (IL)-4 (stage 1), followed by expression of interferon (IFN)-γ and upregulation of CD44 (stage 2).[101] Notably, those developmental stages of iNKT cells, including antigenic activation, expansion, and effector differentiation, strongly resemble the dynamics of classical T lymphocytes primed in peripheral lymphatic tissue. Stage 2 iNKT cells upregulate the receptor for sphingosine-1-phosphate that facilitates thymic exit and seeding of peripheral tissues, such as liver and spleen.[102] Subsequently, the cells express receptors typical for the NK cell lineage, such as the activating receptors NK1.1 and NKG2D, as well as the inhibitory

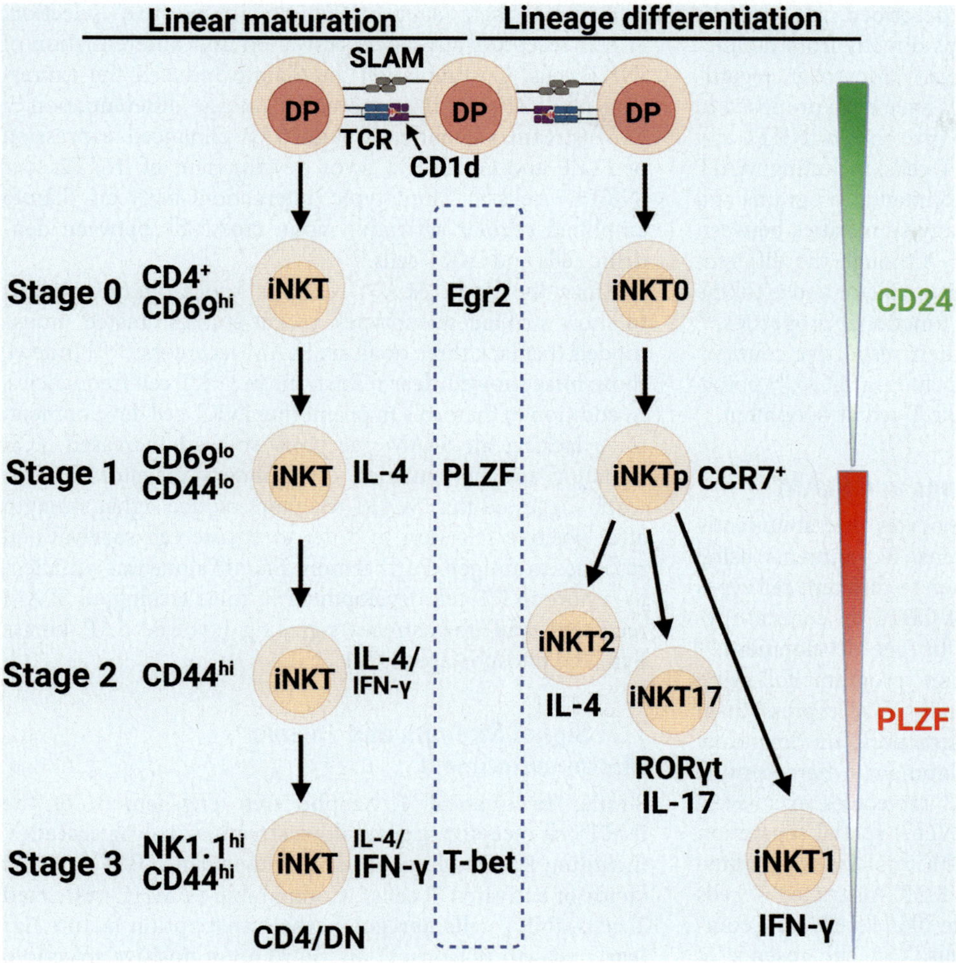

FIG. 36.3. Thymic development of iNKT cells. Shown is the integration of the two models explaining iNKT cell generation. *Left:* The linear maturation model describes iNKT cell development following the depicted stages. *Right:* The lineage differentiation model shows a common iNKT cell precursor that gives rise to separate iNKT cell subsets, including iNKT1, iNKT2, and iNKT17 cells. Key transcription factors are encircled by the dashed line. DP, double-positive thymocytes.

receptors CD94/NKG2A and members of the Ly49 family. This terminal differentiation program is dependent on IL-15 signaling and regulated by the transcription factor T-bet.[103] Alternatively, iNKT cells can also remain in the thymus and undergo terminal maturation as permanent residents in the medulla.

According to the lineage differentiation concept,[99] freshly selected DP thymocytes expressing the invariant T cell receptor develop into iNKT cell precursors characterized by the chemokine receptor CCR7.[104] This common progenitor is able to give rise to three different iNKT cell subsets, called iNKT1, iNKT2, and iNKT17, each characterized by different transcription factors and cytokine profiles. Accordingly, iNKT2 cells express the highest level of PLZF and predominantly produce IL-4, whereas iNKT17 cells show medium levels of PLZF and express the transcription factor RORγt to allow for IL-17 production. By contrast, the iNKT1 cell subset is low in PLZF and mainly expresses IFN-γ under control of the transcription factor T-bet.

Current studies exploring iNKT cell biology often analyze iNKT cells with regard to stages as well as specific subsets. Therefore, we aimed to combine both developmental models in a unifying synopsis. The parallel depiction of the two pathways shows correspondent features at the different levels (Fig. 36.3).

The abundance of the different iNKT cell subsets varies between diverse experimental mouse strains, with the most widely used C57BL/6 mice mainly harboring iNKT1 cells. However, Balb/c mice exhibit an increased frequency of iNKT2 cells that display a steady-state production of IL-4. Notably, this iNKT2-derived IL-4 mediates the generation of MHC-I–restricted CD8 T cells that show a memory-like phenotype.[99] Specifically, those CD8 T cells express the memory marker CD44, the IL-2 receptor β-chain (CD122), and the transcription factor eomesodermin, and they are rapidly secreting cytokines without prior antigenic priming.[105] Several mouse strains that have mutations associated with TCR signaling, including the Tec kinases Itk and Rlk,[106] the inhibitor of E proteins Id3,[107] and the transcription factor KLF2,[105] show a similar presence of memory-like CD8 T cells based on increased frequencies of IL-4–producing iNKT2 cells (or alternatively PLZF-expressing γδ T cells). Taken together, the abundance and cytokine production by iNKT cells has a strong impact on the development of MHC-I–restricted CD8 T cells and underlines the important crosstalk between innate-like and adaptive T cells in the thymus.

Recent genomic, epigenetic, and single-cell analyses corroborated the lineage specification into iNKT cell subsets.[108-112] Like often in biology, different findings from different groups can lead to different conclusions. In this spirit, an alternative developmental pathway has been described, in which a minor fraction of iNKT cells develops directly from double-negative cells, bypassing the DP stage.[113] Moreover, regarding subset differentiation, iNKT2 cells have been proposed to represent precursors that are able to give rise to iNKT1 and iNKT17 cells.[111] Overall, innate-like T cells, including MAIT and γδ T cells, show common developmental programs and shared differentiation paths with many similarities between those PLZF-expressing lineages.[110,112] Although the different subsets clearly differ in their transcription factor and cytokine make-up, and consequently in their functional properties,[108] iNKT1, iNKT2, and iNKT17 have their respective counterparts among other innate lineages (including ILCs),[109] pointing to a general principle of innate-like T cell development.

Homotypic Thymocyte Interactions and SLAM

iNKT cells are selected by DP thymocytes that abundantly express CD1d to present self-antigens. Experiments using genetic tools to target CD1d expression to different cell types revealed that exclusive expression of CD1d on cortical thymocytes is sufficient for iNKT cell lineage development.[114] Notably, the terminal differentiation program following thymic emigration required additional CD1d expression in peripheral tissues. These cellular interactions are drastically different from classical T cell development, where peptide self-ligands are presented on MHC molecules exclusively expressed on thymic epithelial cells. This spatial separation seems to be critical, because when MHC molecules are transgenically expressed on thymocytes, MHC-restricted T cells develop into PLZF-expressing innate-like T cells and compete with iNKT cells and MAIT cells.[115,116] The absence of MHC/peptide complexes on thymocytes facilitates the generation of innate-like T cells. Thus, the exclusive expression of CD1d on thymocytes creates a specific niche for iNKT cell development.

The thymocyte-thymocyte interactions are supported by the homophilic binding of a family of receptors called SLAM (Signaling Lymphocytic Activation Molecule).[117] The SLAM family contains six members, with Slamf1 and Slamf6 (also called Ly108) having a prominent role in iNKT cell development.[117] Downstream signaling is mediated by the adaptor SLAM-associated protein (SAP) and the kinase Fyn. Consequently, SAP- and Fyn-deficient mice showed hampered iNKT cell generation, characterized by a very early developmental arrest that suggests that the SLAM/SAP/Fyn axis supports positive selection.[118,119] In an experimental system in which iNKT precursors lacking CD1d and Slamf1 were selected by Slamf6-deficient thymocytes, iNKT cell numbers were drastically reduced, indicating that SLAM receptors are crucial when the iNKT cell receptor engages its CD1d/lipid ligands.[117] Moreover, the SLAM family receptor genetic locus exhibits pronounced polymorphism, which might explain different support for positive selection and ultimately iNKT cell numbers in various mouse strains. For example, the nonobese diabetic (NOD) strain lacks Slamf1 expression in thymocytes and exhibits iNKT cell deficiency.[120]

Beyond their supportive role in positive selection, SLAM receptors also affect activation and differentiation of iNKT cells. Costimulation via Slamf6 induced the expression of PLZF and thus promoted lineage differentiation.[121] Downstream signaling through SAP enhanced expression of PLZF and GATA3 to favor development of iNKT2 over iNKT17 cells.[122] Homotypic interaction between Slamf6 amplified cellular activation upon cross-talk between dendritic cells and iNKT cells.[123]

Since the single SLAM receptor knock-out mice failed to show striking phenotypes, recent studies created mouse models that lack three or all six SLAM receptors.[124,125] Indeed, those mice showed clear reductions in iNKT cell frequencies. In addition to their role in promoting iNKT cell development, mice lacking all SLAM receptors showed increased TCR signaling and upregulation of inhibitory receptors.[125] This work suggested that SLAM receptors reduce signal strength after positive selection in order to ensure cell survival and response to antigen. Furthermore, Slamf6 alone was sufficient to restore iNKT cell development in mice lacking all SLAM receptors, and downstream signaling involved SAP, kinase Fyn, and phosphatase SHP-1.[125]

TCR Signal Strength and Thymic Microenvironment

Similar to classical T lymphocytes, engagement of the iNKT cell receptor triggers downstream signaling cascades, including Ras/MAP kinase and calcineurin/NFAT (nuclear factor of activated T cells). Comparable to MHC-restricted T cells, iNKT cells upregulate the transcription factors Egr (early growth response)1 and Egr2 upon positive selection, with Egr1 positioned in the Ras/MAP kinase pathway and Egr2 mainly induced by NFAT.[126,127] However, iNKT cells display an elevated and sustained expression of Egr1 and Egr2 that is central to lineage differentiation.[127] Especially Egr2 directly binds to the promoters of lineage-specific genes like *Zbtb16*, coding for PLZF, and *IL2rb* that encodes the β-chain of the IL-2 receptor that also facilitates IL-15 binding.[127] Thus, Egr2-mediated gene expression controls early (PLZF) as well as late (IL-15 signaling) stages of iNKT cell differentiation.

The strong induction of vital regulators, such as Egr2 and PLZF, is dependent on TCR signaling, and therefore, TCR signal strength emerges as critical determinant of iNKT cell lineage regulation. Nur77 is a transcription factor that integrates TCR signaling in lymphocytes and also plays a role in iNKT cell differentiation and effector functions.[128] Since its expression intensity correlates with TCR signal strength, Nur77 levels can be used as reporter for the magnitude of TCR engagement.[129] Indeed, thymic iNKT cells show higher Nur77 expression than MHC-restricted T lymphocytes, reflecting strong TCR signaling. Furthermore, emerging evidence demonstrates that differentiation of the various iNKT subsets depends on TCR density.[130,131] In this context, iNKT subsets showed different avidity for α-GalCer–loaded CD1d tetramers, with iNKT2 cells as the strongest binders, followed by iNKT17 cells in a medium range, and iNKT1 cells staining with tetramers in the weakest manner.[131] This phenomenon correlated well with the strength in Egr2 induction, again with iNKT2 cells showing the highest expression, iNKT17 in the middle, and iNKT1 cells with the lowest Egr2 abundance. Moreover, this TCR signal strength-induced differentiation mechanism was also represented on the genomic level, where iNKT2 cells exhibited the most pronounced chromatin accessibility of NFAT- and Egr2-binding elements.[131] Notably, mice expressing a mutant ZAP70, an important factor in TCR downstream signaling, showed a shift to iNKT1 cells, suggesting that a loss in TCR signal strength causes iNKT cell amplification.[132] Interestingly, this altered thymic differentiation had an impact on disease outcome, as an increase in iNKT1 cells proved to be protective in an arthritis model.

After receiving a strong TCR stimulus upon positive selection in the thymic cortex, iNKT cells subsequently migrate to the medulla, where they interact with further cell types to affect their lineage commitment and/or survival.[133] Medullary thymic epithelial cells (mTECs) come in different flavors, and their surprising heterogeneity is involved in stimulation of the different iNKT cell subsets. Thymic tuft cells, a cell type usually located at mucosal barriers and characterized by the expression of taste receptors, express the protein kinase DCLK1 as marker as well as IL-25 to promote the expansion of iNKT2 cells.[134] Additionally, CD11c$^+$ macrophages have been described to activate medullary iNKT2 cells and to provide strong CD1d-mediated stimulation.[135] Moreover, mTECs expressing CD104 (integrin β4) and CCL21 (the ligand for chemokine receptor CCR7 produced by iNKT precursors) are capable of IL-15 transpresentation to stimulate iNKT1 and to a lesser extent iNKT17 cells.[136] Both, tuft cells and CD104$^+$CCL21$^+$ mTECs, express low levels of MHC-II and the costimulatory molecule CD80 and are dependent on lymphotoxin β receptor for their development.[136] Furthermore, the presence of iNKT cells promotes the development of mTECs,[137] and mTECs in turn can also influence iNKT cell-mediated dendritic cell activation,[136] indicating that different antigen-presenting cells are functionally connected in the thymus. Furthermore, the thymic residence of iNKT2 cells is ensured by CD69 expression on iNKT cell precursors, which suppresses sphingosine-1-phosphate expression and thus precludes a precocious thymic exit.[138] On the other hand, iNKT2 cells secrete IL-4 and IL-13 to coordinate thymic emigration of classical MHC-restricted T lymphocytes.[139] Taken together, the microenvironment of the thymus, including cortex, medulla, and their different cell types, emerges as a sophisticated network for iNKT cell development. There appears to be a division of labor, with positive selection in the cortex based on lipid presentation by DP thymocytes, and activation in the medulla, a process involving various cell types that potentially also affect lineage differentiation, survival, and additional mechanisms of selection. However, it remains currently unclear which CD1d-presented lipid antigens facilitate those different processes. Ongoing investigations try to elucidate the different self-lipid entities that govern the differential functions of thymic cortex and medulla in the development of iNKT cells.

The Master Regulator PLZF

The protein PLZF, encoded by the *Zbtb16* gene, is the master transcription factor of the iNKT cell lineage.[140,141] Of note, it is also expressed in other innate-like T lymphocytes, including γδ T and MAIT cells, as well as ILCs. PLZF is a member of the BTB domain-containing zinc finger proteins, in which the BTB domain homodimerizes and binds a corepressor complex containing NCoR (nuclear receptor corepressor), and the zinc fingers mediate DNA binding. PLZF is induced in developing iNKT cells early after TCR engagement. The iNKT cell subsets have different levels of PLZF abundance, with iNKT2 cells showing the highest expression, which is then descending via iNKT17 to iNKT1 cells. Mice deficient for PLZF showed an early block in iNKT cell generation.[140,141] Most notably, PLZF-deficient iNKT cells failed to acquire their typical effector functions. By contrast, they rather behaved like classical naïve MHC-restricted T lymphocytes, recirculating between blood and lymphatic tissue instead of homing to the liver and producing IL-2 rather than IFN-γ and IL-4 following antigenic stimulation. Strikingly, ectopic expression of PLZF in classical CD4 T cells triggered the characteristic iNKT effector program, including upregulation of CD44, homing to liver and lung, and dual production of IFN-γ and IL-4.[142,143] Thus, while PLZF is induced in naturally developing iNKT cells upon TCR engagement, PLZF protein alone is able to turn on the cell effector program independently of iNKT cell-related signals.

PLZF represents a specific trait of innate-like T cells, because its expression is stably suppressed in classical T cells and not inducible by TCR signaling as well as inflammatory cues.[144] Moreover, PLZF exhibits dosage effects on iNKT cell development since 50% reduction in PLZF protein expression cut the number of generated iNKT cells in half.[145] In addition, this dosage effect had an impact on lineage differentiation, with reduced iNKT2 and iNKT17 subsets and biased development toward iNKT1 cells. Global genomic analyses investigated the genes that are bound and regulated by PLZF using ChIP-Seq.[146] This approach revealed an astonishingly complex gene network controlled by a single transcription factor. Accordingly, PLZF bound genes coding for cytokine receptors, including the ones for IL-12 and IL-18. Furthermore, several transcription factor genes involved in T helper cell differentiation were bound by PLZF, such as the loci coding for c-Maf, T-bet, GATA-3, and RORγt.[146] Interestingly, PLZF did not directly control the effector cytokine genes for IFN-γ, IL-4, and IL-17. However, PLZF suppressed the transcription of Bach2, a powerful repressor of effector differentiation in naïve T lymphocytes, thus facilitating the establishment of the iNKT cell signature program.[146] Finally, exploration of the PLZF enhancer landscape identified a Runx1-bound enhancer element critical for the kinetic regulation of PLZF expression and the linked generation of iNKT cells.[147]

Transcription Factors

The transcription factor T-bet is crucial in many cell types, such as memory CD8 T cells and natural killer cells. In iNKT cell development, T-bet is relatively late expressed and mediates the terminal maturation program.[103] Concomitant with decrease in PLZF, T-bet allows for the induction of IL-2 receptor β-chain (CD122), which enables cellular IL-15 signaling. Thus, T-bet is essential for differentiation of iNKT1 cells.

Mice lacking GATA-3 showed a complex phenotype, with reduction of iNKT cell numbers mainly affecting peripheral tissues, such as liver and spleen.[148] By contrast, thymic iNKT cells were not reduced and exhibited normal maturation toward the $CD44^+NK1.1^+$ stage. Moreover, iNKT cells from GATA-3–deficient animals were not able to produce Th2 cytokines, such as IL-4 and IL-13, while their propensity to produce IFN-γ was preserved.[148] From the perspective of the lineage differentiation model, the effects of GATA-3 deficiency can be explained as mainly impacting the iNKT2 cell subset, leading to lack in Th2 cytokine production and maintenance of iNKT1 cell maturation. Furthermore, since GATA-3 induces ThPOK, which is required for CD4 expression, iNKT cells in GATA-3–deficient mice were devoid of CD4 molecules.

c-Myc is a proto-oncogene and important regulator of cell proliferation. Early in iNKT cell generation, c-Myc is induced to drive an extensive wave of cell division.[100] Thus, c-Myc is central to facilitate cell cycle entry of iNKT cells following positive selection.

NF-κB has many different functions in various cell types and signals downstream of external triggers like cytokines or TCR activation. The versatility of NF-κB is largely due to the fact that this group of transcription factors contains five different proteins, including NF-κB1 (p105/p50) and NF-κB2 (p100/p52) in different dimer combinations with RelA (p65), RelB, and c-Rel. In the canonical pathway, the enzyme IκB kinase (IKK) becomes activated to phosphorylate the inhibitor IκBα, which subsequently releases the RelA/p50 dimer for nuclear translocation to regulate transcription. By contrast, in the noncanonical pathway, NF-κB–inducing kinase (NIK) leads to activation of IKKα, subsequent processing of p100 and translocation of the RelB/p52 complex to the nucleus. The study of mice lacking IKKβ in T cells revealed a complete block of iNKT cell development, indicating an essential function of the canonical NF-κB pathway.[149] Considering that activation of NF-κB downstream of TCR signaling involves protein kinase C theta (PKC-θ) and Bcl10, the respective gene-deficient mice showed that PKC-θ is required for thymic development and Bcl10 for peripheral homeostasis of iNKT cells.[149] Moreover, reduction of iNKT cells was largely based on increased apoptosis in the absence of NF-κB.[150] However, when mice were tested that lacked a single protein of the NF-κB family, such as NF-κB1, RelA, or c-Rel, iNKT cell development proved to be normal, highlighting the redundancy in the NF-κB system.[151] The single factors had though an effect on late-stage maturation and cytokine function (except for IL-17 production). Lastly, the noncanonical NF-κB pathway involved NIK-mediated activation of RelB in thymic stromal cells to promote the development of early iNKT cell precursors.[152]

Smad molecules are important transcriptional regulators in iNKT cell development located downstream of TGF-β signaling. Upon ligand binding, TGF-β receptor recruits signal transducers called Smads and phosphorylates them. In turn, receptor-regulated Smads, such as Smad2 and 3, dissociate from the receptor, and associate with co-Smad4 to form a trimeric complex. Subsequently, the trimer translocates to the nucleus and acts as a transcriptional regulator through interaction with Smad-responsive elements. Following an alternative pathway, the transcriptional intermediary factor Tif-1γ competes with Smad4 for binding to Smad2/3, thus mediating Smad4-independent effects on gene expression. Initial studies using mice lacking TGF-β receptor II in T cells demonstrated that TGF-β signaling is essential for iNKT cell generation in the thymus, whereas the development of classical T lymphocytes proved to be largely normal.[153,154] Follow-up work dissected the TGF-β signaling cascade further and investigated mice lacking Tif-1γ or Smad4.[155] These experiments showed that Tif-1γ maintains IL-7 receptor expression and prevents apoptosis early in iNKT cell development to allow for lineage expansion. By contrast, Smad4 helped to maintain T-bet and IL-2 receptor β expression to promote lineage maturation.[155]

Th-POK is the master transcription factor of the CD4 T helper cell lineage, is induced by GATA-3, and is required to facilitate CD4 and to repress CD8 expression in iNKT cells.[156] Therefore, Th-POK expression explains the absence of $CD8^+$ iNKT cells in the murine system rather than negative selection of CD8-expressing cells. Furthermore, Th-POK–deficient iNKT cells showed impaired functional maturity based on reduced responses to antigen and less production of effector cytokines, such as IFN-γ and IL-4. In sharp contrast, Th-POK repressed the differentiation of iNKT17 cells and IL-17 production.[157,158]

The absence of the transcription factor Bcl11b led to the lack of iNKT cells in thymus and periphery due to impaired self-glycolipid processing and presentation in DP thymocytes.[159] The deletion of Bcl11b in a PLZF-dependent fashion following positive selection showed that this transcription factor favored the development of iNKT1 and iNKT2 cells, while suppressing the iNKT17 program.[159] The overall reduction of iNKT cell frequencies was due to Bcl11b-regulated apoptosis.

Hobit is a transcriptional repressor homologous to Blimp-1 and specifically expressed in iNKT cells. Hobit-deficient mice exhibited reduced thymic iNKT cell frequency as well as maturation.[160] In peripheral tissues, the number of immature iNKT cells was rather increased. Functionally, Hobit repressed IFN-γ secretion but promoted granzyme B expression.[160] Taken together, similar to the functions of Blimp-1 in other cell types, Hobit controlled the abundance of mature iNKT cells and governed their effector functions.

The transcription factor Bcl6 plays important roles in immune cells like follicular helper T cells (Tfh) and germinal center B cells. With regard to innate-like T cells, Bcl6 deficiency led to lack of iNKT cells as well as MAIT cells.[161] For iNKT cell generation, Bcl6 was transiently expressed at an early stage of development after positive selection. Mechanistically, Bcl6 promoted the expression of PLZF and PLZF-targeted genes. The chromatin landscape regulated by Bcl6 favored suppression of a gene expression program characteristic of naïve, classical T lymphocytes, thus driving the effector program of innate-like T cells.[161]

MicroRNAs

MicroRNAs are small noncoding RNAs that regulate gene expression on a posttranscriptional level through RNA interference with protein-coding mRNAs. In the microRNA processing pathway, the RNAse III enzyme Dicer plays a central role in the generation of mature, functional microRNAs. Genetic deletion of Dicer in cortical thymocytes led to a severe reduction in iNKT cell numbers, and the few remaining cells showed impaired differentiation.[162] In a complementary study, Dicer had been deleted in hematopoietic cells, showing that in addition to intrinsic effects in iNKT cells, microRNA deficiency in other cell types also affects iNKT cell development extrinsically.[163] Follow-up studies revealed the miR-17~92 family clusters as specific microRNAs regulated in iNKT cells in a Dicer-dependent fashion.[164] Mechanistically, absence of those microRNAs mediated increased expression of TGF-β receptor II and associated TGF-β downstream signaling responsible for impaired iNKT cell differentiation. Of note, additional ablation of TGF-β receptor II in Dicer-KO mice rescued cellular maturation.[164]

Other groups focused on the function of miR-181, which is the most abundant microRNA in DP thymocytes.[165,166] Lack of miR-181 caused severe reduction in iNKT cell frequencies due to an early block in development. Functionally, miR-181 proved to be critical to promote a sufficiently strong TCR signal upon agonist selection.[165] Furthermore, miR-181 deficiency triggered an upregulation of the phosphatase PTEN, which reduced PI3-kinase activity and subsequent Akt activation. This mechanism affected metabolic fitness and early iNKT cell generation.[166]

NOD mice represent a mouse strain characterized by an increase in iNKT17 cells. Reduced expression of Th-POK facilitates the amplified iNKT17 program. Notably, microRNA-133b was found to be upregulated in NOD mice to suppress Th-POK expression, providing a microRNA-mediated mechanism for iNKT17 cell differentiation.[167]

Let-7 is the most abundant microRNA in the entire genome. In the thymus, let-7 specifically targets the mRNA for PLZF. In this context, the dynamic upregulation of let-7 during iNKT cell development ensures the subsequent downregulation of PLZF at later stages that promotes the differentiation of iNKT1 cells.[168] Accordingly, absence of let-7 caused sustained expression of PLZF and the subsequent accumulation of iNKT2 and iNKT17 cells. The induction of let-7 was induced by IL-15, vitamin D, and retinoic acid.[168] Overall, this represents an interesting example of a microRNA regulating a lineage-specific master transcription factor in the thymus.

EPIGENETICS

The field of epigenetics studies inherited gene expression and phenotypes that are independent of changes in the DNA sequence. Classical epigenetic regulation involves DNA methylation and a diverse set of histone modifications, including methylation, acetylation, and ubiquitination, for example. In this context, iNKT cell development and function is also epigenetically regulated and involves several enzymes and chromatin modifiers that either add or remove the respective epigenetic DNA or histone marks.

Among the different histone modifications, trimethylation of histone 3 at lysine 4 (H3K4me3) represents and active mark that usually facilitates gene expression, whereas H3K27me3 is a repressive mark that hinders transcription. Generation of H3K27me3 is mediated by the polycomb repressive complex 2 (PRC2) that contains the methyltransferase Ezh2 in association with the proteins Suz12 and Eed. Studies investigating histone marks revealed that the PLZF (*Zbtb16*) locus contains a bivalent promoter in DP thymocytes, harboring active H3K4me3 as well as repressive H3K27me3, and that the transition to the iNKT cell lineage was characterized by loss of the repressive mark.[169] This led to the idea that a chromatin-based mechanism is required to couple TCR specificity with lineage commitment. Accordingly, Ezh2-deficient mice showed less H3K27me3 in iNKT cells, which presumably led to increased PLZF expression. As a consequence, Ezh2-KO mice harbored increased numbers of iNKT cells mainly of the iNKT2 subset, and iNKT2-derived IL-4 promoted the generation of memory-like CD8 T cells in the thymus.[169] Notably, MHC-restricted, peptide-specific CD4 T cells exhibited PLZF expression and an innate-like phenotype in the absence of Ezh2, underlining the concept that the bivalent PLZF promoter ensures the developmental separation of iNKT cells and classical T lymphocytes. Furthermore, when the H3K27 demethylases UTX or Jmjd3 were lacking, iNKT cell generation was severely impaired, basically showing the opposite effects of Ezh2 deficiency.[169]

However, a separate study on Ezh2 provided an alternative explanation, showing a chromatin-independent effect in iNKT cell development.[170] Of note, mice lacking the other proteins of PRC2, such as Suz12 and Eed, exhibited strongly impaired iNKT cell numbers, suggesting an additional function of Ezh2. Indeed, PLZF turned out to be a substrate for Ezh2-mediated methylation, which subsequently led to its ubiquitination and proteasomal degradation.[170] Therefore, degradation of PLZF was hampered in the absence of Ezh2 and consequently triggered its increased abundance, associated with the described phenotype of amplified iNKT cell numbers and a shift to iNKT2 cells.[170]

Jarid2 is yet another component in a methyltransferase complex that allows for the generation of the repressive histone mark H3K9me3. In thymocytes, Jarid2 is induced by the strength of TCR signaling upon positive selection.[171] Functionally, Jarid2 binds to the *Zbtb16* locus, facilitates H3K9 trimethylation, and consequently suppresses PLZF expression. In turn, when Jarid2 is lacking, iNKT cell frequencies are increased with a bias toward IL-4–producing iNKT2 cells, a phenotype similar to Ezh2 deficiency.[171]

Several reports demonstrated that the histone demethylases UTX and Jmjd3, which remove the repressive mark H3K27me3, are absolutely required for iNKT cell development.[169,172,173] Deeper genomic analysis of UTX revealed multiple mechanisms of epigenetic regulation.[172] First, UTX binds to the promoter regions of key signature genes, such as

Tbx21 (T-bet) and *IL2rb*, and promotes H3K27 demethylation at those sites to allow for gene expression. Second, UTX physically interacts with central transcriptional regulators like PLZF as well as the AP-1 transcription factor JunB. Third, the chromatin accessibility of superenhancers is facilitated by UTX. Superenhancers are strong enhancer elements that are usually abundant in H3K27 acetylation and play critical roles in cell lineage specification. In this context, many key iNKT cell factors, including the PLZF and JunB gene, harbor superenhancers, underscoring their critical functions in iNKT cell identity.[172]

In addition to histone modifications, changes in DNA methylation have an important impact on gene expression. The family of TET proteins are dioxygenases that convert 5-methylcytosine in DNA to 5-hydroxymethylcytosine (5hmC) and other oxidation products, subsequently leading to DNA demethylation. Mice lacking Tet2 and Tet3 in DP thymocytes showed augmented iNKT cell numbers, with a reduction in iNKT1 cells and an amplified iNKT17 subset.[174] Interestingly, in wild-type iNKT cells, the gene bodies of the crucial transcription factors T-bet and Th-POK contained abundant 5hmC marks as a consequence of TET action. In Tet2/3-deficient animals, however, DNA methylation was increased and responsible for repressing T-bet and Th-POK.[174]

Epigenetic regulation can also affect other T cell functions, including TCR rearrangement and cytokine effector responses. The chromatin-modifying factor TRIM28 has been shown to bind to the TCRα enhancer to induce H3K4me3 and associated recruitment of the recombination-activating proteins Rag1 and Rag2.[175] Consequently, ablation of TRIM28 abrogated TCR rearrangement and iNKT cell development. Regarding cytokine regulation, iNKT cells are able to produce effector cytokines in response to IL-12 and IL-18 signaling, independently of TCR triggering. This innate responsiveness was mechanistically mediated by histone H4 acetylation at the locus for the IFN-γ gene.[176]

The genome organizer SATB1 is essential for T cell lineage specification in the thymus by binding and activation of regulatory elements of key transcription factors. In this context, SATB1 was shown to bind to enhancer elements of Th-POK as well as CD4. As a result, absence of SATB1 in thymocytes fully abolished iNKT cell generation.[177] Finally, modulation of histone ubiquitination is also able to alter iNKT cell development.[178] Accordingly, the deubiquitinating enzyme ubiquitin-specific peptidase (USP) 22 bound to the transcriptional coactivator called mediator complex subunit 1 (MED1) and concomitantly removed ubiquitin from histone H2A.[178] This deubiquitination step empowered MED1 to drive IL2rb and T-bet expression.

MICROBIOTA

Early studies used germ-free mice to investigate whether putative antigenic ligands from microbiota are required for iNKT cell development. Since the frequencies of iNKT cells in thymus, spleen, liver, and bone marrow were not altered in germ-free animals, this suggested that rather endogenous antigens are necessary for iNKT cell generation.[179] However, more recent experiments revealed that the microbiota do affect iNKT cell numbers in the gut and the lung, organs closely associated with microbial exposure. In this context, germ-free mice showed increased iNKT cell numbers in colonic tissue and lung early in life when compared to control (specific pathogen-free) animals, which subsequently led to pronounced morbidity when challenged using models of colitis and allergic lung inflammation.[180] Functionally, lack of microbes was associated with increased expression of the chemokine ligand CXCL16 in the gut and lung, potentially attracting iNKT cells to those mucosal sites. Notably, when germ-free mice were colonized with microbiota early in life, colitis and lung inflammation were prevented. In sharp contrast, colonization in adult mice had no effect on disease occurrence.[180] These results support ideas formulated in the hygiene hypothesis, which proposed that exposure to microbes early in life reduces the risk for allergic and other diseases later on. Follow-up work then demonstrated that *Bacteroides fragilis* among the microbiome was responsible for suppressing colonic iNKT cell expansion early in life in order to prevent subsequent colitis.[181] Interestingly, colonization with a *B. fragilis* species lacking sphingolipids failed to induce iNKT cell tolerance. Moreover, a specific glycosphingolipid isolated from *B. fragilis* was sufficient to inhibit iNKT cell activation and to serve as a treatment to prevent gut inflammation.[181]

The microbiota are abundant in the intestinal lumen and produce a lot of compounds that might affect host physiology. Among those molecules are potential lipid antigens that could influence iNKT cell functions. Indeed, *B. fragilis* in humans often occupies large parts of the intestinal community and is characterized by a membrane rich in sphingolipids. The creation of a mutant strain that lacks a key enzyme in the sphingolipid pathway facilitated comparative mass spectrometry analysis, which revealed a novel lipid species, namely an α-galactosylceramide.[182] This *B. fragilis*-specific α-GalCer potently stimulated iNKT cells. Furthermore, microbial α-GalCer species have been isolated from the large intestine of mice and showed quantitative changes upon induction of dysbiosis.[183] Taken together, the archetypical antigen α-GalCer is not only from sponge origin or endogenous in nature but also produced by the gut microflora, which likely has an impact on host immunity.

In addition to the effects of the microbiota on lipid presentation and T cell function, iNKT cells in turn also influence the gut flora.[184] Accordingly, iNKT cell activation regulated the IgA repertoire as well as the composition and compartmentalization of intestinal bacteria.[185] Consequently, the lack of CD1d on $CD11c^+$ dendritic cells and macrophages affected immune homeostasis and led to dysbiosis. Moreover, the altered microflora in iNKT cell–deficient mice predisposed animals to more pronounced inflammation in a chemically induced colitis model.[186]

METABOLISM

Adipose tissue is central to the regulation of metabolic homeostasis. Interestingly, although generally thought to be rare in the human system, iNKT cells are quite abundant in adipose tissue in humans as well as mice.[187] Animals lacking iNKT cells were prone to develop obesity upon a high-fat diet and showed metabolic syndrome, including glucose intolerance and insulin resistance. Transfer of iNKT cells into obese mice or treatment with the antigen α-GalCer prevented obesity, fatty liver, and insulin resistance via anti-inflammatory cytokine production by adipose iNKT cells.[187] Corroborating these results, a separate study demonstrated that iNKT cells in adipose tissue prevent insulin resistance also in lean mice.[188] The anti-inflammatory functions of adipose iNKT cells are based on their particular properties. Accordingly, they lack the master regulator PLZF but express the transcription factor E4BP4 (*Nfil3*) that facilitates IL-10 production.[189] IL-10 in turn activates anti-inflammatory macrophages of the M2 phenotype. Moreover, adipose iNKT cells produce IL-2 that drives the frequencies and suppressive functions of regulatory T cells (Treg).[189] With regard to weight control in obesity, activated adipose iNKT cells triggered a conversion from white fat toward brown fat tissue, which uses energy for heat production.[190] This thermogenic switch was mediated by induction of fibroblast growth factor 21 (FGF21) in fat tissue. Therapy with a glucagon-like peptide 1 (GLP-1) receptor agonist also required iNKT cell-mediated FGF21 induction to achieve weight loss.[190] Lastly, two adipose iNKT cell subsets were identified with complementary functions in adipose tissue homeostasis.[191] The first, characterized as NK1.1$^-$, mainly produced IL-10 due to activation of the ER stress response, involving IRE1α and XBP1, in response to free fatty acids. This subtype improved metabolic functions in obese animals. By contrast, NK1.1$^+$ iNKT cells largely secreted IFN-γ to promote NK cell–mediated killing of pathogenic macrophages in lean mice.[191] Taken together, these results demonstrate the importance of adipose iNKT cells to govern and maintain physiological metabolism.

However, in contrast to the abovementioned body of work, there are reports that showed that high-fat diet and iNKT cell activation with α-GalCer led to increased tissue inflammation and obesity-associated metabolic syndrome.[192,193] Furthermore, specific ablation of CD1d in adipocytes reduced iNKT cell activation in adipose tissue and improved high-fat diet-induced obesity and insulin resistance, pointing rather to a detrimental role of adipose iNKT cells in obesity and metabolic disease.[193] It is puzzling how such similar approaches can lead to opposite results.

Visceral fat contains fat-associated lymphoid clusters (FALCs), which are known to contain ILCs and support activation of innate T and B cells, such as B-1 cells. Notably, the generation of FALCs proved to be dependent on CD1d expression and iNKT cell activation, suggesting a role of iNKT cells in determining the immunological architecture of fat tissue.[194] Furthermore, the tissue tropism of adipose iNKT cells can also be influenced in the thymus. The partial disruption of a hydrophobic patch that is formed in the iNKT cell receptor upon pairing of TCR-α with TCR-β caused an accumulation of E4BP4$^+$ adipose iNKT cells upon thymic development.[195]

Autophagy is the organized process of degradation and recycling of cellular components. Initially, a targeted organelle, such as a mitochondrium, is isolated in a double-membrane vesicle called an autophagosome, which subsequently fuses with a lysosome for degradation. Autophagy is connected to metabolism since lack of nutrients and starvation are able to trigger autophagy in order to recover cellular energy and building blocks. The class III PI3-kinase called Vps34 is implicated in regulation of autophagy, and consequently, its specific ablation in T cells led to accumulation of cellular organelles and subsequent apoptosis.[196] While the development of classical T cells in the thymus was not affected, iNKT cells were virtually absent in Vps34-deficient mice. Furthermore, experiments deleting the critical autophagy genes Atg5 and Atg7 in T cells showed that iNKT cell development was severely impaired and arrested at an early stage.[197,198] Defects in autophagy did not alter CD1d expression or antigen presentation.[197] Mechanistically, accumulation of mitochondria and reactive oxygen species (ROS) in Atg7-deficient animals triggered apoptotic cell death.[198] Altogether, these results highlight that autophagy is essential for proper development of iNKT cells.

The mechanistic target of rapamycin (mTOR) is a protein kinase that is central to regulation of cell growth and metabolism. In principle, mTOR builds two complexes with different functions. In combination with Raptor and other binding partners, mTOR builds mTOR complex 1 (mTORC1), which has a multitude of functions in regulating cell growth, proliferation, transcription/translation, autophagy, and metabolism, leaving the impression that it is basically involved in all cell processes. By contrast, when mTOR partners with Rictor and other binding elements, it generates mTORC2 that has a much more restricted functional scope by mainly regulating insulin receptor signaling and the associated glucose metabolism. Mice lacking Raptor in T cells, and thus deficient for mTORC1, showed drastic impairment of iNKT cell development.[199,200] The few remaining iNKT cells exhibited a reduced response to antigen stimulation. Moreover, the nuclear localization of PLZF was hampered in iNKT cells lacking mTORC1.[199] Mice deficient in mTORC2 due to ablation of Rictor in T cells also showed a reduction of iNKT cells, although their loss was less extensive when compared to mTORC1 deficiency. Of note, RORγt and IL-17 expression were impaired and led to a drastic decrease in iNKT17 cells in Rictor-KO animals.[201] Interestingly, lack of the phosphatase PTEN increased mTORC2 activity and iNKT17 cell numbers, an effect abolished again in mice lacking Rictor in addition. Lastly, Fnip1 is an adaptor protein of AMP-activated protein kinase (AMPK), which is an energy-sensing kinase that becomes activated when ATP is low. Subsequently, AMPK stimulates glucose and fatty acid uptake and oxidation, triggers autophagy, and inhibits mTOR to limit energy consumption. Fnip1-deficient mice showed reduced iNKT cell numbers and failed to downregulate PLZF, which could explain their block in maturation.[202] Notably, mTOR activity was increased and ultimately predisposed iNKT cells to cell death in the context of low ATP availability. This study provides an example that increased mTOR activation can also negatively affect iNKT cell generation. Taken together, the data indicate that a fine balance of mTOR activation is critical to ensure metabolic homeostasis in iNKT cells.

Finally, the redox system is another regulatory entity that is tightly associated with metabolism, as various metabolic pathways influence oxidative phosphorylation. Compared to conventional CD4 T cells, iNKT cells showed less propensity for glucose uptake, and the following glucose metabolization mainly involved the pentose phosphate pathway and oxidative phosphorylation.[203] By contrast, an environment enriched in lactate rather inhibited iNKT cell functions. However, the enhanced oxidative phosphorylation characteristic for iNKT cells leads to byproducts of oxidative stress, namely the ROS superoxide and hydrogen peroxide. In this context, the redox system consisting of Keap1 and Nrf2 evolved in order to control ROS levels and avoid toxicity. Nrf2 is a master transcription factor to trigger the antioxidative response. However, under normal conditions, Nrf2 is sequestered in the cytosol by binding to Keap1. Upon oxidative stress, Nrf2 is released from Keap1 and translocates to the nucleus to regulate antioxidative gene expression. In mice lacking Keap1, iNKT cell development was perturbed and caused increased glucose uptake and reduced ROS levels.[204] Additional lack of Nrf2 restored redox balance, metabolism, and iNKT cell generation. Thus, a fine-tuned redox balance based on the Keap1/Nrf2 axis is crucial to maintain proper iNKT cell functions.

TISSUE iNKT CELLS

Following development in the thymus, iNKT cells join the bloodstream to migrate to peripheral lymphoid as well as nonlymphoid tissues. In the most widely used experimental mouse strain C57BL/6, iNKT cells represent 0.2% to 0.5% of T cells in the thymus, 2.5% in the spleen, up to 40% in the liver, 0.2% in blood and peripheral lymph nodes, 8% in bone marrow, and 5% of T cells in the lung. The different iNKT cell subsets show a certain tissue tropism, with iNKT1 cells abundant in thymus, liver, and spleen, iNKT2 cells mainly located in thymic medulla, blood, and mesenteric lymph node, and iNKT17 cells resident in skin, skin-draining lymph nodes, and lung.[133,205] These proportions exhibit interstrain variations, with Balb/c animals showing a relative increase in iNKT2 cells, and NOD mice harboring a larger fraction of iNKT17 cells.[133]

Interestingly, after performing thymectomy in adult mice, iNKT cells persist as long-lived cells in the periphery and show striking features of tissue residency.[206] In the liver, iNKT cells remain intravascular in steady-state and continuously patrol the liver sinusoids.[207] Only upon antigenic stimulation, crawling iNKT cells come to a halt. Liver iNKT cells express the chemokine receptor CXCR6 that binds to its ligand CXCL16 produced by liver sinusoidal endothelial cells. This interaction was vital for iNKT cell survival but dispensable for their hepatic accumulation and crawling behavior in liver sinusoids.[207] The use of parabiotic mice, where the vascular system of two individual animals is joined, revealed that iNKT cells do not recirculate.[208] The tissue-resident phenotype was mediated by interactions between LFA-1 and ICAM-1 and regulated by PLZF, since expression of PLZF in conventional CD4 T cells induced a similar intravascular surveillance program. Of note, the transcriptional signature that determines tissue residency was already acquired in the thymus, controlled by the transcription factors Hobit and Blimp1, and proved to be generally operative in other tissue-resident lymphocytes.[209,210]

The study of peripheral iNKT cells not only examines their effect on tissue function but also, vice versa, the possible imprint of tissue on iNKT cells. In this context, analysis of transcriptome and chromatin landscape demonstrated that each of the iNKT cell subsets (iNKT1, 2, and 17) was similar when compared between tissues.[211] Moreover, upon antigenic stimulation, chromatin and transcriptional profiles resembled rather follicular helper T cells or NK- or effector-like cells, suggesting that iNKT cell gene programs show a certain degree of plasticity.[211] Furthermore, it has been reported that iNKT cells in different lymphoid tissues show different TCR repertoires.[212] Modification of the TCR repertoire also pertained to the comparison between recent thymic emigrants and mature cells as well as iNKT cells after immunization. Thus, these findings suggest tissue-specific modulation of the iNKT cell receptor repertoire. Using somatic cell nuclear transfer to generate mice cloned from iNKT cell nuclei, this system demonstrated that monoclonal iNKT cells can give rise to all three iNKT subsets.[213] Notably, the tissue of origin was a stronger predictor of iNKT cell function than TCR specificity, as skin-draining lymph nodes were abundant in iNKT17 cells irrespective of the TCR expressed. Lastly, the exact anatomic localization of iNKT cells influences their cytokine responses, as intravascular iNKT1 cells responded to intravenous α-GalCer injection and iNKT2 cells in mesenteric lymph nodes rather to oral antigen challenge, for example.[133]

With regard to homeostatic regulation of T lymphocytes, antigen presentation and cytokine milieu are vital influences. Although CD1d-mediated selection in the thymus is central to their development, maintenance of peripheral iNKT cells is independent of CD1d and thus contact with self-antigens.[214,215] In addition to NK cells and memory CD8 T cells, iNKT cells proved to rely on IL-15 because lack of IL-15 or IL-15 receptor impaired their survival.[216] The provision of IL-15 is special since the cytokine is not directly binding to the IL-15 receptor on the iNKT cell surface but instead is transpresented by the IL-15 receptor α-chain on the surface of accessory cell types. Genetic ablation and tracing experiments revealed that mTECs present IL-15 to iNKT cells during development, and that dendritic cells as well as stromal cells in lymphoid tissues were required for IL-15–mediated homeostasis in the periphery.[217,218] Notably, it appeared that the iNKT cell subsets have different homeostatic requirements, as iNKT17 cells were independent of IL-15 but instead their survival was dramatically affected by lack of IL-7.[219]

In addition to the cytokine profiles characteristic of the described subsets, tissue iNKT cells also produce other cytokine flavors depending on the context. Accordingly, iNKT cells pretreated with α-GalCer behave like regulatory T cells and secrete IL-10 (iNKT10 cells).[220] These Treg-like cells show a distinct predilection for adipose tissue. Moreover, PLZF controls the effector cytokine program in iNKT cells and is assisted by the transcription factor Yin Yang 1 (YY1), which is essential to help PLZF deploy its full functions.[221] In this line, iNKT cells low in YY1 failed to express ThPOK and T-bet and were dedicated IL-10-producers with an impact on immunosuppression.[222]

When iNKT cells become activated in tissues, they are able to rapidly produce copious amounts of cytokines. This immediate response is facilitated by a preformed, constitutive mRNA pool that predisposes iNKT cells for rapid cytokine production and release.[101] Additionally, TCR engagement was shown to activate the ER stress sensor IRE1α and the downstream p38 MAP-kinase pathway in order to promote the stabilization of preformed cytokine mRNAs.[223] As a result, deletion of IRE1α in iNKT cells abrogated cytokine responses. With regard to IFN-γ secretion, the transcription factor Bhlhe40 cooperated with Tbx21 (T-bet) at the Ifng locus concerning promoter activation.[224] Interestingly, Bhlhe40 is one of the few examples of transcription factors whose deficiency does not affect iNKT cell development but rather shows functional consequences on cytokine production. Finally, addressing the killer in iNKT cells, tissue iNKT cells also need to exercise their cytotoxic functions for clearing infected or transformed target cells. The use of an inducible system for the ablation of the adaptor molecule SAP demonstrated that the interaction of SAP with the tyrosine kinase Fyn was required for iNKT cells to gain their cytotoxic properties, form lytic synapses, and kill T-cell and B-cell leukemia targets.[225]

ANERGY/TOLERANCE

Peripheral iNKT cells can exhibit a phenomenon of unresponsiveness to strong antigenic stimulation. In this context, systemic challenge with α-GalCer induced rapid activation of iNKT cells, which was accompanied by cytokine burst and parallel TCR downregulation, the latter rendering iNKT cells undetectable with CD1d tetramers.[226] Several days after stimulation, iNKT cells massively expanded, prior to a phase of contraction based on Bim-mediated apoptosis.[227] Subsequently, the contracted iNKT cell population proved to be anergic in response to α-GalCer rechallenge, characterized by a failure to produce IFN-γ. This state of anergy could be overcome by exogenous IL-2 administration or blockade of PD-1.[228] Furthermore, compared to naked α-GalCer, stimulation with α-GalCer–loaded dendritic cells (DCs) was able to trigger a sustained expansion and prolonged IFN-γ secretion by iNKT cells.[229]

Functionally, the strength of stimulation was crucial for inducing hyporesponsiveness, since lipid antigens of lower avidity or cytokine-mediated activation failed to induce iNKT cell anergy.[230] And although DCs and B cells have been implicated in the process, mice lacking DCs or B cells were still capable of causing iNKT cell hyporesponsiveness, hinting at a different bone marrow–derived cell type involved in anergy induction.[230] Mice lacking Id2, a negative regulator of the E protein transcription factors, harbored anergic iNKT cells that produced high amounts of PD-1 and IL-10.[231] Engagement of the TCR by cognate lipid antigen caused Id2 downregulation in wild-type iNKT cells, suggesting an Id2-mediated mechanism of iNKT cell anergy. Lastly, other molecules play a role in antigen- and TCR-independent regulation of iNKT cell responsiveness. The surface glycoprotein CD160, a receptor for herpes virus entry mediator (HVEM), controls an overreaching cytokine response. Accordingly, animals deficient for CD160 suffered from acute hepatic injury following systemic α-GalCer challenge due to overstimulated iNKT cells.[232] These results suggest CD160 as a novel checkpoint to control iNKT cell hyperactivation.

CROSS-TALK WITH OTHER CELL TYPES

iNKT cells interact with a multitude of other immune cells and are centrally positioned with regard to the regulation of innate as well as adaptive immunity. In this context, iNKT cells become activated by antigen-presenting cells, including dendritic cells (DCs) and B lymphocytes, following lipid presentation on CD1d. Activated iNKT cells in turn modulate DC and B cell functions, overall boosting adaptive immune responses, such as MHC-restricted CD8 and CD4 T cell activation as well as antibody production, respectively. In parallel, activated iNKT cells influence the downstream functions of innate immune cells like NK cells and neutrophils (Fig. 36.4).

Dendritic Cells

The exogenous administration of α-GalCer serves as a paradigm for the classical iNKT cell–mediated cascade of activation events. Initially, DCs stimulate iNKT cells via α-GalCer/CD1d presentation. Subsequently, activated iNKT cells produce copious amounts of cytokines and upregulate CD40-ligand, which engages CD40 on DCs. This leads to DC activation, which is reflected by IL-12 production and upregulation of MHC and costimulatory molecules (B7.1/B7.2).[233,234] The secreted IL-12 further stimulates iNKT cells, and MHC and B7 expression promote priming of peptide-specific CD8 and CD4 T cells (Fig. 36.4). With regard to activation of CD8 T cells reactive with extracellular antigen, also called cross-priming, it has been shown that iNKT cells license CD8α$^+$ DCs to produce the chemokine ligand CCL17 in order to attract naïve CCR4$^+$ CD8 T cells.[235] Furthermore, protein immunization in the presence of α-GalCer induced the full maturation of DCs, facilitating augmented CD8 and CD4 T cell immune responses.[236,237] These experiments revealed the extraordinary nature of α-GalCer to serve as a powerful adjuvant.

Several synthetic lipid antigen agonists have been produced with the goal to bias the cytokine response of iNKT cells toward IFN-γ or IL-4. A classic example is an α-GalCer variant with a truncated phytosphingosine chain (psC9:0), called OCH.[238] Experiments using OCH stimulation *in vivo* demonstrated preferential IL-4 production by iNKT cells, which allowed the suppression of CNS inflammation in a model of experimental autoimmune encephalomyelitis (EAE).[238] Various studies aimed at elucidating the underlying biology of the cytokine bias induced by the lipid variants. In this context, the lower TCR affinity of OCH has been suggested,[239] although other short or unsaturated variants displayed affinities comparable to α-GalCer. Moreover, analysis of CD1d loading showed that α-GalCer requires lysosomal

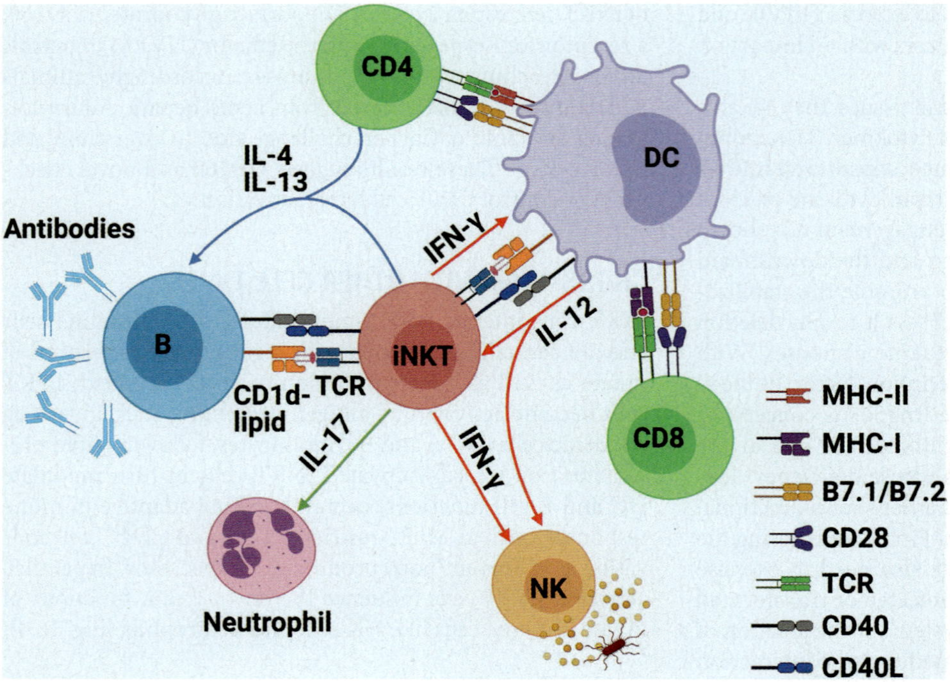

FIG. 36.4. Cross-talk between iNKT cells and other cell types. iNKT cells are activated by antigen-presenting cells, such as DCs and B cells. Subsequently, activated iNKT cells stimulate DCs to increase their antigen presentation functions toward classical CD8 and CD4 T cells. Moreover, iNKT cells provide help for B cells to promote antibody responses and influence other innate immune cells like NK cells. Red arrows represent iNKT1, blue arrows iNKT2, and the green arrow iNKT17 cytokines.

loading and surface presentation associated with lipid rafts, whereas short or unsaturated lipid antigens were able to be loaded directly on the cell surface and presented outside cholesterol-rich domains.[240] Another explanation is based on the type of antigen-presenting cell involved in lipid display. Accordingly, α-GalCer was mainly presented by DCs and macrophages, which also produced IL-12, whereas the IL-4–inducing lipid variants were broadly presented by all CD1d-expressing cells that largely lacked IL-12 secretion,[241] underscoring again the important role of IL-12 in full iNKT cell activation. Lastly, a specific subset of antigen-presenting cell, the CD8α$^+$ DEC-205$^+$ DC, has been shown to mainly capture and present a variety of exogenously administered glycolipid antigens, including variants that trigger divergent cytokine responses.[242] Upon lipid presentation, these DCs modulated their expression of costimulatory and inhibitory molecules, which proved to be dependent on the glycolipid entity. Taken together, the cytokine bias of iNKT cells is determined by the function of the antigen-presenting cell that is instructed by the structure of the lipid antigen.

B Cells

Activated iNKT cells are able to provide help for B cell responses, fostering antibody production, affinity maturation, and immunoglobulin class switching. Help can be indirect (noncognate) when stimulated iNKT cells upregulate CD40-ligand in order to promote antigen presentation by DCs, which subsequently prime MHC-II–restricted CD4 T cells that in turn support B cell activation[243] (Fig. 36.4). Moreover, many studies demonstrated that iNKT cells can also provide cognate help in case of direct lipid antigen presentation by B lymphocytes.[244-246] Upon differentiation, iNKT cells exhibited a follicular helper T cell phenotype, regulated by the transcription factor Bcl-6 and characterized by expression of CXCR5, PD-1, and IL-21.[247,248] Notably, immunization with α-GalCer coupled to the hapten nitrophenyl (NP) induced a faster B cell and germinal center response than immunization with NP coupled to protein antigen. Strikingly though, α-GalCer-NP stimulation led to less sustained germinal center reactions, fewer long-lived plasma cells, and inferior B cell memory. Moreover, cognate interaction preferentially expanded IL-10–producing regulatory B cells.[249] These results suggested that cognate iNKT cell help, though quick, was overall less efficient than conventional CD4 T cell help. However, an independent study performed immunization of mice with nanoparticles containing α-GalCer and a defined B cell antigen from *S. pneumoniae*, namely capsular polysaccharide. Using this system, iNKT cell–mediated cognate B cell help facilitated prolonged antibody responses and robust B cell memory, with rather limited differentiation of iNKT cells into follicular helper cells.[250] Thus, the complexity and intrinsic nature of the B cell antigen appeared to affect the efficacy of cognate iNKT cell help.

Marginal zone (MZ) B cells in the spleen expressed higher amounts of CD1d and proved to be more potent APCs toward iNKT cell activation *in vitro* than other B cell subsets.[245] Therefore, MZ B cells have been implicated in splenic iNKT cell stimulation, and indeed, humoral immunity was impaired upon α-GalCer-NP immunization when MZ B cells were depleted.[251] Microscopy studies demonstrated that splenic iNKT cells patrol the marginal zone located between the blood circulation (red pulp) and the lymphoid tissue (white pulp), well positioned to rapidly respond to blood-borne antigen.[252,253] Systemic lipid antigen was mainly captured and presented by DCs and macrophages in the marginal zone. And although IL-4 production by iNKT cells was not altered when CD1d was lacking on B cells,[253] a sizable population of

IFN-γ–secreting iNKT cells remained after DC and macrophage depletion,[252] suggesting an additional role of MZ B cells in splenic iNKT cell function.

The anatomical location, in which iNKT cell–mediated B cell help occurs, is decisive for the ensuing physiological outcome. The generation of transnuclear mice caused an increased abundance of tissue-resident iNKT cells and facilitated the analysis of rare populations. Accordingly, iNKT cells in Peyer patches of the gut were found to be prominent IL-4 producers that helped B cells in class switching to IgG1, thus potentially impacting on mucosal immune responses.[254] In the human system, lipid antigen presentation by B cells proved to be essential to maintain sufficient iNKT cell numbers. In patients with systemic lupus erythematosus, an autoimmune disease that involves dysregulation of B cells and autoantibody production, iNKT cell frequencies and functions were reduced.[255] This defect was due to impaired CD1d recycling in B cells, a mechanism that could be triggered by type I interferon. Consequently, normalization of CD1d expression reconstituted physiological iNKT cell homeostasis.[255] Furthermore, regulatory B cells (Bregs) induced iNKT cells to produce IFN-γ, which suppressed Th17 cell–mediated adaptive responses and ameliorated experimental arthritis,[256] thus providing an example how the cross-talk between B and iNKT cells controls inflammation.

NK Cells

Innate-like T lymphocytes and innate immune cells closely cooperate with each other, as exemplified by the relationship between iNKT and NK cells. In this context, stimulation of mice using α-GalCer revealed that, in addition to iNKT cells, NK cells were activated rapidly after *in vivo* challenge.[257,258] Accordingly, lipid stimulation triggered an activation sequence, in which iNKT cells consecutively triggered NK cells to amplify their IFN-γ production as well as cytotoxic activity. This direct effect was predominantly carried out by iNKT cell–derived IFN-γ.[257] Additionally, activated iNKT cells upregulated CD40-ligand to license IL-12 production by DCs. IL-12 in turn was able to augment NK cell functions as well, following a mechanism of transactivation.[258] Taken together, iNKT cells swiftly activate NK cells either directly via IFN-γ or indirectly through IL-12 (Fig. 36.4).

Neutrophils

iNKT17 cells, by definition, are well equipped to produce the cytokine IL-17, which attracts neutrophilic granulocytes to tissue sites in order to drive inflammatory processes. Experiments showed that IL-18, a product of inflammasome activation, specifically induced CD1d expression on neutrophils.[259] Subsequently, CD1d-positive neutrophils activated iNKT cells and enabled them to express Fas-ligand to potentially kill death receptor Fas-expressing B lymphocytes. This mechanism could explain how the cross-talk between neutrophils and iNKT cells prevents autoreactive antibody responses. However, in chronic IL-18–mediated inflammation, lipid antigen-stimulated iNKT cells vetoed the licensing mechanism of neutrophils.[260] As a result, iNKT cells differentiated into follicular helper cells to promote anti-DNA antibody production and autoimmunity.

Neutrophils are also involved in iNKT cell trafficking. Intravital microscopy studies of the lung demonstrated that iNKT cells mainly resided inside the vasculature in the steady-state.[261] After α-GalCer challenge or *S. pneumoniae* infection, neutrophils rapidly extravasated and secreted CCL17, which animated iNKT cells to leave the intravascular space and enter the lung interstitium. Blockade of CCL17 increased susceptibility to *S. pneumoniae* infection,[261] highlighting the importance of iNKT cell interaction with neutrophils in the context of host defense.

Macrophages

When antigen is injected subcutaneously, it next reaches draining lymph nodes to trigger an immune response. The same principle holds true for lipid antigen presentation, as $CD169^+$ subcapsular sinus macrophages proved to be efficient antigen-presenting cells for early CD1d-restricted iNKT cell activation in lymph nodes.[262] Following priming by $CD169^+$ macrophages, iNKT cells localized in interfollicular areas produced an early wave of secreted IL-4 that facilitated the recruitment of B lymphocytes into the germinal center reaction.[263] Consequently, CD1d-deficient mice failed to generate efficient germinal center responses and antiviral B cell immunity.

DISEASES

Due to their extensive cross-talk with other cell types and regulation of innate and adaptive immunity, iNKT cells are involved in various pathological conditions and diseases. This impact on pathologies spans infectious diseases caused by a multitude of pathogens, inflammatory conditions, allergic as well as metabolic disease, and cancer.

Infections

In microbial infections, one can discern two basic mechanisms of iNKT cell induction (Fig. 36.5). If the infecting microbe contains antigenic lipid entities, those microbial CD1d ligands are able to directly activate iNKT cells, following the processing and display on antigen-presenting cells. Alternatively, assuming that certain microbes are devoid of veritable lipid antigens, they can still stimulate iNKT cells via activation of antigen-presenting cells. In this context, microbes trigger the production of self-antigens and cytokines like IL-12 through engagement of pattern recognition receptors, such as toll-like receptors (TLRs) (Fig. 36.5). These two mechanisms are though not mutually exclusive, since innate and cytokine-driven signals are also able to support iNKT cell responses to defined microbial lipid antigens.

Bacterial Infections

Sphingomonas spp. are gram-negative, LPS-negative α-proteobacteria, which generate α-glucuronosylceramide and α-galacturonosylceramide as powerful iNKT cell antigens.[34,35]

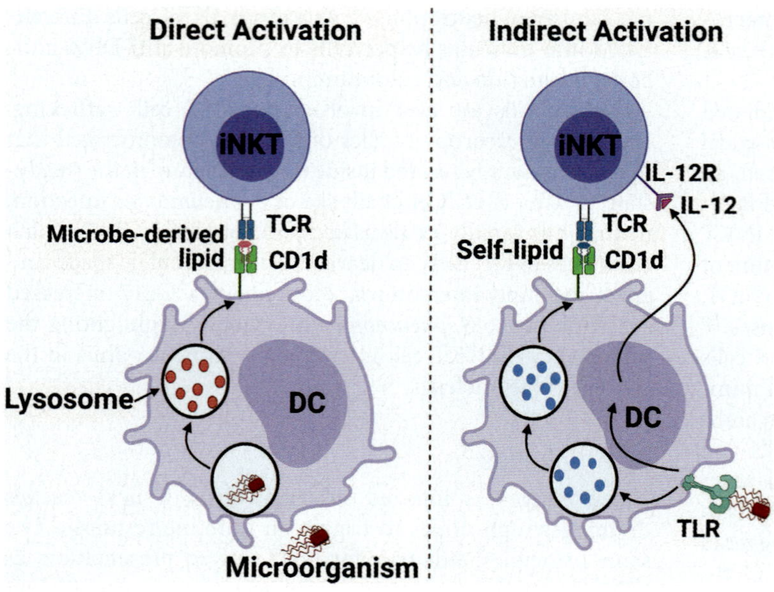

FIG. 36.5. Activation mode of iNKT cells in infection. *Left*: Microbes that harbor iNKT cell lipid antigens can directly activate iNKT cells. *Right*: In the absence of microbial lipid antigens, infections can cause the induction of self-lipid antigens for subsequent iNKT cell stimulation. In this scenario, iNKT cell activation is promoted by IL-12, which can be induced by engagement of toll-like receptors (TLRs).

Upon infection, *Sphingomonas* was phagocytosed by dendritic cells and macrophages and triggered an activation cascade comparable to exogenous α-GalCer stimulation. iNKT cells proved to be required for protection because their absence led to strikingly increased bacterial burden.[34,35] *Sphingomonas* infection can have a severe course of disease in immunocompromised patients but generally does not represent a dangerous pathogen. Another member of the class of α-proteobacteria, namely *Ehrlichia*, has a much higher pathogenicity, at least in certain regions of Africa and East Asia. In the mouse model, *Ehrlichia muris* strongly activated iNKT cells, and bacterial clearance of *Ehrlichia* was massively impaired in the absence of CD1d or iNKT cells.[34]

S. pneumoniae are gram-positive diplococci able to cause severe pneumonia, sepsis, and meningitis in children and the elderly. *Streptococcus* produces α-glucosyldiacylglycerol as potent antigen, and mice lacking CD1d or iNKT cells exhibited higher bacterial burden upon infection.[37] *S. pneumoniae* is a prime example for production of potent lipid antigens, direct activation of iNKT cells, as well as a critical role of iNKT cell–mediated protection. This protective function is highlighted by the development of an experimental vaccine, which used a conjugate of *S. pneumoniae* capsule polysaccharide linked to α-GalCer and proved to be effective against pneumococcal disease.[264] Furthermore, although α-glucosyldiacylglycerol from *S. pneumoniae* directly stimulates iNKT cells, it has been shown that IL-12 from TLR-stimulated antigen-presenting cells supported iNKT cell activation in infection, regardless of the presence of defined lipid antigens.[265] However, an independent study performed a variety of infections and stimulations with TLR ligands, using Nur77-reporter mice that indicate lipid antigen-induced TCR signaling in iNKT cells. As a result, whereas *Streptococcus* and *Sphingomonas* infection induced Nur77, *Salmonella typhimurium* infection and TLR stimulation, although previously described to trigger self-lipid antigens, did not lead to TCR-mediated Nur77 expression.[266] These results demonstrated that several microbes can induce iNKT cell activation in the absence of antigenic stimulation.

B. burgdorferi generates α-galactosyldiacylglycerol to potently activate iNKT cells and is the etiologic agent of Lyme disease, a multisystem inflammatory disorder that affects the skin, joints, heart, and eventually the brain.[36] Upon systemic challenge with *B. burgdorferi*, the bacteria reaching the liver were predominantly taken up by Kupffer cells, prior to interaction with liver iNKT cells.[267] Although iNKT cells were not crucial in the acute phase of infection, they played a vital role to prevent chronic joint inflammation and were also capable of eliminating bacteria from infected tissue.[268] Mechanistically, iNKT cells in joints were located in the extravascular space lining the blood vessels and responded quickly in response to *Borrelia* invaders in a cytotoxic manner.[269]

Pseudomonas aeruginosa is a gram-negative bacterium that can cause severe pneumonia, often in the context of hospital-acquired opportunistic infections. CD1d-deficient mice showed an increase in bacterial burden in the lung when compared to wild-type controls. Administration of α-GalCer promoted iNKT cell–mediated activation of alveolar macrophages, which subsequently effectively cleared *P. aeruginosa* via enhanced phagocytosis.[270]

Another important lung pathogen is *Mycobacterium tuberculosis*, the causative agent of tuberculosis. Previous studies demonstrated that iNKT cells recognize infected macrophages, respond with IFN-γ secretion, and protect mice against aerosol challenge with *M. tuberculosis*.[271] Moreover, iNKT cells were able to directly control mycobacterial growth by granulocyte-macrophage colony-stimulating factor (GM-CSF) production, which acted as antimicrobial factor.[272] Translating the importance of innate-like T cells to human disease, a patient with hereditary T-bet deficiency suffered from mycobacterial disease after the systemic spread of the *Mycobacterium bovis* BCG strain following vaccination.[273] Interestingly, NK cells and innate-like T cells, including iNKT cells, MAIT, and γδ T cells, were strikingly reduced in that patient and produced very little IFN-γ. In sharp contrast, the numbers of mycobacteria-specific conventional CD4 and CD8 T cells as well as their IFN-γ production proved to be normal. These findings suggested that the function of innate-like T lymphocytes is essential for controlling mycobacterial disease.

H. pylori is usually a resident of the stomach and associated with the development of gastritis, peptic ulcer, and gastric cancer. *H. pylori* synthesizes cholesteryl α-glucosides, which functioned as lipid antigens and activated iNKT cells *in vivo*.[39,274] Mice devoid of iNKT cells exhibited a reduced inflammatory response to *H. pylori* and increased bacterial counts in the stomach.[274]

Lastly, if the infectious disease process is not contained, infections can overwhelm the host immune defense and become systemic, leading to bacteremia and ultimately septic shock. When sepsis is driven by gram-negative bacteria, the shock syndrome is usually caused by LPS. Experiments mimicking septic shock, using LPS for intravenous low-dose priming and subsequent challenge, revealed that iNKT cell–deficient mice were less susceptible to shock mortality and that IFN-γ played a central pathogenic role.[275] Furthermore, the acute phase of bacteremia can be followed by postsepsis immunosuppression, predisposing the host for secondary, opportunistic infections. Studies on sepsis patients showed an early IFN-γ signature in the blood that was correlated with secondary infections with the fungus *Candida albicans*.[276] In mouse models, activated iNKT cells caused IFN-γ production by NK cells in an mTORC1-dependent fashion. Accordingly, IFN-γ blockade or inhibition of mTOR using rapamycin was able to reduce immunosuppression and prevent secondary *Candida* infection.[276] Moreover, iNKT cell modulation was experimentally used for sepsis immunotherapy in a stepwise approach, in which administration of OCH biased iNKT cells toward IL-4 production to reduce immunopathology, and a second shot using α-GalCer promoted iNKT cell–derived IFN-γ in order to relieve immunosuppression.[277] Finally, other serious health conditions, such as stroke, can cause immunosuppression and subsequent infections. Notably, upon experimental stroke, sympathetic innervation of hepatic iNKT cells suppressed their functions. Conversely, blocking noradrenergic neurotransmission and stimulation with α-GalCer were able to prevent stroke-associated infections.[278]

Fungal and Parasitic Infections

A. fumigatus is a fungus that can cause allergic asthma and severe infection in patients with immunodeficiency. Intratracheal infection of CD1d-deficient mice caused a markedly increased fungal burden in the lungs when compared to wild-type controls.[279] Functionally, binding of the fungal cell wall polysaccharides, β-1,3 glucans, to Dectin-1 on antigen-presenting cells triggered IL-12 production. Thus, *A. fumigatus* elicited indirect activation of iNKT cells. Subsequent work identified an *A. fumigatus*–derived glycosphingolipid, called asperamide B, which potently stimulated iNKT cells.[38] Intranasal treatment with asperamide B caused airway hyperreactivity in an IL-33–dependent manner. This suggests direct lipid antigen recognition by iNKT cells as a mechanism how *A. fumigatus* triggers asthma.

Plasmodium spp. is a parasitic protozoan and the etiologic agent of malaria. When rodent malaria strains, such as *Plasmodium berghei* and *yoelii*, were used for infection and vaccination, CD1d-deficient mice revealed that iNKT cells were not required for protection.[280,281] However, since iNKT cells are abundant in the liver, it has been speculated that iNKT cell activation might influence the liver stages in the malaria parasitic life cycle. Indeed, stimulation with α-GalCer specifically protected from the liver stages in *Plasmodium* infection.[282] Furthermore, an experimental vaccine consisting of a *Plasmodium* peptide antigen conjugated to α-GalCer was able to efficiently induce protection against liver-stage malaria in mice based on single-shot immunization.[283] Of note, this approach facilitated the substantial generation of liver-resident memory CD8 T cells. These experiments provided yet another example of how iNKT cells, although not primarily involved in the natural course of infection, can be nevertheless activated to fulfill important protective functions against disease.

Viral Infections

Although viruses usually do not code for the synthesis of lipid antigens, viral infections are still capable of leading to indirect iNKT cell activation. In this context, viral nucleic acids and other danger molecules are able to engage TLRs on dendritic cells to subsequently induce IL-12 production. IL-12 and potentially other cytokines then stimulate iNKT cells to deploy their effector functions, and lipid antigen presentation through CD1d is not required for this mechanism.[284] This cytokine-mediated, antigen-independent iNKT cell activation has been shown for murine cytomegalovirus (MCMV) infection, in which IL-12 proved to be essential for IFN-γ secretion by iNKT cells.[285] Moreover, NK cells are generally also important contributors to antiviral protection and showed a stronger receptiveness to IL-18 than to IL-12 in MCMV infection. Of note, mice lacking iNKT cells exhibited higher viral titers especially in the liver, highlighting their protective impact on MCMV infection.[285]

Influenza virus causes highly contagious pulmonary disease especially in the young and elderly and represents a major burden on global health. Using a mouse-adapted influenza A strain for intranasal inoculation, iNKT cells did not affect the overall viral load.[286] However, lung pathology was exacerbated and CD8 T cell responses reduced when iNKT cells were lacking. Functionally, iNKT cells produced IL-22, which could function to preserve epithelial and mucosal integrity of the lung.[287] Furthermore, iNKT cells were shown to suppress the influx of myeloid-derived suppressor cells (MDSCs) and thus facilitated a productive immune response.[288] Intranasal α-GalCer treatment of piglets infected with a pandemic influenza strain drastically reduced viral burden and lung pathology,[289] suggesting possible options for iNKT cell–based therapy.

Hepatitis B virus (HBV) primarily infects hepatocytes and leads in most cases to swift and durable immunity, while a minor percentage can also develop into chronic disease. HBV-expressing hepatocytes showed alterations of their endogenous lipid profile and produced self-lipid antigens that stimulated NKT cells.[8] Lack of iNKT cells, CD1d, or the lipid transfer protein MTP led to reduced HBV-specific T- and B-cell immunity as well as impaired viral control.

Kaposi sarcoma-associated herpes virus (KSHV) codes for two modulator of immune recognition (MIR) proteins, which facilitated downregulation of surface CD1d following ubiquitination of its cytoplasmic tail.[290] KSHV-mediated impairment of CD1d caused abrogated iNKT cell responses. Similarly, herpes simplex virus 1 (HSV-1), the etiologic agent of viral hepatitis and keratitis in immunocompromised patients, is able to manipulate CD1d function. Initial work demonstrated that HSV-1 interfered with CD1d recycling from lysosomal compartments to the plasma membrane.[291,292] Follow-up studies revealed that HSV-1's protein kinase US3 phosphorylates the host kinesin motor protein KIF3A to drive CD1d downregulation.[293] Moreover, iNKT cells helped to generate a virus-specific IgG antibody response.[294] Taken together, these data indicate that CD1d-mediated antigen presentation is a target of viral immune evasion mechanisms.

Inflammation

The inflammasome is central to driving IL-1β and IL-18 secretion, and thus plays an important role in the regulation of inflammatory responses. iNKT cells have the ability to influence innate inflammatory activation in other cell types. Accordingly, this cross-talk revealed that human iNKT cells induce inflammasome-mediated IL-1β activation in monocytes.[295] Remarkably, this phenomenon did not depend on microbial danger signals or alarmins as indicators of cell damage, such as extracellular ATP and its signaling via P2X7 receptors. Moreover, studies demonstrated that iNKT cells are capable of triggering IL-1β in dendritic cells (DCs) in a largely inflammasome-independent fashion.[296] In this context, activated iNKT cells translocated intracellular Fas ligand to the cell surface to trigger caspase 8 in DCs, subsequently shifting to a pyroptotic program activating the pore-forming protein gasdermin D that allows for IL-1β secretion. Thus, iNKT cells could help to kick-start an inflammatory response even in the absence of cytosolic inflammasome triggers. Furthermore, tumor necrosis factor (TNF) derived from iNKT cells was shown to facilitate the NLRP3 inflammasome priming step, by amplifying the production of IL-1β and IL-18.[297] Lastly, on the iNKT cell side, engagement of the TNF superfamily receptor OX40 on hepatic iNKT cells induced a strong pyroptotic program, leading to pyroptosis of iNKT cells and hepatic inflammation.[298] Mechanistically, OX40 ligation promoted caspase-1 activation, processing of pro–IL-1β, and cleavage of gasdermin D.

Purinergic signaling involves the recognition of purines and purine nucleotides, involving adenosine and ATP receptors, surface ATPases like CD39, and intracellular processing enzymes. In this context, human iNKT cells have been shown to secrete extracellular ATP to bind to its respective receptor P2X7 to induce a protective inflammatory response in DCs.[299] Alternatively, purinergic signaling in iNKT cells is also able to counter inflammation. Accordingly, exposure of iNKT cells to adenosine A2 receptor agonists reduced IFN-γ production and concomitantly increased expression of IL-13 and the ecto-ATPase CD39.[300] Taken together, this purinergic Th2-type cytokine bias functioned as an anti-inflammatory program in iNKT cells.

Once an inflammatory cascade is elicited, it is also vital to contain and resolve the inflammation at a certain point in order to avoid chronic inflammatory processes and disease. iNKT cells also have critical functions in helping the tissue to switch from inflammation to resolution and repair. In the liver, where iNKT cells are abundant, sterile injury caused a characteristic triphasic behavior, including repulsion shortly after initial injury, followed by retention of iNKT cells at the border of the tissue damage, prior to infiltration of the inflamed site.[301] Functionally, iNKT cells were stimulated by Kupffer cells and endothelial cells to produce IL-4, which proved to be essential for affecting inflammatory monocytes, cause hepatocyte division, and eventually lead to tissue restitution.

The cytokine IL-17 is a crucial factor in inflammation, especially of the joints. In patients with spondyloarthritis, the inflamed joints contained a high frequency of RORγt-expressing iNKT cells that were the most pronounced IL-17 producers.[302] Of note, RORγt inhibition abrogated iNKT17 functions, indicating a possible therapeutic option against arthritis.

Metabolic syndrome affects glucose tolerance as well as liver functions. Initial fatty liver disease (steatosis) can lead to chronic inflammation (steatohepatitis). In experimental models that mimic nonalcoholic steatohepatitis (NASH) by feeding mice a methionine-choline-deficient diet, iNKT cells strongly promoted liver fibrogenesis through osteopontin-mediated activation of hepatic stellate cells, the main culprits and collagen producers in liver fibrosis.[303] Moreover, iNKT cell–derived cytokines also augmented the extent of steatosis and liver inflammation.[304] Mechanistically, iNKT cells secreted the TNF superfamily member LIGHT that enabled lipid uptake by hepatocytes. The impact of liver CD8 T cells and iNKT cells on hepatocytes ultimately drove the transition from NASH to hepatocellular carcinoma.[305]

CD1d-mediated antigen presentation also plays a role in inflammation of the bile ducts. Intrabiliary administration of oxazolone in mice caused acute cholangitis with bilirubin accumulation and severe weight loss. The use of iNKT cell–deficient mice or a CD1d blocking antibody abrogated inflammation of the bile ducts, suggesting CD1d-based therapeutic possibilities for cholangiopathies.[306]

In the context of allogeneic bone-marrow transplantation (BMT), donor T cells are able to attack host tissue, causing organ damage and vigorous inflammation, called graft-versus-host disease (GVHD). In murine models of allogeneic BMT, adoptive transfer of iNKT cells promoted the expansion of donor Tregs and suppressed inflammation.[307] Moreover, application of α-GalCer proved to be effective in prophylactic and therapeutic treatment of GVHD in mice.[308]

Asthma

Asthma is a clinical syndrome characterized by airway hyperreactivity and chronic lung inflammation. Allergic asthma can be caused by inhaled allergens that drive a Th2-based immune response, involving amplification of eosinophils and IgE production. In addition to classical allergy, asthma can also be provoked by nonallergic causes, such as environmental pollutants. Initial studies exploring the role of iNKT cells in asthma used a sensitization and challenge model based on administration of ovalbumin and the adjuvant alum. iNKT cell–deficient mice showed less airway hyperreactivity, which could be reversed by adoptive transfer of iNKT cells producing IL-4 and IL-13.[309] A follow-up study in asthma patients stated a high abundance of iNKT cells in bronchoalveolar lavage fluid, but those results could not be reproduced by independent investigations. However, the initial results in the murine system were corroborated using a refined mouse model of iNKT cell deficiency (Traj18$^{-/-}$ mice), which showed that eosinophilia and airway hyperreactivity were reduced upon challenge with the allergens ovalbumin or cockroach antigen in the absence of iNKT cells.[310]

Other studies used direct iNKT cell stimulation via administration of lipid antigens through the airways of mice. Intratracheal instillation of α-GalCer led to rapid egress of intravascular iNKT cells into lung tissue, subsequently facilitating the formation of eosinophilic granuloma.[311] Concomitant exposure to α-GalCer and the protein antigen ovalbumin caused strong allergization as indicated by massive airway inflammation upon antigen challenge. Likewise, extracts from house dust contained iNKT cell antigens, which served as adjuvants in the ovalbumin-induced asthma model, promoting strong sensitization to airborne antigen in an iNKT cell–dependent fashion.[312] Moreover, intranasal application of the lipid antigen asperamide B derived from *A. fumigatus*, a fungus associated with asthma in humans, was able to trigger airway hyperreactivity by activating pulmonary iNKT cells in an IL-33–dependent manner.[38]

Nonallergic asthma can be elicited by ozone, a prominent air pollutant. Repeated exposure to ozone in wild-type mice caused airway hyperreactivity and lung inflammation, which was abrogated in animals lacking iNKT cells.[313] Additionally, IL-17 production proved to be essential for the asthmatic phenotype.

Furthermore, different subsets of iNKT cells appear to have a different impact on the outcome of allergic asthma in mice. In this context, infection of young mice with influenza A virus expanded double-negative (DN) iNKT cells, which prevented airway hyperreactivity in the ovalbumin-induced asthma model when the mice were adult.[39] Functionally, those DN iNKT cells preferentially produced IFN-γ, thus potentially creating a cytokine bias that suppressed the allergic phenotype.[314]

Taken together, several mouse models of allergic and nonallergic asthma support a function of iNKT cells in the regulation of this disease complex. However, the translation to the human system and the role of iNKT cells in asthma patients remain elusive.

Diabetes

Type 1 diabetes, or insulin-dependent diabetes mellitus, is an autoimmune disease based on the T cell–mediated destruction of insulin-secreting β-cells in the pancreas. As a result, loss of β-cells leads to lack of insulin, which causes uncontrolled hyperglycemia and its clinical consequences. In the murine system, the NOD strain is predisposed to develop spontaneous diabetes that is thought to be due to functional impairment of regulatory cells and an immune bias toward Th1 responses. NOD mice exhibited reduced numbers of iNKT cells that produced less IL-4, suggesting a potential role of iNKT cells in disease development. Indeed, adoptive transfer of iNKT cells or Vα14-Jα18 transgenesis reduced the frequency of diabetes in NOD mice.[315,316] Complementarily, NOD animals deficient for CD1d demonstrated exacerbated diabetes pathogenesis.[317] Moreover, treatment with α-GalCer prevented autoimmune diabetes in NOD mice.[318,319] The antidiabetic function of iNKT cells appeared to rely on an intricate network of cellular interactions, in which tolerogenic plasmacytoid DCs facilitated the generation of Tregs in an iNKT cell–dependent fashion.[320] On the other hand, NOD mice contained an increased abundance in iNKT17 cells, which were able to cause diabetes development in adoptive transfer experiments,[321] indicating that different iNKT subsets have differential pathogenic functions.

With regard to human type 1 diabetes, an initial study concluded that iNKT cell numbers were reduced and their function strongly biased toward IFN-γ production in diabetes patients.[322] However, subsequent work, using improved detection methods based on CD1d-tetramer staining, failed to corroborate those findings. Rather, the frequency of iNKT cells in peripheral blood of diabetes patients proved to be variable between donors, but not different compared to controls.[323] Furthermore, beyond the question of cell numbers, clonal analysis of the iNKT cell repertoire in diabetes patients revealed a loss of clones expressing high-affinity TCRs,[324] pointing to possible qualitative alterations of iNKT cells in human diabetes.

EAE/Multiple Sclerosis

Multiple sclerosis (MS) is a chronic inflammatory disease that involves demyelination of the neurons in the brain and spinal cord, which eventually can lead to paralysis. The mouse model for MS is EAE, which uses immunization with a self-peptide from myelin oligodendrocyte glycoprotein together with adjuvant to induce inflammation of the CNS. Stimulation of iNKT cells by exogenous α-GalCer administration was able to suppress EAE in mice.[325] This effect required IL-4 and IL-10 in order to drive a Th2-based immune response that countered cell-mediated inflammation. Moreover, treatment with the α-GalCer derivative OCH, which harbors a truncated sphingosine base and is known to trigger a Th2 bias, proved to be very efficient in preventing EAE in an IL-4–dependent manner.[238] Functionally, treatment with lipid antigen promoted the expansion of MDSCs that also infiltrated the CNS. Since MDSC depletion aggravated EAE, MDSC-mediated inhibition of inflammation proved to be a key function of iNKT cell activation.[326]

Primary Biliary Cirrhosis

Primary biliary cirrhosis (PBC) is characterized by the destruction of the bile duct epithelium, lymphocytic infiltrates in the liver, and antimitochondrial antibodies, ultimately leading to cirrhosis. The autoantibodies recognize a mitochondrial enzyme, namely pyruvate dehydrogenase complex E2, which is also conserved in the *Sphingomonas* strain *Novosphingobium aromaticivorans*. Lipid chemistry identified α-glucuronosylceramide derived from *N. aromaticivorans* as potent iNKT cell antigen.[327] Additionally, patients with PBC showed an accumulation of iNKT cells in liver biopsies when compared to peripheral blood and livers from healthy donors.[328] Notably, infection of mice with *N. aromaticivorans* induced antibodies against microbial as well as mammalian pyruvate dehydrogenase E2 and triggered T cell–mediated inflammation and pathology of bile ducts.[327] Antibody responses and disease induction depended on the presence of iNKT cells. These results suggested that a liver-tropic bacterium containing lipid ligands activated iNKT cells, which in turn facilitated an immune response to shared bacterial/mitochondrial antigens, leading to liver autoimmunity.

Cancer

iNKT cells have important functions in tumor immunity. At the time of the discovery of α-GalCer as lipid antigen, a seminal study generated Jα18-KO mice to demonstrate that iNKT cells are required for protection against tumor growth using the B16 melanoma metastasis model.[329] Accordingly, activated iNKT cells prevented the outgrowth of lung metastasis following intravenous injection of B16 melanoma cells, a system that subsequently became a prototypical model for testing iNKT cell–mediated cancer immunity. Of note, stimulation of iNKT cells with lipid antigen or cytokines is a prerequisite for tumor protection against B16 melanoma. Following studies explored tumor immunity against spontaneous fibrosarcoma, triggered by the treatment with the chemical carcinogen methylcholanthrene (MCA), and showed that iNKT cell–deficient mice were less protected from cancer development.[330] However, this impact on MCA-induced fibrosarcoma could not be corroborated in subsequent independent experiments, which highlighted variations in tumor development dependent on mouse colony source and environmental factors.[331]

iNKT cells are abundant in the liver, an organ well positioned to receive input signals from the gut via the portal vein. Treatment with antibiotics altered the commensal gut flora and led to increased liver iNKT cell numbers and function that protected mice in liver cancer and metastasis models.[332] Mechanistically, altered intestinal microbiota caused a reduced conversion of primary bile acids that reached the gut from the liver. Consequently, primary bile acids could recirculate back to the liver and induce upregulation of CXCL16 on liver sinusoidal endothelial cells, which in turn facilitated the accumulation of $CXCR6^+$ iNKT cells and their antitumor immunosurveillance.[332] Furthermore, the tumor microenvironment is able to render iNKT cells dysfunctional. In this context, lactic acid enriched in the tumor microenvironment was able to inhibit the expression of the nuclear receptor PPAR-γ, which under normal circumstances upregulated lipid biosynthesis and cholesterol necessary for optimal IFN-γ production.[333] Interestingly, treatment with the PPAR-γ agonist pioglitazone, an established antidiabetic drug, restored effector functions in tumor-infiltrating iNKT cells from liver cancer patients and in melanoma mouse models.[333] On the other hand, iNKT cells themselves can undergo malignant transformation. In p53-deficient mice, developing peripheral T cell lymphoma originated from iNKT cells and showed signs of chronic TCR stimulation.[334] Since CD1d proved to be required for iNKT lymphoma survival, treatments with CD1d blocking antibody might represent a therapeutic modality.

In addition to their intrinsic roles in different tumor types, iNKT cells are attractive candidates for targeting to improve immunotherapies. Their antitumor effector functions and CD1d restriction, which helps to avoid anti-MHC T cell responses and associated GVHD, make them suitable for therapeutic modulation. One approach to increase the available number of iNKT cells used engineering of human hematopoietic stem cells by lentiviral transduction with the human Vα24 iNKT cell receptor, prior to reconstitution of humanized mice. This set-up allowed for long-term generation of HSC-iNKT cells *in vivo*, which protected mice from multiple human tumor xenografts.[335] A further possibility to increase the availability of iNKT cells has been shown by reprogramming of human $V\alpha24^+$ iNKT cells into induced pluripotent stem cells, which could be redifferentiated into iNKT cells on demand.[336] Those iPS-iNKT cells proved to be effective in various *in vivo* cancer models. Moreover, a prerequisite for direct iNKT cell recognition is the expression of CD1d on tumor cells. However, many tumors fail to express or downregulate CD1d. To overcome this obstacle, a fusion protein composed of CD1d and a single-chain antibody against CD19 was generated in order to target iNKT cells to $CD19^+$ B cell lymphoma. After loading of CD1d with α-GalCer, the fusion protein proved to be effective against $CD19^+CD1d^-$ tumors in mice.[337] A similar approach employing a bispecific T-cell engager (BiTE), which consisted of soluble CD1d and a single-chain antibody against a tumor antigen, was refined by adding covalently linked α-GalCer to CD1d. This BiTE was special since repeated administration avoided anergy induction of iNKT cells and thus preserved lasting antitumor immunity.[338] Lastly, the concept of chimeric antigen receptor (CAR) T cell therapy has been expanded to the iNKT cell field. CAR-iNKT cells directed against CD19 proved to be more efficacious than conventional CAR-T cells with regard to control of CD19-expressing lymphoma in experimental mouse models.[339]

Notably, iNKT cell–based therapies find their way into the clinic. After establishing a preclinical model with CAR iNKT cells specific for ganglioside GD2,[340] which is abundantly present in neuroblastoma, patients have been infused with autologous GD2-specific CAR iNKT cells after *in vitro* expansion and showed no adverse events nor signs of toxicity.[341] Importantly, CAR iNKT cells expanded in neuroblastoma patients, localized to the tumor sites, and were able to mediate regression of metastatic bone lesions, demonstrating that iNKT cells can be applied to human cancer therapy.[341] Furthermore, iNKT cell biology can support cancer prognosis. In patients with chronic lymphocytic leukemia (CLL), disease progression correlated with iNKT cell impairments, including numbers and function. Multivariable analysis identified low iNKT cell frequencies as independent prognostic marker for CLL progression.[342]

TYPE II NKT CELLS

In contrast to iNKT cells, type II NKT cells do not express the invariant Vα14-Jα18 chain, but rather produce a diverse TCR repertoire (dNKT). Although restricted by CD1d, dNKT cells fail to recognize α-GalCer and instead respond to a different set of lipid antigens. The classical lipid antigen for type II NKT cells is sulfatide (sulfated β-GalCer), which is abundant in nervous tissue of the brain and spinal cord. Notably, immunization with sulfatide was able to abrogate EAE based on dNKT-mediated inhibition of autoaggressive T lymphocytes.[343] Moreover, dNKT hybridoma reacted with phospholipids of microbial and mammalian origin, including phosphatidylglycerol, cardiolipin, and phosphatidylinositol.[344] Follow-up work revealed that the bacterial pathogen *Listeria monocytogenes* produced a phosphatidylglycerol with shorter, fully saturated fatty acids, which proved to be highly antigenic compared to its mammalian counterpart,[345] suggesting a possible differential recognition of lipid antigens by type II NKT cells upon infection. Another phospholipid antigen identified was LPC, which stimulated dNKT cells to subsequently suppress iNKT cell activation, thus alleviating autoimmune hepatitis.[346] Furthermore, β-glucosylceramide has been discovered as type II NKT cell antigen.[347,348] It accumulates upon induced or inherited lipid alterations, such as the lysosomal storage disorder called Gaucher disease, and stimulates dNKT cells with a T_{FH} phenotype, promoting downstream B cell activation and chronic inflammation.[348]

Structural biology resolved the ternary complex between a type II NKT cell receptor and sulfatide or lysosulfatide presented on CD1d.[349,350] The dNKT cell receptor engaged CD1d above the A′ roof and in an orthogonal orientation. This binding mode was very different compared to iNKT cells, which bound the CD1d/lipid complex in a parallel fashion and predominantly over the F′ pocket. However, the crystal structure of another dNKT cell receptor recognizing the microbial lipid antigen α-glucuronosyldiacylglycerol on CD1d demonstrated that TCR engagement occurred above the F′ pocket in a parallel fashion, resembling the iNKT cell binding mode.[351] This suggests that type II NKT cells pursue different recognition strategies with regard to CD1d, thus potentially increasing their ability to respond to a multitude of different lipid ligands.

Regarding the development of type II NKT cells, they required PLZF and SAP like their invariant counterparts.[352] Moreover, activated dNKT cells contributed to the antitumor immune response in the B16 melanoma model. Lastly, cytokines, including IL-12, IL-18, or IL-15, were able to stimulate IFN-γ production by dNKT cells in a CD1d-independent manner. By contrast, TCR triggering rather induced proliferation of dNKT cells.[353] Thus, type II NKT cells are equipped with innate and adaptive activation mechanisms.

GROUP 1 CD1-RESTRICTED T LYMPHOCYTES

In contrast to iNKT cells, T lymphocytes restricted by CD1a-c rather represent polyclonal populations with a diverse TCR repertoire. Since CD1a-c are predominantly expressed in humans and absent in normal experimental mice, our knowledge on group 1 CD1-restricted T cells is mainly based on experiments in the human system or using human CD1-transgenic animals. The CD1 isoforms have various biological functions that are the consequence of their differential expression profile on multiple cell types as well as their localization in different cellular compartments. The cellular trafficking pathways of CD1 molecules facilitate the loading of different lipid entities for the subsequent presentation to CD1-responsive T lymphocytes. Indeed, a massive variety of exogenous or self-lipid ligands and antigens has been described with regard to CD1-mediated presentation. Thus, depending on the CD1 isoform, the type of antigen-presenting cell plus its tissue localization, and the CD1/lipid complex presented, group 1 CD1-restricted T cells play important roles in infectious disease, inflammation, and cancer.

CD1 EXPRESSION AND SUBCELLULAR LOCALIZATION

With the advent of monoclonal antibody technology, CD1 molecules were the first proteins characterized when human thymocytes were used for immunization of mice.[354] Accordingly, thymocytes abundantly express CD1a-c,[3] which is required for thymic selection of CD1-restricted T cells. Whereas CD1d is expressed on various cell types including epithelial cells, group 1 CD1 molecules are mainly produced by antigen-presenting cells.[7] Bone marrow–derived dendritic cells (DCs) from CD1-transgenic mice highly express CD1a, CD1b, and CD1c.[355] In peripheral tissues, CD1a is abundant on Langerhans cells, a special type of DC in the skin. CD1b is expressed on lymph node DCs, whereas CD1c is produced by Langerhans cells and classical type 2 DCs (cDC2). Moreover, B lymphocytes express high levels of CD1c.[7] In addition, all three CD1 molecules can be induced on monocytes in the human system upon treatment with GM-CSF and IL-4.

With regard to regulation of gene expression, human serum lipids, such as cardiolipin, suppressed CD1a-c transcription in DCs, involving a mechanism based on activation of nuclear receptor PPAR-γ.[356] Furthermore, a single nucleotide polymorphism was identified in the 5′ untranslated region of the CD1a locus, which drastically reduced CD1a expression in DCs,[357] highlighting an important genomic region for CD1 regulation. Otherwise, genomic and epigenetic control of group 1 CD1 expression remains unclear and provides fertile ground for future study.

CD1a, CD1b, and CD1c differ in their subcellular trafficking pathways. CD1a is peculiar among the different isoforms, as it does not contain a tyrosine sorting motif in its cytoplasmic tail. Accordingly, CD1a was found in early

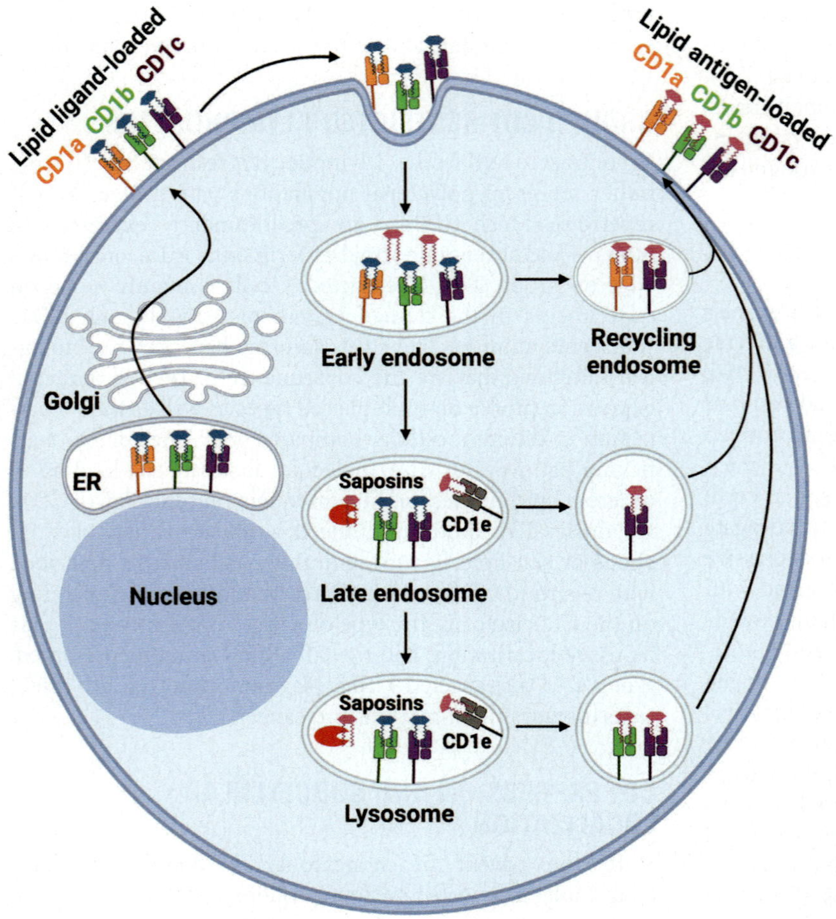

FIG. 36.6. Subcellular localization of group 1 CD1 molecules. Shown are the trafficking pathways of CD1a, CD1b, and CD1c along the endosomal route. In late endosomal compartments, saposins and CD1e assist in lipid antigen processing and loading. Note that the differential localization of CD1a-c ensures a broad surveillance for lipid antigens.

and recycling endosomes and failed to reach deeper endosomal compartments.[358] Moreover, this trafficking pathway proved to be required for the intersection with lipid antigen to allow for efficient presentation, as experimental targeting of CD1a to the late endosomal compartment abolished CD1a-restricted T cell activation.[359] Alternatively, another study reported that CD1a could also be loaded with antigen while on the cell surface.[360] Further cell biological analysis revealed that CD1a is endocytosed in a clathrin- and dynamin-independent manner and follows a recycling pathway defined by Rab22a and Arf6.[361] Mechanistically, the Arf-like small GTPase Arl13b colocalized with Rab22a and Arf6 and interacted with the actin cytoskeleton,[362] thus representing a major regulator of endocytic recycling traffic of CD1a.

CD1b contains a tyrosine-based sorting motif that allows it to traffic deep into late endosomes and lysosomes, comparable to MHC-II.[363] Endocytosis of CD1b was facilitated by interaction with the adaptor protein AP-2.[364] The subsequent transition of CD1b to lysosomes was mediated by AP-3, and among the CD1 isoforms, only CD1b was able to interact with AP-3, which explained its deep trafficking along the endosomal gradient.[365] Furthermore, the lipid length proved to be decisive for the location of antigen loading of CD1b, since lipids with long (C-80) alkyl chains needed lysosomal trafficking, while shorter (C-32) lipid tails were able to load CD1b on the cell surface.[366]

CD1c turned out to be more promiscuous regarding its choice of endosomal residence. In this context, CD1c was located in early recycling endosomes like CD1a, lysosomal compartments like CD1b, and in vesicles in between, facilitating a broad surveillance of endosomal structures.[367] However, alternative results showed that CD1c could also be loaded with antigen on the cell surface, owing to the fact that CD1c demonstrated striking surface expression.[368]

Taken together, the differential subcellular localization pattern indicates a certain division of labor between group 1 CD1 molecules, which facilitates broad monitoring of lipid antigens on cell surface and in endosomal compartments (Fig. 36.6).

LIPID LIGANDS AND ANTIGEN PRESENTATION

Seminal work discovered human T cells that recognized CD1a or CD1c, thus expanding the realm of antigen-presenting molecules beyond MHC proteins.[369] Following studies then established that CD1b was mediating presentation of an antigen derived from *M. tuberculosis*,[370] to eventually demonstrate that a mycobacterial lipid antigen, namely mycolic acid, mediated CD1b-restricted T cell activation.[371] Since that

TABLE 36.2 Lipid Ligands of Group 1 CD1 Molecules

	Ligands	Origin
CD1a	Sulfatides	*Self*
	Phosphoglycerolipids PI, PC, PE, PG, DPG	*Self* *Bacteria* *Plants*
	Lysophospholipids, LPC	*Self* *Bacteria*
	Sphingomyelin	*Self*
	Squalene, wax esters, triacylglycerides	*Self*
	Didehydroxymycobactin	*Mycobacterium*
	Urushiol	*Plants (Poison ivy)*
	Farnesol	*Skin creams*
CD1b	Gangliosides GM1, GD1a, GD1b, GT1b, GQ1b, GD3	*Self*
	Sulfatides	*Self*
	Sphingomyelin	*Self*
	Mycolic acid	*Mycobacterium*
	Glucose and glycerol monomycolates	*Mycobacterium* *Corynebacterium* *Rhodococcus* *Nocardia*
	Lipoarabinomannan	*Mycobacterium*
	Phosphatidylinositol mannoside	*Mycobacterium*
	Diacylsulfoglycolipid	*Mycobacterium*
	Phosphoglycerolipids PI, PC, PE, and PG	*Self* *Staphylococcus aureus* *Salmonella typhimurium*
CD1c	Sulfatides	*Self*
	Lysophospholipids LPC, LSM, mLPA	*Self* *Bacteria*
	Phospholipids, PC	*Self*
	Monoacylglycerol	*Self*
	Sphingomyelin	*Self*
	Mannosyl-β-1-phosphomycoketide, phosphomycoketide	*Mycobacterium*

initial discovery, a plethora of microbial and self-lipid ligands has been added to the ever-growing list of lipid entities presented by group 1 CD1 molecules (Table 36.2).

The processing of lipid ligands and antigens is assisted by several chaperones and helper molecules. In late endosomal compartments, saposins helped loading of lipid antigens onto CD1 molecules.[372] Due to their amphiphilic nature, saposins were able to bridge the biophysical gap between hydrophilic CD1 proteins and hydrophobic lipids. Saposin B directly bound lipids and CD1 in a soluble state, whereas saposin C facilitated lipid loading based on its insertion into lipid bilayers associated with membrane perturbation.[373] Moreover, the ER-resident chaperone MTP promoted activation of group 1 CD1-restricted T cells since its silencing hampered lipid presentation.[374] Another CD1 isoform, CD1e, is not expressed on the surface of antigen-presenting cells and therefore does not participate in T cell engagement. By contrast, CD1e is confined to lysosomes in a soluble form, binds glycolipids, and assists in loading of other CD1 molecules.[375] In this line, CD1e was shown to help editing hexamannosylated phosphatidylinositol (PIM6) for subsequent loading onto CD1b. The crystal structure of CD1e revealed an antigen-binding groove with significantly wider portals,[376] suggesting a propensity for rapid lipid exchange and editing processes, which pertained to all CD1 proteins present in lysosomes, including CD1c and CD1d.[377]

With regard to CD1a-restricted antigens and presentation pathways, Langerhans cells isolated from skin proved to be powerful antigen-presenting cells via CD1a after pulsing with lipid extract from *M. tuberculosis* or *Mycobacterium leprae*.[378,379] In addition, langerin, a C-type lectin located on the cell surface and in special endosomal structures, called Birbeck granules, functionally supported CD1a-mediated lipid presentation.[379] In terms of precise lipid species, CD1a has been shown to present the self-glycolipid sulfatide, which is also promiscuously presented by CD1b and CD1c.[380] The first microbial CD1a-restricted antigen was the lipopeptide named didehydroxymycobactin derived from *M. tuberculosis*, a new entity that combined antigenic acyl and peptide moieties.[381] In addition to microbial antigens, CD1a also presented lipids from plants. Accordingly, phospholipids (PC and PE) from pollen activated CD1a-restricted T cells and drove allergic responses, including IgE production.[382]

Beyond the recognition of exogenous antigens, the CD1a system exhibited pronounced autoreactivity. In this context, peripheral blood from normal healthy donors contained a large proportion of polyclonal T cells reactive with CD1a-expressing cells in the absence of additional antigen.[383,384] Notably, those CD1a-autoreactive T cells expressed skin-homing receptors and produced IL-22 in response to stimulation by Langerhans cells. When CD1a was eluted, hundreds of different lipids were recovered and analyzed by mass spectrometry. Lipids with polar headgroups proved to be inhibitory, whereas apolar, headless ligands facilitated CD1a autoreactivity.[385] Prominent species among those lipids were identified as squalene, wax esters, and fatty acids, which accumulate in skin and sebum, thus suggesting a role of those antigens in skin immunity.[385] This concept of CD1a autoreactivity was also valid when studying human skin. Using CD1a tetramer staining without added antigen, a large CD1a-autoreactive T cell pool was detected in normal human skin, demonstrating that CD1a-autoreactive T cells belong to the physiological repertoire.[386] Notably, lipidomics analysis indicated that CD1a preferentially captured sphingomyelin, a lipid that interferes with autoreactive T cell activation.[387] This supports the idea that CD1a autoreactivity represents a default mode of stimulation that can be controlled by inhibitory ligands. Lastly, an additional player in CD1a biology was identified with group 2 ILCs from human skin capable of presenting endogenous antigens via CD1a.[388]

Structural biology elucidated the precise molecular underpinnings of CD1a presentation. The crystal structure of the CD1a/sulfatide complex revealed that the headgroup was anchored in the junction between A′ and F′ pocket, therefore exposing the carbohydrate head into the F′ pocket for possible recognition by T cells.[389] The binary crystal structure of lipopeptide and CD1a demonstrated that the alkyl chain is deeply buried in the A′ pocket, thereby positioning the protruding peptide moiety for potential TCR activation.[390] Lastly, the first ternary structure of the CD1a/lipid complex with a prototypical autoreactive TCR revealed surprising molecular relationships between the three components. The TCR docked over the lateral side of the A′ roof and had no direct contact with the ligand LPC.[391] Whereas LPC was described as a permissive ligand allowing TCR docking, other lipids, such as sphingomyelin, interfered with TCR binding and were thus nonpermissive. This mechanism demonstrated true autoreactivity of the TCR with CD1a itself.

CD1b was described to present a number of mycobacterial antigens, including mycolic acid,[371] lipoarabinomannan (LAM),[392] and glucose monomycolate (GMM).[393] Notably, the GMM antigen could also be generated based on teamwork between mycobacteria and the host, when the pathogen used host glucose for coupling to mycolic acid.[394] Similar to the concept of CD1a autoreactivity, CD1b-restricted T cells have been isolated that recognize phospholipids, predominantly phosphatidylglycerol (PG).[395]

Antigen processing for CD1b-mediated presentation involved the mannose receptor for uptake of LAM and subsequent transport to late endosomes.[396] Furthermore, saposin C proved to be critical for antigen loading onto CD1b with regard to mycolic acid, GMM, and LAM.[372] Moreover, CD1b was able to accommodate long lipid tails using a mechanism based on acidification, independently of enzymatic cleavage.[397]

The initial structure of CD1b showed that it has the largest groove size compared to CD1a, CD1b, and CD1d. Compared to CD1a and CD1c with the usual A′ and F′ pockets, CD1b possessed four antigen-binding pockets and additional channels.[398] Thus, CD1b has been described as a maze for alkyl chains, which could explain how long lipid antigens, such as mycolic acid and GMM, are accommodated within CD1b. Using samples from tuberculosis patients, analysis of the TCR repertoire of lymphocytes reactive with CD1b tetramers led to the identification of a highly conserved T cell population, which was characterized as bearing a germline-encoded, mycolyl lipid-reactive TCR.[399] The ternary structure of the GEM TCR recognizing GMM presented by CD1b showed that the TCR docked centrally above CD1b.[400] Interestingly, the GEM TCR used both α- and β-chains to grab the glucose portion of GMM like a pair of tweezers. Furthermore, the resolved structure of an autoreactive TCR bound to the CD1b/phosphatidylglycerol (PG) complex demonstrated that TCR engagement changed the conformation of the lipid in order to properly fit TCR binding.[401] Whereas PG represents a rare self-lipid, TCRs also recognized common membrane phospholipids, such as phosphatidylcholine (PC). The ternary structure revealed that the TCR was able to recognize several phospholipids by shunting the different headgroups along a lateral escape channel in the TCR, which allowed the recognition of the central phosphate group.[402]

An initial report describing the first mycobacterial antigen presented by CD1c showed that isoprenoid glycolipids were differentially recognized by peripheral blood T cells from subjects recently infected with *M. tuberculosis* when compared to healthy controls.[403] Follow-up work refined the chemical analysis of those antigens, identified the lipid backbone as mycoketide, specifically produced by *M. tuberculosis*, and defined the respective CD1c antigen as mannosyl-β-1-phosphomycoketide.[404] Subsequent studies demonstrated that mannosyl-β-1-phosphomycoketide was deglycosylated to phosphomycoketide as additional CD1c-restricted antigen.[405] Furthermore, CD1c also presents self-lipids, such as methyl-lysophosphatidic acid (mLPA).[406] Interestingly, mLPA was increased in leukemia cells that additionally expressed CD1c. Consequently, mLPA-specific, CD1c-restricted T cells were able to kill CD1c$^+$ leukemia cells, while sparing normal healthy cells.

The structural analysis of the binary complex of CD1c presenting mannosyl-β-1-phosphomycoketide indicated that the alkyl chain was buried in the A′ pocket. Of note, the F′ pocket displayed an open conformation rather resembling the peptide-binding grooves of MHC molecules.[407] The crystal structure of CD1c and phosphomycoketide (PM) was similar with regard to the A′ pocket architecture, whereas the phosphate group strongly shifted compared to mannosyl-β-1-PM.[408] Finally, the first ternary structure of CD1c/lipid complexes and an autoreactive TCR demonstrated that the TCR docked over the portals of CD1c, not contacting any self-lipids, because they were fully buried in the CD1c pockets.[409] Thus, the CD1c crystal structure suggests a mechanism of veritable autoreactivity toward CD1c itself.

DISEASES

Infectious Disease

M. tuberculosis, the etiologic agent of tuberculosis, has a lipid-rich cell wall that renders it resilient in harsh environments. On the other hand, many of those mycobacterial lipids have been described as antigens to stimulate CD1-restricted T lymphocytes. This evidently suggested the hypothesis that T cells recognizing CD1-presented lipids might play a role in protection against tuberculosis. However, with group 1 CD1 molecules lacking in normal mice, establishing the proof of protection in the human system is challenging. Standard T cell experiments used peripheral blood T cells from humans infected with *M. tuberculosis* in coculture with CD1-expressing cells pulsed with lipid antigen.[403] Increased CD1-mediated T cell responses from infected individuals were interpreted as a function of the CD1 system in tuberculosis.

Improved techniques for detection of lipid antigen-specific T cells using CD1 tetramers refined subsequent analysis of T cell frequencies and function. In this context, tetramers for CD1a, CD1b, and CD1c were loaded with the respective lipid antigens lipopeptide, GMM, and phosphomycoketide, respectively, to measure antigen-specific T cell frequencies.[405,410-412] Overall, the frequencies of tetramer-positive T cells in *M. tuberculosis*–infected donors were consistently low, but sorting of specific T cells and subsequent *in vitro* expansion allowed analysis of reactivity and cytokine production. However, when a large cohort of humans with latent infection, or active tuberculosis, as well as healthy controls were tested with CD1b tetramers loaded with mycolic acid or GMM, the antigen-specific T cell frequencies did not correlate with disease state.[413] Rather, comparable numbers of CD1-specific T cells were present in the normal human immune repertoire. On a technical note, tetramer staining of T cells requires additional controls because other TCR-independent ligands have been described to bind to CD1, including fatty acid translocase and other members of the CD36 family.[414]

Another approach to achieve *in vivo* evidence of protection against infection is the use of CD1-transgenic mice. Accordingly, animals transgenic for human CD1a-c were infected with mycobacteria and showed adaptive CD1-restricted T cell responses recognizing the described lipid antigens.[355] In an additional mouse model, consisting of human CD1a-c transgene and a transgenic TCR specific for CD1b-restricted mycolic acid, mice showed lower bacterial burden after *M. tuberculosis* aerosol infection than controls.[415] This experiment provided the proof of principle that group 1 CD1-restricted T lymphocytes can mediate protection against tuberculosis.

In a different infectious disease model involving *Staphylococcus aureus*, human CD1-transgenic mice displayed T cell responses specific for *S. aureus*–derived lipids. Importantly, CD1-transgenic animals proved to be protected from pathology induced by systemic *S. aureus* infection.[416]

Inflammation

CD1a as a hallmark of Langerhans cells is known since several decades. However, the functional importance of Langerhans cell–mediated CD1a presentation in skin immunity and inflammation only recently emerged. In this context, bee and wasp venom was shown to induce CD1a-restricted T cell activation due to the action of venom-derived PLA2 that generated lysophospholipids and free fatty acids, which acted as lipid antigens in CD1a-mediated inflammatory responses.[417] Indeed, individuals allergic to bee and wasp venom exhibited elevated CD1a-mediated T cell responses.[418] Atopic dermatitis is a chronic inflammatory skin disease characterized by barrier defects and a certain allergic component driven or exacerbated by triggers, such as house dust mite (HDM). Challenge of human skin with HDM amplified HDM-derived PLA2 that generated lipid antigens presented by CD1a.[419] Accordingly, CD1a-mediated inflammatory T cell responses were increased in atopic dermatitis patients. Moreover, the protein filaggrin, which increases skin integrity and is often dysfunctional in atopic dermatitis patients, blocked PLA2 function, thus reducing CD1a-mediated inflammation.[419] Furthermore, CD1a-transgenic mice showed increased inflammatory responses in a contact dermatitis model using the lipidic compound urushiol from the sap of the poison ivy plant.[420] The crystal structure of CD1a in complex with urushiol demonstrated CD1a-mediated lipid display. Of note, treatment of mice with a CD1a blocking antibody abrogated skin inflammation.[420] Usually, allergens like urushiol are considered to work as haptens, which by themselves are not immunogenic, but have to covalently bind to host carrier proteins to induce inflammation. The above findings introduced a novel concept in allergy, in which a classical allergen did not function via haptenization, but instead acted as CD1a-presented lipid antigen. These results were supported by independent studies that showed that contact sensitizers were able to cause CD1a-restricted T cell responses.[421] Subsequent studies corroborated this concept and demonstrated that contact allergens from botanical extracts and skin care products, such as farnesol, bound to CD1a and triggered CD1a-mediated T cell activation.[422] Lastly, CD1a-based lipid presentation also played an important role in psoriasis,[420,423] another chronic skin condition characterized by inflammatory plaque formation and Th17 cell responses. Interestingly, the search for protein autoantigens in psoriasis proved not to be fruitful. Thus, lipid antigens might represent dominant triggers in psoriatic inflammation. Experimentally, CD1a-transgenic mice exhibited amplified skin inflammation in a murine psoriasis model, which was treatable by administration of anti-CD1a antibodies.[420] Additionally, psoriasis patients showed increased PLA2 expression as well as CD1a-mediated T cell stimulation associated with IL-17 and IL-22 production.[423] Taken together, CD1a-restricted T cell activation by Langerhans cells emerged as important mechanism in various inflammatory skin diseases (Fig. 36.7). CD1a-based therapies might represent powerful options to treat skin inflammation.

Using a double-transgenic murine model expressing human CD1 molecules as well as a CD1b-specific TCR, mice were additionally crossed to apolipoprotein E-deficient animals to induce lipid accumulation. As a result, hyperlipidemia led to spontaneous psoriasis-like skin inflammation driven by IL-17 production.[424] Given that psoriasis patients often have dyslipidemia, the CD1 system might provide a causative link between lipid accumulation and skin inflammation.

In inflammatory disease of the thyroid, group 1 CD1 molecules were expressed on thyroid-resident DCs and B cells. CD1-restricted T cells were absent from blood but seemed to preferentially home to inflamed thyroid tissue. Moreover, thyroid-derived T cells were reactive to CD1a and CD1c, suggesting a role of CD1 presentation in autoimmune thyroiditis.[425]

Cancer

In a mouse model transgenic for group 1 CD1 molecules and an autoreactive CD1b-restricted TCR, T lymphocytes reactive with phospholipids presented by CD1b demonstrated antitumor effects against murine T cell lymphoma.[426] Therefore, CD1b-restricted T cells could be promising for cancer immunotherapy.

Finally, treatment of leukemia often involves hematopoietic stem cell transplantation, which requires MHC matching of donor and recipient and frequently leads to GVHD. Therefore, a CD1-mediated approach would be able to avoid possible mismatch situations due to the lack of polymorphism in CD1. Notably, malignant B cells express CD1c and present the tumor-associated lipid antigen mLPA. Human T cells transduced with the mLPA-specific TCR killed CD1c-expressing tumor cells and protected against leukemia xenografts *in vivo*.[427] To conclude, lipid-specific TCR transfer into human T cells could represent a novel strategy for treatment of cancer.

ACKNOWLEDGMENTS

We would like to thank Jamie Rossjohn for providing the crystal structure image shown in Figure 36.2. The other Figures were created using BioRender.com.

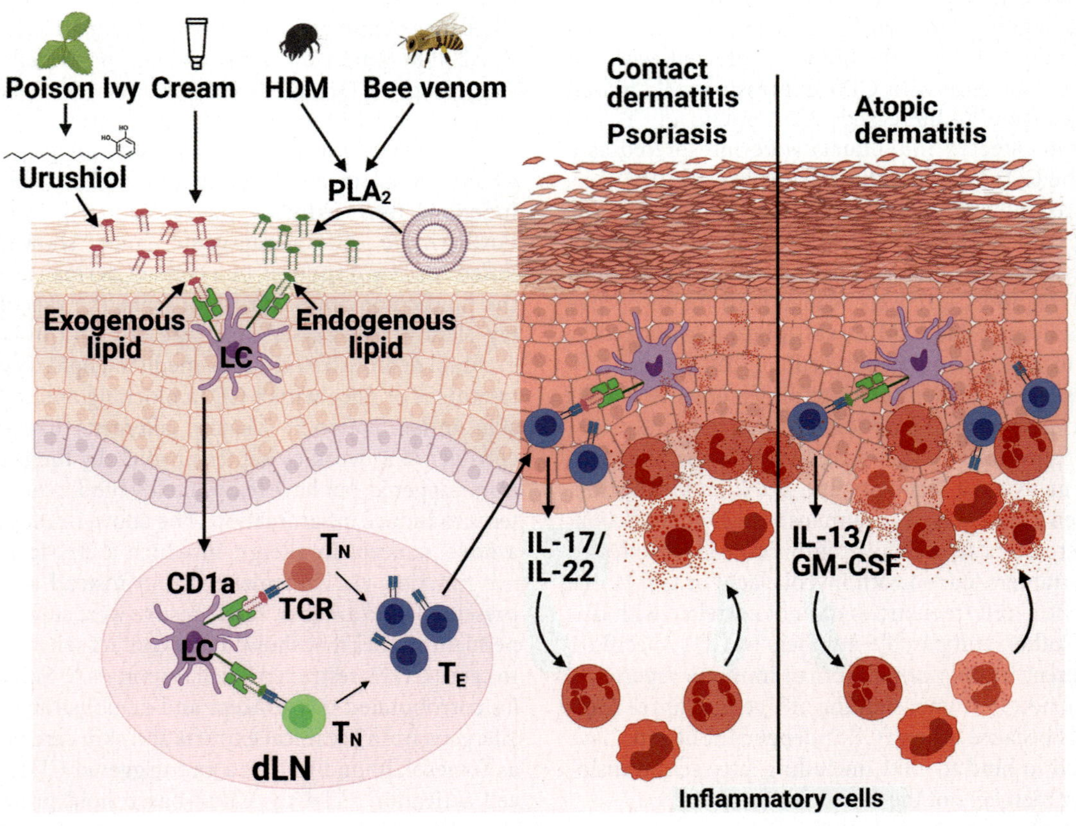

FIG. 36.7. The role of CD1a on Langerhans cells in skin inflammation. Depicted are several external triggers that can lead to provision of lipid antigens that are captured by Langerhans cells (LC) in the skin. Activated LCs migrate to draining lymph nodes to present lipid antigens via CD1a for the priming of CD1a-restricted T cells. Activated lymphocytes home to the skin and deploy their effector functions, causing skin inflammation. Depending on the initial trigger and the quality of the immune response, CD1a-mediated T cell activation is important in various skin diseases. dLN, draining lymph node; HDM, house dust mite; PLA2, phospholipase A2; T_E, effector T cell; T_N, naïve T cell.

REFERENCES

1. Lantz O, Bendelac A. An invariant T cell receptor alpha chain is used by a unique subset of major histocompatibility complex class I-specific CD4+ and CD4−8− T cells in mice and humans. *J Exp Med*. 1994;180:1097-1106.
2. Benlagha K, Kyin T, Beavis A, Teyton L, Bendelac A. A thymic precursor to the NK T cell lineage. *Science*. 2002;296:553-555.
3. Calabi F, Jarvis JM, Martin L, Milstein C. Two classes of CD1 genes. *Eur J Immunol*. 1989;19:285-292.
4. Park SH, Roark JH, Bendelac A. Tissue-specific recognition of mouse CD1 molecules. *J Immunol*. 1998;160:3128-3134.
5. Bradbury A, Calabi F, Milstein C. Expression of CD1 in the mouse thymus. *Eur J Immunol*. 1990;20:1831-1836.
6. Roark JH, Park SH, Jayawardena J, Kavita U, Shannon M, Bendelac A. CD1.1 expression by mouse antigen-presenting cells and marginal zone B cells. *J Immunol*. 1998;160:3121-3127.
7. Dougan SK, Kaser A, Blumberg RS. CD1 expression on antigen-presenting cells. *Curr Top Microbiol Immunol*. 2007;314:113-141.
8. Zeissig S, Murata K, Sweet L, et al. Hepatitis B virus-induced lipid alterations contribute to natural killer T cell-dependent protective immunity. *Nat Med*. 2012;18:1060-1068.
9. Schrumpf E, Tan C, Karlsen TH, et al. The biliary epithelium presents antigens to and activates natural killer T cells. *Hepatology*. 2015;62:1249-1259.
10. Olszak T, Neves JF, Dowds CM, et al. Protective mucosal immunity mediated by epithelial CD1d and IL-10. *Nature*. 2014;509:497-502.
11. Kang SJ, Cresswell P. Calnexin, calreticulin, and ERp57 cooperate in disulfide bond formation in human CD1d heavy chain. *J Biol Chem*. 2002;277:44838-44844.
12. Zhu Y, Zhang W, Veerapen N, Besra G, Cresswell P. Calreticulin controls the rate of assembly of CD1d molecules in the endoplasmic reticulum. *J Biol Chem*. 2010;285:38283-38292.
13. Brozovic S, Nagaishi T, Yoshida M, et al. CD1d function is regulated by microsomal triglyceride transfer protein. *Nat Med*. 2004;10:535-539.
14. Dougan SK, Salas A, Rava P, et al. Microsomal triglyceride transfer protein lipidation and control of CD1d on antigen-presenting cells. *J Exp Med*. 2005;202:529-539.
15. Yuan W, Kang SJ, Evans JE, Cresswell P. Natural lipid ligands associated with human CD1d targeted to different subcellular compartments. *J Immunol*. 2009;182:4784-4791.
16. Jayawardena-Wolf J, Benlagha K, Chiu YH, Mehr R, Bendelac A. CD1d endosomal trafficking is independently regulated by an intrinsic CD1d-encoded tyrosine motif and by the invariant chain. *Immunity*. 2001;15:897-908.
17. Cernadas M, Sugita M, van der Wel N, et al. Lysosomal localization of murine CD1d mediated by AP-3 is necessary for NK T cell development. *J Immunol*. 2003;171:4149-4155.
18. Elewaut D, Lawton AP, Nagarajan NA, et al. The adaptor protein AP-3 is required for CD1d-mediated antigen presentation of glycosphingolipids and development of Valpha14i NKT cells. *J Exp Med*. 2003;198:1133-1146.
19. Lawton AP, Prigozy TI, Brossay L, et al. The mouse CD1d cytoplasmic tail mediates CD1d trafficking and antigen presentation by adaptor protein 3-dependent and -independent mechanisms. *J Immunol*. 2005;174:3179-3186.
20. Prigozy TI, Naidenko O, Qasba P, et al. Glycolipid antigen processing for presentation by CD1d molecules. *Science*. 2001;291:664-667.
21. Darmoise A, Teneberg S, Bouzonville L, et al. Lysosomal alpha-galactosidase controls the generation of self lipid antigens for natural killer T cells. *Immunity*. 2010;33:216-228.
22. Zhou D, Cantu C, Sagiv Y, et al. Editing of CD1d-bound lipid antigens by endosomal lipid transfer proteins. *Science*. 2004;303:523-527.
23. Kang SJ, Cresswell P. Saposins facilitate CD1d-restricted presentation of an exogenous lipid antigen to T cells. *Nat Immunol*. 2004;5:175-181.
24. Yuan W, Qi X, Tsang P, et al. Saposin B is the dominant saposin that facilitates lipid binding to human CD1d molecules. *Proc Natl Acad Sci U S A*. 2007;104:5551-5556.
25. Schrantz N, Sagiv Y, Liu Y, Savage PB, Bendelac A, Teyton L. The Niemann-Pick type C2 protein loads isoglobotrihexosylceramide onto CD1d molecules and contributes to the thymic selection of NKT cells. *J Exp Med*. 2007;204:841-852.
26. van den Elzen P, Garg S, León L, et al. Apolipoprotein-mediated pathways of lipid antigen presentation. *Nature*. 2005;437:906-910.
27. Freigang S, Landais E, Zadorozhny V, et al. Scavenger receptors target glycolipids for natural killer T cell activation. *J Clin Invest*. 2012;122:3943-3954.
28. Torreno-Pina JA, Manzo C, Salio M, et al. The actin cytoskeleton modulates the activation of iNKT cells by segregating CD1d nanoclusters on antigen-presenting cells. *Proc Natl Acad Sci U S A*. 2016;113:E772-E781.
29. Keller CW, Loi M, Ewert S, et al. The autophagy machinery restrains iNKT cell activation through CD1D1 internalization. *Autophagy*. 2017;13:1025-1036.
30. Kawano T, Cui J, Koezuka Y, et al. CD1d-restricted and TCR-mediated activation of valpha14 NKT cells by glycosylceramides. *Science*. 1997;278:1626-1629.
31. Benlagha K, Weiss A, Beavis A, Teyton L, Bendelac A. *In vivo* identification of glycolipid antigen-specific T cells using fluorescent CD1d tetramers. *J Exp Med*. 2000;191:1895-1903.
32. Matsuda JL, Naidenko OV, Gapin L, et al. Tracking the response of natural killer T cells to a glycolipid antigen using CD1d tetramers. *J Exp Med*. 2000;192:741-754.
33. Kawahara K, Moll H, Knirel YA, Seydel U, Zahringer U. Structural analysis of two glycosphingolipids from the lipopolysaccharide-lacking bacterium Sphingomonas capsulata. *Eur J Biochem*. 2000;267:1837-1846.
34. Mattner J, DeBord KL, Ismail N, et al. Exogenous and endogenous glycolipid antigens activate NKT cells during microbial infections. *Nature*. 2005;434:525-529.
35. Kinjo Y, Wu D, Kim G, et al. Recognition of bacterial glycosphingolipids by natural killer T cells. *Nature*. 2005;434:520-525.
36. Kinjo Y, Tupin E, Wu D, et al. Natural killer T cells recognize diacylglycerol antigens from pathogenic bacteria. *Nat Immunol*. 2006;7:978-986.
37. Kinjo Y, Illarionov P, Vela JL, et al. Invariant natural killer T cells recognize glycolipids from pathogenic Gram-positive bacteria. *Nat Immunol*. 2011;12:966-974.
38. Albacker LA, Chaudhary V, Chang Y-J, et al. Invariant natural killer T cells recognize a fungal glycosphingolipid that can induce airway hyperreactivity. *Nat Med*. 2013;19:1297-1304.
39. Chang YJ, Kim HY, Albacker LA, et al. Influenza infection in suckling mice expands an NKT cell subset that protects against airway hyperreactivity. *J Clin Invest*. 2011;121:57-69.
40. Bendelac A, Lantz O, Quimby ME, Yewdell JW, Bennink JR, Brutkiewicz RR. CD1 recognition by mouse NK1+ T lymphocytes. *Science*. 1995;268:863-865.
41. Chiu YH, Park S-H, Benlagha K, et al. Multiple defects in antigen presentation and T cell development by mice expressing cytoplasmic tail-truncated CD1d. *Nat Immunol*. 2002;3:55-60.
42. Brigl M, Bry L, Kent SC, Gumperz JE, Brenner MB. Mechanism of CD1d-restricted natural killer T cell activation during microbial infection. *Nat Immunol*. 2003;4:1230-1237.
43. Zhou D, Mattner J, Cantu C, et al. Lysosomal glycosphingolipid recognition by NKT cells. *Science*. 2004;306:1786-1789.
44. Porubsky S, Speak AO, Luckow B, et al. Normal development and function of invariant natural killer T cells in mice with isoglobotrihexosylceramide (iGb3) deficiency. *Proc Natl Acad Sci U S A*. 2007;104:5977-5982.
45. Speak AO, Salio M, Neville DCA, et al. Implications for invariant natural killer T cell ligands due to the restricted presence of isoglobotrihexosylceramide in mammals. *Proc Natl Acad Sci U S A*. 2007;104:5971-5976.
46. Porubsky S, Speak AO, Salio M, et al. Globosides but not isoglobosides can impact the development of invariant NKT cells and their interaction with dendritic cells. *J Immunol*. 2012;189:3007-3017.
47. Li Y, Teneberg S, Thapa P, Bendelac A, Levery SB, Zhou D. Sensitive detection of isoglobo and globo series tetraglycosylceramides in human thymus by ion trap mass spectrometry. *Glycobiology*. 2008;18:158-165.
48. Christiansen D, Milland J, Mouhtouris E, et al. Humans lack iGb3 due to the absence of functional iGb3-synthase: implications for NKT cell development and transplantation. *PLoS Biol*. 2008;6:e172.
49. Sanderson JP, Brennan PJ, Mansour S, et al. CD1d protein structure determines species-selective antigenicity of isoglobotrihexosylceramide (iGb3) to invariant NKT cells. *Eur J Immunol*. 2013;43:815-825.
50. Fox LM, Cox DG, Lockridge JL, et al. Recognition of lysophospholipids by human natural killer T lymphocytes. *PLoS Biol*. 2009;7:e1000228.
51. Chang DH, Deng H, Matthews P, et al. Inflammation-associated lysophospholipids as ligands for CD1d-restricted T cells in human cancer. *Blood*. 2008;112:1308-1316.
52. Facciotti F, Ramanjaneyulu GS, Lepore M, et al. Peroxisome-derived lipids are self antigens that stimulate invariant natural killer T cells in the thymus. *Nat Immunol*. 2012;13:474-480.
53. Kain L, Webb B, Anderson BL, et al. The identification of the endogenous ligands of natural killer T cells reveals the presence of mammalian alpha-linked glycosylceramides. *Immunity*. 2014;41:543-554.
54. Brennan PJ, Tatituri RVV, Heiss C, et al. Activation of iNKT cells by a distinct constituent of the endogenous glucosylceramide fraction. *Proc Natl Acad Sci U S A*. 2014;111:13433-13438.
55. Brennan PJ, Cheng T-Y, Pellicci DG, et al. Structural determination of lipid antigens captured at the CD1d-T-cell receptor interface. *Proc Natl Acad Sci U S A*. 2017;114:8348-8353.
56. Kirschbaum C, Greis K, Mucha E, et al. Unravelling the structural complexity of glycolipids with cryogenic infrared spectroscopy. *Nat Commun*. 2021;12:1201.
57. Salio M, Speak AO, Shepherd D, et al. Modulation of human natural killer T cell ligands on TLR-mediated antigen-presenting cell activation. *Proc Natl Acad Sci U S A*. 2007;104:20490-20495.
58. Paget C, Mallevaey T, Speak AO, et al. Activation of invariant NKT cells by toll-like receptor 9-stimulated dendritic cells requires type I interferon and charged glycosphingolipids. *Immunity*. 2007;27:597-609.
59. Paget C, Deng S, Soulard D, et al. TLR9-mediated dendritic cell activation uncovers mammalian ganglioside species with specific ceramide backbones that activate invariant natural killer T cells. *PLoS Biol*. 2019;17:e3000169.
60. Bedard M, Shrestha D, Priestman DA, et al. Sterile activation of invariant natural killer T cells by ER-stressed antigen-presenting cells. *Proc Natl Acad Sci U S A*. 2019;116:23671-23681.
61. Govindarajan S, Verheugen E, Venken K, et al. ER stress in antigen-presenting cells promotes NKT cell activation through endogenous neutral lipids. *EMBO Rep*. 2020;21:e48927.
62. Melum E, Jiang X, Baker KD, et al. Control of CD1d-restricted antigen presentation and inflammation by sphingomyelin. *Nat Immunol*. 2019;20:1644-1655.
63. Halder RC, Tran C, Prasad P, et al. Self-glycerophospholipids activate murine phospholipid-reactive T cells and inhibit iNKT cell activation by competing with ligands for CD1d loading. *Eur J Immunol*. 2019;49:242-254.
64. Paduraru C, Bezbradica JS, Kunte A, et al. Role for lysosomal phospholipase A2 in iNKT cell-mediated CD1d recognition. *Proc Natl Acad Sci U S A*. 2013;110:5097-5102.
65. Garg S, Sharma M, Ung C, et al. Lysosomal trafficking, antigen presentation, and microbial killing are controlled by the Arf-like GTPase Arl8b. *Immunity*. 2011;35:182-193.
66. Zajonc DM, Cantu C, Mattner J, et al. Structure and function of a potent agonist for the semi-invariant natural killer T cell receptor. *Nat Immunol*. 2005;6:810-818.
67. Borg NA, Wun KS, Kjer-Nielsen L, et al. CD1d-lipid-antigen recognition by the semi-invariant NKT T-cell receptor. *Nature*. 2007;448:44-49.
68. Li Y, Girardi E, Wang J, et al. The Valpha14 invariant natural killer T cell TCR forces microbial glycolipids and CD1d into a conserved binding mode. *J Exp Med*. 2010;207:2383-2393.
69. Pellicci DG, Clarke AJ, Patel O, et al. Recognition of beta-linked self glycolipids mediated by natural killer T cell antigen receptors. *Nat Immunol*. 2011;12:827-833.
70. Lopez-Sagaseta J, Sibener LV, Kung JE, Gumperz J, Adams EJ. Lysophospholipid presentation by CD1d and recognition by a human Natural Killer T-cell receptor. *EMBO J*. 2012;31:2047-2059.
71. Uldrich AP, Patel O, Cameron G, et al. A semi-invariant Valpha10+ T cell antigen receptor defines a population of natural killer T cells with distinct glycolipid antigen-recognition properties. *Nat Immunol*. 2011;12:616-623.
72. Le Nours J, Praveena T, Pellicci DG, et al. Atypical natural killer T-cell receptor recognition of CD1d-lipid antigens. *Nat Commun*. 2016;7:10570.
73. Pellicci DG, Uldrich AP, Le Nours J, et al. The molecular bases of delta/alphabeta T cell-mediated antigen recognition. *J Exp Med*. 2014;211:2599-2615.
74. Reinink P, Van Rhijn I. Mammalian CD1 and MR1 genes. *Immunogenetics*. 2016;68:515-523.
75. Salomonsen J, Sorensen MR, Marston DA, et al. Two CD1 genes map to the chicken MHC, indicating that CD1 genes are ancient and likely to have been present in the primordial MHC. *Proc Natl Acad Sci U S A*. 2005;102:8668-8673.
76. Miller MM, Wang C, Parisini E, et al. Characterization of two avian MHC-like genes reveals an ancient origin of the CD1 family. *Proc Natl Acad Sci U S A*. 2005;102:8674-8679.
77. Zajonc DM, Striegl H, Dascher CC, Wilson IA. The crystal structure of avian CD1 reveals a smaller, more primordial antigen-binding pocket compared to mammalian CD1. *Proc Natl Acad Sci U S A*. 2008;105:17925-17930.
78. Dvir H, Wang J, Ly N, Dascher CC, Zajonc DM. Structural basis for lipid-antigen recognition in avian immunity. *J Immunol*. 2010;184:2504-2511.
79. Ly N, Danzl NM, Wang J, Zajonc DM, Dascher CC. Conservation of CD1 protein expression patterns in the chicken. *Dev Comp Immunol*. 2010;34:123-132.
80. Yang Z, Wang C, Wang T, et al. Analysis of the reptile CD1 genes: evolutionary implications. *Immunogenetics*. 2015;67:337-346.
81. Rogers SL, Kaufman J. Location, location, location: the evolutionary history of CD1 genes and the NKR-P1/ligand systems. *Immunogenetics*. 2016;68:499-513.
82. Ohta Y, Kasahara M, O'Connor TD, Flajnik MF. Inferring the "primordial immune complex": origins of MHC class I and antigen receptors revealed by comparative genomics. *J Immunol*. 2019;203:1882-1896.
83. Wang C, Perera TV, Ford HL, Dascher CC. Characterization of a divergent non-classical MHC class I gene in sharks. *Immunogenetics*. 2003;55:57-61.
84. Almeida T, Ohta Y, Gaigher A, et al. A highly complex, MHC-linked, 350 million-year-old shark nonclassical class I lineage. *J Immunol*. 2021;207:824-836.
85. Flajnik MF, Kasahara M, Shum BP, Salter-Cid L, Taylor E, Du Pasquier L. A novel type of class I gene organization in vertebrates: a large family of non-MHC-linked class I genes is expressed at the RNA level in the amphibian Xenopus. *EMBO J*. 1993;12:4385-4396.

86. Edholm ES, Albertorio Saez L-M, Gill AL, et al. Nonclassical MHC class I-dependent invariant T cells are evolutionarily conserved and prominent from early development in amphibians. *Proc Natl Acad Sci U S A*. 2013;110:14342-14347.
87. Edholm ES, Grayfer L, De Jesus Andino F, Robert J. Nonclassical MHC-restricted invariant valpha6 T cells are critical for efficient early innate antiviral immunity in the Amphibian *Xenopus laevis*. *J Immunol*. 2015;195:576-586.
88. Edholm ES, Banach M, Hyoe Rhoo K, Pavelka MSJr, Robert J. Distinct MHC class I-like interacting invariant T cell lineage at the forefront of mycobacterial immunity uncovered in Xenopus. *Proc Natl Acad Sci U S A*. 2018;115:E4023-E4031.
89. Rhoo KH, Edholm E-S, Forzán MJ, et al. Distinct host-mycobacterial pathogen interactions between resistant adult and tolerant tadpole life stages of *Xenopus laevis*. *J Immunol*. 2019;203:2679-2688.
90. Banach M, Edholm ES, Gonzalez X, Benraiss A, Robert J. Impacts of the MHC class I-like XNC10 and innate-like T cells on tumor tolerance and rejection in the amphibian Xenopus. *Carcinogenesis*. 2019;40:924-935.
91. Gapin L, Matsuda JL, Surh CD, Kronenberg M. NKT cells derive from double-positive thymocytes that are positively selected by CD1d. *Nat Immunol*. 2001;2:971-978.
92. Bezbradica JS, Hill T, Stanic AK, Van Kaer L, Joyce S. Commitment toward the natural T (iNKT) cell lineage occurs at the CD4+8+ stage of thymic ontogeny. *Proc Natl Acad Sci U S A*. 2005;102:5114-5119.
93. Egawa T, Eberl G, Taniuchi I, et al. Genetic evidence supporting selection of the Valpha14i NKT cell lineage from double-positive thymocyte precursors. *Immunity*. 2005;22:705-716.
94. D'Cruz LM, Knell J, Fujimoto JK, Goldrath AW. An essential role for the transcription factor HEB in thymocyte survival, Tcra rearrangement and the development of natural killer T cells. *Nat Immunol*. 2010;11:240-249.
95. Wei DG, Curran SA, Savage PB, Teyton L, Bendelac A. Mechanisms imposing the Vbeta bias of Valpha14 natural killer T cells and consequences for microbial glycolipid recognition. *J Exp Med*. 2006;203:1197-1207.
96. Chun T, Page MJ, Gapin L, et al. CD1d-expressing dendritic cells but not thymic epithelial cells can mediate negative selection of NKT cells. *J Exp Med*. 2003;197:907-918.
97. Bedel R, Berry R, Mallevaey T, et al. Effective functional maturation of invariant natural killer T cells is constrained by negative selection and T-cell antigen receptor affinity. *Proc Natl Acad Sci U S A*. 2014;111:E119-E128.
98. Benlagha K, Wei DG, Veiga J, Teyton L, Bendelac A. Characterization of the early stages of thymic NKT cell development. *J Exp Med*. 2005;202:485-492.
99. Lee YJ, Holzapfel KL, Zhu J, Jameson SC, Hogquist KA. Steady-state production of IL-4 modulates immunity in mouse strains and is determined by lineage diversity of iNKT cells. *Nat Immunol*. 2013;14:1146-1154.
100. Dose M, Sleckman BP, Han J, Bredemeyer AL, Bendelac A, Gounari F. Intrathymic proliferation wave essential for Valpha14+ natural killer T cell development depends on c-Myc. *Proc Natl Acad Sci U S A*. 2009;106:8641-8646.
101. Stetson DB, Mohrs M, Reinhardt RL, et al. Constitutive cytokine mRNAs mark natural killer (NK) and NK T cells poised for rapid effector function. *J Exp Med*. 2003;198:1069-1076.
102. Allende ML, Zhou D, Kalkofen DN, et al. S1P1 receptor expression regulates emergence of NKT cells in peripheral tissues. *FASEB J*. 2008;22:307-315.
103. Townsend MJ, Weinmann AS, Matsuda JL, et al. T-bet regulates the terminal maturation and homeostasis of NK and Valpha14i NKT cells. *Immunity*. 2004;20:477-494.
104. Wang H, Hogquist KA. CCR7 defines a precursor for murine iNKT cells in thymus and periphery. *Elife*. 2018;7.
105. Weinreich MA, Odumade OA, Jameson SC, Hogquist KA. T cells expressing the transcription factor PLZF regulate the development of memory-like CD8+ T cells. *Nat Immunol*. 2010;11:709-716.
106. Atherly LO, Lucas JA, Felices M, Yin CC, Reiner SL, Berg LJ. The Tec family tyrosine kinases Itk and Rlk regulate the development of conventional CD8+ T cells. *Immunity*. 2006;25:79-91.
107. Verykokakis M, Boos MD, Bendelac A, Kee BL. SAP protein-dependent natural killer T-like cells regulate the development of CD8(+) T cells with innate lymphocyte characteristics. *Immunity*. 2010;33:203-215.
108. Engel I, Seumois G, Chavez L, et al. Innate-like functions of natural killer T cell subsets result from highly divergent gene programs. *Nat Immunol*. 2016;17:728-739.
109. Lee YJ, Starrett GJ, Lee ST, et al. Lineage-specific effector signatures of invariant NKT cells are shared amongst gammadelta T, innate lymphoid, and Th cells. *J Immunol*. 2016;197:1460-1470.
110. Lee M, Lee E, Han SK, et al. Single-cell RNA sequencing identifies shared differentiation paths of mouse thymic innate T cells. *Nat Commun*. 2020;11:4367.
111. Baranek T, Lebrigand K, de Amat Herbozo C, et al. High dimensional single-cell analysis reveals iNKT cell developmental trajectories and effector fate decision. *Cell Rep*. 2020;32:108116.
112. Harsha Krovi S, Zhang J, Michaels-Foster MJ, et al. Thymic iNKT single cell analyses unmask the common developmental program of mouse innate T cells. *Nat Commun*. 2020;11:6238.
113. Dashtsoodol N, Shigeura T, Aihara M, et al. Alternative pathway for the development of Valpha14(+) NKT cells directly from CD4(−)CD8(−) thymocytes that bypasses the CD4(+) CD8(+) stage. *Nat Immunol*. 2017;18:274-282.
114. Wei DG, Lee H, Park S-H, et al. Expansion and long-range differentiation of the NKT cell lineage in mice expressing CD1d exclusively on cortical thymocytes. *J Exp Med*. 2005;202:239-248.
115. Lee YJ, Jeon YK, Kang BH, et al. Generation of PLZF+ CD4+ T cells via MHC class II-dependent thymocyte-thymocyte interaction is a physiological process in humans. *J Exp Med*. 2010;207:237-246.
116. Georgiev H, Peng C, Huggins MA, Jameson SC, Hogquist KA. Classical MHC expression by DP thymocytes impairs the selection of non-classical MHC restricted innate-like T cells. *Nat Commun*. 2021;12:2308.
117. Griewank K, Borowski C, Rietdijk S, et al. Homotypic interactions mediated by Slamf1 and Slamf6 receptors control NKT cell lineage development. *Immunity*. 2007;27:751-762.
118. Nichols KE, Hom J, Gong S-Y, et al. Regulation of NKT cell development by SAP, the protein defective in XLP. *Nat Med*. 2005;11:340-345.
119. Eberl G, Lowin-Kropf B, MacDonald HR. Cutting edge: NKT cell development is selectively impaired in Fyn- deficient mice. *J Immunol*. 1999;163:4091-4094.
120. Jordan MA, Fletcher JM, Jose R, et al. Role of SLAM in NKT cell development revealed by transgenic complementation in NOD mice. *J Immunol*. 2011;186:3953-3965.
121. Dutta M, Kraus ZJ, Gomez-Rodriguez J, et al. A role for Ly108 in the induction of promyelocytic zinc finger transcription factor in developing thymocytes. *J Immunol*. 2013;190:2121-2128.
122. Michel ML, Lenoir C, Massot B, et al. SLAM-associated protein favors the development of iNKT2 over iNKT17 cells. *Eur J Immunol*. 2016;46:2162-2174.
123. Baglaenko Y, Cruz Tleugabulova M, Gracey E, et al. Invariant NKT cell activation is potentiated by homotypic trans-ly108 interactions. *J Immunol*. 2017;198:3949-3962.
124. Huang B, Gomez-Rodriguez J, Preite S, Garrett LJ, Harper UL, Schwartzberg PL. CRISPR-mediated triple knockout of SLAMF1, SLAMF5 and SLAMF6 supports positive signaling roles in NKT cell development. *PLoS One*. 2016;11:e0156072.
125. Lu Y, Zhong M-C, Qian J, et al. SLAM receptors foster iNKT cell development by reducing TCR signal strength after positive selection. *Nat Immunol*. 2019;20:447-457.
126. Lazarevic V, Zullo AJ, Schweitzer MN, et al. The gene encoding early growth response 2, a target of the transcription factor NFAT, is required for the development and maturation of natural killer T cells. *Nat Immunol*. 2009;10:306-313.
127. Seiler MP, Mathew R, Liszewski MK, et al. Elevated and sustained expression of the transcription factors Egr1 and Egr2 controls NKT lineage differentiation in response to TCR signaling. *Nat Immunol*. 2012;13:264-271.
128. Kumar A, Hill TM, Gordy LE, et al. Nur77 controls tolerance induction, terminal differentiation, and effector functions in semi-invariant natural killer T cells. *Proc Natl Acad Sci U S A*. 2020;117:17156-17165.
129. Moran AE, Holzapfel KL, Xing Y, et al. T cell receptor signal strength in Treg and iNKT cell development demonstrated by a novel fluorescent reporter mouse. *J Exp Med*. 2011;208:1279-1289.
130. Joseph C, Klibi J, Amable L, et al. TCR density in early iNKT cell precursors regulates agonist selection and subset differentiation in mice. *Eur J Immunol*. 2019;49:894-910.
131. Tuttle KD, Krovi SH, Zhang J, et al. TCR signal strength controls thymic differentiation of iNKT cell subsets. *Nat Commun*. 2018;9:2650.
132. Zhao M, Svensson MND, Venken K, et al. Altered thymic differentiation and modulation of arthritis by invariant NKT cells expressing mutant ZAP70. *Nat Commun*. 2018;9:2627.
133. Lee YJ, Wang H, Starrett GJ, Phuong V, Jameson SC, Hogquist KA. Tissue-specific distribution of iNKT cells impacts their cytokine response. *Immunity*. 2015;43:566-578.
134. Miller CN, Proekt I, von Moltke J, et al. Thymic tuft cells promote an IL-4-enriched medulla and shape thymocyte development. *Nature*. 2018;559:627-631.
135. Wang H, Breed ER, Lee YJ, Qian LJ, Jameson SC, Hogquist KA. Myeloid cells activate iNKT cells to produce IL-4 in the thymic medulla. *Proc Natl Acad Sci U S A*. 2019;116:22262-22268.
136. Lucas B, White AJ, Cosway EJ, et al. Diversity in medullary thymic epithelial cells controls the activity and availability of iNKT cells. *Nat Commun*. 2020;11:2198.
137. White AJ, Jenkinson WE, Cowan JE, et al. An essential role for medullary thymic epithelial cells during the intrathymic development of invariant NKT cells. *J Immunol*. 2014;192:2659-2666.
138. Kimura MY, Igi A, Hayashizaki K, et al. CD69 prevents PLZF(hi) innate precursors from prematurely exiting the thymus and aborting NKT2 cell differentiation. *Nat Commun*. 2018;9:3749.
139. White AJ, Baik S, Parnell SM, et al. A type 2 cytokine axis for thymus emigration. *J Exp Med*. 2017;214:2205-2216.
140. Kovalovsky D, Uche OU, Eladad S, et al. The BTB-zinc finger transcriptional regulator PLZF controls the development of invariant natural killer T cell effector functions. *Nat Immunol*. 2008;9:1055-1064.
141. Savage AK, Constantinides MG, Han J, et al. The transcription factor PLZF directs the effector program of the NKT cell lineage. *Immunity*. 2008;29:391-403.
142. Kovalovsky D, Alonzo ES, Uche OU, Eidson M, Nichols KE, Sant'Angelo DB. PLZF induces the spontaneous acquisition of memory/effector functions in T cells independently of NKT cell-related signals. *J Immunol*. 2010;184:6746-6755.
143. Savage AK, Constantinides MG, Bendelac A. Promyelocytic leukemia zinc finger turns on the effector T cell program without requirement for agonist TCR signaling. *J Immunol*. 2011;186:5801-5806.
144. Zhang S, Laouar A, Denzin LK, Sant'Angelo DB. Zbtb16 (PLZF) is stably suppressed and not inducible in non-innate T cells via T cell receptor-mediated signaling. *Sci Rep*. 2015;5:12113.
145. Park JY, DiPalma DT, Kwon J, Fink J, Park JH. Quantitative difference in PLZF protein expression determines iNKT lineage fate and controls innate CD8 T cell generation. *Cell Rep*. 2019;27:2548-2557 e2544.
146. Mao AP, Constantinides MG, Mathew R, et al. Multiple layers of transcriptional regulation by PLZF in NKT-cell development. *Proc Natl Acad Sci U S A*. 2016;113:7602-7607.
147. Mao AP, Ishizuka IE, Kasal DN, Mandal M, Bendelac A. A shared Runx1-bound Zbtb16 enhancer directs innate and innate-like lymphoid lineage development. *Nat Commun*. 2017;8:863.
148. Kim PJ, Pai S-Y, Brigl M, Besra GS, Gumperz J, Ho I-C. GATA-3 regulates the development and function of invariant NKT cells. *J Immunol*. 2006;177:6650-6659.
149. Schmidt-Supprian M, Tian J, Grant EP, et al. Differential dependence of CD4+CD25+ regulatory and natural killer-like T cells on signals leading to NF-kappaB activation. *Proc Natl Acad Sci U S A*. 2004;101:4566-4571.
150. Stanic AK, Bezbradica JS, Park J-J, et al. NF-kappa B controls cell fate specification, survival, and molecular differentiation of immunoregulatory natural T lymphocytes. *J Immunol*. 2004;172:2265-2273.
151. Stankovic S, Gugasyan R, Kyparissoudis K, et al. Distinct roles in NKT cell maturation and function for the different transcription factors in the classical NF-kappaB pathway. *Immunol Cell Biol*. 2011;89:294-303.
152. Sivakumar V, Hammond KJ, Howells N, Pfeffer K, Weih F. Differential requirement for Rel/nuclear factor kappa B family members in natural killer T cell development. *J Exp Med*. 2003;197:1613-1621.
153. Li MO, Sanjabi S, Flavell RA. Transforming growth factor-beta controls development, homeostasis, and tolerance of T cells by regulatory T cell-dependent and -independent mechanisms. *Immunity*. 2006;25:455-471.
154. Marie JC, Liggitt D, Rudensky AY. Cellular mechanisms of fatal early-onset autoimmunity in mice with the T cell-specific targeting of transforming growth factor-beta receptor. *Immunity*. 2006;25:441-454.
155. Doisne JM, Bartholin L, Yan K-P, et al. iNKT cell development is orchestrated by different branches of TGF-beta signaling. *J Exp Med*. 2009;206:1365-1378.
156. Engel I, Hammond K, Sullivan BA, et al. Co-receptor choice by V alpha14i NKT cells is driven by Th-POK expression rather than avoidance of CD8-mediated negative selection. *J Exp Med*. 2010;207:1015-1029.
157. Engel I, Zhao M, Kappes D, Taniuchi I, Kronenberg M. The transcription factor Th-POK negatively regulates Th17 differentiation in Valpha14i NKT cells. *Blood*. 2012;120:4524-4532.
158. Enders A, Stankovic S, Teh C, et al. ZBTB7B (Th-POK) regulates the development of IL-17-producing CD1d-restricted mouse NKT cells. *J Immunol*. 2012;189:5240-5249.
159. Uddin MN, Sultana DA, Lorentsen KJ, et al. Transcription factor Bcl11b sustains iNKT1 and iNKT2 cell programs, restricts iNKT17 cell program, and governs iNKT cell survival. *Proc Natl Acad Sci U S A*. 2016;113:7608-7613.
160. van Gisbergen KP, Kragten NAM, Hertoghs KML, et al. Mouse Hobit is a homolog of the transcriptional repressor Blimp-1 that regulates NKT cell effector differentiation. *Nat Immunol*. 2012;13:864-871.
161. Gioulbasani M, Galaras A, Grammenoudi S, et al. The transcription factor BCL-6 controls early development of innate-like T cells. *Nat Immunol*. 2020;21:1058-1069.
162. Fedeli M, Napolitano A, Wong MPM, et al. Dicer-dependent microRNA pathway controls invariant NKT cell development. *J Immunol*. 2009;183:2506-2512.
163. Zhou L, Seo K-H, He H-Z, et al. Tie2cre-induced inactivation of the miRNA-processing enzyme Dicer disrupts invariant NKT cell development. *Proc Natl Acad Sci U S A*. 2009;106:10266-10271.
164. Fedeli M, Riba M.; Garcia Manteiga J. M., et al. miR-17 approximately 92 family clusters control iNKT cell ontogenesis via modulation of TGF-beta signaling. *Proc Natl Acad Sci U S A*. 2016;113:E8286-E8295.

165. Zietara N, Lyszkiewicz M, Witzlau K, et al. Critical role for miR-181a/b-1 in agonist selection of invariant natural killer T cells. *Proc Natl Acad Sci U S A*. 2013;110:7407-7412.
166. Henao-Mejia J, Williams A, Goff LA, et al. The microRNA miR-181 is a critical cellular metabolic rheostat essential for NKT cell ontogenesis and lymphocyte development and homeostasis. *Immunity*. 2013;38:984-997.
167. Di Pietro C, De Giorgi L, Cosorich I, Sorini C, Fedeli M, Falcone M. MicroRNA-133b regulation of Th-pok expression and dendritic cell signals affect NKT17 cell differentiation in the thymus. *J Immunol*. 2016;197:3271-3280.
168. Pobezinsky LA, Etzensperger R, Jeurling S, et al. Let-7 microRNAs target the lineage-specific transcription factor PLZF to regulate terminal NKT cell differentiation and effector function. *Nat Immunol*. 2015;16:517-524.
169. Dobenecker MW, Kim JK, Marcello J, et al. Coupling of T cell receptor specificity to natural killer T cell development by bivalent histone H3 methylation. *J Exp Med*. 2015;212:297-306.
170. Vasanthakumar A, Xu D, Lun AT, et al. A non-canonical function of Ezh2 preserves immune homeostasis. *EMBO Rep*. 2017;18:619-631.
171. Pereira RM, Martinez GJ, Engel I, et al. Jarid2 is induced by TCR signalling and controls iNKT cell maturation. *Nat Commun*. 2014;5:4540.
172. Beyaz S, Kim JH, Pinello L, et al. The histone demethylase UTX regulates the lineage-specific epigenetic program of invariant natural killer T cells. *Nat Immunol*. 2017;18:184-195.
173. Northrup D, Yagi R, Cui K, et al. Histone demethylases UTX and JMJD3 are required for NKT cell development in mice. *Cell Biosci*. 2017;7:25.
174. Tsagaratou A, González-Avalos E, Rautio S, et al. TET proteins regulate the lineage specification and TCR-mediated expansion of iNKT cells. *Nat Immunol*. 2017;18:45-53.
175. Zhou XF, Yu J, Chang M, et al. TRIM28 mediates chromatin modifications at the TCRalpha enhancer and regulates the development of T and natural killer T cells. *Proc Natl Acad Sci U S A*. 2012;109:20083-20088.
176. Wang X, Bishop KA, Hegde S, Rodenkirch LA, Pike JW, Gumperz JE. Human invariant natural killer T cells acquire transient innate responsiveness via histone H4 acetylation induced by weak TCR stimulation. *J Exp Med*. 2012;209:987-1000.
177. Kakugawa K, Kojo S, Tanaka H, et al. Essential roles of SATB1 in specifying T lymphocyte subsets. *Cell Rep*. 2017;19:1176-1188.
178. Zhang Y, Wang Y, Gao B, et al. USP22 controls iNKT immunity through MED1 suppression of histone H2A monoubiquitination. *J Exp Med*. 2020;217.
179. Park SH, Benlagha K, Lee D, Balish E, Bendelac A. Unaltered phenotype, tissue distribution and function of Valpha14(+) NKT cells in germ-free mice. *Eur J Immunol*. 2000;30:620-625.
180. Olszak T, An D, Zeissig S, et al. Microbial exposure during early life has persistent effects on natural killer T cell function. *Science*. 2012;336:489-493.
181. An D, Oh SF, Olszak T, et al. Sphingolipids from a symbiotic microbe regulate homeostasis of host intestinal natural killer T cells. *Cell*. 2014;156:123-133.
182. Wieland Brown LC, Penaranda C, Kashyap PC, et al. Production of alpha-galactosylceramide by a prominent member of the human gut microbiota. *PLoS Biol*. 2013;11:e1001610.
183. von Gerichten J, Lamprecht D, Opálka L, et al. Bacterial immunogenic alpha-galactosylceramide identified in the murine large intestine: dependency on diet and inflammation. *J Lipid Res*. 2019;60:1892-1904.
184. Nieuwenhuis EE, Matsumoto T, Lindenbergh D, et al. Cd1d-dependent regulation of bacterial colonization in the intestine of mice. *J Clin Invest*. 2009;119:1241-1250.
185. Saez de Guinoa J, Jimeno R, Gaya M, et al. CD1d-mediated lipid presentation by CD11c(+) cells regulates intestinal homeostasis. *EMBO J*. 2018;37.
186. Selvanantham T, Lin Q, Guo CX, et al. NKT cell-deficient mice harbor an altered microbiota that fuels intestinal inflammation during chemically induced colitis. *J Immunol*. 2016;197:4464-4472.
187. Lynch L, Nowak M, Varghese B, et al. Adipose tissue invariant NKT cells protect against diet-induced obesity and metabolic disorder through regulatory cytokine production. *Immunity*. 2012;37:574-587.
188. Schipper HS, Rakhshandehroo M, van de Graaf SFJ, et al. Natural killer T cells in adipose tissue prevent insulin resistance. *J Clin Invest*. 2012;122:3343-3354.
189. Lynch L, Michelet X, Zhang S, et al. Regulatory iNKT cells lack expression of the transcription factor PLZF and control the homeostasis of T(reg) cells and macrophages in adipose tissue. *Nat Immunol*. 2015;16:85-95.
190. Lynch L, Hogan AE, Duquette D, et al. iNKT cells induce FGF21 for thermogenesis and are required for maximal weight loss in GLP1 therapy. *Cell Metab*. 2016;24:510-519.
191. LaMarche NM, Kane H, Kohlgruber AC, Dong H, Lynch L, Brenner MB. Distinct iNKT cell populations use IFNgamma or ER stress-induced IL-10 to control adipose tissue homeostasis. *Cell Metabol*. 2020;32:243-258.e246.
192. Wu L, Parekh VV, Gabriel CL, et al. Activation of invariant natural killer T cells by lipid excess promotes tissue inflammation, insulin resistance, and hepatic steatosis in obese mice. *Proc Natl Acad Sci U S A*. 2012;109:E1143-E1152.
193. Satoh M, Hoshino M, Fujita K, et al. Adipocyte-specific CD1d-deficiency mitigates diet-induced obesity and insulin resistance in mice. *Sci Rep*. 2016;6:28473.
194. Benezech C, Luu N-T, Walker JA, et al. Inflammation-induced formation of fat-associated lymphoid clusters. *Nat Immunol*. 2015;16:819-828.
195. Vieth JA, Das J, Ranaivoson FM, et al. TCRalpha-TCRbeta pairing controls recognition of CD1d and directs the development of adipose NKT cells. *Nat Immunol*. 2017;18:36-44.
196. Parekh VV, Wu L, Boyd KL, et al. Impaired autophagy, defective T cell homeostasis, and a wasting syndrome in mice with a T cell-specific deletion of Vps34. *J Immunol*. 2013;190:5086-5101.
197. Salio M, Puleston DJ, Mathan TSM, et al. Essential role for autophagy during invariant NKT cell development. *Proc Natl Acad Sci U S A*. 2014;111:E5678-E5687.
198. Pei B, Zhao M, Miller BC, et al. Invariant NKT cells require autophagy to coordinate proliferation and survival signals during differentiation. *J Immunol*. 2015;194:5872-5884.
199. Shin J, Wang S, Deng W, Wu J, Gao J, Zhong X-P. Mechanistic target of rapamycin complex 1 is critical for invariant natural killer T-cell development and effector function. *Proc Natl Acad Sci U S A*. 2014;111:E776-E783.
200. Zhang L, Tschumi BO, Corgnac S, et al. Mammalian target of rapamycin complex 1 orchestrates invariant NKT cell differentiation and effector function. *J Immunol*. 2014;193:1759-1765.
201. Wei J, Yang K, Chi H. Cutting edge: discrete functions of mTOR signaling in invariant NKT cell development and NKT17 fate decision. *J Immunol*. 2014;193:4297-4301.
202. Park H, Tsang M, Iritani BM, Bevan MJ. Metabolic regulator Fnip1 is crucial for iNKT lymphocyte development. *Proc Natl Acad Sci U S A*. 2014;111:7066-7071.
203. Kumar A, Pyaram K, Yarosz EL, et al. Enhanced oxidative phosphorylation in NKT cells is essential for their survival and function. *Proc Natl Acad Sci U S A*. 2019;116:7439-7448.
204. Pyaram K, Kumar A, Kim Y-H, et al. Keap1-Nrf2 system plays an important role in invariant natural killer T cell development and homeostasis. *Cell Rep*. 2019;27:699-707.e694.
205. Doisne JM, Becourt C, Amniai L, et al. Skin and peripheral lymph node invariant NKT cells are mainly retinoic acid receptor-related orphan receptor (gamma)t+ and respond preferentially under inflammatory conditions. *J Immunol*. 2009;183:2142-2149.
206. McNab FW, Pellicci DG, Field K, et al. Peripheral NK1.1 NKT cells are mature and functionally distinct from their thymic counterparts. *J Immunol*. 2007;179:6630-6637.
207. Geissmann F, Cameron TO, Sidobre S, et al. Intravascular immune surveillance by CXCR6+ NKT cells patrolling liver sinusoids. *PLoS Biol*. 2005;3:e113.
208. Thomas SY, Scanlon ST, Griewank KG, et al. PLZF induces an intravascular surveillance program mediated by long-lived LFA-1-ICAM-1 interactions. *J Exp Med*. 2011;208:1179-1188.
209. Salou M, Legoux F, Gilet J, et al. A common transcriptomic program acquired in the thymus defines tissue residency of MAIT and NKT subsets. *J Exp Med*. 2019;216:133-151.
210. Mackay LK, Minnich M, Kragten NAM, et al. Hobit and Blimp1 instruct a universal transcriptional program of tissue residency in lymphocytes. *Science*. 2016;352:459-463.
211. Murray MP, Engel I, Seumois G, et al. Transcriptome and chromatin landscape of iNKT cells are shaped by subset differentiation and antigen exposure. *Nat Commun*. 2021;12:1446.
212. Jimeno R, Lebrusant-Fernandez M, Margreitter C, et al. Tissue-specific shaping of the TCR repertoire and antigen specificity of iNKT cells. *Elife*. 2019;8:e51663.
213. Clancy-Thompson E, Chen GZ, Tyler PM, et al. Monoclonal invariant NKT (iNKT) cell mice reveal a role for both tissue of origin and the TCR in development of iNKT functional subsets. *J Immunol*. 2017;199:159-171.
214. Matsuda JL, Gapin L, Sidobre S, et al. Homeostasis of V alpha 14i NKT cells. *Nat Immunol*. 2002;3:966-974.
215. Vahl JC, Heger K, Knies N, et al. NKT cell-TCR expression activates conventional T cells *in vivo*, but is largely dispensable for mature NKT cell biology. *PLoS Biol*. 2013;11:e1001589.
216. Ohteki T, Ho S, Suzuki H, Mak TW, Ohashi PS. Role for IL-15/IL-15 receptor beta-chain in natural killer 1.1+ T cell receptor-alpha beta+ cell development. *J Immunol*. 1997;159:5931-5935.
217. Castillo EF, Acero LF, Stonier SW, Zhou D, Schluns KS. Thymic and peripheral microenvironments differentially mediate development and maturation of iNKT cells by IL-15 transpresentation. *Blood*. 2010;116:2494-2503.
218. Cui G, Hara T, Simmons S, et al. Characterization of the IL-15 niche in primary and secondary lymphoid organs *in vivo*. *Proc Natl Acad Sci U S A*. 2014;111:1915-1920.
219. Webster KE, Kim H-O, Kyparissoudis K, et al. IL-17-producing NKT cells depend exclusively on IL-7 for homeostasis and survival. *Mucosal Immunol*. 2014;7:1058-1067.
220. Sag D, Krause P, Hedrick CC, Kronenberg M, Wingender G. IL-10-producing NKT10 cells are a distinct regulatory invariant NKT cell subset. *J Clin Invest*. 2014;124:3725-3740.
221. Darcy PW, Jin K, Osorio L, Denzin LK, Sant'Angelo DB. Coexpression of YY1 is required to elaborate the effector functions controlled by PLZF in NKT cells. *J Immunol*. 2019;203:627-638.
222. Darcy PW, Denzin LK, Sant'Angelo DB. YY1(lo) NKT cells are dedicated IL-10 producers. *Sci Rep*. 2020;10:3897.
223. Govindarajan S, Gaublomme D, Van der Cruyssen R, et al. Stabilization of cytokine mRNAs in iNKT cells requires the serine-threonine kinase IRE1alpha. *Nat Commun*. 2018;9:5340.
224. Kanda M, Yamanaka H, Kojo S, et al. Transcriptional regulator Bhlhe40 works as a cofactor of T-bet in the regulation of IFN-gamma production in iNKT cells. *Proc Natl Acad Sci U S A*. 2016;113:E3394-E3402.
225. Das R, Bassiri H, Guan P, et al. The adaptor molecule SAP plays essential roles during invariant NKT cell cytotoxicity and lytic synapse formation. *Blood*. 2013;121:3386-3395.
226. Parekh VV, Wilson MT, Olivares-Villagómez D, et al. Glycolipid antigen induces long-term natural killer T cell anergy in mice. *J Clin Invest*. 2005;115:2572-2583.
227. Uldrich AP, Crowe NY, Kyparissoudis K, et al. NKT cell stimulation with glycolipid antigen in vivo: costimulation-dependent expansion, Bim-dependent contraction, and hyporesponsiveness to further antigenic challenge. *J Immunol*. 2005;175:3092-3101.
228. Parekh VV, Lalani S, Kim S, et al. PD-1/PD-L blockade prevents anergy induction and enhances the anti-tumor activities of glycolipid-activated invariant NKT cells. *J Immunol*. 2009;182:2816-2826.
229. Fujii S, Shimizu K, Kronenberg M, Steinman RM. Prolonged IFN-gamma-producing NKT response induced with alpha-galactosylceramide-loaded DCs. *Nat Immunol*. 2002;3:867-874.
230. Wingender G, Birkholz AM, Sag D, et al. Selective conditions are required for the induction of invariant NKT cell hyporesponsiveness by antigenic stimulation. *J Immunol*. 2015;195:3838-3848.
231. Stradner MH, Cheung KP, Lasorella A, Goldrath AW, D'Cruz LM. Id2 regulates hyporesponsive invariant natural killer T cells. *Immunol Cell Biol*. 2016;94:640-645.
232. Kim TJ, Park G, Kim J, et al. CD160 serves as a negative regulator of NKT cells in acute hepatic injury. *Nat Commun*. 2019;10:3258.
233. Kitamura H, Iwakabe K, Yahata T, et al. The natural killer T (NKT) cell ligand alpha-galactosylceramide demonstrates its immunopotentiating effect by inducing interleukin (IL)-12 production by dendritic cells and IL-12 receptor expression on NKT cells. *J Exp Med*. 1999;189:1121-1128.
234. Tomura M, Yu WG, Ahn HJ, et al. A novel function of Valpha14+CD4+NKT cells: stimulation of IL-12 production by antigen-presenting cells in the innate immune system. *J Immunol*. 1999;163:93-101.
235. Semmling V, Lukacs-Kornek V, Thaiss CA, et al. Alternative cross-priming through CCL17-CCR4-mediated attraction of CTLs toward NKT cell-licensed DCs. *Nat Immunol*. 2010;11:313-320.
236. Fujii S, Shimizu K, Smith C, Bonifaz L, Steinman RM. Activation of natural killer T cells by alpha-galactosylceramide rapidly induces the full maturation of dendritic cells *in vivo* and thereby acts as an adjuvant for combined CD4 and CD8 T cell immunity to a coadministered protein. *J Exp Med*. 2003;198:267-279.
237. Fujii S, Liu K, Smith C, Bonito AJ, Steinman RM. The linkage of innate to adaptive immunity via maturing dendritic cells *in vivo* requires CD40 ligation in addition to antigen presentation and CD80/86 costimulation. *J Exp Med*. 2004;199:1607-1618.
238. Miyamoto K, Miyake S, Yamamura T. A synthetic glycolipid prevents autoimmune encephalomyelitis by inducing TH2 bias of natural killer T cells. *Nature*. 2001;413:531-534.
239. Oki S, Chiba A, Yamamura T, Miyake S. The clinical implication and molecular mechanism of preferential IL-4 production by modified glycolipid-stimulated NKT cells. *J Clin Invest*. 2004;113:1631-1640.
240. Im JS, Arora P, Bricard G, et al. Kinetics and cellular site of glycolipid loading control the outcome of natural killer T cell activation. *Immunity*. 2009;30:888-898.
241. Bai L, Constantinides MG, Thomas SY, et al. Distinct APCs explain the cytokine bias of alpha-galactosylceramide variants *in vivo*. *J Immunol*. 2012;188:3053-3061.
242. Arora P, Baena A, Yu KOA, et al. A single subset of dendritic cells controls the cytokine bias of natural killer T cell responses to diverse glycolipid antigens. *Immunity*. 2014;40:105-116.
243. Tonti E, Galli G, Malzone C, Abrignani S, Casorati G, Dellabona P. NKT-cell help to B lymphocytes can occur independently of cognate interaction. *Blood*. 2009;113:370-376.
244. Galli G, Nuti S, Tavarini S, et al. CD1d-restricted help to B cells by human invariant natural killer T lymphocytes. *J Exp Med*. 2003;197:1051-1057.
245. Leadbetter EA, Brigl M, Illarionov P, et al. NK T cells provide lipid antigen-specific cognate help for B cells. *Proc Natl Acad Sci U S A*. 2008;105:8339-8344.
246. Barral P, Eckl-Dorna J, Harwood NE, et al. B cell receptor-mediated uptake of CD1d-restricted antigen augments

antibody responses by recruiting invariant NKT cell help *in vivo*. *Proc Natl Acad Sci U S A*. 2008;105:8345-8350.

247. Chang PP, Barral P, Fitch J, et al. Identification of Bcl-6-dependent follicular helper NKT cells that provide cognate help for B cell responses. *Nat Immunol*. 2011;13:35-43.
248. King IL, Fortier A, Tighe M, et al. Invariant natural killer T cells direct B cell responses to cognate lipid antigen in an IL-21-dependent manner. *Nat Immunol*. 2011;13:44-50.
249. Vomhof-DeKrey EE, Yates J, Hägglöf T, et al. Cognate interaction with iNKT cells expands IL-10-producing B regulatory cells. *Proc Natl Acad Sci U S A*. 2015;112:12474-12479.
250. Bai L, Deng S, Reboulet R, et al. Natural killer T (NKT)-B-cell interactions promote prolonged antibody responses and long-term memory to pneumococcal capsular polysaccharides. *Proc Natl Acad Sci U S A*. 2013;110:16097-16102.
251. Muppidi JR, Arnon TI, Bronevetsky Y, et al. Cannabinoid receptor 2 positions and retains marginal zone B cells within the splenic marginal zone. *J Exp Med*. 2011;208:1941-1948.
252. Barral P, Sanchez-Nino MD, van Rooijen N, Cerundolo V, Batista FD. The location of splenic NKT cells favours their rapid activation by blood-borne antigen. *EMBO J*. 2012;31:2378-2390.
253. King IL, Amiel E, Tighe M, et al. The mechanism of splenic invariant NKT cell activation dictates localization *in vivo*. *J Immunol*. 2013;191:572-582.
254. Clancy-Thompson E, Chen GZ, LaMarche NM, et al. Transnuclear mice reveal Peyer's patch iNKT cells that regulate B-cell class switching to IgG1. *EMBO J*. 2019;38:e101260.
255. Bosma A, Abdel-Gadir A, Isenberg DA, Jury EC, Mauri C. Lipid-antigen presentation by CD1d(+) B cells is essential for the maintenance of invariant natural killer T cells. *Immunity*. 2012;36:477-490.
256. Oleinika K, Rosser EC, Matei DE, et al. CD1d-dependent immune suppression mediated by regulatory B cells through modulations of iNKT cells. *Nat Commun*. 2018;9:684.
257. Carnaud C, Lee D, Donnars O, et al. Cutting edge: cross-talk between cells of the innate immune system. NKT cells rapidly activate NK cells. *J Immunol*. 1999;163:4647-4650.
258. Eberl G, MacDonald HR. Selective induction of NK cell proliferation and cytotoxicity by activated NKT cells. *Eur J Immunol*. 2000;30:985-992.
259. Hagglof T, Sedimbi SK, Yates JL, et al. Neutrophils license iNKT cells to regulate self-reactive mouse B cell responses. *Nat Immunol*. 2016;17:1407-1414.
260. Sedimbi SK, Hägglöf T, Garimella MG, et al. Combined proinflammatory cytokine and cognate activation of invariant natural killer T cells enhances anti-DNA antibody responses. *Proc Natl Acad Sci U S A*. 2020;117:9054-9063.
261. Thanabalasuriar A, Neupane AS, Wang J, Krummel MF, Kubes P. iNKT cell emigration out of the lung vasculature requires neutrophils and monocyte-derived dendritic cells in inflammation. *Cell Rep*. 2016;16:3260-3272.
262. Barral P, Polzella P, Bruckbauer A, et al. CD169(+) macrophages present lipid antigens to mediate early activation of iNKT cells in lymph nodes. *Nat Immunol*. 2010;11:303-312.
263. Gaya M, Barral P, Burbage M, et al. Initiation of antiviral B cell immunity relies on innate signals from spatially positioned NKT cells. *Cell*. 2018;172:517-533.e520.
264. Cavallari M, Stallforth P, Kalinichenko A, et al. A semisynthetic carbohydrate-lipid vaccine that protects against S. pneumoniae in mice. *Nat Chem Biol*. 2014;10:950-956.
265. Brigl M, Tatituri RVV, Watts GFM, et al. Innate and cytokine-driven signals, rather than microbial antigens, dominate in natural killer T cell activation during microbial infection. *J Exp Med*. 2011;208:1163-1177.
266. Holzapfel KL, Tyznik AJ, Kronenberg M, Hogquist KA. Antigen-dependent versus -independent activation of invariant NKT cells during infection. *J Immunol*. 2014;192:5490-5498.
267. Lee WY, Moriarty TJ, Wong CHY, et al. An intravascular immune response to Borrelia burgdorferi involves Kupffer cells and iNKT cells. *Nat Immunol*. 2010;11:295-302.
268. Tupin E, Benhnia MR-E-I, Kinjo Y, et al. NKT cells prevent chronic joint inflammation after infection with Borrelia burgdorferi. *Proc Natl Acad Sci U S A*. 2008;105:19863-19868.
269. Lee WY, Sanz M-J, Wong CHY, et al. Invariant natural killer T cells act as an extravascular cytotoxic barrier for joint-invading Lyme Borrelia. *Proc Natl Acad Sci U S A*. 2014;111:13936-13941.
270. Nieuwenhuis EE, Matsumoto T, Exley M, et al. CD1d-dependent macrophage-mediated clearance of *Pseudomonas aeruginosa* from lung. *Nat Med*. 2002;8:588-593.
271. Sada-Ovalle I, Chiba A, Gonzales A, Brenner MB, Behar SM. Innate invariant NKT cells recognize Mycobacterium tuberculosis-infected macrophages, produce interferon-gamma, and kill intracellular bacteria. *PLoS Pathog*. 2008;4:e1000239.
272. Rothchild AC, Jayaraman P, Nunes-Alves C, Behar SM. iNKT cell production of GM-CSF controls *Mycobacterium tuberculosis*. *PLoS Pathog*. 2014;10:e1003805.
273. Yang R, Mele F, Worley L, et al. Human T-bet governs innate and innate-like adaptive IFN-gamma immunity against mycobacteria. *Cell*. 2020;183:1826-1847.e1831.
274. Ito Y, Vela JL, Matsumura F, et al. *Helicobacter pylori* cholesteryl alpha-glucosides contribute to its pathogenicity and immune response by natural killer T cells. *PLoS One*. 2013;8:e78191.
275. Dieli F, Sireci G, Russo D, et al. Resistance of natural killer T cell-deficient mice to systemic Shwartzman reaction. *J Exp Med*. 2000;192:1645-1652.
276. Kim EY, Ner-Gaon H, Varon J, et al. Post-sepsis immunosuppression depends on NKT cell regulation of mTOR/IFN-gamma in NK cells. *J Clin Invest*. 2020;130:3238-3252.
277. Choi J, Mele TS, Porcelli SA, Savage PB, Haeryfar SMM. Harnessing the versatility of invariant NKT cells in a stepwise approach to sepsis immunotherapy. *J Immunol*. 2021;206:386-397.
278. Wong CH, Jenne CN, Lee WY, Leger C, Kubes P. Functional innervation of hepatic iNKT cells is immunosuppressive following stroke. *Science*. 2011;334:101-105.
279. Cohen NR, Tatituri RVV, Rivera A, et al. Innate recognition of cell wall beta-glucans drives invariant natural killer T cell responses against fungi. *Cell Host Microbe*. 2011;10:437-450.
280. Molano A, Park S-H, Chiu Y-H, Nosseir S, Bendelac A, Tsuji M. Cutting edge: the IgG response to the circumsporozoite protein is MHC class II-dependent and CD1d-independent. Exploring the role of GPIs in NK T cell activation and antimalarial responses. *J Immunol*. 2000;164:5005-5009.
281. Romero JF, Eberl G, MacDonald HR, Corradin G. CD1d-restricted NK T cells are dispensable for specific antibody responses and protective immunity against liver stage malaria infection in mice. *Parasite Immunol*. 2001;23:267-269.
282. Gonzalez-Aseguinolaza G, de Oliveira C, Tomaska M, et al. Alpha -galactosylceramide-activated Valpha 14 natural killer T cells mediate protection against murine malaria. *Proc Natl Acad Sci U S A*. 2000;97:8461-8466.
283. Holz LE, Chua YC, de Menezes MN, et al. Glycolipid-peptide vaccination induces liver-resident memory CD8(+) T cells that protect against rodent malaria. *Sci Immunol*. 2020;5(48):eaaz8035.
284. Tyznik AJ, Tupin E, Nagarajan NA, Her MJ, Benedict CA, Kronenberg M. Cutting edge: the mechanism of invariant NKT cell responses to viral danger signals. *J Immunol*. 2008;181:4452-4456.
285. Tyznik AJ, Verma S, Wang Q, Kronenberg M, Benedict CA. Distinct requirements for activation of NKT and NK cells during viral infection. *J Immunol*. 2014;192:3676-3685.
286. Paget C, Ivanov S, Fontaine J, et al. Potential role of invariant NKT cells in the control of pulmonary inflammation and CD8+ T cell response during acute influenza A virus H3N2 pneumonia. *J Immunol*. 2011;186:5590-5602.
287. Paget C, Ivanov S, Fontaine J, et al. Interleukin-22 is produced by invariant natural killer T lymphocytes during influenza A virus infection: potential role in protection against lung epithelial damages. *J Biol Chem*. 2012;287:8816-8829.
288. De Santo C, Salio M, Masri SH, et al. Invariant NKT cells reduce the immunosuppressive activity of influenza A virus-induced myeloid-derived suppressor cells in mice and humans. *J Clin Invest*. 2008;118:4036-4048.
289. Artiaga BL, Yang G, Hutchinson TE, et al. Rapid control of pandemic H1N1 influenza by targeting NKT-cells. *Sci Rep*. 2016;6:37999.
290. Sanchez DJ, Gumperz JE, Ganem D. Regulation of CD1d expression and function by a herpesvirus infection. *J Clin Invest*. 2005;115:1369-1378.
291. Yuan W, Dasgupta A, Cresswell P. Herpes simplex virus evades natural killer T cell recognition by suppressing CD1d recycling. *Nat Immunol*. 2006;7:835-842.
292. Raftery MJ, Winau F, Kaufmann SH, Schaible UE, Schonrich G. CD1 antigen presentation by human dendritic cells as a target for herpes simplex virus immune evasion. *J Immunol*. 2006;177:6207-6214.
293. Xiong R, Rao P, Kim S, Li M, Wen X, Yuan W. Herpes simplex virus 1 US3 phosphorylates cellular KIF3A to downregulate CD1d expression. *J Virol*. 2015;89:6646-6655.
294. Raftery MJ, Wolter E, Fillatreau S, Meisel H, Kaufmann SHE, Schönrich G. NKT cells determine titer and subtype profile of virus-specific IgG antibodies during herpes simplex virus Infection. *J Immunol*. 2014;192:4294-4302.
295. Felley LE, Sharma A, Theisen E, Romero-Masters JC, Sauer J-D, Gumperz JE. Human invariant NKT cells induce IL-1beta secretion by peripheral blood monocytes via a P2X7-independent pathway. *J Immunol*. 2016;197:2455-2464.
296. Donado CA, Cao AB, Simmons DP, Croker BA, Brennan PJ, Brenner MB. A two-cell model for IL-1beta release mediated by death-receptor signaling. *Cell Rep*. 2020;31:107466.
297. Chow MT, Duret H, Andrews DM, et al. Type I NKT-cell-mediated TNF-alpha is a positive regulator of NLRP3 inflammasome priming. *Eur J Immunol*. 2014;44:2111-2120.
298. Lan P, Fan Y, Zhao Y, et al. TNF superfamily receptor OX40 triggers invariant NKT cell pyroptosis and liver injury. *J Clin Invest*. 2017;127:2222-2234.
299. Xu X, Pocock GM, Sharma A, et al. Human iNKT cells promote protective inflammation by inducing oscillating purinergic signaling in monocyte-derived DCs. *Cell Rep*. 2016;16:3273-3285.
300. Yu JC, Lin G, Field JJ, Linden J. Induction of antiinflammatory purinergic signaling in activated human iNKT cells. *JCI Insight*. 2018;3.
301. Liew PX, Lee WY, Kubes P. iNKT cells orchestrate a switch from inflammation to resolution of sterile liver injury. *Immunity*. 2017;47:752-765.e755.
302. Venken K, Jacques P, Mortier C, et al. RORgammat inhibition selectively targets IL-17 producing iNKT and gammadelta-T cells enriched in Spondyloarthritis patients. *Nat Commun*. 2019;10:9.
303. Syn WK, Agboola KM, Swiderska M, et al. NKT-associated hedgehog and osteopontin drive fibrogenesis in non-alcoholic fatty liver disease. *Gut*. 2012;61:1323-1329.
304. Maricic I, Marrero I, Eguchi A, et al. Differential activation of hepatic invariant NKT cell subsets plays a key role in progression of nonalcoholic steatohepatitis. *J Immunol*. 2018;201:3017-3035.
305. Wolf MJ, Adili A, Piotrowitz K, et al. Metabolic activation of intrahepatic CD8+ T cells and NKT cells causes nonalcoholic steatohepatitis and liver cancer via cross-talk with hepatocytes. *Cancer Cell*. 2014;26:549-564.
306. Berntsen NL, Fosby B, Tan C, et al. Natural killer T cells mediate inflammation in the bile ducts. *Mucosal Immunol*. 2018;11:1582-1590.
307. Schneidawind D, Pierini A, Alvarez M, et al. CD4+ invariant natural killer T cells protect from murine GVHD lethality through expansion of donor CD4+CD25+FoxP3+ regulatory T cells. *Blood*. 2014;124:3320-3328.
308. Du J, Paz K, Thangavelu G, et al. Invariant natural killer T cells ameliorate murine chronic GVHD by expanding donor regulatory T cells. *Blood*. 2017;129:3121-3125.
309. Akbari O, Stock P, Meyer E, et al. Essential role of NKT cells producing IL-4 and IL-13 in the development of allergen-induced airway hyperreactivity. *Nat Med*. 2003;9:582-588.
310. Chandra S, Zhao M, Budelsky A, et al. A new mouse strain for the analysis of invariant NKT cell function. *Nat Immunol*. 2015;16:799-800.
311. Scanlon ST, Thomas SY, Ferreira CM, et al. Airborne lipid antigens mobilize resident intravascular NKT cells to induce allergic airway inflammation. *J Exp Med*. 2011;208:2113-2124.
312. Wingender G, Rogers P, Batzer G, et al. Invariant NKT cells are required for airway inflammation induced by environmental antigens. *J Exp Med*. 2011;208:1151-1162.
313. Pichavant M, Goya S, Meyer EH, et al. Ozone exposure in a mouse model induces airway hyperreactivity that requires the presence of natural killer T cells and IL-17. *J Exp Med*. 2008;205:385-393.
314. Chuang YT, Leung K, Chang Y-J, et al. A natural killer T-cell subset that protects against airway hyperreactivity. *J Allergy Clin Immunol*. 2019;143:565-576.e567.
315. Hammond KJ, Poulton LD, Palmisano LJ, Silveira PA, Godfrey DI, Baxter AG. alpha/beta-T cell receptor (TCR)+CD4−CD8− (NKT) thymocytes prevent insulin-dependent diabetes mellitus in nonobese diabetic (NOD)/Lt mice by the influence of interleukin (IL)-4 and/or IL-10. *J Exp Med*. 1998;187:1047-1056.
316. Lehuen A, Lantz O, Beaudoin L, et al. Overexpression of natural killer T cells protects Valpha14- Jalpha281 transgenic nonobese diabetic mice against diabetes. *J Exp Med*. 1998;188:1831-1839.
317. Wang B, Geng YB, Wang CR. CD1-restricted NK T cells protect nonobese diabetic mice from developing diabetes. *J Exp Med*. 2001;194:313-320.
318. Hong S, Wilson MT, Serizawa I, et al. The natural killer T-cell ligand alpha-galactosylceramide prevents autoimmune diabetes in non-obese diabetic mice. *Nat Med*. 2001;7:1052-1056.
319. Sharif S, Arreaza GA, Zucker P, et al. Activation of natural killer T cells by alpha-galactosylceramide treatment prevents the onset and recurrence of autoimmune Type 1 diabetes. *Nat Med*. 2001;7:1057-1062.
320. Beaudoin L, Diana J, Ghazarian L, Simoni Y, Boitard C, Lehuen A. Plasmacytoid dendritic cells license regulatory T cells, upon iNKT-cell stimulation, to prevent autoimmune diabetes. *Eur J Immunol*. 2014;44:1454-1466.
321. Simoni Y, Gautron A-S, Beaudoin L, et al. NOD mice contain an elevated frequency of iNKT17 cells that exacerbate diabetes. *Eur J Immunol*. 2011;41:3574-3585.
322. Wilson SB, Kent SC, Patton KT, et al. Extreme Th1 bias of invariant Valpha24JalphaQ T cells in type 1 diabetes. *Nature*. 1998;391:177-181.
323. Lee PT, Putnam A, Benlagha K, Teyton L, Gottlieb PA, Bendelac A. Testing the NKT cell hypothesis of human IDDM pathogenesis. *J Clin Invest*. 2002;110:793-800.
324. Tocheva AS, Mansour S, Holt TGH, et al. The clonal invariant NKT cell repertoire in people with type 1 diabetes is characterized by a loss of clones expressing high-affinity TCRs. *J Immunol*. 2017;198:1452-1459.
325. Singh AK, Wilson MT, Hong S, et al. Natural killer T cell activation protects mice against experimental autoimmune encephalomyelitis. *J Exp Med*. 2001;194:1801-1811.
326. Parekh VV, Wu L, Olivares-Villagomez D, Wilson KT, Van Kaer L. Activated invariant NKT cells control central nervous system autoimmunity in a mechanism that involves myeloid-derived suppressor cells. *J Immunol*. 2013;190:1948-1960.

327. Mattner J, Savage PB, Leung P, et al. Liver autoimmunity triggered by microbial activation of natural killer T cells. *Cell Host Microbe.* 2008;3:304-315.
328. Kita H, Naidenko OV, Kronenberg M, et al. Quantitation and phenotypic analysis of natural killer T cells in primary biliary cirrhosis using a human CD1d tetramer. *Gastroenterology.* 2002;123:1031-1043.
329. Cui J, Shin T, Kawano T, et al. Requirement for Valpha14 NKT cells in IL-12-mediated rejection of tumors. *Science.* 1997;278:1623-1626.
330. Smyth MJ, Thia KYT, Street SEA, et al. Differential tumor surveillance by natural killer (NK) and NKT cells. *J Exp Med.* 2000;191:661-668.
331. Kammertoens T, Qin Z, Briesemeister D, Bendelac A, Blankenstein T. B-cells and IL-4 promote methylcholanthrene-induced carcinogenesis but there is no evidence for a role of T/NKT-cells and their effector molecules (Fas-ligand, TNF-alpha, perforin). *Int J Cancer.* 2012;131:1499-1508.
332. Ma C, Han M, Heinrich B, et al. Gut microbiome-mediated bile acid metabolism regulates liver cancer via NKT cells. *Science.* 2018;360(6391):eaan5931.
333. Fu S, He K, Tian C, et al. Impaired lipid biosynthesis hinders anti-tumor efficacy of intratumoral iNKT cells. *Nat Commun.* 2020;11:438.
334. Bachy E, Urb M, Chandra S, et al. CD1d-restricted peripheral T cell lymphoma in mice and humans. *J Exp Med.* 2016;213:841-857.
335. Zhu Y, Smith DJ, Zhou Y, et al. Development of hematopoietic stem cell-engineered invariant natural killer T cell therapy for cancer. *Cell Stem Cell.* 2019;25:542-557 e549.
336. Yamada D, Iyoda T, Vizcardo R, et al. Efficient regeneration of human Valpha24(+) invariant natural killer T cells and their anti-tumor activity in vivo. *Stem Cell.* 2016;34:2852-2860.
337. Das R, Guan P, Wiener SJ, et al. Enhancing the antitumor functions of invariant natural killer T cells using a soluble CD1d-CD19 fusion protein. *Blood Adv.* 2019;3:813-824.
338. Kharkwal SS, Johndrow CT, Veerapen N, et al. Serial stimulation of invariant natural killer T cells with covalently stabilized bispecific T-cell engagers generates antitumor immunity while avoiding anergy. *Cancer Res.* 2021;81:1788-1801.
339. Rotolo A, Caputo VS, Holubova M, et al. Enhanced anti-lymphoma activity of CAR19-iNKT cells underpinned by dual CD19 and CD1d targeting. *Cancer Cell.* 2018;34:596-610.e511.
340. Heczey A, Liu D, Tian G, et al. Invariant NKT cells with chimeric antigen receptor provide a novel platform for safe and effective cancer immunotherapy. *Blood.* 2014;124:2824-2833.
341. Heczey A, Courtney AN, Montalbano A, et al. Anti-GD2 CAR-NKT cells in patients with relapsed or refractory neuroblastoma: an interim analysis. *Nat Med.* 2020;26:1686-1690.
342. Gorini F, Azzimonti L, Delfanti G, et al. Invariant NKT cells contribute to chronic lymphocytic leukemia surveillance and prognosis. *Blood.* 2017;129:3440-3451.
343. Jahng A, Maricic I, Aguilera C, Cardell S, Halder RC, Kumar V. Prevention of autoimmunity by targeting a distinct, noninvariant CD1d-reactive T cell population reactive to sulfatide. *J Exp Med.* 2004;199:947-957.
344. Tatituri RV, Watts GFM, Bhowruth V, et al. Recognition of microbial and mammalian phospholipid antigens by NKT cells with diverse TCRs. *Proc Natl Acad Sci U S A.* 2013;110:1827-1832.
345. Wolf BJ, Tatituri RVV, Almeida CF, et al. Identification of a potent microbial lipid antigen for diverse NKT cells. *J Immunol.* 2015;195:2540-2551.
346. Maricic I, Girardi E, Zajonc DM, Kumar V. Recognition of lysophosphatidylcholine by type II NKT cells and protection from an inflammatory liver disease. *J Immunol.* 2014;193:4580-4589.
347. Rhost S, Löfbom L, Rynmark B-M, et al. Identification of novel glycolipid ligands activating a sulfatide-reactive, CD1d-restricted, type II natural killer T lymphocyte. *Eur J Immunol.* 2012;42:2851-2860.
348. Nair S, Boddupalli CS, Verma R, et al. Type II NKT-TFH cells against Gaucher lipids regulate B-cell immunity and inflammation. *Blood.* 2015;125:1256-1271.
349. Patel O, Pellicci DG, Gras S, et al. Recognition of CD1d-sulfatide mediated by a type II natural killer T cell antigen receptor. *Nat Immunol.* 2012;13:857-863.
350. Girardi E, Maricic I, Wang J, et al. Type II natural killer T cells use features of both innate-like and conventional T cells to recognize sulfatide self antigens. *Nat Immunol.* 2012;13:851-856.
351. Almeida CF, Sundararaj S, Le Nours J, et al. Distinct CD1d docking strategies exhibited by diverse Type II NKT cell receptors. *Nat Commun.* 2019;10:5242.
352. Zhao J, Weng X, Bagchi S, Wang CR. Polyclonal type II natural killer T cells require PLZF and SAP for their development and contribute to CpG-mediated antitumor response. *Proc Natl Acad Sci U S A.* 2014;111:2674-2679.
353. Tripathi P, Sedimbi SK, Singh AK, et al. Innate and adaptive stimulation of murine diverse NKT cells result in distinct cellular responses. *Eur J Immunol.* 2019;49:443-453.
354. McMichael AJ, Pilch JR, Fabre JW, Mason DY, Galfré G, Milstein C. A human thymocyte antigen defined by a hybrid myeloma monoclonal antibody. *Eur J Immunol.* 1979;9:205-210.
355. Felio K, Nguyen H, Dascher CC, et al. CD1-restricted adaptive immune responses to Mycobacteria in human group 1 CD1 transgenic mice. *J Exp Med.* 2009;206:2497-2509.
356. Leslie DS, Dascher CC, Cembrola K, et al. Serum lipids regulate dendritic cell CD1 expression and function. *Immunology.* 2008;125:289-301.
357. Seshadri C, Shenoy M, Wells RD, et al. Human CD1a deficiency is common and genetically regulated. *J Immunol.* 2013;191:1586-1593.
358. Sugita M, Grant EP, van Donselaar E, et al. Separate pathways for antigen presentation by CD1 molecules. *Immunity.* 1999;11:743-752.
359. Cernadas M, Cavallari M, Watts G, Mori L, De Libero G, Brenner MB. Early recycling compartment trafficking of CD1a is essential for its intersection and presentation of lipid antigens. *J Immunol.* 2010;184:1235-1241.
360. Manolova V, Kistowska M, Paoletti S, et al. Functional CD1a is stabilized by exogenous lipids. *Eur J Immunol.* 2006;36:1083-1092.
361. Barral DC, Cavallari M, McCormick PJ, et al. CD1a and MHC class I follow a similar endocytic recycling pathway. *Traffic.* 2008;9:1446-1457.
362. Barral DC, Garg S, Casalou C, et al. Arl13b regulates endocytic recycling traffic. *Proc Natl Acad Sci U S A.* 2012;109:21354-21359.
363. Sugita M, Jackman RM, van Donselaar E, et al. Cytoplasmic tail-dependent localization of CD1b antigen-presenting molecules to MIICs. *Science.* 1996;273:349-352.
364. Briken V, Jackman RM, Dasgupta S, Hoening S, Porcelli SA. Intracellular trafficking pathway of newly synthesized CD1b molecules. *EMBO J.* 2002;21:825-834.
365. Sugita M, Cao X, Watts GFM, Rogers RA, Bonifacino JS, Brenner MB. Failure of trafficking and antigen presentation by CD1 in AP-3-deficient cells. *Immunity.* 2002;16:697-706.
366. Moody DB, Briken V, Cheng T-Y, et al. Lipid length controls antigen entry into endosomal and nonendosomal pathways for CD1b presentation. *Nat Immunol.* 2002;3:435-442.
367. Sugita M, van Der Wel N, Rogers RA, Peters PJ, Brenner MB. CD1c molecules broadly survey the endocytic system. *Proc Natl Acad Sci U S A.* 2000;97:8445-8450.
368. Briken V, Jackman RM, Watts GF, Rogers RA, Porcelli SA. Human CD1b and CD1c isoforms survey different intracellular compartments for the presentation of microbial lipid antigens. *J Exp Med.* 2000;192:281-288.
369. Porcelli S, Brenner MB, Greenstein JL, Terhorst C, Balk SP, Bleicher PA. Recognition of cluster of differentiation 1 antigens by human CD4−CD8−cytolytic T lymphocytes. *Nature.* 1989;341:447-450.
370. Porcelli S, Morita CT, Brenner MB. CD1b restricts the response of human CD4−8− T lymphocytes to a microbial antigen. *Nature.* 1992;360:593-597.
371. Beckman EM, Porcelli SA, Morita CT, Behar SM, Furlong ST, Brenner MB. Recognition of a lipid antigen by CD1-restricted alpha beta+ T cells. *Nature.* 1994;372:691-694.
372. Winau F, Schwierzeck V, Hurwitz R, et al. Saposin C is required for lipid presentation by human CD1b. *Nat Immunol.* 2004;5:169-174.
373. Leon L, Tatituri RVV, Grenha R, et al. Saposins utilize two strategies for lipid transfer and CD1 antigen presentation. *Proc Natl Acad Sci U S A.* 2012;109:4357-4364.
374. Kaser A, Hava DL, Dougan SK, et al. Microsomal triglyceride transfer protein regulates endogenous and exogenous antigen presentation by group 1 CD1 molecules. *Eur J Immunol.* 2008;38:2351-2359.
375. de la Salle H, Mariotti S, Angenieux C, et al. Assistance of microbial glycolipid antigen processing by CD1e. *Science.* 2005;310:1321-1324.
376. Garcia-Alles LF, Giacometti G, Versluis C, et al. Crystal structure of human CD1e reveals a groove suited for lipid-exchange processes. *Proc Natl Acad Sci U S A.* 2011;108:13230-13235.
377. Facciotti F, Cavallari M, Angenieux C, et al. Fine tuning by human CD1e of lipid-specific immune responses. *Proc Natl Acad Sci U S A.* 2011;108:14228-14233.
378. Pena-Cruz V, Ito S, Dascher CC, Brenner MB, Sugita M. Epidermal Langerhans cells efficiently mediate CD1a-dependent presentation of microbial lipid antigens to T cells. *J Invest Dermatol.* 2003;121:517-521.
379. Hunger RE, Sieling PA, Ochoa MT, et al. Langerhans cells utilize CD1a and langerin to efficiently present nonpeptide antigens to T cells. *J Clin Invest.* 2004;113:701-708.
380. Shamshiev A, Gober H-J, Donda A, Mazorra Z, Mori L, De Libero G. Presentation of the same glycolipid by different CD1 molecules. *J Exp Med.* 2002;195:1013-1021.
381. Moody DB, Young DC, Cheng T-Y, et al. T cell activation by lipopeptide antigens. *Science.* 2004;303:527-531.
382. Agea E, Russano A, Bistoni O, et al. Human CD1-restricted T cell recognition of lipids from pollens. *J Exp Med.* 2005;202:295-308.
383. de Jong A, Peña-Cruz V, Cheng T-Y, Clark RA, Van Rhijn I, Moody DB. CD1a-autoreactive T cells are a normal component of the human alphabeta T cell repertoire. *Nat Immunol.* 2010;11:1102-1109.
384. de Lalla C, Lepore M, Piccolo FM, et al. High-frequency and adaptive-like dynamics of human CD1 self-reactive T cells. *Eur J Immunol.* 2011;41:602-610.
385. de Jong A, Cheng T-Y, Huang S, et al. CD1a-autoreactive T cells recognize natural skin oils that function as headless antigens. *Nat Immunol.* 2014;15:177-185.
386. Cotton RN, Cheng TY, Wegrecki M, et al. Human skin is colonized by T cells that recognize CD1a independently of lipid. *J Clin Invest.* 2021;131(1):e140706.
387. Cotton RN, Wegrecki M, Cheng TY, et al. CD1a selectively captures endogenous cellular lipids that broadly block T cell response. *J Exp Med.* 2021;218(7):e20202699.
388. Hardman CS, Chen YL, Salimi M, et al. CD1a presentation of endogenous antigens by group 2 innate lymphoid cells. *Sci Immunol.* 2017;2(18):eaan5918.
389. Zajonc DM, Elsliger MA, Teyton L, Wilson IA. Crystal structure of CD1a in complex with a sulfatide self antigen at a resolution of 2.15 Å. *Nat Immunol.* 2003;4:808-815.
390. Zajonc DM, Crispin MDM, Bowden TA, et al. Molecular mechanism of lipopeptide presentation by CD1a. *Immunity.* 2005;22:209-219.
391. Birkinshaw RW, Pellicci DG, Cheng T-Y, et al. Alphabeta T cell antigen receptor recognition of CD1a presenting self lipid ligands. *Nat Immunol.* 2015;16:258-266.
392. Sieling PA, Chatterjee D, Porcelli SA, et al. CD1-restricted T cell recognition of microbial lipoglycan antigens. *Science.* 1995;269:227-230.
393. Moody DB, Reinhold BB, Guy MR, et al. Structural requirements for glycolipid antigen recognition by CD1b-restricted T cells. *Science.* 1997;278:283-286.
394. Moody DB, Guy MR, Grant E, et al. CD1b-mediated T cell recognition of a glycolipid antigen generated from mycobacterial lipid and host carbohydrate during infection. *J Exp Med.* 2000;192:965-976.
395. Van Rhijn I, van Berlo T, Hilmenyuk T, et al. Human autoreactive T cells recognize CD1b and phospholipids. *Proc Natl Acad Sci U S A.* 2016;113:380-385.
396. Prigozy TI, Sieling PA, Clemens D, et al. The mannose receptor delivers lipoglycan antigens to endosomes for presentation to T cells by CD1b molecules. *Immunity.* 1997;6:187-197.
397. Cheng TY, Relloso M, Van Rhijn I, et al. Role of lipid trimming and CD1 groove size in cellular antigen presentation. *EMBO J.* 2006;25:2989-2999.
398. Gadola SD, Zaccai NR, Harlos K, et al. Structure of human CD1b with bound ligands at 2.3 A, a maze for alkyl chains. *Nat Immunol.* 2002;3:721-726.
399. Van Rhijn I, Kasmar A, de Jong A, et al. A conserved human T cell population targets mycobacterial antigens presented by CD1b. *Nat Immunol.* 2013;14:706-713.
400. Gras S, Van Rhijn I, Shahine A, et al. T cell receptor recognition of CD1b presenting a mycobacterial glycolipid. *Nat Commun.* 2016;7:13257.
401. Shahine A, Van Rhijn I, Cheng TY, et al. A molecular basis of human T cell receptor autoreactivity toward self-phospholipids. *Sci Immunol.* 2017;2(16):eaao1384.
402. Shahine A, Reinink P, Reijneveld JF, et al. A T-cell receptor escape channel allows broad T-cell response to CD1b and membrane phospholipids. *Nat Commun.* 2019;10:56.
403. Moody DB, Ulrichs T, Mühlecker W, et al. CD1c-mediated T-cell recognition of isoprenoid glycolipids in *Mycobacterium tuberculosis* infection. *Nature.* 2000;404:884-888.
404. Matsunaga I, Bhatt A, Young DC, et al. *Mycobacterium tuberculosis* pks12 produces a novel polyketide presented by CD1c to T cells. *J Exp Med.* 2004;200:1559-1569.
405. Ly D, Kasmar AG, Cheng T-Y, et al. CD1c tetramers detect ex vivo T cell responses to processed phosphomycoketide antigens. *J Exp Med.* 2013;210:729-741.
406. Lepore M, de Lalla C, Gundimeda SR, et al. A novel self-lipid antigen targets human T cells against CD1c(+) leukemias. *J Exp Med.* 2014;211:1363-1377.
407. Scharf L, Li N-S, Hawk AJ, et al. The 2.5 Å structure of CD1c in complex with a mycobacterial lipid reveals an open groove ideally suited for diverse antigen presentation. *Immunity.* 2010;33:853-862.
408. Roy S, Ly D, Li N-S, et al. Molecular basis of mycobacterial lipid antigen presentation by CD1c and its recognition by alphabeta T cells. *Proc Natl Acad Sci U S A.* 2014;111:E4648-E4657.
409. Wun KS, Reijneveld JF, Cheng T-Y, et al. T cell autoreactivity directed toward CD1c itself rather than toward carried self lipids. *Nat Immunol.* 2018;19:397-406.
410. Kasmar AG, Van Rhijn I, Magalhaes KG, et al. Cutting Edge: CD1a tetramers and dextramers identify human lipopeptide-specific T cells ex vivo. *J Immunol.* 2013;191:4499-4503.
411. Kasmar AG, van Rhijn I, Cheng T-Y, et al. CD1b tetramers bind alphabeta T cell receptors to identify a mycobacterial glycolipid-reactive T cell repertoire in humans. *J Exp Med.* 2011;208:1741-1747.
412. James CA, Yu KKQ, Gilleron M, et al. CD1b tetramers identify T cells that recognize natural and synthetic diacylated sulfoglycolipids from *Mycobacterium tuberculosis. Cell Chem Biol.* 2018;25:392-402 e314.
413. Lopez K, Iwany SK, Suliman S, et al. CD1b tetramers broadly detect T cells that correlate with mycobacterial exposure but not tuberculosis disease state. *Front Immunol.* 2020;11:199.
414. Gherardin NA, Redmond SJ, McWilliam HEG, et al. CD36 family members are TCR-independent ligands for CD1 antigen-presenting molecules. *Sci Immunol.* 2021;6:eabg4176.

415. Zhao J, Siddiqui S, Shang S, et al. Mycolic acid-specific T cells protect against *Mycobacterium tuberculosis* infection in a humanized transgenic mouse model. *Elife*. 2015;4:e08525.
416. Visvabharathy L, Genardi S, Cao L, et al. Group 1 CD1-restricted T cells contribute to control of systemic *Staphylococcus aureus* infection. *PLoS Pathog*. 2020;16:e1008443.
417. Bourgeois EA, Subramaniam S, Cheng T-Y, et al. Bee venom processes human skin lipids for presentation by CD1a. *J Exp Med*. 2015;212:149-163.
418. Subramaniam S, Aslam A, Misbah SA, et al. Elevated and cross-responsive CD1a-reactive T cells in bee and wasp venom allergic individuals. *Eur J Immunol*. 2016;46:242-252.
419. Jarrett R, Salio M, Lloyd-Lavery A, et al. Filaggrin inhibits generation of CD1a neolipid antigens by house dust mite-derived phospholipase. *Sci Transl Med*. 2016;8:325ra318.
420. Kim JH, Hu Y, Yongqing T, et al. CD1a on Langerhans cells controls inflammatory skin disease. *Nat Immunol*. 2016;17:1159-1166.
421. Betts RJ, Perkovic A, Mahapatra S, et al. Contact sensitizers trigger human CD1-autoreactive T-cell responses. *Eur J Immunol*. 2017;47:1171-1180.
422. Nicolai S, Wegrecki M, Cheng TY, et al. Human T cell response to CD1a and contact dermatitis allergens in botanical extracts and commercial skin care products. *Sci Immunol*. 2020;5.
423. Cheung KL, Jarrett R., Subramaniam S, et al. Psoriatic T cells recognize neolipid antigens generated by mast cell phospholipase delivered by exosomes and presented by CD1a. *J Exp Med*. 2016;213:2399-2412.
424. Bagchi S, He Y, Zhang H, et al. CD1b-autoreactive T cells contribute to hyperlipidemia-induced skin inflammation in mice. *J Clin Invest*. 2017;127:2339-2352.
425. Roura-Mir C, Catálfamo M, Cheng T-Y, et al. CD1a and CD1c activate intrathyroidal T cells during Graves' disease and Hashimoto's thyroiditis. *J Immunol*. 2005;174:3773-3780.
426. Bagchi S, Li S, Wang CR. CD1b-autoreactive T cells recognize phospholipid antigens and contribute to antitumor immunity against a CD1b(+) T cell lymphoma. *OncoImmunology*. 2016;5:e1213932.
427. Consonni M, Garavaglia C, Grilli A, et al. Human T cells engineered with a leukemia lipid-specific TCR enables donor-unrestricted recognition of CD1c-expressing leukemia. *Nat Commun*. 2021;12:4844.

CHAPTER 37

Mucosal-Associated Invariant T Cells and Other MR1-Reactive T Cells

Sidonia B. G. Eckle, BSc, MSc, PhD • Wael Awad, PhD • Jamie Rossjohn, FRS • James McCluskey, FAA • Alexandra J. Corbett, BSc (Hons), PhD

INTRODUCTION: HISTORY OF THE DISCOVERY OF MAIT CELLS

Identification of a Population of T Cells With Semi-Invariant TCR Usage

In 1993, about 10 years after the discovery of the T cell receptor (TCR) in the early 1980s, a number of different laboratories identified a discrete subset of αβ T cells now known as mucosal-associated invariant T (MAIT) cells (Fig. 37.1). Their initial discovery arose when characterizing the TCR usage among CD4−CD8− (double negative, DN) αβ TCR+ T cells in human peripheral blood from several donors. Porcelli et al noticed preferential expression of two TCR α-chain rearrangements that were subsequently shown to be the hallmark of Type I natural killer T (NKT) cells and MAIT cells. The invariant *TRAV10-TRAJ18* (IMGT nomenclature, or *Vα24-Jα18* Arden nomenclature) was later assigned to Type I NKT cells, and the essentially invariant *TRAV1-2-TRAJ33* (IMGT nomenclature, or *Vα7.2-Jα33* Arden nomenclature) is found in the majority of MAIT cells (Table 37.1). The TRAJ33 CDR3α-loop was of constant length featuring two variable amino acids at the V-J junction (CAXXDSNYQLIW).[1] The two variable amino acids of the human *TRAV1-2-TRAJ33* rearrangement usually involve the residues VR (CAVRDSNYQLIW) and are located at the base of the CDR3α-loop rather than at the sites of direct antigen recognition by the MAIT TCR.[2,3] While nongermline nucleotide additions can be involved in the generation of CAVRDSNYQLIW, the CDR3α-loop can also be entirely germline-encoded, whereby an overlap of the *TRAV1-2* and *TRAJ33* gene enables nucleotides from either gene to contribute to the formation of the codons at the V-J junction[4] (Fig. 37.2A, B). Thus, the most dominant *TRAV1-2-TRAJ33* amino acid sequence can arise from multiple recombination events, a process referred to as convergent recombination.[4] This semi-invariant pattern of αβ-TCR expression helped define MAIT cells along with other phenotypic characteristics.

Later, *TRAJ12* and *TRAJ20* segments rearranged to *TRAV1-2* were also observed to be frequently used by human MAIT TCRs.[5-7] While there was a marked preference for the three TCR α-chain rearrangements in the human DN population, these rearrangements were also found among T cells that are not DN,[1] including strong enrichment among CD8αα+ and persistence among CD4+ T cells.[8] Later, it was established that human MAIT cells in the peripheral blood are predominantly CD8αβ+ (median 44.7%) or CD8αα+ (median 35%) but also frequently DN while minor proportions are CD4+ (median 1.3%) and CD4+CD8+ (double positive, DP) (median 1.3%).[7,9] In humans, MAIT cells represent on average 3% of T cells in the blood[9] and are found throughout peripheral tissues, where they are particularly enriched in the liver (up to 50% of T cells) and at mucosal sites such as the gut and lungs (Fig. 37.3A).[6,7,10-22] MAIT cells are less frequent in mice, where they represent 0.5% to 2% of T cells in a broad range of tissues, and 0.1% of T cells in the circulation of naïve C57BL/6 and BALB/c mice (Fig. 37.3A).[8,22-24] The early observation of the presence of MAIT cells at mucosal sites such as the gut lamina propria (LP) is what prompted the naming of MAIT cells,[22] which on reflection may be a simplification of the distribution and function of these cells.

Among DN αβ T cells, Porcelli et al also noticed a bias in *TRBV* gene segments, including *TRBV6* and *TRBV20* (or *Vβ13* and *Vβ2* based on Arden nomenclature),[1] and Tilloy et al confirmed that the repertoire of TCRβ-chains paired with the invariant TRAV1-2 α-chain was skewed toward these gene segments.[8] Although Lepore et al noted later a bias in the CDR3β-loop length of 11 to 14 amino acids,[6] there was no obvious restriction in *TRBJ* usage[8] or any obvious amino acid motif.[6] Thus, the *TRAV1-2-TRAJ33+* TCRs have been described as "semi-invariant."[8] Tilloy et al further observed a marked oligoclonality of the TCR β-chain repertoire,[8] which was later confirmed,[6] suggesting peripheral expansion of MAIT cells, which was also consistent with the memory phenotype (ie, CD45R0+CD45RA−) observed among *TRAV1-2-TRAJ33+* T cells.[8]

MAIT cells are evolutionarily conserved in mammals with the canonical TCR α-chain composed of homologous *TRAV* and *TRAJ* elements having a CDR3α-loop of constant length also in mice (*TRAV1-TRAJ33* or *Vα19-Jα33* based on Arden nomenclature),[8] cattle[8,25] sheep,[26] macaques,[27-29] and pigs[30] (Fig. 37.2C). Akin to the human *TRAV1-2-TRAJ33* rearrangement, MAIT TCRs can arise from convergent recombination.[4] Based on analysis of available genomic data, homologs of the *TRAV1-2* and *TRAJ33* genes are present in most mammals, including marsupials.[31] In comparison to other *TRAV* genes, the CDR1- and CDR2-regions of *TRAV1-2* are more conserved across species; *TRAJ33* is also highly conserved in

TABLE 37.1 Comparison of the Properties of MAIT Cells and Type I NKT Cells

	MAIT Cells	Type I NKT Cells
Restriction element	MR1	CD1d
TCR usage	Human: *TRAV1-2-TRAJ33/12/20 TRBV6/20* Mouse: *TRAV1 (Vα19)-TRAJ33 (Jα33) TRBV19 (Vβ6.1)/13 (Vβ8)*	Human: *TRAV10-TRAJ18 TRBV25* Mouse: *TRAV11 (Vα14)-TRAJ18 (Jα18) TRBV13 (Vβ8)/29 (Vβ7)/1(Vβ2)*
Intradonor conservation of TCR sequences	High	High
Interdonor conservation of TCR sequences	High	High
Distribution	Human: present in blood and lymphoid and nonlymphoid tissues; enriched at mucosal sites and in the liver Mouse: similar to human, but at lower frequencies, and enriched in the skin	Human: present in blood and lymphoid and nonlymphoid tissues; enriched in liver, lungs, adipose tissue, and intestine Mouse: similar to human, but at higher frequencies
Development	In thymus, dependent on MR1	In thymus, dependent on CD1d
Subsets	MAIT-1 (PLZF+ T-bet+), MAIT-17 (PLZF+ RORγt+), which preferentially localize to different tissues	NKT-1 (T-bet+), NKT-2 (GATA3+ $PLZF^{hi}$), NKT-17 (RORγt+), NKT-10 ($Id2^{lo}$), NKTreg, NKTfh, which preferentially localize to different tissues

evolution.[32] However, the *TRAV1-2* gene segment is missing from the genome of some placental mammals including carnivores, lagomorphs as well as in the armadillo among xenarthrans.[32] Similar to the human, the TCR β-chain of mouse MAIT cells is biased to selection of *TRBV19* and *TRBV13* (or *Vβ6* and *Vβ8* based on Arden nomenclature), both being the murine orthologous segments of human *TRBV6*.[8]

Initial speculation considered the existence of the dominant *TRAV1-2-TRAJ33* rearrangement and selected *TRBV* gene segments in multiple unrelated donors to potentially originate from either preferential TCR rearrangements, particular requirements for stable assembly of TCRα- and β-chain pairs or structural requirements tuned to the recognition of a nonpolymorphic antigen-presenting molecule and a limited spectrum of antigens.[1] Indeed, the presence of *TRAV1-2-TRAJ33+* T cells was found to be independent of the expression of major histocompatibility complex (MHC)-I, MHC-II, CD1d, and TAP but was dependent on β_2-microglobulin (β_2m), suggesting their probable selection by an MHC-I-like molecule distinct from CD1.[8] The constrained TCR usage, their high conservation between mammalian species, and the abundance of these cells pointed early on to MAIT cells playing an ancient and important physiological role.[8]

Identification of MR1, the Restriction Element of MAIT Cells

Shortly after the identification of the semi-invariant MAIT TCR, in 1995 when actively screening the human genome for novel MHC-related genes by polymerase chain reaction methods, Hashimoto et al described such a novel gene, located on chromosome 1 (1q25.3) in close proximity to *CD1d* (*1q21–23*)[33,34] and outside the MHC located on chromosome 6 (6p21.3).[35] Sharing similarities with classical human leukocyte antigens (HLA)-I genes, the gene is generally referred to as the MHC-I related gene 1 (*MR1*), and the matching protein, MHC-I related protein 1 (MR1) (Table 37.2). The predicted amino acid sequence of MR1 suggested an architecture similar to MHC-I molecules, including a leader sequence, three extracellular domains, with the α1-and α2-domains forming a putative antigen-binding cleft delimited by two α-helices contributed from each domain, and the α3-domain adopting an immunoglobulin (Ig) fold, as well as a transmembrane region and a cytoplasmic tail, each encoded by separate exons,[35-38] with the exon/intron structure similar to that of MHC-I genes.[36,38] Of note, among MHC-I-like molecules, the MR1 antigen-binding domains are the most closely related to those of human, mouse, and nonmammalian classical MHC-I molecules.[35-37] Altogether, most of the invariant or conserved residues thought to maintain the MHC-I scaffold were conserved in MR1.[36,37] While some authors were less convinced,[39] others correctly predicted that MR1 was complexed to β_2m,[37] and this was later formally demonstrated.[39,40] This further implied that MR1 possessed an MHC-I-like structure and suggested that it might function as an antigen-presenting molecule, as also suggested by structure-based computational studies.[41] Several authors have also speculated on whether MR1 interacts with the coreceptor CD8, based on the presence or absence of conserved or homologous residues in the key CD8-binding site within the α3-domain of the MHC-I molecule.[35,37,39] Although it had been proposed that lipid-based antigens might bind to MR1,[42,43] the residues lining the putative ligand binding groove were noted to be markedly different from both MHC-I and CD1d,[39,41] suggesting that MR1 did not bind peptide or glycolipid antigens. In particular, of the eight residues critical for anchoring peptide termini in MHC-I molecules only two in the A′-pocket (Tyr7 and Trp167) and one in the F′-pocket (Trp147) were conserved in MR1.[31,35,36] Thus, MR1 was suspected to bind a unique ligand with an anchoring scheme unlike that of other MHC molecules.[41]

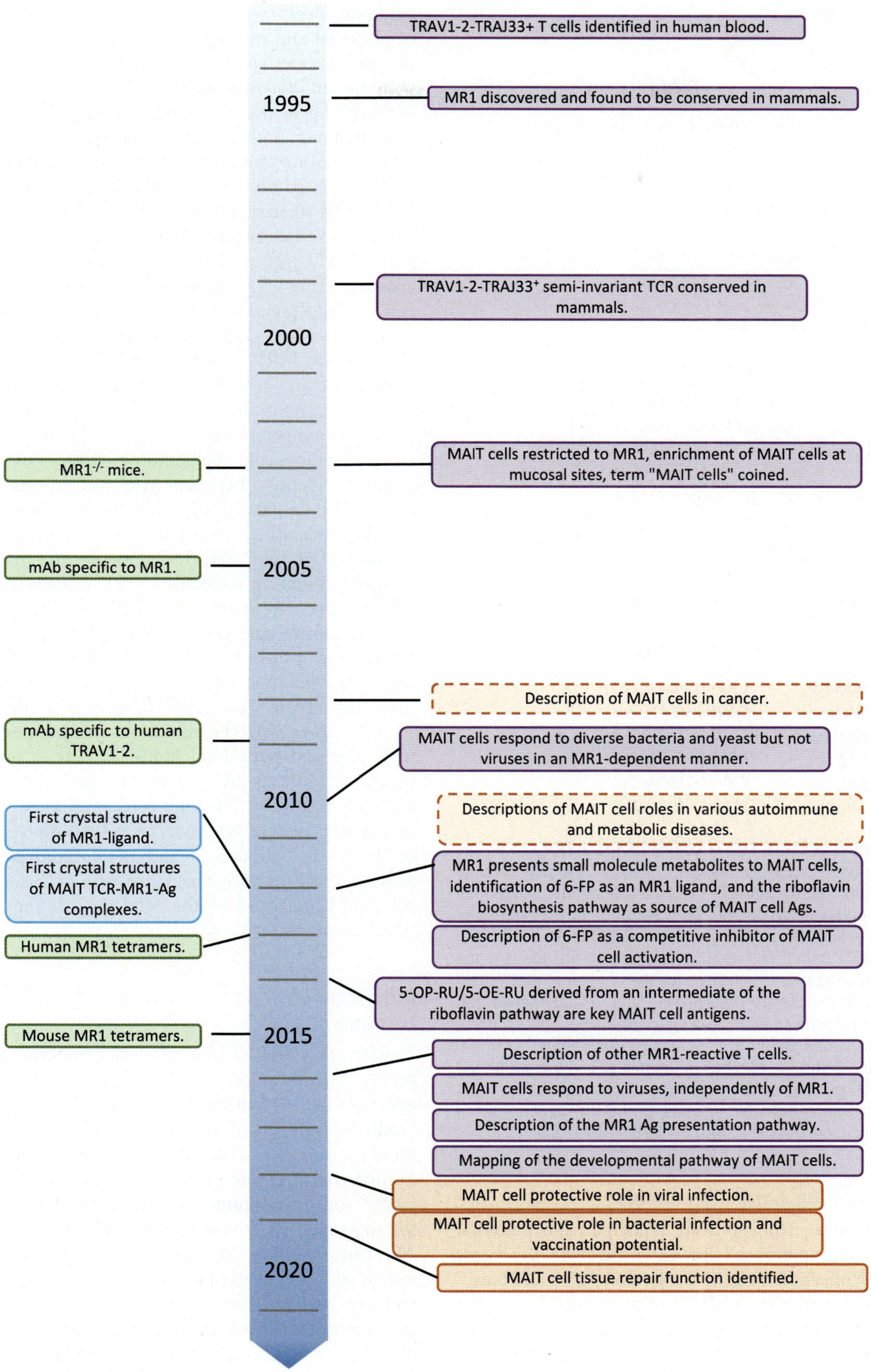

FIG. 37.1. **A timeline highlighting some of the key discoveries relevant to the MR1-MAIT cell field.** Key definitive discoveries are highlighted in *purple*, crystal structures in *blue*, and analytical tools in *green*. *Orange boxes* describe functional roles in disease contexts. Position on the timeline indicates the original description for each event.

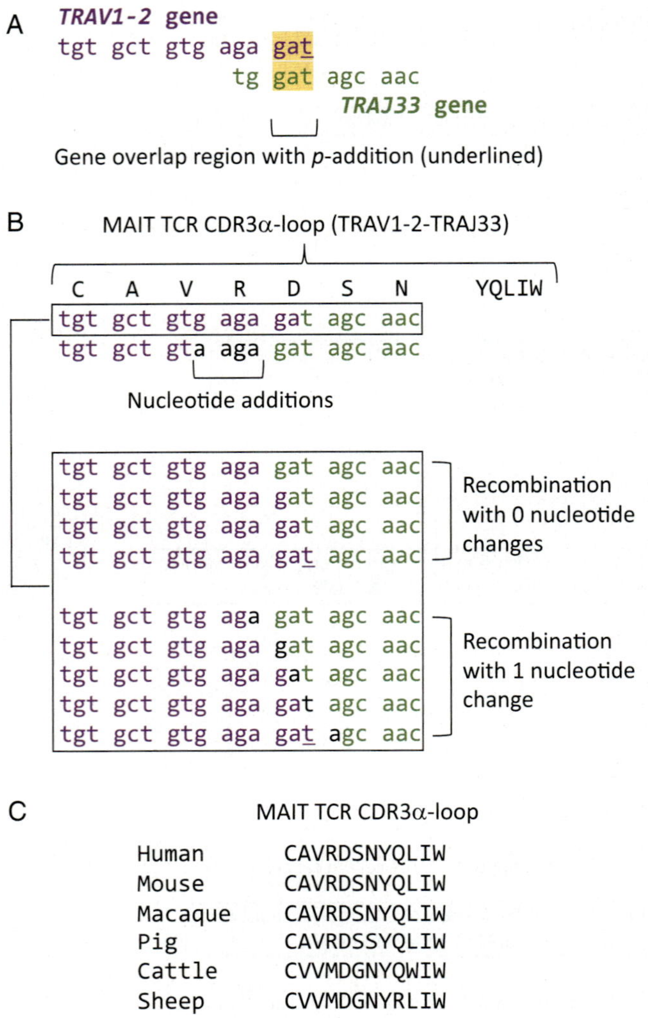

FIG. 37.2. Canonical MAIT TCR α-chain, *TRAV1-2-TRAJ33*. A: The relevant portions of the *TRAV1-2 gene* (*purple*) and the *TRAJ33* gene (*green*) are shown at the top with possible p-additions underlined. The overlap region between the *TRAV1-2* and *TRAJ33* genes is highlighted in *orange*. **B:** Convergent recombination in the human *TRAV1-2-TRAJ33+* MAIT TCR α-chain. Shown at the top are the two nucleotide sequences that code for the canonical amino acid sequence and alignments to the *TRAV1-2* and *TRAJ33* genes that give minimal nucleotide additions (*black*); possible p-addition contributions are underlined. Below are the multiple recombination mechanisms requiring 0 and 1 nucleotide addition that could produce the most frequently seen nucleotide sequence (*black box*). **C:** Alignment showing conservation between species of the CDR3α-loop of the most commonly used *TRAV1-2-TRAJ33+* MAIT TCR α-chain.

MR1 Conservation Across Mammals

Human MR1 gene orthologs were reported from placental mammals, including mouse (Chr. 1),[35,36] rat (Chr. 13),[37] cow (Chr. 16),[26,44,45] sheep (Chr. 12),[26,45] pig (Chr. 9),[45,46] and bats,[47] as well as nonhuman primates,[36,48] and marsupials,[31] together with their absence from other lower jawed vertebrates, including cartilaginous fish (eg, shark), teleosts (eg, zebrafish), and monotremes (eg, platypus).[31] Displaying also conserved synteny within the gene organization and chromosomal location,[26,31,36,38,45,46] MR1 might thus have been established in a common ancestor of mammalian species[35] and more precisely of placental and marsupial mammals after divergence from the monotremes about 170 million years ago.[31,32] Consistent with presentation of a putative invariant ligand, a single copy of the *MR1* gene was identified, and depending on the species, there was only limited polymorphism.[31,36,38,45,49-52] In humans, a patient homozygous for an MR1 mutation Arg9His (α1-domain) has been described. This MR1 variant did not activate MAIT cells when incubated with stimulating antigen and the patient lacked circulating MAIT cells.[51] A distinct polymorphism His17Arg (α1-domain) in human MR1 was shown to create an HLA-DR–restricted minor histocompatibility antigen.[52] In addition, three intronic single-nucleotide polymorphisms have been described in human *MR1*, one of which impacted on MR1 expression and correlated with susceptibility to tuberculosis.[53] None of the allelic variants of *MR1* described in nonhuman primates and opossum are known to impact on antigen presentation or TCR reactivity.[31,48,49]

Furthermore, an outstanding degree of amino acid sequence identity between MR1 of various species was noted. This is particularly high in the ligand-binding α1-/α2-domains, which is not seen for any other MHC-I like or classical or nonclassical MHC molecules that function as antigen-presenting molecules,[31,35,37,45] suggestive of a strong selective pressure maintaining MR1 structure.[36] For example, the amino acid sequences of phylogenetically distant human and mouse MR1 molecules are overall 81% identical, in the putative ligand-binding domains 90% identical (as compared to 70% in MHC-I) and in the α3-domain 73%. Together, with the ubiquitous gene expression of MR1 in tissues and peripheral blood as well as numerous cell types,[31,35,37,45,46] this favors a general and important physiological function.[36,45] The high conservation among mammalian species, where synonymous substitutions predominate in the α1-/α2-domains,[44] and the limited polymorphism indicate that a purifying selection process (ie, against diversification) has acted on the *MR1* gene as the evolutionary selection pressure.[32]

MR1 Transcription and Alternative Splicing-Derived Isoforms

As with other MHC-I-like, classical, and nonclassical MHC-I genes,[54,55] in addition to full length *MR1* transcripts (also referred to as *MR1a* in most species), other transcripts encode predicted isoforms of MR1 through alternative splicing and these transcripts can be expressed at high levels. So far, MR1 isoforms have been described in humans and mice,[35,36,45] rats,[37] cows,[26] sheep,[26] nonhuman primates,[48] and pigs.[46] In contrast, for marsupial *MR1*, no spliced variants were identified.[31] Alternatively spliced *MR1* variants are expressed in numerous tissues and peripheral blood, as well as different cell lineages, and can be highly abundant.[35,36,56] However, unlike *MR1a*, they are not conserved between species and many are not predicted to be functional. For instance, in humans, three other isoforms exist, designated *hMR1B*, *hMR1C*, and *hMR1D*. All lack the α3-domain, and while *hMR1C* encodes a putative soluble form (which thus might be secreted) involving the

TABLE 37.2 Comparison of the Properties of the MHC-I Like Molecules MR1 and CD1d

	MR1	CD1d
Gene location (human)	Chromosome 1.	Chromosome 1.
Polymorphism	Nonpolymorphic.	Nonpolymorphic.
Biochemical properties	Standard MHC-I-fold, including noncovalent association with β_2m. Antigen-binding cleft (~760 A^3) is mostly solvent exposed, consisting of a mixture of charged and hydrophobic residues, with a preponderance of aromatic residues within the α1- and α2-helices. Pockets: deep A′- and a shallower F′-pocket with biochemical properties suited to present small aromatic molecules. All known MR1 ligands bind in the A′ pocket, where they are sequestered in an "aromatic cradle," formed by aromatic and charged amino acid residues. Most MR1 ligands form a covalent bond (Schiff base) with Lys43 of MR1, located in the A′ pocket.	Standard MHC-I-fold, including noncovalent association with β_2m. Antigen-binding cleft (~1690 A^3) is shielded from solvent and lined by hydrophobic amino acids. Pockets: a deep A′- and a shallower F′-pocket, with biochemical properties suited to present lipid-based antigens. Each pocket can bind one alkyl chain from the antigen and spacer molecules, such as fatty acids, are recruited when the alkyl chain of the antigen is too short to fill either pocket.
Ligands	Metabolite-based, eg: • agonist: 5-OP-RU (bacterial-, fungal-, plant-derived). • nonagonist: 6-FP (photodegradation product of folic acid produced by plants and bacteria). • undefined self-ligands (including tumor-derived) recognized by MAIT or other MR1-reactive T cells.	Lipid-based, eg: • agonist: αGalCer (marine sponge-derived) and structural variant $\alpha GalCer_{Bf}$ (bacteria-derived), GD3 (tumor-derived). • nonagonist: GPI (self-ligand).
Distribution	Ubiquitous in tissues (based on gene expression). Ubiquitous in nucleated cell types (based on gene expression).	Wide range of tissues. Wide range of cell types, including hemopoietic and nonhemopoietic cell types.
Dominant steady-state cellular localization	ER (small amounts in the endosomes and at the cell surface).	Endosomes (small amounts at the cell surface).
Dominant route of antigen presentation	During assembly in the ER.	Exchange for endogenous ligand during recycling in endosomal compartments.
Dependence on the MHC-I peptide loading complex (PLC), PLC-binding partners, and ER-resident chaperones	No role in antigen loading but tapasin and TAPBPR stabilize and increase the half-life of the ligand-free ER pool of MR1.	No role in antigen loading but calnexin, calreticulin, and ERp57 are important for folding.
Dependence on MHC-II chaperones (Ii, HLA-DM)	No known role.	No known role.

α1- and α2-domains, *hMR1B* and *hMR1D* remain putative integral transmembrane proteins[36] (Fig. 37.4). In mice, there are five other transcripts, *mMR1-B* to *-F*. *MR1B*, *-C*, *-E*, and *-F* all have aborted open reading frames and *mMR1D* is a truncated, partially spliced transcript carrying part of an intron (intron 3).[36] Although it is curious that some of the MR1 splice variants can be highly expressed (in contrast to those of other MHC molecules), there is only marginal evidence for a potential biological role.[56,57]

Coevolution and Interdependence Between MAIT TCRs and MR1

In early studies, the canonical transcript *MR1a* (hereafter referred to as *MR1*) was considered the most likely to encode the restriction element of MAIT TCRs. MR1 was first proposed[8] and later proven to be the restriction element for MAIT cells.[22] Treiner et al demonstrated that *TRAV1-TRAJ33*+ T cells are essentially absent in mice lacking the α1- and α2-domains of MR1[22] demonstrating the dependence of MAIT cells on MR1 for their selection and expansion. Analyzing *TRAV1-TRAJ33* sequences revealed that 80% of the TCR sequences corresponded to the invariant TCR α-chain and only 6% were out of frame in MR1-proficient mice. In contrast, in MR1-deficient mice, only 7% of the TCR sequences encoded the invariant chain and 47% were out of frame. Overall, 92% of the *TRAV1-TRAJ33* sequences from MR1-proficient mice were either canonical or of identical CDR3 length, whereas only 12% of the sequences from MR1-deficient mice fit these criteria.[22] Interestingly, in mammals in which *TRAV1* is absent (carnivores, lagomorphs, and the armadillo), the *MR1* gene is lost or inactivated, suggesting that the *MR1* and *TRAV1* genes coevolved in mammals.[32] The independent loss of the MR1-MAIT cell system in several mammals suggests that in these

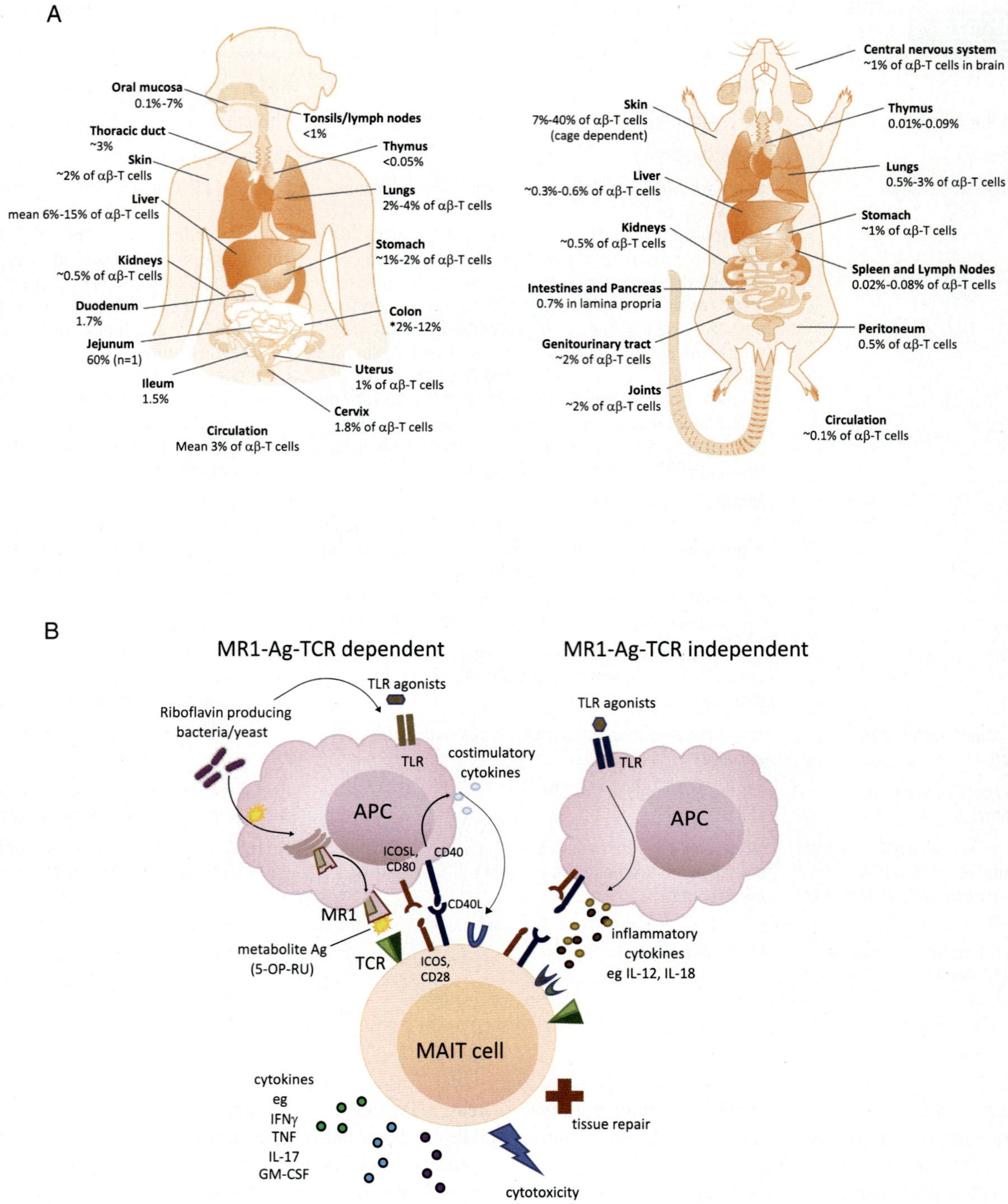

FIG. 37.3. Distribution and activation of mucosal-associated invariant T (MAIT) cells. A: Frequencies of MAIT cells in healthy humans (*left*) and in mice (*right*—based on the C57BL/6 inbred strain). * indicates the range between different studies. **B:** MR1-antigen (Ag)-TCR dependent and independent activation and functional responses. (A, Adapted from Wang H, Chen Z, McCluskey J, et al. Mouse models illuminate MAIT cell biology. *Mol Immunol.* 2021;130:55-63; with permission from Elsevier and with permission of Annual Reviews, Inc, Provine NM, Klenerman P. MAIT Cells in Health and Disease. *Annu Rev Immunol.* 2020; permission conveyed through Copyright Clearance Center, Inc. B, Modified by permission from Springer Nature: Godfrey DI, Koay HF, McCluskey J, et al. The biology and functional importance of MAIT cells. *Nat Immunol.* 2019;20(9):1110-1128.)

species MAIT cells are dispensable, or that another T cell subset has taken over their role, or that they may have had a deleterious role.[32] The constrained nature of the semi-invariant MAIT TCR and mono-/oligomorphic nature of MR1 as well as their extraordinary conservation among most mammalian species is, however, consistent with an important physiological function shared among most mammalian species.[31,32] Further illustrating the phylogenetic conservation of the MAIT-MR1 axis is the recognition of MR1 orthologs by species-disparate MAIT cells.[8,25,27,44,47,58-62]

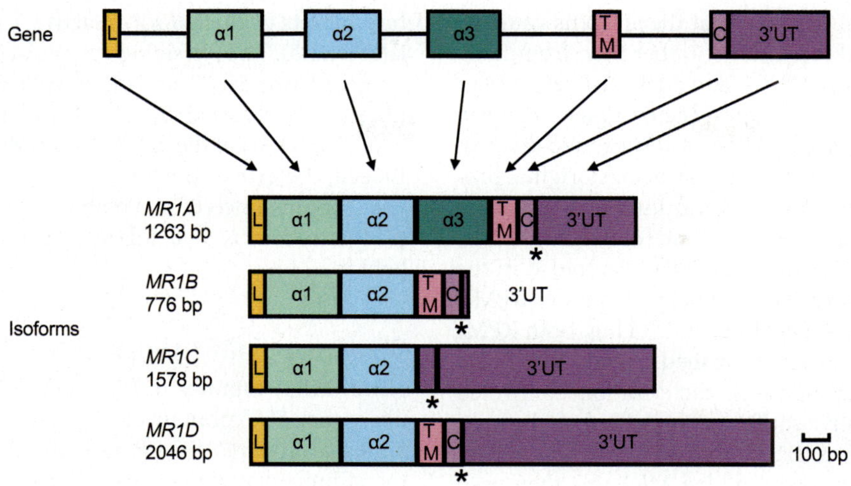

FIG. 37.4. Schematic representation of human MR1 isoforms. Each box corresponds to an exon, coding for the leader peptide (L), the α1-, α2-, and α3-domains, the transmembrane domain (TM), cytoplasmic tail (C), and 3′ untranslated region (3′UT); stop codons are indicated with a star. The cDNA length of each isoform in base pairs (bp) is indicated.

MAIT Cells Are Stimulated, in an MR1-Dependent Manner, by a Wide Range of Bacteria and Fungi

Whether MAIT cells bind MR1 alone or in association with a ligand remained an open question for a long time.[22] However, circumstantial evidence by Miley et al suggested the need for a ligand or an MR1-associated molecule to drive both MR1 folding and efficient cell surface expression, as well as the production of recombinant, soluble MR1 protein *in vitro*.[39] It was further hypothesized that such a ligand or associated molecule could be derived from or induced by microbes and might be directly recognized by MAIT cells but that is otherwise in limited supply.[22,41,44] Indeed, in steady-state, MR1 accumulates largely in the ER, in association with the peptide-loading complex (PLC) consistent with limiting antigen in the absence of infection.[39,40] Huang et al provided further evidence that MR1 captures microbial antigen by showing that MR1-restricted MAIT cell activation was disrupted by changes in single residues in the A′- but not the F′-pocket of the MR1 groove, that the TCR mutagenesis footprint on MR1 resembled that of peptide-MHC-I molecules, and that acid eluates of recombinant MR1 enhanced MAIT cell activation in an MR1-specific manner.[44]

Importantly, Gold et al and Bourhis et al then independently demonstrated that cells infected by a wide range of Gram-positive and Gram-negative bacteria and fungi or coincubated with fixed bacteria, and to some extent, heat-killed bacteria, but not viruses, are capable of stimulating MAIT cells in an MR1-dependent manner.[60,63] In fact, some studies suggested that individual MAIT cell clones recognized cells infected with different bacteria.[60,63] This strongly suggested that MR1 presented conserved components (ligands) that were either microbe-derived or microbe-induced, thereby activating MAIT cells.[60,63] Interestingly, some bacteria, such as *Enterococcus faecalis*, did not induce MAIT cell activation. This suggested that they either lack the necessary trigger for activation or actively inhibit activation.[60] The latter consideration however was refuted as they did not impact on MAIT cell activation by other MAIT-cell–stimulating bacteria.[60] The origin and chemical identity of the MR1 ligand became a key question in the field, as understanding this would help unlock the biology of MAIT cells. Of note, it was later also shown that MAIT cells can sense viral infections,[64] described in detail in section *Viral Infections*. This activation is mediated by inflammatory cytokines present in the context of viral infections, which activate MAIT cells independently of MR1-antigen and TCR interaction.

MR1 ANTIGEN DISCOVERY

Overview of the Antigen Discovery Process

In 2009, more than 15 years after the discovery of MAIT cells, the identification of MAIT cell antigens presented by MR1 emerged as a road block to the further understanding of their biological role and function. The ultimate discovery involved a multidisciplinary approach, including protein chemistry, mass spectrometry, cellular biology, genetics, structural biology, and small molecule chemistry. A detailed recount of the events that led to the discovery of the vitamin B–related compounds as MAIT cell antigens was previously published,[65] and a summary is provided here. After it had been established that individual MAIT cell clones could be stimulated with cells infected with a number of bacteria and fungi or by antigen-presenting cells (APCs) coincubated with heat-killed or-fixed versions of these,[60,63] *Salmonella enterica* serovar Typhimurium (*S.* Typhimurium) was chosen as a model organism for antigen discovery. Importantly, it was noticed that the culture supernatant of *S.* Typhimurium was stimulatory to MAIT cells in an MR1-dependent manner, suggesting that the antigen(s) of interest were likely to be secreted or shed *in vitro*, and hence, they were probably soluble.[66] From a practical perspective, this implied that the antigens could be recovered from bacterial supernatant rather than from more complex bacterial lysates. MHC- and MHC-I-like proteins are generally stable only in the presence of a given antigen, and circumstantial evidence by Miley et al suggested that this also applied to MR1.[39] Indeed, using Garboczi's original protocol of MHC-I folding,[67] supernatant from *S.* Typhimurium sponsored *in vitro* folding of recombinant, soluble MR1 in the presence of β_2m.[66] Small amounts of MR1 also folded in the presence of control Roswell Park Memorial Institute (RPMI) medium, used to culture *S.* Typhimurium.[66] Thus, both RPMI medium and *S.* Typhimurium furnished ligands that could bind to MR1 and biochemical characterization confirmed that MR1 adopted its native conformation with these ligands, associated with β_2m in a 1:1 stoichiometry.[66] Mass spectrometry of eluates from ligand-folded MR1 was then used to identify the nature of the ligands captured by MR1 from RPMI medium and supernatant of *S.* Typhimurium culture.

Discovery of an MR1 Ligand That Acts as Competitive Inhibitor of MAIT Cell Activation

RPMI medium contains defined components, including members of the vitamin B family, which are yellow in color. Indeed, MR1 folded in the presence of a solution of pharmacy-grade vitamin B tablets. Among individual vitamins, folic acid (vitamin B9) yielded significant amounts of native MR1, which could be analyzed by mass spectrometry.[66] This analysis revealed that MR1 did not capture folic acid itself, but a photodegradation product of folic acid termed 6-formylpterin (6-FP, mass-to-charge (*m/z*) 190.03 in negative ion mode, $C_7H_5N_5O_2$) (Fig. 37.5A). In line with its ability to fold with MR1, exogenous 6-FP also induced a substantial increase in MR1 levels on the cell surface of APCs, which in the absence of MR1 ligand are very low.[66] The crystal structure of MR1 folded in the presence of 6-FP definitively showed that MR1 was noncovalently associated with β_2m and had adopted an MHC-I-like fold, whereby 6-FP was sequestered deep within the base of the MR1 antigen-binding cleft, which appeared ideally suited to present small aromatic molecules (Fig. 37.5B). Here, the pterin ring of 6-FP was sandwiched between aromatic residues (Tyr7, Tyr62, Trp69, and Trp156) and formed a hydrogen bond with Arg9 of the cleft. Most remarkably, 6-FP was covalently bonded to MR1 through a Schiff base (chemically known as an imine, a carbon-nitrogen double bond), formed between the formyl group of 6-FP and the side chain ε-amine of lysine 43 of MR1 (Fig. 37.5C). Such a covalent interaction between an antigen and antigen-presenting molecule remains unique to MR1 among antigen-presenting molecules and probably partly explains why such a small ligand is stably bound in the otherwise large cleft of MR1.

While 6-FP was a bona fide MR1 ligand,[66] like RPMI medium, it did not stimulate MAIT cells. Instead 6-FP acted as a competitive inhibitor of MAIT cell activation by microbes and microbial supernatant, containing stimulatory or agonist MR1 ligands.[2,66] As outlined in section *Recognition of MR1-Ligand Complexes by MAIT and Other MR1-Reactive T Cells*, rare subsets of MR1-6-FP–reactive T cells were identified in later studies. The physiological role of 6-FP as an MR1-ligand in modulating MAIT and other MR1-reactive T cells remains unclear. Synthesized by plants and microbes, folic acid is taken up as part of the diet and is produced by the microbiota. Indeed, the extent to which folic acid is degraded to 6-FP *in vivo* remains uncertain. Notably, degradation of folic acid to 6-FP can occur *in vivo*, following skin exposure to ultraviolet light.[68]

Discovery of MR1 Ligands That Activate MAIT Cells

Since MR1 captures 6-FP from RPMI medium, it was likely that *S.* Typhimurium culture in RPMI medium contained a mixture of 6-FP and stimulatory antigen(s). Thus, for stimulatory antigen identification by mass spectrometry, MR1 was instead folded in the presence of supernatant of *S.* Typhimurium culture in folate-free medium, thereby enhancing the yield and purity of MR1 that had captured stimulatory antigen(s). Mass spectrometry analysis of a ligand recovered from MR1 did not match the chemical composition of any known compounds (*m/z* 329.11 in negative ion mode, $C_{12}H_{18}N_4O_7$).[66] However, some ribityllumazine compounds from the riboflavin (vitamin B2) biosynthetic pathway represented close matches for this *m/z* (Fig. 37.6). They included 7-hydroxy-6-methyl-8-D-ribityllumazine (RL-6-Me-7-OH, $C_{12}H_{16}N_4O_7$, *m/z* 327.0946 in negative ion mode) and 6,7-dimethyl-8-D-ribityllumazine (RL-6,7-diMe, $C_{13}H_{18}N_4O_6$, *m/z* 325.1153 in negative ion mode).[66] Synthetic versions of RL-6-Me-7-OH and RL-6,7-diMe turned out to be stimulatory to MAIT cells *in vitro*[66] and crystallographic analysis of the crystal structure involving RL-6-Me-7-OH unambiguously confirmed this antigen in the MR1 binding cleft and also revealed the molecular basis of its stimulatory capacity toward MAIT cells compared to the nonstimulatory 6-FP antigen,[2] described in detail in section *Recognition of MR1-Ligand Complexes by MAIT and Other MR1-Reactive T Cells*. RL-6-Me-7-OH and RL-6,7-diMe were, however, much less potent in stimulating MAIT cells than *S.* Typhimurium supernatant, suggesting that the potently stimulatory antigen with the composition $C_{12}H_{18}N_4O_7$ was yet to be fully identified.[66] The observation that MAIT cell stimulatory microbes identified in previous studies[60,63] possessed a complete riboflavin biosynthetic pathway[66] suggested that the potently activating MAIT cell ligand might be a distinct metabolite derived from riboflavin biosynthesis.

Three compounds, not previously described in riboflavin biosynthesis, could be attributed to the formula $C_{12}H_{18}N_4O_7$. They were a dihydrogen-reduced form of RL-6-Me-7-OH (abbreviated rRL-6-Me-7-OH); a dihydrogen-reduced form of 6-hydroxymethyl-8-D-ribityllumazine (called rRL-6-CH_2OH or rRL-6HM); or 5-(2-oxopropylideneamino)-6-D-ribitylaminouracil (abbreviated 5-OP-RU), the latter expected to be far too unstable in aqueous conditions (Fig. 37.7). rRL-6-Me-7-OH was too unstable, but rRL-6-CH_2OH could be chemically synthesized from the key riboflavin biosynthesis intermediate 5-amino-6-D-ribitylaminouracil (5-A-RU), and this synthesis product potently activated MAIT cells in an

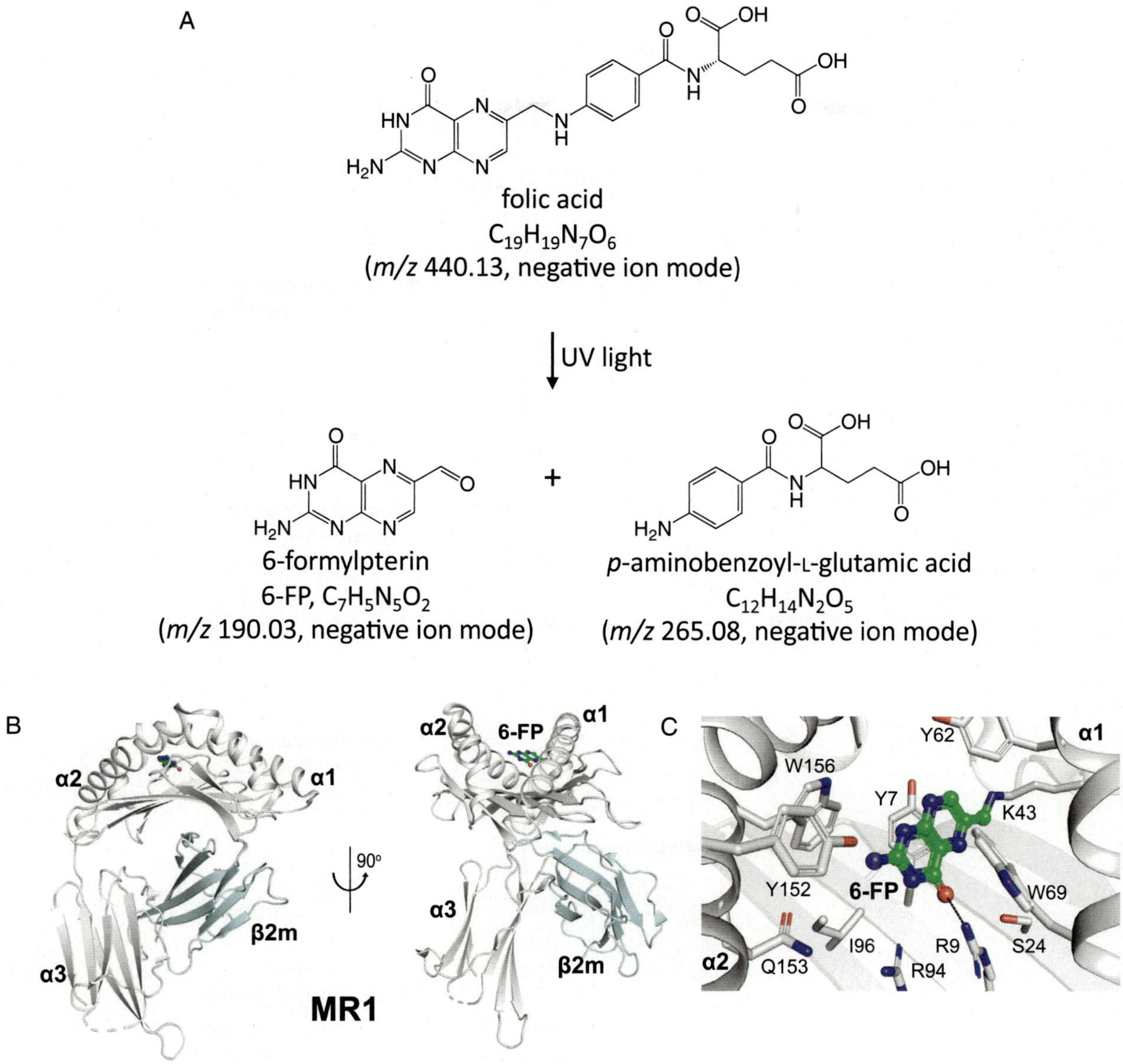

FIG. 37.5. The MR1 ligand 6-formylpterin. A: Photodegradation of folic acid in ultraviolet light to 6-formylpterin and *p*-aminobenzoyl-l-glutamic acid. **B:** Cartoon presentation of the crystal structure of the human MR1-6-FP complex (PDB; 4GUP). MR1 and β_2m are colored *white* and *pale cyan*, respectively. **C:** The interactions of 6-FP within the A′-pocket of the MR1 binding cleft. MR1 α-helices and β-sheets are presented as *white* cartoon. The MR1-antigen interacting residues and the 6-FP ligand are displayed as *white* and *green sticks*, respectively. H-bond is presented as *black dashed line.*

MR1-dependent manner.[66] Limiting availability of synthetic rRL-6-CH_2OH combined with a low MR1 folding efficiency in the presence of rRL-6-CH_2OH prevented determining the crystal structure of MR1 loaded with synthetic rRL-6-CH_2OH and complexed to a MAIT TCR. However, a version of MR1 in which lysine was mutated to alanine (K43A), MR1-K43A, could be folded with β_2m in the absence of ligand (designated "empty" MR1-K43A) and subsequently loaded with synthetic rRL-6-CH_2OH.[7] Intriguingly, the electron density of the crystal structure of MR1-K43A loaded with synthetic rRL-6-CH_2OH and complexed with a MAIT TCR indicated that a single-ring structure was present in the MR1 antigen-binding cleft, instead of the expected two-ring structure of rRL-6-CH_2OH.[69] Importantly, fluorochrome-coupled tetramers of MR1-K43A loaded with a synthetic preparation of rRL-6-CH_2OH stained human MAIT cells,[7] suggesting that the single-ring structure captured by MR1 was a MAIT cell-stimulating antigen. A genetic approach was then taken to understand whether this single-ring structure captured from rRL-6-CH_2OH synthesis originated from natural riboflavin biosynthesis.[69]

Confirmation of the Riboflavin Synthesis Pathway as the Source of MAIT Cell Antigens

Riboflavin (vitamin B2) biosynthesis is an essential metabolic process for many bacteria and fungi, whereas the pathway

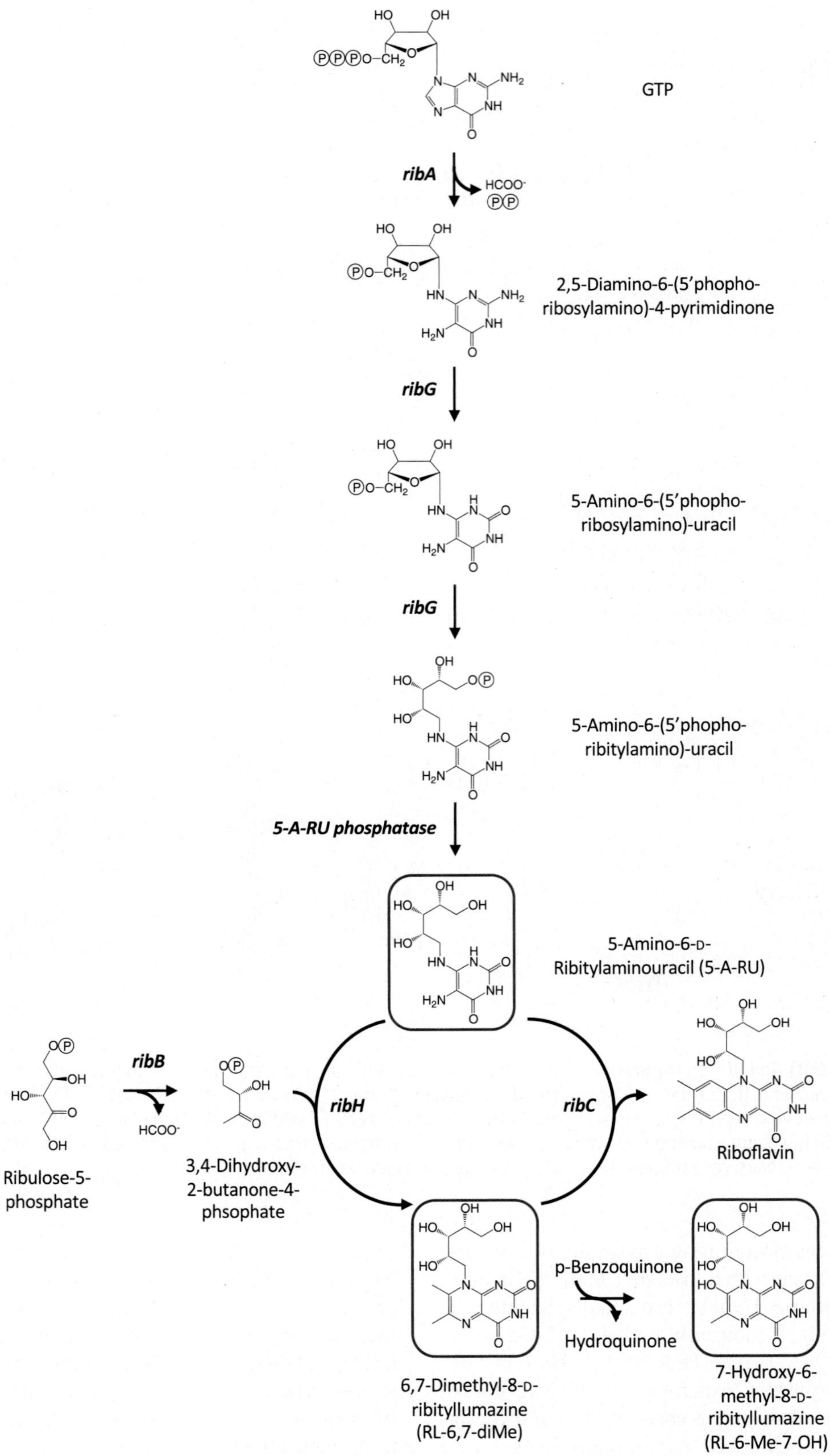

FIG. 37.6. **The riboflavin biosynthesis pathway.** The biosynthesis of one riboflavin molecule requires one molecule of guanosine 5′-triphosphate (GTP) and two molecules of ribulose 5-phosphate as substrates. Molecules that represent mucosal-associated invariant T (MAIT) cell antigens (RL-6,7-diMe, RL-6-Me-7-OH) or are required for their genesis (5-A-RU) are highlighted.

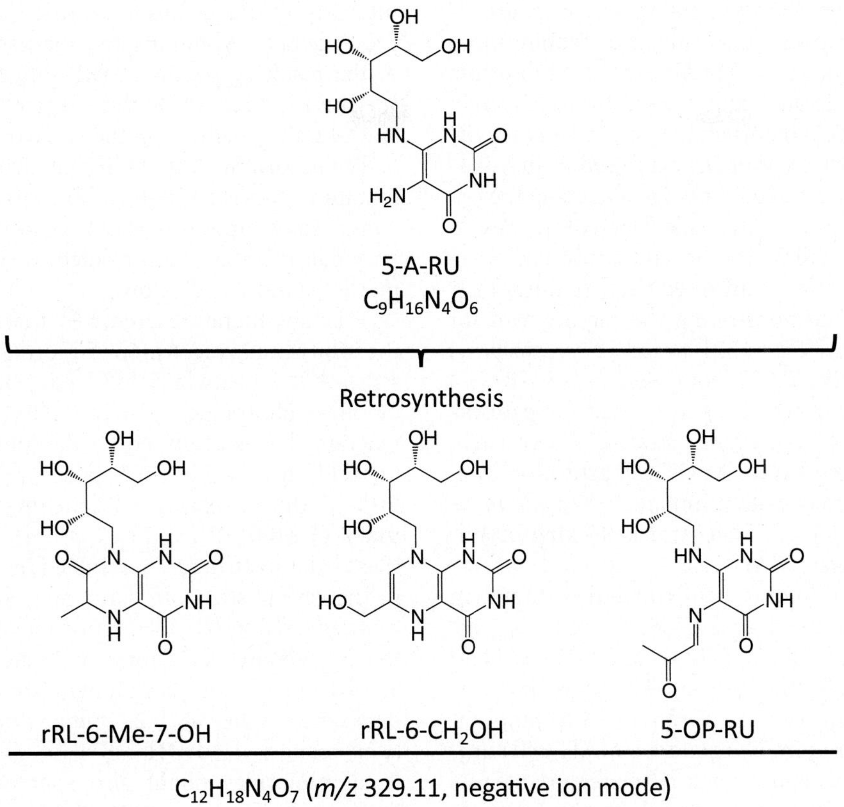

FIG. 37.7. rRL-6-Me-7-OH, rRL-6-CH$_2$OH, and 5-OP-RU are regioisomers ($C_{12}H_{18}N_4O_7$, *m/z* 329.11, negative ion mode). The three regioisomers can be derived nonenzymatically from the key vitamin B2 biosynthesis intermediate 5-A-RU.

is absent from higher organisms, including mammals, and from viruses. Biosynthesis of riboflavin involves multiple reactions catalyzed by a series of enzymes[70] (Fig. 37.6). Firstly, 3,4-dihydroxy-2-butanone-4-phosphate synthase (*rib B*) converts ribulose-5-phosphate into 3,4-dihydroxy-2-butanone 4-phosphate. In a second arm of the pathway, guanosine 5′-triphosphate (GTP) is converted by GTP cyclohydrolase II (*rib A*) into 2,5-diamino-6-ribosylamino-4(3H)-pyrimidinone 5′-phosphate. Subsequent steps modify this to 5-amino-6-ribosylamino-2,4(1H,3H)-pyrimidine 5′-phosphate, 5-amino-6-ribitylamino-2,4(1H,3H)-pyrimidinedione 5′-phosphate, and 5-amino-6-ribitylamino-2,4(1H,3H)-pyrimidinedione (5-amino-6-D-ribitylaminouracil [5-A-RU]), which is then dephosphorylated either by a specific or nonspecific phosphatase,[71] and condensed with 3,4-dihydroxy-2-butanone 4-phosphate to form 6,7-dimethyl-8-ribityllumazine (catalyzed by lumazine synthase, *rib H*). Lastly, riboflavin synthase (*rib E*) catalyzes a dismutation reaction of two molecules of 6,7-dimethyl-8-ribityllumazine to form riboflavin, with 5-amino-6-ribitylamino-2,4(1H,3H)-pyrimidinedione, which then again acts as a substrate for lumazine synthase.

In some bacteria (such as the Gram-positive *Bacillus* and *Lactococcus* spp.), the enzymes driving the individual steps are organized in a single operon, the *rib* operon,[72] whereas in others (eg, *Salmonella* and *Escherichia* spp.), these genes are separated (although in *S.* Typhimurium, *ribD* and *ribH* are grouped together). There are several other genetic nuances to maintaining the essential riboflavin pathway in different organisms. In some bacteria and plants, *ribA* and *ribB* are combined into a bifunctional enzyme, *ribAB*. In fungi and in plants, there is a variation in the early steps of the pathway, upstream of 5-A-RU, the key intermediate relevant to MAIT cell activation.[73-75] Partly as a consequence of these variations, the use of enzyme names differs between different bacteria such that, for example, *ribD* in *Salmonella* spp. is named *ribG* in *Bacillus* and *Lactococcus* spp.

Like other metabolic pathways, riboflavin synthesis is highly regulated. The regulation of riboflavin synthesis is complex and differs between different microorganisms. In *B. subtilis* and other Gram-positive bacteria, the Rib operon, containing genes *ribG, ribB, ribA,* and *ribH*, is under transcriptional negative feedback regulation, so that high concentrations of riboflavin or flavin mononucleotides downregulate further riboflavin production.[70,76] Riboflavin production is also regulated by a range of mechanisms in other species.[76] In line with this negative feedback regulation, MAIT cell activation is reduced when bacterial cultures are grown in high concentrations of riboflavin.[69,77,78] Bacterial species also differ in their expression of a range of transporters that import riboflavin into cells, a factor not yet fully explored in understanding the antigen generation and modulation of MAIT cells.[79,80]

In order to confirm the riboflavin pathway as the source of MAIT cell-activating compounds, a mutant *S.* Typhimurium strain, with a deletion of *ribD-ribH* (*S.* Typhimurium SL1344Δ*ribDH*), was generated. Supernatant from this strain failed to activate MAIT TCR reporter cell lines, and this activity was reconstituted by addition of *ribDH* expressed from a plasmid.[69] A *S.* Typhimurium Δ*ribDH strain* (generated on the less virulent BRD509 background)[23] also had a markedly decreased ability to trigger MAIT cell activation and accumulation *in vivo*. Similar results have now confirmed these findings in *E. coli* and *S. pneumoniae* demonstrating the requirement for *ribD/G* in these bacteria.[77,78,81] In 571 strains of *Streptococcus pneumoniae* the genes "*rib DEAH*" (equivalent to "*rib GBAH*") were found in a highly conserved operon,[78] but the pathway was missing from 2400 *S. pyogenes* strains,[77,78] consistent with a previously reported lack of MAIT cell activation by *S. pyogenes*.[60] Infection of mice with a mutant of *S. epidermidis* (lacking *ribD*)[82] or injection of *E. coli* extracts (lacking *ribA* or *ribD*)[81] also failed to activate MAIT cells *in vivo*.

Although these experiments confirmed the riboflavin biosynthesis pathway as the source of MAIT cell antigens, the candidate compounds rRL-6-CH_2OH and 5-OP-RU had not been described within this pathway. Therefore, individual gene deletion mutant bacteria were needed to pinpoint the step in the pathway required to generate MAIT cell stimulatory molecules. To this end, a set of *rib* gene mutants was generated using *Lactococcus lactis*, in which the four key genes are expressed as a single operon (*ribGBAH*). These mutants were created on the background of the roseoflavin-resistant mutant CB013, which constitutively expresses riboflavin due to a mutation in the negative feedback regulation upstream of the *rib* operon.[83] Stop codons were introduced into each of the four genes, separately. When culture supernatant from these mutants was tested in MAIT TCR reporter activation assays,[69] deletion of *ribA* and *ribG* (the latter equivalent to *ribD* in *Salmonella* spp.) abrogated activation, whereas *ribB* and *ribH* deletions did not. These results pointed to 5-A-RU as being necessary for production of stimulatory MAIT cell antigen. Consistent with this, *L. lactis* Δ*ribA*, *L. lactis* Δ*ribG*, *E. faecalis* (which lacks riboflavin synthesis), and *S.* Typhimurium Δ*ribDH* supernatants all failed to support *in vitro* folding of soluble, recombinant MR1 protein, contrasting the equivalent supernatants from cultures of wild-type bacteria. A molecular species (m/z 329.11) was identified in MR1 eluates by mass spectrometry, and this correlated with their capacity to activate MAIT TCR reporter cell lines.[69] Taken together, these experiments defined 5-A-RU as the key metabolite necessary for generating activating MAIT cell antigens in bacteria.

Formation of Pyrimidine Antigens From the Riboflavin Pathway Intermediate, 5-A-RU

While synthetic 5-A-RU alone stimulated MAIT cells to variable degrees, the mass spectrometry features of this molecule (5-A-RU, $C_9H_{16}N_4O_6$, *m/z* 275.09) were very different from the antigen derived from *S.* Typhimurium ($C_{12}H_{18}N_4O_7$, *m/z* 329.11). Similarly, although recombinant MR1 could be folded in the presence of 5-A-RU, and the folded product was intensely yellow as observed when MR1 was folded in the presence of *S.* Typhimurium supernatant, mass spectrometry of the resulting purified MR1-antigen, gave an *m/z* 315.09 (negative ion mode), 40 units larger than the *m/z* of 5-A-RU and 14 units smaller than the *m/z* of the antigen derived from *S.* Typhimurium, 329.11. Insights from the old biochemical literature, X-ray crystallographic structures of MR1-antigen complexes bound to a MAIT TCR and fundamental chemistry considerations contributed to solving this biochemical puzzle, as described below.

Previous literature predicted that *en route* to producing the ribityllumazine RL-6,7-diMe during riboflavin synthesis, 5-A-RU forms a Schiff base with 3,4-dihydroxy-2-butanone-4-phosphate (3,4-DH-2-B-4-P), a condensation reaction that is catalyzed by the lumazine synthase (*ribH*), to yield an unstable ring-opened pyrimidine intermediate, 5-(1-methyl-2-oxopropylideneamino)-6-D-ribitylaminouracil (5-MOP-RU, $C_{13}H_{20}N_4O_7$, *m/z* 343.13).[84] Though not identical, this structure resembled the single-ring ligand seen in the crystal structure containing MR1-K43A loaded with synthetic rRL-6-CH_2OH.[69] Previous literature also showed that 5-MOP-RU could form spontaneously, in the absence of any enzymatic catalysis, namely from Schiff base formation between 5-A-RU and 2,3-butanedione, the dephosphorylated version of 3,4-DH-2-B-4-P.[70] This suggested that other small molecules might also spontaneously undergo similar reactions during *in vitro* folding of MR1, one of which might generate the dominant *m/z* 329.11 species derived from *S.* Typhimurium culture supernatant and another, the *m/z* 315.09 species observed following folding of MR1 in the presence of 5-A-RU. Indeed, methylglyoxal, lacking a methyl group compared with 2,3-butanedione, was predicted to form a Schiff base with 5-A-RU yielding 5-OP-RU ($C_{12}H_{18}N_4O_7$), matching the mass of the antigen eluted from MR1 (*m/z* 329.11), and a compound theoretically generated *en route* to creating 7-methyl-8-D-ribityllumazine (RL-7-Me) (Fig. 37.8A). Similarly, glyoxal, lacking two methyl groups compared to 2,3-butanedione, would be expected to form a Schiff base with 5-A-RU to yield 5-(2-oxoethylideneamino)-6-D-ribitylaminouracil (5-OE-RU, $C_{11}H_{16}N_4O_7$). By analogy with 5-OP-RU, 5-OE-RU was theoretically generated *en route* to creating 8-D-ribityllumazine (RL).[69]

Equimolar amounts of 5-A-RU and methylglyoxal in MR1 refolds yielded MR1-5-OP-RU in addition to MR1-5-OE-RU, while a molar excess of methylglyoxal solely yielded MR1-5-OP-RU, as determined by mass spectrometry.[69] Notably, both glyoxal and methylglyoxal are formed from a number of metabolic pathways, including mammalian and bacterial glycolysis,[85] and glyoxal is widely abundant and is even produced industrially as a precursor to many products. The almost ubiquitous presence of glyoxal explained the presence of the *m/z* 315.09 species in MR1 folding experiments with 5-A-RU alone or 5-A-RU together with methylglyoxal.[69] Importantly, MR1 tetramers loaded with 5-OP-RU or 5-OE-RU were very stable and stained MAIT cells with high specificity and sensitivity.[69] Moreover, mass spectrometry of MR1 folded in the presence of *S.* Typhimurium supernatant revealed a minor *m/z* 315.09 species along with the dominant *m/z* 329.11

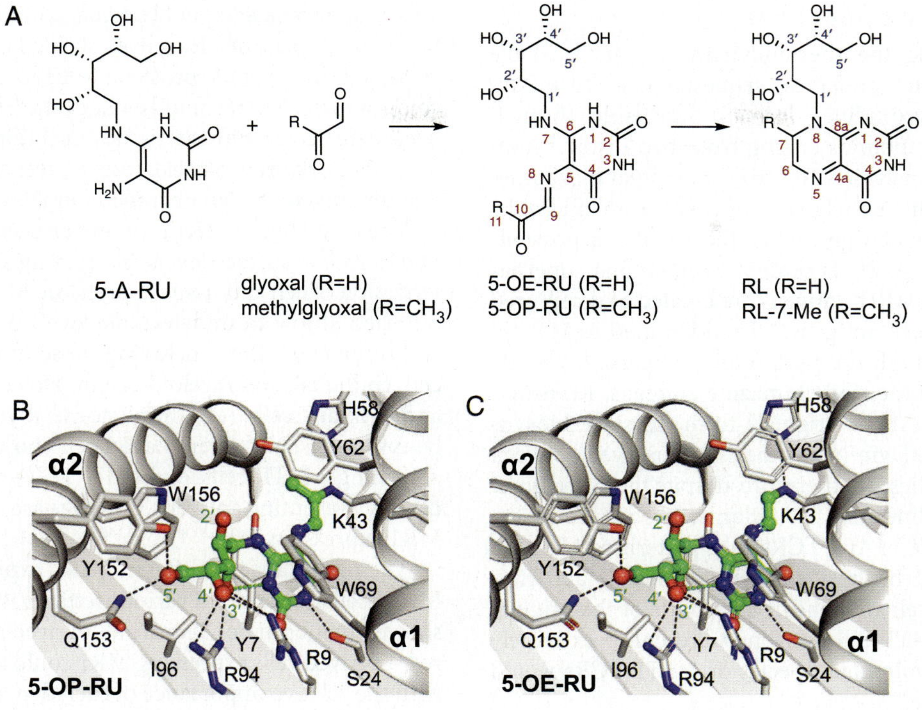

FIG. 37.8. Formation and MR1 capture of the potent mucosal-associated invariant T (MAIT) cell antigens. A: Formation of pyrimidines and ribityllumazines from condensation of small metabolites with 5-A-RU. **B:** H-bonding interactions between the MR1-residues (*white sticks*) and 5-OP-RU ligand (*green sticks*) or (C) 5-OE-RU (*green sticks*). All H-bonds are displayed as *black dashed lines* except of the intramolecular interaction between the 3′OH and the uracil ring of the ligand, which is shown as *green dashed line*.

species. Also, folding of MR1 in the presence of *E. coli* (DH5α) supernatant yielded a distinct, more abundant, *m/z* 315.09 in comparison to a *m/z* 329.11 species.[69] In both cases, the mass spectrometry profiles matched those observed from *in vitro* folded MR1, yielding MR1-5-OP-RU and MR1-5-OE-RU. Thus, both 5-OP-RU and 5-OE-RU were likely acting as potent, physiological MAIT cell antigens. Described in detail in section *The Molecular Basis of MAIT TCR Recognition of MR1-Ligand*, crystal structures of a MAIT TCR in complex with MR1 folded in the presence of a mixture of 5-A-RU and either methylglyoxal or glyoxal, confirmed that 5-OP-RU and 5-OE-RU, respectively, were present in the MR1 antigen-binding cleft, covalently bound via Schiff base to the lysine 43 residue of MR1 (Fig. 37.8B, C). Furthermore, the above mentioned single-ring molecule present in the antigen-binding cleft of MR1-K43A loaded with synthetic rRL-6-CH_2OH appeared to be 5-OP-RU,[69] consistent with 5-OP-RU being generated during the synthesis of rRL-6-CH_2OH.[86] Thus, on balance, we consider that rRL-6-CH_2OH is itself not a physiological agonist of MR1, and it is quite possible that it cannot stimulate MAIT cells but its relative instability means this has not been formally tested.

Neither 5-OP-RU nor 5-OE-RU were previously known to exist naturally, and they are the first examples of organic heterocycles identified as T cell antigens. However, they are also chemically relatively unstable under physiological conditions where they readily cyclize to ribityllumazines.[87] 5-OP-RU presentation by MR1-K43A highlighted that Schiff base formation was not obligatory for MR1-antigen presentation. Indicating that other interactions with MR1-K43A were evidently sufficient to ligate 5-OP-RU and prevent its cyclization to a ribityllumazine. However, the lack of a Schiff base with MR1 impacted profoundly on the stability of the antigen-MR1-K43A complex, which was markedly reduced.[88]

5-MOP-RU is highly unstable and unlikely to form an MR1 ligand, unlike 5-OP-RU and 5-OE-RU, which are predicted to assume similar conformations in solution as compared to those when bound to MR1.[87] Indeed MR1-MOP-RU has not been observed to form upon folding of MR1 in the presence of 5-A-RU and 2,3-butanedione and, unlike methylglyoxal and glyoxal, 2,3-butanedione in conjunction with 5-A-RU has not been observed to induce MR1-upregulation or MAIT cell activation above that observed with 5-A-RU alone.[69,87] 5-OP-RU, on the other hand, has a half-life of 88 min in PBS, pH 7.4, 37°C, and its increased stability compared to 5-OE-RU is likely to account for its slightly more potent agonist activity.[87] A synthesis was therefore devised to produce 5-OP-RU and 5-OE-RU in dimethylsulfoxide, where they were stable and could be extensively characterized in solution by nuclear magnetic resonance spectroscopy.[69] Later, this method was further optimized to provide a reliable source of pure MAIT cell antigens for studying their chemical reactivity and MAIT cell activation *in vitro* and *in vivo*.[87]

Summary and Future Directions

Instead of capturing the thermodynamically more stable ribityllumazines, MR1 prefers to sequester less thermodynamically stable pyrimidines formed transiently through condensation of compounds arising from two distinct metabolic pathways, namely 5-A-RU from microbial biosynthesis of riboflavin and methylglyoxal/glyoxal from microbial and perhaps mammalian glycolysis; the latter is a possibility given that free 5-A-RU is present in microbial supernatant.[69] Thus, the way MR1 antigens are created and captured is unique among T cell antigens.[69] The identified MAIT cell antigens, including both the pyrimidine antigens, 5-OP-RU and 5-OE-RU,[69] and the ribityllumazine antigens, RL-6-Me-7-OH and RL-6,7-diMe,[2,66] originate from the same biosynthetic pathway (riboflavin biosynthesis), conserved between many microbial species. Furthermore, despite their structural differences, pyrimidines and ribityllumazines are both recognized by the same MAIT TCRs in a pattern-recognition like manner.[88] As such, the identified pyrimidine and ribityllumazine MAIT cell antigens fit all aspects of the original hypothesis that "MAIT cell antigens were either extremely conserved between microbial species or MAIT TCRs would be nondiscriminative."[60]

Given the various versions of the riboflavin pathway, variants of 5-A-RU and ribityllumazines might exist. Also, physiological byproducts other than methylglyoxal and glyoxal might be capable of forming a Schiff base with both MR1 and 5-A-RU, thus forming antigens around a generic scaffold that could differ based on the microbe as previously described for 5-OP-RU versus 5-OE-RU[69] or the host context. The current state of emerging knowledge on additional MR1 ligand diversity is described in detail in section *Diversity of MR1 Ligands and Their Presentation by MR1*.

The full extent of MR1 polymorphism and its functional consequences are yet to be investigated. This may be of clinical importance in the context of transplantation and MR1-reactive T cell–based immunotherapies.

The discovery of 5-OP-RU and 5-OE-RU as very potent antigens and the development of MR1-tetramers[7,24,69,89-91] have opened up the field to start elucidating the roles of MAIT cells in immunity, as well as in immunopathology, as described in section *Physiological and Pathophysiological Roles of MAIT Cells*. MR1 tetramers have also allowed further characterization of the phenotypic identity of MAIT cells and their development, as described in sections *Development of MAIT Cells and MAIT Cell Phenotype and Functional Capacity*, respectively. The technology for refolding MR1, *in vitro*, and advanced mass spectrometry techniques have opened the door to a search for novel MR1 ligands, some of which are recognized by non-MAIT cells (MR1-T cells) in malignancy and possibly other scenarios.

MR1 ANTIGEN-PRESENTATION PATHWAY

MR1 Expression and Tissue Distribution

MR1 transcripts were reported to be present ubiquitously among diverse human cell types, such as T and B cell lymphomas, fibrosarcoma, vulval sarcoma, epithelial carcinoma, adenocarcinoma and glioblastoma, and tissues, including the heart, brain, placenta, lung liver, skeletal muscle, kidney, pancreas, spleen, thymus, prostate, testis, ovary, small intestine, colon, and peripheral blood leukocytes.[31,35,37,45,46] MR1 has also been detected in murine tissues, including the heart, brain, lung, liver, skeletal muscle, kidney, testis, spleen, intestines, and thymus, with high expression in DP thymocytes.[35,37,45,92]

The detection of MR1 protein has been more problematic due to its low surface levels, hampering a systematic characterization of cell and tissue expression. MR1 protein has been reported at low to undetectable levels on a range of mouse and human cell lines including those of fibroblast, B cell, T cell, epithelial, and myeloid origin. However, many cell lines and primary cells (eg, from mouse thymus, spleen, lungs, lymph nodes, and liver) have been shown to be capable of activating MAIT cells or MAIT TCR–expressing reporter cells by presenting antigens, and thus are assumed to sponsor MR1 expression.[39,63,66,81,88,92-105,106]

The importance of different MR1-expressing cell types in MAIT cell activation in various settings is unclear. One study showed that both bone marrow–derived and non-bone marrow–derived cells expressing MR1 could activate MAIT cells, with the relative importance of each depending on the bacterial pathogen, such that infection with *Legionella longbeachae*, an intracellular bacteria which resides in phagocytes, resulted in MAIT cell expansion *in vivo* when bone marrow–derived cells expressed MR1, whereas for *S.* Typhimurium, which is capable of actively invading epithelial cells, a potent MAIT cell response required MR1 expression by non-bone marrow–derived cells.[107]

During Steady State the Majority of MR1 Molecules Reside in the ER

MAIT cell development, expansion, and TCR-mediated activation are dependent on MR1 antigen-presentation. MR1 is mostly known for presenting exogenous antigens, that is antigens that are not synthesized by the APC itself, including extracellular antigens that are synthesized by microbes and released to the extracellular environment, as well as intracellular antigens that are synthesized by microbes and released to the lumen of phagosomal and endosomal compartments harboring endocytosed bacteria, or to the cytosol of cells infected by cytosolic bacteria. Some data suggests that MR1 also presents undefined endogenous antigens, such as those produced by cancer cells[108,109] that may stimulate a range of non-MAIT cells, referred to as other MR1-reactive T cells, described in section *Recognition of MR1-Ligand Complexes by MR1-Reactive T Cells Other Than MAIT Cells*. While *MR1* mRNA is broadly expressed in all nucleated cells,[31,35,37,45,46] during steady state, the majority of MR1 proteins reside in the ER, with only a small portion present in endosomes and at the cell surface.[39,40,63,69,88,92,93,98,100,101,103,105,110,111] In the ER, in the absence of exogenous antigens, MR1 primarily adopts a ligand-receptive, incompletely folded, "open" conformation that lacks β_2m, also referred to as the "off" state of MR1 presentation[39,94,103,104] (Fig. 37.9A). While an open conformation is also known for "empty" MHC-I molecules,[113] ligand is not generally limiting at this stage of antigen-presentation for classical MHC molecules which are able to constitutively bind endogenous ligands in the ER shortly after synthesis (cytosolic peptides for MHC-I, invariant chain Ii for MHC-II), triggering their release from the ER and trafficking to the cell surface where they are expressed at high levels, even in the absence of infection.[114-117]

The ligand-free ER pool of MR1 molecules is stabilized and the half-life is increased by the same chaperones that stabilize empty MHC-I in the ER, tapasin (TPN), a member of the MHC-I PLC, and TPN-related protein (TAPBPR), an ER-resident chaperone. TPN and TAPBPR probably use a conserved binding mode established for MHC-I molecules.[118,119] MR1 was also found to be associated with all other members of the PLC, as well as PLC-binding partners and ER-resident chaperones. However, those did not appear to impact on MR1-antigen presentation.[8,22,39,40,104] Similarly, chaperones specific to professional APCs and relevant to MHC-II presentation, such as Ii and HLA-DM, are not essential for MR1-antigen presentation,[22,120] consistent with the broad types of "professional" and "nonprofessional" cells acting as MR1 APCs. Curiously, and not yet explained, Huang et al found that MAIT cell activation was enhanced by overexpression of the MHC class II chaperones Ii and HLA-DM and was strikingly diminished by silencing expression of endogenous Ii.[110]

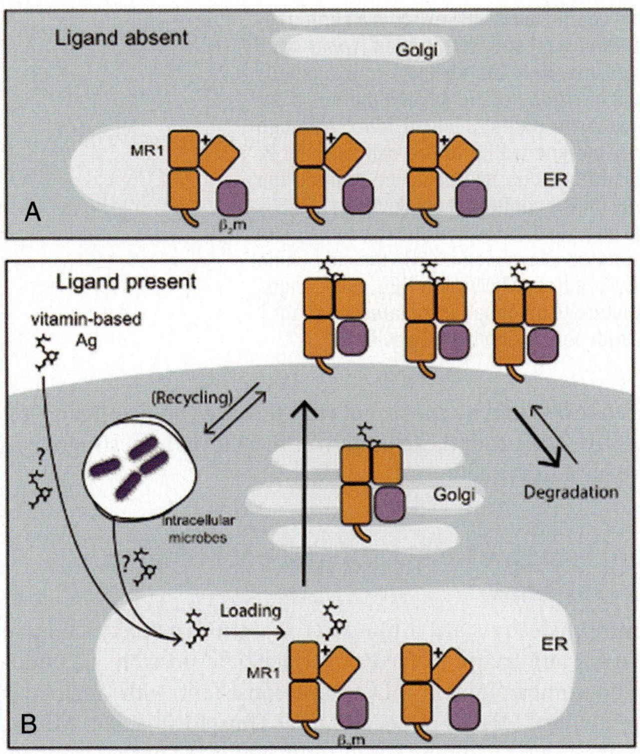

FIG. 37.9. The intracellular pathway of MR1 antigen presentation. A: In the absence of exogenous antigen (Ag) (ie, during steady state), MR1 predominantly resides in the endoplastic reticulum (ER) in an unfolded conformation. **B:** Ag derived from exogenous sources or from intracellular microbes enter the ER via an unknown mechanism and load onto immature MR1, where the charged Lys43 residue (+) is neutralized, triggering a conformational change that includes a strong β_2m association. These complexes traffic through the Golgi apparatus to the cell surface where they can present Ag for mucosal-associated invariant T (MAIT) cell recognition. MR1-Ag complexes remain at the cell surface for several hours before being internalized and delivered to the endocytic route for degradation. Additionally, a small number of MR1 molecules are recycled and returned to the surface, potentially after acquiring new ligands in endosomes. (Modified from McWilliam HEG, Villadangos JA. How MR1 presents a pathogen metabolic signature to mucosal-associated invariant T (MAIT) cells. *Trends Immunol.* 2017;38(9):679-689, with permission from Elsevier.)

Exogenous Antigens Are Loaded in the ER

In the presence of exogenous MAIT cell antigens, the intracellular reservoir of presynthesized MR1 molecules can dramatically translocate to the cell surface, proportionally to antigen dose,[39,63,69,88,101,103,105,111] referred to as the "on" state of MR1 antigen presentation. In detail, this occurs for exogenous antigens primarily as follows: exogenous antigens reach the ER lumen via unknown mechanisms that might include active transport or pinocytosis[104] and are captured there by ligand-receptive MR1 molecules within 10 minutes.[103-105] In turning the MR1 presentation pathway "on," antigen ligation triggers a "molecular switch" in MR1 that leads to its release from the PLC in association with TAPBPR and allows egress from the ER and trafficking through the Golgi apparatus to the surface for presentation to MAIT cells[103,104] (Fig. 37.9B). Complexes are formed over 1 hour, and leave the ER within 2 hours.[104] Unique to MR1 antigen-presentation, and akin to pathogen-driven posttranslational modification, this "molecular switch" involves covalent binding of the ligand to the Lys43 side chain of the MR1 cleft via a Schiff base, and thereby stably neutralizing the positive charge of the lysine sidechain.[66,69,103] This in turn triggers a conformational change in MR1 to adopt the completely folded, "closed" state that includes stable association with β_2m.[103,104] The Lys43 residue of MR1 thus functions as a "destabilizing" motif that inhibits complete folding of MR1 in the absence of ligand resulting in its retention in the ER.[103] Consistent with this notion, MR1 molecules with Lys43 mutated to Ala lack the need for neutralization, and thus spontaneously fold and egress to the cell surface in the absence of antigen.[103] Conversely, MR1 molecules, with Lys43 mutated to Arg, feature a positively charged sidechain with a severely diminished capacity to form a Schiff base, thus preventing folding and ER escape even in the presence of antigen because these Arg43 MR1 molecules cannot bind antigen[103] (Fig. 37.10). Notably, while some ligands are known to bind to MR1 without forming a Schiff base, the resulting complexes are not efficiently presented on the cell surface,[66,122,123] and it is possible that they follow other pathways of presentation, such as surface or endosomal loading, described below.

Following surface expression, like other antigen-presenting molecules, MR1 molecules are internalized by endocytosis and most are then delivered to lysosomes and degraded[103] (Fig. 37.9B). Interestingly, regardless of the type of ligand bound to MR1, the half-life of MR1 surface expression remains 2 to 4 hours.[103] This is in stark contrast to the half-life of MHC-I molecules, which is strongly influenced by the affinity of their

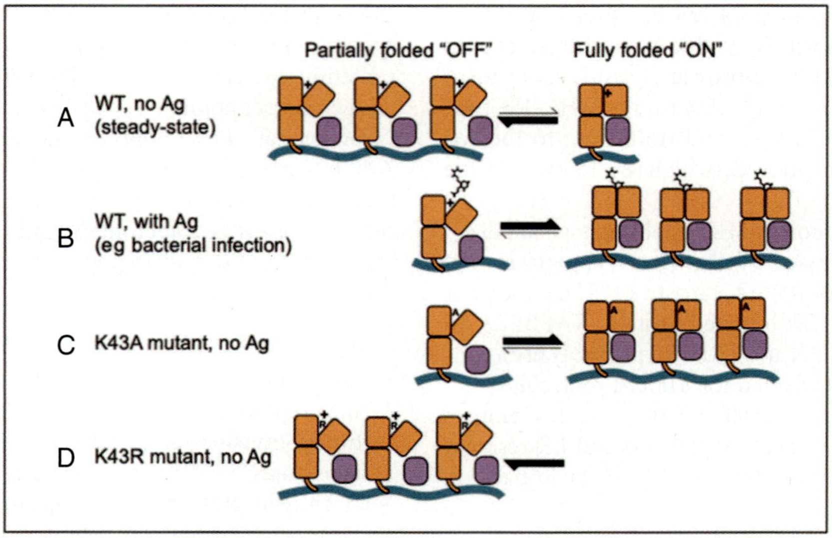

FIG. 37.10. The MR1 molecular switch and folding equilibrium. MR1 exists in a conformational equilibrium between "partially folded" (*left-hand side*) and "fully folded" (*right-hand side*) that is dependent on a destabilizing Lys43 residue. **A:** In the absence of exogenous antigen (Ag) (steady state), the Lys43 side chain is charged (+) and inhibits the complete folding of MR1, and the equilibrium strongly favors the partially folded form, although a small number of molecules may spontaneously fold and traffic to the cell surface. **B:** Binding of Ag (eg, during infection with riboflavin-producing microbes) neutralizes the Lys43 side chain and the equilibrium favors the fully folded conformation. **C:** When Lys43 is mutated to alanine (K43A), there is no charged side chain in this position and the equilibrium shifts toward the folded form. **D:** When Lys43 is mutated to arginine (K43R), the side chain remains charged and folding does not occur. (Modified from McWilliam HEG, Villadangos JA. How MR1 presents a pathogen metabolic signature to mucosal-associated invariant T (MAIT) cells. *Trends Immunol.* 2017;38(9):679-689, with permission from Elsevier.)

peptide ligands.[115] Instead, MR1 surface expression is highly dependent on the pool of ligand-receptive MR1 in the ER and is tightly regulated by the availability of ligand in the ER and characterized by "presentation on demand" and rapid "off-on-off" mechanisms. Hence, availability of ligand directly drives prolonged MR1 surface expression and ligand absence leads to a steady decline in MR1 surface expression[103,104] (Fig. 37.10). Thus, MR1 senses MR1–antigen-producing microbes in real-time, as soon as they generate ligand which is calibrated to the growth phase of riboflavin-dependent bacteria. This allows for rapid MAIT cell activation in the early stages of an infection.[103]

These principles of MR1 antigen-presentation appear to be true for MR1 antigen-presentation in the context of intracellular *S.* Typhimurium infection, where bacteria reside in vacuoles.[103] In contrast, in the context of other intracellular bacteria, such as *Mycobacterium tuberculosis* (Mtb), which reside in an Mtb-specific phagosomal compartment, the pathway of MR1 presentation is less clear with evidence for obligatory endosomal trafficking for antigen presentation. Data at hand[101,105] support the following theoretical possibilities: MR1-antigen loading may occur in the phagosome, whereby MR1 molecules might be recruited directly from the ER via an unknown mechanism or from the cell surface in a fashion similar to endosomal recycling, described below, or Mtb-derived MR1 antigens traffic from the phagosomes to the ER and follow the ER loading pathway.[112,121] The latter scenario would be consistent with the large pool of ligand-receptive MR1 found exclusively in the ER in the resting APCs we have examined.

Origin of Low Levels of Surface MR1 During Steady-State

Even in steady-state, a few MR1 molecules actually do leave the ER and traffic to the cell surface,[103,110] possibly via endocytic compartments.[110] Data are consistent with analogous behavior of MHC-I molecules.[124,125] Thus, ER-resident MR1 is maintained in a folding equilibrium between a partially folded, predominant open conformer that cannot leave the ER in the absence of ligand, and a minor one that is fully folded in a closed conformation and can traffic to the cell surface without ligand (Fig. 37.10).[104,112] Unlike MHC-I molecules, unloaded empty MR1 molecules may be as stable on the cell surface as their ligand-loaded counterparts.[103] In steady-state, MR1 cell surface expression in both wild type and MR1-overexpressing cell lines occurred in proportion to their total cellular pool of MR1.[103,112] These "escapee" MR1 molecules are considered to be either empty or loaded with a low-affinity ligand, and unlike 6-FP loaded molecules, they can be loaded with 5-OP-RU on the cell surface.[103] Furthermore, MR1 surface expression increases in cells cultured at low temperature (26°C),[98] which would favor the stabilization and egress of empty molecules from the ER, thereby resembling MHC-I.[95,126] It is possible that while the physicochemical conditions of the ER destabilize empty MR1 molecules, conditions at the cell surface are more favourable.[103] In summary, during steady-state endogenous MR1 ligands are in limited supply, and cells do not appear to obtain ligands from the extracellular environment. See the next section for the potential significance of surface-expressed empty MR1.

Extracellular Ligand Capture and Exchange in Endosomal Compartments and Surface Loading

While most MR1 molecules are targeted for degradation in the lysosomes following surface presentation, a small percentage (5%) escape transfer to the lysosomes and return to the cell surface, that is they recycle[103] (Fig. 37.9B), a pathway that is important for MHC-II and CD1 antigen-presentation[114,127] and can occur for MHC-I molecules.[128] When transiting through the endosomes before recycling back to the surface, ligand exchange with other extracellular antigens present in endosomal compartments may occur, noting that Schiff bases are labile at the moderate to low pH (5-6) found in late endosomes,[103,105] potentially facilitating exchange. The low levels of empty MR1 on the surface may also capture extracellular antigen via the recycling route.[129] While the recycling route may allow presentation of antigens that cannot efficiently reach the ER or endosomes, or that require endosomal processing, it is likely to depend on accumulation of surface MR1-antigen that has originated from the ER.[129] For example, MR1 loading of a prodrug of 5-A-RU, which requires cleavage by enzymes abundant in the endolysosomal compartments, occurs predominantly as part of the recycling route, ultimately loading ER-derived MR1.[130]

Summary and Future Directions

Most knowledge on MR1 trafficking comes from experiments involving extracellular, synthetic antigens. Because of the much lower abundance of antigen generated during infection, it has been much more difficult to analyze MR1 trafficking in this context. All the more difficult is unraveling the relative role of ligand loading in the ER versus surface loading and endosomal compartments in different infections. One of the key questions regarding MR1-ligand loading in the ER is how ligands are transported into the lumen of the ER, where MHC-I molecules primarily rely on TAP to import peptides into the ER from the cytosol.[112] By analogy, it is speculated that specific transporters might be dedicated to importing Vit B molecules and related antigens from the extracellular space or the lumen of endosomes to the cytosol and from there to the ER.[112] Alternatively, the metabolites might follow a retrograde vesicular transport pathway. In any event, defining this novel pathway of antigen handling and presentation is critical to understanding the MR1–MAIT cell system.

Both these loading options might be influenced by TLR2 and TLR9 agonists as well as inflammatory cytokines which were shown to increase surface expression of MR1 in the absence of exogenous ligand in some APCs although the mechanisms are unclear.[111,131,132] Furthermore, the mechanisms of internalization and degradation of MR1-antigen complexes from the plasma membrane are unclear. Importantly, a detailed understanding of all aspects of MR1 trafficking might reveal potential targets for immune modulation as well as immune escape. Another aspect, about which little is known, is the cell type that is relevant for antigen-presentation to induce MAIT cell activation *in vivo* which might differ on the context of homeostasis and infection.[107]

DIVERSITY OF MR1 LIGANDS AND THEIR PRESENTATION BY MR1

Overview of MR1 Ligand Binding

MR1 forms a heterodimer with β_2m which adopts a standard MHC-I-fold, as revealed by X-ray crystallography.[66] The α1- and α2-domains form the antigen-binding cleft, which comprises two long α-helices sitting atop an antiparallel β-sheet[66] (Fig. 37.11A). MR1 features two pockets. The A′ pocket is equivalent to the MHC-I pocket that binds the N-terminal peptide residue. The shallower F′ pocket is equivalent to the MHC-I pocket that binds the C-terminal residue of peptides.[66] Thus far, all known ligands were found to bind in a similar location within the A′ pocket of MR1 (Fig. 37.11B), where they are sequestered in an "aromatic cradle," formed by aromatic and charged amino acid residues.[2,66,69,88,122,123] Most MR1 ligands, including the potent antigens 5-OP-RU and 5-OE-RU as well as 6-FP, are covalently bound to MR1, namely via a Schiff base between a residual carbonyl group of the MR1 ligand and the side chain ε-amine of lysine 43 of MR1[65,69,88] (Figs. 37.5C, 37.8B, C, 37.11B). All MR1 ligands, regardless of Schiff base formation, are small (140-400 Da) relative to the binding cleft (~760 A^3) and buried deeply within the cleft so that their surface exposure to TCRs is <1%,[2,66,69,88,122,123] much less solvent-exposure than that of peptides presented by MHC-I molecules (10%-30%).[133]

Pyrimidine and Lumazine MR1 Ligands

The well-defined riboflavin biosynthesis-derived MR1-ligands (5-OP-RU and 5-OE-RU) are chemically classified as pyrimidines and are presented in an identical manner by MR1.[69] The α-amino carbonyl of 5-OP-RU/5-OE-RU forms a Schiff base with MR1-Lys43, and the uracil ring is sandwiched between Tyr7 and Tyr62, with its amine and carbonyl groups hydrogen bonding to Ser24 and Arg9, respectively (Fig. 37.8B, C).[69] The ribityl moiety of 5-OP-RU and 5-OE-RU, respectively, is stabilized by a network of H-bonds with MR1 residues Arg9, Arg94, Tyr152, and Gln153 (Fig. 37.8B, C).[69] Moreover, the ribityl 3′-OH group noncovalently interacts with the amine group of the uracil ring.[69] These intra- and inter-molecular interactions limit the flexibility and movement of the ribityl moiety of 5-OP-RU/5-OE-RU within the MR1-binding cleft, perhaps facilitating engagement by MAIT TCRs.[69,134]

RL-6-Me-7-OH and RL-6,7-diMe, classified chemically as ribityllumazines, are substantially less potent MAIT cell antigens than either 5-OP-RU or 5-OE-RU.[2,66,69] Structural data for RL-6-Me-7-OH bound to MR1 revealed that it did not form a Schiff base with MR1-Lys43 (Fig. 37.11C).[2]

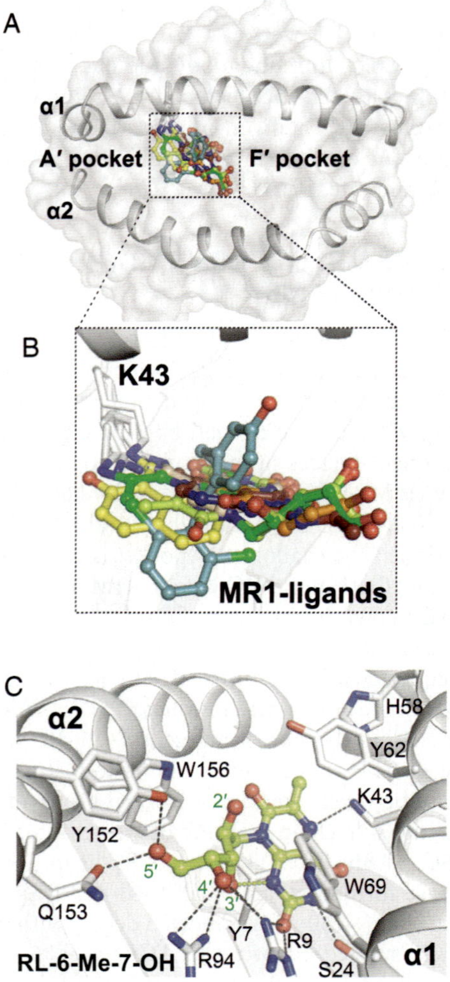

FIG. 37.11. Diverse chemical scaffolds can bind in the A′ pocket of MR1. A: Surface presentation of the MR1-antigen binding A′ and F′ pockets that are formed between MR1 α1 and α2 helices. All recognized MR1-ligands, to date, dock in the A′ pocket as shown in **(B)**. **C:** Interaction of MR1 with the lumazine ligand RL-6-Me-7-OH (*lemon sticks*). All H-bonds are displayed as *black dashed lines* except of the intramolecular H-bond between the 3′OH and the uracil ring which is shown as *lemon dashed line*.

However, Lys43 maintained a similar orientation within the MR1 cleft, as when covalently bound to Schiff base forming ligands, and formed a hydrogen bond with RL-6-Me-7-OH (Fig. 37.11C).[2,69] Also, the ribityl chain of RL-6-Me-7-OH was located in an identical position and formed conserved interactions with the antigen-binding pocket of MR1 as in the case of 5-OP-RU/5-OE-RU (Figs. 37.8B, C and 37.11C).[2,69]

Other microbial MR1-ligands that are structurally similar to the ribityllumazines RL-6-Me-7-OH and RL-6,7-diMe have been identified.[135] Namely, when soluble recombinant MR1 was expressed in the context of live bacterial infection, the lumazines 6-(2-carboxyethyl)-7-hydroxy-8-ribityllumazine (photolumazine I; PLI) and 6-(1H-indol-3-yl)-7-hydroxy-8-ribityllumazine (photolumazine III; PLIII) as well as the riboflavin analog 7,8-didemethyl-8-hydroxy-5-deazariboflavin (FO) were isolated (Fig. 37.12A).[135] None of the three ligands would be predicted to form a Schiff base with MR1 Lys43, and although all three ligands featured ribityl moieties, only PLI and PLIII, possible adducts of 5-A-RU, stimulated selected TRAV1-2+ and TRAV1-2− MR1-reactive T cell clones, while FO was antagonistic.[135]

Furthermore, nonmicrobial MR1 ligands have been identified through *in silico* screens of a range of small organic molecules, including drugs, drug metabolites, and drug-like molecules and commercial compound libraries.[123] Based on chemical substructures, they included the pyrimidines mercaptopurine, floxuridine, and doxofylline, as putative MR1 ligands.[122] These molecules were not predicted to form a Schiff base, and in cellular assays, floxuridine and mercaptopurine did not upregulate MR1 to detectable levels but nonetheless acted as weak agonists of a MAIT TCR reporter cell line.[122] Doxofylline weakly upregulated MR1 but was not agonistic for MAIT cell activation.[122] Another nine molecules whose chemical substructures classified as pyrimidines and heterocycles were identified as MR1 ligands, including 3-[(2,6-dioxo-1,2,3,6-tetrahydropyrimidin-4-yl) formamido] propanoic acid (DB28, which also featured a carboxylate as chemical substructure) and its methyl ester analogue, methyl 3-[(2,6-dioxo-1,2,3,6-tetrahydropyrimidin-4-yl) formamido] propanoate (NV18.1) (Fig. 37.12B).[123] None of the nine molecules upregulated MR1 to detectable levels, likely due to the inability to form a Schiff base with MR1 Lys43 (Fig. 37.12C).[123] Except for DB28 and NV18.1, they weakly stimulated some peripheral blood mononuclear cell (PBMC)-derived TRAV1-2+ MAIT cells (described in more detail in section *Recognition of MR1-Ligand Complexes by MAIT and Other MR1-Reactive T Cells*).[123] Indeed, DB28 and NV18.1 caused MR1 to be retained in the ER/early Golgi in its immature form and thus prevented MR1 presentation at the cell surface.[123] They also prevented presentation of stimulatory ligands *in vitro* but not *in vivo*, in a mechanism distinct from competitive inhibition by pterins.[123] The mechanism by which these molecules retain MR1 in the ER remains unclear; notably, K43A MR1 surface expression was not affected.[123]

Pterin (Vit B_9)-Related MR1 Ligands

MR1 can also present a range of small molecule pterins, namely the photodegradation product of folic acid (vitamin B9), known as 6-FP,[66] the synthetically acetylated form of 6-FP, acetyl-6-formylpterin (Ac-6-FP),[88] the photodegradation product 2,4-diamino-6-formylpteridine (2,4-DA-6-FP) of the drugs aminopterin and methotrexate which are synthetic derivatives of folic acid,[122] and a synthetic derivative of Ac-6-FP, 2-acetylamino-4-hydroxy-6-formylpteridine dimethyl acetal (Fig. 37.12D).[81] 6-FP, Ac-6-FP, and 2,4-DA-6-FP form a Schiff base with MR1-Lys43[66,88,122] (Figs. 37.5C and 37.12E). While the relevant formyl group of the synthetic derivative of Ac-6-FP, 2-acetylamino-4-hydroxy-6-formylpteridine dimethyl acetal, is blocked thereby preventing Schiff base formation,[81] partial hydrolysis of the acetal might generate an aldehyde capable of reacting with Lys43 of MR1. Indeed, the latter and all other listed pterins could induce cell surface

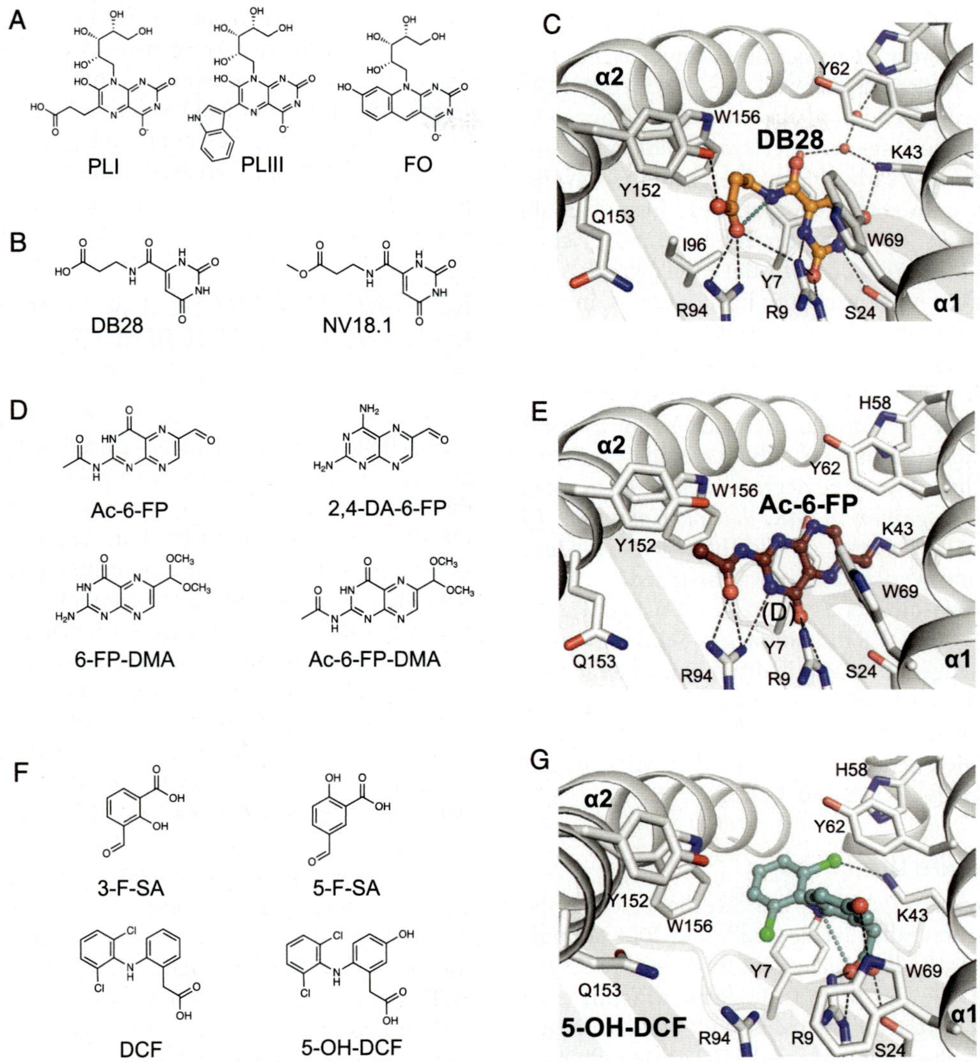

FIG. 37.12. MR1 presentation of diverse MR1 ligands. A, B, D, F: Schematics of the various categories of MR1 ligands, including: **(A)** Riboflavin-related ligands: 6-(2-carboxyethyl)-7-hydroxy-8-ribityl-lumazine (photolumazine I; PLI) and 6-(1H-indol-3-yl)-7-hydroxy-8-ribityllumazine (photolumazine III; PLIII) ribityl-scaffolds: 7,8-didemethyl-8-hydroxy-5-deazariboflavin (FO); **(B)** MR1 downregulating compounds: 3-[(2,6-dioxo-1,2,3,6-tetrahydropyrimidin-4-yl) formamido] propanoic acid (DB28) and methyl 3-[(2,6-dioxo-1,2,3,6-tetrahydropyrimidin-4-yl) formamido] propanoate (NV18.1); **(D)** Folate-based metabolites including: acetyl-6-formyl pterin (Ac-6-FP), 2,4-diamino-6-formylpteridine (2,4-DA-6-FP), 6-formylpterin dimethyl acetal (6-FP-DMA) and acetyl-6-formyl pterin dimethyl acetal (Ac-6-FP-DMA), and **(F)** drug-related compounds: 3-formyl-salicylic acid (3-F-SA), 5-formyl-salicylic acid (5-F-SA), diclofenac (DCF), and 5-OH-diclofenac (5-OH-DCF). **C, E, G:** The interaction between **(C)** DB28 (*orange sticks*), **(E)** Ac-6-FP (*ruby sticks*), and **(G)** 5-OH-DCF (*light teal*) metabolites within the MR1-binding groove. The ligand intermolecular and intramolecular H-bonds are colored *black* and *teal*, respectively.

upregulation of MR1,[66,81,88,122] while an equivalently modified version of 6-FP, 2-amino-4-hydroxy-6-formylpteridine dimethyl acetal, did not.[81] Apart from the exceptions as detailed in section *Recognition of MR1-Ligand Complexes by MAIT and Other MR1-Reactive T Cells,* 6-FP and Ac-6-FP typically do not stimulate MAIT cells.[66,88] Instead, 6-FP and Ac-6-FP, as well as 2-acetylamino-4-hydroxy-6-formylpteridine dimethyl acetal, act as potent competitive inhibitors of activation by riboflavin biosynthesis-derived antigens such as 5-OP-RU.[2,81,88] So far, Ac-6-FP remains the most potent competitive inhibitor of MAIT cell activation by 5-OP-RU and other pyrimidine antigens, both *in vitro* in human and mouse cell line assays and *in vivo* in mice.[122] The acetyl group in Ac-6-FP which is absent in 6-FP forms additional Van der Waals interactions with Tyr152, Trp156, and Ile96 as well as hydrogen bonds to Arg94 of the base of the MR1 cleft (Fig. 37.12E), correlating with an 8°C increase in the stability of MR1-Ac-6-FP in comparison to MR1-6-FP.[88] This, as well as other factors, such as solubility, molecular stability, and capacity for cellular uptake probably also contribute to the potency of Ac-6-FP as a competitive inhibitor.

Other MR1 Ligands

The same *in silico* screen that identified three pyrimidine drugs as MR1 ligands (described above) also identified 19 MR1 ligands which based on chemical substructures belonged to one or more of the following diverse classes of molecules: aromatic aldehyde, aromatic carboxylate, phenol aniline, flavone, isoflavone, enone, quinone.[122] Most of these ligands could not stimulate a MAIT TCR reporter cell line, and in crystal structures, no TCR contacts with the MR1 ligands were observed.[122] All of the nonstimulatory ligands tested, including the salicylic acid analogs 3-formyl-salicylic acid (3-F-SA) and 5-formyl-salicylic acid (5-F-SA) (Fig. 37.12F), formed a Schiff base with MR1-Lys43 and competitively inhibited MAIT cell activation to a similar or lesser extent as 6-FP. Some nonpyrimidine ligands were able to stimulate a MAIT TCR reporter cell line to varying degrees (described in more detail in section *Recognition of MR1-Ligand Complexes by MAIT and Other MR1-Reactive T Cells*), of which diclofenac was most potent and whose activity was assigned to one of its metabolites, 5-OH-diclofenac (5-OH-DCF) (Fig. 37.12F, G).[122] Diclofenac/5-OH-DCF did not form a Schiff base with MR1 and were wedged through hydrophobic interactions between Tyr7 and Tyr62 of the aromatic cradle orienting the central ring of diclofenac/5-OH-diclofenac essentially perpendicular to the plane of any of the other MR1 ligands described to date (Figs. 37.11B and 37.12G).[122]

Summary and Future Directions

Accumulated data thus far[2,66,69,122,123] indicate that MR1 has the capacity to bind structurally diverse small (150-400 Da) molecules. Most of the MR1 ligands (except for RL-6-Me-7-OH, diclofenac/5-OH-DCF, DB28/NV18.1) form a Schiff base with Lys43 of MR1. In the absence of a Schiff base, hydrogen bonds are formed between ligands and MR1 with additional compensatory bonds supporting the energetics of binding.[123] However, regardless of the Schiff base, all ligands to date are located broadly within a similar location in the A′-pocket of MR1, although some are displaced; or oriented differently, involving mostly rotations with broadly similar depositions of the planes of the aromatic rings (Fig. 37.11A, B).[2,66,69,122,123] Indeed, the precise chemical structure of the ligand and its capacity to form a Schiff base with MR1 has a profound effect on ligand propensity to upregulate MR1 cell surface expression. Here, the efficiency of MR1 cell surface upregulation inversely correlates with ligand polarity and associated hydrophilicity of the MR1-antigen interface. Notably, there is no apparent correlation between ligand propensity to upregulate MR1 cell surface expression and potency to stimulate T cells.[134]

In the contexts of bacterial infections,[135,136] cancer and steady-state,[108,109] there is evidence for the existence of additional MR1 ligands of unknown chemical identity, most of which appear to be recognized by other MR1-reactive T cells rather than MAIT cells, detailed in section *Recognition of MR1-Ligand Complexes by MAIT and Other MR1-Reactive T Cells*. There is also the potential that additional drugs binding to MR1 may be involved in inducing hypersensitivities mediated by MAIT or other MR1-reactive T cells.[137] Furthermore, two distinct mechanisms for MR1-ligand dependent competitive inhibition for activation of MAIT cells and other MR1-reactive T cells have been descri bed.[66,88,123] These could be of physiological importance, representing potential mechanisms of immune evasion by pathogens and cancers as well as immune-response dampening effects of drugs.

RECOGNITION OF MR1-LIGAND COMPLEXES BY MAIT AND OTHER MR1-REACTIVE T CELLS

Working Characteristics of MAIT Cells

There is no agreed definition of a MAIT cell, but there are well defined characteristics recognized by the field. We operationally define MAIT cells by a range of phenotypic features discussed below and by their specificity for MR1-5-OP-RU/5-OE-RU (and less potent ribityllumazines). This specificity is generally a function of their semi-invariant TCR and MR1-restriction, with only unusual examples of noncanonical TCR usage by cells that retain MAIT specificity and functionally behavior. This definition generally assumes MAIT cells do not react to MR1 loaded with other ligands (nonribityllumazines), such as 6-FP and Ac-6-FP, and MR1 bearing the Lys43 to Ala point mutation (MR1-K43A). Nonetheless, as described in sections *Reactivity to MR1-5-OP-RU/5-OE-RU (and Less Potent Ribityllumazines) and Cross-Recognition of Other MR1-Ligands and MR1-Centric/AutoReactivity*, a small fraction of MAIT cells can cross-recognize MR1 loaded with other ligands, such as 6-FP or Ac-6-FP or be MR1-autoreactive. Accordingly, the phenotypic features of MAIT cells most widely used in their characterization are the relatively fixed TCR α-chain (TRAV1-2-TRAJ33/12/20), the expression of surface markers (eg, CD45RO and $CD161^{hi}$) noting the latter can be downregulated,[17,138,139] and transcription factors (for example, promyelocytic leukemia zinc finger [PLZF], T-bet, RORγt). The structural features of the MAIT TCR recognition of MR1-antigen complexes explain the extreme bias in the use of the fixed TCR α-chain as described below.

The Molecular Basis of MAIT TCR Recognition of MR1-Ligand

All classical MAIT TCRs studied to date recognize MR1-presenting 5-OP-RU, 5-OE-RU,[69,88] and the ribityllumazine RL-6-Me-7-OH[2] using a consensus docking mode, as revealed by crystal structures: the MAIT TCR docks centrally in an orthogonal position atop the A′ pocket of MR1, and the α- and β-chains of the TCR sit over the α2-and α1-helices of MR1, respectively, contributing almost equally to the buried surface area at the interface between the TCR- and MR1-ligand (Fig. 37.13). Crucial to antigen recognition is a single hydrogen bond between the evolutionary conserved (*TRAJ33/12/20*-encoded) Tyr95 residue from the CDR3-loop of the MAIT TCR α-chain and the 2′-OH group on the antigen ribityl moiety of 5-OP-RU, 5-OE-RU, and RL-6-Me-7-OH, respectively (Fig. 37.14A, B).[2,69,88,134] The MAIT TCR Tyr95α residue also forms a hydrogen bond with Tyr152 of the α2-helix of MR1

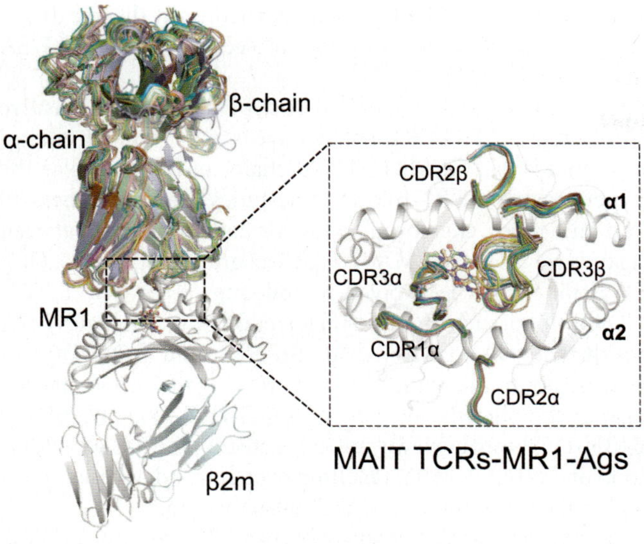

FIG. 37.13. Consensus docking of classical MAIT TCRs atop MR1 molecule. Superimposition of the MAIT TRAV1-2+ TCR-MR1-antigen structures and zoomed-in top view of the antigen (Ag)-binding cleft of the respective MR1 molecule displaying similar docking of the TCR CDR loops atop the MR1 molecule.

(Fig. 37.14A, B). This network of hydrogen bonds, involving the MAIT TCR Tyr95α residue, the Tyr152 residue of MR1, and the 2′-OH group on the antigen ribityl moiety of 5-OP-RU, 5-OE-RU, and RL-6-Me-7-OH, respectively, has been referred to the "*interaction triad*" and represents a limiting molecular feature for MAIT TCR MR1-ligand recognition, with its partial or complete disruption resulting in reduced or absence of MR1-antigen recognition.[134] Indeed, this interaction triad appeared conserved in crystal structures of various MAIT TCRs (featuring diverse TCRβ-chains) in complex with MR1-5-OP-RU, which explained their similar affinities for MR1-5-OP-RU (Fig. 37.14C). The importance of this interaction triad was demonstrated by site-directed mutagenesis of the TCR Tyr95α[2,3,69,88,134] and the MR1 Tyr152 residue,[140] either of which virtually abolished MAIT cell recognition and activation. Similarly, synthetic analogues of 5-OP-RU which lacked the 2′-OH group of the ribityl moiety or the entire ribityl moiety significantly impaired MAIT cell recognition.[134] Surprisingly, while the ribityl chain of MR1 ligands is vital for potent MAIT cell stimulation, it is not essential and may even be detrimental to MR1 cell surface upregulation.[134] In the case of the MAIT TCR β-chain, no particular residue appeared crucial for MR1-ligand recognition.[3,140] Instead, the TCR β-chain, and in particular its hypervariable CDR3-loop, appears to play a role in fine-tuning MR1-antigen recognition,[88] whereby the affinities between various MAIT TCRs and MR1-5-OP-RU are similar (K_D 2-8 μM).[2,88]

Recognition of MR1-Ligand Complexes by MR1-Reactive T Cells Other Than MAIT Cells

Overview of Non-MAIT MR1-Reactive T Cells

MR1-antigen reactivities distinct from MR1-5-OP-RU specificity have recently been described, reviewed previously in detail.[141] For instance, these can be observed when MR1-antigen reactivity of T cell clones or TCR reporter cell lines is assessed by binding to MR1 tetramers or stimulation by APCs expressing physiological levels of MR1 or overexpressing MR1. These reactivities are mediated by "other MR1-reactive T cells," distinct from MAIT cells.[58,59,108,109] Other MR1-reactive T cells can possess some features of a MAIT cell phenotype and are therefore referred to as "MAIT-like" cells. However, non-MAIT MR1-reactive T cells are generally phenotypically heterogenous and distinct from typical MAIT cells. Most non-MAIT cells that are MR1-reactive do not express a TRAV1-2+ TCR[58,59,108,109,135] and some express a γδ TCR rather than an αβ TCR.[142] Together with MAIT cells, other MR1-reactive T cells can be classified as MR1-reactive T cells. Subsets of other MR1-reactive T cells that have been characterized thus far indicate that they are more than 10- to 100-fold less frequent in human blood than MAIT cells.

MAIT-Like Cells

A small fraction of human MR1-5-OP-RU tetramer$^+$ cells (0.78%) is TRAV1-2−, which in total makes up a median of 0.02% of circulating T cells.[9,58,59] Among total TRAV1-2− T cells, the frequency of MR1-5-OP-RU tetramer$^+$ cells is similar to that of MR1-pterin tetramer+ cells (median <0.2%).[58] Most

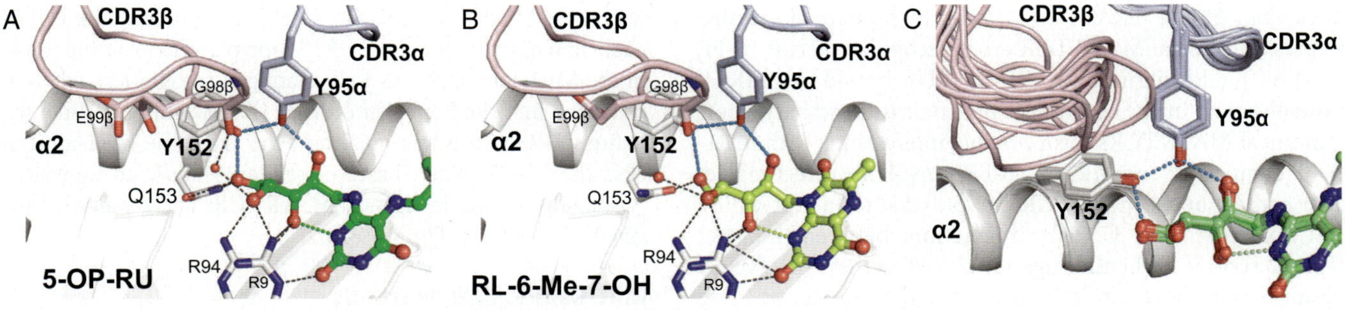

FIG. 37.14. TCR-MR1-antigen "interaction triad" formation. The interfaces between the CDR3α- (*light blue*) and CDR3β- (*pink*) loops of MAIT A-F7 TCR and MR1-5-OP-RU **(A)** and MR1-RL-6-Me-7-OH **(B)**. 5-OP-RU and RL-6-Me-7-OH and their relevant intramolecular H-bonds are colored *green* and *lemon*, respectively. C, Conservation of the MAIT TCR–MR1–antigen interaction triad (encompassing the 5-OP-RU, Tyr95α of CDR3α- loop, and Tyr152 of MR1) among various structures of classical MAIT TCRs in complex with MR1-5-OP-RU despite the usage of different *TRAJ* and *TRBV* genes as well as variations in the CDR3β-loop lengths and sequences.

of the MR1-reactive TRAV1-2− T cells (>80%) are very heterogenous in terms of TCR usage and lack MAIT cell markers and as such are referred to as non-MAIT−like, while the residual fraction features a MAIT-like phenotype (CD161+, IL-18Rα+, CD218a+, CD26+, PLZF+, CD8α+, or DN).[59] In fact, based on *ex vivo* tetramer staining, in most donors all of the MAIT-like TRAV1-2− MR1-reactive T cells are specific for 5-OP-RU akin to MAIT cells, while in a few donors, some MAIT-like TRAV1-2− cells also reacted with 6-FP, with most of the 6-FP−reactive T cells otherwise found among human non-MAIT−like TRAV1-2− cells.[59] Reporter cell lines overexpressing some TRAV36+ TCRs showed weak reactivity to 6-FP,[59] perhaps an artifact of their over expression.

Interestingly, 5-OP-RU−specific TRAV1-2− MAIT-like cells were enriched for a second public TCR α-chain usage next to the classical MAIT TCR α-chain, involving *TRAV36* recombined primarily with *TRAJ34* or otherwise *TRAJ37,* forming an entirely or highly germline-encoded CDR3α-loop of constant length (11 amino acids).[58,59] This germline arrangement is only possible with *TRAJ34*. This *TRAV36-TRAJ34* α-chain further paired with a nearly invariant β-chain, composed mostly of *TRBV28-TRBJ2-5* and a CDR3β-loop of constant length (14 amino acids) with semi-invariant sequence motifs.[59] This TCR was also featured by a clonally expanded population in the blood and bone marrow of a multiple myeloma cancer patient.[58] The crystal structure of a TRAV36+ TCR in complex with MR1-5-OP-RU revealed a recognition mechanism convergent with that by classical MAIT TCRs although the TCR docked more centrally on MR1 compared to the orthogonal docking by classical MAIT TCRs (Fig. 37.15A-D).[58] In the absence of a Tyr95α, the CDR1α-loop of the TRAV36+ TCRα-chain was dominant in contacting MR1.[58] Asn29α, positioned similarly to Tyr95α of classical MAIT TCRs, directly contacted the 2′-OH group of the ribityl moiety of 5-OP-RU (in an H-bond) (Fig. 37.16A).[58] This "deformed" interaction triad also lacked direct or water-mediated interactions between Tyr152 of MR1 and the ribityl moiety of 5-OP-RU, associated with a >fivefold reduced affinity between the TCR and MR1-5-OP-RU.[58]

Reactivity to MR1-5-OP-RU/5-OE-RU (and Less Potent Ribityllumazines) and Cross-Recognition of Other MR1-Ligands

Less than 5% of TRAV1-2+ T cells in human blood that express a classical MAIT TCR cross-recognize pterins (6-FP, Ac-6-FP) and pyrimidines (5-OP-RU) bound to MR1.[58] Notably, even in the absence of pterin cross-reactivity, some canonical MAIT TCRs form notable interactions with MR1-6-FP complexes[2] and MR1-Ac-6-FP,[88] involving direct molecular contact between the TCR Tyr95α residue and Ac-6-FP[88] but not 6-FP[2] (Fig. 37.16C). Interactions between the MR1 Tyr152 residue and either ligand, as well as the TCR Tyr95α residue, were however lacking so that the interaction triad was distorted. Furthermore, selected MAIT TCR expressing reporter cell lines and subsets of primary TRAV1-2+ MAIT cells present in some but not all donors have been described to cross-react with MR1-presented, chemically diverse drugs, drug metabolites, and drug-like molecules, such as 5-F-SA and 5-OH-DCF.[122,123]

That MAIT TCR responses are dependent on, or sensitive to, discrete MAIT TCR β-chain usage is in line with an emerging concept that the MAIT TCRβ-chain, and in particular, the hypervariable CDR3 β-loop "fine-tunes" responsiveness to certain ligands, thereby allowing for discrimination between ligands.[5,58,59,88,123,135,136] MAIT TCR recognition of MR1-5-OH-DCF, for instance, involved remodeling of the MAIT TCR CDR3β-loop, such that Glu99β formed salt-bridges with MR1 residues Arg9 and Arg94 and also contacted 5-OH-DCF in a hydrogen bond, while the aromatic ring of Tyr95α stacked against the phenylacetic ring of 5-OH-DCF (Fig. 37.16D).[122] 5-OH-DCF was however not contacted by the TCR CDR3α-loop or Tyr152 of MR1, resulting overall in a deformed interaction triad and modest MAIT cell activation.[122]

Also, a very small fraction of human MR1-reactive TRAV1-2− T cells in human blood were capable of recognizing MR1-5-OP-RU/5-OE-RU and cross-recognizing other MR1 ligands,[58,59] all of which expressed diverse TCRs and only one TCR possessed the Tyr95α.[58] Both TRAV1-2+ and TRAV1-2− cross-reactive T cells typically differentially recognized antigen, whereby some TCRs preferred antigens of a specific class (riboflavin-based or folate-derived), while others were capable of distinguishing between antigens of the same class (6-FP or Ac-6-FP).[58] TRAV1−, MR1-reactive T cells were also identified in C57BL/6 mice deficient in *TRAJ33*.[59] Together with TRAJ33− MAIT cells, they comprised 50-fold less MR1-5-OP-RU−reactive T cells than in wild-type mice.[59] Gene sequencing revealed that they used a broad range of *TRAV* and *TRAJ* genes, with variable CDR3α length and no tyrosine at position 95 and TCR reporter cell lines expressing these TCRs were MR1-5-OP-RU and MR1-Ac-6-FP−reactive.[59]

Furthermore, a CD8α+ CD161− TRAV1-2− T cell clone has been identified that responded to 5-OP-RU, RL-6-Me-7-OH, and PLI but not RL-6,7-diMe and 6-FP.[135,136,143] The same clone (TRAV12−2+) also responded to infection with *Streptococcus pyogenes*, which lacks the riboflavin biosynthetic pathway, suggesting its capacity to cross-react with an undefined riboflavin-independent MR1 antigen/s.[136] The crystal structure of the relevant TRAV12-2+ TCR in complex with 5-OP-RU revealed a distinct docking mode compared to classical MAIT TCRs as well as the TRAV36+ TCR (Fig. 37.15E, F).[143] Indeed, the TCRβ-chain and in particular the CDR3β-loop played a prominent role in the MR1-interaction as well as antigen recognition, where it projected into the F′ pocket of MR1 (Fig. 37.16B).[143] While the entire CDR3α-loop was quite far away from the MR1-antigen cleft, the CDR3α Asn97 based on sequence was in an equivalent position as Tyr95α in classical MAIT TCRs and mediated antigen-MR1 contacts (Fig. 37.16B).[51]

MR1-Centric/AutoReactivity

MR1-autoreactivity is a hallmark of MR1-reactive γδ TCR expressing T cells,[142] a phenotypically diverse population of T cells that resembles other cells of the γδ T cell lineage.[142] Some

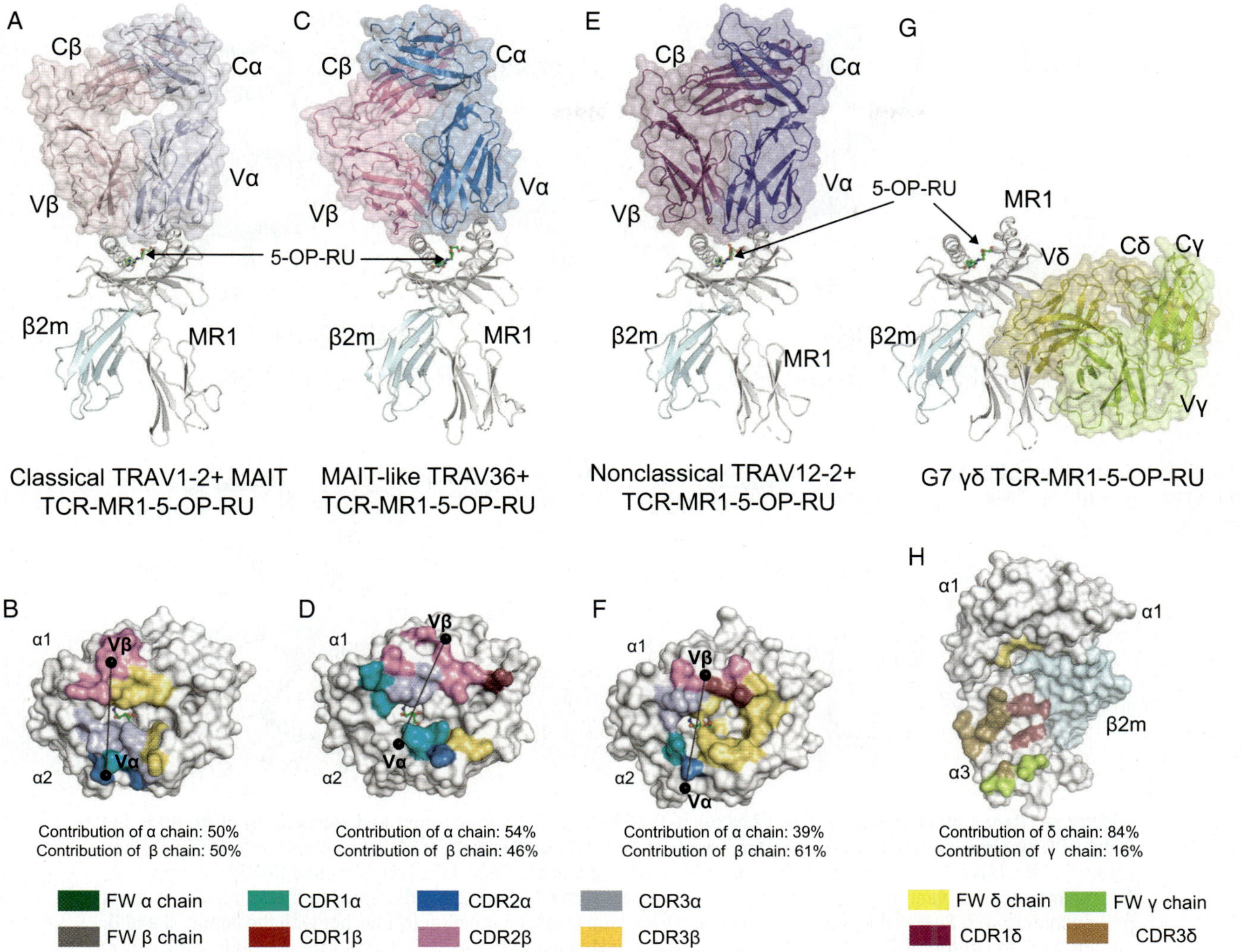

FIG. 37.15. Crystal structures of the ternary complexes of classical TRAV1-2+ MAIT TCR-MR1-5-OP-RU (PDB; 4NQC) **(A, B)**, mucosal-associated invariant T (MAIT)-like TRAV36+ TCR-MR1-5-OP-RU (PDB; 5D7L) **(C, D)**, nonclassical TRAV12-2+ TCR-MR1-5-OP-RU (PDB; 6XQP) **(E and F)**, and G7 γδ TCR-MR1-5-OP-RU (PDB; 6MWR) (G, H). Top panels **(A, C, E, and G)** depict ribbon diagrams of the ternary complexes. The MR1 and β_2m molecules are colored *white* and *pale cyan*, respectively, and 5-OP-RU is presented as *green sticks*. A-F,7 TCRα, light blue; A-F7 TCRβ, pink; MAV36 TCRα, sky blue; MAV36 TCRβ, violet; TRAV12-2 TCRα, slate; TRAV12 to 2 TCRβ, violet-purple; TCRγ, lemon; G7 TCRδ, yellow-orange. The lower panels **(B, D, F, and H)** illustrate the T cell receptor (TCR) footprints on the molecular surface of MR1-5-OP-RU. The centers of mass of the TCR variable domains are shown as *black spheres*. The atomic footprints of CDR loops are colored as follows: CDR1α, teal; CDR2α, sky blue; CDR3α, light blue; framework region of α-chain, dark green; CDR1β, maroon; CD2β, violet; CDR3β, yellow-orange; framework region of β-chain, dark gray; framework region of δ chain, yellow-orange; framework region of γ chain, lemon; CDR1δ, raspberry; CDR3δ, deep olive.

MR1-reactive γδ TCRs were found to bind underneath the antigen-binding cleft (contacting primarily the α3-domain of MR1), while others interacted with the antigen-binding cleft of MR1, akin to the classical MAIT αβ TCRs (Fig. 37.15G, H).[142] Accordingly, activation of MR1-reactive γδ T cells was primarily dependent on MR1, with some reactivity potentially modulated by MR1-bound antigen.[142] MR1-reactive γδ T cells were found in human PBMCs (<0.001%-0.1% of CD3+ T cells and from <0.1%-5% of γδ T cells) and in tissues (liver, stomach, lung, duodenum, and in the skin in the case of a patient with Merkel cell carcinoma).[142] Although MR1-reactive γδ T cells were DN or CD8α+ akin to MAIT cells, they expressed variable levels of CD161.[142] *TRDV1* usage dominates among TCR δ-chains (72%) and most other TCRs feature *TRDV3+* δ-chains, both of which mostly paired with *TRGV8+* γ-chains, although all other *TRGV* genes were represented.[142]

Unusual, and presumably, rare patterns of MR1-centric-/auto-reactivity were also observed both among some canonical MAIT TCR expressing human T cells (frequency unknown)[58] and T cell hybridomas derived from mice.[8,22,44,94,110,140]

In line with the high level of MR1 and MAIT TCR α-chain conservation across species, xenoreactivity between

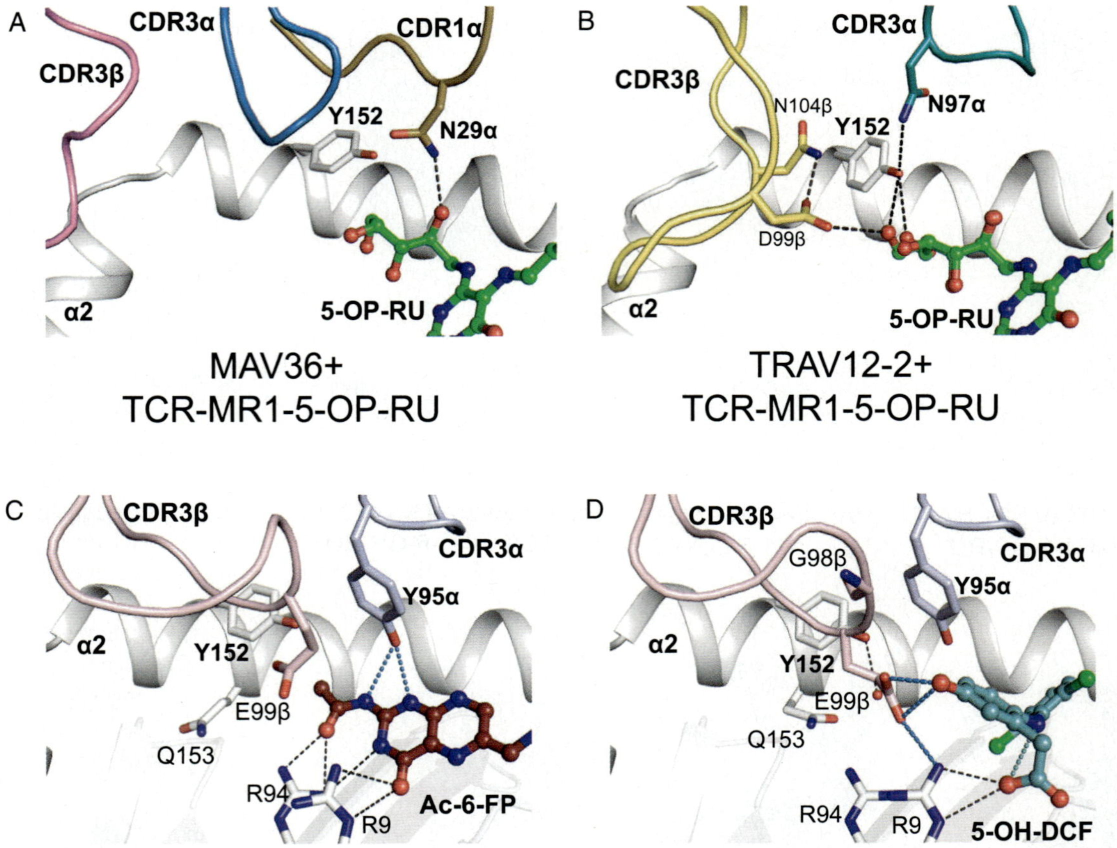

FIG. 37.16. Recognition of 5-OP-RU by TRAV36+ and TRAV12-2+ T cell receptors and recognition of Ac-6-FP and 5-OH-DCF by classical mucosal-associated invariant T (MAIT) cell receptors. A: Interactions of the CDR1α-loop (*brown*) of the MAV36+ TCR with 5-OP-RU, including the locations of CDR3α-loop (*sky blue*) and CDR3β-loops (*violet*). **B**, The interface between the CDR3β-loop (*orange-yellow*) of the TRAV12-2+ TCR, depicting the direct and indirect polar contacts with 5-OP-RU ligand, whereby the CDR3α-loop (*teal*) did not directly interact with the ligand. **C and D:** The interactions of the CDR3α- (*light blue*) and CDR3β (*pink*)-loops of the A-F7 TCR with MR1-Ac-6-FP (*ruby sticks*) **(C)** and MR1-5-OH-DCF (*light teal*) **(D)**. The ligand intermolecular and intramolecular H-bonds are colored *black* and *teal*, respectively.

MAIT cells and MR1 from different species has been described.[25,27,44,58-62] TCR xenoreactivity can occur in an autoreactive manner, independently of whether MR1 is loaded with exogenous antigen, or can be dependent on MR1 being loaded with exogenous antigen and can sometimes be stronger than species-matched reactivity.[47]

Reactivity to Other MR1-Ligands But Not MR1-5-OP-RU/5-OE-RU (and Less Potent Ribityllumazines)

Examples of T cell reactivity to MR1-ligands other than 5-OP-RU/5-OE-RU have mostly been identified among TRAV1-2− MR1-reactive T cells in human peripheral blood. For instance, the TRAV1-2− MAV21 TCR (*TRAV21+*) specifically recognized pterins but not 5-OP-RU.[58] Similarly, the MC.7.G5 clone (*TRAV38.2/DV8-TRAJ31*, *TRBV25.1-TRBJ2.3*), specifically recognized undefined antigen/s expressed or upregulated in cancerous cells but not noncancerous cells (resting, activated, stressed, or infected). The MC.7.G5 clone did not recognize bacterial riboflavin-based antigens or Ac-6-FP which instead acted as competitive inhibitors.[108] The reactivity of the MC.7.G5 occurred in the context of physiological levels of MR1 and high effector T cell to target ratios.[108] Importantly, the reactivity was also donor-unrestricted, with multiple HLA-mismatched tumor cells recognized in an MR1-dependent manner, with no apparent alloreactivity.[108] This T cell clone has raised interest in potential adoptive immunotherapy for cancers.[108] While another such clone was identified from a different donor, the prevalence of such clones within and between individuals is unknown and so is their phenotype. Previously, however, a somewhat similar population of T cells had been identified that comprised 0.02% to 0.04% of circulating T cells in humans, designated "MR1-T" cells.[109] In contrast to the MC.7.G5 clone, they were not cancer-specific, recognizing endogenous cellular antigens from various tissues, including cancer tissues, in the context of physiological levels of MR1 and dependent on MR1. Like the MC.7.G5 clone, they did not recognize pterins which instead acted as competitive inhibitors. However, while the MC.7.G5 clone did not bind MR1-K43A tetramer or respond to MR1-K43A overexpressing cells,[108] MR1-T cell clones recognized MR1 wild-type and MR1-K43A overexpressing cells equally.[109] Thus, recognition of relevant antigens by MR1-T cells either did not depend on Schiff base formation with MR1 or the reactivity was MR1-centric with pterins representing nonpermissive antigens and 5-OP-RU/ribityllumazines permissive antigens. At least two clones did however display antigen specificity by differentially recognizing fractions of cell/tumor lysates.[109] Except for one clone, which had a MAIT phenotype (TRAV1-2+, CD161+, CD8–), all were TRAV1-2–, expressed diverse TCR α- and β-chains, were CD161– and mostly CD8α+.[109] MR1-T cell clones displayed multiple chemokine receptor expression profiles and secretion of diverse effector molecules and thus appeared phenotypically and functionally heterogenous and mostly distinct from MAIT cells. Aside from a potential antitumour/cancer function, MR1-T cells might also contribute to maturation of monocyte-derived dendritic cells in a T-helper–like function and intestinal epithelial barrier homeostasis.[109]

Summary and Future Directions

The study of non-MAIT MR1-reactive T cells is still emerging and the physiological or pathophysiological relevance of these diverse populations of cells is unknown, notwithstanding the antitumour/cancer effect of some clones.[108] Identification of the chemical identities and biosynthetic origins of the tumour-associated and endogenous antigens recognized by some of the other MR1-reactive T cells will allow for the generation of tetramers which in turn will help in determining their prevalence (within an individual and at the population level), their phenotype, and their physiological roles. It is also of interest to better understand the developmental origin of other MR1-reactive T cells, particularly of those that do not feature a MAIT cell phenotype. One might speculate that they have been selected on other MHC molecules and cross-react on MR1 in line with their phenotypic profile being distinct from MAIT cells.

While the classical MAIT TCR α-chain usage correlates well with specific recognition of riboflavin biosynthesis-derived antigens, the existence of other MR1-reactive T cells that are TRAV1-2– highlights that the classical MAIT TCR α-chain is not the only solution that creates MR1-reactivity, particularly for reactivities beyond 5-OP-RU/5-OE-RU (and less potent ribityllumazines). Crystal structures demonstrate that diverse TCR usage among other MR1-reactive T cells enable varied docking strategies which translate into divergent mechanisms of MR1-antigen reactivity. Furthermore, in discrete subsets of classical MAIT TCRs, CDR3β-hypervariability was shown to influence antigen preference and to allow for cross-reactivity as well as MR1-centered reactivity.[5,58,59,88,123,135,136] This is akin to the observation that the CDR3β-loop of type I NKT cells can play an important role in autoreactivity.[144,145]

DEVELOPMENT OF MAIT CELLS

Thymic Development

Requirement for MR1

MAIT cell development has been most extensively studied in mice where they develop from common T cell progenitors in the thymus followed by a broadly similar, but distinct, differentiation process to other T cell subsets. A functional thymus is required for their development, since invariant *TRAV1-TRAJ33* (*Vα19-Jα33*) rearrangements were not detected in nude mice.[8] Early studies demonstrated a requirement for MR1 for positive selection of MAIT cells (see section *Identification of MR1, the Restriction Element of MAIT Cells*).[8,22,94] Consistent with this finding, MR1-deficient mice are also deficient in MAIT cells.[22,146]

Development of conventional αβ-T cells depends on TCR interactions with classical MHC-peptide complexes on thymic epithelial cells, whereas that of innate-like T cells depends on selection by hematopoietic cells. For instance, NKT cell development requires the expression of CD1d on cortical thymocytes.[147] Interestingly, CD1d-deficient mice have greater numbers of MAIT cells in both the thymus and spleen,[148] suggesting these cell types may have a shared niche.

The low surface expression of MR1 on most cells hampered early identification of the selecting element for MAIT cells. MAIT cells were reported to be absent in the intestines of β_2m–/– mice.[8] β_2m–sufficient B cells could reconstitute MAIT cells in these mice suggesting they were responsible for MAIT cell selection or expansion.[22] It was reported that expansion of MAIT cells in the periphery, but not thymic selection (fetal thymic organ culture), was heavily dependent on B lymphocytes.[146] In B cell–deficient Ig Heavy chain joining region (J_H–/–) mice, these transcripts were severely reduced in mesenteric lymph nodes and lamina propria lymphocytes (LPL), but in μMT–/– mice (which lack the transmembrane region of IgM), normal numbers of MAIT cells were found in LPL, showing that residual gut IgA+ B cells in these mice are sufficient for the accumulation of MAIT cells. However, a recent study showed that MAIT cell frequency in the thymus and peripheral organs (spleen and lungs) was not affected in B6-MAITCast Jh–/– mice, which lack B cells. Thus, B cells are not required for MAIT development.[149]

Subsequently, MR1 expression on DP cortical thymocytes was shown to be important for the development of MAIT cells in reaggregated thymic organ culture with TRAV1*i* (Vα19*i*) or TRBV19 (Vβ6) Tg mouse thymocytes and, *in vivo,* using bone marrow transplantation assays.[150] However, both hematopoietic and epithelial cells, including single-positive (SP) thymocytes, DP thymocytes, dendritic cells, and medullary and cortical thymic epithelial cells, may be able to promote MAIT cell thymic development. These cells could each activate MAIT cells, *in vitro,* in response to 5-OP-RU, with DP thymocytes being most efficient.[149]

Stages of Thymic Development

Koay et al described three developmental stages of MAIT cell lineage development (Fig. 37.17), which were roughly parallel in mice and humans. Stage 1 cells, which were only found in the thymus, were identified in mice as CD24+CD44–, stage 2 as CD24–CD44–, and stage 3, which most closely resembled peripheral MAIT cells, as CD24–CD44+.[148] Similarly, three stages were described in human thymic MAIT cells, defined as CD27–CD161–, CD27+CD161–, and CD27+CD161+ for

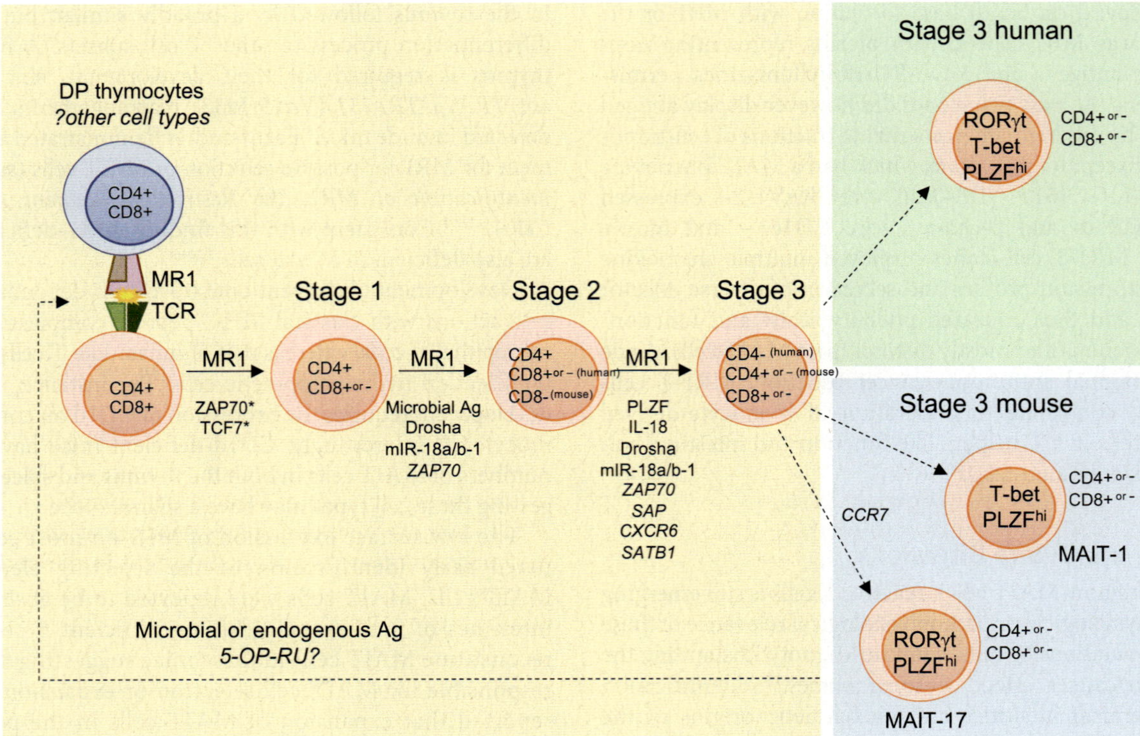

FIG. 37.17. Model of three-stage thymic development of mucosal-associated invariant T (MAIT) cells. Three stages of thymic development, common to mice and humans, are shown. Factors that are important for progression through each stage are shown below arrows. In mice, two lineages (T-bet–expressing MAIT-1 cells and RORγt-expressing MAIT-17 cells) exist, whereas human MAIT cells express both transcription factors. (Adapted with permission of American Association for the Advancement of Science, from Koay HF, Su S, Amann-Zalcenstein D, et al. A divergent transcriptional landscape underpins the development and functional branching of MAIT cells. *Sci Immunol.* 2019;4(41); permission conveyed through Copyright Clearance Center, Inc, Godfrey DI, Koay HF, McCluskey J, et al. The biology and functional importance of MAIT cells. *Nat Immunol.* 2019;20(9):1110-1128 and Pellicci DG, Koay HF, Berzins SP. Thymic development of unconventional T cells: how NKT cells, MAIT cells and gammadelta T cells emerge. *Nat Rev Immunol.* 2020;20(12):756-770.)

stages 1, 2, and 3, respectively. By analyzing thymic MAIT cells in mice at 2, 4, and 8 weeks, stage 1 cells represented the major subset early in life, with stage 2, and later, stage 3 cells found to increase in relative abundance over time. Analysis of transcription factor expression and the production of cytokines in response to phorbol-12-myristate-13-acetate (PMA) and ionomycin stimulation showed that stage 3 cells were functionally mature, whereas stage 1 and 2 were not, consistent with the proposed progression of cells through these developmental stages. This progression was confirmed using a coculture system of OP9 mouse bone marrow stromal cells and thymocytes, in which stage 1 cells differentiated into the later stages over 5 days of culture, whereas stage 3 cells maintained their phenotype over time in coculture.

Next, Koay et al analyzed factors that were important for progression from stage 1 to 3 in the same coculture system. Addition of MR1 blocking antibody showed that, consistent with previous reports, MR1 was necessary for positive selection and for transition through each checkpoint. Maturation of MAIT cells from stage 2 to stage 3 depended on the transcription factor PLZF (ZBTB16), which is also expressed in other innate-like T cells. It was also reported that interleukin-18 (IL-18) was necessary for this transition, consistent with the expression of IL-18Rα detected in cord blood,[11] but a subsequent report showed essentially normal MAIT cell numbers in the periphery in IL-18–deficient mice, which also responded normally by expansion following infection with *S.* Typhimurium.[107] The development of murine MAIT cells also required micro RNAs; short 18 to 20 nt noncoding RNAs, that regulate gene expression. The microRNA (miRNA) processing enzyme Drosha was required for progression from stage 1 to stage 2 as shown by a block of cells at stage 1 in *Drosha*$^{fl/fl}$*Cd4*-Cre. Loss of the micro RNA pair miR-181a/b-1, which is most highly expressed in DP thymocytes and acts as a rheostat for TCR signal strength, mirrored the loss of all miRNAs; in miR-181a/b-1–deficient mice, thymic MAIT cells did not mature, as defined by expression of transcription factors PLZF, T-bet, and RORγt,[154] thus showing miR-181a/b-1 plays a similar role in MAIT cell thymic maturation to that previously determined for other T cells such as iNKT cells[153] and Tregs.[155] A recent study demonstrated a lack of both iNKT and MAIT cells in the thymus of *Cd4*^{Cre}*Bcl6*$^{F/F}$ mice, indicating a role for the transcription factor BCL-6[156] in thymic development of these cells.

Koay et al then performed transcriptional analysis of MAIT cells in thymi from humans and mice.[157] In mice, bulk sequencing (100 cells) of immature stage 1 and mature stage 3 MAIT cells was used to analyze the MAIT cell transcriptomes at different stages of thymic development, with over 300 differentially expressed genes identified, including confirmation of previous findings at the protein level for several molecules, such as CD218 (IL18R), CD161 (KLRB1), CD27, and PLZF (ZBTB16), which have higher expression at stage 3. For human MAIT cells, a comparison of stage 1 to stage 3 thymic MAIT cells revealed 625 differentially expressed genes, with most changes occurring during transition from stage 2 to 3. Many of these genes were overlapping between human and mouse MAIT cells. Additional genes, such as SLAM-associated protein (SAP) and SATB1, were important in regulating stage 1 to 2 and 2 to 3 transitions in mice, respectively, as supported by the lack of these cells in the corresponding gene-deficient mice.[157,158] This is in contrast to human MAIT cells, which were reported to develop independent from SAP.[146] Expression of Zap70 which is critical for T cell signaling was also important for maturation to stage 3, indicating a requirement for ongoing MR1-dependent signaling.[157]

At stage 3, mouse MAIT cells branch into the two main functional subsets that are also found in the periphery. In mice, most MAIT cells are MAIT-17 cells. These express high levels of the transcription factor RORγt consistent with the capacity to produce IL-17 and intermediate levels of PLZF. MAIT-1 cells, which express T-bet and produce IFNγ upon activation, are less abundant. Nur77, a transcription factor correlated with TCR signaling strength, was expressed preferentially in MAIT-17, but not MAIT-1 cells in the thymus. Thus, MAIT-17 cells may receive stronger TCR signals in the thymus.[149] In contrast to mouse MAIT cells, human thymic MAIT cells coexpress T-bet and RORγt at stage 3.[148] The different functions of these MAIT cell subsets (described in section *Physiological and Pathophysiological Roles of Mait Cells*) in mice and humans are currently unclear.

Peripheral Expansion of MAIT Cells

Early studies showed that MAIT cell development is a stepwise process involving thymic selection followed by expansion in the periphery.[146] MAIT cell numbers are low in the thymus and cord blood. In human donors ranging from 5 days to 14 years of age, thymic MAIT cells remain at low numbers (<0.05% of T cells), while there was an increase in numbers in the peripheral blood over this same period.[148] This pattern is consistent with the idea of peripheral MAIT cell expansion reflected by their increased numbers after birth.[9,146,148]

There are several changes to the MAIT cell population following thymic egress and upon peripheral expansion. After exiting the thymus, MAIT cells acquire their effector/memory phenotype (eg, CD44^{hi}, CD62^{lo}), along with a gradual increase in PLZF and ability to produce cytokines in response to bacterial stimuli.[159] In stage 1, mouse MAIT cells are predominantly CD4+ or DP cells, with stage 2 being mostly CD4+, and stage 3 cells are reminiscent of those in the periphery (mostly DN or CD8+). The vast majority of CD8+ cells in the thymus expressed CD8αβ heterodimers in both mouse and human, whereas in the periphery, CD8αα is most prevalent, suggesting further extrathymic maturation, and consistent with earlier studies that suggested MAIT cells developed from precommitted CD161++CD8αβ+ T cells.[159,160] The vast majority of MAIT cells in blood from adult donors coexpressed CD218 (IL-18Rα) and CD161, while most (up to 97%) thymic MAIT cells were CD218−CD161−.[148] RNA-seq analysis on sorted stage 3 MAIT cells from paired samples of human thymus and blood revealed gene differences consistent with a further maturation in the periphery. This was less clear in mice where stage 3 thymic MAIT cells had a very similar transcriptome to MAIT cells in the spleen.[157] An oligoclonal expansion of the TCR repertoire is also consistent with peripheral expansions.[6-9]

Requirement for Microbial Antigens in MAIT Cell Development

Early studies reported the absence of MAIT cells in germ-free mice.[22,146] It is now clear that low numbers of MAIT cells are present in the absence of commensal microbiota. However, MAIT cells are developmentally blocked at stage 1 of thymic maturation and are virtually absent in the periphery of germ-free mice, indicating that commensal microbiota is required for their full development as well as expansion in the gut lamina propria.[22,148,149] MAIT cells could be reconstituted in germ-free mice by microbial colonization or by the introduction of riboflavin-pathway sufficient bacteria,[82] whereas *ribD*-deficient *E. coli* could not reconstitute MAIT cell development or migration to the periphery,[149] indicating that 5-OP-RU or related antigens were required. Recently, Legoux et al showed that 5-OP-RU injected intraperitoneally, or painted onto the skin above the thymus, could be captured by thymocytes and presented to MAIT reporter cells *in vitro* and activated endogenous MAIT cells in the thymus.[149] This study also showed a 5-OP-RU dose-dependent depletion of thymic MAIT cells, which was more pronounced in germ-free mice than in mice housed under specific pathogen-free (SPF) conditions, suggesting a role in negative selection, and that microbial factors other than 5-OP-RU may also influence MAIT cell thymic development. In agreement with this observation, topical application of 5-OP-RU to SPF-housed, but not germ-free, mice reconstituted MAIT cells in the skin, supporting the hypothesis that both 5-OP-RU and other microbial factors were important for their development.[82]

Constantinides et al went on to show that MAIT cell development depends on exposure to riboflavin-synthesizing microbes during a critical time window.[82] MAIT cells appeared in mucosal tissues of SPF-housed mice at 3 weeks of age, and colonization of germ-free mice within this period could reconstitute MAIT cell development, whereas colonization of adult germ-free mice by cohousing with SPF mice failed to induce MAIT cell development. MAIT cells in the skin were similar in frequency in mice cohoused, but differed between cages (ranging from <5%-40% of αβ-T cells) further suggesting that their peripheral expansion differed depending on the microbiota.[82]

Summary and Future Directions

Early studies indicated that MR1 and microbiota were required for the thymic selection of MAIT cells. In the thymus, like for NKT cells, double-positive thymocytes appear most important for their selection, whereas in the periphery, where MAIT cells undergo further expansion, many different cell types can act as APCs. Positive selection of MAIT cells can be achieved experimentally by exposure to 5-OP-RU or by the presence of microbiota during early life. The implications of these findings for early-life exposure to microbiota or factors that may alter microbiota (eg, antibiotics) in humans are yet to be explored.

The identification of 5-OP-RU itself as the selecting antigen would be consistent with a recent report of MAIT cell deficiency in an individual with a point mutation in MR1 ($MR1^{R9H}$), which abrogated binding of activating ligand.[51] However, the details of trafficking of antigen to the thymus and factors controlling antigen concentration requirements in the thymus for positive selection remain unclear.

From studies of mouse and human MAIT cells, a picture is emerging of the key cell surface proteins and transcription factors that control the progression through distinct developmental stages. It is clear that, despite similarities, these are somewhat distinct from other T cells, including the related NKT cells. As discussed in the next section, functional differentiation of MAIT cells into MAIT-1 and MAIT-17 cells is a further step in their development, controlled by transcriptional signals (principally T-bet and RORγt, respectively) and providing qualitatively distinct patterns of immunity in response to pathogen challenge.

MAIT CELL PHENOTYPE AND FUNCTIONAL CAPACITY

Identification of MAIT Cells

Identification of Human MAIT Cells Using Phenotypic Markers

Human MAIT cells can be identified through their expression of the TRAV1-2 (Vα7.2) TCRα chain in conjunction with other markers such as high levels of CD161 (KLRB1). Combinations of these and other surrogate markers, including CD26 and IL18Rα, have been used to define MAIT cells in several studies. MAIT cells also comprise CD4+, CD8+, and DN cells, as well as small proportions of DP cells[9] (see also section *Identification of a Population of T Cells With Semi-Invariant TCR Usage*). It is worth noting that the expression of some markers can change in different settings, particularly CD161 expression which was shown to decrease in rheumatoid arthritis[139] and HIV infection.[161] Additionally, the proportion of MAIT cells (defined with MR1-tetramers, described below) expressing CD161 was reduced in kidney compared with blood.[162] Hence, phenotypic definition of MAIT cells by surface markers should include MR1-antigen tetramers for accuracy.

Assessment of MAIT Cells in Mice

The frequency of MAIT cells in mice is relatively low, representing only 0.1% to 2% of αβ T cells in different tissues and in blood (Fig. 37.3A). Additionally, no antibodies specific for TRAV1 are currently available to detect the invariant MAIT mouse TCR. These factors hampered the precise identification of MAIT cells in laboratory mice prior to the generation of MR1-tetramer reagents.[7,23,24,69,90]

To date, several mouse models have been used to study the immune function of MAIT cells. These include mice that are deficient in MAIT cells as well as those where MAIT cells are increased in abundance. Consistent with the role of MR1 in thymic selection of MAIT cells, MR1−/− mice are deficient in MAIT cells but otherwise lack a clear immune phenotype.[22] MAIT cells are also not found in TCR Jα18 knockout mice[163] or in Jα33 knockout mice.[59] However, the latter had a residual population of MAIT-like TRAV1+ or TRAV1− cells that could respond to antigen or bacterial infection.[59] Collectively, these mice have been valuable tools in the study of the effect of MAIT cell deficiency in infection or disease settings (see section *Physiological and Pathophysiological Roles of MAIT Cells*).

Mice transgenically expressing the TRAV1 invariant (Vα19*i*) TCRα chain have also been developed by several groups.[22,146,164] These mice have much higher numbers of MAIT cells than wild-type mice. However, it should be noted that approximately one-third of their MR1-5-OP-RU-reactive cells do not express the PLZF, a hallmark of classical MAIT cells. Moreover, in these TRAV1*i*Tg mice, MR1-tetramer−reactive cells could be detected even in the absence of MR1 (TRAV1*i*Tg.MR1−/− mice).[90] This suggests that MR1-reactive T cells can be selected by classical MHC molecules at least in the context of the overexpression of the invariant, MAIT TCR alpha chain. The advent of MR1-5-OP-RU-tetramers[7,69] has enabled the detection of MAIT cells in naïve SPF-housed mice[23,24,89] reducing the need to use transgenic models to study MAIT cell function. In addition, the CAST/EiJ mouse strain[165] has approximately 20-fold more MAIT cells than C57BL/6 mice. These mice have now been used in several settings to enable easier assessment of MAIT cell immune responses and role in disease. Additionally, methods have been developed to boost MAIT cell numbers in wild-type laboratory strains to aid in studying their impact on infection or disease outcome. For instance, challenge of wild-type mice with 5-OP-RU,[23,107] premixed 5-A-RU and methylglyoxal[81,130] or a prodrug designed to release 5-A-RU,[130] have all been shown to boost MAIT cell numbers in mice, in the presence of TLR agonists or other costimuli.

Development of MR1-Tetramers

The development of MR1-antigen tetramer reagents[7] (Fig. 37.18), the equivalent of MHC-peptide[166] or CD1d-lipid antigen[167] tetramers, was a breakthrough in the field since these reagents enabled the definitive identification of MAIT cells from human and mouse samples. The first reagents were generated using a modified version of human MR1, containing a Lys to Ala mutation at position 43 at the base of the MR1 A′ binding pocket. Unlike wild-type MR1, this mutant protein was able to form stably in the absence of ligand, enabling the folding of "empty" MR1-β_2m molecules, which could then be loaded with antigen, biotinylated and tetramerized using standard procedures. These K43A tetramers proved useful for the staining of MAIT cells in human blood and intestinal cell preparations[7] and were capable of detecting MAIT cells

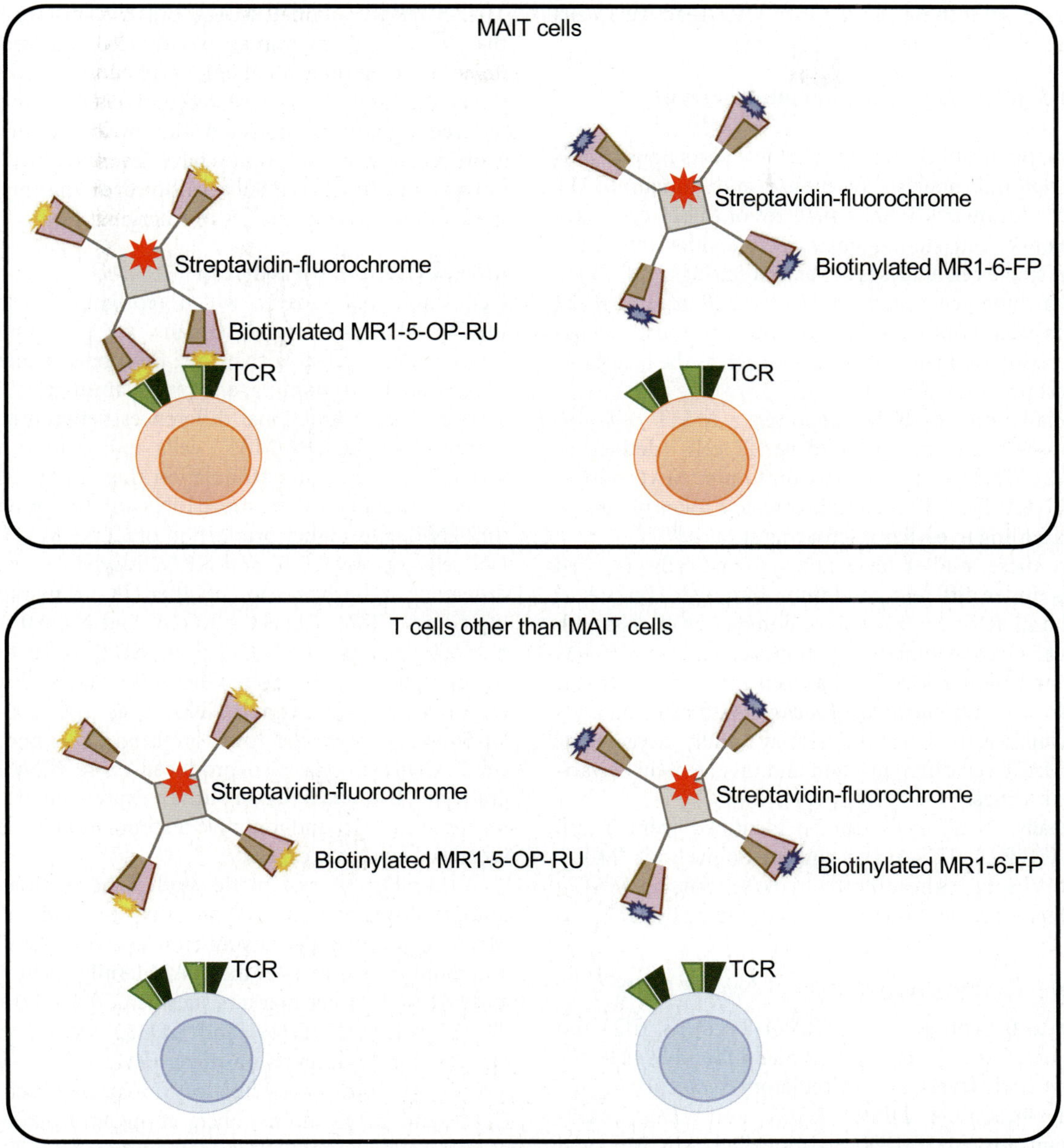

FIG. 37.18. MR1 tetramers as analytical tools to identify MAIT cells. Biotinylated soluble versions of ligand-loaded soluble MR1 are tetramerized via streptavidin, coupled to a fluorochrome, thereby allowing for flow cytometric detection. 5-OP-RU-loaded MR1 tetramers specifically bind to, and this way identify, MAIT cells, while 6-FP-loaded tetramers do not.

in mice.[90] However, these tetramers were relatively unstable and challenging to work with. The production of wild-type MR1 tetramers required the understanding of the formation of the potent pyrimidine antigens, 5-OP-RU and 5-OE-RU from their precursor molecules 5-A-RU and methylglyoxal or glyoxal, respectively. Generation of mouse and human,[23,24,69,90] and later macaque,[28] wild-type MR1 tetramers with each of these molecules, as well as with 6-FP or Ac-6-FP, resulted in more stable reagents. These are now available through the NIH Tetramer core facility (https://tetramer.yerkes.emory.edu/reagents/mr1) and have enabled the detection and study of MAIT cells in many contexts.

It should be noted that while MR1-tetramers containing 6-FP or Ac-6-FP are typically used as negative control reagents; these also detect a small number of specifically reactive or MR1-autoreactive cells,[58] although the biological relevance of these cells is unclear (see section *Recognition of MR1-Ligand Complexes by MAIT and Other MR1-Reactive T Cells*). MAIT cells detected using either MR1-5-OP-RU tetramers or monoclonal antibody to TRAV1-2 showed >95% correlation in human blood.[7] MR1-5-OP-RU and MR1-5-OE-RU also appeared to stain identical populations.[7] More recently MR1-tetramers have been generated loaded with different sources of metabolite molecules.[135] It is yet to be shown if these can identify MR1-reactive T cells other than 5-OP-RU–specific MAIT cells.

MAIT Cell Signatures and Identification in Large Datasets

With several publications over the past few years now including transcriptomic analysis of mouse and human MAIT cells,[157,158,168-173] transcriptional signatures of MAIT cells have been developed and subsequently used to identify MAIT cells within single cell analyses in various settings.[174,175] These signatures include genes such as *TRAV1-2, Rorc,* and *Klrb1 (CD161)*, as well as other molecules not previously recognized to be associated with MAIT cells, such as the bicarbonate transport protein SLC4A10.

Additionally, one study has reported a MAIT cell proteome analysis from human blood MAIT cells (defined as CD3+Vα7.2+CD161+), in comparison to non-MAIT CD8+ T cells and NK cells.[176] These studies create a fuller picture of MAIT cells adding to earlier microarray studies.[160,177]

To date, these studies have not systematically assessed MAIT cells under different conditions, although Hinks et al performed bulk RNA seq analysis of human MAIT cells with and without *in vitro* stimulation (6 hours culture with 10 nM 5-OP-RU) and mouse MAIT cells from mice after intranasal infection with *Legionella longbeachae*. As further datasets are made available, these techniques may rapidly increase our picture of MAIT cell differentiation and involvement in various disease settings.

Additionally, MAIT cells can be identified from T cell sequences via their TCR usage and an online tool "MAIT match" (https://services.healthtech.dtu.dk/service.php?MAIT_Match-1.0) has been developed to assist in this analysis.[178]

Phenotypic Markers and Subsets of MAIT Cells

In addition to the expression of TRAV1-2+ TCRs, CD3 and high levels of CD161 (KLRB1), in humans, blood MAIT cells also express high levels of costimulatory molecules CD26 (dipeptidyl peptidase-4, DPP4), CD27, and CD28, multidrug-resistant protein (MDR1, also called ABCB1), and receptors for cytokines, including for IL-7, IL-12, IL-18, and IL-23.[11,21,91,136,179-181] Mouse MAIT cells express costimulatory molecules, including CD27, CD28, low levels of CD154 (CD40L), and high levels of ICOS,[24,107] as well as receptors for IL-7 and IL-18.[24,182] NK1.1 (KLRB1) was only expressed on 30% to 50% in C57BL/6 mice,[24] although it should be noted that this is not completely orthologous to the human CD161.[183] MAIT cells have been described as "effector-memory" cells, due to their expression of high levels of CD44 and low levels of CD62L.[11,24,182]

MAIT cells were originally considered as functionally homogeneous,[11] but more recently, increasing levels of functional diversity have been reported. Dias et al screened MAIT cells (defined on the basis of CD161 and Vα7.2 expression) in human PBMCs from healthy individuals by flow cytometry for a panel of 332 immunoreceptor surface proteins. MAIT cells clustered separately from other T cells expressing either TRAV1-2 or CD161, and the phenotypes were consistent with MR1-5-OP-RU defined MAIT cells.[182] This study revealed that MAIT cells in human blood were relatively homogeneous in their effector-memory phenotype, costimulatory molecules, and chemokine receptor expression profile but differed in markers associated with innate immune function. More recently, several studies have described MAIT cell phenotypic and functional subsets; however, the importance of these in various contexts is still emerging.

CD8+, CD4+, and DN Subsets

CD8+ and DN subsets, which represent the majority of MAIT cells in mice and humans, are phenotypically and functionally similar, in that they both have a memory phenotype, are tissue homing, and proinflammatory.[184] However, there are some functional differences between subsets. For example, human DN MAIT cells appear more capable of producing IL-17 upon stimulation than CD8+ MAIT cells, which conversely secrete more IFNγ and tumor necrosis factor (TNF).[185] A higher proportion of CD8+ MAIT cells than DN cells express CD16 and NKG2D, whereas there was no difference in the expression of other NK cell markers, including CD56, CD94, CD244, NKG2A, and NKp80 or in T cell markers such as CD25, CD26, or CD27 or in tissue-homing receptors.[184] However, Gherardin and colleagues later reported that NKG2D and NKG2A, as well as CXCR6 and CD56 were expressed at lower levels on CD4+ and DP MAIT cells.[9] CD8+ T cells also produced more IFNγ, TNF, and granzyme B and had higher surface expression of CD107A in response to PMA and ionomycin stimulation,[184] reflecting an increased cytotoxic capacity.

A number of systematic studies have broadened our understanding of MAIT cell phenotypic variation.[9,59,184] Thus, MAIT cells vary in their expression levels of NKp80, NKG2D, and multiple members of the SLAM family, including CD150 and CD244.[11] Other markers such as CD63, CD195, CD196, CD218, CD244, CD319, and CD352 are reported to be expressed at high levels on most MAIT cells.[182] MAIT cells also express other costimulatory molecules such as CD26, CD27, and CD28 and members of the tetraspanins, a family of costimulatory molecules including CD53, CD63, CD81, and CD9. Immunoproteome analysis of MAIT cells showed that CD95 (Fas), CD172γ (SIRPγ), CD52, and CD43 were expressed at moderately high levels.[176]

MAIT-1 and MAIT-17 Functional Subsets

Both human and mouse MAIT cells express the transcription factor PLZF, encoded by zinc finger BTB domain containing 16 (ZBTB16), the CTL-associated transcription factors Eomesodermin (EOMES) and Blimp-1 (PRDM1) and either T-bet (TBX21), which drives type 1 function (MAIT-1 cells) or RORγt (RORC) and STAT3, which drive type 17 responses (MAIT-17 cells).[185-189] The proportions of these subsets of MAIT cells differ between mice and humans. In humans, RORγt and T-bet are coexpressed, whereas in mice, it was reported that their expression can be mutually exclusive, with MAIT-17 cells dominating.[24,165] However, the expression of these transcription factors appears to change upon activation with the proportion of RORγt+T-bet+ double-positive

MAIT cells increasing following infection.[107,190] The plasticity of MAIT cells and the relationship between these functional subsets is currently unknown.

Tissue Distribution and Tissue-Specific MAIT Cell Phenotypes

MAIT cells are present in humans and, at lower frequency, in mice in the circulation and throughout peripheral tissues, with higher abundance in nonlymphoid tissues, including at mucosal sites such as the lungs, female reproductive tract and intestines, and in the liver (Fig. 37.3A).[191,192] In the intestines, MAIT cells are mainly found in the lamina propria and are less abundant among intraepithelial lymphocytes and in Peyer patches.[22] MAIT cells are also present in the skin, at similar frequency to the circulation, and are more readily detected in the healthy dermis, although they can also be detected in the epidermis in inflammatory conditions.[193-195] Their location within lymphoid organs is yet to be described, although small numbers of MAIT cells can be isolated from mouse spleen and lymph nodes.[23,24,89]

Consistent with their relative abundance in blood and peripheral organs (Fig. 37.3A) and scarcity in lymphoid tissues, human MAIT cells express CCR2, CCR3, CCR5, CCR6, CXCR1, CXCR3, and CXCR6 are low or negative for CCR4, CCR7, CXCR3 and CX3CR1, CXCR2, and CXCR7, as well as having heterogenous levels of CXCR4, CCR4, and intermediate expression of CCR9.[11,21,182,184,196] MAIT cells have also been found to recirculate from tissues via the lymph.[197]

The phenotype of MAIT cells differs in different tissue sites. In skin, most MAIT cells express cutaneous lymphocyte-associated antigen (CLA).[193,194] In mucosal tissues, MAIT cells have a phenotype associated with tissue residency including expression of CD69 and CLA, whereas CD103 had bimodal expression on MAIT cells in the thymus, spleen, and lungs of naïve mice.[24] MAIT cells are also present in the LP of the stomach where they express a memory phenotype similar to blood MAIT cells but are more likely to have an activated phenotype expressing CD69 and CD103.[198,199]

MAIT cells are also biased toward particular functions in a tissue-specific manner. For example, in the kidneys, human MAIT cells predominantly produced IL-2, GM-CSF, and IL-17A upon stimulation with PMA and ionomycin.[162] MAIT cells in the female genital tract were skewed toward production of IL-17 and IL-22,[13] and lung MAIT cells were also skewed toward IL-17 production.[172] Liver MAIT cells produced higher levels of both Type 1 and Type 17 cytokines than blood MAIT cells.[21] Similarly, MAIT cells in the oral mucosa appear to be educated to the tissue environment, such that they express CD69, CD103, as well as HLA-DR and PD-1 and were biased toward IL-17 production and had lower perforin expression than MAIT cells from the blood.[20] MAIT cells in the human gut can express markers including PD-1, TIGIT, LAG-3, and CTLA-4, possibly due to increased exposure to antigens resulting in an altered activation state.[200,201]

There is an increase in the proportion of CD4−CD8− (DN) MAIT cells in the kidneys[162] and in the oral mucosa[20] compared to blood. Terpstra et al. also reported a distinct population of MAIT cells in the kidneys (CD27+/−CD45RA−CCR7−CD28− MAIT cells) that were absent in the circulation.

MAIT-1 cells have been reported as less "tissue resident" than MAIT-17 cells, which have higher expression of CD103 and CCR6.[172] CCR7 is selectively required for the development of the MAIT-17 subset.[157] The reduction in MAIT cells was more marked for MAIT-17 than MAIT-1 cells in germ-free mice in the thymus and spleen, suggesting differences in the requirements of these cells for development.[149] Thus, the local tissue microenvironment may drive functional differences in MAIT cells, and functional subsets of MAIT cells appear to exhibit different tissue-homing properties, which are yet to be understood.

Cytokine Production

Along with their "effector-memory" phenotype ($CD44^{hi}$, $CD62L^{lo}$), the ability of MAIT cells to rapidly produce cytokines upon stimulation afforded them the description as "innate-like" T cells. In response to infection or TCR-stimulation, mouse MAIT cells can produce several cytokines including type 1 cytokines IFNγ and TNF as well as IL-17 and GM-CSF[23,24,107,165,190,202] (Fig. 37.3B). These cytokines are made by MAIT-1 and MAIT-17 subsets, respectively, in mice, whereas in humans, the majority of MAIT cells express both T-bet and RORγt and predominantly secrete IFNγ and TNF as well as IL-2,[11,60,63,120] although IL-17A–producing cells have been detected in the lungs[172,203] and small proportions of MAIT cells from peripheral blood of ankylosing spondylitis patients produced IL-17A and IL-22.[204] Peripheral blood MAIT cells are also capable of IL-17 production following PMA and ionomycin stimulation.[11] Another study reported that MAIT cells in the female genital tract preferentially produce IL-17 and IL-22.[13] Since the majority of studies on human MAIT cells have analyzed peripheral blood, whereas most mouse studies use tissues such as the lungs, some of the apparent species differences may be due to differences between circulating and tissue-resident MAIT cells. It remains unclear if distinct MAIT-2, MAIT-9, MAITfh, MAITreg, or other subsets equivalent to their conventional T cell counterparts exist. Transcriptomic analyses revealed the presence of only very small numbers of MAIT-2 cells.[173] However, there is a growing number of reports of the production by MAIT cells of type 2 cytokines including IL-4, IL-5,[165] and IL-13,[205] as well as IL-6, IL-10, and IL-21[165,206] along with several reports of MAIT cells providing help for antibody production.[206-209] MR1-restricted TRAV1*i* (Vα19*i*) T cells from transgenic mice were also found to secrete IFNγ, IL-4, IL-5, and IL-10[164] and MAIT cells could increase type 2 cytokine production in mixed cultures with mouse bone marrow–derived macrophages[210] or human monocytes.[11] Pulmonary MAIT cells in mice produce TGFβ in response to *M. tuberculosis* infection,[211] and thus may play a role in regulating other arms of the immune response.

Cytotoxic Capacity

Human MAIT cells have been shown in several studies to express cytotoxic molecules and to degranulate upon stimulation, suggesting cytotoxic capacity[120,188,212,213] (Fig. 37.3B). The expression of granzymes (Gzm) by MAIT cells differs from that of conventional T cells, with human peripheral blood MAIT cells (defined as Vα7.2+CD161++CD8+) expressing only low levels of granzyme B (GzmB) or perforin, but with high levels of GzmA and GzmK.[176,188,189,214] GzmA and GzmK are only weakly cytotoxic but can induce production of proinflammatory cytokines and may be important in control of bacterial pathogens through other mechanisms.[215] Interestingly, Vα7.2−CD161++CD8+ cells also had high GzmA and low perforin but differed from Vα7.2+ MAIT cells in their levels of GzmB and GzmK.

Upon TCR-dependent activation, the expression of GzmB increased after at least 18 hours of stimulation. Perforin expression could also be enhanced by either TCR-dependent or indirect stimulation of MAIT cells. In contrast, GzmA and GzmK decreased upon stimulation with *Escherichia coli*, but not indirect stimulation by cytokines, such as IL-12 and IL-18.[188]

The cytotoxic capacity of MAIT cells has been demonstrated *in vitro* against *E. coli*- or *Shigella flexneri*-exposed cell lines,[120,188,213] myeloma cell lines pulsed with antigen,[216] and other cell lines.[18,217]

Neither the recognition events nor the mechanisms of killing by MAIT cells have been fully explored, but may include both MR1-dependent and -independent recognition, and various effector molecules. For example, IL-7 was reported to enhance cytotoxic function.[187] Rha et al[18] also examined the impact of cytokines on the cytotoxic capacity of MAIT cells collected from donor liver perfusates prior to transplantation and showed that exposure to IL-15 increased the expression of activating receptors such as NKG2D, NKp30, NKp44, and DNAM-1 as well as GzmB–mediated killing, independent of MR1/TCR recognition. In this setting, killing of K562 cells by liver MAIT cells could be blocked by Gzm inhibitors, but not by anti-Fas Ligand (FasL) or anti-TRAIL antibodies. Thus, conditions that result in IL-15 upregulation, such as hepatitis A infection, may also drive MAIT cell cytotoxic function in an NK-like manner, independent of antigen recognition,[18] whereas the killing of *E. coli*–pulsed cell lines was MR1-dependent, but NKG2D-independent.[120,188]

It is likely that different subsets of MAIT cells differ in their cytotoxic capacity and mode of killing. For example, granulysin and FasL were expressed on a small percentage of PBMC MAIT cells.[185,188] The expression of cytotoxic molecules, including GzmB and perforin, was higher in CD8+ MAIT cells than the DN MAIT cell population in human blood.[185] Expression of GzmB was higher on T-bet+ and Blimp-1+ MAIT cells than MAIT cells lacking these markers.[188] The CD8+ MAIT cell subset also expressed higher levels of NK cell receptors such as CD94/NKG2A and NKG2D.[185] Thus, the control of MAIT cell cytotoxicity, and the role of this function in various settings, such as infection and cancer remains to be fully understood.

Summary and Future Directions

It is now clear that MAIT cells are heterogenous, exiting the thymus predominantly as MAIT-1 and MAIT-17 subsets. However, there is also great diversity in their phenotypic characteristics and expression of the CD4 and CD8 coreceptors, NK cell receptors, and activation markers. The functional implications of this heterogeneity remain unclear. Additionally, the functional capacity for production of a broad array of cytokines and acquisition of cytotoxic activity suggests functional differentiation is part of MAIT cell biology. The recent explosion in large dataset analysis including RNA-seq analysis of various mouse and human MAIT cell subsets/activation states has, and will continue to, yield further information leading to a clearer picture of MAIT cell phenotype and functional capacity.

PHYSIOLOGICAL AND PATHOPHYSIOLOGICAL ROLES OF MAIT CELLS

Activation of MAIT Cells

Antigen-Specific Activation

MAIT cells recognize and respond to riboflavin-based antigens, with the most potent being 5-OP-RU. Human peripheral blood MAIT cells produce cytokines, including IFNγ and TNF, in response to a range of riboflavin-synthesizing bacteria[60,63,66,78,218] or to 5-OP-RU antigen at subnanomolar concentrations.[69,122,134]

Like conventional T cells, full activation of MAIT cells requires additional costimulatory signals in addition to TCR triggering. MAIT cells express high levels of ICOS, and this plays an important role in their response to bacteria *in vivo*.[107] In contrast, despite their expression of other costimulatory molecules, including CD28,[24,107] loss of CD80/86 in mice does not appear to affect the expansion of MAIT cells in response to infection. It has also been reported that, relative to other T cells, MAIT cells are poorly stimulated by anti-CD3/anti-CD28.[21] Nonetheless, costimulation through CD28 was shown to contribute to MAIT cell upregulation of GzmB in human MAIT cells.[188] Human MAIT cells from pleural effusions showed increased expression of the costimulatory molecule 4-1BB following *M. tuberculosis* stimulation, and thus, this may also be important for their activation.[219]

The *in vitro* addition of agonists of TLR 1, 2, and 6 to MR1+ THP-1 cells, was found to increase their ability to stimulate MAIT cell IFNγ production in the presence of fixed *E. coli*.[111] TLR agonists may also enhance antigen-specific activation of MAIT cells by direct effects on MR1 expression. Both a B-lymphoblastoid cell line and human peripheral blood B cells showed increased MR1 expression following stimulation through TLR9.[131] Heat-killed *Listeria monocytogenes*, lacking MAIT stimulatory antigens (Rib pathway deficient), but possessing TLR2 agonist activity, only weakly enhanced MR1 expression in cell lines. Differences in the upregulation of MR1 in different cell lines suggest different mechanisms of costimulation are likely to operate.[111,131]

In vivo, MAIT cells expand to high numbers in mice in response to 5-OP-RU coadministered with a range of TLR agonists including Pam2Cys (TLR2/6), Poly I:C (TLR3), CpG (TLR9), and LPS (TLR2/4),[23,168] whereas no expansion was seen with heat-killed *Listeria monocytogenes* (TLR2), flagellin (TLR5), or imiquimod (TLR7).[168] 5-OP-RU administered alone induced CD69 upregulation on mouse MAIT cells, but without provoking proliferation of these cells.[23] Recently, Wang et al demonstrated expansion of the MAIT cell population in response to 5-OP-RU and IL-23 and showed that this cytokine was also important for accumulation of MAIT-17 cells following bacterial infection.[107] *In vitro*, IL-12 and IL-18 were required, in addition to MR1, for stimulation of MAIT cells by *E. coli*.[220] IL-18 was also important for MAIT cell production of IFNγ in response to *F. tularensis, in vitro*, but this cytokine was not required *in vivo*.[221] IL-15 can also enhance TCR-dependent MAIT cell responses as well as increasing their expression of cytotoxic molecules.[222] MAIT cell responses to *Haemophilus influenzae* were influenced by the synergy of IL-12 and IL-7,[223] and IL-7 was found to drive a MAIT-17 cell response.[224]

Activation by Microorganisms and Interactions With the Microbiome

In 2010, two papers reported MAIT cell reactivity against a range of bacteria and fungi, but not to viruses.[60,63] MAIT cells have since been shown to respond to many microorganisms including both Gram-positive and Gram-negative bacteria and fungal species (reviewed in Ref. 65). The antigen-specific nature of this response has been confirmed, using *rib* gene mutant strains of S. Typhimurium, *E. coli*, and *L. lactis* (see section *Confirmation of the Riboflavin Synthesis Pathway as the Source of MAIT Cell Antigens*).[23,69,81] Given the conservation of the riboflavin biosynthesis pathway, it is likely that many commensal microorganisms also produce MAIT cell antigens. Indeed, the microbiome is required for MAIT cell development (see section *Development of MAIT Cells*). MAIT cells may also be maintained by the presence of the microbiota since antibiotic treatment reduced the MAIT cells response to subsequent *M. tuberculosis* infection in mice.[225] Disruption of the mucosal barrier during viral infections may also increase the presentation of microbial antigens.[226] Conversely, MAIT cells may play a role in shaping the microbiota, which is altered in *Mr1−/−* mice, impacting on the resistance of these mice to *Clostridium difficile* infection.[227] The effect of MR1 and MAIT cells on gut integrity may also affect autoimmune diseases such as diabetes[132] (see section *Autoimmune and Inflammatory Diseases*). Thus, MAIT cells may play both direct and indirect roles in both infectious and noninfectious diseases via their interactions with the microbiota.[228,229]

MR1/TCR-Independent Activation

In the absence of specific antigen, the activation of MAIT cells can occur in a TCR and MR1-independent manner through cytokines, including type I IFNs, IL-12, and IL-18.[230,231] IL-15 in mixed cell cultures also drives MAIT cell production of IFNγ, as well as increasing GzmB and perforin expression, although this response may be mediated through IL-18 produced by CD14+ cells.[222] MAIT cells were also shown to be a major source of proinflammatory cytokines following exposure to bacterial superantigens.[232]

Although viruses do not produce riboflavin-based antigens, both human and mouse MAIT cells have been shown to respond to infection with several viruses, including influenza, hepatitis C, dengue, Puumala orthohantavirus, and zika viru ses.[161,214,230,231,233-237] This response is thought to be driven by the inflammatory cytokine response, with TLR agonists such as TLR8 agonists playing a role.[220,230,238]

The TCR-MR1−independent activation of MAIT cells, including during viral infections, has been shown in several studies to result in an impaired functional phenotype and the expression of exhaustion markers.[214,231,233,234,239-242] Following superantigen exposure, this anergic phenotype prevented cognate responses to subsequent bacterial stimuli.[232] Dysfunctional MAIT cells have also been reported in chronic viral infection,[17] cancer,[171,174,216,243-245] and in patients with alcoholic liver disease,[246] where MAIT cell activation is likely to be antigen-independent. Patients with viral infections, including human immunodeficiency virus, hepatitis B, C, and D, dengue, measles, influenza A viruses, as well as severe COVID-19 patients[239,247-249] have reduced numbers of circulating MAIT cells. A similar finding has been reported in some studies in nonhuman primates following infection with Simian immunodeficiency virus (SIV) or Simian HIV.[29,250] Interestingly, MR1 expression has been shown to be downregulated in the context of infections by viruses, including human cytomegalovirus and herpes simplex virus-1, potentially reducing TCR-MR1-antigen dependent MAIT cell activation.[251] Conversely, activation of MAIT cells, and not depletion, was reported in pigtail macaques[28] and in rhesus macaques, and the MAIT cell frequency correlated with lower viral loads,[226] suggesting a protective immune function.

Protective Immunity to Infection

Bacterial Infections

Despite the conservation of the riboflavin biosynthesis pathway and thus, the capacity of many microbes to produce MAIT stimulatory antigens including 5-OP-RU and 5-OE-RU, a protective role for MAIT cells in infection has only been demonstrated for a handful of bacterial species. However, there is growing evidence that MAIT cells are part of the immune response to many microbes including both pathogens and commensal microbiota.

Mouse models of bacterial infections have demonstrated a protective role in reducing bacterial load or increasing survival following infection with *Francisella tularensis*, *Klebsiella pneumoniae*, *Mycobacterium bovis* BCG, *M. tuberculosis*, *E. coli*, and *Legionella longbeachae*.[60,90,100,190,202,252] In these models, MAIT cell−deficient mice showed impaired control of infection. Importantly, when MAIT cells were increased in number by adoptive transfer into immune deficient recipients or primed and expanded prior to infection (vaccination) enhanced immune control of infection with *Legionella longbeachae*, or *M. bovis* BCG was reported.[107,190,210] Conflicting results for the role of MAIT cells in *M. tb* infections point to the possibility that the timing of MAIT activation is critical and highlight the complexities of immune cell cross talk.[90,210,211] Interaction with other cell types is increasingly the subject of study, with recruitment of other immune cells (including neutrophils, macrophages, dendritic cells, eosinophils, and other T cells) by MAIT cells not only mediating their protective effects[202] but also driving immune pathology.[199] While IFNγ and IL-17 production by MAIT cells was critical to their ability to provide protective immunity in immune-compromised mice infected with *L. longbeachae* and *M. tuberculosis*, respectively,[190,211] the mechanisms behind their protective effect in immune-sufficient hosts remains unclear. Indeed, MAIT cell production of TGFβ delayed the priming of conventional CD4+ T cells during *M. tuberculosis* infection,[211] while their role in driving differentiation of monocyte-derived dendritic cells and subsequent recruitment of CD4+ T cells appeared important for their protective effects against *F. tularensis*.[253] MAIT cells can play an indirect role in many immune reactions, such as driving dendritic cell maturation and production of IL-12 through a CD40L-dependent mechanism.[253,254] Thus, results from the study of MAIT cells in isolation in experimental models need to be further validated before extrapolating to more physiological contexts.

In human infection, the inability to conduct experimental manipulation of MAIT cells means that most studies have focused on enumerating and characterizing MAIT cells in peripheral blood. These studies have not generally proven a protective role one way or the other for MAIT cells in human infections.

Viral Infections

As outlined in section *Activation of MAIT Cells*, many viral infections can activate MAIT cells impacting on their numbers and functional capacity. A study by van Wilgenburg and colleagues demonstrated experimentally that MAIT cells are activated, as indicated by expression of CD25, CD69, and GzmB, and transiently accumulate in the lungs in a mouse model of influenza A virus infection (PR8 strain). Mice which are deficient in MAIT cells (MR1−/−) showed greater weight loss and more severe symptoms of infection than wild-type MR1-positive mice, and adoptive transfer of MAIT cells into recipient mice could ameliorate disease.[235] The MAIT cell response in this setting was dependent on IL-18 and was independent of MR1, as shown by the accumulation of adoptively transferred wild-type MAIT cells in *Mr1−/−* mice following infection. MAIT cell numbers have been correlated with lower viral loads and recovery from viral infections in humans[214,231] and are also activated during SIV infection in nonhuman primates (Ref. 28 and reviewed in Ref. 255), with one study showing an effect of MAIT cells in providing help for B cell Ab production following SIV infection in macaques.[208]

Fungal Infections

Due to the presence of the riboflavin biosynthesis pathway, including the precursor 5-A-RU, in fungi as well as bacteria (https://www.genome.jp/kegg/pathway.html), it is likely that many fungi are capable of generating MAIT antigens. However, research on the activation and response of MAIT cells during fungal infection has been limited. MAIT cell activation by *Candida albicans*, *C. glabrata*, *Saccharomyces cerevisiae*, and *Aspergillus fumigatus* has been demonstrated *in vitro*.[60,63,256] There was also a modest increase of MAIT cells producing IL-17 reported, following lung infection by *Blastomyces dermatitidis*.[257] MAIT cells are present at mucosal sites, including the lungs, reproductive tract, and in the skin, which are often sites of fungal infections, and thus, their response could provide protection against invasive disease. It is highly plausible that they play an important protective role against fungal infection, which is yet to be elucidated.

Pathological Role in Chronic Bacterial Infection

MAIT cells are potent producers of proinflammatory cytokines.[11,60,63] Thus, there is potential for immune pathology if their responses are uncontrolled during chronic infection. *Helicobacter pylori*, which causes gastritis and gastric ulcers, and can lead to gastric cancer in humans, represents a chronic infection where the bacteria are not cleared by the immune system without antibiotic intervention.[258] MAIT cells become activated in an MR1-dependent manner when exposed to *Helicobacter pylori*.[198,199] Indeed, both MAIT-1 and MAIT-17 cells accumulate in gastric mucosa infected with *H. pylori*, and their presence was associated with accelerated gastritis and increased recruitment of neutrophils, macrophages, dendritic cells, eosinophils, non-MAIT T cells, and marked gastric atrophy.[199] MAIT cells in this model were skewed toward a MAIT-1 (T-bet+, RORγt−) phenotype.

Autoimmune and Inflammatory Diseases

Many studies have shown a reduction in MAIT cell numbers in the circulation of patients with various autoimmune or inflammatory conditions including rheumatoid arthritis, type I diabetes, cardiovascular disease, autoimmune hepatitis, spondyloarthropathies, Sjogren syndrome, systemic lupus erythematosus, asthma, chronic obstructive pulmonary disease, celiac disease, and inflammatory bowel diseases (reviewed in Ref. 151). In multiple sclerosis (MS) both an increase and a decrease in circulating MAIT cells have been reported.[259-263] MAIT cells were also reported to be activated in psoriasis,[195] increased in number in skin during dermatitis herpetiformis,[193] and found in multiple sclerosis brain lesions.[259]

An increase in MAIT cell expression of exhaustion markers has been observed in several studies, and there are changes in their homing receptors and antiapoptotic molecules such as BCL-2.[132] These observations initially suggested the activation of MAIT cells in these conditions. However, their role in autoimmune responses is unclear. In the corresponding mouse models, MAIT cells have been reported as either protective (experimental autoimmune encephalomyelitis—model of MS,[264] nonobese diabetic [NOD] mouse model of type I diabetes,[132] the TNBS model of inflammatory bowel disease,[265] nonalcoholic steatohepatitis,[266] graft-versus-host disease)[267] or deleterious (systemic lupus erythematosus,[209] collagen-induced arthritis,[268] type 2 diabetes,[269] liver fibrosis[270]). Mouse models revealing MAIT cell effects in disease are reviewed in Ref. 192.

Although possible mechanisms of modulation of autoimmunity are indicated by the production of cytokines such as IL-17 and IL-22, and the expression of cytotoxic molecules such as GzmB, there is little definitive direct evidence to indicate mechanisms of activation or function in these conditions. The limitations with mouse models make the extrapolation to humans a challenge, where the results are not always consistent with observations in human disease cohorts. In type I diabetes, the NOD mouse model showed a clear reduction in disease in the presence of MAIT cells, compared to NOD. MR1−/− mice, which lack MAIT cells. The apparent protective effects are thought to arise from the role of MAIT cells in preserving gut mucosal integrity.[132] Intriguingly in the same study, a direct role of MAIT cells in cytotoxicity of pancreatic B cells was reported.

MAIT cells isolated from obese donors or from patients with type 2 diabetes secrete higher levels of IL-17 than MAIT cells from healthy donors and are increased in numbers in adipose tissue,[271] suggesting metabolic changes in tissues may have an effect. It has been speculated that MAIT cells might regulate the metabolic effects of obesity in coordination with the gut microbiota.[269]

Wound Repair

The recent description of a MAIT cell tissue repair gene signature, characterised by the expression of tissue repair and angiogenesis genes (*Tnf, Csf2, Hif1a, Furin, Vegfb, Ptges2, Pdgfb, Tgfb1, Mmp25, Hmgb1*)[168-170] by human and mouse MAIT cells and the demonstration of a MAIT cell role in wound healing[82] has added another dimension to the immune function of MAIT cells. Being located in the skin and mucosal sites, MAIT cells may play important, and as yet unexplored, function in tissue repair in various settings. The loss of gut mucosal integrity in mice lacking MAIT cells[132] also suggested that they are important for maintaining normal tissue homeostasis at barrier sites. Thus, their immune function may extend well beyond pathogen recognition and control.

MAIT Cells and Other MR1-Reactive T Cells in Cancer

Recently, several studies have investigated the potential role of MAIT cells in cancer immunology, although to date there is little direct evidence of their role. In human cancer patients, a reduction in circulating MAIT cells has been reported in mucosal, but not nonmucosal, tumor types.[217] MAIT cells also have been observed to accumulate in or around tumors, including in colorectal carcinoma and liver metastases from colorectal tum ors[171,201,243,272-274] as well as esophageal adenocarcinoma.[275]

As MR1 has been shown to present a range of ligands with diverse chemical scaffolds,[122,133,276] it is plausible that novel tumor antigens generated during tumorigenesis can also be presented by MR1. Whether these would be recognized by MAIT cells or by other MR1-reactive T cells[59,108,109] is currently unclear.

MAIT cells from patients with cancer are reported to have impaired effector function and increased expression of activation and exhaustion markers,[171,174,216,243-245] suggestive of antigen-independent or chronic activation. It is possible that cytokine-mediated stimulation may be driving this activation, similar to that during viral infection (see sections *MR1/TCR-Independent Activation and Viral Infections*). Alternatively, since many tumors contain microorganisms,[277,278] it is possible that antigen-specific MAIT cell activation may occur, but this has not been studied to date in the context of tumors.

MAIT cells express cytotoxic molecules, including perforin and Gzm's (see section *Cytotoxic Capacity*). Direct killing of tumor cells by human MAIT cells has been demonstrated for myeloma and esophageal adenocarcinoma cell lines *in vitro*, following antigen stimulation.[216,275] In a recent publication, an MR1-reactive T cell clone was observed to kill tumor cells but spare healthy primary cells, and this appeared to be MR1-dependent.[108] These cells could also kill Jurkat T cells when transferred into profoundly immunodeficient, NOD *scid* gamma mice. Thus, the therapeutic potential of harnessing MR1-reactive T cells in cancer is currently an area of great interest. However, MAIT cells have been associated with both positive and negative effects in the context of cancer. Recently, Haeryfar and colleagues studied human MAIT cell transcriptomic signatures, showing their association with lower survival in a range of tumors, but with a positive prognosis in esophageal carcinoma.[174]

Two studies have assessed MAIT cells in mouse models of metastatic melanoma. MR1−/− or B6-MAITcast.MR1−/− were more resistant to the metastatic tumor growth than their MR1-sufficient counterparts (B6 or B6-CAST mice, respectively),[175,279] suggesting a role for MAIT cells in driving tumor growth or in inhibiting antitumor immunity. Evaluation of MR1 antibody blocking, MAIT cell adoptive transfers, and the survival of MR1 deficient melanoma cells yielded results consistent with the direct recognition of tumor cells via an MR1-dependent mechanism and presumably an undefined antigen.[279] Both studies used 5-OP-RU to boost the MAIT cell effect but differed in their findings. When tumors were pulsed with 5-OP-RU, the Yan study showed a further increase in tumor metastases,[279] whereas the Petley study showed a decrease, and also showed a decrease in tumor burden when MAIT cells were primed with 5-OP-RU, *in vivo*. In both studies, the effects appeared to be partly indirect and mediated through modulating NK cells.[175] Thus, the role and therapeutic potential of MAIT and other MR1-reactive T cells in tumor immunology is an evolving field.

Future Directions: MAIT Cell Immune Roles and Potential as Vaccination or Therapeutic Targets

The role of MAIT cells in contributing to host protection against bacterial pathogens is now well established. Additionally, there is increasing interest in MAIT cells as players in both protective and aberrant immunity in viral infections, autoimmune diseases, and cancer, with their precise function in these settings less well defined. Due to their unique characteristics, recognition of conserved microbial small-molecule antigens, rapid response and restriction to monomorphic MR1 protein, MAIT cells are potentially important targets for both vaccination and immune therapy.

MAIT cell "vaccination" proof of principle studies show enhanced clearance of the bacterial pathogens *Legionella longbeachae* and *Francisella tularensis* in mouse models and long-term persistence of MAIT cells after vaccination or infection.[107,190,252] Although only one component of the immune response, with a relatively small contribution to direct bacterial clearance, MAIT cell boosting may result in substantial increase in immunity in immunocompromised individuals, and there is interest in boosting MAIT cells in the elderly where their numbers naturally decline. Notably, there is potent MAIT cell activation and accumulation, *in vivo*, in response to infection of mice with *S.* Typhimurium, but the presence or absence of MAIT cells had no effect on bacterial load.[23] Nonetheless, avoidance of MAIT cell recognition by a sub-Saharan strain of *S.* Typhimurium, ST313, is associated with nontyphoidal invasive *Salmonella* disease in Africa.[280] This observation suggests that MAIT cells may be an important component of T cell *Salmonellae* in immunosuppressed individuals in sub-Saharan Africa.[280]

MAIT cells are also being explored for their potential in boosting other immune responses—for example in recruiting other immune cells including T cells, monocytes and neutrophils,[199,202,253] and in activation of APCs and providing T cell help for Ab production.[207,254] Thus, targeting MAIT cells for their adjuvant effects may be beneficial.[281] Indeed, it was recently shown that MAIT cell activation by the adenoviral vector vaccine ChAdOx1 correlated with enhanced T cell responses in mice and humans.[282]

The therapeutic inhibition of MAIT cells has also been demonstrated in a mouse model, whereby the nonstimulatory MR1 ligands Ac-6-FP or 3-F-SA were able to competitively inhibit MAIT cell activation by 5-OP-RU in combination with riboflavin-deficient bacteria.[122] In this study, MAIT cell enrichment was prevented. *In vivo* delivery of Ac-6-FP also reduced MAIT cell activation and decreased insulin sensitivity in a high fat diet model of obesity and type 2 diabetes.[269] Further definition of the precise role of MAIT cells in various contexts, described above, will no doubt be followed by additional attempts to boost or inhibit their function.

FINAL REMARKS/CONCLUSION

There is striking conservation of MAIT TCRs and their receptor MR1 across most mammals, yet the evolutionary pressure that has maintained MAIT cells remains something of a conundrum. MAIT cells are functionally highly potent and sit at the interface between innate and adaptive immunity. A clear-cut MAIT cell phenotype does not seem to exist for MR1−/− or other MAIT cell–deficient models, either in the context of infection or homeostasis.

Instead, protective immunity by MAIT cells is only observed in response to a handful but not all riboflavin-producing pathogens. For the former, their role appears incremental, hard to demonstrate, and quantitatively swamped by the impact of conventional T and B cell immunity. Additionally, the anticipated model of universal protection against pathogens, which was hypothesized based on conserved recognition of a shared antigen common to diverse riboflavin synthesizing bacteria and fungi, does not appear to play out. Thus, pathogen and host factors in addition to the pathogen's capacity to produce riboflavin appear important in triggering protective MAIT cell immunity, most of which remain to be understood.

A compelling possibility would be a role for MAIT cells in the health of the reproductive tract where natural selection could act very directly and potently to preserve these immune genes across mammals. However, there is scant evidence of such a function. MAIT cell numbers in the reproductive tract look normal and model infections have yielded very little to support their special role in these organs.

A role for MAIT cells in regulating susceptibility to autoimmunity is plausible but the evidence for their role in the common autoimmune diseases is patchy and inconsistent with a highly selectable characteristic across all mammals through evolution. Similarly, the importance of MAIT cells in cancer is still being evaluated but evidence to date suggests this is not a major function that could support evolutionary selection of reproductive fitness where by and large, in mammals cancer is a postreproduction affliction. Some tantalizing evidence suggests that MAIT cells contribute to maintaining barrier integrity and can accelerate wound healing, but this is not particularly dramatic. MR1-deficient mice and patients with MAIT cell deficiency do not have an obvious barrier integrity or wound healing defect.

The messiness of evolution as a process can frustrate our understanding of the forces leading to its contemporary design. It is possible that the MAIT-MR1 axis is an adjunct to the more flexible adaptive arms of T and B cell immunity, perhaps added around the time of the mammalian expansion and maintained because of its incremental contribution to protecting against a range of pathogens. It is also possible that this axis of immunity has been shaped by a single pathogen or limited subset of pathogens, not yet studied sufficiently or no longer important in the era of vaccines and antibiotics.

We cannot rule out any of these possibilities. However, to us, the most plausible explanation is that MAIT cell immunity provides an incremental contribution to protection against a range of bacterial and fungal pathogens, which is masked by a strong and specific adaptive immune response. Thus, the importance of MAIT cell immunity lies in its influence when the immune system is on the brink of losing control of an infection and where every little bit of protection matters. This is not widely understood in the antibiotic era but may include settings in immune compromised hosts, very young children and sometimes where secondary infection arises from a primary infectious challenge like measles. In these situations, a half to one log control in bacterial counts can make the difference, either in controlling the pathogen at the initial stage of infection when pathogen load is first set, or toward the end of the infection where resolution requires bacterial control sufficient to prevent septicaemia or to allow other arms of immunity to manage the fall-out.

Thus, the field of MAIT cell biology has considerable distance to go before we have a complete picture of their biology and the forces that have maintained them throughout mammalian evolution. Nonetheless, their high conservation, including conserved antigen recognition, and rapid, potent, and polyfunctional response, indicate there may be an untapped potential for harnessing MAIT cells for vaccination and therapeutic applications against pathogens, cancer, and aberrant immunity (Fig. 37.19). In particular, a deeper understanding of the heterogeneity within the MAIT cell population and their cross-talk with other cell types will ultimately enable this, regardless of having necessarily

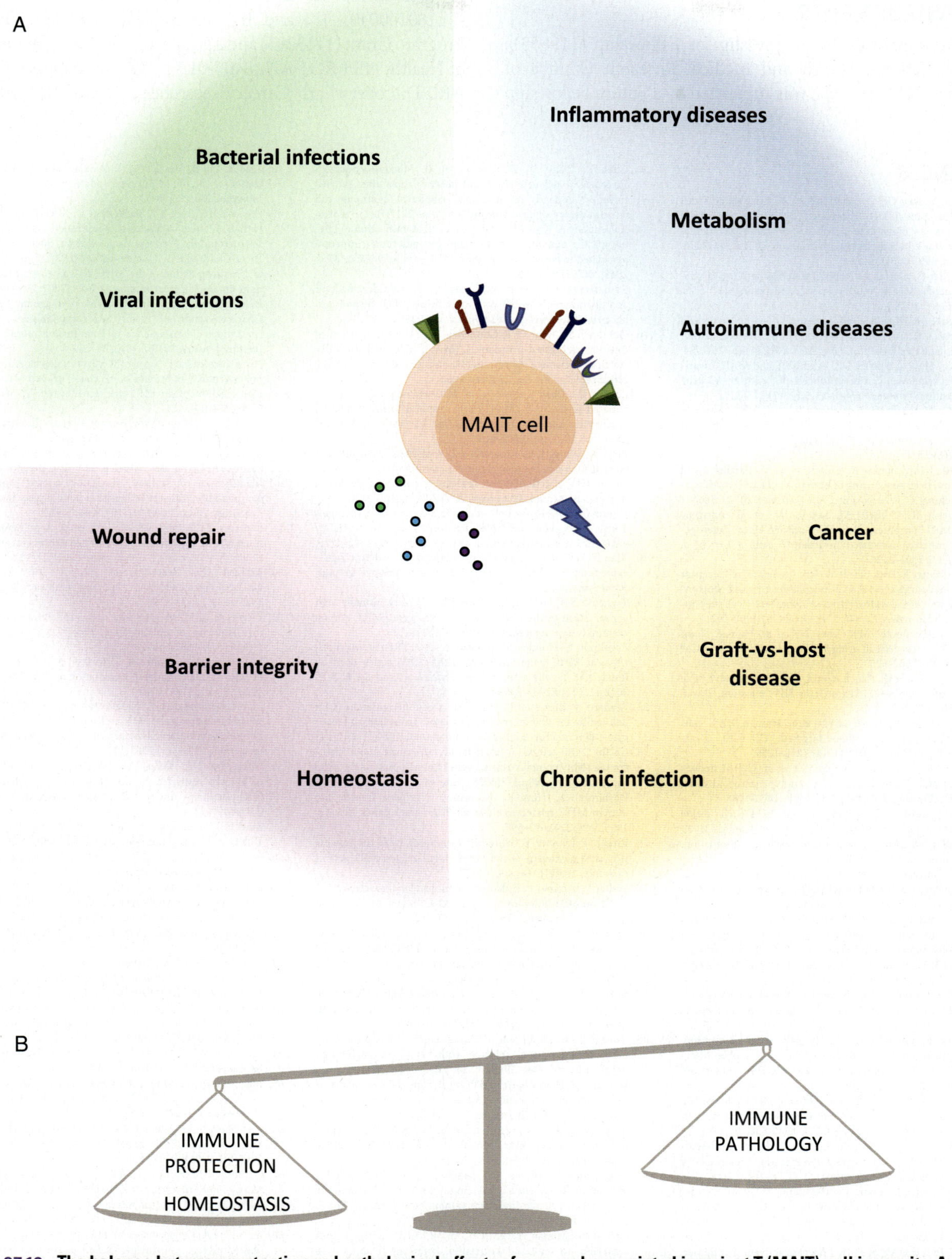

FIG. 37.19. The balance between protective and pathological effects of mucosal-associated invariant T (MAIT) cell immunity. A: MAIT cell contributions have been studied in many areas including infection, autoimmunity, homeostasis, and cancer. **B:** Factors that drive MAIT cell immune roles in tipping the balance between protection, wound healing, and pathological effects are emerging.

understood the underlying evolutionary pressures driving MAIT cell biology.

ACKNOWLEDGMENTS

SBGE was supported by an Investigator Fellowship (1196881) from the National Health and Medical Research Council of Australia (NHMRC). AJC was supported by a Future Fellowship (FT160100083) from the Australian Research Council (ARC), an Investigator Fellowship (1193745) from the NHMRC, and a Dame Kate Campbell Research Fellowship from the University of Melbourne. JR is an Australian ARC Laureate Fellow (FL160100049). JM and JR were supported by an NHMRC Program Grant (1113293) and a grant from the National Institutes of Health, NIH RO1 AI148407-01A1. WA was supported by an ARC Discovery Early Career Researcher Award (DE220101491).

REFERENCES

1. Porcelli S, Yockey CE, Brenner MB, et al. Analysis of T cell antigen receptor (TCR) expression by human peripheral blood CD4−8− alpha/beta T cells demonstrates preferential use of several V beta genes and an invariant TCR alpha chain. *J Exp Med.* 1993;178(1):1-16.
2. Patel O, Kjer-Nielsen L, Le Nours J, et al. Recognition of vitamin B metabolites by mucosal-associated invariant T cells. *Nat Commun.* 2013;4:2142.
3. Reantragoon R, Kjer-Nielsen L, Patel O, et al. Structural insight into MR1-mediated recognition of the mucosal associated invariant T cell receptor. *J Exp Med.* 2012;209(4):761-774.
4. Greenaway HY, Ng B, Price DA, et al. NKT and MAIT invariant TCRα sequences can be produced efficiently by VJ gene recombination. *Immunobiology.* 2013;218(2):213-224.
5. Gold MC, McLaren JE, Reistetter JA, et al. MR1-restricted MAIT cells display ligand discrimination and pathogen selectivity through distinct T cell receptor usage. *J Exp Med.* 2014;211(8):1601-1610.
6. Lepore M, Kalinichenko A, Colone A, et al. Parallel T-cell cloning and deep sequencing of human MAIT cells reveal stable oligoclonal TCRβ repertoire. *Nat Commun.* 2014;5:3866.
7. Reantragoon R, Corbett AJ, Sakala IG, et al. Antigen-loaded MR1 tetramers define T cell receptor heterogeneity in mucosal-associated invariant T cells. *J Exp Med.* 2013;210(11):2305-2320.
8. Tilloy F, Treiner E, Park SH, et al. An invariant T cell receptor alpha chain defines a novel TAP-independent major histocompatibility complex class Ib-restricted alpha/beta T cell subpopulation in mammals. *J Exp Med.* 1999;189(12):1907-1921.
9. Gherardin NA, Souter MN, Koay HF, et al. Human blood MAIT cell subsets defined using MR1 tetramers. *Immunol Cell Biol.* 2018;96(5):507-525.
10. Cosgrove C, Ussher JE, Rauch A, et al. Early and nonreversible decrease of CD161++/MAIT cells in HIV infection. *Blood.* 2013;121(6):951-961.
11. Dusseaux M, Martin E, Serriari N, et al. Human MAIT cells are xenobiotic-resistant, tissue-targeted, CD161hi IL-17-secreting T cells. *Blood.* 2011;117(4):1250-1259.
12. Fergusson JR, Smith KE, Fleming VM, et al. CD161 defines a transcriptional and functional phenotype across distinct human T cell lineages. *Cell Rep.* 2014;9(3):1075-1088.
13. Gibbs A, Leeansyah E, Introini A, et al. MAIT cells reside in the female genital mucosa and are biased towards IL-17 and IL-22 production in response to bacterial stimulation. *Mucosal Immunol.* 2017;10(1):35-45.
14. Hama I, Tominaga K, Yamagiwa S, et al. Different distribution of mucosal-associated invariant T cells within the human cecum and colon. *Cent Eur J Immunol.* 2019;44(1):75-83.
15. Hinks TS, Wallington JC, Williams AP, et al. Steroid-induced deficiency of mucosal-associated invariant T cells in the chronic obstructive pulmonary disease lung. Implications for nontypeable Haemophilus influenzae infection. *Am J Respir Crit Care Med.* 2016;194(10):1208-1218.
16. Jeffery HC, van Wilgenburg B, Kurioka A, et al. Biliary epithelium and liver B cells exposed to bacteria activate intrahepatic MAIT cells through MR1. *J Hepatol.* 2016;64(5):1118-1127.
17. Leeansyah E, Ganesh A, Quigley MF, et al. Activation, exhaustion, and persistent decline of the antimicrobial MR1-restricted MAIT-cell population in chronic HIV-1 infection. *Blood.* 2013;121(7):1124-1135.
18. Rha MS, Han JW, Kim JH, et al. Human liver CD8+ MAIT cells exert TCR/MR1-independent innate-like cytotoxicity in response to IL-15. *J Hepatol.* 2020;73(3):640-650.
19. Serriari NE, Eoche M, Lamotte L, et al. Innate mucosal-associated invariant T (MAIT) cells are activated in inflammatory bowel diseases. *Clin Exp Immunol.* 2014;176(2):266-274.
20. Sobkowiak MJ, Davanian H, Heymann R, et al. Tissue-resident MAIT cell populations in human oral mucosa exhibit an activated profile and produce IL-17. *Eur J Immunol.* 2019;49(1):133-143.
21. Tang XZ, Jo J, Tan AT, et al. IL-7 licenses activation of human liver intrasinusoidal mucosal-associated invariant T cells. *J Immunol.* 2013;190(7):3142-3152.
22. Treiner E, Duban L, Bahram S, et al. Selection of evolutionarily conserved mucosal-associated invariant T cells by MR1. *Nature.* 2003;422(6928):164-169.
23. Chen Z, Wang H, D'Souza C, et al. Mucosal-associated Invariant T cell activation and accumulation after *in vivo* infection depends on microbial riboflavin synthesis and co-stimulatory signals. *Mucosal Immunol.* 2017;10(1):58-68.
24. Rahimpour A, Koay HF, Enders A, et al. Identification of phenotypically and functionally heterogeneous mouse mucosal associated invariant T cells using MR1 tetramers. *J Exp Med.* 2015;20(7):1095-1108.
25. Edmans MD, Connelley TK, Jayaraman S, et al. Identification and phenotype of MAIT cells in cattle and their response to bacterial infections. *Front Immunol.* 2021;12:627173.
26. Goldfinch N, Reinink P, Connelley T, et al. Conservation of mucosal associated invariant T (MAIT) cells and the MR1 restriction element in ruminants, and abundance of MAIT cells in spleen. *Vet Res.* 2010;41(5):62.
27. Greene JM, Dash P, Roy S, et al. MR1-restricted mucosal-associated invariant T (MAIT) cells respond to mycobacterial vaccination and infection in nonhuman primates. *Mucosal Immunol.* 2017;10(3):802-813.
28. Juno JA, Wragg KM, Amarasena T, et al. MAIT cells upregulate α4β7 in response to acute simian immunodeficiency virus/simian HIV infection but are resistant to peripheral depletion in pigtail macaques. *J Immunol.* 2019;202(7):2105-2120.
29. Vinton C, Wu F, Rossjohn J, et al. Mucosa-associated invariant T cells are systemically depleted in simian immunodeficiency virus-infected rhesus macaques. *J Virol.* 2016;90(9):4520-4529.
30. Xiao X, Li K, Ma X, et al. Mucosal-associated invariant T cells expressing the TRAV1-TRAJ33 chain are present in pigs. *Front Immunol.* 2019;10:2070.
31. Tsukamoto K, Deakin JE, Graves JA, et al. Exceptionally high conservation of the MHC class I-related gene, MR1, among mammals. *Immunogenetics.* 2013;65(2):115-124.
32. Boudinot P, Mondot S, Jouneau L, et al. Restricting non-classical MHC genes coevolve with TRAV genes used by innate-like T cells in mammals. *Proc Natl Acad Sci U S A.* 2016;113(21):E2983-E2992.
33. Bernard A, Boumsell L. The clusters of differentiation (CD) defined by the first International workshop on human leucocyte differentiation antigens. *Hum Immunol.* 1984;11(1):1-10.
34. Calabi F, Milstein C. A novel family of human major histocompatibility complex-related genes not mapping to chromosome 6. *Nature.* 1986;323(6088):540-543.
35. Hashimoto K, Hirai M, Kurosawa Y. A gene outside the human MHC related to classical HLA class I genes. *Science.* 1995;269(5224):693-695.
36. Riegert P, Wanner V, Bahram S. Genomics, isoforms, expression, and phylogeny of the MHC class I-related MR1 gene. *J Immunol.* 1998;161(8):4066-4077.
37. Walter L, Günther E. Isolation and molecular characterization of the rat MR1 homologue, a non-MHC-linked class I-related gene. *Immunogenetics.* 1998;47(6):477-482.
38. Yamaguchi H, Kurosawa Y, Hashimoto K. Expanded genomic organization of conserved mammalian MHC class I-related genes, human MR1 and its murine ortholog. *Biochem Biophys Res Commun.* 1998;250(3):558-564.
39. Miley MJ, Truscott SM, Yu YY, et al. Biochemical features of the MHC-related protein 1 consistent with an immunological function. *J Immunol.* 2003;170(12):6090-6098.
40. Yamaguchi H, Hashimoto K. Association of MR1 protein, an MHC class I-related molecule, with beta(2)-microglobulin. *Biochem Biophys Res Commun.* 2002;290(2):722-729.
41. Hansen TH, Huang S, Arnold PL, et al. Patterns of nonclassical MHC antigen presentation. *Nat Immunol.* 2007;8(6):563-568.
42. Okamoto N, Kanie O, Huang YY, et al. Synthetic alpha-mannosyl ceramide as a potent stimulant for an NKT cell repertoire bearing the invariant Valpha19-Jalpha26 TCR alpha chain. *Chem Biol.* 2005;12(6):677-683.
43. Shimamura M, Huang YY, Okamoto N, et al. Glycolipids with nonreducing end alpha-mannosyl residues that have the potential to activate invariant Valpha19 NKT cells. *FEBS J.* 2007;274(11):2921-2932.
44. Huang S, Martin E, Kim S, et al. MR1 antigen presentation to mucosal-associated invariant T cells was highly conserved in evolution. *Proc Natl Acad Sci U S A.* 2009;106(20):8290-8295.
45. Yamaguchi H, Hirai M, Kurosawa Y, et al. A highly conserved major histocompatibility complex class I-related gene in mammals. *Biochem Biophys Res Commun.* 1997;238(3):697-702.
46. Xiao X, Liu B, Ma X, et al. Molecular cloning and characterization of the pig MHC class -related MR1 gene. *Dev Comp Immunol.* 2019;96:58-67.
47. Leeansyah E, Hey YY, Sia WR, et al. MR1-Restricted T cells with MAIT-like characteristics are functionally conserved in the pteropid bat *Pteropus alecto. iScience.* 2020;23(12):101876.
48. Parra-Cuadrado JF, Gómez del Moral M, García-Pavía P, et al. Characterization of the MHC class I-related MR1 locus in nonhuman primates. *Immunogenetics.* 2001;53(8):643-648.
49. Ellis-Connell AL, Kannal NM, Balgeman AJ, et al. Characterization of major histocompatibility complex-related molecule 1 sequence variants in non-human primates. *Immunogenetics.* 2019;71(2):109-121.
50. Parra-Cuadrado JF, Navarro P, Mirones I, et al. A study on the polymorphism of human MHC class I-related MR1 gene and identification of an MR1-like pseudogene. *Tissue Antigens.* 2000;56(2):170-172.
51. Howson LJ, Awad W, von Borstel A, et al. Absence of mucosal-associated invariant T cells in a person with a homozygous point mutation in MR1. *Sci Immunol.* 2020;5(49):eabc9492.
52. Stumpf AN, van der Meijden ED, van Bergen CA, et al. Identification of 4 new HLA-DR-restricted minor histocompatibility antigens as hematopoietic targets in antitumor immunity. *Blood.* 2009;114(17):3684-3692.
53. Seshadri C, Thuong NT, Mai NT, et al. A polymorphism in human MR1 is associated with mRNA expression and susceptibility to tuberculosis. *Gene Immun.* 2017;18(1):8-14.
54. Krangel MS. Secretion of HLA-A and -B antigens via an alternative RNA splicing pathway. *J Exp Med.* 1986;163(5):1173-1190.
55. Woolfson A, Milstein C. Alternative splicing generates secretory isoforms of human CD1. *Proc Natl Acad Sci U S A.* 1994;91(14):6683-6687.
56. Narayanan GA, Nellore A, Tran J, et al. Alternative splicing of MR1 regulates antigen presentation to MAIT cells. *Sci Rep.* 2020;10(1):15429.
57. Lion J, Debuysscher V, Wlodarczyk A, et al. MR1B, a natural spliced isoform of the MHC-related 1 protein, is expressed as homodimers at the cell surface and activates MAIT cells. *Eur J Immunol.* 2013;43(5):1363-1373.
58. Gherardin NA, Keller AN, Woolley RE, et al. Diversity of T Cells restricted by the MHC class I-related molecule MR1 facilitates differential antigen recognition. *Immunity.* 2016;44(1):32-45.
59. Koay HF, Gherardin NA, Xu C, et al. Diverse MR1-restricted T cells in mice and humans. *Nat Commun.* 2019;10(1):2243.
60. Le Bourhis L, Martin E, Peguillet I, et al. Antimicrobial activity of mucosal-associated invariant T cells. *Nat Immunol.* 2010;11(8):701-708.
61. Lopez-Sagaseta J, Dulberger CL, Crooks JE, et al. The molecular basis for Mucosal-Associated Invariant T cell recognition of MR1 proteins. *Proc Natl Acad Sci U S A.* 2013;110(19):E1771-E1778.
62. López-Sagaseta J, Dulberger CL, McFedries A, et al. MAIT recognition of a stimulatory bacterial antigen bound to MR1. *J Immunol.* 2013;191(10):5268-5277.
63. Gold MC, Cerri S, Smyk-Pearson S, et al. Human mucosal associated invariant T cells detect bacterially infected cells. *PLoS Biol.* 2010;8(6):e1000407.
64. Ussher JE, Willberg CB, Klenerman P. MAIT cells and viruses. *Immunol Cell Biol.* 2018;96(6):630-641.
65. Kjer-Nielsen L, Corbett AJ, Chen Z, et al. An overview on the identification of MAIT cell antigens. *Immunol Cell Biol.* 2018;96(6):573-587.
66. Kjer-Nielsen L, Patel O, Corbett AJ, et al. MR1 presents microbial vitamin B metabolites to MAIT cells. *Nature.* 2012;491(7426):717-723.
67. Garboczi DN, Hung DT, Wiley DC. HLA-A2-peptide complexes: refolding and crystallization of molecules expressed in Escherichia coli and complexed with single antigenic peptides. *Proc Natl Acad Sci U S A.* 1992;89(8):3429-3433.
68. Juzeniene A, Grigalavicius M, Ma LW, et al. Folic acid and its photoproducts, 6-formylpterin and pterin-6-carboxylic acid, as generators of reactive oxygen species in skin cells during UVA exposure. *J Photochem Photobiol B.* 2016;155:116-121.
69. Corbett AJ, Eckle SB, Birkinshaw RW, et al. T-cell activation by transitory neo-antigens derived from distinct microbial pathways. *Nature.* 2014;509(7500):361-365.

70. Bacher A, Eberhardt S, Fischer M, et al. Biosynthesis of vitamin b2 (riboflavin). *Annu Rev Nutr*. 2000;20:153-167.
71. Sa N, Rawat R, Thornburg C, et al. Identification and characterization of the missing phosphatase on the riboflavin biosynthesis pathway in *Arabidopsis thaliana*. *Plant J*. 2016;88(5):705-716.
72. Mironov VN, Kraev AS, Chikindas ML, et al. Functional organization of the riboflavin biosynthesis operon from *Bacillus subtilis* SHgw. *Mol Gen Genet*. 1994;242(2):201-208.
73. Herz S, Eberhardt S, Bacher A. Biosynthesis of riboflavin in plants. The ribA gene of *Arabidopsis thaliana* specifies a bifunctional GTP cyclohydrolase II/3,4-dihydroxy-2-butanone 4-phosphate synthase. *Phytochemistry*. 2000;53(7):723-731.
74. Fischer M, Bacher A. Biosynthesis of vitamin B2 in plants. *Physiol Plantarum*. 2006;126:304-318.
75. Brutinel ED, Dean AM, Gralnick JA. Description of a riboflavin biosynthetic gene variant prevalent in the phylum Proteobacteria. *J Bacteriol*. 2013;195(24):5479-5486.
76. Vitreschak AG, Rodionov DA, Mironov AA, et al. Regulation of riboflavin biosynthesis and transport genes in bacteria by transcriptional and translational attenuation. *Nucleic Acids Res*. 2002;30(14):3141-3151.
77. Hartmann N, McMurtrey C, Sorensen ML, et al. Riboflavin metabolism variation among clinical isolates of *Streptococcus pneumoniae* results in differential activation of MAIT cells. *Am J Respir Cell Mol Biol*. 2018;58(6): 767–776.
78. Kurioka A, van Wilgenburg B, Javan RR, et al. Diverse *Streptococcus pneumoniae* strains drive a mucosal-associated invariant T-cell response through major histocompatibility complex class I-related molecule-dependent and cytokine-driven pathways. *J Infect Dis*. 2018;217(6):988-999.
79. Garcia-Angulo VA. Overlapping riboflavin supply pathways in bacteria. *Crit Rev Microbiol*. 2017;43(2):196-209.
80. Gutierrez-Preciado A, Torres AG, Merino E, et al. Extensive identification of bacterial riboflavin transporters and their distribution across bacterial species. *PLoS One*. 2015;10(5):e0126124.
81. Soudais C, Samassa F, Sarkis M, et al. *In vitro* and *in vivo* analysis of the gram-negative bacteria-derived riboflavin precursor derivatives activating mouse MAIT cells. *J Immunol*. 2015;194(10):4641-4649.
82. Constantinides MG, Link VM, Tamoutounour S, et al. MAIT cells are imprinted by the microbiota in early life and promote tissue repair. *Science*. 2019;366(6464):eaax6624.
83. Burgess C, O'Connell-Motherway M, Sybesma W, et al. Riboflavin production in Lactococcus lactis: potential for in situ production of vitamin-enriched foods. *Appl Environ Microbiol*. 2004;70(10):5769-5777.
84. Cushman M, Yang D, Gerhardt S, et al. Design, synthesis, and evaluation of 6-carboxyalkyl and 6-phosphonoxyalkyl derivatives of 7-oxo-8-ribitylaminolumazines as inhibitors of riboflavin synthase and lumazine synthase. *J Org Chem*. 2002;67(16):5807-5816.
85. Wang Y, Ho CT. Flavour chemistry of methylglyoxal and glyoxal. *Chem Soc Rev*. 2012;41(11):4140-4149.
86. Eckle SB, Corbett AJ, Keller AN, et al. Recognition of vitamin B precursors and byproducts by mucosal associated invariant T cells. *J Biol Chem*. 2015;290(51):30204-30211.
87. Mak JY, Xu W, Reid RC, et al. Stabilizing short-lived Schiff base derivatives of 5-aminouracils that activate mucosal-associated invariant T cells. *Nat Commun*. 2017;8:14599.
88. Eckle SB, Birkinshaw RW, Kostenko L, et al. A molecular basis underpinning the T cell receptor heterogeneity of mucosal-associated invariant T cells. *J Exp Med*. 2014;211(8):1585-1600.
89. Chen Z, Wang H, D'Souza C, et al. Characterization and purification of mouse mucosal-associated invariant T (MAIT) cells. *Curr Protoc Immunol*. 2019;127(1):e89.
90. Sakala IG, Kjer-Nielsen L, Eickhoff CS, et al. Functional heterogeneity and antimycobacterial effects of mouse mucosal-associated invariant T cells specific for riboflavin metabolites. *J Immunol*. 2015;195(2):587-601.
91. Souter MNT, Loh L, Li S, et al. Characterization of human mucosal-associated invariant T (MAIT) cells. *Curr Protoc Immunol*. 2019;127(1):e90.
92. Chua WJ, Kim S, Myers N, et al. Endogenous MHC-related protein 1 is transiently expressed on the plasma membrane in a conformation that activates mucosal-associated invariant T cells. *J Immunol*. 2011;186(8):4744-4750.
93. Aldemir H. Novel MHC class I-related molecule MR1 affects MHC class I expression in 293T cells. *Biochem Biophys Res Commun*. 2008;366(2):328-334.
94. Huang S, Gilfillan S, Cella M, et al. Evidence for MR1 antigen presentation to mucosal-associated invariant T cells. *J Biol Chem*. 2005;280(22):21183-21193.
95. Ljunggren HG, Stam NJ, Ohlen C, et al. Empty MHC class I molecules come out in the cold. *Nature*. 1990;346(6283):476-480.
96. Gozalbo-Lopez B, Gomez del Moral M, Campos-Martin Y, et al. The MHC-related protein 1 (MR1) is expressed by a subpopulation of CD38+, IgA+ cells in the human intestinal mucosa. *Histol Histopathol*. 2009;24(11):1439-1449.
97. Braganza CD, Motozono C, Sonoda KH, et al. Agonistic or antagonistic mucosal-associated invariant T (MAIT) cell activity is determined by the 6-alkylamino substituent on uracil MR1 ligands. *Chem Commun (Camb)*. 2020;56(39):5291-5294.
98. Abos B, Gomez Del Moral M, Gozalbo-Lopez B, et al. Human MR1 expression on the cell surface is acid sensitive, proteasome independent and increases after culturing at 26°C. *Biochem Biophys Res Commun*. 2011;411(3):632-636.
99. Harriff MJ, Cansler ME, Toren KG, et al. Human lung epithelial cells contain *Mycobacterium tuberculosis* in a late endosomal vacuole and are efficiently recognized by CD8(+) T cells. *PLoS One*. 2014;9(5):e97515.
100. Georgel P, Radosavljevic M, Macquin C, et al. The non-conventional MHC class I MR1 molecule controls infection by *Klebsiella pneumoniae* in mice. *Mol Immunol*. 2011;48(5):769-775.
101. Harriff MJ, Karamooz E, Burr A, et al. Endosomal MR1 trafficking plays a key role in presentation of *Mycobacterium tuberculosis* ligands to MAIT cells. *PLoS Pathog*. 2016;12(3):e1005524.
102. Laugel B, Lloyd A, Meermeier EW, et al. Engineering of Isogenic cells deficient for MR1 with a CRISPR/Cas9 lentiviral system: tools to study microbial antigen processing and presentation to human MR1-restricted T cells. *J Immunol*. 2016;197(3):971-982.
103. McWilliam HE, Eckle SB, Theodossis A, et al. The intracellular pathway for the presentation of vitamin B-related antigens by the antigen-presenting molecule MR1. *Nat Immunol*. 2016;17(5):531-537.
104. McWilliam HEG, Mak JYW, Awad W, et al. Endoplasmic reticulum chaperones stabilize ligand-receptive MR1 molecules for efficient presentation of metabolite antigens. *Proc Natl Acad Sci U S A*. 2020;117(40):24974-24985.
105. Karamooz E, Harriff MJ, Narayanan GA, et al. MR1 recycling and blockade of endosomal trafficking reveal distinguishable antigen presentation pathways between *Mycobacterium tuberculosis* infection and exogenously delivered antigens. *Sci Rep*. 2019;9(1):4797.
106. Lamichhane R, Ussher JE. Expression and trafficking of MR1. *Immunology*. 2017;151(3):270-279.
107. Wang H, Kjer-Nielsen L, Shi M, et al. IL-23 costimulates antigen-specific MAIT cell activation and enables vaccination against bacterial infection. *Sci Immunol*. 2019;4(41):eaaw0402.
108. Crowther MD, Dolton G, Legut M, et al. Genome-wide CRISPR-Cas9 screening reveals ubiquitous T cell cancer targeting via the monomorphic MHC class I-related protein MR1. *Nat Immunol*. 2020;21(2):178-185.
109. Lepore M, Kalinichenko A, Calogero S, et al. Functionally diverse human T cells recognize non-microbial antigens presented by MR1. *Elife*. 2017;6:e24476.
110. Huang S, Gilfillan S, Kim S, et al. MR1 uses an endocytic pathway to activate mucosal-associated invariant T cells. *J Exp Med*. 2008;205(5):1201-1211.
111. Ussher JE, van Wilgenburg B, Hannaway RF, et al. TLR signaling in human antigen-presenting cells regulates MR1-dependent activation of MAIT cells. *Eur J Immunol*. 2016;46(7):1600-1614.
112. McWilliam HEG, Villadangos JA. How MR1 presents a pathogen metabolic signature to mucosal-associated invariant T (MAIT) cells. *Trends Immunol*. 2017;38(9):679-689.
113. Smith JD, Lie WR, Gorka J, et al. Disparate interaction of peptide ligand with nascent versus mature class I major histocompatibility complex molecules: comparisons of peptide binding to alternative forms of Ld in cell lysates and the cell surface. *J Exp Med*. 1992;175(1):191-202.
114. Barral DC, Brenner MB. CD1 antigen presentation: how it works. *Nat Rev Immunol*. 2007;7(12):929-941.
115. Heemels MT, Ploegh H. Generation, translocation, and presentation of MHC class I-restricted peptides. *Annu Rev Biochem*. 1995;64:463-491.
116. Pamer E, Cresswell P. Mechanisms of MHC class I – restricted antigen processing. *Annu Rev Immunol*. 1998;16:323-358.
117. Villadangos JA, Cardoso M, Steptoe RJ, et al. MHC class II expression is regulated in dendritic cells independently of invariant chain degradation. *Immunity*. 2001;14(6):739-749.
118. Jiang J, Natarajan K, Boyd LF, et al. Crystal structure of a TAPBPR-MHC I complex reveals the mechanism of peptide editing in antigen presentation. *Science*. 2017;358(6366):1064-1068.
119. Thomas C, Tampé R. Structure of the TAPBPR-MHC I complex defines the mechanism of peptide loading and editing. *Science*. 2017;358(6366):1060-1064.
120. Le Bourhis L, Dusseaux M, Bohineust A, et al. MAIT cells detect and efficiently lyse bacterially-infected epithelial cells. *PLoS Pathog*. 2013;9(10):e1003681.
121. Kulicke C, Karamooz E, Lewinsohn D, et al. Covering all the bases: complementary MR1 antigen presentation pathways sample diverse antigens and intracellular compartments. *Front Immunol*. 2020;11:2034.
122. Keller AN, Eckle SB, Xu W, et al. Drugs and drug-like molecules can modulate the function of mucosal-associated invariant T cells. *Nat Immunol*. 2017;18(4):402-411.
123. Salio M, Awad W, Veerapen N, et al. Ligand-dependent downregulation of MR1 cell surface expression. *Proc Natl Acad Sci U S A*. 2020;117(19):10465-10475.
124. Elliott T How do peptides associate with MHC class I molecules? *Immunol Today*. 1991;12(11):386-388.
125. Townsend A, Elliott T, Cerundolo V, et al. Assembly of MHC class I molecules analyzed *in vitro*. *Cell*. 1990;62(2):285-295.
126. Schumacher TN, Heemels MT, Neefjes JJ, et al. Direct binding of peptide to empty MHC class I molecules on intact cells and *in vitro*. *Cell*. 1990;62(3):563-567.
127. Neefjes J, Jongsma ML, Paul P, et al. Towards a systems understanding of MHC class I and MHC class II antigen presentation. *Nat Rev Immunol*. 2011;11(12):823-836.
128. Blander JM. The comings and goings of MHC class I molecules herald a new dawn in cross-presentation. *Immunol Rev*. 2016;272(1):65-79.
129. McWilliam HE, Villadangos JA. MR1 antigen presentation to MAIT cells: new ligands, diverse pathways? *Curr Opin Immunol*. 2018;52:108-113.
130. Lange J, Anderson RJ, Marshall AJ, et al. The chemical synthesis, stability, and activity of MAIT cell prodrug agonists that access MR1 in recycling endosomes. *ACS Chem Biol*. 2020;15(2):437-445.
131. Liu J, Brutkiewicz RR. The Toll-like receptor 9 signalling pathway regulates MR1-mediated bacterial antigen presentation in B cells. *Immunology*. 2017;152(2):232-242.
132. Rouxel O, Da Silva J, Beaudoin L, et al. Cytotoxic and regulatory roles of mucosal-associated invariant T cells in type 1 diabetes. *Nat Immunol*. 2017;18(12):1321-1331.
133. Awad W, Le Nours J, Kjer-Nielsen L, et al. Mucosal-associated invariant T cell receptor recognition of small molecules presented by MR1. *Immunol Cell Biol*. 2018;96(6):588-597.
134. Awad W, Ler GJM, Xu W, et al. The molecular basis underpinning the potency and specificity of MAIT cell antigens. *Nat Immunol*. 2020;21(4):400-411.
135. Harriff MJ, McMurtrey C, Froyd CA, et al. MR1 displays the microbial metabolome driving selective MR1-restricted T cell receptor usage. *Sci Immunol*. 2018;3(25):eaao2556.
136. Meermeier EW, Laugel BF, Sewell AK, et al. Human TRAV1-2-negative MR1-restricted T cells detect *S. pyogenes* and alternatives to MAIT riboflavin-based antigens. *Nat Commun*. 2016;7:12506.
137. de Lima Moreira M, Souter MNT, Chen Z, et al. Hypersensitivities following allergen antigen recognition by unconventional T cells. *Allergy*. 2020;75(10):2477-2490.
138. Freeman ML, Morris SR, Lederman MM. CD161 expression on mucosa-associated invariant T cells is reduced in HIV-infected subjects undergoing antiretroviral therapy who do not recover CD4(+) T cells. *Pathog Immun*. 2017;2(3):335-351.
139. Koppejan H, Jansen D, Hameetman M, et al. Altered composition and phenotype of mucosal-associated invariant T cells in early untreated rheumatoid arthritis. *Arthritis Res Ther*. 2019;21(1):3.
140. Young MH, U'Ren L, Huang S, et al. MAIT cell recognition of MR1 on bacterially infected and uninfected cells. *PLoS One*. 2013;8(1):e53789.
141. Souter MNT, Eckle SBG. Biased MAIT TCR usage poised for limited antigen diversity? *Front Immunol*. 2020;11:1845.
142. Le Nours J, Gherardin NA, Ramarathinam SH, et al. A class of γδ T cell receptors recognize the underside of the antigen-presenting molecule MR1. *Science*. 2019;366(6472):1522-1527.
143. Awad W, Meermeier EW, Sandoval-Romero ML, et al. Atypical TRAV1-2(−) T cell receptor recognition of the antigen-presenting molecule MR1. *J Biol Chem*. 2020;295(42):14445-14457.
144. Mallevaey T, Clarke AJ, Scott-Browne JP, et al. A molecular basis for NKT cell recognition of CD1d-self-antigen. *Immunity*. 2011;34(3):315-326.
145. Matulis G, Sanderson JP, Lissin NM, et al. Innate-like control of human iNKT cell autoreactivity via the hypervariable CDR3beta loop. *PLoS Biol*. 2010;8(6):e1000402.
146. Martin E, Treiner E, Duban L, et al. Stepwise development of MAIT cells in mouse and human. *PLoS Biol*. 2009;7(3):e54.
147. Wei DG, Lee H, Park SH, et al. Expansion and long-range differentiation of the NKT cell lineage in mice expressing CD1d exclusively on cortical thymocytes. *J Exp Med*. 2005;202(2):239-248.
148. Koay HF, Gherardin NA, Enders A, et al. A three-stage intrathymic development pathway for the mucosal-associated invariant T cell lineage. *Nat Immunol*. 2016;17(11):1300-1311.
149. Legoux F, Bellet D, Daviaud C, et al. Microbial metabolites control the thymic development of mucosal-associated invariant T cells. *Science*. 2019;366(6464):494-499.
150. Seach N, Guerri L, Le Bourhis L, et al. Double-positive thymocytes select mucosal-associated invariant T cells. *J Immunol*. 2013;191(12):6002-6009.
151. Godfrey DI, Koay HF, McCluskey J, et al. The biology and functional importance of MAIT cells. *Nat Immunol*. 2019;20(9):1110-1128.
152. Pellicci DG, Koay HF, Berzins SP. Thymic development of unconventional T cells: how NKT cells, MAIT cells and γδ T cells emerge. *Nat Rev Immunol*. 2020;20(12):756-770.
153. Zietara N, Lyszkiewicz M, Witzlau K, et al. Critical role for miR-181a/b-1 in agonist selection of invariant natural killer T cells. *Proc Natl Acad Sci U S A*. 2013;110(18):7407-7412.
154. Winter SJ, Kunze-Schumacher H, Imelmann E, et al. MicroRNA miR-181a/b-1 controls MAIT cell development. *Immunol Cell Biol*. 2019;97(2):190-202.
155. Lyszkiewicz M, Winter SJ, Witzlau K, et al. miR-181a/b-1 controls thymic selection of T_{reg} cells and tunes their suppressive capacity. *PLoS Biol*. 2019;17(3):e2006716.

156. Gioulbasani M, Galaras A, Grammenoudi S, et al. The transcription factor BCL-6 controls early development of innate-like T cells. *Nat Immunol.* 2020;21(9):1058-1069.
157. Koay HF, Su S, Amann-Zalcenstein D, et al. A divergent transcriptional landscape underpins the development and functional branching of MAIT cells. *Sci Immunol.* 2019;4(41):eaay6039.
158. Legoux F, Gilet J, Procopio E, et al. Molecular mechanisms of lineage decisions in metabolite-specific T cells. *Nat Immunol.* 2019;20(9):1244-1255.
159. Leeansyah E, Loh L, Nixon DF, et al. Acquisition of innate-like microbial reactivity in mucosal tissues during human fetal MAIT-cell development. *Nat Commun.* 2014;5:3143.
160. Walker LJ, Kang YH, Smith MO, et al. Human MAIT and CD8αα cells develop from a pool of type-17 precommitted CD8+ T cells. *Blood.* 2012;119(2):422-433.
161. Tang X, Zhang S, Peng Q, et al. Sustained IFN-I stimulation impairs MAIT cell responses to bacteria by inducing IL-10 during chronic HIV-1 infection. *Sci Adv.* 2020;6(8):eaaz0374.
162. Terpstra ML, Remmerswaal EBM, van der Bom-Baylon ND, et al. Tissue-resident mucosal-associated invariant T (MAIT) cells in the human kidney represent a functionally distinct subset. *Eur J Immunol.* 2020;50(11):1783-1797.
163. Xie J, Pan Y, Tao H, et al. Deficiency of mucosal-associated invariant T cells in TCRJα18 germline knockout mice. *Immunohorizons.* 2019;3(6):203-207.
164. Kawachi I, Maldonado J, Strader C, et al. MR1-restricted V alpha 19i mucosal-associated invariant T cells are innate T cells in the gut lamina propria that provide a rapid and diverse cytokine response. *J Immunol.* 2006;176(3):1618-1627.
165. Cui Y, Franciszkiewicz K, Mburu YK, et al. Mucosal-associated invariant T cell-rich congenic mouse strain allows functional evaluation. *J Clin Invest.* 2015;125(11):4171-4185.
166. Altman JD, Moss PA, Goulder PJ, et al. Phenotypic analysis of antigen-specific T lymphocytes. *Science.* 1996;274(5284):94-96.
167. Benlagha K, Weiss A, Beavis A, et al. *In vivo* identification of glycolipid antigen-specific T cells using fluorescent CD1d tetramers. *J Exp Med.* 2000;191(11):1895-1903.
168. Hinks TSC, Marchi E, Jabeen M, et al. Activation and *in vivo* evolution of the MAIT cell transcriptome in mice and humans reveals tissue repair functionality. *Cell Rep.* 2019;28(12):3249-3262 e5.
169. Lamichhane R, Schneider M, de la Harpe SM, et al. TCR- or cytokine-activated CD8(+) mucosal-associated invariant T cells are rapid polyfunctional effectors that can coordinate immune responses. *Cell Rep.* 2019;28(12):3061.e5-3076.e5.
170. Leng T, Akther HD, Hackstein CP, et al. TCR and inflammatory signals tune human MAIT cells to exert specific tissue repair and effector functions. *Cell Rep.* 2019;28(12):3077.e5-3091.e5.
171. Duan M, Goswami S, Shi JY, et al. Activated and exhausted MAIT cells foster disease progression and indicate poor outcome in hepatocellular carcinoma. *Clin Cancer Res.* 2019;25(11):3304-3316.
172. Lu B, Liu M, Wang J, et al. IL-17 production by tissue-resident MAIT cells is locally induced in children with pneumonia. *Mucosal Immunol.* 2020;13:824-835.
173. Salou M, Legoux F, Gilet J, et al. A common transcriptomic program acquired in the thymus defines tissue residency of MAIT and NKT subsets. *J Exp Med.* 2019;216(1):133-151.
174. Yao T, Shooshtari P, Haeryfar SMM. Leveraging public single-cell and bulk transcriptomic datasets to delineate MAIT cell roles and phenotypic characteristics in human malignancies. *Front Immunol.* 2020;11:1691.
175. Petley E, Koay H-F, Henderson M, et al. MAIT cells regulate NK cell mediated tumor immunity. *Nat Commun.* 2021;12(1):4746.
176. Bulitta B, Zuschratter W, Bernal I, et al. Proteomic definition of human mucosal-associated invariant T cells determines their unique molecular effector phenotype. *Eur J Immunol.* 2018;48(8):1336-1349.
177. Turtle CJ, Delrow J, Joslyn RC, et al. Innate signals overcome acquired TCR signaling pathway regulation and govern the fate of human CD161(hi) CD8alpha(+) semi-invariant T cells. *Blood.* 2011;118(10):2752-2762.
178. Wong EB, Gold MC, Meermeier EW, et al. TRAV1-2(+) CD8(+) T-cells including oligoconal expansions of MAIT cells are enriched in the airways in human tuberculosis. *Commun Biol.* 2019;2:203.
179. Cossarizza A, Chang HD, Radbruch A, et al. Guidelines for the use of flow cytometry and cell sorting in immunological studies (second edition). *Eur J Immunol.* 2019;49(10):1457-1973.
180. Sharma PK, Wong EB, Napier RJ, et al. High expression of CD26 accurately identifies human bacteria-reactive MR1-restricted MAIT cells. *Immunology.* 2015;145(3):443-453.
181. Eberhard JM, Hartjen P, Kummer S, et al. CD161+ MAIT cells are severely reduced in peripheral blood and lymph nodes of HIV-infected individuals independently of disease progression. *PLoS One.* 2014;9(11):e111323.
182. Dias J, Leeansyah E, Sandberg JK. Multiple layers of heterogeneity and subset diversity in human MAIT cell responses to distinct microorganisms and to innate cytokines. *Proc Natl Acad Sci U S A.* 2017;114(27):E5434-E5443.
183. Kirkham CL, Carlyle JR. Complexity and diversity of the NKR-P1:clr (Klrb1:Clec2) recognition systems. *Front Immunol.* 2014;5:214.
184. Brozova J, Karlova I, Novak J. Analysis of the phenotype and function of the subpopulations of mucosal-associated invariant T cells. *Scand J Immunol.* 2016;84(4):245-251.
185. Dias J, Boulouis C, Gorin JB, et al. The CD4(−)CD8(−) MAIT cell subpopulation is a functionally distinct subset developmentally related to the main CD8(+) MAIT cell pool. *Proc Natl Acad Sci U S A.* 2018;115(49):E11513-E11522.
186. Billerbeck E, Kang YH, Walker L, et al. Analysis of CD161 expression on human CD8+ T cells defines a distinct functional subset with tissue-homing properties. *Proc Natl Acad Sci U S A.* 2010;107(7):3006-3011.
187. Leeansyah E, Svard J, Dias J, et al. Arming of MAIT cell cytolytic antimicrobial activity is induced by IL-7 and defective in HIV-1 infection. *PLoS Pathog.* 2015;11(8):e1005072.
188. Kurioka A, Ussher JE, Cosgrove C, et al. MAIT cells are licensed through granzyme exchange to kill bacterially sensitized targets. *Mucosal Immunol.* 2015;8(2):429-440.
189. Kurioka A, Jahun AS, Hannaway RF, et al. Shared and distinct phenotypes and functions of human CD161++ Vα7.2+ T cell subsets. *Front Immunol.* 2017;8:1031.
190. Wang H, D'Souza C, Lim XY, et al. MAIT cells protect against pulmonary Legionella longbeachae infection. *Nat Commun.* 2018;9(1):3350.
191. Provine NM, Klenerman P. MAIT cells in health and disease. *Annu Rev Immunol.* 2020;38:203-228.
192. Wang H, Chen Z, McCluskey J, et al. Mouse models illuminate MAIT cell biology. *Mol Immunol.* 2021;130:55-63.
193. Li J, Reantragoon R, Kostenko L, et al. The frequency of mucosal-associated invariant T cells is selectively increased in dermatitis herpetiformis. *Australas J Dermatol.* 2017;58(3):200-204.
194. Nel I, Bertrand L, Toubal A, et al. MAIT cells, guardians of skin and mucosa? *Mucosal Immunol.* 2021;14(4):803-814.
195. Teunissen MBM, Yeremenko NG, Baeten DLP, et al. The IL-17A-producing CD8+ T-cell population in psoriatic lesional skin comprises mucosa-associated invariant T cells and conventional T cells. *J Invest Dermatol.* 2014;134(12):2898-2907.
196. Lee CH, Zhang HH, Singh SP, et al. C/EBPδ drives interactions between human MAIT cells and endothelial cells that are important for extravasation. *Elife.* 2018;7:e32532.
197. Voillet V, Buggert M, Slichter CK, et al. Human MAIT cells exit peripheral tissues and recirculate via lymph in steady state conditions. *JCI Insight.* 2018;3(7):e98487.
198. Booth JS, Salerno-Goncalves R, Blanchard TG, et al. Mucosal-associated invariant T cells in the human gastric mucosa and blood: role in Helicobacter pylori infection. *Front Immunol.* 2015;6:466.
199. D'Souza C, Pediongco T, Wang H, et al. Mucosal-associated invariant T cells augment immunopathology and gastritis in chronic *Helicobacter pylori* infection. *J Immunol.* 2018;200(5):1901-1916.
200. Schmaler M, Colone A, Spagnuolo J, et al. Modulation of bacterial metabolism by the microenvironment controls MAIT cell stimulation. *Mucosal Immunol.* 2018;11(4):1060-1070.
201. Sundstrom P, Ahlmanner F, Akeus P, et al. Human mucosa-associated invariant T cells accumulate in colon adenocarcinomas but produce reduced amounts of IFN-gamma. *J Immunol.* 2015;195(7):3472-3481.
202. Meierovics A, Yankelevich WJ, Cowley SC. MAIT cells are critical for optimal mucosal immune responses during *in vivo* pulmonary bacterial infection. *Proc Natl Acad Sci U S.A.* 2013;110(33):E3119-E3128.
203. Jiang J, Chen X, An H, et al. Enhanced immune response of MAIT cells in tuberculous pleural effusions depends on cytokine signaling. *Sci Rep.* 2016;6:32320.
204. Toussirot E, Laheurte C, Gaugler B, et al. Increased IL-22- and IL-17a-producing mucosal-associated invariant T cells in the peripheral blood of patients with ankylosing spondylitis. *Front Immunol.* 2018;9:1610.
205. Kelly J, Minoda Y, Meredith T, et al. Chronically stimulated human MAIT cells are unexpectedly potent IL-13 producers. *Immunol Cell Biol.* 2019;97(8):689-699.
206. Bennett MS, Trivedi S, Iyer AS, et al. Human mucosal-associated invariant T (MAIT) cells possess capacity for B cell help. *J Leukoc Biol.* 2017;102(5):1261-1269.
207. Leung DT, Bhuiyan TR, Nishat NS, et al. Circulating mucosal associated invariant T cells are activated in *Vibrio cholerae* O1 infection and associated with lipopolysaccharide antibody responses. *PLoS Neglected Trop Dis.* 2014;8(8):e3076.
208. Rahman MA, Ko EJ, Bhuyan F, et al. Mucosal-associated invariant T (MAIT) cells provide B-cell help in vaccinated and subsequently SIV-infected Rhesus Macaques. *Sci Rep.* 2020;10(1):10060.
209. Murayama G, Chiba A, Suzuki H, et al. A critical role for mucosal-associated invariant T cells as regulators and therapeutic targets in systemic lupus erythematosus. *Front Immunol.* 2019;10:2681.
210. Yu H, Yang A, Derrick S, et al. Artificially induced MAIT cells inhibit M. bovis BCG but not *M. tuberculosis* during *in vivo* pulmonary infection. *Sci Rep.* 2020;10(1):13579.
211. Sakai S, Kauffman KD, Oh S, et al. MAIT cell-directed therapy of *Mycobacterium tuberculosis* infection. *Mucosal Immunol.* 2021;14(1):199-208.
212. Sundstrom P, Szeponik L, Ahlmanner F, et al. Tumor-infiltrating mucosal-associated invariant T (MAIT) cells retain expression of cytotoxic effector molecules. *Oncotarget.* 2019;10(29):2810-2823.
213. Dias J, Sobkowiak MJ, Sandberg JK, et al. Human MAIT-cell responses to Escherichia coli: activation, cytokine production, proliferation, and cytotoxicity. *J Leukoc Biol.* 2016;100(1):233-240.
214. Loh L, Wang Z, Sant S, et al. Human mucosal-associated invariant T cells contribute to antiviral influenza immunity via IL-18-dependent activation. *Proc Natl Acad Sci U S A.* 2016;113(36):10133-10138.
215. Voskoboinik I, Whisstock JC, Trapani JA. Perforin and granzymes: function, dysfunction and human pathology. *Nat Rev Immunol.* 2015;15(6):388-400.
216. Gherardin NA, Loh L, Admojo L, et al. Enumeration, functional responses and cytotoxic capacity of MAIT cells in newly diagnosed and relapsed multiple myeloma. *Sci Rep.* 2018;8(1):4159.
217. Won EJ, Ju JK, Cho YN, et al. Clinical relevance of circulating mucosal-associated invariant T cell levels and their anti-cancer activity in patients with mucosal-associated cancer. *Oncotarget.* 2016;7(46):76274-76290.
218. Loh L, Gherardin NA, Sant S, et al. Human mucosal-associated invariant T cells in older individuals display expanded TCRαβ clonotypes with potent antimicrobial responses. *J Immunol.* 2020;204(5):1119-1133.
219. Jiang J, Cao Z, Shan W, et al. 4-1BB expression on MAIT cells is associated with enhanced IFN-gamma production and depends on IL-2. *Cell Immunol.* 2018;328:58-69.
220. Jo J, Tan AT, Ussher JE, et al. Toll-like receptor 8 agonist and bacteria trigger potent activation of innate immune cells in human liver. *PLoS Pathog.* 2014;10(6):e1004210.
221. Jesteadt E, Zhang I, Yu H, et al. Interleukin-18 is critical for mucosa-associated invariant T cell gamma interferon responses to *Francisella* species *In vitro* but not *In vivo*. *Infect Immun.* 2018;86(5):e00117-e00118.
222. Sattler A, Dang-Heine C, Reinke P, et al. IL-15 dependent induction of IL-18 secretion as a feedback mechanism controlling human MAIT-cell effector functions. *Eur J Immunol.* 2015;45(8):2286-2298.
223. Wallington JC, Williams AP, Staples KJ, et al. IL-12 and IL-7 synergize to control mucosal-associated invariant T-cell cytotoxic responses to bacterial infection. *J Allergy Clin Immunol.* 2018;141(6):2182.e6-2195.e6.
224. Gracey E, Qaiyum Z, Almaghlouth I, et al. IL-7 primes IL-17 in mucosal-associated invariant T (MAIT) cells, which contribute to the Th17-axis in ankylosing spondylitis. *Ann Rheum Dis.* 2016;75(12):2124-2132.
225. Dumas A, Corral D, Colom A, et al. The host microbiota contributes to early protection against lung colonization by *Mycobacterium tuberculosis. Front Immunol.* 2018;9:2656.
226. Murugesan A, Ibegbu C, Styles TM, et al. Functional MAIT cells are associated with reduced simian-human immunodeficiency virus infection. *Front Immunol.* 2019;10:3053.
227. Smith AD, Foss ED, Zhang I, et al. Microbiota of MR1 deficient mice confer resistance against *Clostridium difficile* infection. *PLoS One.* 2019;14(9):e0223025.
228. Tastan C, Karhan E, Zhou W, et al. Tuning of human MAIT cell activation by commensal bacteria species and MR1-dependent T-cell presentation. *Mucosal Immunol.* 2018;11(6):1591-1605.
229. Krause JL, Schape SS, Schattenberg F, et al. The activation of mucosal-associated invariant T (MAIT) cells is affected by microbial diversity and riboflavin utilization *in vitro*. *Front Microbiol.* 2020;11:755.
230. Ussher JE, Bilton M, Attwod E, et al. CD161(++) CD8(+) T cells, including the MAIT cell subset, are specifically activated by IL-12+IL-18 in a TCR-independent manner. *Eur J Immunol.* 2014;44(1):195-203.
231. van Wilgenburg B, Scherwitzl I, Hutchinson EC, et al. MAIT cells are activated during human viral infections. *Nat Commun.* 2016;7:11653.
232. Shaler CR, Choi J, Rudak PT, et al. MAIT cells launch a rapid, robust and distinct hyperinflammatory response to bacterial superantigens and quickly acquire an anergic phenotype that impedes their cognate antimicrobial function: defining a novel mechanism of superantigen-induced immunopathology and immunosuppression. *PLoS Biol.* 2017;15(6):e2001930.
233. Paquin-Proulx D, Avelino-Silva VI, Santos BAN, et al. MAIT cells are activated in acute Dengue virus infection and after *in vitro* Zika virus infection. *PLoS Negl Trop Dis.* 2018;12(1):e0006154.
234. Yong YK, Saeidi A, Tan HY, et al. Hyper-expression of PD-1 is associated with the levels of exhausted and dysfunctional phenotypes of circulating CD161(++)TCR iVα7.2(+) mucosal-associated invariant T cells in chronic hepatitis B virus infection. *Front Immunol.* 2018;9:472.
235. van Wilgenburg B, Loh L, Chen Z, et al. MAIT cells contribute to protection against lethal influenza infection *in vivo*. *Nat Commun.* 2018;9(1):4706.

236. Maleki KT, Tauriainen J, Garcia M, et al. MAIT cell activation is associated with disease severity markers in acute hantavirus infection. *Cell Rep Med.* 2021;2(3):100220.
237. Preeyaa SU, Murugesan A, Sopnajothi S, et al. Peripheral follicular T helper cells and mucosal-associated invariant T cells represent activated phenotypes during the febrile phase of acute dengue virus infection. *Viral Immunol.* 2020;33(9):610-615.
238. Chua WJ, Truscott SM, Eickhoff CS, et al. Polyclonal mucosa-associated invariant T cells have unique innate functions in bacterial infection. *Infect Immun.* 2012;80(9):3256-3267.
239. Jouan Y, Guillon A, Gonzalez L, et al. Phenotypical and functional alteration of unconventional T cells in severe COVID-19 patients. *J Exp Med.* 2020;217(12):e20200872.
240. Dias J, Hengst J, Parrot T, et al. Chronic hepatitis delta virus infection leads to functional impairment and severe loss of MAIT cells. *J Hepatol.* 2019;71(2):301-312.
241. Huang W, He W, Shi X, et al. Mucosal-associated invariant T-cells are severely reduced and exhausted in humans with chronic HBV infection. *J Viral Hepat.* 2020;27(11):1096-1107.
242. Rudak PT, Yao T, Richardson CD, et al. Measles virus Infects and programs MAIT cells for apoptosis. *J Infect Dis.* 2021;223(4):667-672.
243. Shaler CR, Tun-Abraham ME, Skaro AI, et al. Mucosa-associated invariant T cells infiltrate hepatic metastases in patients with colorectal carcinoma but are rendered dysfunctional within and adjacent to tumor microenvironment. *Cancer Immunol Immunother.* 2017;66(12):1563-1575.
244. Li S, Simoni Y, Becht E, et al. Human tumor-infiltrating MAIT cells display hallmarks of bacterial antigen recognition in colorectal cancer. *Cell Rep Med.* 2020;1(3):100039.
245. Cogswell DT, Gapin L, Tobin HM, et al. MAIT cells: partners or enemies in cancer immunotherapy? *Cancers (Basel).* 2021;13(7):1502.
246. Riva A, Patel V, Kurioka A, et al. Mucosa-associated invariant T cells link intestinal immunity with antibacterial immune defects in alcoholic liver disease. *Gut.* 2018;67(5):918-930.
247. Parrot T, Gorin JB, Ponzetta A, et al. MAIT cell activation and dynamics associated with COVID-19 disease severity. *Sci Immunol.* 2020;5(51):eabe1670.
248. Ussher JE, Phalora P, Cosgrove C, et al. Molecular analyses define Vα7.2-Jα33+ MAIT cell depletion in HIV infection: a case-control study. *Medicine (Baltimore).* 2015;94(29):e1134.
249. Flament H, Rouland M, Beaudoin L, et al. Outcome of SARS-CoV-2 infection is linked to MAIT cell activation and cytotoxicity. *Nat Immunol.* 2021;22(3):322-335.
250. Ellis AL, Balgeman AJ, Larson EC, et al. MAIT cells are functionally impaired in a Mauritian cynomolgus macaque model of SIV and Mtb co-infection. *PLoS Pathog.* 2020;16(5):e1008585.
251. McSharry BP, Samer C, McWilliam HEG, et al. Virus-mediated suppression of the antigen presentation molecule MR1. *Cell Rep.* 2020;30(9):2948.e4-2962.e4.
252. Zhao Z, Wang H, Shi M, et al. Francisella tularensis induces Th1 like MAIT cells conferring protection against systemic and local infection. *Nat Commun.* 2021;12(1):4355.
253. Meierovics AI, Cowley SC. MAIT cells promote inflammatory monocyte differentiation into dendritic cells during pulmonary intracellular infection. *J Exp Med.* 2016;213(12):2793-2809.
254. Salio M, Gasser O, Gonzalez-Lopez C, et al. Activation of human mucosal-associated invariant T cells induces CD40L-dependent maturation of monocyte-derived and primary dendritic cells. *J Immunol.* 2017;199(8):2631-2638.
255. Barber-Axthelm IM, Kent SJ, Juno JA. Understanding the role of mucosal-associated invariant T-cells in non-human primate models of HIV infection. *Front Immunol.* 2020;11:2038.
256. Jahreis S, Bottcher S, Hartung S, et al. Human MAIT cells are rapidly activated by *Aspergillus* spp. in an APC-dependent manner. *Eur J Immunol.* 2018;48(10):1698-1706.
257. Hernandez-Santos N, Wiesner DL, Fites JS, et al. Lung epithelial cells coordinate innate lymphocytes and immunity against pulmonary fungal infection. *Cell Host Microbe.* 2018;23(4):511.e5-522.e5.
258. Everhart JE. Recent developments in the epidemiology of *Helicobacter pylori. Gastroenterol Clin North Am.* 2000;29(3):559-578.
259. Held K, Bhonsle-Deeng L, Siewert K, et al. αβ T-cell receptors from multiple sclerosis brain lesions show MAIT cell-related features. *Neurol Neuroimmunol Neuroinflamm.* 2015;2(4):e107.
260. Miyazaki Y, Miyake S, Chiba A, et al. Mucosal-associated invariant T cells regulate Th1 response in multiple sclerosis. *Int Immunol.* 2011;23(9):529-535.
261. Negrotto L, Canto E, Rio J, et al. Peripheral blood non-MAIT CD8+CD161hi cells are decreased in relapsing-remitting multiple sclerosis patients treated with interferon beta. *J Neuroimmunol.* 2015;288:98-101.
262. Salou M, Nicol B, Garcia A, et al. Neuropathologic, phenotypic and functional analyses of mucosal associated invariant T cells in multiple sclerosis. *Clin Immunol.* 2016;166-167:1-11.
263. Carnero Contentti E, Farez MF, Correale J. Mucosal-associated invariant T cell features and TCR repertoire characteristics during the course of multiple sclerosis. *Front Immunol.* 2019;10:2690.
264. Croxford JL, Miyake S, Huang YY, et al. Invariant V(alpha)19i T cells regulate autoimmune inflammation. *Nat Immunol.* 2006;7(9):987-994.
265. Ruijing X, Mengjun W, Xiaoling Z, et al. Jα33+ MAIT cells play a protective role in TNBS induced intestinal inflammation. *Hepatogastroenterology.* 2012;59(115):762-767.
266. Li Y, Huang B, Jiang X, et al. Mucosal-associated invariant T cells Improve nonalcoholic fatty liver disease through regulating macrophage polarization. *Front Immunol.* 2018;9:1994.
267. Varelias A, Bunting MD, Ormerod KL, et al. Recipient mucosal-associated invariant T cells control GVHD within the colon. *J Clin Invest.* 2018;128(5):1919-1936.
268. Chiba A, Tajima R, Tomi C, et al. Mucosal-associated invariant T cells promote inflammation and exacerbate disease in murine models of arthritis. *Arthritis Rheum.* 2012;64(1):153-161.
269. Toubal A, Kiaf B, Beaudoin L, et al. Mucosal-associated invariant T cells promote inflammation and intestinal dysbiosis leading to metabolic dysfunction during obesity. *Nat Commun.* 2020;11(1):3755.
270. Hegde P, Weiss E, Paradis V, et al. Mucosal-associated invariant T cells are a profibrogenic immune cell population in the liver. *Nat Commun.* 2018;9(1):2146.
271. Magalhaes I, Pingris K, Poitou C, et al. Mucosal-associated invariant T cell alterations in obese and type 2 diabetic patients. *J Clin Invest.* 2015;125(4):1752-1762.
272. Ling L, Lin Y, Zheng W, et al. Circulating and tumor-infiltrating mucosal associated invariant T (MAIT) cells in colorectal cancer patients. *Sci Rep.* 2016;6:20358.
273. Zheng C, Zheng L, Yoo JK, et al. Landscape of infiltrating T cells in liver cancer revealed by single-cell sequencing. *Cell.* 2017;169(7):1342.e16-1356.e16.
274. Zabijak L, Attencourt C, Guignant C, et al. Increased tumor infiltration by mucosal-associated invariant T cells correlates with poor survival in colorectal cancer patients. *Cancer Immunol Immunother.* 2015;64(12):1601-1608.
275. Melo AM, O'Brien AM, Phelan JJ, et al. Mucosal-associated invariant T cells display diminished effector capacity in oesophageal adenocarcinoma. *Front Immunol.* 2019;10:1580.
276. Keller AN, Corbett AJ, Wubben JM, et al. MAIT cells and MR1-antigen recognition. *Curr Opin Immunol.* 2017;46:66-74.
277. Livyatan I, Nejman D, Shental N, et al. Characterization of the human tumor microbiome reveals tumor-type specific intra-cellular bacteria. *OncoImmunology.* 2020;9(1):1800957.
278. Nejman D, Livyatan I, Fuks G, et al. The human tumor microbiome is composed of tumor type-specific intracellular bacteria. *Science.* 2020;368(6494):973-980.
279. Yan J, Allen S, McDonald E, et al. MAIT cells promote tumor Initiation, growth, and metastases via tumor MR1. *Cancer Discov.* 2020;10(1):124-141.
280. Preciado-Llanes L, Aulicino A, Canals R, et al. Evasion of MAIT cell recognition by the African *Salmonella* Typhimurium ST313 pathovar that causes invasive disease. *Proc Natl Acad Sci U S A.* 2020;117(34):20717-20728.
281. Downey AM, Kaplonek P, Seeberger PH. MAIT cells as attractive vaccine targets. *FEBS Lett.* 2019;593(13):1627-1640.
282. Provine NM, Amini A, Garner LC, et al. MAIT cell activation augments adenovirus vector vaccine immunogenicity. *Science.* 2021;371(6528):521-526.

CHAPTER 38

The Mucosal Immune System

Yasmine Belkaid, PhD • Brian L. Kelsall, MD, PhD • Warren Strober, MD

INTRODUCTION

Most antigens encountered by the immune system enter the body through the mucosal surfaces of the respiratory, gastrointestinal (GI), and urogenital tract and the vast majority of infectious microorganisms utilize the mucosae as portals of entry. These sites are, therefore, charged with the formidable task of protecting the host from environmental and pathogenic challenges while preserving vital physiologic tissue functions. The GI tract represents the most reactive and complex immune environment of the host. Along with its constant exposure to food antigens and primary role in acquisition of metabolites, gut mucosal surfaces host complex microbial communities whose combined membership outnumbers host somatic cells. The microbiota control numerous aspects of host physiology, including the host immune system.[1] These microbial communities also present a significant challenge for the host immune system. However, when uncontrolled, reactivity against innocuous antigens, such as those derived from food and intestinal flora, poses a substantial risk that can lead to tissue damage and severe inflammatory disorders. Therefore, multiple highly specialized innate and adaptive immune cell types, as well as structural features, have been put in place to prevent overt reactivity and favor the induction of tolerogenic responses. However, a certain degree of constitutive activation and inflammation is beneficial for the host, not only to reinforce barrier functions but also to allow for the development of protective responses in the context of infectious challenges. The majority of infectious diseases either target or are acquired through mucosal surfaces of the GI, respiratory, and genital tracts. The complex and highly dynamic maintenance of immune tolerance to mucosal antigens concomitantly with the induction of protective responses to pathogens requires an arsenal of unique cells specifically conditioned by mucosal environments.

Mucosal tissues have been traditionally divided into three major components: the mucosa-associated lymphoid tissues (MALT), such as Peyer patches; the nonlymphoid tissue associated tissues, such as the absorptive intestinal villus; and the draining lymph nodes (LNs).[2,3] Tissues lacking MALT, such as the oral and vaginal mucosa, are called type 1 mucosal tissues and covered by stratified squamous epithelial cells, while those that have MALT are called type 2 mucosal tissues and are covered by columnar epithelial cells, goblet cells, M cells, tuft cells, other enteroendocrine cells, and stem cells.[4] Draining LNs are present in both type 1 and type 2 mucosal tissues and are the only major site for the induction of adaptive immune responses for type 1 tissues, whereas somewhat different immune responses are generated in MALT and draining LNs for type 2 tissues.[4]

Type 2 mucosal barriers are unique from more secure type 1 mucosal barriers, as in type 2 mucosal tissues there is a need to adsorb nutrients and fluids, but exclude the vast majority of exogenous antigens and bacteria.[4,5] At the same time, selective uptake of bacterial and exogenous antigens, as well as bacterial metabolites, continuously occurs to allow for the development and maintenance of immune tissues and for the generation of factors that in turn protect against pathogen invasion and maintain homeostasis with our commensal bacteria.

Owing to limitations in time and space, in this chapter we will focus on studies of the mucosal immune system of the GI tract to illustrate major principles of immune regulation at mucosal sites.

COMPARTMENTALIZATION OF THE MUCOSAL IMMUNE SYSTEM

The mucosal immune system of the GI tract contains unique cell subsets that are structured in such a way as to favor the coexistence of the host with its flora and support food absorption as well as induction of effector responses to pathogens.[6] The primary barrier preventing transit of microorganisms and large molecules from the environment into the body is a simple epithelium made up of a single layer of cells overlaid with mucus and connected by junctional complexes. Furthermore, the intestinal immune system is organized into inductive and effector sites to allow for appropriate antigen sampling and immune responses. Inductive sites include both draining mesenteric lymph nodes (MLNs) and gut-associated lymphoid tissues (GALTs) consisting of organized lymphoid structures embedded directly in the submucosa within the bowel wall. GALT includes Peyer's, cecal, and colonic patches of the small intestine, colon, and cecum (human appendix), respectively, and isolated lymphoid follicles (ILFs) present throughout the intestine. Effector sites include the intestinal lamina propria (LP), which is the diffuse area beneath the epithelium that is not associated with GALT, and the intraepithelial cell compartment in which cells are located above the basement membrane of the epithelial layer and in direct contact with the basal epithelial cell wall.[7]

Inductive sites are precisely organized to allow for optimum antigen exposure and cellular interactions for the induction of T- and B-cell responses. MLNs develop in utero, have a typical LN structure, and drain lymph from the LP and GALT. Conventional dendritic cells (cDC) from the intestinal LP and GALT capture luminal antigens by a variety of mechanisms and from the LP migrate to the MLNs to initiate immune responses similar to LNs draining nonmucosal tissues. Peyer's, cecal, and colonic patches are macroscopic lymphoid aggregates consisting of large B-cell follicles with interspersed T-cell areas (interfollicular regions, IFRs). Peyer's and colonic (and likely cecal) patches appear during the later stages of gestation and are present in germ-free mice but appear to require exposure to microbiota for full functional development. ILFs are single follicular structures that develop from rudimentary collections of cells called cryptopatches after birth in response to environmental signals, such as those derived from the microbiota.[8-10] All of the GALT tissues are separated from the intestinal lumen by a single layer of epithelial cells known as the follicle-associated epithelium (FAE) and a more diffuse area below the epithelium known as the subepithelial dome (SED). Antigens are sampled directly from the intestinal lumen, largely across specialized epithelial cells, called Microfold (M) cells within the FAE that have the capacity to transport organisms and particles from the gut lumen across the epithelial barrier.[11] These cells also represent a major point of pathogen entry.[12] Of note, DCs (and macrophages) are enriched in the SED and are thought to largely capture antigens and induce T- and B-cell responses within the follicular structure of the Peyer's patch (PP) rather than migrate to the MLNs.

An important aspect of mucosal immune responses concerns the trafficking of cells primed in mucosal inductive sites to mucosal effector sites. Adaptive T- and B-cells primed in mucosal inductive sites home preferentially back to mucosal tissues including the LP, the IEL compartment, as well as GALT tissues, where they are localized and maintained to promote barrier integrity, provide host defense, and contribute to the induction of new and ongoing adaptive immunity. As in other tissues, such as the skin, this occurs using tissue-specific addressins that include chemokines expressed in tissues, chemokine receptors and integrins by immune cells, and integrin-binding molecules expressed on high-endothelial venules.

INNATE FUNCTIONS OF THE MUCOSAL BARRIER

The intestinal epithelial barrier consists of different intestinal epithelial cell (IEC) subtypes. Mature enterocytes and colonocytes are polarized epithelial cells that are lined by tight junctions. The majority of cells in the intestine epithelium turn over every few days in the adult and the rate of epithelium renewal is controlled by environmental factors including the microbiota.[13] IECs' identity and function are highly dependent on their localization along the GI tract.[13] Epithelial cells coordinate and orchestrate immune responses within the GI tract via diverse mechanisms including the expression of factors with pleiotropic effects on mucosal immunity. Epithelial cells have the ability to metabolize vitamin A and synthesize its metabolite retinoic acid (RA).[14-18] RA plays a central role in the control of various aspects of innate and adaptive immune responses at mucosal sites including the promotion of cell migration to mucosal sites, the generation of IgA, and the induction of regulatory and effector T cells.[2,19,20]

Epithelial cells are also a source of serum amyloids (SAAs), the release of which is heightened by encounter with adherent microbes. Local secretion of SAA1 and SAA2 by epithelial cells can license T cells and innate lymphoid cells for cytokine production.[21,22] SAAs are also carriers of high-density lipoprotein and retinol (152), allowing the delivery of these molecules to antigen-presenting cells.[23] Epithelial cells also release numerous cytokines in response to the microbiota or inflammatory triggers including IL-18, IL-25, IL-33, and thymic stromal lymphopoietin (TSLP). Epithelial-derived alarmins can promote innate lymphoid cell (ILC) responses that in turn can contribute to protective immunity, tissue repair, or the induction and control of commensal-specific T cells (see chapter 12 for further discussion about ILCs).[22,24] Epithelial cells can also coordinate immune responses in the gut via local release of chemokines involved in cell tropism, such as CXCL9, CXCL10, and CCL20.[25] IECs can express numerous antigen presentation molecules including MHC class II. Expression of MHCII by epithelial cells contributes to the accumulation of CD4+CD8αα+ intraepithelial lymphocytes (IELs).[26]

Goblet Cells

Goblet cells are scattered between enterocytes and are responsible for the secretion of high-molecular-weight glycoprotein mucins that represent the backbone of the mucous layer. The mucus limits commensal and pathogenic bacterial contact with the epithelium preventing translocation of bacteria to distal sites.[27] To date, 17 mucin (MUC) genes have been described, seven of them coding for secreted mucins (MUC2, MUC5A, MUC5B, MUC6, MUCC19, MUC7, and MUC8) that are differentially expressed in various mucosal compartments.[27] Mucus is found in the entire GI tract but varies in thickness ranging from 700 μm in the stomach and large intestine to 150 to 300 μm in the small intestine.[28] The inner layer of the colonic mucus is attached to the epithelium and has a stratified appearance, ranging in thickness from 50 to 100 μm. Based on its structural capacity, this layer acts as a filter and is impermeable to commensal bacteria.[29-31] The inner mucous layer is renewed by goblet cell secretion and at its luminal border is converted into an outer mucous layer that expands in volume allowing commensal bacteria access to this zone.[30] Contrasting to the large intestine, the small intestine is only covered by a single layer of mucus that is not attached to the epithelium and is more permeable to bacteria.[28,32] Commensal bacteria have a high proportion of their genome devoted to production of enzymes involved in glycan degradation and are able to utilize the released mucin monosaccharides as energy sources.[33,34] Goblet cells also contribute to lumen to LP antigen transfer.[35]

Microfold Cells

Microfold (M) cells are a rare specialized population of epithelial cells that overlay intestinal lymphoid tissue, such as Peyer patches. M cells play a role in transepithelial antigen transport and can deliver foreign antigens and microbes to lymphoid tissues. M cells also express the IgA receptor on the apical surface allowing them to bind and retrotransport luminal secretory IgA-bound antigens back into Peyer patches.[36]

Paneth Cells

Paneth cells are localized at the base of the intestinal crypt and are specialist secretors of antimicrobial peptides (AMPs) such as secretory phospholipase A2 (sPLA2), lysozyme, C-type lectin regenerating islet-derived protein IIIγ (RegIIIγ), angiogenin4, cathelicidins, and alpha defensins. AMPs protect the epithelial barrier and maintain a homeostatic relationship with the microbiota.[37] A central strategy utilized by the mucosal immune system to maintain its homeostatic relationship with the microbiota and limit pathogen exposure is to minimize contact between luminal microorganisms and the epithelial cell surface, and this process is highly dependent on antimicrobial proteins. Some of these molecules, such as β-defensins, are constitutively expressed. In other cases, engagement of pattern recognition receptors (PRRs) by commensally derived products induces the expression of a variety of AMPs.[38] These proteins can exert antimicrobial functions by direct killing resulting from enzymatic attack of the bacterial cell wall or by disrupting the bacterial inner membrane.[38] In addition, some antimicrobial proteins can function by depriving bacteria of essential heavy metals, such as iron.[39] One of the best-characterized mucosal AMPs is RegIIIγ. This lectin is expressed soon after birth or following colonization of germ-free mice.[40] Accumulation of RegIIIγ in the mucus contributes to the maintenance of the segregation between the microbiota and the host intestine.[41] Production of RegIIIγ is tightly controlled by the flora in a MyD88-dependent manner and has a direct microbicidal effect on gram-positive bacteria.[40,42,43] The PRR nucleotide-binding oligomerization domain-containing protein 2 (NOD2) controls expression of a subset of α-defensins and cryptidins by Paneth cells.[44] Although Paneth cells play a major role in the production of AMPs, all intestinal cell lineages including enterocytes and goblet cells can produce these factors.

Enteroendocrine Cells

At least 10 subsets of enteroendocrine cells have been described with diverse roles in the control of host physiology such as appetite and serotonin and glucose levels. These cells act via the release of peptide hormones and neuropeptides including somatostatin, neurotensin, vasoactive intestinal peptide, and enteroglucagon.[45] In the small intestine, release of hormones is highly controlled by dietary nutrients while colonic enteroendocrine cells may be preferentially responding to microbial stimulation.[45]

Tuft Cells

Tuft cells are rare chemosensory epithelial cells that contribute to a specific class of immune responses and in particular those associated with type 2 immunity. Tuft cells are the exclusive source of IL-25 within the gut, a factor known to promote type 2 immune responses in various contexts including exposure to parasitic helminths and protozoans.[46-48] Tuft cells promote the expansion and activation of innate lymphoid cells (ILC2s) and eosinophilia within the gut. In the Crypt, IL-13 induces the differentiation of tuft cells from stem progenitor cells, a phenomenon that contributes to the reinforcement of local type 2 immunity.[48]

MUCOSAL ANTIBODIES

IgA antibodies represent a key aspect of immunity at mucosal barriers.[38,49] About three-quarters of total antibody production is composed of immunoglobulin A (IgA); on the order of 3 to 5 g is produced per day in humans, with the vast majority secreted across barrier surfaces.[50-52] Intestinal IgA+ plasma cells (PCs) constitute 80% of all PCs in humans.[53] In contrast, IgG is dominant in urine, bile, and genital and bronchoalveolar secretions, whereas IgD is detected in nasal lacrimal as well as bronchoalveolar secretions[54-56] and IgE in nasal bronchoalveolar and intestinal secretions in the context of allergy.[57] Human B cells produce two IgA subclasses, IgA1 and IgA2, whereas mice only produce one kind of IgA.[58,59] IgA1 can be detected at systemic and mucosal sites, whereas IgA2 is essentially present at mucosal sites.[60] IgA can exist in three main forms, monomeric, dimeric, and secretory (sIgA), which is the main immunoglobulin found in mucous secretions, including tears, saliva, colostrum, and secretions from the genitourinary tract, GI tract, prostate, and respiratory epithelium.[50-52] IgA plays a major role in the control of host-microbe dialogue at mucosal tissues. It is critically important for shaping the microbiota, mediating pathogen clearance, neutralizing toxins and inflammatory microbial-derived molecules, and preventing adhesion of commensal bacteria to the epithelial surfaces by generating steric hindrance.[58,59] Moreover, IgA can also induce bacterial agglutination and interact with the mucous layer through secretory component.[61-66] One remarkable feature of IgA is its capacity to disseminate between mucosa-associated lymphoid structures. Such a feature allows specificities of IgA present in breast milk to be matched to the dominant microbes resident in the maternal microbiota.[67,68]

IgA Generation and Secretion

More than 50 years ago, Peyer patches were identified as the major source of IgA-producing B cells (16). IgA can also be generated in regional LNs as well as in ILFs.[9] The diffuse tissue of the LP can also support IgA production in the absence of follicular structures.[69-71] In order to express IgA, B cells must undergo class switch recombination (CSR) in which DNA is spliced out of the heavy chain constant region loci to replace the μ/δ genes with the α gene segment downstream of the recombined VDJ segment. The GC-rich splice sites upstream of the gene segment for each Ig constant region are the targets for enzymes such as AID, which catalyze the reaction sequence. In humans, there are two Cα regions, generating the two IgA isotypes, IgA1 and IgA2.[58-60] Following induction in the intestinal lymphoid structures, B cells recirculate through the mesentery, lymphatic systems, and the bloodstream to home back to the intestinal mucosa where they develop into IgA-secreting PCs. Secretory IgA (SIgA) is produced as a result of cooperation between PCs in the LP that secrete dimeric IgA (dIgA) and epithelial cells that transport IgA into the lumen. This mechanism of transport involves covalent binding of dIgA to the polymeric immunoglobulin receptor (pIgR), which then carries it within vesicles through IECs to the luminal surface. At this point, IgA is released by proteolytic cleavage of the pIgR leaving a fragment of the transport protein (termed secretory component) that remains associated with the joining chain of the IgA dimer forming a secretory IgA complex with noninflammatory protective functions.[72-74] As such, secretory IgA can bind to bacteria without activating complement or stimulating the release of inflammatory mediators by innate cells.[75,76]

IgA Regulation of Host-Microbe Interactions

At mucosal sites, secretory IgA and IgM (SIgA and SIgM) not only provide protection against toxins and invading pathogens but also control host-commensal interactions by controlling microbiota composition, localization, and function.[62,77] IgAs can express broad reactivity toward diverse members of the microbiota, supporting the idea that IgAs may contribute to a canonical form of immunity to overcome the extraordinary antigenic diversity contained within the microbiota. Within the gut, a significant fraction of the microbiota is coated with IgA antibodies[68,78] and intestinal IgA+ PCs are largely dependent on microbial colonization.[79] Seminal studies uncovered a T cell–independent arm of the IgA response to commensal bacteria[80] and revealed that lack of activation-induced cytidine deaminase (AID)-dependent antibodies leads to aberrant expansion of anaerobic bacteria and ILF hyperplasia.[81] Since then, much work has been devoted to characterizing the ontogeny, specificity, and function of microbiota-reactive antibodies (Fig. 38.1).

Mucosal IgA responses to the microbiota are thought to be induced locally by B1 and B2 precursors, via both T-cell independent (TI) and T-cell dependent (TD) pathways.[82-85] However, the exact contribution of each of these pathways in the control of host-microbiota interactions remains an object of active investigation. TI, in situ CSR to IgA can occur in the LP and ILFs and involves interactions with the epithelial cell– or DC-derived ligands BAFF and APRIL.[83,86,87] Additional factors present in the intestinal microenvironment, such as RA and TGF-β, have also been implicated in gut homing and IgA CSR.[85,88,89] Furthermore, IEC endoplasmic reticulum stress induces the activation and proliferation of peritoneal B1 cells and recruitment of microbiota-reactive, TI PCs to the intestine.[90] TD IgA responses in Peyer patches, including those of microbiota-reactive IgA, require B cell CD40-CD40L interactions with T cells, which lead to upregulation of CCR6 in B cells, allowing their migration to the SED of the activated B cell and local DC-mediated activation of TGF-β and IgA class switching.

Sequencing of members of the microbiota bound to IgA revealed that IgA predominantly targets bacteria that reside in the small intestine.[91,92] T cells are dispensable for most commensal IgA binding, with select taxa requiring TD responses for IgA coating.[91] A large fraction of IgA antibodies have natural polyreactive specificities and bind multiple structurally diverse antigens and commensal bacteria with low affinity.[82] Furthermore, the presence of somatic hypermutation (SHM) in some of these sequences does not confer increased affinity, arguing that TD affinity maturation may not play a major role in all microbiota-reactive IgA.[82,91,93] However, TD examples of high-affinity antibody responses to commensal microbiota do occur,[92,94,95] and the extent to which TI and TD responses contribute to commensal-specific IgA is still debated.[92,93] Regardless, in the absence of T cells, the levels of intestinal IgA PCs and luminal IgA are severely reduced,[80,91] supporting the idea that T cells do play important roles in enabling IgA responses to the microbiota. Abundant IgA coating in the context of inflammatory bowel disease (IBD) identifies specific members of the microbiota with the potential to drive intestinal inflammation,[95] and IgA binding also identifies microbial taxa capable of inducing enteropathy in undernourished children.[96] Experimental evidence supports the idea that mucosal IgA responses may lack classical memory characteristics with established mucosal IgA specificities outcompeted by novel antibacterial responses allowing the mucosal immune system to adapt alteration in microbial composition.[97,98] Whether this phenomenon is true for TI and TD responses remains to be addressed. Thus, the current model is that IgA responses to the microbiota include both TI and TD specificities that play complementary roles in the control of host-microbiota responses.

Selective IgA deficiency is the most common primary immunodeficiency in humans.[99] However, a large percentage of individuals are asymptomatic, and compensatory mechanisms such as increased secretory IgM have been reported.[99] Indeed, IgM and IgG antibodies also play important roles in protection of many barrier sites,[100] and although IgM+ and IgG+ PCs are mostly absent from the murine intestine, they are abundant in the human gut.[93,101,102] Early work suggested that IgG antibodies do not recognize most commensal bacteria during homeostasis and that commensal IgG recognition is primarily induced against microbes able to breach the intestinal barrier.[103] Indeed, invasive members of the *Enterobacteriaceae* family that are able to reach systemic sites induce IgG antibodies that can protect against related pathogens.[104] However, during homeostasis, mice mount systemic IgG antibody responses to a large, phylogenetically diverse fraction of the intestinal microbiota.[105] The bulk of this response consists of TI, Toll-like receptor (TLR)-dependent IgG2b and IgG3 antibodies, and the large overlap between IgG2b/3- and IgA-targeted commensals suggests that they may share similar specificities and ontogeny.[105] In addition, a small subset of commensal bacteria are targeted by high-affinity TD IgG1 and induce cognate T follicular helper (Tfh)

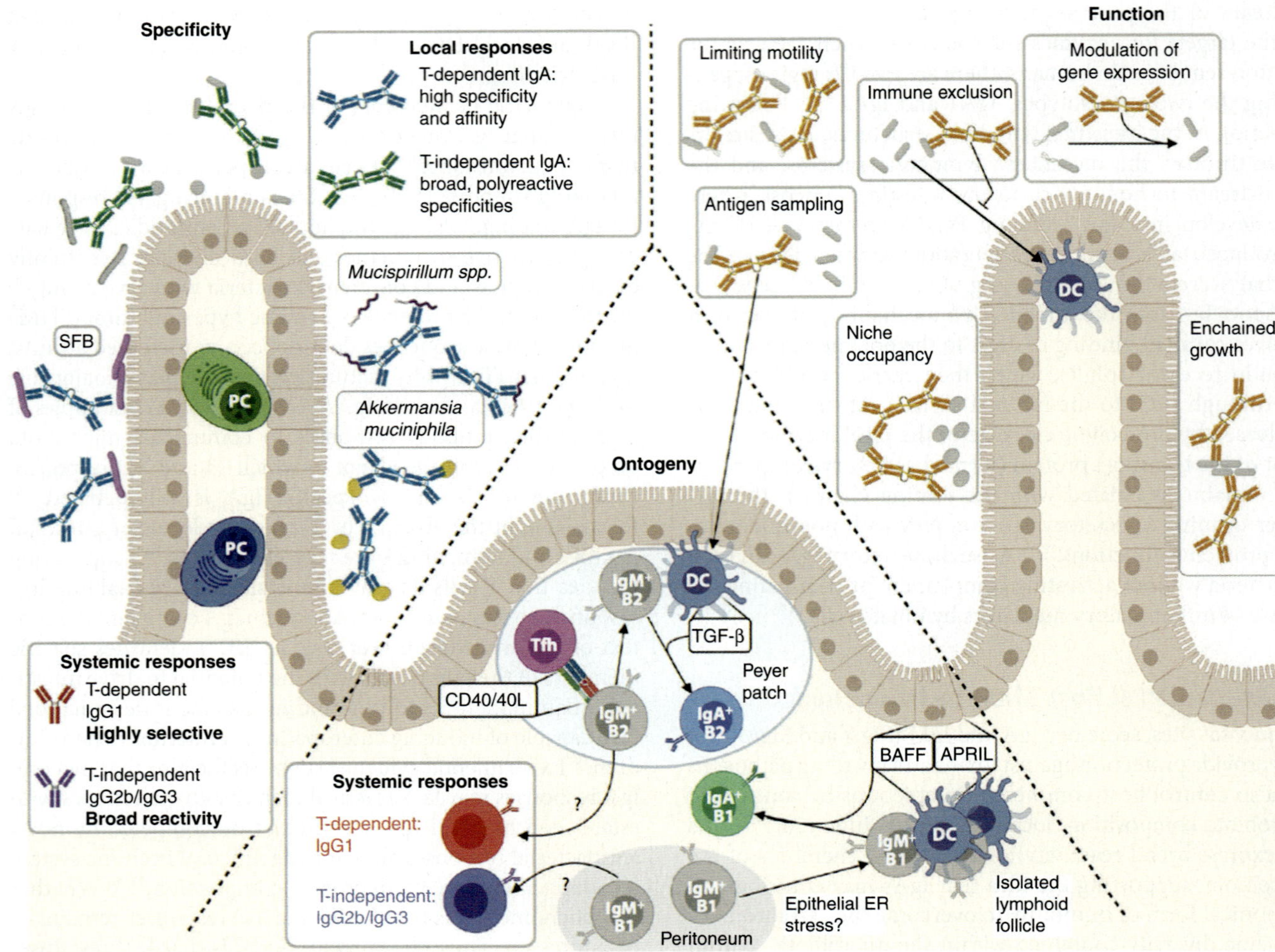

FIG. 38.1. **T-dependent and T-independent antibody responses control host-microbiota mutualism.** Specificity: T-dependent and T-independent IgAs coat a large and diverse fraction of the microbiota at homeostasis. Systemic, T-independent IgG2b and IgG3 responses also target a diverse and overlapping fraction of commensal bacteria, whereas IgG1 and IgA responses that are highly selective to immunostimulatory members of the microbiota are induced in a T-dependent manner. Ontogeny: B1 cells are a major contributor to T-independent IgA, and potentially to IgG2b and IgG3. B1 cells are thought to migrate from the peritoneum to the intestine in response to poorly characterized signals, including intestinal epithelial ER stress, and undergo class-switch recombination in response to local factors including epithelial- and dendritic cell–derived BAFF and APRIL. T-dependent IgA responses occur locally in lymphoid structures such as the Peyer patches from naive B2 cells. Interactions with antigen-specific Tfh cells induce migration of B cells to the subepithelial dome, where interactions with dendritic cells induce the TGF-β-mediated class-switch recombination to IgA. The ontogeny of systemic IgG1 and IgG2b/3 remains unclear but is thought to originate from similar B-cell precursors as IgA responses. Function: Limiting bacterial motility, enabling occupancy of mucosal-associated niches, preventing antigen uptake and immune activation, enchainment of growing bacteria limiting invasion, and modulation of bacterial gene expression have all been proposed as functions of microbiota-reactive antibodies. Abbreviations: DC, dendritic cell; ER, endoplasmic reticulum; PC, plasma cell; Tfh, T follicular helper. (Figure adapted from image created with BioRender.com.)

responses.[94] Thus, similar to the case of IgA, both TI and TD IgG responses to the microbiota are induced and target different subsets of commensal bacteria.

Recent efforts have been devoted to elucidating the role of antibodies, and in particular IgA, in the control of host-microbiota mutualism. Study of the divergent adaptive immune system of teleosts revealed that their secretory immunoglobulin (SIgT) is involved in both protection from mucosal infections and establishment of homeostasis with the microbiota[106]; IgA antibodies regulate the colonization, diversity, and composition of bacterial communities in the intestine via diverse mechanisms.[107-109] IgA can enable colonization of a privileged mucus-associated niche for some commensal species,[110] and enchainment of dividing bacteria may limit invasiveness.[111] Furthermore, TLR5-dependent, flagellin-binding IgA can limit bacterial motility and overall flagellin expression by the microbiota.[112] IgA binding can also modulate bacterial gene expression.[77] Microbiota-specific IgA can also mediate immune exclusion of its targets and prevent further priming of immune responses.[113] Polyreactive

IgA antibodies can enhance bacterial delivery to the Peyer patches in the small intestine.[107] Finally, commensal-reactive antibodies can also positively and negatively impact disease outcomes: systemic IgA, which unlike mucosal IgA is mostly TD, can protect against sepsis after barrier breach and bacteremia.[114] On the other hand, microbiota-specific IgG antibodies, which are increased in patients with ulcerative colitis (UC), can contribute to pathogenesis by activating intestinal macrophages through Fc receptors and fueling inflammation.[115]

Large amounts of IgA are secreted into breast milk by mammary gland B cells in order to protect the neonate gut. sIgA is the most abundant antibody in breast milk.[116] The delivery of IgA to the infant gut is of fundamental importance as children only start to produce IgA by 4 weeks of age.[117] Breast milk–derived IgA plays a fundamental role in the development of the microbiome in infant and protects against necrotizing enterocolitis in preterm infa nts.[109,117,118] Furthermore, maternal antibodies, which have broad commensal-reactive specificities, also limit adaptive immune reactivity to the microbiota in early life[105] and impact microbiota composition into adulthood.[109] Furthermore, in mice, the homeostatic set point of colonic RORγt+ T_{reg} is vertically transmitted during a short postnatal window by maternal IgA. RORγt+ T_{reg}, in turn, controls intestinal and milk IgA levels, thus transmitting this set point throughout multiple generations.[119]

INTRAEPITHELIAL LYMPHOCYTES

IELs reside within the epithelial layer and represent a highly heterogeneous group of cells distributed along the GI tract.[120-122] IELs represent one of the most abundant T cell populations, and it has been estimated that 1 IEL is present for every 4 to 10 epithelial cells in the small intestine and 30 to 50 in the large intestine.[123] IELs constitutively patrol the space between epithelial cells above the basement membrane where they are poised for rapid activation of cytolytic and type 1 cytokine responses directed at infected or stressed epithelial cells.[122] Although these cells share properties and localization, IELs arise from highly diverse lineages and are composed of both conventional and unconventional lymphocyte populations[122] as well as group 1 innate lymphoid cells (ILC1) and ILC1-like cells (Table 38.1). The intraepithelial T-cell compartment in both humans and mice is composed of TCRγδ+, TCRβ+CD8αβ+, and TCRαβ+CD4+T cells, and TCRαβ+CD8αα+ IELs (in human TCRαβ+CD8aa+) also express CD8αβ.[124] The human small and large intestines are dominated by TCRαβ+ IELs. In contrast, the murine small intestine has an equal distribution of TCRγδ+ and TCRαβ+ IEL with a shift toward TCRαβ+ in the colon.[121,125]

Conventional IELs

Conventional CD4+TCRαβ+ and CD8+TCRαβ+ IELs are dominantly tissue resident and derived from encounter of naive T cell with antigens in Peyer patches or in the periphery

TABLE 38.1 Characteristics of IEL Subsets

IEL Subset	Repertoire	Ligands	Effector Molecules
Mouse			
Conventional TCRαβ+	Diverse	MHC class I MHC class II	TNF-α, granzyme B, IFN-γ
Unconventional TCRαβ+	Diverse	MHC class I MHC class II Nonclassical MHC class, RAE1, MULT1, H60a, Qa-1	TNF-α, granzyme B, IFN-γ
Unconventional TCRγδ+	TCRVγ7 enriched	BTNL1, BTNL6, RAE1, MULT1, H60a, Qa-1	TNF-α, granzyme B, IFN-γ
ILC1		RAE-1, MULT1, H60a, Qa-1	TNF-α, granzyme B, IFN-γ
CD8αα innate IEL		Thymus leukemia antigen (TL)	IFN-γ
Human			
Conventional TCRαβ+	Diverse	MHC class I MHC class II MICA, MICB ULBP	TNF-α, granzyme B, IFN-γ
Unconventional TCRαβ+	ND	ND	ND
Unconventional TCRγδ+	TCRVγ1-enriched	CD1, MIA, MICB, ULBP	TNF-α, granzyme B, IFN-γ
ILC1	None	MICA, MICB, ULBP	TNF-α, granzyme B, IFN-γ
Intracellular CD3+IEL	None	ND	

Abbreviations: BTNL, butyrophilin-like protein; H60a, histocompatibility antigen 60; IEL, intraepithelial lymphocyte; IFN-γ, interferon γ; ILC1, group 1 innate lymphoid cell; MIC, MHC class I polypeptide-related sequence; MULT1, mouse ULBP- like transcript 1; ND, not determined; RAE1, mRNA export factor; TCR, T-cell receptor; TCRV, TCR variable chain; TL, thymus leukemia antigen; TNF, tumor necrosis factor; ULBP, UL16-binding protein.

in a context favoring the induction of gut homing molecules such as α4β7.[17,126] The nature of the antigens driving the activation and accumulation of conventional T cells remains largely unknown. However, the significant decrease of this compartment in germ-free and protein antigen–free mice supports the idea that these cells may be enriched in microbiota or dietary antigens specificities.[127] Under defined conditions, such as in the presence of TGF-β and RA in the context of strong TCR signaling, conventional IELs can acquire expression of CD8αα. Converted CD4+CD8αα+TCRαβ+ lose expression of the T-helper inducing POZ/Kruppel-like factor (THPOK) and concomitantly turn on the expression of Runt-related transcription factor3 (RUNX3) as well as T-bet and IEL-associated genes such as natural killer receptor 2B4 and granzyme B.[120,128-130] Local antigen recognition is an essential signal for differentiation of CD4+ T cells at the epithelium, yet differentiated IELs are able to preserve an effector program in the absence of TCR signaling.[26] LP T_{reg} cells can also undergo similar conversion in a microbiota-dependent manner upon migration to the epithelium, a process associated with the loss of FOXP3.[131]

Unconventional IELs

Unconventional IELs include TCRαβ and TCRγδ+ cells that can be double negative for CD4 or CD8ab coreceptor or CD8aa positive.[122] Different from conventional T cells, unconventional IELs recognize self-ligands rather than foreign antigens and already express a cytolytic effector program prior to stimulation.[122] The vast majority of unconventional IELs are thymically derived, although few IELs can be detected in mice devoid of thymus.[132] The current model for the development of unconventional TCRαβ+ IEL proposes that thymic IEL precursors escape thymic negative selection in a process referred to as agonist selection, whereby elevated TCR signals promote clonal deviation rather than clonal deletion.[122] The TCRαβ+ IEL repertoire is polyclonal at birth, and in response to exogenous stimuli, including luminal antigens and microbial composition, these cells can expand and become oligoclonal in the adult.[133] TCRαβ+ IEL TCRs are broadly MHC cross-reactive and can recognize many MHC class I and MHC class II molecules across multiple MHC haplotypes.[134] The maturation and expansion of IELs in the periphery is highly controlled by the cytokine interleukin 15 and by dietary metabolites.[120,135-137]

TCRγδ IELs seed the intestine early in life, irrespective of microbial colonization or exposure to dietary antigen.[138] In mice, unlike unconventional TCRαβ+IELs, TCRVγ7+ IELs exit the thymus as naïve cells and undergo activation and maturation in the gut epithelium.[134,138] Butyrophilin and butyrophilin-like family members that are structurally related to the B7 family are expressed on the surface of various populations of epithelial cells and can serve as ligands for various populations of TCRγδ T cells. Butyrophilin-like protein (BTNL1), BTNL4, and BTNL6 are expressed in the intestinal epithelium, and in the absence of BTNL1 TCRVγ7 IELs are selectively lost.[138] In humans, the tissue-resident TCRγδ+ IEL compartment is dominated by a unique subset of innate-like, semi-invariant Vγ4+/Vδ1+ and Vγ4+/Vδ1− IELs that recognize BTNL3/BTNL8, constitutively express natural cytotoxicity receptors, and persist throughout life.[139]

All IEL lineages share canonical functional properties such as their ability to patrol the epithelium and exhibit potent cytolytic and Th1 effector function.[122] Within the gut IELs have complex functions that include the promotion of local immunity and maintenance of barrier integrity. A role for IEL in promoting host defense has been revealed in various models of mucosal infection. For instance, TCRαβ IEL can promote effector responses to a variety of pathogens such as *Eimeria vermiformis*, *Encephalitozoon cuniculi*, and *Toxoplasma gondii*.[140-143] In the context of bacterial or parasitic infections, IELs can mediate pathogen clearance via their capacity to produce various inflammatory mediators such as IFN-γ and TNF-α.[144] IEL can also promote immunity to nematodes by releasing cytokines such as interleukin-13 that can promote mucus formation.[145] Moreover, IELs express various NK receptors endowing them with a strong cytolytic activity allowing for direct elimination of damaged or infected cells.[120] Their protective role can result from the capacity of these cells to maintain the integrity of tight junctions between epithelial cells therefore preventing epithelial transmigration of pathogens.[146] In this regard, IELs have been associated with the maintenance of barrier function in the context of simian immunodeficiency virus or human immunodeficiency virus.[147,148] IEL can also exert regulatory function at mucosal sites. In particular, TCRγδ IELs play a major anti-inflammatory role via their capacity to promote tissue repair upon injury and to release an array of factors with regulatory properties. In particular, γδ T cells produce large amounts of keratinocyte growth factor that plays an important role in supporting epithelial integrity.[149,150] IELs can also produce the immunoregulatory cytokines IL-10 and TGF-β that can both limit inflammatory responses at mucosal sites.[151-153] Various lines of evidence suggest that, in defined settings, IEL can also contribute to tissue damage and inflammation. Experimentally, these cells have been shown to promote mucosal inflammation.[154-157] Furthermore, CD8αα+ IELs can also contribute to the initiation of experimental colitis. More particularly, IELs have been associated with the exacerbation of celiac disease, an inflammatory disorder in genetically susceptible individuals associated with the induction of inflammatory T-cell responses against dietary gluten.[139,158] Celiac disease is characterized by a dramatic increase of the IEL compartment.[158] TCR-activated CD8ααTCRαβ IELs cause severe villous atrophy by targeting IECs that express stress-induced MHC class I polypeptide related sequence (MIC) antigen in an NKG2D-dependent manner.[159,160] In addition, IL-15, a factor that is overexpressed in various mucosal disorders including celiac and Crohn diseases[158] can trigger potent cytotoxic responses by CD8αα+ IEL via a NKG2D-DAP10-dependent signaling pathway.[160] Supporting a role for IEL in celiac disease pathogenesis, genome-wide association studies (GWASs) have identified a number of genes involved in the development, migration, and cytotoxic/NK function of IEL.[161] Intestinal γδ IELs produce fibroblast growth factors, which promote the proliferation and differentiation of epithelial cells, returning these tissues to homeostasis following damage.[150,162] In addition to mediating epithelial repair directly via the expression of growth factors, γδ T cells can also promote tissue homeostasis through the recruitment of regulatory T (T_{reg}) cells. Within the gut, γδ T cells have been shown to also contribute to the adaptation of gut epithelial cells to nutrient sensing.[163]

UNCONVENTIONAL T CELL NETWORK

In addition to unconventional IEL within the epithelium, a large fraction of T cells residing at mucosal sites including the LP are specific for modified peptides and small molecules including defined lipids and metabolites produced by microbes.[164] These unconventional T cells are restricted by monomorphic MHC class Ib (MHC-Ib) molecules, and some populations, such as invariant natural killer T (iNKT) cells, mucosal-associated invariant T (MAIT) cells, and subsets of γδ T cells, express a semi-invariant T-cell receptor (TCR) that limits their antigenic range analogously to an innate immune receptor.[164] Also referred to as innate-like T cells, these populations developmentally acquire effector characteristics prior to thymic egress, including rapid cytokine release and expression of chemokine receptors and integrins, which results in their accumulation in tissues before conventional effector T cells, during the second trimester in humans and within 3 weeks of birth in mice.[165-170] Recent evidence supports the idea that unconventional T cells compete for niches within tissues and may exert overlapping functions associated with the preservation of tissue homeostasis (Fig. 38.2).

Although iNKT cells can be selected on self-ligands,[171] they are hyporesponsive in the absence of commensals.[172,173] However, iNKT cells accumulate within the colonic LP and lungs of GF mice due to elevated expression of the chemokine CXCL16 by epithelial cells, a phenomenon associated with exacerbation of colonic inflammation and allergic airway responses.[174] The microbiota also provide antigens that can directly modulate the development and or accumulation of iNKT cells during the first 2 weeks of life in mice.[165] As such, early-life exposure to *Bacteroides fragilis* prevents the accumulation of iNKT cells within the colon, but not the lungs, of their progeny, indicating that the inhibitory effects of the sphingolipids are localized to the predominant site of microbial colonization.[165] Thus, microbial colonization during the neonatal period imprints the abundance of iNKT cells in barrier tissues, which can have long-lasting effects on host physiology.

iNKT cells are abundant in mice, whereas the predominant unconventional αβ T cells in humans are MAIT cells, which recognize intermediates of vitamin B2 (riboflavin) synthesis presented by MR1.[175] Because riboflavin synthesis is broadly conserved among bacteria and fungi, MAIT cells respond to a diverse array of microbes, including known commensals.[166,176] Consequently, MAIT cells are nearly absent from peripheral tissues in animals that are either GF or colonized with commensals that do not synthesize riboflavin.[166,177-179] In conventionally housed mice, MAIT cells accumulate in barrier tissues prior to weaning and neonatal colonization of

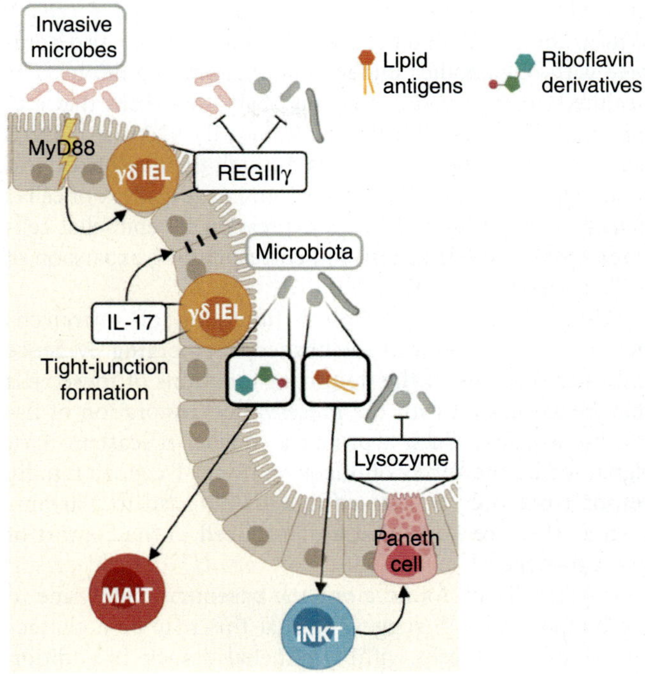

FIG. 38.2. Mucosal unconventional T cells. Unconventional T cells engage in a lifelong dialogue with the microbiota and control tissue homeostasis. Unconventional T cells, including γδ T cells, MAIT cells, and iNKT cells, seed the gut, where they control numerous physiological tissue responses. These include regulation of the microbiota composition and promotion of tissue repair. In response to epithelial MyD88 signaling, gut epithelium–resident γδ T cells produce the AMP REGIIIγ, which in turn regulates microbial composition and localization. Colonization can also be regulated indirectly in response to microbial antigens: Microbiota-derived lipids induce iNKT-dependent Paneth cell secretion of lysozyme. γδ T cell–derived IL-17A contributes to tight-junction formation between enterocytes, and following tissue injury these cells are potent producers of the epithelial mitogen KGF. Abbreviations: AMP, antimicrobial peptide; IEL, intraepithelial lymphocyte; iNKT, invariant natural killer T; KGF, keratinocyte growth factor; MAIT, mucosal-associated invariant T. (Figure adapted from image created with BioRender.com.)

GF animals with riboflavin-synthesizing intestinal commensals results in a comparable number of MAIT cells at these sites.[166,178]

Recent evidence supports the idea that, within mucosal tissues, unconventional T cells should be assessed as a network of cells with potentially overlapping and/or redundant functions. Indeed, although unconventional T cells respond to disparate antigens, they express remarkably similar transcriptional programs,[180] resulting in an inherent competition that results in the maintenance of a relatively stable level of effector cells within tissues.[166,181,182] Experimentally, cytokines such as IL-23 and IL-1 have been proposed to contribute to this phenomenon, but for the most part, the nature of the trophic/survival factors responsible for the tissue-specific composition and frequency of unconventional T cells remains poorly understood. The intrinsic ability of these populations to compete was confirmed in a patient harboring a homozygous point mutation in MR1 associated with an absence of MAIT cells and concomitant expansion in γδ T cells.[183] Within the gut, defined subsets of γδ T cells can also compete with one another under inflammatory conditions. The healthy human intestine contains Vγ4+Vδ1+ IELs that recognize the butyrophilin-like molecules BTNL3 and BTNL8 and promote tissue homeostasis through their cytotoxic and tissue repair programs.[138,184] However, in the context of celiac disease, abrogation of BTNL8 expression by epithelial cells leads Vγ4+ Vδ1+ IELs loss and corresponding expansion of inflammatory Vδ1+ IELs.[139]

Although the antimicrobial functions of unconventional T cells have been established,[164] emerging evidence indicates that one of the dominant functions of these cells may be associated with the preservation/restoration of tissue homeostasis. Such a property can be, at least in part, explained by the ability of unconventional T cells to rapidly respond not only to defined ligands but also to alarmins such as IL-18 or IL-33 that are produced in the context of tissue damage.[185]

MAIT cells are found along the basement membrane in the human colon,[186] suggesting that this may be a characteristic of MAIT cells within epithelial tissues. In addition to their unique localization, MAIT cells express a tissue repair transcriptional program, which is enhanced by TCR-mediated stimulation alone.[166,186,187] However, when activated in the presence of stimulatory cytokines, MAIT cells acquire a distinct transcriptional profile characterized by effector functions, with diminished expression of tissue repair genes.[186]

Growing evidence indicates that the role of unconventional T cells is highly dependent on their tissue localization, host developmental stage, and disease context. Indeed, although the ability of unconventional T cells to promote epithelial growth and angiogenesis is beneficial in the context of tissue repair, these functions can also exacerbate tumorigenesis.[188] Conversely, γδ T cells and other unconventional T cells also exhibit antitumor responses.[189-191]

MUCOSAL MONONUCLEAR PHAGOCYTES

Immune reactivity against nonpathogenic gut elements is not only wasteful but is also known to lead to severe tissue damage. However, the development of active immunity is required to protect the host against invasive pathogens. This tight equilibrium is under the control of a complex network of mucosal mononuclear phagocytes (MP) with unique properties that form an organized cellular network in the LP and are distributed throughout the GALT and draining LNs.[192]

Origin and Identification of Mucosal Mononuclear Phagocytes

Within mucosal tissues there is considerable developmental and functional heterogeneity of MPs, which includes populations of conventional and plasmacytoid dendritic cells (cDC and pDC), as well as monocytes (Mo) and macrophages (Mph). The identification of these populations depends on both common and unique surface markers that derive from their unique cellular origins and differentiation trajectories (Fig. 38.3). Of MHCII+ lineage cells in the LP, Mphs are by far the dominant population in all regions of the small intestine and colon, making up 70% to 90% of these cells, the other 10% to 30% made up of monocytes, developing Mphs, and cDCs. Furthermore, unique Mph populations are present in the submucosal, serosal, and muscularis layers of the intestine. GALT tissues contain unique Mph and cDC populations that are differentially distributed to specific regions of these tissues.

cDCs originate from circulating pre-cDC1 and pre-cDC2 that differentiate in the bone marrow in response to Flt3L and mature into cDC populations in tissues and lymphoid organs under the influence of the hematopoietic cytokines Flt3L, GM-CSF (Csf2), and others, and unique tissue signals and transcription factors.[193,194] In general, cDCs can be identified based on their expression of high levels of CD11c, and MHCII, and their lack of expression of lymphocyte and granulocyte markers, as well as low-level expression of well-established monocyte/macrophage surface proteins, such as FcγR1 (CD64), EMR1 (recognized by the F4/80 antibody), and MerTK. Furthermore, these cells express the zinc finger and BTB domain–containing protein 46 (ZBTB46) not found in Mphs.[195,196]

In contrast to cDCs, Mph develop either from embryonic precursor cells from the fetal liver and/or yolk sac or from circulating bone marrow–derived monocytes[197] under the influence of M-CSF and other tissue-specific factors that result in tissue-specific resident Mph including the presence of microbial flora.[192,198-200] Compared with other tissues, in the intestine, a larger proportion of Mphs are derived from circulating monocytes that replace embryonic precursors starting after weaning. As a whole, tissue Mphs in the intestine can be identified by their lack of lymphocyte and granulocyte markers and their high expression of MHCII, CD64, EMR1 (F4/80), and MerTK.[195,196]

MP Within Mucosa-Associated Lymphoid Tissues

Populations of cDCs that migrate from the tissues to draining LNs can be distinguished from blood-derived resident cDC by their higher expression of MHCII and lower expression of CD11c. Migratory cDCs in the MLN consist of the main populations of cDC found in the LP and reflect those of the specific tissues from which they drain.[203] For example, the proximal MLNs draining the small intestine have more CD103+ cDC2 and the most distal MLNs more 103$^-$ cDC2 from the colon, which is also true for colon-draining caudal and iliac LNs. Furthermore, CD103+ cDC2 and CD103+ cDC1 make up the largest proportions of migratory cDCs in the MLN.[203]

GALT tissues contain several populations of pre-cDC–derived cDCs and monocyte-derived cells. Within PPs Batf3 and IRF-8-dependent CD8+ cDC1 are present predominantly in the T cell–rich IFR, similar to peripheral LNs, whereas cDC2 populations are present in the SED region as well as in the IFR.[204,205] In the SED, CD103−CD11c+CD11b−CD8-CD172α+ cDC2 (originally

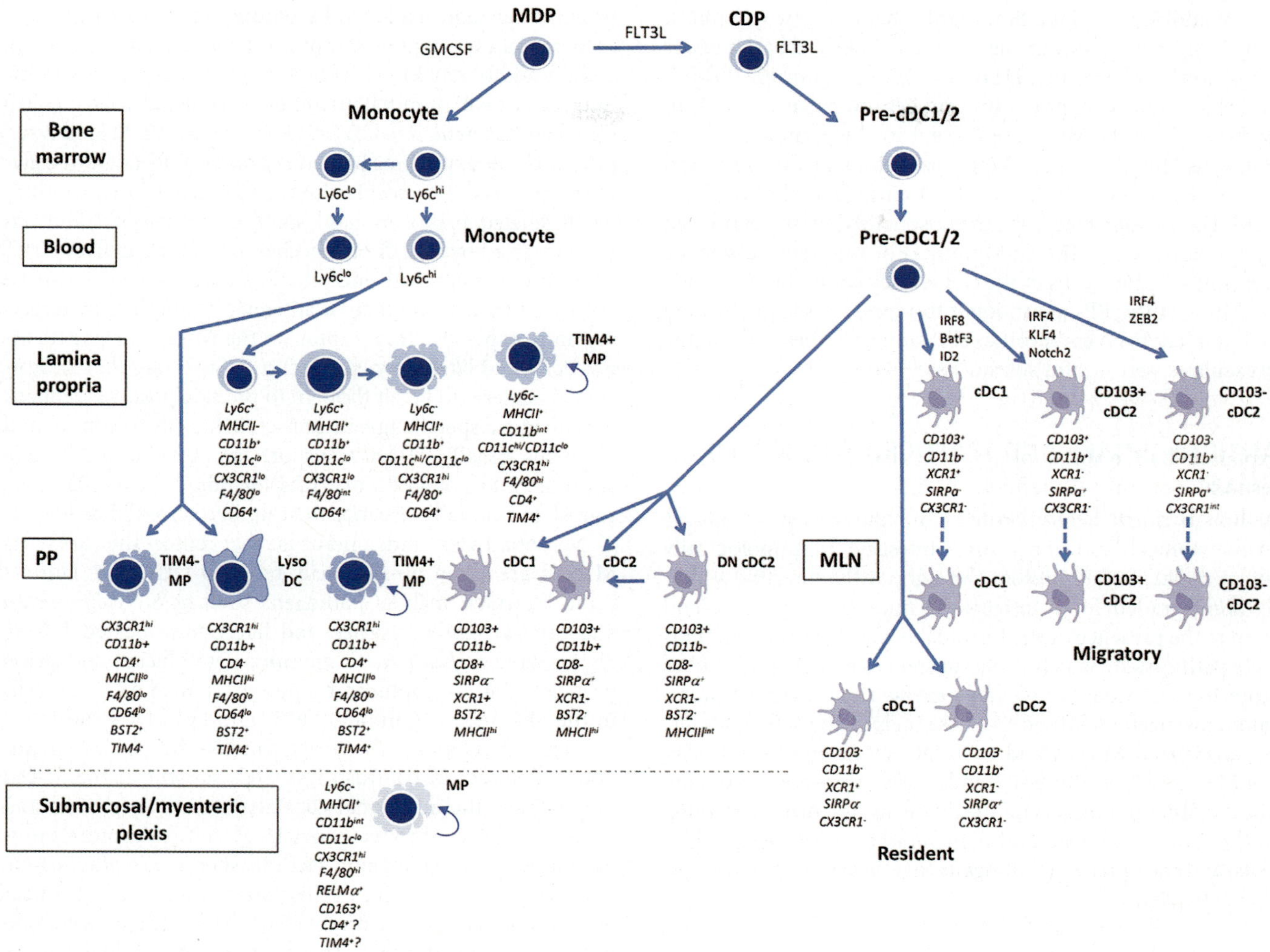

FIG. 38.3. **Origin and identification of mucosal mononuclear phagocytes.**

termed "double negative" (DN) because of their lack of CD8 and CD11b expression) are thought to further differentiate into CD103−CD11c+CD11b−CD8−CD11b+CD172α+ cDC2 present in the SED through as yet unclear signals.[206]

cDC are also present within ILFs of the small intestine and colon. ILFs develop from rudimentary structures called cryptopatches, collections of lymphoid-tissue inducer (LTi) ILC3 cells and lymphoid tissue organizing (LTo) stromal cells that are present at birth. Upon bacterial colonization, CD11c+ cells are recruited to cryptopatches and produce CXCL13, which contributes to attracting B cells to form an ILF.[207] ILF CD11c+ cells appear to be largely CD103− cells[208] and have recently been shown to be a specialized population of cDC2 that constitutively expresses IL-23, IL-6, and uniquely LysM, IL-22 binding protein (IL-22BP) and IL1RN, called cryptopatch and ILF associated cDCs (CIA-DCs). However, others have found that these cells are more consistent with a Mph phenotype[209-211] and curiously are very similar in function to Lyso-DCs present in PPs (see below).

In addition to cDCs, blood monocyte-derived cells have been identified in Peyer patches.[204] Two populations express CD11c, CX3CR1, merTK, CD172a (SIRPa), BST2, and high levels of lysozyme but uniquely do not express F4/80 or CD64. One population has a high turnover rate, expresses high levels of CD11b and genes of the MHC II presentation pathways, and has been called "Lyso-DCs" because of their ability to present antigens to naïve T cells.[212] Lyso-DCs are primarily located in the SED, where they differentiate further as they get closer to the FAE where they can localize in M-cell pockets and extend dendrites into the intestinal lumen to capture antigens and microbes through M cell–specific transcellular pores.[212-214] These Lyso-DCs express CD24, embigin, and JAM-A, the latter of which may be involved in their interaction with epithelial cells or cell trafficking. Another developmentally distinct Lyso-DC population is present in the B-cell follicle that lacks SED lyso-DC markers and expresses many genes distinct from SED lyso-DCs and cDC2.[214] Owing to their location, these cells likely have an as yet unclear functional relationship with B-cell differentiation.

An additional population that is characteristic of Mph in other tissues is found in the SED and follicle that expresses CD4, and low levels of CD11b and MHC II, but not TIM-4, and is not capable of presenting antigens to naïve T cells.[204] In addition, TIM-4+ Mphs are present in the germinal centers (GCs) as "tingible-body" Mphs, as well as in the IFRs, and at the base of the follicle near the draining lymphatics. Since TIM-4 is a major receptor for phosphatidyl serine expressed by apoptotic cells, TIM-4+ Mphs may be more specialized for clearing apoptotic cells in the IFR as they are in the GC. TIM-4+ Mphs in the PPs are thought to represent self-replicating cells derived from embryonic cells, similar to these cells in the muscularis, serosa, and submucosa.[215]

ANTIGEN UPTAKE AND TRAFFICKING OF MYELOID CELLS

Although most dietary proteins are digested by proteases in the stomach and upper small intestine, immunologically intact molecules are present throughout the intestine, either as nondegraded food antigens or microbial products, and despite the presence of the formidable epithelial barrier, multiple pathways by which these substances as well as microbes from the intestinal lumen can traverse the intestinal epithelium have been identified.[216] These include the active passage by specialized M cells and Lyso-DC across the FAE overlying MALT tissues, the active and passive transport across the villus epithelium by several mechanisms, the direct sampling of the lumen and dead and dying epithelial cells by myeloid cells, and the uptake of pathogens that have breached the epithelial barrier.

Mucosa-Associated Lymphoid Tissues

Luminal antigens, including macromolecules, bacteria, and viruses, gain access to the cells of MALT tissues, like the PPs and ILFs through active transport by microfold (M) cells present in the FAE.[216-218] The FAE lacks goblet cells and has poorly developed surface mucins so the mucous coat is significantly more permissive to antigen and microbe uptake and invasion. M cells are specialized epithelial cells that develop in MALT tissues in response to nonhematopoietic stromal cell expressed receptor activator of nuclear factor kB ligand (RANKL), which activates RANK on epithelial cells resulting in lineage commitment dependent on the ets-transcription factor SpiB. Subsequently, CCR6+ CD11c^{int} B cells drive M-cell maturation due to CD137 interaction with M-cell CD137L, resulting in the formation of M-cell pockets, invaginations in the expanded basolateral membrane and full functional maturation. M-cell development can be enhanced by systemic administration of RANKL, which has allowed detailed analysis of the M-cell differentiation program.[218]

M-cells take up particulate matter through clathrin-mediated endocytosis, actin-dependent phagocytosis, or micropinocytosis and transport their cargo through the cell to be released in M-cell pockets. M-cell pockets contain B cells, T cells, cDC2, and lyso-DCs. M-cell transport is relatively promiscuous and mediated by binding to surface-expressed carbohydrates in regions free of overlying mucus. In addition, several additional M-cell receptors allow for more specific interactions with luminal microbes. This includes β1-integrin (*Y. pseudotuberculosis*), CD155 (Polio virus), PrP^{c} (*B. abortus*, prions, GP60-expressing bacteria), and the GPI-anchored glycoprotein 2 (GP2) a receptor for type 1 pili, that enhance entry by flagellated pathogens and some commensals[219] and has been targeted by specific antibodies for vaccine delivery.[216,217] In addition, microbial entry to PPs across M cells can be enhanced by the presence of antigen-specific IgA by targeting an isotype-specific receptor on the M-cell surface that is not FcRa and likely involves dectin-1. Such targeting of IgA-coated bacteria to PPs it thought to provide positive feedback for enhanced specific IgA responses, primarily to commensal microbes. Finally, M-cell transport can be enhanced by oral administration of TLR2 or TLR4 ligands[220] so that bacterial signals can actually enhance their uptake across M cells.

Microparticles and microbes that enter the SED via M cells are taken up primarily by Lyso-DCs, and Tim-4− Mphs,[204] which include pathogens such as *S. typhimurium* and *Listeria monocytogenes*, and likely translocated *E. coli*, SIgA-coated *S. flexneri*, *C. albicans*, beta-glucan, and prion particles that are captured by poorly defined CD11c+ cells in the SED. Although both cDC2s and Lyso-DCs can interact with M cells directly, antigen uptake by cDC2 remains poorly defined. In contrast, lyso-DCs present in the M-cell pocket have the additional capacity to actively extend and retract dendrites through pores in M cells to capture luminal antigens and pathogenic bacteria, such as *S. typhimurium*. However, the importance of this pathway is unclear as a near majority of bacteria are present in TIM-4− Mphs, which do not extend dendrites.[212,213] The ability to extend dendrites by Lyso-DCs correlates with their expression of E-cadherin and JAM-A, the latter likely contributing to interactions between the M cell and dendrite membranes. Finally, the SED CD11c+ cells can capture apoptotic epithelial cells, including M cells, and apoptotic bodies in steady-state conditions, and Lyso-DCs may be most involved in this setting.[212,221] Following reovirus infection of the FAE, CD11c+CD8−CD11b−cells take up apoptotic infected FAE cells in the SED, which are likely Mphs, due to their poor ability to present viral antigens to primed T cells ex vivo.[221]

Finally, it is clear that soluble antigens like ovalbumin can gain access to PP, as evidenced by the efficient de novo differentiation of Foxp3 regulatory T cells in PPs by orally administered ovalbumin[222]; however, the mechanism of soluble antigen uptake in the PPs has not been described, as it has for the non-FAE epithelium (see below). In addition, intestinal bacteria enter the ILFs via M cells similar to PPs and induce the production of IgA in a largely T cell–independent manner (see below). After activation, both lyso-DCs and cDC2 in the SED appear to migrate to the IFR to initiate T-cell responses. This occurs experimentally with infectious agents, such as *T. gondii*, or following oral administration of cholera toxin or the TLR7/8 agonist R848.[205,214] For Lyso-DCs, this involves the upregulation of CD40, IL-12p40, and CCR7, the latter of which may be responsible for migration to CCL19 and CCL21 expressed in the IFR, where they closely associate with proliferating CD4+ T cells.[214] In addition, cDC2s can directly interact with B cells in the SED to initiate B cell switching to IgA,[85] and CCR6+ CD11c+ cells, presumably cDCs, have been shown to form clusters with CD4+ T cells in the SED following *Salmonella* infection,[223] suggesting that T cells may also be activated at this site. Although it has been suggested that cDCs from the PPs and ILFs can migrate to the MLNs, there are few data to support this possibility.

Villus Epithelium, Lamina Propria, and Draining Lymph Nodes

An additional site for antigen entry into the intestine is the nonfollicular absorptive epithelium, where despite well-developed mucous and epithelial barriers both soluble antigens and microbes can gain access to cDCs and Mphs in the LP. This occurs via several mechanisms that include paracellular transport ("leak") across epithelial cell junctions; active endocytosis and transcellular transport across epithelial cells, in particular goblet cells; transport of antibody-bound antigens by Fc receptors; and active uptake by populations of Mphs and cDCs that extend dendrites or migrate into the epithelial layers. In addition, pathogens have developed sophisticated mechanisms to invade and destroy epithelial barriers to allow for invasion into the LP and spread to systemic tissues.

Epithelial cohesion and polarity are maintained by the apical junctional complex, which is composed of tight and adherens junctions, and by the subjacent desmosomes.[224] Transport of molecules across these adhesions are largely controlled at the level of tight junctions, which are rate limiting for the uptake of luminal molecules, owing to their small pore size compared with deeper adhesions.[225,226] Studies using live confocal imaging after oral gavage of fluorescent 10-kDa dextran to normal mice demonstrated that some degree of paracellular leakage of molecules occurs in the small intestine, which is transported to the draining lymphatics but interestingly not avidly taken up by underlying cDCs in the LP.[227] Furthermore, epithelial cell junctions are altered by inflammatory signals to enhance paracellular transport. This includes the inflammatory cytokines IFN-γ, TNF-α, IL-1β, and IL-12 as well as direct activation of NFkb by bacterial products, as seen with *Listeria monocytogenes*.[228] Activation of NFkb in epithelial cells results in activation of myosin light-chain kinase and subsequent redistribution or internalization of junctional proteins claudin, occludins, and E-cadherin that can increase permeability. Increased permeability can facilitate uptake of bacterial pathogens through exposure of ligands, such as E-cadherins for *Listeria*,[228] and likely results in enhanced uptake of soluble food and bacterial products as well.

In addition to paracellular "leakage," transcellular transport of antigens across IECs by active endocytosis was first demonstrated with horseradish peroxidase in rats.[229] Transcellular transport across epithelial cells has also been shown with nanoparticles (in the range of 20–40 nm) and bacterial debris,[230] as well as with intestinal pathogens such as adherent-invasive *E. coli* that can induce micropinocytosis of junctional proteins together with adherent bacteria for transepithelial cell transport into the LP[231] and with antibody-bound antigens or immune complexes mediated by Fc receptors.[232,233]

The polymeric-immunoglobulin receptor (PIgR) is well known for its role in binding and transporting polymeric IgA and immune complexes from the LP to the luminal surface, and the expression and function of which is highly upregulated by inflammatory signals. In contrast to PIgR, the neonatal Fc receptor (FcRn) is an MHC I–like molecule associated with β2 microglobulin that acts to transport IgG and associated immune complexes in a bidirectional manner. In rodents, FcRn is expressed on epithelia of the intestine only in neonates before weaning, thus its name, but in humans it is continually expressed into adulthood, being present at higher levels in the proximal small intestine.[234] FcRn facilitates transport of IgG and immune complexes from maternal milk to an infant's blood stream across the villus epithelium and transport of IgG or immune complexes into the intestinal lumen. FcRn expression has been found more ubiquitously in the liver, kidneys, and lung and has broader roles in controlling the half-life of IgG as well as albumin.[233]

The role of FcRn in antigen uptake in the intestine has been explored in mice with genetically induced constitutive expression of human FcRn on IECs.[235-237] In such mice, oral administration of IgG-antigen immune complexes was shown to be transported to underlying cDCs, which migrate to LNs, where they cross-present the associated antigen to induce CD8+ T-cell responses.[238] FcRn is also expressed by CD11c+ cells, likely cDCs, where after uptake by FcγRs, intracellular FcRn helps direct Ag associated with IgG to MHCII loading compartments for enhancing CD4+ T-cell responses.[233] In this manner, IgG-antigen immune complexes present in the milk of systemically immunized mother mice can be transferred to neonatal mice via FcRn to induce CD4+FoxP3+ regulatory T cells that can prevent subsequent allergic responses to the complexed antigen when weaned mice are subject to allergic sensitization protocols,[239,240] an effect also dependent on expression of FcRn on CD11c+ cells.[235] As FcRn is also expressed highly by Mph, FcRn may facilitate one response over another depending on the capturing cell type or additional cosignals, including those from additional FcRs on cDCs or local macrophages.[241]

Specialized antigen sampling M cells normally present in the FAE have also been identified in the villus epithelium that mediates transcellular transport of bacteria and particulates,[242] which can be induced by systemic administration of RANKL similar to those in lymphoid follicles.[243] Villus M cells can bind bacteria,[242] and RANKL administration induced M cell differentiation on small intestinal villi, which were able to avidly take up *S. typhimurium* and *Y. enterocolitica*, as well as 200 nm fluorescent microbeads.[243] Furthermore, villus M cells may be induced by infection with pathogenic bacteria[244] or stimulation with bacterial toxins, such as cholera toxin.[245]

Importantly, a significant role for constitutive transcellular transport of antigens by endocytosis and transport across goblet cells has been recently demonstrated. These so-called goblet cell–associated passages, or GAPs, were identified by microscopy following intraluminal administration of fluorescent dextran, bovine serum albumin, and ovalbumin in both the small intestine[227] and colon.[246] GAPs were shown to be "columns" of antigen that are actively passaged to underlying CD103+ cDCs.[227] GAPs were also associated with goblet cells at other mucosal surfaces and shown to be present in the human jejunum but have been primarily studied in the mouse intestine.

GAPs are largely present in the small intestine and distal, but not proximal, colon and develop in these regions at different stages of life. GAPs are induced by acetylcholine acting on goblet cell–expressed muscarinic ACh receptor 4 (mAChR4), which also induces compound but not primary mucous granule exocytosis and inhibited by EGFR signaling.[247] In the murine small intestine GAPs are formed just before weaning at day 18 of life, are present through adulthood, and are not dependent on the presence of commensal bacteria. Interesting, *Salmonella typhimurium* infection resulted in both GAP-dependent bacterial uptake early in infection and the inhibition of GAP formation at later time points secondary to IL-1β signaling, resulting in less dissemination of bacteria and dietary antigens to the MLNs, as did elimination of goblet cells, suggesting that, in addition to dietary antigens, intestinal pathogens may use GAPs for bacterial entry.[248]

GAPs are expressed in the proximal colon for only a short window, from day 10 of life to 21, near weaning, and are inhibited by intestinal bacteria signaling through TLRs and EGFR pathways. Following treatment with antibiotics, proximal colon GAPs were not inhibited and allowed for pathogenic immune responses to commensal and pathogens resulting in intestinal inflammation.[246-248] In the distal colon GAPs are present post weaning and are not inhibited by commensal bacteria. It has been speculated that the suppression of GAPs after weaning in the proximal colon where the mucous layer may not be as protective is critical for limiting harmful immune responses to intestinal bacteria, but exposure during the critical period of GAP formation during weaning may be harmful. The distal colon may be protected from exposure throughout life due a thicker mucous coat; however, the functions of persistent GAPs at this site are not yet known.[249]

In addition to para- and transcellular transport, studies have also demonstrated direct uptake of luminal antigens by myeloid cells. CX3CR1+ Mph associated with the basal surface of the villus epithelium can extend protrusions directly into the lumen across interepithelial tight junctions to sample luminal antigens and bacteria. In now classic studies, Rescigno and colleagues demonstrated that the development of these transepithelial dendrites (TEDs) involves coordinated expression of the tight junction proteins occludin, claudin-1, and junctional adhesion molecule (JAM) by the myeloid cells that allows for extended dendrites to form tight-junction-like connections with epithelial cells that prevent loss of barrier function.[250] TEDs are present in the proximal jejunum and distal ilium, the later induced in response to pathogenic and nonpathogenic *S. typhimurium* infection and by orally administered TLR-2, -4, and -9 ligands, and were capable of capturing both invasive and noninvasive *S. typhimurium* and nonpathogenic *E. coli*.[251,252] TED induction was dependent on MyD88 signaling in nonhematopoietic cells, was generated without the disruption of barrier function, and was absent in mice deficient for CX3CR1 or CX3CL1 that also displayed defects in the internalization of noninvasive bacteria,[251,253,254] suggesting that epithelial activation by bacteria drives TED formation, possibly by the elaboration of CX3CL1. These findings are also consistent with the observation that depletion of CD11c+ cells reduced bacterial colonization in the colonic LP after oral administration of noninvasive *S. typhimurium*.[253] Of interest, TEDs are not found in the colon, suggesting a unique mechanism of small bowel antigen uptake.

Furthermore, CX3CR1+ cells in the small intestine were shown to be superior to CD103+ DCs in the uptake of soluble antigens, with active accumulation of orally administered ovalbumin into vacuoles in CX3CR1+ cells in the LP[255] this occurred in the absence of TEDs,[256] suggesting these cells may capture soluble antigens transported to the LP by other routes (*for example,* across goblet cells or via trans- or paracellular transport across epithelial cells).

In contrast to CX3CR1+ Mph, CD103+ LP cDCs have been found more deeply within the LP, or more superficially in the epithelium,[256,257] and similar to intestinal Mph, LP CD103+ CD11b+ cDCs capture soluble antigens and bacteria given orally, and also have the capacity to migrate to the MLN where they activate T cells.[256] Furthermore, CD103+ DCs were also found to carry bacteria as well as fragments of IECs, possibly in the form of apoptotic bodies, to the MLN,[257,258] suggesting that luminal antigens taken up by, or bound to, epithelial cells may be carried to the MLN together with epithelial cell fragments by CD103+ cDCs.

How LP CD103+ cDCs acquire bacterial, soluble, and epithelial-associated antigens has been addressed in several studies, suggesting that this can occur by several mechanisms. Firstly, small numbers of highly motile CX3CR1−CD11c+CD103+ cDCs were found to migrate into the epithelium above the basement membrane where they moved between epithelial cells, before returning to the LP.[255] Within the epithelium, these cells captured soluble Alexa 594–labeled ovalbumin, but not as efficiently as CX3CR1+ cells in the LP. In addition, CD103+ cDCs in the LP actively migrated into the epithelium following *Salmonella* exposure, which depended on MyD88 and/or TRIF signaling, where they sent TEDs into the lumen to capture invasive and noninvasive *S. typhimurium*.[255] Furthermore, similar to CX3CR1+ cells, CD103+ cDCs expressed the tight junction proteins claudin 4 and ZO-2 that may allow for tight junction formation with epithelial cells.[255] Secondly, CD11c+ CD103+ CX3CR1−cDCs were found to directly capture antigens transported by GAPs. Antigen capture by CD103+ cDCs occurred throughout the length of the small intestine, was limited to small molecules and not intact bacteria or particulates, and appeared to be a major pathway for the uptake of soluble proteins as LP CD103+ cDCs from mice fed antigens without goblet cells were not able to stimulate T cells ex vivo, and soluble antigen uptake by TEDs was not observed in these studies.[227] Thirdly, it was recently shown that CX3CR1+ cells in the small intestine, in particular in the duodenum, acquired antigen in ligated intestinal loops and were able to transfer antigen to CD103+

cells in a connexin-43–dependent manner. Connexin-43 is a major component of the gap junctions that were formed between the CX3CR1+ cells and CD103+cDC. Furthermore, oral tolerance to soluble antigen was ablated in connexin-43 or CX3CR1-deficient mice.[259]

cDCs leave the intestinal wall and migrate to draining LNs in steady-state conditions with a half-life of 7–10 days, and this turnover is increased with inflammation.[196,260] The mesenteric, duodenopancreatic, iliac, and caudal LNs drain different segments of the intestine.[7] The MLNs drain the small intestine, cecum, and the ascending colon, while the remainder of the colon drains to caudal and iliac LNs.[7,201,202] Populations of cDCs that migrate from the tissues can be distinguished from blood-derived resident cDCs by their higher expression of MHCII and lower expression of CD11c. Migratory cDCs in the MLN consist of the main populations of cDCs found in the LP and reflect those of the specific tissues from which they drain.[203] For example, the proximal MLNs draining the small intestine have more CD103+ cDC2 and the most distal MLNs more 103^- cDC2 from the colon, which is also true for the caudal and iliac LNs. Furthermore, CD103+ cDC2s and CD103+ cDC1 make up the largest proportions of migratory cDCs in the MLN.[203]

SPECIALIZED FUNCTIONS OF MUCOSAL MYELOID CELL POPULATIONS

In addition to roles shared with similar cell types from other tissues, Mph and cDC populations in the intestine have evolved specialized functions that play essential roles in tissue and immune homeostasis, as well as host defense, and that are highly influenced by local microenvironmental conditions (Table 38.2).

TABLE 38.2 Specialized Functions of Intestinal Myeloid Cells

Tissue	Cell Type	Cellular Origin	Major Reported Functions	Selected References
Lamina propria	Monocyte (Ly6C^{hi}/MHCII^{lo})	BMo	Precursor to inflammatory and regulatory Mo and Mph	200,261,262
	Transitional/inflammatory monocyte (Ly6C^{int}/MHCII^{int})	BMo	Contribution to inflammatory colitis	200,261-263
			Inhibition of PMNs via PGE_2	264
	Macrophage (Ly6C^{lo}/MHCII^{hi})			
	General/nonlocalized	BMo > SR	Anti-inflammatory cytokine production (via IL-10 signaling)	265-270
			Support for T_{reg} differentiation via IL-1b/ILC3/Csf2/cDC	271
			Support for T_{reg} survival/expansion by IL-10 production	272-274
			Source of IL-23 production for support of ILC3 production of IL22	210
			Pathogenic T-cell expansion in chronic colitis	275
			Phagocytosis and killing of bacteria	276-278
			Tissue repair following sterile injury or colitis	279
	Epithelial cell associated	?	Luminal antigen uptake and transfer to cDCs for oral tolerance	259
			Uptake of dying epithelial cells	280
			Uptake of luminal bacteria and fungal products	250-252,281
			Epithelial stem cell survival and replication	282,283
	Blood vessel associated	SR + BMo	Preservation of blood vessel integrity and permeability	215,279
			Prevention of bacterial translocation to the blood	279
			Uptake of antigens from the blood	198

TABLE 38.2 Specialized Functions of Intestinal Myeloid Cells (Continued)

Tissue	Cell Type	Cellular Origin	Major Reported Functions	Selected References
			Induction of colitis via CCL8	284
	Neuron associated (submucosal plexis)	SR > BMo	Survival of neurons	215
	Muscle associated (myenteric plexis/ muscularis externa)	SR > BMo	Survival of neurons	215,285
			Regulation of peristalsis	215,286
			Tissue protection/repair	287,288
	Conventional dendritic cells			
	cDCs (CD103+; both cDC1 and cDC2)	pre-cDCs	Antigen uptake and trafficking to MLN	256
			Induction of T_{reg}	222,289-291
			Induction of oral tolerance	292
			Induction of intestinal homing receptors	208,293-295
	cDC1 (CD103+)	pre-cDC1	Cross-presentation to CD8+ T cells in MLNs	296
			Priming of T_{reg} in MLN	297
	cDC2 (CD103+)	pre-cDCs (pre-cDC2?)	Development intestinal Th17 cells	298-300
			Generation of Th2 responses to helminths	301
			IL-23 production for host defense	302,303
	cDC2 (CD103-)	pre-cDC2 (pre-cDC2?)	Induction of Th1 and Th17 cell differentiation	203
			Induction of T_{reg} in the caudal and iliac LNs	202
			Generation of Th2 responses to helminths	301
Peyer patch	Macrophage			
	TIM-4+ Mph	SR > BMo	Clearance of apoptotic cells	304
	TIM-4- Mph	BMo	Contribution to microbial defense	304
			Antigen and apoptotic cell uptake	304
	Lyso-DC	BMo	Antigen uptake across M cells	213
			Uptake of pathogenic bacteria and dead epithelial cells	212
			Migration and Th1 cell priming	214,304
			Induction of IL-17 production via IL-6 and IL-23	305
			Contribution to microbial defense	304
	Conventional dendritic cells			
	cDC1	pre-cDC1	Differentiation of Th1 cells and CTLs	304,306
	DNcDC2 (CD11b-CD8-)	pre-cDCs (pre-cDC2?)	Induction of IgA B-cell differentiation	85,307
			Uptake of apoptotic epithelial cells	221
	cDC2	pre-cDCs (pre-cDC2?)	IL-10 production and differentiation of Th2 cells	306,308
			Induction of IL-17 production via IL-6 and IL-23	305
ILF	CIA-DCs	?	Regulation of IL-22 production and function	209,211

Abbreviations: BMo, blood monocyte; pre-cDC, blood precursors to cDCs; SR, self-replicating.

Unique Functions of Mucosal Monocytes and Macrophages

Mph populations are determined by cues from surrounding cells that define specific anatomical, developmental, and functional "macrophage niches".[197,309,311] In addition, Mph can be influenced by the amount of time spent exposed to these environmental factors, and the presence of inflammation that can result in both loss of resident Mph, thus opening of the niche for circulating precursors, and specific signals that guide their development. Thus, local conditions can affect cell phenotypes and functions, as well as the relative proportions of long-lived embryonically derived or bone marrow–derived cells that are present, due to rates of cell death, and replacement by monocyte precursors into a defined niche.[197]

Maintenance of the Enteric Nervous System and Blood-Gut Barrier

Several such niches have been described in the intestine that are more or less organized in different layers and structural components of the intestinal wall. Thus, for example, Mph associated with nerves in the muscularis layers (MMphs), primarily in the myenteric plexus, are largely derived from embryonic precursors, have a unique morphology, and are in close association with nerves that produce CSF-1 to promote their turnover. In addition, adrenergic and cholinergic neurotransmitters influence their function by signals through the β2-adrenergic receptor (b2AR) and the α7 nicotinic acetylcholine receptor (α7nAChR), respectively.[286,287,312,313] In steady-state conditions in response to microbiota CSF-1 produced by muscularis neurons stimulates MMph production of bone morphogenic protein 2 (BMP2) that sustains the development of the enteric nervous system and also constitutively activates enteric neurons, thus regulating colonic contractibility.[286] During infection and inflammation, signaling via b2AR and α7nAChR dampens the inflammatory potential of MMphs.[287,288,313] Furthermore, a population of largely self-renewing Mph is associated with larger vessels in the submucosa that regulate blood vessel integrity.[215] Of importance, loss of self-maintaining Mph results in neuro- and vascular degeneration, delayed intestinal transit, and vascular leakage, as well as decreased intestinal secretion due to loss of nervous control of anion secretion by epithelial cells.[215]

Contributions to Structural and Immune Homeostasis in the Villus Epithelium

In addition to these self-renewing populations, Mphs in the LP (and submucosal layers) primarily differentiate from blood monocytes along the "waterfall pathway" described above and make up the vast majority of monocyte-derived cells in the colon and small intestine. Important factors for intestinal Mph development include TGF-β, short-chain fatty acids, IL-10, CXCL1, TSLP, aryl-hydrocarbon receptor (AHR), and Notch ligands.[314]

Intestinal LP Mphs function to both protect against microbial invasion and ensure structural and immunological homeostasis. They are highly phagocytic and express associated genes for uptake of microbes and apoptotic cells, including *Cd206* (MRC-1), *Mertk, Axl, CD36, Marco, Clec7a, Itgav*, and *Itga5*.[315] They are also highly capable of killing internalized microbes that involves enhanced phagolysosomal killing by AMPs and proteins, including cathepsins, calprotectin, and lysozyme, together with active xenophagy.[196,276] Enhanced killing functions may be due to exposure to TGFβ, and particularly to the short-chain fatty acid butyrate produced by intestinal bacteria, which acts to enhance xenophagy and AMP and lysozyme production in Mphs in vitro through a mechanism involving HDAC3 inhibition.[277,315,316] In contrast, PGE_2 appears to oppose intracellular killing mechanisms in Mphs, especially when it is overproduced in the absence of IL-10 signaling.[276]

Intestinal Mphs also support the growth and repair of the intestinal epithelial barrier by silently eliminating apoptotic epithelial cells and providing essential epithelial growth factors, such as PGE_2, hepatocyte growth factor, and WNT ligands, to support crypt stem cell niches and Paneth cell development.[196,283] They also protect the epithelium against toxic metabolites from enteric fungi.[281] Furthermore, in the steady state, they play an important role in soluble antigen uptake and passage to CD103+ cDCs, as well as expansion of Foxp3+ T_{reg} for the generation of oral tolerance to soluble antigens.[216,259,317] For the latter, IL-10 has been implicated, but also IL-1b produced at low levels may stimulate ILC3 cells to make GM-CSF, which both sustains the development of Mphs in the tissues and drives RA production from cDCs, which promotes the generation of Foxp3+T_{reg}.[271]

Although intestinal Mphs are highly active cells, they are poorly responsive to external stimulation, such as TLR ligands and other PRRs, with the lack of production of inflammatory cytokines.[314] Instead, LP Mphs are a source of anti-inflammatory cytokines, such as IL-10, IL1rl1, and IL1RN, as well as low levels of IL-1β and TNF-α.[200,261,265,266,278] Such poor responsiveness is thought to be largely due to exposure of Mphs to IL-10 that is produced by many sources, but critically from CD4+ Foxp3+ regulatory T cells. Thus, humans with *IL10Ra* or *IL10Rb* mutations develop early-onset IBD,[318] and in the absence of IL-10R on $CX3CR1^-$ or lysozyme 2–expressing cells, mice develop spontaneous colitis,[268,270] similar to total IL-10-deficient mice.[319] However, if IL-10 is deleted in the same populations there is no spontaneous inflammation.[268,270] Furthermore, Mphs from humans with *IL10Ra* or *IL10Rb* mutations and global and myeloid cell–specific IL-10R-deficient mice, as well as STAT3-deficient myeloid cells,[320] are highly sensitive to TLR stimulation, with the production of inflammatory cytokines, have an altered metabolic profile[268,270,276], and paradoxically have a decreased capacity to kill intracellular organisms due to secondary effects of overproduced PGE_2.[270,276] Mice also develop spontaneous colitis when IL-10 is deleted from Foxp3+ cells,[321] indicating that IL-10 from T_{reg} but not Mphs themselves is essential to prevent untoward inflammatory responses by intestinal monocyte/Mphs. TGF-β has also been implicated in the suppressive function of Mph in humans.[278,322]

IL-10 signaling has several functional mechanisms to inhibit inflammatory responses by monocyte and Mph populations. IL-10 drives metabolic reprograming in Mphs in vitro, inducing a specific mTORC1 inhibitor (DDIT4) that both blocks a switch from oxidative phosphorylation to glycolysis necessary for proinflammatory cytokine production in response to LPS and induces mitophagy that limits reactive oxygen species important for inflammasome activation.[267] IL-10 also inhibits accessibility of inflammatory genes at the level of chromatin remodeling,[323,324] can induce inhibitors of NFkB signaling,[325] and may enhance ubiquitin-mediated degradation of inflammasome components.[269,326] How these latter mechanisms relate to metabolic changes is not yet clear. The degree to which IL-10 affects the maintenance of an anti-inflammatory state in resident Mphs and/or inhibits early monocyte differentiation into an inflammatory phenotype is also not yet clear; however, it is likely primarily important in the latter, as in colitis models and *C. rodentium* infection, IL-10R deficiency enhances the number of inflammatory monocytes, as well as their production of IL-1β, NO, TNFa, and IL-23, the latter which consequently drives local IL-17 production from Th17 cells primed in MLNs and IL-22 produced from ILC3s.[327]

In contrast to IL-10 responsiveness, IL-10 production from Mphs may be induced by bacterial signals such as TLR ligands, interactions of Mph CX3CR1 with CX3CL1 or TSLPR with TSLP (IL-50) expressed by epithelial cells, or exposure to local type I IFNs or PGE_2.[327] Mph-derived IL-10 can have autocrine effects in vitro and can enhance T_{reg} suppressive function, possibly by maintaining Foxp3 expression by these cells[274] or by inducing their production of IL-10 to act as amplifiers of negative regulatory mechanisms.[272] IL-10 production by Mphs also contributes to epithelial repair and possibly direct suppression of T-cell differentiation in some models of colitis.[327] Regardless of these functions, IL-10 deficiency in intestinal Mphs does not lead to overt spontaneous intestinal inflammation, but more work needs to be done to decipher the functions of Mph-derived IL-10 in other settings.

Recent single-cell studies have shown additional developmental and functional heterogeneity in intestinal MPs populations.[198,215,328] Studies in adult mice supported the development of CCR2+ precursors into mature macrophages along two major pathways, possibly similar to other tissues.[329] Both developmental pathways are inhibited in the relative absence of complex microbiota in mice before weaning or in germ-free adult mice or those given antibiotics,[192,198-200] whereas embryonically derived cells are dependent on bacterial flora for their maintenance in tissues.[330]

One population expresses high levels of CD11c, low levels of CD206 (MRC1), and a unique marker, CD121b (IL1R2) an IL1-decoy receptor, resides in the more distal regions of the colon crypt LP closer to the villus tips, some in close association with small vessels in the villi, others in close proximity to the epithelial cell basement membrane where they may be particularly influenced by microbial products or apoptotic epithelial cells.[198] Owing to their proximity to the surface epithelium,[198,200] and their presence in both the small intestine and colon, $CD11c^{hi}$ $CD206^{int}CD121b+$ Mphs likely perform many functions attributed to intestinal Mph in prior studies discussed above. This population also expresses several intestinal Mph genes that are dependent on TGF-β signaling,[198,200,315] suggesting that this is one factor driving their differentiation.

In contrast, $CD11c^{lo}CD206^{hi}$, CD169+ Mphs in the colon LP and submucosal tissues appear to be derived from monocytes in adult mice and closely associated with blood vessels.[198,284] These cells express many typical Mph markers, including scavenger receptors MRC1 (CD206), CD36, CD163, immunoregulatory factors ATF3 and IL-10, clotting and repair factors F13a, and chemokines CCL2, CCL6, CCL7, CCL8, and CXCL4 (Pf4).[198] In addition, a small population of these Mphs expresses Lyve1 that may guide interactions with the lymphatic epithelium.[198] These CD169+ Mphs constitutively produce chemokines for bone marrow–derived cells, particularly for monocytes.[198,331] During experimental dextran sodium sulfate (DSS) colitis enhanced production of CCL8 by CD169+ Mph contributes to disease severity by attracting inflammatory monocytes into the tissues.[284] CD169+ Mphs in the colon also avidly capture intravascular antigens,[198] similar to their function in other organs,[332] the biological significance of which in the intestine is unclear. Specific factors important for the differentiation of this population of Mphs versus $CD11c^{hi}$ cells discussed above are not yet clear; however, vitamin A may be important.[333]

In the PPs, in addition to their ability to phagocytose microbes and dead cells, the high levels of lysozyme produced by lyso-DCs, as well as Tim-4 Mphs together with their location near the FAE, indicate their function in host defense. Both populations also produce IL-22-binding protein (IL-22BP), a natural inhibitor of IL-22, suggesting that they act to limit AMP production by the FAE and consequently contribute to easier uptake and passage of bacteria by M cells into the SED. Lyso-DCs can also present antigens to T cells, although not as efficiently as PP cDC populations, and upon stimulation with a TLR7/8 agonist, R848, migrate to IFR regions where they cluster with CD4+ T cells; however, the contribution to adaptive immunity by these cells vis-à-vis PP cDCs is not yet clear.[204,212] In contrast, TIM-4+ PP Mphs are essential for clearing apoptotic B cells in the GCs and may have a similar function in the IFRs for the clearance of T or other cells.[204]

Roles in Mucosal Inflammation

In experimental colitis and Crohn disease (CD) and UC in humans, blood monocytes are recruited to the intestine, where they differentiate into inflammatory monocyte-derived cells. In the mouse, these cells are largely Ly6c+MHCII+ cells (falling into "P2" of the waterfall model in mice), and also mature over time to $Ly6c\text{-}MHCII\text{+}CX3CR1^{int}$ Mphs ("P3"). In humans these cells are largely $CD14^{hi}CD1c^{hi}$ cells. Inflammatory monocyte-derived cells are essential for colitis development in many IBD models, as shown by cell deletion and prevention of recruitment via CCR2 in global and cell-specific CCR2 deficiency.[196] These inflammatory cells play a primary role in driving inflammation by the production of inflammatory cytokines, including TNF-α, and IL-1β, for which they are the primary source, as well as IL-6, NO, and IL-12, IL-23, CCL11, and other monocyte and neutrophil attracting chemokines. These cells also express high levels of TREM-1 that acts to amplify inflammatory signals.[334] A similar accumulation of inflammatory monocytes has been shown in multiple colitis models, including T cell transfer models, IL-10R deficient, including the anti-IL-10R+ *H. hepaticus*, and DSS colitis models, as well as in intestinal infection, such as those caused by *C. rodentium* and *T. gondii*.[314,327] During the latter infections, however, inflammatory monocytes are protective, by producing IL-23 and IL1β that drives IL-22 production from ILC3s and subsequent production of AMPs from IELs, as well as IL-17 production by Th17 cells, and by limiting immunopathology via the production of PGE_2 that inhibits neutrophil-mediate inflammation.[264]

Ly6C^{hi} monocytes are recruited to intestinal inflammation in the setting of an altered environmental milieu, so are likely influenced by local inflammatory factors, such as IFN-γ, TNF, and IL-1β, as well as increased exposure to bacterial products that affect their differentiation into inflammatory cells rather than regulatory tissue Mphs.[314] In this regard, IFN-γ signaling in particular has been highlighted, acting either locally in tissues or systemically in the bone marrow, the latter resulting in circulating activated monocytes predisposed to an inflammatory phenotype.[261,263,335]

In contrast to these detrimental and protective proinflammatory effects of monocytes during inflammation and infection, evidence supports the likelihood that resident CX3CR1^{hi} Mphs persist during inflammation and can retain their anti-inflammatory signature, suggesting they may continue to play an immunoregulatory role even during inflammation.[261,262,336,337] In fact, when they are deleted, colitis in some models, particularly those with strong neutrophil involvement, like DSS colitis, is worse due to enhanced neutrophil infiltration.[338] In addition, inflammatory monocytes can produce IL-10 and PGE_2 to help control neutrophilic inflammation during *T. gondii* infection and DSS colitis.[264]

Unique Functions of Mucosal Dendritic Cells

Unique functions of intestinal cDC populations include the imprinting of lymphocytes with unique homing receptors that allow for their recirculation to intestinal tissues, the capacity to drive the differentiation of regulatory T cells that are involved in tolerance to soluble oral antigens, the provision of signals for the differentiation of IgA-producing B cells, and the ability to induce appropriate T-cell responses both to commensal bacteria to maintain homeostasis and to intestinal pathogens for host defense.

Induction of Homing Receptors on Lymphocytes

Intestinal T and B cells are initially activated in LNs, PPs, and ILFs, where they are induced to express tissue-specific homing receptors that allow them to return to mucosal tissues. CD103+ cDCs in PPs and MLNs induce on both CD4+ and CD8+ T cells, as well as IgA+ B cells, the expression of CCR9, the receptor for CCL25, which is highly expressed in the intestinal mucosa, and α4β7, an integrin that binds to mucosal vascular addressin cell adhesion molecule 1 (MAdCAM-1) expressed on high endothelial venules of intestinal tissue.[126,208,293,294] Such imprinting is dependent on RA released by cDCs.[17,295] CD103−cDC2 are also capable of inducing α4β7 and CCR9 on CD8 T cells via production of RA in vitro; however, since CD103+ cDCs constitute the majority of cDCs migrating to the MLN, the effect of RA on homing in vivo has been ascribed to CD103+ cDCs.[203] RA signaling itself confers on CD103+ DCs the ability to produce RA, thereby providing autocrine signals for RA production by cDCs. RA also is produced by IECs from retinol present in intestinal bile, and by stromal cells in the MLN, and the importance of these sources, as well as from cDCs in vivo, is not yet clear.[339]

Generation of IgA B-Cell Responses

B cells are induced to differentiate into IgA PCs through successive stages of B-cell CSR, SHM, affinity maturation, migration, and PC differentiation.[53,317] cDCs have also been implicated in IgA-B cell switching in Peyer patches. In recent studies, initial signals for IgA CSR in response to commensal bacteria and cholera toxin, a highly T cell–dependent response, occurred in the PP SED. This required CCR6 expression on pregerminal center B cells induced by CD40/CD40L interactions to position the B cells in the SED.[85] Here CD11b+ and CD11b−CD8− cDCs have intimate prolonged contact with CCR6+ naive B cells, which express low levels of AID, which promotes IgA B-cell CSR together with TGF-β activated by αvβ8-expression by the cDCs. B cells thus activated in the SED may also be dependent on T-follicular helper cells (Tfh) to provide CD40, IL-21, or other signals for full B-cell differentiation, SHM, and affinity maturation.

Furthermore, cDCs appear to have a major role in driving T follicular helper cell (TfH) differentiation required for T cell–dependent IgA B-cell responses to commensal bacteria. Thus, *Zbtb46*^{cre}+x MHCII$^{fl/fl}$ mice with loss of MHCII expression specifically on cDCs had significantly fewer Tfh (by 50%) as well as GC B cells (by 70%) compared with *Cre*- littermates, and when made germ-free, did not generate these cells when reconstituted with commensal bacteria.[340] Finally, MHCII+CD11c+ PP cells produce RA that contributes to IgA B-cell differentiation, likely by enhancing TGF-β signaling, and induces α4β7 expression on B-cell blasts that allow for homing to the intestinal LP. Together, these data indicate an important role for cDC2s, or possibly CD103−CD11b+ cDCs, in the induction of Tfh and TD responses in the PPs.

In contrast to T cell–dependent responses, T cell–independent IgA responses are also generated in the PPs, as well as in the ILFs and LP, although the overall data for the latter suggests this plays a minor role.[53] IgA B-cell CSR relies on TGF-β signaling together with the B-cell activators APRIL, and to a lesser degree BAFF substitution for CD40 signaling, to drive AID expression by binding to transmembrane activator and CAML interactor (TACI) on B cells. In PPs, follicular dendritic cells respond to TLR signals and RA to produce TGF-β, as well as BAFF. BAFF and APRIL in PPs also are produced by pDCs and CD11c+ cells, which may include cDCs and Mphs. CD11c+ TLR5+ cells release RA, which contributes to IgA CSR. The role of cDCs versus macrophages is even less clear in T cell–independent B-cell CSR in the ILFs, which occurs in response to stimulation with commensal bacteria, as these populations have been poorly defined.

Induction of Foxp3+ T_{reg} for Oral Tolerance

As further discussed, oral tolerance has been shown to be primarily mediated by the de novo differentiation of dietary antigen-specific CD4+Foxp3+T_{reg} (iT_{reg}) in the MLN, with the expansion in the LP before further dissemination to systemic sites.[317] T_{reg} induction in the MLNs is mediated by cDCs, primarily CD103+ cDCs that have migrated from the intestinal LP in a CCR7-dependent manner, and is dependent on cDC production of RA and TGF-β, as well as the expression by cDCs of αvβ8 integrin, which functions to activate latent TGF-β.[290,291,339] TGF-β is essential for de novo induction of iT_{reg}, whereas RA is not and in the absence of TGF-β signaling can contribute to the induction of Th17 cells in the steady state, and in the induction of Th1 and Th17 cells in inflammatory or infectious conditions likely due to its ability to contribute directly to TCR activation.[2,341] Of interest, recent studies suggest that CD103+CD11b−cDC1s are important for the optimal generation of iT_{reg} in MLN following oral OVA administration, possibly due to their high expression of RA and $\alpha_v\beta_8$ integrin, or to enhanced expression of TGF-β2, a lesser studied type of TGF, compared with other cDC populations.[297] cDC1 are not required, however, for oral tolerance induction. CD103+ cDC also express indoleamine 2,3 dioxygenase, an enzyme that participates in tryptophan catabolism and contributes to iT_{reg} induction. Its immunoregulatory effects may be due to either the reduction of local tryptophan concentration or the production of immunomodulatory tryptophan metabolites, like kynurenin that activates the AHR to suppress Th17 development, drive T_{reg} homing to the intestine via CCL20/CCR6 induction, and enhance T_{reg} functions by suppressing their expression of IFN-γ.[342] Others factors expressed by cDCs in the MLNs implicated in iT_{reg} induction are PDL1 and PDL2 and TRAF6.[343,344] Finally, soluble antigen administration to the colon also results in T_{reg} induction and systemic tolerance. However, in this case, RA and MLNs are not required, and it is likely mediated by colon CD103−CD11b+ cDCs draining to caudal and inguinal LNs, which represent primary sites for lymphatic drainage from the colon, but not small intestine.[202]

Generation of T-Cell Responses to Commensal and Pathogenic Intestinal Microbes

Specific cDC populations have unique capacities to drive T-cell responses, including those to commensal bacteria. In early mouse studies, different PP CD11c+ cells were shown to induce unique T-cell responses ex vivo: MHCII^{hi}CD11c+CD11b− cells induced Th1 cells, whereas MHCII^{hi}CD11c+CD11b+ cells induced Th2 and IL-10-producing cells.[306] Since then, additional functions have been identified for intestinal cDC populations that have been more well defined, and in mice lacking specific cDC populations.[196,339]

In vitro T-cell differentiation assays with purified cDC populations have indicated that cDC1s are specialized in driving Th1 responses and are the most capable cDC population for cross-presenting soluble antigens to CD8+ T cells, similar to these cells in other tissues and lymphoid organs, whereas cDC2s can drive Th2 and Th17, and differentiation, and CD103−CD11b+ cDC1s Th1 and CD8+ cell differentiation. In addition, as mentioned above, both CD103+ populations as well as CD103-cDC2s can induce T_{reg} differentiation as well as the intestinal homing receptor CCR9. These functions are correlated with the production of IL-12, IL-6, IL-12, and IL-23; TGF-β; and RA, respectively.

The in vivo functions of cDC1s have been evaluated in *Batf3*$^{-/-}$, *Cd11c-cre.Irf8*$^{fl/fl}$, *Zbtb46-cre.Irf8*$^{fl/fl}$, *Xcr1*-DTA mice, and *Clec9a*-DTR mice, which all have substantially reduced cDC1 populations.[339] cDC1s have specific effects on the differentiation and/or survival of Th1 cells in vivo, as seen in vitro, likely due to their ability to secrete IL-12. Thus, in these mice lacking cDC1, overall, the induction of Th1 cells in the MLN and their presence in the LP and colon are reduced. In addition, IL-12 production by cDC1s is important both for driving intestinal Th1 responses during *T. gondii* infection and contributing to constitutive release of IFN-γ by intestinal T cells, which protects epithelial cells against damage following DSS exposure, resulting in less colitis in this experimental model.

cDC1s also have a broader effect on intestinal T-cell homeostasis as *Cd11c-cre.Irf8*$^{fl/fl}$ and *Xcr1*-DTA mice treated with DT have reduced numbers of CD4+ and CD8+ LP T cells and conventional CD8αβ+TCRαβ+ and unconventional CD8αα+TCRαβ+ and CD8αα+TCRγδ+ intraepithelial lymphocytes in the small intestine. These effects may be due to poor homing of T cells to the small intestine, as cDC1s express the highest level of RA in the MLN, compared with other cDCs, and their elimination resulted in lower numbers of CCR9+$\alpha_4\beta_7$+ small intestinal homing T cells in the MLNs. In addition, low numbers of CD8+ T cells in the LP in these mice may be due to decreased priming in the MLN, as cDC1s are specialized at cross-presenting antigens, including those derived from epithelial cells. Effects of cDC1 on IEL populations are less clear.

In contrast to cDC1s, in vivo functions of CD103+ and CD103−cDC2 have been evaluated in *Cd11c-cre.Irf4*$^{fl/fl}$, *Cd11c-cre.Notch2*$^{fl/fl}$, *Cd11c-cre.Klf4*$^{fl/fl}$, *Clec4a4*-DTR, *Sirpα*$^{-/-}$, as well as *huLangerin*-DTA mice.[339] CD103+ cDC2s have a specific role in Th17 responses in vivo. Thus, mice that have a significant, but not complete, reduction of SIRPα+ cDC2s, but not cDC1s, have a selective loss of Th17 cells in the small intestine and colon LP in the steady state.[299,300] The lack of Th17 cells in the steady state in these cDC-deficient mice implies their role in Th17 responses to commensal bacteria. Furthermore, *Cd11c-cre.Irf4*$^{fl/fl}$ mice also fail to generate colonic Th17 responses to infection with *Tritrichomonas musculis*, an asymptomatic protozoan parasite that infects many mouse colonies.[298] The loss of Th17 cells in the absence of cDC2s may be due to poor T-cell priming in the MLN, as both cDC2 numbers and Th17 cells are reduced in the steady state and generate reduced Th17 responses in the MLN following systemic immunization with OVA, αCD40, and LPS.[299]

Furthermore, the induction of Th17 responses in the MLN, as well as Th17 differentiation by cDC2s, was dependent on IL-6, and cDC2s produce IL-6 in vitro, implying a role for this cytokine in Th17 differentiation to commensal bacteria.[299] In addition to IL-6, IL23, and TGF-β, which drive Th17 induction in vitro and in many other systems, may also contribute to cDC-driven Th17 responses to commensal bacteria, as small intestine LP CD103−CD11b+ cDCs express IL12/23p40 mRNA and LP cDC2s express mRNA for IL-23 and TGF-β in addition to IL-6 in the steady state.[300] Therefore, these studies provide strong evidence that cDC2s are important for driving Th17 cells in the intestine; however, the relative contributions of CD103+ and CD103−cDC2s, as well as the cytokines and mechanisms involved, require further study.

cDC2s have also been shown to be important for Th2 cell responses in the intestine. Although little is known about Th2 responses to commensal bacteria, *Cd11c-cre.Irf4*$^{fl/fl}$ mice do not develop a protective Th2 response to intestinal helminth infections in mice, including those caused by *Trichuris muris* and *Nippostrongylus brasiliensis*, as well as following challenge with *Schistosoma mansoni* eggs injected into the subserosal intestinal tissue.[301] Furthermore, CD103+ cDC2s are important for Th2 responses in the small intestine following *S. mansoni* egg challenge, whereas CD103− cDC2s are important following challenge in the colon, indicating that different cDC populations are involved in the induction of Th2 cells in different sites. This is consistent with the role for these cells in driving oral tolerance and colonic tolerance as discussed above. Finally, consistent with studies of oral tolerance mentioned above, CD103+ cDCs are also implicated in driving T_{reg} induction to commensal bacteria, as *huLangerin-DTA x BatF3*$^{-/-}$ mice deficient in all CD103+ DC populations have fewer T_{reg} in the intestinal LP. However, data regarding the specific CD 103+, as well as CD103− cDC populations involved in iT_{reg} induction in vivo are not at all clear, as data in different cDC-deficient mice provide conflicting results. Overall, however, they suggest that both cDC1 and cDC2s may have redundant functions for driving T_{reg} to commensal bacteria, possible in response to different composition and localization of commensal bacteria.

IMMUNE REGULATION AT THE MUCOSA

Failure to regulate responses to the microbiota or food antigens can have catastrophic consequences for the host as exemplified by the numerous inflammatory disorders including IBDs.[345] The mucosal environment requires a complex network of immunoregulatory mediators to maintain its integrity at steady state and in the face of infection or inflammation. Because of the extraordinary antigenic pressure at these sites, many mucosal cell types can in a context-dependent manner exert regulatory activities. Under homeostatic conditions, the coordinated action of these cells results in the induction of regulatory responses toward dietary or commensal-derived antigens. The acquisition of oral tolerance—the active suppression of inflammatory responses against the myriad of antigens derived from food and the microbiota—is a fundamental aspect of the mucosal immune system. Multiple mechanisms are involved in the induction of oral tolerance.[346] Experimentally, oral administration of soluble proteins results in systemic tolerance that prevents the initiation of potentially damaging inflammatory responses upon rechallenge with the antigen in the periphery.[347-349] A complex regulatory network including specialized populations of antigen presenting cells (APCs), lymphocytes, and innate cytokines controls GI tract homeostasis and converges to favor the induction of regulatory responses toward antigens present at mucosal sites. Early reports suggested that commensals played a role in this process as oral tolerance could not be induced in the absence of gut flora or gut flora–derived signals.[346,350,351,351-353] Over the last few years, various populations of regulatory T cells have been ascribed a central role in the induction of oral tolerance. As further discussed, Foxp3-expressing CD4+ T_{reg} play an essential role in maintaining peripheral immune tolerance as well as preventing autoimmunity and chronic inflammation with several subsets described on the basis of their origin, generation, and mechanism of action.

Foxp3+ T_{reg} cells play a nonredundant role in the maintenance of immunological homeostasis.[354] Within the gut, induced T_{reg} cells constrain responses to orally acquired antigens. This phenomenon is promoted by a complex network of cells including specialized antigen-presenting cells as well as a milieu enriched in TGF-β, RA, as well as microbiota-derived metabolites.[222,289,355-357] As previously mentioned, it has been long recognized that "oral tolerance," a phenomenon associated with the induction of T_{reg} cells,[356] cannot be induced in the absence of the microbiota.[351] Indeed, defined subsets of T_{reg} cells are reduced under germ-free conditions or in the context of antibiotic treatment.[358-361]

At barrier sites, experimental evidence supports the idea that T_{reg} may be enriched for TCR specific for microbiota-derived antigens[362-365] and a large number of isolates or microbiota-derived consortia can promote T_{reg} cells induction and/or function.[94,358,366,367] In contrast, microbes that dominate in the context of inflammatory states such as IBDs fail to promote T_{reg} cells.[95,368] Although IL-10 production by T_{reg} plays a dominant role in the GI tract, T_{reg} cells employ numerous redundant and complementary mechanisms to maintain mucosal homeostasis.[354,369]

The capacity of T_{reg} to adapt to specific environments and to respond to tissue-specific cues contributes to their ability to regulate mucosal tissues.[354] Under steady state, T_{reg} within the gut also express the transcription factor GATA3 or RORγT.[354] Expression of the canonical Th2 transcription factor GATA3 promotes T_{reg} fitness during inflammation and limits their acquisition of effector cytokine production.[370,371] T_{reg} cells expressing RORγt are particularly enriched in the colon.[359,360,371] In mice, the accumulation of RORgT-expressing T_{reg} cells occurs between week 2 and 3 post birth[359,360] at a time that coincides with weaning and as such massive immune reprograming and microbiota alteration.[372] This specific subset of T_{reg} cells has been associated with the control of aberrant immune responses at mucosal sites.[359,360,365]

Numerous microbiota-associated factors converge to promote optimal T_{reg} cell induction and function. For instance, short-chain fatty acids (SCFAs), metabolites produced in the colon by bacterial fermentation of dietary fibers and resistant starch, can promote numerous biological functions including T_{reg}-cell induction.[359,373-377] Mechanistically, SCFAs have been proposed to act either both directly on T cells and indirectly on macrophages or DCs via HDAC inhibition and/or G protein–coupled receptors such as Gpr43.[359,373,376,377]

Recent work revealed that diet alteration impacted the level of colonic RORγt T_{reg} cells and proposed a role for microbial bile metabolites in this phenomenon.[378-380] Secondary bile acids can also enhance the ability of cDCs to promote T_{reg} cell as in the case of isodeoxycholic acid.[378] Because of the fundamental importance of regulatory T cells in constraining reactivity to the microbiota, each of these factors are likely to act in a synergistic and redundant manner.

As previously mentioned in the context of SCFA and secondary bile acids, the microbiota can promote T_{reg} induction or function via its action on the antigen-presenting cell network. For instance, microbiota control luminal antigen sampling by mucosal cDCs and promote the induction or function of mucosal macrophages or cDCs associated with local expansion of T_{reg} cells.[199,251,381] The microbiota can also activate gut resident macrophages, and as such enhance the production of GM-CSF by ILC3.[271] This response, in turn, reinforces the ability of mucosal antigen-presenting cells to produce both IL-10 and RA, thereby promoting local immune regulation and T_{reg} induction.[271] Commensal-derived products can also act by controlling directly or indirectly the function of inflammatory cells.[264,382] Although most of what is known today about the regulatory properties of the microbiota arose from the exploration of the bacterial components of the flora, other microbes such as fungi and viruses are likely to promote similar or complementary aspects of the regulatory network. For instance, in the GI tract, experimentally, interaction of commensal fungi with the C-type lectin receptor Dectin-1 was able to prevent inflammation in the context of acute mucosal injury.[383]

INFLAMMATORY BOWEL DISEASE

The IBDs, UC and CD, are chronic and/or recurrent inflammations of the large intestine (UC) or small and/or large intestine (CD), commonly thought to be due to an unrestrained (and pathologic) mucosal immune response to elements of the intestinal (commensal) microflora or (less likely) to antigenic components in food. Two observations are the main supports of this view. First, the highly diverse types of experimental murine models of intestinal inflammation have in common the fact that they are dependent on the presence of a normal intestinal microflora and humans with these inflammations can sometimes be treated with bowel sterilization or specialized diets. Second, the majority of genetic factors associated with IBD have in common the fact that they are in one way or another the cause of mucosal immune dysregulation due mainly (if not solely) to either a defect intrinsic to the mucosal immune system or a defect of the mucosal immune system that is accompanied by and to some extent dependent on a concomitant defect in IEC function.

This formulation of IBD pathogenesis accommodates current concepts of the main forms of IBD. The inflammation in CD, for example, is dominated by a mucosal immune system abnormality unassociated with an independent epithelial cell defect, and the latter occurs as a secondary manifestation. This accords with the pathologic picture that characterizes CD, in that CD is characterized by a penetrating and often granulomatous inflammation leading to bowel wall thickening and possible intestinal obstruction or fistula formation associated with occasional epithelial layer ulceration. The inflammation in UC, in contrast, originates from a mucosal immune system abnormality that is either unique to UC or partially shared with CD and that is dependent on or causes an abnormality of intestinal epithelial layer integrity. This accords with the main pathologic features of UC in that it is a relatively superficial inflammation marked by extensive ulceration, intestinal crypt distortion, and pericrypt neutrophil infiltration that is accompanied by an LP infiltration enriched in NKT cells.

Environmental and Genetic Factors Underlying IBD

UC and CD are driven by both environmental and genetic factors, the former accounting for the fact that they occur more frequently in the more urbanized, economically advanced countries. This epidemiologic distribution has been attributed to the "hygiene" hypothesis, which holds that, in the relatively sanitized environments characterizing such countries, there is reduced generation of suppressor cells that ordinarily downregulate immune responses and are of particular importance in the GI tract, an immunologic domain constantly exposed to a potentially inflammatory microflora.[384] It should be noted, however, that other environmental factors have also been identified such as perinatal infection-associated antibiotic administration.[385]

Genetic factors in IBD, also a prominent feature of the disease, were noted early on by the fact that IBD not infrequently clusters within families and is more common in genetically related family members than in spouses or in monozygotic twin pairs than in dizygotic twin pairs. In some cases, the clusters consist of both patients with UC and patients with CD, arguing that the two forms of IBD are genetically interrelated. More recently, the genetic basis of IBD has gained strong molecular footing with the advent of GWASs in which large populations are examined to determine associations between single nucleotide polymorphisms and increased (or decreased) occurrence of disease. These studies have led to the identification of over 200 disease "risk polymorphisms" that result in a variable (and usually low) increase (or decrease) in risk for disease development in individuals bearing the polymorphism.[386] In general, these polymorphisms are incapable of causing disease when present alone and, indeed, quite frequently occur in normal individuals who never develop IBD. In theory, they cause disease when accompanied by one or more additional risk polymorphisms or perhaps certain environmental factors. This implies that, from a genetic point of view, in the vast majority of cases of UC and CD, the genetic basis of the disease is both complex and variable. This is not to say that IBD is never due to a solitary gene mutation. On the contrary, with the recent application of genetic sequencing methodology, it has become apparent that, in rare instances, IBD can result from mutations in a single gene that causes disease in a classical Mendelian pattern. The latter usually occurs in very young patients with very-early-onset IBD (termed VEO-IBD) and usually involves a gene that plays a key role in the regulation of mucosal immune responses.[387] The best example of this are the rare instances of severe CD due to inactivating mutations of the IL-10- or IL-10R-encoding genes. In some cases, single gene mutations cause only (or mainly) GI disease and are thus *bone fide* causes of monogenic IBD; in other, more frequent cases, the GI disease is one component of a spectrum of abnormalities involving multiple organs that thus does not truly mimic ordinary IBD.

An overall survey of the risk polymorphisms associated with IBD discloses that polymorphisms affecting genes involved in immune function constitute more than half of the polymorphisms. Furthermore, these immune function–associated polymorphisms occur almost as frequently in UC as in CD, except for polymorphisms affecting genes targeting macrophage/DC function, which occurs mainly in CD. This indicates that UC as well as CD has a primary immune cell basis, at least in part. On the other hand, polymorphisms affecting genes involved in epithelial barrier function occur mainly in UC, emphasizing the greater input of primary epithelial cell abnormalities to this disease. A full discussion of the possible mechanisms by which the various polymorphisms cause IBD is not possible here; nevertheless, it is useful to consider several of the best studied polymorphisms because of the insight this provides into IBD pathogenesis in particular and to mucosal immune function in general.

The first IBD-associated gene polymorphism (or rather set of polymorphisms) discovered and the one giving rise to the greatest increased risk for IBD development (~17-fold) are the polymorphisms in the Leucine-Rich Region (LRR domain) of the NOD2/CARD15 gene.[388] NOD2 is a member of the NLR family of innate intracellular sensors and is usually stimulated by muramyl dipeptide (MDP), a component of the walls of intestinal commensal bacteria; its dysfunction is therefore well positioned to be a cause of increased intestinal responses to commensal bacteria. Since the NOD2 polymorphisms cause a loss in NOD2 function the dominant theory of how the polymorphisms result in disease is that they cause a defect in the ability of the mucosal immune system to contain (ie, eliminate) particular microbial organisms that penetrate the epithelial barrier and enter the subjacent LP where they have the potential to induce inflammation.[389] Evidence that NOD2 mediates host defense responses to various pathogens supports this idea, as do reports showing that NOD2 responses are necessary to maintain homeostatic responses to components of the gut microbiome. With respect to the latter, it has been shown that defects in NOD2 function within epithelial cells disrupt Paneth cell defensin production and thus cause a colitogenic microflora in the lumen of the terminal ileum, the site of CD most associated with NOD2 polymorphisms. In addition, NOD2 defects have been shown to interfere with efficient autophagic destruction of intestinal organisms by intestinal phagocytic cells. Finally, NOD2 deletion has been shown to cause an altered intestinal microbiome that is associated in some as yet undefined way with increased severity of dextran sulfate (DSS)-induced colitis. Despite these data, the idea that NOD2 loss-of-function polymorphisms cause CD by causing a kind of mucosal immunodeficiency has serious difficulties. The problem is that such immunodeficiency predicts that NOD2 dysfunction would be accompanied by frank evidence of bacterial invasion of the intestinal mucosa; in addition, it predicts that NOD2 dysfunction would lead to increased susceptibility to pathogenic intestinal infection, particularly in the face of successful therapy that uniformly undermines immune responses. In the absence of these expected manifestations one has to conclude that the view that NOD2 polymorphisms cause CD by its detrimental effect on host defense is not yet proven.

A second theory of how NOD2 polymorphisms lead to CD is based on the observation that NOD2 host defense function is accompanied (or followed in time) by a regulatory function that affects not only its ability to induce proinflammatory effects but also those induced by other innate factors, namely, TLR-induced responses. The first evidence that this might be the case is that TLR-induced in vitro responses are enhanced in the absence of NOD2 and pre-exposure of cells in vitro to MDP dampens subsequent proinflammatory responses to TLR ligands.[390,391] This immunoregulatory effect was then demonstrated in vivo in studies showing that intraperitoneal administration of MDP downregulates DSS-colitis, presumably by amplifying NOD2 immunoregulatory function. In addition, mice lacking NOD2 and administered T cells responsive to OVA-peptide were susceptible to colitis induced by subsequent intrarectal administration of OVA-expressing *E. coli,* whereas wild-type littermates subjected to the same regimen were not susceptible; however, this susceptibility was lost in mice with both NOD2 and TLR2 deletion, indicating that the colitis was dependent on both a response to a mucosal antigen as well as to a TLR2 adjuvant effect that was enhanced in the absence of NOD2. This evidence of immunoregulation of TLR responses resulting from pre-exposure to MDP was eventually demonstrated in cells from humans, and in this context the fact that this was found to be defective in those bearing a NOD2 polymorphism completed the circle of evidence that the polymorphism was acting through its immunoregulatory effect to cause increased risk of CD, at least in part.[392] The knowledge that NOD2 activation has immunoregulatory consequences introduced the question of its mechanism. The best evidence available so far is that NOD2 activation leads to induction of IRF4, a multifunctional molecule that causes deubiquitination of key inducers of NF-κB activation such as TRAF6 and thus attenuates NF-κB-mediated inflammation.[393]

The concept that excessive mucosal responses to enteric bacteria is the cause of CD finds strong support in the evidence discussed above showing that the proximate cause of increased risk of disease in individuals bearing the NOD2 polymorphism is the loss of the ability to regulate responses to a peptide derived from gut bacteria. However, this conclusion is also supported by studies of the mechanisms of disease underlying other polymorphisms. An instructive example of this is the increased risk of CD associated with a polymorphism in the Atg16L1 gene giving rise to a threonine to alanine substitution that causes impairment of Atg16L1 protein function.[394] This polymorphism occurs in 65% of patients with CD; it also occurs in 52% of control individuals without disease; thus, the increased risk is significant but not high (about twofold). The function that is potentially impaired by the presence of the polymorphism is the ability to generate autophagy, an intracellular vesicular/lysosomal process, whereby the cell clears itself of potentially toxic materials including damaged mitochondria, aggregates resulting from cellular stress and ingested pathogenic bacteria. This follows from the fact that Atg16L1 complexed with Atg5 and Atg12 is necessary to the lipidation of the LC3 component of the autophagic vesicle and thus facilitates both the initial formation of the autophagic membrane and the subsequent ability of the fully formed autophagosome to take up "cargo" for degradation.

The mechanism whereby defective autophagy due to the Atg16L1 polymorphism could predispose to CD is potentially multifactorial given the protean reach of autophagic function.[395] Early on, studies of mechanism focused on the effect of the Atg16L1 polymorphism on the function of Paneth cells, that is, cells located in the terminal ileum known to secrete antibacterial agents such as lysozyme, that regulate luminal bacterial composition. It was found that such cells exhibited morphologic and functional abnormalities in the presence of autophagy defects due to Atg16L1 loss or dysfunction.[396] Later, studies focused on IECs more generally, since it could be shown that if such cells bear the polymorphism they exhibit increased apoptosis and loss of barrier function in the presence of factors that induce cellular stress or when exposed to pathologic organisms. Evidence that the polymorphism predisposed mice to epithelial barrier disruption and *Salmonella* invasion supported this possibility.[397]

Extensive studies relating autophagy to epithelial cell endoplasmic reticulum (ER) stress expanded on this theme.[398,399] It was shown in these studies that deletion of Atg16L1 in epithelial cells led to late-onset terminal ileitis similar to that in Crohn ileitis; however, such ileitis occurred earlier in mice with epithelial deletion of both Atg16L1 and Xbp1, the deletion of the latter causing impaired ER stress resolution and resultant accumulation of cytotoxic IRE1a-containing clusters. Since IRE1a clusters could be shown to be cleared by selective autophagy (via optineurin), this led to the idea that intact autophagy protects the cell from the untoward effects of the ER stress response. In turn, this supported the notion that individuals bearing the Atg16L1 polymorphism are subject to Crohn ileitis because of reduced autophagic clearance of IRE1a resulting from ER stress even in the absence an Xbp1 mutation causing poor regulation of the stress response. This explanation of how Atg16L1 dysfunction causes CD is elegant but not yet convincingly proven. This is evident from studies that show that, although there are increased numbers of IRE1a+ crypt epithelial cells in individuals bearing the Atg16L1 polymorphism, this also occurs in the absence of CD and positive cells have not been identified as a marker specific to Atg16L1-associated CD. In addition, there is no evidence that patients with CD bearing an Atg16L1 polymorphism also bear a Xbp1 polymorphism or indeed another polymorphism affecting the ER stress response. Finally, inasmuch as the mechanisms proposed now apply to ER stress occurring in all secretory epithelial cells one would expect the polymorphism to be a risk factor by this mechanism in UC as well as in CD, yet its effect is only manifest in the latter form of IBD.

Although the effects of the Atg16L1 polymorphism on epithelial cell function may be significant, they may nevertheless be secondary to a more important effect on mucosal immune cells. This is suggested by the fact that Atg16L1 bearing this polymorphism functions normally until cleaved by caspase 3, a factor whose production by epithelial cells is likely dependent on preexisting proinflammatory cytokine production by mucosal immune cells.[400] This would explain the lack of spontaneous intestinal inflammation in mice bearing this polymorphism (as opposed to mice with total Atg16L1 deletion) and the lack of CD in the vast majority of individuals with this polymorphism. The most obvious connection between the polymorphism and mucosal immune function is that defective autophagy in macrophages or DCs leads to reduced clearance of damaged mitochondria and ROS accumulation and this, in turn, leads to activation of the NLRP3 inflammasome and increased IL-1β and IL-18 secretion.[401] However, although both of these cytokines have been linked to intestinal inflammation, it is not yet clear that inflammasome activation is a sufficiently robust mechanism in itself to sustain CD in most patients. Another possible mechanism through which the Atg16L1 polymorphism could be acting on mucosal immune cells is via a positive effect of impaired autophagy on NF-κB activation. Evidence that this is in fact the case has recently appeared: impaired autophagy, including that occurring in association with the Atg16L1 polymorphism, leads to accumulation of the p62 and p62-mediated facilitation of NF-κB-activating factors such as TRAF6.

Another significant polymorphism associated with IBD, in this case affecting both UC and CD (as well as psoriasis and ankylosing spondylitis), is in the gene encoding the IL-23 receptor and thus affecting the signaling pathway regulating the generation of IL-17 and IL-22, key cytokines mediating IBD.[402] Interesting, this polymorphism was shown to provide decreased rather than increased risk for the development of IBD, indicating that genetic polymorphisms can also be protective. In the several studies in which the effect of one of these polymorphisms (R381Q) on immune function was evaluated it was found that stimulated T cells bearing the R381Q IL-23R polymorphism exhibited reduced IL-23R expression, reduced STAT3 activation, and reduced CD4+ and CD8+ T-cell production of IL-17 and IL-22.[403,404] Thus, in keeping with its protective effect, the presence of this polymorphism lead to reduced production of some of the cytokines causing CD. The discovery of this polymorphism had clinical implications in that it suggested that biologic agents that suppress IL-12R signaling, such as anti-IL-12p40 (which inhibits both IL-12 and IL-23), would be a useful IBD treatment, as already suggested by studies of animal models. The successful treatment of CD with this antibody validates this idea.

To conclude this brief survey of the mechanisms by which risk polymorphisms cause IBD it is useful to discuss the risk polymorphism affecting the hepatocyte nuclear factor 4-alpha (*HNF4A*) gene, which is associated mainly with UC but has also been found to occur in pediatric CD (in one study) and is one of several polymorphisms that affect epithelial cell function in UC. The HNF4-alpha protein is a factor shown to regulate both hepatocyte and IEC function.[405] In the latter cell it is upregulated by exposure to various cytokines (eg, IL-1β and TGF-β) and has positive effects as a transcription factor involved in the expression of the epithelial junctional protein, claudin 15, and that of other junctional proteins; in addition, it has transcriptional effects that promote stem cell proliferation, Paneth cell differentiation, and goblet cell production of mucin. Individuals bearing this polymorphism express lower levels of HNF4-α and are thus presumed to have diminished epithelial functions that are important in the prevention of inflammation, particularly as this relates to barrier function. In this way, this and other risk polymorphisms affecting epithelial functions trigger or abet mucosal immune cells with defects causing inflammation. Finally, it should be noted that, although the HNF4-α risk polymorphism (or other epithelial cell-related polymorphisms) causes only minor increases in risk, their importance is nevertheless supported by the fact that mice with deletions of genes encoding mucins develop spontaneous colitis[406]; thus, this kind of abnormality has the potential to cause IBD by this mechanism alone.

Immunologic Mechanisms Operative in IBD: Cytokines

The understanding of the pathogenesis of IBD took a great leap forward some 30 years ago with studies elucidating the main proinflammatory cytokines mediating these diseases (Fig. 38.4). The insights in this area that were developed initially and that still hold today are first that the complex and variable genetic factors driving these forms of IBD (partially reviewed above) are ultimately "channeled" into a more or less homogeneous and distinctive set of cytokines characteristic of either CD or UC and second that the IBD-defining cytokines are accompanied by any overarching set of "general-purpose" cytokines such as TNF-α, IL-1β, and IL-6 that are produced in both forms of IBD at a somewhat variable extent and that are, in fact, the most frequent targets of therapy.

The most important early studies in this area led to the concept that CD is mediated by a Th1-type set of cytokines including IL-12 and IFN-γ,[407] but later this was modified by

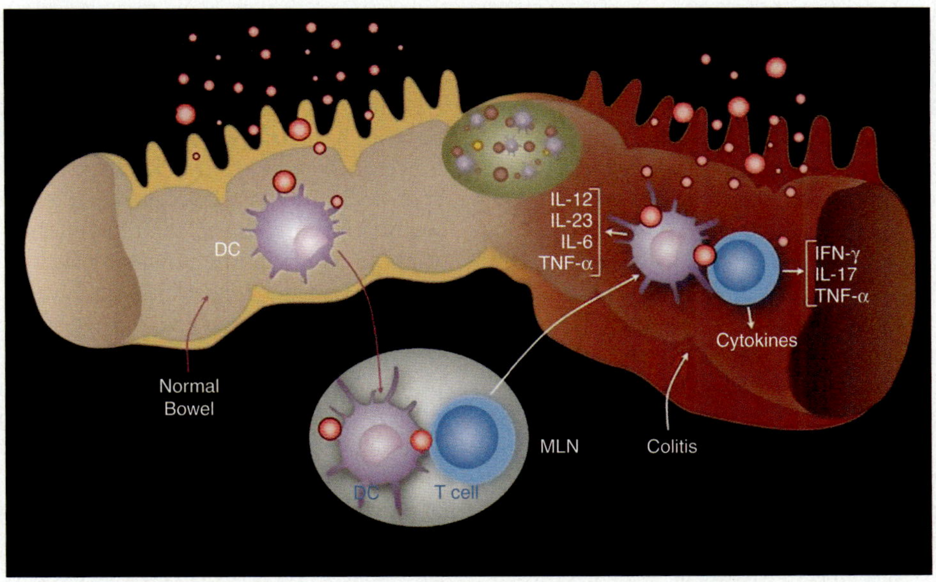

FIG. 38.4. The major proinflammatory cytokines causing Crohn's disease. In the normal gastrointestinal tract (depicted on the left side of the diagram) a small number of bacteria (or products of such bacteria) gain access to the gut mucosal tissue (lamina propria) because of intact epithelial cell barrier function. Immune responses to these bacteria are limited and do not cause significant inflammation because of various immunoregulatory mechanisms. In contrast, in Crohn disease (depicted on the right side of the diagram) bacteria or bacterial products elicit a massive proinflammatory Th1/Th17 response driven by dendritic cell production of IL-23 and IL-12. This results in CD4+ T-cell production of IFN-γ and IL-17 as well as dendritic cell (or macrophage) production of TNF-α, IL-1β, and IL-6. Dendritic cells inducing disease migrate to the mesenteric lymph node (MLN) for initial encounter with T cells and then migrate back to the lamina propria. Bacterial entry into the lamina propria is increased because of cytokine degradation of barrier function. B cells developing in Peyer patches (green structure) do not contribute to the inflammatory response.

the fact that Th17-type cytokines including IL-23 and IL-17 were also being produced and the idea emerged that this form of IBD is a hybrid Th1/Th17-driven inflammation dependent on both IL-12 and IL-23.[408] However, although there is evidence from murine models that CD-like inflammation can begin either as a Th1- or Th17-type inflammatory response, the end result is a response quantitatively dominated by T cells producing IFN-γ. This fact conforms with the observation that, even when intestinal inflammation originates largely as an IL-23-driven process, as it does in the cell-transfer model of CD-like inflammation, the Th17 T cells initially induced are "plastic" and differentiate into T cells that respond to IL-12 with the production of IFN-γ exclusively or both IL-17 and IFN-γ. Such Th17 T-cell transformation could explain the paradox by which a genetic polymorphism associated with reduced IL-23 signaling reduces risk of CD occurrence (as discussed above), the majority of T cells in CD lesional tissue are producing IFN-γ rather than IL-17, and treatment of patients with CD with anti-IL-17 sometimes exacerbates disease, most likely because such treatment undermines IL-17 downregulation of Th1 T-cell proliferation or IL-17 support of epithelial barrier function.[409,410] In addition, the transformation concept explains the fact that the defining characteristic of CD inflammation is its transmural granulomatous character presumably due to IFN-γ activation of macrophages and subsequent macrophage production of TNF-α Finally, the transformation concept fits nicely with the fact that anti-IL-12p40, an antibody that blocks both IL-23 and IL-12 signaling, is a remarkably effective treatment of CD even in patients resistant to anti-TNF-α treatment.[411] It should be noted, however, that IFN-γ dominance does not mean that IL-17 has no pathologic effects in CD inflammation; on the contrary, it can cause increased pathology by any of several mechanisms, including its capacity to facilitate neutrophil infiltration and the latter's untoward effect on epithelial barrier integrity.

Another cytokine playing a role in CD, IL-22, is usually considered a Th17 cytokine since it is mainly induced by IL-23. However, it is also produced by Th1 cells, NK cells, ILC3 cells, and even neutrophils, suggesting that its induction can also occur under non-Th17 conditions.[412] IL-22 has been shown to have multiple positive effects on epithelial cell function including enhancing effects on cell growth, barrier function, mucus production, and AMP production. These positive effects suggest that the increased levels of IL-22 observed in CD constitute an anti-inflammatory component of the increased Th17 response occurring in CD; in addition, it could explain the fact that amelioration of IBD by administration of the aryl hydrocarbon receptor ligand is accompanied by and presumably due to IL-22 induction. This said, it should be noted that increased IL-22 production in CD is accompanied by increased production of antagonistic IL-22-binding protein.[413] Thus, the positive effects of IL-22 production may be self-limited.

The group of cytokines driving UC are different from those driving CD and are also somewhat less homogeneous. Early on, studies of patient LP T cells established that the latter were producing increased amounts of Th2 cytokines such as IL-5 (but not IL-4) rather than Th1 cytokines.[407] Later studies of an influential murine model of colitis induced by intrarectal administration of oxazolone (oxazolone-colitis) provided evidence that experimental UC-like colitis is driven by NKT cells restricted by CD1d and producing IL-13[414,415]; in addition, the latter cytokine was shown to have detrimental effects on epithelial cells via activation of caspase 3.[416] Studies of patients corroborated these findings by showing that UC LP populations contained a substantial percentage of type II NKT cells (NKT cells with noninvariant TCRs) recognizing lyso-sulfatide glycolipid antigen and producing IL-13; these cells bore IL-13Rα2 receptors and exhibited epithelial cell cytotoxicity that was augmented by IL-13.[417] More recently, these findings were at least partially supported by a study of a large number of patients with UC showing that the patients comprised two rather sharply divided groups: a larger group (containing about two-thirds of the total) with high tissue IL-13 mRNA levels and a smaller group (containing about one-third of the total) with normal tissue IL-13 mRNA levels.[418] Of importance, the high-IL-13 group exhibited more widespread and more severe inflammation than their low IL-13 counterparts. Overall, these and related studies have established that UC is driven in most (but possibly not in all) cases by cells producing IL-13; moreover, there is evidence that these cells are NKT cells that are responding to epithelial cell-derived self-antigens, as exemplified by their ability to recognize lyso-sulfatide, a possible epithelial cell-derived antigen. However, proof of the validity of this conclusion by demonstration that patients respond clinically to antibodies blocking IL-13 or its relevant receptors is currently not at hand.[419] In one study of a large patient group, an antibody specific for an IL-13 receptor (IL-4Rα) not likely to be involved in either oxazolone-colitis or UC not surprisingly showed no capacity to ameliorate UC; however, in another study of a large patient group an antibody specific for IL-13 itself resulted in a remission rate (18%) that was significantly higher than that of the placebo (6%). These studies leave open the possibility that an antibody targeting the IL-13 receptor most likely involved in UC pathogenesis, the IL-13Rα2 receptor, could prove efficacious in UC.

When it became apparent that Th17 responses could be involved in IBD, numerous studies (almost exclusively mRNA-based studies) were conducted that showed that IL-17 levels are also increased in UC, albeit not to the extent it is increased in CD. Indeed, in the study of the large group of patients referred to above in which IL-13 was investigated, assessment of IL-17 showed that the great majority of patients exhibited elevated IL-17A mRNA levels, regardless of their IL-13 mRNA levels.[418] However, this elevation was associated with equally elevated IFN-γ mRNA levels, strongly suggesting that these increases were a secondary effect consequent to a more primary process resulting in loss of epithelial barrier function and increased exposure of the mucosal immune system to commensal microbial organisms. This is not to say, however, that this proinflammatory IL-17/IFN-γ response in UC is trivial and does not influence the level of inflammation to some degree. This is supported by the fact that treatment of patients with anti-IL-12p40 has been shown to have at least some ameliorative effect in UC.[420] Along somewhat similar lines, levels of IL-9, a cytokine shown to degrade epithelial barrier function in several ways, has also been shown to be increased in UC.[421] Since IL-9 synthesis is induced by IL-4 in combination with TGF-β it seems likely this cytokine is secondarily induced in the Th2-like microenvironment in UC tissue. Again, its synthesis is not unimportant given the role of barrier dysfunction in UC.

As mentioned above, several "all-purpose" or generic proinflammatory cytokines are produced in excess in both CD and UC such as TNF-α, IL-1β, and IL-6. In contrast to most of the more disease-specific proinflammatory cytokines already discussed, these cytokines are produced mainly by macrophages or DCs (and perhaps other innate cells) rather than activated T cells and their production is secondary to stimulation by cytokines produced by activated T cells. TNF-α has the first position in the ranking of the importance of the all-purpose cytokines instigating IBD inflammation given the fact that it is the cytokine whose neutralization has been most successful in the control of IBD inflammation. Its high potency as a proinflammatory cytokine is due to the fact that, in the form of a homotrimer, it can signal via two different cellular receptors (TNFR1 and TNFR2), both of which give rise to NF-κB and MAPK activation and to the cell proliferation/survival and cytokine synthesis that ensues from such signaling.[422] A second important effect of TNF-α signaling, in this case via TNFR1 exclusively, is the induction of complexes that lead to the activation of necroptosis and apoptosis mechanisms dependent on MLKL and caspase 8, respectively.[423,424] The latter process affects epithelial cell survival in IBD and thereby affects epithelial barrier function.

IL-1β is a second major NF-κB-inducing proinflammatory cytokine produced in excess in IBD but one whose importance is perhaps overlooked due to the fact that blockade of its function has so far not proven useful as a treatment of the usual case of IBD. This cytokine, along with IL-18, is synthesized in a "pro" form and must be cleaved in one of the inflammasome complexes (usually NLRP3) prior to secretion. This, in turn, requires inflammasome activation by a TLR ligand plus one of several coactivators such as ATP. In murine models of colitis inhibition of inflammasome function has had mixed effects on colitis induction, whereas enhancement of inflammasome function has led repeatedly to enhanced colitis induction.[425] This is mirrored in the fact that several monogenic mutations leading to increased NLRP3 inflammasome activation is accompanied by CD.[426] Thus, IL-1β becomes a more important proinflammatory cytokine when its production exceeds that of the already increased production ordinarily accompanying IBD.

The third member of the all-purpose proinflammatory triad, IL-6, is also an important proinflammatory actor in IBD. Its proinflammatory capacity stems from its capacity to induce NF-κB and NF-κB-dependent effector cytokines, as well as its role in the induction of Th17 differentiation.[427] IL-6 can signal via its receptor either directly or by first binding to soluble IL-6R to form a complex that interacts with membrane-bound signal transducer (gp130), thus by-passing membrane-bound receptor (trans-signaling). Such IL-6 trans-signaling of hematopoietic cells is thought to convey the majority of its proinflammatory effects.[428] A diagrammatic representation of the cytokines causing CD is presented in Fig. 38.4.

Immunologic Mechanisms Operative in IBD: T Cells

In addition to their capacities to produce the CD- and UC-defining cytokines discussed above, the T cells that drive IBD inflammation share qualities that facilitate sustained inflammation, namely, the capacity to home to and remain in the inflamed mucosal tissue. These cells, termed tissue-resident memory cells, are a somewhat heterogeneous group of T cells (usually CD4+ T cells) that express CD69, a receptor that opposes the activity of tissue egress-inducing sphingosine-1 phosphate receptor-1; in addition, they frequently also express other molecules that favor mucosal tissue retention such as CD103, the αE subunit of the αEβ7 integrin that binds to E-cadherin and CXCR6, the latter a chemokine associated with mucosal homing.[429] Other defining features of resident memory cells are expression of the CD49a and CD101 surface markers as well as various transcription factors such as Blimp-1, Hobit, and Runx3. CD8+ CD103+ resident memory T cells have also been described in CD, but the extent of their contribution to disease is poorly defined. Studies of IBD patient tissue disclose that levels of CD4+ tissue resident T cells, as indicated by mean CD69 mRNA levels, are increased in both UC and CD, but control individuals and patients overlap in this regard; this suggests that the activity rather than the number of resident memory cells is a distinguishing disease factor.[430] In any case, these studies show that it is in fact the CD69+ T cells that are, overwhelmingly, the origin of the effector cytokines responsible for disease. This correlates with the fact that CD69 mRNA levels associate with disease recurrence and that depletion of these cells by knockout of key transcription factors eliminates colitis in murine models. The fact that these cells are found in both UC and CD indicates that in both diseases an adaptive T-cell response to an antigen, most likely a microbial- or self-antigen, is a key initiating factor, albeit one that depends on amplification by one or another innate cell or factor; it is in fact the latter that guides the tissue resident T cell into a pathway of differentiation that defines the form of IBD.

Immunologic Mechanisms Operative in IBD: ILCs

The discovery and delineation of various ILCs lacking TCRs has drawn attention to their possible role in IBD. In the mucosal immune system these cells occupy a position close to the epithelium and respond rapidly to various cytokines; thus, with the caveat in mind that their responses are dependent on cytokine signals from already activated macrophages or DCs, these cells have properties that allow them to act as early initiators (or inhibitors) of disease.[431] Unfortunately, however, most studies of ILCs in IBD have necessarily been conducted with murine models of colitis employing RAG-deficient or SCID mice so as to distinguish effects of ILCs from the vastly larger population of T cells that usually characterizes an inflammation; this makes it difficult to evaluate the role of these cells in nonlymphopenic (non-T cell–deficient) humans with chronic IBD. This problem is highlighted by the fact that, although changes in the fraction of the total number of ILCs attributable to an individual ILC subset have been shown to occur in IBD, these changes have not been shown to be accompanied by changes in the total number of ILCs; since the latter number is less than 1% of the number of T cells, this implies that the fractional changes do not cause a dramatic increase in ILC subset cells relative to the number of conventional T cells.[432]

Studies of ILC in IBD have to some extent focused on the ILC3 subsets that are IL-23/IL-1β-responsive (RORγt-expressing) cells producing IL-17, IL-22, Csf-2 (GM-CSF), and IFN-γ, that is, cells resembling Th17/Th22 T cells. The majority of these cells bear NK-cell markers (such as NKp44) and produce more IL-22 than IL-17, whereas a minority do not bear these markers and produce more IL-17 than IL-22. In studies of experimental colitis induced in RAG1-deficient mice these cells can have protective or destructive effects depending on the murine model in which they have been studied[433]; this probably reflects the fact that these cytokines have been shown to exert both positive and negative effects on epithelial cell and/or macrophage functions and do so under different conditions. In studies of humans with IBD the ILC3 (NKp44+) fraction of total ILCs falls, whereas the ILC3 (NKp44−) fraction increases, indicating a shift of ILC3 production of IL-22 to IL-17.[434] The importance of this shift to the overall IBD inflammation is unclear given the ambiguities relating to IL-22 and IL-17 function and is in any case mitigated by the fact that in the developing IBD inflammation ILC3 cells are rapidly outnumbered by Th17/Th22 conventional T cells.

The latter concern may be less applicable to ILC3 production of Csf2 (GM-CSF), a factor that plays a role in an inflammatory process via its ability to recruit/expand proinflammatory myeloid cell populations and could be produced in sufficient quantity by a small population of cells. Relating to this is the demonstration that ILC3 production of Csf2 is necessary for optimal accumulation of inflammatory monocytes in RAG2−/− mice with anti-CD40-induced colitis and the latter cells produce IL-23, which further stimulates ILC3 production of Csf-2 in a positive feedback loop; in addition, the Csf-2 produced during inflammation also mobilizes ILC3 from cryptopatches and thereby enhances the distribution of ILC3 cells.[435] With respect to the critical question of whether these findings apply to humans with IBD there are data showing that the fraction of colonic ILCs expressing Csf-2 is about twice that of the fraction of T cells expressing Csf-2; however, this does not establish the essentiality of ILC3 versus T cells as a source of Csf-2 in IBD because activated Th17 cells, presumably capable of producing Csf-2, are present in far greater numbers in the inflamed gut. Finally, it has been shown that, in the uninflamed gut, RORγt+ ILC3 cells (including both LTi cells and NKp44+ ILC3 cells) are a major source of Csf-2 and the latter is essential for recruitment of DCs that promote Foxp3+T_{reg} cell development and oral tolerance; such ILC3 Csf-2 production is enhanced by microbe-stimulated DCs/macrophages producing IL-10.[271] Whether such Csf-2-mediated T_{reg} induction is increased in IBD (and in this case driven by Th17 T cells mainly) has not been determined; if it is, it would presumably lead to decreased, rather than increased, inflammation.

ILC1 cells, ILCs that are responsive to IL-12, IL-15, or IL-18 and produce IFN-γ, are increased as a fraction of the total ILC population in CD.[436] This increase is accompanied by a decrease in the fraction of NKp44+ILC3 cells and is in fact most likely due to conversion of NKp44+ILC3 cells into ILC1 cells, which is known to occur under the influence of an IL-12 signal.[432,437] Thus, the overall profile of ILC may shift away from cells producing IL-22/IL-17 toward cells producing IFN-γ, a more proinflammatory stance. However, whether this has a significant effect on the course of IBD seems questionable.

Overall, although changes occurring in ILC subsets in established human IBD are of interest, the significance of these changes to disease pathogenesis are yet to be proven given the very small number of ILCs actually present and the redundancy in their function with respect to conventional T cells.

Immunologic Mechanisms Operative in IBD: Regulatory T Cells

An important issue relating to the understanding of IBD concerns the question of whether the excessive mucosal responsiveness generally thought to underlie this disease is due a hyperactive immunologic process intrinsic to the responding mucosal cells, or, alternatively, due to a normoactive immunologic process also intrinsic to responding mucosal cells that is, however, being provided insufficient extrinsic negative regulation by regulatory T cells. This question arises from the well-known fact (reviewed elsewhere in this chapter) that responses to orally administered antigens or to components of the endogenous microbiome are kept in check in the antigen-rich mucosal milieu by oral tolerance, an immunosuppressive mechanism that for the most part is mediated by DC induction of Foxp3+ regulatory T cells (T_{reg}s) expressing immunosuppressive factors such as TGF-β and IL-10. Studies providing initial evidence supporting the view that regulatory T-cell defects might be responsible for IBD came from murine models of gut inflammation. It was shown, for instance, that colitis in mice induced by intrarectal trinitrobenzenesulfonic acid (TNBS) is prevented by prefeeding TNBS-haptenated colonic protein; moreover, the suppressive effect of feeding was blocked by coadministration of anti-TGF-β, presumably due to blockade of the suppressive factor produced by the induced suppressor cells.[438] In addition, it was shown that colitis induced in RAG2−/− mice by transfer of $CD45RB^{hi}$ cells in the cell transfer model of colitis could be prevented by cotransfer of $CD45RB^{lo}$ cells, which contained a subset dependent on TGF-β secretion and in later studies identified as Foxp3+ T_{reg}s.[439] The cell transfer model ultimately proved to be the dominant model for studying suppressor cells in colitis because this model allowed exquisite characterization of the transferred suppressor cells and their effects on the inflammation mediated by the $CD45RB^{hi}$ cotransferred effector cells. In addition, with use of the cell transfer model it could be shown that transfer of sufficient numbers of purified Foxp3+ T_{reg}s not only prevented nascent colitis but also ameliorated established colitis.[440] Thus, these studies suggested that IBD could be due to a regulatory cell defect that could be treated by regulatory cell replacement or by targeted regulatory cell activation.

If indeed the latter hypothesis offers a valid way of explaining IBD, one might expect that the population of Foxp3+ T_{reg}s would be deficient in IBD patient intestinal tissue either with respect to cell number or cell function. However, in the many studies of this possibility conducted over the past 2 decades, no convincing evidence has emerged supporting the idea that either of these types of T_{reg} deficiencies occurs in either of the two forms of IBD.[441] In fact, in the most convincing studies such as those conducted in children prior to treatment who therefore are free of secondary conditions that could bias the outcome, the percentage of T_{reg}s in the CD4+ T cells of the inflamed CD or UC intestine is actually significantly higher than in tissue from control individuals and likewise there is no change in subsets bearing markers purporting to show whether the T_{reg}s have a thymic or a peripheral origin that might suggest altered T_{reg} activity.[442] In CD this increase in T_{reg} numbers could be due to the fact that a mucosal DC subpopulation in CD expresses increased amounts of αVβ8, an integrin known to activate TGF-β, a major component of DC-mediated Foxp3+ T_{reg} induction.[443]

The above findings cast strong doubt on the possibility that T_{reg} function is decreased in IBD due to decreased T_{reg} survival or persistence in inflamed mucosal tissue since this would inevitably result in reduced T_{reg} abundance. However, it is still possible that T_{reg} function is impaired in IBD in spite of normal or even increased T_{reg} levels because effector cells in IBD are resistant to T_{reg} cell suppression. This possibility is supported by the finding that CD effector cells express high levels of Smad7 due to the high and persistent TGF-β levels and signaling present in the intestine and, as a result, are resistant to suppressive TGF-β-mediated suppression by T_{reg}s.[444] However, Foxp3+ T_{reg}s have no difficulty suppressing effector cells in the cell transfer model and in several studies CD effector cells have been shown to have been found to be normally suppressed.[445] Overall, one must conclude that, although the concept that IBD is due to a Foxp3+ suppressor cell defect is a sustainable concept in relation to the mechanism of disease in models of intestinal inflammation, it does not seem to be applicable to human IBD. It should be noted, however, that this conclusion does not detract from the possibility that IBD can be treated by enhancing suppressor cell function.

The likelihood that deficient Foxp3+ suppressor cell activity does not account for the hyperresponsiveness occurring in IBD still leaves open the possibility that defects in other suppressor cells or mechanisms might cause disease. There is evidence, for instance, that Foxp3-negative suppressor cells exist that bear surface TGF-β in association with latency-associated peptide and that regulate the extent of disease in UC.[446] In addition, an ILC3 suppressor cell has been described that bears MHC class II antigen and, when stimulated by commensal organisms, downregulates T-cell responses via production of IL-10 and TGF-β.[447] If, in fact, abnormalities of these suppressor cells are significant causes of IBD, they would have to have unique functions not already displayed by Foxp3+ T_{reg}s. An abnormality of regulation that clearly does lead to IBD inflammation is not a regulatory cell but rather a regulatory cytokine. This is evident by the fact that mice and humans with genetically determined IL-10 (or IL-10R) deficiency develop severe intestinal inflammation due to lack of IL-10-mediated suppressor function. As already mentioned, in humans, this defect causes VEO-IBD that is extremely resistant to conventional treatment.[448]

Immunologic Mechanisms Operative in IBD: The Gut Microbiome

Given the likelihood that much, if not all, of the mucosal hyperresponsivity associated with IBD is driven by antigens expressed by organisms in the gut microbiome, it was reasonable to conduct studies to determine if the microbiome is altered in IBD and, if so, if this alteration (often called dysbiosis) is a contributor to disease.[449] To address this question, many studies have been conducted defining the full spectrum of bacterial species in the large and small intestine of patients and controls using sophisticated 16S rRNA sequence analyses of the bacterial microbiome. In the best and most comprehensive of these studies it was found that about a quarter to a third of patients harbored microbiomes that differed from that of controls. The changes consisted mainly of decreases in bacterial species in the *Firmacute* and *Bacteroidete* phyla and, in patients with CD, increases in the *Proteobacteria* phyla.[450] The decreased *Firmacutes* are of interest because this phylum includes organisms belonging to various clostridial strains that have been shown to induce Foxp3+ T cells and decrease the severity of induced colitis when administered to mice.[375] One clostridial organism, *Faecalibacterium prausnitzii*, has been subjected to particular attention because this organism has been shown to produce a soluble substance with immunosuppressive properties that affects Th17/T_{reg} balance; this may consist (at least in part) of butyrate, an SCFA.[451] However, in a longitudinal study of patients with CD, the correlation between calprotectin (a marker of disease activity) and *F. prausnitzii* carriage was marginal at best and could easily have been an effect rather than a cause of disease improvement.[452] The decrease in bacteria in the *Bacteroidete* phylum is also of interest because this phylum harbors *B. fragilis*, an organism that can induce Foxp3+ regulatory cells in mice via its production of a particular polysaccharide (PSA) that interacts with either regulatory cells or DCs via TLR2, and colonization of mice with this organism inhibits induced colitis.[453] However, the significance of these findings is unclear because this particular organism has not been associated with IBD.

The above studies, for the most part conducted in humans, are mirrored in studies of murine models of intestinal inflammation, which frequently show that experimental gut inflammation is also accompanied by various changes in the prevailing gut microflora.[389] In some cases the changed microflora can be shown to be "colitogenic" in that it causes colitis in normal mice; however, this colitis rarely, if ever, persists.[454] The latter correlates with the fact that dysbiosis in mice with colitis or humans with CD disappears with effective therapy addressing immunologic abnormalities. These facts argue that the altered intestinal microbiome is secondary to a more primary immune abnormality. As such, it can aggravate but not cause IBD.

In summary, the above discussion of IBD provides ample evidence that this disease is a multifaceted inflammation whose pathogenesis involves many key aspects of mucosal immune function. Chief among the latter is the interplay between innate and adaptive mucosal responses to the resident microflora and the way these are regulated in health and disease.

ACKNOWLEDGMENTS

This work was supported by the Intramural Research program of the National Institute of Allergy and Infectious diseases.

REFERENCES

1. Ansaldo E, Farley TK, Belkaid Y. Control of immunity by the microbiota. *Annu Rev Immunol*. 2021;39:449-479.
2. Hall JA, Grainger JR, Spencer SP, et al. The role of retinoic acid in tolerance and immunity. *Immunity*. 2011b;35(1):13-22.
3. Mayer L. Mucosal immunity. *Immunol Rev*. 2005;206:5.
4. Iwasaki A. Mucosal dendritic cells. *Annu Rev Immunol*. 2007;25:381-418.
5. France MM, Turner JR. The mucosal barrier at a glance. *J Cell Sci*. 2017;130(2):307-314.
6. Mowat AM. Anatomical basis of tolerance and immunity to intestinal antigens. *Nat Rev Immunol*. 2003;3(4):331-341.
7. Mowat AM, Agace WW. Regional specialization within the intestinal immune system. *Nat Rev Immunol*. 2014;14(10):667-685.
8. Eberl G, Lochner M. The development of intestinal lymphoid tissues at the interface of self and microbiota. *Mucosal Immunol*. 2009;2(6):478-485.
9. Hamada H, Hiroi T, Nishiyama Y, et al. Identification of multiple isolated lymphoid follicles on the antimesenteric wall of the mouse small intestine. *J Immunol*. 2002;168(1):57-64.
10. Moghaddami M, Cummins A, Mayrhofer G. Lymphocyte-filled villi: comparison with other lymphoid aggregations in the mucosa of the human small intestine. *Gastroenterology*. 1998;115(6):1414-1425.
11. Miller H, Zhang J, Kuolee R, Patel GB, Chen W. Intestinal M cells: the fallible sentinels? *World J Gastroenterol*. 2007;13(10):1477-1486.
12. Ohno H. Intestinal M cells. *J Biochem*. 2016;159(2):151-160.
13. Heppert JK, Davison JM, Kelly C, Mercado GP, Lickwar CR, Rawls JF. Transcriptional programmes underlying cellular identity and microbial responsiveness in the intestinal epithelium. *Nat Rev Gastroenterol Hepatol*. 2021;18(1):7-23.
14. Edele F, Molenaar R, Gütle D, et al. Cutting edge: instructive role of peripheral tissue cells in the imprinting of T cell homing receptor patterns. *J Immunol*. 2008;181(6):3745-3749.
15. Hammerschmidt SI, Ahrendt M, Bode U, et al. Stromal mesenteric lymph node cells are essential for the generation of gut-homing T cells in vivo. *J Exp Med*. 2008;205(11):2483-2490.
16. Iliev ID, Mileti E, Matteoli G, Chieppa M, Rescigno M. Intestinal epithelial cells promote colitis-protective regulatory T-cell differentiation through dendritic cell conditioning. *Mucosal Immunol*. 2009;2(4):340-350.
17. Iwata M, Hirakiyama A, Eshima Y, Kagechika H, Kato C, Song S-Y. Retinoic acid imprints gut-homing specificity on T cells. *Immunity*. 2004;21(4):527-538.
18. Molenaar R, Knippenberg M, Goverse G. et al. Expression of retinaldehyde dehydrogenase enzymes in mucosal dendritic cells and gut-draining lymph node stromal cells is controlled by dietary vitamin A. *J Immunol*. 2011;186(4):1934-1942.
19. Cantorna MT, Nashold FE, Hayes CE. Vitamin A deficiency results in a priming environment conducive for Th1 cell development. *Eur J Immunol*. 1995;25(6):1673-1679.
20. Chun TY, Carman JA, Hayes CE. Retinoid repletion of vitamin A-deficient mice restores IgG responses. *J Nutr*. 1992;122(5):1062-1069.
21. Ivanov II, Atarashi K, Manel N, et al. Induction of intestinal Th17 cells by segmented filamentous bacteria. *Cell*. 2009;139(3):485-498.
22. Sano T, Huang W, Hall JA et al. An IL-23R/IL-22 circuit regulates epithelial serum amyloid A to promote local effector Th17 responses. *Cell*. 2015;163(2):381-393.
23. Derebe MG, Zlatkov CM, Gattu S et al. Serum amyloid A is a retinol binding protein that transports retinol during bacterial infection. *Elife*. 2014;3:e03206.
24. Thaiss CA, Zmora N, Levy M, Elinav E. The microbiome and innate immunity. *Nature*. 2016;535(7610):65-74.
25. Tanoue T, Morita S, Plichta DR et al. A defined commensal consortium elicits CD8 T cells and anti-cancer immunity. *Nature*. 2019;565(7741):600-605.
26. Bilate AM, London M, Castro TBR et al. T cell receptor is required for differentiation, but not maintenance, of intestinal CD4(+) intraepithelial lymphocytes. *Immunity*. 2020;53(5):1001-1014.e20.
27. McGuckin MA, Lindén SK, Sutton P, Florin TH. Mucin dynamics and enteric pathogens. *Nat Rev Microbiol*. 2011;9(4):265-278.
28. Atuma C, Strugala V, Allen A, Holm L. The adherent gastrointestinal mucus gel layer: thickness and physical state in vivo. *Am J Physiol Gastrointest Liver Physiol*. 2001;280(5):G922-G929.
29. Ambort D, van der Post S, Johansson MEV, et al. Function of the CysD domain of the gel-forming MUC2 mucin. *Biochem J*. 2011;436(1):61-70.
30. Johansson ME, Larsson JM, Hansson GC. The two mucus layers of colon are organized by the MUC2 mucin, whereas the outer layer is a legislator of host-microbial interactions. *Proc Natl Acad Sci U S A*. 2011;108(suppl 1):4659-4665.
31. Johansson ME, Phillipson M, Petersson J, Velcich A, Holm L, Hansson GC. The inner of the two Muc2 mucin-dependent mucus layers in colon is devoid of bacteria. *Proc Natl Acad Sci U S A*. 2008;105(39):15064-15069.
32. Hansson GC, Baeckstrom D, Carlstedt I, Klingalevan K. Molecular cloning of a cDNA coding for a region of an apoprotein from the 'insoluble' mucin complex of rat small intestine. *Biochem Biophys Res Commun*. 1994;198(1):181-190.
33. Qin J, Li R, Raes J, et al. A human gut microbial gene catalogue established by metagenomic sequencing. *Nature*. 2010;464(7285):59-65.
34. van Passel MW, Kant R, Zoetendal EG, et al. The genome of *Akkermansia muciniphila*, a dedicated intestinal mucin degrader, and its use in exploring intestinal metagenomes. *PLoS One*. 2011;6(3):e16876.
35. Kulkarni DH, Gustafsson JK, Knoop KA, et al. Goblet cell associated antigen passages support the induction and maintenance of oral tolerance. *Mucosal Immunol*. 2020;13(2):271-282.
36. Hand TW, Reboldi A. Production and function of immunoglobulin A. *Annu Rev Immunol*. 2021;39:695-718.
37. Gallo RL, Hooper LV. Epithelial antimicrobial defence of the skin and intestine. *Nat Rev Immunol*. 2012;12(7):503-516.
38. Hooper LV, Macpherson AJ. Immune adaptations that maintain homeostasis with the intestinal microbiota. *Nat Rev Immunol*. 2010;10(3):159-169.
39. Flo TH, Smith KD, Sato S, et al. Lipocalin 2 mediates an innate immune response to bacterial infection by sequestrating iron. *Nature*. 2004;432(7019):917-921.
40. Cash HL, Whitham CV, Behrendt CL, Hooper LV. Symbiotic bacteria direct expression of an intestinal bactericidal lectin. *Science*. 2006;313(5790):1126-1130.
41. Vaishnava S, Yamamoto M, Severson KM, et al. The antibacterial lectin RegIIIgamma promotes the spatial segregation of microbiota and host in the intestine. *Science*. 2011;334(6053):255-258.
42. Ismail AS, Severson KM, Vaishnava S, et al. Gammadelta intraepithelial lymphocytes are essential mediators of host-microbial homeostasis at the intestinal mucosal surface. *Proc Natl Acad Sci U S A*. 2011;108(21):8743-8748.
43. Mukherjee S, Partch CL, Lehotzky RE, et al. Regulation of C-type lectin antimicrobial activity by a flexible N-terminal prosegment. *J Biol Chem*. 2009;284(8):4881-4888.
44. Kobayashi KS, Chamaillard M, Ogura Y, et al. Nod2-dependent regulation of innate and adaptive immunity in the intestinal tract. *Science*. 2005;307(5710):731-734.
45. Gribble FM, Reimann F. Function and mechanisms of enteroendocrine cells and gut hormones in metabolism. *Nat Rev Endocrinol*. 2019;15(4):226-237.
46. Gerbe F, Sidot E, Smyth DJ, et al. Intestinal epithelial tuft cells initiate type 2 mucosal immunity to helminth parasites. *Nature*. 2016;529(7585):226-230.
47. Howitt MR, Lavoie S, Michaud M, et al. Tuft cells, taste-chemosensory cells, orchestrate parasite type 2 immunity in the gut. *Science*. 2016;351(6279):1329-1333.
48. von Moltke J, Ji M, Liang H-E Locksley RM. Tuft-cell-derived IL-25 regulates an intestinal ILC2-epithelial response circuit. *Nature*. 2016;529(7585):221-225.
49. Endt K, Stecher B, Chaffron S, et al. The microbiota mediates pathogen clearance from the gut lumen after non-typhoidal *Salmonella* diarrhea. *PLoS Pathog*. 2010;6(9):e1001097.
50. Chodirker WB, Tomasi TBJr. Gamma-globulins: quantitative relationships in human serum and nonvascular fluids. *Science*. 1963;142(3595):1080-1081.
51. Delacroix DL, Hodgson HJ, McPherson A, Dive C, Vaerman JP. Selective transport of polymeric immunoglobulin A in bile. Quantitative relationships of monomeric and polymeric immunoglobulin A, immunoglobulin M, and other proteins in serum, bile, and saliva. *J Clin Invest*. 1982;70(2):230-241.
52. Tomasi TB Jr, Tan EM, Solomon A, Prendergast RA. Characteristics of an immune system common to certain external secretions. *J Exp Med*. 1965;121:101-124.
53. Fagarasan S, Kawamoto S, Kanagawa O, Suzuki K. Adaptive immune regulation in the gut: T cell-dependent and T cell-independent IgA synthesis. *Annu Rev Immunol*. 2010;28:243-273.
54. Brandtzaeg P, Farstad IN, Johansen F-E, et al. The B-cell system of human mucosae and exocrine glands. *Immunol Rev*. 1999;171:45-87.
55. Chen K, Cerutti A. New insights into the enigma of immunoglobulin D. *Immunol Rev*. 2010;237(1):160-179.
56. Preud'homme JL, Petit I, Barra A, et al. Structural and functional properties of membrane and secreted IgD. *Mol Immunol*. 2000;37(15):871-887.
57. Gould HJ, Sutton BJ. IgE in allergy and asthma today. *Nat Rev Immunol*. 2008;8(3):205-217.
58. Cerutti A, Cols M, Gentile M, et al. Regulation of mucosal IgA responses: lessons from primary immunodeficiencies. *Ann N Y Acad Sci*. 2011;1238:132-144.
59. Chen K, Cerutti A. AIDing the pursuit of IgA diversity. *Nat Immunol*. 2011;12(3):197-198.
60. Cerutti A. 2008. The regulation of IgA class switching. *Nat Rev Immunol* 8(6):421-434.
61. Boullier S, Tanguy M, Kadaoui KA, et al. Secretory IgA-mediated neutralization of *Shigella flexneri* prevents intestinal tissue destruction by down-regulating inflammatory circuits. *J Immunol*. 2009;183(9):5879-5885.
62. Fernandez MI, Pedron T, Tournebize R, Olivo-Marin J-C, Sansonetti PJ, Phalipon A. Anti-inflammatory role for intracellular dimeric immunoglobulin a by neutralization of lipopolysaccharide in epithelial cells. *Immunity*. 2003;18(6):739-749.
63. Hornquist E, Lycke N. Cholera toxin increases T lymphocyte responses to unrelated antigens. *Adv Exp Med Biol*. 1995;371B:1507-1512.
64. Martinoli C, Chiavelli A, Rescigno M. Entry route of *Salmonella typhimurium* directs the type of induced immune response. *Immunity*. 2007;27(6):975-984.
65. Wade TK, Wade WF. Variable gene family usage of protective and non-protective anti-*Vibrio cholerae* O1 LPS antibody heavy chains. *Microbiol Immunol*. 2008;52(12):611-620.
66. Wei M, Shinkura R, Doi Y, Maruya M, Fagarasan S, Honjo T. Mice carrying a knock-in mutation of *Aicda* resulting in a defect in somatic hypermutation have impaired gut homeostasis and compromised mucosal defense. *Nat Immunol*. 2011;12(3):264-270.
67. Brandtzaeg P. Mucosal immunity: integration between mother and the breast-fed infant. *Vaccine*. 2003;21(24):3382-3388.
68. Kramer DR, Cebra JJ. Early appearance of "natural" mucosal IgA responses and germinal centers in suckling mice developing in the absence of maternal antibodies. *J Immunol*. 1995a;154(5):2051-2062.
69. Crouch EE, Li Z, Takizawa M, et al. Regulation of AID expression in the immune response. *J Exp Med*. 2007;204(5):1145-1156.
70. Tsuji M, Suzuki K, Kitamura H, et al. Requirement for lymphoid tissue-inducer cells in isolated follicle formation and T cell-independent immunoglobulin A generation in the gut. *Immunity*. 2008;29(2):261-271.
71. Uematsu S, Fujimoto K, Jang MH, et al. Regulation of humoral and cellular gut immunity by lamina propria dendritic cells expressing Toll-like receptor 5. *Nat Immunol*. 2008;9(7):769-776.
72. Favre L, Spertini F, Corthesy B. Secretory IgA possesses intrinsic modulatory properties stimulating mucosal and systemic immune responses. *J Immunol*. 2005;175(5):2793-2800.

73. Mostov KE. Transepithelial transport of immunoglobulins. *Annu Rev Immunol.* 1994;12:63-84.
74. Phalipon A, Corthesy B. Novel functions of the polymeric Ig receptor: well beyond transport of immunoglobulins. *Trends Immunol.* 2003;24(2):55-58.
75. Monteiro RC, Van De Winkel JG. IgA Fc receptors. *Annu Rev Immunol.* 2003;21:177-204.
76. Pasquier B, Launay P, Kanamaru Y, et al. Identification of FcalphaRI as an inhibitory receptor that controls inflammation: dual role of FcRgamma ITAM. *Immunity.* 2005;22(1):31-42.
77. Peterson DA, McNulty NP, Guruge JL, Gordon JI. IgA response to symbiotic bacteria as a mediator of gut homeostasis. *Cell Host Microbe.* 2007;2(5):328-339.
78. van der Waaij LA, Limburg PC, Mesander G, van der Waaij D. In vivo IgA coating of anaerobic bacteria in human faeces. *Gut.* 1996;38(3):348-354.
79. Macpherson AJ, Hunziker L, McCoy K, Lamarre A. IgA responses in the intestinal mucosa against pathogenic and non-pathogenic microorganisms. *Microbes Infect.* 2001;3(12):1021-1035.
80. Macpherson AJ, Gatto D, Sainsbury E, Harriman GR, Hengartner H, Zinkernagel RM. A primitive T cell-independent mechanism of intestinal mucosal IgA responses to commensal bacteria. *Science.* 2000;288(5474):2222-2226.
81. Fagarasan S, Muramatsu M, Suzuki K, Nagaoka H, Hiai H, Honjo T. Critical roles of activation-induced cytidine deaminase in the homeostasis of gut flora. *Science.* 2002;298(5597):1424-1427.
82. Bunker JJ, Erickson SA, Flynn TM, et al. Natural polyreactive IgA antibodies coat the intestinal microbiota. *Science.* 2017;358(6361):eaan6619.
83. Fagarasan S, Kinoshita K, Muramatsu M, Ikuta K, Honjo T. *In situ* class switching and differentiation to IgA-producing cells in the gut lamina propria. *Nature.* 2001;413(6856):639-643.
84. Macpherson AJ, Uhr T. Induction of protective IgA by intestinal dendritic cells carrying commensal bacteria. *Science.* 2004;303(5664):1662-1665.
85. Reboldi A, Arnon TI, Rodda LB, Atakilit A, Sheppard D, Cyster JG. IgA production requires B cell interaction with subepithelial dendritic cells in Peyer's patches. *Science.* 2016;352(6287):aaf4822.
86. He B, Xu W, Santini PA, et al. Intestinal bacteria trigger T cell-independent immunoglobulin A(2) class switching by inducing epithelial-cell secretion of the cytokine APRIL. *Immunity.* 2007;26(6):812-826.
87. Litinskiy MB, Nardelli B, Hilbert DM, et al. DCs induce CD40-independent immunoglobulin class switching through BLyS and APRIL. *Nat Immunol.* 2002;3(9):822-829.
88. Cazac BB, Roes J. TGF-beta receptor controls B cell responsiveness and induction of IgA in vivo. *Immunity.* 2000;13(4):443-451.
89. Mora JR, Iwata M, Eksteen B, et al. Generation of gut-homing IgA-secreting B cells by intestinal dendritic cells. *Science.* 2006;314(5802):1157-1160.
90. Grootjans J, Krupka N, Hosomi S, et al. Epithelial endoplasmic reticulum stress orchestrates a protective IgA response. *Science.* 2019;363(6430):993-998.
91. Bunker JJ, Flynn TM, Koval JC, et al. Innate and adaptive humoral responses coat distinct commensal bacteria with immunoglobulin A. *Immunity.* 2015;43(3):541-553.
92. Macpherson AJ, Yilmaz B, Limenitakis JP, Ganal-Vonarburg SC. IgA function in relation to the intestinal microbiota. *Annu Rev Immunol.* 2018;36:359-381.
93. Bunker JJ, Bendelac A. IgA responses to microbiota. *Immunity.* 2018;49(2):211-224.
94. Ansaldo E, Slayden LC, Ching KL, et al. *Akkermansia muciniphila* induces intestinal adaptive immune responses during homeostasis. *Science.* 2019;364(6446):1179-1184.
95. Palm NW, de Zoete MR, Cullen TW, et al. Immunoglobulin A coating identifies colitogenic bacteria in inflammatory bowel disease. *Cell.* 2014;158(5):1000-1010.
96. Kau AL, Planer JD, Liu J, et al. Functional characterization of IgA-targeted bacterial taxa from undernourished Malawian children that produce diet-dependent enteropathy. *Sci Transl Med.* 2015;7(276):276ra24.
97. Hapfelmeier S, Lawson MAE, Slack E, et al. Reversible microbial colonization of germ-free mice reveals the dynamics of IgA immune responses. *Science.* 2010;328(5986):1705-1709.
98. Li H, Limenitakis JP, Greiff V, et al. Mucosal or systemic microbiota exposures shape the B cell repertoire. *Nature.* 2020;584(7820):274-278.
99. Yel L. Selective IgA deficiency. *J Clin Immunol.* 2010;30(1):10-16.
100. Chen K, Magri G, Grasset EK, Cerutti A. Rethinking mucosal antibody responses: IgM, IgG and IgD join IgA. *Nat Rev Immunol.* 2020;20(7):427-441.
101. Benckert J, Schmolka N, Kreschel C, et al. The majority of intestinal IgA+ and IgG+ plasmablasts in the human gut are antigen-specific. *J Clin Invest.* 2011;121(5):1946-1955.
102. Magri G, Comerma L, Pybus M, et al. Human secretory IgM emerges from plasma cells clonally related to gut memory B cells and targets highly diverse commensals. *Immunity.* 2017;47(1):118-134.e8.
103. Slack E, Hapfelmeier S, Stecher B, et al. Innate and adaptive immunity cooperate flexibly to maintain host-microbiota mutualism. *Science.* 2009;325(5940):617-620.
104. Zeng MY, Cisalpino D, Varadarajan S, et al. Gut microbiota-induced immunoglobulin G controls systemic infection by symbiotic bacteria and pathogens. *Immunity.* 2016;44(3):647-658.
105. Koch MA, Reiner GL, Lugo KA, et al. Maternal IgG and IgA antibodies dampen mucosal T helper cell responses in early life. *Cell.* 2016;165(4):827-841.
106. Xu Z, Takizawa F, Casadei E, et al. Specialization of mucosal immunoglobulins in pathogen control and microbiota homeostasis occurred early in vertebrate evolution. *Sci Immunol.* 2020;5(44):eaay3254.
107. Fransen F, Zagato E, Mazzini E, et al. BALB/c and C57BL/6 mice differ in polyreactive IgA abundance, which impacts the generation of antigen-specific IgA and microbiota diversity. *Immunity.* 2015;43(3):527-540.
108. Kawamoto S, Tran TH, Maruya M, et al. The inhibitory receptor PD-1 regulates IgA selection and bacterial composition in the gut. *Science.* 2012;336(6080):485-489.
109. Rogier EW, Frantz AL, Bruno MEC, et al. Secretory antibodies in breast milk promote long-term intestinal homeostasis by regulating the gut microbiota and host gene expression. *Proc Natl Acad Sci U S A.* 2014;111(8):3074-3079.
110. Donaldson GP, Ladinsky MS, Yu KB, et al. Gut microbiota utilize immunoglobulin A for mucosal colonization. *Science.* 2018;360(6390):795-800.
111. Moor K, Diard M, Sellin ME, et al. High-avidity IgA protects the intestine by enchaining growing bacteria. *Nature.* 2017;544(7651):498-502.
112. Cullender TC, Chassaing B, Janzon A, et al. Innate and adaptive immunity interact to quench microbiome flagellar motility in the gut. *Cell Host Microbe.* 2013;14(5):571-581.
113. Cong Y, Feng T, Fujihashi K, Schoeb TR, Elson CO. A dominant, coordinated T regulatory cell-IgA response to the intestinal microbiota. *Proc Natl Acad Sci U S A.* 2009;106(46):19256-19261.
114. Wilmore JR, Gaudette BT, Gomez Atria D, et al. Commensal microbes induce serum IgA responses that protect against polymicrobial sepsis. *Cell Host Microbe.* 2018;23(3):302-311 e3.
115. Castro-Dopico T, Dennison TW, Ferdinand JR, et al. Anti-commensal IgG drives intestinal inflammation and type 17 immunity in ulcerative colitis. *Immunity.* 2019;50(4):1099-1114.e10.
116. Gutierrez-Merino J, Isla B, Combes T, Martinez-Estrada F, Maluquer De Motes C. Beneficial bacteria activate type-I interferon production via the intracellular cytosolic sensors STING and MAVS. *Gut Microbes.* 2020;11(4):771-788.
117. Rognum TO, Thrane PS, Stoltenberg L, Vege Å, Brandtzaeg P. Development of intestinal mucosal immunity in fetal life and the first postnatal months. *Pediatr Res.* 1992;32(2):145-149.
118. Gopalakrishna KP, Macadangdang BR, Rogers MB, et al. Maternal IgA protects against the development of necrotizing enterocolitis in preterm infants. *Nat Med.* 2019;25(7):1110-1115.
119. Ramanan D, Sefik E, Galván-Peña S, et al. An immunologic mode of multigenerational transmission governs a gut Treg setpoint. *Cell.* 2020;181(6):1276-1290.e13.
120. Cheroutre H, Lambolez F, Mucida D. The light and dark sides of intestinal intraepithelial lymphocytes. *Nat Rev Immunol.* 2011a;11(7):445-456.
121. Mayassi T, Jabri B. Human intraepithelial lymphocytes. *Mucosal Immunol.* 2018;11(5):1281-1289.
122. McDonald BD, Jabri B, Bendelac A. Diverse developmental pathways of intestinal intraepithelial lymphocytes. *Nat Rev Immunol.* 2018;18(8):514-525.
123. Beagley KW, Fujihashi K, Lagoo AS, et al. Differences in intraepithelial lymphocyte T cell subsets isolated from murine small versus large intestine. *J Immunol.* 1995;154(11):5611-5619.
124. Verstichel G, Vermijlen D, Martens L, et al. The checkpoint for agonist selection precedes conventional selection in human thymus. *Sci Immunol.* 2017;2(8):eaah4232.
125. Hendricks DW, Fink PJ. Uneven colonization of the lymphoid periphery by T cells that undergo early TCR{alpha} rearrangements. *J Immunol.* 2009;182(7):4267-4274.
126. Mora JR, Bono MR, Manjunath N, et al. Selective imprinting of gut-homing T cells by Peyer's patch dendritic cells. *Nature.* 2003;424(6944):88-93.
127. Jung J, Surh CD, Lee YJ. Microbial colonization at early life promotes the development of diet-induced CD8alphabeta intraepithelial T cells. *Mol Cells.* 2019;42(4):313-320.
128. Konkel JE, Maruyama T, Carpenter AC, et al. Control of the development of CD8alphaalpha+ intestinal intraepithelial lymphocytes by TGF-beta. *Nat Immunol.* 2011;12(4):312-319.
129. Mucida D, Husain MM, Muroi S, et al. Transcriptional reprogramming of mature CD4(+) helper T cells generates distinct MHC class II-restricted cytotoxic T lymphocytes. *Nat Immunol.* 2013;14(3):281-289.
130. Reis BS, Rogoz A, Costa-Pinto FA, Taniuchi I, Mucida D. Mutual expression of the transcription factors Runx3 and ThPOK regulates intestinal CD4(+) T cell immunity. *Nat Immunol.* 2013;14(3):271-280.
131. Bilate AM, Bousbaine D, Mesin L, et al. Tissue-specific emergence of regulatory and intraepithelial T cells from a clonal T cell precursor. *Sci Immunol.* 2016;1(2):eaaf7471.
132. Cheroutre H, Lambolez F. The thymus chapter in the life of gut-specific intra epithelial lymphocytes. *Curr Opin Immunol.* 2008;20(2):185-191.
133. Probert CS, Saubermann LJ, Balk S, Blumberg RS. Repertoire of the alpha beta T-cell receptor in the intestine. *Immunol Rev.* 2007;215:215-225.
134. McDonald BD, Bunker JJ, Erickson SA, Oh-Hora M, Bendelac A. Crossreactive alphabeta T Cell receptors are the predominant targets of Thymocyte negative selection. *Immunity.* 2015;43(5):859-869.
135. Bruce D, Cantorna MT. Intrinsic requirement for the vitamin D receptor in the development of CD8alphaalpha-expressing T cells. *J Immunol.* 2011;186(5):2819-2825.
136. Klose CS, Blatz K, d'Hargues Y, et al. The transcription factor T-bet is induced by IL-15 and thymic agonist selection and controls CD8alphaalpha(+) intraepithelial lymphocyte development. *Immunity.* 2014;41(2):230-243.
137. Li Y, nnocentin S, Withers DR, et al. Exogenous stimuli maintain intraepithelial lymphocytes via aryl hydrocarbon receptor activation. *Cell.* 2011;147(3):629-640.
138. Di Marco Barros R, Roberts NA, Dart RJ, et al. Epithelia use butyrophilin-like molecules to shape organ-specific gammadelta T cell compartments. *Cell.* 2016;167(1):203-218.e17.
139. Mayassi T, Ladell K, Gudjonson H, et al. Chronic inflammation permanently reshapes tissue-resident immunity in celiac disease. *Cell.* 2019;176(5):967-981.e19.
140. Chardes T, Buzoni-Gatel D, Lepage A, Bernard F, Bout D. *Toxoplasma gondii* oral infection induces specific cytotoxic CD8 alpha/beta+ Thy-1+ gut intraepithelial lymphocytes, lytic for parasite-infected enterocytes. *J Immunol.* 1994;153(10):4596-4603.
141. Lepage AC, Buzoni-Gatel D, Bout DT, Kasper LH. Gut-derived intraepithelial lymphocytes induce long term immunity against *Toxoplasma gondii*. *J Immunol.* 1998;161(9):4902-4908.
142. Moretto M, Weiss LM, Khan IA. Induction of a rapid and strong antigen-specific intraepithelial lymphocyte response during oral *Encephalitozoon cuniculi* infection. *J Immunol.* 2004;172(7):4402-4409.
143. Roberts SJ, Smith AL, West AB, Wen L, et al. T-cell alpha beta + and gamma delta + deficient mice display abnormal but distinct phenotypes toward a natural, widespread infection of the intestinal epithelium. *Proc Natl Acad Sci U S A.* 1996;93(21):11774-11779.
144. Lefrancois L, Goodman T. In vivo modulation of cytolytic activity and Thy-1 expression in TCR-gamma delta+ intraepithelial lymphocytes. *Science.* 1989;243(4899):1716-1718.
145. Inagaki-Ohara K, Sakamoto Y, Dohi T, Smith AL. Gammadelta T cells play a protective role during infection with *Nippostrongylus brasiliensis* by promoting goblet cell function in the small intestine. *Immunology.* 2011;134(4):448-458.
146. Dalton JE, Cruickshank SM, Egan CE, et al. Intraepithelial gammadelta+ lymphocytes maintain the integrity of intestinal epithelial tight junctions in response to infection. *Gastroenterology.* 2006;131(3):818-829.
147. Epple HJ, Allers K, Tröger H, et al. Acute HIV infection induces mucosal infiltration with CD4+ and CD8+ T cells, epithelial apoptosis, and a mucosal barrier defect. *Gastroenterology.* 2010;139(4):1289-1300.
148. Raffatellu M, Santos RL, Verhoeven DE, et al. Simian immunodeficiency virus-induced mucosal interleukin-17 deficiency promotes Salmonella dissemination from the gut. *Nat Med.* 2008;14(4):421-428.
149. Boismenu R, Havran WL. Modulation of epithelial cell growth by intraepithelial gamma delta T cells. *Science.* 1994;266(5188):1253-1255.
150. Chen Y, Chou K, Fuchs E, Havran WL, Boismenu R. Protection of the intestinal mucosa by intraepithelial gamma delta T cells. *Proc Natl Acad Sci U S A.* 2002;99(22):14338-14343.
151. Bhagat G, Naiyer AJ, Shah JG, et al. Small intestinal CD8+TCRgammadelta+NKG2A+ intraepithelial lymphocytes have attributes of regulatory cells in patients with celiac disease. *J Clin Invest.* 2008;118(1):281-293.
152. Inagaki-Ohara K, Dewi FN, Hisaeda H, et al. Intestinal intraepithelial lymphocytes sustain the epithelial barrier function against *Eimeria vermiformis* infection. *Infect Immun.* 2006;74(9):5292-5301.
153. Kamanaka M, Kim ST, Wan YY, et al. Expression of interleukin-10 in intestinal lymphocytes detected by an interleukin-10 reporter knockin tiger mouse. *Immunity.* 2006;25(6):941-952.
154. Giacomelli R, Parzanese I, Frieri G, et al. Increase of circulating gamma/delta T lymphocytes in the peripheral blood of patients affected by active inflammatory bowel disease. *Clin Exp Immunol.* 1994;98(1):83-88.
155. Kanazawa H, Ishiguro Y, Munakata A, Morita T. Multiple accumulation of Vdelta2+ gammadelta T-cell clonotypes in intestinal mucosa from patients with Crohn's disease. *Dig Dis Sci.* 2001;46(2):410-416.
156. Kawaguchi-Miyashita M, Shimada S, Kurosu H, et al. An accessory role of TCRgammadelta (+) cells in the exacerbation of inflammatory bowel disease in TCRalpha mutant mice. *Eur J Immunol.* 2001;31(4):980-988.

157. Simpson SJ, Holländer GA, Mizoguchi E, et al. Expression of pro-inflammatory cytokines by TCR alpha beta+ and TCR gamma delta+ T cells in an experimental model of colitis. *Eur J Immunol.* 1997;27(1):17-25.
158. Abadie V, Discepolo V, Jabri B. Intraepithelial lymphocytes in celiac disease immunopathology. *Semin Immunopathol.* 2012;34(4):551-566.
159. Hue S, Mention J-J, Monteiro RC, et al. A direct role for NKG2D/MICA interaction in villous atrophy during celiac disease. *Immunity.* 2004;21(3):367-377.
160. Meresse B, Chen Z, Ciszewski C, et al. Coordinated induction by IL15 of a TCR-independent NKG2D signaling pathway converts CTL into lymphokine-activated killer cells in celiac disease. *Immunity.* 2004;21(3):357-366.
161. Abadie V, Sollid LM, Barreiro LB, Jabri B. Integration of genetic and immunological insights into a model of celiac disease pathogenesis. *Annu Rev Immunol.* 2011;29:493-525.
162. Jameson J, Ugarte K, Chen N, et al. A role for skin gammadelta T cells in wound repair. *Science.* 2002;296(5568):747-749.
163. Sullivan ZA, Khoury-Hanold W, Lim J, et al. Gammadelta T cells regulate the intestinal response to nutrient sensing. *Science.* 2021;371(6535):eaba8310.
164. Godfrey DI, Uldrich AP, McCluskey J, Rossjohn J, Moody DB. The burgeoning family of unconventional T cells. *Nat Immunol.* 2015;16(11):1114-1123.
165. An D, Oh SF, Olszak T, et al. Sphingolipids from a symbiotic microbe regulate homeostasis of host intestinal natural killer T cells. *Cell.* 2014;156(1-2):123-133.
166. Constantinides MG, Link VM, Tamoutounour S, et al. MAIT cells are imprinted by the microbiota in early life and promote tissue repair. *Science.* 2019;366(6464):eaax6624.
167. Haas JD, Ravens S, Düber S, et al. Development of interleukin-17-producing gammadelta T cells is restricted to a functional embryonic wave. *Immunity.* 2012;37(1):48-59.
168. Leeansyah E, Loh L, Nixon DF, Sandberg JK. Acquisition of innate-like microbial reactivity in mucosal tissues during human fetal MAIT-cell development. *Nat Commun.* 2014;5:3143.
169. Narayan K, Sylvia KE, Malhotra N, et al. Intrathymic programming of effector fates in three molecularly distinct gammadelta T cell subtypes. *Nat Immunol.* 2012;13(5):511-518.
170. Tieppo P, Papadopoulou M, Gatti D, et al. The human fetal thymus generates invariant effector gammadelta T cells. *J Exp Med.* 2020;217(3).
171. Bendelac A, Savage PB, Teyton L. The biology of NKT cells. *Annu Rev Immunol.* 2007;25:297-336.
172. Wei B, Wingender G, Fujiwara D, et al. Commensal microbiota and CD8+ T cells shape the formation of invariant NKT cells. *J Immunol.* 2010;184(3):1218-1226.
173. Wingender G, Stepniak D, Krebs P, et al. Intestinal microbes affect phenotypes and functions of invariant natural killer T cells in mice. *Gastroenterology.* 2012;143(2):418-428.
174. Olszak T, An D, Zeissig S, et al. Microbial exposure during early life has persistent effects on natural killer T cell function. *Science.* 2012;336(6080):489-493.
175. Constantinides MG. Interactions between the microbiota and innate and innate-like lymphocytes. *J Leukoc Biol.* 2018;103(3):409-419.
176. Le Bourhis L, Martin E, Péguillet I, et al. Antimicrobial activity of mucosal-associated invariant T cells. *Nat Immunol.* 2010;11(8):701-708.
177. Koay HF, Gherardin NA, Enders A, et al. A three-stage intrathymic development pathway for the mucosal-associated invariant T cell lineage. *Nat Immunol.* 2016;17(11):1300-1311.
178. Legoux F, Bellet D, Daviaud C, et al. Microbial metabolites control the thymic development of mucosal-associated invariant T cells. *Science.* 2019;366(6464):494-499.
179. Treiner E, Duban L, Bahram S, et al. Selection of evolutionarily conserved mucosal-associated invariant T cells by MR1. *Nature.* 2003;422(6928):164-169.
180. Gutierrez-Arcelus M, Teslovich N, Mola AR, et al. Lymphocyte innateness defined by transcriptional states reflects a balance between proliferation and effector functions. *Nat Commun.* 2019;10(1):687.
181. Hatano S, Murakami T, Noguchi N, Yamada H, Yoshikai Y. CD5(−)NK1.1(+) gammadelta T cells that develop in a Bcl11b-independent manner participate in early protection against infection. *Cell Rep.* 2017;21(5):1191-1202.
182. Sandrock I, Ziętara N, Łyszkiewicz M, et al. MicroRNA-181a/b-1 is not required for innate gammadelta NKT effector cell development. *PLoS One.* 2015;10(12):e0145010.
183. Howson LJ, Awad W, von Borstel A, et al. Absence of mucosal-associated invariant T cells in a person with a homozygous point mutation in *MR1*. *Sci Immunol.* 2020;5(49).
184. Melandri D, Zlatareva I, Chaleil RAG, et al. The gammadeltaTCR combines innate immunity with adaptive immunity by utilizing spatially distinct regions for agonist selection and antigen responsiveness. *Nat Immunol.* 2018;19(12):1352-1365.
185. Rider P, Voronov E, Dinarello CA, Apte RN, Cohen I. Alarmins: feel the stress. *J Immunol.* 2017;198(4):1395-1402.
186. Leng T, Akther HD, Hackstein C-P, et al. TCR and inflammatory signals tune human MAIT cells to exert specific tissue repair and effector functions. *Cell Rep.* 2019;28(12):3077-3091.e5.
187. Hinks TSC, Marchi E, Jabeen M, et al. Activation and in vivo evolution of the MAIT cell transcriptome in mice and humans reveals tissue repair functionality. *Cell Rep.* 2019;28(12):3249-3262.e5.
188. Jin C, Lagoudas GK, Zhao C, et al. Commensal microbiota promote lung cancer development via gammadelta T cells. *Cell.* 2019;176(5):998-1013.e16.
189. Crowther MD, Dolton G, Legut M, et al. Genome-wide CRISPR-Cas9 screening reveals ubiquitous T cell cancer targeting via the monomorphic MHC class I-related protein MR1. *Nat Immunol.* 2020;21(2):178-185.
190. Gherardin NA, Loh L, Admojo L, et al. Enumeration, functional responses and cytotoxic capacity of MAIT cells in newly diagnosed and relapsed multiple myeloma. *Sci Rep.* 2018;8(1):4159.
191. Godfrey DI, Le Nours J, Andrews DM, Uldrich AP, Rossjohn J. Unconventional T cell targets for cancer immunotherapy. *Immunity.* 2018;48(3):453-473.
192. Bain CC, Bravo-Blas A, Scott CL, et al. Constant replenishment from circulating monocytes maintains the macrophage pool in the intestine of adult mice. *Nat Immunol.* 2014;15(10):929-937.
193. Anderson DA3rd, Dutertre C-A, Ginhoux F, Murphy KM. Genetic models of human and mouse dendritic cell development and function. *Nat Rev Immunol.* 2021;21(2):101-115.
194. Sichien D, Lambrecht BN, Guilliams M, Scott CL. Development of conventional dendritic cells: from common bone marrow progenitors to multiple subsets in peripheral tissues. *Mucosal Immunol.* 2017;10(4):831-844.
195. Cerovic V, Bain CC, Mowat AM, Milling SWF. Intestinal macrophages and dendritic cells: what's the difference? *Trends Immunol.* 2014;35(6):270-277.
196. Joeris T, Müller-Luda K, Agace WW, Mowat AM. Diversity and functions of intestinal mononuclear phagocytes. *Mucosal Immunol.* 2017;10(4):845-864.
197. Bleriot C, Chakarov S, Ginhoux F. Determinants of resident tissue macrophage identity and function. *Immunity.* 2020;52(6):957-970.
198. Kang B, Alvarado LJ, Kim T, et al. Commensal microbiota drive the functional diversification of colon macrophages. *Mucosal Immunol.* 2020;13(2):216-229.
199. Niess JH, Adler G. Enteric flora expands gut lamina propria CX3CR1+ dendritic cells supporting inflammatory immune responses under normal and inflammatory conditions. *J Immunol.* 2010;184(4):2026-2037.
200. Rivollier A, He J, Kole A, Valatas V, Kelsall BL. Inflammation switches the differentiation program of Ly6Chi monocytes from antiinflammatory macrophages to inflammatory dendritic cells in the colon. *J Exp Med.* 2012;209(1):139-155.
201. Houston SA, Cerovic V, Thomson C, Brewer J, Mowat AM, Milling S. The lymph nodes draining the small intestine and colon are anatomically separate and immunologically distinct. *Mucosal Immunol.* 2016;9(2):468-478.
202. Veenbergen S, van Berkel LA, du Pré MF, et al. Colonic tolerance develops in the iliac lymph nodes and can be established independent of CD103(+) dendritic cells. *Mucosal Immunol.* 2016;9(4):894-906.
203. Cerovic V, Houston SA, Scott CL, et al. Intestinal CD103(−) dendritic cells migrate in lymph and prime effector T cells. *Mucosal Immunol.* 2013;6(1):104-113.
204. Da Silva C, Wagner C, Bonnardel J, Gorvel J-P, Lelouard H. The peyer's patch mononuclear phagocyte system at steady state and during infection. *Front Immunol.* 2017;8:1254.
205. Iwasaki A, Kelsall BL. Localization of distinct Peyer's patch dendritic cell subsets and their recruitment by chemokines macrophage inflammatory protein (MIP)-3alpha, MIP-3beta, and secondary lymphoid organ chemokine. *J Exp Med.* 2000;191(8):1381-1394.
206. Bonnardel J, Da Silva C, Wagner C, et al. Distribution, location, and transcriptional profile of Peyer's patch conventional DC subsets at steady state and under TLR7 ligand stimulation. *Mucosal Immunol.* 2017;10(6):1412-1430.
207. McDonald KG, McDonough JS, Dieckgraefe BK, Newberry RD. Dendritic cells produce CXCL13 and participate in the development of murine small intestine lymphoid tissues. *Am J Pathol.* 2010;176(5):2367-2377.
208. Jaensson E, Uronen-Hansson H, Pabst O, et al. Small intestinal CD103+ dendritic cells display unique functional properties that are conserved between mice and humans. *J Exp Med.* 2008;205(9):2139-2149.
209. Guendel F, Kofoed-Branzk M, Gronke K, et al. Group 3 innate lymphoid cells program a distinct subset of IL-22BP-producing dendritic cells demarcating solitary intestinal lymphoid tissues. *Immunity.* 2020;53(5):1015-1032 e8.
210. Longman RS, Diehl GE, Victorio DA, et al. CX(3)CR1(+) mononuclear phagocytes support colitis-associated innate lymphoid cell production of IL-22. *J Exp Med.* 2014;211(8):1571-1583.
211. Savage AK, Liang HE, Locksley RM. The development of steady-state activation hubs between adult LTi ILC3s and primed macrophages in small intestine. *J Immunol.* 2017;199(5):1912-1922.
212. Lelouard H, Henri S, De Bovis B, et al. Pathogenic bacteria and dead cells are internalized by a unique subset of Peyer's patch dendritic cells that express lysozyme. *Gastroenterology.* 2010;138(1):173-184.e1-3.
213. Lelouard H, Fallet M, de Bovis B, Méresse S, Gorvel JP. Peyer's patch dendritic cells sample antigens by extending dendrites through M cell-specific transcellular pores. *Gastroenterology.* 2012;142(3):592-601 e3.
214. Wagner C, Bonnardel J, Da Silva C, et al. Differentiation paths of Peyer's patch LysoDCs are linked to sampling site positioning, migration, and T cell priming. *Cell Rep.* 2020;31(1):107479.
215. De Schepper S, Verheijden S, Aguilera-Lizarraga J, et al. Self-maintaining gut macrophages are essential for intestinal homeostasis. *Cell.* 2018;175(2):400-415.e13.
216. Schulz O, Pabst O. Antigen sampling in the small intestine. *Trends Immunol.* 2013;34(4):155-161.
217. Dillon A, Lo DD. M cells: intelligent engineering of mucosal immune surveillance. *Front Immunol.* 2019;10:1499.
218. Kanaya T, Williams IR, Ohno H. Intestinal M cells: tireless samplers of enteric microbiota. *Traffic.* 2020;21(1):34-44.
219. Hase K, Kawano K, Nochi T. et al. Uptake through glycoprotein 2 of FimH(+) bacteria by M cells initiates mucosal immune response. *Nature.* 2009;462(7270):226-230.
220. Chabot S, Wagner JS, Farrant S, Neutra MR. TLRs regulate the gatekeeping functions of the intestinal follicle-associated epithelium. *J Immunol.* 2006;176(7):4275-4283.
221. Fleeton MN, Contractor N, Leon F, Wetzel JD, Dermody TS, Kelsall BL. Peyer's patch dendritic cells process viral antigen from apoptotic epithelial cells in the intestine of reovirus-infected mice. *J Exp Med.* 2004;200(2):235-245.
222. Sun CM, Hall JA, Blank RB, et al. Small intestine lamina propria dendritic cells promote de novo generation of Foxp3 T reg cells via retinoic acid. *J Exp Med.* 2007a;204(8):1775-1785.
223. Salazar-Gonzalez RM, Niess JH, Zammit DJ, et al. CCR6-mediated dendritic cell activation of pathogen-specific T cells in Peyer's patches. *Immunity.* 2006;24(5):623-632.
224. Buckley A, Turner JR. Cell biology of tight junction barrier regulation and mucosal disease. *Cold Spring Harb Perspect Biol.* 2018;10(1):a029314.
225. Adson A, Raub TJ, Burton PS, et al. Quantitative approaches to delineate paracellular diffusion in cultured epithelial cell monolayers. *J Pharm Sci.* 1994;83(11):1529-1536.
226. Menard S, Cerf-Bensussan N, Heyman M. Multiple facets of intestinal permeability and epithelial handling of dietary antigens. *Mucosal Immunol.* 2010;3(3):247-259.
227. McDole JR, Wheeler LW, McDonald KG, et al. Goblet cells deliver luminal antigen to CD103+ dendritic cells in the small intestine. *Nature.* 2012;483(7389):345-349.
228. Drolia R, Tenguria S, Durkes AC, Turner JR, Bhunia AK. Listeria adhesion protein induces intestinal epithelial barrier dysfunction for bacterial translocation. *Cell Host Microbe.* 2018;23(4):470-484 e7.
229. Walker WA, Cornell R, Davenport LM, Isselbacher KJ. Macromolecular absorption. Mechanism of horseradish peroxidase uptake and transport in adult and neonatal rat intestine. *J Cell Biol.* 1972;54(2):195-205.
230. Howe SE, Lickteig DJ, Plunkett KN, Ryerse JS, Konjufca V. The uptake of soluble and particulate antigens by epithelial cells in the mouse small intestine. *PLoS One.* 2014;9(1):e86656.
231. Shawki A, McCole DF. Mechanisms of intestinal epithelial barrier dysfunction by adherent-invasive *Escherichia coli*. *Cell Mol Gastroenterol Hepatol.* 2017;3(1):41-50.
232. Johansen FE, Kaetzel CS. Regulation of the polymeric immunoglobulin receptor and IgA transport: new advances in environmental factors that stimulate pIgR expression and its role in mucosal immunity. *Mucosal Immunol.* 2011;4(6):598-602.
233. Pyzik M, Sand KMK, Hubbard JJ, Andersen JT, Sandlie I, Blumberg RS. The neonatal Fc receptor (FcRn): a misnomer? *Front Immunol.* 2019;10:1540.
234. Rodewald R, Kraehenbuhl JP. Receptor-mediated transport of IgG. *J Cell Biol.* 1984;99(1 pt 2):159s-164s.
235. Ohsaki A, Venturelli N, Buccigrosso TM, et al. Maternal IgG immune complexes induce food allergen-specific tolerance in offspring. *J Exp Med.* 2018;215(1):91-113.
236. Yoshida M, Claypool SM, Wagner JS, et al. Human neonatal Fc receptor mediates transport of IgG into luminal secretions for delivery of antigens to mucosal dendritic cells. *Immunity.* 2004;20(6):769-783.
237. Yoshida M, Kobayashi K, Kuo TT, et al. Neonatal Fc receptor for IgG regulates mucosal immune responses to luminal bacteria. *J Clin Invest.* 2006;116(8):2142-2151.
238. Baker K, Qiao S-W, Kuo TT, et al. Neonatal Fc receptor for IgG (FcRn) regulates cross-presentation of IgG immune complexes by CD8−CD11b+ dendritic cells. *Proc Natl Acad Sci U S A.* 2011;108(24):9927-9932.
239. Mosconi E, Rekima A, Seitz-Polski B, et al. Breast milk immune complexes are potent inducers of oral tolerance in neonates and prevent asthma development. *Mucosal Immunol.* 2010;3(5):461-474.
240. Nakata K, Kobayashi K, Ishikawa Y, et al. The transfer of maternal antigen-specific IgG regulates the development of allergic airway inflammation early in life in an FcRn-dependent manner. *Biochem Biophys Res Commun.* 2010;395(2):238-243.
241. Lambrecht BN. FcRn is mother's milk to allergen tolerance. *J Exp Med.* 2018;215(1):1-3.
242. Jang MH, Kweon M-N, Iwatani K, et al. Intestinal villous M cells: an antigen entry site in the mucosal epithelium. *Proc Natl Acad Sci U S A.* 2004;101(16):6110-6115.

243. Knoop KA, Kumar N, Butler BR, et al. RANKL is necessary and sufficient to initiate development of antigen-sampling M cells in the intestinal epithelium. *J Immunol.* 2009;183(9):5738-5747.
244. Valpotic H, Kovšca Janjatović A, Lacković G, et al. Increased number of intestinal villous M cells in levamisole -pretreated weaned pigs experimentally infected with F4ac(+) enterotoxigenic *Escherichia coli* strain. *Eur J Histochem.* 2010;54(2):e18.
245. Wang J, Gusti V, Saraswati A, Lo DD. Convergent and divergent development among M cell lineages in mouse mucosal epithelium. *J Immunol.* 2011a;187(10):5277-5285.
246. Knoop KA, McDonald KG, Kulkarni DH, Newberry RD. Antibiotics promote inflammation through the translocation of native commensal colonic bacteria. *Gut.* 2016;65(7):1100-1109.
247. Knoop KA, Gustafsson JK, McDonald KG, et al. Microbial antigen encounter during a preweaning interval is critical for tolerance to gut bacteria. *Sci Immunol.* 2017;2(18):eaao1314.
248. Kulkarni DH, McDonald KG, Knoop KA, et al. Goblet cell associated antigen passages are inhibited during *Salmonella typhimurium* infection to prevent pathogen dissemination and limit responses to dietary antigens. *Mucosal Immunol.* 2018;11(4):1103-1113.
249. Knoop KA, Newberry RD. Goblet cells: multifaceted players in immunity at mucosal surfaces. *Mucosal Immunol.* 2018;11:1551-1557.
250. Rescigno M, Urbano M, Valzasina B, et al. Dendritic cells express tight junction proteins and penetrate gut epithelial monolayers to sample bacteria. *Nat Immunol.* 2001;2(4):361-367.
251. Chieppa M, Rescigno M, Huang AYC, Germain RN. Dynamic imaging of dendritic cell extension into the small bowel lumen in response to epithelial cell TLR engagement. *J Exp Med.* 2006;203(13):2841-2852.
252. Niess JH, Brand S, Gu X, et al. CX3CR1-mediated dendritic cell access to the intestinal lumen and bacterial clearance. *Science.* 2005;307(5707):254-258.
253. Hapfelmeier S, Müller AJ, Stecher B, et al. Microbe sampling by mucosal dendritic cells is a discrete, MyD88-independent step in DeltainvG S. *Typhimurium colitis. J Exp Med.* 2008;205(2):437-450.
254. Rescigno M. Dendritic cells in bacteria handling in the gut. *J Leukoc Biol.* 2011;90(4):669-672.
255. Farache J, Koren I, Milo I, et al. Luminal bacteria recruit CD103+ dendritic cells into the intestinal epithelium to sample bacterial antigens for presentation. *Immunity.* 2013;38(3):581-595.
256. Schulz O, Jaensson E, Persson EK, et al. Intestinal CD103+, but not CX3CR1+, antigen sampling cells migrate in lymph and serve classical dendritic cell functions. *J Exp Med.* 2009;206(13):3101-3114.
257. Bogunovic M, Ginhoux F, Helft J, et al. Origin of the lamina propria dendritic cell network. *Immunity.* 2009;31(3):513-525.
258. Huang FP, Platt N, Wykes M, et al. A discrete subpopulation of dendritic cells transports apoptotic intestinal epithelial cells to T cell areas of mesenteric lymph nodes. *J Exp Med.* 2000;191(3):435-444.
259. Mazzini E, Massimiliano L, Penna G, Rescigno M. Oral tolerance can be established via gap junction transfer of fed antigens from CX3CR1(+) macrophages to CD103(+) dendritic cells. *Immunity.* 2014;40(2):248-261.
260. Scott CL, Bain CC, Wright PB, et al. CCR2(+)CD103(−) intestinal dendritic cells develop from DC-committed precursors and induce interleukin-17 production by T cells. *Mucosal Immunol.* 2015;8(2):327-339.
261. Bain CC, Scott CL, Uronen-Hansson H, et al. Resident and pro-inflammatory macrophages in the colon represent alternative context-dependent fates of the same Ly6Chi monocyte precursors. *Mucosal Immunol.* 2013;6(3):498-510.
262. Zigmond E, Varol C, Farache J, et al. Ly6C hi monocytes in the inflamed colon give rise to proinflammatory effector cells and migratory antigen-presenting cells. *Immunity.* 2012;37(6):1076-1090.
263. Kamada N, Hisamatsu T, Okamoto S, et al. Unique CD14 intestinal macrophages contribute to the pathogenesis of Crohn disease via IL-23/IFN-gamma axis. *J Clin Invest.* 2008;118(6):2269-2280.
264. Grainger JR, Wohlfert EA, Fuss IJ, et al. Inflammatory monocytes regulate pathologic responses to commensals during acute gastrointestinal infection. *Nat Med.* 2013;19(6):713-721.
265. Bernardo D, Marin AC, Fernández-Tomé S, et al. Human intestinal pro-inflammatory CD11c(high)CCR2(+)CX3CR1(+) macrophages, but not their tolerogenic CD11c(−)CCR2(−) CX3CR1(−) counterparts, are expanded in inflammatory bowel disease. *Mucosal Immunol.* 2018;11(4):1114-1126.
266. Bujko A, Atlasy N, Landsverk OJB, et al. Transcriptional and functional profiling defines human small intestinal macrophage subsets. *J Exp Med.* 2018;215(2):441-458.
267. Ip WKE, Hoshi N, Shouval DS, Snapper S, Medzhitov R. Anti-inflammatory effect of IL-10 mediated by metabolic reprogramming of macrophages. *Science.* 2017;356(6337):513-519.
268. Shouval DS, Biswas A, Goettel JA, et al. Interleukin-10 receptor signaling in innate immune cells regulates mucosal immune tolerance and anti-inflammatory macrophage function. *Immunity.* 2014;40(5):706-719.
269. Shouval DS, Biswas A, Kang YH, et al. Interleukin 1beta mediates intestinal inflammation in mice and patients with interleukin 10 receptor deficiency. *Gastroenterology.* 2016;151(6):1100-1104.
270. Zigmond E, Bernshtein B, Friedlander G, et al. Macrophage-restricted interleukin-10 receptor deficiency, but not IL-10 deficiency, causes severe spontaneous colitis. *Immunity.* 2014;40(5):720-733.
271. Mortha A, Chudnovskiy A, Hashimoto D, et al. Microbiota-dependent crosstalk between macrophages and ILC3 promotes intestinal homeostasis. *Science.* 2014;343(6178):1249288.
272. Chaudhry A, Samstein RM, Treuting P, et al. Interleukin-10 signaling in regulatory T cells is required for suppression of Th17 cell-mediated inflammation. *Immunity.* 2011;34(4):566-578.
273. Hadis U, Wahl B, Schulz O, et al. Intestinal tolerance requires gut homing and expansion of FoxP3+ regulatory T cells in the lamina propria. *Immunity.* 2011;34(2):237-246.
274. Murai M, Turovskaya O, Kim G, et al. Interleukin 10 acts on regulatory T cells to maintain expression of the transcription factor Foxp3 and suppressive function in mice with colitis. *Nat Immunol.* 2009;10(11):1178-1184.
275. Arnold IC, Mathisen S, Schulthess J, Danne C, Hegazy AN, Powrie F. CD11c(+) monocyte/macrophages promote chronic *Helicobacter hepaticus*-induced intestinal inflammation through the production of IL-23. *Mucosal Immunol.* 2016;9(2):352-363.
276. Mukhopadhyay S, Heinz E, Porreca I, et al. Loss of IL-10 signaling in macrophages limits bacterial killing driven by prostaglandin E2. *J Exp Med.* 2020;217(2).
277. Schulthess J, Pandey S, Capitani M, et al. The short chain fatty acid butyrate imprints an antimicrobial program in macrophages. *Immunity.* 2019;50(2):432-445.e7.
278. Smythies LE, Sellers M, Clements RH, et al. Human intestinal macrophages display profound inflammatory anergy despite avid phagocytic and bacteriocidal activity. *J Clin Invest.* 2005;115(1):66-75.
279. Honda M, Surewaard BGJ, Watanabe M, et al. Perivascular localization of macrophages in the intestinal mucosa is regulated by Nr4a1 and the microbiome. *Nat Commun.* 2020;11(1):1329.
280. Cummings RJ, Barbet G, Bongers G, et al. Different tissue phagocytes sample apoptotic cells to direct distinct homeostasis programs. *Nature.* 2016;539(7630):565-569.
281. Chikina AS, Nadalin F, Maurin M, et al. Macrophages maintain epithelium integrity by limiting fungal product absorption. *Cell.* 2020;183(2):411-428.e16.
282. Pull SL, Doherty JM, Mills JC, Gordon JI, Stappenbeck TS. Activated macrophages are an adaptive element of the colonic epithelial progenitor niche necessary for regenerative responses to injury. *Proc Natl Acad Sci U S A.* 2005;102(1):99-104.
283. Sehgal A, Donaldson DS, Pridans C, Sauter KA, Hume DA, Mabbott NA. The role of CSF1R-dependent macrophages in control of the intestinal stem-cell niche. *Nat Commun.* 2018;9(1):1272.
284. Asano K, Takahashi N, Ushiki M, et al. Intestinal CD169(+) macrophages initiate mucosal inflammation by secreting CCL8 that recruits inflammatory monocytes. *Nat Commun.* 2015;6:7802.
285. Matheis F, Muller PA, Graves CL, et al. Adrenergic signaling in muscularis macrophages limits infection-induced neuronal loss. *Cell.* 2020;180(1):64-78.e16.
286. Muller PA, Koscsó B, Rajani GM, et al. Crosstalk between muscularis macrophages and enteric neurons regulates gastrointestinal motility. *Cell.* 2014;158(2):300-313.
287. Gabanyi I, Muller PA, Feighery L, et al. Neuro-immune interactions drive tissue programming in intestinal macrophages. *Cell.* 2016;164(3):378-391.
288. Matteoli G, Gomez-Pinilla PJ, Nemethova A, et al. A distinct vagal anti-inflammatory pathway modulates intestinal muscularis resident macrophages independent of the spleen. *Gut.* 2014;63(6):938-948.
289. Coombes JL, Siddiqui KRR, Arancibia-Cárcamo CV, et al. A functionally specialized population of mucosal CD103+ DCs induces Foxp3+ regulatory T cells via a TGF-beta and retinoic acid-dependent mechanism. *J Exp Med.* 2007;204(8):1757-1764.
290. Paidassi H, Acharya M, Zhang A, et al. Preferential expression of integrin alphavbeta8 promotes generation of regulatory T cells by mouse CD103+ dendritic cells. *Gastroenterology.* 2011;141(5):1813-1820.
291. Worthington JJ, Klementowicz JE, Rahman S, et al. Loss of the TGFbeta-activating integrin alphavbeta8 on dendritic cells protects mice from chronic intestinal parasitic infection via control of type 2 immunity. *PLoS Pathog.* 2013;9(10):e1003675.
292. Worbs T, Bode U, Yan S, et al. Oral tolerance originates in the intestinal immune system and relies on antigen carriage by dendritic cells. *J Exp Med.* 2006;203(3):519-527.
293. Annacker O, Coombes JL, Malmstrom V, et al. Essential role for CD103 in the T cell-mediated regulation of experimental colitis. *J Exp Med.* 2005;202(8):1051-1061.
294. Johansson-Lindbom B, Svensson M, Pabst O, et al. Functional specialization of gut CD103+ dendritic cells in the regulation of tissue-selective T cell homing. *J Exp Med.* 2005;202(8):1063-1073.
295. Svensson M, Johansson-Lindbom B, Zapata F, et al. Retinoic acid receptor signaling levels and antigen dose regulate gut homing receptor expression on CD8+ T cells. *Mucosal Immunol.* 2008;1(1):38-48.
296. Cerovic V, Houston SA, Westlund J, et al. Lymph-borne CD8alpha+ dendritic cells are uniquely able to cross-prime CD8+ T cells with antigen acquired from intestinal epithelial cells. *Mucosal Immunol.* 2015;8(1):38-48.
297. Huang G, Wang Y, Chi H. Control of T cell fates and immune tolerance by p38alpha signaling in mucosal CD103+ dendritic cells. *J Immunol.* 2013;191(2):650-659.
298. Chudnovskiy A, Mortha A, Kana V, et al. Host-protozoan interactions protect from mucosal infections through activation of the inflammasome. *Cell.* 2016;167(2):444-456.e14.
299. Persson EK, Uronen-Hansson H, Semmrich M, et al. IRF4 transcription-factor-dependent CD103(+)CD11b(+) dendritic cells drive mucosal T helper 17 cell differentiation. *Immunity.* 2013;38(5):958-969.
300. Schlitzer A, McGovern N, Teo P, et al. IRF4 transcription factor-dependent CD11b+ dendritic cells in human and mouse control mucosal IL-17 cytokine responses. *Immunity.* 2013;38(5):970-983.
301. Demiri M, Müller-Luda K, Agace WW, Svensson-Frej M. Distinct DC subsets regulate adaptive Th1 and 2 responses during *Trichuris muris* infection. *Parasite Immunol.* 2017;39e12458:1-10.
302. Kinnebrew MA, Buffie CG, Diehl GE. et al. Interleukin 23 production by intestinal CD103(+)CD11b(+) dendritic cells in response to bacterial flagellin enhances mucosal innate immune defense. *Immunity.* 2012;36(2):276-287.
303. Satpathy AT, Briseño CG, Lee JS, et al. Notch2-dependent classical dendritic cells orchestrate intestinal immunity to attaching-and-effacing bacterial pathogens. *Nat Immunol.* 2013;14(9):937-948.
304. Bonnardel J, Da Silva C, Henri S, et al. Innate and adaptive immune functions of Peyer's patch monocyte-derived cells. *Cell Rep.* 2015;11(5):770-784.
305. Martinez-Lopez M, Iborra S, Conde-Garrosa R, et al. Microbiota sensing by Mincle-Syk Axis in dendritic cells regulates interleukin-17 and -22 production and promotes intestinal barrier integrity. *Immunity.* 2019;50(2):446-461.e9.
306. Iwasaki A, Kelsall BL. Freshly isolated Peyer's patch, but not spleen, dendritic cells produce interleukin 10 and induce the differentiation of T helper type 2 cells. *J Exp Med.* 1999;190(2):229-239.
307. Sato A, Hashiguchi M, Toda E, Iwasaki A, Hachimura S, Kaminogawa S. CD11b+ Peyer's patch dendritic cells secrete IL-6 and induce IgA secretion from naive B cells. *J Immunol.* 2003;171(7):3684-3690.
308. Iwasaki A, Kelsall BL. Unique functions of CD11b+, CD8 alpha+, and double-negative Peyer's patch dendritic cells. *J Immunol.* 2001;166(8):4884-4890.
309. Amit I, Winter DR, Jung S. The role of the local environment and epigenetics in shaping macrophage identity and their effect on tissue homeostasis. *Nat Immunol.* 2016;17(1):18-25.
310. Mucida D, Esterhazy D. SnapShot: gut immune niches. *Cell.* 2018;174(6):1600-1600.e1.
311. Viola MF, Boeckxstaens G. Niche-specific functional heterogeneity of intestinal resident macrophages. *Gut.* 2021;70(7):1383-1395.
312. Kulkarni S, Kurapati S, Bogunovic M. Neuro-innate immune interactions in gut mucosal immunity. *Curr Opin Immunol.* 2021;68:64-71.
313. Matteoli G, Boeckxstaens GE. The vagal innervation of the gut and immune homeostasis. *Gut.* 2013;62(8):1214-1222.
314. Bain CC, Schridde A. Origin, differentiation, and function of intestinal macrophages. *Front Immunol.* 2018;9:2733.
315. Schridde A, Bain CC, Mayer JU, et al. Tissue-specific differentiation of colonic macrophages requires TGFbeta receptor-mediated signaling. *Mucosal Immunol.* 2017;10(6):1387-1399.
316. Chang PV, Hao L, Offermanns S, Medzhitov R. The microbial metabolite butyrate regulates intestinal macrophage function via histone deacetylase inhibition. *Proc Natl Acad Sci U S A.* 2014;111(6):2247-2252.
317. Pabst O, Mowat AM. Oral tolerance to food protein. *Mucosal Immunol.* 2012;5(3):232-239.
318. Glocker EO, Kotlarz D, Boztug K, et al. Inflammatory bowel disease and mutations affecting the interleukin-10 receptor. *N Engl J Med.* 2009;361(21):2033-2045.
319. Kuhn R, Löhler J, Rennick D, Rajewsky K, Müller W. Interleukin-10-deficient mice develop chronic enterocolitis. *Cell.* 1993;75(2):263-274.
320. Takeda K, Clausen BE, Kaisho T, et al. Enhanced Th1 activity and development of chronic enterocolitis in mice devoid of Stat3 in macrophages and neutrophils. *Immunity.* 1999;10(1):39-49.
321. Rubtsov YP, Rasmussen JP, Chi EY, et al. Regulatory T cell-derived interleukin-10 limits inflammation at environmental interfaces. *Immunity.* 2008;28(4):546-558.
322. Maheshwari A, Kelly DR, Nicola T, et al. TGF-beta2 suppresses macrophage cytokine production and mucosal inflammatory responses in the developing intestine. *Gastroenterology.* 2011;140(1):242-253.
323. Kobayashi T, Matsuoka K, Sheikh SZ, et al. IL-10 regulates Il12b expression via histone deacetylation: implications for intestinal macrophage homeostasis. *J Immunol.* 2012;189(4):1792-1799.

324. Simon JM, Davis JP, Lee SE, et al. Alterations to chromatin in intestinal macrophages link IL-10 deficiency to inappropriate inflammatory responses. *Eur J Immunol.* 2016;46(8):1912-1925.
325. Hirotani T, Lee PY, Kuwata H, et al. The nuclear IkappaB protein IkappaBNS selectively inhibits lipopolysaccharide-induced IL-6 production in macrophages of the colonic lamina propria. *J Immunol.* 2005;174(6):3650-3657.
326. Filardy AA, He J, Bennink J, Yewdell J, Kelsall BL. Posttranscriptional control of NLRP3 inflammasome activation in colonic macrophages. *Mucosal Immunol.* 2016;9(4):850-858.
327. Kole A, Maloy KJ. Control of intestinal inflammation by interleukin-10. *Curr Top Microbiol Immunol.* 2014;380:19-38.
328. Corbin AL, Gomez-Vazquez M, Berthold DL, et al. IRF5 guides monocytes toward an inflammatory CD11c(+) macrophage phenotype and promotes intestinal inflammation. *Sci Immunol.* 2020;5(47):eaax6085.
329. Chakarov S, Lim HY, Tan L, et al. Two distinct interstitial macrophage populations coexist across tissues in specific subtissular niches. *Science.* 2019;363(6432):eaau0964.
330. Shaw TN, Houston SA, Wemyss K, et al. Tissue-resident macrophages in the intestine are long lived and defined by Tim-4 and CD4 expression. *J Exp Med.* 2018;215(6):1507-1518.
331. Jones GR, Bain CC, Fenton TM, et al. Dynamics of colon monocyte and macrophage activation during colitis. *Front Immunol.* 2018;9:2764.
332. Asano K, Kikuchi K, Tanaka M. CD169 macrophages regulate immune responses toward particulate materials in the circulating fluid. *J Biochem.* 2018;164(2):77-85.
333. Hiemstra IH, Beijer MR, Veninga H, et al. The identification and developmental requirements of colonic CD169(+) macrophages. *Immunology.* 2014;142(2):269-278.
334. Schenk M, Bouchon A, Seibold F, Mueller C. TREM-1--expressing intestinal macrophages crucially amplify chronic inflammation in experimental colitis and inflammatory bowel diseases. *J Clin Invest.* 2007;117(10):3097-3106.
335. Nakanishi Y, Sato T, Takahashi K, Ohteki T. IFN-gamma-dependent epigenetic regulation instructs colitogenic monocyte/macrophage lineage differentiation in vivo. *Mucosal Immunol.* 2018;11(3):871-880.
336. Tamoutounour S, Henri S, Lelouard H, et al. CD64 distinguishes macrophages from dendritic cells in the gut and reveals the Th1-inducing role of mesenteric lymph node macrophages during colitis. *Eur J Immunol.* 2012;42(12):3150-3166.
337. Weber B, Saurer L, Schenk M, Dickgreber N, Mueller C. CX3CR1 defines functionally distinct intestinal mononuclear phagocyte subsets which maintain their respective functions during homeostatic and inflammatory conditions. *Eur J Immunol.* 2011;41(3):773-779.
338. Qualls JE, Kaplan AM, Van Rooijen N, Cohen DA. Suppression of experimental colitis by intestinal mononuclear phagocytes. *J Leukoc Biol.* 2006;80(4):802-815.
339. Bekiaris V, Persson EK, Agace WW. Intestinal dendritic cells in the regulation of mucosal immunity. *Immunol Rev.* 2014;260(1):86-101.
340. Loschko J, Schreiber HA, Rieke GJ, et al. Absence of MHC class II on cDCs results in microbial-dependent intestinal inflammation. *J Exp Med.* 2016;213(4):517-534.
341. Hall JA, Cannons JL, Grainger JR, et al. Essential role for retinoic acid in the promotion of CD4(+) T cell effector responses via retinoic acid receptor alpha. *Immunity.* 2011a;34(3):435-447.
342. Matteoli G, Mazzini E, Iliev ID, et al. Gut CD103+ dendritic cells express indoleamine 2,3-dioxygenase which influences T regulatory/T effector cell balance and oral tolerance induction. *Gut.* 2010;59(5):595-604.
343. Fukaya T, Takagi H, Sato Y, et al. Crucial roles of B7-H1 and B7-DC expressed on mesenteric lymph node dendritic cells in the generation of antigen-specific CD4+Foxp3+ regulatory T cells in the establishment of oral tolerance. *Blood.* 2010;116(13):2266-2276.
344. Han D, Walsh MC, Cejas PJ, et al. Dendritic cell expression of the signaling molecule TRAF6 is critical for gut microbiota-dependent immune tolerance. *Immunity.* 2013;38(6):1211-1222.
345. Friedrich M, Pohin M, Powrie F. Cytokine networks in the pathophysiology of inflammatory bowel disease. *Immunity.* 2019;50(4):992-1006.
346. Weiner HL, da Cunha AP, Quintana F, Wu H. Oral tolerance. *Immunol Rev.* 2011;241(1):241-259.
347. Higgins PJ, Weiner HL. Suppression of experimental autoimmune encephalomyelitis by oral administration of myelin basic protein and its fragments. *J Immunol.* 1988;140(2):440-445.
348. Kato H, Fujihashi K, Kato R, Yuki Y, McGhee JR. Oral tolerance revisited: prior oral tolerization abrogates cholera toxin-induced mucosal IgA responses. *J Immunol.* 2001;166(5):3114-3121.
349. Smith DW, Nagler-Anderson C. Preventing intolerance: the induction of nonresponsiveness to dietary and microbial antigens in the intestinal mucosa. *J Immunol.* 2005;174(7):3851-3857.
350. Kiyono H, McGhee JR, Wannemuehler MJ, Michalek SM. Lack of oral tolerance in C3H/HeJ mice. *J Exp Med.* 1982;155(2):605-610.
351. Wannemuehler MJ, Kiyono H, Babb JL, Michalek SM, McGhee JR. Lipopolysaccharide (LPS) regulation of the immune response: LPS converts germfree mice to sensitivity to oral tolerance induction. *J Immunol.* 1982;129(3):959-965.
352. Bashir ME, Louie S, Shi HN, Nagler-Anderson C. Toll-like receptor 4 signaling by intestinal microbes influences susceptibility to food allergy. *J Immunol.* 2004;172(11):6978-6987.
353. Michalek SM, Kiyono H, Wannemuehler MJ, Mosteller LM, McGhee JR. Lipopolysaccharide (LPS) regulation of the immune response: LPS influence on oral tolerance induction. *J Immunol.* 1982;128(5):1992-1998.
354. Whibley N, Tucci A, Powrie F. Regulatory T cell adaptation in the intestine and skin. *Nat Immunol.* 2019;20(4):386-396.
355. Esterhazy D, Canesso MCC, Mesin L, et al. Compartmentalized gut lymph node drainage dictates adaptive immune responses. *Nature.* 2019;569(7754):126-130.
356. Mucida D, Kutchukhidze N, Erazo A, Russo M, Lafaille JJ, Curotto de Lafaille MA, Oral tolerance in the absence of naturally occurring Tregs. *J Clin Invest.* 2005;115(7):1923-1933.
357. Mucida D, Park Y, Kim G, et al. Reciprocal TH17 and regulatory T cell differentiation mediated by retinoic acid. *Science.* 2007;317(5835):256-260.
358. Geuking MB, Cahenzli J, Lawson MAE, et al. Intestinal bacterial colonization induces mutualistic regulatory T cell responses. *Immunity.* 2011;34(5):794-806.
359. Ohnmacht C, Park J-H, Cording S, et al. MUCOSAL IMMUNOLOGY. The microbiota regulates type 2 immunity through RORgammat(+) T cells. *Science.* 2015;349(6251):989-993.
360. Sefik E, Geva-Zatorsky N, Oh S, et al. MUCOSAL IMMUNOLOGY. Individual intestinal symbionts induce a distinct population of RORgamma(+) regulatory T cells. *Science.* 2015;349(6251):993-997.
361. Yang BH, Hagemann S, Mamareli P, et al. Foxp3(+) T cells expressing RORgammat represent a stable regulatory T-cell effector lineage with enhanced suppressive capacity during intestinal inflammation. *Mucosal Immunol.* 2016;9(2):444-457.
362. Lathrop SK, Bloom SM, Rao SM, et al. Peripheral education of the immune system by colonic commensal microbiota. *Nature.* 2011;478:250-254.
363. Nutsch K, Chai JN, Ai TL, et al. Rapid and efficient generation of regulatory T cells to commensal antigens in the periphery. *Cell Rep.* 2016;17(1):206-220.
364. Scharschmidt TC, Vasquez KS, Truong H-A, et al. A wave of regulatory T cells into neonatal skin mediates tolerance to commensal microbes. *Immunity.* 2015;43(5):1011-1021.
365. Xu M, Pokrovskii M, Ding Y, et al. c-MAF-dependent regulatory T cells mediate immunological tolerance to a gut pathobiont. *Nature.* 2018;554(7692):373-377.
366. Faith JJ, Ahern PP, Ridaura VK, Cheng J, Gordon JI. Identifying gut microbe-host phenotype relationships using combinatorial communities in gnotobiotic mice. *Sci. Transl Med.* 2014;6(220):220ra11.
367. Geva-Zatorsky N, Sefik E, Kua L, et al. Mining the human gut microbiota for immunomodulatory organisms. *Cell.* 2017;168(5):928-943.e11.
368. Britton GJ, Contijoch EJ, Mogno I, et al. Microbiotas from humans with inflammatory bowel disease alter the balance of gut Th17 and RORgammat(+) regulatory T cells and exacerbate colitis in mice. *Immunity.* 2019;50(1):212-224 e4.
369. Izcue A, Coombes JL, Powrie F. Regulatory T cells suppress systemic and mucosal immune activation to control intestinal inflammation. *Immunol Rev.* 2006;212:256-271.
370. Wang Y, Su MA, Wan YY. An essential role of the transcription factor GATA-3 for the function of regulatory T cells. *Immunity.* 2011b;35(3):337-348.
371. Wohlfert EA, Grainger JR, Bouladoux N, et al. GATA3 controls Foxp3(+) regulatory T cell fate during inflammation in mice. *J Clin Invest.* 2011a;121(11):4503-4515.
372. Al Nabhani Z, Dulauroy S, Marques R, et al. A weaning reaction to microbiota is required for resistance to immunopathologies in the adult. *Immunity.* 2019;50(5):1276-1288 e5.
373. Arpaia N, Campbell C, Fan X, et al. Metabolites produced by commensal bacteria promote peripheral regulatory T-cell generation. *Nature.* 2013;504:451-455.
374. Atarashi K, Tanoue T, Oshima K, et al. Treg induction by a rationally selected mixture of *Clostridia* strains from the human microbiota. *Nature.* 2013;500(7461):232-236.
375. Atarashi K, Tanoue T, Shima T, et al. Induction of colonic regulatory T cells by indigenous *Clostridium* species. *Science.* 2011;331(6015):337-341.
376. Furusawa Y, Obata Y, Fukuda S, et al. Commensal microbe-derived butyrate induces the differentiation of colonic regulatory T cells. *Nature.* 2013.
377. Smith PM, Howitt MR, Panikov N, et al. The microbial metabolites, short-chain fatty acids, regulate colonic Treg cell homeostasis. *Science.* 2013;341(6145):569-573.
378. Campbell C, McKenney PT, Konstantinovsky D, et al. Bacterial metabolism of bile acids promotes generation of peripheral regulatory T cells. *Nature.* 2020;581(7809):475-479.
379. Hang S, Paik D, Yao L, et al. Bile acid metabolites control TH17 and Treg cell differentiation. *Nature.* 2019;576(7785):143-148.
380. Song X, Sun X, Oh SF, et al. Microbial bile acid metabolites modulate gut RORgamma(+) regulatory T cell homeostasis. *Nature.* 2020;577(7790):410-415.
381. Kim M, Galan C, Hill AA, et al. Critical role for the microbiota in CX3CR1(+) intestinal mononuclear phagocyte regulation of intestinal T cell responses. *Immunity.* 2018;49(1):151-163.e5.
382. Maslowski KM, Vieira AT, Ng A, et al. Regulation of inflammatory responses by gut microbiota and chemoattractant receptor GPR43. *Nature.* 2009;461(7268):1282-1286.
383. Iliev ID, Funari VA, Taylor KD, et al. Interactions between commensal fungi and the C-type lectin receptor Dectin-1 influence colitis. *Science.* 2012;336(6086):1314-1317.
384. Panelli S, Epis S, Cococcioni L, et al. Inflammatory bowel diseases, the hygiene hypothesis and the other side of the microbiota: parasites and fungi. *Pharmacol Res.* 2020;159:104962.
385. Ananthakrishnan AN, Bernstein CN, Iliopoulos D, et al. Environmental triggers in IBD: a review of progress and evidence. *Nat Rev Gastroenterol Hepatol.* 2018;15(1):39-49.
386. Uniken Venema WT, Voskuil MD, Dijkstra G, Weersma RK, Festen EA. The genetic background of inflammatory bowel disease: from correlation to causality. *J Pathol.* 2017;241(2):146-158.
387. Ouahed J, Spencer E, Kotlarz D, et al. Very early onset inflammatory bowel disease: a clinical approach with a focus on the role of genetics and underlying immune deficiencies. *Inflamm Bowel Dis.* 2020;26(6):820-842.
388. Ogura Y, Bonen DK, Inohara N, et al. A frameshift mutation in NOD2 associated with susceptibility to Crohn's disease. *Nature.* 2001;411(6837):603-606.
389. Strober W, Asano N, Fuss I, Kitani A, Watanabe T. Cellular and molecular mechanisms underlying NOD2 risk-associated polymorphisms in Crohn's disease. *Immunol Rev.* 2014;260(1):249-260.
390. Strober W, Watanabe T. NOD2, an intracellular innate immune sensor involved in host defense and Crohn's disease. *Mucosal Immunol.* 2011;4(5):484-495.
391. Watanabe T, Kitani A, Murray PJ, Strober W. NOD2 is a negative regulator of Toll-like receptor 2-mediated T helper type 1 responses. *Nat Immunol.* 2004;5(8):800-808.
392. Hedl M, Li J, Cho JH, Abraham C. Chronic stimulation of Nod2 mediates tolerance to bacterial products. *Proc Natl Acad Sci U S A.* 2007;104(49):19440-19445.
393. Watanabe T, Asano N, Meng G, et al. NOD2 downregulates colonic inflammation by IRF4-mediated inhibition of K63-linked polyubiquitination of RICK and TRAF6. *Mucosal Immunol.* 2014;7(6):1312-1325.
394. Hampe J, Franke A, Rosenstiel P, et al. A genome-wide association scan of nonsynonymous SNPs identifies a susceptibility variant for Crohn disease in ATG16L1. *Nat Genet.* 2007;39(2):207-211.
395. Larabi A, Barnich N, Nguyen HTT. New insights into the interplay between autophagy, gut microbiota and inflammatory responses in IBD. *Autophagy.* 2020;16(1):38-51.
396. Cadwell K, Liu JY, Brown SL, et al. A key role for autophagy and the autophagy gene Atg16l1 in mouse and human intestinal Paneth cells. *Nature.* 2008;456(7219):259-263.
397. Conway KL, Kuballa P, Song JH, et al. Atg16l1 is required for autophagy in intestinal epithelial cells and protection of mice from Salmonella infection. *Gastroenterology.* 2013;145(6):1347-1357.
398. Kaser A, Tomczak M, Blumberg RS. "ER stress(ed out)!": Paneth cells and ischemia-reperfusion injury of the small intestine. *Gastroenterology.* 2011;140(2):393-396.
399. Tschurtschenthaler M, Adolph TE, Ashcroft JW, et al. Defective ATG16L1-mediated removal of IRE1alpha drives Crohn's disease-like ileitis. *J Exp Med.* 2017;214(2):401-422.
400. Murthy A, Li Y, Peng I, et al. A Crohn's disease variant in Atg16l1 enhances its degradation by caspase 3. *Nature.* 2014;506(7489):456-462.
401. Gao P, Liu H, Huang H, Zhang Q, Strober W, Zhang F. The inflammatory bowel disease-associated autophagy gene Atg16L1T300A acts as a dominant negative variant in mice. *J Immunol.* 2017;198(6):2457-2467.
402. Duerr RH, Taylor KD, Brant SR, et al. A genome-wide association study identifies IL23R as an inflammatory bowel disease gene. *Science.* 2006;314(5804):1461-1463.
403. Pidasheva S, Trifari S, Phillips A, et al. Functional studies on the IBD susceptibility gene IL23R implicate reduced receptor function in the protective genetic variant R381Q. *PLoS One.* 2011;6(10):e25038.
404. Sarin R, Wu X, Abraham C. Inflammatory disease protective R381Q IL23 receptor polymorphism results in decreased primary CD4+ and CD8+ human T-cell functional responses. *Proc Natl Acad Sci U S A.* 2011;108(23):9560-9565.
405. Babeu JP, Boudreau F. Hepatocyte nuclear factor 4-alpha involvement in liver and intestinal inflammatory networks. *World J Gastroenterol.* 2014;20(1):22-30.
406. Fu J, Wei B, Wen T, et al. Loss of intestinal core 1-derived O-glycans causes spontaneous colitis in mice. *J Clin Invest.* 2011;121(4):1657-1666.
407. Fuss IJ, Neurath M, Boirivant M, et al. Disparate CD4+ lamina propria (LP) lymphokine secretion profiles in inflammatory bowel disease. Crohn's disease LP cells manifest increased secretion of IFN-gamma, whereas ulcerative colitis LP cells manifest increased secretion of IL-5. *J Immunol.* 1996;157(3):1261-1270.

408. Strober W, Zhang F, Kitani A, Fuss I, Fichtner-Feigl S. Proinflammatory cytokines underlying the inflammation of Crohn's disease. *Curr Opin Gastroenterol.* 2010;26(4):310-317.
409. Lee JS, Tato CM, Joyce-Shaikh B, et al. Interleukin-23-independent IL-17 production regulates intestinal epithelial permeability. *Immunity.* 2015;43(4):727-738.
410. O'Connor WJr, Kamanaka M, Booth CJ, et al. A protective function for interleukin 17A in T cell-mediated intestinal inflammation. *Nat Immunol.* 2009;10(6):603-609.
411. Mannon PJ, Fuss IJ, Mayer L, et al. Anti-interleukin-12 antibody for active Crohn's disease. *N Engl J Med.* 2004;351(20):2069-2079.
412. Dudakov JA, Hanash AM, van den Brink MR. Interleukin-22: immunobiology and pathology. *Annu Rev Immunol.* 2015;33:747-785.
413. Muhl H, Bachmann M. IL-18/IL-18BP and IL-22/IL-22BP: two interrelated couples with therapeutic potential. *Cell Signal.* 2019;63:109388.
414. Boirivant M, Fuss IJ, Chu A, Strober W. Oxazolone colitis: a murine model of T helper cell type 2 colitis treatable with antibodies to interleukin 4. *J Exp Med.* 1998;188(10):1929-1939.
415. Fuss IJ, Heller F, Boirivant M, et al. Nonclassical CD1d-restricted NK T cells that produce IL-13 characterize an atypical Th2 response in ulcerative colitis. *J Clin Invest.* 2004;113(10):1490-1497.
416. Kawashima R, Kawamura YI, Oshio T, et al. Interleukin-13 damages intestinal mucosa via TWEAK and Fn14 in mice-a pathway associated with ulcerative colitis. *Gastroenterology.* 2011;141(6):2119-2129 e8.
417. Fuss IJ, Joshi B, Yang Z, et al. IL-13Ralpha2-bearing, type II NKT cells reactive to sulfatide self-antigen populate the mucosa of ulcerative colitis. *Gut.* 2014;63(11):1728-1736.
418. Butera A, Di Paola M, Vitali F, et al. IL-13 mRNA tissue content identifies two subsets of adult ulcerative colitis patients with different clinical and mucosa-associated microbiota profiles. *J Crohns Colitis.* 2020;14(3):369-380.
419. Hoving JC. Targeting IL-13 as a host-directed therapy against ulcerative colitis. *Front Cell Infect Microbiol.* 2018;8:395.
420. Sands BE, Sandborn WJ, Panaccione R, et al. Ustekinumab as induction and maintenance therapy for ulcerative colitis. *N Engl J Med.* 2019;381(13):1201-1214.
421. Gerlach K, McKenzie AN, Neurath MF, Weigmann B. IL-9 regulates intestinal barrier function in experimental T cell-mediated colitis. *Tissue Barriers.* 2015;3(1–2):e983777.
422. Kalliolias GD, Ivashkiv LB. TNF biology, pathogenic mechanisms and emerging therapeutic strategies. *Nat Rev Rheumatol.* 2016;12(1):49-62.
423. Chen J, Kos R, Garssen J, Redegeld F. Molecular insights into the mechanism of necroptosis: the necrosome as a potential therapeutic target. *Cells.* 2019;8(12):1486.
424. Dannappel M, Vlantis K, Kumari S, et al. RIPK1 maintains epithelial homeostasis by inhibiting apoptosis and necroptosis. *Nature.* 2014;513(7516):90-94.
425. Mao L, Kitani A, Strober W, Fuss IJ. The role of NLRP3 and IL-1beta in the pathogenesis of inflammatory bowel disease. *Front Immunol.* 2018b;9:2566.
426. Mao L, Kitani A, Similuk M, et al. Loss-of-function CARD8 mutation causes NLRP3 inflammasome activation and Crohn's disease. *J Clin Invest.* 2018a;128(5):1793-1806.
427. Waldner MJ, Neurath MF. Master regulator of intestinal disease: IL-6 in chronic inflammation and cancer development. *Semin Immunol.* 2014;26(1):75-79.
428. Rose-John S. IL-6 trans-signaling via the soluble IL-6 receptor: importance for the pro-inflammatory activities of IL-6. *Int J Biol Sci.* 2012;8(9):1237-1247.
429. Samat AAK, van der Geest J, Vastert SJ, van Loosdregt J, van Wijk F. Tissue-resident memory T cells in chronic inflammation-local cells with systemic effects? *Cells.* 2021;10(2):409.
430. Zundler S, Becker E, Spocinska M, et al. Hobit- and Blimp-1-driven CD4(+) tissue-resident memory T cells control chronic intestinal inflammation. *Nat Immunol.* 2019;20(3):288-300.
431. Geremia A, Arancibia-Carcamo CV. Innate lymphoid cells in intestinal inflammation. *Front Immunol.* 2017;8:1296.
432. Li J, Doty AL, Tang Y, et al. Enrichment of IL-17A(+) IFN-gamma(+) and IL-22(+) IFN-gamma(+) T cell subsets is associated with reduction of NKp44(+) ILC3s in the terminal ileum of Crohn's disease patients. *Clin Exp Immunol.* 2017;190(1):143-153.
433. Eken A, Singh AK, Treuting PM, Oukka M. IL-23R+ innate lymphoid cells induce colitis via interleukin-22-dependent mechanism. *Mucosal Immunol.* 2014;7(1):143-154.
434. Forkel M, van Tol S, Höög C, Michaëlsson J, Almer S, Mjösberg J. Distinct alterations in the composition of mucosal innate lymphoid cells in newly diagnosed and established Crohn's disease and ulcerative colitis. *J Crohns Colitis.* 2019;13(1):67-78.
435. Pearson C, Thornton EE, McKenzie B, et al. ILC3 GM-CSF production and mobilisation orchestrate acute intestinal inflammation. *Elife.* 2016;5:e10066.
436. Bernink JH, Peters CP, Munneke M, et al. Human type 1 innate lymphoid cells accumulate in inflamed mucosal tissues. *Nat Immunol.* 2013;14(3):221-229.
437. Forkel M, Mjosberg J. Dysregulation of group 3 innate lymphoid cells in the pathogenesis of inflammatory bowel disease. *Curr Allergy Asthma Rep.* 2016;16(10):73.
438. Neurath MF, Fuss I, Kelsall BL, Presky DH, Waegell W, Strober W. Experimental granulomatous colitis in mice is abrogated by induction of TGF-beta-mediated oral tolerance. *J Exp Med.* 1996;183(6):2605-2616.
439. Powrie F, Carlino J, Leach MW, Mauze S, Coffman RL. A critical role for transforming growth factor-beta but not interleukin 4 in the suppression of T helper type 1-mediated colitis by CD45RB^{low} CD4+ T cells. *J Exp Med.* 1996;183(6):2669-2674.
440. Mottet C, Uhlig HH, Powrie F. Cutting edge: cure of colitis by CD4+CD25+ regulatory T cells. *J Immunol.* 2003;170(8):3939-3943.
441. Lord JD. Promises and paradoxes of regulatory T cells in inflammatory bowel disease. *World J Gastroenterol.* 2015;21(40):11236-11245.
442. Sznurkowska K, Boćkowska M, Zieliński M, et al. Peripheral regulatory T cells and anti-inflammatory cytokines in children with juvenile idiopathic arthritis. *Acta Biochim Pol.* 2018;65(1):119-123.
443. Fenton TM, Kelly A, Shuttleworth EE, et al. Inflammatory cues enhance TGFbeta activation by distinct subsets of human intestinal dendritic cells via integrin alphavbeta8. *Mucosal Immunol.* 2017;10(3):624-634.
444. Fantini MC, Rizzo A, Fina D, et al. Smad7 controls resistance of colitogenic T cells to regulatory T cell-mediated suppression. *Gastroenterology.* 2009;136(4):1308-1316, e1-3.
445. Lord JD, Shows DM, Chen J, Thirlby RC. Human blood and mucosal regulatory T cells express activation markers and inhibitory receptors in inflammatory bowel disease. *PLoS One.* 2015;10(8):e0136485.
446. Butera A, Sanchez M, Pronio A, et al. CD3+CD4+LAP+Foxp3-regulatory cells of the colonic lamina propria limit disease extension in ulcerative colitis. *Front Immunol.* 2018;9:2511.
447. Wang S, Xia P, Chen Y, et al. Regulatory innate lymphoid cells control innate intestinal inflammation. *Cell.* 2017;171(1):201-216.e18.
448. Zhu L, Shi T, Zhong C, Wang Y, Chang M, Liu X. IL-10 and IL-10 receptor mutations in very early onset inflammatory bowel disease. *Gastroenterology Res.* 2017;10(2):65-69.
449. Belkaid Y, Hand TW. Role of the microbiota in immunity and inflammation. *Cell.* 2014;157(1):121-141.
450. Frank DN, StAmand AL, Feldman RA, Boedeker EC, Harpaz N, Pace NR. Molecular-phylogenetic characterization of microbial community imbalances in human inflammatory bowel diseases. *Proc Natl Acad Sci U S A.* 2007;104(34):13780-13785.
451. Zhou L, Zhang M, Wang Y, et al. Faecalibacterium prausnitzii produces butyrate to maintain Th17/Treg balance and to ameliorate colorectal colitis by inhibiting histone deacetylase 1. *Inflamm Bowel Dis.* 2018;24(9):1926-1940.
452. Bjorkqvist O, Repsilber D, Seifert M, et al. Alterations in the relative abundance of *Faecalibacterium prausnitzii* correlate with changes in fecal calprotectin in patients with ileal Crohn's disease: a longitudinal study. *Scand J Gastroenterol.* 2019;54(5):577-585.
453. Round JL, Lee SM, Li J, et al. The toll-like receptor 2 pathway establishes colonization by a commensal of the human microbiota. *Science.* 2011;332(6032):974-977.
454. Garrett WS, Lord GM, Punit S, et al. Communicable ulcerative colitis induced by T-bet deficiency in the innate immune system. *Cell.* 2007;131(1):33-45.

CHAPTER 39

Introduction to Neuroimmunology

Andrea Francesca Salvador, MSc, PhD • Jonathan Kipnis, PhD

INTRODUCTION

Several chapters in this book describe immune responses that occur in different tissues. In this respect, as in many others, the central nervous system (CNS) is unique; the field of knowledge concerning the function of the immune system within the CNS is called neuroimmunology. Today, this term is applied more widely so that it also covers immune system interactions with the peripheral nervous system—but that intriguing topic[1,2] deserves a chapter of its own.

The CNS consists of the brain, the spinal cord, some of the cranial nerves (optic, trigeminal, olfactory), and the retina. Historically, it was accepted as dogma that the uniqueness of the CNS was attributable to the "fact" that under physiological conditions the CNS is walled off by a blood–brain barrier (BBB) from the periphery, and hence that the immune system can gain access to the CNS only under conditions of pathology. This notion was reinforced in the mid-1940s by the transplantation experiments of Sir Peter Medawar, who was subsequently awarded a Nobel prize for transplantation tolerance mechanisms. Medawar and his coworkers found that skin grafts are rejected much more slowly from the CNS than from a peripheral tissue such as the skin. Transplantation of an autologous (same or identical donor) skin graft into the mouse skin remained intact, whereas transplantation of an allogeneic skin graft (from a similar but nonidentical donor, such as a different mouse strain) into the skin was rejected within days. Yet, compared to the allogeneic graft in the skin, an allogeneic graft transplanted into the mouse brain remained unrejected for a much longer time.[3-5] This suggested that unique mechanisms are at play within the brain that limit immune response, a notion that gave rise to the brain's designation as an "immune-privileged" organ. However, in another set of experiments, Medawar's group concurrently transplanted allogeneic skin grafts on the skin and in the brain of the same mouse. Both grafts were rejected. Since the brain graft was rejected after immune cells (T cells) had been activated on the peripheral (skin-to-skin) graft, those peripheral T cells had clearly been able to reach the brain and reject the transplanted brain graft. This last set of experiments is critical for our understanding of what "immune privilege" is and what it is not.[3-5] Unfortunately, however, those seminal findings did not carry enough weight for contemporaries to question the concept of immune privilege as it was then generally understood or to recognize the true complexity of the immune uniqueness of the CNS. It seems likely that because Medawar's findings were largely ignored (or simply because of failure to explain their true implications), the development of the fledgling field of neuroimmunology slowed down for many decades. Thanks to fundamentally important later discoveries,[6-19] neuroimmunology was able to make some critical strides in addressing multiple sclerosis (MS). However, our knowledge of the neuroimmunological relationship for many years relied solely on the pathological interactions between the two systems as studied in MS and other inflammatory conditions. MS patients had the good fortune to have benefited from the great minds working on this disease and learning how T cells and macrophages attack neural tissue (resulting in the development of 13 Food and Drug Administration [FDA]-approved drugs for MS to date), but knowledge of homeostatic interactions between the two systems stagnated for many years before taking off.

In this chapter, several subareas of neuroimmunology will be covered, while (for reasons of space) other subareas of this growing and exciting field will be tackled at another time. First, we will talk about microglia, the brain's only true parenchyma-resident immune cells. Other immune cells may reside in specialized niches of the CNS, whereas microglia reside throughout the brain parenchyma. We will then discuss the roles of other immune cells and processes in neurological diseases and conditions. Starting with MS, where the picture is clearest owing to many years of collective studies and progress, we will then proceed to discuss the immune response to acute CNS injury, and finally, to chronic neurodegenerative diseases such as Alzheimer disease (AD). Shifting gears, we will discuss the role of immune cells and their derived molecules (cytokines) on homeostatic brain function, and speculate on the reason, why the immune system would need to exert an effect on brain function. Lastly, we will discuss CNS drainage and the role of newly characterized meningeal lymphatic vessels in the maintenance of CNS immune surveillance and possibly also of "immune privilege."

IMMUNE PRIVILEGE

The CNS has been regarded as an immune-privileged organ as it can tolerate the administration of an antigen without provoking an inflammatory immune response,[20] thereby allowing the survival of the antigen/graft for an extended period of time. This appears to be an evolutionary adaption to insulate and protect vital organs that have low tolerance to inflammation and low capacity to regenerate. This status is not entirely unique to the CNS, but it is shared by other organs such as the eye chamber, uterus, and gonads.[21] While the concept of immune privilege is continuously being refined in accordance with new discoveries, it is important to consider the mechanisms that maintain this status.

Initially, the existence of physical blood-tissue barriers (eg, BBB, blood-testis barrier, blood-ocular barrier) that exclude the infiltration of leukocytes in nonpathological conditions was believed to underlie immune privilege. The BBB (Fig. 39.1), for example, is formed by specialized nonfenestrated endothelial cells that are joined together by adherens and tight junctions with high-electrical resistance that not only restrict immune cell entry but also regulate the entry of potentially toxic substances and pathogens from the blood. They also express transporters that regulate the efflux of waste from the CNS or influx of nutrients.[22] The endothelial cells are then covered by mural cells (ie, pericytes and vascular smooth muscle cells), basement membranes, and these are then ensheathed by astrocytic end feet. While the BBB is crucial for restricting immune cell entry and breaks are associated with disease, its disruption does not appear to

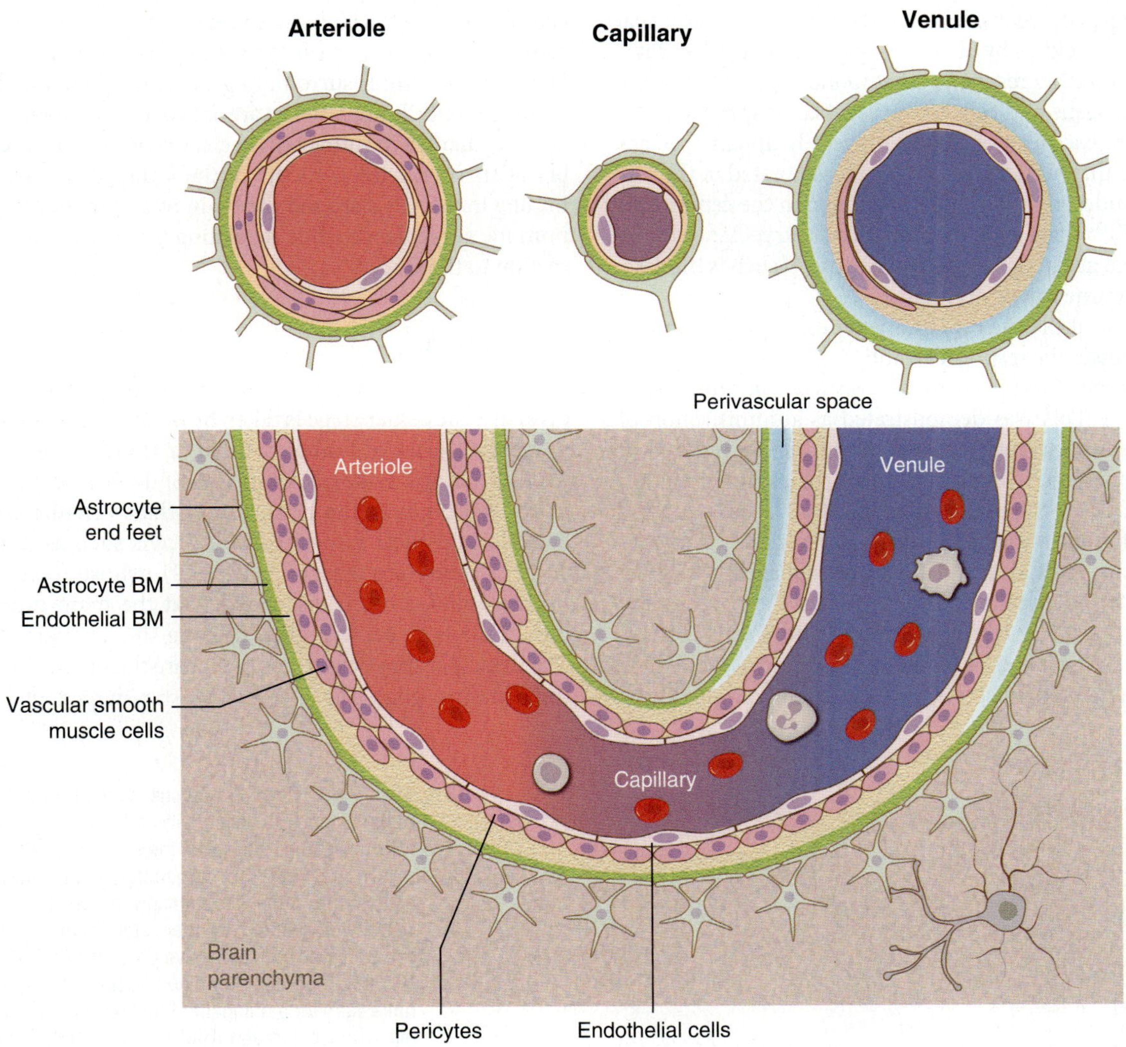

FIG. 39.1. The blood-brain barrier. The blood-brain barrier refers to the specialized structure and function of central nervous system (CNS) microvasculature that allows it to restrict the entry of toxins, pathogens, and immune cells to the brain. The blood vessels are composed of endothelial cells held together by adherent junctions as well as tight junctions that control paracellular solute movement.[22] They also express lower levels of adhesion molecules that limit the entry and surveillance of immune cells.[23] Endothelial cells also express efflux and nutrient transporters enabling the removal of waste and the transport of specific nutrients to the CNS. Covering the blood endothelial cells are mural cells, among which are pericytes located at the abluminal surface of the endothelial tube. These are then ensheathed by endothelial basement membranes and the astrocytic end feet or glia limitans. Because astrocytes are also in contact with neurons, they relay neuronal activity-related information to blood vessels, which then respond by regulating blood flow.[22,24] Arteries are specifically made of endothelial cells, surrounded by multiple layers of vascular smooth muscle cells and the astrocytic end feet. Capillaries, on the other hand, are composed of endothelial cells covered by a single layer of pericytes on the vascular basement membrane. Postcapillary venules are also covered by pericytes that have an expanded perivascular space.[20] Altogether, these structures help maintain CNS homeostasis and support neuronal function.

be sufficient to cause autoimmunity.[25] As such, it is thought that the role of the BBB is to maintain conditions that enable the proper functioning of neurons and other CNS-resident cells.[22]

If the breach of blood-tissue barriers is not sufficient to induce autoimmunity, perhaps restriction of antigen encounter in secondary lymphoid organs underlies immune privilege. Antigens are carried by lymphatic vessels, and it was a long-held assumption that immune-privileged organs lack these structures rendering the immune system "ignorant" of them. This belief has been reversed, following crucial discoveries, among which is the characterization of lymphatic vessels that drain the CNS,[26,27] the meningeal lymphatic vessels, and other immune-privileged organs.[28] Manipulation of the meningeal lymphatic vessels either by ablation, ligation, or supplementation of growth factors results to corresponding changes in the drainage of antigens and cells to the cervical lymph nodes.[29,30] Prior to the rediscovery of the meningeal lymphatics, it has been reported that humoral responses are generated in the cervical lymph nodes upon injection of antigen in the cerebrospinal fluid (CSF),[31] suggesting that functional surveillance of the CNS is taking place in the absence of disease. If such is the case, what does immune privilege entail?

A closer look at the eye reveals one type of mechanism that actively suppresses the response to antigen in the tissue. This is referred to as the anterior chamber-associated immune deviation (ACAID). This was demonstrated as administration of foreign antigen to the anterior chamber of the eye resulted in immune suppression,[32] which was measured as the induction of CD8+ and CD4+ suppressor or regulatory T cells and the generation of noncomplement fixing antibodies.[33] Antigen-presenting cells (APCs) (eg, F4/80 macrophages)[34] migrate from the eye to the thymus and spleen where they present the antigen from the ocular region to B cells[35] and T cells thereby, promoting the generation of regulatory T cells.[36] If ACAID is successfully generated, injection of the antigen typically in the ear will result in reduced swelling compared to hosts that did not undergo ocular inoculation.[33] Analysis of the ocular microenvironment revealed that it harbors immunosuppressive molecules such as the neuropeptides vasoactive intestinal peptide (VIP), somatostatin (SOM), alpha-melanocyte-stimulating hormone (a-MSH), the cytokine TGF-B2, and other molecules such as indoleamine-2,3-dioxygenase (IDO).[37-41] These molecules can reduce T cell proliferation, cytotoxic T-cell activity, production of IFN-γ, and they can also promote anti-inflammatory cytokine production of macrophages. The expression of FasL in resident cells of the eye can also suppress T cell activation.[42] At homeostatic conditions, expression of major histocompatibility complex (MHC) molecules on ocular cells is minimal,[43] which is also reported for the CNS-resident cells[25] and may be critical for maintaining immune privilege. However, structures surrounding the parenchyma such as the meninges and choroid plexus are active immunological niches, and they harbor immune cells such as professional APCs capable of surveying the contents of the brain.[44,45] Perhaps, positioning these APCs around the brain was a strategy to provide immune surveillance while regulating the detrimental inflammation to the CNS.[46]

MICROGLIA

The brain is a complex organ, and among its rich cellular diversity, the center stage is taken by neurons known from the pioneering work of Ramón y Cajal (a Nobel prize winner in 1906). Neurons make up about 50% of the brain's cells. Other major cell types in the brain (other than vascular endothelial cells and their associated mural cells such as fibroblasts and pericytes) are the glia (Fig. 39.2), whose name derives from a belief, since transcended, that the main function of these cells is to act like glue, holding the neurons together. The principal subpopulations of glia are the oligodendrocytes (myelinating glia), the astrocytes (star-shaped glia whose

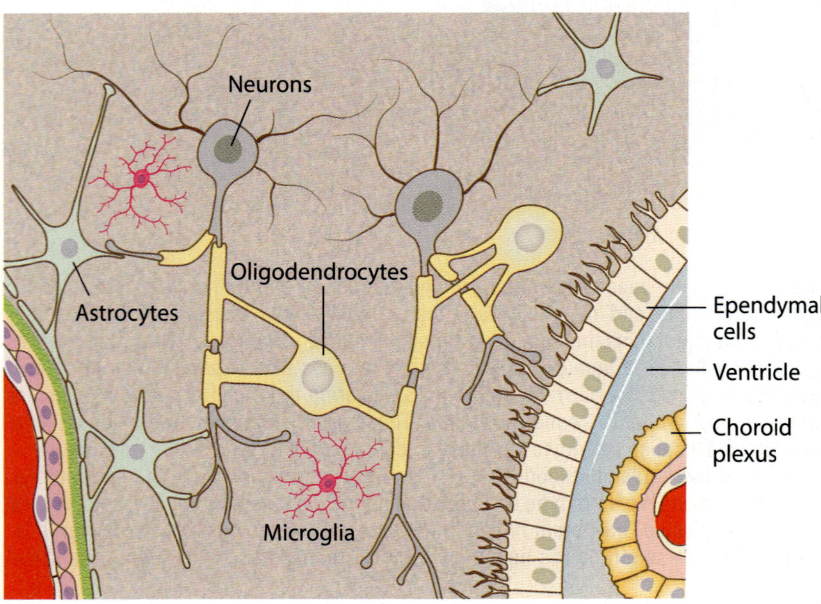

FIG. 39.2. Central nervous system (CNS)-resident cells. Among the major cells of the healthy CNS parenchyma are neurons with their axons ensheathed by myelin derived from oligodendrocytes, which then allows the rapid propagation of electrical signals. Astrocytes marked by their star-shaped morphology are endowed with extensive processes that make contact with neurons and blood vessels. These cells are necessary for a number of functions in the CNS including synapse formation and release of trophic factors to neurons.[47] Microglia are highly motile-resident myeloid cells of the CNS that interact with the rest of the CNS cells. They have been implicated in supporting normal brain development and tissue surveillance.[47] Ependymal cells form the lining of the brain's ventricles, the choroid plexus, and the central canal of the spinal cord. Through the uptake of water and molecules from the blood, the ependymal cells help facilitate the production and circulation of cerebrospinal fluid (CSF). (Reprinted from Liddelow SA, Marsh SE, Stevens B. Microglia and astrocytes in disease: dynamic duo or partners in crime? *Trends Immunol.* 2020;41:820-835. doi:10.1016/j.it.2020.07.006, with permission from Elsevier.)

many functions include secretion of neurotrophic factors and controlling water balance), and the microglia (the brain's resident immune cells). Other cells have recently been identified and proposed as a standalone subclass of glia. Neurons attracted intense research by neuroscientists over the last century, while research on glial cells lagged behind. Of all the glia, probably the least attention was devoted to the microglia which, being "immune cells," were of less interest to neuroscientists. That perception, however, has since undergone a fundamental change, and numerous leading neuroscience laboratories are turning their attention to glia most especially microglia.

Microglia were first described in 1920s by Pío del Rio Hortega, a student of Ramón y Cajal. In his works published in 20 to 30s, Hortega described the many different shapes and forms taken by microglia based on their activation states (injury, chronic neurodegeneration, infection, and others).[48-50] Following his works, researchers began referring to microglia as "activated" (meaning in a pathological state) or "resting." The perception of "resting" microglia was changed, when live imaging of *CX3CR1*^{GFP} mice was performed using a multiphoton intravital microscope through a window in the mouse skull.[51,52] Live imaging of GFP-labeled microglia revealed that the processes of those so-called "resting" microglia are constantly moving, a property that was attributed to the function of microglia in tissue surveillance. Interestingly, however, such movements of other, non–brain-tissue resident macrophages that also reportedly survey the tissues in which they reside are not as active as those of the microglia. It is possible that the observed movement of microglial processes is indeed a unique feature of these brain-resident macrophages but the true reason for their movements has yet to be revealed.

In 1908, the Russian-French scientist Élie Metchnikoff was awarded a Nobel Prize for his work on innate immune cells and their mediation of phagocytosis. Little was known at that time about macrophages and their role in tissue homeostasis. Today, we know that every tissue houses its specialized macrophages which, though similar in their cellular properties—maintain their specificity to the tissue in which they reside.[53-55]

Because microglia are the main phagocytic cells in the brain, their principal function is likely to be phagocytosis of cellular debris and overall maintenance of brain homeostasis.[56-62] Microglia have been shown to play an important role in neurogenesis by phagocytosing apoptotic neural progenitors. Among the best-described functions of microglia is their ability to phagocytose dysfunctional synapses, an essential activity, both for neurodevelopment and for neurodegeneration. In the case of neurodevelopment, most (if not all) animals, including humans, are born with more neurons and neuronal connections than they need, and the surplus are eliminated during the postnatal process of neural-circuit refinement.[63-68] In addition, microglia can recognize weak synapses and eliminate them, thereby ensuring that the synapses which remain are strong and can enhance the circuit's functioning.[66] With aging, however, this ability comes at a price, as microglia have been found to strip synapses from neurons, possibly advancing disease progression.[69,70] In both mouse and human late-stage AD microglia have been shown to ingest synapses, thus presumably contributing to disease pathology.[70] It is also feasible, however, that such elimination of dysfunctional synapses from a diseased adult brain may in fact be beneficial because in pruning the connection it might leave a weaker but better-functioning circuit. It has also been suggested that microglia in AD are neuroprotective. TREM2, a molecule that in the brain parenchyma is exclusively expressed in microglia, reportedly mediates a change in microglial phenotype into neuroprotective, disease-associated microglia (DAM).[71] While the exact mechanism of the neuroprotective function of these cells has yet to be elucidated, a *TREM2* mutation associated with increased AD prevalence has been shown in mice to interfere with generation of the DAM signature hence, worsening disease outcome. These results might suggest that microglia may have two phases in AD, with a neuroprotective role in the early stages of the disease and an increasingly neurotoxic phenotype as the disease progresses.[71-74]

Recent studies demonstrated a new myeloablative drug that resulted in efficient microglia depletion within 2 to 3 weeks of continuous drug application.[75] Upon removal of the drug, however, the microglia proliferated, regaining control over the niche. A most interesting finding was that upon microglial depletion, the mice suffered no deterioration or any other effects on behavioral tests.[75] In a fanciful military analogy, it seems possible that as with soldiers in an army, whose activity may seem irrelevant during peacetime but is vital in times of war, so it is with microglia, which are quiet in healthy individuals but, like soldiers, become energized for protection in the event of disease, injury, or inflammation.

Questions about the origin of microglia and the source of their precursors have been perplexing neurobiologists for decades. Early investigations revealed that upon whole-body irradiation and reconstitution with bone marrow from GFP mice, most of the macrophages in peripheral tissues are replenished by monocytes derived from the GFP-labeled bone marrow. Similarly, in the brain, by about 6 weeks postirradiation, peripheral cells could be seen to take the residence.[76] This phenomenon was attributed to a dysfunctional BBB, although testing with tracers indicated that the BBB at 5 to 6 weeks postirradiation is not compromised.

How, then, are the microglia being replenished and why only after 6 weeks? Interestingly, peripheral cells will enter the brain only if they bear functional CCR2 (a chemokine receptor for CCL2) and only if the head is irradiated. If the entire body except for the brain undergoes irradiation, replenishment of microglia by peripheral cells will not take place.[76] Although some studies have refuted the ability of monocytes to enter the brain even upon irradiation,[77] the current dogma—based on results from multiple groups—is that microglia can be replaced by peripheral cells upon irradiation. Such replacement does not seem to alter mouse behavior, but these cells may alter the rheostat during neuroinflammation, injury, and other pathological deviations from homeostasis.[78,79]

It was shown in 2010 that the precursors of microglia, as well as of all other tissue-resident macrophages, originate from primordial yolk-sac macrophages and populate all bodily tissues, including the CNS, at the time of blood vessel formation (around E8.5 in mice).[80] Later studies showed that, unlike in the brain, macrophages in peripheral tissues are replaced throughout life by monocyte precursors, meaning that the prevalence of original yolk-sac–derived cells reduces with postnatal time.[53] Microglia, on the other hand, to the best of our knowledge, are not replaced by peripheral blood monocytes, at least not under healthy conditions. Instead, they reside in the brain and maintain their pool by continuous proliferation. According to the lineage-tracing studies, the original progenitor population gives rise to the entire microglial pool and maintains it throughout life. In the case of injury or disease, however, the niche opens and allows peripheral monocytes to take up temporary residence, some of them dwelling in the tissue long enough to be called "long-lived resident monocyte-derived macrophages."[77] Interestingly, it seems microglia elimination and creation of the space is what allows monocyte entry. If microglia are depleted and their reconstitution of the niche by proliferation is prevented, peripheral monocytes invade and occupy the empty niche. Replacement of microglia by monocyte-derived macrophages does not appear to affect either gross anatomy or behavior in mice.[81] However, once an animal with replaced microglia is faced with a challenge such as peripheral or central infection, injury, or inflammation, the behavior of the new "immigrant" monocyte-derived macrophages differs from that of the original microglia.[81,82] Despite the fact that in healthy mice the monocyte-derived macrophages do not alter brain function, they are not identical with microglia, and their RNA-sequencing (RNAseq) signature reveals their origin.[79] Although there is no single gene that can predict whether a cell is of yolk sac or of peripheral monocyte origin, there is a distinct gene signature enriched in classical microglia compared to brain-engrafting macrophages. Therefore, single-cell sequencing (scRNA-seq) can facilitate prediction of the cell type being analyzed. Despite advances in RNA sequencing that allow us to identify potentially novel cell subsets, the specific contributions of microglia and monocyte-derived macrophages to brain physiology and pathology have yet to be determined.

NEUROINFLAMMATION: MS AND EAE

The field of neuroimmunology has evolved largely from studies on the human disease of MS. First described in 1868 by Jean-Martin Charcot, MS is a disease in which autoimmune T cells against myelin antigens trigger the disease cascade.[83,84] The human disease is complex and multifaceted, with different manifestations of inflammatory lesions in the CNS ([85-87] for further reading). These lesions are present in the white matter with perivascular infiltrates of myeloid and lymphoid cells. Lesions in some patients also have astrocyte proliferation, scarring, and axonal damage.[88,89] Patients with highly severe disease also develop ectopic lymphoid follicle-like structures in the leptomeninges.[90] The clinical manifestation of the disease is highly diverse, and the symptoms include increased reflexes, spasticity, weakness, ataxia, sensory loss, extensor plantar responses, and visual impairment.[88] MS is an autoimmune disease, and some decades ago, it was demonstrated that besides the recognized death of neurons in MS, degeneration of synapses occurs as well, hence its current definition as an autoimmune neurodegenerative disease. In addition, stripping of synapses by microglia has been demonstrated in MS and may also underlie the neurodegenerative feature of the disease.[91] The incidence of the disease and its progression also varies, and they are influenced by factors such as age,[92] gender,[93] season,[93] and gut microbiome.[94] Because of the complexity of MS, modeling the disease in animals has been invaluable in deciphering the mechanisms that contribute to its pathogenesis.

There are several mouse models of MS, and one of which is experimental autoimmune encephalomyelitis (EAE) typified by immune cell infiltration, inflammation, and CNS demyelination. This can be induced by the active immunization with myelin-derived peptides combined with an adjuvant[95] or the adoptive transfer of preactivated myelin-specific T cells[96] into naïve, susceptible mice or rats[96] (and even marmosets[97]). During the effector phase of the disease, the self-reactive cells (either generated in draining lymph nodes or adoptively transferred) circulate in the periphery and eventually migrate into CNS, creating local inflammation with or without destruction of myelin sheaths and with or without relapses/remissions. Typical EAE disease in rodents is characterized by ascending flaccid paralysis that starts as tail limping and further progresses to the hind and forelimbs.[96] These symptoms are accompanied by weight loss,[98] and inflammation and infiltration of immune cells to the spinal cord,[96] which do not fully recapitulate the human disease where inflammation extends to the brain. Aside from typical EAE symptoms, atypical EAE symptoms associated with brain inflammation have been reported, which include leaning or circling, ataxia, hyperreflexivity, and proprioception defects.[99] The development of atypical EAE appears to be dependent on antigen combination and the mouse strain[96,100,101] and has been linked to IFN-γ deficiency.[102] Induction of EAE results to different EAE disease outcomes depending on the mouse strain. For example, SJL (H-2^s) mice exhibit a relapsing-remitting disease course, C57BL/6 (H-2^b) have a chronic-progressive outcome, and in PL/J (H-2^u), the disease is acute monophasic.[98]

EAE has been useful to characterize the mechanism of inflammation and degeneration in the CNS, but it is limited in modeling how the disease is triggered.[98] To address this concern, genetically engineered mice that express myelin-specific T cell receptors (TCRs) have been used as these develop spontaneous EAE with variable incidence. These genetic models bypass the need for priming and injection of strong adjuvants, which are considered as nonphysiological stimuli.[96] Some of these transgenic mice include humanized mice that carry human leukocyte antigen (HLA) class II molecules linked to MS predisposition[103-105] and mice that express myelin oligodendrocyte glycoprotein (MOG)-specific TCRs (2D2 mice).[106] They exhibit a spectrum of symptoms that range from classical hind limb paralysis, inflammatory demyelination in the brainstem, and optic neuritis.[98] Despite the limitations of EAE models, a wealth of information has been generated related to disease pathogenesis, which have been exploited to uncover possible therapeutic targets.

Among the cells implicated in EAE, CD4+ T cells have been deemed as the primary effectors, due in part to the induction protocol that significantly enriches for them. These cells through their expression of VLA-4 (integrin α4) infiltrate the CNS parenchyma.[107] They also require presentation by local APCs to achieve full reactivation. Antigen presentation by CD11c+ dendritic cells (DCs) and not by CNS-resident cells was reported to be sufficient to facilitate disease development.[108] Upon reactivation, the T cells then release cytokines and chemokines that recruit macrophages and monocytes to the site of inflammation. These myeloid cells secrete cytokines that further the inflammation and then become the main purveyors of tissue damage.[109,110] The role of microglia in antigen presentation is still unclear but inhibiting microglia activation has been reported to dampen the severity of EAE.[111] Astrocytes have been reported to produce cytokines, chemokines, and neurotoxic substances that also promote more inflammation and damage.[112,113] The perpetuation of the damage in the CNS leads to epitope spreading as degraded myelin epitopes are released, processed, and presented by APCs to myelin-specific T cells that mobilize nonspecifically to the inflamed CNS.[99]

Initially, CD4+ Th1 cells have been implicated in driving the disease pathology, and this is through their ability to secrete IFN-γ and IL-12.[99,114] However, multiple lines of evidence have suggested that Th17 cells substantially contribute to the disease pathology. Studies have shown that mice deficient for IL-23, which is crucial for the differentiation of Th17 cells, are resistant to EAE.[115] Additionally, increased expression of IL-17 has been detected in MS lesions,[116] and transfer of Th17 cells induces greater disease compared to Th1 cells. Other reports have also suggested otherwise as *Il17a* and *Il17f*--deficient animals were not completely resistant to EAE,[117] and both Th1 and Th17 cells can induce EAE.[118] It is worth noting that the granulocyte-macrophage colony-stimulating factor (GM-CSF) produced by helper T cells in response to IL-23 has a nonredundant role in EAE development unlike IL-17 and IFN-γ.[112,119] GM-CSF–deficient animals are resistant to EAE,[120] and transfer of T cells deficient for GM-CSF to naïve C57BL/6 animals do not result to CNS inflammation unlike T cells deficient for IL-17a or IFN-γ.[119] GM-CSF can specifically signal to Ly6C^{hi} monocytes to acquire a pathogenic signature.[121] Reports also indicate that a feedback loop occurs between myeloid cells and pathogenic T cells as myeloid cells produce IL-1b that signals to T cells to produce GM-CSF.[121]

Aside from the aforementioned cytokines, tumor necrosis factor (TNF) has also been shown to contribute to disease pathology and protection.[112] It is produced by numerous cells, and studies have demonstrated that it can promote apoptosis of oligodendrocytes,[122] it can stimulate the upregulation of adhesion molecules on the BBB to allow the infiltration of T cells,[123] and it can induce the expression of MHC molecules on glial cells that can predispose them to cytotoxicity by CD8+ T cells.[123,124] These results underlie the complexity of the disease making it challenging to attribute the pathogenicity to single cytokine or effector cell.

As was mentioned, the induction protocol of EAE biases investigation to the contribution of CD4+ T cells to the disease. However, CD8+ T cells have been increasingly recognized as effectors to disease pathology as adoptive transfer experiments of myelin-specific CD8+ T cells lead to severe CNS inflammation and more lesions present in the brain compared to the spinal cord.[125] In addition, MS lesions and blood samples display an enrichment for CD8+ T cells compared with CD4+ T cells suggesting clonal expansion.[126,127] Models to study the role of CD8+ T cells have been developed, which allow the expression of TCRs that recognize peptides associated with MHC class I molecules.[128] The relevant APCs for CD8+ T cells are not well-characterized, but a recent paper suggested that oligodendrocyte precursor cells (OPCs) have the ability to cross-present antigens to CD8+ T cells in an IFN-γ–dependent manner.[129] Activated CD8+ T cells are then able to exert their effector functions by releasing soluble factors (ie, proinflammatory cytokines, IFN-γ and TNF)[130] and by engaging cell contact–mediated lysis pathways.[99] B cells may also contribute to disease pathology as their depletion using monoclonal antibody therapy (Rituximab) in MS patients can reduce inflammatory brain lesions,[131] and the presence of meningeal B cell follicles correlates with worsened pathology.[132] In mouse models of EAE, MOG-specific antibodies have been shown to promote more demyelination.[133] In these cases, B cells may function as APCs and sources of antibodies that target myelin and neurons. Some reports, however, have suggested a protective role for B cells[134] as they recruit Tregs to the CNS[135] and secrete IL-10.[136] A recent study demonstrated that IgA+–producing plasmablasts and plasma cells from the gut migrate to the CNS in EAE models to suppress inflammation.[137]

Given the considerable involvement of immune cells in the development of disease, targeting them is a viable strategy for therapeutic intervention. In fact, MS is among very few neurological diseases in which therapies target the underlying cause. One of the most promising current therapies is based on an inhibitory antibody that targets the integrin α4β1 (VLA4) needed by T cells for their migration into the brain parenchyma.[138-140] Inhibition of the integrin results in inability of T cells to invade the brain and spinal cord, causing the disease to recede. The caveat of using this drug, however, is that while efficient at fighting off inflammatory events in MS patients, it also effectively eliminates many cells of meningeal immunity, thus exposing the meninges to attack by pathogens. Therefore, patients on this treatment have increased risk of developing progressive multifocal leukoencephalopathy (PML), a fatal demyelinating disease caused by the John Cunningham (JC) virus.[141,142]

What drives brain-reactive autoimmune T cells to attack the brain? This is the major question in MS. Autoimmune clones should have undergone central tolerance and would therefore either be well controlled by Treg cells or would not exist. Studies from the late 1990s indicate that the numbers of autoimmune T cells recognizing brain antigens in MS patients do not differ from their numbers in healthy controls, suggesting that the disease is determined not by the number of cells but rather by their phenotype and their ability to invade and attack the CNS.

VIRAL INFECTION IN THE CNS

Despite the presence of specialized and elaborate barriers that protect the CNS, a wide array of viruses (ie, DNA, RNA, naked, and enveloped viruses) can manage to infiltrate and cause disease. The incidence of CNS infections is continually rising, most prominently in immunocompromised individuals, and the emergence of zoonotic viruses is increasing. CNS virus infections, therefore, represent a significant burden to human health. It is critical that once a virus enters the CNS, it is cleared effectively, but at the same time, inflammation and other immune-mediated pathologies must be kept at bay to prevent secondary damage and ensure the survival of CNS-resident cells. Therefore, understanding the mechanisms that maintain this delicate balance is a fundamental aspect of neuroimmunology.

Viral inoculation typically occurs in the periphery, and generally, the host is equipped to restrict the entry to the CNS. When viruses gain access to the CNS, they employ distinct strategies as there is no representative virus that spreads and infects the CNS.[143] One of the main strategies is the hematogenous route. Some viruses including the West Nile virus (WNV), Hepatitis C virus (HCV), Epstein-Barr virus (EBV), and human cytomegalovirus (HCMV) can directly infect human microvascular endothelial cells *in vitro*.[144] The human T-lymphotropic virus type I, for example, binds to glucose transporter type I (GLUT1) on endothelial cells allowing the latter's infection.[144] Upon infection, alterations in the properties of the barrier-forming cells have been observed, which include increase in chemokine production that mobilizes immune cells, induction of matrix metalloproteinase expression that degrades the basement membrane, and loss of tight junction proteins.[145-147] Virus infection also leads to the upregulation of proinflammatory cytokines, IL-1b, TNF-α, and IL-6.[148] Altogether, these compromise the integrity of the BBB, allowing viruses and infected cells to enter the CNS. In the "Trojan horse" strategy, the pathogen concealed within immune cells participating in tissue surveillance is carried across the BBB.[143,149] Lentiviruses, the human T-cell leukemia virus, the simian immunodeficiency virus (SIV), and human immunodeficiency virus (HIV) infect monocytes and macrophages to traverse the barrier.[150-152] Detection of the JC virus in peripheral B cells has been reported, and the B cells might move the virus to glial cells *in vivo*.[153,154] However, conclusive evidence for this is still lacking.

Some viruses invade the CNS through direct infection of nerve endings in the periphery effectively bypassing the BBB. The rabies virus, for example, is introduced to the host's neuromuscular junction, a synapse between muscles and motor neurons, after being bitten by an infected animal. The virus then hijacks the axonal transport system and retrogradely moves to the spinal cord and to the motor centers in the brain.[143] The olfactory epithelium can also be infected by viruses[155] such as the herpes simplex type-1 (HSV), vesicular stomatitis virus (VSV), influenza A virus, Borna disease virus, and severe acute respiratory syndrome coronavirus 2 (SARS-CoV-2 or COVID19).[156] The viruses in the olfactory epithelium can then cross the cribriform plate, or they can then migrate to the innervating-olfactory nerve, which then terminates in the olfactory bulb of the CNS.[143] Once the virus enters the CNS, they spread through the parenchyma by infecting neighboring cells depending on their tropism. For example, neurotropic viruses can be released into synaptic clefts and be taken up by membrane fusion by synaptically connected neurons. Some viruses infect glial cells (eg, HIV infects microglia, JC virus infects oligodendrocytes), the meninges, or the ependyma.[157]

Depending on the location of the infection, the immune response can lead to meningitis, encephalitis, or myelitis. Meningitis is the inflammation of the membranes that surround the brain, and it can be caused by St. Louis encephalitis virus (SLEV),[158] bunyaviruses,[159] lymphocytic choriomeningitis virus (LCMV),[160] and HIV.[161] This aseptic inflammation is accompanied by fever, headache, and neck stiffness that may be attributed to the increase cell count and cytokines in the CSF.[162] To determine the underlying cause of the pathology, mouse models of viral infection have been utilized. In a mouse model of intracerebral infection of LCMV, fatal meningitis develops, and this is driven by substantial infiltration of innate immune cells, particularly monocytes and neutrophils along with vascular injury.[163] Antiviral CD8+ T cells induce a vigorous response that includes secretion of chemokines that support immune cell trafficking and type I interferons that signal to myeloid cells to initiate their antiviral program.[164]

Encephalitis refers to the inflammation of brain parenchyma, and it can be caused by infection of cytopathic or lysis-causing viruses such as the human enterovirus (HEV) and alphaviruses.[165] It can also be indirect damage resulting from the robust immune response following infection as in the case for insect-transmitted flaviviruses (eg, WNV, Japanese encephalitis virus [JEV]).[166] Postmortem analysis of brain tissue from WNV-infected humans and animals revealed the presence of lesions characterized by neuronal cell death, activation of microglia and infiltrating macrophages, and accumulation of T cells and plasma cells in perivascular spaces and in the parenchyma.[167,168] These lesions were typically localized in the cortex, hippocampus, thalamus, cerebellum, and brainstem.[169] Neuronal infection can induce release of reactive oxygen species (ROS) and cytokines that further the pathology[170] and infected CD11b+ cells can also induce more neurotoxicity.[166] HIV-1 infection may also lead to encephalitis and neuronal death, which underlie cognitive and motor impairments of HIV-associated dementia.[149,171] This may be attributed to the activation of infected perivascular macrophages and microglia and their subsequent release of free radicals, inflammatory cytokines, and excitatory amino acids, which promotes neurotoxicity.[150] Rabies encephalopathy, on the other hand, is not induced by immune-related pathology.[149] The symptoms related to this disease such as sporadic edema and gliosis[172] are attributed instead to immune suppression and disruption of neuronal function.[173-175] Lastly, reactivation of latent viruses such as HSV-1 (in the trigeminal ganglia) can lead to encephalitis.[176] This pathology is associated with increased infiltration in perivascular spaces, neuronal infection, engulfment of neurons by microglia, and the recruitment of eosinophils.[177]

Myelitis, on the other hand, impacts the spinal cord rather than the brain tissue.[144] Poliovirus causes poliomyelitis or inflammation of the gray matter of the spinal cord where the neuronal cell bodies in the anterior horn of the spinal cord harbor the viral antigen. This leads to significant perivascular inflammation and motor neuron loss, which manifests as flaccid limb paralysis.[178] Rabies virus, WNV, and JEV are also capable of causing myelitis.[179,180] Similar to polio, paralytic rabies infection also causes motor neuron loss due to spinal cord inflammation. However, peripheral nerve demyelination is also a contributing factor to the observed paralysis.[181]

There is no consensus on how viruses invade the CNS, and how they cause pathology. However, the essential components of the immune system that mediate clearance of viruses have been identified. The innate immune response to the viral infection is primarily achieved through the local production of type I interferons.[182] This cytokine can be produced by infected cells, and it can induce the upregulation of IFN-stimulated genes (ISGs) that block viral replication.[183] Some reports have also suggested that type I interferons can have barrier-tightening functions, preventing further entry of virus and infected cells.[148] Inflammasome signaling has also been implicated in controlling CNS viral infections, particularly flaviviruses and VSV.[184,185] Autophagy has also been shown as an antiviral mechanism of infected neurons that would protect them from undergoing apoptosis. Deletion of key components of the autophagy pathway (eg, Atg6) resulted in poor clearance of the Sindbis virus (SINV), and Atg6 overexpression reduced cell death upon infection.[186,187]

Antiviral CD8+ T cells enter the CNS parenchyma, and they employ multiple strategies against CNS viruses.[182] They can secrete antiviral cytokines such as IFN-γ and trigger cell death of infected cells through perforin, Fas-FasL, or TRAIL-dependent pathways.[188] This has been reported for the clearance of WNV infection,[189] but it is also important for clearing the mouse hepatitis virus (MHV) from glial cells and macrophages.[190] IFN-γ, in particular, has been also shown to support the clearance of measles, LCMV, and VSV from neurons.[191] Application of this cytokine to infected neuronal cells *in vitro* blocks viral replication by engaging the JAK/STAT pathway.[192] Virus-specific CD4+ T cells also participate in viral clearance as they accumulate near blood vessels, secrete IFN-γ, and provide factors that support the survival and function of CD8+ T cells.[193,194] Passively transferred antibodies can also mediate the clearance of viruses, for example, rabies, as the antibody binds to the infected cell and prevents virus budding.[195] Despite these efforts, some viruses are capable of establishing persistent or latent infections (eg, HIV, HSV, and JC virus) effectively evading complete clearance by the immune system. It is possible that the continuous viral replication causes chronic immune activation that disrupts CNS homeostasis.[144] How does the immune system control these viruses at their latent state? Is it possible that the immune system was regulated by the CNS to induce this state to prevent further damage? How do other immune cells exert their antiviral effect? Why are there specific hubs for virus infection and response? These are all exciting questions that merit further mechanistic investigation.

CNS INJURY

The immune system is indeed powerfully equipped for destruction, especially when the target is a pathogen. It can also target cancers, as discussed elsewhere in this book. When out of control, it can also destroy healthy tissues—hardly a physiological role for a system whose aim is to protect its host organism. In spite of its propensity for destroying healthy tissue along with pathogens, the immune system is the main healing system of the body. Immune responses are initiated as a reaction to damage associated with injury or infection, and the later stage of an immune response to such injury or infection is usually a healing response. Such responses may, however, become uncontrolled and wreak havoc. Peripheral nerves, for example, are close to the CNS, and peripheral nerve injury results in an immune response required for optimal tissue recovery; however, if there is interference with the immune response to the peripheral nerve injury, the outcome in terms of neuronal survival is less favorable.[196-200]

What happens in the CNS after an injury? Usually, as outlined above, an injury is followed by an immune response.[201] However, the widespread belief in the unique status of the CNS as an immune-privileged organ and given that most of our early knowledge about neuroimmune interactions came from EAE/MS, where CNS tissue was destroyed by the immune system, the longstanding dogma was that the immune response after injury to the CNS contributes to further neurodegeneration. Therefore, for the sake of tissue protection, the immune response must be blocked or at least attenuated. Many approved treatments were indeed aimed at immune suppression, although their therapeutic efficacy has not stood the test of time, and both clinicians and scientists are now questioning the usefulness of that approach. Notably, experimental evidence in rodent models of CNS injury does not support a detrimental role for the immune system. On the contrary, immune system suppression has been found to result in further tissue degeneration, whereas boosting of immune cells was shown to improve neuronal survival.[196-200]

The mechanism underlying such immune neuroprotection is not yet fully understood, although some insights have clearly emerged. Injury to the CNS differs from injury to other tissues because neurons are terminally differentiated cells with little or no regenerative capacity. Following the initial CNS insult and death of injured neurons, destructive injury-triggered processes initiate a cascade of events, leading to the death of nearby neurons that were undamaged by the primary injury. The final infarct zone is composed mostly of the damage done through such secondary neuronal death. It is, therefore, reasonable to assume that the therapeutic approach to any CNS injury should incorporate two beneficial treatment modalities: *regeneration* (replacement of dead and damaged neurons) and *neuroprotection* (prevention of secondary death, thereby sparing healthy neurons from dying) (Fig. 39.3).

The immune response triggered after CNS injury is not essentially different (albeit with attenuated intensity) from the immune response after injuries to other tissues. The first wave of responders consists of the myeloid cells (neutrophils, monocytes,

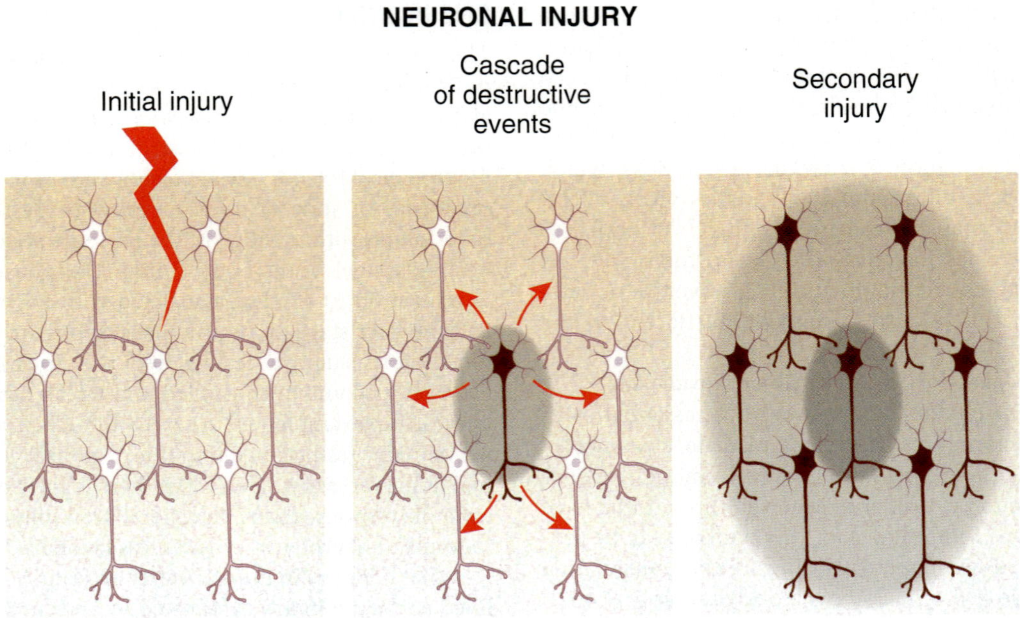

FIG. 39.3. Secondary neurodegeneration. After initial injury to the central nervous system (CNS), the damage spreads in part due to a limited immune support, amounting for a much larger loss of the neural tissue. The most efficient therapies would both target the initially damaged area and prevent the subsequent loss of the tissue due to secondary neurodegeneration.

and others) that comprise the organism's innate immunity, and this is followed by a second wave comprising the apparatus of the adaptive immunity.[201] Myelin, the oligodendroglial sheath wrapped around healthy axons for their efficient propagation of electrical signals, contains lipid and proteins. Some of these are growth inhibitory and will prevent the regrowth of axons unless myelin debris is removed from the injured site or receptors recognizing these growth-inhibitory molecules are blocked. For reasons not well understood, the phagocytic activity of the resident macrophages of the CNS, the microglia, is poor, especially against myelin, and cannot cope with the copious local amounts of myelin debris that exists after a CNS injury, whereas the monocyte-derived macrophages that invade injured CNS tissues are much better at eliminating such myelin debris. These monocytes, however, are prevented from gaining easy access to the core of the injury site because of the astrocyte-induced formation of an obstructive glial scar.[198] Notably, an exogenous supply of appropriately stimulated macrophages was found to benefit the outcome of CNS injury and even to induce neuronal regeneration.[196-200] Likewise, boosting of T-cell responses after CNS injury was shown to offer neuroprotection.

In particular, autoimmune T cells have been shown to offer beneficial effects upon CNS injury, which is coined as protective autoimmunity.[202,203] Seminal studies have reported that injection of activated T cells specific to myelin basic protein (MBP) or T cells specific to a cryptic MBP epitope (p51-70) increase the recovery and protect the secondary degeneration of retinal ganglion cells (RGCs) after partial optic nerve injury.[204] The anti-MBP T cells were observed in the injured optic nerve, and interestingly, the ability of the injected autoimmune cells to induce autoimmune disease was not hampered by the injury.[204] Administration of DCs pulsed with MBP also improve functional recovery after spinal cord injury, which was lost upon depletion of mature T cells.[205] Follow-up studies have since demonstrated that CNS injury elicits a systemic neuroprotective response in a defined time frame that can be transferred.[202] More specifically, rats subjected to spinal cord contusion injury and were subsequently subjected to unilateral optic nerve injury at specific time points (7-, 11-, 14-, or 17-days post injury) after the first injury displayed improved survival of RGCs. Splenocytes obtained from rats that have undergone spinal cord injury and activated with MBP *in vitro* also raise a neuroprotective response after transfer to another animal, and this beneficial effect is not evoked after passive transfer of antibodies.[202] The systemic neuroprotective response evoked after CNS injury appears to be physiological and T-cell dependent as it is lost after thymectomy. In support of these findings, transgenic mice that express the TCR for MBP had better survival of RGCs after optic nerve injury. These results indicate that there is an endogenous protective immune response after injury, and autoimmune T cells underlie the effect. The mechanism as to how these autoimmune T cells exert their neuroprotection has not been clearly elucidated, but it may be through regulation of the myeloid cells[203,204,206-210] or through the production of neurotrophic factors.[211,212] However, in other reports, bystander natural killer and T cells appear to be producing similar or higher levels of neurotrophic factors compared to autoimmune T cells that can reduce the loss of mechanically transected neurons.[213]

The fact that CNS cells such as astrocytes and oligodendrocytes contain alarmins—for example, interleukin (IL)-33—which stand ready to be released to summon peripheral immune assistance after injury to the CNS[201,214] suggests that the nervous and the immune systems are not as isolated as initially thought. Not only does IL-33 induce monocyte recruitment through induction of chemokine release by astrocytes, but this cytokine also acts on T cells to skew them toward neuroprotective IL-4–producing cells. In addition, it also triggers a meningeal-resident group 2 innate lymphoid cells (ILC2s) response that produces more IL-5/13, which keeps this immune response type 2 skewed. Type 2 immune responses, and particularly, production of IL-4, have been shown to directly benefit neuronal survival by potentiating the signaling of protective neurotrophic factors on injured neurons.[201,214-216]

After CNS injury, CNS-draining lymph nodes (such as deep cervical lymph nodes, dCLNs) expand as a result of T-cell activation and proliferation. As with antitumor immunity, partial alleviation of immune suppression through the ablation of Treg cells results in a more efficient immune response and better neuroprotection. In particular, mice depleted of Tregs using thymectomy 3 days after birth or nude mice that received splenocytes devoid of Tregs had significantly better survival of RGCs after optic nerve injury. Conversely, BALB/C mice that received activated Tregs had worsened RGC survival.[203] Complete Treg ablation, however, unleashes a destructive immune response, further demonstrating the delicate nature of the balance between protection and destruction offered by the immune system.[206,217-221] The signals that are necessary to instruct the beneficial or destructive immune responses and how these might imprint leukocytes are currently unknown. The peripheral nervous system might contribute to this function as injury results to overall immune suppression.[222] It is also important to note that aside from T cells and monocyte-derived macrophages, other immune cells such as neutrophils activated by fungal cell wall products[223,224] have been shown to be critical players in CNS injury, as they release a host of growth factors that promote neuronal survival and regeneration.[225] These findings also indicate that there are plenty of therapeutic strategies that can be explored to mitigate damage after CNS injury.

Given the tight control exerted on the immune response to CNS injury and the potential harm that that an uncontrolled immune response could mediate, it is plausible that immune cells may play physiological roles in noninjured CNS. This raises the question: what might these putative functions be? Moreover, the fact that injured CNS drains into the lymph nodes raises new questions: what is the pathway for such drainage? And if the drainage is lymphatic in nature, why and in what way should the CNS be considered to be immunologically unique?

THE ROLE OF IMMUNE CELLS AND CYTOKINES IN CNS FUNCTION

The CNS parenchyma, according to the classical view, is devoid of immune cells except for microglia and perivascular macrophages. Several recent studies indicate, however, that adaptive immune cells are found within the CNS parenchyma and may play a role in the maturation of microglia as well as in regulation of the neurogenic niche.[226-230] Surrounding the brain are the meninges, consisting of three membranous layers (Fig. 39.4). The innermost layer attached to the brain surface is the pia mater, the outermost layer attached to the skull is the dura mater, and between them lies the arachnoid mater. Between the arachnoid and the pia is the subarachnoid space, occupied by CSF that is produced through the filtration of blood by the epithelial cells of the ventricular choroid plexus and fills the ventricles of the brain and the central canal of the spinal cord.

The blood vessels of the brain are composed of unique endothelial cells whose specialized tight junctions prevent soluble and cellular (immune) contents of the blood from reaching the brain. The meningeal vasculature, however, is more permissive to molecular and cellular exchange with the blood, especially the vasculature of the dura mater, which is fenestrated, and discontinuous. Therefore, it is not unlike that of peripheral organs.[235]

The meninges are characterized by their content of a large and diverse population of immune. cells.[236] Recent reports have described the source of meningeal immune cells under homeostasis. Meningeal T cells are mainly repopulated from the blood, and a substantial portion of B cells, monocytes, and neutrophils arise from the adjacent skull bone marrow.[232-234] The question, then, is: what are these cells doing in that area, and how important are they for the function of the brain?

When we consider characteristic animal behaviors in the contexts of, for example, foraging, parental, social, or mating, we realize that such behaviors predispose individuals to a risk of infection. It could be argued, therefore, that a functional immune system would be an essential requirement for those behavioral traits to develop. If the two systems involved have—as some claim—indeed codeveloped, they would out of necessity depend on one another and work in synchrony. Thus, if one system malfunctions, the other may be affected as well. Indeed, mice that are deficient in adaptive immunity

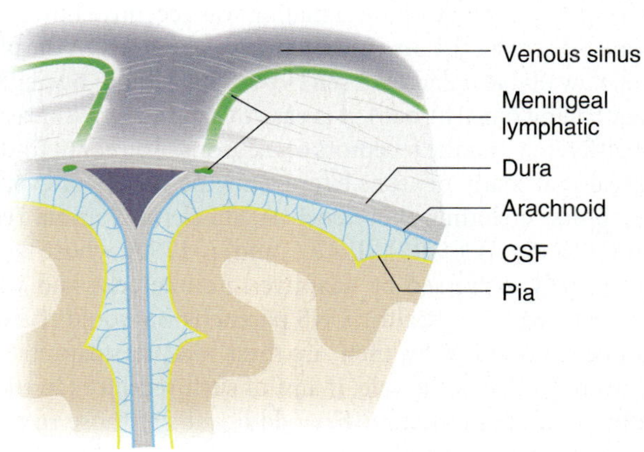

FIG. 39.4. Meningeal coverings of the central nervous system (CNS). Meninges is composed of three layers—pia, arachnoid, and dura maters. Pia mater is attached to the brain parenchyma and dura mater to the skull. Underneath the dura mater is the arachnoid mater, which creates a subarachnoid space allowing for cerebrospinal fluid (CSF) to flow around the brain. The meninges also harbor a diverse array of immune cells, some of which were derived from the local skull bone marrow upon traversing through special vascular channels.[231-234]

exhibit a range of abnormal behaviors, such as processing of fear memories, aberrant social behavior, and anxiety.[237-245] How these behaviors are affected by immune dysfunction is a matter of active investigation.

One of the oldest examples used to illustrate immune-mediated impacts on behavior is sickness behavior.[246-253] The well-known malaise that we experience when we have a common cold or a more debilitating infection can be mimicked in animals simply by injecting them with the endotoxin lipopolysaccharide (LPS) or even with a single inflammatory cytokine such as IL-1b or IL-6. Notably, sickness behavior will develop if such molecules are injected either into the bloodstream or into the CSF. The assumption was (and to some extent still is) that the effect of these inflammatory cytokines is mediated by microglia, which presumably trigger the cascade of molecular events that culminate in sickness behavior.[246-253] Some studies, however, have linked other cell types to transducing the inflammatory signal from the periphery to the CNS. Among the cell types that contribute to this function are brain endothelial cells,[254-256] epithelial cells,[254] and blood vessel–related mural cells.[257] Endothelial cells, in particular, have abundant expression of proinflammatory cytokine receptors such as IL-1R1, and upon engagement, they produce secondary signals such as prostaglandins[258,259] that facilitate some of the symptoms of sickness behavior such as fever and prolonged sleep.[260] Endothelial cell–specific knockout of IL-1R1 or selective overexpression of the receptor in endothelial cells suggest that these cells are necessary[253] and sufficient to induce sickness symptoms upon injection of IL-1.[257] More recently, chemokine signals derived from brain endothelial, epithelial, and mural cells have been shown to signal directly to neurons to change their excitability and synaptic properties in response to cytokines released upon viral infection (eg, type I interferon) or LPS administration (eg, IL-1b).[254,257]

One of the most unexpected group of target cells capable of responding to cytokine signaling was recently shown to be none other than the neurons themselves.[237,244] Given the plethora of available datasets of scRNA-seq and single-nuclei RNA sequencing (snRNA-seq), it is easy to examine the expression of different cytokine receptors on neurons. Abundant findings reveal that many neurons express a wide variety of cytokine receptors, including those such as interferon-gamma receptor (IFNGR1, IFNGR2), IL-4RA, and IL-17RA, whose ligands are usually expressed by adaptive immune cells and whose presence within a healthy CNS parenchyma would therefore not be expected. Why, then, are these receptors expressed on neurons? What is the role, if any, of such receptors in modulating neuronal function? How do ligands of those receptors reach the parenchyma?

An intriguing body of evidence regarding some of these molecules has recently accrued, as described in the following example. A previous study had shown that IFNγ-knockout (IFNγ-ko) and IFNγR-knockout (IFNGR1-ko) mice exhibit social deficits. Surprisingly, conditional deletion of IFNγR from microglia (the immediate suspect in mediating this immune signal), in mice did not lead to behavioral changes, suggesting that cells other than microglia were responding to IFNγ signaling. Indeed, mice with conditional deletion of IFNGR1 from neurons were found to exhibit changes in social behavior reminiscent of those observed in IFNγ-ko and IFNGR1-ko mice. Furthermore, conditional deletion of Stat1 (a signaling molecule downstream of IFNGR1) from inhibitory neurons also resulted in impaired social behavior.[244] Collectively, the results of that study suggested that meningeal T cells produce IFNγ, which acts on inhibitory neurons and, through IFNGR1 signaling, affects social behavior. The stimuli activating the meningeal T cells leading to production of the cytokine have yet to be investigated.

Another example is IL-4, a cytokine that interacts directly with neurons. Both IL-4 and IL-13 can signal through an IL-4/IL-13R heterodimer (while each cytokine also has its unique receptor composed of a specific alpha chain and a common gamma chain). Mice deficient in IL-4 exhibit cognitive deficits. The mechanism involved was initially proposed to be mediated by astrocytes, although recent works have demonstrated the expression of IL-4R on neurons in the context of CNS injury or neuroinflammation, which may underlie the effect of Il-4 on memory. IL-4R expression on neurons is increased after mechanical or inflammatory CNS injury, and IL-4 has been shown to promote the expression of neurotrophic factors through IL-4R expressed on neurons. A similar mechanism may also underlie the role of IL-4 in cognition, but this has yet to be demonstrated.[240,241,261,262]

A more widely studied cytokine that has an abundance of behavioral effects through its neuronally expressed receptors is IL-17. Initial studies have demonstrated that maternal IL-17 can cross the placenta and impact neurons, causing autism-like behaviors in offspring.[263,264] An interesting feature of IL-17 is that its effect on neurons is evolutionarily conserved, as indicated by the fact that, in the primitive roundworm *Caenorhabditis elegans*, an IL-17 orthologue mediates behavioral changes through its direct impact on sensory neurons.[265] Recent studies in mice have demonstrated that meningeal γδT cells can serve as the closest source of IL-17a for the brain. Mice lacking γδT cells or deficient in IL-17a exhibit cognitive, social, and anxiety behaviors, mediated in part via glial cells and in part via neurons.[245,266,267] Notably, the expression pattern of IL-17R is predominantly cortical.[245] It is likely that many other cytokine receptors are expressed on neurons and other CNS cells. Defining cytokine receptor expression in the CNS is, therefore, key in predicting which cytokines can influence behaviors.

It is worth mentioning that apart from immune cells, glial cells also produce cytokines that have critical functions in the CNS. Glial-derived TNF-a, for example, is involved in a process that strengthens synapses at homeostatic conditions.[268,269] Other cytokines (ie, chemokines and TGF-b) have also been associated with CNS development acting as neurotrophic factors and influencing the differentiation of progenitors to distinct neuronal subsets.[270,271] Proinflammatory cytokine IL-6, in particular, has been shown to support the survival of oligodendrocytes *in vitro*,[272] regulate neuronal activity, and induce behavioral abnormalities in the offspring upon injection in pregnant mice.[273] These findings suggest that the context and perhaps the cellular source of cytokine can lead to profoundly different effects on the CNS. A systematic way to characterize and interrogate these cellular sources would further our comprehension of how cytokines affect CNS function. More importantly, understanding how cytokines affect behaviors, and what combination of cytokines can affect which behaviors, is the focus of attention in systems neuroimmunology, a research field that is currently in its infancy.

NEURAL REGULATION OF THE IMMUNE RESPONSE

While the impact of the immune system on CNS function has been well-appreciated in the context of disease and increasingly, in physiology, our understanding of the neural regulation of immunity is more limited. In this section, we will discuss the studies related to this matter, particularly how the manipulation of the CNS, innervation of lymphoid organs, and neurotransmitters can alter peripheral immunity.

To evaluate how specific regions of the CNS can regulate immune function, investigators as early as the 1960s have conducted lesion and stimulation studies on various brain regions and measured cellular and humoral immune responses as well as changes in lymphoid tissues.[274] Bilateral lesions in specific nuclei of the hypothalamus have resulted in decreased T cell function and number,[275] decreased antibody titer against ovalbumin,[276] decreased phagocytic activity of neutrophils,[277] and changes in thymocyte and splenocyte numbers.[278] Manipulation of the brain's limbic structures such as the hippocampus and the amygdala can also modulate the immune response. Lesions of the rodent hippocampus resulted in the increased proliferative capacity of thymocytes and splenocytes, which is in contrast to lesions preoptic-anterior hypothalamus[279] and it also increased antibody titers against ovalbumin.[280] Bilateral ablation of the cortex resulted in changes in the number of T cells as well as in natural killer cytotoxic activity.[281] Unilateral manipulation of the cortex may indicate that there is lateralization effect on the immune response; that is, splenocytes of animals that have a lesion on the right cortex were able to proliferate more upon exposure to a mitogen compared to animals that have a lesion on the left cortex.[282] While informative, these lesion studies introduce substantial injury to the brain that can confound the interpretation of results.

The development of optogenetics and chemogenetics allowed the targeting of specific cell types and cell projections and the functional interrogation of specific neuronal circuits by either stimulating or inhibiting neuronal activity at a given time frame and correlating it to a defined phenotype.[283,284] These methods can be used to more precisely study the contributions of different brain regions on the immune responses. Optogenetics relies on the expression of light-sensitive channels on a predetermined neuron (or cell type) and its subsequent activation through laser diodes connected to optic fibers.[283,285] Chemogenetics, on the other hand, requires the expression of a mutated muscarinic receptor (ie, designer receptors exclusively activated by designer drugs or DREADDs) on the target cell population and the delivery of a synthetic, pharmacologically inert drug.[286]

Chemogenetics has been used to activate the brain's reward system to determine its impact on immunity. Expression of DREADDs in tyrosine hydroxylase (Th)-positive cells of the ventral tegmental area (VTA) and injection of an inert ligand (clozapine-N-oxide; CNO) improved the immune response to bacterial infection[287] and solid tumors.[288] They observed enhanced activation status of immune cells, enhanced phagocytic activity of macrophages, increased class-switching of B cells, and increased interferon-gamma production of T cells upon *E. coli* infection of VTA-stimulated animals.[287] The boost in immune response in the tumor model was attributed to the decreased immunosuppression of myeloid-derived suppressor cells (MDSCs) found in the bone marrow.[288] They show that the effect is dependent on the sympathetic nervous system (SNS) because administration of a neurotoxic substance (6-hydroxydopamine) that targets the catecholaminergic neurons in the periphery ablates the enhanced immune response. Interestingly, it has been previously reported that self-stimulation of the VTA of rats resulted in an increase in the humoral response and increase hypersensitivity to bovine serum albumin (BSA).[289]

Aside from direct manipulation of neuronal circuits, immune conditioning akin to the Pavlovian conditioning paradigm has also been reported. In one of the first studies demonstrating this phenomenon, rats were repeatedly exposed to a conditioned, neutral stimulus, saccharin, and to an immunosuppressive drug, cyclosporin A in order to establish their coupling.[290,291] After training, the administration of saccharin by itself can induce immunosuppression measured by a decrease in antibody titers. This "learned immune response" has been demonstrated in animal models of inflammatory diseases such as lupus and arthritis,[292,293] and it has been linked to multiple brain regions, but most prominently, in the insular cortex.[294] More recently, it has been reported that behavioral paradigms specifically activating brain regions that innervate the spleen (eg, central nucleus of the amygdala and paraventricular nucleus [PVN]) result to the enhancement of humoral immunity.[295] These studies suggest that conditioning or behavioral intervention can be used to prepare the organism in anticipation of an immune insult.[291]

Stress has also been linked with changes in systemic immune response, and this is primarily mediated by the activation of the hypothalamus-pituitary-adrenal (HPA) axis. Cytokines and the neuroendocrine input from the periphery can activate either the adrenals, pituitary, or hypothalamus. Upon integration of the signals, the PVN of the hypothalamus secretes corticotropin-releasing hormone (CRH) into the hypophysial-portal circulation.[296] The hormone then binds to its corresponding receptor on the anterior pituitary corticotropic cells, and this leads to the release of adrenocorticotropic hormone (ACTH) into circulation. ACTH allows the release of glucocorticoids from the adrenal cortex. Glucocorticoids bind to their receptors thereby, allowing their translocation into the nucleus, where they interact with transcription factors to modulate the expression of glucocorticoid-sensitive genes. Generally, the effect of glucocorticoids is anti-inflammatory or immunosuppressive. Some of the ways that glucocorticoids exert their immunosuppressive effects are through the inhibition of vasodilation and subsequent reduction of leukocyte trafficking,[297] modulation of NF-kB and other transcription factors,[298,299] and alteration of cellular differentiation.[300,301] It is important to note, however, that the glucocorticoid receptor is widely expressed in multiple cell types; therefore, its effects extend to metabolic, cardiovascular, and endocrine functions.

Neurotransmitters and neuropeptides have also been shown to regulate peripheral immunity. The presence of adrenergic receptors necessary for the signaling of norepinephrine and epinephrine on both innate and adaptive immune cells has been well-documented.[302] The β2-adrenergic receptor (β2AR), in particular, is predominantly expressed in immune cells.[303] However, the level of expression is dynamic and is dependent on the activation status of the cell, the inflammatory milieu, and differentiation status. For example, T cells that have their TCR engaged results in the downregulation of β2AR suggesting that naïve T cells have higher expression of the receptor making them more susceptible to regulation by the SNS.[302] Moreover, Th1 cells and regulatory cells express β2AR but not Th2 cells[304] because the gene for the receptor is epigenetically silenced. Inflammation induced by LPS also promotes the downregulation of β2AR and the upregulation of α2AR in macrophages and DCs.[302] The effect of adrenergic signaling on immune cells is multifaceted, but it has been reported to influence proliferation, differentiation, migration, cytokine production, and antibody production.[303]

Neuropeptide receptors have also been reported to be expressed on immune cells. The receptor for substance P has been observed in monocytes, DCs, and lymphocytes.[305,306] Receptors for calcitonin gene-related peptide (CGRP) and VIP are expressed on murine DC lines and T cells, respectively. DCs that engage CGRP will downregulate the expression of MHC class II and IL-12.[307] It can also inhibit neutrophils' bactericidal capacity as well their recruitment in bacterial skin infection[308] and the recruitment of γδT cells in bacterial lung infection.[309] The effect of VIP is primarily immunosuppressive as it promotes the differentiation of tolerogenic DCs endowed with the ability to generate Treg cells,[310,311] it enhances Th2 responses,[312] and it reduces the expression of inflammatory cytokines[313] and costimulatory molecules.[314] ILC2s also express receptors (NMUR1) for the neuropeptide, neuromedin U (Nmu) produced by closely associated cholinergic neurons in the intestinal mucosa[315,316] and in the lungs.[317] Upon engagement of the receptor, ILC2s are activated to proliferate and to produce type 2 cytokines such as IL-5 and IL-13 critical for the mounting of immune responses against parasitic infections (eg, *Nippostrongylus brasiliensis*). This is evidenced by the decrease of infection burden upon *in vivo* administration of the neuropeptide and the increase in worm burden and impairment of type 2 responses in mice deficient for NMUR1 (Nmur1$^{-/-}$).[315,316] In the lungs, however, Nmu acts together with IL-25 to activate ILC2s, which leads to an exacerbated allergic lung inflammation.[317]

Innervation of lymphoid organs has also been recognized as a means for the nervous system to influence peripheral immunity. Innervation of the bone marrow has been reported as early as the 1960s,[318] and it modulates hematopoiesis, bone formation, lymphocyte maturation,[319] mobilization of cells,[320,321] and the maintenance of the quiescent state of hematopoietic stem cells (HSCs).[322] The role of splenic innervation, on the other hand, has been the most well-studied and evaluated in the context of the *inflammatory reflex*, which was initially described as the suppression of peripheral inflammation induced by endotoxemia through the parasympathetic nervous system.[323,324] Sensory fibers receive information about the inflammatory state, and they are relayed by the afferent vagus nerve to the nucleus tractus solitarius (NTS) and to the other higher order brain regions. Efferent fibers of the vagus nerve then synapse at the celiac ganglion, where fibers that innervate the spleen arise from.[325] In the initial study, stimulation of the vagus nerve resulted in the secretion of acetylcholine in the spleen, which inhibited the release of TNF-α from splenic macrophages.[323] This was controversial because innervation of the spleen is primarily mediated by sympathetic, norepinephrine-producing fibers, and there is an absence of histological evidence for cholinergic innervation.[326,327] However, recent studies proposed that the splenic nerve, which originates from the celiac ganglion, releases noradrenaline to the spleen. Choline acetyltransferase (ChAT)-positive T cells shown to be in close proximity with the nerve endings through immunofluorescence analysis bind norepinephrine through their β2ARs to produce acetylcholine.[328,329] The acetylcholine acts on the α7 nicotinic acetylcholine receptor (α7nAChR), most notably on macrophages[330] to dampen NF-κB nuclear translocation and proinflammatory cytokine production.[331] The model for this process is still being refined, but the autonomic nervous system undoubtedly plays an important role in regulating the inflammatory state of the spleen and the whole organism.[319]

MENINGEAL IMMUNITY AND LYMPHATICS

Instead of the old dogma that the immune and the nervous systems act as separate entities, interacting only under conditions of pathology, it now appears that the two systems also communicate with each other physiologically. The longstanding dogma of immune privilege, however, is still widely prevalent. One of the claims for the lack of communication between the two systems is based on the presumed absence of lymphatic drainage that could connect the brain with the immune system. This is a claim that has recently been challenged. Although the phenomenon of meningeal immunity has been touched on above, here we need to take a closer look.[27,332-334]

The brain parenchyma, as already mentioned, is devoid of immune cells except for microglia and perivascular macrophages. That does not mean, however, that the brain is not surveyed by the immune system. Also described above is the triple layer of membranes, the meninges, surrounding the brain and the spinal cord. The CSF that fills the subarachnoid space and perfuses the brain and spinal cord is replenished around three times a day. Where does the "old" CSF go? The CSF, during its perfusion of brain and spinal cord parenchyma, removes all toxins and metabolites produced by and accumulating in those tissues. The paths and the forces influencing the CSF during such perfusion are still under investigation.

A plausible current theory posits the "glymphatic" route (Fig. 39.5), suggesting that the CSF travels within perivascular spaces, originating at the Virchow-Robin periarterial spaces at the brain borders and continuing throughout the periarterial space. Arterial pulsation pushes the CSF, along with its content of small molecules (including cytokines produced by meningeal immune cells and released into the CSF) through astrocytic end feet into the parenchyma, washing it and collecting the metabolites. From there, the fluid is recollected at perivenous sites and removed from the parenchyma. The flow of fluid is also facilitated by water channels and by aquaporin-4 (AQP4) molecules expressed mostly by astrocytes.[335-340]

The interstitial fluid (ISF), which is formed in part by blood endothelial cells, perfuses all tissues in the body including the CNS, toward its removal in the lymph. On the way, it collects the accumulating tissue-derived contents of metabolites and toxins for disposal via the lymphatic system. Such perfusion necessitates the cooperative action of the two vascular networks, the blood vessels, and the vascular lymphatics. Since the brain parenchyma is known to be devoid of lymphatic vasculature, a critical question remains: how is the brain perfused?

A few years ago, a network of lymphatic vessels was discovered or rather rediscovered in the brain's vicinity—not within the parenchyma but in the meningeal membrane system surrounding the brain.[26,27,30,332-334,341-347] This system drains the CSF, which through the glymphatic system also contains brain ISF and antigens from the CNS (Fig. 39.6). Via these meningeal lymphatic vessels, CNS antigens reach the draining lymph nodes, either directly as soluble molecules or through their capture and presentation by meningeal APCs and their conveyance on the meningeal APCs[235] to the draining lymph nodes. This latter route, although indirect, allows the immune system to survey the brain and sample its antigens. Several experimental methods have been devised to ablate or otherwise interfere with meningeal lymphatic vessel function. These include pharmacological approaches for its ablation, surgical approaches involving ligation of afferent meningeal lymphatics before they drain into the dCLNs, and genetic approaches such as the use of Prox1^{het} mice, whose lymphatics are leaky and therefore dysfunctional. Using several of these approaches, it was shown that the meningeal lymphatic route is the main (and possibly the only) pathway for drainage of CNS-derived molecules. A nasal route, previously proposed[348] to serve as the drainage route for the CSF/CNS, does not have significant contribution to drainage. Recent data from investigations in humans contradict claims that the nasal route may function physiologically,[349,350] thus supporting the candidacy of meningeal lymphatic vessels as the main, if not the only lymphatic drainage pathway. Further evidence for exclusion of the nasal lymphatics theory comes from the finding that ablation of meningeal lymphatic vessels, but not of nasal lymphatics, attenuates EAE in mice.[29] Autoimmune T cells isolated from the draining lymph nodes (dCLNs) of EAE mice with and without functional meningeal lymphatics showed transcriptional differences, suggesting that CNS drainage is required for autoimmune T cells to acquire a pathological phenotype, which is probably a prerequisite

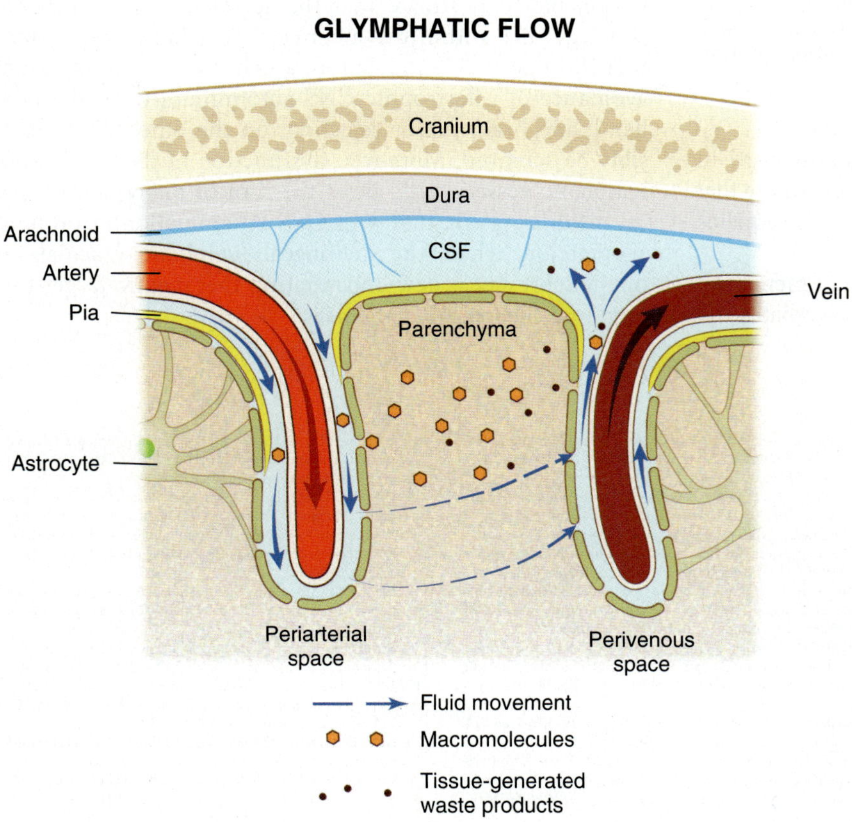

FIG. 39.5. Glymphatic system. Glymphatic system is a conceptual framework underlying the cerebrospinal fluid (CSF) flow through the brain parenchyma. The CSF enters the brain through Virchow-Robin spaces and continues through the perivascular space throughout the brain parenchyma. Due to arterial pulsation, the CSF and its soluble molecules exit the vasculature and create a flow through the brain parenchyma and are being recollected at the perivenous space through which it is removed from the brain.

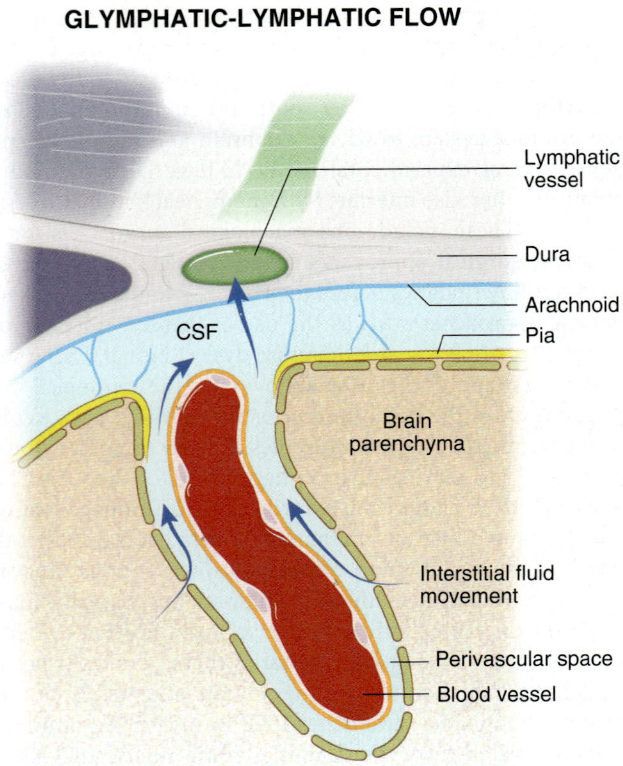

FIG. 39.6. Lymphatic-glymphatic connection. After leaving the brain, the cerebrospinal fluid (CSF) that is carrying the waste products from the brain is drained through meningeal lymphatics. Ablation of meningeal lymphatics or their deterioration with aging interfere with glymphatic flow, whereas expansion of meningeal lymphatic vessels allows a more substantial fluid movement through the brain. The glymphatic system could thus be referred as a prelymphatic brain-fluid dynamics.

for their invasion of the parenchyma.[29] These findings are in agreement with the abovementioned data demonstrating that lymphatic ablation by surgical removal of dCLNs also ameliorates the course of EAE.

Unlike with EAE, and as expected, ablation of meningeal lymphatic vessels in animal models of AD exacerbates the pathology, with more (and larger) plaques being detected both in the meninges and in the parenchyma.[30,333,351] In AD, removal of amyloid beta would be needed in order to avoid its aggregation. As such clogging or ablation of the "brain sink," that is, of meningeal lymphatic vessels exacerbates the disease. Moreover, studies in mice have shown that enhancing the meningeal lymphatic vessels by using recombinant vascular endothelial growth factor (VEGF)-c, a growth factor for lymphatic endothelial cells that signals through their expressed VEGFR3, alleviates pathologies associated with aging and AD.[30,351,351]

If these meningeal vessels are so important, why have they been overlooked for so long? As a matter of fact, they were first described as early as about 250 years ago by an Italian physician, Paulo Mascagni, but his work on meningeal lymphatics, which included the claim that they could drain the brain, was discredited by both his contemporaries and his successors.[352]

Coming full circle back to the beginning of this chapter, if the CNS is endowed with functional lymphatic vessels through which molecules can be drained into the dCLNs, why is the immune response to brain antigens insufficient to reject grafts (or tumors) in the brain? Recent studies have demonstrated that while meningeal lymphatic vessels are present around the brain, their function may be suboptimal, and therefore likely to be incapable of developing autoimmunity against the brain. Enhancement of meningeal lymphatic vessels by means of VEGFc therapy has proven sufficient for mounting a more efficient immune response against brain tumors, resulting in their almost complete rejection.[353,353] Based on these and other results, it seems reasonable to postulate that the secret of "CNS immune privilege" lies within the meningeal lymphatic vessels and that through their modulation, a delicate balance between avoidance of catastrophic CNS autoimmunity, on the one hand, and brain tumor rejection, on the other, could feasibly be achieved. Moreover, dysfunction of the meningeal lymphatic vessels could be at the root of many neurological disorders associated with CNS inflammation or protein aggregation. Thus, the meningeal lymphatic vasculature might represent a new and exciting therapeutic target for numerous neurological diseases.

REFERENCES

1. Chu C, Artis D, Chiu IM. Neuro-immune interactions in the tissues. *Immunity*. 2020;52:464-474. doi:10.1016/j.immuni.2020.02.017.
2. Baral P, Udit S, Chiu IM. Pain and immunity: implications for host defence. *Nat Rev Immunol*. 2019;19:433-447. doi:10.1038/s41577-019-0147-2.
3. Medawar PB. Immunity to homologous grafted skin; the fate of skin homografts transplanted to the brain, to subcutaneous tissue, and to the anterior chamber of the eye. *Br J Exp Pathol*. 1948;29:58-69.
4. Ridge JP, Fuchs EJ, Matzinger P. Neonatal tolerance revisited: turning on newborn T cells with dendritic cells. *Science*. 1996;271:1723-1726.
5. Brent L. The discovery of immunologic tolerance. *Hum Immunol*. 1997;52:75-81.
6. Bhat R, Steinman L. Innate and adaptive autoimmunity directed to the central nervous system. *Neuron*. 2009;64:123-132. doi:10.1016/j.neuron.2009.09.015.
7. Feldmann M, Steinman L. Design of effective immunotherapy for human autoimmunity. *Nature*. 2005;435:612-619. doi:10.1038/nature03727.
8. Steinman L. Immune therapy for autoimmune diseases. *Science*. 2004;305:212-216. doi:10.1126/science.1099896.
9. Steinman L. Assessment of animal models for MS and demyelinating disease in the design of rational therapy. *Neuron*. 1999;24:511-514. doi:10.1016/s0896-6273(00)81107-1.
10. Steinman L. Multiple sclerosis: a coordinated immunological attack against myelin in the central nervous system. *Cell*. 1996;85:299-302. doi:10.1016/s0092-8674(00)81107-1.
11. Hafler DA, Saadeh MG, Kuchroo VK, Milford E, Steinman L. TCR usage in human and experimental demyelinating disease. *Immunol Today*. 1996;17:152-159. doi:10.1016/0167-5699(96)80611-6.
12. Steinman L. Autoimmune disease. *Sci Am*. 1993;269:106-114. doi:10.1038/scientificamerican0993-106.
13. Wraith DC, McDevitt HO, Steinman L, Acha-Orbea H. T cell recognition as the target for immune intervention in autoimmune disease. *Cell*. 1989;57:709-715. doi:10.1016/0092-8674(89)90786-1.
14. Baecher-Allan C, Kaskow BJ, Weiner HL. Multiple sclerosis: mechanisms and immunotherapy. *Neuron*. 2018;97:742-768. doi:10.1016/j.neuron.2018.01.021.
15. Chitnis T, Weiner HL. CNS inflammation and neurodegeneration. *J Clin Invest*. 2017;127:3577-3587. doi:10.1172/JCI90609.
16. Monsonego A, Weiner HL. Immunotherapeutic approaches to alzheimer's disease. *Science*. 2003;302:834-838. doi:10.1126/science.1088469.
17. Weiner HL, Selkoe DJ. Inflammation and therapeutic vaccination in CNS diseases. *Nature*. 2002;420:879-884. doi:10.1038/nature01325.
18. Smith DR, Weiner HL. Immunologic aspects of neurologic and neuromuscular diseases. *J Am Med Assoc*. 1997;278:1956-1961.
19. Weiner HL. Oral tolerance: immune mechanisms and treatment of autoimmune diseases. *Immunol Today*. 1997;18:335-343. doi:10.1016/s0167-5699(97)01053-0.
20. Alves de Lima K, Rustenhoven J, Kipnis J. Meningeal immunity and its function in maintenance of the central nervous system in health and disease. *Annu Rev Immunol*. 2020;38:597-620. doi:10.1146/annurev-immunol-102319-103410.

21. Medzhitov R, Schneider DS, Soares MP. Disease tolerance as a defense strategy. *Science*. 2012;335:936-941. doi:10.1126/science.1214935.
22. Daneman R, Prat A. The blood-brain barrier. *Cold Spring Harb Perspect Biol*. 2015;7:a020412. doi:10.1101/cshperspect.a020412.
23. Daneman R, et al. The mouse blood-brain barrier transcriptome: a new resource for understanding the development and function of brain endothelial cells. *PloS One*. 2010;5:e13741. doi:10.1371/journal.pone.0013741.
24. Pulido RS, et al. Neuronal activity regulates blood-brain barrier efflux transport through endothelial circadian genes. *Neuron*. 2020;108(5):937-952.e7. doi:10.1016/j.neuron.2020.09.002.
25. Carson MJ, Doose JM, Melchior B, Schmid CD, Ploix CC. CNS immune privilege: hiding in plain sight. *Immunol Rev*. 2006;213:48-65. doi:10.1111/j.1600-065X.2006.00441.x.
26. Aspelund A, et al. A dural lymphatic vascular system that drains brain interstitial fluid and macromolecules. *J Exp Med*. 2015;212:991-999. doi:10.1084/jem.20142290.
27. Louveau A, Harris TH, Kipnis J. Revisiting the mechanisms of CNS immune privilege. *Trends Immunol*. 2015;36:569-577. doi:10.1016/j.it.2015.08.006.
28. Zhao S, Zhu W, Xue S, Han D. Testicular defense systems: immune privilege and innate immunity. *Cell Mol Immunol*. 2014;11:428-437, doi:10.1038/cmi.2014.38.
29. Louveau A, et al. CNS lymphatic drainage and neuroinflammation are regulated by meningeal lymphatic vasculature. *Nat Neurosci*. 2018;21:1380-1391. doi:10.1038/s41593-018-0227-9.
30. Da Mesquita S, Fu Z, Kipnis J. The meningeal lymphatic system: a new player in neurophysiology. *Neuron*. 2018;100:375-388. doi:10.1016/j.neuron.2018.09.022.
31. Cserr HF, Harling-Berg CJ, Knopf PM. Drainage of brain extracellular fluid into blood and deep cervical lymph and its immunological significance. *Brain Pathol*. 1992;2:269-276. doi:10.1111/j.1750-3639.1992.tb00703.x.
32. Kaplan HJ, Streilein JW. Immune response to immunization via the anterior chamber of the eye. II. An analysis of F1 lymphocyte-induced immune deviation. *J Immunol*. 1978;120:689-693.
33. Taylor AW. Ocular immune privilege. *Eye*. 2009;23:1885-1889. doi:10.1038/eye.2008.382.
34. Lin H-H, et al. The macrophage F4/80 receptor is required for the induction of antigen-specific efferent regulatory T cells in peripheral tolerance. *J Exp Med*. 2005;201:1615-1625. doi:10.1084/jem.20042307.
35. D'Orazio TJ, Niederkorn JY. Splenic B cells are required for tolerogenic antigen presentation in the induction of anterior chamber-associated immune deviation (ACAID). *Immunology*. 1998;95:47-55. doi:10.1046/j.1365-2567.1998.00581.x.
36. Zhou R, et al. The living eye "disarms" uncommitted autoreactive T cells by converting them to Foxp3(+) regulatory cells following local antigen recognition. *J Immunol*. 2012;188:1742-1750. doi:10.4049/jimmunol.1102415.
37. Ryu Y-H, Kim J-C. Expression of indoleamine 2,3-dioxygenase in human corneal cells as a local immunosuppressive factor. *Invest Ophthalmol Vis Sci*. 2007;48:4148-4152. doi:10.1167/iovs.05-1336.
38. Taylor AW, Streilein JW, Cousins SW. Identification of alpha-melanocyte stimulating hormone as a potential immunosuppressive factor in aqueous humor. *Curr Eye Res*. 1992;11:1199-1206. doi:10.3109/02713689208999545.
39. Taylor AW, Streilein JW, Cousins SW. Immunoreactive vasoactive intestinal peptide contributes to the immunosuppressive activity of normal aqueous humor. *J Immunol*. 1994;153:1080-1086.
40. Taylor AW, Yee DG. Somatostatin is an immunosuppressive factor in aqueous humor. *Invest Ophthalmol Vis Sci*. 2003;44:2644-2649. doi:10.1167/iovs.02-1216.
41. Masli S, Turpie B, Hecker KH, Streilein JW. Expression of thrombospondin in TGFbeta-treated APCs and its relevance to their immune deviation-promoting properties. *J Immunol*. 2002;168:2264-2273. doi:10.4049/jimmunol.168.5.2264.
42. Griffith TS, Brunner T, Fletcher SM, Green DR, Ferguson TA. Fas ligand-induced apoptosis as a mechanism of immune privilege. *Science*. 1995;270:1189-1192. doi:10.1126/science.270.5239.1189.
43. Benhar I, London A, Schwartz M. The privileged immunity of immune privileged organs: the case of the eye. *Front Immunol*. 2012;3:296. doi:10.3389/fimmu.2012.00296.
44. Mrdjen D, et al. High-dimensional single-cell mapping of central nervous system immune cells reveals distinct myeloid subsets in health, aging, and disease. *Immunity*. 2018;48:380-395.e386. doi:10.1016/j.immuni.2018.01.011.
45. Van Hove H, et al. A single-cell atlas of mouse brain macrophages reveals unique transcriptional identities shaped by ontogeny and tissue environment. *Nat Neurosci*. 2019;22(6):1021-1035.
46. Papadopoulos Z, Herz J, Kipnis J. Meningeal lymphatics: from anatomy to central nervous system immune surveillance. *J Immunol*. 2020;204:286-293. doi:10.4049/jimmunol.1900838.
47. Liddelow SA, Marsh SE, Stevens B. Microglia and astrocytes in disease: dynamic duo or partners in crime? *Trends Immunol*. 2020;41:820-835. doi:10.1016/j.it.2020.07.006.
48. Kaur C, Hao AJ, Wu CH, Ling EA. Origin of microglia. *Microsc Res Tech*. 2001;54:2-9. doi:10.1002/jemt.1114 [pii]10.1002/jemt.1114.
49. Alliot F, Godin I, Pessac B. Microglia derive from progenitors, originating from the yolk sac, and which proliferate in the brain. *Brain Res Dev Brain Res*. 1999;117:145-152. doi:S0165380699001133 [pii].
50. Rio-Hortega D. *Cytology and Cellular Pathology of the Nervous System*. Microglia P. B. Hoebaer; 1932:483-534.
51. Davalos D, et al. ATP mediates rapid microglial response to local brain injury in vivo. *Nat Neurosci*. 2005;8:752-758. doi:nn1472 [pii]10.1038/nn1472.
52. Nimmerjahn A, Kirchhoff F, Helmchen F. Resting microglial cells are highly dynamic surveillants of brain parenchyma in vivo. *Science*. 2005;308:1314-1318. doi:1110647 [pii]10.1126/science.1110647.
53. Lavin Y, Mortha A, Rahman A, Merad M. Regulation of macrophage development and function in peripheral tissues. *Nat Rev Immunol*. 2015;15:731-744. doi:10.1038/nri3920.
54. Lavin Y, Merad M. Macrophages: gatekeepers of tissue integrity. *Cancer Immunol Res*. 2013;1:201-209. doi:10.1158/2326-6066.CIR-13-0117.
55. Hashimoto D, et al. Tissue resident macrophages self-maintain locally throughout adult life with minimal contribution from circulating monocytes. *Immunity*. 2013;38:792-804. doi:10.1016/j.immuni.2013.04.004.
56. Butovsky O, Weiner HL. Microglial signatures and their role in health and disease. *Nat Rev Neurosci*. 2018;19:622-635. doi:10.1038/s41583-018-0057-5.
57. Kierdorf K, Prinz M. Microglia in steady state. *J Clin Invest*. 2017;127(9):3201-3209. doi:10.1172/JCI90602.
58. Derecki NC, Katzmarski N, Kipnis J, Meyer-Luehmann M. Microglia as a critical player in both developmental and late-life CNS pathologies. *Acta Neuropathol*. 2014;128:333-345. doi:10.1007/s00401-014-1321-z.
59. Schwartz M, Kipnis J, Rivest S, Prat A. How do immune cells support and shape the brain in health, disease, and aging? *J Neurosci*. 2013;33:17587-17596. doi:10.1523/JNEUROSCI.3241-13.2013.
60. Norden DM, Godbout JP. Review: microglia of the aged brain – primed to be activated and resistant to regulation. *Neuropathol Appl Neurobiol*. 2013;39:19-34. doi:10.1111/j.1365-2990.2012.01306.x.
61. Derecki NC, Cronk JC, Kipnis J. The role of microglia in brain maintenance: implications for Rett syndrome. *Trends Immunol*. 2013;34:144-150. doi:10.1016/j.it.2012.10.002.
62. Cronk JC, Kipnis J. Microglia – the brain's busy bees. *F1000Prime Rep*. 2013;5:53. doi:10.12703/P5-53.
63. Bilbo S, Stevens B. Microglia: the brain's first responders. *Cerebrum*. 2017;2017:cer-14-17.
64. Wu LJ, Stevens B, Duan S, MacVicar BA. Microglia in neuronal circuits. *Neural Plast*. 2013;2013:586426. doi:10.1155/2013/586426.
65. Stephan AH, Barres BA, Stevens B. The complement system: an unexpected role in synaptic pruning during development and disease. *Annu Rev Neurosci*. 2012;35:369-389. doi:10.1146/annurev-neuro-061010-113810.
66. Schafer DP, et al. Microglia sculpt postnatal neural circuits in an activity and complement-dependent manner. *Neuron*. 2012;74:691-705. doi:10.1016/j.neuron.2012.03.026S0896-6273(12)00334-0 [pii].
67. Tremblay ME, et al. The role of microglia in the healthy brain. *J Neurosci*. 2011;31:16064-16069. doi:10.1523/JNEUROSCI.4158-11.2011.
68. Sierra A, et al. Microglia shape adult hippocampal neurogenesis through apoptosis-coupled phagocytosis. *Cell Stem Cell*. 2010;7:483-495. doi:10.1016/j.stem.2010.08.014.
69. Shi Y, et al. ApoE4 markedly exacerbates tau-mediated neurodegeneration in a mouse model of tauopathy. *Nature*. 2017;549:523-527. doi:10.1038/nature24016.
70. Hong S, et al. Complement and microglia mediate early synapse loss in alzheimer mouse models. *Science*. 2016;352:712-716. doi:10.1126/science.aad8373.
71. Keren-Shaul H, et al. A unique microglia type associated with restricting development of alzheimer's disease. *Cell*. 2017;169:1276-1290.e17. doi:10.1016/j.cell.2017.05.018.
72. Ulrich JD, Ulland TK, Colonna M, Holtzman DM. Elucidating the role of TREM2 in alzheimer's disease. *Neuron*. 2017;94:237-248. doi:10.1016/j.neuron.2017.02.042.
73. Ulland TK, et al. TREM2 maintains microglial metabolic fitness in alzheimer's disease. *Cell*. 2017;170:649-663.e13. doi:10.1016/j.cell.2017.07.023.
74. Wang Y, et al. TREM2 lipid sensing sustains the microglial response in an alzheimer's disease model. *Cell*. 2015;160:1061-1071. doi:10.1016/j.cell.2015.01.049.
75. Elmore MR, et al. Colony-stimulating factor 1 receptor signaling is necessary for microglia viability, unmasking a microglia progenitor cell in the adult brain. *Neuron*. 2014;82:380-397. doi:10.1016/j.neuron.2014.02.040.
76. Mildner A, et al. Microglia in the adult brain arise from Ly-6ChiCCR2+ monocytes only under defined host conditions. *Nat Neurosci*. 2007;10:1544-1553. doi:nn2015 [pii]10.1038/nn2015.
77. Ajami B, Bennett JL, Krieger C, Tetzlaff W, Rossi FMV. Local self-renewal can sustain CNS microglia maintenance and function throughout adult life. *Nat Neurosci*. 2007;10:1538-1543. doi:10.1038/nn2014.
78. Norris GT, et al. Neuronal integrity and complement control synaptic material clearance by microglia after CNS injury. *J Exp Med*. 2018;215:1789-1801. doi:10.1084/jem.20172244.
79. Cronk JC, et al. Peripherally derived macrophages can engraft the brain independent of irradiation and maintain an identity distinct from microglia. *J Exp Med*. 2018;215:1627-1647. doi:10.1084/jem.20180247.
80. Ginhoux F, et al. Fate mapping analysis reveals that adult microglia derive from primitive macrophages. *Science*. 2010;330:841-845. doi:10.1126/science.1194637.
81. Bennett FC, et al. A combination of ontogeny and CNS environment establishes microglial identity. *Neuron*. 2018;98:1170-1183.e78. doi:10.1016/j.neuron.2018.05.014.
82. Schafer DP, Stevens B. Brains, blood, and guts: MeCP2 regulates microglia, monocytes, and peripheral macrophages. *Immunity*. 2015;42:600-602. doi:10.1016/j.immuni.2015.04.002.
83. Steinman L. New targets for treatment of multiple sclerosis. *J Neurol Sci*. 2008;274:1-4. doi:S0022-510X(08)00323-7 [pii]10.1016/j.jns.2008.06.040.
84. Cleveland DW, Rothstein JD. From Charcot to Lou Gehrig: deciphering selective motor neuron death in ALS. *Nat Rev Neurosci*. 2001;2:806-819. doi:10.1038/3509756535097565 [pii].
85. Steinman L, Zamvil SS. Beginning of the end of two-stage theory purporting that inflammation then degeneration explains pathogenesis of progressive multiple sclerosis. *Curr Opin Neurol*. 2016;29:340-344. doi:10.1097/WCO.0000000000000317.
86. Steinman L. No quiet surrender: molecular guardians in multiple sclerosis brain. *J Clin Invest*. 2015;125:1371-1378. doi:10.1172/JCI74255.
87. Steinman L. A molecular trio in relapse and remission in multiple sclerosis. *Nat Rev Immunol*. 2009;9:440-447. doi:10.1038/nri2548.
88. Seamons A, Perchellet A, Goverman J. Immune tolerance to myelin proteins. *Immunol Res*. 2003;28:201-221. doi:10.1385/IR:28:3:201.
89. Robertson N. Multiple Sclerosis. Clinical and pathogenetic basis. *J Neurol Neurosurg Psychiatry*. 1997;63(6):815.
90. Serafini B, Rosicarelli B, Magliozzi R, Stigliano E, Aloisi F. Detection of ectopic B-cell follicles with germinal centers in the meninges of patients with secondary progressive multiple sclerosis. *Brain Pathol*. 2004;14:164-174. doi:10.1111/j.1750-3639.2004.tb00049.x.
91. Mahad DH, Trapp BD, Lassmann H. Pathological mechanisms in progressive multiple sclerosis. *Lancet Neurol*. 2015;14:183-193. doi:10.1016/S1474-4422(14)70256-X.
92. Smith ME, Eller NL, McFarland HF, Racke MK, Raine CS. Age dependence of clinical and pathological manifestations of autoimmune demyelination. Implications for multiple sclerosis. *Am J Pathol*. 1999;155:1147-1161. doi:10.1016/S0002-9440(10)65218-2.
93. Teuscher C, et al. Gender, age, and season at immunization uniquely influence the genetic control of susceptibility to histopathological lesions and clinical signs of experimental allergic encephalomyelitis: implications for the genetics of multiple sclerosis. *Am J Pathol*. 2004;165:1593-1602. doi:10.1016/S0002-9440(10)63416-5.
94. Berer K, et al. Gut microbiota from multiple sclerosis patients enables spontaneous autoimmune encephalomyelitis in mice. *Proc Natl Acad Sci USA*. 2017;114:10719-10724. doi:10.1073/pnas.1711233114.
95. Stromnes IM, Goverman JM. Active induction of experimental allergic encephalomyelitis. *Nat Protoc*. 2006;1:1810-1819. doi:10.1038/nprot.2006.285.
96. Stromnes IM, Goverman JM. Passive induction of experimental allergic encephalomyelitis. *Nat Protoc*. 2006;1:1952-1960. doi:10.1038/nprot.2006.284.
97. Brok HP, et al. Myelin/oligodendrocyte glycoprotein-induced autoimmune encephalomyelitis in common marmosets: the encephalitogenic T cell epitope pMOG24-36 is presented by a monomorphic MHC class II molecule. *J Immunol*. 2000;165:1093-1101. doi:10.4049/jimmunol.165.2.1093.
98. Schreiner B, Heppner FL, Becher B. Modeling multiple sclerosis in laboratory animals. *Semin Immunopathol*. 2009;31:479-495. doi:10.1007/s00281-009-0181-4.
99. Goverman J. Autoimmune T cell responses in the central nervous system. *Nat Rev Immunol*. 2009;9:393-407. doi:10.1038/nri2550.
100. Storch MK, et al. Autoimmunity to myelin oligodendrocyte glycoprotein in rats mimics the spectrum of multiple sclerosis pathology. *Brain Pathol*. 1998;8:681-694.
101. Muller DM, Pender MP, Greer JM. A neuropathological analysis of experimental autoimmune encephalomyelitis with predominant brain stem and cerebellar involvement and differences between active and passive induction. *Acta Neuropathol*. 2000;100:174-182. doi:10.1007/s004019900163.
102. Wensky AK, et al. IFN-gamma determines distinct clinical outcomes in autoimmune encephalomyelitis. *J Immunol*. 2005;174:1416-1423. doi:10.4049/jimmunol.174.3.1416.
103. Madsen LS, et al. A humanized model for multiple sclerosis using HLA-DR2 and a human T-cell receptor. *Nat Genet*. 1999;23:343-347. doi:10.1038/15525.
104. Quandt JA, et al. Unique clinical and pathological features in HLA-DRB1*0401-restricted MBP 111-129-specific humanized TCR transgenic mice. *J Exp Med*. 2004;200:223-234. doi:10.1084/jem.20030994.

105. Ellmerich S, et al. High incidence of spontaneous disease in an HLA-DR15 and TCR transgenic multiple sclerosis model. *J Immunol.* 2005;174:1938-1946. doi:10.4049/jimmunol.174.4.1938.
106. Bettelli E, Baeten D, Jäger A, Sobel RA, Kuchroo VK. Myelin oligodendrocyte glycoprotein-specific T and B cells cooperate to induce a Devic-like disease in mice. *J Clin Invest.* 2006;116:2393-2402. doi:10.1172/JCI28334.
107. Baron JL, Madri JA, Ruddle NH, Hashim G, Janeway CA. Surface expression of alpha 4 integrin by CD4 T cells is required for their entry into brain parenchyma. *J Exp Med.* 1993;177:57-68. doi:10.1084/jem.177.1.57.
108. Greter M, et al. Dendritic cells permit immune invasion of the CNS in an animal model of multiple sclerosis. *Nat Med.* 2005;11:328-334. doi:10.1038/nm1197.
109. Fife BT, Huffnagle GB, Kuziel WA, Karpus WJ. CC chemokine receptor 2 is critical for induction of experimental autoimmune encephalomyelitis. *J Exp Med.* 2000;192:899-905. doi:10.1084/jem.192.6.899.
110. Izikson L, Klein RS, Charo IF, Weiner HL, Luster AD. Resistance to experimental autoimmune encephalomyelitis in mice lacking the CC chemokine receptor (CCR)2. *J Exp Med.* 2000;192:1075-1080. doi:10.1084/jem.192.7.1075.
111. Rubino SJ, et al. Acute microglia ablation induces neurodegeneration in the somatosensory system. *Nat Commun.* 2018;9:4578. doi:10.1038/s41467-018-05929-4.
112. Becher B, Spath S, Goverman J. Cytokine networks in neuroinflammation. *Nat Rev Immunol.* 2017;17:49-59. doi:10.1038/nri.2016.123.
113. De Keyser J, Zeinstra E, Frohman E. Are astrocytes central players in the pathophysiology of multiple sclerosis? *Arch Neurol.* 2003;60:132-136. doi:10.1001/archneur.60.1.132.
114. Panitch HS, Hirsch RL, Haley AS, Johnson KP. Exacerbations of multiple sclerosis in patients treated with gamma interferon. *Lancet.* 1987;1:893-895. doi:10.1016/s0140-6736(87)92863-7.
115. Cua DJ, et al. Interleukin-23 rather than interleukin-12 is the critical cytokine for autoimmune inflammation of the brain. *Nature.* 2003;421:744-748. doi:10.1038/nature01355.
116. Lock C, et al. Gene-microarray analysis of multiple sclerosis lesions yields new targets validated in autoimmune encephalomyelitis. *Nat Med.* 2002;8:500-508. doi:10.1038/nm0502-500.
117. Haak S, et al. IL-17A and IL-17F do not contribute vitally to autoimmune neuro-inflammation in mice. *J Clin Invest.* 2009;119:61-69. doi:10.1172/JCI35997.
118. Luger D, et al. Either a Th17 or a Th1 effector response can drive autoimmunity: conditions of disease induction affect dominant effector category. *J Exp Med.* 2008;205:799-810. doi:10.1084/jem.20071258.
119. Codarri L, et al. RORγt drives production of the cytokine GM-CSF in helper T cells, which is essential for the effector phase of autoimmune neuroinflammation. *Nat Immunol.* 2011;12:560-567. doi:10.1038/ni.2027.
120. McQualter JL, et al. Granulocyte macrophage colony-stimulating factor: a new putative therapeutic target in multiple sclerosis. *J Exp Med.* 2001;194:873-882. doi:10.1084/jem.194.7.873.
121. Croxford AL, et al. The cytokine GM-CSF drives the inflammatory signature of CCR2+ monocytes and licenses autoimmunity. *Immunity.* 2015;43:502-514. doi:10.1016/j.immuni.2015.08.010.
122. Hövelmeyer N, et al. Apoptosis of oligodendrocytes via Fas and TNF-R1 is a key event in the induction of experimental autoimmune encephalomyelitis. *J Immunol.* 2005;175:5875-5884. doi:10.4049/jimmunol.175.9.5875.
123. Rosenman SJ, Shrikant P, Dubb L, Benveniste EN, Ransohoff RM. Cytokine-induced expression of vascular cell adhesion molecule-1 (VCAM-1) by astrocytes and astrocytoma cell lines. *J Immunol.* 1995;154:1888-1899.
124. Benveniste EN, Sparacio SM, Bethea JR. Tumor necrosis factor-alpha enhances interferon-gamma-mediated class II antigen expression on astrocytes. *J Neuroimmunol.* 1989;25:209-219. doi:10.1016/0165-5728(89)90139-2.
125. Huseby ES, et al. A pathogenic role for myelin-specific CD8(+) T cells in a model for multiple sclerosis. *J Exp Med.* 2001;194:669-676. doi:10.1084/jem.194.5.669.
126. Goverman J, Perchellet A, Huseby ES. The role of CD8(+) T cells in multiple sclerosis and its animal models. *Curr Drug Targets Inflamm Allergy.* 2005;4:239-245.
127. Friese MA, Fugger L. Autoreactive CD8+ T cells in multiple sclerosis: a new target for therapy? *Brain.* 2005;128:1747-1763. doi:10.1093/brain/awh578.
128. Friese MA, et al. Opposing effects of HLA class I molecules in tuning autoreactive CD8+ T cells in multiple sclerosis. *Nat Med.* 2008;14:1227-1235. doi:10.1038/nm.1881.
129. Kirby L, et al. Oligodendrocyte precursor cells present antigen and are cytotoxic targets in inflammatory demyelination. *Nat Commun.* 2019;10:3887. doi:10.1038/s41467-019-11638-3.
130. Zang YCQ, et al. Increased CD8+ cytotoxic T cell responses to myelin basic protein in multiple sclerosis. *J Immunol.* 2004;172:5120-5127. doi:10.4049/jimmunol.172.8.5120.
131. Hauser SL, et al. B-cell depletion with rituximab in relapsing-remitting multiple sclerosis. *N Engl J Med.* 2008;358:676-688. doi:10.1056/NEJMoa0706383.
132. Magliozzi R, et al. Meningeal B-cell follicles in secondary progressive multiple sclerosis associate with early onset of disease and severe cortical pathology. *Brain.* 2007;130:1089-1104. doi:10.1093/brain/awm038.
133. Iglesias A, Bauer J, Litzenburger T, Schubart A, Linington C. T-and B-cell responses to myelin oligodendrocyte glycoprotein in experimental autoimmune encephalomyelitis and multiple sclerosis. *Glia.* 2001;36:220-234. doi:10.1002/glia.1111.
134. Matsushita T, Yanaba K, Bouaziz J-D, Fujimoto M, Tedder TF. Regulatory B cells inhibit EAE initiation in mice while other B cells promote disease progression. *J Clin Invest.* 2008;118:3420-3430. doi:10.1172/JCI36030.
135. Mann MK, Maresz K, Shriver LP, Tan Y, Dittel BN. B cell regulation of CD4+CD25+ T regulatory cells and IL-10 via B7 is essential for recovery from experimental autoimmune encephalomyelitis. *J Immunol.* 2007;178:3447-3456. doi:10.4049/jimmunol.178.6.3447.
136. Fillatreau S, Sweenie CH, McGeachy MJ, Gray D, Anderton SM. B cells regulate autoimmunity by provision of IL-10. *Nat Immunol.* 2002;3:944-950. doi:10.1038/ni833.
137. Rojas OL, et al. Recirculating intestinal IgA-producing cells regulate neuroinflammation via IL-10. *Cell.* 2019;176:610-624. e18. doi:10.1016/j.cell.2018.11.035.
138. Steinman L. Immunology of relapse and remission in multiple sclerosis. *Annu Rev Immunol.* 2014;32:257-281. doi:10.1146/annurev-immunol-032713-120227.
139. Steinman L. Development of therapies for autoimmune disease at Stanford: a tale of multiple shots and one goal. *Immunol Res.* 2014;58:307-314. doi:10.1007/s12026-014-8509-0.
140. Rudick R, Polman C, Clifford D, Miller D, Steinman L. Natalizumab: bench to bedside and beyond. *JAMA Neurol.* 2013;70:172-182. doi:10.1001/jamaneurol.2013.598.
141. Langer-Gould A, Atlas SW, Green AJ, Bollen AW, Pelletier D. Progressive multifocal leukoencephalopathy in a patient treated with natalizumab. *N Engl J Med.* 2005;353:375-381. doi:10.1056/NEJMoa051847.
142. Major EO. Progressive multifocal leukoencephalopathy in patients on immunomodulatory therapies. *Annu Rev Med.* 2010;61:35-47. doi:10.1146/annurev.med.080708.082655.
143. Koyuncu OO, Hogue IB, Enquist LW. Virus infections in the nervous system. *Cell Host Microbe.* 2013;13:379-393. doi:10.1016/j.chom.2013.03.010.
144. McGavern DB, Kang SS. Illuminating viral infections in the nervous system. *Nat Rev Immunol.* 2011;11:318-329. doi:10.1038/nri2971.
145. Gralinski LE, Ashley SL, Dixon SD, Spindler KR. Mouse adenovirus type 1-induced breakdown of the blood-brain barrier. *J Virol.* 2009;83:9398-9410. doi:10.1128/JVI.00954-09.
146. Verma S, Kumar M, Gurjav U, Lum S, Nerurkar VR. Reversal of West Nile virus-induced blood-brain barrier disruption and tight junction proteins degradation by matrix metalloproteinases inhibitor. *Virology.* 2010;397:130-138. doi:10.1016/j.virol.2009.10.036.
147. Xu Z, Waeckerlin R, Urbanowski MD, van Marle G, Hobman TC. West Nile virus infection causes endocytosis of a specific subset of tight junction membrane proteins. *PloS One.* 2012;7:e37886. doi:10.1371/journal.pone.0037886.
148. Daniels BP, et al. Viral pathogen-associated molecular patterns regulate blood-brain barrier integrity via competing innate cytokine signals. *mBio.* 2014;5:e01476. doi:10.1128/mBio.01476-14.
149. Swanson PA, McGavern DB. Viral diseases of the central nervous system. *Curr Opin Virol.* 2015;11:44-54. doi:10.1016/j.coviro.2014.12.009.
150. Kaul M, Garden GA, Lipton SA. Pathways to neuronal injury and apoptosis in HIV-associated dementia. *Nature.* 2001;410:988-994. doi:10.1038/35073667.
151. Clay CC, et al. Neuroinvasion of fluorescein-positive monocytes in acute simian immunodeficiency virus infection. *J Virol.* 2007;81:12040-12048. doi:10.1128/JVI.00133-07.
152. Alexaki A, Wigdahl B. HIV-1 infection of bone marrow hematopoietic progenitor cells and their role in trafficking and viral dissemination. *PLoS Pathog.* 2008;4:e1000215. doi:10.1371/journal.ppat.1000215.
153. Monaco MC, Atwood WJ, Gravell M, Tornatore CS, Major EO. JC virus infection of hematopoietic progenitor cells, primary B lymphocytes, and tonsillar stromal cells: implications for viral latency. *J Virol.* 1996;70:7004-7012. doi:10.1128/JVI.70.10.7004-7012.1996.
154. Chapagain ML, Nerurkar VR. Human polyomavirus JC (JCV) infection of human B lymphocytes: a possible mechanism for JCV transmigration across the blood-brain barrier. *J Infect Dis.* 2010;202:184-191. doi:10.1086/653823.
155. Mori I, Nishiyama Y, Yokochi T, Kimura Y. Olfactory transmission of neurotropic viruses. *J Neurovirol.* 2005;11:129-137. doi:10.1080/13550280590922793.
156. Meinhardt J, et al. Olfactory transmucosal SARS-CoV-2 invasion as a port of central nervous system entry in individuals with COVID-19. *Nat Neurosci.* 2020;24(2):168-175. doi:10.1038/s41593-020-00758-5.
157. Griffin DE. Immune responses to RNA-virus infections of the CNS. *Nat Rev Immunol.* 2003;3:493-502. doi:10.1038/nri1105.
158. Tsai TF, et al. Epidemiological aspects of a St. Louis encephalitis outbreak in Harris county, Texas, 1986. *J Infect Dis.* 1988;157:351-356. doi:10.1093/infdis/157.2.351.
159. Balfour HH, Siem RA, Bauer H, Quie PG. California arbovirus (La Crosse) infections. I. Clinical and laboratory findings in 66 children with meningoencephalitis. *Pediatrics.* 1973;52:680-691.
160. Skogland JE. An unusual form of lymphocytic choriomeningitis. *Arch NeurPsych.* 1939;42:507. doi:10.1001/archneurpsyc.1939.02270210145007.
161. Hollander H, Stringari S. Human immunodeficiency virus-associated meningitis. Clinical course and correlations. *Am J Med.* 1987;83:813-816. doi:10.1016/0002-9343(87)90635-8.
162. Liu S-C, et al. Different cytokine levels in enterovirus meningitis and encephalitis. *Infect Dis Clin Pract.* 2005;13:241-246. doi:10.1097/01.idc.0000175819.56238.7e.
163. Kim JV, Kang SS, Dustin ML, McGavern DB. Myelomonocytic cell recruitment causes fatal CNS vascular injury during acute viral meningitis. *Nature.* 2009;457:191-195. doi:10.1038/nature07591.
164. Nayak D, et al. Type I interferon programs innate myeloid dynamics and gene expression in the virally infected nervous system. *PLoS Pathog.* 2013;9:e1003395. doi:10.1371/journal.ppat.1003395.
165. Romero JR, Newland JG. Viral meningitis and encephalitis: traditional and emerging viral agents. *Semin Pediatr Infect Dis.* 2003;14:72-82. doi:10.1053/spid.2003.127223.
166. Chambers TJ, Diamond MS. Pathogenesis of flavivirus encephalitis. *Adv Virus Res.* 2003;60:273-342. doi:10.1016/s0065-3527(03)60008-4.
167. Eldadah AH, Nathanson N. Pathogenesis of West Nile Virus encepahlitis in mice and rats. II. Virus multiplication, evolution of immunofluorescence, and development of histological lesions in the brain. *Am J Epidemiol.* 1967;86:776-790. doi:10.1093/oxfordjournals.aje.a120786.
168. Cho H, Diamond MS. Immune responses to West Nile virus infection in the central nervous system. *Viruses.* 2012;4:3812-3830. doi:10.3390/v4123812.
169. Omalu BI, Shakir AA, Wang G, Lipkin WI, Wiley CA. Fatal fulminant pan-meningo-polioencephalitis due to West Nile virus. *Brain Pathol.* 2003;13:465-472. doi:10.1111/j.1750-3639.2003.tb00477.x.
170. Kumar M, Verma S, Nerurkar VR. Pro-inflammatory cytokines derived from West Nile virus (WNV)-infected SK-N-SH cells mediate neuroinflammatory markers and neuronal death. *J Neuroinflammation.* 2010;7:73. doi:10.1186/1742-2094-7-73.
171. Grill MF, Price RW. Central nervous system HIV-1 infection. *Handb Clin Neurol.* 2014;123:487-505. doi:10.1016/B978-0-444-53488-0.00023-7.
172. Murphy FA. Rabies pathogenesis. *Arch Virol.* 1977;54:279-297. doi:10.1007/BF01314774.
173. Fu ZF, et al. Differential effects of rabies and borna disease viruses on immediate-early-and late-response gene expression in brain tissues. *J Virol.* 1993;67:6674-6681. doi:10.1128/JVI.67.11.6674-6681.1993.
174. Ceccaldi PE, Fillion MP, Ermine A, Tsiang H, Fillion G. Rabies virus selectively alters 5-HT1 receptor subtypes in rat brain. *Eur J Pharmacol.* 1993;245:129-138. doi:10.1016/0922-4106(93)90120-x.
175. Vidy A, Chelbi-Alix M, Blondel D. Rabies virus P protein interacts with STAT1 and inhibits interferon signal transduction pathways. *J Virol.* 2005;79:14411-14420. doi:10.1128/JVI.79.22.14411-14420.2005.
176. Aurelius E, Johansson B, Sköldenberg B, Forsgren M. Encephalitis in immunocompetent patients due to herpes simplex virus type 1 or 2 as determined by type-specific polymerase chain reaction and antibody assays of cerebrospinal fluid. *J Med Virol.* 1993;39:179-186. doi:10.1002/jmv.1890390302.
177. Kleinschmidt-DeMasters BK, Gilden DH. The expanding spectrum of herpesvirus infections of the nervous system. *Brain Pathol.* 2001;11:440-451. doi:10.1111/j.1750-3639.2001.tb00413.x.
178. Bodian D. Histopathologic basis of clinical findings in poliomyelitis. *Am J Med.* 1949;6:563-578. doi:10.1016/0002-9343(49)90130-8.
179. Solomon T, et al. Poliomyelitis-like illness due to Japanese encephalitis virus. *Lancet.* 1998;351:1094-1097. doi:10.1016/S0140-6736(97)07509-0.
180. Samuel MA, Wang H, Siddharthan V, Morrey JD, Diamond MS. Axonal transport mediates West Nile virus entry into the central nervous system and induces acute flaccid paralysis. *Proc Natl Acad Sci USA.* 2007;104:17140-17145. doi:10.1073/pnas.0705837104.
181. Mitrabhakdi E, et al. Difference in neuropathogenetic mechanisms in human furious and paralytic rabies. *J Neurol Sci.* 2005;238:3-10. doi:10.1016/j.jns.2005.05.004.
182. Griffin DE, Metcalf T. Clearance of virus infection from the CNS. *Curr Opin Virol.* 2011;1:216-221. doi:10.1016/j.coviro.2011.05.021.
183. Nair S, Diamond MS. Innate immune interactions within the central nervous system modulate pathogenesis of viral infections. *Curr Opin Immunol.* 2015;36:47-53. doi:10.1016/j.coi.2015.06.011.
184. Ramos HJ, et al. IL-1β signaling promotes CNS-intrinsic immune control of West Nile virus infection. *PLoS Pathog.* 2012;8:e1003039. doi:10.1371/journal.ppat.1003039.
185. Nair S, et al. Interferon regulatory factor-1 protects from fatal neurotropic infection with vesicular stomatitis virus by specific inhibition of viral replication in neurons. *PLoS Pathog.* 2014;10:e1003999. doi:10.1371/journal.ppat.1003999.

186. Orvedahl A, et al. Autophagy protects against Sindbis virus infection of the central nervous system. *Cell Host Microbe.* 2010;7:115-127. doi:10.1016/j.chom.2010.01.007.
187. Liang XH, et al. Protection against fatal Sindbis virus encephalitis by beclin, a novel Bcl-2-interacting protein. *J Virol.* 1998;72:8586-8596. doi:10.1128/JVI.72.11.8586-8596.1998.
188. Richner JM, et al. Age-dependent cell trafficking defects in draining lymph nodes impair adaptive immunity and control of West Nile virus infection. *PLoS Pathog.* 2015;11:e1005027. doi:10.1371/journal.ppat.1005027.
189. Shrestha B, Samuel MA, Diamond MS. CD8+ T cells require perforin to clear West Nile virus from infected neurons. *J Virol.* 2006;80:119-129. doi:10.1128/JVI.80.1.119-129.2006.
190. Lin MT, Stohlman SA, Hinton DR. Mouse hepatitis virus is cleared from the central nervous systems of mice lacking perforin-mediated cytolysis. *J Virol.* 1997;71:383-391. doi:10.1128/JVI.71.1.383-391.1997.
191. Rottenberg M, Kristensson K. Effects of interferon-gamma on neuronal infections. *Viral Immunol.* 2002;15:247-260. doi:10.1089/08828240260066206.
192. Burdeinick-Kerr R, Govindarajan D, Griffin DE. Noncytolytic clearance of sindbis virus infection from neurons by gamma interferon is dependent on Jak/STAT signaling. *J Virol.* 2009;83:3429-3435. doi:10.1128/JVI.02381-08.
193. Stohlman SA, Bergmann CC, Lin MT, Cua DJ, Hinton DR. CTL effector function within the central nervous system requires CD4+ T cells. *J Immunol.* 1998;160:2896-2904.
194. Zhou J, et al. Maintenance of CD8+ T cells during acute viral infection of the central nervous system requires CD4+ T cells but not interleukin-2. *Viral Immunol.* 2005;18:162-169. doi:10.1089/vim.2005.18.162.
195. Hooper DC, Phares TW, Fabis MJ, Roy A. The production of antibody by invading B cells is required for the clearance of rabies virus from the central nervous system. *PLoS Negl Trop Dis.* 2009;3:e535. doi:10.1371/journal.pntd.0000535.
196. Wattananit S, et al. Monocyte-derived macrophages contribute to spontaneous long-term functional recovery after stroke in mice. *J Neurosci.* 2016;36:4182-4195. doi:10.1523/jneurosci.4317-15.2016.
197. Shechter R, et al. Recruitment of beneficial M2 macrophages to injured spinal cord is orchestrated by remote brain choroid plexus. *Immunity.* 2013;38:555-569. doi:10.1016/j.immuni.2013.02.012.
198. Shechter R, Raposo C, London A, Sagi I, Schwartz M. The glial scar-monocyte interplay: a pivotal resolution phase in spinal cord repair. *PloS One.* 2011;6:e27969. doi:10.1371/journal.pone.0027969.
199. Schwartz M. "Tissue-repairing" blood-derived macrophages are essential for healing of the injured spinal cord: from skin-activated macrophages to infiltrating blood-derived cells? *Brain Behav Immun.* 2010;24:1054-1057. doi:S0889-1591(10)00037-1 [pii]10.1016/j.bbi.2010.01.010.
200. Rossignol S, Schwab M, Schwartz M, Fehlings MG. Spinal cord injury: time to move? *J Neurosci.* 2007;27:11782-11792. doi:10.1523/jneurosci.3444-07.2007.
201. Gadani SP, Walsh JT, Lukens JR, Kipnis J. Dealing with danger in the CNS: the response of the immune system to injury. *Neuron.* 2015;87:47-62. doi:10.1016/j.neuron.2015.05.019.
202. Yoles E, et al. Protective autoimmunity is a physiological response to CNS trauma. *J Neurosci.* 2001;21:3740-3748.
203. Kipnis J, et al. Neuroprotective autoimmunity: naturally occurring CD4+CD25+ regulatory T cells suppress the ability to withstand injury to the central nervous system. *Proc Natl Acad Sci U S A.* 2002;99:15620-15625. doi:10.1073/pnas.232565399232565399 [pii].
204. Moalem G, et al. Autoimmune T cells protect neurons from secondary degeneration after central nervous system axotomy. *Nat Med.* 1999;5:49-55. doi:10.1038/4734.
205. Hauben E, et al. Vaccination with dendritic cells pulsed with peptides of myelin basic protein promotes functional recovery from spinal cord injury. *J Neurosci.* 2003;23:8808-8819.
206. Kipnis J, Avidan H, Caspi RR, Schwartz M. Dual effect of CD4+CD25+ regulatory T cells in neurodegeneration: a dialogue with microglia. *Proc Natl Acad Sci U S A.* 2004;101(Suppl 2):14663-14669. doi:10.1073/pnas.0404842101040484210l [pii].
207. Kipnis J, et al. Therapeutic vaccination for closed head injury. *J Neurotrauma.* 2003;20:559-569. doi:10.1089/0897715037671 68483.
208. Kipnis J, Mizrahi T, Yoles E, Ben-Nun A, Schwartz M. Myelin specific Th1 cells are necessary for post-traumatic protective autoimmunity. *J Neuroimmunol.* 2002;130:78-85. doi:S0165572802002199 [pii].
209. Kipnis J, et al. Neuronal survival after CNS insult is determined by a genetically encoded autoimmune response. *J Neurosci.* 2001;21:4564-4571. doi:21/13/4564 [pii].
210. Hauben E, et al. Posttraumatic therapeutic vaccination with modified myelin self-antigen prevents complete paralysis while avoiding autoimmune disease. *J Clin Invest.* 2001;108:591-599.
211. Ehrhard PB, Erb P, Graumann U, Otten U. Expression of nerve growth factor and nerve growth factor receptor tyrosine kinase Trk in activated CD4-positive T-cell clones. *Proc Natl Acad Sci USA.* 1993;90:10984-10988. doi:10.1073/pnas.90.23.10984.
212. Kerschensteiner M, et al. Activated human T cells, B cells, and monocytes produce brain-derived neurotrophic factor in vitro and in inflammatory brain lesions: a neuroprotective role of inflammation? *J Exp Med.* 1999;189:865-870. doi:10.1084/jem.189.5.865.
213. Hammarberg H, et al. Neuroprotection by encephalomyelitis: rescue of mechanically injured neurons and neurotrophin production by CNS-infiltrating T and natural killer cells. *J Neurosci.* 2000;20:5283-5291.
214. Gadani SP, Walsh JT, Smirnov I, Zheng J, Kipnis J. The glia-derived alarmin IL-33 orchestrates the immune response and promotes recovery following CNS injury. *Neuron.* 2015;85:703-709. doi:10.1016/j.neuron.2015.01.013.
215. Vogelaar CF, et al. Fast direct neuronal signaling via the IL-4 receptor as therapeutic target in neuroinflammation. *Sci Transl Med.* 2018;10:eaao2304. doi:10.1126/scitranslmed.aao2304.
216. Walsh JT, et al. MHCII-independent CD4+ T cells protect injured CNS neurons via IL-4. *J Clin Invest.* 2015;125:699-714. doi:10.1172/JCI76210.
217. Bodhankar S, et al. Regulatory CD8(+)CD122 (+) T-cells predominate in CNS after treatment of experimental stroke in male mice with IL-10-secreting B-cells. *Metab Brain Dis.* 2015;30:911-924. doi:10.1007/s11011-014-9639-8.
218. Walsh JT, et al. Regulatory T cells in central nervous system injury: a double-edged sword. *J Immunol.* 2014;193:5013-5022. doi:10.4049/jimmunol.1302401.
219. Walsh JT, Kipnis J. Regulatory T cells in CNS injury: the simple, the complex and the confused. *Trends Mol Med.* 2011;17:541-547. doi:10.1016/j.molmed.2011.05.012.
220. Schwartz M, Kipnis J. Therapeutic T cell-based vaccination for neurodegenerative disorders: the role of CD4+CD25+ regulatory T cells. *Ann N Y Acad Sci.* 2005;1051:701-708. doi:1051/1/701 [pii]10.1196/annals.1361.114.
221. Kipnis J, et al. Dopamine, through the extracellular signal-regulated kinase pathway, downregulates CD4+CD25+ regulatory T-cell activity: implications for neurodegeneration. *J Neurosci.* 2004;24:6133-6143. doi:10.1523/JNEUROSCI.0600-04.200424/27/6133 [pii].
222. Meisel C, Schwab JM, Prass K, Meisel A, Dirnagl U. Central nervous system injury-induced immune deficiency syndrome. *Nat Rev Neurosci.* 2005;6:775-786. doi:10.1038/nrn1765.
223. Yin Y, et al. Macrophage-derived factors stimulate optic nerve regeneration. *J Neurosci.* 2003;23:2284-2293.
224. Neirinckx V, et al. Neutrophil contribution to spinal cord injury and repair. *J Neuroinflammation.* 2014;11:150. doi:10.1186/s12974-014-0150-2.
225. Sas AR, et al. A new neutrophil subset promotes CNS neuron survival and axon regeneration. *Nat Immunol.* 2020;21:1496-1505. doi:10.1038/s41590-020-00813-0.
226. Pasciuto E, et al. Microglia require CD4 T cells to complete the fetal-to-adult transition. *Cell.* 2020;182:625-640.e24. doi:10.1016/j.cell.2020.06.026.
227. Dulken BW, et al. Single-cell analysis reveals T cell infiltration in old neurogenic niches. *Nature.* 2019;571:205-210. doi:10.1038/s41586-019-1362-5.
228. Baruch K, et al. Aging. Aging-induced type I interferon response at the choroid plexus negatively affects brain function. *Science.* 2014;346:89-93. doi:10.1126/science.1252945.
229. Ziv Y, et al. Immune cells contribute to the maintenance of neurogenesis and spatial learning abilities in adulthood. *Nat Neurosci.* 2006;9:268-275. doi:10.1038/nn1629.
230. Butovsky O, et al. Microglia activated by IL-4 or IFN-gamma differentially induce neurogenesis and oligodendrogenesis from adult stem/progenitor cells. *Mol Cell Neurosci.* 2006;31(1):149-160.
231. Herisson F, et al. Direct vascular channels connect skull bone marrow and the brain surface enabling myeloid cell migration. *Nat Neurosci.* 2018;21:1209-1217. doi:10.1038/s41593-018-0213-2.
232. Herisson F, et al. Direct vascular channels connect skull bone marrow and the brain surface enabling myeloid cell migration. *Nat Neurosci.* 2018;21:1209-1217. doi:10.1038/s41593-018-0213-2.
233. Cugurra A, et al. Skull and vertebral bone marrow are myeloid cell reservoirs for the meninges and CNS parenchyma. *Science.* 2021;373:eabf7844. doi:10.1126/science.abf7844.
234. Brioschi S, et al. Heterogeneity of meningeal B cells reveals a lymphopoietic niche at the CNS borders. *Science.* 2021;373:eabf9277. doi:10.1126/science.abf9277.
235. Wang Y, et al. Early developing B cells undergo negative selection by central nervous system-specific antigens in the meninges. *Immunity.* 2021;54(12):2784-2794.e6. doi:10.1016/j.immuni.2021.09.016.
236. Rustenhoven J, et al. Functional characterization of the dural sinuses as a neuroimmune interface. *Cell.* 2021;184:1000-1016.e1027. doi:10.1016/j.cell.2020.12.040.
237. Mrdjen D, et al. High-dimensional single-cell mapping of central nervous system immune cells reveals distinct myeloid subsets in health, aging, and disease. *Immunity.* 2018;48:599. doi:10.1016/j.immuni.2018.02.014.
238. Filiano AJ, Gadani SP, Kipnis J. How and why do T cells and their derived cytokines affect the injured and healthy brain? *Nat Rev Neurosci.* 2017;18:375-384. doi:10.1038/nrn.2017.39.
239. Marin I, Kipnis J. Learning and memory ... and the immune system. *Learn Mem.* 2013;20:601-606. doi:10.1101/lm.028357.112.
240. Kipnis J, Gadani S, Derecki NC. Pro-cognitive properties of T cells. *Nat Rev Immunol.* 2012;12:663-669. doi:10.1038/nri3280.
241. Gadani SP, Cronk JC, Norris GT, Kipnis J. IL-4 in the brain: a cytokine to remember. *J Immunol.* 2012;189:4213-4219. doi:10.4049/jimmunol.1202246.
242. Derecki NC, et al. Regulation of learning and memory by meningeal immunity: a key role for IL-4. *J Exp Med.* 2010;207:1067-1080. doi:10.1084/jem.20091419.
243. Kipnis J, Derecki NC, Yang C, Scrable H. Immunity and cognition: what do age-related dementia, HIV-dementia and "chemo-brain" have in common? *Trends Immunol.* 2008;29:455-463. doi:10.1016/j.it.2008.07.007.
244. Lewitus GM, Zhu J, Xiong H, Hallworth R, Kipnis J. CD4(+)CD25(−) effector T-cells inhibit hippocampal long-term potentiation in vitro. *Eur J Neurosci.* 2007;26:1399-1406. doi:EJN5788 [pii]10.1111/j.1460-9568.2007.05788.x.
245. Filiano AJ, et al. Unexpected role of interferon-γ in regulating neuronal connectivity and social behaviour. *Nature.* 2016;535:425-429. doi:10.1038/nature18626.
246. Alves de Lima K, et al. Meningeal gd T cells regulate anxiety-like behavior via IL-17a signaling in neurons. *Nat Immunol.* 2020;21(11):1421-1429.
247. McCusker RH, Kelley KW. Immune-neural connections: how the immune system's response to infectious agents influences behavior. *J Exp Biol.* 2013;216:84-98. doi:10.1242/jeb.073411.
248. Kelley KW, Dantzer R. Alcoholism and inflammation: neuroimmunology of behavioral and mood disorders. *Brain Behav Immun.* 2011;25(Suppl 1):S13-S20. doi:10.1016/j.bbi.2010.12.013.
249. Goehler LE, Lyte M, Gaykema RP. Infection-induced viscerosensory signals from the gut enhance anxiety: implications for psychoneuroimmunology. *Brain Behav Immun.* 2007;21:721-726. doi:S0889-1591(07)00041-4 [pii]10.1016/j.bbi.2007.02.005.
250. Kelley KW, et al. Cytokine-induced sickness behavior. *Brain Behav Immun.* 2003;17(Suppl 1):S112-S118.
251. Dantzer R, Konsman JP, Bluthe RM, Kelley KW. Neural and humoral pathways of communication from the immune system to the brain: parallel or convergent? *Auton Neurosci.* 2000;85:60-65. doi:10.1016/S1566-0702(00)00220-4.
252. Dantzer R, et al. Cytokines and sickness behavior. *Ann N Y Acad Sci.* 1998;840:586-590.
253. Dantzer R, et al. Molecular basis of sickness behavior. *Ann N Y Acad Sci.* 1998;856:132-138.
254. Kent S, Bluthe RM, Kelley KW, Dantzer R. Sickness behavior as a new target for drug development. *Trends Pharmacol Sci.* 1992;13:24-28.
255. Blank T, et al. Brain endothelial-and epithelial-specific interferon receptor chain 1 drives virus-induced sickness behavior and cognitive impairment. *Immunity.* 2016;44:901-912. doi:10.1016/j.immuni.2016.04.005.
256. Ching S, et al. Endothelial-specific knockdown of interleukin-1 (IL-1) type 1 receptor differentially alters CNS responses to IL-1 depending on its route of administration. *J Neurosci.* 2007;27:10476-10486. doi:10.1523/JNEUROSCI.3357-07.2007.
257. Liu X, et al. Cell-type-specific interleukin 1 receptor 1 signaling in the brain regulates distinct neuroimmune activities. *Immunity.* 2019;50:317-333.e16. doi:10.1016/j.immuni.2018.12.012.
258. Duan L, et al. PDGFRβ cells rapidly relay inflammatory signal from the circulatory system to neurons via chemokine CCL2. *Neuron.* 2018;100:183-200.e188. doi:10.1016/j.neuron.2018.08.030.
259. Matsumura K, et al. Brain endothelial cells express cyclooxygenase-2 during lipopolysaccharide-induced fever: light and electron microscopic immunocytochemical studies. *J Neurosci.* 1998;18:6279-6289.
260. Schiltz JC, Sawchenko PE. Distinct brain vascular cell types manifest inducible cyclooxygenase expression as a function of the strength and nature of immune insults. *J Neurosci.* 2002;22:5606-5618. doi:10.1523/JNEUROSCI.22-13-05606.2002.
261. Saper CB, Romanovsky AA, Scammell TE. Neural circuitry engaged by prostaglandins during the sickness syndrome. *Nat Neurosci.* 2012;15:1088-1095. doi:10.1038/nn.3159.
262. Brombacher TM, et al. IL-4R alpha deficiency influences hippocampal-BDNF signaling pathway to impair reference memory. *Sci Rep.* 2020;10:16506. doi:10.1038/s41598-020-73574-3.
263. Brombacher TM, et al. IL-13-Mediated regulation of learning and memory. *J Immunol.* 2017;198:2681-2688. doi:10.4049/jimmunol.1601546.
264. Kim S, et al. Maternal gut bacteria promote neurodevelopmental abnormalities in mouse offspring. *Nature.* 2017;549:528-532. doi:10.1038/nature23910.
265. Choi GB, et al. The maternal interleukin-17a pathway in mice promotes autism-like phenotypes in offspring. *Science.* 2016;351:933-939. doi:10.1126/science.aad0314.
266. Chen C, et al. IL-17 is a neuromodulator of Caenorhabditis elegans sensory responses. *Nature.* 2017;542:43-48. doi:10.1038/nature20818.
267. Reed MD, et al. IL-17a promotes sociability in mouse models of neurodevelopmental disorders. *Nature.* 2020;577:249-253. doi:10.1038/s41586-019-1843-6.
268. Ribeiro M, et al. Meningeal γδ T cell-derived IL-17 controls synaptic plasticity and short-term memory. *Sci Immunol.* 2019;4:eaay5199. doi:10.1126/sciimmunol.aay5199.
269. Beattie EC, et al. Control of synaptic strength by glial TNFalpha. *Science.* 2002;295:2282-2285. doi:10.1126/science.1067859.

270. Stellwagen D, Malenka RC. Synaptic scaling mediated by glial TNF-alpha. *Nature.* 2006;440:1054-1059. doi:10.1038/nature04671.
271. Roussa E, et al. Transforming growth factor beta is required for differentiation of mouse mesencephalic progenitors into dopaminergic neurons in vitro and in vivo: ectopic induction in dorsal mesencephalon. *Stem Cell.* 2006;24:2120-2129. doi:10.1634/stemcells.2005-0514.
272. Deverman BE, Patterson PH. Cytokines and CNS development. *Neuron.* 2009;64:61-78. doi:10.1016/j.neuron.2009.09.002.
273. Barres BA, Schmid R, Sendtner M, Raff MC. Multiple extracellular signals are required for long-term oligodendrocyte survival. *Development.* 1993;118:283-295.
274. Smith SEP, Li J, Garbett K, Mirnics K, Patterson PH. Maternal immune activation alters fetal brain development through interleukin-6. *J Neurosci.* 2007;27:10695-10702. doi:10.1523/JNEUROSCI.2178-07.2007.
275. Wrona D. Neural-immune interactions: an integrative view of the bidirectional relationship between the brain and immune systems. *J Neuroimmunol.* 2006;172:38-58. doi:10.1016/j.jneuroim.2005.10.017.
276. Katayama M, Kobayashi S, Kuramoto N, Yokoyama MM. Effects of hypothalamic lesions on lymphocyte subsets in mice. *Ann N Y Acad Sci.* 1987;496:366-376. doi:10.1111/j.1749-6632.1987.tb35789.x.
277. Tyrey L, Nalbandov AV. Influence of anterior hypothalamic lesions on circulating antibody titers in the rat. *Am J Physiol.* 1972;222:179-185. doi:10.1152/ajplegacy.1972.222.1.179.
278. Hefco V, Olariu A, Hefco A, Nabeshima T. The modulator role of the hypothalamic paraventricular nucleus on immune responsiveness. *Brain Behav Immun.* 2004;18:158-165. doi:10.1016/S0889-1591(03)00114-4.
279. Cross RJ, Markesbery WR, Brooks WH, Roszman TL. Hypothalamic-immune interactions. I. The acute effect of anterior hypothalamic lesions on the immune response. *Brain Res.* 1980;196:79-87. doi:10.1016/0006-8993(80)90717-9.
280. Cross RJ, Brooks WH, Roszman TL, Markesbery WR. Hypothalamic-immune interactions. Effect of hypophysectomy on neuroimmunomodulation. *J Neurol Sci.* 1982;53:557-566.
281. Nance DM, Rayson D, Carr RI. The effects of lesions in the lateral septal and hippocampal areas on the humoral immune response of adult female rats. *Brain Behav Immun.* 1987;1:292-305. doi:10.1016/0889-1591(87)90032-8.
282. Renoux G, Biziere K, Renoux M, Bardos P, Degenne D. Consequences of bilateral brain neocortical ablation on imuthiol-induced immunostimulation in mice. *Ann N Y Acad Sci.* 1987;496:346-353. doi:10.1111/j.1749-6632.1987.tb35786.x.
283. Neveu PJ, Taghzouti K, Dantzer R, Simon H, Le Moal M. Modulation of mitogen-induced lymphoproliferation by cerebral neocortex. *Life Sci.* 1986;38:1907-1913.
284. Tye KM, Deisseroth K. Optogenetic investigation of neural circuits underlying brain disease in animal models. *Nat Rev Neurosci.* 2012;13:251-266. doi:10.1038/nrn3171.
285. Ben-Shaanan T, Schiller M, Rolls A. Studying brain-regulation of immunity with optogenetics and chemogenetics; a new experimental platform. *Brain Behav Immun.* 2017;65:1-8. doi:10.1016/j.bbi.2016.11.024.
286. Boyden ES, Zhang F, Bamberg E, Nagel G, Deisseroth K. Millisecond-timescale, genetically targeted optical control of neural activity. *Nat Neurosci.* 2005;8:1263-1268. doi:10.1038/nn1525.
287. Armbruster BN, Li X, Pausch MH, Herlitze S, Roth BL. Evolving the lock to fit the key to create a family of G protein-coupled receptors potently activated by an inert ligand. *Proc Natl Acad Sci USA.* 2007;104:5163-5168. doi:10.1073/pnas.0700293104.
288. Ben-Shaanan TL, et al. Activation of the reward system boosts innate and adaptive immunity. *Nat Med.* 2016;22:940-944. doi:10.1038/nm.4133.
289. Ben-Shaanan TL, et al. Modulation of anti-tumor immunity by the brain's reward system. *Nat Commun.* 2018;9:2723. doi:10.1038/s41467-018-05283-5.
290. Vlajković S, Dugandzija-Novaković S, Milanović S, Janković BD. Brain self-stimulation and immunity: effect on humoral and cell-mediated immune responses. *Int J Neurosci.* 1993;69:235-250.
291. Ader R, Cohen N. Behaviorally conditioned immunosuppression. *Psychosom Med.* 1975;37:333-340. doi:10.1097/00006842-197507000-00007.
292. Schiller M, Ben-Shaanan TL, Rolls A. Neuronal regulation of immunity: why, how and where? *Nat Rev Immunol.* 2021;21(1):20-36. doi:10.1038/s41577-020-0387-1.
293. Ader R, Cohen N. Behaviorally conditioned immunosuppression and murine systemic lupus erythematosus. *Science.* 1982;215:1534-1536. doi:10.1126/science.7063864.
294. Lysle DT, Luecken LJ, Maslonek KA. Suppression of the development of adjuvant arthritis by a conditioned aversive stimulus. *Brain Behav Immun.* 1992;6:64-73. doi:10.1016/0889-1591(92)90060-2.
295. Ramírez-Amaya V, et al. Insular cortex lesions impair the acquisition of conditioned immunosuppression. *Brain Behav Immun.* 1996;10:103-114. doi:10.1006/brbi.1996.0011.
296. Zhang X, et al. Brain control of humoral immune responses amenable to behavioural modulation. *Nature.* 2020;581:204-208. doi:10.1038/s41586-020-2235-7.
297. Silverman MN, Pearce BD, Biron CA, Miller AH. Immune modulation of the hypothalamic-pituitary-adrenal (HPA) axis during viral infection. *Viral Immunol.* 2005;18:41-78. doi:10.1089/vim.2005.18.41.
298. Perretti M, Ahluwalia A. The microcirculation and inflammation: site of action for glucocorticoids. *Microcirculation.* 2000;7:147-161.
299. Clark AR. Anti-inflammatory functions of glucocorticoid-induced genes. *Mol Cell Endocrinol.* 2007;275:79-97. doi:10.1016/j.mce.2007.04.013.
300. Coutinho AE, Chapman KE. The anti-inflammatory and immunosuppressive effects of glucocorticoids, recent developments and mechanistic insights. *Mol Cell Endocrinol.* 2011;335:2-13. doi:10.1016/j.mce.2010.04.005.
301. Rea D, et al. Glucocorticoids transform CD40-triggering of dendritic cells into an alternative activation pathway resulting in antigen-presenting cells that secrete IL-10. *Blood.* 2000;95:3162-3167.
302. Hong JY, et al. Long-term programming of CD8 T cell immunity by perinatal exposure to glucocorticoids. *Cell.* 2020;180:847-861.e15. doi:10.1016/j.cell.2020.02.018.
303. Takenaka MC, Guereschi MG, Basso AS. Neuroimmune interactions: dendritic cell modulation by the sympathetic nervous system. *Semin Immunopathol.* 2017;39:165-176. doi:10.1007/s00281-016-0590-0.
304. Elenkov IJ, Wilder RL, Chrousos GP, Vizi ES. The sympathetic nerve – an integrative interface between two supersystems: the brain and the immune system. *Pharmacol Rev.* 2000;52:595-638.
305. Sanders VM, et al. Differential expression of the beta2-adrenergic receptor by Th1 and Th2 clones: implications for cytokine production and B cell help. *J Immunol.* 1997;158:4200-4210.
306. Lai JP, Douglas SD, Ho WZ. Human lymphocytes express substance P and its receptor. *J Neuroimmunol.* 1998;86:80-86.
307. Lambrecht BN. Immunologists getting nervous: neuropeptides, dendritic cells and T cell activation. *Respir Res.* 2001;2:133-138. doi:10.1186/rr49.
308. Carucci JA, et al. Calcitonin gene-related peptide decreases expression of HLA-DR and CD86 by human dendritic cells and dampens dendritic cell-driven T cell-proliferative responses via the type I calcitonin gene-related peptide receptor. *J Immunol.* 2000;164:3494-3499.
309. Pinho-Ribeiro FA, et al. Blocking neuronal signaling to immune cells treats streptococcal invasive infection. *Cell.* 2018;173:1083-1097.e22. doi:10.1016/j.cell.2018.04.006.
310. Baral P, et al. Nociceptor sensory neurons suppress neutrophil and γδ T cell responses in bacterial lung infections and lethal pneumonia. *Nat Med.* 2018;24:417-426. doi:10.1038/nm.4501.
311. Delgado M, Gonzalez-Rey E, Ganea D. The neuropeptide vasoactive intestinal peptide generates tolerogenic dendritic cells. *J Immunol.* 2005;175:7311-7324. doi:10.4049/jimmunol.175.11.7311.
312. Gonzalez-Rey E, Chorny A, Fernandez-Martin A, Ganea D, Delgado M. Vasoactive intestinal peptide generates human tolerogenic dendritic cells that induce CD4 and CD8 regulatory T cells. *Blood.* 2006;107:3632-3638. doi:10.1182/blood-2005-11-4497.
313. Gonzalez-Rey E, Chorny A, Delgado M. Regulation of immune tolerance by anti-inflammatory neuropeptides. *Nat Rev Immunol.* 2007;7:52-63. doi:10.1038/nri1984.
314. Delgado M, Abad C, Martinez C, Leceta J, Gomariz RP. Vasoactive intestinal peptide prevents experimental arthritis by downregulating both autoimmune and inflammatory components of the disease. *Nat Med.* 2001;7:563-568. doi:10.1038/87887.
315. Delgado M, Reduta A, Sharma V, Ganea D. VIP/PACAP oppositely affects immature and mature dendritic cell expression of CD80/CD86 and the stimulatory activity for CD4(+) T cells. *J Leukoc Biol.* 2004;75:1122-1130. doi:10.1189/jlb.1203626.
316. Cardoso V, et al. Neuronal regulation of type 2 innate lymphoid cells via neuromedin U. *Nature.* 2017;549:277-281. doi:10.1038/nature23469.
317. Klose CSN, et al. The neuropeptide neuromedin U stimulates innate lymphoid cells and type 2 inflammation. *Nature.* 2017;549:282-286. doi:10.1038/nature23676.
318. Wallrapp A, et al. The neuropeptide NMU amplifies ILC2-driven allergic lung inflammation. *Nature.* 2017;549:351-356. doi:10.1038/nature24029.
319. Calvo W. The innervation of the bone marrow in laboratory animals. *Am J Anat.* 1968;123:315-328. doi:10.1002/aja.1001230206.
320. Jung W-C, Levesque J-P, Ruitenberg MJ. It takes nerve to fight back: the significance of neural innervation of the bone marrow and spleen for immune function. *Semin Cell Dev Biol.* 2017;61:60-70. doi:10.1016/j.semcdb.2016.08.010.
321. Afan AM, Broome CS, Nicholls SE, Whetton AD, Miyan JA. Bone marrow innervation regulates cellular retention in the murine haemopoietic system. *Br J Haematol.* 1997;98:569-577. doi:10.1046/j.1365-2141.1997.2733092.x.
322. Katayama Y, et al. Signals from the sympathetic nervous system regulate hematopoietic stem cell egress from bone marrow. *Cell.* 2006;124:407-421. doi:10.1016/j.cell.2005.10.041.
323. Yamazaki S, et al. Nonmyelinating Schwann cells maintain hematopoietic stem cell hibernation in the bone marrow niche. *Cell.* 2011;147:1146-1158. doi:10.1016/j.cell.2011.09.053.
324. Borovikova LV, et al. Vagus nerve stimulation attenuates the systemic inflammatory response to endotoxin. *Nature.* 2000;405:458-462. doi:10.1038/35013070.
325. Tracey KJ. The inflammatory reflex. *Nature.* 2002;420:853-859. doi:10.1038/nature01321.
326. Pavlov VA, Tracey KJ. The vagus nerve and the inflammatory reflex—linking immunity and metabolism. *Nat Rev Endocrinol.* 2012;8:743-754. doi:10.1038/nrendo.2012.189.
327. Nance DM, Sanders VM. Autonomic innervation and regulation of the immune system (1987-2007). *Brain Behav Immun.* 2007;21:736-745. doi:10.1016/j.bbi.2007.03.008.
328. Bratton BO, et al. Neural regulation of inflammation: no neural connection from the vagus to splenic sympathetic neurons. *Exp Physiol.* 2012;97:1180-1185. doi:10.1113/expphysiol.2011.061531.
329. Rosas-Ballina M, et al. Splenic nerve is required for cholinergic antiinflammatory pathway control of TNF in endotoxemia. *Proc Natl Acad Sci USA.* 2008;105:11008-11013. doi:10.1073/pnas.0803237105.
330. Rosas-Ballina M, et al. Acetylcholine-synthesizing T cells relay neural signals in a vagus nerve circuit. *Science.* 2011;334:98-101. doi:10.1126/science.1209985.
331. Wang H, et al. Nicotinic acetylcholine receptor alpha7 subunit is an essential regulator of inflammation. *Nature.* 2003;421:384-388. doi:10.1038/nature01339.
332. Pavlov VA, et al. Selective alpha7-nicotinic acetylcholine receptor agonist GTS-21 improves survival in murine endotoxemia and severe sepsis. *Crit Care Med.* 2007;35:1139-1144. doi:10.1097/01.CCM.0000259381.56526.96.
333. Raper D, Louveau A, Kipnis J. How do meningeal lymphatic vessels drain the CNS? *Trends Neurosci.* 2016;39:581-586. doi:10.1016/j.tins.2016.07.001.
334. Louveau A, Da Mesquita S, Kipnis J. Lymphatics in neurological disorders: a neuro-lympho-vascular component of multiple sclerosis and alzheimer's disease? *Neuron.* 2016;91:957-973. doi:10.1016/j.neuron.2016.08.027.
335. Louveau A, et al. Structural and functional features of central nervous system lymphatic vessels. *Nature.* 2015;523:337-341. doi:10.1038/nature14432.
336. Mestre H, et al. Aquaporin-4-dependent glymphatic solute transport in the rodent brain. *Elife.* 2018;7:e40070. doi:10.7554/eLife.40070.
337. Louveau A, et al. Understanding the functions and relationships of the glymphatic system and meningeal lymphatics. *J Clin Invest.* 2017;127:3210-3219. doi:10.1172/JCI90603.
338. Lee H, et al. The effect of body posture on brain glymphatic transport. *J Neurosci.* 2015;35:11034-11044. doi:10.1523/JNEUROSCI.1625-15.2015.
339. Lee H, et al. The effect of body posture on brain glymphatic transport. *J Neurosci.* 2015;35:11034-11044. doi:10.1523/JNEUROSCI.1625-15.2015.
340. Thrane AS, Rangroo Thrane V, Nedergaard M. Drowning stars: reassessing the role of astrocytes in brain edema. *Trends Neurosci.* 2014;37:620-628. doi:10.1016/j.tins.2014.08.010.
341. Iliff JJ, et al. Brain-wide pathway for waste clearance captured by contrast-enhanced MRI. *J Clin Invest.* 2013;123:1299-1309. doi:10.1172/JCI67677.
342. Jung E, et al. Development and characterization of A novel prox1-EGFP lymphatic and Schlemm's canal reporter rat. *Sci Rep.* 2017;7:5577. doi:10.1038/s41598-017-06031-3.
343. Absinta M, et al. Human and nonhuman primate meninges harbor lymphatic vessels that can be visualized noninvasively by MRI. *Elife.* 2017;6:e29738. doi:10.7554/eLife.29738.
344. Patel TK, et al. Dural lymphatics regulate clearance of extracellular tau from the CNS. *Mol Neurodegener.* 2019;14:11. doi:10.1186/s13024-019-0312-x.
345. van Lessen M, et al. Intracellular uptake of macromolecules by brain lymphatic endothelial cells during zebrafish embryonic development. *Elife.* 2017;6:e25932. doi:10.7554/eLife.25932.
346. Bower NI, et al. Mural lymphatic endothelial cells regulate meningeal angiogenesis in the zebrafish. *Nat Neurosci.* 2017;20:774-783. doi:10.1038/nn.4558.
347. Venero Galanternik M, et al. A novel perivascular cell population in the zebrafish brain. *Elife.* 2017;6:e24369. doi:10.7554/eLife.24369.
348. Ahn JH, et al. Meningeal lymphatic vessels at the skull base drain cerebrospinal fluid. *Nature.* 2019;572:62-66. doi:10.1038/s41586-019-1419-5.
349. Ransohoff RM, Engelhardt B. The anatomical and cellular basis of immune surveillance in the central nervous system. *Nat Rev Immunol.* 2012;12:623-635. doi:10.1038/nri3265.
350. Ringstad G, Eide PK. Cerebrospinal fluid tracer efflux to parasagittal dura in humans. *Nat Commun.* 2020;11:354. doi:10.1038/s41467-019-14195-x.
351. Melin E, Eide PK, Ringstad G. In vivo assessment of cerebrospinal fluid efflux to nasal mucosa in humans. *Sci Rep.* 2020;10:14974. doi:10.1038/s41598-020-72031-5.
352. Da Mesquita S, et al. Functional aspects of meningeal lymphatics in ageing and Alzheimer's disease. *Nature.* 2018;560:185-191. doi:10.1038/s41586-018-0368-8.
353. Sandrone S, Moreno-Zambrano D, Kipnis J, van Gijn J. A (delayed) history of the brain lymphatic system. *Nat Med.* 2019;25:538-540. doi:10.1038/s41591-019-0417-3.
354. Hu X, et al. Meningeal lymphatic vessels regulate brain tumor drainage and immunity. *Cell Res.* 2020;30:229-243. doi:10.1038/s41422-020-0287-8.
355. Song E, et al. VEGF-C-driven lymphatic drainage enables immunosurveillance of brain tumours. *Nature.* 2020;577:689-694. doi:10.1038/s41586-019-1912-x.

CHAPTER 40

Vaccines

Bali Pulendran, PhD

INTRODUCTION

These are extraordinary times for immunology, vaccinology, and the world. In the few short months since its emergence, the devastating impact of the SARS-CoV-2 virus has radically transformed virtually every dimension of our lives and fueled an urgent global imperative to develop safe and effective vaccines. At the time of writing this chapter, in the fall of 2020, an intense race is on to develop vaccines against COVID-19. Since the emergence of SARS-CoV-2 in China in December 2019, nearly 185 vaccine candidates, encompassing a multitude of platform technologies, are being developed. No fewer than 55 of these have already been tested in phase I and II clinical trials in humans; 22 candidates have been, or are currently being, tested in large phase III efficacy trials; and encouraging interim efficacy data from these trials have already been released for at least 6 of the candidate vaccines. On December 11, 2020, less than a year since the emergence of SARS-CoV-2, Pfizer/BioNTech's mRNA vaccine was approved for emergency use authorization in the United States. This was followed close on the heels by Moderna's mRNA vaccine, which was granted emergency use authorization just 7 days later, on December 18, 2020. The stunning pace at which these vaccines are being developed is unprecedented in the history of vaccinology. As the world expectantly awaits a vaccine, vaccinology and immunology have suddenly been thrust into global center stage. Once arcane terms such as "neutralizing antibodies," "T cells," and "herd immunity," previously relegated to the pages of scholarly textbooks such as "*Fundamental Immunology*," have now gained commonplace acceptance in the public's lexicon. Equally important has been the world's focus on the economics, politics, and ethics of how to ensure global equitable access to COVID-19 vaccines. This virus has certainly shone a spotlight on the field of vaccinology, positioned as it is, at the focal point of immunology, microbiology, public health, clinical medicine, economics, sociology, ethics, and international diplomacy.

Yet, it is important to remember that this is not the first moment in human history that vaccines have saved lives. Indeed, vaccination represents one of the most effective public health tools in history and has led to the eradication of smallpox, the near eradication of polio, and dramatic reductions in global mortality and morbidity rates for a multitude of infectious diseases. In fact, so great has been its success that vaccination has led to a dangerous sense of complacency that has driven down childhood vaccine uptake in many communities. Indeed, the deadliest adverse reaction to vaccination has been complacency!

The dawn of vaccinology can be traced to 1796 with Jenner's now famous experiment in human immunology. The evolution of vaccinology since that time has been punctuated by successive waves of conceptual and technological innovations. These include Pasteur's invention of the method of attenuation; Koch's germ theory of disease; von Behring and Ehrlich's discovery of antitoxins; invention of inactivated vaccines and live attenuated vaccines such as bacille Calmette-Guérin (BCG) and yellow fever vaccine 17D (YF-17D); innovation in cell culture methods that led to the development of vaccines against polio, mumps, measles, and rubella (MMR); conjugate chemistry that led to conjugate vaccines against meningococcal and pneumococcal infections; and the revolution in recombinant DNA in the 1970s that led to vaccines against hepatitis B, human papilloma virus (HPV), and other pathogens. This spirit of technological innovation has continued well into the 21st century. In the late 1990s, the ability to sequence the genomes of microorganisms led to the identification of novel vaccine antigens, a process termed "reverse vaccinology," that led to the development of a vaccine against the theretofore intractable meningococcal B (MenB) pathogen. Just in the past decade, spectacular progress in structural biology has led to delineation of atomic-level structures of candidate antigens for vaccines against respiratory syncytial virus (RSV), influenza, human immunodeficiency virus (HIV), and other pathogens that are currently being tested in humans.

Ironically, however, despite their common origins in the work of Edward Jenner and Louis Pasteur in the 18th and 19th centuries, immunology and vaccinology have evolved along such divergent trajectories that until a decade or so ago, immunologists and vaccinologists existed in parallel universes! Vaccinologists have developed many successful vaccines largely through empirical methods, with little insight into the immunological mechanisms by which they stimulate protective immunity. However, the failure to develop effective vaccines against pandemics such as HIV, malaria, tuberculosis (TB), and dengue, despite decades of effort, has highlighted the paramount importance of understanding the immunology of vaccination and infection. In this context, a revolution in our understanding of the molecular mechanisms of innate immunity and pathogen sensing via pattern recognition receptors (PRRs) that occurred over the past 2 decades is facilitating a deeper, mechanistic understanding of how vaccines stimulate immunity. In addition, in the past decade, advances in the use of systems biology approaches to probe the molecular networks driving vaccine immunity in humans (termed "systems vaccinology") have enhanced our understanding of the mechanisms of vaccine immunity and helped identify early signatures that can predict the later vaccine response. What has also emerged over the past decade is the realization that many vaccines such as the oral polio vaccine (OPV) or candidate malaria vaccines seem to have much lower efficacies in populations in certain regions of the developing world, compared to their efficacies in populations in the developed world. The reasons for this are unclear, but systems-based approaches are being used to analyze potential differences in the underlying immune response.

In this chapter, I attempt to discuss these exciting new advances in a historic context. The present COVID-19 pandemic offers a unique vantage point from which to survey vaccinology as it was for the past two centuries since Jenner's initial experiment (vaccinology BC [before COVID-19]), as it now is, during the time of COVID-19, and to contemplate what it could be, in the post-COVID era (vaccinology AC). This chapter is not meant to be an exhaustive compendium of past, current, and emerging vaccines. The reader is referred to many excellent reference articles that provide such information. Rather, in this chapter, I wish to introduce the students of fundamental immunology to the triumphant contributions that vaccines have made to public health; the formidable challenges in developing effective vaccines against pathogens such as HIV, malaria, and TB; and the exciting advances that have occurred over the past decade. These advances illuminate vast vistas of the promise and challenge of vaccinology in the 21st century.

THE HISTORY OF VACCINES

Smallpox and the Origins of Vaccination

The history of vaccines begins with smallpox. Although the concept of immunity was first documented by the ancient Greek historian Thucydides, who wrote "*the same man was never attacked twice—never at least fatally*"[1] when he described the plague of Athens in 430 BC, it is with smallpox that the dawn of vaccinology lies. Smallpox is a severe viral disease that is transmitted by inhalation of *Variola virus*, of the genus *Orthopoxvirus*, and causes often fatal disease killing more than 30% of the infected people. The earliest credible evidence of smallpox can be found in ancient Egypt.[2] Pharaoh Ramses V, who died as a young man in 1157 BC, might have been the earliest known victim of smallpox. In 1979, the American physician Donald Hopkins[3] was able to examine the mummy of Ramses V, housed in the mummy room of the Cairo Museum, and noted that "*it is one of the best preserved royal mummies in the [Cairo] museum. Inspection of the mummy revealed a rash of elevated 'pustules,' each about two to four millimetres in diameter, that was most distinct in the lower face, neck and shoulders, but also visible in the arms....*"

The earliest recorded attempt to induce protection against an infectious disease dates back to 1000 AD in China, where smallpox was endemic. Chinese physicians obtained dried scabs from the lesions of infected people, ground them into a powder, and blew them into the nostrils of the person to be protected.[4,5] By the 17th century, this practice of "variolation" had been introduced in the Ottoman Empire. Lady Mary Wortley Montagu, the wife of the British Ambassador in Constantinople, had observed this practice and was considerably impressed by it, and she had her infant son inoculated. Upon her return to England in 1721, she promoted the technique and had her 5-year-old daughter inoculated in front of King George I. This helped the practice gain wider acceptance.[4,5]

The Dawn of Vaccinology: Edward Jenner and Vaccination

Cows develop a disease similar to smallpox that is caused by the cowpox virus, part of the genus *Orthopoxvirus*, that is closely related to the *vaccinia* virus. The virus is zoonotic, that is, it can be transmitted between species, such as cows and humans. It was known that milkmaids who touched the lesion-filled udders of infected cows developed cowpox, which caused a mild disease with sores on their hands and fingers. By 1768, it was known that prior infection with cowpox rendered a person immune to smallpox. The first known use of cowpox to induce protective immunity against smallpox occurred in 1774. Benjamin Jesty, who was a farmer, inoculated his wife and two sons with cowpox.[2,6] None of them developed smallpox. Fifteen years later, when the two sons were inoculated with pus from smallpox, there was no disease, suggesting that the prior cowpox inoculation had protected them from smallpox. However, it was Edward Jenner's landmark experiment in human immunology in 1796 that proved the concept of variolation. On May 14, 1796, Jenner inoculated 8-year-old James Phipps with pus taken from cowpox blisters on the hands of Sarah Nelmes, a milkmaid[2,6] (Fig. 40.1). Phipps developed a fever, but no disease. Six weeks later, when Jenner injected Phipps with variolous material from smallpox, no disease followed. (He would never be able to convince local Institutional Review Board authorities to do this experiment today)! Five years later, the boy was again challenged with variolous material and showed no sign of infection, demonstrating that immunity was persistent. Jenner submitted his research for publication to the *Proceedings of the Royal Society*, but the interpretation of the study was met with skepticism and the paper was rejected. After extensive revisions, with inclusion of additional data from further experiments in 23 cases (including his 11-month-old son Robert!), Jenner privately published the results in a small booklet.[7] Although other people before Jenner had inoculated people with cowpox with the aim of protecting against smallpox, Jenner's unique contribution was to prove by experimental challenge with smallpox that cowpox-inoculated people were immune to smallpox.[3] Jenner coined the term "vaccination," which is derived from the Latin word *vacca* for cow. It took a while before vaccination was accepted by the scientific establishment, but in 1840, the British government banned the use of smallpox to induce protective immunity (ie, the practice of variolation) and instead offered vaccination free of charge.

Several countries in Europe made it compulsory, and mortality from smallpox dropped precipitously. Spain organized a successful 3-year-long mission to the Americas, Philippines, Macao, and China, led by Francisco Javier de Balmis, with the aim of vaccinating populations *en masse* with smallpox vaccine. Napoleon had all his French troops vaccinated and awarded a medal to Jenner, and even though France and Britain were at war at that time, he released two British prisoners at the request of Jenner.[8] He is said to have remarked that he could *"not refuse anything to one of the greatest benefactors of mankind."*[9] Jenner was famous in his lifetime but gained reverential status posthumously and is widely regarded as the father of vaccinology. Monuments to Edward Jenner exist in several countries. A prominent statue of Jenner is located in Kensington Gardens in London. A descriptive bronze plaque on the ground in front of the statue reads: *"Edward Jenner, MD, FRS, 1749–1823, country doctor who benefited mankind."*

Global Eradication of Smallpox (Fig. 40.1)

In 1958, Viktor Zhdanov, the Deputy Minister of Health for Soviet Union called on the World Health Assembly, the governing body of the World Health Organization (WHO), to initiate a plan to rid the world of smallpox. The proposal (Resolution WHA11.54) was accepted in 1959. Nearly 1 billion people lived in smallpox endemic areas, and global eradication was set as a new WHO goal.[10,11] Unfortunately, because of a lack of funds, personnel, and commitment from countries, as well as a shortage of vaccine supply, smallpox was still widespread in 1966, with regular outbreaks in countries across South America, Africa, and Asia. The Intensified Eradication Program began in 1967 with a promise of renewed efforts. Dr Donald A. Henderson ("D.A. Henderson") was appointed as head of the Smallpox Eradication Unit, with a pledge for more technical and financial support from the campaign's largest donor, the United States. At that time, smallpox had already been eliminated in North America (1952) and Europe (1953), but not in Asia, Africa, and South America. The 1967 plan was to vaccinate 200 million people, and Henderson proposed a 10-year mass vaccination campaign to vaccinate every human living in endemic areas.[10,11] It soon became apparent that this was not practical. When it was shown in West Africa by Dr. Bill Foege, a Centers for Disease Control and Prevention (CDC)–trained missionary living in Eastern Nigeria, that a targeted vaccination effort to immunize known or suspected contacts of infected individuals could be an alternative to mass vaccination, Henderson embraced and promoted this strategy.[12] This strategy known as "ring vaccination" was accomplished by a combination of

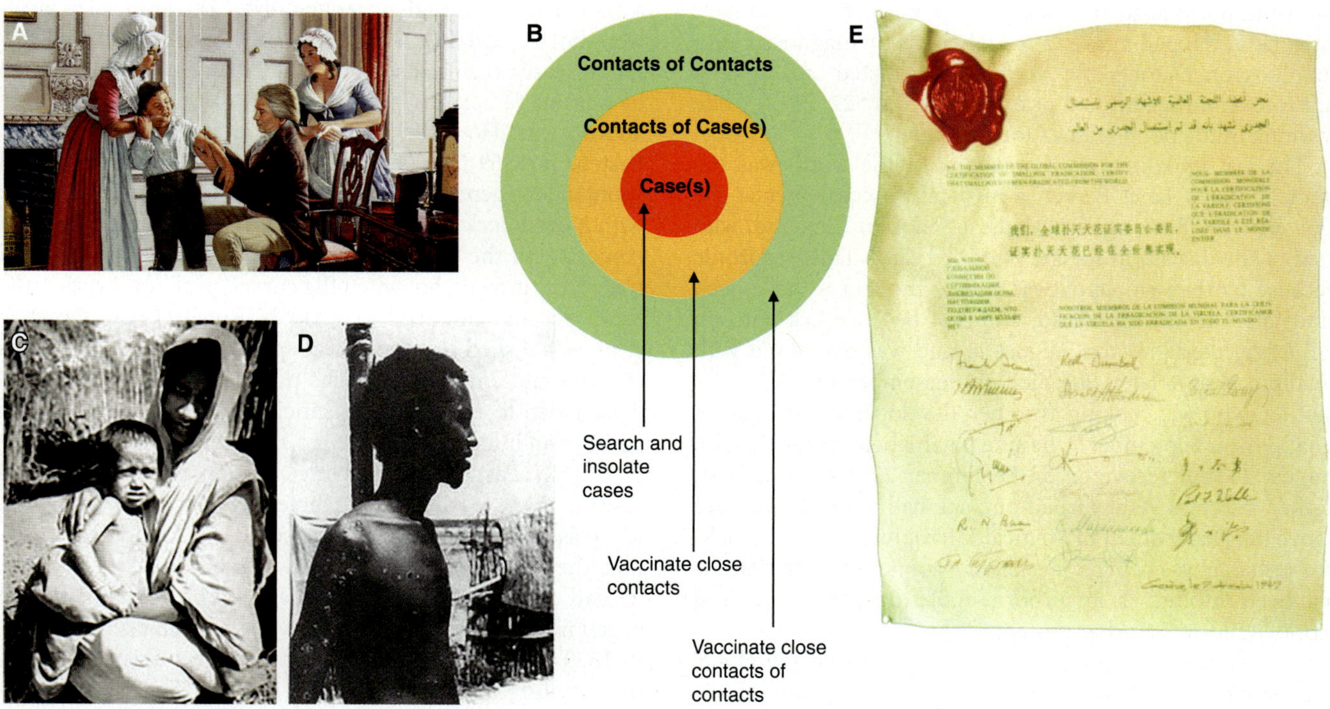

FIG. 40.1. The dawn of vaccinology: from Edward Jenner to the eradication of smallpox. A: Edward Jenner's landmark experiment in human immunology. On May 14, 1796, Jenner put cowpox fluid into James Phipps' arm and later challenged him with smallpox. **B:** The concept of ring vaccination, involving surveillance of rapidly identifying new smallpox cases, and vaccinating known or suspected contacts around an infected person, effectively creating a buffer or "ring" of protection around the infected foci to prevent the spread of the disease. **C:** Rahima Banu, a 3-year-old girl living in the district of Barisal in Bangladesh, who was the last person in Asia to have active smallpox. **D:** Ali Maow Maalin, a hospital cook in Somalia, was the last person to have naturally acquired smallpox in Africa. **E:** On December 9, 1979, a World Health Organization commission certified that smallpox had been eradicated from the world, and this certification was officially accepted by the 33rd World Health Assembly in 1980. (A, University of Michigan, Gift of Pfizer Inc UMHS 23; C, Centers for Disease Control and Prevention; D, Wikipedia; E, World Health Organization.)

outstanding surveillance of rapidly identifying new smallpox cases and vaccinating known or suspected contacts around an infected person, effectively creating a buffer or "ring" of protection around the infected foci to prevent the spread of the disease (Fig. 40.1). Another factor that helped was the invention of the bifurcated needle, which allowed a more rapid and efficient delivery of the smallpox vaccine. The great success of ring vaccination was illustrated dramatically in late 1975 with Rahima Banu, a 3-year-old girl living in the district of Barisal in Bangladesh, who was the last person in Asia to have active smallpox.[13] She was quarantined in her family's home with 24-hour armed guards until she was no longer infectious. A house-to-house ring vaccination campaign was immediately undertaken to vaccinate every person living within a 1.5-mile radius of Rahima's home. Furthermore, a team from the Smallpox Eradication Program visited every house, school, and public meeting place within a 5-mile radius of Rahima's home to ensure that the disease did not spread. A reward was also offered to anyone for reporting a smallpox case. This was a classic example of the success of ring vaccination.[13]

Ali Maow Maalin, a hospital cook in Somalia, was the last person to have naturally acquired smallpox in Africa (Fig. 40.1). Although he had previously worked in the smallpox eradication campaign, he himself had not been vaccinated, as he was afraid of the needle. On October 12, 1977, Maalin accompanied two smallpox-infected children in a vehicle from the hospital to the local smallpox office. Several days later, he developed a fever and was correctly diagnosed with smallpox on October 30. Maalin was isolated and made a full recovery. Maalin died of malaria on July 22, 2013, while working in the polio eradication campaign.[13,14]

The last known case of smallpox in the world was the tragic story of Janet Parker, a medical photographer at the University of Birmingham Medical School in England, who died of the disease in the summer of 1978.[15] This was ironic because by then smallpox was widely thought to have been eradicated across the world. She first became sick with an influenza-like illness on August 11 and was diagnosed with smallpox 9 days later. She died on September 11, 1978. Her mother, who was looking after her, developed a mild case of smallpox on September 7, despite having been vaccinated on August 24, but recovered on September 22. How was Janet Parker infected? As it turned out, she had worked one floor above the Department of Medical Microbiology where smallpox research was being done, and it was later thought that she had been infected via an airborne route through the medical school building's air duct system.[15]

Jenner's work culminated on May 8, 1980, in the global eradication of smallpox from the planet with an announcement by the World Health Assembly: "*The world and all its people have won freedom from smallpox, which was the most devastating disease sweeping in epidemic form through many countries since earliest times, leaving death, blindness and disfigurement in its wake*."[6] Jenner's pioneering work had won the war against a disease that had plagued humanity for 3000 years, killing 300 million people in the 20th century alone.

The Dawn of Immunology: Pasteur, Koch, von Behring, and Ehrlich (1875–1910)

The Jennerian revolution heralded the birth of vaccinology, but exactly how vaccination worked and how to extend this practice to other diseases remained a mystery. Indeed, the concept that infectious disease was caused by microbes was unknown. It was Louis Pasteur (1822–1895)[16] and Robert Koch (1843–1910)[17] who established the etiological cause of infectious disease, in what was described as the "germ theory of disease." It is hard to overstate the great revolution that germ theory was, as it represented a transformational shift in consciousness in how people viewed disease. The Greek philosopher Aristotle (384–322 BC) had articulated the theory of "spontaneous generation," the idea that life spontaneously arose from nonliving matter. People believed that foul odor and evil spirits could cause a person to become ill. Ideas that were contrarian to spontaneous generation, and which, in hindsight, seem vaguely to be the first intimations of the germ theory, can be traced back to Lucretius (c. 99 BC–c. 55 BC) who wrote that the world contained various "seeds," some of which could sicken a person if they were inhaled or ingested.[1] A basic form of germ theory was proposed in 1025 by Ibn Sina in *The Canon of Medicine*, which became the most authoritative medical textbook in Europe till the 16th century, in which he proposed that people can transmit disease to others by breath, noted contagion with TB, and discussed transmission of disease through water and dirt.[18] However, such views were not accepted in Europe, where Galen's miasma theory that diseases such as cholera or the Black Death were caused by miasma (a noxious form of bad air) ruled among scientists and doctors. However, it was an experiment done by Louis Pasteur in 1859 that sounded the death knell for the idea of spontaneous generation. Pasteur boiled meat broth in a flask with a swan neck, whose curved neck isolated the broth from contact with the external environment. The broth remained free of any growth, but when the curved neck was broken, the broth became clouded[16] (Fig. 40.2). Robert Koch, Pasteur's rival, established the etiological agents of specific infectious diseases and enunciated Koch's postulates, which needed to be satisfied to establish a specific pathogen as the causative agent of an infectious disease.[17]

In 1877, Pasteur was studying chicken cholera, a diarrheal disease that was destroying breeding chicken population. He reasoned that if a vaccine could be found for smallpox, then vaccines could be found for all diseases. By 1878, he had cultured the causative agent of chicken cholera and began inoculating chicken. However, many chicken died, but in 1879, he instructed his assistant to inject chickens with a fresh culture of bacteria before a holiday. The assistant forgot to do this and went off on a holiday. When he returned a month later, he injected the month-old culture into chicken. Surprisingly, the chicken showed mild signs of disease and survived. Pasteur was puzzled by this result, so when the chicken became healthy, he injected them with fresh bacteria. The chicken did not develop cholera. Pasteur reasoned that the month-long culture had attenuated the bacteria, which, without causing disease, had taught the immune system to

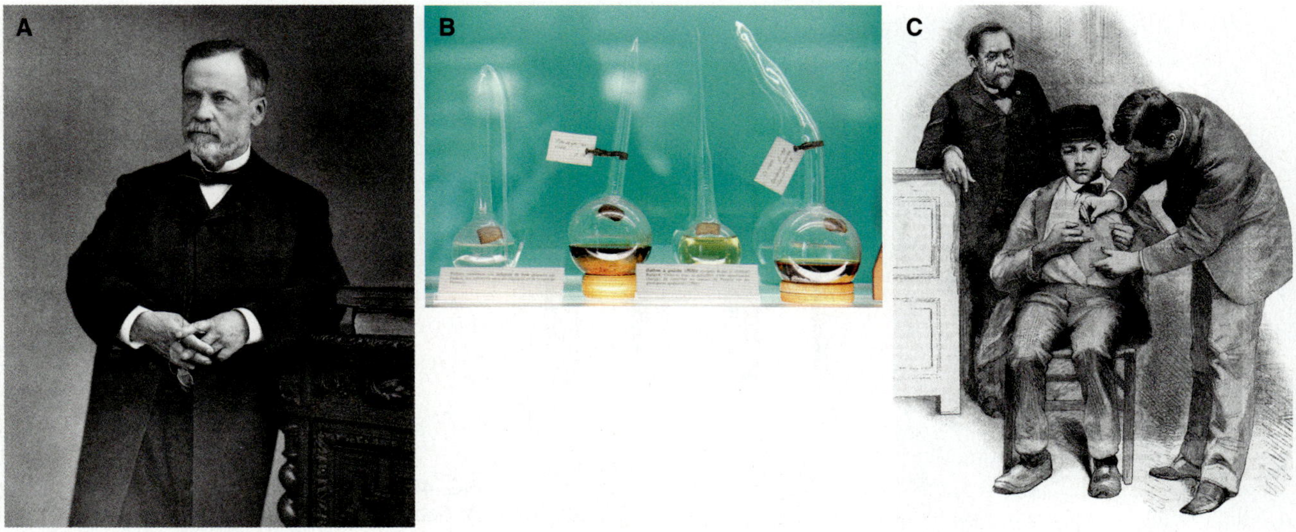

FIG. 40.2. Louis Pasteur's revolution in vaccinology. A: Louis Pasteur. **B:** Flasks used by Pasteur to disprove spontaneous generation are housed at the room of scientific souvenirs at the Musée Pasteur, Paris, France. **C:** Pasteur looks on as his assistant inoculates Joseph Meister, a shepherd boy who had been bitten by a rabid dog. (Alarmy.)

respond to the infection. Pasteur had in fact discovered live attenuated vaccines! This discovery of the chicken cholera vaccine revolutionized vaccinology and is considered the birth of immunology. As Pasteur is said to have remarked: "*Dans les champs de l'observation le hasard ne favorise que les esprits prepares!" "In the field of observation, chance favors only the prepared mind!*"[16] How many of us would have simply ignored the "failed" experiment and started a new experiment with new cultures! Pasteur thought that this approach of attenuation could be extended to other diseases. So in 1881, a mere 2 years after the "failed" chicken cholera experiment, he developed a live attenuated vaccine for anthrax, and in 1882, with surprising speed, 85,000 sheep were vaccinated. ("Warp Speed vaccines" existed even back then!) In 1885, nearly a century after Jenner's first experiment, Pasteur decided to treat his first human patient, 9-year-old Jacob Meister, who had been bitten by rabid dogs. Pasteur injected the boy with a daily series of progressively more virulent doses of the attenuated virus from rabies-infected rabbits (Fig. 40.2). The boy survived, and Pasteur became an international hero. Until Pasteur developed the rabies vaccine, the term "vaccines" had only referred to cowpox inoculation for smallpox. Pasteur's procedure was originally called "Pasteur's treatment," but Pasteur decided to honor Edward Jenner and decided to use the term vaccines for all inoculations. Royalty from all over the world came to honor Pasteur, and he built the famous Pasteur Institute in Paris entirely from benefactions. Founded in 1888, the Institut Pasteur is one of the world's premier biomedical research institutes and one of Pasteur's enduring legacies. The Musée Pasteur, located within the grounds of the Institut Pasteur at 25 rue du Docteur Roux, in Paris, was established in honor of Pasteur. The memory of this great scientist and benefactor of humanity is preserved in the apartment on the Institut's ground where he and his family lived. Personal momentos from heads of state adorn the walls. It also preserves his laboratory equipment and notebooks, including the flasks that he used to disprove the spontaneous generation experiments (Fig. 40.2). Below the museum is the Neo-Byzantine funeral chapel, where Pasteur and his wife are entombed. Louis Pasteur's work represented one of the greatest revolutions in vaccinology, and as we will see below, in subsequent decades, several live attenuated vaccines were developed against some of the world's deadliest diseases, including diphtheria (1888), TB (1927), yellow fever (1936), measles (1963), mumps (1967), rubella (1969), varicella (1995), and rotavirus (1998).[19]

Nineteenth-Century Vaccinology

It was Robert Koch who in the last decades of the 19th century established that infectious diseases are caused by specific microorganisms, thus putting the final nail in the coffin for spontaneous generation theory.[17] Koch is known as one of the founders of bacteriology and identified the etiological agents of TB, cholera, and anthrax (Fig. 40.3). In a makeshift laboratory in his own house, he examined under the microscope the blood of cows that had died of anthrax and observed rod-shaped bacteria. He hypothesized that they caused anthrax and infected mice with blood from anthrax-infected cows and saw that the mice also developed anthrax. This research led to the development of four principles of Koch's postulates that need to be satisfied to conclusively prove that a certain organism caused a specific disease: (1) the bacteria must be present in every case of the disease; (2) the bacteria must be isolated from the host with the disease and grown in culture; (3) the specific disease must be reproduced when the bacteria is inoculated into a susceptible host; and (4) the bacteria must be reisolated from the inoculated animal and matched to the original bacteria. For his discovery of *Mycobacterium tuberculosis* as the causative agent of TB, and subsequent work on

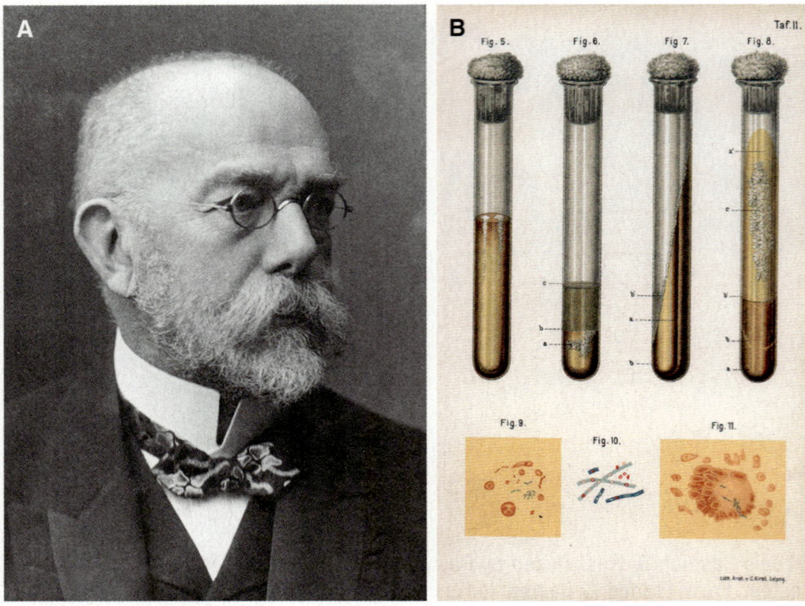

FIG. 40.3. Robert Koch, the father of microbiology. **A:** Robert Koch, German physician and pioneering microbiologist. **B:** Various types of bacteria identified by Koch, in 1889. 5: Pneumonia culture; 6: albumen from a rotten egg; 7 and 8: tuberculosis. (Alarmy.)

TB, Koch won the Nobel Prize in Physiology or Medicine in 1905. Pasteur had originally thought that vaccines could only be made with live attenuated organisms. It was Koch's discovery of *Vibrio cholera* that led Daniel Elmer Salomon and Theobald Smith to show that killed vibrio from infected pigs protected pigeons from disease. This led to the development of killed, whole organism vaccines against typhoid fever (1896), cholera (1896), and plague (1897).[19]

The last 2 decades of the 19th century represented a heady time for immunology. As if the achievements of Pasteur and Koch were not enough, Metchnikoff discovered the role of phagocytes in immunity in 1883[20] and von Behring and Kitasato discovered antibodies (antitoxin) in 1890 and pioneered serum immunotherapy against diphtheria and tetanus by passive transfer of antidiphtheria toxin–specific antibodies from horses, leading to the first Nobel Prize in Medicine to von Behring in 1901.[21] However, the results of serum therapy were variable because of poor quantitation. Robert Koch, who was the director of the Institute for Infectious Diseases (now the Robert Koch Institute) in Berlin, where von Behring and Paul Ehrlich worked, suggested that the two cooperate on the project. The collaboration was successful, and clinical tests with diphtheria serum early in 1894 were successful and the chemical company Hoechst started to market the diphtheria remedy of Behring-Ehrlich.[22] Paul Ehrlich made many seminal contributions, including a major conceptual paradigm in immunology. He postulated that cells express side chains (receptors) that bound toxins, resulting in the secretion of "immune bodies" (antibodies).[23] In 1908, Ehrlich was awarded the Nobel Prize in Physiology or Medicine, jointly with Metchnikoff, for his work on immunology.

Twentieth-Century Vaccinology

If the 18th century has heralded the birth of vaccinology with Jenner's smallpox vaccine, and the 19th century had yielded vaccines against rabies, typhoid, cholera, and plague, the 20th century was a golden age for vaccinology, with approximately 30 vaccines being licensed. Successive waves of technological innovations, including live attenuated vaccines, inactivated toxoid vaccines, polysaccharide vaccines, genetically toxified vaccines, and recombinant vaccines, have advanced vaccine development greatly (Table 1).[19]

Inactivated Pathogens and Toxoids (1910–1930)

Toward the end of the 19th century, it was discovered that killing of bacteria by heat or chemical treatment still retained their immunogenicity. Salmon and Smith in the United States and Roux and Chamberland at the Pasteur Institute developed the first inactivated vaccines.[24,25] The first inactivated vaccines to be made were against typhoid, plague, and cholera. Following the work of Kitasato and von Behring on serum therapy, during the 1920s, Alexander Glenny at the Wellcome Research Laboratories in London[26] and Gaston Ramon, a veterinarian at the Pasteur Institute, working independently developed an inactivated vaccine for diphtheria and tetanus by treating the toxoids with formalin.[27]

Live Attenuated Vaccines (1910–1950)

During the same period, major advances were being made with live attenuated vaccines. At the Pasteur Institute in Lille, France, following in the footsteps of Pasteur, Calmette and Guérin developed the live attenuated BCG vaccine against TB.[28] They isolated a virulent strain of *Mycobacterium bovis* that causes TB in cattle in 1902 and cultured it in glycerine potato medium an amazing 239 times over a period of 13 years to attenuate it. By 1919, they had a live, but highly attenuated strain that was unable to cause disease in a variety of animals, including calves, horses, monkeys, and guinea pigs. This strain was known as the BCG strain; human trials did not start immediately because of safety concerns. However, in 1921, it was given to a newborn infant whose mother had

TABLE 40.1 The History of Development of Human Vaccines

Year	Live or Live Attenuated	Killed Whole Organisms	Inactivated Toxoids	Polysaccharide or Conjugate	Recombinant Protein	Recombinant Vector	mRNA
18th century							
1798	Smallpox (1798)						
19th century							
1896		Typhoid(1896) Cholera(1896)					
1897		Plague (1897)					
20th century							
1923–1927	Tuberculosis (bacille Calmette-Guérin) (1927)	Pertussis (whole cell) (1926)	Diphtheria (1923) Tetanus (1926)				
1935–1938	Yellow fever (1935)	Influenza (1936) Rickettsia (1938)					
1955–1957		Inactivated polio (1955) Oral polio (1957)					
1969–1977	Rubella (1969)		Anthrax toxin (1970)	Meningococcal polysaccharide (1974) Pneumococcal polysaccharide (14-valent) (1977)			
1980–1987	Adenovirus (1980)	Rabies (cell culture) (1980) Tick-borne encephalitis (1981)		Pneumococcal polysaccharide (23-valent) (1983) *Haemophilus influenzae* type B polysaccharide (1985) and conjugate (1987)	Hepatitis B surface antigen (1986)		
1989–1992	Typhoid (Salmonella Ty21a) (1989)	Japanese encephalitis (mouse brain) (1992)					
1993–1999	Cholera (attenuated) (1994) Varicella (1995) Rotavirus reassortants (1999) Cold-adapted influenza (1999)		Acellular pertussis (1996)	Typhoid (Vi) polysaccharide (1994)	Cholera recombinant toxin B (1993) Lyme OspA (1998)		

(Continued)

TABLE 40.1 The History of Development of Human Vaccines (Continued)

Year	Live or Live Attenuated	Killed Whole Organisms	Inactivated Toxoids	Polysaccharide or Conjugate	Recombinant Protein	Recombinant Vector	mRNA
21st century							
2000				Pneumococcal conjugate (heptavalent) (2000)			
2005–2010	Zoster (2006) Rotavirus (attenuated and new reassortants) (2006)	Cholera (whole cell only) (2009) Japanese encephalitis (Vero cell) (2009)		Meningococcal conjugate (quadrivalent) (2005) Pneumococcal conjugate (13-valent) (2010)	Human papilloma virus—GARDASIL (quadrivalent) (2006) Recombinant shingles (2006) Human papilloma virus—CERVARIX (bivalent) (2009)		
2013					Meningococcal B proteins (2013) Recombinant seasonal influenza vaccine (Flublok) (2013) GARDASIL—human papillomavirus (9-valent) (2014)		
2019–2020						Vesicular stomatitis vector expressing Ebola antigen (2019) Ad26 vector expressing Ebola antigen (2020)	COVID-19 mRNA vaccines from Pfizer/BioNech and Moderna (2020)

Adapted from *Plotkin's Vaccines*, 7th ed, with permission from Elsevier.

died of TB, and who was to be raised by his grandmother who was also suffering from TB. During the next 3 years, a further 317 infants were vaccinated.[28] Today, the BCG vaccine is the most widely used of all current vaccines, having being administered to more than 4 billion people, with 100 million children annually. The true efficacy of the BCG vaccine has been a matter of considerable debate for decades. Multiple clinical trials have been performed with highly variable results. One reasonably consistent finding is that vaccination with BCG is 70% to 80% effective in infants and toddlers against the disseminated forms of TB, such as TB meningitis and miliary TB in children.[29] Protection lasts for 10 to 15 years, and a few studies suggest that it can even last for 50 years. It does not prevent primary infection and does not prevent reactivation of latent TB. In adults, BCG vaccination provides variable efficacy from 0% to 80%, which appears to correlate with geographical latitude and may be related to immunological sensitization to atypical, nontuberculous mycobacteria, which are particularly abundant in tropical environments.

Around the same time as the development of the BCG vaccine, another landmark event was the development of the live attenuated YF-17D in 1936. Yellow fever is a deadly disease caused by the mosquito-borne yellow fever virus and is characterized by multiorgan failure and up to 50% fatality.[30-32] The disease occurred mainly in sub-Saharan Africa and the tropical regions of South America. In the late 18th century, there were major epidemics of yellow fever in many cities in America. Refugees from a yellow fever epidemic in the Caribbean fled to Philadelphia, causing a major epidemic in the city that killed 10% of its population. In 1927, the yellow fever virus was isolated from an infected person in Ghana called Mr. Asibi (hence the name "Asibi strain"), by Stokes, Bauer, and Hudson, together with the Rockefeller Foundation's West African Yellow Fever Commission laboratory. The yellow fever vaccine was developed by Max Theiler, a South African–born virologist working at the Rockefeller University in New York,[31] who passaged the virus 176 times in mouse brain and mouse embryo tissue culture and then in chicken embryo and chicken embryo tissue culture, with the brain and spinal cord removed to prevent the emergence of neurotropic virus. The work resulted in a publication in the *Journal of Experimental Medicine* in 1937, in which the live attenuated YF-17D was first tested in humans and was shown to induce neutralizing antibodies.[32] In 1945, the WHO granted the use of two substrain seed lots of this vaccine in humans: 17DD, which is used in South America, and 17D-204, which is used in most of the rest of the world.[31] Both substrains of the vaccine are currently manufactured in embryonated chicken eggs. Theiler was awarded the Nobel Prize in 1951 for this work, and this remains the only Nobel Prize to be awarded for the development of a viral vaccine. YF-17D is one of the most successful vaccines ever made, having being administered to nearly 600 million people with an efficacy of 99%.[30,31] In 1936, the same year that YF-17D was developed, embryonated eggs were used to develop the first two human vaccines against influenza type A.

Cell Culture Revolution: Polio, Measles, Mumps, and Rubella (1950–1970)

To mass produce vaccines, the viruses or bacteria with which to develop these vaccines had to be grown in large quantities. Unlike bacteria which could be grown in culture medium, viruses needed to be grown in cell culture. In 1936, Albert Sabin and Peter Olitsky at the Rockefeller Institute successfully grew poliovirus in a culture of brain tissue from a human embryo. This was the first demonstration of the growth of the poliovirus outside the human body, but despite this success, Sabin and Olitsky were unwilling to use this method to grow virus to make a vaccine because they feared potential damage to the nervous system of vaccinees.[33] They tried to grow poliovirus in cultures using tissues from other organs without success. In 1949, John Enders and his colleagues at the Children Hospital in Boston successfully grew the Lansing strain of the poliomyelitis virus in human embryonic tissues derived from skin and muscle.[34] This discovery was the breakthrough needed to develop a polio vaccine, and Enders and his colleagues Thomas Weller and Frederick Robbins were awarded the 1954 Nobel Prize in Physiology or Medicine for this work. The first effective polio vaccine was developed in 1952 by Jonas Salk and a team at the University of Pittsburgh, who grew three wild-type strains of poliovirus—Mahoney (type 1 poliovirus), MEF-1 (type 2 poliovirus), and Saukett (type 3 poliovirus)—in monkey kidney tissue culture (Vero cell line) and then inactivated with formalin.[35] The vaccine was initially tested in a small group of adults in 1953. It was subsequently tested in the Francis Field Trial led by Thomas Francis, the largest medical experiment in history involving 1.8 million children in 44 states within the United States, in which 440,000 children received one or more injections of the vaccine, 210,000 received a placebo, and 1.2 million children received no vaccination and served as a control.[36] The results of the field trial were announced on April 12, 1955, on the 10th anniversary of the death of Franklin D. Roosevelt, perhaps the most famous victim of polio. In the United States, following a mass immunization campaign, the annual number of polio cases fell from 35,000 in 1953 to 5600 by 1957. By 1961, only 161 cases were recorded in the United States. An unfortunate tragedy was the Cutter incident, in which faulty formalin inactivation techniques resulted in 149 cases of polio.[37] In 1957, Albert Sabin developed a live attenuated OPV, consisting of the three serotypes of poliovirus.[33,38] Once Sabin's OPV became widely available, by 1965, it had essentially supplanted Salk's inactivated vaccine, which had suffered negative publicity because of the Cutter incident.[39] The advantage of the OPV is that it is easy to administer, induces mucosal immunity, and can spread by shedding to contacts, thus inducing immunity in nonvaccinees. However, OPV has also been associated with adverse reactions, including vaccine-associated paralytic polio (VAPP) or circulating vaccine-derived poliovirus (cVDPV). On September 20, 2015, the Global Commission for the Certification of Poliomyelitis Eradication concluded that wild poliovirus type 2 has been eradicated worldwide. Ninety-seven percent of cVDPV was caused by the type 2 serotype, and as a result in 2016, all 155 countries using OPV switched trivalent OPV (tOPV) with bivalent OPV (bOPV) during a synchronized 2-week period—April 17 to May 1, 2016. tOPV will not be used after this period. As of April 14, 2020, at the time of writing this review, polio has been eradicated in all countries, except in 3 countries (Afghanistan, Pakistan, and Nigeria) where a total of 41 cases of wild poliovirus type 1 serotype have been detected. The success of the live attenuated yellow fever vaccine and OPV paved the way for the development of several other live attenuated vaccines, including the MMR vaccines. In 1963, John Enders and colleagues transformed the Edmonston B strain of measles virus into a vaccine, but this was quite reactogenic, so Maurice Hilleman developed a more attenuated measles vaccine. That was followed by an attenuated mumps and rubella vaccine, and today, the measles vaccine is combined with MMR or with MMR and varicella (MMRV).

Reassortment: Seasonal Influenza and Rotavirus Vaccines

Another technological innovation that occurred in the first half of the 20th century was the technique of reassortment. Many RNA viruses have what is known as "segmented genomes," in which their genome is divided into physically distinct segments, each of which encodes one or more open reading frames. A segmented genome can result in genetic reassortment of different viral strains that have infected the same cell. This property of reassortment has been used to make seasonal influenza vaccines,[40,41] as well one of the two rotavirus vaccines.[42]

Polysaccharide Vaccines and Conjugate Vaccines

One of the earliest discoveries in the history of bacteriology was that many bacterial species are covered by polysaccharide capsules, and that antibodies against the capsule could enhance the phagocytosis of bacteria by macrophages. Gotschlich and Artenstein were the first to make a polysaccharide vaccine against meningococcus,[43] which controlled disease in military recruits. It was also discovered that there were multiple serotypes of pneumococcal polysaccharides, and subsequently, combinations of multiple polysaccharides were used in vaccines.[44,45] Bacterial capsular polysaccharide antigens were used to make vaccines against *Haemophilus influenzae* type B (Hib)[46,47] and typhoid.[48] It was subsequently discovered that although capsular polysaccharide vaccines generated serum antibodies that prevented bacteremia and organ disease in adults, they were not immunogenic in infants. This issue was solved by conjugate vaccines, in which polysaccharides were conjugated to carrier proteins, to stimulate T helper cell generation to provide help to B cells. Furthermore, while polysaccharide vaccines did not prevent nasopharyngeal carriage of bacteria, conjugate vaccines did so, thus preventing bacterial shedding and enhancing herd immunity. Although the advantage of protein conjugation to polysaccharides had been discovered in 1929 by Avery and Goebel,[49] it was only in 1954 that this concept was applied to make the conjugate vaccine against Hib.[48] Conjugate vaccines have also been developed against meningococcal and pneumococcal bacteria, and these vaccines stimulate immune responses that control invasive infections as well as the spread of organisms.

Recombinant Protein Vaccines

The advent of the recombinant DNA revolution in the 1970s ushered in a new molecular era in vaccine development. In 1986, the hepatitis B vaccine was the first vaccine to be manufactured using recombinant DNA technology and consisted of the surface antigen of the hepatitis B virus (HBsAg) that could self-assemble into 22 nM virus-like particles (VLPs), which improved its immunogenicity. The development of this vaccine represented a landmark event in vaccinology for several reasons. Not only was it the first recombinant vaccine to be made but it was also the first time that the vaccine was initially developed for use in special populations who were at high risk including doctors, nurses, and blood bank workers, thus foreshadowing the era of precision vaccines. Also, this was the first anticancer vaccine to be developed. It is estimated that there are 300 million people worldwide who are chronically infected with hepatitis B,[50] and over time, approximately 25% develop serious liver problems, including cirrhosis, liver failure, and even hepatocellular carcinoma. Therefore, the hepatitis B vaccine was the first anticancer vaccine to be developed in history.

The history of its development starts with Baruch Blumberg, an American physician and geneticist, who identified the relevant antigen from hepatitis B virus (HBV) from an Australian Aboriginal person (Australia antigen) in 1964. Blumberg, known to everyone as Barry, was an unlikely person to have made the discovery that had such a huge impact on public health. He was not a virologist, nor a hepatologist, nor an infectious disease specialist, nor an epidemiologist, and had very little clinical training in medicine.[51] During his graduate studies at Oxford, he was influenced by Anthony Allison, who had shown that people with the sickle cell trait were resistant to the deadly falciparum malaria. This cemented his interest in studying how genetic polymorphisms affected susceptibility to disease. He traveled the world collecting blood samples from different populations, and using an immunoassay known as the Ouchterlony method, he showed that the serum of a hemophiliac patient from the United States was reactive with a protein in the blood of an Australian Aborigine. He initially thought that he had discovered a new lipoprotein antigen (caused by a genetic polymorphism present in the aboriginal person), but it soon became apparent that it may have something to do with hepatitis.[52] At around the same time, Alfred Prince at the New York Blood Center was specifically looking for serum hepatitis antigen in the blood of patients with hepatitis B and realized that it was identical to the Australia antigen.[53] Under the electron microscope (EM), Blumberg had observed that purified Australia antigen appeared as small round particles that, unlike viruses, were of variable sizes between 17 and 25 nm.[53] In 1970, using an EM, David S. Dane (London) discovered that this antigen appeared not only on the small pleomorphic particles but also on larger, VLPs of about 42 nm in size, with a clearly visible inner core.[53] The Australia antigen in fact turned out to be the HBV surface protein. This antigen was the basis for the hepatitis B vaccines, developed initially as an inactivated vaccine, but later using recombinant DNA technology, as will be discussed below. Today the hepatitis B vaccine is the most effective way of preventing infection, and the WHO recommends that all infants receive the vaccine within 24 hours of birth, followed by two or three booster immunizations at least 4 weeks apart. In 1976, Blumberg was awarded the Nobel Prize in Physiology or Medicine for his work on hepatitis B (together with Carleton Gajdusek for his work on kuru) for "*their discoveries concerning new mechanisms for the origin and dissemination of infectious diseases.*" Jonathan Chernoff, the scientific director of the Fox Chase Cancer Center in Philadelphia, where Blumberg spent most of his professional life, said *"I think it's fair to say that Barry prevented more cancer deaths than any person who's ever lived."*[54]

In the 4 decades that have elapsed since its licensure in 1986, the hepatitis B vaccine has saved tens of millions of lives and transformed the way in which vaccines are made. Today, the recombinant DNA revolution has paved the way for the development of several other vaccines, including the licensed HPV vaccine, the recombinant shingles vaccine, the recombinant Ebola vaccine (in which the gene for the Ebola glycoprotein is expressed in the vesicular stomatitis viral vector or the adenovirus 26 vector), the recombinant MenB vaccine, and several other vaccines that are in advanced stages of development and testing such as the RTS,S malaria vaccine and several candidate vaccines against SARS-CoV-2. These will be discussed in detail later.

Reverse Vaccinology

Toward the end of the 20th century, the Human Genome Project was well underway, and Craig Venter published the genome of the first living organism, the *H. influenzae* bacterium. This technological *tour de force* suddenly opened a new horizon for vaccinology—that of accessing the genomes of pathogens directly, without the need to grow or attenuate microorganisms as Pasteur had done. This new approach was termed "reverse vaccinology" by Rino Rappuoli. Reverse vaccinology was first used to identify candidate vaccine antigens in MenB, a serogroup of *Neisseria meningitidis*, that caused over 50% of meningococcal meningitis, and that had been difficult to make a vaccine against using conventional approaches, because its capsular polysaccharide was identical to human self-antigens and bacterial cell surface proteins were highly variable. Rino Rappuoli and colleagues at Chiron in Siena, in collaboration with Craig Venter and his team at The Institute for Genomic Research in Rockville, Maryland, began a project to sequence the MenB genome, which culminated in 2000 with two landmark papers in *Science* that annotated the genome of the MenB strain MC58, and used these data to identify candidate vaccine antigens.[55,56] Analysis of over 600 potential antigens in mice led to the identification of some 90 novel surface proteins, of which 29 were antigenic in inducing antibodies that could kill bacteria *in vitro*, in the presence of complement. In the years that followed, several of these candidate antigens were tested in mice and later in humans, and the vaccine BEXSERO was approved for human use in Europe in 2013, and then in the United States in 2015. Today, BEXSERO is routinely used for immunization in newborns younger than 12 months in the United Kingdom and also licensed for immunization in adolescents. Currently, reverse vaccinology approaches are being applied to other pathogens, including group A and group B *Streptococcus* species, and antibiotic-resistant strains of *Staphylococcus aureus* and *Streptococcus pneumoniae.*

CLASSIFICATION OF VACCINES (FIG. 40.4)

As discussed above, successive waves of technological innovation have resulted in the generation of new types of vaccines. Today, the portfolio of licensed vaccines and those under development encompasses a diverse range of platform technologies. The historical development of many of these has been discussed above (Table 1), but here I will briefly describe these.

Live Attenuated Vaccines

Live attenuated vaccines consist of a live pathogen that has been deliberately weakened in the laboratory such that it does not cause disease but can stimulate the immune system to launch robust protective immunity against a subsequent exposure with that pathogen. Live attenuated vaccines include the most widely used vaccine in the world, such as the BCG vaccine that has an efficacy of 70% to 80% in protecting infants and toddlers against the most severe forms of TB as well as several other live viral vaccines. Other licensed live attenuated vaccines include smallpox, YF-17D, OPV, MMR, chickenpox, and the chickenpox (varicella) vaccine. These live vaccines mimic the potent and durable immunity that typically results from infection with many live viral infections and stimulate strong and very durable immune responses that can often protect against infection for several decades, and even a lifetime. As discussed above, it was Louis Pasteur who invented attenuation, with his work on the chicken cholera vaccines and his development of the live attenuated anthrax and rabies vaccines.

A common way of developing live attenuated vaccines is to culture the pathogenic virus through a series of tissue culture steps on animal tissues, such as mouse brain and mouse embryo tissues followed by culture in chicken embryo, as was the case with the live attenuated YF-17D. With each passage, the virus becomes adapted to replicating in mouse or chicken cells, but gradually loses its capacity to replicate in human cells. When the resulting attenuated vaccine is given to a human, it can evoke a potent immune response, but has greatly attenuated replication capacity compared to the pathogen from which it was derived. Therefore, the live attenuated virus is highly immunogenic, but not pathogenic. Currently, live attenuated vaccines that are recommended for use include the vaccines for MMR, varicella (chickenpox), influenza (nasal spray), rotavirus, OPV, shingles (Zostavax), and YF-17D. Such vaccines are generally highly efficacious and safe and can induce very durable immunity that can last a lifetime.

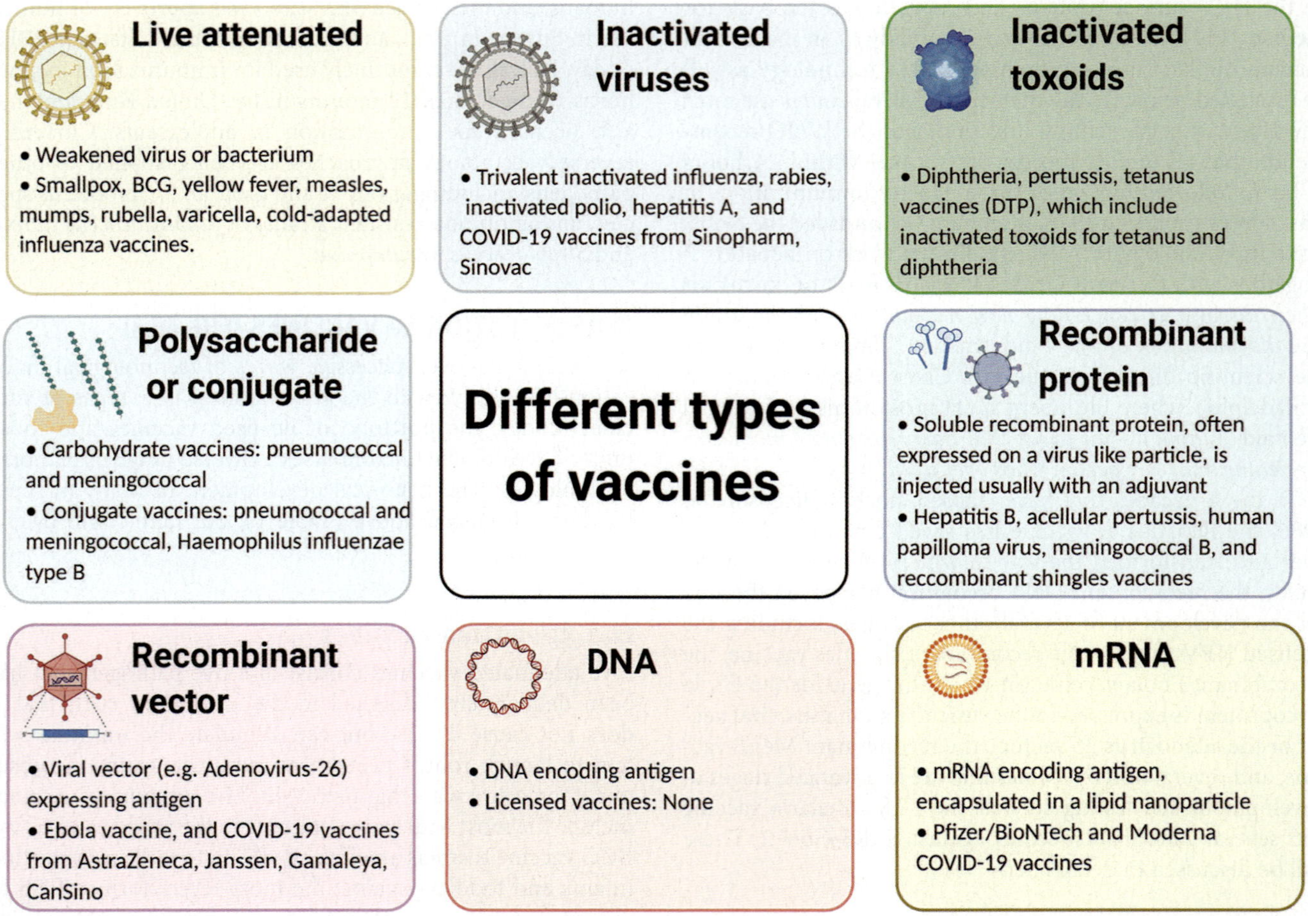

FIG. 40.4. Different types of vaccines. (Created with Biorender.com.)

Inactivated or Killed Pathogen Vaccines

Inactivated or killed whole organism vaccines are developed by inactivating the pathogen using heat or chemicals such as formaldehyde or formalin. Examples of licensed inactivated vaccines include the inactivated seasonal influenza vaccine, Salk's inactivated polio vaccine (IPV), inactivated rabies vaccine, and the hepatitis A vaccine developed by chemical inactivation of the virus by Provost and colleagues. Such vaccines cannot replicate, but can induce immunity, albeit not as strong and durable as the immune responses typically induced by live viruses. As such, they are often administered with an adjuvant, which is a substance that can enhance the strength and durability of the immune response induced by such vaccines.

Inactivated Toxoid Vaccines

The pathogenicity of many bacteria is mediated by toxins they produce. For example, tetanus is caused by a neurotoxin produced by the bacterium *Clostridium tetani,* and diphtheria is caused by a toxin produced by *Corynebacterium*. As discussed above, Gaston Ramon and Alexander Glenny were the first to produce vaccines against tetanus and diphtheria by inactivating their toxins. Similar to inactivated whole organisms, inactivation of toxins can be done by heat or formalin or other chemical treatment. Such inactivated toxins are called toxoids, and the tetanus toxoid and diphtheria toxoid are included in the combined diphtheria, tetanus, and pertussis (DTP) vaccines.

Polysaccharide and Conjugate Vaccines

Polysaccharide vaccines consist of capsular polysaccharides from bacteria and induce antibody responses that recognize these carbohydrate antigens on the surface of the bacteria and promote their phagocytosis by macrophages. Licensed carbohydrate vaccines include the pneumococcal and meningococcal vaccines, which include carbohydrates from multiple serotypes of bacteria. Although such vaccines are immunogenic in adults, they are weakly immunogenic in infants. This problem is overcome by conjugating a carrier protein to the carbohydrate to provide antigenic epitopes for T helper cells, which provide help to B cells, thus causing enhanced antibody responses to the carbohydrate epitopes. The carrier proteins that have been included in conjugate vaccines thus far are tetanus toxoid (T), diphtheria toxoid (D), a genetically modified cross-reacting material (CRM) of diphtheria toxin, meningococcal outer membrane protein complex, and *H. influenzae* protein D.[57] Today, licensed conjugate vaccines include the vaccines against Hib and pneumococcal and meningococcal infections.

Recombinant Vaccines

The genetic engineering revolution in the 1970s led to the development of recombinant vaccines, in which the gene encoding a protein from the pathogen is expressed in cells in culture, to produce purified protein that can then be used as an antigen in a vaccine. Examples of licensed recombinant vaccines include the hepatitis B vaccine, the HPV vaccine, and the MenB vaccine. In addition, recombinant DNA technology can also be used to insert a gene encoding a protein from the pathogen into another harmless carrier virus (called a "viral vector"), which can be injected into a person to induce an immune response. When the viral vector enters the body and replicates *in vivo*, it expressed the protein from the pathogen, thus stimulating an immune response against that protein. Licensed recombinant viral vector vaccines include the vesicular stomatitis virus–Ebola vaccine licensed by Merck and the Ad26.ZEBOV/MVA-BN-Filo developed by Janssen.

DNA Vaccines

DNA plasmid vaccines consist of a plasmid DNA that expressed genes encoding a protein from a given pathogen. Although currently there are no licensed DNA vaccines, their well-established manufacturing process allows them to be developed quickly during a pandemic. Furthermore, innovations in delivery methods, which include electroporation, gene gun delivery, and jet delivery, facilitate a more robust antibody and T-cell response than that observed with the traditional needle and syringe immunization.

mRNA Vaccines

The recently developed mRNA vaccines represent a new era in vaccinology. They offer many advantages over conventional vaccine approaches because of their capacity for rapid development, low cost, and immunogenicity. Many of the COVID-19 vaccines that are currently in development have used the mRNA vaccine approach and have been generated with unprecedented speed. Indeed, in November 2020, Pfizer/BioNTech and Moderna, vaccine manufacturers who are testing the candidate mRNA COVID-19 vaccines in phase III clinical trials, announced their interim analysis results of a 90% to 95% efficacy against symptomatic disease. On December 11 and 13, 2020, respectively, the Pfizer/BioNTech and Moderna mRNA vaccines were granted emergency use authorization in humans. These will be discussed in detail below.

THE IMMUNOLOGY OF VACCINES

Here, I will briefly review some general immunological principles that form the basis of vaccination. The following topics will be reviewed in great depth in the other chapters in this book and will thus be only dealt with briefly here.

Immunological Memory

The capacity of the adaptive immune system to remember an encounter with a microbe, such that it launches a faster, enhanced response upon reencounter with the microbe, is called immunological memory. This phenomenon was noted by Thucydides during the plague of Athens in ancient Greece but strikingly illustrated by the story of the Danish physician Peter Panum, who was sent to investigate a measles outbreak on the Faroe Islands in 1846. The islands had not had an outbreak for 65 years since the previous outbreak in 1781. Panum noted: "*It is quite remarkable that of the many aged people still living on the Faroes who had had measles in 1781, not one, as far as I could find out by careful inquiry, was attacked the second time.*"[58] This led to the notion that immunological memory could be maintained for a lifetime, even in the absence of antigen, and that it could be harnessed for vaccination. Immunological memory is mediated through the induction of antigen-specific memory B and T cells, as well as a durable antibody response.

Antibody Responses (Fig. 40.5)

Antibodies have long been known to be a major mechanism of immunological defense, and the magnitude of the antigen-specific antibody response is considered to be the primary correlate of protection against many pathogens. Antibodies that can "neutralize" the pathogen or toxin, by binding and preventing it from interacting with its target receptor on a host cell, are called neutralizing antibodies. Neutralizing antibodies are considered to be the main mechanism by which antibodies prevent many infections such as yellow fever, smallpox, measles, mumps, and chickenpox and are the basis on which the vast majority of vaccines against COVID-19 are being developed. In addition to neutralizing antibodies, in the case of bacterial vaccines, bactericidal antibodies and opsonophagocytic antibodies, which help phagocytic cells to phagocytose and kill bacteria, are considered to be the main mechanism of protection against bacteria.

Many viral infections such as measles, mumps, and chickenpox induce neutralizing antibody titers that are remarkably stable, with estimated half-lives ranging from 50 years for varicella-zoster virus (VZV, chickenpox) to 92 years for vaccinia to greater than 500 and 3000 years (!) for mumps and measles, respectively.[59] Consistent with this, live viral vaccines induce antibody responses that can last a lifetime. However, waning antibody titers has been documented with inactivated subunit vaccines such as those against seasonal influenza, *Bordetella pertussis*, *Salmonella enterica serovar Typhi* and *N. meningitidis*, and candidate vaccines against HIV and malaria. This lack of durability of the antibody response or "waning immunity" represents a significant shortcoming of these vaccines. Experiments in mice and humans have shown that long-lived antibody secreting plasma cells (LLPCs) in the bone marrow play a key role in the maintenance of antibody titers.[60] Therefore, learning how to design vaccines that can efficiently stimulate the generation of LLPCs is a major challenge in vaccinology and a topic of much current discussion with regard to COVID-19 vaccines.

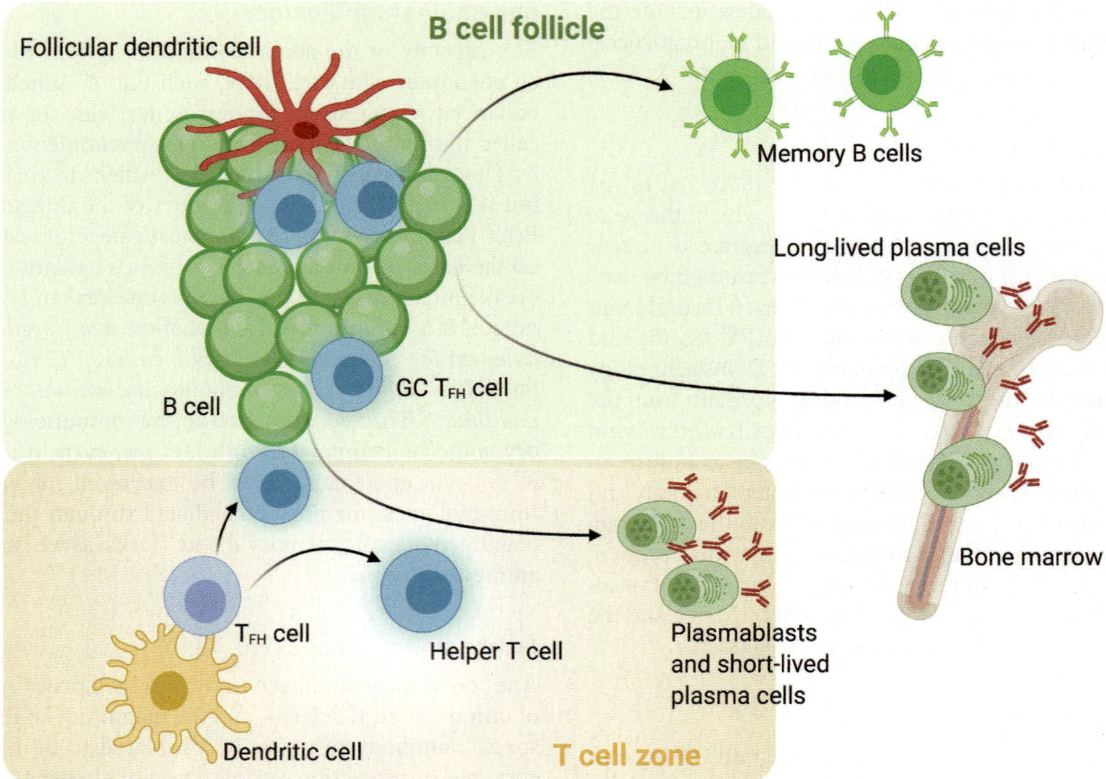

FIG. 40.5. T cell–dependent antibody responses. The germinal center response generates long-lived plasma cells in the bone marrow that maintain the durability of antibody responses. GC, germinal center; T_{FH}, T follicular helper. (Created with Biorender.com.)

The key to understanding the mechanisms that control the generation of LLPCs lies in dynamic microenvironments called germinal centers (GCs) in lymphoid organs.[61] It is in GCs that the antigen-driven selection of B cells that have acquired a higher affinity of antigen binding through the process of somatic hypermutation of their immunoglobulin genes occurs, eventually leading to affinity maturation of the antibody response. The GC response begins when, in response to vaccination, activated antigen-specific B cells migrate into primary B-cell follicles and proliferate rapidly and form GCs (Fig. 40.5). In addition, some activated antigen-specific CD4+ T helper cells express the transcription factor Bcl6, which stimulates the expression of the chemokine receptors CXCR5 and CCR7, which facilitates the migration of the T helper cells to the T-B border. Here, they are optimally positioned to interact with antigen-specific B cells and provide signals such as IL-21 and CD40L that are essential for the proliferation and differentiation of GC cells.[62] As the GC response matures, B cell–derived signals such as IL-6, IL-23, and IL-12 induce the migration of T helper cells at the T-B border and their migration into GCs, where they differentiate into GC T follicular helper (T_{FH}) cells. There, these T_{FH} cells orchestrate the process of affinity maturation and differentiation of GC B cells into memory B cells and LLPCs. Therefore, learning how to induce T_{FH} and GC responses is a key step in the generation of LLPCs, and as has been shown recently and discussed below, recent evidence points to a role for adjuvants in this process[62] (reviewed in the chapter on B-cell immunity).

Finally, antibodies can also protect against infection by certain other mechanisms that are independent of their neutralization capacity. These include a range of mechanisms, all of which involve the interaction of the Fγc receptor portion of the antibody molecules with Fγc receptors on immune cells such as natural killer (NK) cells, monocytes, or macrophages.[63] For example, one such mechanism is called antibody-dependent cellular cytotoxicity (ADCC), in which antibodies that have bound viral antigens on the surface of an infected cell also bind and cross-link Fγc receptors on NK cells, triggering NK cell degranulation and release of perforin and other molecules that target and can kill the virally infected cell.[63] Another such mechanism is antibody-dependent cellular phagocytosis (ADCP), in which antibodies that bind viral antigens expressed on an infected cell also bind Fγc receptors on macrophages, thus triggering internalization and degradation of the target cell through acidification of the phagosome. As will be discussed below, emerging evidence suggests that such neutralization-independent effector mechanisms could play a role in controlling infections against pathogens.

CD8+ T-Cell Immunity

Although antibodies are thought to be a major mechanism of immune defense against pathogens, it has long been known that T cells also play vital roles in protective immunity. As discussed above, CD4+ T helper cells are critical for the generation of antibody responses, but CD8+ T cells can kill virally infected cells directly by release of perforin or granzyme. Tissue-resident memory T (TRM) cells are a recently discovered subset of T lymphocytes that are widely distributed in tissues, and emerging data reveal a vital role for them in protective immunity.[64] In tissues, TRM cells are well positioned to respond rapidly to infections and kill virally infected target cells directly. In addition to direct killing of infected cells, TRM cells also recruit other immune cells to the site of infection, thus mobilizing multiple host defense mechanisms of tissues.[65] Recent studies have highlighted the importance or synergistic interactions between TRM cells and antibodies in vaccination-induced protection.[66] In recent studies in nonhuman primates (NHPs), it was shown that vaccination-induced CD8+ TRM cells lowered the threshold of neutralizing antibodies required for protection against mucosal challenge with a virus, indicating a potential synergy between TRM cells and antibody responses.[66] These immunological features make TRM cells an attractive target for vaccinologists, and learning how to induce TRM cells and maintain them in tissues represents an important goal in vaccinology.

Original Antigenic Sin

Original antigenic sin (OAS) refers to the propensity of the immune system to respond preferentially to the antigen encountered in a previous exposure to a vaccine or infection, when a subsequent encounter with a slightly different cross-reactive antigen occurs, either through a subsequent infection or vaccination. The immune system thus has memory against the original antigen, which confers an advantage for it to respond more efficiently to the original antigen, even when a slightly different, but cross-reactive antigen is encountered. It is thought that OAS occurs by memory B cells being triggered by weakly cross-reactive antigens to produce OAS antibodies. OAS can occur in both antibodies produced by B cells, as well as T cells. OAS was first described in 1960 by Thomas Francis Jr., in the article "On the Doctrine of the Original Antigenic Sin,"[67] and is named by analogy to the theological concept of original sin. This phenomenon has been described in mice, rabbits, ferrets, and humans, in relation to several viruses, including influenza and dengue infection and vaccination.[68] Although OAS has been described for decades, the immunological mechanism(s) that result in this outcome are incompletely understood. Understanding the mechanistic basis of OAS is therefore a critical challenge in the development of improved vaccines against influenza, dengue, and possibly SARS-CoV-2.

ADJUVANTS

Brief History of Adjuvants

An adjuvant is a substance that is added to a vaccine to enhance its immunogenicity. It is difficult to overstate the critical role of adjuvants in vaccinology. Without adjuvants, many of our licensed vaccines would not exist (Fig. 40.6). While live vaccines such as the yellow fever vaccine or BCG contain their own stimuli (including endogenous toll-like receptor (TLR) ligands such as viral RNA or bacterial DNA), which stimulate the innate immune system and in effect act as their endogenous adjuvants, subunit vaccines such as recombinant protein vaccines need to be supplemented with exogenous substances to enhance their immunogenicity. The French veterinarian Gaston Ramon, while working at the Pasteur Institute in the 1920s, showed that coadministration of inactivated diphtheria toxin with other substances such as tapioca, lecithin, agar, oil, saponin, or breadcrumbs increased antitoxin antibody responses to diphtheria.[69,70] Curiously, these substances also caused enhanced inflammation at the site of injection, and Gaston invented the term adjuvant, derived from the Latin word "to help," to denote the capacity of these additives to help enhance the antibody response. Apart from being famous for the first person to observe the adjuvant effects of substances, Gaston is also famous for being the world record holder of Nobel Prize nominations (155 nominations), without ever winning one. At around the same time, the British immunologist Alexander Glenny and colleagues, working at the Wellcome Physiological Laboratories in London, in an attempt to purify inactivated diphtheria toxin (toxoid) to make a vaccine, noticed that when aluminum sulfate was used in the purification process, the toxoid induced stronger antibody response in guinea pigs, compared to soluble toxoid.[26,71] This was the first demonstration of the adjuvant properties of alum, which has since been used in billions of doses of vaccines and is included in vaccines given to millions of people annually. However, despite its widespread use, for the 7 decades from its licensure in the 1920s till the late 1990s, alum remained the only adjuvant included in licensed vaccines such as the DTP or HPV vaccines. For this reason, the development of vaccine adjuvants has been described as one of the slowest processes in the history of medicine.[72] A major obstacle to the development of adjuvants is that despite the dozens of candidates that have shown promising results in preclinical models, only a small fraction has the safety profile required for licensure in humans.

In the 1990s, the oil-in-water emulsion adjuvant MF59 was developed by Chiron Corporation.[73,74] The first clinical trial of an MF59-adjuvanted influenza vaccine was conducted nearly 30 years ago in 1992. In 1997, MF59 was first included in a licensed product in Europe known as Fluad, which is a trivalent inactivated vaccine against seasonal influenza, licensed for adults older than 65 years. Since then, it has been administered to more than 100 million people in more than 30 countries, in both seasonal and pandemic influenza vaccines. In addition, MF59 has been used in the H1N1 pandemic influenza vaccines (Focetria and Celtura) that have been administered to diverse populations including pregnant women and young children.[73,74] In the nearly quarter century that has elapsed since the licensure of MF59, only four other adjuvants have been included in licensed vaccines. These include the AS0 adjuvant systems developed by GlaxoSmithKline (GSK) (AS01 included in the recombinant shingles vaccine Shingrix and the RTS,S malaria vaccine Mosquirix; AS04 included in the hepatitis B vaccine Fendrix, and the human papilloma vaccine CERVARIX; AS03 included in the pandemic influenza vaccines Pandemrix and Arepanrix) and cytosine phosphoguanosine (CpG) 1018 (hepatitis B vaccine, Heplisav-B, developed by Dynavax) vaccines.[75]

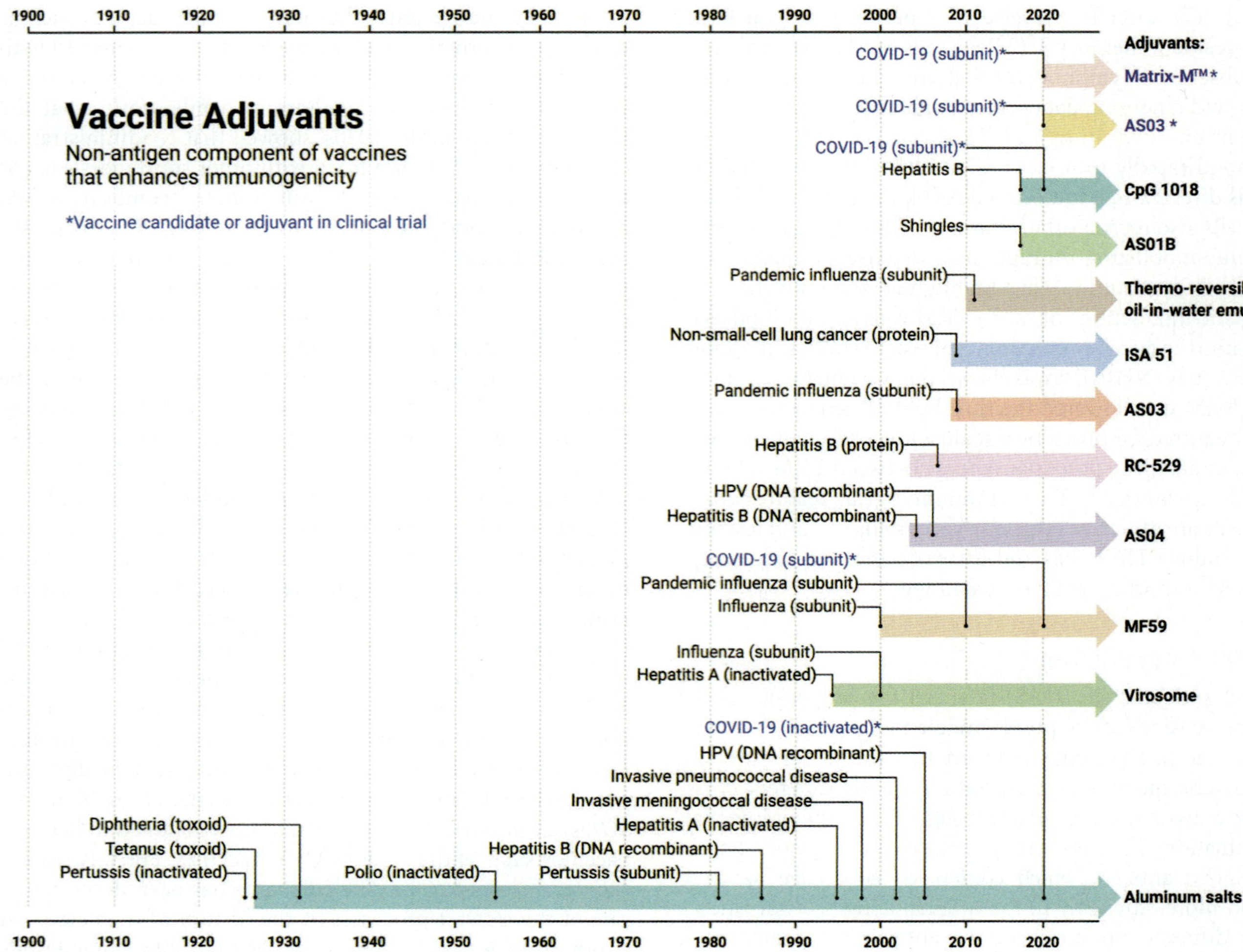

FIG. 40.6. Timeline of vaccine adjuvants. (Created with Biorender.com.)

The Science of Adjuvants

One of the ironies of the adjuvant field is that despite the widespread use of such adjuvants, their molecular mechanisms of action are not well understood. Indeed, adjuvants have remained somewhat of a "witches brew" of substances. The mechanisms through which they stimulated immune responses were not known, and it was thought that adjuvants such as alum had an "antigen depot effect," in which the antigen was released slowly over a period, thus contributing to enhanced immune responses. While it was well established that the antigen in a vaccine could be recognized by B-cell receptors and T-cell receptors on lymphocytes and trigger their activation, the notion that adjuvants themselves could activate the immune system directly was not appreciated. However, a revolution in our understanding of how the innate immune system senses microbes has provided a conceptual framework for understanding the mechanism of action of adjuvants. Innate immunity has, in effect, become the "science of adjuvants."[76,77] Charles Janeway had proposed in 1989 that the evolutionarily ancient innate immune system was endowed with the so-called pathogen recognition receptors, which could recognize highly conserved pathogen-associated molecular patterns (PAMPs) expressed by pathogens, leading to the activation of innate immune cells and the ensuing adaptive response.[78] Experimental evidence for this concept came in the 1990s with the discovery of TLRs,[79,80] which were shown to be expressed on dendritic cells (DCs),[81] and play a critical role in sensing pathogens[82] and activating DCs to stimulate antigen-specific T- and B-cell responses.[83,84] These discoveries represented a watershed moment in the history of immunology, as they provided a new paradigm of how the immune system senses microbes through specific PRRs that recognize conserved PAMPs and modulates adaptive immunity[78,85,86] (Fig. 40.7). This has fundamentally transformed our understanding of how the immune system senses microbes and launches protective immunity and was recognized by the awarding of the 2011 Nobel Prize in Physiology or Medicine, one half jointly to Bruce Beutler and Jules Hoffmann *"for their discoveries concerning the activation of innate immunity"* and the other half to Ralph Steinman *"for his discovery of the dendritic cell and its role in adaptive immunity."*[87] In the decade following the discovery of TLRs, several other families of PRRs, including RIG-I–like receptors,[88] C-type lectin-like receptors,[89] NOD-like receptors (NLRs),[90] and cytosolic DNA

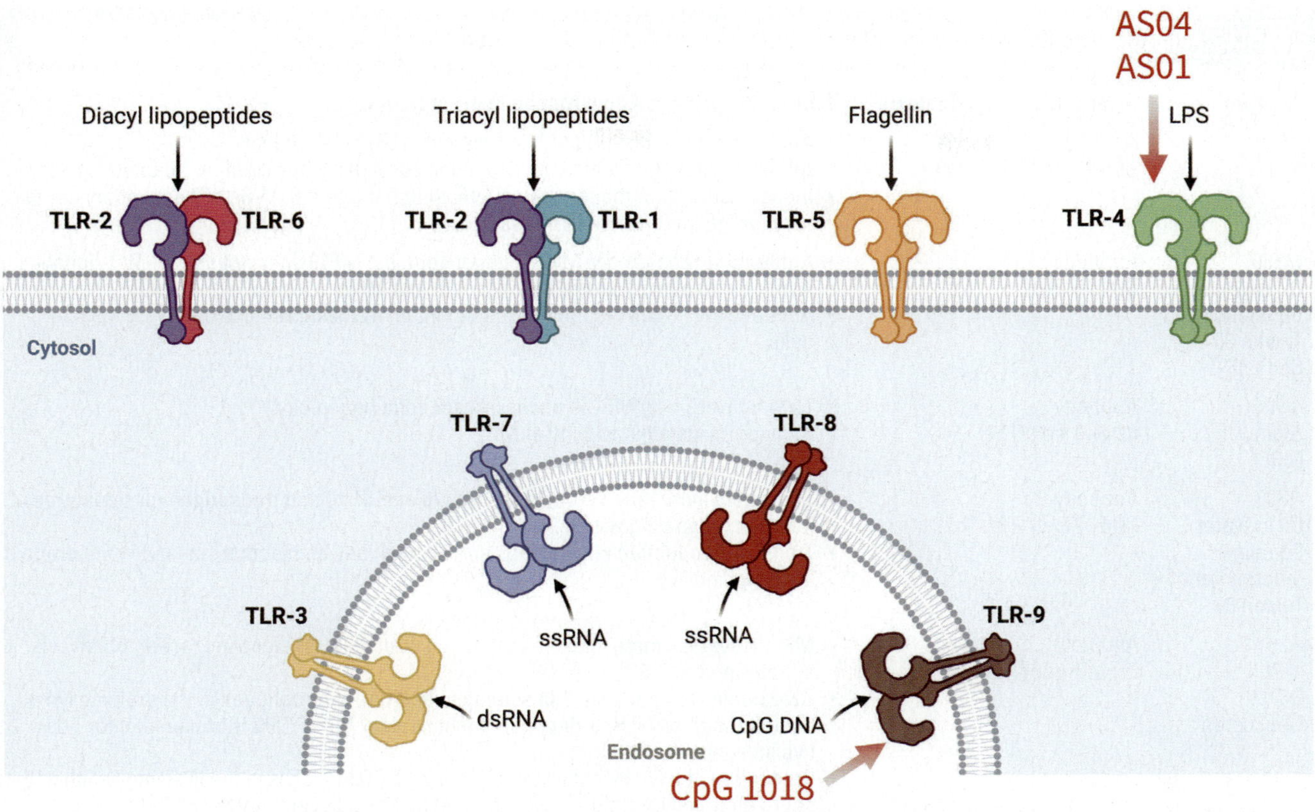

FIG. 40.7. Toll-like receptors (TLRs) represent target of vaccine adjuvants. The licensed adjuvants AS04 and AS01 target TLR4 and CpG 1018 targets TLR9. (Created with Biorender.com.)

sensing receptors[91] were identified, and these were also shown to be expressed on DCs and to stimulate adaptive immunity. These discoveries have also provided a conceptual framework for understanding how adjuvants stimulate the immune response, and in the past decade, several novel adjuvants that target specific TLRs are being developed (Fig. 40.7). These will be discussed in detail below.

Despite these seminal advances, however, our understanding of the molecular mechanisms by which adjuvants that are used in licensed vaccines (eg, alum, MF59, AS03) stimulate immune responses remains poor. What is known about the mechanisms of action of these adjuvants is summarized below, but neither alum nor MF59 nor AS03 depends on TLR signaling to mediate its immunogenicity. Interestingly, our knowledge of how the innate immune system is stimulated has evolved beyond the pathogen recognition paradigm proposed by Charles Janeway, and evidence has emerged that other physiological events such as cell death via necroptosis or pyroptosis, or amino acid starvation (that can be sensed by ancient nutrient sensing pathways),[76] or tissue damage that leads to the release of damage-associated molecular patterns (DAMPs) can also lead to activation of the innate immune system, and ensuing adaptive immunity.[92,93] The extent to which adjuvants mediate their immunogenicity through these pathways remains to be determined and will be discussed below. In the following section, I will first review our current state of knowledge about the mechanism of action of adjuvants used in licensed vaccines and then discuss the status of emerging adjuvants. (Please see chapter on pathogen recognition receptors.)

Mechanism of Action of Endogenous Adjuvants in Licensed Live Vaccines (Table 2)

Although adjuvants are only added to inactivated vaccines such as the recombinant hepatitis B vaccine, over the last decade, it has become apparent that many live vaccines that have been used safely in billions of humans contain their own "endogenous" adjuvants. For example, the live attenuated YF-17D, which is one of the most successful vaccines ever developed, having being administered to 600 million people globally, and inducing neutralizing antibody responses that last for several decades, and robust CD8+ T-cell responses, was developed by attenuation of the pathogen.[30] The mechanisms by which such potent and durable immune responses are stimulated were not known. In 2006, it was shown that YF-17D activates myeloid DCs and plasmacytoid DCs by signaling through multiple TLRs (TLR2, 3, 7, 8, and 9).[94] Experiments using mice deficient in TLR signaling revealed a key role for TLRs in mediating the immunogenicity of YF-17D.[94] In addition, later work showed that YF-17D also signals via RIG-I and melanoma differentiation-associated protein 5

TABLE 40.2 The Known Knowns of the Mechanism of Action of Licensed Human Adjuvants

Adjuvant	Immune Response Induced	Current Knowledge About Mechanisms of Action
Alum	Antibody CD4+ T cell (Th1/Th2)	• Antibody responses are toll-like receptor (TLR) independent • Activates NLRP3 inflammasome, but unclear if this is required for adaptive immunity • Induces cell death that releases DNA, which triggers STING-IRF3 activation, necessary for IgE antibody and Th2 response
MF59 Oil in water Squalene Tween 80 Span 85	Antibody CD4+ T cell	• Antibody responses are MyD88 dependent, but MF59 does not induce TLR signaling • Antibody responses are ASC dependent, but MF59 does not induce NLRP3 signaling • Antibody and T-cell responses dependent on transient release of ATP by muscle cells
AS04 Alum MPL	Antibody CD4+ T cell (Th1)	• TLR4 activation by MPL is necessary for immunogenicity • Alum prolongs adjuvanticity of MPL
AS03 Oil in water Squalene α-Tocopherol Tween 80	Antibody CD4+ T cell (Th1/Th2)	• Stimulates increased serum levels of IL-6 and IP-10 and transcriptional signatures of interferon signaling within 24 h • Transient production of cytokines and chemokines at injection site and in the draining lymph nodes
AS01 MPL QS-21 Liposomes	Antibody CD4+ T cell (Th1)	• MPL activates innate immune responses through TLR4 signaling, acting chiefly via adaptor protein TRIF • QS-21 activates caspase-1 in subcapsular sinus macrophages in the draining lymph node through an NLRP3-dependent mechanism, but NLRP3 is not needed for adaptive immunity • Production of IFNγ by natural killer cells and CD8+ T cells in the draining lymph node, hours after immunization, seems critical for the adjuvant activity of AS01
CpG 1018	Antibody CD4+ T cell (Th1)	• Signals through TLR9 and stimulates strong B-cell and NK cell activation

(MDA5).[95] In addition, BCG, which has been administered in over 4 billion doses, activates innate immunity by signaling via TLR2, TLR4, TLR9, and DC-SIGN.[96] Furthermore, in mice signaling via TLR7 was shown to be essential for the immunogenicity of the 2009 pandemic split influenza vaccine,[97] and the inactivated influenza vaccine.[98]

These findings have provided a strong rationale for the development of TLR ligand–based adjuvants, for recombinant or inactivated subunit vaccines that do not contain their own adjuvants. Thus, many studies have evaluated synthetic TLR ligands as vaccine adjuvants in mice and in NHPs.[99-106] Based on our observation that YF-17D stimulates multiple DC subsets with multiple TLR ligands,[94] we reasoned that activating the innate immune system with multiple TLR ligands could induce synergistic innate activation that promotes robust and durable antibody responses. To test this hypothesis, we developed synthetic polylactic glycolic acid (PLGA) nanoparticles containing a combination of a TLR7/8 ligand (the synthetic imidazoquinolinone compound R848, also known as resiquimod, which signals via MyD88, a critical adaptor protein in TLR signaling) and a TLR4 ligand (MPL, which signals via MyD88 and TRIF, a second adaptor protein involved in TLR4 signaling) and antigen.[104] These nanoparticle vaccines induced remarkably durable and robust antibody responses and GC and long-lived plasma cell responses that lasted for almost the lifetime of the mice.[104] Strikingly, these responses were considerably greater and durable than those induced by nanoparticles containing either TLR ligand alone. Interestingly, the adjuvant effect of these nanoparticles containing TLR ligands was dependent on activation of DCs via TLRs, as well as direct activation of B cells via TLRs. These PLGA nanoparticles containing TLR ligands were also shown to be highly immunogenic in NHPs.[66,104,105,107] In particular, we observed that immunization of rhesus macaques with nanoparticles containing a combination of synthetic TLR7/8 and TLR4 ligands induced a robust and durable antibody response and long-lived bone marrow plasma cell responses that persisted for as long as the animals were followed (~70 weeks).[105] In particular, encapsulation of the novel synthetic imidazoquinolinone 3M-052 into PLGA nanoparticle induced a very high magnitude and durable long-lived bone marrow plasma cell responses that persisted for longer than a year. 3M-052 is structurally similar to resiquimod (R848), with the exception of an 18-C fatty acyl chain that renders 3M-052 more hydrophobic, thus augmenting bioavailability at the site of immunization and in draining lymphoid organs, and reducing systemic dissemination.[105] Interestingly, we also observed that 3M-052 formulated in alum could also stimulate similar robust and durable antibody responses and long-lived bone marrow plasma cell responses, arguing that the clinically relevant alum formulation was as effective as PLGA in inducing durable antibody responses.[105] Based on these encouraging results in NHPs, 3M-052/alum is currently being evaluated as a vaccine adjuvant in phase I clinical trials (see http://clinicaltrials.gov NCT04177355). Since previous work has demonstrated that bone marrow long-lived plasma cells play a critical role in the maintenance of durable antibody responses,[108] the development of adjuvants that induce such cells is a major goal in vaccinology. In this context, recent work has shown that delivery systems that can release the antigen and/or adjuvant *in vivo* with delayed release kinetics, such as osmotic pumps[109] or slow-release gels,[110] can enhance the magnitude and persistence of antibody response, compared with the antigen plus adjuvant delivered as a bolus. While it is thought that the delayed kinetics of antigen release is the determinant of this enhanced and persistent antibody response, it is also plausible that sustained activation of innate cells by slow release of the adjuvants could also have promoted stronger and more durable antibody responses.

Mechanism of Action of Exogenous Adjuvants in Licensed Subunit Vaccines (Table 2)

Alum

Alum is the oldest adjuvant, having been incorporated into billions of doses of vaccines, and administered annually to hundreds of millions of people. Alum-adjuvanted vaccines induce antibody responses and CD4+ T-cell responses[111-113] that are Th2 polarized in mice[114-117] but have a mixed Th1/Th2 profile in humans.[118] For decades, it was thought that alum mediated its adjuvanticity by acting as an antigen depot, where antigen was released with delayed kinetics from the site of injection. Consistent with this depot effect, recent work has shown that chemical modification of antigens with short peptides composed of repeating phosphoserine residues enhanced binding to alum, leading to prolonged release kinetics and enhanced GC and antibody responses, compared to conventional alum-adsorbed antigens.[119] However, work during the past decade has revealed that in addition to its depot effects, alum also exerts potent immune activation effects on the innate immune system. Thus, experiments in mice have provided some mechanistic insights into the adjuvant effects of alum. For example, conditional deletion of DCs in mice results in impaired immune responses to alum-adjuvanted vaccines.[120] The adjuvant effects of alum was not dependent on TLR signaling, as immune responses to alum-adjuvanted vaccines in mice deficient in MyD88 and TRIF were normal.[121] Subsequent studies showed that alum activated the NLRP3 inflammasome,[122-125] although whether NLRP3 inflammasome is necessary for adaptive immunity induced by alum-adjuvanted vaccines is unclear, as some studies show impaired responses in mice deficient in NLRP3,[122,125] whereas others do not.[124,126] Interestingly, alum was also shown to induce tissue damage, and immunization with alum was shown to induce inflammatory DCs, by inducing uric acid a DAMP.[120] In addition, immunization of mice with antigen plus alum induced cell death and release of host cell DNA, which triggered type I interferon (IFN) secretion and a canonical Th2 response (characterized by T cell–mediated IL-4, IL-5, and IL-13 production) and antigen-specific IgE responses, via a mechanism dependent on the TANK-binding kinase 1 (TBK1) and the transcription factor IRF3, which have been shown to be activated by the cytosolic DNA sensor STING.[127] Thus, Th2 responses and antigen-specific IgE responses (but not antigen-specific IgG1 responses) were impaired in mice deficient in TBK1 or IRF3.[127] Subsequent work showed that STING-deficient mice were impaired in their capacity to produce antigen-specific IgE response, and CD4+ T-cell priming.[128] Taken together, these studies in mice demonstrate that alum mediates its adjuvant activity, partly by an antigen depot effect and partly by inducing cell death and release of DNA and the resulting STING-TBK1-IRF3–dependent activation of Th2 responses and antigen-specific IgE responses.[127,128] How alum mediates antigen-specific IgG1 response remains to be determined.

MF59

MF59 is an oil-in-water emulsion adjuvant that has been included in licensed influenza vaccines in Europe since 1997 and has now been administered to more than 100 million people in over 30 countries.[73] It has also recently been approved for use with Fluad and Fluad Quadrivalent, two inactivated seasonal influenza vaccines for people older than 65 years. MF59 comprises squalene oil, stabilized in an aqueous buffer by the surfactants Tween 80 and Span 85.[73] Intramuscular immunization of mice with MF59 plus antigen results in activation of DCs and macrophages that produce chemokines such as CCL2, CCL4, CCL5, and CXCL8 that recruit neutrophils and eosinophils.[74,129] The adjuvant effects of MF59 were shown to be dependent on MyD88, but independent of NLRP3 inflammasome activation.[130] Thus, immunization of mice deficient in MyD88 with an MF59-adjuvanted vaccine resulted in lower antigen-specific antibody titers compared to the titers induced in littermate control mice,[130] but this appears to occur via a TLR-independent pathway as MF59 did not stimulate TLR signaling in a cell line *in vitro*.[130] Furthermore, immunization of mice deficient in the apoptosis-associated, speck-like protein containing caspase recruitment domain (ASC), an adaptor protein within the NLRP3 inflammasome, with an MF59-adjuvanted vaccine resulted in reduced antigen-specific IgG antibody titers.[131] Interestingly, the response was intact in NLRP3 and knockout mice, suggesting that the adjuvant effects of MF59 are mediated via an ASC-dependent, but inflammasome-independent pathway.[131] Finally, immunization of mice with MF59 plus antigen stimulates antigen-specific CD8+ T-cell responses in tissues such as the lung and liver, via a mechanism dependent on the receptor-interacting serine/threonine kinase (RIP3 kinase), a key mediator of necroptosis.[132] However, this RIP3K-dependent effect on CD8+ T-cell response was independent of necroptosis because mice deficient in MLKL (a molecule that is essential for necroptosis) had normal CD8+ T-cell responses.[132] Furthermore, the antibody response induced by immunization of mice with antigen plus MF59 was not impaired in RIP3K knockout mice. However, antibody responses did seem to be dependent on apoptosis, since they were impaired by administration of the pan caspase inhibitor Z-VAD-FMK, but normal in caspase-1–deficient mice.[132]

AS04

The AS0 adjuvant systems developed by GSK over the past 3 decades include a combination of immune stimulatory molecules as well as alum and emulsions and liposomes. AS04 consists of 3-O-desacyl-4′-monophosphoryl lipid A (MPLA), a detoxified form of lipopolysaccharide (LPS) extracted from *Salmonella minnesota*, adsorbed on aluminum hydroxide.[133] AS04 is used in two vaccines: (1) the HPV vaccine that consists of the L1 protein from the HPV 16 and 18 oncogenic strains and prevents cervical cancer and (2) the hepatitis B vaccine (HBV) that consists of the recombinant HbsAg.[134-136] The adjuvanticity of AS04 is mediated by MPL-mediated TLR4 activation, as well as by the intrinsic adjuvant properties of alum. Thus, AS04-adjuvanted HBV and HPV vaccines had higher magnitude and durability of antibody responses, compared to the titers induced by the corresponding alum adjuvanted-vaccines, demonstrating the adjuvant effects of MPL in humans.[134-136] Remarkably, the efficacy of the AS04 HPV 16/18 vaccine is nearly 100% more than 9 years after primary immunization.[137]

AS03

AS03 is an oil-in-water emulsion adjuvant and, like MF59, contains squalene, α-tocopherol (vitamin E), and the surfactant Tween 80 (also known as polysorbate 80) as additional components and is included in the H1N1 pandemic influenza vaccines (Pandemrix and Arepanrix) developed by GSK.[138] The mechanisms of action of AS03 and the molecular pathways through which it stimulates innate and adaptive immune responses remain largely unknown. In humans, vaccination with AS03-adjuvanted H5N1 avian influenza vaccine stimulates increased serum levels of IL-6 and IP-10, as well as transcriptional signatures of INF signaling and antigen processing and presentation in DCs, monocytes, and neutrophils, within 24 hours.[139,140] Notably, this INF response was observed to correlate with the ensuing antibody response. In mice, immunization with AS03 plus antigen stimulates a transient production of cytokines and chemokines at the site of injection and in the draining lymph nodes[141] and enhanced antigen uptake by monocytes.[142] Despite these insights, much further work is needed to understand the molecular mechanism of action of these adjuvants. GSK has committed to manufacturing 1 billion doses of AS03 in 2021 to support multiple COVID-19 vaccine collaborations.[143] AS03 is currently being evaluated in clinical trials as a vaccine adjuvant for a number of different COVID-19 vaccines, including a recombinant spike protein vaccine being developed by Sanofi,[144] Medicago,[145] and other vaccine manufacturers.

AS01

AS01 is a liposome-based adjuvant developed by GSK more than 20 years ago and is included in a licensed vaccine against VZV (Shingrix), which is highly efficacious (97.2% overall efficacy in all age groups, and an unprecedently high efficacy in the elderly population).[146] AS01 comprises two immunostimulatory agents: MPL (the TLR4 ligand also included in AS04) and the saponin QS-21, quenched with cholesterol.[147] QS-21 is a saponin molecule isolated and purified (fraction 21) from the bark of the evergreen, soap bark tree *Quillaja saponaria* Molina, which is native to warm, temperate central Chile. Preclinical studies showed that QS-21 itself was a potent adjuvant and could enhance both antibody and cell-mediated immune responses.[148,149] However, although the QS-21 alone was potent, there were concerns about its reactogenicity when used in human vaccines. Cholesterol was included in the liposomes to quench (detoxify) QS-21, which like other saponins can punch holes in the lipid bilayer of cell membranes. Cholesterol provides a substrate such that QS-21 punches holes in the liposomes, rather than in cells. AS01 is also used in the RTS,S malaria vaccine (Mosquirix), which is currently being implemented in Africa in a limited campaign.[150]

What is known about the molecular mechanism of action of AS01 is that both MPL and QS-21 exert potent immune stimulatory effects. MPL activates innate immune responses through TLR4 signaling, acting chiefly via adaptor protein TRIF. Experiments in mice show that QS-21 activates caspase-1 in subcapsular sinus macrophages (SSMs) in the draining lymph node through an NLRP3-dependent mechanism,[151] NLRP3 does not seem to play a role in adjuvanticity *in vivo*.[152] When formulated in liposomes, QS-21 enters the cells through cholesterol-dependent endocytosis and induces lysosomal destabilization, followed by Syk kinase activation.[153] Collectively, the combination of the specific activation of different innate pathways by the two individual molecules is critical for the full adjuvant effect of AS01, as depletion of TLR4, caspase-1, or SSMs individually impairs the adjuvant effect in mouse models. Hence, a striking feature of AS01 is the synergy that occurs to induce novel pathways that are not triggered by either component alone.[151,154] Early production of interferon gamma (IFNγ) by NK cells and CD8+ T cells in the draining lymph node, hours after immunization, seems critical for the adjuvant activity of AS01, as blockade of this cytokine *in vivo* abrogates the adjuvant activity AS01.[155] Thus, AS01 is a chimeric adjuvant in which two well-established adjuvant molecules (MPL and saponin) were co in a novel delivery system (liposome), resulting in synergistic pathways of activation of the innate immune system. Its extraordinary success with the Shingrix vaccine, particularly in promoting an unprecedentedly high degree of efficacy in the elderly population, has raised its profile as a kind of aspirational target for other adjuvants. The one caveat is that AS01 has been reported to have some reactogenicity, almost always nonserious reactions such as local inflammation at the site of injection, fever, chills, fatigue, and headache.[155] Therefore, a detailed molecular understanding of the immunogenicity and reactogenicity of AS01 will guide the development of new adjuvants that recapitulate the potent immunogenicity of this molecule, but devoid of its reactogenicity.

CpG 1018

Unlike mammalian DNA, bacterial DNA has a high frequency of unmethylated CpG dinucleotides.[156,157] Like bacterial DNA, synthetic oligonucleotides with unmethylated CpG dinucleotides signal through TLR9 to induce potent innate activation.[158,159] In humans, TLR9 is expressed on B cells and plasmacytoid DCs, but in mice, TLR9 is also expressed on myeloid cells.[160] TLR9 is expressed in the endosomal compartments, thus CpG DNA needs internalization and endosomal maturation to activate TLR9. Different classes of CpG DNA have been described. The CpG DNA of the class B is a very potent Th1 adjuvant and stimulates strong B-cell and NK cell activation.[161] In contrast, CpG DNAs of the class A have phosphorothioate-modified ends and a phosphodiester palindromic center portion, which facilitates the spontaneous formation of nanoparticle-like complexes and their localization to transferrin receptor+ early endosomes in plasmacytoid DCs, where they stimulate IFNα production, via a TLR9-dependent mechanism.[162,163] The class C CpG DNA has functional properties of both class A and B and is able to stimulate Th1 responses, and B cells as well as type 1 IFN secretion.

CpG 1018, a 22-mer unmethylated CpG class B oligonucleotide developed by Dynavax, is the adjuvant that is used in Heplisav-B, one of the vaccines against hepatitis B, licensed for use in adults (older than 18 years).[164] A major advantage of Heplisav-B is the simplified two-dose regimen, unlike the three-dose regimen used with the other HBV vaccines that were already licensed and widely prior to the development of Heplisav-B. CpG nucleotides enhance the immunogenicity and durability of antibody responses in both immune-competent and vaccine-hyporesponsive populations such as the elderly.[165] CpG 1018 is currently being evaluated in clinical trials as a potential vaccine adjuvant for COVID-19 vaccines.[145]

Liposomes and Virosomes

Liposomes consist of amphipathic lipid molecules such as phospholipids that self-assemble as lipid bilayers in the form of closed vesicles that encapsulate an aqueous volume. Liposomes are critical components of two of the six types of licensed adjuvants. As discussed above, AS01 consists of liposomes containing MPLA and QS-21 and has been used with the Shingrix and Mosquirix vaccines.[146,147,150] A specialized form of liposomes called virosomes (VLPs that contain functional viral fusion proteins and viral antigens and are efficiently delivered to antigen-presenting cells, thus enhancing the immunogenicity of the vaccine) were approved for use with a hepatitis A vaccine[166] and an influenza vaccine.[167] The first virosomal vaccine to be approved for human use was the hepatitis A vaccine Epaxal, which was approved in Switzerland in 1994.[168] It was followed a few years later by Inflexal V, an influenza vaccine.[169] Both of these products are now owned by Crucell, which is part of Janssen Pharmaceuticals, but both products were discontinued by Crucell in 2014.[170] Mymetics, a United States–registered biotechnology company based in Switzerland, also has a pipeline of potential virosomal vaccines to influenza, RSV, malaria, and chikungunya (https://www.mymetics.com). Medicago, a Canadian vaccine company, is developing VLP vaccines against influenza[171] and COVID-19,[145] using living plants as bioreactors to produce noninfectious VLPs that express the antigens contained in the target virus, but lack the core genetic material that makes them noninfectious and unable to replicate. Therefore, liposomes and virosomes hold promise as vaccine adjuvant and delivery technologies.

Novel Adjuvants in Clinical Development

The discovery of TLRs and other PRRs in the late 1990s and early 2000 has led to intense interest in evaluating various TLRs as targets for adjuvant development. Extensive preclinical studies in mice and NHPs have revealed the potent adjuvant effects of various TLR ligands,[99-106] and numerous clinical trials are underway to test them in humans. The TLR9 ligand CpG DNA 1018 and the TLR4 ligand MPL, included in AS04 used in HBV and HPV vaccines[134-136] and AS01 used in Shingrix and Mosquirix,[146,150] have already been used in licensed vaccines. Other adjuvants that target TLR3, TLR7, TLR8, or TLR7/8, or RIG-I are being developed.[172] In addition, a plethora of TLR agonists are being evaluated in the immunotherapy of cancers.[173-175] Here I briefly summarize these ligands that are currently in clinical testing as vaccine adjuvants against infectious diseases.

TLR7 and 8 Ligands

In humans, TLR7 is expressed constitutively in the endosomal compartments of plasmacytoid DCs and B cells. TLR8 is also expressed in endosomal compartments, but unlike TLR7, it is expressed in human myeloid DCs and monocytes, but not in plasmacytoid DCs. TLR7 ligands activate plasmacytoid DCs to produce IFNα, whereas TLR8 ligands stimulate IL-12 production from myeloid DCs and monocytes.[81,176] Ligands that stimulate signaling via both TLR7 and TLR8 induce both IFNα in plasmacytoid DCs and IL-12 in myeloid DCs and monocytes and are potent adjuvants at stimulating T- and B-cell responses. Notably, our previous work showed that the highly successful live attenuated YF-17D signaled via TLR7 and 8.[94] Several companies including 3M and GSK are developing synthetic small molecules that target TLR7/8.[177] A TLR7 ligand developed by 3M known as Aldara is licensed for use as a 5% imiquimod cream, for topical applications against actinic keratosis, basal cell carcinoma, and for anogenital HPV warts.[178] Furthermore, TLR7 or TLR7/8 agonists have been extensively investigated in the immunotherapy of various malignancies, either alone or in combination with peptide-based DCs or tumor cell lysate vaccines.[179] Yet, despite encouraging enhancement of tumor-specific immunity, there has been limited success in terms of therapeutic clinical success. The reasons for this are many, including immunosuppressive circuits in solid tumors and acquired immune resistance mechanisms in tumors, such as loss of critical antigenic tumor epitopes or loss of key proteins that mediate tumor cells' sensitivity to immune effectors.[173-175,179] However, first-generation TLR7/8 ligands such as the TLR7 agonist imiquimod and the TLR7/8 agonist resiquimod are being actively explored as potential vaccine adjuvants for infectious diseases.[99,172,180,181] Several studies in mice and in NHPs have assessed the adjuvant effects of TLR7 ligands such as R837 or TLR8 ligands such as R848, used as a soluble form mixed with the vaccine antigen, or covalently linked to the antigen, or encapsulated in nanoparticles and codelivered with the antigen, or coadministered with DNA vaccines.[66,101-105,107,182-188] Such studies have demonstrated significant adjuvant effects of these synthetic TLR agonists, and in general, encapsulation in nanoparticles or covalent coupling with the antigen enhances their adjuvant effects, resulting in a higher magnitude of antigen-specific T- and B-cell responses. However, a problem with such adjuvants has been the release of systemic cytokines leading to systemic side effects.[189] Therefore, significant recent effort has focused on the design of second-generation TLR7/8 ligands that exert their adjuvant effects at the local site of immunization and draining lymph nodes, without causing systemic side effects. For example, the novel TLR7/8 ligand known as 3M-052 developed by 3M contains an 18-carbon fatty acyl chain added to imidazoquinoline ring, which makes the molecule more hydrophobic and thus facilitates its sustained, slow release from the immunization site and easier to incorporate into oil-in-water emulsions or liposomes.[178,190] In mice, immunization with 3M-052 encapsulated in liposomes with recombinant hemagglutinin (HA) surface antigen from influenza,[178] or hepatitis B surface antigen, or the recombinant LecA antigen from *Entamoeba histolytica*,[191] was shown to drive potent antibody responses and a Th1 response, in the absence of systemic TNFa or Th1 cytokines. Further studies showed that adsorption of 3M-052 in aluminum hydroxide enhanced the adjuvant activity of 3M-052.[192] Experiments in macaques showed potent adjuvant effects of 3M-052 in the context of pneumococcal[193] and HIV[66,105,107,194] antigens. A striking feature of 3M-052 encapsulated in PLGA nanoparticles or formulated in alum is its capacity to induce high magnitude and remarkably durable antibody responses and long-lived bone marrow plasma cell responses.[105] In addition, 3M-052 encapsulated in PLGA nanoparticles or formulated in alum induced robust activation of blood monocytes, and a strong plasmablast response, GC B cells and T_{FH} cells, and persistent antigen-specific plasma cells in draining lymph nodes. These results demonstrate the superior efficacy of 3M-052 in stimulating robust innate immunity leading to a high magnitude and durability of antibody responses and long-lived bone marrow plasma cells.[105] The research data from this study have prompted a phase I clinical trial to assess the potential of 3M-052 in the context of HIV Env antigens; see http://clinicaltrials.gov NCT04177355. Furthermore, 3M-052 is being evaluated for its adjuvant potential in the context of malaria antigens. Although no serious adverse reactions were observed in any of these studies, in some studies, mild to moderate local erythema (redness), which resolved spontaneously, was observed in some of the animals.[193] Thus, a key goal in the translation of this adjuvant to the clinic will be to optimize the correct dose that yields strong immunogenicity, without any systemic or local adverse reactions. There are also several other TLR7 agonists under development as vaccine adjuvants. A TLR7 agonist of the benzonaphthyridines series adsorbed to aluminum hydroxide (Alum-TLR7), and in mice, it was shown to have potent adjuvant effects in enhancing immune responses to glycoconjugate vaccines, and a recombinant candidate vaccine against the *N. meningitidis* B (MenB), as well as vaccines against *B. pertussis* and anthrax, with inflammation that was spatially restricted to the site of immunization.[195,196]

TLR3 Ligands

TLR3 recognizes double-stranded RNA associated with viral infections and induces activation of IRF3 and NF-kB. The synthetic TLR3 agonist polyinosinic:polycytidylic acid (poly(I:C)) is expressed in the endosomal compartments of B cells, macrophages, and DCs.[81,176] A wealth of preclinical data generated in preclinical and clinical studies have demonstrated potent adjuvant effects of poly(I:C). Poly-ICLC (Hiltonol, Oncovir, Inc) is a synthetic double-stranded RNA that signals through both TLR3 and MDA5 and stimulates potent activation of dendritic and NK cells.[197-199] Hiltonol is being tested as an adjuvant in the context of peptide-based cancer vaccines, as well as in situ tumor vaccination by mimicking natural immune responses to infection, and has been shown to enhance the effectiveness of cancer immunotherapy by promoting T-cell tumor infiltration.[200,201] There are significant clinical data with the use of poly-ICLC in cancer immunotherapy, demonstrating enhanced local and systemic immune responses and clinical benefit in some of the patients.[200-202] However, there are currently no approved products containing this class of molecules as vaccine adjuvants against infectious diseases. A challenge has been with consistent manufacturing of RNA-based TLR3 ligands, but improvements in agonist design and manufacturing have led to the development of Riboxxol, a 50 bp developed by Riboxx, which has the advantage of molecular homogeneity and manufacturing advantage. Furthermore, Riboxxol binds exclusively to TLR3 and not to RIG-I or MDA5, thus there may be subtle but important differences in the types of innate and adaptive immune responses to that induced by poly(I:C).[203]

TLR4 Ligands

TLR4 ligands have the longest history of clinical development as adjuvants. Edgar Ribi, while he was working at the Rocky Mountain Laboratories of the National Institutes of Health (NIH) in Hamilton, Montana, and later at his company known as Ribi ImmunoChem, developed MPLA, when he was trying to separate the adjuvant activity of bacterial LPSs from its toxic effects.[204] Ribi ImmunoChem developed various MPL-based adjuvants but was subsequently bought by the Seattle based company Corixa Corporation, which was later sold to GSK, which continues to produce MPL at the Hamilton site. GSK coupled formulation technology with MPL to generate adjuvants such as AS04 (used in the HBV vaccine Fendrix and the HPV vaccine CERVARIX) and AS01 (used the recombinant shingles vaccine Shingrix and the RTS,S malaria vaccine Mosquirix). The Infectious Disease Research Institute (IDRI) in Seattle has developed several formulations of synthetic TLR4 agonists known as GLA, including the squalene oil in water emulsion adjuvant (GLA-SE), liposomal QS-21 (GLA-LS), and nanoalum. Currently, GLS-SE is being tested in several clinical trials in vaccines against TB, leprosy, schistosomiasis, and cancer.[205-207]

Saponin-based Adjuvants

Saponins (also called triterpene glycosides) are plant-derived organic chemicals that form foamy bubbles when agitated in water.[208] They are widely distributed but found particularly in the soapwort plant and the soapbark tree (*Q. saponaria*) and used in soaps. Saponins obtained from *Q. saponaria* are potent adjuvants and used in AS01b. Saponins have also been used in ISCOMs (immunostimulating complex). ISCOMs are nanoparticles of 40 nM diameter and consist of stable complexes composed of saponin, cholesterol, phospholipid, and incorporated antigen(s). In mice, ISCOMs induce long-lasting biologically active antibodies and cellular immune responses including cytotoxic T cells in mice.[209] A major challenge with manufacturing ISCOMs was with the efficient incorporation of antigens, but it was subsequently learned that physical incorporation of the antigen was not as crucial as initially thought. Therefore, matrix formulations that are homogenous populations of physically stable nanoparticles consisting of saponin, cholesterol, and phospholipid were developed and could be mixed with antigens. The novel formulation Matrix M consists of two separate matrix particles, each with a different saponin fraction, resulting in improved adjuvanticity and safety. One type of particle consists of highly adjuvant active saponin molecules (Fraction C saponin), and the other of a weaker but very well-tolerated saponin (Fraction A). By mixing the two matrix particles, the dose of the highly active but slightly reactogenic Matrix C component may be lowered by a factor of 5 to 10 while keeping the high-dose adjuvant activity.[208,210-212] Hence, the safety margin is substantially increased, which is essential for certain sensitive animal species. Matrix M has recently entered into a human clinical trial with promising results in terms of both efficacy and safety.[213] Matrix M is currently being used as an adjuvant in a phase III clinical trial to evaluate the efficacy of the recombinant SARS-CoV-2 spike (S) protein vaccine against COVID-19 being developed by Novavax. The results of the phase I and II studies indicate that the adjuvant induces potent neutralizing antibodies.[214]

The Critical Role of Formulations

As will be evident from the discussion above, formulation of adjuvants can have a major impact on their potency. Thus, adjuvants such as AS04 and AS01 included in licensed vaccines or those under development such as 3M-052/alum or GLA-SE include TLR ligands formulated in alum or saponin or squalene. Indeed, optimal formulation in delivery systems including nanoparticles has been shown to greatly enhance the adjuvant properties of synthetic molecules such as TLR agonists.[215,216] Furthermore, recent work in mice and macaques have shown that formulations can alter the release kinetics of the antigens, thus modulating the magnitude of the antibody responses.[109,110] In addition, the physical properties of nanoparticles such as size, antigen density, or the glycosylation of antigens on the surface of nanoparticles can affect their immunogenicity.[217] To the latter point, glycosylated antigens displayed in a multimeric form on the surface of nanoparticles can interact with mannose-binding lectins, which facilitate their rapid shuttling to follicular DCs in GCs, leading to enhanced antibody responses.[218]

SOME EXAMPLES OF LICENSED VACCINES

The development of viral vaccines has led monumental and historic accomplishments in public health, such as the eradication of smallpox, the near eradication of polio, and dramatic impact in controlling MMRV and hepatitis B and A. However, despite decades of effort, the development of vaccines against pathogens such as HIV, malaria, and TB has proved challenging, and although the seasonal influenza vaccine we have can be reasonably effective, the threat of strain of pandemic influenza has reenergized the quest for a universal influenza vaccine. While it is not possible to provide a comprehensive survey of all successful vaccines, and discuss all emerging vaccines, the author wishes to highlight some of the triumphs and challenges we have faced. For an excellent comprehensive discussion of vaccines against the entire gamut of vaccines, the reader is referred to the *Vaccines* text book by Stanley Plotkin, Walt Orenstein and Paul Offit.[219] Rather, here I wish to highlight some of the triumphs of the past and challenges we face in the quest to develop viral vaccines against old recalcitrant pathogens and insidious new pathogens such as SARS-CoV-2. Indeed, the emergence of COVID-19 has underscored the global imperative to develop vaccines against emerging pathogens.

Polio Vaccines and the Global Eradication of Polio

The eradication of smallpox is one of the greatest achievements in human history, but smallpox remains the only human disease to be eradicated in human history. Another disease that has almost been eradicated from the planet is polio.[220,221] Polio, or poliomyelitis, is a crippling, potentially fatal disease that has been known since ancient times. Polio is caused by the poliovirus, which is a human enterovirus of the *Picornaviridae* family, composed of a single-stranded, positive-sense RNA genome and a protein capsid.[222,223] It spreads from person to person through fecal-oral transmission or by consumption of contaminated food. It mainly affects children younger than 5 years, and in most people, it causes asymptomatic infection that is spontaneously resolved. However, about 24% of those infected develop minor clinical symptoms such as sore throat, fever, headache, and cold. In less than 1% of infections, it can lead to paralytic polio when the virus enters the central nervous system and replicates in the motor neurons of the spinal cord, and 5% to 10% of these people can die when their breathing muscles become paralyzed.[223] There are three antigenically distinct serotypes of the poliovirus, and although infection can provide lifelong immunity, infection with one serotype does not confer protection against the other two serotypes. For much of the first half of the 20th century, the specter of polio cast a shadow of terror in the mind of every parent. In the 1940s, the frequency of polio outbreaks in the United States increased with an average of more than 35,000 disabled people per year. Parents faced the very real prospect of their children becoming crippled or being imprisoned in large tank-like respirators called iron lungs as they labored to catch every breath, or even dying. In the summer months, when the virus was raging, parents were frightened to let their children play outside, travel was restricted, and quarantines imposed. As discussed above in the section on the History of Vaccines, the development of Salk's IPV in 1955 and Sabin's OPV in 1961 had a dramatic impact on polio. Briefly, Salk produced his formalin-inactivated trivalent IPV (consisting of the three strains of poliovirus) by growing the virus on monkey kidney cells and inactivating with formalin.[35,38] In 1954, the Salk vaccine was tested in the Francis field trial with 1.6 million children and found to be effective.[36] Within 2 weeks of announcement of the results of that trial on April 12, 1955, the vaccine was administered to school children. Sabin developed his live attenuated OPV consisting of three strains, by sequential *in vitro* and *in vivo* passages of the wild strains.[33,38] He reported the results of his trivalent oral vaccine given to 26,033 children from a city of 100,000 people in South America. The immunogenicity and efficaciousness of the vaccine, and with the tragedy of the Cutter incident caused by faulty inactivation with Salk's IPV, led to diminished confidence in that vaccine and its gradual replacement by OPV.[39]

Polio Endgame

The stunning success of polio vaccines has led to a dramatic decline in the number of polio infections throughout the world. In 1988, the World Health Assembly of the WHO adopted a resolution for the worldwide eradication of polio. It launched the Global Polio Eradication Initiative (GPEI), a public-private partnership consisting of the WHO, Rotary International, the U.S. CDC in Atlanta, and the UNICEF.[224] Two other key stakeholders established in the late 1990s, the Bill & Melinda Gates Foundation (BMGF) and the GAVI Alliance, later joined GPEI. Its goal was to eradicate polio worldwide. But why is polio such a good target for eradication? Eradication is defined as "the *worldwide absence of a specific disease agent in nature as a result of deliberate control efforts that may be discontinued where the agent is judged no longer to present a significant risk from extrinsic sources.*"[225] The absence of a nonhuman reservoir so that the virus does not replicate in the environment, the availability of sensitive and practical tools to diagnose polio (clinical symptoms, polymerase chain reaction) and effective vaccines, and proof of eradication in all but two countries in the world, all make polio an excellent target for eradication.

Because of vaccination, two of the three strains of wild poliovirus have been eradicated throughout the world, and only the wild poliovirus type 1 strain is circulating in two countries: Pakistan and Afghanistan. In August 2020, Africa was the last WHO region to be certified as wild poliovirus free. In Pakistan and Afghanistan, a total of 33 cases were reported in 2018, and 94 cases in 2019. Despite the great strides made by GEPI, as of September 7, 2020, there were 61 cases in Afghanistan, and as of September 16, 2020, there were 142 cases in Pakistan. This increase in 2020 has been due to the COVID-19 pandemic mitigation efforts that have halted the polio vaccination campaign for several months.

A major challenge with polio eradication is that the live attenuated strains used in the OPV, which replicate in the gut and are excreted for several weeks, can, on very rare occasions, regain virulence through reversion of key attenuating mutations. In areas of low vaccine coverage, this genetic instability can lead to cVDPVs, which are transmissible and cause polio in the community.[226,227] This has led to outbreaks of polio caused by cVDPV, which is rapidly spreading in outbreaks reported from nearly 20 countries of four different WHO regions (African, Eastern Mediterranean, Southeast Asian, and Western Pacific) (between 2 and 364 cases per year between 2008 and 2019).[220] In addition to cVDPV, very rarely there can VAPP, which is an adverse event following exposure to OPV. There is very little evidence that the vaccine virus circulates from VAPP cases, and there are no outbreaks associated with VAPP.[226,227] Interestingly, 97% of cVDPV cases and nearly a third of the VAPP cases were caused by serotype 2, and therefore, the risks associated with this serotype in the tOPV vaccine were thought to outweigh the benefits of vaccination. Thus, in 2016, all 155 countries that used OPV switched to a bOPV (which excludes serotype 2), during a synchronized 2-week period between April 17 and May 1, 2016. In addition, countries started to introduce at least one dose of IPV (containing all three strains), to ensure that immunity to serotype 2 is maintained even after that serotype is removed from the OPV vaccine. The polio eradication campaign is at a critical juncture today. There is no doubt that it has succeeded in eradicating this disease that has afflicted humanity since ancient times, from almost every part of the world. Yet, significant challenges remain, and a sustained commitment to vaccinating populations throughout the world will be imperative to achieve global eradication of this disease.

Seasonal Influenza Vaccine (Fig. 40.8)

Influenza represents a serious global health issue. It is estimated that, annually, there are more than 1 billion cases, of which 3 to 5 million are severe cases, and as many as 290,000 to 650,000 influenza-related respiratory deaths worldwide.[228] The name "influenza" comes from a medieval Latin word meaning "influence," an epidemic attributed to "influence of the stars," that originated in 15th-century Italy. Influenza-like illnesses have been described in Europe at least since the time of Charlemagne, but the earliest record of a pandemic that clearly documents influenza-like illness can be traced back to the pandemics of 1510 and 1580.[229,230] Since then, numerous influenza pandemics have occurred throughout history. In the 19th century, there have been at least four influenza pandemics, three have occurred in the 20th century (with the 1918 "Spanish flu" killing an estimated 50 million people), and the 2009 H1N1 Swine flu pandemic represents the first to occur in the 21st century.[231]

The influenza virus was isolated in 1933 by Smith, Andrewes, and Laidlaw at the National Institute for Medical Research in Mill Hill, London, who filtered throat washings from infected individuals through a membrane impenetrable to bacteria and were able to transfer the filtrate to ferrets that developed high fever and influenza-like symptoms.[232] They were further able to show transmissibility of the virus between animals by placing normal ferrets in the same cage as infected ferrets, and they further showed that infected ferrets that recovered from the disease had acquired immunity to the same strain of the virus, and that the serum from a recovered animal could neutralize the virus.[232] This was a momentous landmark, and is said to have prompted one of the scientists to exclaim: "*The ferrets are sneezing!*" Later it was observed

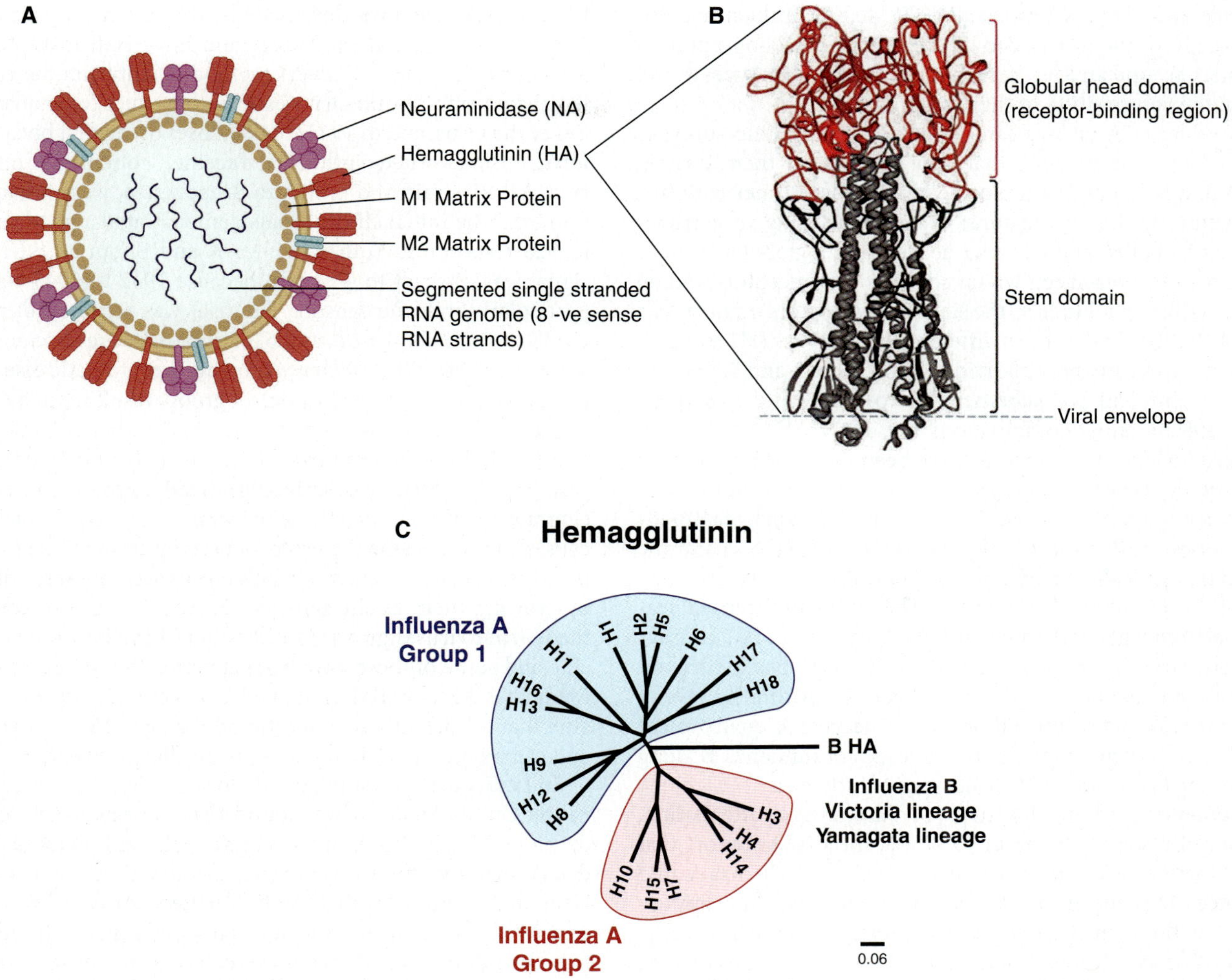

FIG. 40.8. Influenza. A: Structure of influenza A virus. **B:** The hemagglutinin (HA) protein mediates entry of influenza virus particles into host cells. The globular head domain (*red*) is variable between subtypes and strains and undergoes changes. The stem region (*gray*) is highly conserved and is the target of effort to generate broadly cross-reactive neutralizing antibodies. **C:** Phylogenetic tree of influenza virus hemagglutinins. There are two phylogenetically distinct groups of influenza virus A. These are group 1 (*blue*) and group 2 (*red*). Scale bar indicates a 6% change at the amino acid level. (Source: A, Created with Biorender.com and B, From Epidermal Electronics; Ekiert DC, Friesen RH, Bhabha G, et al. A highly conserved neutralizing epitope on group 2 influenza A viruses. *Science.* 2011;333(6044):834-835, Reprinted with permission from AAAS.)

that sera from some infected patients did not react with the influenza virus isolated by Smith, Andrewes, and Laidlaw, suspecting many to believe that different strains of influenza existed. In 1940, Thomas Francis at the Rockefeller University in New York obtained throat washings and blood samples from ill children and infected ferrets with this and isolated another strain of influenza that he named influenza B.[233]

Influenza virus is a single-stranded RNA virus of the orthomyxovirus family. Unusually, the influenza virus genome does not exist as a single stretch of RNA, but rather as segmented pieces of negative-sense RNA, each piece encoding one or two gene products (Fig. 40.8A). There are three genera (types) of influenza viruses that infect humans: Influenza A, B, and C. These are distinct, based on antigenic differences in the viral nucleoprotein (NP) and matrix protein (MP).[234] Influenza A and B viruses can infect humans and cause seasonal influenza epidemics. Influenza A can infect many other vertebrates such as birds, pigs, and horses and can have a reservoir in birds, mainly wild birds such as waterfowl. Influenza A is the only type that is known to cause pandemics.[234] The main host for influenza B is humans, and this is possibly the reason why they have not caused pandemics. Influenza type C infections generally cause mild illness and are not thought to cause human flu epidemics.

Influenza A and B virions are indistinguishable by electron microscopy and their lipid envelopes are studded with glycoprotein spikes of HA and neuraminidase (NA). In influenza A, a smaller number of M2 ion channels traverse the lipid membrane, which overlays the M1 protein that encases the RNA core (Fig. 40.8A). Influenza B is structurally similar to influenza A, except that in addition to HA and NA, instead of M2 protein, it has two proteins called NB and BM2. The organization of the genome structure of influenza A and B is similar, with each virus containing eight negative-sense, single-stranded RNA molecules (Fig. 40.8A). Influenza C also has a glycoprotein-studded lipid envelope overlying a protein matrix around an RNA core, but is different from types A and B in the composition of the glycoproteins.

Influenza A viruses can be further divided into subtypes based on two proteins that they express on their surface: HA and NA. The HA molecule consists of a trimer with two structurally distinct regions: a stem consisting of a triple-stranded coiled coil of alpha helices, and a globular head of antiparallel beta sheet, which contains the "receptor-binding site," (RBS) that binds to the sialic acid receptors on host cells (Fig. 40.8B). There are 18 different HA subtypes (H1 to H18), which can be further subdivided into groups 1 and 2.[234] There are 11 different NA subtypes (N1 to N11) (Fig. 40.8C). Of the 198 potential combinations (eg, H1N1, H3N2, H2N2), approximately 130 subtypes have been detected in nature.[235] There are two serologically distinct lineages of influenza B HA subtypes (B/Victoria/2/87-like and B/Yamagata/16/88-like). Two influenza A subtypes (H1N1 and H3N2) and the two antigenically distinct lineages of influenza B viruses currently cocirculate in humans.[234] The currently licensed seasonal trivalent inactivated influenza vaccine (TIV) consists of one strain each from the H1 and H3 subtypes (which are the dominant subtypes that have been circulating for several years) and one of the influenza B lineages. A quadrivalent influenza vaccine includes two lineages of influenza B along with the H1N1 and H3N2 subtypes of influenza.

Following in the footsteps of Pasteur and attenuation, Smorodintsev in Russia in 1936 was the first to report that vaccination with an attenuated strain of influenza could protect humans from reinfection.[236] However, the evaluation methods in these trials were largely dependent on self-reporting by subjects, so it is unclear how effective this vaccine might have been. Still, this type of vaccine was widely used in the Union of Soviet Socialist Republics for over 50 years till the end of the 20th century, with a billion people having been vaccinated with it. Since then, there have been several attempts to develop attenuated seasonal influenza vaccines with varying degrees of success, but today, the live attenuated seasonal influenza virus (sold under the brand name FluMist Quadrivalent in the United States) in the form of the nasal spray is recommended by the CDC for use in healthy children older than 24 months.[236] FluMist was initially developed by Hunein Maassab at the University of Michigan, and later by the company Aviron in Mountain View, California, which was subsequently acquired by MedImmune. The FDA approved the use of FluMist in 2003.[236] The cold-adapted influenza vaccine version (which is stable in a refrigerator and does not require freezing for storage) was approved by the FDA in 2007 to 2008.

The first effort to develop an inactivated seasonal influenza vaccine outside of the Union of Soviet Socialist Republics began in the United States and England, driven largely by the objective of protecting military soldiers.[236] In 1941, Thomas Francis (who previously isolated influenza B) was appointed the director of the Commission on Influenza of the US Army to develop and conduct field trials to evaluate influenza vaccines. Today, the influenza vaccine is grown in chicken eggs, but this technique was developed by the Australian scientist Macfarlane Burnet at the Walter and Eliza Hall Institute in Melbourne in 1936.[237] Burnet's method of culturing the virus in chicken embryos made it possible to produce large quantities of the virus, and in 1942, 10,000 doses of the first bivalent, formalin-inactivated influenza vaccine, containing influenza A/PR8 strain and influenza B/Lee strain, were tested in humans. The initial clinical trials demonstrated a good serological response to both influenza A and B, but an efficacy signal was difficult to ascertain because 1942 to 1943 was a particularly mild flu season. The trial was repeated during the 1942–1943 flu season, and 6263 subjects were vaccinated, with only 2.2% of the vaccinees showing symptomatic disease, compared with 7.7% in the control group, resulting in a 69% efficacy.[236] These trials led to the conclusion that vaccination with the IIV resulted in protection against symptomatic disease. In 1944, Stanley described in detail a protocol to concentrate, purify, and inactivate the virus produced in chicken eggs, and this became the protocol used by commercial manufacturers. Until recently, all influenza vaccines were made by growing them in chicken eggs. Since 2007, new vaccines made from viruses grown in cell culture have become available and some of these have been approved by the European Medicines Agency (EMA). Flublok is a recombinant HA vaccine that is currently licensed for adults aged 18 to 49 years and can be used in individuals who are allergic to eggs.[238]

In 1947, during the annual influenza vaccination campaign in the United States, it was found that the existing vaccine developed by the US Army was ineffective, and investigators determined that the antigenic composition of the circulating virus had changed (antigenically "drifted" away) from that seen in the virus in the vaccine. This highlighted the need for continuous surveillance for circulating influenza viruses to detect new variants resulting from antigenic drift, and the WHO Influenza Center was established in 1948, whose primary goals were to "*collect and characterize influenza viruses, develop methods for the laboratory diagnosis of influenza virus infections, establish a network of laboratories, and disseminate data accumulated from their investigations.*[239]" Another seasonal influenza vaccine failure occurred in 1958, due to the presence of a new virus, which had two surface antigens that were totally different from the ones present in the previous strains. Since the new virus was so antigenically different from the previous one, it had undergone an "antigenic shift." This led to the realization that influenza viruses were constantly changing, either gradually and subtly or abruptly and dramatically through antigenic shift. Subsequent work showed that antigenic drift was caused by accumulation of point mutations in the viral genome, leading to changes in the antigenic epitopes of HA and NA. In contrast, antigenic shift can occur when two or more different strains of the virus that infects a cell can recombine resulting in exchange of the HAs (eg, H1 replaced by H5), resulting in the formation of chimeric virus.[240] Since influenza A viruses can infect not only humans but also birds and pigs and since pigs can be infected by not just Swine but also avian and human viruses, pigs are thought to be incubators from which pandemic can emerge.

It thus became clear that it was necessary to regularly change the antigenic composition of the seasonal influenza vaccine, based on surveillance data on viral strains that were likely to be recirculating in the coming season. Since 1973, the WHO has issued recommendations to vaccine manufacturers on what influenza strains should be included in the seasonal flu vaccines. This is done by the WHO Collaborating Centers for Reference and Research on Influenza (located at the CDC in Atlanta, St. Jude Children's Research Hospital in Memphis, Francis Crick Institute in London, and in Melbourne, Tokyo, and Beijing) that receive surveillance data from over 100 countries that do year-round testing of thousands of influenza samples from patients.[241] Twice a year, the WHO convenes a meeting with the six directors of the WHO Collaborating Centers, and other key regulatory laboratories, in which they review the surveillance data and make recommendations on the composition of the influenza vaccine. These meetings occur in February for selection of the upcoming Northern Hemisphere's seasonal influenza vaccine, and in September for the Southern Hemisphere's vaccine.[241] When there is a mismatch between the circulating strains and those included in the vaccine, then the efficacy of the vaccine can be low, as was the case in the 2014 to 2015 flu season.[242-244] However, even when there is a good match between the circulating and vaccine strains, the vaccine efficacy is usually around 40% to 60%. This is due to several reasons, including suboptimal immunity in the elderly and waning immunity. Previous studies have documented that antibody responses to the seasonal influenza vaccine decline, often to baseline levels, within a few months, thus necessitating annual revaccination. Consistent with this, a recent study demonstrates bone marrow plasma cells (which are long-lived, antibody-producing cells that reside in the bone marrow and maintain persistent antibody titers) were generated by seasonal influenza vaccination of humans, but that most were short-lived and lost within 1 year.[245] Although currently licensed seasonal influenza vaccines do confer significant protection in the healthy adult population, the efficacy of these vaccines in the elderly population, who are most vulnerable to complication from influenza, is very poor because of immune senescence.[246] Two strategies to enhance immunogenicity of TIV in the elderly are to increase the antigen dose or include an adjuvant.[247] A high-dose IIV with four times the normal amount of HA antigen (Fluzone High Dose) has been introduced for use in the elderly. Also, an MF59-adjuvanted flu vaccine (Fluad) has been licensed for use in the elderly.[246,247]

Correlates and Mechanisms of Protective Immunity

Influenza vaccination induces antibodies that bind to HA and NA. Neutralizing antibodies directed against HA is the primary mechanism of protection against infection with influenza. Neutralization assays and HA inhibition (HI) have both been used to measure strain-specific antibodies, but the HI assay is used more commonly because of its simplicity. The HI assay is based on the principle that HA binds to red blood cells (RBCs) and that antibodies specific to HA inhibit binding of RBCs to HA. Normally, RBCs in solution will sink to the bottom of a tube or assay well and form a red dot; but when an influenza virus is added to the RBC solution, the HA proteins will bind to multiple RBCs, facilitating the formation of an RBC lattice-like structure ("hemagglutination"), which prevents the RBCs from sinking to the bottom and forming the red dot. However, when sera that contain antibodies specific to HA are added to the well, the antibodies compete with RBC for binding to HA, and thus inhibit hemagglutination. Currently, HI titers in the serum are the most widely used correlate of protection against inactivated seasonal influenza vaccines. An HI titer of 40 is often referred to as the "protective titer," but this is misleading because the titer represents the level at which approximately 50% of individuals will be protected, and there is no titer value that can guarantee protection from infection.[248] Furthermore, other immune parameters include antibodies against NA (which are known to reduce the disease severity), as well as T-cell responses against the more conserved internal viral proteins. Thus, in the elderly, antibody titers to the inactivated trivalent seasonal influenza were unreliable predictors of the risk of influenza illness.[249] Instead, elderly individuals who developed strong influenza-specific T-cell responses were less likely to develop flu regardless of postvaccination antibody titers.[249] Interestingly, those elderly individuals who developed a high IFNγ:IL-10 ratio following ex vivo peripheral blood mononuclear cell (PBMC) stimulation with live influenza were more likely to be protected from influenza.[249]

Finally, for live attenuated influenza vaccines (LAIV), a recent study demonstrates that antibody responses, including HI titers, did not correlate with detection or quantitation of LAIV on subsequent challenge.[250] This is likely due to the multifaceted immune response elicited by LAIV, with induction of local IgA and mucosal T-cell responses, including TRM cells. These studies highlight our lack of detailed understanding about the diverse immunological mechanisms by which seasonal influenza vaccines protect against influenza. Therefore, future work should be aimed at a detailed analysis of vaccination-induced immune responses in humans, with a view to obtain a detailed mechanistic understanding. In this regard, as discussed below, the use of systems biological approaches to study mechanisms of vaccine-induced protective immunity in controlled human infection models (CHIM) of influenza infection will be especially valuable.[251]

Hepatitis A and B

Viral hepatitis remains a serious global health issue and is caused by five main strains of the hepatitis virus—A, B, C, D, and E. The WHO estimates that hepatitis B and C affect about 325 million people worldwide, leading to 1.4 million deaths from viral hepatitis-related cirrhosis and liver cancer,[252,253] more than that with either HIV or TB alone. It is the second major killer infectious disease after TB, and nine times more people are infected with hepatitis than HIV. Deaths from hepatitis have been on the rise over the past decade.[254,255] Hepatitis has been dubbed "the silent killer" because the disease is usually asymptomatic and most of the 325 million people who are infected do not know that they are infected, until the virus has caused severe liver damage. Nearly 2 billion people have been infected with hepatitis B at some point in their lives.

Although "epidemic jaundice" has been documented since the time of the Sumerians in 3000 BC, and around 400 BC Hippocrates described the first clinical features of a transmissible, epidemic jaundice,[256,257] it is only in the last few decades that the viral etiologies of this disease have been described.[256] A major milestone was the discovery of the HBV, which started with the identification of the Australia antigen by Blumberg, and has been described above, and which led to the development of the hepatitis B vaccine.

Although both hepatitis A and B viruses can cause hepatitis, they differ in many important respects. Hepatitis A is caused by the hepatitis A virus, which is classified as a hepatovirus and is a small, nonenveloped RNA virus similar to members of the picornavirus family.[258] In contrast, HBV is a double-stranded DNA virus and a member of the hepadnavirus group. Hepatitis A is transmitted from person to person via the fecal-oral route with contaminated food or water.[253] It generally causes asymptomatic infection in children younger than 5 years but can lead to acute liver failure in about 1% of children. In adults, it can cause symptomatic disease. Hepatitis A can be spread through food prepared by an infected person and is mainly caused by poor sanitation and hygiene. It causes an acute infection, with an incubation period of about 15 to 45 days, and most people recover after several weeks. This is unlike hepatitis B, which has a longer incubation period of about 2 to 6 months, and is transmitted through serum or blood, and can result in chronic infection in 90% of infants following acute infection at birth, and in about 5% of newly infected adults.[259] This led to the concept of "short incubation" (hepatitis A) versus "long incubation" (hepatitis B) infections. In 1977, Finestone at the NIH used immune electron microscopy and identified a new agent in stool specimens of patients with acute hepatitis A outbreaks. This was followed by the development of serological assays for detection of hepatitis A antigens in stools and measurement of hepatitis A–specific serum antibody titers, and the development of hepatitis A vaccines. Today, several inactivated or live attenuated hepatitis A vaccines are available. Live hepatitis A vaccine has not received FDA approval in the United States but is in use in India and China.

The first to be developed by Maurice Hilleman and colleagues at Merck starting in the 1970s was a formalin-inactivated, alum-adjuvanted vaccine that was licensed for use in children in 2005 as VAQTA. Results of a formalin-inactivated hepatitis A vaccine in humans were first published in 1991 by a group at the Walter Reed Army Institute of Research and showed that inactivated vaccine could stimulate antibodies against hepatitis A.[260] Also in 1991, results were published on the safety and immunogenicity of a live attenuated hepatitis A vaccine in humans.[260] Merck and GSK independently developed their vaccines, and in 1992, Merck scientists published results of a clinical study with their formalin-inactivated, alum-adjuvanted vaccine (VAQTA), showing efficacy against symptomatic disease, in 1000 children in upstate New York between the ages of 2 and 16 years. In 1994, GSK published results of their formalin-inactivated, alum-adjuvanted vaccine HAVRIX, reporting a 94% efficacy in a trial done in Thailand with 40,000 children aged between 1 and 16 years. In 1995, HAVRIX was approved by the FDA for use in adults and children 2 years and older. A year later, the Merck vaccine VAQTA was approved for use in persons aged 2 years and older.[260] In 2001, GSK's Twinrix, an alum-adjuvanted bivalent hepatitis A and hepatitis B vaccine, was approved for use in adults 18 years and older. Vaccination had resulted in a dramatic reduction in hepatitis A infection rates, with a 93.7% drop in the number of reported cases of acute hepatitis A from 1990 to 2009.[261] The durability of the antibody response has been shown to persist for at least 6 to 10 years in most adults receiving the vaccine.[262]

In the case of hepatitis B vaccine, following the identification of the Australia antigen by Blumberg in 1964, and its connection to hepatitis B by Prince (discussed above), Maurice Hilleman at Merck developed an inactivated vaccine, by inactivating purified HbsAg from the blood of infected individuals with pepsin (that destroyed most proteins, but not HbsAg), urea, and formaldehyde (the same method used to inactivate Salk's polio vaccine).[263] Hilleman's vaccine was licensed by the FDA in 1981 but was taken off the market in 1986 because doctors were reluctant to use it because of fears concerning vaccinating people with human blood and because of a compelling alternative. The advent of the genetic engineering revolution by the pioneering work or Herbert Boyer at University of California, San Francisco, and Stanley Cohen at Stanford paved the way for a new generation of recombinant DNA vaccines and transformed the landscape of vaccinology forever. In 1986, the vaccine RECOMBIVAX HB, manufactured by Merck, and consisting of the recombinant surface antigen of hepatitis B (Australia antigen, expressed in yeast, that self-assembles into a VLPs), adjuvanted with alum was approved for marketing in Germany and in the United States. This marked a major milestone not just in vaccinology but also in biotechnology, as this was the first vaccine to be produced with recombinant DNA technology, and only the third recombinant product to be licensed for human use, after human insulin in 1982, human growth hormone in 1985, and preceding INF-α in 1986.[264] Two other recombinant hepatitis B vaccines quickly followed, Engerix-B (also alum adjuvanted) marketed by SmithKline Biologicals (now GSK) approved in Belgium in 1986 and in the United States in 1989 and GenHevac B by Pasteur Vaccines approved in France in May 1989.[264] In 2005, another GSK hepatitis B vaccine, Fendrix, was approved for sale in Europe. It is identical to Engerix-B, except that it is adjuvanted with AS04, being the first vaccine to be approved for use in humans with an adjuvant other than alum.[265] In 2017, some 25 years since the previous approval of a hepatitis B vaccine, the FDA announced the approval of Heplisav-B, a recombinant hepatitis B vaccine adjuvanted with CpG 1018 from Dynavax. Unlike the other vaccines, which require at least a three-dose immunization regimen, Heplisav-B is recommended for use as a two-dose regimen. These vaccines are highly immunogenic and the response rates to primary vaccination are as much as 85%

to 100% of people developing antibody titers ≥10 mIU/mL, a level that is considered protective.[266] Long-term follow-up studies in vaccinated infants have documented an absence of antibody titers in 50% to 70% of people, 15 to 30 years later.[267,268] In adults, studies suggest that antibody responses to hepatitis B vaccination decrease after 10 to 31 years and fall below a level considered protective in 25% of the cases. However, these subjects responded rapidly and robustly to a booster immunization, demonstrating that hepatitis B vaccination stimulates long-lived memory responses.[266]

MMRV Vaccine

Measles Vaccine

Measles is a highly contagious disease caused by the measles virus, a single-stranded RNA virus of the *Paramyxoviridae* family. It is spread by contact with nasal or throat droplets from infected people or by airborne spread when an infected person sneezes or coughs. The R_0 (the average number of people infected by every infected person) is thought to be 12 to 18. In most cases, the symptoms appear within 7 to 14 days postinfection and include high fever, cough, and runny nose, but there can be a range of more serious complications, with 1 to 3 of every 1000 children dying from respiratory or neurological complications.[269] Natural infection results in an acute infection that stimulates durable antibody response with an estimated half-life of 3000 years!

Prior to the introduction of the measles vaccine 1963, there were an estimated 2.6 million deaths per year.[270] Vaccination has had a dramatic impact, with an estimated 23.2 million deaths being prevented in 2000 to 2018. However, measles surged in 2019, when the number of cases worldwide increased to 869,770, the highest number reported for 23 years, since 1996. The number of deaths worldwide from measles increased 50% from 2016, with 207,500 deaths in 2019 alone.[271] The reasons for this are complex, but all are connected with low vaccination rates, either due to poor health infrastructure in developing countries, civil war, or vaccine hesitancy. The latter is an especially challenging issue, caused by low awareness about the benefits of vaccination, or misinformation and mistrust, or complacency.

The starting point for the development of the measles vaccine was the isolation of the measles virus in cell culture from the infected blood of a 13-year-old boy called David Edmonston, by John Enders and Thomas Peebles. This so-called Edmonston B strain was attenuated by serial passage in cell culture and adapted to grow in chicken embryo and was used to develop the live attenuated vaccine in 1963. However, this vaccine was quite reactogenic and had to be coadministered with immunoglobulins to reduce its side effects. In 1968, Maurice Hilleman and colleagues developed a more attenuated version of the vaccine, which was as immunogenic as the original strain, but did not require coadministration of immunoglobulin. Today, measles vaccine is usually combined with MMR or MMRV vaccine.[272] MMR is recommended at 12 to 15 months, followed by a booster shot at 4 to 6 years. A single dose of MMR is 93% efficacious against measles, while two doses confer a 97% efficacy. Vaccination stimulates a similar immune response to that found in natural infection, with IgG, IgM, and IgA antibodies in serum and nasal secretion, and durable protection that can last several decades.

Mumps Vaccine

Mumps is a highly contagious, acute viral disease caused by an enveloped, nonsegmented, negative-sense RNA virus of the *Paramyxoviridae* family. Like measles, mumps is an airborne virus that can be spread by an infected person by droplets via sneezing and coughing. Typically, infection causes parotitis, but can also cause inflammation of other organs including orchitis, pancreatitis, and meningitis, which is caused as the result of the high neurotropism of the virus.[273] Before the introduction of the live attenuated mumps vaccine in December 1967, 95% of adults developed seropositivity for mumps, acquired mainly during childhood. Like that observed with measles, antibody responses to mumps infection are long lived, with a half-life of 542 years.[59]

Following the introduction of mumps vaccine, in the United States, the cases fell sharply by 99.9%, with similar results in other countries.[273] The mumps vaccine that we use today was developed in 1963 by Maurice Hilleman at Merck, who isolated the mumps virus from a throat swab of his 5-year-old daughter and developed a new vaccine, which was licensed in 1967. The mumps strain used in the vaccine was named Jeryl Lynn, after Hilleman's daughter, and was included in the MMR vaccine that was licensed in 1971 and used worldwide. The mumps vaccine is highly efficacious, with an efficacy of about 88% after two doses and 78% after a single dose.

Rubella Vaccine

Rubella (also known as German measles) is a contagious viral infection that often results in asymptomatic disease, but can result in mild, self-limiting illness characterized by maculopaular rash, lymphadenopathy, and low-grade fever.[274] However, infection of pregnant mothers, especially during the first trimester, can result in miscarriage, congenital defects, or death. In particular, many infants can develop congenital rubella syndrome characterized by pathology in multiple organs such as the brain, lungs, and liver, leading to encephalitis and mental retardation. Prior to the introduction of the rubella vaccine in 1969, rubella was a common infection. For example, during the 1962 to 1965 rubella epidemic, in the United States alone, an estimated 12.5 million people got rubella, 11,000 pregnant women lost their babies, 2100 newborns died, and 20,000 babies were born with congenital defects. Although rubella is called German measles, it is not caused by the measles virus. Rather, rubella virus is a positive-sense, single-stranded RNA virus, encapsulated in an icosahedral capsid, and a member of the genus *Rubivirus* and belongs to the family of *Matonaviridae*.[275]

Following the isolation of the rubella virus in 1962 by Parkman and Weller,[275] several groups set about trying to make a vaccine. Their common strategy was to develop an attenuated vaccine by cell culture of the rubella virus, but each group used cells from different species.[274] The resulting live attenuated vaccines were licenses for use in the United States in 1969 to 1907, but one strain, the RA 27/3 strain developed by Stanley Plotkin at the Wistar Institute using a human diploid fibroblast cell, was only licensed in Europe, because of concerns by the American regulatory authorities about using a vaccine developed in human cells.[274] However, based on better immunogenicity and improved safety profiles, in 1979, the RA 27/3 strain replaced the all other vaccines, which were discontinued. The RA 27/3 rubella vaccine is available as the MMR vaccine or the MMRV vaccine and is safe and more immunogenic than rubella vaccines used previously. Greater than 95% of vaccinees aged 12 months and older developed seroconversion after a single dose, and greater than 90% have protection from clinical rubella and viremia for at least 15 years.[276]

Varicella Vaccine

Varicella, commonly known as "chickenpox," is a highly contagious acute viral infection caused by the VZV. VZV is a double-stranded DNA virus of the *Alphaherpesvirinae* family, which can also cause herpes zoster, hence the term VZV.[277] Like other herpes viruses, following a primary infection, VZV can persist in the body (in sensory ganglia) as a latent infection usually for decades. Primary infection with VZV causes chickenpox, while reactivation results in herpes zoster (shingles). VZV can be spread by direct contact, or infected respiratory secretions via aerosol, or direct contact with fluid from the blisters. Although VZV was isolated in 1952, it was only in 1974 that a sufficiently attenuated strain of the virus was generated to develop a vaccine. Michiaki Takahashi in Japan isolated varicella from a 3-year-old boy with mild chickenpox and attenuated it by passaging it in human embryonic lung fibroblasts and in guinea pig embryo fibroblasts and developed a live attenuated varicella vaccine.[278-280] The strain was named Oka after the boy. The vaccine was initially tested in healthy children and then in immunocompromised children and was shown to be safe and effective. Japan was among the first countries to vaccinate against chickenpox. The vaccine was further developed by pharmaceutical companies such as Merck (VARIVAX) and GSK (VARILRIX). The vaccine was first licensed for use in the United States in 1995 for routine in healthy individuals older than 1 year. The efficacy of varicella vaccine is estimated to be 70% to 90% effective against varicella infection, and 90% to 100% effective against moderate and severe disease.[281] Several studies indicate that varicella vaccination induces durable antibody responses that last for several years after vaccination.[282]

Human Papillomavirus

HPV is the most common sexually transmitted infection. There are at least 100 types of HPV, and most do not cause disease and are spontaneously cleared within 2 years of infection.[283] However, at least 14 of these strains are thought to be "high-risk types" and can cause persistent infection that progresses to cervical cancer. HPV is a nonenveloped DNA virus with a double-stranded viral genome of approximately 8 kb in length, consisting of six early proteins and two late, structural proteins L1 and L2.[284] HPV 16 and 18 are the two most common high-risk genotypes and cause approximately 70% of all cervical cancers. Two "low-risk" genotypes (HPV 6 and 11) cause genital warts, a common benign condition of the external genitalia that causes significant morbidity.[284] In 2018, there were approximately 570,000 cases of cervical cancer and 310,000 deaths, making this the fourth most common cancer in women, after breast, colorectal, and lung cancers.[285] The link between HPV and cervical cancer was first demonstrated in the early 1980s by Harald zur Hausen, a German virologist, who proposed the idea of an HPV vaccine against cervical cancer. zur Hausen was awarded the Nobel Prize in Physiology or Medicine in 2008 for this accomplishment.[286] The HPV vaccine was first developed by Ian Frazer and Jian Zhou at the University of Queensland in Brisbane, Australia. In 1990, Frazer and Zhou showed that recombinant L1 capsid proteins of HPV 16 self-assembled into VLPs, even without the DNA of the virus.[287] Immunization with L1-based VLPs were able to induce antibody titers 100-fold greater than that seen with natural infection. In 2006, GARDASIL, a quadrivalent vaccine comprising VLPs derived from the L1 capsid proteins of HPV types 6, 11, 16, and 18, adjuvanted with alum, developed by Merck was approved for use in females aged 9 to 26 years. In 2009, a second HPV vaccine, CERVARIX, a bivalent vaccine consisting of VLPs for HVP types 16 and 18, adjuvanted with AS04, developed by GSK was licensed for use in females aged between 10 and 25 years.[288] Both GARDASIL and CERVARIX were shown to be efficacious in protecting against the genotypes they contained. In 2014, the US FDA approved GARDASIL 9, a 9 valent HPV vaccine indicated in females aged between 9 and 45 years for prevention of cancers caused by HPV types 6, 11, 16, 18, 31, 33, 45, 52, and 58.[288]

Diphtheria, Tetanus, and Pertussis

The "DTP" vaccine, like the MMR vaccine discussed above, is a combination vaccine against three different infectious diseases: diphtheria, pertussis (whooping cough), and tetanus. The vaccine comprises inactivated toxoids for tetanus and pertussis combined with either killed whole cells of the bacteria that cause pertussis (called "DTwP") or pertussis antigens (termed "DTaP").

Tetanus

Tetanus is a noncontagious disease caused by the gram-positive bacterium *C. tetani*, which is found in soil and which produces a powerful neurotoxin that causes the symptoms of tetanus. The toxin affects the inhibitory neurons, leading to nervous system disorders including muscle spasms, seizures, and lockjaw. The fatality rate is high and can range from 30% overall to 52% in patients older than 60 years.[289] The serum immunotherapy against tetanus and diphtheria pioneered by von Behring and Kitasato in 1890 provided proof of concept that antibodies could protect against tetanus. In the 1920s, Gaston Ramon at the Pasteur Institute and Alexander Glenny and Barbara Hopkins at the Wellcome Research Laboratories in London, working independently, developed vaccines against diphtheria and tetanus, by using formaldehyde-inactivated toxins from these pathogens. Addition of alum improved the efficacy of these vaccines.[290]

Diphtheria

Diphtheria is caused by the toxin produced by an aerobic gram-positive bacillus called *Corynebacterium diphtheriae.* It was a major cause of death in children around the world until a vaccine was introduced in the 1920s. The bacteria can only produce this toxin when they are infected by bacteriophage that carries the genetic information for the toxin (TOX gene). Only toxigenic strains can cause severe disease. Diphtheria infects the upper respiratory tract and is contagious and can spread through respiratory droplets from coughing and sneezing. The toxin inhibits cellular protein synthesis and causes cell death and tissue destruction. A vaccine made by inactivating the toxin by formaldehyde was developed in 1921, but not widely used until the early 1930s. In the 1940s, it was incorporated with tetanus and pertussis vaccine and became routinely used, even to this day.[291] After a primary series of three immunizations in adults, or four doses in infants, a protective level of antitoxin antibody of greater than 0.1 IU/mL is reached in more than 95% of vaccinees. The vaccine is estimated to have a clinical efficacy of 97%. However, childhood immunity wanes, requiring a booster dose every 10 years. Finally, an enzymatically inactive, nontoxic form of diphtheria toxin containing a single amino acid substitution, called CRM197, has been used as a carrier protein for conjugate vaccines against encapsulated bacteria and is currently used to vaccinate children globally against *H. influenzae*, pneumococcus, and meningococcus.[292]

Pertussis

Pertussis, or whooping cough, is an acute respiratory disease caused by the gram-negative bacterium *B. pertussis.* The main clinical symptoms are a protracted cough that can last for several weeks, characterized by the high-pitched whooping cough. In the early part of the 20th century, before the implementation of the vaccine, pertussis was a major cause of childhood mortality. There are an estimated 16 million cases of pertussis, resulting in 195,000 pediatric deaths around the world annually.[293] In the United States, the introduction of vaccination in the 1940s resulted in a reduction in the incidence of disease from 150 per 100,000 to 1 per 100,000 in 1970.[294]

The isolation of the bacterium in 1906 was the starting point for the development of the whole cell pertussis (wP) vaccine, which was made by inactivating a suspension of *B. pertussis* by formaldehyde. The vaccine was licensed in the United States in 1914. It was later combined with the diphtheria and tetanus toxoid vaccines in 1948 to create the DTP vaccine.[295] The widespread use of the vaccine has reduced the incidence of pertussis by more than 80% compared with the prevaccine era. Clinical trials done in the Faroe Islands during a pertussis epidemics and later observational studies showed an efficacy of 70% to 90% in preventing serious pertussis disease, after the primary series of four immunizations.

The common incidence of minor adverse reactions such as local redness, swelling, and pain at the injection site and the much rarer occurrence of more serious adverse reactions fueled public anxiety, much of it misguided, about the safety of this vaccine. This was especially so in developed countries, where the success of vaccination in reducing the incidence of pertussis to such low levels resulted in complacency and a lack of appreciation to the benefits of vaccination. So strong was vaccine hesitancy that it resulted in declining vaccination rates and a resurgence of the disease in many parts of the world. This prompted the search of a less reactogenic vaccine and led to the development of the first acellular pertussis (aP) vaccine by Sato and colleagues in Japan in 1981, as a DTaP combination vaccine.[296] Acellular pertussis vaccines are subunit vaccines that contain purified, inactivated components of *B. pertussis*, including pertussis toxin, a protein complex of the ADP-ribosyl transferase activity, and filamentous HA, which is a cell surface protein. Since then, several acellular pertussis vaccines have been developed, and acellular pertussis vaccines have completely replaced pertussis vaccines in the United States, Canada, Australia, and many European and Asian countries, but whole cell pertussis vaccines are still used in many developing countries.

The routine vaccination schedule for DTaP to children is a three-dose primary series at ages 2, 4, and 6 months, followed by boosters at age 15 to 18 months and 4 to 6 years. There have been many clinical studies to evaluate the efficacy of the different acellular pertussis vaccines, and although comparisons between the different acellular vaccines cannot be made because of differences in the clinical trial design, what is consistent is the fact that the acellular pertussis vaccines are less reactogenic, and at least as efficacious as whole cell pertussis vaccines, with short-term efficacy of 80% to 85% for populations in the United States. However, in the past several years, there has been a resurgence of pertussis in several industrialized countries, including the United States and Australia.[297] In the United States, 4000 cases were reported annually in the 1980s, but 25,827, 25,616, 27,500, and 48,277 cases were reported in 2004, 2005, 2010, and 2012, respectively.[297-299] Although this increase may be partly due to improved diagnostic methods, there is concern that waning immunity to acellular pertussis vaccination could also be a major contributing factor. The durability of protective immunity caused by natural infection with *B. pertussis* has been estimated to range from 7 to 20 years,[300-302] although one study suggests that it can be as short as 3.5 years.[303] The durability of protection induced by vaccination with whole cell pertussis has been estimated to last for several years.[304,305] In contrast, the duration of immunity after immunization with aPVs appears to be shorter.[306,307] Another factor could be a lack of mucosal immunity induced by the whole cell pertussis vaccine. Natural infection stimulates both mucosal and systemic immune responses, while acellular pertussis only induces systemic immune responses.[308-310] Since *B. pertussis* is a mucosal pathogen, the induction of mucosal immunity is critical for its control.

SOME EXAMPLES OF EMERGING VACCINES

Malaria (Fig. 40.9)

Malaria is an infection caused by *Plasmodium* parasites that is transmitted to humans by the female *Anopheles* mosquito. There are five species of the *Plasmodium* parasite that can cause malaria in humans—*Plasmodium falciparum, Plasmodium vivax, Plasmodium ovale, Plasmodium malariae,* and *Plasmodium knowlesi. P. falciparum* is responsible for the majority of malaria deaths and is the most prevalent in sub-Saharan Africa. Every 2 minutes, a child dies of malaria. Nearly half the world's population, or about 3.2 billion people, live in areas of the world vulnerable to malaria.[311] In 2018, malaria afflicted 228 million people and killed an estimated 405,000 people, mostly in sub-Saharan Africa.[312] During the past 15 years, increased diagnostic testing, invention of preventive measures such as insecticide-impregnated bed nets, and early treatment with artemisinin-based combination therapy have reduced the annual incidence substantially.[313] However, resistance to insecticides and antimicrobials has been an emerging problem highlighting the need for a vaccine.

The fact that individuals in endemic areas do develop natural immunity to malaria (although this immunity is not sterilizing and wanes when the individual migrates to a malaria-free area),[314] as well as the efficacy of therapeutic antibodies to malaria,[315] provides a rational basis for the concept of vaccination against malaria. However, despite considerable effort over the past several decades, no effective malaria vaccine exists. Several major challenges have posed serious obstacles to developing a malaria vaccine. The *Plasmodium* parasite is extremely complex, with a 23-megabase genome consisting of 14 chromosomes and about 5300 genes[316] and a complex, multistage, multihost lifecycle, and extensive genetic polymorphisms in genes encoding candidate antigens (Fig. 40.9A). Furthermore, *P. falciparum* and *P. vivax* are difficult to grow *in vitro*, and no animal model truly reflects the disease. However, there is a CHIM, which has been used to test the efficacy of vaccine candidates.

The *Plasmodium* parasite's lifecycle consists of three stages, each of which presents a unique challenge to vaccine developers. In the first stage, the pre-erythrocytic stage, also known

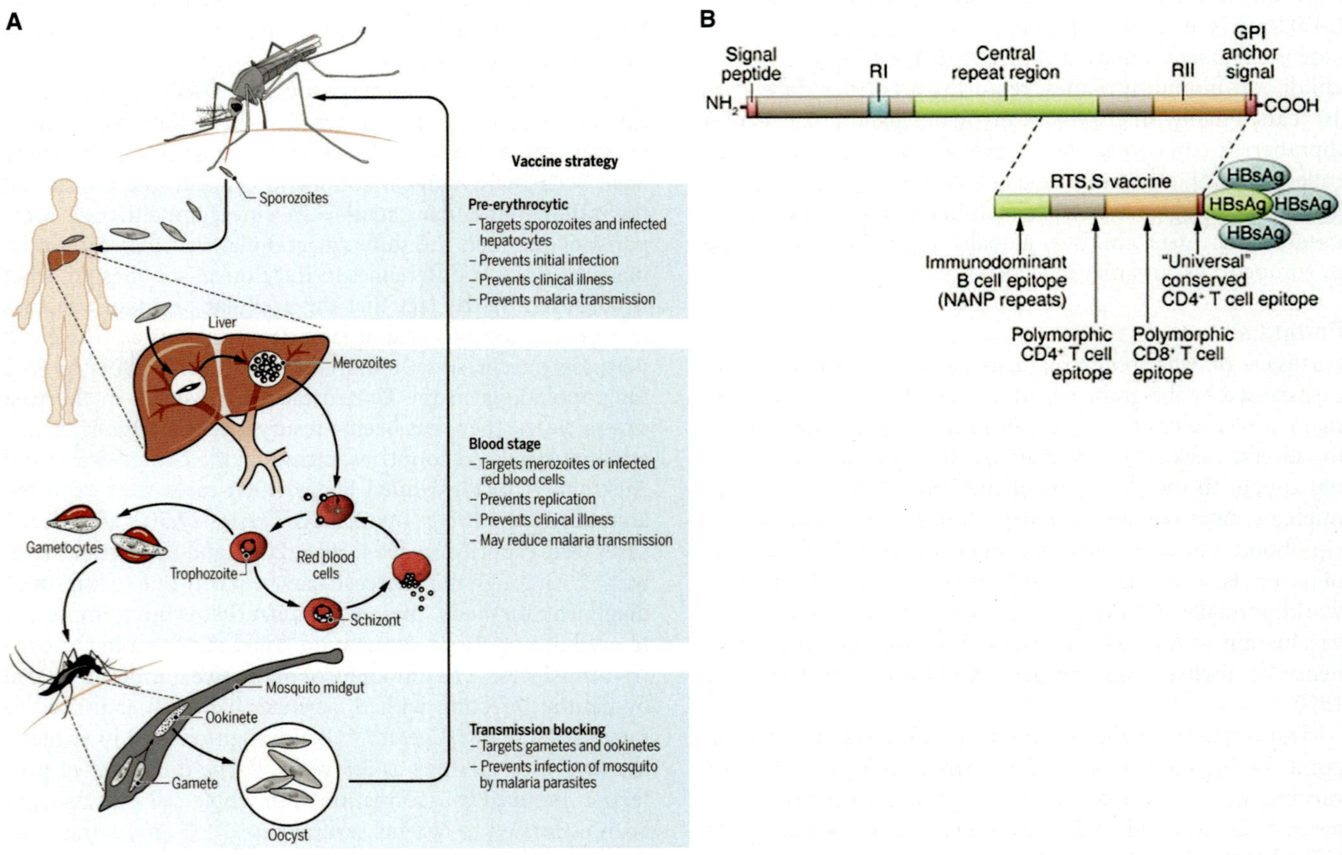

FIG. 40.9. **A: The lifecycle of *Plasmodium falciparum*.** *Plasmodium* parasite's lifecycle consists of three stages, each of which presents a unique challenge to vaccine developers. In the first stage, the pre-erythrocytic stage, also known as the liver stage, occurs when an infected female mosquito bites a human and injects a form of the *Plasmodium* parasite known as the sporozoite into the blood. The dominant surface antigen of the sporozoite is the circumsporozoite protein (CSP) and is the antigen included in the RTS,S malaria vaccine. The second erythrocytic stage occurs when the liver schizont ruptures and the merozoites are released and enter the bloodstream and infect erythrocytes. Within erythrocytes, the merozoites reproduce asexually and release hundreds of new merozoites. Finally, in the sexual stage, the parasite is taken up by a mosquito, where it sexually reproduces in the mosquito gut. Transmission blocking vaccines, which aim to interrupt parasite transmission, induce antibodies that target surface antigens of parasites within mosquitos, during the sexual stage of their lifecycle. **B:** The structure of the CSP protein included in the RTS,S vaccine. (From KITTERMAN/SCIENCE TRANSLATIONAL MEDICINE, Beeson J, Kurtovic L, Dobano C, Opi DH, et al. Challenges and strategies for developing efficacious and long-lasting malaria vaccines. *Sci Transl Med.* 2019;11(474):eaau1458. doi:10.1126/scitranslmed.aau1458. Reprinted with permission from AAAS; From Advances and challenges in malaria vaccine development. *J Clin Invest.* 2010;120(12):4168-4178.

as the liver stage, occurs when an infected female mosquito bites a human and injects a form of the *Plasmodium* parasite known as the sporozoite into the blood. The dominant surface antigen of the sporozoite is the circumsporozoite protein (CSP), first identified and cloned by the Nussenzweigs in the early 1980s.[317] This antigen is the basis for the development of the RTS,S malaria vaccine by GSK, which as described below represented a turning point in the history of malaria vaccines, when in 2015, it was favorably reviewed by the EMA and introduced into national pilot implementation programs. Thus, it was the first human antiparasite vaccine to pass regulatory scrutiny.[318] Three other sporozoite antigens known as thrombospondin-related adhesion protein, liver stage antigens 1 and 3 (LSA1 and 3), have also been tested in clinical trials as candidate malaria vaccine antigens. The sporozoite rapidly and efficiently infects liver hepatocytes (Fig. 40.9A). As few as 2 to 10 sporozoites can initiate infection in hepatocytes in 5 to 30 minutes. Within the hepatocytes, each sporozoite undergoes multiple rounds of asexual reproduction to generate as many as 50,000 merozoites, which are contained in a schizont. Antigenic peptides derived from merozoites are displayed in the context of class I major histocompatibility complex (MHC) molecules on the surface of hepatocytes, and thus represent targets for CD8+ T-cell immunity.

The second erythrocytic stage occurs when the liver schizont ruptures and the merozoites are released and enter the bloodstream and infect erythrocytes (Fig. 40.9A). Within erythrocytes, the merozoites reproduce asexually and release hundreds of new merozoites. Over the past 2 decades, several clinical trials have tested vaccines against many blood-stage antigens, including MSP1, MSP3, AMA1, and EBA-175, with modest or little efficacy. Finally, in the sexual stage, the parasite is taken up by a mosquito, where it sexually reproduces in the mosquito gut. The so called transmission blocking vaccines, which induce antibodies that target surface antigens of parasites within mosquitoes (during the sexual stage of their lifecycle), have also been tested. The four leading candidates are the gamete surface proteins Pfs230 and Pfs48/45 of *P. falciparum* and the zygote surface proteins expressed only postfertilization in the mosquito host such as Pfs25 and Pfs28.[318,319]

Whole Organism Malaria Vaccines

In the 1970s, immunization of humans with whole sporozoite vaccines was shown to induce sterilizing immunity against a sporozoite challenge.[318,320] However, whole sporozoite vaccines were thought to be impractical and thus not pursued. In 2010, scientists at Sanaria Inc., in Rockville, Maryland, developed a method to harvest, purify, and cryopreserve *P. falciparum* sporozoites (PfSPZ) from the salivary glands of aseptic mosquitoes and attenuated these sporozoites using radiation. Intravenous injection of these radiation-attenuated sporozoites into humans resulted in sterile immunity against CHIM.[321] However, this protection was dependent on the dose and regimen of the vaccine.[321-323] Importantly, the intravenous route of immunization was essential to induce protective immunity.[324] A clinical trial conducted in Mali in an area with intense malaria transmission showed that five administrations of the irradiated sporozoite vaccine (2.7×10^5 dosage) to adults reduced the risk of new *P. falciparum* infection by 52% in a time-to-event analysis over 24 weeks after last dose, and reduced the proportion infected across the transmission season by 29%.[323] These results, while promising, highlight the enormous challenge in translating vaccine candidates that show high efficacy in populations living in developed countries to those living in disease-endemic areas in low-income countries. Improvement of these promising results will likely require the use of multiple circulating *P. falciparum* strains to induce broad immunity and detailed analysis of the immune response to better define correlates of protection. Immunization with irradiated sporozoites (which express thousands of antigens) stimulates a spectrum of immune responses, including CD4, CD8, and γδ T cells, as well as antibody responses. In particular, CSP-specific antibody responses and Vδ2 T cells have been associated with protection.[325,326]

Adjuvanted Malaria Subunit Vaccines

The CPS antigen expressed in sporozoites was the first malaria gene to be cloned and is the antigen in the RTS,S vaccine, developed by GSK and the Walter Reed Army Institute of Research. RTS,S consists of B-cell epitopes from the central NANP **r**epeat region (r for "R") and C-terminal regions containing T-cell epitopes ("T"), fused to the hepatitis B surface antigen ("S"). RTS is expressed in yeasts that also express hepatitis B "S" genes, thus synthesizing S and RTS polypeptides that self-assemble into VLPs (RTS,S)[327] (Fig. 40.9B). RTS,S is administered with GSK's proprietary AS01 adjuvant and, in clinical trials in CHIM models, has shown approximately 50% protection.[328,329] A phase III trial in Africa showed immunization of children with three doses at 1-month intervals, followed by a booster dose at 18 months post third dose, and resulted in a 36% and 18% reduction in symptomatic malaria in children and infants, respectively. Statistically significant protection against severe disease was only seen in children.[330] RTS,S (brand name Mosquirix) was favorably reviewed by EMA in 2015. This represented a milestone in malaria vaccine development, as it represented the first human parasite vaccine to have passed the highest level of regulatory scrutiny. In 2019, a pilot implementation program was launched in Malawi, Kenya, and Ghana to assess efficacy and safety in real-life rural settings in over 1 million children. The development of RTS,S represents a landmark in malaria vaccine research, but its partial efficacy and low durability of immune responses highlight the need for improvement. An ongoing trial in West Africa is assessing annual booster shots with RTS,S, and other trials are using fractional dosing regimen that delivers one-fifth the full dosage, a regimen that has been shown to enhance efficacy in a CHIM model.[331] A variation of the RTS,S vaccine is the R21 vaccine developed at the Jenner Institute at Oxford, which does not contain "S," and thus has a higher ratio of CSP to hepatitis B protein. In mice, R21 adjuvanted with the saponin-based adjuvant Matrix M was shown to induce sterile protection upon challenge with transgenic sporozoites expressing homologous CSP. R21 has advanced to phase I/II testing in Africa (ClinicalTrials.gov

IDs NCT02925403; NCT03580824; NCT02925403). In addition, structural vaccinology approaches are being applied to design improved CSP-based vaccines by defining epitopes of functional monoclonal antibodies.

Blood-Stage Antigen Vaccines

The blood stage of the parasite lifecycle, when merozoites replicate within erythrocytes, is when disease symptoms appear. Six decades ago, it was shown that passive transfer of IgG purified from immune African adults cleared parasitemia from African children.[318] This suggests that vaccines that stimulate immunity to blood-stage antigens could be effective. Over the past 20 years, many blood stage–derived vaccines (several containing MSP1 and AMA1, and some containing EBA-175 and MSP3) were evaluated in clinical trials, but provided little evidence for protection in CHIM or against natural infection. Novel blood-stage antigens, such as PhRF5 and AMA1-RON2, are currently being tested.

Transmission Blocking Vaccines

Transmission blocking vaccines stimulate antibodies against antigens in the sexual stages of the parasites (gametes or zygotes) that are present in mosquitoes. These antibodies can kill parasites taken up by mosquitoes in their blood meal and prevent transmission via the vector. The four leading candidates for transmission blocking vaccines include Pfs230 and Pfs48/45, which are surface proteins expressed by gametocytes in humans, and zygote proteins such as Pfs25 and Pfs28. Ongoing studies are evaluating the immunogenicity of these candidates in clinical trials.[319]

Human Immunodeficiency Virus

HIV/AIDS is a major global health challenge with 38 million infected people globally in 2019 and an estimated 1.7 million newly infected people in 2019.[332] Despite the tremendous public health success of antiretroviral therapy in maximally suppressing HIV and stopping the progression of HIV in infected people, the development of a safe and effective vaccine is the most effective way of preventing infection. Yet, despite almost 40 years of intense effort, there is no licensed vaccine against HIV. A major challenge in developing an HIV vaccine is that the virus has a high rate of mutation, which results in a great diversity of antigenic epitopes on the viral envelope protein (Env), which is also the main target of neutralizing antibodies. Env is a heterodimer of the glycoproteins gp120 and gp41 that form trimers on the virion surface and mediate entry by binding to CD4 and CCR5 or CXCR4 on the target cell surface.[333] The high mutation rate of Env is caused by the error-prone viral reverse transcriptase (1-10 mutations per genome per viral replication cycle). This high mutation rate, coupled with a dense canopy of glycans that shroud critical neutralizing epitopes on the Env spike protein, enables the virus to evade neutralizing antibodies.[334] In addition, most primary HIV-1 particles tend to express immunodominant, nonfunctional forms of Env on their surface, and these envelopes lead to the preferential elicitation of nonneutralizing antibody responses.[335]

However, approximately 10% to 30% of HIV-1–infected people develop neutralizing antibodies that can bind to Env and neutralize a diverse range of viral isolates.[336] The fact that such broadly neutralizing antibodies (bnAbs) can develop during natural infection,[337-342] and that when passively transferred, they can prevent simian human immunodeficiency virus (SHIV) infection in macaques[343-345] and maintain viral suppression in chronically infected humans,[346,347] raises the possibility that such bnAbs could be generated through vaccination.

Fig. 40.10 shows the efficacy trials that have been performed to date. Early attempts to develop an HIV vaccine used the monomeric Env gp120 immunogen, but unfortunately, the results of the clinical trial revealed that gp120 failed to neutralize HIV primary isolates, prevent HIV infection, and reduce viral loads or delay disease progression.[348,349] Following these trials, two phase IIb trials evaluated the replication-defective adenovirus type 5 vaccine expressing the HIV genes Gag, Pol, and Nef produced by Merck, to prevent infection or reduce early plasma viral loads in populations in the Americas, Caribbean, Australia (Step trial), and South Africa (Phambili trial).[350] Both trials were stopped after the first interim efficacy analysis of the Step trial revealed no efficacy. Post hoc analysis of the men enrolled in the Step study showed enhanced frequency of HIV infection in the subgroup of vaccinated men who were Ad5 seropositive and uncircumcised, compared to a placebo group. Circumcision is known to lower HIV transmission in heterosexual and homosexual men.[351] This was a setback for the field as it showed enhanced disease in a subset of the vaccinees, possibly as a result of preexisting antibodies to the Ad5 virus used to develop the vector, which is endemic in most populations in the world. Ironically, this vaccine had shown protective efficacy against a humanized SHIV and a modest degree of protection against the pathogenic SIV in NHPs.[352] The first glimmer of hope in the HIV vaccine landscape came from the RV144 "Thai trial." The vaccine tested in RV144 trial in Thailand consisted of a prime-boost vaccination strategy over a period of 6 months, involving four consecutive immunizations with a canarypox vector (ALVAC expressing the genes for HIV Env, Gag, and Pol), followed by two booster immunizations with a recombinant gp120 adjuvanted with alum. This trial provided the first clinical evidence of efficacy for any HIV-1 vaccine.[353] The results demonstrated 31.2% efficacy and that protection correlated with nonneutralizing IgG1 and IgG3 antibodies specific to the V1/V2 region of Env, as well ADCC activity, and decreased IgA responses.[354] Of note, while vaccine efficacy in the first year was nearly 60%, it rapidly waned over time, and over the 3.5 years of the trial, the efficacy was 31.2%.[354,355] After decades of failure, the results of this trial raised optimism that an HIV vaccine could be possible, and that even nonneutralizing antibodies could have an impact on protection, and there is some evidence from NHP studies that nonneutralizing antibodies can confer a degree of protection.[103,107,356-358] The results of the RV144 trial suggested that (1) nonneutralizing antibodies could mediate protection and (2) the durability of vaccine responses needed to be enhanced. Thus, the Uhambo study (HVTN 702) was conducted in South Africa to confirm and extend the results of the RV144

FIG. 40.10. Timeline for HIV vaccine clinical efficacy trials. (Adapted from Robinson HL. HIV/AIDS Vaccines 2018. *Clin Pharmacol Ther.* 2018;104(6):1062-1073. doi:10.1002/cpt.1208.)

trial, which was conducted with clade B and CRF01_AE antigens that matched the prevailing HIV-1 strains in Thailand. The HVTN 702 trial used clade C antigens prevalent in South Africa, and the more potent MF59 adjuvant, as well as an additional booster immunization. Unfortunately, the results of this study that were announced in February 2020 showed no efficacy in preventing HIV-1 infection.[359]

Another candidate vaccine platform that is currently being tested in clinical trials is the Ad26 vector engineered to express HIV antigens. A prototype Ad26 vector expressing a single clade A HIV-1 Env immunogen (Ad26.ENVA.01) was shown to be safe, well-tolerated, and immunogenic.[360] A follow-up clinical trial to test the Ad26 vector engineered to express bioinformatically optimized HIV-1 mosaic Env/Gag/Pol antigens, followed by a booster immunization with alum-adjuvanted gp140, is being tested in a phase IIb clinical efficacy trial in 2600 women in five countries in sub-Saharan Africa called Imbokodo (HVTN 705, HPX 2008). Preliminary efficacy results are expected in 2021. A phase III trial called Mosaico (HVTN 706, HPX 3002) is currently being launched with a similar vaccine involving the mosaic Ad26-Env/Gag/Pol with bivalent mosaic/clade C gp140 in 3800 men and transgender individuals who have sex with men/transgender individuals in the Americas and Europe.

One of the most challenging problems, perhaps the most challenging problem, is the development of immunogens capable of inducing bnAbs. Such antibodies have proven extremely difficult to elicit by vaccination, and despite large research efforts, no such immunogens yet exist. The extremely high mutability of the virus, the degree of Env sequence diversity, and the extensive glycan shield that protects the Env trimer surface from antibody attack pose major obstacles to the elicitation of bnAbs.[361]

Soluble trimers that mimic the native Env spike have recently been developed. BG505 SOSIP.664 is a soluble protein that links Env gp120 and gp41 subunits with a disulfide bond and is stabilized with an I559P substitution in the prefusion state.[362,363] Recent studies in rabbits and macaques demonstrate that BG505 SOSIP.664 gp140 immunogen elicits significant autologous neutralizing antibody (ie, antibodies that specifically neutralize the virus strain from which the immunogen was derived) responses but with very limited neutralization against other HIV strains.[66,363-366] These autologous BG505-specific NAbs target an immunodominant hole in the glycan shield of this virus,[367,368] although other epitopes are also targeted.[368]

The development of SOSIP Env trimers as immunogens that are capable of eliciting significant titers of autologous neutralization antibodies represents an important first step in the quest to develop immunogens that elicit bnAbs. There is currently intense research in many laboratories to use structural biology to design improved SOSIP immunogens. The fact that some chronically infected people who have been sequentially exposed to multiple variants of HIV do develop bnAbs[337-342] raises the possibility that such bnAbs could also be generated through clever vaccination strategies. One concept that is being actively investigated is the idea that through a series of consecutive immunizations with different HIV Env immunogens, the evolution of the antigen-specific B-cell response could be "shepherded" through a series of steps that culminate in the generation of B-cell clones that make bnAbs.[369-371] This approach involves priming with an immunogen that can bind to the B-cell receptor expressed on a naïve, antigen-specific B cells (so-called "germline targeting") followed by booster immunizations with a series of immunogens, each slightly different from the preceding one, to coax and cajole the B-cell clone to develop along a developmental pathway that eventually generates bnAb producing B cells (370–372). An example of this strategy is the use of the eOD-GT8 nanoparticle–based immunogen that was designed to stimulate precursor naïve B cells for the VRC01 class of bnAbs, which target the CD4-binding site, and it is currently being tested in a phase I trial.[372,373] Sequential immunogens have also been designed that aim to induce V3-specific bnAbs.[374] So far, none of the immunogens have elicited bnAbs in NHPs or humans.

TB Vaccines

TB kills more people than any other infectious disease.[375] This disease has existed since ancient times, and *M. tuberculosis*, the bacterium that causes TB, has coevolved with humankind since about 70,000 years ago, when *Homo sapiens* walked out of Africa.[376] In 2019, an estimated 10 million people developed active TB, and 1.4 million people died.[377] It is estimated that about 2 billion people, or one third of the world's population, are infected with TB, which remains latent with no obvious clinical symptoms.[378] However, only around 10% of infected persons become ill with active TB, but in such cases, the disease can manifest a wide spectrum of clinical symptoms, ranging from mild to severe disease with lung and thoracic lymph node involvement or even extrapulmonary manifestations.[379] People with immunocompromised immune systems, including those with HIV, have an increased risk of acquiring active TB. Moreover, the emergence of multidrug-resistant TB is a major public health crisis.

TB is different from HIV and malaria, in that a licensed vaccine already exists. BCG, the only TB vaccine approved for use in humans, was developed by Calmette and Guérin more than 100 years ago, by serial passaging of *M. bovis* in media, amazingly, for a period of 13 years. A seed stock of the BCG was distributed worldwide to other vaccine developers, who attenuated it further, and as a result, there are now more than 16 strains of BCG.[380] BCG is the most widely used vaccine in the world, with 100 million children receiving it every year. There have been many clinical trials performed over the past several decades to measure the efficacy of BCG vaccination, and these have yielded highly variable results. The reasons for this high variability are unclear, but could be connected with geographical regions, genetic differences in the different strains of BCG used, nutritional status of people, or exposure to TB or other environmental mycobacteria that could interfere with BCG vaccination–induced immunity.[381] However, one consistent finding is that BCG vaccination at birth is very effective (70%-80%) at protecting against disseminated TB such as TB meningitis or military TB that spreads outside the lung. It also protects against leprosy. However, it fails to provide consistent protection against lung disease, especially in people living in TB-endemic areas of the world. A meta-analysis of the literature showed that people with previous exposure to *M. tuberculosis* or environmental mycobacteria had lower efficacy to BCG.[382] Consistent with this finding, the efficacy of BCG is highest in unexposed children (average efficacy about 50%) and lowest in adults living in TB-endemic areas (typically no efficacy). Thus, although improved diagnostics and novel antibiotics (for drug-resistant TB) would greatly help reduce the global burden of disease, the development of a safe and highly effective TB vaccine remains the only way of preventing TB and is therefore a major global health priority. Yet, since the development of BCG by Calmette and Guérin more than 100 years ago till the turn of the 21st century, there was little progress in making a TB vaccine. Major challenges to the development of an effective TB vaccine include the poor predictive value of preclinical animal models, the highly variable vaccine efficacies observed in different populations (such as those with pre-exposure to TB, or those coinfected with HIV), and a paucity of understanding of the immunological mechanisms and correlates of protection.

With regard to correlates and mechanisms of protection, early work in mice demonstrated the importance of the T helper 1 (Th1) response in mediating protective immunity against TB.[383] Th1 cells secrete IFNγ and activate antibacterial responses in macrophages, which are the target of *M. tuberculosis* infection.[383] Consistent with this, mice deficient in IFNγ exhibit impaired immunological control of *M. tuberculosis* growth and dissemination,[384] and humans with genetic deficiencies in the IFNγ receptor have been shown to develop fatal, disseminated BCG infection (which occurs in very rare cases after BCG vaccination)[385] or disseminated *Mycobacterium tuberculosis* infection.[386] Furthermore, in a study in South African infants, the frequency of BCG-specific, IFNγ-producing CD4+ T cells was associated with reduced TB risk during the first 1 to 3 years of life.[387] However, work during the past 2 decades has revealed diverse immunological mechanisms that could play a role in protection. Thus, experiments in animal models provide evidence for a role for CD8+ T cells in the control of *M. tuberculosis* infection, and in humans, infection results in antigen-specific CD8+ T cells.[388] Furthermore, studies in animal models have revealed roles for γδ T cells and T-cell effector mechanisms that can occur independently of IFNγ or stimulate activation of the innate immune system.[384] Thus, a major challenge for TB vaccinology is to acquire a deeper, mechanistic understanding of the complex immunological mechanisms that mediate protection. Interestingly, recent studies in animal models show that the route of administration of BCG could result in enhanced protective immunity against *M. tuberculosis*. In NHPs, intravenous immunization with BCG stimulated robust antigen-specific CD4+ and CD8+ T-cell responses in the lung and other sites, as well as significantly enhanced protection against challenge with *M. tuberculosis*, with 60% of the animals showing no detectable infection.[389] In addition, in mice, intranasal immunization with BCG stimulated increased frequencies of antigen-specific, tissue-resident CD4+ T cells and enhanced protection, compared to intradermal immunization.[390] Although an intravenous or intranasal route of vaccination is unlikely to be deployable as a global strategy for vaccination, this preclinical result enables the mechanistic dissection of what immune responses may be necessary for protection against *M. tuberculosis*.

Substantial investment in research and development over the past 2 decades has reenergized the TB vaccine field, and the first rays of hope are now beginning to emerge. In contrast to the situation in 2000 when the TB vaccine landscape was barren, in 2019, more than 20 TB vaccine candidates have been evaluated in clinical trials, including 7 in phase IIb or phase III clinical trials. These candidates include subunit vaccines consisting of recombinant proteins plus adjuvants, recombinant viral vectors expressing TB antigens, and recombinant and attenuated strains of BCG.

The "M72/AS01$_E$" vaccine candidate developed by GSK contains a recombinant fusion protein derived from two *M. tuberculosis* antigens (Mtb32A and Mtb39A), adjuvanted with AS01$_E$. A phase IIb efficacy trial involving 3575 HIV-uninfected adults aged between 18 to 50 years with latent TB was performed in Africa. In 2018, an interim after all participants had completed at least 2 years of follow-up showed that two doses of the vaccine administered 1 month apart provided 54% protection against bacteriologically confirmed, symptomatic, active pulmonary TB.[391] The final analysis showed that the vaccine efficacy at 36 months was 49.7%, and the antigen-specific antibody response and CD4+ T-cell response was increased after the first dose and sustained throughout the follow-up period.[392] Thus, vaccine efficacy was sustained for at least 3 years. This encouraging result has invigorated the TB vaccine field, but these results need to be confirmed in larger and longer studies in diverse populations in different regions of the world and of different ages, and in subjects who have no prior history of exposure to TB. Importantly, blood samples banked from this trial must be analyzed to define immunological correlates of protection, and efforts to perform such analyses are beginning. Such correlates will greatly help the evaluation and advancement of other vaccine candidates in clinical trials.

The capacity of a vaccine to prevent active disease remains the main efficacy readout for licensure of a new TB vaccine. However, clinical trials designed to evaluate the efficacy of vaccines to prevent active disease require a very large number of subjects, and long follow-up periods, because of the small fraction of people who will develop active disease. Therefore, alternative efficacy endpoints that would require smaller trials (eg, prevention of *M. tuberculosis* infection) have been considered. In 2018, the results of a phase II trial in South Africa to evaluate the capacity of BCG or a candidate subunit vaccine to prevent infection were reported.[393] The subunit vaccine known as H4:IC31 consisted of a recombinant fusion protein H4 (mycobacterial antigens Ag85B and TB10.4) and an adjuvant known as IC31, which consists of an oligodeoxynucleotide that does not contain CpG DNA nucleotides, but signals via TLR9.[394] In this study, 990 adolescents in a high-risk setting, who had received neonatal BCG vaccination, were randomly assigned to receive BCG or H4:IC31 or placebo. The subjects did not have HIV or TB (as judged by the QuantiFERON-TB assay [QFT], which involves stimulation of blood cells *in vitro* with antigens from *M. tuberculosis*, followed by assessment of IFNγ production). The results of this study showed that there was no significant reduction in *M. tuberculosis* infection, as defined by the QFT assay. However, BCG vaccination significantly reduced the rate of sustained QFT conversion, with an efficacy of 45.4%. Thus, the rate of sustained QFT, which could reflect sustained *M. tuberculosis* infection, was reduced by BCG vaccination.[393] In contrast to this study, a previous BCG revaccination study in Brazil, involving more than 200,000 children, found no significant protection induced by BCG revaccination on prevention of disease.[395] These discrepant results could be due to differences in the levels of nontuberculous mycobacteria between the sites, which could have impacted on the efficacy.

In addition to these trials, other TB vaccine candidates that are in advanced stages of testing include a live attenuated, recombinant BCG vaccine originating from the Max Planck Institute for Infection Biology, and developed by Vakzine Projekt Management GmbH, in collaboration with Serum Institute of India,[396] and the live attenuated *M. tuberculosis* MTBVAC that was rationally attenuated to include many genetic regions encoding immunodominant antigens that were absent in BCG and to genetically delete virulence genes, thus improving safety.[397-399] It is hoped that these "BCG replacement vaccines" would induce immunity that provides consistent and durable protection against TB. In addition, there are other recombinant protein antigens plus adjuvant candidate vaccines,[400] such as ID93 + GLA-SE, a TB fusion protein antigen combined with the TLR4 agonist adjuvant, and GLA-SE developed by IDRI.[207] As will be evident, the past 2 decades have witnessed considerable progress in TB vaccinology, with encouraging results observed in two efficacy trials, and numerous candidate vaccines being assessed in preclinical and early clinical trials.

Universal Influenza Vaccine (Fig. 40.11)

As discussed above, seasonal influenza vaccines are developed annually against the specific strains that are predicted by the WHO Collaborating Centers, to be circulating in the upcoming flu season. Thus, annual vaccinations are required for efficient protection against the specific influenza strains each year, which continuously acquire mutations in HA (antigenic drift). When there is an antigenic mismatch between the predicted strains and the strains that actually circulate, then the vaccine efficacy is very low, as was the case in the 2014 to 2015 season.[242-244] In addition, pandemics can occur due to antigenic shift in HA molecules, as was recently the case with the 2009 H1N1 pandemic, where it has been estimated that the vaccine efficacy was less than 10%.[240,401] Another shortcoming of the seasonal influenza vaccines is their failure to stimulate durable antibody responses[245] unlike that seen with live attenuated vaccines such as measles, mumps, smallpox, or yellow fever, which can induce lifelong protection. Clearly, a vaccine that would stimulate broad and durable protective immunity against most or all seasonal influenza strains, as well as against pandemic strains would be of enormous value. Thus, there are several major efforts underway to develop a "universal influenza vaccine." The current thinking is that a vaccine that is greater than 90% efficacious (as is the case with several licensed vaccines) and stimulates greater than 75% protection against symptomatic influenza disease, for a

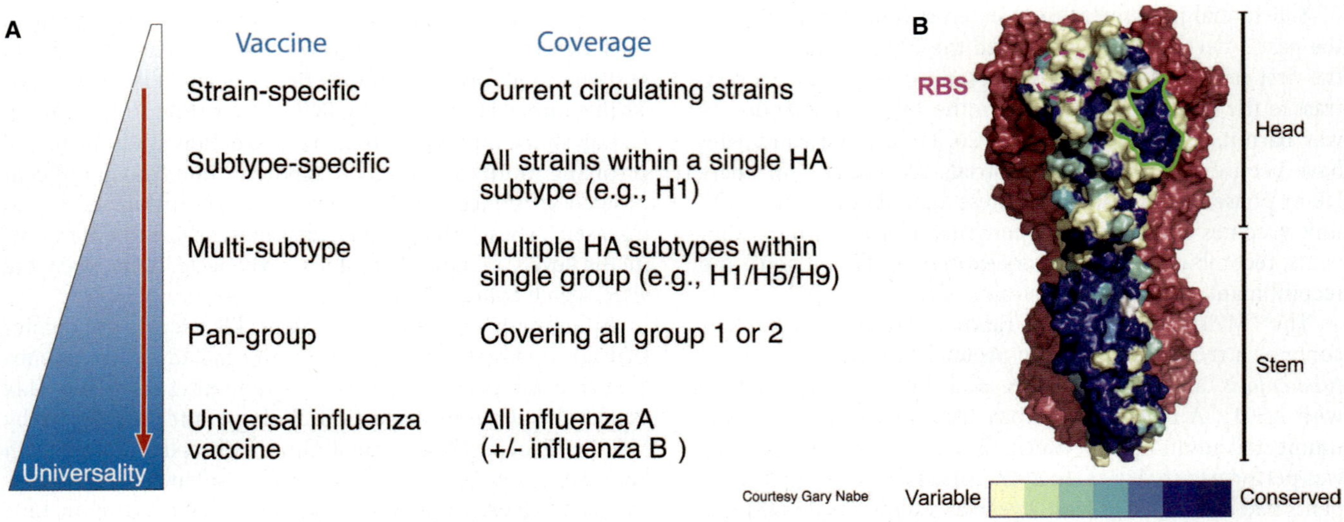

FIG. 40.11. A: Steps toward a universal influenza vaccine. **B:** Sequence conservation on the surface of a hemagglutinin monomer on a scale from beige (most variable) to dark blue (most conserved). The receptor-binding site (RBS) and the conserved lateral patch are outlines in pink and green, respectively. (A, Reprinted from Paules CI, Marston HD, Eisinger RW, et al. The pathway to a universal influenza vaccine. *Immunity.* 2017;47(4):599-603, with permission from Elsevier. B, From Raymond DD, Bajic G, Ferdman J, et al. Conserved epitope on influenza-virus hemagglutinin head defined by a vaccine-induced antibody. *PNAS.* 2018;115(1):168-173.)

period 12 months or longer in all populations, is considered to be a reasonable target.[402] There is reasonable consensus in the field that a universal flu vaccine should include antigens that stimulate protection against both influenza A (which can cause pandemics as well as significant seasonal morbidity and mortality) and influenza B (that can also cause severe seasonal disease).[402] With regard to influenza A, a vaccine could ideally stimulate broad protection against all HA subtypes in groups 1 and 2, although progress toward this ideal is likely to occur in incremental steps (Fig. 40.11A). The universal influenza vaccine platforms being developed include technologies such as DNA and mRNA, viral vectors, and recombinant proteins.[403] In addition to HA, a range of other influenza proteins including NA, as well as the M2 protein and internal viral NP and M1 protein, which contain cross-reactive T-cell epitopes, are also attractive targets.

HA-specific antibodies are induced robustly by natural infection with influenza or vaccination, and the HI assay that is used to quantitate HA-specific antibodies is frequently used as a correlate of vaccine efficacy. Recent advances in structural biology have revealed deep insights into certain highly conserved antigenic epitopes within HA that could represent attractive targets for a universal influenza vaccine. HA is a type 1 membrane glycoprotein that forms a homotrimer and consists of a variable globular head domain that mediates for viral entry when its RBS binds to sialic acid expressed on respiratory epithelial cells[404,405] (Fig. 40.11B) and the highly conserved stem/stalk domain that mediates membrane fusion and entry into cells.[406] Molecular structural analysis of HAs from different strains reveals that although the global architecture is conserved, they display differences in the surface sequence composition especially at regions at or proximal to the RBS, as well as in the glycosylation patterns[404,405] (Fig. 40.11B). Antibodies directed against the head region of HA can be induced by infection or vaccination[407,408] and block viral entry into cells. However, these antibodies are usually strain specific and are unable to efficiently neutralize drifted or shifted variants. In contrast, the stem region of HA is more highly conserved between strains, and bnAbs against this region have been identified[409-416] (Fig. 40.10B).

Although most influenza vaccines elicit strain-specific responses, some present highly conserved epitopes contained within the stem or head region of HA and elicit cross-reactive antibodies. A major thrust toward developing a universal influenza vaccine has been to devise vaccine strategies aimed at stimulating antibodies directed against the highly conserved but immunosubdominant stem region of HA.[406] These approaches include engineering stabilized headless HA[417-419] or sequentially immunizing with so-called "chimeric HAs" that all contain an identical stalk domain but different head domains.[420,421] Experiments in mice and ferrets show that these vaccines induce broadly cross-reactive antibodies that can protect against heterologous viruses. Unlike HA head-specific antibodies (which mediate hemagglutination activity and neutralize virus by blocking viral entry into cells), broadly protective antibodies against the stem region have been shown to confer protection *in vivo* via Fc receptor–mediated effector functions such as ADCC or ADCP.[422,423] As a result of these studies, inactivated and live attenuated vaccines based on the chimeric HA technology have advanced to phase I clinical trials in humans (ClinicalTrials.gov Identifiers: NCT03275389 and NCT03300050). Current efforts are underway to optimize stem HA antigens to ensure that the natural tertiary and quaternary epitopes are preserved. A potential concern is that constantly drifting influenza virus could acquire mutations to evade recognition by stem-specific antibodies. Although this is a possibility, studies in animal models indicate that the mutational tolerance of the HA stem domain is very restricted.[424,425] These results suggest that a polyclonal response that recognized multiple epitopes in the HA stem region would pose major structural and fitness constraints for the virus to evade the immune system. In addition, a vaccine that stimulates such a polyclonal response, coupled with immune responses against NA and the conserved NP, would likely pose major challenges for a virus to evade immunity without compromising fitness.

Another approach to design vaccines that confer broader protection is to use computational algorithms to optimize broadly reactive antigens (COBRA) that induce cross-reactive antibodies.[426] Such COBRA HA antigens have been included in a "cocktail" of divergent H1 HAs expressed in HA-ferritin nanoparticles to induce broad antibody responses in mice.[427] Immunization with a cocktail of HAs induces the expansion of multiple lineages of HA-specific B cells, each of which is capable of producing antibodies to a particular HA or its close relatives, but which collectively provide broadly cross-reactive antibody responses at a population level. This concept of a polyvalent antigen design is not new in vaccinology. For example, the pneumococcal and meningococcal vaccines contain carbohydrates from multiple serotypes, with vaccine PNEUMOVAX containing no fewer than 23 serotypes. The seasonal influenza vaccine itself contains HAs from type A and B viruses and groups 1 and 2 within type A. Thus, in mice, "mosaic nanoparticles" that display receptor-binding domains (RBDs) of viral HAs from multiple strains on the same nanoparticle have been shown to elicit broad antibody responses.[428] Immunization with mosaic RBD nanoparticles elicited broader antibody responses than that induced by a mixture of nanoparticles, each containing a different RBD. Importantly, a monoclonal antibody with broadly neutralizing activity could be isolated from mice immunized with RBD nanoparticles.

Finally, in the case of highly conserved RBD, investigators have isolated monoclonal antibodies from humans vaccinated the seasonal influenza vaccine, which bind to conserved regions of the H1 HA such as the RBS, and which are capable of neutralizing diverse H1 viruses.[429] Interestingly, the unmutated common ancestor of six distinct RBS-antibody lineages from a single individual could bind the HA of a virus circulating at the time the subject was born, but not the HAs of viruses circulating 5 years later.[430] Importantly, mature antibodies in these lineages bound strains from the entire lifetime of the participant. These exciting results suggest that immunological imprinting through vaccination or infection can drive the evolution of germline precursors of limited breadth to mature antibody lineages with broad cross-reactivity.

RSV Vaccines

RSV is a common respiratory virus that is a major cause of severe respiratory tract infection worldwide. It usually causes mild, cold-like symptoms, but can result in serious, sometimes fatal infection in infants and elderly adults. It is a negative-sense, single-stranded RNA virus belonging to the *Paramyxoviridae* family of viruses. RSV is the most common cause of bronchiolitis and pneumonia in children younger than 1 year and is responsible for 7% of deaths in children aged 1 month to 1 year.[431,432] Most children will have had RSV by the age of 2 years. RSV is also increasingly recognized as significant cause of respiratory illness in adults older than 65 years. The development of an RSV vaccine has been a global health priority for decades, but several factors have thwarted efforts to develop a vaccine.[431,432] In infants, severe RSV typically occurs between the second and third months of life, a narrow developmental period characterized by a nascent immune system and the presence of high titers of maternal antibodies. Both of these factors make it more challenging to induce robust immune response to vaccination. In the elderly or immunocompromised, the risk of severe disease is exacerbated by immune senescence and comorbidities.[431,432] These distinct immunological factors that underpin enhanced susceptibility to severe disease in the different demographic subgroups pose major obstacles to vaccine design.

In 1966, a formalin-inactivated RSV vaccine candidate was tested in infants and toddlers in the United States with disastrous consequences. Most of the children experienced enhanced RSV disease, when they became infected with wild-type RSV in the community, and two immunized toddlers died upon infection.[431,432] Postmortem analysis found evidence of eosinophilia and extensive bronchopneumonia. Initial studies indicated the children vaccinated with this vaccine failed to develop neutralizing titers comparable to that observed to the titers developed in age-matched children with RSV infection. It was thus postulated that the suboptimal neutralizing antibody titers were ineffective at clearing the virus or, alternatively, that nonneutralizing antibodies caused enhanced lung disease by formation of immune complexes. Later studies showed that formalin inactivation induced a Th2 response in children and that enhancement of disease was the result of hyperinflammation.[431,432] More recent work has shown that formalin inactivation resulted in loss of neutralization epitopes in the F viral glycoprotein trimer, which is involved in viral cell entry. This sad tragedy dampened RSV vaccine efforts for several years, and other approaches such as passive immunization with monoclonal antibodies were tried and led to the development in the 1990s of palivizumab, a humanized monoclonal antibody directed against the RSV F protein. Palivizumab has been shown to be 80% effective in preventing severe RSV infection in neonates.[431] The success of palivizumab has revived interest in developing vaccines that induce neutralizing antibody responses against the RSV F protein. A key issue appeared to be related to the confirmation of the F glycoprotein trimer, which transitions from a metastable prefusion conformation to a highly stable postfusion conformation. The metastable conformation of the F protein makes it especially challenging as a vaccine target antigen. In 2013, Kwong et al. determined the atomic-level crystal structure of the prefusion F protein by using potent neutralizing antibodies (AM14 and D25) that stabilized its prefusion conformation.[433-435] Based on this structure, they came up with a strategy to stabilize the metastable prefusion state by addition of disulfide bond and cavity-filling mutations (Cav1). Immunization with this prefusion-stabilized F

"DC-Cav1" variant induced a high magnitude of neutralizing antibodies in mice and macaques. The recent results of a phase I trial in humans to test the DC-Cav1 antigen, with or without an alum adjuvant, showed that immunization induced a greater than 10-fold increase in the magnitude of neutralizing antibody titers that targeted the prefusion-specific surface of RSV-F protein.[433-435]

The tragedy of the F1-RSV vaccine trial, as well as the narrow developmental window of opportunity (the first 2-3 months of life) to evoke a robust immune response, has dampened enthusiasm to develop RSV vaccines for the seronegative infant population.[431] Instead, effort has focused on vaccinating pregnant mothers with a view to providing protection to infants via transplacental transfer of maternal antibodies. An advanced maternal vaccine candidate is being developed by Novavax and consists of a recombinant RSV F nanoparticle vaccine, containing a highly conserved epitope which is the target of palivizumab, adjuvanted with alum. In 2016, a clinical trial showed that immunization of women resulted in enhanced antibody titers that reduced the risk of RSV infections.[436] However, the results of a phase III clinical trial (ClinicalTrials.gov Identifier: NCT02624947) showed that the vaccine just failed to reach its primary end point of preventing RSV lower respiratory tract infection. The results showed 44% efficacy of the vaccine against RSV lower respiratory tract infection hospitalizations and 48% efficacy against RSV lower respiratory tract infection with severe hypoxemia.

In addition, there are efforts to develop vaccines against older children. Although the highest burden of disease is in infants and the elderly, there are significant healthcare costs with RSV in older children, and it is thought that reducing circulation of the virus in older children may mitigate the impact of RSV on the vulnerable infant and elderly populations by reducing viral shedding. At the present time, there are 10 RSV vaccines targeting toddlers as well as older adults, including a replication-deficient adenovirus vector RSV vaccine (Clinicaltrials.gov Identifier: NCT03303625).

Finally, several vaccine candidates are being tested in the elderly population. The aforementioned Novavax nanoparticle RSV F candidate was assessed in a phase III clinical trial, and unfortunately, there was no evidence or protection against a lower respiratory tract infection.[437] Despite this setback, there is a robust pipeline of candidate RSV vaccines. For example, prefusion RSV F proteins are being evaluated in older adults (Clinicaltrials.gov Identifier: NCT03572062). Pfizer and GSK are in phase III studies with prefusion vaccines. Viral vector vaccines expressing both T- and B-cell antigens are being tested in older adults. A recent study has reported the successful development of self-assembling nanoparticle formulations presenting prefusion-stabilized F proteins in a polymeric array on a nanoparticle scaffold. In mice and NHPs, the prefusion-stabilized F protein nanoparticles induced >10-fold higher neutralizing antibodies than did previous trimeric formulations of prefusion-stabilized F protein.[438] These developments offer encouragement that a vaccine against RSV in older adults may be achievable.

COVID-19 VACCINES

The development of safe and effective COVID-19 vaccines, in sufficient quantities to vaccinate the entire planet, is an urgent global imperative. On December 31, 2019, the Chinese authorities alerted the WHO China Country Office of a mysterious pneumonia-like disease in a cluster of patients in the city of Wuhan in Hubei province. On January 7, 2020, the causative agent was identified as a new coronavirus (2019-nCoV) by the Chinese authorities, and the disease was later named as COVID-19 ("CO" stands for corona, "VI" stands for virus, "D" stands for disease and "19" for 2019) by the WHO.[439] Coronaviruses are members of the genus *Coronaviridae*, a group of RNA viruses that contain crown-shaped peplomers. SARS-CoV-2 is phylogenetically 80% identical to the severe acute respiratory syndrome coronavirus (SARS-CoV) and 50% similar to the Middle East respiratory syndrome coronavirus (MERS-CoV), which caused global outbreaks in 2002 to 2003 and 2011, respectively. COVID-19 was declared a global pandemic by the WHO on March 11, 2020. As of today, December 21, 2020, COVID-19 has infected 77,184,964 cases and killed 1,699,878 people worldwide.[440]

As of today, there are over 165 vaccine candidates being developed, with 54 vaccines in clinical trials and 13 vaccines in large-scale phase III efficacy trials.[441] These vaccines include a diverse range of platform technologies including mRNA from Pfizer/BioNTech[442] and Moderna,[443] adenoviral vector–based vaccines from Johnson & Johnson/Beth Israel Deaconess Medical Center[444] and AstraZeneca/Oxford,[445] and recombinant protein plus adjuvant from Novavax[446] and Sanofi/GSK (Fig. 40.12). Earlier in November 2020, Pfizer/BioNTech and Moderna announced the interim results of the phase III efficacy trials, indicating that their vaccines were over 94% effective,[447] and on November 23, 2020, AstraZeneca announced the results of their phase III study indicating an average efficacy of 70%.[448] On December 11 and 18, 2020, respectively, Pfizer/BioNTech and Moderna were given approval by the US FDA for emergency use authorization of their mRNA vaccines, these being the first mRNA vaccines to be approved for use in humans (Fig. 40.11). What is remarkable about this effort is the speed with which these vaccines have been developed, in a matter of 11 months from identification of the virus in January 2020 to their anticipated licensure in December 2020. As the world anxiously awaits several billions of doses of safe and effective COVID-19 vaccines, these encouraging results provide hope that we will indeed have a portfolio of vaccines that immunize the planet against this scourge. A portfolio of other vaccine candidates is expected to follow suit.

The antigen being used in the current COVID-19 vaccines is the SARS-CoV-2 spike protein, and the majority of the current vaccine candidates are being evaluated for their capacity to induce neutralizing antibodies to the spike protein. A critical antigenic target in the spike protein is the RBD that binds to the ACE2 receptor in cells to mediate viral entry. In some cases, the vaccines may also induce CD8+ T-cell responses that kill virally infected cells. However, most emphasis is being placed on generating neutralizing antibodies directed against the spike protein. Fig. 40.12 summarizes the leading vaccines

Type of vaccine	mRNA vaccine 	Recombinant viral vector vaccine 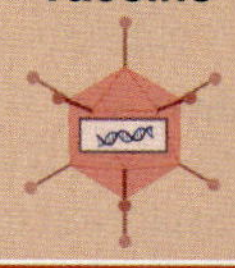	Inactivated virus vaccine 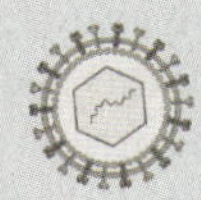	Recombinant protein vaccine 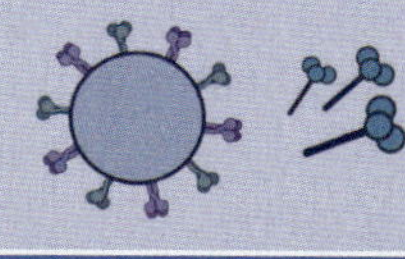
Status	Pfizer/BioNTech vaccine: approved in several countries. Emergency use in U.S., elsewhere Moderna vaccine: approved for use in Switzerland. Emergency use in U.S., European Union, elsewhere	Oxford/AstraZeneca: Emergency use in Britain, European Union, India and elsewhere CanSinoBIO: Ad5 vaccine approved in China. Emergency use in Mexico and Pakistan Gamaleya: Ad5/Ad26 vaccine early use in Russia; emergency use in certain other countries Johnson and Johnson (Ad26): approved for emergency use in US, European Union and elsewhere	Sinopharm: approved in China; limited in UAE, Bahrain Sinovac: inactivated vaccine approved in China, emergency use elsewhere	In phase 3 trials: Novavax: Spike protein with Matrix M adjuvant Sanofi/GSK: Spike protein with AS03 adjuvant Clover/GSK: Spike protein with CpG-1018 adjuvant Medicago: Spike protein on virus like particle with AS03 adjuvant
Vaccine efficacy	Pfizer/BioNTech: 95% Moderna: 90%	Gamaleya: 92%	Not known	Not known
Regimen	2 doses (21 to 28 days apart)	2 doses (21 to 28 days apart) Johnson and Johnson: 1 dose	2 doses (21 to 28 days apart)	2 doses (21 to 28 days apart)
Storage	Pfizer/BioNTech: -75°C Moderna: 22°C	Gamaleya: -20°C Oxford/AstraZeneca; CanSino; Johnson and Johnson: refrigerator	Refrigerator	Refrigerator

FIG. 40.12. Leading COVID-19 vaccine candidates, licensed or in advanced stages of efficacy trials.

that are currently in phase III trial. These include mRNA vaccines; viral vector vaccines comprising the ChmpAd, Ad26, and Ad5 vectors; and recombinant protein vaccines administered with adjuvants. All but one of these candidate vaccines are administered as a two-dose regimen, each dose being about a month apart. The Ad26 vaccine from Johnson &Johnson is a one-dose regimen. The vaccines induce neutralizing antibodies, albeit at varying levels, although strict comparisons are difficult because some studies are performed using pseudoviruses, while others are performed using wild-type SARS-CoV-2 viruses.

Moderna's mRNA-1273, COVID-19 vaccine, developed jointly with scientists at the Vaccine Research Center at the NIH, is a lipid nanoparticle–encapsulated, mRNA vaccine that encodes the spike glycoprotein stabilized in its prefusion conformation.[443] Pfizer/BioNTech's BNT162b1 vaccine is a lipid nanoparticle–formulated mRNA vaccine encoding the trimeric SARS-CoV-2 spike protein RBD.[442,449] Both vaccines are given as a two-dose regimen and are currently in phase III trials, and interim analyses have indicated a 90% to 95% efficacy in protecting against symptomatic disease. Previous phase I/II clinical trials have revealed significant titers of neutralization antibodies observed after the booster immunization. T-cell responses have been reported for the Pfizer/BioNTech vaccine, which showed significant frequencies of RBD-specific CD4+ and CD8+ T-cell responses. A challenge with the mRNA vaccines is that they need to be stored at temperatures of −70 °C (Pfizer) or −20 °C (Moderna), thus posing challenges to their distribution in rural areas and in developing countries.

Viral vector candidates are being developed by several companies. The ChAdOx1 COVID-19 vaccine (AZD1222) was developed jointly by scientists at Oxford University and AstraZeneca and consists of two doses, given a month apart, of a chimpanzee adenoviral vector encoding the sequence of the spike protein.[445] Humans have a low seroprevalence of the chimpanzee adenoviral vector; therefore, this vector does not have the complications of preexisting immunity. The results of the interim analysis of a phase III trial announced today indicate that two different dosing regimens demonstrated efficacy, with one showing better efficacy. Thus, one regimen involving half dose of the ChAdOx1 followed by full dose in 2741 people showed an efficacy of 90%, whereas the other regimen involving 8895 people given two full doses showed an efficacy of 62%, resulting in an average efficacy of 70%.[448] The reasons for this difference in efficacies between the two regimens is at present unclear, but the publication of the data from this and the other phase III trials are eagerly awaited, in order to gain further insights. The results of a previous phase I study with this vaccine had shown detectable neutralization titers after the booster immunization, and substantial T-cell responses, measured by IFNγ ELISPOT assays, still detectable at late time points postvaccination. An advantage of the AstraZeneca/Oxford vaccine is that it can be stored and transported at normal refrigerated temperatures of 2 °C to 8 °C, thus facilitating its distribution with the existing immunization infrastructure in developing countries. Another viral vector approach developed by Janssen/Beth Israel Deaconess Medical Center uses the Ad26 vector expressing the spike protein.[444] A phase I/IIa clinical trial showed that a single dose of the vaccine induced neutralizing antibodies and CD4+ and CD8+ T-cell responses. The results after the booster immunization are yet to be released. In addition to these vectors, a single-dose Ad5 vector vaccine (CanSino Biologics, China) and Ad26/Ad5 prime-boost vaccine (Gamaleya National Research Center for Epidemiology and Microbiology in Russia) are being developed and have been approved for limited or early use in China and Russia, respectively.

Lastly, several companies are developing recombinant protein vaccines administered with adjuvants. The first recombinant protein vaccine candidate (NVX-CoV2373, Novavax) adjuvanted with Matrix M, a saponin-based adjuvant, is being tested in a phase III trial. The results of the phase I clinical trial have been reported.[446] Two immunizations spaced about a month apart induced a high-magnitude (>3000) neutralization titers.[446] Sanofi and GSK are testing a recombinant spike protein adjuvanted with AS03 in phase I and II clinical trials and intend to make available 200 million doses of this vaccine, if approved.[450] Medicago, a Canadian company, is developing a recombinant VLP expressing the stabilized prefusion form of the SARS-CoV-2 spike protein produced by transient expression in the plant *Nicotiana benthamiana*.[145] This candidate vaccine was administered alone or with AS03 or CpG 1018 adjuvants and was evaluated in a Phase I trial in healthy adults. Both AS03 and CpG 1018 elicited strong humoral and T-cell responses after the second dose. Even at the lowest antigen dose, AS03 induced potent neutralizing antibody titers.[145]

The development of so many candidate vaccines from concept to anticipated licensure in 11 months is a remarkable achievement that is unprecedented in the history of vaccinology. Typically, vaccine development from concept to preclinical testing in animals to phase I, II, and III clinical trials can take a decade or even 15 years and cost up to $500 million.[451,452] In the case of COVID-19 vaccines, however, on January 13, 2020, just 2 days after the Chinese authorities released the genomic sequence of SARS-CoV-2, scientists at Moderna and the NIH finalized the sequence of the mRNA-1273 vaccine, and amazingly in just 63 days on March 16, the vaccine had been injected into the first human in a phase I trial! In the United States, the government launched Operation Warp Speed, Manhattan Project style project, with the ambitious goal of delivering tens of millions of doses of safe and effective SARS-CoV-2 vaccines starting at the end of 2020, with the ultimate goal of deploying 300 million doses by the middle of 2021.[453,454] Led by the United States Department of Health and Human Services and the Department of Defense in partnership with several commercial vaccine manufacturers, Operation Warp Speed has had a galvanic effect on the development of COVID-19 vaccines. What made Operation Warp Speed possible? A major factor was the mRNA technology that enabled the candidate vaccine to be synthesized within weeks of the viral genomic sequence becoming known. However, perhaps the biggest contributing factor was the fact that the government had provided financial resources for the companies to allow the financial risk of manufacturing the candidate vaccines at scale, even before the results of the phase III trial became known.

Very early in the pandemic, it was realized that in order to end to the pandemic, not only was it necessary to develop vaccines but also to ensure global access of these vaccines. In other words, we needed to vaccinate the planet! The WHO, in partnership with the Coalition for Epidemic Preparedness Innovations (CEPI) and the Gavi, the Vaccine Alliance, started COVAX, a global initiative aimed at working with vaccine manufacturers to provide countries worldwide equitable access to safe and effective vaccines. One hundred and seventy two countries are now engaged in discussions to participate in COVAX, and the goal of COVAX is to deliver two billion doses of vaccines by the end of 2021.[455]

THE CHALLENGE AND PROMISE OF 21ST-CENTURY VACCINOLOGY

As will be evident from the preceding pages, despite the many triumphs of vaccinology, major challenges remain. A notable failure to date is that, despite several decades of effort, there are no efficacious vaccines for pathogens such as TB, malaria, and HIV. A major obstacle has been that there is an incomplete understanding of what the immunological mechanisms of protection are, and how to efficiently stimulate them through vaccination. Although research in animal models has been very useful in providing mechanistic insights, it has increasingly become clear that many vaccine concepts that work well in animal models simply fail to translate into humans.[456] Furthermore, vaccine efficacies can vary greatly between human populations in different regions of the world, or in different age groups. This highlights the imperative to harness the "human model" at a relatively early stage of the vaccine development pipeline.[456]

Systems Vaccinology

For nearly a century, vaccine developers have relied on a single measurement, typically the magnitude of antibody titers in the serum or plasma, as measured by assays such as ELISA or neutralization assays, to assess the response to a vaccine. Vaccinologists have used such serological measurements to define "correlates of protection," which is a threshold of antibody titers that represents a signature of protective immunization. Indeed, the magnitude of antigen-specific antibody titers is considered to be the primary correlate of protection against most pathogens.[457] However, despite their widespread use, such measurements do not capture the full spectrum of different immunological responses that may mediate protection against different pathogens or even against a single pathogen. For example, as stated above, antibody responses to influenza vaccination is not a reliable correlate of protection in the elderly population, in whom T-cell responses have been shown to be better correlates.[249] Furthermore, CHIM studies with malaria indicate two different vaccination regimens (involving either three consecutive immunizations with RTS,S [RRR] or two consecutive RTS,S immunization preceded by a primary immunization with adenovirus 35 [Ad35] expressing the CSP protein [ARR]), both yielded about 50% protection, albeit with different correlates.[328] The correlate of protection for the RRR regimen was the magnitude of CSP-specific antibody titers on the day of challenge with *P. falciparum*, while that for the ARR regimen was the magnitude of the CSP-specific polyfunctional CD4+ T cells, 2 weeks after priming with Ad35.[328]

Recent advances in the use of systems biological approaches to vaccinology have allowed a comprehensive analysis of the myriad of innate and adaptive immune responses stimulated by vaccination. The use of omics technologies, such as transcriptomics, metabolomics, mass cytometry, and epigenomics, offers an exquisitely penetrating look at the immune response to vaccination, even at the single cell resolution. This coupled with computational approaches and machine learning tools has been used to define "molecular signatures" that predict vaccine immunogenicity and efficacy.[30,458] The first example of this approach came from studies using the YF-17D, where transcriptional profiling of PBMCs isolated 3 to 7 days postvaccination revealed a signature of gene expression involving antiviral sensing and viral immunity including the type 1 IFN pathway.[95,459] Computational analysis and machine learning approaches revealed gene signatures that correlated with, and predicted in an independent study, the ensuing neutralizing antibody and T-cell responses that occurred weeks to months later (Fig. 40.13A). Subsequently, this "systems vaccinology" approach has been applied to study immune responses to several other vaccines, including seasonal influenza,[460-464] meningococcal,[465] shingles,[466] malaria,[328,467] Ebola,[468] and other vaccines, and helped identify early molecular signatures of immunogenicity. In addition, "baseline signatures" (ie, signatures prior to infection) that correlate with immune responses to vaccination have also been identified.[463,469] Furthermore, recent studies have performed integrated analysis of orthogonal datasets (transcriptomics, metabolomics, multiplex analysis of serum cytokines, multiparameter analysis of cell populations) to generate an integrated, multifactorial, response network (MMRN) of immunity to vaccination.[466] This analysis has revealed novel correlates of vaccine immunity, including signatures of metabolic pathways such as sterol metabolism that correlate with the adaptive immune response.[466]

Systems vaccinology approaches are yielding several important insights about the molecular mechanisms that regulate innate and adaptive responses to vaccination. For example, systems analysis of the immune response to yellow fever YF-17D vaccination in humans revealed that YF-17D–induced early expression of the general control nonderepressible 2 kinase (GCN2) in the blood strongly correlated with the magnitude of the later CD8+ T-cell response.[30] Further experiments in knockout mice that were deficient in GCN2 revealed a key role for YF-17D–induced GCN2 activation in programming DCs to initiate autophagy and enhance antigen presentation to both CD4+ and CD8+ T cells.[470] Later, GCN2 was also known to modulate intestinal DCs and macrophages to suppress intestinal inflammation.[471] Thus, these findings reveal an unappreciated connection between the evolutionarily ancient amino acid sensing pathway mediated by GCN2 and immunity to the YF-17D.

Another example of a mechanistic insight that emerged was on the impact of the gut microbiota on vaccine-induced immunity. Analysis of the transcriptional signatures to vaccination with the seasonal influenza vaccine revealed that expression of the gene encoding TLR5, a receptor for bacterial flagellin, was induced within a few days of vaccination and strongly correlated with the HA titers measured 28 days later.[460] Experiments in mice showed that TLR5 signaling by flagellin from the gut microbiota was critical for induction of antibody responses to influenza vaccination, highlighting a role for microbiota on vaccine immunity.[472] These experiments prompted the question of whether the microbiota could have an impact on vaccine responses in humans and led to a clinical trial in humans in which broad-spectrum antibiotics were administered to healthy adults for a few days before and after vaccination against seasonal influenza. Antibiotics treatment led to a 10,000-fold, albeit transient, reduction in the total quantity of gut bacteria, as well as a profound and long-lasting restriction in bacterial diversity that took several months to regain its baseline state.[473] Yet, despite these major changes in the gut microbiota, the magnitude and kinetics of the antibody response to vaccination was unaffected. This raised the possibility that prior exposure to influenza infection or vaccination had imprinted immunological memory that was able to withstand even the most severe dysbiosis caused by antibiotics. To test this, a second trial was performed by recruiting subjects who had not received influenza vaccination or had influenza infection in the 3 years preceding enrollment. These subjects were also prescreened to select those with the lowest baseline antibody titers against influenza.[473] In this second trial, the same antibiotics regimen as was used in the initial trial was observed to have a significant impairment in the H1N1-specific neutralization titers as well as the binding IgG1 and IgA response induced by vaccination. In addition to these effects on the adaptive immune response, antibiotics administration also enhanced inflammatory signatures (including transcriptional signatures of AP-1/NR4A expression). In addition, there was a striking perturbation in the serum metabolic profile, with a 1000-fold reduction in serum secondary bile acids, which was highly correlated with AP-1/NR4A signaling and inflammasome activation. Integrative MMRN analysis revealed significant associations between bacterial species and metabolic phenotypes, highlighting a key role for the microbiome in modulating immunity to vaccination in humans.[473]

In addition to multiomics measurements of vaccine responses, another approach that has been developed recently is an unbiased and comprehensive survey of the diverse effector functions of antibody responses to vaccination.[474,475] Coupled with machine learning tools, this approach is aimed at capturing broadly the compendium of antibody effector phenotypes (eg, glycosylation profiles of vaccine-induced

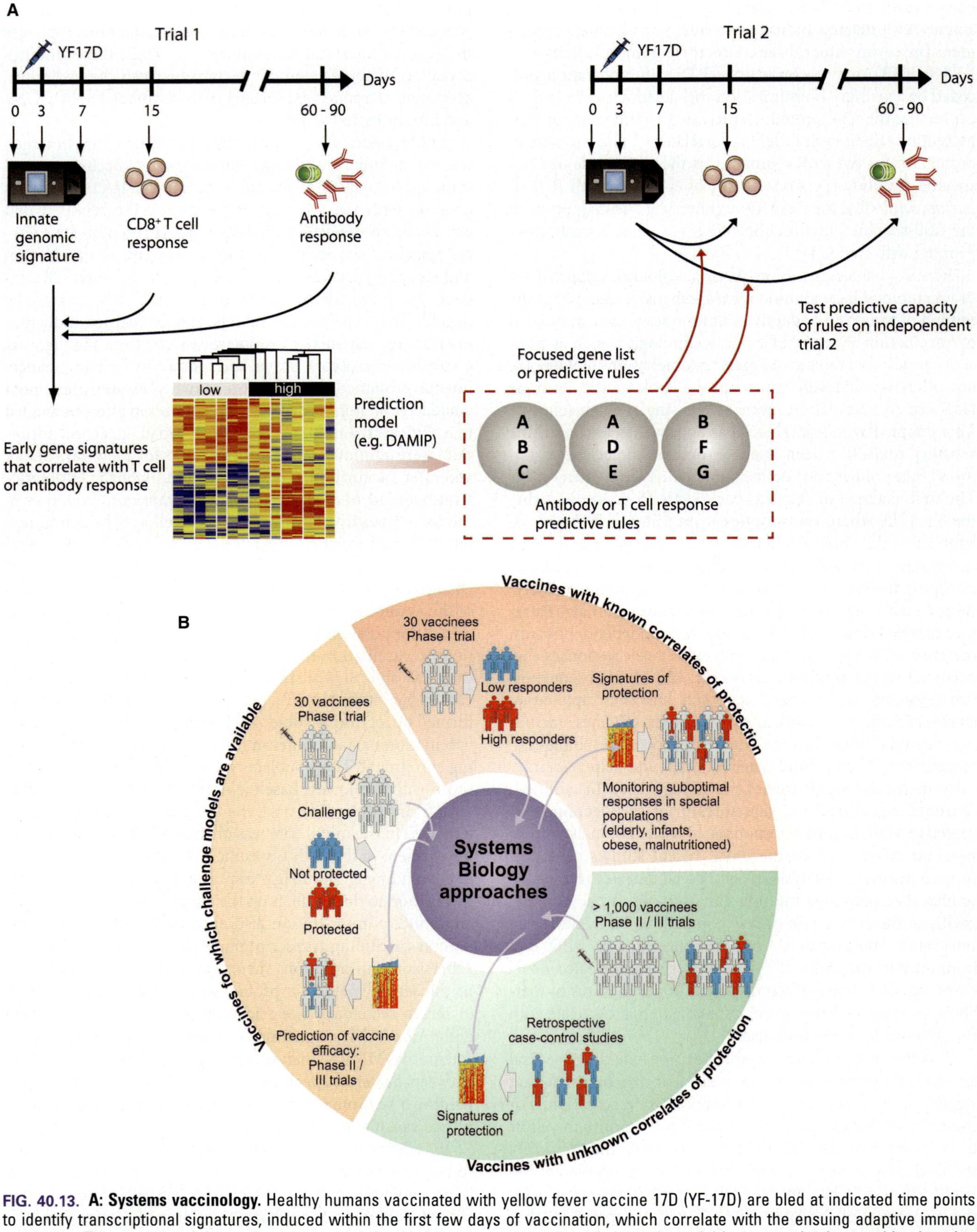

FIG. 40.13. **A: Systems vaccinology.** Healthy humans vaccinated with yellow fever vaccine 17D (YF-17D) are bled at indicated time points to identify transcriptional signatures, induced within the first few days of vaccination, which correlate with the ensuing adaptive immune responses. The capacity of these signatures to predict the immune response in an independent trial is assessed using machine learning techniques. **B:** The application of systems vaccinology approaches in clinical trials. (A, Reprinted from Pulendran B, Li S, Nakaya HI. Systems vaccinology. *Immunity*. 2010;33(4):516-529, with permission from Elsevier and B, From The application of systems vaccinology approaches in clinical trials. *PNAS*. 2014;111(34):12300-12306.)

antibody molecules) and functions (eg, ADCC, ADCP), with a view to obtaining a global picture of the architecture of the antibody response to vaccination.[474,476-478]

How can systems vaccinology approaches be used to define correlates of vaccine efficacy in clinical trials? One approach is the use of CHIMs, in which vaccinated humans can be challenged with the corresponding pathogens to define correlates and mechanisms of vaccine efficacy. Such models exist for pathogens such as influenza, malaria, typhoid, and cholera, and discussions are underway to establish models for other pathogens such as *B. pertussis*, Zika, and more recently SARS-CoV-2, although clearly the serious ethical issues involved the need for very serious consideration.[476] Systems biological analysis of immunity to vaccination in CHIM has already been done in the context of RTS,S malaria vaccination and helped identify signatures of vaccine efficacy.[328,467,479] As mentioned above, these studies reveal that protective immunity to malaria can be mediated by multiple mechanisms, including the magnitude of the circumsporozoite-specific antibody response, as well as the magnitude of the CSP-specific CD4+ T-cell polyfunctional response.[328] Another approach to defining correlates of efficacy is to use animal challenge models, such as the NHPs model for HIV, which can be used to define signatures of vaccine immunogenicity and efficacy that can then be translated into human trials. However, there are varied opinions on the extent to which animal models reflect infection in humans. In the case of phase IIb or III efficacy trials, which involve several thousand subjects, retrospective, nested, case-controlled studies can be performed, in which "banked" blood samples can be analyzed using systems-based approaches to identify correlates of vaccine efficacy. Thus, systems vaccinology can help not only test current candidates but also guide the design of novel vaccine candidates and inform basic mechanistic insights regarding the immune response to SARS-CoV-2. Mechanistic validation of the omics signatures identified in vaccine trials can be facilitated by *in vitro* studies with human organoids such as tonsils, which represent a functional organotypic system that recapitulates key elements of the immune responses *in vitro*.[480]

Vaccinating Diverse Populations

Finally, it is increasingly clear that vaccine efficacies can vary enormously between different populations. Thus, as stated above, the efficacy of the BCG vaccine varies from 0% to 80%, with higher efficacies in Europe compared to Africa.[481,482] Also, vaccines against poliomyelitis, rotavirus, and malaria are less efficacious in Africa and Asia as compared with Europe or the United States.[323,483,484] The reasons for this are unclear but have been speculated to be due to differences in nutritional status, host genetics, and the microbiome.[485] Systems vaccinology approaches hold much promise in defining the mechanisms underpinning these population differences in vaccine efficacy.

Another major variable in human populations is age. During the past century, economic development and advances in public health and medicine have led to improved life expectancy. Thus, this aging population demographics poses new public health challenges on how to vaccinate the elderly, a population characterized by a progressively dysfunctional innate and adaptive immune system. Indeed, vaccine responses in the elderly are impaired in ways that differ, depending on the type of vaccine (eg, live attenuated, polysaccharide, or conjugate) and the mechanisms of protection. At the other extreme of the demographic spectrum, vaccinating the very young poses equally important public health challenges. During the first year of life, the immature state of a developing immune system, together with a nascent microbiome, represents hurdles that need to be overcome in vaccination.[486] Another important issue in the very young is safety. As vividly illustrated by recent events in the development of COVID-19 vaccines, vaccines are almost always tested in healthy, adult populations, to determine their safety, before they are assessed in the infant populations. Thus, the precious very young members of our species have a very high safety threshold, when it comes to vaccinology. Nevertheless, neonatal immunization has proven to be safe and effective against many diseases such as hepatitis B, TB, pertussis, pneumococcus, Hib, and rotavirus, illustrating the enormous power of vaccines in protecting the very young. Applying systems vaccinology approaches to the neonatal and infant populations could help identify ways to effectively and safely stimulate the nascent immune system.

CONCLUSION: PER ASPERA AD IMMUNITAS

The recent emergence of COVID-19 has provided a stark reminder of humanity's sheer dependency on vaccines. The international and collaborative response of governments, industry, and academia to rapidly develop and deploy a portfolio of vaccines worldwide is remarkable and unprecedented. Yet, SARS-CoV-2 has held a mirror on humanity. To immunologists and vaccinologists, this mirror provides an opportunity to reflect on how vaccinology once was, how it currently is during the time of the COVID-19 pandemic, and how it could be in the future. Vaccinology BC (before corona) consisted of the traditional vaccine development paradigm that typically took 10 to 15 years and $500 million to develop a vaccine, involving preclinical testing in animal models, followed by clinical testing in phase I, II, and III clinical trials. Clearly that paradigm has been highly successful in enabling the development of all the licensed vaccines that we use today. But what would vaccinology AC (after corona) look like? Clearly, an intense effort such as Operation Warp Speed would not be financially sustainable in the "post-COVID era," in developing vaccines against other pathogens. Would vaccinology return to "business as usual," (ie, up to 15 years and $500 million per vaccine)? But what lessons can we learn from the COVID-19 experience? This author believes that there are profound lessons to be learnt for every dimension of vaccinology. The silver lining of the COVID-19 tragedy is that it has forced us to redefine the way that vaccines are made, tested, and made globally accessible against a host of diseases and against pandemics that will emerge in the future. New technologies such as mRNA vaccines, which have already shown promising results in phase III efficacy trials, need to be rigorously tested in clinical trials to assess their long-term safety and mechanisms of action. Clearly, their early performance in the COVID-19 vaccine trials is exciting, and they hold much promise in revolutionizing the way in which vaccines are made. The way in which vaccines are tested has much to gain from the use of systems vaccinology approaches that will allow the rapid screening of multiple new vaccine candidates in relatively small phase I trials, to define early molecular signatures of immunogenicity and durability of vaccine responses, thereby negating the need to wait several weeks or months. Equally important will be ensuring global access to vaccines. Gavi, the Vaccine Alliance (previously the GAVI Alliance, and before that the Global Alliance for Vaccines and Immunization), is having a major impact. Gavi represents a public-private partnership that comprises the WHO, UNICEF, the World Bank, and the BMGF and brings together the governments of developing countries and donor governments, as well as the vaccine industry. As part of its mission is *"to save lives, reduce poverty and protect the world against the threat of epidemics, Gavi has helped vaccinate more than 822 million children in the world's poorest countries, preventing more than 14 million future* deaths."[487] In the context of COVID-19, Gavi is coleading, together with the CEPI and WHO, a consortium known as COVAX, to accelerate the development and manufacture of COVID-19 vaccines and to ensure fair and equitable access to every country in the world. The goal is to vaccinate the planet!

Finally, it is important to remember that COVID-19 is not the first pandemic, nor will it be the last. Pathogens have emerged as far back as history can remember, and they will continue to emerge for thousands of years to come. Unfortunately, past economic models of vaccine development have provided little financial incentive for vaccine manufacturers to invest in developing vaccines against emerging pathogens or neglected diseases such as Ebola that seemed geographically remote in regions thousands of miles away in Africa. However, the increasing impact of climate change and growing trends in urbanization provide a spark that could trigger the explosion of the next pandemic. Therefore, the emergence of pandemics poses one of the gravest existentialist threats to humanity. The reactive approach to vaccine development—to start developing a vaccine, once a pandemic has gotten underway—is now increasingly recognized as archaic and an impediment to global security. Research initiatives such as CEPI, a public-private partnership alliance funded by the Wellcome Trust and the BMGF aim to fund early stage trials of neglected vaccine candidates against known threats with the goal of establishing investigational vaccine stockpiles before epidemics begin—"just in case."[488] Clearly, there is much work for immunologists and vaccinologists to do. To channel Winston Churchill, never has the future of humanity been so dependent on so much, owed to so many, by so few!

POSTSCRIPT: In the few short weeks that have elapsed between the initial submission of this chapter on December 22, 2020, and the resubmission of the revised version on March 14, 2021, much has happened. On December 22, 2020, the cumulative confirmed COVID-19 deaths in the world were 1.72 million. On March 13, 2021, it was 2.65 million. Clearly, the virus continues to inflict a heavy toll on humanity. Yet, a handful of new vaccines that were in phase III efficacy trials have now been approved for use in certain countries. These include the vaccines developed by Oxford/AstraZeneca, Janssen, and Gamaleya, as well as the inactivated vaccines developed by Sinopharm and Sinovac. The rapid development of so many excellent vaccines is unprecedented and will no doubt be viewed by historians in the future as one of the shining moments in the history of vaccinology. Yet, as of March 14, 2021, only some 355 million vaccine doses have been administered worldwide—roughly 4.6 doses for every 100 people on the planet. Sadly, there has been a stark gap between the vaccination rates in different countries, with most countries in Africa, and several countries in the developing world, yet to report a single dose. Furthermore, the emergence of several variants of concern, including the B.1.351 lineage first identified in South Africa, has raised concerns that the immune responses induced by the current COVID-19 vaccines may confer only partial or limited immunity against such variants. Thus, many vaccine manufacturers are beginning to tweak their vaccines to include spike proteins from the B.1.351 lineage. Clearly, in our war against this virus, and many others that will surely emerge in the future, there is much for immunologists to do.

ACKNOWLEDGMENTS

I thank my mentor Gustav Nossal for his reading of this chapter. I also thank Rino Rappuoli, Stanley Plotkin, and Barry Rouse for their comments on the chapter, and Chunfeng Li for his assistance with endnote. I thank the National Institutes of Health, the Bill & Melinda Gates Foundation, and Open Philanthropy for their generous support of my work.

REFERENCES

1. Thucydides. *History of the Peloponnesian War*. Open Road Media; 2020.
2. Thein MM, Goh LG, Phua KH. The smallpox story: from variolation to victory. *Asia Pac J Publ Health*. 1988;2(3):203-210.
3. Hopkins DR. *Princes and Peasants: Smallpox and History*. University of Chicago Press; 1983.
4. Boylston A. The origins of inoculation. *J R Soc Med*. 2012;105(7):309-313.
5. Doherty M, Robertson MJ. Some early trends in immunology. *Trends Immunol*. 2004;25(12):623-631.
6. Riedel S. Edward Jenner and the history of smallpox and vaccination. *Proc (Bayl Univ Med Cent)*. 2005;18(1):21-25.
7. Jenner E. *An Inquiry Into the Causes and Effects of the Variolae Vaccinae*. Dodo Press; 2009.
8. De Beer GR. The relations between fellows of the Royal Society and French men of science when France and Britain were at war. *Notes and Records of the Royal Society of London*. 1952;9(2):244-299.
9. Morgan AJ, Poland GA. The Jenner society and the edward Jenner Museum: tributes to a physician-scientist. *Vaccine*. 2011;29(suppl 4):D152-D154.
10. Henderson DA. The eradication of smallpox--an overview of the past, present, and future. *Vaccine*. 2011;29(suppl 4): D7-D9.
11. Fenner F. *Smallpox and its Eradication*. World Health Organization; 1988.
12. Foege WH, Milbank Memorial F. *House on Fire: The Fight to Eradicate Smallpox*. University of California Press: Milbank Memorial Fund; 2012.
13. Centers for Disease Control and Prevention. *History of Smallpox*. 2016. Updated August 30, 2016. https://www.cdc.gov/smallpox/history/history.html
14. Polio Global Eradication Initiative. *Remembering Ali Maalin, the Last Man With Smallpox was an Advocate for Polio Eradication*. 2018.
15. Khan F, Ali M. The last case of smallpox. *Lancet Infect Dis*. 2018;18(12):1318.
16. Debrè P. *Louis Pasteur*. Johns Hopkins University Press; 1998.
17. Brock TD. *Robert Koch: A Life in Medicine and Bacteriology*. ASM Press; 1999.

18. Masic I. Thousand-year anniversary of the historical book: "Kitab al-Qanun fit-Tibb"- the Canon of Medicine, written by Abdullah ibn Sina. *J Res Med Sci*. 2012;17(11):993-1000.
19. Plotkin S. History of vaccination. *Proc Natl Acad Sci U S A*. 2014;111(34):12283-12287.
20. Gordon S. Elie Metchnikoff: father of natural immunity. *Eur J Immunol*. 2008;38(12):3257-3264.
21. Kaufmann SHE. Emil von Behring: translational medicine at the dawn of immunology. *Nat Rev Immunol*. 2017;17(6): 341-343.
22. Ehrlich and von behring. *J Am Med Assoc*. 2017;317(13):1381.
23. Valent P, Groner B, Schumacher U, et al. Paul Ehrlich (1854-1915) and his contributions to the foundation and birth of translational medicine. *J Innate Immun*. 2016;8(2):111-120.
24. Salmon D, Smith T. On a new method of producing immunity from contagious diseases. *Proc Biol Soc Wash*. 1880;3:29-33.
25. Roux E, Chamberland CE. Immunite contre la septicemie conferee par des substances solubles. *Ann Inst Pasteur*. 1987;1:561-572.
26. Matsiko A. Alum adjuvant discovery and potency. *Nature Research Journals*. September 28, 2020. https://www.nature.com/articles/d42859-020-00011-w
27. Breuer T. *Celebrating Gaston Ramon—the Vet Who Discovered Vaccinology's Secret Weapon*. World Economic Forum; 2019.
28. Luca S, Mihaescu T. History of BCG vaccine. *Maedica (Bucur)*. 2013;8(1):53-58.
29. Dockrell HM, Smith SG. What have we learnt about BCG vaccination in the last 20 years? *Front Immunol*. 2017;8:1134.
30. Pulendran B. Learning immunology from the yellow fever vaccine: innate immunity to systems vaccinology. *Nat Rev Immunol*. 2009;9(10):741-747.
31. Querec TD, Pulendran B. Understanding the role of innate immunity in the mechanism of action of the live attenuated Yellow Fever Vaccine 17D. *Adv Exp Med Biol*. 2007;590:43-53.
32. Norrby E. Yellow fever and Max Theiler: the only Nobel prize for a virus vaccine. *J Exp Med*. 2007;204(12):2779-2784.
33. Shampo MA, Kyle RA, Steensma DP. Albert Sabin—Conqueror of poliomyelitis. *Mayo Clin Proc*. 2011;86(7):e44.
34. Rosen FS. Isolation of poliovirus—John Enders and the Nobel prize. *N Engl J Med*. 2004;351(15):1481-1483.
35. Tan SY, Ponstein N. Jonas Salk (1914-1995): a vaccine against polio. *Singapore Med J*. 2019;60(1):9-10.
36. Lambert SM, Markel H. Making history: Thomas Francis, Jr, MD, and the 1954 Salk poliomyelitis vaccine field trial. *Arch Pediatr Adolesc Med*. 2000;154(5):512-517.
37. Fitzpatrick M. The cutter incident: how America's first polio vaccine led to a growing vaccine crisis. *J R Soc Med*. 2006;99(3):156.
38. Baicus A. History of polio vaccination. *World J Virol*. 2012;1(4):108-114.
39. Offit PA. The Cutter incident, 50 years later. *N Engl J Med*. 2005;352(14):1411-1412.
40. Francis T Jr, Salk JE, Brace WM. The protective effect of vaccination against epidemic influenza B. *J Am Med Assoc*. 1946;131:275-278.
41. Maassab HF, DeBorde DC. Development and characterization of cold-adapted viruses for use as live virus vaccines. *Vaccine*. 1985;3(5):355-369.
42. Clark HF, Offit PA, Plotkin SA, Heaton PM. The new pentavalent rotavirus vaccine composed of bovine (strain WC3) -human rotavirus reassortants. *Pediatr Infect Dis*. 2006;25(7):577-583.
43. Gotschlich EC, Liu TY, Artenstein MS. Human immunity to the meningococcus. 3. Preparation and immunochemical properties of the group A, group B, and group C meningococcal polysaccharides. *J Exp Med*. 1969;129(6):1349-1365.
44. Heidelberger M, Mac LC, Di Lapi MM. The human antibody response to simultaneous injection of six specific polysaccharides of pneumococcus. *J Exp Med*. 1948;88(3):369-372.
45. Austrian R. Pneumococcal polysaccharide vaccines. *Rev Infect Dis*. 1989;11(suppl 3):S598-S602.
46. Schneerson R, Barrera O, Sutton A, Robbins JB. Preparation, characterization, and immunogenicity of Haemophilus influenzae type b polysaccharide-protein conjugates. *J Exp Med*. 1980;152(2):361-376.
47. Anderson P, Peter G, Johnston RB Jr, et al. Immunization of humans with polyribophosphate, the capsular antigen of Hemophilus influenzae, type b. *J Clinical Invest*. 1972;51(1):39-44.
48. Landy M, Gaines S, Seal JR, et al. Antibody responses of man to three types of antityphoid immunizing agents: heat-phenol fluid vaccine, acetone-dehydrated vaccine, and isolated Vi and O antigens. *Am J Public Health Nation's Health*. 1954;44(12):1572-1579.
49. Avery OT, Goebel WF. Chemo-immunological studies on conjugated carbohydrate-proteins: II. Immunological specificity of synthetic sugar-protein antigens. *J Exp Med*. 1929;50(4):533-550.
50. Hepatitis B Foundation. Hepatitis B Facts and Figures. 2021. https://www.hepb.org/what-is-hepatitis-b/what-is-hepb/facts-and-figures/
51. London WT. *Baruch S. Blumberg, 1925-2011. Biographical Memoirs*. 2014.
52. Blumberg BS. *Hepatitis B: The Hunt for a Killer Virus*. Princeton University Press; 2002.
53. Gerlich WH. Medical virology of hepatitis B: how it began and where we are now. *Virol J*. 2013;10:239.
54. Brown E. Nobelist Baruch Blumberg, Who Discovered Hepatitis B, Dies at 85. *Washington Post*. April 6, 2011.
55. Pizza M, Scarlato V, Masignani V, et al. Identification of vaccine candidates against serogroup B meningococcus by whole-genome sequencing. *Science*. 2000;287(5459):1816-1820.
56. Tettelin H, Saunders NJ, Heidelberg J, et al. Complete genome sequence of Neisseria meningitidis serogroup B strain MC58. *Science*. 2000;287(5459):1809-1815.
57. Pichichero ME. Protein carriers of conjugate vaccines: characteristics, development, and clinical trials. *Hum Vaccin Immunother*. 2013;9(12):2505-2523.
58. Moss W. Measles in vaccinated individuals and the future of measles elimination. *Clin Infect Dis*. 2018;67(9):1320-1321.
59. Amanna IJ, Carlson NE, Slifka MK. Duration of humoral immunity to common viral and vaccine antigens. *N Engl J Med*. 2007;357(19):1903-1915.
60. Nutt SL, Hodgkin PD, Tarlinton DM, et al. The generation of antibody-secreting plasma cells. *Nat Rev Immunol*. 2015;15(3):160-171.
61. Victora GD, Nussenzweig MC. Germinal centers. *Annu Rev Immunol*. 2012;30:429-457.
62. Crotty ST. Follicular helper cell biology: a decade of discovery and diseases. *Immunity*. 2019;50(5):1132-1148.
63. Bournazos S, Gupta A, Ravetch JV. The role of IgG Fc receptors in antibody-dependent enhancement. *Nat Rev Immunol*. 2020;20(10):633-643.
64. Parga-Vidal L, van Gisbergen K. Area under immunosurveillance: dedicated roles of memory CD8 T-cell subsets. *Cold Spring Harbor Perspect Biol*. 2020;12(11):a037796.
65. Masopust D, Soerens AG. Tissue-resident T cells and other resident leukocytes. *Annu Rev Immunol*. 2019;37:521-546.
66. Arunachalam PS, Charles TP, Joag V, et al. T cell-inducing vaccine durably prevents mucosal SHIV infection even with lower neutralizing antibody titers. *Nat Med*. 2020;26(6):932-940.
67. Thomas Francis J. On the doctrine of original antigenic sin. *Proc Am Phil Soc*. 1960;104(6):572-578.
68. Yewdell JW, Santos JJS. Original antigenic sin: how original? How sinful? *Cold Spring Harb Perspect Med*. 2020;11(5):a038786.
69. Ramon G. Sur la toxine et sur l'anatoxine diphtheriques. *Ann Inst Pasteur*. 1924;38(1):1-10.
70. Di Pasquale A, Preiss S, Tavares Da Silva F, et al. Vaccine adjuvants: from 1920 to 2015 and beyond. *Vaccines (Basel)*. 2015;3(2):320-343.
71. Oakley CL. *Alexander Thomas Glenny, 1882-1965*. Biographical Memoirs of Fellows of the Royal Society; 1966.
72. De Gregorio E, Tritto E, Rappuoli R. Alum adjuvanticity: unraveling a century old mystery. *Eur J Immunol*. 2008;38(8):2068-2071.
73. O'Hagan DT, Ott GS, Nest GV, et al. The history of MF59((R)) adjuvant: a phoenix that arose from the ashes. *Expert Rev Vaccine*. 2013;12(1):13-30.
74. O'Hagan DT, Ott GS, De Gregorio E, et al. The mechanism of action of MF59—an innately attractive adjuvant formulation. *Vaccine*. 2012;30(29):4341-4348.
75. O'Hagan DT, Lodaya RN, Lofano G. The continued advance of vaccine adjuvants—"we can work it out." *Semin Immunol*. 2020;50:101426.
76. Pulendran B. The varieties of immunological experience: of pathogens, stress, and dendritic cells. *Annu Rev Immunol*. 2015;33:563-606.
77. Kwissa M, Kasturi SP, Pulendran B. The science of adjuvants. *Expert Rev Vaccine*. 2007;6(5):673-684.
78. Janeway CA Jr. Approaching the asymptote? Evolution and revolution in immunology. *Cold Spring Harbor Symp Quant Biol*. 1989;54(pt 1):1-13.
79. Lemaitre B, Nicolas E, Michaut L, et al. The dorsoventral regulatory gene cassette spatzle/Toll/cactus controls the potent antifungal response in Drosophila adults. *Cell*. 1996;86(6):973-983.
80. Poltorak A, He X, Smirnova I, et al. Defective LPS signaling in C3H/HeJ and C57BL/10ScCr mice: mutations in Tlr4 gene. *Science*. 1998;282(5396):2085-2088.
81. Iwasaki A, Medzhitov R. Toll-like receptor control of the adaptive immune responses. *Nat Immunol*. 2004;5(10):987-995.
82. Takeuchi O, Hoshino K, Kawai T, et al. Differential roles of TLR2 and TLR4 in recognition of gram-negative and gram-positive bacterial cell wall components. *Immunity*. 1999;11(4):443-451.
83. Schnare M, Barton GM, Holt AC, et al. Toll-like receptors control activation of adaptive immune responses. *Nat Immunol*. 2001;2(10):947-950.
84. Steinman RM. Decisions about dendritic cells: past, present, and future. *Annu Rev Immunol*. 2012;30:1-22.
85. Beutler B. Inferences, questions and possibilities in Toll-like receptor signalling. *Nature*. 2004;430(6996):257-263.
86. Kawai T, Akira S. The role of pattern-recognition receptors in innate immunity: update on Toll-like receptors. *Nat Immunol*. 2010;11(5):373-384.
87. Nobel Media AB. *The Nobel Prize in Physiology or Medicine 2011*. NobelPrizeorg; 2011.
88. Chow KT, Gale M Jr, Loo YM. RIG-I and other RNA sensors in antiviral immunity. *Annu Rev Immunol*. 2018;36:667-694.
89. Brown GD, Willment JA, Whitehead L. C-type lectins in immunity and homeostasis. *Nat Rev Immunol*. 2018;18(6):374-389.
90. Mangan MSJ, Olhava EJ, Roush WR, et al. Targeting the NLRP3 inflammasome in inflammatory diseases. *Nat Rev Drug Discov*. 2018;17(9):688.
91. Hu MM, Shu HB. Innate immune response to cytoplasmic DNA: mechanisms and diseases. *Annu Rev Immunol*. 2020;38:79-98.
92. Gong T, Liu L, Jiang W, et al. DAMP-sensing receptors in sterile inflammation and inflammatory diseases. *Nat Rev Immunol*. 2020;20(2):95-112.
93. Zindel J, Kubes P. DAMPs, PAMPs, and LAMPs in immunity and sterile inflammation. *Annu Rev Pathol*. 2020;15:493-518.
94. Querec T, Bennouna S, Alkan S, et al. Yellow fever vaccine YF-17D activates multiple dendritic cell subsets via TLR2, 7, 8, and 9 to stimulate polyvalent immunity. *J Exp Med*. 2006;203(2):413-424.
95. Querec TD, Akondy RS, Lee EK, et al. Systems biology approach predicts immunogenicity of the yellow fever vaccine in humans. *Nat Immunol*. 2009;10(1):116-125.
96. Tsuji S, Matsumoto M, Takeuchi O, et al. Maturation of human dendritic cells by cell wall skeleton of Mycobacterium bovis bacillus Calmette-Guerin: involvement of toll-like receptors. *Infect Immun*. 2000;68(12):6883-6890.
97. Jeisy-Scott V, Kim JH, Davis WG, et al. TLR7 recognition is dispensable for influenza virus A infection but important for the induction of hemagglutinin-specific antibodies in response to the 2009 pandemic split vaccine in mice. *J Virol*. 2012;86(20):10988-10998.
98. Koyama S, Ishii KJ, Kumar H, et al. Differential role of TLR- and RLR-signaling in the immune responses to influenza A virus infection and vaccination. *J Immunol*. 2007;179(7):4711-4720.
99. Coffman RL, Sher A, Seder RA. Vaccine adjuvants: putting innate immunity to work. *Immunity*. 2010;33(4):492-503.
100. Reed SG, Orr MT, Fox CB. Key roles of adjuvants in modern vaccines. *Nat Med*. 2013;19(12):1597-1608.
101. Wille-Reece U, Flynn BJ, Lore K, et al. HIV Gag protein conjugated to a Toll-like receptor 7/8 agonist improves the magnitude and quality of Th1 and CD8+ T cell responses in nonhuman primates. *Proc Natl Acad Sci U S A*. 2005;102(42):15190-15194.
102. Wille-Reece U, Flynn BJ, Lore K, et al. Toll-like receptor agonists influence the magnitude and quality of memory T cell responses after prime-boost immunization in nonhuman primates. *J Exp Med*. 2006;203(5):1249-58.
103. Kasturi SP, Kozlowski PA, Nakaya HI, et al. Adjuvanting a simian immunodeficiency virus vaccine with toll-like receptor ligands encapsulated in nanoparticles induces persistent antibody responses and enhanced protection in TRIM5alpha restrictive macaques. *J Virol*. 2017;91(4):e01844.
104. Kasturi SP, Skountzou I, Albrecht RA, et al. Programming the magnitude and persistence of antibody responses with innate immunity. *Nature*. 2011;470(7335):543-547.
105. Kasturi SP, Rasheed MAU, Havenar-Daughton C, et al. 3M-052, a synthetic TLR-7/8 agonist, induces durable HIV-1 envelope-specific plasma cells and humoral immunity in non-human primates. *Sci Immunol*. 2020;5(48):eabb1025.
106. Francica JR, Zak DE, Linde C, et al. Innate transcriptional effects by adjuvants on the magnitude, quality, and durability of HIV envelope responses in NHPs. *Blood Adv*. 2017;1(25):2329-2342.
107. Petitdemange C, Kasturi SP, Kozlowski PA, et al. Vaccine induction of antibodies and tissue-resident CD8+ T cells enhances protection against mucosal SHIV-infection in young macaques. *JCI Insight*. 2019;4(4):e126047.
108. Slifka MK, Antia R, Whitmire JK, et al. Humoral immunity due to long-lived plasma cells. *Immunity*. 1998;8(3):363-372.
109. Cirelli KM, Carnathan DG, Nogal B, et al. Slow delivery immunization enhances HIV neutralizing antibody and germinal center responses via modulation of immunodominance. *Cell*. 2020;180(1):206.
110. Roth GA, Gale EC, Alcantara-Hernandez M, et al. Injectable hydrogels for sustained codelivery of subunit vaccines enhance humoral immunity. *ACS Cent Sci*. 2020;6(10):1800-1812.
111. Burny W, Callegaro A, Bechtold V, et al. Different adjuvants induce common innate pathways that are associated with enhanced adaptive responses against a model antigen in humans. *Front Immunol*. 2017;8:943.
112. Nardin EH, Oliveira GA, Calvo-Calle JM, et al. Phase I testing of a malaria vaccine composed of hepatitis B virus core particles expressing Plasmodium falciparum circumsporozoite epitopes. *Infect Immun*. 2004;72(11):6519-6527.
113. Salmon-Ceron D, Excler JL, Sicard D, et al. Safety and immunogenicity of a recombinant HIV type 1 glycoprotein 160 boosted by a V3 synthetic peptide in HIV-negative volunteers. *AIDS Res Hum Retroviruses*. 1995;11(12):1479-1486.
114. Brewer JM, Conacher M, Hunter CA, et al. Aluminium hydroxide adjuvant initiates strong antigen-specific Th2 responses in the absence of IL-4- or IL-13-mediated signaling. *J Immunol*. 1999;163(12):6448-6454.
115. Brewer JM, Conacher M, Satoskar A, et al: In interleukin-4-deficient mice, alum not only generates T helper 1 responses equivalent to freund's complete adjuvant, but continues to induce T helper 2 cytokine production. *Eur J Immunol*. 1996;26(9):2062-2066.

116. Serre K, Mohr E, Benezech C, et al. Selective effects of NF-kappaB1 deficiency in CD4(+) T cells on Th2 and TFh induction by alum-precipitated protein vaccines. *Eur J Immunol.* 2011;41(6):1573-1582.
117. MacLeod MK, David A, McKee AS, et al. Memory CD4 T cells that express CXCR5 provide accelerated help to B cells. *J Immunol.* 2011;186(5):2889-2896.
118. Zhou C, Li C, Gong GZ, et al: Analysis of immunological mechanisms exerted by HBsAg-HBIG therapeutic vaccine combined with Adefovir in chronic hepatitis B patients. *Hum Vaccin Immunother.* 2017;13(9):1989-1996.
119. Moyer TJ, Kato Y, Abraham W, et al. Engineered immunogen binding to alum adjuvant enhances humoral immunity. *Nat Med.* 2020;26(3):430-440.
120. Kool M, Soullie T, van Nimwegen M, et al. Alum adjuvant boosts adaptive immunity by inducing uric acid and activating inflammatory dendritic cells. *J Exp Med.* 2008;205(4):869-882.
121. Gavin AL, Hoebe K, Duong B, et al. Adjuvant-enhanced antibody responses in the absence of toll-like receptor signaling. *Science.* 2006;314(5807):1936-1938.
122. Eisenbarth SC, Colegio OR, O'Connor W, et al. Crucial role for the Nalp3 inflammasome in the immunostimulatory properties of aluminium adjuvants. *Nature.* 2008;453(7198):1122-1126.
123. Li H, Nookala S, Re F. Aluminum hydroxide adjuvants activate caspase-1 and induce IL-1beta and IL-18 release. *J Immunol.* 2007;178(8):5271-5276.
124. McKee AS, Munks MW, MacLeod MK, et al. Alum induces innate immune responses through macrophage and mast cell sensors, but these sensors are not required for alum to act as an adjuvant for specific immunity. *J Immunol.* 2009;183(7):4403-4414.
125. Kool M, Petrilli V, De Smedt T, et al. Cutting edge: alum adjuvant stimulates inflammatory dendritic cells through activation of the NALP3 inflammasome. *J Immunol.* 2008;181(6):3755-3759.
126. Franchi L, Nunez G. The Nlrp3 inflammasome is critical for aluminium hydroxide-mediated IL-1beta secretion but dispensable for adjuvant activity. *Eur J Immunol.* 2008;38(8):2085-2089.
127. Marichal T, Ohata K, Bedoret D, et al. DNA released from dying host cells mediates aluminum adjuvant activity. *Nat Med.* 2011;17(8):996-1002.
128. McKee AS, Burchill MA, Munks MW, et al. Host DNA released in response to aluminum adjuvant enhances MHC class II-mediated antigen presentation and prolongs CD4 T-cell interactions with dendritic cells. *Proc Natl Acad Sci U S A.* 2013;110(12):E1122-E1131.
129. Calabro S, Tortoli M, Baudner BC, et al. Vaccine adjuvants alum and MF59 induce rapid recruitment of neutrophils and monocytes that participate in antigen transport to draining lymph nodes. *Vaccine.* 2011;29(9):1812-1823.
130. Seubert A, Calabro S, Santini L, et al. Adjuvanticity of the oil-in-water emulsion MF59 is independent of Nlrp3 inflammasome but requires the adaptor protein MyD88. *Proc Natl Acad Sci U S A.* 2011;108(27):11169-11174.
131. Ellebedy AH, Lupfer C, Ghoneim HE, et al. Inflammasome-independent role of the apoptosis-associated speck-like protein containing CARD (ASC) in the adjuvant effect of MF59. *Proc Natl Acad Sci U S A.* 2011;108(7):2927-2932.
132. Kim EH, Woodruff MC, Grigoryan L, et al. Squalene-based adjuvants stimulate CD8 T cell, but not antibody responses, through a RIPK3-dependent pathway. *Elife.* 2020;9:e52687.
133. Didierlaurent AM, Morel S, Lockman L, et al. AS04, an aluminum salt- and TLR4 agonist-based adjuvant system, induces a transient localized innate immune response leading to enhanced adaptive immunity. *J Immunol.* 2009;183(10):6186-6197.
134. Garcon N, Morel S, Didierlaurent A, et al. Development of an AS04-adjuvanted HPV vaccine with the adjuvant system approach. *BioDrugs.* 2011;25(4):217-226.
135. Kundi M. New hepatitis B vaccine formulated with an improved adjuvant system. *Expert Rev Vaccine.* 2007;6(2):133-140.
136. Skinner SR, Apter D, De Carvalho N, et al. Human papillomavirus (HPV)-16/18 AS04-adjuvanted vaccine for the prevention of cervical cancer and HPV-related diseases. *Expert Rev Vaccine.* 2016;15(3):367-387.
137. Naud PS, Roteli-Martins CM, De Carvalho NS, et al. Sustained efficacy, immunogenicity, and safety of the HPV-16/18 AS04-adjuvanted vaccine: final analysis of a long-term follow-up study up to 9.4 years post-vaccination. *Hum Vaccines Immunother.* 2014;10(8):2147-2162.
138. Garcon N, Di Pasquale A. From discovery to licensure, the Adjuvant System story. *Hum Vaccines Immunother.* 2017;13(1):19-33.
139. Howard LM, Hoek KL, Goll JB, et al. Cell-based systems biology analysis of human AS03-adjuvanted H5N1 avian influenza vaccine responses: a phase I randomized controlled trial. *PLoS One.* 2017;12(1):e0167488.
140. Sobolev O, Binda E, O'Farrell S, et al. Adjuvanted influenza-H1N1 vaccination reveals lymphoid signatures of age-dependent early responses and of clinical adverse events. *Nat Immunol.* 2016;17(2):204-213.
141. Orr MT, Fox CB. AS03 stresses out macrophages: commentary on 'Activation of the endoplasmic reticulum stress sensor IRE1alpha by the vaccine adjuvant AS03 contributes to its immunostimulatory properties'. *NPJ Vaccines.* 2018;3:27.
142. Morel S, Didierlaurent A, Bourguignon P, et al. Adjuvant System AS03 containing alpha-tocopherol modulates innate immune response and leads to improved adaptive immunity. *Vaccine.* 2011;29(13):2461-2473.
143. *GSK Announces Intention to Produce 1 Billion Doses of Pandemic Vaccine Adjuvant in 2021 to Support Multiple COVID-19 Vaccine Collaborations*[press release]. GSK; May 28, 2020. https://www.gsk.com/en-gb/media/press-releases/gsk-announces-intention-to-produce-1-billion-doses-of-pandemic-vaccine-adjuvant/
144. *Sanofi and GSK Initiate Phase 1/2 Clinical Trial of COVID-19 Adjuvanted Recombinant Protein-Based Vaccine Candidate*[press release]. Sanofi; September 3, 2020. https://www.sanofi.com/en/media-room/press-releases/2020/2020-09-03-07-00-00
145. Ward BJ, Gobeil P, Séguin A, et al. *Phase 1 Trial of a Candidate Recombinant Virus-like Particle Vaccine for Covid-19 Disease Produced in Plants.* medRxiv; 2020. doi:10.1101/2020.11.04.20226282.
146. Lal H, Cunningham AL, Godeaux O, et al. Efficacy of an adjuvanted herpes zoster subunit vaccine in older adults. *N Engl J Med.* 2015;372(22):2087-2096.
147. Kensil CR, Patel U, Lennick M, et al. Separation and characterization of saponins with adjuvant activity from Quillaja saponaria Molina cortex. *J Immunol.* 1991;146(2):431-437.
148. Newman MJ, Wu JY, Gardner BH, et al. Saponin adjuvant induction of ovalbumin-specific CD8+ cytotoxic T lymphocyte responses. *J Immunol.* 1992;148(8):2357-2362.
149. Wu JY, Gardner BH, Murphy CI, et al. Saponin adjuvant enhancement of antigen-specific immune responses to an experimental HIV-1 vaccine. *J Immunol.* 1992;148(5):1519-1525.
150. Olotu A, Fegan G, Wambua J, et al. Seven-year efficacy of RTS,S/AS01 malaria vaccine among young african children. *N Engl J Med.* 2016;374(26):2519-2529.
151. Detienne S, Welsby I, Collignon C, et al. Central role of CD169+ lymph node resident macrophages in the adjuvanticity of the QS-21 component of AS01. *Sci Rep.* 2016;6:39475.
152. Marty-Roix R, Vladimer GI, Pouliot K, et al. Identification of QS-21 as an inflammasome-activating molecular component of saponin adjuvants. *J Biol Chem.* 2016;291(3):1123-1136.
153. Welsby I, Detienne S, N'Kuli F, et al. Lysosome-dependent activation of human dendritic cells by the vaccine adjuvant QS-21. *Front Immunol.* 2016;7:663.
154. Coccia M, Collignon C, Herve C, et al. Cellular and molecular synergy in AS01-adjuvanted vaccines results in an early IFNgamma response promoting vaccine immunogenicity. *NPJ Vaccines.* 2017;2:25.
155. Tavares-Da-Silva F, Co MM, Dessart C, et al. Review of the initial post-marketing safety surveillance for the recombinant zoster vaccine. *Vaccine.* 2020;38(18):3489-3500.
156. Krieg AM. CpG motifs in bacterial DNA and their immune effects. *Annu Rev Immunol.* 2002;20:709-760.
157. Krieg AM, Yi AK, Matson S, et al. CpG motifs in bacterial DNA trigger direct B-cell activation. *Nature.* 1995;374(6522):546-549.
158. Hemmi H, Takeuchi O, Kawai T, et al. A Toll-like receptor recognizes bacterial DNA. *Nature.* 2000;408(6813):740-745.
159. Bauer S, Kirschning CJ, Hacker H, et al. Human TLR9 confers responsiveness to bacterial DNA via species-specific CpG motif recognition. *Proc Natl Acad Sci U S A.* 2001;98(16):9237-9242.
160. Hornung V, Rothenfusser S, Britsch S, et al. Quantitative expression of toll-like receptor 1-10 mRNA in cellular subsets of human peripheral blood mononuclear cells and sensitivity to CpG oligodeoxynucleotides. *J Immunol.* 2002;168(9):4531-4537.
161. Hartmann G, Weeratna RD, Ballas ZK, et al. Delineation of a CpG phosphorothioate oligodeoxynucleotide for activating primate immune responses in vitro and in vivo. *J Immunol.* 2000;164(3):1617-1624.
162. Krug A, Rothenfusser S, Hornung V, et al. Identification of CpG oligonucleotide sequences with high induction of IFN-alpha/beta in plasmacytoid dendritic cells. *Eur J Immunol.* 2001;31(7):2154-2163.
163. Verthelyi D, Ishii KJ, Gursel M, et al. Human peripheral blood cells differentially recognize and respond to two distinct CPG motifs. *J Immunol.* 2001;166(4):2372-2377.
164. A two-dose hepatitis B vaccine for adults (Heplisav-B). *J Am Med Assoc.* 2018;319(8):822-823.
165. Cooper C, Mackie D. Hepatitis B surface antigen-1018 ISS adjuvant-containing vaccine: a review of HEPLISAV safety and efficacy. *Expert Rev Vaccine.* 2011;10(4):417-427.
166. Bovier PA. Epaxal: a virosomal vaccine to prevent hepatitis A infection. *Expert Rev Vaccine.* 2008;7(8):1141-1150.
167. Herzog C, Hartmann K, Kunzi V, et al. Eleven years of Inflexal V-a virosomal adjuvanted influenza vaccine. *Vaccine.* 2009;27(33):4381-4387.
168. Loutan L, Bovier P, Althaus B, et al. Inactivated virosome hepatitis A vaccine. *Lancet.* 1994;343(8893):322-324.
169. Mischler R, Metcalfe IC. Inflexal V a trivalent virosome subunit influenza vaccine: production. *Vaccine.* 2002;20(suppl 5):B17-B23.
170. *Discontinued Products: April 2014.* The Pharmaceutical Journal; 2014:292.
171. Ward BJ, Makarkov A, Seguin A, et al. Efficacy, immunogenicity, and safety of a plant-derived, quadrivalent, virus-like particle influenza vaccine in adults (18-64 years) and older adults (≥65 years): two multicentre, randomised phase 3 trials. *Lancet.* 2020;396(10261):1491-1503.
172. Reed SG, Tomai M, Gale MJ Jr. New horizons in adjuvants for vaccine development. *Curr Opin Immunol.* 2020;65:97-101.
173. Pavlick A, Blazquez AB, Meseck M, et al. Combined vaccination with NY-ESO-1 protein, poly-ICLC, and montanide improves humoral and cellular immune responses in patients with high-risk melanoma. *Cancer Immunol Res.* 2020;8(1):70-80.
174. Anwar MA, Shah M, Kim J, et al. Recent clinical trends in Toll-like receptor targeting therapeutics. *Med Res Rev.* 2019;39(3):1053-1090.
175. Smith M, Garcia-Martinez E, Pitter MR, et al. Trial Watch: Toll-like receptor agonists in cancer immunotherapy. *Oncoimmunology.* 2018;7(12):e1526250.
176. Takeda K, Kaisho T, Akira S. Toll-like receptors. *Annu Rev Immunol.* 2003;21:335-376.
177. Reed SG, Bertholet S, Coler RN, et al. New horizons in adjuvants for vaccine development. *Trends Immunol.* 2009;30(1):23-32.
178. Smirnov D, Schmidt JJ, Capecchi JT, et al. Vaccine adjuvant activity of 3M-052: an imidazoquinoline designed for local activity without systemic cytokine induction. *Vaccine.* 2011;29(33):5434-5442.
179. Frega G, Wu Q, Le Naour J, et al. Trial Watch: experimental TLR7/TLR8 agonists for oncological indications. *OncoImmunology.* 2020;9(1):1796002.
180. Reed SG, Hsu FC, Carter D, et al. The science of vaccine adjuvants: advances in TLR4 ligand adjuvants. *Curr Opin Immunol.* 2016;41:85-90.
181. Carter D, Duthie MS, Reed SG. Adjuvants. *Curr Top Microbiol Immunol.* 2020;428:103-127.
182. Pulendran B, Kumar P, Cutler CW, et al. Lipopolysaccharides from distinct pathogens induce different classes of immune responses in vivo. *J Immunol.* 2001;167(9):5067-5076.
183. Kwissa M, Amara RR, Robinson HL, et al. Adjuvanting a DNA vaccine with a TLR9 ligand plus Flt3 ligand results in enhanced cellular immunity against the simian immunodeficiency virus. *J Exp Med.* 2007;204(11):2733-2746.
184. Wille-Reece U, Wu CY, Flynn BJ, et al. Immunization with HIV-1 Gag protein conjugated to a TLR7/8 agonist results in the generation of HIV-1 Gag-specific Th1 and CD8+ T cell responses. *J Immunol.* 2005;174(12):7676-7683.
185. Ahonen CL, Doxsee CL, McGurran SM, et al. Combined TLR and CD40 triggering induces potent CD8+ T cell expansion with variable dependence on type I IFN. *J Exp Med.* 2004;199(6):775-784.
186. Zaks K, Jordan M, Guth A, et al. Efficient immunization and cross-priming by vaccine adjuvants containing TLR3 or TLR9 agonists complexed to cationic liposomes. *J Immunol.* 2006;176(12):7335-7345.
187. Park H, Adamson L, Ha T, et al. Polyinosinic-polycytidylic acid is the most effective TLR adjuvant for SIV Gag protein-induced T cell responses in nonhuman primates. *J Immunol.* 2013;190(8):4103-4115.
188. Moody MA, Santra S, Vandergrift NA, et al. Toll-like receptor 7/8 (TLR7/8) and TLR9 agonists cooperate to enhance HIV-1 envelope antibody responses in rhesus macaques. *J Virol.* 2014;88(6):3329-3339.
189. Kwissa M, Nakaya HI, Oluoch H, et al. Distinct TLR adjuvants differentially stimulate systemic and local innate immune responses in nonhuman primates. *Blood.* 2012;119(9):2044-2055.
190. Singh M, Khong H, Dai Z, et al. Effective innate and adaptive antimelanoma immunity through localized TLR7/8 activation. *J Immunol.* 2014;193(9):4722-4731.
191. Abhyankar MM, Orr MT, Lin S, et al. Adjuvant composition and delivery route shape immune response quality and protective efficacy of a recombinant vaccine for Entamoeba histolytica. *NPJ Vaccines.* 2018;3:22.
192. Fox CB, Orr MT, Van Hoeven N, et al. Adsorption of a synthetic TLR7/8 ligand to aluminum oxyhydroxide for enhanced vaccine adjuvant activity: a formulation approach. *J Control Release.* 2016;244(pt A):98-107.
193. Dowling DJ, van Haren SD, Scheid A, et al. TLR7/8 adjuvant overcomes newborn hyporesponsiveness to pneumococcal conjugate vaccine at birth. *JCI Insight.* 2017;2(6):e91020.
194. Phillips B, Van Rompay KKA, Rodriguez-Nieves J, et al. Adjuvant-dependent enhancement of HIV env-specific antibody responses in infant rhesus macaques. *J Virol.* 2018;92(20):e01051.
195. Wu TY, Singh M, Miller AT, et al. Rational design of small molecules as vaccine adjuvants. *Sci Transl Med.* 2014;6(263):263ra160.
196. Buonsanti C, Balocchi C, Harfouche C, et al. Novel adjuvant Alum-TLR7 significantly potentiates immune response to glycoconjugate vaccines. *Sci Rep.* 2016;6:29063.

197. Martins KA, Bavari S, Salazar AM. Vaccine adjuvant uses of poly-IC and derivatives. *Expert Rev Vaccine.* 2015;14(3):447-459.
198. Zhu Q, Egelston C, Gagnon S, et al. Using 3 TLR ligands as a combination adjuvant induces qualitative changes in T cell responses needed for antiviral protection in mice. *J Clin Invest.* 2010;120(2):607-616.
199. Zhu Q, Egelston C, Vivekanandhan A, et al. Toll-like receptor ligands synergize through distinct dendritic cell pathways to induce T cell responses: implications for vaccines. *Proc Natl Acad Sci U S A.* 2008;105(42):16260-16265.
200. Sultan H, Wu J, Fesenkova VI, et al. Poly-IC enhances the effectiveness of cancer immunotherapy by promoting T cell tumor infiltration. *J Immunother Cancer.* 2020;8(2):e001224.
201. Sultan H, Salazar AM, Celis E. Poly-ICLC, a multi-functional immune modulator for treating cancer. *Semin Immunol.* 2020:101414.
202. Kyi C, Roudko V, Sabado R, et al. Therapeutic immune modulation against solid cancers with intratumoral poly-ICLC: a pilot trial. *Clin Cancer Res.* 2018;24(20):4937-4948.
203. Schau I, Michen S, Hagstotz A, et al. Targeted delivery of TLR3 agonist to tumor cells with single chain antibody fragment-conjugated nanoparticles induces type I-interferon response and apoptosis. *Sci Rep.* 2019;9(1):3299.
204. Reed SG, Carter D, Casper C, et al. Correlates of GLA family adjuvants' activities. *Semin Immunol.* 2018;39:22-29.
205. Mahipal A, Ejadi S, Gnjatic S, et al. First-in-human phase 1 dose-escalating trial of G305 in patients with advanced solid tumors expressing NY-ESO-1. Cancer immunology, immunotherapy: CII. 2019;68(7):1211-1222.
206. Penn-Nicholson A, Tameris M, Smit E, et al. Safety and immunogenicity of the novel tuberculosis vaccine ID93 + GLA-SE in BCG-vaccinated healthy adults in South Africa: a randomised, double-blind, placebo-controlled phase 1 trial. *Lancet Respir Med.* 2018;6(4):287-298.
207. Coler RN, Day TA, Ellis R, et al. The TLR-4 agonist adjuvant, GLA-SE, improves magnitude and quality of immune responses elicited by the ID93 tuberculosis vaccine: first-in-human trial. *NPJ Vaccines.* 2018;3:34.
208. Lovgren Bengtsson K, Morein B, Osterhaus AD. ISCOM technology-based Matrix M adjuvant: success in future vaccines relies on formulation. *Expert Rev Vaccines.* 2011;10(4):401-403.
209. Takahashi H, Takeshita T, Morein B, et al. Induction of CD8+ cytotoxic T cells by immunization with purified HIV-1 envelope protein in ISCOMs. *Nature.* 1990;344(6269):873-875.
210. Radosevic K, Rodriguez A, Mintardjo R, et al. Antibody and T-cell responses to a virosomal adjuvanted H9N2 avian influenza vaccine: impact of distinct additional adjuvants. *Vaccine.* 2008;26(29-30):3640-3646.
211. Madhun AS, Haaheim LR, Nilsen MV, et al. Intramuscular Matrix-M-adjuvanted virosomal H5N1 vaccine induces high frequencies of multifunctional Th1 CD4+ cells and strong antibody responses in mice. *Vaccine.* 2009;27(52):7367-7376.
212. Cox RJ, Pedersen G, Madhun AS, et al. Evaluation of a virosomal H5N1 vaccine formulated with Matrix M adjuvant in a phase I clinical trial. *Vaccine.* 2011;29(45):8049-8059.
213. Drane D, Gittleson C, Boyle J, et al. ISCOMATRIX adjuvant for prophylactic and therapeutic vaccines. *Expert Rev Vaccine.* 2007;6(5):761-772.
214. Keech C, Albert G, Reed P, et al. First-in-Human trial of a SARS CoV 2 recombinant spike protein nanoparticle vaccine. *medRxiv.* 2020. doi:10.1101/2020.08.05.20168435.
215. Irvine DJ, Aung A, Silva M. Controlling timing and location in vaccines. *Adv Drug Deliv Rev.* 2020;158:91-115.
216. Irvine DJ, Read BJ. Shaping humoral immunity to vaccines through antigen-displaying nanoparticles. *Curr Opin Immunol.* 2020;65:1-6.
217. Walls AC, Fiala B, Schafer A, et al. Elicitation of potent neutralizing antibody responses by designed protein nanoparticle vaccines for SARS-CoV-2. *Cell.* 2020;183(5):1367-1382.e17.
218. Tokatlian T, Read BJ, Jones CA, et al. Innate immune recognition of glycans targets HIV nanoparticle immunogens to germinal centers. *Science.* 2019;363(6427):649-654.
219. Plotkin S, Orenstein W, Offit P, et al. *Plotkins Vaccines.* 2017.
220. Bandyopadhyay AS, Macklin GR. Final frontiers of the polio eradication endgame. *Curr Opin Infect Dis.* 2020;33(5):404-410.
221. Grassly NC, Orenstein WA. Securing the eradication of all polioviruses. *Clin Infect Dis.* 2018;67(suppl 1):S1-S3.
222. Wolbert JG, Higginbotham K. *Poliomyelitis. StatPearls.* Treasure Island; 2020.
223. World Health Organization. Poliomyelitis. Accessed August 17, 2021. https://www.who.int/health-topics/poliomyelitis#tab=tab_1
224. Polio Global Eradication Initiative. Who we are. Accessed August 17, 2021. https://polioeradication.org/who-we-are/
225. Cochi S, Dowdle WR. The eradication of infectious diseases: understanding the lessons and advancing experience. *Strungman Forum Reports.* 2012;5.
226. Minor P. Vaccine-derived poliovirus (VDPV): impact on poliomyelitis eradication. *Vaccine.* 2009;27(20):2649-2652.
227. Kew O, Morris-Glasgow V, Landaverde M, et al. Outbreak of poliomyelitis in Hispaniola associated with circulating type 1 vaccine-derived poliovirus. *Science.* 2002;296(5566):356-359.
228. World Health Organization. Global Influenza Strategy 2019-2030. https://www.who.int/influenza/Global_Influenza_Strategy_2019_2030_Summary_English.pdf?ua=1
229. Knobler S. *The Threat of Pandemic Influenza. Are We Ready?: Workshop Summary.* National Academies Press (US); 2005.
230. Morens DM, North M, Taubenberger JK. Eyewitness accounts of the 1510 influenza pandemic in Europe. *Lancet.* 2010;376(9756):1894-1895.
231. Centers for Disease Control and Prevention. *Influenza.* Updated April 15, 2019. https://www.cdc.gov/vaccines/pubs/pinkbook/flu.html
232. Smith W, Andrewes CH, Laidlaw PP. A virus obtained from influenza patients. *Lancet.* 1933;2(5732):66-68.
233. Thomas Francis Jr. *The history of Vaccines.* https://www.historyofvaccines.org/node/2382
234. Houser K, Subbarao K. Influenza vaccines: challenges and solutions. *Cell Host Microbe.* 2015;17(3):295-300.
235. Centers for Disease Control and Prevention. *Types of Influenza Viruses.* Updated November 18, 2019. https://www.cdc.gov/flu/about/viruses/types.htm
236. Hannoun C. The evolving history of influenza viruses and influenza vaccines. *Expert Rev Vaccines.* 2013;12(9):1085-1094.
237. Walter+Eliza Hall Institute of Medical Research. *Studying Infectious Diseases.* https://discovery.wehi.edu.au/timeline#diseases
238. Grohskopf LA, Olsen SJ, Sokolow LZ, et al. Prevention and control of seasonal influenza with vaccines: recommendations of the Advisory Committee on Immunization Practices (ACIP) – United States, 2014-15 influenza season. *MMWR Morb Mortal Wkly Rep.* 2014;63(32):691-697.
239. Centers for Disease Control and Prevention. Influenza Historic Timeline. Accessed August 17, 2021. https://www.cdc.gov/flu/pandemic-resources/pandemic-timeline-1930-and-beyond.htm
240. Centers for Disease Control and Prevention. Past Pandemics. Accessed August 17, 2021. https://www.cdc.gov/flu/pandemic-resources/basics/past-pandemics.html
241. Centers for Disease Control and Prevention. CDC's World Health Organization (WHO) Collaborating Center for Surveillance, Epidemiology and Control of Influenza. Updated December 6, 2019. https://www.cdc.gov/flu/weekly/who-collaboration.htm
242. Chambers BS, Parkhouse K, Ross TM, et al. Identification of hemagglutinin residues responsible for H3N2 antigenic drift during the 2014-2015 influenza season. *Cell Rep.* 2015;12(1):1-6.
243. Flannery B, Clippard J, Zimmerman RK, et al. Early estimates of seasonal influenza vaccine effectiveness – United States, January 2015. *MMWR Morb Mortal Wkly Rep.* 2015;64(1):10-15.
244. Pebody RG, Warburton F, Ellis J, et al. Low effectiveness of seasonal influenza vaccine in preventing laboratory-confirmed influenza in primary care in the United Kingdom: 2014/15 mid-season results. *Euro Surveill.* 2015;20(5):21025.
245. Davis CW, Jackson KJL, McCausland MM, et al. Influenza vaccine-induced human bone marrow plasma cells decline within a year after vaccination. *Science.* 2020;370(6513):237-241.
246. Haq K, McElhaney JE. Immunosenescence: influenza vaccination and the elderly. *Curr Opin Immunol.* 2014;29:38-42.
247. Tregoning JS, Russell RF, Kinnear E. Adjuvanted influenza vaccines. *Hum Vaccines Immunother.* 2018;14(3):550-564.
248. Plotkin SA. Correlates of protection induced by vaccination. *Clin Vaccine Immunol.* 2010;17(7):1055-1065.
249. McElhaney JE, Xie D, Hager WD, et al. T cell responses are better correlates of vaccine protection in the elderly. *J Immunol.* 2006;176(10):6333-6339.
250. Wright PF, Hoen AG, Ilyushina NA, et al. Correlates of immunity to influenza as determined by challenge of children with live, attenuated influenza vaccine. *Open Forum Infect Dis.* 2016;3(2):ofw108.
251. Cortese M, Sherman AC, Rouphael NG, et al. Systems biological analysis of immune response to influenza vaccination. *Cold Spring Harb Perspect Med.* 2021;11(6):a038596.
252. Jefferies M, Rauff B, Rashid H, et al. Update on global epidemiology of viral hepatitis and preventive strategies. *World J Clin Cases.* 2018;6(13):589-599.
253. Ogholikhan S, Schwarz KB. Hepatitis vaccines. *Vaccines (Basel).* 2016;4(1):6.
254. World Health Organization. Hepatitis. https://www.who.int/hepatitis/about/en/ Accessed August 17, 2021. https://www.who.int/health-topics/hepatitis#tab=tab_1
255. Stanaway JD, Flaxman AD, Naghavi M, et al. The global burden of viral hepatitis from 1990 to 2013: findings from the Global Burden of Disease Study 2013. *Lancet.* 2016;388(10049):1081-1088.
256. Trepo C. A brief history of hepatitis milestones. *Liver Int.* 2014;34(suppl 1):29-37.
257. Wong DT, Mihm MC, Boyer JL, et al. Historical path of discovery of viral hepatitis. *Harvard Medical Student Review.* 2015;(3).
258. Zuckerman AJ. Hepatitis viruses. In: Baron S, ed. *Medical Microbiology.* 4th ed. University of Texas Medical Branch at Galveston; 1996.
259. Centers for Disease Control and Prevention. The ABCs of Hepatitis—for Health Professionals. Accessed August 17, 2021. https://www.cdc.gov/hepatitis/resources/professionals/pdfs/abctable.pdf
260. National Vaccine Information Center. What is the history of Hepatitis A vaccine use in America? https://www.nvic.org/vaccines-and-diseases/hepatitis-a/vaccine-history.aspx#_edn5
261. Centers for Disease Control and Prevention. Hepatitis A. Updated April 15, 2019. https://www.cdc.gov/vaccines/pubs/pinkbook/hepa.html
262. Crum-Cianflone NF, Wilkins K, Lee AW, et al. Long-term durability of immune responses after hepatitis A vaccination among HIV-infected adults. *J Infect Dis.* 2011;203(12):1815-1823.
263. The Vaccine Makers Project. Unit 2: Lesson 4 – Vaccine History and Research. Accessed August 17, 2021. https://vaccinemakers.org/sites/default/files/resources/HS.reading%20passage-Blood.unit2_.lesson4.FINAL_.pdf
264. Huzair F, Sturdy S. Biotechnology and the transformation of vaccine innovation: the case of the hepatitis B vaccines 1968-2000. *Stud Hist Philos Biol Biomed Sci.* 2017;64:11-21.
265. Eng NF, Bhardwaj N, Mulligan R, et al. The potential of 1018 ISS adjuvant in hepatitis B vaccines: HEPLISAV review. *Hum Vaccines Immunother.* 2013;9(8):1661-1672.
266. Gara N, Abdalla A, Rivera E, et al. Durability of antibody response against hepatitis B virus in healthcare workers vaccinated as adults. *Clin Infect Dis.* 2015;60(4):505-513.
267. Lu CY, Chiang BL, Chi WK, et al. Waning immunity to plasma-derived hepatitis B vaccine and the need for boosters 15 years after neonatal vaccination. *Hepatology.* 2004;40(6):1415-1420.
268. Hammitt LL, Hennessy TW, Fiore AE, et al. Hepatitis B immunity in children vaccinated with recombinant hepatitis B vaccine beginning at birth: a follow-up study at 15 years. *Vaccine.* 2007;25(39-40):6958-6964.
269. Centers for Disease Control and Prevention. Complications of Measles. Accessed August 17, 2021. https://www.cdc.gov/measles/symptoms/complications.html
270. World Health Organization. *Measles.* 2019. https://www.who.int/news-room/fact-sheets/detail/measles
271. *Worldwide Measles Deaths Climb 50% from 2016 to 2019 Claiming over 207 500 Lives in 2019* [press release]. World Health Organization; 2020. https://www.who.int/news/item/12-11-2020-worldwide-measles-deaths-climb-50-from-2016-to-2019-claiming-over-207-500-lives-in-2019
272. Centers for Disease Control and Prevention. Measles History. Accessed August 17, 2021. https://www.cdc.gov/measles/about/history.html
273. Rubin S, Eckhaus M, Rennick LJ, et al. Molecular biology, pathogenesis and pathology of mumps virus. *J Pathol.* 2015;235(2):242-252.
274. Plotkin SA. The history of rubella and rubella vaccination leading to elimination. *Clin Infect Dis.* 2006;43(suppl 3):S164-S168.
275. Parkman PD. Togaviruses: rubella virus. In: Baron S, ed. *Medical Microbiology.* 4th ed. University of Texas Medical Branch at Galveston; 1996.
276. Centers for Disease Control and Prevention. *Rubella.* Updated April 15, 2019. https://www.cdc.gov/vaccines/pubs/pinkbook/rubella.html#vaccines
277. Gershon AA. Varicella-zoster vaccine. In: Arvin A, ed. *Human Herpesviruses: Biology, Therapy, and Immunoprophylaxis.* Cambridge University Press; 2007.
278. Takahashi M, Asano Y, Kamiya H, et al. Development of varicella vaccine. *J Infect Dis.* 2008;197(suppl 2):S41-S44.
279. Ozaki T, Asano Y. Development of varicella vaccine in Japan and future prospects. *Vaccine.* 2016;34(29):3427-3433.
280. Gershon AA, Katz SL. Perspective on live varicella vaccine. *J Infect Dis.* 2008;197(suppl 2):S242-S245.
281. Krause PR, Klinman DM. Efficacy, immunogenicity, safety, and use of live attenuated chickenpox vaccine. *J Pediatr.* 1995;127(4):518-525.
282. Centers for Disease Control and Prevention. *Varicella Vaccine Effectiveness and Duration of Protection.* Updated August 30, 2012. https://www.cdc.gov/vaccines/vpd-vac/varicella/hcp-effective-duration.htm
283. World Health Organization. *Human Papillomavirus (HPV) and Cervical Cancer 2020.* Updated November 11, 2020. https://www.who.int/news-room/fact-sheets/detail/human-papillomavirus-(hpv)-and-cervical-cancer
284. World Health Organization. *Human Papillomavirus.* Accessed August 17, 2021. https://www.who.int/teams/immunization-vaccines-and-biologicals/diseases/human-papillomavirus-vaccines-(HPV)
285. Arbyn M, Weiderpass E, Bruni L, et al. Estimates of incidence and mortality of cervical cancer in 2018: a worldwide analysis. *Lancet Glob Health.* 2020;8(2):e191-e203.
286. Hampton T. Nobel Prize honors HIV, HPV discoveries. *J Am Med Assoc.* 2008;300(18):2109.
287. Scolnick EM. A vaccine to prevent cervical cancer: academic and industrial collaboration and a Lasker award. *Clin Transl Immunology.* 2018;7(1):e1002.
288. Harper DM, DeMars LR. HPV vaccines—a review of the first decade. *Gynecol Oncol.* 2017;146(1):196-204.
289. Medscape. *What Is the Mortality Rate for Tetanus (Lockjaw)?* 2019. Updated January 18, 2019. https://www.medscape.com/answers/229594-5925/what-is-the-mortality-rate-for-tetanus-lockjaw

290. Kaufmann SH. Remembering Emil von Behring: from Tetanus Treatment to Antibody Cooperation with Phagocytes. *mBio*. 2017;8(1):e00117.
291. Vitek C, Wharton M. Diphtheria toxoid. In: Plotkin S, ed. *Vaccines*. 5th ed. Saunders; 2008.
292. Malito E, Bursulaya B, Chen C, et al. Structural basis for lack of toxicity of the diphtheria toxin mutant CRM197. *Proc Natl Acad Sci U S A*. 2012;109(14):5229-5234.
293. Centers for Disease Control and Prevention. *Pertussis (Whooping Cough) Fast Facts*. 2017. Updated August 7, 2017. https://www.cdc.gov/pertussis/fast-facts.html
294. Clark TA. Changing pertussis epidemiology: everything old is new again. *J Infect Dis*. 2014;209(7):978-981.
295. Robbins JB, Schneerson R, Kubler-Kielb J, et al. Toward a new vaccine for pertussis. *Proc Natl Acad Sci U S A*. 2014;111(9):3213-3216.
296. Sato Y, Sato H. Development of acellular pertussis vaccines. *Biologicals*. 1999;27(2):61-69.
297. Esposito S, Stefanelli P, Fry NK, et al. Pertussis prevention: reasons for resurgence, and differences in the current acellular pertussis vaccines. *Front Immunol*. 2019;10:1344.
298. Jajosky RA, Hall PA, Adams DA, et al; Centers for Disease Control and Prevention. Summary of notifiable diseases – United States, 2004. *MMWR Morb Mortal Wkly Rep*. 2006;53(53):1-79.
299. McNabb SJ, Jajosky RA, Hall-Baker PA, et al. Summary of notifiable diseases – United States, 2005. *MMWR Morb Mortal Wkly Rep*. 2007;54(53):1-92.
300. Miller E, Gay NJ. Epidemiological determinants of pertussis. *Dev Biol Stand*. 1997;89:15-23.
301. Broutin H, Rohani P, Guegan JF, et al. Loss of immunity to pertussis in a rural community in Senegal. *Vaccine*. 2004;22(5-6):594-596.
302. Wirsing von Konig CH, Postels-Multani S, Bock HL, et al. Pertussis in adults: frequency of transmission after household exposure. *Lancet*. 1995;346(8986):1326-1329.
303. Wendelboe AM, Van Rie A, Salmaso S, et al. Duration of immunity against pertussis after natural infection or vaccination. *Pediatr Infect Dis J*. 2005;24(5 suppl):S58-S61.
304. Lambert HJ. Epidemiology of a small pertussis outbreak in Kent County, Michigan. *Public Health Rep*. 1965;80:365-369.
305. Jenkinson D. Duration of effectiveness of pertussis vaccine: evidence from a 10 year community study. *Br Med J (Clin Res Ed)*. 1988;296(6622):612-614.
306. Clark TA, Messonnier NE, Hadler SC. Pertussis control: time for something new? *Trends Microbiol*. 2012;20(5):211-213.
307. Vickers D, Ross AG, Mainar-Jaime RC, et al. Whole-cell and acellular pertussis vaccination programs and rates of pertussis among infants and young children. *Can Med Assoc J*. 2006;175(10):1213-1217.
308. Esposito S, Agliardi T, Giammanco A, et al. Long-term pertussis-specific immunity after primary vaccination with a combined diphtheria, tetanus, tricomponent acellular pertussis, and hepatitis B vaccine in comparison with that after natural infection. *Infect Immun*. 2001;69(7):4516-4520.
309. Esposito S, Principi N. Prevention of pertussis: an unresolved problem. *Hum Vaccines Immunother*. 2018;14(10):2452-2459.
310. Consortium P. PERISCOPE: road towards effective control of pertussis. *Lancet Infect Dis*. 2019;19(5):e179-e186.
311. *Ten Things You Didn't Know about Malaria*[press release]. Unicef; 2018. https://www.unicef.org/press-releases/ten-things-you-didnt-know-about-malaria
312. *The "World Malaria Report 2019" at a Glance*[press release]. World Health Organization; 2019. https://www.who.int/news-room/feature-stories/detail/world-malaria-report-2019
313. Rossberg F, Tiedt N. Influence of right atrium transmural pressure on heart rate. Article in German. *Z Gesamte Inn Med*. 1974;29(2):52-56.
314. Doolan DL, Dobano C, Baird JK. Acquired immunity to malaria. *Clin Microbiol Rev*. 2009;22(1):13-36, Table of Contents.
315. Kisalu NK, Idris AH, Weidle C, et al. A human monoclonal antibody prevents malaria infection by targeting a new site of vulnerability on the parasite. *Nat Med*. 2018;24(4):408-416.
316. Gardner MJ, Hall N, Fung E, et al. Genome sequence of the human malaria parasite *Plasmodium falciparum*. *Nature*. 2002;419(6906):498-511.
317. Cohen J, Nussenzweig V, Nussenzweig R, et al. From the circumsporozoite protein to the RTS, S/AS candidate vaccine. *Hum Vaccine*. 2010;6(1):90-96.
318. Duffy PE, Patrick Gorres J. Malaria vaccines since 2000: progress, priorities, products. *NPJ Vaccines*. 2020;5(1):48.
319. Talaat KR, Ellis RD, Hurd J, et al. Safety and immunogenicity of pfs25-EPA/alhydrogel(R), a transmission blocking vaccine against plasmodium falciparum: an open label study in malaria naive adults. *PLoS One*. 2016;11(10):e0163144.
320. Goh YS, McGuire D, Renia L. Vaccination with sporozoites: models and correlates of protection. *Front Immunol*. 2019;10:1227.
321. Seder RA, Chang LJ, Enama ME, et al. Protection against malaria by intravenous immunization with a nonreplicating sporozoite vaccine. *Science*. 2013;341(6152):1359-1365.
322. Mordmuller B, Surat G, Lagler H, et al. Sterile protection against human malaria by chemoattenuated PfSPZ vaccine. *Nature*. 2017.
323. Sissoko MS, Healy SA, Katile A, et al. Safety and efficacy of PfSPZ Vaccine against Plasmodium falciparum via direct venous inoculation in healthy malaria-exposed adults in Mali: a randomised, double-blind phase 1 trial. *Lancet Infect Dis*. 2017;17(5):498-509.
324. Epstein JE, Tewari K, Lyke KE, et al. Live attenuated malaria vaccine designed to protect through hepatic CD8(+) T cell immunity. *Science*. 2011;334(6055):475-480.
325. Ishizuka AS, Lyke KE, DeZure A, et al. Protection against malaria at 1 year and immune correlates following PfSPZ vaccination. *Nat Med*. 2016;22(6):614-623.
326. Zaidi I, Diallo H, Conteh S, et al. Gammadelta T cells are required for the induction of sterile immunity during irradiated sporozoite vaccinations. *J Immunol*. 2017;199(11):3781-3788.
327. Ballou WR. The development of the RTS,S malaria vaccine candidate: challenges and lessons. *Parasite Immunol*. 2009;31(9):492-500.
328. Kazmin D, Nakaya HI, Lee EK, et al. Systems analysis of protective immune responses to RTS,S malaria vaccination in humans. *Proc Natl Acad Sci U S A*. 2017;114(9):2425-2430.
329. Kester KE, Cummings JF, Ofori-Anyinam O, et al. Randomized, double-blind, phase 2a trial of falciparum malaria vaccines RTS,S/AS01B and RTS,S/AS02A in malaria-naive adults: safety, efficacy, and immunologic associates of protection. *J Infect Dis*. 2009;200(3):337-346.
330. The RTS,S Clinical Trials Partnership. A phase 3 trial of RTS,S/AS01 malaria vaccine in African infants. *N Engl J Med*. 2012;367(24):2284-2295.
331. Regules JA, Cicatelli SB, Bennett JW, et al. Fractional third and fourth dose of RTS,S/AS01 malaria candidate vaccine: a phase 2a controlled human malaria parasite infection and immunogenicity study. *J Infect Dis*. 2016;214(5):762-771.
332. *Global HIV & AIDS Statistics—2020 Fact Sheet*[press release]. UNAIDS; 2020. https://www.unaids.org/en/resources/fact-sheet
333. Burton DR, Ahmed R, Barouch DH, et al. A blueprint for HIV vaccine discovery. *Cell Host Microbe*. 2012;12(4):396-407.
334. Ng'uni T, Chasara C, Ndhlovu ZM. Major scientific hurdles in HIV vaccine development: historical perspective and future directions. *Front Immunol*. 2020;11:590780.
335. Poignard P, Moulard M, Golez E, et al. Heterogeneity of envelope molecules expressed on primary human immunodeficiency virus type 1 particles as probed by the binding of neutralizing and nonneutralizing antibodies. *J Virol*. 2003;77(1):353-365.
336. Escolano A, Dosenovic P, Nussenzweig MC. Progress toward active or passive HIV-1 vaccination. *J Exp Med*. 2017;214(1):3-16.
337. Binley JM, Lybarger EA, Crooks ET, et al. Profiling the specificity of neutralizing antibodies in a large panel of plasmas from patients chronically infected with human immunodeficiency virus type 1 subtypes B and C. *J Virol*. 2008;82(23):11651-11668.
338. Doria-Rose NA, Klein RM, Manion MM, et al. Frequency and phenotype of human immunodeficiency virus envelope-specific B cells from patients with broadly cross-neutralizing antibodies. *J Virol*. 2009;83(1):188-199.
339. Li Y, Svehla K, Louder MK, et al. Analysis of neutralization specificities in polyclonal sera derived from human immunodeficiency virus type 1-infected individuals. *J Virol*. 2009;83(2):1045-1059.
340. Rusert P, Kouyos RD, Kadelka C, et al. Determinants of HIV-1 broadly neutralizing antibody induction. *Nat Med*. 2016;22(11):1260-1267.
341. Sather DN, Armann J, Ching LK, et al. Factors associated with the development of cross-reactive neutralizing antibodies during human immunodeficiency virus type 1 infection. *J Virol*. 2009;83(2):757-769.
342. Simek MD, Rida W, Priddy FH, et al. Human immunodeficiency virus type 1 elite neutralizers: individuals with broad and potent neutralizing activity identified by using a high-throughput neutralization assay together with an analytical selection algorithm. *J Virol*. 2009;83(14):7337-7348.
343. Gautam R, Nishimura Y, Pegu A, et al. A single injection of anti-HIV-1 antibodies protects against repeated SHIV challenges. *Nature*. 2016;533(7601):105-109.
344. Shingai M, Donau OK, Plishka RJ, et al. Passive transfer of modest titers of potent and broadly neutralizing anti-HIV monoclonal antibodies block SHIV infection in macaques. *J Exp Med*. 2014;211(10):2061-2074.
345. Pegu A, Borate B, Huang Y, et al. A meta-analysis of passive immunization studies shows that serum-neutralizing antibody titer associates with protection against SHIV challenge. *Cell Host Microbe*. 2019;26(3):336-346.e3.
346. Bar KJ, Sneller MC, Harrison LJ, et al. Effect of HIV antibody VRC01 on viral rebound after treatment interruption. *N Engl J Med*. 2016;375(21):2037-2050.
347. Mendoza P, Gruell H, Nogueira L, et al. Combination therapy with anti-HIV-1 antibodies maintains viral suppression. *Nature*. 2018;561(7724):479-484.
348. Flynn NM, Forthal DN, Harro CD, et al. Placebo-controlled phase 3 trial of a recombinant glycoprotein 120 vaccine to prevent HIV-1 infection. *J Infect Dis*. 2005;191(5):654-665.
349. Pitisuttithum P, Gilbert P, Gurwith M, et al. Randomized, double-blind, placebo-controlled efficacy trial of a bivalent recombinant glycoprotein 120 HIV-1 vaccine among injection drug users in Bangkok, Thailand. *J Infect Dis*. 2006;194(12):1661-1671.
350. Gray G, Buchbinder S, Duerr A. Overview of STEP and Phambili trial results: two phase IIb test-of-concept studies investigating the efficacy of MRK adenovirus type 5 gag/pol/nef subtype B HIV vaccine. *Curr Opin HIV AIDS*. 2010;5(5):357-361.
351. Yuan T, Fitzpatrick T, Ko N-Y, et al. Circumcision to prevent HIV and other sexually transmitted infections in men who have sex with men: a systematic review and meta-analysis of global data. *Lancet Global Health*. 2019;7(4):e436-e447.
352. Sekaly RP. The failed HIV Merck vaccine study: a step back or a launching point for future vaccine development? *J Exp Med*. 2008;205(1):7-12.
353. Rerks-Ngarm S, Pitisuttithum P, Nitayaphan S, et al. Vaccination with ALVAC and AIDSVAX to prevent HIV-1 infection in Thailand. *N Engl J Med*. 2009;361(23):2209-2220.
354. Haynes BF, Gilbert PB, McElrath MJ, et al. Immune-correlates analysis of an HIV-1 vaccine efficacy trial. *N Engl J Med*. 2012;366(14):1275-1286.
355. Corey L, Gilbert PB, Tomaras GD, et al. Immune correlates of vaccine protection against HIV-1 acquisition. *Sci Transl Med*. 2015;7(310):310rv7.
356. Barouch DH, Alter G, Broge T, et al. Protective efficacy of adenovirus/protein vaccines against SIV challenges in rhesus monkeys. *Science*. 2015;349(6245):320-324.
357. Barouch DH, Liu J, Li H, et al. Vaccine protection against acquisition of neutralization-resistant SIV challenges in rhesus monkeys. *Nature*. 2012;482(7383):89-93.
358. Roederer M, Keele BF, Schmidt SD, et al. Immunological and virological mechanisms of vaccine-mediated protection against SIV and HIV. *Nature*. 2014;505(7484):502-508.
359. *Experimental HIV Vaccine Regimen Ineffective in Preventing HIV*[press release]. National Institutes of Health; 2020. https://www.nih.gov/news-events/news-releases/experimental-hiv-vaccine-regimen-ineffective-preventing-hiv
360. Baden LR, Karita E, Mutua G, et al. Assessment of the safety and immunogenicity of 2 novel vaccine platforms for HIV-1 prevention: a randomized trial. *Ann Intern Med*. 2016;164(5):313-322.
361. Ward AB, Wilson IA. The HIV-1 envelope glycoprotein structure: nailing down a moving target. *Immunol Rev*. 2017;275(1):21-32.
362. Sanders RW, van Gils MJ, Derking R, et al. HIV-1 VACCINES. HIV-1 neutralizing antibodies induced by native-like envelope trimers. *Science*. 2015;349(6244):aac4223.
363. de Taeye SW, Ozorowski G, Torrents de la Pena A, et al. Immunogenicity of stabilized HIV-1 envelope trimers with reduced exposure of non-neutralizing epitopes. *Cell*. 2015;163(7):1702-1715.
364. Voss JE, Andrabi R, McCoy LE, et al. Elicitation of neutralizing antibodies targeting the V2 apex of the HIV envelope trimer in a wild-type Animal model. *Cell Rep*. 2017;21(1):222-235.
365. Pauthner MG, Nkolola JP, Havenar-Daughton C, et al. Vaccine-induced protection from homologous tier 2 SHIV challenge in nonhuman primates depends on serum-neutralizing antibody titers. *Immunity*. 2019;50(1):241-252.e6.
366. Pauthner M, Havenar-Daughton C, Sok D, et al. Elicitation of robust tier 2 neutralizing antibody responses in nonhuman primates by HIV envelope trimer immunization using optimized approaches. *Immunity*. 2017;46(6):1073-1088.e6.
367. McCoy LE, van Gils MJ, Ozorowski G, et al. Holes in the glycan shield of the native HIV envelope are a target of trimer-elicited neutralizing antibodies. *Cell Rep*. 2016;16(9):2327-2338.
368. Klasse PJ, LaBranche CC, Ketas TJ, et al. Sequential and simultaneous immunization of rabbits with HIV-1 envelope glycoprotein SOSIP.664 trimers from clades A, B and C. *PLoS Pathog*. 2016;12(9):e1005864.
369. Haynes BF, Burton DR, Mascola JR. Multiple roles for HIV broadly neutralizing antibodies. *Sci Transl Med*. 2019;11(516).
370. Burton DR. Advancing an HIV vaccine; advancing vaccinology. *Nat Rev Immunol*. 2019;19(2):77-78.
371. Sok D, Burton DR. Recent progress in broadly neutralizing antibodies to HIV. *Nat Immunol*. 2018;19(11):1179-1188.
372. Sok D, Briney B, Jardine JG, et al. Priming HIV-1 broadly neutralizing antibody precursors in human Ig loci transgenic mice. *Science*. 2016;353(6307):1557-1560.
373. Jardine JG, Ota T, Sok D, et al. HIV-1 VACCINES. Priming a broadly neutralizing antibody response to HIV-1 using a germline-targeting immunogen. *Science*. 2015;349(6244):156-161.
374. Briney B, Sok D, Jardine JG, et al. Tailored immunogens direct affinity maturation toward HIV neutralizing antibodies. *Cell*. 2016;166(6):1459-1470.e11.
375. TB Alliance. *TB is a Pandemic*. Accessed December 21, 2020. https://www.tballiance.org/why-new-tb-drugs/global-pandemic
376. Comas I, Coscolla M, Luo T, et al. Out-of-Africa migration and Neolithic coexpansion of *Mycobacterium tuberculosis* with modern humans. *Nat Genet*. 2013;45(10):1176-1182.
377. *Tuberculosis*[press release]. World Health Organization; 2020. https://www.who.int/news-room/fact-sheets/detail/tuberculosis

378. *2008 Tuberculosis Facts*[press release]. World Health Organization; 2008. https://www.who.int/tb/publications/2008/factsheet_april08.pdf
379. Lin PL, Flynn JL. The end of the binary era: revisiting the spectrum of tuberculosis. *J Immunol.* 2018;201(9):2541-2548.
380. Fletcher HA. Sleeping beauty and the story of the Bacille Calmette-Guerin Vaccine. *mBio.* 2016;7(4):e01370.
381. McShane H. Insights and challenges in tuberculosis vaccine development. *Lancet Respir Med.* 2019;7(9):810-819.
382. Mangtani P, Abubakar I, Ariti C, et al. Protection by BCG vaccine against tuberculosis: a systematic review of randomized controlled trials. *Clin Infect Dis.* 2014;58(4):470-480.
383. O'Garra A, Redford PS, McNab FW, et al. The immune response in tuberculosis. *Annu Rev Immunol.* 2013;31:475-527.
384. Bhatt K, Verma S, Ellner JJ, et al. Quest for correlates of protection against tuberculosis. *Clin Vaccine Immunol.* 2015;22(3):258-266.
385. Jouanguy E, Altare F, Lamhamedi S, et al. Interferon-gamma-receptor deficiency in an infant with fatal bacille Calmette-Guerin infection. *N Engl J Med.* 1996;335(26):1956-1961.
386. Seneviratne SL, Doffinger R, Macfarlane J, et al. Disseminated *Mycobacterium tuberculosis* infection due to interferon gamma deficiency. Response to replacement therapy. *Thorax.* 2007;62(1):97-99.
387. Fletcher HA, Snowden MA, Landry B, et al. T-cell activation is an immune correlate of risk in BCG vaccinated infants. *Nat Commun.* 2016;7:11290.
388. Lin PL, Flynn JL. CD8 T cells and *Mycobacterium tuberculosis* infection. *Semin Immunopathol.* 2015;37(3):239-249.
389. Darrah PA, Zeppa JJ, Maiello P, et al. Prevention of tuberculosis in macaques after intravenous BCG immunization. *Nature.* 2020;577(7788):95-102.
390. Bull NC, Stylianou E, Kaveh DA, et al. Enhanced protection conferred by mucosal BCG vaccination associates with presence of antigen-specific lung tissue-resident PD-1(+) KLRG1(-) CD4(+) T cells. *Mucosal Immunology.* 2019;12(2):555-564.
391. Van Der Meeren O, Hatherill M, Nduba V, et al. Phase 2b controlled trial of M72/as01e vaccine to prevent tuberculosis. *N Engl J Med.* 2018;379(17):1621-1634.
392. Tait DR, Hatherill M, Van Der Meeren O, et al. Final analysis of a trial of M72/AS01E vaccine to prevent tuberculosis. *N Engl J Med.* 2019;381(25):2429-2439.
393. Nemes E, Geldenhuys H, Rozot V, et al. Prevention of *M. tuberculosis* infection with H4:IC31 vaccine or BCG revaccination. *N Engl J Med.* 2018;379(2):138-149.
394. Schellack C, Prinz K, Egyed A, et al. IC31, a novel adjuvant signaling via TLR9, induces potent cellular and humoral immune responses. *Vaccine.* 2006;24(26):5461-5472.
395. Rodrigues LC, Pereira SM, Cunha SS, et al. Effect of BCG revaccination on incidence of tuberculosis in school-aged children in Brazil: the BCG-REVAC cluster-randomised trial. *Lancet.* 2005;366(9493):1290-1295.
396. Nieuwenhuizen NE, Kulkarni PS, Shaligram U, et al. The recombinant bacille calmette-guerin vaccine VPM1002: ready for clinical efficacy testing. *Front Immunol.* 2017;8:1147.
397. Arbues A, Aguilo JI, Gonzalo-Asensio J, et al. Construction, characterization and preclinical evaluation of MTBVAC, the first live-attenuated M. tuberculosis-based vaccine to enter clinical trials. *Vaccine.* 2013;31(42):4867-4873.
398. Scriba TJ, Kaufmann SH, Henri Lambert P, et al. Vaccination against tuberculosis with whole-cell mycobacterial vaccines. *J Infect Dis.* 2016;214(5):659-664.
399. Marinova D, Gonzalo-Asensio J, Aguilo N, et al. MTBVAC from discovery to clinical trials in tuberculosis-endemic countries. *Expert Rev Vaccine.* 2017;16(6):565-576.
400. Suliman S, Luabeya AKK, Geldenhuys H, et al. Dose optimization of H56:IC31 vaccine for tuberculosis-endemic populations. A double-blind, placebo-controlled, dose-selection trial. *Am J Respir Crit Care Med.* 2019;199(2):220-231.
401. Centers for Disease Control and Prevention. *Seasonal Influenza Vaccine Effectiveness.* 2014-2015. Updated October 3, 2019. https://www.cdc.gov/flu/vaccines-work/2014-2015.html
402. Paules CI, Marston HD, Eisinger RW, et al. The pathway to a universal influenza vaccine. *Immunity.* 2017;47(4):599-603.
403. Wei CJ, Crank MC, Shiver J, et al. Next-generation influenza vaccines: opportunities and challenges. *Nat Rev Drug Discov.* 2020;19(4):239-252.
404. Wiley DC, Skehel JJ. The structure and function of the hemagglutinin membrane glycoprotein of influenza virus. *Annu Rev Biochem.* 1987;56:365-394.
405. Wu NC, Wilson IA. A perspective on the structural and functional constraints for immune evasion: insights from influenza virus. *J Mol Biol.* 2017;429(17):2694-2709.
406. Coughlan L, Palese P. Overcoming barriers in the path to a universal influenza virus vaccine. *Cell Host Microbe.* 2018;24(1):18-24.
407. Schmidt AG, Therkelsen MD, Stewart S, et al. Viral receptor-binding site antibodies with diverse germline origins. *Cell.* 2015;161(5):1026-1034.
408. Lee PS, Wilson IA. Structural characterization of viral epitopes recognized by broadly cross-reactive antibodies. *Curr Top Microbiol Immunol.* 2015;386:323-341.
409. Corti D, Voss J, Gamblin SJ, et al. A neutralizing antibody selected from plasma cells that binds to group 1 and group 2 influenza A hemagglutinins. *Science.* 2011;333(6044):850-856.
410. Dreyfus C, Ekiert DC, Wilson IA. Structure of a classical broadly neutralizing stem antibody in complex with a pandemic H2 influenza virus hemagglutinin. *J Virol.* 2013;87(12):7149-7154.
411. Dreyfus C, Laursen NS, Kwaks T, et al. Highly conserved protective epitopes on influenza B viruses. *Science.* 2012;337(6100):1343-1348.
412. Ekiert DC, Bhabha G, Elsliger MA, et al. Antibody recognition of a highly conserved influenza virus epitope. *Science.* 2009;324(5924):246-251.
413. Ekiert DC, Friesen RH, Bhabha G, et al. A highly conserved neutralizing epitope on group 2 influenza A viruses. *Science.* 2011;333(6044):843-850.
414. Friesen RH, Lee PS, Stoop EJ, et al. A common solution to group 2 influenza virus neutralization. *Proc Natl Acad Sci U S A.* 2014;111(1):445-450.
415. Joyce MG, Wheatley AK, Thomas PV, et al. Vaccine-induced antibodies that neutralize group 1 and group 2 influenza A viruses. *Cell.* 2016;166(3):609-623.
416. Kallewaard NL, Corti D, Collins PJ, et al. Structure and function analysis of an antibody recognizing all influenza A subtypes. *Cell.* 2016;166(3):596-608.
417. Steel J, Lowen AC, Wang TT, et al. Influenza virus vaccine based on the conserved hemagglutinin stalk domain. *mBio.* 2010;1(1):e00018.
418. Yassine HM, Boyington JC, McTamney PM, et al. Hemagglutinin-stem nanoparticles generate heterosubtypic influenza protection. *Nat Med.* 2015;21(9):1065-1070.
419. Impagliazzo A, Milder F, Kuipers H, et al. A stable trimeric influenza hemagglutinin stem as a broadly protective immunogen. *Science.* 2015;349(6254):1301-1306.
420. Hai R, Krammer F, Tan GS, et al. Influenza viruses expressing chimeric hemagglutinins: globular head and stalk domains derived from different subtypes. *J Virol.* 2012;86(10):5774-5781.
421. Nachbagauer R, Liu WC, Choi A, et al. A universal influenza virus vaccine candidate confers protection against pandemic H1N1 infection in preclinical ferret studies. *NPJ Vaccines.* 2017;2:26.
422. DiLillo DJ, Palese P, Wilson PC, et al. Broadly neutralizing anti-influenza antibodies require Fc receptor engagement for in vivo protection. *J Clin Invest.* 2016;126(2):605-610.
423. DiLillo DJ, Tan GS, Palese P, et al. Broadly neutralizing hemagglutinin stalk-specific antibodies require FcγR interactions for protection against influenza virus in vivo. *Nat Med.* 2014;20(2):143-151.
424. Heaton NS, Sachs D, Chen CJ, et al. Genome-wide mutagenesis of influenza virus reveals unique plasticity of the hemagglutinin and NS1 proteins. *Proc Natl Acad Sci U S A.* 2013;110(50):20248-20253.
425. Doud MB, Bloom JD. Accurate measurement of the effects of all amino-acid mutations on influenza hemagglutinin. *Viruses.* 2016;8(6):155.
426. Giles BM, Ross TM. A computationally optimized broadly reactive antigen (COBRA) based H5N1 VLP vaccine elicits broadly reactive antibodies in mice and ferrets. *Vaccine.* 2011;29(16):3043-3054.
427. Darricarrere N, Pougatcheva S, Duan X, et al. Development of a pan-H1 influenza vaccine. *J Virol.* 2018;92(22):e01349.
428. Kanekiyo M, Joyce MG, Gillespie RA, et al. Mosaic nanoparticle display of diverse influenza virus hemagglutinins elicits broad B cell responses. *Nat Immunol.* 2019;20(3):362-372.
429. Schmidt AG, Xu H, Khan AR, et al. Preconfiguration of the antigen-binding site during affinity maturation of a broadly neutralizing influenza virus antibody. *Proc Natl Acad Sci U S A.* 2013;110(1):264-269.
430. Schmidt AG, Do KT, McCarthy KR, et al. Immunogenic stimulus for germline precursors of antibodies that engage the influenza hemagglutinin receptor-binding site. *Cell Rep.* 2015;13(12):2842-2850.
431. Drysdale SB, Barr RS, Rollier CS, et al. Priorities for developing respiratory syncytial virus vaccines in different target populations. *Sci Transl Med.* 2020;12(535):eaax2466.
432. Anderson LJ, Dormitzer PR, Nokes DJ, et al. Strategic priorities for respiratory syncytial virus (RSV) vaccine development. *Vaccine.* 2013;31(suppl 2):B209-B215.
433. Kwong PD, DeKosky BJ, Ulmer JB. Antibody-guided structure-based vaccines. *Semin Immunol.* 2020:101428.
434. McLellan JS, Chen M, Joyce MG, et al. Structure-based design of a fusion glycoprotein vaccine for respiratory syncytial virus. *Science.* 2013;342(6158):592-598.
435. McLellan JS, Chen M, Leung S, et al. Structure of RSV fusion glycoprotein trimer bound to a prefusion-specific neutralizing antibody. *Science.* 2013;340(6136):1113-1117.
436. Glenn GM, Fries LF, Thomas DN, et al. A randomized, blinded, controlled, dose-ranging study of a respiratory syncytial virus recombinant fusion (F) nanoparticle vaccine in healthy women of childbearing age. *J Infect Dis.* 2016;213(3):411-422.
437. *Novavax Announces Topline RSV F Vaccine Data From Two Clinical Trials in Older Adults*[press release]. Novavax; 2016. https://ir.novavax.com/news-releases/news-release-details/novavax-announces-topline-rsv-f-vaccine-data-two-clinical-trials
438. Marcandalli J, Fiala B, Ols S, et al. Induction of potent neutralizing antibody responses by a designed protein nanoparticle vaccine for respiratory syncytial virus. *Cell.* 2019;176(6):1420-1431.e17.
439. Muralidar S, Ambi SV, Sekaran S, et al. The emergence of COVID-19 as a global pandemic: understanding the epidemiology, immune response and potential therapeutic targets of SARS-CoV-2. *Biochimie.* 2020;179:85-100.
440. *COVID-19 Coronavirus Pandemic [press release].* Worldometers. Accessed December 21, 2020. https://www.worldometers.info/coronavirus/
441. *Coronavirus Vaccine Tracker*[press release]. Nytimes; 2020. https://www.nytimes.com/interactive/2020/science/coronavirus-vaccine-tracker.html?auth=login-email&login=email
442. Mulligan MJ, Lyke KE, Kitchin N, et al. Phase I/II study of COVID-19 RNA vaccine BNT162b1 in adults. *Nature.* 2020;586(7830):589-593.
443. Jackson LA, Anderson EJ, Rouphael NG, et al. An mRNA vaccine against SARS-CoV-2—preliminary report. *N Engl J Med.* 2020;383(20):1920-1931.
444. Sadoff J, Le Gars M, Shukarev G, et al. Safety and immunogenicity of the Ad26.COV2.S COVID-19 vaccine candidate: interim results of a phase 1/2a, double-blind, randomized, placebo-controlled trial. *medRxiv.* 2020. doi:10.1101/2020.09.23.20199604.
445. Folegatti PM, Ewer KJ, Aley PK, et al. Safety and immunogenicity of the ChAdOx1 nCoV-19 vaccine against SARS-CoV-2: a preliminary report of a phase 1/2, single-blind, randomised controlled trial. *Lancet.* 2020;396(10249):467-478.
446. Keech C, Albert G, Cho I, et al. Phase 1-2 trial of a SARS-CoV-2 recombinant spike protein nanoparticle vaccine. *N Engl J Med.* 2020;383:2320-2332.
447. *Pfizer and Moderna Covid Vaccines 95% Effective in Clinical Trials*[press release]. PBS; 2020. https://www.pbs.org/wgbh/nova/article/pfizer-moderna-covid-vaccines-clinical-trials/
448. *AZD1222 Vaccine Met Primary Efficacy Endpoint in Preventing COVID-19*[press release]. AstraZeneca; 2020. https://www.astrazeneca.com/media-centre/press-releases/2020/azd1222hlr.html
449. Sahin U, Muik A, Derhovanessian E, et al. COVID-19 vaccine BNT162b1 elicits human antibody and TH1 T-cell responses. *Nature.* 2020;586:594-599.
450. *Sanofi and GSK to Support COVAX with 200 Million Doses of Adjuvanted, Recombinant Protein-Based COVID-19 Vaccine*[press release]. Sanofi; 2020. https://www.sanofi.com/en/media-room/press-releases/2020/2020-10-28-07-00-00
451. Gouglas D, Thanh Le T, Henderson K, et al. Estimating the cost of vaccine development against epidemic infectious diseases: a cost minimisation study. *Lancet Glob Health.* 2018;6(12):e1386-e1396.
452. Serdobova I, Kieny MP. Assembling a global vaccine development pipeline for infectious diseases in the developing world. *Am J Public Health.* 2006;96(9):1554-1559.
453. Slaoui M, Hepburn M. Developing safe and effective covid vaccines—operation Warp speed's strategy and approach. *N Engl J Med.* 2020;383(18):1701-1703.
454. Slaoui M, Greene SE, Woodcock J. Bridging the gap at Warp speed—delivering options for preventing and treating covid-19. *N Engl J Med.* 2020;383(20):1899-1901.
455. *172 Countries and Multiple Candidate Vaccines Engaged in COVID-19 Vaccine Global Access Facility*[press release]. World Health Organization; 2020. https://www.who.int/news/item/24-08-2020-172-countries-and-multiple-candidate-vaccines-engaged-in-covid-19-vaccine-global-access-facility
456. Pulendran B, Davis MM. The science and medicine of human immunology. *Science.* 2020;369(6511):eaay4014.
457. Plotkin SA. Updates on immunologic correlates of vaccine-induced protection. *Vaccine.* 2020;38(9):2250-2257.
458. Pulendran B, Li S, Nakaya HI. Systems vaccinology. *Immunity.* 2010;33(4):516-529.
459. Gaucher D, Therrien R, Kettaf N, et al. Yellow fever vaccine induces integrated multilineage and polyfunctional immune responses. *J Exp Med.* 2008;205(13):3119-3131.
460. Nakaya HI, Wrammert J, Lee EK, et al. Systems biology of vaccination for seasonal influenza in humans. *Nat Immunol.* 2011;12(8):786-795.
461. Nakaya HI, Hagan T, Duraisingham SS, et al. Systems analysis of immunity to influenza vaccination across multiple years and in diverse populations reveals shared molecular signatures. *Immunity.* 2015;43(6):1186-1198.
462. Nakaya HI, Clutterbuck E, Kazmin D, et al. Systems biology of immunity to MF59-adjuvanted versus nonadjuvanted trivalent seasonal influenza vaccines in early childhood. *Proc Natl Acad Sci U S A.* 2016;113(7):1853-1858.
463. Tsang JS, Schwartzberg PL, Kotliarov Y, et al. Global analyses of human immune variation reveal baseline predictors of post-vaccination responses. *Cell.* 2014;157(2):499-513.
464. HIPC-CHI Signatures Project Team; HIPC-I Consortium. Multicohort analysis reveals baseline transcriptional predictors of influenza vaccination responses. *Sci Immunol.* 2017;2(14):eaal4656.
465. Li S, Rouphael N, Duraisingham S, et al. Molecular signatures of antibody responses derived from a systems biology study of five human vaccines. *Nat Immunol.* 2014;15(2):195-204.

466. Li S, Sullivan NL, Rouphael N, et al. Metabolic phenotypes of response to vaccination in humans. *Cell*. 2017;169(5):862-877.e17.
467. van den Berg RA, Coccia M, Ballou WR, et al. Predicting RTS,S vaccine-mediated protection from transcriptomes in a malaria-challenge clinical trial. *Front Immunol*. 2017;8:557.
468. Rechtien A, Richert L, Lorenzo H, et al. Systems vaccinology identifies an early innate immune signature as a correlate of antibody responses to the Ebola vaccine rVSV-ZEBOV. *Cell Rep*. 2017;20(9):2251-2261.
469. Kotliarov Y, Sparks R, Martins AJ, et al. Broad immune activation underlies shared set point signatures for vaccine responsiveness in healthy individuals and disease activity in patients with lupus. *Nat Med*. 2020;26(4):618-629.
470. Ravindran R, Khan N, Nakaya HI, et al. Vaccine activation of the nutrient sensor GCN2 in dendritic cells enhances antigen presentation. *Science*. 2014;343(6168):313-317.
471. Ravindran R, Loebbermann J, Nakaya HI, et al. The amino acid sensor GCN2 controls gut inflammation by inhibiting inflammasome activation. *Nature*. 2016;531(7595):523-527.
472. Oh JZ, Ravindran R, Chassaing B, et al. TLR5-mediated sensing of gut microbiota is necessary for antibody responses to seasonal influenza vaccination. *Immunity*. 2014;41(3):478-492.
473. Hagan T, Cortese M, Rouphael N, et al. Antibiotics-driven gut microbiome perturbation alters immunity to vaccines in humans. *Cell*. 2019;178(6):1313-1328.e13.
474. Ackerman ME, Barouch DH, Alter G. Systems serology for evaluation of HIV vaccine trials. *Immunol Rev*. 2017;275(1):262-270.
475. Wang TT, Ravetch JV. Functional diversification of IgGs through Fc glycosylation. *J Clin Invest*. 2019;129(9):3492-3498.
476. Suscovich TJ, Fallon JK, Das J, et al. Mapping functional humoral correlates of protection against malaria challenge following RTS,S/AS01 vaccination. *Sci Transl Med*. 2020;12(553):eabb4757.
477. Chung AW, Kumar MP, Arnold KB, et al. Dissecting polyclonal vaccine-induced humoral immunity against HIV using systems serology. *Cell*. 2015;163(4):988-998.
478. Boudreau CM, Yu WH, Suscovich TJ, et al. Selective induction of antibody effector functional responses using MF59-adjuvanted vaccination. *J Clin Invest*. 2020;130(2):662-672.
479. Moncunill G, Scholzen A, Mpina M, et al. Antigen-stimulated PBMC transcriptional protective signatures for malaria immunization. *Sci Transl Med*. 2020;12(543):eaay8924.
480. Wagar L, Salahudeen A, Constantz CM, et al. Modeling human adaptive immune responses with tonsil organoids. *Nat Med*. 2021;27(1):125-135.
481. Hur YG, Gorak-Stolinska P, Lalor MK, et al. Factors affecting immunogenicity of BCG in infants, a study in Malawi, the Gambia and the UK. *BMC Infect Dis*. 2014;14:184.
482. Fine PE. Variation in protection by BCG: implications of and for heterologous immunity. *Lancet*. 1995;346(8986):1339-1345.
483. Hanlon P, Hanlon L, Marsh V, et al. Trial of an attenuated bovine rotavirus vaccine (RIT 4237) in Gambian infants. *Lancet*. 1987;1(8546):1342-1345.
484. Tate JE, Burton AH, Boschi-Pinto C, et al. 2008 estimate of worldwide rotavirus-associated mortality in children younger than 5 years before the introduction of universal rotavirus vaccination programmes: a systematic review and meta-analysis. *Lancet Infect Dis*. 2012;12(2):136-141.
485. de Jong SE, Olin A, Pulendran B. The impact of the microbiome on immunity to vaccination in humans. *Cell Host Microbe*. 2020;28(2):169-179.
486. Whittaker E, Goldblatt D, McIntyre P, et al. Neonatal immunization: rationale, current state, and future prospects. *Front Immunol*. 2018;9:532.
487. GAVI. *Gavi, the Vaccine Alliance, Helps Vaccinate Almost Half the World's Children Against Deadly and Debilitating Infectious Diseases*. Accessed December 21, 2020. https://www.gavi.org/our-alliance/about
488. CEPI. Why We Exist? Creating a World in Which Epidemics are no Longer a Threat to Humanity. Accessed December 21, 2020. https://cepi.net/about/whyweexist/

CHAPTER 41 Maternal-Fetal Immunology

Gabrielle Rizzuto, MD, PhD • Adrian Erlebacher, MD, PhD

INTRODUCTION

Reproductive immunology is the study of phenomena that lie at the intersection of reproductive biology and immunology. It encompasses the immunology of male and (nonpregnant) female reproductive organs, which we do not discuss here, as well as the immunology of pregnancy, which is the topic of this chapter. Historically, the field has been dominated by interest in the paradox of "fetomaternal tolerance"—that is, how the fetus and placenta avoid being rejected by the maternal immune system—and this question will also be the focus of much of our discussion. Indeed, there is the sense that immunologists' failure to have "solved" this problem, despite its articulation almost 70 years ago, means that there is something fundamental to the nature of peripheral tolerance that we still do not have a handle on. Insight into fetomaternal tolerance, thus, in turn, is felt to have the potential of suggesting new ways to modulate the immune system in the clinic, with obvious application to autoimmune disease, transplantation, and cancer.

In addition to fetomaternal tolerance, we will discuss the influence of the maternal immune system over key developmental processes of pregnancy, including embryo implantation, placental development, and parturition (the act of giving birth). This influence, which has been increasingly appreciated over the last 30 years,[1] normally fosters pregnancy success but can also cause pregnancy complications with attendant detrimental effects on maternal and neonatal health. For example, intrauterine inflammation is now thought to be a significant component of the pathogenesis of preterm birth (PTB), whose adverse effects on neonatal health can extend into adulthood, while systemic maternal inflammation is thought to negatively affect fetal brain development and increase the risk of schizophrenia and autism spectrum disorders. We will also discuss infectious disease control during pregnancy, the pathogens that take advantage of the unique immunological environment of the maternal-fetal interface, and the specialized defense mechanisms that have evolved to counteract these pathogens and thus minimize vertical transmission to the fetus. In fact, reproductive immunology becomes a fascinating case study in evolutionary biology when considering the many adaptations that must be present to reconcile the competing demands of reproduction and host defense.

THE MATERNAL-FETAL INTERFACE

Anatomy, Development, and Microbiology

Maternal-fetal immunology is a challenging field since female reproductive systems are highly divergent across mammalian species. Fortunately, the basic topological arrangement between mother and conceptus (ie, all tissues derived from the fertilized egg, which includes the placenta) is always the same. The trophectoderm (ie, outer, epithelial) layer of the blastocyst gives rise to trophoblast cells, which comprise much of the placenta, including its outermost layer. Trophoblasts also comprise the outermost layer of the chorioamniotic membranes, which enclose the amniotic cavity where the fetus develops suspended in amniotic fluid.[1,2] Accordingly, trophoblasts constitute the entire "fetal" component of the maternal-fetal interface. Fortuitously, mouse and human placentas are also both "hemochorial," which means that the trophoblasts that mediate nutrient and gas exchange between mother and conceptus are directly bathed in maternal blood. However, there are key differences between mice and humans in placental substructure and trophoblast subtypes, as well as in the development of the decidua, which is the uterine tissue layer that encases the conceptus and thus forms the maternal component of the maternal-fetal interface. Given that the mouse is the primary model organism for work in maternal-fetal immunology, we will highlight these differences as we present some necessary background on reproductive biology.

The Decidua

The nonpregnant uterus of all mammals is a three-layered tube (Fig. 41.1A [left] and C). Its outer serosal layer surrounds the myometrium, which is the layer of thick smooth muscle that contracts during labor to expel the conceptus. The endometrium—the uterus's inner layer that abuts the uterine lumen and gives rise to the decidua—is essentially a dense fibroblastic stroma, lined by a simple cuboidal epithelium that contains glands and torturous "spiral" arteries. In mice, the endometrium becomes receptive to embryo implantation following copulation, an act that sets in motion a stereotypical endocrinological response that culminates in the ovary producing increasing amounts of progesterone overlaid with a small spike in estrogen production on gestation day (gd) 3.5 (see section *Implantation, Decidualization, and Early Pregnancy Failure*; N.B.: given that mice typically copulate at night, we will use a standard nomenclature that counts noon of the following day as gd0.5). Together, these hormones alter gene expression in endometrial stromal and epithelial cells to allow implantation to occur on gd4.5. In humans, the endometrium becomes receptive to embryo implantation on a monthly basis during the secretory phase of the menstrual cycle (Fig. 41.1C), a period similarly associated with increased progesterone production by the ovary.

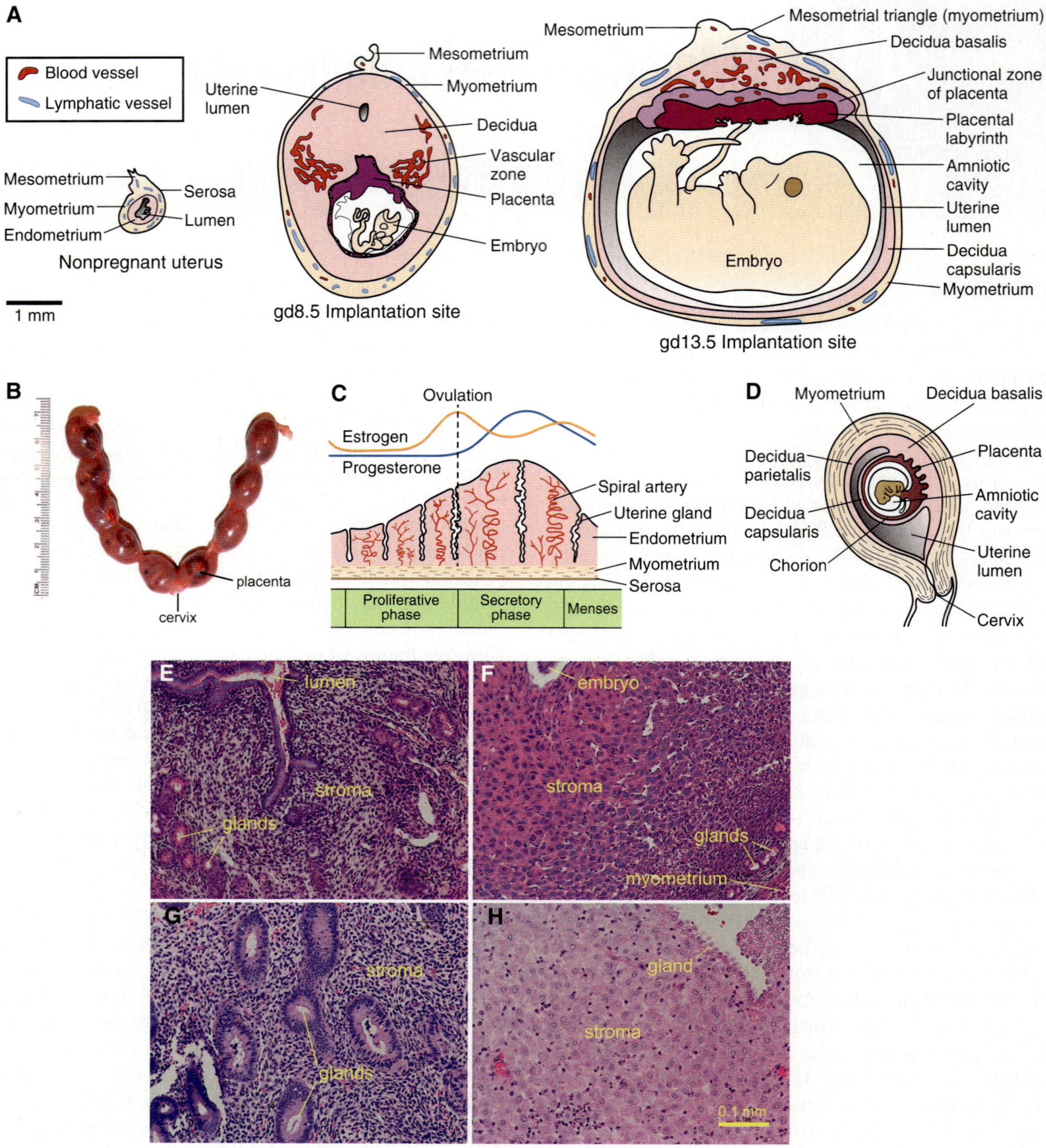

FIG. 41.1. Anatomic and histologic overview of the pregnant and nonpregnant uterus. A: Schematic of the pregnant and nonpregnant mouse uterus. The three images are shown to scale and all depict cross-sections that are perpendicular to the long axis of the uterine horn. The uterine serosa covers the myometrium in all three cases but is only indicated for the nonpregnant uterus. The middle and right images encompass all the tissues and tissue layers that constitute what we refer to as an "implantation site"—that is, the conceptus (embryo/fetus, placenta, and ancillary structures like the chorion and amnion), the decidua, and the overlying segment of myometrium. The mesometrium is the membrane that runs along the length of the uterus and harbors all the blood and lymphatic vessels that supply the uterus. Its position defines the mesometrial and antimesometrial poles of the implantation site. **B:** Whole-mount photograph of a pregnant mouse uterus on gd17.5. Note that the mouse uterus is bicornuate (ie, it possesses two horns). **C:** Schematic of the human endometrium as it appears over the course of the menstrual cycle. The associated fluctuations in serum levels of estrogen and progesterone, produced by the ovary, are also shown. **D:** Schematic of the pregnant human uterus in the first trimester. At ~17 weeks gestation (not shown), the chorioamniotic membranes (which arise from the fusion of the chorion and amnion) and associated decidua capsularis fuses with the decidua parietalis on the far side of the uterus. The chorioamniotic membranes then detach again prior to delivery, still attached to a layer of decidual tissue (now called the decidua parietalis). There is no equivalent layer of decidua parietalis in mice; rather, the entire antimesometrial pole of the decidua is comprised of decidua capsularis, and this tissue regresses around midgestation, leaving behind a layer of undecidualized endometrium covered with luminal epithelial cells that are not attached to the chorion (not depicted). **E-H:** H&E, stained sections of the undecidualized mouse **(E)** and human **(G)** endometrium and the decidualized mouse **(F)** and human **(H)** decidua. Panel E is a section of a segment of undecidualized uterus between implantation sites on gd6.5; Panel F is from one such implantation site. Panel G is from the human uterus during its proliferative phase; Panel H is from a first-trimester elective termination specimen.

The transformation of the endometrium into the decidua, that is, decidualization, entails the transdifferentiation of endometrial stromal cells (ESCs) into decidual stromal cells (DSCs) (for a review, see Ref. 3). The former cells are small and fibroblastic, whereas the latter are large and epithelioid (Fig. 41.1E-H). In mice, decidualization is triggered by implantation and unfolds in a highly stereotypical fashion around each embryo, both temporally and spatially. Thus, since mice are multiparous and all embryos implant at the same time, decidualization occurs synchronously at multiple locations along the length of each uterine horn, giving rise to a set of synchronously growing and developing "implantation sites"—a term we will use to refer to the conceptus, its associated decidua, and the segment of overlying myometrium (Fig. 41.1A [middle and right] and B). In contrast to mice, the human endometrium spontaneously decidualizes to a limited extent during the secretory phase of the menstrual cycle in anticipation of implantation; however, the full transformation of ESCs into DSCs is not apparent unless implantation occurs.

Importantly, the decidua shows regional specialization. The decidua basalis anchors the placenta to the uterus and thus serves as the physical substratum for placental development and the support structure for the uterine spiral arteries that supply the placenta with maternal blood[3] (Fig. 41.1A [middle and right] and D). By contrast, the decidua capsularis lies at the opposite pole of the embryo, which means that in humans, it surrounds the chorioamniotic membranes (see Fig. 41.1 legend for additional detail). Commonly referred to as the "fetal membranes" by researchers and pathologists, the chorioamniotic membranes and associated decidua are delivered along with the rest of the conceptus and provide much of the material for the published analyses of the human decidua in late gestation.

From an immunological perspective, the decidua is important because it is a prime location for infection and harbors all the maternal leukocytes that come into proximity with trophoblasts under normal circumstances. How its immunological milieu is established thus has major implications for mechanisms of fetomaternal tolerance (see section *Mechanisms of Fetomaternal Tolerance*), host defense (see section *Infectious Diseases and Pregnancy*), and the immunological control of placental development (see section *Developmental Influences of the Maternal Immune System Over Pregnancy*). At present, there are indications that trophoblasts themselves can modulate decidual leukocyte phenotypes, as will be discussed immediately below. However, a much more fundamental role has been ascribed to DSCs and the gene expression changes that occur as ESCs transform into DSCs given the pervasive effects of these changes over decidual immune cell phenotypes, trafficking, and population dynamics. For example, differentiating mouse and humans DSCs upregulate genes encoding the cytokine IL-15 and the chemokine CXCL14. These factors are thought to promote the massive accumulation of decidual natural killer (dNK) cells necessary for uterine spiral artery remodeling, the process that ensures sufficient placental perfusion with maternal blood (see section *Spiral Artery Remodeling, IUGR, and Preeclampsia*). Conversely, mouse DSCs transcriptionally silence *Cxcl9*, *Cxcl10*, *Cxcl12*, and *Cxcl16*, which encode chemokines that recruit type-1 polarized T cells (Th1 cells and cytotoxic T-lymphocytes [CTLs]) from the blood to peripheral tissues, as well as *Csf1*, which encodes the prototypical macrophage growth factor CSF-1.[4,5] Thus, and as discussed further in sections *Immune Cells of the Pregnant Uterus* and *Mechanisms of Effector Phase Tolerance to the Conceptus*, the mouse decidua cannot accumulate activated T cells, in contrast to the endometrium, nor can it homeostatically expand its resident macrophage population as it grows in the early gestation.[5,6]

In mice, the gene signature of experimentally induced "deciduomas" that lack embryos are nearly identical to true pregnancy decidua.[7] Moreover, cultured human ESCs upregulate IL-15 and CXCL14 when treated with progesterone.[8] These data suggest that DSCs establish their unique transcriptional state through an endocrine-regulated developmental program without much input from the conceptus. Specific regulatory circuits with relevance to immune cell control, however, have only recently started coming into view, and considerable challenges face attempts to identify pathways that are conserved across species. For example, *Cxcl9*, *10*, *12*, *16*, and *Csf1* silencing represents a small part of a much broader epigenetic program activated in mouse DSCs that affect ~800 protein-coding genes.[4] The program is mediated via polycomb repressive complex-2 (PRC2), which catalyzes generation of the repressive histone mark H3K27me3 on gene promoters. However, it is unknown what induces PRC2 activity during the early stages of decidualization or how PRC2 is targeted to these 800 specific loci. It is also unclear whether an analogous pathway applies to human DSCs, even though human T cells, like mouse T cells, appear unable to accumulate at the maternal-fetal interface (see sections *Immune Cells of the Pregnant Uterus* and *Mechanisms of Effector Phase Tolerance to the Conceptus*). The role of progesterone also remains incompletely defined—even though ESCs express its nuclear receptor, progesterone exposure *in vivo* is insufficient to induce the DSC phenotype in mice (since decidualization only occurs at implantation sites) nor do secretory phase ESCs in humans fully manifest the DSC phenotype, as mentioned above. Further complicating the picture is the likelihood that the DSC transcriptome evolves over course of gestation to meet the changing needs of pregnancy, including the onset of labor.[4] Thus, the regulation of decidual leukocytes is also thought to change over time.

The Placenta

The placenta is the organ of nutrient and gas exchange between fetal and maternal blood.[2] This takes place in its central region—the labyrinth in mice and the villous tree in humans—where a vast fetal capillary network sheathed by layers of specialized trophoblast subtypes receives deoxygenated fetal blood from umbilical cord arteries and sends oxygenated and nutrient-rich blood back to the fetus via the umbilical cord vein. The outermost cell layer of the human villous tree is a fused, syncytial layer comprised of syncytiotrophoblasts. These cells are bathed in maternal blood but are not a site of leukocyte-trophoblast interaction under normal circumstances. Indeed, routine microscopic examination of the human placenta rarely, if ever, reveals leukocytes adherent to syncytiotrophoblasts even though the latter cells' apical surface is under low shear stress. Why this is the case has not been mechanistically explored in mice, but in humans, it likely relates to the inability of syncytiotrophoblasts to express many of the adhesion molecules typically used by endothelial cells to recruit blood-borne leukocytes, including E-selectin, VCAM-1, and ICAM-2 even when the cells are exposed to inflammatory cytokines.[9,10] Only ICAM-1 is inducible, and this is apparent *in vivo* only in focal pathological lesions of placental villi that show leukocyte infiltration (eg, see section *Are There Any Examples of Placental "Rejection"?*).[9-12] In addition, syncytiotrophoblasts express high levels of ACKR2, a decoy receptor for CC chemokines, which thus potentially mitigates firm adhesion for all responsive leukocytes.[13,14] By whatever mechanism, its upshot is that maternal immune cells only interact with trophoblasts within the decidua under normal circumstances, as alluded to above. These trophoblasts, which in humans are called extravillous trophoblasts (EVTs) because they originate from the villi that anchor the placenta to the decidua, populate the decidua interstitially as well as migrate into uterine spiral arteries to reroute maternal blood flow to the placenta. Importantly, syncytiotrophoblasts express the neonatal Fc receptor, FcRn, which demonstrates low pH-dependent binding to maternal IgG in endosomes and then releases the IgG at the syncytiotrophoblast's basolateral surface into the villous core.[15,16] Maternal IgG is then transferred across the villous capillary endothelium, likely via FcγRIIb2,[17] to reach the fetal circulation. This passive transfer of maternal antibodies occurs predominantly in the third trimester and provides the neonate with a certain level of preexisting immunity to the spectrum of pathogens that will be present in its new environment.

Like DSCs, trophoblasts within the decidua are also thought to regulate decidual immune cells. Most clearly, classical major histocompatibility complex (MHC) class I (MHC-I) molecules expressed by both mouse and human trophoblasts modulate the phenotypes of dNK cells and their influence over spiral artery remodeling (see sections *Immune Cells of the Pregnant Uterus* and *Spiral Artery Remodeling, IUGR, and Preeclampsia*). Other than this example, however, trophoblast-leukocyte interactions are poorly understood and starkly illustrate the mouse-human dichotomy in reproductive systems. For example, human EVTs and syncytiotrophoblasts express the T cell inhibitory molecule PD-L1, in contrast to mouse trophoblasts.[18-22] Although not a uniform finding,[20] human EVTs have also been reported to produce the macrophage growth factor CSF-1 and the anti-inflammatory cytokine IL-10[23] in contrast to mouse trophoblasts.[19,24] Perhaps, most famously, human EVTs express human leukocyte antigen (HLA)-G, a nonclassical MHC class I molecule that does not even exist in mice. Extensive work on HLA-G since its discovery in the 1980s[25] has revealed multiple transmembrane and secreted isoforms, a limited degree of polymorphism, and an unusual mode of transfer between cells known as trogocytosis, and its provision through its leader sequence of a high-affinity peptide for HLA-E, another nonclassical MHC class I molecule.[25,26] There is also agreement that HLA-G does not directly interact with the T-cell receptor (TCR) complex on CD8 T cells but rather engages two inhibitory receptors known as LILRB1 and LILRB2 expressed primarily by myeloid cells. LILRB1 is also expressed by subsets of NK cells, including dNK cells. However, there is considerable controversy over whether HLA-G performs an important function at the maternal-fetal interface, such as the attenuation of local immune responses. Given that it lacks a mouse homolog, all functional studies to date have been inferential in nature and indirect. For contrasting opinions, the reader is referred to references 25-28, with key points of contention being whether HLA-G is ever actually expressed by cells other than EVTs, whether it meaningfully engages KIR2DL4, a killer Ig-like receptor expressed by NK cells, and whether correlations between HLA-G alleles and clinical outcomes in the setting of various human diseases have true significance. Of note, some people are homozygous for null alleles for HLA-G, indicating that the molecule is not singularly required for pregnancy.

Does the Maternal-Fetal Interface Have a Microbiome?

For over a century, microbiology experiments supported the "sterile womb paradigm," which posits that the healthy placenta and pregnant uterus do not harbor commensal organisms. Indeed, the fact that germ-free mammals can be generated by cesarean delivery provides support for fetal sterility *in utero*.[29] Modern advances in sequencing technology have led investigators to revisit this paradigm, however, and some studies have instead concluded that healthy human and mouse placentas are colonized by a low abundance, metabolically rich microbiome,[30-33] and that the fetal gastrointestinal tract may even become colonized *in utero*[34,35] Although intriguing, these conclusions are currently controversial. Thus, with the exception of the ~5% of placentas colonized with Group B *Streptococcus (GBS)* organisms, which are common and well-appreciated causes of "ascending" placental infection (see section *Immune Defenses at the Maternal-Fetal Interface*), other work finds that the bacteria detected in healthy placental samples are instead environmental contaminants.[36-39] Of note, recent sequencing studies have identified a predominance of *Lactobacillus* bacteria in the nonpregnant human uterus,[40,41] and there is some evidence that altered colonization may be associated with reproductive failure.[42,43] The field thus awaits a consensus understanding of whether a microbiome is present in the nonpregnant and pregnant uterus, the placenta, and/or the fetus; and if so, how this impacts fertility, pregnancy, and fetal immune development.

Immune Cells of the Pregnant Uterus

Although it would inherently make sense that the maternal-fetal interface would be an immunological desert in order to avoid any possibility of causing immunological damage to the conceptus, research over the last 30 years has revealed this to be only partially true. Thus, while certain immune cell types in certain species (eg, Th1 cells and CTLs in mice) are actively excluded from the decidua, others are present in large numbers and contribute to some of the key developmental events of pregnancy. In this section, we provide an overview of how immune cell dynamics and phenotypes are controlled within the decidua and myometrium. Endometrial immune cell dynamics in response to copulation and semen exposure are discussed, along with the role of the maternal immune system in implantation in section *Implantation, Decidualization, and Early Pregnancy Failure*.

NK Cells

NK cells are the most prevalent immune cell type at the maternal-fetal interface and have characteristics that distinguish them from NK cells elsewhere in the body (for reviews, see Refs. 44-47). In humans, they were first visualized histologically as large granulated lymphocytes that appear in the late secretory endometrium and then become prominent in the first-trimester decidua, where they constitute ~70% of all leukocytes.[48,49] Subsequent flow cytometric studies revealed the cells to have a characteristic lin$^-$ CD56^{superbright} CD16$^-$ surface phenotype that contrasts with peripheral blood NK cells.[46] Thus, these CD56^{superbright} cells have historically been referred to as dNK cells. We will also refer to NK cells in mouse implantation sites as dNK cells, although we note that in this species, the cells also accumulate in portions of the myometrium (see below). Mouse dNK cells are CD45$^+$ CD3$^-$ CD19$^-$ NKp46$^+$ NK1.1$^+$ (strain-dependent) by flow cytometry, uniformly express the transcription factor T-bet (Tbx21), and are distinguished from rare uterine group 1 innate lymphoid cells (ILC1s) by their expression of the transcription factor Eomes.[50-54] They comprise ~50% of all decidual leukocytes as early as gd5.5,[55] and, like human dNK cells, become large and granular but only after implantation. This phenotypic shift is evident histologically, as the cells become positive for periodic acid Shiff (PAS) reagent and the lectin *Dolichos biflorus* agglutinin (DBA), a unique feature of dNK cells in mice. NK cells decline in frequency in both the mouse and human decidua after their peak in early-mid gestation, but in humans, they remain a sizable population even at term gestation.[56-62]

A key function for dNK cells is to provide developmental support to the conceptus via the elaboration of cytokines, chemokines, and growth factors. Most importantly, the cells control uterine spiral artery remodeling, a critical developmental process that ensures adequate perfusion of the placenta with maternal blood[44-46] and that will be discussed in section *Spiral Artery Remodeling, IUGR, and Preeclampsia*. In addition, recent work on human cytomegalovirus (HCMV) and *Listeria monocytogenes* has revealed that they contribute to host defense at the maternal-fetal interface. This function will be discussed in section *Immune Defenses at the Maternal-Fetal Interface*. As researchers dissect these functions, they are starting to incorporate recent data indicating that dNK cells fall into distinct subsets. In mice, dNK cells are comprised of CD49a$^+$ DX5$^-$ and CD49a$^-$ DX5$^+$ cells.[50-53] Parabiosis experiments have revealed the CD49a$^+$ cells to be noncirculating tissue-resident (tr)NK cells, consistent with the CD49a$^+$ DX5$^-$ phenotype being a general indicator of tissue residency, and the CD49a$^-$ DX5$^+$ cells to be derived from circulating, conventional (c)NK cells.[52,62] trNK cells are ~threefold more abundant than cNK cells in the nonpregnant uterus,[50-52,54] but their ratio changes in a highly dynamic fashion as both populations expand over the first half of gestation within the growing implantation site.[54] Thus, trNK cells become by far the dominant subset prior to ~gd8.5 as a result of a rapid burst of proliferation, after which point cNK cells become the dominant subset, presumably as a result of their recruitment from the blood since they do not proliferate in situ.[53,54,62] The two subsets also appear to have distinct functions, with cNK cells being the primary drivers of spiral artery remodeling, a process that commences with onset of their influx on gd8.5.[51] This function is consistent with their higher expression of interferon (IFNγ), a cytokine long known to be critical for spiral artery remodeling in mice.[53,54,63] The functions of trNK cells are less well understood but potentially involve restricting trophoblast invasion into the decidua and controlling fetal growth.[64,65]

In the first-trimester human decidua, almost all dNK cells are CD49a$^+$ tissue-resident cells, with single-cell (sc) RNA-seq and CyTOF analyses dividing them into two subsets, both of which are proliferating to some extent: CD49^{hi} EOMES^{hi} TBX21^{lo} dNK1 cells and CD49^{int} EOMES^{int} TBX21^{int} dNK2 cells.[20,53,66] A third population, called dNK3 cells in the scRNA-Seq study,[20] are likely intraepithelial ILC1s.[66] Intriguingly, dNK1 cells, which have a low capacity to produce cytokines and chemokines following nonspecific stimulation,[66] resemble a population of "pregnancy-trained" dNK cells that increase in prevalence with repeat pregnancies, which suggests that dNK cells can demonstrate features of immunological memory as described for NK cells elsewhere in the body.[66,67] Whether they contribute to any of the pregnancy complications that decline in incidence with increasing gravidity, such as preeclampsia,[68] is currently unknown.

Aside from their division into subsets, dNK cells are instructed to assume their final, unique phenotype by locally produced factors. In humans, these include TGF-β and IL-15, which together cause peripheral blood NK cells to display certain features of dNK cells, including their distinct patterns of surface marker and activating/inhibitory receptor expression.[53,66,69-72] While IL-15 in the decidua is produced by DSCs (see below), the source of TGF-β may be either DSCs or decidual macrophages.[70,73] Activating/inhibitory receptor expression patterns differ between dNK subsets and are influenced by an NK cell education process that takes place partly within the endometrium, prior to pregnancy, and partly within the decidua.[20,66,74-76] For example, dNK1 cells express high levels of the HLA-G receptor LILRB1, suggesting direct interactions with EVTs.[20,66] Once established, these patterns allow EVTs and other cell types to modulate dNK cell secretory phenotypes in ways that impact upon pregnancy success.[65,72,77] Importantly, dNK cells show substantially reduced cytotoxicity toward MHC-I–deficient cells, despite their granules containing perforin and granzymes, and an extremely limited ability to kill EVTs (which express a limited repertoire of MHC-I; see section *Introduction to Fetomaternal Tolerance – History and Overview*). These findings have implications for our understanding of mechanisms of NK cell tolerance toward the conceptus and are discussed further in section *NK Cell Tolerance Toward the Conceptus*.

The factors that drive dNK cell expansion and localization during pregnancy remain incompletely defined. In the human decidua, dNK cells are somewhat enriched at sites of trophoblast invasion and vascular remodeling but are otherwise relatively homogeneously distributed.[78] In contrast, the distribution of dNK cells in mouse implantation sites is highly polarized, as they accumulate in the decidua basalis (the portion of the decidua described in section *Anatomy, Development, and Microbiology; The Decidua* that anchors the placenta; Fig. 41.2A) and overlying

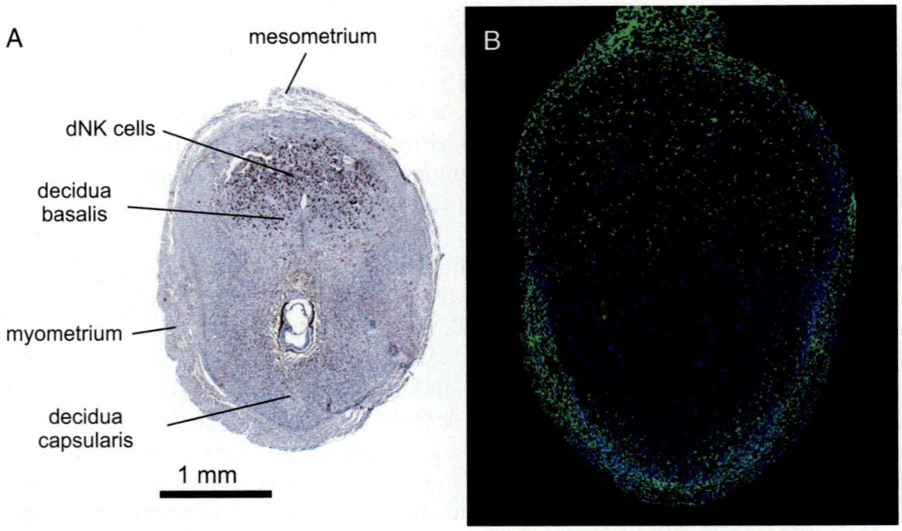

FIG. 41.2. Distribution of natural killer (NK) cells and macrophages in mouse implantation sites. A: A gd7.5 implantation site was stained with *Dolichos biflorus* agglutinin (DBA) lectin (*brown*) to reveal the accumulating population of decidual (d)NK cells in the decidua basalis. As gestation progresses, NK cells also accumulate in the mesometrial triangle above the decidua basalis. **B:** A gd8.5 implantation site was stained with antibodies to F4/80 (*green*) to reveal macrophages. Notice the clearing of macrophages from the decidua, in particular the decidua capsularis. The density of macrophages in the myometrium is similar to what is seen in the myometrium of the nonpregnant uterus.[6,79] DAPI (4′,6-diamidino-2-phenylindole) counterstain. (A, Image courtesy of Patricia Lima and B. Anne Croy (Queen's University).)

segment of myometrium (variously called the mesometrial triangle, metrial gland, or mesometrial lymphoid aggregate of pregnancy), but are absent from the decidua capsularis.[46] This distribution makes physiological sense given that the decidua basalis contains the spiral arteries that supply the placenta with maternal blood and that will be remodeled by dNK cells.

One cytokine that likely drives dNK cell accumulation is IL-15. IL-15 required for the generation of dNK cells in mice[80,81] is highly expressed by both mouse and human DSCs and is detectable on the surface of mouse DSCs consistent with their concomitant expression of IL-15Rα.[4,20,82] IL-15 induction in the mouse decidua likely promotes the expansion of both trNK cells and cNK cells by prolonging their survival, but the extent to which it drives dNK cell proliferation is unclear since trNK but not cNK cells proliferate within the pregnant uterus.[62] cNK cell recruitment to the decidua is likely mediated by the chemokine CXCL14, which is highly expressed by DSCs of both mice and humans[4,20,83] and is likely further promoted in mice by the endothelial cell adhesion molecule VCAM-1, which is expressed by vessels in the decidua basalis.[84] Members of the TGF-β superfamily also locally control dNK cell numbers, directly or indirectly, as does adrenomedullin produced by the murine placenta.[85,86] Importantly, mice with artificial deciduomas lacking implanted embryos contain high numbers of dNK cells distributed with a pattern essentially the same as that seen with true implantation sites,[87] demonstrating that the pattern is established independently of embryonic/placental signals.

Innate Lymphoid Cells

In addition to NK cells, the pregnant uterus harbors other kinds of ILCs. As mentioned above, mouse implantation sites contain a few ILC1s,[50-52,54] and the first-trimester human decidua contains ILC1s that are likely to be intraepithelial.[66,88] The first-trimester human decidua also contains ILC3s, but these cells are very rare in the mouse decidua.[50,53,66,88] ILC2s are completely absent from both the first-trimester human and mouse decidua[50,88] but are present in the mouse myometrium during pregnancy where they constitutively produce their signature cytokine, IL-5.[50] This tissue distribution may explain the parallel uterine distribution of eosinophils, as ILC2s locally control tissue eosinophil numbers throughout the body through their expression of this key eosinophil survival factor[89] (see below). Otherwise, the function of uterine ILCs during pregnancy is unknown. Interestingly, recent work suggests that decidual ILC2s become quite abundant in the late gestation human decidua, suggesting that they are dynamically regulated.[90]

T Cells

After NK cells, T cells are the best studied cell type in the pregnant uterus, given the field's interest in mechanisms of fetomaternal tolerance. In the first-trimester human decidua, they comprise ~10% of all leukocytes,[59,60,91] a frequency that progressively increases to ~40% to 50% by term gestation.[60,91,92] The cells divide roughly evenly between CD4$^+$ and CD8$^+$ cells, with the two subsets constituting ~40% each of all CD3$^+$ cells at term gestation and the remaining ~20% being CD4$^-$ CD8$^-$ T cells in turn comprised of γδ T cells and poorly understood αβ T cells.[92,93] Importantly, the increase in T cell frequencies with advancing gestation likely reflects a relative loss of NK cells rather than change in true histological tissue densities, which remain relatively constant.[58,59] Indeed, the most dramatic shift in endometrial/decidual T cell tissue densities, assessed histologically, occurs with the onset of pregnancy, as T cells are ~50% less dense in the decidua as compared to the secretory endometrium.[94] An analogous pattern is seen in mice with respect to the ability of the endometrium but not the decidua to recruit, and thus accumulate, blood-borne effector T cells.[5] This phenomenon, which is relevant to mechanisms of fetomaternal tolerance, was discussed in sections *Anatomy, Development, and Microbiology* and will be discussed further in *Mechanisms of Effector Phase Tolerance to the Conceptus*. In mice, T cells comprise ~5% of all leukocytes in both the decidua and myometrium on gd9.5.[95] Like in humans, CD4$^-$ CD8$^-$ T cells with unknown function[96] comprise ~20% to 40% of all decidual TCRαβ$^+$ T cells.[97]

CD8 T cells in the human decidua are mainly CD45RO$^+$, indicative of prior antigen encounter, and have been historically considered to be effector memory cells given their predominant CD45RA⁻ CCR7⁻ surface phenotype.[56,98-100] However, the cells also uniformly express high levels of CD69, which is now recognized as the signature marker of tissue-resident memory (Trm) cells in humans,[101] and many also express CD103, another Trm marker.[56,102-106] Thus, decidual CD8 T cells might largely be Trm cells, as noted explicitly in recent work,[107] and thus likely the same Trm cells that comprise half of all endometrial CD45RO⁺ CD8 T cells at the time of implantation.[108] Decidual CD8 T cells moreover show an activated surface phenotype as well as a robust capacity to proliferate, kill target cells, and produce cytolytic molecules and cytokines (IFNγ, TNFα, IL-2, IL-10) upon mitogen stimulation.[99,100,103-105,109,110] Thus, they might also be "effector-like" Trm cells that are able, without chronic antigenic stimulation, to persist in peripheral tissues in a semiactivated state, poised to mount rapid recall responses.[111] The cells also express PD-1 and other exhaustion markers,[100,103,110] a phenotype that may be relevant to mechanisms of fetomaternal tolerance, but it is worth noting that nonexhausted Trm cells in other locations, including the nonpregnant endometrium, also express these markers.[101,108]

Like decidual CD8 T cells, decidual CD4 T cells in humans are primarily CD45RO⁺ CD45RA⁻ CCR7⁻ CD69⁺ and robust cytokine producers. Thus, they may also largely be Trm cells and/or effector memory cells.[56,98,100,105,112] Like elsewhere in the body, they divide into the various T-helper subsets, with 5% to 40%, 1% to 5%, 1% to 7%, and 5% to 10% being Th1, Th2, Th17, and FoxP3⁺ CD25^{hi} T_{reg} cells, respectively, in both the first and third trimester.[91,100,112-115] Like their CD8 counterparts, many of the cells express PD-1,[100] while the T_{reg} cells are predominantly HELIOS⁺ and thus thymically derived.[116] In the mouse decidua, CD4 T cell subset proportions remain poorly defined, and the extent to which mouse decidual T cells, in general, are Trm cells has not been addressed. T_{reg} cells comprise ~2% to 4% of the CD4 T cells; however, surveys performed to date need to be interpreted with caution since mouse decidual T cell preparations contain large numbers of naïve cells, indicating extensive blood contamination.[97]

The antigenic specificity of decidual T cells remains largely unknown. To the extent that the cells are Trm cells, we would expect their specificity to be determined by events prior to pregnancy onset. In particular, prior infections, not only of the uterus itself but also systemically given that circulating Trm precursors widely seed peripheral tissues, would be expected to define the decidual T cell repertoire. Accordingly, HCMV- and EBV-specific CD8 T cells can be detected at nontrivial frequencies of 1% to 3% in first and third trimester decidual specimens.[103,117] Given that pregnancy systemically generates memory T_{reg} cells and quasimemory CD8 cells specific for trophoblast antigens (see section *The Nonimmunogenic Nature of Trophoblast Antigens*), it is likely that prior pregnancies would generate decidual T cells with these specificities as well. Indeed, paternal HLA-C alloreactivity and clonal expansion of human decidual T cells have been documented to a limited extent.[20,93,100] Importantly, however, the apparent activated state of decidual T cells does not necessarily imply antigen reengagement since, as mentioned, effector-like Trm cells show the same phenotype at baseline. Moreover, IL-15, which as mentioned is highly expressed by human and mouse DSCs, strongly induces surface activation marker expression on human CD8 T cells in the absence of concurrent TCR stimulation.[118] In a similarly antigen-independent fashion, human decidual macrophages and EVTs each mildly increase T_{reg} cell frequencies in bulk CD4 T cell cultures,[23,116,119] although paternal alloantigen-specific T_{reg} cell expansion in the pregnant mouse uterus has also been documented in certain experimental systems.[120,121]

Given their potential to produce proinflammatory cytokines and, in principle, even kill trophoblasts, decidual Th1 cells, and Th17 cells, and CTLs are assumed to pose latent threats to pregnancy success. In contrast, decidual T_{reg} cells are thought to promote successful pregnancy via effects on implantation, decidual vascular remodeling, fetomaternal tolerance, and the maintenance of uterine quiescence prior to labor onset. These functions will be described in sections *Mechanisms of Effector Phase Tolerance to the Conceptus, Implantation, Decidualization, and Early Pregnancy Failure, Spiral Artery Remodeling, IUGR, and Preeclampsia*, and *Parturition and PTL*. Importantly, Rag-deficient mice show no reproductive defects, demonstrating that B and T cells are not required for pregnancy success in this species and that the proposed functions for decidual T_{reg} cells all involve the suppression of inflammation generated by other adaptive immune cell types and not the direct regulation of developmental processes.

B Cells

B cells comprise only ~2% of decidual leukocytes, a percentage that remains constant over gestation.[60] The population is largely comprised of naïve and transitional B cells, suggesting that they might be mainly intravascular.[122]

Macrophages and Monocytes

Macrophages constitute ~20% of all leukocytes in the first-trimester human decidua are identified by their CD14⁺ surface phenotype and uniformly express CD68, consistent with tissue residence.[123] Although subsets are coming into view,[124,125] the cells in aggregate express high levels of CD163, CD206, CD209 (DC-SIGN), and IL-10, indicating a specialized M2-like phenotype with the potential to suppress inflammation and promote tissue repair.[123,126-128] Indeed, macrophages appear to be a major source of IL-10 in the first-trimester decidua and secrete this cytokine spontaneously.[129]

Macrophages in mouse implantation sites divide evenly into two subsets, F4/80⁺ MHCII^{hi} and F4/80⁺ MHCII^{lo}, with the MHCII^{lo} subset roughly corresponding to M2 cells.[6] As with dNK cells, their population dynamics in early gestation demonstrate how the developmental changes undertaken by the pregnant uterus determine resident immune cell behavior. Accordingly, uterine macrophage densities are controlled in a tissue layer-specific fashion by local levels of CSF-1, which, as mentioned, is the prototypical macrophage growth and survival factor and whose encoding gene *Csf1* is transcriptionally silenced in DSCs (see section *Anatomy, Development, and Microbiology*). Thus, while *Csf1* expression increases in the myometrium as it rapidly grows, it remains low in the decidua, and this means that only the myometrium maintains relatively constant macrophage tissue densities[6] (through a combination of in situ proliferation and Ly6C^{hi} monocyte precursor recruitment from the blood). In contrast, macrophage tissue densities, which are high in the undecidualized endometrium, drop off in the decidua[4,6,79,130-132] (Fig. 41.2B). Remarkably, Ly6C^{hi} monocytes are abundant in the early mouse decidua, but the cells remain intravascular in association with the endothelial cells of the vascular zone.[6,133] CSF-1 also prevents myometrial macrophages from converting into MHCII^{hi} cells, thus fostering the M2-like phenotype.[6]

It is unlikely that analogous pathways control human uterine macrophage dynamics since macrophages robustly proliferate in the first-trimester human decidua, and their histological tissue density in this tissue layer is similar to or even modestly higher than that of endometrial macrophages in the nonpregnant uterus.[48,126,127,134] Moreover, CSF-1 exposure *in vitro*, particularly in combination with IL-10, causes human blood monocytes to assume the M2-like phenotype of first-trimester decidual macrophages, a phenotype that contrasts with their more M1-like phenotype in the endometrium.[122,124,126,127,134] Together, these data suggest that first-trimester decidual macrophages are under the influence of CSF-1. Interestingly, *CSF1*, while expressed by human DSCs at relatively low levels, like murine DSCs, is highly expressed by human decidual pericytes.[20] This observation might explain not only why macrophages modestly accumulate around spiral arteries in humans[78] but also since pericytes are largely absent from the mouse decidua,[135] why decidual macrophages do not expand in mice. Additionally, CSF-1 is produced by first trimester human EVTs but not by mouse trophoblasts.[23,24] Decidual macrophage tissue densities modestly decline after the first trimester,[58] suggesting that CSF-1 production wanes with advancing gestation.

While macrophages likely clear the endometrial debris generated by menstruation,[136] their role in normal pregnancy remains unclear. Human decidual macrophages might contribute to spiral artery remodeling (see section *Spiral Artery Remodeling, IUGR, and Preeclampsia*),[78,126,137] and their production of IL-10 might help ensure that the uterus remains in a noninflamed, quiescent state until term gestation (see sections *Implantation, Decidualization, and Early Pregnancy Failure* and *Parturition and PTL*). This latter function might involve perpetuation of the M2-like phenotype and the induction of decidual T_{reg} cells.[23,116] Human decidual macrophages also form conjugates with dNK cells in situ and inhibit dNK cell-mediated killing of EVTs *in vitro*, in this case via a TGF-β–dependent mechanism.[73,127] Macrophages, perhaps with a more M1-phenotype, may also promote normal human parturition since they accumulate within decidua and myometrium upon labor onset (see section *Parturition and PTL*).[138-141] Similar, but mild, accumulations of monocytes and macrophages are also apparent in the late gestation rodent uterus.[138,142]

However, pending a close examination of possible roles in spiral artery remodeling, uterine macrophages, and monocytes in mice appear to play little role in normal pregnancy and parturition. Thus, CSF-1–deficient *op/op* mice, which bear very few uterine macrophages, do not have reported reproductive phenotypes aside from reduced ovulation rates.[130,143,144] Similarly, *Ccr2*$^{-/-}$ mice, which are unable to recruit Ly6C^{hi} monocytes from the blood to the uterus, have no reported reproductive defects and display normal parturition timing, as do mice depleted of circulating monocytes.[6,145,146] Rather, the existence of mechanisms to actively prevent macrophage accumulation within the mouse decidua, and to induce the unique M2-like phenotype of decidual macrophages in humans, highlights the potential negative impact of these cells. Such a net-negative likely reflects their ability to produce inflammatory mediators and is consistent with the cells, in humans, being more M1-like in spontaneous preterm labor (PTL) and in cases of spontaneous abortion.[128,141]

Macrophages also comprise almost all the immune cells that reside within the body of the placenta, that is, within the villous tree in humans (where they are called Hofbauer cells) and the placental labyrinth in mice.[147-149] These populations are fetal in origin. While mouse placental macrophages have barely been studied, Hofbauer cells assume an M2-like phenotype like their decidual counterparts, presumably in order to attenuate placental inflammation under normal conditions. Recent work in the rhesus macaque has also identified multiple subsets.[148] As will be described in sections *Are There Any Examples of Placental "Rejection"?* and *Immune Defenses at the Maternal-Fetal Interface*, the cells have roles in placental inflammation and infection,[147] but their functions during normal gestation are unknown.

Dendritic Cells

Like elsewhere in the body, conventional CD11c$^+$ dendritic cells (DCs) (ie, both cDC1s and cDC2s) populate the mouse endometrium and are able to migrate to the draining LN, where they presumably initiate T cell responses to uterine pathogens.[79] But like their macrophage counterparts, their tissue density in the growing decidua drops off dramatically in early gestation due to their failure to undergo homeostatic expansion, leaving very few DCs near the conceptus.[6,79] Remarkably, the mouse decidua also lacks lymphatic vessels, which means that the few decidual DCs that do exist are unable to migrate to the uterine-draining LN.[79] Together, these observations mean that there is minimal DC surveillance of the maternal-fetal interface, a feature of mouse pregnancy that has implications for mechanisms of fetomaternal tolerance and infectious disease control and that will be discussed further in sections *How Maternal T Cells Come to Engage Trophoblast Antigens* and *Immune Defenses at the Maternal-Fetal Interface*. Plasmacytoid DCs are largely absent from the mouse decidua.[97]

Identified as CD83$^+$ cells, human DCs are exceedingly rare in the decidua, but in contrast to mice, they are also quite rare in the endometrium.[127,134,150-152] Moreover, DC and macrophage population dynamics in the human uterus appear unlinked since DC tissue densities, if anything, are lower in the first-trimester decidua compared to the late secretory endometrium, whereas macrophage tissue densities are similar or higher, as mentioned above.[48,127,128,134] Provocatively, the first-trimester decidual DCs form conjugates with T cells in situ and have potent stimulatory capacity when cultured with allogeneic T cells *ex vivo*, which together suggest that they might activate decidual T cells.[150] However, the data on whether the human decidua contains lymphatic vessels and thus could support decidual DC trafficking to the uterine LN is contradictory.[150,153-155] Interestingly, the fetal membranes have recently been shown to contain lymphatic vessels at term gestation, but the route of drainage is not yet established, nor did the single-cell RNA-Seq study that identified these vessels uncover the presence of membrane-resident DCs.[57]

Potential developmental functions for decidual DCs have remained unclear.[152] There have been suggestions that they are required for implantation and decidualization and to tolerize maternal T cells toward fetoplacental antigens, but the lack of overt reproductive deficits in DC-deficient *Flt3l*$^{-/-}$ mice, even in strain disparate (allogeneic) matings, has called these ideas into question.[6,152,156-159] Similarly, indications that decidual DCs might induce decidual T_{reg} cells, or bidirectionally interact with dNK cells to promote decidual angiogenesis, currently lack direct *in vivo* experimental support.[157,160-163] Instead, their low density and tissue entrapment, together with their ability to mature in response to inflammatory stimuli[79] and activate T cells,[150] highlights their potential detrimental effects on pregnancy outcome.

Neutrophils and Other Myeloid Cells

In humans, neutrophils comprise a negligible fraction of first-trimester decidual leukocytes and accumulate in the decidua only in cases of infection (see sections *Immune Defenses at the Maternal-Fetal Interface* and *Infection-Mediated Pregnancy Complications*). They are similarly scarce in the myometrium for most of gestation but then become abundant in this tissue layer upon labor onset, similar to macrophages.[139,140,164] In mice, they comprise ~20% and ~1% of all leukocytes in the decidua and myometrium on gd9.5, respectively, and accumulate in the decidua at the leading edge of invasive trophoblasts and in the decidual vascular zone, where they remain intravascular.[95,133] Antibody-mediated neutrophil depletion during this period impairs placental development and causes intrauterine growth restriction (IUGR), suggesting that neutrophils regulate placentation;[165] however, these results have not been confirmed using genetically engineered mice with low-neutrophil numbers. At the end of mouse gestation, neutrophils represent ~1% to 2% of all uterine leukocytes, and their tissue densities mildly increase in both the decidua and myometrium as a consequence of labor onset.[138,142,146,166]

Strikingly, eosinophils are virtually absent from the mouse decidua but are present in the endometrium and myometrium.[95,167] As this distribution parallels that of ILC2s, which control tissue eosinophil numbers (see above), the mouse decidua, thus, assumes a decidedly nontype 2 immune flavor. Eosinophils cyclically infiltrate the nonpregnant mouse uterus via estrogen-induced CCL11 expression, but CCL11-deficient mice show no obvious reproductive defects.[167] The numbers and location of mast cells, which have recently been implicated in implantation and spiral artery remodeling in mice, are unclear.[168] Mast cells are detectable in the human endometrium throughout the menstrual cycle, while eosinophils appear only with the onset of menses.[169] Both cell types remain poorly characterized in the human decidua. Their potential role in parturition is discussed in section *Parturition and PTL*.

SYSTEMIC CHANGES TO THE MATERNAL IMMUNE SYSTEM DURING PREGNANCY

In addition to the work described above on uterine immune cells, many studies have used human peripheral blood specimens to characterize the systemic changes to the immune system that occur with pregnancy. These studies have been motivated by desires to gain insights into mechanisms of fetomaternal tolerance and the potentially altered responses of pregnant women to infection and vaccination,[170,171] and, more recently, by the goal of developing blood tests that might be able to predict PTL and other pregnancy complications.[172,173]

Perhaps, the most straightforward interpretation of this work is just how little the maternal immune system is systemically altered during pregnancy. At the cellular level, every alteration so far detected has been at most twofold, and many are far subtler. Moreover, certain parameters have shown remarkable variability between studies—for example, the question of whether blood T cell numbers decline during pregnancy, as suggested by some studies, remains unsettled.[174-177] One consistent finding has been a mild increase (<twofold) in blood neutrophil numbers,[173,174,177,178] which is presumably due to increased granulopoiesis given that serum G-CSF levels are also elevated during pregnancy.[179,180] Plasmacytoid DCs also increase (~twofold) in number.[174,181,182] Observations on the numbers of monocytes and NK cells have been very inconsistent, although there might be some subset-dependent alterations in functionality.[174-177,181-189] There are also slight reductions in Th1 and Th17 cell frequencies in the blood of women in their second and third trimesters, as measured by surface markers as well as IFNγ and TNFα production, but very little evidence of a Th2-skewing, in contradiction to the historical suggestion that pregnancy is a Th2-like phenomenon.[174,182,184,188,190,191] As will be discussed in more detail below (see section *Limits to the Treg Cell Paradigm*), studies on T_{reg} cell frequencies in pregnant women have also yielded inconsistent results,[192] in contrast to the relatively uniform observation that T_{reg} cells systemically expand in mice by midgestation (<twofold), particularly following strain disparate (ie, allogeneic) as opposed to strain identical (ie, syngeneic) mating.[156,193-197]

Although not a universal finding, several studies have documented small reductions in B cell numbers in the maternal blood during the third trimester of human pregnancy, particularly affecting transitional B cells.[174-178,188,198,199] This observation is consistent with work in mice showing that pregnancy levels of estrogen impair B cell development by inhibiting IL-7 production by bone marrow stromal cells.[200-202] Total splenic B cells in mice also decrease in absolute number starting at midgestation, in association with decreased splenic expression of the B cell survival factor BAFF/Blys.[203] In contrast, splenic marginal zone B cells increase in number, as do total B cells in the uterine LN and peritoneal cavity.[203] These increase parallel augmented production of "natural antibodies" of the IgA, IgM, and IgG3 subclasses in both mice and humans[199,204] and together suggest that B cell responses during pregnancy become somewhat more innate-like. Pregnancy also alters the properties of newly produced antibodies. First, there is an increased proportion of IgG with galactosylated and sialylated Fc regions, which are more efficiently transported across the placenta into the fetal bloodstream,[205] and which also happen to exert an anti-inflammatory effector function.[206,207] Second, the proportion of IgG antibodies bearing mannose-rich oligosaccharides attached to the Fd fragment of one Fab chain, which renders them univalent and less effective at mediating antigen precipitation, complement fixation, and clearance, rises from ~10% in nonpregnant humans to 30% to 40% during pregnancy.[208] The frequency of such "asymmetric antibodies" is reduced with recurrent spontaneous abortion (RSA).[209,210]

Together with the aforementioned alterations in T cell responses, these observations suggest that adaptive immunity is impaired or suppressed to some extent during pregnancy, at least quantitatively. This possibility is consistent with analyses of human immune responses to Hepatitis C virus (HCV) and influenza virus, as discussed further in section *Systemic Immunity to Infection and Vaccines During Pregnancy*, with studies in mice showing systemic reductions in the magnitude of adaptive immune responses to live attenuated Zika virus (ZIKV) vaccination,[211] and with observations that antibody titers generated against certain but not all influenza virus vaccine strains are upward of twofold diminished in women vaccinated in the third trimester (even though these titers are still effective at preventing infection).[212] Moreover, women with MS and RA show symptomatic improvement during pregnancy, a phenomenon linked to increased IgG galactosylation/sialylation as well to increased T_{reg} cell functionality (see section *Pregnancy and Maternal Autoimmune Disease*). As the most simple example of a pregnancy-associated systemic immune deficit, the expansion of endogenous ovalbumin (OVA)-specific CD8 T cells following the intravenous injection of soluble chicken egg OVA plus adjuvants is fivefold lower in pregnant versus nonpregnant mice (although still robust in absolute terms).[213] Not all adaptive immune responses are impaired, however. For example, immune monitoring studies have failed to detect differences following primary HCMV infection or following *Listeria monocytogenes*, *Fusobacterium nucleatum*, or lymphocytic choriomeningitis virus (LCMV) infection in mice; similarly, systemic infections other than that by HCV and influenza virus do not show worsened clinical severity during pregnancy. These responses are discussed further in section *Immune Defenses at the Maternal-Fetal Interface*. In addition, baseline frequencies of IFNγ- and granzyme B-producing T cells specific for tetanus toxoid and antigens derived from a variety of pathogens, including HCMV, influenza virus, measles virus, herpes simplex virus-1 (HSV-1), and Epstein-Barr virus, are largely unaltered during human pregnancy.[182] Perhaps, most strikingly, pregnant women can mount robust immune responses to paternal alloantigens expressed by fetal blood cells, with the most important example being Rh(D) antigen, the inciting antigen of Rh disease (see section *Humoral Fetomaternal Tolerance*). Together, these observations suggest that pregnancy generates a select set of systemic "holes" in the adaptive immune system, but the exact nature of these holes, their causes, and why they are relevant to only certain pathogens is unclear.

Proximal and ultimate causes for these alterations also remain incompletely defined, although the dominant steroid hormones of pregnancy, progesterone, and estrogen, play key roles.[214] Serum levels of both hormones are elevated during pregnancy, which in the case of progesterone allows it to activate the broadly expressed nuclear glucocorticoid receptor (GR). This cross-reactivity contributes to the pregnancy-associated expansion of T_{reg} cells in mice mentioned above.[215] To the extent that progesterone acts through its own nuclear receptor (PR) to modulate immune cell behavior, such effects are presumably induced indirectly given that immune cells barely express PR.[216,217] For example, pregnancy-induced thymic involution, a universal phenomenon across species that might contribute toward the reductions in blood T cell numbers sometimes seen in late gestation, requires PR expression by thymic stromal cells.[218] Various immune cell types also express membrane progesterone receptors, but their functions have remained poorly characterized. In contrast to progesterone, nuclear estrogen receptors (ERα and ERβ) are broadly expressed across immune cell types, and estrogen is known to have many direct and indirect immunological effects.[219] Of known relevance to pregnancy, these include the inhibition of B cell lymphopoiesis mentioned above and the fostering of CD4 T cell conversion to T_{reg} cells.[220] Immune modulatory function has also been ascribed to chorionic gonadotropin, a primate-specific protein produced by trophoblasts,[214] as well as to nonhormonal material released by the placenta into maternal blood, including pregnancy-specific glycoproteins and exosomes.[221,222] Systems biology approaches might help shed light on the relative importance of these various factors in altering immune cell behavior; for example, recent work has revealed increased STAT5 signaling in blood T cells during human pregnancy.[173]

MECHANISMS OF FETOMATERNAL TOLERANCE

Introduction to Fetomaternal Tolerance—History and Overview

The conceptus is not genetically identical to the mother yet fails to elicit a traditional allograft rejection response. Pregnancy, therefore, is a striking exception to the self/nonself paradigm of immune tolerance, an exception that was obvious to the original leaders in the field. Accordingly, in 1953, Peter Medawar hypothesized three mechanisms that could ensure survival of the fetoplacental allograft: (1) "antigenic immaturity" of the conceptus, meaning that it does not express rejection-inducing antigens; (2) anatomic separation of the conceptus from the mother; and (3) generalized immunosuppression of the mother.[223]

Research performed over the decades since Medawar proposal has mostly disqualified his three ideas. First, the conceptus is not antigenically "immature" since trophoblasts alone express many protein species that could in theory mediate immune rejection. As will be discussed further below, these include classical MHC class I molecules (albeit to limited deg ree),[149,224-226] nonclassical class Ib molecules with limited polymorphism (ie, HLA-E and –G in humans[226]), trophoblast-specific proteins not otherwise expressed in normal adult tissues such as placenta specific glycoproteins[222] and cancer testes antigens,[227] and lastly ubiquitously expressed minor histocompatibility antigens including H-Y antigen encoded on the Y-chromosome.[228,229]

Second, and as we described in section *Anatomy, Development, and Microbiology*, the conceptus is not walled off from the maternal immune system. Rather, trophoblasts come into contact with decidual immune cells, while endovascular trophoblasts and trophoblasts in the villous tree and labyrinth are bathed in maternal blood. In fact, this latter trophoblast-blood interface, with its massive surface area, is the site where the human placenta releases trophoblast-derived multinucleated "syncytial knots," microvesicles, exosomes, soluble protein, lipids, and nucleic acids directly into the maternal bloodstream, with quantity estimates reaching several grams each day in late gestation.[221,230,231] Analogous processes are likely to occur in mice with trophoblast-derived material also known to reach the uterine-draining LNs in this species by cell-free transport through the uterine lymphatics.[79] Lastly, occult placental hemorrhage, a frequent occurrence in humans, exposes the mother to fetal blood cells.[232] Thus, there is ample opportunity for the maternal immune system to become aware of fetal and placental cells. However, only fetal blood cell exposure induces productive immune responses, as we discuss below.

The third and final idea of Medawar proposal—that is, systemic immunosuppression—must also largely be discounted. Clearly, pregnant women bear little resemblance to nonpregnant transplant recipients who, as a consequence of pharmacologically induced immunosuppression, suffer from opportunistic infections and show worse clinical courses with common infections. Rather, systemic antipathogen defenses and vaccine responses remain strong during gestation, as we will describe in section *Systemic Immunity to Infection and Vaccines During Pregnancy*, while exposure to fetal blood cells leads to productive immunity, as mentioned immediately above. This is not to say that systemic immune alterations are not apparent during pregnancy—we have discussed these already in section *Systemic Changes to the Maternal Immune System During Pregnancy*—but they are not large enough in magnitude to explain fetomaternal tolerance and the ones that affect the adaptive immune system are seemingly quite selective.

In retrospect, a 1979 paper by Thomas Wegmann and colleagues provided prescient insight into the nature of the maternal T cell response to the fetoplacental allograft. Their work showed that maternal T cells in mice are not spontaneously primed to paternal alloantigens during gestation but *are* primed to these antigens when the mice are injected with paternal splenocytes.[233] Such intragestational priming could be induced without any adverse effects on pregnancy outcome. Together, these findings suggest a candidate fetomaternal tolerance paradigm that explains many of the observations thus far in the field. We state this paradigm in the form of three tenets, which, given the unique positioning of trophoblasts as the cells that directly contact maternal blood and tissue, focus on trophoblast antigens rather than paternal alloantigens more generally. Indeed, as will emerge from our discussion below, "paternal alloantigens" becomes less of a useful concept since it glosses over the cellular sources of these antigens, which current data suggest is critical.

Three tenets of fetomaternal tolerance are as follows:

1. Pregnancy does not trigger the generation of T cell effectors specific for trophoblast antigens.
2. Pregnancy does not elicit T cell tolerance to trophoblast antigens.
3. Pregnancy proceeds unharmed even in the presence of trophoblast antigen-specific effector T cells.

It is perhaps not so surprising that Wegmann findings were not universally appreciated at the time of their publication. Other analyses of the responses to antigens expressed by the conceptus in some cases suggested effector T cell priming and in other cases tolerance induction,[234,235] while it was already well-established that productive humoral responses could be mounted to fetal blood group antigens and paternal HLA molecules (see section *Humoral Fetomaternal Tolerance* below). Furthermore, muddying the water was the inferential nature of all immune tolerance studies performed in this era. The field thus awaited direct visualization of antigen-specific T cell responses to defined trophoblast antigens. Initially, this came about in the 1990s with descriptions of systemic T cell phenotypes in pregnant TCR-transgenic female mice bearing concepti that expressed the cognate antigen (paternal strain [ie, allogeneic] H-2K or male H-Y). These experiments unfortunately also yielded a variety of conflicting findings, including no obvious differences,[224,225] clonal deletion, and transient anergy,[236,237] or mild increases in T cell numbers.[238] Their inherent relevance was also called into question with the realization that peripheral tolerance studies on intact TCR-transgenic mice could yield aberrant responses given the exceedingly high T cell precursor frequencies involved.[239]

The more contemporary era in the study of fetomaternal tolerance started in the mid-2000s with the applications of TCR-transgenic T cell adoptive transfers and MHC-peptide tetramer technologies, which together have allowed researchers to directly visualize the antigen-driven responses of maternal T cells when the cells are present at more physiologically relevant, if not endogenous, precursor frequencies. Thus armed, researchers have been able to dissect the precise anatomic and cellular pathways that mediate T cells awareness to trophoblast antigens and to directly explore how T cells with defined specificities respond to trophoblast antigens under controlled experimental conditions. Remarkably, the conclusions of this work are similar to those of Wegmann et al, namely that trophoblast antigens induce neither T cell immunity nor tolerance—an outcome that leads us to refer to them as being "nonimmunogenic"—and that mechanisms also exist to protect the fetus from rejection even if trophoblast antigen-specific maternal T cells do become activated. The sections below (*How Maternal T Cells Come to Engage Trophoblast Antigens* to *Mechanisms of Effector Phase Tolerance to the Conceptus*; see also Fig. 41.3) will discuss the key data that support these conclusions, as well as potential underlying mechanisms. We emphasize that unanswered questions and inconsistencies persist. Critically, a different picture emerges when we consider nontrophoblast antigens, such as those expressed by fetal blood cells, since these antigens can clearly drive class-switched B cell responses during pregnancy, and thus also, by inference, helper T cell responses. These responses are addressed in their own section (see section *Humoral Fetomaternal Tolerance*). We then discuss various examples of fetal loss and placental damage in mice and humans and whether or not they represent true rejection (see section *Are There Any Examples of Placental "Rejection"?*). Lastly, we discuss NK cell tolerance to the fetus (see section *NK Cell Tolerance Toward the Conceptus*).

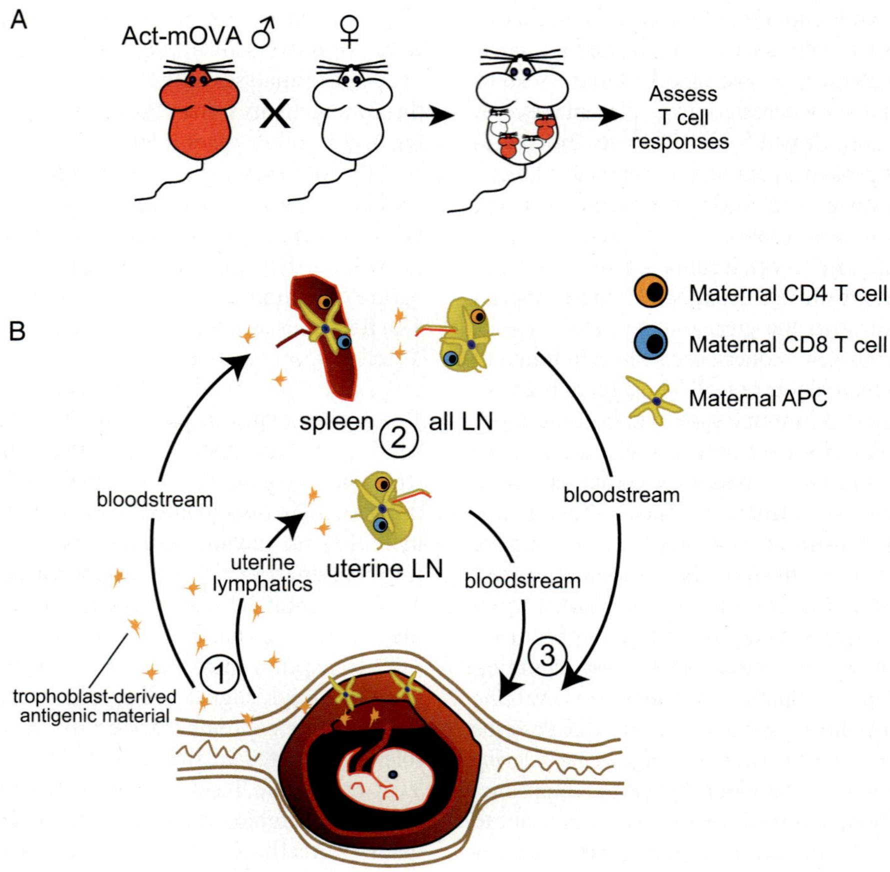

FIG. 41.3. Responses of T cells to trophoblast antigens based upon evidence from the mOVA mouse model. A: The membrane-bound form of ovalbumin (mOVA) model. Use of a transgenically expressed mOVA as a surrogate fetal/placental antigen has allowed researchers to identify the antigen presentation pathways that govern maternal T cell recognition of trophoblast antigens. In this model, nontransgenic females are mated to males that are hemizygous for the Act-mOVA transgene,[240] which directs mOVA expression, under control of the ubiquitously active β-actin promoter, throughout the conceptus. Endovascular trophoblasts, which are bathed in maternal blood, show particularly high mOVA expression.[241] OVA-specific T cell responses can be monitored via the use of OVA-specific peptide-MHC (major histocompatibility complex) tetramers or via the adoptive transfer of OVA-specific T-cell receptor (TCR) transgenic T cells. Since, on average, only 50% of the concepti within a litter express mOVA, antigen-specific fetal loss can be assessed as significant deviations from this Mendelian proportion. **B:** Responses of T cells to mOVA: key points of regulation. (1) Due to the low tissue density of dendritic cells (DCs) and absence of lymphatic vessels within the decidua, maternal T cells become aware of mOVA exclusively as a result of its release from trophoblasts and cell-free transport within the uterine lymphatic vessels and the maternal bloodstream, followed by its uptake and presentation by spleen- and LN-resident antigen-presenting cells (APCs). The conceptus starts releasing mOVA at midgestation (~gd10.5), coincident with the onset of placental perfusion, but mOVA is robustly expressed prior to this time. (2) Presentation of mOVA is "nonimmunogenic" in that it induces an abortive proliferative response and fails to elicit effector T cell priming even when the mice are injected with adjuvants to induce costimulation. The causes of this response are currently incompletely understood but in part involves the action of T_{reg} cells. (3) Even if OVA-specific T cells are activated via experimental means, both mOVA-expressing and nontransgenic concepti remain completely viable up through midgestation because the decidua excludes the cells from the maternal-fetal interface. In late gestation, experimentally activated OVA-specific T cells induce antigen nonspecific fetal demise, likely, through their elaboration of circulating inflammatory mediators that nonspecifically impair placental function. See text for further details.

How Maternal T Cells Come to Engage Trophoblast Antigens

Due to their 10-fold greater prevalence in otherwise naïve hosts, the T cells that recognize intact, nonself MHC molecules collectively pose a much greater threat to organ transplants than the T cells that recognize minor histocompatibility antigens, that is, antigens ingested by host antigen-presenting cells (APCs) and presented by self-MHC molecules. These two pathways have thus been given different names, the direct and indirect allorecognition pathways.[242] A third allorecognition pathway, called the semidirect pathway, involves the transfer of intact donor MHC/peptide complexes to host APCs in a process termed "cross-dressing." This pathway might be of even greater operational relevance than the direct pathway but activates the same large set of T cells.[242]

Accordingly, any discussion of why trophoblasts do not activate maternal T cells needs to start with a consideration of the nature of trophoblast antigens and the relevant pathways of their recognition. As mentioned above, trophoblasts express tissue-specific antigens, ubiquitously expressed minor histocompatibility antigens, and MHC class Ib molecules with limited polymorphism (which in the case of HLA-G are also trophoblast-specific). All of these could provide peptides to activate maternal T cells via the indirect pathway. On the other hand, trophoblasts express a very limited repertoire of classical MHC molecules: they do not express MHC-II whatsoever and show restricted, trophoblast subtype-specific expression of classical MHC-I. In humans, only HLA-C is expressed, and only by EVTs,[226,243,244] while in mice the only trophoblasts that express high MHC-I levels are glycogen cells, a subtype that appears in late gestation and invades interstitially into the decidua, like EVTs.[149,224,225] Endovascular trophoblasts that line remodeled decidual spiral arteries in mice express low levels of H-2K/D,[241] while the trophoblasts of the human villous tree and mouse labyrinth are completely negative for classical MHC molecules.[149,224,225,241] These observations predict less of an opportunity for trophoblasts to trigger T cell activation via the direct and semidirect pathways or for them to serve as targets for directly alloreactive CTLs. Moreover, they predict less T cell activation via the indirect pathway as well, given that MHC molecules themselves provide allopeptides that drive indirect allorecognition.[242] However, it does not entirely remove the threat posed by these pathways: for example, HLA-C mismatch alone compromises graft acceptance in the bone marrow transplant setting.[245]

The extent of direct/semidirect versus indirect allorecognition of trophoblasts has been resolved in mice using a surrogate trophoblast antigen system in which wild-type females are mated to males hemizygous for a transgene that directs ubiquitous expression of a transmembrane form of chicken ovalbumin (mOVA)[241] (Fig. 41.3A). In transgenic concepti, mOVA is highly expressed by trophoblasts of the developing placenta as early as ~gd7.5, and then by the endovascular trophoblasts that are bathed in maternal blood. Beginning at midgestation, when the placenta starts being perfused with maternal blood, adoptively transferred OVA-specific TCR-transgenic CD4 and CD8 T cells start proliferating in the spleen and all lymph nodes, indicating systemic immune recognition of mOVA$^+$ concepti.[241,246-248] Using this system in conjunction with congenic maternal strains unable to present mOVA-derived peptides to the transgenic T cells, it could be demonstrated that T cells engage mOVA exclusively via the indirect allorecognition pathway, that is, following the antigen's uptake by maternal APCs and its presentation in the context of maternal MHC.[241] The lack of a proliferative response in adoptively transferred TCR-transgenic CD4 and CD8 T cells directly alloreactive against paternal MHC molecules confirmed these results,[241,249] and also refuted data from the 1990s using intact TCR-transgenic mice that suggested direct recognition of paternal MHC molecules.[237] Together, these observations also demonstrated the absence of professional APCs of fetal origin that could somehow reach the secondary lymphoid organs of the mother. Rather, since large amounts of mOVA can be detected in maternal blood in both cell-free and exosome-associated forms starting at midgestation,[213,250] immune recognition of the mouse conceptus appears mediated by the hematogenous release and systemic circulation of trophoblast-derived antigenic material, which then gets taken up by spleen and LN-resident APCs (Fig. 41.3B).

While the lack of direct allorecognition of trophoblasts thus greatly reduces the numerical threat to pregnancy posed by maternal T cells and could therefore be considered a "mechanism" of fetomaternal tolerance, T cell activation via the indirect pathway alone can still elicit graft rejection in the transplant setting.[242] In fact, expression of mOVA as a sole minor histocompatibility is sufficient to trigger skin graft rejection in mice.[240] In this kind of MHC-matched context, rejection is thought to be initiated by tissue-resident DCs of host origin that interact with the transplant and then migrate to its draining LN loaded with minor histocompatibility antigens. Remarkably, in the case of the fetoplacental allograft, an analogous process does not take place in mice since, as discussed in section *Immune Cells of the Pregnant Uterus*, DCs are scant within the decidua, and those that reside there are unable to migrate to the uterine-draining LNs due to the decidua's lack of lymphatic vessels.[79] Rather, trophoblast antigens are presented by uterine LN-resident APCs only after their transport in cell-free form via the myometrial lymphatics and bloodstream (Fig. 41.3B). As also discussed in section *Immune Cells of the Pregnant Uterus*, the human decidua contains exceedingly few DCs and might also lack lymphatic vessels.

The Nonimmunogenic Nature of Trophoblast Antigens

Thus far, we have described how the placenta differs from a surgical organ transplant in two critical ways: first, there is an absence of direct and semidirect allorecognition, and second, immunogenic DCs do not survey the maternal-fetal interface. However, we have not yet described the behavior of maternal T cells that do encounter trophoblast antigens within secondary lymphoid organs. Again using the mOVA model antigen system, it was found that OVA-specific CD8 T cells proliferate and alter their activation marker profile but fail to numerically accumulate to a substantial degree (indicating concurrent deletion) and fail to acquire effector functions (IFNγ production and cytotoxicity).[213,241,246,247,250] This nonimmunogenic outcome is perhaps unsurprising given that T cell priming to foreign antigens typically requires the presence of danger signals. However, systemic administration of strong adjuvants such as the TLR3 agonist poly(I:C), even together with agonistic anti-CD40 antibodies to simulate CD4 T cell help, still fails to elicit robust CD8 T cell priming to trophoblast mOVA.[213,241,250] Perhaps, even more remarkably, pregnant females bearing mOVA concepti nonetheless manifest robust CD8 T cell immunity when intravenously injected with soluble OVA (from chicken eggs) plus adjuvants.[213] These results thus mirror those obtained by Wegmann in the 1970s: during pregnancy, CD8 T cells are not primed to trophoblast antigens, but they can be primed to these same antigens when their source is not a trophoblast.

The above findings also demonstrate that exposure to trophoblast antigens does not induce antigen-specific tolerance, a failure that is perhaps even more evident from analyses of the long-term fate of CD8 T cells that encountered trophoblast mOVA. After undergoing the abortive proliferative response mentioned above, these cells are not completely deleted but persist for up to 6 months postpartum and show a surface phenotype indicating prior antigen exposure.[247,250,251] Moreover, mice bearing these cells demonstrate accelerated rejection of mOVA-expressing skin grafts and an increased ability to clear OVA-expressing tumor cells, although the T cells themselves show some defects in cytokine production and cytotoxicity, as well as quantitative impairments in their ability to expand following antigen reengagement.[213,247,250,251] Similar findings have emerged from studies on maternal T cell responses to male H-Y antigen. These studies are less conclusive, given the near impossibility of generating pregnant mice with all-female litters as antigen-specificity negative controls, nor have adoptive transfer studies using H-Y–specific TCR-transgenic T cell been performed. Nonetheless, tetramer staining experiments have revealed expanded (albeit minute) numbers of H-Y–specific CD8 T cells during mouse pregnancy, suggesting antigen exposure and persistence of the cells into the postpartum period, at which point they demonstrate a memory phenotype and the capacity to produce IFNγ and kill male cells *ex vivo*.[252,253] On the other hand, parous mice are partially impaired in their ability to reject male skin grafts.[252] Together with the results from the mOVA system, these data provide evidence that at least some trophoblast antigens do not induce long-term tolerance but rather a quasimemory phenotype. Although much less thoroughly studied, CD4 T cells also display the same kind of response to trophoblast antigens as CD8 T cells—abortive expansion and lack of differentiation into effector cells, with some conversion to T_{reg} cells as described further below.[197,241,254,255]

Due to technical limitations and ethical considerations, there has been much less analysis of how maternal T cells respond to paternal antigens during human pregnancy. In one study, the use of peptide-HLA dextramers revealed expansion of a minute population of functional, H-Y–specific CD8 T cells in the blood of pregnant women bearing male fetuses.[256] In other studies, nonpregnant multiparous women were found to sometimes bear expanded numbers of functional CD8 T cells specific to minor histocompatibility antigens, including H-Y, and to have developed CTL reactivity to cells expressing paternal class I HLA.[252,257-259] However, it needs to be emphasized that priming to these antigens could have resulted from exposure to fetal blood cells given that prenatal placental hemorrhage is a significant feature of human but not mouse pregnancy[232,253,260] and likely the reason antibodies to RBC alloantigens and paternal HLA molecules sometimes develop before delivery (see section *Humoral Fetomaternal Tolerance* below). Antibodies to paternal HLA, in fact, arise in the same women who develop paternal HLA-specific CTLs.[258,259] Tetramer studies that evaluate maternal T cell responses to oncofetal antigens, which are frequently expressed by trophoblasts,[227] might thus provide better insight into true trophoblast antigen-induced responses. Provocatively, exposure to oncofetal antigens during human pregnancy does not significantly raise or lower the risk of subsequent nonhormonally driven cancers,[261] consistent with the aforementioned mouse studies demonstrating the lack of strong outcomes. Of note, direct contact between maternal T cells and fetal leukocytes, which is expected following placental hemorrhage, predicts that a large fraction of the paternal HLA-specific CTLs will be directly alloreactive to paternal HLA, in contrast to the T cells that respond to trophoblast antigens, which occurs solely via the indirect pathway.

T_{reg} Cells and the Attenuation of T Cell Priming to Trophoblast Antigens

Given their general importance in peripheral tolerance, many studies have focused on the potential role of T_{reg} cells in attenuating effector T cell priming to trophoblast antigens. Indeed, the notion that pregnancy could generate dominantly suppressive lymphocytes was originally raised in the 1970s[262] but then fell out of fashion along with the field of "suppressor T cells." In 2004, however, the idea was reignited with the observation of midgestation fetal loss following strain-disparate (ie, allogeneic) mating but not strain-identical (ie, syngeneic) mating when T cell-deficient females were reconstituted with CD25-depleted T cells within 4 days prior to copulation.[193] Similar results soon followed[196,263] along with observations that T_{reg} cell transfer could attenuate fetal loss in a model of early gestation spontaneous abortion.[264,265] T_{reg} cell frequencies were also found to increase in the uterine LN of mice soon after mating, as well as systemically at midgestation, an effect more consistently observed following allogeneic (as opposed to syngeneic) mating.[120,156,193-195,197,266-269]

Importantly, interpretation of these studies in light of subsequent ones using the Foxp3-DTR system, which allows for timed FOXP3$^+$ T_{reg} cell depletion via diphtheria toxin injection, suggested that T_{reg} cells might contribute to pregnancy success in part via paternal antigen-independent effects on implantation and decidualization (see section *Implantation, Decidualization, and Early Pregnancy Failure*). Thus, to better address the role of T_{regs} cells in fetomaternal tolerance, researchers have also interfered with their function after midgestation, that is, when maternal T cells become aware of trophoblast antigens. Strikingly, partial T_{reg} cell depletion at midgestation using the Foxp3-DTR system was found to induce a high rate of fetal loss following allogeneic but not syngeneic mating, an effect that could be prevented by either CD8 T cell depletion or blockade of the CXCR3 chemokine receptor that partially mediates CTL homing to peripheral tissues.[255,270] A similar but much more subtle pattern of late gestation fetal loss was evident in female mice bearing a deletion of the "conserved noncoding sequence 1" (CNS1) enhancer element of the *Foxp3* locus, but this observation was provocative in its own right since the CNS1 element is required to induce *Foxp3* expression in naïve CD4 T cells and thus convert them to "induced" iT_{reg} cells in the periphery.[121] Together, these results suggested that trophoblast antigens induce antigen-specific peripheral iT_{reg} conversion and expansion, which then prevents the generation of effector Th1 cells and CTLs. Indeed, using MHC-II tetramers and a variation of the mOVA mating system in which a peptide, termed 2W1S, has been incorporated into the mOVA construct, it was shown that 2W1S-mOVA antigen release from the placenta induces the expansion of 2W1S-specific iT_{reg} cells as well as the appearance of anergic 2W1S-specific CD4 T cells on a path to iT_{reg} conversion.[255,271] The 2W1S-specific

iT_{reg} cells, moreover, persist postpartum as memory cells and help attenuate immunological threats to second pregnancies.[255] These observations provided a clear antigen-specific example of the aforementioned finding that T_{reg} cells mildly expand at midgestation in allogeneically mated mice. The idea of placental antigen-driven T_{reg} conversion/expansion is also appealing because placental mammals are the only animals bearing the CNS1 element and appear to be experiencing purifying selection in the functional domains of FOXP3 protein.[121,272] Thus, the evolution of iT_{reg} cells appears linked to the evolution of placentation, and thus, the need to prevent fetal rejection.

Limits to the T_{reg} Cell Paradigm

Despite its many attractions, several findings have prevented universal acceptance of the idea that T_{reg} cells are central to fetomaternal tolerance. One set of findings concerns the extent to which T_{regs} actually alter the response of maternal T cells specific for trophoblast antigens. Thus, while Foxp3-DTR–mediated T_{reg} cell depletion induces OVA-specific T cell expansion and effector functions in allogeneically mated mice bearing mOVA concepti, the effect is associated with massive fetal demise and so could be due to the release of large amounts of mOVA from the fetus proper or from nontrophoblastic placental cells.[197] Indeed, when T_{reg} cells are depleted at midgestation using anti-CD25 antibodies, OVA-specific CD8 T cells do not expand even in mice given adjuvants, nor does this treatment induce fetal demise.[196,213] Other findings do not align with standard views on how T_{reg} cells maintain peripheral tolerance. These include the existence of the nontolerant, quasimemory response described above for OVA-specific CD8 T cells persisting into the postpartum period,[247,250] the lack of tolerance induction to H-Y antigen during murine pregnancy,[252,253,273] and the observation that the injection of exogenous OVA plus adjuvants induces OVA-specific effector CD8 T cells in pregnant mice bearing mOVA concepti.[213] In addition, 2W1S-specific T_{regs} do not inhibit the expansion of 2W1S-specific CD4 T effectors or their upregulation of T-bet, but instead, only limit their production of IFNγ.[255] Lastly, the role of T_{reg} cells in human pregnancy is controversial—while some studies have uncovered correlations between pregnancy complications and reduced T_{reg} cell function (see sections *Are There Any Examples of Placental "Rejection"?; Implantation, Decidualization, and Early Pregnancy Failure* and *Spiral Artery Remodeling, IUGR, and Preeclampsia*), it is unclear whether blood T_{reg} cell frequencies increase during normal human gestation (reviewed in Ref. 192).

Thus, we would argue that both the specific functions and ultimate importance of T_{reg} cells in fetomaternal tolerance remain unclear. Indeed, various possibilities have been put forth to explain why Foxp3-DTR–mediated depletion induces fetal loss only in allogeneically mated mice that do not invoke suppression of T cells specific for trophoblast antigens and direct T cell assault on the conceptus.[274] Unfortunately, alternative suggestions to account for the unique T cell response to trophoblast antigens have remained controversial and/or have lacked mechanistic detail.[18,213,248,275] One idea is that trophoblasts render their set of expressed proteins less immunogenic via posttranslational carbohydrate modifications. This possibility has a controversial history[276-278] but is consistent with older work on the immunogenicity of ectopically transplanted trophoblasts, the unique biology of protein glycosylation by trophoblasts, and recent observations on how protein glycosylation alters antigen immunogenicity.[274,279-282] It is also consistent with the aforementioned inability of trophoblast mOVA to prevent CD8 T cell priming to exogenously injected OVA.[213]

Even if their role in determining trophoblast antigen-specific responses turns out to be limited, T_{reg} cells might nonetheless contribute to the other curtailments of adaptive immunity evident during pregnancy and discussed in section *Systemic Changes to the Maternal Immune System During Pregnancy*. Expanded T_{reg} cell function has also been implicated in the increased susceptibility of pregnant mice to *Listeria monocytogenes* and *Salmonella Typhimurium* infection;[197,283] however, the experimental demonstration of this idea is confounded by the fact that the immune-privileged state of the pregnant uterus (see below) allows these same organisms to establish intrauterine infectious reservoirs. Thus, the improvement in bacterial clearance seen following T_{reg} cell depletion might be due to loss of these reservoirs, given that T_{reg} cell depletion also causes fetal demise and resorption of implantation sites. More clearly, expanded T_{reg} cell function during mouse pregnancy prevents the onset of experimental autoimmune encephalomyelitis, an animal model of multiple sclerosis (MS) (see section *Pregnancy and Maternal Autoimmune Disease*).[215] There has also been recent interest in the possible role of IL-10- or IL-35–producing regulatory B cell and B-1a B cell subsets in dampening inflammation during gestation.[284-289] Importantly, the extent to which fetomaternal tolerance relies upon antigen-nonspecific immune suppression, whether induced by T_{reg} cells, regulatory B cells, or any other mechanism, remains unknown.

Mechanisms of Effector Phase Tolerance to the Conceptus

In addition to the obstacles to effector T cell priming outlined above, it is clear that uterus-specific mechanisms also prevent activated T cells from harming the conceptus—that is, the third tenet of fetomaternal tolerance listed at the beginning of this section. The contemporary evidence for existence of such mechanisms is the inability of Th1 cells and CTLs with established trophoblast specificity to induce antigen-specific fetal loss in mice even after they are systemically activated during pregnancy through experimental means.[5,241,273] A similar kind of failure is evident when solid tumors persist despite the successful induction of systemic Th1/CTL responses toward tumor antigens. Importantly, many of the mechanisms of this added layer of protection are expected to apply not only to T cells with specificity toward trophoblast antigens but also to T cells with other specificities. For example, activated T cells generated during pregnancy in response to a nonuterine viral or bacterial infection might negatively impact upon reproductive outcome, if not cause frank fetal demise, if the cells could accumulate at the maternal-fetal interface and produce proinflammatory cytokines there. This is a significant concern given the ability of activated T cells to widely disseminate throughout the body. But, as will be discussed in section *Systemic Immunity to Infection and Vaccines During Pregnancy*, gestational infection by pathogens that do not directly infect the maternal-fetal interface can stimulate robust systemic immune responses without causing pregnancy complications.

One major reason effector Th1 cells and CTLs cannot attack the fetus in mice is that these cells cannot access the decidua from the bloodstream.[5] In contrast, the myometrium and segments of undecidualized endometrium between each implantation site are competent for T cell recruitment (Fig. 41.4). Mechanistically, the decidua excludes Th1 cells and CTLs because DSCs are unable to express the key chemokines that otherwise would recruit these T cell subsets to peripheral tissues. As alluded to in section *Anatomy, Development, and Microbiology*, these include CXCL9, CXCL10, CXCL12, and CXCL16, whose encoding genes are rendered transcriptionally silent in DSCs through the actions of the repressive histone mark H3K27me3.[4] In addition, DSCs are unable to express CCL5, the last of the currently appreciated set of Th1/CTL-attracting chemokines, but this apparently occurs in an H3K27me3-independent fashion.[4,5] It is unknown whether T cell exclusion from the decidua is a feature of human pregnancy, although the aforementioned reduction in T cell tissue densities in the first-trimester decidua compared to the secretory endometrium is suggestive, as is the failure of T cells to accumulate around foci of decidual HCMV infection (see Section *Immune Defenses at the Maternal-Fetal Interface*). There are also indications of H3K27me3-mediated *CXCL10* silencing in human ESCs exposed to chorionic gonadotropin.[290] It is currently unknown whether mechanisms exist to exclude Th17 cells from the decidua, as well as the point in late gestation when (and if) loss of the H3K27me3 mark across the DSC genome might again allow for T cell recruitment.[4]

Even if Th1 cells and CTLs are unable to access the human decidua from the blood, this tissue still hosts many Th1, CD8, Th17, and T_{reg} cells, all with broad functional capabilities (see section *Immune Cells of the Pregnant Uterus*). A large fraction of these cells are likely tissue-resident memory (Trm) cells that had seeded the endometrium prior to pregnancy and that lack specificity for trophoblast antigens. However, given that multiparous women systemically bear memory CD8 T cells (and presumably CD4 T cells) with specificities to paternal HLA and minor histocompatibility antigens such as H-Y (see above), the decidual memory T cell population of multiparous women likely contains cells with these specificities as well. Indeed, fetal alloreactivity and clonal expansion of human decidual T cells have been documented to a limited extent.[20,93,100] Thus, the fact that decidual T cells fail to attack the conceptus under normal circumstances implies yet additional mechanisms of fetomaternal tolerance that operate within the decidua itself, including potentially at the level of direct T cell-trophoblast interactions. The existence of such mechanisms also explains the lack of significant fetal loss in mice in reports in which Trm cells with known trophoblast specificity are expected to populate the uterus.[5,224,273,291-293]

Accordingly, and as guided by human decidual leukocyte phenotypes, many mechanisms of intradecidual fetomaternal tolerance have been proposed over the last few decades. Unfortunately, the experiments that have attempted to functionally evaluate them in mice are difficult to interpret because they were not designed to take into account our more recent appreciation of the limited extent of T cell priming to trophoblast antigens and the inability of effector T cells to access the decidua. For example, the fertility of PD-1- and PD-L1–deficient mice, even when mated in allogeneic fashion,[248] ostensibly rules out the idea that decidual T cells are inactivated by PD-L1 expressed at the maternal-fetal interface.[294] However, this interpretation is complicated by the absence of trophoblast antigen-specific T cell activation in PD-1- and PD-L1–deficient mice,[248] nor would PD-1/PD-L1 deficiency be expected to overcome the inability of blood-borne T cells to access the decidua. Similar concerns apply to the literature that ostensibly rules out IL-10–mediated suppression of decidual T cells.[295] It has also been hard to understand how examples of fetal loss resulting from attempts to

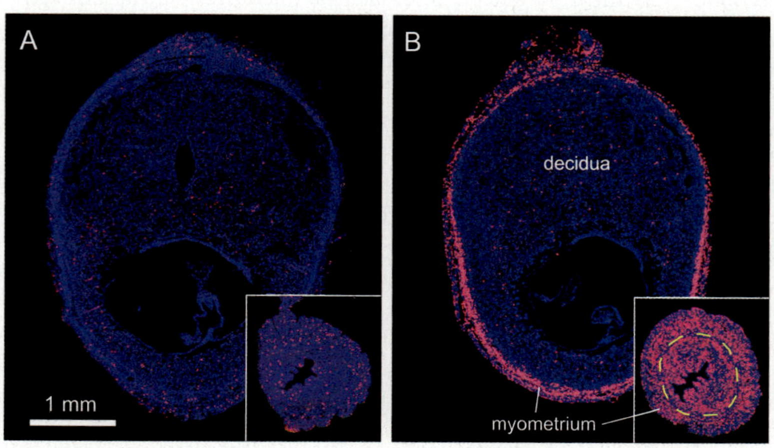

FIG. 41.4. Exclusion of activated Th1 cells and cytotoxic T-lymphocytes (CTLs) from the mouse decidua. Mice were given ovalbumin (OVA)-specific T-cell receptor (TCR) transgenic CD8 T cells and immunized with OVA plus adjuvants 2 to 3 weeks prior to mating. On gd5.5, the mice were either given no additional treatment **(A)** or were rechallenged via intravenous injection with OVA and adjuvants to reactivate the OVA-specific T cells **(B)**. The mice were sacrificed on gd8.5 and implantation sites, and segments of undecidualized uterus between implantation sites (insets) were stained with antibodies to CD3 to visualize all T cells (*red*, with DAPI [4′,6-diamidino-2-phenylindole] counterstain). For the rechallenged mouse, T cells are recruited to the myometrium of implantation sites and to both the myometrium and endometrium (within *yellow dashed line*) of undecidualized uterine segments. In contrast, T cells are excluded from the decidua. (From Nancy P, Tagliani E, Tay CS, et al. Chemokine gene silencing in decidual stromal cells limits T cell access to the maternal-fetal interface. *Science*. 2012;336:1317-1321. Reprinted with permission from AAAS.)

disrupt proposed mechanisms of intradecidual fetomaternal tolerance[159,296] could represent true T cell-mediated rejection without claiming that the experimental manipulation in question also somehow induced T cell priming to trophoblast antigens and allowed for T cell influx into the decidua.

T_{reg} cells have also been proposed to act within the decidua to help maintain fetomaternal tolerance.[297] This idea grew out of the work described above on the detrimental effects of depleting T_{reg} cells during mouse pregnancy and was bolstered by the finding of decreased decidual T_{reg} cells frequencies in patients with spontaneous abortion.[298-302] However, the specific contribution of decidual T_{reg} cells to pregnancy success remains unclear since we currently lack the experimental means to selectively interfere with their function. Thus, fetal loss following T_{reg} ablation, even if trophoblast antigen-driven, could reflect systemic rather than intrauterine processes. This possibility is discussed further below in sections *Are There Any Examples of Placental "Rejection"?* and *Implantation, Decidualization, and Early Pregnancy Failure.* These sections also discuss the possibility that decidual T_{regs} foster pregnancy success not by promoting fetomaternal tolerance per se but rather by maintaining general immune homeostasis at the maternal-fetal interface.

Mechanisms of intradecidual fetomaternal tolerance thus emerge as another open area in reproductive immunology. It even remains unclear whether mouse and human trophoblasts are intrinsically susceptible to being killed by CTLs in an antigen-specific fashion. We anticipate that experiments in mice that start with the seeding of the endometrium with Trm cells bearing defined specificities will provide a rigorous way to probe key questions. Indeed, this approach, although not framed as such, was taken by investigators who tested whether fetomaternal tolerance depended upon the restricted placental expression pattern of MHC-I or trophoblast expression of FasL, a T cell apoptosis-factor implicated in immune privilege at other anatomic sites. In these experiments, virgin mice were immunized with paternal strain allogeneic splenocytes to generate a pool of paternal alloantigen-specific memory T cells, most likely including uterine Trm cells. Following mating, however, there was no evidence of fetal loss even when the placenta was transgenically directed to express high levels of a paternal MHC-I molecule, when the conceptus was rendered FasL-deficient, or when the mother was rendered Fas deficient.[224,292] These observations make a strong case that neither low MHC-I expression by the placenta nor placental FasL expression are key intradecidual mechanisms of fetomaternal tolerance. Considering that tumor cell expression of PD-L1 attenuates the activity of tumor-infiltrating lymphocytes and the high expression of PD-L1 on human syncytiotrophoblasts and EVTs,[20-22] the role of this molecule, in particular, seems ripe for reassessment. Roles for other immune checkpoint molecules are also of significant interest.[303] Intriguingly, there are now three case reports of pregnant women with cancer who received checkpoint blockade reagents in their first trimester, including anti-PD-L1 antibodies, and all three gestations were successful with no evidence of placental inflammation.[304]

Humoral Fetomaternal Tolerance

Currently, there is the presumption that trophoblast antigens are nonimmunogenic for maternal B cells, like they are for T cells, since there is no known example of a mouse or woman who has mounted an antibody response against an antigen exclusively expressed by trophoblasts. Moreover, the inability of trophoblast antigens to prime CD4 T cells predicts that any trophoblast-directed humoral response would generate only lower affinity antibodies that have not undergone affinity maturation. However, our appreciation of the maternal B cells response to trophoblast antigens is far more rudimentary than our appreciation of the T cell response, as experiments to directly address this aspect of pregnancy immunology have not been reported. Thus, our discussion here instead emphasizes the dichotomy we encountered above for T cells, namely that strong responses can be generated to paternal alloantigens—HLA and ubiquitously expressed minor antigens in humans; OVA in the mOVA mouse model—so long as sources of these antigens are *not* trophoblasts. In fact, an even more striking example of this dichotomy comes from the B cell side of the story, as maternal B cells are robustly activated during pregnancy to alloantigens expressed by fetal blood cells. In the most severe example of this situation, sensitization to alloantigens expressed by fetal red blood cells (RBCs) causes the clinical condition of hemolytic disease of the fetus and newborn (HDFN). Prenatal sensitization to the entire set of paternal HLA molecules, including all those not expressed by trophoblasts, also occurs. We will discuss both entities below. Importantly, the antibodies in both cases are class-switched,[305,306] implying the induction of T cell help. Thus, the existence of RBC and HLA alloimmunization provides yet additional evidence that T cells (in this case, CD4 T cells) can be primed to paternal alloantigens during pregnancy, so long as their source is not a trophoblast.

B Cell Responses to Nontrophoblast Paternal Alloantigens

Although many different RBC alloantigens can cause HDFN, most cases involve antibodies against the integral RBC membrane protein Rhesus (Rh)D.[306,307] We will thus discuss "Rh disease" as representative of the larger disease entity. Rh disease occurs when anti-Rh(D) antibodies present in $Rh(D)^-$ pregnant women cross the placenta, bind to the RBCs of a $Rh(D)^+$ fetus, and induce their lysis (Fig. 41.5). Outcomes range from mild neonatal jaundice to heart failure, generalized edema, and fetal or neonatal death. Since most sensitization events occur during delivery of a $Rh(D)^+$ baby, Rh disease is typically a problem for subsequent pregnancies. However, sensitization can also occur from unrecognized or therapeutic abortions or, with greatest relevance to the present discussion, from prenatal placental hemorrhages that are often occult but nonetheless allow antigenically relevant quantities of fetal RBCs to access the maternal bloodstream. Rh(D) alloimmunization and full-blown Rh disease can thus occur during an otherwise normal first pregnancy. Prior to the advent of therapy, fetal and neonatal death due to Rh disease occurred in ~0.1% of all pregnancies and Rh incompatibility had a ~16% chance of inducing alloimmunization.[306,308] Happily, disease incidence was dramatically reduced following the advent of prophylactic therapy using anti-Rh(D) IgG, which successfully blocks alloimmunization when given to $Rh(D)^-$ women carrying a $Rh(D)^+$ fetus,[307] but the global

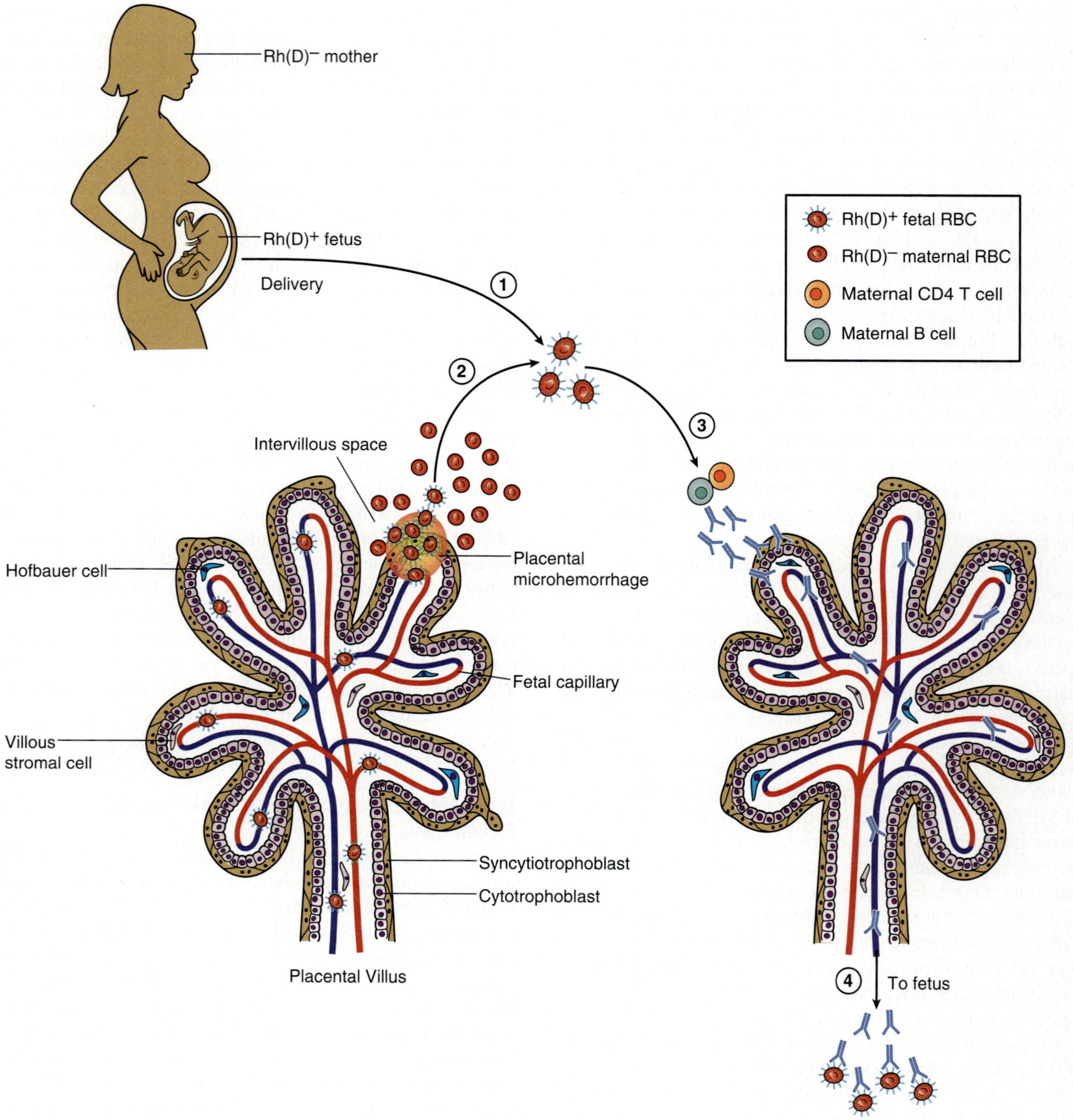

FIG. 41.5. Pathogenesis of Rhesus (Rh) disease. Rh disease occurs in Rh(D)⁻ women bearing Rh(D)⁺ fetuses. Sensitization typically occurs during delivery, when the mother is exposed to fetal blood (1); however, it can also occur as the result of a placental microhemorrhage during an otherwise normal pregnancy (2). In both cases, Rh(D) expressed by fetal red blood cells (RBCs) primes maternal B cells and CD4 T cells, leading to a class-switched antibody response (3). These Rh(D)-specific antibodies are transported across the placenta (when priming occurs during delivery, it is a second pregnancy that is affected), where they reach the fetal circulation and lyse fetal RBCs (4).

burden of Rh disease remains quite high given that this therapy is not employed in many low resource settings. Although rare, alloimmunization can also occur to fetal platelet antigens, causing neonatal alloimmune thrombocytopenia, or to neutrophil antigens, causing fetal neutropenia.[309]

How the administration of anti-Rh(D) IgG prevents immunization to Rh(D) is not entirely clear. Antigen masking that blocks maternal B cell receptor (BCR) recognition of Rh(D) is unlikely a major mechanism since effective suppression occurs with anti-Rh(D) IgG doses far below the point of antigen saturation.[310] Other mechanisms have therefore been proposed, including antibody-mediated clearance of fetal RBCs before an anti-Rh(D) response can be generated, antibody-mediated Rh(D)-antigen-loss ("antigen modulation"), and anergy/tolerance induction via engagement of the inhibitory FcγRIIb receptor expressed by B cells.[311-314]

The generation of antipaternal HLA antibodies provides the second obvious example of how maternal B cell responses appear intact when directed toward nontrophoblast antigens. These antibodies have been known for decades to be present in parous women[315] and are clinically significant since they can mediate postpartum transfusion reactions and preclude a woman from receiving an organ transplant from her male partner.[316] They are also extremely common—a recent high-sensitivity study performed on samples obtained within several months of delivery revealed incidences of 48% after a first pregnancy and 74% after a third or more.[305] Critically, they are directed not only toward HLA-C, the sole classical HLA expressed by human trophoblasts but also toward HLA-A, HLA-B, and HLA-DR.[305,317] Thus, as with RBC alloantigens, they are likely generated as a result fetal blood exposure. Such exposure could occur during delivery or as a result of prenatal placental microhemorrhage, given that anti-HLA antibodies can be detected as early as 28 weeks into a first pregnancy.[318]

Maternal B Cell Responses During Pregnancy

At present, analyses of antigen-driven B cell responses toward any cellular component of the mouse conceptus have been rudimentary. Mice lack Rh(D) and related antigens, but a transgenic model of HDFN has been developed in which the human RBC protein KEL is expressed by fetal RBCs. This leads to the progressive generation of pathogenic antibodies by maternal B cells over multiple pregnancies and can cause the intrauterine death of KEL$^+$ concepti.[260] Unlike with humans, however, sensitization occurs only during delivery and not prenatally since delivery is necessary in mice for fetal RBCs to access the maternal bloodstream.[260] Parous mice, like humans, also bear antipaternal alloantibodies, but their specificity and the circumstances of their generation remain poorly defined. They arise only in some strain-mating combinations and only after multiple pregnancies, similar to the anti-KEL response, but unlike the anti-KEL response they do not appear to cause disease.[319] Lastly, evidence that maternal B cells are even aware of trophoblast antigens during pregnancy comes solely from experiments that employed female mice bearing BCR transgenes specific for H-2K molecules, an approach that is inherently difficult to interpret due to the high precursor frequencies involved. These experiments revealed a 60% to 70% deletion of the transgenic B cells at midgestation in the spleen, blood, and bone marrow following mating with males expressing the cognate MHC molecule but not following syngeneic or third party mating.[320] A lower level of deletion was observed when allo-H-2K expression was restricted via a transgene to trophoblast giant cells and occurred only with developing B cells in the bone marrow and not with mature B cells in the periphery.[321]

Are Anti-trophoblast Antibodies Pathogenic?

Aside from HDFN and the rare alloimmunization conditions mentioned above, the presence of paternal alloantibodies in maternal blood does not typically cause pregnancy complications or even significant placental pathology. This is true not only for the multitude of women bearing anti-HLA antibodies but also for mice hyperimmunized prior to pregnancy with paternal cells so that cytotoxic antipaternal H-2 antibodies are expected to exist.[279] At present, there is no obvious explanation for these divergent outcomes. One possibility is that anti-Rh(D) antibodies have an inherently greater potential to cause disease given the high density of Rh(D) molecules on the RBC plasma membrane. Accordingly, antibodies to other paternal alloantigens (eg, paternal HLA) might also bind fetal and placental cells but cause only clinically insignificant damage. In support of this idea, anti-HLA antibodies may contribute to the pathogenesis of villitis of unknown etiology (VUE), a placental lesion described further below that is typically an incidental histological finding. Similarly, the role of anti-HLA antibodies in RSA appears minor.[322-324] Another possibility, suggested by mouse studies in the 1970s, is that the placenta functions as an immunoabsorbent that prevents antibodies with certain specificities, such as for H-2 molecules, from reaching the fetal circulation and thus causing widespread pathology.[325] Human syncytiotrophoblasts are also thought to be somewhat resistant to complement-mediated lysis since they express high levels of the complement inhibitors CD46, CD55, and CD59.[326,327] In mouse embryos, inactivation of the complement inhibitor Crry causes C3-dependent fetal demise in early gestation associated with complement deposition on trophoblasts.[328] Death occurs without prior immunization, thus revealing a role for Crry in inhibiting antibody-independent complement activation, but it is unknown whether Crry also minimizes damage caused by the antibody-triggered classical pathway. Similar C3-dependent death is evident in embryos lacking the sialic acid activating enzyme CMP-sialic acid synthase, suggesting that protein sialylation by trophoblasts also protects them from complement-mediated lysis.[329]

Importantly, trophoblasts are not completely impervious to the potentially pathogenic effects of antibody binding. This susceptibility is evident from the case of the antiphospholipid antibody syndrome, an autoimmune condition frequently associated with systemic lupus erythematosus (SLE) in which autoantibodies develop against phospholipid-binding proteins expressed on the surfaces of endothelial cells.[330,331] Whether preexisting or arising during pregnancy, these maternal antibodies bind the same proteins on syncytiotrophoblast membranes, where they activate both the complement and coagulation cascades and cause a variety of pregnancy complications, including preeclampsia, IUGR, and miscarriage.[330,331] These complications are thought in part to be due to placental thrombosis given the moderate therapeutic efficacy of antithrombotic agents, but animal models have also implicated roles for complement activation and inflammation.[330,332] Perhaps not surprisingly, given the ability of maternal alloantibodies to cross the placenta and cause HDFN, the transplacental transport of maternal autoantibodies can also be pathogenic. This is best exemplified by women with autoimmune diseases such as SLE or Sjogren syndrome who bear autoantibodies toward the ribonuclear proteins Ro/SSA. These antibodies cross the placenta to cause congenital heart block and other manifestations of neonatal lupus.[333]

Are There Any Examples of Placental "Rejection"?

Given that multiple, redundant mechanisms appear to enforce fetomaternal tolerance, the question arises of whether rejection of the fetus and/or placenta could ever occur, either spontaneously or when only one mechanism at a time is experimentally targeted. Clearly, HDFN is an example of alloimmune-mediated fetal loss, but its pathogenesis involves bypassing the placenta entirely—both as antigen source and target end-organ. Here, in this last section of fetomaternal tolerance, we discuss whether there exists any example of the maternal immune system causing fetal loss because it reacted to the placenta as if it were a classical allograft.

We begin by emphasizing that many immunological manipulations cause fetal loss in mice. Many of these do not show an alloantigen-specific component and instead just illustrate the potentially deleterious effects of nonspecific inflammation on pregnancy outcome. Most classically, the administration of low-dose *lipopolysaccharide* (LPS) to pregnant mice at midgestation rapidly causes isolated fetal resorptions, most likely secondary to focal, nonspecific uteroplacental inflammation and hemorrhage.[334,335] A second example is the complete fetal loss seen in early pregnancy following systemic activation of maternal B cells and DCs via the CD40 signaling pathway. In this case, pregnancy failure is due to the induction of systemic inflammation, and in particular, TNFα production that, in turn, inhibits progesterone production by the ovary.[335] A third example is the fetal loss induced by injection of antiphospholipid antibodies, which as mentioned above induces complement deposition on trophoblasts.[332]

Accordingly, examples of fetal loss that either require an adaptive immune response or that show paternal alloantigen-specificity—typically scored as fetal loss in allogeneic but not syngeneic mating combinations—are more suggestive of true fetal rejection. These examples include systemic T_{reg} cell depletion (see section Treg *Cells and the Attenuation of T Cell Priming to Trophoblast Antigens*), administration of anti-PD-L1- antibodies,[18] administration of the indoleamine 2,3-dioxygenase substrate 5-methyl-tryptophan,[296] and depletion of myeloid-derived suppressor cells.[336] Another example comes from work on the abortion-prone CBA/J(female) × DBA/2(male) mouse mating combination, a model of spontaneous early pregnancy failure that occurs in a pattern suggestive of T cell involvement.[337] However, in some of these cases, fetal loss was no longer evident when genetic approaches were employed,[248,338] and in all of these cases, the extent of maternal T cell accumulation at the maternal-fetal interface was very subtle (if at all evident) and in no way comparable to the level of infiltration seen during acute organ transplant rejection or tumor rejection.[121,270] Thus, even if T cell activation toward trophoblast antigens truly occurred, fetal loss in these examples was ultimately driven by circulating mediators rather than by direct T cell attack on the conceptus. Indeed, a clear example of this kind of situation is the partial, late-gestation fetal loss observed when large numbers of preactivated OVA-specific CD8 T cells are transferred at gd3.5 into mOVA-mated females, as both $mOVA^+$ and $mOVA^-$ concepti are equally affected.[339] A similar set of considerations applies to the view that spontaneous abortion in humans, either isolated or recurrent, represents fetal or placental rejection. Although associated with various alterations in maternal T cells, including decreased T_{reg} cell and increased Th17 cell frequencies within both the blood and uterus, it is not associated with overt T cell infiltration into the uterus.[297,298,301,340-345] Similarly, and as mentioned above, antipaternal HLA antibodies are not a major cause of spontaneous abortion in humans.[322-324]

Villitis of Unknown Etiology

VUE is a histologically identified lesion of the third-trimester placenta characterized by maternal CD8, and to a lesser extent, CD4 T cell infiltration into the chorionic villi in the absence of underlying infection.[346-348] Infiltration usually occurs multifocally, with the affected areas showing features of inflammation including expression of the T cell-attracting chemokines CXCL9, CXCL10, CXCL11, and CCL5, aggregation of activated Hofbauer cells, and upregulation of ICAM-1 by syncytiotrophoblasts.[11,12,349-352] Recent TCR-sequencing data[353] have moreover revealed not only that the infiltrating T cells are clonally expanded but that the specificities of the clones vary between individuals, as would be expected for a response toward a polymorphic target like HLA. In contrast, the expanded clones seen in cases of placental HCMV infection are shared between individuals and are largely absent from cases of VUE. VUE also has a high risk of recurrence with subsequent pregnancies, consistent with immunological memory and a noninfectious etiology, and is strongly associated with the presence of anti-HLA antibodies in maternal serum and the deposition of complement C4d and membrane attack complex components on and around the syncytiotrophoblasts of the inflamed villi.[352,354-356] Together, these observations make a compelling case that VUE represents a true form of placental rejection (Fig. 41.6). Although exact pathogenic mechanisms have not been defined, a possible scenario is that maternal CD4 and CD8 T cells become primed to paternal HLA molecules following maternal exposure to fetal blood cells, all via the canonical allorecognition pathways available to standard blood and organ transplants. Some of the CD4 T cells provide help for a humoral response to paternal HLA, while both CD4 and CD8 T cells, after reaching the villi via the bloodstream and breaching its trophoblastic layers, manifest effector function after being stimulated again by any of the fetal cell types that constitute the villi. Alloantibody binding and complement activation may both trigger villous inflammation and thus initial T cell adhesion to the syncytiotrophoblast layer, as well as fuel ongoing inflammation. Chronic chorioamnionitis and chronic deciduitis, two additional but much more subtle examples of lymphocyte accumulation at the maternal-fetal interface, may also reflect adaptive immune responses toward the conceptus.[346] These lesions are much less well studied but frequently coexist with VUE. Of note, even though villous trophoblasts show signs of apoptosis in the affected areas of VUE,[352] it is unclear whether they are being directly killed by the invading T cells or by complement. It is also possible that trophoblasts die secondarily to damage to the nontrophoblastic cell components of the villi.

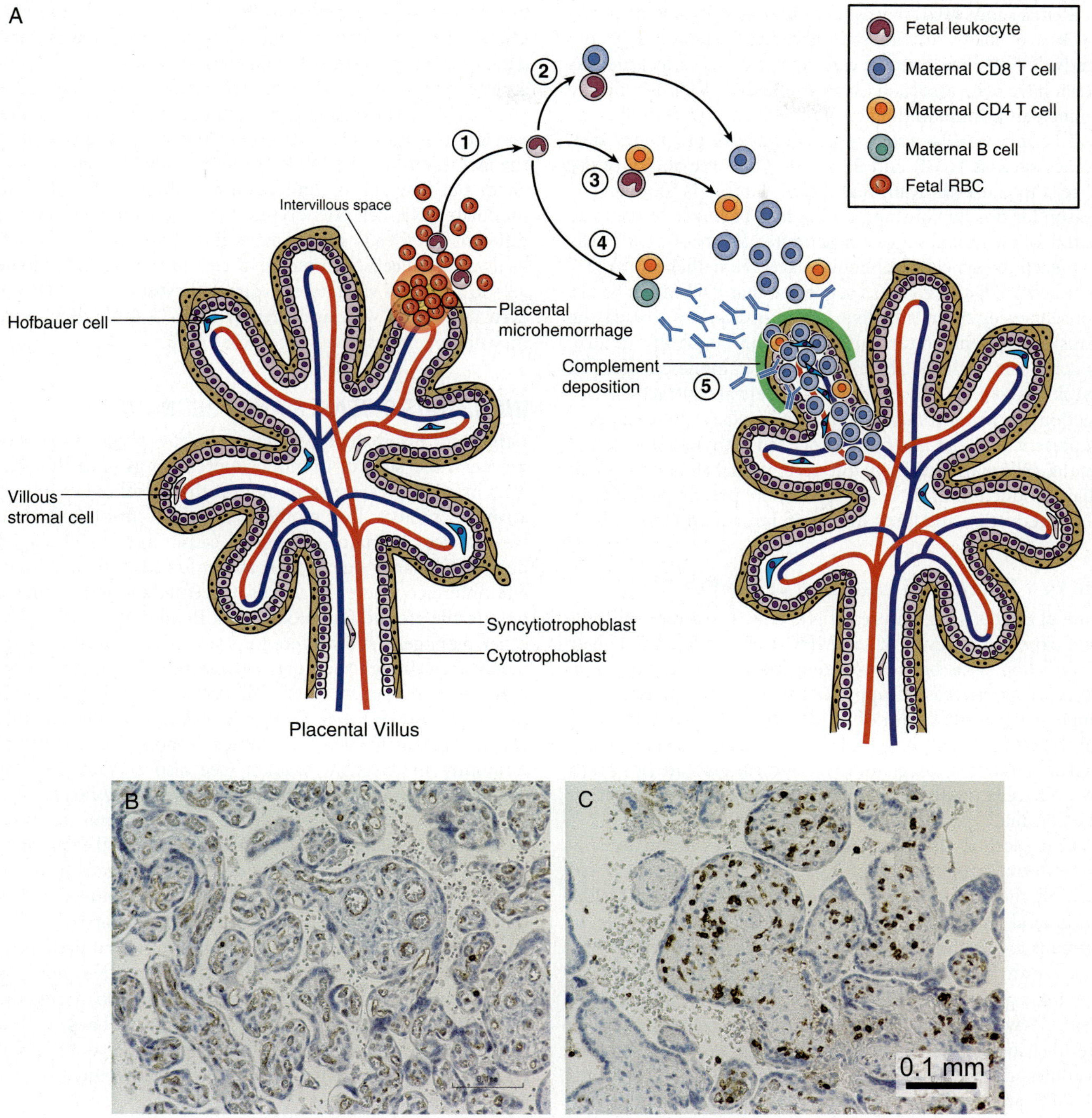

FIG. 41.6. Villitis of unknown etiology (VUE). VUE, a multifocal inflammatory lesion of the placental villi, likely represents a true example of placental "rejection." **A:** Proposed pathogenesis. We speculate that the series of events that cause VUE starts with placental microhemorrhage (or a prior delivery, like with Rh disease) that exposes the mother to fetal white blood cells (1). Priming of maternal T and B cells then proceeds according to the tenets of classical transplantation immunology. Thus, CD8 (2) and CD4 T (3) cells are primarily activated via the direct and semidirect (not depicted) allorecognition pathways, that is, by their T-cell receptor (TCRs), respectively, binding to paternal class I and class II human leukocyte antigen (HLA) molecules expressed by the fetal cells, while B cells are activated to these same molecules with help from CD4 T cells specific for paternal HLA-derived peptides (4). The antipaternal HLA antibodies so produced cross the syncytiotrophoblast layer of the placental villus, whereupon they bind to its nontrophoblastic cellular constituents, namely Hofbauer cells, stromal cells, and endothelial cells. This induces focal complement deposition, inflammation (including chemokine expression), and damage to the syncytium, which in turn fosters recruitment of the activated T cells and thus further damage (5). Minor histocompatibility antigens expressed by the fetal blood cells are also expected to activate additional T and B cells, which will also contribute to inflammation and damage (not depicted). **B and C:** Anti-CD3 immunostaining (*brown*) demonstrating T cell infiltration into villi affected by VUE **(C)** and the absence of T cells from normal villi **(B)**. The slight staining of fetal capillaries is nonspecific.

Remarkably, VUE is very common as it is present in 2% to 34% of all third-trimester placentas, depending on the study.[346,348] Although severe cases that diffusely affect the placenta have been associated with stillbirth,[357] VUE is typically an incidental, mild finding in an otherwise normal pregnancy and is only weakly associated with adverse pregnancy outcomes such as IUGR and PTL.[346,348] Thus, the ability of the T cells to cause clinically significant damage is limited. The reason for this remains unclear but may be linked to the focal nature of the lesion, suggesting a limited ability of CD8 T cells to breach the syncytiotrophoblast layer. Nonetheless, the link between VUE and PTB has suggested that PTB might be the consequences of an adaptive immune response toward the fetus and placenta in certain cases.[346] This possibility, in turn, is consistent with the ability of certain immune cell-derived cytokines to indirectly promote myometrial contractility (see section *Parturition and PTL*), but the idea lacks direct support at present. PTB has also been associated with the activation of fetal T cells with specificity toward maternal alloantigens and their production of IFNγ and TNFα, but precise pathways of T cell activation and labor induction remain undefined.[358]

NK Cell Tolerance Toward the Conceptus

One of the defining features of NK cells is that they kill cells that express low levels of self MHC-I molecules. While best understood as a barrier to bone marrow transplantation, there is an increasing appreciation that this activity also impacts upon the health of solid organ transplants.[359] Thus, given that EVTs only express HLA-C, an additional immunological paradox of pregnancy is that these cells are not killed by dNK cells despite the 2 cell types' close proximity within the decidua.

One piece of the puzzle likely relates to the intrinsically low cytotoxic potential of resting dNK cells and the restricted fashion that cytotoxicity can be induced in them. Unlike peripheral blood NK cells, freshly isolated dNK cells in humans are unable to kill MHC-I–deficient NK cell targets, a defect attributed to an inability of the cells to polarize their cytotoxic granules to the immune synapse.[360-362] Similar lytic defects have been observed for mouse dNK cells.[363] However, freshly isolated human dNK cells polarize their cytotoxic granules, degranulate, and kill cultured DSCs infected with HCMV, an effect that is initiated through ligation of the activating receptors NKG2D and CD94/NKG2C/E.[362,364,365] Increased cytotoxicity can also be induced via ligation of the NKp46 natural cytotoxicity receptor.[72] Thus, the engagement of activating receptors, and not just decreased inhibitory receptor signaling, appears critical for the ability of dNK cells to manifest cytotoxic responses. What induces the dNK cell's state of altered responsiveness *in vivo* is not entirely clear but has been linked to exposure to TGF-β produced by either DSCs or decidual macrophages.[69-71,73]

Superimposed upon this baseline reduction in dNK cytotoxicity, EVTs are intrinsically resistant to dNK cell-mediated lysis even when the cells are HCMV-infected.[362,364,365] This in part may be due to the ability of HLA-G expressed by EVTs to inhibit dNK cells via engagement of the inhibitory receptor LILRB1, as well as to the ability of HLA-E, presenting a HLA-G leader sequence-derived peptide, to inhibit dNK cells via engagement of CD94/NKG2A, another inhibitory receptor.[26] IL-15, which is produced by DSCs, also influences dNK cell cytotoxicity,[362,365,366] but its effects are complex and interpretations of *in vitro* experiments are complicated by the fact that this cytokine is not only a key NK cell survival factor but may act in combinatorial fashion with TGF-β to modulate dNK-cell phenotypes.[69] It is also unclear whether different dNK cells subsets have different capacities to kill virus-infected cells. Of note, dNK cells are not expected to be able to kill EVTs via antibody-directed cytotoxicity (ADCC), even in women with antipaternal HLA-C antibodies, since they do not express the Fc receptor CD16.

INFECTIOUS DISEASES AND PREGNANCY

Earlier, in this chapter, we described how pregnancy is not a generalized state of immunosuppression as once thought. This makes logical sense in light of the evolutionary imperative to protect the mother and fetus from infection. Nevertheless, certain pathogens cause novel pathologies during pregnancy due to their ability to infect the maternal-fetal interface where the uterus, placenta, and fetus provide new, fertile ground for colonization. In addition, a select few other pathogens show altered disease courses during pregnancy even though their organ tropism is the same in pregnant and nonpregnant hosts. We discuss this latter, select set first in section *Systemic Immunity to Infection and Vaccines During Pregnancy* due to the clinical importance of systemic immunity in pregnant women (see also section *Systemic Changes to the Maternal Immune System During Pregnancy*). In section *Immune Defenses at the Maternal-Fetal Interface*, we discuss the pathogens that infect the maternal-fetal interface. In one respect, the ability of these pathogens to do so reflects the diminished nature of cellular immunity within the decidua that is necessary to maintain fetomaternal tolerance. However, it is clear that many of the relevant pathogens employ virulence mechanisms that allow them to exploit this unique immunological environment. We will also discuss the distinctive tissue-specific countermeasures that have evolved to protect the placenta from infection and prevent vertical transmission to the fetus, in particular those mechanisms active within trophoblasts[367-369] (Fig. 41.7).

The systemic and local infections that do occur during pregnancy affect maternal, fetal, and neonatal health in many ways. Congenital infection, which occurs when pathogens gain access to the fetus, has the strongest negative impact on fetal outcome. However, placental infection, and even maternal or fetal inflammation, can impair placental function and thus indirectly affects fetal development. Infection is also the most commonly identified cause of preterm birth (PTB) and can lead to fetal loss. In section *Infection-Mediated Pregnancy Complications*, we review these infection-induced complications and describe treatment strategies. Lastly, in section *The Role of Infection in Infertility and Ectopic Pregnancy*, we discuss pathogens that prevent implantation or lead to ectopic pregnancy (ie, implantation outside of the uterus).

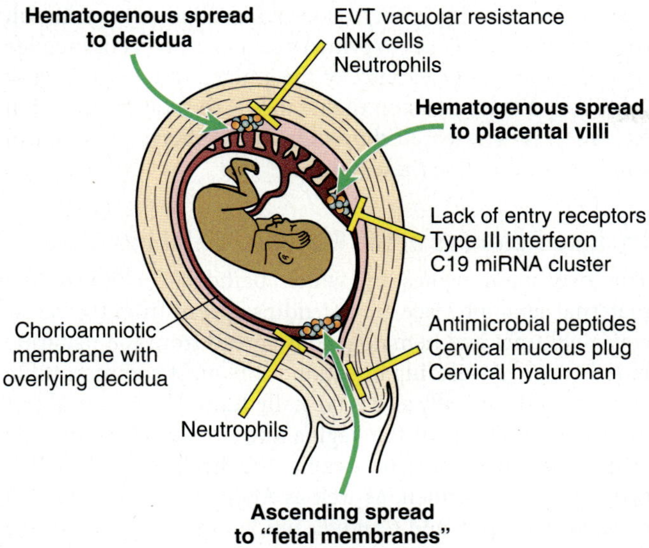

FIG. 41.7. Host defense mechanisms and routes of uterine infection during pregnancy. The three main routes of infection are depicted. For simplicity, infecting organisms are illustrated in a uniform way, but we note that different pathogens take advantage of different routes of spread (see text for details). The host defense lists are only partial and mainly serve to highlight that different locations within the pregnant uterus employ different mechanisms. Of note, infection of the decidua can be a launching point for spread to the placenta. Also, of note, this diagram is representative of the late second trimester up until term gestation, and thus shows the chorioamniotic membrane with overlying decidua capsularis fused with the decidua parietalis (see also Fig. 41.1D).

Systemic Immunity to Infection and Vaccines During Pregnancy

The influence of pregnancy on immunity to pathogens that do not infect the maternal-fetal interface has a controversial history,[370] but the consensus opinion at present is that the effect is subtle except in select cases of chronic HCV and influenza virus, which we discuss in detail here. Chronic HCV infection in nonpregnant, healthy individuals is characterized by stable viremia and CD8 T cell responses that exert selective pressure on the virus, leading to the evolution of viral quasispecies expressing escape mutations in class I HLA epitopes.[371] By contrast, women in the latter half of pregnancy do not show stable viremia but rather a small-to-moderate elevation in viral load and the appearance of quasispecies that lack escape mutations.[372,373] These changes are hypothesized to arise because of a loss of selective pressure, in turn, due to the suppression of virus-specific CD8 T cells. Remarkably, ~30% of women experience dramatic reductions in viral load in the postpartum period, with complete viral clearance occurring in a small number of patients. This phenotypic reversal is associated with the emergence of polyfunctional HCV-specific CD4 effector T cells[374] and shows genetic linkage with polymorphisms in *IFNL3*, *IL-28B*, and certain *HLA-DPB1* alleles, supporting an immune mechanism.[375,376] Provocatively, the third trimester and postpartum waning and waxing of maternal T cell responses to HCV parallel a similar waning and waxing of rheumatoid arthritis (RA) and MS disease severity during the same respective time periods (see section *Developmental Influences of the Maternal Immune System Over Pregnancy*), suggesting common underlying mechanisms.

Fortunately, the third-trimester elevation in HCV viremia does not pose a major threat to maternal health. Influenza, by contrast, can be very severe during the third trimester, particularly for women with comorbidities such as asthma. Indeed, during the influenza pandemic of 1918, physicians noted a 10-fold higher fatality rate among pregnant (27%) versus nonpregnant (2%) women.[377] Increased influenza disease severity has also been observed in pregnant mice, whose lungs show more severe histopathology when compared to nonpregnant controls. These mice also show exaggerated innate responses with increased neutrophil and myeloid cell recruitment and increased production of inflammatory mediators,[378-380] whose adverse effects on the lung are likely compounded by the fact that pregnancy is associated with a reduced capacity for respiratory epithelial cell regeneration.[379] Additional immune deficits in infected pregnant mice include mild (<twofold) reductions in lung concentrations of type I IFN and other proinflammatory cytokines, costimulatory marker expression by lung APCs, lung *Cxcl9* and *Cxcl10* expression, with diminished CD8 T cell recruitment, CD8 T cell killing capacity, and anti-influenza antibody titers.[381] Similarly, influenza-infected pregnant women show reduced anti-influenza IgG2 titers[382] and diminished neutralization capacity of third-trimester serum,[383] while PBMCs from uninfected pregnant women show less types I and III IFN production following exposure to influenza virus particles *in vitro*.[384] To the extent they involve adaptive immune cells, these deficits are in-line with the more general defects in adaptive immunity seen in late pregnancy (see section *Systemic Changes to the Maternal Immune System During Pregnancy*) and likely foster further tissue damage by delaying viral clearance. Importantly, respiratory and cardiovascular adaptations of late pregnancy, including decreased pulmonary volumes, less effective clearing of pulmonary secretions, and increased oxygen demand predispose pregnant women to severe respiratory infection.[385]

Why Are Not All Systemic Infections Worse With Pregnancy?

As we discussed in section *Systemic Changes to the Maternal Immune System During Pregnancy*, work on mice and humans has suggested that systemic adaptive immunity is quantitatively depressed in certain ways during pregnancy. However, only in the cases of HCV and influenza have these deficits been consistently linked to altered disease course. Indeed, a comprehensive study on primary HCMV infection did not reveal appreciable differences in systemic T cell and antibody responses between pregnant and nonpregnant women.[386] Even pathogens such as *Listeria monocytogenes* (*L. monocytogenes*; discussed further below), LCMV, and *Fusobacterium nucleatum (F. nucleatum)*, which all robustly grow in the mouse decidua and then placenta, are effectively cleared from maternal liver, spleen, and lungs in the same hosts.[387-389] At present, there is no clear explanation for this pathogen-specific pattern, but it is important to note that a complete dissection of immune responses to all pathogens over the course of gestation has not been undertaken. Thus, it remains possible that immune responses are modestly impaired to a larger number of pathogens than currently appreciated but that such deficits are clinically relevant in only certain cases.

It also bears emphasis that the nonimmunological physiologic changes of pregnancy might impact upon disease course for many pathogens even if the immune response is unaltered. In particular, respiratory infection might be more severe given the aforementioned changes in pulmonary function seen in late gestation.[390] The virulent and highly transmissible SARS-CoV, MERS-CoV, and SARS-CoV-2 viruses, which cause severe acute respiratory syndrome (SARS), Middle East respiratory syndrome (MERS), and COVID-19, respectively, are thus potentially of great concern for the pregnant population. Unfortunately, our understanding of the pathogenesis of these viruses during pregnancy is still immature and only a limited number of cases have been reported. Vertical transmission appears to be rare for all three, and so far, data suggest adverse pregnancy outcomes for SARS and MERS are likely secondary to severe maternal illness.[391,392] The high case fatality rate for SARS in pregnant women suggests that illness might be more severe for this population. MERS, on the other hand, shows a similar fatality rate in pregnant and nonpregnant women.[393] Fortunately, thus far, COVID-19 does not appear to be more severe for pregnant women;[392] however, it remains to be determined if SARS-CoV-2 elicits adverse pregnancy outcomes.

Vaccines

Consistent with the quantitative depression of systemic immunity seen in late gestation, antibody titers in third trimester pregnant women given certain (but not all) inactivated influenza vaccine strains are somewhat diminished compared to nonpregnant women.[212,394] The elicited titers are nonetheless protective, as pregnant and nonpregnant women achieve similar levels of risk reduction.[395-397] Moreover, vaccination with inactivated influenza vaccine during pregnancy is safe,[398] and thus most guidelines recommend that pregnant women receive inactivated influenza vaccines as well as boosters against tetanus, diphtheria, and pertussis. Novel vaccines against other pregnancy and neonatal pathogens, including GBS, respiratory syncytial virus, HSV, and HCMV are currently under development (as reviewed in Ref. 399). Because of the risk of vertical transmission, live attenuated vaccines are generally not acceptable, with exceptions for endemic viral outbreaks.[400] Finally, because ZIKV is closely related to dengue virus, there is concern that antidengue antibodies could worsen ZIKV pathology via antibody-dependent enhancement (see Chapter 33). Although this has been seen in mouse models,[401] epidemiology studies suggest that preexisting immunity to dengue virus actually lowers the risk of ZIKV infection.[402]

Vaccination not only prevents infection during pregnancy, but the passive transfer of maternal IgG across the placenta to the fetus confers significant protection against influenza, rubella, tetanus, and pertussis in newborns and young infa nts.[403-407] A potential disadvantage of passively transferred maternal immunity is that maternal antibodies may inhibit the responses of infants to primary vaccination via immediate neutralization and clearance of the antigen or engagement of inhibitory Fc receptors on the infant's B cells (as discussed in Ref. 408). New research shows that novel vaccine design can somewhat overcome these barriers.[409]

Immune Defenses at the Maternal-Fetal Interface

The two main routes by which pathogens colonize the maternal-fetal interface are ascending spread from the cervicovaginal tract and hematogenous spread from the maternal bloodstream. Ascending infection is the most common route encountered clinically and is typically caused by bacterial and fungal commensals of the vagina and cervix. These include GBS,[410] which is present among the healthy rectovaginal flora in 30% of women, as well as *Mycoplasma hominis* and *Ureaplasma* spp.[411] and *Candida* spp.[412] Once they transverse the cervix, these organisms infect the abutting fetal membranes. By contrast, a wide variety of bacteria, fungi, protozoa, and viruses reach the maternal-fetal interface via the blood. These include *L. monocytogenes*,[413] *Salmonella typhi*,[414,415] *Treponema palladium*,[416] *F. nucleatum*,[417] *Toxoplasma gondii*,[418] *Trypanosoma cruzi*,[419] Hepatitis E virus,[420] ZIKV, and rubella virus,[421] which all primarily seed and multiply within the decidua before spreading to the placenta via anchoring villi. Notably, most of these pathogens have intracellular life cycles and so access the decidua via trafficking within maternal blood-borne monocytes.[422] In addition, *Plasmodium falciparum* (*P. falciparum*)[423] accumulates at the syncytiotrophoblast layer of the villous tree, while HCMV[421] can directly infect the villous tree as well as the decidua.

Ascending and hematogenous pathogens possess virulence mechanisms that promote seeding and growth at the maternal-fetal interface.[415,424-427] However, vertical transmission to the fetus is the exception rather than the rule, and in the case of viruses occurs only for a subset of quasispecies present in the mother.[428,429] These observations support the existence of uteroplacental defense mechanisms,[430] which are currently thought to be lower reproductive tract defenses for ascending pathogens and the syncytial barrier of the placental villi and innate immunity within the decidua for hematogenous pathogens. We, therefore, focus the following discussion on these defenses and highlight important facets of certain pathogens as they relate to the immune response, but we refer the reader to the comprehensive reviews cited above for more details on the pathogenesis of each organism.

Lower Reproductive Tract Defenses

Physical and molecular barriers prevent vaginal and cervical organisms from traveling upward to the uterus.[431] An important first line of defense is provided by the numerous, short, amphipathic antimicrobial peptides (AMPs) that are produced by cervicovaginal immune and epithelial cells and that interact with and specifically disrupt bacterial membranes.[432] Antiviral protection is provided by epithelial-derived type III IFN.[433] In addition, the "cervical mucous plug" (CMP), which is a dense viscoelastic substance comprised mainly of mucinous glycoproteins produced by secretory cervical glandular cells, sterically excludes bacteria and may have antiviral properties.[434] The mouse CMP, produced upon copulation, falls out within 1-day, while the human CMP, produced upon fertilization, remains over the course of gestation. The CMP also contains immunoglobulins, complement proteins, live phagocytes, the glycosaminoglycan hyaluronan, and various AMPs including lysozyme, lactoferrin, and calprotectin.[434] The importance of hyaluronan, in particular, is evident from the exacerbation of ascending infection-induced PTB in mice lacking cervical hyaluronan[435] and the association between ascending infection in humans and GBS strains that express the virulence factor hyaluronidase, which digests hyaluronan.[436]

Villous Syncytiotrophoblast and Cytotrophoblast Defenses

The villous tree is very resistant to direct bacterial and protozoal infection in large part due to the lack of pathogen entry receptor expression by syncytiotrophoblasts and this layer's dense actin cytoskeleton that physically resists the distortion needed for bacterial entry.[437-440] The syncytium also resists viral infection due to its constitutive expression of type III IFN that contrasts with the inducible nature of type III IFN expression by gastrointestinal and respiratory tract epithelia.[441,442] Type III IFN protects the mature syncytium from ZIKV infection, but primitive, undifferentiated trophoblasts do not produce type III IFN and are thus vulnerable to ZIKV.[443] As a consequence, ZIKV infection during early gestation shows a high risk of vertical transmission and severe congenital infection.[444] The layer of mononuclear cytotrophoblasts that lies immediately beneath the syncytiotrophoblast layer also possesses antiviral defense properties and can resist infection by a variety of DNA and RNA viruses.[440,445,446] Although underlying mechanisms have yet to be defined, this ability is at least in part conferred by expression of C19MC, a primate-specific microRNA cluster.[445] C19MC can also confer protection upon other villous cells via its packaging into exosomes.[446] Importantly, once pathogens breach the cytotrophoblast layer and reach the villous stroma, the main immune cells they will encounter before reaching the fetal circulation are Hofbauer cells. As compared to monocyte-derived macrophages, the M2-polarization of Hofbauer cells and their tendency to elaborate regulatory cytokines like IL-10 and TGF-β is believed to underlie their ability to limit HIV replication.[447] However, it is not clear how well these cells resist the spread of other viruses, as vertical transmission of ZIKV, in particular, is thought to occur through infection of Hofbauer cells.[147]

Despite its many defenses, placental villi are important direct sites of infection for at least two pathogens. First, HCMV can gain access to the villous stroma during a primary infection because the initial set of induced antibodies, while too low avidity to effect viral neutralization, are nonetheless capable of transferring HCMV across the syncytiotrophoblast layer via the neonatal Fc receptor.[448] Second, *P. falciparum* specifically targets syncytiotrophoblasts during malaria infection by altering the expression of a family of *P. falciparum*-encoded clonally variant adhesion proteins (*var* genes, PfEMP1 protein family) in infected RBCs. In nonpregnant hosts, these PfEMP1 variants mediate adhesion of parasitized RBCs to endothelial cells. During pregnancy, however, an atypical variant (VAR2CSA) is induced that specifically binds to chondroitin sulfate A on the syncytial surface and thus allows infected RBCs to sequester within the placenta.[427,449] Unfortunately, the signals that trigger expression of VAR2CSA during pregnancy are not yet known. Multiparous women living in malarial-endemic regions usually do not suffer severe placental malaria because they maintain protective antibodies against VAR2CSA.[450] However, placental malaria in women lacking anti-VAR2CSA antibodies progresses to a chronic stage in which monocyte, macrophage, B cell, and fibrin accumulation in the intervillous space impairs nutrient transport and leads to poor outcomes.[451-453]

Decidual and Extravillous Trophoblast Defenses

Much of our knowledge about the immune response to decidual infection comes from work on *L. monocytogenes.* This facultative intracellular gram-positive bacterium causes a foodborne self-limited gastrointestinal illness in healthy nonpregnant individuals but can cause severe disease in pregnant women because of its ability to colonize and grow rampantly in the decidua. Consistent with this notion, work on *L. monocytogenes* infection during mouse pregnancy indicates that systemic disease is driven by the reseeding of bacteria from the decidua back to other maternal organs.[454]

L. monocytogenes grows rampantly within the decidua because of the decidua's deficits in cellular immunity. Thus, as studied in mice, the infected decidua is unable to accumulate macrophages and CD8 T cells at foci of bacterial growth, nor can it recruit Ly6C^{hi} inflammatory monocytes from the blood.[80,87,389,455,456] These deficits are consistent with the unique limits DSCs place over the trafficking and population dynamics of macrophages, monocytes, and T cells described in sections *Immune Cells of the Pregnant Uterus* and *Mechanisms of Effector Phase Tolerance to the Conceptus* and stand in stark contrast to the ease by which macrophages, monocytes, and T cells accumulate around infectious foci and control bacterial growth in nondecidual tissues, including the undecidualized endometrium.[389,456-458] As a result, the decidual immune response to *L. monocytogenes* falls to neutrophils, dNK cells, and trophoblasts.[389,456,459,460] The function of neutrophils in this regard is canonical (ie, production of reactive oxygen species etc. and NET generation),[461-463] but dNK cells show an unusual ability to kill *L. monocytogenes* organisms within infected EVTs via nanotube-mediated granulysin transfer in a way that keeps the EVTs alive.[460] Moreover, EVTs have an intrinsic capacity to restrain *L. monocytogenes* intravacuolar replication.[464] Neutrophils are also the main inflammatory cell recruited to the human amniotic cavity, fetal membranes, intervillous space, and villi once infection spreads to the placenta.[465] However, despite the ability of *L. monocytogenes* to grow rapidly within the decidua, infection during pregnancy is rather rare, and this correlates with the difficulty faced by the organism in initially colonizing the decidua.[87] Since blood-borne *L. monocytogenes* dissemination occurs within infected mononuclear cells, one important bottleneck is the inability of the decidua to recruit Ly6C^{hi} monocytes.[466]

The decidual immune response to HCMV provides a second example of how pathogens take advantage of the unique immunologic environment of this tissue. In the absence of high anti-HCMV neutralizing antibody titers, the decidua becomes a local reservoir of active infection[467] likely exacerbated by the local constraints on T cells, which would otherwise control the virus.[468] Indeed, the stroma of infected placental villi shows prominent T cell accumulation,[469] unlike areas of decidual HCMV infection.[470] In the absence of T cells, antiviral defense within the decidual falls to other cell types, including dNK cells, which eliminate HCMV-infected decidual fibroblasts.[362,365] dNK and macrophages also protect against decidual HIV-1 infection. Thus, while decidual macrophages and Hofbauer cells are the main local targets of HIV-1 infection,[123] they are less permissive to infection and better at restricting replication when compared to blood monocyte-derived macrophages.[123,447,471] Moreover, dNK cells augment the resistance of decidual macrophages to HIV infection in an IFNγ- and contact-dependent manner.[472] Although not yet addressed directly, it seems unlikely that decidual DCs would be able to initiate adaptive immune responses to decidual pathogens given their low tissue densities in mice and humans and inability to migrate to the uterus-draining LN in mice (see section *Immune Cells of the Pregnant Uterus*). Lastly, the extent to which decidual infection is fostered by local immune suppression, potentially mediated by T_{reg} cells or IL-10 (see section *Immune Cells of the Pregnant Uterus*), is currently unknown.

Infection-Mediated Pregnancy Complications

Effects on Fetal Development

Pathogens that affect fetal development are easily recalled using the "TORCH" pneumonic, which stands for *Toxoplasma gondii,* Others (*Treponema pallidum,* Parvovirus B19, Varicella virus, ZIKV, *P. falciparum*), *Rubella, (H)CMV,* and *HSV*). Infection during early gestation, when organogenesis is occurring, is generally more damaging than infection later in gestation. For instance, ZIKV-induced microcephaly is caused by infection of neural progenitor cells in the developing CNS, which inhibits their growth and differentiation.[473-475] Other important clinical manifestations caused by direct infection include hydrops fetalis elicited by parvovirus B19-mediated destruction of fetal erythroid progenitors,[476] bone deformities in congenital syphilis,[416] and cataracts, deafness, and heart defects in congenital rubella syndrome.[477]

Transplacental transport of pathogen-associated molecular patterns (PAMPs) and the maternal immune response to infection can also affect fetal development even if the fetus is itself not infected. The fetal brain is particularly sensitive to inflammation since inflammatory mediators both damage neurons and activate microglia, which then produce more cytokines, glutamate, and oxidative free radicals. Ascending infection, modeled in mice by administration of either LPS or *E. coli,* is associated with TLR4-dependent fetal brain inflammation,[478,479] while bacterial peptidoglycan alters postnatal behavior after it crosses the placenta and engages TLR2 in the fetal brain.[480] In humans, there is a correlation between severe placental inflammation and brain injury,[481,482] while infection can affect oxygenation and cerebral blood flood to exacerbate neonatal hypoxic-ischemic injury.[483,484] *In utero* inflammation also affects the fetal lungs and immune system.[485]

Intriguingly, epidemiologic and experimental evidence supports a role for pregnancy infections in the development of schizophrenia and autism spectrum disorders. The "neurodevelopmental theory of schizophrenia" emerged from the observation that this disease is more common in babies born during the annual influenza season and from mother infected during the 1957 influenza pandemic.[486,487] Subsequent studies confirmed that maternal infection slightly elevates the risk of offspring developing schizophrenia or autism.[488-490] Proposed mechanisms include (1) transplacental transport of cytokines that directly bind to receptors on brain cells; (2) transplacental transport of antibodies that cross-react with proteins expressed in the brain; (3) pathologic reprogramming of the fetal adrenal axis secondary to maternal stress hormone exposure; and (4) infection-induced placental secretion of neurotransmitters such as serotonin.[491] Mouse experiments show that maternal Th17 cell-derived IL-17a can alter fetal brain development to elicit autism-like behaviors.[492]

Preterm Birth

Infection is the most commonly identified cause of PTB and accounts for an estimated 30% to 40% of cases. The most common pathogenic organisms are aerobic and anaerobic bacteria and fungi that typically colonize the genitourinary tract, oral cavity, and skin—such as GBS, *Mycoplasma hominis, Gardnerella vaginalis, Candida* spp., and *F. nucleatum*[493]. Altered microbiomes or infections at distant sites have also been suggested to increase the risk of PTB due to trafficking of these organisms to the maternal-fetal interface. Thus, there is an association between PTB and rectovaginal colonization with GBS, which is a commensal in 30% of healthy women,[494] between PTB and vaginal dysbiosis,[495] and a controversial association between PTB and oral cavity infections.[496]

The general paradigm for infectious PTB is that organisms gain access to the amniotic cavity and the local inflammatory response culminates in labor via mechanisms described in section *Parturition and PTL.*[497] GBS, for example, invades the amniotic epithelial cells of the fetal membranes via the action of the virulence toxin β-hemolysin/cytolysin and activates NF-κB to drive production of IL-6, IL-1β, and IL-8 leading to neutrophil recruitment.[426] In other cases, it is believed that GBS-derived toxin-containing membrane vesicles promote PTB in the absence of uterine colonization,[498] raising the possibility that infection underlies cases of PTB secondary to so-called "sterile" amniotic fluid inflammation.[499] Concurrent viral infection is thought to lower the threshold for an inflammatory response to bacteria.[500-502] For example, hematogenous infection with gammaherpesvirus 68 and vaginal infection with HSV-2 both sensitize the mouse uterus to bacterial-driven PTB, possibly through upregulation of TLRs, inhibition of type 1 IFN-β production, and promotion of cervical remodeling.[500,502] This "two-hit" model helps to explain why vaginal infections with HSV-2 and *human papillomavirus* (HPV) are risk factors for PTB.

Fetal Loss

Infection underlies 10% to 25% of fetal losses in high-income settings and upward of 50% of losses in low-to-medium income settings.[503] A major immediate antecedent is congenital malformation, which as discussed above is a common consequence of TORCH organism infection. Influenza infection is also associated with higher rates of fetal loss, as well as PTL, fetal growth restriction, and perinatal mortality, but these outcomes are thought to be secondary to systemic inflammation since influenza infection of the maternal-fetal interface and vertical transmission are very rare in mice and humans.[504] In the cases of *L. monocytogenes* and *S. typhimurium* infection in mice, fetal loss is caused by both infection of the maternal-fetal interface as well as by systemic impairments in maternal T_{reg} cells, leading to systemic inflammation.[197] Finally, fetal loss can also occur secondary to disruptions in placental development and hence placental insufficiency. For example, HCMV dysregulates Wnt/β-catenin signaling in trophoblasts, interferes with trophoblast migration, and elicits increased collagen deposition around endothelial cells, thus impairing nutrient/oxygen exchange.[505-508] More generally, viral infections are associated with elevations in type I IFN and IFN-induced transmembrane proteins (IFITMs), whose normal role is to block viral entry into cells. But IFITMs also happen to block cytotrophoblast fusion and therefore syncytiotrophoblast formation, thereby disrupting placental architecture and function. Thus, type I IFN driven fetal loss is observed in WT but not *Ifitm*$^{-/-}$ mice,[509] and type I IFN underlies fetal loss in ZIKV-infection.[510]

Treatment Strategies

The incidence of several congenital infections, such as rubella and varicella, has decreased because of maternal vaccination. Unfortunately, vaccination against all pregnancy pathogens is not feasible, as some are healthy commensals. When administered to GBS-positive mothers, intrapartum antibiotics successfully prevent neonatal sepsis and meningitis and increase the time to delivery in cases of premature preterm rupture of the fetal membranes, a prelude to PTB.[511] Moreover, screening and treatment of pregnant woman during gestation for asymptomatic bacteriuria, bacterial vaginosis, *Trichomonas vaginalis*, and candida reduces the incidence of PTB.[512,513] However, global prophylactic antibiotic treatment of women already in preterm labor does not reduce adverse outcomes,[514] while nonsteroidal anti-inflammatory drugs (NSAIDs), which show efficacy in mouse models of PTB by preventing the generation of procontractile prostaglandins, are contraindicated during human pregnancy because of adverse fetal effects.[515] Newer targeted immune strategies, such as TLR4 antagonists, may be of use since these suppress uterine and placental inflammation in infection-induced PTB mouse models.[516,517]

The Role of Infection in Infertility and Ectopic Pregnancy

Sexually transmitted *Chlamydia trachomatis* and *Neisseria gonorrhea* both target epithelial cells of the female reproductive tract and are both associated with increased risk of infertility and ectopic pregnancy.[518,519] These organisms initially infect the vaginal canal, where they often fail to elicit symptoms. Left untreated, however, they may ascend into the upper reproductive tract, enter the fallopian tubes, and elicit local inflammation and scaring.[520,521] The consequent losses of beating ciliated epithelial cells, epithelial secretion, and muscular tubal contractions are thought to compromise the transit of both male and female gametes to the tube for fertilization, as well as the transit of the fertilized ovum into the uterus. The molecular mechanism for ovum attachment to the tubal epithelial in cases of ectopic pregnancy remains unknown. Infertility is also associated with plasma cell-rich chronic inflammation of the endometrium (ie, chronic endometritis) that likely has an infectious etiology since polymicrobial bacterial species are identified on biopsy, and antibiotics lead to not only resolution of inflammation but also improved fertility outcomes.[522]

DEVELOPMENTAL INFLUENCES OF THE MATERNAL IMMUNE SYSTEM OVER PREGNANCY

In addition to combatting infection, the maternal immune system also contributes to pregnancy outcome by regulating developmental events within the uterus. Here, we describe these contributions and how their disruption leads to pregnancy pathologies in mice and humans.

Implantation, Decidualization, and Early Pregnancy Failure

The maternal immune system first impacts upon pregnancy by influencing the events associated with embryo implantation. These events include the establishment of the receptive uterine state, the attachment reaction itself, and the subsequent generation of a fully decidualized endometrial stroma that can support embryonic and placental development. Importantly, implantation-associated events are primarily controlled by the endocrine system and uterus-intrinsic developmental pathways rather than by the immune system.[3] Indeed, virtually every mouse strain with an immunological deficit is able to become pregnant. For example, *Rag2*$^{-/-}$ *Il2rg*$^{-/-}$, *Csf1*$^{-/-}$ (*op/op*), *Csf3r*$^{-/-}$, and *Flt3l*$^{-/-}$ mice are all overtly fertile, thus ruling out absolute reproductive requirements for lymphocytes, ILCs (including NK cells), macrophages, monocytes, neutrophils, and DCs. However, a few specific immune pathways are required for the establishment of early pregnancy in mice, as will be mentioned. Moreover, implantation in both mice and humans is associated with a transient wave of uterine inflammation that is thought, via ripple effects extending throughout gestation and beyond, to have profound negative consequences if over-exuberant or uncontrolled.[523,524]

Establishment of the receptive uterine state is driven by the ovarian hormones, estrogen and progesterone. In mice, copulation triggers a programmed hypothalamic-pituitary-ovarian response leading to tonically increased progesterone production that is overlaid, 3 days after mating, with a transient surge in estrogen production ("nidatory," ie, implantation estrogen). Together, these hormones induce gene expression changes in the endometrial stroma and epithelium that generate a "window of implantation" lasting for about 1 day. Importantly, nidatory estrogen induces uterine glandular epithelial cells to express leukemia inhibitory factor (LIF), a IL-6 cytokine family member that also fosters subsequent decidualization.[525] Nidatory estrogen also modestly elevates uterine levels of IL-1α/β, IL-6, and TNFα,[526,527] but this follows upon a more full-blown wave of inflammation that is induced by endometrial exposure to non-LPS TLR4 ligands present in semen and that becomes evident immediately after copulation.[528] This earlier wave is characterized by elevated uterine expression of multiple cytokines and chemokines together with endometrial accumulation of macrophages, lymphocytes, and neutrophil s.[143,528-531] In humans, rising progesterone levels in the secretory phase of the menstrual cycle create a similar window of implantation lasting ~2 days, also associated with increased endometrial LIF expression. Although inflammation is not an overt feature of the secretory phase, this period shows mild endometrial IL-1 and IL-6 upregulation together with mild increases in macrophage tissue densities.[48,127,134,136,532-534] These findings have raised the possibility that inflammation may promote implantation, an idea furthered by the observation that endometrial biopsies can increase implantation rates for *in vitro* fertilization patients in proportion to the extent of the induced inflammation.[523,535] Preimplantation inflammation must be under strict control, however, as even subtle but generalized endometrial inflammation (ie, chronic endometritis) is an established cause of repeated implantation failure.[536]

Blastocyst attachment to the uterine luminal epithelium and the immediate response of the epithelium and its underlying stroma induces further inflammation, in this case highly localized to the attachment site. This response cannot be studied in humans due to ethical considerations, but in Rhesus monkeys, it is associated with macrophage accumulation,[537] while in mice, it induces abutting uterine epithelial and stromal cells to express COX-2, which leads to the localized production of inflammatory prostaglandins, in particular, PGE_2 and PGI_2. These prostaglandins promote implantation and are critical for the formation of decidual tissue.[538-540] In addition, ESCs produce IL-11, another IL-6 family member, which drives decidualization in autocrine fashion.[541]

It is not surprising that embryo implantation would induce an inflammatory response in the endometrium, given that the process essentially generates a small wound. What is remarkable, however, is the short-lived nature of the response given that the wound never heals but rather keeps increasing in size as the conceptus grows and trophoblasts invade into the uterine stroma. In fact, the decidua assumes a decidedly noninflammatory state once the initial stages of implantation are complete. In mice, this state is characterized by an inability to recruit T cells and monocytes from the blood, to expand its resident macrophage pool, and to express various proinflammatory genes. We have already described these phenomena when we discussed DSCs, decidual leukocytes, and intrauterine mechanisms of fetomaternal tolerance (see sections *Anatomy, Development, and Microbiology*; *Immune Cells of the Pregnant Uterus* and *Mechanisms of Effector Phase Tolerance to the Conceptus*). Even prior to implantation, the copulation-induced accumulation of macrophages, lymphocytes, and neutrophils evident throughout the mouse uterus recedes within 2 days and does not reemerge in response to nidatory estrogen or the attachment reaction. In humans, the noninflammatory state of the early decidua is characterized by reduced T cell tissue densities and the polarization of resident macrophages to an IL-10-producing M2-like phenotype lacking the M1-like inflammatory features that are instead associated with spontaneous abortion.[94,123,125,127,128,134] In both species, the stereotypical inflammatory reaction to wounding, including the recruitment of neutrophils and monocytes followed by the formation of granulation tissue, is not apparent.

The uterus thus appears to actively transition to a noninflammatory state as it proceeds across the peri-implantation period. The mechanistic basis of this transition remains incompletely understood but is a subject of intense interest given its obvious importance for reproductive success. In mice, one likely mechanism specific to decidual tissue is the targeted accumulation of the repressive histone mark H3K27me3 on select genes within DSCs, which prevents them from expressing CSF-1 and T cell-attracting chemokines (see section *Anatomy, Development, and Microbiology*).[4,5] A second mechanism is IL-10 production by decidual macrophages (see section *Immune Cells of the Pregnant Uterus*), although it should be noted that IL-10-deficient mice are fertile.[126,295,542] Third, T_{reg} cells are thought to suppress inflammation throughout the uterus, with effect starting even before implantation.[524,543] In mice, T_{reg} cells increase in frequency in the uterine-draining LN immediately after copulation as a consequence of uterine exposure to seminal plasma components, and then increase in tissue density in the uterus by gd3.5, presumably as a result of their homing there from the LN.[120,266-269,297,544,545] Most of them are thymus-derived.[545] Their role in promoting early pregnancy is evident from the reproductive deficits caused by their experimental ablation during this period, namely implantation failure and the resorption of otherwise successfully implanted early-stage concepti.[156,193,196,264,267,546] Conversely, T_{reg} cell transfer attenuates early fetal loss in abortion-prone mating combinations.[264,265] In humans, RSA, a condition formally defined by repeated fetal loss prior to 20 weeks gestation for no otherwise identifiable reason is associated with decreased decidual T_{reg} cell and increased decidual Th17 cell frequencies.[297-302,341,342]

Together, these observations suggest that uterine T_{reg} cells tamp down inflammatory responses at multiple points across the peri-implantation period, following copulation and the nidatory estrogen surge (mice), during the secretory phase of the menstrual cycle (humans), and during the attachment reaction and its immediate aftermath. Importantly, uterine

T_{reg} cells, at least in mice, suppress the components of these inflammatory responses that are generated by effector T cells (and possibly B cells) since Rag-deficient mice, lacking both T_{reg} cells and effector T cells, do not show implantation defects or early resorptions. In other words, T_{reg} cells do not support early mouse pregnancy by suppressing inflammation caused by myeloid cells, innate immune cells, the parenchymal components of the uterus, or by complement. A major question thus remains concerning the antigenic specificity of both the T_{reg} cells and their effector T cell targets, including the extent to which either population recognizes paternal alloantigens in semen, semen-specific "tissue" antigens, and/or antigens expressed by the early conceptus. Complex results have come from the many studies that pertain to this issue[120,193,196,264,267-269,544,546]; indeed, it remains possible that many of the effector T cells that cause implantation or early pregnancy failure in mice following T_{reg} cell depletion or in women with implantation failure or RSA are merely uterus-resident T cells with specificities toward tissue antigens or commensal microbes that must be kept in check, as everywhere else in the body, by the constant actions of similarly self- or commensal-specific T_{reg} cells.

Spiral Artery Remodeling, IUGR, and Preeclampsia

Spiral artery remodeling is the next major event in pregnancy under the influence of the maternal immune system (Fig. 41.8). The process takes place late in the first trimester of human pregnancy and commences with apoptosis of the arteries' smooth muscle cells. Trophoblasts (EVTs in humans) then undergo a process termed endovascular invasion whereby they migrate into the vessels' lumens and replace the vessels' endothelial cell linings.[68,548] High resistance, low flow vessels thus transform into low resistance, high flow vessels that are capable of supplying the placenta with the large volumes of blood required for fetal growth and development. Critically, perfusion of the human placenta is inadequate if spiral artery remodeling does not extend deeply enough into the uterus, a situation considered to be the root cause of many important pregnancy complications, including preeclampsia, IUGR, and others of the so-called "great obstetrical syndromes" that manifest themselves in the second and third trimester.[549] Impaired spiral artery remodeling, if severe enough and associated with a general defect in EVT invasion into the decidua, has also been linked to pregnancy loss in the first and early second trimester.[345]

Unlike the case with implantation, where the role of the maternal immune system is less well defined, there is direct evidence that the maternal immune system contributes to spiral artery remodeling. In mice, the process occurs between gd8.5 to gd12.5 and is initiated by dNK cells, which closely associate with uterine spiral arteries. In their absence (eg, in *Rag2*$^{-/-}$ *Il2rg*$^{-/-}$ or *Il15*$^{-/-}$ mice), the vessels retain their smooth muscle cells, display thick walls and narrow lumens, and fail to undergo the hemodynamic adaptations described above.[63,80,81,550] In humans, dNK cells also accumulate around uterine spiral arteries as the vessels shed their smooth muscle coating,[78] with functional evidence for dNK cells driving such shedding based upon immunogenetic studies that link the expectation of dNK cell activation deficits with impaired spiral artery remodeling and its clinical consequences. Understanding these studies requires us to remind the reader that NK cell activation state is determined, in general, by the aggregate level of signaling induced by an NK cell's set of activating and inhibitory receptors (see Chapter 18). In humans, these include activating and inhibitory killer Ig-like receptors (KIRs), which engage class I HLA ligands. Thus, by taking advantage of the polymorphic nature of the KIR locus, which determines the set of KIR molecules that can be expressed by each of the mother's dNK cells, together with the polymorphic nature of HLA-C molecules (and thus KIR ligands) expressed by trophoblasts, researchers were able to link increased incidences of preeclampsia, IUGR, and RSA with combinations of the maternal KIR haplotype and the paternally inherited HLA-C haplotype of the conceptus that predicts lower aggregate dNK cell activation.[77,243,551-554] This kind of combinatorial influence is further supported by data in mice showing that spiral artery remodeling is impaired when paternal MHC-I expression is experimentally manipulated to inhibit maternal dNK cell activation.[555] Presumably, relevant dNK cell activation states are established when dNK cells directly interact with invading trophoblasts, although whether this interaction occurs in areas of incipient vascular remodeling remains controversial.[78,243,556]

How dNK cells actually induce spiral artery remodeling has remained unclear. As alluded to in section *Immune Cells of the Pregnant Uterus*, induction of vascular smooth muscle cell apoptosis in mice is thought to be more a function of the conventional NK cell subset recruited to the decidua starting at gd8.5 rather than the tissue-resident subset.[51] The process has also long been known to depend upon the cells' production of IFNγ although target cell types and downstream events have remained undefined.[63] A recent study has demonstrated an additional requirement for dNK cell production of VEGF-C and its action on decidual endothelial cells.[86] First trimester human dNK cells also disrupt the smooth muscle cell lining of heterologous vessels *in vitro*, similar to what is seen with remodeling spiral arteries *in vivo*, and these effects can be mimicked by purified IFNγ and VEGF-C.[557]

Preeclampsia is the obstetrical complication in humans most classically associated with inadequate spiral artery remodeling. It affects ~5% of all pregnancies and occurs when the underperfused placenta systemically releases substances in the late second and third trimesters that are toxic to maternal endothelial cells.[68,558] These substances include the VEGF decoy receptor soluble Flt1. Ensuing systemic vascular dysfunction causes the two cardinal characteristics of the disease—severe hypertension and proteinuria—but the syndrome can also affect other organ systems and even cause death. By contrast, inadequate spiral artery remodeling does not cause these phenotypes in mice but rather only IUGR, with decreased fetal weights seen following manipulations that either deplete dNK cells or cause deficits in their activation, including alterations to paternal MHC class I expression.[80,86,555,559,560] The phenotypic disparity between mice and

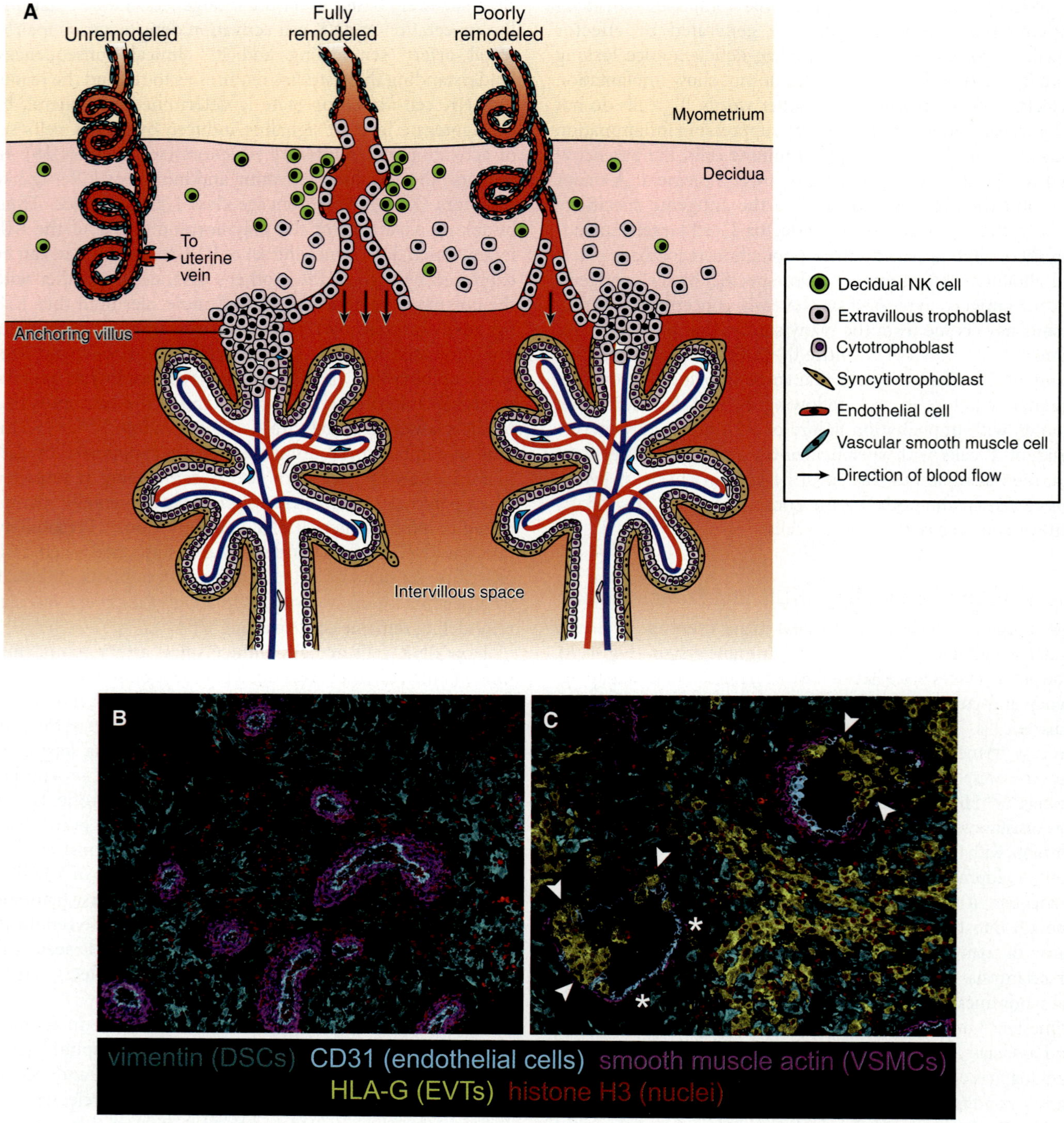

FIG. 41.8. Spiral artery remodeling. A: Schematic of normal and pathological remodeling in the pregnant uterus. The process occurs in the first trimester and early second trimester. The vessel on the left shows a normal spiral artery, replete with full endothelial cell and vascular smooth muscle cell lining, prior to being remodeled. Blood flow returns via a uterine vein (not shown). The vessel in the middle shows a fully remodeled spiral artery. The remodeled vessel has lost its smooth muscle cell lining, and its endothelial cells have been replaced by extravillous trophoblasts (EVTs) that had migrated out from anchoring placental villi into the decidua. The remodeling process extends all the way through the decidua and into the superficial layers of the myometrium, thus greatly reducing vascular resistance and increasing blood flow to the placenta. Decidual natural killer (dNK) cells, which aggregate at areas of ongoing remodeling, are also depicted. EVTs also invade into the uterine veins, thus providing an outlet for placental blood flow (not shown). The vessel on the right is a poorly remodeled spiral artery. It retains much of its own endothelial cell and vascular smooth muscle cell lining, and EVT invasion does not reach the myometrium. As a result, blood flow to the placenta is insufficient to meet the demands of the growing conceptus, which is thought to lead to preeclampsia, intrauterine growth restriction, and other maternal and fetal pathologies in the third trimester. Poor spiral artery remodeling is associated with inadequate dNK cell activation. In mice, the spiral arteries similarly lose their vascular smooth muscle cell lining, which by itself increases blood flow, but trophoblast invasion into the vessels is normally "shallow" and restricted to the decidua. **B and C:** Histology of human spiral arteries before **(B)** and during remodeling **(C)**. Both specimens were from second-trimester elective terminations. Sections were stained with antibodies toward the indicated proteins, and images were acquired

humans might reflect differences in the nature of endovascular invasion between the two species. In humans, endovascular invasion employs EVTs and normally spans the full thickness of the decidua up through the superficial layers of the myometrium. In mice, endovascular invasion employs a subtype of trophoblast giant cell and is much more shallow. Intriguingly, endovascular invasion in rats extends all the way to myometrium and thus is more similar to humans than to mice.[561] In all three species, and again in a species-specific fashion, dNK cells control the extent of endovascular invasion, a function seemingly independent of their function in inducing vascular smooth muscle cell apoptosis. In humans, dNK cells promote EVT migration via their production of various chemokines and cytokines, as visualized *in vitro*,[77,562,563] whereas mouse and rat dNK cells (and more likely the tissue-resident subset in mice) appear to inhibit trophoblast invasion, as evidenced by the over-invasion of the spiral arteries *in vivo* when the cells are absent or functionally deficient.[64,564] Importantly, trophoblast over-invasion is just as problematic for pregnancy as under-invasion since it adheres the placenta too tightly to the uterine wall. This pathology, termed placenta accreta, increta, or percreta, depending upon the extent of over-invasion, can cause severe hemorrhage during delivery. Thus, dNK cells could be considered to be critical for normal pregnancy by determining the overall "set-point" of trophoblast invasion into the uterus.

Macrophages, mast cells, T_{reg} cells, and complement have also been implicated in spiral artery remodeling,[78,121,137,168,524,565,566] but specific pathways have remained elusive. Akin to their contributions to implantation discussed in the prior section, T_{reg} cells might foster the process by dampening excessive inflammation.[567] Consistent with this possibility, systemic inflammation and decreased blood T_{reg} cell frequencies are evident in first-trimester pregnant women destined to develop preeclampsia.[568] Decreased decidual T_{reg} cell frequencies and altered decidual macrophage and APC phenotypes have also been observed in the third trimester in women with preeclampsia, but the causal relevance of these observations to the remodeling events of the first trimester is unclear.[566,569-571] Another idea is that spiral artery remodeling might be impaired if there is an occult placental infection although recent deep analysis of the placenta microbiome argues against this possibility.[78]

Given the clinical importance of preeclampsia and the relative ease of obtaining peripheral blood, many studies have sought to identify systemic immune alterations that would help explain elements of the end-organ damage seen in affected women.[572-574] These studies have uncovered decreased T_{reg} cell frequencies, increased effector T cell frequencies (in particular of Th17 cells), increased proinflammatory cytokine levels, and increased complement and innate immune cell activation.[566,569,572,573,575-577] However, these findings have been inconsistent and relatively subtle; moreover, the positive associations so far uncovered might only reflect additional manifestations of underlying disease processes rather than underlying causes since they virtually all come from analyses of women in their third trimester who already have the disease. Indeed, women in the second trimester either experiencing early-onset preeclampsia or destined to develop preeclampsia in the third trimester do not display decreased T_{reg} cell frequencies.[172,578] End-organ damage in preeclampsia has also been linked to the generation of agonistic autoantibodies toward the angiotensin II type 1 receptor, a receptor that is broadly expressed by multiple cell types, including endothelial cells, vascular smooth muscle cells, and syncytiotrophoblasts.[579] These antibodies are apparent not only in human pregnancies with placental underperfusion but also in a rat model of preeclampsia in which uterine arteries are surgically occluded, thus reducing placental perfusion.[580-582] When injected into nonpregnant mice, the antibodies cause hypertension, and when injected into pregnant mice cause hypertension, kidney and other end-organ damage, and the release of endothelial cell modulators from the placenta.[583] Several inflammatory mediators have been identified that connect activation of the angiotensin II type 1 receptor to placental dysfunction[579]; in fact, systemically induced type 1 inflammation in mice, by itself, can cause an otherwise normal placenta to adversely impact maternal endothelial cell and renal function.[584] Together, these results suggest a complex interplay between placental under-perfusion, immune activation, autoantibody production, and end-organ damage. Interestingly, the antiangiotensin receptor antibodies are natural antibodies produced by B-1a cells but why they are produced in pregnancies with placental under-perfusion remains unknown.[585]

Parturition and PTL

The molecular basis of parturition (ie, the act of giving birth) remains one of the great mysteries of reproductive biology. Not only is it unknown how gestation length is timed, but it is unknown what, exactly, is being timed. Presumably, the approach of term gestation (37 weeks in humans) sets in motion a sequence of events that ultimately causes the uterus, which has otherwise remained in a quiescent, noncontractile state for the entirety of gestation, to abruptly shift toward a state of high contractility. Similarly, the cervix undergoes a "ripening" process that opens it up to allow passage of the conceptus. In recent years, there has been a great deal of interest in how the maternal immune system might contribute toward or even drive these processes.[586] This interest has been further fostered by the clinical problem of PTL. PTL currently has

FIG. 41.8. (Cont'd) using a custom-designed MIBI-TOF mass spectrometer (Ion Path), as previously described.[547] The images were pseudocolored and further manipulated in Adobe Photoshop for instructional purposes. Note that the unremodeled vessels are covered with thick, continuous layers of vascular smooth muscle cells (VSMCs) and have narrow lumens, whereas the vessels undergoing remodeling have largely lost these cells and have wide lumens. VSMC loss is apparent in areas of the vessels that still possess endothelial cells (*asterisks*). EVTs can be seen invading interstitially throughout the decidua and into the vessels (*arrowheads*), where they replace the vessels' endothelial cells. (Courtesy of Shirley Greenbaum, Erin Soon, and Michael Angelo (Stanford University).)

no effective treatment yet affects over 10% of all pregnancies around the world and is showing a steady increase in incidence.[587] PTB, the consequence of PTL, not only increases the risk of neonatal infection and other short-term complications stemming from the immaturity of multiple organ systems but also has long-term sequelae, including neurocognitive, visual and auditory impairments.

The possibility that the immune system is involved in human parturition grew out of the long-recognized importance of prostaglandins in driving parturition across all mammalian species.[588] These molecules, in particular PGE_2 and PGF2α, are produced at high levels by decidual tissues upon labor onset and are potent and direct inducers of myometrial contraction. It has also long been known that ascending infection, and hence inflammation of the fetal membranes with associated neutrophil influx (ie, chorioamnionitis), is a frequent trigger for PTL in women (see section *Infection-Mediated Pregnancy Complications*). This link led to the discovery that amniotic fluid concentrations of inflammatory cytokines typically made by immune cells (IL-1β, IL-6, and TNFα) are frequently elevated not only in women with chorioamnionitis-triggered PTL, as might be expected, but also in women experiencing normal labor.[589,590] These descriptive studies have been robustly extended over the last 20 years to document increased expression of inflammatory chemokines and cytokines by various uterine tissue layers and cells in both term and PTL.[92,140,591-594] A recent single-cell RNA-Seq analysis has suggested that many of the gene expression changes occur in macrophages[57] and indeed decidual macrophages show a small degree of CD80 upregulation with both spontaneous term and PTL onset.[141] At the histological level, monocytes accumulate within the cervix during its ripening,[595] while after labor onset macrophages and neutrophils accumulate within the myometrium and macrophages accumulate in the fetal membranes.[138-140] Lastly, immune cells not only produce agents that are directly contractile for the myometrium (eg, PGE_2, histamine, serotonin)[588,596] but also cytokines (eg, IL-1β, IL-6) can indirectly promote myometrial contractility by a variety of mechanisms, including inducing higher myometrial smooth muscle cell expression of connexin-43, a gap junction protein that promotes cell-cell electrical coupling, as well as higher expression of receptors for both PGF2α and oxytocin, another contraction-inducing hormone that is released by the pituitary gland.[597,598]

Thus, although many nonimmunological processes also contribute to labor induction, the emerging consensus is that natural parturition at term gestation in part involves the activation of uterine immune cells (particularly within the decidua and fetal membranes) and their production of inflammatory mediators. In the case of PTL secondary to chorioamnionitis, which accounts for ~30% to 40% of all PTB,[599] the inflammatory response to the infection presumably feeds into the same inflammatory pathways that contribute toward normal parturition; in the case of "idiopathic" PTL, which accounts for ~45% of PTB,[599] inflammation of the fetal membranes or decidual tissue somehow spontaneously occurs, and this also feeds into the pathways that drive normal parturition. Such spontaneous inflammation might be associated with loss of T_{reg} cell function and the assumption of a more M1-like phenotype in macrophages.[141,600]

Despite this consensus, however, the precise ways that the maternal immune system and uterine inflammation contribute to labor induction remain unclear. The main difficulty in tackling this question lies in the inadequacy of current rodent models. With regards to natural parturition, rodents show features of uterine inflammation with approaching labor and labor onset, and IL-6–deficient mice show a 24 hours parturition delay.[145,601-603] However, parturition timing is unaltered in mice bearing complete or significant losses in T cells, B cells, NK cells, macrophages, monocytes, neutrophils, mast cells, and eosinophils.[130,145,146,559,604,605] Most likely, these negative results stem from the dominant role of the endocrine system in controlling parturition onset in rodents, for which the immediate antecedent to labor induction is cessation of progesterone production by the ovary (ie, "luteolysis").[606] In contrast, both term and preterm parturition in humans occurs without a decline in serum progesterone levels, instead emphasizing the primacy of uterus-intrinsic pathways. The upshot of this dichotomy is that the aforementioned studies that ostensibly rule out various immune cell populations in the control of parturition timing only rule out contributions to events that lie upstream or downstream of luteolysis and not necessarily contributions to the pathways that drive labor induction in humans. That such uterus-intrinsic pathways exist in mice is nonetheless clear from the fact that labor induction still occurs (albeit with delayed onset) when progesterone levels are kept experimentally high,[4] but roles for immune cells in this context have not been explored.

Rodent models have only been modestly more helpful in mechanistically dissecting the immunological and inflammatory pathways that potentially drive PTL. These models are typically based upon either inducing chorioamnionitis via direct uterine infection (typically using *GBS* or *E. coli*, sometimes in combination with local or systemic viral infection as we discussed in section *Infection-Mediated Pregnancy Complications*), or by simulating the inflammatory response to such infection via systemic or intrauterine injection of LPS.[607,608] Here too the difficulty has been that systemic inflammation, whether induced directly or secondary to intrauterine manipulations, is capable of inducing luteolysis and thus labor onset secondary to progesterone withdrawal.[609] Luteolysis and labor induction can occur as quickly as 24 hours following such manipulations, and unless the issue is directly addressed, render many studies difficult to interpret. Thus, studies demonstrating requirements for proinflammatory cytokines, such as IL-1, and the counter-regulatory influences of IL-10, T_{regs}, and B cells[610-613] could be seen more as confirming the general features of inflammation than revealing something specific to the pathways of labor induction. Importantly, studies that involve careful titration of LPS-dosing and direct assessments of serum progesterone levels have provided evidence that intrauterine inflammation per se can cause PTL in rodents.[614] Such results mimic observations with nonhuman primates that direct intra-amniotic injection of proinflammatory mediators such as IL-1β and TNFα induce PTL,[615] and together provide evidence that intrauterine inflammatory processes are sufficient to induce parturition across many species. Key downstream pathways that link inflammation to increased myometrial contractility, including ones that might be targeted to prevent or forestall PTL, however, have yet to be determined.

PREGNANCY AND MATERNAL AUTOIMMUNE DISEASE

Pregnancy dramatically impacts preexisting maternal autoimmune disease, as many autoimmune conditions improve during gestation and then flare in the postpartum period. This pattern is perhaps most famously seen with RA, as ~60% of patients experience clinical improvement during pregnancy while ~50% experience a disease flare at roughly 6 weeks postpartum.[616] Similarly, the relapse rate for MS significantly decreases during pregnancy to below that seen even in patients on currently available treatments but then rises during the postpartum year above the prepregnancy rate.[617] The pattern is not uniform, however, as pregnancy does not have a beneficial effect on SLE.[219] Here, we will discuss these three autoimmune diseases and how their divergent responses might be explained by the intersection of respective pathogenic mechanisms with pregnancy-specific processes.

Rheumatoid Arthritis

One possible explanation for the effect of pregnancy on RA relates to the transient, pregnancy-induced increase in decorated (ie, highly galactosylated and sialylated) IgG species that possess less proinflammatory and more anti-inflammatory effector functions than their undecorated counterparts (see section *Systemic Changes to the Maternal Immune System During Pregnancy*). These undecorated counterparts, however, are known to increase in frequency in RA patients[618] (see Chapter 46) and have been implicated as a causal determinant of RA disease severity in mice.[619] Thus, pregnancy-induced IgG modifications run counter to the requirements for strong disease induction. Indeed, the proportion of highly galactosylated and sialylated IgG1 and IgG2 positively correlates with RA amelioration during pregnancy,[620,621] while the postpartum rise in agalactosylated IgG occurs in synchrony with the postpartum disease flare.[622]

Expanded T_{reg} cell function during pregnancy, as discussed in section *Treg Cells and the Attenuation of T Cell Priming to Trophoblast Antigens*, provides a second potential explanation for RA amelioration during pregnancy. This possibility emerged from the observation that disease activity inversely correlates with CD4⁺ CD25⁺ T cell frequencies in the blood of pregnant RA patients,[623] taken together with studies on the collagen-induced arthritis (CIA) mouse model of RA, in which T_{reg} cells are established suppressors of synovial inflammation.[624] This model recapitulates the human pattern of disease (ie, amelioration during pregnancy and increased severity postpartum)[625,626] and shows greater disease improvement with lower proinflammatory blood cytokines following allogeneic versus syngeneic mating.[627,628] Greater disease improvement thus parallels the greater expansion of T_{reg} cells in allogeneic pregnancies that we described in section *T_{reg} Cells and the Attenuation of T Cell Priming to Trophoblast Antigens*. In humans, the degree of mother/fetus HLA class II disparity also correlates with disease amelioration, thus somewhat paralleling the findings in mice, and further suggesting the involvement of a maternal immune response to paternal alloantigens.[629,630] Intriguingly, $CD4^+$ $CD25^+$ T cell transfer from pregnant mice into nonpregnant recipients decreases disease severity but only if the cell donors were collagen-immunized.[631] This observation supports a role for antigen-specific T_{reg} cells.

Mechanistically, pregnancy hormones and/or placenta-derived soluble factors appear as upstream mediators of pregnancy-associated RA amelioration. Thus, pregnancy-like doses of estrogen and progesterone attenuate disease severity in the Zap70-mutant ("SKG" strain) mouse model of arthritis,[632] while exogenous estrogen protects against disease onset and the postpartum flare in the CIA model.[626,633] Moreover, estrogens promote IgG galactosylation in nonpregnant women and men.[634] Unfortunately, however, estrogen induces only minimal improvement in nonpregnant RA patients.[635] Other studies demonstrate decreased CIA severity in rats injected with placental supernatant,[636] while treatment with human pregnancy-specific glycoprotein 1a decreases CIA severity in mice with an associated decrease in systemic proinflammatory cytokines and expansion of splenic T_{reg} cells.[637] Finally, less common murine arthritis models, such as proteoglycan (PG)-induced and pristane-induced arthritis, also show improvement during pregnancy.[638,639] However, PG arthritis is unlike the other models in that it is independent of IL-17 and does not flare in the postpartum period, thus indirectly suggesting a role for Th17 cells in the postpartum flare.[639,640]

Multiple Sclerosis

In contrast to RA, efforts to understand why MS remits during pregnancy have focused more exclusively on active T cell suppression. Accordingly, TCR sequencing has revealed that the MS patient repertoire becomes less dominated by antigen-experienced (and possibly CNS antigen-reactive) $CD45RO^+$ CD4 and CD8 clones during the third trimester and that clonality reemerges in the postpartum period.[641] Additionally, decreased disease activity in pregnancy correlates with fewer IFNγ-producing CD4 cells and decreased inflammation-related transcripts in PBMCs.[642,643] Importantly, the experimental autoimmune encephalitis (EAE) animal model of MS (see Chapter 46 for basic details) recapitulates much of the human pregnancy phenotype, namely a delay in disease onset with decreased disease severity in animals administered CNS antigens before or soon after mating, but then severe disease in the postpartum period.[215,644] Using the EAE model in combination with mice bearing T cell-specific deficiencies in the GR, it was found that EAE protection requires the pregnancy-induced expansion/conversion of T_{reg} cells (see also section *Systemic Changes to the Maternal Immune System During Pregnancy*), in turn, the result of GR activation within T cells by pregnancy levels of progesterone.[215]

Although its relevance to disease improvement during pregnancy has yet to be directly established, estrogen likely provides additional protection since the administration of pregnancy-like doses of 17β-estradiol (the form of estrogen produced by the ovary) ameliorates EAE severity in nonpregnant mice.[219,633] The effect requires B cells (likely B_{reg} cells), PD-L1, and direct activation of ERα in CD4 T cells to inhibit Th1 and Th17 differentiation.[645,646] Estrogen also promotes T_{reg} cell differentiation, but this effect appears not to be required for EAE amelioration.[220,647] Lastly, ERα signaling in astrocytes is necessary for protection during the effector phase of EAE.[648] Consistent with estrogen's many effects, estriol, an estrogen that is uniquely produced by the conceptus, shows promise as an MS treatment in an early clinical study.[649]

EAE protection in nonpregnant mice can also be elicited by pregnancy-like doses of other factors, including glucocorticoids,[650] and conceptus-derived soluble factors;[644,651,652] however, their relative importance is not yet known. Additional intriguing but controversial clinical observations, such as whether breastfeeding decreases the risk of postpartum relapse[653] and whether parity decreases the risk of an initial MS diagnosis or severe disease later in life,[654] remain unconfirmed.

Systemic Lupus Erythematosis

In contrast to RA and MS, SLE does not show clinical improvement during pregnancy. Instead, pregnancy is a weak determinant of SLE disease flare, while having SLE reciprocally increases rates of pregnancy complications, including preeclampsia, PTB, fetal loss, and the antiphospholipid antibody syndrome described in section *Humoral Fetomaternal Tolerance*.[655] Importantly, estrogens are established positive modulators of SLE risk, and multiple studies have shown that estrogen enhances the type I IFN response (as reviewed in Ref. 656) and autoantibody production by B cells,[657] which are central to SLE pathogenesis. These effects likely explain why SLE patients do not show clinical improvement during pregnancy, unlike patients with RA or MS.[219]

THE LONG-TERM EFFECTS OF PREGNANCY ON THE IMMUNOLOGICAL HEALTH OF THE OFFSPRING

As discussed in section *Mechanisms of Fetomaternal Tolerance*, the maternal immune system responds to fetal blood cell alloantigens in an immunogenic fashion and to trophoblast antigens in a nonimmunogenic fashion. In contrast, tolerance is uniformly induced when the fetal immune system encounters noninherited maternal antigens (NIMAs). This latter effect underlies the improved immunologic acceptance of NIMA-matched organ grafts in humans[658-660] and mice[661-663] as well as the partial tolerance to Rh(D) antigen that is exhibited by the $Rh(D)^-$ offspring of $Rh(D)^+$ mothers.[664]

How fetal and neonatal lymphocytes come to engage NIMAs is not fully understood. Tolerance in murine systems appears to require both *in utero* and breast milk exposure.[663,665,666] In turn, *in utero* exposure involves the transplacental transport of both soluble maternal protein and rare maternal cells,[667-670] while breast milk exposure allows maternal protein and cells to access the neonatal gut.[671,672] Because they can live for long periods of time (a state known as "maternal microchimerism") and thus serve as persistent sources of antigens, the maternal cells that have crossed into the offspring appear more important for inducing tolerance to NIMAs. Mechanistically, NIMA exposure elicits antigen-specific T_{reg} cells in human fetuses[673] and neonatal mice,[661,662,666] as well as deletional tolerance of alloreactive T and B cells in neonatal mice.[663,674] Furthermore, the fact that 50% of "highly sensitized" human patients fail to develop antibodies against noninherited maternal HLA[675] suggests that B cell tolerance to NIMAs occurs in humans as well.

Investigators have also addressed the reciprocal situation, that is, whether the long-term persistence of fetal cells in the mother ("fetal microchimerism") influences the maternal immune system.[676] In particular, microchimeric fetal cells have been hypothesized to promote autoimmune diseases such as systemic sclerosis that arise in postreproductive years.[677,678] As of now, however, pregnancy is not an established risk factor for systemic sclerosis, and the biological relevance of microchimeric fetal cells to gravid women remains unclear.[679,680]

Importantly, the phenomenon of NIMA-induced tolerance adds to the ways researchers are beginning to appreciate how events during pregnancy influence the long-term immunological health of the offspring. Clearly, these also include the protection afforded the newborn as the result of transplacental maternal antibody transport and the fact that the offspring's first encounter with pathogens might occur during pregnancy.[681] In addition, recent work suggests that the occurrence of maternal allergic disease during pregnancy influences the offspring's risk of developing allergies.[682,683] Most generally, maternal health affects the developing fetus in its entirety, and this notion forms the basis of the "fetal/developmental origins of adult disease hypothesis," also called the "Barker hypothesis," for its initial proponent.[684] The extent to which such long-term effects, which include adult-onset hypertension and ischemic heart disease, might involve long-term alterations to the immune system instigated *in utero* is currently unknown.

ACKNOWLEDGMENTS

We thank Drs. Francesco Colucci and Dorothy Sojka for valuable insights, and members of the Erlebacher lab past and present for many useful discussions. Work in the Erlebacher lab is currently supported by grants from the NIH (R01AI143187, R01AI150191), the March of Dimes Foundation (#6-FY18-798) and the Burroughs Wellcome Fund (2019 Preterm Birth Research Grant).

REFERENCES

1. Moffett A, Loke C. Immunology of placentation in eutherian mammals. *Nat Rev Immunol.* 2006;6:584-594.
2. Maltepe E, Bakardjiev AI, Fisher SJ. The placenta: transcriptional, epigenetic, and physiological integration during development. *J Clin Invest.* 2010;120:1016-1025.
3. Cha J, Sun X, Dey SK. Mechanisms of implantation: strategies for successful pregnancy. *Nat Med.* 2012;18:1754-1767.
4. Nancy P, Siewiera J, Rizzuto G, et al. H3K27me3 dynamics dictate evolving uterine states in pregnancy and parturition. *J Clin Invest.* 2018;128:233-247.
5. Nancy P, Tagliani E, Tay CS, et al. Chemokine gene silencing in decidual stromal cells limits T cell access to the maternal-fetal interface. *Science.* 2012;336:1317-1321.
6. Tagliani E, Shi C, Nancy P, et al. Coordinate regulation of tissue macrophage and dendritic cell population dynamics by CSF-1. *J Exp Med.* 2011;208:1901-1916.
7. McConaha ME, Eckstrum K, An J, et al. Microarray assessment of the influence of the conceptus on gene expression in the mouse uterus during decidualization. *Reproduction.* 2011;141:511-527.
8. Katoh N, Kuroda K, Tomikawa J, et al. Reciprocal changes of H3K27ac and H3K27me3 at the promoter regions of the critical genes for endometrial decidualization. *Epigenomics.* 2018;10:1243-1257.
9. Maubert B, Guilbert LJ, Deloron P. Cytoadherence of *Plasmodium falciparum* to intercellular adhesion molecule 1 and chondroitin-4-sulfate expressed by the syncytiotrophoblast in the human placenta. *Infect Immun.* 1997;65:1251-1257.
10. Xiao J, Garcia-Lloret M, Winkler-Lowen B, et al. ICAM-1-mediated adhesion of peripheral blood monocytes to the maternal surface of placental syncytiotrophoblasts: implications for placental villitis. *Am J Pathol.* 1997;150:1845-1860.
11. Labarrere CA, Ortiz MA, Sosa MJ, et al. Syncytiotrophoblast intercellular adhesion molecule-1 expression in placental villitis of unknown cause. *Am J Obstet Gynecol.* 2005;193:483-488.
12. Juliano PB, Blotta MH, Altemani AM. ICAM-1 is overexpressed by villous trophoblasts in placentitis. *Placenta.* 2006;27:750-757.
13. Madigan J, Freeman DJ, Menzies F, et al. Chemokine scavenger D6 is expressed by trophoblasts and aids the survival of mouse embryos transferred into allogeneic recipients. *J Immunol.* 2010;184:3202-3212.
14. Martinez de la Torre Y, Buracchi C, Borroni EM, et al. Protection against inflammation-and autoantibody-caused fetal loss by the chemokine decoy receptor D6. *Proc Natl Acad Sci U S A.* 2007;104:2319-2324.
15. Simister NE, Mostov KE. An Fc receptor structurally related to MHC class I antigens. *Nature.* 1989;337:184-187.
16. Roopenian DC, Akilesh S. FcRn: the neonatal Fc receptor comes of age. *Nat Rev Immunol.* 2007;7:715-725.
17. Takizawa T, Anderson CL, Robinson JM. A novel Fc gamma R-defined, IgG-containing organelle in placental endothelium. *J Immunol.* 2005;175:2331-2339.
18. Guleria I, Khosroshahi A, Ansari MJ, et al. A critical role for the programmed death ligand 1 in fetomaternal tolerance. *J Exp Med.* 2005;202:231-237.
19. Nelson AC, Mould AW, Bikoff EK, et al. Single-cell RNA-seq reveals cell type-specific transcriptional signatures at the maternal-foetal interface during pregnancy. *Nat Commun.* 2016;7:11414.
20. Vento-Tormo R, Efremova M, Botting RA, et al. Single-cell reconstruction of the early maternal-fetal interface in humans. *Nature.* 2018;563:347-353.
21. Brown JA, Dorfman DM, Ma FR, et al. Blockade of programmed death-1 ligands on dendritic cells enhances T cell activation and cytokine production. *J Immunol.* 2003;170:1257-1266.
22. Petroff MG, Chen L, Phillips TA, et al. B7 family molecules: novel immunomodulators at the maternal-fetal interface. *Placenta.* 2002;23(suppl A):S95-S101.
23. Svensson-Arvelund J, Mehta RB, Lindau R, et al. The human fetal placenta promotes tolerance against the semiallogeneic fetus by inducing regulatory T cells and homeostatic M2 macrophages. *J Immunol.* 2015;194:1534-1544.
24. Pollard JW, Bartocci A, Arceci R, et al. Apparent role of the macrophage growth factor, CSF-1, in placental development. *Nature.* 1987;330:484-486.
25. Ferreira LMR, Meissner TB, Tilburgs T, et al. HLA-G: at the interface of maternal-fetal tolerance. *Trends Immunol.* 2017;38:272-286.
26. Apps R, Gardner L, Moffett A. A critical look at HLA-G. *Trends Immunol.* 2008;29:313-321.
27. Morandi F, Rizzo R, Fainardi E, et al. Recent advances in our understanding of HLA-G biology: lessons from a wide spectrum of human diseases. *J Immunol Res.* 2016;2016:4326495.
28. Rajagopalan S. HLA-G-mediated NK cell senescence promotes vascular remodeling: implications for reproduction. *Cell Mol Immunol.* 2014;11:460-466.
29. Perez-Munoz ME, Arrieta MC, Ramer-Tait AE, et al. A critical assessment of the "sterile womb" and "in utero colonization" hypotheses: implications for research on the pioneer infant microbiome. *Microbiome.* 2017;5:48.
30. Aagaard K, Ma J, Antony KM, et al. The placenta harbors a unique microbiome. *Sci Transl Med.* 2014;6:237ra65.
31. Collado MC, Rautava S, Aakko J, et al. Human gut colonisation may be initiated in utero by distinct microbial communities in the placenta and amniotic fluid. *Sci Rep.* 2016;6:23129.
32. Prince AL, Ma J, Kannan PS, et al. The placental membrane microbiome is altered among subjects with spontaneous preterm birth with and without chorioamnionitis. *Am J Obstet Gynecol.* 2016;214(627):e1-e16.
33. Leoni C, Ceci O, Manzari C, et al. *Human Endometrial Microbiota at Term of Normal Pregnancies.* Genes; 2019:10.
34. Rackaityte E, Halkias J, Fukui EM, et al. Viable bacterial colonization is highly limited in the human intestine in utero. *Nat Med.* 2020;26:599-607.
35. Younge N, McCann JR, Ballard J, et al. Fetal exposure to the maternal microbiota in humans and mice. *JCI Insight.* 2019;4:e127806.
36. Lauder AP, Roche AM, Sherrill-Mix S, et al. Comparison of placenta samples with contamination controls does not provide evidence for a distinct placenta microbiota. *Microbiome.* 2016;4:29.
37. Leiby JS, McCormick K, Sherrill-Mix S, et al. Lack of detection of a human placenta microbiome in samples from preterm and term deliveries. *Microbiome.* 2018;6:196.
38. Theis KR, Romero R, Winters AD, et al. Does the human placenta delivered at term have a microbiota? Results of cultivation, quantitative real-time PCR, 16S rRNA gene sequencing, and metagenomics. *Am J Obstet Gynecol.* 2019;220(267):e1-e39.
39. de Goffau MC, Lager S, Sovio U, et al. Human placenta has no microbiome but can contain potential pathogens. *Nature.* 2019;572:329-334.
40. Chen C, Song X, Wei W, et al. The microbiota continuum along the female reproductive tract and its relation to uterine-related diseases. *Nat Commun.* 2017;8:875.
41. Mitchell CM, Haick A, Nkwopara E, et al. Colonization of the upper genital tract by vaginal bacterial species in nonpregnant women. *Am J Obstet Gynecol.* 2015;212(611):e1-e9.
42. Moreno I, Codoner FM, Vilella F, et al. Evidence that the endometrial microbiota has an effect on implantation success or failure. *Am J Obstet Gynecol.* 2016;215:684-703.
43. Verstraelen H, Vilchez-Vargas R, Desimpel F, et al. Characterisation of the human uterine microbiome in non-pregnant women through deep sequencing of the V1-2 region of the 16S rRNA gene. *PeerJ.* 2016;4:e1602.
44. Moffett A, Colucci F. Uterine NK cells: active regulators at the maternal-fetal interface. *J Clin Invest.* 2014;124:1872-1879.
45. Sojka DK, Yang L, Yokoyama WM. Uterine natural killer cells. *Front Immunol.* 2019;10:960.
46. Gaynor LM, Colucci F. Uterine natural killer cells: functional distinctions and influence on pregnancy in humans and mice. *Front Immunol.* 2017;8:467.
47. Bjorkstrom NK, Ljunggren HG, Michaelsson J. Emerging insights into natural killer cells in human peripheral tissues. *Nat Rev Immunol.* 2016;16:310-320.
48. Bulmer JN, Morrison L, Longfellow M, et al. Granulated lymphocytes in human endometrium: histochemical and immunohistochemical studies. *Hum Reprod.* 1991;6:791-798.
49. King A, Wellings V, Gardner L, et al. Immunocytochemical characterization of the unusual large granular lymphocytes in human endometrium throughout the menstrual cycle. *Hum Immunol.* 1989;24:195-205.
50. Doisne JM, Balmas E, Boulenouar S, et al. Composition, development, and function of uterine innate lymphoid cells. *J Immunol.* 2015;195:3937-3945.
51. Boulenouar S, Doisne JM, Sferruzzi-Perri A, et al. The residual innate lymphoid cells in NFIL3-deficient mice support suboptimal maternal adaptations to pregnancy. *Front Immunol.* 2016;7:43.
52. Sojka DK, Plougastel-Douglas B, Yang L, et al. Tissue-resident natural killer (NK) cells are cell lineages distinct from thymic and conventional splenic NK cells. *Elife.* 2014;3:e01659.
53. Montaldo E, Vacca P, Chiossone L, et al. Unique Eomes(+) NK cell subsets are present in uterus and decidua during early pregnancy. *Front Immunol.* 2015;6:646.
54. Filipovic I, Chiossone L, Vacca P, et al. Molecular definition of group 1 innate lymphoid cells in the mouse uterus. *Nat Commun.* 2018;9:4492.
55. Croy BA, Chen Z, Hofmann AP, et al. Imaging of vascular development in early mouse decidua and its association with leukocytes and trophoblasts. *Biol Reprod.* 2012;87:125.
56. Feyaerts D, Benner M, van Cranenbroek B, et al. Human uterine lymphocytes acquire a more experienced and tolerogenic phenotype during pregnancy. *Sci Rep.* 2017;7:2884.
57. Pique-Regi R, Romero R, Tarca AL, et al. Single cell transcriptional signatures of the human placenta in term and preterm parturition. *Elife.* 2019;8:e52004.
58. Williams PJ, Searle RF, Robson SC, et al. Decidual leucocyte populations in early to late gestation normal human pregnancy. *J Reprod Immunol.* 2009;82:24-31.
59. Kwan M, Hazan A, Zhang J, et al. Dynamic changes in maternal decidual leukocyte populations from first to second trimester gestation. *Placenta.* 2014;35:1027-1034.
60. Bartmann C, Segerer SE, Rieger L, et al. Quantification of the predominant immune cell populations in decidua throughout human pregnancy. *Am J Reprod Immunol.* 2014;71:109-119.
61. Paffaro VA Jr, Bizinotto MC, Joazeiro PP, et al. Subset classification of mouse uterine natural killer cells by DBA lectin reactivity. *Placenta.* 2003;24:479-488.
62. Sojka DK, Yang L, Plougastel-Douglas B, et al. Cutting edge: local proliferation of uterine tissue-resident NK cells during decidualization in mice. *J Immunol.* 2018;201:2551-2556.
63. Ashkar AA, Di Santo JP, Croy BA. Interferon gamma contributes to initiation of uterine vascular modification, decidual integrity, and uterine natural killer cell maturation during normal murine pregnancy. *J Exp Med.* 2000;192:259-270.
64. Sliz A, Locker KCS, Lampe K, et al. Gab3 is required for IL-2- and IL-15-induced NK cell expansion and limits trophoblast invasion during pregnancy. *Sci Immunol.* 2019;4:eaav3866.
65. Fu B, Zhou Y, Ni X, et al. Natural killer cells promote fetal development through the secretion of growth-promoting factors. *Immunity.* 2017;47:1100-1113.e6.
66. Huhn O, Ivarsson MA, Gardner L, et al. Distinctive phenotypes and functions of innate lymphoid cells in human decidua during early pregnancy. *Nat Commun.* 2020;11:381.
67. Gamliel M, Goldman-Wohl D, Isaacson B, et al. Trained memory of human uterine NK cells enhances their function in subsequent pregnancies. *Immunity.* 2018;48:951-962.e5.
68. Burton GJ, Redman CW, Roberts JM, et al. Pre-eclampsia: pathophysiology and clinical implications. *Br Med J.* 2019;366:l2381.
69. Siewiera J, Gouilly J, Hocine HR, et al. Natural cytotoxicity receptor splice variants orchestrate the distinct functions of human natural killer cell subtypes. *Nat Commun.* 2015;6:10183.
70. Keskin DB, Allan DS, Rybalov B, et al. TGFbeta promotes conversion of CD16+ peripheral blood NK cells into CD16-NK cells with similarities to decidual NK cells. *Proc Natl Acad Sci U S A.* 2007;104:3378-3383.
71. Cerdeira AS, Rajakumar A, Royle CM, et al. Conversion of peripheral blood NK cells to a decidual NK-like phenotype by a cocktail of defined factors. *J Immunol.* 2013;190:3939-3948.
72. El Costa H, Casemayou A, Aguerre-Girr M, et al. Critical and differential roles of NKp46-and NKp30-activating receptors expressed by uterine NK cells in early pregnancy. *J Immunol.* 2008;181:3009-3017.
73. Co EC, Gormley M, Kapidzic M, et al. Maternal decidual macrophages inhibit NK cell killing of invasive cytotrophoblasts during human pregnancy. *Biol Reprod.* 2013;88:155.
74. Manaster I, Mizrahi S, Goldman-Wohl D, et al. Endometrial NK cells are special immature cells that await pregnancy. *J Immunol.* 2008;181:1869-1876.
75. Male V, Sharkey A, Masters L, et al. The effect of pregnancy on the uterine NK cell KIR repertoire. *Eur J Immunol.* 2011;41:3017-3027.
76. Yadi H, Burke S, Madeja Z, et al. Unique receptor repertoire in mouse uterine NK cells. *J Immunol.* 2008;181:6140-6147.
77. Xiong S, Sharkey AM, Kennedy PR, et al. Maternal uterine NK cell-activating receptor KIR2DS1 enhances placentation. *J Clin Invest.* 2013;123:4264-4272.
78. Smith SD, Dunk CE, Aplin JD, et al. Evidence for immune cell involvement in decidual spiral arteriole remodeling in early human pregnancy. *Am J Pathol.* 2009;174:1959-1971.
79. Collins MK, Tay CS, Erlebacher A. Dendritic cell entrapment within the pregnant uterus inhibits immune surveillance of the maternal/fetal interface in mice. *J Clin Invest.* 2009;119:2062-2073.
80. Barber EM, Pollard JW. The uterine NK cell population requires IL-15 but these cells are not required for pregnancy nor the resolution of a Listeria monocytogenes infection. *J Immunol.* 2003;171:37-46.
81. Ashkar AA, Black GP, Wei Q, et al. Assessment of requirements for IL-15 and IFN regulatory factors in uterine NK cell differentiation and function during pregnancy. *J Immunol.* 2003;171:2937-2944.
82. Chiossone L, Vacca P, Orecchia P, et al. In vivo generation of decidual natural killer cells from resident hematopoietic progenitors. *Haematologica.* 2014;99:448-457.
83. Cao Q, Chen H, Deng Z, et al. Genetic deletion of Cxcl14 in mice alters uterine NK cells. *Biochem Biophys Res Commun.* 2013;435:664-670.
84. Kruse A, Martens N, Fernekorn U, et al. Alterations in the expression of homing-associated molecules at the maternal/fetal interface during the course of pregnancy. *Biol Reprod.* 2002;66:333-345.
85. Peng J, Monsivais D, You R, et al. Uterine activin receptor-like kinase 5 is crucial for blastocyst implantation and placental development. *Proc Natl Acad Sci U S A.* 2015;112:E5098-E5107.
86. Pawlak JB, Balint L, Lim L, et al. Lymphatic mimicry in maternal endothelial cells promotes placental spiral artery remodeling. *J Clin Invest.* 2019;129:4912-4921.
87. Rizzuto G, Tagliani E, Manandhar P, et al. Limited colonization undermined by inadequate early immune responses defines the dynamics of decidual listeriosis. *Infect Immun.* 2017;85;e00153-17.
88. Vacca P, Montaldo E, Croxatto D, et al. Identification of diverse innate lymphoid cells in human decidua. *Mucosal Immunol.* 2015;8:254-264.

89. Nussbaum JC, Van Dyken SJ, von Moltke J, et al. Type 2 innate lymphoid cells control eosinophil homeostasis. *Nature*. 2013;502:245-248.
90. Xu Y, Romero R, Miller D, et al. Innate lymphoid cells at the human maternal-fetal interface in spontaneous preterm labor. *Am J Reprod Immunol*. 2018;79:e12820.
91. Tilburgs T, Claas FH, Scherjon SA. Elsevier Trophoblast Research Award Lecture: unique properties of decidual T cells and their role in immune regulation during human pregnancy. *Placenta*. 2010;31(suppl):S82-S86.
92. Rinaldi SF, Makieva S, Saunders PT, et al. Immune cell and transcriptomic analysis of the human decidua in term and preterm parturition. *Mol Hum Reprod*. 2017;23:708-724.
93. Tilburgs T, Scherjon SA, van der Mast BJ, et al. Fetal-maternal HLA-C mismatch is associated with decidual T cell activation and induction of functional T regulatory cells. *J Reprod Immunol*. 2009;82:148-157.
94. Vassiliadou N, Bulmer JN. Quantitative analysis of T lymphocyte subsets in pregnant and nonpregnant human endometrium. *Biol Reprod*. 1996;55:1017-1022.
95. Blaisdell A, Erlebacher A. Flow cytometric analysis of myometrial and decidual cell suspension. In: Croy BA, Yamada AT, DeMayo FJ, Adamson SL, eds. *The Guide to Investigation of Mouse Pregnancy*. Academic Press; 2014:619-628.
96. Johansson M, Lycke N. A unique population of extrathymically derived alpha beta TCR+CD4-CD8-T cells with regulatory functions dominates the mouse female genital tract. *J Immunol*. 2003;170:1659-1666.
97. Li Y, Lopez GE, Vazquez J, et al. Decidual-placental immune landscape during syngeneic murine pregnancy. *Front Immunol*. 2018;9:2087.
98. Saito S, Nishikawa K, Morii T, et al. A study of CD45RO, CD45RA and CD29 antigen expression on human decidual T cells in an early stage of pregnancy. *Immunol Lett*. 1994;40:193-197.
99. Tilburgs T, Schonkeren D, Eikmans M, et al. Human decidual tissue contains differentiated CD8+ effector-memory T cells with unique properties. *J Immunol*. 2010;185:4470-4477.
100. Powell RM, Lissauer D, Tamblyn J, et al. Decidual T cells exhibit a highly differentiated phenotype and demonstrate potential fetal specificity and a strong transcriptional response to IFN. *J Immunol*. 2017;199:3406-3417.
101. Kumar BV, Ma W, Miron M, et al. Human tissue-resident memory T cells are defined by core transcriptional and functional signatures in lymphoid and mucosal sites. *Cell Rep*. 2017;20:2921-2934.
102. Tilburgs T, Scherjon SA, Roelen DL, et al. Decidual CD8+CD28-T cells express CD103 but not perforin. *Hum Immunol*. 2009;70:96-100.
103. van der Zwan A, Bi K, Norwitz ER, et al. Mixed signature of activation and dysfunction allows human decidual CD8(+) T cells to provide both tolerance and immunity. *Proc Natl Acad Sci U S A*. 2018;115:385-390.
104. Zeng W, Liu X, Liu Z, et al. Deep surveying of the transcriptional and alternative splicing signatures for decidual CD8(+) T cells at the first trimester of human healthy pregnancy. *Front Immunol*. 2018;9:937.
105. Saito S, Nishikawa K, Morii T, et al. Expression of activation antigens CD69, HLA-DR, interleukin-2 receptor-alpha (IL-2R alpha) and IL-2R beta on T cells of human decidua at an early stage of pregnancy. *Immunology*. 1992;75:710-712.
106. Kieffer TEC, Laskewitz A, Scherjon SA, et al. Memory T cells in pregnancy. *Front Immunol*. 2019;10:625.
107. Huang X, Liu L, Xu C, et al. Tissue-resident CD8(+) T memory cells with unique properties are present in human decidua during early pregnancy. *Am J Reprod Immunol*. 2020;84:e13254.
108. Southcombe JH, Mounce G, McGee K, et al. An altered endometrial CD8 tissue resident memory T cell population in recurrent miscarriage. *Sci Rep*. 2017;7:41335.
109. Scaife PJ, Bulmer JN, Robson SC, et al. Effector activity of decidual CD8+ T lymphocytes in early human pregnancy. *Biol Reprod*. 2006;75:562-567.
110. Wang SC, Li YH, Piao HL, et al. PD-1 and Tim-3 pathways are associated with regulatory CD8+ T-cell function in decidua and maintenance of normal pregnancy. *Cell Death Dis*. 2015;6:e1738.
111. Casey KA, Fraser KA, Schenkel JM, et al. Antigen-independent differentiation and maintenance of effector-like resident memory T cells in tissues. *J Immunol*. 2012;188:4866-4875.
112. Zeng W, Liu Z, Liu X, et al. Distinct transcriptional and alternative splicing signatures of decidual CD4(+) T cells in early human pregnancy. *Front Immunol*. 2017;8:682.
113. Mjosberg J, Berg G, Jenmalm MC, et al. FOXP3+ regulatory T cells and T helper 1, T helper 2, and T helper 17 cells in human early pregnancy decidua. *Biol Reprod*. 2010;82:698-705.
114. Wu HX, Jin LP, Xu B, et al. Decidual stromal cells recruit Th17 cells into decidua to promote proliferation and invasion of human trophoblast cells by secreting IL-17. *Cell Mol Immunol*. 2014;11:253-262.
115. Tilburgs T, Roelen DL, van der Mast BJ, et al. Evidence for a selective migration of fetus-specific CD4+CD25bright regulatory T cells from the peripheral blood to the decidua in human pregnancy. *J Immunol*. 2008;180:5737-5745.
116. Salvany-Celades M, van der Zwan A, Benner M, et al. Three types of functional regulatory T cells control T cell responses at the human maternal-fetal interface. *Cell Rep*. 2019;27:2537-2547.e5.
117. van Egmond A, van der Keur C, Swings GM, et al. The possible role of virus-specific CD8(+) memory T cells in decidual tissue. *J Reprod Immunol*. 2016;113:1-8.
118. Bastidas S, Graw F, Smith MZ, et al. CD8+ T cells are activated in an antigen-independent manner in HIV-infected individuals. *J Immunol*. 2014;192:1732-1744.
119. Tilburgs T, Crespo AC, van der Zwan A, et al. Human HLA-G+ extravillous trophoblasts: immune-activating cells that interact with decidual leukocytes. *Proc Natl Acad Sci U S A*. 2015;112:7219-7224.
120. Shima T, Inada K, Nakashima A, et al. Paternal antigen-specific proliferating regulatory T cells are increased in uterine-draining lymph nodes just before implantation and in pregnant uterus just after implantation by seminal plasma-priming in allogeneic mouse pregnancy. *J Reprod Immunol*. 2015;108:72-82.
121. Samstein RM, Josefowicz SZ, Arvey A, et al. Extrathymic generation of regulatory T cells in placental mammals mitigates maternal-fetal conflict. *Cell*. 2012;150:29-38.
122. Leng Y, Romero R, Xu Y, et al. Are B cells altered in the decidua of women with preterm or term labor? *Am J Reprod Immunol*. 2019;81:e13102.
123. Quillay H, El Costa H, Marlin R, et al. Distinct characteristics of endometrial and decidual macrophages and regulation of their permissivity to HIV-1 infection by SAMHD1. *J Virol*. 2015;89:1329-1339.
124. Houser BL, Tilburgs T, Hill J, et al. Two unique human decidual macrophage populations. *J Immunol*. 2011;186:2633-2642.
125. Svensson J, Jenmalm MC, Matussek A, et al. Macrophages at the fetal-maternal interface express markers of alternative activation and are induced by M-CSF and IL-10. *J Immunol*. 2011;187:3671-3682.
126. Svensson-Arvelund J, Ernerudh J. The role of macrophages in promoting and maintaining homeostasis at the fetal-maternal interface. *Am J Reprod Immunol*. 2015;74:100-109.
127. Kämmerer U, Eggert AO, Kapp M, et al. Unique appearance of proliferating antigen-presenting cells expressing DC-SIGN (CD209) in the decidua of early human pregnancy. *Am J Pathol*. 2003;162:887-896.
128. Tsao FY, Wu MY, Chang YL, et al. M1 macrophages decrease in the deciduae from normal pregnancies but not from spontaneous abortions or unexplained recurrent spontaneous abortions. *J Formos Med Assoc*. 2018;117:204-211.
129. Lidstrom C, Matthiesen L, Berg G, et al. Cytokine secretion patterns of NK cells and macrophages in early human pregnancy decidua and blood: implications for suppressor macrophages in decidua. *Am J Reprod Immunol*. 2003;50:444-452.
130. Pollard JW, Hunt JS, Wiktor-Jedrzejczak W, et al. A pregnancy defect in the osteopetrotic (op/op) mouse demonstrates the requirement for CSF-1 in female fertility. *Dev Biol*. 1991;148:273-283.
131. Stewart IJ, Mitchell BS. The distribution of uterine macrophages in virgin and early pregnant mice. *J Anat*. 1991;179:183-196.
132. Brandon JM. Macrophage distribution in decidual tissue from early implantation to the periparturient period in mice as defined by the macrophage differentiation antigens F4/80, macrosialin and the type 3 complement receptor. *J Reprod Fertil*. 1995;103:9-16.
133. Kruse A, Merchant MJ, Hallmann R, et al. Evidence of specialized leukocyte-vascular homing interactions at the maternal/fetal interface. *Eur J Immunol*. 1999;29:1116-1126.
134. Rieger L, Honig A, Sutterlin M, et al. Antigen-presenting cells in human endometrium during the menstrual cycle compared to early pregnancy. *J Soc Gynecol Investig*. 2004;11:488-493.
135. Shawber CJ, Lin L, Gnarra M, et al. Vascular Notch proteins and Notch signaling in the peri-implantation mouse uterus. *Vasc Cell*. 2015;7:9.
136. Thiruchelvam U, Dransfield I, Saunders PT, et al. The importance of the macrophage within the human endometrium. *J Leukoc Biol*. 2013;93:217-225.
137. Lash GE, Pitman H, Morgan HL, et al. Decidual macrophages: key regulators of vascular remodeling in human pregnancy. *J Leukoc Biol*. 2016;100:315-325.
138. Hamilton S, Oomomian Y, Stephen G, et al. Macrophages infiltrate the human and rat decidua during term and preterm labor: evidence that decidual inflammation precedes labor. *Biol Reprod*. 2012;86:39.
139. Thomson AJ, Telfer JF, Young A, et al. Leukocytes infiltrate the myometrium during human parturition: further evidence that labour is an inflammatory process. *Hum Reprod*. 1999;14:229-236.
140. Singh N, Herbert B, Sooranna GR, et al. Is myometrial inflammation a cause or a consequence of term human labour? *J Endocrinol*. 2017;235:69-83.
141. Xu Y, Romero R, Miller D, et al. An M1-like macrophage polarization in decidual tissue during spontaneous preterm labor that is attenuated by Rosiglitazone treatment. *J Immunol*. 2016;196:2476-2491.
142. Edey LF, Georgiou H, O'Dea KP, et al. Progesterone, the maternal immune system and the onset of parturition in the mouse. *Biol Reprod*. 2018;98:376-395.
143. Pollard JW, Lin EY, Zhu L. Complexity in uterine macrophage responses to cytokines in mice. *Biol Reprod*. 1998;58:1469-1475.
144. Cohen PE, Zhu L, Pollard JW. Absence of colony stimulating factor-1 in osteopetrotic (csfmop/csfmop) mice disrupts estrous cycles and ovulation. *Biol Reprod*. 1997;56:110-118.
145. Menzies FM, Khan AH, Higgins CA, et al. The chemokine receptor CCR2 is not required for successful initiation of labor in mice. *Biol Reprod*. 2012;86:118.
146. Timmons BC, Mahendroo MS. Timing of neutrophil activation and expression of proinflammatory markers do not support a role for neutrophils in cervical ripening in the mouse. *Biol Reprod*. 2006;74:236-245.
147. Zulu MZ, Martinez FO, Gordon S, et al. The elusive role of placental macrophages: the Hofbauer cell. *J Innate Immun*. 2019;11:447-456.
148. Reyes L, Golos TG. Hofbauer cells: their role in healthy and complicated pregnancy. *Front Immunol*. 2018;9:2628.
149. Redline RW, Lu CY. Localization of fetal major histocompatibility complex antigens and maternal leukocytes in murine placenta. Implications for maternal-fetal immunological relationship. *Lab Invest*. 1989;61:27-36.
150. Kämmerer U, Schoppet M, McLellan AD, et al. Human decidua contains potent Immunostimulatory CD83+ dendritic cells. *Am J Pathol*. 2000;157:159-169.
151. Kemp B, Schmitz S, Krusche CA, et al. Dendritic cells are equally distributed in intrauterine and tubal ectopic pregnancies. *Fertil Steril*. 2011;95:28-32.
152. Tagliani E, Erlebacher A. Dendritic cell function at the maternal-fetal interface. *Expert Rev Clin Immunol*. 2011;7:593-602.
153. Volchek M, Girling JE, Lash GE, et al. Lymphatics in the human endometrium disappear during decidualization. *Hum Reprod*. 2010;25:2455-2464.
154. Windsperger K, Dekan S, Pils S, et al. Extravillous trophoblast invasion of venous as well as lymphatic vessels is altered in idiopathic, recurrent, spontaneous abortions. *Hum Reprod*. 2017;32:1208-1217.
155. He N, van Iperen L, de Jong D, et al. Human extravillous trophoblasts penetrate decidual veins and lymphatics before remodeling spiral arteries during early pregnancy. *PLoS One*. 2017;12:e0169849.
156. Chen T, Darrasse-Jeze G, Bergot AS, et al. Self-specific memory regulatory T cells protect embryos at implantation in mice. *J Immunol*. 2013;191:2273-2281.
157. Krey G, Frank P, Shaikly V, et al. In vivo dendritic cell depletion reduces breeding efficiency, affecting implantation and early placental development in mice. *J Mol Med (Berl)*. 2008;86:999-1011.
158. Plaks V, Birnberg T, Berkutzki T, et al. Uterine DCs are crucial for decidua formation during embryo implantation in mice. *J Clin Invest*. 2008;118:3954-3965.
159. Blois SM, Ilarregui JM, Tometten M, et al. A pivotal role for galectin-1 in fetomaternal tolerance. *Nat Med*. 2007;13:1450-1457.
160. Karsten CM, Behrends J, Wagner AK, et al. DC within the pregnant mouse uterus influence growth and functional properties of uterine NK cells. *Eur J Immunol*. 2009;39:2203-2214.
161. Du MR, Guo PF, Piao HL, et al. Embryonic trophoblasts induce decidual regulatory T cell differentiation and maternal-fetal tolerance through thymic stromal lymphopoietin instructing dendritic cells. *J Immunol*. 2014;192:1502-1511.
162. Laskarin G, Redzovic A, Rubesa Z, et al. Decidual natural killer cell tuning by autologous dendritic cells. *Am J Reprod Immunol*. 2008;59:433-445.
163. Leno-Duran E, Munoz-Fernandez R, Olivares EG, et al. Liaison between natural killer cells and dendritic cells in human gestation. *Cell Mol Immunol*. 2014;11:449-455.
164. Milne SA, Henderson TA, Kelly RW, et al. Leukocyte populations and steroid receptor expression in human first-trimester decidua; regulation by antiprogestin and prostaglandin E analog. *J Clin Endocrinol Metab*. 2005;90:4315-4321.
165. Nadkarni S, Smith J, Sferruzzi-Perri AN, et al. Neutrophils induce proangiogenic T cells with a regulatory phenotype in pregnancy. *Proc Natl Acad Sci U S A*. 2016;113:E8415-E8424.
166. Rinaldi SF, Catalano RD, Wade J, et al. Decidual neutrophil infiltration is not required for preterm birth in a mouse model of infection-induced preterm labor. *J Immunol*. 2014;192:2315-2325.
167. Gouon-Evans V, Pollard JW. Eotaxin is required for eosinophil homing into the stroma of the pubertal and cycling uterus. *Endocrinology*. 2001;142:4515-4521.
168. Meyer N, Zenclussen AC. Mast cells-Good guys with a bad image? *Am J Reprod Immunol*. 2018;80:e13002.
169. Jeziorska M, Salamonsen LA, Woolley DE. Mast cell and eosinophil distribution and activation in human endometrium throughout the menstrual cycle. *Biol Reprod*. 1995;53:312-320.

170. Sacks G, Sargent I, Redman C. An innate view of human pregnancy. *Immunol Today*. 1999;20:114-118.
171. Kourtis AP, Read JS, Jamieson DJ. Pregnancy and infection. *N Engl J Med*. 2014;370:2211-2218.
172. Han X, Ghaemi MS, Ando K, et al. Differential dynamics of the maternal immune system in healthy pregnancy and preeclampsia. *Front Immunol*. 2019;10:1305.
173. Aghaeepour N, Ganio EA, McIlwain D, et al. An immune clock of human pregnancy. *Sci Immunol*. 2017;2:eaan2946.
174. Kraus TA, Engel SM, Sperling RS, et al. Characterizing the pregnancy immune phenotype: results of the viral immunity and pregnancy (VIP) study. *J Clin Immunol*. 2012;32:300-311.
175. Kuhnert M, Strohmeier R, Stegmuller M, et al. Changes in lymphocyte subsets during normal pregnancy. *Eur J Obstet Gynecol Reprod Biol*. 1998;76:147-151.
176. Watanabe M, Iwatani Y, Kaneda T, et al. Changes in T, B, and NK lymphocyte subsets during and after normal pregnancy. *Am J Reprod Immunol*. 1997;37:368-377.
177. Zhang J, Shynlova O, Sabra S, et al. Immunophenotyping and activation status of maternal peripheral blood leukocytes during pregnancy and labour, both term and preterm. *J Cell Mol Med*. 2017;21:2386-2402.
178. Hong S, Banchereau R, Maslow BL, et al. Longitudinal profiling of human blood transcriptome in healthy and lupus pregnancy. *J Exp Med*. 2019;216:1154-1169.
179. Makinoda S, Mikuni M, Furuta I, et al. Serum concentration of endogenous G-CSF in women during the menstrual cycle and pregnancy. *Eur J Clin Invest*. 1995;25:877-879.
180. Kraus TA, Sperling RS, Engel SM, et al. Peripheral blood cytokine profiling during pregnancy and post-partum periods. *Am J Reprod Immunol*. 2010;64:411-426.
181. Le Gars M, Kay AW, Bayless NL, et al. Increased proinflammatory responses of monocytes and plasmacytoid dendritic cells to influenza A virus infection during pregnancy. *J Infect Dis*. 2016;214:1666-1671.
182. Shah NM, Imami N, Johnson MR. Progesterone modulation of pregnancy-related immune responses. *Front Immunol*. 2018;9:1293.
183. Sacks GP, Studena K, Sargent K, et al. Normal pregnancy and preeclampsia both produce inflammatory changes in peripheral blood leukocytes akin to those of sepsis. *Am J Obstet Gynecol*. 1998;179:80-86.
184. Sacks GP, Redman CW, Sargent IL. Monocytes are primed to produce the Th1 type cytokine IL-12 in normal human pregnancy: an intracellular flow cytometric analysis of peripheral blood mononuclear cells. *Clin Exp Immunol*. 2003;131:490-497.
185. Faas MM, de Vos P. Maternal monocytes in pregnancy and preeclampsia in humans and in rats. *J Reprod Immunol*. 2017;119:91-97.
186. Pflitsch C, Feldmann CN, Richert L, et al. In-depth characterization of monocyte subsets during the course of healthy pregnancy. *J Reprod Immunol*. 2020;141:103151.
187. Kay AW, Fukuyama J, Aziz N, et al. Enhanced natural killer-cell and T-cell responses to influenza A virus during pregnancy. *Proc Natl Acad Sci U S A*. 2014;111:14506-14511.
188. Apps R, Kotliarov Y, Cheung F, et al. Multimodal immune phenotyping of maternal peripheral blood in normal human pregnancy. *JCI Insight*. 2020;5:e134838.
189. Le Gars M, Seiler C, Kay AW, et al. Pregnancy-induced alterations in NK cell phenotype and function. *Front Immunol*. 2019;10:2469.
190. Wegmann TG, Lin H, Guilbert L, et al. Bidirectional cytokine interactions in the maternal-fetal relationship: is successful pregnancy a TH2 phenomenon? *Immunol Today*. 1993;14:353-356.
191. Saito S, Sakai M, Sasaki Y, et al. Quantitative analysis of peripheral blood Th0, Th1, Th2 and the Th1:Th2 cell ratio during normal human pregnancy and preeclampsia. *Clin Exp Immunol*. 1999;117:550-555.
192. Jiang TT, Chaturvedi V, Ertelt JM, et al. Regulatory T cells: new keys for further unlocking the enigma of fetal tolerance and pregnancy complications. *J Immunol*. 2014;192:4949-4956.
193. Aluvihare VR, Kallikourdis M, Betz AG. Regulatory T cells mediate maternal tolerance to the fetus. *Nat Immunol*. 2004;5:266-271.
194. Zhao JX, Zeng YY, Liu Y. Fetal alloantigen is responsible for the expansion of the CD4(+)CD25(+) regulatory T cell pool during pregnancy. *J Reprod Immunol*. 2007;75:71-81.
195. Thuere C, Zenclussen ML, Schumacher A, et al. Kinetics of regulatory T cells during murine pregnancy. *Am J Reprod Immunol*. 2007;58:514-523.
196. Shima T, Sasaki Y, Itoh M, et al. Regulatory T cells are necessary for implantation and maintenance of early pregnancy but not late pregnancy in allogeneic mice. *J Reprod Immunol*. 2010;85:121-129.
197. Rowe JH, Ertelt JM, Aguilera MN, et al. Foxp3(+) regulatory T cell expansion required for sustaining pregnancy compromises host defense against prenatal bacterial pathogens. *Cell Host Microbe*. 2011;10:54-64.
198. Lima J, Martins C, Leandro MJ, et al. Characterization of B cells in healthy pregnant women from late pregnancy to postpartum: a prospective observational study. *BMC Pregnancy Childbirth*. 2016;16:139.
199. Ziegler KB, Muzzio DO, Matzner F, et al. Human pregnancy is accompanied by modifications in B cell development and immunoglobulin profile. *J Reprod Immunol*. 2018;129:40-47.
200. Medina KL, Smithson G, Kincade PW. Suppression of B lymphopoiesis during normal pregnancy. *J Exp Med*. 1993;178:1507-1515.
201. Medina KL, Kincade PW. Pregnancy-related steroids are potential negative regulators of B lymphopoiesis. *Proc Natl Acad Sci U S A*. 1994;91:5382-5386.
202. Bosco N, Ceredig R, Rolink A. Transient decrease in interleukin-7 availability arrests B lymphopoiesis during pregnancy. *Eur J Immunol*. 2008;38:381-390.
203. Muzzio DO, Soldati R, Ehrhardt J, et al. B cell development undergoes profound modifications and adaptations during pregnancy in mice. *Biol Reprod*. 2014;91:115.
204. Muzzio DO, Ziegler KB, Ehrhardt J, et al. Marginal zone B cells emerge as a critical component of pregnancy well-being. *Reproduction*. 2016;151:29-37.
205. Jennewein MF, Goldfarb I, Dolatshahi S, et al. Fc glycan-mediated regulation of placental antibody transfer. *Cell*. 2019;178:202-215.e14.
206. Anthony RM, Wermeling F, Karlsson MC, et al. Identification of a receptor required for the anti-inflammatory activity of IVIG. *Proc Natl Acad Sci U S A*. 2008;105:19571-19578.
207. Karsten CM, Pandey MK, Figge J, et al. Anti-inflammatory activity of IgG1 mediated by Fc galactosylation and association of FcgammaRIIB and dectin-1. *Nat Med*. 2012;18:1401-1406.
208. Margni RA, Malan Borel I. Paradoxical behavior of asymmetric IgG antibodies. *Immunol Rev*. 1998;163:77-87.
209. Zenclussen AC, Gentile T, Kortebani G, et al. Asymmetric antibodies and pregnancy. *Am J Reprod Immunol*. 2001;45:289-294.
210. Barrientos G, Fuchs D, Schrocksnadel K, et al. Low levels of serum asymmetric antibodies as a marker of threatened pregnancy. *J Reprod Immunol*. 2009;79:201-210.
211. Shan C, Xie X, Luo H, et al. Maternal vaccination and protective immunity against Zika virus vertical transmission. *Nat Commun*. 2019;10:5677.
212. Schlaudecker EP, Ambroggio L, McNeal MM, et al. Declining responsiveness to influenza vaccination with progression of human pregnancy. *Vaccine*. 2018;36:4734-4741.
213. Tay CS, Tagliani E, Collins MK, et al. Cis-acting pathways selectively enforce the non-immunogenicity of shed placental antigen for maternal CD8 T cells. *PLoS One*. 2013;8:e84064.
214. Schumacher A, Costa SD, Zenclussen AC. Endocrine factors modulating immune responses in pregnancy. *Front Immunol*. 2014;5:196.
215. Engler JB, Kursawe N, Solano ME, et al. Glucocorticoid receptor in T cells mediates protection from autoimmunity in pregnancy. *Proc Natl Acad Sci U S A*. 2017;114:E181-E190.
216. Wu C, Jin X, Tsueng G, et al. BioGPS: building your own mash-up of gene annotations and expression profiles. *Nucleic Acids Res*. 2016;44:D313-D316.
217. Heng TS, Painter MW, Immunological Genome Project C. The Immunological Genome Project: networks of gene expression in immune cells. *Nat Immunol*. 2008;9:1091-1094.
218. Tibbetts TA, DeMayo F, Rich S, et al. Progesterone receptors in the thymus are required for thymic involution during pregnancy and for normal fertility. *Proc Natl Acad Sci U S A*. 1999;96:12021-12026.
219. Khan D, Ansar Ahmed S. The immune system is a natural target for estrogen action: opposing effects of estrogen in two prototypical autoimmune diseases. *Front Immunol*. 2015;6:635.
220. Polanczyk MJ, Carson BD, Subramanian S, et al. Cutting edge: estrogen drives expansion of the CD4+CD25+ regulatory T cell compartment. *J Immunol*. 2004;173:2227-2230.
221. Mincheva-Nilsson L, Baranov V. The role of placental exosomes in reproduction. *Am J Reprod Immunol*. 2010;63:520-533.
222. Moore T, Dveksler GS. Pregnancy-specific glycoproteins: complex gene families regulating maternal-fetal interactions. *Int J Dev Biol*. 2014;58:273-280.
223. Medawar PB. Some immunological and endocrinological problems raised by the evolution of viviparity in vertebrates. *Symp Soc Exp Biol*. 1953;7:320-338.
224. Rogers AM, Boime I, Connolly J, et al. Maternal-fetal tolerance is maintained despite transgene-driven trophoblast expression of MHC class I, and defects in Fas and its ligand. *Eur J Immunol*. 1998;28:3479-3487.
225. Shomer B, Toder V, Egorov I, et al. Expression of allogeneic MHC class I antigens by transgenic mouse trophoblast does not interfere with the normal course of pregnancy. *Transgenic Res*. 1998;7:343-355.
226. Apps R, Murphy SP, Fernando R, et al. Human leucocyte antigen (HLA) expression of primary trophoblast cells and placental cell lines, determined using single antigen beads to characterize allotype specificities of anti-HLA antibodies. *Immunology*. 2009;127:26-39.
227. Jungbluth AA, Silva WA Jr, Iversen K, et al. Expression of cancer-testis (CT) antigens in placenta. *Cancer Immun*. 2007;7:15.
228. Linscheid C, Petroff MG. Minor histocompatibility antigens and the maternal immune response to the fetus during pregnancy. *Am J Reprod Immunol*. 2013;69:304-314.
229. Holland OJ, Linscheid C, Hodes HC, et al. Minor histocompatibility antigens are expressed in syncytiotrophoblast and trophoblast debris: implications for maternal alloreactivity to the fetus. *Am J Pathol*. 2012;180:256-266.
230. Chamley LW, Chen Q, Ding J, et al. Trophoblast deportation: just a waste disposal system or antigen sharing? *J Reprod Immunol*. 2011;88:99-105.
231. Huppertz B, Frank HG, Kingdom JC, et al. Villous cytotrophoblast regulation of the syncytial apoptotic cascade in the human placenta. *Histochem Cell Biol*. 1998;110:495-508.
232. Wylie BJ, D'Alton ME. Fetomaternal hemorrhage. *Obstet Gynecol*. 2010;115:1039-1051.
233. Wegmann TG, Waters CA, Drell DW, et al. Pregnant mice are not primed but can be primed to fetal alloantigens. *Proc Natl Acad Sci U S A*. 1979;76:2410-2414.
234. Jacoby DR, Olding LB, Oldstone MB. Immunologic regulation of fetal-maternal balance. *Adv Immunol*. 1984;35:157-208.
235. Clark DA. Materno-fetal relations. *Immunol Lett*. 1985;9:239-247.
236. Jiang SP, Vacchio MS. Multiple mechanisms of peripheral T cell tolerance to the fetal "allograft". *J Immunol*. 1998;160:3086-3090.
237. Tafuri A, Alferink J, Moller P, et al. T cell awareness of paternal alloantigens during pregnancy. *Science*. 1995;270:630-633.
238. Zhou M, Mellor AL. Expanded cohorts of maternal CD8+ T-cells specific for paternal MHC class I accumulate during pregnancy. *J Reprod Immunol*. 1998;40:47-62.
239. Kearney ER, Pape KA, Loh DY, et al. Visualization of peptide-specific T cell immunity and peripheral tolerance induction in vivo. *Immunity*. 1994;1:327-339.
240. Ehst BD, Ingulli E, Jenkins MK. Development of a novel transgenic mouse for the study of interactions between CD4 and CD8 T cells during graft rejection. *Am J Transplant*. 2003;3:1355-1362.
241. Erlebacher A, Vencato D, Price KA, et al. Constraints in antigen presentation severely restrict T cell recognition of the allogeneic fetus. *J Clin Invest*. 2007;117:1399-1411.
242. Marino J, Paster J, Benichou G. Allorecognition by T Lymphocytes and allograft rejection. *Front Immunol*. 2016;7:582.
243. Hiby SE, Apps R, Sharkey AM, et al. Maternal activating KIRs protect against human reproductive failure mediated by fetal HLA-C2. *J Clin Invest*. 2010;120:4102-4110.
244. Proll J, Blaschitz A, Hutter H, et al. First trimester human endovascular trophoblast cells express both HLA-C and HLA-G. *Am J Reprod Immunol*. 1999;42:30-36.
245. Petersdorf EW, Longton GM, Anasetti C, et al. Association of HLA-C disparity with graft failure after marrow transplantation from unrelated donors. *Blood*. 1997;89:1818-1823.
246. Moldenhauer LM, Diener KR, Thring DM, et al. Cross-presentation of male seminal fluid antigens elicits T cell activation to initiate the female immune response to pregnancy. *J Immunol*. 2009;182:8080-8093.
247. Barton BM, Xu R, Wherry EJ, et al. Pregnancy promotes tolerance to future offspring by programming selective dysfunction in long-lived maternal T cells. *J Leukoc Biol*. 2017;101:975-987.
248. Taglauer ES, Yankee TM, Petroff MG. Maternal PD-1 regulates accumulation of fetal antigen-specific CD8+ T cells in pregnancy. *J Reprod Immunol*. 2009;80:12-21.
249. Wegorzewska M, Nijagal A, Wong CM, et al. Fetal intervention increases maternal T cell awareness of the foreign conceptus and can lead to immune-mediated fetal demise. *J Immunol*. 2014;192:1938-1945.
250. Jasti S, Farahbakhsh M, Nguyen S, et al. Immune response to a model shared placenta/tumor-associated antigen reduces cancer risk in parous mice. *Biol Reprod*. 2017;96:134-144.
251. Kinder JM, Turner LH, Stelzer IA, et al. CD8(+) T cell functional exhaustion overrides pregnancy-induced fetal antigen alloimmunization. *Cell Rep*. 2020;31:107784.
252. James E, Chai JG, Dewchand H, et al. Multiparity induces priming to male-specific minor histocompatibility antigen, HY, in mice and humans. *Blood*. 2003;102:388-393.
253. Bonney EA, Matzinger P. The maternal immune system's interaction with circulating fetal cells. *J Immunol*. 1997;158:40-47.
254. Kahn DA, Baltimore D. Pregnancy induces a fetal antigen-specific maternal T regulatory cell response that contributes to tolerance. *Proc Natl Acad Sci U S A*. 2010;107:9299-9304.
255. Rowe JH, Ertelt JM, Xin L, et al. Pregnancy imprints regulatory memory that sustains anergy to fetal antigen. *Nature*. 2012;490:102-106.
256. Lissauer D, Piper K, Goodyear O, et al. Fetal-specific CD8+ cytotoxic T cell responses develop during normal human pregnancy and exhibit broad functional capacity. *J Immunol*. 2012;189:1072-1080.
257. Verdijk RM, Kloosterman A, Pool J, et al. Pregnancy induces minor histocompatibility antigen-specific cytotoxic T cells: implications for stem cell transplantation and immunotherapy. *Blood*. 2004;103:1961-1964.
258. van Kampen CA, Versteeg-van der Voort Maarschalk MF, Langerak-Langerak J, et al. Pregnancy can induce long-persisting primed CTLs specific for inherited paternal HLA antigens. *Hum Immunol*. 2001;62:201-207.
259. Bouma GJ, van Caubergh P, van Bree SP, et al. Pregnancy can induce priming of cytotoxic T lymphocytes specific for

paternal HLA antigens that is associated with antibody formation. *Transplantation.* 1996;62:672-678.

260. Stowell SR, Henry KL, Smith NH, et al. Alloantibodies to a paternally derived RBC KEL antigen lead to hemolytic disease of the fetus/newborn in a murine model. *Blood.* 2013;122:1494-1504.
261. Hognas E, Kauppila A, Hinkula M, et al. Incidence of cancer among grand multiparous women in Finland with special focus on non-gynaecological cancers: a population-based cohort study. *Acta Oncol.* 2016;55:370-376.
262. Ruocco MG, Chaouat G, Florez L, et al. Regulatory T-cells in pregnancy: historical perspective, state of the art, and burning questions. *Front Immunol.* 2014;5:389.
263. Darrasse-Jeze G, Klatzmann D, Charlotte F, et al. CD4+CD25+ regulatory/suppressor T cells prevent allogeneic fetus rejection in mice. *Immunol Lett.* 2006;102:106-109.
264. Zenclussen AC, Gerlof K, Zenclussen ML, et al. Abnormal T-cell reactivity against paternal antigens in spontaneous abortion. *Am J Pathol.* 2005;166:811-822.
265. Zenclussen AC, Gerlof K, Zenclussen ML, et al. Regulatory T cells induce a privileged tolerant microenvironment at the fetal-maternal interface. *Eur J Immunol.* 2006;36:82-94.
266. Guerin LR, Moldenhauer LM, Prins JR, et al. Seminal fluid regulates accumulation of FOXP3+ regulatory T cells in the preimplantation mouse uterus through expanding the FOXP3+ cell pool and CCL19-mediated recruitment. *Biol Reprod.* 2011;85:397-408.
267. Teles A, Schumacher A, Kuhnle MC, et al. Control of uterine microenvironment by foxp3(+) cells facilitates embryo implantation. *Front Immunol.* 2013;4:158.
268. Teles A, Thuere C, Wafula PO, et al. Origin of Foxp3(+) cells during pregnancy. *Am J Clin Exp Immunol.* 2013;2:222-233.
269. Robertson SA, Guerin LR, Bromfield JJ, et al. Seminal fluid drives expansion of the CD4+CD25+ T regulatory cell pool and induces tolerance to paternal alloantigens in mice. *Biol Reprod.* 2009;80:1036-1045.
270. Chaturvedi V, Ertelt JM, Jiang TT, et al. CXCR3 blockade protects against Listeria monocytogenes infection-induced fetal wastage. *J Clin Invest.* 2015;125:1713-1725.
271. Kalekar LA, Schmiel SE, Nandiwada SL, et al. CD4(+) T cell anergy prevents autoimmunity and generates regulatory T cell precursors. *Nat Immunol.* 2016;17:304-314.
272. Andersen KG, Nissen JK, Betz AG. Comparative genomics reveals key gain-of-function events in Foxp3 during regulatory T cell evolution. *Front Immunol.* 2012;3:113.
273. Bonney EA, Onyekwuluje J. The H-Y response in mid-gestation and long after delivery in mice primed before pregnancy. *Immunol Invest.* 2003;32:71-81.
274. Fisher SJ, Laine RA. Accumulation of malto-oligosaccharides in the syncytiotrophoblastic cells of first-trimester human placentas. *Biochem J.* 1981;200:93-98.
275. Kshirsagar SK, Alam SM, Jasti S, et al. Immunomodulatory molecules are released from the first trimester and term placenta via exosomes. *Placenta.* 2012;33:982-990.
276. Currie GA, Van Doorninck W, Bagshawe KD. Effect of neuraminidase on the immunogenicity of early mouse trophoblast. *Nature.* 1968;219:191-192.
277. Taylor PV, Hancock KW, Gowland G. Effect of neuraminidase on immunogenicity of early mouse trophoblast. *Transplantation.* 1979;28:256-257.
278. Simmons RL, Lipschultz ML, Ray PK. Failure of neuraminidase to unmask histocompatibility antigens on trophoblast. *Nat New Biol.* 1971;231:111-112.
279. Beer AE, Billingham RE. Immunobiology of mammalian reproduction. *Adv Immunol.* 1971;14:1-84.
280. Whyte A, Loke YW. Increased sialylation of surface glycopeptides of human trophoblast compared with fetal cells from the same conceptus. *J Exp Med.* 1978;148:1087-1092.
281. Clark GF. The role of glycans in immune evasion: the human fetoembryonic defence system hypothesis revisited. *Mol Hum Reprod.* 2014;20:185-199.
282. Wolfert MA, Boons GJ. Adaptive immune activation: glycosylation does matter. *Nat Chem Biol.* 2013;9:776-784.
283. Pejcic-Karapetrovic B, Gurnani K, Russell MS, et al. Pregnancy impairs the innate immune resistance to *Salmonella typhimurium* leading to rapid fatal infection. *J Immunol.* 2007;179:6088-6096.
284. Rolle L, Memarzadeh Tehran M, Morell-Garcia A, et al. Cutting edge: IL-10-producing regulatory B cells in early human pregnancy. *Am J Reprod Immunol.* 2013;70:448-453.
285. Liu J, Chen X, Hao S, et al. Human chorionic gonadotropin and IL-35 contribute to the maintenance of peripheral immune tolerance during pregnancy through mediating the generation of IL-10(+) or IL-35(+) Breg cells. *Exp Cell Res.* 2019;383:111513.
286. Muzzio DO, Soldati R, Rolle L, et al. B-1a B cells regulate T cell differentiation associated with pregnancy disturbances. *Front Immunol.* 2014;5:6.
287. Guzman-Genuino RM, Eldi P, Garcia-Valtanen P, et al. Uterine B cells exhibit regulatory properties during the peri-implantation stage of murine pregnancy. *Front Immunol.* 2019;10:2899.
288. Jensen F, Muzzio D, Soldati R, et al. Regulatory B10 cells restore pregnancy tolerance in a mouse model. *Biol Reprod.* 2013;89:90.
289. Busse M, Campe KJ, Nowak D, et al. IL-10 producing B cells rescue mouse fetuses from inflammation-driven fetal death and are able to modulate T cell immune responses. *Sci Rep.* 2019;9:9335.
290. Silasi M, You Y, Simpson S, et al. Human chorionic gonadotropin modulates CXCL10 expression through histone methylation in human decidua. *Sci Rep.* 2020;10:5785.
291. Xin L, Ertelt JM, Rowe JH, et al. Cutting edge: committed Th1 CD4+ T cell differentiation blocks pregnancy-induced Foxp3 expression with antigen-specific fetal loss. *J Immunol.* 2014;192:2970-2974.
292. Chaouat G, Clark DA. FAS/FAS ligand interaction at the placental interface is not required for the success of allogeneic pregnancy in anti-paternal MHC preimmunized mice. *Am J Reprod Immunol.* 2001;45:108-115.
293. Norton MT, Fortner KA, Oppenheimer KH, et al. Evidence that CD8 T-cell homeostasis and function remain intact during murine pregnancy. *Immunology.* 2010;131:426-437.
294. Petroff MG, Perchellet A. B7 family molecules as regulators of the maternal immune system in pregnancy. *Am J Reprod Immunol.* 2010;63:506-519.
295. Svensson L, Arvola M, Sallstrom MA, et al. The Th2 cytokines IL-4 and IL-10 are not crucial for the completion of allogeneic pregnancy in mice. *J Reprod Immunol.* 2001;51:3-7.
296. Munn DH, Zhou M, Attwood JT, et al. Prevention of allogeneic fetal rejection by tryptophan catabolism. *Science.* 1998;281:1191-1193.
297. Tsuda S, Nakashima A, Shima T, et al. New paradigm in the role of regulatory T cells during pregnancy. *Front Immunol.* 2019;10:573.
298. Sasaki Y, Sakai M, Miyazaki S, et al. Decidual and peripheral blood CD4+CD25+ regulatory T cells in early pregnancy subjects and spontaneous abortion cases. *Mol Hum Reprod.* 2004;10:347-353.
299. Mei S, Tan J, Chen H, et al. Changes of CD4+CD25high regulatory T cells and FOXP3 expression in unexplained recurrent spontaneous abortion patients. *Fertil Steril.* 2010;94:2244-2247.
300. Jin LP, Chen QY, Zhang T, et al. The CD4+CD25 bright regulatory T cells and CTLA-4 expression in peripheral and decidual lymphocytes are down-regulated in human miscarriage. *Clin Immunol.* 2009;133:402-410.
301. Lee SK, Kim JY, Lee M, et al. Th17 and regulatory T cells in women with recurrent pregnancy loss. *Am J Reprod Immunol.* 2012;67:311-318.
302. Yang H, Qiu L, Chen G, et al. Proportional change of CD4+CD25+ regulatory T cells in decidua and peripheral blood in unexplained recurrent spontaneous abortion patients. *Fertil Steril.* 2008;89:656-661.
303. Miko E, Meggyes M, Doba K, et al. Immune checkpoint molecules in reproductive immunology. *Front Immunol.* 2019;10:846.
304. Xu W, Moor RJ, Walpole ET, et al. Pregnancy with successful foetal and maternal outcome in a melanoma patient treated with nivolumab in the first trimester: case report and review of the literature. *Melanoma Res.* 2019;29:333-337.
305. Masson E, Vidal C, Deschamps M, et al. Incidence and risk factors of anti-HLA immunization after pregnancy. *Hum Immunol.* 2013;74:946-951.
306. Urbaniak SJ, Greiss MA. RhD haemolytic disease of the fetus and the newborn. *Blood Rev.* 2000;14:44-61.
307. Webb J, Delaney M. Red blood cell alloimmunization in the pregnant patient. *Transfus Med Rev.* 2018;32:213-219.
308. Tovey LA. Oliver memorial lecture. Towards the conquest of Rh haemolytic disease: Britain's contribution and the role of serendipity. *Transfus Med.* 1992;2:99-109.
309. Arneth B. Neonatal immune incompatibilities between newborn and mother. *J Clin Med.* 2020;9:1470.
310. Kumpel BM, Elson CJ. Mechanism of anti-D-mediated immune suppression – a paradox awaiting resolution? *Trends Immunol.* 2001;22:26-31.
311. Stowell SR, Liepkalns JS, Hendrickson JE, et al. Antigen modulation confers protection to red blood cells from antibody through Fcgamma receptor ligation. *J Immunol.* 2013;191:5013-5025.
312. Girard-Pierce KR, Stowell SR, Smith NH, et al. A novel role for C3 in antibody-induced red blood cell clearance and antigen modulation. *Blood.* 2013;122:1793-1801.
313. Mener A, Patel SR, Arthur CM, et al. Antibody-mediated immunosuppression can result from RBC antigen loss independent of Fcgamma receptors in mice. *Transfusion.* 2019;59:371-384.
314. Bernardo L, Yu H, Amash A, et al. IgG-mediated immune suppression to erythrocytes by polyclonal antibodies can occur in the absence of activating or inhibitory Fcgamma receptors in a full mouse model. *J Immunol.* 2015;195:2224-2230.
315. Van Rood JJ, Eernisse JG, Van Leeuwen A. Leucocyte antibodies in sera from pregnant women. *Nature.* 1958;181:1735-1736.
316. Porrett PM. Biologic mechanisms and clinical consequences of pregnancy alloimmunization. *Am J Transplant.* 2018;18:1059-1067.
317. Honger G, Fornaro I, Granado C, et al. Frequency and determinants of pregnancy-induced child-specific sensitization. *Am J Transplant.* 2013;13:746-753.
318. Regan L, Braude PR, Hill DP. A prospective study of the incidence, time of appearance and significance of anti-paternal lymphocytotoxic antibodies in human pregnancy. *Hum Reprod.* 1991;6:294-298.
319. Bell SC, Billington WD. Anti-fetal allo-antibody in the pregnant female. *Immunol Rev.* 1983;75:5-30.
320. Ait-Azzouzene D, Gendron MC, Houdayer M, et al. Maternal B lymphocytes specific for paternal histocompatibility antigens are partially deleted during pregnancy. *J Immunol.* 1998;161:2677-2683.
321. Ait-Azzouzene D, Caucheteux S, Tchang F, et al. Transgenic major histocompatibility complex class I antigen expressed in mouse trophoblast affects maternal immature B cells. *Biol Reprod.* 2001;65:337-344.
322. Meuleman T, van Beelen E, Kaaja RJ, et al. HLA-C antibodies in women with recurrent miscarriage suggests that antibody mediated rejection is one of the mechanisms leading to recurrent miscarriage. *J Reprod Immunol.* 2016;116:28-34.
323. Lashley EE, Meuleman T, Claas FH. Beneficial or harmful effect of antipaternal human leukocyte antibodies on pregnancy outcome? A systematic review and meta-analysis. *Am J Reprod Immunol.* 2013;70:87-103.
324. Bartel G, Walch K, Wahrmann M, et al. Prevalence and qualitative properties of circulating anti-human leukocyte antigen alloantibodies after pregnancy: no association with unexplained recurrent miscarriage. *Hum Immunol.* 2011;72:187-192.
325. Wegmann TG. The presence of class I MHC antigens at the maternal-fetal interface and hypotheses concerning the survival of the murine fetal allograft. *J Reprod Immunol.* 1981;3:267-270.
326. Girardi G, Bulla R, Salmon JE, et al. The complement system in the pathophysiology of pregnancy. *Mol Immunol.* 2006;43:68-77.
327. Tedesco F, Narchi G, Radillo O, et al. Susceptibility of human trophoblast to killing by human complement and the role of the complement regulatory proteins. *J Immunol.* 1993;151:1562-1570.
328. Xu C, Mao D, Holers VM, et al. A critical role for murine complement regulator Crry in fetomaternal tolerance. *Science.* 2000;287:498-501.
329. Abeln M, Albers I, Peters-Bernard U, et al. Sialic acid is a critical fetal defense against maternal complement attack. *J Clin Invest.* 2019;129:422-436.
330. Salmon JE, Mineo C, Giles I, et al. *Antiphospholipid Syndrome*; 2017:117-143.
331. Tedesco F, Borghi MO, Gerosa M, et al. Pathogenic role of complement in antiphospholipid syndrome and therapeutic implications. *Front Immunol.* 2018;9:1388.
332. Holers VM, Girardi G, Mo L, et al. Complement C3 activation is required for antiphospholipid antibody-induced fetal loss. *J Exp Med.* 2002;195:211-220.
333. Brucato A, Cimaz R, Caporali R, et al. Pregnancy outcomes in patients with autoimmune diseases and anti-Ro/SSA antibodies. *Clin Rev Allergy Immunol.* 2011;40:27-41.
334. Gendron RL, Nestel FP, Lapp WS, et al. Lipopolysaccharide-induced fetal resorption in mice is associated with the intra-uterine production of tumour necrosis factor-alpha. *J Reprod Fertil.* 1990;90:395-402.
335. Erlebacher A, Zhang D, Parlow AF, et al. Ovarian insufficiency and early pregnancy loss induced by activation of the innate immune system. *J Clin Invest.* 2004;114:39-48.
336. Ostrand-Rosenberg S, Sinha P, Figley C, et al. Frontline Science: myeloid-derived suppressor cells (MDSCs) facilitate maternal-fetal tolerance in mice. *J Leukoc Biol.* 2017;101:1091-1101.
337. Bonney EA, Brown SA. To drive or be driven: the path of a mouse model of recurrent pregnancy loss. *Reproduction.* 2014;147:R153-R167.
338. Baban B, Chandler P, McCool D, et al. Indoleamine 2,3-dioxygenase expression is restricted to fetal trophoblast giant cells during murine gestation and is maternal genome specific. *J Reprod Immunol.* 2004;61:67-77.
339. Moldenhauer LM, Diener KR, Hayball JD, et al. An immunogenic phenotype in paternal antigen-specific CD8(+) T cells at embryo implantation elicits later fetal loss in mice. *Immunol Cell Biol.* 2017;95:705-715.
340. Arruvito L, Sanz M, Banham AH, et al. Expansion of CD4+CD25+and FOXP3+ regulatory T cells during the follicular phase of the menstrual cycle: implications for human reproduction. *J Immunol.* 2007;178:2572-2578.
341. Wang WJ, Hao CF, Yi L, et al. Increased prevalence of T helper 17 (Th17) cells in peripheral blood and decidua in unexplained recurrent spontaneous abortion patients. *J Reprod Immunol.* 2010;84:164-170.
342. Nakashima A, Ito M, Shima T, et al. Accumulation of IL-17-positive cells in decidua of inevitable abortion cases. *Am J Reprod Immunol.* 2010;64:4-11.
343. Ji J, Zhai H, Zhou H, et al. The role and mechanism of vitamin D-mediated regulation of Treg/Th17 balance in recurrent pregnancy loss. *Am J Reprod Immunol.* 2019;81:e13112.
344. Lee SK, Kim JY, Hur SE, et al. An imbalance in interleukin-17-producing T and Foxp3(+) regulatory T cells in women with idiopathic recurrent pregnancy loss. *Hum Reprod.* 2011;26:2964-2971.

345. Jauniaux E, Burton GJ. Pathophysiology of histological changes in early pregnancy loss. *Placenta*. 2005;26:114-123.
346. Kim CJ, Romero R, Chaemsaithong P, et al. Chronic inflammation of the placenta: definition, classification, pathogenesis, and clinical significance. *Am J Obstet Gynecol*. 2015;213:S53-S69.
347. Redline RW, Patterson P. Villitis of unknown etiology is associated with major infiltration of fetal tissue by maternal inflammatory cells. *Am J Pathol*. 1993;143:473-479.
348. Tamblyn JA, Lissauer DM, Powell R, et al. The immunological basis of villitis of unknown etiology – review. *Placenta*. 2013;34:846-855.
349. Kim MJ, Romero R, Kim CJ, et al. Villitis of unknown etiology is associated with a distinct pattern of chemokine up-regulation in the feto-maternal and placental compartments: implications for conjoint maternal allograft rejection and maternal anti-fetal graft-versus-host disease. *J Immunol*. 2009;182:3919-3927.
350. Kim JS, Romero R, Kim MR, et al. Involvement of Hofbauer cells and maternal T cells in villitis of unknown aetiology. *Histopathology*. 2008;52:457-464.
351. Enninga EAL, Leontovich AA, Fedyshyn B, et al. Upregulation of HLA-class I and II in placentas diagnosed with villitis of unknown etiology. *Reprod Sci*. 2020;27:1129-1138.
352. Ito Y, Matsuoka K, Uesato T, et al. Increased expression of perforin, granzyme B, and C5b-9 in villitis of unknown etiology. *Placenta*. 2015;36:531-537.
353. Enninga EAL, Raber P, Quinton RA, et al. Maternal T cells in the human placental villi support an allograft response during noninfectious villitis. *J Immunol*. 2020;204:2931-2939.
354. Lee J, Romero R, Xu Y, et al. Maternal HLA panel-reactive antibodies in early gestation positively correlate with chronic chorioamnionitis: evidence in support of the chronic nature of maternal anti-fetal rejection. *Am J Reprod Immunol*. 2011;66:510-526.
355. K AL, Kim YW, Shim JY, et al. Distinct patterns of C4d immunoreactivity in placentas with villitis of unknown etiology, cytomegaloviral placentitis, and infarct. *Placenta*. 2013;34:432-435.
356. Rudzinski E, Gilroy M, Newbill C, et al. Positive C4d immunostaining of placental villous syncytiotrophoblasts supports host-versus-graft rejection in villitis of unknown etiology. *Pediatr Dev Pathol*. 2013;16:7-13.
357. Derricott H, Jones RL, Greenwood SL, et al. Characterizing villitis of unknown etiology and inflammation in stillbirth. *Am J Pathol*. 2016;186:952-961.
358. Frascoli M, Coniglio L, Witt R, et al. Alloreactive fetal T cells promote uterine contractility in preterm labor via IFN-gamma and TNF-alpha. *Sci Transl Med*. 2018;10:eaan2263.
359. Benichou G, Yamada Y, Aoyama A, et al. Natural killer cells in rejection and tolerance of solid organ allografts. *Curr Opin Organ Transplant*. 2011;16:47-53.
360. Kopcow HD, Allan DS, Chen X, et al. Human decidual NK cells form immature activating synapses and are not cytotoxic. *Proc Natl Acad Sci U S A*. 2005;102:15563-15568.
361. King A, Kalra P, Loke YW. Human trophoblast cell resistance to decidual NK lysis is due to lack of NK target structure. *Cell Immunol*. 1990;127:230-237.
362. Siewiera J, El Costa H, Tabiasco J, et al. Human cytomegalovirus infection elicits new decidual natural killer cell effector functions. *PLoS Pathog*. 2013;9:e1003257.
363. Stewart IJ, Peel S. Mouse metrial gland cells do not kill Yac-1 myeloma cells. *J Reprod Immunol*. 1993;24:165-171.
364. Crespo AC, Strominger JL, Tilburgs T. Expression of KIR2DS1 by decidual natural killer cells increases their ability to control placental HCMV infection. *Proc Natl Acad Sci U S A*. 2016;113:15072-15077.
365. Tilburgs T, Evans JH, Crespo AC, et al. The HLA-G cycle provides for both NK tolerance and immunity at the maternal-fetal interface. *Proc Natl Acad Sci U S A*. 2015;112:13312-13317.
366. Apps R, Sharkey A, Gardner L, et al. Ex vivo functional responses to HLA-G differ between blood and decidual NK cells. *Mol Hum Reprod*. 2011;17:577-586.
367. Robbins JR, Bakardjiev AI. Pathogens and the placental fortress. *Curr Opin Microbiol*. 2012;15:36-43.
368. Arora N, Sadovsky Y, Dermody TS, et al. Microbial vertical transmission during human pregnancy. *Cell Host Microbe*. 2017;21:561-567.
369. Zeldovich VB, Bakardjiev AI. Host defense and tolerance: unique challenges in the placenta. *PLoS Pathog*. 2012;8:e1002804.
370. Hamadeh MA, Glassroth J. Tuberculosis and pregnancy. *Chest*. 1992;101:1114-1120.
371. Cox AL, Mosbruger T, Mao Q, et al. Cellular immune selection with hepatitis C virus persistence in humans. *J Exp Med*. 2005;201:1741-1752.
372. Honegger JR, Kim S, Price AA, et al. Loss of immune escape mutations during persistent HCV infection in pregnancy enhances replication of vertically transmitted viruses. *Nat Med*. 2013;19:1529-1533.
373. Gervais A, Bacq Y, Bernuau J, et al. Decrease in serum ALT and increase in serum HCV RNA during pregnancy in women with chronic hepatitis C. *J Hepatol*. 2000;32:293-299.
374. Coss SL, Torres-Cornejo A, Prasad MR, et al. CD4+ T cell restoration and control of hepatitis C virus replication after childbirth. *J Clin Invest*. 2020;130:748-753.
375. Honegger JR, Tedesco D, Kohout JA, et al. Influence of IFNL3 and HLA-DPB1 genotype on postpartum control of hepatitis C virus replication and T-cell recovery. *Proc Natl Acad Sci U S A*. 2016;113:10684-10689.
376. Hashem M, Jhaveri R, Saleh DA, et al. Spontaneous viral load decline and subsequent clearance of chronic hepatitis C virus in postpartum women correlates with favorable interleukin-28B gene allele. *Clin Infect Dis*. 2017;65:999-1005.
377. Harris JW. Influenza occurring in pregnant women. *J Am Med Assoc*. 1919;72:970-980.
378. Chan KH, Zhang AJ, To KK, et al. Wild type and mutant 2009 pandemic influenza A (H1N1) viruses cause more severe disease and higher mortality in pregnant BALB/c mice. *PloS One*. 2010;5:e13757.
379. Marcelin G, Aldridge JR, Duan S, et al. Fatal outcome of pandemic H1N1 2009 influenza virus infection is associated with immunopathology and impaired lung repair, not enhanced viral burden, in pregnant mice. *J Virol*. 2011;85:11208-11219.
380. Lauzon-Joset JF, Scott NM, Mincham KT, et al. Pregnancy induces a steady-state shift in alveolar macrophage M1/M2 phenotype that is associated with a heightened severity of influenza virus infection: mechanistic insight using mouse models. *J Infect Dis*. 2019;219:1823-1831.
381. Engels G, Hierweger AM, Hoffmann J, et al. Pregnancy-related immune adaptation promotes the emergence of highly virulent H1N1 influenza virus strains in allogenically pregnant mice. *Cell Host Microbe*. 2017;21:321-333.
382. Zheng R, Qin X, Li Y, et al. Imbalanced anti-H1N1 immunoglobulin subclasses and dysregulated cytokines in hospitalized pregnant women with 2009 H1N1 influenza and pneumonia in Shenyang, China. *Hum Immunol*. 2012;73:906-911.
383. Mayer AE, Parks GD. An AGM model for changes in complement during pregnancy: neutralization of influenza virus by serum is diminished in late third trimester. *PLoS One*. 2014;9:e112749.
384. Forbes RL, Wark PA, Murphy VE, et al. Pregnant women have attenuated innate interferon responses to 2009 pandemic influenza A virus subtype H1N1. *J Infect Dis*. 2012;206:646-653.
385. Wise RA, Polito AJ, Krishnan V. Respiratory physiologic changes in pregnancy. *Immunol Allergy Clin North Am*.. 2006;26:1-12.
386. Revello MG, Lilleri D, Zavattoni M, et al. Lymphoproliferative response in primary human cytomegalovirus (HCMV) infection is delayed in HCMV transmitter mothers. *J Infect Dis*. 2006;193:269-276.
387. Constantin CM, Masopust D, Gourley T, et al. Normal establishment of virus-specific memory CD8 T cell pool following primary infection during pregnancy. *J Immunol*. 2007;179:4383-4389.
388. Han YW, Redline RW, Li M, et al. Fusobacterium nucleatum induces premature and term stillbirths in pregnant mice: implication of oral bacteria in preterm birth. *Infect Immun*. 2004;72:2272-2279.
389. Redline RW, Lu CY. Role of local immunosuppression in murine fetoplacental listeriosis. *J Clin Invest*. 1987;79:1234-1241.
390. Longman RE, Johnson TR. Viral respiratory disease in pregnancy. *Curr Opin Obstet Gynecol*. 2007;19:120-125.
391. Wong SF, Chow KM, Leung TN, et al. Pregnancy and perinatal outcomes of women with severe acute respiratory syndrome. *Am J Obstet Gynecol*. 2004;191:292-297.
392. Breslin N, Baptiste C, Gyamfi-Bannerman C, et al. COVID-19 infection among asymptomatic and symptomatic pregnant women: two weeks of confirmed presentations to an affiliated pair of New York City hospitals. *Am J Obstet Gynecol MFM*. 2020;2:100118.
393. Dashraath P, Wong JLJ, Lim MXK, et al. Coronavirus disease 2019 (COVID-19) pandemic and pregnancy. *Am J Obstet Gynecol*. 2020;222:521-531.
394. Schlaudecker EP, McNeal MM, Dodd CN, et al. Pregnancy modifies the antibody response to trivalent influenza immunization. *J Infect Dis*. 2012;206:1670-1673.
395. Misra RS, Nayak JL. The importance of vaccinating children and pregnant women against influenza virus infection. *Pathogens*. 2019;8:265.
396. Thompson MG, Li DK, Shifflett P, et al. Effectiveness of seasonal trivalent influenza vaccine for preventing influenza virus illness among pregnant women: a population-based case-control study during the 2010-2011 and 2011-2012 influenza seasons. *Clin Infect Dis*. 2014;58:449-457.
397. Jackson LA, Patel SM, Swamy GK, et al. Immunogenicity of an inactivated monovalent 2009 H1N1 influenza vaccine in pregnant women. *J Infect Dis*. 2011;204:854-863.
398. Tamma PD, Ault KA, del Rio C, et al. Safety of influenza vaccination during pregnancy. *Am J Obstet Gynecol*. 2009;201:547-552.
399. Chu HY, Englund JA. Maternal immunization. *Clin Infect Dis*. 2014;59:560-568.
400. da Silva FC, Magaldi FM, Sato HK, et al. Yellow fever vaccination in a mouse model is associated with uninterrupted pregnancies and viable neonates except when administered at implantation period. *Front Microbiol*. 2020;11:245.
401. Brown JA, Singh G, Acklin JA, et al. Dengue virus immunity increases Zika virus-induced damage during pregnancy. *Immunity*. 2019;50:751-762.e5.
402. Rodriguez-Barraquer I, Costa F, Nascimento EJM, et al. Impact of preexisting dengue immunity on Zika virus emergence in a dengue endemic region. *Science*. 2019;363:607-610.
403. Zaman K, Roy E, Arifeen SE, et al. Effectiveness of maternal influenza immunization in mothers and infants. *N Engl J Med*. 2008;359:1555-1564.
404. Nunes MC, Cutland CL, Jones S, et al. Duration of infant protection against influenza illness conferred by maternal immunization: secondary analysis of a randomized clinical trial. *JAMA Pediatr*. 2016;170:840-847.
405. Roper MH, Vandelaer JH, Gasse FL. Maternal and neonatal tetanus. *Lancet*. 2007;370:1947-1959.
406. Achievements in public health: elimination of rubella and congenital rubella syndrome-US, 1969-2004. *Ann Pharmacother*. 2005;39:1151-1152.
407. Moniz MH, Beigi RH. Maternal immunization. Clinical experiences, challenges, and opportunities in vaccine acceptance. *Hum Vaccin Immunother*. 2014;10:2562-2570.
408. Englund J, Glezen WP, Piedra PA. Maternal immunization against viral disease. *Vaccine*. 1998;16:1456-1463.
409. Willis E, Pardi N, Parkhouse K, et al. Nucleoside-modified mRNA vaccination partially overcomes maternal antibody inhibition of de novo immune responses in mice. *Sci Transl Med*. 2020;12:eaav5701.
410. Vornhagen J, Adams Waldorf KM, Rajagopal L. Perinatal group B Streptococcal infections: virulence factors, immunity, and prevention strategies. *Trends Microbiol*. 2017;25:919-931.
411. Waites KB, Schelonka RL, Xiao L, et al. Congenital and opportunistic infections: ureaplasma species and Mycoplasma hominis. *Semin Fetal Neonatal Med*. 2009;14:190-199.
412. Garcia-Flores J, Cruceyra M, Canamares M, et al. Candida chorioamnionitis: report of two cases and review of literature. *J Obstet Gynaecol*. 2016;36:843-844.
413. Charlier C, Disson O, Lecuit M. Maternal-neonatal listeriosis. *Virulence*. 2020;11:391-397.
414. Vigliani MB, Bakardjiev AI. First trimester typhoid Fever with vertical transmission of salmonella typhi, an intracellular organism. *Case Rep Med*. 2013;2013:973297.
415. Chattopadhyay A, Robinson N, Sandhu JK, et al. *Salmonella enterica* Serovar typhimurium-induced placental inflammation and not bacterial burden correlates with pathology and fatal maternal disease. *Infect Immun*. 2010;78:2292-2301.
416. Rac MW, Revell PA, Eppes CS. Syphilis during pregnancy: a preventable threat to maternal-fetal health. *Am J Obstet Gynecol*. 2017;216:352-363.
417. Han YW. Fusobacterium nucleatum: a commensal-turned pathogen. *Curr Opin Microbiol*. 2015;23:141-147.
418. Borges M, Magalhães Silva T, Brito C, et al. How does toxoplasmosis affect the maternal-foetal immune interface and pregnancy? *Parasite Immunol*. 2019;41:e12606.
419. Kemmerling U, Osuna A, Schijman AG, et al. Congenital transmission of Trypanosoma cruzi: a review about the interactions between the parasite, the placenta, the maternal and the fetal/neonatal immune responses. *Front Microbiol*. 2019;10:1854.
420. Gouilly J, Chen Q, Siewiera J, et al. Genotype specific pathogenicity of hepatitis E virus at the human maternal-fetal interface. *Nat Commun*. 2018;9:4748.
421. Pereira L. Congenital viral infection: traversing the uterine-placental interface. *Annu Rev Virol*. 2018;5:273-299.
422. Vigliani MB, Bakardjiev AI. Intracellular organisms as placental invaders. *Fetal Matern Med Rev*. 2014;25:332-338.
423. Desai M, Hill J, Fernandes S, et al. Prevention of malaria in pregnancy. *Lancet Infect Dis*. 2018;18:e119-e132.
424. Faralla C, Rizzuto GA, Lowe DE, et al. InlP, a new virulence factor with strong placental tropism. *Infect Immun*. 2016;84:3584-3596.
425. Coppenhagen-Glazer S, Sol A, Abed J, et al. Fap2 of Fusobacterium nucleatum is a galactose-inhibitable adhesin involved in coaggregation, cell adhesion, and preterm birth. *Infect Immun*. 2015;83:1104-1113.
426. Whidbey C, Harrell MI, Burnside K, et al. A hemolytic pigment of Group B *Streptococcus* allows bacterial penetration of human placenta. *J Exp Med*. 2013;210:1265-1281.
427. Salanti A, Staalsoe T, Lavstsen T, et al. Selective upregulation of a single distinctly structured var gene in chondroitin sulphate A-adhering Plasmodium falciparum involved in pregnancy-associated malaria. *Mol Microbiol*. 2003;49:179-191.
428. Fauteux-Daniel S, Larouche A, Calderon V, et al. Vertical transmission of hepatitis C virus: variable transmission bottleneck and evidence of midgestation In Utero Infection. *J Virol*. 2017;91:e01372-17.
429. Wolinsky SM, Wike CM, Korber BT, et al. Selective transmission of human immunodeficiency virus type-1 variants from mothers to infants. *Science*. 1992;255:1134-1137.
430. Yockey LJ, Lucas C, Iwasaki A. Contributions of maternal and fetal antiviral immunity in congenital disease. *Science*. 2020;368:608-612.
431. Wira CR, Rodriguez-Garcia M, Patel MV. The role of sex hormones in immune protection of the female reproductive tract. *Nat Rev Immunol*. 2015;15:217-230.
432. Yarbrough VL, Winkle S, Herbst-Kralovetz MM. Antimicrobial peptides in the female reproductive tract: a critical component of the mucosal immune barrier with

physiological and clinical implications. *Hum Reprod Update.* 2015;21:353-377.

433. Caine EA, Scheaffer SM, Arora N, et al. Interferon lambda protects the female reproductive tract against Zika virus infection. *Nat Commun.* 2019;10:280.
434. Becher N, Adams Waldorf K, Hein M, et al. The cervical mucus plug: structured review of the literature. *Acta Obstet Gynecol Scand.* 2009;88:502-513.
435. Akgul Y, Word RA, Ensign LM, et al. Hyaluronan in cervical epithelia protects against infection-mediated preterm birth. *J Clin Invest.* 2014;124:5481-5489.
436. Vornhagen J, Quach P, Boldenow E, et al. Bacterial hyaluronidase promotes ascending GBS infection and preterm birth. *mBio.* 2016;7:e00781.
437. Robbins JR, Skrzypczynska KM, Zeldovich VB, et al. Placental syncytiotrophoblast constitutes a major barrier to vertical transmission of Listeria monocytogenes. *PLoS Pathog.* 2010;6:e1000732.
438. Zeldovich VB, Clausen CH, Bradford E, et al. Placental syncytium forms a biophysical barrier against pathogen invasion. *PLoS Pathog.* 2013;9:e1003821.
439. Robbins JR, Zeldovich VB, Poukchanski A, et al. Tissue barriers of the human placenta to infection with Toxoplasma gondii. *Infect Immun.* 2012;80:418-428.
440. Koi H, Zhang J, Makrigiannakis A, et al. Syncytiotrophoblast is a barrier to maternal-fetal transmission of herpes simplex virus. *Biol Reprod.* 2002;67:1572-1579.
441. Wells AI, Coyne CB. Type III interferons in antiviral defenses at barrier surfaces. *Trends Immunol.* 2018;39:848-858.
442. Corry J, Arora N, Good CA, et al. Organotypic models of type III interferon-mediated protection from Zika virus infections at the maternal-fetal interface. *Proc Natl Acad Sci U S A.* 2017;114:9433-9438.
443. Bayer A, Lennemann NJ, Ouyang Y, et al. Type III interferons produced by human placental trophoblasts confer protection against Zika virus infection. *Cell Host Microbe.* 2016;19:705-712.
444. Sheridan MA, Yunusov D, Balaraman V, et al. Vulnerability of primitive human placental trophoblast to Zika virus. *Proc Natl Acad Sci U S A.* 2017;114:E1587-E1596.
445. Mouillet JF, Ouyang Y, Bayer A, et al. The role of trophoblastic microRNAs in placental viral infection. *Int J Dev Biol.* 2014;58:281-289.
446. Delorme-Axford E, Donker RB, Mouillet JF, et al. Human placental trophoblasts confer viral resistance to recipient cells. *Proc Natl Acad Sci U S A.* 2013;110:12048-12053.
447. Johnson EL, Chakraborty R. Placental Hofbauer cells limit HIV-1 replication and potentially offset mother to child transmission (MTCT) by induction of immunoregulatory cytokines. *Retrovirology.* 2012;9:101.
448. Maidji E, McDonagh S, Genbacev O, et al. Maternal antibodies enhance or prevent cytomegalovirus infection in the placenta by neonatal Fc receptor-mediated transcytosis. *Am J Pathol.* 2006;168:1210-1226.
449. Fried M, Duffy PE. Adherence of Plasmodium falciparum to chondroitin sulfate A in the human placenta. *Science.* 1996;272:1502-1504.
450. Fried M, Nosten F, Brockman A, et al. Maternal antibodies block malaria. *Nature.* 1998;395:851-852.
451. Fried M, Muga RO, Misore AO, et al. Malaria elicits type 1 cytokines in the human placenta: IFN-gamma and TNF-alpha associated with pregnancy outcomes. *J Immunol.* 1998;160:2523-2530.
452. Muehlenbachs A, Fried M, Lachowitzer J, et al. Genome-wide expression analysis of placental malaria reveals features of lymphoid neogenesis during chronic infection. *J Immunol.* 2007;179:557-565.
453. Bauserman M, Conroy AL, North K, et al. An overview of malaria in pregnancy. *Semin Perinatol.* 2019;43:282-290.
454. Bakardjiev AI, Theriot JA, Portnoy DA. Listeria monocytogenes traffics from maternal organs to the placenta and back. *PLoS Pathog.* 2006;2:e66.
455. Redline RW, McKay DB, Vazquez MA, et al. Macrophage functions are regulated by the substratum of murine decidual stromal cells. *J Clin Invest.* 1990;85:1951-1958.
456. Redline RW, Lu CY. Specific defects in the anti-listerial immune response in discrete regions of the murine uterus and placenta account for susceptibility to infection. *J Immunol.* 1988;140:3947-3955.
457. Redline RW, Shea CM, Papaioannou VE, et al. Defective anti-listerial responses in deciduoma of pseudopregnant mice. *Am J Pathol.* 1988;133:485-497.
458. Pamer EG. Immune responses to Listeria monocytogenes. *Nat Rev Immunol.* 2004;4:812-823.
459. Guleria I, Pollard JW. The trophoblast is a component of the innate immune system during pregnancy. *Nat Med.* 2000;6:589-593.
460. Crespo AC, Mulik S, Dotiwala F, et al. Decidual NK cells transfer granulysin to selectively kill bacteria in trophoblasts. *Cell.* 2020;82:1125-1139.e18.
461. Gomez-Lopez N, Romero R, Garcia-Flores V, et al. Amniotic fluid neutrophils can phagocytize bacteria: a mechanism for microbial killing in the amniotic cavity. *Am J Reprod Immunol.* 2017;78:e12723.
462. Tong M, Potter JA, Mor G, et al. Lipopolysaccharide-stimulated human fetal membranes induce neutrophil activation and release of vital neutrophil extracellular traps. *J Immunol.* 2019;203:500-510.
463. Gomez-Lopez N, Romero R, Leng Y, et al. Neutrophil extracellular traps in acute chorioamnionitis: a mechanism of host defense. *Am J Reprod Immunol.* 2017;77:e12617.
464. Zeldovich VB, Robbins JR, Kapidzic M, et al. Invasive extravillous trophoblasts restrict intracellular growth and spread of Listeria monocytogenes. *PLoS Pathog.* 2011;7:e1002005.
465. Benirschke K, Burton GJ, Baergen RN. *Pathology of the Human Placenta.* Springer; 2006.
466. Drevets DA. Dissemination of Listeria monocytogenes by infected phagocytes. *Infect Immun.* 1999;67:3512-3517.
467. Pereira L, Maidji E, McDonagh S, et al. Human cytomegalovirus transmission from the uterus to the placenta correlates with the presence of pathogenic bacteria and maternal immunity. *J Virol.* 2003;77:13301-13314.
468. Klenerman P, Oxenius A. T cell responses to cytomegalovirus. *Nat Rev Immunol.* 2016;16:367-377.
469. Aronoff DM, Correa H, Rogers LM, et al. Placental pericytes and cytomegalovirus infectivity: implications for HCMV placental pathology and congenital disease. *Am J Reprod Immunol.* 2017;78:e12728.
470. Tamiolakis D, Venizelos I, Lambropoulou M, et al. Human decidual cells activity in women with spontaneous abortions of probable CMV aetiology during the first trimester of gestation. An immunohistochemical study with CMV-associated antigen. *Acta Med.* 2004;47:195-199.
471. El Costa H, Quillay H, Marlin R, et al. The local environment orchestrates mucosal decidual macrophage differentiation and substantially inhibits HIV-1 replication. *Mucosal Immunol.* 2016;9:634-646.
472. Quillay H, El Costa H, Duriez M, et al. NK cells control HIV-1 infection of macrophages through soluble factors and cellular contacts in the human decidua. *Retrovirology.* 2016;13:39.
473. Garcez PP, Loiola EC, Madeiro da Costa R, et al. Zika virus impairs growth in human neurospheres and brain organoids. *Science.* 2016;352:816-818.
474. Li C, Xu D, Ye Q, et al. Zika virus disrupts neural progenitor development and leads to microcephaly in mice. *Cell Stem Cell.* 2016;19:120-126.
475. de Vries LS. Viral infections and the neonatal brain. *Semin Pediatr Neurol.* 2019;32:100769.
476. Bonvicini F, Bua G, Gallinella G. Parvovirus B19 infection in pregnancy-awareness and opportunities. *Curr Opin Virol.* 2017;27:8-14.
477. George S, Viswanathan R, Sapkal GN. Molecular aspects of the teratogenesis of rubella virus. *Biol Res.* 2019;52:47.
478. Tulina NM, Brown AG, Barila GO, et al. The absence of TLR4 prevents fetal brain injury in the setting of intrauterine inflammation. *Reprod Sci.* 2019;26:1082-1093.
479. Suff N, Karda R, Diaz JA, et al. Ascending vaginal infection using bioluminescent bacteria evokes intrauterine inflammation, preterm birth, and neonatal brain injury in pregnant mice. *Am J Pathol.* 2018;188:2164-2176.
480. Humann J, Mann B, Gao G, et al. Bacterial peptidoglycan traverses the placenta to induce fetal neuroproliferation and aberrant postnatal behavior. *Cell Host Microbe.* 2016;19:388-399.
481. Dollner H, Vatten L, Halgunset J, et al. Histologic chorioamnionitis and umbilical serum levels of pro-inflammatory cytokines and cytokine inhibitors. *Br J Obstet Gynaecol.* 2002;109:534-539.
482. Redline RW, Wilson-Costello D, Borawski E, et al. Placental lesions associated with neurologic impairment and cerebral palsy in very low-birth-weight infants. *Arch Pathol Lab Med.* 1998;122:1091-1098.
483. Roescher AM, Timmer A, van der Laan ME, et al. In preterm infants, ascending intrauterine infection is associated with lower cerebral tissue oxygen saturation and higher oxygen extraction. *Pediatr Res.* 2015;77:688-695.
484. Yang D, Sun YY, Bhaumik SK, et al. Blocking lymphocyte trafficking with FTY720 prevents inflammation-sensitized hypoxic-ischemic brain injury in newborns. *J Neurosci.* 2014;34:16467-16481.
485. Kramer BW, Kallapur S, Newnham J, et al. Prenatal inflammation and lung development. *Semin Fetal Neonatal Med.* 2009;14:2-7.
486. Barry H III, Barry H Jr. Season of birth. An epidemiological study in psychiatry. *Arch Gen Psychiatry.* 1961;5:292-300.
487. Brown AS, Derkits EJ. Prenatal infection and schizophrenia: a review of epidemiologic and translational studies. *Am J Psychiatry.* 2010;167:261-280.
488. Khandaker GM, Zimbron J, Lewis G, et al. Prenatal maternal infection, neurodevelopment and adult schizophrenia: a systematic review of population-based studies. *Psychol Med.* 2013;43:239-257.
489. Jiang HY, Xu LL, Shao L, et al. Maternal infection during pregnancy and risk of autism spectrum disorders: a systematic review and meta-analysis. *Brain Behav Immun.* 2016;58:165-172.
490. Al-Haddad BJS, Jacobsson B, Chabra S, et al. Long-term risk of neuropsychiatric disease after exposure to infection in utero. *JAMA Psychiatry.* 2019;76:594-602.
491. Al-Haddad BJS, Oler E, Armistead B, et al. The fetal origins of mental illness. *Am J Obstet Gynecol.* 2019;221:549-562.
492. Choi GB, Yim YS, Wong H, et al. The maternal interleukin-17a pathway in mice promotes autism-like phenotypes in offspring. *Science.* 2016;351:933-939.
493. Doyle RM, Alber DG, Jones HE, et al. Term and preterm labour are associated with distinct microbial community structures in placental membranes which are independent of mode of delivery. *Placenta.* 2014;35:1099-1101.
494. Bianchi-Jassir F, Seale AC, Kohli-Lynch M, et al. Preterm birth associated with group B Streptococcus maternal colonization Worldwide: systematic review and meta-analyses. *Clin Infect Dis.* 2017;65:S133-S42.
495. Brown RG, Marchesi JR, Lee YS, et al. Vaginal dysbiosis increases risk of preterm fetal membrane rupture, neonatal sepsis and is exacerbated by erythromycin. *BMC Med.* 2018;16:9.
496. McCuaig R, Wong D, Gardiner FW, et al. Periodontal pathogens in the placenta and membranes in term and preterm birth. *Placenta.* 2018;68:40-43.
497. Adams Waldorf KM, McAdams RM. Influence of infection during pregnancy on fetal development. *Reproduction.* 2013;146:R151-R162.
498. Surve MV, Anil A, Kamath KG, et al. Membrane vesicles of group B Streptococcus disrupt feto-maternal barrier leading to preterm birth. *PLoS Pathog.* 2016;12:e1005816.
499. Romero R, Miranda J, Chaiworapongsa T, et al. Prevalence and clinical significance of sterile intra-amniotic inflammation in patients with preterm labor and intact membranes. *Am J Reprod Immunol.* 2014;72:458-474.
500. Cardenas I, Means RE, Aldo P, et al. Viral infection of the placenta leads to fetal inflammation and sensitization to bacterial products predisposing to preterm labor. *J Immunol.* 2010;185:1248-1257.
501. McGee D, Smith A, Poncil S, et al. Cervical HSV-2 infection causes cervical remodeling and increases risk for ascending infection and preterm birth. *PLoS One.* 2017;12:e0188645.
502. Racicot K, Kwon JY, Aldo P, et al. Type I interferon regulates the placental inflammatory response to bacteria and is targeted by virus: mechanism of polymicrobial infection-induced preterm birth. *Am J Reprod Immunol.* 2016;75:451-460.
503. Goldenberg RL, McClure EM, Saleem S, et al. Infection-related stillbirths. *Lancet.* 2010;375:1482-1490.
504. Komine-Aizawa S, Suzaki A, Trinh QD, et al. H1N1/09 influenza A virus infection of immortalized first trimester human trophoblast cell lines. *Am J Reprod Immunol.* 2012;68:226-232.
505. Angelova M, Zwezdaryk K, Ferris M, et al. Human cytomegalovirus infection dysregulates the canonical Wnt/beta-catenin signaling pathway. *PLoS Pathog.* 2012;8:e1002959.
506. van Zuylen WJ, Ford CE, Wong DD, et al. Human cytomegalovirus modulates expression of noncanonical Wnt receptor ROR2 to alter trophoblast migration. *J Virol.* 2016;90:1108-1115.
507. Tabata T, Petitt M, Fang-Hoover J, et al. Cytomegalovirus impairs cytotrophoblast-induced lymphangiogenesis and vascular remodeling in an in vivo human placentation model. *Am J Pathol.* 2012;181:1540-1559.
508. Tabata T, Kawakatsu H, Maidji E, et al. Induction of an epithelial integrin alphavbeta6 in human cytomegalovirus-infected endothelial cells leads to activation of transforming growth factor-beta1 and increased collagen production. *Am J Pathol.* 2008;172:1127-1140.
509. Buchrieser J, Degrelle SA, Couderc T, et al. IFITM proteins inhibit placental syncytiotrophoblast formation and promote fetal demise. *Science.* 2019;365:176-180.
510. Yockey LJ, Iwasaki A. Interferons and proinflammatory cytokines in pregnancy and fetal development. *Immunity.* 2018;49:397-412.
511. Stinson LF, Payne MS. Infection-mediated preterm birth: bacterial origins and avenues for intervention. *Aust N Z J Obstet Gynaecol.* 2019;59:781-790.
512. Kiss H, Petricevic L, Husslein P. Prospective randomised controlled trial of an infection screening programme to reduce the rate of preterm delivery. *Br Med J.* 2004;329:371.
513. Roberts CL, Algert CS, Rickard KL, et al. Treatment of vaginal candidiasis for the prevention of preterm birth: a systematic review and meta-analysis. *Syst Rev.* 2015;4:31.
514. Kenyon S, Pike K, Jones DR, et al. Childhood outcomes after prescription of antibiotics to pregnant women with preterm rupture of the membranes: 7-year follow-up of the ORACLE I trial. *Lancet.* 2008;372:1310-1318.
515. Antonucci R, Zaffanello M, Puxeddu E, et al. Use of non-steroidal anti-inflammatory drugs in pregnancy: impact on the fetus and newborn. *Curr Drug Metab.* 2012;13:474-490.
516. Chin PY, Dorian CL, Hutchinson MR, et al. Novel Toll-like receptor-4 antagonist (+)-naloxone protects mice from inflammation-induced preterm birth. *Sci Rep.* 2016;6:36112.
517. Liu H, Redline RW, Han YW. Fusobacterium nucleatum induces fetal death in mice via stimulation of TLR4-mediated placental inflammatory response. *J Immunol.* 2007;179:2501-2508.
518. Reekie J, Donovan B, Guy R, et al. Risk of ectopic pregnancy and tubal infertility following Gonorrhea and Chlamydia infections. *Clin Infect Dis.* 2019;69:1621-1623.

519. Davies B, Turner KME, Frølund M, et al. Risk of reproductive complications following chlamydia testing: a population-based retrospective cohort study in Denmark. *Lancet Infect Dis.* 2016;16:1057-1064.
520. Lujan AL, Croci DO, Gambarte Tudela JA, et al. Glycosylation-dependent galectin-receptor interactions promote *Chlamydia trachomatis* infection. *Proc Natl Acad Sci U S A.* 2018;115:E6000-E6009.
521. Lenz JD, Dillard JP. Pathogenesis of *Neisseria gonorrhoeae* and the host defense in ascending infections of human fallopian tube. *Front Immunol.* 2018;9:2710.
522. Kitaya K, Matsubayashi H, Yamaguchi K, et al. Chronic endometritis: potential cause of infertility and obstetric and neonatal complications. *Am J Reprod Immunol.* 2016;75:13-22.
523. Dekel N, Gnainsky Y, Granot I, et al. The role of inflammation for a successful implantation. *Am J Reprod Immunol.* 2014;72:141-147.
524. Robertson SA, Care AS, Moldenhauer LM. Regulatory T cells in embryo implantation and the immune response to pregnancy. *J Clin Invest.* 2018;128:4224-4235.
525. Kimber SJ. Leukaemia inhibitory factor in implantation and uterine biology. *Reproduction.* 2005;130:131-145.
526. De M, Sanford TR, Wood GW. Interleukin-1, interleukin-6, and tumor necrosis factor alpha are produced in the mouse uterus during the estrous cycle and are induced by estrogen and progesterone. *Dev Biol.* 1992;151:297-305.
527. De M, Sanford TR, Wood GW. Expression of interleukin 1, interleukin 6 and tumour necrosis factor alpha in mouse uterus during the peri-implantation period of pregnancy. *J Reprod Fertil.* 1993;97:83-89.
528. Schjenken JE, Glynn DJ, Sharkey DJ, et al. TLR4 signaling is a major mediator of the female tract response to seminal fluid in mice. *Biol Reprod.* 2015;93:68.
529. Sanford TR, De M, Wood GW. Expression of colony-stimulating factors and inflammatory cytokines in the uterus of CD1 mice during days 1 to 3 of pregnancy. *J Reprod Fertil.* 1992;94:213-220.
530. De M, Choudhuri R, Wood GW. Determination of the number and distribution of macrophages, lymphocytes, and granulocytes in the mouse uterus from mating through implantation. *J Leukocyte Biol.* 1991;50:252-262.
531. McMaster MT, Newton RC, Dey SK, et al. Activation and distribution of inflammatory cells in the mouse uterus during the preimplantation period. *J Immunol.* 1992;148:1699-1705.
532. Messaoudi S, El Kasmi I, Bourdiec A, et al. 15 years of transcriptomic analysis on endometrial receptivity: what have we learnt? *Fertil Res Pract.* 2019;5:9.
533. Dimitriadis E, White CA, Jones RL, et al. Cytokines, chemokines and growth factors in endometrium related to implantation. *Hum Reprod Update.* 2005;11:613-630.
534. Evans J, Salamonsen LA, Winship A, et al. Fertile ground: human endometrial programming and lessons in health and disease. *Nat Rev Endocrinol.* 2016;12:654-667.
535. Gnainsky Y, Granot I, Aldo PB, et al. Local injury of the endometrium induces an inflammatory response that promotes successful implantation. *Fertil Steril.* 2010;94:2030-2036.
536. Wang WJ, Zhang H, Chen ZQ, et al. Endometrial TGF-beta, IL-10, IL-17 and autophagy are dysregulated in women with recurrent implantation failure with chronic endometritis. *Reprod Biol Endocrinol.* 2019;17:2.
537. Slukvin, Breburda EE, Golos TG. Dynamic changes in primate endometrial leukocyte populations: differential distribution of macrophages and natural killer cells at the rhesus monkey implantation site and in early pregnancy. *Placenta.* 2004;25:297-307.
538. Lim H, Paria BC, Das SK, et al. Multiple female reproductive failures in cyclooxygenase 2-deficient mice. *Cell.* 1997;91:197-208.
539. Lim H, Gupta RA, Ma WG, et al. Cyclo-oxygenase-2-derived prostacyclin mediates embryo implantation in the mouse via PPARdelta. *Genes Dev.* 1999;13:1561-1574.
540. Cheng JG, Stewart CL. Loss of cyclooxygenase-2 retards decidual growth but does not inhibit embryo implantation or development to term. *Biol Reprod.* 2003;68:401-404.
541. Robb L, Li R, Hartley L, et al. Infertility in female mice lacking the receptor for interleukin 11 is due to a defective uterine response to implantation. *Nat Med.* 1998;4:303-308.
542. Thaxton JE, Sharma S. Interleukin-10: a multi-faceted agent of pregnancy. *Am J Reprod Immunol.* 2010;63:482-491.
543. Saito S, Shima T, Nakashima A, et al. Role of paternal antigen-specific Treg cells in successful implantation. *Am J Reprod Immunol.* 2016;75:310-316.
544. Robertson SA, Prins JR, Sharkey DJ, et al. Seminal fluid and the generation of regulatory T cells for embryo implantation. *Am J Reprod Immunol.* 2013;69:315-330.
545. Moldenhauer LM, Schjenken JE, Hope CM, et al. Thymus-derived regulatory T cells exhibit Foxp3 epigenetic modification and phenotype Attenuation after mating in mice. *J Immunol.* 2019;203:647-657.
546. Heitmann RJ, Weitzel RP, Feng Y, et al. Maternal T regulatory cell depletion impairs embryo implantation which can be corrected with adoptive T regulatory cell transfer. *Reprod Sci.* 2017;24:1014-1024.
547. Keren L, Bosse M, Marquez D, et al. A structured tumor-immune microenvironment in triple negative breast cancer revealed by multiplexed ion beam imaging. *Cell.* 2018;174:1373-1387.e19.
548. Cartwright JE, Fraser R, Leslie K, et al. Remodelling at the maternal-fetal interface: relevance to human pregnancy disorders. *Reproduction.* 2010;140:803-813.
549. Brosens I, Pijnenborg R, Vercruysse L, et al. The "Great Obstetrical Syndromes" are associated with disorders of deep placentation. *Am J Obstet Gynecol.* 2011;204:193-201.
550. Zhang J, Adams MA, Croy BA. Alterations in maternal and fetal heart functions accompany failed spiral arterial remodeling in pregnant mice. *Am J Obstet Gynecol.* 2011;205(485):e1-16.
551. Hiby SE, Walker JJ, O'Shaughnessy KM, et al. Combinations of maternal KIR and fetal HLA-C genes influence the risk of preeclampsia and reproductive success. *J Exp Med.* 2004;200:957-965.
552. Colucci F. The role of KIR and HLA interactions in pregnancy complications. *Immunogenetics.* 2017;69:557-565.
553. Hiby SE, Apps R, Chazara O, et al. Maternal KIR in combination with paternal HLA-C2 regulate human birth weight. *J Immunol.* 2014;192:5069-5073.
554. Nakimuli A, Chazara O, Hiby SE, et al. A KIR B centromeric region present in Africans but not Europeans protects pregnant women from pre-eclampsia. *Proc Natl Acad Sci U S A.* 2015;112:845-850.
555. Kieckbusch J, Gaynor LM, Moffett A, et al. MHC-dependent inhibition of uterine NK cells impedes fetal growth and decidual vascular remodelling. *Nat Commun.* 2014;5:3359.
556. Madeja Z, Yadi H, Apps R, et al. Paternal MHC expression on mouse trophoblast affects uterine vascularization and fetal growth. *Proc Natl Acad Sci U S A.* 2011;108:4012-4017.
557. Robson A, Harris LK, Innes BA, et al. Uterine natural killer cells initiate spiral artery remodeling in human pregnancy. *FASEB J.* 2012;26:4876-4885.
558. Phipps EA, Thadhani R, Benzing T, et al. Pre-eclampsia: pathogenesis, novel diagnostics and therapies. *Nat Rev Nephrol.* 2019;15:275-289.
559. Burke SD, Barrette VF, Bianco J, et al. Spiral arterial remodeling is not essential for normal blood pressure regulation in pregnant mice. *Hypertension.* 2010;55:729-737.
560. Croy BA, Burke SD, Barrette VF, et al. Identification of the primary outcomes that result from deficient spiral arterial modification in pregnant mice. *Pregnancy Hypertens.* 2011;1:87-94.
561. Ain R, Canham LN, Soares MJ. Gestation stage-dependent intrauterine trophoblast cell invasion in the rat and mouse: novel endocrine phenotype and regulation. *Dev Biol.* 2003;260:176-190.
562. Hanna J, Goldman-Wohl D, Hamani Y, et al. Decidual NK cells regulate key developmental processes at the human fetal-maternal interface. *Nat Med.* 2006;12:1065-1074.
563. Kennedy PR, Chazara O, Gardner L, et al. Activating KIR2DS4 is expressed by uterine NK cells and contributes to successful pregnancy. *J Immunol.* 2016;197:4292-4300.
564. Renaud SJ, Scott RL, Chakraborty D, et al. Natural killer-cell deficiency alters placental development in rats. *Biol Reprod.* 2017;96:145-158.
565. Singh J, Ahmed A, Girardi G. Role of complement component C1q in the onset of preeclampsia in mice. *Hypertension.* 2011;58:716-724.
566. Hsu P, Santner-Nanan B, Dahlstrom JE, et al. Altered decidual DC-SIGN+ antigen-presenting cells and impaired regulatory T-cell induction in preeclampsia. *Am J Pathol.* 2012;181:2149-2160.
567. Robertson SA, Green ES, Care AS, et al. Therapeutic potential of regulatory T cells in preeclampsia-opportunities and challenges. *Front Immunol.* 2019;10:478.
568. Salazar Garcia MD, Mobley Y, Henson J, et al. Early pregnancy immune biomarkers in peripheral blood may predict preeclampsia. *J Reprod Immunol.* 2018;125:25-31.
569. Sasaki Y, Darmochwal-Kolarz D, Suzuki D, et al. Proportion of peripheral blood and decidual CD4(+) CD25(bright) regulatory T cells in pre-eclampsia. *Clin Exp Immunol.* 2007;149:139-145.
570. Schonkeren D, van der Hoorn ML, Khedoe P, et al. Differential distribution and phenotype of decidual macrophages in preeclamptic versus control pregnancies. *Am J Pathol.* 2011;178:709-717.
571. Quinn KH, Lacoursiere DY, Cui L, et al. The unique pathophysiology of early-onset severe preeclampsia: role of decidual T regulatory cells. *J Reprod Immunol.* 2011;91:76-82.
572. Rahimzadeh M, Norouzian M, Arabpour F, et al. Regulatory T-cells and preeclampsia: an overview of literature. *Expert Rev Clin Immunol.* 2016;12:209-227.
573. Bellos I, Karageorgiou V, Kapnias D, et al. The role of interleukins in preeclampsia: a comprehensive review. *Am J Reprod Immunol.* 2018;80:e13055.
574. Saito S, Nakashima A, Shima T, et al. Th1/Th2/Th17 and regulatory T-cell paradigm in pregnancy. *Am J Reprod Immunol.* 2010;63:601-610.
575. Santner-Nanan B, Peek MJ, Khanam R, et al. Systemic increase in the ratio between Foxp3+ and IL-17-producing CD4+ T cells in healthy pregnancy but not in preeclampsia. *J Immunol.* 2009;183:7023-7030.
576. Toldi G, Saito S, Shima T, et al. The frequency of peripheral blood CD4+ CD25high FoxP3+ and CD4+ CD25-FoxP3+ regulatory T cells in normal pregnancy and pre-eclampsia. *Am J Reprod Immunol.* 2012;68:175-180.
577. Pierik E, Prins JR, van Goor H, et al. Dysregulation of complement activation and placental dysfunction: a potential target to treat preeclampsia? *Front Immunol.* 2019;10:3098.
578. Boij R, Mjosberg J, Svensson-Arvelund J, et al. Regulatory T-cell subpopulations in severe or early-onset preeclampsia. *Am J Reprod Immunol.* 2015;74:368-378.
579. Xia Y, Kellems RE. Angiotensin receptor agonistic autoantibodies and hypertension: preeclampsia and beyond. *Circ Res.* 2013;113:78-87.
580. Wallukat G, Homuth V, Fischer T, et al. Patients with preeclampsia develop agonistic autoantibodies against the angiotensin AT1 receptor. *J Clin Invest.* 1999;103:945-952.
581. Walther T, Wallukat G, Jank A, et al. Angiotensin II type 1 receptor agonistic antibodies reflect fundamental alterations in the uteroplacental vasculature. *Hypertension.* 2005;46:1275-1279.
582. LaMarca B, Wallukat G, Llinas M, et al. Autoantibodies to the angiotensin type I receptor in response to placental ischemia and tumor necrosis factor alpha in pregnant rats. *Hypertension.* 2008;52:1168-1172.
583. Zhou CC, Zhang Y, Irani RA, et al. Angiotensin receptor agonistic autoantibodies induce pre-eclampsia in pregnant mice. *Nat Med.* 2008;14:855-862.
584. Zenclussen AC, Fest S, Joachim R, et al. Introducing a mouse model for pre-eclampsia: adoptive transfer of activated Th1 cells leads to pre-eclampsia-like symptoms exclusively in pregnant mice. *Eur J Immunol.* 2004;34:377-387.
585. Jensen F, Wallukat G, Herse F, et al. CD19+CD5+ cells as indicators of preeclampsia. *Hypertension.* 2012;59:861-868.
586. Gomez-Lopez N, StLouis D, Lehr MA, et al. Immune cells in term and preterm labor. *Cell Mol Immunol.* 2014;11:571-581.
587. Romero R, Dey SK, Fisher SJ. Preterm labor: one syndrome, many causes. *Science.* 2014;345:760-765.
588. Challis JR, Sloboda DM, Alfaidy N, et al. Prostaglandins and mechanisms of preterm birth. *Reproduction.* 2002;124:1-17.
589. Romero R, Avila C, Santhanam U, et al. Amniotic fluid interleukin 6 in preterm labor. Association with infection. *J Clin Invest.* 1990;85:1392-1400.
590. Opsjøn S-L, Wathen NC, Tingulstad S, et al. Tumor necrosis factor, interleukin-1, and interleukin-6 in normal human pregnancy. *Am J Obstet Gynecol.* 1993;169:397-404.
591. Haddad R, Tromp G, Kuivaniemi H, et al. Human spontaneous labor without histologic chorioamnionitis is characterized by an acute inflammation gene expression signature. *Am J Obstet Gynecol.* 2006;195(394):e1-e24.
592. Lui S, Duval C, Farrokhnia F, et al. Delineating differential regulatory signatures of the human transcriptome in the choriodecidua and myometrium at term labor. *Biol Reprod.* 2018;98:422-436.
593. Stanfield Z, Lai PF, Lei K, et al. Myometrial transcriptional signatures of human parturition. *Front Genet.* 2019;10:185.
594. Bollapragada S, Youssef R, Jordan F, et al. Term labor is associated with a core inflammatory response in human fetal membranes, myometrium, and cervix. *Am J Obstet Gynecol.* 2009;200(104):e1-e11.
595. Timmons B, Akins M, Mahendroo M. Cervical remodeling during pregnancy and parturition. *Trends Endocrinol Metab.* 2010;21:353-361.
596. Menzies FM, Shepherd MC, Nibbs RJ, et al. The role of mast cells and their mediators in reproduction, pregnancy and labour. *Hum Reprod Update.* 2011;17:383-396.
597. Nadeau-Vallee M, Quiniou C, Palacios J, et al. Novel noncompetitive IL-1 receptor-biased ligand prevents infection- and inflammation-induced preterm birth. *J Immunol.* 2015;195:3402-3415.
598. Christiaens I, Zaragoza DB, Guilbert L, et al. Inflammatory processes in preterm and term parturition. *J Reprod Immunol.* 2008;79:50-57.
599. Goldenberg RL, Culhane JF, Iams JD, et al. Epidemiology and causes of preterm birth. *Lancet.* 2008;371:75-84.
600. Schober L, Radnai D, Schmitt E, et al. Term and preterm labor: decreased suppressive activity and changes in composition of the regulatory T-cell pool. *Immunol Cell Biol.* 2012;90:935-944.
601. Migale R, MacIntyre DA, Cacciatore S, et al. Modeling hormonal and inflammatory contributions to preterm and term labor using uterine temporal transcriptomics. *BMC Med.* 2016;14:86.
602. Robertson SA, Christiaens I, Dorian CL, et al. Interleukin-6 is an essential determinant of on-time parturition in the mouse. *Endocrinology.* 2010;151:3996-4006.
603. Wahid HH, Dorian CL, Chin PY, et al. Toll-like receptor 4 is an essential upstream regulator of on-time parturition and perinatal viability in mice. *Endocrinology.* 2015;156:3828-3841.
604. Menzies FM, Higgins CA, Shepherd MC, et al. Mast cells reside in myometrium and cervix, but are dispensable in mice for successful pregnancy and labor. *Immunol Cell Biol.* 2012;90:321-329.
605. Robertson SA, Mau VJ, Young IG, et al. Uterine eosinophils and reproductive performance in interleukin 5-deficient mice. *J Reprod Fertil.* 2000;120:423-432.

606. Ratajczak CK, Muglia LJ. Insights into parturition biology from genetically altered mice. *Pediatr Res.* 2008;64:581-589.
607. Elovitz MA, Mrinalini C. Animal models of preterm birth. *Trends Endocrinol Metab.* 2004;15:479-487.
608. McCarthy R, Martin-Fairey C, Sojka DK, et al. Mouse models of preterm birth: suggested assessment and reporting guidelines. *Biol Reprod.* 2018;99:922-937.
609. Gross G, Imamura T, Vogt SK, et al. Inhibition of cyclooxygenase-2 prevents inflammation-mediated preterm labor in the mouse. *Am J Physiol Regul Integr Comp Physiol.* 2000;278:R1415-R1423.
610. Huang B, Faucette AN, Pawlitz MD, et al. Interleukin-33-induced expression of PIBF1 by decidual B cells protects against preterm labor. *Nat Med.* 2017;23:128-135.
611. Bizargity P, Del Rio R, Phillippe M, et al. Resistance to lipopolysaccharide-induced preterm delivery mediated by regulatory T cell function in mice. *Biol Reprod.* 2009;80:874-881.
612. Nadeau-Vallee M, Chin PY, Belarbi L, et al. Antenatal suppression of IL-1 protects against inflammation-induced fetal injury and improves neonatal and developmental outcomes in mice. *J Immunol.* 2017;198:2047-2062.
613. Thaxton JE, Romero R, Sharma S. TLR9 activation coupled to IL-10 deficiency induces adverse pregnancy outcomes. *J Immunol.* 2009;183:1144-1154.
614. Deng W, Yuan J, Cha J, et al. Endothelial cells in the decidual bed are potential therapeutic targets for preterm birth prevention. *Cell Rep.* 2019;27:1755-1768.e4.
615. Sadowsky DW, Adams KM, Gravett MG, et al. Preterm labor is induced by intraamniotic infusions of interleukin-1beta and tumor necrosis factor-alpha but not by interleukin-6 or interleukin-8 in a nonhuman primate model. *Am J Obstet Gynecol.* 2006;195:1578-1589.
616. Jethwa H, Lam S, Smith C, et al. Does rheumatoid arthritis really improve during pregnancy? A systematic review and metaanalysis. *J Rheumatol.* 2019;46:245-250.
617. Finkelsztejn A, Brooks JB, Paschoal FM Jr, et al. What can we really tell women with multiple sclerosis regarding pregnancy? A systematic review and meta-analysis of the literature. *Br J Obstet Gynaecol.* 2011;118:790-797.
618. Parekh RB, Dwek RA, Sutton BJ, et al. Association of rheumatoid arthritis and primary osteoarthritis with changes in the glycosylation pattern of total serum IgG. *Nature.* 1985;316:452-457.
619. Ohmi Y, Ise W, Harazono A, et al. Sialylation converts arthritogenic IgG into inhibitors of collagen-induced arthritis. *Nat Commun.* 2016;7:11205.
620. van de Geijn FE, Wuhrer M, Selman MH, et al. Immunoglobulin G galactosylation and sialylation are associated with pregnancy-induced improvement of rheumatoid arthritis and the postpartum flare: results from a large prospective cohort study. *Arthritis Res Ther.* 2009;11:R193.
621. Bondt A, Hafkenscheid L, Falck D, et al. ACPA IgG galactosylation associates with disease activity in pregnant patients with rheumatoid arthritis. *Ann Rheum Dis.* 2018;77:1130-1136.
622. Rook GA, Steele J, Brealey R, et al. Changes in IgG glycoform levels are associated with remission of arthritis during pregnancy. *J Autoimmun.* 1991;4:779-794.
623. Forger F, Marcoli N, Gadola S, et al. Pregnancy induces numerical and functional changes of CD4+CD25 high regulatory T cells in patients with rheumatoid arthritis. *Ann Rheum Dis.* 2008;67:984-990.
624. Komatsu N, Okamoto K, Sawa S, et al. Pathogenic conversion of Foxp3+ T cells into TH17 cells in autoimmune arthritis. *Nat Med.* 2014;20:62-68.
625. Waites GT, Whyte A. Effect of pregnancy on collagen-induced arthritis in mice. *Clin Exp Immunol.* 1987;67:467-476.
626. Mattsson R, Mattsson A, Holmdahl R, et al. Maintained pregnancy levels of oestrogen afford complete protection from post-partum exacerbation of collagen-induced arthritis. *Clin Exp Immunol.* 1991;85:41-47.
627. Gonzalez DA, de Leon AC, Moncholi CV, et al. Arthritis in mice: allogeneic pregnancy protects more than syngeneic by attenuating cellular immune response. *J Rheumatol.* 2004;31:30-34.
628. Gonzalez DA, de Leon AC, Moncholi CV, et al. Cytokine profile in collagen-induced arthritis: differences between syngeneic and allogeneic pregnancy. *Inflamm Res.* 2008;57:266-271.
629. Nelson JL, Hughes KA, Smith AG, et al. Maternal-fetal disparity in HLA class II alloantigens and the pregnancy-induced amelioration of rheumatoid arthritis. *N Engl J Med.* 1993;329:466-471.
630. van der Horst-Bruinsma IE, de Vries RR, de Buck PD, et al. Influence of HLA-class II incompatibility between mother and fetus on the development and course of rheumatoid arthritis of the mother. *Ann Rheum Dis.* 1998;57:286-290.
631. Munoz-Suano A, Kallikourdis M, Sarris M, et al. Regulatory T cells protect from autoimmune arthritis during pregnancy. *J Autoimmun.* 2012;38:J103-J108.
632. Inoue K, Inoue E, Imai Y. Female sex hormones ameliorate arthritis in SKG mice. *Biochem Biophys Res Commun.* 2013;434:740-745.
633. Jansson L, Olsson T, Holmdahl R. Estrogen induces a potent suppression of experimental autoimmune encephalomyelitis and collagen-induced arthritis in mice. *J Neuroimmunol.* 1994;53:203-207.
634. Ercan A, Kohrt WM, Cui J, et al. Estrogens regulate glycosylation of IgG in women and men. *JCI Insight.* 2017;2:e89703.
635. Bijlsma JW, Huber-Bruning O, Thijssen JH. Effect of oestrogen treatment on clinical and laboratory manifestations of rheumatoid arthritis. *Ann Rheum Dis.* 1987;46:777-779.
636. Cortes M, Canellada A, Miranda S, et al. Placental secreted factors: their role in the regulation of anti-CII antibodies and amelioration of collagen induced arthritis in rats. *Immunol Lett.* 2008;119:42-48.
637. Falcon CR, Martinez FF, Carranza F, et al. In vivo expression of recombinant pregnancy-specific glycoprotein 1a inhibits the symptoms of collagen-induced arthritis. *Am J Reprod Immunol.* 2014;72:527-533.
638. Thompson SJ, Hitsumoto Y, Zhang YW, et al. Agalactosyl IgG in pristane-induced arthritis. Pregnancy affects the incidence and severity of arthritis and the glycosylation status of IgG. *Clin Exp Immunol.* 1992;89:434-438.
639. Buzas EI, Hollo K, Rubliczky L, et al. Effect of pregnancy on proteoglycan-induced progressive polyarthritis in BALB/c mice: remission of disease activity. *Clin Exp Immunol.* 1993;94:252-260.
640. Doodes PD, Cao Y, Hamel KM, et al. Development of proteoglycan-induced arthritis is independent of IL-17. *J Immunol.* 2008;181:329-337.
641. Ramien C, Yusko EC, Engler JB, et al. T cell repertoire dynamics during pregnancy in multiple sclerosis. *Cell Rep.* 2019;29:810-815.e4.
642. Gilli F, Lindberg RL, Valentino P, et al. Learning from nature: pregnancy changes the expression of inflammation-related genes in patients with multiple sclerosis. *PLoS One.* 2010;5:e8962.
643. Airas L, Saraste M, Rinta S, et al. Immunoregulatory factors in multiple sclerosis patients during and after pregnancy: relevance of natural killer cells. *Clin Exp Immunol.* 2008;151:235-243.
644. Gatson NN, Williams JL, Powell ND, et al. Induction of pregnancy during established EAE halts progression of CNS autoimmune injury via pregnancy-specific serum factors. *J Neuroimmunol.* 2011;230:105-113.
645. Lelu K, Laffont S, Delpy L, et al. Estrogen receptor alpha signaling in T lymphocytes is required for estradiol-mediated inhibition of Th1 and Th17 cell differentiation and protection against experimental autoimmune encephalomyelitis. *J Immunol.* 2011;187:2386-2393.
646. Bodhankar S, Wang C, Vandenbark AA, et al. Estrogen-induced protection against experimental autoimmune encephalomyelitis is abrogated in the absence of B cells. *Eur J Immunol.* 2011;41:1165-1175.
647. Subramanian S, Yates M, Vandenbark AA, et al. Oestrogen-mediated protection of experimental autoimmune encephalomyelitis in the absence of Foxp3+ regulatory T cells implicates compensatory pathways including regulatory B cells. *Immunology.* 2011;132:340-347.
648. Spence RD, Hamby ME, Umeda E, et al. Neuroprotection mediated through estrogen receptor-alpha in astrocytes. *Proc Natl Acad Sci U S A.* 2011;108:8867-8872.
649. Voskuhl RR, Wang H, Wu TC, et al. Estriol combined with glatiramer acetate for women with relapsing-remitting multiple sclerosis: a randomised, placebo-controlled, phase 2 trial. *Lancet Neurol.* 2016;15:35-46.
650. Gold SM, Voskuhl RR. Pregnancy and multiple sclerosis: from molecular mechanisms to clinical application. *Semin Immunopathol.* 2016;38:709-718.
651. Hayrabedyan S, Shainer R, Yekhtin Z, et al. Synthetic PreImplantation Factor (sPIF) induces posttranslational protein modification and reverses paralysis in EAE mice. *Sci Rep.* 2019;9:12876.
652. Zhang B, Harness J, Somodevilla-Torres MJ, et al. Early pregnancy factor suppresses experimental autoimmune encephalomyelitis induced in Lewis rats with myelin basic protein and in SJL/J mice with myelin proteolipid protein peptide 139-151. *J Neurol Sci.* 2000;182:5-15.
653. Hellwig K, Rockhoff M, Herbstritt S, et al. Exclusive breastfeeding and the effect on postpartum multiple sclerosis relapses. *JAMA Neurol.* 2015;72:1132-1138.
654. Hutchinson M. Multiparity in women with multiple sclerosis causes less long-term disability: commentary. *Mult Scler.* 2014;20:1437-1438.
655. Buyon JP, Kim MY, Guerra MM, et al. Predictors of pregnancy outcomes in patients with lupus: a cohort study. *Ann Intern Med.* 2015;163:153-163.
656. Hughes GC, Choubey D. Modulation of autoimmune rheumatic diseases by oestrogen and progesterone. *Nat Rev Rheumatol.* 2014;10:740-751.
657. Bynoe MS, Grimaldi CM, Diamond B. Estrogen up-regulates Bcl-2 and blocks tolerance induction of naive B cells. *Proc Natl Acad Sci U S A.* 2000;97:2703-2708.
658. Nijagal A, Fleck S, Hills NK, et al. Decreased risk of graft failure with maternal liver transplantation in patients with biliary atresia. *Am J Transplant.* 2012;12:409-419.
659. van Rood JJ, Loberiza FR Jr, Zhang MJ, et al. Effect of tolerance to noninherited maternal antigens on the occurrence of graft-versus-host disease after bone marrow transplantation from a parent or an HLA-haploidentical sibling. *Blood.* 2002;99:1572-1577.
660. Burlingham WJ, Grailer AP, Heisey DM, et al. The effect of tolerance to noninherited maternal HLA antigens on the survival of renal transplants from sibling donors. *N Engl J Med.* 1998;339:1657-1664.
661. Molitor-Dart ML, Andrassy J, Kwun J, et al. Developmental exposure to noninherited maternal antigens induces CD4+ T regulatory cells: relevance to mechanism of heart allograft tolerance. *J Immunol.* 2007;179:6749-6761.
662. Matsuoka K, Ichinohe T, Hashimoto D, et al. Fetal tolerance to maternal antigens improves the outcome of allogeneic bone marrow transplantation by a CD4+ CD25+ T-cell-dependent mechanism. *Blood.* 2006;107:404-409.
663. Andrassy J, Kusaka S, Jankowska-Gan E, et al. Tolerance to noninherited maternal MHC antigens in mice. *J Immunol.* 2003;171:5554-5561.
664. Owen RD, Wood HR, Foord AG, et al. Evidence for actively acquired tolerance to Rh antigens. *Proc Natl Acad Sci U S A.* 1954;40:420-424.
665. Dutta P, Molitor-Dart M, Bobadilla JL, et al. Microchimerism is strongly correlated with tolerance to noninherited maternal antigens in mice. *Blood.* 2009;114:3578-3587.
666. Kinder JM, Jiang TT, Ertelt JM, et al. Cross-generational reproductive fitness enforced by microchimeric maternal cells. *Cell.* 2015;162:505-515.
667. Nijagal A, Wegorzewska M, Jarvis E, et al. Maternal T cells limit engraftment after in utero hematopoietic cell transplantation in mice. *J Clin Invest.* 2011;121:582-592.
668. Jonsson AM, Uzunel M, Gotherstrom C, et al. Maternal microchimerism in human fetal tissues. *Am J Obstet Gynecol.* 2008;198(325):e1-e6.
669. Piotrowski P, Croy BA. Maternal cells are widely distributed in murine fetuses in utero. *Biol Reprod.* 1996;54:1103-1110.
670. Stevens AM, Hermes HM, Kiefer MM, et al. Chimeric maternal cells with tissue-specific antigen expression and morphology are common in infant tissues. *Pediatr Dev Pathol.* 2009;12:337-346.
671. Moles JP, Tuaillon E, Kankasa C, et al. Breastmilk cell trafficking induces microchimerism-mediated immune system maturation in the infant. *Pediatr Allergy Immunol.* 2018;29:133-143.
672. Molitor ML, Haynes LD, Jankowska-Gan E, et al. HLA class I noninherited maternal antigens in cord blood and breast milk. *Hum Immunol.* 2004;65:231-239.
673. Mold JE, Michaelsson J, Burt TD, et al. Maternal alloantigens promote the development of tolerogenic fetal regulatory T cells in utero. *Science.* 2008;322:1562-1565.
674. Vernochet C, Caucheteux SM, Gendron MC, et al. Affinity-dependent alterations of mouse B cell development by noninherited maternal antigen. *Biol Reprod.* 2005;72:460-469.
675. Claas FH, Gijbels Y, van der Velden-de Munck J, et al. Induction of B cell unresponsiveness to noninherited maternal HLA antigens during fetal life. *Science.* 1988;241:1815-1817.
676. Adams Waldorf KM, Nelson JL. Autoimmune disease during pregnancy and the microchimerism legacy of pregnancy. *Immunol Invest.* 2008;37:631-644.
677. Lissauer DM, Piper KP, Moss PA, et al. Fetal microchimerism: the cellular and immunological legacy of pregnancy. *Expert Rev Mol Med.* 2009;11:e33.
678. Nelson JL. Microchimerism and autoimmune disease. *N Engl J Med.* 1998;338:1224-1225.
679. Abbot S, Bossingham D, Proudman S, et al. Risk factors for the development of systemic sclerosis: a systematic review of the literature. *Rheumatol Adv Pract.* 2018;2:rky041.
680. Kinder JM, Stelzer IA, Arck PC, et al. Immunological implications of pregnancy-induced microchimerism. *Nat Rev Immunol.* 2017;17:483-494.
681. Odorizzi PM, Feeney ME. Impact of in utero exposure to malaria on fetal T cell immunity. *Trends Mol Med.* 2016;22:877-888.
682. Hamada K, Suzaki Y, Goldman A, et al. Allergen-independent maternal transmission of asthma susceptibility. *J Immunol.* 2003;170:1683-1689.
683. Gatford KL, Wooldridge AL, Kind KL, et al. Pre-birth origins of allergy and asthma. *J Reprod Immunol.* 2017;123:88-93.
684. Barker DJ. The fetal and infant origins of adult disease. *BMJ.* 1990;301:1111.

SECTION VII

Immune Responses to Pathogens

CHAPTER 42

Response to Parasites

David Sacks, PhD • Marion Pepper, PhD • Mary F. Fontana, PhD • P'ng Loke, PhD

PARASITES AND THE IMMUNE SYSTEM

Global Health Importance of Parasitic Diseases

The term "parasite" is said to be derived from a Greek word referring to an "unwanted dinner guest". The term is formally used as a designation for eukaryotic protozoan and metazoan pathogens residing on or within their hosts. Although the taxonomic basis of their classification into a single group is under question, parasites as infectious agents do share many biologic characteristics. They frequently display complex life cycles consisting of morphologically and antigenically distinct stages and produce long-lived or chronic infections to ensure transmission between their hosts. The induction of severe morbidity or mortality is an atypical outcome. However, in the tropical and subtropical regions where transmission is high, the low frequency of disease translates into a major global health and economic problem because of the sheer numbers of people exposed and because of the confounding issues of malnutrition and coinfection. Much of the global burden from parasitic disease is due to malaria and a group of tropical infectious diseases commonly referred to as "neglected." Together, these diseases affect more than 1 billion people and cause more than 1 million deaths annually. They are primarily diseases of poverty, causing disabilities that impair physical and cognitive development, disrupt productivity, care of families, and school attendance, all of which contribute to the cycle of poverty and disease.[1] We also poorly understand the interindividual heterogeneity of immune responses against the parasites, which is directly related to the pathogenesis of infections. The reason for some individuals having asymptomatic or mild infections, whereas others suffer from severe disease and mortality, is often related to variation in immune responses between these individuals. How genetic and environmental factors contribute toward variation in disease pathogenesis and magnitude of responses against the parasites is poorly understood and likely important for the development of vaccines against these organisms. There is often a fine balance between parasite clearance and immunopathology that has to be successfully navigated.

A list of the most important parasitic infections of humans, along with estimates of their prevalence, annual mortality, and current control methods, is presented in Table 42.1. The table includes a metric for the health impact of each disease, quantified as the disability-adjusted life-year (DALY), based on years of life lost from premature death and years of life lived in less than full health. A striking point in the table is the complete absence of effective vaccines to protect human populations. The measures currently available to reduce the burden of tropical parasitic diseases are confined to drug-treatment programs and/or to vector control. These interventions have resulted in selection for both resistant parasites and vectors, which, along with the high cost and low sustainability of the interventions, has reinforced the need for preventive vaccines.

Clearly, the development of vaccines to prevent human parasitic diseases remains one of the major unachieved goals of modern immunology and one of its most difficult challenges. While the absence of strong market incentives remains a barrier to the development of so-called "antipoverty" vaccines,[1] the greater impediments may be the complexity of parasites as immunologic targets, their remarkable adaptability to immunologic pressure, and alterations in host immune functionality caused by prior infection in endemic regions. The field of immunoparasitology is focused on developing a basic understanding of the host-pathogen interface for the ultimate purpose of developing interventions to prevent or moderate the intensity of infection and disease. While these primary objectives remain largely unfulfilled, the studies have provided immunologists with major insights into the structure and functioning of the vertebrate immune system. There is much to be learned from parasites about the fundamental nature of immune responses. Unfortunately, while

TABLE 42.1 Global Impact of Parasitic Diseases and Current Control Measures

	Annual Deaths (000s)		Disease Burden (DALYs,[a] 000s)		Current Control Method Used
	2016	2000	2016	2000	
Malaria	446	767	37,369	67,855	Vector control, chemotherapy
Soil-transmitted helminths[b]	6	12	4536	5899	Chemotherapy, hygiene
Schistosomiasis	24	55	2,543	4,292	Chemotherapy, hygiene
Cysticercosis	26	42	1,912	3,362	Chemotherapy, hygiene
Lymphatic filariasis	0	0	1,186	1,937	Vector control, chemotherapy
Leishmaniasis	14	16	1,069	1,146	Vector control, chemotherapy
Onchocerciasis	0	0	962	1,371	Vector control, chemotherapy
Chagas' disease	8	9	252	329	Vector control
African trypanosomiasis	3	33	203	2,127	Vector control

Data compiled from World Health Organization. Disease burden and mortality estimates (2000-2016). https://www.who.int/healthinfo/global_burden_disease/estimates/en/
[a]Disability-adjusted life years, a measure of the time lived with a disability, combined with the time lost due to premature mortality.
[b]Hookworm, ascariasis, and trichuriasis. Data are the combined values for all three infections.

immunologists have benefited tremendously from studying the response against parasites, we have yet to be able to fully utilize that knowledge to reduce the morbidity and global burden that is caused by these infections.

Some Hallmarks of the Immune Response to Parasites

The hallmark of parasitic infections is their chronicity, which implies a certain capacity to avoid or delay sterilizing immunity. The relatively low transmissibility of blood and tissue parasites that depend on an invertebrate host for cyclical transmission requires persisting infection in order to prolong the time available to the parasite for transmission to the vector or intermediate host. Parasites have evolved numerous mechanisms of immune escape that underlie the slow or incomplete development of immunity. These mechanisms are as varied as the organisms themselves and are commonly associated with the ability of the parasite to evade protective immunity or to regulate its induction or expression. Immune evasion can be achieved in multiple ways, including clonal antigenic variation (eg, African trypanosomes and *Plasmodium falciparum*) or immune sequestration (eg, infection of erythrocytes by malaria or of muscle cells by *Trypanosoma cruzi*). Immunoregulation refers to the suppression of protective immunity. It is especially relevant to many helminth pathogens, which, because they do not replicate in their mammalian hosts, cannot employ rapid antigenic variation as a mechanism to avoid complete elimination during primary or subsequent infection. Immunomodulatory pathways can serve both to prevent parasite elimination, suppress host immunopathology, and impact the development of immune memory. From an evolutionary perspective, in order to minimize the impact that a pathogen has on health, there is the option to confer "resistance" by eliminating the pathogen or to minimize the damage inflicted, conferring "tolerance."[2] The study of tolerance pathways in parasitic infections have yielded important insights into the mechanisms by which regulatory cells and cytokines control immune effector functions.

Another important immunologic concept promoted by the study of parasites, related to chronicity, is concomitant immunity, which refers to the state of acquired resistance against reinfection that coincides with the persistence of the primary infection. This concept was first described in the context of tumor immunity and was extended to infection immunity in experimental models of schistosomiasis.[3] Concomitant immunity is observed in many other parasitic infections, including Leishmaniasis and Toxoplasmosis, and is thought to underlie the durability of acquired resistance against reinfection. From a coevolutionary perspective, concomitant immunity concedes benefits to both parasite and host in the inability to achieve sterile cure.

Perhaps, the most important lesson that parasites have taught immunologists pertains to the dichotomy in the functional subsets of CD4+ T cells elicited by different classes of parasites. Parasitic infections often induce CD4+ T cell responses that are highly polarized in terms of their Th1/Th2 lymphokine profiles. Following on the initial description of functionally distinct CD4+ T cell clones *in vitro*,[4] *Leishmania major* models of cutaneous leishmaniasis in different inbred mouse strains were the first to establish the relevance of these subsets to infection outcome *in vivo*, with Th1 and Th2 responses responsible for healing and nonhealing infections, respectively.[5,6] Th1-dominant responses mediating immunity against intracellular pathogens like *Leishmania* and *Toxoplasma* remain a guiding principle for the development of vaccines against this class of organisms. By contrast, extracellular helminths induce highly polarized Th2 responses that contribute to the mast cell, eosinophil, IgE and mucosal responses that are typically associated with these infections. The study of intestinal nematodes firmly established the role of Th2 responses in acquired resistance to infection, and they remain widely used in clinical and experimental models to define the effector components of type 2 immunity. Because parasites can elicit such polarized responses, they have contributed in important ways to understanding the mechanisms underlying Th subset selection. It is also important to acknowledge that under most infection conditions, both type 1 and type 2 cytokines and responses are concurrently active, and it is the proportion of cell types and cytokines that are skewed in one direction or the other. Recent studies have begun to delve into the details of how both type 1 and type 2 responses function together, even if the overall response is more polarized in a certain direction. These contributions to immunology and others will be detailed below in the context of a selected number of parasitic infections chosen because of their global importance and the relatively advanced state of knowledge regarding the basic immunoprotective and immunopathogenic mechanisms that continue to inform the quest for preventive and therapeutic vaccines.

IMMUNE RESPONSE TO TRYPANOSOMATID PATHOGENS

The Basic Biology and Global Health Importance of Trypanosomatid Pathogens

The Trypanosomatidae family, order Kinetoplastida, includes the genera *Trypanosoma* and *Leishmania,* protozoan parasites that have digenetic life cycles involving a vertebrate host and an insect vector. *Trypanosoma brucei* spp., *Trypanosoma cruzi,* and *Leishmania* spp. are transmitted by tsetse flies, triatomine bugs, or sand flies, respectively (Fig. 42.1). They are responsible for African trypanosomiasis (sleeping sickness), Chagas' disease, and different clinical forms of leishmaniasis in their human hosts. While human African trypanosomiasis (HAT) is approaching elimination in many parts of sub-Saharan Africa, other species of African trypanosomes, primarily *T. congolense,* remain important parasites of domestic livestock, placing at least one-third of the African cattle herd at risk of fatal infection. Chagas' disease affects nearly 20 million people in the Americas, with infections progressing to a symptomatic chronic phase in which 30% of patients develop cardiomyopathy or neuropathies and dilatations of the colon or esophagus at some point during their lifetimes. Leishmaniasis affects over 12 million people annually in tropical and subtropical regions, with over 20 different species of *Leishmania* producing a spectrum of clinical outcomes ranging from self-limiting cutaneous lesions to tissue destructive mucosal involvement, to visceral dissemination that is fatal in the absence of treatment.

Throughout their life cycles, these parasites progress through several developmental stages, for which the insect stages are in each case exclusively extracellular, flagellated forms that replicate in the midgut or salivary glands of their respective vectors (Fig. 42.1). Their differentiation into a nondividing infectious, metacyclic stage is required for successful transmission to a

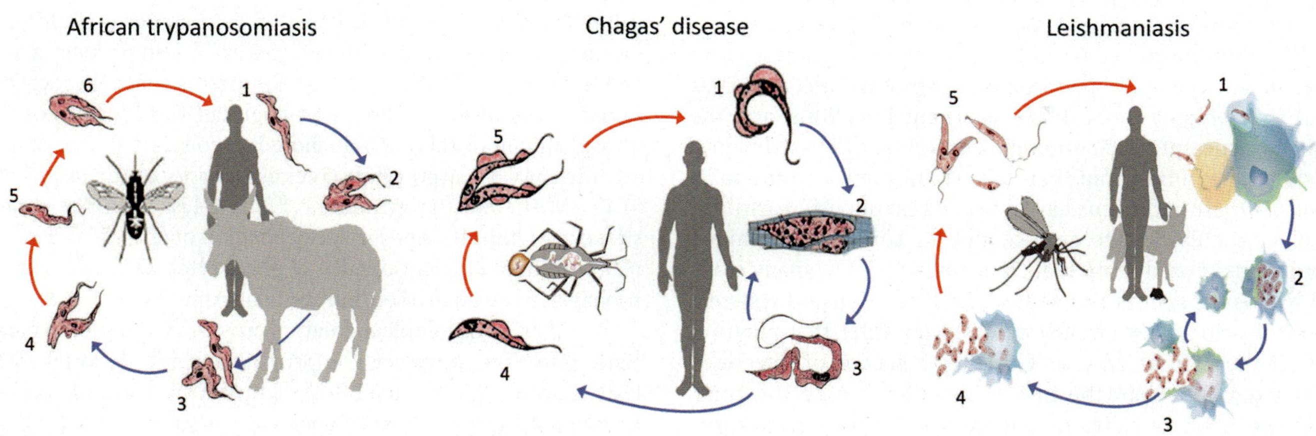

FIG. 42.1. Life cycles of Trypanosomatid pathogens. African trypanosomiasis: (1) Tsetse fly takes a blood meal, principally from cattle, wild ungulates, and humans, and injects metacyclic trypomastigotes, (2) Metacyclic trypomastigotes transform into bloodstream trypomastigotes which multiply as long slender forms by binary fission and are carried to other sites, (3) As parasite numbers increase in the bloodstream, there is differentiation to nondividing stumpy forms that are preadapted for transmission to tsetse flies, (4) Bloodstream trypomastigotes transform into procyclic trypomastigotes in the fly midgut where they multiply by binary fission, (5) Procyclic trypomastigotes migrate from the midgut to the salivary gland where they differentiate into epimastigotes, (6) Epimastigotes multiply by binary fission and transform into nondividing metacyclic trypomastigotes that are preadapted for survival in the vertebrate host. **Chagas' disease:** (1) Metacyclic trypomastigotes that are deposited in the triatomine feces at the bite site, or via contamination at mucosal sites such as the oral mucosa or conjunctiva, penetrate various nucleated cells where they transform into amastigotes, (2) Amastigotes multiply by binary fission in cells of infected tissues, (3) Intracellular amastigotes transform into trypomastigotes, burst out of the cell, and after release into the bloodstream can be carried to new infection sites such as muscle where they can invade other cells or be transmitted to the vector when the triatomine bug takes a blood meal, (4) Trypomastigotes transform into epimastigotes in the midgut, (5) Epimastigotes divide by binary fission, migrate to the hindgut, and transform into nondividing metacyclic trypomastigotes. **Leishmaniasis:** (1) Sand fly takes a blood meal from humans and other reservoir hosts such as dogs and rodents and regurgitates metacyclic promastigotes into the skin, which are taken up by phagocytes, mainly macrophages and neutrophils, which can themselves be captured by macrophages, (2) Promastigotes transform into intracellular amastigotes which multiply by binary fission, (3) Intracellular amastigotes burst out of the infected cell to infect other cells, including in other tissues, or are transmitted to sand flies during blood feeding, (4) Amastigotes are released by infected cells in the sand fly midgut, (5) Amastigotes transform into promastigotes in the midgut, where they multiply by binary fission, and migrate to the anterior midgut, where they differentiate into nondividing metacyclic promastigotes. Life cycles are modified from those available on the CDC Website, www.cdc.gov/dpdx/az.html. All images were acquired from the public domain.

vertebrate host, which occurs by delivery of tsetse saliva in the case of African trypanosomes, by regurgitation of sand fly midgut contents in the case of *Leishmania*, or by release with the bug feces that may contaminate the injured skin in the case of *T. cruzi*. Chagas' disease can also be transmitted orally in contaminated food, with trypomastigotes gaining entrance through mucous membranes. In the vertebrate, African trypanosomes maintain a purely extracellular lifestyle in blood and tissues, with infections marked by immune dysfunction and immunopathology related to prolonged, systemic inflammatory reactions. By contrast, *T. cruzi* metacyclic trypomastigotes parasitize many types of nucleated cells, escaping to the cytoplasm and replicating as intracellular amastigotes prior to transforming back to nondividing trypomastigotes that are released into the blood and disseminate to different tissues. *T. cruzi* can persist at low level in muscle or in peripheral neuronal cells in the gastrointestinal tract, and the cardiac and gut pathologies that are associated with Chagas' disease are due to the ongoing, localized immune responses in these parasitized tissues. *Leishmania* are obligate intracellular organisms in their vertebrate hosts, where they can be found as replicating amastigotes in the phagolysosomes of phagocytic cells, mainly macrophages. The different disease forms, including cutaneous, mucocutaneous, and visceral leishmaniasis, are associated with the tissue tropisms of different *Leishmania* species and strains. Pathogenesis is related to uncontrolled infections, as with the fatal forms of visceral leishmaniasis, or to the immunopathology driven by chronic, localized infections in the skin or mucosa.

The immunoprotective and immunopathologic responses that are elicited by Trypanosomatid pathogens, the strategies that they employ to establish persistent infections, and the status of current vaccination efforts will be discussed below. It should be pointed out that most studies of the immunobiology of these pathogens have been carried out using artificial needle inocula that do not account for some of the unique conditions related to natural transmission. Thus, many studies bypass the skin as a route for initial infection and systemic dissemination, they employ cultured parasites that may not reflect the natural dose or virulence behavior of infectious stages developing in the insect, and they ignore the components of vector saliva or microbiota that may be cotransmitted and have immunomodulatory properties. Lastly, it is important to keep in mind that animal models, particularly murine models, may in some cases be a poor proxy for studying human immune responses.

Innate Recognition and Host Defense

Humoral Mechanisms

Innate resistance can be mediated by preexisting soluble factors that recognize and directly damage parasites within a short time after their transmission or target them for subsequent killing by effector cells. The alternative and lectin pathways of complement activation (see Chapter 14), comprised of dozens of plasma-circulating proteins, provide a first line of defense against many extracellular parasites. Complement recognizes "nonself" as a primitive surveillance system independent of antibodies, triggered by repetitive polysaccharide structures in the case of the lectin pathway and other surface molecules such as sialic acid, glycoproteins, and glycolipids, in the case of the alternative pathway. Complement recognition sets off a proteolytic cascade, which for both pathways converges at an early step in the formation of a C3 convertase complex and the subsequent generation of C3 cleavage fragments C3a and C3b, which have anaphylatoxin and opsonic properties, respectively. C3b is also necessary for the generation of the C5 convertase, which gives rise to C5a, another powerful inflammatory mediator, and to the assemblage of the terminal complement components into the membrane attack complex (MAC), which can produce lytic pores in the pathogen membrane. The ability of the transmitted, metacyclic stages of Trypanosomatids to withstand complement-mediated damage is essential to their successful parasitism and is achieved by different surface molecules that inhibit, dampen, or resist the effects of complement activation. *T. cruzi* metacyclics and bloodstream trypomastigotes rely on members of the large family of inactive transialidases to function as analogs of human decay accelerating factor, which interferes with assembly of C3 convertases by binding to C3b.[7] Another analog of an endogenous human complement regulatory protein is *T. cruzi* glycoprotein gp58/68, which inhibits alternative pathway C3 convertase formation by preventing the binding of factor B to parasite-bound C3b.[8] Finally, *T. cruzi* calreticulin inhibits the human lectin pathway by binding to the collagenous portion of L-Ficolin, preventing the oligomerization required for complement activation.[9] The sand fly transmitted, metacyclic promastigotes of *Leishmania* activate C3 but resist the lytic effects of alternative and lectin pathway activation by expression of a dense, elongated surface glycocalyx structure comprised of lipophosphoglycan (LPG), a glycosylphosphatidylinositol (GPI)-anchored glycolipid. LPG serves as an efficient C3 acceptor but serves as a barrier to the insertion of the MAC into the membrane.[10] These lysis-resistant parasites exploit the efficient surface deposition of opsonic C3b and iC3b for more efficient uptake by phagocytes, especially macrophages, to facilitate their intracellular parasitism.

Another set of soluble mediators providing a barrier to parasitic infection comprises the primate-specific trypanosome lysis factors (TLF-1 and TLF-2). Humans are naturally resistant to pathogenic animal trypanosomes such as *T.b. brucei*, *T. vivax*, and *T. congolense* due to the activity of proteins present within the high-density lipid (HDL) fraction of normal serum, apolipoprotein L1 (APOL1), and haptoglobin-related protein (HRP).[11,12] The HRP is thought to act primarily as a ligand which binds to hemoglobin and facilitates uptake of the lytic complex by a parasite haptoglobin-hemoglobin receptor (HpHbR). After being endocytosed, APOL1 is released, forming anionic pores in the endosomal and mitochondrial membranes, triggering apoptotic-like death. Primates evolved this innate defense mechanism by exploiting the requirement of African trypanosomes for large amounts of lipid, which they obtain in the form of HDL from their hosts. HAT is caused by *T. brucei* subspecies such as *T.b. rhodesiense* that have coevolved resistance to normal human serum by expression the *SRA* gene, encoding a truncated variant surface glycoprotein (VSG) that blocks the membrane-forming capacity of APOL1.[13] Another strain that causes human disease, *T.b. gambiense,* employs an additional resistance mechanism, involving a specific point mutation in HpHbR, reducing TLF1 uptake.[14] These mechanisms offer possible targets for inhibitors or antibodies that could render human trypanosomes susceptible to normal innate resistance.

Cellular Mechanisms

Cells of the macrophage lineage represent an innate first line of defense against protozoan pathogens, and along with dendritic cells (DCs), influence the development of adaptive immunity. Macrophages are present in all tissues, and depending on the tissue and inflammatory conditions, may be seeded embryonically to establish long-term residence or be continuously replaced by monocyte precursors. Macrophages possess primary defense mechanisms that are induced by the attachment and engulfment of microbes, including activation of oxidative metabolism, and the development of phagosomes into digestive organelles via fusion with lysosomes. *Leishmania* have evolved multiple strategies to avoid or withstand these assaults. In the early stage of infection in the skin, infiltrating CCR2+ inflammatory monocytes, which have a less developed microbicidal arsenal, serve as an initial replicative niche for the parasite.[15] In mature macrophages, *Leishmania* can remodel the phagosome and engage deactivating signaling pathways. For example, LPG released by promastigotes incorporates into the phagosome membrane and delays phagosome fusion with the endosome, in part, by promoting the accumulation of periphagosomal F-actin.[16] The delayed phagosome maturation buys the time needed for differentiation into the more hydrolase resistant, amastigote stage. Another well-described *Leishmania* virulence factor, the major surface protease GP63, is involved in the activation of SHP1, a cellular phosphotyrosine phosphatase (PTP), which serves as an endogenous negative regulator of JAK2 activation and its downstream upregulation of inducible nitric oxide synthase (iNOS) in macrophages.[17] *Leishmania* can also engage, either directly or indirectly, phagocytic receptors that are unlinked from activation processes, including complement and mannose receptors and receptors for apoptotic cells. C3 opsonization of infective promastigotes via the alternative and lectin pathways, mentioned above, efficiently targets the parasite to macrophage CR1 and CR3 receptors, promoting their silent uptake and subsequent intracellular survival.[18] Receptors for apoptotic cells, including the TAM family of receptor tyrosine kinases, are relevant to an intriguing "Trojan Horse" model of infection,[19] in which neutrophils, which are among the first cells to take up metacyclic promastigotes in the skin following transmission by infected sand flies,[20] are themselves captured by tissue macrophages. Such a process of silent parasite transfer requires that the organisms remain viable following their initial uptake by the neutrophils, which has been confirmed,[20] as has the ability of *Leishmania* to inhibit or resist neutrophil microbicidal machinery, including the oxidative burst and release of neutrophil extracellular traps (NETs).[21]

Distinct from *Leishmania*, *T. cruzi* can gain entry into both professional phagocytes and nonphagocytic cells, such as myocytes, which are the main infected cells responsible for chronic Chagas' disease. Following binding of trypomastigote infective forms to the plasma membrane, invasion of nonphagocytic cells is triggered by a rise in intracellular Ca^{2+} which signals the recruitment of host cell lysosomes to the attachment site, similar to the cAMP-regulated events that are associated with plasma membrane repair.[22] Lysosome recruitment is thought to provide membrane for the nascent parasitophorous vacuole (PV) and/or to facilitate membrane invaginations by releasing lysosomal acid sphingomyelinase, which converts sphingomyelin to ceramide. For entry into macrophages, *T. cruzi* utilizes conventional actin-mediated phagocytosis, which upon ingestion activates NADPH oxidase and iNOS, producing superoxide and NO that are potentially lethal to the parasite. However, *T. cruzi* possesses powerful antioxidant enzymes, including trypanothione, glutathione, and ascorbate.[23] Regardless of cell type or cell entry mechanism, all parasites are found within a lysosome-fused vacuole within 1 hour of entry, with acidification of the vacuole thought to be essential for the activity of a porin-like protein, TcTOX, responsible for escape into the cytoplasm where differentiation into the replicative, amastigote stage can occur.[24]

Although African trypanosome spp. maintain an exclusive extracellular lifestyle in their vertebrate hosts, there is evidence that the numbers and activation states of macrophages are substantially altered in trypanosome-infected animals. In *T. brucei*–infected mice, a soluble form of VSG released from the parasite contributed to the induction of proinflammatory programs in the liver and splenic macrophages, prominently IFN-α/β, and mice lacking the receptor for type I IFNs exhibited delayed control of parasites during the first week of infection.[25] Macrophages also contribute to parasite clearance from the circulation during the adaptive phase of immunity, which relies on antibody-mediated phagocytosis by liver macrophages. Activation of tissue macrophages by IFN-γ for release of trypanocidal molecules, such as TNF-α and NO, mediates killing of extravascular parasites. At the same time, the hyperactivation of tissue macrophages, resulting in the release of excessive amounts of these molecules, contribute to the characteristic sequelae of sleeping sickness, namely cachexia and anemia.[26]

Role of Pattern Recognition Receptors

Innate cells express germline-encoded, membrane-bound, and cytoplasmic pattern recognition receptors (PRRs) to recognize pathogen-associated molecular patterns (PAMPs), as well as host-derived, damage-associated molecular patterns (DAMPs) (see Chapter 13). Activation of PRRs initiates signaling cascades, typically resulting in a proinflammatory response that shapes innate host defense against infection, instructs the development of acquired resistance, and/or contributes to immunopathologic responses. The principle PRRs that have been investigated in the context of infection with Trypanosomatid pathogens include C-type lectin receptors (CLR), NOD-like receptors (NLR), and toll-like receptors (TLR) (see Table 42.2). TLR signaling contributes to immunity against all three classes of Trypanosmatid diseases since genetic ablation of MyD88, which is the major adaptor protein involved in most TLR-signaling pathways, resulted in increased susceptibility to *T. brucei*, *T. cruzi*, and *L. major* infections in mice.[44-46] In each case, the susceptibility

TABLE 42.2 Innate Recognition of Protozoan Pathogens

PAMPs	Parasite	Structure	Innate Receptors/ Sensors	References
GPI anchors	*Leishmania* spp.	LPG	TLR2, TLR4	27
	Trypanosoma cruzi	GPI-mucins GIPL-ceramide	TLR2 TLR4	27,28
	African trypanosomes	VSG	TLR2	29,30
	Plasmodium spp.		TLR2, TLR4	31
Genomic DNA	*Leishmania* spp.	Unmethylated CpG	TLR 9	32
	Trypanosoma cruzi	Unmethylated CpG	TLR9	28
	African trypanosomes		TLR 9	30
	Plasmodium spp.		TLR9, AIM2, cGAS	33,34
ssRNA	*Trypanosoma cruzi* *Leishmania* spp. *Plasmodium* spp.		TLR7 TLR7 MDA-5	35 32 36
dsRNA	*Leishmania* spp.	Endosymbiotic virus (LRV1)	TLR3	37
Damage Signal	**Parasite Infection Initiating the Damage Signal**		**Innate Sensors**	**References**
K^+ efflux, ROS	*Leishmania* spp.		NLRP3	38,39
	Trypanosoma cruzi		NLRP3, NLRP1	40,41
Hemazoin/Uric acid	*Plasmodium* spp.		NLRP3	42,43

GIPL, glycoinositolphospholipid; GPI, glycosylphosphatidylinositol; LPG, lipophosphoglycan; PAMPs, pathogen-associated molecular patterns; ROS, reactive oxygen species; VSG, variant surface glycoprotein.

was associated with defective IL-12 and IFN-γ production. In the mouse model of African trypanosomiasis due to *T. congolense*, signaling through TLR2 was shown to be essential for resistance and for parasite-driven macrophage MAPK/STAT1 phosphorylation and IL-6/IL-12/TNF-α secretion.[29] In a *T. brucei* infection model of cerebral involvement, both TLR2 and TLR9 were found to control T cell penetration into the brain as well as microglial activation.[30] For *T. cruzi*, the protection conferred by TLR engagement was accounted for in part by the cooperative roles of TLR2 and TLR9 and correlated with defective Th1 responses *in vivo*.[28] Studies conducted in mice that contain a point mutation in the chaperone protein UNC93B1, resulting in the inability to translocate all of the nucleic acid–sensing TLRs (TLR3, TLR7, and TLR9) from the endoplasmic reticulum to the endolysosomes, were extremely susceptible to *T. cruzi* and revealed an essential role for TLR7 in resistance.[35] For cutaneous leishmaniasis due to *L. major*, the UNC93B1-mutant mice also showed an extreme susceptibility that could be reversed by treatment using recombinant IL-12 or anti–IL-10 receptor antibody.[32] In the case of cutaneous leishmaniasis due to *L. guyanensis*, TLR3 stimulation promoted not resistance but pathology, driven by the genomic dsRNA of an endosymbiotic virus (LRV1).[37] The disease-exacerbating role of LRV1 in mice relied on type 1 IFN and is thought to be relevant to the inflammatory pathology associated with mucocutaneous leishmaniasis in humans.

The abundant GPI-anchored glycoconjugates found on the surface or released by each of the Trypanosomatid parasites have also been identified as TLR agonists (Table 42.2). Thus, the LPG from *Leishmania* spp., which shares some structural similarities with bacterial lipopolysaccharide (LPS), and the GPI anchor fractions from either *T. brucei* VSG or the mucin-like molecules from *T. cruzi*, have each been shown to stimulate macrophages to upregulate TNF-α, iNOS, and IL-12, primarily through engagement of TLR2 and TLR4.[27] In the case of *Leishmania* and *T. cruzi* infection models in mice, however, these TLRs appear to play a limited role in controlling infection outcome *in vivo*.

In addition to the cell membrane–associated PRRs, there are cytosolic sensors present in macrophages and other innate immune cells. The inflammasomes are multiprotein complexes that are assembled in the cytoplasm when intracellular PRR detect microbial or damage signals. Inflammasome assembly regulates the activation of the cysteine protease caspase-1, leading to proteolytic processing and secretion of the proinflammatory cytokines IL-1β and IL-18. The most studied inflammasomes in the context of protozoan pathogens are the nucleotide-binding domain leucine-rich repeat protein (NLR) family members, NLRP3 and NLRP1, which have been shown to respond to different kinds of cellular perturbations, such as the efflux of K^+, the production of reactive oxygen species (ROS) and the leakage of lysosomal cathepsins (Table 42.2). For *Leishmania*, infection of bone marrow–derived macrophages triggered Caspase-1 activation and IL-1β production that was abolished in the absence of NLRP3.[38] The IL-1β secretion led to NO-mediated containment of *L. amazonensis* growth *in vitro* and *in vivo*. The virulence factor GP63 was shown to limit the magnitude of NLRP3 activation in macrophages, either by inhibiting ROS production or by directly degrading inflammasome components.[47] Not surprisingly, the NLRP3 inflammasome/IL-1β axis has also been shown to contribute to the pathogenesis of cutaneous lesions in mice due in large part to the persistent neutrophilic infiltrate that is driven by this innate response.[39]

In *T. cruzi* infection, both NLRP1 and NLRP3 have been shown to be activated and to control various aspects of resistance in mice or in human macrophages.[40,41] Potassium efflux, ROS production, and lysosomal damage were variously implicated in the *T. cruzi* induced activation of these platforms. Despite the fact that *T. cruzi* gains access to the host cell cytoplasm, there is no evidence that the inflammasome sensors that have been shown to directly bind microbial ligands, such as the binding of viral dsDNA to AIM2 or bacterial flagellin to NAIP, are activated by *T. cruzi*.

DCs and Mechanisms Underlying Th1/Th2 Response Selection

Because parasites often stimulate CD4+ T cell responses that are highly polarized in either the Th1 or Th2 direction, parasitic infection models have become important tools for studying the mechanisms underlying Th1/Th2 response selection. DCs are professional antigen-presenting cells (APCs) that are believed to be an important source of the signals that determine CD4+ T cell subset development. Their role is best understood for Th1 responses, which are associated with upregulated expression on DCs of MHC II and costimulatory molecules, including CD80, CD86, and CD40, and the release of instructional cytokines, principally IL-12 although IL-1α/β, IL-18, and IFN-α have also been shown to condition a Th1 cell fate. Signal-dependent activation of NF-κB, IFN regulatory factors, and STATs control the timing and duration of the transcriptional programs leading to Th1 development. During *L. major* infection, Th1 effector choice is critically determined at the level of the DCs because the absence of STAT1 only in the DC and not the T cells impaired Th1 cell priming.[48] IL-12 production by DCs infected with *L. major* is TLR9-dependent and driven by parasite DNA.[49] TLR9-signaling in DC was also shown to be critical for early IFN-γ expression by NK cells, which contribute to immunity during the early phase of infection with *L. major in vivo*.[50] In cutaneous leishmaniasis, the DCs that are critical to Th1 development are differentiated locally in the skin from monocyte-derived cells, where they become activated to produce iNOS (referred to as Tip- DC), and subsequently, migrate into the draining lymph nodes.[51] While all *Leishmania* infections are initiated in the skin, there is evidence that different strains and species have very different effects on DC activation that may explain the severity or chronicity of infection. For instance, in contrast to *L. major*, *L. mexicana* produces nonhealing lesions in C57Bl/6 mice, and unlike *L. major*, the uptake of *L. mexicana* amastigotes by bone marrow–derived DCs is silent.[52] Even for healing strains of *Leishmania*, the delayed onset of Th1 immunity is crucial to the early establishment of infection. The early capture of infected, apoptotic neutrophils, mentioned in the context of tissue macrophages and mediated in part by TAM receptor tyrosine kinases, has also been described for DCs at the site of *L. major* inoculation in the skin, with powerful inhibitory effects on the ability of the DCs to initiate T cell priming.[53] Engagement of another receptor, the CLR receptor Mincle, was triggered by a leishmanial ligand to target an immunoreceptor tyrosine-based inhibitory motif (ITIM) signaling pathway that suppressed DC activation.[54] The absence of Mincle resulted in enhanced early immunity that correlated with stronger DC activation and migration to draining lymph nodes.

With regard to the classical susceptibility model involving *L. major* infection in BALB/c mice, there remains little evidence that Th2 polarization in these mice is driven by intrinsic differences in DCs and the manner in which they interact with the parasite or its products. For example, skin-derived DCs that can produce IL-12p40 in response to *L. major in vitro* and prime for protective T cell responses *in vivo* are readily obtainable from BALB/c mice.[55] Other differences that better account for the dichotomy in Th response selection relate to the manner in which the parasite disseminates from the inoculation site, with dissemination to the viscera occurring rapidly in BALB/c mice compared to containment in the footpad and draining nodes in resistant mice.[56] These differences suggest that distinct populations of DCs, modulated by specific tissue environments and with the capacity to induce preferential priming of either Th1 or Th2 cells, become activated in resistant and susceptible mouse strains. That the route of challenge can influence Th response selection has been clearly shown in normally resistant C57Bl/6 mice that developed strong Th2 responses and nonhealing infections following intravenous or intranasal inoculation with *L. major*.[57,58]

In the case of Chagas' disease, the fact that nucleic acid–sensing TLRs are required for control of the acute stage of *T. cruzi* infection in the mouse suggests that these TLRs operate at the levels of DCs to promote the Th1 response. In fact, TLR9 and TLR7 engagement by *T. cruzi* DNA and RNA, respectively, were shown to contribute to IL-12 production by DCs.[35] This response is also likely responsible for the early activation of IFN-γ–producing NK cells that play a role in biasing early Th1 differentiation.[59] It is interesting that *T. cruzi* encounters with DCs have more consistently been found to impart negative or tolerogenic signals. *In vivo*, splenic DCs from *T. cruzi* infected mice displayed low expression of MHCII and costimulatory molecules,[60] and DCs with tolerogenic properties were induced that promoted Treg cells and parasite persistence.[61] The tolerogenic pathway may involve the abundant sialylated ligands on *T. cruzi* that interact with the inhibitory sialic acid-binding Ig-like lectin Siglec-E on the DC, which like Mincle, contains ITIMs in its cytoplasmic tail.[62] Whether DC dysfunction contributes to the persistence of infection and the resulting pathology during the chronic stage has not been directly addressed.

Adaptive Immunity and Immunopathology

African Trypanosomiasis

The mechanisms underlying adaptive immunity to Trypanosomatid pathogens and the strategies they employ to avoid elimination are dictated by their differing extracellular or intracellular lifestyles and the different tissue environments in which they are found. For African trypanosomes, which are inoculated as metacyclic trypomastigotes into the skin and remain as extracellular proliferating slender or nondividing stumpy trypomastigote forms in blood and tissues, their survival relies on a remarkable system of antigenic variation to avoid antibody-mediated clearance. Best understood for *T. brucei*, these parasites are able to turn on and off VSG-encoding genes from a genomic repertoire of approximately 2000 genes.[63] Each parasite expresses one gene at a time from one of 15 to 20 short (30-60 Kb) telomeric expression sites, while all other nonexpressed VSG genes are located in subtelomeric arrays and on minichromosomes. The VSGs are expressed as GPI-anchored homodimers with approximately 5 million dimers per cell forming a dense surface structure that effectively shields the underlying invariant proteins from antibodies. VSG expression switches stochastically, such that new serotypes continuously arise to produce waves of parasitemia as prior variants are controlled and new ones emerge. In addition to changes in amino-acid sequence, posttranslational modifications, such as *O*-glycosylation of the VSG, can change the antigenic character of the surface coat.[64] Passive transfer of variant-specific antibodies or B cells was shown to confer variant-specific protection in immunocompromised mice.[65] T cells are also needed to promote a stronger and higher affinity antibody response, and they contribute to other control mechanisms. In particular, VSG-specific Th1 cells are a primary source of INF-γ that activates tissue macrophages for production of NO and other proinflammatory mediators that can promote parasite killing in extravascular tissues.[45,66]

As mentioned above, the release of excessive amounts of these molecules can have immunopathologic consequences. In HAT, the late encephalitic stage of infection is characterized by the presence of parasites and/or increased numbers of lymphocytes in cerebrospinal fluid. In a mouse model of cerebral involvement, migration of *T. brucei* and T cells from cerebral blood to the brain parenchyma was impeded in IFN-γ–deficient animals, with IFN-γ thought to promote transmigration by modulating endothelial basement membrane laminin expression.[67] In the parenchyma, the T cells can interact with resident microglia, astrocytes, or additional lymphocytes that can each produce additional inflammatory mediators. In humans, increased frequency of progression to and severity of the encephalitic stage was found to be associated with higher plasma IFN-γ concentration.[68] Importantly, the neuroinflammation in the mouse model was shown to be ameliorated by systemic administration of IL-10.[69]

In addition to antigenic variation, another hallmark of infection with African trypanosomes is immunosuppression. The suppression can be profound during the chronic stage of infection, and while it further compromises the ability of the host to eliminate the parasite, it also serves to moderate the deleterious effects of sustained inflammation. A switch from M1 to M2-like macrophage activation programs is a prominent feature of trypanotolerant in mice. Thus, *T. congolense* infected mice with myeloid cells making a sustained M1-type response exhibit increased liver damage and lethality, while mouse strains that switch to a M2-like response exhibit increased survival, with IL-10 production being a critical component of their infection tolerance.[70] A released, *T. brucei*–derived immunomodulatory factor that promotes IL-10 production in myeloid cells was identified as a kinesin heavy chain isoform, with the IL-10 functioning in an autocrine fashion to upregulate arginase activity, which fueled the synthesis of polyamines that are essential to the growth of extracellular trypanosomes in the host.[71] Other sources of IL-10 that contributed to a trypanotolerant phenotype in mice include hepatocytes and an expanded population of Foxp3+ T regulatory cells in the spleen and liver.[72,73] IL-27, a member of the IL-12 super-family produced by macrophages and DCs, was also shown to inhibit hyperactivation of Th1 cells during infection, with mice deficient in IL-27R showing severe liver pathology and reduced survival.[74]

Recent studies identifying adipose tissue and skin as major reservoirs of infection suggest that colonization of these anatomical compartments is an important feature of chronicity that favors cyclical transmission. Skin from experimentally infected mice as well as human skin biopsies from undiagnosed individuals living in an endemic region were found to harbor substantial numbers of trypanosomes which could be transmitted to the tsetse vector.[75] In the mouse, high concentrations of parasites were also found in most fatty deposits, where they rewired their gene expression programs to utilize the rich carbon source of lipids/fatty acids via β-oxidation.[76] In addition to offering a nutrient-rich niche for parasite growth, the adipose tissue may represent an immunologically privileged site, enhancing its role as an infection reservoir able to seed persistent numbers of parasites to the blood and skin for cyclical transmission.

Chagas' Disease

In both mice and humans, the acute stage of *T. cruzi* infection is a self-limited febrile illness and is effectively controlled by adaptive immune mechanisms. Critically, the immunity is not sterile, and parasites can persist for years in host tissue. The importance of anti–*T. cruzi* antibodies in protection was demonstrated by the ability of serum from chronically infected humans or mice to passively transfer immunity against acute infection,[77] and by mice lacking functional B cells (muMT−/−) which succumbed to uncontrolled parasite growth during the acute phase.[78] The mucin-like, GPI-anchored glycoproteins that cover the surface of *T. cruzi* mammalian stages are thought to be targets of the protective response since lytic antibodies directed against α-galactosyl epitopes present in the mucins were described in sera from Chagas' patients.[79] Interestingly, lysis is decreased if the mucin coat is heavily sialylated, with sialic acid acquired from the host due to the presence on the parasite surface of a unique parasite-encoded transsialidase (TS), which catalyzes the transfer of sialic acid from host sialyl donor glycoconjugates to the terminal β-galactoses present on the parasite mucins.[80] The *T. cruzi* surface mucins are encoded by over 800 genes, and the simultaneous expression of a large repertoire of the mucin core polypeptides has been observed in blood-stage trypomastigotes.[81] Simultaneous presentation to the immune system of so many peptide variants is believed to provide another mechanism of immune escape, possibly by affecting antibody affinity maturation, or via the well-described ability of altered peptide ligands to induce a T-cell receptor-mediated anergic state in CD4+ Th cells. Accordingly, there is a poor- and low-affinity antibody response to the mucin core polypeptides in experimental and clinical studies.[82]

As the parasites spend the majority of their time replicating as amastigotes in the cytoplasm of many different types of nucleated cells, cell-mediated immunity is an important component of the protective response. *In vitro* generated CD4+ Th1 but not Th2 cells can adoptively transfer protection against a systemic *T. cruzi* challenge, and recombinant IL-12 and anti–IL-4 treatment-induced type 1 polarized responses that were highly protective against lethal challenge in mice.[83] CD8+ T cells are an essential component of the protective response, as might be expected for a pathogen that can infect diverse nucleated cells lacking in expression of MHC II molecules required for cognate antigen recognition by CD4+ T cells. Genetic deletion or antibody depletion of CD8+ T cells results in uncontrolled infection during the acute stage and a greater number of persisting parasites during the chronic stage.[84] The maintenance of a robust parasite-specific CD8+ T cell response during infection relies on cross-talk between different cell subsets, including B cells and Th17 cells, able to secrete different mediators, including IL-17 and IL-21.[85] There is evidence in the mouse that the CD8+ T cells control intracellular infection through perforin/granzyme-mediated killing of infected cells, with an intriguing cooperative association proposed involving the action of perforin, which allows the granzymes to enter the infected cell and lyse the amastigotes.[86] The antigen targets of the CD8+ T cell response in C57Bl/6 mice are biased against a small set of epitopes present in the enzyme active transsialidases (TS), encoded by approximately 20 genes, which are part of a much larger family of inactive TS family members encoded by thousands of genes. TS antigens also appear to comprise a significant proportion of the CD8+ T cell–specific response in humans, absent the epitope dominance observed in the inbred mice.[87]

A number of mechanisms have been suggested to dampen the Ag-specific CD8+ T cell response during infection and to favor *T. cruzi* persistence. One mechanism relates to the flood of TS variants that get presented to the immune system, as >3000 full and partial length TS genes appear to be transcriptionally active during infection. Another intriguing mechanism involves the TS modifying an acceptor on the CD8+ T cells themselves. Upon activation, cytotoxic CD8+ T cells are normally desialylated, increasing the avidity of CD8αβ-MHC class I interactions and enhancing their effector activity. *T. cruzi* was shown to use its TS enzyme to resialylate the T cell surface, inhibiting their cytotoxic response *in vitro* and promoting parasite persistence *in vivo*.[88] The possibility that the CD8+ T cells are driven to an exhausted state that might further comprise their function is suggested by the elevated coexpression of inhibitory receptors, such as PD-1, CTLA-4, and TIM-3, in CD8+ T cells from patients with severe disease compared with asymptomatic patients.[89] However, when the issue of CD8+ T cell exhaustion was addressed in the mouse, the cells retained their ability to produce IFN-γ and granzyme B throughout the course of the acute and chronic infection, and blocking PD-1 or IL-10 failed to enhance CD8+ T cell effector activity.[90] Immunosuppression of the CD4+ T cell compartment following its initial expansion during the acute infection in mice has also been described, associated with activation-induced cell death detected by apoptosis of splenic T cells *ex vivo*, and susceptibility to Fas-mediated death *in vitro*.[91] The removal of the apoptotic T cells by splenic macrophages induced their production of prostaglandins (PGE_2) and TGF-β and was coupled with a remarkable enhancement in the ability of the macrophages to support *T. cruzi* growth.

The persistence of parasites following control of the acute phase infection can give rise, sometimes decades later, to the pathologic features of chronic Chagas' disease, of which cardiomyopathy is the most frequent and severe manifestation. While the pathogenesis of Chagas' heart disease was argued for years to be driven by an underlying autoimmune mechanism, the current consensus, based on immunohistochemical and PCR-based methods applied in clinical and experimental studies, is that parasite persistence in the cardiac tissue drives the inflammatory response.[92,93] The inflammatory lesions in patients are typically composed of macrophages, a predominance of CD8+ over CD4+ Th1 cells, a proinflammatory cytokine profile, and low levels of IL-10.[94] In comparison to patients with cardiac or gastrointestinal forms of disease, individuals in the asymptomatic, indeterminant group were found to have a higher frequency of T cells with a regulatory phenotype (Foxp3+, $CD25^{high}$), and elevated levels of IL-10 produced by peripheral blood cells, suggesting that Treg and IL-10 might protect against chronic Chagas' disease.[95] In the mouse, the use of blocking antibodies and *IL-10*–deficient animals demonstrated the requirement for IL-10 to restrain pathology, which was driven by IL-12 and CD4+ T cell–derived IFN-γ.[96] In addition, IL-10 production from T cells was shown to confer protection against fatal acute myocarditis in the mouse.[97] While some of the IL-10 producing T cells coexpressed IFN-γ, others appeared to be type 1 regulatory cells (Tr1) cells, based on the requirement for IL-27, a known inducer of IL-10–producing Tr1 cells, to dampen cardiac inflammation.[98]

Leishmaniasis

For acquired resistance against phagosomal pathogens such as *Leishmania*, the need to recognize peptides that are generated within endocytic compartments dictates that CD4+ T cells will be a critical component of the protective response. *L. major* infection in C57Bl/6 mice that develop self-limiting cutaneous lesions remains a classical model to address mechanisms underlying Th1 polarization and its counterregulation. IFN-γ is the key lymphokine-mediating control of virtually all *Leishmania* infections, as demonstrated by the extreme susceptibility of IFN-γ–deficient mouse strains. In addition, genetic ablation of other cytokines (IL-12, TNF-α), receptors (IFN-γR, IL-12R, TNFR), transcription factors (T-bet, STAT4), or costimulatory molecules (CD40-CD40L) that are involved in the development or function of Th1 cells, have all been shown to compromise acquired resistance to *L. major*.[99] The mechanism of action of IFN-γ on infected cells in the mouse is the upregulation of iNOS and the subsequent generation of NO. Simultaneous exposure of infected cells to TNF-α provides especially powerful signals to promote killing by inhibiting the expression of arginase 1, which can compete with iNOS for the common substrate L-arginine.[100] Despite being confined to the phagosomal compartment, parasite antigens can access the MHC Class I processing machinery to elicit CD8+ T cells that contribute to acquired resistance against low-dose challenge with *L. major*.[101] In mouse models of visceral leishmaniasis, the protective response in the liver is also associated with infiltrating CD4+ and CD8+ T cells, with the later observed within hepatic granulomas to directly engage infected Kupffer cells that generate ROS and NO to eliminate the parasite.[102]

Clinically, the largely descriptive studies conducted on patients with cutaneous and visceral forms of disease have generally supported an important role for Th1 responses in acquired resistance or moderation of disease. In situ studies of localized, self-limiting skin lesions have revealed a granulomatous organization, variable distribution of neutrophils, activated CD4+ and CD8+ T cells, with CD4+ cells representing the majority of IFN-γ–producing cells.[103] While the role of NO in controlling human infection is not so well established as in mice, several studies have shown an inverse correlation between the level of iNOS and the parasite burden.[104] As IFN-γ can enhance the respiratory burst in human monocytes and macrophages for ROS-dependent killing of *Leishmania*, the relative roles of ROS and NO for resistance in human infection remain unclear. CD8+ T cell function in resistance is also not understood. While the CD8+ T cell present in peripheral blood of *L. major* patients can produce antigen-specific IFN-γ, their production of granzyme B is a strong marker for protection against the intermediate or severe forms of cutaneous disease.[105] It should be noted that immunostaining or transcriptional profiling of self-limiting cutaneous lesions have almost always revealed a mixed Th1/Th2 response, with type 2 cytokines especially evident during the active stage of lesion development. In human VL, active cases almost always progress to a fatal outcome in the absence of treatment, so it is not possible to associate an immunological profile with self-cure. However, drug-cured patients who are believed to be immune to reinfection, reconstitute an antigen-specific IFN-γ response in their peripheral blood cells, and asymptomatic cases can be identified based on their long-term maintenance of an antigen-specific IFN-γ response.[106] In addition to Th1 cytokines, a prospective study of an *L. donovani* exposed cohort in Sudan identified IL-17 and IL-22 produced by PBMCs as a strong correlate of protection against progression to visceral disease.[107] The role of these cytokines in increasing the production of IL-6, which has proinflammatory effects, and in recruitment of neutrophils to the target organs, were suggested as ways that they contributed to protective immunity.

The manner in which *Leishmania* evades adaptive immune mechanisms, either to prevent complete elimination or to promote nonhealing, chronic forms of disease, has been intensively explored in experimental and clinical studies. The type 2 polarized response that is induced in *L. major*–infected BALB/c mice for years provided the main conceptual framework to understand non-healing forms of clinical disease. It has not been possible, however, to clearly associate type 2 dominance with any severe infection outcome in humans. IFN-γ–producing cells or mRNA remain readily detectable in patients with VL or chronic cutaneous lesions, and the main opposing cytokine most commonly found in these clinical settings is not IL-4 or IL-13, but IL-10.[108,109] The association of IL-10 with the pathogenesis of human VL is considered a hallmark of the disease, with active cases showing highly elevated levels of plasma IL-10 and IL-10 mRNA in lesional tissues.[110] Considering other chronic infection models in mice, an essential role for IL-10 in *Leishmania* persistence is typically observed, including the persistence of low numbers of *L. major* in the skin or *L. donovani* in the liver and spleen following containment in resistant mice.[111,112] IL-10 dampens MHC class II expression on APCs and downregulates TNF-α and NO production by infected macrophages. Th2 cells, Foxp3+ regulatory T cells, B cells, NK cells, macrophages, and DCs, have all been shown to be a source of IL-10 in *Leishmania*-infected mice or humans,[110] reflecting the centrality of this cytokine in the anti-inflammatory response. A critical source of IL-10 in C57Bl/6 mice infected with nonhealing substrains of *L. major*, and the predominant IL-10 expressing cells in splenic biopsies of VL patients, are conventional CD4+ T cells that in some cases coexpress IFN-γ, designated Tr1. These cells have been increasingly recognized for their role in localized control of tissue inflammation. Consistent with an important regulatory role of IL-27 in promoting T cell IL-10 production, VL patients show elevated expression IL-27 subunit genes, and mice with defective IL-27 signaling phenocopy IL-10–deficient mice, showing reduced parasite burdens but exacerbated liver or skin pathology in visceral and cutaneous infection models.[113,114]

IL-10 can also suppress immunity by promoting T cell exhaustion. Immune checkpoint molecules, principally PD-1 and CTLA-4, indicative of functionally exhausted cells, have been identified on CD4+ and CD8+ T cells from chronically infected mice and humans, and in cutaneous and visceral forms of disease.[115] Their exhausted phenotypes can be rescued in mice using antibody blockade of the interactions of PD-1 and CTL-4 with their respective ligands, CD80/CD86 and PDL-1, which in each case increased proinflammatory cytokine production and promoted more effective clearance of parasites from the viscera or skin.[116,117]

While regulatory cells and cytokines contribute to parasite persistence, their main function is to control tissue damage in sites of strong inflammation associated with the antimicrobial response. Thus, patients with severe, tissue destructive forms of mucocutaneous leishmaniasis have a deficit in IL-10 and TGF-β and an overexuberant production of proinflammatory cytokines, especially TNF-α.[118] High levels of TNF-α are also associated with lesion size and chronicity in patients with localized cutaneous lesions, and a combination of antiparasitic drugs and TNF-α inhibitors have shown therapeutic benefit in clinical trials.[119] IL-17 levels, likely due to its involvement in neutrophil recruitment, also correlate with the severity of lesions in patients with cutaneous and mucocutaneous disease.[120] As a central mediator of the pro-inflammatory response, IL-1 levels along with higher expression of genes associated with inflammasome activation, are also associated with pathology in these patients. In mice, while IL-1 appears to contribute to the priming phase of the protective T cell response, its excessive production during the chronic phase has been shown to promote the severity of cutaneous lesions, in part by mediating the expansion of Th17 cells.[121] Finally, cytotoxic CD8+ T cells have been associated with ulcerative lesions in *L. braziliensis* patients, with intralesional cells exhibiting a cytotoxicity transcriptional signature marked by granzyme expression.[122] Infection of recombination activating gene *(Rag)*-deficient mice leads to little pathology, but when reconstituted with CD8+ T cells, they produce nonhealing, ulcerative lesions that are perforin-dependent, thought to be driven by the ability of the killed target cells to promote inflammasome activation.[121]

Vaccines Against Trypanosomatid Pathogens

Owing to the extraordinary diversity of the VSGs against which protective B and T cell responses must be induced, the prospects for developing an effective vaccine against either human or animal African trypanosomiases are considered slim. Antigenically conserved sequences have been identified in the membrane-proximal, VSG C-terminal domain which, despite being inaccessible to antibodies, could serve as a hypothetical vaccine target if they could be made immunogenic for Th1 cells, especially as IFN-γ released from these cells have been shown to activate macrophages in tissues to kill trypanosomes in a variant nonspecific manner.[66] This raises a risk of intensifying immunopathologic outcomes during the chronic stage should the vaccine fail to confer sterilizing immunity. Perhaps, more realistic are therapeutic vaccination strategies, for example, antibodies that target proteins such as SRA that would render human-infective *T. brucei* subspecies susceptible to normal innate resistance and APOL1.

Several defined, experimental vaccines for Chagas' disease have been developed using a variety of delivery systems, including plasmids, adenoviruses, and recombinant proteins/peptides, and heterologous prime-boost vaccination.[123] Many of the vaccines successfully induced protective immunity against *T. cruzi* challenge infection in mice, typically associated with activation of a Th1 immune profile and measured as a CD8+ T cell–dependent reduction in acute-phase parasitemia or associated mortality. Some of the formulations were also shown to be efficacious as therapeutic vaccines, tested in mice with ongoing *T. cruzi* infection to reduce cardiac inflammation while increasing survival times. The protective antigens tested include cruzipain, GP82, and several proteins from the transialidase (TS) superfamily. While responses directed against the immunodominant CD8+ T cell epitopes in the TS family of proteins conferred especially strong immunity, the protection provided was shown to be strain-specific.[124] Furthermore, no regimens, including TS vaccination, have been able to induce sterilizing immunity in mice, raising concerns that the chronic cardiac and intestinal pathologies that are driven by inflammatory reactions to persistent infection may not be adequately prevented and might even be exacerbated. It should be stressed that these concerns have not been borne out in the mouse models, insofar as every vaccine that decreased acute-phase parasitemia and for which parasite burdens in heart or skeletal muscle were also evaluated, ameliorated infection and inflammation in these tissues.[123] Nonetheless, the rationale for the development of a safe and effective vaccine against human Chagas' disease remains suspect, particularly as vector-control methods have proven to be so highly effective.

The failure to develop a safe, effective vaccine against any form of Leishmanial disease in humans, the cutaneous forms, in particular, represent one of the more frustrating efforts in the parasite vaccinology field. This is because people often develop immunity to reinfection following a single primary exposure, an experience that is similar to the naturally acquired immunity induced by most of the infectious pathogens for which licensed vaccines exist. The naturally acquired immunity against zoonotic cutaneous leishmaniasis is the basis for a practice referred to as "leishmanization," which involves the needle inoculation with viable *L. major* parasites in a selected site, usually the arm, and has been used for generations by people living in regions endemic for cutaneous leishmaniasis in the old world to provide protection against natural exposure.[125] This practice has been discontinued because of safety concerns regarding the severity of the lesions in some individuals. In mice, as in people, a primary infection with *L. major* that heals confers strong and long-lived protection against reinfection and represents the gold standard of acquired resistance against any form of Leishmaniasis. Reproducing this protective response is a key objective of anti-*Leishmania* vaccines. A variety of vaccine formulations, including killed-whole cells, defined recombinant protein- and DNA-based vaccines targeting many different antigens, recombinant viral vectors, and live-attenuated *Leishmania*, have been developed that confer some level of protection in animal models of cutaneous and visceral disease.[126] Effective immunization, in many cases, requires antigen delivery with Th1-inducing adjuvants, including TLR ligands such as CpG or monophosphoryl lipid A (MPL). A novel approach to eliciting a protective response is to target the components of sand fly saliva that have been shown to enhance infections following sand fly transmission.[127] In no case, however, have any of the nonliving vaccines tested in mice been shown to confer the same degree of protective potency and durability as a primary, healed infection, and in some cases, the vaccines failed entirely when evaluated under the more stringent conditions of challenge using infected sand flies. Similarly, in humans, attempts to reproduce the protection conferred by leishmanization using whole-cell killed vaccines plus BCG as adjuvant have failed in multiple phase III clinical trials.[128] The acquired resistance observed in healed mice is sometimes referred to as concomitant immunity, meaning that the protection against reinfection coincides with the persistence of the primary infection. While central memory and tissue-resident memory CD4+ T cells contribute to the protective response against reinfection,[129] the rapid recruitment to the challenge site of IFN-γ^+ CD4+ T cells–bearing markers of terminally differentiated, short-lived Th1 effector cells is required to fully reconstitute the protective response in adoptive transfer studies.[130] Thus, concomitant immunity in this model is mediated by a population of preexisting effector T cells, which is maintained as a stable population in the healed mice by the persisting infection. These findings argue for live-attenuated vaccines that have the potential to provide a continuity of exposure to a multiplicity of antigens to generate and maintain the protective response.

Currently, there are three anti-*Leishmania* vaccines in the clinical pipeline against human disease. One is a recombinant fusion protein containing two *Leishmania* antigens, nucleoside hydrolase (NH) and sterol 24-c-methyltransferase (SMT), that when formulated with a synthetic TLR4 agonist glucopyranosyl lipid A, confers strong protection against experimental VL.[131] Another is a live-attenuated vaccine that uses *centrin* gene-deficient *L. donovani* parasites that do not proliferate in macrophages and are highly efficacious against VL in animal models.[132] The third is a simian adenovirus, ChAd63, encoding a synthetic gene for the coexpression of two *Leishmania* antigens, KMP-11 and HASPB, which demonstrated CD8+ T cell–dependent protection against VL in animal models.[133] Its clinical use is intended as a therapeutic vaccine to potentiate protective CD8+ T cell responses in patients with post-kala-azar dermal leishmaniasis (PKDL). The application of vaccines as therapeutics rather than preventives may obviate the requirement for antigen persistence since efficacy may require only a transient enhancement or skewing of an ongoing immune response. Indeed, the killed whole-cell vaccines that failed in prophylactic trials against VL in Sudan or CL in Venezuela provided therapeutic benefit when used in conjunction with antimony in Sundanese patients with PKDL[134] and is the current recommended treatment for CL in Venezuela.[135]

THE IMMUNE RESPONSE TO *PLASMODIUM*

The Basic Biology of *Plasmodium* Infections

Parasites of the genus *Plasmodium* (Phylum: Apicomplexa) are the most significant protozoan pathogens of humans, causing 228 million cases of malaria disease and over 400,000 deaths in 2018 (Global Malaria Program, 2019). Like the Trypanosomatids discussed above, *Plasmodium* parasites are digenetic, alternating between infection of vertebrate and invertebrate hosts to complete their life cycle (Fig. 42.2). For the five *Plasmodium* species that infect humans, the relevant insect vector is the female Anopheles mosquito. *P. falciparum* (written hereafter as Pf) is notorious for causing the majority of severe illness and death in humans, largely in sub-Saharan Africa. *P. vivax* is also a major cause of illness worldwide, particularly in Central and South America, India, and Southeast Asia. Although less frequently lethal than Pf, *P. vivax* is notable for its ability to develop into dormant liver-stage parasites, known as hypnozoites, that can cause relapse of disease weeks or months after an initial infection is treated and thought to be cleared. Three other species—*P. ovale*, *P. malariae*, and *P. knowlesi*—cause infrequent disease in humans, with the latter species primarily infecting primates but occasionally spilling over into humans. Like *P. vivax*, *P. ovale* can also establish dormant liver infection.[136]

Infection of a human host begins with a bite from an infected mosquito, which deposits an average of 18 to 20 haploid *Plasmodium* sporozoites (SPZ) in the skin.[137,138] Some SPZ

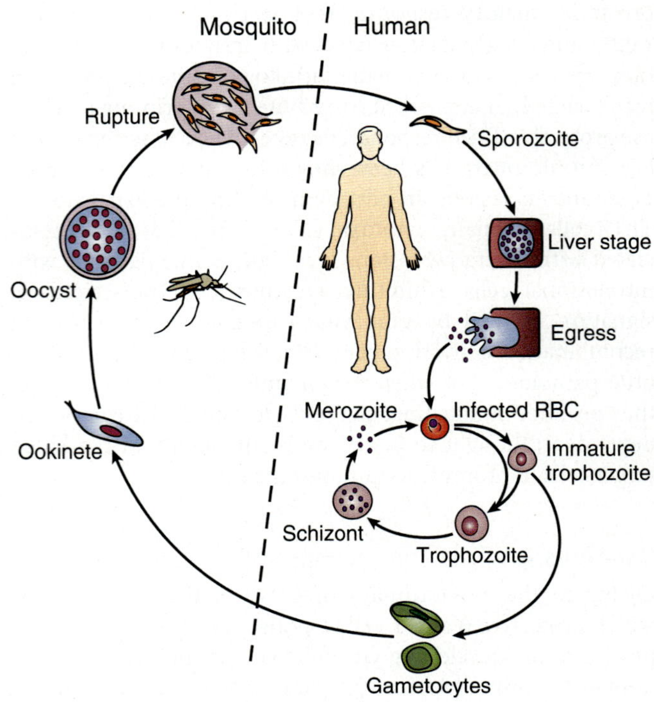

FIG. 42.2. Malaria life cycle. (1) A female *Anopheles* mosquito injects a small number of haploid *Plasmodium* sporozoites into a human host during a blood meal. From the skin, they migrate through blood vessels to the liver, (2) Preerythrocytic or liver-stage parasites undergo schizogony (nuclear replication without cell division) within hepatocytes, (3) Merozoites emerge from infected hepatocytes into the bloodstream, (4) The erythrocytic or blood-stage parasites undergo repeated rounds of replication in red blood cells, developing from infectious merozoites to trophozoites to schizonts, which rupture and release new merozoites, (5) A fraction of immature trophozoites undergo alternative differentiation into male and female gametocytes, which can be taken up by a mosquito during a subsequent blood meal, (6) Within the mosquito, the gametocytes fuse to form the diploid ookinete, (7) The ookinete invades the wall of the mosquito's midgut and develops into an oocyst, (8) The mature oocyst ruptures to release sporozoites, which migrate to the mosquito's salivary glands, from which they can be inoculated into a future human host.

remain in the skin, and others drain through the lymphatics to nearby lymph nodes, where they are degraded. But about one-third of the deposited SPZ use gliding motility to penetrate blood vessels and travel to the liver,[137] reaching it within minutes of accessing the bloodstream[139] After arresting in the liver sinusoids, they traverse across the barrier of endothelial and Kupffer cells and establish infection in hepatocytes within a specialized PV. Over the next 2 to 12 days, depending on the species, the liver-stage parasites (also called exoerythrocytic or preerythrocytic parasites) replicate through the process of schizogony, in which many nuclei are synthesized in the absence of cell division. Eventually, each infected hepatocyte releases many thousands of merozoites into the bloodstream,[140] where they initiate repeated rounds of infection and schizogony within red blood cells (RBCs), destroying their host cell upon completion of each round of replication. During this erythrocytic or blood stage of infection, a proportion of parasites activate a developmental switch that drives formation of male and female gametocytes. If ingested by a mosquito during a blood meal, these gametocytes can go on to complete the sexual stages of the parasite life cycle within the insect vector and may eventually be transmitted to a future human host.

The infected human host remains asymptomatic for several days to weeks after an infectious mosquito bite; the preerythrocytic stage of infection is both clinically "silent" and inaccessible to diagnosis. It is only during the erythrocytic stage, with its repeated infection and destruction of RBCs, that the host experiences the cardinal symptoms of malaria; most prominently, fever, often accompanied by rigors, malaise, headache, muscle pain, and nausea. The fever may be periodic, especially with *P. vivax* and *P. ovale* infection, corresponding to the synchronized rupture of RBCs by emerging blood-stage parasites. Although the majority of malaria episodes in humans are uncomplicated and eventually resolve even without treatment, a small fraction of sick people develops very serious symptoms, including severe anemia, respiratory distress, renal failure, and cerebral malaria.[141] These complications are most common in children under the age of five, who account for nearly 70% of all malaria deaths (Global Malaria Program, 2019).

Individual *Plasmodium* strains exhibit strong preferences or requirements for specific host species, and the species that infect humans do not infect model organisms (with the exception of some nonhuman primates). Thus, most animal research on antimalarial immunity has been performed using *Plasmodium* spp. that do not infect humans. In more recent decades, animal work has focused largely on laboratory mice, using a small cluster of *Plasmodium* spp., originally isolated from rodents such as the African thicket rat. Among these, *P. chabaudi* (Pch) and nonlethal *P. yoelii* (Py) strains are commonly used to study uncomplicated infection; a lethal strain of Py is used to study severe anemia; and the lethal parasite *P. berghei* ANKA serves as a model for cerebral malaria.[142] As with all model systems, important differences exist between mouse and human; nevertheless, studies in mice have significantly advanced our knowledge of immune responses to malaria.

Innate Immune Responses to Malaria

Humoral Mechanisms

Plasmodium parasites are exposed to soluble immune factors during several phases of their life cycle within the mammalian host. In immune individuals, SPZ-specific antibodies can fix complement on the parasite surface before it reaches the liver, inhibiting its traversal through host cells and perhaps causing parasite lysis.[143,144] High levels of complement-fixing SPZ-specific antibodies are rarely found in naturally exposed humans, but when present, they are associated with protection from malaria.[143] The merozoites that initiate blood-stage infection are also susceptible to complement-mediated lysis and must rapidly infect a new cell within a few minutes of egress from their previous host cell. Several studies have reported a strong correlation between the abundance of merozoite-specific complement-fixing antibodies and protection from malaria, suggesting that this may be an important mechanism of restriction.[145,146] To evade complement-mediated killing, merozoites recruit the host proteins Factor H and Factor H-like Protein 1, which bind to the parasite surface and suppress activation of the alternative complement pathway, reducing merozoite lysis.[147,148] Another host protein, the soluble complement regulator C1 esterase inhibitor also binds the merozoite through interactions with the parasite surface protein MSP3 to further inhibit complement fixation and lysis.[149]

Cellular Mechanisms

Some of the earliest studies of the immune response to malaria noted the dramatic expansion of myeloid cells that accompanies this disease— not only in humans but also in primates, birds, and rodents.[150-152] But the importance of myeloid cells in control of blood-stage malaria has only been demonstrated in the last few decades by studies showing increased parasite burden and mortality in *Plasmodium*-infected mice following pharmacological depletion of phagocytes.[151,153,154] Several papers have particularly implicated CD169+ macrophages, found in primary and secondary lymphoid organs, in protection from malaria[154,155]; these macrophages specialize in antigen capture and may prime DCs for both cytokine secretion and antigen presentation, promoting parasite control.[156] Additionally, other tissue macrophages and recruited monocytes can phagocytose and degrade infected RBCs, directly reducing parasite burden.[152-155,157] The relevant mechanism of killing is not well established: activated macrophages and monocytes produce ROS that are capable of killing parasites *in vitro*,[158,159] but mice lacking NADPH oxidase—the primary enzymatic source of ROS in activated macrophages—have little to no defect in control of *Plasmodium* infection, suggesting other killing mechanisms act redundantly *in vivo*.[160] One possible such mechanism is production of ROS by xanthine oxidase,[160,161] which also drives inflammatory cytokine production.[162]

In addition to directly capturing and killing parasites, macrophages and monocytes are early sentinels of the immune response and produce cytokines that may both shape protective responses and contribute to immunopathology.[163] Along with plasmacytoid dendritic cells (pDCs), splenic red pulp macrophages produce type I IFN early in infection.[164] The role of this cytokine in malaria is likely context-dependent; it has been shown to restrict parasitemia in several nonlethal malaria models, though conflicting data exist,[157,165] whereas in lethal models, it contributes to immunopathology and mortality.[166] TNF-α and IL-1β, two other macrophage-derived cytokines, are classical drivers of fever, a hallmark of malaria pathogenesis; TNF, in particular, is associated both with parasite clearance and clinical illness.[167,168] IL-12 from macrophages elicits IFN-γ production from NK and T cells and is associated with protection in mice.[169] Meanwhile, secretion of immunoregulatory IL-10 serves to dampen excess production of inflammatory cytokines, such as TNF and IFN-γ, and limit pathology.[170] Neutrophils are not required for control of *Plasmodium* infection but may contribute to tissue damage through activation of NETs, release of IL-1α, and promotion of excessive inflammation.[171-173] Altogether, an appropriate balance of pro- and anti-inflammatory cytokine production is critical during the antimalarial response to enable pathogen clearance without causing excessive damage to the host.

In addition to macrophages and monocytes, several other innate or innate-like cells are important for control of *Plasmodium*. Myeloid production of IL-12 serves to activate NK cells,[174] which contribute to parasite control most likely by secreting IFN-γ early in infection that activates macrophages to take up and kill infected RBCs.[174,175] The cytotoxic function of NK cells is not required for control of blood-stage infection in mice but has been linked to lysis of infected RBCs in humans and may help limit growth of liver-stage parasites within hepatocytes.[174,176,177] Another innate source of IFN-γ during malaria may be γδ T cells. These innate-like T cells, especially those expressing the Vγ9 and Vδ2 TCR chains, proliferate and secrete IFN-γ in response to both viable and irradiated *Plasmodium* SPZ; they can persist long-term at elevated frequencies in repeatedly exposed patients and have been associated with protection in human vaccine trials.[178,179] Human $V\gamma9^+V\delta2^+$ T cells respond to nonpeptidic phosphoantigens such as (E)-4-hydroxy-3-methyl-but-enyl-pyrophosphate (HMBPP), a molecule used in synthesis of the ancient and diverse class of five-carbon molecules known as isoprenoids.[180] In malaria, soluble phosphoantigens are released from infected RBCs upon parasite egress and can activate γδ T cells without cell-cell contact. Production of these phosphoantigens requires an intact isoprenoid biosynthesis pathway, suggesting that HMBPP may be the (or a) relevant activating ligand in malaria.[181] Like other apicomplexans, *Plasmodium* utilizes an isoprenoid synthesis pathway distinct from the one used by mammals, making it a logical target for immune surveillance.

In mice, γδ T cell–deficient mice experience higher liver-stage parasite burdens than wild-type mice,[182] and depletion of γδ T cells during vaccination with irradiated SPZ-impaired CD8+ T cell responses and ablated sterile protection although depletion during challenge with live parasites did not impair immunity. IFN-γ can mediate killing of liver-stage parasites, suggesting a mechanism through which gd T cells may contribute to protection.[183] Like SPZ, *Plasmodium*-infected RBCs also stimulate robust proliferation and IFN-γ secretion from $V\gamma9^+V\delta2^+$ γδ T cells.[179,184,185] γδ T cells can kill Pf merozoites *in vitro* through a mechanism that requires granulysin and cell-cell contact, though they do not affect growth of intraerythrocytic parasites.[186,187] Studies in mice provide conflicting data on whether γδ T cells are important for protection from malaria,[188-190] but some studies in humans have found correlations between γδ T cell responses and protection.[179,191,192] In children with repeated malaria, $V\delta2^+$ T cell frequency is associated with protection from infection, but frequent exposure drives these cells to become progressively more dysfunctional and eventually be depleted.[193] Cytokines secreted by γδ T cells may also contribute to pathology.[163,191]

DCs and Shaping of Adaptive Responses

pDCs are early sensors of blood-stage parasites and secrete proinflammatory cytokines, especially Type I IFN, to activate the immune system.[157,194] Along with macrophages, conventional $CD11c^{hi}$ DCs secrete IL-12 to elicit IFN-γ from NK cells and Th1 cells; this IFN-γ response is protective in nonlethal malaria models but promotes pathology in a mouse model of cerebral malaria.[195,196] Whether conventional DCs play a central role in antigen presentation during malaria is less clear: although $CD11c^{hi}$ DCs from infected mice are capable of activating *Plasmodium*-specific T cells *ex vivo*,[157] a recent study found that selective depletion of DCs did *not* alter antigen-specific CD4+ T cell activation or polarization during blood-stage infection *in vivo*. Interestingly, depletion of B cells did,[197] indicating that B cells may act as the most critical APCs in this context. Earlier studies had reported a requirement for $CD11c^{hi}$ DCs in both T cell activation and parasite uptake; however, these studies utilized a less selective mouse model involving depletion of all $CD11c^+$ cells, which includes activated B cells and CD8+ T cells, making the specific role of DCs difficult to disentangle.[198,199]

The possibility that B cells, and not conventional DCs, may be the primary orchestrators of T cell activation during malaria is made more compelling by several studies, suggesting that *Plasmodium* parasites drive DCs to become dysfunctional.[200-203] *Plasmodium*-infected RBCs are poor inducers of DC maturation compared to canonical stimuli such as TLR ligands, and *in vitro* incubation with infected RBCs can render DCs hypofunctional and/or refractory to further signaling downstream of PRRs.[200,202] Suppression of DC function can be recapitulated by incubation with hemozoin.[204] *In vivo*, human patients with uncomplicated malaria have decreased numbers of circulating DCs, and remaining DCs exhibit impaired maturation and antigen presentation and increased apoptosis relative to DCs from healthy controls.[201] The extent to which *Plasmodium* suppresses DC function *in vivo* may vary between genetically distinct human populations; intact DC activation correlates with protection from malaria in one study from a malaria-endemic region of Africa.[205] Inhibition or downregulation of DC function may be a parasite immune evasion mechanism or alternatively may reflect an immunoregulatory program that serves to limit excess activation and pathology for the host. In summary, DCs orchestrate proinflammatory innate and adaptive responses during acute infection and may also modulate the adaptive response in ways that both limit immunopathology and inhibit pathogen clearance.

Innate Immune Sensing

TLRs are likely some of the first and most critical host sensors to detect *Plasmodium* infection. Evidence for the importance of TLRs comes from the study of mice deficient in MYD88, a common signaling adaptor for most TLRs as well as for the IL-1 and IL-18 receptors. *Myd88−/−* mice infected with rodent-specific malaria parasites exhibit diminished production of key cytokines, including TNF and IFN-γ, fail to initiate cell-mediated immune responses, and are unable to control parasitemia.[206,207] In addition to TLR signaling, these defects may also reflect contributions from IL-1 and IL-18 or a role for MYD88 in T cell priming and polarization.[207] Despite its importance in initiating immune responses, MYD88-dependent signaling is not exclusively protective in malaria; it can also drive pathological inflammation. Mice lacking MYD88 are protected from cerebral malaria during infection with the normally lethal parasite *P. berghei* ANKA.[208]

Several individual TLRs have been implicated in sensing of *Plasmodium*. As with other protozoan parasites, *Plasmodium* membrane proteins possess abundant GPI lipid anchors that can trigger signaling through TLRs 2 and 4.[31] Parasite RNA and DNA are sensed by endosomally localized TLR7 and TLR9, respectively, likely after phagocytosis of parasites or infected RBCs by macrophages and plasmacytoid DCs. Hemozoin, a crystalline byproduct of the parasite's hemoglobin metabolization, also stimulates TLR9, but this likely reflects contamination with parasite DNA.[33] In general, mice lacking single TLRs have some selective defects in cytokine production and immune activation but display minimal difficulty in controlling infection, suggesting that multiple TLRs contribute redundantly to parasite sensing and host resistance.[206]

In addition to TLRs, which monitor the extracellular and endosomal compartments, several intracellular innate immune detection pathways contribute to detection of *Plasmodium* parasites. RNA from liver-stage parasites can activate the cytosolic sensor MDA-5 in infected hepatocytes, leading to production of type I IFN. This, in turn, induces expansion of natural killer T (NKT) cells expressing IFN-γ in the liver, which are critical for controlling subsequent infections.[36,183]

During the blood stage, *Plasmodium* parasites infect RBCs, which lack cytosolic immune sensors. However, cytosolic parasite sensing still occurs in phagocytes that engulf infected RBCs. After uptake, particles of hemozoin in the phagolysosome may destabilize both parasite and host membranes, allowing some contents to reach the cytosol, where they engage cytosolic PRRs.[34] The most important of these is likely cGAS, which recognizes genomic parasite DNA and activates a signaling pathway involving the downstream adaptor STING, the kinase TBK1, and the transcription factors IRF3 and IRF7, leading to a transcriptional program typified by production of type I IFN.[209,210] The effect of this intracellular sensing pathway on host resistance to malaria is complex and may depend on the infection model: cGAS-deficient mice infected with a nonlethal *Plasmodium* strain displayed increased parasite burden, indicating that the pathway is protective(159). In contrast, elimination of cytosolic nucleic acid sensing pathways during a normally lethal *Plasmodium* infection in mice led to *increased* TLR7-dependent type I interferon in pDCs, resulting in decreased parasite burden and improved survival.[165] Thus, cGAS-mediated signaling can promote both pathogen control and pathology, depending on the context, and can cross-regulate other innate sensing pathways.

Plasmodium-derived signals can engage several additional cytosolic PRRs that lead to activation of inflammasomes. Synthetic hemozoin potently activates the NLRP3 inflammasome, likely by destabilizing the phagolysosome and causing release of DAMPs such as uric acid.[42,43] Additionally, hemozoin facilitates entry of parasite genomic DNA into the cytosol, where it can activate the AIM2 inflammasome.[34] Both inflammasomes mediate release of the inflammatory cytokines IL-1β and IL-18, the first of which promotes fever. Thus, it is likely that inflammasome activation is largely harmful in malaria, exacerbating clinical symptoms without contributing much to restriction of parasite burden. *Nlrp3−/−* mice are partially protected from cerebral malaria[211] and another lethal malaria model,[43] but mice lacking downstream components of the inflammasome (such as Caspase-1 or IL-1β) are not protected, leaving the precise mechanism of protection undefined.[208,211] In humans, the intensity of placental malaria correlates positively with inflammasome activation and IL-1β production within placental tissue and with low birth weight, again suggesting that activation of this immune pathway increases pathology.[212] In addition to the direct effects of elevating inflammatory cytokines, inflammasome activation may antagonize production of type I interferons, dampening its protective effects.[213]

Adaptive Responses and the Challenges of Vaccine-Induced Immune Protection

Plasmodium parasites exhibit unique biology in each stage of their life cycle, including cellular and tissue tropisms, expression of ligands of innate cell sensors, replicative capacity, and kinetics. The perception of antigens under these different conditions invokes distinct responses from adaptive immune cells. Over the past several decades, key mechanisms of adaptive immune responses primarily induced by immunization have been identified that contribute to protection against distinct life stages of the parasite during experimental infection in mice and humans. The trafficking of SPZ from the skin to the liver can be inhibited by antibodies expressed by B cells.[137,214,215] Antigen-experienced CD8+ T cells induced by immunization with attenuated SPZ can recognize parasite-peptide: MHCI complexes on infected hepatocytes to provide sterile protection in mice and nonhuman primates.[177,216] CD4+ T cells and B cells play important roles in controlling parasite replication during the blood stage of the response in mice and humans.[217] CD4+ T cells also modulate inflammation through the production of suppressive cytokines like IL-10, which prevents immune pathology induced by the blood stage.[218]

Yet while each of these immune mechanisms contributes to *Plasmodium*-specific immunity in response to vaccination, their role in immunity acquired after natural infection is less clear. Whereas some immunization strategies can elicit sterilizing immunity, natural infection results in clinical immunity: the development of immune responses that limit parasite burden, reduce mortality, and protect individuals from serious clinical symptoms but do not prevent reinfection.[219] The mechanisms that confer clinical immunity following natural infection may be very different from those that generate sterile immunity in response to immunization.[220] Thus far, the immune correlates of protection that result in naturally acquired clinical immunity have been difficult to identify.[221]

In this section, we highlight the similarities and differences in acquired resistance to natural infection versus immunization. We will not present a broad overview of all *Plasmodium* vaccine strategies, as those have been recently reviewed elsewhere.[222,223] Instead, we highlight recent discoveries elucidating the mechanisms of *Plasmodium*-specific immune protection, the vaccines that specifically induce these responses, and the implications for our understanding of naturally acquired immunity.

Immunization, But Rarely Natural Infection, Induces SPZ-specific Humoral Immunity

SPZ-specific antibodies are capable of inhibiting infection of hepatocytes both *in vitro* and *in vivo* and reduce or prevent the parasite progression into symptomatic blood-stage infection in both humans and mouse models. The major target of anti-Pf SPZ antibodies is the circumsporozoite protein (PfCSP). PfCSP is the most abundant surface antigen on the SPZ and is essential for SPZ motility and hepatocyte invasion.[224,225] It is also the immunodominant antigen on the sporozoite surface[226] and a major focus of vaccine strategies. In malaria-endemic areas, however, there is an inverse correlation between levels of *Plasmodium* exposure and the presence of CSP-specific antibodies and memory B cells, suggesting that the parasite employs profound immunomodulatory strategies that must be overcome by immunization.[227-229]

Natural infection with *Plasmodium* SPZ presents the immune system with multiple challenges that together prevent the efficient generation of protective humoral immunity against antigens expressed in the preerythrocytic stages. Very few SPZ are deposited during the mosquito blood meal, which may contribute to the weak induction of CSP-specific memory B cells.[229] In addition, SPZ-specific antigen availability is transient, as SPZ migrate to the liver within hours, and then develop into merozoites that express an entirely different genetic program and antigenic profile.[226] To further complicate matters, inflammation associated with the emergence of the blood-stage disrupts development of the liver stage–specific germinal center response, decreasing the subsequent development of memory B cells and long-lived plasma cells (LLPCs).[230] These challenges may explain why immunization with high numbers (thousands to millions) of whole attenuated parasites, which cannot mature into the blood-stage, results in sterile immune responses in both human and mouse models, but natural infection does not.

Indeed, multiple immunization strategies can provide sterile immune protection through strong induction of liver stage–specific humoral immune responses. Immunization with parasites attenuated by irradiation, drug chemoprophylaxis, or gene deletion engenders complete sterile protection that is mediated by CD8+ T cells and B cells.[231] Attenuated parasite vaccines elicit robust antibody responses against preerythrocytic antigens in early phase controlled human malaria infection (CHMI) studies with malaria-naïve human volunteers, as well as animal studies.[178,232] Robust antibody responses also form in response to the most advanced malaria vaccine, RTS,S. RTS,S is a virus-like particle consisting of a portion of the central repeat region and C terminal domain of CSP fused to the Hepatitis B Surface antigen. RTS,S, in formulation with the adjuvant AS01, is currently undergoing pilot implementation in sub-Saharan Africa after completion of Phase III clinical trials[233] (WHO, Global Malaria Program, 2020). The efficacy of RTS,S seen in malaria-naïve individuals (around 50% of volunteers) drops significantly in malaria-experienced individuals.[234,235] RTS,S elicited 56% protection against clinical malaria among 5- to 17-month-old children[236] and 30% efficacy among younger infants in Phase III clinical trials in endemic areas. These differences in efficacy raise questions about how either previous or continuous exposure to the parasite may alter the CSP-specific immune repertoire or function of the CSP-specific immune memory cells.

The intense focus on understanding how CSP-specific humoral immunity is associated with immune protection has led to numerous advances in our understanding of the development and function of CSP-specific B cells and their antibodies. The isolation of hundreds of CSP-specific monoclonal antibodies from RTS,S- and attenuated SPZ-immunized individuals has allowed for molecular and functional assays of CSP-specific antibodies to determine how they inhibit liver-stage disease.[226] Antibodies against SPZ can prevent migration to the liver and block hepatocyte invasion by inhibiting parasite egress from the sinusoids or hepatocyte traversal or through direct killing.[237] However, the short time it takes for SPZ to migrate from the skin to the liver requires that high titers of antibody are present in both the skin and blood at the time of infection. Successful antibody-mediated immunization strategies will, therefore, need to generate germinal center–derived LLPCs to continuously maintain anti-SPZ antibody titers. Understanding how to generate strong preerythrocytic humoral immunity in preexposed individuals is an ongoing area of research that will be critical for developing better vaccination strategies.

CD8+ T Cells in the Liver Kill Infected Hepatocytes

SPZ that migrate to the liver rapidly infect hepatocytes, in which they spend several days developing into erythrocyte-infective merozoites. The liver stage of infection is an important vaccine target since it presents the immune system with a final opportunity to eliminate the parasite prior to the emergence of blood-stage infection and clinical disease. Liver stage–directed vaccine strategies include viral vector, subunit DNA, and whole SPZ vaccines. Studies in animals immunized with radiation-attenuated SPZ (RAS) have demonstrated that liver-stage protection is mediated primarily by *Plasmodium*-specific CD8+ T cells that recognize infected hepatocytes, although CD4+ T cells also appear to contribute to protection and the maintenance of parasite-specific CD8+ T cells in the liver.[231,238,239] CD8+ T cells that recognize parasite antigens cluster around infected hepatocytes and eliminate them via both direct killing and cytokine secretion (IFN-γ and TNF).[240-242] A variety of parasite-specific memory CD8+ T cells can develop, including both circulating and tissue-resident memory cells.[177] While the relative importance of the two subsets in immune protection is unclear, studies in a murine immunization model have shown that a very high number of circulating memory CD8+ T cells (10% of total CD8+ T cells) must be specific for CSP to protect from reinfection.[243] More recent studies have highlighted the potential protective role of liver-resident CD8+ memory T cells, which are not measurable in the blood. Liver tissue-resident memory cells (Trm) that express high levels of the surface markers CD69, CXCR6, CXCR3, and CD101, but lack expression of CD62L and KLRG1, are maintained in the liver and are required for sterilizing immunity in response to attenuated parasite vaccination or viral vector strategies.[244,245] Liver Trm cells patrol the liver sinusoids using a crawling motion that enables scanning of infected hepatocytes for cognate antigen.[246] Recent evidence has shown that CXCL9-dependent CD8+ T cell positioning in the periportal regions of the liver lobules may position these immune sentinels at the site of parasite entry.[247] Understanding how to generate and maintain long-lived CD8+ memory T cells in this important site of immune surveillance will be an important next step.

B Cells and Their Antibodies are Critical for Control of Blood Stage Parasitemia

Despite the absence of sterile immunity in naturally infected individuals, repeat *Plasmodium* infection does slowly lead to the production of antibodies that can control blood-stage parasitemia, prevent clinical disease, and reduce mortality.[248,249] Much evidence supports the importance of blood stage–specific antibodies in controlling infection and providing immunity to reinfection in humans and animal models.[250-252] Antibodies that recognize either merozoites or antigens expressed on the surface of infected RBCs function through several mechanisms, including the inhibition of cell invasion, opsonization of merozoites and infected RBCs, and complement fixation.[226,253-255]

Yet while antibodies induced by a blood-stage infection can suppress disease, they are rarely capable of preventing the reemergence of blood-stage parasites in subsequent infections, suggesting that B cell responses will need to be optimized by vaccination.[256] Recent work has, therefore, focused on detailed analyses of merozoite-expressed antigens associated with cross-strain immune protection as well as specific B cell subsets and their antibodies that can contribute to control of parasitemia.

Leading blood-stage vaccines against antigens like apical membrane antigen 1 (AMA-1) have demonstrated partial efficacy in part due to strain-specific antigenic polymorphisms.[257] A major focus has therefore been the identification of merozoite-expressed antigens that can induce strain-transcendent responses. A variety of approaches have been taken to identify novel vaccine antigens, including antigen arrays[258] and direct testing of antibodies generated after immunization with viral vectors expressing blood-stage antigens.[259] While many new and interesting antigens have been revealed, one of the most exciting developments is the discovery that immunization with the PfRH5 protein, required for erythrocyte invasion, induces cross-strain neutralizing antibodies that can inhibit parasite growth, making this a leading vaccine candidate that is currently in Phase I trials.[260]

Understanding the functional contributions of B cells has also been a major focus for learning how to generate optimal humoral immunity to blood-stage parasites. In both humans and animal models, infection induces merozoite-specific memory B cells that can be rapidly reactivated to make antibody. Importantly, memory B cells can display significant phenotypic and functional heterogeneity. Subsets of merozoite-specific B cells express unique surface markers associated with development, activation, and function, including CD21 and CD27 in humans and CD80 and CD73 in mice.[261] A population termed "atypical memory B cells" which were originally thought to be dysfunctional and to arise specifically in chronic infections forms in both humans and animal models of infection; however, recently it was shown that these cells are part of a "typical" immune response, developing in acute as well as chronic settings and displaying functional responsiveness.[262,263] Additionally, *Plasmodium*-specific MBCs express BCRs of a variety of isotypes that appear to have different functional properties and correlate with protection from natural infection.[230,264,265] Understanding the significance of these various populations will be essential for generating long-lived immune protection from blood-stage parasitemia and hence, malaria disease.

CD4+ T Cells Contribute to the Control of Blood-Stage Infection

CD4+ T cells contribute to immune protection from *Plasmodium* after both natural exposure and vaccination. These cells function as central immunomodulators in the response to blood-stage infection: they secrete cytokines that can activate innate cells,[266] provide signals to B cells to class-switch to specific isotypes and enter the memory pool,[267] and quell inflammation through their expression of suppressive cytokines.[218] Evidence that CD4+ T cells contribute to control of malaria comes from studies of HIV+ humans in malaria-endemic regions, who experience more frequent and severe episodes of malaria relative to HIV- individuals.[268] The CD4+ T cell response to *Plasmodium* is predominantly a Th1 response, as denoted by the expression of Tbet and the cytokine IFN-γ. Yet within this Th1 response, there is a division of labor among *Plasmodium*-specific CD4+ T cells that has been previously described for bacterial and viral infections, in which some cytokine-producing CD4+ T cells become innate cell activators (Teff) while others become B cell-helping T follicular helper cells (T_{FH}).[269] Recent insights from both humans and animal models of *Plasmodium* infection suggest that both populations contribute to protection and may be important targets of vaccines.

Cytokines expressed by CD4+ T cells, including IFN-γ, TNF, IL-4, and IL-10, play important roles in controlling parasite replication,[188] and IFN-γ production is a hallmark of both human and murine malaria.[267] In a murine model of *Plasmodium* infection, however, macrophages can be activated in an IFN-γ–independent manner,[153] suggesting alternative roles for IFN-γ in mediating protection, including activating other innate or nonhematopoietic cells or inducing class-switch recombination in B cells. CD4+ T cells can also express additional cytokines that activate innate cells, including MCSF, shown to be necessary for the proliferation and activation of myeloid subsets that could contribute to control of parasitemia.[153,154] It is also important to note that CD4+ T cell–derived IFN-γ is a primary mediator of pathology in experimental cerebral malaria models, although the specific target of the cytokine has not been elucidated. The complex role of CD4+ T cell–expressed IFN-γ suggests the need for careful consideration in vaccine design.

In addition to promoting immune activation through secretion of IFN-γ and other cytokines, CD4+ T cells also can suppress inflammation through production of IL-10.[270,271] Notably, malaria-experienced humans possess elevated frequencies of *Plasmodium*-specific, T-Bet+ CD4+ T cells that coproduce IFN-γ and IL-10.[191,271] These are called T regulatory Type 1 (TR_1) cells, although they lack FOXP3, the master transcription factor for natural regulatory T cells. It is suggested that TR_1 cells limit immunopathology by dampening excessive inflammation, but they may also restrict the development of antiparasitic responses, slowing clearance of the infection.[191,270-272] More recently, it has been shown that IL-10 production is required for *Plasmodium*-specific germinal center formation.[273] A distinct, non-TR_1 subset of antigen-specific CD4+ T cells serves an immunoregulatory role in malaria by producing IL-27, which suppresses clonal expansion and cytokine secretion by CD4+ T cells.[274]

Plasmodium-specific FOXP3+ regulatory T cells do develop in malaria, but their role in control of both immunopathology and infection is not yet well understood.[275]

Plasmodium-specific CD4+ T cells also play critical roles in directing the differentiation of B cells. CD4+ Th1 T_{FH} cells play an important role in inducing B cell responses as well as suppressing inflammation. CD4+ T_{FH} cells emerge early in response to murine *Plasmodium* infection and express the chemokine receptor CXCR5, which directs their migration to the B cell follicle. CD4+ T_{FH} cells responding to *Plasmodium* express the canonical T_{FH} transcription factor BCL6, yet also express the master Th1 regulator Tbet, reflecting the inflammatory environment in which they were primed.[197] This endows these cells with the capacity to produce key cytokines associated with B cell differentiation, including IL-4 and IL-21, as well as IFN-γ, required for B cell class-switching to specific isotypes. Interactions between T_{FH} cells and B cells are required for the generation of germinal centers and the formation of humoral memory. T_{FH} cells and circulating T_{FH} cells that can downregulate expression of BCL6 yet retain expression of CXCR5 have now been described extensively in both murine and human *Plasmodium* infection.[276,277] An urgent need for the field is to more clearly define the specific roles of the various CD4+ T cell subsets.

The immune effector mechanisms that contribute to control of *Plasmodium* infections are summarized in Figure 42.3.

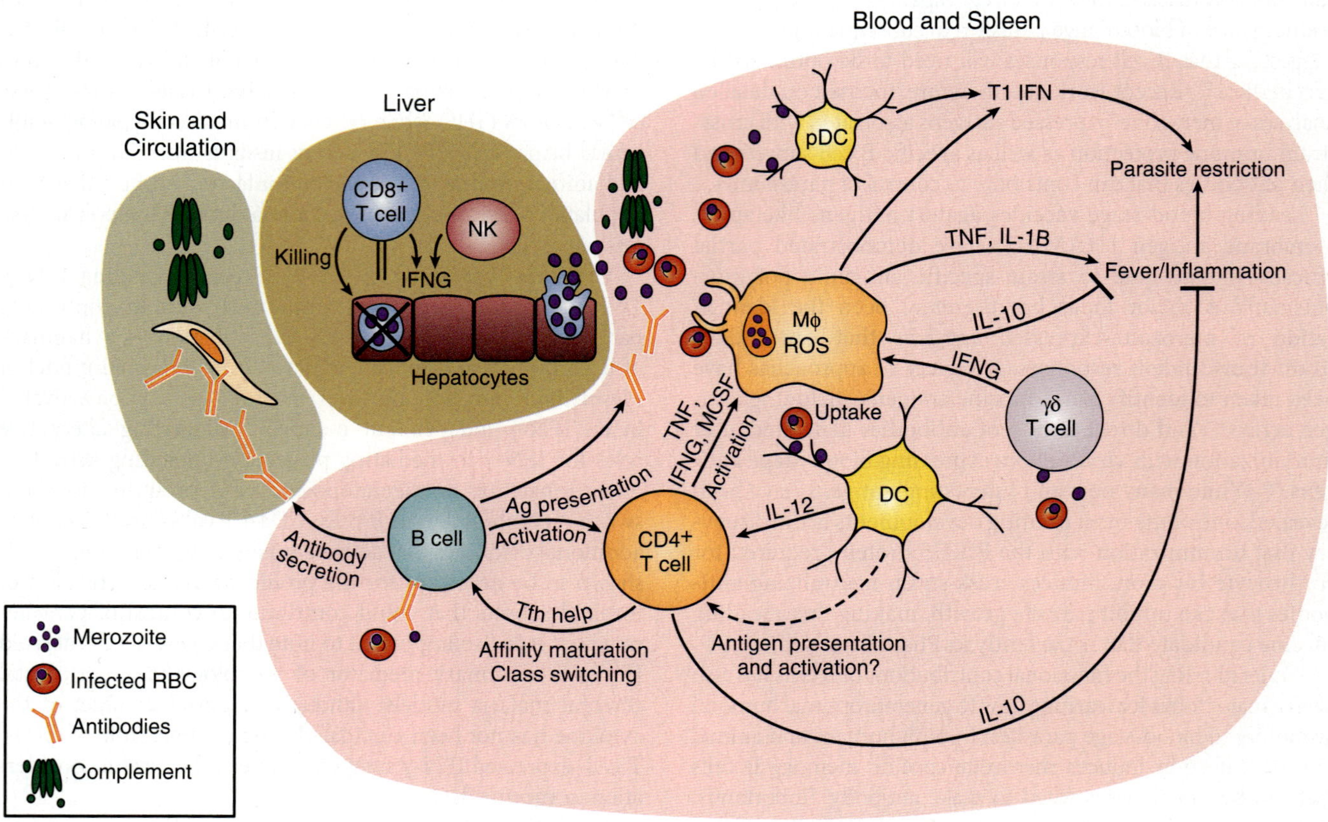

FIG. 42.3. Immune effector mechanisms that restrict *Plasmodium* infection. When infection is initiated in an immune host, preexisting *Plasmodium*-specific antibodies can bind sporozoites in the skin and during transit to the liver. Antibody binding both inhibits motility and induces complement deposition, leading to sporozoite lysis. During liver-stage infection, tissue-resident CD8+ T cells and NK cells can kill infected hepatocytes and the parasites they harbor. After infectious merozoites egress from the liver into the bloodstream, they are vulnerable to complement-mediated lysis, again assisted by parasite-specific antibodies. These effector mechanisms likely also function against infected red blood cells (iRBCs). In the spleen, macrophages and conventional dendritic cells (DCs) phagocytose and degrade iRBCs, perhaps through production of reactive oxygen species (ROS), directly limiting parasite burden. They also make proinflammatory cytokines, including TNF, IL-1β, and IL-12, that help orchestrate the broader protective immune response: TNF and IL-1β induce fever and widespread immune activation, while IL-12 from DCs polarizes the CD4+ T cell response toward Th1. In addition to macrophages, plasmacytoid dendritic cells (pDCs) sense iRBCs and make large amounts of T1IFN, which helps restrict parasite burden through poorly understood mechanisms. Whether conventional DCs are critical for antigen presentation and CD4+ T cell activation during blood-stage malaria is still an object of active study; meanwhile, B cells have been shown to serve as key antigen-presenting cells in this context. In turn, CD4+ T follicular helper (T_{FH}) cells provide activating signals to B cells to promote class-switching and affinity maturation, leading to optimal production of *Plasmodium*-specific antibodies. These antibodies aid in elimination of blood-stage parasites not only through neutralization and complement activation but also through opsonization, which facilitates uptake by phagocytes. Effector CD4+ T cells produce cytokines such as TNF, IFN-γ, and MCSF that synergistically activate macrophages and monocytes (not pictured) to take up and degrade iRBCs. Along with myeloid cells, CD4+ T cells also secrete the anti-inflammatory cytokine IL-10, which limits immunopathology. γδ T cells can sense parasites directly, and like CD4+ T cells, produce IFN-γ to aid in macrophage activation.

IMMUNE RESPONSE TO HELMINTHS

The Basic Biology of Helminth Infections

The word "helminth" is used as a nonphylogenetic term referring to multicellular animals (or metazoans) that have adopted a parasitic lifestyle in mammalian hosts. Most people refer to this group of organisms as parasitic worms and group them together despite the large phylogenetic distance between the Nematodes and the Platyhelminthes, which constitute the two main Phyla within the animal kingdom from which these parasites belong. The anatomy and physiology of these two groups of parasitic organisms are completely different. For example, while the nematodes have a through gut, the cestodes and trematodes lack an alimentary canal for digestion. Whereas the nematodes are covered in a tough exoskeleton called the cuticle, cestodes and trematodes are covered in a cellular structure called the tegument that has the capacity to absorb as well as to secrete materials into the host. Despite these differences in physiology, all of these worms induce remarkably similar immune responses. Features that are shared between the worms are their large size compared to other microorganisms, as well as the complex life cycles whereby an infective stage establishes an infection before growing into long-lasting adults to produce eggs or offspring for transmission to another host. Adult helminths can live for decades as a foreign organism within their hosts, much like a successful organ transplant, and hence can teach us about mechanisms for inducing tolerance within a host.

Most helminths do not replicate within the host (with the exception of *Strongyloides stercoralis*), which is unlike the protozoans described above, as well as viruses and bacteria. Hence, they cannot evade the immune system through the process of antigenic variation. Whereas protozoans may undergo genomic alterations that lead to alterations of the cell surface proteins, the infective stage through to the adult helminths carries the same genomic content within their cells. Immune evasion strategies have thus been selected through the process of long-term coevolution rather than within-host selection during replication. Helminths have thus evolved distinct methods to coexist with their host through the activation of an immune regulatory network.

During the course of vertebrate evolution, most species have harbored parasitic helminths. From fish to mammals, intestinal nematodes that live as mature adults in the intestines for prolonged periods of time are the most common, but other worms can also live within the tissues of their vertebrate hosts. This constant presence during evolution must have exerted a powerful force of selection on our immune system. Just like the gut microbiota, the immune response has coevolved with these organisms in order to tolerate their presence in our bodies. However, unlike the many bacterial communities in the gut microbiota that have evolved as commensals, helminths are still parasites that cause considerable pathology in some of their infected hosts.

There is considerable heterogeneity between individual host responses to helminth infections. This variability in our ability to mount different magnitude of immune responses against the worms is associated with the degree of disease that the worms can cause. Over one billion people worldwide, mainly children, are infected by helminths.[278] While the majority of people carry low-level infections and don't have any symptoms, pathology can occur in some individuals that have very high worm burdens. Because so many people are infected, disease for even a small minority of individuals results in a lot of worldwide morbidity. In terms of a loss of DALYs, which is a measure of morbidity rather than mortality, the combined effects of helminth infections are greater than that estimated for malaria and approaching that attributed to tuberculosis.[279] Hence, helminths cause low mortality but high morbidity in many parts of the world without adequate fecal waste sanitation and contribute toward the spiral of poverty and reduced productivity in many underresourced parts of the world.

In addition to pathology associated with high worm burdens, immunopathology can be associated with an overexuberant unregulated response in some individuals resulting in collateral tissue damage.[280] For the filarial nematode infections, in particular, this can result in debilitating tissue damage with disease manifestations such as elephantiasis and river blindness.[281] In asymptomatic individuals with helminth infections, there is an appropriate balance between "parasite resistance" and "disease tolerance" mechanisms, which results in most infected individuals retaining a small number of worms that are tolerated by the host without causing significant pathology.[282] This situation is beneficial for both the host and the parasite, as transmission persists without overt impact on host health (see Fig. 42.4).

While immunologists have made significant advances in characterizing and understanding the immune responses to these parasites, we have yet to make significant contributions toward utilizing that knowledge toward developing measures (such as vaccines) that can reduce the morbidity and global burden that is caused by these infections.

Hallmarks of the Type 2 Immune Response to Helminths

As a result of the unique biological features of these organisms (discussed above), helminths evoke an immune response that is almost entirely distinct from other microbial pathogens. Remarkably, despite the vast phylogenetic distances and the very distinct and complex life cycles, parasitic helminths induce an immune response that is quite similar to each other. This response is referred to as a Type 2 immune response.[283] In the past, this was referred to as a Th2 response because the cytokines that are associated with this response (eg, IL-4, IL-5, and IL-13) were thought to be produced predominantly by the adaptive arm of the immune response and by CD4+ effector T cells. It is now clear that the innate immune system is both an important initiator of the adaptive Th2 response but also an important early source of many Type 2 cytokines. Understanding the immune responses to helminths is to a large extent about understanding the role that Type 2 immunity plays in host-pathogen relationships. However, under most infection conditions, both Type 1 and Type 2 cytokines and responses are active; it is merely the proportion of cell types and cytokines that are skewed in one direction or the

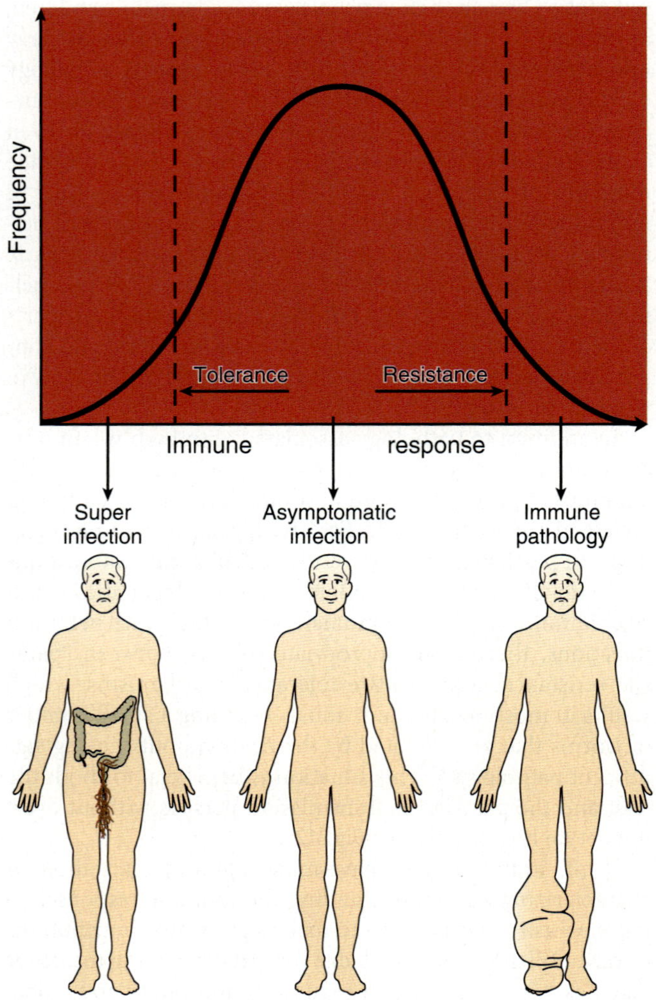

FIG. 42.4. Interindividual variation in immune response to helminths determines pathogenesis. There is considerable variation in the magnitude of the immune response that different individuals will mount against helminth infections. If we assume that this characteristic may follow a normal distribution (as shown), there will be a small minority of individuals that have a very weak immune response to the worms and can become superinfected. These individuals suffer morbidity from having excessively large worm burdens. Individuals with an immune response that is too great may suffer from immune pathology resulting from the collateral damage against the host tissue. For the majority of asymptomatic individuals, a balance of "resistance" mechanisms to expel the worms to prevent excess worm burdens and "tolerance" mechanisms to repair tissue damage and regulate inflammation enables the host and the parasites to coexist while maintaining host fitness, as well as transmitting the parasites to other hosts.

other, and neither Type 1 nor Type 2 responses occur in isolation. Typical features of the Type 2 response include the production of IgE, increased abundance of eosinophils and basophils, activation of mast cells, increased mucus production, and the differentiation of CD4+Th2 cells that produce cytokines such as IL-4, IL-5, and IL-13.

While the primary role of the Type 2 response is likely to be protection against helminth infections, it is now clear that this class of responses is also important in the maintenance of tissue homeostasis, especially by promoting tissue repair[284] and maintaining appropriate metabolic function in adipose tissues.[285] Dysregulated Type 2 responses are pathogenic in the context of allergic diseases[286] and may inhibit an effective antitumor immune response promoting tumorigenesis.[287] Hence, understanding antihelminth immune responses can often lead to immunological discoveries that are of considerable importance for other diseases. An excellent example of this is the discovery of innate lymphoid type 2 cells (ILC2s).[288] Because of the robust immune responses triggered by helminth infections in mucosal surfaces, it became clear that a population of lymphocytes exhibited properties of the innate immune system and rapidly produced large quantities of Type 2 cytokines (IL-13 and IL-5, in particular) in intestinal tissues before the CD4+ Th2 cells have expanded. Because ILC2s are among the first immune cells to respond to helminths, they play a critical role in shaping the adaptive response that follows. As tissue-resident cells, ILC2s are poised to respond to disruptions in the tissue and the release of alarmins or neuropeptides. ILC2s do not respond directly to PRRs or rearranged antigen-specific receptors but rather to the cytokine milieu (eg, IL-25, IL-33, and thymic stromal lymphopoietin [TSLP]) that results from barrier disruption. While these cells were identified and characterized in the context of helminth infection, it is now clear that they are important in a wide range of inflammatory settings.

Mouse Models of Helminth Infection for Immunology Studies

Our understanding of the many immunological features of helminth infections has been greatly facilitated by a number of different mouse models for helminth infection. While all of these worms will induce Type 2 immune responses, the nature of the response varies as befitting the different tissue location of residence for the larval and adult stages of these parasites. *Trichuris muris* is closely related to the human whipworm, *Trichuris trichiura*, with the larval stage hatching from the eggs in the cecum and the adult stage residing in the large intestine.[289] The life cycle for this helminth is completely restricted to the intestinal system, and the mouse model has been used to characterize mucosal immune responses in the gut associated with resistance and susceptibility to infection.[290] Depending on the strain of the mouse and the dosage of the infection, *T. muris* can establish either a chronic infection or be rapidly expelled during an acute infection.[290] *Nippostrongylus brasiliensis* and *Heligmosomoides polygyrus* are two widely used mouse models of small intestinal infection.[291] Both belong to the *Strongylida* order, which includes the human hookworm parasites *Ancylostoma duodenale* and *Necator americanus*, hence they could be considered models of human hookworm infection. However, there are important differences between these two organisms. *N. brasiliensis* in mice is an acute model for the life cycle, including skin penetration and lung migration prior to entrance into the intestinal lumen. However, unlike human hookworm, this parasite does not persist normally and is expelled from immune-competent animals. This model has been useful for defining components of the Type 2 response necessary for rapid expulsion of parasites (eg, IL-13) and is also a useful model for studying tissue repair mechanisms in the lungs, which undergo considerable damage during the tissue migration stages of the infection.[284] On the other hand, primary infections with *H. polygyrus* is a useful model for chronic hookworm infection because it can persist for many months in the mouse. As such, this model has been particularly useful for characterizing immunoregulatory mechanisms and molecules in host-pathogen interactions.[292] Mouse models of schistosomiasis infection (especially *S. mansoni* and *S. japonicum*) have been good models of pathogenesis[293] for human disease.[294] In particular, it has helped gain insight into multiple aspects of macrophage function, granuloma formation, mechanisms regulating fibrosis, and resolution of inflammation. Finally, a mouse model of the filarial nematode, *Litomosoides sigmodontis* has also been particularly useful in characterization of immunomodulation and tissue responses to infection, especially in the context of myeloid cell responses.[295]

Type 2 Cytokine Responses

Of all the Type 2 cytokines, signaling through the IL-4 receptor is perhaps the most important pathway to understand in the context of helminth infections. Dr. William E. Paul (the long-time editor of this textbook) and his laboratory made fundamental discoveries on multiple aspects of IL-4 and this pathway.[296] While the ability of IL-4 to stimulate B cell proliferation was the first function to be discovered, it is now clear that the function of this molecule (as well as IL-13) is extremely broad, and its biological effects on many different cell types remain to be fully understood. The production of IgE during helminth infections is important for immunity against the parasites,[297] and IL-4 is the principle cytokine driving IgE production during infection.[298] IL-13 shares the IL-4Rα–binding chain with IL-4, signaling through a complex with IL-13Rα1. This complex is known as the type II IL-4 receptor and can bind both IL-4 and IL-13, whereas the type I IL-4 receptor complex only binds to IL-4 and has the common γc as the partner to IL-4Rα. Because the expression of type I and type II receptors is quite distinct, IL-4 and IL-13 can act quite differently in the context of helminth infections, with IL-13, in particular, playing a more important role on nonhematopoietic cells such as epithelial cells. IL-4, in contrast, is important for the proper differentiation of CD4+ Th2 cells in a positive feedback loop.[299] The Th2 cells, in turn, become the major producers of Type 2 cytokines during helminth infections. Innate immune cells can be important sources of IL-4 and IL-13 before the CD4+ Th2 cells become fully differentiated and activated, as well as during chronic infection. Granulocytes such as eosinophils and basophils have both been described to be important sources of IL-4, acting, in particular, on macrophages in the tissues. ILC2s rapidly produce IL-13 in response to stimulus, which could be IL-25 produced by tuft cells in the intestinal epithelium, or IL-33 from various cell types.

Downstream of IL-4Rα signaling is activation of the Jak-Stat–signaling pathway, with Stat-6 being the primary Stat activated by IL-4 and IL-13. Stat-6 becomes phosphorylated after IL-4R engagement from its interaction with Jak1 and Jak3 and homodimerizes before translocating into the nucleus to function as a signal-dependent transcription factor (SDTFs). Most of the IL-4 responsive genes are regulated by Stat-6 in a wide range of cell types, including macrophages, epithelial cells mediating the effector functions of the Type 2 response (described below).

Initiation of Type 2 Responses during Helminth Infections

Whereas the initiation of Type 1 responses for viruses and bacteria through the engagement of PAMPs with their corresponding PRRs is well defined, the mechanisms that initiate Type 2 immune responses are less well understood. While PAMPs from helminths are not clearly defined, they can release a heterogeneous mix of excretory/secretory (ES) products that can promote type 2 responses. For example, Omega-1 secreted by schistosome eggs is a potent inducer of type 2 immunity,[300,301] as it binds to the mannose receptor on DCs and conditions them to prime Th2 cell responses.[302] The large size of these multicellular parasites is also important for the initiation of the Type 2 response. Helminths will cause extensive tissue damage as they cross epithelial barriers and migrate through host tissues, through physical disruption, as well as the proteases that they secrete. Danger-associated molecular patterns (DAMPs) released from dying cells will then trigger signaling pathways that help to initiate and amplify the Type 2 response. Examples of such DAMPS include extracellular adenosine triphosphate (ATP),[303] trefoil factors (TFF),[304,305] chitinase-like proteins (CLPs).[306,307] IL-1β production downstream of the NLRP3 inflammasome may also be an important early signal.[308]

TSLP, IL-33, and IL-25 are three important cytokines for the initiation of type 2 immunity. IL-33 is an alarmin released during cell injury or death expressed by hematopoietic and nonhematopoietic cell types, including epithelial cells, endothelial cells, fibroblasts, macrophages, and mast cells. IL-33 binds its receptor ST2, which interacts with the coreceptor IL-1 receptor accessory protein (IL-1RAcP) for cell signaling. IL-33 is stored in the nucleus until cell damage or death, whereby it can be cleaved by inflammatory proteases secreted by neutrophils and mast cells into highly active shorter mature forms.[309,310] In apoptotic cells, caspase 3 and caspase 7 can cleave IL-33 into inactive forms, limiting its inflammatory effects. During intestinal helminth infection, IL-33 is important for the host protective Type 2 response.[311,312]

TSLP is expressed at barrier surfaces like the skin, lungs, and intestines and is upregulated in response to inflammation and injury. TSLP will increase the expression of costimulatory ligands on DCs and enable them to enhance Th2 cell development. It may also directly stimulate Th2 cell development and can directly activate mast cells, basophils, and ILC2s. In mouse models of *Trichuris* infection, TSLP has been shown to enable resistance by blocking IL-12 production from DCs and thus biasing toward Th2 cell differentiation.[313,314] However, for the helminths that live in the small intestine, *H. polygyrus* and *N. brasiliensis*, TSLP does not seem to be required for the proper development of Type 2 responses. IL-25 is a member of the IL-17 cytokine family produced by epithelial cells and is a potent inducer of type 2 immunity. Recently, Tuft cells in both the intestine and the lungs were identified as an important source of IL-25.[315-317] IL-25 is particularly important in the intestine, where it is required for host resistance against intestinal nematodes, including

T. muris,[318] *H. polygyrus*, and *N. brasiliensis*.[319] In addition to tuft cells, IL-25 can also be produced by Th2 cells and myeloid cells. IL-25 will then, in turn, activate ILC2s to produce Type 2 cytokines but also act on other immune cells including Th2 cells, but also epithelial cells, endothelial cells, and fibroblasts.[320,321] IL-25 is required for type 2 immunity and associated host protection against intestinal nematodes, including *T. muris*,[318] *H. polygyrus*, and *N. brasiliensis*.[319]

While IL-33, TSLP, and IL-25 are the best understood cytokines at initiating Type 2 responses, other cytokines including type 1 IFNs and IL-17 have also been implicated in this process, but these observations remain at a relatively early stage of development.

Effector Components of Type 2 Responses During Helminth Infections

Macrophages are Critical Mediators of Pathogenesis and Protection

The functional state of macrophages is determined by its environment and exposure to PAMPs such as bacterial products or viral nucleic acids. In the context of helminth infections, the Type 2 cytokines IL-4 and IL-13 induces an alternative activation state that is quite distinct from the classical response to pattern recognition receptor (PRR) stimulation, most clearly, exemplified by the response to LPS. The alternatively activated macrophage (AAM) phenotype appears to be a general response to tissue damage and can be observed in various settings that involve wound healing. Because of the large size of helminths, their colonization can cause considerable tissue damage,[284] which can induce the production of IL-4 and IL-13 by innate cells such as ILC2s, eosinophils, neutrophils, and basophils.[283] This initial source of IL-4 and IL-13 is then bolstered and eventually replaced by the CD4+ Th2 response as the infection progresses, which instructs and maintains macrophages in the AAM phenotype.[322] Other signals from the tissues such as the structurally related defense collagens SP-A and C1q can act as local amplifiers of AAM activation.[323] Similarly, the phagocytosis of apoptotic cells is often necessary for the appropriate activation of the tissue repair program in AAM.[306]

However, it is important to note that the distinct macrophage phenotypes that are described after IL-4/IL-13 and LPS stimulation is often a feature of *in vitro* stimulation with very high concentrations of ligand, which may not be physiological. Single-cell RNA sequencing (scRNAseq) experiments are beginning to identify a wide spectrum of macrophage activation phenotypes that are much more complex than the M1/CAM versus M2/AAM paradigm. Understanding the heterogeneity of macrophage activation states during infection is an active area of current research. Nonetheless, it is clear that a key difference between the IL-4 driven alternatively activated state and the LPS/IFN activated state is the usage of arginine in the macrophages.[324] IFNs and LPS induce nitric oxide (NO) synthase, which metabolizes arginine to NO and citrulline, whereas IL-4 induces arginase-1 (Arg1), which instead hydrolyzes arginine to ornithine and urea.[325] Arg1 activity from AAMs can deplete arginine availability, thus restraining T cell function by amino acid starvation.[326] During *S. mansoni* infection, macrophage Arg1 is important in preventing the unregulated expansion of CD4+ Th2 cells from causing too much fibrosis.[327] However, in addition to these regulatory functions of AAMs, these cells are also important for killing helminths as part of the effectors of Type 2 resistance mechanisms.[283]

Protective Effects of the Type 2 Response

The Type 2 response is protective to the host because it is important for both "resistance" mechanisms of expelling the parasites and preventing reinfection, as well as "tolerance" mechanisms for minimizing the tissue damage and inflammatory responses that are caused by the presence of chronic infection. The expulsion of helminths from the intestines involves the effect of IL-4 and IL-13 on different aspects of host physiology. These cytokines can induce increased epithelial cell turnover and electrolyte secretion,[328] as well as smooth muscle hypercontractility.[328] IL-4 and or IL-13 stimulation also drives goblet cell hyperplasia and alters both the quantity and type of mucus produced during helminth infection.[329] Muc2 and Muc5Ac both contributes to the expulsion of intestinal nematodes, such as *T. muris*.[329] IL-4 and/or IL-13 activation of goblet cells also leads to increased production of resistin-like molecule β (Relm-β), which can directly interfere with parasite feeding of some intestinal worms.[330]

For killing helminths in the tissues, IL-4 and/or IL-13 AAM have emerged as the critical effector cells, especially during a secondary infection.[283] For the tissue migrating larvae of *N. brasiliensis* and *H. polygyrus*, AAM are involved in trapping and killing these parasites, and the molecule Relmα has been implicated in this process.[331] Arg1 expression also appears important for this function and in response to IL-13 secretion by Th2 cells and ILC2s, AAM will trap *N. brasiliensis* larvae as they enter the lung after reinfection.[307] *N. brasiliensis* larvae can also be trapped in the skin by AAM, with basophils providing a source of IL-4 for their activation there. In the small intestine, *H. polygyrus* larvae can also be trapped by AAM in the submucosa through an IL-4/IL-13 dependent mechanism, which also involves IgG-mediated induction of Arg1.

Eosinophils, the increased abundance being an immunological characteristic of helminth infections, are also important for killing parasites in the tissues (see Chapter 18). Originally described in an *in vitro* setting, antibodies were shown to activate eosinophils to kill various helminths, including *Ascaris suum*, *Schistosoma manson*, and *Brugia malayi*, for the tissue migrating larval stages in particular. However, the *in vivo* situation is more complicated because these cells play an immunoregulatory role also. During a primary infection, eosinophils can actually protect the muscle-encysted *Trichinella spiralis* larvae in mice,[332] while during secondary infection, these cells are important for killing the parasites.[333] Other helminths where eosinophils have been shown to mediate resistance include the larval stages of the mouse filarial nematode *Litomosoides sigmodontis*[334] as well as *N. brasiliensis*.[335] Hence, many cells involved in Type 2 responses have complex roles in parasite killing as well as immune regulation.

Although antibodies are usually required for primary resistance against helminths, they play an important role in conferring protective immunity after repeated exposure. Helminth-specific IgG and IgE antibodies can activate macrophages and eosinophils to attack larval stages of these parasites when they migrate through the tissues. Transfer of parasite-specific antibodies from mothers to their offspring may also be important for reducing worm burden in suckling neonates.[283] Concomitant immunity in schistosomiasis, whereby chronic infection is associated with antilarval immunity, has also been associated with high levels of parasite-specific antibodies. In summary, because of the large size of these multicellular parasites, the resistance mechanisms required to limit the number of helminths in an infected individual are quite distinct from those responsible for clearing other pathogens. Extracellular mechanisms such as encasement, expulsion, and tissue remodeling are more critical features of these responses. However, many important questions remain unanswered; for example, how antibodies and eosinophils directly kill the worms is unclear. How encasement and tissue remodeling mediated by macrophages promote worm killing is also unclear.

The innate and adaptive immune mechanisms that mediate resistance to helminth infections are summarized in Figure 42.5.

Chronic Type 2 Responses and Fibrotic Pathology

There exists a fine balance between a protective anti-infective immune response and collateral damage-causing host pathology. An example of this is schistosomiasis, whereby the

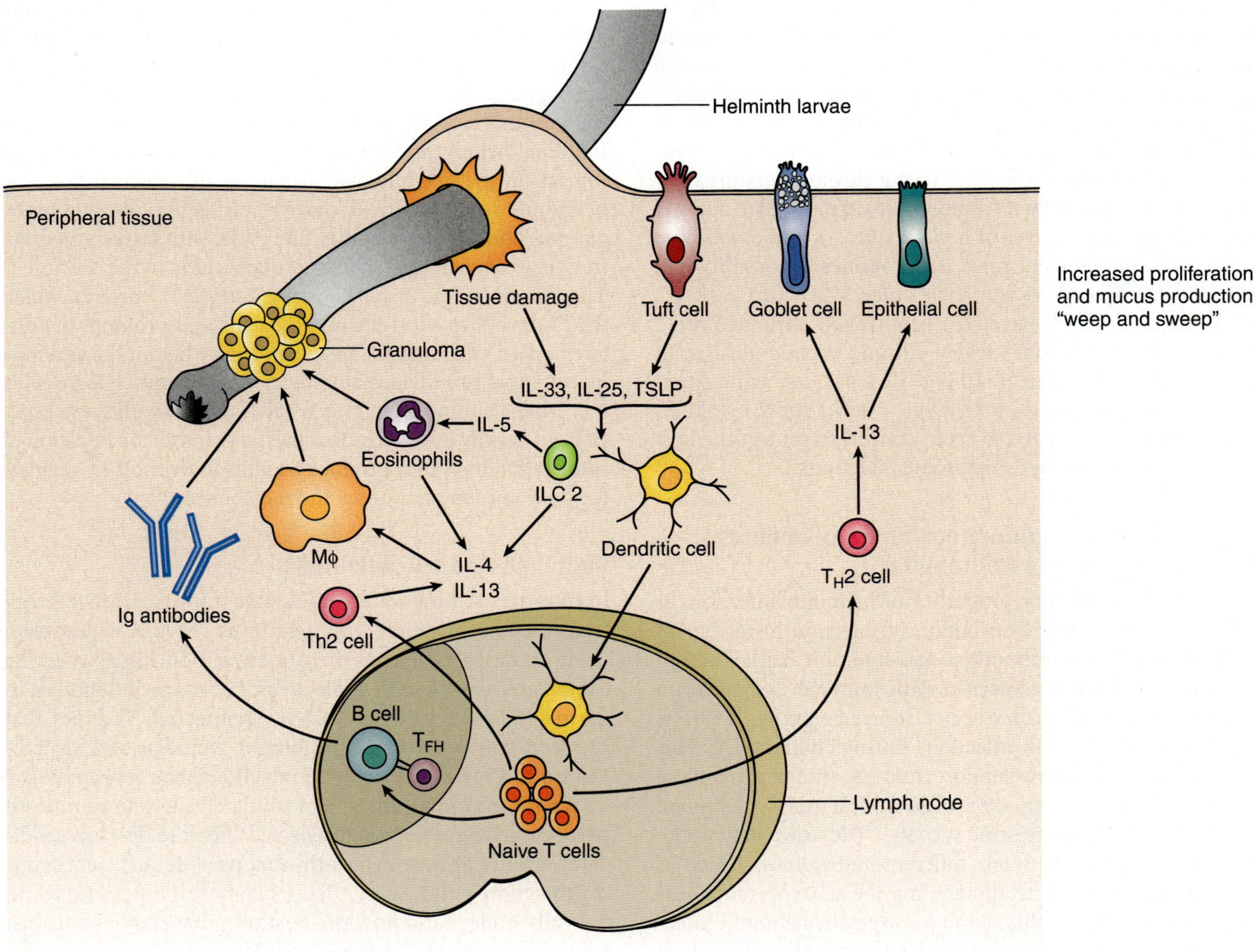

FIG. 42.5. Immune response to helminth infections. As helminth larvae migrate through the peripheral tissues of the host, they cause tissue damage that triggers the release of alarmins IL-25, IL-33, and TSLP. Tuft cells in the intestines can also produce IL-25 in response to infection. IL-25, IL-33, and TSLP activate innate immune cells, including innate lymphoid type 2 cells (ILC2s) and dendritic cells. ILC2s can rapidly produce IL-13, which acts on recruited macrophages, as well as IL-5 to recruit eosinophils into the tissues. Activated dendritic cells migrate to the lymph nodes carrying antigens from the helminths. In the lymph nodes, dendritic cells activate naïve T cells specific for helminth antigens, and they differentiate into T helper 2 cells (Th2) cells as well as T follicular helper cells (T_{FH}). T_{FH} migrate to the follicles to interact with B cells in the germinal center reaction and promote class-switching to IgE antibody production. IgE antibodies bind helminth parasites promoting granuloma formation by macrophages and eosinophils. Th2 migrate from the lymph nodes to the peripheral tissues, where they produce IL-4 and IL-13 to stimulate macrophages to adopt an alternatively activated M2 phenotype. These IL-4 activated macrophages can repair tissue damage, as well as form granulomas to kill helminth larvae with help from eosinophils. Th2 cells that migrate to the intestines will stimulate intestinal epithelial cells and goblet cells mainly through the production of IL-13, which increases proliferation as well as mucus production. This is known as the "weep and sweep" mechanism to promote expulsion of intestinal helminths from the intestines.

granulomatous response around the schistosome eggs causes chronic inflammation and fibrotic pathology that eventually leads to hepatic fibrosis, portal hypertension, and bleeding from esophageal and gastric varices.[293] In the experimental mouse model of schistosomiasis, the CD4+ Th2 cells are essential for the formation of these granulomas and the resulting pathologies. However, in the absence of granuloma formation when Type 2 cytokines or CD4+ T cells are depleted, the eggs are hepatotoxic and can reduce barrier function in the intestine. Mice that cannot mount an appropriate Type 2 response all die at a relatively early stage of infection, associated with intestinal and liver damage from the toxic egg products.[293] Safe passage of eggs across the intestinal walls into the lumen and excretion into the environment is dependent on Type 2 cytokine signaling. Hence, the Type 2 granulomas are critical in the short term for enabling a successful host-pathogen relationship, whereby the host can tolerate the presence of *S. mansoni* parasites without severe symptoms. However, in the long-term, this Type 2 response will eventually lead to fibrosis as a result of the chronic inflammatory response that cannot be resolved. IL-13, in particular, appears to be an important driver of hepatic fibrosis, and *S. mansoni* infected IL-13–deficient mice suffer reduced morbidity and mortality compared to wild-type mice. Mice that lack IL-10, IL-12, and IFN-γ have an uncontrolled chronic Type 2 response and much more severe chronic hepatic fibrosis. In contrast, mice that lack IL-10 and IL-4 have an acute hepatotoxic response. Hence, both Type 1 and Type 2 cytokines and responses are needed to act in concert to protect the host from pathology during Schistosome infections.

Effects of Helminth Infections on Autoimmunity, Metabolism, and the Microbiota

The immune regulatory properties of helminth infection, as well as the geographic distribution of infection in regions of the world where autoimmunity is less common, has led to suggestions that diseases associated with immune dysregulation in urbanized, helminth-free societies may be partly driven by the absence of helminth infections during childhood.[336] This concept has considerable support from experiments in animal models; however, placebo-controlled clinical trials with human subjects has not had the same success.[337] Heterogeneity in the immune responses between different individuals, which is poorly reflected in animal models, may be an important limiting factor. Notably, in animal models whereby helminth infections (most commonly, involving *H. polygyrus*) have improved inflammatory pathologies driven by Type 1/Type 17 conditions (eg, experimental autoimmune encephalomyelitis or type 1 diabetes), as well as diseases driven by Type 2 conditions (eg, airway inflammation) can be improved via regulatory cytokines such as IL-10.[338] Hence, the anti-inflammatory effects of helminth infection is not simply a result of the Type 2 bias. It is possible that persistent childhood exposure to helminths, but also other bacterial, viral, and fungal infections, results in a more developed immunoregulatory response from the repeated exercise of activating and downregulating responses appropriately during infection.[339] The effects of Type 2 responses during helminth infections may also have implications for metabolic diseases since it is now clear that there is a strong relationship between Type 2 cellular immunity and improved adipose tissue metabolism. There is evidence from both experimental mouse models and a few clinical studies that helminth infection can improve glucose tolerance and is protective against weight gain and insulin resistance.[283] As a result, there is considerable interest in the identification of helminth-derived molecules that can improve metabolism, as well as reduce inflammation and autoimmunity, through a "drugs from bugs" concept.[340] Alterations to the gut microbiota from helminth infection could also play an important role in modulating metabolism and immune responses.[341] Intestinal helminths, in particular, occupy the same environmental niche as the gut microbiota and must have coevolved in this environment. The Type 2 cytokine-driven epithelial cell response alters the microbiota during infection by changing the environmental conditions in the gut and favoring certain bacterial communities.[283] Large intestinal helminths such as *Trichuris* sp. might have a bigger impact on the gut microbiota than small intestinal dwelling helminths. Interestingly, *Trichuris* parasites also respond to the gut microbiota as part of their life cycle, with larvae emerging from eggs when triggered by specific bacteria in the caecum.[342] There is also some evidence that *Clostridiales* bacteria, which tend to produce short-chain fatty acids and promote the differentiation of Treg cells, are expanded by helminths and may inhibit more proinflammatory bacteria.[283] Hence, it is possible that while helminths can clearly cause disease pathology, especially in heavily infected individuals, asymptomatic infection in some individuals may produce benefits to the host by regulating both immune responses and metabolism.[343]

Vaccines Against Helminths

In countries lacking water and sewage infrastructure to break transmission, infection with helminths remains widespread. Control measures such as drug treatment during deworming campaigns are active in some areas, but many individuals in these endemic regions are quickly reinfected. Vaccines that can reduce infection and transmission would be a more practical strategy to eradicate these infections than repeated drug treatment in communities. Unfortunately, despite significant advances in technology, it has been difficult to develop candidate vaccines against helminths that provide sufficient levels of protection without side effects. Likely, this is because we do not fully understand how the host immune response against helminths mediates resistance to infection.

The only successful examples of vaccines against helminths are in veterinary medicine. Live irradiated larvae of the lungworm *Dictyocaulus viviparus* is in commercial use for cattle. Indeed, irradiated schistosomes in animal models can achieve up to 60% to 80% immunity and are considered the gold standard for schistosome vaccines. However, batch variability and short-shelf life makes this approach impossible for a human vaccine. Native antigen has also been successfully used against *Haemonchus contortus* in sheep. However, to produce this vaccine, adult *H. contortus* has to obtained from infected sheep in order to extract the antigens for the vaccine. The only successful example of a commercial recombinant vaccine against helminths is the EG95 vaccine that is used to control the spread of *Echinococcus granulosus* in sheep and goats. A better understanding of why this antigen is sufficiently immunogenic to be used as a recombinant vaccine may be helpful for the development of other helminth vaccines. Studying human populations that are exposed to high transmission rates of infection with no clinical manifestations of disease (referred to as putatively immune individuals) may also be useful for the identification of such immunogenic antigens.

Vaccines for human helminth infections are badly needed, even if only partially protective, to reduce transmission within communities, thus reducing disease burden. For the schistosome parasites, targeting the larval migration stages of the parasite may be the best strategy. Early studies immunizing mice with irradiated cercariae showed that high levels of protection were possible. There are currently four main schistosomiasis vaccine candidates in or near clinical development, targeting different parasite proteins.[344] These include the glutathione S-transferase of *S. haematobium* (rSh28GST), the fatty acid-binding protein (FABP) of *S. mansoni* (Sm14), a tetraspanin protein (Sm-TSP2) important for maintaining tegument integrity, and finally Sm-p80, a subunit of calpain also expressed on the parasite surface. Hookworm is another major target for human helminth vaccine development. In clinical studies testing a recombinant protein Na-ASP-2 on healthy adults, it was found that previous hookworm infection had already induced IgE antibodies against this antigen and that the vaccination caused an allergic reaction which resulted in the suspension of these studies.[345] Hence, it is now clear that screening for preexisting IgE antibodies specific for the vaccine antigen in endemic regions needs to be performed before moving a recombinant protein into clinical testing, to avoid similar allergic reactions to vaccinations. Currently, Na-GST-1 and Na-APR-1 are two candidates proceeding toward clinical trials.[346] Testing of some of these candidates will be further facilitated by the development of controlled human hookworm infections, which is now possible because of the development of cGMP grade *Necator americanus* infective larvae that can be safely inoculated and shown to establish a reliable patent infection in healthy human volunteers.[347]

An alternative strategy to vaccine development has emerged recently targeting immune evasion molecules produced by these parasites, instead of targeting the parasites themselves. This may shift the balance of the immune response in favor of the host to eliminate the parasites by essentially disarming the parasite immune evasion weapons. In addition to secreted proteins produced by helminths (eg, TGF-β like molecules that can induce Tregs[292]), extracellular vesicles and their contents (including nucleic acids) have been identified to be a new front in the host-parasite battle and may also be attractive targets for vaccines.[348] However, none of these approaches have yet to reach clinical development.

CONCLUSION

The period since the publication of the last edition of Fundamental Immunology has seen intensified research to better define the innate and adaptive determinants of the immune response to parasites. The important role of endosomal and cytosolic nucleic acid sensing pathways in innate responses to protozoan pathogens has been a notable advance, as has the importance of ILC2s in the development of type 2 responses in helminth infection. The heterogeneity of macrophages as tissue-resident or monocyte-derived cells, and their respective antimicrobial and/or tissue-remodeling functions, has been explored in far greater detail for each of the parasitic infections discussed. The nature of immune evasion strategies has been more fully developed, particularly with respect to how intracellular parasites disrupt signaling pathways in innate cells and how helminth-derived molecules can act on immune cells. The dichotomy of Th1/Th2 response selection that seemed inherent to different classes of parasites has been refined to accommodate the evidence that these responses are typically found to be concurrent, though in different proportions. Studies identifying the role of regulatory cells and cytokines, as well as T cell exhaustion mechanisms that dampen the immune response, have continued to showcase how parasitic infections can reveal factors controlling the balance between effective immunity and pathology. Despite these advances, the goal of developing a safe and effective vaccine has still not been met for malaria nor any of the neglected tropical diseases. Live, whole-organism vaccines against malaria and cutaneous leishmaniasis remain the most uniformly effective vaccines against parasitic diseases in humans, arguing for the development of live-attenuated vaccines. Moving forward, a systems biology approach will be needed to reveal the unique features of the network of interacting cells and cytokines that condition immunity, persistence, or pathology. How these networks differ between individuals and the contribution of genetic and environmental factors toward this variation should be better defined. Host-parasite genomic and transcriptomic databases, including single-cell sequencing, are likely to dominate the future landscape in immunoparasitology research. Priority should be given to the generation of these databases in the temporal and localized tissue environments of human hosts.

REFERENCES

1. Hotez PJ, Bottazzi ME, Strych U. New vaccines for the world's poorest people. *Annu Rev Med.* 2016;67:405-417.
2. Schneider DS, Ayres JS. Two ways to survive infection: what resistance and tolerance can teach us about treating infectious diseases. *Nat Rev Immunol.* 2008;8(11):889-895.
3. Clegg JA, Smithers SR, Terry RJ. Concomitant immunity and host antigens associated with schistosomiasis. *Int J Parasitol.* 1971;1(1):43-49.
4. Mosmann TR, Cherwinski H, Bond MW, Giedlin MA, Coffman RL. Two types of murine helper T cell clone. I. Definition according to profiles of lymphokine activities and secreted proteins. *J Immunol.* 1986;136(7):2348-2357.
5. Heinzel FP, Sadick MD, Holaday BJ, Coffman RL, Locksley RM. Reciprocal expression of interferon gamma or interleukin 4 during the resolution or progression of murine Leishmaniasis. Evidence for expansion of distinct helper T cell subsets. *J Exp Med.* 1989;169(1):59-72.
6. Scott P, Natovitz P, Coffman RL, Pearce E, Sher A. Immunoregulation of cutaneous Leishmaniasis. T cell lines that transfer protective immunity or exacerbation belong to different T helper subsets and respond to distinct parasite antigens. *J Exp Med.* 1988;168(5):1675-1684.
7. Rimoldi MT, Sher A, Heiny S, Lituchy A, Hammer CH, Joiner K. Developmentally regulated expression by *Trypanosoma cruzi* of molecules that accelerate the decay of complement C3 convertases. *Proc Natl Acad Sci U S A.* 1988;85(1):193-197.
8. Fischer E, Ouaissi MA, Velge P, Cornette J, Kazatchkine MD. gp 58/68, a parasite component that contributes to the escape of the trypomastigote form of *T. cruzi* from damage by the human alternative complement pathway. *Immunology.* 1988;65(2):299-303.
9. Sosoniuk E, Vallejos G, Kenawy H, et al. *Trypanosoma cruzi* calreticulin inhibits the complement lectin pathway activation by direct interaction with L-Ficolin. *Mol Immunol.* 2014;60(1):80-85.
10. Puentes SM, Da Silva RP, Sacks DL, Hammer CH, Joiner KA. Serum resistance of metacyclic stage *Leishmania major* promastigotes is due to release of C5b-9. *J Immunol.* 1990;145(12):4311-4316.
11. Raper J, Portela MP, Lugli E, Frevert U, Tomlinson S. Trypanosome lytic factors: novel mediators of human innate immunity. *Curr Opin Microbiol.* 2001;4(4):402-408.
12. Vanhamme L, Pays E. The trypanosome lytic factor of human serum and the molecular basis of sleeping sickness. *Int J Parasitol.* 2004;34(8):887-898.

13. Oli MW, Cotlin LF, Shiflett AM, Hajduk SL. Serum resistance-associated protein blocks lysosomal targeting of trypanosome lytic factor in *Trypanosoma brucei*. *Eukaryot Cell.* 2006;5(1):132-139.
14. Kieft R, Capewell P, Turner CM, Veitch NJ, MacLeod A, Hajduk S. Mechanism of *Trypanosoma brucei* gambiense (group 1) resistance to human trypanosome lytic factor. *Proc Natl Acad Sci U S A.* 2010;107(37):16137-16141.
15. Carneiro MB, Lopes ME, Hohman LS, et al. Th1-Th2 cross-regulation controls early *Leishmania* infection in the skin by modulating the size of the permissive monocytic host cell reservoir. *Cell Host Microbe.* 2020;27(5):752-768 e7.
16. Arango Duque G, Descoteaux A. Leishmania survival in the macrophage: where the ends justify the means. *Curr Opin Microbiol.* 2015;26:32-40.
17. Isnard A, Shio MT, Olivier M. Impact of *Leishmania* metalloprotease GP63 on macrophage signaling. *Front Cell Infect Microbiol.* 2012;2:72.
18. Mosser DM, Edelson PJ. The third component of complement (C3) is responsible for the intracellular survival of *Leishmania major*. *Nature.* 1987;327(6120):329-331.
19. Laskay T, van Zandbergen G, Solbach W. Neutrophil granulocytes as host cells and transport vehicles for intracellular pathogens: apoptosis as infection-promoting factor. *Immunobiology.* 2008;213(3-4):183-191.
20. Peters NC, Egen JG, Secundino N, et al. *In vivo* imaging reveals an essential role for neutrophils in Leishmaniasis transmitted by sand flies. *Science.* 2008;321(5891):970-974.
21. Martinez-Lopez M, Soto M, Iborra S, Sancho D. *Leishmania* hijacks myeloid cells for immune escape. *Front Microbiol.* 2018;9:883.
22. Caradonna KL, Burleigh BA. Mechanisms of host cell invasion by *Trypanosoma cruzi*. *Adv Parasitol.* 2011;76:33-61.
23. Piacenza L, Zago MP, Peluffo G, Alvarez MN, Basombrio MA, Radi R. Enzymes of the antioxidant network as novel determiners of *Trypanosoma cruzi* virulence. *Int J Parasitol.* 2009;39(13):1455-1464.
24. Andrews NW, Abrams CK, Slatin SL, Griffiths G. A T. cruzi-secreted protein immunologically related to the complement component C9: evidence for membrane pore-forming activity at low pH. *Cell.* 1990;61(7):1277-1287.
25. Lopez R, Demick KP, Mansfield JM, Paulnock DM. Type I IFNs play a role in early resistance, but subsequent susceptibility, to the African trypanosomes. *J Immunol.* 2008;181(7):4908-4917.
26. Onyilagha C, Uzonna JE. Host immune responses and immune evasion strategies in African trypanosomiasis. *Front Immunol.* 2019;10:2738.
27. Gazzinelli RT, Denkers EY. Protozoan encounters with Toll-like receptor signalling pathways: implications for host parasitism. *Nat Rev Immunol.* 2006;6(12):895-906.
28. Bafica A, Santiago HC, Goldszmid R, Ropert C, Gazzinelli RT, Sher A. Cutting edge: TLR9 and TLR2 signaling together account for MyD88-dependent control of parasitemia in Trypanosoma cruzi infection. *J Immunol.* 2006;177(6):3515-3519.
29. Kuriakose S, Onyilagha C, Singh R, Olayinka-Adefemi F, Jia P, Uzonna JE. TLR-2 and MyD88-dependent activation of MAPK and STAT proteins regulates proinflammatory cytokine response and immunity to experimental trypanosoma congolense infection. *Front Immunol.* 2019;10:2673.
30. Amin DN, Vodnala SK, Masocha W, Sun B, Kristensson K, Rottenberg ME. Distinct Toll-like receptor signals regulate cerebral parasite load and interferon α/β and tumor necrosis factor α-dependent T-cell infiltration in the brains of *Trypanosoma brucei*-infected mice. *J Infect Dis.* 2012;205(2):320-332.
31. Krishnegowda G, Hajjar AM, Zhu J, et al. Induction of proinflammatory responses in macrophages by the glycosylphosphatidylinositols of Plasmodium falciparum: cell signaling receptors, glycosylphosphatidylinositol (GPI) structural requirement, and regulation of GPI activity. *J Biol Chem.* 2005;280(9):8606-8616.
32. Schamber-Reis BL, Petritus PM, Caetano BC, et al. UNC93B1 and nucleic acid-sensing Toll-like receptors mediate host resistance to infection with *Leishmania major*. *J Biol Chem.* 2013;288(10):7127-7136.
33. Parroche P, Lauw FN, Goutagny N, et al. Malaria hemozoin is immunologically inert but radically enhances innate responses by presenting malaria DNA to Toll-like receptor 9. *Proc Natl Acad Sci U S A.* 2007;104(6):1919-1924.
34. Kalantari P, DeOliveira RB, Chan J, et al. Dual engagement of the NLRP3 and AIM2 inflammasomes by *Plasmodium*-derived hemozoin and DNA during malaria. *Cell Rep.* 2014;6(1):196-210.
35. Caetano BC, Carmo BB, Melo MB, et al. Requirement of UNC93B1 reveals a critical role for TLR7 in host resistance to primary infection with *Trypanosoma cruzi*. *J Immunol.* 2011;187(4):1903-1911.
36. Liehl P, Zuzarte-Luís V, Chan J, et al. Host-cell sensors for *Plasmodium* activate innate immunity against liver-stage infection. *Nat Med.* 2014;20(1):47-53.
37. Ives A, Ronet C, Prevel F, et al. *Leishmania* RNA virus controls the severity of mucocutaneous Leishmaniasis. *Science.* 2011;331(6018):775-778.
38. Lima-Junior DS, Costa DL, Carregaro V, et al. Inflammasome-derived IL-1β production induces nitric oxide-mediated resistance to *Leishmania*. *Nat Med.* 2013;19(7):909-915.
39. Charmoy M, Hurrell BP, Romano A, et al. The Nlrp3 inflammasome, IL-1β, and neutrophil recruitment are required for susceptibility to a nonhealing strain of *Leishmania major* in C57BL/6 mice. *Eur J Immunol.* 2016;46(4):897-911.
40. Paroli AF, Gonzalez PV, Diaz-Lujan C, et al. NLRP3 inflammasome and caspase-1/11 pathway orchestrate different outcomes in the host protection against *Trypanosoma cruzi* acute infection. *Front Immunol.* 2018;9:913.
41. Silva GK, Gutierrez FR, Guedes PM, et al. Cutting edge: nucleotide-binding oligomerization domain 1-dependent responses account for murine resistance against Trypanosoma cruzi infection. *J Immunol.* 2010;184(3):1148-1152.
42. Griffith JW, Sun T, McIntosh MT, Bucala R. Pure hemozoin is inflammatory *in vivo* and activates the NALP3 inflammasome via release of uric acid. *J Immunol.* 2009;183(8):5208-5220.
43. Shio MT, Tiemi Shio M, Eisenbarth SC, et al. Malarial hemozoin activates the NLRP3 inflammasome through Lyn and Syk kinases. *PLoS Pathog.* 2009;5(8):e1000559.
44. Campos MA, Closel M, Valente EP, et al. Impaired production of proinflammatory cytokines and host resistance to acute infection with *Trypanosoma cruzi* in mice lacking functional myeloid differentiation factor 88. *J Immunol.* 2004;172(3):1711-1718.
45. Drennan MB, Stijlemans B, Van den Abbeele J, et al. The induction of a type 1 immune response following a *Trypanosoma brucei* infection is MyD88 dependent. *J Immunol.* 2005;175(4):2501-2509.
46. Muraille E, De Trez C, Brait M, De Baetselier P, Leo O, Carlier Y. Genetically resistant mice lacking MyD88-adapter protein display a high susceptibility to *Leishmania major* infection associated with a polarized Th2 response. *J Immunol.* 2003;170(8):4237-4241.
47. Shio MT, Christian JG, Jung JY, Chang KP, Olivier M. PKC/ROS-Mediated NLRP3 inflammasome activation is attenuated by *Leishmania* zinc-metalloprotease during infection. *PLoS Negl Trop Dis.* 2015;9(6):e0003868.
48. Johnson LM, Scott P. STAT1 expression in dendritic cells, but not T cells, is required for immunity to *Leishmania major*. *J Immunol.* 2007;178(11):7259-7266.
49. Liese J, Schleicher U, Bogdan C. TLR9 signaling is essential for the innate NK cell response in murine cutaneous Leishmaniasis. *Eur J Immunol.* 2007;37(12):3424-3434.
50. Scharton TM, Scott P. Natural killer cells are a source of interferon gamma that drives differentiation of CD4+ T cell subsets and induces early resistance to *Leishmania major* in mice. *J Exp Med.* 1993;178(2):567-577.
51. Leon B, Lopez-Bravo M, Ardavin C. Monocyte-derived dendritic cells formed at the infection site control the induction of protective T helper 1 responses against *Leishmania*. *Immunity.* 2007;26(4):519-531.
52. Bennett CL, Misslitz A, Colledge L, Aebischer T, Blackburn CC. Silent infection of bone marrow-derived dendritic cells by *Leishmania mexicana* amastigotes. *Eur J Immunol.* 2001;31(3):876-883.
53. Ribeiro-Gomes FL, Peters NC, Debrabant A, Sacks DL. Efficient capture of infected neutrophils by dendritic cells in the skin inhibits the early anti-*Leishmania* response. *PLoS Pathog.* 2012;8(2):e1002536.
54. Iborra S, Martinez-Lopez M, Cueto FJ, et al. *Leishmania* uses Mincle to target an inhibitory ITAM signaling pathway in dendritic cells that dampens adaptive immunity to infection. *Immunity.* 2016;45(4):788-801.
55. von Stebut E, Belkaid Y, Nguyen BV, Cushing M, Sacks DL, Udey MC. *Leishmania major*-infected murine Langerhans cell-like dendritic cells from susceptible mice release IL-12 after infection and vaccinate against experimental cutaneous Leishmaniasis. *Eur J Immunol.* 2000;30(12):3498-3506.
56. Laskay T, Diefenbach A, Rollinghoff M, Solbach W. Early parasite containment is decisive for resistance to *Leishmania major* infection. *Eur J Immunol.* 1995;25(8):2220-2227.
57. Constant SL, Lee KS, Bottomly K. Site of antigen delivery can influence T cell priming: pulmonary environment promotes preferential Th2-type differentiation. *Eur J Immunol.* 2000;30(3):840-847.
58. Nabors GS, Nolan T, Croop W, Li J, Farrell JP. The influence of the site of parasite inoculation on the development of Th1 and Th2 type immune responses in (BALB/c × C57BL/6) F1 mice infected with *Leishmania major*. *Parasite Immunol.* 1995;17(11):569-579.
59. Cardillo F, Voltarelli JC, Reed SG, Silva JS. Regulation of Trypanosoma cruzi infection in mice by gamma interferon and interleukin 10: role of NK cells. *Infect Immun.* 1996;64(1):128-134.
60. Soto CDA, Mirkin GA, Solana ME, Cappa SMG. *Trypanosoma cruzi* infection modulates *in vivo* expression of major histocompatibility complex class II molecules on antigen-presenting cells and T-cell stimulatory activity of dendritic cells strain-dependent manner. *Infect Immun.* 2003;71(3):1194-1199.
61. Poncini CV, Gimenez G, Pontillo CA, et al. Central role of extracellular signal-regulated kinase and Toll-like receptor 4 in IL-10 production in regulatory dendritic cells induced by *Trypanosoma cruzi*. *Mol Immunol.* 2010;47(11-12):1981-1988.
62. Erdmann H, Steeg C, Koch-Nolte F, Fleischer B, Jacobs T. Sialylated ligands on pathogenic *Trypanosoma cruzi* interact with Siglec-E (sialic acid-binding Ig-like lectin-E). *Cell Microbiol.* 2009;11(11):1600-1611.
63. Mugnier MR, Cross GA, Papavasiliou FN. The *in vivo* dynamics of antigenic variation in *Trypanosoma brucei*. *Science.* 2015;347(6229):1470-1473.
64. Pinger J, Nesic D, Ali L, et al. African trypanosomes evade immune clearance by O-glycosylation of the VSG surface coat. *Nat Microbiol.* 2018;3(8):932-938.
65. Campbell GH, Phillips SM. Adoptive transfer of variant-specific resistance to Trypanosoma rhodesiense with B lymphocytes and serum. *Infect Immun.* 1976;14(5):1144-1150.
66. Hertz CJ, Filutowicz H, Mansfield JM. Resistance to the African trypanosomes is IFN-γ dependent. *J Immunol.* 1998;161(12):6775-6783.
67. Masocha W, Robertson B, Rottenberg ME, Mhlanga J, Sorokin L, Kristensson K. Cerebral vessel laminins and IFN-γ define *Trypanosoma brucei* brucei penetration of the blood-brain barrier. *J Clin Invest.* 2004;114(5):689-694.
68. Maclean L, Odiit M, Macleod A, et al. Spatially and genetically distinct African Trypanosome virulence variants defined by host interferon-gamma response. *J Infect Dis.* 2007;196(11):1620-1628.
69. Rodgers J, Bradley B, Kennedy PG, Sternberg JM. Central nervous system parasitosis and neuroinflammation ameliorated by systemic IL-10 administration in *Trypanosoma brucei*-infected mice. *PLoS Negl Trop Dis.* 2015;9(10):e0004201.
70. Bosschaerts T, Guilliams M, Noel W, et al. Alternatively activated myeloid cells limit pathogenicity associated with African trypanosomiasis through the IL-10 inducible gene selenoprotein P. *J Immunol.* 2008;180(9):6168-6175.
71. De Muylder G, Daulouede S, Lecordier L, et al. A *Trypanosoma brucei* kinesin heavy chain promotes parasite growth by triggering host arginase activity. *PLoS Pathog.* 2013;9(10):e1003731.
72. Guilliams M, Oldenhove G, Noel W, et al. African trypanosomiasis: naturally occurring regulatory T cells favor trypanotolerance by limiting pathology associated with sustained type 1 inflammation. *J Immunol.* 2007;179(5):2748-2757.
73. Stijlemans B, Korf H, De Baetselier P, et al. Hepatocyte-derived IL-10 plays a crucial role in attenuating pathogenicity during the chronic phase of *T. congolense* infection. *PLoS Pathog.* 2020;16(2):e1008170.
74. Liu G, Xu J, Wu H, et al. IL-27 signaling is crucial for survival of mice infected with African trypanosomes via preventing lethal effects of CD4+ T cells and IFN-γ. *PLoS Pathog.* 2015;11(7):e1005065.
75. Capewell P, Cren-Travaille C, Marchesi F, et al. The skin is a significant but overlooked anatomical reservoir for vector-borne African trypanosomes. *Elife.* 2016;5:e17716.
76. Trindade S, Rijo-Ferreira F, Carvalho T, et al. *Trypanosoma brucei* parasites occupy and functionally adapt to the adipose tissue in mice. *Cell Host Microbe.* 2016;19(6):837-848.
77. Krettli AU, Brener Z. Protective effects of specific antibodies in *Trypanosoma cruzi* infections. *J Immunol.* 1976;116(3):755-760.
78. Kumar S, Tarleton RL. The relative contribution of antibody production and CD8+ T cell function to immune control of *Trypanosoma cruzi*. *Parasite Immunol.* 1998;20(5):207-216.
79. Almeida IC, Ferguson MA, Schenkman S, Travassos LR. Lytic anti-α-galactosyl antibodies from patients with chronic Chagas' disease recognize novel O-linked oligosaccharides on mucin-like glycosyl-phosphatidylinositol-anchored glycoproteins of *Trypanosoma cruzi*. *Biochem J.* 1994;304(pt 3):793-802.
80. Pereira-Chioccola VL, Acosta-Serrano A, Correia de Almeida I, et al. Mucin-like molecules form a negatively charged coat that protects *Trypanosoma cruzi* trypomastigotes from killing by human anti-α-galactosyl antibodies. *J Cell Sci.* 2000;113(pt 7):1299-1307.
81. Buscaglia CA, Campo VA, Frasch AC, Di Noia JM. Trypanosoma cruzi surface mucins: host-dependent coat diversity. *Nat Rev Microbiol.* 2006;4(3):229-236.
82. Campo VA, Buscaglia CA, Di Noia JM, Frasch AC. Immunocharacterization of the mucin-type proteins from the intracellular stage of *Trypanosoma cruzi*. *Microbes Infect.* 2006;8(2):401-409.
83. Hoft DF, Schnapp AR, Eickhoff CS, Roodman ST. Involvement of $CD4^+$ Th1 cells in systemic immunity protective against primary and secondary challenges with *Trypanosoma cruzi*. *Infect Immun.* 2000;68(1):197-204.
84. Tarleton RL. CD8+ T cells in *Trypanosoma cruzi* infection. *Semin Immunopathol.* 2015;37(3):233-238.
85. Rodriguez EVA, Furlan CLA, Vernengo FF, Montes CL, Gruppi A. Understanding $CD8^+$ T cell immunity to *Trypanosoma cruzi* and how to improve it. *Trends Parasitol.* 2019;35(11):899-917.
86. Dotiwala F, Mulik S, Polidoro RB, et al. Killer lymphocytes use granulysin, perforin and granzymes to kill intracellular parasites. *Nat Med.* 2016;22(2):210-216.
87. Alvarez MG, Postan M, Weatherly DB, et al. HLA Class I-T cell epitopes from trans-sialidase proteins reveal functionally

distinct subsets of CD8+ T cells in chronic Chagas disease. *PLoS Negl Trop Dis.* 2008;2(9):e288.
88. Freire-de-Lima L, Alisson-Silva F, Carvalho ST, et al. *Trypanosoma cruzi* subverts host cell sialylation and may compromise antigen-specific CD8+ T cell responses. *J Biol Chem.* 2010;285(18):13388-13396.
89. Lasso P, Mateus J, Pavia P, et al. Inhibitory receptor expression on CD8+ T cells is linked to functional responses against *Trypanosoma cruzi* antigens in chronic chagasic patients. *J Immunol.* 2015;195(8):3748-3758.
90. Pack AD, Collins MH, Rosenberg CS, Tarleton RL. Highly competent, non-exhausted CD8+ T cells continue to tightly control pathogen load throughout chronic *Trypanosoma cruzi* infection. *PLoS Pathog.* 2018;14(11):e1007410.
91. Lopes MF, Freire-de-Lima CG, DosReis GA. The macrophage haunted by cell ghosts: a pathogen grows. *Immunol Today.* 2000;21(10):489-494.
92. Bellotti G, Bocchi EA, de Moraes AV, et al. *In vivo* detection of *Trypanosoma cruzi* antigens in hearts of patients with chronic Chagas' heart disease. *Am Heart J.* 1996;131(2):301-307.
93. Zhang L, Tarleton RL. Parasite persistence correlates with disease severity and localization in chronic Chagas' disease. *J Infect Dis.* 1999;180(2):480-486.
94. Dutra WO, Gollob KJ. Current concepts in immunoregulation and pathology of human Chagas disease. *Curr Opin Infect Dis.* 2008;21(3):287-292.
95. de Araujo FF, Vitelli-Avelar DM, Teixeira-Carvalho A, et al. Regulatory T cells phenotype in different clinical forms of Chagas' disease. *PLoS Negl Trop Dis.* 2011;5(5):e992.
96. Hunter CA, Ellis-Neyes LA, Slifer T, et al. IL-10 is required to prevent immune hyperactivity during infection with *Trypanosoma cruzi. J Immunol.* 1997;158(7):3311-3316.
97. Roffe E, Rothfuchs AG, Santiago HC, et al. IL-10 limits parasite burden and protects against fatal myocarditis in a mouse model of *Trypanosoma cruzi* infection. *J Immunol.* 2012;188(2):649-660.
98. Medina TS, Oliveira GG, Silva MC, et al. Ebi3 prevents *Trypanosoma cruzi*-induced myocarditis by dampening IFN-γ-driven inflammation. *Front Immunol.* 2017;8:1213.
99. Sacks D, Noben-Trauth N. The immunology of susceptibility and resistance to *Leishmania major* in mice. *Nat Rev Immunol.* 2002;2(11):845-858.
100. Schleicher U, Paduch K, Debus A, et al. TNF-mediated restriction of arginase 1 expression in myeloid cells triggers type 2 NO synthase activity at the site of infection. *Cell Rep.* 2016;15(5):1062-1075.
101. Belkaid Y, Von Stebut E, Mendez S, et al. CD8+ T cells are required for primary immunity in C57BL/6 mice following low-dose, intradermal challenge with *Leishmania major. J Immunol.* 2002;168(8):3992-4000.
102. Beattie L, Peltan A, Maroof A, et al. Dynamic imaging of experimental *Leishmania donovani*-induced hepatic granulomas detects Kupffer cell-restricted antigen presentation to antigen-specific CD8 T cells. *PLoS Pathog.* 2010;6(3):e1000805.
103. Santos CD, Brodskyn CI. The role of CD4 and CD8 T cells in human cutaneous Leishmaniasis. *Front Public Health.* 2014;2:165.
104. Qadoumi M, Becker I, Donhauser N, Rollinghoff M, Bogdan C. Expression of inducible nitric oxide synthase in skin lesions of patients with American cutaneous Leishmaniasis. *Infect Immun.* 2002;70(8):4638-4642.
105. Boussoffara T, Chelif S, Ben Ahmed M, et al. Immunity against Leishmania major infection: parasite-specific granzyme B induction as a correlate of protection. *Front Cell Infect Microbiol.* 2018;8:397.
106. Singh OP, Gidwani K, Kumar R, et al. Reassessment of immune correlates in human visceral Leishmaniasis as defined by cytokine release in whole blood. *Clin Vaccine Immunol.* 2012;19(6):961-966.
107. Pitta MG, Romano A, Cabantous S, et al. IL-17 and IL-22 are associated with protection against human kala azar caused by *Leishmania donovani. J Clin Invest.* 2009;119(8):2379-2387.
108. Karp CL, el-Safi SH, Wynn TA, et al. *In vivo* cytokine profiles in patients with kala-azar. Marked elevation of both interleukin-10 and interferon-gamma. *J Clin Invest.* 1993;91(4):1644-1648.
109. Melby PC, Andrade-Narvaez FJ, Darnell BJ, Valencia-Pacheco G, Tryon VV, Palomo-Cetina A. Increased expression of proinflammatory cytokines in chronic lesions of human cutaneous Leishmaniasis. *Infect Immun.* 1994;62(3):837-842.
110. Nylen S, Sacks D. Interleukin-10 and the pathogenesis of human visceral Leishmaniasis. *Trends Immunol.* 2007;28(9):378-384.
111. Belkaid Y, Hoffmann KF, Mendez S, et al. The role of interleukin (IL)-10 in the persistence of *Leishmania major* in the skin after healing and the therapeutic potential of anti-IL-10 receptor antibody for sterile cure. *J Exp Med.* 2001;194(10):1497-1506.
112. Murray HW, Lu CM, Mauze S, et al. Interleukin-10 (IL-10) in experimental visceral Leishmaniasis and IL-10 receptor blockade as immunotherapy. *Infect Immun.* 2002;70(11):6284-6293.
113. Anderson CF, Stumhofer JS, Hunter CA, Sacks D. IL-27 regulates IL-10 and IL-17 from CD4+ cells in nonhealing *Leishmania major* infection. *J Immunol.* 2009;183(7):4619-4627.
114. Rosas LE, Satoskar AA, Roth KM, et al. Interleukin-27R (WSX-1/T-cell cytokine receptor) gene-deficient mice display enhanced resistance to *Leishmania donovani* infection but develop severe liver immunopathology. *Am J Pathol.* 2006;168(1):158-169.
115. Kumar R, Chauhan SB, Ng SS, Sundar S, Engwerda CR. Immune checkpoint targets for host-directed therapy to prevent and treat Leishmaniasis. *Front Immunol.* 2017;8:1492.
116. da Fonseca-Martins AM, Ramos TD, Pratti JES, et al. Immunotherapy using anti-PD-1 and anti-PD-L1 in *Leishmania amazonensis*-infected BALB/c mice reduce parasite load. *Sci Rep.* 2019;9:20275.
117. Joshi T, Rodriguez S, Perovic V, Cockburn IA, Stager S. B7-H1 blockade increases survival of dysfunctional CD8⁺ T cells and confers protection against *Leishmania donovani* infections. *PLoS Pathog.* 2009;5(5):e1000431.
118. Silveira FT, Lainson R, De Castro Gomes CM, Laurenti MD, Corbett CE. Immunopathogenic competences of *Leishmania* (*V.*) *braziliensis* and *L.* (*L.*) *amazonensis* in American cutaneous Leishmaniasis. *Parasite Immunol.* 2009;31(8):423-431.
119. Ribeiro de Jesus A, Luna T, Pacheco de Almeida R, Machado PR, Carvalho EM. Pentoxifylline down modulate *in vitro* T cell responses and attenuate pathology in *Leishmania* and HTLV-I infections. *Int Immunopharmacol.* 2008;8(10):1344-1353.
120. Boaventura VS, Santos CS, Cardoso CR, et al. Human mucosal Leishmaniasis: neutrophils infiltrate areas of tissue damage that express high levels of Th17-related cytokines. *Eur J Immunol.* 2010;40(10):2830-2836.
121. Scott P, Novais FO. Cutaneous Leishmaniasis: immune responses in protection and pathogenesis. *Nat Rev Immunol.* 2016;16(9):581-592.
122. Santos Cda S, Boaventura V, Ribeiro Cardoso C, et al. CD8⁺ granzyme B⁺-mediated tissue injury vs. CD4⁺IFNγ⁺-mediated parasite killing in human cutaneous Leishmaniasis. *J Invest Dermatol.* 2013;133(6):1533-1540.
123. Quijano-Hernandez I, Dumonteil E. Advances and challenges towards a vaccine against Chagas disease. *Hum Vaccin.* 2011;7(11):1184-1191.
124. Haolla FA, Claser C, de Alencar BC, et al. Strain-specific protective immunity following vaccination against experimental *Trypanosoma cruzi* infection. *Vaccine.* 2009;27(41):5644-5653.
125. Khamesipour A, Dowlati Y, Asilian A, et al. Leishmanization: use of an old method for evaluation of candidate vaccines against Leishmaniasis. *Vaccine.* 2005;23(28):3642-3648.
126. Alvar J, Croft SL, Kaye P, Khamesipour A, Sundar S, Reed SG. Case study for a vaccine against Leishmaniasis. *Vaccine.* 2013;31(suppl 2):B244-B249.
127. Kamhawi S, Aslan H, Valenzuela JG. Vector saliva in vaccines for visceral Leishmaniasis: a brief encounter of high consequence? *Front Public Health.* 2014;2:99.
128. Noazin S, Modabber F, Khamesipour A, et al. First generation Leishmaniasis vaccines: a review of field efficacy trials. *Vaccine.* 2008;26(52):6759-6767.
129. Glennie ND, Scott P. Memory T cells in cutaneous Leishmaniasis. *Cell Immunol.* 2016;309:50-54.
130. Peters NC, Pagan AJ, Lawyer PG, et al. Chronic parasitic infection maintains high frequencies of short-lived Ly6C+CD4+ effector T cells that are required for protection against re-infection. *PLoS Pathog.* 2014;10(12):e1004538.
131. Coler RN, Duthie MS, Hofmeyer KA, et al. From mouse to man: safety, immunogenicity and efficacy of a candidate Leishmaniasis vaccine LEISH-F3+GLA-SE. *Clin Transl Immunol.* 2015;4(4):e35.
132. Zhang WW, Karmakar S, Gannavaram S, et al. A second generation leishmanization vaccine with a markerless attenuated *Leishmania major* strain using CRISPR gene editing. *Nat Commun.* 2020;11(1):3461.
133. Osman M, Mistry A, Keding A, et al. A third generation vaccine for human visceral Leishmaniasis and post kala azar dermal Leishmaniasis: first-in-human trial of ChAd63-KH. *PLoS Negl Trop Dis.* 2017;11(5):e0005527.
134. Musa AM, Khalil EA, Mahgoub FA, et al. Immunochemotherapy of persistent post-kala-azar dermal Leishmaniasis: a novel approach to treatment. *Trans R Soc Trop Med Hyg.* 2008;102(1):58-63.
135. Convit J, Ulrich M, Zerpa O, et al. Immunotherapy of American cutaneous Leishmaniasis in Venezuela during the period 1990-99. *Trans R Soc Trop Med Hyg.* 2003;97(4):469-472.
136. Howes RE, Battle KE, Mendis KN, et al. Global epidemiology of *Plasmodium vivax. Am J Trop Med Hyg.* 2016;95(6 suppl):15-34.
137. Amino R, Thiberge S, Martin B, et al. Quantitative imaging of *Plasmodium* transmission from mosquito to mammal. *Nat Med.* 2006;12(2):220-224.
138. Medica DL, Sinnis P. Quantitative dynamics of *Plasmodium yoelii* sporozoite transmission by infected anopheline mosquitoes. *Infect Immun.* 2005;73(7):4363-4369.
139. Shin SC, Vanderberg JP, Terzakis JA. Direct infection of hepatocytes by sporozoites of *Plasmodium berghei. J Protozool.* 1982;29(3):448-454.
140. Vaughan AM, Kappe SHI. Malaria parasite liver infection and exoerythrocytic biology. *Cold Spring Harb Persp Med.* 2017;7(6):a025486.
141. Bartoloni A, Zammarchi L. Clinical aspects of uncomplicated and severe malaria. *Mediterr J Hematol Infect Dis.* 2012;4(1):e2012026.
142. Stephens R, Culleton RL, Lamb TJ. The contribution of *Plasmodium chabaudi* to our understanding of malaria. *Trends Parasitol.* 2012;28(2):73-82.
143. Kurtovic L, Behet MC, Feng G, et al. Human antibodies activate complement against *Plasmodium falciparum* sporozoites, and are associated with protection against malaria in children. *BMC Med.* 2018;16(1):61.
144. Zenklusen I, Jongo S, Abdulla S, et al. Immunization of malaria-preexposed volunteers with PfSPZ vaccine elicits long-lived IgM invasion-inhibitory and complement-fixing antibodies. *J Infect Dis.* 2018;217(10):1569-1578.
145. Boyle Michelle J, Reiling L, Feng G, et al. Human antibodies fix complement to inhibit *Plasmodium falciparum* invasion of erythrocytes and are associated with protection against malaria. *Immunity.* 2015;42(3):580-590.
146. Richards JS, Arumugam TU, Reiling L, et al. Identification and prioritization of merozoite antigens as targets of protective human immunity to *Plasmodium falciparum* malaria for vaccine and biomarker development. *J Immunol.* 2013;191(2):795-809.
147. Kennedy AT, Schmidt CQ, Thompson JK, et al. Recruitment of factor H as a novel complement evasion strategy for blood-stage *Plasmodium falciparum* infection. *J Immunol.* 2016;196(3):1239-1248.
148. Rosa TFA, Flammersfeld A, Ngwa CJ, et al. The *Plasmodium falciparum* blood stages acquire factor H family proteins to evade destruction by human complement. *Cell Microbiol.* 2016;18(4):573-590.
149. Kennedy AT, Wijeyewickrema LC, Huglo A, et al. Recruitment of human C1 esterase inhibitor controls complement activation on blood stage *Plasmodium falciparum* merozoites. *J Immunol.* 2017;198(12):4728-4737.
150. Ben-Harel S. Studies of bird malaria in relation to the mechanism of relapse. *Am J Epidemiol.* 1923;3(6):652-685.
151. Stevenson MM, Kraal G. Histological changes in the spleen and liver of C57BL/6 and A/J mice during *Plasmodium chabaudi* AS infection. *Exp Mol Pathol.* 1989;51(1):80-95.
152. Taliaferro WH, Cannon PR. The cellular reactions during primary infections and superinfections of *Plasmodium brasilianum* in Panamanian monkeys. *J Infect Dis.* 1936;59(1):72-125.
153. Couper KN, Blount DG, Hafalla JC, van Rooijen N, de Souza JB, Riley EM. Macrophage-mediated but gamma interferon-independent innate immune responses control the primary wave of *Plasmodium yoelii* parasitemia. *Infect Immun.* 2007;75(12):5806-5818.
154. Fontana MF, de Melo GL, Anidi C, et al. Macrophage colony stimulating factor derived from CD4+ T cells contributes to control of a blood-borne infection. *PLoS Pathog.* 2016;12(12):e1006046.
155. Gupta P, Lai SM, Sheng J, et al. Tissue-resident CD169⁺ macrophages form a crucial front line against *Plasmodium* infection. *Cell Rep.* 2016;16(6):1749-1761.
156. Grabowska J, Lopez-Venegas MA, Affandi AJ, den Haan JMM. CD169+ macrophages capture and dendritic cells instruct: the interplay of the gatekeeper and the general of the immune system. *Front Immunol.* 2018;9:2472.
157. Voisine C, Mastelic B, Sponaas A-M, Langhorne J. Classical CD11c+ dendritic cells, not plasmacytoid dendritic cells, induce T cell responses to *Plasmodium chabaudi* malaria. *Int J Parasitol.* 2010;40(6):711-719.
158. Dockrell HM, Playfair JH. Killing of *Plasmodium yoelii* by enzyme-induced products of the oxidative burst. *Infect Immun.* 1984;43(2):451-456.
159. Wozencraft AO, Dockrell HM, Taverne J, Targett GA, Playfair JH. Killing of human malaria parasites by macrophage secretory products. *Infect Immun.* 1984;43(2):664-669.
160. Gillman BM, Batchelder J, Flaherty P, Weidanz WP. Suppression of *Plasmodium chabaudi* parasitemia is independent of the action of reactive oxygen intermediates and/or nitric oxide. *Infect Immun.* 2004;72(11):6359-6366.
161. Berman PA, Human L, Freese JA. Xanthine oxidase inhibits growth of *Plasmodium falciparum* in human erythrocytes *in vitro. J Clin Invest.* 1991;88(6):1848-1855.
162. Ty MC, Zuniga M, Götz A, et al. Malaria inflammation by xanthine oxidase-produced reactive oxygen species. *EMBO Mol Med.* 2019;11(8):e9903.
163. Stanisic DI, Cutts J, Eriksson E, et al. γδ T cells and CD14+ monocytes are predominant cellular sources of cytokines and chemokines associated with severe malaria. *J Infect Dis.* 2014;210(2):295-305.
164. Baccarella A, Fontana MF, Chen EC, Kim CC. Toll-like receptor 7 mediates early innate immune responses to malaria. *Infect Immun.* 2013;81(12):4431-4442.
165. Hahn WO, Butler NS, Lindner SE, et al. cGAS-mediated control of blood-stage malaria promotes *Plasmodium*-specific germinal center responses. *JCI Insight.* 2018;3(2):e94142.
166. Ball EA, Sambo MR, Martins M, et al. IFNAR1 controls progression to cerebral malaria in children and CD8+ T cell brain pathology in *Plasmodium berghei*-infected mice. *J Immunol.* 2013;190(10):5118-5127.
167. Mordmüller BG, Metzger WG, Juillard P, et al. Tumor necrosis factor in Plasmodium falciparum malaria: high plasma level is associated with fever, but high production capacity is associated with rapid fever clearance. *Eur Cytokine Netw.* 1997;8(1):29-35.

168. Haque A, Best SE, Ammerdorffer A, et al. Type I interferons suppress CD4+ T-cell-dependent parasite control during blood-stage *Plasmodium* infection. *Eur J Immunol.* 2011;41(9):2688-2698.
169. Sam H, Stevenson MM. Early IL-12 p70, but not p40, production by splenic macrophages correlates with host resistance to blood-stage *Plasmodium chabaudi* AS malaria. *Clin Exp Immunol.* 1999;117(2):343-349.
170. Li C, Corraliza I, Langhorne J. A defect in interleukin-10 leads to enhanced malarial disease in *Plasmodium chabaudi chabaudi* infection in mice. *Infect Immun.* 1999;67(9):4435-4442.
171. Chen L, Zhang Z, Sendo F. Neutrophils play a critical role in the pathogenesis of experimental cerebral malaria. *Clin Exp Immunol.* 2000;120(1):125-133.
172. Kho S, Minigo G, Andries B, et al. Circulating neutrophil extracellular traps and neutrophil activation are increased in proportion to disease severity in human malaria. *J Infect Dis.* 2019;219(12):1994-2004.
173. Knackstedt SL, Georgiadou A, Apel F, et al. Neutrophil extracellular traps drive inflammatory pathogenesis in malaria. *Sci Immunol.* 2019;4(40):eaaw0336.
174. Mohan K, Moulin P, Stevenson MM. Natural killer cell cytokine production, not cytotoxicity, contributes to resistance against blood-stage *Plasmodium chabaudi* AS infection. *J Immunol.* 1997;159(10):4990-4998.
175. De Souza JB, Williamson KH, Otani T, Playfair JH. Early gamma interferon responses in lethal and nonlethal murine blood-stage malaria. *Infect Immun.* 1997;65(5):1593-1598.
176. Roland J, Soulard V, Sellier C, et al. NK cell responses to *Plasmodium* infection and control of intrahepatic parasite development. *J Immunol.* 2006;177(2):1229-1239.
177. Lefebvre MN, Harty JT. You shall not pass: memory CD8 T cells in liver-stage malaria. *Trends Parasitol.* 2020;36(2):147-157.
178. Seder RA, Chang L-J, Enama ME, et al. Protection against malaria by intravenous immunization with a nonreplicating sporozoite vaccine. *Science.* 2013;341(6152):1359-1365.
179. Teirlinck AC, McCall MB, Roestenberg M, et al. Longevity and composition of cellular immune responses following experimental *Plasmodium falciparum* malaria infection in humans. *PLoS Pathog.* 2011;7(12):e1002389.
180. Morita CT, Jin C, Sarikonda G, Wang H. Nonpeptide antigens, presentation mechanisms, and immunological memory of human Vγ2Vδ2 T cells: discriminating friend from foe through the recognition of prenyl pyrophosphate antigens. *Immunol Rev.* 2007;215:59-76.
181. Guenot M, Loizon S, Howard J, et al. Phosphoantigen burst upon *Plasmodium falciparum* schizont rupture can distantly activate Vγ9Vδ2 T cells. *Infect Immun.* 2015;83(10):3816-3824.
182. McKenna KC, Tsuji M, Sarzotti M, Sacci JB, Witney AA, Azad AF. γδ T cells are a component of early immunity against preerythrocytic malaria parasites. *Infect Immun.* 2000;68(4):2224-2230.
183. Miller JL, Sack BK, Baldwin M, Vaughan AM, Kappe SHI. Interferon-mediated innate immune responses against malaria parasite liver stages. *Cell Rep.* 2014;7(2):436-447.
184. Goodier M, Fey P, Eichmann K, Langhorne J. Human peripheral blood γδ T cells respond to antigens of *Plasmodium falciparum*. *Int Immunol.* 1992;4(1):33-41.
185. Ho M, Tongtawe P, Kriangkum J, et al. Polyclonal expansion of peripheral γδ T cells in human *Plasmodium falciparum* malaria. *Infect Immun.* 1994;62(3):855-862.
186. Elloso MM, van der Heyde HC, vande Waa JA, Manning DD, Weidanz WP. Inhibition of *Plasmodium falciparum in vitro* by human γδ T cells. *J Immunol.* 1994;153(3):1187-1194.
187. Troye-Blomberg M, Worku S, Tangteerawatana P, et al. Human γδ T cells that inhibit the *in vitro* growth of the asexual blood stages of the *Plasmodium falciparum* parasite express cytolytic and proinflammatory molecules. *Scand J Immunol.* 1999;50(6):642-650.
188. van der Heyde HC, Pepper B, Batchelder J, Cigel F, Weidanz WP. The time course of selected malarial infections in cytokine-deficient mice. *Exp Parasitol.* 1997;85(2):206-213.
189. Sayles PC, Rakhmilevich L. Exacerbation of *Plasmodium chabaudi* malaria in mice by depletion of TCR alpha beta+ T cells, but not TCR gamma delta+ T cells. *Immunology.* 1996;87(1):29-33.
190. Seixas EM, Langhorne J. γδ T cells contribute to control of chronic parasitemia in *Plasmodium chabaudi* infections in mice. *J Immunol.* 1999;162(5):2837-2841.
191. Jagannathan P, Eccles-James I, Bowen K, et al. IFNγ/IL-10 co-producing cells dominate the CD4 response to malaria in highly exposed children. *PLoS Pathog.* 2014;10(1):e1003864.
192. Mamedov MR, Scholzen A, Nair RV, et al. A macrophage colony-stimulating-factor-producing γδ T cell subset prevents malarial parasitemic recurrence. *Immunity.* 2018;48(2):350-363.e7.
193. Farrington LA, Jagannathan P, McIntyre TI, et al. Frequent malaria drives progressive Vδ2 T-cell loss, dysfunction, and CD16 up-regulation during early childhood. *J Infect Dis.* 2016;213(9):1483-1490.
194. Yu X, Cai B, Wang M, et al. Cross-regulation of two type I interferon signaling pathways in plasmacytoid dendritic cells controls anti-malaria immunity and host mortality. *Immunity.* 2016;45(5):1093-1107.
195. Ing R, Stevenson MM. Dendritic cell and NK cell reciprocal cross talk promotes gamma interferon-dependent immunity to blood-stage *Plasmodium chabaudi* AS infection in mice. *Infect Immun.* 2009;77(2):770-782.
196. Ryg-Cornejo V, Nie CQ, Bernard NJ, et al. NK cells and conventional dendritic cells engage in reciprocal activation for the induction of inflammatory responses during *Plasmodium berghei* ANKA infection. *Immunobiology.* 2013;218(2):263-271.
197. Arroyo EN, Pepper M. B cells are sufficient to prime the dominant CD4+ Tfh response to *Plasmodium* infection. *J Exp Med.* 2020;217(2):e20190849.
198. Borges da Silva H, Fonseca R, Cassado ADA, et al. *In vivo* approaches reveal a key role for DCs in CD4+ T cell activation and parasite clearance during the acute phase of experimental blood-stage malaria. *PLoS Pathog.* 2015;11(2):e1004598.
199. Ueffing K, Abberger H, Westendorf AM, Matuschewski K, Buer J, Hansen W. Conventional CD11c^{high} dendritic cells are important for T cell priming during the initial phase of *Plasmodium yoelii* infection, but are dispensable at later time points. *Front Immunol.* 2017;8:1333.
200. Götz A, Tang MS, Ty MC, et al. Atypical activation of dendritic cells by *Plasmodium falciparum*. *Proc Natl Acad Sci U S A.* 2017;114(49):E10568-E10577.
201. Pinzon-Charry A, Woodberry T, Kienzle V, et al. Apoptosis and dysfunction of blood dendritic cells in patients with falciparum and vivax malaria. *J Exp Med.* 2013;210(8):1635-1646.
202. Urban BC, Ferguson DJ, Pain A, et al. *Plasmodium falciparum*-infected erythrocytes modulate the maturation of dendritic cells. *Nature.* 1999;400(6739):73-77.
203. Woodberry T, Minigo G, Piera KA, et al. Low-level *Plasmodium falciparum* blood-stage infection causes dendritic cell apoptosis and dysfunction in healthy volunteers. *J Infect Dis.* 2012;206(3):333-340.
204. Millington OR, Di Lorenzo C, Phillips RS, Garside P, Brewer JM. Suppression of adaptive immunity to heterologous antigens during *Plasmodium* infection through hemozoin-induced failure of dendritic cell function. *J Biol.* 2006;5(2):5.
205. Arama C, Giusti P, Boström S, et al. Interethnic differences in antigen-presenting cell activation and TLR responses in Malian children during *Plasmodium falciparum* malaria. *PLoS One.* 2011;6(3):e18319.
206. Franklin BS, Rodrigues SO, Antonelli LR, et al. MyD88-dependent activation of dendritic cells and CD4+ T lymphocytes mediates symptoms, but is not required for the immunological control of parasites during rodent malaria. *Microbes Infect.* 2007;9(7):881-890.
207. Gowda NM, Wu X, Gowda DC. TLR9 and MyD88 are crucial for the development of protective immunity to malaria. *J Immunol.* 2012;188(10):5073-5085.
208. Kordes M, Matuschewski K, Hafalla JC. Caspase-1 activation of interleukin-1β (IL-1β) and IL-18 is dispensable for induction of experimental cerebral malaria. *Infect Immun.* 2011;79(9):3633-3641.
209. Gallego-Marin C, Schrum JE, Andrade WA, et al. Cyclic GMP-AMP synthase is the cytosolic sensor of *Plasmodium falciparum* genomic DNA and activates type I IFN in malaria. *J Immunol.* 2018;200(2):768-774.
210. Sharma S, DeOliveira RB, Kalantari P, et al. Innate immune recognition of an AT-rich stem-loop DNA motif in the *Plasmodium falciparum* genome. *Immunity.* 2011;35(2):194-207.
211. Reimer T, Shaw MH, Franchi L, et al. Experimental cerebral malaria progresses independently of the Nlrp3 inflammasome. *Eur J Immunol.* 2010;40(3):764-769.
212. Reis AS, Barboza R, Murillo O, et al. Inflammasome activation and IL-1 signaling during placental malaria induce poor pregnancy outcomes. *Sci Adv.* 2020;6(10):eaax6346.
213. Yu X, Du Y, Cai C, et al. Inflammasome activation negatively regulates MyD88-IRF7 type I IFN signaling and anti-malaria immunity. *Nat Commun.* 2018;9(1):4964.
214. Nardin EH, Nussenzweig V, Nussenzweig RS, et al. Circumsporozoite proteins of human malaria parasites *Plasmodium falciparum* and *Plasmodium vivax*. *J Exp Med.* 1982;156(1):20-30.
215. Hopp CS, Chiou K, Ragheb DR, et al. Longitudinal analysis of *Plasmodium* sporozoite motility in the dermis reveals component of blood vessel recognition. *Elife.* 2015;4:e07789.
216. Chakravarty S, Cockburn IA, Kuk S, Overstreet MG, Sacci JB, Zavala F. CD8+ T lymphocytes protective against malaria liver stages are primed in skin-draining lymph nodes. *Nat Med.* 2007;13(9):1035-1041.
217. Doolan DL, Dobano C, Baird JK. Acquired immunity to malaria. *Clin Microbiol Rev.* 2009;22(1):13-36.
218. Kumar R, Ng S, Engwerda C. The role of IL-10 in malaria: a double edged sword. *Front Immunol.* 2019;10:229.
219. Gupta S, Snow RW, Donnelly CA, Marsh K, Newbold C. Immunity to non-cerebral severe malaria is acquired after one or two infections. *Nat Med.* 1999;5(3):340-343.
220. Langhorne J, Ndungu FM, Sponaas AM, Marsh K. Immunity to malaria: more questions than answers. *Nat Immunol.* 2008;9(7):725-732.
221. McCall MBB, Kremsner PG, Mordmuller B. Correlating efficacy and immunogenicity in malaria vaccine trials. *Semin Immunol.* 2018;39:52-64.
222. Cockburn IA, Seder RA. Malaria prevention: from immunological concepts to effective vaccines and protective antibodies. *Nat Immunol.* 2018;19(11):1199-1211.
223. Laurens MB. The promise of a malaria vaccine-are we closer? *Annu Rev Microbiol.* 2018;72:273-292.
224. Cerami C, Frevert U, Sinnis P, et al. The basolateral domain of the hepatocyte plasma membrane bears receptors for the circumsporozoite protein of *Plasmodium falciparum* sporozoites. *Cell.* 1992;70(6):1021-1033.
225. Tewari R, Spaccapelo R, Bistoni F, Holder AA, Crisanti A. Function of region I and II adhesive motifs of *Plasmodium falciparum* circumsporozoite protein in sporozoite motility and infectivity. *J Biol Chem.* 2002;277(49):47613-47618.
226. Tan J, Piccoli L, Lanzavecchia A. The antibody response to Plasmodium falciparum: cues for vaccine design and the discovery of receptor-based antibodies. *Annu Rev Immunol.* 2019;37(1):225-246.
227. Del Giudice G, Engers HD, Tougne C, et al. Antibodies to the repetitive epitope of Plasmodium falciparum circumsporozoite protein in a rural Tanzanian community: a longitudinal study of 132 children. *Am J Trop Med Hyg.* 1987;36(2):203-212.
228. Wipasa J, Suphavilai C, Okell LC, et al. Long-lived antibody and B cell memory responses to the human malaria parasites, *Plasmodium falciparum* and *Plasmodium vivax*. *PLoS Pathog.* 2010;6(2):e1000770.
229. Triller G, Scally SW, Costa G, et al. Natural parasite exposure induces protective human anti-malarial antibodies. *Immunity.* 2017;47(6):1197-1209.e10.
230. Keitany GJ, Kim KS, Krishnamurty AT, et al. Blood stage malaria disrupts humoral immunity to the pre-erythrocytic stage circumsporozoite protein. *Cell Rep.* 2016;17(12):3193-3205.
231. Schofield L, Villaquiran J, Ferreira A, Schellekens H, Nussenzweig R, Nussenzweig V. γ interferon, CD8+ T cells and antibodies required for immunity to malaria sporozoites. *Nature.* 1987;330(6149):664-666.
232. Keitany GJ, Sack B, Smithers H, et al. Immunization of mice with live-attenuated late liver stage-arresting *Plasmodium yoelii* parasites generates protective antibody responses to preerythrocytic stages of malaria. *Infect Immun.* 2014;82(12):5143-5153.
233. Casares S, Brumeanu T-D, Richie TL. The RTS,S malaria vaccine. *Vaccine.* 2010;28(31):4880-4894.
234. RTS,S Clinical Trials Partnership. Efficacy and safety of RTS,S/AS01 malaria vaccine with or without a booster dose in infants and children in Africa: final results of a phase 3, individually randomised, controlled trial. *Lancet.* 2015;386(9988):31-45.
235. White MT, Verity R, Griffin JT, et al. Immunogenicity of the RTS,S/AS01 malaria vaccine and implications for duration of vaccine efficacy: secondary analysis of data from a phase 3 randomised controlled trial. *Lancet Infect Dis.* 2015;15(12):1450-1458.
236. Agnandji ST, Lell B, Soulanoudjingar SS, et al. First results of phase 3 trial of RTS,S/AS01 malaria vaccine in African children. *N Engl J Med.* 2011;365(20):1863-1875.
237. Julien JP, Wardemann H. Antibodies against *Plasmodium falciparum* malaria at the molecular level. *Nat Rev Immunol.* 2019;19(12):761-775.
238. Weiss WR, Sedegah M, Beaudoin RL, Miller LH, Good MF. CD8+ T cells (cytotoxic/suppressors) are required for protection in mice immunized with malaria sporozoites. *Proc Natl Acad Sci U S A.* 1988;85(2):573-576.
239. Carvalho LH, Sano G, Hafalla JC, Morrot A, Curotto de Lafaille MA, Zavala F. IL-4-secreting CD4+ T cells are crucial to the development of CD8+ T-cell responses against malaria liver stages. *Nat Med.* 2002;8(2):166-170.
240. Cockburn IA, Amino R, Kelemen RK, et al. *In vivo* imaging of CD8+ T cell-mediated elimination of malaria liver stages. *Proc Natl Acad Sci U S A.* 2013;110(22):9090-9095.
241. Kimura K, Kimura D, Matsushima Y, et al. CD8+ T cells specific for a malaria cytoplasmic antigen form clusters around infected hepatocytes and are protective at the liver stage of infection. *Infect Immun.* 2013;81(10):3825-3834.
242. Butler NS, Schmidt NW, Harty JT. Differential effector pathways regulate memory CD8 T cell immunity against *Plasmodium berghei* versus *P. yoelii* sporozoites. *J Immunol.* 2010;184(5):2528-2538.
243. Schmidt NW, Podyminogin RL, Butler NS, et al. Memory CD8 T cell responses exceeding a large but definable threshold provide long-term immunity to malaria. *Proc Natl Acad Sci U S A.* 2008;105(37):14017-14022.
244. Fernandez-Ruiz D, Ng WY, Holz LE, et al. Liver-resident memory CD8+ T cells form a front-line defense against malaria liver-stage infection. *Immunity.* 2016;45(4):889-902.
245. McNamara HA, Cai Y, Wagle MV, et al. Up-regulation of LFA-1 allows liver-resident memory T cells to patrol and remain in the hepatic sinusoids. *Sci Immunol.* 2017;2(9):eaaj1996.
246. Holz LE, Prier JE, Freestone D, et al. CD8+ T cell activation leads to constitutive formation of liver tissue-resident memory T cells that seed a large and flexible niche in the liver. *Cell Rep.* 2018;25(1):68-79.e4.
247. Gola A, Dorrington MG, Speranza E, et al. Commensal-driven immune zonation of the liver promotes host defence. *Nature.* 2021;589(7840):131-136.
248. Cohen S, Mc GI, Carrington S. Gamma-globulin and acquired immunity to human malaria. *Nature.* 1961;192:733-737.
249. Edozien JC. The effect of malaria on the serum gamma globulin concentration in Africans. *West Afr Med J.* 1961;10:304-310.

250. Sabchareon A, Burnouf T, Ouattara D, et al. Parasitologic and clinical human response to immunoglobulin administration in falciparum malaria. *Am J Trop Med Hyg*. 1991;45(3):297-308.
251. Osier FHA, Fegan G, Polley SD, et al. Breadth and magnitude of antibody responses to multiple *Plasmodium falciparum* merozoite antigens are associated with protection from clinical malaria. *Infect Immun*. 2008;76(5):2240-2248.
252. Burns JM Jr, Dunn PD, Russo DM. Protective immunity against *Plasmodium yoelii* malaria induced by immunization with particulate blood-stage antigens. *Infect Immun*. 1997;65(8):3138-3145.
253. Wahlgren M, Goel S, Akhouri RR. Variant surface antigens of *Plasmodium falciparum* and their roles in severe malaria. *Nat Rev Microbiol*. 2017;15(8):479-491.
254. Marsh K, Howard RJ. Antigens induced on erythrocytes by P. falciparum: expression of diverse and conserved determinants. *Science*. 1986;231(4734):150-153.
255. Bull PC, Lowe BS, Kortok M, Molyneux CS, Newbold CI, Marsh K. Parasite antigens on the infected red cell surface are targets for naturally acquired immunity to malaria. *Nat Med*. 1998;4(3):358-360.
256. Tran TM, Li S, Doumbo S, et al. An intensive longitudinal cohort study of Malian children and adults reveals no evidence of acquired immunity to *Plasmodium falciparum* infection. *Clin Infect Dis*. 2013;57(1):40-47.
257. Thera MA, Doumbo OK, Coulibaly D, et al. A field trial to assess a blood-stage malaria vaccine. *N Engl J Med*. 2011;365(11):1004-1013.
258. Osier FH, Mackinnon MJ, Crosnier C, et al. New antigens for a multicomponent blood-stage malaria vaccine. *Sci Transl Med*. 2014;6(247):247ra102.
259. Douglas AD, Williams AR, Illingworth JJ, et al. The blood-stage malaria antigen PfRH5 is susceptible to vaccine-inducible cross-strain neutralizing antibody. *Nat Commun*. 2011;2:601.
260. Ragotte RJ, Higgins MK, Draper SJ. The RH5-CyRPA-Ripr complex as a malaria vaccine target. *Trends Parasitol*. 2020;36(6):545-559.
261. Harms Pritchard G, Pepper M. Memory B cell heterogeneity: remembrance of things past. *J Leukoc Biol*. 2018;103(2):269-274.
262. Perez-Mazliah D, Gardner PJ, Schweighoffer E, et al. *Plasmodium*-specific atypical memory B cells are short-lived activated B cells. *Elife*. 2018;7:e39800.
263. Kim CC, Baccarella AM, Bayat A, Pepper M, Fontana MF. FCRL5[+] memory B cells exhibit robust recall responses. *Cell Rep*. 2019;27(5):1446-1460 e4.
264. Boyle MJ, Chan JA, Handayuni I, et al. IgM in human immunity to *Plasmodium falciparum* malaria. *Sci Adv*. 2019;5(9):eaax4489.
265. Arama C, Skinner J, Doumtabe D, et al. Genetic resistance to malaria is associated with greater enhancement of immunoglobulin (Ig)M than IgG responses to a broad array of *Plasmodium falciparum* antigens. *Open Forum Infect Dis*. 2015;2(3):ofv118.
266. von der Weid T, Langhorne J. The roles of cytokines produced in the immune response to the erythrocytic stages of mouse malarias. *Immunobiology*. 1993;189(3-4):397-418.
267. Su Z, Stevenson MM. Central role of endogenous gamma interferon in protective immunity against blood-stage *Plasmodium chabaudi* AS infection. *Infect Immun*. 2000;68(8):4399-4406.
268. Hochman S, Kim K. The impact of HIV and malaria coinfection: what is known and suggested venues for further study. *Interdiscip Perspect Infect Dis*. 2009;2009:617954.
269. Pepper M, Pagan AJ, Igyarto BZ, Taylor JJ, Jenkins MK. Opposing signals from the Bcl6 transcription factor and the interleukin-2 receptor generate T helper 1 central and effector memory cells. *Immunity*. 2011;35(4):583-595.
270. Couper KN, Blount DG, Wilson MS, et al. IL-10 from CD4CD25Foxp3CD127 adaptive regulatory T cells modulates parasite clearance and pathology during malaria infection. *PLoS Pathog*. 2008;4(2):e1000004.
271. Boyle MJ, Jagannathan P, Bowen K, et al. The development of *Plasmodium falciparum*-specific IL10 CD4 T cells and protection from malaria in children in an area of high malaria transmission. *Front Immunol*. 2017;8:1329.
272. Zander RA, Guthmiller JJ, Graham AC, et al. Type I interferons induce T regulatory 1 responses and restrict humoral immunity during experimental malaria. *PLoS Pathog*. 2016;12(10):e1005945.
273. Guthmiller JJ, Graham AC, Zander RA, Pope RL, Butler NS. Cutting edge: IL-10 is essential for the generation of germinal center B cell responses and anti-Plasmodium humoral immunity. *J Immunol*. 2017;198(2):617-622.
274. Kimura D, Miyakoda M, Kimura K, et al. Interleukin-27-Producing CD4[+] T cells regulate protective immunity during malaria parasite infection. *Immunity*. 2016;44(3):672-682.
275. Kurup SP, Butler NS, Harty JT. T cell-mediated immunity to malaria. *Nat Rev Immunol*. 2019;19(7):457-471.
276. Perez-Mazliah D, Nguyen MP, Hosking C, et al. Follicular helper T cells are essential for the elimination of *Plasmodium* infection. *EBioMedicine*. 2017;24:216-230.
277. Figueiredo MM, Costa PAC, Diniz SQ, et al. T follicular helper cells regulate the activation of B lymphocytes and antibody production during *Plasmodium vivax* infection. *PLoS Pathog*. 2017;13(7):e1006484.
278. Jourdan PM, Lamberton PHL, Fenwick A, Addiss DG. Soil-transmitted helminth infections. *Lancet*. 2018;391(10117):252-265.
279. Hotez PJ, Brindley PJ, Bethony JM, King CH, Pearce EJ, Jacobson J. Helminth infections: the great neglected tropical diseases. *J Clin Invest*. 2008;118(4):1311-1321.
280. Allen JE, Maizels RM. Diversity and dialogue in immunity to helminths. *Nat Rev Immunol*. 2011;11(6):375-388.
281. Babu S, Nutman TB. Immunopathogenesis of lymphatic filarial disease. *Semin Immunopathol*. 2012;34(6):847-861.
282. Girgis NM, Gundra UM, Loke P. Immune regulation during helminth infections. *PLoS Pathog*. 2013;9(4):e1003250.
283. Harris NL, Loke P. Recent advances in type-2-cell-mediated immunity: insights from helminth infection. *Immunity*. 2017;47(6):1024-1036.
284. Gause WC, Wynn TA, Allen JE. Type 2 immunity and wound healing: evolutionary refinement of adaptive immunity by helminths. *Nat Rev Immunol*. 2013;13(8):607-614.
285. Odegaard JI, Chawla A. Alternative macrophage activation and metabolism. *Annu Rev Pathol*. 2011;6:275-297.
286. Umetsu DT, Dekruyff RH. Immune dysregulation in asthma. *Curr Opin Immunol*. 2006;18(6):727-732.
287. Wynn TA, Chawla A, Pollard JW. Macrophage biology in development, homeostasis and disease. *Nature*. 2013;496(7446):445-455.
288. Klose CS, Artis D. Innate lymphoid cells as regulators of immunity, inflammation and tissue homeostasis. *Nat Immunol*. 2016;17(7):765-774.
289. Klementowicz JE, Travis MA, Grencis RK. Trichuris muris: a model of gastrointestinal parasite infection. *Semin Immunopathol*. 2012;34(6):815-828.
290. Grencis RK. Immunity to helminths: resistance, regulation, and susceptibility to gastrointestinal nematodes. *Annu Rev Immunol*. 2015;33:201-225.
291. Camberis M, Le Gros G, Urban J Jr. Animal model of *Nippostrongylus brasiliensis* and *Heligmosomoides polygyrus*. *Curr Protoc Immunol*. 2003;Chapter 19:Unit 19 2.
292. McSorley HJ, Hewitson JP, Maizels RM. Immunomodulation by helminth parasites: defining mechanisms and mediators. *Int J Parasitol*. 2013;43(3-4):301-310.
293. Wilson MS, Mentink-Kane MM, Pesce JT, Ramalingam TR, Thompson R, Wynn TA. Immunopathology of schistosomiasis. *Immunol Cell Biol*. 2007;85(2):148-154.
294. Colley DG, Bustinduy AL, Secor WE, King CH. Human schistosomiasis. *Lancet*. 2014;383(9936):2253-2264.
295. Finlay CM, Allen JE. The immune response of inbred laboratory mice to Litomosoides sigmodontis: a route to discovery in myeloid cell biology. *Parasite Immunol*. 2020;42(7):e12708.
296. Paul WE. History of interleukin-4. *Cytokine*. 2015;75(1):3-7.
297. Fitzsimmons CM, Falcone FH, Dunne DW. Helminth allergens, parasite-specific IgE, and its protective role in human immunity. *Front Immunol*. 2014;5:61.
298. Kuhn R, Rajewsky K, Muller W. Generation and analysis of interleukin-4 deficient mice. *Science*. 1991;254(5032):707-710.
299. Paul WE, Zhu J. How are Th2-type immune responses initiated and amplified? *Nat Rev Immunol*. 2010;10(4):225-235.
300. Everts B, Perona-Wright G, Smits HH, et al. Omega-1, a glycoprotein secreted by *Schistosoma mansoni* eggs, drives Th2 responses. *J Exp Med*. 2009;206(8):1673-1680.
301. Steinfelder S, Andersen JF, Cannons JL, et al. The major component in schistosome eggs responsible for conditioning dendritic cells for Th2 polarization is a T2 ribonuclease (omega-1). *J Exp Med*. 2009;206(8):1681-1690.
302. Everts B, Hussaarts L, Driessen NN, et al. Schistosome-derived omega-1 drives Th2 polarization by suppressing protein synthesis following internalization by the mannose receptor. *J Exp Med*. 2012;209(10):1753-1767, S1.
303. Kouzaki H, Iijima K, Kobayashi T, O'Grady SM, Kita H. The danger signal, extracellular ATP, is a sensor for an airborne allergen and triggers IL-33 release and innate Th2-type responses. *J Immunol*. 2011;186(7):4375-4387.
304. Wills-Karp M, Rani R, Dienger K, et al. Trefoil factor 2 rapidly induces interleukin 33 to promote type 2 immunity during allergic asthma and hookworm infection. *J Exp Med*. 2012;209(3):607-622.
305. Belle NM, Ji Y, Herbine K, et al. TFF3 interacts with LINGO2 to regulate EGFR activation for protection against colitis and gastrointestinal helminths. *Nat Commun*. 2019;10(1):4408.
306. Bosurgi L, Cao YG, Cabeza-Cabrerizo M, et al. Macrophage function in tissue repair and remodeling requires IL-4 or IL-13 with apoptotic cells. *Science*. 2017;356(6342):1072-1076.
307. Chen F, Wu W, Millman A, et al. Neutrophils prime a long-lived effector macrophage phenotype that mediates accelerated helminth expulsion. *Nat Immunol*. 2014;15(10):938-946.
308. Ritter M, Gross O, Kays S, et al. *Schistosoma mansoni* triggers Dectin-2, which activates the Nlrp3 inflammasome and alters adaptive immune responses. *Proc Natl Acad Sci U S A*. 2010;107(47):20459-20464.
309. Han H, Roan F, Ziegler SF. The atopic march: current insights into skin barrier dysfunction and epithelial cell-derived cytokines. *Immunol Rev*. 2017;278(1):116-130.
310. Afferni C, Buccione C, Andreone S, et al. The pleiotropic immunomodulatory functions of IL-33 and its implications in tumor immunity. *Front Immunol*. 2018;9:2601.
311. Humphreys NE, Xu D, Hepworth MR, Liew FY, Grencis RK. IL-33, a potent inducer of adaptive immunity to intestinal nematodes. *J Immunol*. 2008;180(4):2443-2449.
312. Hung LY, Lewkowich IP, Dawson LA, et al. IL-33 drives biphasic IL-13 production for noncanonical Type 2 immunity against hookworms. *Proc Natl Acad Sci U S A*. 2013;110(1):282-287.
313. Massacand JC, Stettler RC, Meier R, et al. Helminth products bypass the need for TSLP in Th2 immune responses by directly modulating dendritic cell function. *Proc Natl Acad Sci U S A*. 2009;106(33):13968-13973.
314. Taylor BC, Zaph C, Troy AE, et al. TSLP regulates intestinal immunity and inflammation in mouse models of helminth infection and colitis. *J Exp Med*. 2009;206(3):655-667.
315. von Moltke J, Ji M, Liang HE, Locksley RM. Tuft-cell-derived IL-25 regulates an intestinal ILC2-epithelial response circuit. *Nature*. 2016;529(7585):221-225.
316. Howitt MR, Lavoie S, Michaud M, et al. Tuft cells, taste-chemosensory cells, orchestrate parasite type 2 immunity in the gut. *Science*. 2016;351(6279):1329-1333.
317. Gerbe F, Sidot E, Smyth DJ, et al. Intestinal epithelial tuft cells initiate type 2 mucosal immunity to helminth parasites. *Nature*. 2016;529(7585):226-230.
318. Owyang AM, Zaph C, Wilson EH, et al. Interleukin 25 regulates type 2 cytokine-dependent immunity and limits chronic inflammation in the gastrointestinal tract. *J Exp Med*. 2006;203(4):843-849.
319. Fallon PG, Ballantyne SJ, Mangan NE, et al. Identification of an interleukin (IL)-25-dependent cell population that provides IL-4, IL-5, and IL-13 at the onset of helminth expulsion. *J Exp Med*. 2006;203(4):1105-1116.
320. Fort MM, Cheung J, Yen D, et al. IL-25 induces IL-4, IL-5, and IL-13 and Th2-associated pathologies *in vivo*. *Immunity*. 2001;15(6):985-995.
321. Angkasekwinai P, Park H, Wang YH, et al. Interleukin 25 promotes the initiation of proallergic type 2 responses. *J Exp Med*. 2007;204(7):1509-1517.
322. Girgis NM, Gundra UM, Ward LN, Cabrera M, Frevert U, Loke P. Ly6C[high] monocytes become alternatively activated macrophages in schistosome granulomas with help from CD4+ cells. *PLoS Pathog*. 2014;10(6):e1004080.
323. Minutti CM, Jackson-Jones LH, Garcia-Fojeda B, et al. Local amplifiers of IL-4Rα-mediated macrophage activation promote repair in lung and liver. *Science*. 2017;356(6342):1076-1080.
324. Qualls JE, Murray PJ. Immunometabolism within the tuberculosis granuloma: amino acids, hypoxia, and cellular respiration. *Semin Immunopathol*. 2016;38(2):139-152.
325. Rath M, Muller I, Kropf P, Closs EI, Munder M. Metabolism via arginase or nitric oxide synthase: two competing arginine pathways in macrophages. *Front Immunol*. 2014;5:532.
326. Van de Velde LA, Subramanian C, Smith AM, et al. T cells encountering myeloid cells programmed for amino acid-dependent immunosuppression use Rictor/mTORC2 protein for proliferative checkpoint decisions. *J Biol Chem*. 2017;292(1):15-30.
327. Pesce JT, Ramalingam TR, Mentink-Kane MM, et al. Arginase-1-expressing macrophages suppress Th2 cytokine-driven inflammation and fibrosis. *PLoS Pathog*. 2009;5(4):e1000371.
328. Finkelman FD, Shea-Donohue T, Morris SC, et al. Interleukin-4-and interleukin-13-mediated host protection against intestinal nematode parasites. *Immunol Rev*. 2004;201:139-155.
329. Hasnain SZ, Gallagher AL, Grencis RK, Thornton DJ. A new role for mucins in immunity: insights from gastrointestinal nematode infection. *Int J Biochem Cell Biol*. 2013;45(2):364-374.
330. Herbert DR, Yang JQ, Hogan SP, et al. Intestinal epithelial cell secretion of RELM-β protects against gastrointestinal worm infection. *J Exp Med*. 2009;206(13):2947-2957.
331. Batugedara HM, Li J, Chen G, et al. Hematopoietic cell-derived RELMα regulates hookworm immunity through effects on macrophages. *J Leukoc Biol*. 2018;104(4):855-869.
332. Huang L, Beiting DP, Gebreselassie NG, et al. Eosinophils and IL-4 support nematode growth coincident with an innate response to tissue injury. *PLoS Pathog*. 2015;11(12):e1005347.
333. Huang L, Gebreselassie NG, Gagliardo LF, et al. Eosinophils mediate protective immunity against secondary nematode infection. *J Immunol*. 2015;194(1):283-290.
334. Specht S, Saeftel M, Arndt M, et al. Lack of eosinophil peroxidase or major basic protein impairs defense against murine filarial infection. *Infect Immun*. 2006;74(9):5236-5243.
335. Knott ML, Matthaei KI, Giacomin PR, Wang H, Foster PS, Dent LA. Impaired resistance in early secondary *Nippostrongylus brasiliensis* infections in mice with defective eosinophilopoeisis. *Int J Parasitol*. 2007;37(12):1367-1378.
336. Guarner F, Bourdet-Sicard R, Brandtzaeg P, et al. Mechanisms of disease: the hygiene hypothesis revisited. *Nat Clin Pract Gastroenterol Hepatol*. 2006;3(5):275-284.
337. Scholmerich J, Fellermann K, Seibold FW, et al. A randomised, double-blind, placebo-controlled trial of *Trichuris suis* ova in active Crohn's disease. *J Crohns Colitis*. 2017;11(4):390-399.
338. Smits HH, Everts B, Hartgers FC, Yazdanbakhsh M. Chronic helminth infections protect against allergic diseases by active regulatory processes. *Curr Allergy Asthma Rep*. 2010;10(1):3-12.

339. Bach JF. The hygiene hypothesis in autoimmunity: the role of pathogens and commensals. *Nat Rev Immunol.* 2018;18(2):105-120.
340. Smallwood TB, Giacomin PR, Loukas A, Mulvenna JP, Clark RJ, Miles JJ. Helminth immunomodulation in autoimmune disease. *Front Immunol.* 2017;8:453.
341. Loke P, Lim YA. Helminths and the microbiota: parts of the hygiene hypothesis. *Parasite Immunol.* 2015;37(6):314-323.
342. Hayes KS, Bancroft AJ, Goldrick M, Portsmouth C, Roberts IS, Grencis RK. Exploitation of the intestinal microflora by the parasitic nematode Trichuris muris. *Science.* 2010;328(5984):1391-1394.
343. Wammes LJ, Mpairwe H, Elliott AM, Yazdanbakhsh M. Helminth therapy or elimination: epidemiological, immunological, and clinical considerations. *Lancet Infect Dis.* 2014;14(11):1150-1162.
344. Hotez PJ, Bottazzi ME, Bethony J, Diemert DD. Advancing the development of a human schistosomiasis vaccine. *Trends Parasitol.* 2019;35(2):104-108.
345. Diemert DJ, Pinto AG, Freire J, et al. Generalized urticaria induced by the Na-ASP-2 hookworm vaccine: implications for the development of vaccines against helminths. *J Allergy Clin Immunol.* 2012;130(1):169-176 e6.
346. Hotez PJ, Diemert D, Bacon KM, et al. The human hookworm vaccine. *Vaccine.* 2013;31(suppl 2):B227-B232.
347. Diemert D, Campbell D, Brelsford J, et al. Controlled human hookworm infection: accelerating human hookworm vaccine development. *Open Forum Infect Dis.* 2018;5(5):ofy083.
348. Drurey C, Coakley G, Maizels RM. Extracellular vesicles: new targets for vaccines against helminth parasites. *Int J Parasitol.* 2020;50(9):623-633.

CHAPTER 43

Immune Responses to Bacteria

S. Celeste Morley, MD, PhD • Moon H. Nahm, MD

INTRODUCTION

Humans, along with all other vertebrates, have co-evolved with microbes (bacteria, viruses, and fungi) over thousands of years in a complex dance of immune modulation and microbial adaptation. Thousands of microbial species comfortably co-exist with the human host, comprising the microbiome. Specific microbiome components drive development of immunocompetence, and effective host immunity permits development of niches for microbial symbionts. When this truce is broken, however, bacterial invasion and pathogenicity can drive acute inflammation, sometimes with sufficient severity to kill the human host. Pathogenic bacterial infection remains one of the greatest threats to human health worldwide. This chapter will outline and illustrate key concepts in considering the various stages of bacterial infections, and how a multilayered, comprehensive immune response is generated to prevent and/or clear infections. While a complete and highly detailed review of each infection is beyond the scope of the chapter, pathogens have been selected to highlight specific essential aspects of anti-bacterial immunity. The goal of this chapter is to provide a conceptual framework for analyzing interactions between hosts and pathogens, with a focus on microbes important to human health and society.

REVIEW OF MICROBIOLOGICAL CONCEPTS

We will first review fundamental concepts of microbiology to facilitate a discussion of how different bacterial pathogens elicit and evade complex, flexible, and multilayered host immune responses. Understanding the microbial components that distinguish prokaryotic structures from host cell structures is essential to understanding how the host immune response distinguishes invading microbes from surrounding host tissue. Prokaryotic bacteria have simple cell structures (Fig. 43.1). They are enveloped by a cell membrane and cell wall, with no internal organelles defined by enclosed lipid bilayers. There is usually one chromosome (often circular) containing the majority of essential genetic material, with plasmids containing some swappable elements. Ribosomes contain 30S and 50S subunits. There are no endoplasmic reticulum, no Golgi, no defined nucleus, and no mitochondria. Despite the relative simplicity of organization, bacteria are metabolically complex and can present a variety of challenges to the host immune system.

Bacterial Classification Systems

Bacteria are classified in multiple ways. First, they can be classified by their shape. Round organisms are termed "cocci," while rod-shaped organisms are termed "bacilli" or "rods." Some of the more common pathogens observed in clinical medicine include both cocci (Fig. 43.2; eg, *Streptococcus pyogenes*, *Staphylococcus aureus*, and *Streptococcus pneumoniae*) and rods (eg, *Escherichia coli, Enterobacter cloacae, Pseudomonas aeruginosa*). These shapes are determined by the cell wall synthesis mechanisms. Another, more central classification system for bacteria was developed in 1884 by Hans Christian Gram. The Gram method relies upon the structural difference in the peptidoglycan layer of the bacterial cells walls. The peptidoglycan layer is formed by cross-linking the peptide stem of the glycan with the interpeptide bridge.[1] The cell wall of "Gram-positive" organisms contain a thick layer of peptidoglycan that surrounds the cell membrane and that retains the purple dye crystal violet after exposure to a decolorizing agent such as ethanol (Figs. 43.1A and 43.2). These organisms thus appear purple under a microscope. "Gram-negative" organisms have a much thinner layer of peptidoglycan sandwiched between two lipid bilayers (the outer membrane and an inner membrane) and therefore do not retain the crystal violet during decolorization (Fig. 43.1B). The area between the outer cell membrane and the plasma (inner) membrane is termed the "periplasmic space" in Gram-negative organisms. After decolorization, organisms are counterstained with a red dye (either safranin or fuchsine), and Gram-negative bacteria thus appear pink under a microscope ("Gram-positive" organisms also retain safranin, but the pink color cannot be appreciated because of the intense purple staining by the crystal violet.). Conserved elements of peptidoglycan are used by host defense receptors to distinguish bacteria from host cells. Muramyl dipeptide (MDP) is an essential structure found in almost all Gram-positive and Gram-negative bacteria and, as discussed later, is recognized by the pattern recognition receptor (PRR) nucleotide-binding oligomerization domain–containing domain 2 (NOD2).[2] In contrast, Gram-negative bacteria (and a few Gram-positive bacteria, such as *Listeria* spp.) contain meso-diaminopimelic acid (meso-DAP), which is recognized by the PRR NOD1.[2] The structural differences in the cell walls of Gram-positive and Gram-negative organisms thus determine much of their biology, including elicitation of immune responses.

While the Gram stain of many bacteria provides for rapid identification in a microbiological laboratory, some pathogens defy this system. "Gram-variable" organisms have an intermediate degree of retention of the crystal violet and, therefore, may appear as a mixture of pink and purple, depending on multiple factors (the age of the culture, the time for which the organisms were exposed to the decolorizing agent, etc.).

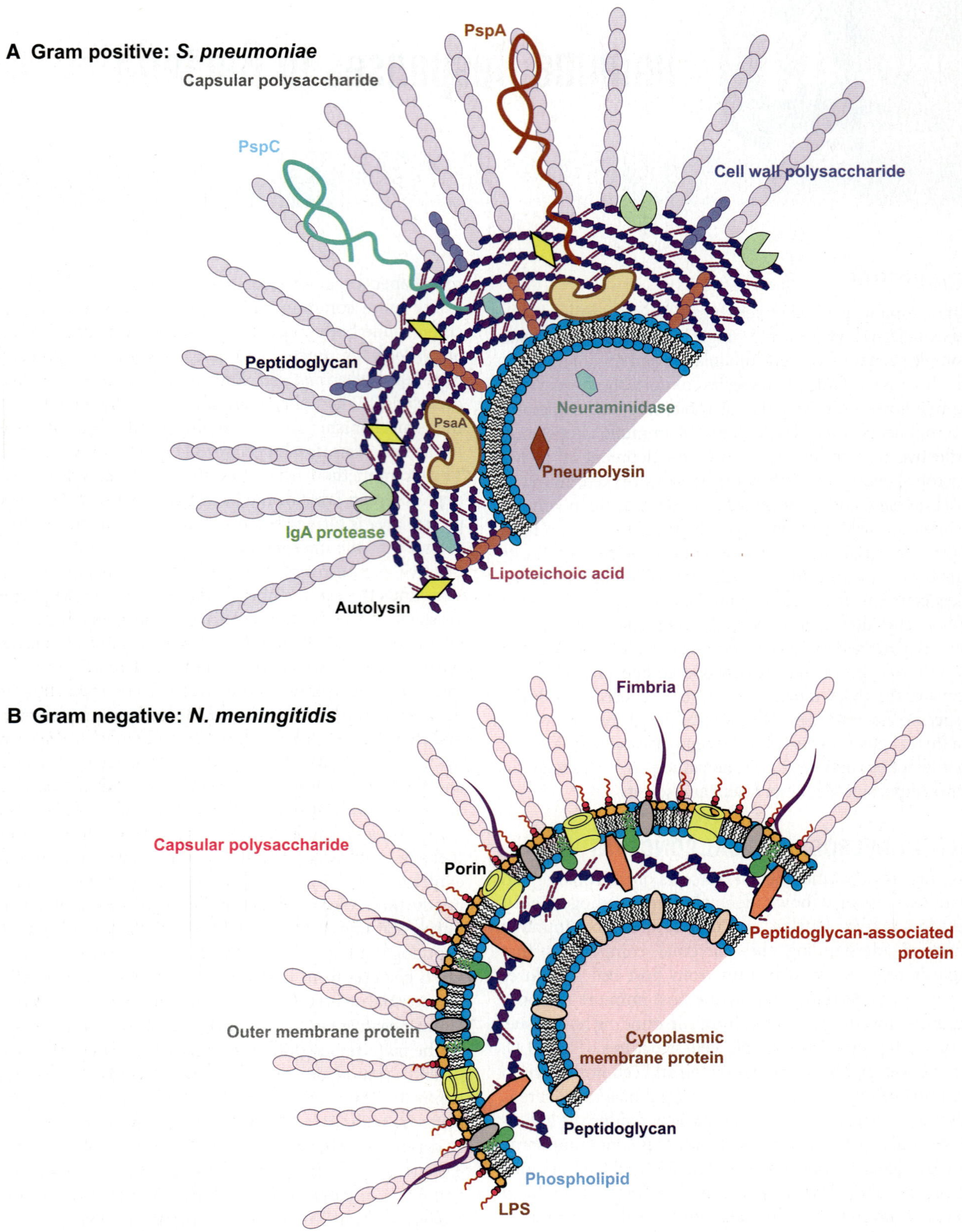

FIG. 43.1. **Structural differences in the cell walls of Gram-positive and Gram-negative organisms.** Retention of crystal violet, enabled by the thick peptidoglycan layer of the cell wall, differentiates Gram-positive bacteria from Gram-negative bacteria. Both Gram-positive and Gram-negative organisms can produce polysaccharide capsules that inhibit phagocytosis, and both contain structural elements (pathogen-associated microbial patterns) recognized by host microbial sensors. **A:** *Streptococcus pneumoniae* is a Gram-positive organism that causes pneumonia, sepsis, and meningitis. Virulence factors include the polysaccharide capsule (over 100 serotypes), lipoteichoic acid, IgA protease, pneumolysin, neuraminidase, autolysin, and adhesion proteins CbpA (PspC) and PspA. **B:** *Neisseria meningitidis* is a Gram-negative microbe that colonizes young teens and adults and can cause meningitis. It also produces a polysaccharide capsule. Like all Gram-negative organisms, the outer cell membrane contains lipopolysaccharides (LPSs). Fimbriae mediate attachment to host cells. Porins are outer membrane proteins that activate TLR2 and act as immune adjuvants. Corresponding labels of structural elements are similar colors.

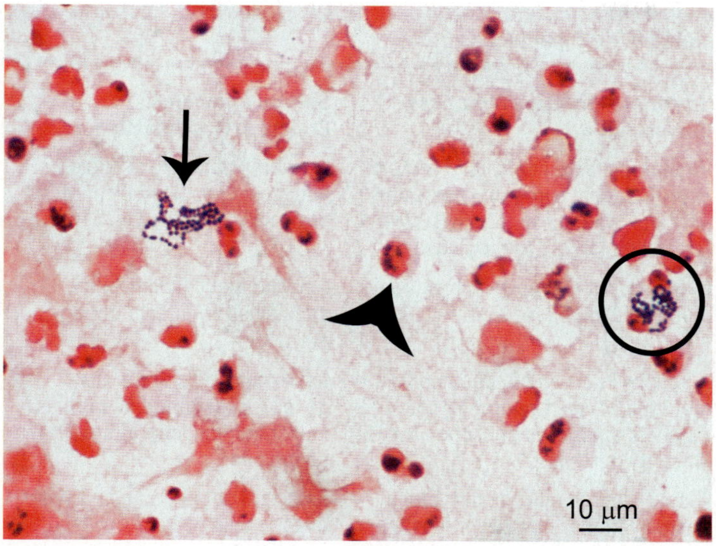

FIG. 43.2. Gram stain of clinical specimen shows abundant neutrophils and Gram-positive cocci in chains. Purulent material was drained from an abscess by a surgical team. The sample was Gram stained and imaged in the clinical microbiology laboratory. The image shows abundant neutrophils (*arrowhead*) and Gram-positive cocci in chains (*thin black arrow*). Bacteria that have been phagocytosed by a neutrophil can be seen (*circle*). Microscopy is insufficient to identify the microorganism; the culture generated from this specimen grew *Streptococcus pyogenes* (group A streptococcus). The patient recovered completely after appropriate therapy. (Image courtesy St. Louis Children's Hospital Clinical Microbiology Laboratory.)

Neisseria meningitidis are Gram-negative bacteria, but may appear Gram variable in practice. *Nocardia* species are only weakly Gram positive and therefore can be considered Gram variable. Some organisms do not stain with either crystal violet or the counterstain and are therefore classed as "Gram indeterminate" and require other staining protocols. For instance, *Mycobacterium tuberculosis* (Mtb) and other atypical mycobacteria species require acid-fast stain and are called "acid-fast" bacilli. The inability to Gram stain Mtb is due to its thick, waxy coating incorporating long-chain mycolic acids. The lipid-rich cell envelope is a major virulence factor that challenges the host immune system and effective antimicrobial therapy.

Finally, bacteria can be classified by their dependence upon oxygen for growth. "Obligate aerobes," such as *Nocardia asteroides*, must have direct access to oxygen at all times. "Obligate anaerobes," such as *Actinomyces* and *Clostridium* species, cannot handle reactive oxygen species (ROS) and die upon exposure to oxygen. Organisms that can survive under aerobic or anaerobic conditions are termed "facultative anaerobes." *S. aureus*, for example, produces high levels of catalase and easily resists the damaging effects of ROS encountered when grown in atmospheric oxygen. However, *S. aureus* can also grow in oxygen-poor environments, such as in the middle of a necrotic abscess. Once again, the host immune response must employ different strategies to effectively clear aerobic and anaerobic bacteria. For instance, the ability of neutrophils to generate cytoplasmic ROS from the NADPH oxidative complex is essential to clearing catalase-positive organisms, such as *S. aureus*.

Bacterial species can be described or characterized in terms of their relationships with human hosts: commensals, symbionts, pathogens, and pathobionts. In a state of health, the bacteria residing on host barriers, without eliciting noticeable immune responses, are termed "commensals." Commensals may simply coexist with the host, neither deriving a benefit from nor granting any benefits to the host. Many disease-causing bacteria can live on barrier surfaces as "commensals." Usually, either a change in the bacteria (eg, upregulation of a virulence factor) or a break in host immunity converts the "commensal" to a "pathogen." (A "commensal" becomes a "pathogen" when it invades and damages the host.) A "symbiont" implies that the microbe both derives and provides a benefit from its coexistence with the host. For instance, some gastrointestinal (GI) microbes may secrete enzymes that aid the host in digestion while also securing nutrients from the host GI tract. A "pathobiont" is a term applied to a bacterial species that can reside in harmony with its host under some circumstances, but causes disease under others. An "opportunistic pathogen" can be part of the host microbiome or can be acquired from an external source and does not cause disease in hosts with intact immune systems, but is virulent in immunodeficient hosts. Thus, the condition of the host immune system may be as important in determining the pathogenicity of a microbe as the intrinsic nature of the bacteria themselves.

Virulence Factors

Virulence factors are loosely defined as anything produced by a bacterium—for example, polysaccharide, lipid, or protein—that promotes survival of the organism within the host and causes tissue damage or disease. A virulence factor may enable escape from an immune response or may kill attacking immune cells. Virulence factors may also permit growth in a nutrient-poor environment, such as iron-scavenging systems. Virulence factors may also promote invasion across a barrier and enable an organism to gain access to a space that is usually sterile (eg, into a joint capsule or into the pleural space). Examples of virulence factors are listed in Table 43.1. A well-known example of a virulence factor is the thick polysaccharide capsule produced by bacteria. Capsules coat organisms, hide many immunologically recognizable cell wall structures, and inhibit phagocytosis. Another example is enzymes, such as catalase. Catalase enables bacteria to survive in atmospheric oxygen, but it can also neutralize host defenses that rely upon generation of ROS. Other examples include "secretion systems" that are produced by multiple Gram-negative species. These "secretion systems"

TABLE 43.1 Examples of Bacteria Commonly Associated With Diseases

Species	Disease	Important Virulence Structures/ Molecules	Special Adaptations Critical to Host Infection
Streptococcus pyogenes (group A streptococcus)	Acute pharyngitis, scarlet fever, necrotizing fasciitis, streptococcal toxic shock syndrome, rheumatic fever, glomerulonephritis	Hyaluronic acid capsule, M protein, streptococcal pyrogenic exotoxins, streptolysin O, streptolysin S, NAD-glycohydrolase, C5a peptidase	Molecular mimicry of human antigens, high diversity of M proteins
Streptococcus agalactiae (group B streptococcus)	Bacteremia, neonatal pneumonia, meningitis	Capsule, beta hemolysin, hyaluronidase, C5a peptidase	Asymptomatic colonization, acquisition by infants during parturition
Streptococcus pneumoniae	Pneumonia, otitis media, meningitis, sinusitis	Capsule, PspA and C, pneumolysin, neuraminidase, hyaluronidase, teichoic acids with phosphorylcholine, IgA1 protease	Asymptomatic colonization, genetic transformation permitting continual generation of new genotypes
Staphylococcus aureus	Impetigo, folliculitis, boils, cellulitis, wound infections, toxic shock, osteomyelitis, endocarditis, bacteremia, pneumonia, food poisoning	Tissue-degrading enzymes, alpha toxin and other membrane-damaging toxins, epidermolytic toxins, enterotoxins, capsule, protein A, TSST-1, pigment	Resistant to dehydration, asymptomatic colonization
Haemophilus influenzae	Otitis media bronchitis, pneumonia, sepsis, meningitis (encapsulated strains)	Lipopolysaccharide with phosphorylcholine, fimbria, high-molecular-weight adhesions, IgA1 protease	Phase and antigenic variation, molecular mimicry of human antigens
Neisseria gonorrhoeae	Urethritis, cervicitis, salpingitis, endometritis, prostatitis, arthritis, proctitis, pharyngitis	Lipopolysaccharide, fimbria, peptidoglycan, Opa protein adhesin, IgA1 protease	Phase and antigenic variation, molecular mimicry of human antigens
Neisseria meningitidis	Bacteremia, meningitis, septic arthritis	Capsule, lipopolysaccharide, fimbria, IgA1 protease	Phase and antigenic variation, molecular mimicry of human antigens, asymptomatic carriage common
Helicobacter pylori	Peptic ulcer disease	Urease, flagella, CagA	Ability to survive at low pH provides a niche lacking bacterial competition or efficient immune surveillance
Escherichia coli	Urinary tract infection, sepsis, traveler's diarrhea, dysentery, neonatal meningitis, hemolytic-uremic syndrome	Capsular polysaccharide, lipopolysaccharide, fimbria, toxins, siderophores	Antigenic heterogeneity of capsule and lipopolysaccharide
Pseudomonas aeruginosa	Infections in compromised hosts, pneumonia, sepsis	Lipopolysaccharide, proteases, lipases, lecithinases, exotoxin A, elastase, flagella	Relatively large genomic size (~6 megabases) allows considerable adaptability to changes in environmental conditions and biofilms
Clostridium perfringens	Gas gangrene, anaerobic cellulitis, endometritis, food poisoning	More than 10 exotoxins	Opportunistic infection of wounds

are multimolecular transmembrane proteins that form either adhesive or motility structures, such as pili or flagella. Type III secretion systems (T3SS) can function as "injectisomes," which form pores in the target host cell membranes and "inject" bacterial effector proteins. Bacterial effector proteins may then subvert host immune defense, induce bacterial internalization, or enhance bacterial intracellular survival.[3,4] The T3SS is a major virulence factor in promoting infection by *P. aeruginosa* and *Shigella* spp.

A subset of virulence factors are toxins. Toxins are any soluble small molecule (lipid or protein) released from pathogenic bacteria that act directly to disrupt or poison host cells. Bacterial toxins include endotoxins, which are intrinsic components of the bacteria that are released upon bacteria lysis, and exotoxins, which are released from live bacteria. Both endotoxins and exotoxins can act directly upon the immune system to cause disease. All Gram-negative organisms contain lipopolysaccharides (LPSs) in their outer cell membranes (Fig. 43.1). LPS, which is an endotoxin, is one of the most potent activators of Toll-like receptors (TLRs) and inflammasomes. Direct administration of isolated LPS to animals is sufficient to induce massive systemic inflammation and death due to shock.

Examples of exotoxins include cholera toxin (CTx), produced by *Vibrio cholerae*, and toxic shock syndrome toxin 1 (TSST-1), produced by *S. aureus*.[5] CTx induces host cells to secrete fluid and causes diarrhea. TSST-1 binds to the side of the $V_{\beta}2$ T cell receptor (TCR) chain and cross-links the TCR to major histocompatibility complex (MHC) II in the absence of cognate antigen.[6] By cross-linking TCR and MHC II, up to 5% to 30% of all T cells are simultaneously activated, leading to systemic release of large amounts of inflammatory cytokines such as IL-1β, IL-2, and TNFα. In a normal host T cell response, less than 0.01% of T cells are activated.[6] Because T cells are activated in the absence of cognate peptide through a TCR-MHC interaction, TSST-1 is also termed a "superantigen." Toxic shock syndrome produces a classic constellation of symptoms, including high fevers, generalized erythroderma, conjunctivitis, and dangerously low blood pressures and can result in multisystem organ failure. Toxic shock can be rapidly fatal (within 24–48 hours of symptom onset) if not recognized and reversed. Toxic shock syndrome can also be induced by *S. pyogenes* (group A streptococci [GAS]) expressing streptococcal pyrogenic exotoxin A (SpeA) or SpeC.[7] Like TSST-1, SpeA and SpeC cross-link MHC class II and TCR independently of cognate peptide and nonspecifically activate up to 30% of host T cells, generating a massive cytokine storm that can be rapidly lethal. Additional toxins and virulence factors are listed in Table 43.1, presented in Section 5 of this chapter, and also discussed in the context of individual infections.

REVIEW OF HOST DEFENSE MECHANISMS

Having reviewed key microbiological concepts for understanding microbial diversity, we will now briefly discuss essential aspects of effective antibacterial host immunity. Detailed discussion of each concept can be found in other chapters.

Tissue Barriers

The first lines of defense against bacteria are tissue barriers. The most obvious host barrier is the skin, although every mucosal surface (eg, the nasal mucosa, the oropharynx mucosa, the vaginal mucosa, the epithelia of the respiratory tract, the colonic mucosa) is considered a barrier between the host and the external world. Internal tissue barriers can limit bacterial invasion and spread as well. For instance, the lungs are wrapped in dual pleural membranes. Under homeostatic conditions, the two pleural membranes are in close approximation and slide along each other during breathing, lubricating the easy expansion and contraction of the lungs within the rib cage. The pleural membranes also serve as a barrier to pathogenic spread outside of the lungs. Another example is blood-brain barrier, which isolates and protects the brain from the other parts of the body.

When an exterior barrier, such as the skin, is intact, otherwise pathogenic bacteria may reside on the "outside" of the host without eliciting a significant immune response. Interior barriers, such as the pleural membranes, guard normally sterile sites. Invasion across an internal barrier, such as migration of bacteria from the airways in the lung into the space between the pleural membranes, can enable bacteria to find a niche that permits disease progression.

Barriers restrict bacterial growth through a variety of mechanisms. First, tight cell junctions physically prevent the microorganisms from crossing into protected spaces. The stratum corneum of the skin acts as a physical barrier to prevent bacterial translocation. Abrasions, cuts, or other damage to the epithelium results in disruption of this key host defense and permits microbes access to host tissues. Additionally, some bacteria, such as *S. aureus*, produce virulence factors to cleave tight junctions and disrupt the barriers.[8] Epidermal keratinocytes act as sentinel cells, releasing cytokines and other inflammatory mediators upon engagement of PRRs that bind conserved structural elements of bacteria, such as lipoteichoic acid or LPS (Fig. 43.1).[9] These conserved structural elements are termed "pathogen-associated microbial patterns" or PAMPs (see Chapter 12: Pattern Recognition Receptors and the IL-1 Family). Keratinocytes can initiate a nonspecific innate immune response whenever a microbe crosses the epithelial barrier by expressing receptors that recognize these PAMPs.

Barrier defenses in the respiratory tract include generation of mucus and ciliary beating. Mucus is composed of a mucopolysaccharide gel heavily enriched with antimicrobial peptides (AMPs), lysozyme, lactoferrin and immunoglobulins.[10] The mucus layer is about 10 to 12 microns thick and composed of two layers; the inner layer is relatively thin, permitting movement of cilia, while the outer layer is thicker for more effective trapping of microbes.[10] Microbes become mechanically trapped in the mucus layer. The cilia, projecting from the apical surface of respiratory epithelia, continually beat unidirectionally and rapidly (10–15 Hz) to sweep the mucus and trapped microbes up and out of the lower airways into the oropharynx, after which it is expectorated or swallowed. Cilia are complex structures, formed from a ring of nine pairs of microtubules surrounding a central microtubule pair, and are about 6 to 7 μm long and 0.2 to 0.3 μm in diameter. The ciliary "power strokes" are ATP-dependent and powered by the molecular motor dynein. The cilia bend backwards during recovery strokes to prevent retrograde mucus flow. In addition to mucus clearance, cilia are sensing organelles, and contribute to innate immune defenses by inducing production of AMPs.[11] Proper synchronization and effective power stroke generation prevent airway infections. Defects in ciliary motility (as in primary cilia dyskinesia) result in recurrent bacterial pneumonias and sinus infections.[11] Several pathogenic organisms, such as *S. pneumoniae* and *P. aeruginosa*, target and disrupt ciliary beating through elaboration of virulence factors to enable colonization and disease promotion.[11] The importance of normal mucus layer can be shown in patients with cystic fibrosis (CF). CF is a severe lung disease caused by impaired mucus secretion, which disrupts barrier function of the respiratory tract and permits growth of lung pathogens.[12]

Immediate innate defenses generated by keratinocytes and other barrier cells include production of AMPs, which are small peptides that bind bacterial cell walls and kill the organisms. Because AMPs exert pleiotropic effects upon host immune responses in addition to their direct bactericidal activity, they have also been termed "host defense peptides."[13] Beyond their direct microbicidal activity, AMPs can serve as chemoattractants, activate cellular inflammation, and promote wound healing.[13] Examples of AMPs include cathelicidins and defensins.

The human cathelicidin LL-37 directly lyses pathogenic bacteria, such as GAS. Mice lacking the murine ortholog mCRAMP are more susceptible to GAS infection.[14] Upon cleavage in the extracellular milieu, the positively charged LL-37 directly associates with the negatively charged microbial membranes, and the alpha-helical structure of LL-37 enables pore formation within microbial cell walls. Pore formation induces microbial cell lysis.[15] Cholesterol in mammalian cell membranes protect host cells from LL-37–mediated damage. LL-37 also acts as a neutrophil chemoattractant, activates the NLRP3 inflammasome, and stimulates neutrophil netosis.[15]

Defensins are small (3–5 kDa) cationic peptides that contain three disulfide bridges. They are divided into alpha-defensins and beta-defensins based on their structure.[16] Neutrophils produce alpha-defensins (also called human neutrophil peptides).[17] Human beta-defensin 1 (HBD1) is produced constitutively at all mucosal surfaces by epithelial cells, including in the intestine, urinary tract, and airways.[17] HBD1 is also produced by hematopoietic cells such as dendritic cells (DCs) and monocytes. Human beta-defensins 2 and 3 are upregulated during inflammation by mediators including TNFα, IL-1β, and IFNγ.[18] Defensins bind the membranes of bacteria, fungi, and some enveloped viruses, killing through a variety of mechanisms including induction of lysis and action on intracellular targets.[16,19]

The critical role of tissue barriers in the host immune response to bacterial infections is revealed in the increased susceptibility to infections observed when these barriers are disrupted or penetrated. Atopic dermatitis, or eczema, is a skin disease characterized by disrupted integrity of the skin, dysbiosis (pathological changes in the microbiome), and inappropriate T cell immune responses.[20] Patients suffering from CF or third-degree burns, with complete loss of skin integrity, are at extraordinarily high risk of systemic infection with pathogens such as *P. aeruginosa* and *S. aureus*.[21] *N. meningitidis* bloodstream infections commonly result in meningitis because the bacteria can bind to specific proteins expressed on brain vascular endothelial cells, enabling transmigration across the blood-brain barrier.[22] Thus, intact barriers serve as the first line of host immune defense against bacterial pathogens.

Innate Immunity: PRRs

PRRs recognize common, conserved molecular elements that are present at significant levels during pathogen encounters. These common molecular elements may be derived from microbes and are then termed PAMPs. As previously discussed, PAMPs include bacterial components, such as LPS derived from Gram-negative cell membranes, or fragments of peptidoglycan derived from bacterial cell walls.[23] Molecular elements may also be host derived and are then termed "damage-associated molecular patterns" (DAMPs). Examples of DAMPs include ATP, cellular DNA, and uric acid.[23] PRRs are divided into two categories: membrane bound receptors (like TLRs) and cytosolic receptors (like NOD-like receptors [NLRs]). We illustrate their functions using TLR and NLR as examples (see Chapter 12: Pattern Recognition Receptors and the IL-1 Family for details).

TLRs. TLRs are dimeric receptors that structurally fall into the IL-1R/TLR superfamily. These type 1 transmembrane proteins contain an extracellular portion comprising multiple leucine-rich repeats, a single transmembrane domain, and an intracellular signaling domain. The leucine-rich repeats interact with PAMPs or DAMPs. The intracellular signaling domain is conserved between TLRs and IL-1Rs, creating the superfamily. TLRs must dimerize to commence signaling (Fig. 43.3A). While most TLRs function as homodimers, some may also function as heterodimers, broadening the range of PAMPs and DAMPs that are recognized. There are at least 10 human TLRs now identified. Some TLRs, such as TLR1 and TLR2, are positioned at the extracellular host cell membrane and engage extracellular PAMPs and DAMPs, such as LPS or flagellin. Other TLRs, such as TLR3 and TLR8, are localized to endosomal cell membranes and recognize intracellular DAMPs and PAMPs, such as double-stranded DNA. Pathogens normally thought of as "extracellular," such as *S. pneumoniae*, may also activate these endosomal TLRs through release of bacterial nucleic acids.[27]

After dimerization, the intracellular tails of TLRs engage with adaptor proteins, such as MyD88 or TRIF. Engagement with adaptor proteins triggers downstream signaling through the mitogen-activated protein kinase (MAPK), I-kappa kinase, and type 1 IFN response pathways. Activation of these pathways results in the activation of the transcription factors NF-κB, AP-1, and IRF7. These transcription factors then drive upregulation of proinflammatory cytokines, such as TNFα, IL-6, IFNγ, and type 1 IFNs. Production of these proinflammatory cytokines constitute induction of the host immune response.

NLRs. NLRs include the cytosolic receptors NOD1 and NOD2. Both NOD1 and NOD2 respond to the conserved peptidoglycan structure of bacterial cells walls. NOD1 recognizes the structural element of DAP peptidoglycan of Gram-negative bacteria, while NOD2 recognizes MDP, a component of peptidoglycan found in both Gram-positive and Gram-negative bacteria (peptidoglycan thus serves as a "PAMP"). NOD1 is ubiquitously expressed, while NOD2 is most highly expressed by monocytes and macrophages.[28] Because NOD1 and NOD2 are intracellular receptors, recognition of peptidoglycan of extracellular bacteria happens after its transport into intracellular compartments. This may occur through phagocytosis and bacterial escape from lysosomes or from active transport of bacterial cell wall components across a cell membrane.[9] As mentioned previously, NOD1 recognizes DAP, a component of all Gram-negative and a few Gram-positive bacteria. NOD2 recognizes MDP, a structure found in the peptidoglycan of Gram-positive and Gram-negative bacteria. NOD1 and NOD2 are maintained in an inactive conformation by binding to HSP70/90. Similar to the TLRs, NOD1 and NOD2 oligomerize after ligand engagement. Signaling through the NLRs occurs via at least three distinct pathways (Fig. 43.3A) to activate the transcription factors NF-κB, AP-1, and IRF7. Activation of these transcription factors elicits proinflammatory cytokines (eg, IL-6, TNFα), chemokines, and AMPs.[29] As discussed later in the chapter, bacterial virulence factors may act upon these signaling pathways to impair the host immune response (Fig. 43.3A).

Inflammasomes, a subset of NLRs, are multimolecular complexes that activate caspase-1 and cleave the immature pro-IL-1β to the bioactive IL-1β (Fig. 43.3B). All inflammasomes contain an inflammasome receptor (or sensor) and caspase-1; some also require the adaptor ASC for activation. There are at least five known inflammasome sensors: NLRP3, NLRC4, AIM2, NLRP1, and pyrin. NLRP3 is the most widely expressed inflammasome.[30] NLRP3 activation requires two steps *in vitro*. In the first step, "priming" occurs when proinflammatory signaling (eg, LPS engagement of TLR4) induces NF-κB–mediated upregulation of the NLRP3 inflammasome components, including the NLRP3 receptor, the adaptor ASC, pro-caspase-1, and pro-IL-1β. The

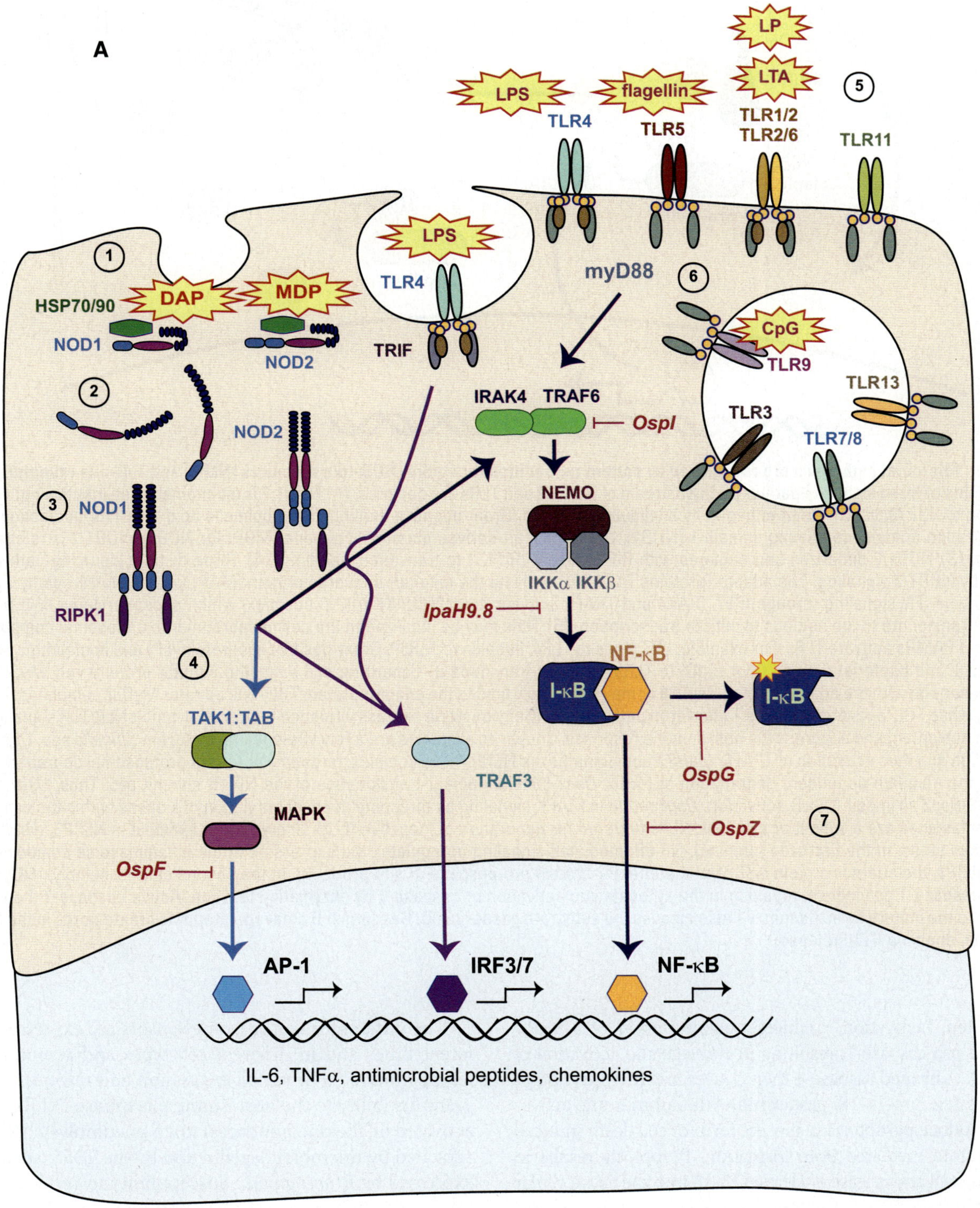

FIG. 44.3 Continued

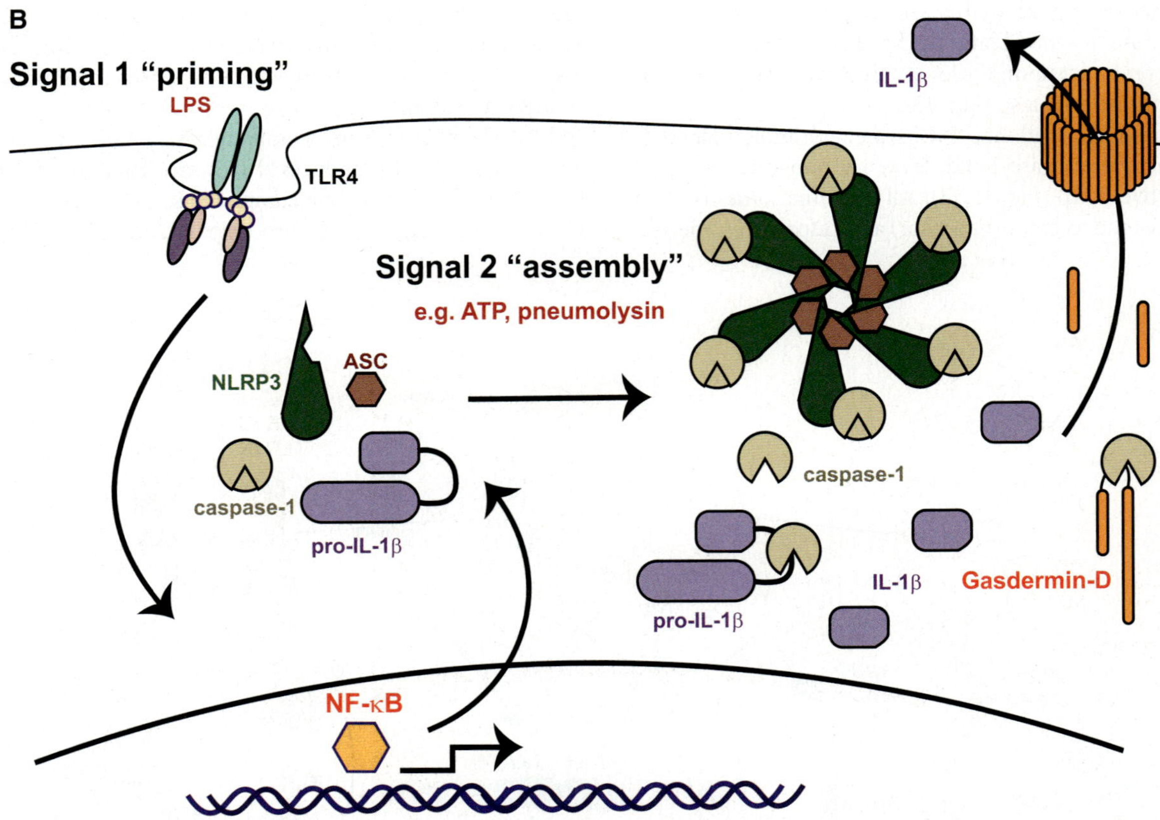

FIG. 43.3. Microbial pathogens are recognized by pattern recognition receptors (NOD-like receptors [NLRs] and Toll-like receptors [TLRs]). **A:** Highlights of three signaling pathways downstream of NOD1/2 and TLRs are depicted. **(1)** NOD1/2 is maintained in an inactive conformation by binding to HSP70/90. **(2)** When engaged by a stimulatory ligand (Gram-negative derived diaminopimelic acid [DAP] for nucleotide-binding oligomerization domain–containing domain 1 [NOD1], peptidoglycan-derived muramyl dipeptide [MDP] for NOD2), NOD1/2 is released from inhibition. **(3)** NOD1/2 dimerizes and engages with RIP kinase 1 (RIPK1) to transduce signaling. **(4)** Three distinct signaling pathways are induced by NOD1/2 signaling. The adaptor proteins TAK1:TAB activate the mitogen-activated protein (MAP) kinase (MAPK) pathway. TRAF3 activates type IFN signaling through IRF7. IRAK4 and TRAF6 activate the NEMO/IKKα/IKKβ complex, which phosphorylates Iκ-B, releasing NF-κB to translocate to the nucleus to initiate transcription. **(5)** TLRs may be localized to the cell membrane or to endosomal compartments. A variety of PAMPs activate TLRs. For example, LPS binds to TLR4; flagellin to TLR5[24]; Gram-positive lipoprotein (LP) and lipoteichoic acid (LTA) to TLR 1/2/6; and bacterial-derived DNA (CpG) to TLR9. **(6)** TLRs transduce by dimerizing and engaging adaptor proteins (eg, MyD88, TRIF). TLR4 may engage with the adaptor protein MyD88 or may alternately bind to the adaptor protein TRIF. The adaptor MyD88 stimulates the NF-κB pathway, while TRIF activates the type 1 interferon pathway. **(7)** Many bacteria express virulence factors that inhibit NLR/TLR signaling (such as NF-κB activation) and suppress the host immune response. Shown as examples are a few virulence factors from *Shigella* spp. *OspI* inhibits ubiquination and thus activation of TRAF6. *IpaH9.8* ubiquitinates the NEMO/IκKα complex, targeting NEMO for proteosomal degradation. *OspG* prevents the ubiquitin-dependent degradation of I-κBα. *OspZ* inhibits nuclear translocation of the NF-κB subunit p65. Thus, *Osp1*, *IpaH9.8*, *OspG*, and *OspZ* all impair NF-κB activation. *OspF* reduces MAPK signaling by threonine dephosphorylation of kinases within the cascade.[25,26] **B:** Inflammasomes are a subset of cytosolic NLRs required for generation of bioactive IL-1β, Shown as an example is NLRP3, which is activated in two steps. In the first step (priming), NF-κB–mediated signaling upregulates the expression of the inflammasome components (the sensor NLRP3, the adaptor protein ASC, and the effector protein pro-caspase-1) and pro-IL-1β. In the second step (assembly), NLRP3, ASC, and pro-caspase-1 polymerize, resulting in the autocatalytic activation of caspase-1 by enzymatic cleavage. Active caspase-1 then cleaves pro-IL-1β to the mature form. Caspase-1 also cleaves the substrate gasdermin D. Gasdermin D cleavage products create pores in the host cell membrane, enabling IL-1β release.

second step, "activation," induces polymerization of NLRP3, ASC, and pro-caspase-1, resulting in cleavage and activation of caspase-1. Activated caspase-1 then cleaves pro-IL-1β and other substrates (eg, pro-IL-18, gasdermin).[30] Inflammasome activation can induce pyroptosis, a specific form of cell death induced by gasdermin-mediated pore formation. Pyroptosis results in lysis of the cell and release of cleaved IL-1β into the extracellular milieu. Multiple bacterial pathogens induce pyroptosis of host immune cells, as discussed below.

The different inflammasome sensors are expressed at different times and in different cell types and respond to different stimuli.[31] The varied expression and response patterns grant flexibility to the host immune response. NLRP3 can be activated by the pneumococcal toxin pneumolysin. NLRC4 is activated by microbial flagellin and by the T3SS, such as that expressed by *P. aeruginosa*. Susceptibility to certain bacterial infections can be explained by impairment in inflammasome activation.[32]

Innate Immunity: Complement

Complement is an extracellular network of more than 60 proteins that provide immediate, innate responses to kill invading pathogens and stimulate a rapid immune response. Complement components are synthesized by the liver and by endothelial, epithelial, and some hematopoietic cells.[33] Proinflammatory mediators such as IL-6 can rapidly induce increased production of complement; complement is thus an "acute phase reactant."[34] Three pathways of complement activation are currently defined: the classical pathway, the alternative activation pathway, and the lectin activation pathway. All three pathways converge on the activation of C3, which leads to opsonized phagocytosis and the formation of the membrane attack complex (MAC). The MAC is formed by complement components C5-C9 and creates pores in the cell membrane of Gram-negative bacteria. Pore creation causes microbial lysis (see Chapter 13: The Complement System).[35,36]

Complement can enhance killing by serving as chemoattractants for innate immune cells and by promoting opsonophagocytosis. C3a and particularly C5a are potent chemoattractants for neutrophils. C3b deposition on the surface of microbes opsonizes the bacteria for phagocytosis by cells expressing the CR3 receptors (eg, neutrophils, alveolar macrophages). Complement deficiencies are associated with increased susceptibility to invasive meningococcal and pneumococcal infections (sepsis, septic shock, and meningitis).

C-reactive protein (CRP) is an acute phase reactant synthesized by hepatocytes and is also induced by IL-6, IL-1β, or TNFα.[37] CRP binds to the phosphorylcholine component of bacterial membranes and recruits and activates the classical complement pathway.[37] Because CRP is rapidly and frequently induced to very high levels in the blood during acute infections, clinicians often use CRP as a marker of disease. Notably, any process or tissue damage, such as trauma, burns, or surgery, will induce CRP; CRP is thus useful as a marker of inflammation, but not specific for bacterial infection. For additional information on the complement cascade, please see Chapter 13: The Complement System.

Innate Immunity: Effector Cells

Professional phagocytes such as neutrophils and macrophages are the major hematopoietically-derived cells that respond during acute bacterial infections (see Chapters 14: Neutrophils and 15: Macrophages and Phagocytosis). The tissue-resident macrophages are often the first immune cell to engage with invading microbes. Macrophages phagocytose the pathogen for lysosomal degradation. Phagocytosis of bacteria is enhanced by deposition of complement or by binding of specific antipathogen antibodies. Macrophages then engage and actively endocytose these opsonized pathogens through complement receptors (eg, CR3) or through FcγR. Lysosomal killing of bacteria occurs through a variety of mechanisms, some of which are still not fully elucidated.[38,39] One such mechanism relies upon phagolysosome acidification and enzymatic degradation. Acidification enhances the enzymatic activity of lysosomal hydrolases. Bacteria are degraded in these acidified intracellular compartments.[39] Activated macrophages can also produce nitric oxide (NO) and other ROS that exert oxidative stress upon microbes, killing them through disruption of membrane lipids, DNA, and metabolic enzymes.[39] In addition to actively killing invading organisms, macrophages express multiple PRRs and initiate an inflammatory response through the production of IL-6, TNFα, IL-1β and other mediators. Production of these mediators will recruit additional effector cells, such as neutrophils.

Macrophages have often been characterized as polarizing toward an M1 ("proinflammatory") or toward an M2 ("prohealing") state.[40] This binary system oversimplifies the plasticity and heterogeneity of macrophages *in vivo* since in-depth studies have revealed a much broader spectrum of macrophage activation states.[40,41] However, because the terminology and concept of M1 and M2 macrophages are widely used, we will describe it here. In general, the M1 state, also called "classical activation," is induced by LPS and IFNγ and characterized by production of proinflammatory mediators, such as IL-1β, IL-6, IL-12, and inducible nitric oxide synthase (iNOS).[41] The prohealing, M2 state ("alternatively activated") is induced by IL-4 and is associated with production of mediators required for wound healing, such as transforming growth factor beta (TGF-β) and arginase.[40-42] Generally, acute bacterial infections trigger M1 polarization.[43] As pathogens are eliminated and the immune system shifts from "eradicating pathogens" to "tissue healing," macrophages will shift from an M1 state toward the M2 state. Prolonged or inappropriate M1 activation can prevent wound healing.[43]

Neutrophil infiltration is the hallmark of an ongoing, acute bacterial infection. In overwhelming systemic infections, neutrophils and neutrophil precursors will demarginate from the bone marrow, flooding the bloodstream with an influx of these antibacterial effectors. Neutrophils possess an arsenal of antibacterial effectors. Like macrophages, neutrophils can engulf and kill bacteria through phagocytosis and lysosomal-mediated degradation. Neutrophils express the NADPH oxidase and generate cellular ROS that are required for killing of multiple microbes, including *S. aureus*. Neutrophils also contain three varieties of preformed granules that contain antibacterial proteins: primary azurophilic, secondary specific, and tertiary gelatinase granules.[39] Azurophilic and specific granules contain elastase, cathepsins, lactoferrin, defensins, and myeloperoxidase. Gelatinase granules carry matrix metalloproteinase 9 (MMP9), collagenase, and gelatinase, all of which can degrade extracellular matrix (ECM). Killing of *S. pneumoniae* is primarily dependent on enzymes, but not ROS. These granules fuse with phagosomes to deliver antimicrobial effector peptides, killing the engulfed bacteria.[39,44,45]

Neutrophils can extrude their nucleic acid and cytoplasmic contents in a process termed "netosis." Netosis results in the creation of "neutrophil extracellular traps" (NETs). NETs consist of sticky aggregates of decondensed chromatin and antibacterial enzymes.[46] Because of their delicate structure, they are difficult to preserve and stain for imaging, which had previously delayed recognition of this antibacterial mechanism.[47] NETs trap and kill microbes and prevent dissemination. However, NET production is nonspecific, and excessive or inappropriate NET formation can lead to damage of surrounding host tissues and contribute to pathologic autoinflammatory processes.[48]

Together, macrophages and neutrophils comprise the major innate immune response to initial bacterial infections. For additional information on these essential antibacterial effector cells, please see Chapters 14: Neutrophils and Chapter 15: Macrophages and Phagocytosis.

Adaptive Immunity

Relative to innate immunity, the adaptive immune system plays minor roles in the immediate response to acute bacterial infection. During the initial infection, T cells may help by producing IFNγ and promoting M1 macrophage polarization. Early B cell production of IgM antibodies may assist in opsonophagocytosis prior to the development of high-affinity IgG responses. In cases of neonates, maternal antibodies can provide protection for many months. In contrast to the initial infection, the adaptive immune response is critically important in protecting the host from recurrent bacterial infections. Vaccination, which is designed to induce adaptive immunity without true infections, clearly illustrates the importance of adaptive immunity. Currently, children in the United States are vaccinated against the bacterial pathogens *S. pneumoniae, Haemophilus influenzae* type B, meningococcus, and pertussis and against the toxins produced by diphtheria and tetanus. Vaccination strategies usually rely on generation of microbe-specific antibodies that facilitate phagocytosis. The specific contributions of adaptive immunity in antibacterial responses will be detailed by pathogen. More information on generation of humoral immunity can be found in Chapter 26: B Cell Responses.

INTERACTIONS BETWEEN PATHOGENIC BACTERIA AND HOST

A Peaceful Coexistence With the Microbiome

The advent of inexpensive high-throughput sequencing platforms in the early 2000s drove a revolution in microbiology.[49] Prior to large-scale sequencing, characterization of a population of bacteria required months of careful subculturing and pain-staking biochemical identification of individual bacterial species. Culture-based analyses of mixed microbial populations are also severely limited by the inability to culture a significant portion of microorganisms *in vitro*. However, development of high throughput sequencing–based techniques overcame these limitations, and rapid analysis of whole populations is now possible. By using degenerate primers, PCR-based amplification of bacterial DNA, and shotgun sequencing through the 16S ribosomal subunit, the "microbiomes" of host organisms are being characterized.[49] A "microbiome" is defined as the whole population of microorganisms (often used to refer to bacteria, although viruses, parasites, and fungi may be included) that are residing at a given site at a given time. For instance, swabbing the nasopharynx, extracting all the DNA, and amplifying and sequencing all bacterial sequences found in the sample reveals the "nasopharyngeal" microbiome. A microbiome is described by the diversity of the bacteria (eg, the different bacterial species) and by the relative quantities of each. A host's microbiome encompasses all the bacteria residing in or on the host at a given time—all bacteria on the skin and at all mucosa barriers (eg, from the GI tract).

The microbiome affects interactions between bacteria and the immune response in multiple ways. First, bacterial species within the microbiome compete with each other, and intermicrobial competition can drive the expression of virulence factors or suppress overgrowth of pathogenic bacteria. Secondly, interactions between specific species and mucosal sites can affect the development of the host immune system (eg, Th17 cells). Finally, interactions between the microbiome and host immunity can affect immune responses to vaccines[50] and drive some disease processes (eg, atopic dermatitis and inflammatory bowel disease [IBD]).

The ability of the microbiome to drive development of host immune populations is well illustrated by the role of segmented filamentous bacteria (SFB) in promoting Th17 cell development in murine GI mucosa.[51] Investigators noted that the intestines of C57Bl/6 (B6) mice from one commercial supplier (Taconic Laboratories) contained significant populations of Th17 cells, while the intestines of B6 mice purchased from another (Jackson Laboratory [JAX]) did not. Transfer of cecal contents from Taconic B6 mice to JAX B6 mice induced development of Th17 cells in the lamina propria of the JAX B6 recipients. Likewise, transfer of cecal contents from Taconic B6 mice into germ-free B6 mice purchased from either Taconic or Jackson also induced Th17 cell development. The ability of cecal contents to induce Th17 development implied a microbiota-driven process. When the gut microbiota of the two B6 mice were compared, *Lactobacillus murinus* and the SFB were the two taxa most overrepresented in Taconic B6 gut microbiota. Oral gavage of JAX B6 mice with SFB (but not with *L. murinus*) induced Th17 population increases similar to Taconic B6 cecal contents.[51] SFB adhere tightly to the epithelial cell surface of the disteal ileum in mice and induce production of serum amyloid A (SAA) by intestinal epithelial cells. SAA in the presence of DCs induces Th17 differentiation. Thus, by binding to intestinal epithelial cells, SFB in gut microbiota induced the acute phase reactant SAA. SAA acted upon gut DCs to induce the cytokine milieu essential to promoting Th17 cells.[51] Significantly, the presence of SFB-driven Th17 cells subsequently protected animals from infection with the gut pathogen *Citrobacter rodentium*. Thus, commensal microbiota can drive immune cell development and determine the outcome of a host-pathogen interaction.

Another excellent illustration of how commensal microbiota can alter host-pathogen interactions is the effect of gut microbiota producing the compound desaminotyrosine (DAT).[52] The gut commensal organism *Clostridium orbiscindens* was found to protect animals from pulmonary influenza challenge. Antibiotic treatment that eliminated *C. orbiscindens* from the gut rendered mice more susceptible to lung influenza infection, and reconstitution of *C. orbiscindens* by oral gavage into previously antibiotic-treated animals restored protection. The mechanism was found to be the metabolite, DAT, generated by *C. orbiscindens*. DAT in turn induced type I IFNs in pulmonary innate immune cells. Upregulation of type I IFN in mice then rendered the mice more resistant to viral infection. Thus, a single commensal bacterium can determine the efficacy of the host immune response by altering the "set point" of basal IFN signaling through the metabolites produced.

Commensal organisms may also drive disease processes by affecting host immunity. An area of intense investigation seeks to define the role of gut microbiota on T cells involved in IBD. IBD, which includes Crohn disease and ulcerative colitis, is considered primarily an autoimmune disease, and its therapies incorporate agents that inhibit T cells. However, IBD flares are also controlled with broad-spectrum antibiotics, implying a causative role for gut bacteria that can be suppressed with those agents.[49] IBD is associated with "dysbiosis" or an unhealthy imbalance in relative microbial species within the GI tract. A current working model posits that IBD is triggered by an aberrant mucosal immune reaction to gut microbiota, which disrupts the homeostatic balance required to maintain healthy intestinal barriers.[49] The dysbiosis results in a relative loss of Firmicutes, Bacteroidetes, and Bifidobacteria and a relative increase in Enterobacterales and Proteobacteria. Pathobionts breach the mucosal barrier and trigger inflammation. Chronic inflammation results in increasing dysbiosis, creating a pathogenic loop of aberrant mucosal immunity and dysbiosis.[49] Thus, a break in the harmonious coexistence between the microbiome and host defense can lead to chronic disease.

Carriage or Colonization

Bacterial carriage, or microbial colonization, implies that bacteria reside on a barrier surface (such as skin or nasal mucosa) without significant induction of inflammation. Colonization may induce a subclinical immune response, meaning that the host immune system recognizes and responds to the bacteria being carried on a mucosal surface, but there is insufficient generation of inflammatory mediators to generate detectable signs or symptoms. Colonization does not always result in infection, but may be a prerequisite. Several key human pathogens are common colonizers. For instance, *S. pneumoniae* colonizes the nasopharynx of approximately 30% of US two-year-olds, and *S. aureus* colonizes about 30% of US adults. *E. coli* and other enteric Gram-negative rods are carried in the stool. With some GI pathogens, such as *Shigella* spp. or *Salmonella enterocolitis*, a prolonged carrier state may occur after an infection is resolved. The majority of human infections occur with bacteria that are carried on a barrier surface.

Invasion and Infection

Invasion occurs when the bacteria move across what should be an intact barrier. For instance, *S. aureus* colonizing on the skin may enter, or invade, the skin through a cut or abrasion. Bacteria naturally residing in the oropharynx may invade the bloodstream through a scratch or abrasion on the gums when brushing teeth. "Invasion" does not necessarily result in infection, since rapid, effective immune responses may clear the invading organisms prior to the bacteria establishing a niche, or triggering sufficient inflammation to produce symptoms. *Clearance* generally implies that all invading or pathogenic bacteria have been removed from the host by the immune system. Clearance may occur after invasion, but before a symptomatic infection has occurred.

An *infection* occurs when bacteria breach a surface barrier, escape immediate clearance, and establish a sufficient hold in the host to begin dividing rapidly and triggering inflammation. In clinical practice, acute "inflammation" is a syndrome generally characterized by warmth, swelling, redness, pain, and loss of function.[53] "Inflammation" has a different meaning in immunology, usually implying the rapid, nonspecific immune response generated through PRR signaling and inflammasome activation. Inflammation is usually characterized by production of cytokines such as IL-6, IL-12, TNFα, and IL-1β and the recruitment and influx of inflammatory cells, such as neutrophils, via chemokine production. Cytokine generation and neutrophil infiltration produce the clinical signs and symptoms of inflammation.[54] During inflammation, production of iNOS, and thus generation of NO, relaxes vascular smooth muscle cells, dilating the blood vessels to increase blood flow to the infected area. Histamine production causes contraction of endothelial cells, generates gaps between cells, and increases vascular permeability. IL-1β and TNFα stimulate endothelial cells to express selectins and integrins that enhance transmigration of neutrophils and other leukocytes.[54] Increased permeability permits extravasation of fluid and proteins into the inflamed site; the proteins that extravasate into inflamed sites increase the ECM for trafficking of leukocytes. Increased blood flow enhances the trafficking of leukocytes and increases the oxygen tension of the infected area. Increased blood flow is often essential to bacterial clearance; areas of the body without blood flow (eg, due to trauma, pressure, or vascular disease) are at high risk for infection. Increased blood flow induces the symptoms of redness and warmth associated with localized skin and soft-tissue infections; increased fluid and cellular accumulation (release of cellular and bacterial components and host and pathogen cellular death) at these sites induce the pain and swelling.

In systemic infections (for instance, bacteremia, when bacteria are actively circulating in the bloodstream) or in severe localized infections (for instance, pyomyositis, a large muscle abscess), the high levels of cytokines circulating in the bloodstream can produce the clinical syndrome of sepsis. The clinical definition of "sepsis" was refined in 2016 to specifically mean life-threatening organ dysfunction caused by a dysregulated host response to infection.[55] Sepsis can be marked by high fevers, sweating, and changes in mental status. The high levels of circulating cytokines may cause systemic blood vessel dilation, dropping the blood pressure of the host. "Septic shock" is a subset of sepsis and is characterized by decreased blood pressure (termed "hypotension"). Hypotension reduces delivery of blood, and thus of oxygen, to critical organs, which can result in rapid tissue necrosis. Tissue necrosis releases the metabolite lactate from dying cells. Physicians will administer fluid to expand the blood volume and, if needed, administer drugs called "vasopressors" to counteract the vasodilatory effect of the inflammatory cytokines, inducing blood vessel constriction and increasing blood pressure. "Septic shock" is now defined as hypotension of sufficient severity to require vasopressors, plus elevated lactate (reflecting organ damage). Hospital mortality rates of patients meeting this definition are above 40%, even with appropriate therapy.[55] The administration of antibiotics to kill infecting organisms will also be given during sepsis. Effective therapy will eventually reduce burden of pathogenic organisms, and thus the PAMP production triggering systemic inflammation. Paradoxically, however, initial bactericidal antibiotic administration may lyse a large number of microbes simultaneously, releasing a burst of PAMPs that temporarily enhances the proinflammatory process. The host factors driving septic shock are still not fully elucidated. Despite decades of research, therapeutic agents directed at the host inflammatory mediators (such as steroids, monoclonal antibodies against TNFα, inhibition of IL-1R) have not been shown to reduce morbidity or mortality associated with septic shock.[56-59]

Clearance of Infections

Complete pathogen clearance almost always relies on an effective host immune response. As detailed above, pathogen clearance in acute infections relies heavily upon macrophages and neutrophils, supported by complement and microbial-specific antibodies. Antibiotic use can greatly reduce pathogen burden, but complete clearance is extremely difficult in patients without functional immune systems. For instance, neutrophils are absolutely required to prevent or clear infections with the Gram-negative rod *P. aeruginosa.* While antibiotic therapy can reduce the burden of organisms, complete resolution of a *P. aeruginosa* infection is all but impossible in patients who are temporarily without neutrophils (eg, "neutropenic" patients). Neutropenic patients include those who have recently undergone treatment with bone marrow suppressive therapies, such as those with leukemia or those in preparation for a bone marrow transplant. Neutropenic patients are extremely susceptible to *P. aeruginosa* during the few weeks it takes for the bone marrow to either recover or engraft and begin generating neutrophils. Neutropenic patients suffering from severe infections who cannot clear pathogens despite appropriate antibiotic therapy may receive infusions of granulocytes (ie, neutrophils) in attempts to resolve the infections. Granulocyte infusions are highly inflammatory, however, and are reserved for life-threatening infections.

The efficacy of a host immune response can determine the pathogenicity of a microbe. In many host-pathogen encounters, the healthy host immune defense is sufficient to clear an invading pathogen before acute inflammation is ever induced. Every infection may therefore be characterized, to some degree, as a failure of the immunological layer critical to the host defense. Some bacteria, such as *S. aureus,* become pathogenic in people with normal immune systems if there is a breach in the protective barrier such as skin. Other bacteria, such as *P. aeruginosa,* are uncommon causes of infections in people with normal immunity, even with a barrier breach, as they are easily killed by active neutrophils. However, *Pseudomonas* is highly virulent in patients without neutrophils. Some organisms are so unlikely to cause infections in immunocompetent individuals that infections with these organisms will raise a concern for a clinical immunodeficiency. For instance, *Serratia marcescens* is a Gram-negative rod commonly found in the environment and in tap water. *S. marcescens* is easily cleared by neutrophils and therefore almost never causes infections in healthy hosts, although everyone is routinely exposed. Therefore, a child presenting with an active *S. marcescens* infection may be evaluated for defects in neutrophils, such as chronic granulomatous disease.

Bacterial clearance can also induce profound host tissue damage and abscess formation. Neutrophils, as part of the host response, may produce ROS that can kill some bacteria but also damages host tissue. Neutrophils may die by netosis, resulting in accumulation of extracellular DNA and microbial debris. Dying bacteria may leave virulence factors or toxins that actively kill host cells. *S. aureus* releases multiple toxins that cause host cell lysis.[8] An abscess forms when fibroblasts and fibrin deposition enclose an area of necrotic tissue and uncleared infection. The fibrin "wall" contains the infection and prevents pathogen spread to other tissues.[60] Thus, an abscess is a walled-off pocket of pus, which is the purulent fluid that accumulates in a local tissue site of infection, and it contains live inflammatory cells (usually neutrophils), bacteria, dead tissue, and debris.[60] Because there is no blood flow to pus pockets, there can be no delivery of oxygen, fresh inflammatory cells, or antibiotics. Sometimes, it is beneficial (if not necessary) for surgeons to drain pus out of tissues for the body to clear and heal from the infection.

Healing/Resolution

The proinflammatory response that initiates bacterial clearance also begins the healing response.[61] IL-1β and TNFα recruit and activate fibroblasts, which migrate to the site of infection and proliferate. Neutrophils produce matrix metalloproteases that digest necrotic tissue, in addition to their primary role in phagocytic elimination of pathogens. Macrophages clear necrotic debris through phagocytosis and produce growth factors such as the platelet-derived growth factor, fibroblast growth factor, and TGF-β. After bacterial clearance, macrophages and other immune components must then shift from the proinflammatory response, required for effective pathogen killing, to a prohealing response. Macrophages increase production of IL-4 and IL-13 and reduce generation of IL-1β, TNFα, and IL-6. Activated fibroblasts proliferate and produce collagen and fibrinogen, which reconstitute the ECM. Depending on the affected organs, tissue progenitor cells may proliferate and differentiate to replace the cells destroyed during the infection, either by direct pathogen-mediated killing or through the bystander effects of exuberant inflammation. New blood vessel growth (angiogenesis) will occur. Over time, the deposited ECM is cross-linked and remodeled to produce scar tissue. Inability to clear infection or failure to downregulate the proinflammatory response can lead to chronic, nonhealing wounds.[61,62]

Secondary Versus Recurrent Infections

Acute infection (eg, viral infection) and inflammation may, somewhat paradoxically, result in temporary host immunosuppression, which increases the risk for a secondary bacterial infection. A *secondary* bacterial infection usually does not refer to a second exposure to the same pathogen, but to a bacterial infection occurring after a prior viral infection, trauma, or other tissue injury. For instance, *S. pneumoniae* and *S. aureus* are significant causes of severe secondary bacterial pneumonias during recovery from preceding viral lung infections, such as influenza or respiratory syncytial virus.[63,64] Preceding varicella infection was identified as a risk factor for invasive GAS infections in children in the mid-1990s.[65,66] Introduction of the varicella-zoster virus vaccine has profoundly reduced pediatric incidence of GAS necrotizing fasciitis and other invasive GAS diseases.[67,68] Severe burns may be complicated by secondary bacterial infections with *S. aureus* or *P. aeruginosa*.[69] How antiviral immune responses or tissue damage increases the susceptibility of human hosts to secondary bacterial infections is poorly understood.

"Recurrent" bacterial infection describes a second infection with the same species. Secondary immune responses to recurrent bacterial infections depend upon the adaptive immune response. Production of long-lasting, high-affinity IgG is often essential to rapid host defense against recurrent challenges with bacteria, and function by opsonizing bacteria and increasing the efficiency of phagocytosis by macrophages and neutrophils. However, immune protection is often specific to a capsule type or LPS O-antigen type, and one may be reinfected by the same species of bacteria expressing different capsule or LPS types. All current vaccination strategies currently rely upon generation of adaptive immunological memory.

SPECIFIC BACTERIAL INFECTIONS

The World Health Organization released a statement in 2017 prioritizing twelve global pathogens in addition to Mtb for focused study[70] because of increasing antimicrobial resistance and continued high impact on human health. While research into microbiological virulence factors has led to novel therapeutic agents, discovery of new classes of antibiotics is becoming increasingly difficult. Meanwhile, research into the multilayered host immune responses that ultimately control and clear these pathogens has lagged. Determining the precise role of immune cells and molecules that control and clear bacterial infections is extremely challenging, as experimental manipulation of any part alters the response of the whole, clouding interpretation of what occurs "normally." The essential elements of effective antimicrobial response are often best elucidated by the study of patients uniquely susceptible to specific organisms. For instance, complement deficiencies result in extreme susceptibility to *N. meningitidis*, revealing the importance of the MAC to meningococcal clearance. To provide examples of how key effector mechanisms vary with microbe, we summarize the host immune responses toward four of the twelve pathogens listed in the 2017 WHO statement: *Streptococcus pneumoniae, Staphylococcus aureus, Shigella spp,* and *Pseudomonas aeruginosa,* in addition to *Neisseria meningitidis* and *Mycobacterium tuberculosis. N. meningitidis* is selected for inclusion because of ongoing outbreaks in sub-Saharan Africa and the severity of disease that profoundly affect health outcomes. Each organism was selected to highlight a key aspect of effective host immunity (Table 43.2). The response to each organism will be reviewed using the framework of the "life cycle" of an infection, from colonization to clearance.

S. pneumoniae

Introduction, Epidemiology, and Diseases/Conditions

S. pneumoniae, also called pneumococcus, is a Gram-positive coccus that often appears in pairs or chains under the microscope.[71] It is facultatively anaerobic and an extracellular pathogen. There is no known animal reservoir for pneumococcus. Pneumococcus is spread to new individuals by contact with a carrier, who carries it in the nasopharynx often without any symptoms. Up to 30% of US toddlers may be asymptomatic carriers at any given time. It can cause infections like otitis media and pneumonia as well as invasive pneumococcal diseases (IPDs). IPD is defined as a pneumococcal infection in a normally sterile site and includes sepsis and meningitis. Pneumococcus is notorious as a causative agent of the lethal secondary bacterial pneumonia following viral infections such as influenza[72] and even with SARS-CoV-2.[73] Pneumococcal infections kill over 800,000 children younger than 5 years annually, despite vaccines and effective antibiotics. Pneumococcal sepsis can be rapidly fatal in adults, again despite initiation of appropriate antimicrobial therapy.

TABLE 43.2 Specific Aspects of Host Immunity Essential in Response to Selected Pathogens

Organism	Highlights of Host Defense
Streptococcus pneumoniae	Humoral immunity against polysaccharide capsule
Staphylococcus aureus	NADPH-oxidative burst, neutrophils, Th17 cells
Neisseria meningitidis	Complement-mediated killing
Shigella species	Subverts host immunity during infections
Pseudomonas aeruginosa	Intact epithelial barriers (eg, skin, respiratory tract)
Mycobacterium tuberculosis	T cell–mediated, granulomatous response

Pneumococcus expresses thick and diverse polysaccharide capsule. So far, at least 100 different types (serotypes) of polysaccharides are identified.[74,75] The capsule shields pneumococci from innate immunity and allows them to evade the host phagocytes. The capsule is the most significant virulence factor of pneumococcus: capsule can increase virulence of pneumococcus more than a million fold in an experimental infection,[76] and almost all IPD isolates express capsule. Consistent with the importance of the capsule, pneumococcal virulence for mice is starkly dependent on its serotypes,[77] and invasiveness of a pneumococcal strain to humans varies more than 100-fold among serotypes.[78] Antibodies to capsule are highly protective, and infection with one serotype leads to type-specific immunity. A noted immunologist C. Janeway Jr. stated that each serotype of *S. pneumoniae* should be considered as a distinct organism.[79]

Pathogenic Mechanisms

Since there are no animal reservoirs, nasopharyngeal colonization is the only source of *S. pneumoniae* and is an important step in pathogenesis (Fig. 43.4A, B). As pneumococci newly arrive in the host nasopharynx, they encounter the first defense barrier, which is the thick layer of mucin coating the nasopharyngeal epithelia. The sialic acid-rich mucin is enriched for AMPs and antibodies and can trap bacteria for expulsion. Pneumococci counter these barrier defenses through production of the IgA1-cleaving zinc metalloprotease, ZmpA, and their capsular polysaccharides, which repel the mucin. Pneumococci also produce sialidases and glycosidases, which are useful for acquiring nutrients in this environment. In addition, the newly arriving pneumococcus needs to compete with other bacteria (other preexisting pneumococcus or other bacterial species) in the nasopharynx. Once the new pneumococci

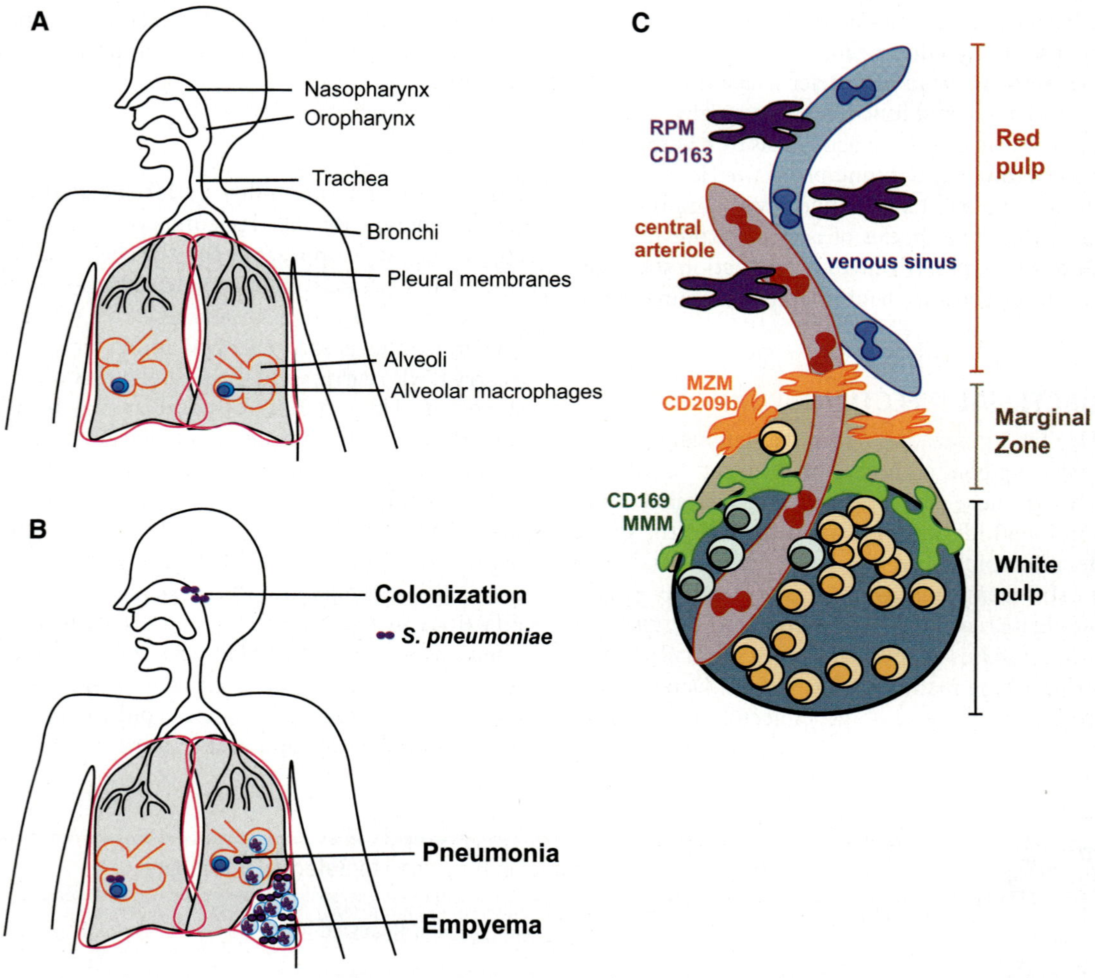

FIG. 43.4. *Streptococcus pneumoniae* colonizes the nasopharynx and is a common cause of pneumonia. A: In health, the dual pleural membranes encase the lungs. Tissue-resident alveolar macrophages adhere to alveolar epithelial cells in the airways as sentinel cells. **B:** *S. pneumoniae* asymptomatically resides in the nasopharynx of about 30% of US toddlers. Pneumococcus in the lungs is first ingested by alveolar macrophages, which phagocytose and kill the organisms and initiate recruitment of other inflammatory cells, such as neutrophils, through secretion of proinflammatory cytokines and chemokines. If pneumococcus escapes the lung and crosses a pleural membrane, it can cause an empyema, an infected fluid collection in the pleural space. Large empyemas compress the lungs and compromise breathing. C, Specialized splenic macrophages and marginal zone B cells are essential to host defense against encapsulated bacteria, such as *S. pneumoniae* and *Neisseria meningitidis*. Asplenic patients have a >70-fold risk of pneumococcal sepsis. MMM, metallophilic macrophages; MZM, marginal zone macrophages; RPM, red pulp macrophages.

reach the host epithelial cells, they attach to the host cells using various adhesion molecules. For instance, pneumococcal phosphorylcholine in teichoic acid targets the host platelet activating factor receptor, and the pneumococcal choline binding protein A (CbpA, alternative name PspC) targets the host pIgR.

Successful colonization involves mild inflammation in the host mucosa,[80,81] which facilitates adhesion of pneumococci to the host as well as secretion and spreading of pneumococci to new hosts.[81] Inflammation may be triggered by pneumococcal toxins (eg, pneumolysin), a viral infection, or vaccination with live virus (eg, live attenuated influenza vaccine).[82,83] Inflammation may involve various inflammatory mechanisms including TLR2-mediated signaling and macrophage recruitment.

Colonization induces anticolonization adaptive immune responses such as Th17 cells or antibody responses against pneumococcal proteins and polysaccharides (eg, capsule). A role for Th17 cells was shown with mouse models,[84] but their importance in humans is still unclear. These immune responses protect against continuing colonization and invasive disease. A recent study of human experimental colonization implicated roles for both mucosal-associated invariant T (MAIT) cells and B cell production of serotype-specific antibodies in the immune response.[85] Antibodies to the capsule may agglutinate pneumococci and facilitate their expulsion. A role for anticapsule antibodies against colonization was clearly shown with the use of pneumococcal conjugate vaccines (PCVs) among children. With the use of PCV, carriage of vaccine-type pneumococcus is significantly reduced, but carriage of nonvaccine serotypes including noncapsulated pneumococcus has increased. While the replacing serotypes tend to be less invasive, the overall prevalence of pneumococcal colonization has remained largely the same. This shift in serotype is termed "serotype replacement."

Otitis media (middle ear infections) and *pneumonia* develop when pneumococci extend into the respective anatomical locations. Young children are prone to otitis media for anatomical reasons and may develop chronic otitis with development of biofilms.[86] Pneumococci may also gain access to the lower respiratory tract without crossing a tissue barrier and cause pneumonia.[72] This is common among individuals who have viral infections such as influenza. Pneumococcal pneumonia is a significant cause of worldwide mortality in young children and adults older than 65 years. Pneumococci overcome the innate mucociliary defense through degradation and disruption of the mucus layer and direct toxicity toward cilia.[87] Pneumolysin induces apoptosis of respiratory epithelial cells and impairs the ciliary beating required for clearance.[11,88]

In the lungs, the tissue-resident alveolar macrophages are the first professional immune cell encountered by invading pneumococci (Fig. 43.4A). CR3 on alveolar macrophages will bind C3b, a complement component, deposited on the surface of pneumococci. Alveolar macrophages express FcγRs, which may be useful to bind to IgG targeting the pneumococcal capsule. Following phagocytosis, pneumococci are trafficked to the phagolysosome and likely killed. Apoptosis of the alveolar macrophages that have engulfed pneumococci appears to be essential in preventing progression to pneumonia. The major virulence factor protecting pneumococci from phagocytosis is the polysaccharide capsule, which sterically inhibits adhesion of the macrophage and microbe.

Pneumococcal pneumonia is highly proinflammatory, due to host immune responses and pneumococcal virulence factors (Fig. 43.4B). CXCL1 and CXCL2 produced by infected alveolar macrophages and epithelial cells cause a rapid influx of neutrophils to the airspaces.[89] Neutrophils kill pneumococci through phagocytosis and the release of granules containing serine proteases.[90] Notably, production of cellular ROS through the NADPH oxidative burst is relatively dispensable for the elimination of pneumococci. Pneumococci produce neuraminidases, hyaluronate lyase, and metalloproteases that directly damage host cells. Plasminogen binding proteins, such as enolase and glyceraldehyde 3-phosphate dehydrogenase, sequester host plasminogen on the microbial surface, supporting adhesion. Pneumolysin activates the NLRP3 inflammasome, resulting in production of IL-1β.

Invasion of pneumococcus to the blood begins with "shedding" of the capsule with the help of LytA[91,92] and adherence of the bacteria to the surface of respiratory epithelial cells. There are many adhesive molecules expressed by the surface of pneumococci. Although the relative importance of these is yet to be determined, major adhesive molecules include phosphorylcholine, CbpA, pneumococcal adherence and virulence protein A (PavA), and the pilus subunit RrgA. After adhering to the host cell surface, pneumococci can subvert host recycling mechanisms to undergo transcellular invasion. For instance, the pneumococcal protein CbpA can bind to the polymeric Ig receptor (pIgR) on the apical surface. Binding of pIgR stimulates endocytosis of pneumococci and subsequent transport across the basolateral membrane, using the endogenous host pathway of pIgR recycling.[93] Pneumococci may also invade by passing between cells in paracellular invasion, induced when pneumococcal toxins, such as pneumolysin, damage the epithelium and pneumococcal enzymes, such as hyaluronate lyase, degrade ECM to permit microbe migration.

Once a barrier is breached and invasion has occurred, resident immune cells sense pneumococci through the aforementioned microbial sensors (PAMPs). Pneumolysin binds and activates TLR4.[94] TLR7, TLR9, and TLR13 all appear to be activated during pneumococcal infection, and triple knockout mice lacking TLR7, TLR9, and TLR13 rapidly succumb to IPD.[27] IRAK-4 is important for host defense against pneumococci early in life but its importance wanes as IRAK-4–deficient patients become older.[95] TLR signaling, mediated by MyD88 and IRAK-4, activates NF-κB and subsequent production of CXCL1 and CXCL2, which recruit neutrophils.[27] Pneumolysin also activates the NLRP3 inflammasome, triggering IL-1β release.[96]

Pneumococci evade these early innate responses through elaboration of additional virulence factors. For instance, pneumococci suppress neutrophil recruitment through expression of proteins such as CbpA. Pneumolysin has been recently shown to bind to the mannose receptor C type 1 on human DCs and murine alveolar macrophages, which results in suppression of bacterial killing in lysosomes and in T cell polarization away from an effective antibacterial Th1 phenotype.[97]

Pneumococci may subsequently translocate from the airways into the bloodstream, causing sepsis. Elimination of pneumococci circulating in the blood relies upon opsonization by antibodies to the capsule. When antibodies bind the capsule, they activate complement via the classical pathway and opsonize pneumococci for phagocytes such as neutrophils or macrophages to clear. Pneumococcal clearance from the bloodstream also relies on splenic anatomy; asplenic patients have over 70-fold greater chance of suffering pneumococcal sepsis than those with functional spleens. The spleen is also thought to harbor multiple subsets of neutrophils, which can be mobilized in response to pneumococcal or other bacterial sepsis.[98]

Splenic tissue is roughly divided into the vascular "red pulp" and the immunological "white pulp." The border between the red pulp and the white pulp constitutes the marginal zones (Fig. 43.4C). Highly specialized leukocytes within the marginal zones express receptors that specifically bind to pneumococcal components and "clear" pneumococci from the bloodstream. These specialized leukocytes include the marginal zone B (MZB) cells and the marginal zone macrophages (MZM). MZM are one of the three tissue-resident macrophage lineages identified in the spleen, the others being metallophilic macrophages and red pulp macrophages.

MZM express the C-type lectin SIGN-R1, which binds to certain serotypes of pneumococcal capsule.[99] Serotype 14 that binds SIGN-R1 is not virulent in mice,[77] although it is highly virulent in humans. High levels of SIGN-R1 expression enable MZM to rapidly bind and remove circulating pneumococci from the bloodstream. Mice lacking SIGN-R1 are much more susceptible to pneumococcal infection than are wild-type mice.[100] MZB cells can also engage directly with pneumococcus and secrete low-affinity IgM against the phosphorylcholine expressed on pneumococcal teichoic acid or lipoteichoic acid. Because MZB can begin secretion of antipneumococcal IgM within 5 days of infection, their rapid "innate" response may provide some humoral protection prior to generation of the high-affinity IgG antibodies manufactured during a germinal center response. Antiphosphorylcholine antibodies, either preexisting (made by B-1 cells) or induced rapidly after infection, were found to be important in protecting mice in the early phase of bacteremia.[101]

Once in the blood, pneumococci may breach additional tissue barriers and gain access to the pleural space or the brain causing an empyema (Fig. 43.4B) or meningitis, respectively. Recently, it was found that pneumococci in the blood may invade the heart or other organs and weaken their function.[102] This may explain why patients who recovered from pneumococcal sepsis have a high probability of heart failure.

Critical Protective Immunity

Since humans are commonly exposed to pneumococcus, persons with a defect in the immune system that is critical and nonredundant for the defense against pneumococcus suffer from pneumococcal infections early in life. For instance, the critical need for antibodies in defense against IPD has been shown with patients with Bruton agammaglobulinemia or with common variable immune deficiency, who often suffer from IPDs early in their lives.[95] Moreover, ability to make antibodies against polysaccharide is important since children younger than 2 years, who cannot make antibodies to polysaccharide antigens, are susceptible to IPDs and other encapsulated organisms. Thus, these observations suggest antibodies, particularly antibodies to polysaccharides, are critical for protective immunity against pneumococci.

Next, importance of the early components (C1, C2, C3, and C4) of complement cascade in the host defense against IPD can be demonstrated. Patients with complement deficiency affecting C3 activation are susceptible to pneumococcal infections. C2 deficiency is relatively common, and a study of 40 C2-deficient patients in Sweden found more than half of the patients had a history of invasive infections, with pneumococcus being the most common etiologic agent.[103] In contrast, use of a monoclonal antibody to C5 (eculizumab), or patients' deficiencies of the complement components necessary for the formation of MAC, are primarily associated with susceptibility to meningococci but not pneumococci.

Neutrophils are necessary for effective protection, as shown when noninvasive pneumococci become invasive and kill neutrophil-deficient mice.[80] However, their role needs to be examined further since agranulocytic patients suffer from staphylococcal or pseudomonas infections more than pneumococcal infections.[95] Further, splenectomy dramatically increases the risk of pneumococcal sepsis.[104] Taken together, protection against IPD requires antibodies, complement, and phagocytes. These findings taken together support that antibody-mediated opsonophagocytosis is critical in the defense against IPD.

Immunotherapy (Vaccine)

Serotype-specific protective properties of anti-capsule antibodies have been clearly established with antiserum treatment used before antibiotics were available. Based on this observation, early attempts were made to elicit anticapsule antibodies by using purified capsular polysaccharide as a vaccine. Such attempts led to the development of a vaccine containing capsular polysaccharide from 23 different capsule types that are commonly responsible for IPDs (trade name Pneumovax). The vaccine has been found to reduce IPD death among adults but was not immunogenic in young children, who are vulnerable to IPD and need effective vaccines.

To induce anticapsule antibodies in young children, capsular polysaccharide has been converted to a T cell–dependent antigen by conjugating the polysaccharide to a carrier protein. Commonly used carrier proteins are a nontoxic variant of diphtheria toxin (DTx; CRM) or tetanus toxoid. The resulting vaccine, which is called PCV, has been found to be highly effective against IPD. In addition, it was effective against nonbacteremic pneumonia and NP colonization and provided strong herd immunity to the adults who were not vaccinated. However, the PCVs are only effective against the serotypes included in the vaccine (and cross-reactive serotypes). Further, its use has resulted in serotype shifts to non–vaccine-type IPDs as described above, even though the total incidence of IPDs has stayed low. Thus, serotypes included in the PCVs have steadily increased: the first PCV had seven serotypes, current PCVs have 10 to 13 serotypes, and PCVs containing 15 to 31 serotypes are at various stages of development.[105] Also, improvements in conjugation technologies have led to the introduction of a very low cost conjugate vaccine in 2019.[106] Taken together, the PCVs have greatly reduced the burden of pneumococcal diseases and improved public health throughout the world.

Since manufacturing PCVs containing many capsule types is technologically challenging, there are efforts to produce vaccines based on pneumococcal proteins and T cell–mediated immunity. The protein antigens that have been investigated include PspA, CbpA, or pneumolysin.[107,108] While these proteins showed promising results in animal studies, their use in vaccines for humans is unclear at the moment.

S. aureus

Introduction, Epidemiology, and Diseases/Conditions

S. aureus is a Gram-positive, facultatively anaerobic coccus that appears in grape-like clusters upon Gram staining. When cultured on blood agar plates, *S. aureus* colonies can appear "golden" in color, from which the species derives its name (in Latin, *aureus* = golden). *S. aureus* commonly colonizes the anterior nares, axillae, inguinal folds, and other moist areas,[109] with a population prevalence of carriage around 30%.[110] *S. aureus* can persist on high-touch environmental surfaces in the homes of people colonized and can be transferred from carriers to noncarriers through sharing of domestic items (eg, bath towels).[109] There is no animal reservoir for *S. aureus*, although domestic pets may acquire *S. aureus* colonization from their owners.[109]

S. aureus was first identified as a pus-forming pathogen in 1880 and remains the most common cause of skin and soft-tissue infections.[60] *S. aureus* produces a panoply of toxins, including proteases, pore-forming cytotoxic toxins, exfoliative toxins (ETs), and T cell superantigens, all of which may induce tissue damage and severe disease.[111] *S. aureus* causes severe, invasive disease such as postviral pneumonias and sepsis. Sepsis with *S. aureus* can have mortality rates as high as 50%,[112,113] and overall mortality of *S. aureus* infections remains 16% to 30%,[114] in part because of increasing antimicrobial resistance. The interactions between *S. aureus*, the microbiome, and the host immune system are extraordinarily complex and have not yet been fully elucidated. Despite many years of research, all attempts at developing effective vaccines have failed.[114]

Pathogenic Mechanisms

Colonization and Barrier Disruption. As noted, *S. aureus* is a common colonizer of skin, preferring warmer and moister areas, such as the axillae.[115-117] *S. aureus* expresses adhesion proteins that bind to host proteins such as loricrin, cytokeratin 10, cytokeratin 8, desmoglein, and fibronectin.[118] Loricrin is the most abundant protein in stratified corneum and is essential to proper formation of keratinized layers.[118] Damage or inflammation of the epithelial barrier can enhance the likelihood or density of colonization by *S. aureus*. Increased type 1 IFN expression may enhance the colonization of patients with systemic lupus erythematous by *S. aureus*.[116] Inflammation induced by physical disruption (tape stripping) is sufficient to induce neutrophil recruitment and netosis, which also enhance *S. aureus* colonization in a mouse model. This inflammatory process was dependent upon MyD88, TLR2, and TLR9. Thus, inflammatory cells essential to clear *S. aureus* infections can paradoxically enhance carriage.[117] Colonization by *S. aureus* can also promote skin disease, such as atopic dermatitis, through induction of cytokines such as IL-33.[20,119] Presence of other staphylococcal species, such as *S. epidermidis*, can reduce *S. aureus* carriage, an example of microbial intercompetition.

S. aureus does not commonly colonize respiratory epithelium beyond the anterior nares. However, it is a major pathogen in patients with CF, a respiratory disease marked by barrier dysfunction. CF is caused by mutations in the CF transmembrane conductance regulator (CFTR). CFTR is a chloride/bicarbonate ion exchange channel present at most barrier surfaces, including the airways, the GI tract, and the skin. CF mutations in CFTR result in excessive chloride secretion and disrupted osmotic balance. The skin of patients with CF classically tastes "salty" due to this osmotic imbalance, and a primary diagnostic modality for CF remains the "CF sweat test," in which the chloride concentration of subject's sweat is assayed. In the GI tract, CFTR mutations disrupt gut motility and pancreatic secretion. It is currently believed that the CFTR mutations are maintained in the human populations (eg, Caucasian) historically ravaged by cholera epidemics; heterozygosity for CFTR mutations reduces the extreme GI fluid losses found during cholera.[120] In the respiratory tract, homozygosity for CFTR mutations results in highly viscous mucus that impairs ciliary clearance, and lungs become clogged with thick mucus entrapping inflammatory cells and microbes. Along with *P. aeruginosa* (reviewed below), *S. aureus* is a major pathogen in CF.[121] Recently, medications restoring the CFTR function in patients with CF have been introduced,[122] and their use alters the classical microbiology of patients' respiratory tracks.

Colonization of CF airways by *S. aureus* relies on modulation of normal immune defenses and production of biofilms.[123] Biofilms protect bacteria from antibiotics, oxidative stress, phagocytosis, complement deposition, and antibody attachment. Biofilms are composed of polysaccharides, DNA, and bacterial proteins. *S. aureus* contains two gene clusters, one that encodes the enzymatic pathway for production of polysaccharide intercellular adhesin and the other for production of capsular polysaccharide.[123] Sessile *S. aureus* residing in biofilms is resistant to neutrophil-mediated killing. Toxin production (Panton-Valentine leukocidin [PVL] and γ-hemolysin AB) by sessile *S. aureus* induces neutrophil netosis and death, preventing bacterial killing by the neutrophils.[123] Staphylococcal biofilms also induce M2 polarization of responding macrophages.[124] Because M1 activation and production of ROS is key to macrophage-mediated killing of *S. aureus*, the induction of M2 phenotypes permits staphylococcal escape.[125]

Critical host defense and virulence factors during invasion and infection. Keratinocytes, which form the epidermis bound by the two cell layers (stratum corneum and stratum basale), provide the first layer of antistaphylococcal defense (Fig. 43.5). First, the stratum corneum and cell junctions physically prevent translocation. Secondly, keratinocytes produce human beta-defensins and cathelicidin LL-37, which kill *S. aureus*. While intact epidermis presents a physical barrier to *S. aureus* invasion, disruption of the barrier by cuts or burns permits translocation of bacteria into the deeper layers of skin, and infection may ensue. Keratinocytes then drive the initial host response through release of proinflammatory factors. Lipoteichoic acid, lipoprotein, and peptidoglycan in staphylococcal cell walls activate TLR2/6 and TLR1/2.[126] TLR2$^{-/-}$ mice exhibit reduced ability to clear staphylococcal skin infections.[28] After internalization, MDP induces NOD2 signaling. Induction of proinflammatory mediators, such as IL-8, by keratinocytes then recruits neutrophils.

Neutrophils, with the ability to generate the NADPH-mediated oxidative burst, are a key effector cell in clearing *S. aureus*. Neutrophils will "swarm" to the site of infections, engulfing and eliminating *S. aureus* in phagolysosomes. Evidence for the essential role for neutrophils is provided by the dramatic increase in susceptibility to *S. aureus* infections in patients with neutrophil deficiencies or dysfunction. For instance, chronic granulomatous disease, which is a genetic deficiency in components required to generate NADPH-mediated ROS, increases susceptibility to severe staphylococcal infections in the skin and other locations. Defects in

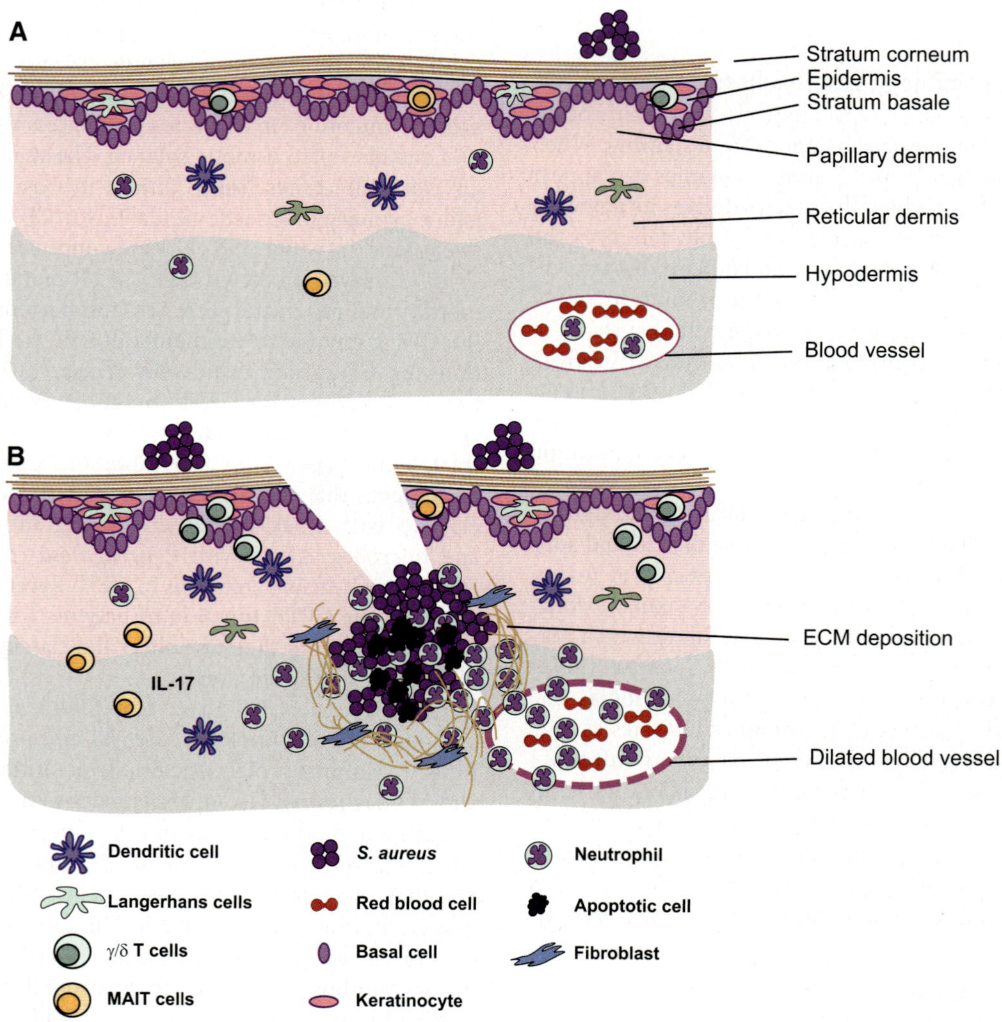

FIG. 43.5. *Staphylococcus aureus* colonizes the skin and is the most common cause of purulent skin and soft tissue infections. A: In health, *S. aureus* resides on the outer layer of skin, the stratum corneum. The thick layer of cross-linked extracellular matrix (ECM) proteins physically prevents translocation of microbes. **B:** Breaks in the outer layer of skin permit bacterial translocation. Keratinocytes initiate an inflammatory response, recruiting neutrophils. Tissue-resident hematopoietic cells, such as Langerhans cells and dermal γδ T cells, also induce IL-17 production and neutrophil recruitment. *S. aureus* produces multiple virulence factors that cause tissue destruction, including proteases that cleave ECM and cell-cell junctions, and toxins that directly lyse and kill host cells. Neutrophils engulf and kill bacteria, releasing products such as reactive oxygen species that are also toxic to surrounding host tissue. Sites of staphylococcal infections can therefore become rapidly necrotic and filled with host cell and microbial debris. Proinflammatory cytokines induce local blood vessel dilation and porosity, increasing oxygen and leukocyte delivery to the site of infection. Fibroblasts are recruited to the area of necrosis and infection and deposit fibrinogen and collagen to create a "wall" to isolate the infection and prevent spread to other tissues. Deposition of extracellular matrix around the infection forms an abscess.

neutrophil migration, such as those found in leukocyte adhesion deficiency or Rac2 deficiency, also increase susceptibility to staphylococcal infections.[8,114] While the NADPH-mediated oxidative burst is required for efficient staphylococcal killing, neutrophil granules also produce antimicrobial peptides and proteases that kill bacteria following phagosome and granule fusion.[8]

Neutrophils, monocytes, and macrophages also express TLR1/2, TLR2/6, and NOD2 and also participate in the PRR-mediated generation of proinflammatory mediators such as TNFα. Furthermore, pore-forming toxins and other virulence factors expressed by *S. aureus* activate the NLRP3 inflammasome, IL-1β production, and pyroptosis. In skin infections, NLRP3 activation promotes bacterial clearance.[123]

All three pathways of complement activation are induced by *S. aureus* during infection.[126] While the MAC cannot kill Gram-positive organisms due to the thick peptidoglycan layer, complement nevertheless plays a significant role in clearing *S. aureus*. First, deposition of C3b opsonizes *S. aureus* for phagocytosis by macrophages and neutrophils. Complement components, such as C3a, serve as chemoattractants for neutrophils, recruiting these essential effectors to the sites of infection. The importance of complement to clearance is reflected by the multiple virulence factors expressed by *S. aureus* that counter activation and deposition of complement (details below).[126]

While *S. aureus* is classically thought of as an extracellular pathogen, it adapts to intracellular niches by avoiding phagolysosome-mediated bacterial death.[127,128] *S. aureus* residing in intracellular components alters basal metabolism and is more resistant to antimicrobial therapy. Intracellular reservoirs of *S. aureus* are thought to contribute to both chronic and recurrent infections.[127,129] Recently, a novel antibody-antibiotic conjugate was developed to specifically target these intracellular reservoirs, although the conjugate has not yet moved to clinical practice.[129]

S. aureus is a formidable pathogen because of the many virulence factors elaborated to counteract neutrophil-mediated killing and suppress effective immune responses. Pore-forming toxins, such as PVL, directly bind and lyse immune cells. Factors such as staphylococcal superantigen-like protein 3 and Toll/Interleukin-1 receptor domain protein inhibit TLR2 signaling, impairing NF-κB activation.[114] Staphylococcal protein A (SpA), which is on the microbial cell surface, binds the Fc portion of IgG, resulting in the "wrong" orientation of IgG to bind FcγR on phagocytes. SpA binds to IgG with such high affinity that it is sold commercially to facilitate IgG purification. SpA also serves as a B cell directed superantigen by binding VH3-containing Fab portions of B cell receptors, inducing apoptosis.[130] Staphylococcal complement inhibitor (SCIN) inhibits C3 convertase activity, reducing production of bioactive C3a, C3b, and C5a. By degrading or binding C3b, the virulence factors complement factors I and H prevent deposition of C3b on staphylococcal membranes, impairing opsonization. Thus, SpA suppresses both innate and adaptive immune responses.[114,126,131]

S. aureus infection causes tissue necrosis due to production of high levels of cytotoxic pore-forming toxins and proteases that cleave ECM and cell junctions.[132] Staphylococcal α-toxin, also called α-hemolysin (Hla), is the archetypal pore-forming toxin. Staphopain A (ScpA) degrades collagen, fibrinogen, fibronectin, and elastin.[132] ETs cleave desmoglein, damaging tight junctions between keratinocytes and disrupting the barrier. ETs produce the clinical syndrome of staphylococcal scalded skin syndrome. Thus, sites of *S. aureus* infection rapidly fill with products of bacterial and host cell death, releasing PAMPs and DAMPs and triggering a rapid and profound proinflammatory milieu. Some of these proinflammatory mediators recruit fibroblasts. Fibroblasts deposit fibrin to create a barrier surrounding the necrotic area of inflammation and microbial proliferation, generating a walled-off, purulent abscess. Because blood flow to the necrotic abscess is absent, resolution can be delayed due to impaired oxygen delivery.

If *S. aureus* disseminates in the bloodstream, septic shock can rapidly occur. The virulence factors, leukocidin ED and γ-hemolysin AB, bind to the Duffy antigen receptor for chemokines (DARC) on endothelial cells to cause vascular leakage. DARC is a chemokine receptor that is upregulated specifically on endothelial cells in postcapillary venules and is thought to promote the adhesion and transmigration of leukoctyes responding to infections.[133] Both leukocidin ED and γ-hemolysin AB are extracellular toxins and form pores in host cell membranes to kill them. Injection of these purified leukocidins can kill mice.[134] Death of endothelial cells leads to breakdown of the integrity of blood vessel walls, resulting in leakage of plasma into the interstitial spaces. Plasma leakage from the bloodstream decreases blood pressure, resulting in shock.

The severity of staphylococcal septic shock has also been linked to α-hemolysin.[113] In addition to hypotension and vascular leak compromising oxygen delivery, sepsis can also be accompanied by dysregulated thrombosis (clotting) of capillary beds in tissues. The hypercoagulability associated with sepsis leads to inappropriate clot formation, again depriving tissues of much-needed blood flow and delivery of oxygen and leukocytes. In some cases, the rapid onset of hypercoagulability consumes available clotting factors, leading to a seemingly paradoxical state of uncontrolled bleeding. When a patient in septic shock develops hypercoagulation in capillary beds leading to consumption of all available clotting factors, this is termed "disseminated intravascular coagulation," a condition that can be rapidly fatal if not reversed by the delivery of blood components to replace the missing factors. As previously mentioned, activation of PRRs generates systemic production of IL-6, TNFα, and IL-1β, which also contribute to shock and tissue damage.

Adaptive immunity. Both T and B cells contribute to protective immunity. Humoral immunity includes antibody production against many staphylococcal cell wall components and toxins, including capsular polysaccharides, LTA, peptidoglycan, and TSST-1. Although many of these antibodies assist opsonophagocytosis, few correlate with protection from subsequent infection. The exceptions are antibodies against TSST-1; IgG responses are likely developed in response to colonization with TSST-1 producing strains. Inability or failure to generate anti-TSST-1 Ig has been identified as a major risk factor for susceptibility to toxic shock.[135,136] Additionally, severe invasive staphylococcal infections give rise to development of long-lasting anti-Hla antibodies that have correlated with reduced subsequent infections. However, superficial (skin) infections with *S. aureus* induced only short-lived antibody responses to Hla and PVL and provided no protection.[137] All attempts to elicit effective humoral protection through vaccination against staphylococcal toxins or components have thus far failed.[114]

Effective T cell–mediated responses in skin infections rely upon ROR-γt–induced Th17 polarization and IL-17 production.[138] Hyper-IgE syndrome is a clinical immunodeficiency, resulting from deficiencies in the transcription factor Stat3, characterized by failure to produce Th17 cells, hyperproduction of IgE, and susceptibility to staphylococcal skin infections. The reoccurrence of severe, poorly healing staphylococcal skin infections in hyper-IgE syndrome explains the original name for the disease, Job syndrome.[139] Production of IL-17 at the site of staphylococcal skin infections recruits neutrophils and upregulates production of antimicrobial peptides.[138] Additional unconventional T cells that assist in developing protective antistaphylococcal dermal immunity include clonally expanded γδ T cells[138,140] and MAIT cells.[114] These unconventional T cells induce IL-17 production to recruit neutrophils to sites of infection. In contrast to superficial infections, bacteremia elicits IFNγ-dominant T cell responses and T cell memory.[141] *S. aureus* suppresses effective T cell responses through production of Hla, which kills DCs and thus reduces antigen presentation to T cells.[142]

S. aureus thus presents many challenges to human immunity. The immunosuppressive effects of Hla and other toxins, combined with production of cytotoxins and matrix-cleaving proteases, cause acute, life-threatening inflammation. The ability of *S. aureus* to survive inside phagolysosomes permits the bacteria to persist in chronic infections. No immunotherapeutic directed against proinflammatory mediators such as TNFα has improved mortality during septic shock, and all attempts to generate a protective antistaphylococcal vaccine have failed so far.[114] Much research remains to be done to define effective immune responses that can be harnessed for therapeutic benefit.

N. meningitidis

Introduction, Epidemiology, and Diseases/Conditions

N. meningitidis is a Gram-negative, aerobic diplococcus that usually resides as a nasopharyngeal commensal, but may cause invasive disease in humans. *N. meningitidis* can cause acute, life-threatening meningitis in children and young adults, with a case fatality rate of 80% when untreated and as high as 10% to 40% with appropriate therapy.[143] As of 2016, meningococcus was reported to infect up to 1.2 million people annually and to be responsible for as many as 135,000 deaths.[143] In the United States, meningococcal meningitis cases are usually sporadic. However, large outbreaks of meningococcal meningitis have occurred in sub-Saharan Africa and during the Hajj.[143] As with pneumococcus, the polysaccharide capsule of meningococcus is a major virulence factor and the basis for serogrouping. While 13 total serotypes of meningococcus have been identified, only 6 are associated with invasive disease: A, B, C, W135, X, and Y.

Pathogenic Mechanisms

Colonization. Up to 10% of a population may be asymptomatically colonized with meningococcus in the oropharynx or nasopharynx at any given time.[144] Meningococcus is carried primarily by adolescents and young adults, with peak carriage occurring around age 19 years.[145] Carriage can last from a few weeks to over a year.[145] It is estimated that less than 1% of people who carry meningococcus will suffer infection. Meningococcus only exists in the human population (eg, there is no "animal reservoir"), and transmission occurs by direct contact with oropharyngeal secretions from a current carrier. Because meningococcus is a strict human pathogen, there are few good animal models, and thus the determinants of protective immunity remain incompletely described. Most of our understanding of meningococcal pathogenesis and immunity are derived from cell culture models or from humanized mice.[146]

Barrier function is critical to protection against meningococcal infection (Fig. 43.6). Airway mucus traps and restricts meningococcus, preventing contact between the bacteria and host epithelial cells.[10] When cultured in a mucus layer generated by respiratory epithelial cells, meningococci behaved as a commensal, with reduced expression of virulence factors compared to meningococci cultured in broth. Furthermore, meningococci residing in the mucus layer did not elicit detectable inflammatory markers in the cultured epithelial cells.[10] The relative avirulence of meningococci residing in an airway mucus layer may explain why there is a high meningococcal colonization prevalence, but low infection rate.

Colonization with meningococcus induces humoral immunity. Bactericidal activity in the serum is thought to be the major effective immune response against colonization and infection with meningococcus.[147] The meningococcal antigens that elicit protective Ig during colonization are a constellation of outer membrane proteins. In a study of military recruits, titers to outer membrane proteins of meningococcus rose at least 10-fold in the weeks following colonization with a novel strain. The antibodies were strain-specific and elicited against membrane proteins class 1 protein, Opa, and class 3 protein.[144] A dramatic (16-fold) increase in total bactericidal activity in serum from colonized individuals was also detected.[144] Meningococci undergo significant variation in expression of surface adhesion molecules during asymptomatic colonization,[145] suggesting that humoral immunity elicited to these surface molecules is driving selection of variants in the bacterial population.

Host Defense and Virulence Factors in Invasion/Infection. Meningococcus is most commonly associated with the clinical syndromes of septic shock and meningitis. While meningococcus is exquisitely sensitive to antibiotics, the rapidity and highly inflammatory nature of meningococcal infections causes significant mortality in infected individuals, and survivors often suffer permanent sequelae, such as deafness or loss of limbs. The virulence factors that contribute to the pathogenesis of meningococcal infections include the polysaccharide capsule, outer membrane surface proteins, IgA1 protease, and LPS. The polysaccharide capsule is a major virulence factor, protecting the bacterium against complement-mediated

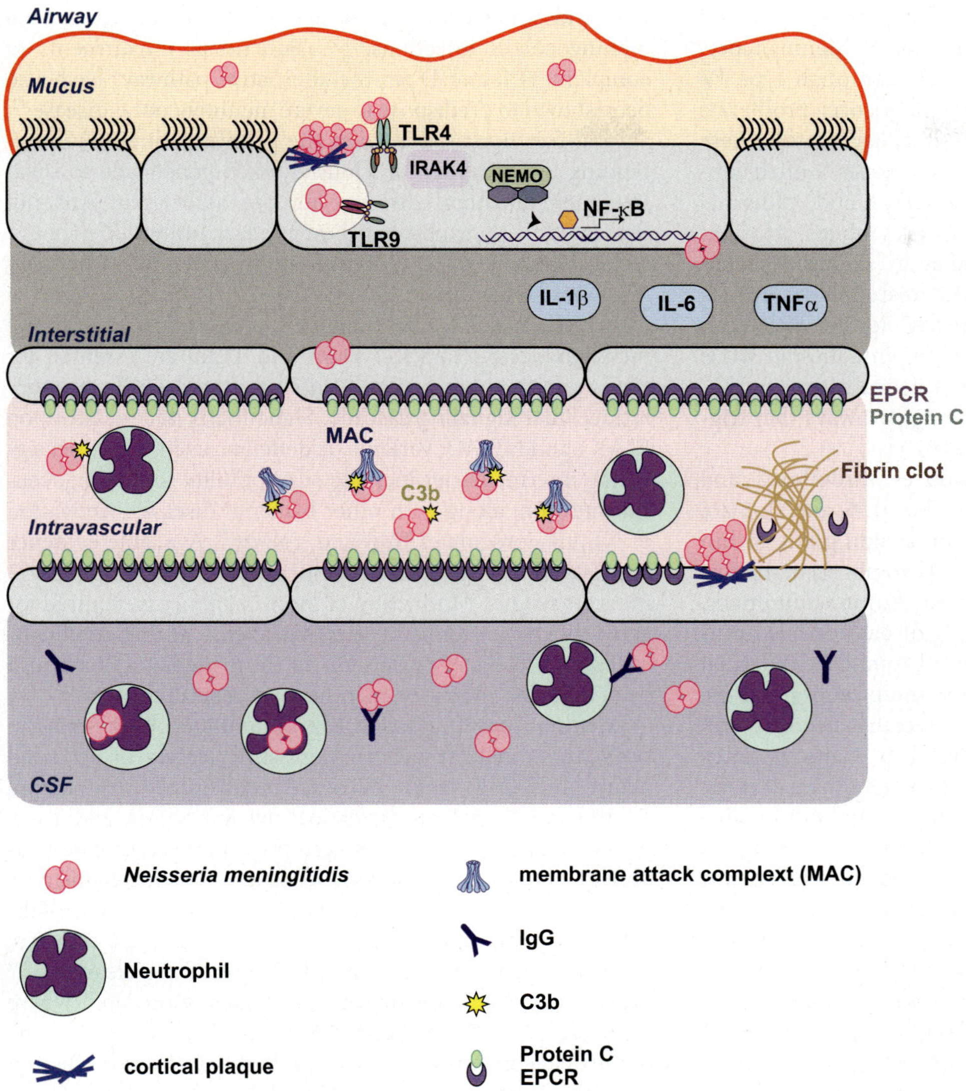

FIG. 43.6. Pathogenesis of *Neisseria meningitidis* infection. Meningococcus colonizes the nasopharynx of teens and young adults. The thick mucus layer may contribute to the avirulence of colonizing strains. Pathogenic strains adhere to the respiratory epithelium and induce cortical plaques that stimulate internalization of the microbes. TLR4 and TLR9 can sense meningococcal pathogen-associated microbial patterns and activate the NF-κB pathway through IRAK4 and NEMO, inducing production of IL-1β, IL-6, TNFα, and other proinflammatory mediators. If meningococcus translocates into the bloodstream, septic shock may result. Because meningococcemia results in loss of EPCR from endothelial cells, fibrin clots in small vessels can form rapidly and cause necrosis of digits and limbs. The major defense against meningococcus is the complement pathway. Neutrophils are also recruited, but their requirement for effectively clearing infection is not yet established. Meningococcus has a predilection for crossing the endothelial cells of the blood-brain barrier, which permits invasion of the meninges and resulting meningitis.

lysis and phagocytosis.[143] Like all Gram-negative organisms, meningococcal cell membranes contain LPS. LPS induces massive inflammation through activation of TLR4[143] in epithelial and endothelial cells, as well as in "professional" immune cells. Type IV pili are one of the major adhesive structures and also a meningococcal virulence factor. Type IV pili are filamentous glycoproteins up to 6 microns long and can extend through the capsule to contact and adhere to host cells.[143] Two classes of Opacity proteins, *Opa* and *Opc*, are expressed in the outer membrane of meningococci.[143] *Opa* binds CEACAMs, adhesion molecules expressed by epithelial cells, and may be upregulated on vascular endothelial cells via TNFα.[143] Genetic variants in CEACAM3 and CEACAM6 have been linked to increased susceptibility to invasive meningococcal disease.[147] *Opc* binds to ECM components vitronectin and fibronectin. Bacteria adhered to vitronectin or fibronectin enter cells via binding to integrin receptors that adhere to these ECM proteins, αVβ3 and α5β1, respectively. The integrins αVβ3 and α5β1 are both expressed on human brain microvascular endothelial cells. Minor adhesion proteins include Neisserial adhesin A (NadA), a trimeric protein that supports adhesion to and invasion of host cells. Other minor adhesive proteins include Neisseria hia homologue A protein and the autotransporter meningococcal serine protease A (MspA).[143]

Invasion by meningococcus begins when bacteria contact respiratory epithelial cells in the nasopharynx (Fig. 43.6). Meningococci adhere to respiratory epithelial cells through type IV pili[10,148] and stimulate internalization. Invasion of epithelial cells induces NF-κB signaling.[149] Some virulent strains secrete a serine protease (IgA protease) that can traffic to the cell nucleus, where it cleaves NF-κB and terminates signaling. Cells in which NF-κB was cleaved demonstrated sustained activation of Jun kinase (Jnk) and increased apoptosis.[149] Strains in which the IgA1 protease could not traffic to the nucleus did not cleave NF-κB, and sustained NF-κB signaling induced cytoprotective responses that prevented cellular apoptosis.[149] Cell death may then disrupt the barrier, permitting meningococcal invasion. Meningococcal invasion through epithelial cells may also occur through a tubulin-dependent, transcellular transport process that does not disrupt barrier integrity.[148,150]

Meningococcal bacteremia almost always precedes or accompanies meningococcal meningitis.[151] Meningococci bind to human vascular endothelial cells through type IV pili, *Opa* and *Opc*. Once bound, meningococci proliferate and form microcolonies or small bacterial aggregates. These aggregates induce formation of "cortical plaques," which contain ezrin, moesin, ICAM-1/ICAM-2, CD44, and epidermal growth factor receptor. These plaques accumulate tyrosine-phosphorylated proteins and cortical actin. Endothelial cells undergo significant shape change and create cellular projections that reach into the bacterial microcolonies. These cellular protrusions help the bacteria maintain adhesion while under shear stress in the bloodstream. Cortical plaque formation also activates intracellular signaling pathways that trigger active endocytosis of the bacteria.[143]

Invasion of endothelial cells induces thrombosis and inflammation.[152] Human cytokines TNFα, IL-6, and IL-8 are readily detectable in mice bearing human skin grafts within 4 hours of meningococcal infection.[152] Thrombosis in meningococcemia can be severe, resulting in purpura fulminans. "Purpura" are skin findings consisting of dark purple, non-blanching, flat lesions and result from thrombosis of dermal microvessels, loss of vascular integrity, and bleeding into the skin. With disease progression, tissue becomes necrotic, and patients with purpura fulminans may lose digits or limbs to complete thrombosis and necrosis of the affected areas. Purpuric lesions are commonly seen in patients with meningococcemia, and as many as 25% of infections may progress to purpura fulminans. The induction of purpura fulminans during meningococcemia may be due to its predilection to adhere specifically to endothelial cells in microvessels. Meningococcus has also been shown to induce shedding of endothelial protein C receptor (EPCR). In health, EPCR retains the anticoagulation protein C on the surface of vascular endothelial cells. Protein C maintains a negative feedback loop, preventing uncontrolled activation of the coagulation cascade. During meningococcemia, microbe-induced loss of EPCR results in protein C inactivation, leading to rapid and uncontrolled thrombosis.[153]

Clearance. Complement is the most important component of anti-meningococcal host defense (Fig. 43.6).[147,154,155] The complement component C3b directly binds to meningococcus, facilitating opsonphagocytosis. C3b can also initiate formation of the MAC, comprising components C5-C9. The MAC lyses and eliminates meningococci. Effective clearance of meningococci is so dependent upon the MAC that any functional impairment of any MAC component increases susceptibility to invasive meningococcal disease.[147] Patients with deficiencies of C5-C9 often present with recurrent invasive meningococcal infections.[147] Treatment with eculizumab, an anti-C5 monoclonal antibody used in patients with complement-driven paroxysmal nocturnal hemoglobinuria and atypical hemolytic-uremic syndrome (HUS), increases the risk of invasive meningococcal disease by over 1000-fold.[156]

In addition to MAC formation, functional deficiencies in components of the alternative pathway of activation have also been associated with increased susceptibility to meningococcus. Genetic deficiency of properdin, a component that binds and stabilizes C3b, has been linked to rapidly fatal meningococcal infections.[147,157] Functional impairments of complement factor D (in the alternative pathway) have also been shown to predispose to severe meningococcal disease.[147] Several functional variants in the gene encoding mannose-binding lectin 2 were previously linked genetically to invasive meningococcal disease, but the association was not replicated.[147] An expert panel on primary immunodeficiency recently removed "MBL deficiency" from the list of primary immunodeficiencies.[158]

Initiation of NF-κB signaling is also critical to effective meningococcal clearance. TLR4 and TLR9 are essential for sensing meningococcus and initiating the proinflammatory intracellular signaling cascade. Signaling is dependent upon IRAK4 and NEMO; variants or deficiencies in these two signaling intermediaries increase susceptibility to meningococcal infection, along with other bacterial diseases. Activation of NF-kB upregulates IL-6 and TNFα.[159] Available evidence suggests IL-1β is generated by activation of the NLRP3 inflammasome.[160] Induction of these cytokines is required for generation of a rapid and effective defense against meningococcus; deficiencies or variants in these mediators have also been linked to increased meningococcal susceptibility.[147]

Meningococcal infection is profoundly proinflammatory, activating a cellular defense consisting largely of activated neutrophils (Fig. 43.6). Activated neutrophils phagocytose meningococci that are opsonized by antibodies and complement. Complement is also essential for activating neutrophils to effectively kill meningococcus.[161] The mechanism by which neutrophils kill meningococcus has not been established. Meningococcal infection induces NET formation, but netosis does not appear to be required for clearance.[162] Trafficking of meningococcus to phagolysosomes, where exposure to AMPs, serine proteases, and ROS occurs, may be the primary means by which engulfed bacteria are killed.[162] Functional deficiency of the NADPH oxidative burst (as in CGD) has not been associated with increased susceptibility to meningococcal disease. However, neutrophils alone may be insufficient to control or clear meningococci, since severe meningococcal infections can progress despite robust neutrophil responses.[162]

Immunotherapy (Vaccines)

Since meningococci are clearly susceptible to complement mediated killing, meningococcal vaccine development strategies induce antibodies that enhance complement deposition on the bacterial surface. In the past, purified polysaccharide capsules were used as vaccines with limited success, although the polysaccharides are weakly immunogenic and not useful for children. Consequently, the polysaccharide vaccines have been replaced with two types of vaccines that are currently available. One type is a conjugate vaccine that induces Ig to the polysaccharide capsules of groups A, C, W, and /or Y. As described for PCVs, the meningococcal conjugate vaccines elicit T cell help and strong Ig responses even in young children. The other type (Bexsero and Trumenba) contains membrane proteins found in serogroup B strains such as factor H binding protein (FHBP), neisserial heparin binding antigen,

PorA, and NadA.[163] NadA is a trimeric protein that supports adhesion to and invasion of host cells.[164] FHBP normally binds the host's factor H protein and reduces complement deposition on meningococci. The fact that a serogroup B vaccine (Bexsero) contains FHBP only clearly reflects the critical importance of complement deposition for meningococcal killing.

Shigella spp.

Introduction, Epidemiology, and Diseases/Conditions

Shigella spp. are Gram-negative, facultatively anaerobic bacilli that cause shigellosis, otherwise known as dysentery.[25] Bacterial dysentery is a clinical syndrome of fever, abdominal pain, and profuse diarrhea that often contains blood and mucus. *Shigella* are highly virulent, with an estimated infectious dose of 10 to 100 organisms sufficient to cause disease. *Shigella* was estimated to be the second most common cause of diarrheal death worldwide in 2016, affecting all age groups, but with the largest impact upon children younger than 5 years of age in low-middle income countries.[165] Over 900 million children younger than 5 years are estimated to suffer diarrheal disease annually, of whom about 500,000 die.[166] Four species of *Shigella* have been identified: *Shigella flexneri*, *Shigella sonnei*, *Shigella dysenteriae*, and *Shigella boydii*. Each species comprises multiple serotypes, such that over 50 serotypes of *Shigella* spp. have been defined.

Shigella can be found in contaminated water and soil, and transmission is via the fecal-oral route.[167] Transmission of *Shigella* can also be person to person,[167] and flies may carry organisms from feces to food.[168] Those who are recovering from *Shigella* infection may shed bacteria in the stool for some weeks to months after becoming asymptomatic;[168,169] these asymptomatic carriers likely contribute to disease persistence and spread within the community.

Pathogenic Mechanisms

Host defense and virulence factors during invasion and infection. After ingestion, *Shigella* invades across the colonic epithelium to cause acute inflammation and disease (Fig. 43.7). The major virulence factors include T3SS, LPS, and inhibitors of the NF-κB signaling pathway.[25,170] *S. flexneri* can induce epithelial cells to produce factors that favor its survival and invasion. First, *S. flexneri* increases the production of specific mucins to alter the composition of the mucus layer and enhance its ability to invade colonic epithelium.[171,172] *Shigella* also expresses enzymes that cleave host mucin, facilitating access to host epithelium.[25] Secondly, *S. flexneri* induces downregulation of human beta-defensins and LL-37, to which it is otherwise highly susceptible.[172]

Shigella translocates across the epithelial barrier via one of two mechanisms (Fig. 43.7). The primary means is attachment to M (microfold) cells, which are specialized epithelial cells that overlie gut-associated lymphoid tissue (GALT), such as Peyer patches or lymphoid follicles. M cells "sample" lumen contents through endocytosis and transcytosis of luminal particulate matter. Luminal samples are then transferred to the mononuclear phagocytes localized in close approximation to the M cells.[173] *Shigella* hijacks this system to promote its uptake and transfer into the tissue-resident mononuclear phagocytes. After infection, *Shigella* promotes the apical secretion of IL-8, a neutrophil chemoattractant. Neutrophils transmigrate across the epithelial barrier, disrupting the epithelium. Neutrophil-mediated barrier disruption permits the second mechanism of bacterial translocation. Thus, the initial innate immune response actually enables barrier invasion by *Shigella*.[25,173]

The T3SS drives invasion into host cells. T3SS, present on the bacterial surface, delivers *Shigella* virulence factors directly into the host cell cytoplasm. Injected effector proteins (such as IcsA) induce actin-based ruffling of the epithelial cell membrane to prompt bacterial internalization through macropinocytosis.[25] Inside the epithelial cell, additional effectors (eg, VirA, IcsB) inhibit autophagy (see Mtb, below) and disrupt host cellular endosomal trafficking pathways to prevent bacterial destruction in autophagolysosomes. *Shigella* is transcytosed to the mononuclear phagocytes that reside in the basolateral pocket of the M cell. After invading the macrophages, *Shigella* escapes from the phagolysosome and induces pyroptosis. Macrophage pyroptotic death releases *Shigella*, which then invades the basolateral aspect of colonic epithelial cells through binding of basolateral receptors such as CD43. Infection of colonic epithelium results in epithelial cell death and destruction of tight junctions between cells. Loss of barrier integrity permits additional translocation of bacteria and drives the major symptom of bacillary dysentery: profuse, watery diarrhea, often containing blood and mucus.

Several *Shigella* virulence factors directly suppress the innate immune system. For instance, the proteins Osp1 and OspG inhibit NF-κB signaling through modulation of ubiquitin-mediated degradation pathways (Fig. 43.3A). OspZ and IpaH9.8 impair NF-κB through other mechanisms. Reduced NF-κB signaling results in downregulation of the antimicrobial peptides and other proinflammatory mediators.[25] OspF inhibits the MAPK signaling pathway, reducing activation of the transcription factor AP-1. One of the downstream effects is to reduce production of the chemokine CCL20. Loss of CCL20 reduces the recruitment of DCs, thus impeding the development of an adaptive immune response. Despite these inhibitors, however, a neutrophil-mediated innate response is usually able to clear the initial infection within 5 to 7 days.[25] NETs effectively trap extracellular *Shigella*, and neutrophil elastase cleaves virulence factors such as IpaB.[47]

Shigella spp. may elaborate Shiga toxin (STx), named after Kiyoshi Shiga, who first described *Shigella* as a cause of bacillary dysentery. STx-like toxins are also produced by enterogenic and enterohemorrhagic *E. coli* and can cause HUS. A more detailed description of the actions of STx is provided in the section below, "Additional Select Toxin-mediated Diseases."

Adaptive Responses. *Shigella* produces multiple virulence factors that impair and delay effective adaptive immune responses. Reduction of CCL20 production impairs effective recruitment of DCs. Bacterial effector IpgD dephosphorylates PIP2 in T cell membranes, interfering with the cytoskeletal reorganization required for T cell migration. IpaD cooperates with additional bacterial cosignals to induce B cell apoptosis.

Natural humoral immunity to *Shigella* targets mostly carbohydrate O-antigen of LPS, which is a weakly immunogenic polysaccharide antigen. Further, the natural immunity is O-antigen serotype specific and *Shigella* species express diverse

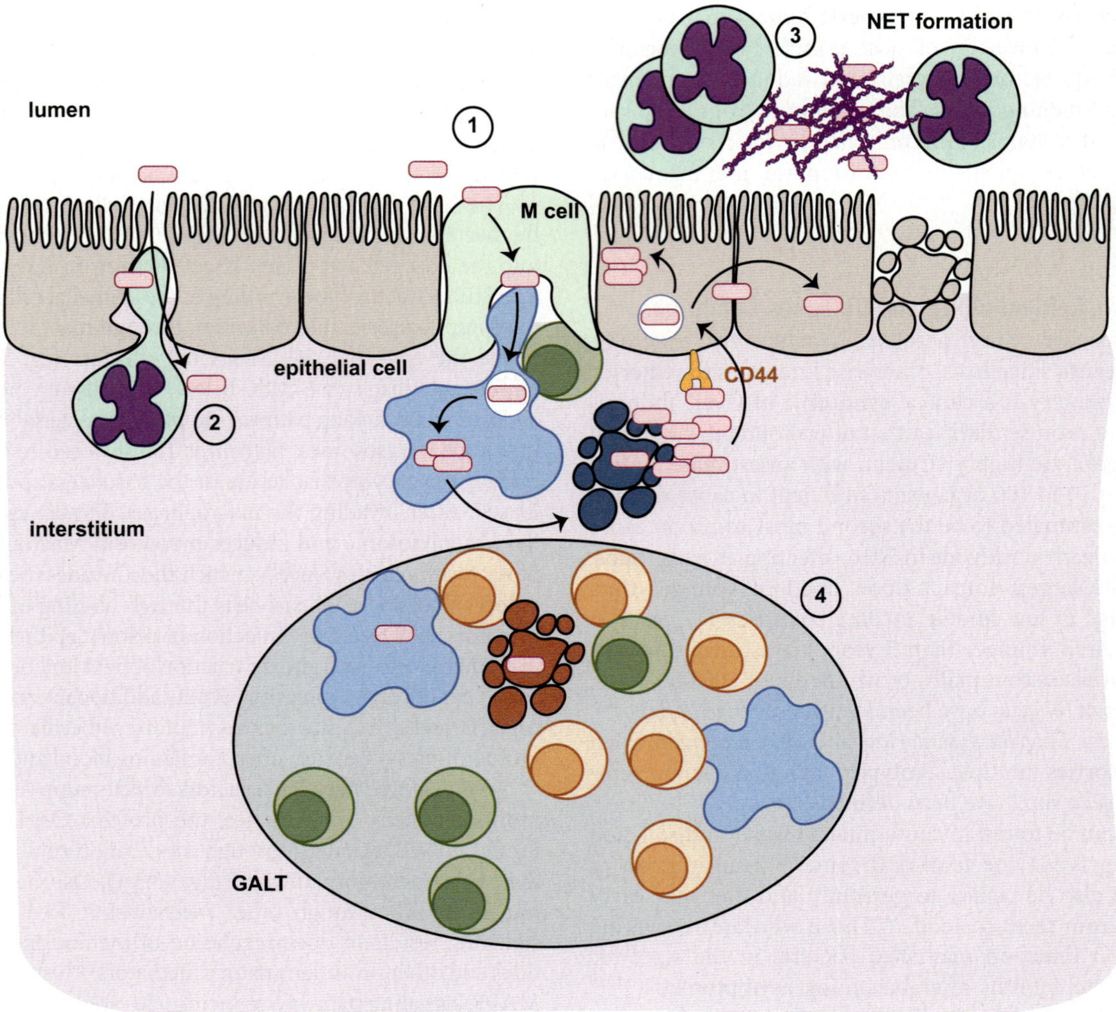

FIG. 43.7. Pathogenesis and host defense in *Shigella* spp. infection. *Shigella* is transmitted through ingestion of contaminated food or water. **(1)** Microbes first gain access by translocating across M cells in the intestinal epithelium, trafficking into the myeloid cells that reside on the basolateral side. Once internalized, *Shigella* escapes the phagolysosome, proliferates, and induces apoptosis of the myeloid cell. Cell death releases live bacteria into the interstitium. Bacteria may then enter intestinal epithelial cells on the basolateral side, again escaping from phagosomes to proliferate and kill the host cell. **(2)** After infection, intestinal epithelial cells release proinflammatory mediators, such as IL-8, that recruit neutrophils. Inflammation induced by cytokines and translocation of neutrophils across the epithelium into the luminal space permits additional translocation of *Shigella* across the epithelial barrier **(3)** Neutrophils in the luminal space eliminate bacteria, in part through netosis. The significant inflammatory response and epithelial damage result in the classic symptoms of bacterial dysentery: profuse, watery diarrhea, often with mucus or blood. **(4)** Trafficking of infected myeloid cells to the gut-associated lymphoid tissue (GALT) can initiate an adaptive response. *Shigella* can also induce apoptosis of B cells, impairing an effective host response. Initial infections with *Shigella* only induce T-independent humoral responses, which are serotype-specific and short-lived. Repeated *Shigella* infections can induce T-dependent humoral responses, which eventually provide protection for the host against further infection. NET, neutrophil extracellular trap

O-antigens. Consequently, reinfections with other serotypes are frequent. In the wake of dramatic successes with PCVs, there are active efforts to develop a multivalent O-antigen conjugate vaccine, which should provide longer lasting immunity to multiple prevalent serotypes. An interesting approach to develop a *Shigella* conjugate vaccine is to genetically engineer a bacterial strain that naturally produces a conjugate vaccine expressing an O-antigen. Such an engineered strain would simplify the production of conjugate vaccines and drastically reduce the cost, making *Shigella* conjugate vaccines available to lower- and middle-income countries. Proof of principle studies have found this bioconjugate approach possible.[174]

P. aeruginosa

Introduction, Epidemiology, and Diseases/condition

P. aeruginosa is a Gram-negative, aerobic/facultative, oxidase-positive bacillus and an opportunistic pathogen. *P. aeruginosa* is ubiquitously found in water and in soil. As an environmental pathogen, *P. aeruginosa* is not normally carried by humans. However, patients with certain disease processes, such as CF, or with indwelling foreign material (such as those with plastic tracheostomy collars or with indwelling central lines) may become colonized by *P. aeruginosa*. Once colonized, *P. aeruginosa* forms biofilms, enabling significant microbial resistance against host immunity and antibiotic therapies, and eradication of carriage becomes extraordinarily difficult. The biofilms formed by *P. aeruginosa* are one of its greatest virulence factors.

Host Defense and Virulence Factors During Colonization and Invasion/Infection

Colonization. Like *S. aureus, P. aeruginosa* is a major pathogen colonizing the respiratory tract of patients with CF. The ubiquity of *P. aeruginosa* in the environment ensures universal airway exposure. Under normal circumstances, *P. aeruginosa* is cleared by mucociliary action. In CF, *P. aeruginosa* colonizes the thick secretions of the respiratory tract and forms biofilms, generating chronic inflammation. *P. aeruginosa* produces three exopolysaccharides to create the biofilm: alginate, Pel, and Psl.[175,176] Alginate protects bacteria from macrophage and neutrophil-mediated phagocytosis, and alginate scavenges ROS produced by immune cells.

Host AMPs have activity against *P. aeruginosa* and specifically can disperse the biofilm.[177] Human cathelicidin LL-37 has been shown to inhibit biofilms formed by *P. aeruginosa*[177] at concentrations of 0.5 μg/mL, much less than the MIC needed to kill microbes in planktonic growth.[177]

Invasion. The flagellum and pili of *P. aeruginosa* mediate adherence to host cells. The T3SS of *P. aeruginosa* directly injects toxins into host cells, damaging or eliminating the epithelial barrier.[178] Injection of the microbial protein VgrG2b recruits host microtubules to promote bacterial uptake into the host cell. *P. aeruginosa* is profoundly proinflammatory, triggering TLR4 through LPS in the cell walls and TLR5 through flagellin. Internal DNA motifs can also induce TLR3 signaling. Flagellin is also a potent stimulator of the NLRC4 inflammasome. Thus, intrinsic components of *P. aeruginosa* are sufficient to induce a rapid, systemic cytokine storm of IL-1β, TNFα, IL-6, and other proinflammatory mediators. Translocation of *P. aeruginosa* to the bloodstream and hyperproduction of these proinflammatory mediators can produce septic shock.

Clearance of *P. aeruginosa* from the bloodstream or other distal sites is largely dependent upon phagocytosis by neutrophils.[179] Uptake of *P. aeruginosa* by PMNs is facilitated by opsonization by C3b (engaged by CR3, expressed on neutrophils) and antibodies (engaged by FcγRII). After phagocytosis, engulfed bacteria are trafficked into the phagolysosome. Phagolysosomes fuse with several classes of granules to expose the bacteria to lethal peptides, such as bacterial/permeability-increasing protein (BPI), neutrophil elastase, proteinase 3, cathepsins, defensins LL-37, and lysozyme. BPI neutralizes LPS. Exposure to ROS generated by the NADPH oxidase system in phagolysosomes can also be bactericidal; however, *P. aeruginosa* produces catalase and may resist ROS-mediated killing. However, antipseudomonal activity of neutrophils is robust even in the absence of NADPH-generated ROS, so this mechanism is dispensable for antipseudomonal host def ense.[177,179,180] Neutrophils also undergo netosis in response to *P. aeruginosa* infections. In CF, ongoing netosis in response to persistent *P. aeruginosa* colonization may contribute to the chronic inflammation and tissue damage that characterizes the disease process.

Adaptive Responses

Patients chronically exposed to *P. aeruginosa*, such as patients with CF, develop a humoral antibody response. T cell–mediated responses appear to be skewed toward Th2, with diminished IFNγ and elevated IL-10 production. Antipseudomonal immunoglobulins form immune complexes with the bacteria, opsonizing for neutrophil-mediated phagocytosis. In sections from removed CF lungs, aggregates of *P. aeruginosa* surrounded by neutrophils can be observed.[181] However, humoral immunity is insufficient to promote effective clearance of bacteria in isolation. There is no vaccine available for *P. aeruginosa*. There have been dramatic improvements recently in treatments of CF, which directly target the defective transporter, and these treatments will reduce *P. aeruginosa* disease burden.

M. tuberculosis

Introduction, Epidemiology, and Diseases/Conditions

M. tuberculosis (Mtb) is a major global health threat, causing over 1.5 million deaths annually.[182,183] Mtb cannot be Gram stained because of the thick, waxy cell wall consisting of mycolic acids. The mycolic acids, along with other lipids, in the cell wall are a major virulence factor. The lipid-rich cell envelope defends against antibiotics and against the host immune system, and prevents routine Gram staining.[184] Instead, Mtb must be illuminated using the alternative acid-fast staining method. It is predominantly, but not always, aerobic and facultatively intracellular. Mtb is extraordinarily slow growing, with a doubling time of approximately 20 hours.[185] The prolonged doubling time creates treatment challenges, as infected patients must be treated for extraordinarily long durations with combinations of drugs. Insufficient treatment permits escape of drug-resistant mutants, and the human population is now challenged by the emergence of both multidrug resistant and extensively drug-resistant strains of Mtb.[186] Mtb pathogenesis and elements of host immunity have recently been reviewed in detail.[187]

Host Defense and Virulence Factors in Invasion/Infection

Although humans are the primary reservoir for Mtb, ***colonization*** with Mtb has not been found. Once exposed, some people seem able to resist and clear the infection, some develop a lifelong coexistence with latent tuberculosis infection (LTBI), while about 5% to 10% develop active pulmonary disease.[182] Those with LTBI have a 10% chance of developing active disease later in life.[188] Most cases of active Mtb are pulmonary. Rarely (<2% of active cases), hematogenous spread can result in disseminated Mtb infection. Disseminated Mtb is more likely to occur in young children and in immunocompromised individuals.[189]

Invasion. Mtb is transmitted by aerosol dispersion, which occurs when an infected individual coughs. Respiratory droplets expelled from the airways carry the infective Mtb bacteria into the surrounding air. Contacts who then breathe in the droplets become infected. Because the droplets that carry the bacteria are relatively small, they can remain suspended in the air for hours. After bacteria are inhaled into the airways, they are first engaged by alveolar macrophages. After engulfment, Mtb is trafficked into the phagolysosomes for bacterial killing. While effective AM responses may be sufficient to clear the organisms and end infection, Mtb has developed an array of mechanisms to subvert the host response, often converting the macrophage phagolysosome into comfortable living quarters. Inability to kill the intracellular Mtb results in eventual granuloma formation (Fig. 43.8A).

Mtb subverts the cellular process of autophagy in escaping host defense against Mtb.[190,191] ***Autophagy*** ("self-eating") was first described as a recycling pathway found in almost all eukaryotic cells that enables cells to survive conditions of starvation by digesting and reusing intracellular organelles and proteins. Autophagy begins with creation of the autophagosome, a cellular compartment with a double lipid membrane and expressing LC3 on the surface. The exact intracellular origin of the autophagosomal membrane is unclear; it has been postulated to derive from the plasma membrane, endoplasmic reticulum, and/or Golgi apparatus, along with other intracellular membranes.[192] After engulfment, a phagosome may fuse with the inner membrane of the autophagosome to deliver contents to the autophagy pathway. The autophagosome then fuses with the lysosome. Lysosomes, marked by LAMP1/2, contain hydrolytic enzymes and an ATP-dependent proton pump. After fusion, the acidic environment of the newly created autophagolysosome, along with the hydrolytic enzymes, can kill the engulfed bacteria.[193] Engulfed Mtb produces virulence factors that can prevent phagolysosome maturation. Failure to acidify the autophagolysosome prevents maturation, and viable Mtb persist in immature phagosomes.[193]

Induction of autophagy may enhance macrophage-mediated clearance of Mtb.[191] Autophagy is induced by TLR and NLR signaling, and by the cytokines IFNγ and TNFα. For effective control of Mtb, macrophages must adopt the proinflammatory M1 phenotype, with production of IFNγ and TNFα. Production of IL-4 or IL-10, associated with adoption of the M2 phenotype, correlates with increased risk of Mtb infection.[194] Production of the M1-associated mediators TNFα and IFNγ promotes phagosome maturation (eg, fusion with the lysosome) and enhances Mtb killing. Patients who receive anti-TNFα monoclonal antibodies (eg, infliximab) as therapy for autoimmune disease such as Crohn or rheumatoid arthritis are at great risk of reactivating Mtb if they are latently infected. If alveolar macrophages are able to eliminate invading bacteria through autophagolysosome-mediated killing, then patients do not become infected with Mtb. It is not clear how many exposed people can "resist" Mtb infection in this way.[182] Humoral immunity may also contribute to the ability to resist Mtb infection.[195]

Mtb evades host immunity by inhibiting phagosome maturation, preventing fusion of the autophagososome with the lysosome (Fig. 43.8A).[193,196] Mtb may achieve inhibition of autophagy through dysregulation of the serine/threonine kinase mTOR, which is a master regulator of autophagy.[196] Mtb can escape the phagosome and exit into the cytosol, where it replicates. After Mtb replication and host cell death, Mtb is released into the host to infect surrounding macrophages and other antigen-presenting cells. The subversion of autophagy by Mtb provides the bacteria with an intracellular niche, shielding them from immune-mediated clearance.

The hallmark of Mtb infection is ***granuloma formation*** (Fig. 43.8B). Granulomas are highly organized collections of myeloid cells and lymphocytes that restrict dissemination of Mtb. Granulomas may also benefit Mtb, as many harbor live microbes.[197] Granulomas that undergo caseating necrosis may break and release live Mtb into the airways, which are then coughed into the air to propagate infection.[197] Granulomas undergo significant changes in cellular composition during the course of an Mtb infection, and granulomas in various stages of maturation can be found concurrently in infected specimens. Granulomas can contain activated macrophages, multinucleated giant cells, foamy macrophages, DCs, neutrophils, B cells, and αβ and γδT cells. Because of the complexity and chronicity of granuloma formation, the precise mechanisms that guide effective granuloma formation remain poorly defined. Additionally, while mice are susceptible to Mtb infection, the mouse model has limitations: mice always develop active tuberculosis (ie, mice have no equivalent LTBI), and the more commonly used low-dose models of infection in C57/Bl6 mice generate granulomas that are histologically different from those of human and nonhuman primates. A new advance, recognizing that ultra-low-dose infection (1–3 microbes) of mice may create granulomas closer to those seen in human disease, may now accelerate research in this area.[198]

Granulomas aid in host defense by trapping Mtb within the granuloma and preventing dissemination. However, granulomas may also delay clearance, as the wall protecting the host from the bacteria may also protect the bacteria from host defense. Granuloma formation may also protect Mtb from antimicrobial therapy. Granulomas begin when alveolar macrophages ingest but cannot kill Mtb. Activation of the alveolar macrophage induces production of chemoattractants that recruit additional inflammatory cells, such as neutrophils, monocyte-derived macrophages, and γδT cells. Mtb may promote granuloma formation by stimulating epithelial cells to secrete MMP9. MMP9 further recruits resting macrophages.[197] Early granulomas therefore consist of a loose aggregate of innate myeloid cells, in which viable Mtb replicate in immature phagosomes.[197]

Maturation of the granuloma requires activation of T cells. DCs migrate to pulmonary or mediastinal lymph nodes and present Mtb antigens to conventional αβ T cells, promoting differentiation or activation of Th1, Th17, and cytotoxic CD8+ T cells. T cells then migrate to and surround the nascent granulomas, activating the macrophages and enhancing Mtb killing. Mature granulomas therefore appear to consist of lymphocyte "rims" surrounding cores of infected myeloid cells. Cell turnover in granulomas appears to be quite dynamic; infected macrophages die and are replaced by newly recruited myeloid cells. Lymphocytes traffic in and out of the aggregate, and the surrounding vasculature and tissue are remodeled in response to inflammation. Macrophages in the center of the granuloma undergo morphological changes, possibly accumulating lipids and becoming "foamy," or differentiating into "epithelioid" macrophages. Epithelioid macrophages, also called histiocytes, are activated macrophages differentiated to form tight interdigitations with neighboring cells.[199] Epithelioid macrophages histologically define the granulomas found in

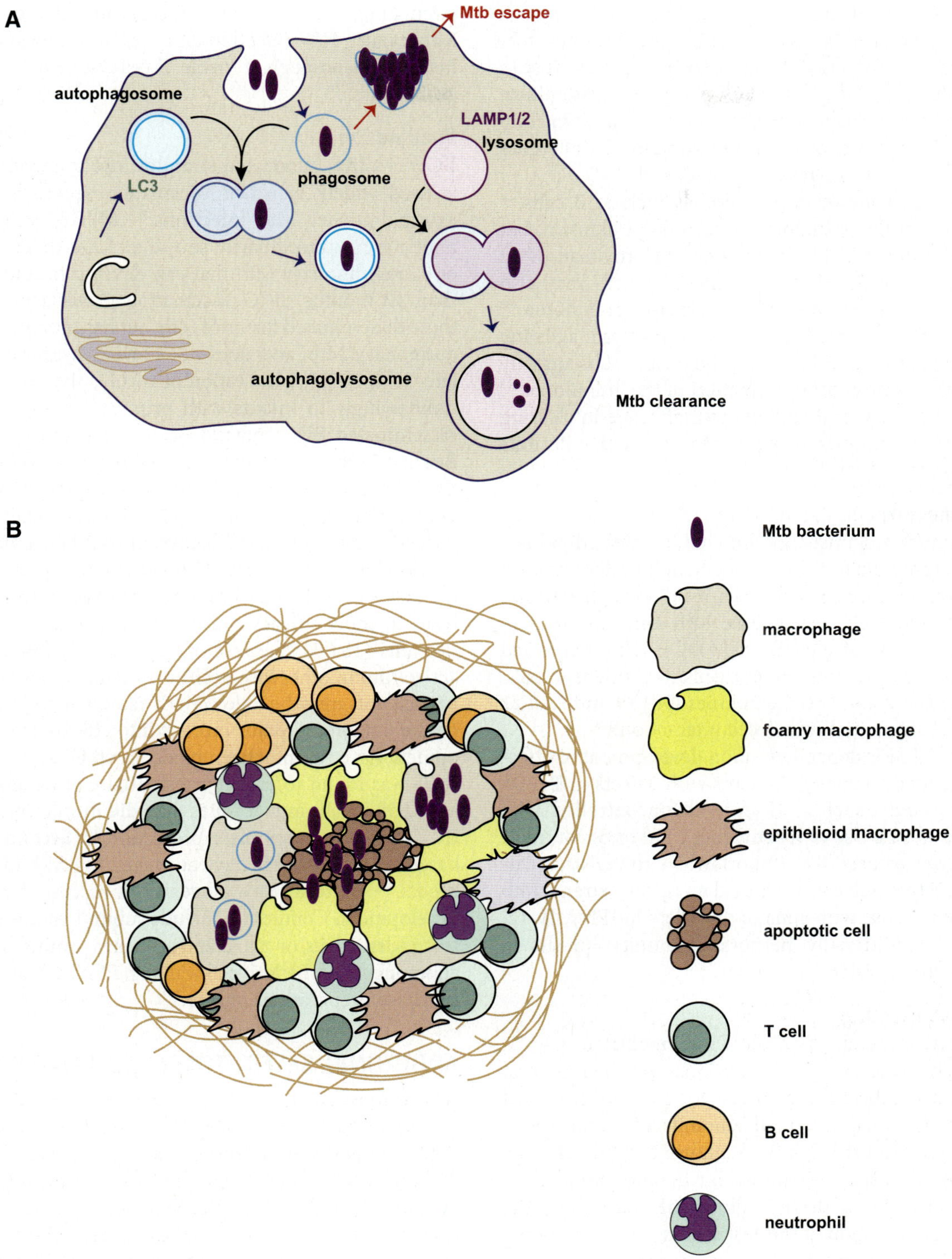

FIG. 43.8. Restriction of *Mycobacterium tuberculosis* (Mtb) through autophagolysosomal degradation and granuloma formation. A: Autophagy is a cellular process upregulated during starvation to recycle cellular contents. Autophagosomes are dual-membrane intracellular vesicles tagged with LC3 that target contents for degradation in lysosomes. After phagocytosis, fusion of an Mtb-laden phagosome with an autophagosome promotes fusion with a LAMP1/2-positive lysosome. Lysosomal fusion acidifies the environment and creates a microbicidal compartment, eliminating Mtb. Virulent Mtb expresses factors that actively inhibit phagolysosomal maturation and acidification. Mtb may then proliferate in the phagosome, eventually escaping and killing the host cell. Cell death then releases viable microbes and infection spreads. **B:** Granulomas are a complex multicellular histological structure than restricts spread of Mtb to tissues, but may also protect Mtb from killing. At the center of granulomas are necrotic tissue and infected macrophages. Macrophages that have become engorged with components of the tuberculosis cell walls (mycolic acids) are lipid laden and appear "foamy." Macrophages within the granuloma may differentiate into an "epithelioid" appearance, forming tight interdigitations with neighboring epithelioid cells. Activated T cells form a rim around the granuloma, delivering IFNγ and TNFα and promoting the M phenotype necessary for Mtb control. B cells are also recruited into an organized granuloma, though the function is currently unclear. Neutrophils are often present in granulomas; a high neutrophil predominance is unhelpful and correlates with loss of Mtb control. The granuloma is vascularized and surrounded by collagen and fibrinogen deposition to restrict spread of infection. The granuloma may undergo necrosis and caseate, liquefying and releasing viable Mtb to continue infection. Alternatively, Mtb may be effectively cleared from the granuloma, and the granuloma may undergo calcification and healing.

tuberculosis. Mtb produces a number of lipid products, such as trehalose dimycolate (a mycolic acid in the cell wall), that engorge foamy macrophages.[200] Macrophages may also fuse to generate multinucleated giant cells.[201] Mtb may disseminate when infected macrophages migrate from the granuloma into surrounding host tissue, or when the granuloma undergoes necrosis and breaks down, releasing viable Mtb.[202,203]

Additional cells found in granulomas include B cells.[204] B cell function in the granuloma is currently unknown, but a possible role in CXCL13 production and provision of an organizing structure has been speculated. B cells may also serve as antigen-presenting cells within the granuloma.[203] Group 3 ILCs are activated early in Mtb infection, assisting in the recruitment of alveolar macrophages and subsequently lymphocytes, thus promoting formation of the granuloma.[205] Neutrophils may be recruited, but are ineffective in clearing Mtb, and in some contexts, may provide a niche that permits ongoing Mtb replication.[206,207]

Adaptive Immunity

Adaptive immune responses are initiated by infected pulmonary DCs that migrate to the draining lymph nodes. Infected DCs may present antigen to T cells residing in the lymph node, or infected DCs may release Mtb antigens that are phagocytosed by noninfected, resident DCs. The noninfected resident DCs then present Mtb antigens to resident T cells. The cytokine IL-12 and the chemokines CCL19 and CCL21 are essential to this initial DC+T cell interaction.[208]

Control of LTBI is dependent upon development of effective Th1-mediated responses. Patients with defects in T cell–mediated immunity, such as those with uncontrolled HIV infection, are at high risk of reactivating LTBI or of contracting Mtb if newly exposed. The importance of the IFNγ signaling axis to effective Mtb control is evident by the extreme Mtb susceptibility of those with genetic disorders in IFNγ, IL-12, and STAT1.[209] Additionally, humoral immunity appears to contribute to the "resister" phenotype.[187,195]

Immune-Based Testing

Testing for Mtb depends upon the T cell–mediated type IV hypersensitivity reaction. The traditional purified protein derivative (PPD) tuberculin skin test for Mtb is performed by intradermally injecting small amounts of sterile (noninfectious) protein purified from a culture of Mtb. After 48 hours, the injection site is examined for the induration associated with dense T cell–driven cellular infiltration into the injected area. A limitation of this test is cross-reactivity of the T cells specific for either atypical mycobacterial species or the vaccine strain (bacille Calmette-Guérin [BCG]). Newer tests now rely on a modified IFNγ ELISPOT assay. In brief, T cells are isolated from the blood and exposed to specific peptide sequences found in the Mtb cell surface proteins CFP-10, ESAT-6, and TB7.7. The number of T cells that produce the Th1 signature cytokine IFNγ is then determined. A large number of IFNγ-producing memory T cells in response to exposure to specific Mtb antigen indicate prior or current LTBI infection. The benefit of the ELISPOT test is its relative specificity compared to the tuberculin skin test. Both tests are limited in their utility by dependence upon sufficient circulating, active memory T cells. In patients with severe Mtb, cancer, HIV infection, or other immunosuppression, an insufficient number of viable T cells limit the sensitivity of both assays.[210]

Vaccination

BCG is a live, attenuated strain of *Mycobacterium bovis* and is used widely in infant vaccination schedules to prevent extrapulmonary Mtb infections.[211] Miliary and Mtb meningitis are extraordinarily severe and life-threatening forms of extrapulmonary Mtb that can develop in infants younger than 12 months. BCG vaccination protects infants against these disseminated forms of Mtb, but has little efficacy against pulmonary Mtb, and does not prevent reactivation of LTBI later in life.[183,209] A limitation of BCG is that, if administered unknowingly to infants with primary immunodeficiencies, disseminated BCG infection can occur and can be fatal.[212] Because of the relatively low prevalence of Mtb infection and the limitations of the BCG vaccine, the control of Mtb in the United States depends upon rapid contact tracing of infected individuals and prompt effective antimicrobial therapy.

Development of a new Mtb vaccine has been hindered by the lack of a validated immunoprotective correlate. Clinical trials to show efficacy must include a large number of participants (10,000) because of the relatively low incidence of active Mtb in otherwise healthy individuals, and clinical trial participants must be followed for extended periods of time to define outcomes. Major inroads into effective vaccine development have recently been made. First, IV administration of BCG was found to induce substantially more anti-Mtb CD4 and CD8 T cells in macaques than did intradermal or aerosol BCG, providing complete protection.[213] Secondly, a CMV-based vaccine was effective at preventing Mtb in a macaque model.[214] There are now currently 14 vaccines for Mtb under development. Candidates include subunit vaccines, viral vectors (adenovirus or avirulent influenza strains) that express components of the Mtb cell wall (eg, ESAT-6), or avirulent or sterilized preparations of Mtb.[183]

ADDITIONAL SELECT TOXIN-MEDIATED DISEASES

As previously mentioned, multiple clinically and historically important bacterial diseases are driven by toxin production, not bacterial invasion per se. Some of these infections, such as diphtheria and tetanus, have been so successfully controlled by routine childhood vaccination that they are at risk of being regarded as insignificant. However, these diseases have not disappeared, as they reappear in undervaccinated communities because of vaccine hesitancy (eg, incorrect perception that the risk of the vaccine outweighs the risk of the disease) or poor socioeconomic conditions. While other infections, such as anthrax, occur far more rarely than in past history, they could be used as potent bioweapons. Finally, some toxin-mediated illnesses, such as *Clostridioides difficile*, are currently ongoing challenges to medical practice. We briefly review the pathophysiology of these select toxin-mediated bacterial diseases, with an emphasis on immunomodulation, development of protective immunity, and/or the historical importance to vaccine development and public health. We discuss the respiratory infections (pertussis and diphtheria) first and then the systemic syndromes (tetanus, botulism, and anthrax) and the GI diseases (cholera, STx, and *C. difficile*) (Table 43.3).

Bordetella pertussis: Pertussis (Whooping Cough)

Pertussis, or "whooping cough," is caused by infection with the Gram-negative organism *Bordetella pertussis*. While thought to be primarily an extracellular pathogen, *B. pertussis* may invade alveolar macrophages and respiratory epithelial cells, evade phagolysosomes, and establish an intracellular reservoir.[215] Humans are the only known reservoir,[216] and transmission is via respiratory droplets. In older children and adults, pertussis causes an upper respiratory infection with a prolonged, paroxysmal cough. In young infants, the paroxysmal cough may be followed by an inspiratory gasp, producing the "whooping" sound for which the disease is named. Pertussis may also induce apnea, cyanosis, pulmonary hypertension, and respiratory failure in young infants (younger than 4 months).[216,217] In the first half of the 20th century, pertussis was the primary cause of infant mortality.[218] Pertussis may also be accompanied by profound lymphocytosis, in contrast to many other bacterial infections.[216]

Pertussis toxin (PTx) drives pertussis pathogenesis. *B. pertussis* attaches to ciliated cells of the respiratory epithelium and does not disseminate systemically. PTx is an ADP-ribosyltransferase that irreversibly inhibits the Gα_i subunit of G protein–coupled receptors (GPCRs) through ribosylation.[216,219] As chemokine receptors are GPCRs, PTx-mediated

TABLE 43.3 Summary of Select Toxin-Mediated Bacterial Diseases

Toxin(s)	Organism(s)	Disease/Syndrome(s)	Key Points
PTx	*Bordetella pertussis*	Pertussis (whooping cough)	• PTx is used experimentally to inhibit GPCRs, which includes many chemokine receptors. • Use of acellular pertussis vaccine has resulted in waning immunity; whole cell pertussis vaccine had too many adverse events.
DTx	*Corynebacterium diphtheria*	Diphtheria	• Equine antisera were developed in the late 1800s to treat diphtheria.
TeNT	*Clostridium tetani*	Tetanus (lockjaw)	• Equine antisera were developed early in the 20th century to treat tetanus. It was the first passive immunotherapy to be used prophylactically.
BTx	*Clostridium botulinum*	Botulism	• The only directed therapeutic for infant botulism is BabyBIG, which consists of pooled sera from human volunteers immunized with botulinum toxin. One course of BabyBIG costs $50,000.
LT ET	*Bacillus anthracis*	Cutaneous ulcers GI anthrax Inhalation anthrax	• Anthrax was used as a model organism to derive Koch's postulates, in generation of the first live attenuated vaccine, and to first describe phagocytosis. • Anthrax toxins act directly on immune cells to shut down the early innate responses, permitting disease progression.
CTx	*Vibrio cholerae*	Cholera—profuse, watery diarrhea	• Mucosal immunity to cholera is poorly understood; while natural infection seems to provide lifelong immunity, current vaccination strategies seem to provide only short-term protection. • CTx itself has profound immunomodulatory effects and is being used in adjuvant strategies and as platforms for other vaccine efforts.
STx	*Shigella* spp. EHEC ETEC	Bacillary dysentery Bloody diarrhea Hemorrhagic colitis HUS	• STx-producing organisms (such as ETEC and EHEC) cause up to 1 billion cases of diarrhea annually and cause significant childhood stunting. • No vaccine is currently available.
TcdA TcdB	*Clostridium difficile*	Antibiotic-associated diarrhea Fulminant colitis	• *C. difficile* infection is characterized by derangements in the host GI microbiome; fecal microbiota transplant is a recognized therapy for recurrent CDI. • Antitoxin immunoglobulins can provide protective immunity, and vaccines are currently under development.

BTx, botulism toxin; CDI, *Clostridium difficile* infection; CTx, cholera toxin; DTx, diphtheria toxin; EHEC, enterohemorrhagic *Escherichia coli*; ET, edema toxin; ETEC, enterotoxigenic *Escherichia coli*; GI, gastrointestinal; GPCRs, G protein–coupled receptors; HUS, hemolytic-uremic syndrome; LT, lethal toxin; PTx, pertussis toxin; STx, Shiga toxin; TeNT, tetanus neurotoxin.

GPCR inactivation profoundly affects host immunity. For instance, CXCR4, the receptor for CXCL12, is PTx sensitive and PTx treatment of T cells prevents CXCL12-induced migration.[219] In fact, PTx can be purchased from commercial scientific providers as an experimental reagent to inhibit GPCRs and to alter leukocyte migration. During infection, pertussis may be marked by profound leukocytosis (total white blood cell count, ≥30,000 cells/μL; normal ≤12,000 cells/μL) and lymphocytosis, impaired macrophage activation, and impaired leukocyte trafficking.[216] Pulmonary circulation can become hyperviscous, and pulmonary hypertension can develop. Leukocytosis and pulmonary hypertension are significantly associated with likelihood of death.[220] Despite ongoing infant mortality due to pertussis, the pathogenesis leading to death remains unclear and vaccine developments remain empirical. All organ systems, including the cardiovascular system, use GPCRs (including chemokine and chemokine receptors) to maintain homeostasis. PTx may have direct cardiotoxic effects that lead to cardiovascular collapse, independent of the effects of PTx on hematopoietic cells.

B. pertussis remains susceptible to macrolide antibiotics, such as azithromycin. However, because the disease is driven primarily by PTx, severe morbidity and mortality can occur even with appropriate antimicrobial therapy. Vaccination remains the most important strategy for combatting pertussis. The first vaccine, consisting of formalin-inactivated "whole cell" *B. pertussis,* was developed in the mid-1940s.[218,221] While effective, the so-called "whole cell" vaccine produced a number of adverse events, including high fevers, seizures, and (rarely) encephalopathy.[221] A subunit or "acellular" pertussis vaccine was therefore developed and introduced into the vaccine schedule in the 1990s. Acellular pertussis vaccine includes up to five different protein antigens, one of which is inactivated PTx, and effectively protects pediatric recipients from disease.[221] However, immunity of a population wanes with time. Waning herd immunity has led to sporadic outbreaks of disease in presumably vaccinated populations—largely among young infants aged 2 months or younger, who are too young to have received the first dose of vaccine. To counteract waning immunity, a booster dose is now recommended between ages 11 and 18 years, and pregnant women are given a booster to protect their infants.

The waning immunity afforded by the acellular pertussis vaccine has raised multiple questions. First, it is unclear if humoral immunity is sufficient to provide protection against pertussis infection. Natural immunity and immunity induced by the whole cell vaccine may generate cell-mediated responses, including those of Th1 and Th17 cell subsets that may be important in antipertussis immunity. Secondly, it is unclear if different adjuvants provided during acellular pertussis vaccination might be sufficient to provide longer lasting immunity.

Corynebacterium diphtheriae: Diphtheria

Diphtheria is a toxin-mediated disease caused by colonization or infection with *Corynebacterium diphtheriae. C. diphtheriae* is an anaerobic, Gram-positive coccobacillus,[222] with humans the only known reservoir.[223] DTx is an exotoxin that ADP-ribosylates elongation factor 2, inhibiting protein synthesis, and causing cell death.[224] The lethal dose of DTx is 100 to 150 ng/kg.[225] Infection with *C. diphtheriae* usually occurs in the nasopharynx, with localized tissue necrosis induced by toxin and generation of a grayish "pseudomembrane" visible retropharyngeally. Systemic release of toxin is cardiotoxic, and death occurs from respiratory or cardiac failure.[224] In unimmunized populations, diphtheria currently carries a case fatality rate ranging from 5% to 17%.[224] In the prevaccination, preantibiotic era, diphtheria was a major cause of childhood mortality, with case fatality rates as high as 40%.[224,226,227] Diphtheria continues to be endemic or to break out in sporadic epidemics in areas characterized by poor vaccine coverage and population instability.[227]

Immunity against diphtheria is provided by neutralizing antitoxin antibodies. Along with tetanus, diphtheria was one of the first two diseases to be successfully treated with equine antisera (passive immunization). The Iditarod, an annual sled dog race in Alaska run from Anchorage to Nome, commemorates a heroic sled dog relay in 1925 that carried equine anti-DTx antisera from an airstrip in Anchorage to the village of Nome in 5.5 days. At the time, Nome was threatened with an outbreak of diphtheria that could have killed all children in the village. Active immunization against diphtheria, using diphtheria toxoid, began in the 1940s and has dramatically reduced morbidity and mortality.[226,227] Currently, diphtheria vaccine is included in the recommended vaccine schedule for all children.

Clostridium tetani: Tetanus

Tetanus (once called "lockjaw") is a toxin-mediated disease characterized by spastic paralysis. The name "lockjaw" derives from the symptoms of uncontrolled muscle contractures preventing mouth opening (trismus). Tetanus may have been first described around 3000 BC and was definitively recognized as a syndrome by Hippocrates in 400 BC.[228] The tetanus neurotoxin (TeNT) is produced by the anaerobe *Clostridium tetani. C. tetani* is ubiquitously present in soil and frequently contaminated battlefield and surgical wounds in the preantibiotic and prevaccine era.[229,230] TeNT is taken up by peripheral motor nerves and undergoes retroaxonal transport to the spinal cord. Within the spinal cord, TeNT is trafficked from peripheral motor neurons to inhibitory interneurons. Inhibitory interneurons are GABAergic neurons that inhibit activation of opposing muscles during contraction.[231] Because TeNT blocks neurotransmitter release from inhibitory neurons, TeNT results in uncontrolled spastic contractures (activation of muscles)[232] that can break bones. Similar to botulism, tetanus is fatal because of respiratory or cardiovascular paralysis.[232] The common occurrence of tetanus among wounded soldiers drove interest in antitetanus therapies in the first half of the 20th century.

The history of tetanus immunization is important to vaccinology. Tetanus was first recognized as toxin mediated in 1890, and, along with diphtheria, was one of the first two diseases against which passive immunization therapies were developed using equine antisera.[229,230] Antisera were generated by injecting horses every 8 days with increasing doses of purified tetanus toxoid, and equine antisera were then administered to patients. With recognition that passive immunization against tetanus was largely ineffective after symptom onset (now known to be due the localization of TeNT to the spinal cord in symptomatic disease), prophylactic passive immunization was begun during WWI by administering equine antisera to all wounded soldiers.[230] While effective, equine antisera also carried significant risk of adverse events, ranging from local symptoms to anaphylactic shock and death. Active immunization strategies were therefore developed in the 1920s, attaining widespread use among soldiers in World War II.[233] Active immunization strategies rely on administration of chemically inactivated toxin, termed "toxoid."[229] These medical examples showed that antitoxin antibodies are sufficient to provide protective immunity.

Active immunization using tetanus toxoid is now standard in most high-income countries and is effective at preventing tetanus. Incidentally, chemically inactivated tetanus toxoid is also used as a "carrier protein" for polysaccharide conjugate vaccines against encapsulated organisms, such as the anti-PCVs.[234] Tetanus immunization is included in the routine pediatric immunization schedule recommended by the CDC. While neutralizing immunoglobulins are generated by vaccination, immunity is not lifelong. Booster shots are required to maintain effective levels of circulating antitoxin antibodies.[234] While effectively controlled by vaccination in high-income countries, hundreds of thousands of infants and mothers still die of tetanus in undervaccinated low- and middle-income countries.[228]

Clostridium botulinum: Botulism

Botulism is a severe disease characterized by flaccid paralysis caused by toxin-mediated blockade of acetylcholine (a neurotransmitter) at neuromuscular junctions. Acetylcholine is released by nerves to induce muscular contraction; in the absence of acetylcholine release, muscles cannot contract and flaccid paralysis results. Botulism is fatal when neuromuscular blockade results in diaphragmatic paralysis and respiratory failure.[235] Botulism toxin (BTx) is produced primarily by *Clostridium botulinum*, but may also be produced by other *Clostridium* species.[236] BTx is an "AB" toxin, with protomer A serving as the "activator" and protomer B serving as the "binding" domain.[237] *C. botulinum* is a spore-forming anaerobe found in the soil. The lethal dose of BTx is estimated at 1 ng/kg, and botulism is classified as a category A bioweapon.[238] Commercially produced BTx (Botox) is now used therapeutically to treat migraines and to reduce contractures in children with disorders such as cerebral palsy, as well as cosmetically to reduce facial wrinkling.[239-241]

Three clinical syndromes of botulism intoxication have been described: food-borne botulism, infant botulism, and wound-associated botulism. Food-associated botulism is the most commonly observed form and occurs when a patient ingests improperly handled food that has been contaminated with *C. botulinum* spores (eg, canned goods that were insufficiently sterilized). During storage, spores germinate and organisms produce toxins that are then ingested. In infant botulism, the immature infant gut permits colonization by *C. botulinum*. BTx produced by colonizing *C. botulinum* then results in the clinical illness. Because the gut is often paralyzed in infant botulism, constipation and poor feeding are frequent presenting symptoms.[242] Finally, necrotic or contaminated wounds may provide a niche for *C. botulinum* colonization, with subsequent toxin production inducing disease. In general, treatment of botulism is supportive care. Because botulism results in respiratory paralysis, patients may require intubation in an intensive care unit.

Protective immunity against botulism is provided by antitoxin immunoglobulins.[243] Infant botulism is treated with passive immunization, using human antitoxin immunoglobulin licensed under the name "BabyBIG." BabyBIG is generated by immunizing human volunteers with a pentavalent toxin, and then pooling the plasma to create a blood product containing high titers of polyclonal, antitoxin immunoglobulin. While a single course of BabyBIG currently costs around $50,000, administration is cost effective, as it significantly reduces an otherwise prolonged stay in the hospital. In a single center retrospective study, BabyBIG reduced the average duration of hospitalization from 23.5 to 13.5 days.[242]

Bacillus anthracis: Anthrax

Anthrax is caused by infection with *Bacillus anthracis*, an aerobic, Gram-positive, spore-forming rod found ubiquitously in soil. As cows, sheep, and other animals are susceptible to infection, anthrax persists in areas where humans live in close proximity to their livestock, and it is estimated that 2000 to 20,000 cases occur annually worldwide. There are three forms of anthrax: cutaneous, GI, and inhalational. Cutaneous anthrax, the most common form, occurs with localized infection causing tissue necrosis, painless ulceration, and the classic black eschar. An eschar results from full-thickness skin necrosis, generating a layer of thick, hard, adherent dead tissue at the area of injury. The name "anthrax" derives from the Greek word *anthrakites*, meaning "coal-like," in reference to this cutaneous eschar. GI anthrax occurs after ingestion of contaminated meat, such as beef or mutton.[244] Ingested spores vegetate, causing gastroenteritis and systemic symptoms such as fever and malaise. In severe cases, GI anthrax may result in intestinal perforation, sepsis, and death. Untreated, mortality of GI anthrax may be as high as 40%.[244] Inhalational anthrax occurs after spores are inhaled into the lungs. After germination, vegetative organisms invade the airways, traffic to draining lymph nodes, and proliferate rapidly. Inhalational anthrax is profoundly inflammatory, rapidly progressive, and inevitably fatal when untreated. Even with appropriate therapy, mortality rate is 50%.[245] Because of the small inoculum required to produce lethal disease, the ability of anthrax spores to persist in the environment, and the ease with which spores could theoretically be rapidly disseminated in the air, anthrax is a considered a Category A (highest priority) agent of bioterrorism.[246]

Anthrax pathogenesis is driven by production of the two toxins, lethal toxin (LT) and edema toxin (ET). Three polypeptide chains combine to form LT and ET. The polypeptide named "protective antigen" (PA) combines with another polypeptide termed "lethal factor" (LF) to form LT, while it forms ET by combining with a polypeptide termed "edema factor" (EF). After binding to cell surface receptors, PA generates a pore in the cell membrane and enables transport of LF or EF into the cytosol. EF is a calmodulin-dependent adenylate cyclase, and ET significantly increases intracellular cyclic AMP levels.[247] In contrast, LF is a zinc metalloprotease,[248] and LT cleaves MAPK kinases (such as MEK1), ablating all downstream MAPK.[249,250] Early in infection, the actions of LT and ET act directly on myeloid cells to impair the early innate response.

During inhalational anthrax, dormant spores are engulfed by alveolar macrophages and/or DCs.[250,251] Within hours, spores will have germinated and vegetative bacteria begun replicating in the draining mediastinal lymph nodes. Whether spores and/or bacteria are carried to the nodes inside phagocytic cells or if local alveolar replication induces breaks in the epithelial barrier that then permit tissue invasion by bacteria is currently unclear.[250] Regardless, rapid production of LT inhibits MAPK signaling in innate immune cells, like alveolar macrophages, and prevents generation of proinflammatory cytokines such as IL-1β, TNFα, and IL-6. Bactericidal function may also be impaired by LT.[251] LT also reduces chemotaxis and neutrophil recruitment.[248] LT and ET render DCs and T cells nonresponsive, preventing an adaptive response.[245] ET has pleiotropic effects on innate immunity, including inhibition of macrophage chemotaxis and phagocytosis, impairment of neutrophil priming and motility, and dysregulation of DC cytokine secretion, migration, and activation.[249] ET also cripples adaptive immunity by impairing T cell chemotaxis and activation and by driving helper T cell differentiation toward a Th2 response.[249] By directly suppressing an immune response, ET and LT permit progression of anthrax infection.

Late in infection, the effects of ET and LT on multiple cell types lead to systemic inflammation, growth of *B. anthracis* in the bloodstream (eg, bacteremia), pulmonary hemorrhage, disseminated intravascular coagulation, and circulatory collapse.[252] Morbidity and mortality occur within 2 to 5 days of onset of symptoms, and all cellular and molecular mechanisms leading to sepsis and death are yet to be fully described.[252] The migration of bacteria to draining lymph nodes, followed by rapid replication of the organism, produces the classic "widened mediastinum" observed in chest radiographs obtained from patients with inhalational anthrax.[253] Infection in a macaque model identified severe disruption of cell junctions at the alveolar-capillary barrier, leading to pulmonary hemorrhage and massive pulmonary inflammation.[252] Finally, both LT and ET act directly on vascular smooth muscle cells and cardiomyocytes to induce cardiovascular collapse.[248]

Protective immunity to anthrax is provided by antibodies directed against LT and ET, specifically against PA, the polypeptide shared by LT and ET. The vaccine currently available in the United States is a live attenuated vaccine that consists of a sterile filtrate derived from culture of avirulent, nonencapsulated *B. anthracis*. The vaccine contains multiple bacterial products released during culture, including the immunodominant PA peptide, but no intact organisms (dead or alive). It is delivered by intramuscular injection with aluminum as the adjuvant.[245] The vaccination induced protective humoral immunity directed against PA. Additionally, there are currently three FDA-approved preparations of anti-PA immunoglobulin to ameliorate ongoing anthrax infections.[247]

Anthrax research produced landmark studies in both microbiology and immunology. Robert Koch first demonstrated a causal relationship between a single microbial pathogen and disease by transferring *B. anthracis* (and disease) from one mouse to another.[254,255] Koch later developed his postulates from his work on *B. anthracis* and tuberculosis.[254] The first live attenuated vaccine was developed by Louis Pasteur, for preventing *B. anthracis* infection in sheep.[254] Finally, the field of cellular immunology originated with anthrax: Metchnikoff used *B. anthracis* as the model microbe in his first description of phagocytosis.[248,256]

Vibrio cholerae: Cholera

Vibrio cholerae is a motile, Gram-negative, comma-shaped rod that resides in aquatic environments. Infection with *V. cholerae* causes profuse, watery diarrhea that can rapidly lead to dehydration and death. Contamination of drinking water sources by diarrhea from infected patients can drive human transmission during outbreaks. Cholera epidemics and pandemics have been regular occurrences throughout human history, with six major worldwide pandemics recognized in the 1800s.[257] Notably, modern epidemiology began with the identification of the Broad Street water pump as the origin of the 1854 London cholera epidemic by John Snow.[258] The largest cholera outbreak to date devastated Haiti shortly after the 2010 earthquake; cholera was introduced into the population by UN peacekeepers.[259] Worldwide, an estimated 3 to 5 million people suffer from endemic cholera each year.[260] The O-antigen, a repeating oligosaccharide that comprises the "exposed" part of LPS, defines *V. cholerae* serotypes.[261] While over 200 O group serotypes have been identified, only 2, O1 and O139, have been associated with human disease. The O1 serogroup has been shown to bind to the chitin of marine animals, such as shellfish, from which the epidemiological association with shellfish consumption arises.[262]

The major virulence factor produced by *V. cholerae* is CTx. CTx comprises the invariant subunit A, with enzymatic activity, and five B subunits, which bind the cell surface receptor monosialoganglioside (GM1). After internalization into colonic epithelial cells, the CTx-A polypeptide activates the Gsα subunit of the guanine nucleotide-binding regulatory protein $G\alpha_s$. $G\alpha_s$-mediated signaling leads to inhibition of Na^+/H^+ exchange through Na^+/H^+ membrane transporters (preventing colonic absorption of Na^+ in exchange for a lost H^+) and phosphorylates the CFTR. CFTR phosphorylation causes massive efflux of Cl^- into the gut lumen, which then drives efflux of Na^+, K^+, and $NaHCO_3$. Massive ion efflux causes water loss, leading to the profuse, watery, secretory diarrhea associated with cholera.[263] Death occurs from dehydration and electrolyte imbalances.

Infection with *V. cholerae* begins when microbes penetrate the intestinal mucus layer, adhere to the microvilli of intestinal epithelial cells, proliferate, and upregulate toxin production. Initial infection activates TLRs and NF-κB, inducing acute inflammation marked by production of IL-1β, IL-6, and IL-17.[263] It is estimated that 65% to 100% of infected patients will develop immunity against cholera that lasts at least 3 years.[264] Adaptive immunity is directed primarily against the immunodominant O-antigen and CTx.[265] Even though CTx is the major pathogenic driver, long-lasting immunity against CTx does not appear to develop after natural infection.[266] Anti-CTxB IgG and memory B cells develop, but do not protect against symptomatic disease. Anti-CTxB IgA protects against disease in the short term, but production wanes quickly after acute infection.[263,264] Effective long-term immunity (up to 3 years) correlates best with serum vibriocidal activity, as measured by antibody-dependent complement-mediated killing.[264] Vibriocidal humoral responses are generated primarily against the immunodominant O-antigen, although immunoglobulins against other antigenic targets may play minor roles.[266] Because *V. cholerae* is not a disseminated infection, it is unlikely that serum humoral responses offer direct protection against disease. Instead, it seems likely that protection is associated with local mucosal antibody production, which is mirrored by serum vibriocidal titers.[264]

Several cholera vaccines have been developed. Vaxchora is an oral, live attenuated vaccine that temporarily protects against O1 serotypes. Vaxchora has an estimated 90% efficacy 1 month after administration and 80% efficacy after 3 months. Efficacy after 3 to 6 months is unknown.[267] Dukoral is an oral vaccine comprising killed whole cell *V. cholerae* (three strains, heat- and/or formalin-inactivated) plus recombinant CTxB.[264] Two doses of Dukoral provide 2 years of protection. Shanchol is another orally available vaccine containing four inactivated strains of *V. cholerae*, but without CTxB. Two doses provide 3 years of protection. In all cases, vaccination is provided orally to generate local, mucosal responses, and none seem to provide lifelong immunity. Immunization is provided to those traveling to areas of endemicity or outbreak and to curb ongoing epidemics.[267] A cholera vaccine providing lifelong immunity does not yet exist.

CTxB has recognized immunomodulatory effects that are currently poorly understood.[268] CTxB has been used in experimental models to reduce symptoms of IBD and to suppress experimental asthma.[268] CTxB has also been contemplated as an adjuvant for mucosal vaccination.[269] However, inclusion of CTxB as an adjuvant is limited by prior experience with the closely related *E. coli* heat-labile toxin. Bell's palsy developed in a significant proportion of recipients of a nasal influenza vaccine that incorporated *E. coli* heat-labile toxin as the adjuvant.[270]

S. dysenteriae: Bacillary Dysentery

STx, named for Kiyoshi Shiga, who first described both *Shigella* spp. and the enterotoxin, is produced by *Shigella* spp. and by enterotoxigenic and enterohemorrhagic *E. coli* strains (ETEC and EHEC, respectively).[271] Infection with ETEC is a major cause of childhood stunting in low- and middle-income countries.[272] STx is an RNA glycosidase that poisons ribosomes, inhibits protein synthesis, and causes cell death. STx production by infecting organisms can cause bloody (hemorrhagic) diarrhea and, in 5% to 10% of cases, HUS.[273] In HUS, toxin activity against the vascular endothelial cells of the kidney results in renal failure.[274] While identified first as a toxin associated with Shigella, disease from STx arises more commonly from STx-producing *E. coli* strains, such as such as *E. coli* 0157:H7.

Shigella spp. and STx-producing *E. coli* are found in the environment, and infection occurs with ingestion of contaminated food or water. Sporadic outbreaks in the United States have been associated with contamination of alfalfa sprouts, leafy greens, and insufficiently cooked hamburger meat with *E. coli* 0157.[275] After ingestion, bacteria adhere to intestinal epithelium and release STx. STx directly acts on crypt epithelial cells, resulting in acute damage to the intestinal barrier. Innate immune pathways induce neutrophil migration. Bloody, mucusy diarrhea, or a more severe hemorrhagic colitis, may result.[274,276] Systemic release of STx can damage renal vasculature, causing fibrin clots to form and damaging red blood cells. Damage to red blood cells results in red blood cell lysis (hemolysis) and anemia; fibrin clots in the renal vasculature causes renal failure and a rise in serum urea. The resulting syndrome is thus termed HUS. While STx-induced HUS is relatively rare in the United States, it is the most common cause of acute renal failure in children.[277] In Argentina, HUS is the primary cause of acute renal failure in children, the second most common cause of chronic renal failure, and is responsible for 20% of pediatric renal transplants.[278]

STx may directly modulate the host innate immune response by inducing production of proinflammatory cytokines and chemokines in macrophages.[278] Infiltration of neutrophils and monocytes into affected tissues may contribute to pathogenesis by disrupting tissue barriers, allowing increased dissemination of STx.[277] Recent work in a rabbit model of infection revealed STx-mediated upregulation of genes for IL-23, IL-1α, and the procoagulation tissue factor F3 by colonic epithelial cells, while inhibiting IFNγ signaling.[273] STx may also suppress caspase-11–mediated signaling downstream of LPS recognition, impairing pyroptosis, IL-1β release, and full innate immune activation.[279] In a human challenge model, higher neutrophil infiltration in acute disease correlated with reduced production of anti-ETEC antibodies.[272] Infection with ETEC also resulted in increased IL-17 and IFNγ production, although the degree to which STx specifically contributed to upregulation of these cytokines was not defined.[272] While STx appears to exert immunosuppressive effects in multiple studies, precisely how these immunomodulatory effects promote pathogenesis is an ongoing area of investigation.[272,273]

Infection with STx-producing *E. coli* generate humoral antibody responses to the surface antigens of the infecting microbe, such as the O-antigens of LPS,[272,280] and colonization factors (eg, colonization factor antigen I and heat-labile toxin) that promote adherence of infecting EHEC and ETEC to colonic epithelium.[272] Whether antibodies to STx develop after natural infection is unclear.[280] The multiple serotypes of STx-producing *E. coli* have hampered vaccine development.[280] No vaccine is currently available,[273] although several are under development.[272] Patients developing HUS can be treated with antibody to C5.

Clostridioides difficile: Antibiotic-Associated Diarrhea

Clostridioides difficile (formerly *Clostridium difficile*) is a Gram-positive anaerobe that colonizes the GI tract.[281] The frequency of asymptomatic colonization with *C. difficile* in adults has been estimated at 0.8% to 13%, with increased frequency found in those residing in long-term care facilities.[282,283] *C. difficile* infection (CDI) usually results from dysregulation of the host microbiome. When antibiotic administration suppresses growth of normal flora, overgrowth of *C. difficile* may occur. Disease develops when the secreted toxins TcdA and TcdB are upregulated.[284] TcdA and TcdB enter intestinal epithelial cells via receptor-mediated endocytosis and glycosylate the small GTP-binding proteins Rac and Rho, disrupting enzymatic function. Disrupted Rac and Rho function results in depolymerization of the actin cytoskeleton, causing cell rounding, epithelial cell apoptosis, and loss of barrier integrity.[285] Death of intestinal epithelial cells induces an inflammatory response, with production of inflammatory cytokines and chemokines such as IL-8 and CXCL5 and recruitment of neutrophils. Clinical syndromes induced by *C. difficile* include mild to moderate diarrhea, pseudomembranous colitis, and toxic megacolon. In severe cases of toxic megacolon, a total colectomy may be required to prevent mortality. Despite available therapies, the mortality rate of CDI is estimated at 5%.[283]

CDI exemplifies a disease that is diagnosed based on symptoms induced by the host inflammatory response.[281] Because patients may be asymptomatically colonized with toxin-producing *C. difficile*, detection of the organism and/or the toxins TcdA or TcdB does not define CDI. Criteria for CDI include symptoms of loose, watery stools, fever, and evidence of inflammation (eg, increased white blood cell count). If these symptoms are present and CDI is suspected, stool is then tested for presence of organism and/or toxin. Furthermore, clinical disease is associated with inflammatory markers, not with presence of organism.[286] Detection of fecal CXCL5 and IL-8 mRNA correlated with duration of illness, whereas organism burden did not.[286] Mortality has been associated with increased inflammatory responses, as measured by elevated white blood cell count, but not with toxin concentration.[287] Mild cases may be treated with metronidazole, an antibiotic with directed activity against *C. difficile*. In recurrent CDI that does not resolve with antimicrobial therapy, fecal microbiota transplant has been successfully used to reconstitute a normal gut microbiome,[288] reducing growth of *C. difficile*.

Protective immunity against *C. difficile* is provided by humoral immunity against either microbial surface proteins (protecting against colonization) or against toxins (protecting against symptomatic disease).[289] Higher levels of serum anti-TcdA and anti-TcdB IgG have been reported to correlate with reduced risk of CDI, and higher levels of fecal anti-TcdA IgA correlated with reduced risk of recurrent CDI.[289] Low serum anti-TcdA IgG has also been correlated with increased mortality.[287] In 2017, the fully humanized anti-TcdB monoclonal antibody bezlotoxumab was licensed by the FDA for use in adults to prevent recurrent CDI.[289] Bezlotoxumab significantly reduced recurrence of CDI in clinical trials, although it had no effect on initial disease.[289] Vaccines that elicit either antitoxin IgG or secretory IgA are under development.[283,289]

INTERSECTION OF ANTIBACTERIAL IMMUNITY WITH SCIENCE AND SOCIETY

Immunological research has been shaped by societal events. To understand why scientists pursue certain questions, while leaving others unasked or unanswered, the scientific enterprise must be placed within its historical context. At times, technical advances (such as the invention of the microscope) drive rapid progress in unexpected and novel directions. The societal need for safe foods initiated studies of *Listeria monocytogenes,* a food-borne pathogen. Investigation into the pathogenesis of *L. monocytogenes* infections led, in turn, to studies of cell-mediated immunity and provided much of our understanding of antigen processing and presentation and T cell subset development.[290-296] The scientific findings may also shape society by raising profound philosophical and ethical issues. For instance, the discovery of asymptomatic carriage with *Salmonella enterica* serovar Typhi in the early 1900s resulted in the forcible quarantine of one woman for over 25 years, highlighting the tensions between permitting personal liberty while protecting public health.[297] Finally, scientific inquiry and social advocacy may collide, as exemplified by the political pressure placed on clinicians and scientists studying Lyme disease (infection with *Borrelia burgdorferi*).[298] In this section, we review the immune responses to *L. monocytogenes*, *S.* Typhi, and *B. burgdorferi*, emphasizing the relevance of each to science or society.

L. monocytogenes: The Very Paragon of a Modern Model Pathogen

L. monocytogenes is a Gram-positive bacillus associated with food-borne outbreaks of potentially fatal illness.[299,300] For instance, in 2011, an outbreak of listeriosis (*L. monocytogenes* infection) associated with contaminated cantaloupes resulted in 147 infected patients, 33 of whom died.[299] Outbreaks have also been associated with soft cheeses (eg, queso fresco), ice cream, delicatessen meat, and fish.[301-303] Symptoms may be nonspecific, such as fever and severe muscle aches, or may be associated with the GI tract (eg, nausea and vomiting). *L. monocytogenes* may disseminate, causing meningitis, sepsis, or fetal loss. Infants, pregnant women, those older than 65 years, and immunocompromised patients are at increased risk of serious infection with *L. monocytogenes*, though the mechanisms of increased susceptibility in these populations are unclear. *L. monocytogenes* can survive and replicate at 0°C (temperatures at which refrigerated foods are maintained). It is for this reason that pregnant women are advised to avoid refrigerated deli meats, hot dogs, soft cheeses, and produce of uncertain provenance.

L. monocytogenes is an intracellular pathogen that uses actin "comets" to propel itself inside cells and through cell membranes to propagate infection to adjacent cells. Because of its intracellular motility, actin biologists have used *L. monocytogenes* extensively as a model organism to define the proteins that initiate and regulate *de novo* actin polymerization.[304,305] Infection with *L. monocytogenes* begins after ingestion of contaminated food. *L. monocytogenes* binds directly to intestinal epithelial cells by engaging with the cell surface molecule E-cadherin.[306] After phagocytosis, *L. monocytogenes* escapes the vacuole through lysis of the phagosomal membrane by the microbial toxin listeriolysin (LLO). Polarized expression of the *L. monocytogenes*–derived protein ActA stimulates actin polymerization, propelling the microbe through the cell and into adjacent cells. Deficiency of either LLO or ActA attenuates *L. monocytogenes* infection. The discovery of ActA was immensely important to cell biology, as ActA was the first actin nucleation-promoting factor identified.[304] Using *L. monocytogenes* as a model organism, cell biologists were rapidly able to define the functions of Arp2/3, coronin, VASP, cofilin, and other essential molecular regulators of actin-based motility.[304,305]

L. monocytogenes is easy to culture, reproducibly and robustly infects mice, and is relatively safe to handle under laboratory conditions.[306] For these reasons, *L. monocytogenes* has been used to propel not only actin biology but much of immunology. In fact, cellular immunity was first described in murine infection with *L. monocytogenes.*[296] *L. monocytogenes* was then used as the model pathogen in exhaustive subsequent investigations of key processes such as IFNγ signaling, antigen processing and presentation, and generation of T cell–mediated immunity.[306] *L. monocytogenes* has also been used to probe innate immune components, as early proinflammatory signals are required for induction of T cell responses needed for effective control and clearance of pathogenic infection.[306]

Escape from the phagolysome into the cytosol by *L. monocytogenes* induces a robust innate immune response. The early innate immune response is essential for pathogen control. Mice deficient for the adaptor protein MyD88, the transcription factor NF-κB, IFNγ, or the receptor for TNFα all exhibit increased susceptibility and rapid mortality following challenge with *L. monocytogenes.*[306] Mice deficient for adaptive components (such as SCID mice, lacking both B and T cells) control *L. monocytogenes* early in disease, but express increased susceptibility and mortality at later times after initial challenge due to an inability to clear bacteria.[306] Activation of NF-κB–dependent inflammatory pathways recruit neutrophils and monocytes to the site of infection; mice lacking CCR2, a chemokine receptor essential for $CD11b^+$ monocyte trafficking, are also highly susceptible to infection at early timepoints.[306] The innate immune response is needed to stimulate a protective adaptive response. Although both LLO- and ActA-deficient *L. monocytogenes* are attenuated, they differ in elicitation of T cell responses. LLO-deficient mutants cannot escape the phagolysosome, do not induce innate immunity, and do not elicit protective T cell memory, whereas ActA-deficient mutants can escape the phagolysome, do induce inflammation, and generate protective T cell responses.[306]

Clearance of *L. monocytogenes* relies extensively on α/β T cells.[306] During infection, *L. monocytogenes* is carried into T cell zones of the spleen by macrophages. Uptake of bacteria by splenic DCs enables T cell priming.[292] Intracellular proteasomal processing of *Listeria*-derived antigens, such as LLO, results in antigen presentation on MHC class I.[307] Antigens derived from Listeria are also released and cross-presented on MHC class II.[291] Elicitation of $CD8^+$ effector cytotoxic T cells is required for the eventual clearance of primary *Listeria* infection.[294] CD4+ T cell help during a primary infection is required for development of the robust CD8+ T memory response that provides protective immunity upon *Listeria* rechallenge.[308] Because eradication of *L. monocytogenes* depends upon a robust T cell response, and because immunodominant peptides from LLO were readily identified as key antigens, *L. monocytogenes* infection has been used to delineate antigen processing and presentation,[291,307,309] the function of IFNγ in generating T cell responses,[310] the role of DC subsets,[292,311] identification of Th1 differentiation,[290] differentiation of memory T cells,[293,312] exploration of T cell tonic signaling,[313] and differentiation of T_{fh} cells.[314]

In summary, although a relatively minor human pathogen compared to infections such as pneumococcus or tuberculosis, *L. monocytogenes* has played a major role as a model organism in landmark studies in cell biology and immunology.

S. enterica: Mary Mallon and a Series of Unfortunate Events

S. enterica is a Gram-negative, facultative anaerobic, intracellular bacillus that causes a variety of diseases, including enteric (typhoid) fever, enterocolitis/diarrhea, and sepsis.[315] *S. enterica* is divided into subspecies and serovars based upon differences in the structures of LPS, flagellin, and carbohydrates. Different serovars (serotypes) are associated with different disease presentations, with serovar Typhi associated with typhoid fever and nontyphoidal serovars, most commonly *S.* Typhimurium and *S.* Enteritidis, associated with enterocolitis/diarrhea. Over 1 billion cases of typhoid fever or nontyphoidal enterocolitis occur each year, with significant morbidity and mortality. Here we will discuss the immune response to *S.* Typhi, as humans are the only known reservoirs and the immune mechanisms permitting chronic infection or the asymptomatic carrier state remain under investigation.[315,316] While much of the pathobiology of early, acute infection is similar in nontyphoidal serovar infections, there are key differences, as noted below.

S. Typhi is important historically for providing the first evidence of asymptomatic carriage in the United States, and for raising the significant ethical challenges associated with balancing personal liberty against public health, with the unfortunate events surrounding Mary Mallon.[297] Mallon was a domestic cook who served in the households of wealthy New York families in the early 1900s. Mallon was also an asymptomatic carrier of *S.* Typhi. In 1906, Mallon was hired by a New York banker as a cook at his summer home on Long Island. In 1 week, 6 of his 11 guests became ill with typhoid fever. Epidemiological investigation by a sanitary engineer, hired by the banker, led to the suspicion of Mallon as the source of infection. His suspicions were confirmed when he determined that seven of the eight previous families for whom Mallon had worked had also suffered episodes of typhoid fever. In the early 20th century, typhoid fever carried a mortality rate as high as 10%, and there were no vaccines or effective antibiotics. Mallon, however, refused to be tested for *S.* Typhi and continued to find work as a cook. In 1907, an outbreak of typhoid fever sickening over 3000 New Yorkers was thought to be linked to Mallon. Mallon was arrested, forcibly quarantined, and tested by the New York Department of Health, revealing carriage of S. Typhi in stool. After 2 years of quarantine, Mallon was released with instructions to avoid work as a cook. However, Mallon returned to employment as a cook at Sloane Maternity, soon causing an outbreak that infected 25 staff members and caused 2 fatalities. Tabloid New York newspapers dubbed her "Typhoid Mary." Mallon was once again forcibly quarantined, this time for over 25 years, until her death from a stroke. Despite quarantine and multiple tests, the concept of "asymptomatic carriage" was never made clear to Mallon. Furthermore, while there were multiple outbreaks of typhoid fever, and over 400 asymptomatic carriers later identified, Mallon was singled out for forcible quarantine and public attention. Her case highlights the ethical tension between individual liberty and public health.[297]

Infection with S. Typhi occurs after ingestion of contaminated water or food. Stomach acid and the mucus layer of the intestines present the first barrier challenges to invading microbes.[317] AMPs, such as beta-defensin, and secretory IgA may also protect against invasion.[317] S. Typhi that successfully overcomes these barriers attaches to intestinal epithelial cells using fimbriae. S. Typhi has long been thought to cross the intestinal epithelial barrier through uptake by M cells, like *Shigella* spp., then trafficked to Peyer patches.[315,318] Recent evidence suggests that S. Typhi may also directly infect enterocytes.[317] The T3SS plays an essential role in internalization, injection of virulence factors into host cells, and stimulating uptake. *S. enterica* is an intracellular pathogen, creating for itself protective "*Salmonella*-containing vacuoles (SCVs)." SCVs are distinct from phagosomes or lysosomes, as they prevent assembly of the NADPH oxidase complex on the SCV membrane, allowing microbes to evade ROS-mediated killing.[318] SCVs also avoid fusion with phagosomes (or lysosomes) containing antimicrobial peptides, again permitting *S. enterica* to escape intracellular killing.[317,318]

Early inflammatory responses that induce TNFα and IFNγ are essential for survival during *Salmonella* infection.[319,320] In humans, *Salmonella* activates TLR2 (lipoprotein), TLR4 (LPS), TLR5 (flagellin), and TLR9 (CpG DNA). The most critical receptor appears to be TLR4, as TLR4-deficient animals are profoundly susceptible to *Salmonella* infection, and the contributions of TLR2, TLR5, and TLR9 are apparent only on a TLR4-deficient background.[316] TLR-mediated NF-κB signaling induces proinflammatory mediators such as TNFα. IFNγ and TNFα mobilize the NADPH-oxidative burst in macrophage and neutrophils, which is essential for clearing *Salmonella* from the intracellular compartment. IL-12, derived from T cells or NK cells, is also essential for upregulating IFNγ and enhancing cell-mediated immunity. S. Typhi produces a number of virulence factors that modulate this key host signaling pathway.

The difference between clinical presentation of S. Typhi (enteric fever) and nontyphoidal serovars like S. Typhimurium arises from disparate elicitation of the NF-κB–mediated response. When infected by nontyphoidal *S. enterica*, there is significant activation of TLR5 and the NLRC4 inflammasome by flagellin and induction of NF-κB by T3SS.[317,321] The virulence factor SipA induces IL-8 and thus neutrophil recruitment.[318] The S. Typhimurium virulence factor SipB can bypass inflammasome activation by directly cleaving and activating caspase-1, inducing IL-1β production and pyroptosis of infected cells.[317,318] *S. enterica*–derived LPS also activates TLR4, inducing NF-κB. Thus, nontyphoidal *S. enterica* induces profound inflammation, resulting in neutrophil recruitment, causing enterocolitis and diarrhea. S. Typhi, however, downregulates T3SS and flagellin expression, muting these inflammatory responses.[317] S. Typhi also produces a protective Vi polysaccharide capsule that "masks" LPS from TLR4. By dampening TLR4, TLR5, and NLRC4-generated pro-inflammatory signals, S. Typhi can transverse the epithelial barrier and disseminate without producing diarrhea.

S. Typhi is readily phagocytosed by monocytes, macrophages, and DCs and, as discussed, thus slips across the intestinal barrier without inducing massive inflammation. Myeloid cells that contain viable S. Typhi likely disseminate the pathogen by trafficking through the reticuloendothelial system or the bloodstream.[317,318] During dissemination, S. Typhi may also replicate extracellularly in the bloodstream and may therefore be susceptible to complement and humoral immunity.[316] After systemic dissemination, S. Typhi may be internalized by a variety of professional immune cells, such as splenic or hepatic resident macrophages, or by nonhematopoietic cells such as hepatocytes. Once in the liver, S. Typhi may traffic to the biliary tract, where it can take up residence in biofilms in gallstones.[315] S. Typhi can persist in biofilms complexed with cholesterol-rich gallstones.[315]

Adaptive immune responses are required for long-term control of *S. enterica* infection. Adaptive responses begin with DCs that internalize S. Typhi in the intestine and traffic to mesenteric lymph nodes. Mesenteric lymph nodes are the site of DC antigen presentation to T cells. $CD11c^+CCR6^+$ DCs are required for antigen presentation and effective control, as are mesenteric lymph nodes.[315] Major T cell epitopes derive from *S. enterica* flagellin. Flagellin downregulation by S. Typhi also permits immune escape from cell-mediated immunity. Other *Salmonella* virulence factors ubiquitinate MHC class II molecules, targeting them for degradation and again permitting immune escape. $CD8^+$ effector T cells responses appear most important for clearance, with humoral immunity playing a minor role in primary infection.[315,318] However, B cells are required to generate IgG and IgA and may serve as critical antigen-presenting cells in a secondary immune response against repeat *Salmonella* exposure.[315,318]

Th1 $CD4^+$ T cell response is most effective at clearing *Salmonella*, though a role for Th17 in neutrophil recruitment also seems important. Th2 responses are less effective, and IL-10 production dampens effective cellular immunity and permits disease progression.[315,318] *Salmonella* infection can lead to granuloma formation in livers and spleens, leading to chronic, persistent infection. Like other granulomatous diseases, the granuloma may both protect the host from pathogenic dissemination, while also providing a protected niche for continued pathogen replication. Granulomas induced by S. Typhi appear similar to those induced by Mtb. M1 macrophages constrain

Salmonella while M2 polarization is more permissive for disease progression.[322] TNFα is required for maintenance of restrictive *Salmonella* granulomas.

Immunological determinants for successful control of *Salmonella* have been revealed in studies of patients at increased risk of Salmonella infection. Humans with genetic deficiencies of the IFNγ/IL-12/IL-23 pathway suffer from increased susceptibility to intracellular pathogens, including *S. enterocolitis*.[323,324] Patients with STAT1 deficiencies, necessary for IFNγ signaling, are also at increased risk. Patients with deficiencies of NF-κB signaling and MyD88 are also at increased risk of *S. enterocolitis* disease.[316] Those treated with anti-TNFα antibodies are at risk for *S. enterocolitis* disease, highlighting the requirement for this cytokine in anti-*Salmonella* responses.[316] Finally, patients with CGD, or defects in the NADPH oxidase complex, have presented with S. *enterocolitis* dissemination and complications, suggesting a critical role for neutrophils in restricting S. *enterocolitis* early in disease.[325]

Vaccines against the Vi polysaccharide capsule of S. Typhi are available as T-dependent polysaccharide conjugate vaccines (Vi-TT) and as T-independent polysaccharide formulations (Vi-PS). While efficacious in endemic areas against typhoid fever in adults and older children, the Vi-PS vaccine is insufficiently immunogenic in young children.[326,327] Studies of these vaccines in controlled infection experiments revealed that IgA quantity and avidity correlated with protection from disease, while increased IgG correlated with reduced severity.[327] Thus, while cell-mediated immunity is essential in resolution of natural infection, humoral immunity can be successfully harnessed to prevent or ameliorate infection.

B. burgdorferi: Immune-Mediated Controversies in Lyme Disease

Lyme disease results from infection with the pathogen *B. burgdorferi* (and related *Borrelia* spp.). The disease derives its name from Old Lyme, CT, the town where Dr. Allen Steere, Dr. Stephen Malawista, and others investigated an outbreak of juvenile arthritis, leading to the recognition of Lyme disease as a distinct clinical entity with an infectious origin.[328] The species derives its name from Dr. Willy Burgdorfer, who found that sera from Lyme disease patients reacted with the *Borrelia* spirochetes he had isolated from the GI tracts of *Ixodes* ticks (deer ticks).[329] *Borrelia* spp. are spirochetes: Gram-negative, spiral-shaped, motile bacteria with flagella positioned in between the inner and outer membranes. *B. burgdorferi* are up to 20 to 30 μm in length, longer than many other microbes.[330] *Borrelia* spp. are obligate parasites and cycle between arthropod and vertebrate life stages.[331,332] The usual "reservoir" hosts include small mammals, such as the white-footed mouse,[331,332] and infection of humans is incidental.

Lyme disease has been classified in three phases: early/localized, early/disseminated, and late/disseminated.[333,334] Early/localized Lyme disease consists of the classic "bulls-eye" skin rash at the site of infecting tick bite. The rash, termed *erythema migrans*, has a characteristic appearance, with a slowly spreading red ring surrounding a cleared area. At this stage, a relatively short course (10–14 days) of an appropriate antibiotic (eg, doxycycline) is sufficient to resolve disease.[334] Cardiac manifestations occur within 1 to 2 months of infection and are an example of early/disseminated disease. Lyme arthritis may present months after the initial infection and is therefore considered late/disseminated. Infection of the nervous system may occur in either early or late disseminated disease. The clinical presentation of disseminated Lyme disease therefore varies widely, from monoarticular (one joint) arthritis to meningoencephalitis (infection of the brain and meninges). Disseminated Lyme (early or late) requires up to 28 days of directed antimicrobial therapy for resolution.[334]

The immune response to *Borrelia* has been reviewed in detail.[335] In brief, both innate and adaptive responses are generated, with humoral immunity essential to reducing pathogen burden. In mice, Langerhans cells are the first immune cells to encounter *Borrelia* after intradermal inoculation (used experimentally to mimic transmission via tick bite). Skin-resident macrophages/DCs next encounter and engulf organisms.[335] Macrophages may recognize *Borrelia* through several surface receptors, such as FcγR or CR3. *Borrelia* induces IL-10 production from macrophages and DCs early in infection, suppressing effective immunity and permitting disease progression.[335] Internalization leads to NF-κB–mediated signaling initiated through TLR ligation (eg, TLR2/1[336]), and neutrophils are recruited to infected skin less than 4 hours after injection.[335] *B. burgdorferi* does not produce classical LPS and therefore does not activate TLR4.[335] Interestingly, tick saliva contains elements such as evasin-4, an inhibitor of CCL5 and other chemokines, which impairs inflammatory cell recruitment to the bite site.[337,338] Infiltration of the skin at the site of inoculation (or tick bite) by inflammatory cells, such as monocytes and neutrophils, produces the rash of *erythema migrans*.[335,339] Generation of proinflammatory mediators, like IL-1β and TNFα, induces the systemic symptoms that can accompany *erythema migrans*, such as fever, myalgias (muscle aches), and malaise.[340]

Neutrophils phagocytose spirochetes through "tube" phagocytosis, in which pseudopods wrap around the long spiral-shaped microbes to completely engulf them. Neutrophils kill *Borrelia* through granular components, such as LL-37 and elastin, and through formation of NETs.[335,339] The NADPH-mediated oxidative burst appears to be dispensable for elimination of *Borrelia*. The innate immune response requires TLR2, MyD88, and complement components. C3 enhances neutrophil-mediated capture of spirochetes. Untreated, *Borrelia*-associated *erythema migrans* may persist for weeks to months.[333] Additional components essential early in infection include the complement system and iNKT cells.[335] If the immune response is sufficient to clear all pathogens, the infection will self-resolve.

If not all spirochetes are captured, dissemination may occur rapidly, with detection of *Borrelia* spp. in the hearts and joints of infected mice within 7 to 10 days.[335] Although *Borrelia* spp. are predominantly extracellular, spirochetes may invade vascular endothelial cells as they transmigrate into vessels and lymphatics.[341] While dissemination occurs rapidly, the clinical symptoms of dissemination, such as arthritis, may not appear for 4 days to 4 years after initial infection.[339] Long-term persistence of organisms in the absence of symptoms suggests that immunopathology drives disease, not the organism itself. The adaptive immune system—particularly production of *Borrelia*-specific immunoglobulin—is essential for eventual pathogen clearance. SCID mice eventually succumb to *Borrelia* infection, as they are unable to clear.[335] In reservoir animals, such as mice, production of immunoglobulin is sufficient to protect the host from disease, but insufficient to fully eliminate pathogens. Chronic, persistent infection in reservoir animals increases the likelihood of transmission to ticks, allowing *Borrelia* to move on to the next life stage. In nonreservoir humans, effective immunoglobulin production may completely clear pathogen and eliminate disease. The IgG anti-*Borrelia* seroprevalence in endemic areas can be as high as 5% to 13%, while the incidence of Lyme disease is much lower, suggesting that far more people are exposed to *B. burgdorferi* than are susceptible to symptomatic infection.[335,339]

Antibodies are bactericidal through complement-dependent and complement-independent microbial lysis. Stronger B cell responses correlate with faster resolution of diseases after antibiotic treatment. Seroconversion can take several weeks after infection and may not happen in patients who are treated completely soon after appearance of *erythema migrans*, suggesting that adaptive immunity only occurs after dissemination of organism. IgG may be detectable for years after effective treatment and clinical remission. Intrathecal IgM, IgG, and IgA are found in patients with neurologic borreliosis. In some patients, T-independent anti-*Borrelia* IgM may persist for months, which is unusual. As serum IgM has a half-life of 12 to 48 hours, ongoing IgM production suggests either generation of long-lived IgM secreting plasma cells or ongoing generation of new IgM-secreting plasmablasts. Because IgM production against *Borrelia* is largely T independent, generation of long-lived IgM plasma cells seems unlikely. However, continuous *de novo* production of IgM-secreting plasmablasts would suggest persistent infection with viable *B. burgdorferi* or at least persistence of *Borrelia* antigens. This mystery remains unsolved.[335]

Several lines of evidence indicate that the immune reaction to *Borrelia* drives clinical disease. First, pathogen burden does not correlate with disease severity. For instance, although mice deficient for IL-10 appear to have lower pathogen burden, disease severity is higher. Mice lacking TLR2 and MyD88 have much higher pathogen burden during the initial phase of infection, but clinical severity is unchanged. Finally, mice lacking IFNγ and IFNγR have increased pathogen burden, but improved disease symptoms.[335] Second, the spirochete burden is identical in disease susceptible mouse strains, such as C3H, to those of disease-resistant strains. In C3H mice, susceptibility to *Borrelia*-induced arthritis is due to deficiency of lysosome beta-glucuronidase. Joint invasion by spirochetes damages host tissue and releases glycosaminoglycans. Glycosaminoglycans serve as DAMPs—host-derived molecules that induce inflammation. Because C3H mice lack the enzyme required to degrade glycosaminoglycans, glycosaminoglycan-mediated inflammatory signaling persists, thus causing arthritis.[335]

The immune response during disseminated Lyme depends upon the organ system affected. Disseminated Lyme most often affects joints, the nervous system, or the heart. Most studies have therefore focused on these three organ systems. In general, cellular infiltrates at these sites consist of lymphocytes, indicating a persistent adaptive response.[335] Cardiac Lyme disease, while rare, can be fatal. Lyme carditis usually presents as conduction abnormalities, such as atrioventricular (AV) heart block.[342] AV heart block is a condition in which the electrical signal that starts in the upper chambers of the heart (the atria) is not correctly propagated to the lower chambers (the ventricles), and so the contractions of the upper and lower chambers are uncoordinated. Dysregulation of contractions and electrical impulses can result in dangerous arrhythmias, resulting in sudden cardiac death. Autopsy specimens from five patients who died of acute Lyme carditis revealed significant infiltrations of T cells, histiocytes, and plasma cells, but not of neutrophils, into heart tissue.[335]

Lyme neuroborreliosis presents in a myriad of ways, ranging from cranial nerve palsies (such as Bell's palsy, which causes unilateral or bilateral "drooping" of facial muscles) to meningoencephalitis (causing nonspecific symptoms such as headaches and altered mental status). About 10% of patients with Lyme disease will have neurological complications.[343] The immune response to CNS infection by *Borrelia* provides the diagnostic testing of choice.[334,343] The CSF of patients with Lyme neuroborreliosis is characterized by elevated white blood cell count with a lymphocytic predominance, with >60% T cells and >15% B cells. NK cells and monocytes may also be present, albeit at much lower frequencies.[335] The CSF of patients with Lyme neuroborreliosis also contains anti-*Borrelia* immunoglobulins that are generated locally (eg, intrathecally, within the meninges). The Infectious Diseases Society of America currently recommends simultaneously testing serum and CSF for anti-*Borrelia* IgG to diagnose Lyme neuroborreliosis: the CSF IgG level should be higher than the serum anti-*Borrelia* IgG in patients with Lyme neuroborreliosis, due to intrathecal antibody synthesis.[334] CSF anti-*Borrelia* IgG levels equivalent to that of serum IgG levels suggests that anti-*Borrelia* IgG have simply diffused from serum into CSF, and thus are not indicative of active CNS infection. Per European criteria, where neurologic Lyme disease is more common than in the United States, a definite diagnosis of Lyme neuroborreliosis currently requires (1) symptoms of Lyme neuroborreliosis; (2) elevated lymphocyte infiltration in the CSF; and (3) production of anti-*Borrelia* IgG in the CSF.[343] Tests that attempt to detect the spirochetes in CSF and serum samples are not recommended, as nucleic acid–based tests directed against *Borrelia* have sensitivities as low as 1% to 5%, and culture is worse.[343] The pathology of Lyme neuroborreliosis is not well understood.

Finally, Lyme arthritis develops late in Lyme disease, frequently several months after initial infection, and may present up to 4 years after infection.[339] Arthritis may arise in as many as 50% of patients with untreated *erythema migrans* resulting from *Borrelia* infection.[335] Synovial tissue obtained from infected patients reveals infiltrates of $CD4^+$ T cells, macrophages, B cells, and plasma cells.[335] Currently available evidence suggests that small populations of *Borrelia* may persist within the joint, driving immunopathology. Clearance of organisms through immunity (self-resolution) or through antibiotic administration may lead to complete resolution of disease. However, arthritis may persist for weeks or months in antibiotic-treated patients, for reasons that are as yet unknown. Continued persistence of active *Borrelia* infection is extremely unlikely, as no viable organisms have been detected by culture or PCR from affected synovial tissues (as they have been before antibiotic therapy).[339] It is possible that *Borrelia* antigens (such as *Borrelia*-derived peptidoglycan) may persist in the joint space long after viable organisms have been eliminated, driving ongoing immune responses.[335,339] Molecular mimicry or development of autoantibodies may also result in antibiotic-refractory Lyme arthritis.[339] When symptomatology from confirmed *Borrelia* infection persists beyond the recommended courses of antibiotics, as it may in antibiotic-refractory arthritis, disease may be managed by anti-inflammatory agents.[339]

Innate immunity likely initiates Lyme arthritis, while adaptive immunity perpetuates it.[339] A plethora of immune cells have been recovered from the joints of patients with Lyme arthritis, including α/β and γ/δ T cells. α/β T cells may be Th1, Th2, or Th17. A unique subset of $CD3^+CD4^{neg}CD8^{neg}IL\text{-}23R^+ROR\gamma t^+$ T cells has also been found at the entheses (cartilage and tendon insertion sites at the joints) of Lyme-inflamed joints. However, these specialized $IL\text{-}23R^+ROR\gamma t^+$ T cells are found in other conditions of inflammatory arthritis, and their role in Lyme arthritis is unknown.[335] Production of IFNγ, TNFα, and the T cell chemoattractants CXCL9 and CXCL10 was higher in the joint fluid of patients presenting with antibiotic-refractory Lyme arthritis than in those patients whose Lyme arthritis resolved after antibiotic therapy,[339] again suggesting dysregulated adaptive responses as the primary driver of clinical disease.

Definitive diagnosis of Lyme disease relies upon serological assays that detect the immune response against the organism, because *Borrelia* spp. cannot be routinely cultured in clinical microbiology laboratories, and nucleic acid amplification tests are unfortunately insensitive. However, reliance upon anti-*Borrelia* serologies as the only available diagnostic test has created many challenges. First, the seroprevalence in Lyme-endemic areas has been estimated to be as high as 5% to 13%, complicating the interpretation of a positive test in patients presenting with symptoms (eg, headaches, fatigue) that may or may not be caused by *Borrelia* infection.[339] Interpretation of serologies are also complicated by the temporal development of antibodies—people may be acutely infected with *Borrelia* and symptomatic before sufficient time has passed for detectable, specific immunoglobulins to appear. Finally, as discussed, symptoms of Lyme disease may develop months to years after the initiating tick bite. Diagnosis therefore takes into account the geographic history of the patient (eg, have they ever lived or traveled to a Lyme-endemic region) and the symptoms of the patient, as well as laboratory testing. For instance, patients presenting with *erythema migrans* from Lyme-endemic regions are treated for *Borrelia* infection without undergoing serological testing because *erythema migrans* often develops prior to production of high affinity anti-*Borrelia* anti-IgG.[334]

Two clinical entities linked to Lyme disease continue to cause significant controversy. First, a subset of patients with confirmed Lyme disease continue to suffer nonspecific symptoms, such as fatigue, muscle aches, or difficulty concentrating, after appropriate antibiotic treatment. This constellation of subjective, nonspecific symptoms has been categorized as "posttreatment Lyme disease syndrome (PTLDS)."[344] PTLDS is distinct from clinical presentations such as arthritis or nerve palsies with objective, quantifiable findings, which may persist after antibiotic therapy; these are termed "antibiotic-refractory arthritis" or "antibiotic-refractory neuritis." While it is clear that patients with PTLDS are suffering from one or more debilitating conditions, the etiology (and therefore treatment) remains elusive. It has been suggested that PTLDS may be an immune-mediated pathological condition caused by *Borrelia* exposure, but it may simply be the unfortunate prolonged convalescence some patients suffer after a severe illness. Another possibility is that PTLDS represents an independent disease process, such as clinical depression, that happens to occur in someone who also had *Borrelia* infection. In support of this hypothesis, a recent study of patients presenting with the constellation of PTLDS symptoms of fatigue, muscle aches, and cognitive impairments revealed no significant association with previous Lyme disease or serologies consistent with *Borrelia* exposure.[344] These results suggest that *Borrelia* infection and PTLDS are, in fact, two separate syndromes.[344]

Second, a highly controversial entity termed "chronic Lyme disease" persists.[298,344] Symptoms associated with chronic Lyme disease (CLD) include fatigue, joint and muscle aches, difficulty concentrating, memory loss, and other cognitive impairments (eg, "fogginess"). No consistent or quantifiable objective signs accompany CLD. In contrast to PTLDS, CLD has been diagnosed in patients who have no history of tick bite, no history of travel to *Borrelia*-endemic areas, and negative anti-*Borrelia* serologies.[298] CLD continues to be diagnosed because the subjective symptoms of patients are poorly managed by the current US healthcare system; many CLD patients feel unheard, dismissed and/or ignored by conventional physicians. Advocacy groups and practitioners of alternative medicine offer compassion and support to CLD patients, forming a self-reinforcing network that has been resistant to new data and consensus guidelines.[298,345]

These ongoing controversies present significant challenges for clinicians wishing to improve the care of patients with symptoms classified as PTLDS or CLD, or for immunologists who wish to pursue investigation of post-Lyme immune dysregulation. For both PTLDS and CLD, no scientific evidence exists to suggest that viable Borrelia persist beyond 1 month of antibiotic therapy.[298,334,345] Furthermore, there is no evidence that repeated or prolonged courses of antibiotics decrease suffering in those with CLD.[298,334,345] In fact, prolonged antibiotic treatment of patients with CLD has been associated with severe antibiotic-associated adverse events and death.[345] Patients who pursue treatment for PTLDS or CLD may in fact have serious underlying conditions, such as major depressive disorder, multiple sclerosis, or lupus, that are initially overlooked, delaying needed treatment.[298] Unfortunately, advocacy for CLD has resulted in adverse political pressure upon clinicians and scientists wishing to better understand Lyme disease.[345] Greater definition of the immunopathology caused by *Borrelia* infection, improved diagnostic testing, and better therapies to modulate *Borrelia*-induced immunopathology would dramatically improve care of patients.

An anti-*Borrelia* vaccine was temporarily available after FDA licensing in 1998. The company discontinued the vaccine 3 years later, citing poor sales.[334] Specific, high-affinity IgG are sufficient to protect against reinfection, and serum transfer of IgG is sufficient to protect against initial infection.[335] Several new vaccines for *Borrelia* are currently under development.[340,346]

CONCLUSION

Humans and bacteria have coevolved over thousands of years, and a healthy microbiome is essential for normal human development and immunity. Microbial pathogens disturb this balance, disrupting microbial communities and sickening the human host. The human immune system has developed a complex and multitiered response, incorporating physical barriers, microbial sensors, microbicidal proteins (eg, defensins, complement), immediate effector cells (such as neutrophils and macrophages), and long-lived memory cells to prevent recurrent infections. Despite the many layers of protection, pathogens such as Mtb, pneumococcus, *S. aureus*, and the others described above still cause significant death and disease through expression of an array of virulence factors. Much research in the areas of inflammation, septic shock, effective clearance and healing of infection, and vaccinology remains to be done in antibacterial immunobiology.

REFERENCES

1. Vollmer W, Blanot D, de Pedro MA. Peptidoglycan structure and architecture. *FEMS Microbiol Rev*. 2008;32:149-167.
2. Inohara, Chamaillard, McDonald C, Nuñez G. NOD-LRR proteins: role in host-microbial interactions and inflammatory disease. *Annu Rev Biochem*. 2005;74:355-383.
3. Milne-Davies B, Wimmi S, Diepold A. Adaptivity and dynamics in type III secretion systems. *Mol Microbiol*. 2021;115:395-411.
4. Zhang B, Liu B, Zhou Y, Zhang X, Zou Q, Liu X. Contributions of mass spectrometry-based proteomics to understanding Salmonella-host interactions. *Pathogens*. 2020;9:581. PMC7400052.
5. Schlievert PM, Davis CC. Device-associated menstrual toxic shock syndrome. *Clin Microbiol Rev*. 2020;33:e00032-19. PMC7254860.
6. Moza B, Varma AK, Buonpane RA, et al. Structural basis of T-cell specificity and activation by the bacterial superantigen TSST-1. *Embo J*. 2007;26:1187-1197. PMC1852840.
7. Pandey M, Calcutt A, Ozberk V, et al. Antibodies to the conserved region of the M protein and a streptococcal superantigen cooperatively resolve toxic shock-like syndrome in HLA-humanized mice. *Sci Adv*. 2019;5:eaax3013. PMC6726444.
8. Guerra FE, Borgogna TR, Patel DM, Sward EW, Voyich JM. Epic immune battles of history: neutrophils vs. Staphylococcus aureus. *Front Cell Infect Microbiol*. 2017;7:286. PMC5491559.
9. Trindade BC, Chen GY. NOD1 and NOD2 in inflammatory and infectious diseases. *Immunol Rev*. 2020;297:139-161.
10. Audry M, Robbe-Masselot C, Barnier JP, et al. Airway mucus restricts Neisseria meningitidis away from nasopharyngeal epithelial cells and protects the mucosa from inflammation. *mSphere*. 2019;4:e00494-e00419. PMC6893211.
11. Kuek LE, Lee RJ. First contact: the role of respiratory cilia in host-pathogen interactions in the airways. *Am J Physiol Lung Cell Mol Physiol*. 2020;319:L603-l619. PMC7516383.
12. Deshpande R, Zou C. *Pseudomonas aeruginosa* induced cell death in acute lung injury and acute respiratory distress syndrome. *Int J Mol Sci*. 2020;21:5356. PMC7432812.
13. Prasad SV, Fiedoruk K, Daniluk T, Piktel E, Bucki R. Expression and function of host defense peptides at inflammation sites. *Int J Mol Sci*. 2019;21:104. PMC6982121.
14. Nizet V, Ohtake T, Lauth X, et al. Innate antimicrobial peptide protects the skin from invasive bacterial infection. *Nature*. 2001;414:454-457.
15. Kahlenberg JM, Kaplan MJ. Little peptide, big effects: the role of LL-37 in inflammation and autoimmune disease. *J Immunol*. 2013;191:4895-4901.
16. Wendler J, Schroeder BO, Ehmann D, et al. Proteolytic degradation of reduced human beta defensin 1 generates a novel antibiotic octapeptide. *Sci Rep*.2019;9:3640. PMC6403363 reduced hBD-1.
17. Bonamy C, Sechet E, Amiot A, et al. Expression of the human antimicrobial peptide β-defensin-1 is repressed by the EGFR-ERK-MYC axis in colonic epithelial cells. *Sci Rep*. 2018;8:18043.
18. Chen H, Xu Z, Peng L, et al. Recent advances in the research and development of human defensins. *Peptides*. 2006;27:931-940.
19. Yang M, Zhang C, Zhang MZ, Zhang S. Beta-defensin derived cationic antimicrobial peptides with potent killing activity against gram negative and gram positive bacteria. *BMC Microbiol*. 2018;18:54.
20. Baldwin H, Aguh C, Andriessen A, et al. Atopic dermatitis and the role of the skin microbiome in choosing prevention, treatment, and maintenance options. *J Drugs Dermatol*. 2020;19:935-940.
21. Torres MJM, Peterson JM, Wolf SE. Detection of infection and sepsis in burns. *Surg Infect*. 2020;22:20-27.
22. Miller F, Lécuyer H, Join-Lambert O, et al. Neisseria meningitidis colonization of the brain endothelium and cerebrospinal fluid invasion. *Cell Microbiol*. 2013;15:512-519.
23. Tang D, Kang R, Coyne CB, Zeh HJ, Lotze MT. PAMPs and DAMPs: signal 0s that spur autophagy and immunity. *Immunol Rev*. 2012;249:158-175. PMC3662247.
24. Hayashi F, Smith KD, Ozinsky A, et al. The innate immune response to bacterial flagellin is mediated by Toll-like receptor 5. *Nature*. 2001;410:1099-1103.
25. Mattock E, Blocker AJ. How do the virulence factors of Shigella work together to cause disease? *Front Cell Infect Microbiol*. 2017;7:64. PMC5364150.
26. Sun L, Liu W, Zhang LJ. The role of Toll-like receptors in skin host defense, psoriasis, and atopic dermatitis. *J Immunol Res*. 2019;2019:1824624. PMC6877906.
27. Famà A, Midiri A, Mancuso G, et al. Nucleic acid-sensing Toll-like receptors play a dominant role in innate immune recognition of pneumococci. *mBio*. 2020;11:e00415-20. PMC7157524.
28. Fournier B, Philpott DJ. Recognition of *Staphylococcus aureus* by the innate immune system. *Clin Microbiol Rev*. 2005;18:521-540. PMC1195972.
29. Correa RG, Milutinovic S, Reed JC. Roles of NOD1 (NLRC1) and NOD2 (NLRC2) in innate immunity and inflammatory diseases. *Biosci Rep*. 2012;32:597-608. PMC3497720.
30. Seoane PI, Lee B, Hoyle C, et al. The NLRP3-inflammasome as a sensor of organelle dysfunction. *J Cell Biol*. 2020;219:e202006194. PMC7543090.
31. Alehashemi S, Goldbach-Mansky R. Human autoinflammatory diseases mediated by NLRP3-, pyrin-, NLRP1-, and NLRC4-inflammasome dysregulation updates on diagnosis, treatment, and the respective roles of IL-1 and IL-18. *Front Immunol*. 2020;11:1840. PMC7477077.
32. Vladimer GI, Marty-Roix R, Ghosh S, Weng D, Lien E. Inflammasomes and host defenses against bacterial infections. *Curr Opin Microbiol*. 2013;16:23-31. PMC3697846.
33. Lubbers R, van Essen MF, van Kooten C, Trouw LA. Production of complement components by cells of the immune system. *Clin Exp Immunol*. 2017;188:183-194. PMC5383442.
34. Jain S, Gautam V, Naseem S. Acute-phase proteins: as diagnostic tool. *J Pharm BioAllied Sci*. 2011;3:118-127. PMC3053509.
35. Romano R, Giardino G, Cirillo E, Prencipe R, Pignata C. Complement system network in cell physiology and in human diseases. *Int Rev Immunol*. 2020:1-12.
36. Lewis LA, Ram S. Complement interactions with the pathogenic Neisseriae: clinical features, deficiency states, and evasion mechanisms. *FEBS Lett*. 2020;594:2670-2694.
37. Langereis JD, van der Pasch ES, de Jonge MI. Serum IgM and C-reactive protein binding to phosphorylcholine of nontypeable Haemophilus influenzae increases complement-mediated killing. *Infect Immun*. 2019;87:e00299-e00219. PMC6652778.
38. Wegiel B, Larsen R, Gallo D, et al. Macrophages sense and kill bacteria through carbon monoxide-dependent inflammasome activation. *J Clin Invest*. 2014;124:4926-4940. PMC4347244.
39. Weiss G, Schaible UE. Macrophage defense mechanisms against intracellular bacteria. *Immunol Rev*. 2015;264:182-203. PMC4368383.
40. Murray PJ. Macrophage polarization. *Annu Rev Physiol*. 2017;79:541-566.
41. Mosser DM, Edwards JP. Exploring the full spectrum of macrophage activation. *Nat Rev Immunol*. 2008;8:958-969. PMC2724991.
42. Orecchioni M, Ghosheh Y, Pramod AB, Ley K. Macrophage polarization: different gene signatures in M1(LPS+) vs. classically and M2(LPS–) vs. alternatively activated macrophages. *Front Immunol*. 2019;10:1084.
43. Jablonski KA, Amici SA, Webb LM, et al. Novel markers to delineate murine M1 and M2 macrophages. *PloS One*. 2015;10:e0145342.
44. Teng TS, Ji AL, Ji XY, Li YZ. Neutrophils and immunity: from bactericidal action to being conquered. *J Immunol Res*. 2017;2017:9671604. PMC5337389.
45. Li Y, Cifuentes-Pagano E, DeVallance ER, et al. NADPH oxidase 2 inhibitors CPP11G and CPP11H attenuate endothelial cell inflammation & vessel dysfunction and restore mouse hind-limb flow. *Redox Biol*. 2019;22:101143.
46. Branzk N, Lubojemska A, Hardison SE, et al. Neutrophils sense microbe size and selectively release neutrophil extracellular traps in response to large pathogens. *Nat Immunol*. 2014;15:1017-1025. PMC4236687.
47. Brinkmann V, Reichard U, Goosmann C, et al. Neutrophil extracellular traps kill bacteria. *Science*. 2004;303:1532-1535.
48. Papayannopoulos V. Neutrophil extracellular traps in immunity and disease. *Nat Rev Immunol*. 2018;18:134-147.
49. Cococcioni L, Panelli S, Varotto-Boccazzi I, et al. IBDs and the pediatric age: their peculiarities and the involvement of the microbiota. *Dig Liver Dis*. 2020;53:17-25.
50. de Jong SE, Olin A, Pulendran B. The impact of the microbiome on immunity to vaccination in humans. *Cell Host Microbe*. 2020;28:169-179. PMC7422826.
51. Ivanov II, Atarashi K, Manel N, et al. Induction of intestinal Th17 cells by segmented filamentous bacteria. *Cell*. 2009;139:485-498. PMC2796826.
52. Steed AL, Christophi GP, Kaiko GE, et al. The microbial metabolite desaminotyrosine protects from influenza through type I interferon. *Science*. 2017;357:498-502. PMC5753406.
53. Hannoodee S, Nasuruddin DN. *Acute inflammatory response*. In: *StatPearls* [Internet]. Treasure Island (FL): StatPearls Publishing; 2020.
54. Pober JS, Sessa WC. Inflammation and the blood microvascular system. *Cold Spring Harb Perspect Biol*. 2014;7:a016345. PMC4292166.
55. Singer M, Deutschman CS, Seymour CW, et al. The third international consensus definitions for sepsis and septic shock (Sepsis-3). *J Am Med Assoc*. 2016;315:801-810. PMC4968574.
56. Bone RC. Why sepsis trials fail. *J Am Med Assoc*. 1996;276:565-566.
57. Abraham E, Anzueto A, Gutierrez G, et al. Double-blind randomised controlled trial of monoclonal antibody to human tumour necrosis factor in treatment of septic shock. NORASEPT II Study Group. *Lancet*. 1998;351:929-933.

58. Marshall JC. Why have clinical trials in sepsis failed? *Trends Mol Med*. 2014;20:195-203.
59. Venkatesh B, Finfer S, Cohen J, et al. Hydrocortisone compared with Placebo in patients with septic shock Satisfying the sepsis-3 diagnostic criteria and APROCCHSS study inclusion criteria: a post Hoc analysis of the ADRENAL trial. *Anesthesiology*. 2019;131:1292-1300.
60. Cheng AG, DeDent AC, Schneewind O, Missiakas D. A play in four acts: Staphylococcus aureus abscess formation. *Trends Microbiol*. 2011;19:225-232.
61. Eming SA, Martin P, Tomic-Canic M. Wound repair and regeneration: mechanisms, signaling, and translation. *Sci Transl Med*. 2014;6:265sr266.
62. Goldberg SR, Diegelmann RF. What makes wounds chronic. *Surg Clin North Am*. 2020;100:681-693.
63. Deinhardt-Emmer S, Haupt KF, Garcia-Moreno M, et al. Staphylococcus aureus pneumonia: preceding influenza infection Paves the way for low-virulent strains. *Toxins*. 2019;11:734. PMC6950557.
64. Verma AK, Bansal S, Bauer C, Muralidharan A, Sun K. Influenza infection induces alveolar macrophage dysfunction and thereby enables noninvasive *Streptococcus pneumoniae* to cause deadly pneumonia. *J Immunol*. 2020;205:1601-1607. PMC7484308.
65. Doctor A, Harper MB, Fleisher GR. Group A beta-hemolytic streptococcal bacteremia: historical overview, changing incidence, and recent association with varicella. *Pediatrics*. 1995;96:428-433.
66. Davies HD, McGeer A, Schwartz B, et al. Invasive group A streptococcal infections in Ontario, Canada. Ontario group A streptococcal study group. *N Engl J Med*. 1996;335:547-554.
67. Patel RA, Binns HJ, Shulman ST. Reduction in pediatric hospitalizations for varicella-related invasive group A streptococcal infections in the varicella vaccine era. *J Pediatr*. 2004;144:68-74.
68. Frère J, Bidet P, Tapiéro B, et al. Clinical and microbiological characteristics of invasive group A streptococcal infections before and after Implementation of a Universal Varicella Vaccine Program. *Clin Infect Dis*. 2016;62:75-77.
69. El Hamzaoui N, Barguigua A, Larouz S, Maouloua M. Epidemiology of burn wound bacterial infections at a Meknes hospital, Morocco. *New Microbes New Infect*. 2020;38:100764. PMC7600360.
70. World Health Organization. *2019 Antibacterial Agents in Clinical Development: An Analysis of the Antibacterial Clinical Development Pipeline*. 2019. https://apps.who.int/iris/bitstream/handle/10665/330420/9789240000193-eng.pdf
71. Henriques-Normark B, Tuomanen EI. The pneumococcus: epidemiology, microbiology, and pathogenesis. *Cold Spring Harb Perspect Med*. 2013;3:a010215. PMC3685878.
72. Aguilera ER, Lenz LL. Inflammation as a modulator of host susceptibility to pulmonary influenza, pneumococcal, and co-infections. *Front Immunol*. 2020;11:105. PMC7026256.
73. Zhu X, Ge Y, Wu T, et al. Co-infection with respiratory pathogens among COVID-2019 cases. *Virus Res*. 2020;285:198005. PMC7213959.
74. Ganaie F, Saad JS, McGee L, et al. A new pneumococcal capsule type, 10D, is the 100th serotype and has a large cps fragment from an oral Streptococcus. *mBio*. 2020;11:e00937-20. PMC7240158.
75. Geno KA, Gilbert GL, Song JY, et al. Pneumococcal capsules and their types: past, present, and future. *Clin Microbiol Rev*. 2015;28:871-899. PMC4475641.
76. Avery OT, Dubos R. The protective action of a specific enzyme against type III pneumococcus infection in mice. *J Exp Med*. 1931;54:73-89. PMC2132045.
77. Briles DE, Crain MJ, Gray BM, Forman C, Yother J. Strong association between capsular type and virulence for mice among human isolates of *Streptococcus pneumoniae*. *Infect Immun*. 1992;60:111-116. PMC257510.
78. Brady AM, Calix JJ, Yu J, Geno KA, Cutter GR, Nahm MH. Low invasiveness of pneumococcal serotype 11A is linked to ficolin-2 recognition of O-acetylated capsule epitopes and lectin complement pathway activation. *J Infect Dis*. 2014;210:1155-1165. PMC4215079.
79. Janeway CA Jr. *Immunobiology: The Immune System in Health and Disease*. Garland Science; 2001.
80. Matthias KA, Roche AM, Standish AJ, Shchepetov M, Weiser JN. Neutrophil-toxin interactions promote antigen delivery and mucosal clearance of *Streptococcus pneumoniae*. *J Immunol*. 2008;180:6246-6254.
81. Weiser JN, Ferreira DM, Paton JC. Streptococcus pneumoniae: transmission, colonization and invasion. *Nat Rev Microbiol*. 2018;16:355-367. PMC5949087.
82. Thors V, Christensen H, Morales-Aza B, Vipond I, Muir P, Finn A. The effects of live attenuated influenza vaccine on nasopharyngeal bacteria in healthy 2 to 4 year olds. A randomized controlled trial. *Am J Respir Crit Care Med*. 2016;193:1401-1409. PMC4910891.
83. Mina MJ, McCullers JA, Klugman KP. Live attenuated influenza vaccine enhances colonization of *Streptococcus pneumoniae* and *Staphylococcus aureus* in mice. *mBio*. 2014;5:e01040-13. PMC3944816.
84. Moffitt K, Cheung E, Manis J, Malley R. Evaluation of the role of stat3 in antibody and T(H)17-mediated responses to pneumococcal immunization and infection by use of a mouse model of autosomal dominant hyper-IgE syndrome. *Infect Immun*. 2018;86:e00024-18. PMC5913858.
85. Jochems SP, de Ruiter K, Solórzano C, et al. Innate and adaptive nasal mucosal immune responses following experimental human pneumococcal colonization. *J Clin Invest*. 2019;129:4523-4538. PMC6763269.
86. Yadav MK, Chae SW, Song JJ. In vitro *Streptococcus pneumoniae* biofilm formation and in vivo middle ear mucosal biofilm in a rat model of acute otitis induced by S. Pneumoniae. *Clin Exp Otorhinolaryngol*. 2012;5:139-144. PMC3437414.
87. Fliegauf M, Sonnen AF, Kremer B, Henneke P. Mucociliary clearance defects in a murine in vitro model of pneumococcal airway infection. *PloS One*. 2013;8:e59925. PMC3602288.
88. Nerlich A, von Wunsch Teruel I, Mieth M, et al. Reversion of pneumolysin induced executioner caspase activation redirects cells to survival. *J Infect Dis*. 2020;223:1973-1983.
89. Ishizuka S, Yokoyama R, Sato K, et al. Effect of CARD9 deficiency on neutrophil-mediated host defense against pulmonary infection with *Streptococcus pneumoniae*. *Infect Immun*. 2020;89:e00305-20.
90. Standish AJ, Weiser JN. Human neutrophils kill *Streptococcus pneumoniae* via serine proteases. *J Immunol*. 2009;183:2602-2609.
91. Hammerschmidt S, Wolff S, Hocke A, Rosseau S, Müller E, Rohde M. Illustration of pneumococcal polysaccharide capsule during adherence and invasion of epithelial cells. *Infect Immun*. 2005;73:4653-4667. PMC1201225.
92. Kietzman CC, Gao G, Mann B, Myers L, Tuomanen EI. Dynamic capsule restructuring by the main pneumococcal autolysin LytA in response to the epithelium. *Nat Commun*. 2016;7:10859. PMC4773454.
93. Loughran AJ, Orihuela CJ, Tuomanen EI. Streptococcus pneumoniae: invasion and inflammation. *Microbiol Spectr*. 2019;7. PMC6422050.
94. Wang X, Yuan T, Yuan J, et al. Expression of Toll-like receptor 2 by dendritic cells is essential for the DnaJ-ΔA146Ply-mediated Th1 immune response against *Streptococcus pneumoniae*. *Infect Immun*. 2018;86:e00651-e00617. PMC5820941.
95. Picard C, von Bernuth H, Ghandil P, et al. Clinical features and outcome of patients with IRAK-4 and MyD88 deficiency. *Medicine (Baltimore)*. 2010;89:403-425. PMC3103888.
96. Rabes A, Suttorp N, Opitz B. Inflammasomes in pneumococcal infection: innate immune sensing and bacterial evasion strategies. *Curr Top Microbiol Immunol*. 2016;397:215-227.
97. Subramanian K, Neill DR, Malak HA, et al. Pneumolysin binds to the mannose receptor C type 1 (MRC-1) leading to anti-inflammatory responses and enhanced pneumococcal survival. *Nat Microbiol*. 2019;4:62-70. PMC6298590.
98. Deniset JF, Surewaard BG, Lee WY, Kubes P. Splenic Ly6G(high) mature and Ly6G(int) immature neutrophils contribute to eradication of S. pneumoniae. *J Exp Med*. 2017;214:1333-1350. PMC5413339.
99. Kang YS, Kim JY, Bruening SA, et al. The C-type lectin SIGN-R1 mediates uptake of the capsular polysaccharide of *Streptococcus pneumoniae* in the marginal zone of mouse spleen. *Proc Natl Acad Sci U S A*. 2004;101:215-220. PMC314165.
100. Koppel EA, Wieland CW, van den Berg VC, et al. Specific ICAM-3 grabbing nonintegrin-related 1 (SIGNR1) expressed by marginal zone macrophages is essential for defense against pulmonary *Streptococcus pneumoniae* infection. *Eur J Immunol*. 2005;35:2962-2969.
101. Briles DE, Nahm M, Schroer K, et al. Antiphosphocholine antibodies found in normal mouse serum are protective against intravenous infection with type 3 *streptococcus pneumoniae*. *J Exp Med*. 1981;153:694-705. PMC2186108.
102. Reyes LF, Restrepo MI, Hinojosa CA, et al. Severe pneumococcal pneumonia causes acute cardiac toxicity and subsequent cardiac remodeling. *Am J Respir Crit Care Med*. 2017;196:609-620. PMC5620668.
103. Jönsson G, Truedsson L, Sturfelt G, Oxelius VA, Braconier JH, Sjöholm AG. Hereditary C2 deficiency in Sweden: frequent occurrence of invasive infection, atherosclerosis, and rheumatic disease. *Medicine (Baltimore)*. 2005;84:23-34.
104. Theilacker C, Ludewig K, Serr A, et al. Overwhelming postsplenectomy infection: a prospective multicenter cohort study. *Clin Infect Dis*. 2015;62:871-878.
105. Stacey HL, Rosen J, Peterson JT, et al. Safety and immunogenicity of 15-valent pneumococcal conjugate vaccine (PCV-15) compared to PCV-13 in healthy older adults. *Hum Vaccin Immunother*. 2019;15:530-539. PMC6605726.
106. Feldman C, Anderson R. Recent advances in the epidemiology and prevention of *Streptococcus pneumoniae* infections. *F1000Res*. 2020;9:F1000 Faculty Rev-338. PMC7212261.
107. Chen A, Mann B, Gao G, et al. Multivalent pneumococcal protein vaccines comprising pneumolysoid with epitopes/fragments of CbpA and/or PspA elicit strong and broad protection. *Clin Vaccine Immunol*. 2015;22:1079-1089. PMC4580740.
108. Briles DE, Paton JC, Mukerji R, Swiatlo E, Crain MJ. Pneumococcal vaccines. *Microbiol Spectr*. 2019;7.
109. Mork RL, Hogan PG, Muenks CE, et al. Longitudinal, strain-specific Staphylococcus aureus introduction and transmission events in households of children with community-associated meticillin-resistant S aureus skin and soft tissue infection: a prospective cohort study. *Lancet Infect Dis*. 2020;20:188-198. PMC6995751.
110. Wertheim HF, Melles DC, Vos MC, et al. The role of nasal carriage in *Staphylococcus aureus* infections. *Lancet Infect Dis*. 2005;5:751-762.
111. Starkl P, Watzenboeck ML, Popov LM, et al. IgE effector mechanisms, in concert with mast cells, contribute to acquired host defense against Staphylococcusaureus. *Immunity*. 2020;53:793-804.e799. PMC7572876.
112. van der Poll T, van de Veerdonk FL, Scicluna BP, Netea MG. The immunopathology of sepsis and potential therapeutic targets. *Nat Rev Immunol*. 2017;17:407-420.
113. Surewaard BGJ, Thanabalasuriar A, Zeng Z, et al. α-Toxin induces platelet aggregation and liver injury during *Staphylococcus aureus* sepsis. *Cell Host Microbe*. 2018;24:271-284.e273. PMC6295203.
114. Miller LS, Fowler VG, Shukla SK, Rose WE, Proctor RA. Development of a vaccine against Staphylococcus aureus invasive infections: evidence based on human immunity, genetics and bacterial evasion mechanisms. *FEMS Microbiol Rev*. 2020;44:123-153. PMC7053580.
115. Huitema L, Phillips T, Alexeev V, Tomic-Canic M, Pastar I, Igoucheva O. Intracellular escape strategies of *Staphylococcus aureus* in persistent cutaneous infections. *Exp Dermatol*. 2020;11:10.1111/exd.14235. doi:10.1111/exd.14235
116. Sirobhushanam S, Parsa N, Reed TJ, et al. *Staphylococcus aureus* colonization is increased on lupus skin lesions and is promoted by IFN-mediated barrier disruption. *J Invest Dermatol*. 2020;140:1066-1074.e1064. PMC7183889.
117. Bitschar K, Staudenmaier L, Klink L, et al. *Staphylococcus aureus* skin colonization is enhanced by the interaction of neutrophil extracellular traps with keratinocytes. *J Invest Dermatol*. 2020;140:1054-1065.e1054.
118. Sakr A, Brégeon F, Mège JL, Rolain JM, Blin O. Staphylococcus aureus nasal colonization: an update on mechanisms, epidemiology, risk factors, and subsequent infections. *Front Microbiol*. 2018;9:2419. PMC6186810.
119. Al Kindi A, Williams H, Matsuda K, et al. *Staphylococcus aureus* second immunoglobulin-binding protein drives atopic dermatitis via IL-33. *J Allergy Clin Immunol*. 2020;147:1354-1368.e3.
120. Withrock IC, Anderson SJ, Jefferson MA, et al. Genetic diseases conferring resistance to infectious diseases. *Genes Dis*. 2015;2:247-254. PMC6150079.
121. Wolter DJ, Ramsey BW. Not quite the bully in the Schoolyard: Staphylococcus aureus can survive and co-exist with pseudomonas aeruginosa in the cystic fibrosis lung. *Am J Respir Crit Care Med*. 2020;203:279-281.
122. Hoy SM. Elexacaftor/ivacaftor/tezacaftor: first approval. *Drugs*. 2019;79:2001-2007.
123. Riquelme SA, Wong Fok Lung T, Prince A. Pulmonary pathogens adapt to immune signaling metabolites in the airway. *Front Immunol*. 2020;11:385. PMC7082326.
124. Bosch ME, Bertrand BP, Heim CE, et al. *Staphylococcus aureus* ATP synthase promotes biofilm persistence by influencing innate immunity. *mBio*. 2020;11:e01581-20. PMC7482063.
125. Simonin J, Bille E, Crambert G, et al. Airway surface liquid acidification initiates host defense abnormalities in Cystic Fibrosis. *Sci Rep*. 2019;9:6516. PMC6482305.
126. Thammavongsa V, Kim HK, Missiakas D, Schneewind O. Staphylococcal manipulation of host immune responses. *Nat Rev Microbiol*. 2015;13:529-543. PMC4625792.
127. Garzoni C, Kelley WL. Return of the Trojan horse: intracellular phenotype switching and immune evasion by Staphylococcus aureus. *EMBO Mol Med*. 2011;3:115-117. PMC3395108.
128. Peyrusson F, Varet H, Nguyen TK, et al. Intracellular *Staphylococcus aureus* persisters upon antibiotic exposure. *Nat Commun*. 2020;11:2200.
129. Lehar SM, Pillow T, Xu M, et al. Novel antibody-antibiotic conjugate eliminates intracellular S. aureus. *Nature*. 2015;527:323-328.
130. Pauli NT, Kim HK, Falugi F, et al. *Staphylococcus aureus* infection induces protein A-mediated immune evasion in humans. *J Exp Med*. 2014;211:2331-2339. PMC4235641.
131. Kobayashi SD, DeLeo FR. *Staphylococcus aureus* protein A promotes immune suppression. *mBio*. 2013;4:e00764-00713. PMC3791897.
132. Singh V, Phukan UJ. Interaction of host and *Staphylococcus aureus* protease-system regulates virulence and pathogenicity. *Med Microbiol Immunol*. 2019;208:585-607.
133. Alfano DN, Bubeck Wardenburg J. Another score for the pore: S. aureus leukocidins take a shot on the endothelium. *Cell Host Microbe*. 2019;25:351-353.
134. Lubkin A, Lee WL, Alonzo F 3rd, et al. *Staphylococcus aureus* leukocidins target endothelial DARC to cause lethality in mice. *Cell Host Microbe*. 2019;25:463-470.e469. PMC6468323.
135. Lappin E, Ferguson AJ. Gram-positive toxic shock syndromes. *Lancet Infect Dis*. 2009;9:281-290.
136. Park JY, Kim JS, Woo H. Prevalence of antibody to toxic shock syndrome toxin-1 in burn patients. *Ann Lab Med*. 2015;35:89-93. PMC4272971.

137. Fritz SA, Tiemann KM, Hogan PG, et al. A serologic correlate of protective immunity against community-onset *Staphylococcus aureus* infection. *Clin Infect Dis.* 2013;56:1554-1561. PMC3641868.
138. Cho JS, Pietras EM, Garcia NC, et al. IL-17 is essential for host defense against cutaneous *Staphylococcus aureus* infection in mice. *J Clin Invest.* 2010;120:1762-1773. PMC2860944.
139. Freeman AF, Holland SM. Clinical manifestations, etiology, and pathogenesis of the hyper-IgE syndromes. *Pediatr Res.* 2009;65:32r-37r. PMC2919366.
140. Dillen CA, Pinsker BL, Marusina AI, et al. Clonally expanded γδ T cells protect against *Staphylococcus aureus* skin reinfection. *J Clin Invest.* 2018;128:1026-1042. PMC5824877.
141. Brown AF, Murphy AG, Lalor SJ, et al. Memory Th1 cells are protective in invasive *Staphylococcus aureus* infection. *PLoS Pathog.* 2015;11:e1005226. PMC4634925.
142. Lee B, Olaniyi R, Kwiecinski JM, Wardenburg JB. *Staphylococcus aureus* toxin suppresses antigen-specific T cell responses. *J Clin Invest.* 2020;130:1122-1127. PMC7269593.
143. Simonis A, Schubert-Unkmeir A. Interactions of meningococcal virulence factors with endothelial cells at the human blood-cerebrospinal fluid barrier and their role in pathogenicity. *FEBS Lett.* 2016;590:3854-3867.
144. Jones GR, Christodoulides M, Brooks JL, Miller AR, Cartwright KA, Heckels JE. Dynamics of carriage of Neisseria meningitidis in a group of military recruits: subtype stability and specificity of the immune response following colonization. *J Infect Dis.* 1998;178:451-459.
145. Green LR, Al-Rubaiawi AA, Al-Maeni M, et al. Localized hypermutation is the major driver of meningococcal genetic variability during persistent asymptomatic carriage. *mBio.* 2020;11:e03068-e03019. PMC7157529.
146. Eriksson L, Stenmark B, Deghmane AE, et al. Difference in virulence between Neisseria meningitidis serogroups W and Y in transgenic mice. *BMC Microbiol.* 2020;20:92. PMC7160935.
147. Hodeib S, Herberg JA, Levin M, Sancho-Shimizu V. Human genetics of meningococcal infections. *Hum Genet.* 2020;139:961-980. PMC7272491 of interest.
148. Sutherland TC, Quattroni P, Exley RM, Tang CM. Transcellular passage of Neisseria meningitidis across a polarized respiratory epithelium. *Infect Immun.* 2010;78:3832-3847. PMC2937448.
149. Besbes A, Le Goff S, Antunes A, et al. Hyperinvasive meningococci induce intra-nuclear cleavage of the NF-κB protein p65/RelA by meningococcal IgA protease. *PLoS Pathog.* 2015;11:e1005078. PMC4524725.
150. Barrile R, Kasendra M, Rossi-Paccani S, et al. Neisseria meningitidis subverts the polarized organization and intracellular trafficking of host cells to cross the epithelial barrier. *Cell Microbiol.* 2015;17:1365-1375.
151. Coureuil M, Join-Lambert O, Lécuyer H, Bourdoulous S, Marullo S, Nassif X. Mechanism of meningeal invasion by Neisseria meningitidis. *Virulence.* 2012;3:164-172. PMC3396695.
152. Denis K, Le Bris M, Le Guennec L, et al. Targeting type IV pili as an antivirulence strategy against invasive meningococcal disease. *Nat Microbiol.* 2019;4:972-984.
153. Lécuyer H, Virion Z, Barnier JP, et al. An ADAM-10 dependent EPCR shedding links meningococcal interaction with endothelial cells to purpura fulminans. *PLoS Pathog. e1006981.* 2018;14:e1006981. PMC5908201.
154. Figueroa JE, Densen P. Infectious diseases associated with complement deficiencies. *Clin Microbiol Rev.* 1991;4:359-395. PMC358203.
155. Muenstermann M, Strobel L, Klos A, et al. Distinct roles of the anaphylatoxin receptors C3aR, C5aR1 and C5aR2 in experimental meningococcal infections. *Virulence.* 2019;10:677-694. PMC6650196.
156. Konar M, Granoff DM. Eculizumab treatment and impaired opsonophagocytic killing of meningococci by whole blood from immunized adults. *Blood.* 2017;130:891-899. PMC5561903.
157. Gillet B, Joram N, Bacchi VF, Thomas C, Béné MC, Wuillème S. Neisseria meningitidis inside neutrophils, revealing properdin deficiency. *Int J Infect Dis.* 2020;99:117-118.
158. Brodszki N, Frazer-Abel A, Grumach AS, et al. European society for immunodeficiencies (ESID) and European Reference Network on Rare Primary Immunodeficiency, Autoinflammatory and Autoimmune Diseases (ERN RITA) complement guideline: deficiencies, diagnosis, and management. *J Clin Immunol.* 2020;40:576-591. PMC7253377.
159. Geldhoff M, Mook-Kanamori BB, Brouwer MC, et al. Inflammasome activation mediates inflammation and outcome in humans and mice with pneumococcal meningitis. *BMC Infect Dis.* 2013;13:358. PMC3750264.
160. Idosa BA, Kelly A, Jacobsson S, et al. Neisseria meningitidis-induced caspase-1 activation in human innate immune cells is LOS-dependent. *J Immunol Res.* 2019;2019:6193186. PMC6526529.
161. Krüger S, Eichler E, Strobel L, Schubert-Unkmeir A, Johswich KO. Differential influences of complement on neutrophil responses to Neisseria meningitidis infection. *Pathog Dis.* 2018;76.
162. Lappann M, Danhof S, Guenther F, Olivares-Florez S, Mordhorst IL, Vogel U. In vitro resistance mechanisms of Neisseria meningitidis against neutrophil extracellular traps. *Mol Microbiol.* 2013;89:433-449.
163. Gandhi A, Balmer P, York LJ. Characteristics of a new meningococcal serogroup B vaccine, bivalent rLP2086 (MenB-FHbp; Trumenba®). *Postgrad Med.* 2016;128:548-556.
164. Fiebig T, Cramer JT, Bethe A, et al. Structural and mechanistic basis of capsule O-acetylation in Neisseria meningitidis serogroup A. *Nat Commun.* 2020;11:4723. PMC7501274.
165. Khalil IA, Troeger C, Blacker BF, et al. Morbidity and mortality due to shigella and enterotoxigenic Escherichia coli diarrhoea: the global burden of disease study 1990-2016. *Lancet Infect Dis.* 2018;18:1229-1240. PMC6202441.
166. Baker S, The HC. Recent insights into Shigella. *Curr Opin Infect Dis.* 2018;31:449-454. PMC6143181.
167. Tinevez JY, Arena ET, Anderson M, et al. Shigella-mediated oxygen depletion is essential for intestinal mucosa colonization. *Nat Microbiol.* 2019;4:2001-2009. PMC6817363.
168. Levine MM, DuPont HL, Khodabandelou M, Hornick RB. Long-term shigella-carrier state. *N Engl J Med.* 1973;288:1169-1171.
169. McMurry TL, McQuade ETR, Liu J, et al. Duration of post-diarrheal enteric pathogen carriage in young children in low-resource settings. *Clin Infect Dis.* 2020;72:e806-e814.
170. de Jong MF, Alto NM. Cooperative immune suppression by *Escherichia coli* and Shigella effector proteins. *Infect Immun.* 2018;86:e00560-17. PMC5865018.
171. Sperandio B, Fischer N, Chevalier-Curt MJ, et al. Virulent Shigella flexneri affects secretion, expression, and glycosylation of gel-forming mucins in mucus-producing cells. *Infect Immun.* 2013;81:3632-3643.
172. Ducarmon QR, Zwittink RD, Hornung BVH, van Schaik W, Young VB, Kuijper EJ. Gut microbiota and colonization resistance against bacterial enteric infection. *Microbiol Mol Biol Rev.* 2019;83:e00007-e00019. PMC6710460.
173. Mabbott NA, Donaldson DS, Ohno H, Williams IR, Mahajan A. Microfold (M) cells: important immunosurveillance posts in the intestinal epithelium. *Mucosal Immunol.* 2013;6:666-677. PMC3686595.
174. Riddle MS, Kaminski RW, Di Paolo C, et al. Safety and immunogenicity of a candidate bioconjugate vaccine against Shigella flexneri 2a administered to healthy adults: a single-blind, randomized phase I study. *Clin Vaccine Immunol.* 2016;23:908-917. PMC5139601.
175. Rybtke M, Hultqvist LD, Givskov M, Tolker-Nielsen T. Pseudomonas aeruginosa biofilm infections: community structure, antimicrobial tolerance and immune response. *J Mol Biol.* 2015;427:3628-3645.
176. Rybtke M, Jensen P, Nielsen CH, Tolker-Nielsen T. The extracellular polysaccharide matrix of *Pseudomonas aeruginosa* biofilms is a determinant of polymorphonuclear leukocyte responses. *Infect Immun.* 2020;89:e00631-20.
177. Chen H, Wubbolts RW, Haagsman HP, Veldhuizen EJA. Inhibition and eradication of *Pseudomonas aeruginosa* biofilms by host defence peptides. *Sci Rep.* 2018;8:10446. PMC6041282.
178. Curran CS, Bolig T, Torabi-Parizi P. Mechanisms and targeted therapies for *Pseudomonas aeruginosa* lung infection. *Am J Respir Crit Care Med.* 2018;197:708-727. PMC5855068.
179. Rada B. Interactions between neutrophils and *Pseudomonas aeruginosa* in cystic fibrosis. *Pathogens.* 2017;6. PMC5371898.
180. Marciano BE, Spalding C, Fitzgerald A, et al. Common severe infections in chronic granulomatous disease. *Clin Infect Dis.* 2015;60:1176-1183. PMC4400412.
181. Mauch RM, Jensen P, Moser C, Levy CE, Høiby N. Mechanisms of humoral immune response against *Pseudomonas aeruginosa* biofilm infection in cystic fibrosis. *J Cyst Fibros.* 2018;17:143-152.
182. Simmons JD, Stein CM, Seshadri C, et al. Immunological mechanisms of human resistance to persistent *Mycobacterium tuberculosis* infection. *Nat Rev Immunol.* 2018;18:575-589.
183. Weerasuriya CK, Clark RA, White RG, Harris RC. New tuberculosis vaccines: advances in clinical development and modelling. *J Intern Med.* 2020;288:661-681.
184. Shang S, Kats D, Cao L, et al. Induction of *Mycobacterium tuberculosis* lipid-specific T cell responses by pulmonary delivery of mycolic acid-loaded polymeric micellar nanocarriers. *Front Immunol.* 2018;9:2709.
185. Koch A, Mizrahi V. *Mycobacterium tuberculosis. Trends Microbiol.* 2018;26:555-556.
186. Koch A, Cox H, Mizrahi V. Drug-resistant tuberculosis: challenges and opportunities for diagnosis and treatment. *Curr Opin Pharmacol.* 2018;42:7-15. PMC6219890.
187. Kinsella RL, Zhu DX, Harrison GA, et al. Perspectives and advances in the understanding of tuberculosis. *Annu Rev Pathol.* 2021;16:377-408.
188. Gopal R, Monin L, Slight S, et al. Unexpected role for IL-17 in protective immunity against hypervirulent *Mycobacterium tuberculosis* HN878 infection. *PLoS Pathog.* 2014;10:e1004099.
189. Khan FY. Review of literature on disseminated tuberculosis with emphasis on the focused diagnostic workup. *J Family Community Med.* 2019;26:83-91. PMC6515764.
190. Gutierrez MG, Master SS, Singh SB, Taylor GA, Colombo MI, Deretic V. Autophagy is a defense mechanism inhibiting BCG and *Mycobacterium tuberculosis* survival in infected macrophages. *Cell.* 2004;119:753-766.
191. Songane M, Kleinnijenhuis J, Netea MG, van Crevel R. The role of autophagy in host defence against *Mycobacterium tuberculosis* infection. *Tuberculosis.* 2012;92:388-396.
192. Wei Y, Liu M, Li X, Liu J, Li H. Origin of the autophagosome membrane in mammals. *BioMed Res Int.* 2018;2018:1012789. PMC6174804.
193. Jo EK. Autophagy as an innate defense against mycobacteria. *Pathog Dis.* 2013;67:108-118.
194. Huynh JP, Lin CC, Kimmey JM, et al. Bhlhe40 is an essential repressor of IL-10 during *Mycobacterium tuberculosis* infection. *J Exp Med.* 2018;215:1823-1838. PMC6028511.
195. Lu LL, Smith MT, Yu KKQ, et al. IFN-γ-independent immune markers of *Mycobacterium tuberculosis* exposure. *Nat Med.* 2019;25:977-987. PMC6559862.
196. Garg R, Borbora SM, Bansia H, et al. *Mycobacterium tuberculosis* calcium pump CtpF modulates the autophagosome in an mTOR-dependent manner. *Front Cell Infect Microbiol.* 2020;10:461. PMC7525011.
197. Ehlers S, Schaible UE. The granuloma in tuberculosis: dynamics of a host-pathogen collusion. *Front Immunol.* 2012;3:411. PMC3538277.
198. Plumlee CR, Duffy FJ, Gern BH, et al. Ultra-low dose aerosol infection of mice with *Mycobacterium tuberculosis* more closely models human tuberculosis. *Cell Host Microbe.* 2021;29:68-82.e65. PMC7854984.
199. Turk JL, Narayanan RB. The origin, morphology, and function of epithelioid cells. *Immunobiology.* 1982;161:274-282.
200. Russell DG, Cardona PJ, Kim MJ, Allain S, Altare F. Foamy macrophages and the progression of the human tuberculosis granuloma. *Nat Immunol.* 2009;10:943-948. PMC2759071.
201. McClean CM, Tobin DM. Macrophage form, function, and phenotype in mycobacterial infection: lessons from tuberculosis and other diseases. *Pathog Dis.* 2016;74:ftw068. PMC5985490.
202. Silva Miranda M, Breiman A, Allain S, Deknuydt F, Altare F. The tuberculous granuloma: an unsuccessful host defence mechanism providing a safety shelter for the bacteria? *Clin Dev Immunol.* 2012;2012:139127.
203. Choreño-Parra JA, Bobba S, Rangel-Moreno J, et al. *Mycobacterium tuberculosis* HN878 infection induces human-like B-cell follicles in mice. *J Infect Dis.* 2020;221:1636-1646. PMC7184917.
204. Tsai MC, Chakravarty S, Zhu G, et al. Characterization of the tuberculous granuloma in murine and human lungs: cellular composition and relative tissue oxygen tension. *Cell Microbiol.* 2006;8:218-232.
205. Ardain A, Domingo-Gonzalez R, Das S, et al. Group 3 innate lymphoid cells mediate early protective immunity against tuberculosis. *Nature.* 2019;570:528-532. PMC6626542.
206. Kimmey JM, Huynh JP, Weiss LA, et al. Unique role for ATG5 in neutrophil-mediated immunopathology during *M. tuberculosis* infection. *Nature.* 2015;528:565-569.
207. Nair S, Huynh JP, Lampropoulou V, et al. Irg1 expression in myeloid cells prevents immunopathology during *M. tuberculosis* infection. *J Exp Med.* 2018;215:1035-1045. PMC5881474.
208. Sia JK, Rengarajan J. Immunology of *Mycobacterium tuberculosis* infections. *Microbiol Spectr.* 2019;7. PMC6636855.
209. Boisson-Dupuis S, Bustamante J, El-Baghdadi J, et al. Inherited and acquired immunodeficiencies underlying tuberculosis in childhood. *Immunol Rev.* 2015;264:103-120. PMC4405179.
210. Borgström EW, Fröberg G, Correia-Neves M, et al. CD4(+) T cell proliferative responses to PPD and CFP-10 associate with recent *M. tuberculosis* infection. *Tuberculosis.* 2020;123:101959.
211. Nieuwenhuizen NE, Kaufmann SHE. Next-generation vaccines based on bacille calmette-guérin. *Front Immunol.* 2018;9:121. PMC5807593.
212. Dhar S, Ganjoo S, Dhar S, Srinivas SM. Disseminated cutaneous tuberculosis from BCG vaccination site in an immunocompetent child. *Pediatr Dermatol.* 2020;38:294-295.
213. Darrah PA, Zeppa JJ, Maiello P, et al. Prevention of tuberculosis in macaques after intravenous BCG immunization. *Nature.* 2020;577:95-102. PMC7015856.
214. Hansen SG, Zak DE, Xu G, et al. Prevention of tuberculosis in rhesus macaques by a cytomegalovirus-based vaccine. *Nat Med.* 2018;24:130-143. PMC5909823.
215. Lamberti Y, Gorgojo J, Massillo C, Rodriguez ME. Bordetella pertussis entry into respiratory epithelial cells and intracellular survival. *Pathog Dis.* 2013;69:194-204.
216. Gregg KA, Merkel TJ. Pertussis toxin: a key component in pertussis vaccines? *Toxins.* 2019;11:557. PMC6832755.
217. Liu C, Yang L, Cheng Y, Xu H, Xu F. Risk factors associated with death in infants <120 days old with severe pertussis: a case-control study. *BMC Infect Dis.* 2020;20:852.PMC7668018.
218. Damron FH, Barbier M, Dubey P, et al. Overcoming waning immunity in pertussis vaccines: Workshop of the National Institute of Allergy and Infectious Diseases. *J Immunol.* 2020;205:877-882. PMC7454230.
219. Phillips R, Ager A. Activation of pertussis toxin-sensitive CXCL12 (SDF-1) receptors mediates transendothelial

migration of T lymphocytes across lymph node high endothelial cells. *Eur J Immunol.* 2002;32:837-847.

220. Winter K, Zipprich J, Harriman K, et al. Risk factors associated with infant deaths from pertussis: a case-control study. *Clin Infect Dis.* 2015;61:1099-1106.
221. Control, CfD. Pertussis vaccination: use of acellular pertussis vaccines among infants and YoChildren recommendations of the advisory committee on immunization practic. *MMWR (Morb Mortal Wkly Rep): MMWR (Morb Mortal Wkly Rep).* 1997;46:1-25.
222. Chaudhary A, Pandey S. *Corynebacterium diphtheriae.* In: *StatPearls* [Internet]. Treasure Island (FL): StatPearls Publishing; 2021.
223. Adler NR, Mahony A, Friedman ND. Diphtheria: forgotten, but not gone. *Intern Med J.* 2013;43:206-210.
224. Sharma NC, Efstratiou A, Mokrousov I, Mutreja A, Das B, Ramamurthy T. Diphtheria. *Nat Rev Dis Primers.* 2019;5:81.
225. Gill DM. Bacterial toxins: a table of lethal amounts. *Microbiol Rev.* 1982;46:86-94. PMC373212.
226. Pezzotti P, Bellino S, Prestinaci F, et al. The impact of immunization programs on 10 vaccine preventable diseases in Italy: 1900-2015. *Vaccine.* 2018;36:1435-1443.
227. Clarke KEN, MacNeil A, Hadler S, Scott C, Tiwari TSP, Cherian T. Global epidemiology of diphtheria, 2000-2017(1). *Emerg Infect Dis.* 2019;25:1834-1842. PMC6759252.
228. Megighian A, Pirazzini M, Fabris F, Rossetto O, Montecucco C. Tetanus and tetanus neurotoxin: from peripheral uptake to central nervous tissue targets. *J Neurochem.* 2021. doi:10.1111/jnc.15330
229. Perey BJ. Progress in tetanus prophylaxis: the advent of human antitoxin. *Can Med Assoc J.* 1966;94:437-441. PMC1935316.
230. Aubert N, Brachet-Botineau M, de Olivera Preto GE, Benz-de Bretagne I, Watier H, Brachet G. History, extensive characterization and challenge of anti-tetanus serum from World War I: exciting remnants and deceived hopes – centenarian IgGs lost their neutralization capacity. *Immunol Res.* 2020;68:7-12.
231. Swanson OK, Maffei A. From hiring to firing: activation of inhibitory neurons and their recruitment in behavior. *Front Mol Neurosci.* 2019;12:168.
232. Montecucco C, Schiavo G. Mechanism of action of tetanus and botulinum neurotoxins. *Mol Microbiol.* 1994;13:1-8.
233. Callison C, Nguyen H. *Tetanus prophylaxis.* In: *StatPearls.* StatPearls PublishingCopyright © 2021, StatPearls Publishing LLC; 2021.
234. Przedpelski A, Tepp WH, Pellett S, Johnson EA, Barbieri JT. A novel high-potency tetanus vaccine. *mBio.* 2020;11:e01668-20. PMC7439469.
235. Lonati D, Locatelli CA, Fenicia L, et al. Fatal course of foodborne botulism in an eight-month old infant. *Pediatr Rep.* 2011;3:e31. PMC3283199.
236. Rasetti-Escargueil C, Lemichez E, Popoff MR. Toxemia in human naturally acquired botulism. *Toxins.* 2020;12:716. PMC7697460.
237. Swaminathan S, Eswaramoorthy S. Structural analysis of the catalytic and binding sites of Clostridium botulinum neurotoxin B. *Nat Struct Biol.* 2000;7:693-699.
238. Xiong X, Lv S, Fu C, et al. Production and characterization of a neutralizing antibody against botulinum neurotoxin A. *J Immunol Methods.* 2020;487:112871.
239. León-Valenzuela A, Palacios JS, Del Pino Algarrada R. IncobotulinumtoxinA for the treatment of spasticity in children with cerebral palsy - a retrospective case series focusing on dosing and tolerability. *BMC Neurol.* 2020;20:126. PMC7140581.
240. Frampton JE. OnabotulinumtoxinA in chronic migraine: a profile of its use. *CNS Drugs.* 2020;34:1287-1298. PMC7772159.
241. Park MY, Ahn KY. Scientific review of the aesthetic uses of botulinum toxin type A. *Arch Craniofac Surg.* 2021; 22:1-10.
242. Tseng-Ong L, Mitchell WG. Infant botulism: 20 years' experience at a single institution. *J Child Neurol.* 2007;22:1333-1337.
243. Liu FJ, Shi DY, Mao YY, et al. Immunological characterisation and immunoprotective efficacy of functional domain antigens of botulinum neurotoxin serotype A. *Vaccine.* 2020;38:2978-2983.
244. Nakanwagi M, Ario AR, Kwagonza L, et al. Outbreak of gastrointestinal anthrax following eating beef of suspicious origin: Isingiro District, Uganda, 2017. *PLoS Negl Trop Dis.* 2020;14:e0008026.
245. Savransky V, Ionin B, Reece J. Current status and trends in prophylaxis and management of anthrax disease. *Pathogens.* 2020;9:370. PMC7281134.
246. Das S, Kataria VK. Bioterrorism: a public health perspective. *Med J Armed Forces India.* 2010;66:255-260. PMC4921253.
247. Couse Z, Cui X, Li Y, Moayeri M, Leppla S, Eichacker PQ. A review of the efficacy of FDA-approved B. Anthracis antitoxin agents when combined with antibiotic or hemodynamic support in infection- or toxin-challenged preclinical models. *Toxins.* 2021;13:53. PMC7828353.
248. Moayeri M, Leppla SH, Vrentas C, Pomerantsev AP, Liu S. Anthrax pathogenesis. *Annu Rev Microbiol.* 2015;69:185-208.
249. Pan Z, Dumas EK, Lawrence C, et al. Bacillus anthracis edema toxin inhibits efferocytosis in human macrophages and alters efferocytic receptor signaling. *Int J Mol Sci.* 2019;20:1167. PMC6429438.
250. Patel VI, Booth JL, Dozmorov M, Brown BR, Metcalf JP. Anthrax edema and lethal toxins differentially target human lung and blood phagocytes. *Toxins.* 2020;12:464. PMC7405021.
251. Ribot WJ, Panchal RG, Brittingham KC, et al. Anthrax lethal toxin impairs innate immune functions of alveolar macrophages and facilitates Bacillus anthracis survival. *Infect Immun.* 2006;74:5029-5034. PMC1594833.
252. D'Agnillo F, Zhang X, Williams MC. Structural integrity of the alveolar–capillary barrier in cynomolgus monkeys challenged with fully virulent and toxin-deficient strains of Bacillus anthracis. *Am J Pathol.* 2020;190:2095-2110.
253. Bush LM, Abrams BH, Beall A, Johnson CC. Index case of fatal inhalational anthrax due to bioterrorism in the United States. *N Engl J Med.* 2001;345:1607-1610.
254. Sternbach G. The history of anthrax. *J Emerg Med.* 2003;24:463-467.
255. Grzybowski A, Pietrzak K. Robert Koch (1843-1910) and dermatology on his 171st birthday. *Clin Dermatol.* 2014;32:448-450.
256. Metchnikoff E. Classics in infectious diseases. Concerning the relationship between phagocytes and anthrax bacilli. By Elie Metchnikoff. 1884. (translated by Drs. D. Magasanik and A. Coons). *Rev Infect Dis.* 1984;6:761-770.
257. Kumar A, Das B, Kumar N. Vibrio pathogenicity Island-1: the master determinant of cholera pathogenesis. *Front Cell Infect Microbiol.* 2020;10:561296. PMC7574455.
258. Control, CfD. 150th Anniversary of John Snow and the Pump Handle. *MMWR (Morb Mortal Wkly Rep): MMWR (Morb Mortal Wkly Rep).* 2004;53:783.
259. Fraulin G, Lee S, Bartels SA. 'They came with cholera when they were tired of killing us with bullets': community perceptions of the 2010 origin of Haiti's cholera epidemic. *Glob Public Health.* 2021:1-15.
260. Kuhlmann FM, Santhanam S, Kumar P, Luo Q, Ciorba MA, Fleckenstein JM. Blood group O-dependent cellular responses to cholera toxin: parallel clinical and epidemiological links to severe cholera. *Am J Trop Med Hyg.* 2016;95:440-443. PMC4973196.
261. Lerouge I, Vanderleyden J. O-antigen structural variation: mechanisms and possible roles in animal/plant–microbe interactions. *FEMS (Fed Eur Microbiol Soc) Microbiol Rev.* 2002;26:17-47.
262. Reidl J, Klose KE. Vibrio cholerae and cholera: out of the water and into the host. *FEMS (Fed Eur Microbiol Soc) Microbiol Rev.* 2002;26:125-139.
263. Ramamurthy T, Nandy RK, Mukhopadhyay AK, et al. Virulence regulation and innate host response in the pathogenicity of Vibrio cholerae. *Front Cell Infect Microbiol.* 2020;10:572096. PMC7554612.
264. Gabutti G, Rossanese A, Tomasi A, et al. Cholera, the current status of cholera vaccines and recommendations for travellers. *Vaccines (Basel).* 2020;8:606. PMC7711912.
265. Sit B, Fakoya B, Zhang T, Billings G, Waldor MK. Dissecting serotype-specific contributions to live oral cholera vaccine efficacy. *Proc Natl Acad Sci U S A.* 2021;118:e2018032118. PMC7896348.
266. Harris JB. Cholera: immunity and prospects in vaccine development. *J Infect Dis.* 2018;218:S141-S146. PMC6188552.
267. Ratnayake R, Finger F, Azman AS, et al. Highly targeted spatiotemporal interventions against cholera epidemics, 2000-19: a scoping review. *Lancet Infect Dis.* 2021;21:e37-e48.
268. Royal JM, Matoba N. Therapeutic potential of cholera toxin B subunit for the treatment of inflammatory diseases of the mucosa. *Toxins.* 2017;9:379. PMC5744099.
269. Baldauf KJ, Royal JM, Hamorsky KT, Matoba N. Cholera toxin B: one subunit with many pharmaceutical applications. *Toxins.* 2015;7:974-996. PMC4379537.
270. Mutsch M, Zhou W, Rhodes P, et al. Use of the inactivated intranasal influenza vaccine and the risk of Bell's palsy in Switzerland. *N Engl J Med.* 2004;350:896-903.
271. Forbes JD. Clinically important toxins in bacterial infection: utility of laboratory detection. *Clin Microbiol Newsl.* 2020;42:163-170. PMC7541054.
272. Brubaker J, Zhang X, Bourgeois AL, Harro C, Sack DA, Chakraborty S. Intestinal and systemic inflammation induced by symptomatic and asymptomatic enterotoxigenic *E. coli* infection and impact on intestinal colonization and ETEC specific immune responses in an experimental human challenge model. *Gut Microb.* 2021;13:1-13. PMC7919917.
273. Warr AR, Kuehl CJ, Waldor MK. Shiga toxin remodels the intestinal epithelial transcriptional response to Enterohemorrhagic *Escherichia coli. PLoS Pathog.* 2021;17:e1009290. PMC7880444.
274. Menge C. Molecular biology of *Escherichia coli* shiga toxins' effects on mammalian cells. *Toxins.* 2020;12:345. PMC7290813.
275. CDC. *Reports of Selected E. coli Outbreak Investigations.* 2021. Accessed March 26, 2021. https://www.cdc.gov/ecoli/outbreaks.html
276. Mühlen S, Dersch P. Treatment strategies for infections with shiga toxin-producing *Escherichia coli. Front Cell Infect Microbiol.* 2020;10:169. PMC7218068.
277. Lee MS, Tesh VL. Roles of shiga toxins in immunopathology. *Toxins.* 2019;11:212. PMC6521259.
278. Martorelli L, Garimano N, Fiorentino GA, et al. Efficacy of a recombinant Intimin, EspB and Shiga toxin 2B vaccine in calves experimentally challenged with *Escherichia coli* O157:H7. *Vaccine.* 2018;36:3949-3959.
279. Havira MS, Ta A, Kumari P, et al. Shiga toxin suppresses noncanonical inflammasome responses to cytosolic LPS. *Sci Immunol.* 2020;5:eabc0217.
280. Li Y, Frey E, Mackenzie AM, Finlay BB. Human response to Escherichia coli O157:H7 infection: antibodies to secreted virulence factors. *Infect Immun.* 2000;68:5090-5095. PMC101746.
281. Crobach MJT, Vernon JJ, Loo VG, et al. Understanding *Clostridium difficile* colonization. *Clin Microbiol Rev.* 2018;31:e00021-17. PMC5967689.
282. Solomon K. The host immune response to *Clostridium difficile* infection. *Ther Adv Infect Dis.* 2013;1:19-35. PMC4040718.
283. Karyal C, Hughes J, Kelly ML, et al. Colonisation factor CD0873, an attractive oral vaccine candidate against clostridioides difficile. *Microorganisms.* 2021;9:306. PMC7913071.
284. Fletcher JR, Pike CM, Parsons RJ, et al. Clostridioides difficile exploits toxin-mediated inflammation to alter the host nutritional landscape and exclude competitors from the gut microbiota. *Nat Commun.* 2021;12:462.
285. Korbmacher M, Fischer S, Landenberger M, Papatheodorou P, Aktories K, Barth H. Human α-defensin-5 efficiently neutralizes clostridioides difficile toxins TcdA, TcdB, and CDT. *Front Pharmacol.* 2020;11:1204. PMC7435013.
286. El Feghaly RE, Stauber JL, Deych E, Gonzalez C, Tarr PI, Haslam DB. Markers of intestinal inflammation, not bacterial burden, correlate with clinical outcomes in *Clostridium difficile* infection. *Clin Infect Dis.* 2013;56:1713-1721. PMC3707425.
287. Solomon K, Martin AJ, O'Donoghue C, et al. Mortality in patients with *Clostridium difficile* infection correlates with host pro-inflammatory and humoral immune responses. *J Med Microbiol.* 2013;62:1453-1460.
288. McGill SK. Fecal microbiota transplant for severe clostridioides difficile infection: let's halt the raging fire. *Clin Infect Dis.* 2021. doi:10.1093/cid/ciab047
289. Förster B, Chung PK, Crobach MJT, Kuijper EJ. Application of antibody-mediated therapy for treatment and prevention of *Clostridium difficile* infection. *Front Microbiol.* 2018;9:1382. PMC6027166.
290. Hsieh CS, Macatonia SE, Tripp CS, Wolf SF, O'Garra A, Murphy KM. Development of TH1 CD4+ T cells through IL-12 produced by Listeria-induced macrophages. *Science.* 1993;260:547-549.
291. Janda J, Schöneberger P, Skoberne M, Messerle M, Rüssmann H, Geginat G. Cross-presentation of Listeria-derived CD8 T cell epitopes requires unstable bacterial translation products. *J Immunol.* 2004;173:5644-5651.
292. Edelson BT. Dendritic cells in Listeria monocytogenes infection. *Adv Immunol.* 2012;113:33-49.
293. Khan SH, Badovinac VP. Listeria monocytogenes: a model pathogen to study antigen-specific memory CD8 T cell responses. *Semin Immunopathol.* 2015;37:301-310. PMC4439301.
294. Harty JT, Lenz LL, Bevan MJ. Primary and secondary immune responses to Listeria monocytogenes. *Curr Opin Immunol.* 1996;8:526-530. PMC2784914.
295. Allen PM, Unanue ER. Antigen processing and presentation by macrophages. *Am J Anat.* 1984;170:483-490.
296. Mackaness GB. Cellular resistance to infection. *J Exp Med.* 1962;116:381-406. PMC2137547.
297. Marineli F, Tsoucalas G, Karamanou M, Androutsos G. Mary Mallon (1869-1938) and the history of typhoid fever. *Ann Gastroenterol.* 2013;26:132-134. PMC3959940.
298. Lantos PM. Chronic Lyme disease. *Infect Dis Clin North Am.* 2015;29:325-340. PMC4477530.
299. Ruiz-Llacsahuanga B, Hamilton A, Zaches R, Hanrahan I, Critzer F. Prevalence of Listeria species on food contact surfaces in Washington State apple packinghouses. *Appl Environ Microbiol: AEM.* 2021:02932-02920.
300. Schmitz-Esser S, Anast JM, Cortes BW. A large-scale sequencing-based survey of plasmids in Listeria monocytogenes reveals global dissemination of plasmids. *Front Microbiol.* 2021;12:653155. PMC7994336.
301. Kvistholm Jensen A, Nielsen EM, Björkman JT, et al. Whole-genome sequencing used to investigate a nationwide outbreak of listeriosis caused by ready-to-eat delicatessen meat, Denmark, 2014. *Clin Infect Dis.* 2016;63:64-70.
302. Pouillot R, Klontz KC, Chen Y, et al. Infectious dose of Listeria monocytogenes in outbreak linked to ice cream, United States, 2015. *Emerg Infect Dis.* 2016;22:2113-2119. PMC5189132.
303. Jackson KA, Gould LH, Hunter JC, Kucerova Z, Jackson B. Listeriosis outbreaks associated with soft cheeses, United States, 1998-2014(1). *Emerg Infect Dis.* 2018;24:1116-1118. PMC6004860.

304. Kühn S, Enninga J. The actin comet guides the way: how Listeria actin subversion has impacted cell biology, infection biology and structural biology. *Cell Microbiol.* 2020;22:e13190.
305. Robertin S, Mostowy S. The history of septin biology and bacterial infection. *Cell Microbiol.* 2020;22:e13173.
306. Pamer EG. Immune responses to Listeria monocytogenes. *Nat Rev Immunol.* 2004;4:812-823.
307. Pamer EG, Sijts AJAM, Villanueva MS, Busch DH, Vijh S. MHC class I antigen processing of Listeria monocytogenes proteins: implications for dominant and subdominant CTL responses. *Immunol Rev.* 1997;158:129-136.
308. Sun JC, Bevan MJ. Defective CD8 T cell memory following acute infection without CD4 T cell help. *Science.* 2003;300:339-342. PMC2778341.
309. Allen PM, Beller DI, Braun J, Unanue ER. The handling of Listeria monocytogenes by macrophages: the search for an immunogenic molecule in antigen presentation. *J Immunol.* 1984;132:323-331.
310. Buchmeier NA, Schreiber RD. Requirement of endogenous interferon-gamma production for resolution of Listeria monocytogenes infection. *Proc Natl Acad Sci U S A.* 1985;82:7404-7408. PMC391353.
311. Kc W, Satpathy AT, Rapaport AS, et al. L-Myc expression by dendritic cells is required for optimal T-cell priming. *Nature.* 2014;507:243-247. PMC3954917.
312. Weber KS, Li QJ, Persaud SP, Campbell JD, Davis MM, Allen PM. Distinct CD4+ helper T cells involved in primary and secondary responses to infection. *Proc Natl Acad Sci U S A.* 2012;109:9511-9516. PMC3386110.
313. Persaud SP, Parker CR, Lo WL, Weber KS, Allen PM. Intrinsic CD4+ T cell sensitivity and response to a pathogen are set and sustained by avidity for thymic and peripheral complexes of self peptide and MHC. *Nat Immunol.* 2014;15:266-274. PMC3944141.
314. Bartleson JM, Viehmann Milam AA, Donermeyer DL, et al. Strength of tonic T cell receptor signaling instructs T follicular helper cell-fate decisions. *Nat Immunol.* 2020;21:1384-1396. PMC7578106.
315. Harrell JE, Hahn MM, D'Souza SJ, et al. Salmonella biofilm formation, chronic infection, and immunity within the intestine and hepatobiliary tract. *Front Cell Infect Microbiol.* 2020;10:624622. PMC7885405.
316. Gilchrist JJ, MacLennan CA, Hill AVS. Genetic susceptibility to invasive Salmonella disease. *Nat Rev Immunol.* 2015;15:452-463.
317. Barton AJ, Hill J, Blohmke CJ, Pollard AJ. Host restriction, pathogenesis and chronic carriage of typhoidal Salmonella. *FEMS Microbiol Rev.* 2021. doi:10.1093/femsre/fuab014
318. Coburn B, Grassl GA, Finlay BB. Salmonella, the host and disease: a brief review. *Immunol Cell Biol.* 2007;85:112-118.
319. Nauciel C, Espinasse-Maes F. Role of gamma interferon and tumor necrosis factor alpha in resistance to Salmonella typhimurium infection. *Infect Immun.* 1992;60:450-454. PMC257648.
320. Vázquez-Torres A, Fantuzzi G, Edwards CK, Dinarello CA, Fang FC. Defective localization of the NADPH phagocyte oxidase to Salmonella-containing phagosomes in tumor necrosis factor p55 receptor-deficient macrophages. *Proc Natl Acad Sci U S A.* 2001;98:2561-2565.
321. Gram AM, Wright JA, Pickering RJ, et al. Salmonella flagellin activates NAIP/NLRC4 and canonical NLRP3 inflammasomes in human macrophages. *J Immunol.* 2021;206:631-640. PMC7812056.
322. Pham THM, Brewer SM, Thurston T, et al. Salmonella-driven polarization of granuloma macrophages antagonizes TNF-mediated pathogen restriction during persistent infection. *Cell Host Microbe.* 2020;27:54-67.e55.
323. van de Vosse E, van Dissel JT, Ottenhoff TH. Genetic deficiencies of innate immune signalling in human infectious disease. *Lancet Infect Dis.* 2009;9:688-698.
324. Sanal O, Turul T, De Boer T, et al. Presentation of interleukin-12/-23 receptor beta1 deficiency with various clinical symptoms of Salmonella infections. *J Clin Immunol.* 2006;26:1-6.
325. Song E, Jaishankar GB, Saleh H, Jithpratuck W, Sahni R, Krishnaswamy G. Chronic granulomatous disease: a review of the infectious and inflammatory complications. *Clin Mol Allergy.* 2011;9:10.
326. Jin C, Gibani MM, Moore M, et al. Efficacy and immunogenicity of a Vi-tetanus toxoid conjugate vaccine in the prevention of typhoid fever using a controlled human infection model of Salmonella Typhi: a randomised controlled, phase 2b trial. *Lancet.* 2017;390:2472-2480. PMC5720597.
327. Jin C, Hill J, Gunn BM, et al. Vi-specific serological correlates of protection for typhoid fever. *J Exp Med.* 2021;218:e20201116. PMC7668386.
328. Elbaum-Garfinkle S. Close to home: a history of Yale and Lyme disease. *Yale J Biol Med.* 2011;84:103-108. PMC3117402.
329. Burgdorfer W, Barbour AG, Hayes SF, Benach JL, Grunwaldt E, Davis JP. Lyme disease-a tick-borne spirochetosis? *Science.* 1982;216:1317-1319.
330. Klose M, Scheungrab M, Luckner M, Wanner G, Linder S. FIB-SEM-based analysis of Borrelia intracellular processing by human macrophages. *J Cell Sci.* 2021;134:jcs252320.
331. Tilly K, Rosa PA, Stewart PE. Biology of infection with Borrelia burgdorferi. *Infect Dis Clin North Am.* 2008;22:217-234. PMC2440571.
332. Coburn J, Garcia B, Hu LT, et al. Lyme disease pathogenesis. *Curr Issues Mol Biol.* 2021;42:473-518.
333. Kullberg BJ, Vrijmoeth HD, van de Schoor F, Hovius JW. Lyme borreliosis: diagnosis and management. *Br Med J.* 2020;369:m1041.
334. Lantos PM, Rumbaugh J, Bockenstedt LK, et al. Clinical Practice Guidelines by the Infectious Diseases Society of America (IDSA), American Academy of Neurology (AAN), and American College of Rheumatology (ACR): 2020 guidelines for the prevention, diagnosis and treatment of lyme disease. *Clin Infect Dis.* 2020;72:e1-e48.
335. Bockenstedt LK, Wooten RM, Baumgarth N. Immune response to Borrelia: lessons from Lyme disease spirochetes. *Curr Issues Mol Biol.* 2021;42:145-190.
336. Petnicki-Ocwieja T, Kern A. Mechanisms of Borrelia burgdorferi internalization and intracellular innate immune signaling. *Front Cell Infect Microbiol.* 2014;4:175.
337. Déruaz M, Bonvin P, Severin IC, et al. Evasin-4, a tick-derived chemokine-binding protein with broad selectivity can be modified for use in preclinical disease models. *FEBS J.* 2013;280:4876-4887. PMC4240464.
338. Denisov SS, Ramírez-Escudero M, Heinzmann ACA, et al. Structural characterization of anti-CCL5 activity of the tick salivary protein evasin-4. *J Biol Chem.* 2020;295:14367-14378. PMC7573271.
339. Brouwer MAE, van de Schoor FR, Vrijmoeth HD, Netea MG, Joosten LAB. A joint effort: the interplay between the innate and the adaptive immune system in Lyme arthritis. *Immunol Rev.* 2020;294:63-79. PMC7065069.
340. Rego ROM, Trentelman JJA, Anguita J, et al. Counterattacking the tick bite: towards a rational design of anti-tick vaccines targeting pathogen transmission. *Parasit Vectors.* 2019;12:229. PMC6518728.
341. Ma Y, Sturrock A, Weis JJ. Intracellular localization of Borrelia burgdorferi within human endothelial cells. *Infect Immun.* 1991;59:671-678. PMC257809.
342. Bamgboje A, Akintan FO, Gupta NM, Kaur G, Pekler G, Mushiyev S. Lyme carditis: a reversible cause of acquired third-degree AV block. *Am J Case Rep.* 2021;22:e927885. PMC7797605.
343. Dersch R, Singh AE. Neurosyphilis and lyme neuroborreliosis. *Curr Opin Neurol.* 2021;34:403-409.
344. Nilsson K, Skoog E, Jones V, et al. A comprehensive clinical and laboratory evaluation of 224 patients with persistent symptoms attributed to presumed tick-bite exposure. *PLoS One.* 2021;16:e0247384. PMC7971513.
345. Auwaerter PG, Bakken JS, Dattwyler RJ, et al. Antiscience and ethical concerns associated with advocacy of lyme disease. *Lancet Infect Dis.* 2011;11:713-719. PMC4489928.
346. Kamp HD, Swanson KA, Wei RR, et al. Design of a broadly reactive lyme disease vaccine. *NPJ Vaccines.* 2020;5:33.

CHAPTER 44

Response to Viruses

Hildegund C J. Ertl, MD

INTRODUCTION

Viruses appeared on earth about 1.5 billion years ago during the Mesoproterozoic era. It remains debated which group evolved first, bacteria or viruses. Both may have appeared together, or they may have descended from an ancestral cell in a process during which bacteria became more complex while viruses shed all necessary genes until they were no longer able to replicate outside a cell.[1]

Each year viruses cause more human deaths than some of the most lethal weapons of mankind—cars and guns: 0.011% of the US population die either due to car accidents or gun shots while infections caused by just four out of over 200 human viruses, that is, poliomyelitis, measles, human immunodeficiency, and influenza viruses, kill 0.012% of the population. The picture is even more dire in less developed countries like South Africa where these four viruses take the lives of 0.04% of the population while 0.007% perish due to car accidents or gun-related violence. We have laws to protect us against reckless drivers and gun owners—rules that enforce vaccinations or isolation of sick and quarantine of exposed individuals to limit the spread of viruses exist and have been shown to blunt the impact of childhood infections or the spread of some new viruses with pandemic potential, such as the Ebola virus or the severe acute respiratory syndrome (SARS)-coronavirus of 2003; they have not been successful against others, such as HIV-1 or influenza viruses, and vaccination laws and recommendations are right now failing with nearly 10 million influenza cases in 2018, resulting in a staggering 140,000 deaths that could and should have been prevented.

Viruses are not only harmful parasites—they have been essential to the evolution of complex life forms on the earth and they can benefit their host. Some viruses render plants more resistant to drought.[2] Pepper cryptic virus increases the longevity of pepper seeds and herpesviruses protect mice from the bubonic plague.[3] Integrating viruses shaped evolution. A retrovirus that integrated upstream of the amylase gene relaxed its strict pancreatic tropism and permitted for its expression in saliva and thereby improved digestion of starchy foods. This in turn allowed for a shift from the early nomadic hunter-gather cultures of humans to agriculture and permanent settlements that relied on more carbohydrate-based diets.[4] Mammals would not have evolved without syncytins, which came from a retroviral envelop protein and which are essential for placental development.[5]

Viruses are minute ranging in diameter from 10 to 300 nm with genomes encoding anywhere from 3 to 1000 genes. In their zest to replicate in prokaryotes or eukaryotes, the latter include humans, animals, and plants, they not only raise havoc to the infected being but have even shaped human history. In the 16th century, Europeans conquered the Americas with the help of smallpox, measles, and influenza viruses, which they brought with them and which decimated the native populations thus ensuring an easy victory for the invaders. Smallpox virus was used as a biological weapon during the French and Indian wars (1754-1763) and allegedly again during the American Revolutionary war (1775-1783). Famines were triggered in Europe in the 4th and 13th centuries following the conquests by Huns and then Mongols, who introduced a Rinderpest-causing paramyxovirus into cattle. In the 1880, yellow fever virus prevented the French from building the Panama Canal between the Atlantic and Pacific. Upon instigating sanitation and mosquito eradication programs, the United States was finally able to finish the project in 1914. Currently, HIV-1, which primarily infects young adults and then causes a lifelong infection, is impeding socioeconomic development in African countries. On December 31st, 2019, China reported an outbreak with a new coronavirus, called COVID-2019 that started in the city of Wuhan. The virus most likely was transmitted from bats to an animal that was up for sale at a market and from there to a human. The World Health Organization declared a global health emergency on January 31st, 2020 and by February 19th, 2020 75,204 human cases were confirmed in 25 countries with 74,280 cases in China; 2006 people had died. The entire city of Wuhan was placed under quarantine, stores and factories across China closed, foreigners were evacuated from Wuhan and placed under quarantine in their home countries. Most airlines grounded flights to and from China, and governments imposed travel restrictions. People on a cruise ship were quarantined off the shores of Japan, where the virus spread relentlessly. Comparing the global spread of severe acute respiratory syndrome coronavirus 2 (SARS-CoV-2) with a doubling of confirmed cases from February 9 to 18 to that in the confined space of the cruise ship where during the same time span human cases increased more than sevenfold suggests the draconic quarantine measures may have slowed the spread of the virus (Fig. 44.1).

Viruses can kill rapidly during an acute infection. They can establish a chronic infection that may eventually kill due to accumulating organ or tissue damage. They can kill by causing cancer or by driving autoimmune or degenerative diseases. Individuals that survive a virus infection can suffer

Early stages of the COVID-2019 outbreak

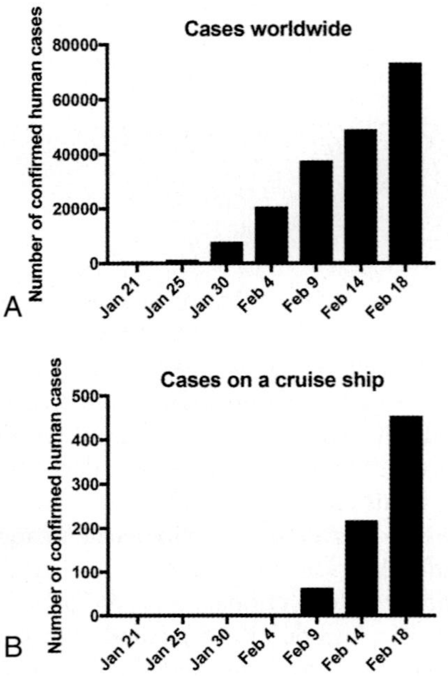

C

Dates	Timeline of COVID-2019 Outbreak
12.31.2019	Wuhan Municipal Health Commission reports on 27 cases of pneumonia caused by an unknown infectious agent
1. 08. 2020	A new coronavirus is identified as the causative agent
1. 13. 2020	First case reported outside China
1. 22. 2020	New SARS virus has spread to other coutries in Asia and to the US
1. 22. 2020	Countries start to impose entry restrictions on Chinese citizens
1. 23. 2020	China imposes strict quarentine on major cities of Hubei
1. 27. 2020	CDC issues a travel warning for China
1. 29. 2020	Airlines start canceling flights to China
1. 31. 2020	WHO declares the virus a public health emergency
2. 02. 2020	US imposes quarentine on inbound travelers from Hubei
2. 13. 2020	China temporarily shuts down nonessential companies in Hubei

FIG. 44.1. The figure shows the spread of the coronavirus, called SARS-CoV-2 that originated in Wuhan in 2019. **A:** shows the number of confirmed cases worldwide as reported by the World Health Organization (WHO) from January 21st-February 18, 2020. **B:** shows the number of confirmed cases on a cruise ship that was quarantined off the shores of Japan. **C:** shows the important dates of the outbreak.

lifelong sequelae, such as disfiguring scars after a small pox-virus infection, paralysis after an infection with poliovirus, or hydrocephalus after an intrauterine Zika virus infection.

Why are viruses so lethal? Viruses require the host cell machinery to replicate. To allow for efficient production of their genomes and proteins, they disrupt the cells' physiological functions, which may eventually cause cell death. Viruses can expand by two logs within hours after cell infection. In theory an infection with 10 virus particles that can expand 100-fold every 24 hours, which is a very conservative estimate, would lead to 10^{16} virus particles within a week, thus greatly outnumbering the 37 trillion cells (3.7×10^{13}) in an adult human body. This calculation is simplistic; far from every cell becomes infected within a week, much of the viral progeny is noninfectious, not every virus infects a unique cell, and many cells are resistant to infections with viruses that require specific receptors for cell entry. Last but not least—the viral invasion triggers an epic battle fought between the virus and the immune system. Cytokines and antibodies produced upon infections can reduce virus replication or prevent viruses from infecting new cells. A plethora of cells of the immune system can either directly kill viruses or disrupt their replication by lysing infected cells. Viruses in turn have acquired skills to circumvent host immunity.

Many viruses establish an acute infection that is either cleared by the immune system or results in death of the host, which ends the virus's ability to spread to the next organism. Other viruses are craftier; after an acute infection that is controlled by the immune system, residual virus particles dodge immune responses and establish chronic infections, which may eventually be cleared or contribute to the death of their host, but in the meantime, over long periods of time, allow the virus to infect new victims. A list of common human viruses that establish acute or chronic infections is shown in Table 44.1. During chronic or persistent infections, some viruses continue to replicate ferociously, such as human immunodeficiency virus (HIV)-1 or hepatitis B or C virus (HBV, HCV), while others, such as herpesviruses, hide from the immune system by going into latency with occasional episodes of reactivation. Neither virus size nor type of genome predicts the type of its infection. Viruses that establish chronic infections tend to have a more restrictive host range with the exception of adeno-associated viruses (AAVs), which are unique by being dependoviruses—a type of virus that requires a helper virus, typically an adenovirus, for its replication. During chronic infection, the genomes of some viruses, such as of oncogenic types of human papilloma viruses (HPV), integrate into the host cell genome, while that of others, such as of herpesviruses are maintained as extrachromosomal circular DNA.

There are millions of different viruses that infected cellular organisms, and it is estimated that at least 320,000 of them can infect mammals. Thus far 219 virus species have been identified in humans and on average an additional 3 to 4 species are detected each year, suggesting a large pool of yet to be discovered human viruses.[6] The pool is constantly enlarged by viruses that successfully cross species barriers. At the beginning of the 20th century, the great ape precursor of HIV-1 crossed into humans[7] and evolved into a virus that has by now infected ~75 million people. In the early 21st century, SARS coronavirus was transmitted from a horseshoe bat to a human and from there to ~8000 other humans.[8] The 2009 influenza pandemic was driven by an H1N1 influenza A virus that evolved by reassortment from human, swine, and avian influenza viruses. The pandemic H1N1 virus infected nearly 61 million people worldwide and killed ~575,000.[9] These numbers are staggering. The number of people that became infected with the pandemic H1N1 influenza virus exceeds the entire population of Italy, the 23rd most populous country on earth, while the number of fatalities is above the numbers of residents of Atlanta, Georgia, the 37th most populous city in the United States. The constant emergence of new viruses, which evolve from those that infect other animals or viral variants that can gain pathogenicity through mutations poses grave threats to humans. Humans have neither immunity nor vaccines to new viruses that can kill large segments of the population as exemplified by the influenza A virus that caused the Spanish Flu pandemic of 1918. This virus infected a third of the world's population, killed 5% to 10% of its victims, and caused widespread socioeconomic disruptions.[10] The threat of large pandemics has increased since then due to enhanced global travel and urbanization as well as climate changes, which expand the geographic ranges of viral transmitters, such as mosquitos.

Antiviral drugs are available for nine different virus infections; most are designed to treat chronic infections (Table 44.1). In the last 50 years, ~90 antiviral drugs have been licensed for chronic infections with HIV-1, HCV, HBV, human cytomegalovirus (HCMV), herpes simplex virus (HSV), varicella-zoster virus (VZV), HPVs that cause anogenital warts, or acute infections with respiratory syncytial virus (RSV), or influenza viruses. Vaccines are available for no more than 15 human viruses (Table 44.1) and only two of

TABLE 44.1 Types of Viruses

Virus	Family	Host(s)	Genome (Size in Base Pairs)	Disease	Vaccine	Antiviral Drug
Acute Infections						
Human adenovirus	Adenoviridae	Humans	dsDNA, (35,937)	Respiratory, eye, or intestinal disease	Live attenuated	None
Hantaan virus	Bunyaviridae	Humans, rodents	sRNA, (18,234)	Renal or respiratory diseases	None	None
Dengue virus	Flaviviridae	Humans, mosquitoes	(+)ss-RNA, (10,745)	Hemorrhagic fever	None	None
Japanese encephalitis virus	Flaviviridae	Humans, horses, birds, mosquitoes	(+)ss-RNA, (10,976)	Encephalitis	Inactivated	None
West Nile virus	Flaviviridae	Humans, birds, ticks, mosquitoes	(+)ss-RNA (10,932)	Encephalitis	None	None
Yellow fever virus	Flaviviridae	Humans, monkeys, mosquitoes	(+)ss-RNA (10,862)	Hemorrhagic fever	Live attenuated	None
Zika virus	Flaviviridae	Humans, monkeys, mosquitoes	(+)ss-RNA (10,794)	Fever, rash, birth defects	None	None
Ebola virus	Filoviridae	Humans, monkeys, bats	(−)ssRNA (18,956)	Hemorrhagic fever	None	None
Influenza A viruses	Orthomyxoviridae	Humans, birds, pigs	(−)ssRNA (13,588)	Respiratory disease	Inactivated or live attenuated	Several
Human parainfluenza	Paramyxoviridae	Humans	(−)ssRNA (15,600)	Respiratory disease	None	None
Measles virus	Paramyxoviridae	Humans	(−)ssRNA 15,894)	Respiratory disease	Live attenuated	None
Mumps virus	Paramyxoviridae	Humans	(−)ssRNA (15,384)	Salivary gland/ Respiratory disease	Live attenuated	None
Human rhinovirus	Picornaviridae	Humans	(+)ss-RNA (7152)	Respiratory disease	None	None
Hepatitis A virus	Picornaviridae	Humans	(+)ss-RNA (7478)	Hepatitis	Inactivated	None
Poliovirus	Picornaviridae	Simians	(+)ss-RNA (7440)	Paralysis	Inactivated or live attenuated	None
Human respiratory syncytial virus	Pneumoviridae	Humans	(−)ssRNA (15,225)	Respiratory disease	None	Ribavarin

(Continued)

TABLE 44.1 Types of Viruses (Continued)

Virus	Family	Host(s)	Genome (Size in Base Pairs)	Disease	Vaccine	Antiviral Drug
Rotavirus	Reoviridae	Humans	dsRNA (23,441)	Gastroenteritis	Live attenuated	None
Rabies virus	Rhabdoviridae	Mammals	(−)ss-RNA (11,932)	Encephalitis	Inactivated	None
Rubella virus	Togaviridae	Humans	(+)ss-RNA (9762)	Fever, rash, birth defects	Live attenuated	None
Smallpox virus	Variola major	Humans	dsDNA (186,103)	Fever, rash	Live attenuated	None
Chronic or Persistant Infections						
Hepatitis C virus	Flaviviridae	Humans	(+)ss-RNA (9646)	Hepatitis	None	Several
Hepatitis B virus	Hepadnaviridae	Humans, Chimpanzees	dsDNA-RT (3182)	Hepatitis	Virus-like particles	Several
Epstein-Barr virus	Herpesviridae	Humans	dsDNA (171,823)	Mononucleosis	None	None
Human cytomegalovirus	Herpesviridae	Humans	dsDNA (235,646)	Mononucleosis	None	Several
Human herpes-virus 1	Herpesviridae	Humans	dsDNA (152,222)	Skin lesions	None	None
Varicella-zoster virus	Herpesviridae	Humans	dsDNA (124,884)	Skin lesions	Live attenuated	Several
Human papillomavirus	Papillomaviridae	Humans	dsDNA (7906)	Cancer	Virus-like particles	None
Adeno-associated virus	Parvoviridae	Vertebrates	ssDNA (4679 bp)	None	None	None
Human immuno-deficiency virus	Retroviridae	Humans	ss-RNA-RT (9181)	Immunodeficiency	None	Several

dsDNA, double-stranded deoxyribonucleic acid; dsDNA-RT, DNA with RNA intermediate; dsRNA, double-stranded ribonucleic acid; ssDNA, single-stranded deoxyribonucleic acid; ss-RNA, single-stranded ribonucleic acid; ss-RNA-RT, single-stranded RNA with DNA intermediate.

those were first licensed in the 21st century although many more potentially deadly viruses were identified during this time. This stresses that the path from pathogen identification to vaccine licensure is not only technically challenging but also costly and time-consuming.

SOME BASICS IN VIROLOGY

Virus Classification

Various methods have been used to group viruses. The Baltimore classification is simple and groups viruses according to their genome and mode of replication into seven groups (Table 44.2). The Holmes classification is based on host range which divides viruses into three groups (Table 44.3). The Lwoff-Horne-Tournier (LHT) system is based on chemical and physical characteristics, that is, (a) nature of the nucleic acid in the virion, (b) symmetry of the protein shell, (c) presence or absence of a lipid membrane and (d) dimensions of the virion and capsid.[11]

The International Committee on Taxonomy of Viruses (ICTV) is formally in charge to group viruses. ICTV's classify viruses into 1: order (virales), 2: family (viridae), 3: subfamily (virinae), 4: genus (virus), and 5: species (virus). For example, rabies virus, the species, belongs to the order of *Mononegavirales*, the family of *Rhabdoviridae*, and the genus of *Lyssavirus*. Measles morbillivirus, another species, is also a member of the order of *Mononegavirales*, the family of *Paramyxoviridae*, the subfamily of *Paramyxovirinae* and the genus of *Morbillivirus* (https://talk.ictvonline.org/taxonomy/). The 2014 ICTV report recognizes 7 orders, 104 families, 23 subfamilies, 505 genera, and 3186 viral species.[12] Only a fraction of families has been assigned to an order. This is a reflection of the viruses' high mutation rates, which make an analyses of ancestral sequence homologies often impossible.

As best shown by the Baltimore classification, viruses differ in their genome, which in turn dictates how a specific virus alerts the innate immune system. Ribonucleic acid (RNA) viruses can carry a single-stranded (ss) or double-stranded (ds)RNA; ssRNA viruses are divided into those with a positive (+) or negative-sense (−) RNA strand. (+)ssRNA serves as a messenger (m) RNA for protein synthesis while (−)ssRNA requires a viral polymerase that converts the (−) into a (+) strand prior to its translation. Most RNA viruses have a linear RNA but hepatitis delta virus, a satellite virus of HBV, as well as some plant viruses have circular RNA. Some viruses, such as influenza viruses, have a segmented, (−)ssRNA, which

TABLE 44.2 Baltimore Classification

Group	Type of Viral Genome	Example
I	Double-stranded (ds)DNA	Human adenovirus
II	Single-stranded (ss)DNA	Adeno-associated virus
III	Double-stranded (ds)RNA	Rotavirus
IV	Positive sense (+) single-stranded (ss)RNA	Human rhinoviruses
V	Negative sense (−) single-stranded (ss)RNA	Influenza viruses
VI	Single-stranded RNA with DNA intermediate (ssRNA-RT)	Rous sarcoma virus
VII	DNA with RNA intermediate (dsDNA-RT)	Hepatitis B virus

allows them to reassort with related viruses into a new subtype in a process called antigenic shift. Humans commonly lack immunity to new viruses that were formed by reassortment, which in turn can cause, especially in the case of influenza A viruses, deadly pandemics like the 1918 Spanish Flu pandemic. Some ssRNA viruses called ssRNA-RT viruses use RNA-dependent deoxyribonucleic acid (DNA) polymerase, ribonuclease H, and DNA-dependent DNA polymerase to reverse transcribe the ssRNA into a complementary (c)DNA, which upon integration into the host cell genome serves as a template for mRNA. DNA viruses can have a ssDNA such as parvoviruses or a dsDNA such as poxviruses. A dsDNA virus can be circular such as in papillomaviruses or linear such as in adenoviruses. ssDNA viruses use a DNA-dependent DNA polymerase to form a dsDNA intermediate, which similar to the dsDNA is transcribed with the help of DNA-dependent RNA polymerases into viral mRNA. dsDNA-RT viruses such as HBV, a hepadnavirus, have a partially ds, partially ssDNA. Within the host cell nucleus, the viral genome is converted to a covalently closed circular (ccc) DNA, which serves as a template for transcription. One of the forms of transcripts is a pregenomic RNA, which is packaged into progeny capsids and with the help of a hairpin structure reverse transcribed into a new viral DNA.

Abbreviations for viral proteins are rather confusing. Hemagglutinin expressed by influenza virus is referred to as HA while neuraminidase is abbreviated to NA; the corresponding genes are called H and N. In other viruses, H refers to the hemagglutinin protein. IE, E, and L reflect immediate early, early, and late gene products, respectively. NS or NSP stands for nonstructural proteins. ORF is an open reading frame, that is, a gene fragment that can be translated into a protein. G or GP stands for glycoprotein, env is an abbreviation for the glycoprotein of HIV-1 while it is called the surface or S antigen for HBV. S can also stand for spike antigen such as in coronaviruses. UL or U are used together with a number for many of the herpesvirus proteins, while in other viruses, viral proteins are abbreviated to VP. CMV proteins have a letter reflecting the virus's species specificity like M for mouse of H for human followed by a number. The poxvirus genome is divided into multiple sections (A-P) and proteins are named accordingly.

TABLE 44.3 Holmes Classification

Group	Name	Host
I	Phaginae	Bacteriae
II	Phytophaginae	Plants
III	Zoophaginae	Animals

Virus Infection

Routes of Infection

Viruses can enter an organism through mucosal surfaces of the respiratory, the intestinal, or the urogenital tract, or the eye. Some viruses need to be injected directly into a body. The route of entry affects the ensuing immune responses as distinct anatomic sites are patrolled by distinct types of immunocytes and carry their unique microbiomes, which also modulate immune reactions.

The most common port of viral entry is the respiratory tract. The estimated total surface area of the respiratory tract of an adult human is somewhere between 50 and 75 m^2 and thereby 50 times larger than the skin surface. Most of the respiratory tract is covered by a single cell epithelium. An average adult takes 12 breath per minute inhaling each time about 500 mL of air that contains not just oxygen, nitrogen, and other gases but also pollutants such as viruses. Mucus, which covers the surface of the upper parts of the airways, traps particles. Cilia, which are tiny hairs that cover cells of these parts of the airways, move mucus upward toward the throat where it is swallowed or expectorated. Macrophages line the alveolar surfaces, which lack ciliated cells; they phagocytose and destroy inhaled viral particles. By the same token, with their massive airflow the airways are a source of viral spread. Viruses that infect and spread through the airways include influenza viruses, RSV, rhinoviruses, most types of adenoviruses, and measles morbillivirus. Other viruses enter cells of the alimentary tract after being ingested with drinks or food. Most viruses are destroyed by the acidic environment of the stomach and the proteolytic enzymes present in the upper intestine. Furthermore, a thick mucus layer protects the intestinal epithelium, which is further covered with protective villi and a surface coat of glycosylated proteins and lipids. Nevertheless, some viruses, such as reoviruses (eg, rotavirus), picornaviruses (eg, poliovirus), and enteric adenoviruses manage to survive the passage through the upper intestinal tract and penetrate intestinal barriers. The female genital tract, which provides the port of entry to sexually transmitted viruses, such as HPVs, HBV, HCV, or HIV-1, is protected by a low pH and a mucus layer. Viruses gain entry through minute lesions. By the same token, the eye is protected by tears and constant wiping by the eyelids, which brush away particulate contaminants. Viruses, such as some types of human adenoviruses, enter through abrasions and, in general, cause only a localized infection. Healthy skin cannot be penetrated by viruses, but again, small wounds can lead to virus infections. Viruses that infect through the skin commonly use insect vectors, such as mosquitos or ticks, to cross the skin barrier; rabies virus is transmitted through the bite of an infected animal.

Route of entry determines ease of transmission, which in turn drives the evolution of viruses toward or away from virulence. In general, a virus that spreads through the airways is transmitted easily, and one infected host can infect multiple new hosts giving the virus ample opportunity to survive. A virus that spreads without trouble can afford to cause a disease that after a few days may kill its host. Ebola virus, which is transmitted through direct contact with infected blood, other bodily fluids or tissues, neither of which is very effective, kills rapidly within 6 to 16 days after the first symptoms appear. One could argue that this very new human virus, which was first detected in 1976 in an infected individual in Africa, will over time evolve to either spread more efficiently or to become less virulent. Rabies virus, which is transmitted by the bite by an infected animal, such as a rabid dog, has a hard time to spread as the dog must find someone to bite, which takes time, and at best, with each bite, one individual becomes infected, while viruses present in the breath of an influenza virus–infected human each time their victim exhales are given a chance to infect multiple new hosts. Some viruses that spread poorly have evolved to optimize their chances to survive. For example, rabies virus, a very successful virus, deals with its low transmission rates by infecting multiple animal species. It also modifies its victim's behavior; a rabid dog is generally very aggressive and therefore more likely to bite than a healthy dog. Viruses that are transmitted from human-to-human by sexual contact also have a more limited choice in new victims and, therefore, in some cases, increase their ability to spread by establishing chronic or persistent infections, such as HIV-1 or herpesviruses. Many of the symptoms that humans develop during an infection also serve the virus to spread to its next host; coughing and sneezing during an infection with influenza virus or rhinovirus expels the pathogen-containing droplets into the air and onto surfaces so that viruses may be inhaled or picked up by their next victim. Diarrhea releases copious amounts of viruses ready for fecal-oral transfer to the next person. Blisters or pustules that form during an HSV or poxvirus infection render the skin contagious to others.

Cell Entry

Once a virus breaches the outer barriers of a body, it has to enter a cell and its genome has to reach a compartment where transcription can commence. Although different types of viruses use unique pathways to enter and transverse cells, there are some commonalities. The first step in the virus life cycle is its attachment to a cell. Attachment is facilitated by viral surface proteins, which bind to cellular receptors. Some viruses use receptors that are expressed by most cells, while others are more selective and require a receptor that is only present on some cell types or in certain organs. Some viruses bind to a receptor, but then need to attach to a second receptor, a coreceptor, to enter a cell. A virus's receptor specificity is one of the factors that influences its tropism and the type of the ensuing disease. Tropism is also affected by the binding of some viruses to serum proteins, which are selectively taken up by cells. For example, the hexon protein of adenovirus binds to coagulation factor (F)X, which promotes infection of hepatocytes.[13]

Upon binding to a receptor on the cell membrane, viruses need to gain access into the cytoplasm. Those that replicate within the nucleus also have to cross the nuclear membrane. Viruses enter the cell either through endocytosis[14] or fusion.[15] During clathrin-mediated endocytosis used by enveloped and nonenveloped viruses, the virus-receptor complexes cluster into a cadherin-coated pit on the cell membrane that becomes invaginated, closes, detaches from the cell membrane and delivers the virus to early endosomes from where it travels to late endosomes. Other viruses use caveolar endocytosis. Caveolae are invaginations in the plasma membrane that are rich in cholesterol, glycosphingolipids, and claveolin, which are used for uptake of macromolecules into endosomes. Some viruses, such as enteroviruses, enter cells through a lipid raft–dependent pathway, others use a cholesterol-and dynamin-dependent but clathrin-and caveolae-independent pathway, or micropinocytosis or phagocytosis. Enveloped viruses can directly fuse into the cell membrane causing release of the viral genome into the cytoplasm.

Once viruses reach the endosomes, they have to penetrate the vesicular membrane to reach the cytoplasm. The decrease in pH between early and late endosomes causes conformational changes of surface proteins of enveloped viruses, which promote fusion of the viral envelope with the endosomal membrane allowing the viral core or the genome to escape into the cytoplasm. Nonenveloped viruses disrupt the endosomal membrane either by a pathway called carpet mechanism or by forming pores. In the carpet-like pathway, viral peptides insert their hydrophobic tails into the membrane causing them to disrupt. Proteins of other viruses form amphipathic α-helices that assemble into a pore within the lipid endosomal. The genomes of some viruses, once they enter the cytoplasm, must reach the nucleus. Viruses may use microtubules to travel to nuclear pores. Small genomes can diffuse through pores into the nucleus, while larger genomes may depend on viral proteins to facilitate nuclear entry.

Different cell-entry pathways are used by viruses. Most adenoviruses bind to the coxsackie-adenovirus receptor (CAR), which is mainly expressed on epithelial cells but is absent on antigen-presenting cells. The upper part of the viral fiber, the so-called fiber knob, attaches to CAR. Arg-Gly-Asp (RGD) motifs present in penton, another viral protein that anchors fiber to the viral capsid, bind to the vitronectin-binding integrins $\alpha_v\beta_3$ and $\alpha_v\beta_5$.[16] Family B adenoviruses, such as human adenovirus of serotype 35, bind to CD46, a complement regulatory protein, which unlike CAR is widely expressed and present on antigen-presenting cells. Binding of penton to integrins triggers virus internalization by endocytosis using the coated pit pathway. The low pH of the late endosomes causes partial uncoating of the viral capsid exposing viral protein VI, which through a not yet fully understood pathway can lyse the endosomal membrane and allow escape of viral particles into the cytoplasm.[17] The adenovirus hexon then binds to the nuclear core complex, which leads to complete disassembly of the viral capsid and transfer of the genome into the nucleus, where virus replication commences.[18] HCMV, a herpesvirus, in general establishes benign persistent infections that can be severe in immunocompromised individuals or in infants infected *in utero*. Glycoproteins (g) B and a second surface protein complex gM/gN initially bind HCMV to heparan sulfate glycosaminoglycans; gH/gL together with gB then allow for membrane fusion. The type of entry varies between cell types; while gH/gL suffices for fusion into fibroblasts, penetration of the membrane of epithelial and endothelial cells requires a complex between gH/gL and UL128-131.[19]

Ebola virus can enter a large variety of target cells including dendritic cells and macrophages, which support early viral replication, as well as endothelial cells, fibroblasts, hepatocytes, and others, which tend to become infected later. The viral glycoprotein (GP) is the only protein expressed on the virus surface. Sugar residues on GP attach to C-type lectins such as dendritic cell–specific intercellular adhesion molecule-3–grabbing nonintegrin (DC-SIGN). Phosphatidylserine (PtdSer), a phospholipid within the viral envelope, binds to PtdSer receptors such as T cell immunoglobulin (Ig) mucin domain (TIM), which allows the virus to enter into endosome by micropinocytosis. Within the endosome, GP is cleaved exposing a receptor-binding domain that allows the virus to attach to an endosomal cholesterol transporter, which facilitates escape into the cytoplasm where the virus replicates.[20]

Influenza viruses have two surface proteins, HA and NA. Subtypes are classified according to the two genes (H and N, respectively) encoding the surface proteins. Viruses have 1 of 18 different H genes referred to as H1-H18, only 3 of which circulate in humans (H1-H3) and 11N genes (N1-N11) with N1 and N2 viruses infecting humans. Influenza A virus HA of human strains binds to sialylated glycan receptors with a terminal α2-6 linked N-acetylneuraminic acid while the HA of avian influenza A viruses bind to α2-3 linked sugar residues, which are not present on human cells. For that reason, avian strains of influenza virus are unable to spread into humans. But, a few mutations with the H gene can switch the virus specificity from bird to human, which could then trigger the next Flu pandemic. Avian strains that are currently of particular concern are H7N9 and H5N1, both of which have caused isolated cases in humans. Upon binding of HA to its receptor, the virus is taken up into cells by endocytosis. The low pH within the endosomes changes the conformation of HA exposing a fusion sequence, which allows the viral ribonucleoprotein (RNP) complex to escape into the cytoplasm. A nuclear localization domain on RNP facilitates transport into the nucleus where the virus replicates.[21] Upon assembly of new progeny, viral NA is essential to cleave sialic acid from host cell glycoproteins, which attach to HA, to allow for viral release.

HBV has a narrow tropism for hepatocytes of humans, chimpanzees, and tupaias, a Malaysian tree shrew. The surface (S) protein of HBV binds to heparan sulfate proteoglycans (HSPGs). This low-affinity interaction is followed by binding of the pre-S1 protein to sodium taurocholate cotransporting polypeptide (NTCP) leading to endocytosis of the virus particles.[22] Fusion of the viral and endosomal membranes releases the nucleocapsid into the cytoplasm. The genome is then transported into the nucleus through nuclear pores.

HSV-1 enters cells through a complex pathway that involves multiple viral and cellular proteins. Initially, the viral gC or gB surface proteins bind to HSGPs. With the help of gB, HSV-1 may then surf along filopodial surfaces. Fusion of the viral membrane occurs with the help of gB, gD, gH, and gL. During this process gD binds to cellular receptors such as nectin-1, herpesvirus entry mediator (HVEM), or 3-O-sulfated HS (3-OS HS) followed by binding of gB to one of its receptors such as plant intracellular ras-group-related LRR (PIRL)1, nonmuscle myosin heavy chain (NMMHC)-IIA or myelin-associated glycoprotein (MAG). Alternatively, the virus is endocytosed upon receptor binding, which is followed by fusion of the viral and vesicular membranes to release the capsid into the cytoplasm.[23] The genome then translocates through nuclear pores into the nucleoplasm where replication occurs.

HIV-1 is also a species-specific virus with a selective tropism for cluster of differentiation (CD) 4+ cells, that is, mainly T helper cells. The HIV-1 envelope protein, which is a trimer composed of three gp120 and gp41 subunits each, initially interacts with HSPGs, integrins, or pathogen recognition receptors (PRR) such as DC-SIGN, which brings the virus into close proximity of CD4, its primary receptors. Gp120 then binds to CD4, which leads to structural changes of the gp120/gp41 complex allowing for its attachment to a coreceptor, which can be C-C chemokine receptor type (CCR) 5 or C-X-C chemokine receptor type (CXCR) 4. Gp41 then promotes fusion and delivery of the viral capsid into the cytoplasm.[24] The capsid is degraded in the cytoplasm and in part at nuclear pores.

Wild-type measles morbillivirus uses signaling lymphocyte-activation molecule (SLAM)-1 as its receptor while the attenuated Edmonston strain can also use CD46. Receptor binding is mediated by the viral hemagglutinin (H) which is followed by entry through the fusion (F) protein.

Poliovirus has four surface virus proteins (VP) 1-VP4. VP1 forms so-called "Mesas," which are surrounded by depressions, so-called "Canyons." "Propellers" are formed by VP2 and VP3. VP4 stabilizes the capsid. The cellular receptor, CD155, binds into the "Canyons." This causes an irreversible conformational change of the virus producing a 135S particle, in which VP4 and the N-terminal end of VP1 are externalized. VP1 potentially together with VP4 then forms a pore in the cellular membrane allowing for passage of the viral RNA.[25]

Viral Replication

Most dsDNA viruses, such as adenoviruses, papillomaviruses, or herpesviruses, replicate in the host cell nucleus. Replication proceeds in two phases. Early regulatory antigens are transcribed first with the help of host cell factors. They produce proteins that alter expression of cellular proteins to ensure viral DNA synthesis, activation of viral genes, or subversion of immune responses. Late protein and DNA synthesis occurs in the second phase. For example, adenovirus has four immediate early/early (E1-E4) and five late (L1-L5) transcriptional units. The genome is flanked by long terminal repeats (LTRs). E1A, an immediate early gene, is transcribed first. The resulting protein is a transactivator needed for transcription of the other early genes. Deletion of this gene from the genome renders adenovirus replication-defective. Translation of E1A is followed by synthesis of E1B resulting in two proteins that inhibit host cell apoptosis. E2 encodes three proteins, that is, a terminal protein that acts as a primer for initiation of DNA synthesis, a DNA-binding protein and a DNA-dependent polymerase; all of these proteins are essential for viral DNA replication. Gene products of E3 are nonessential for the adenoviral life cycle. They interfere with the expression of major histocompatibility complex (MHC) class I antigens to dampen the effects of cytolytic CD8+ T cells, reduce the levels of tumor necrosis factor receptor superfamily, Member 6 (Fas) ligand and tumor necrosis factor (TNF)–related apoptosis-inducing ligand (TRAIL) on the cell surface, and inhibit TNF-mediated inflammation and apoptosis. E4 gene products modulate transcription, facilitate viral DNA synthesis, shut off host cell protein synthesis, and affect cell cycling. Transcription of early genes is followed by viral DNA synthesis, which is initiated at one of the LTRs that contain the origin of replication. Late structural proteins are synthesized in parallel followed by assembly of the virus.

Poxviruses, which also carry a dsDNA, replicate in the cytoplasm. Early during their life cycle these large viruses produce a number of enzymes, such as a DNA polymerase, a helicase, a DNA ligase, and others that are essential for the initial steps of replication. DNA synthesis is then initiated through self-priming at self-complementary regions at the end of the genome followed by synthesis of late structural proteins and virus assembly.

Replication of ssDNA viruses such as parvoviruses occurs in the nucleus and requires as a first step conversion of the ssDNA into a dsDNA. Some of the parvoviruses are dependoviruses, which in addition to host cell factors, such as a DNA polymerase, require helper functions from early gene products of adenovirus.

(+)ssRNA viruses replicate in the cytoplasm. Their genome serves as a template for protein synthesis. Upon attachment of the viral RNA to ribosomes, viral protein synthesis is initiated. Some of the (+)ssRNA viruses produce subgenomic mRNA while the genomic RNA of others produces a polyprotein that is then cleaved into structural and nonstructural proteins. The latter catalyze RNA synthesis for production of viral progeny.

(−)ssRNA viruses first have to convert their genomes into a (+)ssRNA strand. This reaction occurs within the viral capsid and is mediated by viral RNA-dependent RNA polymerases. The resulting mRNA is translated and produces viral proteins. Nonstructural viral proteins initiate conversion of the (−)ssRNA into a complementary antigenomic RNA strand that serves as a template for the production of (−)ssRNA.

ssRNA-RT viruses, such as retroviruses, are reversed transcribed in the cytoplasm by a viral RNA/DNA-dependent DNA polymerase into a dsDNA. The proviral DNA crosses into the nucleus and integrates into the host cell genome. Cellular RNA polymerase II then produces viral mRNAs, which in turn are translated into viral proteins.

dsDNA-RT viruses, such as HBV, have a circular but noncovalently closed partially dsDNA genome. Inside the host cell nucleus, the DNA is converted into a cccDNA. The cccDNA serves as a template for mRNA synthesis and for pregenomic RNA that is packaged into viral progeny capsids where it is reverse transcribed by copackaged enzymes into a new genome. At this stage, the genome-filled capsids can be transported back to the nucleus to amplify cccDNA production, or they are secreted in form of 42 nm infectious virions (Dane particles) or subviral envelope particles (SVPs), which do not contain genomes but carry the S protein and thereby serve as decoys for neutralizing antibodies.

Viruses interfere with host cell metabolism to further their replication. Two important targets are the mechanistic target of rapamycin (mTOR) and 5′ adenosine monophosphate (AMP)-activated protein kinase (AMPK) pathways, which also play crucial roles during activation of lymphocytes. Some viruses enhance while others reduce the activity of the mTOR complex (C)1 pathway, which controls cell proliferation, protein synthesis, and lipogenesis, and through activation of hypoxia-inducible factor (HIF)-α drives cells toward the use of glycolysis for energy production. US3 of HSV-1, Env of HIV-1, NS1 of influenza A virus, and NS5K of HCV enhance mTORC1 activity, while poliovirus, F17 of poxvirus, and dengue virus inhibit mTORC1 activity.[26-31] Dengue virus enhances the activity of AMPK, which is an energy sensor that increases uptake and catalysis of glucose and lipids when cells are starved.[31] Many viruses block mitochondrial functions to prevent apoptosis, inhibit innate immune responses, and modulate production of reactive oxygen species (ROS).[32]

Viral Assembly and Export

After replication of the genome and synthesis of proteins, the virus must assemble, which requires transport of the different components to a single location within the cell for which viruses use intracellular sorting and transport mechanism. Nonenveloped viruses assemble either in the nucleus or the cytoplasm. Proteins of viruses, such as adenoviruses, that assemble in the nucleus need a nuclear localization domain to reach their destiny. Within the nucleus, an empty adenoviral capsid is formed. Specific viral packaging proteins then bind to the packaging domain close to the left-side LTR and insert the genome through a yet to be defined portal within the capsid. The packaging proteins are released. A viral protease cleaves some of the minor capsid proteins causing a conformational change of the capsid into its mature form.[33] The virus is then released upon cell apoptosis.

Cytoplasmic assembly is less well defined. Most viruses, even those that are nonenveloped, assemble along cellular membranes of the secretory pathway. Enveloped viruses must assemble at cell membranes as they depend on their host membranes' liquid bilayer to form their envelope. Different enveloped viruses assemble at distinct membranes; herpesviruses use nuclear membranes, rotaviruses assemble at the membrane of the rough endoplasmic reticulum (ER); vaccinia viruses use an intermediate compartment between the ER and the Golgi complex, the latter is used by bunyavirus. Many viruses including influenza virus, rabies virus, HIV-1, and others assemble at the outer cell membrane.

Viruses use several pathways to be released from their host cells. Viruses that assemble at the outer membrane escape by budding. In this process, the virus inserts its own glycoproteins into the apical surface of the membrane. The viral capsid assembles below and then becomes surrounded by the membrane while pushing through it. As a final step, the membrane stalk is pinched off. Budding slowly causes cell death by membrane depletion. Virus can leave the cells by exocytosis using the cells' regular secretory pathways such as autophagosomes or other types of membrane enclosed vesicles. This type of viral exit does not cause the demise of the infected cells. Alternatively, viruses can lyse the infected cell and escape by breaking down the membrane.

Persistent Virus Infection

Some virus infections such as those with AAVs are silent and cause no symptoms. These viruses fail to initiate an inflammatory reaction and the immune system basically ignores them. During other acute viral infections, the body becomes a battlefield: the viral assailants and the immune system wage a war leaving a plethora of dead cells, unfit viruses, and collateral damage. It is a fight to death of the invader or the defender. Once the battle is won and the virus is defeated, the attacked victim has to mount a major clean-up campaign to get rid of dead cells and repair any damage. If the virus wins and kills its host, it has no cause for celebration either for it will not thrive in a dead host. In some cases, a truce is reached and neither dies. Instead the virus achieves an equilibrium with its host and establishes a chronic infection. It is becoming increasingly appreciated that all humans are chronically infected with somewhere between 8 and 12 different viruses. All humans carry retroviruses and anelloviruses, most are infected with AAV and polyomaviruses; these viruses have in common that they cause no or only trivial symptoms in immunocompetent individuals. Other viruses, such as various herpesviruses and adenoviruses, cause an initial symptomatic infection but then persist asymptomatically at low levels. During latency, herpesviruses do not replicate and produce no or minimal amounts of proteins that could be detected by the immune system. The viral genome remains intact during latency. As exemplified by herpesviruses and adenoviruses, their genomes remain in a stable form within episomes. As cell divisions would over time dilute out the viral genome, these viruses typically persist in nondividing cells. Viruses that assume a state of latency can become reactivated typically under conditions that weaken the immune system. Reactivation of HSV, for example, which can be triggered by minor insults, such as too much sunlight or stress, is rapidly controlled by the immune system. Major insults, such as immunosuppression to allow for an organ transplant, can cause death of the patient due to overwhelming spread of endogenously persisting herpesviruses or adenoviruses.

Other viruses persist by integrating into the host cell genome such as HPVs. In the majority of humans, the infection is transient and rapidly removed by the immune system without any negative outcomes. However, a small percentage of patients are unable to clear the infection and the resulting integration of the viral genome of oncogenic types of HPV, for example, HPV-16 or -18, into host cell DNA can lead to malignancies, such as cervical, anal, vulvar, penile, and head and neck cancer. HPV encodes the accessory proteins E5, E6, and E7, which affect cell cycle regulation. The E6 protein causes degradation pf p53 while the E7 inactivates the retinoblastoma (Rb) tumor suppressor protein. E5 is thought to contribute to oncogenesis through interactions with the epidermal growth factor receptor (EGFR).

In some types of chronic infection, such as those with HBV, HCV, or HIV-1, the virus continues to replicate, and the host continues to eliminate the virus or virus-infected cells, but rather than going into a full-blown attack, the host immune system maintains viral loads at levels that do not cause immediate debilitating harm without achieving viral clearance. This cold war approach can be beneficial for both parties—the host does not immediately face potentially fatal immunopathological damage to essential organs and the virus survives. Viruses promote this very fragile state of balance by subverting immune responses through mutational escape or by interfering with factors that are crucial for optimal stimulation or execution of immune effector function.

Some viruses that typically cause severe symptoms during a primary infection may on rare occasions persist for years before they elicit a fatal or debilitating disease. An infection with rabies virus generally becomes symptomatic with 2 to 4 weeks after the bite by an infected animal and then kills within days. Nevertheless, in rare cases, years pass between viral transmission and onset of symptoms,[34] and it is unknown where and how rabies virus persists during these long-incubation times. Measles morbillivirus in most children results after a 10 to 14-day incubation period in cold-like symptoms, high fever, and a red, blotchy skin rash. Although the acute symptoms disappear after about 2 weeks, it has long-term effects on the immune system causing immune amnesia by replacing memory cells directed against previously encountered pathogens with those specific for measles morbillivirus.[35] About 1 out of 100,000 infected individuals, especially children who are infected early in life, within 5 to 15 years after a primary infection, develop subacute sclerosing panencephalitis caused by persisting measles morbilliviruses.

Nuclear Anti-Viral Factors

All cells, and not just those of the immune system, have developed nuclear defense mechanisms against DNA viruses. The nuclear defense centers around the promyelocytic leukemia nuclear bodies (PML-NBs), also called ND10 nuclear bodies. Most cells have 5 to 10 PML-NBs which are 0.1 to 1 μm in diameter and thereby easily visible by light microscopy.[36] PML-NBs are composed of multimerized PML protein that forms a ring-like structure that binds to the nuclear matrix.

PML-NBs associate with a number of proteins including bloom syndrome protein (BLM), cyclic amp response element (CREB)–binding protein (CBP), death domain–associated protein (Daxx), mouse double minute (Mdm)2 proto-oncoprotein, p53, speckled protein (Sp)100, small ubiquitin-like modifier 1 (SUMO-1), and others.[37] PML-NBs, which are induced by interferon (IFN), are positionally stable within the nucleus, but they become mobile upon cellular stress; their structure disintegrates, and they form microbodies. Most work on the effects of PML-NBs has focused on HSV-1. The DNA of HSV-1 as well as of other viruses such as adenovirus or HCMV associates with PML-NBs in the nucleus[38] where early viral transcription is initiated. Viral proteins that target the DNA to the PML-NBs are involved in the initial transcription of the virus, such as the viral origin of replication (OriS), the viral immediate-early proteins (ICP) 4, and ICP27 of herpesviruses.[39] IPC4 actively recruits PML-NB components so that their numbers initially increase after virus infection. Individual PML-NB components contribute to the silencing of viral gene expression, and HSV-1 and CMV evades these effects by degrading components of PML with the help of viral immediate early proteins ICP10 and IE1, respectively.[40] One of the PML proteins is Daxx, which represses expression of viral genes by recruiting histone deacetylases (HDACs) to targeted promoters. HCMV circumvents this silencing through the viral pp71, which degrades Daxx.

Interferon-inducible protein (IFI) 16 is a member of the pyrin and IFN-inducible, nuclear localization (HIN) domain (PYHIN) protein family.[41] It contains an N-terminal pyrin domain (PYD) for protein-protein interactions, two C-terminal HIN domains, which allow for the protein's nonspecific binding to double-stranded DNA, and a nuclear localization domain. It can sense DNA of HSV-1 and initiate an inflammatory response[42] and lead to silencing of the viral DNA by covering it with a repressive layer of heterochromatin.[43] The virus fights back through ICP10 which targets IFI16 for proteasomal degradation.[44]

Viruses downregulate transcription and translation of host genes to allow for production of their own proteins through a number of mechanisms, such as inhibition of transcription factors, inhibition of processing, polyadenylation, decapping of cellular RNA, by increasing RNA degradation or promoting RNA dissociation from ribosomes in a process called ribosome stalling.[45]

Viruses and RNA Interference

In eukaryotes, gene expression can be silence by RNA interference mediated by short interfering (si)RNA or by micro (mi)RNA. siRNA are 20 to 24 base pair long dsRNAs with a phosphorylated 5′ end and a hydroxylated 3′ end with two overhanging nucleotides. They are produced by the ribonuclease DICER from longer dsRNAs, which are formed during viral replication. Within the cell, the ds siRNA is incorporated into a protein complex to form the RNA-induced silencing complex (RISC). Within this complex, the siRNA unwinds and forms single strands, which can bind to complementary target mRNAs causing their degradation. The inhibitory effects of siRNA on viral replication was first reported for plant viruses.[46] Viruses can inhibit the formation of siRNA by directly binding RNA to prevent its digestion into smaller fragments or by silencing or degrading RISC. Whether siRNAs play a role in antiviral defense in mammals remains debated.[47] miRNAs are generated by RNA transcripts that fold back on themselves to form short hairpins. Similar to siRNAs, they are involved in gene silencing by targeting complementary mRNAs. Several viruses, such as herpesviruses, adenoviruses, and polyoma viruses, which can all establish persistent infections, produce their own miRNAs.[48-50] Other viruses depend on host cell miRNA for their replication, such as HCV, which uses the liver-specific miRNA-122 to replicate and stabilize its genome.[51] Nevertheless, miRNAs can exhibit antiviral function. They play a role in resistance to vesicular stomatitis virus (VSV),[52] retroviruses,[53] HCV,[54] and others. Viruses counterbalance the detrimental effects of cellular miRNAs. For example, RNA encoded by CMV binds and then causes degradation of miRNA-27 and miRNA-17 members.[55,56]

Autophagy

Autophagy is a key degradation process that cells employ to recycle faulty or unwanted proteins. It is used by virus-infected cells to eliminate viruses or their components. Cells use three types of autophagy: microautophagy, macroautophagy, and chaperone-mediated autophagy (CMA). In the latter pathway, heat shock cognate protein 70 binds to a KFERQ motif on the target protein, which diverts it into lysosomes for degradation.[57] During microautophagy, lysosomal membranes engulf their cargo and form a vesicle in which proteins are degraded into nutrients that are released upon hydrolysis of the vesicles' membrane.[58] During macroautophagy, cells form double-membrane vesicles around a portion of the cytoplasm. These vesicles then fuse with lysosomes where their contents are digested.[59] Autophagy not only rids the cell of viral components, but it is also intricately linked to innate activation by regulating the expression of IFNs, while in turn IFNs harness autophagy for viral clearance.[60] Autophagy is regulated by multiple factors such as ER stress, hypoxia, oxidative stress, starvation, deprivation of growth factors, or virus infections. Cells upon virus infections increase the activity of the Eukaryotic Initiation Factor 2 (eIF2α), which is essential for protein synthesis and the IFN-inducible double-stranded RNA-dependent protein kinase R (PKR). Phosphorylation of eIF2α then induces autophagy.[61]

Viruses have evolved to evade autophagy by targeting upstream pathways or by directly inhibiting key components of autophagy. For example, ICP34.5 of HSV-1 antagonized PKR-induced autophagy through dephosphorylation of eIF2α.[62] A number of herpesviruses, including HSV-1 and Kaposi sarcoma-associated herpesvirus (KSHV), inhibits Beclin 1, a crucial protein of autophagy.[63,64] Under normal conditions, Beclin 1 is regulated by the antiapoptotic protein BCL2, which is phosphorylated and then released from Beclin 1 once cells receive an autophagy-inducing signal. Some herpesviruses such as KSHV produce a BCL2 homologue; its binding to Beclin 1 is not affected by such signals.[65] A number of viruses block PKR activation. HCV nonstructural protein (NS)5A and vaccinia virus E3 block PKR dimerization, HIV-1 that blocks the catalytic domain of PKR, influenza virus NS1, HSV-1 US11, reovirus σ3, and rotavirus NSP3 block activation of the PKR pathway by their genomes.[66]

Anti-Viral Innate Immunity

The innate immune system is evolutionary older than the adaptive immune system and can already be found in all living things. The innate immune system responds immediately to an infection and for the first few days after an infection protection against viral spread primarily depends on innate immune defense mechanisms.

PRRs and Downstream Signaling Pathways

Once a virus attaches to or enters into a cell, it alerts the innate immune system through motifs that are not present in a mammalian organism. These so-called pathogen-associated molecular patterns (PAMPs) bind to PRRs that are expressed in different cellular compartments (see Chapter 12).

Toll-Like Receptors

Toll-like receptors (TLRs) were discovered first in the mid-1990s. To date, 13 TLRs have been identified; TLR1-10 are active in humans, TLR11-13 have been deleted from the human genome; in mice, TLR10 is a disrupted pseudogene. TLRs are membrane glycoproteins with an extracellular PAMP-interacting domain that contain leucine-rich repeats, a transmembrane domain and a cytoplasmic tail that signals and shows homology with the interleukin (IL)-1 receptor (R, IL-1R) tail and is therefore called Toll-IL-1R (TIR) domain. The TLR's TIR domain binds to a C-terminal TIR domain on myeloid differentiation primary response 88 (MyD88), which upon activation binds through an N-terminal death domain to IL-1R–associated kinases (IRAK)4 and 1. Upon activation, IRAK associates with TNF receptor–associated factor (TRAF)6, which in turn forms a complex with transforming growth factor (TGF)-ß–activated kinase (TAK)1, TGF-ß–activated kinase 1 binding proteins (TAB)1 and 2, and the E2 ligases Ubc13 and Uev1A. TAK1 becomes activated within the complex and interacts with the inhibitor of nuclear factor kappa-B kinase (IKK) complex, which leads to activation of the nuclear factor kappa-light-chain-enhancer of activated B cell (NF-κB) and activator protein (AP)-1 pathways and induction of cytokine production. An alternative MyD88-independent pathways can be used by TLR3 and 4. Upon binding of the TLR to its ligand, the TLR forms a dimer and its TIR domain binds to TIR-domain containing adapter-inducing IFN-β (TRIF), which in turn through TRAF-interacting protein (TANK) binding kinase (TBK)1 causes activation of IKKε and induction of type I IFNs. In addition, TRIF activates the NK-κB pathway while TBK1 induces the interferon regulatory factor (IRF)3 pathway (Fig. 44.2).

TLRs are expressed by cells of the immune system, such as dendritic cells, macrophages, B cells, and some types of T cells, as well as by fibroblasts, epithelial and endothelial cells, hepatocytes, and others. Crucial for the innate defense against viruses are TLR2, 3, 4, 7, 8, and 9. Of those, TLR2 and 4 are expressed on the cell surface where they can recognize viral envelope proteins during virus attachment. Specifically, TLR2, which dimerizes with other TLRs such as TLR1,[67] is activated by gB and gH of HCMV,[68] gL/H and gB of HSV,[69] UTPase of Epstein-Barr virus (EBV),[70] core and NS3 proteins of HCV,[71] NS4 of rotavirus,[72] or H protein of measles morbillivirus.[73] Vaccinia virus,[74] lymphocytic choriomeningitis virus (LCMV),[75] VZV,[76] and RSV[77] have also been implicated to activate the TLR2 pathway.

TLR4, which can signal through MyD88-dependent and-independent pathways, does not form a dimer with other TLRs but binds to myeloid differentiation protein (MD)-2, also known as lymphocyte antigen 96. It responds to the F protein of RSV,[78] as well as to glycoproteins of VSV,[79] and Ebola virus.[80] TLR4 contributes to protection against influenza viruses,[81] SARS-coronavirus,[82] vaccinia virus,[83] KSHV,[84] and others. TLR3, 7, 8, and 9 are expressed within endosomes where they can recognize nucleic acids upon exposure of genomes following the viruses' uncoating. Endosomal TLRs also react to RNA intermediates, which are part of the virus replication process. TLR3, which only signals though the MyD88-independent pathway, recognizes primarily dsRNA as present in reoviruses but can also be triggered by ssRNA with internal loops as in poliovirus. Intermediate RNAs produced during replication of HSV-1, RSV, influenza A virus, West Nile virus, and others activate TLR3 as well.[85] TLR7 and 9 are primarily expressed in dendritic cells while monocytes/macrophages carry TLR8. TLR7 and 8 recognize ssRNA such as in influenza virus, flavivirus, HIV-1, coxsackievirus, and others. TLR9, which is also present within the ER and lysosomes, responds to unmethylated CpG motifs within viral DNA and is thereby crucial for recognition of herpesviruses, adenoviruses, and poxviruses.

Viruses have evolved to evade recognition by TLRs[86] and nearly every step of TLR signaling is attacked (Fig. 44.2). A comprehensive review is beyond the realm of this chapter, which describes a few examples. TRIF is cleaved or directed toward degradation by herpesviruses[87] and coxsackievirus.[88] IRF3 is steered toward degradation by rotavirus NSP1[89] while herpesviruses target IRF7. Vaccinia virus attacks TIR-containing adaptors such as MyD88 through protein A46R while protein A52R blocks signaling through TRAF6 and IRAK2.[90,91] MyD88 signaling is also blocked by HCV.[92] SARS-coronavirus, RSV, and HSV-1 prevent IRF3 from entering the nucleus[93] or from binding to the IFN promoter.[94,95] HCV NS3A4 inhibits IRF3 phosphorylation.[96] Measles morbillivirus proteins serves as a decoy for IRF7,[97] while the mumps virus protein V can mimic IRF3.[97] The rabies virus matrix (M) protein blocks NF-κB from entering the nucleus.[98] The HBV early (e) Ag binds to MyD88 or the MyD88 adapter-like protein (TIRAP) and thereby inhibits TLR2 signaling. HBV, which has evolved to target multiple steps of activation of the innate immune system,[99] blocks the NF-κB pathway by inducing rubicon, a regulator of autophagy. Rubicon binds to NF-κB–essential activator (NEMO) and prevents its ubiquitination and thereby blocks its functions. Picornavirus 3C is a protease that can cleave TRIF, TAK1, and TAB2.[100] The influenza virus nonstructural protein (NSP) 1 inhibits TRAF3,[101]

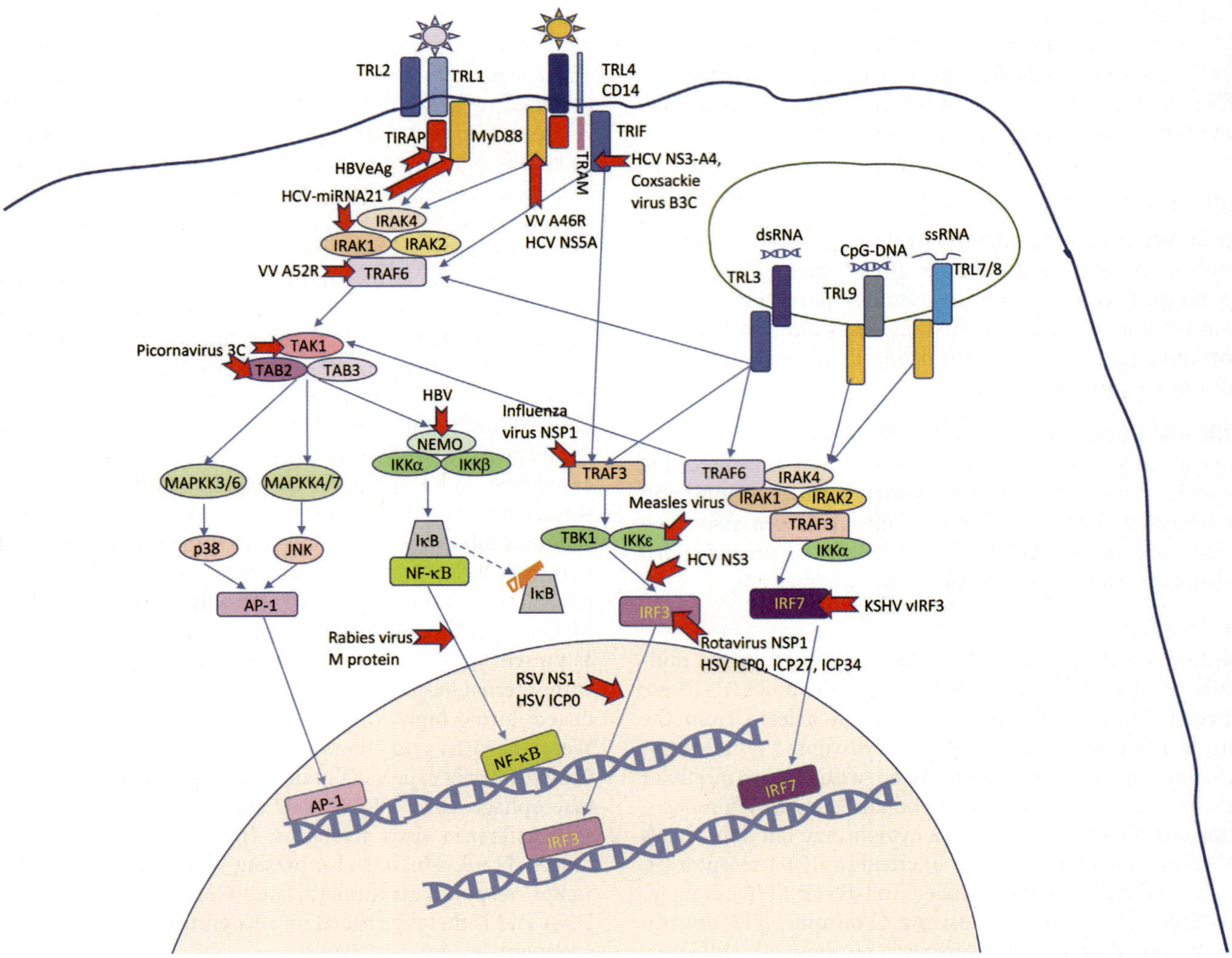

FIG. 44.2. The figure shows the pathways of toll-like receptors (TLR) signaling. *Red lines* indicate parts of the pathways that are affected by viruses. IκB, inhibitor of nuclear factor kappa B; MAPKK, mitogen-activated protein kinase kinase; TRAM, translocation-associated membrane protein. The other abbreviations are explained in the text.

thereby attenuating induction of proinflammatory cytokines. HCV upregulates miR-21, which targets MyD88 and IRAK1.[102]

C-type Lectin Receptors

CLRs are highly conserved molecules with one or more carbohydrate-recognition domains. They form a large family of over 1000 proteins that are involved in pathogen recognition, phagocytosis, antigen uptake, cellular adhesion, migration, and signaling. CLRs are divided into 17 groups (I-XVII). Most are expressed on the cell surface but some are secreted. Myeloid CLRs lectins are expressed by antigen-presenting cells such as dendritic cells and macrophages, where they serve as PRRs specifically recognizing glycans on viral surface proteins. Although viruses use the host cell glycosylation machinery, there are some intrinsic differences in the glycan profiles of host cell proteins and viral proteins, which allow for the recognition of the latter by PRRs. Viral glycoproteins are more heavily glycosylated as exemplified by the envelope protein of HIV-1, which carries over 20 N-glycans. Furthermore, HIV-1 carries mainly oligomannose N-glycans unlike the more complex types expressed by host cells.[103]

Myeloid CLRs belong to groups II, V, and VI. According to their intracellular signaling motifs, CLRs are divided into four classes: CLRs with an immunoreceptor tyrosine-based activation motif (ITAM), which consist of YxxL/I tandem repeats; a hemITAM motif, which consists of a single YxxL/I sequence; an immunoreceptor tyrosine-based inhibition motif (ITIM), that is, a S/I/V/LxYxxI/V/L sequence; and a class that lacks typical signaling motifs. CLRs with ITAM or hemITAM are activating receptors that signal through the spleen-associated tyrosine kinase (Syk).[104] Phosphorylation of Syk activates Vav guanine nucleotide exchange factor 1(Vav) and protein kinase (PK)Cδ, which in turn induce the caspase recruitment domain (CARD)9/mammalian CARD-containing adapter molecule E10 (BCL10)/mucosa-associated lymphoid tissue lymphoma translocation protein (Malt)1 and mitogen-activated protein kinase (MAPK) pathways and production of cytokines and chemokines.[105] Some of the hemITAM-bearing CLRs activate nuclear factor of activated T cells

(NFAT) and MAPK1 through phospholipase (PL)Cγ2 and Syk-independent NF-κB activation through Raf-1 kinase.[105] All of these pathways result in production of ROS and cytokines, such as IL-1β, IL-6, IL-10, IL-12, tumor necrosis factor (TNF)-α, and IL-23 to drive T helper (Th)1 and Th17 cell differentiation. CLRs with ITIM motifs recruit tyrosine phosphatases, such as Src homology region 2 domain-containing phosphatase (SHP)-1 or-2,[106] which may lead to inhibition of TLR8-mediated IL-12 and TNF-α production and TLR9-induced IFN-α production.[107,108] CLRs without an ITAM or ITIM motif such as DC-SIGN, which is carried by human but not mouse myeloid cells, associate with lymphocyte-specific protein (LSP)1, kinase suppressor of ras (KSR)1, connector enhancer of kinase suppressor of ras 1(CNK), and the kinase Raf-1,[109] and modulate signaling by other receptors.

Group II transmembrane CLRs contain a short cytoplasmic tail, a transmembrane domain, an extracellular stalk region, and a single carbohydrate binding domain. DC-SIGN, or its mouse equivalent, dendritic cell–specific intracellular adhesion molecules (SIGN-R1), also called CD209, and the C-type lectin superfamily member 6 (DCIR), are members of this group. DC-SIGN, an hemiITAM containing molecule, is expressed by dendritic cells and macrophages. It acts as an attachment receptor for several viruses such as HIV-1 and-2,[110] Ebola virus,[111] CMV,[112] HCV,[113] measles morbillivirus,[114] West Nile virus,[115] and others.[116,117] Binding of pathogens expressing mannose to DC-SIGN modulates TLR4 activity through a ras-dependent pathway enhancing production of IL-6, IL-10, and IL-12. Binding of fructose-carrying pathogens trigger a LSP1-dependent pathway that increases IL-10 production but reduces secretion of IL-6 and IL-12.[109] It thereby skews T helper cell responses toward the Th2 pathway. In humans, DCIR contains a cytoplasmic ITIM motif, binds HIV-1, and helps the virus to spread to CD4+ T cells.[118] In mice, DCIR assists in the internalization of chikungunya virus, where DCIR limits viral spread.[119] DCIR promotes cross-priming,[120] inhibits TLR8-dependent IL-12 and TNF-α production, and reduces IFN-α responses while sustaining signal transducer and activator of transcription (STAT)1 activation and, thereby, type I IFN production.

Group V contains type II transmembrane proteins with a short cytoplasmic tail, a transmembrane domain, and an extracellular stalk region followed by a single C type lectin like domain without a typical carbohydrate binding motif. Many of the natural killer (NK) cell receptors, such as killer cell lectin-like receptor (KLR)F1, KLRG1, KLRC1, KLRC2, and others, are group V CLRs. Group V also includes myeloid killer–activating receptor-associated protein (DAP-12)-associating lectin (MDL-1). MDL-1, an ITAM containing molecule, binds to the HA of influenza A virus and was shown to contribute to viral pathogenicity by augmenting inflammatory responses.[121] It also interacts with dengue virus and contributes through heightened cytokine release to the dengue toxic shock syndrome.[122]

Group VI are type I transmembrane proteins with an extracellular domain containing an N-terminal ricin-like domain, a fibronectin type 2 domain, followed by a C type lectin-like domain. The proteins have a transmembrane region and a short cytoplasmic domain. Group VI encompasses mannose receptors, which are expressed by most tissue macrophages and dendritic cells and by endothelial cells in lymphatic tissues and livers. Mannose receptors bind HIV-1, influenza, and dengue viruses and result in their endocytosis.[123-125] Binding of HIV-1 to mannose receptors on macrophages facilitates transmission of the virus to CD4+ T cells.[126]

C-type lectin domain containing 9A (DNGR-1), which is not a PRR, is a group IX member with a hemITAM domain. It is mainly expressed on CD8+ dendritic cells. This receptor senses material released by cells dying due to necrosis, such as upon an infection with vaccinia virus,[127] and through recruitment and activation of the tyrosine kinase Syk promotes cross-presentation of antigen derived from the dying cells.[128]

Nucleotide-Binding and Oligomerization Domain (NOD)-Like Receptors (NLRs)

NLRs are located in the cytoplasm where they detect PAMPs including those associated with cell damage upon enhanced production of ROS, increased ion flux, or endosomal disruption.[129] They have been implicated to sense viral RNA or DNA. Overall, there are 22 different NLRs. They have an N-terminal effector binding domain, which is responsible for signaling; a central NOD domain, also called NACHT domain; and a C-terminal leucine-rich repeat (LRR) domain, which is involved in ligand sensing. NLRs are divided into four subfamilies according to their N-terminal domains. NLRA, which only has one member, that is, class II MHC transactivator (CIITA), has an acidic transactivation domain. NLRB, which also has only one member called NLR family apoptosis inhibitory protein (NAIP), carries a baculoviral inhibitory repeat-like domain. NLRC is characterized by a CARD domain while NLRP has a PYD. Members of subfamilies NRLC, which has six members (NLRC1-5, NLRX1), which include NOD1 (NLRC1) and NOD2 (NLRC2), and NRLP, which has 14 members (NLRP1-14), are involved in sensing of viruses. NLRCs are involved in autophagy by recruiting autophagy-related protein 16-like 1 (ATL16L1) to the cell membrane. This may play a role during infection with VSV, which is targeted by NLRX1 to autophagosomes, and then to autolysosomes, where the virus is degraded.[130] NLRs can signal by activating MAPK or NF-κB, thereby inducing an inflammatory reaction; under some circumstances, they deliver inhibitory signals. NLRs can activate transcription factors. Many NLRs, such as NLRP1, NLRP2, NLRP3, NLRP6, NLRP7, NLRP12, NLRC4, and NAIP, can form inflammasomes by activating caspase-1, which converts the procytokines IL-1ß and IL-18 into their active forms. NLRP3, which forms an inflammasome complex after activation, responds to the ssRNA of influenza and sendai viruses and the dsDNA of adenovirus,[131,132] while NOD-2 is triggered by the ssRNA of RSV, influenza A viruses, and parainfluenza viruses.[133] NLRC5 augments cytokine responses to RNA viruses.[134] Its level increases upon HCMV infection due to an autocrine IFN-γ loop. In turn, overexpression results in increased expression of MHC class I antigens as NLRC5 acts as a transactivator on the MHC class I promoter.[135]

Enhanced level of NLRC5 causes activation of antiviral target genes such as IFN-α, the 58 KDa interferon-induced protein kinase-interacting protein (PRKRIR), and 2′-5′-oligoadenylate synthetase (OAS), which upon activating latent RNase L results in viral RNA degradation and the inhibition of viral replication.

Several viruses target NLR pathways to suppress activation of the inflammasome. To name a few examples, NS1 of influenza A virus inhibits activation of caspase 1 through a yet to be defined mechanism.[136] The V protein of measles morbillivirus relocalizes NLRP3 to inhibit IL-1ß secretion.[137] Vaccine virus FL1 protein inhibits activation of inflammasomes by binding to NLRP1.[138] ORF63 of KSHV mimics NLRP1 and thereby blocks activation of caspase 1 and conversion of IL-1ß and IL-18 into their active forms.[139]

Retinoic Acid Inducible Gene-I (RIG-I)–Like Receptors (RLRs)

RLRs are ubiquitously expressed at low levels by all cells. Their expression is upregulated by IFNs. Two RLRs, which function as cytoplasmic RNA sensors, have been identified, that is, RIG-1 and melanoma differentiation–associated protein 5 (MDA5). RIG-1 senses relatively short ss or dsRNA such as found in RNA viruses including influenza virus, VSV, Japanese encephalitis virus, with preferential recognition of 5-triphosphate (5′ppp) sequences, while MDA5 mainly responds to longer stretches of dsRNA as can be found in cells infected with picornavirus.[140,141] RLRs have two N-terminal CARD domains, which are involved in signaling, a central DEAD-box helicase that binds and unwinds the target RNA, and a C-terminal CTC domain that also interacts with viral RNA and determines the specificity of the reaction. Upon ligand binding, one of the RLRs' CARD domains becomes ubiquitinated, which is essential for its translocation to mitochondria for mitochondrial antiviral-signaling protein (MAVS) binding. Signaling through MAVS then results in activation of inositol-3-phosphate synthase isozyme (IPS)-1, which induces the NF-κB and IRF3 pathways and production of type I IFNs.

Several viruses avoid recognition by RIG-1 by modifying triphosphorylated 5-terminus (5′ppp) motifs. Crimean-Congo hemorrhagic fever virus, Borna disease virus, and hantavirus remove the 5′ppp group on their genome post-transcriptionally.[142] The hantavirus genome carries a 5′ monophosphate group, which is not recognized by RIG-1; the virus then uses a prime and align mechanism to initiate RNA synthesis.[143]

Bornavirus trims its genome at the 5′ end and although this reduces infectivity of the virus, it allows it to escape RIG-1 detection.[144] Other viruses cap their RNA by either using cellular enzymes, by snatching cap structures from host cell RNAs, or by encoding proteins that cap their RNA.[145] Examples for the latter are Ebola virus vp35 or polymerase subunit 2 of influenza A virus.[146] Influenza virus with an adaptive mutation in position 627 of the viral polymerase modulates nucleocapsid inhibition by the pathogen sensor RIG-I.[147] Other viruses hide their RNA in compartments that fail to carry RIG-1 sensors such as SARS-coronavirus or dengue virus.[148,149]

Viruses can also inhibit RIG signaling by preventing its ubiquitination either directly, as was shown for influenza A virus[150] or by destroying the E3 ligase Riplet, such as by the NS3/NS4 protease of HCV.[151]

Other viruses such as KSHV or SARS-coronavirus deubiquitinate RIG-1.[151,152] Viruses can directly degrade RIG-1 or MAVS, as was shown for picornavirus,[153] RSV,[154] and enteroviruses.[155] HCV evades signaling through RIG-1, which is highly expressed in hepatocytes, through its NS3/NS4 complex, which has proteolytic activity. It cleaves IPS-1 thereby preventing activation of the IRF3 and NF-κB pathways.[156] Virus evasion through inhibition of other downstream signaling events mirror those described above for NF-κB and IRF activation following TLR signaling.

Absent-in-Melanoma –Like Receptors

ALRs include absent in melanoma 2 (AIM2) and IFI16. AIM2 is located within the cytosol, while IFI16 is a nuclear defense factor described above. Both contain an N-terminal PYD and a C-terminal HIN domain, which senses dsDNA as in poxviruses and herpesviruses. AIM2 in collaboration with IFI16 activates caspase-1, which results in processing of the immature forms of IL-1ß and IL-18 into functional cytokines.

HSV-1 encodes several proteins that block ALR signaling. IPC0 is an E3 ubiquitin ligate, which degrades IFI16.[157] VP22 inhibits caspase 1 activation[158] while UL41 degrades the IFI16 mRNA.[159] The V protein of paramyxoviruses binds to MDA-5, thus preventing binding of its genome.[159] Another cytosolic sensor is cyclic GMP-AMP synthase (cGAS) which through binding to stimulator of interferon genes (STING) triggers TBK1-mediated activation of IRF3.

Inflammasomes

Activation of several of the above described PRRs leads to the formation in inflammasomes. These include NLRP1, NLRP3, NAIP, and AIM2. Upon activation, these sensors oligomerize with adaptor proteins such as apoptosis-associated speck-like protein containing CARD (ASC) forming multiprotein complexes called specks. Once formed, the inflammasome binds to pro–caspase-1 via its own CARD or the CARD of ASC resulting in activation of caspase 1, which proteolytically cleaves pro–IL-1β into IL-1ß and pro–IL-18 in IL-18. Caspase 1 cleaves gasdermin-D which then induces pyroptosis, a form of programmed cell death that allows for release of cytokines and additional danger signals and prevents the misuse of cells for viral replication.

Complement

Recognition of pathogens by PRRs leads to activation of complement through one of three pathways: the classical pathway, the lectin pathway, or the alternative pathway, the latter can activate spontaneously without help from PRRs (see Chapter 13). In the classical pathway, antigen-antibody complexes activate the C1 complex, which is composed of one molecule of C1q, two molecules of C1r, and two molecules of C1s. To avoid spontaneous activation, a protease inhibitor, called C1 inhibitor, binds and inactivates C1s and C1r. C1q upon binding to C1r activates C1r, which cleaves C1s. C1r together with C1s splits C2 and C4 into two component each (a and b). The C4b2b complex, also called C3 convertase, cleaves C3 into

C3a and b, the latter forms together with C4bC2b, the C5 convertase. In the lectin pathway, carbohydrates on the pathogen's surface bind to the mannose-binding lectin (MBL)/ficolin-mannose–associated serine protease 1 (MASP) complex, which binds and activates the C1 complex. The alternative pathway relies on spontaneous hydrolysis of C3, which occurs constantly at low levels. Normally, C3b once formed is inactivated by factors H and I. In presence of a virus, C3b can bind to a hydroxyl or amino group on the virus surface. It then binds factor B, which through factor D is cleaved into Ba and Bb. Bb remains bound to C3b forming the alternative C3 convertase. C3bBb binds P forming C3bBbP and then cleaves more C3, thus amplifying the process. A second C3b binds the complex forming C5 convertase. C5 convertase cleaves C5 into C5a and b, C5b recruits C6-9 to form a membrane attach complex (MAC). To prevent inadvertent activation of the complement cascade, a number of inhibitors control this process. To ensure that the alternative pathway is not activated by components on the surface of normal cells, surface molecules such as CD35 (complement C3b/C4b receptor 1), CD46 (membrane cofactor protein), CD55 (decay accelerating factor [DAF]), and CD59 (protectin) inhibit activation. Soluble complement inhibitors include factor H and C4 binding protein (Fig. 44.3).

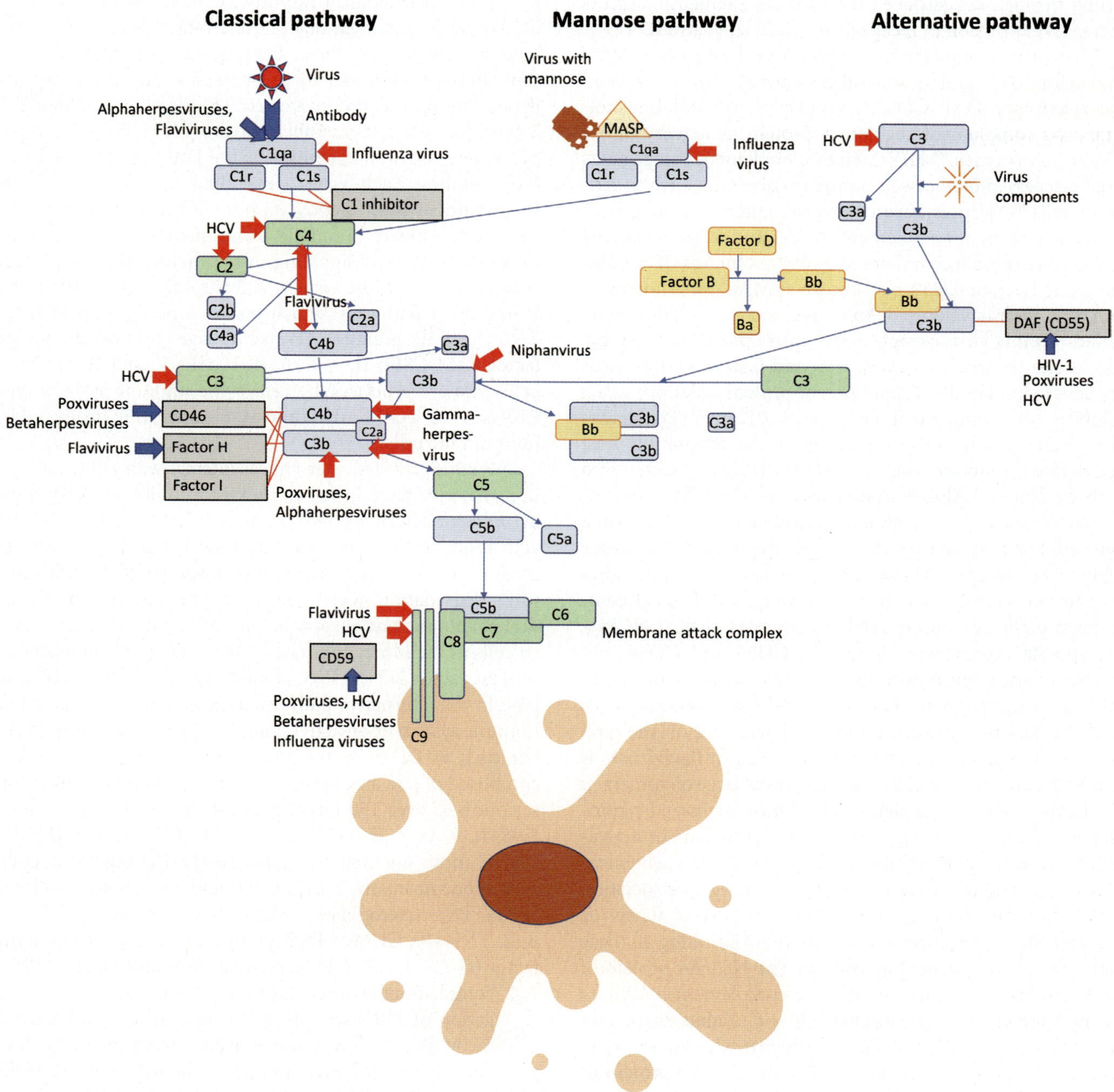

FIG. 44.3. The figure shows the three pathways of complement activation. *Red lines* indicate parts of the pathways that are affected by viruses.

Complement can directly neutralize viruses. Alphaviruses, herpesviruses, coronaviruses, retroviruses, and paramyxoviruses can be lysed directly by MACs.[160] Other viruses, such as SARS-coronavirus, are neutralized by C1q and MBL,[161] while neutralization of, for example, dengue virus requires deposition of additional complement components.[162] Complement can also promote aggregation of viruses,[163] thereby reducing the number of virus particles that can infect individual cells. It can facilitate phagocytosis by leukocytes or monocytes.[164] Upon production of virus-specific antibodies, complement can kill infected cells.[160] Complement further promotes B and T cell responses.[165] Complement signaling within CD4+ T cells increases activation of the mTOR pathway while signaling through cell surface expressed complement receptors increases expression of receptors for IL-2, IL-7, and IL-12.[166]

Viruses use complement receptors for cell entry as was first shown for EBV[167] and then for others, such as adenoviruses and picornaviruses. DAF, CD55, a glycosylphosphatidylinositol-anchored complement regulatory protein, is a receptor for several echoviruses.[168] HIV-1 binds complement components to gain access into otherwise nonpermissive cells, such as dendritic cells.[169] Viruses have developed different strategies to evade activation of complement. HIV-1 and other enveloped viruses incorporate complement inhibitors such as CD55, also known as DAF, or CD59, also known as MAC-inhibitory protein, into their envelope.[170] Poxviruses secrete a viral protein, called vaccinia virus complement control protein (VCP) that accelerates the degradation of C3b and C4b.[171] Variola virus encodes a similar albeit more potent protein called smallpox inhibitor of complement enzymes (SPICE).[172] Poxviruses also evade complement activation by incorporating host cell derived inhibitors such as CD49, CD55, and CD59 into their envelope.[173] Alphaherpesviruses such as HSV-1 use gC to inactivate C3b.[174] The gE and gI proteins of HSV-1 form a heterodimer that mimics an Ig (Ig)G fragment crystalizable region (Fc) receptor (FcR). The complex binds antibodies thereby preventing them from activating the classical complement pathway.[175] Some betaherpesviruses, such as HCMV, incorporate complement inhibitors CD55 and CD46 into their envelope while others, such as HHV-7, induce increased inhibitor expression on host cells.[176,177] Gammaherpesvirus KSHV ORF4 is expressed on infected cells and on the viral envelop. It inactivates C3b and C4b.[178] NS1 of flaviviruses is secreted from infected cells and then forms large oligomers. It blocks the classical complement activation pathway by complexing antibodies.[179] NS1 also inhibits complement activation by recruiting the inhibitory factor H,[180] and it enhances C4 cleavage and binds C4b, thereby preventing its association with other complement component.[181,182] NS1 of flavivirus prevents oligomerization of C9,[183] thereby inhibiting all three pathways of complement activation. The NS3/A4 protein of HCV cleaves C4. It incorporates the complement inhibitors CD59[184] and CD55[185] into its envelope and inhibits expression of C2,[186] C3,[187] C4,[188] and C9.[189] Influenza viruses incorporate CD59 into their envelope.[190] The matrix (M) protein of influenza virus binds and inactivates C1qa. Paramyxoviruses incorporate the complement inhibitors CD55 and CD46 into their envelope or increase their expression in infected cells.[191,192] Niphan virus has a protease that in presence of factor H and the complement receptor degrades C3b.[193]

This list is obviously incomplete, but it shows that many viruses have evolved similar mechanisms, including incorporation of inhibitors into their envelop, upregulation of inhibitors, blockade of antibody binding to cell-bound receptors, and degradation of key proteins of the pathway to evade destruction by complement.

Early Cytokine Responses

Mammals produce numerous cytokines and chemokines within hours after an infection (see Chapter 12). Cytokines are intercellular messengers that upon secretion bind to a receptor to elicit a biological response. Chemokines are a subset of cytokines with chemotactic functions. Functionally cytokines are divided into those that promote or block inflammation although some can act pro-as well as anti-inflammatory depending on the circumstances. There are 33 different ILs, 3 types of IFNs, 16 members of the TNF family, 10 other cytokines, and ~50 chemokines, the latter are grouped into four families. Each virus, depending on its composition, its immunoevasion tactics, its port of entry, and/or its cell specificity activates a unique blend of these factors that are released upon PRR signaling. Nevertheless, there are some communalities. Most viruses activate the NF-κB, IRF3 and 7, and AP-1 pathways, which result in production of IFNs, TNF, TGF-ß, granulocyte-macrophage colony-stimulating factor (GM-CSF), IL-1, IL-6, IL-10, IL-12, and IL-15. Most cytokines are produced by cell of the immune system; type I IFNs, TNF, TGF-ß, IL-1, and IL-6 can also be produced by nonimmune cells.

NF-κB is activated after viruses interact with a PRR. NF-κB in its inactive form is bound to NF-κB inhibitor (IκB) in the cytoplasm. There are two pathways that can cause NF-kB activation. In the canonical pathway that is downstream of most PRRs, IKK upon activation phosphorylates IκB leading to its degradation, which allows for translocation of NF-κB dimers into the nucleus. In the canonical pathway that is used by cells of the innate immune system, NF-κB is composed of nuclear factor NF-κB P65 subunit (RelA), C-Rel, RelB, and p50. In the alternative pathway used by cells of the adaptive immune system upon activation through a TNF family member such as CD40 or lymphotoxin, the NF-κB complex is composed of p52 and RelB. Activated NF-κB drives the transcription of over 150 target genes including ILs (IL-1, IL-2, IL-6, IL-8, IL-9, IL-11, IL-12, IL-15), IFNs (IFN-γ, IFN-ß), C-X-C motif chemokine ligand 10 (IP-10) regulated upon activation, normally T-expressed, and presumably secreted (RANTES) macrophage inflammatory protein (MIP)-1a/ß,-2, TNF-a/ß, CD69, CD95, granulocyte colony-stimulating factor (G-CSF), GM-CFS, interferon regulatory factor (IRF)-1,-2, complement factors B and C4, and others.[194]

Ligation of PRRs such as cGAS, RIG-1, or TLR3 activate STING, MAVS, or TRIF, respectively, which in turn causes phosphorylation of TBK1 leading to the activation of IRF3. MyD88 activates IKKα, which similar to TBK1 activates IRF7.[195] IRF7 is amplified by a positive feedback loop through IFNs.[196] IRF3 and 7 induce expression of multiple genes most notably IFNs and genes that are downstream thereof. TLR signaling through TAK1-mediated activation of MAPK can phosphorylate c-Jun N-terminal kinase (JNK) and ERK, which in turn activate transcription factor AP-1, a sequence-specific transcriptional activator composed of members of the Jun and Fos families. AP-1 interacts with over 2000 target genes including those producing proinflammatory cytokines.

The early cytokine response has three main objectives. Their first goal is to block viral spread. IFNs are the main group of cytokines with potent antiviral activity. IFNs are divided into three major types. IFN-Is produced by monocytes and fibroblasts and include IFN-α, IFN-ß, IFN-ε, IFN-κ, and IFN-ω (see Chapter 9). IFN-II also called immune IFN or IFN-γ is induced by IL-12 and is produced mainly by T cells. IFN-III also called IFN-λ is produced by epithelial cells.

Type I IFNs bind to a common receptor expressed on most cell types. Levels of receptor expression, which are controlled by autophagy, in part regulate the type I IFN response. Receptor engagement activates tyrosine kinase (Tyk)2 and Janus kinase (Jak)1, which phosphorylate and activate STAT proteins, which upon translocation into the nucleus then promote the expression of target genes. STAT1, STAT2, and IRF9 form the IFN-stimulated gene factor 3 (ISGF3) complex, which binds to IFN-stimulated response elements (ISREs) in the promoters of IFN-inducible genes. Type I IFNs also activate the MAPK, phosphoinositide 3-kinase (PI3K), protein kinase B (AKT), and the NF-κB signaling pathways (Fig. 44.4). Type I IFN blocks production of IL-12,[197] a potent inducer of IFN-γ.[198] Type I IFNs also induce autophagy to eliminate viral components.[199]

IFN-γ is the only type II IFN. It binds to the IFN-γ receptor complex causing activation of Jak1 and Jak2, which in turn recruit and phosphorylate STAT1. STAT1 bind to IFN-γ–activated sequence (GAS) sites in the promoters of target genes. In addition, IFN-γ signaling can activate MAPK, PI3K/AKT, and NF-κB signaling pathways (Fig. 44.4). IFN-γ signaling promotes macrophage and natural killer cell activation, upregulates the expression of antigen processing and

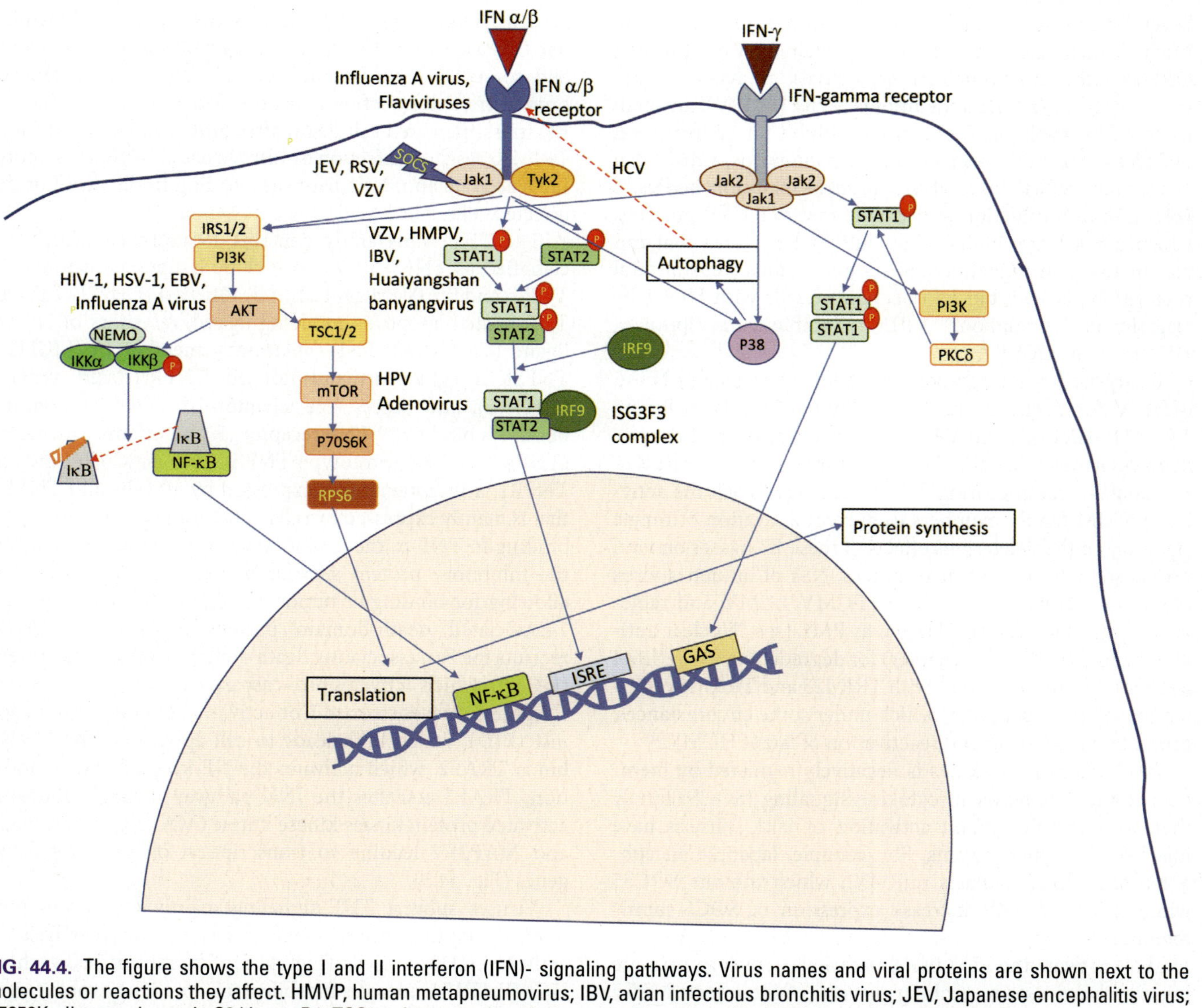

FIG. 44.4. The figure shows the type I and II interferon (IFN)- signaling pathways. Virus names and viral proteins are shown next to the molecules or reactions they affect. HMVP, human metapneumovirus; IBV, avian infectious bronchitis virus; JEV, Japanese encephalitis virus; P7056K, ribosomal protein S6 kinase B1; TSC1, tuberous sclerosis 1 protein. The other abbreviations are explained in the text.

presentation molecules, biases immune responses toward the Th1 pathway, regulates B cell functions, and induces the production of chemokines to attract lymphocytes to the sites of viral infections. IFN-γ signaling also induces through the MAPK pathway autophagy in macrophages.[200]

The type III IFN family consists of IL-29/IFN-λ1, IL-28A/IFN-λ2, IL-28B/IFN-λ3, and IFN-λ4. Type III IFNs bind to a receptor complex formed by the IL-28Rα/IFN-λR1 ligand-binding subunit and the IL-10Rß accessory chain, which is primarily expressed on epithelial cells. Upon receptor ligation, type III IFNs follow signaling pathways that are similar to those induced by as type I IFNs.

Viruses target multiple components of IFN signaling (Fig. 44.4), those involved in NF-κB signaling are addressed in Figure 44.2. The HA of influenza A viruses promotes phosphorylation and degradation of the type I IFN receptor[201] while flaviviruses such as West Nile virus inhibit the receptor's surface expression by blocking its intracellular trafficking.[202] HCV increases autophagy of the IFN-α receptor.[203] Some coronaviruses such as porcine epidemic diarrhea virus induce ubiquination and degradation of STAT1 and STAT2. Several viruses, including the paramyxovirus La Piedad Michoacán Mexico virus through protein V, VZV through ORF63, infectious bronchitis virus (IBV), an avian coronavirus, and human metapneumonia virus (hMPV), through its small hydrophobic (SH) protein inhibit phosphorylation of STAT1 or 2.[204-207] Huaiyangshan banyangvirus, a tick-born virus, traps STAT1 and STAT2 in viral inclusion bodies.[208] IFNs induce production of tripartite motif (TRIM) proteins, which are a large family of ubiquiting E3 ligases that can modulate innate signaling and inhibit various steps of viral replication. TRIM5 blocks infection of cells with HIV-1 by targeting its degradation.[209] TRIM22 inhibits transcription of HIV-1[210] and HBV.[211]

It targets the nucleoprotein of influenza virus and NSA5 of HCV for degradation.[212,213] Additional TRIMs including TRIM11, TRIM37, and TRIM28 interfere with the life cycle of retroviruses. An HIV-1 capsid binding protein TRIM11 accelerates viral uncoating.[214,215] Viruses can block the activity of TRIM family members to prevent activation of innate signaling or the deleterious effects of these E3 ligases on viral propagation. To give some examples, NS1 of influenza virus blocks the activity of TRIM25.[150] ECMV, LCMV, and rabies virus target TRIM19 also known as PML (see "Nuclear antiviral factors" part of this chapter) for degradation.[216,217] UL144 of HCMV forms a complex with TRIM23 and TRAF-6 resulting in NF-kB activation,[218] which under some circumstances, promotes viral spread and reactivation of latent HCMV.[219]

Production of cytokines is negatively regulated by members of the suppressor of cytokine signaling (SOCS) family that, for example, inhibit activation of Jak1. Viruses have hijacked these mechanisms. For example, Japanese encephalitis virus downregulates miR-432, which inhibits SOCS5 while RSV and VZV increase expression of SOCS family members.[77,220,221]

IFNs activate the PI3K/AKT pathways, which in turn are hijacked by viruses to prolong survival of the infected cells. Nef of HIV-1, NS1 of influenza A virus, US3 of HSV-1, and latent membrane protein (LMP)1 of EBV increase the activity of the PI3K/AKT.[222-225] Viruses increase the activity of mTOR through PI3K-dependent and-independent pathways to augment synthesis of their proteins. For example, E6 of HPV and E44ORF1 and 4 of adenoviruses increase AKT/mTOR signaling.[226-228]

IFN-stimulated cells are induced to express new proteins that have antiviral activities, including interferon alpha-inducible protein (ISG)15, ISG20, RNAse-L, PKR, 2′,5′-OAS, myxoma resistance protein (Mx)1, TRIM containing 5α, and others.[229,230] The antiviral activity of IFNs depends on the induction of 2′5′-OAS, which activates RNAse L that in turn degrades preferentially viral dsRNA.[231] IFN-α also induces dsRNA-dependent protein kinases, which phosphorylate numerous proteins including eIF-2, a eukaryotic translation initiation factor, which in turn blocks synthesis of viral proteins.[232] IFNs induce additional antiviral mediators, such as ISG15, which inhibits budding of HIV-1 and Ebola virus. It also blocks assembly of HIV-1 particles. It inhibits formation of the ribonucleocapsid of influenza A virus. ISG20 has antiviral activity against RNA viruses including HCV, hepatitis A virus (HAV), and yellow fever virus. Mx1 is a dynamin-like GTPase with antiviral activity against many RNA viruses and some DNA viruses. It can bind and thereby prevent proper trafficking of viral ribonucleocapsids. It enhances ER stress-mediated cell death after influenza virus infection. TRIM5a restricts infection by retroviruses.[229,230] Furthermore, IFNs activate apoptotic pathways leading to the death of the infected cells.

The TNF superfamily contains 19 members which are classified as TNFSF1-19 and include lymphotoxin α and ß, TNF-α and ß, OX40 ligand (L), CD40L, FasL, CD27L, CD30L, TNF-related apoptosis-inducing ligand, receptors of NF-κB ligand (TRANCE), TNF superfamily member 14 (LIGHT), TNFSK18 (glucocorticoid-induced TNFR-related protein [GITR]L), and others (see Chapter 10). TNF superfamily members bind to 29 TNF receptor (R) superfamily members (TNFR1-29). The prototype TNF is TNF-α, which binds to TNFR1, a receptor that is expressed by all cells and TNFR2 that is mainly expressed by cells of the immune system. Upon binding to TNF-α, the TNFR forms a trimer and dissociates the inhibitory protein silencer of death domains (SODD) allowing for binding of tumor necrosis factor–receptor type 1–associated death-domain protein (TRADD). TRADD recruits the Fas-associating death domain–containing protein (FADD), which activates pro–caspase 8 into active caspase 8. Caspase 8 cleaves caspase 3 or activates apoptotic death agonist (BID), which both leads to cell apoptosis. TADD also binds TRAF2, which activates the NF-κB pathway. In addition, TRAF2 activates the JNK pathway through mitogen-activated protein kinase kinase kinase (MAP3K) 5, MAP3K7, and MAP2K7 leading to transcription of AP-1–regulated genes (Fig. 44.5).

Viruses subvert TNF signaling mainly to prevent premature apoptotic death of infected cells[233] or their lysis by activated CD8+ T cells (Fig. 44.5). Several viruses, such as HCMV, HIV-1, measles morbillivirus, or adenoviruses target death receptors of the TNF family such as Fas (CD95) and

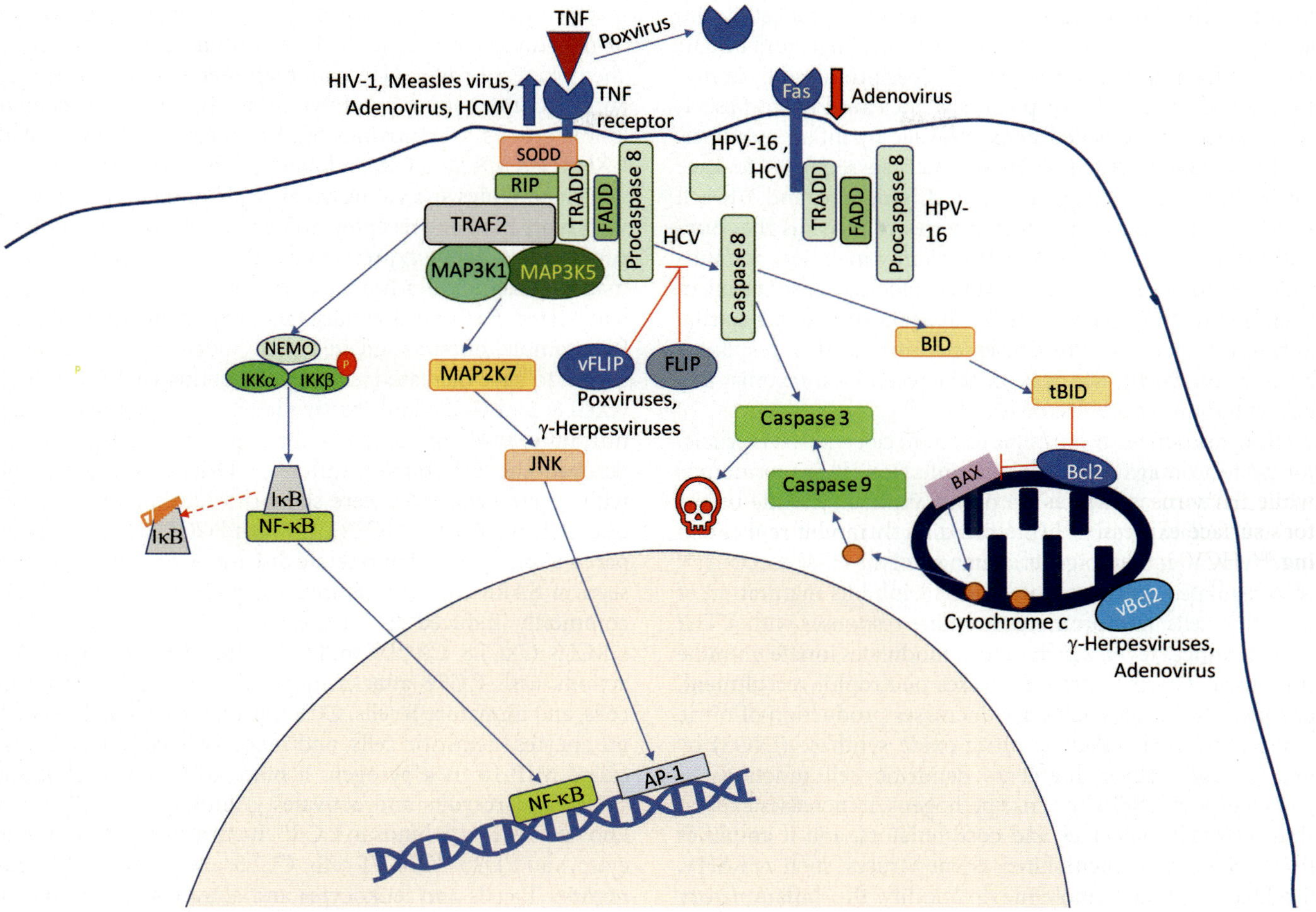

FIG. 44.5. The figure shows tumor necrosis factor (TNF)- signaling pathways. Virus names and viral proteins are shown next to the molecules or reactions they affect. Thin blue lines indicate activation, thin red lines reflect inhibition. Thick blue lines indicate that viruses affect an increase. Thick red lines indicate that viruses affect a decrease. MAP3K, mitogen-activated protein kinase kinase kinase; tBID, C-terminal portion of BID; v-Bcl2, viral BCL2. The other abbreviations are explained in the text.

TRAIL.[234-237] Poxviruses produce TNFR2 orthologues that soak up the corresponding ligands.[238] Polypeptides encoded by E4 of adenovirus downregulate expression of Fas and TRAIL-R1 and 2 from the cell surface to allow for cell survival.[233,239] Poxviruses and gammaherpesviruses encode viral caspase-like apoptosis regulatory protein (FLICE)/caspase-8 inhibitory proteins (vFLIP), which inhibit activation of caspases.[240,241] The E5 protein of HPV-16 decreases formation of the death inducing signaling complex composed of the receptor FADD and caspase 8, while E6 of HPV-15 and NS5A of HCV bind to the cytoplasmic tail of TNFR1, thus preventing binding of TRADD.[242-244] HCMV through UL36 and poxviruses through serpins block activation of caspases.[245,246] Viruses target mitochondrial death pathways that are triggered by members of the TNF superfamily. Ligation of TNFR or other apoptotic signal leads to translocation of proapoptotic BCL2 family members such as BID to the mitochondrial membrane. This blocks BCL2 and allows for the activity of BCL2-associated X (BAX), apoptosis regulator resulting in release of cytochrome c which causes activation of caspase 9, then caspase 3, and apoptosis. Gammaherpesviruses and adenoviruses encode BCL2 orthologues.[247,248]

Type I IFNs and TNF act synergistically to suppress viral spread through a STAT2 and IRF9-dependent pathways.[249] Other cytokines can enhance the antiviral activities of IFNs. Combinations of TNF, IL1β, type-I IFN, and/or type-II IFN inhibit replication of VZV,[250] HSV-1,[251] EBV,[252] and SARS-coronavirus.[253] TNF in addition activates NF-κB and MAPK; it furthermore has albeit weak apoptotic activity.

The second goal of the early cytokine response it to recruit and activate cells of the innate immune system. Type I IFNs and TNF-α promote activation of macrophages, dendritic cells, and NK cells. TGF-β induces chemotaxis of neutrophils[254] and eosinophils.[255] GM-CSF activates myeloid cell differentiation and increases output of neutrophils and monocytes from bone marrow. GM-CSF recruits neutrophils and enhances their functions and survival. It promotes production of natural IgM by B1-B cells.

The IL-1 family includes 11 members, which can bind to 11 different receptors.[256] IL-1 induces fever.[257] IL-1 family members increase expression of adhesion molecules on endothelial cells, thereby facilitating leukocyte attachment. They augment the lifespan and effector functions of neutrophils and macrophages.[258] IL-1α is produced in an immature form by most cells. The IL-1α protein, if released upon cell death, acts similar to a PRR as an alarm signaling danger and initiating an inflammatory response.[259] IL-1β is produced by hematopoietic cells. The IL-1β precursor is cleaved by caspase-1, which becomes activated upon cleavage by inflammasomes. IL-18 is closely related to IL-1β and also requires cleavage by caspase-1. IL-18 plays a role in driving Th1 and NK cell responses. Poxviruses produce a series of proteins that bind and inactivate IL-18.[260] Other IL-1 family members are anti-inflammatory such as IL-37, which blocks transcription of proinflammatory genes.[261] IL-38 appears to act as a partial antagonist of IL-36. Poxviruses circumvent IL-1 responses by secreting B16R, which is an IL-1 receptor antagonist and blocks inflammatory responses.[262]

IL-6 is produced by multiple different cell types. It is crucial for protection against some viral infection. It has numerous functions some of which are proinflammatory while others are anti-inflammatory.[263] IL-6 can directly inhibit replication of HBV by affecting epigenetic changes of the HBV cccDNA.[264] IL-6 regulates expression of GM-CSF, inhibits maturation of dendritic cells, and promotes immune responses with a Th2 bias. It induces fever, which in turn modulates innate immune responses. Thermal stress increases neutrophil recruitment, enhances NK cytotoxicity, and increases production of nitric oxide (NO) and inducible nitric oxide synthase (iNOS) by macrophages. Fever augments dendritic cell functions; it enhances their ability to sense pathogens, it increases expression of MHC molecules and costimulators, and it amplifies responsiveness to chemokines. Some viruses, such as KSHV, produce an IL-6 homologue to modify the inflammatory response.[264]

IL-10 can be part of the early antiviral immune response. It is produced by macrophages and dendritic cells, and it serves to inhibit inflammatory responses to prevent excessive tissue damage.[265] It also suppresses Th1 in favor of Th2 responses. Poxviruses and betaherpesviruses express IL-10 homologues to suppress innate immune responses and inhibit induction of Th1 type adoptive immunity.[266,267]

The IL-12 family consists of five members (IL-12, IL-23, IL-27, IL-35, and IL-39), which are secreted by antigen-presenting cells. All IL-12 family members are composed of two subunits, which can be shared between family members; for example, both IL-12 and IL-23 used the IL-12ß subunit. IL-12 is a potent inducer of IFN-γ while IL-23 inhibits IFN-γ production. IL-12 drives adaptive immune responses toward a Th1 pathway. IL-27 increases IL-10 production and thereby acts as an anti-inflammatory cytokine.[268] IL-15 is crucial for the development, maturation, maintenance, and survival of NK cells. It promotes migration of NK cells to sites of inflammation by increasing expression of adhesion molecules and chemokine receptors, and it augments their cytolytic functions.[269] IL-15 augments the phagocytotic activity of monocytes and macrophages and induces production of chemokines such as IL-8 and monocyte chemotactic protein (MCP)-1.[270] IL-15 increases expression of MHC molecules and costimulators on dendritic cells.[271,272]

Chemokines play a role in recruiting leukocytes to sites of infection (see Chapter 11). According to their structure, chemokines are divided into four families with two major (CC and CXC chemokines) and two minor (CX_3C and C chemokines) groups. Chemokines bind to receptors termed CCR, CXCR, CX_3CR, and CR, and many of the receptors bind several chemokines and *vice versa* most chemokines can interact with more than one receptor. Some chemokines are produced constitutively while expression of other is induced by inflammation. Chemokines function through a gradient to attract leukocytes. Each virus induces its own chemokine mixture; for example, humans infected with Sudan Ebola virus were shown to have increased levels in circulating CCL5, CXCL10, CXCL8, and CCL4 and the two latter chemokines were significantly lower in survivors than in individuals who succumbed the infection.[273] Influenza virus–infected patients with severe symptoms were shown to have increased levels of CXCL10 (IP-10), CXCL8 (IL-8), and CCL2 (MCP-1) compared to patients with moderate disease. A similar profile was seen in SARS-coronavirus infected patients.[273] Chemokines commonly induced by viruses are CCL1-9 and CXCL1, CXCL3, CXCL8, CXCL9, and CXCL10. CCL1 through interactions with CCR8 attacks monocytes, NK cells, dendritic cells, and immature B cells. CCL2, also called MCP-1, recruits monocytes, dendritic cells, and memory T cells. CCL3, also called MIP-1α, is a pyrogen. It binds to CCR1, CCR4, and CCR5 and recruits and activates granulocytes. CCL4, also known as MIP-1ß, binds to CCR5. It attracts NK cells, monocytes, and CD25+CD4+ T cells. CCL5, also termed RANTES, recruits T cells and leukocytes and activates subsets of NK cells. It interacts with CCR1, CCR3, and CCR5. CCL6 is a chemokine only present in rodents; it attracts macrophages, B cells, CD4+ T cells, and eosinophils. CCL7, called MCP3, attracts eosinophils and monocytes and enhances monocyte functions. CCL8, also called MCP-2, attracts monocytes, lymphocytes, basophils, and eosinophils. CCL9, also called MIP-1γ, is a mouse chemokine that attracts dendritic cells through binding to CCR1. CXCL1 and 3, the latter also called MIP-2, recruit neutrophils. CXCL8, called IL-8, recruits neutrophils, basophils, and T-cells and is involved in neutrophil activation. CXCL9, also called monokine, induced by IFN-γ (MIG), recruits NK cells, NKT cells, macrophages, and CD8+ T cells. CXCL10, called IP-10, is a chemoattractant for T cells, monocytes/macrophages, dendritic cells, and NK cells. Several of the chemokines affect angiogenesis by recruiting proangiogenic immunocytes or endothelial progenitors or by directly affecting endothelial cell functions, thereby changing the microenvironment of an inflamed tissue. Proangiogenic chemokines include CXCL1, CXCL2, CXCL3, CXCL5, CXCL6, CXCL8, and CCL2, while CXCL4, CXCL9, and CXCL10 inhibit vessel formation.[274] Antiangiogenic chemokines function through binding to CXCR3, which is expressed by endothelial cells and, upon ligand binding, inhibits cell proliferation and migration.

Large DNA viruses, such as poxviruses and herpesviruses, evade chemokine signaling by mimicking chemokines, chemokine receptors, or by modulating the activity of chemokines through secretion of chemokine-binding proteins or soluble receptors. Antagonist activity reduces recruitment of inflammatory cells that might harm the virus. Viral chemokine homologues can function as agonists or antagonists or as both depending on their target receptor. Agonist activity helps viruses that specifically infect leukocytes by broadening numbers of their potential victims. Poxviruses encode homologues of CC chemokines[275] and of CCR8.[276] They also secrete chemokine-binding proteins that can either bind just to CC chemokines[277] or that through ligation to the heparin-binding domain of chemokines can inhibit C, CC, and CXC chemokines.[278] Poxviruses produce a 35k-Da protein that serves as a receptor for CC chemokines.[279]

An analysis of the herpesvirus genome showed several ORFs with sequence similarities to chemokines, chemokine receptor, or chemokine-binding proteins. This has been studied most intensively for murine gammaherpesvirus (MHV)-68 which produces a protein than can bind CCL3, CCL5, CXCL8, and CX3CL1.[280] The protein is encoded by the M3 gene, which is unique to MHV68. KSHV produces CCL4, CCL2, and CCL20 homologues. HCMV secretes UL21.5, which binds with high affinity to CCL5.[281] Virus chemokine binding proteins described here thus far inhibit the activity of their targets. Alphaherpesvirus gG acts differently: it binds and enhances the activity of CXCL12, CXCL13, and CXCL28.[282] HSV gG modifies the chemokine receptor localization on the cell membrane by promoting clustering, signaling, and internalization. The notion that viruses enhance the activity of chemokines seems counterintuitive unless it allows them to attack cells. Nevertheless, some of the chemokine-signaling pathways change characteristics of the cell membrane or enhance transcription and translation machineries, which can facilitate viral entry or replication, respectively.

The third goal of the early innate cytokine response is to assist in activating antigen-driven adaptive T cell responses, as discussed below in the adaptive immunity sections.

Cells of the Innate Immune System

Cells of the innate immune system, which include neutrophils, eosinophils, basophils, mast cells, macrophages, dendritic cells, NK cells, NKT cells, and γ/δ T cells, provide a first layer of defense against virus infections and promote activation of adaptive immunity. Cells of the innate immune system are activated by PRRs or cytokines, and their response is not antigen-specific. They act rapidly without undergoing the massive proliferation of T and B cells. With the exception of some NK cells and macrophages, cells of the innate immune system fail to establish long-lasting memory.

Granulocytes

There are four types of granulocytes called neutrophils, eosinophils, basophils, and mast cells. Neutrophils constitute 60% to 65% of white blood cells; eosinophils and basophiles are less common in blood with 2% to 3% and 0.4% to 0.5%, respectively. Mast cell reside mainly in tissues. They all contribute to the early antiviral defense response (see Chapters 14, 16, and 17).

Neutrophils are rapidly recruited to sites of infection through proinflammatory cytokines, chemokines, and complement components, and they can limit the viral invasion; animals infected with a highly pathogenic strain of influenza A virus develop significantly higher lung virus titers if neutrophils are depleted from the circulation.[283] Neutrophils are in general short-lived, but their lifespan increases in presence of proinflammatory cytokines such as IFN, TNF, or G-CSF.[284] Neutrophils use several pathways to limit viral spread. They can phagocytose virus particles or infected cells and degrade them with proteolytic enzymes within their cytoplasm or destroy their genomes with ROS.[285-287] Neutrophils release antimicrobial factors, such as defensins, from their granules into the surrounding tissue. Defensins inactivate viruses, such as HIV-1 or herpesviruses, upon binding to lipid components of their membranes.[288,289] Cathelicidins are host defense peptides with direct antiviral activity against rhinoviruses,[290] RSV,[291] Zika virus,[292] and others. They may also damage uninfected cells thus contributing to the early immunopathology of virus infections. Neutrophils with the help of reduced nicotinamide adenine dinucleotide phosphate (NADPH) oxidase produce ROS, which is released and causes damage to viral genomes. Neutrophils secrete neutrophil extracellular traps (NETs), which are composed of cellular DNA, fibers, and antimicrobial proteins such as proteases, defensins, and histones. NFTs can directly kill viruses, or they can prevent them from infecting cells.[293] Overproduction of NFTs may cause pathology; for example, they were shown to contribute to airway obstruction in children infected with RSV.[293] Pathogens, such as herpesviruses or poxviruses, evade NFTs through direct secretion of IL-10 homologues, while HIV-1 induces cellular IL-10 production through binding of its envelope protein to DC-SIGN. IL-10 in turn blocks TLR-induced production of ROS, which is essential for the release of NETs.[294,295] KSHV through LANA1 inhibits the activity of IL-8, which stimulates NET formation. Under some circumstances, neutrophils promote viral spread. Neutrophils recruited to the focal point of a West Nile virus infection become targets for the virus and then transport it to other parts of the body.[296]

Eosinophils play a role in respiratory infections, such as with influenza viruses or RSV. They kill viruses through release of lipid mediators including eicosanoids, leukotrienes, and prostaglandins. They also release ROS and granules containing antimicrobial proteins, such as galectin 10, a lysophospholipase, which hydrolyzes lysophosphatidylcholine to glycerophosphocholine and a free fatty acid, major basic protein, which triggers release of histamine and activates neutrophils and alveolar macrophages, eosinophil cationic protein, and eosinophil-derived neurotoxin, which are two ribonucleases that can directly kill viruses. Eosinophils also release growth factors and cytokines including IL-1, IL-6, IL-8, and IL-13 and TNF-α.

Basophils release histamines and thereby play a role in allergic reactions. Some viruses produce factors that inhibit basophil functions indicating that they play a role in the antiviral defense.[297]

Our knowledge on the role of mast cells during the early stages of a virus infection remains limited. They were shown to be beneficial as well as detrimental to dengue virus–infected hosts; they slow the infection by contributing to the recruitment of other immunocytes, but they also enhance vascular leakage by secreting serotonin, which causes thrombocytopenia and by releasing chymase and leukotrienes which increase vascular permeability; this can be fatal in some patients.[298]

Monocytes/Macrophages

Macrophages are distributed throughout the body (see Chapter 15). They are very abundant, and it is estimated that an adult human has about 10^{10} macrophages. They are named depending on their location; Kupffer cells reside in liver, microglia cells are in the brain, osteoclasts are in bone, alveolar macrophages are in lung, etc. Microglia are derived from yolk sac while other macrophages such as those in skin or liver are derived from fetal liver progenitors; they have the capacity of self-renewal, or they can be replenished from bone marrow–derived cells. At other sites, such as the gut, lymphatic tissues, and blood, macrophages are derived exclusively from bone marrow. Macrophages destroy viruses and remove dead or dying cells by phagocytosis and intracellular degradation. They are also involved in inflammatory reactions and play a role in antigen presentation. Macrophages come in two flavors: proinflammatory M1 cells, also called classical macrophages, and anti-inflammatory M2 cells, called alternative macrophages.[299] Macrophages are plastic and with the appropriate stimulus can switch from an M1 to an M2 phenotype and *vice versa*. For examples, during an acute HCV infection, M1 macrophages secreting IFN-γ and IL-12 are elevated, while in chronically infected patients M2 macrophages become more abundant potentially contributing to the patients' dysfunctional antiviral T cell responses.[300] TNF-α and IFN-γ drive differentiation toward the M1 pathway while IL-4, IL-10, or IL-13 promote the development of M2 macrophages. M1 macrophages kill pathogens and produce the proinflammatory cytokines TNF-α, IL1-β, IL-6, IL-12, and IL-23. They express CD80, CD86, CIITA, MHC class II molecules, cyclooxygenase 2 (COX-2), and iNOS and promote Th1 responses. M2 macrophages express high levels of the mannose receptor CD206. They produce arginase, metalloproteinase, vascular endothelial growth factor A (VEGF-A), endothelial growth factor (EGF), platelet-derived growth factor (PDGF), and IL-8. They thereby promote angiogenesis, lymphogenesis, and wound healing while dampening inflammatory reactions.[301] M2 macrophages express higher levels of the coinhibitor programmed cell death (PD)-1, which was shown to be linked to decreased MHC class II expression and reduced phagocytosis.[302] The metabolism of the two macrophage subsets differs.[303] M1 macrophages use mainly glycolysis while oxidative phosphorylation (OXPHOS) is impaired leading to an accumulation of ROS, which has antimicrobial activity. Due to downregulation of isocitrate dehydrogenase (IDH) and upregulation of the mitochondrial citrate carrier (CIC), citrate is withdrawn from the Krebs cycle and converted in part to acetyl-CoA used for fatty acid synthesis, or it is used to produce itaconate, which through activation of nuclear factor erythroid 2-related factor 2 (Nrf2) dampens inflammation. M2 macrophages rely on OXPHOS for energy production. M1 and M2 macrophages also differ in their amino acid metabolism. In M1 macrophages, arginine is converted to citrulline and NO, the latter reacts with oxygen or ROS to form NO_2, which has antimicrobial activity. M2 macrophages metabolize arginine to ornithine and from there to polyamines, which play a role in tissue repair. They increase the activity of indoleamine 2,3-dioxygenase (IDO), which converts tryptophan to kynurenine, a potent inhibitor of T cell activation. They also increase the activity of glutamine synthetase (GS), which converts glutamate to glutamine, which is then metabolized into uridine diphosphate N-acetylglucosamine (UDP-GlcNAC) needed for increased expression of CD206.[292] Macrophages seem to have memory. Once activated, they can react to a second challenge either with tolerance, when their response is attenuated, or they show increased activity. The mechanisms underlying this innate memory are not yet fully understood; the populations of macrophages may have shifted upon the first insult or macrophages may have undergone lasting epigenetic changes.[304] Although macrophages play a dominant role in the early antiviral immune response, they can serve as reservoirs for some viruses such as Chikungunya virus[305] or HIV-1.[306]

Viruses can affect different stages of macrophage differentiation and functions. They can block macrophage activation by interfering with PRR signaling as described above. EBV through secreted (s) BARF1 inhibits the activity of macrophage colony-stimulating factor (M-CSF), which drives macrophage differentiation. Macrophages activated in presence of sBARF1 show reduced expression of cell surface differentiation markers, produce lower levels of ROS, and their ability to phagocytose dead cells is diminished.[307] HIV-1 drives macrophages toward an M1-like phenotype; nevertheless, these cells differ from typical M1 macrophages by showing reduced antimicrobial activity and by inducing stronger inflammatory responses.[308] HCV core protein through interaction with the complement factor C1q increases the expression of PD-1 and SOSC-1 on/in macrophages, which impairs their ability to produce cytokines such as IL-12, to present antigen to CD4+ T cells or to phagocytose virus particles.[309] Measles morbillivirus drives a Th2 immune response mainly through its effects on antigen-presenting cells. Ligation of measles morbillivirus to the CD46 receptors decreases IL-12 responses[310] driving adaptive immune responses toward the Th2 pathway. HSV-1 infection increases expression of FasL on the surface of infected macrophages potentially allowing for killing of HSV-1–specific T cells.[311] Japanese encephalitis virus inhibits production of cytokines and interferes with antigen presentation by downregulating histone deacetylase 3, thus affecting inflammatory responses through epigenetic changes.[312]

NK Cells

NK cells respond rapidly upon activation by PAMPs or proinflammatory cytokines and are crucial for the defense against mainly DNA viruses as shown by the increased susceptibility of NK-deficient individuals to infections with herpesviruses, papillomaviruses, adenoviruses, HBV, and others (see Chapter 18).[313-317]

NK cells are a very heterogeneous population of innate leukocytes. Their diversity is provided by the expression of combinations of germline-encoded highly variable inhibiting or activating killer Ig-like receptors (KIR). Inhibitory receptors include KIR2DL1, KIR2DL2, KIR2DL3, KIT3DL1, KIR3DL2, KIR3DL3, KIR2DL5A, and B. They have a long cytoplasmic tail with an ITIM motif. Activating KIRs include KIR2DL4, KIT2DS1-5, and KIR3DS1. Activating KIRs with the exception of KIR2DL4, which plays a role in pregnancy, have a short cytoplasmic tail that binds DAP-12. DAP-12 has an ITAM motif, which leads through additional kinases to cell activation. Inhibitory receptors mainly recognize MHC class I molecules so that they can block responses to normal cells. Downregulation of MHC molecules, as observed upon infections with some viruses, results in loss of binding of inhibitory KIRs and thereby triggers activating KIRs. NK cells express Fc receptors that can bind to antibody-coated cells and then kill them in a process called antibody-dependent cell-mediated cytotoxicity (ADCC). CD2 expressed by all NK cells is an essential NK cell costimulator for antibody-mediated NK functions.[318] NK cells inhibit viral spread by directly killing some viruses, such as herpesviruses, influenza viruses, or dengue virus, upon recognition of their surface glycoproteins through natural cytotoxicity receptors (NCRs) such as NKp46, NKp44, and NKp30.[319,320] They lyse infected cells through the release of perforin and granzyme B, or once infected cells are coated by virus-specific antibodies, they can destroy them by ADCC, which plays a major role in reducing viral loads.[321] To name two examples; in the only large-scale HIV-1 vaccine trial that showed any efficacy in preventing virus acquisition, ADCC was identified as the most plausible correlate of protection.[322] Non-neutralizing antibodies to the RSV glycoprotein were shown to protect mice against an RSV infection by triggering clearance of infected cells through ADCC.[323] NK cells secrete proinflammatory cytokines including IFN-γ, TNF, IL-12, IL-15, IL-18, IL-2, CCL5, and others and modulate adaptive immune responses. NK cells can suppress activation of T cell responses by producing IL-10 and by killing activated CD4+ or CD8+ T cells.[324-326] They can accelerate development of a CD8+ T cell response by limiting type I IFN production by plasmacytoid dendritic cells.[327] In other virus infections, such as with LCMV, NK cells inhibit T cell responses through the secretion of type I IFNs.[328] NK cells can kill immature but not mature dendritic cells as the former express only low levels of MHC class I antigens. NK cells, similar to macrophages, may be able to develop memory as was shown in mice infected with mouse (M) CMV. Memory is provided by epigenetic changes in the IFNG locus allowing for increased responses upon a repeat encounter with the virus.

A number of viruses, including poxviruses, herpesviruses, retroviruses, flaviviruses, and papillomaviruses, have developed pathways to evade the activity of NK cells.[329] As NK cells react to cells that express unusually low levels of MHC class I antigens, which can be triggered by viral infections, some viruses can express MHC class I homologues or selectively modulate expression levels of MHC class I molecules on the surface of infected cells to prevent killing by NK cells. Viruses can also inhibit the functions of activating receptor, produce viral cytokine-binding proteins, cytokine receptor antagonists, or directly reduce NK cell responses. MCMV encodes an MHC class I homologue called m14 which renders cell resistant to NK cell–mediated lysis.[330] Some viruses selectively downregulate MHC class I molecules leaving those that are the least suited to present antigen to CD8+ T cells, such as HLA-G and HLA-C, intact to still allow for inhibition of NK cell activity.[331] Other viruses accelerate recycling of MHC class I molecules from the cell surface and again this process seems to be selective. For example, nef of HIV-1 reduces expression of HLA-A and HLA-B but not HLA-C or HLA-E so that cells remain resistant to NK cell–mediated cytolysis.[332] K3 and K5 of KSHV cause rapid endocytosis of MHC class I molecules with E5 being selective while K3 affects all HLA antigens.[333] The HCMV UL40 protein increases expression of HLA-E, which again may block NK cell activity as remains to be investigate in more detail.[334] HCMV downregulates lymphocyte function–associated antigen (LFA)-3, thus preventing activation of the costimulator CD2.[335] MCMV m152 inhibits surface expression of NKG2D.[335]

Viral proteins can block ligand binding to activating receptors such as UL16 of HCMV that binds to NKG2D. HIV-1 uses a different mechanism. Its tat protein binds to a calcium channel blocking calcium influx and thereby NK cell–mediated cytolysis.[336] Viral homologues of cytokines or chemokines described above can affect NK cell activation and migration. Viruses can directly bind or infect NK cells and thereby affect their functions or even survival. HIV-1 can infect and kill NK cells.[337] E2 of HCV binds to CD81 on NK cells, which inhibits NK cell–mediated lysis, proliferation, and cytokine production.[338] Herpesviruses of subfamilies α and ß encode glycoproteins, which are expressed on the cell surface and on the viral membrane, that can bind to the Fc domain of antibodies thus serving as decoy Fc receptors.[339] More specifically, HSV encodes gE and gI which upon forming a complex bind to antibodies; if the antibody is HSV-specific, its variable region will in addition bind to a cell surface expressed antigen. The entire complex is then endocytosed, and upon reaching the late endosomes, which has a low pH, the antibody will dissociate from gE and gI; the latter are recycled back to the surface while the antibody is degraded within the lysosomes.[340] HCMV encodes four viral Fcγ receptors, which differ in their specificity for different IgG subclasses. Their mode of action is similar to that of HSV's gE/gI.[341-343]

Invariant (i)NKT Cells

iNKT cells have characteristics of both T cells and NK cells. They can be positive or negative for NK1.1, CD4, or CD8. They carry an invariant α/β chain T cell receptor (TCR) that recognizes lipids and glycolipids presented by CD1d. Based on expression of transcription factors that are known to play major roles in the differentiation of T helper cells, iNKT cells are divided into three subpopulations. iNKT1 cells are promyelocytic leukemia zinc-finger protein $(PLZF)^{low}$ and T-box transcription factor 21 $(T\text{-}Bet)^{high}$; they produce mainly IFN-γ. iNKT2 cells are $PLZF^{high}$ and transacting T cell–specific transcription factor $(GATA3)^{high}$; they produced IL-4.

iNKT17 cells are PLZF^{intermediate} and retinoid-related orphan receptor (ROR)γ^{high} and produce IL-17.[344] An iNK follicular helper T (Tfh) cell population that expresses BCL6 and produces IL-21 and promotes B cell responses has been identified in lymphatic tissues.[345,346] iNKT cells are further divided into type I and II iNKT cells, depending on their receptor usage and their antigen specificity. Type I iNKT cells recognizes alpha-galactosylceramide (αGalCer), a synthetic glycolipid, while type II iNKT cells respond to sulfatide, lysosulfatide, lysophosphatidylcholine; both are presented by CD1d. Upon activation, iNKT cells secrete a diverse array of cytokines and chemokines, they can mediate cell lysis, and they can promote activation of dendritic cells and T cells.[347] iNKT cells are involved in resistance to a number of viruses including HSV-1 and-2, CMV, HIV-1, HBV, HCV, LCMV, and encephalomyocarditis virus.[348] They may contribute to virus-associated immunopathology upon infections with RSV.[349]

Viruses evade iNKT cells by interfering with antigen presentation by CD1d or by affecting iNKT cells directly. Many viruses upon binding to a PRR induce an IL-12 response, which triggers apoptotic death of iNKT cells.[350] HIV-1 directly infects and kills iNKT cells. Several viruses affect levels of CD1d on antigen-presenting cells. LCMV reduces CD1d expression.[351] Vaccinia virus through B1R and H5R and HSV-1 through VP22 and US3 inhibit antigen presentation by CD1d.[352-354] In addition, HSV-1 increases recycling of CD1d thus reducing its expression on the cell surface.[355] Nef of HIV-1 binds to CD1d, increases its internalization and retention in the trans-Golgi network.[356] E5 of HPV interacts with calnexin, a chaperone that ensures proper folding of CD1d; the incorrectly folded protein is then retained within the ER.[357]

Dendritic Cells

Dendritic cells are essential to activate antiviral T cell responses by presenting antigen in an immunogenic fashion to naïve T cells (see Chapter 34). Dendritic cells are derived from bone marrow precursors. Their differentiation and functions are affected by a number of growth factors such as GM-CSF, M-CFS, and Fms related tyrosine kinase 3 (Flt3). Depending on the surrounding conditions, dendritic cells differentiate into three major subsets: plasmacytoid DC (pDC), myeloid/conventional DC1 (cDC1), and myeloid/conventional DC2 (cDC2). The latter subset is rare and may play a role in fighting wound infections. The different dendritic cell subsets express distinct transcription factors and surface molecules; pDCs carry Ig transcription factor 2 (E2-2), zinc-finger E-box–binding homeobox 2 (ZEB2), IRF8, and IRF4; they express CD123, CD303, and CD304. cDC1 cells are positive for inhibitor of DNA binding 2 (ID2), IRF8, and basic leucine zipper ATF-like transcription factor 3 (BATF3); they express C-type lectin domain containing 9A (CLEC9A), cell adhesion molecule 1 (CADM1), X-C motif chemokine receptor 1 (XCR1), BTLA, CD26, and CD226. cDC2 cells use ID2, ZEB2, IRF4, and notch receptor 2 (Notch2), and express CD2, feline sarcoma oncogene (FCER1), signal regulatory protein alpha (SIRPA), and leukocyte Ig-like receptor A2 (ILT1).

Dendritic cells can also be distinguished based on their anatomic location. Langerhans cells have characteristics of both macrophages and dendritic cells. They are in the epidermis of the skin and the epithelia of the respiratory, digestive, and urogenital tracts. In a quiescent stage, Langerhans cell promote expansion of regulatory T (T_{reg}) cells. Intestinal dendritic cells in a healthy individual constantly monitor the intestinal lumen for pathogens while maintaining tolerance to the intestinal microbiome. Upon activation, they migrate to gut lymphoid tissue where they activate T cell responses. During this process, they secrete alltrans retinoic acid (RA), which induces expression of gut-homing surface molecules such as α4ß7 and CCR9 on responding T cells. Dendritic cells can further be divided into immature and mature cells. Immature dendritic cells are dispersed through an organism. They may undergo homeostatic maturation and migrate to draining lymph nodes. These partially matured dendritic cells induce tolerance rather than T cell activation. Full maturation of dendritic cells is driven by the engagement of PAMPs with PRRs. Dendritic cell subsets differ in their expression of PRRs; mDC express TLR1-8 and 10, pDCs express TLR7, 9, and 10, very low levels of TLR1, but fail to express TLR2, 3, or 8.[358] Maturation can also be induced by proinflammatory cytokines or other cellular stress signals such as by inadvertent shaking of a tissue culture flask harboring immature dendritic cells *in vitro*.[359]

Before maturation, dendritic cells can take up viruses, which facilitates engagement of PAMPs to intracellular PRRs. Upon receiving maturation signals, dendritic cells stop phagocytosis and endocytosis. They start secreting chemokines and thus contribute to the inflammatory response at the site of infection. Cytokine secretion differs between the different types of dendritic cells and is furthermore affected by the type of maturation signal. The differential cytokine production in turn affects the flavor of the ensuing T cell responses; cDCs can produce IL-12 and IL-18 to promote Th1 responses, IL-23 for stimulation of Th17 responses, or TGF-ß for T_{reg} cell activation. pDCs upon maturation release capacious amount of type I IFNs. Maturing dendritic cells upregulate expression of CCR7 and migrate from tissue to the T cell–rich zones of draining lymph nodes. Dendritic cells increase expression of CD40. Signaling through CD40 facilitates full maturation of dendritic cells. CD40 is upregulated on cDCs upon interactions with TLR4 and on pDCs upon TLR7 and 9 signaling.[103] CD40 stimulation causes expression of CD40L, which is induced through TRAF6 and results in activation of MAP and Jun kinases and in production of CD86 and cytokines, such as IL-6 and IL-12. CD40 ligation also results in activation of the NF-κB pathway. Ligation of CD40L promotes the survival of dendritic cells, which, once matured, generally die after a few days.

Once cDCs have matured, they increase synthesis of molecules that are involved in antigen processing and presentation such as proteases, transporters, and MHC class I and II molecules (see Chapter 20). Maturing dendritic cells increase expression of cell surface molecules that regulate immune responses. Such molecules include costimulators such as CD80, CD86, CD40, inducible T-cell costimulator (ICOS) ligand, and ligands for immunoinhibitors such as ligands 1 and 2 for PD-1 (PD-L1, PD-L2), HVEM, and others. Interactions of T and B cells with costimulators in addition to ligation to their cognate antigen is crucial for the induction of potent adaptive immune response while interactions with coinhibitors ensure that immune responses do not become overwhelming. Phagocytosis of virus particles to allow for antigen presentation puts dendritic cells at risk to become infected themselves. A human dendritic cell subset that expresses high levels of CD141 was shown to be resistant to infection by constitutive expression of Ras-related protein Rab-15 (RAB15), a vesicle-trafficking protein that seems to prevent viral escapes from endosomes.[358]

In general, antigen presentation to CD8+ T cells involves proteolytic digestion of newly synthesized proteins that are either misfolded or whose synthesis was terminated prematurely. The resulting peptides are then transported actively by the transporter associated with antigen processing (TAP) into the ER where they associate with MHC class I molecules that are then transported to the cell surface. Cross-presentation is a pathway that allows dendritic cells that cannot be infected by a given virus to nevertheless present viral antigens to CD8+ T cells.[360] Cross-priming is described in more details in the CD8+ T cell section.

Viruses subvert dendritic cell function through various mechanisms. Those that target PRR activation or cytokine responses are described above while those that subvert antigen presentation are described below in the "Adaptive Immunity" sections. Nevertheless, some of the viral evasion strategies are specifically designed to block dendritic cell functions and some examples are described below.

Viruses can affect numbers of circulating dendritic cells by either trapping them at sites of inflammation, by reducing their output from bone marrow, or by causing their death.[361-363] They may affect cytokine production by dendritic cells in part by dysregulating PAMP-mediated activation pathways.[364] Two glycoproteins of human metapneumonia virus, that is, G and SH block uptake of the virus by myeloid dendritic cells.[365] Adenovirus of human serotype 3 was shown to use CD80 and CD86 as cell attachment receptors allowing them to gain entry into otherwise infection-resistant dendritic cells.[366] Several antigen encoded by Ebola virus cooperate to block dendritic cell maturation.[367]

Follicular Dendritic Cells

Follicular dendritic cells (FDCs) although they look like dendritic cells are actually a unique cell subset; they are not derived from bone marrow progenitors, but they are of mesenchymal origin. They can be found in B cell follicles within lymphatic tissues. They express complement receptors CR1 and 2 and the Fc fragment of IgG receptor IIb (FcγRIIB). They secrete CXCL13, which allows them to recruit B cells and then trap B cell-delivered antigen bound to complement or specific antibodies. The immune complexes are displayed by FDCs and serve to activate B cells with complementary receptors. B cells in turn take up the antigen displayed by FDCs and upon digesting and presentation by MHC class II molecules activate Tfh cells. Through CD40L interactions and secretion of IL-21 and IL-4 FDCs trigger formation of germinal centers where B cells undergo affinity maturation and differentiation into antibody-producing cells. FDCs provide survival factors to B cells that undergo antigenic stimulation and help to remove B cells that die due to lack of stimulation by secreting lactadherin (Mfge8), which binds apoptotic cells and allows for their removal by macrophages. FDCs serve as a reservoir for HIV-1 during chronic infection.[368] Bluetonguevirus, a sheep arbovirus, infects and destroys FDCs.[369]

γ/δ T Cells

γ/δ T cells have characteristics of cells of both the innate and adaptive immune system (see Chapter 35). They are mainly found in skin and gut. They express an invariant TCR which may undergo some limited rearrangements to produce junctional diversity. Their receptor recognizes mainly lipids without requiring antigen processing or presentation by MHC molecules. They are activated rapidly by viral components or by other types of danger signals. Upon activation, they can lyse infected cells or secrete proinflammatory cytokines. They can form memory. They can phagocytose pathogens, and they can serve as antigen-presenting cells. They play a role during infections with influenza virus,[370] HIV-1,[371] herpesviruses,[372] HBV,[373] HCV,[374] and others.[375] They limit viral spread during acute infections but may contribute to immunopathology during chronic infections.[376] V∂2 T cells are selectively depleted in HIV-1 infected individuals. It is assumed that this is due to induction of apoptotic pathways upon binding of the viral envelope protein to CCR5 expressed by V∂2 T cells.[371]

Adaptive Immune Responses

T Cells

T cells are able to eliminate virus-infected cells by killing them. Unlike B cells, they do not recognize native viral antigens but rather respond to peptides derived from viral antigens that are presented by MHC antigens on the surface of infected cells (see Chapter 19). This prevents that the TCR is saturated by virus particles, which would inhibit T cell recognition and lysis of infected cells.

The TCR is composed of a heterodimer α/β or γ/δ chains (see Chapter 23). About 95% of T cells carry an α/β chain receptor. Each chain has a V and a C region. The latter is linked to a transmembrane domain and a short cytoplasmic tail. The V region of each chain is composed of three or four hypervariable regions, also called complementarity determining regions (CDRs), for the α and β chains, respectively. The CDRs binds to a complex composed of an MHC molecule and a peptide derived from a foreign protein. CDR2 of the ß chain binds to MHC molecules while CDR3 has specificity for the peptide. CDR1 of the α chain may also bind to the N terminus of a foreign peptide while CDR1 of the ß chain may bind to the peptide's C terminus. Receptor diversity is provided by genetic recombination-activating genes (RAG)1- and RAG2-mediated recombination of V and J gene segments for the α chain and VDJ gene segments for the ß chain. Unlike antibodies, the TCR V regions do not undergo hypermutation upon binding their cognate antigen. The T cells' repertoire is thus more limited than that of B cells. On the cell surface, the α and β chain of the TCR form a complex with three CD3 heterodimers composed of two CD3ζ chains, a CD3γ and ε chain and a CD3δ and ε chain. Signaling is provided through intracellular ITAM motifs on the CD3 chains.

Upon binding of the TCR to an MHC-peptide complex, the lymphocyte protein tyrosine kinase (Lck) phosphorylates the CD3 chains' ITAM motifs (see Chapter 24). CD3ζ chain-associated protein kinase (ZAP70) is recruited to the TCR complex. It phosphorylates linker for activation of T cells (LAT) and lymphocyte cytosolic protein 2 (SLP-76). The latter docks to LAT and forms a complex with Vav, noncatalytic region of tyrosine kinase (NCK), Grb2 family adaptor (GADS), and IL-2 inducible T cell kinase (Itk). Itk becomes activated and phosphorylates PLCγ1, which in turn hydrolyses phosphatidylinositol 4,5-bisphosphate (PIP2) into diacylglycerol (DAG) and inositol trisphosphate (IP3). DAG activates protein kinase C theta (PKCθ) and the MAPK/Erk pathways, resulting in NF-κB activation. IP3 triggers calcium release from the ER resulting in activation of NFAT and production of IL-2 (Fig. 44.6). TCR signaling is controlled by other interacting pathways triggered by binding of costimulators or coinhibitors on T cells to their respective ligands on antigen-presenting cells (Fig. 44.7). Full activation of T cells requires a second signal through binding of CD28 to a costimulator such as CD80 or CD86, which are both expressed by mature, but not by immature, dendritic cells. In unstimulated T cells, the cytoplasmic tail of CD28 binds to the serine/threonine protein phosphatase 2A (PP2A). Upon TCR engagement, CD28 is phosphorylated and PP2A is replaced by PI3K and growth factor receptor–bound protein 2 (GRB2). Activation of PI3K drives activation of AKT and mTORC1. AKT/mTORC1 in turn regulates a number of processes including cell proliferation, survival, migration, autophagy, and metabolism. GRB2 binds to LAT thereby

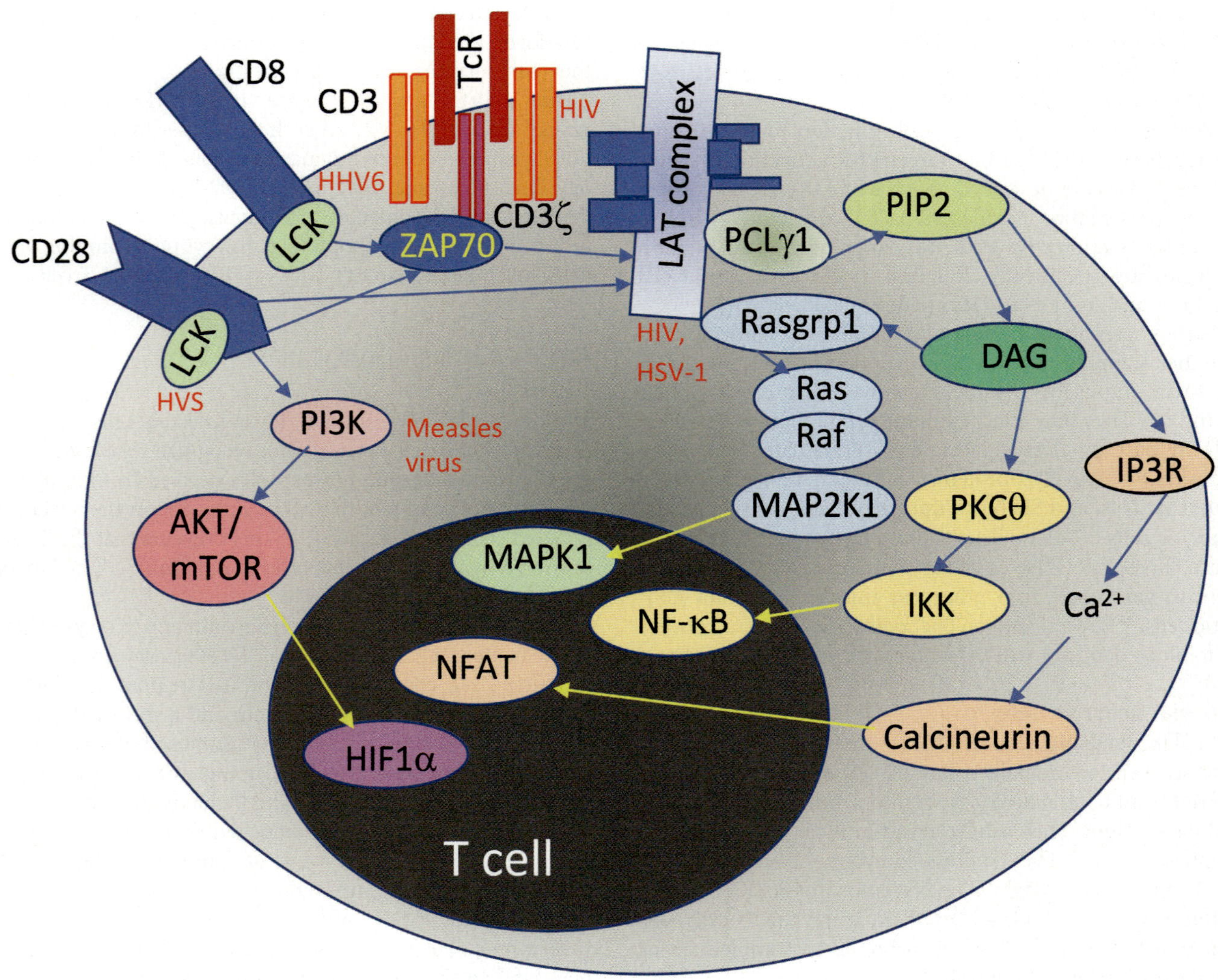

FIG. 44.6. The figure shows the T cell receptor (TCR) signaling pathway and steps that are affected by the indicated viruses. PCLg1, polycomb-like protein; Raf, B-Raf proto-oncogene, serine/threonine kinase; Ras, KRAS proto-oncogene, GTPase; Rasgrp1, RAS guanyl releasing protein 1. The other abbreviations are explained in the text.

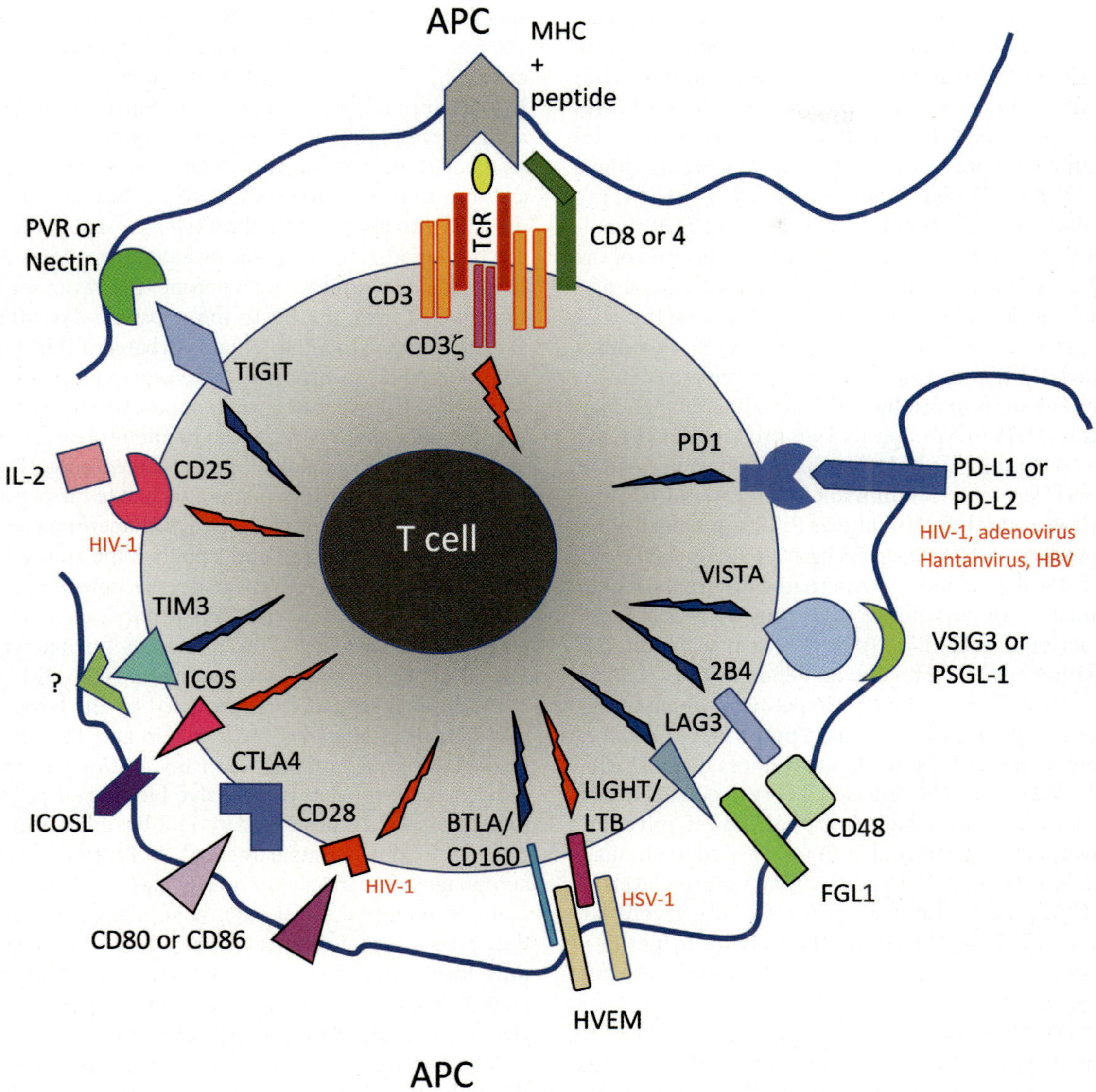

FIG. 44.7. The figure shows costimulators (in *red*) and coinhibitors (in *blue*) on T cells and their respective ligands on antigen-presenting cells (APCs). Virus names that interfere with these pathways are shown next to the molecules they target.

amplifying TCR signaling. A number of coinhibitors dampen TCR or CD28 signaling. BTLA, which binds to HVEM, is already expressed on naïve T cells and reduces activation of T cells through pathways that remain to be characterized. PD-1 is induced upon activation. It binds to PD-L1, which is expressed widely, or PD-L2, which is mainly expressed on antigen-presenting cells. The cytoplasmic tails of PD-1 contain an immunoreceptor tyrosine-based inhibitory motif (ITIM) and an immunoreceptor tyrosine-based switch motif (ITSM).[377] Upon engagement, PD-1 recruits SHP1 and SHP2 phosphatases to its tyrosine phosphorylated ITIM and ITSM motifs; this blocks activation of ZAP70 and PI3K.[378] Once a T cell is activated, it expresses cytotoxic T-lymphocyte–associated protein (CTLA)-4, which outcompetes CD28 binding to CD80 or CD86. Through a YMNM motif in its cytoplasmic tail CTLA-4 recruit SHP2 similar to PD-1 and blocks signaling through CD3ζ and ZAP70. Other costimulators include ICOS, which interacts with ICOS ligand, TNFSF14 also called LIGHT and lymphotoxin B, which both bind to HVEM albeit to a different domain than BTLA thus allowing for formation of a trimolar complex in which inhibition prevails. Coinhibitors include lymphocyte activating 3 (LAG-3), which binds to fibrinogen-like protein 1 (FGL1); T-cell membrane protein 3 (TIM-3), which also regulates functions of NK cells, dendritic cells, and macrophages; CD160 which binds to the BTLA-binding domain of HVEM; 2B4, which binds to CD48; V-Type Ig domain-containing suppressor of T cell activation (VISTA), which binds to Ig superfamily member 11 (VSIG3); and P-selectin glycoprotein ligand-1 (PSGL-1), and T cell immunoreceptor with Ig and ITIM domains (TIGIT), which binds poliovirus receptor (PVR) and nectin cell adhesion molecule 2 (NECTIN2).[379] Coactivators and coinhibitors are expressed at different stages of T cell activation. They serve to ensure an adequate immune response while preventing self-reactive or excessive responses. Viral subversion of costimulatory pathways blunts T cell responses while defects in coinhibitory pathways can result in fatal immunopathology upon a virus infection that would otherwise cause a benign disease.[196]

Viruses disrupt T cell stimulation by affecting TCR signaling (Fig. 44.6). Nef of HIV-1 affects multiple molecules of the TCR/CD3ζ signaling pathway. It binds to the CD3ζ chain,[235] PI3K,[380] Vav,[381] and Lck,[382] and other components of the TCR signaling pathway.[383] Herpesvirus saimiri (HVS) encodes a protein called Tip that interacts with Lck; it remains controversial if this enhances or inhibits TCR signaling.[384,385] Human herpesvirus 6 (HHV6) through its U26 protein blocks recycling of the CD3 molecule.[386] HSV-1 blocks phosphorylation of LAT, which prevents recruitment of PLCγ1, Grb2, and GADS.[387] Measles morbillivirus affects PI3K signaling by preventing degradation of Cbl-b, an E3 ligase that targets the p85 unit of PI3K for degradation.[388] Alternatively, viruses modulate costimulatory or coinhibitory signaling pathways. They can either act on T cells directly or they may act indirectly by manipulating ligand expression on dendritic cells. A main viral target is the CD28—CD80/CD86 pathway, which is crucial for stimulation of naïve T cells. Vpu of HIV-1 reduces surface expression of CD28, while nef reduces overall cellular levels of CD28 in HIV-1–infected CD4+ T cells.[389] CD4+ T cell defects, which are a hallmark of an HIV-1 infection, also influence signaling molecules on CD8+ T cells; circulating CD8+ T cells from HIV-1 patients have reduced levels of CD3ζ and CD28. After stimulation, these CD8+ T cells fail to upregulate CD25, which affects their ability to proliferate in response to IL-2.[390] Herpesviruses also downmodulate expression of CD28.[391]

Another common viral target is the PD-1—PD-L1/2 pathway that dampens T cell responses and thereby prevents fatal excessive immune-mediated damage. This pathway also drives T cell exhaustion during chronic virus infections. A number of viruses increases expression of PD-L1 or PD-L2 on infected cells such as hantavirus,[392] HIV-1,[393] adenovirus,[394] or HBV.[395] HSV gD binds to HVEM and thereby blocks BTLA or CD160 binding without affecting binding of the costimulators LIGHT or lymphotoxin B. Although it seems counterintuitive that HSV-1 would block an immunoinhibitory pathway thus promoting the actions of costimulators, the resultant activation of NF-κB was shown to increase the efficiency of virus replication.[396]

CD4+ T Cells

CD4+ T cells mainly produce cytokines and thereby have regulatory functions (see Chapter 28). CD4+ T cells are divided into different subsets that promote or dampen other immune responses.

Antigen Presentation

CD4+ T cells are stimulated by antigenic peptides displayed by MHC class II molecules, which are only expressed on specialized cells that are involved in antigen presentation such as mature dendritic cells, macrophages, or human B cells (Fig. 44.8) (see Chapter 20). Expression of MHC class II molecules requires CIITA. Virus particles are taken up by antigen-presenting cells through different pathways such as phagocytoses or endocytosis. They then enter endosomes and are transported to lysosomes where they are enzymatically digested into peptides. Viral proteins present in the cytoplasm can be degraded upon uptake by autophagosomes. MHC class II molecules bound to a chaperone, the invariant chain, are transported from the ER to the lysosome-like MHC class II loading compartment. It remains debated if this step involves an intermediate transport to the plasma membrane. Once the MHC class II–invariant chain complex reaches the lysosomes, the invariant chain is degraded by the proteases cathepsins S and L, and only a small fragment called class II–associated invariant chain (CLIP) remains attached to the peptide groove of the MHC class II molecule.[397] A chaperone protein called HLA-DM binds to CLIP and stabilizes the empty MHC class II molecule allowing for repeated replacement of CLIP by different peptides till one with high affinity binds to the peptide groove.[398] In some cells, the process is further regulated by HLA-DO, which can inhibit binding of HLA-DM to CLIP.[399]

Additional regulation is provided by the lysosomal compartment that becomes more acidic and thus more suited to degrade viral proteins upon maturation or activation of antigen-presenting cells. Selective binding of peptides from foreign proteins is promoted by TLR ligands within the MHC class II loading compartment.[400,401] Peptides generated in autophagosomes reach the MHC class II loading compartment by membrane fusion.[402] MHC class II molecules once they bind a peptide are transported to the cell surface where they form microclusters. This spatial distribution in turns facilitates cross-linkage of CD4+ TCRs with several MHC class II molecules carrying the same peptide.[403]

Viruses affect antigen presentation by MHC class II through inhibiting IFN-γ which increases MHC class II expression or by promoting production of IL-10 or secreting IL-10 homologues, which downregulate expression of MHC class II molecules and increase the pH within endosomes, thus diminishing the activity of proteolytic enzymes.[404,405] Other viruses inhibit antigen presentation by MHC class II directly (Fig. 44.8).[406] LANA1 of KSHV and tat of HIV-1 downregulate the transcriptional activity of CIITA.[407] Tat competes with CIITA for the binding to its target promoters and blocks the expression of MHC class II genes in HIV-1–infected cells.[408] The immediate early gene BZLF1 of EBV represses CIITA activity by binding to its promoter.[409] HCMV decreases transcription of CIITA.[410] US3 of HCMV binds to the immature α/β MHC class II dimer reducing binding of the invariant chain.[411] HCV core and NS5A proteins inhibit transcription of cathepsin S, thereby preventing digestion of the invariant chain into CLIP.[411] The US2 protein of HCMV and the late lytic protein BDLF3 of EBV target MHC class II antigens for degradation.[412,413] The EBV lytic protein gp42 binds to MHC class II antigens on the cell surface, thereby blocking binding of the TCR.[414]

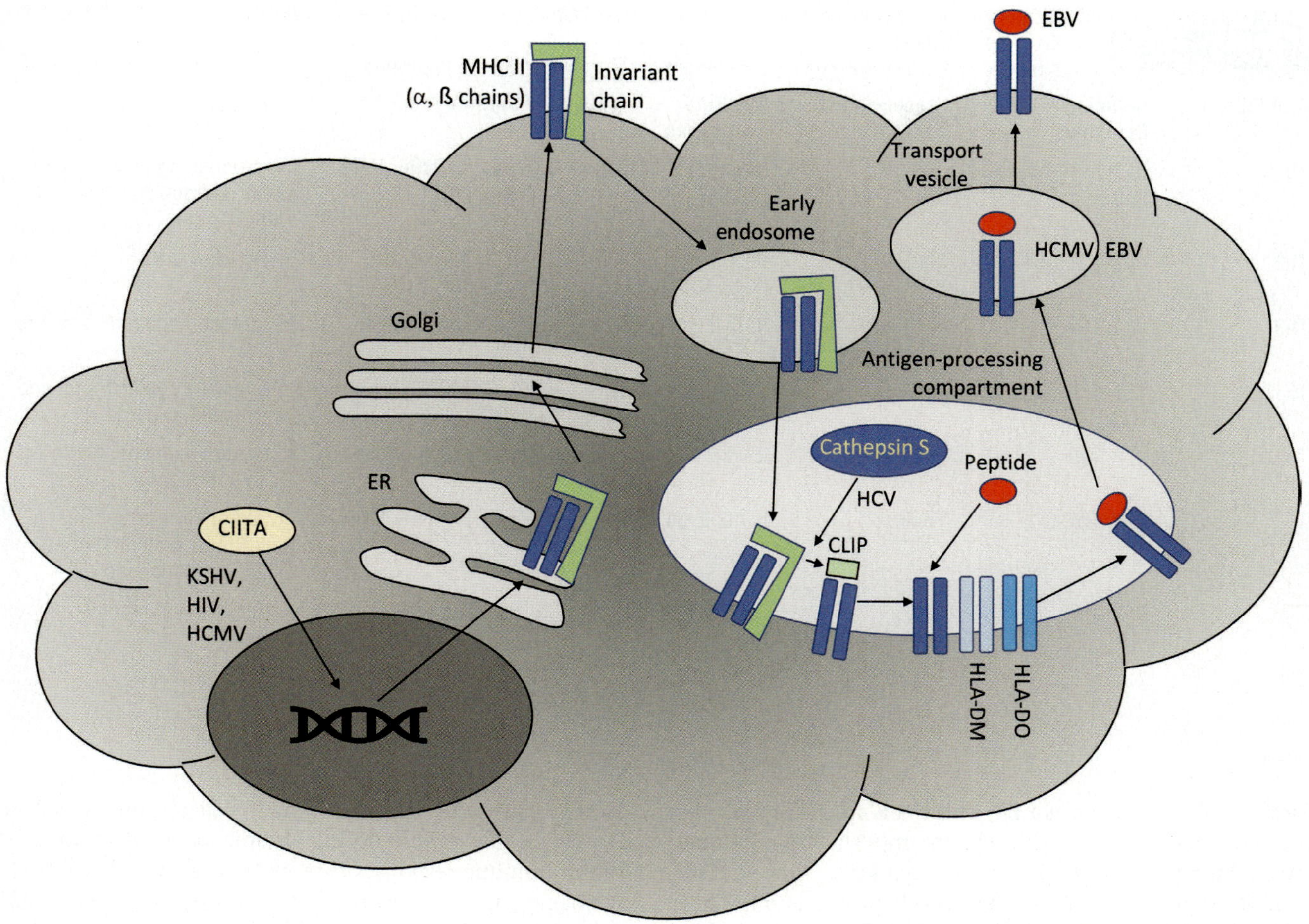

FIG. 44.8. The figure shows the major histocompatibility complex (MHC) class II presentation pathways and viruses and their proteins that interfere with this pathway.

CD4+ T Cell Differentiation

Naïve CD4+ T cells traffic through blood and lymph nodes monitoring MHC class II+ cells for expression of their cognate antigen. The frequency of naïve CD4+ T cells to a given antigen is low but variable.[415] Upon interacting with their cognate antigen, which takes several hours,[416] CD4+ T cells become activated. The cytokine milieu as well as the strength of TCR signaling[417] during stimulation affects CD4+ T cell fate decision, that is, their differentiation into different subsets (Table 44.4).

Upon activation, CD4+ T cells proliferate although their expansion is less pronounced than that of CD8+ T cells. Once antigen has been removed, 90% to 95% of the activated CD4+ T cells undergo apoptotic cell death. The remaining cells differentiate into memory (see Chapter 30). It remains debated if memory formation is imprinted into a given cell during its activation and initial division[418] or if it is determined by the level of signaling during the effector phase.[419] CD4+ T cells with high receptor avidity have a survival advantage and become enriched in the memory pool.[420] Upon reencounter of their antigen, memory CD4+ T cells become activated without the lengthy interaction with their cognate antigens that is needed for stimulation of naïve CD4+ T cells. Recalled memory CD4+ T cells show improved functionality and express higher avidity TCRs than CD4+ T cells initiated during a primary response.[421] In general, Th1 and Th2 cells show fidelity to their original subset designation while Th17, T_{reg}, and Tfh cells have a higher degree of plasticity. It has been reported that prolonged antigenic stimulation during viral persistence causes Th1 cells exhaustion and concomitantly promotes differentiation of naïve CD4+ T cells into Tfh cells.[422]

CD4+ T Cell Function

CD4+ T cell subsets exhibit distinct functions through secretion of cytokines and chemokines.[42] All Th cells can produce IL-2, which promotes their own proliferation as well as proliferation of CD8+ T cells. Th1 cells produce TNF-α and IFN-γ, which promote activation of NK cells, macrophages, and B cells. Th1 cells are essential for activation of CD8+ T cells although some viruses, such as LCMV, Sendai virus, influenza virus, or ectromelia virus, induce CD8+ effector T cells without CD4+ T help. Nevertheless, T helper cell–independent effector CD8+ T cells are fundamentally defective and fail to differentiate efficiently into memory CD8+ T cells.[423] Th1 cells can lyse virus-infected target cells that express MHC

TABLE 44.4 CD4+ T Cell Subsets

Cell Subset	Inducing Cytokines	Transcription Factors	Cytokine Responses	Key Surface Markers	Functions
Th1	IL-2, IFN-γ, IL-18, IL-27	T-Bet	IFN-γ, TNF-α, IL-2	IL-12Rß2, IL-27Rα, IFN-γR2, CCR5, CXCR3	Protection against intracellular pathogens, CD8+ T cell help
Th2	IL-4	Gata3	IL-3, IL-4, IL-5, IL-13, IL-27E/25	ST2/IL-1R4, CXCR4, CCR3, CCR4, CCR8	Help for B cells
Th9	IL-4, TGF-ß	PU.1	IL-9, IL-10	CCR3, CXCR3, CCR6, IL-17RB	Protection against parasites
Th17	TGF-ß, IL-6, IL-21, IL-23	RORγ0	IL-17, IL-21, IL-22, IL-26	BTLA, CD200, CD99, and IL-18Rα	Protection against fungi and extracellular bacteria
Th22	IL-6, TNF-α	?	IL-22	CCR4, CCR10	Protection against microbial skin infections
T_{regs}	IL-2, TGF-ß	FoxP3	IL-10, IL-35, TGF-ß	CD25, CD28, GITR, CTLA-4, ICOS	Maintenance of self-tolerance, inhibition of immune responses
Tfh	IL-6, TNF- α	BCL6	IL-4, IL-10 IL-17, IFN-γ	CXCR5, ICOS, PD-1	Generation of affinity-matured B cells
Tfr	?	FoxP3, Bcl6	IL-10, TGF-ß	PD-1, CXCR5, CTLA4, ICOS, TIGIT, GITR, CD28	Inhibition of Tfh

?, unknown.

class II molecules, such as EBV-infected B cells[424] or HSV-1–infected macrophages.[425] Th2 cells are important to eliminate extracellular pathogens. They secrete cytokines, such as IL-3, IL-4, IL-5, and IL-13 that drive B-cell proliferation. They induce B cell switching to IgG1 and IgE. Furthermore, they can activate eosinophils and macrophages. Th17 cells produce IL-17, IL-21, IL-22, and IL-26. They recruit and activate neutrophils through the secretion of IL-17. They are essential to combat bacteria and fungi, but they can also affect antiviral immune responses. They induce strong inflammatory responses mainly through IL-17–mediated recruitment and activation of neutrophils. In a mouse model of vaginal HSV-2 infection, Th17 cells were shown to be essential for generation of a Th1 response. Upon West Nile virus or adenovirus infection of mice, Th17 cells promote CD8+ T cell responses and are thereby essential to control viral loads.[426,427] Upon infection of mice with an influenza H5N1 virus, Th17 cells were shown to recruit B cells to the infected lungs.[428] They also promote differentiation and migration of B1a-B cells, a subset of B cells that produces natural low avidity IgM antibodies as a first-line defense against pathogens.[429] Upon infection of mice with Theiler's murine encephalomyelitis virus, Th17 cells were found to promote viral persistence and enhance chronic demyelinating disease within the central nervous system by suppressing Th1 and CD8+ T cell responses.[430] Similar results were obtained upon infection of mice with coxsackievirus B3, which causes an acute myocarditis.[431] Th22 cells protect against skin infections, but it remains controversial if they play a role during virus infections. Upon infection with HBV, Th22 cells were shown to contribute to liver damage.[432] T_{regs} suppress immune response to maintain tolerance (see Chapter 29). They play a role in viral infections by preventing damage caused by overwhelming T cell responses. Natural T_{regs} play a role mainly during chronic virus infection. They inhibit immune responses by secretion of TGF-ß and IL-10, by soaking up IL-2 through cell surface–expressed CD25, or by directly killing activated T cells through the release of granzyme or perforin.[433-436] During acute viral infections, T_{regs} increase viral loads and delay clearance mainly by blunting other T cell responses, reducing production of cytokines, and affecting lymphocyte migration patterns. Nevertheless, T_{regs} may also benefit the host by reducing immunopathology. This was shown in numerous viral infections including infections with HSV,[437,438] HIV,[439] HCV,[440] HBV,[441] CMV,[442] EBV,[443] and others.[444] T_{regs} induced by antigen can cross-suppress unrelated responses through secretion of anti-inflammatory cytokines.

Tfh cells, upon initial stimulation through the TCR, CD28, and CD40L, home to lymphatic tissues. They trigger the formation of germinal centers through secretion of IL-4 and IL-21. They then enter germinal centers, which are mainly composed of proliferating and mutating B cells. Within germinal centers, Tfh cells upregulate expression of CD40L, which interacts with CD40 on B cells. Through CD40-CD40L interactions, Tfh cells assist in B cell proliferation, selection, Ig class switching, and plasma cell differentiation. Tfr cells are also found in germinal centers where they suppress the activity of Tfh cells to prevent unduly potent germinal center formation or generation of auto-reactive antibodies.

Many viruses subvert CD4+ T cell responses by modulating antigen-presentation pathways as described above; others target CD4+ T cell directly. HIV-1 and HIV-2 are the best examples of a direct attack; they selectively infect CD4+ cells. The viruses kill most of the infected cells, which are initially replaced by naïve CD4+ T cells from the thymus. Eventually, the pool of naïve cells becomes exhausted and overall numbers of CD4+ T cell decline[445] resulting in immunosuppression and eventual death of the patient due to overwhelming secondary infections. In addition, HIV-1 causes chronic immune stimulation leading to activation-induced death of CD4+ T cells.[446] Ebola virus caused lymphopenia. Although the virus can infect T cells, it fails to replicate due to a cellular restriction factor. Nevertheless, the abortive infection leads to an ER stress response and autophagy followed by CD4+ and CD8+ T cell death.[447] Tfh cells are major reservoir for HIV-1. Their location within B cell follicles, which CD8+ T cells cannot enter, shields virus-infected cells from cell-mediated destruction. Viruses suppress immune responses by promoting T_{reg} cells. Antigen-driven induction of T_{reg} cells depends on dendritic cells with a semimature phenotype. In mice, Japanese encephalitis virus induces upregulation of PD-L1 on dendritic cells, which in turn promotes the development of T_{reg} cells.[448] The same is achieved by HCV through TGF-ß production by hepatocytes.[449] HSV-1 increases expression of HVEM on T_{reg} cells leading to their expansion.[450] By the same token, CD4 binding of HIV-1 increases T_{reg} functions.[451]

CD8+ T Cells

CD8+ T cells also called cytolytic or cytotoxic T cells are the immune system's main weapon to kill virus-infected cells through the release of the toxins perforin and granzyme, thus efficiently stopping virus replication (see Chapter 31).

Antigen Presentation

CD8+ T cells recognize peptides displayed by MHC class I molecules, which are expressed by nearly all nucleated cells of the body (Fig. 44.9) (see Chapter 20). MHC class I molecules are composed of an α chain, also called heavy chain, with three distinct domains (α1-3); the heavy chain binds noncovalently to ß2-microglobin. The heavy chain's α1 and α2 domains form the peptide-binding groove. The binding groove of MHC class I molecules is closed at both ends by conserved tyrosine residues, which imposes a size restriction of 8 to 10 amino acids to the bound peptides. Newly synthesized MHC class I heavy chains are transported into the ER, where they bind to calnexin; this promotes appropriate folding of the protein. Once the heavy chain is folded, calnexin is released and the heavy chain binds to β2-microglobin. The correctly folded MHC class I—ß2-microglobin complex then binds to calreticulin, the thiol oxidoreductase ERp57, and tapasin, the latter two interact with the peptide-binding groove.[452] Tapasin connects MHC class I molecules to the transporter associated with antigen presentation (TAP) and assists in selection and binding of high-affinity peptides. In the classical presentation pathway, newly synthesized proteins, including self-proteins and viral antigens that are incorrectly folded or incompletely synthesized, are initially ubiquitinated and then proteolytically digested into peptides by a proteasome complex. Proteasomes generate the correct carboxyl terminus of most MHC class I binding peptides but commonly leave an extended amino terminus that is trimmed by peptidases either in the cytoplasm or the ER. The peptides are transported from the cytoplasm to the ER by TAP. This transporter is composed of two chains, TAP1 and TAP2, which breach the ER membrane, and within the ER, bind to MHC class I via tapasin to facilitate binding of peptides to the MHC class I groove. Binding of a peptide stabilizes the MHC class I complex, which is then transported via the Golgi to the cell surface.

Some cells, such as dendritic cells, are able to cross-present antigens from viruses that were taken up by endocytosis or phagocytosis (Fig. 44.9). Cross-presentation is essential to allow for stimulation of CD8+ T cell responses to viruses that do not infect and replicate in dendritic cells. It relies on the presence of empty MHC class I molecules in compartments other than the ER. Empty MHC class I molecules are removed from the cell surface by endocytosis. MHC-containing endosomes fuse with sorting endosomes. From there, MHC molecules can be routed back to the cell membrane through a fast pathway, also called vacuolic pathway or the slow cytosolic pathway; the latter goes through the endocytic recycling compartment or to endosome-lysosome compartments from where MHC class I molecules can be diverted to phagosomes. The cytosolic pathway is sensitive to drugs that inhibit the activity of TAP or proteasomes.[453] In the cytosolic pathway, viruses are taken up into phagosomes. Vesicle-trafficking proteins and TAP are recruited from the ER-Golgi intermediate compartment (ERGIC) to the phagosomes. The vesicle trafficking proteins Sec61 allow for exit of polypeptides from the phagosomes into the cytoplasm where they are ubiquinated and further degraded by the proteasome complex. Peptides can then either move into the ER with the help of TAP and enter the classical presentation pathway, or they can diffuse back into endosomes or phagosomes, undergo further trimming by aminopeptidases, and then associate with phagosomal MHC class I molecules.[454] Cross-presentation through the vacuolic pathway is sensitive to inhibitors of lysosomal proteolysis.[455] In this pathway, antigens are degraded into peptides within the endosomes. ROS present in endosomes through the activity of NADPH oxidase disrupts the endosomal membrane, which allows the peptides to escape into the cytoplasm from where they can be transported into the ER and associate there with MHC class I molecules. A number of receptors such as TLRs, mannose receptors, IgGFc receptors, or CD40, as well as heat-shock proteins, influence cross-priming and may provide avenues to improve the immunogenicity of antiviral vaccines.

Numerous viruses interfere with MHC class I presentation pathways (Fig. 44.7). Rhesus (Rh) CMV protein Rh178 blocks translation of the MHC class I heavy chain.[456] The nuclear antigen 1 of EBV and LANA1 of KSHV carry repeat sequences that render the proteins resistant to digestion by proteasomes.[457] Herpesviruses and poxviruses encode proteins that inhibit peptide transport. TAP binds ATP and a peptide and this causes a conformational change that allows the peptide to pass through. ATP is then hydrolyzed, and TAP resumes its initial closed conformation. US6 of HVMC[458] and CPXV012 of cowpox virus[459] block binding of ATP to

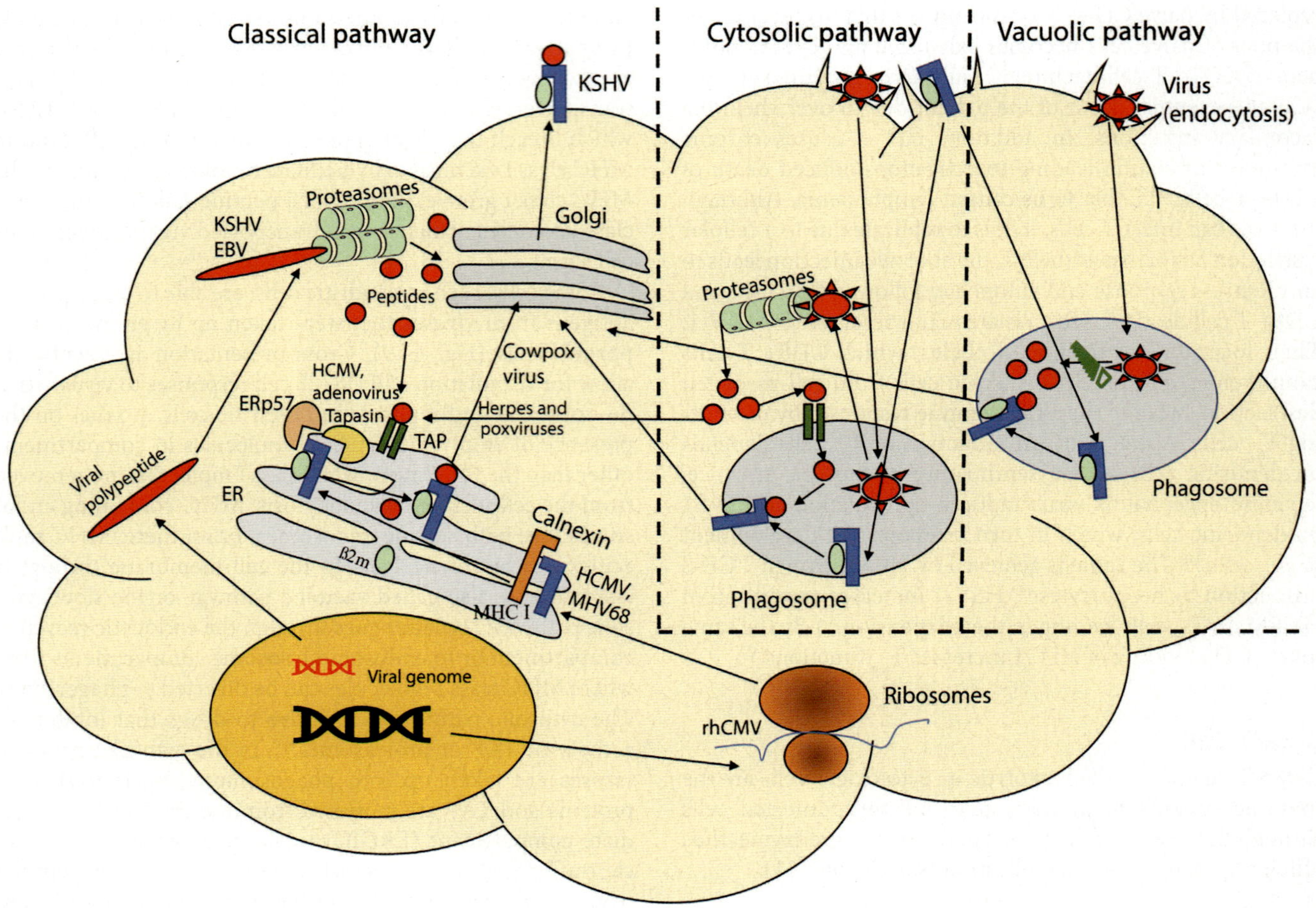

FIG. 44.9. The figure shows the major histocompatibility complex (MHC) class I presentation pathways and viruses and their proteins that interfere with this pathway.

TAP. UL49.5 of bovine herpesvirus 1 blocks conformational changes of TAP.[460] IPC47 of HSV1 inhibits binding of peptides to TAP.[461] BNLF2a of EBV prevents binding of both ATP and peptide to TAP.[462] HCMV blocks binding of tapasin to MHC class I molecules while E3-19K of adenovirus interferes with tapasin's ability to connect MHC class I molecules to TAP, either protein retains MHC class I molecules within the ER.[463,464] US2 and US11 of HCMV drive transfer of newly synthesized MHC class I molecules from the ER into the cytoplasm where they are ubiquinated and degraded.[465,466] mK3 of MHV68 degrades MHC class I once it binds to tapasin.[467] Cowpox virus encodes CPXV203, which binds to MHC class I and prevents its exist from the ER or affects its retrograde transport from the Golgi back to the ER.[468] KSHV through kK3 and kK5 promote endocytosis and lysosomal degradation of surface-expressed MHC class I molecules.[469]

CD8+ T Cell Differentiation

Naïve CD8+ T cells are activated in lymphatic tissues by antigen displayed on mature dendritic cells, which in addition to MHC class I peptide complexes express costimulators such as CD28. CD8+ T cells engage with their cognate antigen through an immunological synapse[470] for about 18 to 24 hours. Then they start to proliferate very rapidly. They expend more vigorously than CD4+ T cells and antiviral CD8+ T cells cycle in 4- to 6-hour intervals, resulting in a more than 10,000 fold expansion within a week. Proliferation is largely driven by IL-2, which is provided by activated CD4+ T cells. CD4+ T cells furthermore promote stimulation of CD8+ T cells through chemokines that recruit naïve CD8+ T cells to lymph nodes or already activated CD8+ T cells to sites of infection.[471,472] Some viruses, such as LCMV or ectromelia virus, can induce a CD8+ T cell response without CD4+ T helper cells while stimulation of CD8+ T cells to others, such as influenza virus or HSV-1, depends on help, although this appears to be dose-dependent with higher doses allowing for CD4+ T cell–independent CD8+ T cell responses.[473-475]

During expansion, CD8+ T cells differentiate into effector cells and then migrate to sites of infection and secrete cytokines and lytic enzymes. Most effector CD8+ T cells are terminally differentiated, and they will die once the antigen has been removed. These end-stage effector cells are characterized by high expression of T-bet and beta-interferon gene positive-regulatory domain I binding factor (Blimp)-1 and by enhanced mTOR activity, which drives glycolytic energy production. About 5% to 10% survive and gradually differentiate into memory CD8+ T cells. These cells have increased levels of BCL6, polycomb complex protein BMI-1, transcription factor (Tcf)7, and they can switch from mTOR-and HIF-1α–driven glycolysis toward mitochondrial energy production. They express CD127, the receptor for IL-7. Effector CD8+ T cells can turn into effector memory (Tem) or central memory T (Tcm) cells. CD8+ Tem cells reside in tissues and maintain functions. Their proliferative potential is low, and over time they die or differentiate into Tcm cells. CD8+ Tcm cells home to lymphatic tissues. They no longer exhibit functions; they proliferate vigorously upon reencounter of their antigen, and they are very long-lived. They are maintained at steady numbers potentially throughout the lifespan of an individual. Naïve, effector, Tem, and Tcm cells can be distinguished by a number of surface markers. Naïve cells are $CD44^{lo}$, $CD62L^{hi}$, $CD127^{hi}$, and $CCR7^{hi}$. Effector and Tem cells are $CD44^{hi}$, $CD62L^{lo}$, $CCR7^{lo}$, and $CD127^{lo}$, while Tcm cells are $CD44^{hi}$, $CD62L^{hi}$, $CCR7^{hi}$, and $CD127^{hi}$. It remains debated at what stage and through what signals CD8+ T cells make the fate decision toward terminally differentiated short-lived effector cells or long-lived memory cells. High TCR signaling strength favors differentiation into short-lived effector cells presumably by inducing transcription factors IRF4 and AP-1 while lower signaling induces BACH2 and NR4A1, which suppress genes associated with effector functions. Other transcription factors, such as T-bet/Eomes, ID2/ID3, and Blimp-1/BCL6, have reciprocal expression patterns in short-lived effector cells and memory precursors and drive differentiation toward divergent cell fates.[476] Presence of the costimulators OX-40 (CD134) and 4-1BB (CD137) during priming seems to predispose CD8+ T cells to differentiate toward memory,[477] while cells with high coinhibitory expression levels are prone to differentiate into short-lived effector cells.[478]

Another costimulator, CD27, was also shown to be required for memory formation.[479] CD4+ T cell help during priming seems to be essential for memory formation[423,480] and helpless CD8+ T cells express elevated levels of T-bet, which represses expression of CD127.[481] Low levels of CD25 expression during the early stages of activation, which decreases a cell's responsiveness to IL-2, favors memory formation. Maintenance of memory is provided by IL-7, which drives homeostatic proliferation of memory CD8+ T cells, and IL-15, which by increasing expression of TNFR9 (4-1BB), TNFR2, and GITR activates the NF-κB pathway and downstream prosurvival signals.[482] Memory CD8+ T cells can be recalled by antigen expressed by cells other than dendritic cells such as B cells and macrophages. They respond to lower amounts of MHC class I peptide complexes than naïve cells. Their activation is less dependent on costimulation or T cell help.[423,475,483] Secondary CD8+ T cell responses come up faster mainly due to increased numbers of precursors although each individual cell undergoes fewer cycles compared to a naïve CD8+ T cell that is being activated. Secondary CD8+ T cells acquire functions faster, secrete higher amounts of factors per cell, and they contract more slowly.[484-486] Metabolism as described above for CD4+ T cells changes during T cell differentiation. While naïve CD8+ T cells use mitochondrial energy production, they switch to the use of glycolysis upon activation, and then they switch back to OXPHOS, which is in part fueled by FAO upon transitioning to memory.

Chronic virus infections drive differentiation of CD8+ T cells toward exhaustion has been observed in mice infected with LCMV, humans infected with HBV, HCV, or HIV, and nonhuman primates infected with simian immunodeficient virus (SIV).[487,488] Exhaustion can also be driven by excessive inflammation and certain cytokines such as IL-10 and TGF-ß[489,490] and by a metabolically challenging microenvironment that lacks nutrients and oxygen.[491] Exhausted CD8+ T cells evolve from a pool of CD127+ T cells that have the potential to differentiate toward memory; KLRG1+CD8+ effector cells that are terminally differentiated do not transition toward exhaustion.[492] Effector CD127+CD8+ T cells that are bombarded for an extended period of time with antigen fail to transition toward memory. They gradually increase expression of coinhibitory markers, lose functions, and change their use of key transcription factors. T cells initially lose the ability to produce IL-2, followed by reduced production of lytic enzymes and loss of TNF-α secretion. Their proliferative capacity declines. During this stage of partial exhaustion, CD8+ T cells increase expression of PD-1, which reduces excessive stimulation through the TCR by blocking ZAP70 and PI3K phosphorylation. This in turn reduces AKT/mTOR signaling and shifts T cell metabolism away from glycolysis toward the use of FAO for energy production. In addition, T cells increase expression of CTLA-4, which blocks further signaling through CD28. This stage of partial exhaustion is followed by a stage of complete exhaustion where CD8+ T cells stop producing IFN-γ and increase expression of additional coinhibitory markers such as LAG-3, TIGIT, TIM3, 2B4, and CD160.[493] Transcription factors that contribute to CD8+ T cell exhaustion include NFAT that in absence of AP-1 activation causes T cell hyporesponsiveness[494] as well as T-bet, Eomes, FoxO1, FoxP1, Blimp-1, NFAT, B cell–activating transcription factor (BATF), IRF4, and von Hippel-Lindau disease tumor suppressor (VHL).[495] Another transcription factor that is induced by calcineurin and NFAT2 is thymocyte selection–associated high-mobility group box (TOX) that drives expression of coinhibitors and induces the epigenetic changes that characterize exhausted CD8+ T cells.[496] It should be pointed out that exhaustion is not a uniform process that affects all CD8+ T cells directed against a given virus simultaneously. Treatment with antibodies that block signaling of coinhibitory receptors can, in some but not all patients, partially restore CD8+ T cell functions. In some cases, these benefits are sustained, while in other patients, T cell recover transiently but then relapse back into exhaustion. Differences in patterns of reversion of exhaustion by blocking antibodies may reflect the heterogeneity of exhausted CD8+ T cells. For example, a subset of PD-1+ CD8+ T cells that expresses coinhibitors as well as costimulators in addition to signal regulatory protein (SIRP)α was found to retain antiviral functions.[497] Also, as exhaustion is driven largely by strong and sustained TCR signaling, CD8+ T cells to subdominant epitopes are more resistant as indicated by a marked shift in the epitope profile of responding CD8+ T cells during chronic virus infections or cancer with a gradual loss of CD8+ T cells to dominant epitopes and proportional increases in CD8+ T cells to subdominant epitopes.[498,499]

Viruses that persist in a latent form, and only periodically upon reactivation produce antigen, fail to drive CD8+ T cell exhaustion. Instead, as best known for HCMV, they gradually expand the pool of effector cells so that in an elderly person up to a quarter of all of CD8+ T cells can be directed to a single epitope of this virus.[500] Other viruses, such as adenoviruses, persist at low levels and can be reactivated in severely immunosuppressed patients. These viruses maintain a large pool of CD8+ Tem cells, which, in addition, may have been generated in part by repeat infections with serologically distinct adenoviruses that carry cross-reactive T cell epitopes.[501]

CD8+ T Cell Functions

The main job of CD8+ T cells is to remove cells that can harm an individual. That includes virus-infected cells, which can spit out thousands of new infectious virus particles. Many antiviral CD8+ T cells recognize epitopes from early viral antigens, which are produced well before a virus completes its life cycle. This serves the organism well as it allows the killing of an infected cell without risking release of infectious virus particles. Once CD8+ effector T cells bind an infected cell, through interactions between the TCR and an MHC class I peptide complex, the cells release the lytic enzymes granzyme B and perforin. Release is directional toward the target cell to avoid death of bystander cells and can be triggered by recognition of as few as 2 to 3 MHC class I peptide complexes.[502] It was initially thought that perforin, with the help of calcium, forms pores in the cell membrane, which allows for diffusion of granzyme B into the cells. Other data suggest perforin and granzyme B are endocytosed by the target cell, and then perforin acts within the endocytic vesicle to release granzyme B into the cytoplasm.[503] Upon entering the cytoplasm, granzyme B can activate caspase 3 directly, which leads to apoptosis. Alternatively, granzyme B activates BID causing activation of BAX which removes BCL2 from the surface of the mitochondria. This results in a change in the mitochondrial membrane potential and causes destabilization of the mitochondria and release of cytochrome C, which can cause necrosis or apoptosis of the cell. Granzyme B also increases release of ROS from mitochondria, which further promotes cell death.[504,505] Killing occurs fast within minutes, and then the CD8+ T cell detaches from the dying cell and moves on to kill its next target,[506] allowing one CD8+ T cell to eliminate multiple infected cells. CD8+ T cells express FasL on their surface and can thereby lyse Fas+ target cells. Binding of FasL on the CD8+ T cells to Fas molecules on the infected target cells causes formation of Fas trimers. The intracellular death domain (DD) of the Fas molecule recruits and activates the death-inducing signaling complex (DISC) through the FADD adaptor molecule. FADD interacts with FLICE, which activates caspase 8, which in turn activates caspase 3. Activated caspase 3 cleave a variety of substrates including DNA repair enzymes, structural proteins, and caspase-activated deoxyribonuclease (CAD) inhibitor (ICAD) as well as other caspases. Activated CAD moves to the nucleus and causes DNA fragmentation, which proceeds apoptotic cell death. As CD8+ T cells express both Fas and FasL, this can lead to fratricide, that is, killing of one CD8+ T cell by another, which may play a role at the end of an immune response during CD8+ T cell contraction. Activated CD8+ T cells in addition secrete cytokines such as IFN-γ, which has antiviral functions and increases the expression of MHC class I molecule, thus facilitating the T cells' detection of their cognate antigen.

Viruses can evade killing of infected cells by subverting apoptotic pathways as described in more detail above in the section discussing TNF-α signaling. Viruses can block lysis upstream of the caspase pathway by directly inhibiting the activity of granzyme B. CrmA of poxviruses and the 100k protein of adenovirus both inhibit granzyme B–mediated proteolysis and thereby CD8+ T cell–mediated cytolysis.[507] Adenovirus L4-100K assembly protein is a granzyme B substrate that potently inhibits granzyme B–mediated cell death.[508]

Regulatory CD8+ T Cells

FoxP3+ regulatory CD8+ T cells can inhibit antiviral immune responses. CD8+ T_{regs} express high levels of CD122, indicating that they proliferate in response to IL-15. They are positive for CD103, an adhesion molecule, CTLA-4 and LAG-3, two coinhibitors, and FasL. They can produce IFN-γ and immunosuppressive cytokines such as IL-10, IL-34, IL-35, and TGF-ß.[509] IL-10 suppresses production of proinflammatory cytokines and expression of MHC molecules and costimulators. IL-34 similar to TGF-ß drives differentiation of macrophages toward the M2 pathway.[510] IL-35 is an anti-inflammatory cytokine that blocks T cell proliferation. Overall, CD8+ T_{regs} can contribute to viral persistence during chronic infections with HBV, HCV, or HIV[511-513] while also preventing excessive T cell–mediated damage.

B Cells

B cells produce antibodies that can neutralize viruses, promote virus opsonization, or allow for cell- or complement-mediated lysis of infected cells (see Chapter 26). They play a minor part during the initial stages of an infection, but they are crucial in providing protection against reinfections. They also secrete cytokines and serve as antigen-presenting cells.

Antibodies

Antibodies are glycoproteins of the Ig family that are composed of two heavy and two light chains, which are linked by disulfide bonds to form a Y-shaped molecule. Light chains are encoded by two different loci, called kappa and lambda. Both heavy and light chains carry a V and a C region. The V domain binds through three hypervariable regions, also called CDRs to antigens, while the C region specifies antibody functions such as binding to complement or Fc receptors. Ig heavy and light chains are encoded by separate gene families and individual C and V domains are encoded by different genes. The light chain V genes are followed by J genes and then by a single C region gene. The heavy chain genes are structured similarly, but they contain additional segments called diversity (D) genes. Rearrangement of the heavy chain genes starts with DJ joining, followed by VDJ joining and then joining to a C gene; for the light chain genes, first VJ joining is followed by joining to the C gene. V(D)J recombination, which is one way to create antibody diversity, requires RAG-1 and-2. Upon activation during B cell proliferation, the variable domain genes undergo somatic hypermutation through activation-induced cytidine deaminase (AID) to increase the antibody's affinity to its antigen.

The C region of Ig heavy chains determines the isotype of an antibody. There are five Ig isotypes, that is, IgM, IgA (IgA1 and IgA2), IgD, IgE, and IgG; the latter is further divided into IgG1, IgG2a and b, IgG3, and IgG4. Most of these antibodies are monomers, that is, they are composed of two heavy and two light chains; IgA can form tetramers and IgM is a pentamer although it is expressed as a monomer on the surface of B cells. IgM is the first antibody that is produced in response to a virus. It is size-wise the largest of all antibodies, which restricts its ability to diffuse into tissues. IgE becomes elevated in response to respiratory viruses, measles morbillivirus, HIV-1, and others and may contribute to viral pathogenesis and the exacerbation of asthma during virus infections.[514,515] IgA is the most widely produced antibody. It is mainly produced at mucosal surfaces where it protects against invading pathogens. It is also present in breast milk and can cross the placenta. IgG is the major Ig class that circulates in blood and that provides protection against systemic viral infections. All IgGs can bind to complement and initiate ADCC upon binding to Fc receptors, but there are quantitative differences. IgG3 is the most potent activator of complement followed by IgG1, while IgG1 is better than IgG3 at mediating ADCC.

B Cell Subsets

Naïve B cells are divided into three subsets, that is, follicular B cells, also called B-2 B cells, B-1 B cells, which are further divided into B-1a and B-1b cells, and marginal zone (MZ) B cells. Follicular B cells upon antigenic stimulation in presence of T cell help differentiate into short-or long-lived plasma cells that produce class-switched affinity matured antibodies, or they can turn into memory B cells. MZ-B cells rapidly produce IgM upon activation, and they deliver antigen to B cell follicles. Their high MHC class II expression levels allow them to present antigen to CD4+ T cells.[516] B1-B cells without stimulation by antigen secrete polyreactive IgM antibodies that provide a first layer of defense against an invading pathogen.

B Cell Activation and Differentiation

Activation of follicular B cells, which are key to the production of affinity-matured antiviral antibodies, requires T cell help. T-B interaction triggers formation of germinal centers, where follicular B cells proliferate, undergo hypermutation, and selection for those that produce high-affinity antibodies. Germinal centers are mainly found in lymphatic tissue, but under some circumstance, such as upon influenza virus infection, can also be formed in tissues such as lungs. B cells proliferate and undergo somatic hypermutation within germinal centers. A number of transcription factors determine if germinal center B cells will differentiate into memory B cells or antibody-secreting B cells.

Long-term B cell memory and long-lived plasma cell, which can maintain antibody titers for years, are maintained in absence of T cell help.[517] Long-lived plasma cell home preferentially to bone marrow, which provides a microenvironment that is highly suited for plasma cell survival. This is best exemplified by long-lived plasma cell induced by vaccinia virus, which could be isolated from bone marrow 35 years after having been induced by a vaccine.[518]

Memory B cells home to the marginal zones of spleens or the subcapsular sinuses of lymph nodes. Alternatively, they can remain in issues as tissue-resident memory B cells. Upon reencounter of their antigen, they can either proliferate and quickly differentiate into plasma cells or they can undergo further affinity maturation within germinal centers. In either case, a secondary B cell response comes up faster and is more potent than a primary response. Although secondary plasma cells produce increased levels of antibodies with higher affinity, they are, in general, short-lived.[519] Regulatory B (B_{regs}) represent a small fraction of all B cells. Upon activation, they suppress B and T cell responses through secretion of IL-10 and TGF-ß.

Chronic exposure to antigen can lead to atypical "exhausted" B cells that show attenuation of BCR signaling, high expression of coinhibitory molecules, such as Fc receptor-like (FcRL)4 and FcRL5, loss of antibody production, and distinct transcriptional profiles.[520] This was first observed in HIV-1 patients where it is thought to contribute to their defective B cell responses. "Exhausted" B cells can also be found in malaria-infected children, where they can represent up to 40% of all circulating B cells. "Exhausted" B cells in malaria-infected patients respond poorly to soluble antigen but show robust BCR signaling to membrane-bound antigen. This in turn suggests a defense mechanism to avoid B cell stimulation to soluble rather harmless antigens and to rather focus responses on antigens expressed by infected cells.

B Cell Effector Functions

The main function of B cells is to produce antibodies. Antibodies can neutralize viruses, cause their demise by promoting their phagocytosis, kill virus factories by ADCC or complement-mediated lysis, and they assist in activation of T cells by presenting antigen. Natural antibodies produced by B1-B cells provide early protection against invading viruses. As it takes time for B cells to differentiate into antibody-secreting plasma cells, high-quality affinity-matured antibodies do not play a role during the early stages of a primary viral infection—they may contribute to viral clearance at later stages by killing infected cells by ADCC or complement-mediated lysis and by neutralizing viruses that are released from infected cells. Neutralizing antibodies are nevertheless crucial for the ability of the human species to survive the constant onslaught with viruses. Maternal IgG and IgA antibodies pass the placenta. Furthermore, they can be transmitted to neonates through breast milk. Most humans are infected with common viruses early in life and the resulting antibody responses are sustained through the reproductive lifespan of a woman. Transfer of maternal antibodies to neonates provides a window of protection during period of immunological immaturity when children are extremely vulnerable to infections. Neutralizing antibodies protect against secondary infections, and they are the main correlate of protection induced by antiviral vaccines. Antibodies cannot only prevent viruses from infecting cells, they can even neutralize viruses within a cell through a process called antibody-dependent intracellular neutralization (ADIN).[521]

A virus that has been coated with antibodies can carry them into the interior of cells. Once inside the cells, TRIM25 attaches to the Fc domain of the antibodies which targets the antibody-virus complex for degradation. ADIN, which only requires about two antibodies to bind to the virus surface, is more efficient than extracellular neutralization, which requires multiple antibodies to prevent virus from binding to its cellular receptor.[522] Viruses evade B cell responses mainly by hypermutation of their surface antigens, which are the main targets for neutralizing antibodies. A classic example is influenza A virus, which constantly creates new B cell epitopes by antigenic drift. By the same token, HIV-1 mutates extremely rapidly; B cells in HIV-1 patients constantly mutate but nevertheless fail to keep up with the more rapid mutations of the virus so that antibodies present in the serum of a patient fail to neutralize the patient's current viruses but rather recognize viruses that were present months ago.[523] Other viruses, such as CMV, EBV, HCV, influenza A virus, and norovirus, infect B cells directly. They can subvert B cell responses at several levels.[524] Measles morbillivirus and EBV prevent B cell activation.[525,526] HCV through binding to CD81, also called target of the antiproliferative antibody 1 (TAPA-1), triggers proliferation of naïve B cells.[527] HCV also induces hypermutations of Ig genes resulting in antibodies with low specificity to the viral antigens.[528] The nucleocapsid of measles morbillivirus binds to FcgRII and thereby inhibits antibody production. Influenza virus can infect specific memory B cells and kill them.[529] HBV produces an excess of noninfectious virus particles expressing the S antigen, which soak up neutralizing antibodies. Ebola virus produces large qualities of truncated secreted glycoprotein that bind antibodies and thereby distract them from attacking the virus.[530] Several viruses, such as Ebola virus,[531] influenza A virus,[532] or arenaviruses,[533] cover their glycoproteins with a glycan shield that prevents binding of neutralizing antibodies.

Other viruses induce B_{regs} such as HBV or HIV-1 during chronic infections.[534] LMP1 of EBV inhibits B cell differentiation into germinal B cells or plasma cells[535] while LMP2A enhances BCR signaling driving premature differentiation of plasma cells, which secrete antibodies that have not yet undergone full affinity maturation.[536]

The Mucosal Immune System

The mucosal immune system provides the first layer of defense against viruses invading through the skin or mucosal surfaces of the airways, the gut, the genital tract, the eye, and surface-linked glands (see Chapter 38). The mucosal immune system thus covers a huge area of nearly 400 m^2 in a human adult. Unlike the systemic immune system, the mucosal immune system faces the challenge to ignore beneficial or innocuous antigens such as those present in food and in the 10^{14} commensal microorganisms that live on mucosal surfaces, where they protect against more pathogenic microorganism and provide nutritional support. At the same time, the mucosal immune system must rapidly react to incoming threats.

Anatomically mucosa-associated lymphoid tissues (MALT) are divided into inductive sites such as the tonsils and adenoids at the back of the oral cavity, Peyer patches along the small intestine and bronchus-associated lymphoid tissues in the respiratory tract. These structures resemble systemic lymph nodes with some notable differences such as a lack of afferent lymphatics, the presence of high-endothelial regions, and lack of a defined capsule. Peyer patches are covered by M cells that take up pathogens and antigens by endocytosis or phagocytosis from the gut lumen. These are then moved by transcytosis through the M cells' basal membranes, which are covered by antigen-presenting cells and lymphocytes. The effector sites are within the mucosal epithelium. For example, the lamina propria of the guts, the multilayered squamous epithelium of the female genital tract, or the ciliated columnar epithelium of the airways contain both T and B cells. During activation at inductive sites, dendritic cells through the release of retinoic acid imprint gut-homing on responding lymphocytes by triggering upregulation of $\alpha4\beta7$ and CCR9. Activated mucosal B and T cells traffic through lymph vessels to lymph nodes such as the mesenteric lymph nodes to the thoracic duct and from there through blood to mucosal sites.

CD8+ T cells within the mucosal immune system can be conventional α/β T cells, which carry the CD8α/ß chains; they are common in the lamina propria. Alternatively, mucosal T cells carry a γ/∂ TCR or a γ/δ TCR with a CD8α homodimer, the latter 2 cell types are more common in the intraepithelial lymphocyte population of the gut epithelium. γ/δ T cells have a restricted T cell repertoire, respond to lipids and glycolipids presented by MHC 1b antigens such as HLA-E. They play a role in maintaining the integrity of the intestinal epithelium by regulating its turnover and repair.[537] α/β T cells with a CD8α homodimer respond to antigen presented by Qa-1. Conventional CD8+ T cell reside in the lamina propria in form of effector memory cells that can respond very rapidly by release of lytic enzymes or cytokines but that have limited proliferative potential. CD4+ T cells are enriched for Th17 cells and natural T_{regs} and iT_{regs}; the latter prevent immune responses against the gut microbiome. The fate decision between T_{reg} and Th17 differentiation is affected by mucosal dendritic cells, which at a resting stage produce retinoic acid, which favors T_{reg} formation, while activated dendritic cells produce IL-6, which drives Th17 formation.

Inductive sites such as Peyer patches contain naïve B cells and FDCs. They can form germinal centers for activation of B2-B cells into affinity-matured plasma cells, which then migrate to mucosal effector sites, where they produce mainly IgA1 and IgA2, the latter is more common in parts of the mucosa that are colonized by microorganisms. IgA can be produced more rapidly by B1-B cells within the mucosa that become activated in response to TLR ligation. IgA that is secreted into the lumen of the gut, the airways, or the genital tract can trap pathogens within the mucus layer that covers the epithelium. Antigen that crosses the epithelial barrier can bind to IgA and then be transported back to the lumen. IgA can bind antigen to M cells and facilitate its transport to underlying antigen-presenting cells. The intestinal barrier is crucial to contain the luminal content while allowing absorption of nutrients. Disruption of the barrier results in local and systemic inflammatory responses. HIV-1 enters through mucosal surfaces of the genital or rectal tract by transcytosis, by entering dendrites of intraepithelial dendritic cells, or by moving through spaces between epithelial cells. The gp120 of HIV-1 then disrupts the mucosal barrier resulting in local cytokine release and T cell infiltration, providing the virus with an ample selection of additional target cells.[538]

Effects of Viruses and Anti-Viral Immune Responses on the Microbiome and *Vice Versa*

The microbiome is shaped by trillions of microorganisms living on mucosal surfaces in harmony with their host organism. The microbiome is influenced by age, region of residence, diet, and lifestyle and in turn influences risks of heart disease, cancer, autism, obesity, inflammatory bowel diseases, etc. It also affects immune responses to pathogens and vaccines. The immune system shapes the microbiome and *vice versa* the microbiome shapes the immune system by activating innate responses. The microbiome ensure integrity of the intestinal barrier by driving proliferation of epithelial cells.[539] It is essential for development of the anatomic structures of the mucosal immune system as was shown with germfree mice that have a paucity of Peyer patches, lower numbers of intraepithelial lymphocytes, and small mesenteric lymph nodes with poorly formed high-endothelial venules.[540] The microbiome also shapes the systemic immune system, which is underdeveloped in germfree mice.[541] Microbes within the intestine increase secretion of antimicrobial peptides from intestinal cells.[542] They modulate expression of PRRs[543] and impact the Th17/T_{reg} balance.[544,545] The microbiota is required for induction of protective adaptive immune responses against viruses; germfree mice develop lower antibody and CD8+ T cell responses to MCMV, and this can be corrected by oral administration of fecal suspension from normal mice.[546] Mice treated with antibiotics to destroy their microbiome are more susceptive to influenza A virus infections, which is linked to poor B and CD8+ T cell responses.[547] Chronic viral infections change the microbiome. HIV-1 patients even upon treatment with antiviral drugs suffer from chronic immune activation; they have a higher content of opportunistic pathogens such *as Pseudomonas aeruginosa* and *Candida albicans* and reduced numbers of commensal bacteria including *Bifidobacteria* and *Lactobacilli* in their intestine.[548]

The microbiome can protect against viral infections. *Saccharomyces cerevisiae* species reduce infections with rotavirus, a virus that causes watery diarrhea mainly in children by inducing chloride secretion and ROS production; this can be blocked by factors secreted by *Saccharomyces boulardii*.[549] Gut bacteria modify the glycosylation patterns of intestinal cells by increasing the content of galactose within glycans, this in turn blocks infection with rotavirus. Glycan-modifying bacteria-derived soluble factors from *Bacteroides thetaiotaomicron* and *Lactobacillus casei* inhibit rotavirus infection of human intestinal cells.[550] *Lactobacillus* species provide protection against sexual transmission of HIV-1 to reproductive-age women by acidification of the vaginal fluid, by trapping HIV-1 virions, and by secretion of lactic acid, which inhibits HIV-1 replication in a pH-independent manner.[551] Desaminotyrosine, a microbiome-derived metabolite, protects from influenza A virus infection by augmenting type I IFN signaling.[552] *Staphylococcus aureus* through TLR2 activation protects against influenza A virus–induced lung pathology by polarizing monocytes toward M2 alveolar macrophages, which reduce inflammatory reactions.

On the other hand, the microbiome can also promote virus infections. For example, the microbiome increases norovirus infection of B cells and increases virus replication.[553] By binding to norovirus, intestinal bacteria promote viral persistence.[554] Components of the microbiome, such as lipopolysaccharide and peptidoglycan, increase infections with poliovirus and reovirus.[555] Through TLR5-mediated activation of the NF-κB pathway, bacterial flagellin facilitates uptake of a number of viruses.[556] Binding of reovirus to lipoteichoic acid and N-acetylglucosamine-containing polysaccharides on gut bacteria enhances its thermostability and therefore its infectivity.[557]

The Aged Immune System

Immune responses become impaired during aging, resulting in an increased susceptibility to viruses and reduced responses to vaccines. Upon immunosenescence, a term that refers to age-related changes in immunity, both innate and adaptive immune responses change. Aging is linked to a decreased ability to fight infections combined with a persistent low-grade inflammation. Bone marrow shows reduced cellularity, and although hematopoietic stem cells increase in numbers, they become more prone to differentiate into myeloid cells rather than lymphocytes, which results in a peripheral decline in naïve B and T cells.[558] As thymus functions are gradually lost during adulthood, human naïve T cells are maintained through homeostatic proliferation rather than by being replenished from bone marrow.[559] This can lead to a reduction in the T cell repertoire, which is also affected by an increase in memory T cells directed to persisting viruses such as HCMV; they become more frequent during aging and outcompete naïve T cells for space and growth factors.[560,561] T cell functions decline, and this has been linked to signaling defects.[562,563] Aged T cells also increase expression of coinhibitory markers that are typically associated with T cell exhaustion.[564] Numbers of B cell precursors decline with age.[565] In the periphery, naïve B cells become scarce while relative proportions of memory B cells increase.[566] The B cell repertoire shrinks, especially in the frail elderly.[567] Aged B cells show reduced class-switch recombination and hypermutation resulting in fewer functional antibodies. This is caused by a decline in transcription factors that induce AID.[568] Upon activation, T and B cells rapidly change their metabolism from mitochondrial energy production toward glycolysis. Aging in turn is associated with a wide range of metabolic defects such as enhanced resistance to insulin, defects in the adsorption of nutrients, and mitochondrial decay, which further affect the ability of T and B cells to quickly respond to a threat.[569-571]

Innate cell functions decline with age. Although their numbers do not decrease in the elderly, neutrophils and macrophages show reduced expression of PRRs, produce less superoxide, and their chemotaxis decreases. Numbers of NK cells tend to increase in the aged, but their functions decline.[572,573] A number of factors contribute to the heightened baseline inflammation of the aged, which is commonly referred to as "inflammaging." Chronic stimulation, by persisting viruses, declines in adaptive responses, increases in exogenous antigens such as cell debris, loss of the integrity of the intestinal barrier in part caused by age-related changes in the microbiome, and increases in adipose tissue may contribute to heightened baseline inflammation.[574-577] Inflammaging is a considerable risk factor for morbidity and mortality of the elderly by increasing degenerative diseases such as Alzheimer, diabetes, osteoporosis, or arteriosclerosis.

Overall declines in immune functions render the aged more susceptible to virus infection. Infections with influenza A virus are by far more severe in the aged and 90% of influenza-related death occur in the elderly. The aged show higher mortality upon infection with West Nile virus or SARS-CoV-2. In addition, the incidence of reactivation of persisting viruses, such as varicella-zoster virus causing shingles, increases in the aged. Exercise and a healthy diet reduce or at least delay the impact of aging on the immune system.[578,579]

Viruses and Autoimmunity

Autoimmune reactions are pathological immune responses that react to self-antigens (see Chapter 46). Although the immune system has multiple control mechanisms to avoid "horror autotoxicus," more than 80 type of different autoimmune diseases have been identified in humans to date. Viruses can trigger autoimmunity through a variety of mechanisms. The inflammatory reaction to a virus infection combined with increased release of self-proteins by damaged or dying cells can trigger "bystander" activation of autoreactive lymphocytes exemplified by West Nile virus–mediated myasthenia gravis.[580] Antigenic mimicry of a self-protein by a viral antigen can lead to responses that cross-react with the virus and a self-component, as has been described for HSV, that can trigger stromal keratitis, coxsackievirus, which can cause an autoimmune myocarditis, and Theiler murine encephalitis virus, which can trigger nervous system demyelination.[581-583]

Antibodies to the EBV nuclear antigen 1 cross-react with a lupus-associated autoantigen.[487,488] In addition, EBV can immortalize B cells including those directed to self-proteins.[585] Comparisons of the human proteome with that of some common viruses revealed massive cross-reactivities suggesting that autoimmunity by molecular mimicry is most likely very common.[586]

Viruses and Cancer

About 15% of all human cancers are linked to persistent virus infections. Cancer-causing viruses include EBV, oncogenic types of HPV, HHV8 (KSHV), HCV, HBV, and HTLV-1. These viruses do not cause cancer directly but rather predispose a cell toward malignant transformation. HPV, a small circular DNA virus, can integrate into the genome. It expresses oncoproteins such as E6, which degrades p53, and E7, which affects Rb. Although most patients are able to eliminate HPV-infected cells through CD8+ T cell–mediated lysis, in some, especially those with immunodeficiencies, the virus persists and can eventually lead to cancer. EBV is linked to B and T cell lymphomas, Hodgkin disease, post-transplant lymphoproliferative disease, leiomyosarcomas, and nasopharyngeal carcinomas, while HHV8 causes Kaposi sarcoma[587] and primary effusion lymphoma. Several of these herpesvirus-associated cancers are rare in immunocompetent patients but are common in patients that are severely immunosuppressed such as AIDS patients or transplant recipients. While HPV, EBV, and HHV8 encode proteins that contribute to cell transformation, sustained inflammation of the liver combined with DNA damage of hepatocytes during chronic HBV and HCV infections can lead to hepatocellular carcinoma.

The Use of Viruses for Prevention or Therapy of Diseases

Genes of viruses can be modified by insertion of foreign sequences such as those encoding the antigen of another pathogen. This approach, which can turn a virus into vaccine carrier, has been tested preclinically for numerous viruses, including DNA and RNA viruses. Needless to say, viruses that serve as vaccine carriers need to be highly attenuated or additional modifications have to be introduced that cripple the virus. One viral vector, a canarypox virus, expressing the rabies virus glycoprotein is now licensed for use in veterinary medicine.[588] A replication-competent adenovirus vector expressing the rabies virus glycoprotein and a vaccinia virus vector expressing the same antigen are used for immunization of wild-life.[589,590]

Viruses have also been harnessed to fix genetic diseases. Adeno-associated vectors, in which the genome but for the inverted terminal repeats (ITRs), has been replaced by an expression cassette for a missing or faulty gene and which are then grown on cell lines that provide the viral genes in trans are licensed for correction of Leber congenital amaurosis, a congenital form of blindness.[591]

Viruses can be used to treat cancer. Lytic viruses, such as influenza virus, rabies virus, and SARS, kill the cells in which they have replicated. Other viruses, such as herpes viruses, HBV, or HIV, are lysogenic, which means they switch from a nonlytic to a lytic cycle. Due to abnormalities in cell signaling, apoptotic pathways, and stress responses, cancer cells are commonly more susceptible to virus infections than normal cells.[592] Cell-lysis induced by virus infections can halt cancer growth at least temporarily as has been known since the late 1800 when it was first reported that some cancer patients spontaneously undergo remission upon a viral infection.[593] In 1949 Moore showed in a rodent model that a virus, in this case Russian Far East encephalitis virus, could completely eradicate a tumor.[594] Other reports use a plethora of viruses, such as yellow fever virus,[595] West Nile virus,[596] vaccinia virus,[597] and adenovirus,[598] in mice or humans. Animal viruses were pursued next under the assumption that they were less likely to cause disease in humans. For example, Newcastle disease virus was first shown to be efficacious in mice[599] followed by its use in human melanoma patients.[600] Vesicular stomatitis virus, a rhabdovirus, was shown to selectively eliminate human tumor cells with an IFN-pathway defect.[601] With advances in molecular biology, methods became available to manipulate virus genomes so that they could be modified to preferentially attack and kill cancer cells. For example, measles virus has been engineered to express single-chain antibodies to carcinoembryonic antigen, which is selectively expressed by adenocarcinomas.[602] Oncolytic viruses not only halt tumor progression by killing infected tumor cells, but during this process, they also cause release of tumor antigens, which can trigger an adaptive antitumor immune response. This can be enhanced by modifying oncolytic viruses to produce immunostimulatory molecules. To name an example, incorporation of a GM-CSF–encoding sequence into HSV-1 increased its ability to cause tumor rejection in mice.[603] Others increased the efficacy of oncolytic viruses by incorporating suicide genes into the viral genome.[604] By now, a number of oncolytic viruses including HSV-1, adenovirus, vaccinia virus, and others are in late-stage clinical trials.[605] In 2015, the first oncolytic virus, an HSV-1 virus, was licensed for use in melanoma patients by the United States Food and Drug Administration.

CONCLUSION

Each year millions of humans die prematurely due to viral infections and many more become temporarily incapacitated. At the beginning of 2020, a pandemic with the new SARS-CoV-2 virus swept the globe and forces governments to close schools and businesses and issue "shelter-in-place" orders to reduce further spread of the virus and prevent a complete breakdown of healthcare services. Vaccines and antiviral drugs licensed for a handful of viruses have lessened the impact of some viral infections, but for most, the immune system remains our only defense. Our knowledge in molecular virology and basic immunology has increased tremendously over the last 20 years. Hopefully, with a better understanding of the intricate pathways that regulate the interactions between viruses and their hosts' immune system, humans will eventually win this eternal battle and lessen human suffering caused by invasive bits of genomes with a few proteins around it. Nevertheless, considering how easily viruses mutate and change their host range, this is far from certain.

REFERENCES

1. Nasir A, Caetano-Anolles G. A phylogenomic data-driven exploration of viral origins and evolution. *Sci Adv*. 2015;1:e1500527.
2. Xu P, Chen F, Mannas JP, Feldman T, Sumner LW, Roossinck MJ. Virus infection improves drought tolerance. *New Phytol*. 2008;180:911-921.
3. Barton ES, White DW, Cathelyn JS, et al. Herpesvirus latency confers symbiotic protection from bacterial infection. *Nature*. 2007;447:326-329.
4. Perry GH, Dominy NJ, Claw KG, et al. Diet and the evolution of human amylase gene copy number variation. *Nat Genet*. 2007;39:1256-1260.
5. Lavialle C, Cornelis G, Dupressoir A, et al. Paleovirology of '*syncytins*', retroviral *env* genes exapted for a role in placentation. *Philos Trans R Soc Lond B Biol Sci*. 2013;368:20120507.
6. Woolhouse M, Scott F, Hudson Z, Howey R, Chase-Topping M. Human viruses: discovery and emergence. *Philos Trans R Soc Lond B Biol Sci*. 2012;367:2864-2871.
7. Sharp PM, Hahn BH. The evolution of HIV-1 and the origin of AIDS. *Philos Trans R Soc Lond B Biol Sci*. 2010;365:2487-2494.
8. Li W. Bats are natural reservoirs of SARS-like coronaviruses. *Science*. 2005;310:676-679.
9. Van Kerkhove MD, Hirve S, Koukounari A, Mounts AW; H1N1pdm Serology Working Group. Estimating age-specific cumulative incidence for the 2009 influenza pandemic: a meta-analysis of A(H1N1)pdm09 serological studies from 19 countries. *Influenza Other Respir Viruses*. 2013;7:872-886.
10. Short KR, Kedzierska K, van de Sandt CE. Back to the future: lessons learned from the 1918 influenza pandemic. *Front Cell Infect Microbiol*. 2018;8:343. https://www.frontiersin.org/article/10.3389/fcimb.2018.00343/full
11. Lwoff A, Horne R, Tournier P. A system of viruses. *Cold Spring Harb Symp Quant Biol*. 1962;27:51-55.
12. Lefkowitz EJ, Dempsey DM, Hendrickson RC, Orton RJ, Siddell SG, Smith DB. Virus taxonomy: the database of the International Committee on Taxonomy of Viruses (ICTV). *Nucleic Acids Res*. 2018;46:D708-D717.
13. Waddington SN, McVey JH, Bhella D, et al. Adenovirus serotype 5 hexon mediates liver gene transfer. *Cell*. 2008;132:397-409.
14. Mercer J, Schelhaas M, Helenius A. Virus entry by endocytosis. *Annu Rev Biochem*. 2010;79:803-833.
15. Poste G. *Mechanisms of virus-induced cell fusion*. In: *International Review of Cytology*. Elsevier; 1972:157-252. [cited February 26, 2020]. https://linkinghub.elsevier.com/retrieve/pii/S0074769608614515
16. Wickham TJ, Mathias P, Cheresh DA, Nemerow GR. Integrins $\alpha_v\beta_3$ and $\alpha_v\beta_5$ promote adenovirus internalization but not virus attachment. *Cell*. 1993;73:309-319.
17. Wiethoff CM, Nemerow GR. Adenovirus membrane penetration: Tickling the tail of a sleeping dragon. *Virology*. 2015;479-480:591-599.
18. Fay N, Panté N. Nuclear entry of DNA viruses. *Front Microbiol*. 2015;6:467. http://www.frontiersin.org/Virology/10.3389/fmicb.2015.00467/abstract
19. Vanarsdall AL, Johnson DC. Human cytomegalovirus entry into cells. *Curr Opin Virol*. 2012;2:37-42.
20. Rhein BA, Maury WJ. Ebola virus entry into host cells: identifying therapeutic strategies. *Curr Clin Microbiol Rep*. 2015;2:115-124.
21. Dou D, Revol R, Östbye H, Wang H, Daniels R. Influenza A virus cell entry, replication, virion assembly and movement. *Front Immunol*. 2018;9:1581. https://www.frontiersin.org/article/10.3389/fimmu.2018.01581/full
22. Watashi K, Wakita T. Hepatitis B virus and hepatitis D virus entry, species specificity, and tissue tropism. *Cold Spring Harb Perspect Med*. 2015;5:a021378.
23. Agelidis AM, Shukla D. Cell entry mechanisms of HSV: what we have learned in recent years. *Future Virol*. 2015;10:1145-1154.
24. Wilen CB, Tilton JC, Doms RW. HIV: cell binding and entry. *Cold Spring Harb Perspect Med*. 2012;2:a006866.
25. Bubeck D, Filman DJ, Cheng N, Steven AC, Hogle JM, Belnap DM. The structure of the poliovirus 135S cell entry intermediate at 10-angstrom resolution reveals the location of an externalized polypeptide that binds to membranes. *J Virol*. 2005;79:7745-7755.
26. Jordan TX, Randall G. Dengue virus activates the AMP kinase-mTOR axis to stimulate a proviral lipophagy. *J Virol*. 2017;91:e02020-e02116. http://jvi.asm.org/lookup/doi/10.1128/JVI.02020-16
27. Kuss-Duerkop SK, Wang J, Mena I, et al. Influenza virus differentially activates mTORC1 and mTORC2 signaling to maximize late stage replication. *PLoS Pathog*. 2017;13:e1006635.
28. Le Sage V, Cinti A, Amorim R, Mouland A. Adapting the stress response: viral subversion of the mTOR signaling pathway. *Viruses*. 2016;8:152.
29. Martin R, Nauta AJ, Ben Amor K, Knippels LMJ, Knol J, Garssen J. Early life: gut microbiota and immune development in infancy. *Benef Microbes*. 2010;1:367-382.
30. Meade N, Furey C, Li H, et al. Poxviruses evade cytosolic sensing through disruption of an mTORC1-mTORC2 regulatory circuit. *Cell*. 2018;174:1143-1157.e17.
31. Stöhr K, Meslin FM. Progress and setbacks in the oral immunisation of foxes against rabies in Europe. *Vet Rec*. 1996;139:32-35.
32. Moreno-Altamirano MMB, Kolstoe SE, Sánchez-García FJ. Virus control of cell metabolism for replication and evasion of host immune responses. *Front Cell Infect Microbiol*. 2019;9:95. https://www.frontiersin.org/article/10.3389/fcimb.2019.00095/full

33. Ahi YS, Mittal SK. Components of adenovirus genome packaging. *Front Microbiol.* 2016;7:1503. http://journal.frontiersin.org/Article/10.3389/fmicb.2016.01503/abstract
34. Johnson N, Fooks A, McColl K. Human rabies case with long incubation, Australia. *Emerg Infect Dis.* 2008;14:1950-1951.
35. Mina MJ, Kula T, Leng Y, et al. Measles virus infection diminishes preexisting antibodies that offer protection from other pathogens. *Science.* 2019;366:599-606.
36. Lallemand-Breitenbach V, de The H. PML nuclear bodies. *Cold Spring Harb Perspect Med.* 2010;2:a000661.
37. Staněk D, Fox AH. Nuclear bodies: news insights into structure and function. *Curr Opin Cell Biol.* 2017;46:94-101.
38. Ishov AM, Maul GG. The periphery of nuclear domain 10 (ND10) as site of DNA virus deposition. *J Cell Biol.* 1996; 134:815-826.
39. Everett RD. DNA viruses and viral proteins that interact with PML nuclear bodies. *Oncogene.* 2001;20:7266-7273.
40. Müller S, Dejean A. Viral immediate-early proteins abrogate the modification by SUMO-1 of PML and Sp100 proteins, correlating with nuclear body disruption. *J Virol.* 1999;73:5137-5143.
41. Unterholzner L, Keating SE, Baran M, et al. IFI16 is an innate immune sensor for intracellular DNA. *Nat Immunol.* 2010;11:997-1004.
42. Orzalli MH, DeLuca NA, Knipe DM. Nuclear IFI16 induction of IRF-3 signaling during herpesviral infection and degradation of IFI16 by the viral ICP0 protein. *Proc Natl Acad Sci U S A.* 2012;109:E3008-E3017.
43. Orzalli MH, Conwell SE, Berrios C, DeCaprio JA, Knipe DM. Nuclear interferon-inducible protein 16 promotes silencing of herpesviral and transfected DNA. *Proc Natl Acad Sci U S A.* 2013;110:E4492-E4501.
44. Tognarelli EI, Palomino TF, Corrales N, Bueno SM, Kalergis AM, González PA. Herpes simplex virus evasion of early host antiviral responses. *Front Cell Infect Microbiol.* 2019;9:127.
45. Herbert K, Nag A. A Tale of two RNAs during viral infection: how viruses antagonize mRNAs and small non-coding RNAs in the host cell. *Viruses.* 2016;8:154.
46. Hamilton AJ. A species of small antisense RNA in posttranscriptional gene silencing in plants. *Science.* 1999;286:950-952.
47. Cullen BR. Viruses and RNA interference: issues and controversies. *J Virol.* 2014;88:12934-12936.
48. Gupta A, Gartner JJ, Sethupathy P, Hatzigeorgiou AG, Fraser NW. Retraction Note: anti-apoptotic function of a microRNA encoded by the HSV-1 latency-associated transcript. *Nature.* 2008;451:600.
49. Kincaid RP, Sullivan CS. Virus-encoded microRNAs: an overview and a look to the future. *PLoS Pathog.* 2012;8:e1003018.
50. Sullivan CS, Grundhoff AT, Tevethia S, Pipas JM, Ganem D. SV40-encoded microRNAs regulate viral gene expression and reduce susceptibility to cytotoxic T cells. *Nature.* 2005; 435:682-686.
51. Jopling CL. Modulation of hepatitis C virus RNA abundance by a liver-specific microRNA. *Science.* 2005;309:1577-1581.
52. Otsuka M, Jing Q, Georgel P, et al. Hypersusceptibility to vesicular stomatitis virus infection in Dicer1-deficient mice is due to impaired miR24 and miR93 expression. *Immunity.* 2007;27:123-134.
53. Triboulet R, Mari B, Lin YL, et al. Suppression of microRNA-silencing pathway by HIV-1 during virus replication. *Science.* 2007;315:1579-1582.
54. Pedersen IM, Cheng G, Wieland S, et al. Interferon modulation of cellular microRNAs as an antiviral mechanism. *Nature.* 2007;449:919-922.
55. Lee S, Song J, Kim S, et al. Selective degradation of host MicroRNAs by an intergenic HCMV noncoding RNA accelerates virus production. *Cell Host Microbe.* 2013;13:678-690.
56. Scheel TKH, Luna JM, Liniger M, et al. A broad RNA virus survey reveals both miRNA dependence and functional sequestration. *Cell Host Microbe.* 2016;19:409-423.
57. Agarraberes FA, Terlecky SR, Dice JF. An intralysosomal hsp70 is required for a selective pathway of lysosomal protein degradation. *J Cell Biol.* 1997;137:825-834.
58. Kunz JB, Schwarz H, Mayer A. Determination of four sequential stages during microautophagy *in vitro*. *J Biol Chem.* 2004;279:9987-9996.
59. Feng Y, He D, Yao Z, Klionsky DJ. The machinery of macroautophagy. *Cell Res.* 2014;24:24-41.
60. Tian Y, Wang ML, Zhao J. Crosstalk between autophagy and type I interferon responses in innate antiviral immunity. *Viruses.* 2019;11:132.
61. Py BF, Boyce M, Yuan J. A critical role of eEF-2K in mediating autophagy in response to multiple cellular stresses. *Autophagy.* 2009;5:393-396.
62. Tallóczy Z, Virgin HW, Levine B. PKR-dependent autophagic degradation of herpes simplex virus type 1. *Autophagy.* 2006;2:24-29.
63. Orvedahl A, Alexander D, Tallóczy Z, et al. HSV-1 ICP34.5 confers neurovirulence by targeting the Beclin 1 autophagy protein. *Cell Host Microbe.* 2007;1:23-35.
64. Pattingre S, Tassa A, Qu X, et al. Bcl-2 antiapoptotic proteins inhibit Beclin 1-dependent autophagy. *Cell.* 2005;122:927-939.
65. Wei Y, Pattingre S, Sinha S, Bassik M, Levine B. JNK1-mediated phosphorylation of Bcl-2 regulates starvation-induced autophagy. *Mol Cell.* 2008;30:678-688.
66. Orvedahl A, Levine B. Viral evasion of autophagy. *Autophagy.* 2008;4:280-285.
67. Sandor F, Latz E, Re F, et al. Importance of extra-and intracellular domains of TLR1 and TLR2 in NFkappa B signaling. *J Cell Biol.* 2003;162:1099-1110.
68. Boehme KW, Guerrero M, Compton T. Human cytomegalovirus envelope glycoproteins B and H are necessary for TLR2 activation in permissive cells. *J Immunol.* 2006;177:7094-7102.
69. Leoni V, Gianni T, Salvioli S, Campadelli-Fiume G. Herpes simplex virus glycoproteins gH/gL and gB bind Toll-like receptor 2, and soluble gH/gL is sufficient to activate NF-κB. *J Virol.* 2012;86:6555-6562.
70. Ariza ME, Glaser R, Kaumaya PTP, Jones C, Williams MV. The EBV-encoded dUTPase activates NF-kappa B through the TLR2 and MyD88-dependent signaling pathway. *J Immunol.* 2009;182:851-859.
71. Chang S, Dolganiuc A, Szabo G. Toll-like receptors 1 and 6 are involved in TLR2-mediated macrophage activation by hepatitis C virus core and NS3 proteins. *J Leukoc Biol.* 2007;82:479-487.
72. Ge Y, Mansell A, Ussher JE, et al. Rotavirus NSP4 triggers secretion of proinflammatory cytokines from macrophages via Toll-like receptor 2. *J Virol.* 2013;87:11160-11167.
73. Bieback K, Lien E, Klagge IM, et al. Hemagglutinin protein of wild-type measles virus activates Toll-like receptor 2 signaling. *J Virol.* 2002;76:8729-8736.
74. Zhu J, Martinez J, Huang X, Yang Y. Innate immunity against vaccinia virus is mediated by TLR2 and requires TLR-independent production of IFN-β. *Blood.* 2007;109:619-625.
75. Zhou S, Kurt-Jones EA, Mandell L, et al. MyD88 is critical for the development of innate and adaptive immunity during acute lymphocytic choriomeningitis virus infection. *Eur J Immunol.* 2005;35:822-830.
76. Wang JP, Kurt-Jones EA, Shin OS, Manchak MD, Levin MJ, Finberg RW. Varicella-zoster virus activates inflammatory cytokines in human monocytes and macrophages via Toll-like receptor 2. *J Virol.* 2005;79:12658-12666.
77. Murawski MR, Bowen GN, Cerny AM, et al. Respiratory syncytial virus activates innate immunity through Toll-like receptor 2. *J Virol.* 2009;83:1492-1500.
78. Kurt-Jones EA, Popova L, Kwinn L, et al. Pattern recognition receptors TLR4 and CD14 mediate response to respiratory syncytial virus. *Nat Immunol.* 2000;1:398-401.
79. Georgel P, Jiang Z, Kunz S, et al. Vesicular stomatitis virus glycoprotein G activates a specific antiviral Toll-like receptor 4-dependent pathway. *Virology.* 2007;362:304-313.
80. Okumura A, Pitha PM, Yoshimura A, Harty RN. Interaction between Ebola virus glycoprotein and host Toll-like receptor 4 leads to induction of proinflammatory cytokines and SOCS1. *J Virol.* 2010;84:27-33.
81. Shinya K, Ito M, Makino A, et al. The TLR4-TRIF pathway protects against H5N1 influenza virus infection. *J Virol.* 2012;86:19-24.
82. Zhao J, Wohlford-Lenane C, Zhao J, et al. Intranasal treatment with poly(I·C) protects aged mice from lethal respiratory virus infections. *J Virol.* 2012;86:11416-11424.
83. Hutchens MA, Luker KE, Sonstein J, Núñez G, Curtis JL, Luker GD. Protective effect of Toll-like receptor 4 in pulmonary vaccinia infection. *PLoS Pathog.* 2008;4:e1000153.
84. Lagos D, Vart RJ, Gratrix F, et al. Toll-like receptor 4 mediates innate immunity to Kaposi sarcoma herpesvirus. *Cell Host Microbe.* 2008;4:470-483.
85. Said EA, Tremblay N, Al-Balushi MS, Al-Jabri AA, Lamarre D. Viruses seen by our cells: the role of viral RNA sensors. *J Immunol Res.* 2018;2018:9480497.
86. Lester SN, Li K. Toll-like receptors in antiviral innate immunity. *J Mol Biol.* 2014;426:1246-1264.
87. Ahmad H, Gubbels R, Ehlers E, et al. Kaposi sarcoma-associated herpesvirus degrades cellular Toll-interleukin-1 receptor domain-containing adaptor-inducing β-interferon (TRIF). *J Biol Chem.* 2011;286:7865-7872.
88. Mukherjee A, Morosky SA, Delorme-Axford E, et al. The coxsackievirus B 3C protease cleaves MAVS and TRIF to attenuate host type I interferon and apoptotic signaling. *PLoS Pathog.* 2011;7:e1001311.
89. Barro M, Patton JT. Rotavirus nonstructural protein 1 subverts innate immune response by inducing degradation of IFN regulatory factor 3. *Proc Natl Acad Sci U S A.* 2005;102:4114-4119.
90. Harte MT, Haga IR, Maloney G, et al. The poxvirus protein A52R targets Toll-like receptor signaling complexes to suppress host defense. *J Exp Med.* 2003;197:343-351.
91. Stack J, Haga IR, Schröder M, et al. Vaccinia virus protein A46R targets multiple Toll-like-interleukin-1 receptor adaptors and contributes to virulence. *J Exp Med.* 2005;201:1007-1018.
92. Abe T, Kaname Y, Hamamoto I, et al. Hepatitis C virus nonstructural protein 5A modulates the Toll-like receptor-MyD88-dependent signaling pathway in macrophage cell lines. *J Virol.* 2007;81:8953-8966.
93. Devaraj SG, Wang N, Chen Z, et al. Regulation of IRF-3-dependent innate immunity by the papain-like protease domain of the severe acute respiratory syndrome coronavirus. *J Biol Chem.* 2007;282:32208-32221.
94. Melroe GT, Silva L, Schaffer PA, Knipe DM. Recruitment of activated IRF-3 and CBP/p300 to herpes simplex virus ICP0 nuclear foci: potential role in blocking IFN-β induction. *Virology.* 2007;360:305-321.
95. Ren J, Liu T, Pang L, et al. A novel mechanism for the inhibition of interferon regulatory factor-3-dependent gene expression by human respiratory syncytial virus NS1 protein. *J Gen Virol.* 2011;92:2153-2159.
96. Otsuka M, Kato N, Moriyama M, et al. Interaction between the HCV NS3 protein and the host TBK1 protein leads to inhibition of cellular antiviral responses. *Hepatology.* 2005;41:1004-1012.
97. Lu LL, Puri M, Horvath CM, Sen GC. Select paramyxoviral V proteins inhibit IRF3 activation by acting as alternative substrates for inhibitor of kappaB kinase epsilon (IKKe)/TBK1. *J Biol Chem.* 2008;283:14269-14276.
98. Ben Khalifa Y, Luco S, Besson B, et al. The matrix protein of rabies virus binds to RelAp43 to modulate NF-κB-dependent gene expression related to innate immunity. *Sci Rep.* 2016;6:39420.
99. Ortega-Prieto AM, Dorner M. Immune evasion strategies during chronic hepatitis B and C virus infection. *Vaccines (Basel).* 2017;5:24.
100. Wang Y, Ma L, Stipkovits L, Szathmary S, Li X, Liu Y. The strategy of picornavirus evading host antiviral responses: non-structural proteins suppress the production of IFNs. *Front Microbiol.* 2018;9:2943.
101. Qian W, Wei X, Guo K, et al. The C-terminal effector domain of non-structural protein 1 of influenza A virus blocks IFN-β production by targeting TNF receptor-associated factor 3. *Front Immunol.* 2017;8:779.
102. Chen Y, Chen J, Wang H, et al. HCV-induced miR-21 contributes to evasion of host immune system by targeting MyD88 and IRAK1. *PLoS Pathog.* 2013;9:e1003248.
103. Doores KJ. The HIV glycan shield as a target for broadly neutralizing antibodies. *FEBS J.* 2015;282:4679-4691.
104. Kerrigan AM, Brown GD. Syk-coupled C-type lectins in immunity. *Trends Immunol.* 2011;32:151-156.
105. Strasser D, Neumann K, Bergmann H, et al. Syk kinase-coupled C-type lectin receptors engage protein kinase C-δ to elicit Card9 adaptor-mediated innate immunity. *Immunity.* 2012;36:32-42.
106. Kanazawa N, Okazaki T, Nishimura H, Tashiro K, Inaba K, Miyachi Y. DCIR acts as an inhibitory receptor depending on its immunoreceptor tyrosine-based inhibitory motif. *J Invest Dermatol.* 2002;118:261-266.
107. Meyer-Wentrup F, Benitez-Ribas D, Tacken PJ, et al. Targeting DCIR on human plasmacytoid dendritic cells results in antigen presentation and inhibits IFN-α production. *Blood.* 2008;111:4245-4253.
108. Meyer-Wentrup F, Cambi A, Joosten B, et al. DCIR is endocytosed into human dendritic cells and inhibits TLR8-mediated cytokine production. *J Leukoc Biol.* 2009;85:518-525.
109. Gringhuis SI, den Dunnen J, Litjens M, van der Vlist M, Geijtenbeek TBH. Carbohydrate-specific signaling through the DC-SIGN signalosome tailors immunity to *Mycobacterium tuberculosis*, HIV-1 and *Helicobacter pylori*. *Nat Immunol.* 2009;10:1081-1088.
110. Curtis BM, Scharnowske S, Watson AJ. Sequence and expression of a membrane-associated C-type lectin that exhibits CD4-independent binding of human immunodeficiency virus envelope glycoprotein gp120. *Proc Natl Acad Sci U S A.* 1992;89:8356-8360.
111. Lin G, Simmons G, Pöhlmann S, et al. Differential N-linked glycosylation of human immunodeficiency virus and Ebola virus envelope glycoproteins modulates interactions with DC-SIGN and DC-SIGNR. *J Virol.* 2003;77:1337-1346.
112. Halary F, Amara A, Lortat-Jacob H, et al. Human cytomegalovirus binding to DC-SIGN is required for dendritic cell infection and target cell trans-infection. *Immunity.* 2002;17:653-664.
113. Cormier EG, Durso RJ, Tsamis F, et al. L-SIGN (CD209L) and DC-SIGN (CD209) mediate transinfection of liver cells by hepatitis C virus. *Proc Natl Acad Sci U S A.* 2004;101: 14067-14072.
114. de Witte L, Abt M, Schneider-Schaulies S, van Kooyk Y, Geijtenbeek TBH. Measles virus targets DC-SIGN to enhance dendritic cell infection. *J Virol.* 2006;80:3477-3486.
115. Davis CW, Nguyen HY, Hanna SL, Sanchez MD, Doms RW, Pierson TC. West Nile virus discriminates between DC-SIGN and DC-SIGNR for cellular attachment and infection. *J Virol.* 2006;80:1290-1301.
116. Londrigan SL, Turville SG, Tate MD, Deng YM, Brooks AG, Reading PC. N-Linked glycosylation facilitates sialic acid-independent attachment and entry of influenza A viruses into cells expressing DC-SIGN or L-SIGN. *J Virol.* 2011;85:2990-3000.
117. Yang ZY, Huang Y, Ganesh L, et al. pH-dependent entry of severe acute respiratory syndrome coronavirus is mediated by the spike glycoprotein and enhanced by dendritic cell transfer through DC-SIGN. *J Virol.* 2004;78:5642-5650.
118. Lambert AA, Gilbert C, Richard M, Beaulieu AD, Tremblay MJ. The C-type lectin surface receptor DCIR acts as a new attachment factor for HIV-1 in dendritic cells and contributes to *trans*-and *cis*-infection pathways. *Blood.* 2008;112:1299-1307.
119. Long KM, Whitmore AC, Ferris MT, et al. Dendritic cell immunoreceptor regulates Chikungunya virus pathogenesis in mice. *J Virol.* 2013;87:5697-5706.

120. Klechevsky E, Flamar AL, Cao Y, et al. Cross-priming CD8+ T cells by targeting antigens to human dendritic cells through DCIR. *Blood*. 2010;116:1685-1697.
121. Teng O, Chen ST, Hsu TL, et al. CLEC5A-mediated enhancement of the inflammatory response in myeloid cells contributes to influenza virus pathogenicity *in vivo*. *J Virol*. 2017;91.
122. Chen ST, Lin YL, Huang MT, et al. CLEC5A is critical for dengue-virus-induced lethal disease. *Nature*. 2008;453:672-676.
123. Azad AK, Rajaram MVS, Schlesinger LS. Exploitation of the macrophage mannose receptor (CD206) in infectious disease diagnostics and therapeutics. *J Cytol Mol Biol*. 2014;1:1000003.
124. Miller JL, de Wet BJM, Martinez-Pomares L, et al. The mannose receptor mediates dengue virus infection of macrophages. *PLoS Pathog*. 2008;4:e17.
125. Upham JP, Pickett D, Irimura T, Anders EM, Reading PC. Macrophage receptors for influenza A virus: role of the macrophage galactose-type lectin and mannose receptor in viral entry. *J Virol*. 2010;84:3730-3737.
126. Nguyen DG, Hildreth JEK. Involvement of macrophage mannose receptor in the binding and transmission of HIV by macrophages. *Eur J Immunol*. 2003;33:483-493.
127. Iborra S, Izquierdo HM, Martínez-López M, Blanco-Menéndez N, Reis e Sousa C, Sancho D. The DC receptor DNGR-1 mediates cross-priming of CTLs during vaccinia virus infection in mice. *J Clin Invest*. 2012;122:1628-1643.
128. Sancho D, Joffre OP, Keller AM, et al. Identification of a dendritic cell receptor that couples sensing of necrosis to immunity. *Nature*. 2009;458:899-903.
129. Kanneganti TD. Central roles of NLRs and inflammasomes in viral infection. *Nat Rev Immunol*. 2010;10:688-698.
130. Lei Y, Wen H, Yu Y, et al. The mitochondrial proteins NLRX1 and TUFM form a complex that regulates type I interferon and autophagy. *Immunity*. 2012;36:933-946.
131. Kanneganti TD, Body-Malapel M, Amer A, et al. Critical role for Cryopyrin/Nalp3 in activation of caspase-1 in response to viral infection and double-stranded RNA. *J Biol Chem*. 2006;281:36560-36568.
132. Muruve DA, Pétrilli V, Zaiss AK, et al. The inflammasome recognizes cytosolic microbial and host DNA and triggers an innate immune response. *Nature*. 2008;452:103-107.
133. Sabbah A, Chang TH, Harnack R, et al. Activation of innate immune antiviral responses by Nod2. *Nat Immunol*. 2009;10:1073-1080.
134. Neerincx A, Lautz K, Menning M, et al. A role for the human nucleotide-binding domain, leucine-rich repeat-containing family member NLRC5 in antiviral responses. *J Biol Chem*. 2010;285:26223-26232.
135. Kobayashi KS, van den Elsen PJ. NLRC5: a key regulator of MHC class I-dependent immune responses. *Nat Rev Immunol*. 2012;12:813-820.
136. Stasakova J, Ferko B, Kittel C, et al. Influenza A mutant viruses with altered NS1 protein function provoke caspase-1 activation in primary human macrophages, resulting in fast apoptosis and release of high levels of interleukins 1β and 18. *J Gen Virol*. 2005;86:185-195.
137. Komune N, Ichinohe T, Ito M, Yanagi Y. Measles virus V protein inhibits NLRP3 inflammasome-mediated interleukin-1β secretion. *J Virol*. 2011;85:13019-13026.
138. Gerlic M, Faustin B, Postigo A, et al. Vaccinia virus F1L protein promotes virulence by inhibiting inflammasome activation. *Proc Natl Acad Sci U S A*. 2013;110:7808-7813.
139. Gregory SM, Davis BK, West JA, et al. Discovery of a viral NLR homolog that inhibits the inflammasome. *Science*. 2011;331:330-334.
140. Kato H, Takeuchi O, Sato S, et al. Differential roles of MDA5 and RIG-I helicases in the recognition of RNA viruses. *Nature*. 2006;441:101-105.
141. Loo YM, Fornek J, Crochet N, et al. Distinct RIG-I and MDA5 signaling by RNA viruses in innate immunity. *J Virol*. 2008;82:335-345.
142. Habjan M, Andersson I, Klingström J, et al. Processing of genome 5′ termini as a strategy of negative-strand RNA viruses to avoid RIG-I-dependent interferon induction. *PLoS One*. 2008;3:e2032.
143. Garcin D, Lezzi M, Dobbs M, et al. The 5′ ends of Hantaan virus (Bunyaviridae) RNAs suggest a prime-and-realign mechanism for the initiation of RNA synthesis. *J Virol*. 1995;69:5754-5762.
144. Schneider U, Schwemmle M, Staeheli P. Genome trimming: a unique strategy for replication control employed by Borna disease virus. *Proc Natl Acad Sci U S A*. 2005;102:3441-3446.
145. Decroly E, Ferron F, Lescar J, Canard B. Conventional and unconventional mechanisms for capping viral mRNA. *Nat Rev Microbiol*. 2011;10:51-65.
146. Cárdenas WB, Loo YM, Gale M, et al. Ebola virus VP35 protein binds double-stranded RNA and inhibits α/α interferon production induced by RIG-I signaling. *J Virol*. 2006;80:5168-5178.
147. Weber M, Sediri H, Felgenhauer U, et al. Influenza virus adaptation PB2-627K modulates nucleocapsid inhibition by the pathogen sensor RIG-I. *Cell Host Microbe*. 2015;17:309-319.
148. Knoops K, Kikkert M, Worm SH, et al. SARS-coronavirus replication is supported by a reticulovesicular network of modified endoplasmic reticulum. *PLoS Biol*. 2008;6:e226.
149. Uchida L, Espada-Murao LA, Takamatsu Y, et al. The dengue virus conceals double-stranded RNA in the intracellular membrane to escape from an interferon response. *Sci Rep*. 2014;4:7395.
150. Gack MU, Albrecht RA, Urano T, et al. Influenza A virus NS1 targets the ubiquitin ligase TRIM25 to evade recognition by the host viral RNA sensor RIG-I. *Cell Host Microbe*. 2009;5:439-449.
151. Oshiumi H, Miyashita M, Matsumoto M, Seya T. A distinct role of Riplet-mediated K63-Linked polyubiquitination of the RIG-I repressor domain in human antiviral innate immune responses. *PLoS Pathog*. 2013;9:e1003533.
152. Chiang JJ, Davis ME, Gack MU. Regulation of RIG-I-like receptor signaling by host and viral proteins. *Cytokine Growth Factor Rev*. 2014;25:491-505.
153. Barral PM, Sarkar D, Fisher PB, Racaniello VR. RIG-I is cleaved during picornavirus infection. *Virology*. 2009;391:171-176.
154. Goswami R, Majumdar T, Dhar J, et al. Viral degradasome hijacks mitochondria to suppress innate immunity. *Cell Res*. 2013;23:1025-1042.
155. Feng Q, Langereis MA, Lork M, et al. Enterovirus 2Apro targets MDA5 and MAVS in infected cells. *J Virol*. 2014;88:3369-3378.
156. Liu HM, Gale M. Hepatitis C virus evasion from RIG-I-dependent hepatic innate immunity. *Gastroenterol Res Pract*. 2010;2010:548390.
157. Johnson KE, Chikoti L, Chandran B. Herpes simplex virus 1 infection induces activation and subsequent inhibition of the IFI16 and NLRP3 inflammasomes. *J Virol*. 2013;87:5005-5018.
158. Maruzuru Y, Ichinohe T, Sato R, et al. Herpes simplex virus 1 VP22 inhibits AIM2-dependent inflammasome activation to enable efficient viral replication. *Cell Host Microbe*. 2018;23:254-265.e7.
159. Orzalli MH, Broekema NM, Knipe DM. Relative contributions of herpes simplex virus 1 ICP0 and vhs to loss of cellular IFI16 vary in different human cell types. *J Virol*. 2016;90:8351-8359.
160. Cooper NR, Nemerow GR. Complement, viruses, and virus-infected cells. *Springer Semin Immunopathol*. 1983;6:327-347.
161. Ip WKE, Chan KH, Law HKW, et al. Mannose-binding lectin in severe acute respiratory syndrome coronavirus infection. *J Infect Dis*. 2005;191:1697-1704.
162. Avirutnan P, Hauhart RE, Marovich MA, Garred P, Atkinson JP, Diamond MS. Complement-mediated neutralization of dengue virus requires mannose-binding lectin. *mBio*. 2011;2:e00276-e00311. https://mbio.asm.org/lookup/doi/10.1128/mBio.00276-11
163. Jayasekera JP, Moseman EA, Carroll MC. Natural antibody and complement mediate neutralization of influenza virus in the absence of prior immunity. *J Virol*. 2007;81:3487-3494.
164. Van Strijp JA, Miltenburg LA, van der Tol ME, Van Kessel KP, Fluit AC, Verhoef J. Degradation of herpes simplex virions by human polymorphonuclear leukocytes and monocytes. *J Gen Virol*. 1990;71(pt 5):1205-1209.
165. Fang Y, Xu C, Fu YX, Holers VM, Molina H. Expression of complement receptors 1 and 2 on follicular dendritic cells is necessary for the generation of a strong antigen-specific IgG response. *J Immunol*. 1998;160:5273-5279.
166. Kolev M, Le Friec G, Kemper C. Complement—Tapping into new sites and effector systems. *Nat Rev Immunol*. 2014;14:811-820.
167. Fingeroth JD, Weis JJ, Tedder TF, Strominger JL, Biro PA, Fearon DT. Epstein-Barr virus receptor of human B lymphocytes is the C3d receptor CR2. *Proc Natl Acad Sci U S A*. 1984;81:4510-4514.
168. Bergelson JM, Chan M, Solomon KR, St John NF, Lin H, Finberg RW. Decay-accelerating factor (CD55), a glycosylphosphatidylinositol-anchored complement regulatory protein, is a receptor for several echoviruses. *Proc Natl Acad Sci U S A*. 1994;91:6245-6248.
169. Posch W, Steger M, Knackmuss U, et al. Complement-opsonized HIV-1 overcomes restriction in dendritic cells. *PLoS Pathog*. 2015;11:e1005005.
170. Saifuddin M, Parker CJ, Peeples ME, et al. Role of virion-associated glycosylphosphatidylinositol-linked proteins CD55 and CD59 in complement resistance of cell line-derived and primary isolates of HIV-1. *J Exp Med*. 1995;182:501-509.
171. Sahu A, Isaacs SN, Soulika AM, Lambris JD. Interaction of vaccinia virus complement control protein with human complement proteins: factor I-mediated degradation of C3b to iC3b1 inactivates the alternative complement pathway. *Mol Immunol*. 1998;35:371.
172. Rosengard AM, Liu Y, Nie Z, Jimenez R. Variola virus immune evasion design: expression of a highly efficient inhibitor of human complement. *Proc Natl Acad Sci U S A*. 2002;99:8808-8813.
173. Vanderplasschen A, Mathew E, Hollinshead M, Sim RB, Smith GL. Extracellular enveloped vaccinia virus is resistant to complement because of incorporation of host complement control proteins into its envelope. *Proc Natl Acad Sci U S A*. 1998;95:7544-7549.
174. Harris SL, Frank I, Yee A, Cohen GH, Eisenberg RJ, Friedman HM. Glycoprotein C of herpes simplex virus type 1 prevents complement-mediated cell lysis and virus neutralization. *J Infect Dis*. 1990;162:331-337.
175. Lubinski JM, Lazear HM, Awasthi S, Wang F, Friedman HM. The herpes simplex virus 1 IgG fc receptor blocks antibody-mediated complement activation and antibody-dependent cellular cytotoxicity *in vivo*. *J Virol*. 2011;85:3239-3249.
176. Nomura M, Kurita-Taniguchi M, Kondo K, et al. Mechanism of host cell protection from complement in murine cytomegalovirus (CMV) infection: identification of a CMV-responsive element in the CD46 promoter region. *Eur J Immunol*. 2002;32:2954-2964.
177. Takemoto M, Yamanishi K, Mori Y. Human herpesvirus 7 infection increases the expression levels of CD46 and CD59 in target cells. *J Gen Virol*. 2007;88:1415-1422.
178. Mullick J, Singh AK, Panse Y, Yadav V, Bernet J, Sahu A. Identification of functional domains in kaposica, the complement control protein homolog of Kaposi's sarcoma-associated herpesvirus (human herpesvirus 8). *J Virol*. 2005;79:5850-5856.
179. Brandt WE, Chiewslip D, Harris DL, Russell PK. Partial purification and characterization of a dengue virus soluble complement-fixing antigen. *J Immunol*. 1970;105:1565-1568.
180. Chung KM, Liszewski MK, Nybakken G, et al. West Nile virus nonstructural protein NS1 inhibits complement activation by binding the regulatory protein factor H. *Proc Natl Acad Sci U S A*. 2006;103:19111-19116.
181. Avirutnan P, Hauhart RE, Somnuke P, Blom AM, Diamond MS, Atkinson JP. Binding of flavivirus nonstructural protein NS1 to C4b binding protein modulates complement activation. *J Immunol*. 2011;187:424-433.
182. Avirutnan P, Fuchs A, Hauhart RE, et al. Antagonism of the complement component C4 by flavivirus nonstructural protein NS1. *J Exp Med*. 2010;207:793-806.
183. Conde JN, da Silva EM, Allonso D, et al. Inhibition of the membrane attack complex by dengue virus NS1 through interaction with vitronectin and terminal complement proteins. *J Virol*. 2016;90:9570-9581.
184. Amet T, Ghabril M, Chalasani N, et al. CD59 incorporation protects hepatitis C virus against complement-mediated destruction. *Hepatology*. 2012;55:354-363.
185. Mazumdar B, Kim H, Meyer K, et al. Hepatitis C virus infection upregulates CD55 expression on the hepatocyte surface and promotes association with virus particles. *J Virol*. 2013;87:7902-7910.
186. Kim H, Meyer K, Di Bisceglie AM, Ray R. Inhibition of C3 convertase activity by hepatitis C virus as an additional lesion in the regulation of complement components. *PLoS One*. 2014;9:e101422.
187. Mazumdar B, Kim H, Meyer K, et al. Hepatitis C virus proteins inhibit C3 complement production. *J Virol*. 2012;86:2221-2228.
188. Banerjee A, Mazumdar B, Meyer K, Di Bisceglie AM, Ray RB, Ray R. Transcriptional repression of C4 complement by hepatitis C virus proteins. *J Virol*. 2011;85:4157-4166.
189. Kim H, Meyer K, Di Bisceglie AM, Ray R. Hepatitis C virus suppresses C9 complement synthesis and impairs membrane attack complex function. *J Virol*. 2013;87:5858-5867.
190. Shaw ML, Stone KL, Colangelo CM, Gulcicek EE, Palese P. Cellular proteins in influenza virus particles. *PLoS Pathog*. 2008;4:e1000085.
191. Johnson JB, Grant K, Parks GD. The paramyxoviruses simian virus 5 and mumps virus recruit host cell CD46 to evade complement-mediated neutralization. *J Virol*. 2009;83:7602-7611.
192. Li Y, Johnson JB, Parks GD. Parainfluenza virus 5 upregulates CD55 expression to produce virions with enhanced resistance to complement-mediated neutralization. *Virology*. 2016;497:305-313.
193. Johnson JB, Borisevich V, Rockx B, Parks GD. A novel factor I activity in Nipah virus inhibits human complement pathways through cleavage of C3b. *J Virol*. 2015;89:989-998.
194. Pahl HL. Activators and target genes of Rel/NF-κB transcription factors. *Oncogene*. 1999;18:6853-6866.
195. Mogensen TH. IRF and STAT transcription factors – From basic biology to roles in infection, protective immunity, and primary immunodeficiencies. *Front Immunol*. 2018;9:3047.
196. Ning S, Pagano JS, Barber GN. IRF7: activation, regulation, modification and function. *Genes Immun*. 2011;12:399-414.
197. Cousens LP, Orange JS, Su HC, Biron CA. Interferon-α/β inhibition of interleukin 12 and interferon-γ production *in vitro* and endogenously during viral infection. *Proc Natl Acad Sci U S A*. 1997;94:634-639.
198. Orange JS, Biron CA. An absolute and restricted requirement for IL-12 in natural killer cell IFN-γ production and antiviral defense. Studies of natural killer and T cell responses in contrasting viral infections. *J Immunol*. 1996;156:1138-1142.
199. Schmeisser H, Bekisz J, Zoon KC. New function of type I IFN: induction of autophagy. *J Interferon Cytokine Res*. 2014;34:71-78.
200. Matsuzawa T, Kim BH, Shenoy AR, Kamitani S, Miyake M, Macmicking JD. IFN-γ elicits macrophage autophagy via the p38 MAPK signaling pathway. *J Immunol*. 2012;189:813-818.
201. Xia C, Vijayan M, Pritzl CJ, Fuchs SY, McDermott AB, Hahm B. Hemagglutinin of influenza A virus antagonizes type I interferon (IFN) responses by inducing degradation of type I IFN receptor 1. *J Virol*. 2015;90:2403-2417.
202. Lubick KJ, Robertson SJ, McNally KL, et al. Flavivirus antagonism of type I interferon signaling reveals prolidase as a regulator of IFNAR1 surface expression. *Cell Host Microbe*. 2015;18:61-74.

203. Gunduz F, Aboulnasr FM, Chandra PK, et al. Free fatty acids induce ER stress and block antiviral activity of interferon α against hepatitis C virus in cell culture. *Virol J*. 2012;9:143.
204. Pisanelli G, Laurent-Rolle M, Manicassamy B, et al. La Piedad Michoacán Mexico Virus V protein antagonizes type I interferon response by binding STAT2 protein and preventing STATs nuclear translocation. *Virus Res*. 2016;213:11-22.
205. Hastings AK, Amato KR, Wen SC, Peterson LS, Williams JV. Human metapneumovirus small hydrophobic (SH) protein downregulates type I IFN pathway signaling by affecting STAT1 expression and phosphorylation. *Virology*. 2016;494:248-256.
206. Verweij MC, Wellish M, Whitmer T, et al. Varicella viruses inhibit interferon-stimulated JAK-STAT signaling through multiple mechanisms. *PLoS Pathog*. 2015;11:e1004901.
207. Kint J, Dickhout A, Kutter J, et al. Infectious bronchitis coronavirus inhibits STAT1 signaling and requires accessory proteins for resistance to type I interferon activity. *J Virol*. 2015;89:12047-12057.
208. Ning YJ, Feng K, Min YQ, et al. Disruption of type I interferon signaling by the nonstructural protein of severe fever with thrombocytopenia syndrome virus via the hijacking of STAT2 and STAT1 into inclusion bodies. *J Virol*. 2015;89:4227-4236.
209. Ribeiro CMS, Sarrami-Forooshani R, Setiawan LC, et al. Receptor usage dictates HIV-1 restriction by human TRIM5α in dendritic cell subsets. *Nature*. 2016;540:448-452.
210. Kajaste-Rudnitski A, Marelli SS, Pultrone C, et al. TRIM22 inhibits HIV-1 transcription independently of its E3 ubiquitin ligase activity, Tat, and NF-kappaB-responsive long terminal repeat elements. *J Virol*. 2011;85:5183-5196.
211. Gao B, Duan Z, Xu W, Xiong S. Tripartite motif-containing 22 inhibits the activity of hepatitis B virus core promoter, which is dependent on nuclear-located RING domain. *Hepatology*. 2009;50:424-433.
212. Di Pietro A, Kajaste-Rudnitski A, Oteiza A, et al. TRIM22 inhibits influenza A virus infection by targeting the viral nucleoprotein for degradation. *J Virol*. 2013;87:4523-4533.
213. Yang C, Zhao X, Sun D, et al. Interferon alpha (IFNα)-induced TRIM22 interrupts HCV replication by ubiquitinating NS5A. *Cell Mol Immunol*. 2016;13:94-102.
214. Yuan T, Yao W, Tokunaga K, Yang R, Sun B. An HIV-1 capsid binding protein TRIM11 accelerates viral uncoating. *Retrovirology*. 2016;13:72.
215. Allouch A, Di Primio C, Alpi E, et al. The TRIM family protein KAP1 inhibits HIV-1 integration. *Cell Host Microbe*. 2011;9:484-495.
216. Geoffroy MC, Chelbi-Alix MK. Role of promyelocytic leukemia protein in host antiviral defense. *J Interferon Cytokine Res*. 2011;31:145-158.
217. El McHichi B, Regad T, Maroui MA, et al. SUMOylation promotes PML degradation during encephalomyocarditis virus infection. *J Virol*. 2010;84:11634-11645.
218. Poole E, Groves I, MacDonald A, Pang Y, Alcami A, Sinclair J. Identification of TRIM23 as a cofactor involved in the regulation of NF-κB by human cytomegalovirus. *J Virol*. 2009;83:3581-3590.
219. Hancock MH, Nelson JA. Modulation of the NFκb signalling pathway by human cytomegalovirus. *Virology (Hyderabad)*. 2017;1:104.
220. Sharma N, Kumawat KL, Rastogi M, Basu A, Singh SK. Japanese encephalitis virus exploits the microRNA-432 to regulate the expression of suppressor of cytokine signaling (SOCS) 5. *Sci Rep*. 2016;6:27685.
221. Choi EJ, Lee CH, Shin OS. Suppressor of cytokine signaling 3 expression induced by varicella-zoster virus infection results in the modulation of virus replication. *Scand J Immunol*. 2015;82:337-344.
222. Wolf D, Witte V, Laffert B, et al. HIV-1 Nef associated PAK and PI3-Kinases stimulate Akt-independent Bad-phosphorylation to induce anti-apoptotic signals. *Nat Med*. 2001;7:1217-1224.
223. Hale BG, Randall RE, Ortín J, Jackson D. The multifunctional NS1 protein of influenza A viruses. *J Gen Virol*. 2008;89:2359-2376.
224. Benetti L, Roizman B. Protein kinase B/Akt is present in activated form throughout the entire replicative cycle of US3 mutant virus but only at early times after infection with wild-type herpes simplex virus 1. *J Virol*. 2006;80:3341-3348.
225. Dawson CW, Tramountanis G, Eliopoulos AG, Young LS. Epstein-Barr virus latent membrane protein 1 (LMP1) activates the phosphatidylinositol 3-kinase/Akt pathway to promote cell survival and induce actin filament remodeling. *J Biol Chem*. 2003;278:3694-3704.
226. Spangle JM, Münger K. The human papillomavirus type 16 E6 oncoprotein activates mTORC1 signaling and increases protein synthesis. *J Virol*. 2010;84:9398-9407.
227. O'Shea C, Klupsch K, Choi S, et al. Adenoviral proteins mimic nutrient/growth signals to activate the mTOR pathway for viral replication. *EMBO J*. 2005;24:1211-1221.
228. Diehl N, Schaal H. Make yourself at home: viral hijacking of the PI3K/Akt signaling pathway. *Viruses*. 2013;5:3192-3212.
229. Borrow P, Martínez-Sobrido L, de la Torre JC. Inhibition of the type I interferon antiviral response during arenavirus infection. *Viruses*. 2010;2:2443-2480.
230. Wilkins C, Gale M. Recognition of viruses by cytoplasmic sensors. *Curr Opin Immunol*. 2010;22:41-47.
231. Baglioni C. *2′,5′-oligo(A) pathway of interferon action*. In: *The Interferon System: A Current Review to 1987*. University of Texas Medical Branch; 1987:365-371.
232. Meurs E, Chong K, Galabru J, et al. Molecular cloning and characterization of the human double-stranded RNA-activated protein kinase induced by interferon. *Cell*. 1990;62:379-390.
233. Benedict CA. Viruses and the TNF-related cytokines, an evolving battle. *Cytokine Growth Factor Rev*. 2003;14:349-357.
234. Raftery MJ, Schwab M, Eibert SM, Samstag Y, Walczak H, Schönrich G. Targeting the function of mature dendritic cells by human cytomegalovirus: a multilayered viral defense strategy. *Immunity*. 2001;15:997-1009.
235. Xu XN, Laffert B, Screaton GR, et al. Induction of Fas ligand expression by HIV involves the interaction of Nef with the T cell receptor zeta chain. *J Exp Med*. 1999;189:1489-1496.
236. Vidalain PO, Azocar O, Lamouille B, Astier A, Rabourdin-Combe C, Servet-Delprat C. Measles virus induces functional TRAIL production by human dendritic cells. *J Virol*. 2000;74:556-559.
237. Hall SJ, Canfield SE, Yan Y, Hassen W, Selleck WA, Chen SH. A novel bystander effect involving tumor cell-derived Fas and FasL interactions following Ad.HSV-tk and Ad.mIL-12 gene therapies in experimental prostate cancer. *Gene Ther*. 2002;9:511-517.
238. Everett H, McFadden G. Poxviruses and apoptosis: a time to die. *Curr Opin Microbiol*. 2002;5:395-402.
239. Shisler J, Yang C, Walter B, Ware CF, Gooding LR. The adenovirus E3-10.4K/14.5K complex mediates loss of cell surface Fas (CD95) and resistance to Fas-induced apoptosis. *J Virol*. 1997;71:8299-8306.
240. Wang GH, Bertin J, Wang Y, et al. Bovine herpesvirus 4 BORFE2 protein inhibits Fas-and tumor necrosis factor receptor 1-induced apoptosis and contains death effector domains shared with other γ-2 herpesviruses. *J Virol*. 1997;71:8928-8932.
241. Garvey TL, Bertin J, Siegel RM, Wang GH, Lenardo MJ, Cohen JI. Binding of FADD and caspase-8 to molluscum contagiosum virus MC159 v-FLIP is not sufficient for its antiapoptotic function. *J Virol*. 2002;76:697-706.
242. Kabsch K, Alonso A. The human papillomavirus type 16 E5 protein impairs TRAIL-and FasL-mediated apoptosis in HaCaT cells by different mechanisms. *J Virol*. 2002;76:12162-12172.
243. Filippova M, Song H, Connolly JL, Dermody TS, Duerksen-Hughes PJ. The human papillomavirus 16 E6 protein binds to tumor necrosis factor (TNF) R1 and protects cells from TNF-induced apoptosis. *J Biol Chem*. 2002;277:21730-21739.
244. Majumder M, Ghosh AK, Steele R, et al. Hepatitis C virus NS5A protein impairs TNF-mediated hepatic apoptosis, but not by an anti-FAS antibody, in transgenic mice. *Virology*. 2002;294:94-105.
245. Skaletskaya A, Bartle LM, Chittenden T, McCormick AL, Mocarski ES, Goldmacher VS. A cytomegalovirus-encoded inhibitor of apoptosis that suppresses caspase-8 activation. *Proc Natl Acad Sci U S A*. 2001;98:7829-7834.
246. Ekert PG, Silke J, Vaux DL. Caspase inhibitors. *Cell Death Differ*. 1999;6:1081-1086.
247. Henderson S, Huen D, Rowe M, Dawson C, Johnson G, Rickinson A. Epstein-Barr virus-coded BHRF1 protein, a viral homologue of Bcl-2, protects human B cells from programmed cell death. *Proc Natl Acad Sci U S A*. 1993;90:8479-8483.
248. Perez D, White E. TNF-α signals apoptosis through a bid-dependent conformational change in Bax that is inhibited by E1B 19K. *Mol Cell*. 2000;6:53-63.
249. Mariani MK, Dasmeh P, Fortin A, et al. The combination of IFN β and TNF induces an antiviral and immunoregulatory program via non-canonical pathways involving STAT2 and IRF9. *Cells*. 2019;8:919.
250. Desloges N, Rahaus M, Wolff MH. Role of the protein kinase PKR in the inhibition of varicella-zoster virus replication by β interferon and γ interferon. *J Gen Virol*. 2005;86:1-6.
251. Wong GH, Goeddel DV. Tumour necrosis factors α and β inhibit virus replication and synergize with interferons. *Nature*. 1986;323:819-822.
252. Lin JC, Zhang ZX, Chou TC, Sim I, Pagano JS. Synergistic inhibition of Epstein-Barr virus: transformation of B lymphocytes by α and γ interferon and by 3′-azido-3′-deoxythymidine. *J Infect Dis*. 1989;159:248-254.
253. Scagnolari C, Trombetti S, Alberelli A, et al. The synergistic interaction of interferon types I and II leads to marked reduction in severe acute respiratory syndrome-associated coronavirus replication and increase in the expression of mRNAs for interferon-induced proteins. *Intervirology*. 2007;50:156-160.
254. Brandes ME, Mai UE, Ohura K, Wahl SM. Type I transforming growth factor-β receptors on neutrophils mediate chemotaxis to transforming growth factor-β. *J Immunol*. 1991;147:1600-1606.
255. Luttmann W, Franz P, Matthys H, Virchow JC. Effects of TGF-β on eosinophil chemotaxis. *Scand J Immunol*. 1998;47:127-130.
256. Garlanda C, Dinarello CA, Mantovani A. The interleukin-1 family: back to the future. *Immunity*. 2013;39:1003-1018.
257. Dinarello CA. Immunological and inflammatory functions of the interleukin-1 family. *Annu Rev Immunol*. 2009;27:519-550.
258. Mantovani A, Cassatella MA, Costantini C, Jaillon S. Neutrophils in the activation and regulation of innate and adaptive immunity. *Nat Rev Immunol*. 2011;11:519-531.
259. Rider P, Carmi Y, Guttman O, et al. IL-1α and IL-1β recruit different myeloid cells and promote different stages of sterile inflammation. *J Immunol*. 2011;187:4835-4843.
260. Xiang Y, Moss B. IL-18 binding and inhibition of interferon γ induction by human poxvirus-encoded proteins. *Proc Natl Acad Sci U S A*. 1999;96:11537-11542.
261. Sharma S, Kulk N, Nold MF, et al. The IL-1 family member 7b translocates to the nucleus and down-regulates proinflammatory cytokines. *J Immunol*. 2008;180:5477-5482.
262. Alcamí A, Smith GL. A soluble receptor for interleukin-1β encoded by vaccinia virus: a novel mechanism of virus modulation of the host response to infection. *Cell*. 1992;71:153-167.
263. Velazquez-Salinas L, Verdugo-Rodriguez A, Rodriguez LL, Borca MV. The role of interleukin 6 during viral infections. *Front Microbiol*. 2019;10:1057.
264. Palumbo GA, Scisciani C, Pediconi N, et al. IL6 inhibits HBV transcription by targeting the epigenetic control of the nuclear cccDNA Minichromosome. *PLoS One*. 2015;10:e0142599.
265. Rojas JM, Avia M, Martín V, Sevilla N. IL-10: a multi-functional cytokine in viral infections. *J Immunol Res*. 2017;2017:6104054.
266. Eberhardt MK, Deshpande A, Fike J, et al. Exploitation of interleukin-10 (IL-10) signaling pathways: Alternate roles of viral and cellular IL-10 in rhesus cytomegalovirus infection. *J Virol*. 2016;90:9920-9930.
267. Chan A, Baird M, Mercer AA, Fleming SB. Maturation and function of human dendritic cells are inhibited by orf virus-encoded interleukin-10. *J Gen Virol*. 2006;87:3177-3181.
268. Guo Y, Cao W, Zhu Y. Immunoregulatory functions of the IL-12 family of cytokines in antiviral systems. *Viruses*. 2019;11.
269. Verbist KC, Klonowski KD. Functions of IL-15 in anti-viral immunity: multiplicity and variety. *Cytokine*. 2012;59:467-478.
270. Carroll HP, Paunovic V, Gadina M. Signalling, inflammation and arthritis: crossed signals. The role of interleukin-15 and-18 in autoimmunity. *Rheumatology (Oxford)*. 2008;47:1269-1277.
271. Pulendran B, Dillon S, Joseph C, Curiel T, Banchereau J, Mohamadzadeh M. Dendritic cells generated in the presence of GM-CSF plus IL-15 prime potent CD8+ Tc1 responses *in vivo*. *Eur J Immunol*. 2004;34:66-73.
272. Omata F, Birkenbach M, Matsuzaki S, Christ AD, Blumberg RS. The expression of IL-12 p40 and its homologue, Epstein-Barr virus-induced gene 3, in inflammatory bowel disease. *Inflamm Bowel Dis*. 2001;7:215-220.
273. McElroy AK, Erickson BR, Flietstra TD, et al. Ebola hemorrhagic fever: novel biomarker correlates of clinical outcome. *J Infect Dis*. 2014;210:558-566.
274. Mehrad B, Keane MP, Strieter RM. Chemokines as mediators of angiogenesis. *Thromb Haemostasis*. 2007;97:755-762.
275. Afonso CL, Tulman ER, Lu Z, Zsak L, Kutish GF, Rock DL. The genome of fowlpox virus. *J Virol*. 2000;74:3815-3831.
276. Lee HJ, Essani K, Smith GL. The genome sequence of Yaba-like disease virus, a yatapoxvirus. *Virology*. 2001;281:170-192.
277. Alcamí A, Symons JA, Collins PD, Williams TJ, Smith GL. Blockade of chemokine activity by a soluble chemokine binding protein from vaccinia virus. *J Immunol*. 1998;160:624-633.
278. Lalani AS, Graham K, Mossman K, et al. The purified myxoma virus γ interferon receptor homolog M-T7 interacts with the heparin-binding domains of chemokines. *J Virol*. 1997;71:4356-4363.
279. Graham KA, Lalani AS, Macen JL, et al. The T1/35kDa family of poxvirus-secreted proteins bind chemokines and modulate leukocyte influx into virus-infected tissues. *Virology*. 1997;229:12-24.
280. Parry CM, Simas JP, Smith VP, et al. A broad spectrum secreted chemokine binding protein encoded by a herpesvirus. *J Exp Med*. 2000;191:573-578.
281. Wang D, Bresnahan W, Shenk T. Human cytomegalovirus encodes a highly specific RANTES decoy receptor. *Proc Natl Acad Sci U S A*. 2004;101:16642-16647.
282. Viejo-Borbolla A, Martinez-Martín N, Nel HJ, et al. Enhancement of chemokine function as an immunomodulatory strategy employed by human herpesviruses. *PLoS Pathog*. 2012;8:e1002497.
283. Tumpey TM, García-Sastre A, Taubenberger JK, et al. Pathogenicity of influenza viruses with genes from the 1918 pandemic virus: functional roles of alveolar macrophages and neutrophils in limiting virus replication and mortality in mice. *J Virol*. 2005;79:14933-14944.
284. Colotta F, Re F, Polentarutti N, Sozzani S, Mantovani A. Modulation of granulocyte survival and programmed cell death by cytokines and bacterial products. *Blood*. 1992;80:2012-2020.
285. Van Strijp JA, Van Kessel KP, van der Tol ME, Fluit AC, Snippe H, Verhoef J. Phagocytosis of herpes simplex virus by human granulocytes and monocytes. *Arch Virol*. 1989;104:287-298.
286. Rothwell SW, Wright DG. Characterization of influenza A virus binding sites on human neutrophils. *J Immunol*. 1994;152:2358-2367.
287. Hashimoto Y, Moki T, Takizawa T, Shiratsuchi A, Nakanishi Y. Evidence for phagocytosis of influenza virus-infected, apoptotic cells by neutrophils and macrophages in mice. *J Immunol*. 2007;178:2448-2457.

288. Wang W, Owen SM, Rudolph DL, et al. Activity of α-and θ-defensins against primary isolates of HIV-1. *J Immunol.* 2004;173:515-520.
289. Hazrati E, Galen B, Lu W, et al. Human α-and β-defensins block multiple steps in herpes simplex virus infection. *J Immunol.* 2006;177:8658-8666.
290. Sousa FH, Casanova V, Findlay F, et al. Cathelicidins display conserved direct antiviral activity towards rhinovirus. *Peptides.* 2017;95:76-83.
291. Currie SM, Gwyer Findlay E, McFarlane AJ, et al. Cathelicidins have direct antiviral activity against respiratory syncytial virus *in vitro* and protective function *in vivo* in mice and humans. *J Immunol.* 2016;196:2699-2710.
292. He M, Zhang H, Li Y, et al. Cathelicidin-derived antimicrobial peptides inhibit Zika virus through direct inactivation and interferon pathway. *Front Immunol.* 2018;9:722.
293. Cortjens B, de Boer OJ, de Jong R, et al. Neutrophil extracellular traps cause airway obstruction during respiratory syncytial virus disease. *J Pathol.* 2016;238:401-411.
294. Saitoh T, Komano J, Saitoh Y, et al. Neutrophil extracellular traps mediate a host defense response to human immunodeficiency virus-1. *Cell Host Microbe.* 2012;12:109-116.
295. Rodríguez-Espinosa O, Rojas-Espinosa O, Moreno-Altamirano MMB, López-Villegas EO, Sánchez-García FJ. Metabolic requirements for neutrophil extracellular traps formation. *Immunology.* 2015;145:213-224.
296. Bai F, Kong KF, Dai J, et al. A paradoxical role for neutrophils in the pathogenesis of West Nile virus. *J Infect Dis.* 2010;202:1804-1812.
297. Shiratori I, Yamaguchi M, Suzukawa M, et al. Down-regulation of basophil function by human CD200 and human herpesvirus-8 CD200. *J Immunol.* 2005;175:4441-4449.
298. Piliponsky AM, Acharya M, Shubin NJ. Mast cells in viral, bacterial, and fungal infection immunity. *Int J Mol Sci.* 2019;20:2851.
299. Benoit M, Desnues B, Mege JL. Macrophage polarization in bacterial infections. *J Immunol.* 2008;181:3733-3739.
300. Ahmed F, Ibrahim A, Cooper CL, Kumar A, Crawley AM. Chronic hepatitis C virus infection impairs m1 macrophage differentiation and contributes to CD8+ T-cell dysfunction. *Cells.* 2019;8:374.
301. Sica A, Mantovani A. Macrophage plasticity and polarization: in vivo veritas. *J Clin Invest.* 2012;122:787-795.
302. Lu D, Ni Z, Liu X, et al. Beyond T cells: understanding the role of PD-1/PD-L1 in tumor-associated macrophages. *J Immunol Res.* 2019;2019:1919082.
303. Viola A, Munari F, Sánchez-Rodríguez R, Scolaro T, Castegna A. The metabolic signature of macrophage responses. *Front Immunol.* 2019;10:1462.
304. Italiani P, Boraschi D. New insights into tissue macrophages: from their origin to the development of memory. *Immune Netw.* 2015;15:167-176.
305. Dupuis-Maguiraga L, Noret M, Brun S, Le Grand R, Gras G, Roques P. Chikungunya disease: infection-associated markers from the acute to the chronic phase of arbovirus-induced arthralgia. *PLoS Negl Trop Dis.* 2012;6:e1446.
306. Koppensteiner H, Brack-Werner R, Schindler M. Macrophages and their relevance in human immunodeficiency virus type I infection. *Retrovirology.* 2012;9:82.
307. Hoebe EK, Le Large TYS, Tarbouriech N, et al. Epstein-Barr virus-encoded BARF1 protein is a decoy receptor for macrophage colony stimulating factor and interferes with macrophage differentiation and activation. *Viral Immunol.* 2012;25:461-470.
308. Cassetta L, Kajaste-Rudnitski A, Coradin T, et al. M1 polarization of human monocyte-derived macrophages restricts pre and postintegration steps of HIV-1 replication. *AIDS.* 2013;27:1847-1856.
309. Zhang Y, Ma CJ, Ni L, et al. Cross-talk between programmed death-1 and suppressor of cytokine signaling-1 in inhibition of IL-12 production by monocytes/macrophages in hepatitis C virus infection. *J Immunol.* 2011;186:3093-3103.
310. Karp CL, Wysocka M, Wahl LM, et al. Mechanism of suppression of cell-mediated immunity by measles virus. *Science.* 1996;273:228-231.
311. Hoves S, Niller HH, Krause SW, et al. Decreased T cell stimulatory capacity of monocyte-derived human macrophages following herpes simplex virus type 1 infection. *Scand J Immunol.* 2001;54:93-99.
312. Adhya D, Dutta K, Kundu K, Basu A. Histone deacetylase inhibition by Japanese encephalitis virus in monocyte/macrophages: a novel viral immune evasion strategy. *Immunobiology.* 2013;218:1235-1247.
313. Biron CA, Byron KS, Sullivan JL. Severe herpesvirus infections in an adolescent without natural killer cells. *N Engl J Med.* 1989;320:1731-1735.
314. Kamili QUA, Seeborg FO, Saxena K, et al. Severe cutaneous human papillomavirus infection associated with natural killer cell deficiency following stem cell transplantation for severe combined immunodeficiency. *J Allergy Clin Immunol.* 2014;134:1451-1453.e1.
315. Bukowski JF, Woda BA, Habu S, Okumura K, Welsh RM. Natural killer cell depletion enhances virus synthesis and virus-induced hepatitis *in vivo*. *J Immunol.* 1983;131:1531-1538.
316. Lunemann S, Malone DFG, Hengst J, et al. Compromised function of natural killer cells in acute and chronic viral hepatitis. *J Infect Dis.* 2014;209:1362-1373.
317. Rehermann B. Pathogenesis of chronic viral hepatitis: differential roles of T cells and NK cells. *Nat Med.* 2013;19:859-868.
318. Liu LL, Landskron J, Ask EH, et al. Critical role of CD2 Co-stimulation in adaptive natural killer cell responses revealed in NKG2C-deficient humans. *Cell Rep.* 2016;15:1088-1099.
319. Arnon TI, Lev M, Katz G, Chernobrov Y, Porgador A, Mandelboim O. Recognition of viral hemagglutinins by NKp44 but not by NKp30. *Eur J Immunol.* 2001;31:2680-2689.
320. Mandelboim O, Lieberman N, Lev M, et al. Recognition of haemagglutinins on virus-infected cells by NKp46 activates lysis by human NK cells. *Nature.* 2001;409:1055-1060.
321. Lu LL, Suscovich TJ, Fortune SM, Alter G. Beyond binding: antibody effector functions in infectious diseases. *Nat Rev Immunol.* 2018;18:46-61.
322. Kim JH, Excler JL, Michael NL. Lessons from the RV144 Thai phase III HIV-1 vaccine trial and the search for correlates of protection. *Annu Rev Med.* 2015;66:423-437.
323. Haynes LM, Caidi H, Radu GU, et al. Therapeutic monoclonal antibody treatment targeting respiratory syncytial virus (RSV) G protein mediates viral clearance and reduces the pathogenesis of RSV infection in BALB/c mice. *J Infect Dis.* 2009;200:439-447.
324. Schuster IS, Wikstrom ME, Brizard G, et al. TRAIL+ NK cells control CD4+ T cell responses during chronic viral infection to limit autoimmunity. *Immunity.* 2014;41:646-656.
325. Rabinovich BA, Li J, Shannon J, et al. Activated, but not resting, T cells can be recognized and killed by syngeneic NK cells. *J Immunol.* 2003;170:3572-3576.
326. Peppa D, Gill US, Reynolds G, et al. Up-regulation of a death receptor renders antiviral T cells susceptible to NK cell-mediated deletion. *J Exp Med.* 2013;210:99-114.
327. Robbins SH, Bessou G, Cornillon A, et al. Natural killer cells promote early CD8 T cell responses against cytomegalovirus. *PLoS Pathog.* 2007;3:e123.
328. Waggoner SN, Cornberg M, Selin LK, Welsh RM. Natural killer cells act as rheostats modulating antiviral T cells. *Nature.* 2011;481:394-398.
329. Orange JS, Fassett MS, Koopman LA, Boyson JE, Strominger JL. Viral evasion of natural killer cells. *Nat Immunol.* 2002;3:1006-1012.
330. Kubota A, Kubota S, Farrell HE, Davis-Poynter N, Takei F. Inhibition of NK cells by murine CMV-encoded class I MHC homologue m144. *Cell Immunol.* 1999;191:145-151.
331. Schust DJ, Tortorella D, Seebach J, Phan C, Ploegh HL. Trophoblast class I major histocompatibility complex (MHC) products are resistant to rapid degradation imposed by the human cytomegalovirus (HCMV) gene products US2 and US11. *J Exp Med.* 1998;188:497-503.
332. Cohen GB, Gandhi RT, Davis DM, et al. The selective down-regulation of class I major histocompatibility complex proteins by HIV-1 protects HIV-infected cells from NK cells. *Immunity.* 1999;10:661-671.
333. Ishido S, Wang C, Lee BS, Cohen GB, Jung JU. Downregulation of major histocompatibility complex class I molecules by Kaposi's sarcoma-associated herpesvirus K3 and K5 proteins. *J Virol.* 2000;74:5300-5309.
334. Wang ECY, McSharry B, Retiere C, et al. UL40-mediated NK evasion during productive infection with human cytomegalovirus. *Proc Natl Acad Sci U S A.* 2002;99:7570-7575.
335. Krmpotić A, Busch DH, Bubić I, et al. MCMV glycoprotein gp40 confers virus resistance to CD8+ T cells and NK cells *in vivo*. *Nat Immunol.* 2002;3:529-535.
336. Poggi A, Carosio R, Spaggiari GM, et al. NK cell activation by dendritic cells is dependent on LFA-1-mediated induction of calcium-calmodulin kinase II: inhibition by HIV-1 Tat C-terminal domain. *J Immunol.* 2002;168:95-101.
337. Chehimi J, Bandyopadhyay S, Prakash K, et al. *In vitro* infection of natural killer cells with different human immunodeficiency virus type 1 isolates. *J Virol.* 1991;65:1812-1822.
338. Crotta S, Stilla A, Wack A, et al. Inhibition of natural killer cells through engagement of CD81 by the major hepatitis C virus envelope protein. *J Exp Med.* 2002;195:35-41.
339. Budt M, Reinhard H, Bigl A, Hengel H. Herpesviral Fcγ receptors: culprits attenuating antiviral IgG? *Int Immunopharm.* 2004;4:1135-1148.
340. Dubin G, Socolof E, Frank I, Friedman HM. Herpes simplex virus type 1 Fc receptor protects infected cells from antibody-dependent cellular cytotoxicity. *J Virol.* 1991;65:7046-7050.
341. Atalay R, Zimmermann A, Wagner M, et al. Identification and expression of human cytomegalovirus transcription units coding for two distinct Fcγ receptor homologs. *J Virol.* 2002;76:8596-8608.
342. Lilley BN, Ploegh HL, Tirabassi RS. Human cytomegalovirus open reading frame TRL11/IRL11 encodes an immunoglobulin G Fc-binding protein. *J Virol.* 2001;75:11218-11221.
343. Sprague ER, Reinhard H, Cheung EJ, et al. The human cytomegalovirus Fc receptor gp68 binds the Fc CH2-CH3 interface of immunoglobulin G. *J Virol.* 2008;82:3490-3499.
344. Constantinides MG, Bendelac A. Transcriptional regulation of the NKT cell lineage. *Curr Opin Immunol.* 2013;25:161-167.
345. Chang PP, Barral P, Fitch J, et al. Identification of Bcl-6-dependent follicular helper NKT cells that provide cognate help for B cell responses. *Nat Immunol.* 2012;13:35-43.
346. Krovi SH, Gapin L. Invariant natural killer T cell subsets-more than just developmental intermediates. *Front Immunol.* 2018;9:1393.
347. Juno JA, Keynan Y, Fowke KR. Invariant NKT cells: regulation and function during viral infection. *PLoS Pathog.* 2012;8:e1002838.
348. Orange JS. Human natural killer cell deficiencies and susceptibility to infection. *Microbes Infect.* 2002;4:1545-1558.
349. van Erp EA, Feyaerts D, Duijst M, et al. Respiratory syncytial virus infects primary neonatal and adult natural killer cells and affects their antiviral effector function. *J Infect Dis.* 2019;219:723-733.
350. Eberl G, MacDonald HR. Rapid death and regeneration of NKT cells in anti-CD3epsilon-or IL-12-treated mice: a major role for bone marrow in NKT cell homeostasis. *Immunity.* 1998;9:345-353.
351. Lin Y, Roberts TJ, Spence PM, Brutkiewicz RR. Reduction in CD1d expression on dendritic cells and macrophages by an acute virus infection. *J Leukoc Biol.* 2005;77:151-158.
352. Webb TJR, Litavecz RA, Khan MA, et al. Inhibition of CD1d1-mediated antigen presentation by the vaccinia virus B1R and H5R molecules. *Eur J Immunol.* 2006;36:2595-2600.
353. Liu J, Gallo RM, Duffy C, Brutkiewicz RR. A VP22-Null HSV-1 is impaired in inhibiting CD1d-mediated antigen presentation. *Viral Immunol.* 2016;29:409-416.
354. Xiong R, Rao P, Kim S, Li M, Wen X, Yuan W. Herpes simplex virus 1 US3 phosphorylates cellular KIF3A to downregulate CD1d expression. *J Virol.* 2015;89:6646-6655.
355. Raftery MJ, Winau F, Kaufmann SHE, Schaible UE, Schönrich G. CD1 antigen presentation by human dendritic cells as a target for herpes simplex virus immune evasion. *J Immunol.* 2006;177:6207-6214.
356. Chen N, McCarthy C, Drakesmith H, et al. HIV-1 down-regulates the expression of CD1d via Nef. *Eur J Immunol.* 2006;36:278-286.
357. Miura S, Kawana K, Schust DJ, et al. CD1d, a sentinel molecule bridging innate and adaptive immunity, is downregulated by the human papillomavirus (HPV) E5 protein: a possible mechanism for immune evasion by HPV. *J Virol.* 2010;84:11614-11623.
358. Dalod M, Chelbi R, Malissen B, Lawrence T. Dendritic cell maturation: functional specialization through signaling specificity and transcriptional programming. *EMBO J.* 2014;33:1104-1116.
359. Reis e Sousa C. Activation of dendritic cells: translating innate into adaptive immunity. *Curr Opin Immunol.* 2004;16:21-25.
360. Blander JM. Regulation of the cell biology of antigen cross-presentation. *Annu Rev Immunol.* 2018;36:717-753.
361. Nattermann J, Zimmermann H, Iwan A, et al. Hepatitis C virus E2 and CD81 interaction may be associated with altered trafficking of dendritic cells in chronic hepatitis C. *Hepatology.* 2006;44:945-954.
362. Donaghy H, Pozniak A, Gazzard B, et al. Loss of blood CD11c(+) myeloid and CD11c(−) plasmacytoid dendritic cells in patients with HIV-1 infection correlates with HIV-1 RNA virus load. *Blood.* 2001;98:2574-2576.
363. Dolganiuc A, Chang S, Kodys K, et al. Hepatitis C virus (HCV) core protein-induced, monocyte-mediated mechanisms of reduced IFN-α and plasmacytoid dendritic cell loss in chronic HCV infection. *J Immunol.* 2006;177:6758-6768.
364. Donaghy H, Gazzard B, Gotch F, Patterson S. Dysfunction and infection of freshly isolated blood myeloid and plasmacytoid dendritic cells in patients infected with HIV-1. *Blood.* 2003;101:4505-4511.
365. Le Nouën C, Hillyer P, Brock LG, et al. Human metapneumovirus SH and G glycoproteins inhibit macropinocytosis-mediated entry into human dendritic cells and reduce CD4+ T cell activation. *J Virol.* 2014;88:6453-6469.
366. Short JJ, Pereboev AV, Kawakami Y, Vasu C, Holterman MJ, Curiel DT. Adenovirus serotype 3 utilizes CD80 (B7.1) and CD86 (B7.2) as cellular attachment receptors. *Virology.* 2004;322:349-359.
367. Lubaki NM, Ilinykh P, Pietzsch C, et al. The lack of maturation of Ebola virus-infected dendritic cells results from the cooperative effect of at least two viral domains. *J Virol.* 2013;87:7471-7485.
368. Dave RS, Jain P, Byrareddy SN. Follicular dendritic cells of lymph nodes as human immunodeficiency virus/simian immunodeficiency virus reservoirs and insights on cervical lymph node. *Front Immunol.* 2018;9:805.
369. Melzi E, Caporale M, Rocchi M, et al. Follicular dendritic cell disruption as a novel mechanism of virus-induced immunosuppression. *Proc Natl Acad Sci U S A.* 2016;113:E6238-E6247.
370. Tu WW, Lau YL, Peiris JSM. Use of humanised mice to study antiviral activity of human γδ-T cells against influenza A viruses. *Hong Kong Med J.* 2014;20(suppl 6):4-6.
371. Pauza CD, Poonia B, Li H, Cairo C, Chaudhry S. γδ T cells in HIV disease: past, present, and future. *Front Immunol.* 2014;5:687.
372. Münz C. Epstein-Barr virus-specific immune control by innate lymphocytes. *Front Immunol.* 2017;8:1658.

373. Chen M, Hu P, Peng H, et al. Enhanced peripheral γδT cells cytotoxicity potential in patients with HBV-associated acute-on-chronic liver failure might contribute to the disease progression. *J Clin Immunol.* 2012;32:877-885.
374. Yin W, Tong S, Zhang Q, et al. Functional dichotomy of Vδ2 γδ T cells in chronic hepatitis C virus infections: role in cytotoxicity but not for IFN-γ production. *Sci Rep.* 2016;6:26296.
375. Zheng J, Liu Y, Lau YL, Tu W. γδ-T cells: an unpolished sword in human anti-infection immunity. *Cell Mol Immunol.* 2013;10:50-57.
376. Zhao Y, Lin L, Xiao Z, et al. Protective role of γδ T cells in different pathogen infections and its potential clinical Application. *J Immunol Res.* 2018;2018:5081634.
377. Sharpe AH, Wherry EJ, Ahmed R, Freeman GJ. The function of programmed cell death 1 and its ligands in regulating autoimmunity and infection. *Nat Immunol.* 2007;8:239-245.
378. Sheppard KA, Fitz LJ, Lee JM, et al. PD-1 inhibits T-cell receptor induced phosphorylation of the ZAP70/CD3zeta signalosome and downstream signaling to PKCtheta. *FEBS Lett.* 2004;574:37-41.
379. Anderson AC, Joller N, Kuchroo VK. Lag-3, Tim-3, and TIGIT: Co-inhibitory receptors with specialized functions in immune regulation. *Immunity.* 2016;44:989-1004.
380. Linnemann T, Zheng YH, Mandic R, Peterlin BM. Interaction between Nef and phosphatidylinositol-3-kinase leads to activation of p21-activated kinase and increased production of HIV. *Virology.* 2002;294:246-255.
381. Fackler OT, Luo W, Geyer M, Alberts AS, Peterlin BM. Activation of Vav by Nef induces cytoskeletal rearrangements and downstream effector functions. *Mol Cell.* 1999;3:729-739.
382. Collette Y, Dutartre H, Benziane A, et al. Physical and functional interaction of Nef with Lck. HIV-1 Nef-induced T-cell signaling defects. *J Biol Chem.* 1996;271:6333-6341.
383. Jerome KR. Viral modulation of T-cell receptor signaling. *J Virol.* 2008;82:4194-4204.
384. Hartley DA, Amdjadi K, Hurley TR, Lund TC, Medveczky PG, Sefton BM. Activation of the Lck tyrosine protein kinase by the Herpesvirus saimiri tip protein involves two binding interactions. *Virology.* 2000;276:339-348.
385. Jung JU, Lang SM, Jun T, Roberts TM, Veillette A, Desrosiers RC. Downregulation of Lck-mediated signal transduction by tip of herpesvirus saimiri. *J Virol.* 1995;69:7814-7822.
386. Sullivan BM, Coscoy L. Downregulation of the T-cell receptor complex and impairment of T-cell activation by human herpesvirus 6 u24 protein. *J Virol.* 2008;82:602-608.
387. Sloan DD, Han JY, Sandifer TK, et al. Inhibition of TCR signaling by herpes simplex virus. *J Immunol.* 2006;176:1825-1833.
388. Avota E, Müller N, Klett M, Schneider-Schaulies S. Measles virus interacts with and alters signal transduction in T-cell lipid rafts. *J Virol.* 2004;78:9552-9559.
389. Pawlak EN, Dirk BS, Jacob RA, Johnson AL, Dikeakos JD. The HIV-1 accessory proteins Nef and Vpu downregulate total and cell surface CD28 in CD4+ T cells. *Retrovirology.* 2018;15:6.
390. Trimble LA, Shankar P, Patterson M, Daily JP, Lieberman J. Human immunodeficiency virus-specific circulating CD8 T lymphocytes have down-modulated CD3zeta and CD28, key signaling molecules for T-cell activation. *J Virol.* 2000;74:7320-7330.
391. Mou D, Espinosa JE, Stempora L, Iwakoshi NN, Kirk AD. Viral-induced CD28 loss evokes costimulation independent alloimmunity. *J Surg Res.* 2015;196:241-246.
392. Raftery MJ, Abdelaziz MO, Hofmann J, Schönrich G. Hantavirus-driven PD-L1/PD-L2 upregulation: an imperfect viral immune evasion mechanism. *Front Immunol.* 2018;9:2560.
393. Boasso A, Hardy AW, Landay AL, et al. PDL-1 upregulation on monocytes and T cells by HIV via type I interferon: restricted expression of type I interferon receptor by CCR5-expressing leukocytes. *Clin Immunol.* 2008;129:132-144.
394. Mühlbauer M, Fleck M, Schütz C, et al. PD-L1 is induced in hepatocytes by viral infection and by interferon-α and-γ and mediates T cell apoptosis. *J Hepatol.* 2006;45:520-528.
395. Maier H, Isogawa M, Freeman GJ, Chisari FV. PD-1:PD-L1 interactions contribute to the functional suppression of virus-specific CD8+ T lymphocytes in the liver. *J Immunol.* 2007;178:2714-2720.
396. Patel A, Hanson J, McLean TI, et al. Herpes simplex type 1 induction of persistent NF-kappa B nuclear translocation increases the efficiency of virus replication. *Virology.* 1998;247:212-222.
397. Hsing LC, Rudensky AY. The lysosomal cysteine proteases in MHC class II antigen presentation. *Immunol Rev.* 2005;207:229-241.
398. Busch R, Rinderknecht CH, Roh S, et al. Achieving stability through editing and chaperoning: regulation of MHC class II peptide binding and expression. *Immunol Rev.* 2005;207:242-260.
399. Denzin LK, Sant'Angelo DB, Hammond C, Surman MJ, Cresswell P. Negative regulation by HLA-DO of MHC class II-restricted antigen processing. *Science.* 1997;278:106-109.
400. Blander JM, Medzhitov R. Toll-dependent selection of microbial antigens for presentation by dendritic cells. *Nature.* 2006;440:808-812.
401. ten Broeke T, Wubbolts R, Stoorvogel W. MHC class II antigen presentation by dendritic cells regulated through endosomal sorting. *Cold Spring Harb Perspect Biol.* 2013;5:a016873.
402. Schmid D, Pypaert M, Münz C. Antigen-loading compartments for major histocompatibility complex class II molecules continuously receive input from autophagosomes. *Immunity.* 2007;26:79-92.
403. Bosch B, Heipertz EL, Drake JR, Roche PA. Major histocompatibility complex (MHC) class II-peptide complexes arrive at the plasma membrane in cholesterol-rich microclusters. *J Biol Chem.* 2013;288:13236-13242.
404. de Waal Malefyt R, Haanen J, Spits H, et al. Interleukin 10 (IL-10) and viral IL-10 strongly reduce antigen-specific human T cell proliferation by diminishing the antigen-presenting capacity of monocytes via downregulation of class II major histocompatibility complex expression. *J Exp Med.* 1991;174:915-924.
405. Fiebiger E, Meraner P, Weber E, et al. Cytokines regulate proteolysis in major histocompatibility complex class II-dependent antigen presentation by dendritic cells. *J Exp Med.* 2001;193:881-892.
406. Forsyth KS, Eisenlohr LC. Giving CD4+ T cells the slip: viral interference with MHC class II-restricted antigen processing and presentation. *Curr Opin Immunol.* 2016;40:123-129.
407. Cai Q, Banerjee S, Cervini A, et al. IRF-4-mediated CIITA transcription is blocked by KSHV encoded LANA to inhibit MHC II presentation. *PLoS Pathog.* 2013;9:e1003751.
408. Kanazawa S, Okamoto T, Peterlin BM. Tat competes with CIITA for the binding to P-TEFb and blocks the expression of MHC class II genes in HIV infection. *Immunity.* 2000;12:61-70.
409. Thakker S, Purushothaman P, Gupta N, Challa S, Cai Q, Verma SC. Kaposi's sarcoma-associated herpesvirus latency-associated nuclear antigen inhibits major histocompatibility complex class II expression by disrupting enhanceosome assembly through binding with the regulatory factor X complex. *J Virol.* 2015;89:5536-5556.
410. Lee AW, Wang N, Hornell TMC, et al. Human cytomegalovirus decreases constitutive transcription of MHC class II genes in mature Langerhans cells by reducing CIITA transcript levels. *Mol Immunol.* 2011;48:1160-1167.
411. Kim H, Mazumdar B, Bose SK, et al. Hepatitis C virus-mediated inhibition of cathepsin S increases invariant-chain expression on hepatocyte surface. *J Virol.* 2012;86:9919-9928.
412. Hegde NR, Tomazin RA, Wisner TW, et al. Inhibition of HLA-DR assembly, transport, and loading by human cytomegalovirus glycoprotein US3: a novel mechanism for evading major histocompatibility complex class II antigen presentation. *J Virol.* 2002;76:10929-10941.
413. Quinn LL, Williams LR, White C, Forrest C, Zuo J, Rowe M. The missing link in Epstein-Barr virus immune evasion: the BDLF3 gene induces ubiquitination and downregulation of major histocompatibility complex class I (MHC-I) and MHC-II. *J Virol.* 2016;90:356-367.
414. Ressing ME, van Leeuwen D, Verreck FAW, et al. Interference with T cell receptor-HLA-DR interactions by Epstein-Barr virus gp42 results in reduced T helper cell recognition. *Proc Natl Acad Sci U S A.* 2003;100:11583-11588.
415. Geiger R, Duhen T, Lanzavecchia A, Sallusto F. Human naive and memory CD4+ T cell repertoires specific for naturally processed antigens analyzed using libraries of amplified T cells. *J Exp Med.* 2009;206:1525-1534.
416. Celli S, Lemaître F, Bousso P. Real-time manipulation of T cell-dendritic cell interactions *in vivo* reveals the importance of prolonged contacts for CD4+ T cell activation. *Immunity.* 2007;27:625-634.
417. Tao X, Constant S, Jorritsma P, Bottomly K. Strength of TCR signal determines the costimulatory requirements for Th1 and Th2 CD4+ T cell differentiation. *J Immunol.* 1997;159:5956-5963.
418. Chang JT. Polarity and lymphocyte fate determination. *Curr Opin Cell Biol.* 2012;24:526-533.
419. Lanzavecchia A, Sallusto F. Progressive differentiation and selection of the fittest in the immune response. *Nat Rev Immunol.* 2002;2:982-987.
420. Malherbe L, Hausl C, Teyton L, McHeyzer-Williams MG. Clonal selection of helper T cells is determined by an affinity threshold with no further skewing of TCR binding properties. *Immunity.* 2004;21:669-679.
421. Berard M, Tough DF. Qualitative differences between naïve and memory T cells. *Immunology.* 2002;106:127-138.
422. Klenerman P, Hill A. T cells and viral persistence: lessons from diverse infections. *Nat Immunol.* 2005;6:873-879.
423. Sun JC, Bevan MJ. Defective CD8 T cell memory following acute infection without CD4 T cell help. *Science.* 2003;300:339-342.
424. Heller KN, Gurer C, Münz C. Virus-specific CD4+ T cells: ready for direct attack. *J Exp Med.* 2006;203:805-808.
425. Torpey DJ, Lindsley MD, Rinaldo CR. HLA-restricted lysis of herpes simplex virus-infected monocytes and macrophages mediated by CD4+ and CD8+ T lymphocytes. *J Immunol.* 1989;142:1325-1332.
426. Acharya D, Wang P, Paul AM, et al. Interleukin-17A promotes CD8+ T cell cytotoxicity to facilitate West Nile virus clearance. *J Virol.* 2017;91.
427. Jie Z, Liang Y, Hou L, et al. Intrahepatic innate lymphoid cells secrete IL-17A and IL-17F that are crucial for T cell priming in viral infection. *J Immunol.* 2014;192:3289-3300.
428. Wang X, Chan CCS, Yang M, et al. A critical role of IL-17 in modulating the B-cell response during H5N1 influenza virus infection. *Cell Mol Immunol.* 2011;8:462-468.
429. Wang X, Ma K, Chen M, Ko KH, Zheng B-J, Lu L. IL-17A promotes pulmonary B-1a cell differentiation via induction of blimp-1 expression during influenza virus infection. *PLoS Pathog.* 2016;12:e1005367.
430. Hou W, Kang HS, Kim BS. Th17 cells enhance viral persistence and inhibit T cell cytotoxicity in a model of chronic virus infection. *J Exp Med.* 2009;206:313-328.
431. Yuan Y, Cruzat VF, Newsholme P, Cheng J, Chen Y, Lu Y. Regulation of SIRT1 in aging: roles in mitochondrial function and biogenesis. *Mech Ageing Dev.* 2016;155:10-21.
432. Mo R, Wang P, Lai R, et al. Persistently elevated circulating Th22 reversely correlates with prognosis in HBV-related acute-on-chronic liver failure. *J Gastroenterol Hepatol.* 2017;32:677-686.
433. Green EA, Gorelik L, McGregor CM, Tran EH, Flavell RA. CD4+CD25+ T regulatory cells control anti-islet CD8+ T cells through TGF-β-TGF-β receptor interactions in type 1 diabetes. *Proc Natl Acad Sci U S A.* 2003;100:10878-10883.
434. Asseman C, Mauze S, Leach MW, Coffman RL, Powrie F. An essential role for interleukin 10 in the function of regulatory T cells that inhibit intestinal inflammation. *J Exp Med.* 1999;190:995-1004.
435. Barthlott T, Kassiotis G, Stockinger B. T cell regulation as a side effect of homeostasis and competition. *J Exp Med.* 2003;197:451-460.
436. Grossman WJ, Verbsky JW, Barchet W, Colonna M, Atkinson JP, Ley TJ. Human T regulatory cells can use the perforin pathway to cause autologous target cell death. *Immunity.* 2004;21:589-601.
437. Suvas S, Azkur AK, Kim BS, Kumaraguru U, Rouse BT. CD4+CD25+ regulatory T cells control the severity of viral immunoinflammatory lesions. *J Immunol.* 2004;172:4123-4132.
438. Suvas S, Kumaraguru U, Pack CD, Lee S, Rouse BT. CD4+CD25+ T cells regulate virus-specific primary and memory CD8+ T cell responses. *J Exp Med.* 2003;198:889-901.
439. Kinter A, McNally J, Riggin L, Jackson R, Roby G, Fauci AS. Suppression of HIV-specific T cell activity by lymph node CD25+ regulatory T cells from HIV-infected individuals. *Proc Natl Acad Sci U S A.* 2007;104:3390-3395.
440. Cabrera R, Tu Z, Xu Y, et al. An immunomodulatory role for CD4(+)CD25(+) regulatory T lymphocytes in hepatitis C virus infection. *Hepatology.* 2004;40:1062-1071.
441. Yang G, Liu A, Xie Q, et al. Association of CD4+CD25+Foxp3+ regulatory T cells with chronic activity and viral clearance in patients with hepatitis B. *Int Immunol.* 2007;19:133-140.
442. Aandahl EM, Michaëlsson J, Moretto WJ, Hecht FM, Nixon DF. Human CD4+ CD25+ regulatory T cells control T-cell responses to human immunodeficiency virus and cytomegalovirus antigens. *J Virol.* 2004;78:2454-2459.
443. Wingate PJ, McAulay KA, Anthony IC, Crawford DH. Regulatory T cell activity in primary and persistent Epstein-Barr virus infection. *J Med Virol.* 2009;81:870-877.
444. Veiga-Parga T, Sehrawat S, Rouse BT. Role of regulatory T cells during virus infection. *Immunol Rev.* 2013;255:182-196.
445. Cooper A, García M, Petrovas C, Yamamoto T, Koup RA, Nabel GJ. HIV-1 causes CD4 cell death through DNA-dependent protein kinase during viral integration. *Nature.* 2013;498:376-379.
446. Deeks SG, Kitchen CMR, Liu L, et al. Immune activation set point during early HIV infection predicts subsequent CD4+ T-cell changes independent of viral load. *Blood.* 2004;104:942-947.
447. Younan P, Santos RI, Ramanathan P, et al. Ebola virus-mediated T-lymphocyte depletion is the result of an abortive infection. *PLoS Pathog.* 2019;15:e1008068.
448. Gupta N, Hegde P, Lecerf M, et al. Japanese encephalitis virus expands regulatory T cells by increasing the expression of PD-L1 on dendritic cells. *Eur J Immunol.* 2014;44:1363-1374.
449. Hall CHT, Kassel R, Tacke RS, Hahn YS. HCV+ hepatocytes induce human regulatory CD4+ T cells through the production of TGF-β. *PLoS One.* 2010;5:e12154.
450. Sharma S, Rajasagi NK, Veiga-Parga T, Rouse BT. Herpes virus entry mediator (HVEM) modulates proliferation and activation of regulatory T cells following HSV-1 infection. *Microbes Infect.* 2014;16:648-660.
451. Ji J, Cloyd MW. HIV-1 binding to CD4 on CD4+CD25+ regulatory T cells enhances their suppressive function and induces them to home to, and accumulate in, peripheral and mucosal lymphoid tissues: an additional mechanism of immunosuppression. *Int Immunol.* 2009;21:283-294.
452. Cresswell P, Bangia N, Dick T, Diedrich G. The nature of the MHC class I peptide loading complex. *Immunol Rev.* 1999;172:21-28.

453. Kovacsovics-Bankowski M, Rock KL. A phagosome-to-cytosol pathway for exogenous antigens presented on MHC class I molecules. *Science.* 1995;267:243-246.
454. Houde M, Bertholet S, Gagnon E, et al. Phagosomes are competent organelles for antigen cross-presentation. *Nature.* 2003;425:402-406.
455. Merzougui N, Kratzer R, Saveanu L, van Endert P. A proteasome-dependent, TAP-independent pathway for cross-presentation of phagocytosed antigen. *EMBO Rep.* 2011;12:1257-1264.
456. Richards R, Scholz I, Powers C, Skach WR, Früh K. The cytoplasmic domain of rhesus cytomegalovirus Rh178 interrupts translation of major histocompatibility class I leader peptide-containing proteins prior to translocation. *J Virol.* 2011;85:8766-8776.
457. Levitskaya J, Coram M, Levitsky V, et al. Inhibition of antigen processing by the internal repeat region of the Epstein-Barr virus nuclear antigen-1. *Nature.* 1995;375:685-688.
458. Ahn K, Gruhler A, Galocha B, et al. The ER-luminal domain of the HCMV glycoprotein US6 inhibits peptide translocation by TAP. *Immunity.* 1997;6:613-621.
459. Alzhanova D, Edwards DM, Hammarlund E, et al. Cowpox virus inhibits the transporter associated with antigen processing to evade T cell recognition. *Cell Host Microbe.* 2009;6:433-445.
460. Koppers-Lalic D, Reits EAJ, Ressing ME, et al. Varicelloviruses avoid T cell recognition by UL49.5-mediated inactivation of the transporter associated with antigen processing. *Proc Natl Acad Sci U S A.* 2005;102:5144-5149.
461. Hill A, Jugovic P, York I, et al. Herpes simplex virus turns off the TAP to evade host immunity. *Nature.* 1995;375:411-415.
462. Horst D, Favaloro V, Vilardi F, et al. EBV protein BNLF2a exploits host tail-anchored protein integration machinery to inhibit TAP. *J Immunol.* 2011;186:3594-3605.
463. Jones TR, Wiertz EJ, Sun L, Fish KN, Nelson JA, Ploegh HL. Human cytomegalovirus US3 impairs transport and maturation of major histocompatibility complex class I heavy chains. *Proc Natl Acad Sci U S A.* 1996;93:11327-11333.
464. Bennett EM, Bennink JR, Yewdell JW, Brodsky FM. Cutting edge: adenovirus E19 has two mechanisms for affecting class I MHC expression. *J Immunol.* 1999;162:5049-5052.
465. Wiertz EJ, Jones TR, Sun L, Bogyo M, Geuze HJ, Ploegh HL. The human cytomegalovirus US11 gene product dislocates MHC class I heavy chains from the endoplasmic reticulum to the cytosol. *Cell.* 1996;84:769-779.
466. Loureiro J, Lilley BN, Spooner E, Noriega V, Tortorella D, Ploegh HL. Signal peptide peptidase is required for dislocation from the endoplasmic reticulum. *Nature.* 2006;441:894-897.
467. Lybarger L, Wang X, Harris MR, Virgin HW, Hansen TH. Virus subversion of the MHC class I peptide-loading complex. *Immunity.* 2003;18:121-130.
468. Byun M, Wang X, Pak M, Hansen TH, Yokoyama WM. Cowpox virus exploits the endoplasmic reticulum retention pathway to inhibit MHC class I transport to the cell surface. *Cell Host Microbe.* 2007;2:306-315.
469. Coscoy L, Ganem D. Kaposi's sarcoma-associated herpesvirus encodes two proteins that block cell surface display of MHC class I chains by enhancing their endocytosis. *Proc Natl Acad Sci U S A.* 2000;97:8051-8056.
470. Germain RN, Miller MJ, Dustin ML, Nussenzweig MC. Dynamic imaging of the immune system: progress, pitfalls and promise. *Nat Rev Immunol.* 2006;6:497-507.
471. Castellino F, Huang AY, Altan-Bonnet G, Stoll S, Scheinecker C, Germain RN. Chemokines enhance immunity by guiding naive CD8+ T cells to sites of CD4+ T cell-dendritic cell interaction. *Nature.* 2006;440:890-895.
472. Nakanishi Y, Lu B, Gerard C, Iwasaki A. CD8(+) T lymphocyte mobilization to virus-infected tissue requires CD4(+) T-cell help. *Nature.* 2009;462:510-513.
473. Bevan MJ. Helping the CD8(+) T-cell response. *Nat Rev Immunol.* 2004;4:595-602.
474. Jennings SR, Bonneau RH, Smith PM, Wolcott RM, Chervenak R. CD4-positive T lymphocytes are required for the generation of the primary but not the secondary CD8-positive cytolytic T lymphocyte response to herpes simplex virus in C57BL/6 mice. *Cell Immunol.* 1991;133:234-252.
475. Riberdy JM, Christensen JP, Branum K, Doherty PC. Diminished primary and secondary influenza virus-specific CD8(+) T-cell responses in CD4-depleted Ig(−/−) mice. *J Virol.* 2000;74:9762-9765.
476. Chen Y, Zander R, Khatun A, Schauder DM, Cui W. Transcriptional and epigenetic regulation of effector and memory CD8 T cell differentiation. *Front Immunol.* 2018;9:2826.
477. Hendriks J, Xiao Y, Rossen JWA, et al. During viral infection of the respiratory tract, CD27, 4-1BB, and OX40 collectively determine formation of CD8+ memory T cells and their capacity for secondary expansion. *J Immunol.* 2005;175:1665-1676.
478. Bally APR, Tang Y, Lee JT, et al. Conserved region C functions to regulate PD-1 expression and subsequent CD8 T cell memory. *J Immunol.* 2017;198:205-217.
479. Hendriks J, Gravestein LA, Tesselaar K, van Lier RA, Schumacher TN, Borst J. CD27 is required for generation and long-term maintenance of T cell immunity. *Nat Immunol.* 2000;1:433-440.
480. Bourgeois C, Tanchot C. Mini-review CD4 T cells are required for CD8 T cell memory generation. *Eur J Immunol.* 2003;33:3225-3231.
481. Intlekofer AM, Takemoto N, Kao C, et al. Requirement for T-bet in the aberrant differentiation of unhelped memory CD8+ T cells. *J Exp Med.* 2007;204:2015-2021.
482. Sabbagh L, Snell LM, Watts TH. TNF family ligands define niches for T cell memory. *Trends Immunol.* 2007;28:333-339.
483. Bachmann MF, Gallimore A, Linkert S, et al. Developmental regulation of Lck targeting to the CD8 coreceptor controls signaling in naive and memory T cells. *J Exp Med.* 1999;189:1521-1530.
484. Veiga-Fernandes H, Walter U, Bourgeois C, McLean A, Rocha B. Response of naïve and memory CD8+ T cells to antigen stimulation *in vivo*. *Nat Immunol.* 2000;1:47-53.
485. Tanchot C, Guillaume S, Delon J, et al. Modifications of CD8+ T cell function during *in vivo* memory or tolerance induction. *Immunity.* 1998;8:581-590.
486. Murali-Krishna K, Lau LL, Sambhara S, Lemonnier F, Altman J, Ahmed R. Persistence of memory CD8 T cells in MHC class I-deficient mice. *Science.* 1999;286:1377-1381.
487. Zajac AJ, Blattman JN, Murali-Krishna K, et al. Viral immune evasion due to persistence of activated T cells without effector function. *J Exp Med.* 1998;188:2205-2213.
488. Fenwick C, Joo V, Jacquier P, Noto A, Banga R, Perreau M, Pantaleo G. T-cell exhaustion in HIV infection. *Immunol Rev.* 2019;292:149-163. doi:10.1111/imr.12823.
489. Brooks DG, Ha SJ, Elsaesser H, Sharpe AH, Freeman GJ, Oldstone MBA. IL-10 and PD-L1 operate through distinct pathways to suppress T-cell activity during persistent viral infection. *Proc Natl Acad Sci U S A.* 2008;105:20428-20433.
490. Tinoco R, Alcalde V, Yang Y, Sauer K, Zuniga EI. Cell-intrinsic transforming growth factor-β signaling mediates virus-specific CD8+ T cell deletion and viral persistence *in vivo*. *Immunity.* 2009;31:145-157.
491. Zhang Y, Kurupati R, Liu L, et al. Enhancing CD8+ T cell fatty acid catabolism within a metabolically challenging tumor microenvironment increases the efficacy of melanoma immunotherapy. *Cancer Cell.* 2017;32:377.e9-391.e9.
492. Xu W, Larbi A. Markers of T cell senescence in humans. *Int J Mol Sci.* 2017;18.
493. Blackburn SD, Shin H, Haining WN, et al. Coregulation of CD8+ T cell exhaustion by multiple inhibitory receptors during chronic viral infection. *Nat Immunol.* 2009;10:29-37.
494. Martinez GJ, Pereira RM, Äijö T, et al. The transcription factor NFAT promotes exhaustion of activated CD8+ T cells. *Immunity.* 2015;42:265-278.
495. Saeidi A, Zandi K, Cheok YY, et al. T-cell exhaustion in chronic infections: Reversing the state of exhaustion and reinvigorating optimal protective immune responses. *Front Immunol.* 2018;9:2569.
496. Khan O, Giles JR, McDonald S, et al. TOX transcriptionally and epigenetically programs CD8+ T cell exhaustion. *Nature.* 2019;571:211-218.
497. Myers LM, Tal MC, Torrez Dulgeroff LB, et al. A functional subset of CD8+ T cells during chronic exhaustion is defined by SIRPα expression. *Nat Commun.* 2019;10:794.
498. Zhang Y, Ertl HCJ. The effect of adjuvanting cancer vaccines with herpes simplex virus glycoprotein D on melanoma-driven CD8+ T cell exhaustion. *J Immunol.* 2014;193:1836-1846.
499. van der Most RG, Murali-Krishna K, Lanier JG, et al. Changing immunodominance patterns in antiviral CD8 T-cell responses after loss of epitope presentation or chronic antigenic stimulation. *Virology.* 2003;315:93-102.
500. Pawelec G, Koch S, Franceschi C, Wikby A. Human immunosenescence: does it have an infectious component? *Ann N Y Acad Sci.* 2006;1067:56-65.
501. Hutnick NA, Carnathan D, Demers K, Makedonas G, Ertl HCJ, Betts MR. Adenovirus-specific human T cells are pervasive, polyfunctional, and cross-reactive. *Vaccine.* 2010;28:1932-1941.
502. Purbhoo MA, Irvine DJ, Huppa JB, Davis MM. T cell killing does not require the formation of a stable mature immunological synapse. *Nat Immunol.* 2004;5:524-530.
503. Tollefson AE, Hermiston TW, Lichtenstein DL, et al. Forced degradation of Fas inhibits apoptosis in adenovirus-infected cells. *Nature.* 1998;392:726-730.
504. Afonina IS, Cullen SP, Martin SJ. Cytotoxic and non-cytotoxic roles of the CTL/NK protease granzyme B. *Immunol Rev.* 2010;235:105-116.
505. Martinvalet D. Mitochondrial entry of cytotoxic proteases: a new insight into the granzyme B cell death pathway. *Oxid Med Cell Longev.* 2019;2019:9165214.
506. Isaaz S, Baetz K, Olsen K, Podack E, Griffiths GM. Serial killing by cytotoxic T lymphocytes: T cell receptor triggers degranulation, re-filling of the lytic granules and secretion of lytic proteins via a non-granule pathway. *Eur J Immunol.* 1995;25:1071-1079.
507. Komiyama T, Ray CA, Pickup DJ, et al. Inhibition of interleukin-1β converting enzyme by the cowpox virus serpin CrmA. An example of cross-class inhibition. *J Biol Chem.* 1994;269:19331-19337.
508. Andrade F, Bull HG, Thornberry NA, Ketner GW, Casciola-Rosen LA, Rosen A. Adenovirus L4-100K assembly protein is a granzyme B substrate that potently inhibits granzyme B-mediated cell death. *Immunity.* 2001;14:751-761.
509. Vieyra-Lobato MR, Vela-Ojeda J, Montiel-Cervantes L, López-Santiago R, Moreno-Lafont MC. Description of CD8+ regulatory t lymphocytes and their specific intervention in graft-versus-host and infectious diseases, autoimmunity, and cancer. *J Immunol Res.* 2018;2018:3758713.
510. Foucher ED, Blanchard S, Preisser L, et al. IL-34 induces the differentiation of human monocytes into immunosuppressive macrophages. Antagonistic effects of GM-CSF and IFNγ. *PLoS One.* 2013;8:e56045.
511. Billerbeck E, Nakamoto N, Seigel B, Blum HE, Chang KM, Thimme R. Determinants of *in vitro* expansion of different human virus-specific FoxP3+ regulatory CD8+ T cells in chronic hepatitis C virus infection. *J Gen Virol.* 2009;90:1692-1701.
512. Franzese O, Kennedy PTF, Gehring AJ, et al. Modulation of the CD8+-T-cell response by CD4+ CD25+ regulatory T cells in patients with hepatitis B virus infection. *J Virol.* 2005;79:3322-3328.
513. Dittmer U, He H, Messer RJ, et al. Functional impairment of CD8(+) T cells by regulatory T cells during persistent retroviral infection. *Immunity.* 2004;20:293-303.
514. Imani F, Proud D, Griffin DE. Measles virus infection synergizes with IL-4 in IgE class switching. *J Immunol.* 1999;162:1597-1602.
515. Suzuki S, Suzuki Y, Yamamoto N, Matsumoto Y, Shirai A, Okubo T. Influenza A virus infection increases IgE production and airway responsiveness in aerosolized antigen-exposed mice. *J Allergy Clin Immunol.* 1998;102:732-740.
516. Allman D, Pillai S. Peripheral B cell subsets. *Curr Opin Immunol.* 2008;20:149-157.
517. Vieira P, Rajewsky K. Persistence of memory B cells in mice deprived of T cell help. *Int Immunol.* 1990;2:487-494.
518. Brynjolfsson SF, Mohaddes M, Kärrholm J, Wick MJ. Long-lived plasma cells in human bone marrow can be either CD19+ or CD19. *Blood Adv.* 2017;1:835-838.
519. Krueger CC, Thoms F, Keller E, Vogel M, Bachmann MF. Virus-specific secondary plasma cells produce elevated levels of high-avidity antibodies but are functionally short lived. *Front Immunol.* 2019;10:1831.
520. Moir S, Ho J, Malaspina A, et al. Evidence for HIV-associated B cell exhaustion in a dysfunctional memory B cell compartment in HIV-infected viremic individuals. *J Exp Med.* 2008;205:1797-1805.
521. Mallery DL, McEwan WA, Bidgood SR, Towers GJ, Johnson CM, James LC. Antibodies mediate intracellular immunity through tripartite motif-containing 21 (TRIM21). *Proc Natl Acad Sci U S A.* 2010;107:19985-19990.
522. McEwan WA, Hauler F, Williams CR, et al. Regulation of virus neutralization and the persistent fraction by TRIM21. *J Virol.* 2012;86:8482-8491.
523. Richman DD, Wrin T, Little SJ, Petropoulos CJ. Rapid evolution of the neutralizing antibody response to HIV type 1 infection. *Proc Natl Acad Sci U S A.* 2003;100:4144-4149.
524. Nothelfer K, Sansonetti PJ, Phalipon A. Pathogen manipulation of B cells: the best defence is a good offence. *Nat Rev Microbiol.* 2015;13:173-184.
525. McChesney MB, Kehrl JH, Valsamakis A, Fauci AS, Oldstone MB. Measles virus infection of B lymphocytes permits cellular activation but blocks progression through the cell cycle. *J Virol.* 1987;61:3441-3447.
526. Babcock GJ, Decker LL, Volk M, Thorley-Lawson DA. EBV persistence in memory B cells *in vivo*. *Immunity.* 1998;9:395-404.
527. Rosa D, Saletti G, De Gregorio E, et al. Activation of naïve B lymphocytes via CD81, a pathogenetic mechanism for hepatitis C virus-associated B lymphocyte disorders. *Proc Natl Acad Sci U S A.* 2005;102:18544-18549.
528. Machida K, Kondo Y, Huang JY, et al. Hepatitis C virus (HCV)-induced immunoglobulin hypermutation reduces the affinity and neutralizing activities of antibodies against HCV envelope protein. *J Virol.* 2008;82:6711-6720.
529. Dougan SK, Ashour J, Karssemeijer RA, et al. Antigen-specific B-cell receptor sensitizes B cells to infection by influenza virus. *Nature.* 2013;503:406-409.
530. Mohan GS, Li W, Ye L, Compans RW, Yang C. Antigenic subversion: a novel mechanism of host immune evasion by Ebola virus. *PLoS Pathog.* 2012;8:e1003065.
531. Cook JD, Lee JE. The secret life of viral entry glycoproteins: Moonlighting in immune evasion. *PLoS Pathog.* 2013;9:e1003258.
532. Tate MD, Job ER, Deng YM, Gunalan V, Maurer-Stroh S, Reading PC. Playing hide and seek: how glycosylation of the influenza virus hemagglutinin can modulate the immune response to infection. *Viruses.* 2014;6:1294-1316.
533. Sommerstein R, Flatz L, Remy MM, et al. Arenavirus glycan shield promotes neutralizing antibody evasion and protracted infection. *PLoS Pathog.* 2015;11:e1005276.
534. Liu J, Zhan W, Kim CJ, et al. IL-10-producing B cells are induced early in HIV-1 infection and suppress HIV-1-specific T cell responses. *PLoS One.* 2014;9:e89236.
535. Tsai CY, Sakakibara S, Yasui T, Minamitani T, Okuzaki D, Kikutani H. Bystander inhibition of humoral immune responses by Epstein-Barr virus LMP1. *Int Immunol.* 2018;30:579-590.

536. Minamitani T, Yasui T, Ma Y, et al. Evasion of affinity-based selection in germinal centers by Epstein-Barr virus LMP2A. *Proc Natl Acad Sci U S A*. 2015;112:11612-11617.
537. Komano H, Fujiura Y, Kawaguchi M, et al. Homeostatic regulation of intestinal epithelia by intraepithelial γ delta T cells. *Proc Natl Acad Sci U S A*. 1995;92:6147-6151.
538. Brenchley JM, Douek DC. The mucosal barrier and immune activation in HIV pathogenesis. *Curr Opin HIV AIDS*. 2008;3:356-361.
539. Rakoff-Nahoum S, Medzhitov R. Regulation of spontaneous intestinal tumorigenesis through the adaptor protein MyD88. *Science*. 2007;317:124-127.
540. Hill DA, Artis D. Intestinal bacteria and the regulation of immune cell homeostasis. *Annu Rev Immunol*. 2010;28:623-667.
541. Mazmanian SK, Liu CH, Tzianabos AO, Kasper DL. An immunomodulatory molecule of symbiotic bacteria directs maturation of the host immune system. *Cell*. 2005;122:107-118.
542. Cash HL, Whitham CV, Behrendt CL, Hooper LV. Symbiotic bacteria direct expression of an intestinal bactericidal lectin. *Science*. 2006;313:1126-1130.
543. Miettinen M, Veckman V, Latvala S, Sareneva T, Matikainen S, Julkunen I. Live *Lactobacillus rhamnosus* and *Streptococcus pyogenes* differentially regulate Toll-like receptor (TLR) gene expression in human primary macrophages. *J Leukoc Biol*. 2008;84:1092-1100.
544. Round JL, Mazmanian SK. Inducible Foxp3+ regulatory T-cell development by a commensal bacterium of the intestinal microbiota. *Proc Natl Acad Sci U S A*. 2010;107:12204-12209.
545. Ivanov II, Frutos Rde L, Manel N, et al. Specific microbiota direct the differentiation of IL-17-producing T-helper cells in the mucosa of the small intestine. *Cell Host Microbe*. 2008;4:337-349.
546. Tanaka K, Sawamura S, Satoh T, Kobayashi K, Noda S. Role of the indigenous microbiota in maintaining the virus-specific CD8 memory T cells in the lung of mice infected with murine cytomegalovirus. *J Immunol*. 2007;178:5209-5216.
547. Ichinohe T, Pang IK, Kumamoto Y, et al. Microbiota regulates immune defense against respiratory tract influenza A virus infection. *Proc Natl Acad Sci U S A*. 2011;108:5354-5359.
548. Gori A, Tincati C, Rizzardini G, et al. Early impairment of gut function and gut flora supporting a role for alteration of gastrointestinal mucosa in human immunodeficiency virus pathogenesis. *J Clin Microbiol*. 2008;46:757-758.
549. Buccigrossi V, Laudiero G, Russo C, et al. Chloride secretion induced by rotavirus is oxidative stress-dependent and inhibited by *Saccharomyces boulardii* in human enterocytes. *PLoS One*. 2014;9:e99830.
550. Varyukhina S, Freitas M, Bardin S, et al. Glycan-modifying bacteria-derived soluble factors from *Bacteroides thetaiotaomicron* and *Lactobacillus casei* inhibit rotavirus infection in human intestinal cells. *Microbes Infect*. 2012;14:273-278.
551. Ñahui Palomino RA, Zicari S, Vanpouille C, Vitali B, Margolis L. Vaginal *Lactobacillus* inhibits HIV-1 replication in human tissues ex vivo. *Front Microbiol*. 2017;8:906.
552. Steed AL, Christophi GP, Kaiko GE, et al. The microbial metabolite desaminotyrosine protects from influenza through type I interferon. *Science*. 2017;357:498-502.
553. Jones MK, Watanabe M, Zhu S, et al. Enteric bacteria promote human and mouse norovirus infection of B cells. *Science*. 2014;346:755-759.
554. Baldridge MT, Nice TJ, McCune BT, et al. Commensal microbes and interferon-λ determine persistence of enteric murine norovirus infection. *Science*. 2015;347:266-269.
555. Kuss SK, Best GT, Etheredge CA, et al. Intestinal microbiota promote enteric virus replication and systemic pathogenesis. *Science*. 2011;334:249-252.
556. Benedikz EK, Bailey D, Cook CNL, et al. Bacterial flagellin promotes viral entry via an NF-kB and Toll like receptor 5 dependent pathway. *Sci Rep*. 2019;9:7903.
557. Berger AK, Yi H, Kearns DB, Mainou BA. Bacteria and bacterial envelope components enhance mammalian reovirus thermostability. *PLoS Pathog*. 2017;13:e1006768.
558. Pang WW, Price EA, Sahoo D, et al. Human bone marrow hematopoietic stem cells are increased in frequency and myeloid-biased with age. *Proc Natl Acad Sci U S A*. 2011;108:20012-20017.
559. den Braber I, Mugwagwa T, Vrisekoop N, et al. Maintenance of peripheral naive T cells is sustained by thymus output in mice but not humans. *Immunity*. 2012;36:288-297.
560. Pourgheysari B, Khan N, Best D, Bruton R, Nayak L, Moss PAH. The cytomegalovirus-specific CD4+ T-cell response expands with age and markedly alters the CD4+ T-cell repertoire. *J Virol*. 2007;81:7759-7765.
561. Hadrup SR, Strindhall J, Køllgaard T, et al. Longitudinal studies of clonally expanded CD8 T cells reveal a repertoire shrinkage predicting mortality and an increased number of dysfunctional cytomegalovirus-specific T cells in the very elderly. *J Immunol*. 2006;176:2645-2653.
562. Li G, Yu M, Lee WW, et al. Decline in miR-181a expression with age impairs T cell receptor sensitivity by increasing DUSP6 activity. *Nat Med*. 2012;18:1518-1524.
563. Li G, Ju J, Weyand CM, Goronzy JJ. Age-associated failure to Adjust type I IFN receptor signaling thresholds after T cell activation. *J Immunol*. 2015;195:865-874.
564. Lee KA, Shin KS, Kim GY, et al. Characterization of age-associated exhausted CD8+ T cells defined by increased expression of Tim-3 and PD-1. *Aging Cell*. 2016;15:291-300.
565. McKenna PM, McGettigan JP, Pomerantz RJ, Dietzschold B, Schnell MJ. Recombinant rhabdoviruses as potential vaccines for HIV-1 and other diseases. *Curr HIV Res*. 2003;1:229-237.
566. Frasca D, Diaz A, Romero M, Landin AM, Blomberg BB. Age effects on B cells and humoral immunity in humans. *Ageing Res Rev*. 2011;10:330-335.
567. Gibson KL, Wu YC, Barnett Y, et al. B-cell diversity decreases in old age and is correlated with poor health status. *Aging Cell*. 2009;8:18-25.
568. Frasca D, Diaz A, Romero M, et al. Intrinsic defects in B cell response to seasonal influenza vaccination in elderly humans. *Vaccine*. 2010;28:8077-8084.
569. Fink RI, Kolterman OG, Griffin J, Olefsky JM. Mechanisms of insulin resistance in aging. *J Clin Invest*. 1983;71:1523-1535.
570. Woudstra T, Thomson ABR. Nutrient absorption and intestinal adaptation with ageing. *Best Pract Res Clin Gastroenterol*. 2002;16:1-15.
571. Sun N, Youle RJ, Finkel T. The mitochondrial basis of aging. *Mol Cell*. 2016;61:654-666.
572. Fulop T, Larbi A, Douziech N, et al. Signal transduction and functional changes in neutrophils with aging. *Aging Cell*. 2004;3:217-226.
573. Solana R, Pawelec G, Tarazona R. Aging and innate immunity. *Immunity*. 2006;24:491-494.
574. Sansoni P, Vescovini R, Fagnoni FF, et al. New advances in CMV and immunosenescence. *Exp Gerontol*. 2014;55:54-62.
575. Deeks SG. HIV infection, inflammation, immunosenescence, and aging. *Annu Rev Med*. 2011;62:141-155.
576. Youm YH, Grant RW, McCabe LR, et al. Canonical Nlrp3 inflammasome links systemic low-grade inflammation to functional decline in aging. *Cell Metabol*. 2013;18:519-532.
577. Meier J, Sturm A. The intestinal epithelial barrier: does it become impaired with age? *Dig Dis*. 2009;27:240-245.
578. Drela N, Kozdron E, Szczypiorski P. Moderate exercise may attenuate some aspects of immunosenescence. *BMC Geriatr*. 2004;4:8.
579. Maijó M, Clements SJ, Ivory K, Nicoletti C, Carding SR. Nutrition, diet and immunosenescence. *Mech Ageing Dev*. 2014;136-137:116-128.
580. Leis AA, Szatmary G, Ross MA, Stokic DS. West Nile virus infection and myasthenia gravis. *Muscle Nerve*. 2014;49:26-29.
581. Zhao ZS, Granucci F, Yeh L, Schaffer PA, Cantor H. Molecular mimicry by herpes simplex virus-type 1: autoimmune disease after viral infection. *Science*. 1998;279:1344-1347.
582. Gauntt CJ, Arizpe HM, Higdon AL, et al. Molecular mimicry, anti-coxsackievirus B3 neutralizing monoclonal antibodies, and myocarditis. *J Immunol*. 1995;154:2983-2995.
583. Croxford JL, Olson JK, Miller SD. Epitope spreading and molecular mimicry as triggers of autoimmunity in the Theiler's virus-induced demyelinating disease model of multiple sclerosis. *Autoimmun Rev*. 2002;1:251-260.
584. Poole BD, Scofield RH, Harley JB, James JA. Epstein-Barr virus and molecular mimicry in systemic lupus erythematosus. *Autoimmunity*. 2006;39:63-70.
585. Nanbo A, Inoue K, Adachi-Takasawa K, Takada K. Epstein-Barr virus RNA confers resistance to interferon-α-induced apoptosis in Burkitt's lymphoma. *EMBO J*. 2002;21:954-965.
586. Kanduc D. The comparative biochemistry of viruses and humans: an evolutionary path towards autoimmunity. *Biol Chem*. 2019;400:629-638.
587. Chang Y, Cesarman E, Pessin MS, et al. Identification of herpesvirus-like DNA sequences in AIDS-associated Kaposi's sarcoma. *Science*. 1994;266:1865-1869.
588. Cadoz M, Strady A, Meignier B, et al. Immunisation with canarypox virus expressing rabies glycoprotein. *Lancet*. 1992;339:1429-1432.
589. Charlton KM, Artois M, Prevec L, et al. Oral rabies vaccination of skunks and foxes with a recombinant human adenovirus vaccine. *Arch Virol*. 1992;123:169-179.
590. Brochier B, Kieny MP, Costy F, et al. Large-scale eradication of rabies using recombinant vaccinia-rabies vaccine. *Nature*. 1991;354:520-522.
591. Bennett J, Wellman J, Marshall KA, et al. Safety and durability of effect of contralateral-eye administration of AAV2 gene therapy in patients with childhood-onset blindness caused by RPE65 mutations: a follow-on phase 1 trial. *Lancet*. 2016;388:661-672.
592. Hanahan D, Weinberg RA. Hallmarks of cancer: the next generation. *Cell*. 2011;144:646-674.
593. Pelner L, Fowler GA, Nauts HC. Effects of concurrent infections and their toxins on the course of leukemia. *Acta Med Scand Suppl*. 1958;338:1-47.
594. Moore AE. The destructive effect of the virus of Russian Far East encephalitis on the transplantable mouse sarcoma 180. *Cancer*. 1949;2:525-534.
595. Moore AE. Viruses with oncolytic properties and their adaptation to tumors. *Ann N Y Acad Sci*. 1952;54:945-952.
596. Southam CM, Moore AE. Clinical studies of viruses as antineoplastic agents with particular reference to Egypt 101 virus. *Cancer*. 1952;5:1025-1034.
597. Newman W, Southam CM. Virus treatment in advanced cancer; a pathological study of fifty-seven cases. *Cancer*. 1954;7:106-118.
598. Huebner RJ, Rowe WP, Schatten WE, Smith RR, Thomas LB. Studies on the use of viruses in the treatment of carcinoma of the cervix. *Cancer*. 1956;9:1211-1218.
599. Cassel WA, Garrett RE. Newcastle disease virus as an antineoplastic agent. *Cancer*. 1965;18:863-868.
600. Cassel WA, Murray DR. A ten-year follow-up on stage II malignant melanoma patients treated postsurgically with Newcastle disease virus oncolysate. *Med Oncol Tumor Pharmacother*. 1992;9:169-171.
601. Stojdl DF, Lichty B, Knowles S, et al. Exploiting tumor-specific defects in the interferon pathway with a previously unknown oncolytic virus. *Nat Med*. 2000;6:821-825.
602. Hammond AL, Plemper RK, Zhang J, Schneider U, Russell SJ, Cattaneo R. Single-chain antibody displayed on a recombinant measles virus confers entry through the tumor-associated carcinoembryonic antigen. *J Virol*. 2001;75:2087-2096.
603. Toda M, Martuza RL, Rabkin SD. Tumor growth inhibition by intratumoral inoculation of defective herpes simplex virus vectors expressing granulocyte-macrophage colony-stimulating factor. *Mol Ther*. 2000;2:324-329.
604. Freytag SO, Rogulski KR, Paielli DL, Gilbert JD, Kim JH. A novel three-pronged approach to kill cancer cells selectively: concomitant viral, double suicide gene, and radiotherapy. *Hum Gene Ther*. 1998;9:1323-1333.
605. Kaufman HL, Kohlhapp FJ, Zloza A. Oncolytic viruses: a new class of immunotherapy drugs. *Nat Rev Drug Discov*. 2015;14:642-662.

CHAPTER 45

Immunology of HIV Infection

Gaurav D. Gaiha, MD, DPhil • Douglas S. Kwon, MD, PhD

INTRODUCTION

In June of 1981, the US Centers for Disease Control (CDC) reported five cases of *Pneumocystis* pneumonia (PCP) in otherwise healthy young gay men in Los Angeles,[1] representing the first published evidence of what was to become the acquired immunodeficiency syndrome (AIDS) epidemic. At that time, PCP was rare, known as an "opportunistic infection" that occurred in the setting of severe immune compromise resulting from cancer treatment. At the same time, there were also reports of an aggressive form of Kaposi Sarcoma (KS), a generally benign cancer, in the same demographic, namely young gay men.[2] Like PCP, KS was infrequent and was typically seen in older men or those who were immunosuppressed following organ transplantation. It made sense in the setting of immune compromise to see PCP or KS, but what was alarming in those initial reports was that it was being observed in otherwise seemingly healthy people.

At that time, pentamidine, the main drug used to treat PCP, was only available through the CDC. Coincident with the first PCP cases reported in Los Angeles, there was a dramatic increase in requests for pentamidine from many other cities across the United States. A consistent and disturbing pattern was quickly noted—the drug was largely being requested for young men who had no known reason to be immunocompromised and at risk for PCP. An emerging epidemic was thus recognized, and already from those first patients, there were clues pointing to the central problem induced by what was ultimately recognized to be infection by a new human retrovirus: that human immunodeficiency virus (HIV) is an infection of the immune system.

When an antibody test for HIV became available in 1985, the kinetics of the expanding epidemic first began to be realized through retrospective examination of stored serum samples. Results from analysis of samples from a hepatitis B vaccine trial that had coincidentally been initiated in 1980 in gay and bisexual men in San Francisco were startling.[3] By 1982, the cumulative incidence of infection in those at-risk men had risen to 42.6%. HIV has since gone on to infect ~79 million persons worldwide and has caused over 36 million deaths.[4] In some of the most affected areas of the world, HIV prevalence exceeds 50% in certain communities and age groups (Fig. 45.1).[5]

Thankfully, the development of effective HIV antiretroviral therapy (ART) has turned HIV infection into a treatable condition for those fortunate to have access to these medications,[6] but viral integration into the host chromosome makes the need for treatment lifelong.[7] Moreover, despite more than 28 million people on treatment, there are still approximately 700,000 deaths from HIV/AIDS each year.[4] The ultimate resolution of the epidemic will almost certainly require an effective vaccine, but despite billions of dollars invested in vaccine development, this has remained an elusive goal.[8]

HIV ORIGIN, TRANSMISSION, AND DISEASE PROGRESSION

Origin of HIV and Relationship to Other Retroviruses

The isolation of HIV in 1983 from a patient in France led to wider sampling of at-risk persons and was paralleled by the discovery of other primate retroviruses. Sequencing of HIV revealed enormous variability among isolates, which were ultimately phylogenetically characterized as belonging to four distinct groups. Group M is the largest and accounts for over 98% of infections worldwide.[9] Based on geographically targeted sequencing, it appears that the epicenter of the global group M epidemic was in Kinshasa in the Democratic Republic of the Congo[10] and that a cross species transmission of simian immunodeficiency virus (SIV) from chimpanzees occurred in the early 1900s.[11,12] In turn, chimpanzees had acquired SIV from cross-species transmission of two viruses in monkeys that recombined. The actual origin of the group M epidemic in humans appears to have been in a remote region of southeastern Cameroon, likely by exposure through butchering of a chimpanzee.[13] Much less frequent are groups N, O, and P, the result of separate cross-species transmissions from chimpanzees (groups N and O) and gorillas (group P), which have remained localized to Cameroon and neighboring countries in West Africa.[9,14,15]

HIV is a typical retrovirus, morphologically characterized by a core containing the viral RNA and two copies of reverse transcriptase (RT), surrounded by an outer envelope that is heavily glycosylated. The error prone RT leads to the rapid generation of new mutations, resulting in enormous viral genetic variation within group M viruses globally. This is largely due to lack of 3′ exonuclease proofreading activity of RT which gives rise to the mutation of approximately 1 in every 10,000 nucleotides during viral genome transcription; in other words, once per replication cycle.[16] In addition, frequent viral recombination and the lifelong nature of infection also contribute to generation of diversity. Within an individual, this produces an enormously diverse quasispecies of viruses whose intraindividual diversity exceeds the global

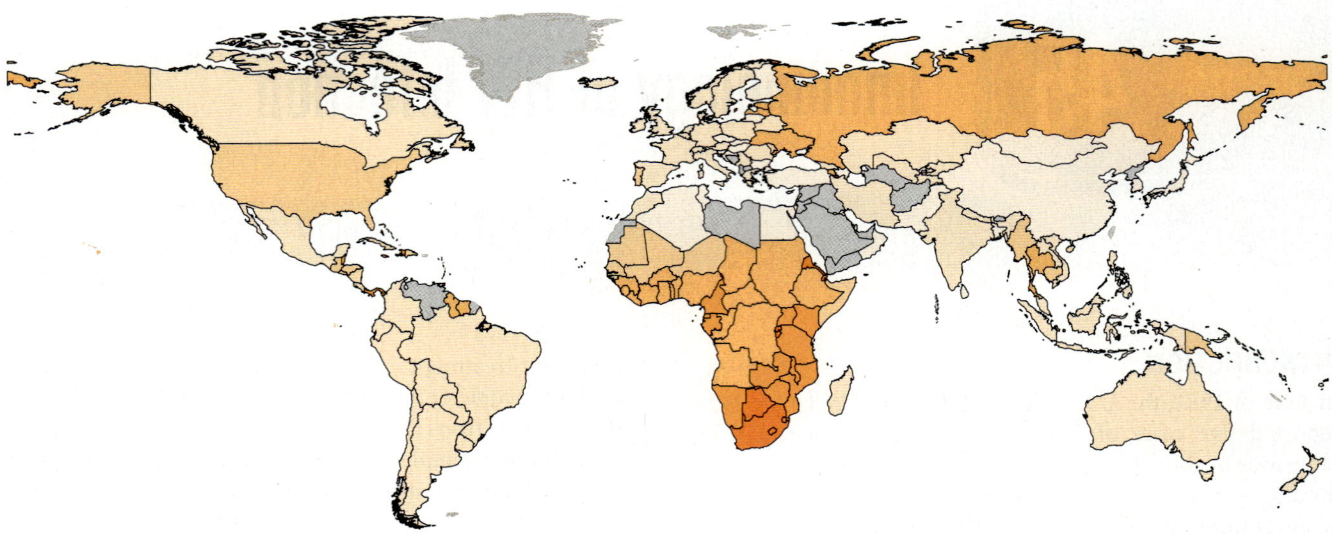

FIG. 45.1. Global human immunodeficiency virus (HIV) prevalence by country.

diversity of influenza in a single season[17] (Fig. 45.2). This represents a significant challenge to the immunologic control of infection and the development of broadly prophylactic vaccines. Currently, there are at least nine distinct clades, or subtypes, of group M HIV-1 worldwide, and many additional subtypes that are the result of recombination events between these major clades.

HIV-1 is closely linked to HIV-2, a less pathogenic human retrovirus that is endemic in West Africa. HIV is also closely related to a number of nonhuman primate (NHP) retroviruses called SIV.[18] Although SIV was initially considered nonpathogenic in primates, infection of certain wild chimpanzee species results in CD4 T cell decline and increased mortality. It is important to note that some wild NHP species tolerate their species-specific SIV infections quite well, including African green monkeys and sooty mangabeys, in whom the infection is characterized by persistent high viral load but a lack of CD4 T cell decline or development of immune deficiency.[19]

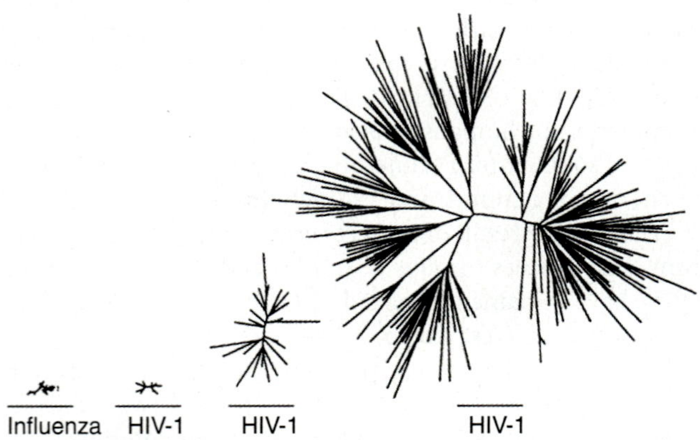

FIG. 45.2. Relative viral diversity of influenza and human immunodeficiency virus (HIV)-1. Comparison of influenza HA1 domain and HIV-1 envelope V2-C5 regions, starting from left with diversity of all H3N2 influenza sequences available from the 1996 influenza season, then HIV-1 subtype B sequence diversity from a single individual, HIV-1 sequences from a subtype B epidemic occurring in Amsterdam in 1990 to 1991, and finally HIV-1 sequences from the Democratic Republic of Congo obtained in 1997. (Adapted from Korber B, Gaschen B, Yusim K, Thakallapally R, Kesmir C, Detours V. Evolutionary and immunological implications of contemporary HIV-1 variation. *Br Med Bull.* 2001;58:19-42. By permission of Oxford University Press.)

Exploiting Knowledge of the Viral Life Cycle to Develop Antiviral Agents

Early in the HIV epidemic, it was realized that expression of CD4 was necessary for cells to become infected—a finding that was assisted by the recognition that CD4+ cells were specifically lost in progressive disease. However, it was also clear that CD4 alone was not sufficient. Expression of recombinant CD4 on murine cells failed to allow viral infection. The discovery of chemokine receptors as necessary coreceptors for HIV entry took over a decade to be realized and was facilitated by a remarkable convergence of clinical observation and basic science. Blood samples from persons identified to be very heavily exposed to HIV, who had never become infected, showed that a group of β chemokines—MIP-1α, MIP-1β, and RANTES—were secreted from CD8 T cells and could inhibit HIV replication.[20] These chemokines were known to be linked through a common receptor, CCR5, a member of a family of G-protein coupled seven transmembrane domain chemokine receptors. In the next year, CCR5 and another chemokine receptor, CXCR4, were identified as the long sought HIV entry coreceptors whose expression along with CD4 were both necessary and sufficient to allow viral entry.[21-23] Shortly thereafter, it was discovered that homozygosity of a naturally occurring 32 base pair deletion of CCR5 (CCR5 Δ32) that abrogated surface expression of the protein and occurred in 1% of Caucasians imparted resistance to HIV acquisition, thus demonstrating the biologic significance of

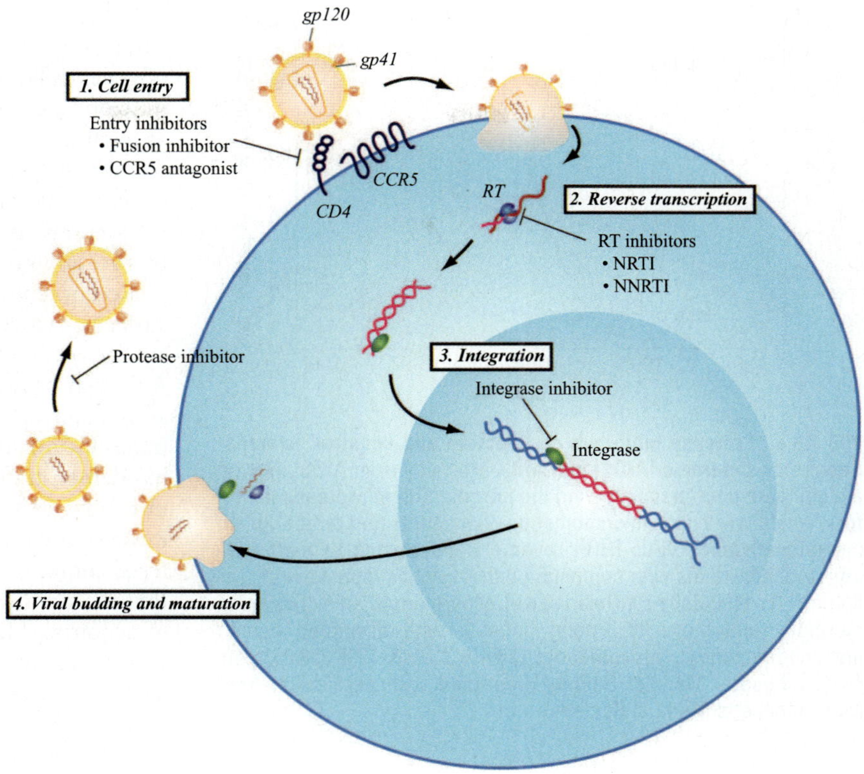

FIG. 45.3. Human immunodeficiency virus (HIV) viral life cycle and targets for antiretroviral therapy. The HIV envelope protein gp120 first contacts CD4 on the target cell, which then induces conformational changes in gp120 that exposes the chemokine receptor binding site. Following binding of chemokine receptor, HIV gp41 mediates fusion of target and viral membranes. The viral single-stranded RNA genome is then reverse transcribed and the viral genome is transported to the nucleus where it is integrated in the host genome. Viral genes are then transcribed and translated and virions then bud from the host membrane. Following budding, virions mature in a process that requires cleavage of HIV polyproteins by the viral protease. Currently, approved antiviral agents target entry, reverse transcription, viral genome integration, and viral protease function.

chemokine receptors for HIV infection.[24-26] Since individuals lacking functional CCR5 were found to be largely immunologically intact, CCR5 became an attractive target for ART, and in 2007, maraviroc, a negative allosteric modulator of the CCR5 receptor, became the first chemokine receptor antagonist approved for treatment of HIV infection.

Even before the development of maraviroc, however, multiple viral proteins had been successfully targeted to treat HIV infection based upon a rapidly developing understanding of the viral life cycle (Fig. 45.3). HIV RT was the first viral protein to be successfully targeted by drug therapy, and the class of RT inhibitors remains the backbone of first line regimens of combination ART strategies used today. Zidovudine, also known as AZT, was the first clinically approved HIV drug. It is a thymidine analog initially developed in 1964 as a potential anticancer agent. Over 20 years later, however, it was found to inhibit replication of HIV as part of a unique collaboration between the National Cancer Institute and the pharmaceutical company Burroughs Wellcome. When AZT was first tested for activity against HIV in 1985, there were over 10,000 people suffering from AIDS, a disease for which there was no available therapy at that time.[27] After proving to be safe in a small phase I trial, AZT quickly went onto efficacy studies and remarkably was approved for clinical use by the Food and Drug Administration just 2 years from its first testing in the laboratory. This would be the first of several drugs that target RT that now constitute one of the mainstays of HIV treatment.

The next major breakthrough in ART was the development of drugs that target the HIV protease. Protease is required for cleavage of viral polyproteins in immature virions released from the infected cell, a step necessary for virion assembly and maturation. The first protease inhibitor, saquinavir, was approved in 1995 and its development marked the age of modern multidrug ART. Up to that point, only four drugs had been developed, all nucleoside reverse transcriptase inhibitors (NRTIs). However, therapy with these NRTIs showed toxicity and none of them alone or in combination could durably control HIV. Combination therapy with agents that inhibited both RT and protease showed strong *in vitro* synergy,[28] could achieve virologic suppression *in vivo*, and resulted in a dramatic decrease in AIDS deaths (Fig. 45.4).[29] This was quickly followed by the development of non-nucleoside reverse transcriptase inhibitors (NNRTIs) which also targeted RT and could also be used effectively with NRTIs to suppress viral replication. This marked the modern area of combination ART and has changed HIV from a nearly uniformly terminal illness to a chronic treatable disease for those with access to medications.

Since the development of RT and protease inhibitors, two other steps in the viral life cycle, entry and viral genome integration, have been successfully targeted by approved drugs. HIV entry is initiated when the viral envelope protein, gp120, contacts CD4 and a chemokine receptor on the target cell membrane. This triggers conformational changes in the transmembrane envelope glycoprotein, gp41, which leads to fusion of viral and host membranes. There are currently two antiretroviral agents that act at this early stage of entry: maraviroc, a CCR5 antagonist that blocks the interaction of gp120 with CCR5,[30] and enfuvirtide (Fuzeon, T-20), approved in 2003, a 36 amino acid peptide that binds to gp41 and prevents formation of a hairpin intermediate necessary for fusion of the

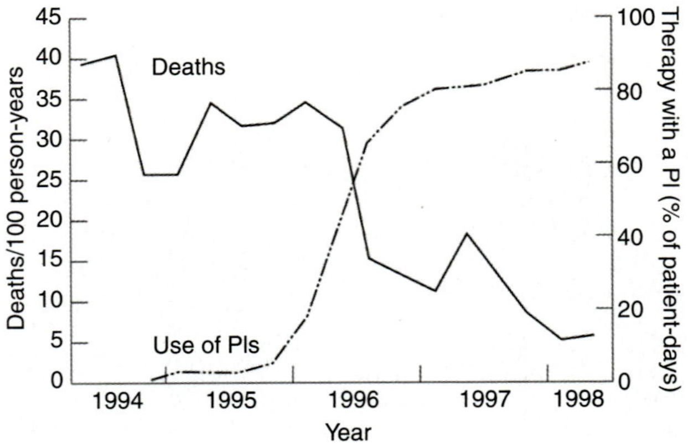

FIG. 45.4. **Protease inhibitors and decline and acquired immunodeficiency syndrome (AIDS) mortality.** Mortality from AIDS showed a significant decrease following the introduction of protease inhibitors, which were used in combination with agents that target human immunodeficiency virus (HIV) reverse transcriptase to treat HIV-infected individuals with combination therapy to suppress viral replication. (From Palella FJ, Delaney KM, Moorman AC, et al. Declining morbidity and mortality among patients with advanced human immunodeficiency virus infection. *N Engl J Med.* 1998;338:853-860. Massachusetts Medical Society. Reprinted with permission from Massachusetts Medical Society.)

viral and target cell membranes.[31] Raltegravir targets HIV integrase and represents a newer class of antiretroviral agents to be developed.[32] Integrase is the viral enzyme that catalyzes the DNA strand transfer reactions necessary for incorporation of the viral genome into the host genome. It represents one of the most potent of currently available HIV drugs, is generally one of the best tolerated therapies, and is now considered a first line agent for the treatment of HIV in many parts of the world.

Although there are currently over 30 approved antiretroviral drugs in various formulations, demand continues for novel agents, especially as drug resistance becomes increasingly common. Studies in Europe have demonstrated that approximately 10% of all newly diagnosed patients harbor virus with at least one major resistance-associated mutation.[33] New drugs remain in development including maturation inhibitors, CD4 receptor inhibitors, and additional agents in existing classes.[34] Additionally, an entirely new class of agents, known as broadly neutralizing antibodies (bNAbs), are rapidly under development for therapeutic purposes. These agents, which target key sites of vulnerability on the HIV envelope protein, are now also being tested for use in preventing infection in individuals at high risk of acquisition and are the subject of intense focus for HIV vaccine development efforts.[35]

Mucosal Transmission and Early Local Viral Replication

Transmission of HIV is relatively inefficient with a risk of approximately 1 in 1000 sexual acts for male-to-female sexual transmission and 1 in 100 for receptive anal intercourse.[36] Over 90% of HIV transmissions worldwide occur across a mucosal barrier, either through sexual exposure or maternal to child transmission, with the remaining cases due to direct blood exposure such as injection drug use. These mucosal surfaces are exposed to an enormously diverse quasispecies of genetically distinct virions, and yet, the majority of transmissions are mediated by a single founder virus,[37] although as many as 16 viruses have been documented to be transmitted simultaneously.[38] Multiple transmitting founder viruses appear to be more prevalent in transmission occurring via intravenous drug use, suggesting that mucosal-specific factors are an important determinant for this restriction.

Recent reports demonstrate that inflammation at mucosal surfaces, particularly in the female genital tract, are an important factor in determining susceptibility to HIV infection.[39] Although sexual transmitted infections such as gonorrhea and chlamydia are known to increase this localized genital inflammation and HIV acquisition risk, more recent evidence has demonstrated the importance of the vaginal microbiome in this process. In a cohort of young South African women, it was found that those harboring high-diversity vaginal bacterial communities dominated by anaerobic bacteria such as *Prevotella* and *Gardnerella* had an over fourfold higher risk of acquiring HIV than women with a *Lactobacillus crispatus*–dominant vaginal community.[40] These high-risk vaginal bacterial communities were also associated with increased production of genital inflammatory cytokines and elevated numbers of activated cervical CD4 T cells,[41] which are likely the first cells to be infected by HIV, following viral exposure in the female genital tract. This increase in local "HIV target cells" within the mucosal tissues of the female genital tract may be one way by which the vaginal microbiome may impact HIV risk.

Following mucosal exposure, a small founder population establishes host infection by initially replicating locally within the exposed mucosal tissue for 7 to 10 days before disseminating to the draining lymph node and then moving quickly into the blood (reviewed in Ref. 42). Once infection has spread to the periphery, a widespread reservoir of latently infected cells is rapidly established at multiple sites and subsequent viral eradication becomes enormously challenging due to this extensive reservoir. The fact that HIV transmission is inherently inefficient and typically initiated by a single virion which requires days of local expansion before systemic infection can take place suggests that these early events represent a vulnerable stage in the process of transmission. Unfortunately, few HIV-specific B or T cell responses are detected locally in the mucosal site of exposure until well after viral dissemination, despite the fact that these sites contain abundant antigen-presenting cells and numerous lymphocytes.[43,44] This lack of early responses likely plays a critical role in the failure of the immune system to contain viral replication within the mucosal site of viral exposure. Although some studies suggest that a subset of persons with multiple exposures will develop adaptive immune responses to HIV[45] and have suggested that these can protect against acquisition; this remains controversial.[46]

Transmission depends on tropism and the affinity of envelope glycoprotein of HIV for CD4 and coreceptor molecules. Acute infection is typically established with R5 viruses that have a gp120 molecule with high affinity for CD4 allowing them to infect T cell and macrophages, which express relatively lower levels of CD4 on their surface. Over time, some viruses within the quasispecies of an infected person can evolve to an X4 tropic phenotype, which is characterized by the use of CXCR4 as its coreceptor, but these viruses somehow are either prevented from establishing the initial infection or rapidly revert to R5 viruses upon transmission.

The mechanism of the transmission "bottleneck" remains incompletely understood. The efficiency of HIV transmission is relatively low and highly linked to the viral load in the donor. A population study of 45 discordant couples in which one partner was HIV positive and the other was not showed a complete lack of transmission in the 15 couples in whom the infected partner had a viral load of less than 1500 RNA copies/μL plasma.[47] Additionally, treatment of couples discordant for HIV infection has confirmed the impact of viral load on transmission: when the HIV-infected partner is on therapy, the risk of HIV transmission to their partner is reduced by 96%.[48] This insight has led to the "Undetectable = Untransmittable" campaign, which advocates that for HIV-infected individuals on ART with consistently undetectable viral loads, the virus cannot be transmitted to a sexual partner. Treatment as Prevention (TasP) has become one of several tools being used to reduce HIV transmission through the world. Unfortunately, periods of highest transmissibility tend to be early in infection before viral load has declined to a steady state and before some people are aware of their diagnosis.[49]

Acute Infection Syndrome and Early Infection

Most persons have had the symptoms that could be consistent with acute HIV infection at some point in their lives, since these are not specific to HIV but rather are typical of many viral infections. Indeed, the diagnosis of acute HIV infection is frequently missed by healthcare providers. Most but not all patients are symptomatic with their initial acute infection, with fever, sore throat, swollen lymph nodes, fatigue, malaise, and often transient oral or genital ulcers.[50] The syndrome looks very much like acute infectious mononucleosis. These nonspecific symptoms eventually resolve, and subsequent disease course is characterized by a gradual CD4 cell decline and slowly rising viral load, with the ultimate development of clinical AIDS when the CD4 count is less than 200 cells/μL or certain opportunistic infections arise. Initial peak viremia in newly infected adults is often 10 million viral particles per mL or more,[51] followed by a decline over the first few months to a quasi-steady-state at a mean of around 30,000 RNA copies/mL (Fig. 45.5).[53]

Current views of the very earliest stages in acute infection stem largely from studies in the nonhuman primate model of AIDS virus infection, in which rhesus macaques are infected with SIV. In high dose challenge models of macaques, virus crosses the mucosal site of exposure within hours and initially establishes a small founder population of infected cells locally (reviewed in42). It is unclear how the virus crosses the mucosal epithelial barrier, but studies have suggested a

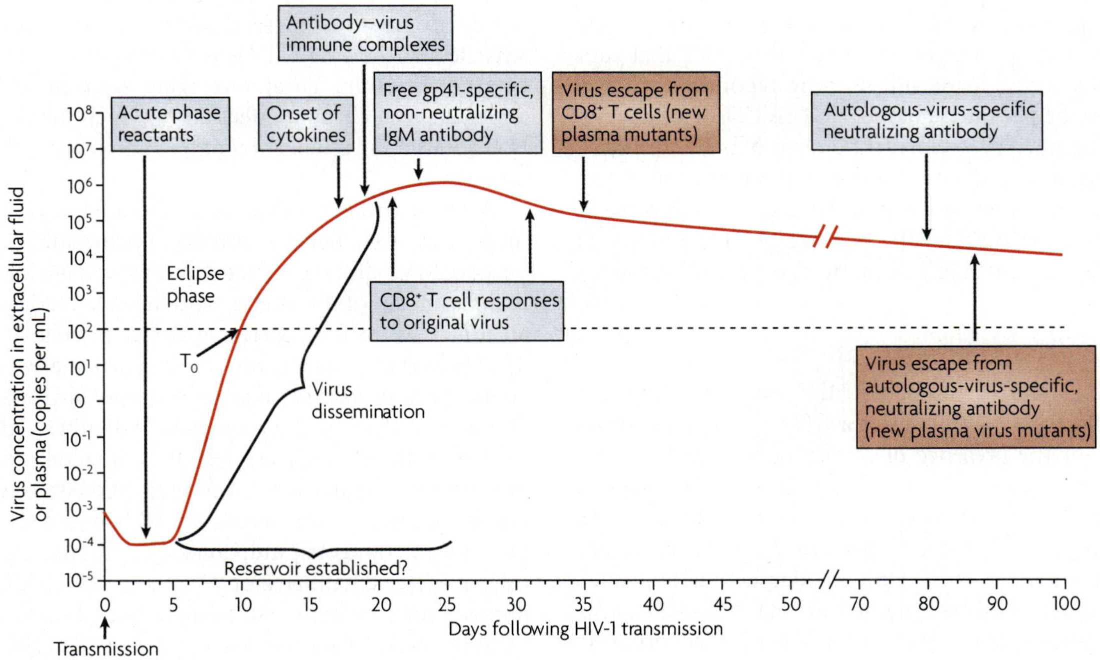

FIG. 45.5. Early immune responses in acute human immunodeficiency virus (HIV)-1 infection. Acute infection is typically characterized by high-viral loads and flu-like symptoms. There is an initial "eclipse phase" in which virus is present but antibody-based HIV tests remain negative. Early adaptive responses are comprised of HIV-specific CD8 T cell responses and non-neutralizing antibody responses. The emergence of CD8 T cell responses coincides with the decline of peak viral loads and the establishment of a viral set point. (Reprinted by permission of Springer Nature: McMichael AJ, Borrow P, Tomaras GD, Goonetilleke N, Haynes BF. The immune response during acute HIV-1 infection: clues for vaccine development. *Nat Rev Immunol.* 2010;10:11-23.)

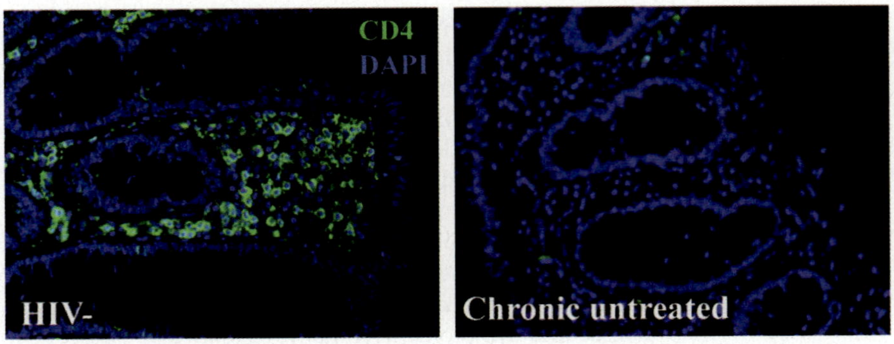

FIG. 45.6. Depletion of CD4 T cells in the intestinal mucosa. Gut-associated lymphoid tissue (GALT) contains the majority of the body's lymphocytes, which reside within both the lamina propria and lymphoid follicles. Lamina proprial CD4 T cells are rapidly and massively depleted early in human immunodeficiency virus (HIV) infection. Shown are representative stains from colonic mucosa from HIV -infected and -uninfected individuals stained by immunofluorescence for CD4 (*green*) and nuclei (DAPI, *blue*).

variety of mechanisms, including transcytosis across epithelial cells, facilitation by mucosal dendritic cells (DCs), and/or entry through mechanical disruption or physiological intercellular spaces in the epithelium. Once across, the first cells to be infected are likely activated mucosal-resident CD4 T cells. Once infection within the tissue is established, there is a period of local expansion before virus is disseminated to regional lymph nodes. This is then followed by rapid progression to the peripheral circulation and seeding of gut-associated lymphoid tissue (GALT). This gut reservoir contains 60% to 80% of the body's CD4 T cells, a large majority relative to the 2% to 3% of lymphocytes present in the peripheral circulation.

Based upon studies initially conducted in rhesus macaques and subsequently corroborated in humans, a hallmark of HIV infection is the rapid and profound depletion of GALT CD4 T cells in the lamina propria within the first weeks of HIV infection (Fig. 45.6). It appears that these cells never fully recover, even with prolonged treatment with ART that suppress peripheral viral loads and result in rebound of CD4 T cell counts in the peripheral blood.[54,55] This CD4 T cell depletion is accompanied by extensive apoptosis of bystander cells, associated immune activation, and disruption of normal epithelial barrier function in the gut. Bystander death actually involves abortive infection, with accumulation of cytoplasmic nucleic acids that elicit proapoptotic and proinflammatory responses.[56]

Natural History of Untreated Infection

The typical course of untreated HIV infection until the development of AIDS, defined as a CD4 count of less than 200 cells/μL or the presence of certain opportunistic infections, is approximately 10 years[57] and involves progressive immunologic decline. As the epidemic has matured, however, it has become clear that not all persons progress. There are now persons with documented untreated infection for greater than 30 years that have maintained normal CD4 cell numbers and an undetectable viral load, whereas others progress from acute infection to AIDS in as little as 6 months without therapy.[58]

The most comprehensive HIV natural history data derive from the Multicenter AIDS Cohort Study (MACS) and other population studies established in the early days of the epidemic. The MACS, which involved recruitment of both infected and at-risk men in Baltimore, Chicago, Los Angeles, and Pittsburgh, was established in 1984 and continues today in conjunction with the Women's Interagency HIV Study. The study currently has more than 7000 participants with an accumulated 93,000 person years of follow-up as of 2013,[59] with a large repository of archived specimens. These multicenter seroprevalent and seroincident cohorts were pivotal in identifying laboratory predictors of prognosis, including CD4 T cell count and viral load, which are both independent predictors of disease progression. Following acute infection, viral load reaches a quasi-set point within approximately 6 months of infection, with a median viral load around 30,000 RNA copies/mL and interquartile ranges from 12,000 to 80,000 copies. There is a marked increase in the development of opportunistic infections when CD4 count drops below 200 cells/mL, which led to the current recommendations that prophylactic treatment of opportunistic infections begin when this level is reached. The sequelae of chronic-untreated HIV infection are the development of these opportunistic diseases, which are what led to the original identification of the new epidemic of HIV infection. What was being seen clinically was the end stage of a chronic infection in which immune dysfunction allowed the establishment of a number of otherwise rare illnesses.

AIDS defining illnesses are Candidiasis of the respiratory tract, *Pneumocystis jirovecii* pneumonia, cytomegalovirus (CMV) disease, extrapulmonary cryptococcal disease, toxoplasmosis of the brain, and mycobacterial infection,[60] all indicative of an underlying defect in cellular immunity. *Mycobacterium tuberculosis* is a particularly devastating coinfection, in part because its widespread prevalence, with almost one-third of HIV-infected individuals globally also infected with tuberculosis (TB). Both infections are far more prevalent in resource limited settings, and coinfection is associated with worse outcomes, with TB being the leading cause of death in coinfected individuals.[61] Although opportunistic infections are a hallmark of progressive HIV, AIDS defining illnesses are not confined to infectious diseases. There are notably several cancers that are also associated with HIV, including KS, lymphoma, and cervical cancer, all of which are likely due to loss of protective immune surveillance and the presence of other viral coinfections. Additionally, there is also evidence that HIV-infected individuals are at increased risk for anal cancer, colonic adenocarcinoma, hepatocellular carcinoma, and lung cancer.[62,63] The risks for these associated infectious diseases decrease with therapy and subsequent immune reconstitution; however, the risk for some cancers, such as cervical and anal cancer, remains elevated.

With the advent of combination ART, HIV has in many ways become a chronic treatable disease. This is evidenced by continued narrowing of the gap in life expectancy between HIV-uninfected and -infected individuals, particularly when individuals have access to care and begin early treatment.[64] The ART Cohort Collaboration study, a group composed of 13 independent studies, found that deaths were due largely due to age-related clinical diseases, such as cardiovascular disease, cancer, fractures, and renal dysfunction, despite immune reconstitution with effective ART. Additionally, treated HIV-infected adults older than 50 years had three times greater risk of death compared to those less than 30 years. The precise mechanism of this age-related morbidity and mortality is unclear, but it is increasingly relevant as the population of HIV-infected individuals ages, a result of the success of ART. One key to understanding this is the observation that HIV-infected individuals have increased systemic immune activation, and that with therapy, the activation level decreases to levels that remain higher than those who are uninfected.[65] Persistent immune activation, even at low levels, may therefore be one mechanism by which HIV creates progressive impairment in untreated individuals.

Chronic Immune Activation and Immune Dysregulation

Peripheral markers of T cell activation predict the rate of disease progression and CD4 T cell decline in untreated infection more accurately than plasma viral load.[66-68] This persistent immune activation is a hallmark of chronic HIV infection, and although it is clear that HIV viremia itself plays a direct role, it does not fully account for the entirety of observed T cell activation seen in infected patients. For example, in sooty mangabeys, a natural host of SIV, high levels of viremia are supported but without a significant increase in immune activation and with the remarkable absence of clinically advanced disease.[69-71] Additionally, patients who are successfully treated with ART and have suppressed viral loads continue to have residual T cell activation relative to HIV-uninfected subjects.[72]

Recent studies have suggested that the gut may be critically involved in chronic immune activation. Individuals with untreated HIV infection have been found to exhibit elevated serum levels of bacterial lipopolysaccharide (LPS), which is thought to be a consequence of increased microbial translocation from the intestinal lumen in the setting of a damaged gut mucosa.[73] Serum LPS levels are closely correlated with markers of systemic T cell activation and in patients who receive allogeneic bone marrow transplants complicated by graft-versus-host disease, it has been shown that plasma LPS levels closely correlate with biochemical and histological measures of mucosal damage.[74] Thus, elevated serum LPS in HIV-infected subjects is believed to reflect the loss of mucosal epithelial integrity concomitant with HIV infection, and it defines a mechanism by which the gut and intestinal microbiota may play a central role in chronic immune activation through increased microbial translocation. There have also been a number of studies that have described specific HIV-associated changes in the gut microbiome, with enrichment of potentially more inflammatory bacterial taxa with HIV that may have contributed to systemic inflammation.[75] The ultimate consequence of this chronic immune activation appears to be higher turnover of CD4 and CD8 T cells, clonal exhaustion, and the progressive impairment of T cell function.

The development of immune dysfunction or "exhaustion" in the presence of ongoing antigen exposure is a cardinal feature of chronic viral infections with persistent viremia.[76] Exhaustion results in the hierarchical impairment of T cell effector functions, with an initial loss of proliferative capacity and IL-2 production and the late loss of the ability to secrete proinflammatory cytokines such as IFNγ.[77] This exhaustion is mediated by several immunoregulatory pathways (reviewed in[78,79]) including programmed death 1 (PD-1), cytotoxic T lymphocyte–associated antigen 4 (CTLA-4), T cell Ig and mucin-3 (TIM-3), and interleukin-10 (IL-10), which are all upregulated *in vivo* in chronic-untreated HIV infection and have been shown to mediate a reversible virus-specific immune dysfunction *in vitro*. These immunoregulatory pathways are therefore being actively explored as targets for therapeutic intervention in chronic HIV infection (reviewed in[78,80-82]).

Latent HIV Reservoir

While combination ART reduces plasma HIV levels below the limit of detection, viremia rebounds rapidly after treatment interruption despite prolonged suppression. The source of this rapid rebound is a latent reservoir of integrated HIV in resting memory CD4 T cells, which is established after acute infection.[83] The development of the viral reservoir is a result of the tropism of HIV for activated CD4 T cells,[84] which transition to a resting memory state that is nonpermissive to viral gene expression during a narrow time window that allows for integration, but not high-level gene expression.[85] Thus, establishment of a latently infected cell is actually quite rare, with a calculated frequency *in vivo* of ~$1/10^6$ infected cells.[7] Given the long half-life of memory CD4 T cell responses (~3.6 years),[86] the latent reservoir remains remarkably stable. Moreover, the homeostatic expansion of latently infected CD4 T cells, particularly those with integration of HIV genomes in genes associated with cell growth, provide an ongoing source of reservoir renewal.[87]

During latency, the virus persists as silent genetic information which is reinforced by epigenetic modifications of the integrated provirus, limiting detection by the immune system. In addition, while most integrated provirus is defective due to hypermutation or deletions, approximately 5% encodes full-length replication-competent viral DNA. When latently infected cells become activated by T cell receptor (TCR) engagement or cytokine signaling, the provirus is transcribed, leading to the synthesis of viral proteins and new viral particles (reviewed in Ref. 88). This persistence and capacity for viral rebound are primary reasons why the latent reservoir is a major obstacle to HIV cure efforts.

From an anatomical perspective, the latent HIV reservoir is located in most secondary lymphoid organs and nonlymphoid tissues such as the gastrointestinal tract, which harbors >60% of latently infected cells.[89] Additional anatomical sites include the spleen, thymus, lymph nodes, bone marrow, lungs, kidneys, liver, adipose tissue, and genitourinary systems. Within lymph nodes, latently infected follicular helper CD4 T cells have received considerable attention as they are preferentially infected by HIV[90] and located in B cell follicles that cannot be readily accessed by CD8 T cells.[91] These follicles represent an immune-privileged site, which are found at other regions of the body such as the CNS, that will require a unique approach to achieve both viral reactivation and immune recognition.

Efforts to achieve an HIV cure have been focused on "shock and kill" strategies to purge the latent HIV reservoir,[92] which utilize both reactivation of viral gene products and subsequent clearance by the immune response, such as cytotoxic T cells (CTLs). These approaches would be carried out on HIV infected patients on treatment in order to mitigate new infection of CD4 T cells. It has been postulated that a reduction of >4 logs of the viral reservoir would be required to prevent viral rebound following ART cessation in ~50% of HIV infected individuals.[93] Additional strategies are focused on the development of therapeutic CTL vaccines to achieve a functional HIV cure by maintaining immune-mediated suppression of viral rebound.[94] The passive administration of bnAbs is also an emerging HIV cure modality by targeting HIV envelope protein that is synthesized and trafficked to the plasma membrane of productively infected cells. Delivery of bnAbs to ART-treated HIV infected individuals led to a delay in viral rebound after ART interruption.[95] BnAb infusion has also been suggested to enhance viral-specific CTL responses though a vaccinal effect that leverages antibody-mediated cross-presentation of viral antigen.[96,97]

IMMUNE RESPONSE TO HIV

Unlike many other viral infections, there does not appear to be natural clearance of HIV infection. The only known cases in which replication-competent virus has been apparently successfully eliminated has occurred in patients who underwent bone marrow transplants from donors who were homozygous for the 32 base pair deletion in CCR5, rendering the newly transplanted cells not susceptible to infection.[98,99] The fact that people do not spontaneously clear virus indicates that natural immunity is insufficient to eliminate HIV. Despite this, there is clear evidence of initial partial control in the early stages of infection, when viral load is decreased from peak levels often in excess of 10 million viral particles per mL of plasma to a set point mean of around 30,000 copies.[51] Moreover, some rare individuals termed elite controllers maintain undetectable viral loads and normal CD4 cell counts, in some cases for over 30 years and counting, suggesting that partial host control is possible. Evidence exists to implicate both innate and adaptive immune responses in situations of partial control in acute and chronic infection, but the actual correlates of protective immunity remain unclear.

NK Cell Response

The potential protective role of natural killer (NK) cells has been well-established in human viral infections. In the setting of HIV infection, NK cells are likely the first line of defense against HIV, given that HIV downregulates HLA-A and HLA-B molecules via its Nef protein[100] and renders infected cells targetable by certain NK cell subsets. NK cell interactions with HIV-infected cells do not depend on expression of an antigen-specific receptor but instead are dependent upon alterations in activating and inhibitory signaling between HIV-infected cells and NK cells. This occurs via interactions with a subset of highly polymorphic killer immunoglobulin-like receptors (KIR) with HLA class I molecules, as well as through indirect mechanisms involving antibody-dependent cellular cytotoxicity (ADCC) and an arsenal of nonspecific inhibitory and activating receptors.[101]

The evidence for a beneficial effect of NK cells in HIV infection is substantial, both in protection from acquisition and in subsequent control of viremia. Persons who have been heavily exposed to HIV and yet remain uninfected have increased NK cell activity[102] and a higher KIR3DS1 to KIR3DL1 transcript ratio,[103] consistent with the genetic studies indicating a protective effect of the compound genotype consisting of the activating KIR KIR3DS1 and Bw4-80I.[104] Moreover, during acute infection, there is an expansion of KIR3DS1+ NK cells in persons expressing the putative ligand Bw4-80I, and these cells are able to lyse HIV-infected cells *in vitro*, in a Bw4-80I–dependent manner.[105]

The mechanism of putative protection by NK cells in HIV infection likely relates to the selective downregulation of both HLA-A and -B alleles, with elevated expression of HLA-A leading to impaired control of HIV.[106] This certainly would be consistent with the role of activating ligands, but inhibitory KIR alleles are also associated with better control of HIV infection.[107] This association with inhibitory KIR alleles is more challenging to explain but has been postulated to relate to the strength of the KIR-HLA interaction during NK cell licensing, where stronger engagement of inhibitory KIRs by HLA molecules can enhance their functional capacity.[108]

There are a number of defects in NK cells that have been identified in HIV-infected persons. There is an inverse correlation between viral load and NK cell–mediated suppression of viral replication, suggesting that NK cell dysfunction may contribute to disease progression.[109] NK cells in HIV-infected persons have impaired cytotoxic capacity associated with decreased levels of granzyme A and perforin,[110] whereas other studies suggest that NK cells may be incompletely activated in chronic HIV infection and that functionally anergic subsets may be expanded.[111] It has also been suggested that NK cells can become HIV-infected and might contribute to the persistent HIV reservoir *in vivo*.[112]

More recently, *in vivo* evidence for antiviral selection pressure mediated by NK cells has been reported. Functional roles for NK cells are suggested by studies in infected persons, showing expansion of a subset of NK cells in acute infection.[105] But the most compelling data implicating NK cells in the host defense against HIV come from the identification of NK cell "footprints" in viral sequences obtained from infected persons. Analysis of KIR genotypes and full-length virus sequencing in a large cohort of persons with chronic HIV infection demonstrated 22 positions throughout the HIV genome in which amino acid polymorphisms were associated with the presence of specific KIR genes.[113] *In vitro* studies showed that these HIV sequence polymorphisms increased the binding of inhibitory KIRs to HIV-infected cells, suggesting a functional consequence favoring the virus. Moreover, these polymorphisms reduced the antiviral activity of KIR-positive NK cells, indicating not only that NK cells exert immune selection pressure on HIV but also that HIV evolves to escape from detection by this response.

CD8 T Cell Response

Systemic spread of HIV infection from an initial localized mucosal infection is associated with a dramatic increase in plasma viremia followed by decline to a quasi-set point level. The appearance of a CTL response during acute infection coincides with the drop in viremia, a month or more before neutralizing antibodies are detectable, suggesting that these cells may be a primary effector mechanism of this decline.[114-116]

The data in support of a critical role of CTL in initial containment of systemic viremia also come from CD8 T cell depletion studies in SIV-infected macaques. Use of a monoclonal antibody to achieve complete CD8 depletion in peripheral blood and near-complete depletion in tissues resulted in persistence of peak viremia in acute infection.[117,118] When CD8 T cells are depleted in chronic infection, well after set point has been established, viral load increases, and then declines again as CD8 cells reappear. CD8 depletion also resulted in more rapid disease progression.[118] These studies also may have depleted CD8–bearing NK cells, so one cannot entirely rule out a contribution of the innate immune response. However, support for a central role for CTL is also implied from large cohort studies, in which the most striking association with viral load and CD4 cell decline is the expression of certain HLA class I alleles[119] (Fig. 45.7). Among these are so-called "protective" alleles associated with natural control, in particular HLA B*57 and B*27,[119,120] and "risk" alleles such certain HLA B*35 subtypes associated with higher viral loads and more rapid disease progression.[121,122]

Because acute HIV infection first becomes symptomatic at the time of peak viremia, very few studies have been able to identify patients early enough to characterize CTL responses during the rapid initial increase in viral load. In the studies that have been done, however, it is clear that the initial CD8 T cell response is narrowly directed, and it remains narrow when initial set point is achieved.[116,123,124] Comprehensive screening of cellular immune responses to sets of 410 overlapping peptides representing the entire HIV proteome at an estimated 28 days after virus exposure revealed an average of only three peptides targeted per person.[124] Since most studies characterizing CD8 T cell responses have relied on IFNγ Elispot assays, and have used a reference strain of virus, it is possible that some responses may be missed due to sequence differences in the infecting strain. Indeed, when autologous peptides have been used, the breadth of responses detectable can be up to 30% higher,

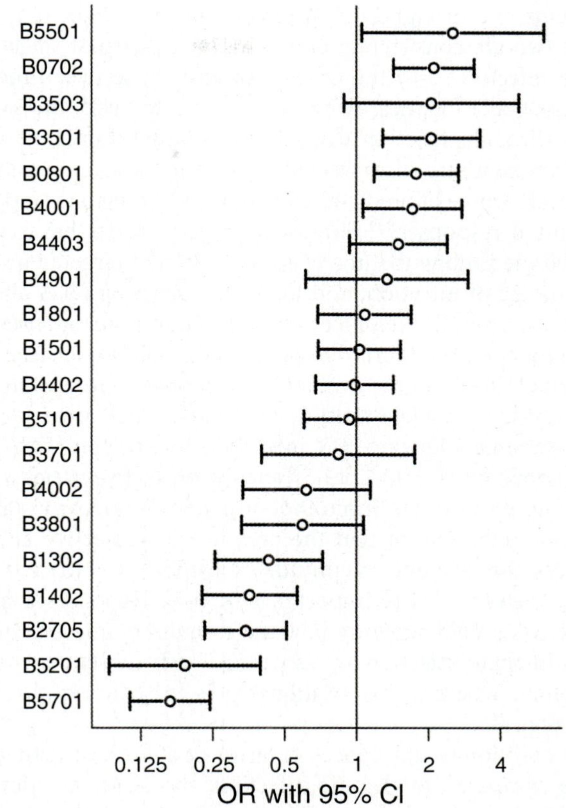

FIG. 45.7. Human leukocyte antigen (HLA)-B alleles and human immunodeficiency virus (HIV) control. Association of HIV immune control and imputed classical HLA-B alleles in a European cohort were calculated using a logistic regression model where all alleles with frequency of 2% or greater were used as covariables in the model. OR <1 indicates association with virologic control of HIV and OR>1 indicated association with progression. (Courtesy of M. Carrington.)

but even with this caveat, most studies suggest that the initial breadth of responses consists of one to three epitope-specific responses.[123,125] Thus, the breadth of responses associated with the dramatic initial decline in viremia is actually quite narrow. The narrowness of these responses is further supported by sequencing of autologous virus in acute infection, where epitopes that subsequently become targeted in the chronic phase of infection are present in wild-type form that should be normally processed and presented, so the lack of detection of responses cannot be ascribed to sequence divergence of the infecting strain from the reference strain used in the assay.[123,124]

The specificity of the initial HIV-specific CD8 T cell response is predominantly targeted to the Nef and Gag proteins during the acute phase of infection[126-128] and subsequently expands toward the Gag, Pol, Env, and accessory proteins during the transition to the chronic phase of infection.[127] During acute infection, certain HLA alleles predominate in terms of presenting HIV epitopes for CD8 T cell recognition, and certain epitopes are preferentially targeted, for reasons that remain unclear. For example, of more than a dozen epitopes that have been identified to be presented and targeted through HLA-A3 in chronic HIV infection, the same two are consistently dominant in the earliest stages of acute infection, and the other responses arise over time.[129] In genetically identical twins acutely infected with the same virus through injection drug use, the initial dominant and subdominant responses were almost identical, again pointing to a high degree of predictability in terms of the major targets for initial responses.[130] Studies in larger cohorts have confirmed the finding that many epitopes are not targeted in the acute stage of infection, and for each restricting HLA allele, there are marked differences in the evolution of responses to different epitopes. Moreover, certain HLA alleles consistently contribute more to the initial CTL response.[131] Interestingly, HLA B*57, the allele most strongly associated with favorable outcome following HIV infection, is associated with less symptomatic acute infection,[120] and the protective alleles were also found to exhibit immunodomination in the early stages of acute infection, in that the presence of protective alleles reduces the absolute magnitude of responses restricted by other alleles.[127,131] The importance of early responses is also reflected in data showing that the immunodominance patterns in acute infection are associated with subsequent viral set point, whereas this relationship is lost during chronic infection.[127]

In addition to differences in terms of epitope targeting in acute compared to chronic infection, the characteristics of the CTL are also different. Functional avidity of responses in acute and chronic infection differential, high-avidity responses are subsequently lost in persons with progressive infection, whereas the original clonotypes persist in those who control spontaneously.[132] Studies in persons who control HIV spontaneously, so-called "elite controllers,"[58] have shown that the ability of CD8 T cells to proliferate to cognate antigen, which leads to increased per cell expression of cytolytic effector molecules and is associated with control of viremia.[133,134] Strong proliferative responses by HIV-specific CD8 T cells are detected in acute infection but decline with the transition to chronic uncontrolled infection.[135] At least part of the proliferative defect in chronic infection is due to a lack of sufficient CD4 T helper cell responses and can be restored *in vitro* by addition of autologous CD4 T cells obtained at the time of acute infection, vaccine-induced CD4 T helper cell responses,[135,136] or by inhibition of pathways of CD8 T cell dysfunction.[137,138]

Although Nef is a primary target in acute infection, data from numerous studies indicate that the targeting of Gag during chronic infection is associated with better outcomes. Studies of persons who control virus spontaneously without medication have shown preferential targeting of Gag, whereas responses are more broadly directed in persons with progressive infection.[139] The magnitude and breadth of CD8 T cell responses to the entire Gag protein or just the p24 protein was found to correlate inversely with viral load and directly with CD4 cell counts; in contrast, there was no association with Env or Nef peptides.[140] In a large study of persons with chronic clade C virus infection in South Africa, the broader the Gag-specific CD8 T cell response, the lower was the viral load, whereas the broader the targeting of Env, the higher the viral load (Fig. 45.8).[142] Moreover, median viral load in those persons in whom Gag was the dominant target of the CD8 T cell response was lower than in those in whom the response was predominantly targeted to non-Gag epitopes.[141]

Based on these observations, there is now an emerging consensus that functional targeting of proliferative CD8 T cells to specific epitopes facilitates immune control of HIV. While the sequence conservation of targeted epitopes does not distinguish the immune responses of elite controllers,[143] multidimensional sequence analysis revealed that epitopes with greater inferred viral fitness cost were preferentially presented by protective HLA alleles.[144] Moreover, a recent study applying network theory to HIV protein structure revealed that epitopes made up of highly interconnected residues located in topologically important regions were preferentially targeted by CD8 T cells in HIV controllers, irrespective of

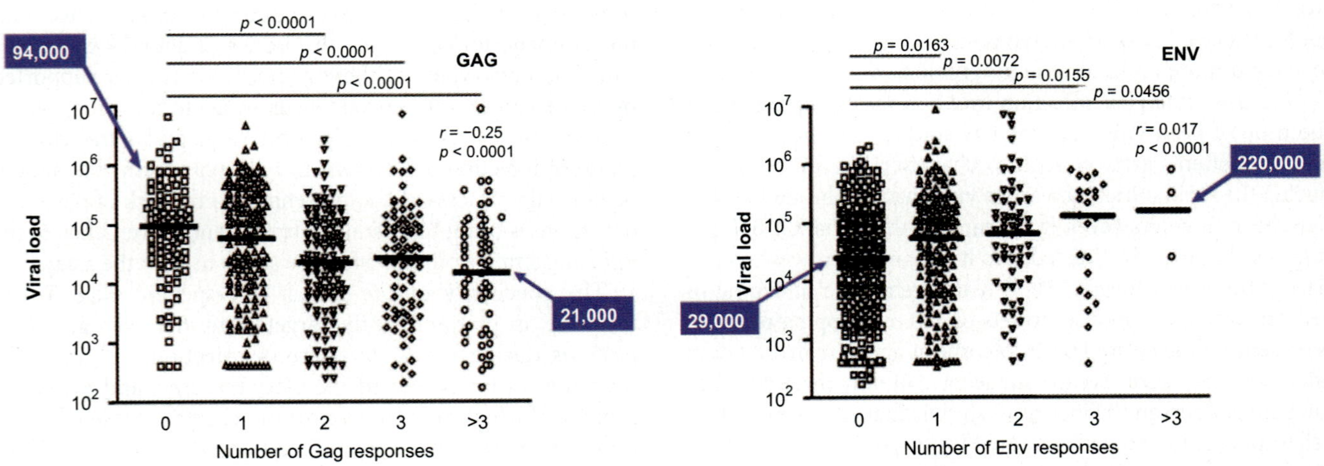

FIG. 45.8. **Breadth of Gag and Env CD8 T cell responses and relation to viral load.** Broader Gag targeting is associated with lower viral loads while dominant Env responses are associated with higher viral loads. (Adapted by permission from Springer Nature: Kiepiela P, Ngumbela K, Thobakgale C, et al. CD8+ T-cell responses to different HIV proteins have discordant associations with viral load. *Nat Med.* 2007;13:46-53.)

HLA allele.[145] Elite controllers have also been found to have a distinct viral reservoir composed of either defective proviruses or intact proviruses integrated at genomic regions with limited transcriptional activity,[146] which likely developed as a result of T cell–mediated clearance of productively infected cells by highly specific CD8 T cell responses. These findings are currently being utilized to generate a therapeutic T cell–based vaccine for HIV in an effort to induce similar responses in the broader population.

CD4 T Cell Response

Immunologic studies performed at the early stages of the HIV epidemic revealed a striking lack of HIV-specific CD4 T cell responses in most infected persons. This has long been seen as a central defect in HIV infection, as virus-specific CD4 cells are central to immune control in every setting in which this has been evaluated. They play a major role in immune induction through the production of IL-2 and the activation of APCs via CD40–CD40 L interactions and are essential for long-term control of viremia due to help provided to CD8 T cells (reviewed in147). HIV preferentially infects CD4 T cells, and during acute infection, there is a dramatic loss of CD4 T cells in the GALT as well as the peripheral blood. Unlike the recovery of peripheral CD4 cell counts when a quasi-set point viremia is achieved, the CD4 cell loss in GALT persists and is not restored even with prolonged antiviral therapy. However, with very early treatment of acute HIV infection, gradual emergence of virus-specific CD4 T cell responses is observed, and similar responses are characteristic of persons who are able to control virus spontaneously without the need for antiviral therapy.[148,149] It has been hypothesized that these antigen-specific CD4 T cells are preferentially deleted at the time of high-level viremia since these cells are being maximally activated at this time. However, the majority of HIV-specific CD4 cells remain uninfected, suggesting that these cells may contribute to the level of relative control that is observed.[150]

Renewed interest in HIV-specific CD4 T cells was sparked in part by results from the RV144 phase III trial of a canarypox prime-recombinant gp120 boost HIV vaccine in which a modest protective effect was correlated with the presence of HIV-specific CD4 T cell responses,[151] although this did not recapitulate in the recent HVTN702 trial which demonstrated a lack of preventive efficacy despite using a similar vaccine strategy. HIV-specific CD4 T cells predominantly target Gag and Env proteins and have been found to display some interesting associations. A study demonstrated that the breadth of total HIV-specific CD4 T cell responses and the breadth and magnitude of Gag-specific responses are both inversely correlated with HIV viral load.[152] Additionally, this study suggested that dominant targeting of Gag was more frequently seen in HIV controllers, while individuals with progressive disease more frequently targeted Env epitopes. Additional data also suggest that the presence of cytolytic CD4 T cells at the time of peak and declining viremia in acute infection is associated with a lower viral set point, suggesting an effector role for CD4 T cells in control of HIV infection.[153]

Immune regulatory networks also appear to play a role and indicate that there is not a total loss of HIV-specific CD4 T cells but rather that these are impaired. In chronic progressive infection, CTLA-4 is upregulated on CD4 T cells but not CD8 T cells and correlates positively with viral load and negatively with the ability of HIV-specific CD4 T cells to produce IFNγ.[154,155] Other studies show impaired IL-2 production by HIV-specific CD4 T cells in persons with persistent viremia[156] and comprehensive, genome-wide transcriptional analyses demonstrate an altered state of differentiation as a hallmark of HIV-specific CD4 T cell dysfunction in chronic disease.[157] In settings of relative control, HIV-specific CD4 T cells consist of both central memory cells capable of producing exclusively IL-2 and effector memory cells producing IL-2 and INFγ. In contrast, in viremic patients, there are elevated levels of effector memory cells that exclusively produce IFNγ.[155] Functional studies are consistent with a defect in these cells in progressive infection, in that HIV-specific CD4 proliferation is negatively associated with viral load,[149] whereas this association is not seen when one measures the ability to produce IFNγ.[156] Higher frequencies of IL-21 expressing CD4 T cells, which represent a unique subset known as T follicular Th cells (Tfh), have also been correlated with lower viral load in viremic individuals.[158]

B Cell Responses

In 1984, Margaret Heckler, the secretary of Health and Human Services, stated that a vaccine for HIV could be expected within 2 years. Unfortunately, the development of an effective vaccine has proved to be much more challenging. For the majority of currently licensed vaccines, neutralizing viral-specific antibodies have been a reliable correlate of protection. In HIV, however, several challenges exist to the induction of protective antibody responses. First and foremost is the enormous diversity of the surface Env protein. This is compounded by the dense glycosylation of the viral Env proteins, which creates a barrier to antibody recognition. This has made it difficult to induce the production of antibodies capable of broad neutralization of an array of viral strains, known as bNAbs. In fact, the initial antibody responses shortly after natural infection are primarily nonneutralizing but are followed by the appearance of responses capable of neutralizing autologous viruses after 12 weeks, but to which escape mutations quickly develop.[159] These antibody responses do not appear to play a significant role in the control of viremia, as they are directed not against contemporaneous viral populations but against viruses from which escape has quickly developed. Nonetheless, it appears that 10% to 25% of chronically infected individuals are able to induce antibodies with narrow neutralization properties, with 2% to 4% that are able to produce broadly nAbs,[160,161] although these appear to arise only after years of infection.

Antibody responses may not be critical for immunological control of viremia, but they are considered essential to achieve protection from acquisition. There is significant evidence from nonhuman primate models of infection that bNAbs are able to prevent HIV acquisition. Passive immunization

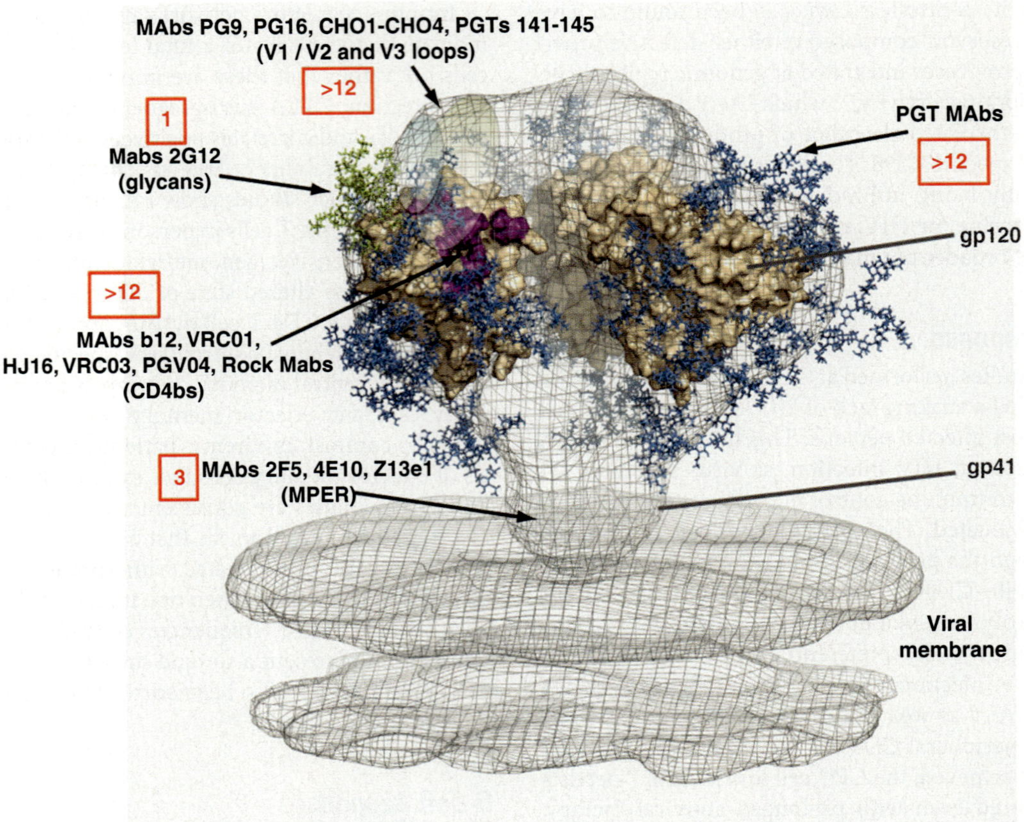

FIG. 45.9. Human immunodeficiency virus (HIV) envelope recognition sites of broadly neutralizing monoclonal antibodies. Identified neutralizing antibodies target include the V2 apex (*blue*), V3 glycan (*purple*), multiple sites on gp120 including the CD4 binding site (CD4bs; *green*), fusion peptide (*red*), and membrane proximal external region (MPER) of gp41. (Courtesy of D. Burton.)

of Rhesus macaques with bNAbs protects against infection via intrarectal challenge with SIV-HIV chimeric virus (SHIV).[162-164] In addition, early treatment of SHIV-infected macaques with bNAbs led to the induction of viral remission in approximately 50% of animals, in part by facilitating the induction of CD8 T cells.[96] New methods of identification and recovery of single antibody-producing cells, however, have allowed for the identification of several other bNAbs. This has revealed several sites of vulnerability on the Env proteins, which include the CD4 binding site on gp120, hypervariable loop 2 (V2) apex, V3 Env glycans, fusion peptide, and the membrane proximal external region of gp41 (Fig. 45.9).[165,166] A major current effort is utilizing bNAbs in both preventive and therapeutic settings by passive delivery, which has been facilitated by the introduction of a mutation in the FcRn binding site which greatly enhances plasma half-life.[167]

Although much of antibody-directed work has been focused on the induction on bNAb responses, more recent interest has developed regarding Fc-mediated antibody effector functions such as antibody-dependent cell-mediated cytotoxicity, in part due to studies that show that alterations to the Fcγ receptor-binding region of the nAb b12 show reduced protection from SHIV challenge in rhesus macaques.[102,168] Additionally, there is some suggestion from the RV144 HIV vaccine trial that non-neutralizing antibody responses to the V1/V2 domains of gp120 are important in mediating modest protection from infection although this was not redemonstrated in the HVTN702 trial. The precise contribution of non-neutralizing antibody responses is likely to be further clarified by ongoing clinical vaccine trials using an adenovirus serotype 26 (Ad26) vector and soluble gp140 envelope protein, which successfully protected ~66% of SHIV-infected rhesus macaques with a polyfunctional antibody response and was highly immunogenic in human subjects.[169]

Dendritic Cells in HIV Infection

A major hallmark of chronic-untreated HIV infection is immune suppression. CD4 T cell loss certainly contributes to this, but additional dysregulation occurs within DCs, which play a key role in providing cross-talk between innate and adaptive immune responses to viral infections (reviewed in[170]). Both plasmacytoid and conventional DCs can harbor HIV, and both are reduced in numbers in HIV infection, and this correlates with increasing viral load and with increased disease progression.[171] This depletion occurs early in acute infection and appears to persist in the chronic phase.

Although DCs express CD4 and conventional coreceptors CCR5 and CXCR4, it remains unclear to what extent these cells are productively infected and subsequently deleted versus rendered dysfunctional. DC infection *ex vivo* is inefficient, in part due to host factors. The host exonuclease, three prime repair exonuclease (TREX1), cleaves ssDNA derived from endogenous retroelements,[172] and TREX1 deletion leads to accumulation of defective HIV viral DNA products and production of type I IFN.[173] In addition to coreceptors, DC also express the C-type lectin DC-SIGN, which virus can use to enter DC through gp120 binding to DC-SIGN. This results in the uptake of virions into early endosomal compartments, where HIV is not degraded but maintained and then released after migration to draining lymph nodes to efficiently infect T cells.[174,175] Clusterin, a glycoprotein found in a number of tissues and blood, specifically blocks HIV-DC-SIGN interactions and prevents transmission of HIV to CD4 T cells when present in semen and may help explain why sexual transmission is less effective than bloodborne exposure, since clusterin in blood is differently glycosylated and does not have this effect.[176]

The HIV single-stranded RNA genome encodes T-like receptor (TLR) 7/9 agonists that are recognized by DCs and lead to the induction of inflammatory cytokines through a mechanism that involves internalization of virions.[177,178] There is also evidence that infected DCs may be able to recognize HIV through interaction of newly transcribed HIV capsid with the cellular protein cyclophilin A, which results in the induction of type I interferons by the transcription factor IRF3.[179] Although the interaction of DCs and HIV is complex and the overall numbers of DCs appear to be reduced in the blood during HIV infection, the degree to which DCs are functionally impaired during infection remains unclear. Plasmacytoid DCs from persons with progressive disease only partially mature upon stimulation and may contribute to chronic immune activation and further CD4 T cell infection by production of copious amounts of IFNα that induces a chronic inflammatory response.[180] There is evidence that myeloid DCs from HIV-infected individuals appear to have reduced responsiveness to TLR agonists and are less efficient at T cell activation.[171,181,182] However, other studies indicate that both myeloid and plasmacytoid DCs from HIV-infected individuals continue to be capable of activating T cells and producing cytokines in response to TLR agonists.[183,184] In fact, myeloid DCs from elite controllers have enhanced cell-intrinsic responses that lead to effective priming of HIV-specific CD8 T cell responses.[185] Transcriptional profiling of these cells in ECs reveals gene expression consistent with profound functional activation and heightened antiviral activity,[186] implicating the importance of this cell type in supporting the development of cellular immune response against the virus.

Immune Selection Pressure and Escape From Cellular Immune Responses

HIV has an enormous ability to escape from immune responses, including NK and antibody responses as discussed above. However, our most detailed understanding of immune escape comes from the study of adaptation to CD8 T cell responses. In most infected persons, virus continues to replicate in the presence of a CD8 T cell response that is insufficient to completely contain viremia. In the same way that incomplete control of viral replication leads to emergence of drug resistant viruses, incomplete immune control is associated with escape mutations that diminish recognition by the established T cell response. Analysis of large datasets of patient-derived sequences reveals a profound impact of CTL responses in driving HIV evolution.[187-189] The clinical significance of this has been shown in infants born to HLA B*27 positive HIV-infected mothers.[190] HLA B*27 positive infants who received the HLA allele from their fathers and the virus from their mothers were able to control viremia through induction of a strong HLA B*27–restricted T cell response to a dominant epitope in Gag. In contrast, in B*27 positive infants whose mothers were also B*27 positive, transmitted viruses were found to have already escaped from the dominant B*27–restricted response in the mother, and thus there was no targeting of this otherwise dominant B*27–restricted Gag epitope in their children.

The kinetics of immune escape are just beginning to be understood, in part because of the paucity of studies in the earliest stages of infection, before the decline in peak viremia. What is clear is that escape can occur within days to weeks of infection. In studies in which autologous virus was sequenced and responses screened with peptides representing the autologous strain, clear evidence of immune escape was evident within 25 to 32 days of infection, coinciding with peak viremia.[119] Longitudinal studies in persons with acute infection have shown selection for CTL escape mutations arise at quite variable rates depending on the epitope and the restricting HLA allele.[189] Deep sequencing has revealed that the virus explores multiple pathways to immune escape, and that there can be numerous variants that arise at low levels within targeted epitopes so low that they are missed by traditional sequencing methods.[191] What remain unclear are the factors that contribute to immunodominance and immunodomination. For a given HLA allele, it has become increasingly clear that the earliest immune responses are predictable, as are the pathways to immune escape.[192] Likewise, it is also clear that certain responses do not arise in the presence of other HLA alleles, a phenomenon called immunodomination.[126]

One of the challenges in defining the role of escape mutations in disease progression is defining mutations that lead to an immunologic impact. Mutations within targeted epitopes that lead to loss of peptide binding to the restricting HLA class I allele are most common,[193] followed by mutations that cause full or partial loss of recognition by the TCR. Mutations within residues flanking epitopes also account for immune escape through alterations in antigen processing.[188,194] Detection of immune escape presents a potential problem using simple interferon gamma Elispot assays, since these may still indicate recognition because they do not incorporate potential confounding effect of antigen processing. More recent focus on assays that incorporate infected target cells, in which antigen-processing events are required for recognition, may provide a better sense of the true measure of immune escape and may be needed in order to finally define the contribution of immune escape to disease progression.

It is important to note that immune escape can have either positive or negative impact on viral control due to the possibility that escape mutations can confer a fitness cost to the virus. In other words, there are competing factors–on the one hand, escape leads to loss of recognition by an established immune response, and on the other hand, it can result in a virus that is markedly impaired in its pathogenic potential. This effect may be most important in the Gag protein, where data suggest multidimensional constraints of HIV evolution due to structural constraints on capsid formation.[144,145,195] Indeed, viruses obtained from elite controllers have been shown to be less fit overall,[196] and selective targeting of regions of Gag under multidimensional constraints on evolution by persons who control HIV spontaneously may contribute to durable control.[195] This has been further confirmed using computational viral fitness landscapes and network analysis of viral protein structures which reveals that epitopes preferential targeted by elite controllers are derived from structurally constrained regions.[144,145] Transmission of escape variants with reduced fitness has been linked to less severe disease in recipients[196,197] and has also been associated with subsequent reversion to wild-type, a clear indication of a fitness cost to the virus that can revert in the absence of ongoing immune selection pressure in a recipient lacking the restricting HLA allele for that epitope.

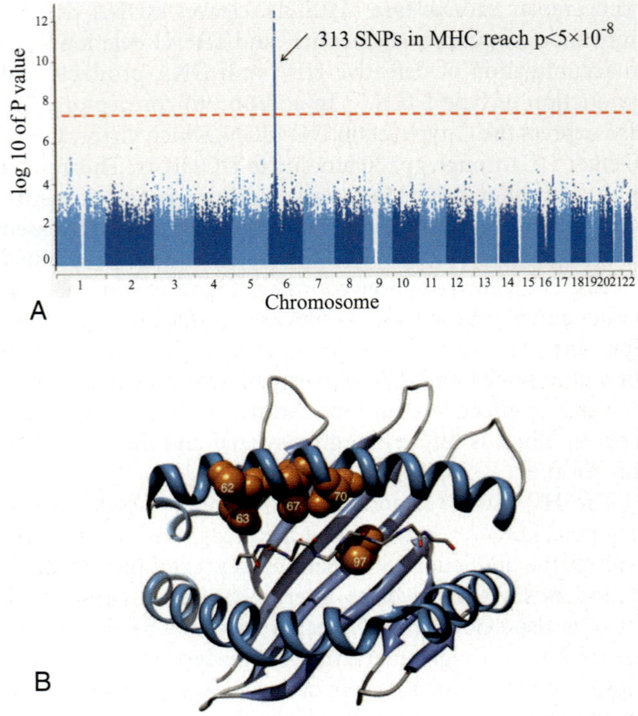

FIG. 45.10. Human leukocyte antigen (HLA) and immune control of human immunodeficiency virus (HIV). A: Genome-wide association study (GWAS) of 1.3 million single-nucleotide polymorphisms (SNPs) show a significant association with a region in chromosome 6 encoding human leukocyte antigens (HLA)-B and -C. **B:** Discrete amino acids within the HLA-B peptide biding groove at positions 67, 70, and 97 are associated with immunologic control of HIV. (Adapted with permission from American Association for the Advancement of Science, from The International HIV Controllers Study. The major genetic determinants of HIV-1 control affect HLA class I peptide presentation. *Science.* 2010;330:1551-1557; permission conveyed through Copyright Clearance Center, Inc.)

HIV IMMUNOGENETICS

The associations between class I alleles and viral control suggest that class I–restricted CTL play a role in limiting HIV replication, and large population studies coupling HLA typing with sequence analysis of the viruses circulating *in vivo* add to this evidence. HLA class I–associated mutational "footprints" have been identified in HIV-1, suggesting that this rapidly mutating virus is adapting to immune selection pressure.[187,198] Given the role of HLA molecules in determining the repertoire and specificity of the immune response, these data are the basis of CD8 T cell responses being involved in immune control.

Considerable additional insight into the immunogenetics of HIV infection has come from genome-wide association studies (GWAS). In a study of persons with acute infection, two single-nucleotide polymorphisms (SNPs) in the major histocompatibility complex (MHC) locus were independently associated with viral control, one associated with HLA-C expression and another that is a proxy for HLA B57,[199] already known to be associated with better outcome. These two plus an additional two independently associated SNPs were subsequently identified in GWAS analyses of persons termed HIV controllers, who maintain viral loads of less than 2000 RNA copies/mL in the absence of therapy.[139,200] GWAS analysis of these HIV controllers also demonstrated that aside from the signal in the region of HLA-B and -C in chromosome 6, most of the previously reported host genetic influences on HIV control could not be replicated (Fig. 45.10A). Indeed, only the chemokine cluster in chromosome 3, which harbors the gene encoding CCR5, the coreceptor associated with HIV entry, was found to be statistically significantly associated with viral control when one controlled for population structure and corrected for multiple comparisons.[122,201]

Further insight into the mechanism of the effects of these SNPs has been revealed by the observations that they "tag" specific features within HLA sequences themselves: specific amino acids within the HLA-B peptide-binding groove, as well as an independent SNP associated with HLA-C expression.[122] Specific amino acids at the HLA-B positions, particularly positions 67, 70, and 97, are associated with differential ability to control HIV viremia, providing a mechanism of action for the SNP associations and suggesting that the nature of the peptide-HLA-B interaction is key to HIV control (Fig. 45.10B). In addition to these HLA-B associations, a SNP 35 kb upstream of the HLA-C gene was also strongly associated with control of viral load and with levels of HLA-C expression. The mechanism of this effect has been linked to variation with the 3′ untranslated region of HLA-C that is posttranscriptionally regulated by a microRNA, resulting in differences in HLA-C expression. The variant that does not bind this microRNA results in high expression of HLA-C and better control of HIV.[202] Unlike HLA-A and -B, which are downregulated during HIV infection, HLA-C is not.[203] Thus, higher levels of HLA-C may prolong the ability of infected cells to be targeted by CTL, after HLA-A and -B have been downregulated, thereby conferring a benefit. However, the antiviral efficacy of HLA-C–restricted responses does not approach the level of immune selection pressure applied through A and B, as evidenced by the lack of HLA-C–associated footprints on the viral genome. Another possibility is that high expression of HLA-C may lead to more effective education and maturation of NK cells due to better engagement of inhibitory NK cell receptors with their HLA-C ligands, resulting in stronger NK cell responses to infected cells.

HLA class II associations with HIV disease outcome have also been reported, but the data are less convincing than for class I. Class II "footprints" have not been convincingly identified, suggesting that responses restricted by HLA class II either exert less selection pressure or that epitopes are presented so promiscuously on different class II alleles that clear signatures cannot be discerned.

In addition to HLA class I alleles, *KIR* locus has also been implicated in modulating control of HIV viremia, suggesting the potential involvement of NK cells in human control of HIV. *KIR*, like *HLA*, is one of the most polymorphic loci in the human genome. The *KIR* genes, located on chromosome 19, encode molecules that regulate NK cell activity, with stimulatory and inhibitory effects. Analysis of AIDS progression rates in untreated persons with HIV infection have shown a protective effect of the activating KIR allele *KIR3DS1* and its putative ligand, HLA-B Bw4 alleles that contain an isoleucine at position 80 (Bw4-80I).[104] In contrast, in persons who express KIR3DS1 in the absence of HLA Bw4-80I progression is more rapid, whereas expression of HLA B Bw4-80I in the absence of KIR3DS1 has no effect. In addition to these effects of an activating KIR on disease progression, other studies have shown that certain alleles of the inhibitory receptor KIR3DL1 also provide protection against disease progression and that protection relates to the relative expression levels of KIR3DL1.[107] Interestingly, this allotype acts synergistically with HLA B*57, which contains the KIR3DS1 ligand Bw4-80I.[101] Whole genome sequencing of HLA B*57 individuals also identified that a single variant encoding an isoleucine-to-valine substitution at position 47 (I47 V) in KIR3DL1 was the only significant modifier of B*57–mediated protection.[204] Thus, collectively this data suggest a protective role of both activating and inhibitory KIRs in immune control of HIV.

HIV PREVENTION

Developing an HIV Vaccine

Despite the success of potent ART, it has been a challenge to deliver medicines to all those in need. Treatment has certainly reached record levels of approximately 21 million HIV-infected individuals,[4] although the prospect of delivering lifelong HIV treatment to individuals living in regions of the world where there is little healthcare infrastructure and unreliable access to food and safe drinking water remains daunting. Thus, the development of an efficacious preventative vaccine remains paramount to ending the AIDS pandemic. It has been estimated that a prophylactic vaccine with only 50% efficacy, delivered to just 30% of the population could reduce annual infections by up to one-third.[205] And yet, the HIV vaccine field has not been able to make significant headway toward this goal. To date, while a number of HIV vaccine candidates have been tested in large scale trials, they have all failed to provide reproducible evidence that an effective vaccine is on the horizon.

The first candidate vaccine to be tested in a large efficacy trial was the VaxGen recombinant gp120 protein, which was used in over 5000 participants in the AIDSVAX trial. The vaccine utilized a recombinant subunit protein to induce antibody responses, an approach that is the basis for the currently licensed hepatitis B vaccine. Initial results in chimpanzees were promising, showing that gp120 vaccination could protect from subsequent HIV challenge with a laboratory isolate.[206] Yet after 3 years of follow-up in the AIDSVAX study, acquisition rates in vaccine recipients and those who received a placebo were statistically indistinguishable.[207,208]

In part because of the difficulty in inducing broadly neutralizing antibody responses against HIV and an appreciation of the important role of CD8 T cell responses in virologic control of infection, interest increased in vaccines that induce T cell responses. Although such vaccines may not protect from HIV acquisition, they may be able to reduce viral set point and delay development of AIDS. Cohort studies have shown that viral set point inversely correlates with disease progression, and thus, such vaccines may be able to delay disease progression.[209] The STEP trial, the second large HIV vaccine trial to be undertaken, tested the T cell vaccine concept in a phase IIb trial of an adenovirus serotype 5 (Ad5) vaccine. The vaccine consisted of an equal mixture of three replication defective adenoviral vectors, each containing a near-consensus clade B gag, pol, or nef gene. The trial was opened in 2005 and cosponsored by Merck and the National Institute of Allergy and Infectious Disease (NIAID), but in September 2007, it was halted early when interim analysis showed that the vaccine failed to prevent HIV infection and had no effect on viral load at 3 months post-infection.[210,211] Although this test-of-concept vaccine failed to show a result in either of the two primary endpoints (HIV acquisition and viral load), it generally has been seen as a first step and not as a condemnation of the T cell vaccine approach per se. In the wake of STEP, a subsequent efficacy trial was the RV144 "Thai vaccine" trial.[151] This approach utilized a recombinant canarypox vector prime with a recombinant gp120 boost strategy and was designed to induce both antibody and T cell responses. Although there were no significant differences in acquisition or viral load using intention-to-treat analysis, a modified intention-to-treat analysis showed a 31% decrease in acquisition in vaccine recipients with no effect on viremia or CD4 T cell count. Interestingly, this effect did not correlate with HIV-specific CD8 T cell responses as measured by ELISpot or intracellular cytokine staining but did correlate with the presence of non-neutralizing antibody responses and Env-directed CD4 T cell responses.

As a follow-up to RV144, the HVTN 702 study or Uhambo trial utilized a similar approach of canary pox vector (ALVAC) encoding HIV genes and two-component subtype C gp120 protein with the adjuvant MF59 as a prime-boost. The vaccine regimen was given to participants in 14 locations in

South Africa and therefore was optimized to clade C virus. While a predecessor trial (HVTN 100) found that this vaccine was safe and induced high titers of antibodies to several HIV strains prevalent in South Africa, the larger HVTN702 trial did not recapitulate the protection seen in RV144 and therefore was stopped early following an interim review by the data safety monitoring board. Two additional trials, known as Imbokodo (HVTN705) and Mosaico (HVTN706), are currently underway at multiple trial sites to examine a vaccine regimen of Ad26 vector and gp140 protein designed to offer protection against a variety of global HIV strains through the induction of polyfunctional antibodies.[169] Results from these studies are expected in the next few years.

The path to an effective vaccine continues to be wrought with challenges, with the suggestion that more focus needs to be placed on understanding basic immunologic mechanisms of vaccine-induced protection. While the apparent partial efficacy of the RV144 trial did not recapitulate in subsequent studies, it did lead to a resurgence in vaccine development. Among the new approaches, currently under investigation are the use of CMV vectors that can induce CTL responses restricted by the nonclassical MHC class I allele HLA-E. While MHC-E–restricted CTLs are rarely seen in natural infections, a rhesus cytomegalovirus (RhCMV) strain 68-1 vectored vaccine-expressing SIV Gag was able to induce CD8 T cells directed against MHC-E–restricted epitopes and MHC class II epitopes in NHPs.[212] These animals were subsequently challenged with pathogenic SIVmac239, and remarkably, approximately half of the animals were able to control and clear the infection.[213,214] Recent studies have now demonstrated MHC class II–restricted T cell responses are not able to mediate protection, indicating the need for MHC-E responses to control and clear the infection.[215]

Given these findings, there is now substantial effort to generate a humanized version of the RhCMV vector and to explore its capacity to induce HLA-E–restricted CD8 T cell responses in humans. This is partly due to the observation that RhCMV68-1 frequently elicits a response to the SIV Gag epitope RMYNPTNIL, which has a close homolog in HIV Gag (RMYSPTSIL; RL9) that was also shown to bind to HLA-E *in vitro*.[216] In addition, a recent study demonstrated that HLA-E–restricted CD8 T cells directed against RL9 can be primed *in vitro* and suppress HIV replication.[217] Thus, efforts are actively underway to enter Phase I clinical trial testing of human CMV vectors in order to offer additional vaccine options to successfully control the epidemic.

Other HIV Prevention Strategies

Globally, over 90% of HIV is transmitted following heterosexual intercourse. Besides abstinence and barrier contraception, the most effective methods of transmission prevention utilize ART (Fig. 45.11). An early method of antiretroviral prevention of transmission is in the setting of prevention of mother-to-child transmission (PMTCT). Peripartum treatment of the mother can reduce the risk of transmission to 5% in breastfeeding children and less than 2% in nonbreastfeeding children, even with short-course therapy.[219] The use of ART for PMTCT has been widely applied and is the major driver of the global reduction in mother-to-child transmission in the developing world. A newer application of "treatment as prevention" was studied in the HPTN 052 study of serodiscordant couples, where ART treatment of the infected partner was shown to reduce transmission by 96%, thus providing compelling evidence that treatment for prevention is effective.[49] A 73% reduction was also seen in the Partners PrEP study of serodiscordant couples in which the uninfected partner was given ART. Clearly, these strategies make sense for individuals at high risk of transmitting within serodiscordant couples who do not use barrier contraception but how widely these strategies should be applied to other populations remain controversial.

The utility of the use of ART in at-risk uninfected individuals as "preexposure prophylaxis" (PrEP) to prevent transmission remains promising. Results in the use of an oral daily fixed-dose combination of tenofovir and emtricitabine (TDF-FTC) as PrEP in HIV-negative men who have sex with men taking part in the iPrEX study showed a 44% decrease in risk of HIV acquisition[220] and 86% reduction in HIV infections in the PROUD study.[221] In the Partners PREP study, use of daily TDF-FTC led to a 75% reduction in HIV infection in heterosexual males and females.

Besides oral therapy, antiretroviral treatment has been incorporated into microbicides to topically delivery drugs at the area of exposure. This approach has the advantage of low toxicity and up to 100× higher levels of drugs within the mucosal site of exposure. The landmark CAPRISA 004 trial of a vaginal microbicide containing tenofovir showed a 39% reduction in HIV acquisition among at-risk heterosexual South African women.[222] However, follow-up studies testing oral tenofovir-emtricitabine (FEM-PrEP)[223] or oral tenofovir-based regimens and vaginal tenofovir (VOICE)[224] for PrEP in sub-Saharan African women failed to show efficacy, likely due to low adherence to study interventions. A subsequent study known an ASPIRE (A Study to Prevent Infection with a RING for Extended Use) utilized an implantable vaginal ring containing the ART drug dapivirine that led to a reduction in HIV infection by 27% in African women,[225] which was recapitulated by The Ring Study which found similar rates of protection.[226] This has generated substantial enthusiasm for newer formulations of antiretroviral medicines for PrEP, including depot injections, long-lasting oral formulations, and implants.

Overall, it is clear that although tremendous advances have been made in the treatment and prevention of HIV, much more remains to be done. Halting the AIDS pandemic will require a multipronged approach, relying on continuing to expand treatment rollout, defining the most effective antiretroviral-based methods for prevention, and implementing structural and behavior interventions such as male circumcision, prevention education, and programs to reduce risk behaviors. Importantly, a safe and effective HIV vaccine will also be needed as a primary component of these efforts. However, to produce such a vaccine will require a more detailed understanding of HIV immunity and host interactions. In the almost 40 years since the discovery of HIV as the causative agent of AIDS, an enormous amount has been

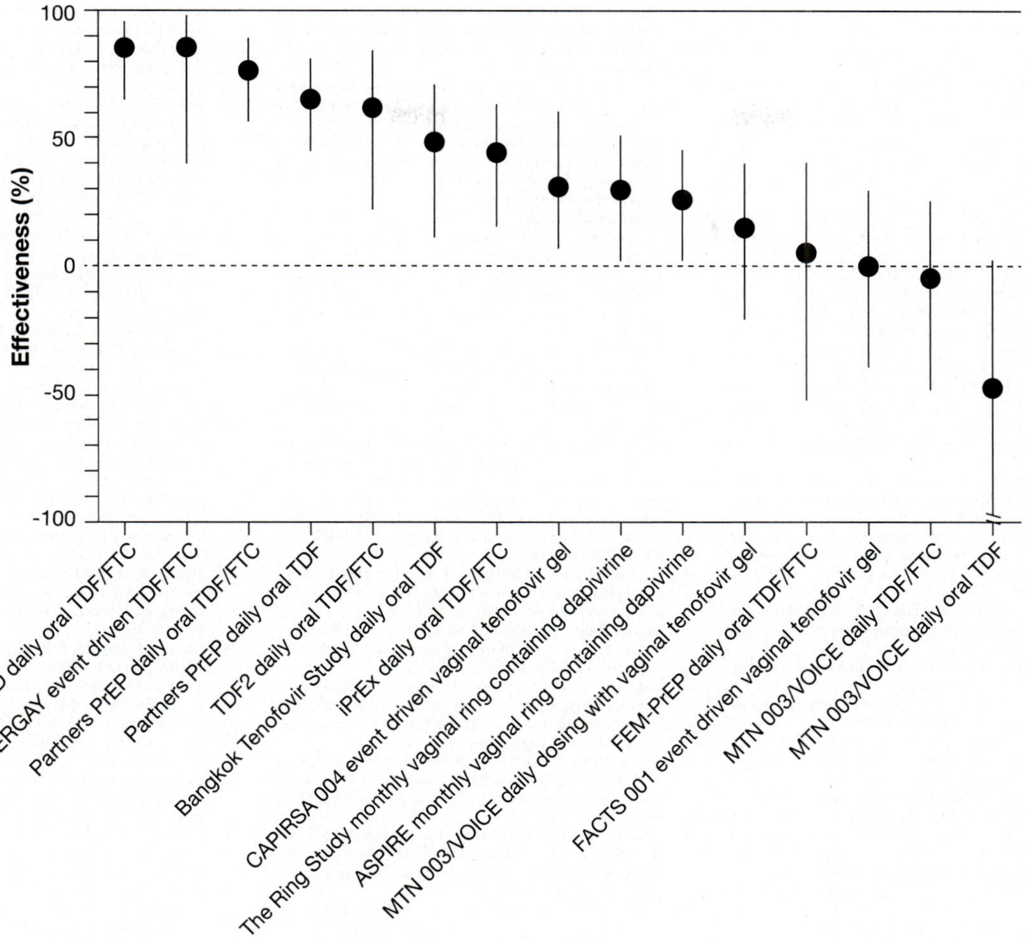

FIG. 45.11. Effectiveness of human immunodeficiency virus (HIV) prevention methods using antiretrovirals. Protection provided by preexposure prophylaxis (PrEP) methods tested in large clinical trials showing efficacy and confident intervals. (Data from Desai M, Field N, Grant R, McCormack S. Recent advances in pre-exposure prophylaxis for HIV. *Br Med J.* 2017;359:j5011.)

learned but significant challenges clearly remain. To overcome these challenges will require the continued dedication and support of clinicians, researchers, patients, funders, and the public at large. But the pace of discover continues to be rapid and hopes high that this is a disease that can be prevented in those not yet infected, and potentially eradicated in those who are.

ACKNOWLEDGMENTS

GDG is supported by an NIH DP2 New Innovators Award, the Bill and Melinda Gates Foundation, a Career Award for Medical Scientists from the Burroughs Wellcome Fund and the Gilead HIV Research Scholars Award. DSK is supported by the NIAID, NIDDK, and the Bill and Melinda Gates Foundation.

REFERENCES

1. Gottlieb MS, Schanker HM, Fan PT, et al. Pneumocystis pneumonia—Los Angeles. *MMWR Morb Mortal Wkly Rep.* 1981;30:250-252.
2. Hymes K, Greene J, Marcus A, et al. Kaposi's sarcoma in homosexual men—a report of eight cases. *Lancet.* 1981;318:598-600.
3. Hessol NA, Koblin BA, Byers RH, et al. Epidemiology of human immunodeficiency virus type 1 infection among homosexual men participating in hepatitis B vaccine trials in Amsterdam, New York City, and San Francisco, 1978-1990. *Am J Epidemiol.* 1993;137:909-915.
4. Joint United Nations Programme on HIV/AIDS (UNAIDS). Geneva: UNAIDS; 2018.
5. Virgin HW, Walker BD. Immunology and the elusive AIDS vaccine. *Nature.* 2010;464:224-231.
6. Walensky RP, Paltiel AD, Losina E, et al. The survival benefits of AIDS treatment in the United States. *J Infect Dis.* 2006;194:11-19.
7. Finzi D, Blankson J, Siliciano JD, et al. Latent infection of CD4+ T cells provides a mechanism for lifelong persistence of HIV-1, even in patients on effective combination therapy. *Nat Med.* 1999;5:512-517.
8. Johnston MI, Fauci AS. HIV vaccine development—improving on natural immunity. *N Engl J Med.* 2011;365:873-875.
9. Sharp PM, Hahn BH. The evolution of HIV-1 and the origin of AIDS. *Philos Trans R Soc Lond B Biol Sci.* 2010;365:2487-2494.
10. Vidal N, Peeters M, Mulanga-Kabeya C, et al. Unprecedented degree of human immunodeficiency virus type 1 (HIV-1) group M genetic diversity in the Democratic Republic of Congo suggests that the HIV-1 pandemic originated in Central Africa. *J Virol.* 2000;74:10498-10507.
11. Korber B, Muldoon M, Theiler J, et al. Timing the ancestor of the HIV-1 pandemic strains. *Science.* 2000;288:1789-1796.
12. Worobey M, Gemmel M, Teuwen DE, et al. Direct evidence of extensive diversity of HIV-1 in Kinshasa by 1960. *Nature.* 2008;455:661-664.
13. Hahn BH, Shaw GM, De Cock KM, Sharp PM. AIDS as a zoonosis: scientific and public health implications. *Science.* 2000;287:607-614.
14. Plantier J-C, Leoz M, Dickerson JE, et al. A new human immunodeficiency virus derived from gorillas. *Nat Med.* 2009;15:871-872.
15. Vallari A, Holzmayer V, Harris B, et al. Confirmation of putative HIV-1 group P in Cameroon. *J Virol.* 2011;85:1403-1407.
16. Roberts JD, Bebenek K, Kunkel TA. The accuracy of reverse transcriptase from HIV-1. *Science.* 1988;242:1171-1173.
17. Korber B, Gaschen B, Yusim K, Thakallapally R, Kesmir C, Detours V. Evolutionary and immunological implications of contemporary HIV-1 variation. *Br Med Bull.* 2001;58:19-42.
18. Gao F, Bailes E, Robertson DL, et al. Origin of HIV-1 in the chimpanzee *Pan troglodytes* troglodytes. *Nature.* 1999;397:436-441.
19. Sodora DL, Allan JS, Apetrei C, et al. Toward an AIDS vaccine: lessons from natural simian immunodeficiency virus infections of African nonhuman primate hosts. *Nat Med.* 2009;15:861-865.

20. Cocchi F, DeVico AL, Garzino-Demo A, Arya SK, Gallo RC, Lusso P. Identification of RANTES, MIP-1 alpha, and MIP-1 beta as the major HIV-suppressive factors produced by CD8+ T cells. *Science.* 1995;270:1811-1815.
21. Alkhatib G, Combadiere C, Broder CC, et al. CC CKR5: a RANTES, MIP-1alpha, MIP-1beta receptor as a fusion cofactor for macrophage-tropic HIV-1. *Science.* 1996;272:1955-1958.
22. Paxton WA, Martin SR, Tse D, et al. Relative resistance to HIV–1 infection of CD4 lymphocytes from persons who remain uninfected despite multiple high–risk sexual exposures. *Nat Med.* 1996;2:412-417.
23. Deng H, Liu R, Ellmeier W, et al. Identification of a major co-receptor for primary isolates of HIV-1. *Nature.* 1996;381:661-666.
24. Samson M, Libert F, Doranz BJ, et al. Resistance to HIV-1 infection in caucasian individuals bearing mutant alleles of the CCR-5 chemokine receptor gene. *Nature.* 1996;382:722-725.
25. Liu R, Paxton WA, Choe S, et al. Homozygous defect in HIV-1 coreceptor accounts for resistance of some multiply-exposed individuals to HIV-1 infection. *Cell.* 1996;86:367-377.
26. Huang Y, Paxton WA, Wolinsky SM, et al. The role of a mutant CCR5 allele in HIV--1 transmission and disease progression. *Nat Med.* 1996;2:1240-1243.
27. Yarchoan R, Mitsuya H, Broder S. AIDS therapies. *Sci Am.* 1988;259:110-119.
28. Chow YK, Hirsch MS, Merrill DP, et al. Use of evolutionary limitations of HIV-1 multidrug resistance to optimize therapy. *Nature.* 1993;361:650-654.
29. Palella FJ, Delaney KM, Moorman AC, et al. Declining morbidity and mortality among patients with advanced human immunodeficiency virus infection. *N Engl J Med.* 1998;338:853-860.
30. Tan Q, Zhu Y, Li J, et al. Structure of the CCR5 chemokine receptor-HIV entry inhibitor maraviroc complex. *Science.* 2013;341:1387-1390.
31. Williams IG. Enfuvirtide (Fuzeon): the first fusion inhibitor. *Int J Clin Pract.* 2003;57:890-897.
32. Hicks C, Gulick RM. Raltegravir: the first HIV type 1 integrase inhibitor. *Clin Infect Dis.* 2009;48:931-939.
33. Vercauteren J, Wensing AMJ, van de Vijver DAMC, et al. Transmission of drug-resistant HIV-1 is stabilizing in Europe. *J Infect Dis.* 2009;200:1503-1508.
34. Ghosh RK, Ghosh SM, Chawla S. Recent advances in antiretroviral drugs. *Expet Opin Pharmacother.* 2011;12:31-46.
35. Okwundu CI, Uthman OA, Okoromah CA. Antiretroviral pre-exposure prophylaxis (PrEP) for preventing HIV in high-risk individuals. *Cochrane Database Syst Rev.* 2012;11(7):CD007189. doi:10.1002/14651858.CD007189.pub3
36. Boily M-C, Baggaley RF, Wang L, et al. Heterosexual risk of HIV-1 infection per sexual act: systematic review and meta-analysis of observational studies. *Lancet Infect Dis.* 2009;9:118-129.
37. Salazar-Gonzalez JF, Salazar MG, Keele BF, et al. Genetic identity, biological phenotype, and evolutionary pathways of transmitted/founder viruses in acute and early HIV-1 infection. *J Exp Med.* 2009;206:1273-1289.
38. Bar KJ, Li H, Chamberland A, et al. Wide variation in the multiplicity of HIV-1 infection among injection drug users. *J Virol.* 2010;84:6241-6247.
39. Anahtar MN, Gootenberg DB, Mitchell CM, Kwon DS. Cervicovaginal microbiota and reproductive health: the virtue of simplicity. *Cell Host Microbe.* 2018;23:159-168.
40. Gosmann C, Anahtar MN, Handley SA, et al. Lactobacillus-deficient cervicovaginal bacterial communities are associated with increased HIV acquisition in young South African women. *Immunity.* 2017;46:29-37.
41. Anahtar MN, Byrne EH, Doherty KE, et al. Cervicovaginal bacteria are a major modulator of host inflammatory responses in the female genital tract. *Immunity.* 2015;42:965-976.
42. Haase AT. Targeting early infection to prevent HIV-1 mucosal transmission. *Nature.* 2010;464:217-223.
43. Haase AT. Perils at mucosal front lines for HIV and SIV and their hosts. *Nat Rev Immunol.* 2005;5:783-792.
44. Reynolds MR, Rakasz E, Skinner PJ, et al. CD8+ T-lymphocyte response to major immunodominant epitopes after vaginal exposure to simian immunodeficiency virus: too late and too little. *J Virol.* 2005;79:9228-9235.
45. Rowland-Jones SL, Dong T, Fowke KR, et al. Cytotoxic T cell responses to multiple conserved HIV epitopes in HIV-resistant prostitutes in Nairobi. *J Clin Invest.* 1998;102:1758-1765.
46. Addo MM, Altfeld M, Brainard DM, et al. Lack of detectable HIV-1–specific CD8+ T cell responses in Zambian HIV-1–exposed seronegative partners of HIV-1–positive individuals. *J Infect Dis.* 2011;203:258-262.
47. Quinn TC, Wawer MJ, Sewankambo N, et al. Viral load and heterosexual transmission of human immunodeficiency virus type 1. *N Engl J Med.* 2000;342:921-929.
48. Cohen MS, Shaw GM, McMichael AJ, Haynes BF. Acute HIV-1 infection. *N Engl J Med.* 2011;364:1943-1954.
49. Cohen MS, Chen YQ, McCauley M, et al. Prevention of HIV-1 infection with early antiretroviral therapy. *N Engl J Med.* 2011;365:493-505.
50. Kahn JO, Walker BD. Acute human immunodeficiency virus type 1 infection. *N Engl J Med.* 1998;339:33-39.
51. Kassutto S, Maghsoudi K, Johnston MN, et al. Longitudinal analysis of clinical markers following antiretroviral therapy initiated during acute or early HIV type 1 infection. *Clin Infect Dis.* 2006;42:1024-1031.
52. McMichael AJ, Borrow P, Tomaras GD, Goonetilleke N, Haynes BF. The immune response during acute HIV-1 infection: clues for vaccine development. *Nat Rev Immunol.* 2010;10:11-23.
53. Lyles RH, Muñoz A, Yamashita TE, et al. Natural history of human immunodeficiency virus type 1 viremia after seroconversion and proximal to AIDS in a large cohort of homosexual men. *J Infect Dis.* 2000;181:872-880.
54. Brenchley JM, Schacker TW, Ruff LE, et al. CD4+ T cell depletion during all stages of HIV disease occurs predominantly in the gastrointestinal tract. *J Exp Med.* 2004;200:749-759.
55. Mattapallil JJ, Douek DC, Hill B, Nishimura Y, Martin M, Roederer M. Massive infection and loss of memory CD4+ T cells in multiple tissues during acute SIV infection. *Nature.* 2005;434:1093-1097.
56. Doitsh G, Cavrois M, Lassen KG, et al. Abortive HIV infection mediates CD4 T cell depletion and inflammation in human lymphoid tissue. *Cell.* 2010;143:789-801.
57. Moss AR, Bacchetti P. Natural history of HIV infection. *AIDS.* 1989;3:55-61.
58. Deeks SG, Walker BD. Human immunodeficiency virus controllers: mechanisms of durable virus control in the absence of antiretroviral therapy. *Immunity.* 2007;27:406-416.
59. Detels R, Jacobson L, Margolick J, et al. The multicenter AIDS cohort study, 1983 to …. *Public Health.* 2012;126, 196-198.
60. Centers for Disease Control and Prevention. *MMWR Morb Mortal Wkly Rep.* 2008.
61. Corbett EL, Watt CJ, Walker N, et al. The growing burden of tuberculosis: global trends and interactions with the HIV epidemic. *Arch Intern Med.* 2003;163:1009-1021.
62. Sigel K, Dubrow R, Silverberg M, Crothers K, Braithwaite S, Justice A. Cancer screening in patients infected with HIV. *Curr HIV/AIDS Rep.* 2011;8:142-152.
63. Chapman C, Aboulafia DM, Dezube BJ, Pantanowitz L. Human immunodeficiency virus–associated adenocarcinoma of the colon: clinicopathologic findings and outcome. *Clin Colorectal Cancer.* 2009;8:215-219.
64. Marcus JL, Chao CR, Leyden WA, et al. Narrowing the gap in life expectancy between HIV-infected and HIV-uninfected individuals with access to care. *J Acquir Immune Defic Syndr.* 2016;73:39.
65. van den Dries L, Claassen MAA, Groothuismink ZMA, van Gorp E, Boonstra A. Immune activation in prolonged cART-suppressed HIV patients is comparable to that of healthy controls. *Virology.* 2017;509:133-139.
66. Fahey JL, Taylor JM, Manna B, et al. Prognostic significance of plasma markers of immune activation, HIV viral load and CD4 T-cell measurements. *AIDS.* 1998;12:1581-1590.
67. Fahey JL, Taylor JM, Detels R, et al. The prognostic value of cellular and serologic markers in infection with human immunodeficiency virus type 1. *N Engl J Med.* 1990;322:166-172.
68. Giorgi JV, Hultin LE, McKeating JA, et al. Shorter survival in advanced human immunodeficiency virus type 1 infection is more closely associated with T lymphocyte activation than with plasma virus burden or virus chemokine coreceptor usage. *J Infect Dis.* 1999;179:859-870.
69. Silvestri G, Sodora DL, Koup RA, et al. Nonpathogenic SIV infection of sooty mangabeys is characterized by limited bystander immunopathology despite chronic high-level viremia. *Immunity.* 2003;18:441-452.
70. Silvestri G, Fedanov A, Germon S, et al. Divergent host responses during primary simian immunodeficiency virus SIVsm infection of natural sooty mangabey and nonnatural rhesus macaque hosts. *J Virol.* 2005;79:4043-4054.
71. Broussard SR, Staprans SI, White R, Whitehead EM, Feinberg MB, Allan JS. Simian immunodeficiency virus replicates to high levels in naturally infected African green monkeys without inducing immunologic or neurologic disease. *J Virol.* 2001;75:2262-2275.
72. Tanko RF, Soares AP, Masson L, et al. Residual T cell activation and skewed CD8+ T cell memory differentiation despite antiretroviral therapy-induced HIV suppression. *Clin Immunol.* 2018;195:127-138.
73. Brenchley JM, Price DA, Schacker TW, et al. Microbial translocation is a cause of systemic immune activation in chronic HIV infection. *Nat Med.* 2006;12:1365-1371.
74. Cooke KR, Hill GR, Crawford JM, et al. Tumor necrosis factor- alpha production to lipopolysaccharide stimulation by donor cells predicts the severity of experimental acute graft-versus-host disease. *J Clin Invest.* 1998;102:1882-1891.
75. Gootenberg DB, Paer JM, Luevano J-M, Kwon DS. HIV-associated changes in the enteric microbial community: potential role in loss of homeostasis and development of systemic inflammation. *Curr Opin Infect Dis.* 2017;30:31-43.
76. Wherry EJ. T cell exhaustion. *Nat Immunol.* 2011;12:492-499.
77. Wherry EJ, Blattman JN, Murali-Krishna K, van der Most R, Ahmed R. Viral persistence alters CD8 T-cell immunodominance and tissue distribution and results in distinct stages of functional impairment. *J Virol.* 2003;77:4911-4927.
78. McLane LM, Abdel-Hakeem MS, Wherry EJ. CD8 T cell exhaustion during chronic viral infection and cancer. *Annu Rev Immunol.* 2019;37:457-495.
79. Virgin HW, Wherry EJ, Ahmed R. Redefining chronic viral infection. *Cell.* 2009;138:30-50.
80. Kaufmann DE, Walker BD. PD-1 and CTLA-4 inhibitory cosignaling pathways in HIV infection and the potential for therapeutic intervention. *J Immunol.* 2009;182:5891-5897.
81. Mylvaganam GH, Chea LS, Tharp GK, et al. Combination anti–PD-1 and antiretroviral therapy provides therapeutic benefit against SIV. *JCI Insight.* 2018;3(18):e122940. https://www.ncbi.nlm.nih.gov/pmc/articles/pmc6237231/
82. Hashimoto M, Kamphorst AO, Im SJ, et al. CD8 T cell exhaustion in chronic infection and cancer: opportunities for interventions. *Annu Rev Med.* 2018;69:301-318.
83. Chun TW, Engel D, Berrey MM, Shea T, Corey L, Fauci AS. Early establishment of a pool of latently infected, resting CD4+ T cells during primary HIV-1 infection. *Proc Natl Acad Sci U S A.* 1998;95(15):8869-8873. https://www.pnas.org/content/95/15/8869.short
84. Zhang Z, Schuler T, Zupancic M, et al. Sexual transmission and propagation of SIV and HIV in resting and activated CD4+ T cells. *Science.* 1999;286:1353-1357.
85. Chun TW, Finzi D, Margolick J, Chadwick K, Schwartz D, Siliciano RF. In vivo fate of HIV-1-infected T cells: quantitative analysis of the transition to stable latency. *Nat Med.* 1995;1:1284-1290.
86. Crooks AM, Bateson R, Cope AB, et al. Precise quantitation of the latent HIV-1 reservoir: implications for eradication strategies. *J Infect Dis.* 2015;212:1361-1365.
87. Maldarelli F, Wu X, Su L, et al. Specific HIV integration sites are linked to clonal expansion and persistence of infected cells. *Science.* 2014;345:179-183.
88. Murray AJ, Kwon KJ, Farber DL, Siliciano RF. The latent reservoir for HIV-1: how immunologic memory and clonal expansion contribute to HIV-1 persistence. *J Immunol.* 2016;197:407-417.
89. Buggert M, Japp AS, Betts MR. Everything in its right place: resident memory CD8+ T cell immunosurveillance of HIV infection. *Curr Opin HIV AIDS.* 2019;14:93.
90. Perreau M, Savoye A-L, De Crignis E, et al. Follicular helper T cells serve as the major CD4 T cell compartment for HIV-1 infection, replication, and production. *J Exp Med.* 2013;210:143-156.
91. Fukazawa Y, Lum R, Okoye AA, et al. B cell follicle sanctuary permits persistent productive simian immunodeficiency virus infection in elite controllers. *Nat Med.* 2015;21:132-139.
92. Archin NM, Sung JM, Garrido C, Soriano-Sarabia N, Margolis DM. Eradicating HIV-1 infection: seeking to clear a persistent pathogen. *Nat Rev Microbiol.* 2014;12:750-764.
93. Ho Y-C, Shan L, Hosmane NN, et al. Replication-competent noninduced proviruses in the latent reservoir increase barrier to HIV-1 cure. *Cell.* 2013;155:540-551.
94. Borducchi EN, Cabral C, Stephenson KE, et al. Ad26/MVA therapeutic vaccination with TLR7 stimulation in SIV-infected rhesus monkeys. *Nature.* 2016;540:284-287.
95. Scheid JF, Horwitz JA, Bar-On Y, et al. HIV-1 antibody 3BNC117 suppresses viral rebound in humans during treatment interruption. *Nature.* 2016;535:556-560.
96. Nishimura Y, Gautam R, Chun T-W, et al. Early antibody therapy can induce long-lasting immunity to SHIV. *Nature.* 2017;543:559-563.
97. Niessl J, Baxter AE, Mendoza P, et al. Combination anti-HIV-1 antibody therapy is associated with increased virus-specific T cell immunity. *Nat Med.* 2020;26:222-227.
98. Hütter G, Nowak D, Mossner M, et al. Long-term control of HIV by CCR5 Delta32/Delta32 stem-cell transplantation. *N Engl J Med.* 2009;360:692-698.
99. Gupta RK, Peppa D, Hill AL, et al. Evidence for HIV-1 cure after CCR5Δ32/Δ32 allogeneic haemopoietic stem-cell transplantation 30 months post analytical treatment interruption: a case report. *Lancet HIV.* 2020;7:e340-e347.
100. Collins KL, Chen BK, Kalams SA, Walker BD, Baltimore D. HIV-1 Nef protein protects infected primary cells against killing by cytotoxic T lymphocytes. *Nature.* 1998;391:397-401.
101. Bottino C, Castriconi R, Moretta L, Moretta A. Cellular ligands of activating NK receptors. *Trends Immunol.* 2005;26:221-226.
102. Scott-Algara D, Truong LX, Versmisse P, et al. Cutting edge: increased NK cell activity in HIV-1-exposed but uninfected Vietnamese intravascular drug users. *J Immunol.* 2003;171:5663-5667.
103. Ravet S, Scott-Algara D, Bonnet E, et al. Distinctive NK-cell receptor repertoires sustain high-level constitutive NK-cell activation in HIV-exposed uninfected individuals. *Blood.* 2007;109:4296-4305.
104. Martin MP, Gao X, Lee J-H, et al. Epistatic interaction between KIR3DS1 and HLA-B delays the progression to AIDS. *Nat Genet.* 2002;31:429-434.
105. Alter G, Martin MP, Teigen N, et al. Differential natural killer cell–mediated inhibition of HIV-1 replication based on distinct KIR/HLA subtypes. *J Exp Med.* 2007;204:3027-3036.
106. Ramsuran V, Naranbhai V, Horowitz A, et al. Elevated HLA-A expression impairs HIV control through inhibition of NKG2A-expressing cells. *Science.* 2018;359:86-90.
107. Martin MP, Qi Y, Gao X, et al. Innate partnership of HLA-B and KIR3DL1 subtypes against HIV-1. *Nat Genet.* 2007;39:733-740.

108. Boudreau JE, Mulrooney TJ, Le Luduec J-B, Barker E, Hsu KC. KIR3DL1 and HLA-B density and binding calibrate NK education and response to HIV. *J Immunol.* 2016;196:3398-3410.
109. Kottilil S, Chun T-W, Moir S, et al. Innate immunity in human immunodeficiency virus infection: effect of viremia on natural killer cell function. *J Infect Dis.* 2003;187:1038-1045.
110. Portales P, Reynes J, Pinet V, et al. Interferon-alpha restores HIV-induced alteration of natural killer cell perforin expression in vivo. *AIDS.* 2003;17:495-504.
111. Fogli M, Costa P, Murdaca G, et al. Significant NK cell activation associated with decreased cytolytic function in peripheral blood of HIV-1-infected patients. *Eur J Immunol.* 2004;34:2313-2321.
112. Bernstein HB, Wang G, Plasterer MC, et al. CD4+ NK cells can be productively infected with HIV, leading to downregulation of CD4 expression and changes in function. *Virology.* 2009;387:59-66.
113. Alter G, Heckerman D, Schneidewind A, et al. HIV-1 adaptation to NK-cell-mediated immune pressure. *Nature.* 2011;476:96-100.
114. Borrow P, Lewicki H, Hahn BH, Shaw GM, Oldstone MB. Virus-specific CD8+ cytotoxic T-lymphocyte activity associated with control of viremia in primary human immunodeficiency virus type 1 infection. *J Virol.* 1994;68:6103-6110.
115. Koup RA, Safrit JT, Cao Y, et al. Temporal association of cellular immune responses with the initial control of viremia in primary human immunodeficiency virus type 1 syndrome. *J Virol.* 1994;68:4650-4655.
116. Ndhlovu ZM, Kamya P, Mewalal N, et al. Magnitude and kinetics of CD8+ T cell activation during hyperacute HIV infection impact viral set point. *Immunity.* 2015;43:591-604.
117. Schmitz JE, Kuroda MJ, Santra S, et al. Control of viremia in simian immunodeficiency virus infection by CD8+ lymphocytes. *Science.* 1999;283:857-860.
118. Jin X, Bauer DE, Tuttleton SE, et al. Dramatic rise in plasma viremia after CD8(+) T cell depletion in simian immunodeficiency virus-infected macaques. *J Exp Med.* 1999;189:991-998.
119. Kaslow RA, Carrington M, Apple R, et al. Influence of combinations of human major histocompatibility complex genes on the course of HIV–1 infection. *Nat Med.* 1996;2:405-411.
120. Migueles SA, Sabbaghian MS, Shupert WL, et al. HLA B*5701 is highly associated with restriction of virus replication in a subgroup of HIV-infected long term nonprogressors. *Proc Natl Acad Sci U S A.* 2000;97:2709-2714.
121. Carrington M, Nelson GW, Martin MP, et al. HLA and HIV-1: heterozygote advantage and B*35-Cw*04 disadvantage. *Science.* 1999;283:1748-1752.
122. The International HIV Controllers Study. The major genetic determinants of HIV-1 control affect HLA class I peptide presentation. *Science.* 2010;330:1551-1557.
123. Goonetilleke N, Liu MKP, Salazar-Gonzalez JF, et al. The first T cell response to transmitted/founder virus contributes to the control of acute viremia in HIV-1 infection. *J Exp Med.* 2009;206:1253-1272.
124. Radebe M, Nair K, Chonco F, et al. Limited immunogenicity of HIV CD8+ T-cell epitopes in acute Clade C virus infection. *J Infect Dis.* 2011;204:768-776.
125. Altfeld M, Addo MM, Shankarappa R, et al. Enhanced detection of human immunodeficiency virus type 1-specific T-cell responses to highly variable regions by using peptides based on autologous virus sequences. *J Virol.* 2003;77:7330-7340.
126. Lichterfeld M, Yu XG, Cohen D, et al. HIV-1 Nef is preferentially recognized by CD8 T cells in primary HIV-1 infection despite a relatively high degree of genetic diversity. *AIDS.* 2004;18:1383-1392.
127. Streeck H, Jolin JS, Qi Y, et al. Human immunodeficiency virus type 1-specific CD8+ T-cell responses during primary infection are major determinants of the viral set point and loss of CD4+ T cells. *J Virol.* 2009;83:7641-7648.
128. Brumme ZL, Brumme CJ, Carlson J, et al. Marked epitope- and allele-specific differences in rates of mutation in human immunodeficiency type 1 (HIV-1) Gag, Pol, and Nef cytotoxic T-lymphocyte epitopes in acute/early HIV-1 infection. *J Virol.* 2008;82:9216-9227.
129. Xu GY, Addo MM, Rosenberg ES, et al. Consistent patterns in the development and immunodominance of human immunodeficiency virus type 1 (HIV-1)-specific CD8+ T-cell responses following acute HIV-1 infection. *J Virol.* 2002;76:8690-8701.
130. Draenert R, Allen TM, Liu Y, et al. Constraints on HIV-1 evolution and immunodominance revealed in monozygotic adult twins infected with the same virus. *J Exp Med.* 2006;203:529-539.
131. Altfeld M, Kalife ET, Qi Y, et al. HLA alleles associated with delayed progression to AIDS contribute strongly to the initial CD8(+) T cell response against HIV-1. *PLoS Med.* 2006;3:e403.
132. Lichterfeld M, Yu XG, Mui SK, et al. Selective depletion of high-avidity human immunodeficiency virus type 1 (HIV-1)-specific CD8+ T cells after early HIV-1 infection. *J Virol.* 2007;81:4199-4214.
133. Migueles SA, Laborico AC, Shupert WL, et al. HIV-specific CD8+ T cell proliferation is coupled to perforin expression and is maintained in nonprogressors. *Nat Immunol.* 2002;3:1061-1068.
134. Migueles SA, Osborne CM, Royce C, et al. Lytic granule loading of CD8+ T cells is required for HIV-infected cell elimination associated with immune control. *Immunity.* 2008;29:1009-1021.
135. Lichterfeld M, Kaufmann DE, Yu XG, et al. Loss of HIV-1–specific CD8+ T cell proliferation after acute HIV-1 infection and restoration by vaccine-induced HIV-1–specific CD4+ T cells. *J Exp Med.* 2004;200:701-712.
136. Robbins GK, Addo MM, Troung H, et al. Augmentation of HIV-1-specific T helper cell responses in chronic HIV-1 infection by therapeutic immunization. *AIDS.* 2003;17:1121-1126.
137. Day CL, Kaufmann DE, Kiepiela P, et al. PD-1 expression on HIV-specific T cells is associated with T-cell exhaustion and disease progression. *Nature.* 2006;443:350-354.
138. Gaiha GD, McKim KJ, Woods M, et al. Dysfunctional HIV-specific CD8+ T cell proliferation is associated with increased caspase-8 activity and mediated by necroptosis. *Immunity.* 2014;41:1001-1012.
139. Pereyra F, Addo MM, Kaufmann DE, et al. Genetic and immunologic heterogeneity among persons who control HIV infection in the absence of therapy. *J Infect Dis.* 2008;197:563-571.
140. Lazaro E, Godfrey SB, Stamegna P, et al. Differential HIV epitope processing in monocytes and CD4 T cells affects cytotoxic T lymphocyte recognition. *J Infect Dis.* 2009;200:236-243.
141. Kiepiela P, Ngumbela K, Thobakgale C, et al. CD8+ T-cell responses to different HIV proteins have discordant associations with viral load. *Nat Med.* 2007;13:46-53.
142. Yang OO, Kalams SA, Trocha A, et al. Suppression of human immunodeficiency virus type 1 replication by CD8+ cells: evidence for HLA class I-restricted triggering of cytolytic and noncytolytic mechanisms. *J Virol.* 1997;71:3120-3128.
143. Migueles SA, Mendoza D, Zimmerman MG, et al. CD8(+) T-cell cytotoxic capacity associated with human immunodeficiency virus-1 control can Be mediated through various epitopes and human leukocyte antigen types. *EBioMedicine.* 2015;2:46-58.
144. Ferguson AL, Mann JK, Omarjee S, Ndung'u T, Walker BD, Chakraborty AK. Translating HIV sequences into quantitative fitness landscapes predicts viral vulnerabilities for rational immunogen design. *Immunity.* 2013;38:606-617.
145. Gaiha GD, Rossin EJ, Urbach J, et al. Structural topology defines protective CD8+ T cell epitopes in the HIV proteome. *Science.* 2019;364:480-484.
146. Jiang C, Lian X, Gao C, et al. Distinct viral reservoirs in individuals with spontaneous control of HIV-1. *Nature.* 2020;585:261-267.
147. Soghoian DZ, Streeck H. Cytolytic CD4+ T cells in viral immunity. *Expert Rev Vaccines.* 2010;9:1453-1463.
148. Rosenberg ES, Altfeld M, Poon SH, et al. Immune control of HIV-1 after early treatment of acute infection. *Nature.* 2000;407:523-526.
149. Rosenberg ES, Billingsley JM, Caliendo AM, et al. Vigorous HIV-1-specific CD4+ T cell responses associated with control of viremia. *Science.* 1997;278:1447-1450.
150. Douek DC, Brenchley JM, Betts MR, et al. HIV preferentially infects HIV-specific CD4+ T cells. *Nature.* 2002;417:95-98.
151. Rerks-Ngarm S, Pitisuttithum P, Nitayaphan S, et al. Vaccination with ALVAC and AIDSVAX to prevent HIV-1 infection in Thailand. *N Engl J Med.* 2009;361:2209-2220.
152. Ranasinghe S, Flanders M, Cutler S, et al. HIV-specific CD4 T cell responses to different viral proteins have discordant associations with viral load and clinical outcome. *J Virol.* 2012;86:277-283.
153. Soghoian DZ, Jessen H, Flanders M, et al. HIV-specific cytolytic CD4 T cell responses during acute HIV infection predict disease outcome. *Sci Transl Med.* 2012;4:123ra25.
154. Kaufmann DE, Kavanagh DG, Pereyra F, et al. Upregulation of CTLA-4 by HIV-specific CD4+ T cells correlates with disease progression and defines a reversible immune dysfunction. *Nat Immunol.* 2007;8:1246-1254.
155. Younes S-A, Yassine-Diab B, Dumont AR, et al. HIV-1 viremia prevents the establishment of interleukin 2–producing HIV-specific memory CD4+ T cells endowed with proliferative capacity. *J Exp Med.* 2003;198:1909-1922.
156. Pitcher CJ, Quittner C, Peterson DM, et al. HIV-1-specific CD4+ T cells are detectable in most individuals with active HIV-1 infection, but decline with prolonged viral suppression. *Nat Med.* 1999;5:518-525.
157. Morou A, Brunet-Ratnasingham E, Dubé M, et al. Altered differentiation is central to HIV-specific CD4+ T cell dysfunction in progressive disease. *Nat Immunol.* 2019;20:1059-1070.
158. Yue FY, Lo C, Sakhdari A, et al. HIV-specific IL-21 producing CD4+ T cells are induced in acute and chronic progressive HIV infection and are associated with relative viral control. *J Immunol.* 2010;185:498-506.
159. Wei X, Decker JM, Wang S, et al. Antibody neutralization and escape by HIV-1. *Nature.* 2003;422:307-312.
160. Stamatatos L, Morris L, Burton DR, Mascola JR. Neutralizing antibodies generated during natural HIV-1 infection: good news for an HIV-1 vaccine? *Nat Med.* 2009;15:866-870.
161. Gray ES, Madiga MC, Moore PL, et al. Broad HIV-1 neutralization mediated by plasma antibodies against the gp41 membrane proximal external region. *J Virol.* 2009;83(21):11265-11274. https://jvi.asm.org/content/jvi/early/2009/08/19/JVI.01359-09.full.pdf
162. Mascola JR, Lewis MG, Stiegler G, et al. Protection of macaques against pathogenic simian/human immunodeficiency virus 89.6 PD by passive transfer of neutralizing antibodies. *J Virol.* 1999;73:4009-4018.
163. Shibata R, Igarashi T, Haigwood N, et al. Neutralizing antibody directed against the HIV–1 envelope glycoprotein can completely block HIV–1/SIV chimeric virus infections of macaque monkeys. *Nat Med.* 1999;5:204-210.
164. Shingai M, Donau OK, Plishka RJ, et al. Passive transfer of modest titers of potent and broadly neutralizing anti-HIV monoclonal antibodies block SHIV infection in macaques. *J Exp Med.* 2014;211:2061-2074.
165. Walker LM, Phogat SK, Chan-Hui P-Y, et al. Broad and potent neutralizing antibodies from an African donor reveal a new HIV-1 vaccine target. *Science.* 2009;326:285-289.
166. Walker LM, Huber M, Doores KJ, et al. Broad neutralization coverage of HIV by multiple highly potent antibodies. *Nature.* 2011;477:466-470.
167. Sievers SA, Scharf L, West AP Jr, Bjorkman PJ. Antibody engineering for increased potency, breadth and half-life. *Curr Opin HIV AIDS.* 2015;10:151.
168. Hessell AJ, Poignard P, Hunter M, et al. Effective, low-titer antibody protection against low-dose repeated mucosal SHIV challenge in macaques. *Nat Med.* 2009;15:951-954.
169. Barouch DH, Tomaka FL, Wegmann F, et al. Evaluation of a mosaic HIV-1 vaccine in a multicentre, randomised, double-blind, placebo-controlled, phase 1/2a clinical trial (APPROACH) and in rhesus monkeys (NHP 13-19). *Lancet.* 2018;392:232-243.
170. Altfeld M, Fadda L, Frleta D, Bhardwaj N. DCs and NK cells: critical effectors in the immune response to HIV-1. *Nat Rev Immunol.* 2011;11:176-186.
171. Donaghy H, Pozniak A, Gazzard B, et al. Loss of blood CD11c+ myeloid and CD11c- plasmacytoid dendritic cells in patients with HIV-1 infection correlates with HIV-1 RNA virus load. *Blood.* 2001;98(8):2574-2576.
172. Stetson DB, Ko JS, Heidmann T, Medzhitov R. Trex1 prevents cell-intrinsic initiation of autoimmunity. *Cell.* 2008;134:587-598.
173. Yan N, Regalado-Magdos AD, Stiggelbout B, Lee-Kirsch MA, Lieberman J. The cytosolic exonuclease TREX1 inhibits the innate immune response to human immunodeficiency virus type 1. *Nat Immunol.* 2010;11:1005-1013.
174. Kwon DS, Gregorio G, Bitton N, Hendrickson WA, Littman DR. DC-SIGN-mediated internalization of HIV is required for trans-enhancement of T cell infection. *Immunity.* 2002;16:135-144.
175. Geijtenbeek TBH, Kwon DS, Torensma R, et al. DC-SIGN, a dendritic cell–specific HIV-1-Binding protein that enhances trans-infection of T cells. *Cell.* 2000;100:587-597.
176. Sabatté J, Ceballos A, Raiden S, et al. Human seminal plasma abrogates the capture and transmission of human immunodeficiency virus type 1 to CD4+ T cells mediated by DC-SIGN. *J Virol.* 2007;81:13723-13734.
177. Meier A, Alter G, Frahm N, et al. MyD88-dependent immune activation mediated by human immunodeficiency virus type 1-encoded Toll-like receptor ligands. *J Virol.* 2007;81:8180-8191.
178. Beignon A-S, McKenna K, Skoberne M, et al. Endocytosis of HIV-1 activates plasmacytoid dendritic cells via Toll-like receptor–viral RNA interactions. *J Clin Invest.* 2005;115:3265-3275.
179. Manel N, Hogstad B, Wang Y, Levy DE, Unutmaz D, Littman DR. A cryptic sensor for HIV-1 activates antiviral innate immunity in dendritic cells. *Nature.* 2010;467:214-217.
180. O'Brien M, Manches O, Sabado RL, et al. Spatiotemporal trafficking of HIV in human plasmacytoid dendritic cells defines a persistently IFN-α–producing and partially matured phenotype. *J Clin Invest.* 2011;121:1088-1101.
181. Yonkers NL, Rodriguez B, Asaad R, Lederman MM, Anthony DD. Systemic immune activation in HIV infection is associated with decreased MDC responsiveness to TLR ligand and inability to activate naive CD4 T-cells. *PLoS One.* 2011;6:e23884.
182. Martinson JA, Roman-Gonzalez A, Tenorio AR, et al. Dendritic cells from HIV-1 infected individuals are less responsive to toll-like receptor (TLR) ligands. *Cell Immunol.* 2007;250:75-84.
183. Sabado RL, O'Brien M, Subedi A, et al. Evidence of dysregulation of dendritic cells in primary HIV infection. *Blood.* 2010;116:3839-3852.
184. Smed-Sörensen A, Loré K, Vasudevan J, et al. Differential susceptibility to human immunodeficiency virus type 1 infection of myeloid and plasmacytoid dendritic cells. *J Virol.* 2005;79:8861-8869.
185. Martin-Gayo E, Buzon MJ, Ouyang Z, et al. Potent cell-intrinsic immune responses in dendritic cells facilitate HIV-1-Specific T cell immunity in HIV-1 elite controllers. *PLoS Pathog.* 2015;11:e1004930.
186. Martin-Gayo E, Cole MB, Kolb KE, et al. A reproducibility-based computational framework identifies an inducible, enhanced antiviral state in dendritic cells from HIV-1 elite controllers. *Genome Biol.* 2018;19:10.
187. Moore CB, John M, James IR, Christiansen FT, Witt CS, Mallal SA. Evidence of HIV-1 adaptation to HLA-restricted immune responses at a population level. *Science.* 2002;296:1439-1443.

188. Allen TM, Altfeld M, Yu XG, et al. Selection, transmission, and reversion of an antigen-processing cytotoxic T-lymphocyte escape mutation in human immunodeficiency virus type 1 infection. *J Virol.* 2004;78:7069-7078.
189. Brumme ZL, Brumme CJ, Heckerman D, et al. Evidence of differential HLA class I-mediated viral evolution in functional and accessory/regulatory genes of HIV-1. *PLoS Pathog.* 2007;3:e94.
190. Goulder PJ, Brander C, Tang Y, et al. Evolution and transmission of stable CTL escape mutations in HIV infection. *Nature.* 2001;412:334-338.
191. Fischer W, Ganusov VV, Giorgi EE, et al. Transmission of single HIV-1 genomes and dynamics of early immune escape revealed by ultra-deep sequencing. *PLoS One.* 2010;5:e12303.
192. Brumme ZL, John M, Carlson JM, et al. HLA-associated immune escape pathways in HIV-1 subtype B Gag, Pol and Nef proteins. *PLoS One.* 2009;4:e6687.
193. Bronke C, Almeida C-AM, McKinnon E, et al. HIV escape mutations occur preferentially at HLA-binding sites of CD8 T-cell epitopes. *AIDS.* 2013;27:899-905.
194. Draenert R, Le Gall S, Pfafferott KJ, et al. Immune selection for altered antigen processing leads to cytotoxic T lymphocyte escape in chronic HIV-1 infection. *J Exp Med.* 2004;199:905-915.
195. Dahirel V, Shekhar K, Pereyra F, et al. Coordinate linkage of HIV evolution reveals regions of immunological vulnerability. *Proc Natl Acad Sci U S A.* 2011;108:11530-11535.
196. Miura T, Brumme ZL, Brockman MA, et al. Impaired replication capacity of acute/early viruses in persons who become HIV controllers. *J Virol.* 2010;84:7581-7591.
197. Crawford H, Lumm W, Leslie A, et al. Evolution of HLA-B* 5703 HIV-1 escape mutations in HLA-B* 5703–positive individuals and their transmission recipients. *J Exp Med.* 2009;206:909-921.
198. Kawashima Y, Pfafferott K, Frater J, et al. Adaptation of HIV-1 to human leukocyte antigen class I. *Nature.* 2009;458:641-645.
199. Fellay J, Shianna KV, Ge D, et al. A whole-genome association study of major determinants for host control of HIV-1. *Science.* 2007;317:944-947.
200. Dalmasso C, Carpentier W, Meyer L, et al. Distinct genetic loci control plasma HIV-RNA and cellular HIV-DNA levels in HIV-1 infection: the ANRS Genome Wide Association 01 study. *PLoS One.* 2008;3:e3907.
201. Fellay J, Ge D, Shianna KV, et al. Common genetic variation and the control of HIV-1 in humans. *PLoS Genet.* 2009;5:e1000791.
202. Kulkarni S, Savan R, Qi Y, et al. Differential microRNA regulation of HLA-C expression and its association with HIV control. *Nature.* 2011;472:495-498.
203. Cohen GB, Gandhi RT, Davis DM, et al. The selective downregulation of class I major histocompatibility complex proteins by HIV-1 protects HIV-infected cells from NK cells. *Immunity.* 1999;10:661-671.
204. Martin MP, Naranbhai V, Shea PR, et al. Killer cell immunoglobulin–like receptor 3DL1 variation modifies HLA-B*57 protection against HIV-1. *J Clin Invest.* 2018;128:1903-1912.
205. Stover J, Bollinger L, Hecht R, Williams C, Roca E. The impact of an AIDS vaccine in developing countries: a new model and initial results. *Health Aff (Millwood).* 2007;26, 1147-1158.
206. Berman PW, Gregory TJ, Riddle L, et al. Protection of chimpanzees from infection by HIV-1 after vaccination with recombinant glycoprotein gp120 but not gp160. *Nature.* 1990;345:622-625.
207. Check E. Back to "plan A". *Nature.* 1990;423:912-914.
208. Flynn NM, Forthal DN, Harro CD, et al; rgp120 HIV Vaccine Study Group. Placebo-controlled phase 3 trial of a recombinant glycoprotein 120 vaccine to prevent HIV-1 infection. *J Infect Dis.* 2005;191:654-665.
209. Gupta SB, Jacobson LP, Margolick JB, et al. Estimating the benefit of an HIV-1 vaccine that reduces viral load set point. *J Infect Dis.* 2007;195:546-550.
210. McElrath MJ, De Rosa SC, Moodie Z, et al. HIV-1 vaccine-induced immunity in the test-of-concept Step Study: a case-cohort analysis. *Lancet.* 2008;372:1894-1905.
211. Buchbinder SP, Mehrotra DV, Duerr A, et al. Efficacy assessment of a cell-mediated immunity HIV-1 vaccine (the Step Study): a double-blind, randomised, placebo-controlled, test-of-concept trial. *Lancet.* 2008;372:1881-1893.
212. Hansen SG, Wu HL, Burwitz BJ, et al. Broadly targeted CD8+ T cell responses restricted by major histocompatibility complex. *Science.* 2016;351:714-720.
213. Hansen SG, Sacha JB, Hughes CM, et al. Cytomegalovirus vectors violate CD8+ T cell epitope recognition paradigms. *Science.* 2013;340:1237874.
214. Hansen SG, Marshall EE, Malouli D, et al. A live-attenuated RhCMV/SIV vaccine shows long-term efficacy against heterologous SIV challenge. *Sci Transl Med.* 2019;11(501):eaaw2607. doi:10.1126/scitranslmed.aaw2607
215. Malouli D, Hansen SG, Hancock MH, et al. Cytomegaloviral determinants of CD8+ T cell programming and RhCMV/SIV vaccine efficacy. *Sci Immunol.* 2021;6(57):eabg5413. doi:10.1126/sciimmunol.abg5413
216. Walters LC, Harlos K, Brackenridge S, et al. Pathogen-derived HLA-E bound epitopes reveal broad primary anchor pocket tolerability and conformationally malleable peptide binding. *Nat Commun.* 2018;9:3137.
217. Yang H, Rei M, Brackenridge S, et al. HLA-E–restricted, Gag-specific CD8+ T cells can suppress HIV-1 infection, offering vaccine opportunities. *Sci Immunol.* 2021;6(57):eabg1703. doi:10.1126/sciimmunol.abg1703
218. Desai M, Field N, Grant R, McCormack S. Recent advances in pre-exposure prophylaxis for HIV. *Br Med J.* 2017;359:j5011.
219. Siegfried N, van der Merwe L, Brocklehurst P, Sint TT. Antiretrovirals for reducing the risk of mother-to-child transmission of HIV infection. *Cochrane Database Syst Rev.* 2011;7:CD003510. doi:10.1002/14651858.CD003510.pub3
220. Grant RM, Lama JR, Anderson PL, et al. Preexposure chemoprophylaxis for HIV prevention in men who have sex with men. *N Engl J Med.* 2010;363:2587-2599.
221. McCormack S, Dunn DT, Desai M, et al. Pre-exposure prophylaxis to prevent the acquisition of HIV-1 infection (PROUD): effectiveness results from the pilot phase of a pragmatic open-label randomised trial. *Lancet.* 2016;387:53-60.
222. Abdool Karim Q, Abdool Karim SS, Frohlich JA, et al. Effectiveness and safety of tenofovir gel, an antiretroviral microbicide, for the prevention of HIV infection in women. *Science.* 2010;329:1168-1174.
223. Van Damme L, Corneli A, Ahmed K, et al. Preexposure prophylaxis for HIV infection among African women. *N Engl J Med.* 2012;367:411-422.
224. Marrazzo JM, Ramjee G, Richardson BA, et al. Tenofovir-based preexposure prophylaxis for HIV infection among African women. *N Engl J Med.* 2015;372:509-518.
225. Baeten JM, Palanee-Phillips T, Brown ER, et al. Use of a vaginal ring containing dapivirine for HIV-1 prevention in women. *N Engl J Med.* 2016;375:2121-2132.
226. Nel A, van Niekerk N, Kapiga S, et al. Safety and efficacy of a dapivirine vaginal ring for HIV prevention in women. *N Engl J Med.* 2016;375:2133-2143.

SECTION VIII

Immunological Disease States and Immunodeficiency

CHAPTER 46

Autoimmunity and Autoimmune Diseases

Gustaf Christoffersson, PhD • Ken T. Coppieters, PhD •
Matthias von Herrath, MD • Dirk Homann, MD, MA

INTRODUCTION

The concept of autoimmunity, that is, the phenomenon of immune reactivity directed against an organism's constituent organs, tissues, cells, and/or extracellular factors, comprises both physiological and pathophysiological components that attain practical relevance in the context of autoimmune diseases, a collection of over 80 disorders with circumscribed pathological features that are thought to result primarily from the induction and perpetuation of aberrant immune responses. The precise etiopathology for most human autoimmune diseases remains incompletely defined but appears to be promoted by the complex interplay of genetic predispositions, environmental insults such as viral infections, and, in an acknowledgment of the many questions yet to be answered, bad luck. However, a role for immune-mediated processes and pathologies has by now been documented in, at times, compelling detail and provides important direction for novel and improved diagnostic, prophylactic, and therapeutic modalities. Herein, we discuss the evolution of major conceptual and practical advances and challenges in autoimmunity and autoimmune diseases in general, and consider individual autoimmune disorders in regard to the contribution of genetic and environmental components, specific pathology and autoimmune features, experimental models, and ongoing research efforts as well as current and potential future therapeutics.

GENERAL CONCEPTS IN AUTOIMMUNITY

Immunity and Autoimmunity

> *"... weil, so schliesst er messerscharf,*
> *nicht sein kann, was nicht sein darf"*
>
> *(" ... for, he reasons pointedly,*
> *that which must not, can not be")*
>
> —Christian Morgenstern *"The Impossible Fact"* (1910)

Autoimmunity is "the Other"[1] of immunity. And in true dialectical fashion, immunity's inception as a scientific discipline encapsulates the conceptual problem that was to shape the immunological debates in the first 15 and last 50 years of the past century. As much as the prefix *auto* assigns a specific place apart from immunity proper, the idea of immunity is not conceivable, for better or worse, without the notion of the self.[2] Immunity's distinct association with the individual self stretches from its etymological roots in the Roman legal concept of an *individual's* exemption from duty, service, or tax, to its official induction into the canon of medical terminology (defined as "*idio*syncratic condition" in the 1878 edition of Littré's *Dictionnaire de Médecine*[3]) and beyond. However, with the dawn of a new century, in the wake of the seminal discoveries of immune protection by active immunization with attenuated pathogens[4] or passive transfer of convalescent serum[5] and the seemingly unstoppable success of the "New Immunology," the notion of the immunological self underwent a dramatic reconfiguration. The price for adaptive immunity, it appeared, was autoimmunity, a concept so problematic that its existence had to be relegated to the realm of the almost unspeakable, that is, the Greco-Latin neologism of a "horror autotoxicus."[6] Paradoxically, the time that witnessed the concept of "horror autotoxicus" acquire quasi-dogmatic status as the "law of immunity research"[7] also became what A. Silverstein has termed the "classical period" of autoimmunity research.[8] Though declared anathema, autoimmune phenomena were reported in a quick succession of widely publicized observations. In 1902, Portier and Richet reported the phenomenon of anaphylaxis (as opposed to *pro*phylaxis),[9] Maurice Arthus characterized the local inflammatory response that was to bear his name in 1903,[10] Donath and Landsteiner described the first human autoimmune disease (paroxysmal cold hemoglobinuria) in 1904,[11] and the term allergy ("altered reactivity") was coined by von Pirquet and Schick in their analysis of serum sickness in 1905.[12] The idea that tissue destruction may lead to expanded immunopathology (nowadays referred to as "determinant-spreading") was proposed by Weil and Braun in 1909: In the course of an infectious disease, "tissue alterations lead to generation of complement-binding factors that, [...] based on their affinity to bodily cells, themselves attack and damage the cells of the organism."[13] Even the concepts of organ-specificity,[14] immunoprivilege, and a breakdown of regulatory mechanisms as a cause for autoimmunity[7] were developed in the early days of autoimmunity research.

This period of extraordinary productivity was followed by an almost 40-year hiatus, "the dark ages of autoimmunity research."[8] The reasons for a generalized disinterest in autoimmune phenomena are manifold and include political reconfigurations after World War I, the death of both Ehrlich (1915) and Metchnikoff (1916), a misconception of "horror autotoxicus" as the immune system's *inability* to generate responses against "self," and a paradigm shift in the field of immunology in favor of immunochemical approaches.[8] The renaissance of autoimmunity research had to await observations about immunological tolerance of mice congenitally infected with lymphocytic choriomeningitis virus (LCMV),[15] description of tolerance in chimeric cattle twins,[16] and Peter Medawar's work on skin transplant rejection[17-20] as well as the integration of these findings into a conceptual framework of self and nonself as determinants for immunological reactivity.[21] Burnet subsequently developed and extended these ideas into the "clonal selection theory of antibody formation,"[22] thereby establishing the conceptual centrality of self, nonself, and immunological tolerance. Although the dogmatic reading of "horror autotoxicus" was still prevalent (Ernest Witebsky and Noel Rose delayed their seminal publication on the discovery of thyroid antibodies[23] for several years assuming an experimental error[8]), Ehrlich's original conception as "regulatory contrievances" that prevent autoimmunity was now validated within the context of the clonal selection theory.

Again, however, the usefulness of self and nonself as distinguishing parameters was challenged at the very time they began their rise to prominence. Ludwik Fleck, in his singular study "Genesis and Development of a Scientific Fact," questioned the capacity of an immune system that only interacts with structures that are strictly nonself: "... it is very doubtful whether an invasion in the old sense is possible, involving as it does an inference by completely foreign organisms in natural conditions. A completely foreign organism could find no receptors capable of reaction and thus could not generate a biological process."[24] This view is echoed and elaborated upon in the work of Jerne[25] and Coutinho,[26] and even the possibility of a beneficial role for autoimmune processes was postulated by concepts such as "physiologic autoimmunity,"[27] "positive autoimmunity,"[28] and "protective autoimmunity."[29] The importance of autoreactivity as an integral aspect of immunity is furthermore demonstrated by the process of "positive selection" of developing T cells in the thymus and the T cell–mediated destruction of transformed or infected tissues that is based on the recognition of "foreign" (eg, viral peptides or even "self" as in the case of some tumor-derived antigens) in the context of "self" (major histocompatibility complexes [MHCs]). In fact, events associated with "danger" or the preservation of "tissue integrity" rather than the discrimination between "self/nonself" have been postulated as a primary driving force that engages the immune system.[30,31] More recently, the notion that immune reactivity is generated by the introduction of strong antigenic discontinuities ("criterion of continuity") has been proposed and may offer an elegant solution to some of the conceptual problems outlined earlier.[32,33] Still, the idea of "recognition"[2,34,35] remains a central element common to all these models as the immune system's principal task is cast as a differentiation between organismal states (self vs nonself, safety vs danger, integrity vs damage, continuity vs discontinuity, etc). Such operational distinction is indeed a powerful tool to conceptualize, with both impressive success and some obvious shortcomings, the functions of the immune system. Yet it may also insinuate a proximity of logical and functional categories that can culminate in the postulate of a principal objective, a veritable *raison d'ètre* for the immune system, and thus reintroduces a tinge of teleology. Rather, we favor an evolutionary perspective that conceives of the adaptive immune system, devoid of a particular purpose,[36] as "the cause of its own necessity."[37]

The notion of autoimmunity as an aberrant phenomenon has informed much of our current understanding about the immune system and its functions, and there appears to be a growing awareness that immunity and autoimmunity are both historically and conceptually inextricably intertwined. An emerging consensus indicates that the anthropomorphisms of "self" and "nonself" should be overcome and that autoimmunity is likely a universal phenomenon in the evolution of the vertebrate immune system. As part of the evolving organism, the immune system processes antigen stimuli in a deterministic fashion restricted by genetics, previous antigenic experience of the host, nature of the antigen, and the conditions of its presentation.[38] However, imbuing the immune system's function with an overriding purpose, no matter how important for our conceptualization and experimentation, has to consider that evolution is ignorant to teleology. In this respect, the remark by the Darwinist Paul Ehrlich that production of autoantibodies is "dysteleological in the extreme"[6] may be extended to the functionality of the immune system as a whole: there is no teleology in autoimmunity nor immunity, just the workings of a complex system under evolutionary constraints. The rules that inform immunity are the same ones that govern autoimmunity; random generation of antigen receptors for adaptive immunity necessitates a tolerance-induction process and the threat of autoimmunity.

The Burden of Autoimmune Diseases

The existence of autoimmune diseases in humans has been known for more than 100 years. By now, autoimmune pathogenesis has been attributed to more than 80 human diseases,[39] yet it is still far from clear which features can conclusively prove an underlying autoimmune pathogenesis. It has been suggested, somewhat provocatively, that with knowledge about an infectious origin, diseases are called immunopathologically mediated, whereas lack of such knowledge results in reference to such diseases as autoimmune.[40] While this argument is akin to the medical taxonomy where diseases of unknown origin are assigned to the domain of the "endogenous," "idiotypic," "essential," or "primary," a "positive" definition for autoimmune diseases is very much needed to provide a specific diagnostic framework that allows for unequivocal identification of distinct autoimmune disorders, yet remains flexible enough to accommodate new insights in etiological and symptomatological processes. A first attempt to provide such a basis for the establishment of the autoimmune origin of human diseases was formulated by Witebsky and associates who modeled their postulates on those of Koch: recognition of an autoimmune response (autoantibody or cell-mediated), identification of a corresponding autoantigen, and induction of an analogous autoimmune response and disease in experimental animals.[41] An update for these criteria has been proposed by Rose and Bona who suggested a combination of direct evidence (transfer of pathogenic antibodies or T cells), indirect evidence (reproduction of disease in experimental animals), and circumstantial evidence (clinical clues) to determine an underlying autoimmune etiology for human diseases.[42] However, it is important to note that any specific guidelines have to be tailored to individual autoimmune disorders. An example for a catalog of diagnostic criteria to be evaluated in a scoring system for identification of patients with a specific autoimmune disease is the report of the International Autoimmune Hepatitis Group.[43] This report also illustrates the importance of distinguishing between an autoimmune and infectious origin for hepatitis[43]: Immunosuppressive therapy has a beneficial effect on the course of autoimmune hepatitis (AIH); responsiveness to such therapy is in fact one of the diagnostic criteria for AIH but may be detrimental when employed for treatment of virus-induced hepatitis.

As a first of its kind, a meta-study by Jacobson et al provided the comprehensive evaluation of prevalence and incidence studies conducted for 24 autoimmune diseases between 1965 and 1995.[44] Overall, 1 in 31 Americans (~8.5 million people, 3.2% prevalence) were estimated to be afflicted by autoimmune diseases, and the most common disorders, divided into organ-specific and systemic conditions, and ranked in order of prevalence, are listed in Table 46.1. An update on these figures came in a 2012 study by Hayter and Cook which estimated the US prevalence to be about 14.7 million people (4.6%). Other epidemiological studies have provided even higher estimates for the contemporary burden of autoimmune diseases. Eaton et al, using national hospitalization registry data in Denmark from 1977 to 2001, arrived at a prevalence of 5.3%,[45] and Cooper et al calculated a global prevalence rate of 7.6% to 9.4% for the period from

TABLE 46.1 Prevalence of Autoimmune Diseases

Autoimmune Disease	Organ	Weighted Mean Prevalence Rate/100,000	Weighted Mean Incidence Rate/100,000
Thyroiditis/hypothyroidism	**Thyroid**	**1323.8**	**21.8**
Graves disease/hyperthyroidism	**Thyroid**	**1151.5**	**13.9**
Rheumatoid arthritis	Joints, lung, heart, other	860.0	23.7
Vitiligo	**Skin**	**400.2**	
Type 1 diabetes	**Pancreatic B cells**	**192.0**	**12.2**
Pernicious anemia	**Stomach**	**150.9**	—[a]
Multiple sclerosis	**Brain/spinal cord**	**58.3**	**3.2**
Glomerulonephritis (primary)	**Kidney**	**40.0**	**3.6**
Systemic lupus erythematosus	Skin, joints, kidney, brain, lung, heart, other	23.8	7.3
Glomerulonephritis (immunoglobulin A)	**Kidney**	**23.2**	**2.4**
Sjögren syndrome		14.4	—[a]
Addison disease	**Adrenal**	**5.0**	—[a]
Myasthenia gravis	**Muscle**	**5.1**	**0.4**
Polymyositis/dermatomyositis	Muscle, lung, heart, joints, other	5.1	1.8
Scleroderma	Skin	4.4	0.8
Primary biliary cirrhosis	**Liver bile ducts**	**3.5**	**0.9**
Uveitis	**Eye**	**1.7**	**18.9**
Chronic active hepatitis	**Liver**	**0.4**	**0.7**

Organ-specific autoimmune diseases in **bold**, systemic autoimmune diseases are in regular type face.
The prevalence/incidence rate from each study within a disease category contributed proportionately to the mean prevalence/incidence rate based on the population size of that study. The proportion or weight was calculated by dividing the study population denominator by the total of all the study population denominators for each disease.
For details see Jacobson et al.
[a]No studies on disease incidence available.
Data taken from Jacobson et al.[44]

1989 to 2008.[46] Though clearly only approximations, it therefore appears that autoimmune diseases are much more frequent than previously thought. Walsh and Rau approached this subject in a different manner by determining the relative ranking of autoimmune diseases in terms of mortality risk among women under the ages of 65. Remarkably, the collection of 24 autoimmune diseases specified by Jacobson et al ranked within the top 10 causes of death.[47] Thus, autoimmune conditions not only decrease the quality of life among afflicted individuals, but due to their high prevalence constitute a major public health burden. In addition, a trend toward rising incidence rates among most autoimmune disease has been noticed over the past few decades. For example, the incidence of multiple sclerosis (MS) in Italy has doubled between 1981 and 2002,[48] and in the United States, the incidence of celiac disease increased fivefold in 15 years.[49] These trends are not likely to abate, and a recent overview of childhood type 1 diabetes (T1D) in Europe found a 3.4% annual increase in incidence rate, which would result in a doubling of incidence by 2038. The economic challenges associated with these developments are indeed staggering as shown by the health care–related expenses in the United States and provided by the American Autoimmune Related Diseases Association (AARDA) for Crohn disease ($10.9-15.5 billion in 2008), rheumatoid arthritis (RA; $19.3 billion in 2005), or psoriasis ($11.2 in 2010). Treatment costs are expected to increase even further due to the nature of some of the more successful therapeutic modalities developed for chronic autoimmune conditions, for example, the introduction of anti-TNF and other biologicals. While clearly an important advance and of great benefit to patients, these drugs do not promote a cure for the underlying disease, and the need for continuous treatment will exacerbate associated health care costs.

The above epidemiological studies also permit several additional, if not entirely unexpected conclusions. Many autoimmune conditions are clearly understudied, and some of the most frequently studied diseases exhibit comparatively low prevalence rates. The cause for the seeming imbalance between the public health burden posed by some autoimmune disorders and their attraction as objects for scientific study remains to be elucidated, but will likely include the presence or absence of effective therapy. Pernicious anemia, the sixth most common autoimmune disease in the United States, can be effectively managed, and therefore elicits only limited epidemiological interest. In contrast, some rare conditions may pose a pronounced burden to afflicted individuals and thus warrant continued efforts to develop more effective prophylactic and therapeutic interventions. Furthermore, the availability of certain models for autoimmune diseases, again not necessarily a reflection of the epidemiological importance of the corresponding human autoimmune disease, will have an impact on choices made by researchers charting their field of study. Additionally, as in other areas of research or clinical medicine, the funds and resources available are the result of multiple factors that may or may not include the public health burden exerted by a particular autoimmune disease. Balancing these aspects to appropriately appreciate and address the burden of autoimmune diseases, based on both the afflicted individual and society at large, is a challenge that will require our continued efforts to identify, investigate, inform, and, hopefully, improve the therapies for many autoimmune diseases.

Spectra and Continua: Organ-Specific and Systemic Autoimmune Disorders, Autoinflammatory Diseases, and the Challenges of Taxonomy

A perennial approach in our quest to make sense of the complex phenomena we encounter is the establishment of dichotomies, however fraught with shortcomings, inconsistencies, and exceptions to the rule. Steeped in clinical traditions and immediately intelligible, the distinction between systemic and organ-specific autoimmune diseases is as useful as it is inadequate. Given that our evolving understanding of autoimmune diseases requires a constant reevaluation of our concepts pertaining to etiopathogenesis and effective treatment modalities, it would be premature to abandon such a simple and still useful classification. Rather, that porous juncture between systemic and organ-specific disorders may reveal hitherto unappreciated aspects of pathogenesis. On the surface, the patterns of pathology result from the distribution of anatomical niches that provide a suitable environment to "interface" antigens and immune effectors. Leaving for the moment aside the difficulties pertaining to the identification of initiating autoantigens in many human autoimmune diseases and the challenging task to correlate markers of immunological activity (eg, autoantibodies) with cause or consequence of tissue destruction, a particularly puzzling phenomenon is the seeming organ-specificity of some disorders in the face of autoimmune responses that target ubiquitous antigens. For instance, the ribonucleoprotein (RNP) antigens implicated in Sjögren syndrome or the transfer RNA synthetases targeted in polymyositis are widely expressed intracellular antigens, yet the pathology of these diseases is relatively circumscribed. Another intriguing example is the K/BxN arthritis model in which pathogenic antibodies recognize the ubiquitous cytoplasmic enzyme glucose-6-phosphate isomerase (GPI). Here, the preferential involvement of the joints apparently results from unique properties of the regional vasculature that allow for an antibody-mediated increase of vasopermeability and amplification of pathology by extracellular GPI deposition in the articular cavities.[50,51] The observation that autoimmune damage is critically dependent on aspects of the local microanatomy emphasizes the importance to consider autoimmune processes in the larger context of interdependent organ systems.

In addition, an examination of some animal models used for the study of particular organ-specific autoimmune disorders further challenges the simple notion of restricted pathology and may provide clues about etiological commonalities of ostensibly disparate clinical autoimmune syndromes. The nonobese diabetic (NOD) mouse is the most widely used animal model for the study of T1D, a severe condition caused by autoimmune destruction of the insulin-producing beta cells in the pancreas.[52,53] However, NOD mice also exhibit aspects of type 2 diabetes[54] and are prone to autoimmune sialitis, thyroiditis, peripheral neuropathy, prostatitis, a lupus-erythematosus-like syndrome that develops after exposure to killed mycobacteria, and, under certain circumstances, exocrine pancreatitis.[53] Similar to the etiology of T1D, specific T cells are involved in the pathogenesis of all these disorders, although antigenic targets and requirements for costimulatory interactions are distinct. Thus, as in human T1D, the NOD mouse combines a generalized genetic susceptibility to multiorgan autoimmunity that is focused on pancreatic beta cells but not limited to endocrine organs.

While the preceding considerations argue against a rigorous opposition of systemic versus organ-specific autoimmune diseases and rather support the notion of a spectrum of clinico-pathological features that variously emphasize prominent organ-specific and systemic features, even the conceptual distinction of autoimmune disorders at large is suffused with certain limitations (eg, absence of MHC or autoantibody associations with some diseases tentatively labeled "autoimmune") that have prompted new taxonomic proposals for immunological diseases in general. Perhaps most prominently, McDermott and colleagues introduced the concept of autoinflammation defined as "self-directed tissue inflammation, where local factors at disease-prone sites determine the activation of the innate immune system."[55] Accordingly, a continuum of immunological diseases is demarcated by rare monogenic disorders that are purely autoimmune and demonstrate preferential involvement of adaptive immune responses (eg, autoimmune polyendocrine syndrome-1 resulting from mutations of the AIRE gene) and purely autoinflammatory conditions defined by mutations in cells or molecules of the innate immune system and localized pathologies that escape the traditional purview of "classical" autoimmune mechanisms (eg, tumor necrosis factor [TNF] receptor–associated periodic syndrome). In between these boundaries, the vast majority of other immunological disorders can be organized along a continuum of pathologies that range from classic polygenic autoimmune diseases to mixed pattern diseases and to polygenic autoinflammatory diseases.[55] The major appeal of this expanded notion of autoimmunity lies in the recognition of innate immune mechanisms as an important component of autoimmune disorders (with inevitable echoes of the Ehrlich-Metchnikoff rivalry informing the current discussions), and in the provision of an inclusive classification of immunological diseases.

Central and Peripheral Tolerance: Implementing an Operational Concept

A detailed historical discussion of the concept of tolerance is beyond the scope of this chapter, but some aspects of the usage of the term tolerance require clarification at the outset. Tolerance in adaptive immunity, *sensu stricto*, is the absence of specific lymphocyte activity, the consequence of physical deletion, or functional silencing of specific T and B cells. Some researchers refer to these tolerance mechanisms as "passive" or "recessive" tolerance to explicitly distinguish them from "active" or "dominant" tolerance. While the latter mechanisms constitute bona fide immune responses (therefore other researchers do not categorize them as a mode of tolerance), their particular nature results in a phenotype that is comparable to that achieved by means of passive/recessive tolerance. Distinct effector mechanisms (eg, immunosuppressive cytokines) and possibly dedicated classes of immune cells (eg, "T regulatory cells, T_{regs}") assure that local or systemic autoimmunity is avoided.

The concept of T cell suppressors, first proposed in the early 1970s by Gershon as CD8+ T cell suppressors,[56] was resurrected in the form of "professional" CD25+FoxP3+CD4+ and "adaptive" regulatory T cells (TR1 and other T_{regs}) and has since attracted considerable attention. However, while there is indeed a CD25+ lineage of T cells committed to regulatory activity in naïve, nonimmunized mice, we wish to underscore that regulatory functions, including those that limit autoimmunity, are a feature of the immune system as a whole and can be exercised by other classes of immune cells as well (eg, CD8+ T cells, γδTCR T cells, natural killer T cells). Thus, while CD25+CD4+ regulatory T cells occupy a distinct and important niche in the complex dynamic network of immune functions, not all T cell regulators are CD25+CD4+ nor do all CD25+CD4+ T cells function as suppressors. Indeed, novel markers might characterize regulatory function better; among these are expression of the transcription factor FoxP3[57-63] and secretion of cytokines with regulatory function such as interleukin (IL)-10[64,65] or tissue growth factor (TGF)-β.[66] The multiplicity of current efforts to understand the nature of CD25+CD4+ regulatory T cells has been expertly reviewed elsewhere,[67,68] and we note, in particular, more recent observations on the transcriptional, phenotypic, and functional instability of the T_{reg} population[69,70] (Chapter 30). This has important repercussions in the context of autoimmunity, as T_{regs} can lose their regulatory capacities when exposed to local conditions of autoimmune inflammation. On the one hand, these results suggest that aberrant conversion from regulatory to effector phenotype may partly underlie progression to overt autoimmune disease in susceptible individuals. On the other hand, such propensity may jeopardize the long-term efficacy of therapeutic strategies based on the induction or adoptive transfer of autoantigen-specific T_{reg}, as these cells could potentially convert to effectors once they arrive in the inflamed tissue. Therefore, these recent insights clearly warrant caution in regard to the popular endeavor of assigning T cells to defined subpopulations based on their "signature" transcription factors and functionalities and emphasize the importance of plasticity and mutability, the mechanistic foundation and relevance of which remains to be explored in further detail. Lastly, we emphasize that changes in T_{reg} functionality rather than a mere numerical decrease appear to constitute a correlate for some human autoimmune diseases.[71] Thus, the simple quantification of peripheral T_{reg} numbers as a measure of risk in autoimmunity-prone individuals or as a biomarker in clinical studies will have to be complemented with functional assays to generate relevant data.

T Cell Tolerance

Autoreactivity, by definition, designates a specific immune response to self-antigens. Antigen nonspecific responses such as inflammatory and innate immune processes should not be considered autoimmune in the strict sense, although they may accompany, enhance, or even trigger autoimmune processes proper. Thus, antigen-specific T- or B-cell immunity will have to underlie a genuine autoimmune disorder. Furthermore, for organ-specific autoimmune diseases, antigen-specificity of primary effector lymphocytes must be largely restricted to autoantigens derived from defined organs or tissues. Once initiated, specific responses that precipitate or "drive" the localized autoimmune reaction may diversify to comprise additional specificities ("determinant spreading") and pathogenic mechanisms.

How does the adaptive immune system restrict generation and activation of autoreactive lymphocytes? The central process by which the generation of T-cell receptor diversity is limited is called thymic selection. Thymic selection is a developmental process that selects T cells with a biased repertoire for export into the periphery.[72-76] T cells that interact at least weakly with self-peptides presented in the context of MHC molecules are chosen in the course of positive selection,[77,78] while those that do not effectively interact with MHC:peptide complexes die "by neglect." However, interactions above a certain avidity threshold result in elimination by negative selection and constitute the basis for "central tolerance"[72,76] (Chapters 28 and 33). Thus, central tolerance prevents widespread autoimmunity as a function of lymphocyte/antigen-complex avidity and preferentially selects T cells with specificity for antigens not expressed in thymic epithelium for export into the periphery. However, central tolerance is not a complete mechanism, and a sizable pool of T cells with intermediate avidity can escape negative selection and constitutes the majority of autoreactive T cells found in the peripheral immune system. The presence of these autoreactive T cells can be considered physiological and they are not noxious for two reasons: Either they are usually not activated and exhibit a "naïve" phenotype or they exhibit regulatory effector functions and act as adaptive T_{regs} following activation. Thus, not all self-reactive lymphocytes need to necessarily exhibit an aggressive phenotype. Depending on their specific effector functions, autoreactive T cells may exhibit regulatory functions and may critically modulate or even abort local autoimmune processes. Such autoreactive regulators might occur physiologically and constitute the majority of autoimmune responses present in healthy individuals. Only an encounter under appropriate stimulatory conditions (ie, presentation of autoantigen-derived peptides presented in the context of MHC class I or II molecules accompanied by antigen-nonspecific costimulatory interactions and strong inflammatory signals) can lead to their full activation and detrimental effector functions in the periphery. As such "armed effectors," autoreactive T cells are now potentially very dangerous and may initiate specific autoimmunity, if they recognize the autoantigens or closely related proteins in a defined tissue. It is thought that a few autoaggressive "driver clones" with highly detrimental effector function can sustain a localized autoimmune process. High receptor/MHC/self-peptide avidity likely but not necessarily predisposes to this phenotype.[79-81]

The presence of autoreactive T cells in the periphery might suggest that detrimental autoimmunity should occur quite frequently if organ-specific autoantigens are not expressed in the thymus, or alternatively or in addition, such physiologically occurring autoimmunity is not of a regulatory nature. Yet, there are several additional mechanisms that maintain tolerance in the periphery. "Peripheral tolerance" involves a set of mechanisms which ensures that autoreactive T lymphocytes are not activated in the periphery. It should be noted that these mechanisms pertain to both autoreactive and "hetero-reactive" T cells and involve the following pathways. First, it has been observed that naïve T cells triggered by a strong signal through the T cell receptor alone may lose the ability to proliferate, and some but not all effector functions become "anergic."[82-84] Presence of certain cytokines or costimulatory interactions can avoid the induction of anergy or may reverse an anergic state. Second, highly activated T cells will eventually undergo activation-induced cell death (AICD).[85] AICD is thought to be essential for the downmodulation of immune responses and the reestablishment of immune homeostasis. Impairment of AICD may lead to continued immune activation and generalized autoreactivity. For CD4 lymphocytes, AICD is FAS/FAS-L dependent[86,87]; it is not clear which interactions precisely control AICD in CD8 cells. Third, molecules that can deliver specific negative signals, such as the B7-binding CTLA-4, are involved in "turning off" of antigen-specific T cells.[88,89] Finally, other factors such as regulatory lymphocytes and regulatory antigen-presenting cells (APCs) might play important roles in maintaining peripheral tolerance.[56,67,90] Last, and maybe most important, it has become recently more apparent that tissue-specific factors decide, whether an organ-specific autoimmune disease develops or not.[91] Along with this notion comes the realization that autoreactivity (ie, the systemic presence of autoreactive T cells) is physiological[92] and that such cells to do harm have to be activated by tissue-specific local factors, as in the majority of cases organ-specific autoimmune disease do not develop together.[93]

B Cell Tolerance

Although they are not selected in the thymus, similar paradigms as those for T cells apply to autoreactive B lymphocytes. Clonal selection occurs after recognizing antigens, is avidity dependent, and allows the B cells to undergo further receptor editing. Both central (ie, bone marrow) as well as peripheral tolerance mechanisms are in place to avoid generating mature B cells with specificity for self-antigens. After B cells mature in the bone marrow, they clonally expand after recognizing antigens in the periphery. T-helper lymphocytes are needed for this process in response to most protein antigens, and these B cell responses are therefore termed "thymus-dependent" or, in other words, require T cells with specificity linked to an epitope on the antigen they are reacting with. T-helper cell–independent B cell responses occur mostly to bacterial and lipid antigens, for example, to lipopolysaccharide, and are therefore rarely autoaggressive. Thus, T cell tolerance controls directly B cell reactivity to autoantigens. In general, systemic autoimmune disorders such as systemic lupus erythematosus (SLE) are B cell–dependent and organ-specific diseases such as MS and T1D are less dependent on autoantibodies, although B cells can play important roles as APCs and antibodies can possibly enhance disease pathogenesis. In both, diabetes and MS, autoantibodies correlate with disease progression. In thyroiditis, autoantibodies are instrumental for causing disease and in AIH their role is thought to be crucial.

In conclusion, central (thymic) and peripheral tolerance mechanisms will effectively control the vast majority of autoreactive lymphocytes, which assures that autoaggressive immune responses are relatively rare (<5% overall population). Some of the considerations described in this paragraph are illustrated in Figure 46.1.

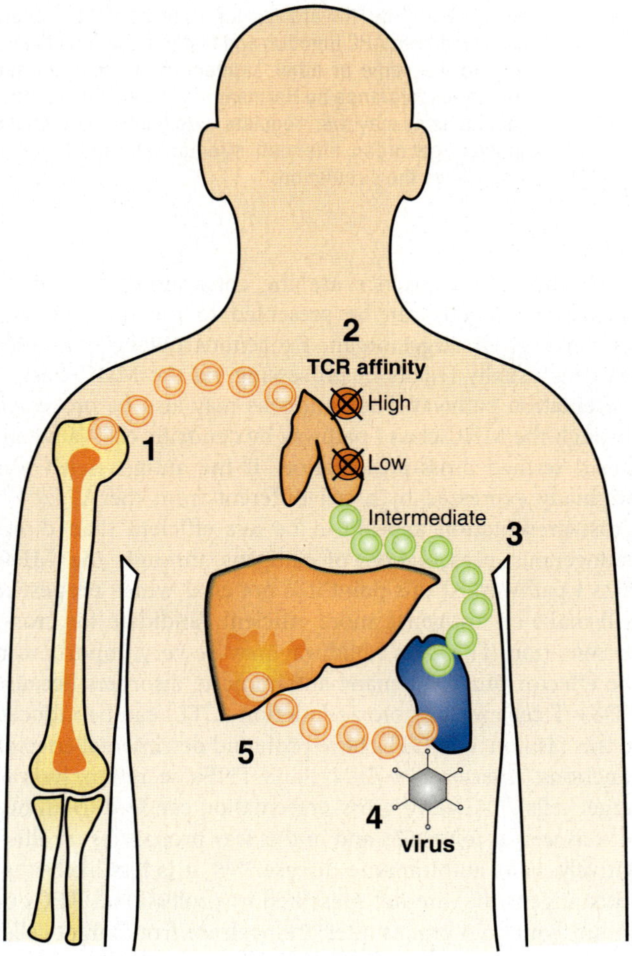

FIG. 46.1. Events important for the pathogenesis of autoimmunity. (1) Lymphocyte precursors migrate from the bone marrow to the thymus to undergo maturation. (2) In the thymus of susceptible individuals, autoreactive lymphocytes can escape thymic selection. Whereas cells with high or low T cell receptor (TCR) affinity are generally deleted, "intermediate" lymphocytes fail to undergo negative selection. (3) Such initially naïve cells migrate to peripheral organs such as the spleen and may remain unresponsive for many years. (4) Certain environmental triggers have the ability to activate these naïve cells by bystander activation, molecular mimicry, or antigen-presenting cell cross-presentation. (5) As a consequence, activated autoreactive lymphocytes migrate to target organs such as the liver and mediate inflammation and tissue damage, leading to clinical autoimmune disease symptoms.

Organ-Specific Tuning—Regulatory and Destructive Autoimmunity

As indicated previously, while autoimmune disorders must conform with a set of general criteria, organ-specific autoimmunity must be considered in the context of the target organ affected. Certain effector functions exerted by autoreactive lymphocytes will be detrimental only to particular cells or tissues. For example, the pancreatic β-cells are more sensitive to the damaging effect of inflammatory cytokines than neighboring α-cells (both cell types are part of the islets of Langerhans) and other cells in their vicinity (eg, fibroblasts or acinar tissue of the pancreas).[94] On the other hand, some organs provide a microenvironment that suppresses inflammatory responses. For example, both the gut and the central nervous system (CNS) contain relatively large amounts of TGF-β, which can have direct anti-inflammatory effector functions on T cells and APCs and activate T_{regs}, unless IL-6 is present.[66,95-98] Also, certain large organs such as the liver may better tolerize lymphocyte responses directed toward them, because they contain a high number of cells that are incapable of costimulation (eg, hepatocytes) and will, therefore, more likely shut down naïve autoreactive lymphocytes that stimulate them.[30] Finally, the precise activation state, phenotype, and effector function of an autoreactive cell are critical in determining their impact on tissues expressing their cognate antigen. Some molecules will exert beneficial functions, for example, IL-4 and probably IL-10 and TGF-β in T1D.[99-101] Antigen-specific adaptive and professional regulatory cells are likely participants of every local autoimmune process and can therefore be essential for delay or prevention of clinical disease altogether. While their induction is a clearly therapeutic goal, phenotype and mechanistic aspects essential for regulatory function are not yet understood in complete detail. Likely, professional and adaptive T_{regs} may exert their function by targeted alteration of APC function. Evidence even suggests that T_{regs} may safeguard against autoimmunity by specifically killing APC that present autoantigens to effector T cells.[102] They may also act as true bystander processors if their suppressive action via APC modulation is extended to aggressive immune responses regardless of specificity found in the microenvironment of the affected organ (Fig. 46.2). Indeed, such cells have been found in several animal models,[100,103-105] but their existence and function in humans is not well defined. Alternatively, regulatory lymphocytes may directly affect activated or naïve aggressive lymphocytes and induce anergy or apoptosis. Lastly, by changing the overall cytokine milieu of a given inflammatory process, the number and function of aggressive cells and cytokines/chemokines may be dampened in a localized area.[106] Probably the best understood balancing circuit employed by autoreactive adaptive T_{regs} are the Th1/Th2 paradigm[107-109] and signaling by IL-10[65] and/or TGF-β.[66] In light of these observations, autoreactivity is not necessarily detrimental and maybe even essential in the sense of T_{regs}.

Last, it might be difficult or even impossible to define a general phenotype of autoaggressive lymphocytes that will cause damage at any site or organ. Rather, autoaggression has to be defined in relation to the target organ or cell that is under attack, lymphocytes detrimental in one organ/disease

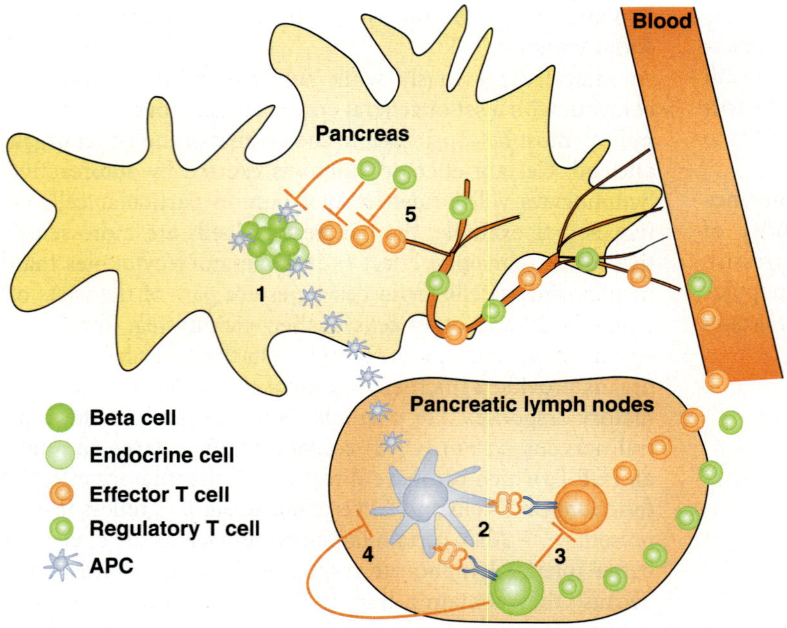

FIG. 46.2. The concepts of bystander suppression and infectious tolerance. (1) Autoantigens are taken up by professional antigen-presenting cells (APCs) in the target organ, here depicted as the pancreas in type 1 diabetes. Upon maturation, APC migrate to the draining lymph nodes. (2) In the lymph nodes, lymphocytes specific for distinct beta cell antigens are activated by the APC. (3) As a consequence, aggressive T cells directed toward one antigen could be regulated by T_{regs} specific for another autoantigen, a process termed "bystander suppression" by Howard Weiner. (4) In theory, T_{regs} could also modulate APC function and foster induction of more T_{regs} to the same or other antigens, a process termed infectious tolerance by Hermann Waldman. (5) Identical mechanisms may then regulate autoimmune responses in the target organ, after both effector cells and T_{regs} gain access via the circulation.

will not necessarily be detrimental to other organs, and ubiquitous autoimmunity will therefore occur very rarely.

Initiating Autoimmunity: Antigens, Genes, and Environment

Antigens

A central challenge in most human autoimmune disorders is the identification of an initiating autoantigen or autoantigens. For example, several candidate autoantigens have been identified in T1D. However, to date it remains unclear and controversial whether any are actually involved in the initiation of the disease. Similar situations exist for MS and arthritis. Thus, distinguishing features that constitute an ideal candidate autoantigen and parameters by which potential candidates can be searched for and identified as the initiating autoantigen for a given organ-specific disease must be determined with much greater precision to yield insights applicable to preventative or therapeutic interventions.

The concept that "processing of antigens determines the self" has been introduced by E. Sercarz.[110,111] Autoantigens that are not expressed in the thymus or are of cryptic nature[110,112,113] may encounter a more extensive T cell repertoire that is likely of higher avidity. Such antigens appear to be better targets. Low levels of thymic antigen expression or limited numbers of thymic medullary cells may lead to partial tolerance.[72,114] In addition, the precise timing of antigen expression during embryonic development may play a major role for the establishment of tolerance. Diversification of epitope specificity from the initial dominant epitope to subdominant or cryptic epitopes is known as epitope spreading. Antigens that are cryptic or not expressed during early embryonic development might become better targets later, because more autoreactive T cells will be present in the periphery. It should be considered that certain "embryonic antigens" are involved in initiating autoimmune diseases.[115]

Of further importance are the constraints with which particular antigens can be presented to the immune system in a given target organ. Exogenous uptake of soluble antigens usually results in processing via the MHC class II presentation pathway. Such antigens may also be processed through the MHC class I pathway by dendritic cells, a mechanism termed cross-presentation, if the antigen itself was originally expressed by a cell different from the APC.[116,117] Cross-presentation appears to be less efficient than direct endogenous presentation of antigens through the MHC class I pathway. At this point it is not clear which properties will make an antigen a more efficient candidate for cross-presentation. The class I pathway may be very important in the effector phase for many autoimmune disorders because CD8+ T cells with cytotoxic function (CTL) can be induced in this fashion, and CTLs have profound detrimental effector functions (interferon-γ [IFNγ] and TNFα secretion, lysis of target cells).[118] Clearly, cross-presentation can lead to immunity, as well as tolerance, and in this way propagates or, alternatively, halts autoimmune disease.[116,119] It is less likely that autoantigens that are not presented by professional APCs or within lymphoid organs after their release from target cells/organs can initiate autoimmunity. Only under special circumstances (ie, extremely high-density presentation of an antigen on a target cell or highly inflammatory conditions including chemotaxis) would this appear possible. This might occur during localized or systemic viral infections and could also be an important event in lymphocyte entry into target organs, especially if chemotaxis is involved and endothelial cells present autoantigens.[120] In addition, experienced or primed autoaggressive lymphocytes will be capable of becoming activated after seeing antigen in the absence of costimulation; thus accumulating a memory pool of autoaggressive T cells might be inherently dangerous to an organism.

Thus, the type of autoantigen will define its own potential "candidacy" in an autoimmune process. Autoantigens that are not expressed during development, or were cryptic for an extended period of time, as well as antigens secreted and cross-presented in lymphoid organs by APCs, appear to be better suited for assuming the role as primary culprits.

Genes

An abundance of empiric evidence indicates the association of many organ-specific autoimmune diseases with certain human leukocyte antigen (HLA) haplotypes as well as other susceptibility or protective genes. The principal specific genetic linkages will be discussed separately for each condition in the section Individual Autoimmune Disorders. In general, MHC class I (such as HLA B 27)[121,122] or class II (such as DR4) genes might predispose to a certain disease by enhanced presentation of pathogenic peptides in the periphery or inefficacious presentation of autoantigen-derived peptides in the thymus. For example, the human HLA-DQ8 molecule has a striking structural similarity to mouse I-A^{g7} class II that predisposes NOD mice to spontaneous autoimmune diabetes.[123,124] However, the link to disease appears not to be as simple as reasoned above, and more complex mechanisms might be in place. Presentation of the pathogenic peptide in gluten-induced celiac disease involves acidic modification of the protein to generate the peptide ultimately presented by MHC class II,[125] and other chemical modifications can be expected to alter peptide binding to MHC and the resulting conformation of the peptide/MHC complex.

Other genes that encode immunoregulatory or inflammatory proteins may be involved in the disease process and, finally, genes that support tissue or wound repair (eg, islet cell regeneration in T1D) may be of help in preventing disease development.[126] For most autoimmune disorders, the genetic links are complex, not absolute, and many susceptibility and resistance genes act in concert to modulate the clinical phenotype. Noteworthy, newly emerging themes in the genetics of autoimmunity are the fact that small nucleotide polymorphisms might result in alternate forms of the same gene and the fact that the character of the full array of MHC region genes is becoming understood.[127] A striking example for the former is the existence of an alternate spliced form of the inhibitor of cellular function, CTLA-4. Kuchroo and colleagues discovered that the presence of a secondary CTLA-4 lacking the B7.1/B7.2-binding motif can be instrumental in the genetic penetrance of diabetes pathogenesis.[128] An important area still largely unexplored is the nature of genes responsible for transcriptional control of other proteins. Experimental models indicate that transcription factors can have profound effects on the development of autoimmune disease, and variations in expression and activity levels may be found between susceptible and protected individuals. In summary, a complex interplay of many genes will predispose for a certain autoimmune disease, but the concordance of clinical manifestation is frequently not higher than 30% to 40% in monozygotic twins.[129-135] For this reason, other factors are to be considered in triggering or propagating the pathogenic autoimmune process.

Our understanding of the genetic component to autoimmunity has greatly benefited from the arrival of the first wave of so-called genome-wide association (GWAS) studies in the 2000s. Prior to the use of this technology, genetic contribution was analyzed by two different approaches, linkage analysis and association studies. Linkage analysis tests for correlation between a genomic region and phenotype in families, whereas association analysis searches for correlations in the entire population.[136] In a genetic association study of autoimmune disease, sequence variants are genotyped in patients and controls and the frequencies of these genotypes are compared. It is then evaluated whether certain alleles are significantly more associated with patients than with controls. Most autoimmune conditions are polygenic diseases, meaning that they are caused by (a range of) combinations of polymorphisms within different genes. As mentioned above, HLA-associated risk was discovered early in many autoimmune disorders and turned out to be well reproducible because of its highly significant association. The vast majority of common genetic variants in autoimmunity, however, display modest effects. Even extremely large linkage studies have usually been unsuccessful in discovering susceptibility genes for polygenic diseases that met stringent statistical thresholds. Likewise, most common variants require large association studies for reliable detection. Advances in the development of high-throughput genotyping, statistical correlation methods, and better insight into the organization of the genome offered the opportunity to design approaches for the unbiased interrogation of the entire genome,[137] and new insights will likely benefit from recent advances in sequencing technology and decreased sequencing costs. Sufficiently large case-control GWAS studies thus have the potential of identifying disease-associated mutations, even if the contribution of that polymorphism is minor. It should be noted here that GWAS studies often cannot unequivocally identify gene associations (at best "candidate genes"), as risk loci contain multiple genes and therefore follow-up fine-mapping or functional studies are generally required. Although GWAS studies have rapidly increased the list of known risk loci for many autoimmune disorders, some have questioned whether this knowledge substantially aids our understanding of their etiology and offers clues for rational intervention. It could be argued that all associations are often eclipsed by the major contributing factor, HLA polymorphisms, and essentially pinpoint genes with key roles in the immune system that could be expected to be involved to some degree. In T1D, for instance, minor associations with important genes encoding molecules involved in immunity such as CTLA-4 and IL-2 were discovered. From a therapeutic perspective, such findings may be hard to translate as, based on the relatively small association rates, these genes and their products may not necessarily represent pivotal targets in a diverse human patient population. In summary, it can be concluded that most likely all major susceptibility genes associated with the most common autoimmune disorders as discussed below are currently known. Future years will tell whether this comprehensive genetic dataset translates to a better understanding of the origins of autoimmunity and if this will open new avenues toward better treatments.

Environment

Viral infections: For many years viral infections have been discussed as potential candidates to trigger autoimmunity in susceptible individuals because of their capacity to directly infect target tissues and induce strong inflammatory responses and immune activation. While the association between viral infections and organ-specific autoimmune disorders is a very intriguing possibility, it has been exceedingly difficult to demonstrate a causative role for specific viruses in human autoimmune diseases. Among the many obstacles are the fact that (1) all individuals undergo a multitude of viral infections during their lifetime, (2) one has to assume that viruses are frequently cleared at the time of diagnosis and viral footprints can be difficult to find in individuals affected by an autoimmune disease ("hit and run" event), (3) the precise viral strain, infection kinetics, and number of T cells and type of effector functions induced may play an instrumental role in determining its effect in an individual genetically at risk, necessitating a very detailed immunological profiling,[138] (4) due to MHC polymorphism, there is a significant variation in specificity of the antiviral response, (5) viral infections might not, per se, trigger autoimmunity, but affect an ongoing autoimmune process in a detrimental way, and (6) there now is increasing evidence that viral and other[139] infections may rather prevent than enhance ongoing autoimmune responses, either by apoptosing aggressive T cells or by augmenting the T_{reg} pool.[140,141] Thus, a successful approach should be to first explore the underlying mechanisms in virally induced autoimmunity and then apply the precise insight and paradigms developed specifically to the human situation.

The most important mechanisms to be considered are summarized in Figure 46.3.

1. Molecular mimicry, which implies the cross-reactivity between viral and self-determinants as a principal cause or mechanism to enhance autoimmunity.[142,143]
2. Bystander activation, which postulates that APCs and autoreactive lymphocytes will become activated indirectly as a consequence of the cytokine/chemokine by virus infection of a particular organ.[144]
3. Virally induced determinant spreading, which involves the presentation of autoantigens (possibly previously cryptic) by virus-activated APCs.[145-147]

The experimental evidence for all of these three scenarios obtained in different mouse models *in vivo* is well documented. However, none of these have been proven for any human autoimmune disorder due to the large size of human trials and the invasive nature of *in vivo* diagnostics required at the present stage. Notable planned studies for this are one Norwegian approach where children with recent-onset T1D will be given antiviral drugs,[148] and a Finnish/Swedish project where a Coxsackie virus vaccine has been developed for trials in preventing T1D.[149] However, in the near future, we will continue to depend on animal models until noninvasive human *in vivo* diagnostic strategies have advanced and allow for imaging of trafficking of antigen-specific lymphocytes and high-resolution definition of immune process present in a specific organ. A final remark has to be made for the existence of a negative association between viral infections and autoimmune disease found in several experimental models.[150] These observations are in support of epidemiological findings that the incidence of many autoimmune disorders is decreased in equatorial countries, where the presence of certain infectious diseases is significantly increased. However, no firm associations have been established to date.

Other environmental causes: Similar to viral infections, other inflammatory stimuli may trigger or enhance autoimmunity. The gut deserves particular attention in this respect. At this site each individual harbors thousands of different bacterial strains and viral infections are common. Furthermore, the mucosal lining is permeable for nutrients and constitutes a very large interactive surface with the environment. Again, the complete absence of all bacteria results in severe immune dysfunction and possibly autoimmunity. However, it would be incorrect to conclude that infections are therefore always protective. Indeed, the commensal flora appears crucial in maintaining proper immune activation and function, but certain pathogens could definitely elicit strong gut immune responses that lead to autoimmune disease.[151,152]

Regulatory Networks in Autoimmune Processes

Cytokines

Cytokines and chemokines are essential regulators of cellular and humoral immune responses and lymphocyte trafficking.[153-155] They play a central role in orchestrating autoimmune processes and constitute a multitude of positive as well as negative feedback loops.[156] It is well-established that certain cytokines can negatively or positively influence the production of other cytokines (ie, the Th1/Th2/Th17 paradigm) and thus determine the balance between pro- and anti-inflammatory factors in the local environment. Furthermore, autocrine production feedback can augment or shut down production of a given cytokine by one cell. Cytokine networks operate with a fair amount of redundancy, and many cytokines and chemokines share common receptors. They are the most likely mediators of "bystander activation" and "suppression" processes and also offer an effective and versatile therapeutic target via the temporally restricted use of cytokine- or chemokine-blocking antibodies. Their precise function can vary quite dramatically in respect to the autoimmune disease under investigation and will be discussed in the disease-oriented portions of the section Individual Autoimmune Disorders. Furthermore, their level and timing of expression during an ongoing disease process will determine whether they have a positive or negative effect (or any at all; see below).

Many autoimmune disorders have traditionally been classified as "Th1-" or "Th2"-driven disorders based on, among other factors, the pathogenic role of cytokines such as IFNγ and IL-4.[107] Recent years have challenged this perspective with of the description the Th17 helper subset, characterized by the production of IL-17.[157-159] This class of T cells is considered highly pro-inflammatory and its discovery was found to be of importance in several autoimmune disorders in both animal models and humans.[160] One example of an autoimmune disease where Th17 cells have taken center stage is RA. The disease's principal animal model was long referred to as a Th1-driven, but the field struggled with the observation that mice treated with neutralizing antibodies to IFNγ and IFNγ receptor–deficient mice developed more severe arthritis.[161,162] Dominance of Th17-directed immunity was put forward

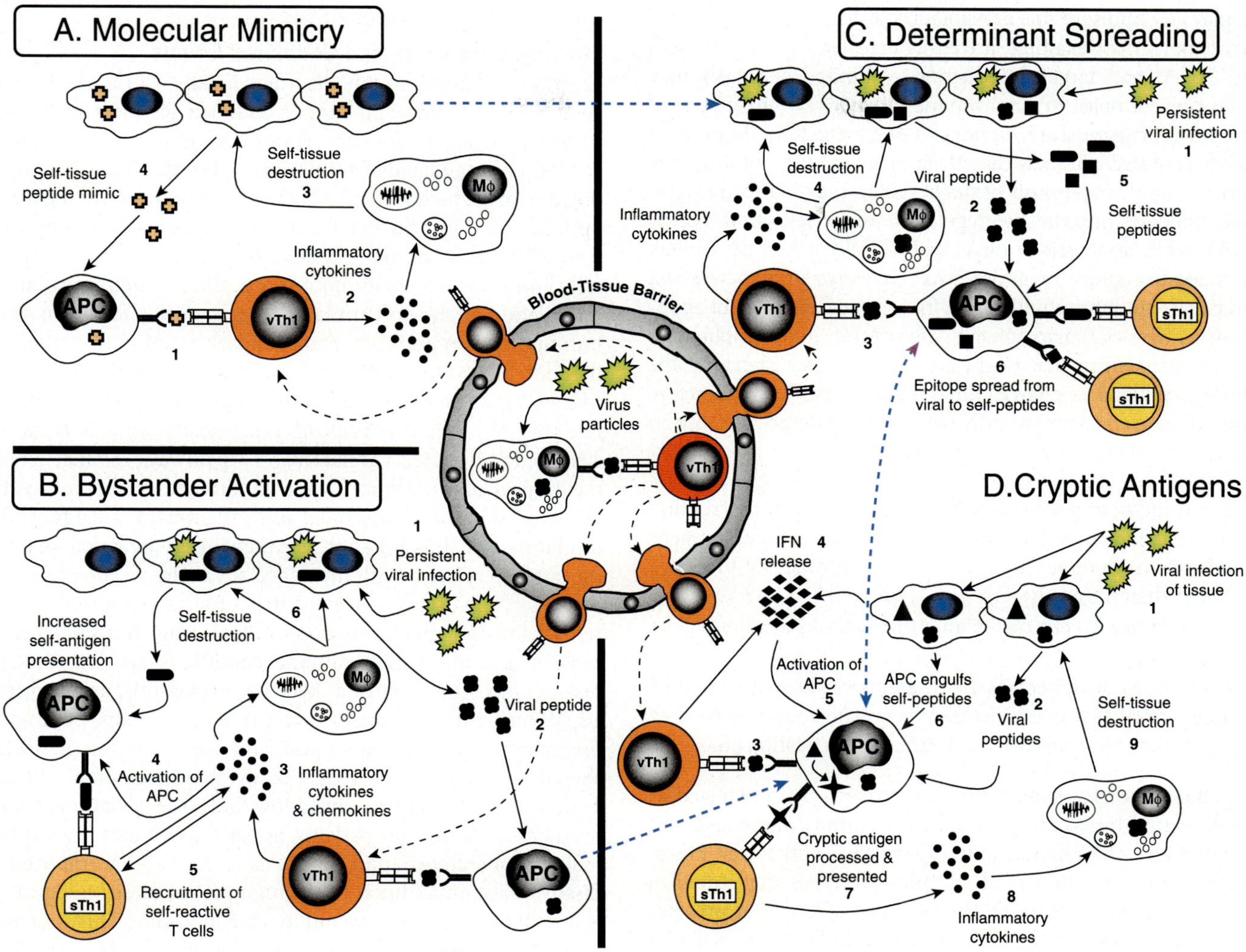

FIG. 46.3. Potential mechanisms of autoimmune disease induction. After a viral infection, activated virus-specific T-helper cells (vTh1) and CD8 CTL will migrate more readily through blood-tissue barriers to the infected and noninfected organs. Molecular mimicry **(A)** describes the activation of cross-reactive lymphocytes that recognize the original viral epitope and a self-epitope (1), which leads to the release of cytokines and chemokines (2), which enhance local inflammation, activate APCs, and indirectly or directly cause tissue destruction (3) and spreading of the autoimmune process (4). In the epitope spread model **(B)**, persistent viral infections (1) could result in the activation of virus-specific T cells (2, 3), which cause tissue damage by killing virally infected cells (4), leading to the release (5) and (cross-)presentation of more autoantigens (6). The bystander activation model **(C)** describes the non-antigen-specific activation of autoreactive T cells. Infiltration by virally specific T cells (1, 2) leads to inflammation and upregulation of immunity throughout the tissue (3, 4), involving the activation of APCs, which now differentially process autoantigens (also de novo or previously cryptoic antigens). This can lead to activation of lymphocytes by TcR-specific or TcR-independent mechanisms (5), which can then cause tissue damage (6). The cryptic antigen model **(D)** describes the initiation of autoimmunity by differential processing of self-antigen/peptides, which can occur under inflammatory conditions. After viral infections, interferons (1) are secreted by antiviral T cells and infected cells (2-4). APCs are activated in this way (5), which enables them to engulf self-peptides (6, *triangle*), or to differentially process endogenous autoantigens. Cytokines can activate proteases more strongly, which might result in the presentation and processing of previously cryptic autoantigens (7-9). The above-displayed pathways are not mutually exclusive and are probably operational at different levels in many autoimmune responses. Currently of high interest is the presentation of neo-antigens and strategies to define them. (Courtesy of Steve Miller and Ludovich Croxford, Northwestern University, Chicago, IL.)

when IL-17 knockout mice were found to develop less arthritis,[163] and treatment with neutralizing antibodies to IL-17 or soluble IL-17 receptor alleviated joint inflammation.[164] In humans, it was found that IL-17 is increased in RA sera and synovial fluid and is present in the T cell–rich areas of the synovium.[165] Furthermore, IL-17 has profound inflammatory effects on a range of cell types within the joint. IL-17 blockade (eg, secukinumab and ixekizumab) is now an approved treatment for psoriasis, ankylosing spondylitis, and psoriatic arthritis, and IL-6 blockade (tocilizumab), a cytokine which governs the generation of Th17 cells, is approved for treatment of RA.

Apoptosis

It appears to be a general paradigm of great functional consequence that activation of the immune system is followed by a process that reverses the activation and reestablishes homeostatic base-line levels of immunity.[85,87] In the absence of such regulatory mechanisms, immune responses will overshoot their goals and excessive immunopathology will occur. Thus, AICD is believed to play an important role in regulating autoimmunity. Apoptotic lymphocytes, for example, are easily detected in islet infiltrates in T1D,[166] and targeted induction of limited apoptosis may even prevent onset of autoimmune disease.[167] While increased apoptosis of aggressive lymphocytes that exceeds the "supply" of newly activated cells may directly limit an ongoing autoimmune process, limited apoptosis of target tissues may indirectly facilitate induction of protective regulatory responses. On the other hand, while apoptosis of target cells should at best be limited, decreased apoptosis of autoreactive aggressive lymphocytes will propagate autoimmunity. In RA, for instance, insufficient apoptosis of synovial macrophages, fibroblasts, and lymphocytes is one mechanism that might contribute to persistence of the disease and lead to synovial lining hyperplasia.[168] It is therefore important to consider precisely which cells undergo apoptosis in order to predict the possible outcome.

- If too many target cells die by apoptosis, organ destruction occurs more rapidly; however, at the same time, antigens releases from apoptotic cells appear to propagate tolerance rather than immunity.[167,169]
- If regulatory T cells die by apoptosis, autoimmunity will be enhanced.[170]
- If aggressive lymphocytes die by apoptosis, disease should be ameliorated. However, since they have to first be activated, they might induce organ damage during their activation phase.

Thus, an ongoing autoimmune process can be viewed as a rather fine-tuned and fragile equilibrium of aggressive and regulatory components, and the precise activation kinetics and survival times of all lymphocyte types implicated in the process will determine the outcome. We are, at present, unable to delineate the precise *in vivo* cellular kinetics, and a more thorough understanding will require improved noninvasive diagnostic techniques.

Kinetics

One of the most important areas for an improved understanding of the pathogenesis of autoimmunity is concerned with the kinetics of immune responses. The pathophysiological or therapeutic effect of a given lymphocyte population not only depends on specificity, activation state, and effector functions, but is also a function of the timing during which phase of an ongoing disease process it is present. Indeed, inflammatory cytokines such as IFNγ or TNFα exhibit opposing effects in T1D depending on the precise time point of generation.[171] Early expression enhances islet destruction and disease development, whereas late expression ameliorates disease by inducing apoptosis of autoaggressive cells. These kinetic issues constitute a major obstacle for successful immune intervention because they preclude the use of specific blocking agents or administration of cytokines without precise knowledge of their kinetically differential role in the disease process. Figure 46.4 illustrates these kinetic considerations in relation to target cell destruction. A better understanding of the underlying "autoimmune-kinetics" is essential and treatments will likely have to be individualized, in particular for antigen-specific immune-based interventions.

Therapeutic Considerations

Efficacy, Specificity, and Undesired Effects

Treatment of autoimmune disorders is not that different conceptually from cancer therapy. A fine balance must be found between efficacy of the intervention and acceptable undesired effects. The main goal of autoimmune disorder therapy is suppression of the pathological autoimmune response. Therapeutic options range in principle from continuous immunosuppression of the entire immune system to specific, targeted, temporally limited, and local immunosuppression. Systemic immune modulation or anti-inflammatory therapy will affect the entire immune system and may compromise the immune status of the individual. One of the more successful examples for effective systemic treatments is the blockade of TNFα to ameliorate RA.[172,173] This class of drugs is especially valuable because TNF blockade affects inflammatory pathways distinct from the targets of conventional anti-inflammatory therapy with corticosteroids or nonsteroidal anti-inflammatory drugs. TNF inhibitors can thus be used in patients that are refractory to the latter treatments with the added advantage of considerably improved outcome in terms of structural bone erosions. The remarkable therapeutic success of these drugs has since been replicated in other autoimmune conditions such as Crohn disease, ulcerative colitis, ankylosing spondylitis, and psoriasis. Whereas these diseases target different organs, common therapeutic improvements upon TNF blockade thus indicate shared mechanisms in terms of cytokine signaling. While undesired effects are relatively low, SLE-like symptoms have been observed in a few patients, as has the enhanced susceptibility to tuberculosis, and certain cancers. Another important disadvantage, in addition to the high cost of these biologicals, is that not all patients respond to TNF blockade. This spurred a search for more potent blockers of inflammation, with the design of Janus kinase (JAK) inhibitors for treatment of RA being a prominent example.[174] This class of small molecular drugs that interrupt signaling downstream of multiple rather than individual cytokines are now approved for treatment of RA. JAK inhibitors have now been proven effective in RA and can also provide symptom relief in some patients who were nonresponders to TNF blockade.

Autoantigen-specific immune interventions, in contrast, bear the promise of lower systemic side effects, since they can be targeted to antigens that are exclusively expressed in the diseased organ.[175,176] However, the efficacy might be lower and suitable target antigens have to be chosen carefully, because enhancement of autoimmunity is an important concern. The goal is either deletion of aggressive autoreactive T cells or induction of regulatory cells.[177] To achieve the latter, response modifiers are probably required at the time of immunization in order to skew the resulting immune response to exert regulatory effector functions. Deletion of autoaggressive lymphocytes or anergy induction is even more risky, since only suboptimal immunization (ie, in the absence of costimulators) will result in this outcome. To control this *in vivo* is rather difficult. Ultimately, antigen-specific therapy will likely have to be individualized due to MHC polymorphism and distinct T cell repertoires between individual patients and should be

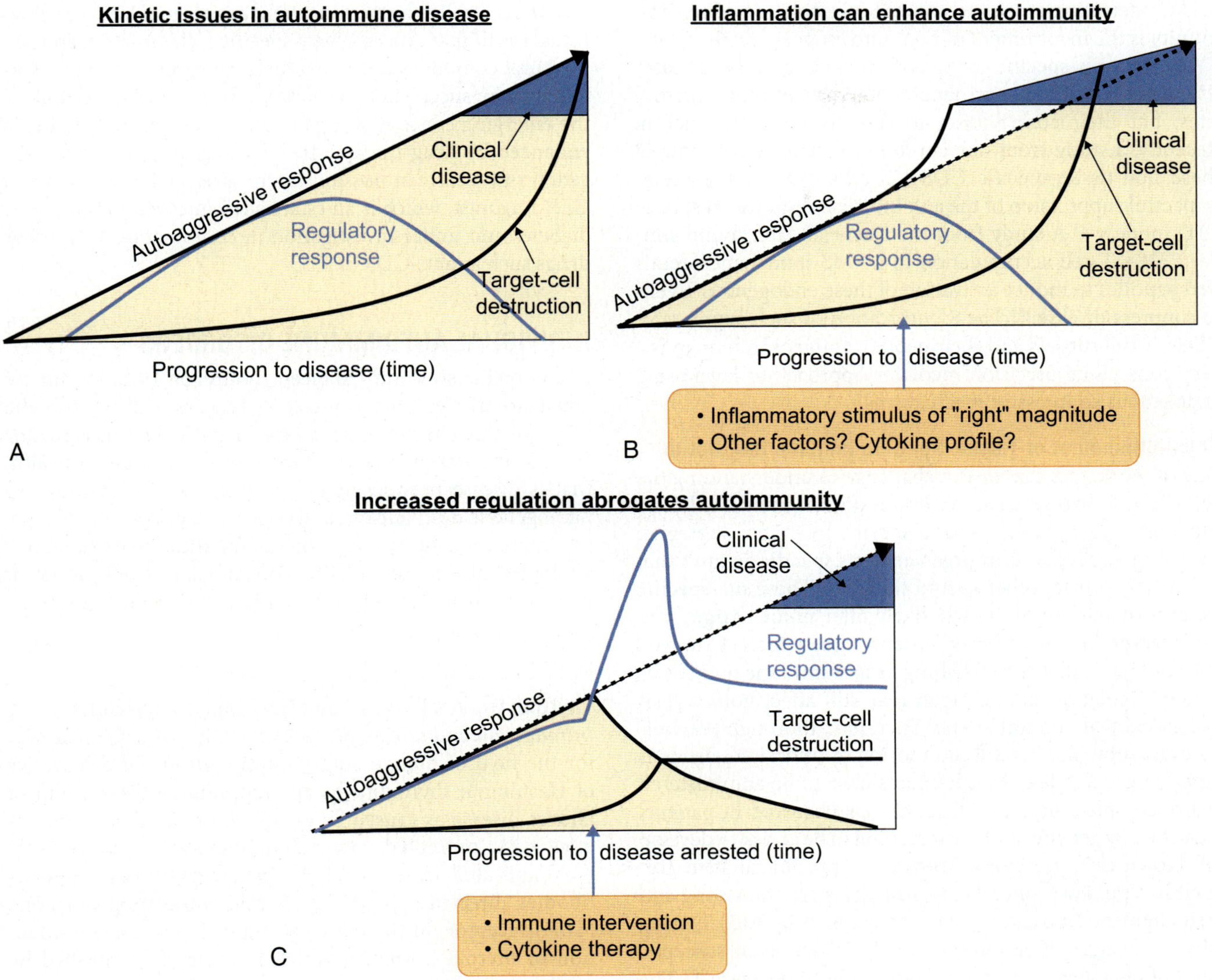

FIG. 46.4. Kinetic issues in autoimmune diseases. Clinical manifestation of autoimmunity is the consequence of a dysequilibrium between protective (regulatory) and aggressive (effector) responses. **A:** An important consideration is that the destruction of the target cell or organ will usually lag somewhat behind the peak of the aggressive responses, because organ regeneration is common. **B:** Inflammatory stimuli will augment the effector arm of the autoreactive response, resulting in more rapid disease development. In contrast, induction of T_{regs} can delay or dampen organ destruction (**C**).

combined with other systemically acting agents, for example, antibodies against CD3 or CD4 or costimulation blockade. Indeed, there is intriguing evidence along these lines in animal models: Combination of a non-Fc-binding anti-CD3 antibody with mucosal immunization of insulin or insulin-derived peptides exhibited clear synergy and enhanced efficacy in reversing recent-onset diabetes in the NOD.[178] In this case, induction of insulin-specific adaptive T_{regs} and their effector function were enhanced and protection by these T_{regs} was transferable and highly effective in that it reversed recent-onset T1D.

Promising Targets

For anti-inflammatory interventions, the factor to be targeted should be as disease-specific as possible. Therefore, blockade of TNF works well for RA but not for diabetes or MS.[179] Experimental evidence supports this observation, because TNF is a crucial mediator found to be elevated in affected joints,[173] but has clearly positive as well as negative effector functions in murine models of MS and diabetes.[180,181] Targeting ubiquitously present chemokines or cytokines will likely not bear much success, because of the resultant generalized immune modulation or suppression.

For antigen-specific interventions, antigens that are already targeted by regulatory autoreactivity are likely to constitute good targets to augment such a preexisting response.[182] In contrast, these antigens should not be selected to delete autoreactive cells, which results in a loss of regulation. For anergy or apoptosis induction, antigens targeted by a primarily aggressive response will be better suited. The fact that auto-antigenic and epitope spreading occurs during each ongoing autoreactive process makes such interventions difficult to design and individualization will likely be necessary.

An interesting emerging concept in regulation of autoimmunity is the involvement of non-autoantigen-specific CD8+ T cells. T cells specific for other antigens than the disease-driving anti-self-antigens generally appear at inflammatory sites, but their contribution to disease has been unclear. Recently, a study from our lab found that large amounts of these non-tissue-specific CD8+ T cells can exert a rather impactful suppression of the autoimmune response in several T1D models.[183] A study from the Davis lab later found similar CD8+ T cell accumulation in an MS model and identified peptides to induce activation of these endogenous T cells to suppress disease.[184] The identification of such suppressive CD8+ T cell subsets and their peptide antigens in human is a very compelling precision medicine approach to harnessing endogenous suppressive mechanisms.

Reestablishment of Tissue-Specific Immune Regulation

One of the factors that pose a challenge to understanding the pathogenesis of distinct autoimmune diseases may also hold a clue to developing effective and specific treatment strategies. Each target cell, tissue, or organ exhibits specific features that distinguish it from other sites of the body. These site-specific features of autoimmunity will likely offer unique target sites for interventions with lower systemic side effects. However, one concern is that reestablishing proper immune homeostasis and regulation in one organ may still affect homeostasis systemically or at another site. Therefore, thorough preclinical evaluation and careful monitoring of undesired effects is urgently needed. Ideally, treatments have to be administered before complete organ destruction has occurred in patients identified by genetic or other screening to be at risk to develop full-blown clinical disease. During the preclinical state, frequently regulatory autoreactive responses are still strong and their augmentation can result in protection (Fig. 46.5). During advanced stages of autoimmunity, T_{reg} function or susceptibility of effector T cells to regulation might decrease.[185] The goal to reestablish homeostasis and proper regulation after an initial insult that caused organ-specific inflammation appears a natural countermeasure to which the specific immune system may be successfully harnessed. An intriguing example is the emerging concept of organ repair that can potentially be enhanced by drug therapy. In T1D, regeneration and replication of β-cells can possibly occur after giving exenatide, a GLP-1 agonist, which is an established intervention in type 2 diabetes and under investigation in combination with other drugs such as anti-CD3 in T1D.

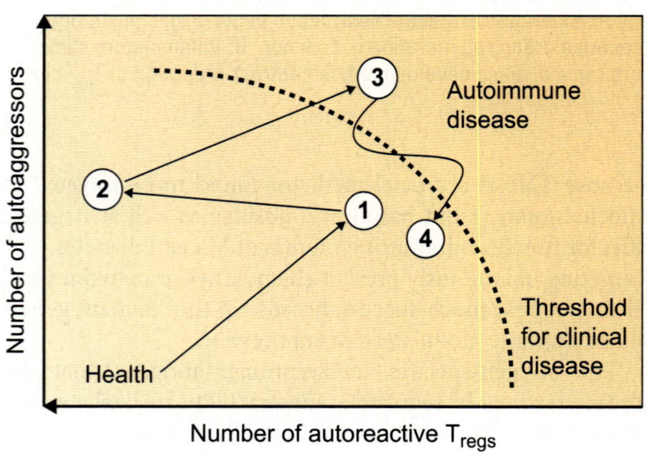

FIG. 46.5. Development of autoimmune disease as a function of aggression versus regulation. Following induction of disease (1), distinct numbers of autoaggressive and autoreactive T_{regs} will be generated and activated. Undulations between protective states and diseases states follow (2, 3), that is, the so-called honeymoon phase in type 1 diabetes, which can eventually lead to an irreversible state of clinical manifestation of disease (4). An additional important component to factor in is the ability of the target organ to regenerate.

INDIVIDUAL AUTOIMMUNE DISORDERS

A comprehensive and balanced discussion of autoimmune and autoinflammatory diseases at large is well beyond the scope of this chapter. Our choice of individual diseases for this section was guided by their relevance to human health, that is disease prevalence and severity, recent insights into pathogenetic mechanisms, as well as the development of novel treatment modalities. Given these limitations, we extend our apologies to all those scientists and clinicians whose work on autoimmune disorders is not mentioned in our discussion.

Endocrine Autoimmune Diseases

Thyroid: Graves Disease and Hashimoto Thyroiditis

Introduction and disease description: 1956 was a seminal year for the field of human autoimmunity given the discoveries of Hashimoto thyroiditis as an autoimmune disease and of Graves disease as caused by an autoantibody. These discoveries have prompted some straightforward and relatively uncomplicated treatments.[186,187] In Graves disease, autoantibodies directed against the thyroid-stimulating hormone (TSH) receptor on thyroid cells stimulate excessive production of thyroid hormone, which is normally controlled by feedback regulation. In contrast, in Hashimoto thyroiditis, autoantibodies to thyroid peroxidase and thyroglobulin are present over years and likely able to fix sublytic doses of complement to cells of the thyroid. The result is an inflammatory reaction, which is also associated with T cell–mediated cytotoxicity. Thyroid damage due to painful Hashimoto thyroiditis may be associated with the development of Graves disease, indicating that there is a tendency for spreading of the autoimmune reaction in humans. Treatment of thyroiditis is relatively straight forward with antithyroid drugs (methimazole) and radioactive iodine.

Autoimmune features: Autoantigens targeted in thyroiditis are thyroid peroxidase, a cell surface protein (Hashimoto thyroiditis), and the TSH receptor (Graves disease). Autoimmune responses to thyroglobulin are also seen in animal models. The B cell epitopes to these autoantigens have been mapped relatively well; however, as it is the case for other human autoimmune disorders, T cell responses, their tracking, and specificity, as well as their eliciting antigens in humans, have remained largely elusive.[186] An interesting genetic link has been established to the CTLA-4 region, which is supported by the finding that NOD mice that also exhibit a link to this gene may present with thyroiditis. The usual suspects, that is APC dysfunction, autoaggressive lymphocytes, and links to viral infections, have been examined, but no conclusive etiological or mechanistic evidence has been obtained to date. It is noteworthy that a role for regulatory T cells has been established in D. Mason's animal model for thyroiditis.[188]

Genetic features and environmental factors: Interestingly, numerous viruses have been implicated in the pathogenesis of different thyroid diseases, but firm evidence for a direct involvement of viruses or virus-induced immune responses leading to clinically manifest disease is scarce. Subacute thyroiditis is a clinical and pathological form of thyroid involvement that appears after infection with viruses such as measles, influenza, adenovirus, Epstein-Barr virus (EBV), and Coxsackie virus.[189] Again, however, a causative role *in vivo* has not been shown for any single infectious agent.[190] Presence of viral material in the thyroid and elevated virus-specific antibody titers were found to correlate with subacute thyroiditis. In other instances, direct virally induced thyroiditis has been documented epidemiologically with thyroid or parathyroid disease.

Retrovirusal, in particular human immunodeficiency virus (HIV), infections have generated much interest. Although HIV infection and acquired immunodeficiency syndrome (AIDS) may affect multiple endocrine organ systems,[191-194] thyroid dysfunction usually reflects weight loss, anorexia, and cachexia of advanced HIV disease, rather than a direct viral effect on the thyroid.[195,196] Thus, direct involvement of the thyroid by HIV or by opportunistic infections is uncommon and may include subclinical hypothyroidism and "euthyroid sick syndrome." The clinical relevance is probably limited, since overt hyper- or hypothyroidism does not occur with greater frequency in HIV-infected and AIDS patients as compared to patients with other nonthyroidal illnesses (HIV and the thyroid gland have been reviewed in Ref. 195). Involvement of the parathyroid in patients with AIDS could be shown by reduced basal and maximal parathyroid hormone levels, but the mechanisms underlying these findings have not yet been elucidated.[197] To account for a possible role of HIV in thyroid autoimmunity, a 66% homology between the HIV-1 nef protein and the human TSH receptor has been noted. However, reactivity of sera of Graves disease patients against a nef peptide showed no significant differences as compared to normal controls.[198] This does not rule out the presence conformationally shared T- or B-cell epitopes with HIV proteins. In analyzing mechanisms of molecular mimicry, studies of potential antigenic surfaces have emerged as an important supplement to analysis of sequence similarity.[199]

In addition, antibodies from a Graves disease patient showed reactivity to the gag proteins of another retrovirus, human foamy virus (HFV).[199] The association of HFV with Graves disease or subacute thyroiditis is controversial. Whereas one study demonstrated HFV-related sequences in the DNA of peripheral blood in two-thirds of the Graves disease patients but none in normal controls,[200] another study could not confirm these findings.[201]

In other studies, an association between human T-lymphotropic virus I and II and the occurrence of autoimmune thyroiditis or Graves disease was reported.[201-204] Furthermore, as in the case of HIV, hepatitis C virus was shown to lead to a wide variety of autoimmune disorders including involvement of the thyroid gland.[205-209] Moreover, treatment of chronic hepatitis B and C with IFNα leads to induction or enhancement of autoimmune disease.[210-212] For congenital rubella infection, which has been associated with diabetes mellitus, Addison disease, growth hormone deficiencies, as well as thyroid disorders,[213] it is not clear whether thyroid involvement is the result of a direct viral effect or a more generalized dysfunction of the immune system.[190]

Animal models: Animal models in mice and chicken have been used to study virus-induced thyroiditis (reviewed in Ref. 190). Mice persistently infected with LCMV showed reductions of thyroglobulin mRNA and circulating thyroid hormones in the absence of thyroid cell destruction.[214] Thyroiditis characterized by focal destruction of the follicular structures, inflammatory infiltration, and generation of antibodies against thyroglobulin and thyroid-peroxidase was observed in a reovirus type 1 mouse model of thyroiditis.[215] The reovirus gene responsible for autoantibody induction was identified and the encoded polypeptide shown to bind to tissue-specific surface receptors.[216] Spontaneous lymphocytic infiltration of the thyroid is observed in the obese strain of chickens. Such chickens express an endogenous retrovirus, avian leukosis virus (ev22), not found in healthy normal inbred strains.[217] Although ev22 appears to be a genetic marker rather than cause for thyroiditis, infection of normal chicken embryos with avian leukosis virus can cause hypothyroidism.[218] Moreover, aberrant MHC class II expression is demonstrated in obese strain chickens and elevated levels of 2,5-oligoadenylate synthetase as well as 2,5-oligoadenylate-polymer levels in the cytosol of thyroid epithelial cells occur, suggesting viral involvement.[219] Again, not all cross-reactivities with self ligands need to increase autoimmunity[220] and regulatory cells also play a major role in modulating autoimmune thyroiditis.[188]

Endocrine Pancreas: Type 1 Diabetes

Introduction and disease description: While the distinct symptoms of diabetes mellitus have been known since antiquity, the underlying pathophysiological processes were only identified in the late 19th and early 20th centuries. The proof of the involvement of the pancreas in diabetes etiology was conducted by von Mering and Minkowski who demonstrated in 1890 that extirpation of the canine pancreas results in the classic symptomatology of hyperglycemia, abnormal hunger, increased thirst, polyuria, and glycosuria.[221] Subsequently, inflammatory changes in the endocrine pancreas, that is the islets of Langerhans, were correlated with diabetes by Schmidt in 1902[222] and 2 decades later, Banting and Best identified insulin as a pancreatic hormone,[223] thereby providing the basis for insulin substitution therapy which remains to this day the cornerstone for T1D management. In a classic 1965 paper, Gepts noted the histopathological similarity between thyroiditis and insulitis and suggested an immune basis for the disease.[224] By 1974, the concept of T1D as an autoimmune syndrome was firmly established by the discovery of islet cell antibodies and an association between T1D and certain HLA genes.[225-227]

Autoimmune features: Today, a quarter century later, the possible autoimmune origin of T1D is understood in much

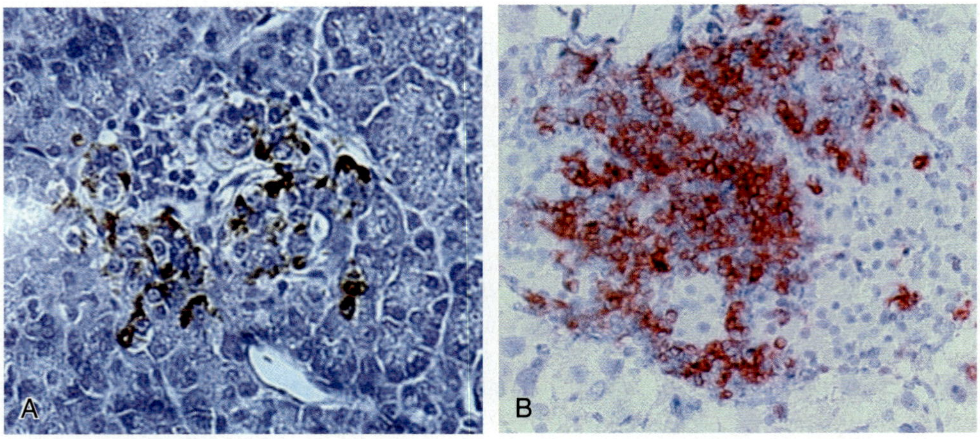

Specific staining is for T cells (CD45RO). Specific staining is for CD8 lymphocytes.

FIG. 46.6. Human and mouse insulitic lesions. Comparison of human insulitis **(A)**, courtesy of Francesco Dotta (University La Sapienzia, Rome, Italy) and mouse insulitis **(B)**, from a RIP-LCMV mouse 14 days post infection with lymphocytic choriomeningitis virus (From von Herrath laboratory, La Jolla Institute for Immunology, La Jolla, CA, USA).

greater detail. However, the lymphocytic infiltration of the islets of Langerhans (Fig. 46.6) and the presence of antibodies specific for β-cell antigens associated with the progressive destruction of insulin-producing β-cells[228,229] still constitute the cardinal evidence for an autoimmune etiology. While there is a reasonably strong genetic linkage to certain HLA molecules, the disease has to be considered polygenic in nature,[129] and a significant discordance of disease among monozygotic twins suggests that environmental factors contribute to trigger and/or exacerbate the disease. Furthermore, it remains unclear which islet antigens are the primary targets. The earliest islet cell–specific antibodies in human individuals at risk are directed to insulin,[230,231] and evidence from relevant animal models and human studies points toward insulin as a primary antigenic target,[232] IGRP (islet-specific glucose-6-phosphatase catalytic subunit related protein),[233,234] and possibly glutamic acid decarboxylase (GAD),[235] but definitive proof for a pathogenic role has only been ascertained in the NOD mouse model but not yet in humans.[236] The cellular infiltrates found in the islets contain both CD4+ and CD8+ T lymphocytes and their irreducible role in β-cell destruction has been documented in several animal models.[237-239] CD8+ T cells can exert direct cytotoxic effects toward MHC class I–expressing β-cells, while CD4+ T cells secreting inflammatory cytokines can provide help to CD8+ T cells as well as B cells. Ultimately, it is important to consider that β-cells constitute about 60% of the islets, which in turn contribute only ~2% to the pancreas mass and demonstrate, unlike many other tissues targeted in autoimmune disorders, an exquisite sensitivity to cytokines such as IL-1, TNF, and interferons that will result in their apoptotic cell death after prolonged exposure.[94]

The detection of islet-antigen-specific antibodies remains an essential tool in identifying prediabetic subjects and monitoring the progression of subclinical and clinical disease. Procedures for autoantibody determination have been substantially refined and standardized worldwide. Data from clinical studies support the notion that with progression of the prediabetic phase, generation of islet antibodies also is increased.[240-242] Usually, antibodies to insulin become discernible first, then to GAD, thereafter to insulinoma antigen 2 (IA2). Individuals with islet antibodies to three or more distinct antigens have a greater than 90% risk of developing T1D.[135] Thus, islet antibodies are an excellent marker for disease risk. However, they appear not to play a role pathogenetically, since transfer of antibodies from mothers to children does not increase the risk for T1D and B cells are not needed for human diabetes.[243-245] In this crucial respect, the NOD mouse appears to provide a paradigm that might not be applicable to human diabetes, since maternal antibodies are an essential factor for diabetes development in NOD offspring.[246]

Measurements of human T cell responses to islet antigens are not yet standardized and can vary considerably between different laboratories. One reason for this may be the source of T cells that are generally subjected to specificity analyses: blood-borne CD4+ or CD8+ T cells may not reflect the specificity distribution and frequencies of islet-specific T cells found in the target organ, that is pancreas and its draining lymph nodes. Even the study of spleen-derived islet-specific T cells readily obtained from NOD mice and analyzed in standardized proliferation assays has shown variations between different NOD colonies.[247] Therefore, measurement of multiple effector functions (ie, cytokine production, etc) in highly standardized assays will likely be required to assess T cell autoreactivity on a routine basis.[248] A study showed that while techniques such as ELISPOT can discriminate between patients and controls, difficulties in detection of low-frequency antigen-specific T cells probably lead to the limited reproducibility of such measurements.[249] Another complicating fact is that T cell responses between individuals are expected to vary and depend on the HLA haplotype and individual trigger(s) that precipitate T1D. A technology that has improved specificity and sensitivity is the use of combinatorial HLA multimers loaded with multiple disease-associated epitopes.[250] A remarkable paper[251] has shown that tracking of immune responses to naturally processed peptide epitopes ("NPPEs") can discern between healthy individuals and those afflicted with T1D: Whereas the number of overall T cells reacting with the NPPEs was similar, individuals with T1D produced more IFNγ relative to IL-10, whereas the ratio was the opposite in healthy individuals.

Genetic features and environmental factors: Data have shown that concordance rates among monozygotic twins are higher than previously thought.[252] Still, the remaining discordance and particularly the major temporal differences in clinical onset indicate that environmental factors have to act in concert with diabetes-susceptibility genes to orchestrate the autoimmune destruction of β-cells. The initial hope that only a few genes would contribute to disease pathogenesis and that genetic links would help to directly understand the mechanistic aspects of T1D pathogenesis has been progressively eroded. Instead, a complex network of susceptibility and resistance genes in both humans and animals (eg, the NOD mouse) has slowly taken shape. A range of GWAS studies has extended the list of gene associations in T1D. So far, more than 60 T1D-associated loci were discovered, some of which encompass important immunity genes (eg, IL-2, IL-2rα [CD25], CTLA-4, or β cell–specific products [insulin]), while other candidate genes do not have an apparent link to immune functions and/or β-cells.[253] The implications of this enhanced genetic characterization have so far not led to the identification of monogenic pathways that directly lead to disease. Rather, consensus is growing that disease susceptibility is the consequence of a global problem of immune regulation, with many genes involved in a variety of combinations in individual patients.[254]

Many of these genes exhibit direct parallels to NOD diabetes susceptibility genes.[255]

The association between particular HLA/MHC class II haplotypes and the occurrence of human diabetes has been of particular interest. DRB1/04-DQA1/0301/B1 and DRB1/03-DQA1/0501/0201 strongly predispose to T1D and more than 80% of patients carry either one or both alleles.[256] In contrast, other MHC class II haplotypes can protect from disease, as evidenced by the sixfold reduced risk to T1D in DRB1/15-DQA1/B1/0602-bearing individuals. Considerable evidence indicates that other genes in the MHC region likely contribute significantly to T1D risk.[127]

An intriguing mechanistic hypothesis was put forth by McDevitt's observation that predisposing HLA class II alleles appear to express small neutral amino acids at position 57 of the DQ allele of Caucasoid populations, whereas an aspartic acid is found in resistant alleles at the same position. Since position 57 is part of the peptide-anchoring pocket, amino acid substitutions in this area will affect peptide binding. Indeed, the susceptibility alleles prefer different peptides, but the contribution to T1D development is not yet clear and a mechanistic link has to be established. Both central and peripheral tolerance mechanisms have been implicated but no direct proof has been obtained. It is important to realize that the human susceptibility MHC class II alleles share amino acids at position 57 with the I-A^{g7} alleles expressed in the NOD mouse and are required for NOD T1D predisposition. However, as previously mentioned, polymorphisms in the MHC class II coding region alone cannot explain diabetes pathogenesis. The amount of complexity involved in the immunogenetics of T1D has been well described by Serreze[257,258]: "Many genes contributing to T1D may contribute to dysregulation of different biochemical steps in a common developmental or metabolic pathway. For example, sequential expression of hundreds, if not thousands, of genes would be expected in the developmental and functional maturation of a macrophage or dendritic cell from stem cell precursors. This process does not occur in a vacuum, but is contingent upon cues provided by the physical environment. In the case for APC development, the microfloral and dietary environments are crucial." Thus, diabetes susceptibility and resistance genes contribute to disease in a polygenetic/multifactorial fashion that appears to gain in complexity as it is being unraveled. The link to environmental factors will be defined to shape gene expression and disease development. Major contributors in this respect appear to be the gut and viral infections.

With more than 400 m^2 of mucosal epithelium, the gut constitutes the largest interactive surface area of the human body connecting us with the environment and its pathogens.[151] Therefore, exposure to antigens or pathogens through the gut, mediated by the largest outpost of the immune system, the gut-associated lymphoid tissue (GALT), will strongly affect specific and general immune functions. It is intriguing that immune tolerance to the numerous foreign protein antigens found in food, as well as bacterial antigens derived from the commensal flora, is generally well maintained.[259] This may be attributable to the high levels of IgA and TGF-β in the gut and to the phenomenon of "oral tolerance."[260] Oral tolerance has been observed in animal models and humans and is characterized by tolerance induction to protein antigens present in the gut. It occurs via two principal mechanisms: Low amounts of antigens will induce a nonaggressive immune-regulatory response while high amounts of antigen can lead to lymphocyte anergy or deletion,[261-263] which is likely achieved via APC modulation. In addition, the profound immune dysregulation found in the absence of a bacterial flora in both animals and humans points to an important physiological role that foreign antigens play in immune homeostasis in the gut.[264] Furthermore, NOD mice only exhibit high levels of autoimmunity when kept in a clean, specific pathogen-free environment; they do not develop T1D when housed under "dirty" conditions.[229] A seminal paper by Wen and colleagues highlighted the pivotal role of the mucosal immune compartment by showing that host recognition of the digestive flora is essential in preventing T1D through engagement of an MyD88-independent signaling pathway.[265] Thus, changes in the intestinal microbiome, integrity of the mucosal barrier, infections of the gut, or certain dietary components can play a role in T1D pathogenesis.

Several reports and studies have attempted to establish a link between the introduction of cow's milk and development of T1D in young infants. This link was not observed in the German, Australian, and US baby-diabetes studies, but was in a Finnish epidemiological study.[266,267] The Finnish study differed from most of the others by an extended observation time involving infant as well as childhood consumption of cow's milk. Therefore, a dietary link between milk feeding and T1D can be considered unlikely but not excluded after long-term exposure to cow albumin or other milk proteins. Similarly, wheat-derived gluten and milk-derived insulin have been implicated as a cause for childhood diabetes. The evidence, however, is not convincing at this point and no firm links have been established.

Some intriguing observations have been published more recently, supporting the concept of a viral etiology for T1D. The mechanistic links between viral infections and autoimmunity can be manifold and have been discussed in detail in the introductory section of this chapter. A significant association between rotavirus infection in young infants and the first occurrence of islet autoantibodies was established by Harrison's group in Australia,[268] but not in Finnish populations.[269] Rotavirus is a double-stranded RNA virus, infects the intestinal mucosa, and is a common cause for seasonal childhood diarrhea. It can polyclonally activate T and B lymphocytes and might possibly harbor antigens that could immunologically mimic islet cell–derived self-proteins. However, it is not clear whether it infects the pancreas or islets directly. The most convincing case can be made for enteroviruses, and there is now robust evidence for a significant association with T1D.[270] Coxsackie B4 virus has been isolated from islets of a child with acute-onset T1D,[271-273] and Coxsackie B3 and 4 strains commonly infect the gut, pancreas, and heart.[274] They lead to profound pancreatitis if they replicate at high enough titers and might harbor a mimicry antigen (P2C protein)[229] cross-reactive on the T cell level with a human GAD epitope. However, this evidence could not be replicated by other laboratories and is still controversial. It has become apparent, however, that Coxsackie B3 viral strains can effectively trigger[275] or prevent[141] T1D in the NOD mouse, depending on the timing of infection. Thus, diabetogenicity of a viral infection may critically depend on the preexistence of insulitis and may either act as an initiating factor or by aggravating ongoing inflammatory processes. Clinical studies have recently provided indirect support for both of these scenarios.[276,277] The finding that infections under certain conditions confer protection might support the "hygiene hypothesis" which suggests that infections protect from rather than enhance autoimmunity. It is still uncertain whether enteroviruses routinely infect or persist in pancreatic islets, although recent reports indicate that viral particles are found more frequently in islets of T1D patients around the time of disease onset.[278] Similar to Coxsackie, other enteroviruses such as polio or echoviruses have been detected in the pancreas and might therefore at least have enhancing effects on ongoing islet destruction in prediabetic individuals at risk.[271] The establishment of a firm association between viral infections and T1D is difficult to confirm, because the underlying mechanistic links established in several animal models allow for the virus to be cleared before autoimmunity develops (ie, in the RIP-LCMV model), viruses need not necessarily directly induce islet-reactive T cell responses but can act as bystander activators and, in many cases, viral infections have been found capable of preventing autoimmunity.

Data suggest that the genetic constitution may (co-) determine whether viral infections will provoke islet autoimmunity in certain individuals. It was found that certain rare polymorphisms in the IFIH1 gene are associated with protection against T1D,[279] while others confer disease risk.[280] This gene encodes a helicase enzyme, IFIH1 (also known as MDA5), that triggers the secretion of interferons in response to viral infection. Most, if not all, viral infections trigger the production of interferon by the host immune system. Being an interferon-response gene, IFIH1 allows the infected cell to sense the RNA genome of enteroviruses and increase interferon production. Interferons then primarily limit viral replication to prevent damage to the infected cell. They also, however, increase the visibility of the infected cell to the immune system by enhancing the expression of MHC molecules. A scenario could be envisioned where mutations leading to decreased IFIH1 activity are associated with milder responses against diabetogenic viruses and thus disease protection. Conversely, mutations leading to IFIH1 hyperactivity could induce inappropriate, exaggerated interferon responses and localized immunity against the infected beta cells. Altogether, such a mechanism would explain why viruses that have presumed diabetogenic effects do not necessarily trigger diabetes in all infected individuals.[281]

Animal models: Since the pancreas and its draining lymphoid organs are notoriously difficult to access, many important insights about diabetes immunology have been gained from suitable animal models that continue to refine our understanding of the pathogenesis and the development of potential prophylactic and therapeutic strategies. There are a multiplicity of animal models for T1D. The most commonly employed models take advantage of natural mutations that give rise to spontaneous diabetes onset or antigen-specific induction of disease using transgenic technology. Other models make use of β-cell damage initiated by treatment with specific chemicals (eg, streptozodozin) or virus infection. Encephalomyocarditis virus is diabetogeneic in mice; the incidence of disease is dependent on both virus and mouse strains used.[143,272,273,282-287] Similarly, Coxsackie virus, associated with diabetes development in humans, causes extensive pancreatic tissue damage and release of sequestered autoantigens that lead to rapid diabetes development in some mouse strains.[288]

a. *Models of spontaneous diabetes onset:* There are several animal models of spontaneous T1D.[289] The two most extensively used are the biobreeding (BB) rat, introduced in 1974 at the Bio Breeding Laboratories in Canada, and the NOD mouse strain established 1974 in Osaka, Japan.[289] Because the BB-rat is associated with leukopenia and other abnormalities,[289] the NOD mouse has been the model of choice due to its genetic linkages that are reminiscent of human T1D.[289] In both models, adoptive transfer of T cells can induce disease.[290] Interestingly, viral infections, first shown with LCMV, can prevent insulin-dependent diabetes mellitus in both BB rats and NOD mice.[291,292] This occurs in the absence of a general immune suppression. While the mechanism involved is unclear, the generation of suppressor T cells has been suggested.[292,293]
b. *Models of antigen-specific diabetes induction:* Expression of influenza virus hemagluttinin (HA) under control of the rat insulin promoter (RIP) resulted in a low incidence (10%-20%) of spontaneous T1D, but no significant enhancement of disease occurred after infection of such RIP-HA mice with influenza virus. However, in RIP-HA

mice expressing a transgenic T cell receptor (TCR) specific for a determinant of the influenza virus HA, the incidence of spontaneous T1D increased to 100%.[294] These studies clearly demonstrate the importance of autoreactive T cells present at sufficient numbers for induction of autoimmune disease. Thus, the inability of influenza strain PR8 to cause T1D in RIP-HA mice likely reflects the low precursor frequency of self-reactive CD8+ T cells. Interestingly, studies from the Sherman laboratory have demonstrated that spontaneous disease in this model is associated with cross-presentation of HA. It has been postulated that cross-presentation constitutes an important mechanism for unmasking of islet-specific antigens to autoaggressive CD8+ T cells. When the model antigen ovalbumin (OVA) was expressed as a self-antigen in β-cells using the RIP, similar observations were made underscoring the likely importance of cross-presentation in autoimmune diabetes. These studies also demonstrated that the clinical outcome correlated with the number of autoreactive CD4+ and CD8+ cells generated and determined a "cut-off level" below which the autoimmune reaction would not take hold of the pancreas.[295-297]

Similar to RIP-HA and RIP-OVA mice, the RIP was also used to express the nucleoprotein (NP) or glycoprotein (GP) of LCMV in the insulin-producing beta cells.[237,298,299] Diabetes develops 2 to 4 weeks after infection with LCMV due to a strong CD8+ or combined CD8+/CD4+ T cell response directed to the viral/self GP or NP in the β-cells. Insulitis is initiated only at a time when the systemic antiviral response reaches its peak and continues well after the LCMV infection has been eliminated.[129,300] Therefore, the localized, islet-specific autoimmune process, although initiated by a response to the viral/self transgene, can be regarded as an autoimmune process that follows kinetics distinctly different from systemic antiviral immunity. Indeed, antigenic spreading to insulin and GAD is observed during the prediabetic phase.[240] Destruction of islet β-cells requires activation of APCs[301] in the islets and is mediated by both, perforin and inflammatory cytokines, predominantly IFNγ.[302] Thus, the RIP-LCMV model reproduces many features found in human diabetes as well as other mouse models. A distinct advantage of RIP-LCMV mice is that the time point for induction of the autoaggressive LCMV NP–specific response can be chosen experimentally and that virus-specific, destructive CD4+ and CD8+ T cells can be enumerated, functionally evaluated, and localized using limiting dilution analyses, MHC tetramers, or intracellular cytokine stains.[303,304] These aspects constituted an advantage in our studies demonstrating that feeding of insulin during the prediabetic period induces insulin B-chain–specific CD4+ regulatory lymphocytes that act as bystander suppressors and can locally downregulate the autoaggressive dibetogenic response in the pancreatic draining lymph node and the islets.[304] Other studies in the RIP-LCMV model have underlined the role of thymic selection in allowing sufficient numbers of low-avidity autoaggressive T cells to emerge in the periphery,[237] the role of non-MHC-linked genes in influencing the kinetics and severity of T1D even in the presence of high numbers of autoaggressive T cells,[304] and importance of APCs in breaking tolerance and sustaining autoaggressive T cell responses.[305-307]

Under some conditions, CD4+ or CD8+ T cells can induce T1D by adoptive immunization.[290,308] In general, T1D develops more slowly or not at all in the absence of CD8+ CTL, MHC I, or perforin.[94,309] Similar considerations are true for MHC class II and costimulatory molecules, unless their elimination affects the generation of regulatory lymphocytes.[310,311] Cytokines, however, frequently play dual roles in T1D pathogenesis in animal models. IFNγ, in general known as a proinflammatory mediator that upregulates MHC molecules, can "unmask" β-cells for immune-recognition by induction of MHC class I expression, but also exert direct antiviral effects and might be beneficial by increasing apoptosis of aggressive T lymphocytes later in the disease.[94] A similar dual role is true for TNFα which appears to enhance early disease (possibly by directly causing β-cell death in conjunction with other cytokines) but ameliorates advanced autoimmunity just prior to onset of clinical disease.[171,312] For these reasons, blockade of such inflammatory mediators might be problematic in T1D. Dual roles were also described for IFNα,[313-315] IL-2,[316] and IL-10.[317,318] In some of these studies, the level of the cytokine might play as important a role as the precise timing of expression. The only cytokine with largely beneficial effects has been IL-4 which ameliorates the disease and therefore might be a good candidate for treating or preventing T1D.[101] Figure 46.7 illustrates the pathogenetic hypotheses generated through research in multiple animal models.

Treatment and prevention: Despite intensive research, no effective prophylaxis, therapy, or cure of T1D is available to date. However, even under optimal disease management with insulin substitution therapy, T1D significantly shortens life-expectancy due to eventual vascular complications in multiple organs. Currently, all strategies that look promising have either been evaluated in animal models or are in early clinical trials in humans (see next paragraphs). Part of the problem in devising such prophylactic or interventive immune-based approaches is that that the only endpoints in human trials are disease prevention and insulin requirements, as well as remaining insulin production (C-peptide levels). No precise interim staging is yet possible, which makes efficient clinical evaluation very difficult compared to MS or RA, where access to the target organ either visually by MRI or directly by sampling fluids is much easier to achieve. Animal models have provided interesting ideas and evidence for a variety of antigen-specific and systemic interventions that bear promise for human diabetes.

a. *Antigen-specific immunoregulation:* Immunization with DNA plasmids that express islet self-antigens, with or without cytokines, that act as response modifiers, can induce autoreactive regulatory CD4+ T cells. These cells are able to suppress autoaggressive CD4 and CD8 cells locally in the pancreatic draining lymph node, where they act as bystander suppressors. Phenotypically, they behave very similarly to the Th2-like regulators[182,320-323] induced after oral antigen administration. In treated mice (using

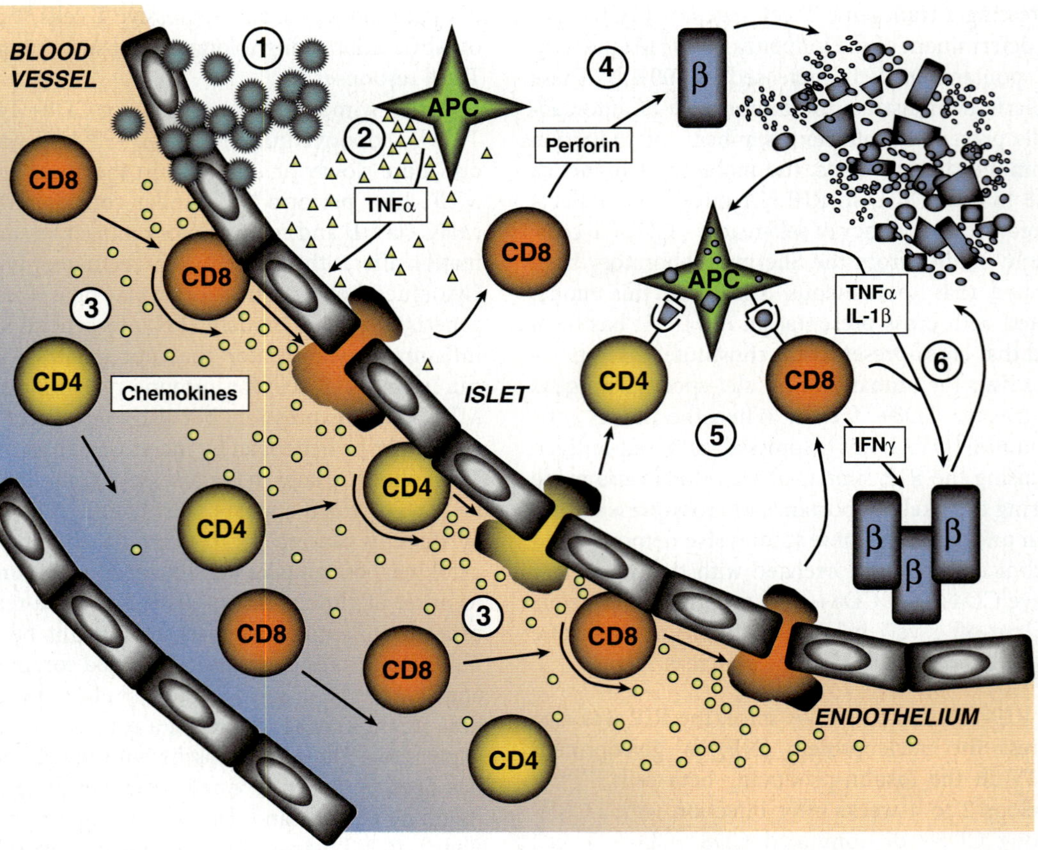

FIG. 47.7. Immunopathogenesis of type 1 diabetes. Viral infections and other inflammatory stimuli (1) can activate antigen-presenting cells (APCs) (2), that secrete chemotactic mediators that entice lymphocytes in blood vessels to "rolling" and entry into the pancreatic islet tissue (3). Entering lymphocytes might directly destroy beta cells (4) or activate APCs locally (5) and also propagate inflammation by secreting pro-inflammatory cytokines (6). Recent intravital imaging data suggest that effectors may remain in a given islet for hours while inducing beta cell death.[319] (Courtesy of Urs Christen, University of Frankfurt, Frankfurt, Germany.)

several distinct diabetes models), insulitis is permanently reduced and progression to clinical diabetes can be prevented in 50% to 80% of the animals.[176,182,320,321] To bring this approach to the clinic as a preventive therapy, a suitable marker will be needed that can predict early outcome, for example, levels and isotypes of autoantibodies. Like other antigen-specific interventions, this strategy will not likely be effective in late stages of the prediabetic phase and should therefore be thought of and tested as an early preventive therapy. The important advantage of this and other antigen-specific approaches is that the risk for systemic side effects is low, since the effector cells will act antigen-specifically only in the area, where autoimmune destruction is ongoing and their cognate antigen (eg, GAD or insulin or heat shock protein) is presented. Discouraging results from recent clinical trials indicate that, while highly effective in animal models, translation of antigen-specific approaches to the clinic remains a challenge.[324] It may thus be possible that successful antigen-specific therapy will require a combination with systemic immune modulators. For example, we have demonstrated that a combination treatment with intranasal proinsulin peptide and anti-CD3-specific antibody can reverse recent-onset diabetes in two murine diabetes models with much higher efficacy as compared to monotherapy with anti-CD3 or antigen alone.[178] Despite the strong rationale to tackle the disease from different angles, drug companies are generally reluctant to include successful drugs in a combinatorial regimen.[325] A specific subtype of antigen-specific approaches is the design of altered peptide ligands: Modified insulin peptides that favor a Th2-like deviation of responder cells have been developed by Neurocrine (San Diego, CA) and showed great promise in the NOD mouse model. These successfully passed preclinical safety studies,[326-328] but efficacy outcomes in the first clinical trial were disappointing.[329] Other approaches, such as using peptides bound to MHC II on nanoparticles have recently shown promise in preclinical models.[330]

Finally, induction of oral tolerance can be seen as a form of mucosally targeted antigen-specific therapy. As for MS and RA, oral antigen trials have failed so far for T1D, except for in a distinct subgroup of patients,[331] and a follow-up study unfortunately did not show any delay of prevention of the disease in at-risk individuals.[332] The reasons for these failures after initially very promising animal data have become increasingly clear from follow-up animal experimentation. Likely, the dose of antigen was significantly too low (presumably by a factor of 10-100). This problem may be overcome by coupling the orally administered proteins to gut response modifiers such as the cholera toxin B subunit. Second, the choice of antigen is very important and even minor amino acid changes can modify the dose-response curves. Without a reliable preclinical marker, it will be very difficult to tune this treatment to the human situation.[264] There is still hope that this is a viable option for primary prevention, and a large European trial with oral insulin therapy for infants with high genetic risk is ongoing.[333]

b. *Therapeutic failures and future perspectives:*

1. Anti-CD3 treatment: One of the more promising drug candidates from the past decade in terms of immune modulation in T1D was anti-CD3. The effector mechanism in mice is systemic and is likely relying on the induction of antigen nonspecific regulators[310] that are $CD62L^{hi}$ and $CD25^{hi}$, as well as anergy or deletion of activated aggressive lymphocytes. From data in animal models and humans, systemic immune suppression is not too profound, indicating reasonable safety for this strategy. Two independent randomized controlled phase 3 trials, however, showed minimal effectiveness of this approach for treatment of individuals recently diagnosed with T1D.[334] These notable and discouraging failures have raised questions concerning the trials' accuracy of dosing, choice of appropriate endpoints, and timing of administration. Recently however, the same drug was trialed in a group of relatives of patients with T1D with high risk of developing the disease. The anti-CD3 treatment in this cohort doubled the time to disease onset compared to placebo (48.4 vs 24.4 months).[335] Still, what we have learned from these previous trials is the inability to translate a therapy that is so potent in mice to humans may indicate that we have in the past relied too heavily on animal models to prioritize drug candidates.
2. CTLA-4Ig, anti-CD20, and anti-cytokine therapy: Encouraging data in mice, likely in combination with the availability of commercial compounds, have triggered a series of trials aimed at tackling islet autoimmunity from diverse angles. CTLA-4Ig (abatacept) modulates costimulation and prevents full T cell activation. A trial showed that the drug initially induces a beneficial effect but also that continued administration did not offer lasting protection. The authors therefore speculated that T cell activation in T1D may lessen with time.[336] Another study found that anti-CD20 therapy (rituximab; depletes B cells) partially preserved beta-cell function over a period of 1 year in patients with T1D.[337] While the outcome was relatively modest, the finding that B lymphocytes contribute to the pathogenesis of T1D is of scientific value. Finally, clinical studies are underway to block cytokines such as TNFα or IL-1β in order to eliminate their known deleterious effects on beta cell viability. It can be concluded from these modest successes that immunomodulatory regimens may hold potential, that is, that tackling the immune component from T1D is indeed a viable option. It is clear, however, that more potent or complementary strategies need to be evaluated in order to achieve significant and lasting protection.
3. Islet transplants: A promising therapeutic advance was made by developing a specific protocol for gentle isolation and purification of human islets ("Edmonton protocol") that appear less prone to rejection after intraportal implantation into diabetic subjects.[338] However, continued immune suppressive treatment is still necessary in recipients of such transplants to prevent renewed autoimmune destruction of islet transplants, emphasizing the need for effective immune-based strategies of T1D prevention. A potential improvement was developed by the company Viacyte in the form of proprietary capsules that are designed to shield islet transplants from the deleterious effects of the immune system. The approach also encompasses a detailed procedure for the generation of viable beta cells from human embryonic stem cells,[339,340] and several other groups are working on producing beta cells from induced pluripotent stem cells. It is hoped that these advanced technologies will one day allow for the safe and lasting transfer of functional islets into T1D patients.

Perspective and conclusions: No single animal model accurately reflects all aspects of human T1D. Consequently, translating the successful T1D prevention and reversion strategies developed in a variety of animal models into clinical applications faces significant obstacles, as is evident from failures in recent clinical trials. For example, to date over 140 therapeutic options have been developed to prevent T1D in the NOD mouse,[228] but not a single one that was tested in clinical trials is as effective in humans. To improve this situation, there is a dire need for preclinical markers that accurately predict the potential success of an intervention being evaluated by a clinical trial. Better tracking reagents for blood autoreactive T cells such as tetramers, noninvasive high-resolution *in vivo* imaging systems, and perhaps isotype profiles of islet antigen autoantibodies and their change after a given preclinical intervention are good candidates.[341] Similar considerations apply to the maintenance of long-term tolerance after islet transplantation, where prolonged systemic immunosuppression might not be feasible.[342-345] In addition, treatment modalities will likely have to be tailored to the specific needs of individual patients. Lastly, one could argue based on clinical studies using cyclosporine or nonmyeloablative stem cell transfer that a robust cure for T1D already exists, be it at a high cost in terms of patient safety and associated side effects. Future progress should therefore be aimed at walking the fine line between profound efficacy and optimal tolerability in order to favorably compare against lifelong insulin replacement therapy.

Central Nervous System and Ocular Autoimmune Diseases

Multiple Sclerosis

Introduction and disease description: Multiple sclerosis (MS) is the most predominant human demyelinating disease of the CNS. An autoimmune etiology is suggested by elevated frequencies of myelin basic protein (MBP)-specific T cells documented in several independent studies of individuals diagnosed with MS.[346-349] Furthermore, demonstrations that adoptive transfer of T cells specific for myelin or other CNS antigens can cause a CNS autoimmune syndrome in experimental animals resembling human MS support the concept of a direct T cell–mediated pathogenesis. In addition, myelin-specific and lipid-specific antibodies have been found in MS cerebrospinal fluid.[350] These antibodies are able to complement-fix and induce antibody-dependent cellular cytotoxicity and may be involved in the demyelinating process. In spite of these clues, the etiology of MS remains unclear and a complex interplay of genetic and environmental factors (similar to T1D etiology) has to be postulated.

MS is a rather heterogeneous disease. However, we are only beginning to understand how clinical and pathological differences may point toward distinct etiologies and, potentially, treatment strategies of differential applicability and efficacy. Different subtypes of MS are somewhat better histologically defined (eg, the distinction between T cell–rich and macrophage-rich lesions as well as the balance between demyelination and remyelination, which can vary dramatically) than those of T1 since many tissues from patients with clinically active disease are available post mortem, whereas pancreatic islets are mostly destroyed in type 1 diabetics and no immunologic correlates can be established at the end-stage.[351,352] MS predominantly affects younger women and its frequency, dependent on geographic location, may approach up to 3/1000.[353] Clinically, the disease can take a mild or very debilitating course and, intriguingly, neuritis of the optical nerve is frequently the first sign for a beginning MS. Diagnosis is usually obtained by EEG and direct imaging that allows for identification of individual lesions, their development over time, and the success of therapeutic interventions. Immune-based approaches such as the use of interferon-β or Copaxone combined with systemic immune-modulatory agents has brought some limited success for certain forms of MS. Thus, a cure for MS appears possibly a little closer than for T1D.

Autoimmune features: Although MBP- and proteolipoprotein (PLP)-specific T cells have been documented in patients with MS, some follow-up studies have reported similar frequencies in normal individuals.[354] More recent studies have found increased numbers of MBP-specific T cells in MS patients reactive to six different MBP epitopes, supporting the concept of determinant spreading as a potential pathogenetic factor. Furthermore, their activation appears to be increased selectively in MS patients and they are less dependent on costimulation.[355] Thus, it appears plausible that resting T cells specific for CNS antigens are present in both, healthy and affected individuals, but their activation is selectively increased under conditions of disease and their epitope/antigen recognition repertoire is expanded. In addition to cellular infiltrations, MS is usually marked by edema and inflammatory signs such as MHC upregulation, cytokine and chemokine induction. A breakdown of the blood-brain barrier facilitates antibody access to the CNS. Antibodies specific for myelin structures and lipid components[356] are commonly found in MS patients and some studies indicate fluctuations, similar to T cell responses of antibody titers in parallel to exacerbations and remissions of disease. These findings still await confirmation in larger patient cohorts. It is important to note that such antibodies can be directed to MBP, myelin oligodendrocyte–associated protein (MOG), and other myelin constituents.[357] The fact that they can fix complement and enhance complement-mediated cytotoxicity implicates them as possible agents of demyelination.

Genetic features and environmental factors: There is a definite genetic link between progressive (HLA-DR3) as well as remitting relapsing MS (DR2) and human MHC class II alleles. However, similar to T1D, there is a significant discordance among monozygotic twins (>55%). A pronounced geographical gradient (higher incidence in northern as compared to southern regions) argues for additional environmental triggers or modulators.[353] Many viruses have been implicated in the pathogenesis of MS, but a causal link has not been established. For example, human herpes virus 6 was discovered around MS lesions, but follow-up studies did not discover any differences comparing healthy, Alzheimer or Parkinson tissues with MS lesions.[358] Furthermore, antibodies to corona virus have been detected in serum of MS patients, which indicates that demyelination might occur similarly to the mouse hepatitis virus (MHV) model. Mechanistically, the same paradigms as for diabetes might apply (described in the introductory section), but a persistent viral infection might be more likely the culprit, based on similar observations in some mouse models (see the TMEV model for example). The difficulty in establishing conclusive proof is the multiplicity of infections during the course of a lifetime and the manifold viral traces that can usually be detected in healthy as well as diseased individuals. Even if the same levels of viral antigen, RNA, or DNA are found comparing healthy patients with MS patients, this still does not rule out the possibility that a certain virus will only induce CNS disease in conjunction with a distinct MHC haplotype. Indeed, animal models indicate that this is in fact the case. With the advent of improved vaccination and antiviral treatment protocols for multiple infectious agents, we may be able to infer a causal relationship, if the incidence of MS should recede after introduction of such an intervention.

Animal models: Experimental allergic encephalitis (EAE) is one of the oldest and most widely studied animal models for a human autoimmune disease.[111,359,360] EAE, a primarily CD4+ T cell–mediated disease, can be induced in susceptible mouse strains (SJL, PL/J) by immunization with MBP, PLP, MOG, or respective MHC class II restricted epitopes together with complete Freund's adjuvant and pertussis toxin. More recently, a role for CD8+ T cells has also been documented.

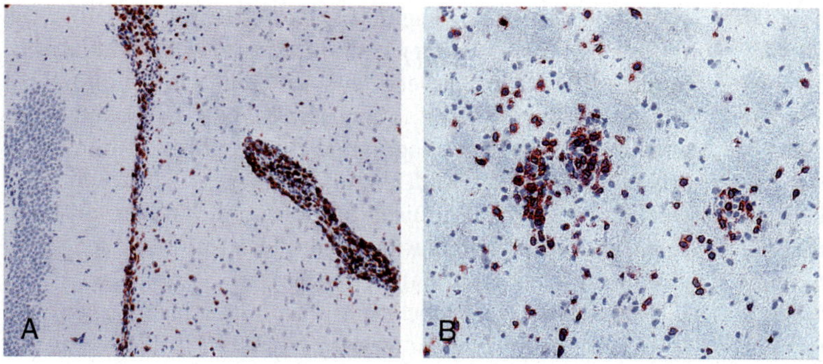

FIG. 46.8. Induction of EAE in SJL (H-2^s) mice with PLP peptide aa 139 to 151 and adjuvant. A: CD4 infiltrate, hippocampus. **B:** CD4 infiltrate, brain stem. (Courtesy of Andreas Holz, Max Planck Institute for Neurobiology, Munich, Germany.)

The clinical picture resembles that of a relapsing/remitting disease with profound neurological symptoms. CNS histological findings include T cell infiltration (mainly CD4+ T cells lymphocytes, Fig. 46.8), APC and microglia activation, as well as disruption of the blood-brain barrier. Some demyelination occurs and histological changes can undergo periods of remission that are accompanied by remyelination. The predominant cytokine profile is Th1-like, and Th2 cytokines have protective function.[361] Epitope spreading occurs and activated APC takes a central role as "drivers" of the autoimmune process.[147] A shortcoming of this model is the major extent of external manipulation required to break tolerance (adjuvant, pertussis). These might be necessary to affect the blood-brain barrier, skew the systemic cytokine profile to the Th1 phenotype, and support prolonged inflammation.

A lesser degree of additional nonspecific inflammatory conditions is required for some virus-induced animal models for MS. Infection with Theiler's murine encephalitis virus (TMEV), which replicates predominantly in neurons (particularly motor neurons during initial stages of infection), results in a poliomyelitis-like syndrome including flaccid paralysis. Certain strains of mice that survive the primary infection become persistently infected and develop a chronic progressive inflammatory demyelinating disease.[147] Demyelination is the consequence of the CD4+ T cell response to persisting TMEV and involves determinant spreading as a consequence of infection and local inflammatory processes.[147] Studies from S. Miller's laboratory have shown that APCs in the CNS are crucial for diversification/epitope-spreading of the initiating anti-TMEV response to other brain self-antigens such as MBP and PLP. Furthermore, when recombinant TMEV expressing PLP epitopes or modified mimic epitopes were employed, infection resulted in enhanced disease. These experiments provided *in vivo* evidence for a possible role of molecular mimicry in CNS autoimmune disease.[362]

Somewhat similar to TMEV, infection with neurotropic strains of MHV induces neurological disease accompanied by demyelination.[363] An interesting difference is that MHV will not persist, and therefore allows for easier differentiation between the antiviral and the autoimmune response. CD4+ as well as CD8+ lymphocytes accompanied by inflammatory cytokines are needed for destruction. These virus-induced models of demyelination suggest that a transient or persisting presence of virus in the CNS may be needed in conjunction with a nontolerized antiviral response for disease

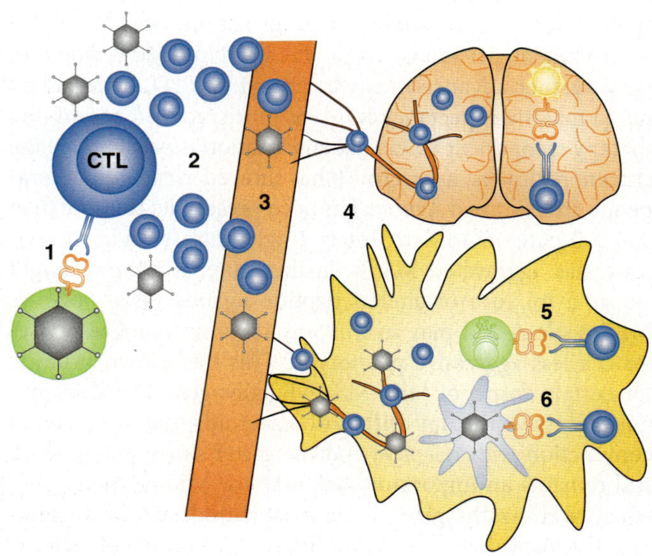

FIG. 46.9. Activated antigen-presenting cells (APCs) might be a crucial local driver once an autoimmune process has been initiated. Induction of profound autoimmune disease of the islets but only mild autoimmune infiltration in the central nervous system in rat insulin- versus myelin basic protein-promoter lymphocytic choriomeningitis virus (LCMV) transgenic mice. (1) In both systems, LCMV infects many peripheral organs which leads to (2) activation and expansion of CTL. (3) Both virus and CTL are able to traffick via the bloodstream but (4) only CTL can cross the blood-brain barrier. In the pancreas, both transgenic beta cells (5) and infected APC (6) drive CTL activation and killing, whereas in the brain only the transgenic oligodendrocytes present antigen (7). The observation that clinical disease only develops in the pancreas but not the brain indicates that APC might be crucial local drivers once an autoimmune process has been initiated.

induction and/or propagation.[363] Indeed, another animal model of virally induced CNS disease, the transgenic MBP-LCMV model developed in analogy to the RIP-LCMV model for T1D only exhibits mild motor dysfunction despite good lymphocytic infiltration.[364] It is possible that activation of APCs is insufficient in this model to locally propagate systemically induced autoaggressive lymphocytes once they have entered the parenchyma of the CNS. These considerations are depicted in Figure 46.9.

Treatment and prevention: Similar to observations in the diabetes models,[322] DNA vaccines that aim to skew the autoreactive response to MBP or PLP from a Th1 to a Th2 phenotype have shown some success in the EAE model. A clinical phase 2 trial of an MBP DNA vaccine indicates that these results at least to some extent can be translated to humans[365]; however this candidate has not moved forward in development. At least for MS, there are preclinical markers that can noninvasively and continuously track the size and appearance of current lesions and in this way indicate success or failure of a given therapy in a more immediate way as compared to T1D.[366,367] Oral tolerance induction had shown much promise in EAE models, but failed in a human trial.[368] Strong placebo effects were observed after feeding irrelevant protein versus MBP, and the outcome was therefore not conclusive, leaving a smaller read-out window. Blockade of TNFα has shown promise in animal models but not humans.[179]

Treatment of MS is, from the immunological point of view, further advanced than treatment of T1D. It should be noted that life expectancy in MS is more reduced because no therapy comparable to insulin substitution is possible. Of the current regimens, interferon-β has showed very strong therapeutic efficiacy and Avonex has been prescribed to more than 500,000 patients to date. Interestingly, interferon-γ, in contrast, has aggravated MS.[369] Furthermore, Copaxone might act as a mixture of altered peptide ligands (as a polymer MBP analog) and may in this way skew immune responses systo a less aggressive phenotype. This has shown promise for certain forms of early MS.[370] Finally, generalized immunosuppressive therapy with corticosteroids that are given in conjunction with relapses, and thereafter slowly phased out, still remains an important backbone for MS treatment. Not effective are azathioprine, cyclophosphamide, or plasmapheresis. Cyclosporine A is helpful but has profound side effects. Altered peptide ligands (APL) to human MBP epitopes have shown strong allergic side effects and have worsened disease in one study indicating that APLs might activate *in vivo* aggressive, as well as regulatory lymphocytes, and that the success can therefore vary from individual to individual.[328,371] Some setbacks have been encountered with monoclonal antibodies. Noteworthy is that anti-CD40L exhibited some efficacy but leads to coagulation disturbances in humans[372] and is currently on hold. In addition, integrin blockers have shown promise, but clinical development has been on hold, in part because combination with interferon-β had led to reemergence of CNS viral infection.

Perspective and conclusions: Therapeutic interventions in MS are becoming more effective, but will likely have to be individualized and fine-tuned in respect to the subtype of MS in a given patient. Antigen-specific therapy is still in development but better understanding of the role of T cells in MS will be needed to translate experimental findings into clinical applications. One interesting strategy that has been trialed involves the intravenous administration of chemically fixed, autologous peripheral blood lymphocytes coupled with five immunodominant myelin peptides.[373] There is also interest in evaluating other therapeutic agents such as CTLA-4Ig with validated benefits in other autoimmune diseases.[374] Combination of systemically acting antibodies shows good promise and will have to be further developed and fine-tuned.

Ocular Autoimmune Diseases

Various inflammatory diseases of the eye accompany other systemic or organ-specific autoimmune disorders and are displayed in Figure 46.10. Some of these diseases will be discussed very briefly since they are described in a separate chapter or section in this book.

M. Bechterew/spondyloarthropathies: Iridocyclitis is frequently associated with Bechterew disease (ankylosing spondylitis) that is characterized by a chronic spondyloarthrosis of the ileosacral joints resulting in profound kyphosis

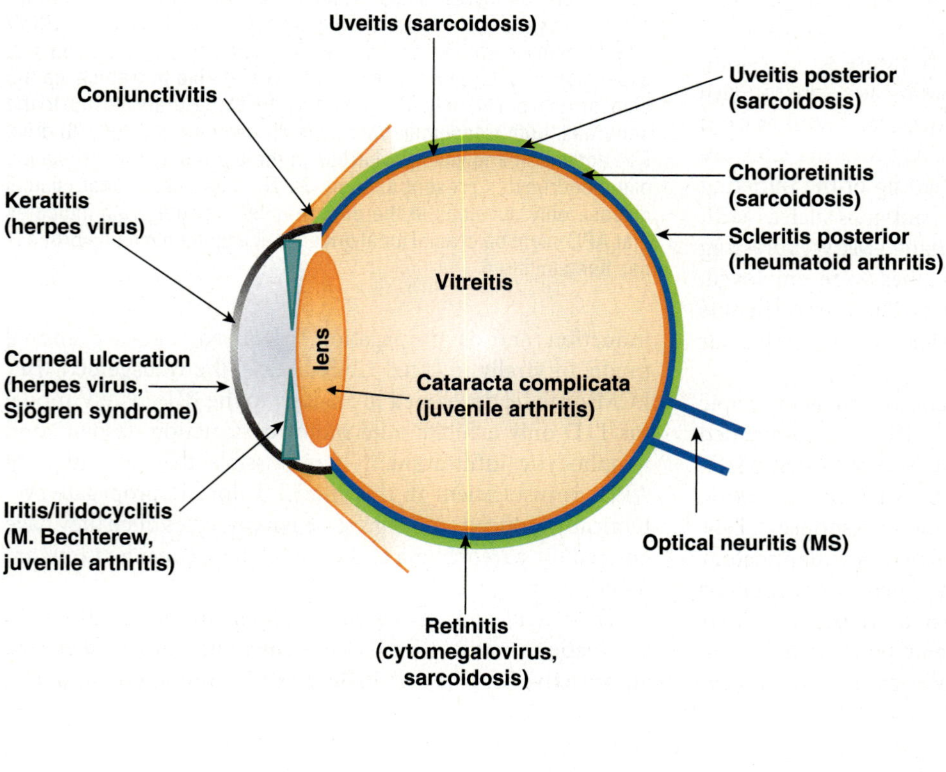

FIG. 46.10. Overview of the structure of the eye and some of its inflammatory diseases and their association with other systemic disorders.

and is associated in 30% to 50% of cases with the HLA B27 allele.[375-377] Interestingly, mucosal inflammation, ulcers, and vasculitis are frequently associated with this disease group. The immunological pathogenesis is not clearly known, but cross-reactive antibodies to bacterial proteins are frequently found in sera of patients with spondyloarthropathies such as Bechterew disease, Reiter syndrome, psoriasis, Crohn disease, or ulcerative colitis. Therefore, molecular mimicry has been hypothesized as a cause for these disorders, but proof has been difficult to obtain. Several animal models have been developed, the most intriguing of which is probably a HLA-B27-transgenic rat that develops joint diseases, genital ulcers, and eye disease. Systemic immunosuppressive interventions are not effective and the treatment of choice is nonsteroidal anti-inflammatory drugs. Local steroids are frequently administered for symptoms affecting the eye.

Rheumatoid arthritis: Rheumatoid arthritis (RA) is an autoimmune joint disease (see below) that is frequently associated with scleritis and episcleritis. Inflammatory infiltrates can be found in the peripheral cornea leading to ulcerations. Local treatments are usually without effect and symptoms are ameliorated in conjunction with systemic immunosuppressive therapy. In contrast, juvenile arthritis is accompanied in about 20% of patients with cataracts and iridocyclitis and ocular symptoms are often sensitive to local and systemic immunosuppression.

Sarcoidosis: Sarcoidosis is a chronic, systemic granulomatous inflammatory disease that involves the eye (uveitis in 20% of cases) and constitutes the cause for uveitis in about 10% of patients. Affected are multiple organs, in particular the lungs. The etiology is unclear and an autoimmune pathogenesis is suspected and multiple other inflammatory changes can occur in the eye (see Fig. 46.10). Local therapy with steroids is usually helpful, and systemic immunosuppression can be considered.

Idiopathic uveitis: About half of the cases of uveitis are not associated with a known primary syndrome and in young adults the association (50%) with HLA B27 is striking. Only few of these patients will ultimately develop Bechterew disease. Again, the etiology is quite unclear, but some evidence suggests that molecular mimicry between ocular and viral antigens, for example herpes virus, could play a role. Peptide therapy is currently being evaluated; otherwise steroids are the only effective choice.

A summary of the eye structures and some of the systemic diseases that are associated with a putative autoimmune affectation of the eye is listed in Figure 46.10.

Animal models: Small rodent models for autoimmune uveitis and herpes virus keratitis have been very useful for understanding the underlying immunopathology of inflammatory eye disorders.

a. *Experimental autoimmune uveoretinitis (EAU)*: EAU is induced, similar to EAE, by immunizing susceptible rodent strains with retinal proteins such as the receptor retinoid binding protein (IRBP).[378] Extensive studies by R. Caspi and colleagues have revealed that EAU follows very similar autoimmune paradigms as other experimental autoimmune diseases including EAE and induced forms of diabetes. For example, a Th1-like response drives the aggressive EAU process,[379] whereas IL-10 in synergy with IL-4 has a pronounced protective function.[99,380] Similarly, disease-resistant mice exhibit a predominantly Th2-like response to IRBP. Experimental evidence indicates that sequestration and local immune privilege of IRBP and possibly other retinal antigens leads to a lack of systemic tolerance to these proteins and, as a consequence, destructive autoimmune responses can be readily induced. Conversely, tolerance can be reestablished by application of tolerizing retinal-Ag/Ig fusion proteins or by inducing peripheral tolerance by systemic expression under an MHC class II promoter in transgenic mice. Interestingly, peptide therapy using an HLA B27–restricted peptide appears to be effective in humans. The immunopathology of this EAU model in comparison to human uveitis, as found in sarcoidosis, is displayed in Figure 46.11. Evidence on the role of Deaf1 has revealed that multiple isoforms of this protein regulate the expression of peripheral tissue antigens, much like the gene AIRE controls antigen expression in the thymus.[381] Remarkably, the only phenotypic defect observed in Deaf1-deficient mice was the development of retina-reactive autoantibodies, which are also found in AIRE knockout mice. These data indicate that somehow the expression of an appropriate range of peripheral tissue antigens is important to maintain tolerance against retinal components.

b. *Herpesvirus-induced keratitis*: Herpesvirus-induced keratitis has been explored in several animal models. The most intriguing studies have come from the laboratory of H. Cantor. Inoculation of a genetically susceptible mouse strain with HSV-1 results in a viral infection of the eye and cornea that is accompanied by an immune response to HSV-1.[142] Interestingly, damage to the eye is only initiated by the virus and strongly depends on CD4 T cells that can cross-react with a self-protein. This animal model is one of the few that strongly indicate a direct immunopathological role for molecular mimicry in an *in vivo* model for autoimmune disease. Studies by Rouse and colleagues have focused on the role of inflammatory bystander effects in the HSV-keratitis model.[382] In these studies, inflammatory cytokines were shown to play a major role in mediating local damage and administration of DNA vaccine constructs expressing certain beneficial cytokines such as IL-10 ameliorated disease.[383] It is important to point out that these studies were performed using a different HSV strain, and that the precise immunopathogenic process (mimicry vs bystander effects) indeed depends on the strain of HSV used.[384] Additional observations have underscored the importance of regulatory cells in down-modulating a virally induced autoimmune disease. It is worthwhile mentioning that Streilein and colleagues have previously demonstrated regulatory cells in ocular autoimmune disease and found an important role for tolerizing APCs.[385,386]

Treatment and prevention: Due to the associated side effects, it is desirable to avoid systemic immunosuppressive

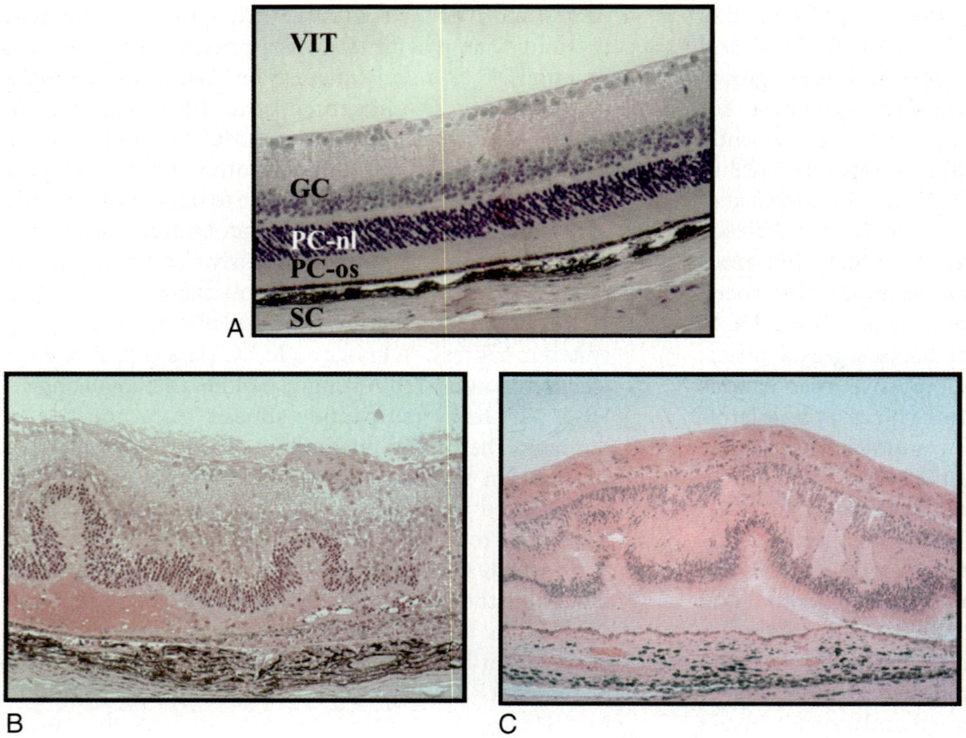

FIG. 46.11. Histopathological comparison of human and mouse uveitis. A: Normal retina, mouse. **B:** Experimental autoimmune uveoretinitis (EAU) mouse. **C:** Ocular sarcoidosis, human. GC, ganglion cell layer; PC-nl, photoreceptor cells—nuclear layer; PC-os: photoreceptor cells—outer segment; SC, sclera; VIT, vitreous. (Courtesy of Rachel Caspi, National Eye Institute, Betheda, MD.)

interventions, as well as the local application of corticosteroids. A Munich group has identified a HLA B27–derived peptide that can function as a mimic to the retinal S antigen and induce EAU in Lewis rats.[387] Using this mimic peptide as an orally administered antigen did produce promising human data, but clinical development seems to have halted. Thus, antigen-specific immune modulation using retinal self-antigens or their peptide mimics may be developed as an effective therapeutic choice. It remains to be seen whether combination of such a treatment with cytokine-DNA vaccines, as demonstrated by Rouse's group for HSV keratitis or by Caspi's group for EAU, both in mouse models, might enhance efficacy. In addition, systemic immune modulation bears promise. Also, a trial using anti-CD25 monoclonal antibody (dacluzimAb) has been conducted at the NIH with promising results. It is noteworthy that in this case the dampening of effector cells rather than any effects on CD25+ T_{regs} appears to explain the efficacy.

Gastrointestinal Autoimmune Diseases

The gut constitutes a unique immunological environment given its large interactive mucosal surface area and the need to maintain tolerance toward food antigens and bacteria normally present within the gut flora. The gut-associated immune system has an important regulatory and barrier function. Immune responses are usually initiated within the Peyer's patches that obtain a significant amount of antigens via the M-cells located within the mucosa and specialized in antigen uptake and transport.[152,388,389] Lymphocytes from the bone marrow will circulate through the Peyer's patches, where B or T cell responses can be initiated. Interestingly, the B cell responses are characteristically IgA high. After antigen encounter, these cells will circulate to the mesenteric lymph nodes and enter the systemic circulation from there. Specifically for the gut, there are other extra-lymphoid locations, where immune cells are found and where immune responses (aggressive or regulatory) can be initiated. One is the intestinal lamina propria, where MHC class II is expressed and bacterial products can be presented. Predominantly CD4+ T and B cells may assume regulatory functions and IgA is secreted with the mucus into the gut lumen. The second are intraepithelial lymphocytes (IELs). A significant proportion of IELs express the γδT cell receptor and they have cytotoxic as well as immune-regulatory functions.[390,391] Certain studies after oral or intranasal feeding of autoantigens have attributed regulatory function to γ/δ lymphocytes. Overall, the hypothesis that the gut environment is ideally suited to induce a tolerizing or regulatory immune response to antigens present in the gut lumen is of high interest and well supported by evidence from animal models (eg, "oral tolerance").[392] The rationale is that the immune system must have evolved to put some unique mechanisms into place to deal with the multitude of "foreign" antigens present in the gut and needed as nutrients and for digestive purposes. Indeed, mice housed within a sterile environment have a shorter life span and multiple immune defects, indicating that the gut's immune system is vital for regular immune development and functions.[265]

Pernicious anemia and autoimmune gastritis: Pernicious anemia is the end-stage of autoimmune gastritis observed in about 10% to 20% of patients[393] and represents the most common cause of vitamin B_{12} deficiency.[394] Autoimmune gastritis is associated with autoantibody production to parietal cells and to its secreted product, intrinsic factor. In addition, a role for CD8 and CD4 T cells has been demonstrated.[395] Animal models have been developed.[396] Standard therapy is parenteral B_{12} administration. Interestingly, this disease also develops in mice after thymectomy and is characterized by a relative lack of CD25+ regulatory CD4+ T cells.[177,397,398]

Ulcerative colitis: Ulcerative colitis is a chronic inflammatory disease of the gut that affects men somewhat more frequently than women. It usually begins in the distal colon and rectum and will spread proximally resulting in severe cases in pancolitis. The hallmark is bloody mucoid diarrhea. The etiology is unclear; evidence points toward a certain genetic predisposition that has to meet triggering environmental factors. Unlike Crohn disease, the inflammatory foci are not granulomatous and not discontinuous. Overall, inflammatory mediators are increased, but systemic symptoms are rare. Treatment is usually achieved by corticosteroids; in severe cases, ulcerative colitis may require removal of the colon.

Crohn disease: As opposed to ulcerative colitis, Crohn disease begins as a discontinuous, granulomatous inflammation of the proximal ileum. Ulcerations are frequent and the disease is more severe than ulcerative colitis. The permeability of the mucosal epithelium is enhanced, which leads to a local breakdown of barrier functions and upregulation of TNF, IL-1, IL-6, and other cytokines. The immune reaction involves the whole mucosa and regional lymph nodes. Interestingly, in addition to standard immunosuppressive treatments, blockade of TNF is a highly efficacious treatment in Crohn disease.[399]

Hepatic Autoimmune Diseases

As an anatomical site for the induction and perpetuation of immune responses, the liver exhibits distinctive immunological features that appear to favor an overall "tolerogenic" microenvironment. This perspective is supported by the apparent absence of liver pathology in spite of the continuous exposure to nutritional antigens and components of commensal gut flora,[400] the ability of viral (hepatitis B and C) and parasitic (malaria falciparum) pathogens to establish persistence and chronic hepatitis,[401,402] as well as the ready acceptance of liver allografts without immunosuppression in experimental models.[403] Moreover, successful liver transplants in humans require comparatively less immunosuppressive therapy,[404] and in some cases, complete weaning of immunosuppressive drugs has been achieved without compromising the functionality of the transplant.[405] The prevailing notion that the liver preferentially promotes abortive immune responses and tolerance was challenged in a meticulous study by Klein and Crispe. Having optimized the considerable surgical feat of orthotopic liver transplants in mice, the authors conclusively demonstrated that the liver effectively initiates the bona fide activation and differentiation of specific T cell responses.[406] Thus, the mechanisms operative in "liver tolerance" are not the consequence of altered antigen presentation, maturation failure, or premature T cell death but rather appear to rely on the activation of poorly defined regulatory circuits by persisting antigens.

Conversely, a breakdown of such tolerance mechanisms, presciently postulated almost 100 years ago as a cause for autoimmune diseases at large,[7] may contribute to pathogenesis of hepatic autoimmune disorders. Here, we will discuss the three major autoimmune diseases of the liver: autoimmune hepatitis (AIH),[407-409] which primarily targets hepatocytes; primary biliary cirrhosis (PBC),[410-412] in which small bile ducts are destroyed; and primary sclerosing cholangitis (PSC),[413,414] in which the damage is mainly focused on medium to large bile ducts. In addition, we have included "halothane hepatitis" as a paradigm for an immune-mediated adverse drug reaction.[415]

Autoimmune Hepatitis

Introduction and disease description: Autoimmunity has been invoked as a cause for hepatitis in the early 1950s,[416-418] and based on the association of AIH and antinuclear antibodies (ANAs), the term lupoid hepatitis was proposed.[419] AIH preferentially affects women but is observed worldwide in children and adults of both sexes in diverse ethnic groups and contributes an estimated 10% to 20% to all cases of chronic hepatitis (which for the most part are due to infections with hepatitis viruses B, C, and D).[407,408] Clinical symptoms range from inapparent to fulminant disease courses, and differential diagnosis is guided by an exemplary scoring system developed by the International Autoimmune Hepatitis Group.[43,420] The presence of autoantibodies targeting different subcellular components is commonly used to differentiate two types of AIH. The etiology of AIH remains undefined, but a contribution of environmental/microbial factors and genetic predisposition, as in so many other autoimmune diseases, is invoked as a trigger for T cell–mediated, destructive autoimmunity. Finally, it is also possible that as of today unidentified hepatitis viruses or other chronic viral infections contribute to some AIH cases.[209,421]

Autoimmune features: The presence of specific autoantibodies in AIH is a key feature in the diagnostic scoring system and the foundation for the clinical distinction of type 1 and type 2 AIH (Table 46.2). Type 1 AIH is characterized by ANAs, a class of antibodies also observed in PBC, PSC, and other forms of chronic hepatitis; smooth muscle and antiactin antibodies; atypical perinuclear anti-neutrophilic antibodies; and antibodies directed against soluble liver antigen and liver-pancreas antigen. None of these antibodies are specific for type 1 AIH, but their combined measurements allow for a reasonable distinction from type 2 AIH and, to a certain extent, other chronic liver diseases. Anti-LKM (anti–liver-kidney mircosomal) antibodies are a serological marker for type 2 AIH, with the majority binding the LKM1 antigen identified as cytochrome monooxygenase P450 IID6 (CYP2D6). In addition, <20% of type 2 AIH patients also carry autoantibodies specific for uridine diphosphate glucuronosyltransferase, designated as LKM3, while 30% to 50% of patients generate antibody responses against formiminotransferase cyclodeaminase (FTCD). Of all these targets in AIH, only one, the hepatocyte-specific asialoglycoprotein receptor (ASGPR), is cell surface expressed and may thus function as a direct immunogen for antibody-mediated

TABLE 46.2 Autoantibodies and Autoantigens in Autoimmune Liver Disease

	Type I AIH	Type II AIH	PBC	PSC
Auto-Abs	ANA: centromere, ribonucleoprotein, cyclin A, histone SMA: F actin, anti-ASGPR anti-SLA/LP pANCA: myeloid-specific nuclear envelop protein	Anti-LKM-1: CYP450, 2D6 Anti-LC-1: formimin-otransferase cyclodeaminase	AMA: PDC-E2, BCOADC-E2, OGDC-E2, PDC-ERBP, PDC-E1a ANA: nuclear pore complex, anti-gp210, anti-p62, nuclear dot pattern, anti-Sp100, anti-PML, anti-SUMOs	pANCA: myeloid-specific nuclear envelope protein ANA: anticardiolipin, thyroperoxidase, rheumatic factor, *Helicobacter pylori* IgG
B cell infiltrate in liver	+with plasma cells	+with plasma cells	+with plasma cells	±

Abs, antibodies; AIH, autoimmune hepatitis; AMA, antimitochondrial autoantibody; ANA, antinuclear antibody; ASGPR, hepatocyte-specific asialoglycoprotein receptor; BCOADC, branched chain 2-oxo-acid dehydrogenase complex; Ig, immunoglobulin; LC, liver cytosol (antigen type 1); LKM, liver-kidney mircosomal; OGDC, 2-oxo-glutarate dehydrogenase complex; pANCA, perinuclear antineutrophilic antibodies; PBC, primary biliary cirrhosis; PDC, pyruvate dehydrogenase complex; PDC-ERBP, ER binding protein of the pyruvate dehydrogenase complex; PML, promyelocytic leukemia protein; PSC, primary sclerosing cholangitis; SLA/LP, soluble liver antigen and liver-pancreas antigen; SMA, smooth muscle antibody; SUMO, small ubiquitin-like modifier.
From Moritoki et al.[422]

hepatocyte damage. ASGPR-reactive antibodies, also not disease-specific, are found in 50% to 75% of all AIH patie nts.[407,408,422] Since the generation of these autoantibodies is T-helper cell dependent, it is not surprising that loss of tolerance also occurs on the T cell level and lymphocytes isolated from livers of patients with AIH type 2 can react *in vitro* and expand after exposure to P4502D6 antigen. This finding supports the concept of a lymphocyte-driven autoaggression and might underlie pathogenesis. Furthermore, mononuclear infiltration of the periportal areas and liver parenchyma is usually present when the disease progresses and causes the AIH-typical "piecemeal" appearance of the liver. Although evidence from animal models and human studies gives some indication that cytotoxic killing of hepatocytes could play a role, further analysis is still hampered by the fact that intrahepatic lymphocytes are difficult to access in humans. The notion that defective suppressor T cell activity contributes to AIH pathogenesis, proposed more than 2 decades ago, is supported by the finding of reduced frequencies of peripheral CD4+CD25+ T_{reg} in AIH patients as compared to healthy controls.[423] Cytokines also appear to play an important role in human AIH. Administration of type 1 interferon led to strong exacerbation of AIH, showing that it likely plays a central role in liver destruction.[424] Interferons upregulate MHC class I and class II molecules, enhance inflammation, and have strong antiviral effects. The outcome of these studies therefore argues against a role of an ongoing chronic (or acute) viral infection in patients with AIH who test negative for hepatitis virus antigen or antibodies. A negative regulatory role can possibly be attributed to IL-6 and TNFα. These two cytokines are reduced in livers of AIH patients and have a negative effect on cytochrome P450 regulation.[425] As a consequence, the lower levels of IL-6 and TNF might enhance CYP2D6 expression and in this way support autoimmunity directed to CYP2D6 in patients with type 2 AIH. Lastly, AIH pathogenesis is not necessarily controlled by the Th1/Th2 paradigm: CD4+ T cells expressing Th1 cytokines are elevated in type 1 AIH, whereas Th2 cytokines are augmented in type 2 AIH in agreement with the high levels of CYP2D6-specific antibodies present in these patients.

Is production of autoantibodies important for AIH pathogenesis or is it just an epiphenomenon signifying the breaking of tolerance to self-antigens? Evidence suggests that autoantibodies in AIH play a role that can be situated somewhere in between T1D, where autoantibodies and B lymphocytes likely play no pathogenetically important role in humans, and SLE, where autoantibodies are complement-fixing and enhance organ destruction. LKM antibodies found in AIH, for example, can inhibit cytochrome P450 function *in vitro* but not *in vivo* and may participate in hepatocyte dysfunction or destruction. Epitope mapping of the CYP2D6 antigen in type 2 AIH revealed two immunodominant regions spanning aa 256 to 269 and aa 181 to 245 that are recognized by most of the AIH type 2 patient's sera. Since there is a significant genetic polymorphism in humans for the CYP2D6 antigen, it is possible that patients with AIH target CYP2D6 protein selectively (LKM1 antibodies), whereas cytochrome P4502D9 (LKM2) is targeted in patients with drug-induced hepatitis, disulfide-isomerase in halothane-induced hepatitis, or UDP-glycosyltransferases in hepatitis D virus–associated autoimmunity. As so often, identification of the major target antigens, clearly a decisive step toward a better understanding of the underlying disease process, has so far not yielded compelling hypotheses as to the precise pathogenetic mechanisms.

Genetic features and environmental factors: No single genetic locus capable of explaining AIH etiology has been identified. HLA-DR3 and HLA-DR4 are associated with type 1 AIH, and the HLA-A1-B8-DR3 haplotype exhibits a particularly strong association with early disease onset, relapse, and the eventual necessity of liver transplantation. In contrast, type 2 AIH is associated with carriage of HLA-DR7. Genes outside the HLA loci potentially conferring disease susceptibility include complement component C4, CTLA-4, vitamin D receptor, Fas receptor, and others.[407,408,422]

The association with hepatitis C, D, and E virus–induced hepatitis and autoimmunity remains very intriguing.

However, at this point, the prevalence of hepatitis C virus (HCV), for example, exhibits drastic variations between studies performed in different countries, and autoantibody titers appear much higher in patients with HCV-negative AIH. In a more recent study, it was found that sera of 38% of chronic hepatitis C patients reacted specifically with CYP2D6, whereas none of the sera obtained from patients with chronic hepatitis B showed CYP2D6 reactivity.[424] Furthermore, it was found that HCV has the potential to induce autoreactive CD8+ T cells that cross-reactively recognize the cytochrome P450 isoforms 2A6 and 2A7 that contain sequence homology to HCV aa 178 to 187. In this context it may be important to emphasize that molecular mimicry also seems to be an important factor in other immune-mediated diseases of the liver. Hence, trifluoroacetyl (TFA)-protein adducts as generated during the metabolism of halothane by CYP2E1 confer molecular mimicry to the lipoic acid prosthetic group of the pyruvate dehydrogenase complex (PDC) and other members of the 2-oxoacid dehydrogenase family,[79,237,299,426] which in turn are major autoantigens in PBC.[72] Consequently, halothane hepatitis and PBC may be linked on the level of cross-reactive autoantibodies that recognize similar target antigens. It remains to be seen, whether novel subtypes of hepatitis viruses can be found in AIH patients and whether further studies will corroborate such an association. It is likely, however, that viruses play a multifactorial role in AIH pathogenesis (similar to T1D) and that no single virus will be identified as a specific cause for liver autoimmunity.

Animal models: Over the past 30 years, many attempts have been made to develop animal models for AIH, however to date, no reliable model is available that reproduces the spontaneously relapsing chronic disease course of type 1 AIH.[427,428]

a. In the HbsAg-transgenic mouse, the hepatitis B virus surface antigen (HBsAg) is expressed in hepatocytes under control of the mouse albumin promoter. Induction of transient hepatitis is possible after adoptive transfer of activated T cells from HBsAg-primed donor mice.[429,430] Although not designed to reproduce features of AIH, this model has been extraordinarily helpful in understanding the role of interferons in inducing liver damage, as well as clearing HBV from the liver. Results have shown that interferons, in the absence of cytotoxicity, can purge virus from infected hepatocytes. Furthermore, induction of HBV-specific CD4+ and CD8+ T cell responses can be evaluated in this model. The transgenic mice exhibit profound liver damage and infiltration after transfer of HBV-specific T lymphocytes.

b. In another model system, the MHC class I molecule H-2K^b is transgenically expressed in the liver of mice that also carry transgenic T cells specific for H-2K^b. Hepatitis induction was only successful when such mice were infected with a liver-specific pathogen, indicating that bystander activation within the liver microenvironment can be very potent in causing autoimmune damage.[431] A similar approach was chosen in yet another model that combines liver-specific expression of a transgene (here, the immunodominant LCMV glycoprotein epitope [GP_{33}]) and transgenic T cells specific for the transgene.[432] These mice remain tolerant to the transgene even in the presence of transferred GP_{33}-specific T cells. However, under conditions of LCMV infection and TCR-transgenic T cell transfer, these mice develop a transient form of hepatitis as evidenced by elevated ALT levels.[432] Together, the above findings emphasize the importance of both autoreactive T cells and generalized inflammatory alterations in the wake of microbial infection.

c. The potential pathogenic contribution of autoantibodies was demonstrated in a new model mouse model. Injection of a monoclonal antibody, derived from a patient with type 1 AIH and specific for an undefined 190 kDa cell surface protein, promotes predominantly perivenular hepatocellular necrosis in mice.[433] However, the validity of this model remains to be critically evaluated as the pathology does not appear to closely resemble that of type 1 AIH.

d. Arguably, more compelling models have been developed for the study of type 2 AIH. DNA immunization of mice with human CYPD6 and FTCD antigens leads to the generation of LMK1- and LC1 autoantibodies (targeting CYPD6 and FTCD, respectively) and specific cytotoxic T cells, pronounced lymphocytic infiltration of the liver, and an elevation of ALT levels indicative of liver damage.[434] Interestingly, work with this model has reproduced some aspects of the genetic predisposition exerted on disease development as C57BL/6 mice demonstrated more pronounced disease than 129/Sv mice and no inflammation was found in Balb/c mice.[435] Lastly, adenoviral infection with recombinant expressing the 2D6 antigen can lead to focal and confluent liver necrosis in mice resembling that of AIH[436] (Fig. 46.12). In summary, the success of most mouse models has only been partial, since hepatitis was transient and induction of chronic disease appeared difficult to achieve. Figure 46.13 illustrates the possible pathogenetic mechanisms implicated in AIH.

Treatment and prevention: Immunosuppression is the therapy of choice for AIH. Since their introduction in 1968,[437] prednisolone and azathioprine have become standard treatment regimens.[408,438] The impressive success rate of this therapeutic approach (over 85% of AIH patients) may have, to a certain extent, removed the urgency to pursue basic AIH research.[428] Immunosuppressive therapy also improves survival of patients with severe AIH,[439] but no guidelines are available for individuals with minimal symptoms. End-stage AIH is an important indication for liver transplantation.[440] Recurrence of AIH has been reported after liver transplantation,[441] but interestingly, is not as frequent as observed for islet transplants in T1D patients. Therapeutic use of interferons, effective for treatment of viral hepatitis, can worsen autoimmune liver disease[424] and challenges the assumption that unknown chronic viral infections of the liver, while possible initiators, maintain the active disease process.

Primary Biliary Cirrhosis

Introduction and disease description: PBC is a slowly progressive hepatic autoimmune disorder that primarily affects

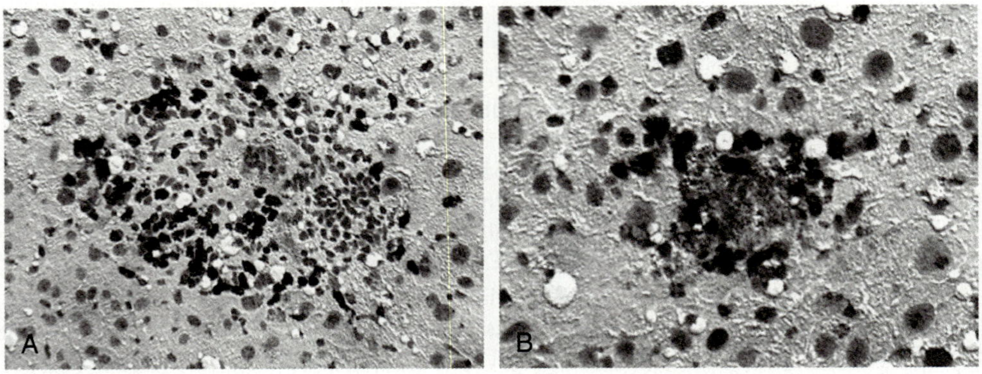

FIG. 46.12. Liver necrosis in 2D6 transgenic mice infected with an adenoviral recombinant expressing 2D6. A: Confluent necrosis with infiltrating CD8 cells. **B:** Focal necrosis with infiltrating CD8 cells. (Courtesy of Urs Christen and Matthias von Herrath, University of Frankfurt, Germany and La Jolla Institute for Allergy and Immunology, La Jolla, CA, USA.)

women. The disease is typically associated with high titers of antimitochondrial autoantibodies (AMAs) that presage onset of clinical disease by 5 to 10 years. Histopathological changes include portal inflammation and destruction of intrahepatic bile ducts that occur at various rates and degrees in different patients. The resulting decrease of bile secretion and retention of toxic substances within the liver leads to further hepatic damage, fibrosis, cirrhosis, and eventual liver failure.[410-412]

Autoimmune features: Although speculation about the autoimmune pathogenesis of PBC centers around the usual suspects of genetic and environmental/microbial factors, PBC may serve as an example how advanced knowledge about the targeted autoantigens can give rise to novel and intriguing hypotheses. The presence of AMAs in PBC were first described in the 1950s, the autoantigens cloned in the 1980s and 1990s, and B cell, CD4+, and CD8+ T cell epitopes of the major autoantigen PDC-E2 mapped in the 1990s and 2000s. Over 90% of PBC patients have high titers of autoantibodies directed against the 2-oxo-acid dehydrogenase complex (2-OADC) E2 subunit, in particular the E2 component of the pyruvate dehydrogenase complex (PDC-E2). Mapping studies have identified determinants within the PDC-E2 lipoyl domains as a primary targets for antibody and T cell reactivity, a motif that is also found in the E2 subunit of other 2-OADC members (OGDC-E2, BCKD-E2; motif is also found in E3-BP).[411] Antibody reactivity to nuclear envelop proteins gp210 and p62 that are found in 10% to 30% of patients is apparently associated with more active disease. The obvious question emerging from the identification of these autoantigens pertains to the fact that they are (1) found in every cell type and (2) located on the inner surface of the inner mitochondrial membrane, thus separated from immune effectors by three membranes. In addition, the existence of a clinical entity known as AMA-negative PBC, a disease very similar to PBC but without detectable AMA, appears to further argue against a pathogenic role of these autoantibodies.[412] Yet, certain particularities of bile duct epithelial cells may predispose them as targets for humoral immune responses. These cells, just as salivary epithelial cells, express receptors involved in the transcytosis of IgA, and high titers of PDC-E2-specific IgA are found in bile, saliva, and urine of PBC patients. AMA IgA was shown to induce apoptosis, which in turn may lead to PDC accumulation in the cytoplasm and cell surface and amplification of the immune response. Further speculation has focused on the possibility that failure of PDC glutathionylation specifically in epithelial cells may promote accumulation of immunogenic forms of PDC, a process that may be initiated by xenobiotic-modified PDC and molecular mimicry.[411] In the "determinant density model,"[442] potentially PDC-reactive T cells survive negative thymic selection due to

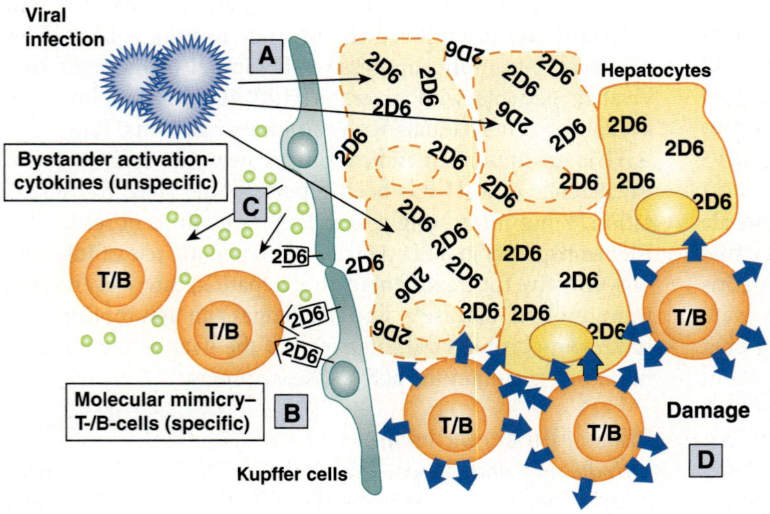

FIG. 46.13. Pathogenesis of liver autoimmunity after viral infection in the 2D6 animal model. Viral infection leads to expression of high cytokine levels in the liver **(A)**. The infection will probably also activate resident macrophages (Kupffer cells) **(B)**. Antiviral T cells react against virally infected cells and destroy them. This leads to release of autoantigens fom hepatocytes, which can be (cross)-presented, leading to pathogenic responses against autoantigens such as cytochrome P450 2D6. Chemokine and cytokine release **(C)** leads to enhanced inflammation and antigen presentation **(D)**.

low TCR avidity and remain "dormant" in spite of sporadic exposure to PDC-derived epitopes. However, strong inflammatory stimuli that allow for improved antigen presentation and activation of toll-like receptors (TLRs; monocytes from PBC patients exhibit enhanced sensitivity to TLR signals), for example through microbial molecular mimicry, may induce the breakdown of tolerance that precipitates the disease. It should be noted that the general validity of this hypothesis, including a key role for low-avidity T cells and TLRs, is fully supported by elegant studies in models for induction of autoimmune diabetes.[114,443] Finally, a model has been proposed in which a primary dysfunction of endothelial cells results in overproduction of endothelins, inflammatory alterations and ischemic damage due to vasoconstriction, and the generation of PDC-specific humoral and cellular responses as a secondary event.[412] These hypotheses are not mutually exclusive and may be critically evaluated in two animal models.

Genetic features and environmental factors: Familial clustering of PBC and a high concordance rate in monozygotic twins (63%) indicate a genetic component to PBC pathogenesis. A flurry of GWAS data has dramatically improved our insight into the genetics landscape of PBC. Before the implementation of GWAS technology, only class II HLA loci had been reproducibly shown to associate with disease. Non-HLA loci were reported on some occasions but replication often proved problematic. GWAS studies confirmed HLA as the strongest association and many additional non-HLA risk loci have been identified, with equivalent effect size to HLA, including IL12A,[444] IL12RB2,[444] STAT4,[445] IRF5-TNPO3,[446] 17q12.21,[446] MMEL1,[446] SPIB,[446] and CTLA-4.[444] As argued before, the list of associated genes comprises many proteins with important roles in innate and adaptive immunity and thus the existence of a monogenic pathway that leads to disease can be excluded. In correspondence with, for instance, T1D and RA, CTLA-4 also confers risk for this condition and many of the candidate genes signal though NF-κB, TLR, and TNF pathways, which are also important in many other autoimmune conditions. The future will tell whether this growing list of gene associations will be instrumental in unraveling the immune pathology that underlies PBC.

Animal models: The absence of a suitable animal model has long been a serious obstacle to PBC research. However, two publications document important progress on this front by describing the spontaneous onset of a PBC-like disease in two mouse strains. NOD.c3c4 mice congenically derived from the NOD strain generate PDC-E2-specific autoantibodies and exhibit biliary pathology characterized by destructive cholangitis, granuloma formation, and eosinophilic infiltration. Importantly, T cells are found in affected areas of the biliary epithelium and transfer of splenocytes or CD4+ T cells induces disease in NOD.c3c4-scid recipients, suggesting a central role for T cells in pathogenesis.[447] In the second model, abolition of TGF-β-signaling in T cells (achieved by T cell–specific expression of a dominant-negative TGF-βRII) results in production of AMAs directed at PDC-E2, BCOADC-E2, and OGDC-E2 as well as lymphocytic liver infiltration, with periportal inflammation analogous to the histological profile in human PBC. These observations implicate the TGF-β pathway in PBC pathogenesis and suggest that the activation of intrinsically self-reactive T cells is a consequence of impaired regulatory T cell function rather than molecular mimicry.[448]

Treatment and prevention: It is noteworthy that despite its presumed autoimmune etiology, PBC is not ameliorated by immunosuppressive therapy. In the absence of optimized therapeutic protocols, treatment is tailored according to individual responsiveness of PBC patients and includes management of disease symptoms (pruritus, osteoporosis, hyperlipidemia, portal hypertension, as well as therapy of the underlying disease by stepwise addition of ursodeoxycholic acid [UDCA], colchicine, and methotrexate). Nevertheless, the use of colchicine and low-dose methotrexate, presumably acting as an immunomodulatory rather than antimetabolic agent, remains controversial. In cases of liver failure, orthotopic liver transplantation is the only effective treatment.[410]

Primary Sclerosing Cholangitis

Introduction and disease description: PSC is a chronic cholestatic liver disease (cholestasis: suppression of biliary flow) frequently associated with inflammatory bowel disease (IBD; >75% of PSC patients). It is characterized by fibrotic inflammation and destruction of large intra- and extrahepatic bile ducts, and may lead to the development of cholangiocarcinoma (in 10%-30% of PSC patients) and ultimately death by liver failure.

Autoimmune features: Although the etiology of PSC remains undefined, a complex multistep process has been delineated and involves cholangiocyte activation through bacterial pathogen-associated molecular patterns (PAMPs), production of proinflammatory cytokines in conjunction with aberrant chemokine expression and endothelial cell adhesion molecules, and the recruitment of T cells presumably specific for enterocyte antigens and primed in GALTs.[414] The generation of autoantibodies is a common feature (Table 46.2), but most antibodies are not specific for PSC and their contribution to disease pathogenesis remains at present unclear.[422]

Genetic features and environmental factors: The relative risk of PSC among first-degree relatives of PSC patients is almost 100-fold increased as compared to the general population, indicating that there is a strong genetic component to the condition.[413] As in many other autoimmune conditions discussed here, GWAS studies have greatly contributed to our understanding of the array of genes that contribute to disease susceptibility. A first report revealed a similar genetic architecture as described in PBC, and it was confirmed that the association with HLA was even more pronounced.[449] The authors also found one non-HLA association at chromosome 13q31 (locus also associated with MS) and next assayed 15 established IBD (which affects 50%-80% of patients with PSC) susceptibility loci. They were able to identify two additional non-HLA susceptibility loci (chromosome 2q35, and chromosome 3p21) for which the G-protein-coupled bile acid receptor 1 (TGR5) and macrophage-stimulating 1 (MST1) were proposed as the likely disease genes. A subsequent study from the same group added two more non-HLA loci, 2q13 and 10p15, with presumable candidate genes BCL2-like

11 (BCL2L11) and interleukin 2 receptor, alpha (IL2RA).[450] Interestingly, IL2ra−/− mice spontaneously develop intestinal and biliary inflammation.[451] Finally, three more IBD susceptibility loci were found to be associated with PSC, encompassing the candidate genes REL, IL2, and CARD9.[452] Collectively, these genetic data offer strong, albeit still indirect, support for an important role for innate and adaptive immune components in PSC. Moreover, the shared genetic patterns between PSC and IBD indicate a certain degree of common immunopathogenesis and once again suggest the potentially shared origin of what first seemed to be distinct autoimmune disorders.

Animal models: None of the animal models developed to date reproduce all pertinent aspects of human PSC.[453] Rat models of small bowel bacterial overgrowth and administration of bacterial cell wall components have provided some support for an etiopathogenetic role of PAMPs. Other models include injury of biliary epithelial or vascular endothelial cells (rodents, dog, monkey) and toxic, infectious or intraluminal biliary injury (rodents) permitting the study of cytokines, chemokines, and inflammation in the development of diffuse bile duct sclerosis.[453]

Treatment and prevention: There is currently no effective medical therapy available for PSC.[454] Treatment with the hydrophilic bile acid UDCA improves liver enzymes, but its effect on liver histology and prognosis remains inconclusive. Orthotopic liver transplant constitutes the only established long-term treatment for PSC with cumulative 5-year survival of about 70%.

Halothane Hepatitis

Halothane hepatitis is a severe, life-threatening form of hepatic damage that affects a small subset of individuals exposed to the anesthetic agent halothane[455] and is thought to have an immunological basis. Sera of afflicted individuals contain autoantibodies directed against the native and the trifluoroacetylated (TFA) form of hepatic proteins. TFA proteins are generated during the oxidative metabolism of halothane (2-bromo-2-chloro-1,1,1-trifluoroethan e) and include cytochrome P450, protein disulfide isomerase, microsomal carboxlesterase, calreticulin, Erp72, GRP78 (BiP), and GRP94. Current evidence suggests that such TFA proteins arise in all individuals exposed to halothane. However, the vast majority of individuals appear to tolerate this covalent protein modification. The lack of immunological responsiveness was suggested to occur due to tolerance induced through the presence of structures in the repertoire of self-determinants that immunochemically and structurally mimic TFA proteins very closely. In fact, lipoic acid, the prosthetic group of the constitutively expressed E2 subunits of members of the 2-oxoacid dehydrogenase complex family, was demonstrated by immunochemical and molecular modeling analysis to perfectly mimic N^6-trifluoroacteyl-L-lysine, the major haptenic group of TFA proteins. Interestingly, a fraction of patients with halothane hepatitis exhibit irregularities in the hepatic expression levels of these cross-reactive proteins. Thus, molecular mimicry of TFA-lysine by lipoic acid, or the impairment thereof, can be considered a susceptibility factor of individuals for the development of halothane hepatitis.[456] A small animal model for chemically induced liver diseases has been described.[457]

Perspective and conclusions: The presence of autoantibodies in AIH, PBC, and PSC is indicative of autoimmune processes, but none of the antibodies described is liver-specific and their contribution to pathogenesis remains unclear. While the complex role of T cells is subject to current investigations, the same caveats of specificity and pathogenicity discussed earlier apply for autoimmune liver diseases.[458] Although the success of immunosuppressive therapy, in fact a criterion for the diagnostic scoring system for AIH,[43,420] further supports autoimmune etiology for AIH, it is conceivable that a viral infection might initiate disease as a "hit-and-run" event. Following elimination of viral antigens, disease may be perpetuated by immune-mediated processes and preclude the identification of a particular virus as causative agent at the time of liver disease diagnosis. Again, the role of "foreign-" and "self-"antigens becomes blurred and their interactions with the immune system in terms of immunity or autoimmunity conceptually problematic. Future investigations should seek to improve animal models, and focus on human autoantigens and the application of contemporary tools for identification and isolation of specific lymphocytes. On the basis of such developments, future *in vivo* tracking of autoaggressive lymphocytes with noninvasive methods should substantially improve our insight into disease pathogenesis.

Renal Autoimmune Diseases

Glomerulonephritis (GN) is the major cause of chronic renal disease and kidney failure and exhibits a wide spectrum of histopathological alterations, disease severity, and clinical outcomes. In most cases, evidence for an immune-mediated pathogenesis, including humoral and cellular responses, has been obtained.[459-462] The precise incidence of defined renal autoimmune disease is difficult to establish since definite diagnosis requires histological evaluation of the kidneys. In this respect, the Italian registry of renal biopsies provides a unique repository, and information about selected glomerular disease obtained from this resource is displayed in Table 46.3.[463,464] The identity of nephritogenic antigens targeted in most renal autoimmune diseases remain poorly defined, with the notable exception of Goodpasture syndrome, a form of anti-glomerular basement membrane (GBM) disease in which autoantibodies bind to α3(IV) collagen in the GBM and are associated with rapidly progressive GN as well as pulmonary hemorrhage. Conformational epitomes have been mapped to two regions within the noncollagenous (NC1) domain of the α3(IV) chain.[465] A role for T cells, both in providing help for autoantibody production and exerting direct effector functions, has also emerged. Cast within the reigning concept of balanced Th1 and Th2 immunity, it appears that Th1 responses are associated with proliferative

TABLE 46.3 Incidence of Glomerular Diseases in 1993

Disease	Incidence per Million Population
Immunoglobulin A nephropathy	8.4
Membranous glomerulonephritis	4.9
Focal segmental glomerulonephritis	2.3
Minimal change disease	1.6
Mesangiocapillary glomerulonephritis	1.4
Poststreptococcal glomerulonephritis	0.7

From Schena et al.[463]

and crescentic forms of GN that cause severe injury while Th2 responses are associated with membranous patterns of injury.[459] These distinctions have also given rise to speculations that have harnessed the "hygiene hypothesis" as a potential explanation for the observation that proliferative GNs have a relatively high incidence in developing countries whereas minimal change disease and IgA nephropathy predominate in developed countries. Therapeutically, immunosuppression with corticosteroids and in combination with cyclophosphamide and plasma exchange (to remove damaging antibodies) is employed to manage diseases. Based on the Th1/Th2 paradigm and its role in developing therapies for other autoimmune diseases that aim to target the Th1-Th2 balance, therapeutic approaches such as oral tolerance, cytokine-based immune modulation (anti-TNFα and IL-12p40 antibodies, recombinant IL-10), anti-adhesion molecules, and chemokine receptor blockade are evaluated for their potential use in the treatment of autoimmune renal diseases.[459]

Cutaneous Autoimmune Diseases

Cutaneous autoimmune disorders are commonly divided into non-organ-specific and organ-specific diseases. The former group contains skin manifestations of systemic autoimmune disorders such as SLE, scleroderma, Crohn disease, and ulcerative colitis. The organ-specific autoimmune skin disorders discussed in this section include several autoimmune bullous dermatoses as well as pigmentation disorder vitiligo. Despite a relative paucity of animal models for human autoimmune skin disorders, the skin is one of the few organs that offer direct access to target antigens and effector cells; consequently, advanced insights into mechanisms of disease induction and perpetuation have been obtained.

Autoimmune Bullous Dermatoses

Autoimmune bullous dermatoses are a group of severe, possibly fatal skin disorders characterized by erosions and blisters of skin and/or mucous membranes (Table 46.4). The term "pemphigus" (from Greek/Latin: pustule, bubble), featured in the name of several of these diseases, dates back to "pemphigoid fever" described by Hippocrates nearly 2400 years ago and has been used to describe chronic blistering skin conditions since the mid-18th century. All forms of pemphigus have distinctive characteristics but exhibit the common feature of epidermal cell-to-cell detachment known as acantholysis. On occasion, pemphigous syndromes are associated with other autoimmune diseases such as SLE, RA, myasthenia gravis, pernicious anemia, and Hashimoto thyroiditis.

Pemphigus vulgaris (PV). PV is the most common and severe form of pemphigus in the United States, constituting more than 80% of reported cases.[466,467] With a worldwide incidence of 1 to 5 cases per million per year in the general population, PV remains a rare disease and preferentially affects individuals in the fourth to sixth decade of life. Clinical features include flaccid, fragile, and noninflammatory bullae (from Latin: bubble) that arise on normal-appearing skin. These lesions tend to coalesce and rupture, leaving larger areas of denuded skin that often heal poorly.[466-468] About half of the patients initially present with lesions confined to the oral mucosa, although any surface with squamous epithelial tissue may be affected. Blister formation begins with edema of epidermal intracellular spaces and progresses to complete loss of cohesion between epidermal cells. In PV, blisters occur in the suprabasilar region of the epidermis while leaving attachment of basal cells to the dermis largely intact.[466-470]

Autoimmune features: The hallmark of PV is autoantibodies generated against the desmosomal glycoprotein desmoglein-3 (Dsg3) (Fig. 46.14). In fact, the unequivocal identification of this antigen makes PV one of just a few human autoimmune diseases for which the relevant autoantigen has been defined. Although the pathogenic contribution of Dsg3 antibodies has been demonstrated in passive transfer experiments, more recent work has emphasized the importance of specific T cells.[471] TCR transgenic mice with autoreactive CD4+ T cells that are specific for Dsg3 were recently generated. It was found that these CD4+ T cells specific for a single skin-associated antigen could induce a pemphigus-like phenotype.[472]

Peripheral CD4+ and occasional CD8+ T cell responses targeting Dsg3 with varying functional profiles (Th1 and Th2) have been described. These autoreactive CD4+ T cells may be involved in the coordination of specific B cell responses as evidenced by Th1-regulated IgG1 and Th2-regulated Dsg3 antibodies in patient sera. Interestingly, IL-10-producing Dsg3-reactive T cells have also been identified in the majority of healthy individuals carrying PV-associated HLA alleles (see below) but were found in less than 20% of PV patients.[473] Given these functional attributes reminiscent of the Tr1 regulatory T cell subset as well as the documentation of CD8+CD28− T cells with potential regulatory function in remittent PV patients (cited in Ref. 471), it has been proposed that clinical disease may emerge as a consequence of an imbalance between autoreactive pathogenic and regulatory T cells.[471]

TABLE 46.4 Autoimmune Bullous Dermatoses

Disorder	Acantholysis	Autoantigen	Autoantibody Class	T Cell Subset
Pemphigus vulgaris	Suprabasila	Desmoglein-3 > Desmoglein-1	IgG4, IgG1, IgA (occasionally)	Th1, Th2, Tr1, CD8+CD28– T_{reg}
Pemphigus foliaceus	Subcorneal	Desmoglein-1	IgG4, IgG1	Th1, Th2
Bullous pemphigoid	Subepidermal	BP180, BP230	IgG1, IgG4, IgA (occasionally)	Th1, Th2
Cicatricial pemphigoid	Subepidermal	BP180 > BP230, Laminin 5	IgG1, IgG4, IgA (occasionally)	Th1
Pemphigus gestationis	Subepidermal	BP180, BP230	IgG1, IgG3	Th1, Th2
Dermatitis herpetiformis (associated with gluten-sensitive enteropathy)	Subepidermal	Epidermal transglutaminase	IgA	Th2
Linear IgA disease	Subepidermal	BP180, BP230	IgA	Th1, Th2
Epidermolysis bullosa acquisita	Subepidermal	Collagen VII	IgG, IgA, and IgM (occasionally)	Th1, Th2, no T_{reg} contribution

CD, cluster of differentiation; Ig, immunoglobulin; Th, helper T; T_{reg}, regulatory T.

Genetic features and environmental factors: Although PV can affect anyone, an enhanced prevalence is observed in people of Mediterranean or Jewish ancestry. Associations with certain HLA alleles (HLA-26, HLA-B38, SC21, HLA-DR4) in Ashkenazy Jews and non-Ashkenazy Jews (HLA-Drw6) have been detected in PV patients, and it has been suggested that the third hypervariable region of the beta-1 chain of the DRB1 allele may constitute a susceptibility locus.[466-468] Recent studies indicate that the H-CDR3 region within anti-Dsg antibodies is critical for disease induction and even one amino acid change can alter their pathogenicity.[474] It was concluded that mutations in these areas can uncouple pathogenicity and binding. Finally, while PV may be associated with other autoimmune diseases,[470] its occurrence with certain malignant diseases has been defined as a separate entity.[475]

Animal models: The pathogenic role of autoantibodies has been demonstrated by passive transfer of PV patient sera into neonatal mice.[476] More recent studies using Dsg3-deficient mice have established that loss of tolerance and induction of autoimmunity required the presence of both autoreactive B and T cells.[477-479] Finally, CD4 T cells from Dsg3-immunized HLA transgenic mice target Dsg3 epitopes identical to those identified in humans and thus constitute a promising model to further analyze the regulation of Dsg3-specific T cells and autoantibody production *in vivo.*[471]

Treatment and prevention: Corticosteroids are the treatment of choice with supplementary immunosuppressive agents (cyclophosphamide, azathioprine, and methotrexate) commonly used to reduce the side effects of steroids. In severe cases, plasmapheresis may be of short-term benefit. Clinical data indicate that B cell depletion, achieved by administrating

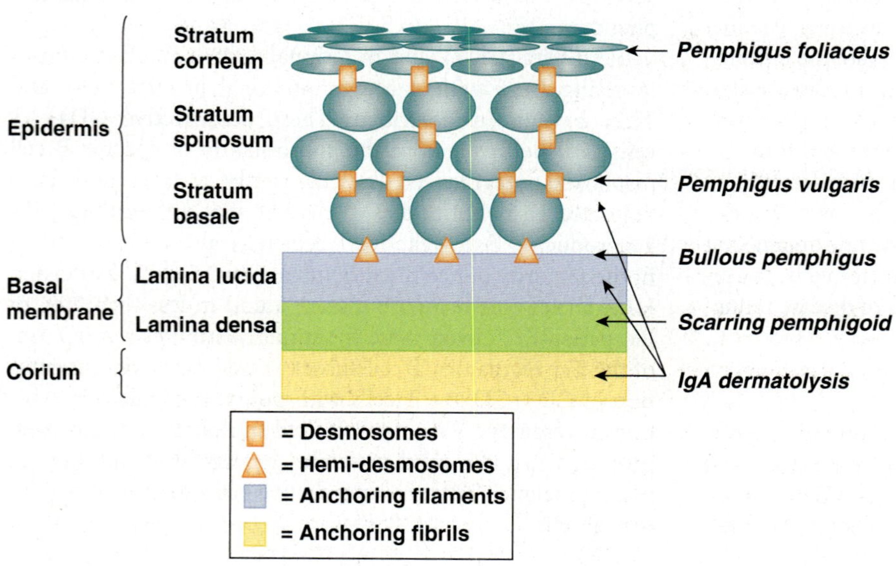

FIG. 46.14. Different layers of the skin and their involvement in autoimmune skin syndromes.

anti-CD20 antibodies (rituximab) is an effective treatment for pemphigus, but it is generally acknowledged that, due to its potentially severe side effects, its use should be limited to the most severe types of the disease.[480]

Other autoimmune bullous disorders include pemphigus foliaceus (PF), bullous pemphigoid (BP), cicatricial (scarring) pemphigoid (CP), epidermolysis bullosa acquisita (EBA), dermatitis herpitiformis (DH), and linear immunoglobulin A disease (LIgAD) (Table 46.4). Two forms of PF are recognized, one affecting individuals irrespective of ethnic background around the world and the other, PF fogo selvagem (Portuguese: savage fire), endemic to certain regions of Brazil. Antibodies found in the circulation of PF patients preferentially target desmoglein-1, are of the IgG4 subclass and, similar to PV, can induce aspects of the disease in a neonatal mouse transfer model[466]; Th1 and Th2 T cells specific for the same antigen have also been described.[471] Information about genetic association remains incomplete due to the rarity of PF. However, PF fogo selvagem demonstrates associations with HLA-DR1 and HLA-DR4. Prior to the advent of corticosteroids, fewer than 10% of patients underwent spontaneous remission and fatalities reached up to 40% within the first 2 years of disease. Compared to PV, PF patients respond more rapidly and to smaller doses of corticosteroids.[466]

Together with PV, BP belongs to the best-characterized bullous disorders with regard to pathogenesis. In BP, IgG autoantibodies are directed against components of the dermoepidermal basement membrane such as BP antigens BP180 and BP230. Again, these antigens are also recognized by CD4+ T cells and comprise both Th1- and Th2-type subsets,[471] and systemic corticosteroids with facultative addition of other immunosuppressive agents are used for treatment. The principal autoimmune targets and basic pathologic features of the remaining bullous dermatoses are listed in Table 46.4.

Perspective and conclusions: The identification of relevant autoantigens, a task that is far from complete in the majority of human autoimmune diseases, is usually considered to be a critical juncture in our understanding of pathogenetic mechanisms and an important step toward the development of targeted immunomodulatory therapies. However, the case of PV and BP is a telling reminder that even with established knowledge about such primary autoantigens, the development of specific immunotherapies faces continuing challenges.

Vitiligo

Introduction and disease description: Vitiligo is an acquired, noncontagious disorder of pigmentation resulting from melanocyte loss (reviewed Refs. 481-483). The disease is characterized by defined areas of depigmentation that range from limited numbers of macules (focal vitiligo) to bilateral, symmetric distribution of the lesions (generalized vitiligo, the most common type) to involvement of the complete body surface area (universal vitiligo). Vitiligo is arguably the most common pigmentary disorder affecting 0.5% to 2% of the population worldwide and demonstrates no predilection for gender or ethnicity. Although commonly regarded as a pathophysiological entity with distinct clinical manifestations, the existence of a spectrum of vitiligo disorders with different etiologies was suggested. The potential autoimmune etiology of vitiligo has been discussed for some time and has gained traction through recent studies on genetic epidemiology. A combination of environmental, genetic, and immunologic factors is thought to contribute to autoimmune melanocyte destruction. This hypothesis is further strengthened by the reported association of vitiligo with other autoimmune diseases such as autoimmune thyroid disease, SLE, Addison disease, adult-onset T1D, and pernicious anemia.[484] Another interesting analogy with T1D was coined by Eisenbarth, who noted that both diseases shared but unexplained focality of melanocytes or beta cells.[485] A lingering question in this regard is what differentiates one melanocyte or beta cell from another in becoming a target for autoimmunity. Nevertheless, the specific causes of vitiligo remain obscure and progress limited due to lack of a clear definition of the disorder as well as a dearth of experimental animal models.[482]

Autoimmune features: T cell infiltrates have been reported in generalized vitiligo and exhibit molecular signatures indicative of CTL and Th1 responses (TNFα and IFNγ).[486] CD8+ T cells tend to predominate in vitiligo lesions, and high frequencies of Melan-A-specific CD8+ T cell subsets apparently correlate with severity of disease.[487,488] Additional infiltrating cells include macrophages and monocytes producing proinflammatory cytokines as well as DCs. The role of DCs in local immune responses remains unclear but the DC-mediated killing of melanocytes has been suggested as a mechanism for depigmentation.[489] Finally, circulating autoantibodies, including those specific for the melanocyte antigen TRP1, have been found in the sera of vitiligo patients, and TRP1-specific antibodies can induce melanocyte loss in mice.[490] Nevertheless, there appears to be an emerging consensus that alterations of humoral immunity are secondary to T cell–mediated autoimmune destruction.

Genetic features and environmental factors: Evidence for genetic factors contributing to onset of generalized vitiligo has been obtained in studies of vitiligo patients' close relatives.[484,491] The largest vitiligo twin study to date has demonstrated concordance for generalized vitiligo in monozygotic twins of 23%, far greater than the risk for nonidentical siblings (6.3%) and the general population risk.[484] At the same time, these data highlight the critical importance of environmental factors. Genetic association and linkage studies initially revealed an association between vitiligo and HLA-DRB1, HLA-DLRB4, and DQB1 alleles as well as other genes.[481,482] It is worth noting that the first vitiligo linkage data excluded a single candidate gene, the microphthalmia-associated transcription factor (MITF). MITF appeared to be a relevant candidate due to numerous mutations at the mouse microphthalmia (mi) locus associated with reduced or absent pigmentation, and with the mivit/vit mutation presenting postnatal depigmentation that resembles human vitiligo.[492] Many of the findings from early genetic studies were difficult to reproduce, and a study systematically tested the association of multiple previously reported candidate genes.[493] It was found that only HLA class I, HLA class II, and PTPN22 were strongly associated, with weaker support for TSLP, XBP1, and FOXD3. It can be concluded that most other previously reported associations most likely represent false-positives. One interesting finding came from association analyses that resulted in the identification of NALP1 (later renamed NLRP1) as a candidate gene.[494] This gene encodes an important regulator of the innate immune system that may monitor bacterial infection of the skin. Importantly, other reports have subsequently found genetic associations of NLRP1 with other autoimmune diseases such as T1D, Addison disease, celiac disease, and systemic sclerosis, pointing toward overlapping disease mechanisms.

The genetic basis of vitiligo has been explored by a series of GWAS studies, which have drawn a more comprehensive and reliable picture of the potential disease genes in (generalized) vitiligo. Almost all susceptibility genes that were found involve candidate genes functional in immune regulation and immune targeting of melanocytes, which has solidified the general consensus that vitiligo is a primary autoimmune disease.[495] The results from three large-scale GWAS have been reported, and major association signals in the MHC (both class I and II genes) but also numerous non-MHC associations were revealed.[496-499] The list is too long to summarize here (refer to Ref. 495) but includes many important immunity genes such as *TYR* (tyrosinase), PTPN22 (lymphoid-specific protein tyrosine phosphatase nonreceptor type 22), IL2RA (interleukin-2-receptor alpha chain), GZMB (granzyme B), and FOXP3 (forkhead box P3). Most of these genes have been discussed in this chapter for other conditions, again highlighting potentially common immune pathways. It is estimated that the current set of known genes accounts for approximately 10% of vitiligo total risk, and it is thus expected that large GWAS studies in the near future will complete this picture.

Animal models: The lack of a tractable experimental animal model has clearly hampered progress of vitiligo research. Nevertheless, the mutant Smyth line of chickens exhibits several relevant clinical features including melanocyte loss as a consequence of T cell infiltration.[500] Furthermore, the potential contribution of herpes virus infection to vitiligo induction in Smyth line chickens[501] emphasizes the role of environmental factors and warrants further investigations into this model.

Treatment and prevention: Routine nonsurgical repigmentation strategies include phototherapy, topical steroids, and inhibitors of the calcineurin pathway. These treatment modalities, while interfering with a multitude of complex pathophysiological pathways, modulate in particular the activity of local T cell responses.

Perspective and conclusions: In addition to genes involved in aberrant immune regulation, more than 120 genes are known to participate in the coordination of mammalian pigmentation. However, the identification of several candidate genes and genetic linkages associated with vitiligo will facilitate the elucidation of biological pathways involved in vitiligo pathogenesis and allow for a better differentiation of the spectrum of vitiligo disorders. As a result, novel therapeutic and prophylactic approaches will likely be developed in the near future.

Cardiac Autoimmune Diseases

The etiology for primary myocarditis is unclear and an autoimmune cause has to be taken into account. The heart muscle is infiltrated mainly by CD4+ T cells and macrophages, and the infiltrates are usually diffuse and nonfocal. The infiltrating cells produce considerable amounts of IFNγ and TNFα. As a consequence of the inflammation, heart muscle cells can swell, the heart can dilate, and severe disturbances of the electrical conduction can occur. It is noteworthy that myocarditis can also occur in conjunction with certain systemic autoimmune diseases such as SLE, scleroderma, RA, polymyositis, and polyarteritis nodosa. Two specific scenarios are noteworthy.

The first is myocarditis occurring after Chagas disease. In this case, cross-reactivity between antigens presented by the parasite causing Chagas disease (*Trypanosoma cruzi*) and cardiac myosin has been found in patients with myocarditis.[502] The scarcity of parasites in the chronic phase of the disease contrasts with the severe cardiac pathology observed in approximately 30% of chronic patients and suggested a role for autoimmunity as the origin of the pathology. In these cases, various *T. cruzi* antigens, such as B13, cruzipain, and Cha, cross-react with host antigens at the B or T cell level, and their role in pathogenesis has been widely studied. Immunization with those antigens and/or passive transfer of autoreactive T lymphocytes in mice leads to clinical disturbances similar to those found in Chagas disease patients. On the other hand, the parasite is becoming increasingly detected in chronically infected hosts and may also be the cause of pathology either directly or through parasite-specific mediated inflammatory responses. Thus, the issue of autoimmunity versus parasite persistence as the cause of Chagas disease pathology is hotly debated among many researchers in the field.

The second is Coxsackie B virus–associated myocarditis. Also in this situation, immunopathology outlasts the direct presence of the virus in heart muscle. This disease has been extensively studied in experimental models by Rose and colleagues.[503] They developed two murine models of myocarditis, one elicited by cardiotropic Coxsackie B3 (CB3) virus infection and the other by cardiac myosin immunization, to better analyze the pathogenetic mechanisms responsible for immune-mediated heart-muscle disease. Both virus infection and myosin immunization produce myocardial inflammation and elicit heart-reactive antibodies which bind to the myocardium *in vivo* and which recognize the cardiac myosin heavy chain. Each model offers unique advantages. The virus-induced disease more closely resembles human myocarditis; myosin immunization isolates the autoimmune components of the disease since no virus infection is involved. They also distinguished strains of mice resistant to autoimmune myocarditis (such as B10.A) from those susceptible to the autoimmune phase of disease (such as A.CA and A/J). Mice from a resistant strain to virus- or myosin-induced autoimmune heart disease develop myocardial inflammation and myosin antibodies if co-treated with TNFα or IL-1 when infected or immunized. Thus, cytokines can modulate the outcome of cardiotropic virus infection and enhance its autoimmune sequela. They also found that blocking IL-1 receptor inhibits autoimmune myocarditis in genetically susceptible mice. Interestingly, there is no regulatory function of Th2 cytokines such as IL-4, illustrating that immune regulation is an organ-specific concept rather than a global paradigm applicable to all autoimmune disease equally.

Rheumatoid Arthritis

Although RA is commonly classified as a systemic autoimmune disease (although it selectively affects joints), we believe that RA offers insights and paradigms that are important for understanding organ-specific diseases such as T1D and MS. A particularly interesting aspect is that diagnostic or therapeutic access to the target organ is comparatively easily achieved, and that this has facilitated the development of rationale and highly effective immunotherapies. In this case, high levels of TNFα were detected in affected joints and TNF blockade has now proven to be a highly effective intervention in RA (see more detailed description below).

Introduction and disease description: RA is a severe debilitating disease with unknown etiology. Epidemiologically, the prevalence is about 1% in the overall population, women are affected three times more frequently than men and the peak of incidence is within the fourth and sixth decade of life.[173,504] Diagnostic criteria are the typical morning stiffness, joint swelling with fluid accumulation, defined radiological changes, subcutaneous rheumatic nodes, and positive rheumatoid factor (autoantibodies; Fig. 46.15). The prognosis is worse in HLA-DR4 positive patients and, similar to diabetes and MS, there is genetic linkage with certain MHC haplotypes. Insights enabled by GWAS studies have added numerous non-HLA-associated disease genes, a list too long to include here (reviewed in Ref. 505). Many important immunity genes, such as CTLA-4, PTPN22, and IL2RB, were identified that have been found in other autoimmune conditions, in addition to genes involved in non-immune-related pathways. One notable example in the latter category is PADI4, coding for a protein that converts the arginine residue (peptidylarginine) to the citrulline residue (peptidylcitrulline) in a process called citullination. Peptidylcitrulline is targeted by a class of autoantibodies in RA that have high diagnostic value and this protein conversion is influenced by smoking. Interaction between PADI4 polymorphisms and smoking was repeatedly reported and thus offers a link between this important environmental determinant and genetic constitution of the host. Without systemic immunosuppressive or anti-inflammatory therapy, the disease will eventually result in destruction of many major joints and immobilize the patient. The major cause of death is infections that easily take a more severe course in immobile individuals.[173]

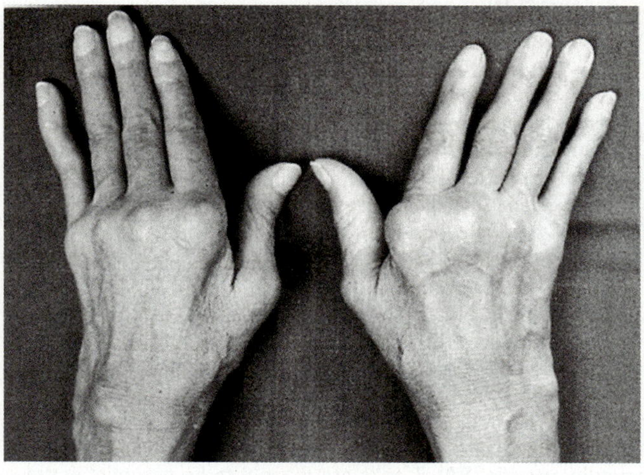

FIG. 46.15. Rheumatoid arthritis. Typical ulnar deviation and swan's neck deformation of the hands in rheumatoid arthritis.

The etiology of RA remains ill-defined. Since there is a possible overlap between infection-associated arthropathies (eg, reactive arthritis and rheumatic fever, where inflammation of joints and tissues occurs following bacterial infection in another part of the body), bacteria have always been good candidates as a cause for arthritis, and molecular mimicry has been suggested as a cause, since cross-reactivities of autoantibodies between self and bacterial proteins have been detected.[122] However, while it is quite possible to demonstrate cross-reactive antibodies and T cells in human blood, proof of their pathogenetic involvement is exceedingly difficult to obtain. The immune system is likely to have developed to tolerate a low-level cross-reactivity, which might even be necessary for its proper function, and it remains very difficult to prove a mechanistic link between autoreactive cells and pathological autoimmunity in humans.

Autoimmune features: The evidence that RA is an autoimmune disease stems, as in the case of MS and T1D, from the observation that T lymphocytes in a patient's blood can react with joint-derived autoantigens and IgM autoantibodies, as well as ANAs, are readily found in humans.[506] Inflammatory signs in the serum include complement activation and increased erythrocyte sedimentation rate, which indicates that the antibodies could possibly play a role in the disease process. Similarly, immunization with collagen can induce arthritis in susceptible mouse strains (Fig. 46.16). Figure 46.17 summarizes the major pathogenetic pathways leading to progressive joint destruction. As in many other autoimmune disease animal models, collagen-induced arthritis models require strong immunization with autoantigens and adjuvant, supporting the concept that breaking of self-tolerance, even on a genetically susceptible background, requires a rather pronounced inflammatory stimulus. In contrast, how disease is initiated in humans is unclear. The immunopathogenesis of RA has been redefined since the discovery of the Th17 lineage, which is now firmly established as a principal instigator of joint inflammation in animal models and has been repeatedly found in joints from affected patients. Current treatment modalities were generally developed based on the classification of RA as a prototypical Th1-driven condition. Potential future treatments, including anti-IL-17 and anti-IL-6 therapy, may be effective by dampening this Th17-driven disease component.

Treatment and prevention: Treatment of RA was historically regarded as a difficult task, because the medications that adequately suppress joint inflammation also had strong systemic side effects. Therefore, a delicate balance

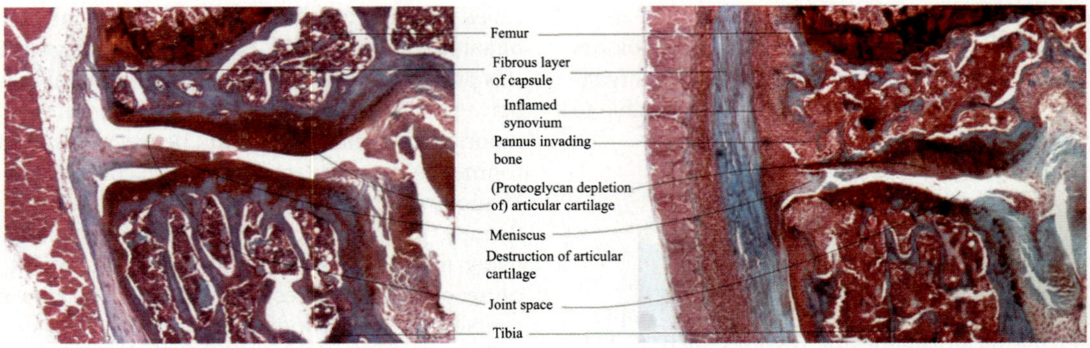

FIG. 46.16. Immunopathology of collagen-induced arthritis in mice. All features that are characteristic to rheumatoid arthritis can be found in sections from knee joints, here stained for proteoglycans with Safranin O. Left is a joint from a healthy control animal, while right is from a mouse with full-blown arthritis. Both inflammation and structural damage to cartilage and bone are evident.

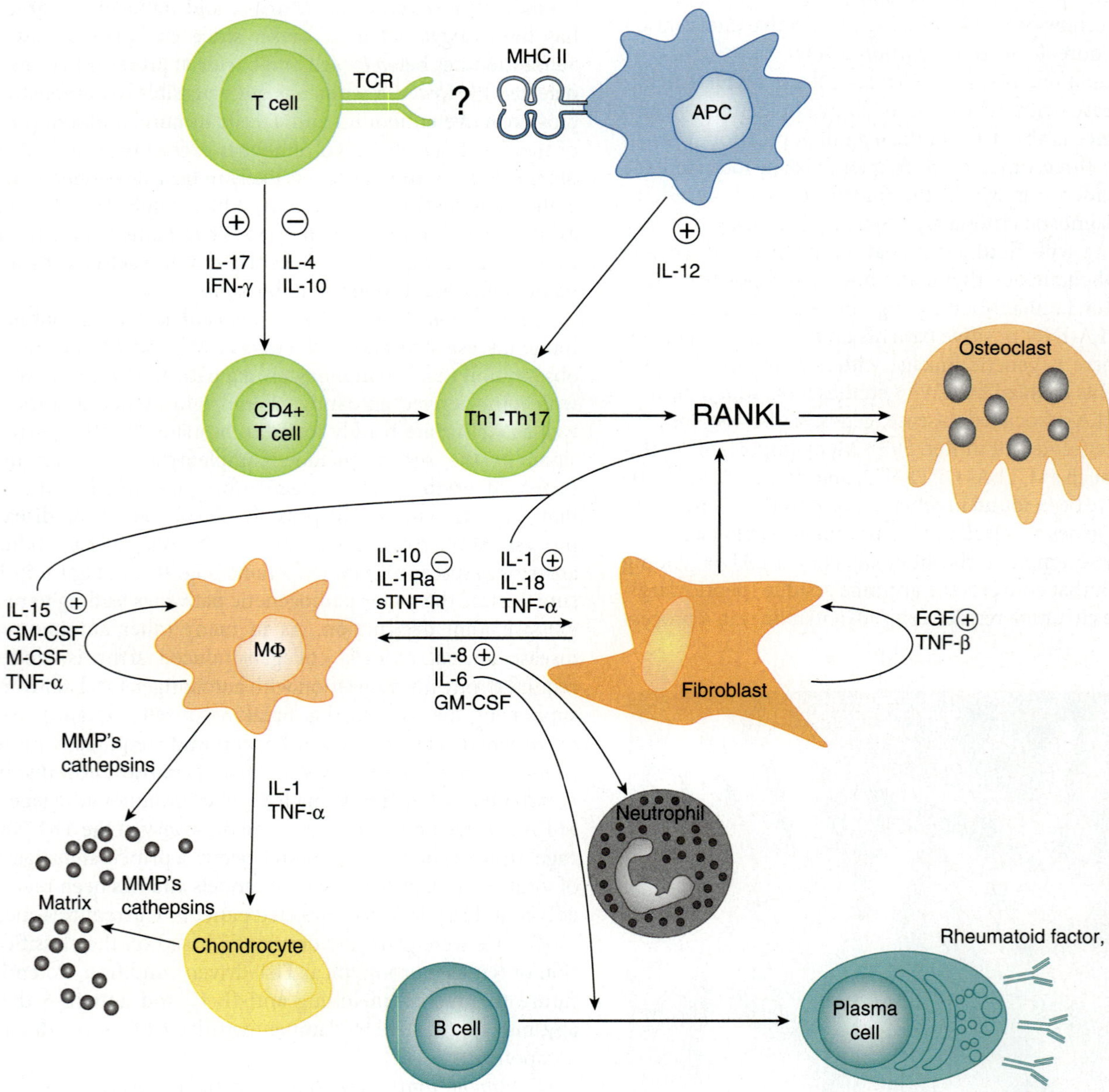

FIG. 46.17. Pathogenic pathways leading to joint destruction in rheumatoid arthritis (RA). T cells are thought to be essential in the etiology of RA, but it is unknown what triggers their activation. Once activated, T cells are skewed to a Th1 or Th17 phenotype and release factors such as RANKL that activate bone resorption through osteoclasts. Macrophages are major effectors and are believed to account for the high levels of TNF found in rheumatoid synovial tissue, resulting in cartilage destruction via effects on chondrocytes. Finally, pathogenic antibodies are produced such as rheumatoid factor as a consequence of B cell activation. Note that the introduction of biologicals such as anti-TNF, CTLA-4Ig, or anti-CD20 intervenes at distinct levels within the pathogenic process.

between different therapeutic approaches that target different stages of the inflammatory process needs to be established. Corticosteroids and nonsteroidal anti-inflammatory drugs can provide some baseline relief but are not able to halt progressive joint destruction. Gold compounds, methotrexate, and cyclophosphamide are more effective but also have profound systemic side effects. For this reason, progress with novel immunomodulatory interventions is of important benefit for RA patients.

Application of cyclosporin A was shown to be effective by suppressing proliferation and activation of T lymphocytes. A more selective intervention has also been established based on research by M. Feldman's group.[172] TNFα was found to play a key role in an *in vitro* model of synovial destruction using human cells, and numerous TNFα blocking agents were subsequently developed. After promising preclinical results in animal models, this intervention was tested in clinical trials and is now licensed for treatment of RA and many other autoimmune conditions, as discussed earlier. Although side effects do occur, as expected, they appear tolerable and clinical improvement of disease is pronounced, although not all patients respond to treatment. In some patients, SLE-like symptoms were observed, indicating that blocking a cytokine beneficial for one autoimmune disorder might be detrimental for another one. Indeed, TNF blockade appears to be less effective in MS, because this cytokine exhibits complex dual functions depending on the disease stage. In T1D, a small trial did suggest some effectiveness, although animal models previously indicated that time of administration crucially dictated outcomes in terms of beta cell preservation.[507] Based on the success story with TNF blockade in RA, and more particularly the approach of determining the dominant cytokine(s) locally at the target site, it may be possible to find similar key cytokines for T1D or MS treatment. It should be stressed that the successful intervention was developed based on human cell cultures and not animal models, underlining the need for direct research on human materials, if and when possible. Other strategies based on therapeutic antibody blockade such as CTLA-4Ig, anti-IL-1 and anti-CD20 administration have expanded the therapeutic arsenal for RA and in combination have led to dramatic improved life quality of RA patients. It also well established that anti-TNF biologics are the first to halt structural joint damage, a feat not achieved by any of the earlier-generation DMARDs. Future improvements likely lie in this aspect of disease, which principally accounts for long-term disability in RA. Finally, the successful introduction of JAK inhibitors, which are small molecular compounds that can be orally administered and act at signaling pathways common to a variety of cytokines, has provided an effective treatment option.

Scleroderma (Systemic Sclerosis)

Introduction and disease description: The term scleroderma, also known as systemic sclerosis (SSc), is derived from the Greek words *skleros* (hard) and *derma* (skin). Apart from thickening and tightening of the skin, this chronic condition often involves damage to internal body organs. The skin thickening in SSc patients typically starts with the fingers and can proceed to affect the whole body. The thickening results from vasculopathy and sclerosis or stiffening of structures due to an increased collagen content of the connective tissue.[508] Depending on the degree of skin thickening, SSc can be divided into limited SSc (lSSc; vascular and/or SSc-associated antibodies but no skin involvement), limited cutaneous SSc (lcSSc; typically involving forearms, hands, legs, feet, and face), or diffuse SSc (dSSc; can affect almost any area of the body).[509,510] SSc is a relatively rare disorder—about 20 new cases are diagnosed annually per million in the United States—which primarily targets adult women.[511] While symptoms may initially be rather benign, serious complications can develop over time and may involve various organs such as the gastrointestinal tract, lung, kidney, heart, and musculoskeletal system.

Autoimmune features: The pathophysiology of SSc is thought to have a vascular, an immunologic, and a fibrotic component.[508] A prominent diagnostic marker in SSc is the presence of ANAs, while also nonnuclear antibodies such as anti-fibrillin and anti-MMP-1/3 are commonly found.[512] B cells in general seem to have a chronically activated phenotype,[513] while an oligoclonal T cell subset has also been found in skin lesions.[514] The fibrotic aspect is caused by the overproduction of cellular matrix by activated fibroblasts, which in turn induces contraction of the affected tissue.[515] The vascular component encompasses reversible vasospasms (as in Raynaud phenomenon) as well as structural changes in vessel architecture sometimes leading to obliteration of the vessel lumen. The elevation of endothelium-derived serum factors in SSc such as von Willebrand factor and endothelin-1 but also circulating endothelial cells point toward the existence of vascular abnormalities.[516-518] Moreover, vasculogenesis appears to be impaired in SSc.[519] It is still largely unresolved how these three mechanisms act in concert to culminate in the characteristic SSc disease phenotype.

Genetic features and environmental factors: Family history constitutes a strong risk factor for SSc, although the low disease concordance rates in twin studies suggest that environmental factors play a role.[520] The advent of GWAS studies has offered more robust datasets in recent years and has added new susceptibility loci.[521,522] In accordance with our present understanding of the pathophysiology of SSc, the associated polymorphisms can be categorized in relation to either immunity, fibrosis, or vascular disease (Table 46.5). Possible environmental factors that may act in concert with a susceptible genetic constitution include viral or bacterial agents,[524] vitamin D levels,[525] and a wide host of occupational factors.[526] The vast range of potential interactions between the many susceptibility genes and environmental factors may account for the disease's notable heterogeneity.

Animal models: Multiple genetic and inducible animal models each cover some of the three aspects of SSc to a varying degree (excellent review in Ref. 527). We mention here three examples that demonstrate the partial spectrum of SSc characteristics mimicked by each model. A widely used inducible model is the bleomycin-induced skin fibrosis model. This compound, originally isolated from *Streptomyces*

TABLE 46.5 Systemic Sclerosis Susceptibility Genes Organized According to Possible Involvement in One of Three Pathophysiologic Pillars of the Disease

Immunity	Vascular Disease	Fibrosis
Transcription factors	*Angiogenesis*	*Fibroblast proliferation, ECM production*
STAT4; TBX21; IRF5	CXCL12; VEGF	CTGF
Lymphocyte (co-)stimulation	*Vascular integrity, survival*	*Platelet aggregation*
CD247 (CD3ξ); CD226; TNFSF4 (OX40L)	CD105; HIF1A	Serotonin 5-HT2A receptor
Phosphatases/kinases	*Vasoconstriction*	*Fibrogenic phenotype*
PTPN22; BLK	ACE; eNOS and iNOS; ET-1 and receptors	IL-1α/β
B cell–specific scaffold protein	*Voltage-gated K+ channel*	*ECM degradation*
BANK1	KCNA5	MMP
Ubiquitin-modifying enzyme	*GPI-anchored surface receptor*	
TNFAIP3	CD87	
Apoptosis	*Blood clotting*	
FAS	Fibrinogen	

CD, cluster of differentiation; CTGF, connective tissue growth factor; ECM, extracellular matrix; FAS, TNF receptor superfamily, member 6 or CD95; GPI, glucose-6-phosphate isomerase; IL, interleukin; MMP, matrix metalloproteinases; TNF, tumor necrosis factor; VEGF, vascular endothelial growth factor.
Principal gene type/function is listed but does not exclude other functions in systemic sclerosis. References to all original studies can be found in Romano et al.[523]
In italics: gene category according to function. Regular text: respective genes within this category related to sclerosis.

verticillus, induces skin fibrosis localized to the injection area after repeated injections.[528] The model mimics the inflammatory changes in SSc that are thought to occur early in the disease course, usually before a patient seeks medical care. It is therefore useful for preclinical evaluation of therapeutics that target inflammation or early-stage cutaneous fibrosis. An obvious disadvantage is that fibrosis is restricted to the injection site, while SSc in humans typically presents as a systemic disease involving both skin and internal organs. Furthermore, vascular phenomena as seen in SSc patients are absent, which is a serious shortcoming.

One of the genetically triggered models is the TSK-1 mouse,[529] where SSc-like skin tightening is caused by a tandem duplication of the fibrillin 1 gene (Fbn1).[530] Whereas the homozygous mouse dies before birth, heterozygous TSK-1 mice develop SSc-like disease, with skin fibrosis and autoimmunity. A remarkable parallel with human disease is the occurrence of autoantibodies against SSc-specific antigens (eg, topoisomerase I). The excessive release of extracellular matrix proteins by activated fibroblasts underlies the fibrosis seen in these animals. These mice do lack a vascular phenotype and are commonly used to study later stages of SSc associated with low inflammation.

One mouse model that does display characteristic SSc-like microvasculopathy but lacks autoimmune phenomena is the *Fra2* (Fos-related AP1 transcription factor) transgenic mouse, which expresses Fra2 under the control of a class I MHC antigen promoter.[531] These examples serve to highlight the notion that a perfect SSc model does not exist, but rather that several models can be chosen to study one or more of the three pillars underlying the disease.

Treatment and prevention: At present most therapeutical approaches are organ-based, while systematic treatment modalities aimed at the vascular and inflammatory component may also be used. Drugs such as methotrexate (skin involvement[532]), cyclophosphamide (interstitial lung disease[533]), or dihydropiridine calcium blockers (Raynaud phenomenon[534]) have proven efficacy but clearly offer incomplete protection. No effective therapies, however, exist today that effectively block the fibrotic aspect.

Systemic Lupus Erythematosus

Introduction and disease description: Systemic lupus erythematosus (SLE) is a systemic autoimmune disease sometimes referred to as the "great imitator." The wide array of clinical manifestations includes skin rashes, photosensitivity, oral ulcers, arthritis, pleuritis, pericarditis, kidney problems, neurological symptoms, and hematologic abnormalities.[535] Clinical course is typically of the relapsing-remitting type. As a consequence, the disease is notoriously difficult to diagnose and adequate treatment ideally involves a team approach due to the number of organ systems involved. SLE affects up to 10 times more women than men while estimated incidence ranges widely, likely due to inconsistencies in diagnosis, from 1.8 to 7.6 cases per 100,000 persons per year in the United States.[536] Hormones are suspected to codetermine disease and thus account for the increased prevalence of SLE among women. Patients commonly present with general complaints such as malaise, fatigue, fever, or weight loss and a correct diagnosis critically depends on careful history and physical examination combined with appropriate laboratory testing.

One of the most characteristic skin manifestations is the "butterfly" rash which shows a malar (across the cheeks) distribution and often develops in response to sunlight exposure. Arthritic symptoms are in fact the most common manifestation of SLE affecting mostly the small joints of the hands, wrists, and knees but may involve any joint. Renal involvement is considered the single most common cause of morbidity and represents a major cause of mortality in SLE. Of particular diagnostic importance in SLE are autoantibodies against intranuclear nucleic acids, proteins, and RNP complexes, commonly known as antinuclear antibodies (see next section). A general ANA test is performed by incubation of the patient's serum with a fixed, permeabilized cell line followed by fluorescently labeled anti-human IgG detection antibody. Specific autoantibodies that are associated with SLE and are used for diagnosis include anti-dsDNA, anti-Smith (part of small nuclear RNPs), anti-U1RNP, anti-Ro/SSA (largely unknown location and function), and anti-La/SSB (a nuclear phosphoprotein). This list is not exhaustive and it has become clear that SLE sera contain antibodies against extranuclear components also, such as antiribosomal antibodies and various cell membrane components. None of these antibodies, however, represents a reliable diagnostic tool that covers the entire spectrum of SLE manifestations. Anti-Smith, for instance, is highly specific but is detected in only 10% of Caucasian SLE patients.

Genetics: As expected from the heterogenous disease phenotype, genetic predisposition in SLE is extremely complex. Before the GWAS era, only a limited set of genes were known to be associated with SLE. As in most autoimmune diseases discussed here, HLA genes contained SLE-associated polymorphisms in addition to genes involved in the complement cascade (C2, C4), Fcγ receptors (FCGR2A, FCGR3A), B cell differentiation (PDC1), a tyrosine phosphatase (PTPN22), and factors involved in cytokine signaling (TNFα, IRF5).

At the time of writing, the combined results from seven GWAS studies are published, drastically adding to the number of SLE genes. While a complete summary of all associations is beyond the scope of this chapter (full list in Ref. 537), it can be concluded that many aspects of SLE genetics show overlap with other autoimmune diseases. For instance, the SLE susceptibility genes BANK1, BLK, IRF5, PTPN22, and STAT4 are all included in Table 46.5 detailing the genetic landscape of systemic sclerosis, while most are well-known autoimmune genes found in several other autoimmune conditions (eg, PTPN22, STAT4, and the HLA genes). Overall, the genetic data indicate that SLE results from interference with a diverse array of pathways such as antigen presentation, clearance of apoptotic material and immune complexes, immune cell function, innate immunity, cytokines, cellular energy machinery and leukocyte/endothelial adhesion, lymphocyte signaling, interferon response, complement activation, and apoptosis. Therefore, it remains to be seen to what extent this wealth of data will translate to conceptual and in particular translational insight. Although it is interesting to see that some of the usual suspects underlie SLE genetics, it is hoped that a thorough genetic understanding will explain which specific gene set provokes the SLE disease phenotype.

Autoimmune features: A myriad of immunological mechanisms is thought to underlie SLE, mirroring the disease's complex genetic and phenotypic profile (Fig. 46.18). How all these individual immune pathways culminate in clinical SLE is still unclear, and it is anticipated that different combinations of mechanisms might affect individual patients. As in many other autoimmune disorders, inadequate thymic selection is hypothesized to be responsible for the release of naïve autoreactive T and B cells into the periphery.[538] What triggers their activation is still elusive, although support for some of the environmental factors mentioned before exists. The EBV, for instance, has received a considerable degree of attention and higher loads were found in patients with SLE in comparison with controls.[539] Molecular similarities between the EBV nuclear antigen 1 and the common lupus autoantigen Ro were identified, fueling speculation about the role of mimicry in SLE.[540] A feature that further confounds our interpretation of the disease is the existence of autoantibodies up to 10 years

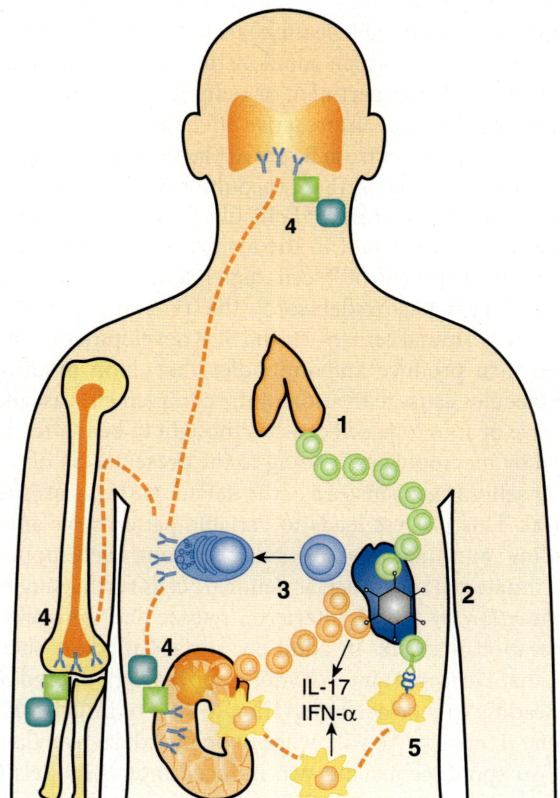

FIG. 46.18. Pathogenic immune mechanisms in systemic lupus erythematosus. (1) As in many autoimmune diseases, defective thymic selection is thought to contribute. (2) Naïve autoreactive cells reach the periphery and are activated by an unknown stimulus, possibly virus. (3) These activated T-helper cells in turn aid in generating plasma cells that produce copious amounts of autoantibodies. (4) These autoantibodies bind components in various organs, here shown as joints, skin, and kidney. Complement activation and tissue inflammation initiates. T cells and other leukocytes may also infiltrate. (5) Due to inadequate processing of apoptotic cells, a new round of autoimmunity is initiated by antigen presentation.

prior to disease onset. Thus, whatever the inciting antigen may be, it is likely that the immune system generally encounters it long before presentation of clinical symptoms.[541]

The recognition of nucleic acid motifs by TLRs may be a reason for activation of B cells in SLE. TLRs 3, 7, and 9 are located in endosomes and may contribute to the distinction between endogenous and foreign nucleic acids. There is mounting evidence to suggest TLRs play a role in the initial activation of a B cell response against nuclear antigens; however, the involvement of specific TLRs is complex. TLR9 has shown protective effects in mouse studies, whereas data from both mice and humans point to a more central role for TLR7 signaling in B cells for mounting an autoreactive response.[542]

Altered T cell signaling of potentially autoreactive T cells may then lead to a lower activation threshold, that is a relatively weak signal from presented self-antigen suffices to trigger activation. T cells are thought to be directly responsible for increased levels of proinflammatory cytokines, notably IL-17 which is found in the blood and kidneys of SLE patients. Another notable feature is a reduced ability of the T cells to secrete IL-2, which could be the cause of an impaired regulatory T cell compartment or reduced AICD followed by increased longevity of autoreactive effectors.

Increased turnover from B cells to plasma cells is a hallmark and is a consequence of their encounter with self-antigen in the presence of T cell help.[543] Follicular helper T cells are specialized subset important in the formation of germinal centers and development of T cell–dependent B cell responses. Emerging evidence underscores the potential pathogenic role in the germinal centers during SLE development. Plasma cells in turn produce autoantibodies that upon binding to self molecules cause activation of the complement cascade. A first wave of apoptotic cell death is thought to be ineffectively cleared by macrophages, leading to the presentation of intracellular self-antigens instead of the default pathway of "silent" removal. This process leads to variable patterns of "antigen spreading" and subsequent rounds of uncontrolled apoptosis. Other innate cell types such as dendritic cells further augment inflammation by type 1 interferon release and upregulation of activation markers. In addition, surface markers such as CD44 that are important in homing appear to be altered on T cells, leading to increased migration to sites of inflammation.

Animal models: We will, without being exhaustive, discuss here two spontaneous and two induced mouse models that have been instrumental in delineating SLE pathophysiology (full review in Refs. 544 and 535). Numerous genetically modified strains were found to exhibit some SLE-like features and are thus informative in the identification of gene function, but are not commonly used to study SLE per se.

One of the most widely adopted models is the NZB/W F1 mouse. This strain originates from a cross between the NZB (New Zealand Black) and NZW (New Zealand White) strains, of which only the former develops mild clinical signs of autoimmunity. The F1 hybrids, however, display a lupus-like phenotype including B cell hyperactivity, development of ANA, and immune complex–mediated glomerulonephritis around 5 to 6 months of age. Mice typically suffer kidney failure and die around 10 to 12 months of age. As in humans, the disease develops more frequently in females and is hormone dependent. Extensive genetic analysis has been performed on related inbred strains that resulted from accidental backcrossing (NZM; New Zealand Mixed strains) and led to the identification of multiple susceptibility loci. Several of these were also identified in human SLE subjects including MHC class II alleles, Fcγ receptors, and components of the complement system.

The MRL lpr/lpr strain is another spontaneous model which develops massive lymphadenopathy; high concentrations of autoantibodies such as ANA, anti-dsDNA, and anti-Sm; and renal disease. In this strain, both males and females are equally affected. The lpr mutation interferes with expression of the Fas receptor (CD95), an important mediator of apoptosis. As discussed above, defective apoptosis may in part underlie SLE in humans and it is known that aberrant Fas signaling can lead to the development of autoimmune lymphoproliferative syndrome which shares many symptoms with SLE.

SLE-like symptoms can be induced by the repeated injection of pristane, a compound found in mineral oil. These features include the development of ANA, immune complex deposition, and severe disease of the kidneys. This model can be reproduced in most mouse strains although the extent of susceptibility varies. In strains such as the SJL/J mouse, females are more prone to disease development.

Alternatively, some aspects of SLE can be mimicked by the experimental induction of graft-versus-host disease. Disease aspects consist of lymphoid hyperplasia, formation of ANA, and development of lupus nephritis leading to severe kidney dysfunction and mortality. Only certain strains are prone to induction, once again illustrating the requirement for a genetic contribution in addition to an environmental stimulus. Disease severity can be tailored by adjusting the number of allografted cells, and onset occurs from a known starting point relative to injection. ANA are usually detectable only 2 weeks after injection which makes it a suitable model for the more rapid assessment of pathogenic pathways and experimental therapies.[545]

Therapy: The multiorgan nature of the condition and its course of relapses and remissions dictates that therapeutic management is in accordance with disease activity. In addition to diverse clinical scoring systems, active disease often correlates with altered levels of immune markers such as increased anti-dsDNA and decreased serum complement. NSAIDs are prescribed to treat musculoskeletal symptoms or SLE-associated fever. Acute symptoms are generally treated by some form of corticosteroid treatment, either topical in case of cutaneous flares or injected when major organ disease develops. Immunosuppressive agents such as azathioprine are used mostly when major organ disease threatens to considerably worsen the patient's condition. The antimalarial agent hydroxychloroquine is known to ameliorate disease by inhibiting the function of TLRs. It is thus clear that, in line with the disease's protean pattern, no disease-specific treatment options exist and that the therapeutic goal often is to prevent where flares where possible (eg, avoid sun exposure) and tackle them in a symptomatic fashion as they occur. Since the approval of an antimalarial and corticosteroids for treatment of SLE in 1955, no new agents were introduced.

After more than 50 years and many failed trials, the FDA finally approved a new drug for SLE. Belimumab (marketed as Benlysta) is a fully human monoclonal antibody that binds to soluble human B-lymphocyte stimulator (BLyS) and inhibits its biological activity.[546] BLyS was shown to be a key survival cytokine for B lymphocytes which, as argued before, are a major pathogenic immune subset in SLE. In SLE patients, BLyS concentrations reflect disease activity and correspond with anti-dsDNA antibody titers.[547,548] It was found that the drug selectively decreases B cell numbers and short-lived plasma cells as well as SLE-associated anti-dsDNA antibody titers.[549] In two phase 3 trials that had enrolled patients with serologically active disease, a modest clinical effect was demonstrated combined with a safety profile similar to that of placebo.[550] It was argued that both inclusion criteria (patients with SLE activity that actually stands to be improved) and optimized disease-assessment indices contributed significantly to the successful outcome of these studies.[551] Initial studies with a humanized anti-CD22 monoclonal also had an encouraging initial outcome, further underscored by a recent meta-analysis,[552] corroborating the idea of B cells as a suitable target cell in SLE.[553] The fact that not just any therapy that targets B cells has therapeutic potential was underscored by two recent phase 3 trials that failed to show efficacy of rituximab (anti-CD20) in SLE.[554,555] A more recent meta-analysis of off-label use of rituximab in SLE did, however, find clinical efficacy which warrants further studies in controlled randomized trials.[556] These advances, however, represents a glimmer of hope for patients affected by this particularly heterogenic and pathophysiologically complex condition.

CONCLUSION

A common trait among diverse animal models for autoimmune diseases and the human conditions they aim to model appears to be the difficulty with which experimental autoimmunity is achieved. Breaking of tolerance to self-antigens requires in most instances strong inflammatory (eg, pertussis toxin and adjuvant in EAE or collagen-induced arthritis) or infectious (eg, RIP-LCMV model for T1D, or TMEV/MHV models for MS) stimuli. In addition, many animal models require a genetically susceptible background or have to rely on artificial autoantigen expression by means of transgenic technology. In contrast to the relatively easy detection of autoreactive lymphocytes in antigen-induced animal models, it has been a daunting endeavor in humans (as well as spontaneous disease models such as the NOD mouse) and data obtained frequently vary considerably between different patients as well as laboratories. What are the implications for our understanding of autoimmune diseases in humans given these challenges?

First, it is unlikely that strong inflammatory stimuli can be provided under natural conditions in the absence of infectious disease. Thus, viral and bacterial infections remain prime candidates for causing secondary autoimmunity (see Miller's TMEV model of APC-mediated determinant spreading to autoantigens). Infectious pathogens can provide the "danger signals" (P. Matzinger) needed for propagation of extended inflammation leading to clinical disease. Nevertheless, proof of a causal relationship is exquisitely difficult (if not impossible), since traces of pathogens detected may have no relation to the underlying disease process, or may be cleared from the system by the time of secondary clinical disease. In addition, disease is likely dependent on individual pathogen strains, making a very detailed immunological profiling in prospective clinical trials necessary. It is possible that the introduction of new antivirals might unmask such an association in the future. Finally, the importance of the gut microbiome has attracted some attention in the context of autoimmune disease.[557] This is not unexpected since the gut harbors the major part of our immune system and immune tolerance can often be induced by passage of an antigen through the gastrointestinal system. We now have gained a more precise view on the vast diversity of commensals inside our bodies[558,559] and it appears that this composition affects the likelihood of a genetically susceptible subject to develop autoimmunity.[265] For instance, generation of the pathogenic Th17 subset described before in many autoimmune conditions critically depends on the composition of microbiota.[560]

Second, a genetically susceptible background will likely be required to provide a "fertile field" for initiating a chronic inflammation involving autoantigens. This probably occurs via a multifaceted network of multiple susceptibility and protective genes, and it will be impossible to treat a respective disease just by analyzing the background genes involved. Last, autoreactive lymphocytes might predominantly be present in the affected organ or site and not in the peripheral blood, which makes their identification and characterization in humans rather difficult.

A word of caution should be devoted to our interpretation of specific findings obtained in individual animal models. Animal models should serve to teach us paradigms of how a disease *could* develop kinetically *in vivo*. The precise parameters, targeted antigens, susceptibility genes, and effector molecules may be considerably different in humans. Thus, for example, if there is indication that insulin is a primary antigen in the NOD mouse, this may or may not have direct relevance to the human disease. Some evidence exists that leads us to believe that insulin, indeed, may also be a primary antigen in human autoimmune diabetes, but other antigens such as GAD might also play a role. The opportunities as well as limits of each animal model have to be delineated. Again, the NOD mouse appears to be prone toward multiorgan autoimmunity (induction of EAE,[561] neuritis,[562] arthritis, and hepatitis have all been observed in NOD or NOD-congenic mouse strains) and exhibits, in addition to diabetes, thyroiditis, sialitis, and orchitis. Thus, diabetes in the NOD mouse is clearly different from typical human T1D. Therefore, treatments capable of correcting the systemic immune dysregulation that predominates in the NOD model may not directly apply to human T1D where no such pronounced systemic dysregulation is present. Based on these considerations, it is not surprising that out of the more than 140 therapeutic strategies that prevent diabetes in the NOD mouse, only a handful have made it to phase 1 human trials. An appealing solution that has shown promise in T1D epitope discovery and also recently treatment is the development of humanized NOD mice expressing HLA class I transgenic molecules for antigen presentation. Remarkably, several of the islet epitopes that become targeted under these conditions correspond to the disease-associated specificities found in PBMC from T1D patients.[563] When those epitopes are administrated in a tolerogenic fashion, the disease course can be significantly altered.[564] While still animal data in principle, these results may have higher translational value when tolerogenic strategies are to be envisioned for humans. Experiments with human cells or materials should be undertaken in order to solidify the choice of molecules or target antigens. The successful story in respect to blockade of TNFα to treat RA underlines the importance of this step. Employment of a multiplicity of models thus becomes imperative to evaluate potential candidate interventions, as does a careful proceeding, objective evaluation, and avoidance of premature conclusions. In addition to the researchers and clinicians, publishers as well as news media will have to share this responsibility. Continued research will undoubtedly provide us eventually with sufficient insight into the complexities of autoimmunity and autoimmune disorders, but patience and perseverance coupled with experimental objectivity will be required.

Therapeutic progress remains variable using monoclonal antibodies to dampen autoimmune processes in humans. Successful examples include anti-TNF, CTLA-4Ig, anti-CD20, and anti-IL-1 treatments, and the promising data of a non-Fc-binding anti-CD3 antibody prolonging time to onset of T1D in at-risk individuals.[335] As discussed higher, many of the successful therapies have also shown efficacy in a diverse array of other autoimmune diseases such as in psoriasis, Crohn disease, RA, MS, and uveitis. Thus, the roadmap to therapeutic success might lie in a well-tuned combination of such therapeutics, likely in conjunction with autoantigen-specific approaches that mediate more long-term and lasting tolerance. Augmentation of adaptive as well as intrinsic T_{regs} might be, in the end, the key for achieving site-specific tolerance (involving bystander suppression of effector responses and "infectious" induction of more T_{regs}) and is being explored with encouraging developments in many preclinical models.

REFERENCES

1. Lacan J. The field of the other and back to the transference. In: Miller J-A, ed. *The Four Fundamental Concepts of Psycho-Analysis*. W.W. Norton & Company; 1981:203-260.
2. Tauber AI. The biological notion of self and non-self. In: Zalta EN, ed. *Stanford Encyclopedia of Philosophy*. The Metaphysics Research Lab; 2006. https://stanford.library.sydney.edu.au/archives/sum2015/entries/biology-self/
3. Littre E, Robin CH. *Dictionnaire de medecine, de chirugie, de pharmacie et del'art veterinaire*. J.-B. Baillière & Fils, Paris; 1878.
4. Pasteur L. De l'attenuation du virus cholera des poules. *Compt Rend Acad Sci*. 1880;91:673-680.
5. von Behring E, Kitasato S. Über das Zustandekommen der Diphtherie-Immunität und der Tetanus-Immunität bei Thieren. *Dtsch Med Wochenschr*. 1890;16:1113-1114.
6. Ehrlich P, Morgenroth J. Zytotoxine als Antikorper. *Berl Klin Wochenschr*. 1901;38:251-260.
7. Römer P, Gebb H. Beiträge zur Frage der Anaphylaxie durch Linseneiweiss und Eiweiss aus andern Geweben des Auges. *Albrecht von Graefes Arch Ophtalmol*. 1912;81:367-402.
8. Silverstein A. *A History of Immunology*. Academic Press; 1989.
9. Portier P, Richet C. De l'action anaphylactique de certain venins. *CR Soc Biol (Paris)*. 1902;54:170-172.
10. Arthus M. Injections repetees de serum du cheval chez le lapin. *C R Soc Biol (Paris)*. 1903;55:817-820.
11. Donath J, Landsteiner K. Uber paroxysmale hamoglobinurie. *Munch Med Wochenschr*. 1904;51:1590-1601.
12. von Pirquet C, Schick B. *Die Serumkrankeit*. Deuticke; 1905.
13. Weil E, Braun H. Über das Wesen der luetischen Erkrankung auf Grund der neueren Forschungen. *Wien Klin Wochenschr*. 1909;22:372.
14. Uhlenhuth P. *Festschrift Zum 60. Jena*. Geburtstag von Robert Koch; 1903.
15. Traub E. Persistence of lymphocytic choriomeningitis virus in immune animals and its relation to immunity. *J Exp Med*. 1936;63:847-861.
16. Owen RD. Immunogenetic consequences of vascular anastomoses between bovine cattle twins. *Science*. 1945;102:400-401.
17. Medawar PB. The behaviour and fate of skin autografts and skin homografts in rabbits. *J Anat*. 1944;78:176-199.
18. Medawar PB. A second study of the behaviour and fate of skin homografts in rabbits. *J Anat*. 1945;79:157-176.
19. Medawar PB. Immunity to homologous grafted skin. I. The suppression of cell division in grafts transplanted to immunized animals. *Br J Exp Pathol*. 1946;27:9-14.
20. Medawar PB. Immunity to homologous grafted skin. II. The immunity relationship between the antigens of blood and skin. *Br J Exp Pathol*. 1946;27:15-24.
21. Burnet FM, Fenner F. *The Production of Antibodies*. Macmillan; 1949.
22. Burnet FM. *The Clonal Selection Theory of Acquired Immunity*. Cambridge University Press; 1957.
23. Witebsky E, Rose NR. Studies on organ specificity. V. Changes in the thyroid glands of rabbits following active immunization with rabbit thyroid extracts. *J Immunol*. 1956;76:417-427.
24. Fleck L. *Entstehung und Entwicklung einer wissenschaftlichen tatsache. Einfurung in die Lehre vom Dekstil und Denkkollektiv*. B. Schwabe und Co. Verlagsbuchhandlung; 1935.
25. Jerne NK. Towards a network theory of the immune system. *Ann Immunol (Paris)*. 1974;125C(1-2):373-389.
26. Coutinho A, Forni L, Holmberg D, Ivars F, Vaz N. From an antigen-centered, clonal perspective of immune responses to an organism-centered, network perspective of autonomous activity in a self-referential immune system. *Immunol Rev*. 1984;79:151-168.
27. Grabar P. Hypothesis. Auto-antibodies and immunological theories: an analytical review. *Clin Immunol Immunopathol*. 1975;4:453-466.
28. Wigzell H. Positive autoimmunity. In: Talal N, ed. *Autoimmunity: Genetic, Immunologic, Virologic and Clinical Aspects*. Academic Press; 1997.
29. Schwartz M, Cohen IR. Autoimmunity can benefit self-maintenance. *Immunol Today*. 2000;21:265-268.
30. Matzinger P. Tolerance, danger, and the extended family. *Annu Rev Immunol*. 1994;12:991-1045.
31. Dembic Z. Immune system protects integrity of tissues. *Mol Immunol*. 2000;37:563-569.
32. Pradeu T, Carosella ED. On the definition of a criterion of immunogenicity. *Proc Natl Acad Sci U S A*. 2006;103:17858-17861.
33. Pradeu T, Carosella ED. Critical analysis of the immunological self/non-self model and of its implicit metaphysical foundations. Article in French. *C R Biol*. 2004;327:481-492.
34. Cohn M. A commentary on the Zinkernagel-Hengartner 'Credo 2004'. *Scand J Immunol*. 2005;62:187-194.
35. Sercarz EE, Maverakis E. Recognition and function in a degenerate immune system. *Mol Immunol*. 2004;40:1003-1008.
36. Tada T. The immune system as a supersystem. *Annu Rev Immunol*. 1997;15:1-13.
37. Hedrick SM. The acquired immune system: a vantage from beneath. *Immunity*. 2004;21:607-615.
38. Rose NR, Mackay IR: Prelude. In: Rose NR, Mackay IR, eds. *The Autoimmune Diseases*. Academic Press; 1998.
39. *Autoimmune Disorders*. US National Library of Medicine and National Institutes of Health. Accessed May 27, 2020. http://www.nlm.nih.gov/medlineplus/ency/article/000816.htm
40. Zinkernagel RM. Anti-infection immunity and autoimmunity. *Ann NY Acad Sci*. 2002;958:3-6.
41. Witebsky E, Rose NR, Terplan K, Paine JR, Egan RW. Chronic thyroiditis and autoimmunization. *J Am Med Assoc*. 1957;164:1439-1447.
42. Rose NR, Bona C. Defining criteria for autoimmune disease (Witebsky's postulates revisited). *Immunol Today*. 1993;114:426-430.
43. Alvarez F, Berg PA, Bianchi FB, et al. International Autoimmune Hepatitis Group Report: review of criteria for diagnosis of autoimmune hepatitis. *J Hepatol*. 1999;31:929-938.
44. Jacobson DL, Gange SJ, Rose NR, Graham NM. Epidemiology and estimated population burden of selected autoimmune diseases in the United States. *Clin Immunol Immunopathol*. 1997;84:223-243.
45. Eaton WW, Rose NR, Kalaydjian A, Pedersen MG, Mortensen PB. Epidemiology of autoimmune diseases in Denmark. *J Autoimmun*. 2007;29:1-9.
46. Cooper GS, Bynum ML, Somers EC. Recent insights in the epidemiology of autoimmune diseases: improved prevalence estimates and understanding of clustering of diseases. *J Autoimmun*. 2009;33:197-207.
47. Walsh SJ, Rau LM. Autoimmune diseases: a leading cause of death among young and middle-aged women in the United States. *Am J Public Health*. 2000;90:1463-1466.
48. Grimaldi LM, Palmeri B, Salemi G, et al. High prevalence and fast rising incidence of multiple sclerosis in Caltanissetta, Sicily, southern Italy. *Neuroepidemiology*. 2007;28:28-32.
49. Catassi C, Kryszak D, Bhatti B, et al. Natural history of celiac disease autoimmunity in a USA cohort followed since 1974. *Ann Med*. 2010;42:530-538.
50. Matsumoto I, Maccioni M, Lee DM, et al. How antibodies to a ubiquitous cytoplasmic enzyme may provoke joint-specific autoimmune disease. *Nat Immunol*. 2002;3:360-365.
51. Binstadt BA, Patel PR, Alencar H, et al. Particularities of the vasculature can promote the organ specificity of autoimmune attack. *Nat Immunol*. 2006;7:284-292.
52. Shoda LK, Young DL, Ramanujan S, et al. A comprehensive review of interventions in the NOD mouse and implications for translation. *Immunity*. 2005;23:115-126.
53. Anderson MS, Bluestone JA. The NOD mouse: a model of immune dysregulation. *Annu Rev Immunol*. 2005;23:447-485.
54. Chaparro RJ, Konigshofer Y, Beilhack GF, Shizuru JA, McDevitt HO, Chien YH. Nonobese diabetic mice express aspects of both type 1 and type 2 diabetes. *Proc Natl Acad Sci U S A*. 2006;103:12475-12480.
55. McGonagle D, McDermott MF. A proposed classification of the immunological diseases. *PLoS Med*. 2006;3:e297.
56. Gershon RK, Kondo K. Infectious immunological tolerance. *Immunology*. 1971;21:903-914.
57. Schubert LA, Jeffery E, Zhang Y, Ramsdell F, Ziegler SF. Scurfin (FOXP3) acts as a repressor of transcription and regulates T cell activation. *J Biol Chem*. 2001;276:37672-37679.
58. Ziegler SF. FOXP3: of mice and men. *Annu Rev Immunol*. 2006;24:209-226.
59. Hori S, Nomura T, Sakaguchi S. Control of regulatory T cell development by the transcription factor Foxp3. *Science*. 2003;299:1057-1061.
60. Khattri R, Cox T, Yasayko SA, Ramsdell F. An essential role for Scurfin in CD4+CD25+ T regulatory cells. *Nat Immunol*. 2003;4:337-342.
61. Fontenot JD, Gavin MA, Rudensky AY. Foxp3 programs the development and function of CD4+CD25+ regulatory T cells. *Nat Immunol*. 2003;4:330-336.

62. Fontenot JD, Rasmussen JP, Williams LM, Dooley JL, Farr AG, Rudensky AY. Regulatory T cell lineage specification by the forkhead transcription factor foxp3. *Immunity.* 2005;22:329-341.
63. Wan YY, Flavell RA. Identifying Foxp3-expressing suppressor T cells with a bicistronic reporter. *Proc Natl Acad Sci U S A.* 2005;102:5126-5131.
64. Haase C, Ejrnaes M, Juedes AE, Wolfe T, Markholst H, von Herrath MG. Immunomodulatory dendritic cells require autologous serum to circumvent nonspecific immunosuppressive activity *in vivo*. *Blood.* 2005;106:4225-4233.
65. Roncarolo MG, Gregori S, Battaglia M, Bacchetta R, Fleischhauer K, Levings MK. Interleukin-10-secreting type 1 regulatory T cells in rodents and humans. *Immunol Rev.* 2006;212:28-50.
66. Li MO, Wan YY, Sanjabi S, Robertson AK, Flavell RA. Transforming growth factor-beta regulation of immune responses. *Annu Rev Immunol.* 2006;24:99-146.
67. Shevach EM. Regulatory T cells in autoimmmunity. *Annu Rev Immunol.* 2000;18:423-449.
68. Shevach EM. CD4+ CD25+ suppressor T cells: more questions than answers. *Nat Rev Immunol.* 2002;2:389-400.
69. Zhou X, Bailey-Bucktrout SL, Jeker LT, et al. Instability of the transcription factor Foxp3 leads to the generation of pathogenic memory T cells *in vivo*. *Nat Immunol.* 2009;10:1000-1007.
70. Murai M, Turovskaya O, Kim G, et al. Interleukin 10 acts on regulatory T cells to maintain expression of the transcription factor Foxp3 and suppressive function in mice with colitis. *Nat Immunol.* 2009;10:1178-1184.
71. Brusko TM, Putnam AL, Bluestone JA. Human regulatory T cells: role in autoimmune disease and therapeutic opportunities. *Immunol Rev.* 2008;223:371-390.
72. Ashton-Rickardt PG, Bandeira A, Delaney J, et al. Evidence for a differential avidity model of T-cell selection in the thymus. *Cell.* 1994;76:651-663.
73. Hoffmann MW, Heath WR, Ruschmeyer D, Miller JF. Deletion of high-avidity T cells by thymic epithelium. *Proc Natl Acad Sci U S A.* 1995;92:9851-9855.
74. Jameson SC, Bevan MJ. T-cell selection. *Curr Opin Immunol.* 1998;10:214-219.
75. Sebzda E, Mariathasan S, Ohteki T, Jones R, Bachmann MF, Ohashi PS. Selection of the T cell repertoire. *Annu Rev Immunol.* 1999;17:829-874.
76. Kishimoto H, Sprent J. The thymus and negative selection. *Immunol Res.* 2000;21:315-323.
77. Nikolic-Zugic J, Bevan MJ. Role of self-peptides in positively selecting the T-cell repertoire. *Nature.* 1990;344:65-67.
78. Surh CD, Lee DS, Fung-Leung WP, Karlsson L, Sprent J. Thymic selection by a single MHC/peptide ligand produces a semidiverse repertoire of CD4+ T cells. *Immunity.* 1997;7:209-219.
79. Steinman L. A few autoreactive cells in an autoimmune infiltrate control a vast population of nonspecific cells: a tale of small bombs and the infantry. *Proc Natl Acad Sci U S A.* 1996;93:2253-2256.
80. van den Elzen P, Menezes JS, Ametani A, et al. Limited clonality in autoimmunity: drivers and regulators. *Autoimmun Rev.* 2004;3:524-529.
81. Han B, Serra P, Amrani A, et al. Prevention of diabetes by manipulation of anti-IGRP autoimmunity: high efficiency of a low-affinity peptide. *Nat Med.* 2005;11:645-652.
82. Allison JP. CD28-B7 interactions in T-cell activation. *Curr Opin Immunol.* 1994;6:414-419.
83. Harding FA, Allison JP. CD28-B7 interactions allow the induction of CD8+ cytotoxic T- lymphocytes in the absence of exogenous help. *J Exp Med.* 1993;177:1791-1796.
84. Schwartz RH. A cell culture model for T lymphocyte clonal anergy. *Science.* 1990;248:1349-1356.
85. Lenardo M, Chan KM, Hornung F, et al. Mature T lymphocyte apoptosis—immune regulation in a dynamic and unpredictable antigenic environment. *Annu Rev Immunol.* 1999;17:221-253.
86. Abbas AK. Die and let live: eliminating dangerous lymphocytes. *Cell.* 1996;84:655-657.
87. Van Parijs L, Abbas AK. Homeostasis and self-tolerance in the immune system: turning lymphocytes off. *Science.* 1998;280:243-248.
88. Chambers CA, Kuhns MS, Egen JG, Allison JP. CTLA-4-mediated inhibition in regulation of T cell responses: mechanisms and manipulation in tumor immunotherapy. *Annu Rev Immunol.* 2001;19:565-594.
89. Salomon B, Bluestone JA. Complexities of CD28/B7: CTLA-4 costimulatory pathways in autoimmunity and transplantation. *Annu Rev Immunol.* 2001;19:225-252.
90. Weiner HL, Friedman A, Miller A, et al. Oral tolerance: immunologic mechanisms and treatment of animal and human organ-specific autoimmune diseases by oral administration of autoantigens. *Annu Rev Immunol.* 1994;12:809-837.
91. Atkinson MA, Bluestone JA, Eisenbarth GS, et al. How does type 1 diabetes develop?: the notion of homicide or beta-cell suicide revisited. *Diabetes.* 2011;60:1370-1379.
92. Yu W, Jiang N, Ebert PJ, et al. Clonal deletion prunes but does not eliminate self-specific alphabeta CD8(+) T lymphocytes. *Immunity.* 2015;42:929-941.
93. Somers EC, Thomas SL, Smeeth L, Hall AJ. Are individuals with an autoimmune disease at higher risk of a second autoimmune disorder? *Am J Epidemiol.* 2009;169:749-755.
94. Seewaldt S, Thomas HE, Ejrnaes M, et al. Virus-induced autoimmune diabetes: most beta-cells die through inflammatory cytokines and not perforin from autoreactive (anti-viral) cytotoxic T-lymphocytes. *Diabetes.* 2000;49:1801-1809.
95. Bettelli E, Carrier Y, Gao W, et al. Reciprocal developmental pathways for the generation of pathogenic effector TH17 and regulatory T cells. *Nature.* 2006;441:235-238.
96. Marth T, Strober W, Seder RA, Kelsall BL. Regulation of transforming growth factor-beta production by interleukin-12. *Eur J Immunol.* 1997;27:1213-1220.
97. Seder RA, Marth T, Sieve MC, et al. Factors involved in the differentiation of TGF-b-producing cells from naive CD4+ T cells: IL-4 and IFN-g have opposing effects, while TGF-b positively regulates its own production. *J Immunol.* 1998;160:5719-5728.
98. King C, Davies J, Mueller R, et al. TGF-b1 alters APC preference, polarizing islet antigen responses toward a Th2 phenotype. *Immunity.* 1998;8:601-613.
99. Rizzo LV, Morawetz RA, Miller-Rivero NE, et al. IL-4 and IL-10 are both required for the induction of oral tolerance. *J Immunol.* 1999;162:2613-2622.
100. Homann D, Holz A, Bot A, et al. Autoreactive CD4+ lymphocytes protect from autoimmune diabetes via bystander suppression using the IL-4/STAT6 pathway. *Immunity.* 1999;11:463-472.
101. King C, Hoenger RM, Cleary MM, et al. Interleukin-4 acts at the locus of the antigen-presenting dendritic cell to counter-regulate cytotoxic CD8+ T-cell responses. *Nat Med.* 2001;7:206-214.
102. Tree TI, Lawson J, Edwards H, et al. Naturally arising human CD4 T-cells that recognize islet autoantigens and secrete interleukin-10 regulate proinflammatory T-cell responses via linked suppression. *Diabetes.* 2010;59:1451-1460.
103. Zhang ZY, Lee CS, Lider O, Weiner HL. Suppression of adjuvant arthritis in Lewis rats by oral administration of type II collagen. *J Immunol.* 1990;145:2489-2493.
104. Maron R, Blogg NS, Polanski M, Hancock W, Weiner HL. Oral tolerance to insulin and the insulin B-chain: cell lines and cytokine patterns. *Ann N Y Acad Sci.* 1996;778:357.
105. Hancock WW, Polanski M, Zhang J, Blogg N, Weiner HL. Suppression of insulitis in non-obese diabetic (NOD) mice by oral insulin administration is associated with selective expression of interleukin-4 and -10, transforming growth factor-beta, and prostaglandin-E. *Am J Pathol.* 1995;147:1193-1199.
106. Bergerot I, Arreaza GA, Cameron MJ, et al. Insulin B-chain reactive CD4+ regulatory T-cells induced by oral insulin treatment protect from type 1 diabetes by blocking the cytokine secretion and pancreatic infiltration of diabetogenic effector T-cells. *Diabetes.* 1999;48:1720-1729.
107. Charlton B, Lafferty KJ. The Th1/Th2 balance in autoimmunity. *Curr Opin Immunol.* 1995;7:793-798.
108. Paul WE. Interleukin-4: a prototypic immunoregulatory lymphokine. *Blood.* 1991;77:1859-1870.
109. Lepault F, Gagnerault MC. Characterization of peripheral regulatory CD4+ T cells that prevent diabetes onset in nonobese diabetic mice. *J Immunol.* 2000;164:240-247.
110. Sercarz EE. Processing creates the self. *Nat Immunol.* 2002;3:110-112.
111. Anderton SM, Viner NJ, Matharu P, Lowrey PA, Wraith DC. Influence of a dominant cryptic epitope on autoimmune T cell tolerance. *Nat Immunol.* 2002;3:175-181.
112. Moudgil KD, Sekiguchi D, Kim SY, Sercarz EE. Immunodominance is independent of structural constraints: each region within hen eggwhite lysozyme is potentially available upon processing of native antigen. *J Immunol.* 1997;159:2574-2579.
113. Bhardwaj V, Kumar V, Geysen HM, Sercarz EE. Degenerate recognition of a dissimilar antigenic peptide by myelin basic protein-reactive T cells. *J Immunol.* 1993;151:5000-5010.
114. Zehn D, Bevan MJ. T cells with low avidity for a tissue-restricted antigen routinely evade central and peripheral tolerance and cause autoimmunity. *Immunity.* 2006;25:261-270.
115. Moudgil KD, Sercarz EE. Understanding crypticity is the key to revealing the pathogenesis of autoimmunity. *Trends Immunol.* 2005;26:355-359.
116. Carbone FR, Kurts C, Bennett SRM, Miller JFAP, Heath WR. Cross-presentation: a general mechanism for CTL immunity and tolerance. *Immunol Today.* 1998;19:368-373.
117. van Stipdonk MJ, Lemmens EE, Schoenberger SP. Naive CTLs require a single brief period of antigenic stimulation for clonal expansion and differentiation. *Nat Immunol.* 2001;2:423-429.
118. Blattman JN, Sourdive DJ, Murali-Krishna K, Ahmed R, Altman JD. Evolution of the T cell repertoire during primary, memory, and recall responses to viral infection. *J Immunol.* 2000;165:6081-6090.
119. Heath WR, Kurts C, Miller JF, Carbone FR. Cross-tolerance: a pathway for inducing tolerance to peripheral tissue antigens. *J Exp Med.* 1998;187:1549-1553.
120. Savinov AY, Wong FS, Stonebraker AC, Chervonsky AV. Presentation of antigen by endothelial cells and chemoattraction are required for homing of insulin-specific CD8+ T cells. *J Exp Med.* 2003;197:643-656.
121. Taurog JD, Maika SD, Satumtira N, et al. Inflammatory disease in HLA-B27 transgenic rats. *Immunol Rev.* 1999;169:209-223.
122. Schwimmbeck P, Yu DT, Oldstone MB. Autoantibodies to HLA-B27 in sera of patients with ankylosing spondylitis and Reiter's syndrome. *J Exp Med.* 1987;166:173-181.
123. McDevitt HO. The role of MHC class II molecules in susceptibility and resistance to autoimmunity. *Curr Opin Immunol.* 1998;10:677-681.
124. Sonderstrup G, McDevitt HO. DR, DQ, and you: MHC alleles and autoimmunity. *J Clin Invest.* 2001;107:795-796.
125. Molberg O, Kett K, Scott H, Thorsby E, Sollid LM, Lundin KE. Gliadin specific, HLA DQ2-restricted T cells are commonly found in small intestine biopsies from coeliac disease patients, but not from controls. *Scand J Immunol.* 1997;46:103-108.
126. von Herrath MG, Wolfe T, Mohrle U, Coon B, Hughes A. MHC non-linked factors can prevent virally induced type 1 diabetesin SV129 (H-2Db) mouse strains. *J Immunol.* 2001;50(12):2700-2708.
127. de Bakker PIW, McVean G, Sabeti PC, et al. A high-resolution HLA and SNP haplotype map for disease association studies in the extended human MHC. *Nat Genet.* 2006;38:1166-1172.
128. Vijayakrishnan L, Slavik JM, Illes Z, et al. An autoimmune disease-associated CTLA-4 splice variant lacking the B7 binding domain signals negatively in T cells. *Immunity.* 2004;20:563-575.
129. Bach J-F. Organ-specific autoimmunity. *Immunol Today.* 1995;16:353-355.
130. Bach J-F. Insulin dependent diabetes mellitus as an autoimmune disease. *Endocr Rev.* 1994;15:516-542.
131. Bach J-F. Predictive medicine in autoimmune diseases: from the identification of genetic predisposition and environmental influence to precocious immunotherapy. *Clin Immunol Immunopathol.* 1994;72:156-161.
132. Todd JA. The role of MHC class II genes in susceptibility to insulin-dependent diabetes mellitus. *Curr Top Microbiol Immunol.* 1990;164:17-40.
133. Vyse TJ, Todd JA. Genetic analysis of autoimmune disease. *Cell.* 1996;85:311-318.
134. Redondo MJ, Rewers M, Yu L, et al. Genetic determination of islet cell autoimmunity in monozygotic twin, dizygotic twin, and non-twin siblings of patients with type 1 diabetes: prospective twin study. *Br Med J.* 1999;318:698-702.
135. Aly TA, Ide A, Jahromi MM, et al. Extreme genetic risk for type 1A diabetes. *Proc Natl Acad Sci U S A.* 2006;103: 14074-14079.
136. Hirschhorn JN, Gajdos ZK. Genome-wide association studies: results from the first few years and potential implications for clinical medicine. *Annu Rev Med.* 2011;62:11-24.
137. Hakonarson H, Grant SF. Genome-wide association studies in type 1 diabetes, inflammatory bowel disease and other immune-mediated disorders. *Semin Immunol.* 2009;21:355-362.
138. Sevilla N, Homann D, von Herrath MG, et al. Virus-induced diabetes in a transgenic model: role of cross-reacting viruses and quantitation of effector T cells needed to cause disease. *J Virol.* 2000;74:3284-3292.
139. Raine T, Zaccone P, Mastroeni P, Cooke A. *Salmonella typhimurium* infection in nonobese diabetic mice generates immunomodulatory dendritic cells able to prevent type 1 diabetes. *J Immunol.* 2006;177:2224-2233.
140. Christen U, Benke D, Wolfe T, et al. Cure of prediabetic mice by viral infections involves lymphocyte recruitment along an IP-10 gradient. *J Clin Invest.* 2004;113:74-84.
141. Filippi CM, Estes EA, Oldham JE, von Herrath MG. Immunoregulatory mechanisms triggered by viral infections protect from type 1 diabetes in mice. *J Clin Invest.* 2009;119:1515-1523.
142. Zhao ZS, Granucci F, Yeh L, Schaffer PA, Cantor H. Molecular mimicry by herpes simplex virus-type 1: autoimmune disease after viral infection. *Science.* 1998;279:1344-1347.
143. Oldstone MB. Molecular mimicry and autoimmune disease. *Cell.* 1987;50:819-820.
144. Horwitz MS, Bradley LM, Harbertson J, Krahl T, Lee J, Sarvetnick N. Diabetes induced by Coxsackie virus: initiation by bystander damage and not molecular mimicry. *Nat Med.* 1998;4:781-785.
145. Olson JK, Girvin AM, Miller SD. Direct activation of innate and antigen-presenting functions of microglia following infection with Theiler's virus. *J Virol.* 2001;75:9780-9789.
146. Miller SD, Vanderlugt CL, Begolka WS, et al. Persistent infection with Theiler's virus leads to CNS autoimmunity via epitope spreading. *Nat Med.* 1997;3:1133-1136.
147. Katz-Levy Y, Neville KL, Girvin AM, et al. Endogenous presentation of self myelin epitopes by CNS-resident APCs in Theiler's virus-infected mice. *J Clin Invest.* 1999;104:599-610.
148. EudraCT Number: 2015-003350-41. EU Clinical Trials Register; 2017. https://eudract.ema.europa.eu/
149. Stone VM, Laitinen OH, Sioofy-Khojine AB, et al. A hexavalent Coxsackievirus B vaccine is highly immunogenic and has a strong protective capacity in mice and nonhuman primates. *Sci Adv.* 2020;6(19):eaaz2433.
150. Oldstone MB. Prevention of type I diabetes in nonobese diabetic mice by virus infection. *Science.* 1988;239:500-502.
151. Blumberg RS, Saubermann LJ, Strober W. Animal models of mucosal inflammation and their relation to human inflammatory bowel disease. *Curr Opin Immunol.* 1999;11:648-656.

152. Kelsall BL, Strober W. Dendritic cells of the gastrointestinal tract. *Springer Semin Immunopathol.* 1997;18:409-420.
153. La Cava A, Sarvetnick N. The role of cytokines in autoimmunity. *Curr Dir Autoimmun.* 1999;1:56-71.
154. von Andrian UH, Mackay CR. T-cell function and migration. Two sides of the same coin. *N Engl J Med.* 2000;343:1020-1034.
155. Godessart N, Kunkel SL. Chemokines in autoimmune disease. *Curr Opin Immunol.* 2001;13:670-675.
156. O'Shea JJ, Ma A, Lipsky P. Cytokines and autoimmunity. *Nat Rev Immunol.* 2002;2:37-45.
157. Langrish CL, Chen Y, Blumenschein WM, et al. IL-23 drives a pathogenic T cell population that induces autoimmune inflammation. *J Exp Med.* 2005;201:233-240.
158. Harrington LE, Hatton RD, Mangan PR, et al. Interleukin 17-producing CD4+ effector T cells develop via a lineage distinct from the T helper type 1 and 2 lineages. *Nat Immunol.* 2005;6:1123-1132.
159. Park H, Li Z, Yang XO, et al. A distinct lineage of CD4 T cells regulates tissue inflammation by producing interleukin 17. *Nat Immunol.* 2005;6:1133-1141.
160. Fouser LA, Wright JF, Dunussi-Joannopoulos K, Collins M. Th17 cytokines and their emerging roles in inflammation and autoimmunity. *Immunol Rev.* 2008;226:87-102.
161. Manoury-Schwartz B, Chiocchia G, Bessis N, et al. High susceptibility to collagen-induced arthritis in mice lacking IFN-gamma receptors. *J Immunol.* 1997;158:5501-5506.
162. Vermeire K, Heremans H, Vandeputte M, Huang S, Billiau A, Matthys P. Accelerated collagen-induced arthritis in IFN-gamma receptor-deficient mice. *J Immunol.* 1997;158:5507-5513.
163. Nakae S, Nambu A, Sudo K, Iwakura Y. Suppression of immune induction of collagen-induced arthritis in IL-17-deficient mice. *J Immunol.* 2003;171:6173-6177.
164. Lubberts E, Koenders MI, Oppers-Walgreen B, et al. Treatment with a neutralizing anti-murine interleukin-17 antibody after the onset of collagen-induced arthritis reduces joint inflammation, cartilage destruction, and bone erosion. *Arthritis Rheum.* 2004;50:650-659.
165. Sarkar S, Fox DA. Targeting IL-17 and Th17 cells in rheumatoid arthritis. *Rheum Dis Clin North Am.* 2010;36:345-366.
166. Kim S, Kim KA, Hwang DY, et al. Inhibition of autoimmune diabetes by Fas ligand: the paradox is solved. *J Immunol.* 2000;164:2931-2936.
167. Hugues S, Mougneau E, Ferlin W, et al. Tolerance to islet antigens and prevention from diabetes induced by limited apoptosis of pancreatic beta cells. *Immunity.* 2002;16:169-181.
168. Pope RM. Apoptosis as a therapeutic tool in rheumatoid arthritis. *Nat Rev Immunol.* 2002;2:527-535.
169. Gallucci S, Lolkema M, Matzinger P. Natural adjuvants: endogenous activators of dendritic cells. *Nat Med.* 1999;5:1249-1255.
170. Delovitch TL, Singh B. The nonobese diabetic mouse as a model of autoimmune diabetes: immune dysregulation gets the NOD. *Immunity.* 1997;7:727-738.
171. Christen U, Wolfe T, Mohrle U, et al. A dual role for TNF-alpha in type 1 diabetes: islet-specific expression abrogates the ongoing autoimmune process when induced late but not early during pathogenesis. *J Immunol.* 2001;166:7023-7032.
172. Maini RN, Elliott M, Brennan FM, Williams RO, Feldmann M. TNF blockade in rheumatoid arthritis: implications for therapy and pathogenesis. *APMIS.* 1997;105:257-263.
173. Feldmann M, Brennan FM, Maini RN. Rheumatoid arthritis. *Cell.* 1996;85:307-310.
174. Garber K. Pfizer's JAK inhibitor sails through phase 3 in rheumatoid arthritis. *Nat Biotechnol.* 2011;29:467-468.
175. Tisch R, McDevitt H. Antigen specific immunotherapy: is it a real possibility to combat T-cell medicated autoimmunity? *Proc Natl Acad Sci U S A.* 1994;91:437-438.
176. Tisch R, Wang B, Weaver DJ, et al. Antigen-specific mediated suppression of beta cell autoimmunity by plasmid DNA vaccination. *J Immunol.* 2001;166:2122-2132.
177. Shevach EM, McHugh RS, Piccirillo CA, Thornton AM. Control of T-cell activation by CD4+ CD25+ suppressor T cells. *Immunol Rev.* 2001;182:58-67.
178. Bresson D, Togher L, Rodrigo E, et al. Anti-CD3 and nasal proinsulin combination therapy enhances remission from recent-onset autoimmune diabetes by inducing Tregs. *J Clin Invest.* 2006;116:1371-1381.
179. Sicotte NL, Voskuhl RR. Onset of multiple sclerosis associated with anti-TNF therapy. *Neurology.* 2001;57:1885-1888.
180. Riminton DS, Korner H, Strickland DH, Lemckert FA, Pollard JD, Sedgwick JD. Challenging cytokine redundance: inflammatory cell movement and clinical course of experimental autoimmune encephalomyelitis are normal in lymphotoxin-deficient, but not tumor necrosis factor-deficient, mice. *J Exp Med.* 1998;187:1517-1528.
181. Sedgwick JD, Riminton DS, Cyster JG, Korner H. Tumor necrosis factor: a master-regulator of leukocyte movement. *Immunol Today.* 2000;21:110-113.
182. Coon B, An L-L, Whitton JL, von Herrath MG. DNA immunization to prevent autoimmune diabetes. *J Clin Invest.* 1999;104:189-194.
183. Christoffersson G, Chodaczek G, Ratliff SS, Coppieters K, von Herrath MG. Suppression of diabetes by accumulation of non-islet-specific CD8+ effector T cells in pancreatic islets. *Sci Immunol.* 2018;3(21):eaam6533.
184. Saligrama N, Zhao F, Sikora MJ, et al. Opposing T cell responses in experimental autoimmune encephalomyelitis. *Nature.* 2019;572:481-487.
185. Tang Q, Henriksen KJ, Bi M, et al. *In vitro*-expanded antigen-specific regulatory T cells suppress autoimmune diabetes. *J Exp Med.* 2004;199:1455-1465.
186. Dayan CM, Daniels GH. Chronic autoimmune thyroiditis. *N Engl J Med.* 1996;335:99-107.
187. Weetman AP. Graves' disease. *N Engl J Med.* 2000;343: 1236-1248.
188. Seddon B, Mason D. Peripheral autoantigen induces regulatory T cells that prevent autoimmunity. *J Exp Med.* 1999;189:877-881.
189. Vrbikova J, Janatkova I, Zamrazil V, Tomiska F, Fucikova T. Epstein-Barr virus serology in patients with autoimmune thyroiditis. *Exp Clin Endocrinol Diabetes.* 1996;104:89-92.
190. Tomer Y, Davies TF. Infections and autoimmune endocrine disease. *Baillieres Clin Endocrinol Metab.* 1995;9:47-70.
191. Sellmeyer DE, Grunfeld C. Endocrine and metabolic disturbances in human immunodeficiency virus infection and the acquired immune deficiency syndrome. *Endocr Rev.* 1996;17:518-532.
192. Grinspoon SK, Bilezikian JP. HIV disease and the endocrine system. *N Engl Med.* 1992;327:1360-1365.
193. Hellerstein MK. Endocrinology abnormalities. In: Cohen PT, Sande M, Volberding P, eds. *The AIDS Knowledge Base.* Massachusetts Medical Society; 1995:1-10.
194. Dluhy RG. The growing spectrum of HIV-related endocrine abnormalities. *J Clin Endocrinol Metab.* 1990;70:563-565.
195. Heufelder AH, Hofbauer LC. Human immunodeficiency virus infection and the thyroid gland. *Eur J Endocrinol.* 1996;134:669-674.
196. Friedman ND, Spelman DW. Subacute thyroiditis presenting as pyrexia of unknown origin in a patient with human immunodeficiency virus infection. *Clin Infect Dis.* 1999;29:1352-1353.
197. Jaeger P, Otto S, Speck RF, et al. Altered parathyroid gland function in severely immunocompromised patients infected with human immunodeficiency virus. *J Clin Endocrinol Metab.* 1994;79:1701-1705.
198. Burch HB, Nagy EV, Lukes YG, et al. Nucleotide and amino acid homology between the human thyrotropin receptor and HIV-1 nef protein: identification and functional analysis. *Biochem Biophys Res Commun.* 1991;181:498-505.
199. Wick G, Trieb K, Aguzzi A, Recheis H, Anderl H, Grubeck-Loebenstein B. Possible role of human foamy virus in Graves' disease. *Intervirology.* 1993;35:101-107.
200. Lagaye S, Vexiau P, Morozov V, et al. Human spumaretrovirus-related sequences in the DNA of leukocytes from patients with Graves disease. *Proc Natl Acad Sci U S A.* 1992;89: 10070-10074.
201. Mine H, Kawai H, Yokoi K, Akaike M, Saito S. High frequencies of human T-lymphotropic virus type I (HTLV-I) infection and presence of HTLV-II proviral DNA in blood donors with anti-thyroid antibodies. *J Mol Med.* 1996;74:471-477.
202. Yokoi K, Kawai H, Akaike M, Mine H, Saito S. Presence of human T-lymphotropic virus type II-related genes in DNA of peripheral leukocytes from patients with autoimmune thyroid diseases. *J Med Virol.* 1995;45:392-398.
203. Mizokami T, Okamura K, Ikenoue H, et al. A high prevalence of human T-lymphotropic virus type I carriers in patients with antithyroid antibodies. *Thyroid.* 1994;4:415-419.
204. Akamine H, Takasu N, Komiya I, et al. Association of HTLV-1 with autoimmune thyroiditis in patients with adult T-cell leukemia (ATL) and in HTLV-I carriers. *Clin Endocrinol (Oxf).* 1996;45:461-466.
205. McMurray RW, Elbourne K. Hepatitis C virus infection and autoimmunity. *Semin Arthritus Rheum.* 1997;26:689-701.
206. Cosserat J, Cacoub P, Bletry O. Immunological disorders in C virus chronic hepatitis. *Nephrol Dial Transplant.* 1996;11:31-35.
207. Degos F. Natural history of hepatitis C virus infection. *Nephrol Dial Transplant.* 1996;4:16-18.
208. Ganne-Carrie N, Medini A, Coderc E, et al. Latent autoimmune thyroiditis in untreated patients with HCV chronic hepatitis: a case-control study. *J Autoimmun.* 2000;14:189-193.
209. Manns MP, Rambusch EG. Autoimmunity and extrahepatic manifestations in hepatitis C virus infection. *J Hepatol.* 1999;1:39-42.
210. Preziati D, La Rosa L, Covini G, et al. Autoimmunity and thyroid function in patients with chronic active hepatitis treated with recombinant interferon alpha-2a. *Eur J Endocrinol.* 1995;132:587-593.
211. Gregorio GV, Jones H, Choudhuri K, et al. Autoantibody prevalence in chronic hepatitis B virus infection: effect in interferon alfa. *Hepatology.* 1996;24:520-523.
212. Fernandez-Soto L, Gonzalez A, Escobar-Jimenez F, et al. Increased risk of autoimmune thyroid disease in hepatitis C vs Hepatitis B before, during, and after discontinuing interferon therapy. *Arch Intern Med.* 1998;158:1445-1448.
213. Sever JL, South MA, Shaver KA. Delayed manifestations of congenital rubella. *Rev Infect Dis.* 1985;7:S164-S169.
214. Klavinskis LS, Notkins AL, Oldstone MB. Persistent viral infection of the thyroid gland: alteration of thyroid function in the absence of tissue injury. *Endocrinology.* 1988;122:567-575.
215. Srinivassapa J, Gazelli C, Onodera T, et al. Virus-induced thyroiditis. *Endocrinology.* 1988;122:563-566.
216. Onodera T, Awaya A. Anti-thyroglobulin antibodies induced with recombinant reovirus infection in BALB/c mice. *Immunology.* 1990;71:581-585.
217. Ziemiecki A, Kromer G, Muller RG, et al. Ev 22, a new endogenous avian leukosis virus locus found in chickens with spontaneous autoimmune thyroiditis. *Arch Virol.* 1988;100:267-271.
218. Carter JK, Smith RE. Rapid induction of hypothyroidism by an avian leukosis virus. *Infect Immun.* 1983;40:795-805.
219. Kuhr T, Hala K, Dietrich H, Herold M, Wick G. Genetically determined target organ susceptibility in the pathogenesis of spontaneous autoimmune thyroiditis: aberrant expression of HMC-class II antigens and the possible role of virus. *J Autoimmun.* 1994;7:13-25.
220. Mason LJ, Timothy LM, Isenberg DA, Kalsi JK. Immunization with a paptide of Sm B/B' results in limited epitope spreading but not autoimmune disease. *J Immunol.* 1999;162:5099-5105.
221. von Mering J, Minkowski O. Diabetes mellitus nach Pankreasextirpation. *Arch Exp Pathol Pharmakol.* 1890;26:371-374.
222. Schmidt MB. Uber die Beziehung der Langerhansschen Insein des Pankreas zum Diabetes Mellitus. *Munch Med Wochenschr.* 1902;49:51.
223. Banting FG, Best CH. The internal secretion of the pancreas. *J Lab Clin Med.* 1922;7:251-266.
224. Gepts W. Pathologic anatomy of the pancreas in juvenile diabetes mellitus. *Diabetes.* 1965;14:619-633.
225. Bottazzo GF, Florin-Christensen A, Doniach D. Islet cell antibodies in diabetes mellitus with autoimmune polyendocrine deficiencies. *Lancet.* 1974;2:1279-1281.
226. MacCuish AC, Irvine WJ, Barnes EW, Duncan LJ. Antibodies to pancreatic islet cells in insulin-dependent diabetics with coexistent autoimmune disease. *Lancet.* 1974;2: 1529-1531.
227. Nerup J, Platz P, Andersen OO, et al. HL-A antigens and diabetes mellitus. *Lancet.* 1974;2:864-866.
228. Atkinson MA, Eisenbarth GS. Type 1 diabetes: new perspectives on disease pathogenesis and treatment. *Lancet.* 2001;358:221-229.
229. Atkinson MA, Maclaren NK. The pathogenesis of insulin-dependent diabetes mellitus. *N Engl J Med.* 1994;331: 1428-1436.
230. Jackson RA, Soeldner JS, Eisenbarth GS. Predicting insulin-dependent diabetes. *Lancet.* 1988;10:627-628.
231. Eisenbarth GS, Gianani R, Yu L, Pietropaolo M, Harrison C, Jackson R. Dual-parameter model for prediction of type I diabetes mellitus. *Proc Assoc Am Phys.* 1998;110:126-135.
232. Wong FS, Karttunen J, Dumont C, et al. Identification of an MHC class I-restricted autoantigen in type 1 diabetes by screening an organ-specific cDNA library. *Nat Med.* 1999;5:1026-1031.
233. Lieberman SM, Evans AM, Han B, et al. Identification of the beta cell antigen targeted by a prevalent population of pathogenic CD8+ T cells in autoimmune diabetes. *Proc Natl Acad Sci U S A.* 2003;100:8384-8388.
234. Trudeau JD, Kelly-Smith C, Verchere CB, et al. Prediction of spontaneous autoimmune diabetes in NOD mice by quantification of autoreactive T cells in peripheral blood. *J Clin Invest.* 2003;111:217-223.
235. Yoon J-W, Yoon C-S, Lim H-W, et al. Control of autoimmune diabetes in NOD mice by GAD expression or suppression in betta cells. *Science.* 1999;284:1183-1187.
236. Nakayama M, Abiru N, Moriyama H, et al. Prime role for an insulin epitope in the development of type 1 diabetes in NOD mice. *Nature.* 2005;435:220-223.
237. von Herrath MG, Dockter J, Oldstone MB. How virus induces a rapid or slow onset insulin-dependent diabetes mellitus in a transgenic model. *Immunity.* 1994;1:231-242.
238. DiLorenzo TP, Graser RT, Ono T, et al. Major histocompatibility complex class I-restricted T cells are required for all but the end stages of diabetes development in nonobese diabetic mice and use a prevalent T cell receptor alpha chain gene rearrangement. *Proc Natl Acad Sci USA.* 1998;95:12538-12543.
239. Graser RT, DiLorenzo TP, Wang F, et al. Identification of a CD8 T cell that can independently mediate autoimmune diabetes development in the complete absence of CD4 T cell helper functions. *J Immunol.* 2000;164:3913-3918.
240. Eisenbarth GA. Type 1 diabetes mellitus. A chronic autoimmune disease. *N Engl J Med.* 1986;314:1360-1368.
241. Durinovic-Bello I, Hummel M, Ziegler AG. Cellular immune response to diverse islet cell antigens in IDDM. *Diabetes.* 1996;45:795-800.
242. Fuchtenbusch M, Kredel K, Bonifacio E, Schnell O, Ziegler A-G. Exposure to exogenous insulin promotes IgG1 and the T-helper 2-associated IgG4 responses to insulin but not to other islet autoantigens. *Diabetes.* 2000;49:918-925.
243. Roep BO, Stobbe I, Duinkerken G, et al. Auto- and alloimmune reactivity to human islet allografts transplanted into type 1 diabetic patients. *Diabetes.* 1999;48:484-490.
244. Martin S, Wolf-Eichbaum D, Duinkerken G, et al. Development of type 1 diabetes despite severe hereditary B-lymphocyte deficiency. *N Engl J Med.* 2001;345:1036-1040.
245. Hiemstra HS, Drijfhout JW, Roep BO. Antigen arrays in T cell immunology. *Curr Opin Immunol.* 2000;12:80-84.

246. Noorchashm H, Lieu YK, Noorchashm N, et al. I-Ag7-mediated antigen presentation by B lymphocytes is critical in overcoming a checkpoint in T cell tolerance to islet beta cells of nonobese diabetic mice. *J Immunol.* 1999;163:743-750.
247. Kaufman DL, Tisch R, Sarvetnick N, et al. Report from the 1st international NOD mouse T-cell workshop and the follow-up mini-workshop. *Diabetes.* 2001;50:2459-2463.
248. van de Linde P, Roep BO. T-cell assays to determine disease activity and clinical efficacy of immune therapy in type 1 diabetes. *Am J Ther.* 2005;12:573-579.
249. Herold KC, Brooks-Worrell B, Palmer J, et al. Validity and reproducibility of measurement of islet autoreactivity by T-cell assays in subjects with early type 1 diabetes. *Diabetes.* 2009;58:2588-2595.
250. Velthuis JH, Unger WW, Abreu JR, et al. Simultaneous detection of circulating autoreactive CD8+ T-cells specific for different islet cell-associated epitopes using combinatorial MHC multimers. *Diabetes.* 2010;59:1721-1730.
251. Arif S, Tree TI, Astill TP, et al. Autoreactive T cell responses show proinflammatory polarization in diabetes but a regulatory phenotype in health. *J Clin Invest.* 2004;113:451-463.
252. Redondo MJ, Jeffrey J, Fain PR, Eisenbarth GS, Orban T. Concordance for islet autoimmunity among monozygotic twins. *N Engl J Med.* 2008;359:2849-2850.
253. Pociot F. Type 1 diabetes genome-wide association studies: not to be lost in translation. *Clin Transl Immunology.* 2017;6:e162.
254. Concannon P, Rich SS, Nepom GT. Genetics of type 1A diabetes. *N Engl J Med.* 2009;360:1646-1654.
255. Wicker LS, Todd JA, Peterson LB. Genetic control of autoimmune diabetes in the NOD mouse. *Annu Rev Immunol.* 1995;13:179-200.
256. McDevitt H, Singer S, Tisch R. The role of MHC class II genes in susceptibility and resistance to type I diabetes mellitus in the NOD mouse. *Horm Metab Res.* 1996;28:287-288.
257. Serreze DV, Leiter EH. Genes and cellular requirements for autoimmune diabetes susceptibility in nonobese diabetic mice. *Curr Dir Autoimmun.* 2001;4:31-67.
258. Von Herrath MG, Theofilopoulos AN, eds. *Molecular Pathology of Type 1 Diabetes Mellitus.* Karger Publishers; 2001.
259. Van Houten N, Blake SF. Direct measurement of anergy of antigen-specific T cells following oral tolerance induction. *J Immunol.* 1996;157:1337-1341.
260. Weiner HL. Oral tolerance: mobilizing the gut. *Hosp Pract.* 1995;30:53-58.
261. Benson JM, Whitacre CC. The role of clonal deletion and anergy in oral tolerance. *Res Immunol.* 1997;148:533-541.
262. Strobel S, Mowat AM. Immune responses to dietary antigens: oral tolerance. *Immunol Today.* 1998;19:173-181.
263. Viney JL, Mowat AM, O'Malley JM, Williamson E, Fanger NA. Expanding dendritic cells *in vivo* enhances the induction of oral tolerance. *J Immunol.* 1998;160:5815-5825.
264. Maeda Y, Noda S, Tanaka K, et al. The failure of oral tolerance induction is functionally coupled to the absence of T cells in Peyer's patches under germfree conditions. *Immunobiology.* 2001;204:442-457.
265. Wen L, Ley RE, Volchkov PY, et al. Innate immunity and intestinal microbiota in the development of Type 1 diabetes. *Nature.* 2008;455:1109-1113.
266. Paronen J, Knip M, Savilahti E, et al. Effect of cow's milk exposure and maternal type 1 diabetes on cellular and humoral immunization to dietary insulin in infants at genetic risk for type 1 diabetes. Finnish Trial to Reduce IDDM in the Genetically at Risk Study Group. *Diabetes.* 2000;49:1657-1665.
267. Karjalainen J, Martin JM, Knip M, et al. A bovine albumin peptide as a possible trigger of insulin-dependent diabetes mellitus. *N Engl J Med.* 1992;327:302-307.
268. Honeyman MC, Coulson BS, Stone NL, et al. Association between rotavirus infection and pancreatic islet autoimmunity in children at risk of developing type 1 diabetes. *Diabetes.* 2000;49:1319-1324.
269. Makela M, Oling V, Marttila J, et al. Rotavirus-specific T cell responses and cytokine mRNA expression in children with diabetes-associated autoantibodies and type 1 diabetes. *Clin Exp Immunol.* 2006;145:261-270.
270. Yeung WC, Rawlinson WD, Craig ME. Enterovirus infection and type 1 diabetes mellitus: systematic review and meta-analysis of observational molecular studies. *Br Med J.* 2011;342:d35.
271. Lonnrot M, Korpela K, Knip M, et al. Enterovirus infection as a risk factor for beta-cell autoimmunity in a prospectively observed birth cohort: the Finnish Diabetes Prediction and Prevention Study. *Diabetes.* 2000;49:1314-1318.
272. Notkins AL, Yoon JW. Virus-induced diabetes mellitus. In: Notkins AL, Oldstone MBA, eds. *Concepts in Viral Pathogenesis.* Springer-Verlag; 1984:241-247.
273. Jenson AB, Rosenberg HS, Notkins AL. Pancreatic islet cell damage in children with fatal viral infections. *Lancet.* 1980;2:354-358.
274. Mena I, Fischer C, Gebhard JR, Perry CM, Harkins S, Whitton JL. Coxsackievirus infection of the pancreas: evaluation of receptor expression, pathogenesis, and immunopathology. *Virology.* 2000;271:276-288.
275. Serreze DV, Ottendorfer EW, Ellis TM, Gauntt CJ, Atkinson MA. Acceleration of type 1 diabetes by a coxsackievirus infection requires a preexisting critical mass of autoreactive T-cells in pancreatic islets. *Diabetes.* 2000;49:708-711.
276. Stene LC, Oikarinen S, Hyoty H, et al. Enterovirus infection and progression from islet autoimmunity to type 1 diabetes: the Diabetes and Autoimmunity Study in the Young (DAISY). *Diabetes.* 2010;59:3174-3180.
277. Oikarinen S, Martiskainen M, Tauriainen S, et al. Enterovirus RNA in blood is linked to the development of type 1 diabetes. *Diabetes.* 2011;60:276-279.
278. Richardson SJ, Willcox A, Bone AJ, Foulis AK, Morgan NG. The prevalence of enteroviral capsid protein vp1 immunostaining in pancreatic islets in human type 1 diabetes. *Diabetologia.* 2009;52:1143-1151.
279. Nejentsev S, Walker N, Riches D, Egholm M, Todd JA. Rare variants of IFIH1, a gene implicated in antiviral responses, protect against type 1 diabetes. *Science.* 2009;324:387-389.
280. Smyth DJ, Cooper JD, Bailey R, et al. A genome-wide association study of nonsynonymous SNPs identifies a type 1 diabetes locus in the interferon-induced helicase (IFIH1) region. *Nat Genet.* 2006;38:617-619.
281. von Herrath M. Diabetes: a virus-gene collaboration. *Nature.* 2009;459:518-519.
282. Babu PG, Huber SA, Craighead JE. Contrasting features of T-lymphocyte-mediated diabetes in encephalomyocarditis virus-infected Balb/cBy and Balb/cCum mice. *Am J Pathol.* 1986;124:193-198.
283. Baum H. Mitochondrial antigens, molecular mimicry and autoimmune disease. *Biochim Biophys Acta.* 1995;1271:111-121.
284. Gianani R, Sarvetnick N. Viruses, cytokines, antigens, and autoimmunity. *Proc Natl Acad Sci U S A.* 1996;93:2257-2259.
285. Foulis AK, McGill M, Farquharson MA, Hilton DA. A search for evidence of viral infection in pancreases of newly diagnosed patients with IDDM. *Diabetologia.* 1997;40:53-61.
286. Shrinivasappa J, Saegusa J, Prabhakar B, et al. Frequency of reactivity of monoclonal antiviral antibodies with normal tissues. *J Virol.* 1986;57:397-401.
287. Oldstone MBA. Molecular mimicry as a mechanism for the cause and as a probe uncovering etiologic agent(s) of autoimmune disease. *Curr Top Microbiol Immunol.* 1989;145:127-135.
288. Horwitz MS, Sarvetnick N. Viruses, host responses, and autoimmunity. *Immunol Rev.* 1999;169:241-253.
289. Buschard K. Diabetic animal models. *APMIS.* 1996;104:609-614.
290. Katz J, Benoist C, Mathis DX. T helper cell subsets in IDDM. *Science.* 1995;268:1185-1188.
291. Wilberz S, Partke HJ, Dagnaes-Hansen F, Herberg L. Persistent MHV (mouse hepatitis virus) infection reduces the incidence of diabetes mellitus in non-obese diabetic mice. *Diabetologia.* 1991;34:2-5.
292. Oldstone MBA. Viruses as therapeutic agents. I. Treatment of nonobese insulin- dependent diabetes mice with virus prevents insulin-dependent diabetes mellitus while maintaining general immune competence. *J Exp Med.* 1990;171:2077-2089.
293. Toshiyuki M, Ogawa M, Kobayashi F, Yoshihiro T, Hirotsugu U. Electron microscope studies on the interaction of pancreatic islet cells and splenic lymphocytes in non-obese diabetic (NOD) mice. *Biomed Res.* 1988;9:67-73.
294. Morgan D, Liblau R, Scott B, et al. CD8+ T cell-mediated spontaneous diabetes in neonatal mice. *J Immunol.* 1996;157:978-984.
295. Kurts C, Cannarile M, Klebba I, Brocker T. Cutting edge: dendritic cells are sufficient to cross-present self-antigens to CD8 T cells in vivo. *J Immunol.* 2001;166:1439-1442.
296. Kurts C, Heath WR, Carbone FR, Kosaka H, Miller JF. Cross-presentation of self antigens to CD8+ T cells: the balance between tolerance and autoimmunity. *Novartis Found Symp.* 1998;215:172-181; discussion 181-190.
297. Kurts C, Heath WR, Kosaka H, Miller JFAP, Carbone FR. The peripheral delection of autoreactive CD8+ T cells induced by cross-presentation of self-antigens involves signaling through CD95 (Fas, Apo-1). *J Exp Med.* 1998;188:415-420.
298. Oldstone MBA, Nerenberg M, Southern P, Price J, Lewicki H. Virus infection triggers insulin-dependent diabetes mellitus in a transgenic model: role of anti-self (virus) immune response. *Cell.* 1991;65:319-331.
299. Ohashi P, Oehen S, Buerki K, et al. Ablation of "tolerance" and induction of diabetes by virus infection in viral antigen transgenic mice. *Cell.* 1991;65:305-317.
300. Berdanier CD. Diet, autoimmunity, and insulin-dependent diabetes mellitus: a controversy. *Proc Soc Exp Biol Med.* 1995;209:223-230.
301. Hiltunen M, Hyoty H, Knip M, et al. Islet cell antibody seroconversion in children is temporally associated with enterovirus infections. Childhood Diabetes in Finland (DiMe) Study Group. *J Infect Dis.* 1997;175:554-560.
302. Peterson JS, Dyrberg T, Karlsen AE, et al. Glutamic acid decarboxylase (GAD_{65}) autoantibodies in prediction of b-cell function and remission in recent onset IDDM after cyclosporin treatment. *Diabetes.* 1994;43:1291-1296.
303. Peterson JS, Hejnaes KR, Moody A, et al. Detection of GAD_{65} antibodies in diabetes and other autoimmune diseases using a simple radioligand assay. *Diabetes.* 1994;43:459-467.
304. Wucherpfennig KW, Strominger JL. Molecular mimicry in T-cell mediated autoimmunity: viral peptides activate human T-cell clones specific for myelin basic protein. *Cell.* 1995;80:695-705.
305. Ludewig B, Odermatt B, Landmann S, Hengartner H, Zinkernagel RM. Dendritic cells induce autoimmune diabetes and maintain disease via de novo formation of local lymphoid tissue. *J Exp Med.* 1998;188:1493-1501.
306. Ludewig B, Odermatt B, Ochsenbein AF, Zinkernagel RM, Hengartner H. Role of dendritic cells in the induction and maintenance of autoimmune diseases. *Immunol Rev.* 1999;169:45-54.
307. Ludewig B, Junt T, Hengartner H, Zinkernagel RM. Dendritic cells in autoimmune diseases. *Curr Opin Immunol.* 2001;13:657-662.
308. von Herrath MG, Guerder S, Lewicki H, Flavell RA, Oldstone MB. Coexpression of B7-1 and viral ("self") transgenes in pancreatic beta cells can break peripheral ignorance and lead to spontaneous autoimmune diabetes. *Immunity.* 1995;3:727-738.
309. Kagi D, Odermatt B, Ohashi PS, Zinkernagel RM, Hengartner H. Development of insulitis without diabetes in transgenic mice lacking perforin-dependent cytotoxicity. *J Exp Med.* 1996;183:2143-2152.
310. Chatenoud L, Salomon B, Bluestone JA. Suppressor T cells – they're back and critical for regulation of autoimmunity! *Immunol Rev.* 2001;182:149-163.
311. Herold KG, Lenschow DJ, Bluestone JA. CD28/B7 regulation of autoimmune diabetes. *Immunol Res.* 1997;16:71-84.
312. Green EA, Flavell RA. The temporal importance of TNF alpha expression in the development of diabetes. *Immunity.* 2000;12:459-469.
313. Brod SA, Atkinson M, Lavis VR, et al. Ingested IFN-alpha preserves residual beta cell function in type 1 diabetes. *J Interferon Cytokine Res.* 2001;21:1021-1030.
314. Stauffer Y, Marguerat S, Meylan F, et al. Interferon-alpha-induced endogenous superantigen. a model linking environment and autoimmunity. *Immunity.* 2001;15:591-601.
315. Bosi E, Minelli R, Bazzigaluppi E, Salvi M. Fulminant autoimmune Type 1 diabetes during interferon-alpha therapy: a case of Th1-mediated disease? *Diabet Med.* 2001;18:329-332.
316. von Herrath MG, Allison J, Miller JF, Oldstone MBA. Focal expression of interleukin-2 does not break unresponsiveness to "self" (viral) antigen expressed in beta cells but enhances development of autoimmune disease (diabetes) after initiation of an anti-self immune response. *J Clin Invest.* 1995;95:477-485.
317. Lee MS, Wogensen L, Shizuru J, Oldstone MBA, Sarvetnick N. Pancreatic islet production of murine interleukin-10 does not inhibit immune-mediated tissue destruction. *J Clin Invest.* 1994;93:1332-1338.
318. Lee M-S, Mueller R, Wicker LS, Peterson LB, Sarvetnick N. IL-10 is necessary and sufficient for autoimmune diabetes in conjunction with NOD MHC homozygosity. *J Exp Med.* 1996;183:2663-2668.
319. Coppieters K, Amirian N, von Herrath M. Intravital imaging of CTLs killing islet cells in diabetic mice. *J Clin Invest.* 2012;122(1):119-131.
320. von Herrath MG, Whitton JL. DNA vaccination to treat autoimmune diabetes. *Ann Med.* 2000;32:285-292.
321. Bot A, Smith D, Bot S, et al. Plasmid vaccination with insulin b chain prevents autoimmune diabetes in nonobese diabetic mice. *J Immunol.* 2001;167:2950-2955.
322. Garren H, Ruiz PJ, Watkins TA, et al. Combination of gene delivery and DNA vaccination to protect from and reverse Th1 autoimmune disease via deviation to the Th2 pathway. *Immunity.* 2001;15:15-22.
323. Tisch R, Wang B, Atkinson MA, Serreze DV, Friedline R. A glutamic acid decarboxylase 65-specific th2 cell clone immunoregulates autoimmune diabetes in nonobese diabetic mice. *J Immunol.* 2001;166:6925-6936.
324. Wherrett DK, Bundy B, Becker DJ, et al. Antigen-based therapy with glutamic acid decarboxylase (GAD) vaccine in patients with recent-onset type 1 diabetes: a randomised double-blind trial. *Lancet.* 2011;378:319-327.
325. von Herrath M. Combination therapies for type 1 diabetes: why not now? *Immunotherapy.* 2010;2:289-291.
326. Gaur A, Boehme SA, Chalmers D, et al. Amelioration of relapsing experimental autoimmune encephalomyelitis with altered myelin basic protein peptides involves different cellular mechanisms. *J Neuroimmunol.* 1997;74:149-158.
327. Pedotti R, Mitchell D, Wedemeyer J, et al. An unexpected version of horror autotoxicus: anaphylactic shock to a self-peptide. *Nat Immunol.* 2001;2:216-222.
328. Kappos L, Comi G, Panitch H, et al. Induction of a non-encephalitogenic type 2 T helper-cell autoimmune response in multiple sclerosis after administration of an altered peptide ligand in a placebo-controlled, randomized phase II trial. The Altered Peptide Ligand in Relapsing MS Study Group. *Nat Med.* 2000;6:1176-1182.
329. Walter M, Philotheou A, Bonnici F, Ziegler AG, Jimenez R. No effect of the altered peptide ligand NBI-6024 on beta-cell residual function and insulin needs in new-onset type 1 diabetes. *Diabetes Care.* 2009;32:2036-2040.

330. Clemente-Casares X, Blanco J, Ambalavanan P, et al. Expanding antigen-specific regulatory networks to treat autoimmunity. *Nature.* 2016;530:434-440.
331. Skyler JS, Krischer JP, Wolfsdorf J, et al. Effects of oral insulin in relatives of patients with type 1 diabetes: the Diabetes Prevention Trial--Type 1. *Diabetes Care.* 2005;28:1068-1076.
332. Writing Committee for the Type 1 Diabetes TrialNet Oral Insulin Study Group; Krischer JP, Schatz DA, Bundy B, Skyler JS, Greenbaum CJ. Effect of oral insulin on prevention of diabetes in relatives of patients with type 1 diabetes: a randomized clinical trial. *J Am Med Assoc.* 2017;318:1891-1902.
333. Ziegler AG, Achenbach P, Berner R, et al. Oral insulin therapy for primary prevention of type 1 diabetes in infants with high genetic risk: the GPPAD-POInT (global platform for the prevention of autoimmune diabetes primary oral insulin trial) study protocol. *BMJ Open.* 2019;9:e028578.
334. Sherry N, Hagopian W, Ludvigsson J, et al. Teplizumab for treatment of type 1 diabetes (Protege study): 1-year results from a randomised, placebo-controlled trial. *Lancet.* 2011;378:487-497.
335. Herold KC, Bundy BN, Long SA, et al. An anti-CD3 antibody, teplizumab, in relatives at risk for type 1 diabetes. *N Engl J Med.* 2019;381:603-613.
336. Orban T, Bundy B, Becker DJ, et al. Co-stimulation modulation with abatacept in patients with recent-onset type 1 diabetes: a randomised, double-blind, placebo-controlled trial. *Lancet.* 2011;378:412-419.
337. Pescovitz MD, Greenbaum CJ, Krause-Steinrauf H, et al. Rituximab, B-lymphocyte depletion, and preservation of beta-cell function. *N Engl J Med.* 2009;361:2143-2152.
338. Shapiro AM, Lakey JR, Ryan EA, et al. Islet transplantation in seven patients with type 1 diabetes mellitus using a glucocorticoid-free immunosuppressive regimen. *N Engl J Med.* 2000;343:230-238.
339. Kroon E, Martinson LA, Kadoya K, et al. Pancreatic endoderm derived from human embryonic stem cells generates glucose-responsive insulin-secreting cells *in vivo*. *Nat Biotechnol.* 2008;26:443-452.
340. D'Amour KA, Bang AG, Eliazer S, et al. Production of pancreatic hormone-expressing endocrine cells from human embryonic stem cells. *Nat Biotechnol.* 2006;24:1392-1401.
341. Ettinger RA, Nepom GT. Molecular aspects of HLA class II alphabeta heterodimers associated with IDDM susceptibility and protection. *Rev Immunogenet.* 2000;2:88-94.
342. Coulombe M, Yang H, Wolf LA, Gill RG. Tolerance to antigen-presenting cell-depleted islet allografts is CD4 T cell dependent. *J Immunol.* 1999;162:2503-2510.
343. Parker DC, Greiner DL, Phillips NE, et al. Survival of mouse pancreatic islet allografts in recipients treated with allogeneic small lymphocytes and antibody to CD40 ligand. *Proc Natl Acad Sci U S A.* 1995;92:9560-9564.
344. Rossini AA, Greiner DL, Mordes JP. Induction of immunologic tolerance for transplantation. *Physiol Rev.* 1999;79:99-141.
345. Welsh RM, Markees TG, Woda BA, et al. Virus-induced abrogation of transplantation tolerance induced by donor-specific transfusion and anti-CD154 antibody. *J Virol.* 2000;74:2210-2218.
346. Wilson SB, Kent SC, Patton KT, et al. Extreme Th1 bias of invariant Valpha24JalphaQ T cells in type 1 diabetes. *Nature.* 1998;391:177-181.
347. Utz U, Biddison WE, McFarland HF, McFarlin DE, Flerlage M, Martin R. Skewed T-cell receptor repertoire in genetically identical twins correlates with MS. *Nature.* 1993;364:243-247.
348. Muraro PA, Vergelli M, Kalbus M, et al. Immunodominance of a low-affinity major histocompatibility complex-binding myelin basic protein epitope (residues 111-129) in HLA-DR4 (B1*0401) subjects is associated with a restricted T cell receptor repertoire. *J Clin Invest.* 1997;100:339-349.
349. Scholz C, Patton KT, Anderson DE, Freeman GJ, Hafler DA. Expansion of autoreactive T cells in multiple sclerosis is independent of exogenous B7 costimulation1. *J Immunol.* 1998;160:1532-1538.
350. de Seze J, Dubucquoi S, Lefranc D, et al. IgG reactivity against citrullinated myelin basic protein in multiple sclerosis. *J Neuroimmunol.* 2001;117:149-155.
351. Mason JL, Suzuki K, Chaplin DD, Matsushima GK. Interleukin-1beta promotes repair of the CNS. *J Neurosci.* 2001;21:7046-7052.
352. Arnett HA, Mason J, Marino M, Suzuki K, Matsushima GK, Ting JP. TNF alpha promotes proliferation of oligodendrocyte progenitors and remyelination. *Nat Neurosci.* 2001;4:1116-1122.
353. Ebers G, Bulman D, Sadovnik A, et al. A population based study of multiple sclerosis in twins. *N Engl J Med.* 1987;315:1638-1642.
354. Vergelli M, Mazzanti B, Traggiai E, et al. Short-term evolution of autoreactive T cell repertoire in multiple sclerosis. *J Neurosci Res.* 2001;66:517-524.
355. Anderson DE, Sharpe AH, Hafler DA. The B7-CD28/CTLA-4 costimulatory pathways in autoimmune disease of the central nervous system. *Curr Opin Immunol.* 1999;11:677-683.
356. Kanter JL, Narayana S, Ho PP, et al. Lipid microarrays identify key mediators of autoimmune brain inflammation. *Nat Med.* 2006;12:138-143.
357. Iglesias A, Bauer J, Litzenburger T, Schubart A, Linington C. T- and B-cell responses to myelin oligodendrocyte glycoprotein in experimental autoimmune encephalomyelitis and multiple sclerosis. *Glia.* 2001;36:220-234.
358. Challoner PB, Smith KT, Parker JD, et al. Plaque-associated expression of human herpes virus 6 in multiple sclerosis. *Proc Natl Acad Sci U S A.* 1995;92:7440-7444.
359. Miller S, Karpus WJ. The immune pathogenesis and regulation of T-cell mediated demyelinating diseases. *Immunol Today.* 1994;15:358-362.
360. Neumann H, Schmidt H, Cavalie A, Jenne D, Wekerle H. Major histocompatibility complex (MHC) Class 1 gene expression in single neurons of the central nervous system: differential regulation by interferon (IFN-gamma) and tumor necrosis factor (TNF)-a. *J Exp Med.* 1997;185:305-316.
361. Ruiz PJ, Garren H, Ruiz IU, et al. Suppressive immunization with DNA encoding a self-peptide prevents autoimmune disease: modulation of T cell costimulation. *J Immunol.* 1999;162:3336-3341.
362. Olson JK, Croxford JL, Calenoff MA, Dal Canto MC, Miller SD. A virus-induced molecular mimicry model of multiple sclerosis. *J Clin Invest.* 2001;108:311-318.
363. Buchmeier MJ, Lane TE. Viral-induced neurodegenerative disease. *Curr Opin Microbiol.* 1999;2:398-402.
364. Evans CF, Horwitz MS, Hobbs MV, Oldstone MBA. Viral infection of transgenic mice expressing a viral protein in oligodendrocytes leads to chronic central nervous system autoimmune disease. *J Exp Med.* 1996;184:2371-2384.
365. Garren H, Robinson WH, Krasulova E, et al. Phase 2 trial of a DNA vaccine encoding myelin basic protein for multiple sclerosis. *Ann Neurol.* 2008;63:611-620.
366. Martin R, Sturzebecher CS, McFarland HF. Immunotherapy of multiple sclerosis: where are we? Where should we go? *Nat Immunol.* 2001;2:785-788.
367. McFarland HF. Complexities in the treatment of autoimmune disease. *Science.* 1996;274:2037-2038.
368. Weiner HL, Mackin GA, Matsui M, et al. Double-blind pilot trial of oral tolerization with myelin antigens in multiple sclerosis. *Science.* 1993;259:1321-1324.
369. Rice GP, Woelfel EL, Talbot PJ, et al. Immunological complications in multiple sclerosis patients receiving interferon. *Ann Neurol.* 1985;18:439-442.
370. Chen M, Gran B, Costello K, Johnson K, Martin R, Dhib-Jalbut S. Glatiramer acetate induces a Th2-biased response and crossreactivity with myelin basic protein in patients with MS. *Mult Scler.* 2001;7:209-219.
371. Bielekova B, Goodwin B, Richert N, et al. Encephalitogenic potential of the myelin basic protein peptide (amino acids 83-99) in multiple sclerosis: results of a phase II clinical trial with an altered peptide ligand. *Nat Med.* 2000;6:1167-1175.
372. Liossis SN, Sfikakis PP. Costimulation blockade in the treatment of rheumatic diseases. *BioDrugs.* 2004;18:95-102.
373. Smith CE, Miller SD. Multi-peptide coupled-cell tolerance ameliorates ongoing relapsing EAE associated with multiple pathogenic autoreactivities. *J Autoimmun.* 2006;27:218-231.
374. Viglietta V, Bourcier K, Buckle GJ, et al. CTLA4Ig treatment in patients with multiple sclerosis: an open-label, phase 1 clinical trial. *Neurology.* 2008;71:917-924.
375. Lopez de Castro JA. The pathogentic role of HLA-B27 in chronic arthritis. *Curr Opin Immunol.* 1998;10:59-66.
376. Wildner G, Diedrichs-Mohring M, Thurau SR. Induction of arthritis and uveitis in Lewis rats by antigenic mimicry of peptides from HLA-B27 and cytokeratin. *Eur J Immunol.* 2002;32:299-306.
377. Lee SJ, Im HY, Schueller WC. HLA-B27 positive juvenile arthritis with cardiac involvement preceding sacroiliac joint changes. *Heart.* 2001;86:E19.
378. Sun B, Rizzo LV, Sun S-H, et al. Genetic susceptibility to experimental autoimmune uveitis involves more than a predisposition to generate a T helper-1-like or a T helper-2-like response. *J Immunol.* 1997;159:1004-1011.
379. Tarrant TK, Silver PB, Chan CC, Wiggert B, Caspi RR. Endogenous IL-12 is required for induction and expression of experimental autoimmune uveitis. *J Immunol.* 1998;161:122-127.
380. Wildner G, Thurau SR. Orally induced bystander suppression in experimental autoimmune uveoretinitis occurs only in the periphery and not in the eye. *Euro J Immunol.* 1995;25:1292-1297.
381. Yip L, Su L, Sheng D, et al. Deaf1 isoforms control the expression of genes encoding peripheral tissue antigens in the pancreatic lymph nodes during type 1 diabetes. *Nat Immunol.* 2009;10:1026-1033.
382. Gangappa S, Sam Babu J, Thomas J, Daheshia M, Rouse BT. Virus-induced immunoinflammatory lesions in the absence of viral antigen recognition. *J Immunol.* 1998;161:4289-4300.
383. Chun S, Daheshia M, Lee S, Rouse BT. Immune modulation by IL-10 gene transfer via viral vector and plasmid DNA: implication for gene therapy. *Cell Immunol.* 1999;194:194-204.
384. Panoutsakopoulou V, Sanchirico ME, Huster KM, et al. Analysis of the relationship between viral infection and autoimmune disease. *Immunity.* 2001;15:137-147.
385. Kosiewicz MM, Alard P, Streilein JW. Alterations in cytokine production following intraocular injection of soluble protein antigen: impairment in IFN-g and induction of TGF-b and IL-4 production. *J Immunol.* 1998;161:5382-5390.
386. Takeuchi M, Alard P, Streilein JW. TGF-b promotes immune deviation by altering accessory signals of antigen-presenting cells. *J Immunol.* 1998;160:1589-1597.
387. Thurau SR, Diedrichs-Mohring M, Fricke H, Burchardi C, Wildner G. Oral tolerance with an HLA-peptide mimicking retinal autoantigen as a treatment of autoimmune uveitis. *Immunol Lett.* 1999;68:205-212.
388. Yamanaka T, Straumfors A, Morton H, Fausa O, Brandtzaeg P, Farstad I. M cell pockets of human Peyer's patches are specialized extensions of germinal centers. *Eur J Immunol.* 2001;31:107-117.
389. Iwasaki A, Kelsall BL. Localization of distinct Peyer's patch dendritic cell subsets and their recruitment by chemokines macrophage inflammatory protein (MIP)-3alpha, MIP-3beta, and secondary lymphoid organ chemokine. *J Exp Med.* 2000;191:1381-1394.
390. Harrison LC, Dempsey-Collier M, Kramer DR, Takahashi K. Aerosol insulin induces regulatory CD8 gamma delta T cells that prevent murine insulin-dependent diabetes. *J Exp Med.* 1996;184:2167-2174.
391. Hanninen A, Harrison LC. Gd T cells as mediators of mucosal tolerance: the autoimmune diabetes model. *Immunol Rev.* 2000;173:109-119.
392. Chen Y, Inobe J, Kuchroo VK, Baron JL, Janeway CAJr, Weiner HL. Oral tolerance in myelin basic protein T-cell receptor transgenic mice: suppression of autoimmune encephalomyelitis and dose-dependent induction of regulatory cells. *Proc Natl Acad Sci U S A.* 1996;93:388-391.
393. Strickland RG, Mackay IR. A reappraisal of the nature and significance of chronic atrophic gastritis. *Am J Dig Dis.* 1973;18:426-440.
394. Carmel R. Prevalence of undiagnosed pernicious anemia in the elderly. *Arch Intern Med.* 1996;156:1097-1100.
395. Toh BH, van Driel IR, Gleeson PA. Pernicious anemia. *N Engl J Med.* 1997;337:1441-1448.
396. Alderuccio F, Sentry JW, Marshall AC, Biondo M, Toh BH. Animal models of human disease: experimental autoimmune gastritis-a model for autoimmune gastritis and pernicious anemia. *Clin Immunol.* 2002;102:48-58.
397. Thornton AM, Shevach EM. Suppressor effector function of CD4+CD25+ immunoregulatory T cells is antigen nonspecific. *J Immunol.* 2000;164:183-190.
398. Itoh M, Takahashi T, Sakaguchi N, et al. Thymus and autoimmunity: production of CD25+CD4+ naturally anergic and suppressive T cells as a key function of the thymus in maintaining immunologic self-tolerance. *J Immunol.* 1999;162:5317-5326.
399. Hanauer SB. Inflammatory bowel disease. *N Engl J Med.* 1996;334:841-848.
400. Crispe IN. Hepatic T cells and liver tolerance. *Nat Rev Immunol.* 2003;3:51-62.
401. Rehermann B, Nascimbeni M. Immunology of hepatitis B virus and hepatitis C virus infection. *Nat Rev Immunol.* 2005;5:215-229.
402. Good MF. Development of immunity to malaria may not be an entirely active process. *Parasite Immunol.* 1995;17:55-59.
403. Calne R, Davies H. Organ graft tolerance: the liver effect. *Lancet.* 1994;343:67-68.
404. Calne RY. Immunological tolerance--the liver effect. *Immunol Rev.* 2000;174:280-282.
405. Mazariegos GV, Reyes J, Marino IR, et al. Weaning of immunosuppression in liver transplant recipients. *Transplantation.* 1997;63:243-249.
406. Klein I, Crispe IN. Complete differentiation of CD8+ T cells activated locally within the transplanted liver. *J Exp Med.* 2006;203:437-447.
407. Manns MP, Vogel A. Autoimmune hepatitis, from mechanisms to therapy. *Hepatology.* 2006;43:S132-S144.
408. Krawitt EL. Autoimmune hepatitis. *N Engl J Med.* 2006;354:54-66.
409. Czaja AJ. Autoimmune liver disease. *Curr Opin Gastroenterol.* 2006;22:234-240.
410. Kaplan MM, Gershwin ME. Primary biliary cirrhosis. *N Engl J Med.* 2005;353:1261-1273.
411. Leung PS, Coppel RL, Gershwin ME. Etiology of primary biliary cirrhosis: the search for the culprit. *Semin Liver Dis.* 2005;25:327-336.
412. Kouroumalis E, Notas G. Pathogenesis of primary biliary cirrhosis: a unifying model. *World J Gastroenterol.* 2006;12:2320-2327.
413. MacFaul GR, Chapman RW. Sclerosing cholangitis. *Curr Opin Gastroenterol.* 2006;22:288-293.
414. O'Mahony CA, Vierling JM. Etiopathogenesis of primary sclerosing cholangitis. *Semin Liver Dis.* 2006;26:3-21.
415. Neuberger J. Halothane hepatitis. *Eur J Gastroenterol Hepatol.* 1998;10:631-633.
416. Waldenström J. Leber, Blutproteine und Nahrungseiweiss. *Dtsch Z Verdau Stoffwechselkrankheiten.* 1950;2:113-119.
417. Kunkel HG, Ahrens EH, Eisenmemger WJ, et al. Extreme hypergammaglobulinemia in young women with liver disease of unknown etiology (abstract). *J Clin Invest.* 1951;30:654.
418. Bearn AG, Kunkel HG, Slater R. The problem of chronic liver disease in young women. *Am J Med.* 1956;21:3-15.

419. Mackay IR, Taft CO, Cowling DS. Lupoid hepatitis. *Lancet*. 1956;2:1323-1326.
420. Johnson PJ, McFarlane IG. Meeting report: international autoimmune hepatitis group. *Hepatology*. 1993;18:998-1005.
421. Manns MP, Obermayer-Straub P. Viral induction of autoimmunity: mechanisms and examples in hepatology. *J Viral Hepat*. 1997;4:42-47.
422. Moritoki Y, Lian ZX, Ohsugi Y, Ueno Y, Gershwin ME. B cells and autoimmune liver diseases. *Autoimmun Rev*. 2006;5:449-457.
423. Longhi MS, Ma Y, Bogdanos DP, Cheeseman P, Mieli-Vergani G, Vergani D. Impairment of CD4+CD25+ regulatory T-cells in autoimmune liver disease. *J Hepatol*. 2004;41:31-37.
424. Ruiz-Moreno M, Rua MJ, Carreno V, Quiroga JA, Manns M, Meyer zum Buschenfelde KH. Autoimmune chronic hepatitis type 2 manifested during interferon therapy in children. *J Hepatol*. 1991;12:265-266.
425. Obermayer-Straub P, Strassburg CP, Manns MP. Target proteins in human autoimmunity: cytochromes P450 and UDP- glucuronosyltransferases. *Can J Gastroenterol*. 2000;14:429-439.
426. Cerasoli DM, McGrath J, Carding SR, Shih FF, Knowles BB, Caton AJ. Low avidity recognition of a class II-restricted neo-self peptide by virus-specific T cells. *Int Immunol*. 1995;7:935-945.
427. Jaeckel E. Animal models of autoimmune hepatitis. *Semin Liver Dis*. 2002;22:325-338.
428. Heneghan MA, McFarlane IG. Of mice and women: toward a mouse model of autoimmune hepatitis. *Hepatology*. 2005;42:17-20.
429. Guidotti LG, Ishikawa T, Hobbs MV, Matzke B, Schreiber R, Chisari FV. Intracellular inactivation of the hepatitis B virus by cytotoxic T lymphocytes. *Immunity*. 1996;4:25-36.
430. Chisari FV, Ferrari C. Hepatitis B virus immunopathogenesis. *Annu Rev Immunol*. 1995;13:29-60.
431. Alferink J, Aigner S, Reibke R, Hammerling GJ, Arnold B. Peripheral T-cell tolerance: the contribution of permissive T-cell migration into parenchymal tissues of the neonate. *Immunol Rev*. 1999;169:255-261.
432. Voehringer D, Blaser C, Grawitz AB, Chisari FV, Buerki K, Pircher H. Break of T cell ignorance to a viral antigen in the liver induces hepatitis. *J Immunol*. 2000;165:2415-2422.
433. Yamauchi K, Yamaguchi N, Furukawa T, et al. A murine model of acute liver injury induced by human monoclonal autoantibody. *Hepatology*. 2005;42:149-155.
434. Lapierre P, Djilali-Saiah I, Vitozzi S, Alvarez F. A murine model of type 2 autoimmune hepatitis: xenoimmunization with human antigens. *Hepatology*. 2004;39:1066-1074.
435. Lapierre P, Beland K, Djilali-Saiah I, Alvarez F. Type 2 autoimmune hepatitis murine model: the influence of genetic background in disease development. *J Autoimmun*. 2006;26:82-89.
436. Holdener M, Hintermann E, Bayer M, et al. Breaking tolerance to the natural human liver autoantigen cytochrome P450 2D6 by virus infection. *J Exp Med*. 2008;205:1409-1422.
437. Mackay IR. Chronic hepatitis: effect of prolonged suppressive treatment and comparison of azathioprine with prednisolone. *Q J Med*. 1968;37:379-392.
438. Al-Khalidi JA, Czaja AJ. Current concepts in the diagnosis, pathogenesis, and treatment of autoimmune hepatitis. *Mayo Clin Proc*. 2001;76:1237-1252.
439. Kirk AP, Jain S, Pocock S, Thomas HC, Sherlock S. Late results of the Royal Free Hospital prospective controlled trial of prednisolone therapy in hepatitis B surface antigen negative chronic active hepatitis. *Gut*. 1980;21:78-83.
440. Sanchez-Urdazpal L, Czaja AJ, van Hoek B, Krom RA, Wiesner RH. Prognostic features and role of liver transplantation in severe corticosteroid-treated autoimmune chronic active hepatitis. *Hepatology*. 1992;15:215-221.
441. Neuberger J, Portmann B, Calne R, Williams R. Recurrence of autoimmune chronic active hepatitis following orthotopic liver grafting. *Transplantation*. 1984;37:363-365.
442. Jones DE. Pathogenesis of primary biliary cirrhosis. *J Hepatol*. 2003;39:639-648.
443. Lang KS, Recher M, Junt T, et al. Toll-like receptor engagement converts T-cell autoreactivity into overt autoimmune disease. *Nat Med*. 2005;11:138-145.
444. Hirschfield GM, Liu X, Xu C, et al. Primary biliary cirrhosis associated with HLA, IL12A, and IL12RB2 variants. *N Engl J Med*. 2009;360:2544-2555.
445. Mells GF, Floyd JA, Morley KI, et al. Genome-wide association study identifies 12 new susceptibility loci for primary biliary cirrhosis. *Nat Genet*. 2011;43:329-332.
446. Hirschfield GM, Liu X, Han Y, et al. Variants at IRF5-TNPO3, 17q12-21 and MMEL1 are associated with primary biliary cirrhosis. *Nat Genet*. 2010;42:655-657.
447. Irie J, Wu Y, Wicker LS, et al. NOD.c3c4 congenic mice develop autoimmune biliary disease that serologically and pathogenetically models human primary biliary cirrhosis. *J Exp Med*. 2006;203:1209-1219.
448. Oertelt S, Lian ZX, Cheng CM, et al. Anti-mitochondrial antibodies and primary biliary cirrhosis in TGF-beta receptor II dominant-negative mice. *J Immunol*. 2006;177:1655-1660.
449. Karlsen TH, Franke A, Melum E, et al. Genome-wide association analysis in primary sclerosing cholangitis. *Gastroenterology*. 2010;138:1102-1111.
450. Melum E, Franke A, Schramm C, et al. Genome-wide association analysis in primary sclerosing cholangitis identifies two non-HLA susceptibility loci. *Nat Genet*. 2011;43:17-19.
451. Hsu W, Zhang W, Tsuneyama K, et al: Differential mechanisms in the pathogenesis of autoimmune cholangitis versus inflammatory bowel disease in interleukin-2Ralpha−/− mice. *Hepatology* 2009;49:133-140.
452. Janse M, Lamberts LE, Franke L, et al. Three ulcerative colitis susceptibility loci are associated with primary sclerosing cholangitis and indicate a role for IL2, REL, and CARD9. *Hepatology*. 2011;53:1977-1985.
453. Vierling JM. Animal models for primary sclerosing cholangitis. *Best Pract Res Clin Gastroenterol*. 2001;15:591-610.
454. Cullen SN, Chapman RW. The medical management of primary sclerosing cholangitis. *Semin Liver Dis*. 2006;26:52-61.
455. Lo SK, Wendon J, Mieli-Vergani G, Williams R. Halothane-induced acute liver failure: continuing occurrence and use of liver transplantation. *Eur J Gastroenterol Hepatol*. 1998;10:635-639.
456. Gut J, Christen U, Frey N, Koch V, Stoffler D. Molecular mimicry in halothane hepatitis: biochemical and structural characterization of lipoylated autoantigens. *Toxicology*. 1995;97:199-224.
457. Kosuda LL, Bigazzi PE. Chemical-induced autoimmunity. In: Smialowicz RJ, Holsapple MP, eds. *Experimental Immunotoxicology*. CRC Press; 1996:419-465.
458. Kita H, Mackay IR, Van De Water J, Gershwin ME. The lymphoid liver: considerations on pathways to autoimmune injury. *Gastroenterology*. 2001;120:1485-1501.
459. Tipping PG, Kitching AR. Glomerulonephritis, Th1 and Th2: what's new? *Clin Exp Immunol*. 2005;142:207-215.
460. Robson MG, Walport MJ. Renal autoimmunity. In: Lahita RG, Chiorazzi N, Reeves WH, eds. *Textbook of the Autoimmune Diseases*. Lippincott Williams & Wilkins; 2000:323-349.
461. Wilson CB, Borza D-B, Hudson BG. Autoimmune renal disease involving basement membrane antigens. In: Theofilopoulos AN, Bona CA, eds. *The Molecular Pathology of Autoimmune Diseases*. 2nd ed. Taylor & Francis; 2002:999-1028.
462. Racusen LC. Autoimmune diseases in the kidney. In: Rose NR, Mackay CR, eds. *The Autoimmune Diseases*. 3rd ed. Academic Press; 1998:603-621.
463. Schena FP. Survey of the Italian registry of renal biopsies. Frequency of the renal diseases for 7 consecutive years. The Italian Group of Renal Immunopathology. *Nephrol Dial Transplant* 1997;12:418-426.
464. Gesualdo L, Di Palma AM, Morrone LF, Strippoli GF, Schena FP. The Italian experience of the national registry of renal biopsies. *Kidney Int*. 2004;66:890-894.
465. Borza DB, Neilson EG, Hudson BG. Pathogenesis of Goodpasture syndrome: a molecular perspective. *Semin Nephrol*. 2003;23:522-531.
466. Lin M-S, Liu Z, Drolet BA, Diaz LA. Cutaneous autoimmune diseases. In: Rose NR, Mackay IR, eds. *The Autoimmune Diseases*. 3rd ed. Academic Press; 1998:545-570.
467. Lin M-S, Arteaga LA, Warren SJP, Diaz LA. Pemphigus. In: Theofilopoulos AN, Bona CA, eds. *The Molecular Pathology of Autoimmune Diseases*. Taylor & Francis; 2002:799-816.
468. Bystryn JC, Rudolph JL. Pemphigus. *Lancet*. 2005;366:61-73.
469. Humbert P, Dupond JL, Vuitton D, Agache P. Dermatological autoimmune diseases and the multiple autoimmune syndromes. *Acta Derm Venereol Suppl (Stockh)*. 1989;148:1-8.
470. Korman N. Pemphigus. *J Am Acad Dermatol*. 1988;18:1219-1238.
471. Hertl M, Eming R, Veldman C. T cell control in autoimmune bullous skin disorders. *J Clin Invest*. 2006;116:1159-1166.
472. Takahashi H, Kouno M, Nagao K, et al. Desmoglein 3-specific CD4+ T cells induce pemphigus vulgaris and interface dermatitis in mice. *J Clin Invest*. 2011;121:3677-3688.
473. Veldman C, Hohne A, Dieckmann D, Schuler G, Hertl M. Type I regulatory T cells specific for desmoglein 3 are more frequently detected in healthy individuals than in patients with pemphigus vulgaris. *J Immunol*. 2004;172:6468-6475.
474. Yamagami J, Payne AS, Kacir S, Ishii K, Siegel DL, Stanley JR. Homologous regions of autoantibody heavy chain complementarity-determining region 3 (H-CDR3) in patients with pemphigus cause pathogenicity. *J Clin Invest*. 2010;120:4111-4117.
475. Camisa C, Helm TN. Paraneoplastic pemphigus is a distinct neoplasia-induced autoimmune disease. *Arch Dermatol*. 1993;129:883-886.
476. Anhalt GJ, Labib RS, Voorhees JJ, Beals TF, Diaz LA. Induction of pemphigus in neonatal mice by passive transfer of IgG from patients with the disease. *N Engl J Med*. 1982;306:1189-1196.
477. Koch PJ, Mahoney MG, Ishikawa H, et al. Targeted disruption of the pemphigus vulgaris antigen (desmoglein 3) gene in mice causes loss of keratinocyte cell adhesion with a phenotype similar to pemphigus vulgaris. *J Cell Biol*. 1997;137:1091-1102.
478. Amagai M, Tsunoda K, Suzuki H, Nishifuji K, Koyasu S, Nishikawa T. Use of autoantigen-knockout mice in developing an active autoimmune disease model for pemphigus. *J Clin Invest*. 2000;105:625-631.
479. Tsunoda K, Ota T, Suzuki H, et al. Pathogenic autoantibody production requires loss of tolerance against desmoglein 3 in both T and B cells in experimental pemphigus vulgaris. *Eur J Immunol*. 2002;32:627-633.
480. Joly P, Mouquet H, Roujeau JC, et al. A single cycle of rituximab for the treatment of severe pemphigus. *N Engl J Med*. 2007;357:545-552.
481. Passeron T, Ortonne JP. Physiopathology and genetics of vitiligo. *J Autoimmun*. 2005;25 suppl:63-68.
482. Spritz RA. The genetics of generalized vitiligo and associated autoimmune diseases. *J Dermatol Sci*. 2006;41:3-10.
483. Boissy RE, Nordlund JJ. Vitiligo: current medical and scientific understanding. *G Ital Dermatol Venereol*. 2011;146:69-75.
484. Alkhateeb A, Fain PR, Thody A, Bennett DC, Spritz RA. Epidemiology of vitiligo and associated autoimmune diseases in Caucasian probands and their families. *Pigment Cell Res*. 2003;16:208-214.
485. Eisenbarth GS. Banting Lecture 2009: an unfinished journey. Molecular pathogenesis to prevention of type 1A diabetes. *Diabetes*. 2010;59:759-774.
486. Le Poole IC, Wankowicz-Kalinska A, van den Wijngaard RM, Nickoloff BJ, Das PK. Autoimmune aspects of depigmentation in vitiligo. *J Investig Dermatol Symp Proc*. 2004;9:68-72.
487. Ogg GS, Rod Dunbar P, Romero P, Chen JL, Cerundolo V. High frequency of skin-homing melanocyte-specific cytotoxic T lymphocytes in autoimmune vitiligo. *J Exp Med*. 1998;188:1203-1208.
488. Lang KS, Caroli CC, Muhm A, et al. HLA-A2 restricted, melanocyte-specific CD8+ T lymphocytes detected in vitiligo patients are related to disease activity and are predominantly directed against MelanA/MART1. *J Invest Dermatol*. 2001;116:891-897.
489. Kroll TM, Bommiasamy H, Boissy RE, et al. 4-Tertiary butyl phenol exposure sensitizes human melanocytes to dendritic cell-mediated killing: relevance to vitiligo. *J Invest Dermatol*. 2005;124:798-806.
490. Hara I, Takechi Y, Houghton AN. Implicating a role for immune recognition of self in tumor rejection: passive immunization against the brown locus protein. *J Exp Med*. 1995;182:1609-1614.
491. Laberge G, Mailloux CM, Gowan K, et al. Early disease onset and increased risk of other autoimmune diseases in familial generalized vitiligo. *Pigment Cell Res*. 2005;18:300-305.
492. Spritz RA, Chiang PW, Oiso N, Alkhateeb A. Human and mouse disorders of pigmentation. *Curr Opin Genet Dev*. 2003;13:284-289.
493. Birlea SA, Jin Y, Bennett DC, et al. Comprehensive association analysis of candidate genes for generalized vitiligo supports XBP1, FOXP3, and TSLP. *J Invest Dermatol*. 2011;131:371-381.
494. Jin Y, Mailloux CM, Gowan K, et al. NALP1 in vitiligo-associated multiple autoimmune disease. *N Engl J Med*. 2007;356:1216-1225.
495. Spritz RA. Recent progress in the genetics of generalized vitiligo. *J Genet Genomics*. 2011;38:271-278.
496. Birlea SA, Gowan K, Fain PR, Spritz RA. Genome-wide association study of generalized vitiligo in an isolated European founder population identifies SMOC2, in close proximity to IDDM8. *J Invest Dermatol*. 2010;130:798-803.
497. Jin Y, Birlea SA, Fain PR, et al. Variant of TYR and autoimmunity susceptibility loci in generalized vitiligo. *N Engl J Med*. 2010;362:1686-1697.
498. Jin Y, Birlea SA, Fain PR, et al. Common variants in FOXP1 are associated with generalized vitiligo. *Nat Genet*. 2010;42:576-578.
499. Quan C, Ren YQ, Xiang LH, et al. Genome-wide association study for vitiligo identifies susceptibility loci at 6q27 and the MHC. *Nat Genet*. 2010;42:614-618.
500. Wang X, Erf GF. Apoptosis in feathers of Smyth line chickens with autoimmune vitiligo. *J Autoimmun*. 2004;22:21-30.
501. Erf GF, Bersi TK, Wang X, Sreekumar GP, Smyth JRJr. Herpesvirus connection in the expression of autoimmune vitiligo in Smyth line chickens. *Pigment Cell Res*. 2001;14:40-46.
502. Cunha-Neto E, Coelho V, Guilherme L, Fiorelli A, Stolf N, Kalil J. Autoimmunity in Chagas' disease. Identification of cardiac myosin-B13 Trypanosoma cruzi protein crossreactive T cell clones in heart lesions of a chronic Chagas' cardiomyopathy patient. *J Clin Invest* 1996;98:1709-1712.
503. Lane JR, Neumann DA, Lafond-Walker A, Herskowitz A, Rose NR. Interleukin 1 or tumor necrosis factor can promote Coxsackie B3-induced myocarditis in resistant B10.A mice. *J Exp Med*. 1992;175(4):1123-1129. doi:10.1084/jem.175.4.1123.
504. Wicks I, McColl G, Harrison L. New perspectives on rheumatoid arthritis. *Immunol Today*. 1994;15:553-556.
505. Suzuki A, Kochi Y, Okada Y, Yamamoto K. Insight from genome-wide association studies in rheumatoid arthritis and multiple sclerosis. *FEBS Lett*. 2011;585(23):3627-3632.
506. Rudolphi U, Hohlbaum A, Lang B, Peter HH, Melchers I. The B cell repertoire of patients with rheumatoid arthritis. Frequencies and specificities of peripheral blood B cells reacting with human IgG, human collagens, a mycobacterial heat shock protein and other antigens. *Clin Exp Immunol*. 1993;92:404-411.
507. Mastrandrea L, Yu J, Behrens T, et al. Etanercept treatment in children with new-onset type 1 diabetes: pilot randomized, placebo-controlled, double-blind study. *Diabetes Care*. 2009;32:1244-1249.

508. Denton CP, Black CM, Abraham DJ. Mechanisms and consequences of fibrosis in systemic sclerosis. *Nat Clin Pract Rheumatol*. 2006;2:134-144.
509. LeRoy EC, Black C, Fleischmajer R, et al. Scleroderma (systemic sclerosis): classification, subsets and pathogenesis. *J Rheumatol*. 1988;15:202-205.
510. LeRoy EC, Medsger TAJr. Criteria for the classification of early systemic sclerosis. *J Rheumatol*. 2001;28:1573-1576.
511. Mayes MD. Scleroderma epidemiology. *Rheum Dis Clin North Am*. 2003;29:239-254.
512. Senecal JL, Henault J, Raymond Y. The pathogenic role of autoantibodies to nuclear autoantigens in systemic sclerosis (scleroderma). *J Rheumatol*. 2005;32:1643-1649.
513. Sato S, Fujimoto M, Hasegawa M, Takehara K. Altered blood B lymphocyte homeostasis in systemic sclerosis: expanded naive B cells and diminished but activated memory B cells. *Arthritis Rheum*. 2004;50:1918-1927.
514. Sakkas LI, Xu B, Artlett CM, Lu S, Jimenez SA, Platsoucas CD. Oligoclonal T cell expansion in the skin of patients with systemic sclerosis. *J Immunol*. 2002;168:3649-3659.
515. Abraham DJ, Eckes B, Rajkumar V, Krieg T. New developments in fibroblast and myofibroblast biology: implications for fibrosis and scleroderma. *Curr Rheumatol Rep*. 2007;9:136-143.
516. Yamane K, Kashiwagi H, Suzuki N, et al. Elevated plasma levels of endothelin-1 in systemic sclerosis. *Arthritis Rheum*. 1991;34:243-244.
517. Del Papa N, Colombo G, Fracchiolla N, et al. Circulating endothelial cells as a marker of ongoing vascular disease in systemic sclerosis. *Arthritis Rheum*. 2004;50:1296-1304.
518. Kahaleh MB, Osborn I, LeRoy EC. Increased factor VIII/von Willebrand factor antigen and von Willebrand factor activity in scleroderma and in Raynaud's phenomenon. *Ann Intern Med*. 1981;94:482-484.
519. Kuwana M, Okazaki Y, Yasuoka H, Kawakami Y, Ikeda Y. Defective vasculogenesis in systemic sclerosis. *Lancet*. 2004;364:603-610.
520. Feghali-Bostwick C, Medsger TAJr, Wright TM. Analysis of systemic sclerosis in twins reveals low concordance for disease and high concordance for the presence of antinuclear antibodies. *Arthritis Rheum*. 2003;48:1956-1963.
521. Zhou X, Lee JE, Arnett FC, et al. HLA-DPB1 and DPB2 are genetic loci for systemic sclerosis: a genome-wide association study in Koreans with replication in North Americans. *Arthritis Rheum*. 2009;60:3807-3814.
522. Radstake TR, Gorlova O, Rueda B, et al. Genome-wide association study of systemic sclerosis identifies CD247 as a new susceptibility locus. *Nat Genet*. 2010;42:426-429.
523. Romano E, Manetti M, Guiducci S, Ceccarelli C, Allanore Y, Matucci-Cerinic M. The genetics of systemic sclerosis: an update. *Clin Exp Rheumatol*. 2011;29(2 Suppl 65):S75-S86.
524. Grossman C, Dovrish Z, Shoenfeld Y, Amital H. Do infections facilitate the emergence of systemic sclerosis? *Autoimmun Rev*. 2011;10:244-247.
525. Caramaschi P, Dalla Gassa A, Ruzzenente O, et al. Very low levels of vitamin D in systemic sclerosis patients. *Clin Rheumatol*. 2010;29:1419-1425.
526. Mora GF. Systemic sclerosis: environmental factors. *J Rheumatol*. 2009;36:2383-2396.
527. Beyer C, Schett G, Distler O, Distler JH. Animal models of systemic sclerosis: prospects and limitations. *Arthritis Rheum*. 2010;62:2831-2844.
528. Yamamoto T, Takagawa S, Katayama I, et al. Animal model of sclerotic skin. I: local injections of bleomycin induce sclerotic skin mimicking scleroderma. *J Invest Dermatol*. 1999;112:456-462.
529. Green MC, Sweet HO, Bunker LE. Tight-skin, a new mutation of the mouse causing excessive growth of connective tissue and skeleton. *Am J Pathol*. 1976;82:493-512.
530. Siracusa LD, McGrath R, Ma Q, et al. A tandem duplication within the fibrillin 1 gene is associated with the mouse tight skin mutation. *Genome Res*. 1996;6:300-313.
531. Eferl R, Hasselblatt P, Rath M, et al. Development of pulmonary fibrosis through a pathway involving the transcription factor Fra-2/AP-1. *Proc Natl Acad Sci U S A*. 2008;105:10525-10530.
532. van den Hoogen FH, Boerbooms AM, Swaak AJ, Rasker JJ, van Lier HJ, van de Putte LB. Comparison of methotrexate with placebo in the treatment of systemic sclerosis: a 24 week randomized double-blind trial, followed by a 24 week observational trial. *Br J Rheumatol*. 1996;35:364-372.
533. Tashkin DP, Elashoff R, Clements PJ, et al. Cyclophosphamide versus placebo in scleroderma lung disease. *N Engl J Med*. 2006;354:2655-2666.
534. Thompson AE, Shea B, Welch V, Fenlon D, Pope JE. Calcium-channel blockers for Raynaud's phenomenon in systemic sclerosis. *Arthritis Rheum*. 2001;44:1841-1847.
535. Klippel JH, Dieppe P. *Rheumatology*. Mosby; 1998.
536. https://www.cdc.gov/lupus/facts/detailed.html. Retrieved May 28 2020.
537. Sestak AL, Furnrohr BG, Harley JB, Merrill JT, Namjou B. The genetics of systemic lupus erythematosus and implications for targeted therapy. *Ann Rheum Dis*. 2011;70(suppl 1):i37-i43.
538. Tsokos GC. Systemic lupus erythematosus. *N Engl J Med*. 2011;365:2110-2121.
539. Kang I, Quan T, Nolasco H, et al. Defective control of latent Epstein-Barr virus infection in systemic lupus erythematosus. *J Immunol*. 2004;172:1287-1294.
540. Poole BD, Scofield RH, Harley JB, James JA. Epstein-Barr virus and molecular mimicry in systemic lupus erythematosus. *Autoimmunity*. 2006;39:63-70.
541. Arbuckle MR, McClain MT, Rubertone MV, et al. Development of autoantibodies before the clinical onset of systemic lupus erythematosus. *N Engl J Med*. 2003;349:1526-1533.
542. Fillatreau S, Manfroi B, Dörner T. Toll-like receptor signalling in B cells during systemic lupus erythematosus. *Nat Rev Rheumatol*. 2021;17(2):98-108. doi:10.1038/s41584-020-00544-4.
543. Carroll MC. A protective role for innate immunity in systemic lupus erythematosus. *Nat Rev Immunol*. 2004;4:825-831.
544. Perry D, Sang A, Yin Y, Zheng YY, Morel L. Murine models of systemic lupus erythematosus. *J Biomed Biotechnol*. 2011;2011:271694.
545. Bruijn JA, van Elven EH, Hogendoorn PC, Corver WE, Hoedemaeker PJ, Fleuren GJ. Murine chronic graft-versus-host disease as a model for lupus nephritis. *Am J Pathol*. 1988;130:639-641.
546. Baker KP, Edwards BM, Main SH, et al. Generation and characterization of LymphoStat-B, a human monoclonal antibody that antagonizes the bioactivities of B lymphocyte stimulator. *Arthritis Rheum*. 2003;48:3253-3265.
547. Cheema GS, Roschke V, Hilbert DM, Stohl W. Elevated serum B lymphocyte stimulator levels in patients with systemic immune-based rheumatic diseases. *Arthritis Rheum*. 2001;44:1313-1319.
548. Petri M, Stohl W, Chatham W, et al. Association of plasma B lymphocyte stimulator levels and disease activity in systemic lupus erythematosus. *Arthritis Rheum*. 2008;58:2453-2459.
549. Wallace DJ, Stohl W, Furie RA, et al. A phase II, randomized, double-blind, placebo-controlled, dose-ranging study of belimumab in patients with active systemic lupus erythematosus. *Arthritis Rheum*. 2009;61:1168-1178.
550. Navarra SV, Guzman RM, Gallacher AE, et al. Efficacy and safety of belimumab in patients with active systemic lupus erythematosus: a randomised, placebo-controlled, phase 3 trial. *Lancet*. 2011;377:721-731.
551. Furie R, Petri M, Zamani O, et al. A phase III, randomized, placebo-controlled study of belimumab, a monoclonal antibody that inhibits B lymphocyte stimulator, in patients with systemic lupus erythematosus. *Arthritis Rheum*. 2011;63:3918-3930.
552. Li J, Wei MM, Song Q, Guo XH, Shao L, Liu Y. Anti-CD22 epratuzumab for systemic lupus erythematosus: a systematic review and meta-analysis of randomized controlled trials. *Exp Ther Med*. 2019;18:1500-1506.
553. Dorner T, Kaufmann J, Wegener WA, Teoh N, Goldenberg DM, Burmester GR. Initial clinical trial of epratuzumab (humanized anti-CD22 antibody) for immunotherapy of systemic lupus erythematosus. *Arthritis Res Ther*. 2006;8:R74.
554. Merrill JT, Neuwelt CM, Wallace DJ, et al. Efficacy and safety of rituximab in moderately-to-severely active systemic lupus erythematosus: the randomized, double-blind, phase II/III systemic lupus erythematosus evaluation of rituximab trial. *Arthritis Rheum*. 2010;62:222-233.
555. Furie R, Looney RJ, Rovin B, et al. Efficacy and safety of rituximab in subjects with active proliferative lupus nephritis (LN): results from the randomized, double-blind phase III LUNAR Study [abstract]. *Arthritis Rheum*. 2009;60:S429.
556. Alshaiki F, Obaid E, Almuallim A, Taha R, El-Haddad H, Almoallim H. Outcomes of rituximab therapy in refractory lupus: a meta-analysis. *Eur J Rheumatol*. 2018;5:118-126.
557. Chervonsky AV. Influence of microbial environment on autoimmunity. *Nat Immunol*. 2010;11:28-35.
558. Gill SR, Pop M, Deboy RT, et al. Metagenomic analysis of the human distal gut microbiome. *Science*. 2006;312:1355-1359.
559. Turnbaugh PJ, Ley RE, Hamady M, Fraser-Liggett CM, Knight R, Gordon JI. The human microbiome project. *Nature*. 2007;449:804-810.
560. Ivanov II, Frutos Rde L, Manel N, et al. Specific microbiota direct the differentiation of IL-17-producing T-helper cells in the mucosa of the small intestine. *Cell Host Microbe*. 2008;4:337-349.
561. Winer S, Astsaturov I, Cheung RK, et al. Type I diabetes and multiple sclerosis patients target islet plus central nervous system autoantigens; nonimmunized nonobese diabetic mice can develop autoimmune encephalitis. *J Immunol*. 2001;166:2831-2841.
562. Salomon B, Rhee L, Bour-Jordan H, et al. Development of spontaneous autoimmune peripheral polyneuropathy in B7-2-deficient NOD mice. *J Exp Med*. 2001;194:677-684.
563. Takaki T, Marron MP, Mathews CE, et al. HLA-A*0201-restricted T cells from humanized NOD mice recognize autoantigens of potential clinical relevance to type 1 diabetes. *J Immunol*. 2006;176:3257-3265.
564. Niens M, Grier AE, Marron M, Kay TW, Greiner DL, Serreze DV. Prevention of "Humanized" diabetogenic CD8 T-cell responses in HLA-transgenic NOD mice by a multipeptide coupled-cell approach. *Diabetes*. 2011;60:1229-1236.

CHAPTER 47 Immunologic Mechanisms of Allergic Disorders

Stéphanie Lejeune, MD, PhD • Kari C. Nadeau, MD, PhD•
David W. Scott, PhD • Joshua D. Milner, MD

INTRODUCTION

The immune system designed to protect against foreign substances or altered self (as in cancer) can also react to many innocuous environmental substances.[1] These aberrant responses to harmless substances can lead to hypersensitivity reactions and/or autoimmunity (*see Chapter 46*). Hypersensitivity reactions are classified into four types based on the mechanism of action[2,3] (Gell and Coombs classification—Fig. 47.1). Type I reactions are immediate IgE-mediated reactions that generally occur within seconds to minutes and can be either local or systemic. Diseases associated with type I allergic diseases are also termed atopic or allergic diseases. Both type II and type III hypersensitivity reactions are mediated by IgG/IgM. Type II reactions are cytotoxic and are mediated against cell surface and extracellular matrix proteins. Immunoglobulin G and IgM damage cells by activating the complement system or by phagocytosis (eg, drug-induced hemolytic anemia). In type III hypersensitivity reactions, IgM and IgG react with soluble antigens forming antigen-antibody complexes. They involve immune complex tissue deposition and complement activation (eg, serum sickness). Type IV hypersensitivity reactions are delayed cell-mediated immune reactions, characterized by antigen–T cell interactions leading to cytotoxicity (eg, contact dermatitis as with poison ivy). Some hypersensitivity reactions may involve more than one type of reaction. For example, eosinophilic esophagitis (EoE) can be induced by both IgE and/or cell-mediated hypersensitivity reaction.

In this chapter, we will focus primarily on type I IgE-mediated hypersensitivity reactions and mechanisms of allergic inflammation. We review the critical role of IgE in allergic disease, the characteristics of common allergens, the various type I allergic disorders, the risk factors for allergic disorders, the cellular and molecular mechanisms of allergic responses and tolerance, and the role of microbiota in allergy. We also discuss systems biology approaches and state-of-the-art high-throughput omics technologies that are enabling further understanding of allergic mechanisms, and the development of novel treatments such as biologics and immunomodulatory strategies for treating allergic diseases.

Allergy and IgE (a Historical and Evolutionary Perspective)

The term "allergy," for altered activity, was coined in the early 1900s by Baron von Pirquet by combining the Greek word "*allos*" (different or strange) with "*ergon*" (activity).[4] Although descriptions of severe and, sometimes, lethal allergic reactions have been recorded prior to the 1900s, the discovery of anaphylaxis by Paul Portier and Charles Richet in 1902 was pivotal in our understanding of hypersensitivity and earned Richet the Nobel prize in 1913.[5] Both are regarded as the first allergologists of the modern age. Further progress was made in 1921 by Prausnitz and Küstner with the discovery that hypersensitivity could be passively transferred through serum from one individual to the other.[6] This groundbreaking discovery sparked the hunt for the serum factor and led to the eventual discovery of IgE in 1966 based on the work by the Ishizakas,[7] Bennich, and Johansson.[8,9]

Evolutionarily, the IgE response, per se, likely evolved from a primitive response to protect against certain toxins (venom) and endoparasites (eg, roundworm), providing individuals with an important adaptive advantage.[10,11] IgE responses may also be critical for defense against ectoparasites, such as ticks.[12,13] Venoms lead to direct activation of host mast cells and IgE. Although they cause immediate deleterious effects mediated by inflammatory chemical mediators found in preformed granules, they also lead to activation of both innate protective mechanisms, such as the release of proteases which inactivate venom, and long-term adaptative mechanisms, providing protection upon reinoculation.[11,14] IgE is also known to be upregulated during infection with parasitic worms, which dampen the host immune system via desensitization of effector cells (mast cells and basophils) to prevent their elimination. While there is no direct evidence yet in humans that IgE plays an essential role in venom detoxification or parasitic immunity, the key observations in animal models strongly suggest its evolutionary role. As stated by the hygiene hypothesis (*see microbes and the hygiene hypothesis*), reduced exposure to microbial loads in the indoor environment during the second half of the 20th century, possibly associated with modifications of the microbiota, may also have led to subsequent lack of early training of the immune system to function properly, hypersensitization, and increases in the incidence of allergic disorders and immune-mediated disorders.[15,16]

Some key effectors of the allergic response may also be implicated in other protective immune mechanisms, such as immune regulation of cancers (*see Chapter 49: Cancer Immunology*), and the study of their interactions constitute the new field of AllergoOncology.[17] Epidemiological studies have suggested that allergic disorders may have a

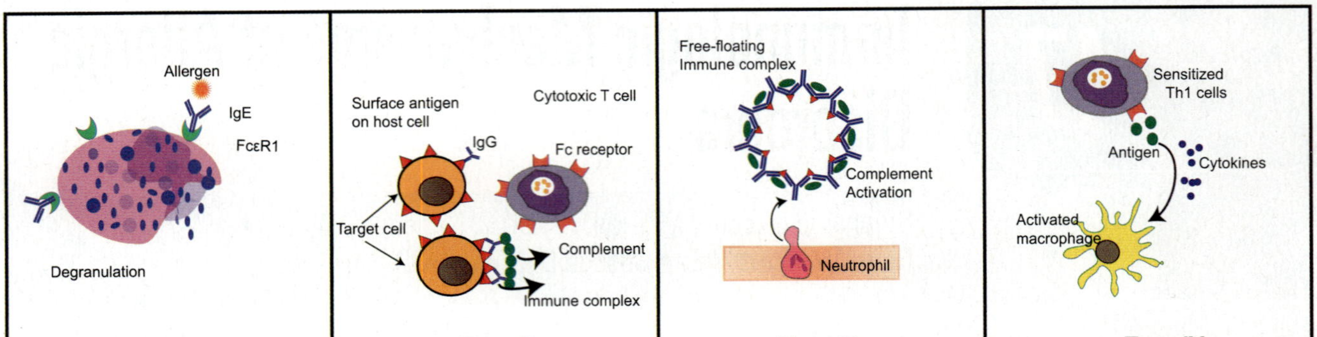

Type I	Type II	Type III	Type IV
They are also called **immediate hypersensitivity.** Initial exposure to the allergen causes sensitization to those genetically predisposed. On subsequent exposure to the allergen, an allergic reaction occurs.	They are also called **cytotoxic hypersensitivity reactions**. In type II hypersensitivities, IgG and IgM antibodies bind to angtigens on cell surfaces. This induces a cascade of reactions that damages cells by activating the complement system or by phagocytsis.	Type III reactions (immune complex reactions) involve circulating antigen-antibody immune complexes that deposit in postcapillary venules, with subsequent complement fixation.	Type IV reactions are delayed hypersensitivity reactions that are mediated by T cells.
Examples: Food allergy, allergic rhinitis	Examples: Hemolytic disease of newborns, hemolytic transfusion reactions.	Example: Serum sickness	Examples: Contact dermatitis, transplant tissue rejection.

FIG. 47.1. Types of hypersensitivities. FcεRI: high-affinity IgE receptor; Ig: Immunoglobulin. Hypersensitivity reactions are classified into four types based on the mechanism of action. This figure represents a modified version of the Gell and Coombs classification (1963).[2,3]

protective effect against the development of cancers.[18,19] Eosinophils, which are evolutionary-conserved cells, are frequently observed in solid tumors, and studies have demonstrated that their presence is a positive prognosis biomarker and/or positive predictive marker of beneficial response to immunotherapy in some cancers.[20,21] They display immune regulatory properties toward other immune cell subsets, cytotoxic, and tissue-repair functions that may be critical in the tumor microenvironment.[21,22] Mast cells can also interact with different regulatory cells, such as immunosuppressive cells like myeloid-derived suppressor cells.[23] Because both eosinophils and mast cells have functional plasticity depending on the tumor microenvironment, there is debate whether they may promote tumor growth or drive immune surveillance and antitumoral effects.[23,24] Increasing research shows the multiple roles of IgE in the tumor microenvironment. They can engage and modulate the function of macrophages and prime them to mediate antitumor functions.[25] They also have a key role in antigen cross-presentation by dendritic cells (DCs), via their high-affinity receptor FcεRI, leading to the generation of cytotoxic $CD8^+$ T lymphocytes with specificity for tumor antigens.[26] Finally, in the skin, IgE are part of the host defense against epithelial damage and tumor development. Stress surveillance by intraepithelial lymphocytes may result in IgE class-switching by B cells leading to autoreactive IgE targeting epithelial carcinogenesis.[27] FcεRI-signalling in basophils can also drive tumor outgrowth of epithelial cells harboring oncogenic mutation.[28]

General Characteristics of Common Allergic Disorders and Their Prevalence

Common allergic disorders include atopic dermatitis (AD), anaphylaxis, allergic asthma, food allergy (FA), and allergic rhinitis (AR). In the past few decades, the prevalence of allergic disorders has increased worldwide,[29] although epidemiological studies have shown high disparities between countries, possibly reflecting environmental and genetic differences.

Atopic Dermatitis

AD is an acute, subacute, or chronic skin disorder characterized by dryness of skin and intense pruritis. Local inflammation and repeated scratching due to itching can induce lichenification over time.[30,31] AD affects approximately 15% to 20% of children and 1% to 3% of adults in the United States and is associated with increased risk of both asthma and FAs.[32,33] Although lesions may vary over time and with external triggers, the disease is characterized by skin barrier disruption, which can be exacerbated by scratching of lesions and the promotion and proliferation of pathogenic commensal skin bacteria, such as *Staphylococcus aureus*.[34] Triggering factors include topical irritants, such as soap, laundry detergents, and clothes, as well as airborne and food allergens. Emotional stress, heat, and sweating can also worsen symptoms.

Anaphylaxis

Anaphylaxis is a rapid and systemic allergic reaction, involving the upper and lower airways, skin, conjunctiva, gastrointestinal tract, and cardiovascular systems.[35,36] Anaphylaxis events are also likely to be underdiagnosed and underreported. It is estimated that 0.3% of the European population and up to 1.6% of US adults may experience anaphylaxis at some point in their lives.[37,38] The most common causes of anaphylaxis are medications, Hymenoptera venom hypersensitivity, and food allergies.[39,40] Untreated, anaphylaxis can be fatal, most often due to upper or lower airways obstruction or cardiovascular collapse.[41,42] In anaphylaxis, biphasic reactions can also occur, with recurrence of symptoms 2 to 24 hours after the initial reaction.[42,43] Cofactors, such as concomitant infection, alcohol intake, medication, or exercise (as seen in food-dependent exercise-induced anaphylaxis to ω5-gliadin in wheat[44,45]) may influence the occurrence of anaphylaxis.

Asthma

Asthma is defined by chronic reversible and diffuse obstruction of the distal airways. Clinical presentations include episodes of acute attacks with wheezing and dyspnea, resulting from bronchial inflammation, mucosal hypersecretion, and bronchoconstriction, when exposed to triggering factors, such as microbes, indoor and outdoor pollutants, or airborne allergens.[46,47] Between episodes, chronic inflammation may persist resulting in chronic cough and bronchial remodeling.[48] Asthma is the most common chronic condition among children. Asthma prevalence, both overall and among adults, has been trending upward in the past 20 years, currently affecting around 7% to 8% of adults and 8% to 11% of children worldwide.[49,50] Allergic and nonallergic asthma are differentiated, although some mechanisms and triggering factors may significantly overlap between these entities. Allergic asthma occurs when asthma symptoms are triggered by exposure to airborne allergens. The prevalence of allergic asthma increases with age with a progression of sensitization during childhood and stabilization in adolescence, as shown by data from neonatal cohorts.[51-53] The importance of sensitizations in the natural history of asthma has been demonstrated, early and multiple sensitization being identified risk factors for persistence of asthma, recurrence of attacks, and long-term alteration of lung function.

Allergic Rhinitis

AR, commonly known as hay fever, is an IgE-mediated inflammation of the nasal mucosa cavity. The prevalence of AR increases during the course of childhood and currently estimated at 10% to 30% of the adult population.[54,55] Its cardinal symptoms include watery rhinorrhea, nasal obstruction, nasal itching and sneezing, often leading to anosmia.[54] Seasonal rhinitis is induced by airborne pollen allergens, such as grass, tree, and weed pollens, whereas perennial rhinitis is induced by perennial allergens, such as house dust mite, animal dander, or mold spores.

Food Allergy

FAs are adverse antigen-specific immune-mediated responses upon exposure to a given food.[56] They have to be distinguished from food intolerance, whose primary mechanism takes place in the digestive system, based on enzyme deficiencies (eg, carbohydrate malassimilation), sensitivity to additives (eg, flavors, dyes, preservatives), or reaction to natural chemicals in foods (eg, histamine).[57] Food allergies affect 8% of children and 5% of adults in the United States and up to more than 10% of young children in Australia. Common allergenic foods include milk, eggs, peanuts, tree nuts (such as almonds, cashews, walnuts), fish, shellfish, soy, and wheat. Among food allergies, peanut allergies tend to be more severe persisting into adulthood. Prevalence of peanut allergy is estimated to be 2% to 3%.[58-60] IgE-mediated FAs are characterized by the rapid onset of symptoms (typically observed within minutes to about 2 hours) after ingestion of the suspected allergen, generally proteins in foods.[42] Other subtypes of IgE-mediated FA include oral allergy syndrome (OAS), also called pollen FA syndrome, alpha-gal allergy, and EoE.

Oral Allergy Syndrome

OAS is a hypersensitivity reaction to plant-based foods due to cross-reactivity between food proteins and inhalant allergens that requires prior sensitization to a cross-reacting inhalant allergen.[42,61] Symptoms are usually mild, such as pruritus of the lips, tongue, and mouth, rarely leading to oral angioedema. The most common example of tree-pollen fruit cross-reactivity is PR-10 syndrome, between birch Bet v1 and other PR-10 in fruits (eg, Mal d 1 in apple), peanuts (Ara h 8), and tree nuts (eg, Cor a 1 in hazelnut).[62]

Alpha-Gal Allergy

Alpha-gal allergy is unusual in that the allergen is an oligosaccharide (galactose-alpha-1,3-galactose, known as α-Gal) rather than a protein. This allergen is present in red meats such as beef, pork, and lamb, and interestingly, also in ticks. Due to cross-reactivity between the two, an individual exposed to α-Gal, a component of tick saliva, from tick bites becomes sensitized to α-Gal in red meats. Ingestion of red meats in these sensitized individuals leads to an allergic reaction with symptoms ranging from mild urticaria to angioedema and severe anaphylaxis.[13,63] Unlike other IgE-mediated reactions, these allergic reactions are also unique in that onset is delayed and effects are only observed 3 to 6 hours after consumption of mammalian meat.

Eosinophilic Esophagitis

EoE has been known for less than 30 years. It is a chronic, local, immune-mediated esophageal disease.[42] The incidence of EoE is possibly underestimated, varying from 1 to 20 new cases per 100,000 inhabitants per year.[42,64] The disease is characterized by clinical symptoms related to esophageal dysfunction, such as dysphagia, food impaction, and gastroesophageal reflux-like symptoms. Endoscopy and identification of esophageal lesions, such as esophageal rings or narrow-caliber esophagus, are indicative of EoE.[64,65] A diagnostic criteria for EoE is eosinophilic infiltration greater than 15 cells/high-power field.[65] Although IgE is associated with the disease, its exact role in the disease is unclear and other cell-mediated reactions are also known to play a role.[66] Unlike other food allergies, patients with EOE appear to be frequently sensitized to both airborne and food allergens, and their individual roles in the pathophysiology of the disease remain to be clarified.[67] FA, however, is known to be a main component of the disease as elimination diets have been fairly successful in treating the disorder.[42,68,69]

Other mixed IgE and cell-based FAs include other eosinophilic gastrointestinal disorders such as eosinophilic colitis, eosinophilic gastroenteritis, and eosinophilic gastritis, but these are fairly rare.

Other Types of Food Allergy

There are a number of non–IgE-mediated FAs such as food protein–induced allergic proctitis, food protein–induced enteropathy, food protein–induced enterocolitis, but these are beyond the scope of this chapter.

Atopic March

Epidemiological studies have recognized a natural progression of allergic diseases from infancy to adulthood. Many individuals with a single allergic disorder eventually develop other atopic comorbidities. This suggests common mechanisms underlying these different diseases or different manifestations of the same disease. Classically, the earliest manifestation of allergic diseases starts with AD in infancy and preschool years. In these individuals, there is increased risk of FA, AR, and/or asthma in subsequent years[70,71] (Fig. 47.2). This natural progression of allergic disorders is termed the allergic or atopic march. Allergen exposure through skin barrier disruption, either due to genetic predisposition or due to environmental causes, is thought to be the primary route by which infants initiate the atopic march (*see dual-allergen hypothesis*). Some allergens, such as mites, cockroaches, fungi, and pollen, produce or contain proteases, which can also induce skin barrier breakdown.[72] These observations suggest that if AD can be prevented or treated early, the development or onset of other allergic disorders could also be prevented.

RISK FACTORS FOR ALLERGIC DISORDERS

Risk factors for allergic diseases are multifactorial, including genetic, epigenetic, and environmental factors. The sudden increase in allergic diseases over the last few decades suggests that modern lifestyle changes (increased hygiene, processed food, antibiotic use, time spent indoors) as well as environmental changes (increased pollution and climate change) have altered the sum total of the exposures that an individual faces before birth (in utero) and during their lifetime (the exposome)[73] (Fig. 47.3). However, the relative contributions of these factors are still not well understood. Epigenetics provides valuable information on how environmental factors alter gene expression, offering important tools for understanding the mechanisms underlying allergic disorders.

Genetic Factors

Twin studies estimate that approximately 25% of the variability in asthma severity is determined by genetic factors, with the remainder determined by environmental influence, psychosocial factors, behavioral traits, and comorbidities.[74] In rare cases, allergy- or atopic effector-related symptoms are associated with monogenic diseases, defining primary atopic disorders which identify critical genes and pathways in allergy pathogenesis.[75,76] These mutations can be classified into (1) defects of specific T and B cell signaling pathways, some of which affect tolerance (eg, mutations in *Caspase Recruitment Domain Family Member CARD11*, *Mucosa-associated lymphoid tissue lymphoma translocation protein MALT1*), leading to high-IgE levels, atopy, autoimmunity and immunodeficiency disorders; (2) those which directly affect immune tolerance machinery such as *FOXP3* mutations, leading to immune dysregulation, polyendocrinopathy, enteropathy, X-linked (IPEX); (3) defects in genes important for cytokine signaling such as *STAT3, IL6R, IL6ST*, and *ZNF341*, leading to markedly elevated IgE, poor inflammatory responses and infection (4) barrier function defects (eg, *Filaggrin, SPINK5, desmoglein* or *corneodesmosin* mutations), leading to chronic mucosal allergic inflammation such as eczema or eosinophilic gut disease; or (5) aberrant-intrinsic mast cell effector functions (eg, *Adhesion G Protein-Coupled Receptor E2 ADGRE2* mutations or *PLCG2*), leading to urticaria or anaphylaxis in the absence of inflammation or antibody-mediated sensitization.[76] There are likely many more monogenic causes of atopy, and the relative contribution of monogenic atopy to the general pool of atopic diseases is not well-known though it is quite possible that it is highly impactful.[77]

Genome-wide association studies (GWAS) have allowed identification of additional determinants associated with allergic disorders, some of which overlap directly or pathophysiologically with those seen in primary atopic disorders.[78] Single nucleotide polymorphisms (SNPs) are the most common variations in the genome. In AD, GWAS studies identified loci involving genes associated with Th2 response (eg, *IL-4, IL-13, TSLP*), skin barrier (eg, *Laminin Subunit Alpha 3 LAMA3, Transmembrane Protein 79 TMEM79*), and vitamin D signaling pathway (eg, *vitamin D receptor [VDR]*).[79] In FA, genetic determinants were found in human leukocyte antigen

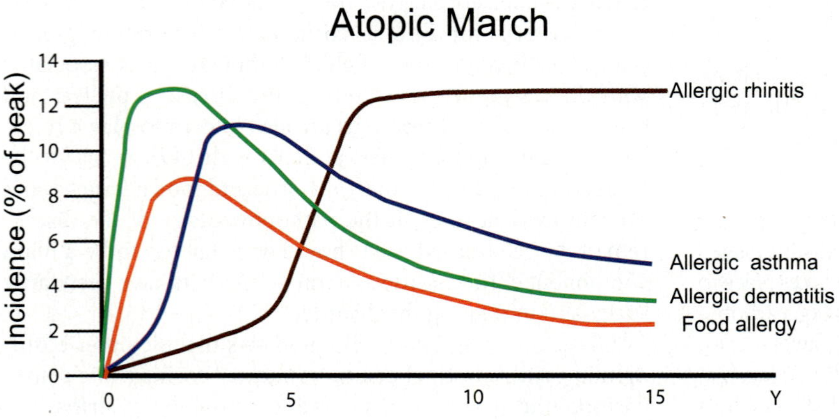

FIG. 48.2. Atopic March. Atopic March: Atopic dermatitis prevalence peaks early in infancy, followed by the development of food allergy, asthma, and allergic rhinitis.

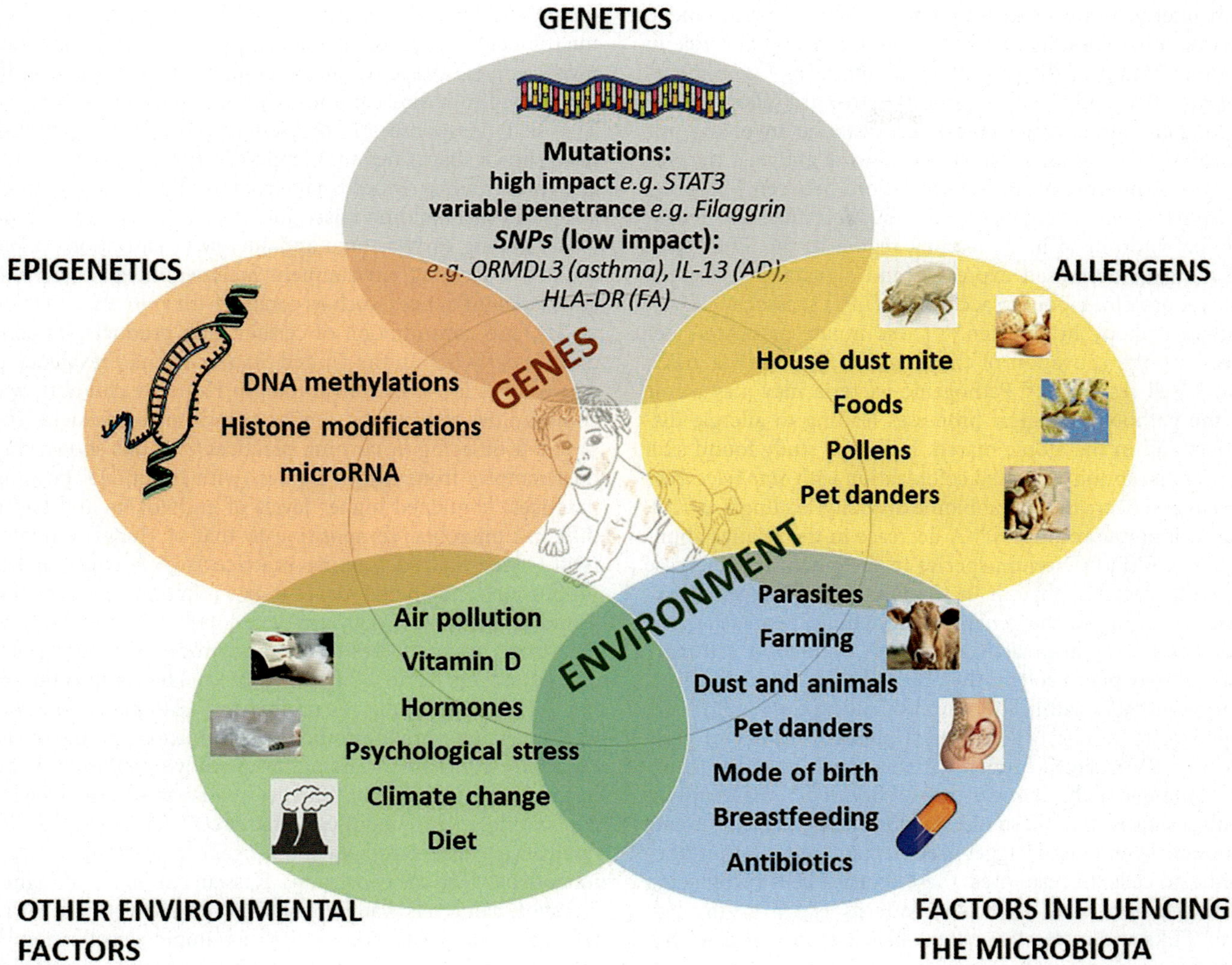

FIG. 47.3. Factors affecting risk of development of allergic disorders. HLA: Human leukocyte antigen; IL: Interleukin; ORMDL3: ORMDL sphingolipid biosynthesis regulator 3; SNP: Single nucleotide polymorphism; STAT3: Signal transducer and activator of transcription 3. Risk factors for allergic disorders include environmental factors (collectively called exposome, ie, the collection of all environmental factors to which an individual is exposed throughout his/her life). Genes affect genetic susceptibility to risk of allergic disease. Epigenetic alterations provide valuable information on how environmental factors alter gene expression.

(HLA) loci (eg, *HLA-DQB1, HLA-DRB1*), and skin barrier genes (eg, genetic polymorphisms of *Filaggrin*), and anti-inflammatory pathways (eg, *IL-10*) and Th2 pathways (eg, *IL-13*).[80,81] *MALT1* variants, in addition to leading primary atopic disorders in extreme form, may more commonly contribute to the development of FA in at-risk individuals who avoid early food introduction.[82] In asthma, association signals are often in or near genes involved in innate immune response pathways including pattern recognition receptor (PRR) recognition (eg, *toll-like receptor TLR1* locus, extracellular receptor on innate immune cells), epithelial cell homeostasis (*e.g. IL-33* locus, alarmin implicated in Th2 response in interaction with innate lymphoid cells [ILC2]), adaptive response (eg, *HLA-DR*, implicated in antigen-presenting cells [APCs] presentation to T cells; *T-cell immunoglobulin mucin-3 TIM3* locus, implicated in Th1 response; *IL-13* locus, implicated in Th2 response; *ORMDL Sphingolipid Biosynthesis Regulator 3 ORMDL3* locus implicated in inhibition of sphingolipid synthesis and modulation of effect of airborne allergens), and airway structural changes (eg, *TEK* locus, involved in airway remodeling).[83,84] In AR, fine mapping of the HLA region has shown an increased risk with some amino acid variants important for antigen binding.[85]

Environmental factors, including infections, diet, and psychological stress, may contribute to the variable expression of allergic disease in primary atopic disorders and to the variability of risk in common allergy-associated variants.[86]

Microbes and the Hygiene Hypothesis

The hygiene hypothesis states that children exposed early in life to microbial diversity as well as helminths are less likely to develop allergic disorders or autoimmune diseases as these infections contribute to the development of the immune system by provoking a strong IgE response leading to desensitization of effector cells like mast cells.[87] The gut, lung, and skin interfaces are in direct contact with the environment, and the microbiota is known to play a fundamental role in the maturation of the cells at these interfaces.[88] The mechanisms underlying a tolerogenic state toward allergens and innocuous microorganisms are likely similar involving primarily innate immune cells that recognize antigens via PRR sensors. A number of strong associations between TLR polymorphisms and allergic disorders support this. For example, polymorphism in the coding region of the *TLR4* gene resulting in decreased expressed and signaling of the cell surface protein. Lipopolysaccharides (LPS) is associated with increased atopy in children.[89] These innate pathways later influence the activation of the adaptive response via APCs and T cell responses.[90] Pathogenic bacteria may play a role in the pathophysiological processes leading to allergic disorders and in the atopic march. Indeed, a study found skin barrier disruption increased colonization with *Staphylococcus aureus* and decreased microbiome diversity, leading to a Th2-biased immune responses.[34] A decrease in the relative abundance of Bifidobacterium species in the gut and an increase in Proteobacteria phylum in the lung (eg, pathogenic bacteria belonging to the genera Haemophilus, Moraxella, and Neisseria) have been associated with childhood asthma.[91] Viruses may play a role in the onset of a Th2-biased inflammation, notably within the airways. During early childhood, exposure to respiratory viruses is intense. History of early RSV- or RV-induced bronchiolitis has been linked with the development of childhood asthma.[92] Both *in vitro* and *in vivo* studies suggest that these viruses closely interact with innate immune defenses and trigger specific responses in respiratory epithelial cells. In responses, these sentinel cells recognizing pathogen-associated molecular patterns (PAMPs) through their TLRs and intracellular RNA helicases mount a defense with the production of antiviral molecules such as type I and III interferon (IFN) and other antimicrobial peptides. Some studies conducted in asthmatic subjects have observed a defect of the production of type I and III IFN in response to respiratory viruses, in particular to RV, in some individuals although results are conflicting.[93-95] Such defects may result in the production of high levels of the alarmins IL-25, IL-33, and TSLP, as seen in some studies, leading to an allergic state.[96] Fungi are other pathogens that may interact with the allergic response.[97] Some subjects may exert an inability to clear fungal pathogens and display a Th2-type response, rather than a protective Th1/Th17 response, leading to a detrimental eosinophilic inflammation. Fungi promote a Th2 reaction via allergens (eg, *Aspergillus fumigatus*) and potent adjuvants (eg, secreted proteases). This may lead ultimately to the modulation of immune response to common allergens.

Microbiota and the "Old Friends" Hypothesis

The "old friends" hypothesis was proposed as an alternative to the hygiene hypothesis.[98] This hypothesis suggests that the increase in allergic disorders is at least partly attributable to immunodysregulation resulting from lack of exposure to certain microorganisms that evolved together with the human organism. It is likely that atopic individuals receive quantitatively and qualitatively different stimulations by microbiota at mucosal surfaces (eg, LPS, peptidoglycans, mycobacterial antigens), and therefore APCs in these individuals fail to redirect immune responses toward protective Th1 or T_{reg} responses.[90,99] In Westernized countries, studies have shown disparities in allergic disorders between urban and rural children. Indeed, data from five European countries suggested that children raised in dairy farms had lower risks of developing early asthma and allergic sensitization.[100] The protective farming environment includes independent and additional influences, such as contact with animals, exposure to hay, and consumption of certain milk products, by children in early life. In the United States, asthma prevalence is 4 to 6 times lower in Amish than in Hutterite children, two communities sharing the similar genetic ancestries and lifestyle but differing in farming practices.[101] It was shown that dust samples from Amish homes (with traditional farming practices) contained higher levels of endotoxins and had a different microbial composition to that of Hutterite homes (with industrialized farming practices). Differences in the proportions and phenotypes of innate immune cells were also found between the two groups of children. Administration of dust from Amish homes in a mouse model of experimental allergic asthma led to a protective effect. The composition of the gut, skin, and lung microbiota may also play a key role in influencing immune response and mucosal homeostasis, especially in the first years of life. A study of children living in the Finnish or Russian Karelia, two adjacent regions with contrasting socioeconomical differences, demonstrated differences in skin microbiota and fungal communities in nasal mucosa between the two groups. Russian children had higher *Acinetobacter* levels than Finnish children and a 10-fold lower risk of asthma, AR, AD, as well as atopic sensitization.[102] Similar results were observed recently when comparing Finnish and Estonian children.[103] It has also been demonstrated that individuals with FA have a distinct gut microbiota composition compared with healthy controls, and dysbiosis precedes the development of FA.[104] Studies suggest that certain bacteria, most notably from the *Clostridia* class, may promote the development of tolerance in the gut.[90] Our understanding of the role of the microbiota in allergic disorders is at its infancy. However, novel high throughput tools are accelerating research in this area. These include the use of humanized microbiota animal models and DNA and RNA sequencing technologies (*see Systemic biology approaches to study allergic disorders*).[91] Further, many factors may influence the development and composition of gut, skin, and lung microbiota such as birth mode (Caesarian or vaginal), breastfeeding, pet exposure, and antibiotic use.[105] Birth by C-section has been associated with allergic outcomes in children, and it has been shown that C-section can induce dysbiosis of the newborn gut microbiome because of lack of exposure to maternal vaginal and fecal flora.[106] Antibiotics use can ultimately alter both the microbiota and the presence of pathogenic bacteria and modify immune mechanisms leading to tolerance and/or allergy.[107] Several studies have shown that early exposure to pets, mainly dogs and cats, is associated with decreased risk of asthma, hay fever, and/or sensitization to airborne allergens.[108] This "protective pet effect" could result from early exposure to allergens or an increase in microbial loads in homes with pets, consistent with the hygiene hypothesis. It is also noteworthy that genetics can play a key role in molding the protective effects of microbial exposure. For instance, variations in a key regulatory of inflammatory NFkB signaling, A20, strongly determine the extent of protection against asthma afforded by being raised on a farm.[109]

Diets

Diets have also been shown to influence the development of inflammatory responses and allergic disorders. Omega-6 polyunsaturated fatty acids induce production of eicosanoid mediators involved in inflammatory mechanisms.[110] Fish and fish oil are sources of long-chain omega-3 polyunsaturated fatty acids, which display opposite biological effects with regard to eicosanoid synthesis. It has been shown that maternal fish oil intake induces immunological changes in cord blood, with lower Th2 cytokines response after allergenic stimulation.[111] Maternal fish and fish oil consumption has been associated with reduced sensitization to common food allergens in the offspring, lower rates of AD in the first year of life, and lower rates of persistent wheeze and asthma at the age of 3 years.[112] In addition, other nutrients and a healthy balanced maternal diet, containing prebiotics and probiotics, may influence immune development in the fetus and thus the development of allergic disorders in the first years of life.[113] As stated earlier, consumption of unprocessed cow's milk contributes to the protective environment associated with living in a farm environment.[100,114] The Learning Early About Peanut Allergy (LEAP) study showed that early introduction of peanuts to the diet of infants at high risk of FA prevents the development of a peanut allergy and indicates that early consumption of food allergens can induce regulatory mechanisms leading to long-term tolerance.[115] These results led to the concept of the dual-allergen exposure hypothesis, which states that tolerance to antigens occurs in the neonate through high-dose oral exposure, whereas allergic sensitization occurs through low-dose cutaneous exposure (see *Dual-allergen hypothesis*).[116,117] Thus, early introduction of peanut and egg in infants' diet and early introduction of a diversity of food may induce tolerance and prevent the development of FAs.[118-120]

There is also increasing research investigating a possible protective effect of *vitamin D* supplementation on the risk of allergic diseases.[121] Vitamin D has numerous immunomodulatory effects on innate and adaptive immunity, modulated by its translocation to the nucleus of target cells and activation of vitamin D responsive elements in promoter regions.[122,123] Notably, in innate immune cells, such as macrophages and epithelial cells, vitamin D can modulate response to microbes by guiding proinflammatory response and production of host-microbial peptides, such as the human cathelicidin antimicrobial peptide LL-37. Numerous studies have reported low vitamin D levels in atopic patients. Research shows a possible positive impact of vitamin D supplementation on AD severity but no demonstrated effect on the prevention of the disease.[124,125] In asthma, studies regarding the preventive effect on the onset of the disease, progression, or improvement of asthma symptoms show conflicting results, pointing to little to no evidence of a beneficial impact.[126,127] In FAs, results are also conflicting but ultraviolet exposure and season of birth have been related to the occurrence of FAs.[128] The levels of vitamin D produced from skin by ultraviolet B exposure might reflect this relationship. Genetic polymorphisms and or epigenetic alterations may also alter vitamin D levels or patients' response to supplementation.[129] Further studies are needed to better determine the conditions of its administration, as well as the phenotypes and risk factors of patients who might benefit from it.

Stress and Hormonal Effects

Among other environmental factors, prenatal and early-life *psychological stress* may increase the risk of childhood allergic disorders. It has long been recognized that stress can lead to flares of allergic disease. In asthmatic children exposed to recurrent maternal distress, it has been proposed that this worsening may be due to a dysfunctional hypothalamic-pituitary-adrenal axis, leading to less production of cortisol in response to stress.[130] Stress may increase mast cell activation, possibly through corticotropin-releasing hormone, which has been shown to be increased under stress in the lungs of asthmatic patients.[48] Additionally, some mast cell–derived mediators, such as histamine and IL-6, are released in response to stress. Recent research has also demonstrated a significant impact of prenatal stress in the development of early wheeze, asthma, AD, and AR, with a possible cumulative effect of stress exposure.[131,132] Several studies have suggested that the children exposed to antenatal stress may display lower levels of gut IgA than their counterparts.[133] Provided through breast milk in early infancy, secretory IgA play a vital role in gut mucosal immunity. Additionally, maternal stress can alter the composition of the vaginal flora and/or alter their ability to breastfeed that may subsequently modify the composition of the child's microbiota and have long-term consequences.[134,135]

In allergic diseases and particularly in asthma, *sex hormonal influence* also impacts the severity of the disease. While childhood asthma is more frequent in boys, there is a switch at the onset of puberty with higher prevalence and severity of the disease in females.[136]

Air Pollution and Climate Change

Climate change has led to increases in heatwaves, thunderstorms, greater variability in temperature, increased air pollution, forest fires, droughts, and floods.[137] These changes have had a significant impact on human health and contribute to the onset and aggravation of allergic disorders, in particular asthma and AR.[138] Exposure to indoor pollutants, such as tobacco smoke, as well as outdoor pollution, derived from traffic and other human activities is associated with allergic disorders. Numerous studies have found particulate matter (PM) as well as increases in the major gaseous air pollutants, NO_2, SO_2, CO, and O_3 to be associated with increased risk of allergic disease.[139,140] PM with diameters ≤ 10 μm or smaller (eg, PM_{10} and $PM_{2.5}$) are particularly harmful as they can penetrate deep into the respiratory tract and skin barrier. PM exposure can exacerbate preexisting asthma and contribute to the development of airborne sensitization, asthma, and AR.[141] Particles can also have a direct effect on airborne allergens, such as pollens, with possible adsorption of particles to the pollen, leading to protein modifications that may enhance pollen allergenicity.[141,142] Increasing evidence also show an effect of antenatal exposure to pollutants on the development of allergic disorders in the offspring.[131,143]

Epigenetic Modifications

Epigenetic modifications play a key role in gene-environment interactions underlying allergic disorders. Unlike genetic variation, changes in the epigenome can occur rapidly and may provide a potential explanation for the rapid rise in prevalence of allergic disorders over the last few decades. These are biochemical changes of the chromatin, affecting DNA or histones/nucleosome positioning but not the nucleotide sequence of the genome.[144,145] DNA methylation occurs at CpG sites, which are typically located in the promoter or enhancer regions of genes, low-DNA methylation levels being associated with a higher transcriptional activity of the gene. Histone modifications include phosphorylation, ubiquitination, acetylation, and methylation, leading to tighter or looser wrapping of DNA, impacting its accessibility to the transcriptional machinery. Other epigenetic changes include microRNA (miRNA), which display silencing effects on their targeted messenger RNAs (mRNAs).

To date, extensive epigenome-wide association studies have mostly been conducted in asthma. In childhood asthma, significant hypomethylation of 14 CpG sites was demonstrated in whole-blood analysis, and these were associated with increased activation of eosinophils, effector/memory CD8 T cells and natural killer (NK) cells, and reduced number of naive T cells.[146] Epigenetic mechanisms could also drive, in part, mechanisms leading to susceptibility to IgE-mediated asthma. A recent study found significant association between lower CpG methylation levels across several loci, involving genes encoding for known eosinophil products, phospholipid inflammatory mediators, specific transcription factors, mitochondrial proteins, and IgE levels in adult and pediatric Caucasian asthma subjects.[147] A direct association between certain environmental factors and epigenetic alterations has been shown by numerous studies. For instance, exposure to a farm environment has been linked with lower risk of asthma and significant hypomethylation of regions containing genes highly associated with asthma (*ORMDL* family) and IgE regulation (DNA repair protein RAD50, IL-13, and IL-4).[148] It has been shown that many SNPs associated with asthma are located in noncoding regions of genes that are subject to DNA methylation, thus suggesting that epigenetic changes may underlie the connection between air pollution, genetic susceptibility, and asthma.[86,149] Further, increased exposure to ambient air pollution has been directly associated with increased asthma and hypermethylation of the *FOXP3* locus, linked with impaired regulatory T cell (T_{reg}) function.[150] The transmission of asthma susceptibility may also be a result of epigenetic variations originating in utero, as suggested by a higher risk of asthma and allergies in children with a maternal history of asthma than in those with a paternal history.[151] These epigenetic differences involve differential DNA methylation profiles of several genes of key regulatory pathways molecules, relating to developmental, metabolic, and inflammatory processes that may contribute to the pathogenesis of asthma.[152]

In FA, epigenetic changes have been associated with molecular and cellular processes leading to tolerance. Compared with age-matched controls, a study demonstrated that children with FA displayed different DNA methylation levels, particularly in several mitogen-activated protein kinases (MAPK) signaling molecules, impacting early CD4+ T cell development.[153] In a large epigenome-wide association study, methylation changes in loci-containing genes associated with Th1/Th2 balance were observed in cow's milk allergic patients subjects but were not observed in nonatopic healthy controls.[154] Another study found a general increase of DNA methylation in Dutch children with cow's milk allergy.[155]

In AD, a small number of studies have explored dysregulation of DNA methylation as a characteristic feature of AD, with conflicting results.[156] Of note, significant differences in methylation profiles were observed within lesioned epidermis compared with control skin.[157] Loci involved genes related to epidermal differentiation and innate immune response, suggesting a role for aberrant DNA methylation in the pathogenesis of AD. Interestingly, those differences were not observed in blood cells, suggesting a tissue-specific methylation profile and a need for more studies in skin cells.

ALLERGENS

Definition and General Characteristics of Allergens

A number of antigens that can trigger inflammatory allergenic immune responses have been identified, originating from plants (trees, grasses), fungi (*Alternaria alternata*), arthropods (mites, cockroaches), and other mammals (cats, dogs, cows).[99] They can be species-specific or molecules with broad biochemical homology among many species (eg, tropomyosin).

From an immunological perspective, allergens are linked by their ability to activate the innate immune system at mucosal interfaces, initiating the influx of APCs that subsequently promote Th2-polarized adaptive immune responses (*see innate immune responses to allergens*). In most cases, an antigen has several epitopes, reacting with different antibodies. Antibodies recognize either a linear sequence of amino acids (linear epitopes) or a three-dimensional motif (conformational epitope).[158] For one antigen source, a major antigenic determinant is defined as an epitope to which >50% of allergic patients' sera react, or in other terms, display specific IgE. Cross-reactivity is defined as the ability of an antigen to bind with an antibody that was raised to a different antigen and is observed in the pollen FA syndrome (PR-10 family).[42,61] It is explained by either linear sequence homology between epitopes from the same biochemical family or by conformational homology, originated from three-dimensional conformation making discontinuous amino acids come together.[158]

Allergen Classification

Allergens can be classified based on their route of exposure and source: aeroallergens, food allergens, latex allergens, pharmaceuticals, and Hymenoptera venom allergens. They can also be classified based on their biochemical group. The allergen nomenclature database (www.allergen.org) currently includes approximately 880 proteins and glycoproteins epitopes from a wide variety of sources.[129,159] Purified allergens are named based on their source, incorporating the first three letters of the genus and the first letter of the species epithet from which the allergen is derived followed by an Arabic numeral. Allergens from the same source are classified by biochemical groups (designated by Arabic numeral that contains two digits), usually according to the order in which they were identified. The main criteria for acceptance of a candidate allergen are descriptions of the allergenic source (including taxonomic name), information on the purification and characterization of the candidate allergenic protein (including amino acid sequence), and immunochemical characterization and demonstration of specific binding by IgE. There are several web databases that archive known allergens and their characteristics, such as their structure, composition, and family.[160] The major determinants of the most common allergens are summarized in Table 47.1.

Aeroallergens

Aeroallergens are those which enter and trigger allergic response in the airways. They can be further classified as perennial or seasonal allergens. The main perennial allergens are house dust mites species (eg, *Dermatophagoides pteronyssinus, Dermatophagoides farinae,* etc), cockroaches, animal danders (eg, cat, dog, horse, rodents), and mold species (eg, *Alternaria alternata, Aspergillus fumigatus*).[161] Seasonal allergens are mostly represented by tree (eg, birch, cypress, oak), grass (eg, Timothy), and weed pollens (eg, ragweed). Another classification distinguishes indoor from outdoor allergens. Indoor allergens are more frequently associated with the development of asthma, whereas outdoor allergens (mostly pollen species) are associated with symptoms of AR. The mechanisms underlying such associations could include the physiochemical nature of allergens (particles size, chemical structure) and pattern of exposure (acute vs chronic).[99] The type of airborne allergens that one is exposed to varies with geographical, climatic, and socioeconomic factors. Indoor allergen concentrations are affected by the presence of factors that favor their accumulation (eg, carpets, high humidity, well-insulated environment),[162] while outdoor allergens are affected by geographical and seasonal factors (eg, precipitation, humidity, and wind). As described earlier, PM can have a direct effect on pollen grains, with possible adsorption of particles to the pollen wall, leading to protein modifications that may enhance pollen allergenicity.[141,142]

Food Allergens

A small number of foods account for the vast majority of FA. The eight common ones are cow's milk, egg, peanut, tree nuts, soy, wheat, fish, and shellfish. Although most reactions are due to ingestion of foods, a small number of patients may react to inhaled food allergens, constituting a specific high-risk group with frequent severe reactions and low-eliciting doses.[163] Depending on the food allergen, the natural history of the FA may vary. Children with cow's milk and hen's egg IgE-mediated allergies usually recover in the first years of life or later in life, particularly to baked milk or egg. In contrast, peanut and tree nut allergies, generally diagnosed in the first year of life, and allergies to fish and selfish, generally diagnosed in adulthood, are often persistent. Allergies to peanuts and tree nuts are among the most severe allergies, often leading to anaphylaxis, and in rare cases, deaths. Most food allergens are water-soluble glycoproteins, of 10 to 40 kDa, that are heat- and acid-stable and resistant to proteolytic degradation. Some proteins, such as those from the PR-10 family, are usually altered by heat, explaining tolerance of baked foods. In alpha-gal allergy, the allergen is an oligosaccharide (α-Gal) rather than a protein.[13]

Latex Allergens

Latex allergy emerged in the 1980s, frequently associated with perioperative anaphylaxis and occupational asthma.[164] High-risk subjects are healthcare workers, rubber industry workers, and subjects undergoing multiple surgical procedures in early infancy, such as spina bifida. Symptoms occur in case of skin and/or mucosal contact with latex and range from contact urticaria and mucosal swelling to severe anaphylaxis and death.[99] Multiple individual latex allergens have been identified. The "latex-fruit syndrome" is a cross-reactivity syndrome, occurring between protein allergens in latex and those in certain plant foods (eg, banana, avocado, kiwi, and chestnut), affecting up to 50% of latex-allergic patients. It is thought to occur as a result of the presence of enzymes such as β-glucanase (Hev b 2 in latex) present in both fruits and rubber.[165]

Pharmaceuticals

Drug hypersensitivity reactions are common clinical problems, and drug hypersensitivity is the most frequent cause of anaphylaxis in adults. They may be classified based on timing (immediate vs delayed), mechanisms, and pattern of clinical manifestations.[166] The most common therapeutics leading to drug hypersensitivity are antibiotics (eg, penicillins, other beta-lactams, and sulfonamides), quaternary ammonium compounds (neuromuscular-blocking agents, such as curare), angiotensin-converting enzyme inhibitors, aspirin and nonsteroidal anti-inflammatory drugs, radiocontrast media, and chemotherapeutic agents. Important risk factors for drug hypersensitivity include chemical properties of the drug and host factors. Drug hypersensitivities are mediated by IgE and other types of hypersensitivity reactions.[167] Specific structural moieties of smaller molecular weight ($\leq$1000 kDa) nonprotein agents and/or their

TABLE 47.1 Molecular Properties of Common Allergens

Allergenic Source	Major Antigenic Determinant	Molecular Weight (kDa)	Biochemical Group
Aeroallergens			
Dermatophagoides pteronyssinus **(House dust mite)**	**Der p 1** (major)	24	Cysteine protease
	Der p 2 (major)	15	Group 2 mite allergen
	Der p 3 (major)	31	Serine protease
	Der p 10 (minor)	36	Tropomyosin
	Der p 11 (major)	103	Paramyosin
	Der p 23 (major)	14	Chitinase class III
Dermatophagoides farinae **(House dust mite)**	**Der f 1** (major)	27	Papain-like cysteine protease
	Der f 2 (major)	15	Group 2 mite allergen
	Der f 10 (major)	37	Tropomyosin
	Der f 11 (major)	98	Paramyosin
Blattella germanica **(German cockroach)**	**Bla g 1** (major)	46	Group 1 cockroach allergen
	Bla g 2 (major)	36	Eukaryotic aspartyl protease
Periplaneta americana **(American cockroach)**	**Per a 1** (major)	45	Group 1 cockroach allergen
	Per a 2 (major)	42	Eukaryotic aspartyl protease
Felis domesticus **(cat)**	**Fel d 1** (major)	38	Uteroglobin
	Fel d 4 (major)	22	Lipocalin
Canis familiaris **(dog)**	**Can f 1** (major)	23-25	Lipocalin
	Can f 2 (major)	19-27	Lipocalin
	Can f 5 (major)	28	Arginine esterase, prostatic kallikrein
Equus caballus **(horse)**	**Equ c 1** (major)	25	Lipocalin
Mus musculus **(mouse)**	**Mus m 1** (major)	17	Lipocalin
Alternaria alternate	**Alt a 1** (major)	15-16	Alt a 1 family
	Alt a 4 (major)	57	Disulfide isomerase
Aspergillus fumigatus	**Asp f 1** (major)	18	Mitogillin family
	Asp f 3 (major)	19	Peroxysomal protein
	Asp f 11 (major)	24	Cyclophilin
Betula verrucosa **(Birch pollen)**	**Bet v 1** (major)	17	PR-10
Phleum pretense **(Timothy grass)**	**Phl p 1** (major)	27	Expansin and expansin-like
	Phl p 2 (major)	10-12	Group 2/3 grass pollen allergen
	Phl p 5 (major)	32	Group 5/6 grass pollen allergen
Ambrosia artemisiifolia **(Ragweed)**	**Amb a 1** (major)	38	Pectate lyase
Food Allergens			
Bos domesticus **(Cow's milk)**	**Bos d 4** (major)	14	Alpha-lactalbumin
	Bos d 5 (major)	18	Beta-lactoglobulin
	Bos d 12 (major)	19	kappa-casein
Gallus domesticus **(Hen's egg)**	**Gal d 1** (major)	28	Ovomucoid
	Gal d 2 (major)	44	Ovalbumin—Egg white
	Gal d 5 (major)	69	Serum albumin—Egg yolk

TABLE 47.1 Molecular Properties of Common Allergens (Continued)

Allergenic Source	Major Antigenic Determinant	Molecular Weight (kDa)	Biochemical Group
***Arachis hypogaea* (Peanut)**	**Ara h 1** (major)	64	Cupin (Vicillin-type, 7S globulin)
	Ara h 2 (major)	17	Conglutin (2S albumin)
	Ara h 3 (minor)	60	Cupin (Legumin-type, 11S globulin, Glycinin)
	Ara h 8 (major)	17	PR-10
	Ara h 9 (minor)	10	Nonspecific lipid-transfer protein type 1
***Corylus avellane* (Hazelnut)**	**Cor a 1** (major)	17	PR-10
	Cor a 2 (minor)	14	Profilin
	Cor a 8 (minor)	9	Nonspecific lipid transfer protein type 1
	Cor a 9 (major)	40	11S seed storage globulin (legumin-like)
	Cor a 14 (minor)	10	2S albumin
***Juglans regia* (Walnut)**	**Jug r 1** (major)	15-16	2S albumin seed storage protein
	Jug r 3 (major)	9	Nonspecific lipid transfer protein type 1 (nsLTP1)
***Anacardium occidentale* (Cashew)**	**Ana o 1** (major)	50	Vicilin-like protein
	Ana o 2 (major)	55	Legumin-like protein
***Pistacia vera* (Pistachio)**	**Pis v 1** (major)	7	2S albumin
	Pis v 2 (major)	32	11S globulin subunit
***Glycine max* (Soybean)**	**Gly m 4** (major)	17	PR-10
	Gly m 8 (major)	28	2S albumin
***Triticum aestivum* (Wheat)**	**Tri a 14** (major)	9	Nonspecific lipid transfer protein type 1 (nsLTP1)
	Tri a 19 (major)	65	Omega-5 gliadin, seed storage protein
***Gadus morhua* (Atlantic cod)**	**Gad m 1** (major)	12	Beta-parvalbumin
***Penaeus indicus* (Shrimp)**	**Pen i 1** (major)	34	Tropomyosin
Meat: beef, pork, lamb	**Galactose-alpha-1,3-galactose**		Carbohydrate
Latex allergens			
***Hevea brasiliensis* (Para rubber tree: latex)**	**Hev b 1** (major)	14	Rubber elongation factor
	Hev b 2 (major)	35-38	beta-1,3-glucanase; glucan endo-1,3-beta-glucosidase, basic vacuolar isoform
	Hev b 6 (major)	20	Hevein precursor
Hymenoptera allergens			
***Apis mellifera* (Honey bee)**	**Api m 1** (major)	16	Phospholipase A2
	Api m 2 (major)	44	Hyaluronidase
	Api m 3 (major)	43	Acid phosphatase
	Api m 5 (major)	100	Dipeptidylpeptidase IV
	Api m 10 (major)	50-55	Icarapin variant 2, carbohydrate-rich protein
***Vespula vulgaris* (Yellow jacket)**	**Ves v 1** (major)	31	Phospholipase A1B
	Ves v 2 (major)	38	Hyaluronidase
	Ves v 5 (major)	23	Antigen 5
***Solenopsis invicta* (Red imported fire ant)**	**Sol i 1** (major)	18	Phospholipase A1B

metabolic products are more likely to form complexes with self-proteins, producing immunogenic hapten-protein conjugates (type II hypersensitivity).[166] Type III hypersensitivity reactions, which involve immune complex tissue deposition and complement activation, may also be involved in some drug hypersensitivity effects.[168] Finally, activation of Mas-related G protein–coupled receptor X2 (MRGPRX2), a mast cell–specific receptor that mediates IgE-independent cell degranulation, has been shown to be involved in some hypersensitivity reactions, possibly leading to anaphylaxis.[169] MRGPRX2 is an atypical receptor for many classes of drugs, including both synthetic and endogenous opioids, possibly mediating drug related to local (eg, itch, urticaria, skin rash, and contact dermatitis) and systemic (eg, anaphylactic shock associated with some heroin fatalities) reactions.[169,170] Hence, neuromuscular-blocking drugs such as atracurium and rocuronium may induce mast cell histamine release mediated by MRGPRX2.[169,171,172] The glycopeptide antibiotic vancomycin, associated with clinical pseudoallergic reactions known as red man syndrome, is also an agonist of MRGPRX2.[173] Murine studies indicate that fluoroquinolone-induced anaphylactic reactions are mediated by mast cells through MRGPRX2 in a dose-dependent manner.[174] Host risk factors for drug hypersensitivity include female gender, history of drug allergy, recurrent exposure to drugs, and genetic factors, such as polymorphisms in drug metabolism genes and some *HLA* alleles, which present the drug in particularly antigenic fashions.[175-177] Beta-lactams hypersensitivity are the most common drug-hypersensitivity, mediated by IgE-allergic responses. The allergic components of penicillin are, in most cases, the R-group side chains (different between penicillins), or rarely, the beta-lactam core ring structure (common to all penicillins, cephalosporins, and carbapenems), leading to cross-reactivity between beta-lactams.[178]

Insect Venom Allergens

Stinging insect allergy is rare. The three common stinging families of the Hymenoptera order are bees (honeybee, bumblebee), vespids (yellow jacket, yellow hornet, white-faced hornet, wasp), and stinging ants (in particular, the imported fire ants).[179] Yellow jackets cause the most frequent insect-sting reactions in North America. Normal reactions to insect stings and bites are localized to the area of the sting, including erythema, pruritus, and mild swelling. IgE-mediated severe reactions include cutaneous severe reactions, such as generalized urticaria, angioedema, or anaphylaxis.[180] Interestingly, severe venom anaphylaxis is not uncommonly associated with the presence of known or indolent-systemic mastocytosis—the clonal expansion of atypical mast cells.[181] Sensitization occurs when high levels of proteins contained in venom are injected subcutaneously during a sting. Many Hymenoptera allergens have proteolytic activity and are cross-reactive. Venom immunotherapy is the most effective preventive treatment for patients with a history of anaphylaxis.

IMMUNE MECHANISMS

Overview of Immune Mechanisms Involved in Allergic Disorders

The immune system is composed of the innate or nonspecific immunity and the adaptive or specific immunity. When the body encounters a novel pathogen, it is initially recognized by the innate immune system, and then the adaptive immune response is activated. During allergic disease, both the innate and adaptive immune systems play a role.

Allergic Response

An allergic response is defined as the immune reaction following exposure to a given allergen in sensitized individuals. In addition to an immediate acute phase allergic reaction, corresponding to a type I hypersensitivity reaction, there is often a subsequent late-phase reaction, leading to chronic inflammation on persistent or repetitive contacts with the allergen[182] (Fig. 47.4).

Allergic sensitization is an initial first step that primes the individual toward an allergic state. Allergic sensitization is facilitated by loss of integrity of barrier interfaces and, in some cases, by endogenous protease activity, enabling allergens to penetrate and interact with innate immune cells and DCs within the epithelium and in the submucosa. The presence of allergens leads to the release of the epithelial-derived inflammatory cytokines, IL-25, IL-33, and TSLP, which have collectively been termed alarmins. These cytokines induce a Th2-prone inflammatory environment in the mucosa via type 2 ILC2 promoting the activation and maturation of DCs, which then migrate to draining lymph nodes. Here, they initiate the adaptive immune response by presenting allergen-derived peptides to naïve T cells using major histocompatibility complex (MHC) class II molecules and driving their differentiation into allergen-specific Th2 cells. This occurs under the influence of various signals, such as IL-4. The initial source of IL-4 remains under debate.[183,184] Some have suggested it is initially produced by basophils, able to act as APCs to prime Th2 cells.[185] However, it has also been demonstrated that Th2 immunity can develop in the absence of basophils and IL-4 in a mouse model of helminth infection.[186] Another model can be proposed in which basophils and DCs cooperate, DCs as key APCs, and basophils as providers of IL-4 to amplify the Th2 response.[187] Following exposure to some allergen proteases, DCs may also release reactive oxygen species leading to CCL7 production, which mediates IL-4$^+$ basophil recruitment into the draining lymph nodes.[188] Other necessary actors in the initiation of the TH2 response might be iNKT cells, as suggested by rodent models of asthma.[99,189] Finally, mouse models and monogenic human diseases associated with atopy caused by mutations in the T cell receptor signaling cascade (eg, *MALT1, CARD11*) suggest that stimulation of naïve CD4+ T cells with weak T cell receptor agonists can result in IL-4 production driving Th2 response.[190]

After being primed, Th2 cells then secrete IL-4 and IL-13 and interact with naïve B cells. Under the influence of these cytokines, costimulatory signals, such as Cluster of differentiation CD40-CD40 Ligand (CD40L) and CD80/86-CD28

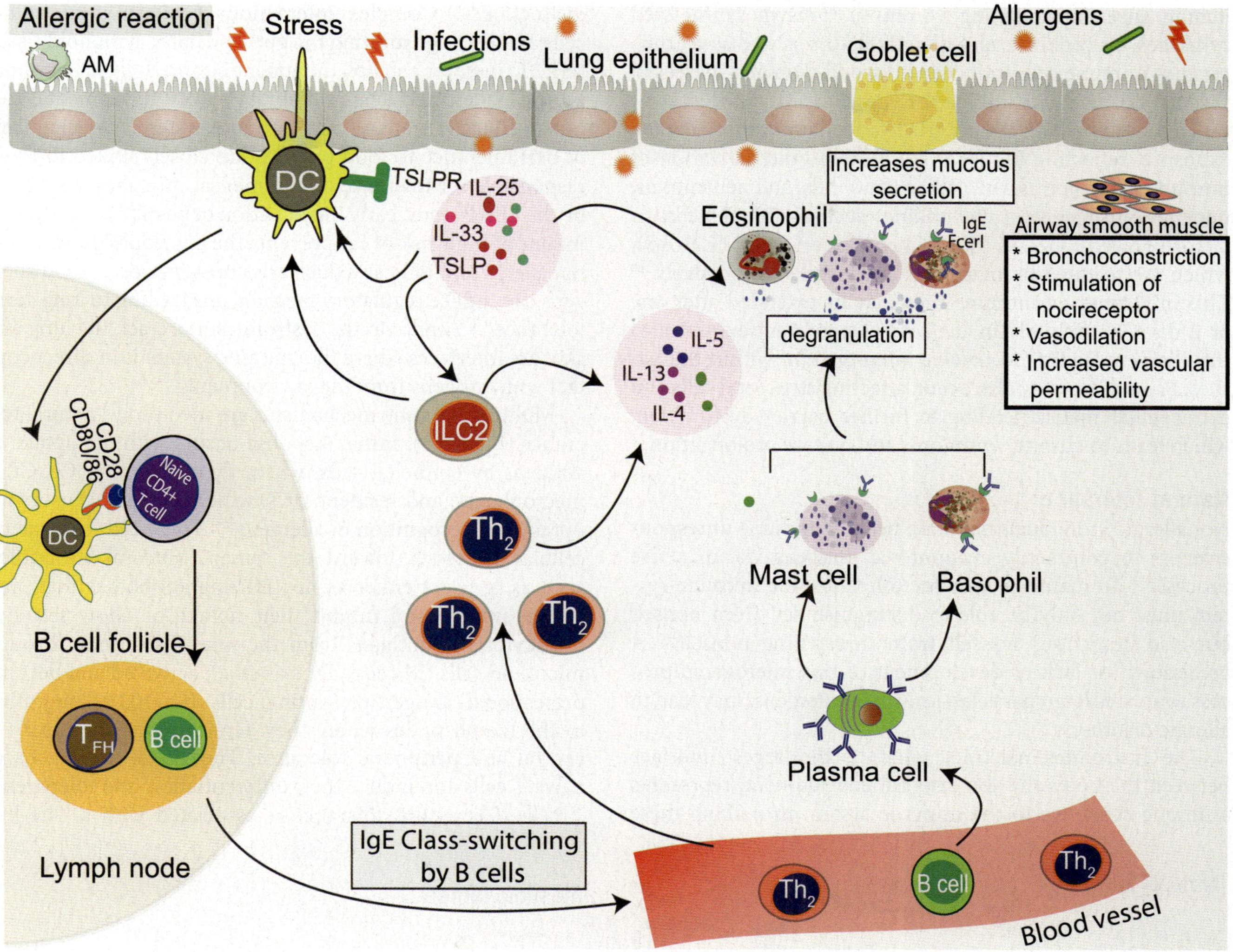

FIG. 47.4. Acute phase of the allergic reaction (lung). AM: Alveolar macrophage; CD: Cluster of differentiation; DC: dendritic cell; FcεRI: high-affinity IgE receptor; IgE: Immunoglobulin E; IL: Interleukin; ILC2: type 2 innate lymphoid cells dendritic cells; MHC: major histocompatibility complex; TFH: T follicular helper cell; TSLP: Thymic stromal lymphopoietin. Alveolar macrophages in the lungs are the first to be exposed to exogenous triggers, including infections and environmental allergens. In the presence of allergens, epithelial cells release the alarmins IL-25, IL-33, and TSLP. These cytokines induce a Th2-inflammatory environment in the mucosa via ILC2, promoting the activation and maturation of DCs, which then migrate to draining lymph nodes. Here, they initiate the adaptive immune response by presenting allergen-derived peptides to naïve T cells using MHC class II molecules and driving their differentiation into allergen-specific Th2 cells under the influence of IL-4. Th2 cells then secrete IL-4 and IL-13 and interact with naïve B cells. Under the influence of these cytokines and costimulatory signals, Th2 cells promote IgE class-switching by B cells. Allergen-specific or nonspecific IgE then bind to high-affinity FcεRI receptors on tissue-resident mast cells and basophils, sensitizing the individual to the allergen. In these sensitized individuals, subsequent exposure and allergen binding to FcεRI-bound IgE attached on the surface of mast cells and basophils activates the release of biologically active products in a process called degranulation. This leads to vasodilation, increased vascular permeability, contraction of bronchial smooth muscle, increased secretion of mucus, stimulation of nociceptors of sensory nerves, all contributing to the symptoms of the allergic reaction.

interactions, Th2 cells promote IgE class switching by B cells. Allergen-specific or nonspecific IgE then bind to high-affinity FcεRI receptors on tissue-resident mast cells, sensitizing individuals and enabling rapidly allergenic response (within minutes) on subsequent exposure to the allergen.[182]

When an allergen binds to FcεRI-bound IgE attached on the surface of mast cells and basophils in sensitized individuals, they activate the release of three classes of biologically active products: (1) those stored in the cytoplasmic granules (eg, histamine, serine proteases such as tryptase, platelet-activating factor [PAF]), (2) lipid-derived mediators (prostaglandin D2 PGD2, leukotrienes), and (3) newly synthesized cytokines, chemokines, and growth factors (*see Chapter 16: Mast Cells, Basophils, and Eosinophils*).[191] Cytoplasmic granules and lipid-derived mediators contribute to the signs and symptoms associated with acute reaction: vasodilation (producing erythema), increased vascular permeability (leading to tissue swelling, extravasation in the gastrointestinal tract), contraction of bronchial smooth muscle (leading to bronchoconstriction within the airways), increased secretion of mucus (leading to bronchial hypersecretion, diarrhea), stimulation of nociceptors of sensory nerves (leading to reaction such as itching, sneezing, coughing, vomiting).[182] Newly synthesized cytokines, chemokines, and growth factors as well as recruitment of immune cells induced by vasodilation and increased vascular permeability initiate the *late-phase reaction*.

This late-phase reaction starts two to 6 hours after allergen exposure, reflecting the local recruitment and activation of immune cells, such as Th2 cells, eosinophils, and neutrophils, under the influence of chemokines such as tumor necrosis factor (TNF-α), C-X-C Motif Chemokine Ligand (CXCL8), which are released by mast cells and T cells, respectively.[182] This infiltration of immune cells may be extended after one or 2 days, particularly in the case of persistent exposure to the allergen, leading to *chronic inflammation* within the tissues. *Remodeling* may also occur, affecting structural cells and extracellular matrix, leading to further barrier dysfunction, which leads to chronic symptoms and disease prolongation.

Natural Tolerance

Nonallergic individuals maintain healthy immune unresponsiveness to commonly encountered antigens via an active process.[192] To maintain immune tolerance, the immune system must not only be able to distinguish self from nonself but also innocuous nonself from threatening nonself.[90] A breakdown or lack of development of this tolerogenic process against foreign particles, known as allergens, may lead to allergic disorders.

The gastrointestinal tract, which is the largest interface between the body and the external environment, represents a unique challenge to the immune system in making these distinctions.[90] Complex interactions between nonimmune cells and cells constituting the gut-associated lymphoid tissue (GALT) in Peyer patches, the appendix, and isolated lymphoid follicles take place. The microbiota plays a fundamental role in the education of the host immune system.[88] Mechanisms of oral tolerance to food proteins are closely linked to those responsible for tolerance to commensal microbes. As shown by the LEAP study, early introduction of peanuts to the diet of infants at high risk of FA prevents the development of a peanut allergy and indicates that early consumption of food allergens can induce regulatory mechanisms leading to long-term tolerance.[115] Similar to the gastrointestinal tract, the lung and skin are interfaces where the immune system is in direct contact with antigens from the environment.

Multiple immune mechanisms are involved in natural tolerance (Fig. 47.5). Initial response occurs with the uptake of antigens by epithelial cells, where they encounter CX_3CR1^+ macrophages, and resident DCs, which are early sensors for uptake and recognition of allergens.[193] These cells can extend cellular processes toward the lumen. Mucosal antibodies, such as IgA, act either as neutralizing antibodies that bind to the antigen and inhibit their uptake or allow selective transcytosis of antigens from the intestinal lumen through microfold cells (M cells). DCs are then activated and become professional antigen-presenting cells ($CD103^+$), migrating to the lymph nodes where they act as critical regulators of central and peripheral tolerance. They present antigens to naïve T cells and induce their differentiation into tolerogenic T cells. Classically, tolerance is associated with a Th1-like

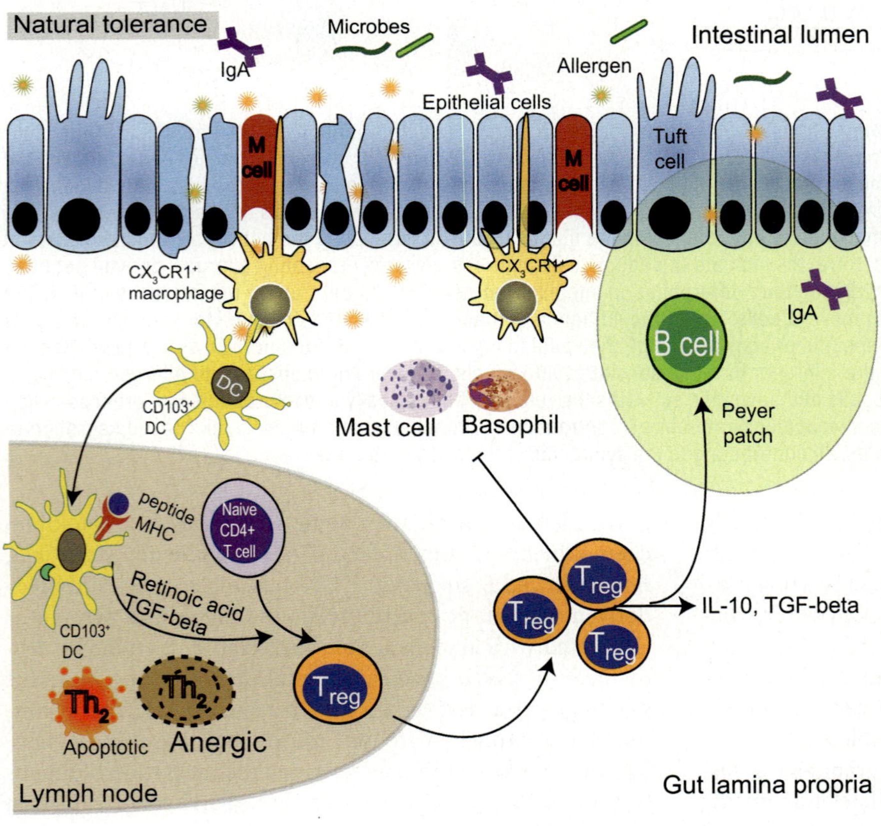

FIG. 47.5. Natural tolerance (gut). CD: Cluster of differentiation; DC: dendritic cell; IgA: Immunoglobulin A; IL: Interleukin; MHC: major histocompatibility complex; TGF: transforming growth factor; T_{reg}: regulatory T cell. Initial response occurs with the uptake of antigens by epithelial cells, where they encounter CX_3CR1^+ macrophages and resident DCs. Mucosal antibodies, such as IgA, act either as neutralizing antibodies that bind to the antigen and inhibit their uptake or allow selective transcytosis of antigens from the intestinal lumen through M cells. Dendritic cells are then activated ($CD103^+$) and migrate to the lymph nodes. They present antigens to naïve T cells and induce their differentiation into T_{reg} cells, expressing tissue-homing receptors. Tregs are then transported to tissues, under the influence of tissue-homing cytokines, such as TGF-β, retinoic acid. They can also inhibit the development and function of other nonregulatory T cells by producing anti-inflammatory cytokines such as IL-10 and TGF-β, leading to apoptosis or anergy of Th2 cells. They can also alter sensitization via downregulation of antigen-presenting functions in DCs and can ultimately inhibit mast cells and basophils degranulation.

predominant response, with production of cytokines such as interferon-gamma (IFN-γ) and TNF-α. In the draining lymph nodes, DCs can also induce differentiation of naïve T cells into Tregs expressing tissue-homing receptors. Tregs are then transported to tissues, under the influence of tissue-homing cytokines, such as TGF-β, retinoic acid, C-C chemokine receptor type 9 (CCR9), and $\alpha_4\beta_7$ integrin (in the gut),[192] CCL20 (in the skin),[194] and CCR4 (in the lung).[195] They can also inhibit the development and function of other nonregulatory T cells by producing anti-inflammatory cytokines such as IL-10 and TGF-β, leading to apoptosis or anergy of Th2 cells. TGF-β is a major immunosuppressive cytokine with critical regulatory functions in allergies. In fact, patients with Loeys-Dietz syndrome, who present mutations in the TGF-β receptor pathway, have a high incidence of allergic disorders, as well as EoE.[196] Tregs can also alter sensitization via downregulation of antigen-presenting functions in DCs and can ultimately inhibit mast cells and basophils degranulation.

Barrier Dysfunction and Dual-Allergen Hypothesis

The skin is the largest "organ" in the body and our first layer of protection against foreign invaders. Antigens can stimulate the immune system if able to penetrate the skin barrier, but some may directly penetrate and bind to APCs (Langerhans or DCs) in the skin. Skin irritation and proteases contained in some allergens can facilitate their passage through the skin.[72] Loss-of-function mutations in the *Filaggrin* gene are frequent in AD, increasing skin permeability.[197] Chronic Th2 inflammation can also enhance this barrier defect by reducing *Filaggrin* gene expression.[198] The dual-allergen exposure hypothesis suggests that tolerance to antigens occurs in the neonate through high-dose oral exposure, whereas allergic sensitization occurs through low-dose cutaneous exposure[116,117] (Fig. 47.6). This hypothesis provides an explanation for the association between early eczema and the development of FAs. It is supported by studies in rodent models showing that allergenic sensitization, characterized by specific IgE responses, occurs after cutaneous exposure to antigens.[199] In human subjects, food allergen–specific T cells have been isolated from lesional skin in patients with eczema.[200] Low-dose exposure to peanut oil applied to inflamed skin in infants has been linked with increased risk of peanut allergy at the age of 5 years.[201] Additionally, it has been shown that food allergens present in household dust may induce sensitization through the skin.[202] The atopic march linking skin barrier dysfunction with allergy sensitization and the development of other allergic disorders also lends support to this hypothesis.[70,71] On the other hand, both mouse and human studies suggest that early consumption of allergenic foods induces tolerance.[117] Thus, low-dose exposure to environmental foods, with antigens penetrating the skin barrier and taken up by Langerhans cells, could lead to Th2 responses and IgE production by B cells. Early high-dose oral consumption may induce Th1 and T_{reg} responses in the gut. This differential response might originate from the immune system's attempt to discriminate danger (eg, skin parasitic infection) from safety (eg, normal diet). Early infancy might provide a window of opportunity to induce tolerance rather than allergy via oral consumption of allergens.

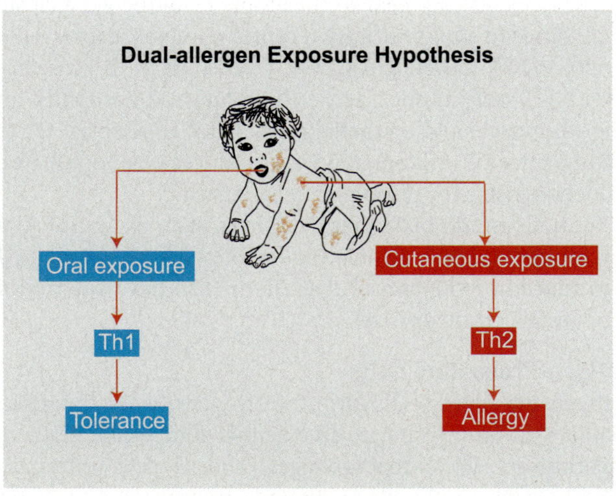

FIG. 47.6. Dual-Allergen Exposure Hypothesis. The dual-allergen exposure hypothesis suggests that tolerance to antigens, mediated by Th1-adaptive immune response, occurs in the neonate through high-dose oral exposure, whereas allergic sensitization, mediated by exaggerated Th2 response occurs through low-dose cutaneous exposure.

Innate Immune Responses to Allergens

The human body is constantly encountering numerous antigens from the environment, including both noxious pathogens that need to be eliminated as well as innocuous materials such as commensal microbes, food proteins, or inhaled particles. As indicated earlier, it is important for the immune system to be able to differentiate the two and mount appropriate responses accordingly. Innate immune cells within the skin and mucosal layers are in the frontline of host defense and are the first to be exposed to environmental allergens. Much of what we know of the innate immune response comes from its response to pathogenic agents, where they play a sentinel role in triggering robust and protective responses (*Chapter 4. Innate Immunity*). Among them, epithelial cells are structural cells with innate immune functions. The main innate immune cells in the mucosa are APCs (eg, DCs), phagocytes (eg, neutrophils and monocytes), and ILCs.

In susceptible individuals, *innocuous allergens can trigger the innate immune system* toward a robust Th2-mediated response that can be detrimental to the body, primarily through three mechanisms. Firstly, allergens may disrupt epithelial barriers to activate the innate immune system through their intrinsic enzymatic activity. Some allergens exhibit cysteine or serine protease activity (eg, allergens from house dust mites, German cockroaches, fungi, Felis domesticus, and ragweed pollens), proteolytic activity in detergents and foods (eg, papain), trypsin inhibition (eg, major peanut allergen Ara h 2), and oxidase activity (eg, ragweed pollen).[99,203] Secondly, allergens and their soluble components may contain PAMPs or danger-associated molecular patterns (DAMPs), such as endotoxin, chitin, β-glycans, and MD-2 like lipid-binding family (eg, Der p 2), which are recognized by PRRs (eg, TLR4) on cells lining mucosal surfaces.[204] It is now recognized that the ability of certain environmental particles to provide these danger signals to innate immune cells likely underlie their ability to be allergens, initiating allergic responses. Finally, many allergens, such as the peanut allergen Ara h 1, and the dust mite allergen Der p 1, also contain carbohydrate moieties that can stimulate the C-type lectin receptors (CLR) on DCs, thereby enhancing their uptake to activate Th2 immune responses.[99]

Epithelial Cells and Epithelium-Derived Cytokines

Epithelial cells are structural cells that form the first line of defense between the environment and the self. Three main types of epithelial cells are involved in allergic disorders: respiratory epithelial cells in AR and asthma (upper and lower airways), skin keratinocytes in AD, and gut epithelial cells in FAs. They constitute a contiguous layer of cells and form a complex physicochemical barrier via the presence of tight intercellular junctions and, in the lungs, by specialized goblet cells, which clears foreign particles through the production of mucus. However, when barrier integrity is lost, allergens from the environment can penetrate the body and increase risk of allergy development through induction of Th2 immune responses. Several genetic defects, as well as mechanical/chemical damage to the skin barrier integrity have been linked to AD and other allergic diseases.[205,206] Beyond their function as a physical barrier, epithelial cells also possess "sensors," known as PRR, such as TLR, CLR, and nucleotide-like oligomerization domain-like receptors (NLR) that are capable of recognizing molecules frequently found in pathogens or molecules released by damaged cells (PAMPs and DAMPs, respectively).[204] Epithelial cells also express a wide range of receptors such as protease-activated receptors and can activate danger sensing pathways in response to extracellular nucleotides or uric acid.

The activation of PRRs on innate immune cells leads to further activation of various signaling pathways. PPR activation leads to assembly and activation of the inflammasome, which initiates transcription and release of epithelial-derived cytokines, IL-25, IL-33, and TSLP. These cytokines, also termed alarmins, activate neighboring cells, DCs, basophils, and ILC2s.[207,208] Tuft cells in the gut are specialized epithelial cells with chemosensory pathways leading to the secretion of IL-25, which has a critical role in the initiation of Th2 response to parasite infections and in the pathophysiology of allergic disorders.[209] It has been recently demonstrated that parasitic helminths are detected by small intestinal tuft cells and that luminal succinate acts as an upstream agonist.[210-212] This chemosensation of luminal ligands, including allergen, involves taste-like signaling molecules such as alpha-gustducin and TRPM5 that induces downstream mediators such as IL-25, eicosanoids, and acetylcholine.[213] IL-25 activates ILC2s to release IL-13 and promote goblet and tuft cell expansion, constituting a tuft cell–ILC2 circuit, promoting a "weep-and-sweep" response.[210,212] In the airway, specialized cells with similar morphology, and possibly function, called brush cells, are present. In the gut, M cells, a type of intestinal epithelial cells overlying the GALT, actively engage in phagocytosis and transcytosis of antigens.[90]

Epithelial-derived cytokines, IL-25, IL-33, and TSLP, also called alarmins, are rapidly released after antigen stimulation or after nonprogrammed cell death to "alarm" the immune system of a potential threat.[207,208] These cytokines activate both the innate and adaptive immune system to external insults and regulate tissue restoration and repair after injury. They are important mediators of inflammation in allergic disease exacerbations, as confirmed by GWAS studies showing their association with asthma and AD.[208] TSLP and IL-33 may preferentially coordinate sensitization, while IL-25 enhances effector functions through ILC2s stimulation.[214] Each of these cytokines may also have tissue-dependent roles.[215]

IL-25 is a member of the IL-17 cytokine family, also known as IL-17E, and is released from the extranuclear cellular compartment of epithelial cells, Th2-polarized cells, activated mast cells, basophils, and eosinophils, upon exposure to proteases.[207] IL-25 influences target cells that express IL-25R, such as APCs, airway smooth muscles, epithelial cells, fibroblasts, eosinophils, iNKT cells, and ILC2s and induces the release of type 2 cytokines IL-4, IL-5, and IL-13. IL-25 is also antagonistic to Th17 cell differentiation.

IL-33 is a member of the IL-1 cytokine family and is produced by epithelial cells and mast cells on detection of substances that they perceive as dangerous. They are also produced, although in lesser amounts, by smooth muscle cells, endothelial cells, macrophages, DCs, and monocytes.[216] There are two isoforms of IL-33, full-length and mature, found in both the nuclear and extracellular space. It binds to a receptor complex consisting of two transmembrane proteins, of which IL-1R1 (called ST2) is the main unit. IL-33 induces production of type 2 cytokines, including IL-4, IL-5, and IL-13, by hematopoietic cells and, more importantly, by ILC2s. It is also characterized by the induction of a sustained deposition of extracellular matrix protein and neoangiogenesis.

TSLP is part of the IL-2 cytokine family produced by epithelial cells, notably in the lung and skin. Its gene is localized on chromosome 5q, next to the atopic cytokine gene cluster. TSLP binds to a high-affinity receptor complex, expressed on myeloid DCs, CD4+ and CD8+ T cells, Tregs, B cells, mast cells, NKT cells, monocytes, eosinophils and basophils, and smooth muscle and epithelial cells. TSLP is essential for the maintenance of inflammation and induces eosinophilia and bronchoconstrictor responses in the airway.

Of note, numerous other cytokines, such as granulocyte-macrophage colony-stimulating factor (GM-CSF) and TGF-β, produced by the barrier epithelium also have key roles in regulating allergic diseases.

Antigen-Presenting Cells

Antigen-presenting cells form the bridge between innate and adaptive immunity and are an essential component of allergy sensitization. DCs, macrophages, and B cells internalize antigens, process them into peptide fragments, and display these peptides bound to a class II MHC molecule on their cell surface. When allergens penetrate the skin, gut, or lung epithelium, they can be taken up by tissue-resident macrophages or DCs.[217] Langerhans cells are tissue-resident DCs that are essential for antigen uptake within the epidermidis.[218]

In the gut, a subset of CD11C+ CXC3CR1+ myeloid DCs in the lamina propria can extend cellular processes into the intestinal lumen and sample antigens without disruption of the epithelial barrier.[90] In the airway, alveolar macrophages are the primary immune cells that clear invading pathogens to maintain homeostasis in the lung.[219] CD103+ CD11b+ β7-integrin conventional DCs and monocyte-derived DCs, stimulated by inflammation, can also sample antigens by extending their dendrites into the airway lumen.[220] Of note, the last subset of DCs present in the lung, plasmacytoid DCs, have a limited capacity to take up and present antigens.[220] Under the influence of antigen-derived danger signals or through indirect sensing of danger by surrounding cells, DCs will then undergo maturation during which they lose their ability to take up antigen and express costimulatory molecules (CD80/86) and chemokines, thus becoming professional APCs. These DCs, loaded with allergens, then migrate to the draining lymph nodes and present peptides on MHC class II molecules to the T cell receptors (TCRs) on naïve CD4 T cells. DCs have the ability to influence T-cell differentiation toward a Th1, Th2, Th17, or T_{reg} cell type by secretion of specific cytokine patterns, depending upon their lineage, their maturation status, and the costimulatory molecules they express.[99] Three signals are required to attract and drive naïve CD4 T cells to become allergen-specific Th2 cells: MHC-peptide–TCR interactions, costimulatory signals, and Th2-polarizing cytokines.[217] In addition, DCs activated in the Th2-promoting environment do not produce IL-12, which is important to drive Th1 response.[221]

As previously stated, some studies suggest basophils, but also eosinophils, mast cells, and ILC2s can act as APCs because they express low level of MHC-class II molecules on their surface under certain conditions. While they might not be sufficient to initiate Th2 polarization from naïve T cells because a lack of costimulatory signals or migration capability, these "nonprofessional" APCs likely collaborate with DCs to maintain Th2 effector and memory cells upon reexposure to the allergens.[183,217]

Phagocytes

Phagocytes, such as neutrophils and monocytes, are important players of innate immunity. They patrol the body and quickly mobilize to the infected site to clear invading pathogens. However, their roles in allergic responses are less clear. Neutrophils are recruited to the airway during asthma and contribute to lung inflammation.[222] Increased levels of neutrophil-attacking cytokines IL-8 and IL-17 in response to inhaled allergens are correlated with the accumulation of neutrophils in severe and uncontrolled asthma.[222] Neutrophils are also associated with remodeling after the late-phase response. Monocytes are also likely to be involved with an allergic reaction as they migrate into tissues and become tissue macrophages or monocyte-derived DCs.[223]

Innate Lymphoid Cells

The importance of the innate lymphoid system in driving and regulating adaptive immune responses has been well recognized. ILCs are primarily localized in barrier and mucosal tissues, with distinct subsets that can be discriminated based on their transcriptional and effector programs.[193] Helper ILCs are implicated in the pathophysiology of allergic disorders and express CD127, CD25, and CD90. Of note, other subsets include cytotoxic ILCs, known as NK cells, and regulatory ILCs. Among helper ILCs, three groups have been described, defined by their expression of lineage-defining transcription factors and with functions mirroring the various subsets of helper T cells (Th1, Th2, and Th17). Of these, ILC2s, which express high levels of the transcription factor Trans-acting T cell–specific transcription factor GATA-3, are important in the response to helminths as well as for the development of allergic response. They are mostly found in the lung, intestine, skin, and adipose tissue. Stimulated by epithelial-derived alarmins TSLP, IL-25, and IL-33, they secrete considerable levels of IL-5, IL-13, as well as IL-4 and IL-9, and thus drive Th2 response.[208,224,225] ILC2s also secret amphiregulin, a ligand for the epithelial growth factor receptor, which mediates epithelial cell repair and contributes to homeostasis after initial inflammation, notably in the gut.[193]

Adaptive Responses Leading to the Allergic Reaction

Adaptive immune responses are initiated a few hours after the allergen breaks in the mucosal interfaces, after a cross-talk with innate immune cells, in particular, DCs. Unlike innate responses, they are specific to an allergen. There are two parts of the adaptive response: the cell-mediated immune response, carried out by T cells, and the humoral immune response, mediated by B cells.[226] In the course of the allergic reaction, the adaptive immune responses upon exposure to an allergen are deleterious and exaggerated, responsible for chronic inflammation. They mainly involve CD4+ Th2 cells that drive the maturation of IgE-secreting B cells. Following initial activation, a subset of memory T cells remain in mucosal surfaces, leading to quick-onset responses upon later exposure to the allergen. In the situation of immune tolerance, adaptive responses are appropriate and involve CD4+ Th1 cells, IL-10–producing cells, and regulatory T cells.

Role of CD4+ T Cells in the Allergic Response

Allergic diseases were initially thought of as imbalance between CD4 Th1 and Th2 cells with Th1 cells promoting tolerogenic and Th2 cells promoting inflammatory allergic responses. With the discovery of additional T cell subsets and a greater understanding of their plasticity, this initial simplistic view that restoring Th1/Th2 balance is being enhanced by a more complex model involving other T cell subsets. CD4 Tregs are now known to be an important cell type for promotion of tolerance. Research have also identified other T cell subtypes that play a role in allergic diseases such as Th17, Th22, follicular helper T cells (Tfh), γδ T Cells, iNKT cell, mucosal-associated invariant T Cells (MAIT), and CD8 T cells.

CD4+ T cells play central and orchestrating roles in the pathophysiology of allergic disorders, as evidenced by increased numbers in the bronchial mucosa, nasal mucosa, and skin of patients with allergic asthma, rhinitis, and allergic dermatitis, respectively, when compared with nonatopic controls.[99] There is indeed a generalized activation in CD4+ T-cells, both locally and systemically, in allergic disorders. Activated T cells produce IL-2 and increase their expression of HLA-DR, IL-2R, and Integrin VLA, promoting their differentiation into effector T cells and memory T cells. These activated T cells rapidly expand into functional subsets, in response to specific stimuli and under the influence of various cytokines. They recruit and activate other immune cells, initiating a complex series of events resulting in the allergic reaction.

Classically, allergic disorders are explained by a Th1/Th2 imbalance resulting in increases in CD4+ Th2 cells and cytokines (IL-4, IL-5, and IL-13) and decreases in CD4+ Th1 cells and cytokines (IFN-γ and TNF-β).

Th2 Cells and Cytokines

Evidence of a Th2-biased response comes from human studies, showing elevated Th2 cytokines and their receptors on T cells at disease sites in asthma,[227] AR,[228] and AD patients.[31] NC/Nga mouse models of AD are also characterized by increased Th2 cytokines associated with infiltration of CD4+ T cells in the dermis.[229] Further proof comes from animal models of asthma and FA. Murine models of experimental asthma and FA sensitized with various allergens (eg, house dust mite in asthma, ovalbumin in FA) have proven consistently the necessary and causal role for Th2-driven CD4+ T cells in the development of the clinical, histological, and immunological features of the diseases.[230,231] Th1 and Th2 cells originate from the same naïve Th cells after being presented the antigen by DCs, their fate being determined by environmental and genetic influences, their cytokine environment, and presence of costimulatory molecules.[232] Th1 differentiation is mediated by STAT4 activation, IL-12 binding to its receptor, and T-bet activation.[233] T-bet gene polymorphisms are associated with asthma phenotypes in adults.[234]

Interleukins IL-4, IL-5, IL-13, and IL-9 are key *Th2 cytokines* associated with allergic response.[99,227] Their genes are found adjacent to one another at the Th2 loci, on chromosome 5, and are activated by transcription factors such as Signal Transducer And Activator Of Transcription STAT6 and GATA-3, the Th2 lineage determining factor.[235] Other pathways include STAT5, Notch, c-Maf, and mechanistic target of rapamycin mTOR.[236] This activation may be alternatively controlled for each cytokine, depending on the cellular environment and on the state of chromatin opening at the locus.

IL-4 and *IL-13* are potent mediators of type 2 inflammation with both overlapping and distinct functions, binding to the same receptor, known as IL-4R.[237] IL-4 is critical in the induction of the Th2 response, whereas IL-13 plays a more decisive role in the effector phase of the allergic response.[99,227,236] IL-4 is secreted during the initial phase of the allergic response by basophils. This plays a critical role in the differentiation of naïve CD4 T cells into Th2 cells, which further orchestrates the allergic response. In B cells, IL-4 and IL-13 drive Ig class switching to IgG1 and IgE. IL-4 and IL-13 also promote FcεRI expression and proliferation on human mast cells, thus inducing many of the manifestations of allergic diseases including airway inflammation, airway hyperresponsiveness, and mucus cell hypersecretion in asthma.[99,237] These cytokines will also activate a population of macrophages, called alternatively activated macrophages, or M2 cells that will promote extracellular matrix repair.

IL-5 primarily promotes bone marrow survival, activation, and systemic recruitment of eosinophils, which later migrates into tissues under the influence of various cytokines, such as eotaxin.[238]

IL-9 plays an important role within the tissues in a number of ways: (1) it targets professional APCs and limits their ability to induce Th1-type immune responses,[239] (2) it activates and induces proliferation of tissue-resident cells such as mast cells within the mucosa, which themselves produce increasing amounts of IL-9 in an autocrine loop,[240,241] (3) it directly affects airway and gut epithelial cells by inducing the expression of mucins, thereby inducing mucus hypersecretion, (4) it facilitates IL-4–mediated production of IgE and IgG in B cells, (5) it is involved in memory B cell differentiation,[239] (6) it acts as a growth factor for other Th subsets, especially for Th17.

Tissue-Resident Memory T Cells

Following initial activation, a subset of T cells can remain in mucosa and associated lymphoid tissues as memory cells, retaining the ability to respond to specific antigens upon reexposure. Hence, tissue-resident memory T cells have been found within the skin in AD, with a predominance of CD8+ T cells.[242] Mouse models of asthma have suggested that CD4+ lung-resident memory T cells persist long-term, following house dust mite challenge, and can be reactivated upon allergen reexposure.[243] In the gut, following prolonged antigen exposure, resident CD4+ T cells accumulate in lymph nodes and Peyer patches.[244] Resident CD4+ T cells constitute up to 50% of all effector/memory cells.

Th17 and Th22 Cells

Beyond the Th1/Th2 paradigm, research advances in the past 20 years have shown an important role for other CD4+ T cells subsets, such as Th17 cells and Tfh cells in allergic disorders. Th17 cells are effector T cells that differentiate under the influence of TGF-β, IL-6, and IL-23 and produce cytokines such as IL17 A/F, IL-21, IL-22, and TNF-α. These cytokines are classically associated with neutrophil responses and protection from bacteria or fungi.[245,246] The role of Th17 cytokines has been predominantly explored in asthma, as they have been association with asthma onset and severity. Patients with severe asthma display elevated levels in plasma, sputum, and bronchial biopsies,[99,247] and polymorphisms in genes and/or promoters for IL17 A/F and p40 subunit of IL-23 in GWAS studies.[248,249] Th17 cytokines induce the accumulation of neutrophils in the airway, can promote mucus hypersecretion, and the production of cytokines and chemokines by epithelial cells.[99,250] In allergic asthma, a mixed neutrophilic-eosinophilic inflammation is common and could be related to a mixed Th17/Th2 activation.[247] In addition, exposure to many known triggers of acute asthma attacks, including airway bacteria, respiratory viruses, inhaled PM, and ozone, have been associated with an increase in Th17 cytokines.[245,251] In the longer term, IL-17A may contribute to the development of airway remodeling during asthma by enhancing the production of profibrotic cytokines, proangiogenic factors, and collagen.[247,252] Th17 inflammation has been far less-studied in other allergic disorders. Evidence suggests that IL-17 contributes in AR, predominantly in mite-induced perennial rhinitis.[228]

In AD skin lesions, Th17 cytokines are expressed, with a predominance of IL-22. A subset of skin-homing T cells producing high amounts of IL-22, rather than IL-17, called *Th22 cells*, has been described.[253-255] Th22 cells may have a potential role in skin homeostasis and inflammation. Although AD is characterized by a strong Th2 phenotype, these different T cell subsets might be involved in different proportions depending on ethnicities. Hence, several studies have shown that Asian patients with AD display stronger Th17/Th22 activation, whereas African American patients lack Th1 and Th17 activation and display higher serum IgE levels.[256-258]

Follicular Helper T Cells

Tfh, which are CXCR5+ CD4+ cells, interact with B cells in germinal centers of lymph nodes, driving them to proliferate into Ig-secreting plasma cells. Under the influence of different cytokine environments, they can either produce IL-21 or IL-4, which exert opposite effects.[259] Most *in vitro* and *in vivo* studies indicate that IL-21 inhibits IgE and potentiates IgG1 class-switching and production by B cells, in the context of limited IL-4 stimulation.[260] This regulatory role of IL-21 is supported by clinical observations in patients with loss-of-function mutations in *IL-21* and *IL-21R* genes, which display increased levels of IgE.[261,262] In addition, mouse models have shown that IL-6, through STAT3 signaling, exerts early roles in Tfh cell development and may restrain the expression of Th2 genes in Tfh cells, ultimately inhibiting IgE class-switch recombination in B cells.[263,264] Loss-of-function mutation in *IL-6R* gene leads to enhanced IgE levels, also suggesting a role for IL-6 in IgE class-switch recombination.[265]

Under certain conditions, Tfh rather produces IL-4 and exerts a pro-Th2 effect on B cells, promoting class-switching toward IgE production.[259] These Th2-like Tfh cells seem to be more predominant in atopic patients, as shown by increased percentages in patients with AR and asthma, and pediatric patients displaying FA symptoms after liver transplant. Another subset of IL-13 producing Tfh cells has also been recently described in mouse and humans, required for production of high- but not low-affinity IgE by B cells in allergen-induced anaphylaxis.[266]

Other T Cells Subsets Involved in the Allergic Reaction

Other T cells, such as CD8+ cells and unconventional T cells may also be involved in the mechanisms of allergic disorders.

CD8+ Cytotoxic T Cells

CD8+ cytotoxic T cells, which are effector cells, may also play a role in IgE-mediated allergic inflammation, as indicated by studies in animal models and humans. In AD, CD4+, and CD8+–resident T cells populations are both present in skin lesions.[99] CD8+ T cells respond primarily to bacterial stimulation, producing a potent and nonpolarized inflammation within the skin, mainly constituted of IFN-γ (Th1 cytokine), IL-13 (Th2 cytokine), and IL-22 (Th17 cytokine).[267] In asthma, studies of postmodern histopathological samples, characterized by infiltrates of CD8+ predominant T cells, suggest that CD8+ T cells play a role in severe acute asthma.[268] CD8+ T cells may be activated by allergens, cross-presented by MHC I complex on airway DCs, and/or by alarmins IL-33 and TSLP.[269] Finally, peanut allergy patients display an increase in allergen-specific CD8+ T cells in the systemic compartment that are specifically activated by peanut peptide in a sequence-dependent manner.[270]

Unconventional T Cells

Unconventional T cells also play a role in allergic disorders. iNKT cells and MAIT share similarities. As opposed to conventional T cells, they both express TCR of limited diversity, constituted of an invariant TCRα chain (Vα24-Jα18 in iNKT, Vα7.2-Jα33 in MAIT), together with a limited set of TCRβ chains.[189] *iNKT* cells are mostly tissue-resident cells. Their primary function is to protect the host from infections. They recognize self and nonself glycosphingolipids and microbial cell wall α-galactosylceramide (α-GalCer) antigens.[271] Three major iNKT cell subsets have been described: iNKT1 (IFN-γ and IL-4 producers), iNKT2 (IL-4 and IL-13 producers), and iNKT17 (IL-17 and IL-22 producers). Rodent models of asthma have shown that iNKT cells are necessary, but not sufficient, to induce Th2-mediated responses and cardinal features of asthma, such as airway responsiveness.[99,189] They are thought to be activated once inflammation is established following a protein-antigen–induced allergic response and serve as adjuvants of the Th2 response. Some studies also suggest that iNKT may be directly activated by some allergen adjuvants that bind lipid ligands, such as those contained in seed storage proteins, nonspecific lipid transfer proteins (Bet v 1-like) from pollens and fruits, house dust mite, and cockroaches' extracts.[272] MAITs are abundant in the gut, lung, and liver and account for up to 10% of T cells compartment in peripheral blood.[189] They recognize microbial-derived vitamin B2 (riboflavin) metabolites, and their expansion in response to commensal flora antigens explains their abundance in the gut. MAIT cells produce IFN-γ, TNF-α, and IL-17 and cytotoxicity through perforin granzyme mechanism. They are able to produce low-to-moderate levels of IL-4 and IL-13 upon stimulation. Asthmatic patients seem to present with a deficiency of MAIT cells in the blood, sputum, and bronchial biopsy samples. It has been shown that an increase in MAIT cell frequency at 1 year of age is associated with a protective role against the development of asthma.[273,274] Because they are associated with immune response against commensal flora, they may also play a role in FA via their impact on gut homeostasis. Finally, *γδ T Cells* are a small subset of T cells that express heterodimeric TCRs, composed of γ and δ chains, as opposed to conventional T cells that express αβ TCRs, and recognize MHC-like molecules and growth factor receptors, triggering both innate and adaptive immune response.[275] Increasing evidence suggest a role for γδ

T cells in allergic disorders. First, human γδ2 T cells are a major source of IL-9, a cytokine mediating Th2-type inflammation.[276] Second, in the gut, they are localized within the epithelium and *lamina propria* and are therefore thought to contribute to immune surveillance.[192] Murine models have shown contradictory results as to their role in FAs. γδ T cell activation have been shown to lead to a breakdown of oral tolerance, while blockade of the γδ TCR has been shown to also result in elevated food-allergic responses upon allergenic sensitization.[277,278] It is now understood that a number of γδ T cell subsets exist, which may likely play different roles in allergic reaction and tolerance.[192] Third, they may specifically contribute to the regulation of the IgE response, as suggested by *in vitro* studies. Human γδ T cells significantly increase IgE synthesis when cultured with B cells in the presence of IL-4.[279] More interestingly, IgE-enhancing human γδ T cells can be antigen-specific, as shown by specific response to phospholipid extracts from tree pollen grain in cell cultures from cypress-sensitive but not healthy subjects.[280]

Role of CD4+ Regulatory T Cells in Immune Tolerance

Tregs represent 5% to 10% of the total $CD4^+$ T cell pool and are associated with tolerance, as shown by their increase after immunotherapy.[281] Two major categories of Tregs are described: natural CD4+ CD25+ FoxP3+ Tregs and induced Tregs (iTregs), which are induced after antigen stimulation.[282]

TCR repertoire diversity is important for T_{reg} expansion and homeostasis.[283] Recent technological advances have enabled our ability to measure TCR repertoire diversity, and these studies have indicated extensive T-cell epitope sharing between the human proteome, gastrointestinal microbiome, and pathogenic bacteria.[192,284] In addition, the TCR repertoires of nTregs and iTregs minimally overlap. While the TCR repertoire of iTregs is directed toward commensal pathogens and environmental allergens, nTregs mainly express anti–self-biased TCR repertoire.[282,285] These findings are consistent with the hygiene hypothesis, linking the lack of allergen-specific tolerance in allergic individuals to reduced infections, alterations in commensal microflora in the intestinal tract, and/or reduced TLR signaling.[99]

The initial induction of iTregs mainly occurs in the lymph nodes, where they interact with DCs, become primed and activated.[195] They exert their suppressive activity in the lymph nodes, and most importantly, within the tissues, where they subsequently migrate, under the influence of tissue-homing cytokines. The generation of iTregs is facilitated by mucosal CD103+ CD11C+ DCs in the gastrointestinal tract and alveolar macrophages in the lungs in the presence of IL-2 and TGF-β.[282] Tregs are also implicated in acquired tolerance as Tregs increase in patients who naturally outgrow allergy.[286] Two main iTreg subtypes have been associated with tolerance: (1) IL-10–secreting FoxP3- Tregs type 1 (Tr1), and (2) TGF-β–secreting induced T helper 3 cells (Th3).[90,281] FoxP3+ natural Tregs secrete moderate amounts of both IL-10 and TGF-β. Tregs suppressive activities involve two main mechanisms.[282] Firstly, they secrete anti-inflammatory cytokines, such as IL-10, and TGF-β, which suppress Th2 cell effector function. They can also induce cytolysis of target cells via granzyme- and perforin-dependent mechanisms. Secondly, they mediate suppressive functions through direct cell-cell contacts, notably with other T cells, thus influencing their activation. They can also influence professional APCs and suppress their ability to present the antigen, via CTLA-4—CD80/86 and lymphocyte-activation gene LAG-3—MHCII interactions. Tregs can also directly inhibit mast cell and basophil degranulation, independent of TGF-β and IL-10.

In atopic individuals, the role of Tregs is controversial. Patients with Treg/*FOXP3* mutations, such as those with IPEX syndrome (Immunodysregulation, Polyendocrinopathy, Enteropathy, X-linked), display multiple autoimmune problems, frank atopic disease, and elevated IgE, suggesting that a quantitative and/or qualitative defect of Tregs is implicated in the pathogenesis of allergic disorders. A reduced frequency of Tregs has indeed been associated with allergic asthma in some studies,[287] and atopic individuals may lack antigen-specific T regs in some conditions,[288] but most evidence show that total Tregs frequencies are overall similar between allergic patients and healthy controls.[289] Most importantly, atopic subjects display an impairment in suppressing effector Th2 cytokines production.[291] Several studies have suggested that this impairment could be partly explained by differences in TCR repertoires in atopic conditions. First, in primary human immunodeficiencies in which limited numbers of T cells are circulating, as seen in treated severe combined immunodeficiency (SCID) children, eosinophilia, elevated levels of serum IgE, and allergic disorders are frequently observed.[290] Second, limited repertoire complexity of Tregs is associated with Th2 phenotype, as demonstrated in SCID mice reconstituted with limited, but not large, repertoire T cells.[291] Third, analysis of CD4+ repertoires in primary atopic disorders has demonstrated an imbalance in allergen-specific Tregs and effectors subsets, suggesting a failure of antigen-specific Tregs to regulate effector cells specific for the same antigens.[292] In atopic subjects, Tregs may display an abnormal phenotype. A study in a murine FA model has suggested that disease-susceptible mice displayed STAT6-dependant impaired allergen-specific T_{reg} generation and function, with a skewed Th2 cell–like phenotype marked by increased expression of IL-4 and transcription factor GATA-3.[293] In addition, aberrations of the T_{reg} cell compartment can also be attributed to a decrease in or failure of their suppressive functions, as demonstrated by *in vitro* study in which atopic Tregs can be distinguished from non-atopic Tregs by limited suppressive functions toward allergen-specific Th2 cytokines–producing CD4+ T cells.[293,294]

Role of B Cells in the Allergic Response

IgE Class-Switch Recombination

Skewing of naïve T cells to a Th2 response is followed by cognate interactions with specific B cells inducing IgE class-switch recombination (*see Chapter 26*). Briefly, the constant-region portion of the Ig heavy chain is modified but the variable region (VDJ cassette) providing antigen specificity does not.[99,295] Initially, all B cells produce IgM antibodies. The genomic DNA of the constant region, at the IgH locus, is spliced, deleted, and rejoined to move the VDJ elements from their location proximal to a position downstream, next to the exons encoding the heavy chains of other isotypes. IgE heavy gene-encoding region is located downstream to IgG3, IgG1, IgA1, IgG2, and IgG4.[295] Thus, IgE-switching can occur directly from IgM or sequentially through one or more of the prior-switched intermediates. This leads to the transcription of a functional IgE heavy chain, called ε.

Regulation of IgE Synthesis

A complex series of secreted signals and cell-surface interactions leads to IgE class switch recombination.[295] The first step includes allergen presentation by APCs to T cells as peptide fragments bound to MHC II molecules and subsequent recognition by TCR on Th2 cells. Two signals are then required to induce isotype switching to IgE. The first signal is mediated by Th2 cytokines IL-4 and IL-13, which initiate the transcription of the IgE ε heavy chain, at the Cε gene locus, via stimulation of the transcription factor STAT6. The second signal is provided by cell-to-cell contact between T and B cells. Specifically, the interaction between CD40 on the surface of B cells with CD40L on the T-cell surface is critical to drive the IgE switch to completion, mediated by proteins belonging to the family of TNF receptor–associated factors (TRAFs) and activation-induced cytidine deaminase (AID). CD40L is encoded on the X chromosome, and individuals with the X-linked hyper-IgM syndrome are deficient in CD40L, leading to the absence of production of IgG, IgA, and IgE by B cells.[296]

In contrast to IL-4 and IL-13, a number of cytokines can negatively regulate IgE synthesis.[99] TGF-β, IFN-α, and IFN-γ can negatively regulate IgE synthesis driven by Th2 cytokines. In addition, IL-10 and IL-12 have been reported to inhibit IgE while enhancing IgG4. As stated earlier, IL-6 may also exert a role in inhibiting IgE class-switch recombination.[263,264] IL-21, mainly produced by Tfh cells, also inhibits IgE and promotes IgG1 class-switch recombination, in the context of limited IL-4 stimulation.[260] Moreover, another mechanism of regulation of IgE is through the competition with IL-4–driven IgG4-blocking antibodies.[297] These Ig compete monovalently with cell-bound IgE for allergen binding if they share the same antigen specificity, without recruiting IgG-dependent effector mechanisms.

Some allergen-specific plasma cells survive in the bone marrow with long-term immunological memory.[295] Some of them might be mucosal IgE plasmablasts, which migrate from local mucosal tissue to the bone marrow where they received survival signals to become memory cells. The constant supply of IgE antibodies from bone marrow plasma cells allows a rapid "recall response" upon reexposure to antigen. Local antigen provocation also stimulates de novo synthesis of IgE by mucosal B cells, preferentially through sequential switching from IgG. Further stimulation may also lead to new gene rearrangements that will allow IgE antibodies to gain high affinity toward the allergen.[298,299]

Biological Effects of IgE

IgE antibodies are responsible for the "early phase" of the allergic reaction. The biological role of IgE is related to its ability to influence the functions of several immune and structural cells via its two specific receptors, *high-affinity (FcεRI)* and *low-affinity CD23 (or FcεRII)* receptors.[300] They are distinguished by their structure and their relative affinities for IgE. Importantly, as opposed to IgG isotypes, IgE antibodies are primarily localized in tissues, where they are tightly bound to their effector cells with a prolonged half-life up to 2 to 3 months, as opposed to a shorter half-life in serum (2-3 days), and thus the plasma levels of IgE are in the low-microgram range, compared to IgG which are 10 mg/mL.[301] This long-term maintenance of IgE-mediated memory in tissue-resident mast cells may explain the delayed efficacy of Omalizumab, in the absence of detectable serum levels of IgE.[301,302]

Role of Effector Cells

Mast cells and basophils both constitutively express high levels of *high-affinity FcεRI* receptors that are important in triggering the release of the mediators of allergy (*see Chapter 16*). For many years, mast cells in tissues were believed to be derived from basophils in the blood. Both are derived from CD34+ hematopoietic stem cell precursors in bone marrow.[303] However, we now know that basophils and mast cells undergo distinct differentiation pathways, as well as location and functions. Mast cells mainly control IgE-dependent acute allergic responses, while basophils may induce IgE-mediated chronic responses and IgE-independent allergic inflammation.

Mast Cells

Mast cells mature in peripheral tissues. The interaction between the c-Kit surface receptor on mast cells with stem cell factor plays a major role in mast cell differentiation and survival.[303] Evidence in mice suggest that mutations in *c-Kit* result in a loss of mast cells. Mast cells are distributed in connective tissues, often near blood or lymphatic vessels and in proximity to nerves, and thus can be exposed to antigens which penetrate natural barriers, including the skin and mucosal tissues. The presence of the high-affinity FcεRI receptor is central to mast cell function. The expression of FcεRI on the surface of mast cells occurs early in their maturation and is regulated by several factors postmaturation, including IgE itself and Th2 cytokines. Mast cell activation is downregulated by inhibitory receptors, including FcγRIIB and CTLA-4.[99] FcεRI signaling occurs upon IgE binding to the receptor. This requires two sequential events: (1) binding of IgE antibody to the FcεRI and (2) cross-linking of IgE antibody by bivalent or multivalent antigen. FcεRI then transduces intracellular signals through its association with cytoplasmic tyrosine kinases that result in degranulation of three classes of biologically active products.[182] The first, immediately released, are those stored in the cytoplasmic granules (histamine, heparin, chondroitin sulfates, serine proteases such as tryptase, PAF). In addition, the activation of mast cells leads to the synthesis and secretion of derivatives of arachidonic acid, known as lipid-derived mediators (prostaglandin D2 PGD2, leukotrienes).[191] Combined, these substances lead to the inflammatory consequences of an allergic reaction, both immediate (acute-phase) within minutes to late-phase reactions in hours. Mast cell granule release is primarily responsible for the acute symptoms and may have local to systemic effects, depending on the route of entry and dose of the "offending" allergen. Prostaglandin D2 binds to its receptor CRTh2, a G protein–coupled receptor expressed by eosinophils, basophils, mast cells, Th2 cells, and ILC2.[304] This attracts and activates these cells and increase their release of Th2 cytokines, enhancing allergic inflammation, despite treatment with inhaled corticosteroids. Mast cells also newly synthesize and release cytokines and chemokines that participate in late-phase reactions, leading to chronic inflammation and remodeling.[182]

Basophils

Basophils are potent IL-4 secreting cells. As such, they play a role in the sensitization phase of the allergic response and drive Th2 cell differentiation and B-cell differentiation to IgE-producing plasma cells.[305] Under the influence of various activation factors, such as IL-3, IL-33, and various proteases, they can degranulate in response to IgE-mediated cross-linking, releasing mediators such as histamine, serotonin, lipid mediators, and various cytokines.[303] Activation by TSLP also induces the release of various chemokines: CCL3, CCL4, CCL12, CXCL2, in an IgE-independent manner. Recent murine and human studies suggest that developmental and functional heterogeneity exists within basophil populations and that basophils can be divided into two categories: IL-3–elicited basophils, which operate in an IgE-dependent manner and may elicit Th2 responses; and TSLP-elicited basophils, which operate in an IgE-independent manner.[305] In EoE, patients frequently display gain-of-function polymorphisms in the gene encoding TSLP and elevated peripheral basophilia.[306]

Eosinophils

Eosinophils are granulocytes that are characterized by their bilobed nucleus and their distinctive cytoplasmic granules that are hematoxylin- and eosin-stained (*see Chapter 17*). Eosinophils accumulate in the bone marrow, tissues, and blood of allergic individuals, under the influence of Th2 cytokines.[99] IL-5 acts as the primary determinant of eosinophil differentiation in the bone marrow, together with GM-CSF and IL-3, inducting pathways such as GATA-1. Th2 cytokines orchestrate their recruitment to tissues by induction of specific adhesion molecules on eosinophils (eg, intercellular adhesion molecule-1, E-selectin, L-selectin, and very late antigen-4/VCAM-1) and upregulation of eosinophil chemoattractants (eg, C-C chemokines such as CCL5, eotaxin, CCL2, CCL13, CCL7, and CCL3). Eotaxin, which binds specifically to CCR3, which is highly expressed on eosinophils, is considered to play a pivotal role in eosinophilic recruitment and subsequent inflammation in the tissues. In the tissues, Th2 cytokines, especially IL-5, are potent inducers of eosinophil granule protein release.[307] This process can also be regulated by other cytokines (eg, GM-CSF), chemokines (eg, eotaxin), Ig (eg, IgE [via its binding on the S-type lectin galectin-3, FcεRII/CD23, and FcεRI], IgA and IgG), lipid mediators (platelet-activating factor, PAF), complement fragments (C5a and C3a), and neuropeptide substance P.[99] Eosinophils granules store four basic, low–molecular-weight proteins: major basic protein (MBP), eosinophil-derived neurotoxin (EDN), eosinophil peroxidase (EPO), and eosinophil cationic protein (ECP).[308] Eosinophils granules also contain varying numbers of lipid bodies that contribute to the formation of eicosanoid mediators (lipoxygenase- and cyclooxygenase-derived eicosanoids) upon stimulation. They are also a major source of cytokines, stored in cytoplasmic-specific granules, including IL-4, IL-5, IL-6, IL-8, IL-10, IL-12, IFN-γ, and eotaxin. All these mediators contribute to the late-phase of the allergic response, chronic inflammation, tissue remodeling, and fibrosis by the release of mediators and immunoregulatory/proinflammatory cytokines. Of note, proteins contained in eosinophil granule (including MBP and EPO) can themselves later stimulate eosinophils degranulation in an autocrine mechanism.[99]

Other Cells Involved in the Allergic Response

Airway Smooth Muscles

Within the lung, IgE also binds to FcεRI on airway smooth muscles, triggering smooth muscle contraction and activating production of IL-4, IL-5, IL-13, TNF-α, TSLP, and chemokines (CCL5, CCL11, CXCL8, CXCL10).[309] The high-affinity receptor FcεRI is also expressed at lower levels on peripheral blood DCs, monocytes, and Langerhans cells.

Cells Expressing the Low-Affinity Receptor FcεRII

The low-affinity receptor FcεRII (CD23) is expressed on B cells, T cells, Langerhans cells, monocytes, macrophages, platelets, airway smooth muscle, and eosinophils.[99] It is a calcium-dependent C-type lectin receptor that exists in two forms, CD23a and CD23b. CD23 expression is upregulated by IgE and IL-4. On the other hand, IFN-γ counteracts the inducing effect of IL-4 on CD23 expression. CD23 is thought to mediate a number of effects including regulation of IgE synthesis, antigen presentation, proliferation and differentiation of B cells, and activation of monocytes. CD23 on B cells serves mainly as a negative regulator of IgE synthesis.[310] Dysregulation of the CD23 pathway in atopic patients might be part of their propensity to develop IgE antibodies and may cause an enhancement of the inflammatory reaction.

The Neuroimmune Axis

Our understanding of the role of the neuroimmune axis in allergic disease is at its infancy. The skin, respiratory tract, and gastrointestinal tract are exposed to allergens and also closely connected to both the peripheral sensory and autonomic nervous systems.[182] Hence, there is intercommunication and cross-talk with neurons and mast cells, eosinophils, DCs, ILC2s, and Th2 cells. Because the skin is highly innervated, mediators like histamine, as well as cytokines and neurotrophins, can activate sensory neurons leading to itching in the skin.[311] In particular, mast cells release algogenic and pruritogenic mediators that directly activate specific nociceptors and sensory nerve fibers.[312] Furthermore, these mediators can lead to sneezing and bronchial constriction in the respiratory tract and other effects in the gastrointestinal tract. Conversely, peripheral neurons release neuropeptides and vasoactive neurotransmitters to modulate the function of mast cells and a variety of cells in the immune system in a feedback mechanism, promoting a vicious cycle. Studying these close interactions could lead to understanding how the allergic response can be aggravated by a range of stimuli, including pruritogens, skin dryness, heat, sweat (in the skin), and psychosocial stress.[48,312]

SYSTEMIC BIOLOGY APPROACHES TO STUDY ALLERGIC DISORDERS

Allergic disorders are manifested via complex dynamic interactions between environmental stimuli and the host immune system. Large variations in endotypes and phenotypes exist, and allergic responses vary between individuals and also within an individual toward different allergens or at different ages of life. Traditionally, physicians and immunologists have studied allergy in a reductionist fashion using hypothesis-driven study designs. However, with the advance of modern high-throughput omics technologies and data analysis approaches. Examples of omic technologies include genomics, epigenomics, transcriptomics, proteomics, metabolomics, microbiomics, and exposomics. The use of the term "omics" suggests a comprehensive high-throughput and systematic investigation of biological parameters.[313] In recent years, system biology approaches have increasingly been used to study the mechanisms of allergic disorders as well as to facilitate the discovery of their diagnosis/prognosis biomarkers.[314]

With the completion of the *Human Genome Project* and the vast genetic information that is made available, there is now growing interest in linking diseases phenotypes with endotypes. Large-scale GWAS compare genetic variants, typically SNPs between healthy and diseases states in large populations to identify genes associated with disease states.[78] As stated earlier (*see Risk factors for allergic disorders*), such studies have identified common risk variants as well as rare mutations associated with the risk of AD,[315] AR,[85] asthma,[84] and FA.[316] Certain gene loci related to immune modulation have been identified. For example, SNPs mapping to the *HLA-DR* and *HLA-DQ* regions, which are receptors on APCs that present antigens to T cells, have been identified in peanut allergy,[80] as well as in asthma.[84] In addition, gene variants related to cutaneous barrier function have been associated with both AD and FA, which support the atopic march and the dual-allergen exposure hypothesis.[117,315] Genetic backgrounds have also been associated with levels of total IgE as well as allergen-specific IgE.[80] Understanding immune repertoire, which refers to gene diversity of B cells (ie, immunoglobulins) and TCRs are important in allergy. As discussed previously in this book, T cells and B cells undergo somatic genomic rearrangements (ie, V[D]J recombination) in regions of Ig and TCR genes. The diversity of both Ig and TCR is of particular interest in allergic disorders because it dictates how an allergen can be recognized by the adaptive immune system. Studies of IgE-producing B cells have resulted in a better understanding of allergenic specific immune response and the mechanism of IgE class switch.[298,299] This area is currently under intense research and could enable more complete characterization of the molecular mechanisms of allergen recognition, leading to the identification of novel biomarkers.[316]

Epigenetic modifications are biochemical modifications that occur on DNA or histones and influence the function of the genome without changing the nucleotide sequence itself.[145] Because they affect the expression profiles of the genes, important roles in the development of allergic disorders have been demonstrated (*see Risk factors for allergic disorders*).[144,145] The results of these studies can provide new hypotheses to which genes/pathways contribute to the early development of allergic disorders.

RNA transcriptomics provides a profile of the total RNA molecules contained in one sample, representing the current state transciptomic profile of the cell. The transcriptome profiling of both circulating leukocytes in the blood and tissue cells specific to allergic disorders has facilitated our understanding of the underlying mechanisms that trigger the allergic response. For example, transcriptome profiling of the lesional and nonlesional skin from AD patients has found increased expression of certain genes related to TREM-1 receptor signaling pathway and IL-36 cytokine.[317] An analysis of esophageal biopsy specimens revealed a set of genes differentially expressed in EoE patients, including several long noncoding RNAs.[318]

Proteomics aims to measure all proteins, which conduct most functions in biological systems. CyTOF is a novel mass cytometry technique, which combines flow cytometry and mass spectrometry and allows for more than 40 protein targets to be detected in a single sampling of cells by labeling with metal isotypes rather than fluorochromes. This method provides lesser signal overlap and offers a platform for a more comprehensive investigation of immune cell phenotypes compared with the traditional flow cytometry method.[319,320] In the field of allergy, proteomic studies are also of particular interest to investigate the allergens, especially in FA. Allergenomics refers to the comprehensive identification, characterization, and quantification of the allergens.[316] As stated earlier, several databases have been established to store the information, nomenclature, and protein structures of allergens (www.allergen.org).[160] Such information is particularly useful to identify allergens contained in the complex products from the modern food industry and to discover cross-reactivity among allergens.

Single-cell omics are a relatively new and quickly evolving field.[321] The technique relies on the measure of multiple molecules in the same individual cell, thus of particular interest in immunology and allergy research, where immune cells are highly heterogeneric and relatively autonomous.[313] More recently, the development and maturation of single-cell RNA-sequencing technologies enable us to access gene expression profiles at a single-cell level. Several studies have since reported the gene expression profiles of important cellular players in the allergic reaction, including Th2 T cells [322] and IgE-producing B cells.[298]

Microbiomics is the measure of DNA, RNA, and proteins of the microbes associated with the host. We have discussed the relationship between microbes and allergy in the previous section (*see Interactions between microbes and atopy*). The major omic technology used to investigate the microbiota is 16s rRNA sequencing.[314] Another approach is to unbiasedly sequence the total DNA present in one ecosystem, and secondly, map genes related to microbes, including viruses and fungi. These methods allow us to investigate the microbiota hosted in the gut, skin, and lung environment and deepen our understanding of their interactions with host immune system.

Metabolomics is the systemic measure of metabolites (generally small molecules) in a living system, including sugar, lipids, amino acids, fatty acids, etc. The identities and quantities of metabolites are measured by nuclear magnetic resonance or liquid chromatography—mass spectrometry. Metabolites from plasma, serum, urine, and exhaled breath condensate have been identified in the context of asthma[314] and/or AD,[315] reflecting various biological processes. Identified metabolites can also come from commensal microbiota and modulate the host immune system. Hence, some microbial metabolites and their corresponding receptor in host immune cells have been identified, although limited studies have been conducted so far.[323]

Allergic disorders are noncommunicable, multifactorial diseases. Their complexity and heterogeneity make it unlikely that a single biomarker or single omic scale is sufficient to serve as unique biomarker. Integrations of data at multiple scales are necessary to build models to facilitate our understanding of the dynamic interactive networks underlying allergic disorders and their different phenotypes.[316] Overall, system biology approaches and omic technologies have great potentials to apply in allergy research to facilitate our understanding of disease mechanism, to identify biomarkers for diagnosis and prognosis, and to assist precision medicine and better treatment.

MODULATING IMMUNOLOGICAL MECHANISMS IN ALLERGIC DISORDERS

Glucocorticosteroids are commonly used as therapy for the treatment of allergic disorders, such as asthma, AR, and AD. They efficiently treat inflammation, targeting the action of phospholipid A2, a mediator released by effector cells in the acute phase of the allergic reaction. They may even induce some tolerance-inducing effects by inducing IL-10 Tregs in certain conditions.[324] However, they do not cure the disease. Furthermore, side effects are common when administered systemically and in the long-term. Until recently, the only effective therapy for FA was strict avoidance of the allergenic food. New approaches with disease-modifying properties that exert significant and durable effect on the mechanisms of allergy have been developed in the past 20 years, leading to a revolution in the field.[325]

Allergen Immunotherapy

Allergen immunotherapy (AIT) has been shown to induce desensitization, which refers to the ability of a subject to be exposed to the allergen without clinical reactivity to it but requiring continued exposure to the allergen to maintain this state of nonreactivity. In some patients, "sustained tolerance," which can be defined as experiencing years of unresponsiveness to the allergen in the absence of intentional persistent exposure, is achieved, but this is less common and a current limitation of immunotherapy.[90,326]

AIT was introduced in clinical practice in the early 20th century for the treatment of AR and Hymenoptera allergy.[327] In the 1980s, subcutaneous immunotherapy (SCIT) was widely prescribed in AR, but patients experienced significant side effects. In the past 15 years, the availability of safer routes of AIT allowed the development of AIT in all allergic disorders, with the use of standardized extracts in AR, asthma, and FA with guidelines regarding their indications, safety, cost, and duration of treatment.[325]

Currently, there are three main routes of delivery: sublingual immunotherapy (SLIT), oral immunotherapy (OIT), and epicutaneous immunotherapy (EPIT), through a patch, to promote the skin's role as a major immunological organ.[192] SLIT, OIT, and EPIT differ in allergen doses, efficacy, and safety. Doses in EPIT and SLIT are usually in the microgram range, whereas doses in OIT range from milligrams to grams.

SLIT is the preferred choice of immunotherapy with aeroallergens, mainly house dust mite or pollen extracts.[325] They have proven to be efficient for AR, but the medical community remains cautious about their efficacy and safety in asthma and in the pediatric population.

In FA, OIT, EPIT, and SLIT have been commonly evaluated in clinical trials, with various allergens, doses, and updosing patterns.[328] SCIT has been shown to enable desensitization but has been largely abandoned due to the high rates of adverse reactions.[326] OIT has higher rates of success than SLIT and EPIT, achieving desensitization rates on average of 80% to 85%.[90] It should be performed under close medical supervision so that adverse reactions can be monitored and treated promptly[192] (Fig. 47.7). Usually, oral food challenges are necessary to determine the eliciting dose before starting therapy and to evaluate the efficacy of the treatment at the end of the protocol and in the longer term. After screening for eligibility, patients undergo an updosing phase with increasing threshold of daily doses, leading to a maintenance phase, with the same daily dose.[192] Subjects may experience reactions with daily dosing during updosing or maintenance phases. The majority of symptoms are mild gastrointestinal complaints, but some cases of EoE have been described.[329] With successful OIT, patients can increase their allergenic threshold dose so that previously allergenic foods can be consumed at normal dietary amounts without allergic reaction and/or with a minor reaction that induces non–life-threatening symptoms. Some protocols may also involve a phase of withdrawal to assess long-term sustained unresponsiveness. This intense

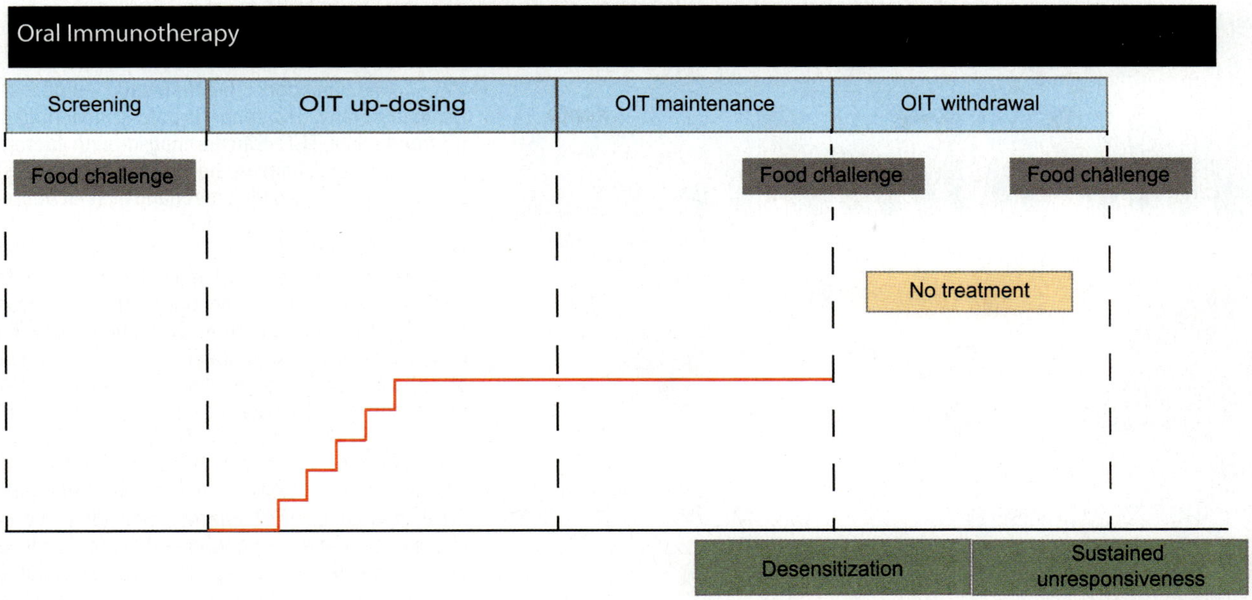

FIG. 47.7. Progression of oral immunotherapy. OIT: Oral immunotherapy. Oral food challenges assist with determination of the eliciting dose before initiation of immunotherapy and to evaluate the efficacy of the treatment during and at the end of the protocol. After screening for eligibility, patients undergo an updosing phase with increasing threshold of daily doses, leading to a maintenance phase where a predetermined daily dose is consumed for a specific amount of time, at the end of which efficacy (desensitization) is determined. In some studies, this is followed by dose discontinuation for a specific time period, after which efficacy (sustained unresponsiveness) is evaluated. OIT should be performed under close medical supervision so that adverse reactions can be monitored and treated promptly.

research has led to the first *Food and Drug Administration (FDA)*-approved OIT treatment, Palforzia, a standardized and characterized peanut powder formulation, for peanut allergy in children aged 4 to 17 years old in 2020.[130] Many questions remain regarding optimization of OIT, including the choice and/or association of allergens in multiple FAs, the optimal doses and time period of maintenance, and the sustainability of the desensitization process.[90] Some novel clinical trials also combine OIT and adjunctive use of biologics, such as anti-IgE (*see Biologics*), with the aim of ensuring safer and faster desensitization.

In parallel with the development of AIT in medical practice, significant research has been performed to understand the *immunological mechanisms leading to desensitization with immunotherapy* (Fig. 47.8). First of all, AIT may induce protection early in its course, in the first 3 to 4 months, due to decreased levels of allergen-specific IgE on the surface of mast cells and basophils, leading to a decreased activation of these effector cells.[90] AIT may also deplete effector cells of some of their mediators, as observed in patients undergoing rush desensitization for Hymenoptera venom allergy.[330] As described earlier (*see Natural tolerance*), innate immune cells are also involved in early tolerogenic responses induced by AIT. Mouse models have shown the induction of APCs, in particular, tolerogenic skin-derived CD103+ DCs after epicutaneous allergen stimulation. AIT may also induce a decrease in ILC2s, which promote pro-Th2 responses in the mucosa. A study has shown that the increase in peripheral ILC2s in patients with grass pollen allergy during the pollen season was inhibited by SCIT.[331]

T cells play a central role in the process of desensitization during OIT. Four main mechanisms are potentially involved, leading to increase of the ratio T_{reg}/Th2.[192] Firstly, there is a deviation of Th2-type responses to a Th1-type response (immune deviation). Secondly, some modifications in Th2 cells will lead to their apoptosis (Th2 deletion). It has also been shown that successful peanut OIT induces allergen-specific CD4+ T cells to expand and shift toward an "anergic" Th2 T-cell phenotype (Th2 anergy) and that the maintenance of this subset is associated with persistent unresponsiveness after withdrawal.[332] Finally, OIT induces the differentiation of T_{reg} cells, secreting IL-10 and TGF-β, leading to a downregulation of Th2-cells (immune regulation). Different subtypes of Tregs may be involved in these processes, possibly with overlapping functions. Interestingly, methylation processes might also be involved, in particular in mechanisms leading to desensitization versus sustained tolerance. Hence, demethylation of CpG sites in the Foxp3 gene has been shown in T_{reg} cells following OIT in peanut allergy patients who achieved sustained unresponsiveness, whereas there was increase methylation in those who lost tolerance after withdrawal.[333] Tregs also mediate the shift in antibody-producing B cells from the IgE to the IgG4 isotype.[334]

AIT also induces changes in B cells and immunoglobulin subsets. During peanut OIT, patients exhibit increased levels of peanut-specific IgG4 and decreased levels of peanut-specific IgE (increased IgG4/IgE ratio).[335] IgG4 have an inhibitory activity on effector cells, mediated by two main mechanisms: (1) IgG4 acts as a "blocking antibody" and binds the allergen before it encounters IgE attached to the surface of basophils

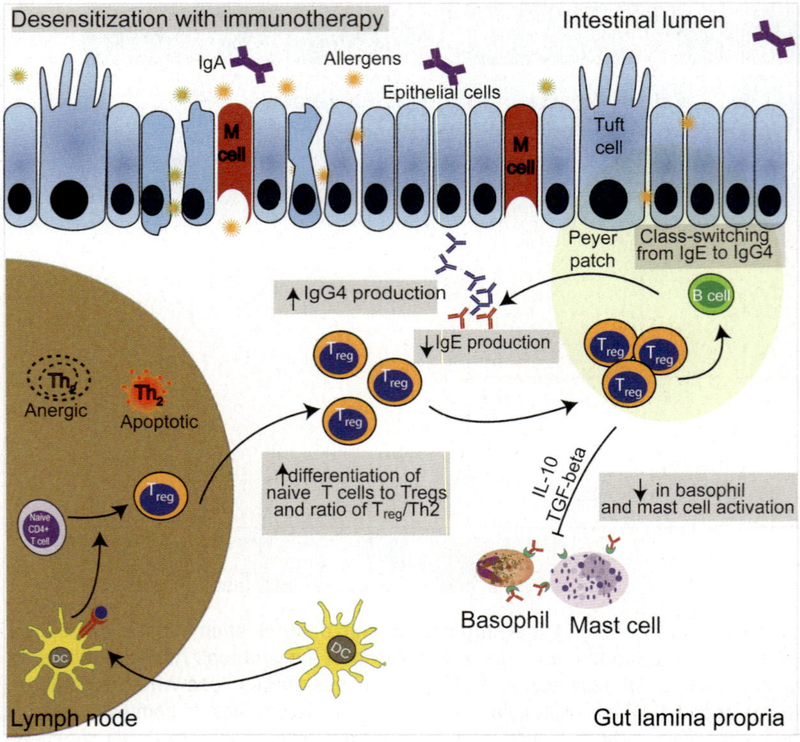

FIG. 47.8. Immune mechanisms induced by Immunotherapy. DC: dendritic cell; Ig: Immunoglobulin; IL: Interleukin; TGF: transforming growth factor; T_{reg}: regulatory T cell. Immunotherapy induces desensitization through the following mechanism: (1) It decreases levels of allergen-specific IgE on the surface of mast cells and basophils, leading to a decreased activation of these effector cells. AIT may also deplete effector cells of some of their mediators, (2) it increases the ratio T_{reg}/Th2 through either promoting naïve T cells differentiation in Tregs, leading to a downregulation of Th2-cells (immune regulation), deviating Th2-type responses to a Th1-type response (immune deviation), or inducing modifications in Th2 cells leading to their apoptosis (Th2 deletion), (4) expanding and shifting allergen-specific CD4+ T cells to an "anergic" Th2 T-cell phenotype (Th2 anergy), and (3) it increases levels of peanut-specific IgG4 and decreases levels of peanut-specific IgE through Ig class-switching, leading to increased IgG4/IgE ratio. IgG4 have an inhibitory activity on effector cells because they act as "blocking antibodies" and bind the allergen before it encounters IgE attached to the surface of basophils or mast cells; and because IgG4 bind to FcγRIIb receptors blocking binding of IgE to these receptors.

or mast cells and (2) IgG4 binds to FcγRIIb receptors blocking binding of IgE to these receptors.[90] Peanut OIT induces an increase of peanut-specific B cells in the blood, mainly of the memory phenotype, with highly mutated IgG4–expressing peanut-specific clones, suggesting that ongoing somatic mutation of IgG4–expressing B cells may increase IgG4 affinity for the allergen.[336] Following venom immunotherapy, a population of IgG4-secreting regulatory B cells producing IL-10 has been isolated.[337] An increase of allergen specific-IgA during the course of egg OIT has also been shown, suggesting a role for IgA in the acquisition of tolerance.[338]

It remains to be determined whether AIT strategies should be directed toward targeting innate immune cells, eliminating effector Th2 cells, or toward increasing antigen-specific T_{reg} responses. Other cells, in particular T cells subsets, may also be involved, such as iNKT, as demonstrated in a study showing increase of iNKT cells with a switch toward a Th1 cytokine profile following cow's milk OIT.[339]

Biologics

All biologics currently approved for use, and most of the biologics under development for allergic diseases are humanized monoclonal antibodies, targeting the Th-2 inflammatory pathway, including anti-IgE, and the upstream mediators, such as anti–IL-5, anti–IL 4, and anti–IL 13.[340] Although most studies have been conducted in severe asthma and AD, new evidence emerge for their use in other allergic disorders, in particular in FA and EoE. Table 47.2 summarizes the main biologics that are *FDA*-approved or in advanced-phase clinical trials.

Omalizumab is a subcutaneous humanized IgG1 monoclonal antibody that binds to the constant region (Fc) of IgE antibodies, thus blocking IgE binding to FcεRI and preventing the activation and degranulation of mast cells and basophils.[341] In 2003, it became the first FDA-approved biologic for moderate-to-severe persistent allergic asthma whose asthma symptoms are not controlled with inhaled corticosteroids, for adults and adolescents ≥12 years, and secondly, for children ≥6 years in 2016. Omalizumab is highly effective to reduce asthma exacerbations and improve asthma control with a good safety profile, as demonstrated in clinical trials and in real-life studies.[342] Improvement of antiviral innate immunity with seasonal treatment, leading to the restoration of type I IFN production in response to RV infection has also been reported, suggesting disease-modifying properties.[343] In FA, omalizumab is now being tested in phase 3 clinical trials, in combination with OIT, to improve the rates and safety of desensitization.[328]

Dupilumab is a monoclonal antibody directed against the common α subunit of IL-4 and IL-13 receptors. Results from clinical trials including adults with persistent moderate-to-severe AD despite treatment with topical corticosteroids and calcineurin inhibitors have shown considerable efficacy, leading to *FDA*-approval in this indication for patients 12 years and older in 2017.[344] In asthma, Dupilumab is indicated as an add-on therapy in the treatment of moderate-to-severe asthma patients with eosinophilic phenotype or with oral corticosteroid-dependent asthma.[345] Clinical trials are ongoing in FA and EoE, testing Dupilumab in monotherapy or in addition to OIT in peanut FA.

Three monoclonal antibodies selectively inhibit IL-5 and thus its downstream action in eosinophils activation and recruitment.[307] *Mepolizumab* is an anti–IL-5 antibody indicated as an add-on maintenance treatment for patients 12 and

TABLE 47.2 Biologics for Treatment of Allergic Disorders

Drug	Target	Disease		Status
Omalizumab (Xolair®)	Anti-IgE	Asthma	FDA approved: 2003	Indicated in moderate-to-severe persistent allergic asthma in patients 6 y of age or older not controlled by inhaled corticosteroids.
		Multiple food allergies	Phase 3: NCT03881696	Tested as monotherapy and as adjunct therapy to multiallergen OIT in food allergic participants.
Dupilumab (Dupixent®)	Anti-IL-4Rα	Atopic Dermatitis	FDA approved: 2017	Indicated for the treatment of patients 12 y and older with moderate-to-severe AD whose disease is not adequately controlled with topical prescription therapies or when those therapies are not advisable.
		Asthma	FDA approved: 2018	Indicated as an add-on therapy in the treatment of moderate-to-severe asthma patients with eosinophilic phenotype or with oral corticosteroid-dependent asthma.
		EoE	Phase 3: NCT03633617	Study to determine the efficacy and safety of dupilumab in adult and adolescent patients with eosinophilic esophagitis.
		Peanut allergy	Phase 2: NCT03793608	Study to evaluate Dupilumab monotherapy in children with peanut allergy.
		Peanut allergy	Phase 2: NCT03682770	Study in pediatric subjects with peanut allergy to evaluate efficacy and safety of Dupilumab as Adjunct to AR101 (peanut oral immunotherapy).
Mepolizumab (Nucala®)	Anti-IL-5	Asthma	FDA approved: 2015	Indicated as an add-on maintenance treatment for patients 12 and older with severe eosinophilic asthma.
Reslizumab (Cinqair®)	Anti-IL-5	Asthma	FDA approved: 2016	Indicated as an add-on maintenance treatment of severe asthma in patients aged 18 y and older.
		EoE	Phase 3: NCT00635089	Open-label extension study of Reslizumab in pediatric subjects with eosinophilic esophagitis.
Benralizumab (Fasenra®)	Anti-IL-5R	Asthma	FDA approved: 2017	Indicated as an add-on maintenance treatment for patients 12 y and older with severe eosinophilic asthma.
Lebrikizumab	Anti-IL-13	Atopic dermatitis	Phase 3: NCT04178967	Evaluation of the efficacy and safety of Lebrikizumab in moderate to severe atopic dermatitis.
		Asthma	Phase 3: NCT01868061	Study of Lebrikizumab in participants with uncontrolled asthma on inhaled corticosteroids and a second controller medication.

(Continued)

TABLE 47.2 Biologics for Treatment of Allergic Disorders (Continued)

Drug	Target	Disease		Status
Tralokinumab	Anti-IL-13	Asthma	Phase 3: NCT02194699	Phase 3 study to evaluate the efficacy and safety of tralokinumab in adults and adolescents with uncontrolled asthma.
		Atopic dermatitis	Phase 3: NCT03363854	Tralokinumab in combination with topical corticosteroids for moderate to severe atopic dermatitis.
Etokimab	Anti-IL-33	Peanut allergy	Phase 2: NCT02920021	Placebo-controlled study to investigate ANB020 activity in adult patients with peanut allergy.
		Atopic dermatitis	Phase 2: NCT03533751	Study investigating the efficacy, safety, and pharmacokinetic profile of ANB020 administered to adult subjects with moderate-to-severe atopic dermatitis (ATLAS).
REGN3500 (SAR440340)	Anti-IL-33	Atopic dermatitis	Phase 2: NCT03738423	Study to investigate efficacy, safety, and pharmacokinetic profiles of REGN3500 administered to adult patients with moderate-to-severe atopic dermatitis.
Tezepelumab	Anti-TSLP	Asthma	Phase 3: NCT03347279	Study to investigate the efficacy and safety of Tezepilumab in adults and adolescents with severe uncontrolled asthma.
		Atopic dermatitis	Phase 2: NCT03809663	Study to investigate the safety and efficacy of Tezepilumab in atopic dermatitis.
Fevipiprant (QAW039)	CRTH2 antagonist	Asthma	Phase 3: NCT03226392	Study of efficacy and safety of QAW039 when added to standard-of-care asthma therapy in patients with uncontrolled asthma.
Omalizumab (Xolair®) and Dupilumab (Dupixent®)	Anti-IgE and Anti-IL-4Rα	Multiple food allergies	Phase 2: NCT03679676	Clinical study using biologics to improve Multi OIT outcomes
Dupilumab (Dupixent®) and REGN3500 (SAR440340)	Anti-IL-4Rα and Anti-IL-33	Asthma	Phase 2: NCT03387852	Evaluation of SAR440340 and as combination therapy with Dupilumab in moderate-to-severe asthma patients

CRTH2, Prostaglandin D2 receptor 2; EoE, Eosinophilic esophagitis; *FDA, Food and Drug Administration*; IgE, Immunoglobulin E; IL, Interleukin; TSLP, thymic stromal lymphopoietin.

older with severe eosinophilic asthma. *Reslizumab* is another anti–IL-5 antibody indicated as an add-on maintenance treatment for patients 18 years and older with severe asthma. It is also tested in phase 3 clinical trial in pediatric subjects with EoE. *Benralizumab* is a monoclonal antibody that binds to IL-5R, consequently blocking the binding of IL-5 to its receptor, indicated as an add-on maintenance treatment for patients 12 years and older with severe eosinophilic asthma.

Several other drugs are being studied in allergic disorders, sometimes in association with immunotherapy or as stand-alone therapy: Lebrikizumab (anti–IL-13), Tralokinumab (anti–IL-13), Etokimab (anti–IL-33), REGN3500 (anti–IL-33), Tezepelumab (anti-TSLP), Fevipiprant (CRTh2 antagonists).

Other Disease-Modifying Therapeutic Perspectives

Other immunomodulatory strategies are currently the object of intense research. Some food-adjuvants, such as unlipidated-outer membrane protein 16 (U-OMP16), chitin plus chitosan, can reduce the Th2-biased inflammation and increase Th1 cytokines, as shown in murine models of cow's milk, peanut, or shrimp FA.[346-348] Another approach is to educate the immune system via a plasmid DNA vaccine that expresses allergen proteins.[328] Indeed, preclinical studies using animal models of allergic diseases have shown promising results with house dust mite allergens-vaccines and peanut allergens-plasmid DNA vaccines, leading to ongoing clinical trials in AR and peanut allergy, respectively.[349,350] TLR agonists have also been studied as an immune deviation therapeutic, to modify innate and adaptive response to allergens. In AR associated with ragweed pollen allergy, studies using a TLR9 agonist as a vaccine/preventive strategy,[351] and a modified allergoid preparation including a TLR4 agonist in treating symptoms[352] have shown clinical efficacy.

Because infectious agents and allergic disorders both interact with the innate and adaptive immune system with common mechanisms, *anti-infectious strategies* might also be used to treat allergic disorders. Bacterial lysates, which contain lysates of 8 pathogenic bacteria including *Haemophilus influenzae* and *Streptococcus pneumoniae*, have shown some efficacy in preventing asthma exacerbations in children, through modulation of PRR recognition by PAMPs contained in lysates, and subsequent induction and maturation of DCs that will promote a Th1 phenotype.[352,353] Antimicrobial peptides are key molecules of the innate immune response against both viruses and bacteria, displaying immunomodulatory activities and altering host cellular responses. Hence, supplementing with antimicrobial peptides, such as LL-37, could enhance anti-infectious innate defenses.[354] Another strategy could rely on inducing its production among epithelial cells through vitamin D supplementation, although limited evidence is available and its impact remains uncertain.[121] In asthma, it has been suggested that IFN supplementation upon virus-induced acute asthma could enhance antiviral responses and modify inflammation in the long-term. Some *in vitro* studies have indeed suggested efficacy of type I IFN in reducing RV replication and modifying the antiviral response of epithelial cells.[355] However, clinical studies have been disappointing regarding the efficacy and safety of such treatment in acute asthma. Finally, strategies targeting the skin, lung, and gut microbiota are developing to treat allergic disorders. Treatments aiming at influencing its composition might have some interest. Transplantation of healthy microbiota seems a promising alternative as demonstrated in murine model and in human gut inflammatory diseases.[356] Several studies have also assessed the effect of probiotics in reducing symptoms of allergic disorders, including AD,[357] asthma,[358] and FA [359] but no final evidence supports their current use as a therapeutic strategy. Another option may be to administer bacteria metabolites, which can mimic the effect of normal microbiota. Supplementation with short-chain fatty acids has shown a reduction of allergic symptoms in murine models of peanut allergy, in combination with OIT.[360] Feeding mice a high-fiber diet that increases the levels of the short-chain fatty acid produced by microbiota leads to a suppression of allergic airway inflammation in mouse models of allergic asthma, through a Treg dependent mechanism.[361,362] Using antibiotics to modify the microbiota to treat allergic disorders also show promise. In particular, macrolides are well-described molecules with known immunomodulatory effects.[363] Although there is still limited evidence, some studies using macrolides have shown promising clinical results in the treatment of acute asthma in preschoolers,[364] and a direct effect on reducing epithelial production of proinflammatory cytokines IL-6, IL-1β, and alarmin IL-33 in adults.[365]

Finally, as stated earlier, *preventive strategies* to limit risk factors and their impact on the mechanism of allergic disorders can be encouraged (*see Risk factors for allergic disorders*). As suggested by a growing body of research in the past 15 years, the most effective preventive strategy might be early introduction of allergens. Both the LEAP study (early introduction of peanut at 4-11 months of age) and the EAT study (early introduction of 6 common allergenic food [peanut, cow's milk, egg, sesame, whitefish, and wheat]) in the diet of breastfed infants at 3 months of age, have shown a protective effect.[115,366] Based on results from these studies, guidelines no longer recommend allergen avoidance but rather support introduction of solid foods at 4 to 6 months of age.[328] Consistent with the dual-allergen hypothesis and atopic march, some researchers have hypothesized that prevention of allergen exposure through maintaining skin barrier integrity could serve as an effective preventive strategy for prevention of AD and other allergic disorders. To date, studies evaluating the effectiveness of application of skin emollients in preventing AD, food sensitization, and FA have shown conflicting results and may be due to the differences in emollients used.[367-369] As stated previously, the role of the microbiota in allergic disorders and immune tolerance is an active area of research.[328] Because probiotics have been reported to modulate immune responses, their supplementation, during pregnancy, breastfeeding, or in the early months of infancy, has been proposed as a preventive intervention.[370] Current evidence does not indicate that probiotic supplementation reduces the risk of developing asthma and FA in children, but there is a possible benefit from using probiotics in the primary prevention of AD.[371]

CONCLUSION AND PERSPECTIVES

As we have shown in this chapter, immune mechanisms leading to the development of allergic disorders are complex and heterogeneous. Some individuals may display a barrier defect and exaggerated innate immune responses, whereas others may display an exaggerated Th2 response and others a lack of regulatory mechanisms. These distinct pathophysiological pathways, called endotypes, may distinguish different types of allergic disorders. These specific endotypes may correspond to different clinical presentations or phenotypes, which can be defined as the observable characteristics of a disease.[372]

In asthma, different phenotypes have been described over the last 2 decades using unbiased methods of clustering. Whereas childhood asthma phenotypes mostly rely on allergenic sensitizations, frequency of exacerbations, and severity [373,374]; adult phenotypes are based on gender, obesity, overlap with chronic obstructive pulmonary disease, and severity.[375] Neonatal cohorts, usually designed with multiple sampling and assessment over time, could provide more insights into the evolution of these phenotypes throughout the years. Hence, this research is constantly progressing, and some subtypes of asthma may be clarified and modified in the coming years.

In other allergic disorders, such as FA, new presentations are also emerging, opposing the "classic" phenotype to "intermittent and cross-reactive phenotype" or "alpha-gal syndrome."[372,376] AD can also be stratified according to different criteria, such as age, severity, or infectious complications.[377]

As stated earlier, digital tools with "big data," "machine learning," and mechanistic studies will provide an innovative way to study and integrate the complexity of the immune response and its implication in the observed phenotypes.[378] Because the development of barrier sites and immunity in the early years may be crucial in these mechanisms, population-based birth cohort studies could provide precious information. A better understanding of the diversity and long-term evolution of patients will make it possible to better target patients who would benefit from new therapies, particularly disease-modifying strategies. It has now become clear that phenotyping and even more so, endotyping patients with allergic disorders is an essential step toward precision medicine.

Finally, in understanding allergic disorders and the attempts to treat them, it is always critical to uncover and understand the fundamental evolutionary purposes of the various elements of the atopic response—be the detoxification, antiparasitic, antineoplastic, or others—and balance those positive roles against the targeted interventions meant to prevent allergic pathology. Clearly, there have been changes in allergy risk over the most recent generation, identifying the root issues (behaviors, antigenic and microbial exposure, and colonization, etc) and targeting those will likely provide the greatest therapeutic window and help better our understanding of allergic inflammation in general.

ACKNOWLEDGMENTS

The authors would like to thank Dr Vanitha Sampath and Dr Ziyuan He (Sean N. Parker Center for Allergy and Asthma Research) for their contribution to this chapter.

REFERENCES

1. Justiz Vaillant AA, Vashisht R, Zito PM. *Immediate hypersensitivity reactions*. [Updated 2020 Jan 9]. In: *StatPearls* [Internet]. StatPearls Publishing; January 2020. Available from: https://www.ncbi.nlm.nih.gov/books/NBK513315/
2. Gell PGH, Coombs RRA. *The classification of allergic reactions underlying disease*. In: *Clinical Aspects of Immunology*. Blackwell; 1963.
3. Rajan TV. The Gell-Coombs classification of hypersensitivity reactions: a re-interpretation. *Trends Immunol*. 2003;24(7):376-379.
4. Bendiner E. Baron von Pirquet: the aristocrat who discovered and defined allergy. *Hosp Pract*. 1981;16(10):137.
5. Bergmann KC, Ring J. *History of Allergy*. S. Karger AG; 2014.
6. Prausnitz C. Studien uber die Ueberempfindichkeit. *Zbl Balst*. 1921;86:160-169.
7. Ishizaka K, Ishizaka T. Identification of gamma-E-antibodies as a carrier of reaginic activity. *J Immunol*. 1967;99(6):1187-1198.
8. Coombs RR, Hunter A, Jonas WE, Bennich H, Johannson SG, Panzani R. Detection of IgE(IgND) specific antibody (probably reagin) to castor-bean allergen by the red-cell-linked antigen-antiglobulin reaction. *Lancet*. 1968;1(7552):1115-1118.
9. Johansson SG, Bennich H. Immunological studies of an atypical (myeloma) immunoglobulin. *Immunology*. 1967;13(4):381-394.
10. Daschner A, González Fernández J. Allergy in an evolutionary framework. *J Mol Evol*. 2020;88(1):66-76.
11. Oettgen HC. Fifty years later: emerging functions of IgE antibodies in host defense, immune regulation, and allergic diseases. *J Allergy Clin Immunol*. 2016;137(6):1631-1645.
12. Karasuyama H, Tabakawa Y, Ohta T, Wada T, Yoshikawa S. Crucial role for basophils in acquired protective immunity to tick infestation. *Front Physiol*. 2018;9:1769.
13. Wilson JM, Schuyler AJ, Schroeder N, Platts-Mills TAE. Galactose-α-1,3-Galactose: atypical food allergen or model IgE hypersensitivity? *Curr Allergy Asthma Rep*. 2017;17(1):8.
14. Galli SJ, Starkl P, Marichal T, Tsai M. Mast cells and IgE can enhance survival during innate and acquired host responses to venoms. *Trans Am Clin Climatol Assoc*. 2017;128:193-221.
15. Maizels RM. Parasitic helminth infections and the control of human allergic and autoimmune disorders. *Clin Microbiol Infect*. 2016;22(6):481-486.
16. Wu Z, Wang L, Tang Y, Sun X. Parasite-derived proteins for the treatment of allergies and autoimmune diseases. *Front Microbiol*. 2017;8:2164.
17. Jensen-Jarolim E, Bax HJ, Bianchini R, et al. AllergoOncology - the impact of allergy in oncology: EAACI position paper. *Allergy*. 2017;72(6):866-887.
18. Bożek A, Jarzab J, Mielnik M, Bogacz A, Kozlowska R, Mangold D. Can atopy have a protective effect against cancer? *PloS One*. 2020;15(2):e0226950.
19. Kozłowska R, Bożek A, Jarząb J. Association between cancer and allergies. *Allergy Asthma Clin Immunol*. 2016;12:39.
20. Gebhardt C, Sevko A, Jiang H, et al. Myeloid cells and related chronic inflammatory factors as novel predictive markers in melanoma treatment with ipilimumab. *Clin Cancer Res*. 2015;21(24):5453-5459.
21. Reichman H, Karo-Atar D, Munitz A. Emerging roles for eosinophils in the tumor microenvironment. *Trends Cancer*. 2016;2(11):664-675.
22. Varricchi G, Galdiero MR, Loffredo S, et al. Eosinophils: the unsung heroes in cancer? *OncoImmunology*. 2018;7(2):e1393134.
23. Rigoni A, Colombo MP, Pucillo C. Mast cells, basophils and eosinophils: from allergy to cancer. *Semin Immunol*. 2018;35:29-34.
24. Dudeck A, Köberle M, Goldmann O, et al. Mast cells as protectors of health. *J Allergy Clin Immunol*. 2019;144(4s):S4-S18.
25. Pellizzari G, Hoskin C, Crescioli S, et al. IgE re-programs alternatively-activated human macrophages towards pro-inflammatory anti-tumoural states. *EBioMedicine*. 2019;43:67-81.
26. Platzer B, Elpek KG, Cremasco V, et al. IgE/FcεRI-mediated antigen cross-presentation by dendritic cells enhances anti-tumor immune responses. *Cell Rep*. 2015;10(9):1487-1495.
27. Crawford G, Hayes MD, Seoane RC, et al. Epithelial damage and tissue γδ T cells promote a unique tumor-protective IgE response. *Nat Immunol*. 2018;19(8):859-870.
28. Hayes MD, Ward S, Crawford G, et al. Inflammation-induced IgE promotes epithelial hyperplasia and tumour growth. *Elife*. 2020;9:e51862.
29. Thomsen SF. Epidemiology and natural history of atopic diseases. *Eur Clin Respir J*. 2015;2.
30. Sidbury R, Kodama S. Atopic dermatitis guidelines: diagnosis, systemic therapy, and adjunctive care. *Clin Dermatol*. 2018;36(5):648-652.
31. Huang E, Ong PY. Severe atopic dermatitis in children. *Curr Allergy Asthma Rep*. 2018;18(6):35.
32. Avena-Woods C. Overview of atopic dermatitis. *Am J Manag Care*. 2017;23(8 Suppl):S115-S123.
33. Shaw TE, Currie GP, Koudelka CW, Simpson EL. Eczema prevalence in the United States: data from the 2003 national survey of children's health. *J Invest Dermatol*. 2011;131(1):67-73.
34. Geoghegan JA, Irvine AD, Foster TJ. Staphylococcus aureus and atopic dermatitis: a complex and evolving relationship. *Trends Microbiol*. 2018;26(6):484-497.
35. Sampson HA, Muñoz-Furlong A, Campbell RL, et al. Second symposium on the definition and management of anaphylaxis: summary report - second national institute of allergy and infectious disease/food allergy and anaphylaxis network symposium. *J Allergy Clin Immunol*. 2006;117(2):391-397.
36. Lieberman P, Nicklas RA, Randolph C, et al. Anaphylaxis —a practice parameter update 2015. *Ann Allergy Asthma Immunol*. 2015;115(5):341-384.
37. Panesar SS, Javad S, de Silva D, et al. The epidemiology of anaphylaxis in Europe: a systematic review. *Allergy*. 2013;68(11):1353-1361.
38. Wood RA, Camargo CA Jr, Lieberman P, et al. Anaphylaxis in America: the prevalence and characteristics of anaphylaxis in the United States. *J Allergy Clin Immunol*. 2014;133(2):461-467.
39. Fischer D, Vander Leek TK, Ellis AK, Kim H. Anaphylaxis. *Allergy Asthma Clin Immunol*. 2018;14(Suppl 2):54.
40. Pouessel G, Antoine M, Lejeune S, et al. The time course of anaphylaxis manifestations in children is diverse and unpredictable. *Clin Exp Allergy*. 2020;50(1):117-120.
41. Pouessel G, Beaudouin E, Tanno LK, et al. Food-related anaphylaxis fatalities: analysis of the allergy Vigilance Network(®) database. *Allergy*. 2019;74(6):1193-1196.
42. Rael E, Sampath V, Nadeau KC. Diagnosis and differential diagnosis of food allergy. In: Gupta RS, ed. *Pediatric Food Allergy: A Clinical Guide*. Springer International Publishing; 2020:31-44.
43. Lee J, Garrett JPD, Brown-Whitehorn T, Spergel JM. Biphasic reactions in children undergoing oral food challenges. *Allergy Asthma Proc*. 2013;34(3):220-226.
44. Christensen MJ, Eller E, Mortz CG, Brockow K, Bindslev-Jensen C. Wheat-dependent cofactor-augmented anaphylaxis: a prospective study of exercise, aspirin, and alcohol efficacy as cofactors. *J Allergy Clin Immunol Pract*. 2019;7(1):114-121.
45. Feldweg AM. Exercise-induced anaphylaxis. *Immunol Allergy Clin*. 2015;35(2):261-275.
46. Castillo JR, Peters SP, Busse WW. Asthma exacerbations: pathogenesis, prevention, and treatment. *J Allergy Clin Immunol Pract*. 2017;5(4):918-927.
47. Saglani S, Fleming L, Sonnappa S, Bush A. Advances in the aetiology, management, and prevention of acute asthma attacks in children. *Lancet Child Adolesc Health*. 2019;3(5):354-364.
48. Theoharides TC, Enakuaa S, Sismanopoulos N, et al. Contribution of stress to asthma worsening through mast cell activation. *Ann Allergy Asthma Immunol*. 2012;109(1):14-19.
49. Dharmage SC, Perret JL, Custovic A. Epidemiology of asthma in children and adults. *Frontiers in Pediatrics*. 2019;7:246.
50. Zahran HS, Bailey CM, Damon SA, Garbe PL, Breysse PN. Vital signs: asthma in children - United States, 2001-2016. *MMWR Morb Mortal Wkly Rep*. 2018;67(5):149-155.
51. Custovic A, Sonntag H-J, Buchan IE, Belgrave D, Simpson A, Prosperi MCF. Evolution pathways of IgE responses to grass and mite allergens throughout childhood. *J Allergy Clin Immunol*. 2015;136(6):1645-1652.e48.
52. Gabet S, Just J, Couderc R, Bousquet J, Seta N, Momas I. Early polysensitization is associated with allergic multimorbidity in PARIS birth cohort infants. *Pediatr Allergy Immunol*. 2016;27(8):831-837.
53. Hose AJ, Depner M, Illi S, et al. Latent class analysis reveals clinically relevant atopy phenotypes in 2 birth cohorts. *J Allergy Clin Immunol*. 2017;139(6):1935-1945.e12.
54. Min Y-G. The pathophysiology, diagnosis and treatment of allergic rhinitis. *Allergy Asthma Immunol Res*. 2010;2(2):65-76.
55. Nathan RA. The burden of allergic rhinitis. *Allergy Asthma Proc*. 2007;28(1):3-9.
56. Nadeau K. Approach to the patient with allergic or immunologic disease. In: Goldman L, Schafer AI, ed. *Goldman-Cecil Medicine*. 26th ed. Elsevier; 2019.
57. Turnbull JL, Adams HN, Gorard DA. Review article: the diagnosis and management of food allergy and food intolerances. *Aliment Pharmacol Ther*. 2015;41(1):3-25.
58. Gupta M, Cox A, Nowak-Węgrzyn A, Wang J. Diagnosis of food allergy. *Immunol Allergy Clin*. 2018;38(1):39-52.
59. Osborne NJ, Koplin JJ, Martin PE, et al. Prevalence of challenge-proven IgE-mediated food allergy using population-based sampling and predetermined challenge criteria in infants. *J Allergy Clin Immunol*. 2011;127(3):668-676.e62.

60. Sicherer SH, Sampson HA. Food allergy: epidemiology, pathogenesis, diagnosis, and treatment. *J Allergy Clin Immunol.* 2014;133(2):291-308.
61. Price A, Ramachandran S, Smith GP, Stevenson ML, Pomeranz MK, Cohen DE. Oral allergy syndrome (pollen-food allergy syndrome). *Dermatitis.* 2015;26(2):78-88.
62. Matricardi PM, Kleine-Tebbe J, Hoffmann HJ, et al. EAACI molecular allergology user's guide. *Pediatr Allergy Immunol.* 2016;27(S23):1-250.
63. Flaherty MG, Kaplan SJ, Jerath MR. Diagnosis of life-threatening alpha-gal food allergy appears to be patient driven. *J Prim Care Community Health.* 2017;8(4):345-348.
64. Lucendo AJ, Molina-Infante J, Arias Á, et al. Guidelines on eosinophilic esophagitis: evidence-based statements and recommendations for diagnosis and management in children and adults. *United European Gastroenterology J.* 2017;5(3):335-358.
65. Liacouras CA, Furuta GT, Hirano I, et al. Eosinophilic esophagitis: updated consensus recommendations for children and adults. *J Allergy Clin Immunol.* 2011;128(1):3-22.
66. Blanchard C, Simon D, Schoepfer A, Straumann A, Simon HU. Eosinophilic esophagitis: unclear roles of IgE and eosinophils. *J Intern Med.* 2017;281(5):448-457.
67. Sugnanam KKN, Collins JT, Smith PK, et al. Dichotomy of food and inhalant allergen sensitization in eosinophilic esophagitis. *Allergy.* 2007;62(11):1257-1260.
68. Lucendo AJ, Arias Á, González-Cervera J, et al. Empiric 6-food elimination diet induced and maintained prolonged remission in patients with adult eosinophilic esophagitis: a prospective study on the food cause of the disease. *J Allergy Clin Immunol.* 2013;131(3):797-804.
69. Molina-Infante J, Arias Á, Alcedo J, et al. Step-up empiric elimination diet for pediatric and adult eosinophilic esophagitis: the 2-4-6 study. *J Allergy Clin Immunol.* 2018;141(4):1365-1372.
70. Hill DA, Spergel JM. The atopic march: critical evidence and clinical relevance. *Ann Allergy Asthma Immunol.* 2018;120(2):131-137.
71. Han H, Roan F, Ziegler SF. The atopic march: current insights into skin barrier dysfunction and epithelial cell-derived cytokines. *Immunol Rev.* 2017;278(1):116-130.
72. Takai T, Ikeda S. Barrier dysfunction caused by environmental proteases in the pathogenesis of allergic diseases. *Allergol Int.* 2011;60(1):25-35.
73. Renz H, Holt PG, Inouye M, Logan AC, Prescott SL, Sly PD. An exposome perspective: early-life events and immune development in a changing world. *J Allergy Clin Immunol.* 2017;140(1):24-40.
74. Raby BA. Asthma severity, nature or nurture: genetic determinants. *Curr Opin Pediatr.* 2019;31(3):340-348.
75. Lyons JJ, Milner JD. Primary atopic disorders. *J Exp Med.* 2018;215(4):1009-1022.
76. Milner JD. Primary atopic disorders. *Annu Rev Immunol.* 2020:38:785-808. doi:10.1146/annurev-immunol-042718-041553.
77. Mucha S, Baurecht H, Novak N, et al. Protein-coding variants contribute to the risk of atopic dermatitis and skin-specific gene expression. *J Allergy Clin Immunol.* 2020;145(4):1208-1218.
78. Bønnelykke K, Sparks R, Waage J, Milner JD. Genetics of allergy and allergic sensitization: common variants, rare mutations. *Curr Opin Immunol.* 2015;36:115-126.
79. Bin L, Leung DYM. Genetic and epigenetic studies of atopic dermatitis. *Allergy Asthma Clin Immunol.* 2016;12:52.
80. Hong X, Hao K, Ladd-Acosta C, et al. Genome-wide association study identifies peanut allergy-specific loci and evidence of epigenetic mediation in US children. *Nat Commun.* 2015;6:6304.
81. Suaini NHA, Wang Y, Soriano VX, et al. Genetic determinants of paediatric food allergy: a systematic review. *Allergy.* 2019;74(9):1631-1648.
82. Winters A, Bahnson HT, Ruczinski I, et al. The MALT1 locus and peanut avoidance in the risk for peanut allergy. *J Allergy Clin Immunol.* 2019;143(6):2326-2329.
83. Hall R, Hall IP, Sayers I. Genetic risk factors for the development of pulmonary disease identified by genome-wide association. *Respirology.* 2019;24(3):204-214.
84. Moffatt MF, Gut IG, Demenais F, et al. A large-scale, consortium-based genomewide association study of asthma. *N Engl J Med.* 2010;363(13):1211-1221.
85. Waage J, Standl M, Curtin JA, et al. Genome-wide association and HLA fine-mapping studies identify risk loci and genetic pathways underlying allergic rhinitis. *Nat Genet.* 2018;50(8):1072-1080.
86. Frischmeyer-Guerrerio PA, Milner JD. Editorial overview: collusion between genes and environment in the pathogenesis of allergic disease. *Curr Opin Immunol.* 2019;60:iii-v.
87. Santiago HC, Nutman TB. Human helminths and allergic disease: the hygiene hypothesis and beyond. *Am J Trop Med Hyg.* 2016;95(4):746-753.
88. Belkaid Y, Harrison OJ. Homeostatic immunity and the microbiota. *Immunity.* 2017;46(4):562-576.
89. Fagerås Böttcher M, Hmani-Aifa M, Lindström A, et al. A TLR4 polymorphism is associated with asthma and reduced lipopolysaccharide-induced interleukin-12(p70) responses in Swedish children. *J Allergy Clin Immunol.* 2004;114(3):561-567.
90. Chinthrajah RS, Hernandez JD, Boyd SD, Galli SJ, Nadeau KC. Molecular and cellular mechanisms of food allergy and food tolerance. *J Allergy Clin Immunol.* 2016;137(4):984-997.
91. Barcik W, Boutin RCT, Sokolowska M, Finlay BB. The role of lung and gut microbiota in the pathology of asthma. *Immunity.* 2020;52(2):241-255.
92. Jartti T, Gern JE. Role of viral infections in the development and exacerbation of asthma in children. *J Allergy Clin Immunol.* 2017;140(4):895-906.
93. Bergauer A, Sopel N, Kroß B, et al. IFN-α/IFN-λ responses to respiratory viruses in paediatric asthma. *Eur Respir J.* 2017;49(2).
94. Lejeune S, Pichavant M, Engelmann I, et al. Severe preschool asthmatics have altered cytokine and anti-viral responses during exacerbation. *Pediatr Allergy Immunol.* 2020;31(6):651-661.
95. Lejeune S, Deschildre A, Le Rouzic O, et al. Childhood asthma heterogeneity at the era of precision medicine: modulating the immune response or the microbiota for the management of asthma attack. *Biochem Pharmacol.* 2020;179:114046.
96. Jackson DJ, Makrinioti H, Rana BM, et al. IL-33-dependent type 2 inflammation during rhinovirus-induced asthma exacerbations in vivo. *Am J Respir Crit Care Med.* 2014;190(12):1373-1382.
97. Bartemes KR, Kita H. Innate and adaptive immune responses to fungi in the airway. *J Allergy Clin Immunol.* 2018;142(2):353-363.
98. Rook GA. Hygiene and other early childhood influences on the subsequent function of the immune system. *Dig Dis.* 2011;29(2):144-153.
99. Wills-Karp M, Lewkowich I. Immunologic mechanisms of allergic disorders. In: Paul WE, ed. *Fundamental Immunology.* 7th ed. Wolters Kluwer; 2012.
100. Waser M, Michels KB, Bieli C, et al. Inverse association of farm milk consumption with asthma and allergy in rural and suburban populations across Europe. *Clin Exp Allergy.* 2007;37(5):661-670.
101. Stein MM, Hrusch CL, Gozdz J, et al. Innate immunity and asthma risk in amish and hutterite farm children. *N Engl J Med.* 2016;375(5):411-421.
102. Ruokolainen L, Paalanen L, Karkman A, et al. Significant disparities in allergy prevalence and microbiota between the young people in Finnish and Russian Karelia. *Clin Exp Allergy.* 2017;47(5):665-674.
103. Ruokolainen L, Parkkola A, Karkman A, et al. Contrasting microbiotas between Finnish and Estonian infants: exposure to acinetobacter may contribute to the allergy gap. *Allergy.* 2020;75(9):2342-2351.
104. Bunyavanich S, Shen N, Grishin A, et al. Early-life gut microbiome composition and milk allergy resolution. *J Allergy Clin Immunol.* 2016;138(4):1122-1130.
105. Kim H, Sitarik AR, Woodcroft K, Johnson CC, Zoratti E. Birth mode, breastfeeding, pet exposure, and antibiotic use: associations with the gut microbiome and sensitization in children. *Curr Allergy Asthma Rep.* 2019;19(4):22.
106. Stokholm J, Thorsen J, Chawes BL, et al. Cesarean section changes neonatal gut colonization. *J Allergy Clin Immunol.* 2016;138(3):881-889.e2.
107. Patrick DM, Sbihi H, Dai DLY, et al. Decreasing antibiotic use, the gut microbiota, and asthma incidence in children: evidence from population-based and prospective cohort studies. *Lancet Respir Med.* 2020;8(11):1094-1105.
108. Ojwang V, Nwaru BI, Takkinen H-M, et al. Early exposure to cats, dogs and farm animals and the risk of childhood asthma and allergy. *Pediatr Allergy Immunol.* 2020;31(3):265-272.
109. Schuijs MJ, Willart MA, Vergote K, et al. Farm dust and endotoxin protect against allergy through A20 induction in lung epithelial cells. *Science.* 2015;349(6252):1106-1110.
110. Miles EA, Calder PC. Can early omega-3 fatty acid exposure reduce risk of childhood allergic disease? *Nutrients.* 2017;9(7):784.
111. Dunstan JA, Mori TA, Barden A, et al. Fish oil supplementation in pregnancy modifies neonatal allergen-specific immune responses and clinical outcomes in infants at high risk of atopy: a randomized, controlled trial. *J Allergy Clin Immunol.* 2003;112(6):1178-1184.
112. Bisgaard H, Stokholm J, Chawes BL, et al. Fish oil-derived fatty acids in pregnancy and wheeze and asthma in offspring. *N Engl J Med.* 2016;375(26):2530-2539.
113. Khan TK, Palmer DJ, Prescott SL. In-utero exposures and the evolving epidemiology of paediatric allergy. *Curr Opin Allergy Clin Immunol.* 2015;15(5):402-408.
114. von Mutius E. Maternal farm exposure/ingestion of unpasteurized cow's milk and allergic disease. *Curr Opin Gastroenterol.* 2012;28(6):570-576.
115. Du Toit G, Roberts G, Sayre PH, et al. Randomized trial of peanut consumption in infants at risk for peanut allergy. *N Engl J Med.* 2015;372(9):803-813.
116. Du Toit G, Sampson HA, Plaut M, Burks AW, Akdis CA, Lack G. Food allergy: update on prevention and tolerance. *J Allergy Clin Immunol.* 2018;141(1):30-40.
117. Lack G. Epidemiologic risks for food allergy. *J Allergy Clin Immunol.* 2008;121(6):1331-1336.
118. Baker MG, Nowak-Wegrzyn A. Food allergy prevention: current evidence. *Curr Opin Clin Nutr Metab Care.* 2020;23(3):196-202.
119. Holl JL, Bilaver LA, Finn DJ, Savio K. A randomized trial of the acceptability of a daily multi-allergen food supplement for infants. *Pediatr Allergy Immunol.* 2020;31(4):418-420.
120. Tran MM, Lefebvre DL, Dai D, et al. Timing of food introduction and development of food sensitization in a prospective birth cohort. *Pediatr Allergy Immunol.* 2017;28(5):471-477.
121. Yepes-Nuñez JJ, Brożek JL, Fiocchi A, et al. Vitamin D supplementation in primary allergy prevention: systematic review of randomized and non-randomized studies. *Allergy.* 2018;73(1):37-49.
122. Wang T-T, Nestel FP, Bourdeau V, et al. Cutting edge: 1,25-dihydroxyvitamin D3 is a direct inducer of antimicrobial peptide gene expression. *J Immunol.* 2004;173(5):2909-2912.
123. Sassi F, Tamone C, D'Amelio P. Vitamin D: nutrient, hormone, and immunomodulator. *Nutrients.* 2018;10(11):1656.
124. Huang CM, Lara-Corrales I, Pope E. Effects of Vitamin D levels and supplementation on atopic dermatitis: a systematic review. *Pediatr Dermatol.* 2018;35(6):754-760.
125. Kim G, Bae J-H. Vitamin D and atopic dermatitis: a systematic review and meta-analysis. *Nutrition.* 2016;32(9):913-920.
126. Litonjua AA. Vitamin D and childhood asthma: causation and contribution to disease activity. *Curr Opin Allergy Clin Immunol.* 2019;19(2):126-131.
127. Hall SC, Agrawal DK. Vitamin D and bronchial asthma: an overview of data from the past 5 years. *Clin Ther.* 2017;39(5):917-929.
128. Matsui T, Tanaka K, Yamashita H, et al. Food allergy is linked to season of birth, sun exposure, and vitamin D deficiency. *Allergol Int.* 2019;68(2):172-177.
129. Koplin JJ, Suaini NHA, Vuillermin P, et al. Polymorphisms affecting vitamin D-binding protein modify the relationship between serum vitamin D (25[OH]D3) and food allergy. *J Allergy Clin Immunol.* 2016;137(2):500-506.e4.
130. Dreger LC, Kozyrskyj AL, HayGlass KT, Becker AB, MacNeil BJ. Lower cortisol levels in children with asthma exposed to recurrent maternal distress from birth. *J Allergy Clin Immunol.* 2010;125(1):116-122.
131. Rosa MJ, Lee AG, Wright RJ. Evidence establishing a link between prenatal and early-life stress and asthma development. *Curr Opin Allergy Clin Immunol.* 2018;18(2):148-158.
132. Flanigan C, Sheikh A, DunnGalvin A, Brew BK, Almqvist C, Nwaru BI. Prenatal maternal psychosocial stress and offspring's asthma and allergic disease: a systematic review and meta-analysis. *Clin Exp Allergy.* 2018;48(4):403-414.
133. Kang LJ, Vu KN, Koleva PT, et al. Maternal psychological distress before birth influences gut immunity in mid-infancy. *Clin Exp Allergy.* 2020;50(2):178-188.
134. Bascom EM, Napolitano MA. Breastfeeding duration and primary reasons for breastfeeding cessation among women with postpartum depressive symptoms. *J Hum Lact.* 2016;32(2):282-291.
135. Jašarević E, Howerton CL, Howard CD, Bale TL. Alterations in the vaginal microbiome by maternal stress are associated with metabolic reprogramming of the offspring gut and brain. *Endocrinology.* 2015;156(9):3265-3276.
136. Laffont S, Blanquart E, Guéry J-C. Sex differences in asthma: a key role of androgen-signaling in group 2 innate lymphoid cells. *Front Immunol.* 2017;8:1069.
137. Intergovernmental panel on climate change. *Climate Change 2013: The Physical Science Basis.* 2013. http://www.ipcc.ch/report/ar5/wg1/
138. Eguiluz-Gracia I, Mathioudakis AG, Bartel S, et al. The need for clean air: the way air pollution and climate change affect allergic rhinitis and asthma. *Allergy.* 2020;75(9):2170-2184.
139. Prunicki M, Stell L, Dinakarpandian D, et al. Exposure to NO(2), CO, and PM(2.5) is linked to regional DNA methylation differences in asthma. *Clin Epigenetics.* 2018;10:2.
140. Rosser F, Forno E, Kurland KS, et al. Annual SO(2) exposure, asthma, atopy, and lung function in Puerto Rican children. *Pediatr Pulmonol.* 2020;55(2):330-337.
141. Sompornrattanaphan M, Thongngarm T, Ratanawatkul P, Wongsa C, Swigris JJ. The contribution of particulate matter to respiratory allergy: a review of current evidence. *Asian Pac J Allergy Immunol.* 2020;38(1):19-28. doi:10.12932/AP-100619-100579.
142. Ribeiro H, Guimarães F, Duque L, Noronha F, Abreu I. Characterisation of particulate matter on airborne pollen grains. *Environ Pollut.* 2015;206:7-16.
143. Wang B, Chen H, Chan YL, Wang G, Oliver BG. Why do intrauterine exposure to air pollution and cigarette smoke increase the risk of asthma? *Front Cell Dev Biol.* 2020;8:38.
144. Portela A, Esteller M. Epigenetic modifications and human disease. *Nat Biotechnol.* 2010;28(10):1057-1068.
145. Potaczek DP, Harb H, Michel S, Alhamwe BA, Renz H, Tost J. Epigenetics and allergy: from basic mechanisms to clinical applications. *Epigenomics.* 2017;9(4):539-571.
146. Xu C-J, Söderhäll C, Bustamante M, et al. DNA methylation in childhood asthma: an epigenome-wide meta-analysis. *Lancet Respir Med.* 2018;6(5):379-388.
147. Liang L, Willis-Owen SAG, Laprise C, et al. An epigenome-wide association study of total serum immunoglobulin E concentration. *Nature.* 2015;520(7549):670-674.
148. Michel S, Busato F, Genuneit J, et al. Farm exposure and time trends in early childhood may influence DNA methylation in genes related to asthma and allergy. *Allergy.* 2013;68(3):355-364.

149. Johansson H, Mersha TB, Brandt EB, Khurana Hershey GK. Interactions between environmental pollutants and genetic susceptibility in asthma risk. *Curr Opin Immunol.* 2019;60:156-162.
150. Nadeau K, McDonald-Hyman C, Noth EM, et al. Ambient air pollution impairs regulatory T-cell function in asthma. *J Allergy Clin Immunol.* 2010;126(4):845-852.e10.
151. Barrett EG. Maternal influence in the transmission of asthma susceptibility. *Pulm Pharmacol Ther.* 2008;21(3):474-484.
152. Gunawardhana LP, Baines KJ, Mattes J, Murphy VE, Simpson JL, Gibson PG. Differential DNA methylation profiles of infants exposed to maternal asthma during pregnancy. *Pediatr Pulmonol.* 2014;49(9):852-862.
153. Martino D, Joo JE, Sexton-Oates A, et al. Epigenome-wide association study reveals longitudinally stable DNA methylation differences in CD4+ T cells from children with IgE-mediated food allergy. *Epigenetics.* 2014;9(7):998-1006.
154. Hong X, Ladd-Acosta C, Hao K, et al. Epigenome-wide association study links site-specific DNA methylation changes with cow's milk allergy. *J Allergy Clin Immunol.* 2016;138(3):908-911.e09.
155. Petrus NCM, Henneman P, Venema A, et al. Cow's milk allergy in Dutch children: an epigenetic pilot survey. *Clin Transl Allergy.* 2016;6:16.
156. Mervis JS, McGee JS. DNA methylation and inflammatory skin diseases. *Arch Dermatol Res.* 2020;312(7):461-466. doi:10.1007/s00403-00019-02005-00409.
157. Rodríguez E, Baurecht H, Wahn AF, et al. An integrated epigenetic and transcriptomic analysis reveals distinct tissue-specific patterns of DNA methylation associated with atopic dermatitis. *J Invest Dermatol.* 2014;134(7):1873-1883.
158. Liu C, Sathe SK. Food allergen epitope mapping. *J Agric Food Chem.* 2018;66(28):7238-7248.
159. Pomés A, Davies JM, Gadermaier G, et al. WHO/IUIS allergen nomenclature: providing a common language. *Mol Immunol.* 2018;100:3-13.
160. Kadam K, Karbhal R, Jayaraman VK, Sawant S, Kulkarni-Kale U. AllerBase: a comprehensive allergen knowledgebase. *Database.* 2017;2017:bax066.
161. Ukhanova O, Bogomolova E. Airborne allergens. In: Pereira C, ed. *Allergic Diseases: New Insights.* 2015.
162. Portnoy J, Miller JD, Williams PB, et al. Environmental assessment and exposure control of dust mites: a practice parameter. *Ann Allergy Asthma Immunol.* 2013;111(6):465-507.
163. Deschildre A, Elegbédé CF, Just J, et al. Peanut-allergic patients in the MIRABEL survey: characteristics, allergists' dietary advice and lessons from real life. *Clin Exp Allergy.* 2016;46(4):610-620.
164. Kelly KJ, Sussman G. Latex allergy: where are we now and how did we get there? *J Allergy Clin Immunol Pract.* 2017;5(5):1212-1216.
165. Barre A, Culerrier R, Granier C, et al. Mapping of IgE-binding epitopes on the major latex allergen Hev b 2 and the cross-reacting 1,3beta-glucanase fruit allergens as a molecular basis for the latex-fruit syndrome. *Mol Immunol.* 2009;46(8-9):1595-1604.
166. Dykewicz MS, Lam JK. Drug hypersensitivity reactions. *Med Clin North Am.* 2020;104(1):109-128.
167. Patton K, Borshoff DC. Adverse drug reactions. *Anaesthesia.* 2018;73(Suppl 1):76-84.
168. Riedl MA, Casillas AM. Adverse drug reactions: types and treatment options. *Am Fam Physician.* 2003;68(9):1781-1790.
169. Ali H. Emerging roles for MAS-related G protein-coupled receptor-X2 in host defense peptide, opioid, and neuropeptide-mediated inflammatory reactions. *Adv Immunol.* 2017;136:123-162.
170. Lansu K, Karpiak J, Liu J, et al. In silico design of novel probes for the atypical opioid receptor MRGPRX2. *Nat Chem Biol.* 2017;13(5):529-536.
171. Spoerl D, D'Incau S, Roux-Lombard P, Harr T, Czarnetzki C. Non-IgE-dependent hypersensitivity to rocuronium reversed by sugammadex: report of three cases and hypothesis on the underlying mechanism. *Int Arch Allergy Immunol.* 2016;169(4):256-262.
172. Navinés-Ferrer A, Serrano-Candelas E, Lafuente A, Muñoz-Cano R, Martín M, Gastaminza G. MRGPRX2-mediated mast cell response to drugs used in perioperative procedures and anaesthesia. *Sci Rep.* 2018;8(1):11628.
173. Grimes J, Desai S, Charter NW, et al. MrgX2 is a promiscuous receptor for basic peptides causing mast cell pseudo-allergic and anaphylactoid reactions. *Pharmacol Res Perspect.* 2019;7(6):e00547.
174. Liu R, Hu S, Zhang Y, et al. Mast cell-mediated hypersensitivity to fluoroquinolone is MRGPRX2 dependent. *Int Immunopharmacol.* 2019;70:417-427.
175. Chung WH, Hung SI, Hong HS, et al. Medical genetics: a marker for Stevens-Johnson syndrome. *Nature.* 2004;428(6982):486.
176. Genin E, Chen DP, Hung SI, et al. HLA-A*31:01 and different types of carbamazepine-induced severe cutaneous adverse reactions: an international study and meta-analysis. *Pharmacogenomics J.* 2014;14(3):281-288.
177. Mallal S, Phillips E, Carosi G, et al. HLA-B*5701 screening for hypersensitivity to abacavir. *N Engl J Med.* 2008;358(6):568-579.
178. Zagursky RJ, Pichichero ME. Cross-reactivity in β-lactam allergy. *J Allergy Clin Immunol Pract.* 2018;6(1):72-81.e71.
179. Golden DB, Demain J, Freeman T, et al. Stinging insect hypersensitivity: a practice parameter update 2016. *Ann Allergy Asthma Immunol.* 2017;118(1):28-54.
180. Abrams EM, Golden DBK. Approach to patients with stinging insect allergy. *Med Clin North Am.* 2020;104(1):129-143.
181. Bonadonna P, Scaffidi L. Hymenoptera anaphylaxis as a clonal mast cell disorder. *Immunol Allergy Clin North Am.* 2018;38(3):455-468.
182. Galli SJ, Tsai M, Piliponsky AM. The development of allergic inflammation. *Nature.* 2008;454(7203):445-454.
183. Min B. Basophils induce Th2 immunity: is this final answer? *Virulence.* 2010;1(5):399-401.
184. van Beek AA, Knol EF, de Vos P, Smelt MJ, Savelkoul HF, van Neerven RJ. Recent developments in basophil research: do basophils initiate and perpetuate type 2 T-helper cell responses? *Int Arch Allergy Immunol.* 2013;160(1):7-17.
185. Sokol CL, Barton GM, Farr AG, Medzhitov R. A mechanism for the initiation of allergen-induced T helper type 2 responses. *Nat Immunol.* 2008;9(3):310-318.
186. Kim S, Prout M, Ramshaw H, Lopez AF, LeGros G, Min B. Cutting edge: basophils are transiently recruited into the draining lymph nodes during helminth infection via IL-3, but infection-induced Th2 immunity can develop without basophil lymph node recruitment or IL-3. *J Immunol.* 2010;184(3):1143-1147.
187. Hammad H, Plantinga M, Deswarte K, et al. Inflammatory dendritic cells – not basophils – are necessary and sufficient for induction of Th2 immunity to inhaled house dust mite allergen. *J Exp Med.* 2010;207(10):2097-2111.
188. Tang H, Cao W, Kasturi SP, et al. The T helper type 2 response to cysteine proteases requires dendritic cell-basophil cooperation via ROS-mediated signaling. *Nat Immunol.* 2010;11(7):608-617.
189. Lezmi G, Leite-de-Moraes M. Invariant natural killer T and mucosal-associated invariant T cells in asthmatic patients. *Front Immunol.* 2018;9:1766.
190. Milner JD. TCR signaling abnormalities in human Th2-associated atopic disease. *Front Immunol.* 2018;9:719.
191. Boyce JA. Mast cells and eicosanoid mediators: a system of reciprocal paracrine and autocrine regulation. *Immunol Rev.* 2007;217:168-185.
192. Sampath V, Nadeau KC. Newly identified T cell subsets in mechanistic studies of food immunotherapy. *J Clin Invest.* 2019;129(4):1431-1440.
193. Branzk N, Gronke K, Diefenbach A. Innate lymphoid cells, mediators of tissue homeostasis, adaptation and disease tolerance. *Immunol Rev.* 2018;286(1):86-101.
194. Ali N, Rosenblum MD. Regulatory T cells in skin. *Immunology.* 2017;152(3):372-381.
195. Ding Y, Xu J, Bromberg JS. Regulatory T cell migration during an immune response. *Trends Immunol.* 2012;33(4):174-180.
196. Williams KW, Milner JD, Freeman AF. Eosinophilia associated with disorders of immune deficiency or immune dysregulation. *Immunol Allergy Clin North Am.* 2015;35(3):523-544.
197. Palmer CN, Irvine AD, Terron-Kwiatkowski A, et al. Common loss-of-function variants of the epidermal barrier protein filaggrin are a major predisposing factor for atopic dermatitis. *Nat Genet.* 2006;38(4):441-446.
198. Howell MD, Kim BE, Gao P, et al. Cytokine modulation of atopic dermatitis filaggrin skin expression. *J Allergy Clin Immunol.* 2007;120(1):150-155.
199. Strid J, Hourihane J, Kimber I, Callard R, Strobel S. Disruption of the stratum corneum allows potent epicutaneous immunization with protein antigens resulting in a dominant systemic Th2 response. *Eur J Immunol.* 2004;34(8):2100-2109.
200. van Reijsen FC, Felius A, Wauters EA, Bruijnzeel-Koomen CA, Koppelman SJ. T-cell reactivity for a peanut-derived epitope in the skin of a young infant with atopic dermatitis. *J Allergy Clin Immunol.* 1998;101(2 Pt 1):207-209.
201. Lack G, Fox D, Northstone K, Golding J. Factors associated with the development of peanut allergy in childhood. *N Engl J Med.* 2003;348(11):977-985.
202. Brough HA, Liu AH, Sicherer S, et al. Atopic dermatitis increases the effect of exposure to peanut antigen in dust on peanut sensitization and likely peanut allergy. *J Allergy Clin Immunol.* 2015;135(1):164-170.
203. Boldogh I, Bacsi A, Choudhury BK, et al. ROS generated by pollen NADPH oxidase provide a signal that augments antigen-induced allergic airway inflammation. *J Clin Invest.* 2005;115(8):2169-2179.
204. Hammad H, Lambrecht BN. Barrier epithelial cells and the control of type 2 immunity. *Immunity.* 2015;43(1):29-40.
205. Kelleher MM, Dunn-Galvin A, Gray C, et al. Skin barrier impairment at birth predicts food allergy at 2 years of age. *J Allergy Clin Immunol.* 2016;137(4):1111-1116.e8.
206. Venkataraman D, Soto-Ramírez N, Kurukulaaratchy RJ, et al. Filaggrin loss-of-function mutations are associated with food allergy in childhood and adolescence. *J Allergy Clin Immunol.* 2014;134(4):876-882.e74.
207. Mitchell PD, O'Byrne PM. Epithelial-derived cytokines in asthma. *Chest.* 2017;151(6):1338-1344.
208. Roan F, Obata-Ninomiya K, Ziegler SF. Epithelial cell-derived cytokines: more than just signaling the alarm. *J Clin Invest.* 2019;129(4):1441-1451.
209. Gerbe F, Jay P. Intestinal tuft cells: epithelial sentinels linking luminal cues to the immune system. *Mucosal Immunol.* 2016;9(6):1353-1359.
210. Nadjsombati MS, McGinty JW, Lyons-Cohen MR, et al. Detection of succinate by intestinal tuft cells triggers a type 2 innate immune circuit. *Immunity.* 2018;49(1):33-41.e37.
211. Schneider C, O'Leary CE, Locksley RM. Regulation of immune responses by tuft cells. *Nat Rev Immunol.* 2019;19(9):584-593.
212. von Moltke J, Ji M, Liang HE, Locksley RM. Tuft-cell-derived IL-25 regulates an intestinal ILC2-epithelial response circuit. *Nature.* 2016;529(7585):221-225.
213. O'Leary CE, Schneider C, Locksley RM. Tuft cells-systemically dispersed sensory epithelia integrating immune and neural circuitry. *Annu Rev Immunol.* 2019;37:47-72.
214. Graham MT, Andorf S, Spergel JM, Chatila TA, Nadeau KC. Temporal regulation by innate type 2 cytokines in food allergies. *Curr Allergy Asthma Rep.* 2016;16(10):75.
215. Ricardo-Gonzalez RR, Van Dyken SJ, Schneider C, et al. Tissue signals imprint ILC2 identity with anticipatory function. *Nat Immunol.* 2018;19(10):1093-1099.
216. Chan BCL, Lam CWK, Tam LS, Wong CK. IL33: roles in allergic inflammation and therapeutic perspectives. *Front Immunol.* 2019;10:364.
217. Schuijs MJ, Hammad H, Lambrecht BN. Professional and "amateur" antigen-presenting cells in type 2 immunity. *Trends Immunol.* 2019;40(1):22-34.
218. Otsuka M, Egawa G, Kabashima K. Uncovering the mysteries of Langerhans cells, inflammatory dendritic epidermal cells, and monocyte-derived Langerhans cell-like cells in the epidermis. *Front Immunol.* 2018;9:1768.
219. Balhara J, Gounni AS. The alveolar macrophages in asthma: a double-edged sword. *Mucosal Immunol.* 2012;5(6):605-609.
220. Vroman H, Hendriks RW, Kool M. Dendritic cell subsets in asthma: impaired tolerance or exaggerated inflammation? *Front Immunol.* 2017;8:941.
221. Everts B, Tussiwand R, Dreesen L, et al. Migratory CD103+ dendritic cells suppress helminth-driven type 2 immunity through constitutive expression of IL-12. *J Exp Med.* 2016;213(1):35-51.
222. Hosoki K, Itazawa T, Boldogh I, Sur S. Neutrophil recruitment by allergens contribute to allergic sensitization and allergic inflammation. *Curr Opin Allergy Clin Immunol.* 2016;16(1):45-50.
223. Plantinga M, Guilliams M, Vanheerswynghels M, et al. Conventional and monocyte-derived CD11b(+) dendritic cells initiate and maintain T helper 2 cell-mediated immunity to house dust mite allergen. *Immunity.* 2013;38(2):322-335.
224. Cosmi L, Liotta F, Maggi L, Annunziato F. Role of type 2 innate lymphoid cells in allergic diseases. *Curr Allergy Asthma Rep.* 2017;17(10):66.
225. Doherty TA. At the bench: understanding group 2 innate lymphoid cells in disease. *J Leukoc Biol.* 2015;97(3):455-467.
226. Marshall JS, Warrington R, Watson W, Kim HL. An introduction to immunology and immunopathology. *Allergy Asthma Clin Immunol.* 2018;14(Suppl 2):49.
227. Lambrecht BN, Hammad H, Fahy JV. The cytokines of asthma. *Immunity.* 2019;50(4):975-991.
228. Scadding G. Cytokine profiles in allergic rhinitis. *Curr Allergy Asthma Rep.* 2014;14(5):435.
229. Zheng T, Zhu Z. Lessons from murine models of atopic dermatitis. *Curr Allergy Asthma Rep.* 2005;5(4):291-297.
230. Leong KP, Huston DP. Understanding the pathogenesis of allergic asthma using mouse models. *Ann Allergy Asthma Immunol.* 2001;87(2):96-109; quiz 110.
231. Schülke S, Albrecht M. Mouse models for food allergies: where do we stand? *Cells.* 2019;8(6):546.
232. Romagnani S. Immunologic influences on allergy and the TH1/TH2 balance. *J Allergy Clin Immunol.* 2004;113(3):395-400.
233. Yin Z, Chen C, Szabo SJ, Glimcher LH, Ray A, Craft J. T-Bet expression and failure of GATA-3 cross-regulation lead to default production of IFN-gamma by gammadelta T cells. *J Immunol.* 2002;168(4):1566-1571.
234. Raby BA, Hwang ES, Van Steen K, et al. T-bet polymorphisms are associated with asthma and airway hyperresponsiveness. *Am J Respir Crit Care Med.* 2006;173(1):64-70.
235. Lavender P, Cousins D, Lee T. Regulation of Th2 cytokine gene transcription. *Chem Immunol.* 2000;78:16-29.
236. Bao K, Reinhardt RL. The differential expression of IL-4 and IL-13 and its impact on type-2 immunity. *Cytokine.* 2015;75(1):25-37.
237. Marone G, Granata F, Pucino V, et al. The intriguing role of interleukin 13 in the pathophysiology of asthma. *Front Pharmacol.* 2019;10:1387.
238. Chusid MJ. Eosinophils: friends or foes? *J Allergy Clin Immunol Pract.* 2018;6(5):1439-1444.
239. Koch S, Sopel N, Finotto S. Th9 and other IL-9-producing cells in allergic asthma. *Semin Immunopathol.* 2017;39(1):55-68.

240. Neurath MF, Finotto S. IL-9 signaling as key driver of chronic inflammation in mucosal immunity. *Cytokine Growth Factor Rev.* 2016;29:93-99.
241. Shik D, Tomar S, Lee JB, Chen CY, Smith A, Wang YH. IL-9-producing cells in the development of IgE-mediated food allergy. *Semin Immunopathol.* 2017;39(1):69-77.
242. Padovan E. Modulation of CD4+ T helper cell memory responses in the human skin. *Int Arch Allergy Immunol.* 2017;173(3):121-137.
243. Turner DL, Goldklang M, Cvetkovski F, et al. Biased generation and in situ activation of lung tissue-resident memory CD4 T cells in the pathogenesis of allergic asthma. *J Immunol.* 2018;200(5):1561-1569.
244. Ugur M, Schulz O, Menon MB, Krueger A, Pabst O. Resident CD4+ T cells accumulate in lymphoid organs after prolonged antigen exposure. *Nat Commun.* 2014;5:4821.
245. Liu D, Tan Y, Bajinka O, Wang L, Tang Z. Th17/IL-17 Axis regulated by airway microbes get involved in the development of asthma. *Curr Allergy Asthma Rep.* 2020;20(4):11.
246. Pelletier M, Maggi L, Micheletti A, et al. Evidence for a cross-talk between human neutrophils and Th17 cells. *Blood.* 2010;115(2):335-343.
247. Chesné J, Braza F, Mahay G, Brouard S, Aronica M, Magnan A. IL-17 in severe asthma. Where do we stand? *Am J Respir Crit Care Med.* 2014;190(10):1094-1101.
248. Randolph AG, Lange C, Silverman EK, et al. The IL12B gene is associated with asthma. *Am J Hum Genet.* 2004;75(4):709-715.
249. Wang JY, Shyur SD, Wang WH, et al. The polymorphisms of interleukin 17A (IL17A) gene and its association with pediatric asthma in Taiwanese population. *Allergy.* 2009;64(7):1056-1060.
250. Hirose K, Ito T, Nakajima H. Roles of IL-22 in allergic airway inflammation in mice and humans. *Int Immunol.* 2018;30(9):413-418.
251. Brandt EB, Kovacic MB, Lee GB, et al. Diesel exhaust particle induction of IL-17A contributes to severe asthma. *J Allergy Clin Immunol.* 2013;132(5):1194-1204.e192.
252. Bellini A, Marini MA, Bianchetti L, Barczyk M, Schmidt M, Mattoli S. Interleukin (IL)-4, IL-13, and IL-17A differentially affect the profibrotic and proinflammatory functions of fibrocytes from asthmatic patients. *Mucosal Immunol.* 2012;5(2):140-149.
253. Czarnowicki T, Gonzalez J, Shemer A, et al. Severe atopic dermatitis is characterized by selective expansion of circulating TH2/TC2 and TH22/TC22, but not TH17/TC17, cells within the skin-homing T-cell population. *J Allergy Clin Immunol.* 2015;136(1):104-115.e07.
254. Sugaya M. The role of Th17-related cytokines in atopic dermatitis. *Int J Mol Sci.* 2020;21(4):1314.
255. Jin M, Yoon J. From bench to clinic: the potential of therapeutic targeting of the IL-22 signaling pathway in atopic dermatitis. *Immune network.* 2018;18(6):e42.
256. Brunner PM, Guttman-Yassky E. Racial differences in atopic dermatitis. *Ann Allergy Asthma Immunol.* 2019;122(5):449-455.
257. Noda S, Suárez-Fariñas M, Ungar B, et al. The Asian atopic dermatitis phenotype combines features of atopic dermatitis and psoriasis with increased TH17 polarization. *J Allergy Clin Immunol.* 2015;136(5):1254-1264.
258. Sanyal RD, Pavel AB, Glickman J, et al. Atopic dermatitis in African American patients is T(H)2/T(H)22-skewed with T(H)1/T(H)17 attenuation. *Ann Allergy Asthma Immunol.* 2019;122(1):99-110.e6.
259. Varricchi G, Harker J, Borriello F, Marone G, Durham SR, Shamji MH. T follicular helper (Tfh) cells in normal immune responses and in allergic disorders. *Allergy.* 2016;71(8):1086-1094.
260. Yang Z, Wu CM, Targ S, Allen CDC. IL-21 is a broad negative regulator of IgE class switch recombination in mouse and human B cells. *J Exp Med.* 2020;217(5):e20190472.
261. Kotlarz D, Ziętara N, Uzel G, et al. Loss-of-function mutations in the IL-21 receptor gene cause a primary immunodeficiency syndrome. *J Exp Med.* 2013;210(3):433-443.
262. Salzer E, Kansu A, Sic H, et al. Early-onset inflammatory bowel disease and common variable immunodeficiency-like disease caused by IL-21 deficiency. *J Allergy Clin Immunol.* 2014;133(6):1651-1659.e12.
263. Dienz O, Eaton SM, Bond JP, et al. The induction of antibody production by IL-6 is indirectly mediated by IL-21 produced by CD4+ T cells. *J Exp Med.* 2009;206(1):69-78.
264. Hercor M, Anciaux M, Denanglaire S, Debuisson D, Leo O, Andris F. Antigen-presenting cell-derived IL-6 restricts the expression of GATA3 and IL-4 by follicular helper T cells. *J Leukoc Biol.* 2017;101(1):5-14.
265. Spencer S, Köstel Bal S, Egner W, et al. Loss of the interleukin-6 receptor causes immunodeficiency, atopy, and abnormal inflammatory responses. *J Exp Med.* 2019;216(9):1986-1998.
266. Gowthaman U, Chen JS, Zhang B, et al. Identification of a T follicular helper cell subset that drives anaphylactic IgE. *Science.* 2019;365(6456):eaaw6433.
267. Hijnen D, Knol EF, Gent YY, et al. CD8(+) T cells in the lesional skin of atopic dermatitis and psoriasis patients are an important source of IFN-γ, IL-13, IL-17, and IL-22. *J Invest Dermatol.* 2013;133(4):973-979.
268. de Magalhães Simões S, dos Santos MA, da Silva Oliveira M, et al. Inflammatory cell mapping of the respiratory tract in fatal asthma. *Clin Exp Allergy.* 2005;35(5):602-611.
269. Hinks TSC, Hoyle RD, Gelfand EW. CD8(+) Tc2 cells: underappreciated contributors to severe asthma. *Eur Respir Rev.* 2019;28(154):190092.
270. Yu W, Zhou X, Dunham D, et al. Allergen-specific CD8(+) T cells in peanut-allergic individuals. *J Allergy Clin Immunol.* 2019;143(5):1948-1952.
271. Zajonc DM, Cantu C 3rd, Mattner J, et al. Structure and function of a potent agonist for the semi-invariant natural killer T cell receptor. *Nat Immunol.* 2005;6(8):810-818.
272. Bublin M, Eiwegger T, Breiteneder H. Do lipids influence the allergic sensitization process? *J Allergy Clin Immunol.* 2014;134(3):521-529.
273. Chandra S, Wingender G, Greenbaum JA, et al. Development of asthma in inner-city children: possible roles of MAIT cells and variation in the home environment. *J Immunol.* 2018;200(6):1995-2003.
274. Iwamura C, Nakayama T. Role of CD1d-and MR1-restricted T cells in asthma. *Front Immunol.* 2018;9:1942.
275. Bonneville M, O'Brien RL, Born WK. Gammadelta T cell effector functions: a blend of innate programming and acquired plasticity. *Nat Rev Immunol.* 2010;10(7):467-478.
276. Peters C, Häsler R, Wesch D, Kabelitz D. Human Vδ2 T cells are a major source of interleukin-9. *Proc Natl Acad Sci U S A.* 2016;113(44):12520-12525.
277. Bol-Schoenmakers M, Marcondes Rezende M, Bleumink R, et al. Regulation by intestinal γδ T cells during establishment of food allergic sensitization in mice. *Allergy.* 2011;66(3):331-340.
278. Frossard CP, Asigbetse KE, Burger D, Eigenmann PA. Gut T cell receptor-γδ(+) intraepithelial lymphocytes are activated selectively by cholera toxin to break oral tolerance in mice. *Clin Exp Immunol.* 2015;180(1):118-130.
279. Huang Y, Yang Z, McGowan J, Huang H, O'Brien RL, Born WK. Regulation of IgE responses by γδ T cells. *Curr Allergy Asthma Rep.* 2015;15(4):13.
280. Russano AM, Agea E, Corazzi L, et al. Recognition of pollen-derived phosphatidyl-ethanolamine by human CD1d-restricted gamma delta T cells. *J Allergy Clin Immunol.* 2006;117(5):1178-1184.
281. Sampath V, Tupa D, Graham MT, Chatila TA, Spergel JM, Nadeau KC. Deciphering the black box of food allergy mechanisms. *Ann Allergy Asthma Immunol.* 2017;118(1):21-27.
282. Georgiev P, Charbonnier LM, Chatila TA. Regulatory T cells: the many faces of Foxp3. *J Clin Immunol.* 2019;39(7):623-640.
283. Föhse L, Suffner J, Suhre K, et al. High TCR diversity ensures optimal function and homeostasis of Foxp3+ regulatory T cells. *Eur J Immunol.* 2011;41(11):3101-3113.
284. Bremel RD, Homan EJ. Extensive T-cell epitope repertoire sharing among human proteome, gastrointestinal microbiome, and pathogenic bacteria: implications for the definition of self. *Front Immunol.* 2015;6:538.
285. Haribhai D, Williams JB, Jia S, et al. A requisite role for induced regulatory T cells in tolerance based on expanding antigen receptor diversity. *Immunity.* 2011;35(1):109-122.
286. Shreffler WG, Wanich N, Moloney M, Nowak-Wegrzyn A, Sampson HA. Association of allergen-specific regulatory T cells with the onset of clinical tolerance to milk protein. *J Allergy Clin Immunol.* 2009;123(1):43-52.e47.
287. Lloyd CM, Hawrylowicz CM. Regulatory T cells in asthma. *Immunity.* 2009;31(3):438-449.
288. Matsumoto K, Inoue H, Fukuyama S, et al. Decrease of interleukin-10-producing T cells in the peripheral blood of severe unstable atopic asthmatics. *Int Arch Allergy Immunol.* 2004;134(4):295-302.
289. Noval Rivas M, Chatila TA. Regulatory T cells in allergic diseases. *J Allergy Clin Immunol.* 2016;138(3):639-652.
290. Lawrence MG, Barber JS, Sokolic RA, et al. Elevated IgE and atopy in patients treated for early-onset ADA-SCID. *J Allergy Clin Immunol.* 2013;132(6):1444-1446.
291. Milner JD, Ward JM, Keane-Myers A, Paul WE. Lymphopenic mice reconstituted with limited repertoire T cells develop severe, multiorgan, Th2-associated inflammatory disease. *Proc Natl Acad Sci U S A.* 2007;104(2):576-581.
292. Barber JS, Yokomizo LK, Sheikh V, et al. Peptide library-based evaluation of T-cell receptor breadth detects defects in global and regulatory activation in human immunologic diseases. *Proc Natl Acad Sci U S A.* 2013;110(20):8164-8169.
293. Noval Rivas M, Burton OT, Wise P, et al. Regulatory T cell reprogramming toward a Th2-cell-like lineage impairs oral tolerance and promotes food allergy. *Immunity.* 2015;42(3):512-523.
294. Ling EM, Smith T, Nguyen XD, et al. Relation of CD4+CD25+ regulatory T-cell suppression of allergen-driven T-cell activation to atopic status and expression of allergic disease. *Lancet.* 2004;363(9409):608-615.
295. Gould HJ, Wu YB. IgE repertoire and immunological memory: compartmental regulation and antibody function. *Int Immunol.* 2018;30(9):403-412.
296. Hirbod-Mobarakeh A, Aghamohammadi A, Rezaei N. Immunoglobulin class switch recombination deficiency type 1 or CD40 ligand deficiency: from bedside to bench and back again. *Expert Rev Clin Immunol.* 2014;10(1):91-105.
297. Lighaam LC, Rispens T. The immunobiology of immunoglobulin G4. *Semin Liver Dis.* 2016;36(3):200-215.
298. Croote D, Darmanis S, Nadeau KC, Quake SR. High-affinity allergen-specific human antibodies cloned from single IgE B cell transcriptomes. *Science.* 2018;362(6420):1306-1309.
299. Hoh RA, Joshi SA, Lee JY, et al. Origins and clonal convergence of gastrointestinal IgE(+) B cells in human peanut allergy. *Sci Immunol.* 2020;5(45):eaay4209.
300. Boonpiyathad T, Sözener ZC, Satitsuksanoa P, Akdis CA. Immunologic mechanisms in asthma. *Semin Immunol.* 2019;46:101333.
301. Lawrence MG, Woodfolk JA, Schuyler AJ, Stillman LC, Chapman MD, Platts-Mills TAE. Half-life of IgE in serum and skin: consequences for anti-IgE therapy in patients with allergic disease. *J Allergy Clin Immunol.* 2017;139(2):422-428.e4.
302. Kubo S, Nakayama T, Matsuoka K, Yonekawa H, Karasuyama H. Long term maintenance of IgE-mediated memory in mast cells in the absence of detectable serum IgE. *J Immunol.* 2003;170(2):775-780.
303. Kubo M. Mast cells and basophils in allergic inflammation. *Curr Opin Immunol.* 2018;54:74-79.
304. Schuligoi R, Sturm E, Luschnig P, et al. CRTH2 and D-type prostanoid receptor antagonists as novel therapeutic agents for inflammatory diseases. *Pharmacology.* 2010;85(6):372-382.
305. Siracusa MC, Kim BS, Spergel JM, Artis D. Basophils and allergic inflammation. *J Allergy Clin Immunol.* 2013;132(4):789-788.
306. Noti M, Wojno ED, Kim BS, et al. Thymic stromal lymphopoietin-elicited basophil responses promote eosinophilic esophagitis. *Nat Med.* 2013;19(8):1005-1013.
307. Caminati M, Menzella F, Guidolin L, Senna G. Targeting eosinophils: severe asthma and beyond. *Drugs Context.* 2019;8:212587.
308. Sastre B, Rodrigo-Muñoz JM, Garcia-Sanchez DA, Cañas JA, Del Pozo V. Eosinophils: old players in a new game. *J Investig Allergol Clin Immunol.* 2018;28(5):289-304.
309. Gounni AS, Wellemans V, Yang J, et al. Human airway smooth muscle cells express the high affinity receptor for IgE (Fc epsilon RI): a critical role of Fc epsilon RI in human airway smooth muscle cell function. *J Immunol.* 2005;175(4):2613-2621.
310. Balbino B, Conde E, Marichal T, Starkl P, Reber LL. Approaches to target IgE antibodies in allergic diseases. *Pharmacol Ther.* 2018;191:50-64.
311. Manti S, Brown P, Perez MK, Piedimonte G. The role of neurotrophins in inflammation and allergy. *Vitam Horm.* 2017;104:313-341.
312. Gupta K, Harvima IT. Mast cell-neural interactions contribute to pain and itch. *Immunol Rev.* 2018;282(1):168-187.
313. Davis MM, Tato CM, Furman D. Systems immunology: just getting started. *Nat Immunol.* 2017;18(7):725-732.
314. Bunyavanich S, Schadt EE. Systems biology of asthma and allergic diseases: a multiscale approach. *J Allergy Clin Immunol.* 2015;135(1):31-42.
315. Ghosh D, Bernstein JA, Khurana Hershey GK, Rothenberg ME, Mersha TB. Leveraging multilayered "omics" data for atopic dermatitis: a road map to precision medicine. *Front Immunol.* 2018;9:2727.
316. Dhondalay GK, Rael E, Acharya S, et al. Food allergy and omics. *J Allergy Clin Immunol.* 2018;141(1):20-29.
317. Suárez-Fariñas M, Ungar B, Correa da Rosa J, et al. RNA sequencing atopic dermatitis transcriptome profiling provides insights into novel disease mechanisms with potential therapeutic implications. *J Allergy Clin Immunol.* 2015;135(5):1218-1227.
318. Sherrill JD, Kiran KC, Blanchard C, et al. Analysis and expansion of the eosinophilic esophagitis transcriptome by RNA sequencing. *Genes Immun.* 2014;15(6):361-369.
319. O'Boyle KC, Ohtani T, Manne S, et al. Exploration of T-cell diversity using mass cytometry. *Methods Mol Biol.* 2020;2111:1-20.
320. Timp W, Timp G. Beyond mass spectrometry, the next step in proteomics. *Sci Adv.* 2020;6(2):eaax8978.
321. Stuart T, Satija R. Integrative single-cell analysis. *Nat Rev Genet.* 2019;20(5):257-272.
322. Chiang D, Chen X, Jones SM, et al. Single-cell profiling of peanut-responsive T cells in patients with peanut allergy reveals heterogeneous effector T(H)2 subsets. *J Allergy Clin Immunol.* 2018;141(6):2107-2120.
323. Kim CH. Immune regulation by microbiome metabolites. *Immunology.* 2018;154(2):220-229.
324. Karaki S, Garcia G, Tcherakian C, et al. Enhanced glucocorticoid-induced leucine zipper in dendritic cells induces allergen-specific regulatory CD4(+) T-cells in respiratory allergies. *Allergy.* 2014;69(5):624-631.
325. Komlósi ZI, Kovács N, Sokolowska M, van de Veen W, Akdis M, Akdis CA. Mechanisms of subcutaneous and sublingual aeroallergen immunotherapy: what is new? *Immunol Allergy Clin North Am.* 2020;40(1):1-14.
326. Sampath V, Sindher SB, Alvarez Pinzon AM, Nadeau KC. Can food allergy be cured? What are the future prospects? *Allergy.* 2020;75(6):1316-1326.
327. Passalacqua G, Canonica GW. Allergen immunotherapy: history and future developments. *Immunol Allergy Clin North Am.* 2016;36(1):1-12.
328. Yu W, Freeland DMH, Nadeau KC. Food allergy: immune mechanisms, diagnosis and immunotherapy. *Nat Rev Immunol.* 2016;16(12):751-765.

329. Cafone J, Capucilli P, Hill DA, Spergel JM. Eosinophilic esophagitis during sublingual and oral allergen immunotherapy. *Curr Opin Allergy Clin Immunol.* 2019;19(4):350-357.
330. Jutel M, Müller UR, Fricker M, Rihs S, Pichler WJ, Dahinden C. Influence of bee venom immunotherapy on degranulation and leukotriene generation in human blood basophils. *Clin Exp Allergy.* 1996;26(10):1112-1118.
331. Lao-Araya M, Steveling E, Scadding GW, Durham SR, Shamji MH. Seasonal increases in peripheral innate lymphoid type 2 cells are inhibited by subcutaneous grass pollen immunotherapy. *J Allergy Clin Immunol.* 2014;134(5):1193-1195.e4.
332. Ryan JF, Hovde R, Glanville J, et al. Successful immunotherapy induces previously unidentified allergen-specific CD4+ T-cell subsets. *Proc Natl Acad Sci U S A.* 2016;113(9):E1286-E1295.
333. Syed A, Garcia MA, Lyu SC, et al. Peanut oral immunotherapy results in increased antigen-induced regulatory T-cell function and hypomethylation of forkhead box protein 3 (FOXP3). *J Allergy Clin Immunol.* 2014;133(2):500-510.
334. Meiler F, Klunker S, Zimmermann M, Akdis CA, Akdis M. Distinct regulation of IgE, IgG4 and IgA by T regulatory cells and toll-like receptors. *Allergy.* 2008;63(11):1455-1463.
335. Vickery BP, Lin J, Kulis M, et al. Peanut oral immunotherapy modifies IgE and IgG4 responses to major peanut allergens. *J Allergy Clin Immunol.* 2013;131(1):128-134.e121-123.
336. Hoh RA, Joshi SA, Liu Y, et al. Single B-cell deconvolution of peanut-specific antibody responses in allergic patients. *J Allergy Clin Immunol.* 2016;137(1):157-167.
337. van de Veen W, Stanic B, Yaman G, et al. IgG4 production is confined to human IL-10-producing regulatory B cells that suppress antigen-specific immune responses. *J Allergy Clin Immunol.* 2013;131(4):1204-1212.
338. Konstantinou GN, Nowak-Węgrzyn A, Bencharitiwong R, Bardina L, Sicherer SH, Sampson HA. Egg-white-specific IgA and IgA2 antibodies in egg-allergic children: is there a role in tolerance induction? *Pediatr Allergy Immunol.* 2014;25(1):64-70.
339. Cianferoni A, Saltzman R, Saretta F, et al. Invariant natural killer cells change after an oral allergy desensitization protocol for cow's milk. *Clin Exp Allergy.* 2017;47(11):1390-1397.
340. Manka LA, Wechsler ME. New biologics for allergic diseases. *Expert Rev Clin Immunol.* 2018;14(4):285-296.
341. Pennington LF, Tarchevskaya S, Brigger D, et al. Structural basis of omalizumab therapy and omalizumab-mediated IgE exchange. *Nat Commun.* 2016;7:11610.
342. Just J, Deschildre A, Lejeune S, Amat F. New perspectives of childhood asthma treatment with biologics. *Pediatr Allergy Immunol.* 2019;30(2):159-171.
343. Teach SJ, Gill MA, Togias A, et al. Preseasonal treatment with either omalizumab or an inhaled corticosteroid boost to prevent fall asthma exacerbations. *J Allergy Clin Immunol.* 2015;136(6):1476-1485.
344. Beck LA, Thaçi D, Hamilton JD, et al. Dupilumab treatment in adults with moderate-to-severe atopic dermatitis. *N Engl J Med.* 2014;371(2):130-139.
345. Castro M, Corren J, Pavord ID, et al. Dupilumab efficacy and safety in moderate-to-severe uncontrolled asthma. *N Engl J Med.* 2018;378(26):2486-2496.
346. Bae MJ, Shin HS, Kim EK, Kim J, Shon DH. Oral administration of chitin and chitosan prevents peanut-induced anaphylaxis in a murine food allergy model. *Int J Biol Macromol.* 2013;61:164-168.
347. Ibañez AE, Smaldini P, Coria LM, et al. Unlipidated outer membrane protein Omp16 (U-Omp16) from Brucella spp. as nasal adjuvant induces a Th1 immune response and modulates the Th2 allergic response to cow's milk proteins. *PloS One.* 2013;8(7):e69438.
348. Jiang T, Ji H, Zhang L, Wang Y, Zhou H. Chitosan oligosaccharide exerts anti-allergic effect against shrimp tropomyosin-induced food allergy by affecting Th1 and Th2 cytokines. *Int Arch Allergy Immunol.* 2019;180(1):10-16.
349. Chua KY, Huangfu T, Liew LN. DNA vaccines and allergic diseases. *Clin Exp Pharmacol Physiol.* 2006;33(5-6):546-550.
350. Zhu Z, Yu J, Niu Y, et al. Enhanced prophylactic and therapeutic effects of polylysine-modified Ara h 2 DNA vaccine in a mouse model of peanut allergy. *Int Arch Allergy Immunol.* 2016;171(3-4):241-250.
351. Creticos PS, Schroeder JT, Hamilton RG, et al. Immunotherapy with a ragweed-toll-like receptor 9 agonist vaccine for allergic rhinitis. *N Engl J Med.* 2006;355(14):1445-1455.
352. Patel P, Holdich T, Fischer von Weikersthal-Drachenberg KJ, Huber B. Efficacy of a short course of specific immunotherapy in patients with allergic rhinoconjunctivitis to ragweed pollen. *J Allergy Clin Immunol.* 2014;133(1):121-129.e121-122.
353. Emeryk A, Bartkowiak-Emeryk M, Raus Z, Braido F, Ferlazzo G, Melioli G. Mechanical bacterial lysate administration prevents exacerbation in allergic asthmatic children-The EOLIA study. *Pediatr Allergy Immunol.* 2018;29(4):394-401.
354. Barlow PG, Findlay EG, Currie SM, Davidson DJ. Antiviral potential of cathelicidins. *Future Microbiol.* 2014;9(1):55-73.
355. Becker TM, Durrani SR, Bochkov YA, Devries MK, Rajamanickam V, Jackson DJ. Effect of exogenous interferons on rhinovirus replication and airway inflammatory responses. *Ann Allergy Asthma Immunol.* 2013;111(5):397-401.
356. Li P, Zhang T, Xiao Y, et al. Timing for the second fecal microbiota transplantation to maintain the long-term benefit from the first treatment for Crohn's disease. *Appl Microbiol Biotechnol.* 2019;103(1):349-360.
357. Makrgeorgou A, Leonardi-Bee J, Bath-Hextall FJ, et al. Probiotics for treating eczema. *Cochrane Database Syst Rev.* 2018;11(11):Cd006135.
358. Wei X, Jiang P, Liu J, Sun R, Zhu L. Association between probiotic supplementation and asthma incidence in infants: a meta-analysis of randomized controlled trials. *J Asthma.* 2020;57(2):167-178.
359. Qamer S, Deshmukh M, Patole S. Probiotics for cow's milk protein allergy: a systematic review of randomized controlled trials. *Eur J Pediatr.* 2019;178(8):1139-1149.
360. Wagenaar L, Bol-Schoenmakers M, Giustarini G, et al. Dietary supplementation with nondigestible oligosaccharides reduces allergic symptoms and supports low dose oral immunotherapy in a peanut allergy mouse model. *Mol Nutr Food Res.* 2018;62(20):e1800369.
361. Cait A, Hughes MR, Antignano F, et al. Microbiome-driven allergic lung inflammation is ameliorated by short-chain fatty acids. *Mucosal Immunol.* 2018;11(3):785-795.
362. Thorburn AN, McKenzie CI, Shen S, et al. Evidence that asthma is a developmental origin disease influenced by maternal diet and bacterial metabolites. *Nat Commun.* 2015;6:7320.
363. Zimmermann P, Ziesenitz VC, Curtis N, Ritz N. The immunomodulatory effects of macrolides-A systematic review of the underlying mechanisms. *Front Immunol.* 2018;9:302.
364. Bacharier LB, Guilbert TW, Mauger DT, et al. Early administration of azithromycin and prevention of severe lower respiratory tract illnesses in preschool children with a history of such illnesses: a randomized clinical trial. *J Am Med Assoc.* 2015;314(19):2034-2044.
365. Yamaya M, Nomura K, Arakawa K, et al. Clarithromycin decreases rhinovirus replication and cytokine production in nasal epithelial cells from subjects with bronchial asthma: effects on IL-6, IL-8 and IL-33. *Arch Pharm Res.* 2020;43(5):526-539.
366. Perkin MR, Logan K, Marrs T, et al. Enquiring about tolerance (EAT) study: feasibility of an early allergenic food introduction regimen. *J Allergy Clin Immunol.* 2016;137(5):1477-1486.e78.
367. Horimukai K, Morita K, Narita M, et al. Application of moisturizer to neonates prevents development of atopic dermatitis. *J Allergy Clin Immunol.* 2014;134(4):824-830.e26.
368. Lowe A, Su J, Tang M, et al. PEBBLES study protocol: a randomised controlled trial to prevent atopic dermatitis, food allergy and sensitisation in infants with a family history of allergic disease using a skin barrier improvement strategy. *BMJ Open.* 2019;9(3):e024594.
369. Sindher S, Alkotob SS, Shojinaga MN, et al. Pilot study measuring transepidermal water loss (TEWL) in children suggests trilipid cream is more effective than a paraffin-based emollient. *Allergy.* 2020;75(10):2662-2664.
370. Fiocchi A, Pawankar R, Cuello-Garcia C, et al. World Allergy Organization-McMaster University Guidelines for allergic disease prevention (GLAD-P): probiotics. *World Allergy Organ J.* 2015;8(1):4.
371. Li L, Han Z, Niu X, et al. Probiotic supplementation for prevention of atopic dermatitis in infants and children: a systematic review and meta-analysis. *Am J Clin Dermatol.* 2019;20(3):367-377.
372. Baker MG, Sampson HA. Phenotypes and endotypes of food allergy: a path to better understanding the pathogenesis and prognosis of food allergy. *Ann Allergy Asthma Immunol.* 2018;120(3):245-253.
373. Just J, Gouvis-Echraghi R, Couderc R, Guillemot-Lambert N, Saint-Pierre P. Novel severe wheezy young children phenotypes: boys atopic multiple-trigger and girls nonatopic uncontrolled wheeze. *J Allergy Clin Immunol.* 2012;130(1):103-110.e08.
374. Fitzpatrick AM, Teague WG, Meyers DA, et al. Heterogeneity of severe asthma in childhood: confirmation by cluster analysis of children in the national institutes of health/national heart, lung, and blood institute severe asthma research program. *J Allergy Clin Immunol.* 2011;127(2):382-389.e381-e313.
375. Chung KF. Precision medicine in asthma: linking phenotypes to targeted treatments. *Curr Opin Pulm Med.* 2018;24(1):4-10.
376. Deschildre A, Lejeune S, Cap M, et al. Food allergy phenotypes: the key to personalized therapy. *Clin Exp Allergy.* 2017;47(9):1125-1137.
377. Cabanillas B, Brehler AC, Novak N. Atopic dermatitis phenotypes and the need for personalized medicine. *Curr Opin Allergy Clin Immunol.* 2017;17(4):309-315.
378. Saglani S, Custovic A. Childhood asthma: advances using machine learning and mechanistic studies. *Am J Respir Crit Care Med.* 2019;199(4):414-422.

CHAPTER 48 Transplantation Immunology

Megan Sykes, MD • Anita S. Chong, PhD • Markus Y. Mapara, MD, PhD

INTRODUCTION

The transplantation of organs and cells between individuals saves or prolongs thousands of lives each year. The growing list of organs transplanted includes corneas, kidneys, livers, hearts, lungs, small intestines, pancreata, hands, faces, ie, uteri and penises. Currently, clinical cellular transplantation includes mainly islets of Langerhans and hematopoietic cells, but the list is likely to expand in the future to include other cell types, such as hepatocytes and myoblasts, which are currently under investigation in experimental models. Additionally, cell therapy with expanded T cells and pluripotent stem cell–derived products from self (autologous) and nonself (allogeneic) donors is a growing area that is beyond the scope of this chapter.

Success of all types of transplants that are genetically nonidentical to the recipient depends on the ability to avoid rejection due to a host-versus-graft immune response. Hematopoietic cell transplantation (HCT) is, in addition, associated with special considerations, as the administration of a donor graft that contains mature T cells to a conditioned, and consequently, immunoincompetent recipient is associated with the risk of rejection in the graft-versus-host direction (ie, graft-versus-host disease [GVHD]). These transplants, performed between different members of the same species, are referred to as allotransplants. However, the success of organ transplantation has presented a new limitation, namely the insufficient supply of human organs. This organ shortage has led to consideration of alternative sources of organs, such as artificial organs, tissue grafts engineered from stem cells, and other species. The latter, referred to as xenografts, are a promising alternative but present even greater immunologic challenges than allografts. This chapter presents an overview of the immunology of organ and cellular allotransplantation and xenotransplantation.

Allogeneic and xenogeneic responses differ from other immunologic responses in at least two fundamental ways. First, they exhibit extraordinary strength that includes multiple, redundant pathways. Second, they involve two different sets of antigen-presenting cells (APCs): those of the donor and those of the recipient. In this chapter, we will emphasize the uniqueness of and challenges presented by the immune response to transplantation.

ORIGINS OF TRANSPLANTATION IMMUNOLOGY

A brief summary of the history and immunogenetic principles underlying transplantation immunology is provided here. These topics are covered in greater detail in Paul's Fundamental Immunology VIIth Edition, to which the reader is referred for further information.

Early History

Although there were sporadic reports of tissue transplants from ancient times, experimental skin grafting did not become widely investigated until the late 1800s. Even then, autografts (where the donor and recipient are the same individual), allografts (where the donor and recipient are of the same species), or even xenografts (where donor and recipient are of different species) were not distinguished, and in fact, those efforts led to a period of confusion in transplantation, without a clear understanding of the processes involved in graft acceptance. Clarification of the immunological basis of allograft rejection is largely attributed to Medawar, who, driven by the need for skin grafts to treat burn victims in World War II, carried out a series of skin grafting experiments in rabbits that led to the conclusion that rejection of skin "belongs to the general category of actively acquired immune reactions."[1]

The ability to transplant solid organs was dependent on the breakthrough in vascular surgery by Alexis Carrel, who was awarded the Nobel Prize in 1912 for his discovery of "vascular suture and the transplantation of blood vessels and organs." By avoiding immune reactions through transplantation between identical twins, the first successful clinical renal transplantation was performed in 1954 by Joseph Murray, who received the 1990 Nobel Prize jointly with Donnall Thomas, who pioneered bone marrow transplantation.

Principles and Early Discoveries of Immunogenetics

Inbred Strains

Rodents are invaluable for the study of the genetic basis for graft rejection due to the availability of a large number of inbred strains. Historically, such strains were produced by sequential pedigreed brother-sister matings for at least 20 generations to ensure that they are homozygous for all autosomal chromosomes and produce identical homozygous progeny. The 20-generation benchmark was based on the probability of fixation of a given autosomal locus at each brother-sister mating, which is 1 in 8; thus, the probability of fixation of each locus is calculated by equation $1 - (7/8)^{n-1}$, where n is the number of generations. For practical purposes, 20 such brother-sister matings would result in a very small chance that any locus will not have reached homozygosity. More recently, the availability of genome sequencing and high-throughput genotyping has permitted the use of polymorphic DNA markers to reduce the number of backcrosses necessary to generate congenic strains. This process, termed speed congenics or marker-assisted accelerated backcrossing, allows for the selection at each generation of offspring that not only retain the desired chromosomal fragment but also "lose" the maximum amount of background genetic information from the donor strain. As a result, a congenic strain can be generated in as few as five generations of backcrosses.

In addition to mice, inbred strains have been produced in several other species, including rats, guinea pigs, rabbits, and swine. Notably, the large litter size, short gestation time, early sexual maturity, and an estrous cycle every 3 weeks have made it possible to develop major histocompatibility complex (MHC) homozygous lines in swine, the only large animal model that readily allows the reproducible investigation of the role of MHC class I and class II genes in allogeneic transplantation.

Genetic Principles Governing Tissue Transplantation

In the early 1900s, tumor biologists noted that tumors arising in inbred animals could frequently be transplanted successfully to other animals of the inbred line but not to outbred animals. Using these models, Little[2] summarized his results in what was considered to be the five laws of transplantation: (1) Transplants within inbred strains will succeed. (2) Transplants between inbred strains will fail. (3) Transplants from a member of an inbred parental strain to an F1 offspring will succeed, but those in the reverse direction will fail. (4) Transplants from F2 and all subsequent generations to F1 animals will succeed. (5) Transplants from inbred parental strains to the F2 generation will usually, but not always, fail. In reconciling his observations with classical Mendelian principles, Little inferred that recipients would reject grafts if the donor expressed a product of any histocompatibility locus that was not expressed by the recipient, that there must be codominant expression of the histocompatibility genes, and that there must be a relatively large number of histocompatibility loci.

Estimating the Number of Histocompatibility Genes

The number of histocompatibility loci was originally determined experimentally by crossing two inbred strains and generating a large cohort of F2 offspring, then transplanting tissues from one of the parental strains to F2 and measuring the fraction of grafts that survived. If the two strains differed by only one histocompatibility locus, then 75% of the grafts would survive. If there were two loci, then $(0.75)^2$ would survive. The end result of those experiments was the identification of 30 to 50 histocompatibility loci, and since there were only 20 chromosome pairs in the mouse genome, they inferred that some chromosomes carried more than one histocompatibility locus.

Producing Congenic Strains and Identifying the Major Histocompatibility Complex

The process to generate strains that differed only at a single histocompatibility locus was pursued by Snell at The Jackson Laboratory[3] and involved the production of congenic strains (inbred strains that differ from one another at only one independently segregating genetic locus) using rejection of the parental skin graft as the trait used to select successive matings. The histocompatibility locus that induced the most vigorous skin rejection was called the MHC, which in mice was called the histocompatibility system 2 or just the H-2, while the other 30 to 50 loci were called minor histocompatibility loci. There are now a large number of congenic mouse strains, including those that isolate the major or minor histocompatibility loci, and some rat congenic strains. In addition, during the generation of congenic strains, it became apparent that the MHC comprised multiple loci, based on recombination within the H-2 locus, and that these encoded two major types of antigens, now referred to as class I and class II MHC antigens.

DONOR ANTIGENS RESPONSIBLE FOR GRAFT REJECTION

Major Histocompatibility Antigens

As discussed above, the genetic analysis of graft rejection indicated that the antigens encoded within the MHC are of particular importance in graft rejection. Table 48.1 summarizes important aspects of the MHC antigens that are especially relevant for transplantation, while a much more detailed description of their structure and function can be found in Chapter 19. Some basic features of MHC antigens are briefly discussed here.

Polymorphism

The MHC antigens exhibit extraordinary polymorphism, providing the species with broad capacity to present the peptides of, and thus respond to, a large number of pathogens. In the human HLA complex, for example, there are currently >4000 alleles at each of the HLA-A and B (class I) and >2500 at the DRB1 (class II) locus. The high degree of polymorphism has important consequences for transplantation. Given that there are three classical class I loci (A, B, and C) and three to four class II loci (DQ, DP, DRB1, ± an expressed DRB 3, 4, or 5 locus present in some haplotypes) on each haplotype, the likelihood of achieving identity for MHC antigens in two unrelated humans is extremely small, though for individuals with two common HLA haplotypes,[4] the likelihood is increased.

TABLE 48.1 Summary of Features of Major Histocompatibility Complex

Class I antigens	Single polymorphic chain. Three domains: alpha 1, 2, and 3 MW: 45,000 Associated with beta 2 Microglobulin A, B, and C loci in humans Expressed on all tissues and cells
Class II antigens	Two polymorphic chains: alpha and beta Each with two domains: alpha 1 and 2, beta 1 and 2 MW: 33,000 and 28,000 DP, DQ, and DR loci in humans Expressed on macrophages, dendritic cells, and B cells; vascular endothelium; activated human T cells

Tissue Distribution

The tissue distribution of the two types of MHC antigens differs. Class I antigens are constitutively expressed on all nucleated cells but at low levels on some types of cells. Class II MHC antigens are more selective in their distribution. They are especially frequent on macrophages, dendritic cells, B lymphocytes, and thymic epithelium. They are present on other lymphoid cells under some circumstances and on vascular endothelium. Their expression on some tissues of the body is regulated, for example, by interferons. One of the important distinctions between rodents and many larger species is the lack of constitutive expression of class II antigens on the vascular endothelium and other cell populations in rodents. In contrast, pigs, monkeys and humans express class II antigens on these tissues,[5] a point to be borne in mind in comparing results in experimental models.

Physiologic Function of MHC Antigens

MHC antigens are called "histocompatibility" antigens because of their powerful role in causing graft rejection; yet, they did not evolve in nature to prevent tissue grafting. While the name serves to emphasize the historical importance of transplantation in the discovery of the MHC, the essential role of MHC antigens is now understood to involve the presentation of peptides of foreign antigens to responding T cells (see Chapters 19 and 20).

The Importance of MHC Antigens in Alloreactivity

MHC antigens are exceptionally important in stimulating T and B cell alloresponses. This section will focus mainly on T cell responses to MHC, as alloantibody responses are discussed later in the chapter.

Vigorous Graft Rejection. Mouse skin grafts differing only in their MHC antigens are typically rejected in eight to 10 days, whereas MHC-matched grafts may be rejected more slowly, depending on the number of minor histocompatibility antigen differences. In pigs, primarily vascularized organs, such as the kidney, may survive indefinitely in some cases, even without immunosuppression, if all of their MHC antigens are matched, whereas MHC-mismatched kidneys are always rejected within 2 weeks.[6]

Strong In Vitro Alloresponses. Allogeneic MHC antigens stimulate an extraordinarily strong T cell response in vitro, whereas responses to non-MHC antigens generally require in vivo priming or stimulation by strong professional APCs. For the most part, CD4 cells recognize class II alloantigens, and CD8 cells recognize class I alloantigens. However, this strong bias is more stringent for CD4 cells than CD8 cells. The standard in vitro assay of T helper function, the mixed lymphocyte response (MLR), measures proliferation of T cells after allogeneic stimulation. This has been adapted into flow cytometric CFSE dye dilution assays, which, along with intracellular cytokine staining, enhances the ability to detect and functionally characterize alloreactive T cells.[7]

Limiting dilution assays can be used to quantify alloreactive proliferating or cytokine-producing cells. Such analyses have led to frequency estimates of approximately 1% to 7% of T cells responding to a particular allogeneic donor,[8,9] while naïve T cells reactive with an exogenous peptide generally represent only approximately 0.01% to 0.001% of the same T cell pool.[10] Recent studies have used a high-throughput T cell receptor (TCR) CDR3β sequencing-based method of measuring divided compared to undivided T cell frequencies in CFSE-MLR to identify and quantify the alloreactive repertoire. These studies estimated alloreactive TCR frequencies in the same range as those determined by the above functional assays.[11,12]

Strong primary direct alloresponses of CD8+ T cells can also be measured in vitro, either in standard cell-mediated lympholysis (CML) assays or in limiting dilution assays measuring CTL precursor frequencies or using the high-throughput TCR-sequencing method described in the preceding paragraph. Prior in vivo activation by alloantigens can be inferred by the detection of cytotoxic activity without further in vitro stimulation and increased alloreactive precursor frequencies using modified CTL assays[13] and ELISpot assays.[14]

Direct Recognition of Allogeneic MHC Antigens. The extraordinary strength of alloreactivity largely reflects the ability of T cells to recognize allogeneic MHC antigens presented on donor APC, referred to as "direct" allorecognition. TCR sequencing-based analyses of human alloreactive T cell repertoires demonstrated the uniqueness of the repertoire for each allogeneic donor which, combined with the measured and extrapolated (for unseen species) frequency of alloreactive T cells for each allogeneic donor, have led to the conclusion that most human TCRs are likely to cross-react on some allogeneic HLA molecule,[11] likely reflecting the absence of allogeneic MHC during intrathymic negative selection, which is induced by strong TCR recognition of MHC-peptide complexes. Three different, but not mutually exclusive hypotheses have been proposed to explain the high frequency of alloreactive T cells:

(I) GENETIC BIAS: Because the thymus only positively selects T cells with some MHC reactivity, a TCR gene pool with intrinsic affinity for MHC molecules would allow for more efficient thymic selection. TCRs indeed have intrinsic affinity for MHC molecules.[15-17] Given the degeneracy in antigens recognized by TCRs, with multiple potential ligands for each, intrinsic allogeneic MHC reactivity thereby characterizes a T cell repertoire that has been negatively selected only by "self" MHC-peptide complexes.

(II) HIGH-DETERMINANT DENSITY: The density of individual peptide determinants presented by an APC should be quite low (since MHC antigens present diverse peptides), whereas the density of a peptide-independent allogeneic MHC determinant on allogeneic APCs would be very high (since every MHC antigen would include the foreign determinant). These abundant allogeneic MHC determinants might activate many T cells that recognize them with relatively low affinities. This hypothesis requires that allogeneic MHC molecules can be recognized at least partly independently of the peptides they present.

While peptide-independent and peptide-"promiscuous" alloreactive T cell clones have been described,[18,19] many peptide-specific alloreactive T cell clones have also been described[20-22] and most allorecognition is likely of this kind (see paragraph below).[23]

(III) MULTIPLE BINARY COMPLEXES: The third explanation for alloreactivity, for which there is strong evidence, is based on the idea that allospecificities include the range of specific peptides presented by allogeneic MHC molecules.[24] Positive selection, which requires lower TCR affinity for self-MHC/peptide complexes than that involved in negative selection,[25] enriches for TCRs capable of seeing self MHC plus peptide, which may cross-react on allogeneic MHC molecules. Indeed, TCRs may be seen as "polyspecific" due to their ability to bind to distinct peptide/MHC ligands in different conformations.[26] T cells strongly responsive to self-peptides on self MHC molecules are eliminated by negative selection, but negative selection would not affect T cells that recognize (self or nonself) peptides presented on allogeneic MHC molecules. Consistent with this hypothesis, many alloreactive T cells have been shown to be peptide-specific or at least peptide-dependent.[18,22,27-30] Delayed rejection of cardiac allografts from H2 DM$^{-/-}$ mice, which lack the capacity to replace invariant chain-derived CLIP peptide with a more diverse array of peptides, provide in vivo evidence for the importance of peptides in direct allorecognition.[31]

Overall, the available information supports the inherent MHC binding capacity of TCR, combined with the multiple binary complex notion, to explain the high frequency of T cells recognizing alloantigens directly.

Minor Histocompatibility Antigens

While initially defined by their ability to cause rapid graft rejection, MHC antigens are defined in part by the location of the genes encoding them and in part by the well-characterized structure of both class I and class II antigens (see Chapter 20). Minor histocompatibility antigens (MiHA), on the other hand, are those capable of eliciting a T cell immune response but that lack the structural characteristics of MHC products. Rather than being allelic cell surface proteins, MiHA are donor-specific peptides presented by MHC molecules that are shared by donor and recipient. Since individuals are tolerant to the peptides derived from their own proteins and presented on their own MHC, they will only respond to the peptides of another individual's proteins on a shared MHC molecule if the peptide is not shared by the recipient. Unlike MHC antigens, MiHC do not readily stimulate primary in vitro cell-mediated responses in MLR and CML assays, reflecting the low frequency of T cells recognizing them in the unprimed T cell repertoire. Nevertheless, they are sufficiently important to necessitate long-term immunosuppressive therapy in recipients of HLA-identical sibling kidney allografts and are the cause of GVHD in the common setting of HLA-identical sibling HCT.

It has been estimated that there may be as many as 720 minor histocompatibility loci in mice,[32] some of which are autosomal and others of which are encoded on the Y chromosome. MiHA can be expressed ubiquitously or in a tissue-selective manner.[33] Many proteins producing MIHA have been identified, some of which are intracellular proteins such as nuclear transcription factors and myosin, while others, like CD31 and CD19, are polymorphic cell surface glycoproteins. Some MiHA are peptides with two allelic variants, both of which can be represented by a particular MHC molecule,[34,35] resulting in bidirectional recognition (eg, the murine H13 locus[35]). Alternatively, allelic variation in MHC binding capacity of a peptide can result in one allele being presented and the other not (eg, the human HA-1 minor antigen, in which only one of two allelic peptides binds effectively to HLA-A2[34]). Minor antigenic determinants can also result from the failure of one allele to be processed to a peptide. An example is HA-8, for which only one allele is effectively transported by the TAP complex, resulting in a null allele despite the presence of the MHC-binding peptide sequence in the molecule.[36]

Both helper determinants, recognized by CD4$^+$ cells, and cytotoxic determinants, recognized by CD8$^+$ cells, are required for effective cytotoxic T cell responses to MiHA.[37,38] When multiple MiHA disparities exist, a phenomenon known as "immunodominance" may occur. Removal of the immunodominant recognition can reveal strong responses to antigens that evoked weak or no responses before. This phenomenon may be due to competition between peptides for presentation by MHC molecules,[32] as well as differing durations of antigen presentation and TCR avidities.[39] An exceptional murine peptide, H60, is derived from an NKG2D-binding class I MHC-like protein, H60a, for which there is a null allele and, in such strains of mice, produces responses that are comparable in potency to those elicited by MHC alloantigens. This phenomenon apparently reflects a high frequency and diversity of naïve T cell TCRs recognizing this peptide. Because of its hematopoietic cell-restricted expression, H60 is a useful target for graft-vs-leukemia effects without GVHD in murine HCT models.[40] It should be borne in mind that immunodominance of CTL responses measured in vitro does not necessarily reflect the immunodominance of the same antigens in vivo,[32,41] possibly reflecting the importance of tissue distribution of minor antigen expression or of helper T cell responses to minor antigens.[42]

The H-Y antigens are encoded on the Y chromosome and therefore expressed only by males.[32,43] They are of biological significance, as they cause rejection of male skin grafts in otherwise syngeneic female mice and, in humans, female to male HCT is associated with increased GVHD rates compared to other combinations in the HLA-identical related donor setting.

Human MiHA have been identified as determinants recognized by CD4+ or CD8+ cells, mainly in the setting of HLA-identical sibling donor HCT, in association with GVHD and marrow graft rejection.[36,42,44-46] Certain MiHA incompatibilities (eg, HA-1) may predispose to GVHD.[47] Immunodominance of CTL responses to particular H-Y and HA determinants as well as expansion of minor antigen-specific CTL detected with tetrameric complexes of HLA molecules and minor antigenic peptides[44,48] have been associated with GVHD and marrow graft rejection.[49]

It is very difficult to detect humoral responses to MiHA, presumably because individual peptide/MHC complexes are too low in abundance on the cell surface to either stimulate an antibody response or be detected on the cell surface with antibodies. However, antibodies against H-Y miHA can be detected in association with chronic GVHD.[50]

Other Antigens of Potential Importance in Transplantation

Superantigens

Superantigens include products of endogenous retroviruses in mice, such as mammary tumor viruses, and bacterial products such as staphylococcal enterotoxin B. Like MHC antigens, superantigens can stimulate primary in vitro T cell proliferative responses and activate a large component of the T cell repertoire. However, these antigens are not presented as peptides in the binding groove of MHC molecules but instead bind to distinct regions of class II MHC molecules, and engage nonvariable portions of Vβ components of the TCR, rather than the hypervariable regions that recognize peptides. Endogenous superantigens are not classical transplantation antigens, perhaps because of their restricted tissue expression patterns, but they have been found to contribute to GVHD in mice.[51]

Tissue-Specific Antigens

Some peptides are derived from proteins with limited tissue distribution which may not include hematopoietic cells used in traditional assays of alloreactivity. One implication is that transplantation tolerance induced by one set of donor cells might not always induce complete tolerance to donor cells of a different sort. For example, skin-specific antigens may be targets of skin graft rejection despite the presence of stable hematopoietic chimerism induced by HCT.[52] Tissue-specific proteins lacking allelic variation may serve as alloantigens when presented by a donor MHC allele that is not shared with the recipient.

Several human MiHA may be expressed only on hematopoietic cells, and these may provide an opportunity to use GVH-alloreactive donor T cells (that recognize such antigens) to achieve graft-versus-leukemia (GVL) effects without GVHD (largely a disease affecting epithelial tissues). However, disparities for some of these minor antigens (eg, HA-1) have in fact been associated with an increased incidence of GVHD.[47,53]

Endothelial Glycoproteins

Blood Group Antigens

Blood group antigens are the products of glycosylation enzymes that are not the same in all individuals. They are expressed on erythrocytes and other cells and, importantly, on vascular endothelium where they may serve as the targets for "natural" antibody-mediated attack on blood vessels of organ grafts. Blood group A and B individuals each express their respective antigen but O individuals have neither. "Natural" antibodies against blood group antigens an individual lacks probably arise due to cross-reactions with common carbohydrate determinants of environmental microorganisms. Type O individuals have antibodies to the antigens of A and B donors, while A and B individuals only have antibodies reactive with antigens from each other, and AB individuals have antibodies to neither. Therefore, O recipients can only receive transfusions from O donors; A and B recipients can receive transfusions from O donors or from individuals sharing their blood type; and AB recipients can receive blood from donors of any blood type. The same rules apply to the transplantation of most organs.[54] Recently, advances have been made in the ability to successfully transplant organs across ABO barriers in infants, where B cell tolerance can be induced, or by adsorbing antibody from the plasma and depleting B cells.[55] Other, non-ABO blood group antigens on erythrocytes are irrelevant to organ transplantation because they are not expressed on vascular endothelium.

Blood group antigens are of lesser importance, but nevertheless significant, in HCT. ABO incompatibility in the host-versus-graft direction ("major" ABO mismatch) can lead to prolonged red cell aplasia following HCT; incompatibility in the GvH direction ("minor" mismatch) can result in initial hemolytic anemia, but this complication can be avoided by washing the donor HCT preparation to rid it of plasma.[56,57]

Species-Specific Carbohydrate Determinants

Closely analogous to the blood group antigens are the carbohydrate determinants expressed on vascular endothelium that show species selectivity. For example, pigs, which are of interest as an organ source for xenotransplantation, have a glycosyl-transferase enzyme not expressed by humans, that glycosylates β-galactosyl N-acetyl glucosamine to form a Galα1-3Galβ1-4GlcNAc (αGal) determinant. This enzyme is present in most species but underwent a loss of function mutation in our nonhuman primate ancestors. Preformed or "natural" antibodies are present in human serum which react to the nonself pig determinant. Like the blood group antibodies, these natural antibodies probably arise from cross-reactions with environmental microorganisms, and they cause hyperacute rejection (HAR) of most primarily vascularized xenogeneic transplants.

"Missing Self" and Natural Killer Cell Recognition

In apparent violation of the laws of transplantation described above, (AxB) F1 mice are capable of rejecting bone marrow from parental donors, a phenomenon termed hybrid resistance. This phenomenon, as well as rapid rejection of fully allogeneic marrow, is mediated by natural killer (NK) cells. NK cells are innate lymphoid cells lacking TCRs that have the ability to mediate cytolysis against certain tumor targets and hematopoietic cells. NK cells also produce a number of proinflammatory, hematopoietic and even anti-inflammatory cytokines and may be divided into subsets on the basis of their cytokine production pattern (see Chapter 18).

The originally puzzling specificity of NK cell–mediated marrow rejection is due to the expression by NK cells of inhibitory receptors that recognize specific groups of class I MHC alleles on target cells and prevent cytolysis by the NK cell. These class I receptors are type II C lectins (Ly49 family) or dimers of CD94 with NKG2 lectins in the mouse and are either immunoglobulin family members (killer inhibitory receptors, KIRs) or CD94/NKG2 molecules in humans.

Recognition by an inhibitory receptor of a class I ligand results in intracellular transmission of an inhibitory signal via an immune receptor tyrosine-based inhibitory motif (ITIM) that interacts with a tyrosine phosphatase and counteracts activating signals transmitted from other cell surface molecules. Recognition of "self" class I inhibitory ligands prevents NK cells from killing normal autologous cells. Other molecules expressed by infected or stressed cells activate NK cells, counterbalancing these inhibitory signals (see Chapter 18).

Inhibitory receptors are clonally distributed on NK cells, each of which may express one or more different inhibitory receptor. For NK cells to be functional yet tolerant of "self," they must express at least one inhibitory receptor for a "self" class I MHC molecule.[58] Thus, as is illustrated in Figure 48.1, an AxB F1 recipient will have subsets of NK cells with inhibitory receptors that recognize MHC of either the A parent, the B parent, or both. The absence of "B" class I molecules on, for example, AA parental hematopoietic cells permits subsets of $(AxB)F_1$ NK cells that have inhibitory receptors only for class I molecules from the B parent to destroy AA cells. Thus, hybrid resistance and NK cell–mediated resistance to fully allogeneic marrow grafts can be explained on the basis of "missing self."[59] The roles of NK cells in transplantation are discussed in later sections.

TRANSPLANTATION OF SPECIFIC ORGANS AND TISSUES

Kidney Transplantation

Kidneys are the most frequently transplanted human organs, with more than 23,000 kidney transplants having been performed in the United States in 2019. Renal allograft survival has increased steadily since its inception and patient and organ survival are >97% and 94%, respectively, at 1 year (https://optn.transplant.hrsa.gov/data/view-data-reports/national-data/#). However, even well-matched recipients of renal transplants must take immunosuppressive medications for the rest of their lives, with attendant risks of infection,

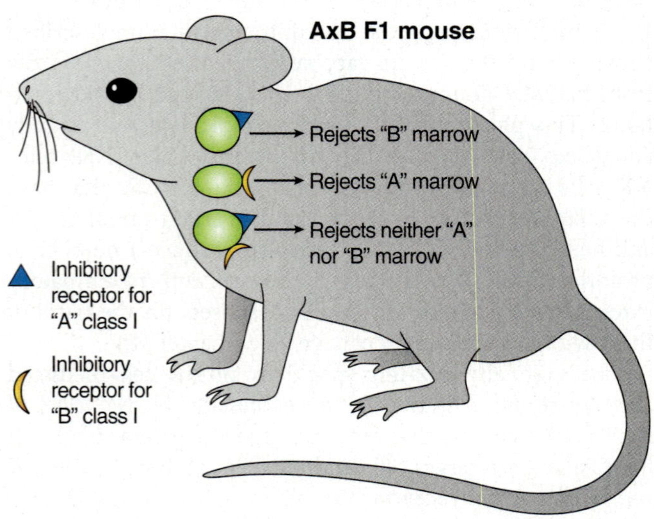

FIG 48.1. An explanation for hybrid resistance. Each filled circle represents a subset of natural killer cells.

cancer, hypertension, nephrotoxicity, diabetes, and metabolic bone disease. As for other organs, a major obstacle in kidney transplantation is the shortage of organs. The waiting time for cadaveric kidneys can be four or more years even for unsensitized candidates and even greater for those with high levels of sensitization. To overcome this graft shortage and broaden donor pools, many patients are now being offered kidneys from so-called "marginal" donors, as defined based on age and/or comorbidities. Further steps to increase the donor pool are based on using hepatitis C positive organ donors, which was recently made possible by the introduction of potent antiviral treatment leading to viral eradication.[60,61] Furthermore, the donor pool has been expanded in recent years by paired living donor exchange programs, ABO-mismatched transplants, and desensitization protocols. Unfortunately, for all kidney transplants, even when immunosuppression is well managed, there is an inexorable loss of kidneys to "chronic rejection." This late graft loss probably involves both immunologic alloreactivity and other factors, including the effects of the early ischemic injury and the ongoing effects of drugs and metabolic abnormalities.[62] Thus, although the 1-year survival rate for kidney transplants has improved from roughly 40% to >90% over the past 40 years, 5-year graft survival is only 71%. Indeed, the half-life for kidney transplants has barely increased during the same period and remains less than 10 years, though recent results suggest that costimulatory blockade could improve these figures.[63]

Liver Transplantation

More than 8000 liver transplants were performed in the US in 2019. Organ and patient survival rates have steadily increased annually, with 1-year graft survivals of about 89% and patient survival rates of more than 90% in 2019. The severe shortage of available organs has recently led to increasing use of living donors, resulting in more than 500 living donor transplants in 2019. Immunologically, the liver is relatively resistant to early antibody-mediated rejection (AMR), allowing successful transplantation even across blood group barriers and in the face of detectable donor-specific antibodies (DSA), although an absence of these antibodies is generally considered preferable. In addition, long-term survival of liver transplants does not appear to be improved by HLA matching between donor and recipient[64] and many centers do not even type recipients and donors. The ability of the liver to tolerate immune insults may relate to its strong regenerative capacity. Long-term results of liver transplantation are almost as good as those for kidney, with 72% graft survival and 76% patient survival at 5 years.

Heart and Lung Transplantation

More than 3000 heart transplants were performed in 2019 and 1-year graft survival is above 90%. A major challenge is the high rate of atherosclerotic disease or "graft vasculopathy" in the coronary arteries of the transplanted hearts, which is considered to be a manifestation of chronic rejection. Nevertheless, patient and graft survivals are close to 80% for heart transplants at 5 years.

The number of lung transplants performed in 2019 in the United States was approximately 2700 and 1-year patient survival is now about 87%. However, these excellent early outcomes are followed by a steep decline, resulting in a 5-year survivals of only about 53% for grafts and 56% for patients (https://optn.transplant.hrsa.gov/data/view-data-reports/national-data/). A major cause of graft loss is a process called bronchiolitis obliterans, thought to be the pulmonary manifestation of chronic rejection. In addition, long-term survival is reduced as a consequence of pulmonary infections.

Pancreas and Islet Transplantation

Successful pancreas transplantation to treat diabetes mellitus has been achieved using new technical approaches, with success rates of about 96% at 1 year when cotransplanted with a donor kidney. There are approximately 850 simultaneous kidney/pancreas transplants per year in the United States. Surprisingly, rejection activity in one organ is not always associated with rejection activity in the other, possibly reflecting tissue-specific antigens or localized inflammatory (or regulatory) events in one but not the other organ. Most pancreas transplantation is carried out as a curative treatment for diabetes mellitus. For this purpose, the potential also exists of transplanting only the insulin-producing islet cells rather than the whole pancreas. In 2000, publication of the "Edmonton Protocol"[65] using a new combination of immunosuppressive drugs that appeared to cause less damage to transplanted islets led to renewed hope for islet transplantation. However, this protocol required use of two or sometimes three pancreata for a single recipient, and by 5 years, in a much larger series of patients, only 10% of patients remained insulin independent.[66] Thus, these results, while representing an advance, remain inferior to the results of whole organ pancreas transplantation.

Hematopoietic Cell Transplantation

A major distinguishing feature of HCT compared to solid organ transplantation is that the recovering immune system in durable hematopoietic chimeras is tolerant of the donor alloantigens, so that in principle there is no requirement for continued immunosuppressive therapy to prevent allograft rejection once the initial immune resistance to the allograft has been overcome. However, development of acute and chronic GVHD are common complications, which may require extended treatment with immunosuppressive agents. Another fundamental difference between HCT for malignant disease and the transplantation of solid organs is that HCT requires initial recipient conditioning to allow eradication of the underlying malignancy and also to prevent graft rejection. This conditioning may result in ablation of the immune and hematopoietic systems prior to transplantation. Originally, hematopoietic cell allografts were administered only as a means of replacing ablated host hematopoiesis. However, experimental and clinical observations revealed that one of the main therapeutic benefits of allogeneic HCT is due to the GVL or graft-versus-malignancy effect of donor lymphocytes. With the recognition of this immunotherapeutic benefit of allogeneic HCT, clinicians began to evaluate less toxic, nonmyeloablative conditioning as a means of allowing allogeneic marrow to engraft so that donor lymphocytes can mediate GVL effects.[67,68] The introduction of these nonmyeloablative conditioning regimens with successful achievement of stable donor cell engraftment has allowed extension of allogeneic transplantation to nonmalignant indications such as sickle cell disease[69] and for the induction of organ allograft tolerance (see below).

The hematopoietic stem and progenitor cells used for HCT can be obtained from the bone marrow or from hematopoietic stem cells and progenitors mobilized from the marrow into peripheral blood by treatment with granulocyte colony–stimulating factor and/or CXCR4/SDF-1 antagonist drugs, or from umbilical cord blood (UCB). HCT is used most commonly for the treatment of otherwise incurable hematologic malignancies, aplastic anemia, hemoglobinopathies, inborn errors of metabolism, and congenital immunodeficiency states. Autologous HCT is used for hematologic rescue following high-dose chemo-/radiotherapy for the treatment of hematologic malignancies, especially plasma cell dyscrasias, and has recently been established as a treatment for certain autoimmune diseases (ie, diffuse systemic sclerosis).

In 2018, more than 8000 allogeneic and 14,000 autologous stem cell transplants were performed in the US. However, autologous HCT will not be considered further here as it does not involve the broaching of any immunologic barriers. Allogeneic HCT can be performed using HLA-matched or mismatched related or unrelated donors. For unrelated donation, high-resolution HLA-matching was imperative until recently, given the extensive risk of graft rejection or GVHD and subsequent excessive treatment-related mortality (TRM) when HLA-mismatched donors were used. With high-resolution HLA matching, finding an appropriately HLA-matched donor has been a significant challenge, especially for patients who are members of minority populations and are underrepresented in the worldwide stem cell donor banks. Expanding the donor pool by utilization of mismatched donors was largely precluded by the associated excessive toxicity. Creation and use of UCB registries were of crucial relevance in helping to alleviate this donor shortage, as HLA-mismatched UCB grafts were better tolerated possibly due to the naivety/immaturity of the donor immune cells. However, UCB cannot solve the donor shortage for a number of reasons, most importantly due to the limited dose of hematopoietic stem cells contained within a cord blood unit. As an alternative, success in traversing extensive HLA-barriers using haploidentical related donors was achieved using the mega-dose concept pioneered by Reisner and Martelli, which involved intensive recipient conditioning to prevent graft rejection, administration of mega-doses of hematopoietic stem cells ($>10 \times 10^6$ CD34 cells/kg recipient body weight), and highly efficient in vivo and ex vivo T cell depletion (TCD) of the graft.[70] This approach to haploidentical transplantation showed initial promise but failed to achieve widespread application due to high rates of infectious and other transplant-related complications.[71] The recent introduction of efficient modulation of host-versus-graft (HvG) and GvH responses by administration of posttransplant cyclophosphamide (PT-Cy)[72] has made it possible to cross HLA barriers safely without the need for myeloablation[73] and has resulted in widespread implementation of haploidentical transplantation across the world. The PT-Cy approach has also been introduced for HLA-mismatched unrelated donor transplantation.[74,75] Thus, this approach has provided a breakthrough in allowing HCT across HLA barriers, thereby overcoming the previous limitation in donor availability when HLA matching was required. Recent studies suggest that haploidentical transplants using PTCy are superior to double UCB transplants in terms of toxicity and overall survival.[76]

The GVL effect of allogeneic lymphocytes, especially T cells, transferred with the donor graft, is due largely to recognition by donor T cells of host alloantigens, which are also expressed on malignant cells. Therefore, while T-cell depletion of the donor graft decreases the incidence of GVHD[77] increased relapse rates[78] and increased risk of engraftment failure due to rejection[79] may result. In the setting of T cell depleted HLA-mismatched transplants, GVL effects can be mediated by NK cells when the host lacks inhibitory MHC ligands recognized by the donor.[80] Such alloreactive NK cells may also help to promote donor engraftment by eliminating host T cells that would otherwise reject donor marrow.[80]

Composite Tissue Allografts

Composite tissue transplantation is an emerging area in transplantation medicine which presents a multitude of challenges related to technical complexity, immunological barriers, and ethical considerations regarding transplants that are needed to improve quality rather than sustain life. Improvements in immunosuppressive treatment options over the last 3 decades have enabled the performance of vascularized composite allografts (VCA), which may include multiple tissues such as skin, bones, muscles, and nerves. Examples of such transplanted grafts are face, limbs, penis, uterus, abdominal wall, and larynx.[81-84] Fifteen VCA procedures were performed in the US as recorded in the UNOS database in 2019.

The simultaneous transplantation of different tissues poses a formidable challenge from an immunological perspective, given the complex cellular composition and associated immunogenicity of these diverse grafts. In accordance with observations in animal studies indicating that skin grafts are highly immunogenic, VCA which frequently include skin are complicated by a significantly higher rate of rejection episodes in the first year compared to solid organ transplants.[85] Split rejection may manifest as acute rejection of skin without affecting other parts of the allografts[86] or chronic rejection in the form of vasculopathy may occur without acute skin changes.

The inclusion of bone marrow in a VCA has been associated with mitigation of rejection, leading to the suggestion that mixed chimerism may be operative in blunting the HVG response, and suggesting that combined VCA and bone marrow transplantation might be advantageous.[87]

Intestinal Transplantation

Intestinal transplantation is relatively uncommon, despite being the only definitive treatment for end-stage intestinal failure from a variety of causes. Its low utilization despite this situation is due in part to outcomes that are inferior to those achieved with other types of solid organ transplants, with 5-year survivals in the 50% range. Intestinal transplantation includes isolated small bowel transplantation as well as multivisceral (MvTx), which includes liver, stomach, and pancreas in addition to the intestine, with removal of the native stomach, pancreatico-duodenal complex, and spleen. Modified MvTx is performed without transplanting the liver. MvTx transplant recipients have demonstrated remarkably high and persistent (for >1 year in some cases) levels of mixed lymphohematopoietic chimerism compared to recipients of isolated intestinal grafts. This chimerism has often occurred without GVHD and is associated with lower rates of rejection.[88,89] The chimerism appears to reflect, in part, the engraftment of donor hematopoietic stem and progenitor cells carried in the allograft[90] and promotion of their engraftment by GvH-reactive T cells that expand in the intestinal mucosa in response to recipient APC entry into the donor tissue.[89,91] The balance of GvH and HvG reactivity is favored by multivisceral compared to isolated intestinal transplantation, possibly reflecting the relative amount of recipient mucosal and lymphoid tissue removed and the amount of donor tissue placed[89] and/or the strong contribution of the donor liver to donor hematopoietic engraftment. Indeed, both chimerism[90] and long-term graft survival[92] are superior in recipients of MvTx transplants compared to isolated intestine or modified MvTx grafts, suggesting that the liver plays a significant role in these outcomes.

MECHANISMS OF REJECTION AND GVHD

Improvements in immunosuppressive protocols in the past 4 decades have resulted in increased 1-year survival rates for most types of organ transplants but have had little impact on rates of late graft loss. While improvements in GVHD prophylaxis have made haploidentical and other alternative HCT sources viable, GVHD remains a common and often severe complication. Multiple, redundant immunological mechanisms involving humoral and cellular responses can cause or contribute to allograft rejection in both the HvG and GvH directions. Many of these processes may be triggered by innate immune mechanisms that are activated by graft ischemia and reperfusion in the case of organ transplants and recipient tissue injury induced by conditioning for HCT. These processes may be mitigated by the same immunoregulatory processes that normally control immune responses to pathogens and injury. Efforts to harness these regulatory responses have led to numerous experimental and clinical efforts at immune modulation in transplantation.

Several different types of graft rejection have been defined by their timing and the contribution of T cells and antibodies. While these types of rejection are not always separate and distinct, the major types are summarized in Table 48.2. While there is considerable interest in and effort aimed at developing noninvasive diagnostic criteria for rejection and GVHD, including identification of biomarkers and, in the case of solid organ grafts, imaging methodologies, the determination

TABLE 48.2 Major Types of Rejection

Type of Rejection	Timing	Mediators/Characteristics
Hyperacute	Immediate (minutes to hours)	Antibodies, complement, coagulation cascade
Antibody-mediated rejection	Days to months	Antibodies, complement, Fc and complement receptors, multiple inflammatory pathways, microvascular injury
Acute cellular rejection	Weeks to months	Cytotoxic T cells, cytokines
Chronic rejection	Years	Vascular neointimal expansion, interstitial fibrosis

of rejection or GVHD currently relies on a pathological diagnosis from a tissue biopsy. A detailed discussion of the pathologic criteria for each form of rejection of each organ is beyond the scope of this chapter. However, we attempt to cover the major mechanisms leading to and mitigating the various types of rejection that are summarized in Table 48.2. These mechanisms are considered in terms of the "afferent" phase culminating in allosensitization and the "efferent" phase of rejection involving a number of redundant effector mechanisms.

The Afferent Phase of Rejection and GVHD: Tissue Injury, Innate Immune Responses, T Cell Sensitization and Help

Role of Ischemia, Tissue Injury, and the Innate Immune Response

The innate immune system can be considered as a "preformed" defense mechanism that is immediately available to defend the host until either the dangerous stimulus is cleared or the adaptive immune system is able to mount an antigen-specific response. Innate immune mechanisms that are activated by graft ischemia and reperfusion in the case of organ transplants and recipient tissue injury induced by conditioning for HCT provide important triggers for the ensuing adaptive immune response that promotes rejection. All forms of organ and tissue transplantation involve ischemic and traumatic injury to the donor tissue. Both warm (while in the body) and cold (once removed from the donor and before being placed in the recipient) ischemia promote ischemia reperfusion injury (IRI), which results from signals within both the graft and the recipient. Pattern recognition receptors exist to detect the unwanted presence of bacterial or viral pathogen-associated molecular patterns, but after transplantation, the TLRs that form part of the pattern recognition receptor family can also be used to detect the molecules produced as a result of implantation of the graft and the associated cell death,[93] so-called damage-associated molecular patterns (DAMPS). These signals include heat shock proteins, reactive oxygen species, high mobility group protein B1, complement breakdown products, nucleic acids (deoxyribonucleic acid and RNA), mitochondrial components, and molecules associated with tissue fibrosis that activate cells of the innate immune system via TLR ligation. These factors result in activation of the coagulation system, endothelial dysfunction, and cell injury and death,[94] with resulting innate immune responses when revascularization occurs. While kidneys from living donors show superior survival compared with those from deceased donors, in part due to shorter ischemic times, the donor pool for several organs has recently been expanded by using organs from donation after circulatory death and extended criteria donors. These practices are associated with increased IRI, which is positively associated with acute and chronic rejection rates,[95-98] reflecting the ability of innate immune responses to trigger adaptive immunity. Not surprisingly, efforts to counteract IRI are being actively pursued as an approach to mitigating allograft rejection.

DAMPs produced by stressed and injured cells[99] play a key role in triggering TLR-4 and the receptor for advanced glycation end products (RAGE), which activate the inflammasome via the NFκB and MAP kinase pathways.[100,101] TLRs are critical sentinels of the innate immune system and contribute to the early response after transplantation. Rejection of skin grafts expressing only a few minor (H-Y) alloantigens required MyD88-dependent TLR signaling[102] and multiple TLR pathways contribute to rejection of fully allogeneic skin grafts.[103] Multiple TLRs have been implicated in ischemia-reperfusion injury in heart, brain, liver, intestine, and kidney.[104,105] Haptoglobin, a serum acute-phase reactant protein released by necrotic tissue, has also been shown to promote graft rejection via a MyD88-dependent pathway.[106] Triggering the above innate immune receptors induces production of inflammatory cytokines such as TNF,[107] IL-1, and IL-6 by activated macrophages and other phagocytic cells. Production of active IL-1β requires proteolytic cleavage of the inactive form, pro–IL-1, in an inflammasome that acts as a platform for caspase 1, the cysteine protease involved in the proteolytic maturation and secretion of IL-1β.

The IRI response may also involve fixation of complement via both the classical (due to natural IgM antibodies) and lectin pathways.[108,109] Activated complement components constitute a proteolytic cascade that generates a range of effector molecules. The anaphylatoxins C5a and C3a are chemoattractant molecules that assist leukocytes to home to the graft while other soluble mediators opsonize cells, targeting them for destruction by phagocytes. Recognition of C3b, C4b, or their fragments covalently bound to target cells by complement receptors on the surface of leukocytes facilitates antigen presentation and T-and B-cell activation.[110,111] Generation of the terminal components of the complement cascade (C5b-9) results in formation of the membrane attack complex and initiation of target cell lysis, playing an important role in IRI.[112]

The induction of chemokine production[113,114] and upregulation of endothelial cell adhesion molecules by the above inflammatory pathways attracts leukocytes, including a rapid influx of neutrophils, macrophages, and APCs,[115] which, using upregulated chemokine receptors, migrate along the chemoattractant gradient to reach the graft.[116] Inflammatory signals affecting blood vessels in the transplant cause vasodilation and endothelial activation. Activated endothelial cells rapidly externalize preformed granules called Weibel- Palade bodies that contain the adhesion molecule P-selectin and rapidly upregulate expression of vascular cell adhesion molecule (VCAM) and CD62E (E-selectin). At the same time, chemokines released from the graft become tethered to the endothelium, and these alterations in endothelial surface markers advertise to passing leukocytes that an inflammatory process is occurring in the neighboring tissue. Leukocytes are usually conveyed within the fast laminar flow at the center of blood vessels, but once activated leukocytes reach postcapillary venules in proximity to the graft, they are able to leave this rapid flow and move toward the edge of the vessel. This occurs in response to the local chemokine gradient and extravasation occurs in a multistep process. Initially, low-affinity interactions develop between endothelial P-selectin and sialyl-Lewis^X moieties that are present on the surface of activated leukocytes. These interactions continually form and break down, and the leukocyte "rolls" along the endothelial surface. Chemokines on the endothelial surface induce conformational changes in leukocyte integrin molecules that allow them to bind other endothelial adhesion molecules such as ICAM-1. These higher affinity interactions cause arrest of the leukocyte on the endothelial surface, allowing it to commence extravasation. Having entered the tissues, the activated leukocytes continue to migrate along chemokine gradients in order to invade the graft.

Ligation of TLRs on DCs induces maturation, as defined by upregulation of costimulatory molecules and MHC class II, enhancing their ability to act as a bridge between the innate and adaptive immune systems. Inflammatory changes and APC activation via the above inflammatory pathways ultimately promotes an influx of T cells[117] sensitized via pathways described below, resulting in adaptive T cells responses to the allograft. In addition, alloreactive memory T cells preexisting due to cross-reactive immunity to pathogens may play an early role in driving the innate immune response,[118] in part by producing additional chemokines,[119] thereby perpetuating rejection.

Similar processes are important in inducing graft-vs-host disease in the epithelial target tissues that include skin, intestine, liver, and lungs. Administration of large numbers of nontolerant donor lymphocytes to established mixed bone marrow chimeras (ie, animals not recently treated with conditioning therapy) leads to a graft-versus-host response that attacks only lymphohematopoietic tissues and does not cause GVHD, which requires T cell migration into the epithelial target tissues.[120,121] In contrast, similar numbers of T cells cause rapidly lethal, severe GVHD in freshly irradiated hosts.[120,122] Conditioning rapidly induces production of chemokines in the GVHD target tissues, promoting immigration of T cells that then elicit a further cascade of chemokines that amplifies the response.[123] Upregulated adhesion molecules also promote leukocyte infiltration through the microvasculature of these tissues. Lethal total body irradiation (TBI) and cyclophosphamide, for example, upregulate the proinflammatory cytokines IL-1, IL-6 and TNF-α,[124] which can upregulate endothelial cell E-selectin, P-selectin, ICAM-1, and VCAM-1. In the absence of such host target tissue inflammation, mature, activated graft-versus-host-reactive effector T cells are unable to traffic into skin and induce injury.[122] Provision of a local toll-like receptor (TLR) stimulus promotes the entry of such cells into the skin and induces localized GVHD,[122] indicating that tissue inflammation provides a critical checkpoint for T cell recruitment to GVHD target tissues. A systemic TLR stimulus in this setting promotes severe, multiorgan GVHD.[122]

However, it should be noted that GVHD can also occur when very large numbers of nontolerant parental T cells are administered to untreated genetically tolerant F1 hosts. Therefore, in the presence of sufficiently powerful graft-versus-host responses, the need for conditioning-induced tissue inflammation to induce epithelial tissue trafficking of T cells that results in GVHD can be bypassed. The requisite tissue inflammation may be induced by very high cytokine levels resulting from the unopposed GVH response. Likewise, skin and cardiac allografts that are allowed to heal before being exposed to alloreactive T cells are rapidly rejected if there is sufficient antigenic disparity between donor and host.[125,126] Similarly, patients with long-standing allografts are rarely able to terminate immunosuppressive therapy without rejection. Therefore, "danger" signals are not a critical requirement for graft rejection, and it is better to picture the antigenic disparity and the recipient's immune responsiveness as the dominant features controlling graft rejection, while danger signals influence the timing, intensity, and character of the rejection response.

T Cell Sensitization

T cells are at the heart of adaptive immune responses to allografts. The mechanisms by which IRI-induced inflammation promotes an influx of leukocytes that ultimately includes T cells is discussed in the preceding section. However, in order to acquire the machinery to enter the graft and attack cells expressing their cognate (donor) alloantigens, T cells must first become activated. This process likely occurs in recipient secondary lymphoid tissues as a result of the rapid emigration of donor-derived passenger leukocytes that, in response to the inflammatory signals that are triggered by ischemia and transplantation, rapidly leave the graft and migrate to these lymphoid tissues.[127]Secondary lymphoid tissues comprise the spleen, lymph nodes, and gut- or mucosal-associated lymphoid tissue, and depending on the location of the graft, the passenger leukocytes will migrate to the tissue that drains the graft site, where they encounter naïve T cells. As a result of transplantation-induced inflammation, both in situ within the tissue and during migration, the passenger leukocytes acquire the phenotypic and functional characteristics of mature DCs, expressing high levels of MHC class I and II molecules as well as other cell surface costimulatory molecules necessary to fully activate naïve CD4+ and CD8+ T cells.

It was originally believed that, once in the secondary lymphoid tissues, these DCs act as professional APCs, presenting antigens expressed in the transplanted tissue to recipient T cells via the direct pathway of allorecognition. Indeed, the failure to reject allografts lacking class II MHC + APCs in certain experimental models emphasized the importance of this pathway in graft rejection.[128,129] However, more recent studies in mouse heart, skin, and islet transplantation have suggested that semidirect presentation in recipient lymphoid tissues may be the principal pathway by which alloreactive T cells are activated.[130-132] As illustrated in Figure 48.2, both direct and semidirect pathways of allorecognition involve recognition by a host T cell of an intact donor MHC-peptide complex. However, the direct pathway involves recognition of these complexes on donor APCs, which are passenger leukocytes from the graft. Semidirect presentation, on the other hand, involves transfer and presentation of intact donor MHC-peptide complexes by recipient APCs. This may occur during direct cell–cell contact in a process known as trogocytosis, via nanotubes, or without cell-cell contact by release of extracellular vesicles such as exosomes. The production of exosomes, mainly by donor passenger leukocytes, can result in sensitization of T cells by recipient APCs picking up the exosomes in secondary lymphoid tissues.[132,133] Endothelial cells may also be a source of donor exosomes,[134] which could be relevant after donor professional APCs have been depleted. Because passenger leukocytes are not self-renewing and are eliminated over time, semidirect presentation is likely to be the more important pathway for long-term maintenance of T cells that recognize donor MHC/peptide complexes. Indeed, transfer of intact donor MHC molecules to recipient DCs within the graft was recently found to drive effector T cell responses within experimental islet and kidney allografts for at least 8 weeks.[135] Nevertheless, responses among circulating T cells to donor HLA tend to subside over time in transplant patients,[136,137] suggesting that semidirect responses may not be sustained long-term, at least outside of the graft.

A third pathway, indirect allorecognition, involves presentation of processed donor antigens (peptides) on recipient APCs. Indirectly alloreactive CD4 T cells have roles in experimental skin, islet, and solid organ graft rejection.[138,139] CD4 cells also contribute to rejection of bone marrow grafts differing only at class I MHC loci, possibly implicating indirect allorecognition.[140,141] Sensitization of indirect CD4 responses to donor MHC-derived peptides has been demonstrated in patients undergoing graft rejection, and these may be correlated with poor outcomes.[139] Moreover, the eventual replacement of donor APCs by recipient APCs implicates the latter in long-term graft recipients.[138] With weak transplantation barriers (eg, class I disparate only or minor HA disparate skin grafts or primarily vascularized allografts in rodents), indirect reactivity is insufficient to cause rejection and donor APC depletion may, by preventing the stronger direct alloresponse, allow the inherent tolerogenity of the organ to prevail. In addition to the more classical CD4 cell–mediated indirect allorecognition, CD8 T cells may also participate in a type of indirect allorecognition following "cross-priming" (see Chapter 20) and contribute to the rejection of bone marrow grafts[142] or skin grafts[143] lacking MHC class I.

T Cell Help

Indirect allorecognition by CD4 T cells provides help for class-switched alloantibody responses.[144] This help requires cognate interactions between recipient class II-restricted indirectly alloreactive CD4 cells and host B cells that recognize donor MHC molecules through their immunoglobulin receptors, process them, and present donor MHC peptides with their class II molecules. Consistently, data suggest a major role for indirect allorecognition in chronic rejection,[139] in which antibody responses are often implicated.[145-147] T cell help through CD40-CD40L interactions and cytokines such as IL-4 and IL-21 results in B cell activation, differentiation to antibody-secreting cells, and class-switching, as detailed in Chapter 26.

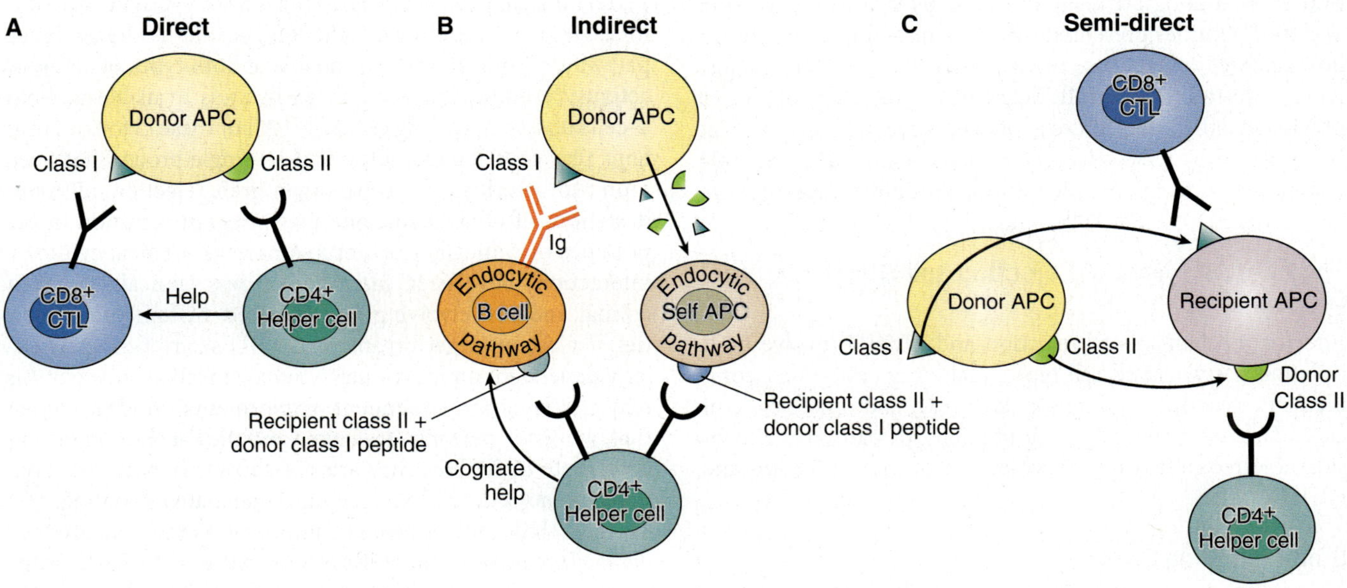

FIG 48.2. Pathways of allorecognition. Direct **(A)**, indirect **(B)**, and semidirect **(C)** pathways of allorecognition.

CD4 T cells also provide "help" for differentiation of effector CD8 T cells that consists of both cytokine (major role for IL-2) production and "conditioning" of the APC, largely by interactions of CD40 on the APC with CD40L on the activated CD4 cell. These interactions upregulate APC expression of MHC, CD80, and CD86 costimulatory molecules and cytokines such as IL-12, making the cell a more effective APC. Studies of antiviral immunity indicate that CD4 help is needed for development of full effector function[148] and for CD8 memory cell survival[149] and function.[150]

Early studies involving very limited (not clinically relevant) antigenic disparities between donors and recipients suggested that a "three-cell cluster" model involving interactions between directly alloreactive helper T cells, cytotoxic effector T cells, and donor APCs was essential for rejection. However, studies involving more extensive, clinically relevant histoincompatibilities[151] suggest that CD4 helper cells sensitized by antigen presented indirectly on recipient APCs can provide help for directly alloreactive CD8+ effector cells. It remains possible that semidirect presentation of donor class I MHC/peptide complexes on recipient APCs allows indirectly alloreactive CD4 cells to provide help to directly alloreactive CD8 cells.

In addition to being typically presented by MHC class II antigens, peptides of exogenous antigens can also be presented by MHC class I molecules via cross-presentation, which is usually carried out by CD8α+ dendritic cells in the mouse and their counterpart in humans defined by expression of CD141[152,153] (see Chapter 20). CD8 cell cross-priming was originally demonstrated in a murine transplantation minor HA-mismatched skin graft model.[154] Cross-primed CD8 cells recognizing donor MiHAs and MHC-derived peptides are theoretically most likely to participate in rejection when there is sharing of class I alleles between the donor and recipient, so that the shared class I/allogeneic peptide epitope can be seen on recipient APCs as well as the parenchymal or endothelial cells of the graft.[155] However, even without class I sharing[156] or expression[143] on donor cells, CD8+ cell cross-priming can lead to skin allograft rejection, perhaps due to recognition of donor peptides presented by recipient endothelial cells on host-derived vessels that revascularize the graft or on host APCs infiltrating the graft. Some of the rejection processes previously attributed to cross-primed CD8 cells may in fact be mediated by CD8 cells seeing intact donor MHC-peptide complexes on recipient APCs due to semidirect presentation.

The Efferent Phase of Rejection and GVHD: Effector Mechanisms

Effector mechanisms of rejection and GVHD involve multiple mechanisms and cell types, including antibodies, complement, cytokines and cytotoxic T cells, and innate effector cells. The net result of this multiplicity of pathways is considerable redundancy of mechanisms of graft rejection and GVHD.

Graft-Infiltrating Cells

Many types of cells infiltrate rejecting grafts, including CD4+ and CD8+ T cells, NK cells, and macrophages. While B cells may be less prominent, their presence has been associated with both acute and chronic rejection, and they are attracting increasing interest for their role not only as producers of antibody effectors of rejection but also as APCs. B cells may be located in tertiary lymphoid organs found in chronically rejecting allografts.[157] B cells producing donor HLA-specific antibodies have been detected in renal and cardiac allografts, raising the possibility that local antibody production may contribute to rejection.[158,159]

The number of invading T cells in a graft is not necessarily correlated with the speed of rejection. This finding suggested that certain critical elements of the graft, such as its blood vessels, are the actual site of graft destruction and, indeed, endothelialitis is an important hallmark of clinically significant acute cellular rejection. Repertoire analysis of graft-infiltrating T cells in acutely rejecting grafts reveals marked polyconality[160] but oligoclonal dominance has been suggested in studies of long-term rejected human kidneys.[161] Using the high-throughput TCR sequencing-based assay mentioned above,[12] the repertoire of HLA-alloreactive T cells was shown to be highly diverse and proportional to the number of HLA disparities.[11] Furthermore, these diverse donor-specific T cell clones were shown to predominate among the recipient-mappable T cells infiltrating the graft mucosa of rejecting human intestinal allografts.[89] Although the human anti-MiHA TCR repertoire is polyclonal,[162] immunodominance for HY specificities has been reported in clinical skin GVHD.[163] In experimental settings, T cells infiltrating xenografts included a broad TCR repertoire.[164] T cells mediating murine GVHD in the setting of multiple minor histoincompatibilities demonstrated a markedly skewed repertoire.[165]

Cytotoxic Mechanisms of Graft Rejection and GVHD

Rejecting organs contain proteins and messenger ribonucleic acid (RNA) encoding perforin, granzymes, and proteases associated with cell-mediated cytotoxicity.[166] Although the perforin/granzyme pathway is the major cytolytic pathway for CD8 T cells and CD4 cells tend to utilize the Fas/FasL pathway,[167] both subsets are capable of both types of cytolytic activity,[168,169] and the perforin pathway is available to both T-cell subsets mediating GVHD.[170] All of these cytotoxic proteins play contributory roles, and no single protein has been found to be critical for solid organ graft rejection, all manifestations of GVHD, or bone marrow graft rejection in the presence of clinically relevant mismatches. Critical cytotoxic interactions have been identified in less clinically relevant animal models involving very limited histoincompatibilities.[171] Certain target organs of GVHD seem to be particularly dependent on individual cytotoxic mechanisms, but this may also be strain-dependent. Experimental models suggest that, while the perforin-granzyme and the Fas/FasL pathways contribute to GVHD, they are not absolutely required, even in recipients lacking TNF receptor 1–mediated signaling.[171,172] Thus, GVHD effector mechanisms demonstrate some redundancy. In contrast, the perforin/granzyme and TRAIL pathways may be especially important for CD8 cell–mediated GVL effects.[173-175]

Noncytotoxic T-Lymphocyte Effector Mechanisms in Graft Rejection and GVHD

T cells can effect rejection of grafts whose parenchymal cells do not express the TCR ligand, indicating the existence of "indirect" effector mechanisms. Such mechanisms target nonparenchymal cells such as APCs and endothelial cells and likely involve release of cytotoxic cytokines and/or impairment of the vascular supply to the graft. Entire skin grafts can be rejected when only the APCs are foreign,[176] indicating that nonselective destruction of grafted tissue can occur. Several studies have implicated indirectly alloreactive CD4 cell–mediated rejection of experimental allografts.[177,178] Replacement of graft endothelium by the host was shown to be needed for rejection through this indirect effector mechanism.[178]

While the relative contribution of cytokine-dependent mechanisms versus direct cell-mediated cytotoxicity to GVHD is still a matter of debate, GVHD of the liver and intestine can be induced by donor T cells in MHC-deficient hosts receiving wild-type host DCs.[179,180] However, CD4-mediated GVHD against MiHAs is markedly attenuated when the target antigens are expressed only on hematopoietic cells.[181] Thus, "indirect" effector mechanisms can destroy transplanted tissue or recipient tissue in the case of GVHD but less efficiently than direct cytotoxic mechanisms. These "indirect" mechanisms of graft rejection and GVHD are likely to include cytokines such as tumor necrosis factor (TNF)-α and IFNγ. Macrophages are activated by lipopolysaccharides translocating from damaged gut epithelium and by IFNγ to release TNF-α, nitric oxide, and other mediators of tissue injury that promote GVHD. In certain models, TNF-α was shown to be critical for wasting disease and intestinal GVHD.

A great redundancy of rejection pathways is suggested by studies detecting Th1, Th2, and Th17 cytokines in rejecting allografts. Experimental models indicate that all three T cell types as well as cytokines produced by activated APCs can promote rejection[182,183] and GVHD.[184] There appears to be strain-dependent tissue specificity to the type of GVHD induced by the various Th subsets.[185-187] Thus, while the concept that Th2 cytokines are anti-inflammatory attracted early interest in the transplantation field, Th2 responses can clearly contribute to both graft rejection and GVHD. With a few special exceptions, studies using various cytokine knockout mice as recipients have failed to reveal any single molecule that is essential for rejection or GVHD.

Antibody-Mediated Effector Mechanisms

Antibodies are powerful mediators of HAR, acute humoral rejection, and chronic rejection through interactions with Fc receptors and, following complement activation, with complement receptors on effector cells such as monocytes and macrophages. Complement activation produces the membrane attack complex, causing endothelial damage and activation of inflammatory and prothrombotic pathways.[188] The timing of appearance and level of DSA influence the various pathologic outcomes of this process, as described below.

Rejection Caused by Preformed Antibodies (HAR)

HAR occurs within minutes to hours after blood flow is established to a transplanted vascularized organ. The phenomenon is visible and dramatic: the organ turns blue and its function ceases. Microscopically, there is extensive evidence of vascular thrombosis and hemorrhage. The important components involved in the mechanism of HAR include (1) donor endothelial antigens; (2) preformed antibodies that can bind these antigens. Antibodies could be directed against ABO blood group antigens or donor HLA antigens in allogeneic transplantation. Additionally, natural antibodies against αGal precluded survival of vascularized pig organs in nonhuman primates for decades due to their ability to initiate HAR; (3) complement; and (4) coagulation cascades, which are activated by the binding of preformed antibodies to the vascular endothelium. The interaction of these components leading to HAR is diagrammed in Figure 48.3.

The role of complement in HAR is inferred both from the accumulation of various complement components in the grafts and from the fact that complement depletion leads to prolonged survival of xenografts. Complement activation leads to production of active protein fragments and complexes of complement components, which cause tissue injury either directly or by recruiting effector cells that mediate destruction of the graft. In allogeneic combinations, this is initiated by antibody-mediated activation of complement through the classical pathway, whereas in xenogeneic combinations, the alternative pathway may also be involved. In both cases, the membrane attack complex, produced by the ordered interaction of several complement components, initiates the destructive pathway.

Complement activation is controlled by several regulatory molecules, including complement receptor 1, decay accelerating factor (DAF; CD55), membrane cofactor protein (CD46), and CD59, which act at different stages along the cascade (see Chapter 13). Many of these molecules are produced by the vascular endothelial cells. Because these regulatory proteins prevent unwanted complement activation in the face of low levels of perturbation to the system, the titer and avidity of the preformed antibodies must be high enough to overcome these downregulating molecules. Thus, preformed antibodies directed at MHC antigens almost always accomplish this activation, whereas the lower affinity blood group antibodies

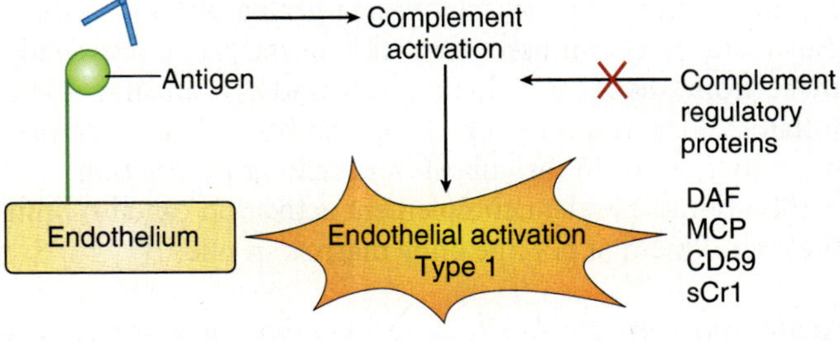

FIG 48.3. Schematic representation of the hyperacute rejection pathway.

lead to HAR in only about 25% of mismatched kidneys. One of the reasons that HAR is such an important feature in xenogeneic transplantation is that the complement regulatory proteins (CRPs) produced by the donor vascular endothelium of one species may not function effectively with complement molecules derived from a different species. Consequently, lower levels of an initial triggering signal lead to explosive complement activation. The role of CRPs in mitigating rejection by low levels of antibodies is illustrated by the dramatic AMR observed for $GalT^{+}$ CD55-deficient heart transplants in $GalT^{-/-}$ mice containing low levels of otherwise innocuous anti-αGal natural antibodies.[189]

Although the membrane attack complex is often thought of as a lytic molecule, its effect on the donor vascular endothelium, even before cell lysis, is to cause endothelial activation. This occurs rapidly, before there is time for new gene transcription or protein synthesis, and results in gaps between endothelial cells and initiation of coagulation pathways due to the loss of antithrombotic molecules from the endothelium.[190] Activated endothelium is also proinflammatory through its chemoattractive and adhesive properties and endothelial activation is responsible for the two principal pathologic findings in HAR: extravascular hemorrhage and edema and intravascular thrombosis.

There are no known treatments that can stop the process of HAR once it has started and thus it is essential to avoid the circumstances that initiate it. Experimentally, this can be accomplished for relatively short periods of time by administration of complement inhibitors, such as cobra venom factor, which depletes complement. In clinical practice, this is accomplished by avoiding transplantation in the face of preformed antibodies, both by avoiding blood-group antigen disparities, by testing recipients before transplantation (ie, a "cross-match") to determine whether they have performed antidonor antibodies. When transplants must be performed in the presence of such antibodies, the application of methods for antibody depletion and recipient "desensitization" is required.[191,192]

Not all tissues are equally susceptible to HAR. Most primarily vascularized organs, such as kidneys and hearts, are very susceptible. Skin grafts are relatively resistant to HAR but high levels of antibody can cause a "white graft" (ie, a failure of blood vessels to communicate with those of the recipient). Free cellular transplants, such as bone marrow cells or hepatocytes, that express some of the antigens recognized by preformed antibodies, are cleared quickly from the circulation by the reticular endothelial system, leading to resistance to engraftment. In the case of HCT, this resistance can be overcome by transplanting larger numbers of cells.[193] Pancreatic islets infused into the portal vein, especially those from xenogeneic donors, can undergo an immediate inflammatory response involving antibodies and complement that results in thrombosis and cellular destruction.[194,195] Antibody-independent complement activation can also limit the engraftment of porcine bone marrow in mice.[196]

Acute and Chronic AMR. A second type of AMR results from the presence of antibodies to HLA or non-HLA donor antigens that develop or increase after transplantation is performed. This type of rejection can be rapid and acute or chronic, causing late allograft loss.[191,192,197] Rapid AMR has also been called "acute humoral" or referred to as "accelerated" rejection when it occurs within the first 5 days after transplant. The process is characterized by vascular injury with fibrinoid necrosis of donor arterioles, intravascular and microvascular inflammation, leukocyte adhesion, and thrombosis, with or without C4d deposition.[198] Accelerated rejection can occur in presensitized, crossmatch-positive recipients, so every effort is made to avoid such transplants. When they are unavoidable, desensitization of recipients is carried out prior to transplantation from donors to whom donor-specific or blood group antibodies are present. Nevertheless, acute AMR remains a significant problem in all forms of organ transplantation, as it is difficult to reverse and is associated with poor outcomes.[147] Vigorous anti–T-cell immunosuppression has little effect on acute AMR. Immunosuppression is typically targeted toward B cells using rituximab, plasma cells using proteasome inhibitors and/or anti–IL-6 receptor antibodies, complement using complement inhibitors, and circulating antibodies using plasmapheresis or the staphylococcal enzyme IdeS.[191] Intravenous immunoglobulin (IvIg) is also commonly used in desensitization and AMR treatment protocols and works through multiple anti-inflammatory mechanisms.[147] While the need for transplantation in presensitized recipients is on the rise in large part due to the increasing need for retransplantation, it should be noted that, despite desensitization protocols, outcomes remain inferior in such patients, with increased rates of acute AMR, chronic rejection, and late graft loss.[147]

Accelerated rejection has been a frequent occurrence in xenotransplantation models when vascularized organ transplants are performed between closely related, concordant xenogeneic species, and between discordant species following adsorption of anti-Gal antibodies.[199] In these cases, the levels of preformed antibodies are not sufficient to cause HAR but antidonor antibodies appear rapidly (within 3–4 days) in association with the onset of rejection. The pathology of acute humoral rejection reveals a paucity of lymphocytes infiltrating the donor graft, antibody binding to donor vascular endothelium, and fibrinoid necrosis of the donor vessels.

As in HAR, the process of acute humoral rejection is initiated by antibody binding to antigens on the donor vascular endothelium. In this case, however, the subsequent endothelial changes occur more slowly, allowing time for gene transcription and new protein synthesis. Many features of endothelial activation appear to be mediated by the transcription factor NF-κB, which generates responses associated with inflammation, including the secretion of inflammatory cytokines such as interleukin (IL)-1 and IL-8, and the expression of adhesion molecules such as E-selectin and intercellular adhesion molecule (ICAM)-1.[200,201] Cytokines, chemokines, and adhesion molecules upregulated by endothelial cells promote the recruitment of monocytes, macrophages, neutrophils, and NK cells, all of which contribute to the inflammatory cascade of AMR.[147,202] In addition, this type of endothelial activation causes the loss of thrombomodulin and other prothrombotic changes.[203] These events explain the intravascular thrombosis and the inflammatory destruction of donor vessels that occurs in the absence of infiltrating lymphocytes in AMR.

Just as there are regulatory processes for complement activation, there are regulatory molecules that counter the tendency toward intravascular coagulation and endothelial activation (eg, tissue factor protein inhibitor expressed by vascular endothelium, which inhibits factor Xa of the clotting cascade) and a number of other protective molecules, including Bcl-x_L, Bcl-2, and A20),[199,204] some of which inhibit activation mediated by NF-κB. Similar to CRPs, some of these regulators may not function across species differences, leading to disordered regulation of the coagulation system.[204]

Although vigorous early antibody responses generate endothelial activation and accelerated rejection, later antibody responses may fail to do so and circulating DSA may be present without AMR.[147] In the context of xenotransplantation, the process that enables transplanted organs to survive in the face of circulating antibodies that can bind endothelial antigens has been called "accommodation." Resistance to endothelial activation has been achieved in vitro by pretreatment with low levels of antiendothelial antibodies that are insufficient to trigger activation. Accommodation is associated with increased expression of the antiapoptotic genes described above and with changes in the isotype of the recipient's antibody responses.[205] The role of different IgG subclasses in determining the pathogenicity of DSA is currently not well-understood.[147]

Chronic AMR is a relentless process caused by repeated vascular injury and uncontrolled repair responses that is characterized by inflammation and interstitial fibrosis.[206] There is no universally accepted effective therapy for chronic AMR though several strategies are under investigation.[206,207] Graft vasculopathy due, at least in part, to chronic AMR is a common final pathway leading ultimately to organ fibrosis and graft loss. In kidney allografts, chronic AMR is also characterized by glomerulopathy and multilamellation of peritubular capillary basement membranes. Graft vasculopathy associated with a history of prior AMR and the presence of donor-specific HLA antibodies affects all graft types, including hearts, kidneys, livers, pancreata, intestines, composite tissue allografts, and lungs.[147] It is characterized by intimal thickening (neointimal expansion) and fibrosis in larger vessels and may not be detected in the small sample provided in a biopsy specimen. HLA antibodies are a major risk factor and probable cause of graft vasculopathy via mechanisms that involve complement, endothelial injury and apoptosis, leukocyte-mediated remodeling, and cytokines such as IL-8, TNF, IFN-γ, and VEGF.[147] Furthermore, graft injury may result in the exposure of cryptic self-antigens in allografts, resulting in the appearance and likely contribution of autoantibodies to the pathology.[209] Chronic AMR is often mixed with T cell–mediated rejection in kidney allografts.

B cells and antibodies also play a role in chronic GVHD (cGVHD), both in animal models and humans.[50,210-213] Antibodies in this setting are directed against alloantigens such as MiHA, including H-Y determinants in male to female transplants, and against autoantigens.[214] T follicular helper cells (Tfh), inadequate regulation by T_{regs} and regulatory B cells, and the cytokines BAFF, IL-4, and IL-21 have all been implicated in the pathophysiology of this disease. Thymic dysfunction and failure of normal T cell negative selection may also contribute to B cell autoreactivity in this setting. B cells in cGVHD demonstrate hyperactivity to stimulation through the BCR. IgG binding to FcRs may contribute to macrophage activation and cytokine secretion. Experimental therapies such as B cell depletion with rituximab showed some efficacy against cGVHD, as have several inhibitors of BCR signaling, one of which, ibrutinib, an inhibitor of BTK/ITK, has been approved for use in steroid-refractory cGVHD.[214]

Role of NK Cells

Although the role of NK cells in bone marrow rejection is well established in mice, the amount of resistance mediated by NK cells to allogeneic hematopoietic stem cells is limited and can be overcome by increasing the dose of donor stem cells administered.[215,216] A role for NK cells in resisting human allogeneic marrow engraftment has not been clearly demonstrated, although they are likely to be important in recipients of reduced toxicity conditioning regimens. Indeed, patients with severe combined (T- and B-cell) immunodeficiency who have functional NK cells require cytotoxic conditioning to permit engraftment of haploidentical marrow, whereas those lacking NK cells do not.[217]

The ability of NK cells to be triggered by "missing self" may have utility in HCT. Donor-derived NK cells with graft-versus-host reactivity due to the lack of donor class I MHC inhibitory ligands in the recipient can kill residual host leukemia cells and alloreactive cells that resist the marrow graft without causing GVHD. The alloreactive donor NK cells may also reduce susceptibility to GVHD by killing recipient APCs needed to activate donor T cells.[80] While striking antileukemic effects of KIR mismatching were detected in heavily conditioned patients with acute myelogenous leukemia who received high doses of haploidentical CD34+ stem cells,[80] the effect of KIR incompatibility has been extensively studied and found to vary in other types of transplants and leukemias.[218,219] Considerable interest and effort has recently been invested in the use of expanded NK cells for antitumor effects in HCT.[219,220]

Unlike mice lacking T cells, NK cell–deficient mice are capable of rejecting solid organ allografts and the role of NK cells in organ rejection is only partially understood. In one experimental model, NK cells were found to be required for acute but not chronic antibody-mediated renal allograft rejection.[221] KIR mismatches have been reported to trigger NK cell recognition of allogeneic endothelial cells and have been associated with antibody-negative chronic renal allograft rejection in humans.[222] The more obvious importance of NK cells for rejecting cellular transplants than for solid allografts is exemplified by the observation that, while bone marrow allografts from class I deficient donors ($\beta 2m^{-/-}$) are subject to potent NK-mediated rejection (because these cells cannot trigger inhibitory receptors on host NK cells),[223] $\beta 2m^{-/-}$ skin grafts are not rejected by syngeneic wild-type recipients.[224] Nevertheless, NK cells are prominent among cells infiltrating rejecting organ allografts and may be early producers of inflammatory cytokines and chemokines and inducers of DC maturation.[225,226] An exception to the T cell–dependence of solid allograft rejection is the ability of NK cells to cause chronic allograft vasculopathy in a murine cardiac allograft model[225] in a manner that may be triggered by viral infection[227] or enhanced by DSA through an Fc receptor-dependent pathway.[228] Indeed, NK cells have been implicated in AMR in humans via mechanisms that involve Fc receptors and production of cytokines such as IFNγ.[147]

Inhibitory receptors on NK cells are quite broad in their class I specificity and fully MHC-mismatched allogeneic bone marrow is less susceptible to NK-mediated marrow destruction compared to class I–deficient marrow,[223,229] presumably because some NK cell subsets transmit inhibitory signals from allogeneic donor MHC molecules through inhibitory receptors. Because of the increased disparity of xenogeneic compared to allogeneic MHC molecules, NK cells may receive fewer inhibitory signals from xenogeneic than allogeneic cells. Indeed, transduction of HLA molecules into porcine endothelial cells reduces NK cell–mediated xenogeneic cell adhesion and cytotoxicity.[230] NK cells may also be activated by interactions of activating receptors with ligands on xenogeneic cells, of which several examples have been identified.[231,232] On balance, activating xenogeneic NK cell–target interactions are likely more effective than inhibitory interactions, explaining the greater prominence of NK cells in resisting xenogeneic than allogeneic bone marrow engraftment.[215,233-235] NK cells have also been implicated in solid organ xenotransplantation, including acute vascular rejection, through mechanisms that likely include antibody-dependent cell-mediated cytotoxicity and proinflammatory cytokine release.[236]

Additional Cell Types in Rejection and GVHD

Two additional types of T cells, namely natural killer T (NKT) cells and gamma delta cells, might have a role in rejection. Although NKT cells have a largely inhibitory role in preventing marrow rejection and GVHD, they have been reported to participate in cellular allograft (and tumor[237]) rejection in mice,[238,239] in part by activating NK cells[239] and by producing IFNγ.[240] Gamma delta T cells play a unique role in xenogeneic bone marrow rejection[234] but have not been widely implicated in allograft rejection or GVHD. They may, however, be important in eliciting GVL responses.

It is likely that proinflammatory cytokines and chemokines produced by activated monocytes and macrophages play a role in endothelial cell activation and lymphocyte recruitment in allograft rejection and macrophages feature prominently in AMR.[147] Additionally, activated macrophages may damage tissue through the production of toxic molecules such as nitric oxide.[241] Recently, a system for monocyte-mediated recognition of allogeneic non-MHC nonself has been described and reported to play a key role in activating alloreactive T cells following transplantation of sterile allografts.[242] One such pathway for activating macrophages involves A-type paired immunoglobulin-like receptors (PIR-As), which confer MHC class I antigen-specific memory on myeloid cells and promote organ graft rejection in mice.[243] Additionally, the balance between activating and inhibitory signals resulting from reciprocal interactions between CD47 and its polymorphic ligand, SIRPα, respectively, may result in recipient macrophage activation to allografts in the absence of adaptive immunity.[244]

In xenotransplantation, macrophages play an especially important role in the rejection of cellular xenografts such as islets[245] and may be activated to do so in a T cell–dependent manner.[246] Macrophages cause almost immediate rejection of xenogeneic bone marrow, even in the absence of adaptive immunity.[247,248] Human macrophages can phagocytose porcine cells in an antibody- and complement-independent manner.[249] Not surprisingly, given the intense inflammation associated with acute xenograft rejection and the role of antibodies and complement activation, macrophages have also been implicated in solid organ and skin xenograft rejection.[250,251] This prominent role for macrophages may reflect the combined ability of certain xenogeneic receptors to activate macrophages,[252] whereas important inhibitory interactions, such as that between CD47 and its macrophage ligand SIRPα, are not effective across certain species barriers.[253,254] The production of pigs expressing human CD47 inhibits phagocytosis by baboon and human macrophages[255] and has markedly improved the survival of porcine bone marrow cells in primates, thereby promoting tolerance induction via mixed chimerism.[256,257]

Eosinophils recruited to allografts by Th2 T-cell responses have been reported to be effectors of graft rejection in some experimental models, and eosinophils are often found clinically in rejecting allografts[258-260] and may contribute to xenograft rejection.[261] Th2-derived eotaxin, IL-4, and IL-5 recruit and activate eosinophils, which release highly cytotoxic substances from granules into the tissue.

Role of the Microbiome in Rejection and GVHD

It has long been recognized that organs colonized with microbiota have poorer clinical outcomes compared to sterile organs. Using rodent models, communities of microorganisms that colonize such organs have been shown to affect alloreactivity and ultimately the outcomes of solid organ transplantation. The microbiota poise dendritic cells for action and modulate their ability to stimulate immune responses directed at the allograft, most strikingly with colonized organs like the skin, lung, and small bowel. Gut microbiota may also affect the metabolism of immunosuppressive drugs, thus affecting their efficacy and side-effects. In mice, the gastrointestinal toxicity of the immunosuppressant mycophenolate mofetil (MMF) is reversed with antibiotics. Furthermore, the dynamic status of the microbiota and their rapid response to the environment has led to the speculation that gut microbiome composition and diversity may be a predictive biomarker for the adequacy of or overdosing of nonspecific immunosuppression. Given that macrolides such as cyclosporine A, tacrolimus and rapamycin, that are used to prevent rejection, were initially screened for their antibiotic and/or antifungal activities, it is reasonable that these immunosuppressants may effect changes in microbiome composition in the gut. Finally, there is a clear potential for microbiota to be harnessed therapeutically, through probiotics, carefully curated antibiotic use, or fecal microbiota transplantation (FMT). Oral antibiotic treatments in rodent models consistently result in prolonged allograft survival and FMT from antibiotic-treated donors can confer protection from rejection. Indeed, certain microbial species have been shown to alter the lymph node microenvironment in a manner that supports immune downregulation.[262] While the near universal use of perioperative and prophylactic antimicrobials demonstrably induces taxonomic changes in the microbiome of transplant patients, whether antibiotic use may be modified to promote changes in gut microbiota toward those that promote the downregulation of alloreactivity require further investigation (reviewed in 263).

For patients undergoing allogeneic HCT for the treatment of hematological cancers, preconditioning regimens result in dramatic shifts in the composition of intestinal flora and loss of gut permeability, leading to increased susceptibility to GVHD. Early studies in the 1970s demonstrated that the incidence of GVHD following HCT was significantly reduced in germ-free mice, leading to broadly adopted antimicrobial prophylaxis, sterilized positive air pressure rooms, low-microbial diet, and use of barrier precautions in patients (reviewed in 264). However, more recently, the use of antibiotics with activity against anaerobes and consequent decreased gut microbial diversity has been associated with worse GVHD and poorer clinical outcome after HSCT. Fecal microbiota transfer is beneficial in restoring a healthy microbiota in patients with *Clostridium difficile* infection and preliminary studies of FMT to prevent GVHD also suggest a beneficial effect (reviewed in 265, 266). Nevertheless, challenges remain in designing FMT formulations, preventing potential risks and implementing application in the clinic for patients with GVHD.

Mechanisms of Chronic Rejection and Chronic GVHD

Chronic Rejection

Clinical survival statistics reveal that even when 1-year graft survival has been achieved, the loss of transplanted organs continues to occur at a rate of about 3% to 5% per year, and a significant proportion of this delayed or late graft failure appears to be due to immunologic mechanisms. The term "chronic rejection" is commonly used to describe this later process of delayed graft destruction, although in kidney transplantation the Banff classification schema has replaced this term with interstitial fibrosis and tubular atrophy.[267] As immunosuppressive reagents have become more effective at controlling acute rejection, chronic rejection has emerged as one of the most important problems in clinical practice. Indeed, while there has been ongoing improvement over the past 30 years in the 1-year graft survival rates for kidney transplants, the half-life for organs that have survived for >1 year has not changed significantly over that entire period of time; as a result of this ongoing loss, only about 50% of transplants are still functioning 10 years later.

Although almost every type of organ transplant suffers from deterioration in function over time, the pathologic manifestations are somewhat different in each case. Kidney biopsies tend to show interstitial fibrosis along with arterial narrowing from hyalinization of the vessels—hence the terminology interstitial fibrosis and tubular atrophy. In the heart, the process is manifested principally as a diffuse myointimal hyperplasia, proceeding to fibrosis of the coronary arteries that has often been referred to as "accelerated atherosclerosis" or "transplant arteriosclerosis." Chronic rejection in lung transplants primarily affects the bronchioles with progressive narrowing of these structures and is referred to as "bronchiolitis obliterans", though vascular lesions similar to those in other types of transplants are also present.[147] While liver was once thought to be relatively resistant to chronic rejection, the progressive destruction of bile ducts referred to as the "vanishing bile duct syndrome" may be another manifestation of this process and obliterative arteriopathy is a feature of chronic fibrosis of liver allografts.[147]

Nonimmunologic factors such as initial ischemic insult, reduced mass of transplanted tissue (especially in the case of kidney transplants with hyperfiltration injury), denervation of the transplanted organ, hyperlipidemia and hypertension associated with immunosuppressive drugs, toxicity of the immunosuppressive drugs themselves, and viral injury, also contribute to late graft loss, which may represent cumulative and incremental damage to the graft from time-dependent nonimmunologic and immunologic causes.[268] Nonetheless, there is almost certainly an important immunologic component in most cases of chronic rejection. Consistently, there is a strong correlation between the onset of chronic rejection and a history of early acute and subclinical rejection episodes.[267] Importantly for long-term outcomes, the clinical data show that the process of chronic rejection is usually refractory to increases in immunosuppressive therapy, in contrast to acute rejection episodes that almost always respond to treatment. As discussed in the section above, the development of chronic kidney graft rejection is frequently, though not always, associated with the presence of antidonor antibodies and the deposition of complement component C4d in the allograft. Components of the innate immune system such as NK cells and macrophages may also contribute[147,202] and a collaboration between NK cells and alloantibodies can produce experimental chronic rejection.[228] In fact, evidence does not support the exclusive involvement of one mechanism. Alloantibody production often reflects activation of indirect pathway T cells and may serve as a marker for other rejection mechanisms as opposed to a cause of chronic rejection. In fact, i-IFTA has been described as the histologic result of active T cell–mediated alloimmunity, including inflammation and tubular injury with fibrotic healing.[267] The presence of alloantibody and C4d may be transient, and there are many examples of chronic changes in the graft in their absence, including in experimental models.[269]

The uncertainties that arise from the interpretation of the clinical data make it important to develop experimental models for studying the mechanisms of chronic rejection. It is difficult in the laboratory, however, to mimic a process that may take 5 or 10 years to develop in patients treated with immunosuppressive drugs. Thus, the effort to study chronic rejection experimentally has depended on surrogate short-term pathologic markers that are thought to predict the long-term changes of chronic rejection. In particular, these studies have concentrated on the development of the myointimal proliferation that is thought to be the precursor of the chronic vascular changes typically observed in patients. In rodents, pigs, and primates, this has often been done with grafts after an initial period of immunosuppression that prevents acute rejection. All of these experimental studies are subject to the caveat that the surrogate pathologic lesion occurs much earlier than the typical changes of chronic rejection in patients. Thus, the process being studied experimentally may not be the same as the clinical process.

The typical pathologic features of the experimental lesion associated with chronic rejection are shown in Figure 48.4. The marked narrowing of the vascular lumen is caused by the substantial proliferation of endothelial and smooth muscle cells, associated with progressive destruction of the media. In time, the cellular proliferation becomes less pronounced and is replaced by concentric fibrosis that narrows the vascular lumen. Immunohistologic staining indicates that there is increased expression of several adhesion molecules, intracellular proteins such as vimentin, nitric oxide synthase, acidic fibroblast growth factor, insulin-like growth factor, IFNγ, and endothelin, each of which may play a role. Ultimately, the ischemia resulting from vascular occlusion results in fibrosis in the parenchyma of the organ and consequent organ dysfunction.

Adoptive transfer experiments into mice with severe combined immunodeficiency have shown that alloantibodies in the absence of T cells can induce the typical pathologic vascular changes, and these vascular lesions can develop in T-cell–deficient mice.[270] However, T cells without B cells have also been shown to cause similar lesions.[145] In various experimental models, CD4+ or CD8+ T cells and class II or class I MHC differences appear to play dominant roles in driving chronic rejection. Consistent with the multiple different pathways

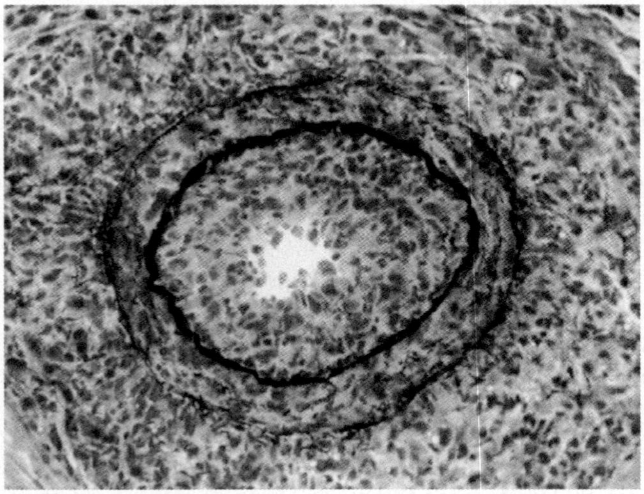

FIG 48.4. Vasculopathy lesion of chronic rejection.

that can lead to typical chronic rejection lesions, murine cardiac allograft recipients with robust T cell tolerance to the donor[271] or completely lacking adaptive immunity may also develop graft vasculopathy, apparently via an NK cell–mediated, viral infection-induced pathway.[227] Basophils, attracted to allografts and activated by IL-3, have also been found to promote fibrosis in murine heart transplants, via production of IL-6.[272] Several additional cytokines, including IFNγ and TGF-β, have been implicated in the development of chronic rejection and fibrosis. The upshot from experimental models is that multiple immunologic mechanisms can create the graft arteriosclerotic lesions that are characteristic of chronic rejection and that T-cell alloreactivity is not essential for their induction. Whether there is a critical final common mediator involved in all of these pathways is not currently known.

Chronic GVHD

cGVHD is the most common and severe complication among patients surviving for more than 100 days after allogeneic BMT and typically afflicts 30% to 70% of HCT recipients. It severely reduces quality of life and increases susceptibility to serious infections. Clinically, acute and cGVHD can be distinguished on the basis of the time of onset, clinical manifestations, and distinct pathobiologic mechanisms. Acute GVHD usually occurs within 2 to 6 weeks following allogeneic BMT and primarily affects the skin, liver, and the gastrointestinal tract with T-cell infiltration of the epithelia of the skin, mucous membranes, bile ducts, and gut. However, acute GVHD has been noted to occur later in recent protocols involving nonmyeloablative conditioning for HCT. In contrast, cGVHD involves a wider range of organs and clinical manifestations that include scleroderma, liver failure, immune complex disease, arthritis and other musculoskeletal manifestations, genital, ocular, oral and esophageal disease, glomerulonephritis, central and peripheral nervous system involvement, and autoantibody formation. Diagnosis and staging is carried out with standardized NIH criteria.[273]

The pathogenesis of cGVHD is incompletely understood. Fibrotic lesions with minimal inflammatory infiltrates are a common feature. However, a recent conceptual framework describes the earliest phase as inflammatory, while the intermediate stage is associated with thymic dysfunction, T and B cell dysregulation, and chronic inflammation, and the third phase is characterized by tissue repair with fibrosis. B cells, IL-17–producing T cells (Th17 cells), T follicular helper (Tfh), and deficient T_{regs} have all been implicated.[274] The disease involves T cell responses to alloantigens or autoantigens and risk is increased in female to male and HLA-mismatched transplants and by the use of mobilized peripheral blood stem cells (PBSC) rather than bone marrow for transplantation, by increased donor and patient age and by a prior episode of acute GVHD. Because most BMT is performed between HLA-identical or closely HLA-matched pairs, alloreactivity may be directed against miHAs presented by shared MHC molecules or against HLA alloantigens when disparities are present. The early phase involves multiple innate pathways triggered by tissue injury and consequent APC activation with amplification of adaptive immunity against the host,[274] including

T cells and hyper-reactive B cells as is discussed in an earlier section. These B cells present antigen to donor T cells, activating Tfh and germinal center reactions in part via IL-21 production, culminating in class-switched high-affinity antibody responses and antibody-mediated injury. T cells developing de novo in a recipient thymus that is damaged due to conditioning, drugs and attack by GVH-reactive T cells undergo disordered negative selection, permitting the emergence of autoreactive T cells into the peripheral repertoire. Furthermore, experimental models demonstrate disruption of peripheral regulatory mechanisms mediated by T_{regs}, Tfr, Tr1 cells, iNKT cells, and B_{regs}. In the later phase of tissue repair, production of TGF-β and PDGF-α by activated macrophages activates fibroblasts that promote tissue fibrosis. Chronic inflammation is maintained by unregulated Th17 cells.[274] Many of these same mechanisms have been implicated in human cGVHD, including induction of Th17 by innate immunity via IL-1 and IL-6–dependent mechanisms. Th1 and Th2 responses, impaired thymic function, and defective regulation by T_{regs} have also been implicated in the clinical disease.[274]

While associated with mortality due to organ damage and infection, cGVHD is also associated with reduced leukemic relapse rates and hence reduced disease-related mortality. cGVHD can be treated with a variety of immunosuppressive approaches, providing it is identified sufficiently early after initiation. Even with treatment, extensive skin involvement, thrombocytopenia, and progression are poor prognostic factors. Recent developments in B cell–targeted therapies have been discussed in the section on antibody-mediated GVHD above. Additionally, T cell targeting with inhibitors of Th2 development and of the JAK1/2 cytokine signaling pathways, as well as inhibitors of IL-6, IL-17, and TGF-β are all being explored in preclinical or clinical studies.[274] These and other modalities are discussed in more detail later in the chapter.

APPROACHES TO PREVENT TRANSPLANT REJECTION

Pharmacological Immunosuppression

A majority of advances in clinical transplantation that have occurred over the past 5 decades have been made possible by the discovery of pharmacological immunosuppressants that nonspecifically blunt immune responses to the allograft. Lifelong maintenance immunosuppression typically comprises a combination of steroids, calcineurin inhibitor (cyclosporine or tacrolimus), and antimetabolite (azathioprine or MMF) (Fig. 48.5). Higher risk recipients additionally receive induction therapy comprising antithymocyte globulin (ATG) to deplete lymphocytes or IL-2 receptor antagonist to blunt early alloreactive T cell expansion. Acute cellular rejection is treated with high-dose steroids, while acute AMR is treated with IVIG, plasmapheresis, and in some cases, anti-CD20 B cell depletion.

Calcineurin Inhibitors

The discovery in the 1970s of cyclosporin[275] revolutionized the field of organ transplantation. Cyclosporine was reported to inhibit lymphocyte proliferation in vitro and in animal models,

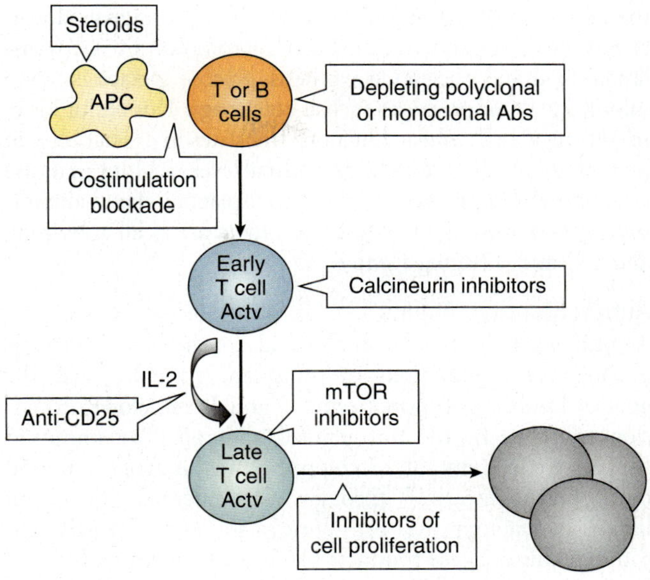

FIG 48.5. Targets of immunosuppression.

and the benefits of using cyclosporin in preventing kidney rejection were initially reported by Roy Calne and colleagues.[276] Almost 10 years later, in the mid-1980s, scientists at Fujisawa Pharmaceuticals isolated FK506, later named tacrolimus, from the soil fungus *Streptomyces tsukubaensis*, that had similar in vitro and in vivo immunosuppressive activities as cyclosporin.[277,278] FK506 has largely replaced cyclosporin as the calcineurin inhibitor of choice for maintenance immunosuppression.

While both cyclosporin and tacrolimus share the same mechanism for suppressing activated T cells, they have distinct molecular structures and bind to different intracellular components, cyclophilin and FK-binding protein, respectively. Nevertheless, both cyclophilin/cyclosporin and FK-binding protein/FK506 complexes directly bind to the calcineurin phosphatase and inhibit its phosphatase activity.[279] Activated calcineurin dephosphorylates the cytoplasmic transcription factor, nuclear factor of activated T cells (NFATc), which permits its translocation into the nucleus where it upregulates the expression of multiple cytokines and costimulatory molecules necessary for full activation of T cells.[280] Thus, inhibiting calcineurin in T cells suppresses their activation, primarily by inhibiting T cell production of IL-2, IFN-gamma, and IL-4, and expression of CD40 ligand.

mTOR Inhibitors

Rapamycin, the inhibitor of the mammalian target of rapamycin (mTOR) pathway was isolated from the bacterium *Streptomyces hygroscopicus* and was found to have antiproliferative and immunosuppressive activity.[281] Loss of mTOR activity results in the inhibition of cell cycle progression, and even though both rapamycin and FK506 bind to FKBP12, the rapamycin-FKBP12 complex mediates its antiproliferative effects through the inhibition of the serine-threonine kinase, target of rapamycin (TOR).[282] mTOR protein is found in two separate complexes, mTOR complex 1 (mTORC1) and mTOR complex 2 (mTORC2), with only mTORC1 being rapamycin-sensitive. Many pathways lead to the activation of the mTOR pathway, including signals downstream of CD28 and IL-2. mTORC1 promotes mRNA translation, lipid and nucleotide synthesis, glucose metabolism and inhibits catabolic processes such as autophagy. This multitude of mTOR-regulated cellular functions translates to crucial roles in physiology and disease at the organismal level, resulting in its use as an immunosuppressive agent in transplantation and autoimmunity and ongoing investigations into its use as an antitumor, antiviral, and antiaging agent.

Antiproliferative Agents

Azathioprine and mycophenolic acid are the two main antiproliferative agents used in transplant patients, with the goal of inhibiting donor-specific T and B cell clonal expansion triggered by the transplanted allograft. The immunosuppressive activity of azathioprine was initially reported by Robert Schwartz[283] through its sequential conversion into 6-mercaptopurine, 6-thiouric acid, 6-methyl-MP, and 6-thioguanine. These products block the de novo pathway of purine synthesis, and 6-thioguanine is incorporated into replicating DNA in place of purine nucleosides, collectively resulting in inhibited cell proliferation.[284] The use of azathioprine in the clinic has largely been supplanted by MMF or mycophenolate sodium,[285] which is rapidly converted by plasma esterases into the active compound, mycophenolic acid. Mycophenolic acid reversibly inhibits inosine monophosphate dehydrogenase (IMPDH), which is the rate-limiting enzyme for the de novo synthesis of guanosine nucleotides necessary for T and B cell proliferation.[286] Most cell types generate guanosine nucleotides by the synthetic IMPDH and salvage pathways; however, lymphocytes are entirely dependent on the IMPDH pathway, thus accounting for the relatively selective blockade by mycophenolic acid. Furthermore, mycophenolic acid preferentially inhibits the type II isoform of IMPDH, which is expressed preferentially on activated lymphocytes, thus explaining its improved efficacy/toxicity profile compared to azathioprine.

Corticosteroids

Three different corticosteroids, prednisolone, prednisone, and methyl prednisolone, have been used in the treatment of acute allograft rejection and in the maintenance posttransplant phase. Corticosteroids have a variety of anti-inflammatory and immunomodulatory effects,[287] including the reduction of histamine and bradykinin release, capillary permeability, and prostaglandin synthesis. Corticosteroids cross into the cytoplasm and bind to glucocorticoid receptors that are bound to cytoplasmic-located heat-shock proteins. The binding to corticosteroids triggers the release of glucocorticoid receptors from heat-shock proteins and their translocation into the nucleus where they inhibit the transcription of many proinflammatory cytokines, including IL-1, IL-2, IL-6, IFN-gamma, and TNF-alpha.

The side effects of steroid treatment are numerous, including diabetogenesis, central obesity, protein loss from skeletal muscle, cataracts and glaucoma, osteoporosis, and impaired wound healing. To avoid long-term side-effects, the use of steroids can be successfully tapered and eliminated in low-risk transplant recipients without affecting long-term graft outcomes.[288]

Biological Agents

Polyclonal Antithymocyte Globulin

Polyclonal antibodies (antithymocyte/lymphocyte globulin or antilymphocyte serum) are widely used during the induction phase of solid organ transplantation or for treating acute allograft rejection. These antibodies are prepared by immunizing rabbits or horses with human lymphocytes or thymocytes, and the serum IgG fraction is purified. Such polyclonal preparations comprise antibodies directed against many different cell-surface molecules expressed on T and B lymphocytes, NK cells, and APCs. They deplete lymphocytes and may block molecules stimulating T cell proliferation. Efficacy of repeated treatment with polyclonal antibodies is reduced by the development of xenospecific neutralizing antibodies, which can be mitigated by the use of preparations from different species.

Monoclonal Antibodies

Anti-CD25 antibodies target the alpha chain that confers high-affinity IL-2 binding to the IL-2R, which also includes the beta and gamma chains. Blockade with anti-CD25 reduces T cell proliferation and is used in induction therapy to inhibit the initial T cell proliferation, thereby reducing the incidence of acute rejection in the first-year posttransplantation. Anti-CD25 cannot be used to treat acute rejection because increased IL-2 levels are sufficient to activate T cells by binding to just the low-affinity IL-2 receptor.

Rituximab recognizes human CD20 and is a chimeric murine monoclonal antibody created by fusing the Fab region of a murine anti-CD20 antibody with the human kappa and gamma 1 constant regions.[289] Because CD20 is expressed by most human B lymphocytes, treatment with rituximab results in B cell depletion and has been used in desensitization, ABO-incompatible transplantation, and acute and chronic AMR. Its efficacy remains to be confirmed in randomized multicenter trials, as anti-CD20 does not target the plasma cells that produce DSA, and there is concern that it could deplete regulatory B cells.

Intravenous Immunoglobulin

IVIG is prepared from the serum of 1000 to 15,000 healthy human donors per batch and is used for the treatment of acute rejection as well as in desensitizing patients with high levels of circulating anti-HLA antibodies to improve their access to transplantation and to prevent acute AMR.[290] Many different mechanisms have been attributed to the immunomodulatory effects of IVIG due to their binding to Fc receptors as well as anti-idiotypic binding of the Fab domains that block binding of circulating anti-HLA antibodies to the allograft. Implicated mechanisms include depletion of T and B cell subsets, inhibition of lymphocyte activation and differentiation, saturation of FcRn recepotrs, resulting in accelerated decay of circulating IgG, inhibition of proinflammatory complement components (C1,C3a,C3b,C4), cytokines (IL-1β, TNF), Fc receptors, and adhesion molecules.

CTLA-4Ig

T cell activation requires signals from the TCR as well as costimulatory molecules including CD28 binding to CD80 and CD86

(B7.1 and B7.2) on activated APCs.[291] Compared to CD28, cytotoxic T-lymphocyte antigen-4 (CTLA4 or CD152) binds with >2500-fold and ~500-fold greater affinity to CD80 and CD86, respectively, and has a coinhibitory function. CTLA4Ig is a fusion protein of CTLA4 and the Fc domain of IgG and was first shown to be effective at preventing rejection in rodent models.[292] Because CTLA4 is less effective at inhibiting CD28 binding to CD86 relative to CD80, two amino acid substitutions were introduced to generate a modified CTLA4Ig, belatacept (LEA29Y), that has ~fourfold and twofold greater affinity to CD80 and CD86[293] and is ~10-fold more potent at inhibiting human T cell activation compared to the parent CTLA4Ig. In renal transplant recipients, belatacept can reduce or replace the use of calcineurin inhibitors, which have significant renal and other toxicities. Belatacept is reported to be particularly effective at preventing the development of DSA and reducing pre- and posttransplant DSA. However, belatacept has been associated with higher rates of early T cell–mediated rejection compared to immunosuppression with calcineurin inhibitors, thus limiting its use.[63]

Additional Costimulatory Blockers

New drugs based on costimulation blockade currently in clinical trials are designed to offer better efficacy and safety compared to belatacept.[294] Two novel anti-CD28 single domain antibody antagonists conjugated to polyethylene glycol are in clinical development, FR104 and lulizumab pegol. Both are anticipated to be superior to belatacept by avoiding the inhibition of coinhibitory CTLA-4 and by also avoiding the proinflammatory effects of an early version of an anti-CD28 mAb agonist. FR104, a pegylated mAb fragment, prevented alloimmunization in nonhuman primate renal allografts and was reported to be well tolerated and efficacious at inhibiting antibody responses in humans.[295] Lulizumab, a single domain pegylated anti-CD28 nanobody, was successfully used in the desensitization of nonhuman primates prior to kidney transplantation and is currently in phase I/II clinical trials for preventing renal transplant rejection.

Anti-CD154 mAbs directed at blocking CD40:CD154 interactions have been reported to be potent inhibitors of T cell activation and to promote prolonged graft acceptance in nonhuman primate models of allogeneic kidney and islet transplantation. However, clinical trials with anti-CD154 were halted by a high incidence of thromboembolic complications due to the activation of platelets through CD154 ligation and the formation of immune complexes with soluble CD40L that resulted in the activation of platelets through the Fc-γ receptor IIA (FCGR2A). Three novel CD40L-binding molecules (letolizumab/BMS-986004; dapirolizumab pegol/CDP7657 and VIB4920) designed to avoid FCGR2A-mediated thrombosis have undergone early phase clinical trials for autoimmune indications, and one has shown efficacy in nonhuman primate kidney transplantation.[294] Letolizumab is an anti-CD40 L domain antibody engineered to express a mutated IgG1 that lacks Fc-binding and complement-fixing effector functions. At high doses, this agent successfully prolonged kidney allograft survival in rhesus macaques without evidence of thromboembolic complications.[296]

A number of antagonistic anti-CD40 antibodies are also under investigation; two have been tested in kidney transplantation, namely CFZ533 (iscalimab) and ASKP1240 (bleselumab).[294] Both are fully humanized, nondepleting anti-CD40 IgG mAbs, that are well tolerated with no thrombotic events reported in clinical trials. A fully Fc-silencing anti-CD40 mAb, iscalimab, prolonged renal allograft survival without B cell depletion in nonhuman primates[297] and phase 2 trials in kidney transplantation are ongoing. Bleselumab, an anti-CD40 recombinant IgG4, demonstrated noninferior efficacy when administered with low-dose tacrolimus compared to standard therapy in kidney transplant patients, but higher rates of acute rejection and increased incidence of CMV and BK virus infections were observed in the bleselumab plus mycophenolate (MMF) group.[298] Thus, research continues to develop efficacious and safe drugs targeting costimulatory molecules on T cells and APCs, including dendritic cells and B cells, with the aim of preventing rejection and potentially promoting transplantation tolerance.

Future Directions

Presensitized Recipients

Presensitized patients are identified by the presence of circulating anti-HLA antibodies and having a history of sensitizing events such as allograft rejection, pregnancy, or blood transfusion. While sensitization would have induced both memory T and B cell responses, assessment of how memory T cells affect graft outcomes is hampered by the lack of a standardized clinical assay able to quantify HLA-specific memory T cells. As a result, sensitized recipients are identified based on the presence of circulating DSA and desensitization has focused on the removal of circulating DSA and the cells that produce them. Transient removal of circulating DSA can be achieved through plasmapheresis or plasma exchange, and more recently, through the use of a novel immunoglobulin G-degrading enzyme of *Streptococcus pyogenes* (IdeS).[299] To achieve a more durable reduction in DSA, IVIG is used in combination with plasmapheresis in desensitization and AMR treatment protocols and works through multiple anti-inflammatory mechanisms as discussed earlier.[147] In addition, the removal of B cells with anti-CD20 monoclonal antibodies has been attempted, although concerns that depletion of regulatory B cells may result in higher rates of acute rejection have tempered the use of this approach. More recently, proteasome inhibitors, such as bortezomib and carfilzomib, which prevent the intracellular degradation of misfolded proteins, thereby inducing the depletion of plasma cells responsible for antibody production, have been evaluated, as well as an anti–IL-6 receptor antibody.[191] Recent studies that used proteasome inhibitors in combination with belatacept show therapeutic promise. Despite the identification of numerous approaches targeting humoral desensitization, the elimination of high-titer DSA and preventing their rebound posttransplantation remain challenging and outcomes remain suboptimal.

Immunosuppression Minimization

Toxicities associated with global immunosuppression, off-target effects of the immunosuppressive drugs, and challenges associated with nonadherence have prompted studies focused on identifying transplant recipients for whom immunosuppression can be successfully reduced or eliminated. The most successful attempts at immunosuppression elimination to reach a state of operational tolerance have been reported for pediatric and adult recipients of allogeneic liver transplants.[300,301] Long-term graft acceptance after cessation of immunosuppression in highly selected patients ranges from 60% in the pediatric group to 50% in adult patients with hepatitis C virus infection.[300,302] Overall, higher rates of successful weaning are observed in older recipients (>60 years), with an absence of autoimmune liver disease and viral infections, with pristine biopsies, and in recipients who have had their transplant for longer periods of time prior to weaning. Currently, prospective clinical trials for weaning off immunosuppression are ongoing and are focused on identifying biomarkers that predict the liver transplant recipients most likely to achieve operational tolerance, as well as biomarkers of a stable operational state of tolerance.

Donor-Specific Tolerance Induction

The Need for Tolerance-Inducing Regimens

Immune tolerance denotes a state in which donor-specific immune nonresponsiveness is maintained without immunosuppressive agents. For the purpose of this discussion, we will define operational tolerance as acceptance of a donor organ by an otherwise intact immune system and will qualify tolerance as "systemic" when the entire immune system fails to respond to antigens from the same donor. Transplantation tolerance is a desirable goal for three major reasons. First, while improvements in immunosuppressive therapy have dramatically increased the success of clinical organ transplantation, these drugs are associated with increased risks of infection and malignancy as well as metabolic and organ toxicities that can result in additional organ failure. Second, despite improved immunosuppression, chronic rejection still contributes to constantly down-sloping long-term survival curves for organ allografts. Third, a critical shortage of allogeneic organs has increased interest in the use of other species as organ and tissue sources. However, immune barriers to xenografts are greater than those to allografts, and the induction of both B-cell and T-cell tolerance may be essential to the ultimate success of xenotransplantation.

Mechanisms of Transplantation Tolerance

Other chapters in this book describe the mechanisms by which tolerance to self-antigens is achieved (see Chapter 32). Rather than reiterate these in detail here, we will mention each and provide examples in which they have been implicated in transplantation. Mechanisms for inducing T- and B-cell tolerance include deletion, anergy, and suppression. In addition, a graft may simply be ignored by recipient lymphocytes, but in most transplant settings, this is unlikely to be a reliable mechanism for sustaining tolerance. The thymus is the central organ for T-cell development. Induction of tolerance among developing thymocytes is referred to as "central," as distinguished from the "peripheral" tolerance that may develop among already-mature T cells when they encounter antigen in the peripheral tissues. The bone marrow is the central organ for B-cell development, and as for T cells, B-cell tolerance may also be thought of as central or peripheral.

Clonal Deletion

B cells: When a developing B cell with autoreactivity encounters self-antigen, it undergoes developmental arrest. Recombination activating gene-dependent light chain receptor editing then takes place. If this results in the formation of a nonautoreactive immunoglobulin receptor, the B-cell survives; if not, the B cell undergoes apoptosis.[303] B cells are susceptible to tolerance at particular stages of development upon recognition of membrane-bound antigen. Several checkpoints for elimination of autoreactive B cells during development in the marrow and after migration to the periphery have been identified.[304,305] Deletion and receptor editing eliminate autoreactive B cells in the bone marrow,[306] while those encountering self-antigen in the periphery become deleted or anergized, depending on the nature of the antigen.[307] In the context of transplantation, B-cell responses to allogeneic MHC molecules are largely T cell–dependent, so an absence of alloantibodies may not necessarily reflect intrinsic B-cell tolerance; an absence of T-cell help would produce a similar outcome. However, MHC-alloreactive B cells may be intrinsically tolerized by encounter with antigens in the absence of T-cell help, apparently by a deletional mechanism[308] or by a nondeletional mechanism associated with the acquisition of B cell-suppressive function by donor-reactive B cells.[309]

In the context of blood group antigens or xenoantigens, tolerance of natural antibody responses must target B cells in addition to T cells, as natural antibody production is largely T cell–independent, though it can be augmented by T-cell help. Cells of the B-1 subset can produce xenoreactive natural antibodies and may be tolerized by an initial anergy mechanism followed by deletion of preexisting and newly developing B cells in mice given hematopoietic cells expressing an antigen for which natural antibody-producing cells preexist in the recipient.[310,311]

T cells. Central T-cell tolerance involves deletion of developing thymocytes that recognize self-antigens presented by hematopoietic cells and thymic epithelial cells. In addition, antigens produced only by nonhematopoietic thymic stromal cells can induce T-cell anergy or drive T_{reg}-cell differentiation. Several hematopoietic cell types, including DCs, B cells, and thymocytes, as well as thymic stromal cells, can induce intrathymic deletion. Medullary thymic epithelial cells expressing tissue-specific antigens can mediate thymocyte deletion by presenting antigens directly or transferring them to DCs. Transplantation tolerance induced by intrathymic deletion should be very reliable, as donor-reactive T cells would be largely absent in the periphery, as discussed in detail below.

Mature peripheral T cells can also be deleted. Peripheral CD8 cells may undergo deletion when antigen is repeatedly encountered on recipient nonhematopoietic cells, as described for GvH-reactive CTLs following HCT.[312,313] Peripheral deletion of donor-reactive T cells, via mechanisms that include exhaustion, activation-induced cell death, "veto" cells, and cytotoxic regulatory cells, have been described in various models, some of which are detailed below.

Anergy

Anergy is a state that may result when lymphocytes recognize antigen without adequate accessory or costimulatory signals or when they encounter ligands for which they have low affinity. Anergy is an important tolerance mechanism for the many self-reactive B cells that escape deletion in the bone marrow, particularly if they recognize abundant but low-avidity antigens. Self-recognizing, anergic B cells may also be repurposed into useful B cells recognizing foreign antigens via BCR mutation in germinal centers in a process termed "clonal redemption."[314] Anergy requires persistent antigen and is characterized by immunoglobulin receptor downregulation, altered signaling patterns, and increased apoptosis upon antigen encounter.[314] T-cell tolerance and the resulting absence of help is important for the maintenance of B-cell anergy, as anergic B cells can be activated in the presence of high-avidity antigen and T-cell help. Anergy is responsible for the early tolerance of natural antibody-producing B-1 cells in mice rendered mixed chimeric with bone marrow cells expressing an antigen recognized by natural antibody-producing cells.[310,311]

A T cell is considered to be "anergic" if it cannot proliferate or produce IL-2 in response to the antigen for which it is specific. T-cell anergy has been associated with altered signaling and tyrosine phosphorylation patterns. T-cell anergy can often be overcome by providing exogenous IL-2. Thymocytes are also susceptible to anergy induction by antigens presented on hematopoietic or nonhematopoietic stromal cells. While deletion may follow the induction of anergy of T cells in the continued presence of antigen,[316] anergic T cells may also persist.[317] Such anergy is reversible and may be overcome by infection[318,319] or by removal of antigen[317] and therefore may not be reliable in transplant recipients, in whom infections may perturb a nondeletional state of tolerance by activating innate immunity.[318] In a transplantation model involving BMT under cover of costimulatory blockade, peripheral donor-reactive CD4 and CD8 T cells are rendered anergic prior to their deletion.[320-322] Anergic T cells may also have a suppressive effect on other T cells,[323] which may involve conditioning of suppressive APCs.

Immunoregulation

Regulatory T cells. The concept of antigen-specific suppression and the existence of a specialized population of T cells, known initially as suppressor cells and more recently as T_{reg} cells, that could control immune responsiveness, originated in the 1970s, when the existence of complex T cell regulatory networks was surmised. In the setting of transplantation, a role for suppressor T cells in tolerance induction was also identified in that period.[324] However, the cellular and molecular characteristics of these suppressor cells remained elusive until 1990, when Hall et al. demonstrated the ability of CD25+CD4+ T cells to adoptively transfer tolerance from rats tolerating cardiac allografts after cyclosporine treatment.[325] Suppressor/T_{reg} cells were further characterized by Sakaguchi and colleagues using a murine neonatal thymectomy model that results in multiorgan autoimmune disease in mice due to insufficiency of CD25+ regulatory cells.[326]

The immunobiology of T_{regs} is covered in Chapter 29. Naturally occurring or thymus-derived CD25+CD4+ T cells with regulatory properties are generated in the thymus through interaction between the developing thymocytes with an intermediate affinity ligand, too low for negative selection, expressed on nonhematopoietic cells of the thymus. Additionally, antigen-induced or adaptive T_{reg} cells are generated in the periphery under certain conditions. The importance of T_{regs} in maintaining normal self-tolerance in humans is well-established, and they have been demonstrated to play a role in many experimental transplant models. Moreover, T_{regs} have been implicated in determining human transplant outcomes and are being explored as a platform for cellular immunotherapy (see below). While CD25 is constitutively expressed by T_{regs}, it is also transiently expressed by activated human T cells. Low levels of CD127 and expression of the Foxp3 forkhead/winged-helix transcription factor are important features of T_{regs}. In humans, mutations in the Foxp3 gene result in the immune dysregulation, polyendocrinopathy, enteropathy, X-linked (IPEX) syndrome, illustrating the physiologic importance of T_{regs}. Other cell surface markers of T_{regs} include CD152 (CTLA4), LAP, and glucocorticoid-induced TNF receptor (GITR)-related protein among many others. The methylation status of the Foxp3 gene is an important marker of the stability of FOXP3 expression that defines functional T_{regs}.

Recipient-derived CD25+CD4+ T cells have been shown to have potent regulatory properties in both the induction and maintenance phases of tolerance to alloantigens in vivo in mice.[327] In HCT, donor CD25+CD4+ T cells present in the bone marrow inoculum protect from GVHD.[328] In humans, low numbers of peripheral blood T_{regs} have been associated with chronic allograft rejection[329] and chronic GVHD.[330] In patients achieving renal allograft tolerance through combined kidney and nonmyeloablative bone marrow transplantation that achieves transient mixed chimerism, T_{regs} have been strongly implicated in tolerance induction, as FoxP3+ cells are prominent in allograft biopsies,[331] T_{regs} are markedly enriched in circulation in the early months posttransplant[332-334] and expansion of donor-specific T_{regs} correlates with successful tolerance induction.[334] In rodent models, natural T_{regs} appear to be much less potent than T_{reg} cells that expand or are induced following exposure to alloantigen, as 10 times fewer antigen-expanded compared to non-specific T_{reg} cells have been able to prevent allograft rejection.[327] The capacity of T_{regs} to prevent rejection as a single modality is limited in the presence of stringent transplant models, including presensitized recipients.[335] In this case, the kinetics of activation of the memory cells is very rapid and, unless very high numbers of T_{regs} are present, the balance between rejection and regulation is overwhelmingly in favor of rejection.[335] Thus, additional measures are needed to overcome the strong memory T cell alloresponse in presensitized hosts.

After transplantation, the ability to expand or generate T_{regs} in vivo may be influenced by the microenvironment. The presence of IL-2, which signals through STAT5 via the CD25-containing high-affinity IL-2 receptor expressed by T_{regs}, is critical for their function.[336] Consequently, numerous experimental therapeutic strategies are being explored that aim to increase the availability of IL-2 to expand T_{regs} in vivo. The balance between the presence of IL-6 and TGF-β in vivo can have a dramatic impact on which T-cell population develops. Importantly, T_{reg} generation can be prevented by inflammation, which is commonly present in vivo after cell or organ transplantation, leading instead to the generation of cells capable of mediating tissue damage, including Th17 cells. Furthermore, T_{regs} have been shown to have plasticity and develop Th17 and Th1 effector function in the presence of inflammatory stimuli.[337-339]

The mechanisms of regulation effected by T_{regs} to prevent allograft rejection involve a variety of different pathways. When T_{regs} are reactivated by their cognate antigen through the TCR, they rapidly and transiently produce IFNγ, which is needed to mediate their regulatory activity and prevent graft rejection.[340] IFNγ triggers the STAT-1 signaling pathway that also influences the functional activity of T_{regs}.[341] The transient production of IFNγ by T_{regs} can have multiple effects on T cells with which they interact, mediated either directly or indirectly through effects on APCs. T_{reg} activity can result in inhibition of cytokine production and secretion, downregulation of costimulatory and/or adhesion molecule expression, inhibition of proliferation, induction of anergy, elimination of the effector population by promoting cell death, or even conversion of naïve and/or effector T cells to a regulatory phenotype, a process known as "infectious tolerance."[342,343] In infectious tolerance, new T cells entering the repertoire after transplantation are converted to induced T_{regs} (iT_{regs}), thereby propagating and reinforcing the tolerant state throughout the posttransplant course. The activity of T_{regs} in transplant models has been shown to be dependent on IL-10,[344] TGF-β,[345] and CTLA-4.[344] CTLA-4 (CD152) is constitutively expressed by T_{regs}, and its engagement of CD80/86 on DCs can induce indoleamine 2,3 dioxygenease (IDO), an enzyme that catalyses the initial and rate-limiting step of the kyneurinine pathway of tryptophan catabolism.[346] IDO is subject to transcriptional regulation by IFNγ in a manner that is dependent on STAT-1 signaling.[347] IFNγ released by T_{regs} can lead to the development of IDO-competent DCs that may acquire the capacity to control T cells locally through tryptophan depletion, thereby inhibiting alloimmune responses.[348] IDO has been implicated in promoting allograft acceptance in several experimental models involving donor-specific cell infusions.[349,350] More recently, IL-35, an immunosuppressive cytokine in the IL-12 family that is produced naturally by T_{regs} and induced in "iTr35's," was found to be transferred to T and B cells via extracellular vesicles that thereby acquire IL-35 on their cell surfaces. This pathway promoted tolerance in a murine donor-specific cell infusion/anti-CD154 tolerance induction model and provided a mechanism for amplification of a suppressive response initially involving a small number of donor-specific T_{regs}.[351]

To elaborate their functional activity, T_{regs} have to be activated through their TCRs, enabling them to suppress the response of other cells present in vitro[352] and in vivo[353] by a process of "bystander regulation." There is evidence that the location in which T_{regs} function in vivo may change with time after transplantation. Indeed, the unique microenvironment in draining lymph nodes, including various lymph node stromal cell populations and the different types of laminins that are present, have a major impact on the effector versus regulatory fate of naïve T cells entering them.[354,355] T_{regs} are found mainly in the draining lymph nodes in the first few days following skin transplantation, whereas they later appear within the allograft itself and can migrate out upon retransplantation to induce suppression in the new host.[356]

Numerous experimental strategies are being explored to exploit T_{regs} to achieve transplantation tolerance and trials in clinical transplantation have begun. These include methods that aim to expand T_{regs} in vivo in the presence of an allograft and others that involve infusion of nonspecific (polyclonal) or donor-specific T_{regs}. Understanding the impact of immunosuppressive drugs on T_{reg} generation and function is an important part of using the potential of T_{regs} in vivo. Data from mouse models suggest that calcineurin inhibitors, by preventing activation-induced cell death, favor effector T cells over T_{regs}, thereby limiting their efficacy.[327] However, initial use of calcineurin inhibitors did not prevent the achievement of renal allograft tolerance via transient mixed chimerism in monkeys[357] or in patients, in which T_{regs} were strongly implicated in the mechanism of tolerance.[331-333] Administration of anti-CD25 monoclonal antibodies after the induction of T_{regs} can prevent long-term graft survival. On the other hand, other classes of immunosuppressive drugs, such as the mTOR inhibitors sirolimus and everolimus, spare T_{regs} and are often used in protocols to expand them in vitro.

Strategies to generate and/or expand T_{regs} ex vivo for therapeutic purposes are being developed with potential application in cell and organ transplantation. A variety of different approaches have been used in both mice and humans, including exposure to rapamycin, TGF-β, IFNγ, and stimulation with anti-CD3/CD28 in the presence of IL-2. Typically, sorting for CD25+CD127low CD4+ T cells is followed by an expansion period.[358] The development, maintenance, and expansion of T_{regs} are critically dependent on IL-2, and trials of low-dose IL-2 have been initiated in autoimmune diseases and GVHD in efforts to expand T_{regs} in vivo.[336]

In proof-of-concept studies using a humanized mouse transplant model, ex vivo expanded T_{regs} were shown to be capable of controlling rejection.[359,360] Clinical trials in recent years have demonstrated the safety of administering expanded polyclonal T_{regs} in numbers greater than one billion in Type 1 diabetic patients[361] and trials are currently underway in solid organ transplantation using both polyclonal[362] and donor-specific T_{regs}.[363] The completed ONE study in Europe documented the safety of expanded T_{regs} in renal transplant recipients.[364] While these studies were aimed at assessing safety and did not involve immunosuppression withdrawal, a small liver transplant study involving infusion of recipient mononuclear cells expanded in the presence of donor antigen and costimulatory blockade, in which T_{regs} were found to be enriched, was associated with successful immunosuppression withdrawal, and hence operational tolerance, in 7 of 10 subjects.[365]

In addition to CD25+CD4+ T_{regs}, several additional types of T cells, including CD8+ cells,[366] iNKT cells[367,368] and novel CD4-CD8-[369] T cells have demonstrated regulatory activity in experimental transplant models.

Antibody-mediated suppression. Idiotypes are unique antigenic determinants that characterize the binding sites of antibody or TCRs. These determinants can be antigenic and induce the production of anti-idiotypic antibodies. Regulatory mechanisms for B-cell responses might include anti-idiotypes, as originally suggested by Jerne. Antibody-mediated suppression could theoretically occur through the recognition of idiotypes of antidonor immunoglobulin receptors, and the development of such antibodies has been suggested as one of the possible benefits of prekidney transplant blood transfusions. Such antibodies have also been suggested to contribute to the apparent hyporesponsiveness to noninherited maternal antigens in renal allograft recipients. While it has been suggested that anti-idiotypic antibodies might inhibit T-cell recognition of antigens, this seems unlikely, given that T cells recognize peptide/MHC complexes, whereas antibodies recognize epitopes of intact molecules. Nevertheless, anti-idiotypic antibodies have been reported to suppress HLA-directed MLRs,[370] perhaps by mimicking HLA and thereby blocking interactions between the TCR and critical HLA components. However, efforts to control transplantation using exogenous anti-idiotypic antibodies to either TCRs or B-cell receptors have been disappointing.[371]

In a process known as enhancement, prolonged graft survival is achieved by antigraft antibodies. Enhancing regimens using anti-MHC antibodies and/or soluble antigen were shown to produce long-term tolerance for rodent kidney allografts through mechanisms that did not simply involve blocking anti-MHC antibodies but also included suppressive cells.[372] Unfortunately, the spectacular success obtained using enhancement for kidney graft survival in rats has not been observed in other species, and this approach has largely been abandoned, at least for the present.

Ignorance and Immune Privilege

Peripheral antigens may be ignored by T cells[373] or B cells[374] that recognize them. This may be due to the presentation of these antigens by nonprofessional APCs, or it may reflect a failure of recipient T cells to migrate to the antigen-bearing tissue, as in murine solid tumor models. The level of peripheral antigen expression, how recently the responding T cell has emerged from the thymus, and the presence of proinflammatory cytokines and costimulatory molecules may all influence the decision of a T cell to ignore or respond to peripheral antigens. However, "ignorance" may be a precarious state that can be upset by inflammatory stimuli that may be induced, for example, by infections[375] or by presentation of antigen on professional APCs, as observed for endocrine allografts that are depleted of APCs prior to transplantation in rodents.[376]

Immunoprotection in specific "immune privileged" niches may protect allografts from immune attack, either by making antigen invisible to the immune system or by providing cell types in the microenvironment that protect the allogeneic cells. An example of the former is corneal transplantation, which is done without systemic immunosuppression due in part to the sequestered status of the anterior chamber of the eye. However, a number of active mechanisms also contribute to this immune privilege.[377] An example of a cell that may be protected largely by other cells in its niche is the hematopoietic stem cell, which can survive in an allogeneic stem cell niche without immunosuppression through mechanisms that involve specialized "niche T_{regs}."[378]

Experimental and Clinical Strategies for Inducing Transplantation Tolerance

Central Tolerance

Hematopoietic chimerism. The pioneering observations made by Owen and later by Medawar in Freemartin cattle 75 years ago revealed that hematopoietic chimerism that develops in utero is associated with transplantation tolerance.[379,380] While intentional chimerism induction with in utero HCT has the potential to cure congenital hematologic disorders and induce donor-specific tolerance, its success in animal models has been impeded by the difficulty in achieving engraftment without GVHD at gestational ages when such disorders can be diagnosed, which unfortunately occurs later than the development of fetal T cell–mediated resistance. Therefore, in utero HCT is limited by many of the same challenges as postnatal HCT, including the requirement for donor T cells to overcome host resistance and the tendency of donor T cells to induce GVHD. Nevertheless, some success in achieving lasting chimerism and donor-specific tolerance has been achieved in mice, dogs, and pigs.[381-383]

Applying the tolerance-inducing capacity of HCT in recipients in which an immune system is already established has been a major goal ever since Owen and Medawar's observations were made in Freemartin cattle. Acquired immune tolerance was first demonstrated by Medawar and colleagues, who injected allogeneic cells and tissue lysates into neonatal mice.[384] Subsequently, tolerance was achieved in untreated neonatal rodents given allogeneic hematopoietic cells shortly after birth. In different models, tolerance was associated with no chimerism, lasting microchimerism and sometimes sufficient levels of chimerism for flow cytometric detection (macrochimerism).[385] Tolerance involved multiple central and peripheral mechanisms,[385,386] with likely roles for regulatory cells,[385] Th2 immunity and noncostimulatory APCs in large donor inocula.[387]

The above studies in neonatally tolerized animals led to the hope that very low levels of chimerism (microchimerism), for which DNA and other evidence was found in human recipients of solid organ allografts who did not receive HCT, might be associated with donor-specific tolerance in adults.[388] Microchimerism, which requires highly sensitive techniques for its detection, should be distinguished from the mixed chimerism discussed below, in which multilineage chimerism is readily measurable by flow cytometry. However, with a few exceptions,[389,390] microchimerism has been shown not to be required for or even associated with tolerance[391] in a variety of models or in humans.[392] In fact, we now know that detection of donor DNA in recipient tissues does not necessarily denote hematopoietic chimerism since cell-free or exosome-derived DNA might be detected by such sensitive PCR-based techniques.

In experimental models involving adult rodents conditioned for allogeneic hematopoietic cell transplants, the capacity of hematopoietic cells to induce intrathymic clonal deletion of donor-reactive thymocytes results in robust tolerance to the most immunogenic allografts, such as fully MHC-mismatched skin and small bowel grafts.[235,393] However, the difficulty in achieving permanent donor hematopoietic engraftment without GVHD or excess toxicity has hampered the use of this approach in humans. HCT for tolerance induction in rodents originally involved recipient treatment with lethal total-body irradiation (TBI) with removal of mature donor T cells from the graft to prevent GVHD. However, MHC-mismatched allogeneic HCT in humans has proved to be less successful and more dangerous than in rodents because of the toxicity associated with myeloablative conditioning, the high risk of GVHD, and the risk of infection and engraftment failure when donor T-cell depletion is used to prevent GVHD. Indeed, alloengraftment across HLA barriers in humans is still somewhat dependent on GVH alloreactivity needed to overcome residual host-vs-graft (HVG) alloreactivity in recipients of all but the heaviest conditioning regimens. In contrast, it is possible to overcome alloreactivity in rodents with various conditioning regimens to achieve permanent engraftment of fully MHC-mismatched bone marrow without any reliance on donor T cells when sufficiently high donor hematopoietic cell doses are given (Fig. 48.6). Thus, more specific and effective methods of overcoming the barriers to marrow engraftment with minimal GVHD risk are needed for the application of HCT for organ allograft tolerance in man.

Achievement of a state of mixed rather than full allogeneic hematopoietic chimerism has several advantages for allograft tolerance induction: (1) Mixed chimerism can be achieved with less toxic (nonmyeloablative) conditioning regimens than those that lead to full donor chimerism. Furthermore, nonmyeloablative regimens preserve some host hematopoiesis so that life-threatening marrow failure is not a risk if donor marrow is rejected; (2) Mixed chimeras, unlike full chimeras, contain abundant recipient APCs in the periphery, allowing optimal antigen presentation to T cells that have developed in the host thymus, and which therefore preferentially recognize peptide antigens presented by host-type MHC molecules. Antiviral CTL responses in mixed chimeras demonstrate exquisite specificity for recipient-derived MHC-restricting elements[394,395]; (3) Mixed chimeras contain hematopoietic cells from both recipient and donor in the thymus that can delete both host-reactive and donor-reactive T cells, resulting in more complete deletion of host-reactive thymocytes in mixed compared to full chimeras.[396] As a result of central T cell deletion, tolerance achieved with durable mixed chimerism is "systemic," as evidenced by specific unresponsiveness to the donor and recipient in MLR and CML assays, and acceptance of donor skin grafted at any time post-BMT.[235]

In moving toward clinical application, several nonmyeloablative approaches have been developed for achieving mixed chimerism and tolerance. An early regimen produced a "clean slate" for donor engraftment by eliminating T cells centrally and peripherally using thymic irradiation and T-cell depleting mAbs and utilized low-dose (3 Gy) TBI to make hematopoietic "space" for marrow engraftment.[235] This regimen showed that overcoming the T cell barrier was sufficient to allow allogeneic hematopoietic engraftment and intrathymic deletion

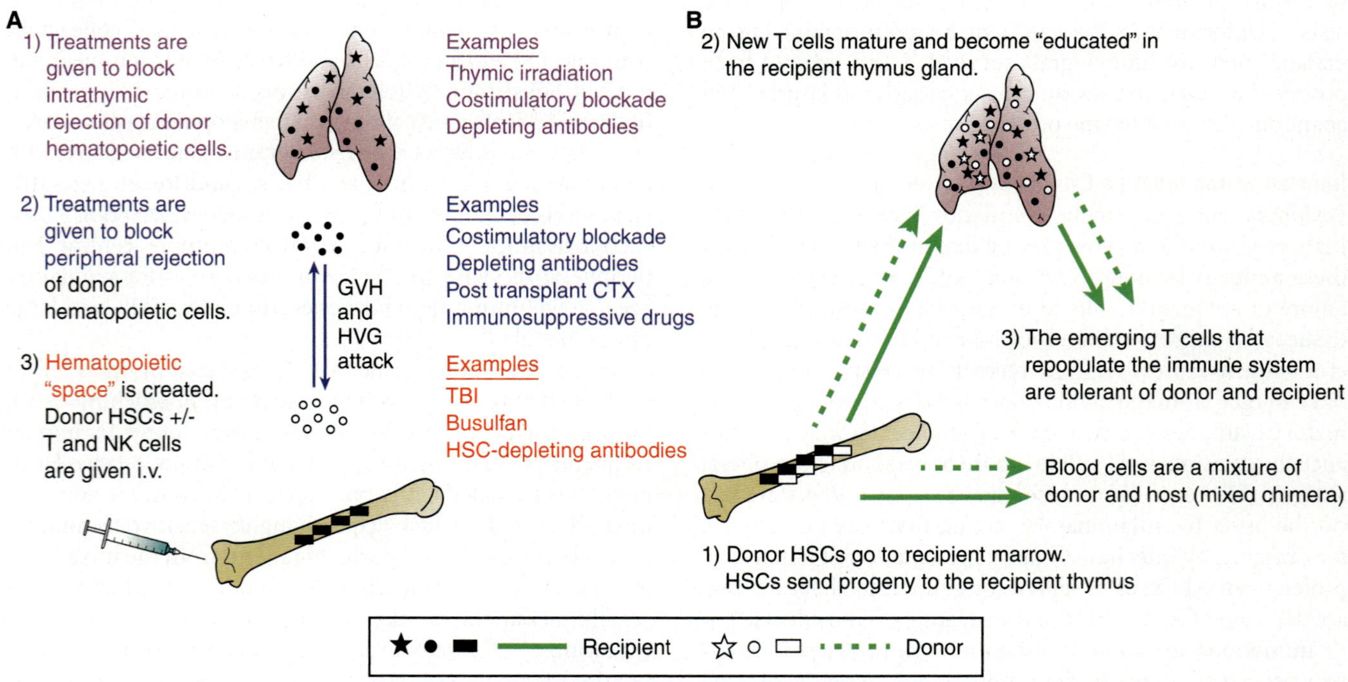

FIG 48.6. Schematic showing methods of overcoming resistance to allogeneic hematopoietic stem cell (HSC) engraftment **(A)** and mechanism by which permanent mixed HSC chimerism achieves central T cell tolerance **(B)**. ☆Dendritic cells, ○ Thymocytes and T cells, ▭ Bone marrow HSCs.

was shown to be the major mechanism maintaining long-term donor-specific tolerance.[397] Administration of donor MHC-specific antibody to eliminate donor chimerism from established mixed chimeras led to loss of tolerance and to the de novo appearance in the periphery of T cells with receptors that recognize donor antigens. However, if the recipient thymus was removed prior to elimination of chimerism, specific skin graft tolerance was preserved and donor-reactive TCR did not appear.[398] Thus, chimerism is needed only in the thymus and not in the periphery in order to ensure persistent tolerance among peripheral T cells that are deleted of donor-reactive cells. In contrast, peripheral anergy and suppression generally require persistent antigen to maintain tolerance. Because thymic APCs are continually turning over, this observation emphasizes the need for hematopoietic stem cell engraftment at sufficient levels to ensure an uninterrupted supply of donor APCs in the recipient thymus for the life of the mixed chimera. The absence of suppressive tolerance mechanisms makes these animals particularly vulnerable to breaking of tolerance when nontolerant T cells emerge from the thymus after intentional depletion of donor chimerism or after exogenous administration of nontolerant host-type T cells.[398]

The requirements for achievement of durable chimerism with central tolerance are summarized in Figure 48.6. T-cell alloreactivity in both the thymus and periphery must be overcome in order to permit allogeneic stem cell engraftment and early seeding of the thymus with allogeneic APCs. Intrathymic alloreactivity can be overcome with irradiation, high doses of T cell-depleting antibodies, or costimulatory blockers. In combination with bone marrow transplantation, GVH alloreactivity, T_{regs}, costimulatory blockers, various immunosuppressive drugs, and total lymphoid irradiation (TLI)[399] can be used to overcome peripheral T cell–mediated alloresistance. TLI establishes tolerance through a complex mechanism involving multiple cells types, including iNKT cells and T_{regs} in rodents. Myelosuppressive preparative treatments (eg, TBI, busulfan) are usually added to promote marrow engraftment, which may occur by creating physical niches due to the destruction of host hematopoietic cells and by upregulating cytokines and chemokines that promote stem cell homing, expansion, and hematopoiesis. Recently, monoclonal antibodies to the hematopoietic stem cell marker CD117 have been used as a more specific, less toxic means of creating such hematopoietic "space."[400] Myelosuppression is required to promote engraftment of syngeneic marrow cells given to mice in conventional doses,[401] but this requirement can be avoided using very high doses of marrow.[402] Engraftment of high doses of allogeneic marrow can be achieved without myelosuppressive treatment in mice and pigs receiving T cell–depleting mAbs[403,404] and in mice receiving costimulatory blockade[405] or expanded recipient T_{reg} cells.[406] Hematopoietic cells themselves have immunosuppressive properties that can help to promote their own engraftment by "veto"[407] and other mechanisms when sufficiently high doses of marrow are given. Therefore, advances in the ability to expand hematopoietic cells in vitro and to utilize Treg therapy may improve the ability to minimize or eliminate host myelosuppression and other toxic conditioning approaches in the future. Peripheral tolerance mechanisms are involved in all of the regimens that do not include measures to globally ablate T cells, though some of these peripheral mechanisms involve specific deletion of donor-reactive T cells (see below).

Some nonmyeloablative conditioning regimens have been adapted and applied with some success for tolerance induction in MHC-mismatched large animals[404,408] though mixed chimerism has been transient in nonhuman primates achieving renal allograft tolerance with one such regimen.[357] In another monkey model, persistent MHC-mismatched chimerism required ongoing immunosuppressive pharmacotherapy, indicating that tolerance was not achieved.[409]

It has been difficult to achieve T-cell depletion with antibodies in large animals and humans that is as exhaustive as that achieved in the previously mentioned rodent models, perhaps due to the use of inadequate doses or suboptimal reagents. A second concern is that, if sufficiently exhaustive T-cell depletion could be achieved in humans, T-cell recovery from the thymus might be dangerously slow, especially in older individuals. The ability to replace globally T cell–depleting antibodies with costimulatory blockade, T_{regs} or other approaches is therefore of considerable interest.

Strategies to Achieve Peripheral Tolerance

Many strategies have been developed for induction of transplantation tolerance among peripheral T cells. The essential difference between this approach and central tolerance is the requirement only in central tolerance to generate a new, tolerant T cell repertoire in the recipient thymus. Instead, peripheral tolerance modifies the repertoire or functions of preexisting T cells in the periphery to achieve allograft tolerance. Provision of donor antigen in a tolerogenic form often involves cell therapy, which may be combined with pharmacologic immunosuppression or biologics such as costimulatory blockers. Thousands of such approaches have been developed in animal models and some have begun clinical trials, most notably cell therapies. These are summarized below.

Cell therapies. The experimental and clinical use of T_{regs} and other types of tolerogenic T cells has already been discussed in the section on mechanisms of tolerance. An extension of this approach that is being actively explored in transplantation and autoimmune diseases involves the engineering of chimeric antigen receptor (CAR)-bearing T_{regs} (CAR-T_{regs}) and other genetic engineering approaches to improve T_{reg} survival, stability, homing, and function.[410] CARs typically consist of an antigen-binding single-chain variable fragment (scFv) antibody domain linked to intracellular components of the TCR and costimulatory signaling pathways. CAR-T_{regs} bearing antidonor class I MHC antibody domains have been shown in humanized mouse models to suppress allograft rejection and GVHD,[411,412] and this modality may soon enter clinical trials.

Many non–T cell types are also being explored for their tolerogenic potential in transplant models and clinical trials. Regulatory B cells (B_{regs}), rather than being a specialized cell type defined by an individual cell surface marker, seem to include several different types of B cells that produce IL-10 and other suppressive cytokines, including transitional B cells, plasmablasts, and, in mice, B-1 cells. While B_{regs} have been more widely studied in autoimmune diseases, their role in transplantation is not well-defined,[413] though B_{regs} have been associated with operational tolerance in renal allograft recipients[414,415] and a paucity of B_{regs} has been associated with clinical chronic GVHD.[416] However, it should be borne in mind that the lack of immunosuppressive therapy in operationally tolerant allograft recipients may account for many of the observed differences between this group and patients receiving conventional therapy,[417] making interpretation of such associations difficult.

Myeloid-derived suppressor cells (MDSCs), which consist of heterogenous immature granulocytic and monocytic subsets that are myeloid precursors with nonspecific T cell suppressive activity, have been shown to play a role in a variety of experimental tolerance models in a network that ultimately involves T_{regs}.[418] They tend to appear in settings of chronic inflammation and produce reactive oxygen and nitrogen species as well as immunosuppressive cytokines, including IL-10 and TGF-β. They are able to promote T cell apoptosis and Treg expansion and can suppress NK cells, B cells, and Tfh. While they have shown an ability to prolong corneal and islet allograft survival in rodents,[418] further studies are needed to better understand and purify tolerogenic subsets for use as immunotherapeutic cells.

"Mesenchymal stromal cells" (MSCs) are heterogeneous cells with the potential to differentiate into tissues of mesodermal lineages in vitro that demonstrate plastic adherence, expression of CD73, CD90, and CD105, and lack expression of several markers, including CD45 and CD34. MSCs also display nonspecific suppressive activity, in large part mediated by a variety of molecules, including indoleamine 2,3-dioxygenase (IDO), nitric oxide, TGF-β, prostaglandin-E_2 (PGE_2), heme-oxygenase 1 (HO-1), galectins, and HLA-G. Their broad suppressive effect on T cell responses involves Treg expansion as well as inhibition of Th1 and Th17 differentiation and enhanced Th2 polarization. Additionally, they directly affect B cell and DC functions. IFN-γ appears to be a key driver of their suppressive function. MSCs have been used to suppress autoimmunity and allograft rejection in numerous animal models and to inhibit GVHD in both animals and clinical trials. Trials of allogeneic and autologous MSCs have begun in organ transplant recipients, with safety having been demonstrated and some early suggestion of efficacy.[419]

Tolerogenic dendritic cells have been found active in a number of experimental models and are now being explored clinically.[420] Immature myeloid DCs that express low levels of MHC class II and costimulatory molecules at the cell surface as well as various classes of mature DCs, including subsets of classical DCs[421] and plasmacytoid DCs,[422] have shown this capacity experimentally. Immature DCs have shown activity alone[423] or in combination with costimulation blockade[424] or sirolimus.[425] Multiple mechanisms have been invoked for tolerance induction by each type of DC, including Treg-mediated and cytokine-dependent pathways.

In addition to infusion of various regulatory cell types, studies in rodents suggest that the mere provision of donor antigen in appropriate quantity and form may be used to promote tolerance. In providing donor antigens prior to organ transplantation, it may not even be necessary to pretreat recipients with every donor major and MiHA they would subsequently encounter on the allograft, as regulatory mechanisms in the host permit spreading of tolerance to additional antigens following exposure to a single donor alloantigen, via linked suppression.[327,426,427] Animal models indicating that pretreatment with donor antigen-bearing cells could promote tolerance were validated by the observation that donor-specific transfusions[428] (DSTs) were associated with improved outcomes in patients receiving renal allografts.[429] However, infusion of cells expressing alloantigens from the organ donor also has the potential to sensitize the recipient to donor alloantigens and the use of DSTs prior to renal transplantation largely ended with the introduction of erythropoietin into clinical practice, which obviated the need for blood transfusions in dialysis patients on transplant waiting lists. Nevertheless, recent rodent and large animal studies demonstrating the capacity of apoptotic donor leukocytes administered with costimulatory blockade and other reagents prior to organ or islet transplantation to promote graft survival[430,431] have suggested a potentially clinically applicable approach to achieving tolerance with donor antigens. However, it is a concern that tolerance induced in this manner was broken by acute murine CMV infection,[432] indicating that this type of nondeletional tolerance can be overcome by inflammatory stimuli. A similar approach, using apoptotic syngeneic lymphocytes, has already entered clinical trials to induce tolerance in patients with autoimmune diseases.

It should be noted that efforts to simply exhaustively T cell deplete a recipient and allow immune reconstitution to occur in the presence of a solid organ allograft[433] or skin allograft (M.Sykes and D.H.Sachs, unpublished data, 1986) have been unsuccessful. These observations highlight the importance of the particular mode of donor antigen provision and the immune regulatory environment needed to achieve allograft tolerance.

Blockade of the CD28/B7 and CD40L/CD40 pathways engendered initial enthusiasm due to their potent ability to prolong skin graft survival in rodents and kidney allografts in monkeys.[434,435] These reagents, however, failed to induce operational tolerance when used alone or in combination. Nevertheless, costimulatory blockade has proved potent in promoting tolerance induction in rodent models of cell therapy. One such combination involves bone marrow transplantation with anti-CD40L, with or without CTLA4Ig.[436,437] Donor bone marrow cells can also provide donor antigen that induces peripheral tolerance when combined with other modalities, most notably costimulatory blockade. Donor BMT has the advantage of then providing an ongoing source of donor antigen to tolerize newly developing T cells in the thymus, thereby combining peripheral and central tolerance mechanisms. While long-term tolerance is maintained by intrathymic deletion in mixed chimeras prepared with costimulatory blockade, initial tolerance of peripheral T cells involves specific deletion of donor-reactive CD4[322] and CD8[438] T cells. CD25-negative CD4 cells are required for CD8 tolerance only during this initial 2-week period before deletion is complete[438] and deletion of peripheral donor-specific CD4 and CD8 cells is preceded by donor-specific unresponsiveness.[321,322,439] Regulatory cells do not appear to play a major role in maintaining the long-term tolerance in this model, which is deletional, both centrally and peripherally.[322] The complete deletion of donor-reactive T cells may preclude the expansion and maintenance of specific regulatory mechanisms, similar to chimeras prepared with global T cell–depleting regimens.[398] Administration of DST with anti-CD40L mAb prevents islet allograft rejection,[440] but in contrast to BMT, does not achieve central deletion and, consequently, is most effective in thymectomized mice.[441] DST promotes peripheral deletion of donor-reactive CD8 T cells in recipients of anti-CD40L.[442]

Peripheral and Central Tolerance Are Interrelated

The thymus can play an active role in tolerizing peripheral T cells. For example, in a pig model involving a short course of CsA with kidney transplantation, the recipient thymus is required to achieve tolerance among preexisting peripheral T cells.[443] In rats, peripheral tolerance can be achieved by intrathymic injection of donor antigens combined with peripheral organ transplantation.[444] The capacity of T cells that are activated in the periphery to migrate to the thymus,[445] migration of donor graft APCs to the thymus, where they induce thymic T-cell deletion,[446] and/or the development of Treg cells that migrate to the periphery and promote tolerance[447] may contribute to this role for the thymus in peripheral tolerance induction. T_{regs} have indeed been implicated in both the pig and rat models above, in which the thymus is required to induce peripheral tolerance.

From Animal Models to Clinical Transplantation Tolerance

Almost every transplant clinician has treated patients who have chosen to withdraw immunosuppressive therapy for a variety of reasons. While the majority of such patients reject their allograft, rare cases achieve "spontaneous" tolerance in this manner, demonstrating that tolerance can be achieved in humans. Blood biomarkers have been described that identify tolerant patients who have been successfully weaned from immunosuppression[414,448] but so far none have been used to identify such patients before immunosuppression withdrawal and thereby shown to have predictive value. In fact, the absence of immunosuppressive drug therapy in these patients may itself account for the differing transcriptional profile of their immune cells compared to conventional transplant recipients.[417] In general, short-term results of most organ allograft transplants are excellent, making it essential to have safe and reliable methods of inducing tolerance in order to ethically justify their use in place of conventional chronic immunosuppressive therapies. Because the risk of rejection due to immunosuppression withdrawal is difficult to accept in patients, it can be argued that extension of results from animal models to humans should only be attempted after: (1) rodent studies have demonstrated robust tolerance in multiple strain combinations using extensively histoincompatible, highly immunogenic grafts such as skin; (2) efficacy has been demonstrated in large animal models; and (3) safety has been demonstrated in large animal models. When complete removal of immunosuppression is likely to be achieved, the level of toxicity accepted in the short-term may be somewhat higher than that which could be acceptable on a long-term basis. Most efforts to achieve peripheral tolerance in larger animals have not been as effective as in rodent models.

Donor bone marrow cell infusions without specific myelosuppressive conditioning have been evaluated in efforts to achieve tolerance in patients receiving organ transplantation with standard chronic immunosuppressive therapy but none of these trials have shown sufficient efficacy to permit immunosuppression withdrawal, despite the achievement of low levels of chimerism in some.[449,450] However, the extensive animal data that have been accumulated on the ability of mixed chimerism induced with reduced intensity conditioning to achieve robust transplantation tolerance, including its demonstration in large animal models,[357] combined with clinical data obtained in patients with a more conventional indication for HCT, have allowed clinical evaluation of this approach for tolerance induction. Mixed chimerism can be achieved with reduced toxicity using nonmyeloablative conditioning in patients with hematologic malignancies, and lymphohematopoietic graft-versus-host reactions induced by donor lymphocyte infusion (DLI) can be used to achieve graft-versus-tumor effects.[451] These observations provided an opportunity to evaluate the potential of this approach to induce transplantation tolerance in patients with a hematologic malignancy, multiple myeloma, and consequent renal failure. Recipients of a simultaneous nonmyeloablative BMT and renal allograft from HLA-identical siblings achieved either transient or durable mixed chimerism or, in a few cases, full allogeneic chimerism (some after DLIs). All three subsets of patients accepted their kidney grafts without any immunosuppression for follow-up periods of up to 20 years, with very good tumor responses.[452] These promising efficacy results, combined with safety data in patients receiving HLA-mismatched transplants with a related regimen for hematologic malignancies,[453] cleared the way for trials of HLA-mismatched haploidentical bone marrow and kidney grafts in patients with no malignant disease (ie, BMT was performed solely for allograft tolerance induction). The regimen chosen had achieved transient mixed chimerism without GVHD in an initial cohort of patients with hematologic malignancies. Donor kidneys were accepted off immunosuppression for >5 years in 7 of 10 patients, with follow-up for as long as 18 years in the first patient at the time of this writing. Although a cytokine release syndrome complicated most of these transplants, these studies provided proof of principle that it is possible to intentionally and safely achieve transplantation tolerance across HLA barriers in humans. Mechanistic studies implicated regulatory cells and the kidney graft itself as possible contributors to long-term, systemic, donor-specific tolerance that may involve eventual peripheral deletion of donor-reactive T cells.[12,334,454] An additional protocol used a TLI/ATG-based nonmyeloablative HCT method to achieve tolerance to HLA-matched kidneys[455] but has not yet successfully achieved tolerance across HLA barriers. A third protocol combines TBI, cyclophosphamide, and fludarabine to achieve full chimerism across extensive HLA barriers and involves administration of bone marrow "facilitating cells" along with donor hematopoietic cells. However, the complication of GVHD has been observed in several recipients of this protocol[456] and animal studies would predict that these full chimeras with extensive HLA mismatches might not regain full immunocompetence.[394,395]

As is discussed in preceding sections, a number of clinical trials are underway to evaluate various cell therapies with the potential to achieve transplantation tolerance without HCT. While some have succeeded in reducing the amount of immunosuppressive therapy in study subjects, none so far have involved complete immunosuppression withdrawal and therefore have not demonstrated induction of operational tolerance. In general, these approaches have not been preceded by large animal studies demonstrating the capacity of the administered cell population to achieve tolerance, making it difficult to ethically justify rapid immunosuppression withdrawal in patients.

MANIPULATIONS TO PREVENT OR TREAT GVHD

In general, approaches to prevent or treat GVHD can be divided into nonspecific immunosuppression and more tailored approaches which aim to minimize effects on global immunocompetence to leave the immune system otherwise intact. Furthermore, based on the distinct clinical features of acute GVHD and chronic GVHD, which have different underlying pathobiologies, interventions may be more aimed at treating or preventing acute versus chronic GVHD.

Immunosuppressive Drugs

Most recipients of allogeneic HCT today receive exogenous immunosuppression in the form of combination therapy with several drugs including a calcineurin inhibitor (cyclosporine or tacrolimus) and an antimetabolite (eg, methotrexate [MTX] or MMF). Sirolimus has been implemented for GVHD prophylaxis in a reduced-intensity/nonmyeloablative conditioning regimen for sickle cell disease based on its favorable side effect profile with regard to neurotoxicity[457] while promoting expansion of T_{reg} cells.[458] In general terms, both standard and experimental immunosuppressive drugs for the prevention or treatment of GVHD in the clinical setting suppress immune responses either by depleting immune cells in vivo (Campath, ATG) or ex vivo, by blocking costimulation (eg, CTLA4-Ig), by inhibiting lymphocyte gene transcription (eg, cyclosporine, tacrolimus), by blocking cytokine function (anti-IL-6 receptor antibodies, tocilizumab) or by blocking cytokine signal transduction (eg, JAK1/2 inhibitors), or by interfering with trafficking of immune cells (CCR5 blockade). A major breakthrough allowing transplantation across HLA barriers was accomplished by administration of the alkylating chemotherapy agent cyclophosphamide at the peak of alloreactive donor and host T cell expansion following transplantation on days 3 and 4 after HCT.

T Cell Depletion

Given that donor T cells are the major cell population initiating GVHD, elimination of donor T cells, either through removal from the graft ex vivo or by depleting T cells in vivo through administration of anti–T cell antibodies, has been used clinically. The major drawback of thorough TCD is impairment of GVL effects, as has been discussed earlier in the chapter and has been most clearly demonstrated in the setting of chronic myelogenous leukemia. However, TCD also results in increased rates of allograft rejection, delayed immune reconstitution, and hence increased risk of infectious complications. Provided that conditioning regimens can be devised to reliably allow engraftment without relying on donor T cells, TCD may be of greatest utility in the setting of nonmalignant indications (eg, sickle cell disease, severe aplastic anemia) and in patients with lower risk of recurrence of malignancy.

Graft Engineering and Ex Vivo TCD

Currently, the only FDA-approved approach for ex vivo TCD relies on positive selection of CD34 cells using the CliniMACS system,[77] which results in near-complete elimination of all non–CD34-expressing cells. Recent experimental approaches aim to allow for better immunocompetence by attempting selective depletion of cell populations responsible for GVHD while retaining T cells that contribute to immune competence. Selective elimination of naïve T cells (T_N) from grafts while retaining T_{CM} and T_{EM} prevents GVHD in murine models,[459] although T_{CM} retained more capacity for GVHD induction than T_{EM}.[460] The clinical translation of this approach in a trial testing this hypothesis using CD45RA-depletion did not mitigate development of aGVHD but showed a significant reduction in cGVHD.[461,462] Alternatively, selective depletion of TCR$\alpha\beta$+ T cells is currently under investigation,[463] especially in conjunction with B cell depletion in the setting of haploidentical transplantation.

In Vivo TCD

In vivo TCD with mAb-based (eg, anti-CD52 mAb Campath) or polyclonal rabbit ATG are quite effective in preventing GVHD and have become important components of conditioning therapies, especially in reduced-intensity conditioning regimens. A recent multicenter randomized GVHD prophylaxis study using the ATG preparation raised against Jurkat cells (ATG-Fresenius) showed significant improvement in cGVHD/relapse-free survival.[464]

Selective In Vivo TCD

Selective depletion of alloantigen-activated and expanding T cells in vivo with the intention to preserve nonalloreactive T cells to prevent global immunosuppression has been a goal for many decades. Although different approaches have been tested over the years, it was not until the introduction of PT-Cy that such a strategy became clinical reality and enabled control of HVG and GVH responses sufficiently to allow transplantation across HLA barriers[465,466] without prohibitive toxicity from infectious complications or GVHD. PT-Cy has led to widespread adoption of haploidentical donor HCT.[467] PT-Cy has been successfully utilized for GVHD prophylaxis not only for haploidentical but also in the HLA-matched and mismatched unrelated donor transplantation settings.[74,468] Two multicenter randomized trials are currently ongoing in the US comparing PT-Cy versus other GVHD prophylaxis regimens following myeloablative and nonmyeloablative conditioning.

Cell Therapies

Treg Cell Therapy

Administration of ex vivo expanded T_{regs} has also been studied for the prevention of GVHD. Clinical proof of concept studies utilized UCB-derived T_{regs} in the setting of double UCB transplantation[469] and peripheral blood-derived Treg cells were tested in the setting of haploidentical transplantation.[470] Although acute GVHD rates were encouraging in both trials, definitive findings from prospective trials and larger patient cohorts are currently not available. As an alternative approach, in vivo expansion of Treg cells by administration of low-dose IL-2 has yielded promising results in refractory cGVHD.[471,472]

Mesenchymal Stromal Cells

MSCs have been extensively studied in various inflammatory conditions for their immunomodulatory properties. There is substantial preclinical and also clinical evidence suggesting that MSCs are effective in the management of GVHD.[473] However, there is a still need for positive prospective randomized trials to definitely prove their utility.

Other Approaches

Costimulatory Blockade

Agents that suppress T-cell costimulation have been tested for prevention of GVHD. CTLA4-immunoglobulin (Abatacept) in particular was shown to be effective in preventing GVHD in HLA-matched and mismatched allogeneic transplants.[474,475]

Targeting Cytokines

A plethora of cytokines have been implicated in the afferent and efferent phases of GVHD. TNF-α was one of the first cytokines identified as a potential target. Clinical trials attempting to block TNF-α have reported mixed results for the treatment of GVHD. Other cytokines which have been targeted in clinical trials include IL-6 through administration of an IL-6–receptor antibody, which was shown to be effective in the treatment of steroid refractory GVHD[476,477] and also as prophylaxis.[478] Of note, the IL-6 receptor antibody Tocilizumab was approved for the treatment of patients developing cytokine release syndrome following infusion of CAR-transgenic T cells.

Small Molecule Inhibitors

Perhaps the most significant progress in terms of developing new treatment strategies against GVHD and resulting in FDA approval has been made through identifying druggable signaling pathways which control certain aspects of the pathobiology of acute and cGVHD. While many approaches have shown encouraging results in smaller studies and others are currently being tested in clinical trials, two drugs have been recently approved for the treatment of acute and chronic GVHD. Based on the crucial role of IFN-γ in the pathobiology of acute GVHD, preclinical studies demonstrated that inhibition of the IFN-γR/JAK1/STAT1 signaling axis mitigated GVHD.[479-481] A subsequent randomized multicenter phase II trial confirmed that treatment with the JAK1/2 inhibitor Ruxolitinib led to superior overall response rates, improved response duration, and better median failure-free survival[482] in patients with steroid-refractory acute GVHD. cGVHD remains one of the most challenging complications following allogeneic HCT, leading to significant morbidity. The emerging evidence that donor B cells are critically involved in some of the autoimmune-like manifestations of cGVHD[483] prompted the evaluation of the BTK inhibitor ibrutinib for patients with treatment-refractory cGVHD. In an open-label single-arm trial, ibrutinib showed activity in all involved organs, with an overall response rate of 67% of patients, providing the basis for achieving FDA approval.

Blocking T Cell Trafficking

The distinct organ tropism of acute GVHD has garnered significant interest to gain understanding of the mechanisms underlying the infiltration of donor immune cells into the GVHD target organs by studying the role of chemokines and their respective receptors in addition to identifying tissue-homing receptors. Targeting the chemotactic response subsequent to conditioning therapy by blocking the chemokine receptor CCR5 showed very promising results in a single-center phase I trial.[484] Alternative strategies to impair T cell migration into GVHD target organs focus on blocking tissue-homing receptors like $\alpha4\beta7$ integrins. mAbs targeting $\alpha4$ (natalizumab) or a combinatorial epitope on $\alpha4\beta7$ complex (vedolizumab) are currently in clinical trials for GVHD prevention or treatment.

Extracorporeal Photopheresis

Extracorporeal photopheresis (ECP) is an immunomodulatory treatment initially established for cutaneous T cell lymphoma which involves extracorporeal exposure of leukapheresis products to a photosensitizing compound 8-methoxypsoralen (8-MOP) and UVA irradiation. Mechanistic studies have hinted at multiple mechanisms of action involving Treg cells and modulating dendritic cell function.[485] ECP remains a dark horse in the field of GVHD management. Although its beneficial effects in acute and chronic GVHD have been well documented in a number of trials, there is a paucity of high-quality data from randomized prospective trials.[486] ECP is generally associated with low levels of adverse events and can frequently allow tapering of steroids. However, its consistent application is hampered by logistical hurdles; for example, the need for frequent apheresis procedures over a long period of time.

XENOTRANSPLANTATION

Over the past 3 decades, the increasing shortage of allogeneic donor organs has evoked a worldwide resurgence of interest in xenotransplantation, that is, the replacement of human organs or tissues with those from a donor of a different species. Routine clinical application of this therapeutic modality is still in the future. However, recent progress, which is reviewed briefly here, offers cause for optimism.

From a phylogenetic viewpoint, nonhuman primate organs would undoubtedly be the most similar to allotransplants immunologically. However, due to considerations of size, availability, ethics and likelihood of transmission of infectious disease, most investigators have decided against the use of primates as a future source of xenogeneic organs. Instead, the discordant species, swine, has been chosen by many as the most suitable xenograft donor. The pig has essentially unlimited availability, as well as favorable breeding characteristics, and many of its organ systems are similar to those of humans. Partially inbred miniature swine are a particularly attractive choice because of their size (adult weights of approximately 120 kg), their physiology (also similar to humans for many organ systems), and their breeding characteristics, which have permitted inbreeding and genetic manipulation.[199]

Mechanisms of Xenograft Rejection

Xenografts are subject to all four of the rejection mechanisms described previously in this chapter and give rise to more powerful immune responses than allografts, probably for each type of rejection. As discussed earlier, primates have high levels of natural antibodies against αGal[487] that react with endothelial cells of pig vessels after xenotransplantation, causing HAR. HAR is particularly vigorous in pig-to-primate transplantation in part because the CRPs expressed by pig endothelium may not be fully functional to control primate complement activation. These factors can also promote acute humoral rejection when HAR is prevented by genetic modification of the pig.[488] Additionally, cell-mediated immune responses to xenografts, including powerful indirect T cell responses, are more powerful than those directed to allografts.[199] Most of the molecular interactions involved in cellular immunity that have been examined appear to be at least partially functional in primate antipig responses. Therefore, human-anti-pig T-cell responses are likely to be as great or greater than those in allogeneic combinations.

Therapeutic Strategies for Xenotransplantation

There are three main strategies that have been pursued to achieve long-term survival of organ xenografts. The first has been to seek nonspecific immunosuppressive drugs that might prove especially effective for xenotransplantation. The second therapeutic approach has been to use genetic engineering of donor animals to lessen the immunologic barriers to xenografts. The third strategy to achieve successful xenotransplantation is the induction of tolerance of donor antigens. Recently, various combinations of these three strategies have resulted in major advances in pig-to-nonhuman primate xenograft survival, resulting in hope that clinical xenotransplantation could soon become a reality.

The heightened immune response to xenografts compared to allografts suggests that larger amounts of exogenous immunosuppression would be required to achieve xenograft survival comparable to that of allografts. Orthotopic, life-sustaining pig heart graft survival >6 months has recently been achieved using GalT KO, human CD46 and human thrombomodulin transgenic pig hearts in baboons treated with rituximab, ATG, anti-CD40/CD40L, MMF, and steroids.[489] Pig kidney graft survival >400 days has been reported in a small number of monkeys with low non-Gal natural antibody levels that received T cell–depleting, B cell–depleting, and costimulatory-blocking antibodies, steroids, and rapamycin or MMF.[199] However, the grafts were eventually rejected. While >6 months' survival of porcine islet xenografts has been achieved in nonhuman primates, again with heavy immunosuppression, the ethics of using such powerful drugs in patients who are treatable with insulin therapy are challenging. Clinical trials have been initiated using encapsulated porcine islets without immunosuppression and, as expected, the grafts did not survive. Nevertheless, important safety data has been obtained from these studies.[199,490]

Because the two features that distinguish xenografts from allografts are the larger number of antigens and the molecular incompatibilities between species, genetic modifications have been aimed primarily at correcting these two disadvantages of xenografts (Fig. 48.7). The first transgenic pigs produced by genetic engineering for xenotransplantation attempted to overcome the species selectivity of CRPs. Transgenic pigs were produced that expressed human genes for several of these CRPs. Organs from animals expressing these are more resistant to HAR than are those from wild-type pigs but are nevertheless susceptible to acute humoral rejection. The alternative genetic engineering strategy to the addition of transgenes is the removal or inactivation of genes through knockout technology. The advent of cloning through nuclear transfer technology, first demonstrated in the famous sheep "Dolly,"[491] made knockout technology possible in several large animal species, including pigs. As described earlier in the chapter, the major stumbling block in pig-to-primate xenotransplantation has been the large amount of natural antibody in primates directed toward the αGal determinant. Nuclear transfer

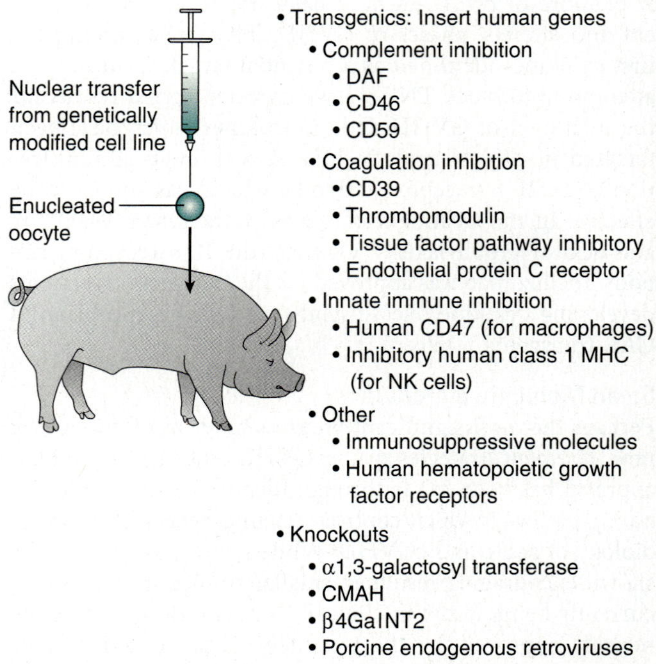

FIG 48.7. Genetic modifications of swine for xenotransplantation.

technology was used to eliminate this enzyme from pigs, resulting in "GalT-KO" pigs,[492,493] which, like humans and Old World primates, do not express αGal on to the surface of their cells. This development was a watershed in the xenotransplantation field that increased survivals of both heart and kidney transplants from pigs to baboons and permitted the advances discussed above to be made. Numerous additional genetic modifications have been permitted by the development of CRISPR/Cas9 technology, including the removal of genes that result in additional carbohydrate natural antibody targets that are present in pigs and missing in humans[494] and the addition of multiple human CRPs, coagulation inhibitors,[495] and CD47 molecules.[256]

The third strategy to achieve successful xenotransplantation is the induction of tolerance of donor antigens. The two major strategies for tolerance induction include mixed chimerism and thymic xenotransplantation.

Mixed Chimerism for Xenograft Tolerance Induction

Host treatment with mAbs to T cells and NK cells along with a low dose (3 Gy) of TBI has permitted rat bone marrow engraftment in mice, resulting in mixed xenogeneic chimerism and donor-specific tolerance.[233] Both $\gamma\delta$ T cells and NK cells play an important role in resisting xenogeneic marrow engraftment. Mixed chimerism is associated with tolerance of both T cells and natural xenoantibody-producing B cells[496,497] and thereby prevents HAR, acute vascular rejection, and cell-mediated rejection of vascularized cardiac xenografts.[497] Relevant to the pig to human combination, both preexisting and newly developing B cells recognizing anti-αGal are tolerized by the induction of mixed chimerism in GalT knockout mice receiving αGal-expressing allogeneic or xenogeneic marrow.[193,496] The relevant B cells are tolerized rapidly by an anergy mechanism that is followed later by clonal deletion and/or receptor editing of αGal-specific B cells.[311]

Studies in mixed allogeneic chimeras show that recipient NK cells are tolerized to the donor,[498] and studies in the rat-to-mouse combination also indicate a state of NK-cell tolerance.[499] While this tolerance is specific for the donor in mixed allogeneic chimeras, whose NK cells retain the ability to reject class I–deficient bone marrow, the NK-cell tolerance in mixed xenogeneic chimeras is associated with global NK-cell unresponsiveness. The dominant tolerance exerted by mixed allogeneic or xenogeneic chimerism favors the "disarming" concept over the "licensing" model,[58] which proposes that inhibitory interactions license NK cells to become functional, as the mechanism for NK-cell tolerance. Such "disarming" of NK cells could also explain dominant hyporesponsiveness in mixed chimeras involving class I–deficient donors or mosaic class I MHC transgenic mice. In general, inhibitory NK receptors are less functional between species than activating receptors. In mixed xenogeneic chimeras, the xenogeneic cells, however minor a population, lack ligands that inhibit recipient NK cells, which become chronically activated and hence anergic. In contrast, allogeneic cells in mixed allogeneic chimeras express MHC molecules that will inhibit some subsets of recipient NK cells expressing relevant inhibitory receptors. Thus, in mixed allogeneic chimeras, only the subset of recipient NK cells lacking an inhibitory ligand for donor cells will chronically receive unopposed stimulation and hence become anergic; the consequence is a functional NK-cell repertoire that is inhibited by recognition of the donor (in addition to recipient) cells. A near-complete lack of inhibitory interactions with a small population of chimeric xenogeneic cells could thereby render all NK cells anergic, perhaps by a mechanism similar to the cross-tolerance of multiple activation pathways described for NK cells receiving sustained stimulation through a DAP10/12-dependent activation molecule.[500]

Despite T-, B-, and NK-cell tolerance in mixed xenogeneic chimeras, the levels of donor chimerism decline gradually over time due to a competitive advantage of recipient marrow over xenogeneic rat marrow as the host recovers from low-dose TBI. Species specificity or selectivity of cytokines, adhesion molecules, and other interactions that regulate hematopoiesis probably account for this competitive advantage of host hematopoiesis.[496] Achievement of xenogeneic hematopoietic repopulation has proved to be an even more formidable challenge in highly disparate (discordant) species combinations. Using immunodeficient mice transgenically expressing porcine hematopoietic cytokines to promote chimerism, proof of principle was obtained that human T cells can be centrally tolerized in a human thymus to porcine xenoantigens via induction of mixed xenogeneic chimerism.[501]

The important role of macrophages in limiting porcine marrow engraftment in primates has been discussed earlier in the chapter. Consistently, porcine mixed chimerism induction has been enhanced by the availability of human CD47 Tg pigs, which greatly improves survival of porcine bone marrow in primates and reduces anti-pig Nab levels,[257] consistent with the induction of B cell tolerance by mixed chimerism as demonstrated in rodent models.[497]

Xenogeneic Thymic Transplantation: An Alternative Method of Achieving Central T Cell Tolerance

Based on the demonstrated capacity of mixed chimerism to induce central, deletional tolerance of xenoreactive T cells and on the difficulty achieving lasting mixed chimerism in NHPs, an alternative approach to T-cell tolerance was developed. In particular, replacement of the recipient thymus with a xenogeneic donor thymus after host T-cell depletion and thymectomy generates de novo an immunocompetent recipient T cell repertoire that is tolerant of the xenogeneic thymus donor.[502-504] Tolerance to both donor and host was shown to involve clonal deletion, reflecting the presence of APCs from both species within the thymic graft.[505] Positive selection in such grafts is mediated only by donor (porcine) thymic MHC, with no influence of the host MHC.[503] Nevertheless, host-restricted immune responses are demonstrable[504] apparently due to the broad TCR repertoire that develops.[506] This approach can also succeed in the pig-human combination, as donor pig-tolerant human T cell repertoires are generated in xenogeneic porcine thymus grafts in humanized mice.[507] This approach has also demonstrated promise in pig-to-primate xenograft models.[508] While the absence of recipient thymic epithelium in the xenogeneic thymus predisposes mice to the development of a multiorgan autoimmune syndrome, this autoimmunity can be avoided by adding recipient thymic epithelial cells to the thymic xenograft, resulting in improved deletion of recipient-reactive T cells and generation of T_{reg} cells that suppress autoimmune disease.[509,510]

Thymic transplantation has shown evidence of an ability to tolerize baboon T cells and prolong pig kidney xenograft survival,[199] and it seems likely that a combination of mixed chimerism and thymic transplantation could optimize the survival of genetically modified pig organs in humans.

Nonimmunologic Barriers to Xenotransplantation

In addition to the immunologic mechanisms that prevent successful xenografting, there are two other potentially important obstacles to clinical application. First, molecular incompatibilities between species may cause physiologic dysfunction of xenogeneic organs. This kind of incompatibility is least likely for the heart, for which the function, albeit vital, is relatively simple. It could certainly become an obstacle, however, for the liver, which produces so many different products, including serum proteins and enzymes, that at least some are likely not to function properly in a primate host. An example is the limitation to liver xenograft survival found to be due in part to failure of porcine coagulation factors to function normally in baboons.[511] On the other hand, if the number of incompatibilities leading to physiologic dysfunction is limited, they could be correctable through genetic modification.

The other nonimmunologic barrier to xenotransplantation is the risk of cross-species transfer of infectious agents, potentially creating a health hazard, not only for the recipient but also for society as a whole. This possibility has gained significant attention, both in the scientific literature and in the lay press, and the issue has become confused by enormous uncertainties about the true risks that are involved. "Zoonosis" is a term that has been used for some time to describe the general process of cross-species infection, and the term "xenozoonosis" has been developed to describe infection transmission that might occur as a result of xenotransplantation. It is important to realize that from the point of view of the individual recipient, the risk of transmitting infection by xenotransplantation is likely to be less than by current clinical allotransplantation, both because of the natural resistance to cross-species transmission of infectious diseases and because it should be possible to screen for and eliminate the presence of known pathogens from the herd of source pigs. The major infectious concern, therefore, is that endogenous retroviral sequences from donor cells might infect the recipient's cells, giving rise to previously unrecognized pathogenic viruses. However, to date there is no evidence that such cross-species transfer after a pig-to-human transplant would generate a virus that would be infectious or pathogenic. Indeed, studies of humans exposed to pig tissues have not revealed any cases of detectable pig endogenous retroviruses.[512] However, because of this unknown risk associated with xenotransplantation, public health agencies and members of the transplant community have designed approaches for oversight of trials and surveillance of xenotransplant recipients for untoward events. Additionally, CRISPR and other genetic modification strategies have been used to markedly reduce the number of potentially infectious PERV loci in the porcine genome.[199,495]

Clinical Progress in Xenotransplantation

Early clinical efforts in xenotransplantation took place in the 1960s and involved organ transplants from nonhuman primates. One of the patients survived for 9 months with normal renal function provided by the kidney of a chimpanzee.[513] Additional clinical trials thereafter, using baboon hearts and livers, were considerably less successful. More recent clinical trials have involved fetal pig cells transplanted into the brains of patients with Parkinson disease. Survival of pig tissue 8 months after the transplant was documented in a patient taking only moderate doses of immunosuppression.[514] These studies suggest that cellular xenotransplantation may be achieved more easily but the promise must be balanced against the risks of immunosuppressive therapy. For organ xenotransplantation, most investigators believe that further clinical trials are close to becoming a reality.

REFERENCES

1. Medawar PB. The behaviour and fate of skin autografts and skin homografts in rabbits: a report to the War Wounds Committee of the Medical Research Council. *J Anat.* 1944;78(Pt 5):176-199.
2. Little CC. The genetics of tissue transplantation in mammals. *J Canc Res.* 1924;8:75-95.
3. Snell GD. Methods for the study of histocompatibility genes. *J Genet.* 1948;49(2):87-108.
4. Baschal EE, Aly TA, Jasinski JM, et al. Defining multiple common "completely" conserved major histocompatibility complex SNP haplotypes. *Clin Immunol.* 2009;132(2):203-214.
5. Houser SL, Benjamin LC, Wain JC, et al. Constitutive expression of major histocompatiblity complex class II antigens in pulmonary epithelium and endothelium varies among different species. *Transplantation.* 2004;77(4):605-607.
6. Pescovitz MD, Sachs DH, Lunney JK, et al. Localization of class II MHC antigens on porcine renal vascular endothelium. *Transplantation.* 1984;37:627-630.
7. Volk H-DKern F. Insights into the specificity and function of (allo)antigen-reactive T cells. *Am J Transplant.* 2001;1(2):109-114.
8. Noorchashm H, Lieu YK, Rostami SY, et al. A direct method for the calculation of alloreactive CD4+ T cell prescursor frequency. *Transplantation.* 1999;67:1281-1284.
9. Fischer-Lindahl K, Wilson DB. Histocompatibility antigen-activated cytotoxic T lymphocytes. I. Estimates of the absolute frequency of killer cells generated in vitro. *J Exp Med.* 1977;145:500-507.
10. Teh HS, Harley E, Phillips RA, et al. Quantitative studies on the precursors of cytotoxic lymphocytes. I. Characterization of a clonal assay and determination of the size of clones derived from single precursors. *J Immunol.* 1977;118(3):1049-1056.
11. DeWolf S, Grinshpun B, Savage T, et al. Quantifying size and diversity of the human T cell alloresponse. *JCI Insight.* 2018;3(15):e121256.
12. Morris H, Susan DW, Robins H, et al. Tracking donor-reactive T cells: evidence for clonal deletion in tolerant kidney transplant patients. *Sci Transl Med.* 2015;7(272):272ra10.
13. Orosz CG, Bishop DK. Limiting dilution analysis of alloreactive T-cell status and distribution during allograft rejection. *Hum Immunol.* 1990;28:72-81.
14. Heeger PS, Greenspan NS, Kuhlenschmidt S, et al. Pretransplant frequency of donor-specific, IFN-g-producing lymphocytes is a manifestation of immunologic memory and correlates with the risk of posttransplant rejection episodes. *J Immunol.* 1999;163:2267-2275.
15. Blackman M, Yague J, Kubo R, et al. The T cell repertoire may be biased in favor of MHC recognition. *Cell.* 1986;47(3):349-357.
16. Huseby E, Kappler J, Marrack P. TCR-MHC/peptide interactions: kissing-cousins or a shotgun wedding? *Eur J Immunol.* 2004;34(5):1243-1250.
17. Scott-Browne JP, Crawford F, Young MH, et al. Evolutionarily conserved features contribute to alphabeta T cell receptor specificity. *Immunity.* 2011;35(4):526-535.
18. Rotzschke O, Falk K, Faath S, et al. On the nature of peptides involved in T cell alloreactivity. *J Exp Med.* 1991;174:1059-1071.
19. Smith PA, Brunmark A, Jackson MR, et al. Peptide-independent recognition by alloreactive cytotoxic T lymphocytes (CTL). *J Exp Med.* 1997;185(6):1023-1033.
20. Felix NJ, Donermeyer DL, Horvath S, et al. Alloreactive T cells respond specifically to multiple distinct peptide-MHC complexes. *Nat Immunol.* 2007;8(4):388-397.
21. Felix NJ, Allen PM. Specificity of T-cell alloreactivity. *Nat Rev Immunol.* 2007;7(12):942-953.
22. Heath WR, Kane KP, Mescher MF, et al. Alloreactive T cells discriminate among a diverse set of endogenous peptides. *Proc Natl Acad Sci U S A.* 1991;88(12):5101-5105.
23. Amir AL, van der Steen DM, Hagedoorn RS, et al. Allo-HLA-reactive T cells inducing graft-versus-host disease are single peptide specific. *Blood.* 2011;118(26):6733-6742.
24. Matzinger P, Bevan MJ. Hypothesis: Why do so many lymphocytes respond to major histocompatibility antigens? *Cell Immunol.* 1977;29:1-5.

25. Huseby ES, White J, Crawford F, et al. How the T cell repertoire becomes peptide and MHC specific. *Cell.* 2005;122(2):247-260.
26. Yin L, Huseby E, Scott-Browne J, et al. A single T cell receptor bound to major histocompatibility complex class I and class II glycoproteins reveals switchable TCR conformers. *Immunity.* 2011;35(1):23-33.
27. Demotz S, Sette A, Sakaguchi K, et al. Self peptide requirement for class II major histocompatibility complex allorecognition. *Proc Natl Acad Sci U S A.* 1991;88(19):8730-8734.
28. Weber DA, Terrell NK, Zhang Y, et al. Requirement for peptide in alloreactive CD4+ T cell recognition of class II MHC molecules. *J Immunol.* 1995;154:5153-5164.
29. Alexander-Miller MA, Burke K, Koszinowski UH, et al. Alloreactive cytotoxic T lymphocytes generated in the presence of viral-derived peptides show exquisite peptide and MHC specificity. *J Immunol.* 1993;151:1-10.
30. Moris A, Teichgraber V, Gauthier L, et al. Cutting edge: characterization of allorestricted and peptide-selective alloreactive T cells using HLA-tetramer selection. *J Immunol.* 2001;166(8):4818-4821.
31. Felix NJ, Brickey WJ, Griffiths R, et al. H2-DMalpha(−/−) mice show the importance of major histocompatibility complex-bound peptide in cardiac allograft rejection. *J Exp Med.* 2000;192(1):31-40.
32. Perreault C, Roy DC, Fortin C. Immunodominant minor histocompatibility antigens: the major ones. *Immunol Today.* 1998;19:69-74.
33. Mutis T, Verdijk R, Schrama E, et al. Feasibility of immunotherapy of relapsed leukemia with ex vivo-generated cytotoxic T lymphocytes specific for hematopoietic system-restricted minor histocompatibility antigens [see comments]. *Blood.* 1999;93(7):2336-2341.
34. den Haan JMM, Meadows LM, Wang W, et al. The minor histocompatibility antigen HA-1: a diallelic gene with a single amino acid polymorphism. *Science.* 1998;279:1054-1057.
35. Mendoza LM, Paz P, Zuberi A, et al. Minors held by major: the H13 minor histocompatibility locus defined as a peptide/MHC class I complex. *Immunity.* 1997;7:461-472.
36. Brickner AG, Warren EH, Caldwell JA, et al. The immunogenicity of a new human minor histocompatibility antigen results from differential antigen processing. *J Exp Med.* 2001;193(2):195-206.
37. Roopenian DC, David AP, Christianson GJ, et al. The functional basis of minor histocompatibility loci. *J Immunol.* 1993;151:4595-4605.
38. Ryu SJ, Jung KM, Yoo HS, et al. Cognate CD4 help is essential for the reactivation and expansion of CD8 memory T cells directed against the hematopoietic cell-specific dominant minor histocompatibility antigen, H60. *Blood.* 2009;113(18):4273-4280.
39. Yoshimura Y, Yadav R, Christianson GJ, et al. Duration of alloantigen presentation and avidity of T cell antigen recognition correlate with immunodominance of CTL response to minor histocompatibility antigens. *J Immunol.* 2004;172(11):6666-6674.
40. Choi EY, Choi K, Nam G, et al. H60: a unique murine hematopoietic cell-restricted minor histocompatibility antigen for graft-versus-leukemia effect. *Front Immunol.* 2020;11:1163.
41. Korngold R, Wettstein PJ. Immunodominance in the graft-vs-host disease T cell response to minor histocompatibility antigens. *J Immunol.* 1990;145:4079-4088.
42. Goulmy E. Human minor histocompatibility antigens. *Curr Opin Immunol.* 1996;8:75-81.
43. Simpson E, Roopenian D, Goulmy E. Much ado about minor histocompatibility antigens. *Immunol Today.* 1998;19:108-111.
44. Vogt MH, Goulmy E, Kloosterboer FM, et al. UTY gene codes for an HLA-B60-restricted human male-specific minor histocompatibility antigen involved in stem cell graft rejection: characterization of the critical polymorphic amino acid residues for T-cell recognition. *Blood.* 2000;96(9):3126-3132.
45. Voogt PJ, Fibbe WE, Marijt WAF, et al. Rejection of bone-marrow graft by recipient-derived cytotoxic T lymphocytes against minor histocompatibility antigens. *Lancet.* 1990;335:131-134.
46. Milosevic S, Bachnick B, Karim K, et al. Identification of MHC II-restricted minor histocompatibility antigens after HLA-identical stem-cell transplantation. *Transplantation.* 2010;90(9):1030-1035.
47. Goulmy E, Schipper R, Pool J, et al. Mismatches of minor histocompatiblity antigens between HLA-idnetical donors and recipients and the development of graft-versus-host disease after bone marrow transplantation. *New Engl J Med.* 1996;334(5):281-285.
48. Mutis T, Gillespie G, Schrama E, et al. Tetrameric HLA class I-minor histocompatibility antigen peptide complexes demonstrate minor histocompatibility antigen-specific cytotoxic T lymphocytes in patients with graft-versus-host disesae. *Nature Med.* 1999;5:839-842.
49. Vogt MH, de Paus RA, Voogt PJ, et al. DFFRY codes for a new human male-specific minor transplantation antigen involved in bone marrow graft rejection. *Blood.* 2000;95(3):1100-1105.
50. Nakasone H, Tian L, Sahaf B, et al. Allogeneic HY antibodies detected 3 months after female-to-male HCT predict chronic GVHD and nonrelapse mortality in humans. *Blood.* 2015;125(20):3193-3201.
51. Jones MS, Riley R, Hamilton BL, et al. Endogenous superantigens in allogeneic bone marrow transplant recipients rapidly and selectively expand donor T cells which can produce IFN-gamma. *Bone Marrow Transplant.* 1994;14:725-735.
52. Ildstad ST, Wren SM, Bluestone JA, et al. Characterization of mixed allogeneic chimeras. Immunocompetence, in vitro reactivity, and genetic specificity of tolerance. *J Exp Med.* 1985;162:231-244.
53. den Haan JMM, Sherman NE, Blokland E, et al. Identification of a graft versus host disease-associated human minor histocompatibility antigen. *Science.* 1995;268:1476-1480.
54. Szulman AE. The histological distribution of the blood group substances in man as disclosed by immunofluorescence: II. The H antigen and its relation to a and B antigens. *J Exp Med.* 1962;115(5):977-996.
55. Montgomery RA, Locke JE, King KE, et al. ABO incompatible renal transplantation: a paradigm ready for broad implementation. *Transplantation.* 2009;87(8):1246-1255.
56. Ludajic K, Balavarca Y, Bickeboller H, et al. Minor ABO-mismatches are risk factors for acute graft-versus-host disease in hematopoietic stem cell transplant patients. *Biol Blood Marrow Transplant.* 2009;15(11):1400-1406.
57. Blin N, Traineau R, Houssin S, et al. Impact of donor-recipient major ABO mismatch on allogeneic transplantation outcome according to stem cell source. *Biol Blood Marrow Transplant.* 2010;16(9):1315-1323.
58. Raulet DH, Vance RE. Self-tolerance of natural killer cells. *Nat Rev Immunol.* 2006;6(7):520-531.
59. Moretta L, Ciccone E, Moretta A, et al. Allorecognition by NK cells: nonself or no self? *Immunol Today.* 1992;13:300-306.
60. Potluri VS, Goldberg DS, Mohan S, et al. National trends in utilization and 1-year outcomes with transplantation of HCV-viremic kidneys. *J Am Soc Nephrol.* 2019;30(10):1939-1951.
61. Sise ME, Goldberg DS, Kort JJ, et al. Multicenter study to transplant hepatitis C-infected kidneys (MYTHIC): an open-label study of combined glecaprevir and pibrentasvir to treat recipients of transplanted kidneys from deceased donors with hepatitis C virus infection. *J Am Soc Nephrol.* 2020:2678-2687.
62. Nankivell BJ, Borrows RJ, Fung CL, et al. The natural history of chronic allograft nephropathy. *N Engl J Med.* 2003;349(24):2326-2333.
63. Vincenti F, Rostaing L, Grinyo J, et al. Belatacept and long-term outcomes in kidney transplantation. *N Engl J Med.* 2016;374(4):333-343.
64. Navarro V, Herrine S, Katopes C, et al. The effect of HLA class I (A and B) and class II (DR) compatibility on liver transplantation outcomes: an analysis of the OPTN database. *Liver Transpl.* 2006;12(4):652-658.
65. Shapiro AM, Lakey JR, Ryan EA, et al. Islet transplantation in seven patients with type 1 diabetes mellitus using a glucocorticoid-free immunosuppressive regimen. *N Engl J Med.* 2000;343(4):230-238.
66. Ryan EA, Paty BW, Senior PA, et al. Five-year follow-up after clinical islet transplantation. *Diabetes.* 2005;54(7):2060-2069.
67. Sykes M, Preffer F, McAffee S, et al. Mixed lymphohematopoietic chimerism and graft-vs-lymphoma effects are achievable in adult humans following non-myeloablative therapy and HLA-mismatched donor bone marrow transplantation. *Lancet.* 1999;353:1755-1759.
68. Giralt S, Thall PF, Khouri I, et al. Melphalan and purine analog-containing preparative regimens: reduced-intensity conditioning for patients with hematologic malignancies undergoing allogeneic progenitor cell transplantation. *Blood.* 2001;97(3):631-637.
69. Hsieh MM, Kang EM, Fitzhugh CD, et al. Allogeneic hematopoietic stem-cell transplantation for sickle cell disease. *N Engl J Med.* 2009;361(24):2309-2317.
70. Aversa F, Tabilio A, Velardi A, et al. Treatment of high-risk acute leukemia with T-cell-depleted stem cells from related donors with one fully mismatched HLA haplotype. *N Engl J Med.* 1998;339(17):1186-1193.
71. Ciceri F, Labopin M, Aversa F, et al. A survey of fully haploidentical hematopoietic stem cell transplantation in adults with high-risk acute leukemia: a risk factor analysis of outcomes for patients in remission at transplantation. *Blood.* 2008;112(9):3574-3581.
72. Luznik L, Jalla S, Engstrom LW, et al. Durable engraftment of major histocompatibility complex-incompatible cells after nonmyeloablative conditioning with fludarabine, low-dose total body irradiation, and posttransplantation cyclophosphamide. *Blood.* 2001;98(12):3456-3464.
73. Luznik L, O'Donnell PV, Symons HJ, et al. HLA-haploidentical bone marrow transplantation for hematologic malignancies using nonmyeloablative conditioning and high-dose, posttransplantation cyclophosphamide. *Biol Blood Marrow Transplant.* 2008;14(6):641-650.
74. Kasamon YL, Ambinder RF, Fuchs EJ, et al. Prospective study of nonmyeloablative, HLA-mismatched unrelated BMT with high-dose posttransplantation cyclophosphamide. *Blood Adv.* 2017;1(4):288-292.
75. Battipaglia G, Labopin M, Kroger N, et al. Posttransplant cyclophosphamide vs antithymocyte globulin in HLA-mismatched unrelated donor transplantation. *Blood.* 2019;134(11):892-899.
76. Fuchs EJ, O'Donnell PV, Eapen M, et al. Double unrelated umbilical cord blood versus HLA-haploidentical bone marrow transplantation (BMT CTN 1101). *Blood.* 2020;137(3):420-428.
77. Pasquini MC, Devine S, Mendizabal A, et al. Comparative outcomes of donor graft CD34+ selection and immune suppressive therapy as graft-versus-host disease prophylaxis for patients with acute myeloid leukemia in complete remission undergoing HLA-matched sibling allogeneic hematopoietic cell transplantation. *J Clin Oncol.* 2012;30(26):3194-3201.
78. Horowitz MM, Gale RP, Sondel PM, et al. Graft-versus-leukemia reactions after bone marrow transplantation. *Blood.* 1990;75(3):555-562.
79. Martin PJ, Hansen JA, Torok-Storb B, et al. Graft failure in patients receiving T cell-depleted HLA-identical allogeneic marrow transplants. *Bone Marrow Transplant.* 1988;3(5):445-456.
80. Ruggeri L, Capanni M, Urbani E, et al. Effectiveness of donor natural killer cell alloreactivity in mismatched hematopoietic transplants. *Science.* 2002;295(5562):2097-2100.
81. Redett RJ III, Etra JW, Brandacher G, et al. Total penis, scrotum, and lower abdominal wall transplantation. *N Engl J Med.* 2019;381(19):1876-1878.
82. Kollar B, Pomahac B, Riella LV. Novel immunological and clinical insights in vascularized composite allotransplantation. *Curr Opin Organ Transplant.* 2019;24(1):42-48.
83. Henderson ML. The landscape of vascularized composite allograft donation in the United States. *Curr Opin Organ Transplant.* 2019;24(6):699-704.
84. Rodrigue JR, Tomich D, Fleishman A, et al. Vascularized composite allograft donation and transplantation: a survey of public attitudes in the United States. *Am J Transplant.* 2017;17(10):2687-2695.
85. Etra JW, Raimondi G, Brandacher G. Mechanisms of rejection in vascular composite allotransplantation. *Curr Opin Organ Transplant.* 2018;23(1):28-33.
86. Cetrulo CL Jr, Torabi R, Scalea JR, et al. Vascularized composite allograft transplant survival in miniature swine: is MHC tolerance sufficient for acceptance of epidermis? *Transplantation.* 2013;96(11):966-974.
87. Cetrulo CL Jr, Drijkoningen T, Sachs DH. Tolerance induction via mixed chimerism in vascularized composite allotransplantation: is it time for clinical application? *Curr Opin Organ Transplant.* 2015;20(6):602-607.
88. Zuber J, Rosen S, Shonts B, et al. Macrochimerism in intestinal transplantation: association with lower rejection rates and multivisceral transplants, without GVHD. *Am J Transplant.* 2015;15(10):2691-2703.
89. Zuber J, Shonts B, Lau SP, et al. Bidirectional intragraft alloreactivity drives the repopulation of human intestinal allografts and correlates with clinical outcome. *Sci Immunol.* 2016;1(eaa h3732):1-10.
90. Fu J, Zuber J, Martinez M, et al. Human intestinal allografts contain functional hematopoietic stem and progenitor cells that are maintained by a circulating pool. *Cell Stem Cell.* 2019;24(2):227-239 e8.
91. Fu J, Zuber J, Shonts B, et al. Lymphohematopoietic graft-versus-host responses promote mixed chimerism in patients receiving intestinal transplantation. *J Clin Invest.* 2021;131(8):e141698.
92. Grant D, Abu-Elmagd K, Mazariegos G, et al. Intestinal transplant registry report: global activity and trends. *Am J Transplant.* 2015;15(1):210-219.
93. Rock KL, Lai JJ, Kono H. Innate and adaptive immune responses to cell death. *Immunol Rev.* 2011;243(1):191-205.
94. Wang X, O'Brien ME, Yu J, et al. Prolonged cold ischemia induces necroptotic cell death in ischemia-reperfusion injury and contributes to primary graft dysfunction after lung transplantation. *Am J Respir Cell Mol Biol.* 2019;61(2):244-256.
95. Mikhalski D, Wissing KM, Ghisdal L, et al. Cold ischemia is a major determinant of acute rejection and renal graft survival in the modern era of immunosuppression. *Transplantation.* 2008;85(7 Suppl):S3-S9.
96. Schneeberger S, Amberger A, Mandl J, et al. Cold ischemia contributes to the development of chronic rejection and mitochondrial injury after cardiac transplantation. *Transpl Int.* 2010;23(12):1282-1292.
97. Solini S, Aiello S, Cassis P, et al. Prolonged cold ischemia accelerates cellular and humoral chronic rejection in a rat model of kidney allotransplantation. *Transpl Int.* 2012;25(3):347-356.
98. Postalcioglu M, Kaze AD, Byun BC, et al. Association of cold ischemia time with acute renal transplant rejection. *Transplantation.* 2018;102(7):1188-1194.
99. Land WG, Agostinis P, Gasser S, et al. Transplantation and damage-associated molecular patterns (DAMPs). *Am J Transplant.* 2016:3338-3361.
100. Weigand K, Brost S, Steinebrunner N, et al. Ischemia/reperfusion injury in liver surgery and transplantation: pathophysiology. *HPB Surg.* 2012;2012:176723.
101. Zhai Y, Petrowsky H, Hong JC, et al. Ischaemia-reperfusion injury in liver transplantation–from bench to bedside. *Nat Rev Gastroenterol Hepatol.* 2013;10(2):79-89.
102. Goldstein DR, Tesar BM, Akira S, et al. Critical role of the Toll-like receptor signal adaptor protein MyD88 in acute allograft rejection. *J Clin Invest.* 2003;111(10):1571-1578.
103. McKay D, Shigeoka A, Rubinstein M, et al. Simultaneous deletion of MyD88 and Trif delays major histocompatibility and minor antigen mismatch allograft rejection. *Eur J Immunol.* 2006;36(8):1994-2002.

104. Arumugam TV, Okun E, Tang SC, et al. Toll-like receptors in ischemia-reperfusion injury. *Shock*. 2009;32(1):4-16.
105. Arslan F, Keogh B, McGuirk P, et al. TLR2 and TLR4 in ischemia reperfusion injury. *Mediators Inflamm*. 2010;2010:704202.
106. Shen H, Heuzey E, Mori DN, et al. Haptoglobin enhances cardiac transplant rejection. *Circ Res*. 2015;116(10):1670-1679.
107. de Vries DK, Lindeman JH, Tsikas D, et al. Early renal ischemia-reperfusion injury in humans is dominated by IL-6 release from the allograft. *Am J Transplant*. 2009;9(7):1574-1584.
108. Chan RK, Ibrahim SI, Takahashi K, et al. The differing roles of the classical and mannose-binding lectin complement pathways in the events following skeletal muscle ischemia-reperfusion. *J Immunol*. 2006;177(11):8080-8085.
109. Zhang M, Takahashi K, Alicot EM, et al. Activation of the lectin pathway by natural IgM in a model of ischemia/reperfusion injury. *J Immunol*. 2006;177(7):4727-4734.
110. Marsh JE, Farmer CK, Jurcevic S, et al. The allogeneic T and B cell response is strongly dependent on complement components C3 and C4. *Transplantation*. 2001;72(7):1310-1318.
111. Carroll MC. The complement system in regulation of adaptive immunity. *Nat Immunol*. 2004;5(10):981-986.
112. Sacks SH, Chowdhury P, Zhou W. Role of the complement system in rejection. *Curr Opin Immunol*. 2003;15(5):487-492.
113. Morita K, Miura M, Paolone DR, et al. Early chemokine cascades in murine cardiac grafts regulate T cell recruitment and progression of acute allograft rejection. *J Immunol*. 2001;167(5):2979-2984.
114. Hancock WW, Gao W, Csizmadia V, et al. Donor-derived IP-10 initiates development of acute allograft rejection. *J Exp Med*. 2001;193(8):975-980.
115. Banchereau J, Briere F, Caux C, et al. Immunobiology of dendritic cells. *Ann Rev Immunol*. 2000;18:767-811.
116. Stein JV, Nombela-Arrieta C. Chemokine control of lymphocyte trafficking: a general overview. *Immunology*. 2005;116(1):1-12.
117. El Sawy T, Belperio JA, Strieter RM, et al. Inhibition of polymorphonuclear leukocyte-mediated graft damage synergizes with short-term costimulatory blockade to prevent cardiac allograft rejection. *Circulation*. 2005;112(3):320-331.
118. Iida S, Miyairi S, Su CA, et al. Peritransplant VLA-4 blockade inhibits endogenous memory CD8 T cell infiltration into high-risk cardiac allografts and CTLA-4Ig resistant rejection. *Am J Transplant*. 2019;19(4):998-1010.
119. Kapoor A, Morita K, Engeman TM, et al. Early expression of interferon-gamma inducible protein 10 and monokine induced by interferon-gamma in cardiac allografts is mediated by CD8+ T cells. *Transplantation*. 2000;69(6):1147-1155.
120. Sykes M, Sheard MA, Sachs DH. Graft-versus-host-related immunosuppression is induced in mixed chimeras by alloresponses against either host or donor lymphohematopoietic cells. *J Exp Med*. 1988;168(6):2391-2396.
121. Pelot MR, Pearson DA, Swenson K, et al. Lymphohematopoietic graft-vs-host reactions can be induced without graft-vs-host disease in murine mixed chimeras established with a cyclophosphamide-based non-myeloablative conditioning regimen. *Biol Blood Marrow Transplant*. 1999;5:133-143.
122. Chakraverty R, Cote D, Buchli J, et al. An inflammatory checkpoint regulates recruitment of graft-versus-host-reactive T cells to peripheral tissues. *J Exp Med*. 2006;203(8):2021-2031.
123. Mapara MY, Leng C, Kim YM, et al. Expression of chemokines in GVHD target organs is influenced by conditioning and genetic factors and amplified by GVHR. *Biol Blood Marrow Transplant*. 2006;12(6):623-634.
124. Xun CQ, Thompson JS, Jennings CD, et al. Effect of total body irradiation, busulfan-cyclophosphamide, or cyclophosphamide conditioning on inflammatory cytokine release and development of acute and chronic graft-versus-host disease in H-2-incompatible transplanted SCID mice. *Blood*. 1994;83:2360-2367.
125. Bingaman AW, Ha J, Waitze S-Y, et al. Vigorous allograft rejection in the absence of danger. *J Immunol*. 2000;164:3065-3071.
126. Anderson CC, Carroll JM, Gallucci S, et al. Testing time-, ignorance-, and danger-based models of tolerance. *J Immunol*. 2001;166(6):3663-3671.
127. Larsen CP, Morris PJ, Austyn JM. Migration of dendritic leukocytes from cardiac allografts into host spleens. *J Exp Med*. 1989;171:307-314.
128. Lafferty KJ, Prowse SJ, Simeonovic CJ, et al. Immunobiology of tissue transplantation: a return to the passenger leukocyte concept. *Ann Rev Immunol*. 1983;1:143-173.
129. Siu JHY, Surendrakumar V, Richards JA, et al. T cell allorecognition pathways in solid organ transplantation. *Front Immunol*. 2018;9:2548.
130. Herrera OB, Golshayan D, Tibbott R, et al. A novel pathway of alloantigen presentation by dendritic cells. *J Immunol*. 2004;173(8):4828-4837.
131. Liu Q, Rojas-Canales DM, Divito SJ, et al. Donor dendritic cell-derived exosomes promote allograft-targeting immune response. *J Clin Invest*. 2016;126(8):2805-2820.
132. Marino J, Babiker-Mohamed MH, Crosby-Bertorini P, et al. Donor exosomes rather than passenger leukocytes initiate alloreactive T cell responses after transplantation. *Sci Immunol*. 2016;1(1):aaf8759.
133. Marino J, Paster J, Benichou G. Allorecognition by T lymphocytes and allograft rejection. *Front Immunol*. 2016;7:582.
134. Smyth LA, Herrera OB, Golshayan D, et al. A novel pathway of antigen presentation by dendritic and endothelial cells: Implications for allorecognition and infectious diseases. *Transplantation*. 2006;82(1 suppl):S15-S18.
135. Hughes AD, Zhao D, Dai H, et al. Cross-dressed dendritic cells sustain effector T cell responses in islet and kidney allografts. *J Clin Invest*. 2020;130(1):287-294.
136. Hornick PI, Mason PD, Yacoub MH, et al. Assessment of the contribution that direct allorecognition makes to the progression of chronic cardiac transplant rejection in humans. *Circulation*. 1998;97(13):1257-1263.
137. Baker RJ, Hernandez-Fuentes MP, Brookes PA, et al. Loss of direct and maintenance of indirect alloresponses in renal allograft recipients: implications for the pathogenesis of chronic allograft nephropathy. *J Immunol*. 2001;167(12):7199-7206.
138. Afzali B, Lombardi G, Lechler RI. Pathways of major histocompatibility complex allorecognition. *Curr Opin Organ Transplant*. 2008;13(4):438-444.
139. Gokmen MR, Lombardi G, Lechler RI. The importance of the indirect pathway of allorecognition in clinical transplantation. *Curr Opin Immunol*. 2008;20(5):568-574.
140. Vallera DA, Taylor PA, Sprent J, et al. The role of host T cell subsets in bone marrow rejection directed to isolated major histocompatibility complex class I versus class II differences of bm1 and bm12 mutant mice. *Transplantation*. 1994;57:249-256.
141. Hayashi H, LeGuern C, Sachs DH, et al. Alloresistance to K locus mismatched bone marrow engraftment is mediated entirely by CD4+ and CD8+ T cells. *Bone Marrow Transplant*. 1996;18:285-292.
142. Haspot F, Li HW, Lucas CL, et al. Allospecific rejection of MHC class I-deficient bone marrow by CD8 T cells. *Am J Transplant*. 2014;14(1):49-58.
143. He C, Heeger PS. CD8 T cells can reject major histocompatibility complex class I-deficient skin allografts. *Am J Transplant*. 2004;4(5):698-704.
144. Steele DJ, Laufer TM, Smiley ST, et al. Two levels of help for B cell alloantibody production. *J Exp Med*. 1996;183(2):699-703.
145. Russell PS, Chase CM, Colvin RB. Alloantibody- and T cell-mediated immunity in the pathogenesis of transplant arteriosclerosis: lack of progression to sclerotic lesions in B cell-deficient mice. *Transplantation*. 1997;64(11):1531-1536.
146. Loupy A, Lefaucheur C. Antibody-mediated rejection of solid-organ allografts. *N Engl J Med*. 2018;379(12):1150-1160.
147. Valenzuela NM, Reed EF. Antibody-mediated rejection across solid organ transplants: manifestations, mechanisms, and therapies. *J Clin Invest*. 2017;127(7):2492-2504.
148. Barber DL, Wherry EJ, Masopust D, et al. Restoring function in exhausted CD8 T cells during chronic viral infection. *Nature*. 2005:682-687.
149. Janssen EM, Lemmens EE, Wolfe T, et al. CD4(+) T cells are required for secondary expansion and memory in CD8(+) T lymphocytes. *Nature*. 2003;421(6925):852-856.
150. Williams MA, Tyznik AJ, Bevan MJ. Interleukin-2 signals during priming are required for secondary expansion of CD8+ memory T cells. *Nature*. 2006;441(7095):890-893.
151. Lee RS, Grusby MJ, Glimcher LH, et al. Indirect recognition by helper cells can induce donor-specific cytotoxic T lymphocytes in vivo. *J Exp Med*. 1994;179(3):865-872.
152. Kurts C, Robinson BW, Knolle PA. Cross-priming in health and disease. *Nat Rev Immunol*. 2010;10(6):403-414.
153. Bachem A, Guttler S, Hartung E, et al. Superior antigen cross-presentation and XCR1 expression define human CD11c+CD141+ cells as homologues of mouse CD8+ dendritic cells. *J Exp Med*. 2010;207(6):1273-1281.
154. Bevan MJ. Cross-priming for a secondary cytotoxic response to minor H antigens with H-2 congenic cells which do not cross-react in the cytotoxic assay. *J Exp Med*. 1976;143(5):1283-1288.
155. Valujskikh A, Zhang Q, Heeger PS. CD8 T cells specific for a donor-derived, self-restricted transplant antigen are nonpathogenic bystanders after vascularized heart transplantation in mice. *J Immunol*. 2006;176(4):2190-2196.
156. Valujskikh A, Lantz O, Celli S, et al. Cross-primed CD8(+) T cells mediate graft rejection via a distinct effector pathway. *Nat Immunol*. 2002;3(9):844-851.
157. Koenig A, Thaunat O. Lymphoid neogenesis and tertiary lymphoid organs in transplanted organs. *Front Immunol*. 2016;7:646.
158. Huibers MM, Gareau AJ, Beerthuijzen JM, et al. Donor-specific antibodies are produced locally in ectopic lymphoid structures in cardiac allografts. *Am J Transplant*. 2017;17(1):246-254.
159. Bachelet T, Couzi L, Lepreux S, et al. Kidney intragraft donor-specific antibodies as determinant of antibody-mediated lesions and poor graft outcome. *Am J Transplant*. 2013;13(11):2855-2864.
160. Douillard P, Cuturi M-C, Brouard S, et al. T cell receptor repertoire usage in allotransplantation: an overview. *Transplantation*. 1999;68:913-921.
161. Gagne K, Brouard S, Giral M, et al. Highly altered Vb repertoire of T cells infiltrating long-term rejected kidney allografts. *J Immunol*. 2000;164:1553-1563.
162. Scheinberg P, Price DA, Ambrozak DR, et al. Alloreactive T cell clonotype recruitment in a mixed lymphocyte reaction: implications for graft engineering. *Exp Hematol*. 2006;34(6):788-795.
163. Dickinson AM, Wang XN, Sviland L, et al. In situ dissection of the graft-versus-host activities of cytotoxic T cells specific for minor histocompatibility antigens. *Nat Med*. 2002;8(4):410-414.
164. Guillet M, Gagne K, Lair D, et al. Different patterns of TCR beta chain regulation following allo- and xeno-transplantation. *Xenotransplantation*. 2004;11(4):315-322.
165. Friedman TM, Statton D, Jones SC, et al. Vbeta spectratype analysis reveals heterogeneity of CD4+ T-cell responses to minor histocompatibility antigens involved in graft-versus-host disease: correlations with epithelial tissue infiltrate. *Biol Blood Marrow Transplant*. 2001;7(1):2-13.
166. Lin CM, Gill RG. Direct and indirect allograft recognition: pathways dictating graft rejection mechanisms. *Curr Opin Organ Transplant*. 2016;21(1):40-44.
167. Graubert TA, Russell JH, Ley TJ. The role of granzyme B in murine models of acute graft-versus-host disease and graft rejection. *Blood*. 1996;87:1232-1237.
168. Williams NS, Engelhard VH. Identification of a population of CD4+ CTL that utilizes a perforin—rather than a Fas ligand-dependent cytotoxic mechanism. *J Immunol*. 1996;156:153-159.
169. Hahn S, Gehri R, Erb P. Mechanism and biological significance of CD4-mediated cytotoxicity. *Immunol Rev*. 1995;146:57-79.
170. Blazar BR, Taylor PA, Vallera DA. CD4+ and CD8+ T cells each can utilize a perforin-dependent pathway to mediate lethal graft-versus-host disease in major histocompatibility complex-disparate recipients. *Transplantation*. 1997;64:571-576.
171. Graubert TA, DiPersio JF, Russell JH, et al. Perforin/granzyme-dependent and independent mechanisms are both important for the development of graft-versus-host disease after murine bone marrow transplantation. *J Clin Invest*. 1997;100:904-911.
172. Jiang Z, Jones M, Levy RB. FasL and perforin cotyotoxically impaired donor T-cells induce GVHD in TNF receptor 1 (p55) knockout MHC I/II mismatched recipients. *Biol Blood Marrow Transplant*. 2001;7:76.
173. Tsukada N, Kobata T, Aizawa Y, et al. Graft-versus-leukemia effect and graft-versus-host disease can be differentiated by cytotoxic mechanisms in a murine model of allogeneic bone marrow transplantation. *Blood*. 1999;93:2738-2747.
174. Schmaltz C, Alpdogan O, Horndasch KJ, et al. Differential use of Fas ligand and perforin cytotoxic pathways by donor T cells in graft-versus-host disease and graft-versus-leukemia effect. *Blood*. 2001;97(9):2886-2895.
175. Schmaltz C, Alpdogan O, Kappel BJ, et al. T cells require TRAIL for optimal graft-versus-tumor activity. *Nat Med*. 2002;8(12):1433-1437.
176. Doody DP, Stenger KS, Winn HJ. Immunologically nonspecific mechanisms of tissue destruction in the rejection of skin grafts. *J Exp Med*. 1994;179:1645-1652.
177. Braun MY, Grandjean I, Feunou P, et al. Acute rejection in the absence of cognate recognition of allograft by T cells. *J Immunol*. 2001;166(8):4879-4883.
178. Kapessidou Y, Habran C, Buonocore S, et al. The replacement of graft endothelium by recipient-type cells conditions allograft rejection mediated by indirect pathway CD4(+) T cells. *Transplantation*. 2006;81(5):726-735.
179. Teshima T, Ordemann R, Reddy P, et al. Acute graft-versus-host disease does not require alloantigen expression on host epithelium. *Nat Med*. 2002;8(6):575-581.
180. Duffner UA, Maeda Y, Cooke KR, et al. Host dendritic cells alone are sufficient to initiate acute graft-versus-host disease. *J Immunol*. 2004;172(12):7393-7398.
181. Jones SC, Murphy GF, Friedman TM, et al. Importance of minor histocompatibility antigen expression by nonhematopoietic tissues in a CD4+ T cell-mediated graft-versus-host disease model. *J Clin Invest*. 2003;112(12):1880-1886.
182. Gorczynski RM. Role of cytokines in allograft rejection. *Curr Pharm Des*. 2001;7(11):1039-1057.
183. Benjamin LC III, Allan JS, Madsen JC. Cytokines in immunity and allograft rejection. *Crit Rev Immunol*. 2002;22(4):269-279.
184. Hill GR, Koyama M. Cytokines and costimulation in acute graft-versus-host disease. *Blood*. 2020;136(4):418-428.
185. Yi T, Chen Y, Wang L, et al. Reciprocal differentiation and tissue-specific pathogenesis of Th1, Th2, and Th17 cells in graft-versus-host disease. *Blood*. 2009;114(14):3101-3112.
186. Nikolic B, Lee S, Bronson RT, et al. Th1 and Th2 cells both contribute to acute GVHD, and each subset has distinct end-organ targets. *J Clin Invest*. 2000;105:1289-1293.
187. Broady R, Yu J, Chow V, et al. Cutaneous GVHD is associated with the expansion of tissue-localized Th1 and not Th17 cells. *Blood*. 2010;116(25):5748-5751.
188. Wasowska BA. Mechanisms involved in antibody- and complement-mediated allograft rejection. *Immunol Res*. 2010;47(1-3):25-44.
189. Shimizu I, Smith NR, Zhao G, et al. Decay-accelerating factor prevents acute humoral rejection induced by low levels of anti-alphaGal natural antibodies. *Transplantation*. 2006;81(1):95-100.
190. Platt JL, Fischel RJ, Matas AJ, et al. Immunopathology of hyperacute xenograft rejection in a swine-to-primate model. *Transplantation*. 1991;52:214-220.

191. Woodle ES, Tremblay S, Rossi A, et al. Plasma cell targeting to prevent antibody-mediated rejection. *Am J Transplant.* 2020;20(4):33-41.
192. Jordan SC, Ammerman N, Choi J, et al. The role of novel therapeutic approaches for prevention of allosensitization and antibody-mediated rejection. *Am J Transplant.* 2020;20(4):42-56.
193. Ohdan H, Swenson KG, Kitamura H, et al. Tolerization of Gala1,3Gal-reactive B cells in presensitized a1,3-galactosyltransferase-deficient mice by nonmyeloablative induction of mixed chimerism. *Xenotransplantation.* 2002;8:227-238.
194. Ozmen L, Ekdahl KN, Elgue G, et al. Inhibition of thrombin abrogates the instant blood-mediated inflammatory reaction triggered by isolated human islets: possible application of the thrombin inhibitor melagatran in clinical islet transplantation. *Diabetes.* 2002;51(6):1779-1784.
195. Hawthorne WJ, Salvaris EJ, Phillips P, et al. Control of IBMIR in neonatal porcine islet xenotransplantation in baboons. *Am J Transplant.* 2014;14(6):1300-1309.
196. Yang Y-G, Chen AM, Sergio JJ, et al. Role of antibody-independent complement activation in rejection of porcine bone marrow cells in mice. *Transplantation.* 1999;69:163-190.
197. Chong AS. Mechanisms of organ transplant injury mediated by B cells and antibodies: implications for antibody-mediated rejection. *Am J Transplant.* 2020;20(4):23-32.
198. Haas M, Sis B, Racusen LC, et al. Banff 2013 meeting report: inclusion of c4d-negative antibody-mediated rejection and antibody-associated arterial lesions. *Am J Transplant.* 2014;14(2):272-283.
199. Sykes MS, Sachs DH. Transplanting organs from pigs to humans. *Sci Immunol.* 2019;4:eaau6298.
200. Millan MT, Geczy C, Stuhlmeier KM, et al. Human monocytes activate porcine endothelial cells, resulting in increased E-selectin, interleukin-8, monocyte chemotactic protein-1, and plasminogen activator inhibitor-type-1 expression. *Transplantation.* 1997;63(3):421-429.
201. Jane-Wit D, Manes TD, Yi T, et al. Alloantibody and complement promote T cell-mediated cardiac allograft vasculopathy through noncanonical nuclear factor-kappaB signaling in endothelial cells. *Circulation.* 2013;128(23):2504-2516.
202. Gill RG, Lin CM. Linking innate immunity and chronic antibody-mediated allograft rejection. *Curr Opin Organ Transplant.* 2019;24(6):694-698.
203. Siegel JB, Grey ST, Lesnikoski B-A, et al. Xenogeneic endothelial cells activate human prothrombin. *Transplantation.* 1997;64:888-896.
204. Schulte am EJ, Rogiers X, Robson SC. Molecular incompatibilities in hemostasis between swine and men–impact on xenografting. *Ann Transplant.* 2001;6(3):12-16.
205. Soares M, Lin Y, Sato K, et al. Accomodation. *Immunol Today.* 1999;20:434-437.
206. Sablik KA, Clahsen-van Groningen MC, Damman J, et al. Banff lesions and renal allograft survival in chronic-active antibody mediated rejection. *Transpl Immunol.* 2019;56:101213.
207. Chiu HF, Wen MC, Wu MJ, et al. Treatment of chronic active antibody-mediated rejection in renal transplant recipients - a single center retrospective study. *BMC Nephrol.* 2020;21(1):6.
208. Shin BH, Everly MJ, Zhang H, et al. Impact of tocilizumab (Anti-IL-6R) treatment on immunoglobulins and anti-HLA antibodies in kidney transplant patients with chronic antibody-mediated rejection. *Transplantation.* 2020;104(4):856-863.
209. Wilkes DS. Autoantibody formation in human and rat studies of chronic rejection and primary graft dysfunction. *Semin Immunol.* 2012;24(2):131-135.
210. Zhang C, Todorov I, Zhang Z, et al. Donor CD4+ T and B cells in transplants induce chronic graft-versus-host disease with autoimmune manifestations. *Blood.* 2006;107(7):2993-3001.
211. Srinivasan M, Flynn R, Price A, et al. Donor B-cell alloantibody deposition and germinal center formation are required for the development of murine chronic GVHD and bronchiolitis obliterans. *Blood.* 2012;119(6):1570-1580.
212. Sarantopoulos S, Ritz J. Aberrant B-cell homeostasis in chronic GVHD. *Blood.* 2015;125(11):1703-1707.
213. Miklos DB, Kim HT, Miller KH, et al. Antibody responses to H-Y minor histocompatibility antigens correlate with chronic graft-versus-host disease and disease remission. *Blood.* 2005;105(7):2973-2978.
214. Zeiser R, Sarantopoulos S, Blazar BR. B-cell targeting in chronic graft-versus-host disease. *Blood.* 2018;131(13):1399-1405.
215. Lee LA, Sergio JJ, Sykes M. Natural killer cells weakly resist engraftment of allogeneic long-term multilineage-repopulating hematopoietic stem cells. *Transplantation.* 1996;61:125-132.
216. Aguila HL, Weissman IL. Hematopoietic stem cells are not direct cytotoxic targets of natural killer cells. *Blood.* 1996;87:1225-1231.
217. O'Reilly RJ, Brochstein J, Collins N, et al. Evaluation of HLA-haplotye disparate parental marrow grafts depleted of T lymphocytes by differential agglutination with a soybean lectin and E-rosette depletion for the treatment of severe combined immunodeficiency. *Vox Sang.* 1986;51(suppl 2):81-86.
218. Vacca P, Montaldo E, Croxatto D, et al. NK cells and other innate lymphoid cells in hematopoietic stem cell transplantation. *Front Immunol.* 2016;7:188.
219. Cichocki F, Verneris MR, Cooley S, et al. The past, present, and future of NK cells in hematopoietic cell transplantation and adoptive transfer. *Curr Top Microbiol Immunol.* 2016;395:225-243.
220. de Witte MA, Kuball J, Miller JS. NK cells and gammadeltaT cells for relapse protection after allogeneic hematopoietic cell transplantation (HCT). *Curr Stem Cell Rep.* 2017;3(4):301-311.
221. Yagisawa T, Tanaka T, Miyairi S, et al. In the absence of natural killer cell activation donor-specific antibody mediates chronic, but not acute, kidney allograft rejection. *Kidney Int.* 2019;95(2):350-362.
222. Koenig A, Chen CC, Marcais A, et al. Missing self triggers NK cell-mediated chronic vascular rejection of solid organ transplants. *Nat Commun.* 2019;10(1):5350.
223. Bix M, Liao N-S, Zijlstra M, et al. Rejection of class I MHC-deficient haemopoietic cells by irradiated MHC-matched mice. *Nature.* 1991;349:329-331.
224. Zijlstra M, Auchincloss HJ, Loring JM, et al. Skin graft rejection by beta2-microglobulin-deficient mice. *J Exp Med.* 1992;175:885-893.
225. Kitchens WH, Uehara S, Chase CM, et al. The changing role of natural killer cells in solid organ rejection and tolerance. *Transplantation.* 2006;81(6):811-817.
226. Fahrner R, Dondorf F, Ardelt M, et al. Role of NK, NKT cells and macrophages in liver transplantation. *World J Gastroenterol.* 2016;22(27):6135-6144.
227. Graham JA, Wilkinson RA, Hirohashi T, et al. Viral infection induces de novo lesions of coronary allograft vasculopathy through a natural killer cell-dependent pathway. *Am J Transplant.* 2009;9(11):2479-2484.
228. Hirohashi T, Chase CM, Della Pelle P, et al. A novel pathway of chronic allograft rejection mediated by NK cells and alloantibody. *Am J Transplant.* 2012;12(2):313-321.
229. Hoglund P, Sundback J, Olsson-Alheim MY, et al. Host MHC class I gene control of NK cell specificity in the mouse. *Immunol Rev.* 1997;155:11-28.
230. del Rio ML, Seebach JD, Fernandez-Renedo C, et al. ITIM-dependent negative signaling pathways for the control of cell-mediated xenogeneic immune responses. *Xenotransplantation.* 2013;20(6):397-406.
231. Nakamura MC, Linnemeyer PA, Niemi EC, et al. Mouse Ly-49D recognizes H-2Dd and activates natural killer cell cytotoxicity. *J Exp Med.* 1999;189:493-500.
232. Forte P, Lilienfeld BG, Baumann BC, et al. Human NK cytotoxicity against porcine cells is triggered by NKp44 and NKG2D. *J Immunol.* 2005;175(8):5463-5470.
233. Sharabi Y, Aksentijevich I, Sundt TM III, et al. Specific tolerance induction across a xenogeneic barrier: production of mixed rat/mouse lymphohematopoietic chimeras using a nonlethal preparative regimen. *J Exp Med.* 1990;172:195-202.
234. Nikolic B, Cooke DT, Zhao G, et al. Both g/d T cells and NK cells inhibit the engraftment of xenogeneic rat bone marrow cells in mice. *J Immunol.* 2001;166:1398-1404.
235. Sharabi Y, Sachs DH. Mixed chimerism and permanent specific transplantation tolerance induced by a non-lethal preparative regimen. *J Exp Med.* 1989;169:493-502.
236. Puga Yung G, Schneider MKJ, Seebach JD. The role of NK cells in pig-to-human xenotransplantation. *J Immunol Res.* 2017;2017:4627384.
237. Berzins SP, Smyth MJ, Baxter AG. Presumed guilty: natural killer T cell defects and human disease. *Nat Rev Immunol.* 2011;11(2):131-142.
238. Toyofuku A, Yasunami Y, Nabeyama K, et al. Natural killer T-cells participate in rejection of islet allografts in the liver of mice. *Diabetes.* 2006;55(1):34-39.
239. Nierlich PN, Klaus C, Bigenzahn S, et al. The role of natural killer T cells in costimulation blockade-based mixed chimerism. *Transpl Int.* 2010;23(11):1179-1189.
240. Mattarollo SR, Yong M, Tan L, et al. Secretion of IFN-gamma but not IL-17 by CD1d-restricted NKT cells enhances rejection of skin grafts expressing epithelial cell-derived antigen. *J Immunol.* 2010;184(10):5663-5669.
241. Bogdan C. Nitric oxide and the immune response. *Nat Immunol.* 2001;2(10):907-916.
242. Oberbarnscheidt MH, Zeng Q, Li Q, et al. Non-self recognition by monocytes initiates allograft rejection. *J Clin Invest.* 2014;124(8):3579-3589.
243. Dai H, Lan P, Zhao D, et al. PIRs mediate innate myeloid cell memory to nonself MHC molecules. *Science.* 2020;368(6495):1122-1127.
244. Dai H, Friday AJ, Abou-Daya KI, et al. Donor SIRPalpha polymorphism modulates the innate immune response to allogeneic grafts. *Sci Immunol.* 2017;2(12):eaam6202.
245. Andres A, Toso C, Morel P, et al. Macrophage depletion prolongs discordant but not concordant islet xenograft survival. *Transplantation.* 2005;79(5):543-549.
246. Yi S, Hawthorne WJ, Lehnert AM, et al. T cell-activated macrophages are capable of both recognition and rejection of pancreatic islet xenografts. *J Immunol.* 2003;170(5):2750-2758.
247. Abe M, Cheng J, Qi J, et al. Elimination of porcine hematopoietic cells by macrophages in mice. *J Immunol.* 2002;168:621-628.
248. Basker M, Alwayn IP, Buhler L, et al. Clearance of mobilized porcine peripheral blood progenitor cells is delayed by depletion of the phagocytic reticuloendothelial system in baboons. *Transplantation.* 2001;72(7):1278-1285.
249. Ide K, Ohdan H, Kobayashi T, et al. Antibody- and complement-independent phagocytotic and cytolytic activities of human macrophages toward porcine cells. *Xenotransplantation.* 2005;12(3):181-188.
250. Candinas D, Belliveau S, Koyomada N, et al. T cell independence of macrophage and natural killer cell infiltration, cytokine production, and endothelial activation during delayed xenograft rejection. *Transplantation.* 1996;62:1920-1927.
251. Xia G, Ji P, Rutgeerts O, et al. Natural killer cell- and macrophage mediated discordant Guinea pig-->rat xenograft rejection in the absence of complement, xenoantibody and T cell immunity. *Transplantation.* 2000;70(1):86-93.
252. Rees MA, Butler AJ, Brons IG, et al. Evidence of macrophage receptors capable of direct recognition of xenogeneic epitopes without opsonization. *Xenotransplantation.* 2005;12(1):13-19.
253. Wang H, Verhalen J, Madariaga ML, et al. Attenuation of phagocytosis of xenogeneic cells by manipulating CD47. *Blood.* 2007;109(2):836-842.
254. Ide K, Wang H, Tahara H, et al. Role for CD47-SIRPalpha signaling in xenograft rejection by macrophages. *Proc Natl Acad Sci U S A.* 2007;104(12):5062-5066.
255. Nomura S, Ariyoshi Y, Watanabe H, et al. Transgenic expression of human CD47 reduces phagocytosis of porcine endothelial cells and podocytes by baboon and human macrophages. *Xenotransplantation.* 2020;27(1):e12549.
256. Tena AA, Sachs DH, Mallard C, et al. Prolonged survival of pig skin on baboons after administration of pig cells expressing human CD47. *Transplantation.* 2017;101(2):316-321.
257. Watanabe H, Ariyoshi Y, Pomposelli T, et al. Intra-bone bone marrow transplantation from hCD47 transgenic pigs to baboons prolongs chimerism to >60 days and promotes increased porcine lung transplant survival. *Xenotransplantation.* 2020;27(1):e12552.
258. Goldman M, Moine AL, Braun M, et al. A role for eosinophils in transplant rejection. *Trends Immunol.* 2001;22(5):247-251.
259. Trull AK, Akhlaghi F, Charman SC, et al. Immunosuppression, eotaxin and the diagnostic changes in eosinophils that precede early acute heart allograft rejection. *Transpl Immunol.* 2004;12(2):159-166.
260. Meleg-Smith S, Gauthier PM. Abundance of interstitial eosinophils in renal allografts is associated with vascular rejection. *Transplantation.* 2005;79(4):444-450.
261. Tanaka K, Yamagami S, Streilein JW. Evidence that T-helper type 2 cell-derived cytokines and eosinophils contribute to acute rejection of orthotopic corneal xenografts in mice. *Transplantation.* 2005;79(10):1317-1323.
262. Bromberg JS, Hittle L, Xiong Y, et al. Gut microbiota-dependent modulation of innate immunity and lymph node remodeling affects cardiac allograft outcomes. *JCI Insight.* 2018;3(19):e121045.
263. Sepulveda M, Pirozzolo IA, legre ML. Impact of the microbiota on solid organ transplant rejection. *Curr Opin Organ Transplant.* 2019;24(6):679-686.
264. Andermann TM, Peled JU, Ho C, et al. The microbiome and hematopoietic cell transplantation: past, present, and future. *Biol Blood Marrow Transplant.* 2018;24(7):1322-1340.
265. Shono Y, van den Brink MRM. Gut microbiota injury in allogeneic haematopoietic stem cell transplantation. *Nat Rev Cancer.* 2018;18(5):283-295.
266. Ouyang J, Isnard S, Lin J, et al. Treating from the inside out: relevance of fecal microbiota transplantation to counteract gut damage in GVHD and HIV infection. *Front Med.* 2020;7:421.
267. Nankivell BJ, Shingde M, Keung KL, et al. The causes, significance and consequences of inflammatory fibrosis in kidney transplantation: the Banff i-IFTA lesion. *Am J Transplant.* 2018;18(2):364-376.
268. Nankivell BJ, Chapman JR. Chronic allograft nephropathy: current concepts and future directions. *Transplantation.* 2006;81(5):643-654.
269. Uehara S, Chase CM, Cornell LD, et al. Chronic cardiac transplant arteriopathy in mice: relationship of alloantibody, C4d deposition and neointimal fibrosis. *Am J Transplant.* 2007;7(1):57-65.
270. Shi C, Lee WS, He Q, et al. Immunologic basis of transplant-associated arteriosclerosis. *Proc Natl Acad Sci U S A.* 1996;93(9):4051-4056.
271. Russell PS, Chase CM, Sykes M, et al. Tolerance, mixed chimerism and chronic transplant arteriopathy. *J Immunol.* 2001;167:5731-5740.
272. Balam S, Schiechl-Brachner G, Buchtler S, et al. IL-3 triggers chronic rejection of cardiac allografts by activation of infiltrating basophils. *J Immunol.* 2019;202(12):3514-3523.
273. Jagasia MH, Greinix HT, Arora M, et al. National Institutes of Health Consensus Development Project on criteria for clinical trials in chronic graft-versus-host disease: I. The 2014 diagnosis and staging working group report. *Biol Blood Marrow Transplant.* 2015;21(3):389-401 e1.
274. Zeiser RB, Blazar BR. Pathophysiology of chronic graft-versus-host disease and therapeutic targets. *N Engl J Med.* 2017;377(26):2565-2579.
275. Borel JF, Feurer C, Gubler HU, et al. Biological effects of cyclosporin A: a new antilymphocytic agent. *Agents Actions.* 1976;6(4):468-475.

276. Calne RY, White DJ, Thiru S, et al. Cyclosporin A in patients receiving renal allografts from cadaver donors. *Lancet.* 1978;2(8104-5):1323-1327.
277. Ochiai T, Nakajima K, Nagata M, et al. Effect of a new immunosuppressive agent, FK 506, on heterotopic cardiac allotransplantation in the rat. *Transplant Proc.* 1987;19(1 pt 2):1284-1286.
278. Kino T, Hatanaka H, Miyata S, et al. FK-506, a novel immunosuppressant isolated from a Streptomyces. II. Immunosuppressive effect of FK-506 in vitro. *J Antibiot (Tokyo).* 1987;40(9):1256-1265.
279. Liu J, Farmer JD Jr, Lane WS, et al. Calcineurin is a common target of cyclophilin-cyclosporin A and FKBP-FK506 complexes. *Cell.* 1991;66(4):807-815.
280. Muller MR, Rao A. NFAT, immunity and cancer: a transcription factor comes of age. *Nat Rev Immunol.* 2010;10(9):645-656.
281. Sehgal SN, Baker H, Vezina C. Rapamycin (AY-22,989), a new antifungal antibiotic. II. Fermentation, isolation and characterization. *J Antibiot (Tokyo).* 1975;28(10):727-732.
282. Heitman J, Movva NR, Hall MN. Targets for cell cycle arrest by the immunosuppressant rapamycin in yeast. *Science.* 1991;253(5022):905-909.
283. Schwartz R, Stack J, Dameshek W. Effect of 6-mercaptopurine on antibody production. *Proc Soc Exp Biol Med.* 1958;99(1):164-167.
284. Elion GB, Callahan SW, Hitchings GH, et al. The metabolism of 2-amino-6-[(1-methyl-4-nitro-5-imidazolyl)thio]purine (B.W. 57-323) in man. *Cancer Chemother Rep.* 1960;8:47-52.
285. Sollinger HW. Mycophenolate mofetil for the prevention of acute rejection in primary cadaveric renal allograft recipients. U.S. Renal Transplant Mycophenolate Mofetil Study Group. *Transplantation.* 1995;60(3):225-232.
286. Allison AC, Kowalski WJ, Muller CD, et al. Mechanisms of action of mycophenolic acid. *Ann N Y Acad Sci.* 1993;696:63-87.
287. Cain DW, Cidlowski JA. Immune regulation by glucocorticoids. *Nat Rev Immunol.* 2017;17(4):233-247.
288. Augustine JJ, Hricik DE. Steroid sparing in kidney transplantation: changing paradigms, improving outcomes, and remaining questions. *Clin J Am Soc Nephrol.* 2006;1(5):1080-1089.
289. Reff ME, Carner K, Chambers KS, et al. Depletion of B cells in vivo by a chimeric mouse human monoclonal antibody to CD20. *Blood.* 1994;83(2):435-445.
290. Tedla FM, Roche-Recinos A, Brar A. Intravenous immunoglobulin in kidney transplantation. *Curr Opin Organ Transplant.* 2015;20(6):630-637.
291. Esensten JH, Helou YA, Chopra G, et al. CD28 costimulation: from mechanism to therapy. *Immunity.* 2016;44(5):973-988.
292. Turka LA, Linsley PS, Lin H, et al. T-cell activation by the CD28 ligand B7 is required for cardiac allograft rejection in vivo. *Proc Natl Acad Sci U S A.* 1992;89(22):11102-11105.
293. Larsen CP, Pearson TC, Adams AB, et al. Rational development of LEA29Y (belatacept), a high-affinity variant of CTLA4-Ig with potent immunosuppressive properties. *Am J Transplant.* 2005;5(3):443-453.
294. Schroder PM, Fitch ZW, Schmitz R, et al. The past, present, and future of costimulation blockade in organ transplantation. *Curr Opin Organ Transplant.* 2019;24(4):391-401.
295. Poirier N, Blancho G, Hiance M, et al. First-in-human study in healthy subjects with FR104, a pegylated monoclonal antibody fragment antagonist of CD28. *J Immunol.* 2016;197(12):4593-4602.
296. Kim SC, Wakwe W, Higginbotham LB, et al. Fc-silent anti-CD154 domain antibody effectively prevents nonhuman primate renal allograft rejection. *Am J Transplant.* 2017;17(5):1182-1192.
297. Cordoba F, Wieczorek G, Audet M, et al. A novel, blocking, Fc-silent anti-CD40 monoclonal antibody prolongs nonhuman primate renal allograft survival in the absence of B cell depletion. *Am J Transplant.* 2015;15(11):2825-2836.
298. Harland RC, Klintmalm G, Jensik S, et al. Efficacy and safety of bleselumab in kidney transplant recipients: a phase 2, randomized, open-label, noninferiority study. *Am J Transplant.* 2020;20(1):159-171.
299. Jordan SC, Lorant T, Choi J. IgG endopeptidase in highly sensitized patients undergoing transplantation. *N Engl J Med.* 2017;377(17):1693-1694.
300. Feng S, Ekong UD, Lobritto SJ, et al. Complete immunosuppression withdrawal and subsequent allograft function among pediatric recipients of parental living donor liver transplants. *J Am Med Assoc.* 2012;307(3):283-293.
301. Benitez C, Londono MC, Miquel R, et al. Prospective multicenter clinical trial of immunosuppressive drug withdrawal in stable adult liver transplant recipients. *Hepatology.* 2013;58(5):1824-1835.
302. Bohne F, Londono MC, Benitez C, et al. HCV-induced immune responses influence the development of operational tolerance after liver transplantation in humans. *Sci Transl Med.* 2014;6(242):242ra81.
303. Nemazee D. Receptor editing in lymphocyte development and central tolerance. *Nat Rev Immunol.* 2006;6(10):728-740.
304. Wardemann H, Yurasov S, Schaefer A, et al. Predominant autoantibody production by early human B cell precursors. *Science.* 2003;301(5638):1374-1377.
305. Meffre EO'Connor KC. Impaired B-cell tolerance checkpoints promote the development of autoimmune diseases and pathogenic autoantibodies. *Immunol Rev.* 2019;292(1):90-101.
306. Ota T, Ota M, Duong BH, et al. Liver-expressed Igkappa superantigen induces tolerance of polyclonal B cells by clonal deletion not kappa to lambda receptor editing. *J Exp Med.* 2011;208(3):617-629.
307. Cambier JC, Gauld SB, Merrell KT, et al. B-cell anergy: from transgenic models to naturally occurring anergic B cells? *Nat Rev Immunol.* 2007;7(8):633-643.
308. Li Y, Ma L, Yin D, et al. Long-term control of alloreactive B cell responses by the suppression of T cell help. *J Immunol.* 2008;180(9):6077-6084.
309. Khiew SH, Jain D, Chen J, et al. Transplantation tolerance modifies donor-specific B cell fate to suppress de novo alloreactive B cells. *J Clin Invest.* 2020;130(7):3453-3466.
310. Ohdan H, Sykes M. B cell tolerance to xenoantigens. *Xenotransplantation.* 2003;10(2):98-106.
311. Kawahara T, Shimizu I, Ohdan H, et al. Differing mechanisms of early and late B cell tolerance induced by mixed chimerism. *Am J Transplant.* 2005;5(12):2821-2829.
312. Flutter B, Edwards N, Fallah-Arani F, et al. Nonhematopoietic antigen blocks memory programming of alloreactive CD8+ T cells and drives their eventual exhaustion in mouse models of bone marrow transplantation. *J Clin Invest.* 2010:3855-3868.
313. Dey B, Yang Y-G, Preffer F, et al. The fate of donor TCR transgenic T cells with known host antigen specificity in a GVHD model. *Transplantation.* 1999;68:141-149.
314. Burnett DL, Reed JH, Christ D, et al. Clonal redemption and clonal anergy as mechanisms to balance B cell tolerance and immunity. *Immunol Rev.* 2019;292(1):61-75.
315. Tan C, Noviski M, Huizar J, et al. Self-reactivity on a spectrum: a sliding scale of peripheral B cell tolerance. *Immunol Rev.* 2019;292(1):37-60.
316. Chai J-G, Bartok I, Scott D, et al. T:T antigen presentaiton by activated murine CD8+ T cells induces anergy and apoptosis. *J Immunol.* 1998;160:3655-3665.
317. Rocha B, Tanchot C, Von Boehmer H. Clonal anergy blocks in vivo growth of mature T cells and can be reversed in the absence of antigen. *J Exp Med.* 1993;177:1517-1521.
318. Young JS, Daniels MD, Miller ML, et al. Erosion of transplantation tolerance after infection. *Am J Transplant.* 2017;17(1):81-90.
319. Rocken M, Urban JF, Shevach EM. Infection breaks T cell tolerance. *Nature.* 1992;359:79-82.
320. Fehr T, Wang S, Haspot F, et al. Rapid deletional peripheral CD8 T cell tolerance induced by allogeneic bone marrow: role of donor class II MHC and B cells. *J Immunol.* 2008;181(6):4371-4380.
321. Haspot F, Fehr T, Gibbons C, et al. Peripheral deletional tolerance of alloreactive CD8 but not CD4 cells is dependent on the PD-1/PD-L1 pathway. *Blood.* 2008;112(5):2149-2155.
322. Kurtz J, Shaffer J, Anosova N, et al. Mechanisms of early peripheral CD4 T cell tolerance induction by anti-CD154 monoclonal antibody and allogeneic bone marrow transplantation: evidence for anergy and deletion, but not regulatory cells. *Blood.* 2004;103:4336-4343.
323. Chai JG, Bartok I, Chandler P, et al. Anergic T cells act as suppressor cells in vitro and in vivo. *Eur J Immunol.* 1999;29(2):686-692.
324. Kilshaw PJ, Brent L, Pinto M. Suppressor T cells in mice made unresponsive to skin allografts. *Nature.* 1975;255(5508):489-491.
325. Hall BM, Pearce NW, Gurley K, et al. Specific unresponsiveness in rats with prolonged cardiac allograft survival after treatment with cyclosporine. *J Exp Med.* 1990;171:141-157.
326. Sakaguchi S, Sakaguchi N, Asano M, et al. Immunologic self-tolerance maintained by activated T cells expressing IL-2 receptor a-chains (CD25). Breakdown of a single mechanism of self-tolerance causes various autoimmune diseases. *J Immunol.* 1995;155:1151-1164.
327. Wood KJ, Bushell A, Jones ND. Immunologic unresponsiveness to alloantigen in vivo: a role for regulatory T cells. *Immunol Rev.* 2011;241(1):119-132.
328. Hoffmann P, Ermann J, Edinger M, et al. Donor-type CD4(+) CD25(+) regulatory T cells suppress lethal acute graft- versus-host disease after allogeneic bone marrow transplantation. *J Exp Med.* 2002;196(3):389-399.
329. Louis S, Braudeau C, Giral M, et al. Contrasting CD25hiCD4+T cells/FOXP3 patterns in chronic rejection and operational drug-free tolerance. *Transplantation.* 2006;81(3):398-407.
330. Zorn E, Kim HT, Lee SJ, et al. Reduced frequency of FOXP3+ CD4+CD25+ regulatory T cells in patients with chronic graft-versus-host disease. *Blood.* 2005;106(8):2903-2911.
331. Kawai T, Cosimi AB, Spitzer TR, et al. HLA-mismatched renal transplantation without maintenance immunosuppression. *N Engl J Med.* 2008;358(4):353-361.
332. Andreola G, Chittenden M, Shaffer J, et al. Mechanisms of donor-specific tolerance in recipients of haploidentical combined bone marrow/kidney transplantation. *Am J Transplant.* 2011;11(6):1236-1247.
333. Sprangers B, DeWolf S, Savage TM, et al. Origin of enriched regulatory T cells in patients receiving combined kidney-bone marrow transplantation to induce transplantation tolerance. *Am J Transplant.* 2017;17(8):2020-2032.
334. Savage TM, Shonts BA, Obradovic A, et al. Early expansion of donor-specific Tregs in tolerant kidney transplant recipients. *JCI Insight.* 2018;3(22):e124086.
335. Yang J, Brook MO, Carvalho-Gaspar M, et al. Allograft rejection mediated by memory T cells is resistant to regulation. *Proc Natl Acad Sci U S A.* 2007;104(50):19954-19959.
336. Abbas AK, Trotta E, Dimitre RS, et al. Revisiting IL-2: Biology and therapeutic prospects. *Sci Immunol.* 2018;3(25):eaat1482.
337. Bailey-Bucktrout SL, Martinez-Llordella M, Zhou X, et al. Self-antigen-driven activation induces instability of regulatory T cells during an inflammatory autoimmune response. *Immunity.* 2013;39(5):949-962.
338. Dominguez-Villar M, Hafler DA. Regulatory T cells in autoimmune disease. *Nat Immunol.* 2018:665-673.
339. Lee YK, Mukasa R, Hatton RD, et al. Developmental plasticity of Th17 and Treg cells. *Curr Opin Immunol.* 2009;21(3):274-280.
340. Wood KJ, Feng G, Wei B, et al. Interferon gamma: friend or foe? *Transplantation.* 2007;84(1 Suppl):S4-S5.
341. Wei B, Baker S, Wieckiewicz J, et al. IFN-gamma triggered STAT1-PKB/AKT signalling pathway influences the function of alloantigen reactive regulatory T cells. *Am J Transplant.* 2010;10(1):69-80.
342. Cobbold S, Waldmann H. Infectious tolerance. *Curr Opin Immunol.* 1998;10(5):518-524.
343. Jonuleit H, Schmitt E, Kakirman H, et al. Infectious tolerance: human CD25(+) regulatory T cells convey suppressor activity to conventional CD4(+) T helper cells. *J Exp Med.* 2002;196(2):255-260.
344. Kingsley CI, Karim M, Bushell AR, et al. CD25+CD4+ regulatory T cells prevent graft rejection: CTLA-4- and IL-10-dependent immunoregulation of alloresponses. *J Immunol.* 2002;168(3):1080-1086.
345. Dieckmann D, Bruett CH, Ploettner H, et al. Human CD4(+) CD25(+) regulatory, contact-dependent T cells induce interleukin 1-producing, contact-independent type 10-like [correction of type 1-like] regulatory T cells. *J Exp Med.* 2002;196(2):247-253.
346. Fallarino F, Grohmann U, Hwang KW, et al. Modulation of tryptophan catabolism by regulatory T cells. *Nat Immunol.* 2003;4(12):1206-1212.
347. Grohmann U, Orabona C, Fallarino F, et al. CTLA-4-Ig regulates tryptophan catabolism in vivo. *Nat Immunol.* 2002;3(11):1097-1101.
348. Beutelspacher SC, Tan PH, McClure MO, et al. Expression of indoleamine 2,3-dioxygenase (Ido) by endothelial cells: implications for the control of alloresponses. *Am J Transplant.* 2006;6(6):1320-1330.
349. Chen G, Kheradmand T, Bryant J, et al. Intragraft CD11b(+) Ido(+) cells mediate cardiac allograft tolerance by ECDI-fixed donor splenocyte infusions. *Am J Transplant.* 2012;12(11):2920-2929.
350. Sucher R, Fischler K, Oberhuber R, et al. Ido and regulatory T cell support are critical for cytotoxic T lymphocyte-associated Ag-4 Ig-mediated long-term solid organ allograft survival. *J Immunol.* 2012;188(1):37-46.
351. Sullivan JA, Tomita Y, Jankowska-Gan E, et al. Treg-cell-derived IL-35-coated extracellular vesicles promote infectious tolerance. *Cell Rep.* 2020;30(4):1039-1051.e5.
352. Thornton AM, Shevach EM. Suppressor effector function of CD4+CD25+ immunoregulatory T cells is antigen nonspecific. *J Immunol.* 2000;164:183-190.
353. Karim M, Feng G, Wood KJ, et al. CD25+CD4+ regulatory T cells generated by exposure to a model protein antigen prevent allograft rejection: antigen-specific reactivation in vivo is critical for bystander regulation. *Blood.* 2005;105(12):4871-4877.
354. Saxena V, Li L, Paluskievicz C, et al. Role of lymph node stroma and microenvironment in T cell tolerance. *Immunol Rev.* 2019;292(1):9-23.
355. Warren KJ, Iwami D, Harris DG, et al. Laminins affect T cell trafficking and allograft fate. *J Clin Invest.* 2014;124(5):2204-2218.
356. Graca L, Cobbold SP, Waldmann H. Identification of regulatory T cells in tolerated allografts. *J Exp Med.* 2002;195(12):1641-1646.
357. Kawai T, Cosimi AB, Colvin RB, et al. Mixed allogeneic chimerism and renal allograft tolerance in cynomologous monkeys. *Transplantation.* 1995;59:256-262.
358. Fuchs A, Gliwinski M, Grageda N, et al. Minimum information about T regulatory cells: a step toward reproducibility and standardization. *Front Immunol.* 2017;8:1844.
359. Nadig SN, Wieckiewicz J, Wu DC, et al. In vivo prevention of transplant arteriosclerosis by ex vivo-expanded human regulatory T cells. *Nat Med.* 2010;16(7):809-813.
360. Sagoo P, Ali N, Garg G, et al. Human regulatory T cells with alloantigen specificity are more potent inhibitors of alloimmune skin graft damage than polyclonal regulatory T cells. *Sci Transl Med.* 2011;3(83):83ra42.
361. Bluestone JA, Buckner JH, Fitch M, et al. Type 1 diabetes immunotherapy using polyclonal regulatory T cells. *Sci Transl Med.* 2015;7(315):315ra189.
362. Chandran S, Tang Q, Sarwal M, et al. Polyclonal regulatory T cell therapy for control of inflammation in kidney transplants. *Am J Transplant.* 2017;17(11):2945-2954.

363. Raffin C, Vo LT, Bluestone JA. Treg cell-based therapies: challenges and perspectives. *Nat Rev Immunol.* 2020;20(3):158-172.
364. Sawitzki B, Harden PN, Reinke P, et al. Regulatory cell therapy in kidney transplantation (The ONE Study): a harmonised design and analysis of seven non-randomised, single-arm, phase 1/2A trials. *Lancet.* 2020;395(10237):1627-1639.
365. Todo S, Yamashita K, Goto R, et al. A pilot study of operational tolerance with a regulatory T cell-based cell therapy in living donor liver transplantation. *Hepatology.* 2016;64(2):632-643.
366. Bezie S, Anegon I, Guillonneau C. Advances on CD8+ Treg cells and their potential in transplantation. *Transplantation.* 2018;102(9):1467-1478.
367. Sykes M, Hoyles KA, Romick ML, et al. In vitro and in vivo analysis of bone marrow derived CD3+, CD4-, CD8-, NK1.1+ cell lines. *Cell Immunol.* 1990;129:478-493.
368. Zeng D, Lewis D, Dejbakhsh-Jones S, et al. Bone marrow NK1.1- and NK1.1+ T cells reciprocally regulate acute graft versus host disease. *J Exp Med.* 1999;189:1073-1081.
369. Zhang ZX, Yang L, Young KJ, et al. Identification of a previously unknown antigen-specific regulatory T cell and its mechanism of suppression [In Process Citation]. *Nat Med.* 2000;6(7):782-789.
370. Hack N, Angra S, McKnight T, et al. HLA class II-like antiidiotypic antibodies from highly sensitized patients inhibit T-cell alloresponses. *Am J Transplant.* 2008;8(1):111-120.
371. Bluestone JA, Leo O, Epstein SL, et al. Idiotypic manipulation of the immune response to transplantation antigens. *Immunol Rev.* 1986;90:5-27.
372. Batchelor JR. The riddle of kidney graft enhancement. *Transplantation.* 1978;26:139-141.
373. Arnold B, Schonrich G, Hammerling GJ. Multiple levels of peripheral tolerance. *Immunol Today.* 1993;14(1):12-14.
374. Goodnow CC. Transgenic mice and analysis of B-cell tolerance. *Ann Rev Immunol.* 1992;10:489-518.
375. Ohashi PS, Oehen S, Buerki K, et al. Ablation of "tolerance" and induction of diabetes by virus infection in viral antigen transgenic mice. *Cell.* 1991;65:305.
376. Lafferty KJ, Babcock SK, Gill RG. Prevention of rejection by treatment of the graft: an overview. *Prog Clin Biol Res.* 1986;224:87-117.
377. George AJ, Larkin DF. Corneal transplantation: the forgotten graft. *Am J Transplant.* 2004;4(5):678-685.
378. Hirata Y, Furuhashi K, Ishii H, et al. CD150(high) bone marrow Tregs maintain hematopoietic stem cell quiescence and immune privilege via adenosine. *Cell Stem Cell.* 2018;22(3):445-453 e5.
379. Owen RD. Immunogenetic consequences of vascular anastomoses between bovine twins. *Science.* 1945;102:400-401.
380. Billingham RE, Lampkin GH, Medawar PB, et al. Tolerance to homografts, twin diagnosis, and the freemartin condition in cattle. *Heredity.* 1952;6:201-212.
381. Riley JS, McClain LE, Stratigis JD, et al. Regulatory T cells promote alloengraftment in a model of late-gestation in utero hematopoietic cell transplantation. *Blood Adv.* 2020;4(6):1102-1114.
382. Vrecenak JD, Pearson EG, Santore MT, et al. Stable long-term mixed chimerism achieved in a canine model of allogeneic in utero hematopoietic cell transplantation. *Blood.* 2014;124(12):1987-1995.
383. Mathes DW, Solari MG, Randolph MA, et al. Long-term acceptance of renal allografts following prenatal inoculation with adult bone marrow. *Transplantation.* 2005;80(9):1300-1308.
384. Billingham RE, Brent L, Medawar PB. "Actively acquired tolerance" of foreign cells. *Nature.* 1953;172:603-606.
385. Streilein JW. Neonatal tolerance of H-2 alloantigens. *Transplantation.* 1991;52:1-10.
386. Borenstein SH, Tao KS, West LJ, et al. Extrathymic deletion of CD8+ alloreactive T cells in a transgenic T cell receptor model of neonatal tolerance. *Transplantation.* 2001;72(11):1807-1816.
387. Ridge JP, Fuchs EJ, Matzinger P. Neonatal tolerance revisited: turning on newborn T cells with dendritic cells. *Science.* 1996;271:1723-1726.
388. Starzl TE, Demetris AJ, Murase N, et al. The lost chord: microchimerism and allograft survival. *Immunol Today.* 1996;17:577-584.
389. Burlingham WJ, Grailer AP, Fechner JH Jr, et al. Microchimerism linked to cytotoxic T lymphocyte functional unresponsiveness (clonal anergy) in a tolerant renal transplant recipient. *Transplantation.* 1995;59:1147-1155.
390. Ko S, Deiwick A, Jager MD, et al. The functional relevance of passenger leukocytes and microchimerism for heart allograft acceptance in the rat. *Nature Med.* 1999;5:1292-1297.
391. Schlitt HJ, Hundreiser J, Ringe B, et al. Systemic microchimerism of donor-type associated with irreversible acute liver graft rejection eight years after transplantation. *New Engl J Med.* 1994;330:646-647.
392. Wood K, Sachs DH. Chimerism and transplantation tolerance: cause and effect. *Immunol Today.* 1996;17:584-588.
393. Orloff MS, Fallon MA, DeMara E, et al. Induction of specific tolerance to small-bowel allografts. *Surgery.* 1994;116:222-228.
394. Ruedi E, Sykes M, Ildstad ST, et al. Antiviral T cell competence and restriction specificty of mixed allogeneic (P1+P2-->P1) irradiation chimeras. *Cell Immuol.* 1989;121:185-195.
395. Koehn BH, Williams MA, Borom K, et al. Fully MHC-disparate mixed hemopoietic chimeras show specific defects in the control of chronic viral infections. *J Immunol.* 2007;179(4):2616-2626.
396. Yoshikai Y, Ogimoto M, Matsuzaki G, et al. Bone marrow-derived cells are essential for intrathymic deletion of self-reactive T cells in both the host- and donor-derived thymocytes of fully allogeneic bone marrow chimeras. *J Immunol.* 1990;145:505-509.
397. Tomita Y, Khan A, Sykes M. Role of intrathymic clonal deletion and peripheral anergy in transplantation tolerance induced by bone marrow transplantion in mice conditioned with a non-myeloablative regimen. *J Immunol.* 1994;153:1087-1098.
398. Khan A, Tomita Y, Sykes M. Thymic dependence of loss of tolerance in mixed allogeneic bone marrow chimeras after depletion of donor antigen. Peripheral mechanisms do not contribute to maintenance of tolerance. *Transplantation.* 1996;62:380-387.
399. Gibbons C, Sykes M. Manipulating the immune system for anti-tumor responses and transplant tolerance via mixed hematopoietic chimerism. *Immunol Rev.* 2008;223:334-360.
400. Kwon HS, Logan AC, Chhabra A, et al. Anti-human CD117 antibody-mediated bone marrow niche clearance in nonhuman primates and humanized NSG mice. *Blood.* 2019;133(19):2104-2108.
401. Tomita Y, Sachs DH, Sykes M. Myelosuppressive conditioning is required to achieve engraftment of pluripotent stem cells contained in moderate doses of syngeneic bone marrow. *Blood.* 1994;83:939-948.
402. Ramshaw HS, Crittenden RB, Dooner M, et al. High levels of engraftment with a single infusion of bone marrow cells into normal unprepared mice. *Biol Blood Marrow Transplant.* 1995;1:74-80.
403. Sykes M, Szot GL, Swenson K, et al. Induction of high levels of allogeneic hematopoietic reconstitution and donor-specific tolerance without myelosuppressive conditioning. *Nature Med.* 1997;3(7):783-787.
404. Fuchimoto Y, Huang CA, Yamada K, et al. Mixed chimerism and tolerance without whole body irradiation in a large animal model. *J Clin Invest.* 2000;105:1779-1789.
405. Wekerle T, Kurtz J, Ito H, et al. Allogeneic bone marrow translantation with costimulatory blockade induces macro-chimerism and tolerance without cytoreductive host treatment. *Nature Med.* 2000;6:464-469.
406. Li HW, Sykes M. Emerging concepts in haematopoietic cell transplantation. *Nat Rev Immunol.* 2012;12(6):403-416.
407. Ophir E, Eidelstein Y, Afik R, et al. Induction of tolerance to bone marrow allografts by donor-derived host nonreactive ex vivo-induced central memory CD8 T cells. *Blood.* 2010;115(10):2095-2104.
408. Huang CA, Fuchimoto Y, Sheir-Dolberg R, et al. Stable mixed chimerism and tolerance using a non-myeloablative preparative regimen in a large animal model. *J Clin Invest.* 2000;105:173-181.
409. Kean LS, Adams AB, Strobert E, et al. Induction of chimerism in rhesus macaques through stem cell transplant and costimulation blockade-based immunosuppression. *Am J Transplant.* 2007;7(2):320-335.
410. Sicard A, Boardman DA, Levings MK. Taking regulatory T-cell therapy one step further. *Curr Opin Organ Transplant.* 2018;23(5):509-515.
411. Sicard A, Lamarche C, Speck M, et al. Donor-specific chimeric antigen receptor Tregs limit rejection in naive but not sensitized allograft recipients. *Am J Transplant.* 2020;20(6):1562-1573.
412. Dawson NA, Lamarche C, Hoeppli RE, et al. Systematic testing and specificity mapping of alloantigen-specific chimeric antigen receptors in regulatory T cells. *JCI Insight.* 2019;4(6):e123672.
413. Wortel CM, Heidt S. Regulatory B cells: phenotype, function and role in transplantation. *Transpl Immunol.* 2017;41:1-9.
414. Newell KA, Asare A, Kirk AD, et al. Identification of a B cell signature associated with renal transplant tolerance in humans. *J Clin Invest.* 2010;120(6):1836-1847.
415. Chesneau M, Pallier A, Braza F, et al. Unique B cell differentiation profile in tolerant kidney transplant patients. *Am J Transplant.* 2014;14(1):144-155.
416. de Masson A, Bouaziz JD, Le Buanec H, et al. CD24(hi)CD27(+) and plasmablast-like regulatory B cells in human chronic graft-versus-host disease. *Blood.* 2015;125(11):1830-1839.
417. Rebollo-Mesa I, Nova-Lamperti E, Mobillo P, et al. Biomarkers of tolerance in kidney transplantation: are we predicting tolerance or response to immunosuppressive treatment? *Am J Transplant.* 2016:3443-3457.
418. Ochando J, Conde P, Utrero-Rico A, et al. Tolerogenic role of myeloid suppressor cells in organ transplantation. *Front Immunol.* 2019;10:374.
419. Podesta MA, Remuzzi G, Casiraghi F. Mesenchymal stromal cells for transplant tolerance. *Front Immunol.* 2019;10:1287.
420. Ochando J, Ordikhani F, Jordan S, et al. Tolerogenic dendritic cells in organ transplantation. *Transpl Int.* 2020;33(2):113-127.
421. Albert ML, Jegathesan M, Darnell RB. Dendritic cell maturation is required for the cross-tolerization of CD8+ T cells. *Nat Immunol.* 2001;2(11):1010-1017.
422. Swiecki M, Colonna M. Unraveling the functions of plasmacytoid dendritic cells during viral infections, autoimmunity, and tolerance. *Immunol Rev.* 2010;234(1):142-162.
423. Lutz MB, Suri RM, Niimi M, et al. Immature dendritic cells generated with low doses of GM-CSF in the absence of IL-4 are maturation resistant and prolong allograft survival in vivo. *Eur J Immunol.* 2000;30(7):1813-1822.
424. Lu L, Li W, Fu F, et al. Blockade of the CD40-CD40 ligand pathway potentiates the capacity of donor-derived dendritic cell progenitors to induce long-term cardiac allograft survival. *Transplantation.* 1997;64:1808-1815.
425. Taner T, Hackstein H, Wang Z, et al. Rapamycin-treated, alloantigen-pulsed host dendritic cells induce ag-specific T cell regulation and prolong graft survival. *Am J Transplant.* 2005;5(2):228-236.
426. Madsen JC, Superina RA, Wood KJ, et al. Immunological unresponsiveness induced by recipient cells transfected with donor MHC genes. *Nature.* 1988;332:161-164.
427. Davies JD, Leong LYW, Mellor A, et al. T cell suppression in transplantation tolerance through linked recognition. *J Immunol.* 1996;156:3602-3607.
428. Peugh WN, Wood KJ, Morris PJ. Genetic aspects of the blood transfusion effect. *Transplantation.* 1988;46(3):438-443.
429. Opelz G, Vanrenterghem Y, Kirste G, et al. Prospective evaluation of pretransplant blood transfusions in cadaver kidney recipients. *Transplantation.* 1997;63(7):964-967.
430. Singh A, Ramachandran S, Graham ML, et al. Long-term tolerance of islet allografts in nonhuman primates induced by apoptotic donor leukocytes. *Nat Commun.* 2019;10(1):3495.
431. McCarthy DP, Bryant J, Galvin JP, et al. Tempering allorecognition to induce transplant tolerance with chemically modified apoptotic donor cells. *Am J Transplant.* 2015;15(6):1475-1483.
432. Yu S, Dangi A, Burnette M, et al. Acute murine cytomegalovirus disrupts established transplantation tolerance and causes recipient allo-sensitization. *Am J Transplant.* 2020;21(2):515-524.
433. Moses RD, Sundeen JT, Orr KS, et al. Cardiac allograft survival across major histocompatibility complex barriers in the rhesus monkey following T lymphocyte- depleted autologous marrow transplantation. III. Late allograft rejection. *Transplantation.* 1989;48:769-773.
434. Larsen CP, Elwood ET, Alexander DZ, et al. Long-term acceptance of skin and cardiac allografts after blocking CD40 and CD28 pathways. *Nature.* 1996;381:434-438.
435. Kirk AD, Harlan DM, Armstrong NN, et al. CTLA4-Ig and anti-CD40 ligand prevent renal allograft rejection in primates. *Proc Natl Acad Sci U S A.* 1997;94(16):8789-8798.
436. Wekerle T, Sayegh MH, Hill J, et al. Extrathymic T cell deletion and allogeneic stem cell engraftment induced with costimulatory blockade is followed by central T cell tolerance. *J Exp Med.* 1998;187(12):2037-2044.
437. Ito H, Kurtz J, Shaffer J, et al. CD4 T cell-mediated alloresistance to fully MHC-mismatched allogeneic bone marrow engraftment is dependent on CD40-CD40L interactions, and lasting T cell tolerance is induced by bone marrow transplantation with initial blockade of this pathway. *J Immunol.* 2001;166:2970-2981.
438. Fehr T, Takeuchi Y, Kurtz J, et al. Early regulation of CD8 T cell alloreactivity by CD4+CD25- T cells in recipients of anti-CD154 antibody and allogeneic BMT is followed by rapid peripheral deletion of donor-reactive CD8+ T cells, precluding a role for sustained regulation. *Eur J Immunol.* 2005;35:2679-2690.
439. Kurtz J, Ito H, Wekerle T, et al. Mechanisms involved in the establishment of tolerance through costimulatory blockade and BMT: lack of requirement for CD40L-mediated signaling for tolerance or deletion of donor-reactive CD4+ cells. *Am J Transplant.* 2001;1:339-349.
440. Parker DC, Greiner DL, Phillips NE, et al. Survival of mouse pancreatic islet allografts in recipients treated with allogeneic small lymphocytes and antibody to CD40 ligand. *Proc Natl Acad Sci U S A.* 1995;92:9560-9564.
441. Markees TG, Phillips NE, Gordon EJ, et al. Long-term survival of skin allografts induced by donor splenocytes and anti-CD154 antibody in thymectomized mice requires CD4+ T cells, Interferon-g, and CTLA4. *J Clin Invest.* 1998;101:2446-2455.
442. Iwakoshi NN, Markees TG, Turgeon N, et al. Skin allograft maintenance in a new synchimeric model system of tolerance. *J Immunol.* 2001;167(11):6623-6630.
443. Yamada K, Gianello PR, Ierenо FL, et al. Role of the thymus in transplantation tolerance in miniature swine: I. Requirement of the thymus for rapid and stable induction of tolerance to class I-mismatched renal allografts. *J Exp Med.* 1997;186:497-506.
444. Oluwole SF, Jin M-X, Chowdhury NC, et al. Effectiveness of intrathymic inoculation of soluble antigens in the induction of specific unresponsiveness to rat islet allografts without transient recipient immunosuppression. *Transplantation.* 1994;58:1077-1081.
445. Agus DB, Surh CD, Sprent J. Reentry of T cells to the adult thymus is restricted to activated T cells. *J Exp Med.* 1991;173:1039-1046.
446. Hadeiba H, Lahl K, Edalati A, et al. Plasmacytoid dendritic cells transport peripheral antigens to the thymus to promote central tolerance. *Immunity.* 2012;36(3):438-450.

447. Pearce NW, Spinelli A, Gurley KE, et al. Specific unresponsiveness in rats with prolonged cardiac allograft survival after treatment with cyclosporine. V. Dependence of CD4+ suppressor cells on the presence of alloantigen and cytokines, including interleukin-2. *Transplantation.* 1993;55:374-379.
448. Brouard S, Mansfield E, Braud C, et al. Identification of a peripheral blood transcriptional biomarker panel associated with operational renal allograft tolerance. *Proc Natl Acad Sci U S A.* 2007;104(39):15448-15453.
449. Ciancio G, Miller J, Garcia-Morales RO, et al. Six-year clinical effect of donor bone marrow infusions in renal transplant patients. *Transplantation.* 2001;71(7):827-835.
450. Mineo D, Ricordi C, Xu X, et al. Combined islet and hematopoietic stem cell allotransplantation: a clinical pilot trial to induce chimerism and graft tolerance. *Am J Transplant.* 2008;8(6):1262-1274.
451. Spitzer TR, MCafee S, Sackstein R, et al. The intentional induction of mixed chimerism and achievement of anti-tumor responses following non-myeloablative conditioning therapy and HLA-matched and mismatched donor bone marrow transplantation for refractory hematologic malignancies. *Biol Blood Marrow Transplant.* 2000;6:309-320.
452. Spitzer TR, Tolkoff-Rubin N, Cosimi AB, et al. Twenty-year follow-up of histocompatibility leukocyte antigen-matched kidney and bone marrow cotransplantation for multiple myeloma with end-stage renal disease: lessons learned. *Transplantation.* 2019;103(11):2366-2372.
453. Spitzer TR, MCafee S, Dey BR, et al. Non-myeloablative haploidentical stem cell transplantation using anti-CD2 monoclonal antibody (MEDI-507)-based conditioning for refractory hematologic malignancies. *Transplantation.* 2003;75:1748-1751.
454. Sykes M. Immune monitoring of transplant patients in transient mixed chimerism tolerance trials. *Hum Immunol.* 2018;79(5):334-342.
455. Scandling JD, Busque S, Lowsky R, et al. Macrochimerism and clinical transplant tolerance. *Hum Immunol.* 2018;79(5):266-271.
456. Leventhal JR, Ildstad ST. Tolerance induction in HLA disparate living donor kidney transplantation by facilitating cell-enriched donor stem cell Infusion: the importance of durable chimerism. *Hum Immunol.* 2018;79(5):272-276.
457. Schmidt V, Prell T, Treschl A, et al. Clinical management of posterior reversible encephalopathy syndrome after allogeneic hematopoietic stem cell transplantation: a case series and review of the literature. *Acta Haematol.* 2016;135(1):1-10.
458. Battaglia M, Stabilini A, Migliavacca B, et al. Rapamycin promotes expansion of functional CD4+CD25+FOXP3+ regulatory T cells of both healthy subjects and type 1 diabetic patients. *J Immunol.* 2006;177(12):8338-8347.
459. Anderson BE, McNiff J, Yan J, et al. Memory CD4+ T cells do not induce graft-versus-host disease. *J Clin Invest.* 2003;112(1):101-108.
460. Zheng H, Matte-Martone C, Jain D, et al. Central memory CD8+ T cells induce graft-versus-host disease and mediate graft-versus-leukemia. *J Immunol.* 2009;182(10):5938-5948.
461. Bleakley M, Heimfeld S, Loeb KR, et al. Outcomes of acute leukemia patients transplanted with naive T cell-depleted stem cell grafts. *J Clin Invest.* 2015;125(7):2677-2689.
462. Bleakley M, Heimfeld S, Jones LA, et al. Engineering human peripheral blood stem cell grafts that are depleted of naive T cells and retain functional pathogen-specific memory T cells. *Biol Blood Marrow Transplant.* 2014;20(5):705-716.
463. Schumm M, Lang P, Bethge W, et al. Depletion of T-cell receptor alpha/beta and CD19 positive cells from apheresis products with the CliniMACS device. *Cytotherapy.* 2013;15(10):1253-1258.
464. Kroger N, Solano C, Wolschke C, et al. Antilymphocyte globulin for prevention of chronic graft-versus-host disease. *N Engl J Med.* 2016;374(1):43-53.
465. Luznik L, Engstrom LW, Iannone R, et al. Posttransplantation cyclophosphamide facilitates engraftment of major histocompatibility complex-identical allogeneic marrow in mice conditioned with low-dose total body irradiation. *Biol Blood Marrow Transplant.* 2002;8(3):131-138.
466. O'Donnell PV, Luznik L, Jones RJ, et al. Nonmyeloablative bone marrow transplantation from partially HLA-mismatched related donors using posttransplantation cyclophosphamide. *Biol Blood Marrow Transplant.* 2002;8(7):377-386.
467. Ruggeri A, Sun Y, Labopin M, et al. Post-transplant cyclophosphamide versus anti-thymocyte globulin as graft-versus-host disease prophylaxis in haploidentical transplant. *Haematologica.* 2017;102(2):401-410.
468. Brissot E, Labopin M, Moiseev I, et al. Post-transplant cyclophosphamide versus antithymocyte globulin in patients with acute myeloid leukemia in first complete remission undergoing allogeneic stem cell transplantation from 10/10 HLA-matched unrelated donors. *J Hematol Oncol.* 2020;13(1):87.
469. Brunstein CG, Miller JS, Cao Q, et al. Infusion of ex vivo expanded T regulatory cells in adults transplanted with umbilical cord blood: safety profile and detection kinetics. *Blood.* 2011;117(3):1061-1070.
470. Di Ianni M, Falzetti F, Carotti A, et al. Tregs prevent GVHD and promote immune reconstitution in HLA-haploidentical transplantation. *Blood.* 2011;117(14):3921-3928.
471. Whangbo JS, Kim HT, Mirkovic N, et al. Dose-escalated interleukin-2 therapy for refractory chronic graft-versus-host disease in adults and children. *Blood Adv.* 2019;3(17):2550-2561.
472. Koreth J, Matsuoka K, Kim HT, et al. Interleukin-2 and regulatory T cells in graft-versus-host disease. *N Engl J Med.* 2011;365(22):2055-2066.
473. Munneke JM, Spruit MJ, Cornelissen AS, et al. The potential of mesenchymal stromal cells as treatment for severe steroid-refractory acute graft-versus-host disease: a critical review of the literature. *Transplantation.* 2016;100(11):2309-2314.
474. Ngwube A, Shah N, Godder K, et al. Abatacept is effective as GVHD prophylaxis in unrelated donor stem cell transplantation for children with severe sickle cell disease. *Blood Adv.* 2020;4(16):3894-3899.
475. Khandelwal P, Yeh RF, Yu L, et al. Graft versus host disease prophylaxis with abatacept reduces severe acute graft versus host disease in allogeneic hematopoietic stem cell transplant for beta thalassemia major with busulfan, fludarabine, and thiotepa. *Transplantation.* 2020;105(4):891-896.
476. Kattner AS, Holler E, Holler B, et al. IL6-receptor antibody tocilizumab as salvage therapy in severe chronic graft-versus-host disease after allogeneic hematopoietic stem cell transplantation: a retrospective analysis. *Ann Hematol.* 2020;99(4):847-853.
477. Drobyski WR, Pasquini M, Kovatovic K, et al. Tocilizumab for the treatment of steroid refractory graft-versus-host disease. *Biol Blood Marrow Transplant.* 2011;17(12):1862-1868.
478. Kennedy GA, Varelias A, Vuckovic S, et al. Addition of interleukin-6 inhibition with tocilizumab to standard graft-versus-host disease prophylaxis after allogeneic stem-cell transplantation: a phase 1/2 trial. *Lancet Oncol.* 2014;15(13):1451-1459.
479. Ma H, Lu C, Ziegler J, et al. Absence of Stat1 in donor CD4(+) T cells promotes the expansion of Tregs and reduces graft-versus-host disease in mice. *J Clin Invest.* 2011;121(7):2554-2569.
480. Choi J, Cooper ML, Alahmari B, et al. Pharmacologic blockade of JAK1/JAK2 reduces GvHD and preserves the graft-versus-leukemia effect. *PloS One.* 2014;9(10):e109799.
481. Choi J, Ziga ED, Ritchey J, et al. IFNgammaR signaling mediates alloreactive T-cell trafficking and GVHD. *Blood.* 2012;120(19):4093-4103.
482. Zeiser R, von Bubnoff N, Butler J, et al. Ruxolitinib for glucocorticoid-refractory acute graft-versus-host disease. *N Engl J Med.* 2020;382(19):1800-1810.
483. Dubovsky JA, Flynn R, Du J, et al. Ibrutinib treatment ameliorates murine chronic graft-versus-host disease. *J Clin Invest.* 2014;124(11):4867-4876.
484. Reshef R, Luger SM, Hexner EO, et al. Blockade of lymphocyte chemotaxis in visceral graft-versus-host disease. *N Engl J Med.* 2012;367(2):135-145.
485. Hart JW, Shiue LH, Shpall EJ, et al. Extracorporeal photopheresis in the treatment of graft-versus-host disease: evidence and opinion. *Ther Adv Hematol.* 2013;4(5):320-334.
486. Jagasia M, Scheid C, Socie G, et al. Randomized controlled study of ECP with methoxsalen as first-line treatment of patients with moderate to severe cGVHD. *Blood Adv.* 2019;3(14):2218-2229.
487. Galili U. The alpha-gal epitope and the anti-Gal antibody in xenotransplantation and in cancer immunotherapy. *Immunol Cell Biol.* 2005;83(6):674-686.
488. Sachs DH, Sykes M, Robson S, et al. Xenotransplantation. *Adv Immunol.* 2001;79:129-223.
489. Langin M, Mayr T, Reichart B, et al. Consistent success in life-supporting porcine cardiac xenotransplantation. *Nature.* 2018;564(7736):430-433.
490. Coe TM, Markmann JF, Rickert CG. Current status of porcine islet xenotransplantation. *Curr Opin Organ Transplant.* 2020;25(5):449-456.
491. Wilmut I, Schnieke AE, McWhir J, et al. Viable offspring derived from fetal and adult mammalian cells. *Nature.* 1997;385(6619):810-813.
492. Lai L, Kolber-Simonds D, Park KW, et al. Production of {alpha}-1,3-Galactosyltransferase knockout pigs by nuclear transfer cloning. *Science.* 2002;295:1089-1092.
493. Dai Y, Vaught TD, Boone J, et al. Targeted disruption of the alpha1,3-galactosyltransferase gene in cloned pigs. *Nat Biotechnol.* 2002;20(3):251-255.
494. Cowan PJ, Tector AJ. The resurgence of xenotransplantation. *Am J Transplant.* 2017;17(10):2531-2536.
495. Yue Y, Xu W, Kan Y, et al. Extensive germline genome engineering in pigs. *Nat Biomed Eng.* 2020;5(2):134-143.
496. Griesemer A, Yamada K, Sykes M. Xenotransplantation: immunological hurdles and progress toward tolerance. *Immunol Revs.* 2014;258(1):241-258.
497. Ohdan H, Yang Y-G, Swenson KG, et al. T cell and B cell tolerance to gal1,3gal-expressing heart xenografts is achieved in a1,3-galactosyltransferase-deficient mice by non-myeloablative induction of mixed chimerism. *Transplantation.* 2001;71:1532-1542.
498. Zhao Y, Ohdan H, Manilay JO, et al. NK cell tolerance in mixed allogeneic chimeras. *J Immunol.* 2003;170:5398-5405.
499. Kawahara T, Rodriguez-Barbosa JI, Zhao Y, et al. Global unresponsiveness as a mechanism of natural killer cell tolerance in mixed xenogeneic chimeras. *Am J Transplant.* 2007;7(9):2090-2097.
500. Coudert JD, Scarpellino L, Gros F, et al. Sustained NKG2D engagement induces cross-tolerance of multiple distinct NK cell activation pathways. *Blood.* 2008;111(7):3571-3578.
501. Lan P, Wang L, Diouf B, et al. Induction of human T cell tolerance to porcine xenoantigens through mixed hematopoietic chimerism. *Blood.* 2004;103:3964-3969.
502. Zhao Y, Swenson K, Sergio JJ, et al. Skin graft tolerance across a discordant xenogeneic barrier. *Nature Med.* 1996;2:1211-1216.
503. Zhao Y, Swenson K, Sergio JJ, et al. Pig MHC mediates positive selection of mouse CD4+ T cells with a mouse MHC-restricted TCR in pig thymus grafts. *J Immunol.* 1998;161:1320-1326.
504. Zhao Y, Fishman JA, Sergio JJ, et al. Immune restoration by fetal pig thymus grafts in T cell-depleted, thymectomized mice. *J Immunol.* 1997;158:1641-1649.
505. Zhao Y, Sergio JJ, Swenson KA, et al. Positive and negative selection of functional mouse CD4 cells by porcine MHC in pig thymus grafts. *J Immunol.* 1997;159:2100-2107.
506. Shimizu I, Fudaba Y, Shimizu A, et al. Comparison of human T cell repertoire generated in xenogeneic porcine and human thymus grafts. *Transplantation.* 2008;86(4):601-610.
507. Kalscheuer H, Onoe T, Dahmani A, et al. Xenograft tolerance and immune function of human T cell developing in pig thymus xenografts. *J Immunol.* 2014;192(7):3442-3450.
508. Yamada K, Yazawa K, Shimizu A, et al. Marked prolongation of porcine renal xenograft survival in baboons through the use of alpha1,3-galactosyltransferase gene-knockout donors and the cotransplantation of vascularized thymic tissue. *Nat Med.* 2005;11(1):32-34.
509. Fudaba Y, Onoe T, Chittenden M, et al. Abnormal regulatory and effector T cell function predispose to autoimmunity following xenogeneic thymic transplantation. *J Immunol.* 2008;181(11):7649-7659.
510. Yan Y, Devos T, Yu L, et al. Pathogenesis of autoimmunity after xenogeneic thymus transplantation. *J Immunol.* 2003;170(12):5936-5946.
511. Navarro-Alvarez N, Shah JA, Zhu A, et al. The effects of exogenous administration of human coagulation factors following pig-to-baboon liver xenotransplantation. *Am J Transplant.* 2016;16(6):1715-1725.
512. Fishman JA. Infectious disease risks in xenotransplantation. *Am J Transplant.* 2018;18(8):1857-1864.
513. Reemtsma K, McCracken BH, Schlegel JU. Renal heterotransplantation in man. *Ann Surg.* 1964;160:384.
514. Deacon T, Schumacher J, Dinsmore J, et al. Histological evidence of fetal pig neural cell survival after transplantation into a patient with Parkinson's disease. *Nature Med.* 1997;3:350-353.

CHAPTER 49

Cancer Immunology and Immunotherapy

Thomas F. Gajewski, MD, PhD • Justin Kline, MD

HISTORICAL AND CONCEPTUAL BACKGROUND

The concept that there may be an interaction between cancer and the immune system is almost as old as the history of the discipline of immunology itself. In the late 19th and early 20th centuries, the battle between two overarching theories of immunity was brewing—the humoral view and the cellular view. Soluble factors (antibodies) could clearly neutralize bacterial toxins and along with complement could lead to killing of bacteria. At the same time, even marine invertebrates were observed to have phagocytes that were described to ingest and destroy bacteria or lead to granuloma formation for containment, and various human pathologic conditions were shown to contain leukocyte reactions. The scientific community moved toward reconciliation of these camps with the awarding of the Nobel Prize in Medicine jointly to Ilya Metchnikoff and (proponent of the cellular model) and Paul Ehrlich (champion of the humoral model) in 1908. It is noteworthy that each of these components of immune responses have been found to have great relevance for the cancer context, both from the perspective of natural endogenous responses and also in terms of therapeutics. Of interest are the early observations regarding immunity to tissues rather than to pathogens. Very few cases of antibody-mediated destruction of tissues could be documented, for example, upon transplantation, with the exception of hemolytic diseases. Importantly, transplanted allogeneic tumors were described to have a proliferative lymphocytic infiltrate as early as 1916. Additionally, allogeneic tumors were found to grow progressively in immunologically privileged sites, such as the brain. However, cotransplantation of lymphoid tissue could promote tumor rejection, which was prevented upon X-irradiation (which kills the radiation-sensitive lymphocytes). Together, these data studying allogeneic tumors in mice from the early 20th century argued that a local lymphocyte-dependent, cell-mediated immune response was capable of tumor rejection. It is interesting to fast forward to the 21st century to the notion of the "T cell–inflamed tumor microenvironment" described in a major subset of human solid tumor patients, which has positive prognostic and predictive value and is consistent with those historical results in mice.

Studies of the rejection of allogeneic tumors were also useful as they contributed to the identification and mapping of the major histocompatibility loci. However, it was questionable whether syngeneic tumors could be handled similarly by the immune system. It was noted that spontaneously arising cancers in mice could frequently be transplanted and passaged in syngeneic wildtype mice without rejection, arguing against the potential for immune-mediated elimination. However, in the 1950s, a distinct result was obtained with fibrosarcomas induced by the carcinogen methylcholanthrene (MCA). MCA-induced tumors were found to induce protection following initial vaccination. In addition, this protection was almost always tumor-specific, without cross-protection between individual tumors. These results argued that putative dominant antigens might often be unique to each tumor. Such concepts were borne out approximately 50 years later with the molecular cloning of tumor antigen genes, which were found often to be represented by nonsynonymous mutations in normal genes, and also with the observed association between tumor mutation load and clinical efficacy of immunotherapy in patients. Numerous other genomic aberrancies have similarly been found to give rise to tumor-associated antigens, which has the potential to provide an ample substrate for adaptive immune recognition.

However, although antigenic tumor cell lines implanted as a cell suspension could be rejected by mice, transplantation of tumor fragments usually did not.[1] Besides cancer cells, fragments grown in vivo contain the additional cellular components of the tumor microenvironment (ie, stroma). These empiric observations were among the first to suggest that the composition of the tumor microenvironment might play a dominant role in preventing immune-mediated tumor elimination in vivo, and in the intervening years, multiple immune regulatory processes have been identified that dominantly act at the level of the tumor site. Such processes likely explain many cases of how cancers replete with antigens nonetheless develop and grow progressively in patients, avoiding immune destruction. Zeroing in on the immune regulatory aspects of the tumor microenvironment has helped to guide development of several contemporary immunotherapies for cancer. The success of immune checkpoint blockade with antibodies targeting CTLA-4 or PD-1/PD-L1 interactions might be highlighted here, which led to the Nobel Prize in Medicine being awarded to James Allison and Tasuku Honjo in 2018.

In general, adaptive immune responses require upstream innate immune activation to generate productive immunity rather than tolerance and to orchestrate dynamic interactions between collaborating immune cells (Fig. 49.1). Like the early studies from Metchnikoff which took advantage of simpler animal models that lack an adaptive immune system, innate immune pathway discovery benefited from a distinct model—in this case, genetic experiments in

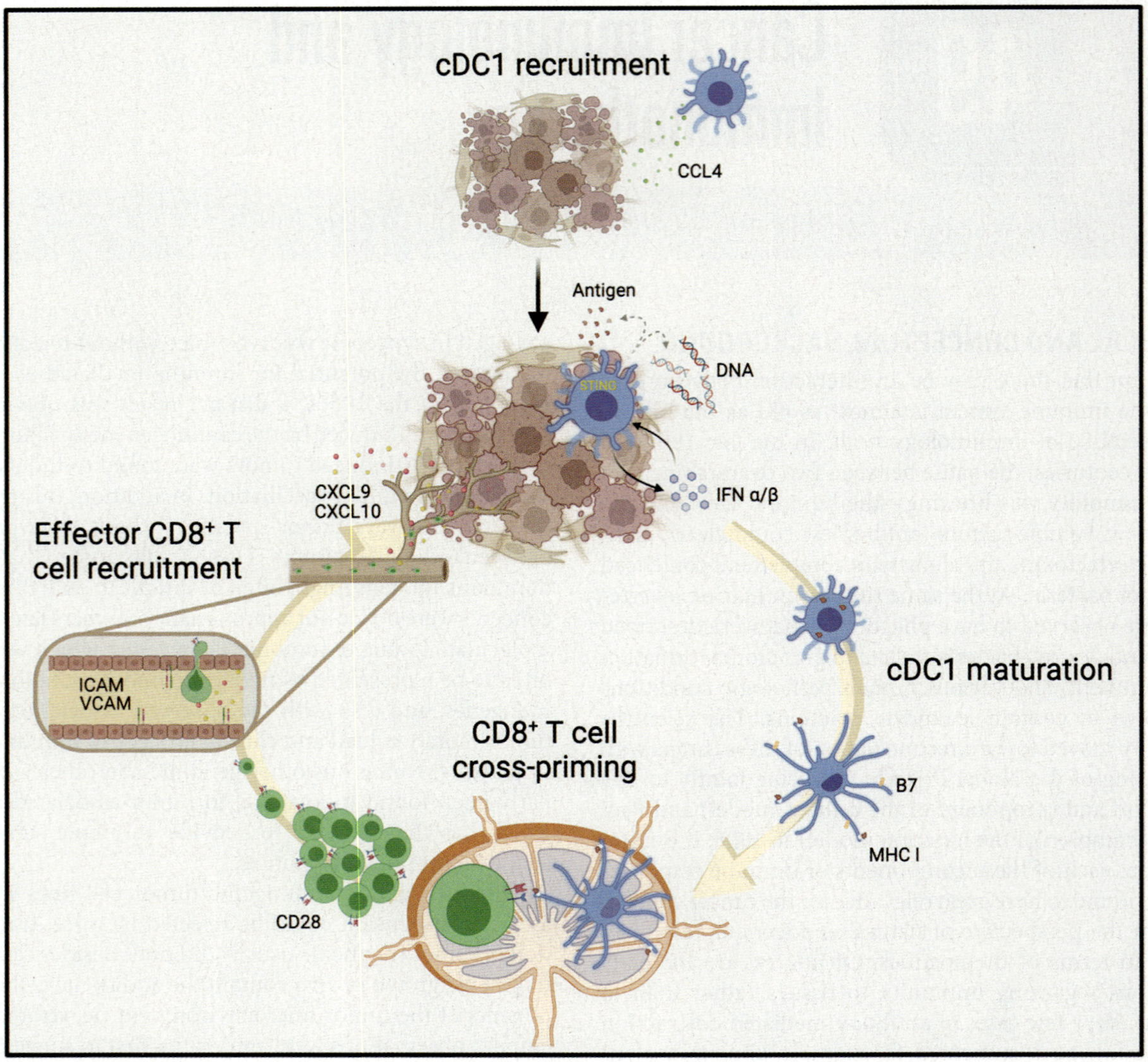

FIG. 49.1. Mechanistic steps in the generation of a spontaneous immune response against cancer.

Drosophila melanogaster. A particular mutant fly was found to have opportunistic fungal infection, and the identified gene was called Toll. Ten human homologs were identified (designated Toll-like receptors or TLRs), and identification of murine TLRs benefited from parallel mutational genetic experiments. In mammals, TLRs were found to be expressed at the highest levels in antigen-presenting cells (APCs), especially dendritic cells (DCs). Collectively, these observations coalesced into a model in which innate immune "sensing" of pathogen-derived products through TLRs would activate DCs, which in turn would productively activate T cells and B cells to generate adaptive immunity. This set of discoveries led to awarding of the Nobel Prize in Medicine to Jules Hoffmann, Bruce Beutler, and Ralph Steinman in 2011. Molecular agonists against specific TLRs have been developed, and one of these directed against TLR7 (Imiquimod) has been FDA-approved and is used clinically, not only to treat genital warts but also to treat basal cell carcinomas of the skin. Thus, innate immune activators can serve as cancer therapeutics. In the context of mouse solid tumor models, host TLR pathways did not appear to be necessary for the spontaneous priming of antitumor T cells. Evaluation of additional candidates led to the discovery that the host STING pathway, a mechanism for sensing of cytosolic DNA originally discovered to be central to responses against DNA viruses, was critically involved. Synthetic STING pathway agonists are similarly being explored for therapeutic activity in cancer.

While a significant body of work in mouse models solidified the concepts that tumor antigens could exist, that spontaneous T cell responses against cancer could develop, and that therapeutic interventions aimed at amplifying such immune responses could cause tumor regression, there remained skepticism by some as to whether similar phenomena occurred in human cancer patients. However, several lines of evidence indeed supported the existence of spontaneous immune responses against cancers in the clinic. First, some cancer patients develop paraneoplastic syndromes, which are immunopathologies that become manifest when antibodies become induced against a tumor protein that are cross-reactive with a normal self-antigen. Second, immunocompromised patients have an increased incidence of certain cancers. This was perhaps most apparent during the early HIV epidemic, when multiple virus-induced cancers were found to emerge rapidly as CD4+ T cell counts dropped. Lymphomas also occur at a greater frequency in these patients, and over time, the incidence of more common epithelial malignancies also modestly increases as patients survive longer with HIV. Third, T cell clones could be identified from the blood of cancer patients that could recognize and kill autologous tumor cell lines in vitro. This rather straightforward result argued that some of the key requisite components are present within many patients (tumor-associated antigen, antigen-specific T cells), but that somehow the endogenous immune response is not properly orchestrated to carry tumor regression to completion. Fourth, in many patients with melanoma, tumor-infiltrating lymphocytes (TILs) could be cultured out ex vivo, many of which were tumor-reactive. The notion that these T cells have the *potential* to mediate tumor regression is supported by the therapeutic activity of ex vivo–grown TIL, administered to patients following lymphodepletion and along with IL-2, which generates a response rate of around 50%. These data argue that even within the tumor microenvironment, there can be T cells capable of tumor recognition and destruction, but that somehow the final effector phase of the immune response has been thwarted, enabling tumor persistence and outgrowth. Together, these clinical observations, built upon the foundational history of mouse tumor immunity models, generated a persistent optimism that immunotherapeutic strategies ultimately could be developed with broad applicability in the treatment of human cancer patients.

EXPERIMENTAL EVIDENCE FOR IMMUNE RECOGNITION OF CANCER

Empirical evidence supporting the notion that the immune system could recognize, and in some cases eliminate, cancers has existed since the late 1800s when William Coley, a surgeon at New York Memorial hospital, treated hundreds of cancer patients with bacteria or bacteria-derived products. Coley hypothesized that the immune system, in response to bacterial infection, could simultaneously mediate tumor regression.[2] While most cancer patients failed to benefit from what later became to be known as "Coley toxins," and in fact some became quite ill and even died, Coley noted that some patients, particularly those with bone and soft tissue sarcomas, benefitted greatly from his approach. As a result of treatment irregularities, inconsistent results, and general disbelief among his peers, along with the advent of chemotherapy and radiation therapy for the treatment of malignant disease, Coley toxins fell from favor, and its use was eventually discontinued. However, a half century later, experimental evidence in animal models supporting the cancer immunosurveillance hypothesis rekindled excitement that the immune system might indeed be harnessed in the fight against cancer. For his initial work in the 1890s, Coley is now regarded by some as the father of cancer immunotherapy.

Mouse Model Evidence

Before much of the direct evidence in support of the cancer immunosurveillance hypothesis had been generated, biologists Lewis Thomas and Macfarlane Burnet hypothesized that a major function of the immune system was to surveil the host for developing tumors based on the expression of unique antigens, ultimately leading to Burnet's immunosurveillance theory.[3,4] He famously wrote, "It is by no means inconceivable that small accumulation of tumor cells may develop and because of their possession of new antigenic potentialities provoke an effective immunological reaction with regression of the tumor and no clinical hint of its existence." Although the cancer immunosurveillance hypothesis would require decades of research to confirm, pioneering biologists believed it to be so based on their rudimentary understanding of the immune system and its functions at the time.

In the late 1950s, a potent chemical carcinogen, MCA, was shown to induce soft tissue fibrosarcomas within months following inoculation into rodents.[5] Over the next several decades, this carcinogen-induced autochthonous cancer model was instrumental in elucidating numerous aspects of cancer immune surveillance. Early experiments indicated that irradiated cell lines derived from primary MCA-induced sarcomas, upon inoculation in isologous mice (those of a similar inbred mouse strain), generated resistance to a subsequent challenge with live sarcoma cells.[5-7] Because in-depth knowledge of the immune system and its components was lacking at the time, it was difficult to conclude with certainty that such resistance was due to a specific antitumor immune response, or rather a result of slight genetic differences between mice in which the primary sarcoma arose and those subsequently challenged with tumor cells derived from it.

George Klein and colleagues in Sweden, extending upon similar results of others,[5,7] definitively demonstrated the existence of immune surveillance against MCA-induced sarcomas in seminal work published in 1960.[6] Here, sarcomas generated in the hindlimbs of inbred mouse strains were surgically resected and cell lines from these tumors were created. When inoculated into naïve mice of the same genetic background as the primary host (isologous mice), tumors developed in most cases. However, when either the primary autologous mouse or secondary isologous mice were repeatedly vaccinated with irradiated sarcoma cells derived from the primary tumor, protection was often conferred against a subsequent live tumor cell challenge.[6] Normal cells from the organs of untreated mice, when irradiated and repeatedly inoculated into isologous animals, provided little or no protection against a subsequent challenge with live sarcoma cells in most cases. Resistance conferred by irradiated tumor cell vaccination in isologous mice was often overcome through a higher live tumor cell inoculum. When pre-immunized mice resistant to a given sarcoma were challenged with sarcoma cells derived from a different primary host, tumors typically grew out similarly as those in untreated isologous mice, indicating that the immune response generated against MCA-induced sarcomas was highly specific and did not provide cross-protection against cancers of a similar histologic origin. In further support of the specificity of the observed resistance, pre-inoculation with sarcoma cells from several mouse strains provided no protection from primary MCA-induced carcinogenesis. Interestingly, pre-immunization with irradiated tumor cells often afforded greater protection to isologous mice than autologous mice from whom the primary sarcomas were derived, which indirectly suggested the existence of a global immunosuppressed state in the latter. The phenomenon of cancer-associated immunosuppression has since been well-characterized, but at the time, this observation was not easily explained. Finally, administration of total body irradiation was associated with more rapid progression of transplanted sarcomas in isologous hosts, suggesting that a radiosensitive tissue or cell compartment was responsible for controlling tumor growth.

In an effort to define the extent to which protection afforded to pre-immunized, isologous mice was dependent on humoral or cellular immunity, tumor cell suspensions were cultured with serum or lymph node cells from untreated or pre-immunized isologous mice prior to inoculation. Here, pre-incubation of tumor cells with lymph node cells from pre-immunized mice but not untreated mice was associated with slower tumor outgrowth upon subsequent inoculation into host mice, all of which had received TBI prior in order to prevent a primary host immune response against the inoculated sarcoma cells. Pre-irradiation of lymph node cells prior to incubation with tumor cells abrogated the effect. Incubation of sarcoma cells with serum, regardless of origin from untreated or pre-immunized isologous mice, had no impact on subsequent tumor outgrowth. This result clearly demonstrated that live lymph node cells from pre-immunized mice were essential in the antitumor immune protection.

The experiments performed by Klein and others were transformational and provided considerable evidence supporting the involvement of the immune system in recognizing and controlling cancer. It is worth noting that these experiments were performed in an era when many immune cells, including T cells and DCs, had not yet been characterized. Furthermore, this body of work highlighted the specific nature of the immune response to cancer that it was mediated by live cells of the lymph node, and that cancer seemed to generate a global immunosuppressive state in the host. Finally, while evidence of cancer immune surveillance was evident in the MCA-induced sarcoma model, it could not always be detected in other spontaneously arising murine tumors,[7] which at the time puzzled scientists in the field. Today, we understand this is likely related to the fact that MCA induces potent mutagenesis in host cells, thereby generating numerous neoantigens that can be strongly targeted by host T cells, whereas in spontaneously arising autochthonous cancers, neoantigens are much less common.

Despite the enthusiasm around the notion of cancer immune surveillance at the time, experimental observations in the 1970s again brought this theory into question. For example, Stutman and colleagues demonstrated that the incidence of primary autochthonous tumors was similar in athymic nude mice and wild-type mice of the identical genetic background, which was in stark contrast to observations noted in the MCA model of carcinogenesis.[8,9] However, it was later shown that nude mice are not completely devoid of T cells[10] and still contain NK cells, which had not been well-characterized at the time. Also, several groups identified interferon gamma (IFN-γ)—a cytokine produced by effector T cells and NK cells—and perforin—a cytolytic molecule secreted by these same cells—as important secreted factors involved in cancer immune surveillance,[11-16] particularly against MCA-induced sarcomas.

Furthermore, in the early 2000s, Robert Schreiber and colleagues at Washington University found a significantly higher incidence of primary MCA-induced sarcomas in recombinase-activating gene (Rag) 2–deficient mice which are entirely devoid of mature B and T lymphocytes, when compared to wildtype animals of the same genetic background.[17] A similar observation was noted in mice lacking the IFN-γ receptor (*Ifngr1*$^{-/-}$) and in mice deficient in Stat1, a transcription factor required for IFN-γ–dependent signaling. *Rag2*$^{-/-}$ and *Rag2*$^{-/-}$*Stat1*$^{-/-}$ mice were also strikingly more prone toward development of spontaneous tumors than wildtype animals when observed over a prolonged period of time.[17] Depletion of CD4+ or CD8+ T cells from wildtype animals was associated with more rapid outgrowth of transplanted cell lines derived from primary MCA tumors, supporting the notion that T cells were particularly critical for mediating antitumor immune responses. While these experiments solidified the role of T cells and IFN-γ in promoting cancer immune surveillance, perhaps the most notable finding was the observation that cell lines derived from primary MCA-induced sarcomas that initially developed in wildtype mice grew similarly when implanted in *Rag2*$^{-/-}$ or wildtype hosts, whereas cell lines from sarcomas that initially developed in *Rag2*$^{-/-}$ mice progressed more slowly in wildtype mice compared with *Rag2*$^{-/-}$ hosts.[17] These compelling data indicated that tumors derived from immunocompetent animals were of lower immunogenicity than those from immunodeficient animals and suggested that the immunogenicity of a tumor could be shaped by the immunologic environment in which it arises.

Evidence Derived From Human Cancer Patients

Technological advances in the 1970s enabled scientists to screen TILs or blood-derived T cell clones for reactivity against autologous cancer cells through in vitro assays, and in some cases, tumor antigen–reactive T cells could be identified.[18] Cell lines established from primary tumors were tested for reactivity by autologous T cells and serum using fibroblasts or other healthy cells as controls.[18] Not infrequently, specific T cell and/or antibody responses against autologous tumor cells could be identified.[18-20] These initial human experiments provided direct evidence for the existence of cancer-specific immune responses, at least in a subset of patients. Further advances in cloning and in cDNA library expression by Boon and colleagues in the 1990s allowed for the identification of shared tumor antigens,[21,22] including the MAGE (melanoma antigen gene) family of cancer-testis antigens, and their recognition by T cell clones, which collectively led to a direct demonstration of cancer-specific immune responses to human tumors. Following this pioneering and incredibly labor-intensive effort, many other shared tumor antigens were characterized, which are further discussed below. With the advent of HLA tetramer technology in the mid-1990s,[23] it became possible to screen patient blood and tumor tissue for T cells reactive to known cancer antigens, which were often identifiable.[24,25] Collectively, the identification of cancer antigens and the frequent detection of expanded antigen-specific T cells in cancer patients provided strong evidence that, as Thomas and Burnet had hypothesized years earlier, T cells could indeed recognize cancers as non-self and mount a response against them.

Parallel work focused on investigating the immune infiltrate that can often be observed within solid tumors. A seminal study published in 2006 by Galon and colleagues sought to identify a correlation between the type, density, and location of T cells in colorectal cancers with clinical outcomes.[26] Immunostaining was utilized to identify numbers of CD3+ T cells, CD8+ T cells, CD45RO+ T cells (memory T cells), and granzyme B+ cells, and their localization (in the tumor center or invasive margin) in tissue microarrays encompassing 415 colorectal carcinomas of various pathological stages. The presence of higher numbers of activated T cells was remarkably effective at predicting improved disease-free and overall survival in patients, regardless of tumor stage. Similar analyses were performed in early-stage melanoma[27] and in advanced-stage ovarian carcinoma,[28] also showing very strong correlations between intratumoral T cell numbers and favorable outcomes. Together, these results suggest that the presence of a baseline antitumor immune response can have positive prognostic value, even if the cancer is not eliminated by the host.

Building on this concept, a consortium led by the Society for the Immunotherapy of Cancer (SITC) was convened to validate the immunoscore in over 3500 colon cancer specimens from patients with TNM stage I-III disease. The investigators found that immunoscore effectively predicted recurrence risk (more effectively than tumor stage, age, and other clinical and pathological variables) suggesting that the immunoscore should be routinely included into conventional staging in colon cancer.[29]

Additional molecular analyses of human cancer tissues also supported a role for the host immune response. First, exome sequencing of human tumors, as collated in The Cancer Genome Atlas (TCGA),[30] identified a wide variation in tumor mutational burden, among individual patients and also between cancer types.[31] While cancers such as melanoma, lung carcinoma, and bladder carcinoma typically carried a higher NSV burden, most pediatric cancers and hematologic malignancies were found to be much lower.[31,32] These results suggested that mutational load (and therefore, predicted neoantigen burden) varied widely between cancer types,[31,33] which was roughly correlating with both the presence of a spontaneously generated antitumor immune response[33] and with the efficacy of new immunotherapy strategies,[34,35] discussed further below. Second, gene expression profiling of human tumors, initially using the Affymetrix gene chip platform but later utilizing RNAseq or the NanoString technology, revealed a tremendous variation in the expression of immune genes in individual tumors. These included T cell transcripts, as well as transcripts associated with DCs, macrophages, and various cytokines and chemokines.[36-39] Similar to the immunoscore approach, tumor-based immune gene signatures were found to have positive prognostic value in different cancer types and also to be associated with the efficacy of emerging immunotherapies.[40] Taken together, these observations argued that, in many cases, cancers were antigenic, but also that they could be immunogenic, culminating in immune cell infiltration and endogenous T cell responses associated with positive prognostic value for the patient.

TUMOR ANTIGENS

The fact that spontaneous T cell responses and immune cell infiltration into the tumor microenvironment could be observed implied that cancer cells likely express antigens that can be recognized by host T cells, even though cancers arise from normal self-tissues. Thousands of antigens have been identified in a broad range of cancers, and an exhaustive listing of these antigens is beyond the scope of this text. The conceptual framework and key principles regarding tumor antigens are developed below.

Methods of Antigen Identification

Multiple strategies have been pursued to identify tumor-associated antigens, both in mouse models and from human cancer patients. In general, these methods can be divided into unbiased, immune response-centered strategies, and biased, candidate antigen-based strategies.

Unbiased Methods for Antigen Identification

The classical approach for identifying tumor antigens begins with a known immune response against the tumor, usually CD8+ T cells. If a tumor cell line can be derived from a patient, then T cells collected either from the blood or from the tumor microenvironment can be expanded in response to tumor cells in vitro and then cloned by limiting dilution or flow cytometric sorting. In this way, clonal populations of T cells specific for the patient's tumor can be studied.[41] To identify the antigen serving as the T cell recognition element, the MHC restriction molecule is evaluated, either using blocking antibodies against specific HLA molecules or using genetic means. Then typically, a cDNA library is generated and co-transfected into third-party recipient cells along with the relevant HLA molecule, the transfected cells successfully recognized by the T cell clone are identified, and the corresponding cDNA is isolated and sequenced. The full-length gene can then be identified, and the type of antigen can be determined.[22]

A related strategy employs not a cDNA library from the tumor cell line but rather naturally processed peptides eluted from immunoprecipitated MHC molecules. Peptide fractions can be pulsed onto cells expressing the corresponding class I molecule, and T cell recognition through cytokine production can be used as a readout.[42]

Rather than using T cell clones as a probe to identify antigens, an alternative approach is to utilize the spontaneous antibody response that is often generated in cancer patients. This technique has been termed "SEREX" and involves screening a cDNA expression library made from a patient's tumor cells with autologous serum to identify the antigen. Once positive transfectants are identified and the gene is characterized, T cell epitopes can be identified from the same protein.[43]

All three of the above methods begin with the knowledge that an immune response was generated against the antigen, so immunogenicity of the putative antigen is guaranteed once it has been identified.

Candidate Approach for Tumor Antigen Identification

The second general strategy for tumor antigen identification is the candidate approach, also called "reverse immunology." In this strategy, information about a molecule expressed by tumor cells is the starting point. This could be a protein that is overexpressed, a mutated oncogene, a chromosomal translocation product, a randomly mutated normal protein, etc. The prior characterization of thousands of epitopes binding specific HLA molecules has led to algorithms that can be employed to predict potential HLA binding, including SYFPEITHI[44] and NetMHC.[45,46] Overlapping peptides can then be synthesized that cover the regions of interest within the protein or that span a mutation or other molecular aberration and then tested for actual HLA binding. The challenge with this strategy is demonstrating immunogenicity and actual T cell recognition, particularly in the human context. If patient-derived T cells are utilized (eg, TIL or expanded oligoclonal T cells from the blood), then autologous T cell recognition can be evaluated. An alternative approach is to load the candidate neoepitopes onto autologous APCs and to prime autologous T cells using multiple stimulations in vitro, either from patient PBMCs or normal donor blood. In general, most candidate epitopes predicted to bind a given HLA molecule do indeed bind, as assessed by stabilized class I MHC expression on TAP-deficient T2 cells. However, only around 20% of these HLA binders can induce a T cell response.[47] Reasons for lack of immunogenicity include a low affinity for peptide-HLA binding, or absence of available TCRs contained within the normal T cell repertoire.

Categories of Tumor-Associated Antigens

Some identified tumor antigens are highly tumor-specific, not being detectably expressed in any normal host cells. Some tumor-specific antigens are also patient-specific, arising from unique mutational events that are not shared by tumors across individual patients. Other tumor antigens are not only enriched in tumor cells but also expressed by normal host cells, and yet the higher levels expressed by tumor cells can be sufficient to cross a threshold for T cell activation and recognition.

Tumor-Specific Antigens

The highest degree of tumor specificity is afforded by mutational events that generate neoepitopes that are exclusively expressed by the cancer cells. Nonsynonymous mutations in any expressed gene can theoretically give rise to a neoepitope that might bind the groove of one of the six expressed class I MHC molecules. Mutational antigens were first identified with mouse tumor cells generated in response to mutagens, either in vitro or in vivo.[48-50] With human cancer, some of the first mutational antigens identified were those found in mutated oncogenes as a candidate approach, such as K-Ras,[51,52] and also with the expression library approach, such as mutant caspase-8.[53] Using modern exome sequencing strategies and autologous T cells, such as those expanded in vivo after TIL-based therapy in melanoma, it has been possible to identify epitopes generated by nonsynonymous mutations in a large fraction of cancer patients.[54] Usually these are passenger mutations and not in driver oncogenes and are unique to individual patients.

Besides neoepitopes generated by nonsynonymous mutations, there are multiple additional molecular mishaps that can result in neoantigen generation. These include intronic sequences that become aberrantly retained to encode new proteins,[55] mutated intron sequences,[56] reverse strand transcripts,[57] translation of alternative open reading frames,[58] chromosomal translocation fusion proteins and frame shift events,[59] post-translational modifications such as phosphorylation,[60] asparagine deamidation,[61] and unusual cases of post-translational peptide splicing.[62] As such, exome sequencing and RNAseq strategies are only capable of detecting a subset of neoantigens present in a given cancer.

A second major category of antigen that is largely tumor-specific is the cancer-testis antigen family. The first of these identified was MAGE-1,[22] which is part of a large family of related genes all encoded on the X chromosome. These genes are generally silent in normal tissues except for male and female germ cells and trophoblasts of the placenta, and they become aberrantly expressed in cancers through epigenetic events.[63] The attractiveness of cancer-testis antigens is their wide expression by a broad range of cancer types and the fact that they are not mutated self-antigens and therefore can enable more of an "off-the-shelf" immunotherapy approach that can be applied to multiple individuals.

A third major category of tumor-specific antigens includes virus-encoded proteins, such as from HPV. Both cervical and vulvar cancers and also head and neck cancers can be caused by persistent HPV infection, and the HPV E6 and E7 oncoproteins have been shown to contain immunogenic determinants.[64] Other virus-derived antigens that have been defined include from the Merkel cell polyomavirus[65] and the EBV LMP1 and LMP2 proteins.[66,67] Since these viral proteins are foreign to the human host, they should be quite immunogenic in naïve individuals.

Tumor-Enriched Antigens

The second major class of tumor antigens shows increased expression in tumor cells over that seen in normal cells, such that immune discrimination is often still preserved. The first antigens of this type uncovered were the melanocyte differentiation antigens, such as Melan-A/MART-1.[68] Melan-A seems to be part of the melanin pigment biochemical machinery, and a candidate approach extending to other such proteins revealed immunogenic epitopes contained within the gp100/HMB45, tyrosinase, Trp1, and Trp2 proteins as well.[69] These antigens are also expressed in normal melanocytes, but immunologic self-tolerance against these proteins appears to be incomplete. Interestingly, melanoma patients with clinical responses to a variety of immunotherapies often develop vitiligo, which is immune-mediated destruction of normal melanocytes leading to depigmentation of the skin and sometimes hair.[70] This observation argues that cross-reactive T cells that are amplified and contribute to destruction of tumor cells also can migrate to the skin and destroy melanocytes.

The second type of antigens enriched in tumors is created by overexpression of normal genes. These include Her2, WT-1, survivin, PRAME, CEA, and mesothelin.[71-77] Candidate overexpressed antigens can be identified by RNAseq or similar techniques, yet it is challenging to predict whether an identified overexpressed protein will be immunogenic. This validation requires detailed definition of epitopes that are tested for immunogenicity with autologous or normal T cells, as well as confirmation of T cell recognition of tumor cells with minimal recognition of normal cells. Despite these challenges, some early clinical successes have been achieved. For example, a TCR-transduced T cell strategy against WT-1 was shown to prevent recurrence of AML in a pilot clinical trial, without apparent toxicity to normal cells.[78]

Considerations of Antigen Quality

Once antigens are identified, there are several qualitative features that need to be considered when transitioning that knowledge into a therapeutic intervention, such as a vaccine or engineered T cell product. Peptide antigens must be expressed by tumor cells and the peptide must be delivered into the groove of the appropriate MHC molecule. The affinity of candidate neoepitopes for binding to MHC molecules appears to be important, as a fast off-rate leads to poor T cell recognition and diminished immunotherapy efficacy.[79] The pool of peptides presented on the surface of a cell is generated via proteasomal processing, and the individual components of the proteasome can be different between cells, particularly cancer cells versus immune cells. As such, one set of peptides might be created from an antigen within DCs through the immunoproteasome, yet a distinct collection of peptides from the same antigen may be presented on the surface of a cancer cell via the conventional proteasome.[80] While a major focus of antigen discovery has been on class I MHC–binding epitopes presented to CD8+ T cells, class II–binding epitopes recognized by CD4+ T cells also can be identified and may be important for the most potent and durable adaptive immune responses.[81-83] Even when each of these details regarding the generation and expression of antigenic peptides is understood in detail, the circulating T cell repertoire in an individual patient may or may not include cells expressing TCRs with specificity for that antigen. Another important consideration is the proportion of cancer cells within a given tumor that expresses a neoantigen. An epitope identified within a driver oncogene is likely to be expressed by all the cancer cells, whereas passenger mutations might be more variable as they are not necessary for the malignant phenotype. Thus, therapeutic targeting of "truncal" mutations might be more effective than targeting of "branch" mutations, and immunotherapy efficacy has indeed been linked to the abundance of "truncal" antigens.[84] Selection for antigen loss variants (ALV) can occur through immunologic editing, particularly with "branch" mutational antigens.

GENERATION OF AN ENDOGENOUS IMMUNE RESPONSE AGAINST CANCER

Since a great number of cancers express antigens that can be recognized by T cells of the host immune system, the question arises as to why spontaneous immune-mediated destruction has generally failed to occur in most patients. In principle, the initial priming of T cell response against tumor-associated antigens would likely occur in a draining lymph node (Fig. 49.1). This would occur when tumor-derived products (including antigen) are taken up by APCs which traffic through afferent lymphatics to be available to present processed antigen to naïve T cells that traffic via the high endothelial venule to the same region of the lymph node. In order for productive T cell activation to occur, those APCs would have to be activated by innate immune pathway engagement during the process, much in the way that pathogens provide PAMPs to provoke productive immune responses rather than peripheral tolerance. Once stimulated by activated APCs, those tumor antigen–specific CD8+ T cells would expand, differentiate into effector cells, and then enter the blood circulation to disseminate throughout the host. At the effector phase, activated T cells would need to home to tumor tissues, traffic within the tumor microenvironment, and remain functional there for eventual killing of antigen-expressing tumor cell targets. It is worth considering experimental evidence supporting this model and the detailed components that have been implicated in the process of an endogenous immune response against cancer.

Cancers can Induce CD8+ T Cell Responses

In mouse models, the fact that allogeneic tumors, as well as syngeneic tumors induced by mutagens or rendered antigenic by treatment of mutagens in vitro, could be rejected in vivo argued that it is feasible for an effective endogenous immune response to be generated against cancer. Modestly immunogenic syngeneic tumors also could generate a detectable endogenous T cell response, yet they still grow progressively. By analyzing human tumor biopsy samples, it became clear that a significant proportion of solid tumors has evidence for a smoldering immune response, as characterized by a T cell–inflamed tumor microenvironment gene signature and/or by immunohistochemistry/immunofluorescence analysis.[26,37,85] The presence of activated CD8+ T cells within the tumor microenvironment has positive prognostic value in multiple cancer types, as well as predictive value for immunotherapy response.[40] More contemporary single-cell RNAseq experiments of human tumor cell suspensions have revealed a complex mixture of subpopulations of immune cells within these infiltrated tumors,[86] and multispectral immunofluorescence analysis has provided visual evidence of cell–cell interactions within solid tumors.[87]

As human tumor antigens became characterized molecularly, it became possible to utilize immune monitoring techniques to measure endogenous CD8+ T cell response against defined class I MHC–binding epitopes in patients. Indeed, detectable circulating T cells specific for defined antigens could be demonstrated using class I MHC/peptide multimers, intracellular flow cytometry cytokine staining, and ELISPOT.[88,89] In addition, antibody responses against defined antigens also could be detected.[90] In some instances, the magnitude of these response increased as tumors grew over time, arguing that increased antigen delivery to the host could drive further T cell expansion.

Analysis of T cells infiltrating human solid tumors has indicated that a proportion of them are indeed specific for tumor-derived antigens, although there clearly are bystander T cells present that are likely recruited nonspecifically via the chemokine gradients present in inflamed tumors.[91,92] The percentage of bystander cells varies widely between patient tumor samples, arguing that the mere presence of T cells within a tumor site does not always imply a tumor-specific immune response. By studying the human TIL from melanoma patients that are expanded ex vivo for therapeutic purposes, combined with exome sequencing of the tumor cells to identify candidate neoantigens, it has been possible to validate at least one T cell–recognized mutated neoantigen in approximately half of the patients who respond to TIL-based therapy in the clinic.[54] Taken together, these collective data argue that spontaneous immune responses against tumor-associated antigens in human cancer patients are not rare events.

Upstream Events Leading to T Cell Priming Against Tumor Antigens

If spontaneous CD8+ T cell responses indeed are induced in mouse models and in many cancer patients, then how does the initial activation of those T cells occur? As mentioned earlier, in principle this should involve a population of APCs, activated through one or more innate immune sensing pathway. Early clues came from interrogating gene expression profiling data comparing T cell–inflamed versus non–T cell–inflamed human melanoma samples, in search for clues suggestive of innate immune pathway activation. Indeed, T cell–inflamed tumors were found to show expression of genes indicative of type I IFN signaling. Since type I IFN (IFN-α/β) production is a key output from several innate immune pathway signaling events, mouse models were utilized to examine whether host type I IFN signaling might function upstream from T cell priming against tumor-associated antigens. In fact, gene-targeted mice lacking the type I IFN receptor (IFNAR) showed defective spontaneous T cell priming and tumor control with multiple transplantable tumors.[93,94] Profiling of immune cells in the tumor microenvironment of these mice revealed diminished accumulation of CD8α+ and CD103+ DCs, which represent the Batf3-lineage cDC1 cells known to be optimized for cross-presentation of antigen to CD8+ T cells. Moreover, conditional knockout mouse and mixed bone marrow chimera approaches revealed that type I IFN signaling on Batf3-lineage DCs was involved.[94] Batf3 knockout mice also showed defects in spontaneous priming of CD8+ T cells against tumor antigens in vivo.[94,95] Together, these data indicated that the presence of a growing tumor in vivo somehow elicits type I IFN production by CD11c+ cells of the host, which leads to optimal activation of tumor antigen–specific CD8+ T cells in vivo. Consistent with these mouse model data, analysis of human solid tumor RNAseq data demonstrated enrichment of a Batf3-DC gene expression profile in T cell–inflamed tumors, suggesting that this biology translates to the human cancer context.[47]

Building off of a demonstrated role for host type I IFN production and signaling upstream from the priming of T cells specific for tumor antigens, it became of interest to understand which innate immune sensing pathway was involved, to drive type I IFN production. Since most solid tumors are not induced by pathogens and arise from self-tissues, the spontaneous generation of an adaptive immune response was in fact somewhat of a conundrum, as TLR agonists and other pathogen-derived molecular patterns would not be expected to be present. Consistent with this notion, tumors implanted into specific TLR gene-targeted mice, as well as into mice deficient in the TLR adapter molecule TRIF, continued to induce host IFN-β production and develop an antitumor T cell response.[96] This led to the investigation of a potential role for the other main category of innate immune sensing pathways, those that detect cytoplasmic nucleic acids. Mice deficient in the cytosolic RNA sensing pathway (MAVS$^{-/-}$ mice) also retained normal T cell priming against tumors. However, mice deficient in cytosolic DNA sensing (STING$^{-/-}$ mice) showed defective host IFN-β production along with defective spontaneous priming of tumor antigen–specific CD8+ T cell responses.[96] Activation of the STING pathway in host APCs appears to occur either through intercellular transfer of tumor-derived DNA or via transfer of the second messenger cGAMP.[97] While these experiments were performed with transplantable tumor models, host STING also was shown to be important in a genetically engineered mouse (GEM) model of melanoma.[98] Although the host STING pathway appears to be the dominant innate immune sensing pathway for a spontaneous adaptive T cell response, this does not preclude the potential utility of pharmacologic agonists of other innate pathways as immunotherapeutics.

Where does the initial priming of tumor antigen–specific CD8+ T cells occur? With protein antigens administered subcutaneously or intramuscularly in mice, this process appears to occur in the vaccine-draining lymph node.[99] Using transplantable tumors implanted subcutaneously, most evidence suggests that the tumor-draining lymph node also is the dominant site of T cell priming. This notion is based on several lines of evidence. First, when TCR Tg T cells specific for a tumor-associated antigen are CFSE labeled and adoptively transferred into tumor-bearing mice in vivo, the first location where CFSE dilution is observed to occur is in the tumor-draining lymph node. Subsequent to that, those activated T cells can be found in the circulation (spleen, distant lymph nodes) and then in the tumor microenvironment.[100-102] Second, DC-mediated T cell priming depends on expression of the chemokine receptor CCR7, which is engaged by the lymph node–based chemokine CCL21.[103] This result argues that DC homing to lymph nodes is important. Third, T cell priming against tumors has been reported to be reduced in genetically altered mice deficient in lymph nodes, although some compensation for lymph nodes is supported by CD4+ T cells.[104] Together, these lines of evidence argue that Batf3-lineage DCs traffic to draining lymph nodes carrying antigen, where tumor antigen–specific CD8+ T cell clones can become engaged for activation, proliferation, and differentiation into effector cells.

In parallel for evidence implicating tumor-draining lymph nodes as a site for initial T cell priming, specialized lymphoid aggregates that resemble a lymph node architecture have been described within tumor sites. These have been termed tertiary lymphoid structures (TLSs) and are driven by lymphotoxin engaging the LTβ receptor. The presence of TLSs in human tumors has been reported to enrich for clinical responders to checkpoint blockade immunotherapy,[105,106] and it has been suggested that initial priming of antitumor T cells might occur in these sites. However, TLSs appear to be driven by chronic inflammation, and T cells themselves are a major source of lymphotoxin. As such, a balanced working model is that that initial priming of antitumor T cells likely occurs in draining lymph nodes, but as T cells accumulate in the tumor microenvironment they have the opportunity to drive TLS generation, which serves as a feedforward mechanism to expand and sustain antitumor immune responses within the tumor site. Nonetheless, engineering tumors to express LIGHT, which engages the LTβR and promotes lymphoid structures within the tumor, can be sufficient to promote potent antitumor immunity.[107]

Effector T Cell Trafficking into the Tumor Microenvironment

Primed effector T cells induced to differentiate and expand in draining lymph nodes would need to find their way into the tumor microenvironment of solid cancers in order to exert their effector function and lead to tumor cell destruction. Effector T cells exit the lymph node via efferent lymphatics, which ultimately converge on the thoracic duct to enter the venous circulation. Such T cells would be expected to circulate through the blood vasculature until they encounter homing signals for exiting the vessels and entering an inflamed tissue, in this case a solid tumor mass. This process is thought to involve the participation of adhesion molecules, including ICAM-1 (binding to LFA-1 on T cells) and VCAM-1 (binding to VLA-4), expressed by inflammation-activated endothelial cells, along with specific chemokines (chemoattracting cytokines) that guide cellular migration along concentration gradients.[108] These signals work in concert to facilitate diapedesis of T cells into the tumor microenvironment. Within the tumor site, if activated T cells re-encounter their cognate antigen, then a "stop" signal is established, which helps to retain specific T cells within the tumor milieu. Antigen nonspecific bystander cells also are found in the tumor microenvironment,[91] and in experimental models these are more likely to traffic through and exit the tumor tissue, having a shorter dwell time.[109]

As important as this process of T cell entry into tumor sites is for the ultimate step of tumor cell killing, it is less well characterized than other aspects of the antitumor immune response. Some experimental models have shown a functional importance of ICAM-1 and VCAM-1 for antitumor immunity, and analysis of human tissues has demonstrated expression of these molecules in T cell–inflamed tumors.[108] In a mouse model giving anti–CTLA-4 mAb, increased T cell entry into tumor sites was associated with ICAM-1 and VCAM-1 expression on the tumor vasculature.[110] However, increased expression of VCAM-1 by tumor cells has been suggested to be a mechanism of T cell trapping and potential immune evasion.[111] Thus, additional experimental evidence is required to clarify the adhesion/homing receptor requirements for T cell trafficking. Inflamed endothelium also can express E-selectin, and activated T cells can express E-selectin ligands,[112] also contributing to T cell entry.

The chemokines involved in effector CD8+ T cell entry into tumors are better characterized. T cell–inflamed tumors contain transcripts for CCL2, -3, -4, and -5, along with CXCL9 and CXCL10, confirmed by protein arrays.[37] While these chemokines sometimes can be produced by tumor cells, they are more abundantly produced by myeloid cells within the tumor microenvironment. CD8+ effector T cells express the chemokine receptors CXCR3 and CCR5, both of which have been shown to be functionally important in careful mouse models.[113-115] In a melanoma genetically GEM model, the critical chemokines CXCL9 and CXCL10 (interacting with CXCR3) were found to be dominantly expressed by Batf3-lineage DCs,[116] arguing for a critical importance of these cells both at the priming phase and the effector phase of the antitumor immune response. Mice lacking Batf3-DCs failed to establish a tumor microenvironment that recruited effector T cells. The chemokine CCL5 has also been shown to be expressed by macrophages and DCs, which interacts with CCR5 and also contributes to T cell homing.[115] In human melanoma samples using the RNAscope technique, expression of the chemokine CXCL10 was found predominantly in immune cells and was sufficient to predict anti–PD-1 efficacy.[87]

Given the functional important of Batf3-DCs in the tumor microenvironment, the question arises as to which factors mediate their recruitment and accumulation in tumor sites. While much less is known about DC migration into tumors than T cell migration, Batf3-DCs express the chemokine receptor XCR1 (which can interact with XCL1) and also CCR5 (interacting with CCL4). In the same melanoma GEM model mentioned earlier, CCL4 was itself made by tumor cells,[98] suggesting that this might be one of the most proximal steps in the establishment of a T cell–inflamed tumor microenvironment. XCL1 can be made by NK cells and other cells, and a role of NK-produced XCL1 has been implicated in Batf3-DC recruitment in tumors as well.[117,118]

CD4+ T Cells and Myeloid Cells in Antitumor Immunity

Because CD8+ effector T cells can be directly cytolytic for class I MHC–expressing tumor cells, much attention has been paid to their detailed biology. However, CD4+ T cells also participate in antitumor immune responses. A traditional view has been that CD4+ T cells provide "help" for CD8+ T cell expansion and differentiation, either through production of IL-2 as a T cell growth factor or through the arming of DCs via CD40L–CD40 interactions for superior CD8+ T cell activation.[119] However, CD4+ T cells also can directly be involved in tumor cell killing. While most solid tumor cell lines lack expression of class II MHC molecules in vitro, class II can be rapidly upregulated in vivo, and IHC of human tumors can reveal class II MHC staining. In some mouse models, the presence of CD4+ T cells was required for optimal immune-mediated tumor control, but not through providing help for CD8+ T cells.[100] Adoptive transfer of a tumor antigen–specific CD4+ T cell clone was sufficient to cause tumor regression in a patient with metastatic melanoma,[120] and some cancer patients responding to immunotherapies can show participation of CD4+ T cells and cDC2-type DCs, as opposed to CD8+ T cells and cDC1 (Batf3-lineage) DCs.[121,122] CD4+ T cells also are important for optimal memory CD8+ T cell responses in vivo,[123] so long-term protection may require CD4+ T cell participation.

Besides T cells, immune-infiltrated tumors also have been found to contain various numbers of NK cells, B cells, and even plasma cells.[26] Myeloid cell composition can include macrophages, DCs, granulocytes, and eosinophils. Within these lineages, distinct differentiation states have been observed, perhaps best characterized by scRNAseq analysis. At least five subsets of macrophages can be found in mouse and human tumors, extending beyond the M1 (inflammatory)–M2 (anti-inflammatory) differentiation spectrum.[124] DCs can be identified corresponding to the cDC1, cDC2, and plasmacytoid lineages.[125] More recently, a DC3 subset has been identified by studying tumor-infiltrating immune cells.[126] Granulocytes can be identified corresponding to classical neutrophils, but also granulocytic myeloid-derived suppressor cells (MDSCs), which are immune regulatory.[127] Different individual tumors contain different proportions of these various cell types and differentiation states, which likely come together to finely tune the effector phase of the antitumor immune response. In as much as those tumors by definition exist and were not spontaneously rejected, it is logical to presume that the net outcome of the dynamic interplay of these cell types is immune regulation.

Non–T cell–inflamed tumors lack most of the cell types and corresponding transcriptional profiles described earlier. Such tumors lack CD8+ T cells, Batf3-lineage DCs, a type I IFN gene signature, and also lack the key chemokines for T cell and DC recruitment. The proximal causes for this failure to generate a T cell–inflamed tumor microenvironment are likely multifactorial and not entirely clear, but this phenotype can be rate-limiting for immunotherapy efficacy as discussed further in subsequent sections. Non–T cell–inflamed tumors still contain macrophages (thought to be mostly M2-lineage), vascular endothelial cells, and fibroblasts, which have the potential to contribute to immune regulation.

Immune-Mediated Tumor Cell Killing

The final critical step in an optimal antitumor T cell response is tumor cell killing by T cells. Using in vitro culture models, two major mechanisms of T cell–mediated killing have been defined experimentally. These are the perforin/granzyme mechanism involving cytotoxic granule release, and the death receptor mechanism, usually through FasL on T cells engaging Fas on tumor cells.[128] Each of these mechanisms is thought to trigger the process of apoptosis in target cells as the final death pathway. However, when examined in vivo, the mechanism of final T cell–mediated tumor cell killing is less clear. Adoptively transferred T cells can still cause tumor regression if they are deficient for either the perforin or FasL genes.[100] Anti–PD-1/PD-L1 immunotherapy also can still be efficacious in perforin knockout mice. Using an in vivo melanoma GEM model, adoptively transferred tumor antigen–specific CD8+ effector T cells were observed to migrate into the tumor microenvironment and make continuous contact with tumor cells, yet tumor regression was not initiated until 5 days later,[116] a delayed kinetic which is not consistent with a rapid cytotoxic granule effect.

Interestingly, recent evidence has suggested that a major mechanism of T cell–mediated tumor cell killing might not be through apoptosis but rather via ferroptosis. Ferroptosis is an iron-dependent process of cell death that is driven by oxidized lipids, which is counterregulated by the glutathione pathway.[129] Effector T cells were found to induce lipid oxidation in target cells, and tumor cells genetically modified so as not to undergo ferroptosis were resistant to T cell–mediated rejection in vivo.[130] While multiple details of this pathway remain undefined, gaining a clear understanding of the major mechanisms of tumor cell killing in vivo may enable development of new treatment strategies to catalyze this process further and improve immunotherapy efficacy.

MECHANISMS FOR LACK OF SPONTANEOUS REJECTION OF CANCERS

Given the cascade of events that can drive tumor antigen–specific T cell priming and differentiation, effector cell trafficking into tumors, and direct tumor cell recognition and killing (Fig. 49.1), it is worth considering why cancers nonetheless grow and progress in immune competent hosts. In general, it is convenient to divide mechanisms of tumor immune escape based on the T cell–inflamed/non–T cell–inflamed tumor microenvironment paradigm.[131] Knowledge of these mechanisms has enabled development of immunotherapeutic interventions that impact on the endogenous immune response.

Tumor Progression despite Spontaneous Immune Cell Activation and Infiltration

T Cell Dysfunction

It was initially an unexpected observation that a major subset of solid tumors contains evidence for a smoldering T cell–based immune response. Analysis of tumor-infiltrating CD8+ T cells when they are indeed present has revealed that a major proportion of them are tumor antigen–specific. This was

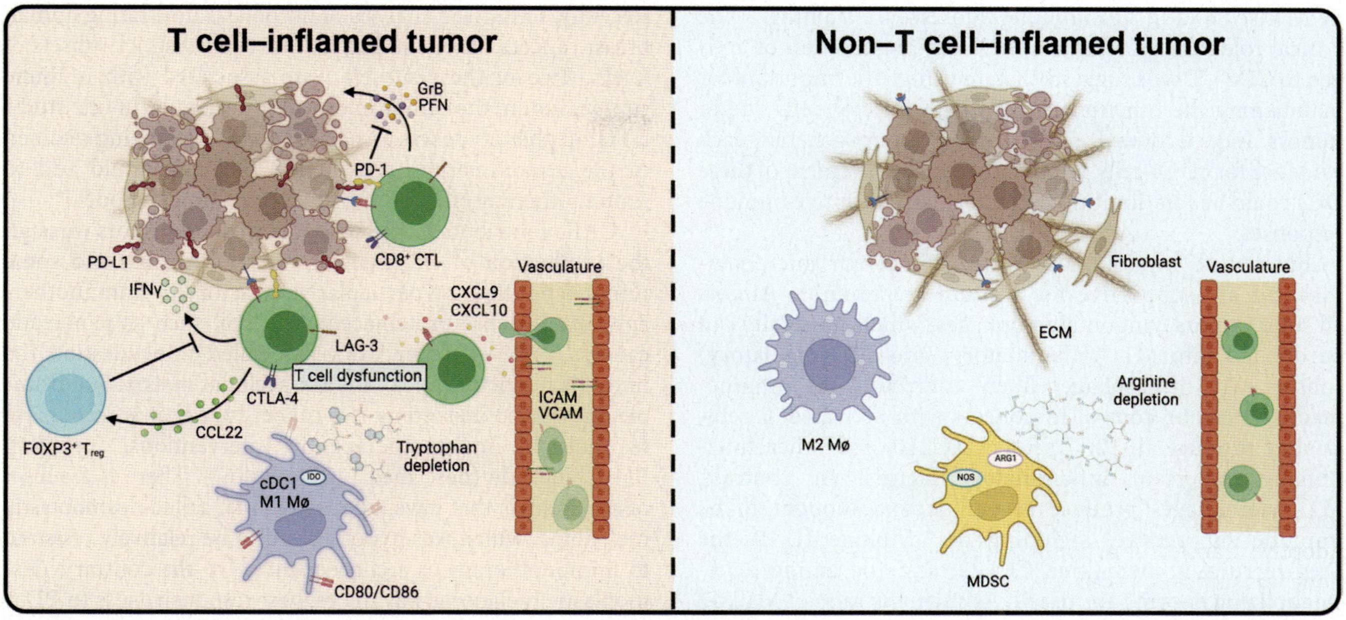

FIG. 49.2. The T cell–inflamed/non–T cell–inflamed tumor microenvironment paradigm.

clinically evident as well, based on the therapeutic efficacy of TIL-based therapy for melanoma, which begins with tumor-infiltrating T cells as starting material, as discussed in more detail below. However, bystander T cells are also present, likely other effector/memory T cells trafficking along chemokine gradients.[91,92] As such, the mere quantitation of T cells within the tumor microenvironment is not always sufficient to argue that tumor antigen specificity is implicated.

The presence of a T cell–inflamed tumor phenotype has been shown to have positive prognostic importance in multiple cancer types,[132,133] arguing that the mere attempt by the host immune response to attack and destroy the tumor is favorable for the patient (Fig. 49.2). But why are T cell–inflamed tumors not rejected completely? Early gene expression profiling studies indicated that T cell–inflamed tumors had the highest abundance of putative negative regulatory processes with the potential for inhibiting T cell function. These include PD-L1 (the ligand for the T cell inhibitory receptor PD-1), the tryptophan catabolizing enzyme indoleamine-2,3-dioxygenase (IDO), and FoxP3+ regulatory T cells (Tregs).[134] Indeed, functional analysis of tumor antigen–specific TIL generally reveled diminished cytokine production and sometimes cytolytic activity within the tumor microenvironment,[135] although it is clear that some functional properties are maintained.[136] IFN-γ produced by tumor antigen–specific TIL contributes to upregulation of PD-L1 and IDO, and chemokines such as CCL22 are produced leading to Treg recruitment.[134] A model has emerged in which chronic antigen stimulation, perhaps without proper costimulatory ligands or in the context of the metabolic limitations of the tumor microenvironment, ultimately leads to T cell dysfunction. Preclinical studies have indicated that PD-L1/PD-1 blockade, IDO inhibition, and Treg depletion each can tip the balance toward immune-mediated tumor control in mouse models in vivo.[137-140]

Perhaps the most convincing evidence for the functional relevance of Treg-mediated suppression has been through the use of Foxp3-DRT mice, a system which enables deletion of Tregs upon diphtheria toxin administration, which has shown tumor regression in multiple tumor models.[141]

Deeper analysis of tumor antigen–specific dysfunctional CD8+ T cells within the tumor microenvironment has revealed expression of multiple inhibitory receptors, including PD-1 but also CTLA-4, LAG-3, and Tim3, Tigit among others.[136,142-144] Unexpectedly, these T cells also express multiple positive costimulatory receptors, including 4-1BB, ICOS, Ox40, and CD27, among others.[136] Thus, not all of the receptors upregulated by dysfunctional TIL are inhibitory. It seems likely that these T cells are poised to receive positive costimulatory signals upon engagement by properly activated APCs, and/or when inhibitory receptors are blocked therapeutically. Additional evidence has indicated that tumor antigen–specific CD8+ TILs are driven to apoptosis, which would limit the numbers of effector T cells as the tumor continues to grow.[145,146]

Recent data have suggested that a subset of tumor antigen–specific CD8+ T cells undergoes an epigenetic shift to the point of irreversible silencing of key T cell effector molecules, a phenotype controlled by Tox.[147-149] Nonetheless, because checkpoint blockade and TIL adoptive transfer strategies can be therapeutically efficacious (as discussed further below), it seems likely that a major subset of tumor antigen–specific T cells is reversibly dysfunctional.

Suppressive Stromal Cells

Besides T cells, multiple additional immune cells and sessile stromal cells are present within the tumor microenvironment. Within the myeloid compartment, T cell–inflamed tumors typically contain Batf3-lineage DCs (cDC1 cells), as indicated by RNAseq and in situ immune fluorescence staining.[47] The critical role for cDC1 cells in the cross-presentation of antigen to CD8+ T cells suggests that they might be important for maintaining the functionality of CD8+ effector cells within tumors. Indeed, non–T cell–inflamed tumors generally lack evidence for cDC1 cells, arguing that the involvement of these DCs could be rate-limiting for endogenous adaptive immune responses.

But not all myeloid cells within the tumor microenvironment are supportive for antitumor immunity. Almost all solid tumors contain macrophages, which classically can be divided into M1 (inflammatory) and M2 (regulatory) subsets. M1 macrophages likely contribute to immune-mediated tumor control in concert with activated T cells, as they produce IL-12, TNF-α, CXCL10, and other functionally important inflammatory factors. In contrast, M2 macrophages produce several factors thought to be immune suppressive, including the cytokine IL-10, the Treg-recruiting chemokine CCL22, and the amino acid–catabolizing enzyme arginase I. As such, the ratio of M1/M2 macrophages has been found to be prognostic in a number of solid tumor histologies.[86,150-152]

More recent scRNAseq studies of immune cells from the tumor microenvironment of both mouse and human systems have called into question the binary division of macrophage phenotypes. At least five subsets of monocyte/macrophage-lineage cells can be observed, which shift in phenotype upon successful immunotherapy.[124] It is not clear whether these are distinct macrophage differentiation states versus activation states, perhaps reflecting cells receiving distinct signals from other nearby immune cells. For example, IFN-γ produced by infiltrating T cells induces expression of iNOS by neighboring macrophages.[153] Based on expression of specific gene sets, these expanded macrophage subtypes also can be inferred to be immune stimulatory or immune suppressive, so the balance is likely to be a determining factor for immune-based tumor control.

Besides monocyte and macrophages, granulocytes also can be found in tumors and can be immune regulatory. Classical neutrophils are usually a component of a productive immune response, and these cells can produce factors that can contribute to tumor control. For example, human neutrophils have been reported to produce neutrophil elastase, which has direct tumoricidal activity.[154] In contrast, granulocytes also can take on an MDSC phenotype that has clear immune suppressive properties.[127] MDSCs can produce arginase I and immune suppressive cytokines and can contribute to the recruitment of FoxP3+ Tregs. A ratio of neutrophils to granulocytic MDSCs, and also neutrophils to lymphocytes, also may be prognostic for cancer outcomes.[155-157]

Besides hematopoietic cell lineages that can be immune regulatory, fibroblasts that compose the tumor stroma also can impact on immune responses. Like M2 macrophages, fibroblasts support the three-dimensional growth of solid tumors. Immune regulation by a subset of cancer-associated fibroblasts (CAFs) has been linked to expression of fibroblast-activating protein[158,159] and also to TGF-β signaling.[160] Recently, scRNAseq analysis of fibroblasts infiltrating human breast cancers has identified at least eight distinct subsets of CAFs. Two of these subsets were associated with immune suppression in the tumor microenvironment.[161] In one study, a TGF-β phenotype was associated with T cells being retained in the tumor margin, and TGF-β blockade could add to immunotherapy efficacy in a mouse preclinical model.[160]

CAFs contribute to the structure of solid tumors through the production of extracellular matrix proteins, and some tumor types that are particularly refractory to immunotherapy have a dense proteinaceous stroma, such as pancreatic cancer. However, other work has called into question the immune restrictive role of the extracellular matrix. For example, metastatic melanoma is a cancer type that is responsive to multiple immunotherapeutic interventions, including PD-1 blockade (described further below). There is a subset of melanomas that have a dense stroma, called desmoplastic melanoma, which was hypothesized to be relatively resistant to immunotherapy as a consequence. On the contrary, desmoplastic melanoma has the highest responsiveness to PD-1 blockade among the melanoma subtypes, with response rates around 70%.[162] These data argue against a general immune suppressive role for dense stroma.

Metabolic Stresses

The tumor microenvironment is metabolically challenging for infiltrating T cells. Tumor cells often utilize glucose uptake and the glycolytic pathway as an energy source for ATP generation. Effector T cells also upregulate the glycolytic machinery, and glucose uptake is required for T cell effector functions including IFN-γ production and cytolysis.[163,164] Thus, there is competition between the tumor cells and T cells for glucose, and the tumor cells might be expected to win because of sheer numbers of cells. Oxidative phosphorylation is also important for sustaining T cell function, and activated T cells contain more mitochondria than resting naïve T cells. In addition, certain costimulatory receptors can increase mitochondrial number during T cell activation.[165] Yet the tumor microenvironment is generally hypoxic compared to normal tissues, creating another stress. Tumor cells also express amino acid transporters and likely compete for other nutrients with activated T cells. In part because of hypoxic conditions and a reliance on the glycolytic pathway, proliferating tumor cells generate an abundance of lactic acid, which creates an acidic microenvironment as well. Ongoing research is identifying proton-sensing receptors on T cells that may detect an acidic microenvironment and impair T cell function. Elevated serum LDH is a strong negative prognostic factor in metastatic melanoma[166] and likely is reflective of a distinct metabolic phenotype. Recent work has indicated that checkpoint blockade therapy is less effective in the setting of tumors that take up greater glucose.[167,168] Added to these metabolic issues is the activity of IDO or arginase I expressed by infiltrating myeloid cells, which deplete tryptophan and arginine, respectively. T cells are particularly sensitive to deprivation of these amino acids, processes which have been associated with T cell dysfunction.[169,170]

Tumor Progression in the Absence of Immune Cell Infiltration

The largest fraction of solid tumor patients has tumors that lack productive immune cell infiltration, sometimes referred to as a non–T cell–inflamed tumor microenvironment.[47,131] This observation suggests a second general mechanism of tumor immune escape, which is immune exclusion. Non–T cell–inflamed tumors lack a type I IFN gene signature, Batf3-lineage cDC1 cells, chemokines for effector T cell recruitment, tumor antigen–specific T cells, and T cell–derived cytokines. Importantly, non–T cell–inflamed tumors also show low expression of key immune regulatory mechanisms, including PD-L1, IDO, and Tregs.[134] This latter result argues that effector T cells themselves are major drivers of immune regulation, as part of a programmed negative feedback loop to control chronic T cell–based inflammation. As expected, non–T cell–inflamed tumors are less likely to respond to current immunotherapy strategies, including checkpoint blockade, vaccination, and adoptive T cell therapy.[116] This may be because of an inability to recruit effector T cells into tumor sites or poor maintenance of T cell function once T cells are recruited due to lack of appropriate DCs.

It is not clear whether the non–T cell–inflamed tumor microenvironment is established as an active process (presence of a dominant negative) or rather is a consequence of deficiencies in early rate-limiting steps important for immune activation, such as STING pathway activation and cDC1 recruitment (absence of a positive). Preclinical data as well as some translational biomarker evidence from cancer patients, thus far, are indicative of the latter, as discussed further below. But it is conceivable that active suppression or repulsion of innate or adaptive immune cells could be operational under some conditions, and it would be important to identify such processes as they could be modulated with therapeutic intent.

Antigenicity Versus Adjuvanticity of Cancers

In order for tumors to generate a spontaneous antitumor T cell response, the first minimal requirement would be expression of neoantigens that can be recognized as targets for T cell–mediated killing. The breadth of antigens includes not only point mutational antigens but also differentiation antigens and cancer-testis antigens, among others. The question arises as to whether non–T cell–inflamed tumors might lack antigens for T cell recognition, thus explaining a paucity of infiltrating T cells. However, the bulk of translational evidence suggests that this is not the case. There is a large variation in the number of nonsynonymous mutations across human cancers, ranging from major carcinogen-induced cancers such as melanoma (driven by UV light) and non–small cell lung cancer (driven by tobacco-induced carcinogenesis) on the high end, down to primary brain tumors and pediatric cancers on the low end. However, within each cancer type, there does not appear to be a correlation between the number of nonsynonymous mutations and the T cell–inflamed/non–T cell–inflamed tumor microenvironment.[47,171] In melanoma, the number of candidate HLA-A2–binding neoepitopes, and also the fraction of peptides that can induce specific CD8+ T cells, also is similar between T cell–inflamed and non–T cell–inflamed tumors.[47] Taken together, these results suggest that tumor *antigenicity* is not the major rate-limiting step for spontaneous antitumor T cell responses, at least among the heavily mutated tumor types. By inference then, these tumors likely have reduced *adjuvanticity* or the ability of the tumor to support the cascade of innate immune activation, T cell priming, and T cell recruitment into the tumor microenvironment.

Based on investigation into molecular mechanisms of immunotherapy resistance (discussed further below), it seems likely that some patients with a non–T cell–inflamed phenotype never primed a de novo T cell response against their tumor, despite an abundance of antigens. However, in other cases, initial T cell priming and trafficking may have occurred, but through a process of immune selection or editing, a tumor cell phenotype emerges that no longer supports T cell recruitment or is resistant to T cell–mediated killing. This latter phenotype fits with observations made in the context of tumor antigen–based vaccines, in which circulating T cells specific for the antigen can be identified, without T cell accumulation within tumor sites. Understanding these mechanistic details in individual patients may be necessary to expand immunotherapy efficacy further.

CANCER IMMUNOTHERAPIES

Cytokines

The first immunotherapy strategies to emerge out of the recombinant DNA biotechnology era were immunomodulatory cytokines (Fig. 49.3). Interferon alpha (IFN-α) was developed as an antiviral drug and was hyped at the time it was featured on the cover of *Time* magazine in 1980. IFN-α can exert direct antiproliferative effects on cancer cells, promote MHC molecule upregulation, and induce production of chemokines for T cell recruitment. Through host modulation, IFN-α can have antiangiogenic effects and promote DC activation/maturation. Ultimately, it was indeed found to have activity in a variety of cancers and gained entry into clinical practice in the treatment of chronic myelogenous leukemia, hairy cell leukemia, Kaposi sarcoma, renal cell carcinoma, and melanoma. A potential immunologic mechanism was suggested by the observation that IFN-α treatment of Kaposi sarcoma in HIV patients was only effective when CD4+ T cell counts were greater than 200/mL.[172] IFN-α became standard of care in the postsurgical adjuvant therapy of patients with stage 3 melanoma in the late 1990s.[173] However, with the development of newer, more effective, and less toxic treatments, the use of IFN-α has fallen out of favor.

Interleukin-2 (IL-2) is a T cell growth factor that has the capability of expanding both T cells and NK cells in vivo. High-dose IL-2 given by bolus intravenous infusion showed clinical activity (response rates around 15%) in both melanoma and renal cell carcinoma and became FDA-approved for these indications.[174] The treatment must be administered as an inpatient and has significant systemic side effects that often require ICU management. For these reasons, the use of IL-2 has diminished with the development and approval of more effective and less toxic immunotherapies. Nonetheless,

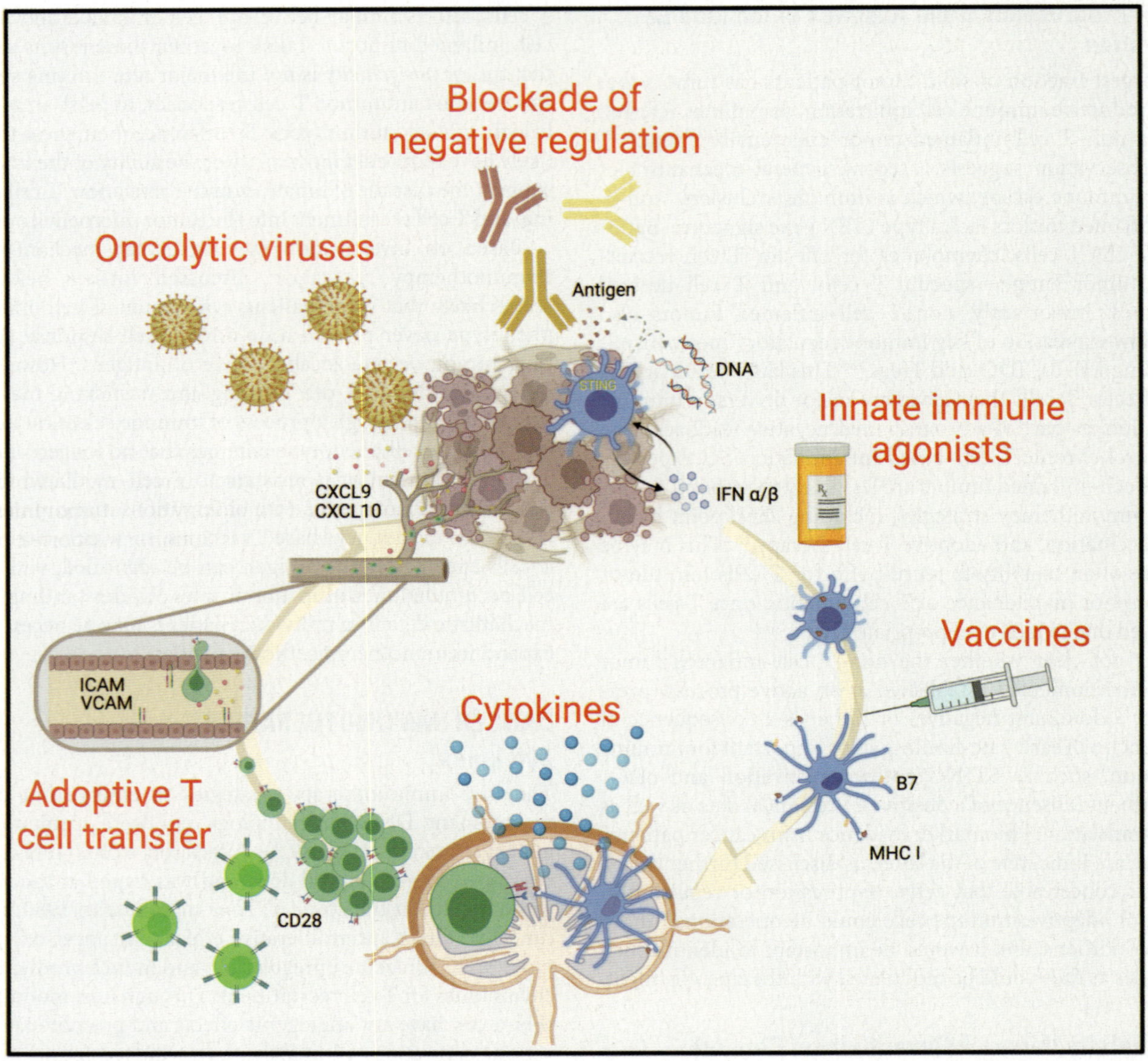

FIG. 49.3. Pictorial representation of immunotherapy strategies.

IL-2 remains an important component of adoptive T cell–type immunotherapy strategies, given to expand and maintain adequate numbers of infused T cell products (discussed further below).

Based on the successful application of IL-2, other cytokines that act to induce proliferation or survival of T cell populations have been developed and tested clinically. These include IL-15 and IL-7, which are two homeostatic cytokines that maintain the survival of naïve and memory T cells in vivo. Endogenous IL-15 and IL-7 become available upon the administration of lymphodepleting chemotherapy, used as a preparative regimen prior to the infusion of engineered T cell products. Intravenously administered recombinant IL-15 and IL-7 has been explored clinically, with the expected biologic effects on T cell and NK cell subsets.[175,176] IL-15 does exhibit some IL-2–like toxicities, but these can be manageable. Because IL-15 is normally "presented" to immune cells in conjunction with the IL-15Rα chain, fusion proteins between IL-15 and IL-15Rα also have been developed and investigated in patients in vivo.[177] Thus far, none of these agents has yet been explored for meaningful clinical activity in cancer settings. These cytokines could have utility as part of T cell adoptive transfer settings, or to support faster T cell recovery following bone marrow transplantation.

A wide range of additional cytokines have been tested, usually by systemic administration, in phase I and phase II clinical trials, without meaningful clinical activity observed and/or showing significant toxicity. These include IL-4, IL-6, IL-12, TNF-α, IFN-γ, M-CSF, GM-CSF, and Flt3L, and none has gained FDA approval for a cancer therapy indication. However, some cytokines may prove useful as adjuncts to other interventions. IL-12 may have properties as a vaccine adjuvant when given subcutaneously, to help promote Th1-type T cell differentiation.[178,179] Flt3L may have the ability to recruit and expand Batf3-lineage DCs, and intratumoral administration has important effects preclinically.[180,181] The GM-CSF cDNA has been included as a component of the oncolytic virus T-Vec, approved for melanoma,[182] with the idea that it may locally recruit DCs (discussed further below).

Therapeutic Vaccines

The molecular identification of tumor antigens catalyzed the development and evaluation of innumerable vaccination approaches with the goal of promoting new T cell responses against a patient's tumor, thereby supporting immune-mediated tumor cell killing (Fig. 49.3). The underlying presumption is that most patients do not spontaneously develop an endogenous T cell response against their tumor, so that inducing one de novo will lead to effector CD8+ T cells that home to tumor sites, migrate into the tumor microenvironment, and carry out the process of tumor cell killing. Note that this process is conceptually distinct from prophylactic vaccines against pathogens, in which the biologic goal is to induce antibody responses that neutralize viruses or bacteria prior to infection. In fact, some virus-specific prophylactic vaccines do indeed decrease the incidence of certain virus-induced cancers. These include the HBV vaccine and prevention of hepatocellular carcinoma[183] and the HPV vaccine and prevention of cervical cancer.[184]

The general principles involved in therapeutic cancer vaccines include the type of antigen selected, the backbone vector/antigen source, the type of adjuvant, and the route of administration. Most categories of antigens have been investigated both preclinically and clinically, including differentiation antigens, cancer-testis antigens, intronic sequences, fusion proteins, viral antigens, and mutated antigens. These have been delivered as recombinant proteins, synthetic peptides, as part of viral vectors (eg, adenovirus) or engineered bacteria (eg, *Listeria monocytogenes*); as DNA plasmids or RNA constructs; as antigens of viruses loaded onto autologous APCs (bulk PBMCs or expanded DC subsets); and also as modified tumor cells containing a range of antigens. The adjuvant components vary with the type of delivery system: with viruses, plasmid DNA, RNA constructs, bacteria, and modified APCs, the natural adjuvanticity of the backbone is relied upon. With peptides or proteins, emulsification in an oil or lipid component is often utilized (eg, Montanide), and admixture with defined TLR agonists has been employed (eg, the TLR4 agonist MPL or the TLR3 agonist poly I:C). Finally, in terms of route of injection, vaccines have been given subcutaneously, intramuscularly, intravenously, and intralymphatically. Clinically, vaccines have been explored in patients with established metastatic disease, or in a postsurgical adjuvant setting, sometimes with rising levels of a cancer biomarker (eg, PSA in the setting of prostate cancer).

Taken as a whole, therapeutic cancer vaccines have generated clinical responses in only a minor subset of patients. An overview of melanoma vaccines published by the NCI Surgery Branch documented a 3% overall response rate, after investigation of a range of vaccine modalities and antigen content,[185] and the best clinical responses have been in the 10% range.[179] In most studies, immunologic monitoring evaluations indeed revealed increased frequencies of T cells specific for the vaccine antigen, suggesting that vaccine itself was having the desired biologic effect. T cell responses are measured clinically using peptide-HLA multimers and flow cytometry, antigen-specific ELISPOT assays, or intracellular/secreted cytokine production upon antigen stimulation. Standardized assay approaches have been developed and in some cases have been commercialized for standard clinical use in other disease settings (eg, the QuantiFERON Gold test for tuberculosis).

It's worth considering mechanisms that may explain why therapeutic cancer vaccines have not been more effective. First, the underlying principle motivating vaccine development might not be entirely valid. It is now known that many patients do indeed generate endogenous T cell responses against their tumor, so T cell priming might not be the rate-limiting barrier in antitumor immunity. Downstream resistance mechanisms and the biology of the tumor microenvironment may dominate.[186] This notion is supported by the observation that clinical benefit with some vaccines has been associated with a pre-existing T cell–inflamed tumor microenvironment, arguing that the inability for a tumor to support effector T cell entry may be a major therapeutic barrier.[39] Second, it is possible that by the time a cancer develops, a new equilibrium dynamic may exist between the tumor cells and the host immune response that is at the threshold of allowing tumor cells to grow. T cells against high-affinity antigens may have become deleted or rendered permanently dysfunctional. Similarly, tumor cells expressing high-affinity antigens may have been eliminated, leaving behind malignant cells with less fit antigens for T cell–based recognition and killing. Third, the magnitude of the T cell response induced by vaccination may not be optimal. Most vaccines have induced only modest increases in the frequencies of vaccine-induced T cells, so the numerical requirements might not have been achieved to overwhelm the number of tumor cells in the host. Fourth, it has been argued that patients with metastatic cancer might be globally immune suppressed, due to bulky tumor, cachexia, poor nutrition, etc., such that immunotherapy might not have the opportunity to be effective. However, this argument is not supported overall by the impressive clinical responses that are often observed with checkpoint blockade therapy, T cell adoptive transfer, and even high-dose IL-2. Fifth, most vaccines have utilized "off-the-shelf" shared antigens that can be applied to many patients, and it has been suggested that individualized mutational neoantigens might induce more effective T cells for tumor shrinkage. The attractiveness of this notion notwithstanding clinical benefit with neoepitope vaccines has, thus far, also been limited.[187] Finally, successfully immune-mediated tumor regression might require a combination of T cells seeing multiple antigens, in part to guard against selection for ALV. It is noteworthy that in-depth studies of exceptional responders have revealed that vaccination against a defined epitope might increase the frequency of antigen-specific T cells circulating in the blood, and yet the dominant expanded T cell population within the microenvironment of a regressing tumor may recognize a distinct antigen not included in the vaccine.[188]

Despite these limitations, one therapeutic cancer vaccine has shown efficacy in phase 3 clinical studies and has been FDA-approved, the Provenge (Sipuleucel-T) vaccine for prostate cancer. The antigen in the vaccine is prostatic acid phosphatase, fused to GM-CSF to target DCs and also contribute to DC activation. This antigen is admixed with PBMCs after autologous leukapheresis and then re-infused intravenously to the patient. In patients with hormone-refractory prostate cancer, a statistically significant though modest improvement in overall survival was observed.[189] However, the mechanism of success versus failure was not elucidated, as patients could show clinical benefit without obvious tumor shrinkage, decline in serum PSA, or induction of CD8+ T cell responses.

Much current work in cancer vaccines is focused on mutational neoantigens, novel delivery vectors (such as RNA constructs), and combination therapies with immune checkpoint blockade.

Blockade of Negative Regulatory Checkpoints

The realization that many patients already have a spontaneous antitumor T cell response that might be held in check by negative regulatory influences spurred investigation into functionally relevant immune inhibitory pathways that might be drugged therapeutically (Fig. 49.3). Checkpoint blockade immunotherapy has become a major therapeutic success, leading to the Nobel Prize in Medicine being awarded to James Allison and Tasuku Honjo in 2018.

CTLA-4

CTLA-4 is an inhibitory receptor upregulated on T cells following initial T cell activation. It is an Ig domain–containing protein with significant homology to the costimulatory receptor CD28 and binds to the same ligands as CD28, namely CD80 and CD86. The proof that CTLA-4 was a negative regulatory receptor came from studies using Fab fragments of anti–CTLA-4 mAbs, which were shown to augment T cell function,[190,191] as well as using CTLA-4 knockout mice, which displayed spontaneous T cell activation and lethal autoimmunity.[192,193] Mechanistically, CTLA-4 competes with CD28 for ligand binding, thus decreasing positive costimulation, based on its markedly higher affinity for CD80/86. But CTLA-4 also directly delivers a negative signal, likely by recruiting protein tyrosine phosphatases.[194-200] In fact, CTLA-4 was found still to inhibit the function of CD28 knockout T cells, arguing that part of the mechanism is direct.[201]

The possibility that CTLA-4 was restraining antitumor immunity came from in vivo mouse studies using transplantable tumor models. Many murine transplantable tumor models are heavily mutated and express antigens, yet grow progressively when implanted in vivo. In this context, a blocking anti–CTLA-4 mAb was found to improve immune-mediated tumor control in vivo, in a T cell–dependent fashion.[202] The striking tumor rejection seen in some models motivated the development of anti-human CTLA-4 mAbs with the hope of similar activity in patients. The first such mAb was ipilimumab, a fully human IgG1 mAb. Several responders were seen in early-phase clinical trials,[203] which led to a randomized study in patients with metastatic melanoma, against an experimental gp100 vaccine as the control arm. A combination arm also was evaluated, with the notion that priming a new T cell response might be needed for optimal therapeutic effects. In fact, both anti–CTLA-4 groups showed similar favorable clinical outcomes compared to the gp100 vaccine control group, with a median overall survival of 10 versus 6.4 months.[204] A second study compared anti–CTLA-4 mAb to the chemotherapy dacarbazine, which also demonstrated improved overall survival.[205] Also of interest was the observation that approximately 20% of treated patients showed durable clinical benefit, even with minimal tumor shrinkage. Based on these data, ipilimumab became the first immune checkpoint therapy approved by the FDA, for metastatic melanoma in 2011.

PD-1/PD-L1

PD-1 is another inhibitory receptor expressed by activated T cells, including tumor antigen–specific CD8+ T cells within the tumor microenvironment. Its ligands are PD-L1 and PD-L2, which are inducibly expressed on cells. While PD-L2 is largely expressed by APCs, PD-L1 can be expressed by a broad array of epithelial cells, including tumor cells. The major inducer of PD-L1 is the cytokine IFN-γ.[134,206] Based on the notion that effector T cells migrate into tumors and produce IFN-γ locally upon antigen re-encounter, a model emerged in which reactivation of T cells within the tumor microenvironment would result in upregulation of PD-L1, which would then dampen the immune response through engagement of PD-1.[134,207] This involvement directly in the site of action at the level of the tumor microenvironment made it extremely attractive to consider pharmacologic blocking of PD-1/PD-L1 interactions as a potential therapeutic. Preclinically, antibodies showed therapeutic effects in transplantable tumor models, either targeting the PD-1 receptor or the PD-L1 ligand.[138,208-210]

The first anti–PD-1 mAb to enter the clinic was nivolumab, which is a fully human IgG4 Ab that blocks PD-1/PD-L1 interactions. The agent already showed activity in phase I clinical studies, with a 28% response rate observed in refractory melanoma patients.[211,212] Similarly exciting phase I clinical trial data were seen with a second anti–PD-1 mAb, pembrolizumab, which became FDA approved in 2014.[213] Several anti–PD-L1 mAbs also showed activity in various cancers, called atezolizumab, durvalumab, and avelumab. It is not clear whether blocking the ligand or the receptor might show superior therapeutic efficacy, and it is unlikely the approaches will be compared head-to-head.

Because the biology of the T cell–inflamed tumor microenvironment is relatively constant across many types of solid tumors, with effector T cells entering and producing IFN-γ which upregulates PD-L1 expression to dampen persistent T cell activation via PD-1, it is perhaps not surprising that Abs targeting the pathway have broad therapeutic activity in multiple cancer types. These include melanoma, Merkel cell carcinoma, squamous cell carcinoma of the skin, basal cell carcinoma, head and neck cancer, non–small cell lung cancer, mesothelioma, triple negative breast cancer, gastric cancer, hepatocellular carcinoma, kidney cancer, and bladder cancer.[214] There also is FDA approval for any cancer showing the highlight-mutated, microsatellite instability phenotype (so-called MSI-hi tumors), regardless of tissue histology.[215,216] While activity in this context is believed to be secondary to a greater number of mutational antigens, it is also conceivable that other biologic properties are associated with tumors having mismatch repair deficiency. Activity has also been seen in Hodgkins disease and in non-Hodgkins lymphoma.[217,218] However, the biological mechanism in Hodgkins disease seems somewhat distinct. The malignant cell in Hodgkins disease is the Reed–Sternberg (RS) cell, which does not express class I MHC molecules. In addition, PD-L1 expression in RS cells is driven not be exogenous IFN-γ exposure but rather through a chromosomal translocation event.[219] Clinical activity with anti–PD-1 mAbs in this disease is best correlated with intratumoral CD4+ T cells.[220] Thus, these data highlight the critical importance of PD-L1/PD-1 interaction in cancer immune evasion, and that there are multiple ways the pathway can be recruited to allow tumor immune escape.

Multiple biomarkers have been evaluated as potential predictors of clinical response to PD-1 blockade therapies. As expected, based on underlying mechanism, patients with a T cell–inflamed tumor microenvironment are enriched for response to single-agent anti–PD-1 Ab treatment. This can be reflected at the level of gene expression profiling, either with RNAseq or the NanoString platform, the latter of which is more forgiving with partially degraded RNA as occurring with paraffin-embedded tissues.[221] While enumeration of CD8+ T cells in the tumor, or assessing PD-L1 expression by IHC, also can enrich for patients likely to respond, these single-parameter assays do not have the dynamic range of a combinatorial transcriptional profile. A second molecular biomarker that enriches for responder patients is mutation load. Tumors with a high mutation count, as typically measured through exome sequencing, are more likely to respond to anti–PD-1 therapy.[35] This is thought to be because heavily mutated tumors are more likely to contain mutational neoantigens. However, there are responder patients with low tumor mutation burdens and nonresponder patients with high mutation loads. Thus, it is hard to predict the presence of functionally useful neoantigens by mutation load itself. The combination of an immune gene signature and mutation load may have slightly better discriminatory value than either assay alone.[222] Nonetheless, there is no baseline predictive biomarker that is sufficiently robust to deny a given patient treatment with anti–PD-1/PD-L1 therapy.

Combination Therapies

Inasmuch as clinical responses to single-agent anti–PD-1 are seen in the minority of patients, the possibility arises that compensation by other negative regulatory pathways could continue to drive T cell inhibition. As such, combinatorial blockade of two inhibitory pathways together could have theoretical advantages. In preclinical models, combining anti–CTLA-4+anti–PD-1/PD-L1 mAb showed superior efficacy than either Ab alone.[138,208] The combination of the anti–CTLA-4 mAb ipilimumab and the anti–PD-1 mAb nivolumab has been tested clinically in patients with metastatic melanoma. Combination therapy generated a higher response rate and progression-free survival rate, along with a modest improvement in overall survival.[223] However, the grade 3/4 toxicity rate also was significantly higher. Side effects with checkpoint blockade therapy reflect inflammation in normal tissues that include skin rash, colitis, hypothyroidism, and others, and the severe adverse events require immune suppressive drug therapy, largely managed with corticosteroids. The combination of ipilimumab and another anti–PD-1 mAb pembrolizumab also has been explored in the second-line setting following anti–PD-1 failure. This study demonstrated approximately a 30% response rate with less toxicity overall,[224] thus establishing the possibility of using combination therapy in patients who fail anti–PD-1 alone.

Other Immune Checkpoints

Based on the success of mAbs targeting CTLA-4 and PD-1, additional inhibitory receptors expressed on activated T cells have been pursued as candidate therapeutic targets. LAG-3 is inducibly expressed on activated T cells and also can be expressed by NK cells. It has sequence homology to CD4, and a major ligand for LAG-3 is in fact class II MHC molecules.[225] LAG-3 is highly expressed by dysfunctional T cells in the tumor microenvironment, along with PD-1 and CTLA-4.[136] LAG-3 knockout mice showed augmented T cell activation, and an anti–LAG-3 mAb augmented antitumor T cell responses and was therapeutic, alone and in combination with anti–PD-1.[144] These preclinical data helped to motivate the development of an anti–LAG-3 mAb for clinical testing. Several clinical responders were observed in a phase I clinical trial with the anti–LAG-3 mAb relatlimab in combination with the anti–PD-1 mAb nivolumab, in patients who failed prior anti–PD-1 therapy alone. A randomized phase 3 trial was then developed in frontline treatment of metastatic melanoma patients, comparing the combo therapy to nivolumab alone. In fact, combination therapy with relatlimab more than doubled progression-free survival, with more manageable side effects than with the anti–PD-1/anti–CTLA-4 combination.[226] Relatlimab is being evaluated for possible approval by the FDA in 2022.

Additional putative negative regulatory receptors expressed on activated T cells have also been explored as potentially druggable immunotherapy targets. These include Tim3 and TIGIT, both of which have demonstrated preclinical supporting evidence.[227] Clinical trials have revealed evidence for clinical activity in some patients, but no studies have advanced toward FDA approval at this point in time.

Agonistic Abs Against Costimulatory Receptors

In addition to inhibitory receptors, dysfunctional T cells in the tumor microenvironment also express costimulatory receptors which, if properly ligated, could potentially augment T cell function and improve immune-mediated tumor control. These include 4-1BB, Ox40, and CD27, which are all TNFR superfamily members.[136] In general, these costimulatory receptors also become upregulated upon T cell activation (CD27 is also expressed by naïve T cells), and the corresponding ligands can be expressed in inflamed tissues to support continued T cell functionality. In preclinical models, agonistic Abs against costimulatory receptors can have profound therapeutic activity in cancer models, and anti–4-1BB mAb is perhaps the most active immunotherapy against mouse tumors. mAbs against each of the corresponding human receptors have entered clinical trial testing. However, limited clinical activity has been observed.[136,228,229] One limiting factor is that the anti–4-1BB Ab first tested revealed liver toxicity that precluded continued dose escalation. However, there are additional pharmacologic considerations that need to be revisited when developing agonistic Abs versus blocking Abs. The optimal dose and schedule of such agents is not clear. For example, engagement of these receptors usually leads to their internalization and sometimes degradation, and receptor re-expression may take significant time. Receptor downmodulation would remove the possibility of positive engagement with endogenous ligands, which could theoretically be detrimental. Thus, lower doses of agonistic Abs might be preferable, with a schedule of administration that gives time for receptor re-expression. Another direction of drug development is creating strategies to preferentially target agonistic Abs such as anti–4-1BB to the tumor microenvironment, to favor engagement with dysfunctional TIL rather than immune cells within the liver.

Adoptive T Cell Therapy

Early History

The ability of lymphocytes isolated from immunized animals and transferred into secondary hosts bearing tumors derived from cells of the original cancer was first demonstrated in the 1960s. These initial experiments were informed through enhanced knowledge of the key role played by T cells in promoting allograft rejection. Here, transfer of syngeneic lymphocytes from tumor-immune rodents was shown to mediate regression of carcinogen-induced or viral-induced sarcomas in tumor-bearing recipient animals,[230,231] and the effect was potentiated when chemotherapy or total body irradiation was administered immediately prior to adoptive lymphocyte transfer.[232,233] The later finding implicated inhibitory cells in the host in suppressing the antitumor efficacy of the transferred lymphocytes.[233,234] The inability to culture or manipulate T cells in vitro at the time precluded modern adoptive cell therapy (ACT) approaches of culturing, selecting, and expanding tumor-infiltrating T cells for reinfusion into cancer-bearing animals or patients. Characterization of the T cell growth factor, interleukin 2 (IL-2), in 1976 enabled successful culture and propagation of T cells, often for months at a time,[235] suitable for use in ACT experiments.

By the 1980s, it was demonstrated that IL-2–expanded lymphocytes from immunized animals could cause complete rejection of established tumors in TBI-conditioned, secondary hosts bearing the same transplanted tumor.[236] Interestingly, in vitro propagation of lymphocytes from immunized animals in IL-2 for prolonged times was associated with a decrease in their efficacy upon adoptive transfer, which was an important observation later confirmed in human ACT studies. Furthermore, purified IL-2 administered to tumor-bearing hosts augmented the ability of IL-2–expanded lymphocytes to mediate tumor regression upon adoptive transfer.[237] These preclinical observations paved the way for subsequent ACT clinical studies in cancer patients (Figs. 49.3 and 49.4). The success of ACT in these early experiences required transferring lymphocytes derived from previously tumor-immunized, syngeneic animals, which limited the potential of ACT in humans.

LAK Cells

Demonstration that human peripheral blood lymphocytes or murine splenocytes expanded and activated through in vitro culture with IL-2—so-called lymphokine-activated killer (LAK) cells as they lacked (or lost) expression of lineage defining cell surface markers—could lyse tumor cell lines in vitro and mediate regression of established tumors in syngeneic mice[238] provided a new opportunity to test ACT in human cancer patients. Investigators at the Surgery Branch of the National Cancer Institute (NCI) were the first to evaluate this approach in humans with advanced cancer. While adoptive transfer of high numbers of LAK cells was relatively well-tolerated, no patients achieved an objective tumor response.[239] Similarly, administration of high doses of IL-2 intravenously also failed to cause tumor regression in most cancer patients and was associated with sometimes severe toxicities.[240,241] However, the combination of LAK cell and IL-2 therapy was shown to mediate tumor regression, occasionally durably, in a small subset of patients with metastatic melanoma, renal cell carcinoma, and colorectal carcinoma.[242] Toxicities were largely attributable to IL-2, including fever, chills, malaise, and significant fluid retention, but abated rapidly upon discontinuation of IL-2.[242] This study was among the first to be demonstrated that the immune system could be harnessed to treat human cancer and put ACT on the map as a new cancer therapeutic modality.

TIL-Based Immunotherapy

A renewed interest in the nature and specificity of lymphocytes that infiltrated tumors, most notably that TILs were potentially enriched for T cells specific for cancer antigens,[243] suggested that T cells isolated directly from tumors, expanded ex vivo, and reinfused back into the patient might be useful as a new type of cancer immunotherapy (Fig. 49.4). Indeed, ACT with autologous TIL expanded in IL-2 was shown to be highly effective in mediating regression of metastatic tumors in mice.[244] In 1988, a group of investigators led by Steven Rosenberg at the NCI found that TIL ACT preceded by a single dose of cyclophosphamide and followed by several days of high-dose IL-2 lead to an impressive overall response rate of 55% in 20 patients with metastatic melanoma resistant to conventional therapies.[245] This small clinical study was instrumental at demonstrating the potential of TIL-based ACT and served as a platform upon which to further develop ACT as an effective cancer immunotherapeutic.

However, a major issue with early TIL-based ACT was the inability of adoptively transferred T cells to persist in the host. Thus, early responses were often followed by disease

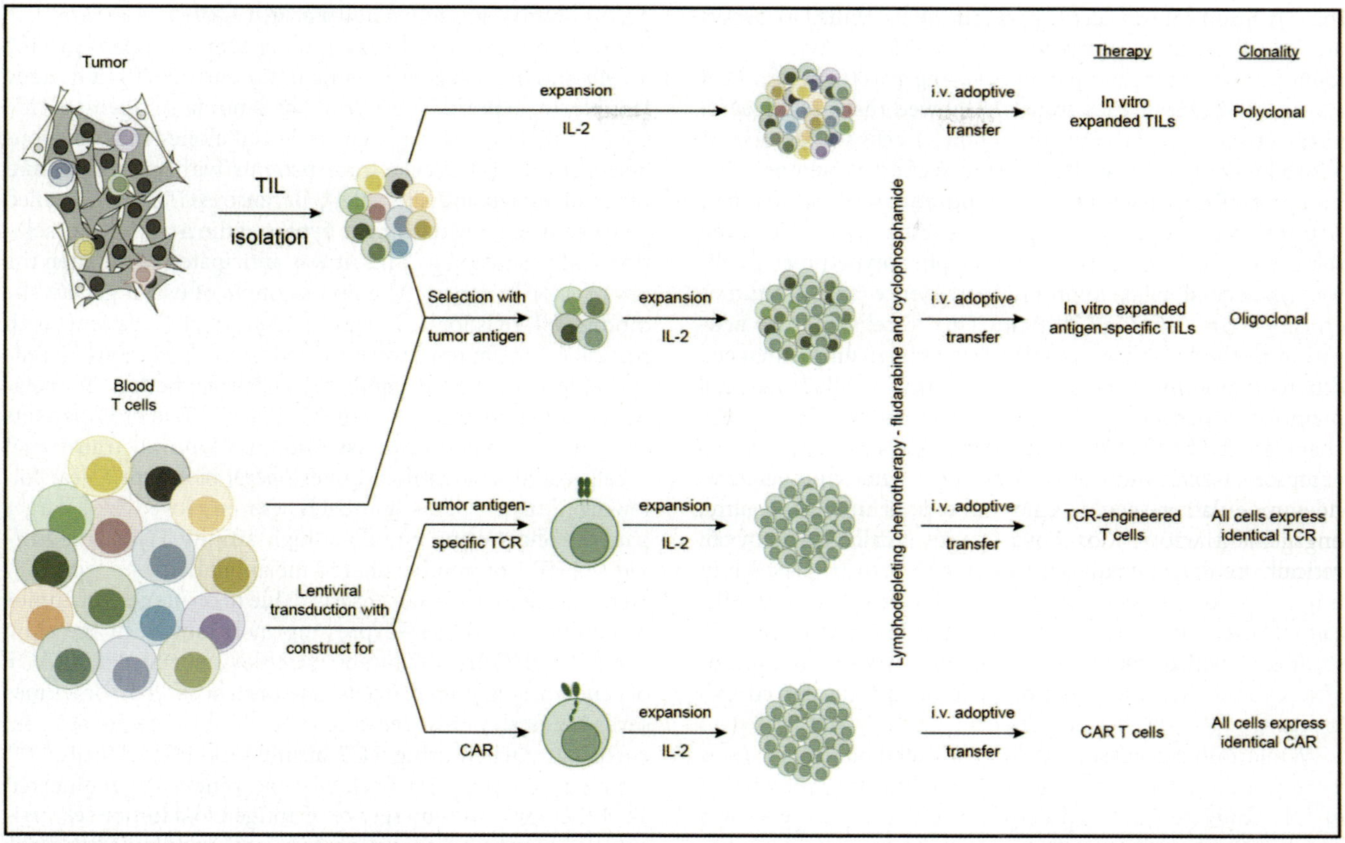

FIG. 49.4. T cell adoptive therapy modalities.

progression using this approach. Preclinical experiments in mouse cancer models indicated that preconditioning of the host either with chemotherapy or TBI was associated with improved activity of ACT.[140,246] First, conditioning therapy led to the depletion of endogenous regulatory cell populations,[246] most notably FoxP3+ regulatory T (Treg) cells, known to effectively impair antitumor responses.[247] Second, conditioning therapy eliminated many endogenous CD4+ T cells, CD8+ T cells, and NK cells that functioned as "cytokine sinks," thereby increasing the availability of T cell homeostatic cytokines, such as IL-7 and IL-15, which facilitated expansion and survival of adoptively transferred T cell populations.[246,248] The intensity of conditioning or "lymphodepleting" therapy with chemotherapy (cyclophosphamide and fludarabine) and/or increasing fractions of TBI was shown to correlate directly with the efficacy of ACT in preclinical melanoma models,[249] and possibly in patients,[250] but also led to more prolonged cytopenias, sometimes requiring reinfusion of autologous stem cells to rescue bone marrow function.

These observations implicated host conditioning as an essential requirement for ACT efficacy and led to a seminal clinical trial performed at the NCI.[251] In this study, 13 HLA-A2+ metastatic melanoma patients were treated with 7 days of cyclophosphamide and fludarabine conditioning therapy prior to receiving TIL therapy and high-dose IL-2. Six of 13 patients achieved objective tumor responses. The adoptively transferred TIL expanded and persisted, and the transferred T cells were detectable for months in some cases. Furthermore, tumor-specific CD8+ T cells could be detected against the MART-1 antigen in selected patient peripheral blood specimens. HLA class I and II expression was significantly increased in posttreatment melanoma biopsies, likely reflecting tumor cell sensing of IFN-γ secreted by adoptively transferred TILs that accumulated in the local environment. Extending this approach to a larger cohort of patients over the next several years revealed that ~20% of those treated with this approach could achieve long-term complete remissions.[252] This was remarkable, as these patients had previously failed treatment with conventional melanoma therapies. The addition of non-myeloablative or myeloablative fractions of TBI along with fludarabine/cyclophosphamide conditioning therapy in single-arm studies appeared to further increase response rates following TIL adoptive therapy,[252] confirming the importance of the conditioning regimen in promoting the effectiveness of ACT for melanoma.

Another important aspect of TIL-based ACT was the characterization of the TIL product itself. In addition to the ability to screen individual cultures from unique fragments of surgically excised primary or metastatic melanomas for tumor specificity, the phenotype and differentiation state of the TIL product was also important in its ability to persist and mediate tumor regression in vivo. Mouse ACT experiments in the B16 melanoma model using gp100-specific TCR transgenic CD8+ T cells (pmel-1) showed that prolonged in vitro activation and expansion of pmel-1 cells was associated with a higher fraction of T cells with an effector memory phenotype and decreased efficacy upon adoptive transfer into B16 melanoma-bearing, lymphodepleted animals. However, the use of naïve or central memory phenotype pmel-1 cells was associated with improved persistence and antitumor efficacy.[253] In a study of human CD8+ T cells, in vitro activation in the presence of active Wnt3A signaling enhanced differentiation into a pool of T cells with so-called stem cell memory properties.[254] These stem cell memory CD8+ T cells, characterized by CD62L positivity and CD44 negativity, also expressed CD95 and the IL-2 receptor beta. Compared to ACT with tumor-specific CD8+ T cells of a naïve, central, or effector phenotype, ACT with CD8+ T cells of a stem cell memory phenotype showed superior in vivo persistence in immunodeficient mice, along with enhanced antitumor efficacy. These murine data were borne out in human TIL ACT studies,[252] which found a correlation between antitumor efficacy and the differentiation state of the transferred TIL product.

While initial analyses of TILs indicated specificity for a variety of shared tumor antigens,[255,256] only a small proportion of TILs could be identified as specifically recognizing known melanoma differentiation or cancer-testis antigens. The ability to enumerate the mutational load of cancers through next-generation sequencing and to predict nonsynonymous mutations that may give rise to neoantigens,[31,47,54,257] along with the finding that the tumor mutational burden correlated with response to checkpoint blockade therapy,[34,35] suggested that cancer neoantigens were key targets of anticancer T cell responses. These observations also implicated neoantigen-specific T cells in driving clinical responses following TIL-based ACT for melanoma, and indeed neoantigen-specific T cells in TILs were identified through WES and in silico HLA binding prediction algorithms to identify candidate neoantigenic epitopes.[54] The hypothesis that neoantigen-specific T cell transfer could mediate tumor regression was proven true when CD4+ TILs from a patient with cholangiocarcinoma enriched for a HLA-DQ bound neoantigen derived from ERBB2IP protein involved in regulating RAS/ERK signaling mediated prolonged tumor regression upon ACT.[258]

Although ACT with TIL was clearly associated with potent clinical activity in metastatic melanoma, several issues precluded its broader use as a cancer immunotherapy. First, although lymphocytes could be readily expanded from tumors of various histologies, only those grown from melanomas regularly demonstrated tumor reactivity.[259] Second, the approach of isolating, expanding, and selecting TILs for tumor reactivity was a complex and expensive task requiring significant laboratory and clinical experience. Therefore, the approach was not feasible at most centers. Importantly, TIL-based therapies are now being pursued by several biotechnology companies, with the hope of commercializing this strategy and bringing it to a larger patient population.

T Cell Clones and TCR-Engineered T Cells

Based on advances in the isolation of tumor antigen–specific T cells and in TCR gene cloning in the early 2000s, it became feasible to express TCRs from melanoma antigen-specific CD8+ and in some cases CD4+ T cell clones in autologous peripheral blood T cells from patients with melanoma and other malignancies (Fig. 49.4). Because ACT with so-called genetically engineered T cells bypassed the need for the isolation and expansion of TILs, it was anticipated to broaden the use of ACT in cancer. The first example of use of genetically modified T cells for ACT was in 2006 when 17 patients with metastatic melanoma were treated with autologous T cells transduced with an unmodified TCR specific for the melanocyte differentiation antigen, MART-1.[260] Two of 17 patients achieved a complete response, and the adoptively transferred T cells could be identified in peripheral blood for a year following infusion. This approach was employed in a larger patient cohort using modified, high-affinity TCRs recognizing MART-1 or another shared melanoma antigen, gp100.[261] Here, severe toxicity was observed due to on-target, off-tumor recognition of antigen-expressing melanocytes in the skin, eye, and ear. Other off-tumor toxicities associated with ACT of genetically modified T cells have been observed, sometimes severe or fatal,[262-265] suggesting the need for extreme caution, particularly when using TCR affinity–modified T cells.[266-268] Another potential drawback of using genetically engineered TCR therapy is the emergence of antigen loss tumor cell variants that may be a more relevant issue when a single antigen is targeted, rather than making use of TILs, which presumably contain T cell clones of broader tumor specificity. Regardless, this ACT using genetically modified T cells can be effective and is an approach that, if delivered safely, could be much more generalizable in treating cancers expressing shared antigens (NY-ESO1, MAGE family antigens). Importantly, the pharmaceutical industry has become involved in developing early-phase clinical studies of such T cells in advanced cancer patients, with the hope of bringing this strategy to the larger marketplace.

CAR-T Cells

Another approach to genetically modify T cells and redirect their specificity toward cell surface tumor antigens in an HLA-independent manner was developed in the 1980s[269] (Fig. 49.4). Here, single-chain variable fragments (scFv) derived from antibodies specific for a given antigen were coupled with the CD3-zeta (z) domain and later with the costimulatory molecule signaling domains, for example from CD28 or CD137 (4-1BB),[270-272] to enable functional T cell signaling. These chimeric antigen receptor (CAR) T cells could be generated through retroviral or lentiviral transduction of peripheral blood T cells to recognize essentially any cell surface cancer antigen. The use of CAR-T cells over other genetically engineered T cells was advantageous because CAR-T cells could recognize antigen in an HLA-independent manner, and because CARs contain an internal costimulatory domain, the need for an APC to provide classical costimulation was bypassed. One important limitation of CAR-T cells is that they require cell surface expression of the target antigen and cannot differentiate between antigens expressed by cancer cells or healthy tissue cells. Thus, CAR-T cells can be used to target malignant cells only where their normal cellular counterparts are dispensable for health or to target cell surface antigens uniquely expressed on cancer cells. To date, CAR-T cells have chiefly been applied to B cell malignancies, including acute lymphoblastic leukemia (ALL),[273-275] chronic lymphocytic leukemia (CLL),[276] B cell non-Hodgkin lymphomas (NHL),[277-279] and more recently multiple myeloma (here, with CAR-T cells targeting B cell maturation antigen; BCMA). Although CARs recognizing numerous B cell surface antigens have been generated,[280] CD19 has been the most widely targeted for several reasons. CD19 is required for normal B cell development in humans and is highly expressed on B cells across developmental stages, with the exception of plasma cells.[281]

First-generation CARs, comprised of an antigen-binding scFv and a CD3ζ signaling domain,[282] generally showed modest proliferation and cytotoxicity.[283,284] It subsequently was determined through preclinical experimentation that second-generation CARs, into which a costimulatory domain was incorporated, most often with CD28 or CD137 (4-1BB), showed enhanced CAR-T cell–intrinsic signaling, culminating in augmented T cell proliferation, cytokine secretion, and target cell cytolysis.[270-272] CD28 costimulated CAR-T cells were particularly proliferative with potent effector functions, but in vivo persistence was typically brief.[277] CAR-T cells containing CD137 costimulatory domains proliferated less vigorously, exhibited less potent effector functions, yet were more capable of durable persistence following adoptive transfer.[278] Given the remarkable potency of CD19-directed CAR-T cell therapy in xenograft models, this approach was moved into the clinic around 2010.[285]

Early-phase studies demonstrated remarkable effectiveness of CAR-T cell therapy in patients with refractory CLL and ALL,[273,276] including the ability to induce durable remissions and disease cures in a fraction of patients, particularly in younger patients with ALL.[274] CD28 costimulated, CD19-directed CAR-T cell therapy was pioneered at the NCI shortly thereafter and was similarly successful in small early-phase studies in patients with relapsed/refractory diffuse large B cell lymphoma (DLBCL),[286] including the ability to mediate prolonged remissions in a fraction of treated patients. Based on the experience of adoptive TIL therapy for melanoma, lymphodepleting chemotherapy was administered several days prior to CAR-T cell infusion.[286]

The early success of this CAR-T cell therapy approach was very exciting, but with the excitement around efficacy, it became clear that CAR-T cell therapy was associated with unique toxicities. For example, cytokine release syndrome (CRS)—a constellation of symptoms, such as fever, hypotension, and hypoxia, which often developed several days following CAR-T cell infusion—developed in a significant fraction of patients.[287] CRS often presented with mild symptoms (fevers), but in some cases, led to life-threatening cardiopulmonary complications, landing patients in the intensive care unit. Neurological symptoms due to the immune effector cell–associated neurotoxicity syndrome (ICANS) were also common[288] and included complaints such as headache, tremor, expressive aphasia, somnolence, and more severe toxicities, including seizures and cerebral edema. Similar to CRS, ICANS was often mild, but could also lead to life-threatening complications. CRS and ICANS, as unique toxicities of CAR-T cell therapy, became well known to centers engaged in CAR-T cell therapy research. Regarding CRS, IL-6 appeared to be a major driver, and IL-6 or IL-6 receptor-blocking antibodies and corticosteroids were shown to be effective antidotes.[288] Regarding ICANS, its etiology is still incompletely understood, although steroids were commonly employed effectively. While these two complications were indeed serious, they were most often completely reversible with effective therapy and supportive care. In addition to CRS and ICANS, other side effects of CD19-directed CAR-T cell therapy were observed, including infections, hypogammaglobulinemia from elimination of normal antibody-producing B cells, and prolonged cytopenias from the lymphodepleting chemotherapy regimen.[289]

Subsequently, larger phase II studies of CD19-directed CAR-T cell therapy were undertaken, driven by pharmaceutical companies. The first FDA approval granted for CD19-directed CAR-T cell therapy was for the indication of B-ALL in patients younger than 25 years, using a 4-1BB costimulated CAR-T construct called tisagenlecleucel.[274] Approvals for tisagenlecleucel, as well as for a CD28 costimulated CAR-T product called axicabtagene ciloleucel, were granted in 2018 after similar phase II studies (Juliet and ZUMA-1) showed that approximately 40% of patients with relapsed/refractory DLBCL achieved durable remissions.[277,278] Given that survival of patients with refractory DLBCL prior to the advent of CAR-T cell therapy was only 6 months,[290] the results seen with CAR-T cell therapy were extremely impressive and led to significant improvements in outcomes for such patients. Currently, CD19-directed CAR-T cell therapy is approved for use in relapsed/refractory DLBCL, follicular lymphoma, mantle cell lymphoma, primary mediastinal B cell lymphoma, and B-ALL.

Although plasma cell malignancies, most notably multiple myeloma, lack CD19 expression, BCMA was identified as a suitable target for CAR-T cell therapy in this disease.[291] Although not considered curative, BCMA-targeted CAR-T cell therapy was shown to induce deep and durable remissions in this otherwise incurable disease.[280,292-294] The first BCMA-specific CAR-T cell product was approved for use in relapsed/refractory myeloma in 2021.[280]

CAR-T Cell Therapy Resistance Mechanisms

Although CD19-directed CAR-T cell therapy was remarkably effective in a fraction of patients with B cell and plasma cell malignancies, a majority of subjects were refractory or experienced a disease relapse within months of this treatment. Interestingly, relapses rarely occurred after 6 months in remitting patients.[278,295] In B-ALL, one common resistance mechanism was the emergence of CD19 ALV that could no longer be targeted by CAR-T cells. CD19 antigen loss was associated with CD19 splice variants in some cases and with loss-of-function mutations in CD19 chaperone proteins in others.[296] Rarely, ALLs underwent a switch to a myeloid phenotype, lacking CD19 cell surface expression, following CAR-T cell therapy.[297] Given that a single antigen is targeted with this ACT approach, it is perhaps not surprising that ALV are a major cause of failure, and early-phase clinical studies of dual-targeting CAR-T cells are underway. In B cell NHLs, ALV have also been described as a failure mechanism associated with CAR-T cell therapy.[298] However, it appears less common than in B-ALL and is rare in the majority of therapy failures that occur within weeks to months after CAR-T cell infusion in patients with lymphoma. Here, resistance mechanisms are less well understood, although upregulation of coinhibitory ligands (PD-L1) and other features of the lymphoma microenvironment have been implicated. Another key factor in CAR-T cell therapy success or failure, regardless of the underlying malignancy, is the CAR-T cell product itself. T cell–intrinsic dysfunction or exhaustion that occurs prior to or following manufacturing of a CAR-T cell product has been correlated with clinical outcomes.[299,300] Through single-cell profiling of patient T cells or CAR-T cell products prior to infusion, the composition of the T cells with regard to differentiation state and expression of genes associated with T cell exhaustion appear to be predictive of CAR-T cell therapy efficacy or failure, and many groups are searching to identify approaches aimed at enhancing the in vivo persistence and function of CAR-T cells, both during the manufacturing process and after infusion in patients.

For all its success in B cell and plasma cell malignancies, CAR-T cell therapy efficacy has not yet been realized in the solid tumor context. Identifying suitable target antigens that are expressed uniquely on cancer cells and not healthy tissues has been one major challenge, where severe on-target, off-tumor toxicity has been observed. Another major obstacle has been to develop strategies to overcome the many immune evasive mechanisms active in the solid tumor microenvironment that can inhibit the infiltration and effector function of adoptively transferred CAR-T cells.

Intratumoral Immune Modulation: Innate Immune Agonists and Oncolytic Viruses

Adaptive immunity begins when critical innate immune cells receive signals that indicate a potential antigenic threat from the host. Innate immune sensing systems likely evolved to respond to invasive pathogens. In that context, specific pathogen-derived molecular patterns engage host-expressed receptors, including TLRs, cytosolic RNA sensing pathways (via RIG-I), and cytosolic DNA sensing pathways (cGAS-STING). It has been thought that provision of innate immune receptor ligands within the tumor microenvironment might trigger de novo adaptive immune responses against tumors that include CD8+ T cell responses against tumor-expressed antigens. While this aspect of the field is in its infancy, there are several agents already in clinical use.

Bacillus Calmette-Guerin (BCG) is an attenuated form of *Mycobacterium bovis* that exerts immune stimulatory effects when applied topically. It has reached clinical practice in the treatment of locally recurrent nonmuscle invasive bladder cancer and was FDA-approved in 1977. BCG is administered intravesicularly and has been shown to decrease recurrence rates by around 60%.[301] Evidence suggests that BCG initiates an immune reaction that includes activation and recruitment of CD4+ and CD8+ T cells.[302] Anecdotal clinical responses have also been seen with intralesional injection in melanoma.[303] The long-term success of BCG has motivated the pursuit of additional molecularly defined agents that might have similar clinical efficacy without bearing the challenges of a live attenuated bacterium as a drug.

During a pharmacologic screen for new compounds that might exert antiviral properties in vitro, researchers at 3M identified what eventually became marketed as imiquimod. The agent was not directly antiviral, but rather triggered an innate immune response in infected cells via the TLR7 pathway. Based on the original viral immunity hypothesis, the drug entered clinical trials to explore efficacy against HPV-driven genital warts and was FDA approved for this indication. Because imiquimod was being applied topically to the skin, interest grew in exploring potential activity against challenging skin cancers, including recurrent basal cell carcinomas. The agent was highly active, with response rates of up to 100% seen in phase 3 trials.[304] Anecdotal responses have been seen with topical application to recurrent melanomas as well.[305] This success has motivated exploration of additional innate immune agonists by intratumoral or topical application. Studies have explored ligands for TLR3, 4, 8, and 9; for RIG-I (RNA sensing); and for STING. Clinical activity has been observed in a range of cancers, both alone and in combination with anti–PD-1 antibody therapies. However, this approach has been hampered by the requirement for intratumor administration for maximal effects and the lack of a biological understanding for treatment success versus failure. Strategies for systemic administration of conjugates that preferentially target the tumor microenvironment are being explored.

Another intratumoral treatment strategy that likely engages innate immune sensing pathways is the use of oncolytic viruses. A wide range of virus backbones have been explored for antitumor activity, including adenovirus, coxsackievirus, measles virus, vaccinia, and herpes simplex virus (HSV-1). The tumor-killing properties of oncolytic viruses likely exploit signaling aberrations in malignant cells, such as defects in the induction of type I IFNs that enable rapid viral replication and oncolysis. However, because the vector itself is a virus, immune activation also occurs, and it is thought that APCs in the tumor microenvironment become activated through innate immune pathways, leading to systemic T cell priming that can exert antitumor activity at distant sites. These characteristics led to the clinical development of talimogene laherparepvec (T-Vec), which is a genetically modified, live replicating HSV-1 variant encoding GM-CSF to recruit DC participation. T-Vec showed single-agent activity in metastatic melanoma patients with lesions amenable to intratumoral injection. In 2015, T-Vec was FDA-approved based on a phase 3 clinical trial, with randomization against GM-CSF alone as a control arm.[182,306] Interestingly, 15% of patients showed regression of visceral lesions that were not injected, supporting the notion that the host immune response is contributing to therapeutic efficacy.

NEXT FRONTIER: MECHANISMS OF FAILURE OF CURRENT IMMUNOTHERAPIES

The therapeutic success of currently approved immunotherapy strategies, combined with mechanistic mouse models, is enabling identification and investigation of pathways explaining primary and secondary resistance. Understanding reasons for failure is highlighting new potential therapeutic interventions to expand the circle of therapeutic efficacy. Some mechanisms of resistance based on immune editing are challenging to address therapeutically. These include selection for tumor cell variants that lack expression of target antigens, class I MHC molecules, or antigen-processing machinery. A convenient framework for considering mechanisms of treatment failure that might be amenable to new therapeutic interventions is based on the T cell–inflamed/non–T cell–inflamed tumor microenvironment paradigm. Hypothetical molecular sources of interpatient heterogeneity that could explain a T cell–inflamed versus non–T cell–inflamed tumor microenvironment include somatic mutational differences between tumor cells, host polymorphisms in immune regulatory genes, and environmental differences, including immunomodulatory effects of the commensal microbiota (Fig. 49.5).

Somatic Tumor Cell Differences

There are multiple oncogenic aberrations that can lead to the development of cancer, and no two cancers of the same histology between individual patients are genetically identical. As such, it was conceivable that activation of certain oncogenic pathways could be immune-evasive, in addition to being tumor-promoting. Most oncogenic events affect signaling pathways, which at minimum could imply a distinct transcriptional output in non–T cell–inflamed tumors that could either be inhibitory for immune system processes or could lack rate-limiting positive factors that are necessary for orchestrating a T cell–centered immune response within tumor sites.

In order to begin to probe candidate somatic cell differences that might be causal for immune exclusion, the melanoma TCGA data set was employed. Importantly, TCGA includes not only exome sequencing data across patient tumor samples but also bulk RNAseq data and germline SNPs. The RNAseq data were used to categorize patients into T cell–inflamed,

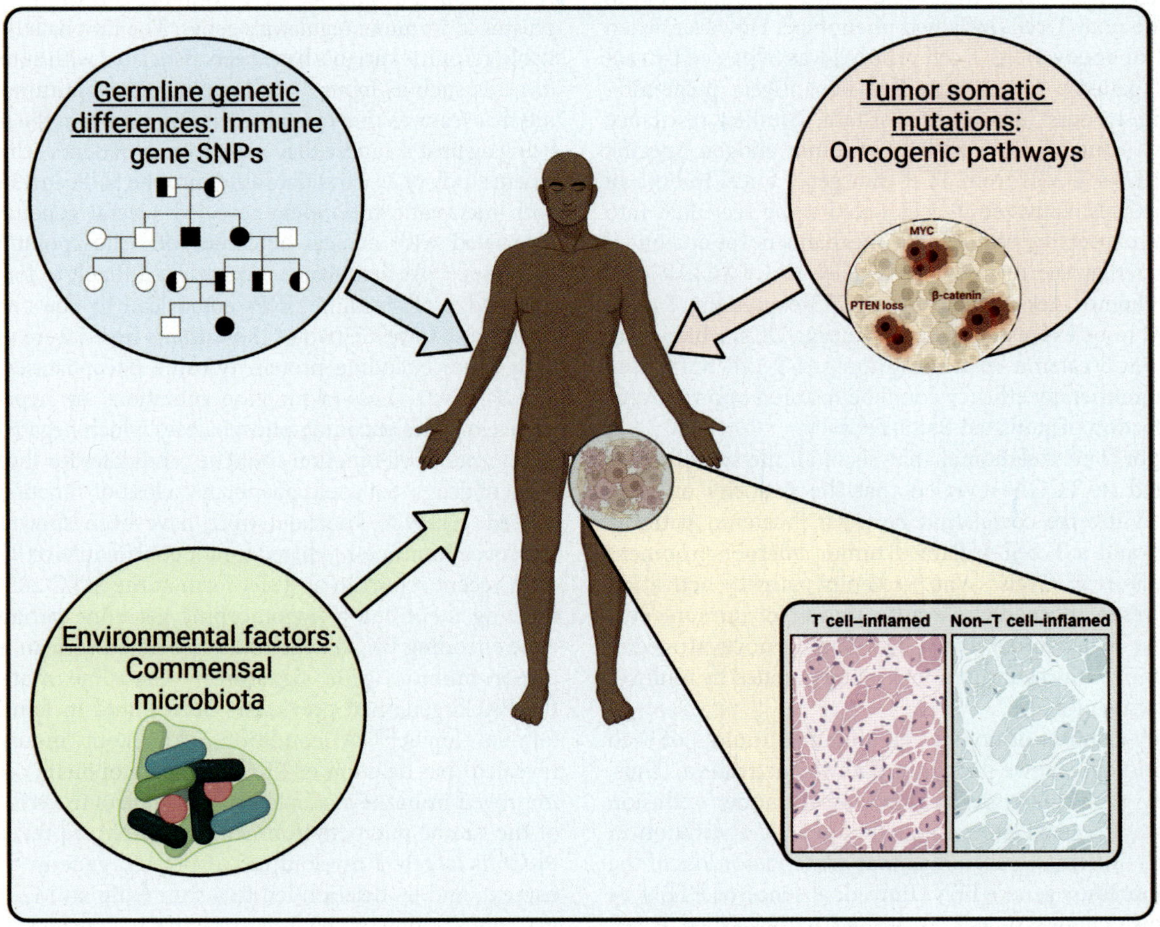

FIG. 49.5. Molecular sources of interpatient heterogeneity that could impact endogenous immune responses against cancer.

non–T cell–inflamed, or intermediate phenotypes. Focusing in on the T cell–inflamed/non–T cell–inflamed extremes, the corresponding exome sequencing data were interrogated for oncogene mutations enriched in the non–T cell–inflamed tumors. This analysis revealed activating mutations in the β-catenin gene, suggesting that Wnt/β-catenin pathway activation was associated with immune exclusion. The analysis was broadened to include other candidate mechanisms for β-catenin pathway activation, and 48% of the non–T cell–inflamed melanomas demonstrated such alterations, including inactivating mutations in negative regulators of the pathway, or overexpression of selected Wnt ligands. In order to determine whether tumor cell–intrinsic β-catenin activation was actually causal for immune cell exclusion and immunotherapy resistance, a GEM model of melanoma was utilized, incorporating Cre-induced expression of stabilized β-catenin. Indeed, tumors induced with the incorporation of active β-catenin lost their immune infiltrate and were resistant to checkpoint blockade immunotherapy. The mechanism was linked to failed recruitment of Batf3-lineage DCs, which led to defective T cell priming in the tumor-draining lymph node. One mechanism explaining poor Batf3-DC recruitment was diminished tumor cell production of the chemokine CCL4.[98]

The ability of tumor cell–intrinsic β-catenin activation to lead to defective T cell priming could have been sufficient to explain the non–T cell–inflamed phenotype. However, when the need for endogenous T cell priming was bypassed by vaccination against a tumor-expressed neoantigen, β-catenin–expressing tumors also were resistant. Similar resistance was found using adoptive transfer of tumor antigen–specific effector CD8+ T cells from TCR transgenic mice. In both of these instances, activated T cells failed to be recruited into β-catenin–expressing tumors. The mechanism was attributed to the fact that the majority of CXCL9 and CXCL10 production, chemokines that recruit CXCR3-expressing T cells, was found to be expressed by Batf3-lineage DCs, which were absent from β-catenin–expressing tumors. T cell trafficking and immunotherapy efficacy could be restored upon intratumoral injection of activated Batf3-DCs.[116]

Going beyond melanoma, analysis of all the solid tumors represented in TCGA revealed that the majority of them showed an inverse correlation between β-catenin pathway activation and a T cell–inflamed tumor microenvironment gene signature.[307] Thus, Wnt/β-catenin pathway activation could represent a broadly relevant pathway for immune evasion across many tumor types. Developing novel strategies for β-catenin pathway inhibition is being pursued by multiple biotech companies.

Yet only a subset of non–T cell–inflamed tumors of each cancer histology shows β-catenin pathway activation. Thus, additional molecular mechanisms of immune exclusion likely exist. The next most frequently genetic aberration in non–T cell–inflamed melanomas was mutation or loss of the tumor suppressor gene PTEN. Indeed, absence of PTEN by immunohistochemistry was associated with lack of T cell infiltration and diminished efficacy of anti–PD-1 therapy in melanoma.[308] These results suggest that developing strategies to block the PI3K pathway preferentially in tumor cells should be considered for improving immunotherapy efficacy. Such an approach would have to take into account the fact that T cell activation also relies on PI3K signaling, so the dose and schedule of such inhibitors, as well as the PI3K isoform being targeted, would need to be considered.

A third tumor cell–intrinsic oncogenic event associated with immunotherapy resistance in the clinic is mutation or loss of the tumor-suppressor gene LKB1. LKB1 loss of function has been identified in several studies to have negative prognostic significance and to be associated with checkpoint blockade resistance in non–small cell lung cancer.[309] A conditional knockout mouse GEM model confirmed decreased T cell infiltration and increased granulocyte-lineage cells, suggesting a shift in immune cell composition that would be predicted to diminish immunotherapy efficacy.[310] The biochemical mechanisms of this effect are being elucidated, which could enable pharmacologic approaches to restore immunogenicity. These and other oncogenic pathways that could contribute to immune evasion have been reviewed.[311]

Germline Polymorphisms in Immune Regulatory Genes

A second hypothetical source of interpatient heterogeneity that could regulate endogenous immune activation and a T cell–inflamed tumor microenvironment is germline polymorphisms in immune regulatory genes. The rationale is that heritable germline variants have been associated with autoimmune diseases such as lupus,[312] and spontaneous antitumor immunity has features that resemble autoimmunity, in the latter case being against a cancer that arose from a patient's self-tissue. A recent study of 25 candidate autoimmune SNPs in 436 patients with metastatic melanoma revealed several genetic variants associated with clinical outcome with checkpoint blockade therapies.[313] At first glance it might be difficult to foresee how knowledge of germline SNPs could lead to new therapeutic strategies. However, two of the variants noted were associated with genes encoding protein tyrosine phosphatases, PTPN2 and PTPN22. Loss-of-function mutations or hypomorphic alleles in genes encoding phosphatases which negatively regulate immune cell function could be candidates for the development of drugs that could phenocopy a loss-of-function variant. Indeed, PTPN22 knockout mice have been shown to have improved immune-mediated tumor control in vivo.[314-316]

A recent report in abstract form at the SITC 2021 annual meeting identified a hypomorphic germline variant in the gene encoding PKCδ which was associated with an increased tumor immune gene signature.[317] Germline mutations in the PKCδ gene had previously been found in families with familial lupus.[318] A conditional knockout mouse model revealed that deletion of PKCδ in hematopoietic cells led to improved immune-mediated tumor control in vivo. Analysis of the tumor microenvironment by scRNAseq revealed that PKCδ is largely a myeloid cell–expressed gene in the tumor context, and its deletion led to a shift from an M2 to an M1 macrophage phenotype. It is attractive to consider that pharmacologic targeting of PKCδ could lead to improved immunotherapy efficacy through the remodeling of myeloid cells within the tumor microenvironment.

Variations in the Commensal Microbiota

A third general category of interpatient heterogeneity that could contribute to the degree of spontaneous immune activation and infiltration into tumors is environmental variability. Many aspects of interactions with the outside environment are difficult to track and study with respect to immunotherapy efficacy, such as chemical exposures and history of previous pathogenic infections. However, one tractable variable that has been pursued with success is the composition of the commensal microbiota. The number of genes contained within commensal gut bacteria is greater than the number in the human genome, and there is tremendous variability in the relative composition of the bacterial communities in the gut across individuals. And importantly, the composition of commensal bacteria has been shown to regulate diverse systemic immune responses, including autoimmunity.[319] With this concept in mind, the potential for gut microbiome composition to regulate spontaneous antitumor immunity was pursued.

Standard specific pathogen-free inbred mice obtained from different vendors were found to have distinct gut microbiota composition, which was found to influence the development of certain autoimmune pathologies.[320,321] Similar experiments performed in transplantable tumor models revealed faster or slower tumor growth that was associated with weaker or stronger antitumor T cell responses, respectively.[322] All of that difference could be explained by the gut microbiota, and poor spontaneous tumor control was improved using fecal transplant from mice with favorable microbiota at baseline. The efficacy of anti–PD-L1 and also of anti–CTLA-4 Ab therapy was improved when favorable gut microbes were present.[322,323] Based on these results, stool samples began to be collected from cancer patients undergoing treatment with anti–PD-1 Ab therapy, and the microbiota was analyzed using 16S ribosomal RNA gene sequencing and also metagenomic shotgun sequencing. Distinct populations of gut bacteria were observed in patients who responded to anti–PD-1 versus nonresponders.[324-326] To determine whether the gut microbiota composition was causal for response versus nonresponse to PD-1 blockade, transfer of patient fecal material into germ-free mice was performed, along with tumor implantation and anti–PD-1 Ab therapy. Microbiota from responder patients supported improved antitumor efficacy in germ-free mice compared to microbiota from nonresponder patients.[325] Together, these data indicate that in some patients, the composition of the gut microbiota can be a major determining factor for immunotherapy efficacy.

Investigations are ongoing to determine the mechanisms by which specific gut microbes might impact on immune responses within the tumor microenvironment and also to identify the collection of favorable versus unfavorable bacterial communities on a species and strain level. Once these details are elucidated, then rationally designed therapies aiming to manipulate the microbiota composition or regulate the downstream effects of the microbiota can be developed. But in the meantime, studies are ongoing evaluating the potential for fecal microbial transplantation (FMT) to restore immunotherapy efficacy. In this strategy, fecal samples are obtained from elected donors with favorable microbiota characteristics, for example, from anti–PD-1 responder patients. Patients who failed anti–PD-1 therapy are eligible to receive FMT followed by resumption of anti–PD-1. In two small pilot trials, several patients were found to regain anti–PD-1 efficacy following FMT. FMT was associated with improved immune cell composition within the tumor microenvironment, as well as stable alteration of the gut microbiota composition.[327] Thus, these early data suggest the feasibility of improving immunotherapy outcomes upon microbiota modulation.

ACKNOWLEDGMENTS

The authors would like to thank Santiago Acero Bedoya and Brendan MacNabb for generating the figures associated with this chapter, several of which were created with BioRender.com, as well as Jack Kline for assistance with literature review and in drafting several sections of the text.

REFERENCES

1. Singh S, Ross SR, Acena M, Rowley DA, Schreiber H. Stroma is critical for preventing or permitting immunological destruction of antigenic cancer cells. *J Exp Med.* 1992;175(1):139-146. doi:10.1084/jem.175.1.139
2. Coley WB. *The Treatment of Inoperable Sarcoma by Bacterial Toxin (The Mixed Toxins of Streptococcus erysipelas and the Bacillus prodigious).* John Bale, Sons and Banielson, Ltd; 1909.
3. Burnet M. Cancer: a biological approach. III. Viruses associated with neoplastic conditions. IV. Practical applications. *Br Med J.* 1957;1(5023):841-847. doi:10.1136/bmj.1.5023.841
4. Burnet FM. The concept of immunological surveillance. *Prog Exp Tumor Res.* 1970;13:1-27. doi:10.1159/000386035
5. Foley EJ. Antigenic properties of methylcholanthrene-induced tumors in mice of the strain of origin. *Cancer Res.* 1953;13(12):835-837.
6. Klein G, Sjogren HO, Klein E, Hellstrom KE. Demonstration of resistance against methylcholanthrene-induced sarcomas in the primary autochthonous host. *Cancer Res.* 1960;20:1561-1572.
7. Prehn RT, Main JM. Immunity to methylcholanthrene-induced sarcomas. *J Natl Cancer Inst.* 1957;18(6):769-778.
8. Stutman O. Tumor development after 3-methylcholanthrene in immunologically deficient athymic-nude mice. *Science.* 1974;183(4124):534-536. doi:10.1126/science.183.4124.534
9. Stutman O. Chemical carcinogenesis in nude mice: comparison between nude mice from homozygous matings and heterozygous matings and effect of age and carcinogen dose. *J Natl Cancer Inst.* 1979;62(2):353-358.
10. Ikehara S, Pahwa RN, Fernandes G, Hansen CT, Good RA. Functional T cells in athymic nude mice. *Proc Natl Acad Sci U S A.* 1984;81(3):886-888. doi:10.1073/pnas.81.3.886
11. Dighe AS, Richards E, Old LJ, Schreiber RD. Enhanced in vivo growth and resistance to rejection of tumor cells expressing dominant negative IFN gamma receptors. *Immunity.* 1994;1(6):447-456.
12. Kaplan DH, Shankaran V, Dighe AS, et al. Demonstration of an interferon gamma-dependent tumor surveillance system in immunocompetent mice. *Proc Natl Acad Sci U S A.* 1998;95(13):7556-7561.
13. Street SE, Cretney E, Smyth MJ. Perforin and interferon-gamma activities independently control tumor initiation, growth, and metastasis. *Blood.* 2001;97(1):192-197. doi:10.1182/blood.v97.1.192
14. Street SE, Trapani JA, MacGregor D, Smyth MJ. Suppression of lymphoma and epithelial malignancies effected by interferon gamma. *J Exp Med.* 2002;196(1):129-134. doi:10.1084/jem.20020063
15. van den Broek ME, Kagi D, Ossendorp F, et al. Decreased tumor surveillance in perforin-deficient mice. *J Exp Med.* 1996;184(5):1781-1790.
16. Smyth MJ, Thia KY, Street SE, MacGregor D, Godfrey DI, Trapani JA. Perforin-mediated cytotoxicity is critical for surveillance of spontaneous lymphoma. *J Exp Med.* 2000;192(5):755-760. doi:10.1084/jem.192.5.755
17. Shankaran V, Ikeda H, Bruce AT, et al. IFNgamma and lymphocytes prevent primary tumour development and shape tumour immunogenicity. *Nature.* 2001;410(6832):1107-1111. doi:10.1038/35074122
18. Knuth A, Danowski B, Oettgen HF, Old LJ. T-cell-mediated cytotoxicity against autologous malignant melanoma: analysis with interleukin 2-dependent T-cell cultures. *Proc Natl Acad Sci U S A.* 1984;81(11):3511-3515. doi:10.1073/pnas.81.11.3511
19. Carey TE, Takahashi T, Resnick LA, Oettgen HF, Old LJ. Cell surface antigens of human malignant melanoma: mixed hemadsorption assays for humoral immunity to cultured autologous melanoma cells. *Proc Natl Acad Sci U S A.* 1976;73(9):3278-3282. doi:10.1073/pnas.73.9.3278
20. Ueda R, Shiku H, Pfreundschuh M, et al. Cell surface antigens of human renal cancer defined by autologous typing. *J Exp Med.* 1979;150(3):564-579. doi:10.1084/jem.150.3.564
21. Traversari C, van der Bruggen P, Van den Eynde B, et al. Transfection and expression of a gene coding for a human melanoma antigen recognized by autologous cytolytic T lymphocytes. *Immunogenetics.* 1992;35(3):145-152.

22. van der Bruggen P, Traversari C, Chomez P, et al. A gene encoding an antigen recognized by cytolytic T lymphocytes on a human melanoma. *Science*. 1991;254(5038):1643-1647.
23. Altman JD, Moss PA, Goulder PJ, et al. Phenotypic analysis of antigen-specific T lymphocytes. *Science*. 1996;274(5284):94-96. doi:10.1126/science.274.5284.94
24. Lee PP, Yee C, Savage PA, et al. Characterization of circulating T cells specific for tumor-associated antigens in melanoma patients. *Nat Med*. 1999;5(6):677-685. doi:10.1038/9525
25. Romero P, Dunbar PR, Valmori D, et al. Ex vivo staining of metastatic lymph nodes by class I major histocompatibility complex tetramers reveals high numbers of antigen-experienced tumor-specific cytolytic T lymphocytes. *J Exp Med*. 1998;188(9):1641-1650. doi:10.1084/jem.188.9.1641
26. Galon J, Costes A, Sanchez-Cabo F, et al. Type, density, and location of immune cells within human colorectal tumors predict clinical outcome. *Science*. 2006;313(5795):1960-1964. doi:10.1126/science.1129139
27. Clark WH Jr, Elder DE, Guerry D, et al. Model predicting survival in stage I melanoma based on tumor progression. *J Natl Cancer Inst*. 1989;81(24):1893-1904. doi:10.1093/jnci/81.24.1893
28. Zhang L, Conejo-Garcia JR, Katsaros D, et al. Intratumoral T cells, recurrence, and survival in epithelial ovarian cancer. *N Engl J Med*. 2003;348(3):203-213. doi:10.1056/NEJMoa020177
29. Pages F, Mlecnik B, Marliot F, et al. International validation of the consensus Immunoscore for the classification of colon cancer: a prognostic and accuracy study. *Lancet*. 2018;391(10135):2128-2139. doi:10.1016/S0140-6736(18)30789-X
30. Cancer Genome Atlas Research N; Weinstein JN, Collisson EA, et al. The cancer genome Atlas pan-cancer analysis project. *Nat Genet*. 2013;45(10):1113-1120. doi:10.1038/ng.2764
31. Alexandrov LB, Nik-Zainal S, Wedge DC, et al. Signatures of mutational processes in human cancer. *Nature*. 2013;500(7463):415-421. doi:10.1038/nature12477
32. Bao R, Spranger S, Hernandez K, et al. Immunogenomic determinants of tumor microenvironment correlate with superior survival in high-risk neuroblastoma. *J Immunother Cancer*. 2021;9(7):e002417. doi:10.1136/jitc-2021-002417
33. Rooney MS, Shukla SA, Wu CJ, Getz G, Hacohen N. Molecular and genetic properties of tumors associated with local immune cytolytic activity. *Cell*. 2015;160(1-2):48-61. doi:10.1016/j.cell.2014.12.033
34. Snyder A, Makarov V, Merghoub T, et al. Genetic basis for clinical response to CTLA-4 blockade in melanoma. *N Engl J Med*. 2014;371(23):2189-2199. doi:10.1056/NEJMoa1406498
35. Rizvi NA, Hellmann MD, Snyder A, et al. Cancer immunology. Mutational landscape determines sensitivity to PD-1 blockade in non-small cell lung cancer. *Science*. 2015;348(6230):124-128. doi:10.1126/science.aaa1348
36. Gajewski TF, Louahed J, Brichard VG. Gene signature in melanoma associated with clinical activity: a potential clue to unlock cancer immunotherapy. *Cancer J*. 2010;16(4):399-403. doi:10.1097/PPO.0b013e3181eacbd8
37. Harlin H, Meng Y, Peterson AC, et al. Chemokine expression in melanoma metastases associated with CD8+ T-cell recruitment. *Cancer Res*. 2009;69(7):3077-3085. doi:10.1158/0008-5472.CAN-08-2281
38. Gajewski TF, Zha Y, Thurner B, Schuler G. Association of gene expression profile in melanoma and survival to a dendritic cell-based vaccine. *J Clin Oncol*. 2009;27(15s):9002.
39. Gajewski TF, Meng Y, Harlin H. Chemokines expressed in melanoma metastases associated with T cell infiltration. Abstract. *J Clin Oncol*. 2007;25(18S):8501.
40. Ayers M, Lunceford J, Nebozhyn M, et al. IFN-gamma-related mRNA profile predicts clinical response to PD-1 blockade. *J Clin Invest*. 2017;127(8):2930-2940. doi:10.1172/JCI91190
41. Degiovanni G, Lahaye T, Herin M, Hainaut P, Boon T. Antigenic heterogeneity of a human melanoma tumor detected by autologous CTL clones. *Eur J Immunol*. 1988;18(5):671-676.
42. Cox AL, Skipper J, Chen Y, et al. Identification of a peptide recognized by five melanoma-specific human cytotoxic T cell lines. *Science*. 1994;264(5159):716-719. doi:10.1126/science.7513441
43. Gnjatic S, Wheeler C, Ebner M, et al. Seromic analysis of antibody responses in non-small cell lung cancer patients and healthy donors using conformational protein arrays. *J Immunol Methods*. 2009;341(1-2):50-58. doi:S0022-1759(08)00324-4 [pii] 10.1016/j.jim.2008.10.016
44. Schuler MM, Nastke MD, Stevanovikc S. SYFPEITHI: database for searching and T-cell epitope prediction. *Methods Mol Biol*. 2007;409:75-93. doi:10.1007/978-1-60327-118-9_5
45. Zhang H, Lundegaard C, Nielsen M. Pan-specific MHC class I predictors: a benchmark of HLA class I pan-specific prediction methods. *Bioinformatics*. 2009;25(1):83-89. doi:10.1093/bioinformatics/btn579
46. Lundegaard C, Lamberth K, Harndahl M, Buus S, Lund O, Nielsen M. NetMHC-3.0: accurate web accessible predictions of human, mouse and monkey MHC class I affinities for peptides of length 8-11. *Nucleic Acids Res*. 2008;36(Web Server issue):W509-W512. doi:10.1093/nar/gkn202
47. Spranger S, Luke JJ, Bao R, et al. Density of immunogenic antigens does not explain the presence or absence of the T-cell-inflamed tumor microenvironment in melanoma. *Proc Natl Acad Sci U S A*. 2016;113(48):E7759–E7768. doi:10.1073/pnas.1609376113
48. De Plaen E, Lurquin C, Van Pel A, et al. Immunogenic (tum-) variants of mouse tumor P815: cloning of the gene of tum-antigen P91A and identification of the tum- mutation. *Proc Natl Acad Sci U S A*. 1988;85(7):2274-2278.
49. Dubey P, Hendrickson RC, Meredith SC, et al. The immunodominant antigen of an ultraviolet-induced regressor tumor is generated by a somatic point mutation in the DEAD box helicase p68. *J Exp Med*. 1997;185(4):695-705.
50. Monach PA, Meredith SC, Siegel CT, Schreiber H. A unique tumor antigen produced by a single amino acid substitution. *Immunity*. 1995;2(1):45-59.
51. Bergmann-Leitner ES, Kantor JA, Shupert WL, Schlom J, Abrams SI. Identification of a human CD8+ T lymphocyte neo-epitope created by a ras codon 12 mutation which is restricted by the HLA-A2 allele. *Cell Immunol*. 1998;187(2):103-116. doi:10.1006/cimm.1998.1325
52. Fossum B, Breivik J, Meling GI, et al. A K-ras 13Gly-->Asp mutation is recognized by HLA-DQ7 restricted T cells in a patient with colorectal cancer. Modifying effect of DQ7 on established cancers harbouring this mutation? *Int J Cancer*. 1994;58(4):506-511. doi:10.1002/ijc.2910580409
53. Mandruzzato S, Brasseur F, Andry G, Boon T, van der Bruggen P. A CASP-8 mutation recognized by cytolytic T lymphocytes on a human head and neck carcinoma. *J Exp Med*. 1997;186(5):785-793.
54. Robbins PF, Lu YC, El-Gamil M, et al. Mining exomic sequencing data to identify mutated antigens recognized by adoptively transferred tumor-reactive T cells. *Nat Med*. 2013;19(6):747-752. doi:10.1038/nm.3161
55. Guilloux Y, Lucas S, Brichard VG, et al. A peptide recognized by human cytolytic T lymphocytes on HLA-A2 melanomas is encoded by an intron sequence of the N-acetylglucosaminyltransferase V gene. *J Exp Med*. 1996;183(3):1173-1183.
56. Coulie PG, Lehmann F, Lethe B, et al. A mutated intron sequence codes for an antigenic peptide recognized by cytolytic T lymphocytes on a human melanoma. *Proc Natl Acad Sci U S A*. 1995;92(17):7976-7980.
57. Van Den Eynde BJ, Gaugler B, Probst-Kepper M, et al. A new antigen recognized by cytolytic T lymphocytes on a human kidney tumor results from reverse strand transcription. *J Exp Med*. 1999;190(12):1793-1800. doi:10.1084/jem.190.12.1793
58. Turajlic S, Litchfield K, Xu H, et al. Insertion-and-deletion-derived tumour-specific neoantigens and the immunogenic phenotype: a pan-cancer analysis. *Lancet Oncol*. 2017;18(8):1009-1021. doi:10.1016/S1470-2045(17)30516-8
59. Bosch GJ, Joosten AM, Kessler JH, Melief CJ, Leeksma OC. Recognition of BCR-ABL positive leukemic blasts by human CD4+ T cells elicited by primary in vitro immunization with a BCR-ABL breakpoint peptide. *Blood*. 1996;88(9):3522-3527.
60. Zarling AL, Obeng RC, Desch AN, et al. MHC-restricted phosphopeptides from insulin receptor substrate-2 and CDC25b offer broad-based immunotherapeutic agents for cancer. *Cancer Res*. 2014;74(23):6784-6795. doi:10.1158/0008-5472.CAN-14-0043
61. Skipper JC, Hendrickson RC, Gulden PH, et al. An HLA-A2-restricted tyrosinase antigen on melanoma cells results from posttranslational modification and suggests a novel pathway for processing of membrane proteins. *J Exp Med*. 1996;183(2):527-534. doi:10.1084/jem.183.2.527
62. Warren EH, Vigneron NJ, Gavin MA, et al. An antigen produced by splicing of noncontiguous peptides in the reverse order. *Science*. 2006;313(5792):1444-1447. doi:10.1126/science.1130660
63. De Smet C, De Backer O, Faraoni I, Lurquin C, Brasseur F, Boon T. The activation of human gene MAGE-1 in tumor cells is correlated with genome-wide demethylation. *Proc Natl Acad Sci U S A*. 1996;93(14):7149-7153.
64. Ressing ME, Sette A, Brandt RM, et al. Human CTL epitopes encoded by human papillomavirus type 16 E6 and E7 identified through in vivo and in vitro immunogenicity studies of HLA- A*0201-binding peptides. *J Immunol*. 1995;154(11):5934-5943.
65. Iyer JG, Afanasiev OK, McClurkan C, et al. Merkel cell polyomavirus-specific CD8(+) and CD4(+) T-cell responses identified in Merkel cell carcinomas and blood. *Clin Cancer Res*. 2011;17(21):6671-6680. doi:10.1158/1078-0432.CCR-11-1513
66. Duraiswamy J, Burrows JM, Bharadwaj M, et al. Ex vivo analysis of T-cell responses to Epstein-Barr virus-encoded oncogene latent membrane protein 1 reveals highly conserved epitope sequences in virus isolates from diverse geographic regions. *J Virol*. 2003;77(13):7401-7410. doi:10.1128/jvi.77.13.7401-7410.2003
67. Lee SP, Thomas WA, Murray RJ, et al. HLA A2.1-restricted cytotoxic T cells recognizing a range of Epstein-Barr virus isolates through a defined epitope in latent membrane protein LMP2. *J Virol*. 1993;67(12):7428-7435. doi:10.1128/JVI.67.12.7428-7435.1993
68. Coulie PG, Brichard V, Van Pel A, et al. A new gene coding for a differentiation antigen recognized by autologous cytolytic T lymphocytes on HLA-A2 melanomas. *J Exp Med*. 1994;180(1):35-42.
69. Brichard V, Van Pel A, Wolfel T, et al. The tyrosinase gene codes for an antigen recognized by autologous cytolytic T lymphocytes on HLA-A2 melanomas. *J Exp Med*. 1993;178(2):489-495.
70. Rosenberg SA, White DE. Vitiligo in patients with melanoma: normal tissue antigens can be targets for cancer immunotherapy. *J Immunother Emphas Tumor Immunol*. 1996;19(1):81-84.
71. Rezvani K, Yong AS, Tawab A, et al. Ex vivo characterization of polyclonal memory CD8+ T-cell responses to PRAME-specific peptides in patients with acute lymphoblastic leukemia and acute and chronic myeloid leukemia. *Blood*. 2009;113(10):2245-2255. doi:10.1182/blood-2008-03-144071
72. Kessler JH, Beekman NJ, Bres-Vloemans SA, et al. Efficient identification of novel HLA-A(*)0201-presented cytotoxic T lymphocyte epitopes in the widely expressed tumor antigen PRAME by proteasome-mediated digestion analysis. *J Exp Med*. 2001;193(1):73-88. doi:10.1084/jem.193.1.73
73. Chen Y, Ayaru L, Mathew S, Morris E, Pereira SP, Behboudi S. Expansion of anti-mesothelin specific CD4+ and CD8+ T cell responses in patients with pancreatic carcinoma. *PLoS One*. 2014;9(2):e88133. doi:10.1371/journal.pone.0088133
74. Rongcun Y, Salazar-Onfray F, Charo J, et al. Identification of new HER2/neu-derived peptide epitopes that can elicit specific CTL against autologous and allogeneic carcinomas and melanomas. *J Immunol*. 1999;163(2):1037-1044.
75. Linehan DC, Goedegebuure PS, Peoples GE, Rogers SO, Eberlein TJ. Tumor-specific and HLA-A2-restricted cytolysis by tumor-associated lymphocytes in human metastatic breast cancer. *J Immunol*. 1995;155(9):4486-4491.
76. Rezvani K, Brenchley JM, Price DA, et al. T-cell responses directed against multiple HLA-A*0201-restricted epitopes derived from Wilms' tumor 1 protein in patients with leukemia and healthy donors: identification, quantification, and characterization. *Clin Cancer Res*. 2005;11(24 pt 1):8799-8807. doi:10.1158/1078-0432.CCR-05-1314
77. Scheibenbogen C, Letsch A, Thiel E, et al. CD8 T-cell responses to Wilms tumor gene product WT1 and proteinase 3 in patients with acute myeloid leukemia. *Blood*. 2002;100(6):2132-2137. doi:10.1182/blood-2002-01-0163
78. Chapuis AG, Egan DN, Bar M, et al. T cell receptor gene therapy targeting WT1 prevents acute myeloid leukemia relapse post-transplant. *Nat Med*. 2019;25(7):1064-1072. doi:10.1038/s41591-019-0472-9
79. Engels B, Engelhard VH, Sidney J, et al. Relapse or eradication of cancer is predicted by peptide-major histocompatibility complex affinity. *Cancer Cell*. 2013;23(4):516-526. doi:10.1016/j.ccr.2013.03.018
80. Morel S, Levy F, Burlet-Schiltz O, et al. Processing of some antigens by the standard proteasome but not by the immunoproteasome results in poor presentation by dendritic cells. *Immunity*. 2000;12(1):107-117.
81. Wong R, Lau R, Chang J, et al. Immune responses to a class II helper peptide epitope in patients with stage III/IV resected melanoma. *Clin Cancer Res*. 2004;10(15):5004-5013. doi:10.1158/1078-0432.CCR-04-0241 10/15/5004 [pii]
82. Wells DK, van Buuren MM, Dang KK, et al. Key parameters of tumor epitope immunogenicity revealed through a consortium approach improve neoantigen prediction. *Cell*. 2020;183(3):818-834 e13. doi:10.1016/j.cell.2020.09.015
83. Alspach E, Lussier DM, Miceli AP, et al. MHC-II neoantigens shape tumour immunity and response to immunotherapy. *Nature*. 2019;574(7780):696-701. doi:10.1038/s41586-019-1671-8
84. McGranahan N, Furness AJ, Rosenthal R, et al. Clonal neoantigens elicit T cell immunoreactivity and sensitivity to immune checkpoint blockade. *Science*. 2016;351(6280):1463-1469. doi:10.1126/science.aaf1490
85. Pages F, Galon J, Dieu-Nosjean MC, Tartour E, Sautes-Fridman C, Fridman WH. Immune infiltration in human tumors: a prognostic factor that should not be ignored. *Oncogene*. 2010;29(8):1093-1102. doi:onc2009416 [pii] 10.1038/onc.2009.416
86. Wang L, Sfakianos JP, Beaumont KG, et al. Myeloid cell-associated resistance to PD-1/PD-L1 blockade in urothelial cancer revealed through bulk and single-cell RNA sequencing. *Clin Cancer Res*. 2021;27(15):4287-4300. doi:10.1158/1078-0432.CCR-20-4574
87. Reschke R, Yu J, Flood B, Higgs EF, Hatogai K, Gajewski TF. Immune cell and tumor cell-derived CXCL10 is indicative of immunotherapy response in metastatic melanoma. *J Immunother Cancer*. 2021;9(9)doi:10.1136/jitc-2021-003521
88. Gajewski TF. Monitoring specific T-cell responses to melanoma vaccines: ELISPOT, tetramers, and beyond. *Clin Diagn Lab Immunol*. 2000;7(2):141-144.
89. Pittet MJ, Valmori D, Dunbar PR, et al. High frequencies of naive Melan-A/MART-1-specific CD8(+) T cells in a large proportion of human histocompatibility leukocyte antigen (HLA)-A2 individuals. *J Exp Med*. 1999;190(5):705-715.

90. Jager E, Nagata Y, Gnjatic S, et al. Monitoring CD8 T cell responses to NY-ESO-1: correlation of humoral and cellular immune responses. *Proc Natl Acad Sci U S A*. 2000;97(9):4760-4765.
91. Harlin H, Kuna TV, Peterson AC, Meng Y, Gajewski TF. Tumor progression despite massive influx of activated CD8(+) T cells in a patient with malignant melanoma ascites. *Cancer Immunol Immunother*. 2006;55(10):1185-1197.
92. Voabil P, de Bruijn M, Roelofsen LM, et al. An ex vivo tumor fragment platform to dissect response to PD-1 blockade in cancer. *Nat Med*. 2021;27(7):1250-1261. doi:10.1038/s41591-021-01398-3
93. Diamond MS, Kinder M, Matsushita H, et al. Type I interferon is selectively required by dendritic cells for immune rejection of tumors. *J Exp Med*. 2011;208(10):1989-2003. doi:10.1084/jem.20101158
94. Fuertes MB, Kacha AK, Kline J, et al. Host type I IFN signals are required for antitumor CD8+ T cell responses through CD8{alpha}+ dendritic cells. *J Exp Med*. 2011;208(10):2005-2016. doi:10.1084/jem.20101159
95. Hildner K, Edelson BT, Purtha WE, et al. Batf3 deficiency reveals a critical role for CD8alpha+ dendritic cells in cytotoxic T cell immunity. *Science*. 2008;322(5904):1097-1100.
96. Woo SR, Fuertes Mercedes B, Corrales L, et al. STING-dependent cytosolic DNA sensing mediates innate immune recognition of immunogenic tumors. *Immunity*. 2014;41(5):830-842. doi:10.1016/j.immuni.2014.10.017
97. Marcus A, Mao AJ, Lensink-Vasan M, Wang L, Vance RE, Raulet DH. Tumor-derived cGAMP triggers a STING-mediated interferon response in non-tumor cells to activate the NK cell response. *Immunity*. 2018;49(4):754-763 e4. doi:10.1016/j.immuni.2018.09.016
98. Spranger S, Bao R, Gajewski TF. Melanoma-intrinsic beta-catenin signalling prevents anti-tumour immunity. *Nature*. 2015;523(7559):231-235. doi:10.1038/nature14404
99. Neeland MR, Shi W, Collignon C, et al. The lymphatic immune response induced by the adjuvant AS01: a comparison of intramuscular and subcutaneous immunization routes. *J Immunol*. 2016;197(7):2704-2714. doi:10.4049/jimmunol.1600817
100. Kline J, Zhang L, Battaglia L, Cohen KS, Gajewski TF. Cellular and molecular requirements for rejection of B16 melanoma in the setting of regulatory T cell depletion and homeostatic proliferation. *J Immunol*. 2012;188(6):2630-2642. doi:10.4049/jimmunol.1100845
101. Spiotto MT, Yu P, Rowley DA, et al. Increasing tumor antigen expression overcomes "ignorance" to solid tumors via crosspresentation by bone marrow-derived stromal cells. *Immunity*. 2002;17(6):737-747.
102. Bai A, Higham E, Eisen HN, Wittrup KD, Chen J. Rapid tolerization of virus-activated tumor-specific CD8+ T cells in prostate tumors of TRAMP mice. *Proc Natl Acad Sci U S A*. 2008;105(35):13003-13008. doi:10.1073/pnas.0805599105
103. Roberts EW, Broz ML, Binnewies M, et al. Critical role for CD103(+)/CD141(+) dendritic cells bearing CCR7 for tumor antigen trafficking and priming of T cell immunity in melanoma. *Cancer Cell*. 2016;30(2):324-336. doi:10.1016/j.ccell.2016.06.003
104. Yu P, Spiotto MT, Lee Y, Schreiber H, Fu YX. Complementary role of CD4+ T cells and secondary lymphoid tissues for cross-presentation of tumor antigen to CD8+ T cells. *J Exp Med*. 2003;197(8):985-995. doi:10.1084/jem.20021804
105. Lauss M, Donia M, Svane IM, Jonsson G. B cells and tertiary lymphoid structures: friends or foes in cancer immunotherapy? *Clin Cancer Res*. 2021;doi:10.1158/1078-0432.CCR-21-1130
106. Yang M, Lu J, Zhang G, et al. CXCL13 shapes immunoactive tumor microenvironment and enhances the efficacy of PD-1 checkpoint blockade in high-grade serous ovarian cancer. *J Immunother Cancer*. 2021;9(1):e001136. doi:10.1136/jitc-2020-001136
107. Yu P, Lee Y, Wang Y, et al. Targeting the primary tumor to generate CTL for the effective eradication of spontaneous metastases. *J Immunol*. 2007;179(3):1960-1968.
108. Harjunpaa H, Llort Asens M, Guenther C, Fagerholm SC. Cell adhesion molecules and their roles and regulation in the immune and tumor microenvironment. *Front Immunol*. 2019;10:1078. doi:10.3389/fimmu.2019.01078
109. Lugade AA, Moran JP, Gerber SA, Rose RC, Frelinger JG, Lord EM. Local radiation therapy of B16 melanoma tumors increases the generation of tumor antigen-specific effector cells that traffic to the tumor. *J Immunol*. 2005;174(12):7516-7523. doi:10.4049/jimmunol.174.12.7516
110. Sackstein R, Schatton T, Barthel SR. T-lymphocyte homing: an underappreciated yet critical hurdle for successful cancer immunotherapy. *Lab Invest*. 2017;97(6):669-697. doi:10.1038/labinvest.2017.25
111. Wu TC. The role of vascular cell adhesion molecule-1 in tumor immune evasion. *Cancer Res*. 2007;67(13):6003-6006. doi:10.1158/0008-5472.CAN-07-1543
112. Slaney CY, Kershaw MH, Darcy PK. Trafficking of T cells into tumors. *Cancer Res*. 2014;74(24):7168-7174. doi:10.1158/0008-5472.CAN-14-2458
113. Mikucki ME, Fisher DT, Matsuzaki J, et al. Non-redundant requirement for CXCR3 signalling during tumoricidal T-cell trafficking across tumour vascular checkpoints. *Nat Commun*. 2015;6:7458. doi:10.1038/ncomms8458
114. Mikucki ME, Skitzki JJ, Frelinger JG, et al. Unlocking tumor vascular barriers with CXCR3: implications for cancer immunotherapy. *OncoImmunology*. 2016;5(5):e1116675. doi:10.1080/2162402X.2015.1116675
115. Dangaj D, Bruand M, Grimm AJ, et al. Cooperation between constitutive and inducible chemokines enables T cell engraftment and immune attack in solid tumors. *Cancer Cell*. 2019;35(6):885-900 e10. doi:10.1016/j.ccell.2019.05.004
116. Spranger S, Dai D, Horton B, Gajewski TF. Tumor-Residing Batf3 dendritic cells are required for effector T cell trafficking and adoptive T cell therapy. *Cancer Cell*. 2017;31(5):711-723 e4. doi:10.1016/j.ccell.2017.04.003
117. de Andrade LF, Lu Y, Luoma A, et al. Discovery of specialized NK cell populations infiltrating human melanoma metastases. *JCI Insight*. 2019;4(23):e133103.doi:10.1172/jci.insight.133103
118. Bottcher JP, Bonavita E, Chakravarty P, et al. NK cells stimulate recruitment of cDC1 into the tumor microenvironment promoting cancer immune control. *Cell*. 2018;172(5):1022-1037 e14. doi:10.1016/j.cell.2018.01.004
119. Cella M, Scheidegger D, Palmer-Lehmann K, Lane P, Lanzavecchia A, Alber G. Ligation of CD40 on dendritic cells triggers production of high levels of interleukin-12 and enhances T cell stimulatory capacity: T-T help via APC activation. *J Exp Med*. 1996;184(2):747-752.
120. Hunder NN, Wallen H, Cao J, et al. Treatment of metastatic melanoma with autologous CD4+ T cells against NY-ESO-1. *N Engl J Med*. 2008;358(25):2698-2703. doi:10.1056/NEJMoa0800251
121. Combes AJ, Samad B, Tsui J, et al. Discovering dominant tumor immune archetypes in a pan-cancer census. *Cell*. 2022;185(1):184-203 e19. doi:10.1016/j.cell.2021.12.004
122. Im K, Combes AJ, Spitzer MH, Satpathy AT, Krummel MF. Archetypes of checkpoint-responsive immunity. *Trends Immunol*. 2021;42(11):960-974. doi:10.1016/j.it.2021.09.007
123. Janssen EM, Lemmens EE, Wolfe T, Christen U, von Herrath MG, Schoenberger SP. CD4+ T cells are required for secondary expansion and memory in CD8+ T lymphocytes. *Nature*. 2003;421(6925):852-856. doi:10.1038/nature01441
124. Gubin MM, Esaulova E, Ward JP, et al. High-dimensional analysis delineates myeloid and lymphoid compartment remodeling during successful immune-checkpoint cancer therapy. *Cell*. 2018;175(5):1443. doi:10.1016/j.cell.2018.11.003
125. Salmon H, Idoyaga J, Rahman A, et al. Expansion and activation of CD103(+) dendritic cell progenitors at the tumor site enhances tumor responses to therapeutic PD-L1 and BRAF inhibition. *Immunity*. 2016;44(4):924-938. doi:10.1016/j.immuni.2016.03.012
126. Maier B, Leader AM, Chen ST, et al. A conserved dendritic-cell regulatory program limits antitumour immunity. *Nature*. 2020;580(7802):257-262. doi:10.1038/s41586-020-2134-y
127. Gabrilovich DI, Nagaraj S. Myeloid-derived suppressor cells as regulators of the immune system. *Nat Rev Immunol*. 2009;9(3):162-174. doi:10.1038/nri2506
128. Kagi D, Vignaux F, Ledermann B, et al. Fas and perforin pathways as major mechanisms of T cell-mediated cytotoxicity. *Science*. 1994;265(5171):528-530.
129. Yan HF, Zou T, Tuo QZ, et al. Ferroptosis: mechanisms and links with diseases. *Signal Transduct Target Ther*. 2021;6(1):49. doi:10.1038/s41392-020-00428-9
130. Wang W, Green M, Choi JE, et al. CD8(+) T cells regulate tumour ferroptosis during cancer immunotherapy. *Nature*. 2019;569(7755):270-274. doi:10.1038/s41586-019-1170-y
131. Gajewski TF, Schreiber H, Fu YX. Innate and adaptive immune cells in the tumor microenvironment. *Nat Immunol*. 2013;14(10):1014-1022. doi:10.1038/ni.2703 ni.2703 [pii]
132. Pages F, Berger A, Camus M, et al. Effector memory T cells, early metastasis, and survival in colorectal cancer. *N Engl J Med*. 2005;353(25):2654-2666. doi:10.1056/NEJMoa051424
133. Ma MW, Medicherla RC, Qian M, et al. Immune response in melanoma: an in-depth analysis of the primary tumor and corresponding sentinel lymph node. *Mod Pathol*. 2012;25(7):1000-1010. doi:10.1038/modpathol.2012.43
134. Spranger S, Spaapen RM, Zha Y, et al. Up-regulation of PD-L1, IDO, and Tregs in the melanoma tumor microenvironment is driven by CD8+ T cells. *Sci Transl Med*. 2013;5(200):200ra116. doi:10.1126/scitranslmed.3006504
135. Mortarini R, Piris A, Maurichi A, et al. Lack of terminally differentiated tumor-specific CD8+ T cells at tumor site in spite of antitumor immunity to self-antigens in human metastatic melanoma. *Cancer Res*. 2003;63(10):2535-2545.
136. Williams JB, Horton BL, Zheng Y, Duan Y, Powell JD, Gajewski TF. The EGR2 targets LAG-3 and 4-1BB describe and regulate dysfunctional antigen-specific CD8+ T cells in the tumor microenvironment. *J Exp Med*. 2017;214(2):381-400. doi:10.1084/jem.20160485
137. Holmgaard RB, Zamarin D, Munn DH, Wolchok JD, Allison JP. Indoleamine 2,3-dioxygenase is a critical resistance mechanism in antitumor T cell immunotherapy targeting CTLA-4. *J Exp Med*. 2013;210(7):1389-1402. doi:10.1084/jem.20130066
138. Spranger S, Koblish HK, Horton B, Scherle PA, Newton R, Gajewski TF. Mechanism of tumor rejection with doublets of CTLA-4, PD-1/PD-L1, or IDO blockade involves restored IL-2 production and proliferation of CD8(+) T cells directly within the tumor microenvironment. *J Immunother Cancer*. 2014;2:3. doi:10.1186/2051-1426-2-3
139. Zhang L, Gajewski TF, Kline J. PD-1/PD-L1 interactions inhibit antitumor immune responses in a murine acute myeloid leukemia model. *Blood*. 2009;114(8):1545-1552.
140. Kline J, Brown IE, Zha YY, et al. Homeostatic proliferation plus regulatory T-cell depletion promotes potent rejection of B16 melanoma. *Clin Cancer Res*. 2008;14(10):3156-3167. doi:10.1158/1078-0432.CCR-07-4696
141. Bos PD, Plitas G, Rudra D, Lee SY, Rudensky AY. Transient regulatory T cell ablation deters oncogene-driven breast cancer and enhances radiotherapy. *J Exp Med*. 2013;210(11):2435-2466. doi:10.1084/jem.20130762
142. Sakuishi K, Apetoh L, Sullivan JM, Blazar BR, Kuchroo VK, Anderson AC. Targeting Tim-3 and PD-1 pathways to reverse T cell exhaustion and restore anti-tumor immunity. *J Exp Med*. 2010;207(10):2187-2194. doi:10.1084/jem.20100643
143. Fourcade J, Sun Z, Benallaoua M, et al. Upregulation of Tim-3 and PD-1 expression is associated with tumor antigen-specific CD8+ T cell dysfunction in melanoma patients. *J Exp Med*. 2010;207(10):2175-2186. doi:10.1084/jem.20100637
144. Woo SR, Turnis ME, Goldberg MV, et al. Immune inhibitory molecules LAG-3 and PD-1 synergistically regulate T-cell function to promote tumoral immune escape. *Cancer Res*. 2012;72(4):917-927. doi:10.1158/0008-5472.CAN-11-1620
145. Horton BL, Williams JB, Cabanov A, Spranger S, Gajewski TF. Intratumoral CD8(+) T-cell apoptosis is a major component of T-cell dysfunction and impedes antitumor immunity. *Cancer Immunol Res*. 2018;6(1):14-24. doi:10.1158/2326-6066.CIR-17-0249
146. Zhu J, Powis de Tenbossche CG, Cane S, et al. Resistance to cancer immunotherapy mediated by apoptosis of tumor-infiltrating lymphocytes. *Nat Commun*. 2017;8(1):1404. doi:10.1038/s41467-017-00784-1
147. Yao C, Sun HW, Lacey NE, et al. Single-cell RNA-seq reveals TOX as a key regulator of CD8(+) T cell persistence in chronic infection. *Nat Immunol*. 2019;20(7):890-901. doi:10.1038/s41590-019-0403-4
148. Khan O, Giles JR, McDonald S, et al. TOX transcriptionally and epigenetically programs CD8(+) T cell exhaustion. *Nature*. 2019;571(7764):211-218. doi:10.1038/s41586-019-1325-x
149. Scott AC, Dundar F, Zumbo P, et al. TOX is a critical regulator of tumour-specific T cell differentiation. *Nature*. 2019;571(7764):270-274. doi:10.1038/s41586-019-1324-y
150. Yao J, Duan L, Huang X, et al. Development and validation of a prognostic gene signature correlated with M2 macrophage infiltration in esophageal squamous cell carcinoma. *Front Oncol*. 2021;11:769727. doi:10.3389/fonc.2021.769727
151. Wu MF, Lin CA, Yuan TH, et al. The M1/M2 spectrum and plasticity of malignant pleural effusion-macrophage in advanced lung cancer. *Cancer Immunol Immunother*. 2021;70(5):1435-1450. doi:10.1007/s00262-020-02781-8
152. Garrido-Martin EM, Mellows TWP, Clarke J, et al. M1(hot) tumor-associated macrophages boost tissue-resident memory T cells infiltration and survival in human lung cancer. *J Immunother Cancer*. 2020;8(2):e000778. doi:10.1136/jitc-2020-000778
153. Coccia EM, Stellacci E, Marziali G, Weiss G, Battistini A. IFN-gamma and IL-4 differently regulate inducible NO synthase gene expression through IRF-1 modulation. *Int Immunol*. 2000;12(7):977-985. doi:10.1093/intimm/12.7.977
154. Cui C, Chakraborty K, Tang XA, et al. Neutrophil elastase selectively kills cancer cells and attenuates tumorigenesis. *Cell*. 2021;184(12):3163-3177 e21. doi:10.1016/j.cell.2021.04.016
155. Passaro A, Mancuso P, Gandini S, et al. Gr-MDSC-linked asset as a potential immune biomarker in pretreated NSCLC receiving nivolumab as second-line therapy. *Clin Transl Oncol*. 2020;22(4):603-611. doi:10.1007/s12094-019-02166-z
156. Vinal D, Gutierrez-Sainz L, Martinez D, et al. Prognostic value of neutrophil-to-lymphocyte ratio in advanced cancer patients receiving immunotherapy. *Clin Transl Oncol*. 2021;23(6):1185-1192. doi:10.1007/s12094-020-02509-1
157. Schalper KA, Carleton M, Zhou M, et al. Elevated serum interleukin-8 is associated with enhanced intratumor neutrophils and reduced clinical benefit of immune-checkpoint inhibitors. *Nat Med*. 2020;26(5):688-692. doi:10.1038/s41591-020-0856-x
158. Arnold JN, Magiera L, Kraman M, Fearon DT. Tumoral immune suppression by macrophages expressing fibroblast activation protein-alpha and heme oxygenase-1. *Cancer Immunol Res*. 2014;2(2):121-126. doi:10.1158/2326-6066.CIR-13-0150
159. Feig C, Jones JO, Kraman M, et al. Targeting CXCL12 from FAP-expressing carcinoma-associated fibroblasts synergizes with anti-PD-L1 immunotherapy in pancreatic cancer. *Proc Natl Acad Sci U S A*. 2013;110(50):20212-20217. doi:10.1073/pnas.1320318110
160. Mariathasan S, Turley SJ, Nickles D, et al. TGFbeta attenuates tumour response to PD-L1 blockade by contributing to exclusion of T cells. *Nature*. 2018;554(7693):544-548. doi:10.1038/nature25501

161. Kieffer Y, Hocine HR, Gentric G, et al. Single-cell analysis reveals fibroblast clusters linked to immunotherapy resistance in cancer. *Cancer Discov*. 2020;10(9):1330-1351. doi:10.1158/2159-8290.CD-19-1384
162. Eroglu Z, Zaretsky JM, Hu-Lieskovan S, et al. High response rate to PD-1 blockade in desmoplastic melanomas. *Nature*. 2018;553(7688):347-350. doi:10.1038/nature25187
163. Cham CM, Driessens G, O'Keefe JP, Gajewski TF. Glucose deprivation inhibits multiple key gene expression events and effector functions in CD8+ T cells. *Eur J Immunol*. 2008;38(9):2438-2450.
164. Cham CM, Gajewski TF. Glucose availability regulates IFN-gamma production and p70S6 kinase activation in CD8+ effector T cells. *J Immunol*. 2005;174(8):4670-4677. doi:174/8/4670
165. Menk AV, Scharping NE, Rivadeneira DB, et al. 4-1BB costimulation induces T cell mitochondrial function and biogenesis enabling cancer immunotherapeutic responses. *J Exp Med*. 2018;215(4):1091-1100. doi:10.1084/jem.20171068
166. Keilholz U, Scheibenbogen C, Sommer M, Pritsch M, Geuke AM. Prognostic factors for response and survival in patients with metastatic melanoma receiving immunotherapy. *Melanoma Res*. 1996;6(2):173-178.
167. Scharping NE, Rivadeneira DB, Menk AV, et al. Mitochondrial stress induced by continuous stimulation under hypoxia rapidly drives T cell exhaustion. *Nat Immunol*. 2021;22(2):205-215. doi:10.1038/s41590-020-00834-9
168. Najjar YG, Menk AV, Sander C, et al. Tumor cell oxidative metabolism as a barrier to PD-1 blockade immunotherapy in melanoma. *JCI Insight*. 2019;4(5):e124989. doi:10.1172/jci.insight.124989
169. Rodriguez PC, Zea AH, Culotta KS, Zabaleta J, Ochoa JB, Ochoa AC. Regulation of T cell receptor CD3zeta chain expression by L-arginine. *J Biol Chem*. 2002;277(24):21123-21129.
170. Munn DH, Mellor AL. IDO and tolerance to tumors. *Trends Mol Med*. 2004;10(1):15-18.
171. Danilova L, Wang H, Sunshine J, et al. Association of PD-1/PD-L axis expression with cytolytic activity, mutational load, and prognosis in melanoma and other solid tumors. *Proc Natl Acad Sci U S A*. 2016;113(48):E7769-E7777. doi:10.1073/pnas.1607836113
172. de Wit R. AIDS-associated Kaposi's sarcoma and the mechanisms of interferon alpha's activity; a riddle within a puzzle. *J Intern Med*. 1992;231(4):321-325. doi:10.1111/j.1365-2796.1992.tb00939.x
173. Kirkwood JM, Strawderman MH, Ernstoff MS, Smith TJ, Borden EC, Blum RH. Interferon alfa-2b adjuvant therapy of high-risk resected cutaneous melanoma: the Eastern Cooperative Oncology Group Trial EST 1684. *J Clin Oncol*. 1996;14(1):7-17.
174. Atkins MB, Lotze MT, Dutcher JP, et al. High-dose recombinant interleukin 2 therapy for patients with metastatic melanoma: analysis of 270 patients treated between 1985 and 1993. *J Clin Oncol*. 1999;17(7):2105-2116.
175. Waldmann TA, Dubois S, Miljkovic MD, Conlon KC. IL-15 in the combination immunotherapy of cancer. *Front Immunol*. 2020;11:868. doi:10.3389/fimmu.2020.00868
176. Mackall CL, Fry TJ, Gress RE. Harnessing the biology of IL-7 for therapeutic application. *Nat Rev Immunol*. 2011;11(5):330-342. doi:10.1038/nri2970
177. Conlon K, Watson DC, Waldmann TA, et al. Phase I study of single agent NIZ985, a recombinant heterodimeric IL-15 agonist, in adult patients with metastatic or unresectable solid tumors. *J Immunother Cancer*. 2021;9(11):e003388. doi:10.1136/jitc-2021-003388
178. Fallarino F, Uyttenhove C, Boon T, Gajewski TF. Improved efficacy of dendritic cell vaccines and successful immunization with tumor antigen peptide-pulsed peripheral blood mononuclear cells by coadministration of recombinant murine interleukin- 12. *Int J Cancer*. 1999;80(2):324-333.
179. Peterson AC, Harlin H, Gajewski TF. Immunization with Melan-A peptide-pulsed peripheral blood mononuclear cells plus recombinant human interleukin-12 induces clinical activity and T-cell responses in advanced melanoma. *J Clin Oncol*. 2003;21(12):2342-2348.
180. Sanchez-Paulete AR, Teijeira A, Quetglas JI, et al. Intratumoral immunotherapy with XCL1 and sFlt3L encoded in recombinant semliki forest virus-derived vectors fosters dendritic cell-mediated T-cell cross-priming. *Cancer Res*. 2018;78(23):6643-6654. doi:10.1158/0008-5472.CAN-18-0933
181. Lynch DH, Andreasen A, Maraskovsky E, Whitmore J, Miller RE, Schuh JC. Flt3 ligand induces tumor regression and antitumor immune responses in vivo. *Nat Med*. 1997;3(6):625-631.
182. Andtbacka RH, Kaufman HL, Collichio F, et al. Talimogene laherparepvec improves durable response rate in patients with advanced melanoma. *J Clin Oncol*. 2015;33(25):2780-2788. doi:10.1200/JCO.2014.58.3377
183. Chang MH, You SL, Chen CJ, et al. Decreased incidence of hepatocellular carcinoma in hepatitis B vaccinees: a 20-year follow-up study. *J Natl Cancer Inst*. 2009;101(19):1348-1355. doi:10.1093/jnci/djp288
184. Lei J, Ploner A, Elfstrom KM, et al. HPV vaccination and the risk of invasive cervical cancer. *N Engl J Med*. 2020;383(14):1340-1348. doi:10.1056/NEJMoa1917338
185. Rosenberg SA, Yang JC, Restifo NP. Cancer immunotherapy: moving beyond current vaccines. *Nat Med*. 2004;10(9):909-915. doi:10.1038/nm1100
186. Gajewski TF. Failure at the effector phase: immune barriers at the level of the melanoma tumor microenvironment. *Clin Cancer Res*. 2007;13(18 pt 1):5256-5261.
187. Ott PA, Hu Z, Keskin DB, et al. An immunogenic personal neoantigen vaccine for patients with melanoma. *Nature*. 2017;547(7662):217-221. doi:10.1038/nature22991
188. Lurquin C, Lethe B, De Plaen E, et al. Contrasting frequencies of antitumor and anti-vaccine T cells in metastases of a melanoma patient vaccinated with a MAGE tumor antigen. *J Exp Med*. 2005;201(2):249-257. doi:10.1084/jem.20041378
189. Kantoff PW, Higano CS, Shore ND, et al. Sipuleucel-T immunotherapy for castration-resistant prostate cancer. *N Engl J Med*. 2010;363(5):411-422. doi:10.1056/NEJMoa1001294
190. Krummel MF, Allison JP. CTLA-4 engagement inhibits IL-2 accumulation and cell cycle progression upon activation of resting T cells. *J Exp Med*. 1996;183(6):2533-2540.
191. Krummel MF, Allison JP. CD28 and CTLA-4 have opposing effects on the response of T cells to stimulation [see comments]. *J Exp Med*. 1995;182(2):459-465.
192. Chambers CA, Sullivan TJ, Allison JP. Lymphoproliferation in CTLA-4-deficient mice is mediated by costimulation-dependent activation of CD4+ T cells. *Immunity*. 1997;7(6):885-895.
193. Waterhouse P, Penninger JM, Timms E, et al. Lymphoproliferative disorders with early lethality in mice deficient in Ctla-4 [see comments]. *Science*. 1995;270(5238):985-988.
194. Schneider H, Rudd CE. Tyrosine phosphatase SHP-2 binding to CTLA-4: absence of direct YVKM/YFIP motif recognition. *Biochem Biophys Res Commun*. 2000;269(1):279-283.
195. Chuang E, Fisher TS, Morgan RW, et al. The CD28 and CTLA-4 receptors associated with the serine/threonine phosphatase PP2A [In Process Citation]. *Immunity*. 2000;13(3):313-322.
196. Nakaseko C, Miyatake S, Iida T, et al. Cytotoxic T lymphocyte antigen 4 (CTLA-4) engagement delivers an inhibitory signal through the membrane-proximal region in the absence of the tyrosine motif in the cytoplasmic tail. *J Exp Med*. 1999;190(6):765-774.
197. Lee KM, Chuang E, Griffin M, et al. Molecular basis of T cell inactivation by CTLA-4. *Science*. 1998;282(5397):2263-2266.
198. Zhang Y, Allison JP. Interaction of CTLA-4 with AP50, a clathrin-coated pit adaptor protein. *Proc Natl Acad Sci U S A*. 1997;94(17):9273-9278.
199. Marengere LE, Waterhouse P, Duncan GS, Mittrucker HW, Feng GS, Mak TW. Regulation of T cell receptor signaling by tyrosine phosphatase SYP association with CTLA-4 [published errata appear in Science 1996 Dec 6;274(5293)1597 and 1997 Apr 4;276(5309):21]. *Science*. 1996;272(5265):1170-1173.
200. Chambers CA, Allison JP. The role of tyrosine phosphorylation and PTP-1C in CTLA-4 signal transduction. *Eur J Immunol*. 1996;26(12):3224-3229.
201. Fallarino F, Fields PE, Gajewski TF. B7-1 engagement of cytotoxic T lymphocyte antigen 4 inhibits T cell activation in the absence of CD28. *J Exp Med*. 1998;188(1):205-210.
202. Leach DR, Krummel MF, Allison JP. Enhancement of antitumor immunity by CTLA-4 blockade [see comments]. *Science*. 1996;271(5256):1734-1736.
203. O'Day SJ, Maio M, Chiarion-Sileni V, et al. Efficacy and safety of ipilimumab monotherapy in patients with pretreated advanced melanoma: a multicenter single-arm phase II study. *Ann Oncol*. 2010;21(8):1712-1717.
204. Hodi FS, O'Day SJ, McDermott DF, et al. Improved survival with ipilimumab in patients with metastatic melanoma. *N Engl J Med*. 2010;363(8):711-723.
205. Robert C, Thomas L, Bondarenko I, et al. Ipilimumab plus dacarbazine for previously untreated metastatic melanoma. *N Engl J Med*. 2011;364(26):2517-2526. doi:10.1056/NEJMoa1104621
206. Sanmamed MF, Chen L. Inducible expression of B7-H1 (PD-L1) and its selective role in tumor site immune modulation. *Cancer J*. 2014;20(4):256-261. doi:10.1097/PPO.0000000000000061
207. Taube JM, Anders RA, Young GD, et al. Colocalization of inflammatory response with B7-h1 expression in human melanocytic lesions supports an adaptive resistance mechanism of immune escape. *Sci Transl Med*. 2012;4(127):127ra37. doi:10.1126/scitranslmed.3003689
208. Curran MA, Montalvo W, Yagita H, Allison JP. PD-1 and CTLA-4 combination blockade expands infiltrating T cells and reduces regulatory T and myeloid cells within B16 melanoma tumors. *Proc Natl Acad Sci U S A*. 2010;107(9):4275-4280. doi:10.1073/pnas.0915174107
209. Blank C, Kuball J, Voelkl S, et al. Blockade of PD-L1 (B7-H1) augments human tumor-specific T cell responses in vitro. *Int J Cancer*. 2006;119(2):317-327.
210. Blank C, Brown I, Peterson AC, et al. PD-L1/B7H-1 inhibits the effector phase of tumor rejection by T cell receptor (TCR) transgenic CD8+ T cells. *Cancer Res*. 2004;64(3):1140-1145.
211. Topalian SL, Hodi FS, Brahmer JR, et al. Safety, activity, and immune correlates of anti-PD-1 antibody in cancer. *N Engl J Med*. 2012;366(26):2443-2454. doi:10.1056/NEJMoa1200690
212. Brahmer JR, Drake CG, Wollner I, et al. Phase I study of single-agent anti-programmed death-1 (MDX-1106) in refractory solid tumors: safety, clinical activity, pharmacodynamics, and immunologic correlates. *J Clin Oncol*. 2010;28(19):3167-3175. doi:10.1200/JCO.2009.26.7609
213. Homet Moreno B, Parisi G, Robert L, Ribas A. Anti-PD-1 therapy in melanoma. *Semin Oncol*. 2015;42(3):466-473. doi:10.1053/j.seminoncol.2015.02.008
214. Gong J, Chehrazi-Raffle A, Reddi S, Salgia R. Development of PD-1 and PD-L1 inhibitors as a form of cancer immunotherapy: a comprehensive review of registration trials and future considerations. *J Immunother Cancer*. 2018;6(1):8. doi:10.1186/s40425-018-0316-z
215. Andre T, Shiu KK, Kim TW, et al. Pembrolizumab in microsatellite-instability-high advanced colorectal cancer. *N Engl J Med*. 2020;383(23):2207-2218. doi:10.1056/NEJMoa2017699
216. Overman MJ, McDermott R, Leach JL, et al. Nivolumab in patients with metastatic DNA mismatch repair-deficient or microsatellite instability-high colorectal cancer (CheckMate 142): an open-label, multicentre, phase 2 study. *Lancet Oncol*. 2017;18(9):1182-1191. doi:10.1016/S1470-2045(17)30422-9
217. Ansell SM, Lesokhin AM, Borrello I, et al. PD-1 blockade with nivolumab in relapsed or refractory Hodgkin's lymphoma. *N Engl J Med*. 2015;372(4):311-319. doi:10.1056/NEJMoa1411087
218. Juarez-Salcedo LM, Sandoval-Sus J, Sokol L, Chavez JC, Dalia S. The role of anti-PD-1 and anti-PD-L1 agents in the treatment of diffuse large B-cell lymphoma: the future is now. *Crit Rev Oncol Hematol*. 2017;113:52-62. doi:10.1016/j.critrevonc.2017.02.027
219. Roemer MG, Advani RH, Ligon AH, et al. PD-L1 and PD-L2 genetic alterations define classical Hodgkin lymphoma and predict outcome. *J Clin Oncol*. 2016;34(23):2690-2697. doi:10.1200/JCO.2016.66.4482
220. Cader FZ, Schackmann RCJ, Hu X, et al. Mass cytometry of Hodgkin lymphoma reveals a CD4(+) regulatory T-cell-rich and exhausted T-effector microenvironment. *Blood*. 2018;132(8):825-836. doi:10.1182/blood-2018-04-843714
221. Salcedo R, Hixon JA, Stauffer JK, et al. Immunologic and therapeutic synergy of IL-27 and IL-2: enhancement of T cell sensitization, tumor-specific CTL reactivity and complete regression of disseminated neuroblastoma metastases in the liver and bone marrow. *J Immunol*. 2009;182(7):4328-4338. doi:10.4049/jimmunol.0800471
222. Newell F, Pires da Silva I, Johansson PA, et al. Multiomic profiling of checkpoint inhibitor-treated melanoma: identifying predictors of response and resistance, and markers of biological discordance. *Cancer Cell*. 2022;40(1):88-102 e7. doi:10.1016/j.ccell.2021.11.012
223. Larkin J, Chiarion-Sileni V, Gonzalez R, et al. Combined nivolumab and ipilimumab or monotherapy in untreated melanoma. *N Engl J Med*. 2015;373(1):23-34. doi:10.1056/NEJMoa1504030
224. Olson DJ, Eroglu Z, Brockstein B, et al. Pembrolizumab plus ipilimumab following anti-PD-1/L1 failure in melanoma. *J Clin Oncol*. 2021;39(24):2647-2655. doi:10.1200/JCO.21.00079
225. Triebel F, Jitsukawa S, Baixeras E, et al. LAG-3, a novel lymphocyte activation gene closely related to CD4. *J Exp Med*. 1990;171(5):1393-1405.
226. Tawbi HA, Schadendorf D, Lipson EJ, et al. Relatlimab and nivolumab versus nivolumab in untreated advanced melanoma. *N Engl J Med*. 2022;386(1):24-34. doi:10.1056/NEJMoa2109970
227. Dougall WC, Kurtulus S, Smyth MJ, Anderson AC. TIGIT and CD96: new checkpoint receptor targets for cancer immunotherapy. *Immunol Rev*. 2017;276(1):112-120. doi:10.1111/imr.12518
228. Curran MA, Kim M, Montalvo W, Al-Shamkhani A, Allison JP. Combination CTLA-4 blockade and 4-1BB activation enhances tumor rejection by increasing T-cell infiltration, proliferation, and cytokine production. *PLoS One*. 2011;6(4):e19499. doi:10.1371/journal.pone.0019499
229. May KFJr, Chen L, Zheng P, Liu Y. Anti-4-1BB monoclonal antibody enhances rejection of large tumor burden by promoting survival but not clonal expansion of tumor-specific CD8+ T cells. *Cancer Res*. 2002;62(12):3459-3465.
230. Delorme EJ, Alexander P. Treatment of primary fibrosarcoma in the rat with immune lymphocytes. *Lancet*. 1964;2(7351):117-120. doi:10.1016/s0140-6736(64)90126-6
231. Fefer A. Immunotherapy and chemotherapy of Moloney sarcoma virus-induced tumors in mice. *Cancer Res*. 1969;29(12):2177-2183.
232. Fernandez-Cruz E, Woda BA, Feldman JD. Elimination of syngeneic sarcomas in rats by a subset of T lymphocytes. *J Exp Med*. 1980;152(4):823-841. doi:10.1084/jem.152.4.823
233. North RJ. Cyclophosphamide-facilitated adoptive immunotherapy of an established tumor depends on elimination of tumor-induced suppressor T cells. *J Exp Med*. 1982;155(4):1063-1074. doi:10.1084/jem.155.4.1063

234. Berendt MJ, North RJ. T-cell-mediated suppression of antitumor immunity. An explanation for progressive growth of an immunogenic tumor. *J Exp Med.* 1980;151(1):69-80. doi:10.1084/jem.151.1.69
235. Morgan DA, Ruscetti FW, Gallo R. Selective in vitro growth of T lymphocytes from normal human bone marrows. *Science.* 1976;193(4257):1007-1008. doi:10.1126/science.181845
236. Eberlein TJ, Rosenstein M, Rosenberg SA. Regression of a disseminated syngeneic solid tumor by systemic transfer of lymphoid cells expanded in interleukin 2. *J Exp Med.* 1982;156(2):385-397. doi:10.1084/jem.156.2.385
237. Donohue JH, Rosenstein M, Chang AE, Lotze MT, Robb RJ, Rosenberg SA. The systemic administration of purified interleukin 2 enhances the ability of sensitized murine lymphocytes to cure a disseminated syngeneic lymphoma. *J Immunol.* 1984;132(4):2123-2128.
238. Mule JJ, Shu S, Schwarz SL, Rosenberg SA. Adoptive immunotherapy of established pulmonary metastases with LAK cells and recombinant interleukin-2. *Science.* 1984;225(4669):1487-1489.
239. Mazumder A, Eberlein TJ, Grimm EA, et al. Phase I study of the adoptive immunotherapy of human cancer with lectin activated autologous mononuclear cells. *Cancer.* 1984;53(4):896-905. doi:10.1002/1097-0142(19840215)53
240. Lotze MT, Matory YL, Ettinghausen SE, et al. In vivo administration of purified human interleukin 2. II. Half life, immunologic effects, and expansion of peripheral lymphoid cells in vivo with recombinant IL 2. *J Immunol.* 1985;135(4):2865-2875.
241. Lotze MT, Frana LW, Sharrow SO, Robb RJ, Rosenberg SA. In vivo administration of purified human interleukin 2. I. Half-life and immunologic effects of the Jurkat cell line-derived interleukin 2. *J Immunol.* 1985;134(1):157-166.
242. Rosenberg SA, Lotze MT, Yang JC, et al. Prospective randomized trial of high-dose interleukin-2 alone or in conjunction with lymphokine-activated killer cells for the treatment of patients with advanced cancer. *J Natl Cancer Inst.* 1993;85(8):622-632. doi:10.1093/jnci/85.8.622
243. Muul LM, Spiess PJ, Director EP, Rosenberg SA. Identification of specific cytolytic immune responses against autologous tumor in humans bearing malignant melanoma. *J Immunol.* 1987;138(3):989-995.
244. Rosenberg SA, Spiess P, Lafreniere R. A new approach to the adoptive immunotherapy of cancer with tumor-infiltrating lymphocytes. *Science.* 1986;233(4770):1318-1321. doi:10.1126/science.3489291
245. Rosenberg SA, Packard BS, Aebersold PM, et al. Use of tumor-infiltrating lymphocytes and interleukin-2 in the immunotherapy of patients with metastatic melanoma. A preliminary report. *N Engl J Med.* 1988;319(25):1676-1680. doi:10.1056/NEJM198812223192527
246. Gattinoni L, Finkelstein SE, Klebanoff CA, et al. Removal of homeostatic cytokine sinks by lymphodepletion enhances the efficacy of adoptively transferred tumor-specific CD8+ T cells. *J Exp Med.* 2005;202(7):907-912. doi:10.1084/jem.20050732
247. Sutmuller RP, van Duivenvoorde LM, van Elsas A, et al. Synergism of cytotoxic T lymphocyte-associated antigen 4 blockade and depletion of CD25(+) regulatory T cells in antitumor therapy reveals alternative pathways for suppression of autoreactive cytotoxic T lymphocyte responses. *J Exp Med.* 2001;194(6):823-832.
248. Klebanoff CA, Khong HT, Antony PA, Palmer DC, Restifo NP. Sinks, suppressors and antigen presenters: how lymphodepletion enhances T cell-mediated tumor immunotherapy. *Trends Immunol.* 2005;26(2):111-117. doi:10.1016/j.it.2004.12.003
249. Wrzesinski C, Paulos CM, Gattinoni L, et al. Hematopoietic stem cells promote the expansion and function of adoptively transferred antitumor CD8 T cells. *J Clin Invest.* 2007;117(2):492-501. doi:10.1172/JCI30414
250. Dudley ME, Yang JC, Sherry R, et al. Adoptive cell therapy for patients with metastatic melanoma: evaluation of intensive myeloablative chemoradiation preparative regimens. *J Clin Oncol.* 2008;26(32):5233-5239. doi:10.1200/JCO.2008.16.5449
251. Dudley ME, Wunderlich JR, Robbins PF, et al. Cancer regression and autoimmunity in patients after clonal repopulation with antitumor lymphocytes. *Science.* 2002;298(5594):850-854. doi:10.1126/science.1076514
252. Rosenberg SA, Yang JC, Sherry RM, et al. Durable complete responses in heavily pretreated patients with metastatic melanoma using T-cell transfer immunotherapy. *Clin Cancer Res.* 2011;17(13):4550-4557. doi:10.1158/1078-0432.CCR-11-0116
253. Gattinoni L, Klebanoff CA, Palmer DC, et al. Acquisition of full effector function in vitro paradoxically impairs the in vivo antitumor efficacy of adoptively transferred CD8+ T cells. *J Clin Invest.* 2005;115(6):1616-1626. doi:10.1172/JCI24480
254. Gattinoni L, Zhong XS, Palmer DC, et al. Wnt signaling arrests effector T cell differentiation and generates CD8+ memory stem cells. *Nat Med.* 2009;15(7):808-813. doi:10.1038/nm.1982
255. Kawakami Y, Eliyahu S, Delgado CH, et al. Identification of a human melanoma antigen recognized by tumor-infiltrating lymphocytes associated with in vivo tumor rejection. *Proc Natl Acad Sci U S A.* 1994;91(14):6458-6462. doi:10.1073/pnas.91.14.6458
256. Kawakami Y, Eliyahu S, Delgado CH, et al. Cloning of the gene coding for a shared human melanoma antigen recognized by autologous T cells infiltrating into tumor. *Proc Natl Acad Sci U S A.* 1994;91(9):3515-3519. doi:10.1073/pnas.91.9.3515
257. van Rooij N, van Buuren MM, Philips D, et al. Tumor exome analysis reveals neoantigen-specific T-cell reactivity in an ipilimumab-responsive melanoma. *J Clin Oncol.* 2013;31(32):e439-42. doi:10.1200/JCO.2012.47.7521
258. Tran E, Turcotte S, Gros A, et al. Cancer immunotherapy based on mutation-specific CD4+ T cells in a patient with epithelial cancer. *Science.* 2014;344(6184):641-645. doi:10.1126/science.1251102
259. Rosenberg SA, Restifo NP. Adoptive cell transfer as personalized immunotherapy for human cancer. *Science.* 2015;348(6230):62-68. doi:10.1126/science.aaa4967
260. Morgan RA, Dudley ME, Wunderlich JR, et al. Cancer regression in patients after transfer of genetically engineered lymphocytes. *Science.* 2006;314(5796):126-129. 10.1126/science.1129003
261. Johnson LA, Morgan RA, Dudley ME, et al. Gene therapy with human and mouse T-cell receptors mediates cancer regression and targets normal tissues expressing cognate antigen. *Blood.* 2009;114(3):535-546. doi:10.1182/blood-2009-03-211714
262. Lamers CH, Sleijfer S, van Steenbergen S, et al. Treatment of metastatic renal cell carcinoma with CAIX CAR-engineered T cells: clinical evaluation and management of on-target toxicity. *Mol Ther.* 2013;21(4):904-912. doi:10.1038/mt.2013.17
263. Parkhurst MR, Yang JC, Langan RC, et al. T cells targeting carcinoembryonic antigen can mediate regression of metastatic colorectal cancer but induce severe transient colitis. *Mol Ther.* 2011;19(3):620-626. doi:10.1038/mt.2010.272
264. Morgan RA, Chinnasamy N, Abate-Daga D, et al. Cancer regression and neurological toxicity following anti-MAGE-A3 TCR gene therapy. *J Immunother.* 2013;36(2):133-151. doi:10.1097/CJI.0b013e3182829903
265. Morgan RA, Yang JC, Kitano M, Dudley ME, Laurencot CM, Rosenberg SA. Case report of a serious adverse event following the administration of T cells transduced with a chimeric antigen receptor recognizing ERBB2. *Mol Ther.* 2010;18(4):843-851. doi:10.1038/mt.2010.24
266. Li Y, Moysey R, Molloy PE, et al. Directed evolution of human T-cell receptors with picomolar affinities by phage display. *Nat Biotechnol.* 2005;23(3):349-354. doi:10.1038/nbt1070
267. Robbins PF, Li YF, El-Gamil M, et al. Single and dual amino acid substitutions in TCR CDRs can enhance antigen-specific T cell functions. *J Immunol.* 2008;180(9):6116-6131. doi:10.4049/jimmunol.180.9.6116
268. Linette GP, Stadtmauer EA, Maus MV, et al. Cardiovascular toxicity and titin cross-reactivity of affinity-enhanced T cells in myeloma and melanoma. *Blood.* 2013;122(6):863-871. doi:10.1182/blood-2013-03-490565
269. Gross G, Waks T, Eshhar Z. Expression of immunoglobulin-T-cell receptor chimeric molecules as functional receptors with antibody-type specificity. *Proc Natl Acad Sci U S A.* 1989;86(24):10024-10028. doi:10.1073/pnas.86.24.10024
270. Maher J, Brentjens RJ, Gunset G, Riviere I, Sadelain M. Human T-lymphocyte cytotoxicity and proliferation directed by a single chimeric TCRzeta/CD28 receptor. *Nat Biotechnol.* 2002;20(1):70-75. doi:10.1038/nbt0102-70
271. Imai C, Mihara K, Andreansky M, et al. Chimeric receptors with 4-1BB signaling capacity provoke potent cytotoxicity against acute lymphoblastic leukemia. *Leukemia.* 2004;18(4):676-684. doi:10.1038/sj.leu.2403302
272. Song DG, Ye Q, Carpenito C, et al. In vivo persistence, tumor localization, and antitumor activity of CAR-engineered T cells is enhanced by costimulatory signaling through CD137 (4-1BB). *Cancer Res.* 2011;71(13):4617-4627. doi:10.1158/0008-5472.CAN-11-0422
273. Grupp SA, Kalos M, Barrett D, et al. Chimeric antigen receptor-modified T cells for acute lymphoid leukemia. *N Engl J Med.* 2013;368(16):1509-1518. doi:10.1056/NEJMoa1215134
274. Maude SL, Laetsch TW, Buechner J, et al. Tisagenlecleucel in children and young adults with B-cell lymphoblastic leukemia. *N Engl J Med.* 2018;378(5):439-448. doi:10.1056/NEJMoa1709866
275. Shah BD, Ghobadi A, Oluwole OO, et al. KTE-X19 for relapsed or refractory adult B-cell acute lymphoblastic leukaemia: phase 2 results of the single-arm, open-label, multicentre ZUMA-3 study. *Lancet.* 2021;398(10299):491-502. doi:10.1016/S0140-6736(21)01222-8
276. Porter DL, Levine BL, Kalos M, Bagg A, June CH. Chimeric antigen receptor-modified T cells in chronic lymphoid leukemia. *N Engl J Med.* 2011;365(8):725-733. doi:10.1056/NEJMoa1103849
277. Neelapu SS, Locke FL, Go WY. CAR T-cell therapy in large B-cell lymphoma. *N Engl J Med.* 2018;378(11):1065. doi:10.1056/NEJMc1800913
278. Schuster SJ, Bishop MR, Tam CS, et al. Tisagenlecleucel in adult relapsed or refractory diffuse large B-cell lymphoma. *N Engl J Med.* 2019;380(1):45-56. doi:10.1056/NEJMoa1804980
279. Abramson JS, Palomba ML, Gordon LI, et al. Lisocabtagene maraleucel for patients with relapsed or refractory large B-cell lymphomas (TRANSCEND NHL 001): a multicentre seamless design study. *Lancet.* 2020;396(10254):839-852. doi:10.1016/S0140-6736(20)31366-0
280. Munshi NC, Anderson LD,Jr, Shah N, et al. Idecabtagene vicleucel in relapsed and refractory multiple myeloma. *N Engl J Med.* 2021;384(8):705-716. doi:10.1056/NEJMoa2024850
281. van Zelm MC, Reisli I, van der Burg M, et al. An antibody-deficiency syndrome due to mutations in the CD19 gene. *N Engl J Med.* 2006;354(18):1901-1912. doi:10.1056/NEJMoa051568
282. Eshhar Z, Waks T, Gross G, Schindler DG. Specific activation and targeting of cytotoxic lymphocytes through chimeric single chains consisting of antibody-binding domains and the gamma or zeta subunits of the immunoglobulin and T-cell receptors. *Proc Natl Acad Sci U S A.* 1993;90(2):720-724. doi:10.1073/pnas.90.2.720
283. Brentjens RJ, Santos E, Nikhamin Y, et al. Genetically targeted T cells eradicate systemic acute lymphoblastic leukemia xenografts. *Clin Cancer Res.* 2007;13(18 pt 1):5426-5435. doi:10.1158/1078-0432.CCR-07-0674
284. Till BG, Jensen MC, Wang J, et al. Adoptive immunotherapy for indolent non-Hodgkin lymphoma and mantle cell lymphoma using genetically modified autologous CD20-specific T cells. *Blood.* 2008;112(6):2261-2271. doi:10.1182/blood-2007-12-128843
285. Kochenderfer JN, Wilson WH, Janik JE, et al. Eradication of B-lineage cells and regression of lymphoma in a patient treated with autologous T cells genetically engineered to recognize CD19. *Blood.* 2010;116(20):4099-4102. doi:10.1182/blood-2010-04-281931
286. Kochenderfer JN, Dudley ME, Kassim SH, et al. Chemotherapy-refractory diffuse large B-cell lymphoma and indolent B-cell malignancies can be effectively treated with autologous T cells expressing an anti-CD19 chimeric antigen receptor. *J Clin Oncol.* 2015;33(6):540-549. doi:10.1200/JCO.2014.56.2025
287. Lee DW, Gardner R, Porter DL, et al. Current concepts in the diagnosis and management of cytokine release syndrome. *Blood.* 2014;124(2):188-195. doi:10.1182/blood-2014-05-552729
288. Neelapu SS, Tummala S, Kebriaei P, et al. Chimeric antigen receptor T-cell therapy - assessment and management of toxicities. *Nat Rev Clin Oncol.* 2018;15(1):47-62. doi:10.1038/nrclinonc.2017.148
289. Brudno JN, Kochenderfer JN. Toxicities of chimeric antigen receptor T cells: recognition and management. *Blood.* 2016;127(26):3321-3330. doi:10.1182/blood-2016-04-703751
290. Crump M, Neelapu SS, Farooq U, et al. Outcomes in refractory diffuse large B-cell lymphoma: results from the international SCHOLAR-1 study. *Blood.* 2017;130(16):1800-1808. doi:10.1182/blood-2017-03-769620
291. Carpenter RO, Evbuomwan MO, Pittaluga S, et al. B-cell maturation antigen is a promising target for adoptive T-cell therapy of multiple myeloma. *Clin Cancer Res.* 2013;19(8):2048-2060. doi:10.1158/1078-0432.CCR-12-2422
292. Ali SA, Shi V, Maric I, et al. T cells expressing an anti-B-cell maturation antigen chimeric antigen receptor cause remissions of multiple myeloma. *Blood.* 2016;128(13):1688-1700. doi:10.1182/blood-2016-04-711903
293. Raje N, Berdeja J, Lin Y, et al. Anti-BCMA CAR T-cell therapy bb2121 in relapsed or refractory multiple myeloma. *N Engl J Med.* 2019;380(18):1726-1737. doi:10.1056/NEJMoa1817226
294. Cohen AD, Garfall AL, Stadtmauer EA, et al. B cell maturation antigen-specific CAR T cells are clinically active in multiple myeloma. *J Clin Invest.* 2019;129(6):2210-2221. doi:10.1172/JCI126397
295. Neelapu SS, Locke FL, Bartlett NL, et al. Axicabtagene ciloleucel CAR T-cell therapy in refractory large B-cell lymphoma. *N Engl J Med.* 2017;377(26):2531-2544. doi:10.1056/NEJMoa1707447
296. Brown CE, Mackall CL. CAR T cell therapy: inroads to response and resistance. *Nat Rev Immunol.* 2019;19(2):73-74. doi:10.1038/s41577-018-0119-y
297. Jacoby E, Nguyen SM, Fountaine TJ, et al. CD19 CAR immune pressure induces B-precursor acute lymphoblastic leukaemia lineage switch exposing inherent leukaemic plasticity. *Nat Commun.* 2016;7:12320. doi:10.1038/ncomms12320
298. Schuster SJ, Svoboda J, Chong EA, et al. Chimeric antigen receptor T cells in refractory B-cell lymphomas. *N Engl J Med.* 2017;377(26):2545-2554. doi:10.1056/NEJMoa1708566
299. Fraietta JA, Lacey SF, Orlando EJ, et al. Determinants of response and resistance to CD19 chimeric antigen receptor (CAR) T cell therapy of chronic lymphocytic leukemia. *Nat Med.* 2018;24(5):563-571. doi:10.1038/s41591-018-0010-1
300. Deng Q, Han G, Puebla-Osorio N, et al. Characteristics of anti-CD19 CAR T cell infusion products associated with efficacy and toxicity in patients with large B cell lymphomas. *Nat Med.* 2020;26(12):1878-1887. doi:10.1038/s41591-020-1061-7
301. Bassi P BCG (Bacillus of Calmette Guerin) therapy of high-risk superficial bladder cancer. *Surg Oncol.* 2002;11(1-2):77-83. doi:10.1016/s0960-7404(02)00008-7
302. Lim CJ, Nguyen PHD, Wasser M, et al. Immunological Hallmarks for clinical response to BCG in bladder cancer. *Front Immunol.* 2020;11:615091. doi:10.3389/fimmu.2020.615091
303. Bornstein RS, Mastrangelo MJ, Sulit H, et al. Immunotherapy of melanoma with intralesional BCG. *Natl Cancer Inst Monogr.* 1973;39:213-220.

304. Geisse JK, Rich P, Pandya A, et al. Imiquimod 5% cream for the treatment of superficial basal cell carcinoma: a double-blind, randomized, vehicle-controlled study. *J Am Acad Dermatol.* 2002;47(3):390-398. doi:S0190962202001263
305. Smyth EC, Flavin M, Pulitzer MP, et al. Treatment of locally recurrent mucosal melanoma with topical imiquimod. *J Clin Oncol.* 2011;29(33):e809-e811. doi:10.1200/JCO.2011.36.8829
306. Dummer R, Gyorki DE, Hyngstrom J, et al. Neoadjuvant talimogene laherparepvec plus surgery versus surgery alone for resectable stage IIIB-IVM1a melanoma: a randomized, open-label, phase 2 trial. *Nat Med.* 2021;27(10):1789-1796. doi:10.1038/s41591-021-01510-7
307. Luke JJ, Bao R, Sweis RF, Spranger S, Gajewski TF. WNT/beta-catenin pathway activation correlates with immune exclusion across human cancers. *Clin Cancer Res.* 2019;25(10):3074-3083. doi:10.1158/1078-0432.CCR-18-1942
308. Peng W, Chen JQ, Liu C, et al. Loss of PTEN promotes resistance to T cell-mediated immunotherapy. *Cancer Discov.* 2016;6(2):202-216. doi:10.1158/2159-8290.CD-15-0283 2159-8290.CD-15-0283 [pii]
309. Skoulidis F, Goldberg ME, Greenawalt DM, et al. STK11/LKB1 mutations and PD-1 inhibitor resistance in KRAS-mutant lung adenocarcinoma. *Cancer Discov.* 2018;8(7):822-835. doi:10.1158/2159-8290.CD-18-0099
310. Koyama S, Akbay EA, Li YY, et al. STK11/LKB1 deficiency promotes neutrophil recruitment and proinflammatory cytokine production to suppress T-cell activity in the lung tumor microenvironment. *Cancer Res.* 2016;76(5):999-1008. doi:10.1158/0008-5472.CAN-15-1439
311. Spranger S, Gajewski TF. Impact of oncogenic pathways on evasion of antitumour immune responses. *Nat Rev Cancer.* 2018;18(3):139-147. doi:10.1038/nrc.2017.117
312. Boackle SA. Advances in lupus genetics. *Curr Opin Rheumatol.* 2013;25(5):561-568. doi:10.1097/BOR.0b013e328363eb4e
313. Chat V, Ferguson R, Simpson D, et al. Autoimmune genetic risk variants as germline biomarkers of response to melanoma immune-checkpoint inhibition. *Cancer Immunol Immunother.* 2019;68(6):897-905. doi:10.1007/s00262-019-02318-8
314. Orozco RC, Marquardt K, Mowen K, Sherman LA. Proautoimmune allele of tyrosine phosphatase, PTPN22, enhances tumor immunity. *J Immunol.* 2021;207(6):1662-1671. doi:10.4049/jimmunol.2100304
315. Cubas R, Khan Z, Gong Q, et al. Autoimmunity linked protein phosphatase PTPN22 as a target for cancer immunotherapy. *J Immunother Cancer.* October 2020;8(2):e001439. doi:10.1136/jitc-2020-001439
316. Brownlie RJ, Wright D, Zamoyska R, Salmond RJ. Deletion of PTPN22 improves effector and memory CD8+ T cell responses to tumors. *JCI Insight.* 2019;5:e127847. doi:10.1172/jci.insight.127847
317. Cron K, Sivan A, Aquino-Michaels K, et al. Decreased host PKC-delta is associated with the T cell-inflamed tumor microenvironment and improves anti-tumor immunity. *J Immunother Cancer.* 2021;9(suppl 2):642.
318. Belot A, Kasher PR, Trotter EW, et al. Protein kinase cdelta deficiency causes mendelian systemic lupus erythematosus with B cell-defective apoptosis and hyperproliferation. *Arthritis Rheum.* 2013;65(8):2161-2171. doi:10.1002/art.38008
319. Dzutsev A, Goldszmid RS, Viaud S, Zitvogel L, Trinchieri G. The role of the microbiota in inflammation, carcinogenesis, and cancer therapy. *Eur J Immunol.* 2014;45(1):17-31. doi:10.1002/eji.201444972
320. Littman DR, Pamer EG. Role of the commensal microbiota in normal and pathogenic host immune responses. *Cell Host Microbe.* 2011;10(4):311-323. doi:10.1016/j.chom.2011.10.004
321. Wu HJ, Ivanov, Darce J, et al. Gut-residing segmented filamentous bacteria drive autoimmune arthritis via T helper 17 cells. *Immunity.* 2010;32(6):815-827. doi:10.1016/j.immuni.2010.06.001
322. Sivan A, Corrales L, Hubert N, et al. Commensal Bifidobacterium promotes antitumor immunity and facilitates anti-PD-L1 efficacy. *Science.* 2015;350(6264):1084-1089. doi:10.1126/science.aac4255
323. Vetizou M, Pitt JM, Daillere R, et al. Anticancer immunotherapy by CTLA-4 blockade relies on the gut microbiota. *Science.* 2015;350(6264):1079-1084. doi:10.1126/science.aad1329
324. Gopalakrishnan V, Spencer CN, Nezi L, et al. Gut microbiome modulates response to anti-PD-1 immunotherapy in melanoma patients. *Science.* 2018;359(6371):97-103. doi:10.1126/science.aan4236
325. Matson V, Fessler J, Bao R, et al. The commensal microbiome is associated with anti-PD-1 efficacy in metastatic melanoma patients. *Science.* 2018;359(6371):104-108. doi:10.1126/science.aao3290
326. Routy B, Le Chatelier E, Derosa L, et al. Gut microbiome influences efficacy of PD-1-based immunotherapy against epithelial tumors. *Science.* 2018;359(6371):91-97. doi:10.1126/science.aan3706
327. Davar D, Dzutsev AK, McCulloch JA, et al. Fecal microbiota transplant overcomes resistance to anti-PD-1 therapy in melanoma patients. *Science.* 2021;371(6529):595-602. doi:10.1126/science.abf3363

CHAPTER 50

Inborn Errors of Immunity

Isabelle Meyts, MD, PhD • Jean-Laurent Casanova, MD, PhD •
Luigi D. Notarangelo, MD, PhD

INTRODUCTION

Protective immunity to infection is carried out by all >400 discernible cell types of the human body. Among them is the immune system, which forms a complex, layered network of tissue-resident and circulating myeloid and lymphoid cells, most of which are of hematopoietic origin. It is geared for identifying and eliminating infectious agents, or perhaps internal threats like cancer cells or tissue damage. When one or more components of the immune system fail, the tissue and organism homeostasis is at stake, and increased susceptibility to infection, autoinflammation, autoimmunity, allergy, and/or tumors occur. This can also result from the derailment of cells other than leukocytes, outside the immune system per se. This immunodeficiency can be secondary (or acquired), for instance, due to chemotherapy, malnutrition, or infection (human immunodeficiency virus), or primary (or inherited). The birth of the field of primary immunodeficiency (PID), also known and further defined as inborn errors of immunity (IEIs), is usually set at 1952 when Colonel Ogden Bruton, a pediatrician, noticed an absent gamma globulin peak on the serum electrophoresis—at that time an innovative technique—of a boy with early-onset recurrent invasive pyogenic infections.[1-3] Bruton described his groundbreaking discovery as "No gammaglobulin, no antibody, ergo infections." He treated the boy successfully with injections of immunoglobulins. Not coincidentally, this unusual presentation with recurrent invasive infection was only noted in the post–World War II era, when infant mortality was decreasing, nutrition was improving, and foremost after the introduction of antibiotics, which allowed for survival of the patient. The landmark finding by Bruton led to the ultimate unraveling of the role and identity of the cells producing these gamma globulins and finally to the description of deleterious mutations in Bruton tyrosine kinase underlying X-linked agammaglobulinemia (XLA) or Bruton disease, more than 40 years later.[4,5] A field of intensive research on B cell immunology had opened up.

Nevertheless, several conditions, now considered classical IEIs, had been described many years before XLA: ataxia-telangiectasia (AT) was described in 1926, Wiskott-Aldrich syndrome (WAS) in 1937 as an atypical form of Werlhof syndrome in three brothers (idiopathic thrombocytopenia), epidermodysplasia verruciformis (EV) in 1946, and hereditary agranulocytosis in 1950.[6-10] Unlike the former two, the latter two conditions were explicitly stated to confer inherited predisposition to infection. Yet, EV was not considered as an IEI because of the lack of detectable anomaly of leukocytes. Agranulocytosis was not considered because in these years hematologists, not immunologists, studied neutrophils. As per AT and WAS, the immunodeficiency got studied from the late 1950s onward, as longer survival of the patients thanks to antibiotics started to unveil a pattern of recurrent infections.[11,12] Hence Bruton disease held stand as the blueprint of conventional PIDs. The term X-linked agammaglobulinemia like other conventional PIDs focused on the immunological aberration as the defining phenotype, not on the clinical expression, the infectious phenotype. The conditions were inherited as Mendelian traits, that is, monogenic disorders with full penetrance. The identification of these conventional PIDs, expressing as multiple and recurrent infections, caused by weakly virulent micro-organisms, and especially, occurring as the consequence of a readily measurable immunological anomaly, has been the adagio until very recently.

The description of more PIDs gradually extended the infectious phenotype to include autoimmunity, allergy, malignancy, and autoinflammation.[13] The first autoinflammatory phenotype reported was probably angioedema in 1965. The increasing recognition of clinical phenotypes other than infectious has led the PID community to embrace the term "inborn error of immunity" (IEI), first introduced in 1966, to so forth denominate inherited defects of the immune system, or of other cells, presenting with increased susceptibility to infection, allergy, autoimmunity, autoinflammation, and/or malignancy.[14-16]

The classification of IEIs based on the predominant immunological defect is still in use. Another concept is the innate/adaptive immunity dichotomy. Although this is an attractive concept for studying immunology at the cellular level, it is less relevant at the genetic and clinical level, as indeed most immune system genes play a role in both arms of the immune system. Moreover, in principle, all cells, not only the adaptive (T and B lymphocytes) and innate leukocytes (NK cells and other innate lymphoid cells, as well as resident and circulating phagocytes and dendritic cells) but all other cells can contribute to host defense, for instance, by lining bodily surfaces (epithelia, mucosa) or by secreting factors that influence development and/or function of immune cells, or that exert direct antimicrobial affects. The best-known example is the liver, which produces most components of the complement cascade. Recently, several conditions have been described in which the disruption of cell-intrinsic immunity underlies a specific phenotype. Within the framework of adaptive and innate immunity and measurable immunological defects, it is also understandable why conditions like EV have been only recently recognized as an IEI. Indeed, EV was first described in 1922 and classified as a congenital dermatosis with persistent and disseminated flat warts and an increased risk of nonmelanoma skin cancer.[9,17,18] An autosomal recessive (AR) mode of inheritance was documented in 1933 and a viral etiology in 1946. Orth identified the first two EV genes through genome-wide linkage analysis with the discovery of homozygous null mutations in *EVER1* and *EVER2*.[19] A third gene, *CIB1*, was recently described.[20] This is a keratinocyte condition, with no consistent leukocyte phenotype, that is, an inborn error of intrinsic immunity to β-human papilloma virus (HPV).

EV is also a beautiful example of an IEI presenting with a narrow infectious phenotype; more specifically it is an example of a Mendelian infection, in which otherwise healthy individuals, with normal resistance to other infections agents, can experience life-threatening illness with a specific pathogen.[21] Other examples include Mendelian susceptibility to mycobacterial diseases (MSMD) with an autosomal recessive (AR), autosomal dominant (AD), or X-linked recessive (XR) predisposition to infection with BCG and environmental mycobacteria, invasive dermatophytosis, chronic mucocutaneous candidiasis AR or AD, or lymphoproliferation caused by Epstein-Barr virus (EBV). Although the condition was described some time ago, the underlying genetic defects for these Mendelian infections have been unraveled from 1996 onward. As explained above, many of these conditions were intuitively not initially regarded as genuine IEI owing to the lack of a readily measurable immunological defect. In fact, the infections were most often regarded as "idiopathic." The dissection at the genetic level of these Mendelian infections offered proof of principle that humans can be selectively susceptible to infection with a single pathogen, while successfully defeating all other pathogens. This paradigm shift led to the discovery of infections that were monogenic yet not Mendelian in the strict sense. In recent years, it was further shown that isolated, severe, rare, or even common infections can be monogenic, at least in some patients, typically with incomplete clinical penetrance. For example, a common variant in TYK2 (P1104A) is responsible for about 1% of tuberculosis cases in populations of European descent.[22-24] P1104A has markedly fluctuated in frequency over the last 10,000 years of European history, with a dramatic decrease in frequency after the Bronze Age, pointing to strong negative selection.[23] Several other examples followed with the study of invasive pyogenic infections, herpes simplex virus encephalitis, and severe influenza pneumonitis.[25-40]

Clinical presentations characterized by increased susceptibility to infections (with a large or a narrow range of pathogens), autoimmunity, autoinflammation, allergy, and malignancy represent a large reservoir of potential phenotypes for the discovery of novel forms of IEI. No current classification of IEI is perfect, not the one based on phenotype (as most IEI are characterized by a blended phenotype), nor the one based on immunological anomalies, nor the one focused on the gene alteration. It mirrors the complexity of IEI and the challenge that lies ahead, to further meticulously describe IEI at the genetic, immunological, and clinical levels. With about 25,000 known protein-coding genes, the >3000 genes closely related biochemically to the currently known >400 loci underlying IEI, and given the differential phenotypes associated with gain or loss of function variants at these loci, we can predict that several thousands of IEI are yet to be unraveled. One of the pioneers in the field, Robert Good, rightfully referred to PIDs/IEIs as "experiments of nature," reviving a line of investigation that was illustrated by Cushing, Osler, Garrod, McQuarrie, and Burnet, among others. Indeed, no matter how rare a disorder, it can provide considerable insight into the fundamental mechanisms governing human immunity.

Scientific experiments suffer from the limitations inherent to their experimental nature, even when performed *in vivo*. Immunity *in natura* can be defined by the careful dissection of IEIs and by other related approaches, such as epidemiologic and evolutionary genetics. For instance, as opposed to experimental infections in mice and other laboratory animals, most human infections are natural, with the notable exception of the inoculation of live vaccines, the study of whose adverse reactions has played an important role. The genetic dissection of IEIs thus provides us with a unique opportunity to cast new light on the function of human genes in a natural ecosystem. Over the years, these observations have provided invaluable immunological insights, sometimes at odds with the corresponding mouse model. One of such striking differences is the greater redundancy displayed by genes in humans underlying Mendelian or genetic infections and even PID, than in mouse models under experimental conditions of infections.[41-44]

The genetic dissection of IEIs first relied on lengthy positional cloning efforts, together with Sanger sequencing and homozygosity mapping approaches. The first IEI to be unraveled at the molecular level was ADA deficiency in 1985 followed by cloning of the *CYBB* gene (responsible for X-linked chronic granulomatous disease) in 1986.[45,46] With the advent of next-generation sequencing (NGS) we have witnessed a rapid rise in the number of described IEIs, and more than 400 distinct defects were included in International Union of Immunological Societies (IUIS) classification of 2019 reflecting an increase of 200 new genes and diseases in 10 years.[15,47,48] A caveat for the use of NGS is the need for meticulous validation of the pathogenicity of any variant deemed causal.[49] Likewise, there is a risk of labeling heterozygous variants as irrelevant to the observed clinical phenotype without functional testing. Indeed, different phenotypes can be allelic at the same locus, and on the other hand different defects at the same locus may govern different phenotypes, by various mechanisms. For instance, different mutations in *RAG1* and *RAG2* result in a large phenotypic variety from severe combined immunodeficiency (SCID) to predominantly autoimmune phenotypes, depending on the residual activity of the mutant protein.[50-53] For several IEI genes, both gain-of-function (GOF) and loss-of-function (LOF) mutations have been described resulting in entirely different phenotypes. For instance, LOF variants in STAT1 result in mycobacterial or viral diseases, whereas GOF variants result in chronic mucocutaneous candidiasis with autoimmunity.[54,55] It is important to emphasize that there is an apparently high level of genetic heterogeneity for the IEI phenotypes (one clinical phenotype, multiple loci). Nevertheless, whatever the level of genetic heterogeneity, there is usually important physiological homogeneity related to the clinical phenotype: thus 33 genetic etiologies are described for MSMD, all resulting in disrupted IFN-γ signaling, instructing us on the essential role of that pathway in host defense against mycobacteria.

Whatever the complexity, thanks to NGS and other molecular advances, including generation of inducible pluripotent stem cells and their differentiation into cell lineages of interest, CRISPR-Cas9 technology, and single cell RNA sequencing, the study of IEI is rich in promises that can teach us human immunology. Of importance, IEIs have also paved the way to the development of innovative therapeutic approaches (the first successful allogeneic hematopoietic stem cell transplantation; the first application of immunoglobulin infusion, of enzyme replacement therapy, of cytokine therapy, and of gene therapy to cure disease); the field of IEI will undoubtedly continue to grow and expand. Even with the 400 or so known IEIs by now, we cannot cover in a reasonable or attractive way the details of these disorders. Of note, for several IEIs, autoimmune phenocopies have been described, where neutralizing autoantibodies directed against an essential cytokine result in a similar infectious susceptibility as has been described for chronic mucocutaneous candidiasis (IL-17A, IL-17F, IL-22), mycobacterial infection (IFN-gamma) and severe viral infection (type I IFN).[50,56-59] Finally, somatic mutations can also result in phenocopies of IEI.[60-62]

We will therefore focus on several IEIs manifesting with one of the five major phenotypes, paying particular attention to the unique immunologic lessons that these experiments of nature continue to teach us.[63]

IEIS GIVING RISE TO SUSCEPTIBILITY TO INFECTION WITH A BROAD OR NARROW RANGE OF PATHOGENS

As explained above, XLA was long regarded as the prototype of the conventional IEI. Despite the clinical description in the 1940s and 1950s of the occurrence of specific forms of infectious disease (such as mycobacterial infections or EV) running in families as Mendelian traits, it was only in the 1990s that a new paradigm emerged with the discovery of interferon (IFN)-γ receptor 1 (IFNGR1) deficiency, shortly thereafter followed by the discovery of other Mendelian defects of the IL-12/IL-23/IFN-γ axis accounting for MSMD. Moreover, in 2002, mutations in *EVER1* and *EVER2* were shown to underlie EV.[9,64,65] Together, these observations demonstrate that susceptibility to specific forms of infectious disease can be sustained by defects of various genes along the same pathway. These discoveries represented a paradigmatic shift in the field of IEI: on one end we have genetic defects that cause susceptibility to a broad range of pathogens (as exemplified by defects of T and B cell development and of granulocyte production and function); on the other hand, other forms of IEI lead to infection sustained by a single pathogen or by a narrow class of pathogens. Furthermore, the observations that *EVER1* and *EVER2* mutations affect the capacity of keratinocytes to resist to HPV infection were followed by other examples of gene defects that affect the capacity of nonhematopoietic cells to resist infection, as exemplified by defects of TLR3 signaling pathway in patients with herpes simplex encephalitis and of genetic defects of type I IFN responses in patients with life-threatening influenza or, more recently, COVID-19.[36,40,66] These forms of IEIs are representative of defects of "cell-intrinsic" (typically in nonhematopoietic cells) immunity.

We will here give an overview of the typical infectious presentation of the main categories of IEIs, as per the main branch of the immune system that is affected.

(Severe) Combined Immunodeficiencies

Combined immunodeficiencies (CIDs) and SCIDs are defects of lymphocyte number and/or function. SCID was first described in 1950 with Glanzmann and Riniker's report on two infants who presented with severe infections, diarrhea, failure to thrive, and disseminated candidiasis, leading to early death. Severe lymphoid depletion was found at the postmortem examination. Similar cases were described in Switzerland in 1958, with early death due to severe infection with often weakly virulent microbes, lymphopenia, and agammaglobulinemia.[67] These were the first descriptions of SCID, at that time called "Swiss type agammaglobulinemia," to distinguish it from Bruton agammaglobulinemia. SCID comprises a group of conditions characterized by a lack of autologous T cells and extreme susceptibility to infections sustained by a broad range of pathogens. Although all forms of SCID are characterized by a severe defect of T cell development, some forms of SCID also include B cell and/or NK cell lymphopenia. Therefore, from an immunological standpoint, SCID can be classified into various subgroups: a) T-B-NK-SCID; b) T-B+NK-SCID; c) T-B+NK+SCID. Even when B cells are not directly affected, the T cell impairment results in a B cell functional defect due to the lack of helper T cell function. The incidence of SCID has remained ill-defined until newborn screening for SCID has become available. It is now estimated to be at approximately 1:50,000 to 1:70,000 in Western countries, but it is much higher in geographical areas characterized by high consanguinity.[68] The genetic causes of CID are multiple and involve receptors and proteins essential for T lymphocyte survival and development. In fact, the study of SCID has played an essential role in unraveling key mechanisms governing human T cell development. In particular, SCID can be caused by gene defects that affect various steps along T cell development:

a. cell survival (*ADA, AK2, RAC2*);
b. thymocyte proliferation (*IL7R, IL2RG, JAK3, CD45*);
c. V(D)J recombination (*RAG1, RAG2, DCLRE1C, PRKDC, LIG4, NHEJ*);
d. Signaling through the pre-TCR and the TCR (*CD3D, CD3E, CD3Z, LAT, SLP76*);
e. Defects of thymus organogenesis (DiGeorge syndrome, *FOXN1, PAX1, CHD7*);
f. Egress from thymus (*CORO1A*)

Because of the severe T cell lymphopenia, patients with SCID are susceptible from early in life to recurrent, often life-threatening infections sustained by a broad range of pathogens, including bacteria, viruses, and fungi. Respiratory tract infections, chronic diarrhea, and failure to thrive are typical clinical manifestations. Opportunistic infections, such as *Pneumocystis jirovecii* pneumonia, are common. Laboratory data reveal profound T cell lymphopenia and lack of *in vitro* T cell proliferation to mitogens and antigens. The number of B and NK cells is variable, depending on the nature of the gene defect. Immunoglobulin levels are not particularly useful in the diagnosis of SCID, because IgG levels at birth reflect transplacental passage of maternally derived immunoglobulins. However, infants with SCID lack specific antibody responses. T cells of maternal origin that have crossed the placenta are detected in the periphery of a significant proportion of SCID babies; however, these T cells are oligoclonal and display an activated (CD45RA-) phenotype. Therefore, it is the absence of autologous naïve (CD45RA+) T cells that is the hallmark of SCID. As compared with infants with typical forms of SCID, those with atypical SCID have a slightly higher number of circulating autologous T cells, yet significantly lower than those seen in healthy babies.[69] In most cases, this phenotype is due to hypomorphic mutations in SCID-associated genes that allows for residual T cell development. Omenn syndrome is a peculiar SCID phenotype characterized by a variable number of autologous, oligoclonal, and activated T cells that infiltrate and damage peripheral organs such as the skin and the liver, causing generalized erythroderma, lymphadenopathy, and hepatosplenomegaly. In most cases, Omenn syndrome is due to hypomorphic mutations of *RAG1* or *RAG2*, which are lymphoid-specific genes involved in the initiation of V(D)J recombination and therefore assembly of immunoglobulin and T cell receptor molecules.[69]

CIDs include a heterogeneous group of conditions that allow residual development of T cells; however, T cell function is compromised, and therefore these patients are also prone to infections since early in life. Infections sustained by cytomegalovirus (CMV) and EBV are common and may lead to complications, including EBV-associated lymphomas. Patients with CID are also at higher risk of developing warts. Moreover, CID often manifests also with immune dysregulation, as a result of impaired T cell homeostasis and immune regulation. Autoimmune cytopenias, other forms of autoimmune disease, and granulomas affecting the skin, lungs, liver and other organs are frequently observed. Presence of the rubella virus vaccine strain has been demonstrated in granulomatous lesions of several patients with CID.[70] Administration of live vaccines, such as BCG, can cause disseminated infection in patients with SCID and is therefore contraindicated.[71] After the widespread introduction of the live attenuated oral rotavirus vaccine, chronic rotavirus diarrhea has become a typical SCID presentation in Western countries.[72] SCIDs are pediatric emergencies and require urgent diagnosis and treatment. Patients need to be aggressively treated for ongoing infections and protected from further exposures with antimicrobial prophylaxis, isolation, and immunoglobulin replacement therapy. However, unless immune reconstitution is achieved (typically with hematopoietic stem cell transplantation, or in selected cases with gene therapy, enzyme replacement therapy, or thymus transplantation), patients with SCID succumb to infections by 1 to 2 years of age. Of importance, newborn screening for SCID has recently become available. This test is based on the polymerase chain reaction–mediated amplification of TCR excision circles (TRECs) from a dried blood spot collected at birth. A low TREC count is indicative of a low number of circulating naïve T cells; eventually, enumeration of T cell subsets by flow cytometry and genetic tests may help obtain a definitive confirmation of SCID. The test has now been implemented in all states of the United States of America and in several other countries worldwide, greatly facilitating the diagnosis of patients with SCID, the implementation of measures to prevent infections, and prompting referral of babies with SCID to transplantation early in life.

Although less severe and not immediately life-threatening, CIDs carry a substantial morbidity and mortality rate owing to infections and require appropriate follow-up and careful consideration of the risks and benefits of a curative approach with hematopoietic stem cell transplantation (HSCT) versus a conservative medical management and follow-up.

Antibody Deficiencies

Disorders of antibody production represent >60% of all IEIs. They form a heterogeneous group of diseases, comprising B cell defects with agammaglobulinemia, such as XLA, class switch recombination defects, hypogammaglobulinemia, common variable immunodeficiency (CVID), immunoglobulin (Ig)-G subclass deficiency, selective IgA deficiency, specific antibody deficiency (SAD) with normal Ig levels, and transient hypogammaglobulinemia of infancy. The incidence varies greatly between the various disorders. For example, X-linked (or Bruton) agammaglobulinemia (XLA) has an estimated prevalence of 1 in 200,000 to 400,000 births, some monogenic forms of hypogammaglobulinemia, such as lipopolysaccharide-responsive beige-like anchor (LRBA) deficiency, are described in a handful of patients, and selective IgA deficiency is as common as 1:400 people, but with variable incidence in different ethnicities.[73]

The main clinical manifestations in patients with antibody deficiency are recurrent bacterial infections, generally occurring after 6 to 12 months of age, when the passive immunity conferred by maternal antibodies decreases below protective levels. The typical sites of infection are the upper and lower airways (otitis, sinusitis, bronchitis, and pneumonia) and the gastrointestinal tract. Invasive infections, such as septic arthritis, meningitis, and sepsis, are common if patients do not receive appropriate treatment. The infections are almost exclusively bacterial, predominantly with streptococci, staphylococci, *Hemophilus influenzae*, *Pseudomonas aeruginosa*, *Mycoplasma*, and *Campylobacter*.[74] Intestinal parasitic infections with *Giardia lamblia* and Cryptosporidium spp. are also reported.[74-76] Patients are not unduly susceptible to viral infections, with two notable exceptions: enteroviruses can cause chronic enteritis and meningoencephalitis in patients with XLA, and patients with activated phosphatidylinositol 3 kinase δ syndrome (APDS) cannot control herpes virus group infections.[77-79]

The mainstay of treatment for humoral IEIs is immunoglobulin replacement therapy, accompanied by prophylactic antibiotics if necessary. Chronic or repeated infections lead to the development of organ damage in the form of hearing loss and bronchiectasis, which are a typical evolution in patients with agammaglobulinemia despite optimal Ig substitution. Patients with pervasive defects of B cell development and function cannot mount an appropriate response to vaccinations, either conjugated or polysaccharide. Of note, patients with milder forms of hypogammaglobulinemia, IgG subclass deficiency, SAD, and selective IgA deficiency can be pauci- or asymptomatic.

Although defects of early B cell development mostly result in B cell lymphopenia, severe hypogammaglobulinemia, and an infectious phenotype, other defects impairing terminal B cell differentiation or function can cause immune dysregulation leading to autoimmune and autoinflammatory manifestations. These features complicate over 30% of CVID and are part of the typical presentation of LRBA deficiency, CTLA4 deficiency, and APDS.[80-82] Moreover, a proportion of patients with selective IgA deficiency have autoimmune disease, celiac disease, allergy, and malignancy, and allergy with asthmatic symptoms is a common manifestation of SAD.[83,84]

Finally, X-linked CD40 ligand (CD40LG) deficiency is the most common form of class-switch recombination defects. These patients have a normal number of circulating B cells; however, they typically manifest dysgammaglobulinemia, with low levels of IgG and IgA and normal to elevated serum IgM.[85] CD40LG is expressed on the surface of activated CD4+ T cells and interacts with CD40 expressed by B cells. Lack of CD40LG expression causes defective germinal center reaction and impaired class switch recombination. Besides recurrent bacterial infections, patients with CD40LG deficiency are also at risk for opportunistic infections sustained by *P. jirovecii*, *Cryptosporidium*, and CMV.[86] These infections are accounted for by impaired interaction between CD40LG-expressing T cells and dendritic cells and monocytes/macrophages, which express CD40. Severe biliary tract disease (sclerosing cholangitis, bile duct carcinoma) is frequently observed.[87,88] Therefore, both X-linked CD40LG deficiency and the rare AR CD40 deficiency are not pure forms of humoral IEI, but rather are combined immune deficiencies, and as such can be cured by allogeneic HSCT.[89-91]

Neutrophil Defects

Neutrophil defects are roughly divided into two categories: defects of neutrophil number and of neutrophil function. In most cases, neutropenia is secondary to infections, malignancies, use of chemotherapeutic or immunosuppressive agents, or autoimmune diseases.[92] Immune-mediated forms include both autoimmune neutropenia and alloimmune neutropenia, a condition of early infancy secondary to transplacental passage of maternal antibodies directed against neutrophils. Besides these acquired conditions, inborn errors of neutrophil number include severe congenital neutropenia (SCN), cyclic neutropenia, and bone marrow failure disorders, in which neutropenia is part of broader defects of hematopoiesis. Some metabolic and glycosylation disorders can also cause neutropenia. Disorders of phagocyte function are heterogeneous, comprising chronic granulomatous disease (CGD), leukocyte adhesion deficiencies (LADs), specific granule deficiency, β-actin deficiency, and several other rare defects. Once again, the in-depth molecular studies of these disorders have provided us with ever increasing insight into granulocyte development, function, and the respiratory burst.

SCN is a rare IEI with an estimated frequency of 1 to 2 cases per 10*6.[93] Rolph Kostmann described this condition already in 1956 in a Swedish family with invasive bacterial infections and severe neutropenia, with a maturation arrest at the promyelocyte-myelocyte stage.[10] More than 50% of the currently described SCN cases are caused by heterozygous mutations in the *ELANE* gene (OMIM*130130).[94] The hallmark manifestations of SCN are early-onset recurrent bacterial infections, usually from birth. Superficial bacterial abscesses, oral ulcers, cutaneous infections, omphalitis, pneumonia, and otitis media are characteristic. Oral ulcers, periodontitis, and gingivitis are common and lead to loss of permanent teeth in childhood. Patients with SCN typically have persistent severe neutropenia with absolute neutrophil count <500/μL.[95] Over 90% of patients affected by SCN respond to treatment with granulocyte colony-stimulating factor (G-CSF), although this therapy is likely associated with an increased risk of myelodysplasia and myeloid leukemia due to acquired mutations in the cytoplasmic tail of the G-CSF receptor (CSF3R).[96-98] HSCT remains the only definitive treatment. Prior to the introduction of G-CSF in 1989, SCN was almost invariably fatal with half of the patients dying from sepsis in the first year of life and the other half dying during early childhood.[99,100]

CGD is the first described, most common, and best characterized condition due to a functional defect of phagocytes. It was first described clinically in individual patients in 1954, and the first series of affected girls was reported in 1965.[101,102] It was first called "fatal granulomatous disease of childhood." A study in 1966 showed the impaired killing of staphylococci by granulocytes is caused by impaired phagocyte respiratory burst due to defects in the NADPH oxidase protein complex, a six-protein complex in its fully assembled state, capable of producing superoxide.[103] This led then to the identification of the first underlying genetic lesion.[46] The majority of CGD cases is accounted for by mutations in the gene encoding the membrane subunit gp91^{phox}, which together with the p22^{phox} constitutes the cytochrome b558. Patients with CGD show a selective susceptibility to staphylococci, *Burkholderia cepacia*, *Klebsiella pneumoniae*, Salmonella spp., *Serratia marcescens*, Nocardia spp., Aspergillus spp., and mycobacteria if the patients are exposed to BCG or tuberculosis. In particular, *Aspergillus nidulans* is a virtually exclusive CGD pathogen causing mostly osteomyelitis. The most frequent sites of infections are the skin, lungs, lymph nodes, and liver, although osteomyelitis and perianal abscesses are also common.[104] Autoinflammatory manifestations are present in patients with CGD, especially in the form of granulomatous lesions especially of the gastrointestinal tract and of the genitourinary tract. In patients with p40 deficiency, the autoinflammation predominates and invasive bacterial or fungal infection is rare. Treatment includes antibiotic and antifungal prophylaxis, immunomodulatory therapy with IFN-γ, and HSCT.[105] Especially since the introduction of reduced intensity conditioning regimens, survival has substantially improved, also for high-risk patients with active infection.

LAD is caused by impaired adhesion of leukocytes to the blood vessel walls, leading to deficient migration of neutrophils to the sites of infection.[106,107] In LAD1, mutations in *ITGB2* (OMIM*600065), the gene encoding the beta-subunit of the integrins, are causative, whereas in LAD2, fucosylation of the selectin ligands is disturbed due to mutations in *SLC35C1* (OMIM*605881). LAD2 is thus a congenital disorder of glycosylation, resulting also in mental retardation and short stature. LAD-III finally is characterized by additional platelet dysfunction and is caused by mutations in *FERMT3* (OMIM*607901) encoding kindlin-3, which regulates beta-integrin conformation in blood cells. Patients have recurrent necrotizing skin and mucosal bacterial infections with high neutrophilia but without pus formation, sepsis, omphalitis, severe gingivitis and periodontitis, and poor wound healing. Patients also develop nonpurulent necrotizing lesions of the skin and mucous membranes, resulting in high mortality at early age. HSCT is the only curative treatment for LAD. Lifelong antibiotic prophylaxis may be envisaged in mild LAD1 with residual function.[107]

Disorders of the Complement System

By 1884, Grohmann had established that cell-free serum *in vitro* is capable of killing micro-organisms.[108] As early as 1923, it was hypothesized that "alexin," the protective substance in serum, killed microorganisms by damaging their membrane.[109] This finding was gradually refined to what we know now as the complement system, a pattern recognition-based system of soluble proteins cross-linking innate and adaptive immunity. It is activated via three pathways: the classical pathway, induced by antigen-antibody complexes; the lectin pathway, induced by microbial mannose residues; and the alternative pathway, induced by exposure to foreign surfaces.[110] Several regulatory factors are present to prevent excessive activation and inflammation, such as C1 inhibitor, factor I, and factor H. Complement is viewed as the first line of defense against microbial particles, but it is now increasingly recognized that complement and the adaptive immune system are intertwined. At least 90% of complement proteins in the serum are synthesized by hepatocytes, thus showing the important role of somatic cells in immunity.[111] Properdin is produced by monocytes/macrophages and granulocytes and lymphocytes, whereas factor D is produced by adipocytes and monocytes/macrophages. Most inherited deficiencies are uncommon, although some may be particularly common in specific populations, for example, C9 deficiency affecting as many as 0.1% of the population in Japan.

Defects of various complement components are typically associated with increased susceptibility to recurrent and invasive bacterial infections and with autoimmune disorders.[110] The most common presentation is recurrent *Neisseria* spp. meningitis, which is caused by defects of factor D and properdin of the alternative pathway and defects of the terminal components (C5-C9). In factor D and properdin deficiencies, the disease course is more severe with fulminant recurrent meningitis and sepsis. Defects of early classical pathway components C1, C2, and C4 cause susceptibility to infection with pyogenic encapsulated bacteria, especially *Streptococcus pneumoniae* and *Haemophilus influenzae*. Moreover, they underlie systemic lupus erythematosus (SLE)-like autoimmune diseases, especially C1q deficiency. C3 deficiency also causes recurrent pyogenic infections, mainly with gram-negative bacteria such as *Neisseria meningitidis*, *Enterobacter aerogenes*, *H. influenzae*, and *Escherichia coli*. It also predisposes to membranoproliferative glomerulonephritis. Factor I and factor H deficiency, causing spontaneous activation of the alternative pathway and subsequent consumption of C3, can cause a similar infectious phenotype as C3 deficiency.

Unconventional IEIs: Inborn Susceptibility to Infection With a Narrow Range of Pathogens

As described in the introduction, the decades-long quest to understand the interindividual variability of susceptibility to infection found its synthesis in the genetic paradigm of infectious diseases, with the description of infections inherited as Mendelian traits and then by monogenic infections (with incomplete penetrance) (Table 50.1). Like most IEIs these also have high genetic heterogeneity, yet physiological homogeneity, in terms of the affected immunological pathway.

We will here describe herpes simplex encephalitis (HSE) as an example of monogenic susceptibility to infection sustained by a single pathogen (Table 50.2).

The herpes simplex virus (HSV) is ubiquitous and infects up to 60% to 70% of the human population. In rare cases, a primary HSV infection manifests as HSE, often with devastating consequences. The study of several patients with HSE has led to the discovery of inborn errors of Toll-like receptor 3 (TLR3) signaling as underlying susceptibility to HSE. TLR3 is an UNC93B-dependent endosomal pattern recognition receptor that binds viral double-stranded RNA. Upon

TABLE 50.1 Mendelian and Monogenic Infections: Nonexhaustive Overview (Excluding Herpes Simplex Virus Encephalitis)

Infectious Agent	Gene(s)	References
Mycobacterium tuberculosis	IL12RB1, TYK2, incl. common variant P1104A	112-117 22,118-121
Isolated staphylococcal and pneumococcal disease	RPSA, IRAK-4, MYD88, NEMO, TIRAP	25-28,122-125
Influenza virus	IRF7, IRF9, TLR3	36,38,40,126,127

TABLE 50.2 Single-Gene Inborn Errors of Immunity Underlying Isolated Herpes Simplex Virus Encephalitis

Gene	Inheritance	Defect	Clinical Presentation	References
TLR3	AD	Partial	Forebrain HSE	30,128
			IAV pneumonia	40
			Zoster ophthalmicus	129
	AR	Partial, complete	Forebrain HSE	128,130
UNC93B1	AR	Complete	Forebrain HSE	29
TRIF	AD	Partial	Forebrain HSE	33
	AR	Complete	Forebrain HSE	33
TRAF3	AD	Partial	Forebrain HSE	32
TBK1	AD	Partial	Forebrain HSE	31
IRF3	AD	Partial	Forebrain HSE	131
SNORA31	AD	Partial	Forebrain HSE	34
IFNAR1	AR	Complete	Forebrain HSE	132
DBR1	AR	Partial	Brainstem HSE	133
			Brainstem encephalitis by Norovirus, IBV virus	35

IAV, influenza A virus; IBV, influenza B virus; HSE, herpes simplex encephalitis.

binding, it recruits adapter molecule TIR domain–containing adapter-inducing IFN-β (TRIF) for the activation of TANK-binding kinase 1 (TBK1) and IκB kinase-ε (IKKε) via TNFR-associated factor 3 (TRAF3). These molecules activate IFN regulatory factor 3 (IRF3) and upregulate production of type I IFNs. Genetic defects in *TLR3*, *UNC93B*, *TRIF*, *TRAF3*, *TBK1*, and *IRF3* have been described to cause forebrain HSE.[2 9-33,131] Studies in induced pluripotent stem cell (iPSC)-derived cortical neurons but not trigeminal ganglion neurons from the patients demonstrated impaired cell intrinsic immunity to HSV-1. Type I IFN was shown to be an essential effector molecule induced by the TLR3 pathway thanks to the identification of a patient with HSE due to AR IFNAR1 deficiency.[132] The mechanism by which TLR3 controls HSV-1 infection in the brain does not seem to involve its recognition of double-stranded (dsRNA) intermediates or by-products in the course of infection, but via its control of the tonic, basal type I IFN levels in cortical neurons.[134] Moreover, the TLR3 pathway is redundant in leukocytes, accounting for the lack of dissemination of HSV-1 in the course of HSE.[135] In all genotypes of forebrain HSE, except for UNC93B deficiency, incomplete penetrance was reported. Of interest, the TLR3 P554S variant found in HSE, was also found in patients with severe influenza pneumonitis. HSE and influenza pneumonitis are thus two severe infections allelic at the TLR3 locus, with incomplete penetrance for both phenotypes accounting for the absence or rarity of dual infections at the individual level.[40] More recently, defects of the TLR3/type I IFN pathway have been identified in 3.5% of patients with life-threatening COVID-19, thus confirming the critical role played by this pathway in antiviral human immune response.[66] AD SNORA31 deficiency underlies forebrain HSE by another mechanism.[34] AR DBR1 deficiency underlies brainstem HSE, the pathophysiology of which is still unknown.[35] Finally HSE can rarely occur in the context of broad defects of type I IFN response, for example, in partial AR STAT1 deficiency.[136,137] The study of HSE has thus given us tremendous insight into the importance of cell-intrinsic immunity and of CNS-intrinsic immunity in particular.

AUTOIMMUNITY AS A PHENOTYPE OF IEI

Autoimmunity is a frequent phenotype of IEI. In fact, in a number of IEIs it may even be the first and/or predominant manifestation. The first IEI with autoimmunity phenotype identified was probably complement deficiency as early as 1972 with manifestations of SLE described in C2 deficiency.[138,139] As autoimmunity is a common trait in the general population, affecting about 1 in 1,000, the question of an underlying IEI, caused by rare variants, is often posed late in the disease course, which renders these patients at risk of complications. A recent French retrospective study focused on the prevalence of autoimmune manifestations among all types of IEIs. One or more autoimmune and inflammatory manifestations were noted in 26.2% of the patients.[140] The study highlighted the particularly increased risk for autoimmune hemolytic anemia (AIHA) (830 times) and immune thrombocytopenia (60 times) in children with IEI, when compared with the general population. Although autoimmune manifestations occurred in patients with all types of IEIs, those with T cell defects or CVID had the highest risk. Moreover, overall survival was significantly worse in patients with IEI and autoimmune/autoinflammatory manifestations. Furthermore, the survival post-HSCT was worse in case of pre-HSCT autoimmunity or inflammation. Vice versa, the frequency of IEI is very high in children presenting with Evans syndrome highlighting the need for expert immunology advice in such patients.[141]

In several IEIs, an increased risk for autoimmune disease is at the center of the clinical phenotype. Patients with autoimmune lymphoproliferative syndrome (ALPS) present mainly with autoimmune cytopenia and generalized lymphoproliferation due to defective apoptosis.[61,142-145] In most cases, ALPS is due to heterozygous dominant-negative mutations of the FAS (CD95) gene, which encodes for a molecule that is expressed on the cell surface of activated T lymphocytes as a trimer. FAS contains a death domain in its intracytoplasmic tail. Following interaction of FAS with its ligand (FASLG), a signaling cascade is triggered that results in apoptosis of T cells, thereby extinguishing T cell–mediated immune responses. Heterozygous, dominant-negative *FAS* mutations generate trimers that are inefficient in delivering apoptosis signals. More rarely, ALPS is due to AR *FASLG* mutations or to defects of other molecules downstream of FAS, such as FADD.[146,147] All these defects affect the extrinsic pathway of apoptosis. More rarely, ALPS may also be due to defects of the intrinsic pathway of apoptosis, which involves release of cytochrome c from the mitochondria.[143]

The study of patients with an ALPS-like phenotype (but without mutations in known ALPS-associated genes) led to the discovery of other IEIs associated with autoimmune disease. The inhibitory cytotoxic T lymphocyte associated antigen 4 (CTLA-4) molecule outcompetes the costimulatory molecule CD28 for the binding to CD80 and CD86 expressed by antigen-presenting cells and mediates their transendocytosis, thereby limiting T cell activation. Furthermore, CTLA-4 also plays an important role in regulatory T (T_{reg}) cell function.[148] Heterozygous LOF mutations of the *CTLA4* gene, resulting in haploinsufficiency, cause an IEI characterized by a high frequency of autoimmune and lymphoproliferative manifestations. In particular, 33% of the patients with CTLA-4 deficiency present with autoimmune cytopenia, which has been reported in even higher frequency (42%) in patients with LRBA deficiency.[149-151] LRBA is an endosomal adaptor molecule that enables recycling of CTLA-4 upon internalization; in the absence of LRBA, CTLA-4 is targeted to lysosomal degradation, thus explaining the similarities of clinical manifestations between CTLA-4 and LRBA deficiencies.[152]

Hyperactivation of the mTOR signaling pathway is another cause of altered T_{reg} function. The activated PI3K delta syndrome (APDS) is due to heterozygous mutations of the *PIK3CD* gene (encoding for the p110δ catalytic subunit of the enzyme) or of the *PIK3R1* gene (encoding the p85α regulatory subunit), resulting in hyperactivation of PI3Kδ and of the mTOR pathway.[153] APDS is characterized by both increased susceptibility to infections and autoimmunity. Autoimmunity is frequently seen also in patients with LOF mutations in nuclear factor kappa B subunit 1 (*NFKB1*) and *NFKB2* genes.[154,155] Finally, GOF mutations of signal transducer and activator of transcription (*STAT1*) and *STAT3* genes represent another group exceptionally affected by autoimmunity.[156,157] The discovery of the molecular defects underlying these forms of IEI has also prompted targeted molecular interventions, including use of anti-IL-6 receptor monoclonal antibody (tocilizumab) and JAK inhibitors in patients with STAT3 GOF, CTLA4-Ig fusion protein (abatacept) in CTLA4 deficiency, and leniolisib (a PI3Kδ inhibitor) in APDS.[152,158,159] These drugs complement other immunosuppressive strategies, such as those based on depletion of B cells and plasma cells (rituximab/proteasome inhibitors) and on T cell inhibition (sirolimus, mycophenolic acid).

Many different mechanisms have been proposed to underlie IEI manifesting with autoimmunity. Cell-autonomous mechanisms are of utmost importance and include (1) impaired negative selection of high-affinity self-reactive T and B cells; (2) defective B cell receptor (BCR) editing in the periphery; (3) impaired peripheral self-antigen-mediated activation-induced cell death; (4) gain of function of molecules involved in B or T cell activation; (5) intrinsic defect in regulatory T cells; and (6) homeostatic expansion of self-reactive T lymphocytes. In a single IEI often several mechanisms are at play to provoke autoimmunity. In CVID, mucosal barrier leakage and ineffective germinal center responses with inadequate clonal redemption have been proposed.[160] In LRBA deficiency and CTLA4 deficiency as well as APDS, expansion of T follicular helper (Tfh) cells has been demonstrated.[160-162] Although infections are classically thought of as an extrinsic trigger to the onset of autoimmunity, prospective studies are needed to definitively confirm this in most IEI. Given the high risk of underlying IEI, patients, and especially children, presenting with autoimmune disease should be meticulously screened for IEI, in collaboration with an immunologist, no matter what the clinical presentation is. In-depth study of rare inherited IEIs associated with autoimmunity as a predominant trait has provided crucial insight into the human immune system's protection against self-directed harmful immune responses. Two IEIs have provided us with a new model for studying autoimmunity: autoimmune polyendocrinopathy-candidiasis-ectodermal dystrophy (APECED) and immunodysregulation polyendocrinopathy enteropathy X-linked (IPEX).

APECED, also known as autoimmune polyglandular syndrome type 1 (APS1, OMIM*240300) is characterized by multiple organ-specific autoimmune manifestations as well as by the production of a variety of autoantibodies.[163-165] It was first reported by Thorpe and Handley in 1929.[166] The predominant clinical features, the so-called Whitaker triad, includes adrenal insufficiency, hypoparathyroidism, and chronic mucocutaneous candidiasis (CMC), along with ectodermal dystrophy manifesting with enamel hypoplasia and keratitis.[163,167-169] Other clinical features are premature ovarian failure, pernicious anemia, vitiligo, alopecia, intestinal malabsorption, and autoimmune pneumonitis.[170] The gene mutated in APECED was unraveled in 1997 as the Autoimmune Regulator (*AIRE*).[171,172] AIRE is expressed by a subset of medullary thymic epithelial cells (mTECs) and plays a key role in central tolerance to self-antigens. In particular, it serves as a transactivator of transcription and regulates the expression of a very large number of self-peptides that are presented in association with MHC molecules expressed by mTECs and by thymic dendritic cells to developing T cells.[173-175] Those T cells that express a T cell receptor with high affinity for the self-peptides are negatively selected in the thymus. In addition, developing T cells with intermediate affinity for self-peptides are poised to become T_{reg} cells.[176,177] AIRE is also expressed in a subset of dendritic cells in the periphery, but its role there is still unclear. The CMC manifestation of APECED was explained by the presence of neutralizing autoantibodies to Th17 cytokines (IL-17A, IL-17F, and IL-22).[56,178] Interestingly, autoantibodies against type I IFN were found in 2006 and thought to be silent until they were shown to underlie critical COVID-19 in these patients.[179] Moreover, no antibodies against Th17 cytokines were found in a variety of other autoimmune diseases.[56] An additional explanation for the CMC was proposed to be excessive type 1 immunity in the patients' mucosae.[180]

In most cases, APECED is inherited as an AR trait with onset in childhood and with 90% of affected patients developing the classical triad of manifestations by age 20 years.[181] The disease occurs more frequently in isolated populations, with a prevalence of 1/9000 in Iranian Jews, 1/25,000 in Finns, and 1/14,000 in Sardinians. In 2015, dominant-negative mutations in *AIRE* were unveiled to cause organ-specific autoimmune disease. The mutations causing AR APECED were mostly located in the CARD domain of AIRE, whereas the mutations acting in an AD pattern were located in the PHD1 domain and a single one in the SAND domain.[182] Patients with dominant-negative *AIRE* mutations manifest with organ-specific autoimmune disease with incomplete penetrance, delayed clinical onset, and slower disease progression.[183] In these patients the presence of autoantibodies against interferon, a hallmark of AR APECED, was less prevalent, as was CMC, probably reflective of some residual AIRE function. The frequency of the dominant-negative PHD1 *AIRE* mutations in various populations proved to be around 0.0009, which is close to the prevalence of organ-specific autoimmune disease in the general population. Treatment hinges on substitution therapy for the failing endocrine organs and on biological treatments to eliminate or suppress the autoantibodies and the autoreactive T cells. Chronic antifungal prophylaxis is often needed.

Optimal organization of the thymic microenvironment and correct maturation of TECs depends on cross talk with developing thymocytes. IEIs characterized by impaired T cell development are often associated with perturbation of thymic architecture and decreased number of AIRE-expressing mTECs. Reduced expression of AIRE in the thymus of these patients may limit expression of self-peptides in the thymus and therefore compromise negative selection of self-reactive T cells, leading to the development of autoimmune manifestations. This phenomenon has been neatly established in patients with hypomorphic *RAG1* or *RAG2* mutations manifesting with Omenn syndrome or with combined immunodeficiency with granulomas and autoimmunity.[184,185]

In 1982, Powell et al described a large kindred with 18 males who died prematurely owing to a syndrome characterized by chronic diarrhea with villous atrophy, eczema, hemolytic anemia, diabetes mellitus, autoimmune thyroiditis, and recurrent infections.[186] The disease was subsequently defined as immune dysregulation, polyendocrinopathy, enteropathy, X-linked (IPEX) syndrome (OMIM*304790). Studies in affected patients and in *scurfy* mice demonstrated that the syndrome is caused by mutations in the gene encoding the forkhead box protein 3 (FOXP3) located on the X chromosome.[187,188] *Scurfy* mice had been described 20 years before, as a spontaneously occurring mutant mouse strain. They have a characteristic skin rash, small-thickened ears, conjunctivitis, and marked runting due to diarrhea and malabsorption, similar to the phenotype of patients with IPEX.[187,189] They die invariably by 3 to 4 weeks of life. Studies in *scurfy* mice and in patients with IPEX aided in unraveling the role of FOXP3 in the development and function of regulatory T cells. FOXP3 is a transcription factor belonging to the forkhead-winged-helix family of transcription factors. It is central to the identity and function of the CD4+ regulatory T cells. T_{regs} can suppress the activation and function of other leukocytes, by secreting anti-inflammatory cytokines, by expressing coinhibitory molecules (CTLA4), by regulation of antigen-presenting cell activity, by starving effector cells, or even by mediating a cytotoxic response. A defect in FOXP3 thus leads to a functional deficiency in T_{regs} and generalized peripheral autoimmunity due to a break of peripheral immunological tolerance. Although FOXP3 has long been seen as the master transcription factor leading to T_{reg} commitment in the thymus, a pioneer factor leading to a T-reg permissive state for gene transcription has been described to precede FOXP3 expression.[190] FOXP3 expression upregulates expression of CD25 (the IL-2 receptor α chain), and signaling through IL-2R is crucial for T_{reg} survival. Moreover, T_{regs} also express CTLA4 at high density. Consistent with these observations, genetic defects of CD25 and of CTLA4 are also associated with autoimmunity.

The classical triad of IPEX consists of early-onset intractable diarrhea, neonatal type 1 diabetes mellitus, and eczema.[191] If undiagnosed, the condition is fatal in early life. The gut disease correlates histopathologically with villous atrophy or graft-versus-host–like disease and is associated with antienterocyte autoantibodies. The dermatitis in IPEX can be eczematiform, psoriatiform, or ichthyosiform. Patients typically also have endocrinopathy, either thyroiditis or insulin-dependent diabetes mellitus (IDDM). Finally, other autoimmune manifestations can also be observed, such as cytopenias, hepatitis, nephritis, and myopathy. Most mutations are in the forkhead domain, followed by the repressor domain and the leucine-zipper domain of FOXP3. A variety of autoantibodies can be found in patients, with the most typical being those against the autoimmune enteropathy related 75-kDA antigen (AIE-75), anti-villin and anti-harmonin antibodies, along with autoantibodies related to endocrine manifestations.[192] In addition to elevated IgE and peripheral blood eosinophilia, these antibodies can be used as a diagnostic tool. CD4+CD25+FOXP3+ T reg cells are usually present, sometimes with reduced FOXP3 expression. Patients with neonatal onset are candidates to HSCT, whereas chronic immunosuppression is reserved to patients with less fulminant onset and less severe clinical course. Overall survival after HSCT was 73.2% and after immunosuppression was 65.1% (95% confidence interval, 62.8-95.8). The pretreatment organ involvement score was the only significant predictor of overall survival after transplant. Important, HSCT resulted in disease resolution with a better quality of life.[193] Gene editing strategies are on the way, and T reg adoptive immunotherapy is also being exploited for common conditions like IDDM and Crohn disease.[194] On the other hand, T_{regs} are being targeted in the tumor microenvironment, where T_{reg} function is not desired.[195] In conclusion, APECED and IPEX are model diseases exemplifying a break in tolerance by impaired elimination of autoreactive T cells and/or impaired T_{reg} function, leading to IEI with predominant autoimmunity.

ALLERGY AS A PHENOTYPE OF INBORN ERRORS OF IMMUNITY

Allergy or atopy-related signs and symptoms are a predominant manifestation in some IEIs. The molecular dissection of these monogenic IEIs with allergy as a predominant presentation has provided us with crucial insights into pathways relevant to the pathophysiology of allergic disease. Severe symptoms of allergy that are refractory to standard-of-care treatment or are accompanied by other manifestations (such as severe and recurrent infections) are suggestive of an underlying IEI. Overall, the pathophysiology of allergy in IEI may reflect disturbances in five major pathways: disturbed cytokine signaling, abnormalities of T cell receptor signaling, perturbed actin skeleton dynamics, failure of tolerance failure, and mast cell dysfunction.[196] Here we describe two groups of disorders in which the allergic phenotype is the predominant feature of an IEI.

The first group of IEIs strongly associated with an allergic phenotype are the so-called hyper-IgE syndromes (HIESs). HIES refers to a group of IEIs associated with recurrent staphylococcal skin abscesses and bacterial pulmonary infection, severe cutaneous allergy, and elevated IgE. The first description of HIES dates back to 1966, when two girls who had recurrent respiratory infection, severe eczema, and recurrent staphylococcal infections with little inflammation were described.[197] This clinical picture was referred to as "Job syndrome" after a biblical passage from the book of Job, where Job is depicted as suffering from widespread boils. A second description and clarification of the phenotype came from Rebecca Buckley who noticed very high levels of serum IgE in two boys with Job syndrome. This finding was subsequently confirmed in several other patients and led to the term HIES (OMIM*147060).[198] Recurrent bacterial infections of skin and lung (with frequent development of pneumatoceles), chronic mucocutaneous candidiasis, and impaired acute phase response (fever, CRP rise, neutrophil rise) are common features. Connective tissue (hyperextensible joints, vascular aneurysms, arterial tortuosity, retention of primary teeth) and skeletal anomalies (scoliosis, recurrent fractures) are additional typical manifestations of the AD subtype of HIES. It was not until 2007 that Minegishi identified dominant-negative mutations in the gene encoding signal transducer and activator of transcription 3 (STAT3) (dominant-negative STAT3 HIES) as the molecular cause of AD-HIES (OMIM*102582).[199,200] As to the allergy phenotype, patients with DN STAT3 HIES have blood and tissue eosinophilia and increased Th2 cell skewing[201,202]; however, other than asthma and severe eczema few other allergy manifestations are seen. As STAT3 controls induction of ROR γT (a transcription factor that plays a critical role in commitment of Th17 cells), DN STAT3 HIES is also characterized by a severe reduction of circulating Th17 cells, which may explain the occurrence of CMC in these patients. Treatment of the disease is largely symptomatic and includes antifungals and antimicrobials, immunoglobulin substitution therapy, and adequate topical management of the eczema and of fungal and bacterial skin infections. The role of HSCT remains controversial.

Genetic studies of patients with previously undefined forms of HIES gave additional insights into the role of the IL-6-gp130-STAT3 signaling pathway (Fig. 50.1). GP130, encoded by *IL6ST*, is a signaling receptor subunit utilized by all IL-6 family cytokines: IL-6, IL-11, IL-27, LIF, OSM, IL-35, cardiotrophin, cardiotrophin-like cytokine, and ciliary neurotrophic factor.[203-206] GP130 contains two boxes that bind JAKs constitutively. Upon cytokine activation, JAKs phosphorylate the four tyrosine residues in the cytoplasmic domain of GP130, leading to recruitment of STAT3, STAT1, and STAT5, ultimately resulting in transcription of downstream target genes. IL-6R and IL-11R cannot signal independently of GP130, whereas IL-27RA, LIF-R, and OSM-R can recruit STATs even in the absence of GP130. These features may help explain similarities and differences of the clinical phenotype observed in patients with distinct genetic bases of HIES. The AR form of HIES can be caused by ZNF341 deficiency,[207,208] whereas another AD form can be caused by DN mutations in IL6ST[209] (Table 50.3; Fig. 50.1).

In 2018, two groups reported independently an AR form of HIES due to biallelic nonsense mutations of a previously uncharacterized gene, the zinc-finger 341 gene, *ZNF341*, which encodes a transcription factor regulating basal and inducible STAT3 expression.[207,208] Clinical manifestations of these patients resembled closely those of patients with DN *STAT3* mutations, including CMC, staphylococcal infections, severe cutaneous allergy, and elevated IgE levels. However, patients with ZNF341 deficiency have stronger inflammatory responses and fewer extrahematopoietic manifestations than patients with DN *STAT3* mutations. ZNF341 is constitutively expressed in the nucleus where it binds a target DNA sequence, also present in the *STAT3* promoter. Patients with ZNF341 deficiency express low basal levels of STAT3 mRNA and protein. Moreover, the autoinduction of STAT3 and the phosphorylation of STAT3 in response to STAT3-activating cytokines are strongly diminished.

Like patients with STAT3 DN mutations, those with ZNF341 deficiency have a low number of Th17 cells, an excess of Th2 cells, and a low proportion of memory B cells. The lack of ZNF341 and the ensuing lack of functional STAT3 and therefore of RORγT results in a selective and severe impairment of differentiation of CD4+ cells into Th17 cells, accounting for the CMC phenotype observed in the patients. In addition, the patients showed higher GATA3 mRNA expression in memory CD4+ T cells, higher frequency of Th2 cells, and increased levels of *IL5* an *IL9* mRNA in CD8+

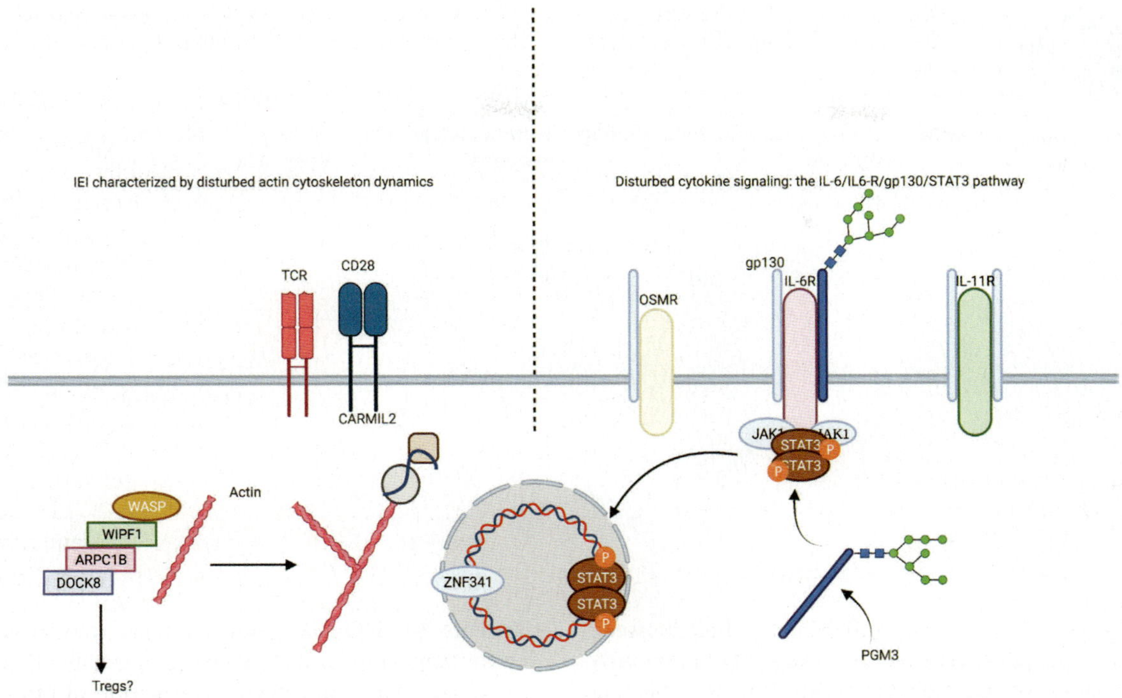

FIG. 50.1. Inborn errors of immunity associated with allergy as a prominent phenotype (Created with BioRender.com.)

T cells, resembling STAT3 DN HIES.[201,202,214] Their stronger inflammatory responses indicate the lower dependence on ZNF341 of the STAT3 function in cell types other than lymphocytes (as shown for monocytes). The milder nonhematopoietic developmental features may be accounted for by the differential requirement for STAT3 function downstream from the receptors responsible for connective tissue and skeletal phenotypes in corresponding cell types.

In 2017, a patient carrying a homozygous missense mutation in *IL6ST* resulting in partial GP130 deficiency was described.[211] The clinical manifestations of this AR subtype of HIES were very similar to the phenotype observed in patients with DN *STAT3* mutations. Another patient harboring another homozygous missense mutation in *IL6ST* was described later.[215] Of note, complete homozygous LOF mutations in IL6ST are associated with skeletal malformations, respiratory failure, and perinatal death resembling Stüve-Wiedemann syndrome, a condition caused by LIF-R deficiency, indicating differential dependence of the different cytokine receptors on GP130 and suggesting that osteopenia and scoliosis in DN STAT3 HIES may be due to perturbed LIF signaling.[216-218] More recently, another clinical phenocopy of DN *STAT3* HIES was described in eight unrelated kindreds with DN mutations in *IL6ST*. The mutant alleles encoded truncated GP130 expressing the transmembrane domain but not the recycling motif and lacking all four STAT3 recruiting tyrosine residues. The overexpressed mutant alleles accumulated at the cell surface and were LOF in terms of STAT3 recruitment in response to IL-6/IL6R-α, LIF, and OSM stimulation but not for IL-27, reflecting IL-27R intrinsic signaling capacity. The patients' leukocytes and fibroblasts also responded poorly to IL-6 and IL-11. These mutants proved DN for IL-6 signaling. The immunophenotype of the patients resembled that of patients with DN STAT3 HIES and AR IL6ST HIES, with increased Th2 cytokine production but with normal frequency of Th17 cells, consistent with the phenotype of the patients.

In 2019, two unrelated patients harboring homozygous mutations in *IL6R* encoding the α-chain of the IL-6R were described. The patients presented the classical HIES triad of skin allergies, elevated IgE concentrations, and cold staphylococcal lesions. This finding provided strong evidence that IL-6 is essential for the regulation of IgE-mediated allergy, notably severe eczema, and for host defense against staphylococci. The patients did not manifest CMC, in line with the decreased yet present Th17 cells in the two patients. Moreover, the patients with IL6R deficiency did not have any skeletal or connective tissue abnormalities, as opposed to patients with IL6ST and DN STAT3 mutations and those with ZNF341 deficiency, thus neatly dissecting the signaling pathways (most likely, LIF and IL-11 signaling, respectively) responsible for these features. Overall, the phenotype of IL-6R-α chain deficiency more closely resembles the phenotype of patients with IL6ST deficiency than that of HIES associated with STAT3 and ZNF341 deficiency. In addition, the report of human IL-6R deficiency shows us that human CRP production is directly dependent on IL-6 signaling (Fig. 50.1; Table 50.3).

Additional evidence for the role of IL-6 in human host defense against staphylococci had come from the description of neutralizing autoantibodies against IL-6.[213] Of interest, these patients did not have eczema or elevated IgE, suggesting incomplete neutralization of IL-6 or alternatively that additional mutations outside the IL-6 pathway may account

TABLE 50.3 Gene Defects and Clinical Characteristics of Diverse Inborn Errors of Immunity Associated With Increased IgE

Gene Defect	High IgE	Eczema	Staphylococcal Disease	Inflammation	Extrahematopoietic Features	IL-6 Response	Mucocutaneous Candidiasis	References
TYK2−/−	+	+	+	−	−	−	+	210
STAT3+/−	+	+	+	−	+	−	+	199
ZNF341−/−	+	+	+	+/−	+/−	−	+	207,208
(AR) IL6ST−/−	+	+	+	−	+	−	+	211
(DN) IL6ST/−	+	+	+	−	+	−	+	209
IL6R−/−	+	+	+	−	−	−	−	212
Anti-IL-6 autoantibodies	−	−	+	−	−	−	+	213

for the allergy phenotype. Nevertheless, the link between IL-6 and an atopic phenotype had been suggested previously by the association of common polymorphisms in *IL6R* that diminish its function and elevated IgE as well as asthma susceptibility.[219,220] Besides illustrating the role of IL-6 cytokine signaling pathways *in natura* in humans, these findings also warn us about the potential long-term side effects of treatment with IL-6 blockers. Indeed, treatment with tocilizumab starting early in life as opposed to adult life is associated with increased atopy. Overall, characterization of the clinical and molecular features of IEI manifesting with allergy allowed one to dissect the specific role played by individual components of the IL-6/IL-6ST/STAT3 signaling pathway. At present, however, the precise etiology of elevated IgE in HIES remains unclear.

Initially classified as HIES, AR DOCK8 deficiency is not a bona fide HIES given the extended phenotype consistent with a combined immunodeficiency.[221-226] Although DOCK8 deficiency (OMIM*243700) shares some features with DN STAT3 HIES including elevated IgE, eczema, recurrent staphylococcal skin abscesses, frequent airway infections, *Candida* infections, and hypereosinophilia, lack of bone and connective tissue abnormalities and normal inflammatory responses are distinctive features when compared with DN STAT3 HIES. Moreover, DOCK8-deficient patients typically have severe cutaneous viral infections due to molluscum contagiosum, HPV, HSV, and varicella-zoster virus. Infections with other viruses, such as CMV, EBV, and rotavirus, and biliary infections with *Cryptosporidium* and ensuing sclerosing cholangitis have been reported as well. DOCK8 deficiency is caused by biallelic LOF mutations in *DOCK8*. The majority of patients reported so far carry a large deletion in the *DOCK8* gene on at least one allele.[221] DOCK8 encodes a guanine nucleotide exchange factor primarily expressed in hematopoietic tissue and necessary for actin cytoskeleton dynamics. DOCK8 activates cell division cycle 42 (CDC42), a Rho GTPase, which in turn activates WASP, necessary for the binding of actin monomers and the elongation of actin filaments.[16,225] DOCK8 plays a crucial role in lymphocyte development and effector function, especially through its role in cell trafficking and survival, formation of immunological synapses, and cell activation.

The immunological phenotype is that of a combined immunodeficiency with lymphopenia, loss of naïve T cells, reduced memory B cells, signs of T cell senescence, and a defect of NK cytotoxicity. There is variable hypogammaglobulinemia with eosinophilia and elevated IgE. Autoimmune cytopenias, solid organ autoimmune manifestations, vasculitis, and atopy are clinical manifestations of immune dysregulation, correlating with autoantibodies and decreased regulatory T cells.[225,226] DOCK8-deficient T cells are preferentially committed to the Th2 lineage at the expense of Th1 and Th17, an additional explanation for the atopic manifestations.[227]

An intriguing feature of DOCK8 deficiency is the specific deficiency in tissue-resident memory T cells in the skin; the dense collagen maze in the skin poses significant stress on the cellular cytoskeleton of the migrating tissue lymphocytes. In the absence of DOCK8, cells cannot resist the shear stresses leading to cell shattering (cytothrypsis), resulting in a tissue-resident memory T cell deficiency, which likely plays a role in the severe susceptibility to cutaneous viral infections observed in these patients.[222] Approximately 17% of patients develop cancer during childhood, with a mortality of 6%.[224] EBV-driven lymphomas, such as diffuse large B cell lymphoma, and HPV-related squamous cell carcinoma of the skin and cervical cancer are the most common malignancies. Other lymphoid and nonlymphoid tumors have been reported, including various other lymphomas, cutaneous T cell leukemia, adnexal carcinoma, and EBV-associated leiomyoma/leiomyosarcoma.[226]

The natural history of the disease is devastating with >90% mortality by age 40 years.[224,228] The only available cure is HSCT, with excellent outcome in terms of survival and resolution of infections and eczema.[228] However, food allergies remain a problem for a number of patients after transplantation, and more follow-up is needed to establish the effectiveness of

HSCT on autoimmune manifestations and the risk of malignancies.[228,229] Somatic reversion partially restoring DOCK8 expression has been reported in several patients, but it is not necessarily associated with improved clinical outcome.[223,230]

DOCK8 deficiency is a model of an actinopathy. In the prototypic actinopathy, the X-linked WAS (OMIM*301000), caused by mutations in the *WAS* gene encoding WASP, allergy is a predominant phenotype: patients have severe eczema and food allergy.[231] The other hallmarks of WAS are microthrombocytopenia with increased risk of bleeding; combined immunodeficiency with recurrent, often life-threatening infections of bacterial, viral, and fungal origin; and predisposition to neoplasms.[232] The clinical spectrum is extensive and there is genotype-phenotype correlation. The occurrence of severe atopy in WAS is incompletely understood. WASP has been shown to be essential for T_{reg} homeostasis in mice, and T_{regs} from patients with WAS show impaired suppressor function *in vitro*, thereby contributing to the autoimmunity of the disease.[233] Other data indicate that WAS T cells are skewed to a Th2 phenotype, with reduced generation of Th1 cells, pointing to the fine balance between different T cell subsets in the development of the clinical atopic phenotype of patients with WAS.[234,235] HSCT is curative, leading to excellent survival and resolution of the main manifestations of the disease.[236] Gene therapy (GT) has been successfully performed in patients lacking a matched donor; however, platelet reconstitution post-GT may not be optimal.[237-239]

Actin cytoskeleton dynamics are complex and involve a multitude of highly conserved actin-binding proteins that deliver actin monomers and facilitate the nucleation, elongation, severing, branching, depolymerization, and cross-linking of F-actin (Fig. 50.2). Disruption of these dynamics has a tremendous impact on mitosis, endocytosis, vesicular trafficking, cell migration, and cytokinesis, and also on lipid metabolism and autophagy, which are crucial functions for most cells but in particular for white blood cells and thrombocytes.[240] Accordingly, mutations in several other key proteins involved in the actin cytoskeleton have been described to underlie IEIs.[16] Deficiencies of WASP-interacting protein (WIP), moesin, ARPC1B, and RLTPR present with severe skin allergy suggesting a common pathophysiology for the development of the atopic trait.[241-252] It is unclear whether this simply reflects default differentiation of T cells into Th2 cells in these actinopathies and in other conditions in which T cell activation is perturbed.[253,254]

Distantly linked to the actin cytoskeleton are heterozygous LOF mutations, in Caspase recruitment domain-containing protein 11 (CARD11). CARD11 is a scaffolding molecule in the CARD-MALT1-BCL2 complex mediating signals from

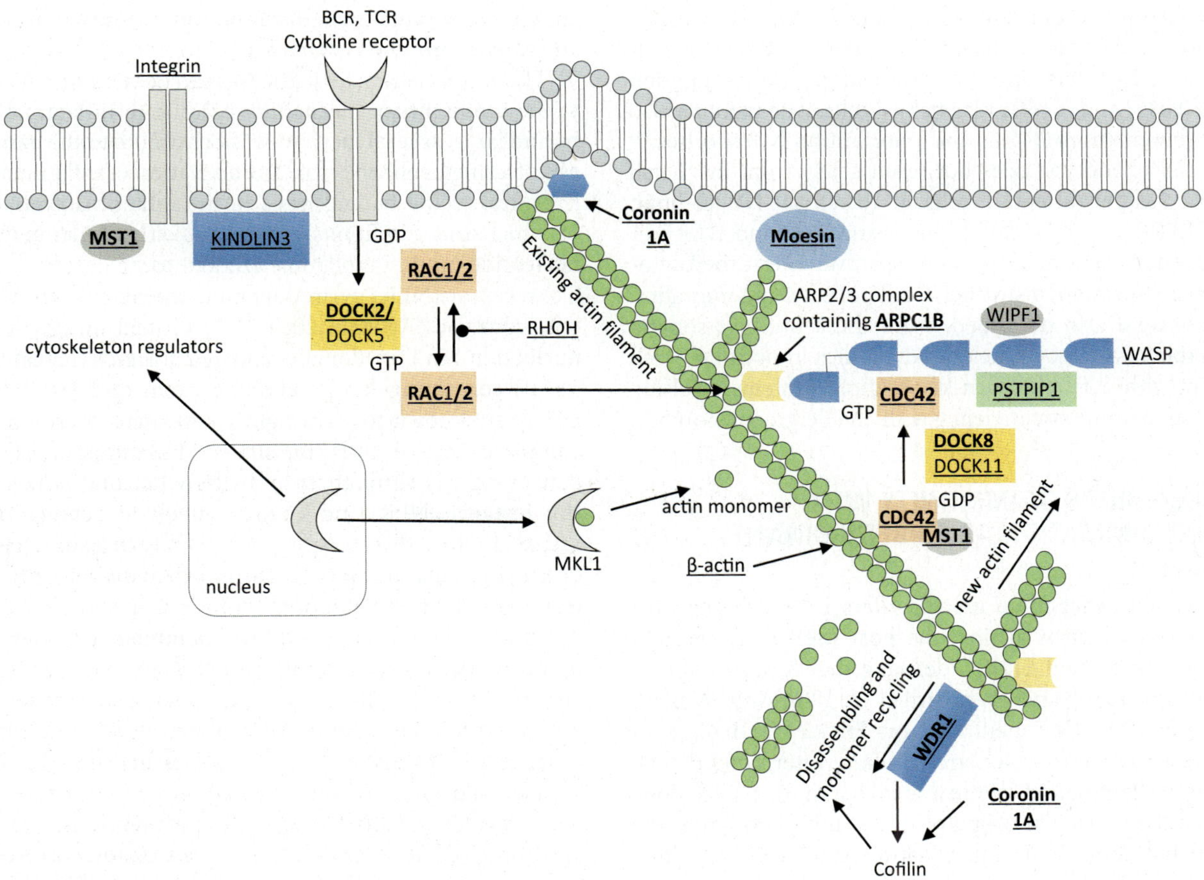

FIG. 50.2. Overview of actin cytoskeleton defects leading to inborn errors of immunity. (Adapted fromTangye SG, Bucciol G, Casas-Martin J, et al. Human inborn errors of the actin cytoskeleton affecting immunity: way beyond WAS and WIP. *Immunol Cell Biol.* 2019;97(4):389-402.)

TCR and BCR to mTORC1 and NF-κB in lymphocytes. Heterozygous LOF mutations in CARD11 were also described to cause a clinical syndrome of atopy, elevated serum IgE, autoimmunity, infection susceptibility, and lymphoma. The patients also had increased susceptibility to viral infection, HPV and molluscum, for instance, resembling DOCK8 deficiency clinically. There is an important phenotypic variability from asymptomatic to combined immunodeficiency with lymphoma.[247,255,256] In line with the central role of CARD11 in activating mTORC1 and NF-κB downstream of TCR, the laboratory findings are those of humoral or common variable immunodeficiency with diminished T cell proliferation, low memory B cells, variable degrees of hypogammaglobulinemia, and neutropenia in some.[255-257] The explanation of atopy is unclear but probably involves TCR signaling alterations, next to metabolic changes as well as effects on actin-mediated trafficking.[257]

Finally, biallelic LOF mutations in RLTPR or capping protein regulator and myosin 1 linker 2 (CARMIL2) have been described in patients with combined immunodeficiency with atopy.[243,244] The patients have cutaneous and pulmonary allergy, as well as a variety of bacterial and fungal infectious diseases, including invasive tuberculosis and mucocutaneous candidiasis. Early-onset inflammatory bowel disease and eosinophilic colitis have also been described.[241-245] Typically, their CD4+ T cells do not respond to CD28 stimulation and the proportions of circulating regulatory T cells and memory CD4+ T cells are reduced. The patients' CD4+ T cells exhibit a Th2 cell bias *in vitro*, contrasting with a reduction of Th1, Th17, and T follicular helper cells. The patients also display few memory B cells and poor antibody responses.[244] This B cell phenotype does not result solely from the T cell deficiency, as the patients' B cells fail to activate NF-κB upon BCR. RLTPR acts as a scaffold between CD28 and CARD11 and more downstream of the NF-κB pathway. Perturbed cytoskeletal organization, disturbed T cell polarity, and migration disorders were also described.[245] All these findings show a crucial role of actin cytoskeleton dynamics in optimal T cell function, with deficiencies in key scaffold proteins resulting in combined immunodeficiency with an allergic phenotype.

INBORN ERRORS OF IMMUNITY WITH AUTOINFLAMMATION AS A PREDOMINANT PHENOTYPE

The possibility that there are disorders in which the host immune system turns against the host itself dates back to early 1900s with Paul Ehrlich defining the "horror autotoxicus" (Wiley, 1910) and even earlier, to 1802 with William Heberden's 1802 Commentaries on History and Care of Disease citing "pains which are regularly intermittent." The term "autoinflammation" per se was coined more than 20 years ago when the molecular origin of an AD inherited syndrome of recurrent fever and inflammation (the TNF receptor-associated periodic syndrome, TRAPS) was unveiled as well as the locus for the gene associated with familial Mediterranean fever (FMF) after a decade of positional-cloning efforts.[258,259] Autoinflammatory syndromes manifest with seemingly unprovoked recurrent episodes of fever, malaise, arthritis, serositis, skin rash, abdominal pain, and central nervous system manifestations yet lack the hallmarks of autoimmunity, that is, autoantibodies and self-reactive T cells. Thus the term "autoinflammation" was born to denote the distinction with diseases traditionally designated as autoimmune diseases (eg, lupus).[260,261] The lack of detectable autoreactive T and B cells is a hallmark of autoinflammation.

Meticulous description of autoinflammatory conditions and unbiased sequencing of patients have indeed taught us that several of the genes underlying autoinflammatory diseases are key for innate immunity. Kastner et al. thus adapted the definition of autoinflammatory diseases into "clinical disorders marked by abnormally increased inflammation, mediated predominantly by the cells and molecules of the innate immune system, with a significant host predisposition."[260] The unprovoked character of the inflammation can be questioned as in most autoinflammatory syndromes, some trigger can be identified, be it sun/cold exposure, clinically silent infection, vaccination, or just simply infection.[262] Also, there is a continuum from autoimmunity to autoinflammation as well as a gigantic overlap between the typical isolated autoinflammatory syndromes and the immunodeficiencies with an infection phenotype, gratifying the term inborn errors of immunity, encompassing the blended phenotype that is usually the patient's reality. This also beautifully demonstrates the entwinement rather than dichotomous separation of adaptive and innate immune systems.[263]

The autoinflammatory diseases can be classified according to the predominant mechanism leading to inflammation as, by and large, several mechanisms contribute to the pathogenesis of a single inborn error. The activation of inflammasomes is central to the pathogenesis in the inflammasomopathies, of which the conditions associated with mutations in the nucleotide-binding oligomerization domain-like (NOD) like receptor (NLR) pyrin domain containing 3 (NLRP3) are the prototype (Agostini 2004).[264,265] Protein members of the nucleotide-binding domain and leucine-rich repeat (LRR) (NLR)-containing family and the pyrin and HIN domain family are known to form tightly regulated macromolecular complexes called inflammasomes. Inflammasomes detect danger signals through their pattern recognition receptors (for instance, NLR) and kindle proteolytic cleavage of pro-caspase1 into active caspase-1, which in turn leads to cleaving of pro-IL-1beta and pro-IL-18 into their biologically active form, and the induction of pyroptosis, a specific form of cell death.[266] IL-1beta is a potent proinflammatory mediator and pyrogen, whereas IL-18 induces IFN-γ and activates NK cells and cytotoxic T cells. Active caspase-1 also induces pyroptosis, a proinflammatory form of cell death. The different inflammasomes distinguish themselves in pathways leading to their activation. Three distinct autoinflammatory disorders, chronic infantile neurologic cutaneous and articular syndrome (CINCA, OMIM*607115 also known as Neonatal onset multisystem inflammatory disease [NOMID]), Muckle Wells syndrome (MWS, OMIM*191900) and familial cold autoinflammatory syndrome (FCAS, OMIM*120100) are linked to mutations in *NLRP3* and are being regarded as a clinical continuum, renamed as cryopyrin-associated periodic syndrome (CAPS).[267-269] Mutations in pyrin (encoded by MEFV *Me*diterranean *FeV*er Gene) have been described as underlying FMF (OMIM*249100).[259] Biallelic mutations in MEFV are found in only two-third cases. Although the pathophysiology of FMF involves increased cleavage of pro-IL-1beta, increased NF-κB signaling is also involved.[270]

As stated above, several inborn errors involving components of the actin cytoskeleton are associated with variable manifestations of autoinflammation. This can be explained by the role of actin in the activation of the pyrin inflammasome, via the interaction of pyrin with the adaptor protein ASC, which is located at sites rich in polymerizing actin,[271] whereas microtubular function is needed for ASC recruitment.[272] Any disturbance of this network can result in autoinflammation. Conditions in which disturbed pyrin-ASC interaction due to disturbed actin cytoskeleton dynamics plays a role are numerous. The first to be described was pyagenic Arthritis, pyoderma gangrenosum, and acne (PAPA) syndrome. In PAPA syndrome, mutations in PSTPIP1, proline-serine-threonine-phosphatase interacting protein 1, lead to hyperphosphorylation, with higher binding affinity to the autoinhibitory domain of pyrin with ensuing constitutive activation of pyrin. PSTPIP1 plays a key role in cytoskeletal organization, especially in podosome and filopodia formation.[273,274] Other inborn errors associated with autoinflammation due to actin cytoskeletal dysregulation include conditions caused by WDR1 deficiency as well as ARPC1B deficiency and more recently by mutations in CDC42.[249-251,275-279] The picture is complex as these defects are intrinsically related to the autophagy defect associated with actin cytoskeleton disturbances.[240,280]

Protein misfolding triggers the unfolded protein response, a cellular process for maintaining homeostasis in situations of endoplasmic reticulum stress. This leads to inadequate cytokine secretion. The prototype condition in which protein misfolding plays a predominant role is TNF-associated periodic syndrome (TRAPS OMIM*142680) in which missense mutations in the p55 subunit ectodomain of the TNF receptor superfamily 1A gene (*TNFRSF1A*) lead to misfolding.[281] TRAPS is the most frequent AD autoinflammatory disorder characterized by prolonged attacks of fevers, peritonitis, and soft tissue inflammation. Other examples include chronic atypical neutrophilic dermatosis with lipodystrophy and elevated temperature (CANDLE) and mutations in POMP. POMP encodes a proteasome maturation protein, and the clinical phenotype caused by heterozygous truncating mutations in POMP lead to a complex phenotype of autoimmunity, immunodeficiency, and autoinflammation,[282] again highlighting the complexity and overlap of phenotypes of IEIs.

Dysregulated signaling in the NF-κB pathway is the key route to autoinflammation in several inborn errors of immunity. NF-κB signaling plays an essential role in almost all cell types, in diverse cellular functions, leading to expression of thousands of genes. The orchestrated action of ubiquitin ligases and deubiquitinases (DUBs) and kinases is crucial for controlled activation of the NF-κB signaling pathway.[283] Any defect in this system can lead to inappropriate and uncontrolled NF-κB signaling and inflammation. Depending on the specifics of the defect, the phenotype can consist of predominant immunodeficiency, autoinflammation, or a mixed phenotype. Finally, the type I IFN-mediated autoinflammatory diseases (so-called type I interferonopathies) encompass a group of disorders hallmarked by inappropriate type I interferon-mediated immune responses.[284] We will here describe the principles of autoinflammation associated with type I interferonopathies and with NF-κB signaling dysregulation.

TYPE I INTERFERONOPATHIES

The concept of type I interferonopathies was posed by Crow in 2011 to refer to a group of novel Mendelian IEIs in which the homeostasis of type I interferon signaling is disturbed. Already 30 years before, Ion Gresser had deduced from experiments in rodents that excess of type I interferon might be harmful to the host.[285] Type I interferons are central to antiviral defense and are induced following detection of nucleic acids by various dedicated innate immune receptors. Intrinsic to this mechanism and to the conserved structure of DNA and RNA is the risk of sensing self-nucleic acids as nonself. Also, the tight regulation of nucleic acid sensing in the (intracellular and extracellular) environment, leading to type I IFN induction, is subject to errors. Thus, both GOF mutations in nucleic acid sensors as well as LOF mutations in regulatory molecules can lead to type I interferonopathies.[286] Intriguing, although there is an association between the diseased status and the excessive type I IFN signaling, there is no definitive proof linking type I interferon directly to the pathogenesis of these conditions. Indeed, some data contradict the direct pathophysiological role of increased type I interferon signaling, for instance, the incomplete penetrance of the phenotype despite the enhanced type I interferon signaling or the absence of interferon signature in patients harboring a relevant mutant.[287] Nevertheless, using increased type I interferon signaling as a diagnostic criterion for type I interferonopathy, over 20 distinct IEIs qualify as type I interferonopathy. The presence of increased type I interferon signaling can be established by a biological assay quantifying antiviral interferon activity or by the measurement of interferon stimulated gene (ISG) expression. Alternatively, serum IFN-α can be measured using a sensitive technological platform.[288]

Several cellular pathways can lead to enhanced type I interferon signaling. First, an abnormal accumulation of endogenous nucleic acids either due to aberrant nuclease activity or due to leakage of and accumulation of nucleic acids can be the culprit. Endogenous nucleic acids have nuclear or mitochondrial sources. A compelling hypothesis says that evolutionarily integrated retroviruses, retroelements, whether or not with retained capacity to replicate and reinvade the host genome, represent an important stimulus for IFN signaling and play a role in proper immune function and also in type I interferonopathy.[289] Indeed, 40% of the human genome is made up of endogenous retroelements, which form a potential trigger for nucleic acid sensors.[290] An example of such a sensor is MDA5, a dsRNA sensor. ADAR1 weakens the dsRNA structure of ALU retroelements, by editing adenosine into inosine, and this modified hybrid is not recognized by wild-type MDA5, whereas it is by mutated MDA5 underlying Aicardi-Goutières syndrome (AGS). In this way, mutated MDA5 becomes constitutively active and leads to increased type I IFN signaling. Likewise, mutations affecting ADAR1 activity fail to mark the Alu hybrids as "self" by editing adenosine into inosine, giving rise to interaction of the hybrids with (wild-type) MDA5, resulting in net enhanced type I IFN signaling.[291] Likewise TREX1, SAMHD1, and ADAR1 are shown to act as LINE-1 restriction factors, whereas RNase H2 promotes LINE-1 retrotransposition.[292-294] Recent evidence suggests that adenosine deaminase type II deficiency is also indirectly due to increased expression of retroelements, caused by increased deoxyinosine levels, leading to increased MAV signaling and type I IFN induction.[295-298] DNase II deficiency is another type I interferonopathy, likely due to a failure of lysosomal clearance of apoptotic DNA and red cell nuclei with ensuing activation of STING, as shown in mice.[299-301] The accumulation of mitochondrial dsRNA in the cytoplasm can also lead to increased type I interferon signaling. Indeed, stressed mitochondria may leak DNA into the cytoplasm, which can then promote type I IFN responses as evident in patients with hypomorphic mutations in PNPT1, encoding PNPase, mutations in which are associated with variable respiratory chain complex deficiencies.[302,303] Likewise nuclear histones are essential in suppressing the immunogenicity of self-DNA, as evidenced by biallelic mutations in LSM11 and RNU7.[304]

Moreover, DNA damage, intrinsic to the mere process of cell division, can result in accumulation of single-stranded DNA, and in case of depletion of DNA damage sensors RPA and Rad51 lead to type I interferon induction, potentially via the STING pathway. This is in line with the finding of increased type I signaling in the microglia of patients with ATM and Artemis deficiency.[305] Clinically, mutations in TREX1 (AD/AR), RNASEH2A/B/C (AR), SAMHD1 (AR), ADAR1 (AR/AD), and IFIH1 (AD) result in AGS, a progressive brain disorder with basal ganglia calcifications and various associated other (neuro)immunologic features. DNase II deficiency leads to a complex phenotype including early-onset anemia, glomerulonephritis, liver fibrosis, arthropathy, and anti-DNA antibodies.[287,306]

Another mechanism of increased type I IFN signaling is the enhanced sensitivity to its ligand or constitutive activation of signaling molecules downstream of the sensors. A noteworthy example is STING-associated vasculitis with onset in infancy (SAVI) caused by heterozygous GOF mutation in TMEM173, encoding STING.[307,308] The SAVI phenotype ranges from early onset with systemic inflammation and severe cutaneous vasculopathy with acral necrosis and interstitial lung disease due to vasculitis to a lupus-like phenotype, depending in part on the specific mutation.[309] Immune cells and fibroblasts of patients with SAVI typically show constitutive upregulation of ISG expression. Recently, homozygous GOF mutations in *STAT2* were described in two unrelated kindreds with severe early-onset severe autoinflammation and elevated IFN signature. The gain of IFN activity was shown to result from impaired function of ubiquitin-specific protease 18, a critical STAT2-dependent negative regulator of type I IFN signaling.[310,311] More specifically, the STAT2 R148Q variant is a GOF because it fails to appropriately traffic USP18 to IFNAR2, thereby preventing USP18 from negatively regulating responses to type I interferons.[311]

This brings us to the third mechanism in which impaired downregulatory mechanisms of the type I IFN response are causal. STAT2 GOF is indeed a phenocopy of inherited USP18 and ISG15 deficiencies in which the regulation of the late responses to IFN-I are impaired. Homozygous LOF of USP18 was identified in five patients of two unrelated families as the cause of a type I interferonopathy manifesting as pseudo-TORCH syndrome (resembling intrauterine infection with toxoplasma, other agents, rubella, CMV, and herpes simplex) with microcephaly and cerebral calcification.[312] The condition is lethal in infancy, although recently another patient was identified and rescued by timely treatment with JAK inhibitor.[313] ISG15 deficiency is defined by two major clinical manifestations: an AGS-like syndrome and susceptibility to mycobacterial infections. Absence of intracellular ISG15 prevents the accumulation of USP18.[314] Free extracellular ISG15 is crucial in IFN-γ-dependent antimycobacterial immunity explaining the increased susceptibility to mycobacterial infection in ISG15 deficiency. Recently, five more patients with ISG15 deficiency were described, presenting with necrotizing skin lesions. Thus a defect in the negative regulation of type I IFN signaling results in an interferonopathy phenotype, suggesting a tonic type I IFN signaling, a topic of further research.[315]

In addition, indirect mechanisms lead to increased type I IFN signaling. For instance, heterozygous mutations in COPA cause a disease resembling SAVI both clinically as well as immunologically. COPA encodes the α-subunit of the coatomer complex I involved in the transport of vesicles from Golgi to the endoplasmic reticulum (ER).[316-318] Of note, STING translocation from ER to the ER-Golgi intermediate compartment is crucial for type I IFN response and requires coatomer complex I function.[318,319] Thus, dysfunctional ER-Golgi trafficking represents an additional mechanism in type I interferonopathy. The unfolded protein response involves activation of IRE-1, which is composed of an ER endoluminal domain that senses the misfolded protein and a cytosolic endo-RNAse that cleaves mRNAs to prevent further overloading of the ER. This RNAse activity leads also to the generation of 3′ cyclic phosphate containing RNAs that are recognized by the cytosolic RNA sensors.[320,321] RNAs are further processed and degraded by RNA exosomes, to which SKIVL2 and TTC37 are associated. This provides a molecular basis to the type I IFN signature seen in patients with trichohepatoenteric syndrome caused by homozygous hypomorphic SKIV2L mutations.[322] In this syndrome, intractable diarrhea, immunodeficiency, and developmental abnormalities coexist.

Finally, the proteasome-associated autoinflammatory syndromes (PRAASs), also referred to as chronic atypical neutrophilic dermatosis with lipodystrophy and elevated temperature (CANDLE), have been linked to type I IFN induction and IFN signature. The proteasome is crucial for proteolysis of ubiquitinated proteins. AR mutations in proteasome subunit b8 (PSMB8) cause a severe autoinflammatory syndrome with panniculitis and lipodystrophy, joint contractures, recurrent fevers, hepatomegaly, and calcification of the basal ganglia. It was further demonstrated that this phenotype is associated with increased ISG expression.[323-325] Mutations in additional proteasome genes result in a similar phenotype, demonstrating a link between proteasomal dysfunction and type I IFN induction.[282] Of note, deficiency of adenosine deaminase type 2, a devastating vasculopathy associated with bone marrow failure and immunodeficiency, leads to increased IFN signaling, potentially due to altered methylation status of retroelements.[297,298,326]

By and large the phenotype of these type I interferonopathies (Table 50.4) is overlapping, for example, the chilblain-like lesions and intracranial calcifications. Some presentations, such as the severe lung disease in SAVI and COPA syndrome stand out. Moreover, there is incomplete penetrance for the typical disease manifestations, sometimes into adulthood. Here differential exposure to environmental triggers such as infection must be considered. Interesting, there is a spillover of type I interferonopathies to autoimmunity. Thus, in several type I interferonopathies, autoantibodies can be found, without ascertaining that these autoantibodies have a pathophysiological importance, as in DNase II deficiency. When it comes to therapy, the pathophysiological hypothesis that increased type I interferon signaling is key has led to the use of JAK1/2 inhibition, such as with ruxolitinib, especially in SAVI[327,328] and recently in a neonate with USP18[329] deficiency. These therapies have also been used with relative success in COPA and in several other IEIs in which a variable type I IFN signature is present.[330]

AUTOINFLAMMATION DUE TO DISTURBED NF-κB SIGNALING

As stated above, posttranslational protein ubiquitination, that is, the attachment of 76 amino acid Ub molecules, is crucial for the regulation of a variety of biological processes, by determining the fate of protein localization, interaction, activity, and degradation. Ubiquitination is initiated by the orchestrated action of Ub-activating enzymes, Ub-conjugating enzymes, and Ub-ligating enzymes. Sequential Ub molecules are added through lysine or methionine residues. Ubiquitination is a continuous and dynamic process thanks to the action of DUBs, of which several like A20 and OTULIN function as negative regulators of NF-κB signaling.[331-334] Linear ubiquitin chain assembly complex (LUBAC) mediates Met1 ubiquitination and is crucial to maintain the stability of the TNF receptor 1, TLRs, IL-1R, CD40, RLR, and inflammasome receptor signaling complexes. Upon triggering, LUBAC is recruited and attaches linear Ub chains to IKK, RIPK1, RIPK2, IRAKs, and MyD88.[335,336] LUBAC depletion results in decreased NF-κB and mitogen-activated protein kinases–mediated signaling. LUBAC consists of HOIP (HOIL-1 interacting protein) and two accessory proteins, HOIL-1 (heme oxidized IRP2 ubiquitin ligase 1) and SHARPIN (SHANK interacting protein like 1). HOIP is the catalytic subunit that is autoinhibited unless the complex is fully assembled. As a consequence, any deficiency of SHARPIN/HOIL-1 results in loss of LUBAC complex stability. Evidencing the need for tight regulation, the DUB OTULIN hydrolyzes the linear Met1-linked Ub chains by interacting with the N-terminal PUB domain of HOIP.[337,338] This connection, HOIP-OTULIN, is crucial for its action. The action of LUBAC is further negatively regulated by TRAF1 (tumor necrosis factor receptor–associated factor 1), which interacts directly with LUBAC to interfere with activation of IKK/NEMO.[339]

Some of the NF-κB pathway defects manifest predominantly with infection, with some features of autoinflammation. For instance, NF-κB essential modulator deficiency (NEMO deficiency) caused by hypomorphic mutations in *IKBKB* results in a complex phenotype with severely increased susceptibility to infection.[340-342] A similar phenotype is caused by heterozygous GOF mutations in IKBa encoded by *NFKBIA*.[26] An important autoinflammatory feature of NEMO deficiency is noninfectious colitis.[343] AR HOIL-1 and HOIP deficiency are the only described deficiencies of LUBAC.[344,345] The patients have immunodeficiency with invasive pyogenic infection and autoinflammation, with amylopectinosis, cardiomyopathy, and death in early childhood.

Deficient function of the ubiquitination system leading to disturbed activation of NF-κB signaling results in phenotypes that are predominantly autoinflammatory. Four patients with AR otulipenia have been described.[346] They had recurrent fevers, joint swelling, gastrointestinal inflammation, and failure to thrive along with skin inflammation: a painful erythematous or pustular rash in another patient, corresponding to small and medium-sized vasculitis. The loss of negative regulation by OTULIN results in the patients' mononuclear cells and fibroblasts mounting a strong inflammatory signature as shown by increased degradation of IKB-α and increased phosphorylation of IKKa, IKKb, and IKB-α, hallmarks of NF-κB pathway activation.[347] A subsequent patient was described with systemic inflammation and fatal pulmonary edema.[347] Treatment with anti-TNF agents has been effective in controlling the inflammation.[346]

Lys(K)63-Ubiquitination was initially regarded as a mechanism for activating the canonical NF-κB pathway. NEMO/IKK, TRAF6, RIPK1 are ubiquitinated both by K63 and linear Ub chains. A20/TNFAIP3 restricts inflammatory responses by hydrolyzing these K63 links and by E3 ligase activity leading to K48 conjugates that are directed for proteasomal degradation.[348] A20 undergoes cleavage by mucosa-associated lymphoid tissue lymphoma translocation protein 1 as well as posttranslational phosphorylation.[349] Common noncoding variants in A20 have been associated with lupus, rheumatoid arthritis, psoriasis, type 1 diabetes mellitus, celiac disease, coronary artery disease, and inflammatory bowel disease. Recently, an AD autoinflammatory disease, haploinsufficiency of A20 (HA20), was demonstrated in over 10 patients. They presented with childhood onset of fevers, arthritis, oral and/or genital ulcers, skin pathergy, and gastrointestinal and ocular inflammation.[350] Additional patients were described presenting with ALPS and AD Behcet disease.[351-353]

TABLE 50.4 Current Overview of Type I Interferonopathies Indicating Protein Function, Link to Interferon Signaling, Proposed Molecular Mechanism, and Currently Recognized Associated Clinical Phenotypes

Gene	Protein Function	Basic Mechanism Underlying Increased Type I Interferon Signaling	Mutation	Clinical Phenotypes
TREX1	Deoxyribonuclease	Cytosolic DNA	LOF (recessive or dominant-negative)	AGS, FCL, SLE
SAMHD1	Control of dNTP pool (±nuclease)	Cytosolic DNA (±cytosolic RNA)	LOF (recessive)	AGS, FCL, CVD
TMEM173	Transduction of cytosolic type I interferon signal	Cytosolic DNA (±cytosolic RNA)	GOF (dominant)	SAVI, FCL
LSM11	Histone pre-mRNA-processing complex	Loss of histone suppression of immunogenicity of self-DNA	LOF (recessive)	AGS
RNU7	Histone pre-mRNA-processing complex	Loss of histone suppression of immunogenicity of self-DNA	LOF (recessive)	AGS
RNASEH2A	Ribonuclease	Cytosolic RNA-DNA hybrids	LOF (recessive)	AGS
RNASEH2B	Ribonuclease	Cytosolic RNA-DNA hybrids	LOF (recessive)	AGS, SP
RNASEH2C	Ribonuclease	Cytosolic RNA-DNA hybrids	LOF (recessive)	AGS
POLA1	Polymerase	Cytosolic RNA-DNA hybrids	X-linked recessive	XLPDR
ADAR1	RNA editing	Cytosolic RNA	LOF (recessive or dominant-negative)	AGS, DSH, BSN, SP
ADA2	Ado deaminase	Retroelements	LOF (recessive)	ADA2 deficiency
IFIH1	dsRNA sensor	Cytosolic RNA	GOF (dominant)	AGS, SP, SMS
RJG-I	dsRNA sensor	Cytosolic RNA	GOF (dominant)	Atypical SMS
SKIV2L	RNA helicase	Cytosolic RNA	LOF (recessive)	THES
UPS18	Inhibition of ISG transcription	IFNAR1 signaling	LOF (recessive)	Pseudo-TORCH
ISG15	Inhibition of ISG transcription	IFNAR1 signaling	LOF (recessive)	MSMD, ICC
STAT2	Negative regulation of responses to IFN-I via trafficking of USP18 to IFNAR2	IFNAR2 signaling	GOF (recessive)	Autoinflammatory syndrome
PSMB8	Proteasome	Unknown	LOF (recessive)	PRAAS
PSMB4	Proteasome	Unknown	LOF (recessive)	PRAAS
PSMA3	Proteasome	Unknown	LOF (recessive)	PRAAS
ACP5	Phosphatase activity related to osteopontin	Unknown	LOF (recessive)	SPENCD, SLE, cytopenias
C1q	Alternative complement pathway activity	Unknown	LOF (recessive)	SLE

AGS, Aicardi-Goutières syndrome; CVD, cerebrovascular disease; DSH, dyschomatosis symmetrica hereditarian; FCL, familial chilblain lupus; ICC, intracranial calcification; PRAAS, proteasome associated autoinflammatory syndrome; SAVI, Sting associated vasculopathy with onset in infancy; SLE, systemic lupus erythematosus; SMS, Singleton-Merten syndrome; SP, spastic paraparesis; SPENCD, spondyloenchondrodysplasia; X-linked reticulate pigmentary disorder.

Adapted with permission of Annual Reviews, Inc, from Uggenti C, Lepelley A, Crow YJ. Self-awareness: nucleic acid-driven inflammation and the type I interferonopathies. *Annu Rev Immunol.* 2019;37:247-267; permission conveyed through Copyright Clearance Center, Inc.

Patient-derived PBMCs and fibroblasts show increased phosphorylation of IKKα/ß and increased degradation of IkB-α. Hydrolyzation of K63 Ub chains from NEMO/IKK, RIPK1, and TNFR1 was impaired. The patients' PBMCs also showed constitutive activation of NLRP3 consistent with the role of A20 to downregulate NLRP3 inflammasome activation, as shown in mice.[354]

Heterozygous mutations in RELA (p65) also lead to a predominantly autoinflammatory phenotype.[355] The first described patient presented with recurrent fevers, mouth ulcerations, fevers, vomiting, and ileitis together with a systemic inflammatory response. Yet another patient had cytopenia and lymphoproliferation and another had neuromyelitis optica associated with Behçet-like disease.[356,357]

The phenotypic heterogeneity, a typical trait for IEIs, stands out in many of the autoinflammatory disorders, and often it is difficult to disentangle which pathway predominantly causes the autoinflammation. An excellent and recently described example is the IEI resulting from mutations in the receptor interacting serine/threonine kinase 1 (RIPK1). RIPK1 acts at the crossroads of inflammation and cell death pathways by inhibiting necroptosis and by serving inflammatory homeostasis. It mediates signal transduction downstream of TNFR1, TLR3, and TLR4 and is involved in caspase-8-mediated apoptosis and in RIPK3-mediated necroptosis.[358-360] Stimulation of these receptors activates the canonical NF-κB pathway and the mitogen-activated protein kinases (MAPKs) signaling pathways.[361-363] This leads to phosphorylation of NF-κB and AP-1 transcription factors, which induces expression of proinflammatory and cell survival genes. RIPK1 function is under strict control through phosphorylation and ubiquitination. It is cleaved by caspase-8 and caspase-6. In 2018, four patients from three unrelated families with complete RIPK1 deficiency caused by rare homozygous mutations of the N-terminal domain, resulting in a premature stop, were reported.[363] The patients had recurrent bacterial and viral infections, early-onset inflammatory bowel disease, and progressive polyarthritis. They were lymphopenic. There was a partial reduction in NF-κB activation with decreased p38 Mitogen activated protein kinase and AP-1 phosphorylation, whereas IL-1beta production was increased, most likely due to increased necroptosis.

More recently, 13 patients from five unrelated families, who had an autoinflammatory disorder, hallmarked by recurrent fevers, oral ulcerations, and lymphadenopathy beginning in early childhood and continuing throughout adulthood, were reported to carry heterozygous mutations (D324N, D324Y, D324H, D324V) in RIPK1 at the evolutionarily conserved aspartate at position 324, crucial for cleavage by caspase-8 and caspase-6.[364,365] Data from both mice as well as the patients showed that the loss of RIPK1 cleavage did not affect TNF-induced NF-κB or MAPK activation but sensitized the cells for TNF-α-induced cell death. PBMCs also produced more TNF-α, IL-6, and IL-1b after LPS or Poly(I:C) treatment. In terms of treatment, tocilizumab but not etanercept proved successful in several patients, in line with upregulation of IL-6 in monocytes compared with controls during fever flares. In all, the RIPK1-deficient patients and the patients presenting with mutations at the cleavage site show the complex intertwinement of signaling pathways in autoinflammatory syndromes, even when caused by mutations in the same gene. Specifically, in RIPK1-mediated disease, NF-κB, MAPK, IL-6, and cell death pathways, in particular necroptosis, contribute to inflammation, broadening the scope of the autoinflammatory spectrum.

MALIGNANCY AS A PHENOTYPE OF IEI

Together with infections, autoimmunity, and allergies, malignancies are a frequent manifestation of IEIs: patients with IEIs have an overall risk of 4% to 25% of developing cancer in their lifetime, and this risk is even higher for selected forms of IEIs. In addition, malignancies represent the second leading cause of death after infection in patients with IEI. More often, they occur in the first and second decade of life, but they can also be the initial manifestation of IEI in early childhood. Because the survival of patients with IEI has been improving, their lifetime risk of malignancy has become more apparent. Cancer occurs more frequently in patients with IEI and at a younger age compared with the general population. Different mechanisms are responsible for this increased susceptibility, depending on the specific IEI.[366] Recent advances in molecular diagnosis of IEI have also shed light on the defects predisposing to cancer. Although the theory of immune surveillance in cancer is an attractive hypothesis for the increased incidence of cancer in IEI, the data plead for a cell-autonomous effect. DNA repair defects are characterized by progressive accumulation of DNA damage, facilitating the onset of cancer.[367] Also, failure to extinguish lymphocyte proliferation (as observed in ALPS and in APDS) may eventually lead to somatic mutations leading to lymphoma. IEIs affecting hematopoiesis (SCN, GATA2 deficiency) are also accompanied by an increased risk of myelodysplasia and overt malignancy. Inability to control oncogenic viruses (such as EBV, HHV8, HCV, HBV, HPV) is key in malignancy development. Especially chronic viral and mucosal cutaneous infections can lead to the development of skin and epithelial cancer. Chronic inflammation at mucosal sites and in the gastrointestinal tract may also facilitate development of cancer. Some forms of IEIs are associated with rare types of malignancies; an example is represented by intestinal neuroectodermal tumors in patients with CD40LG deficiency. It is clear that different mechanisms may be at play in the same IEI, rendering attempts to define more precisely the pathophysiology of cancer in IEI difficult.

International collaborative registry data are and will be crucial to learn more about the incidence and management of cancer in IEIs.[368] Depending on the predominant disease categories represented in the registry, incidences of cancer and types of malignancies differ slightly. Non-Hodgkin and Hodgkin lymphoma account for 50% and 10% of reported cancers in the Immunodeficiency Cancer Registry, respectively. Leukemias and various types of carcinomas represent about 10% each of all cases of cancer in IEI. Lymphomas are often of B cell origin and involve extranodal site; they are related to EBV infection in at least 50% of the cases. In a cohort of >3000 patients with IEI accrued in the USIDNet registry, men with IEI had a 1.91-fold excess relative risk of developing cancer compared with the age-adjusted male population, whereas women had similar overall cancer rates compared with the age-adjusted female population.[369] There were no increases for lung, colon, breast, and prostate cancer, but increases were found for lymphoma in both men (×10) and women (× 8.34). The risk of skin cancer was also 3- to 5-fold higher, and a similar figure was observed for thyroid cancer. Neoplasms of the stomach were found to be the leading cause of death in a cohort of >400 patients with common variable immunodeficiency.[370] As mentioned above, cancer is particularly common in some forms of IEIs. In particular, more than 30% of all cases of IEI-associated cancers occur in patients with AT, and approximately 25% of all cases affect patients with common variable immunodeficiency. Cancer is also common in patients with WAS, various forms of combined immunodeficiency, and IgA deficiency.[366,369]

It is interesting to note that somatic mutations in several IEI-associated genes have been frequently reported in malignancies. Examples include mutations in the *IKZF1*, *TYK2*,

MYD88, STAT3, and *STAT5B* genes.[371] Because malignancies are an integral part of the IEI phenotype, patients with malignancies should be thoroughly investigated for a possible immune defect, especially if the cancer occurs early in life. In a cohort of children presenting with NHL, among all of those with a preexisting condition, 50% had IEI, mostly AT and Nijmegen breakage syndrome.[372] Here, we will focus our attention on AT and other DNA repair defects and on IEI with increased susceptibility to EBV-associated malignancies.

MALIGNANCY IN DNA REPAIR DEFECTS: ATAXIA-TELANGIECTASIA AS A PARADIGM

A host of intertwined cellular pathways are involved in preventing and repairing DNA damage, incurred by cells in response to breaks occurring during normal cell division cycles or by exposure to ionizing or ultraviolet radiation or toxic free radicals and cellular products. Dividing cells and especially cells that exploit the repair of targeted DNA damage, like immune cells, undergo significant DNA repair during the process of lymphocyte receptor repertoire development, isotype class switching of immunoglobulins, and somatic hypermutation and are therefore quite sensitive to any mutation occurring in these DNA repair pathways. Patients with DNA repair disorders accumulate the DNA defects and are usually sensitive even to low doses of toxic radicals or irradiation. The net effect of defects in the DNA repair pathway on the immune system depends on the particular role and expression during the repair mechanism. The central DNA break type is the DNA double-stranded break. Although the precise pathway of sensing DNA damage is still to be unveiled in detail, it is well known that the protein kinases ATM and ATR are central to the DNA damage response.

The first step in the detection of double-stranded DNA breaks is the MRE11-RAD50-Nibrin complex, which in turn activates the ataxia telangiectasia mutated protein (ATM), a ubiquitous nuclear factor that recruits and activates a cascade of proteins around the break to coordinate the action of cell cycle checkpoint and DNA repair by excision repair, nucleotide synthesis, homologous recombination, or nonhomologous end joining.[373,374]

ATAXIA-TELANGIECTASIA

AT is an AR disorder (OMIM*208900) caused by biallelic mutations in the *AT mutated* (*ATM*) gene, encoding for the coordination of cellular responses to genomic damage. ATM belongs to the family of PI3/4-kinases. Upon detection of double-stranded DNA breaks or cell damage, it acts as a DNA damage sensor, activating several proteins dedicated to DNA repair and cell cycle checkpoint.[375] ATM is necessary for the unfolding of heterochromatin, which allows the entry of the DNA repair machinery. It plays a critical role also in VDJ recombination and class switch recombination in lymphocytes. Cells from patients with AT show several cellular abnormalities, including defects in G1 and G2-M arrest, and reduction in DNA synthesis.[376] In response to DNA damage, phosphorylation of key proteins downstream of ATM (eg, p53, Mdm2, BRCA1, Chk2, and Nbs1) is preserved, but with delayed kinetics. The nature of the DNA insult is also important: AT fibroblasts are very sensitive to irradiation; however, they show little sensitivity to ultraviolet radiation, alkylating agents, or inhibitors of DNA replication. Mutations in *ATM* underlying AT cause chromosomal instability, short telomeres, excessive cell damage, and radioresistant DNA synthesis upon exposure to ionizing radiation, accounting for the cancer susceptibility and immunodeficiency shown by affected individuals.[376] Moreover, a minor cytoplasmic proportion of ATM is involved in oxidative stress metabolism, thus explaining the mitochondrial dysfunction seen in these patients.[377]

AT has an estimated prevalence of 1:40,000 to 1:100,000.[378,379] Patients present in infancy with oculomotor apraxia, cerebellar ataxia, followed by appearance of oculocutaneous telangiectasia. The neurological degeneration is progressive, and severe chorea and dystonia affect 90% of patients. The pathophysiology of cerebellar atrophy in patients with *ATM* mutations is not well understood. An immune defect is present in about 70% of patients and is caused by impaired T and B cell receptor recombination and class switch recombination. Laboratory features include lymphopenia, defective cell-mediated immunity, hypogammaglobulinemia, antibody subclass deficiency, and specific antibody deficiency. The most frequent humoral defects are IgG2 deficiency (80%) and IgA deficiency (50%).[380-382] Levels of α-fetoprotein (AFP) are markedly elevated; however, this test cannot be used to identify patients with AT at neonatal age, when AFP levels are physiologically higher. Sinopulmonary infections since childhood are the most common clinical manifestations of this immunodeficiency. The severity of the immunological phenotype correlates with the extent of class switch recombination impairment, and the presence of IgG2 deficiency or a hyper-IgM phenotype is linked to increased mortality.[383] Autoimmune and autoinflammatory manifestations are associated with selective IgA deficiency. Hepatosplenomegaly, lymphadenopathy, and neutropenia are also part of the clinical spectrum of AT. Respiratory failure is a leading cause of death in adolescence, probably due to the contribution of multiple factors such as repeated infections, systemic inflammation, oxidative stress, dysphagia, and abnormal airway clearance due to neurological degeneration. Somatic growth delay is also common. Less prevalent manifestations of AT include cutaneous granulomas, dermatitis, hair graying, insulin resistance, vascular disease, and infertility.[384]

The cancer predisposition shown by patients with AT amounts to 25% lifetime risk. Leukemias and lymphomas are particularly prevalent, appearing in patients with AT at rates that are 70- to 500-fold and 200- to 750-fold higher than in the general population, respectively. Lymphoma and acute leukemia (most notably T cell acute lymphoblastic leukemia) and T cell prolymphocytic leukemia account for 85% of these malignancies, in line with the crucial role of ATM in lymphoma surveillance.[385-387] Cytogenetic analysis may reveal spontaneous translocations (especially involving chromosomes 7 and 14), even prior to cancer development. A high frequency of solid tumors (including breast, gastric, liver, parotid gland,

and esophageal carcinomas) have been reported in older patients with AT.[388] Of interest, ATM directly interacts with BRCA1, and heterozygous AT carriers have an increased risk of breast cancer.[389]

There is no definitive treatment, and current median survival is 25 years. Tumors are often aggressive, and the treatment is hampered by the radiosensitivity of the patients. Allogenic HSCT is used to treat lymphoid malignancies and preemptive HSCT has been attempted; however, overall survival is poor owing to the extreme toxicity experienced by patients in response to chemotherapeutic agents.[390,391] Prophylactic antibiotics and immunoglobulin replacement are indicated in case of immunodeficiency with recurrent pulmonary infections. Neurologic symptoms can benefit from antiepileptic, antidyskinetic, and antioxidant medications, and corticosteroids, the last at the cost of severe side effects on long-term use.[392] Partial improvement of neurologic manifestations without the systemic effects of steroids have been reported with erythrocyte-delivered dexamethasone in an ongoing trial.[393] Finally, antisense oligonucleotides or the use of ribosomal read through molecules to avoid premature stop codons offer hope, as does gene therapy, although a major hurdle is represented by the inability to target the central nervous system.

Mutations leading to defects in several other DNA repair proteins lead to specific primary immunodeficiency conditions, all of which have a particularly increased risk of malignancy. The overall principle is similar to AT: there is increased sensitivity to radiation and other toxic stress due to impaired DNA repair. Depending on the pattern of expression of the protein, these disorders may affect multiple organs and systems, and immunological and hematological abnormalities may or may not be a predominant manifestation; however, these conditions are invariably associated with an increased risk of cancer. These forms of IEIs associated with impairment of the DNA repair machinery include defects of proteins that play a crucial role in the development of T cell and B cell receptors such as DNA Ligase IV (LIG4),[394,395] XLF,[394,396-398] DNA-dependent protein kinase catalytic subunit (DNA-PKcs),[399-401] and Artemis. LIG4-deficient patients may present with a variable degree of combined immunodeficiency (in some cases so severe to fulfill diagnostic criteria for SCID), growth retardation, and microcephaly and are particularly prone to lymphoproliferation and malignancy.[402] AR Cernunnos deficiency resembles SCID, also with microcephaly, facial dysmorphisms, and other developmental defects. Homozygous mutations in *PRKDC* (encoding for DNA-PKcs) lead to defective NHEJ and double-stranded DNA repair, resulting in a radiosensitive form of SCID. Bloom syndrome and Fanconi anemia, respectively, due to mutations in *BLM* (encoding for a RecQL3 DNA helicase critical in suppressing crossover formation between sister chromatids)[403-406] or in 1 of 15 proteins related to DNA interstrand cross-linking repair, lead to variable phenotypes with bone marrow failure as a major manifestation. In a broader concept, these conditions are also related to telomeropathies caused by inherited defects in the telomerase complex, leading to cell senescence.[407,408] Clinically, telomeropathies are characterized by oral leukoplakia, reticular pigmentation, and nail dystrophy, although several other features are also frequently present, such as interstitial lung disease. There is increased risk of myelodysplasia, acute myeloid leukemia, and squamous cell carcinoma.

IEI PREDISPOSING TO EBV-INDUCED MALIGNANCY

EBV is a member of the herpes family of viruses and has been implicated in the pathology of lymphomas as well as epithelial cancers such as gastric cancer and nasopharyngeal carcinoma. This potential for inducing malignancy in the general population is in part attributed to the virus' potential of latency, along with specific environmental factors such as in nasopharyngeal carcinoma.[409] EBV is a ubiquitous virus with up to 95% of population in Western countries infected before adulthood. In infected adolescents and adults, it causes typical symptoms of infectious mononucleosis, whereas infection in younger children is mostly asymptomatic. Several forms of IEIs are associated with increased susceptibility to both severe EBV infection including fatal mononucleosis as well as lymphoproliferative disorders. Examples include mutations in *SH2D1A*, *BIRC4*, *ITK*, *CTPS1*, *MAGT1*, *CD27*, and *CD70* genes.[410] Patients harboring mutations in *RLTPR*, *PRKDC*, *CORO1A*, *STK4*, *MCM4*, *PIK3CD GOF*, *PIK3R1*, *NFKB1*, *GATA2*, and *LRBA* are also prone to severe EBV infections as part of a broader viral susceptibility. Patients with these IEIs are predisposed to developing severe EBV infection, EBV-induced lymphoproliferative disorders, or EBV-induced hemophagocytic lymphohistiocytosis. In a proportion of patients harboring these mutations, lymphoma seems to develop spontaneously.[410,411]

X-linked lymphoproliferative disease (XLP-1) (OMIM #308240) is the most common IEI associated with increased susceptibility to severe EBV infection and was also the first one that was described to be associated with this phenotype.[412-414] XLP-1 is characterized by fulminant infectious mononucleosis, dysgammaglobulinemia, and lymphoma. It represents a prototype of an IEI associated with a very narrow spectrum of infectious vulnerability. The original description reports 13 otherwise healthy males from three unrelated families who developed an often-fatal immune-deficient condition precipitated by EBV infection.[413] Interesting, patients with XLP-1 have normal responses to other childhood viruses including other herpesviruses. In 1998, XLP-1 was unraveled to be caused by mutations in the SH2D1A gene, encoding signaling lymphocytic activation molecule (SLAM)-associated protein (SAP).[415-417] SAP is an SH2 domain–containing intracellular adaptor protein that regulates signaling downstream of the SLAM family of receptors expressed on the surface of hematopoietic cells. Most common clinical features of XLP-1 are EBV-induced fulminant mononucleosis with hemophagocytic lymphohistiocytosis (~45%-70%), B cell lymphoma (~25%), and hypogammaglobulinemia (~50%). Lymphoma and dysgammaglobulinemia occur equally in EBV-naive and EBV-infected patients with XLP.[418,419] The current survival rate in XLP-1 is 71.4%, a huge improvement since the first description.[418,420]

SAP-deficient NK cells are unable to activate 2B4 (CD244) and NTB-A driven cytotoxic responses. Activation of CD8+ T cells by B cells, but not other antigen-presenting cells such as dendritic cells, was also abolished. Indeed, SAP-deficient T cells and NK cells responded normally when activated via SAP-independent pathways.[421-425] Given that B cells are typically infected by EBV, these data explain the flawed activation of CD8+ T cells by B cells in SAP deficiency. Moreover, SAP-deficient CD8+ T cells proliferate more strongly in response to polyclonal stimulation.[426] This, together with a defective reactivation–induced cell death, likely contributes to the EBV-induced HLH. Moreover, SAP expression is essential for the function of Tfh cells and therefore for the development of T cell–dependent humoral immune responses.

Since the description of XLP-1, several other IEIs predisposing to EBV susceptibility and EBV-driven malignancy have been identified (Table 50.5). The molecular mechanisms of EBV susceptibility and severe EBV disease in each of these various disorders intersect at several levels.[410] For example, pathogenic mutations in *ITK* and *MAGT1* abolish the calcium fluxes downstream of PLC-γ1, crucial for T cell activation via the TCR. Next, reduced NKG2D expression and 2B4 on CD8+ T cells is a central mechanism in both CD70 deficiency as well as MAGT1 deficiency and SAP deficiency. Moreover, there is synergy between 2B4 and NKG2D for optimal NK cell activation, which is crucial for optimal anti-EBV cytotoxic activity. Finally, 2B4, NTB-A, and CD27 ligands are expressed more on B cells, the Trojan horse for EBV.[424,457,458] The knowledge gained from studying these conditions has provided a framework for the important role of the CD8+ T cells to control EBV, as neatly used in adoptive anti-EBV-cellular therapies, which have been used in bridging restoration of adequate anti-EBV responses post-HSCT and also in the treatment of posttransplant lymphoproliferative disease developing post solid organ transplant or HSCT.[459]

CONCLUSION AND PERSPECTIVES FOR THE FIELD OF CLINICAL MEDICINE AND FUNDAMENTAL IMMUNOLOGY

We have attempted to provide a glimpse of the expanding diversity of IEIs, by highlighting a selection of disorders resulting primarily in infection, autoimmunity, autoinflammation, allergy, or malignancy. We highlighted how the study of these IEIs has not only been of great medical benefit but has also taught us basic principles of fundamental immunology. The field is so large that we could have easily picked other forms of IEIs to illustrate similar concepts. We also could have highlighted how various IEIs affect different branches of the human host response. Instead, we chose to pick clinically distinct phenotypes with which IEIs may manifest and to provide a framework for suspecting an IEI in these patients, in spite of their variable presentation. For the most anciently studied phenotype of IEI, infection, the field has begun reconciling the germ theory with the genetic theory, providing a framework that explains the gigantic interindividual clinical variability in the course of infections, including common infections, as best exemplified recently with COVID-19.

For patients presenting with autoimmunity, we have discussed APECED and IPEX as prototypic models of monogenic forms of autoimmunity that have unveiled key mechanisms in central and peripheral T cell tolerance. Further research into rare variants in *AIRE* and *FOXP3* may help explain a fraction of autoimmune manifestations in the general population. Intriguing, the frequency of potential pathogenic heterozygous variants in AIRE was around 1:1000, which is the frequency of autoimmunity in the general population. The study of autoimmune cytopenia in children has also unveiled the unexpected high proportion of underlying IEI in this population, calling for a closer collaboration of hematologists and clinical immunologists in the management of these patients. Autoinflammation is another phenotype that has gradually moved into the field. Indeed, several bona fide "classic" IEIs are accompanied by severe autoinflammatory symptoms, for example, chronic granulomatous disease. Although the first described autoinflammatory disorders, the IL-1β inflammasomopathies, are currently the most common in this group, there are still holes in our understanding of the pathophysiology of this growing group of disorders. The same holds true for inflammation associated with mutations in genes involved in type I IFN production and in the NF-κB signaling pathway. Many of these conditions are characterized by a blended phenotype, consisting of autoinflammatory and autoimmune features, reflecting a complex pathophysiology. The description of the important role of necroptosis-induced inflammation may lead to new discoveries in the field. As for allergy, a common and usually mild condition in the general population, several IEIs are typified by severe allergy. HIES was certainly the first one and as we learn more, the actinopathies, with WAS as the prototype, are a group of conditions to look further into. Principles derived from the studies in these IEIs will enable us to better understand allergy in the general population. Finally, the malignancy phenotype manifested by several forms of IEIs is likely to become a field of intensive research. Indeed, especially in cancers occurring in childhood, the risk of an underlying IEI may be very high, although studies using unbiased sequencing of genomic DNA to detect variants underlying IEIs are mostly missing. Nonetheless, data obtained from cancer registries support an intrinsic cause for the increased risk of malignancy in IEI.

Probably more than any other field in medicine, the field of IEI has grown considerably over the past decades. It is probably also a field in which, in part thanks to the fact that the immune system can be readily studied from blood cells, the relationship between genotype, cellular phenotype, immunologic phenotype, and clinical phenotype has been more thoroughly studied than in many other fields of medicine. New phenotypes are continuously being explored, and we can anticipate in all fairness that thousands of IEIs will be deciphered in the future. This meticulous study has taught us, as apparent also from the IEIs highlighted in this chapter, about the tremendous phenotypic variability associated with mutations at a given locus and how GOF or LOF mutations in a given IEI gene can lead to entirely different IEIs, once more highlighting the essence of thorough investigation

TABLE 50.5 Gene Defects Associated With Severe EBV-Related Disease

Mutated Gene	EBV-Related Clinical Features	Mechanism of Disease	References
SH2D1A	EBV viremia FIM/HLH EBV+ B-lymphoma	↓ SLAM/SAP-dependent activation of NK and CD8+ T cells	418-420,423,425,426
XIAP	EBV viremia FIM/HLH	Increased apoptosis	420,427,428
ITK	EBV viremia EBV-lymphoproliferation EBV+ B cell lymphoma	↓ T cell activation, proliferation	429-436
RASGRP1	EBV viremia EBV+ B cell lymphoma, smooth muscle tumor HLH	↓ ERK activation in T cells ↓ T cell proliferation ↓ CTPS1 expression in T cells ↓ NK cell function	437-441
CTPS1	EBV viremia EBV-lymphoproliferation Fulminant mononucleosis	↓ T cell activation/proliferation ↓ IFN production	442-445
MAGT1	Chronic EBV viremia EBV lymphoproliferation (not HLH) EBV+ B-lymphoma	↓ T cell activation ↓ NK/CD8+ T cell cytotoxicity Glycosylation defect	446-455
GATA2	EBV viremia HLH Chronic EBV viremia	Impaired EBV latency? ↓ NK/CD8+ T cell cytotoxicity due to lymphopenia	456
CD27	EBV viremia HLH EBV lymphoproliferation EBV+ B-lymphoma	↓ CD27/CD70-induced T cell proliferation ↓ EBV-induced T cell proliferation/ expansion	437,458,460,461
CD70	EBV viremia EBV-lymphoproliferation EBV+ B-lymphoma	↓ CD8+ T cell cytotoxicity to B cells ↓ NKG2D, 2B4 may impair CD8+ T cell cytotoxicity	458,462,463
TNFRSF9	EBV viremia HLH EBV lymphoproliferation EBV+ B-lymphoma EBV T cell lymphoproliferation	↓ T cell proliferation, ↓ EBV-specific CD8+ T cell cytotoxicity	464-466
PIK3CD	EBV viremia (25%-30%)	↑ CD8+ T cell apoptosis ↓ NK and CD8+ T cell cytotoxicity (immune senescence?)	79,153,467-470
DOCK8	EBV viremia (20%-30%)	Cytoskeletal defects →↓ immune synapse defect ↓ CD8+ T cell proliferation, ↓ cytotoxicity (immune senescence?) ↑ T cell apoptosis	224,226,471
STK4	EBV viremia (~50%)	↓ CD8+ T cell proliferation ↑ T cell apoptosis Cytoskeletal defects →↓ immune synapse defect	472-475
COROA1	EBV viremia (~35%) EBV-lymphoproliferation EBV+ B-lymphoma	Cytoskeletal defects →↓ immune synapse defect ↑ T cell apoptosis	476-480
CARMIL2	EBV viremia (~50%) EBV+ smooth muscle tumors	T cell proliferation, anti-CD28 costimulation, F-actin polymerization	241-243,245

of the relation between the mutation and its immunological pathogenesis at the cellular level and the clinical phenotype at the organism level.

The study of IEIs has on many occasions paved the way to new avenues of immunologic investigation, where fundamental immunology benefits from the insights gained in the study of the role of genes in the human immune system under natural conditions. Vice versa, technological advances like next-generation sequencing and mass cytometry have armed us to better study IEIs at the genetic and cellular levels. Advances in fundamental immunology have provided us the framework to expand our knowledge of IEIs. We believe that IEIs present a new frontier in Fundamental Immunology and that many of the future immunological insights will derive from the investigation of unique patients.

REFERENCES

1. Bruton OC. Agammaglobulinemia. *Pediatrics.* 1952;9(6):722-728.
2. Tiselius A. Electrophoresis of serum globulin. I. *Biochem J.* 1937;31(2):313-317.
3. Tiselius A, Kabat EA. An electrophoretic study of immune sera and purified antibody preparations. *J Exp Med.* 1939;69(1):119-131.
4. Conley ME. B cells in patients with X-linked agammaglobulinemia. *J Immunol.* 1985;134(5):3070-3074.
5. Rawlings DJ, Saffran DC, Tsukada S, et al. Mutation of unique region of Bruton's tyrosine kinase in immunodeficient XID mice. *Science.* 1993;261(5119):358-361.
6. Henner K. Apropos of the description of "Ataxia teleangiectasis" by Mme Louis-Bar. Priority of the description, by Lad. Syllaba and K. Henner in 1926, of the conjunctival vascular network. Article in French. *Rev Neurol (Paris).* 1968;118(1):60-63.
7. Aldrich RA, Steinberg AG, Campbell DC. Pedigree demonstrating a sex-linked recessive condition characterized by draining ears, eczematoid dermatitis and bloody diarrhea. *Pediatrics.* 1954;13(2):133-139.
8. Wiskott A. Familiärer, angeborener Morbus Werlhofii? *Monatsschr KinderHeilkd.* 1937;68:212-216.
9. Lutz W. About verruciform epidermodysplasia. *Dermatologica.* 1946;92(1):30-43.
10. Kostmann R. Infantile genetic agranulocytosis; agranulocytosis infantilis hereditaria. *Acta Paediatr Suppl.* 1956;45(suppl 105):1-78.
11. Bode E, Sedgwick RP. Ataxia-telangiectasia. A familial syndrome of progressive cerebellar ataxia, oculocutaneous telangiectasia and frequent pulmonary infection. A preliminary report on 7 children, an autopsy, and a case history. *USC Med Bull.* 1957;9:15-28.
12. Boder E, Sedgwick RP. Ataxia-telangiectasia; a familial syndrome of progressive cerebellar ataxia, oculocutaneous telangiectasia and frequent pulmonary infection. *Pediatrics.* 1958;21(4):526-554.
13. Cooper MD, Chae HP, Lowman JT, Krivit W, Good RA. Wiskott-Aldrich syndrome. An immunologic deficiency disease involving the afferent limb of immunity. *Am J Med.* 1968;44(4):499-513.
14. Bousfiha A, Jeddane L, Picard C, et al. Human inborn errors of immunity: 2019 update of the IUIS phenotypical classification. *J Clin Immunol.* 2020;40(1):66-81.
15. Tangye SG, Al-Herz W, Bousfiha A, et al. The ever-increasing array of novel inborn errors of immunity: an interim update by the IUIS committee. *J Clin Immunol.* 2021;41(3):666-679.
16. Tangye SG, Bucciol G, Casas-Martin J, et al. Human inborn errors of the actin cytoskeleton affecting immunity: way beyond WAS and WIP. *Immunol Cell Biol.* 2019;97(4):389-402.
17. Lutzner MA. Epidermodysplasia verruciformis. An autosomal recessive disease characterized by viral warts and skin cancer. A model for viral oncogenesis. *Bull Cancer.* 1978;65(2):169-182.
18. Lutzner MA, Blanchet-Bardon C. Epidermodysplasia verruciformis. *Curr Probl Dermatol.* 1985;13:164-185.
19. Ramoz N, Rueda LA, Bouadjar B, Montoya LS, Orth G, Favre M. Mutations in two adjacent novel genes are associated with epidermodysplasia verruciformis. *Nat Genet.* 2002;32(4):579-581.
20. de Jong SJ, Crequer A, Matos I, et al. The human CIB1-EVER1-EVER2 complex governs keratinocyte-intrinsic immunity to beta-papillomaviruses. *J Exp Med.* 2018;215(9):2289-2310.
21. Casanova JL, Abel L. Lethal infectious diseases as inborn errors of immunity: toward a synthesis of the germ and genetic theories. *Annu Rev Pathol.* 2021;16:23-50.
22. Boisson-Dupuis S, Ramirez-Alejo N, Li Z, et al. Tuberculosis and impaired IL-23-dependent IFN-gamma immunity in humans homozygous for a common TYK2 missense variant. *Sci Immunol.* 2018;3(30):eaau8714.
23. Kerner G, Laval G, Patin E, et al. Human ancient DNA analyses reveal the high burden of tuberculosis in Europeans over the last 2,000 years. *Am J Hum Genet.* 2021;108(3):517-524.
24. Kerner G, Ramirez-Alejo N, Seeleuthner Y, et al. Homozygosity for TYK2 P1104A underlies tuberculosis in about 1% of patients in a cohort of European ancestry. *Proc Natl Acad Sci U S A.* 2019;116(21):10430-10434.
25. Bolze A, Mahlaoui N, Byun M, et al. Ribosomal protein SA haploinsufficiency in humans with isolated congenital asplenia. *Science.* 2013;340(6135):976-978.
26. Courtois G, Smahi A, Reichenbach J, et al. A hypermorphic I-kappa-B-alpha mutation is associated with autosomal dominant anhidrotic ectodermal dysplasia and T cell immunodeficiency. *J Clin Invest.* 2003;112:1108-1115.
27. Israel L, Wang Y, Bulek K, et al. Human adaptive immunity rescues an inborn error of innate immunity. *Cell.* 2017;168(5):789-800.
28. Von Bernuth H, Picard C, Jin Z, et al. Pyogenic bacterial infections in humans with MyD88 deficiency. *Science.* 2008;321(5889):691-696.
29. Casrouge A, Zhang SY, Eidenschenk C, et al. Herpes simplex virus encephalitis in human UNC-93B deficiency. *Science.* 2006;314(5797):308-312.
30. Zhang SY, Jouanguy E, Ugolini S, et al. TLR3 deficiency in patients with herpes simplex encephalitis. *Science.* 2007;317(5844):1522-1527.
31. Herman M, Ciancanelli M, Ou YH, et al. Heterozygous TBK1 mutations impair TLR3 immunity and underlie herpes simplex encephalitis of childhood. *J Exp Med.* 2012;209(9):1567-1582.
32. Perez de Diego R, Sancho-Shimizu V, Lorenzo L, et al. Human TRAF3 adaptor molecule deficiency leads to impaired Toll-like receptor 3 response and susceptibility to herpes simplex encephalitis. *Immunity.* 2010;33(3):400-411.
33. Sancho-Shimizu V, Pérez de Diego R, Lorenzo L, et al. Herpes simplex encephalitis in children with autosomal recessive and dominant TRIF deficiency. *J Clin Invest.* 2011;121(12):4889-4902.
34. Lafaille FG, Harschnitz O, Lee YS, et al. Human SNORA31 variations impair cortical neuron-intrinsic immunity to HSV-1 and underlie herpes simplex encephalitis. *Nat Med.* 2019;25(12):1873-1884.
35. Zhang SY, Clark NE, Freije CA, et al. Inborn errors of RNA lariat metabolism in humans with brainstem viral infection. *Cell.* 2018;172(5):952.e18-965.e18.
36. Ciancanelli MJ, Huang SX, Luthra P, et al. Infectious disease. Life-threatening influenza and impaired interferon amplification in human IRF7 deficiency. *Science.* 2015;348(6233):448-453.
37. Hernandez N, Bucciol G, Moens L, et al. Inherited IFNAR1 deficiency in otherwise healthy patients with adverse reaction to measles and yellow fever live vaccines. *J Exp Med.* 2019;216(9):2057-2070.
38. Hernandez N, Melki I, Jing H, et al. Life-threatening influenza pneumonitis in a child with inherited IRF9 deficiency. *J Exp Med.* 2018;215(10):2567-2585.
39. Hambleton S, Goodbourn S, Young DF, et al. STAT2 deficiency and susceptibility to viral illness in humans. *Proc Natl Acad Sci U S A.* 2013;110(8):3053-3058.
40. Lim HK, Huang SXL, Chen J, et al. Severe influenza pneumonitis in children with inherited TLR3 deficiency. *J Exp Med.* 2019;216(9):2038-2056.
41. Cypowyj S, Picard C, Marodi L, Casanova JL, Puel A. Immunity to infection in IL-17-deficient mice and humans. *Eur J Immunol.* 2012;42(9):2246-2254.
42. Fieschi C, Casanova JL. The role of interleukin-12 in human infectious diseases: only a faint signature. *Eur J Immunol.* 2003;33(6):1461-1464.
43. von Bernuth H, Picard C, Puel A, Casanova JL. Experimental and natural infections in MyD88- and IRAK-4-deficient mice and humans. *Eur J Immunol.* 2012;42(12):3126-3135.
44. Meyts I, Casanova JL. Viral infections in humans and mice with genetic deficiencies of the type I IFN response pathway. *Eur J Immunol.* 2021;51(5):1039-1061.
45. Bonthron DT, Markham AF, Ginsburg D, Orkin SH. Identification of a point mutation in the adenosine deaminase gene responsible for immunodeficiency. *J Clin Invest.* 1985;76(2):894-897.
46. Royer-Pokora B, Kunkel LM, Monaco AP, et al. Cloning the gene for an inherited human disorder–Chronic granulomatous disease–on the basis of its chromosomal location. *Nature.* 1986;322(6074):32-38.
47. Tangye SG, Al-Herz W, Bousfiha A, et al. Human inborn errors of immunity: 2019 update on the classification from the international Union of Immunological Societies Expert Committee. *J Clin Immunol.* 2020;40(1):24-64.
48. Meyts I, Bosch B, Bolze A, et al. Exome and genome sequencing for inborn errors of immunity. *J Allergy Clin Immunol.* 2016;138(4):957-969.
49. Casanova JL, Conley ME, Seligman SJ, Abel L, Notarangelo LD. Guidelines for genetic studies in single patients: lessons from primary immunodeficiencies. *J Exp Med.* 2014;211(11):2137-2149.
50. Walter JE, Rosen LB, Csomos K, et al. Broad-spectrum antibodies against self-antigens and cytokines in RAG deficiency. *J Clin Invest.* 2015;125(11):4135-4148.
51. Walter JE, Rucci F, Patrizi L, et al. Expansion of immunoglobulin-secreting cells and defects in B cell tolerance in RAG-dependent immunodeficiency. *J Exp Med.* 2010;207(7):1541-1554.
52. Notarangelo LD, Kim MS, Walter JE, Lee YN. Human RAG mutations: biochemistry and clinical implications. *Nat Rev Immunol.* 2016;16(4):234-246.
53. Lee YN, Frugoni F, Dobbs K, et al. A systematic analysis of recombination activity and genotype-phenotype correlation in human recombination-activating gene 1 deficiency. *J Allergy Clin Immunol.* 2014;133(4):1099-1108.
54. Boisson-Dupuis S, Kong XF, Okada S, et al. Inborn errors of human STAT1: allelic heterogeneity governs the diversity of immunological and infectious phenotypes. *Curr Opin Immunol.* 2012;24(4):364-378.
55. Boisson B, Quartier P, Casanova JL. Immunological loss-of-function due to genetic gain-of-function in humans: autosomal dominance of the third kind. *Curr Opin Immunol.* 2015;32:90-105.
56. Puel A, Doffinger R, Natividad A, et al. Autoantibodies against IL-17A, IL-17F, and IL-22 in patients with chronic mucocutaneous candidiasis and autoimmune polyendocrine syndrome type I. *J Exp Med.* 2010;207(2):291-297.
57. Kampmann B, Hemingway C, Stephens A, et al. Acquired predisposition to mycobacterial disease due to autoantibodies to IFN-gamma. *J Clin Invest.* 2005;115(9):2480-2488.
58. Patel SY, Ding L, Brown MR, et al. Anti-IFN-gamma autoantibodies in disseminated nontuberculous mycobacterial infections. *J Immunol.* 2005;175(7):4769-4776.
59. Bastard P, Michailidis E, Hoffmann HH, et al. Auto-antibodies to type I IFNs can underlie adverse reactions to yellow fever live attenuated vaccine. *J Exp Med.* 2021;218(4):e20202486.
60. Lanzarotti N, Bruneau J, Trinquand A, et al. RAS-associated lymphoproliferative disease evolves into severe juvenile myelo-monocytic leukemia. *Blood.* 2014;123(12):1960-1963.
61. Holzelova E, Vonarbourg C, Stolzenberg MC, et al. Autoimmune lymphoproliferative syndrome with somatic Fas mutations. *N Engl J Med.* 2004;351(14):1409-1418.
62. Aluri J, Cooper MA. Genetic mosaicism as a cause of inborn errors of immunity. *J Clin Immunol.* 2021;41(4):718-728.
63. Good RA. Experiments of nature in immunobiology. *N Engl J Med.* 1968;279(24):1344-1345.
64. Casanova JL. Severe infectious diseases of childhood as monogenic inborn errors of immunity. *Proc Natl Acad Sci U S A.* 2015;112(51):E7128-E7137.
65. Mimouni J. Our experiences in three years of BCG vaccination at the center of the O.P.H.S. at constantine; study of observed cases (25 cases of complications from BCG vaccination). *Alger Medicale.* 1951;55(8):1138-1147.
66. Zhang Q, Bastard P, Liu Z, et al. Inborn errors of type I IFN immunity in patients with life-threatening COVID-19. *Science.* 2020;370(6515):eabd4570.
67. Hitzig WH, Biro Z, Bosch H, Huser HJ. Agammaglobulinemia & alymphocytosis with atrophy of lymphatic tissue. Article in German. *Helv Paediatr Acta.* 1958;13(6):551-585.
68. Kwan A, Abraham RS, Currier R, et al. Newborn screening for severe combined immunodeficiency in 11 screening programs in the United States. *J Am Med Assoc.* 2014;312(7):729-738.
69. Shearer WT, Dunn E, Notarangelo LD, et al. Establishing diagnostic criteria for severe combined immunodeficiency disease (SCID), leaky SCID, and Omenn syndrome: the Primary Immune Deficiency Treatment Consortium experience. *J Allergy Clin Immunol.* 2014;133(4):1092-1098.

70. Neven B, Perot P, Bruneau J, et al. Cutaneous and visceral chronic granulomatous disease triggered by a rubella virus vaccine strain in children with primary immunodeficiencies. *Clin Infect Dis*. 2017;64(1):83-86.
71. Barkai G, Somech R, Stauber T, Barziali A, Greenberger S. Bacille Calmette-Guerin (BCG) complications in children with severe combined immunodeficiency (SCID). *Infect Dis (Lond)*. 2019;51(8):585-592.
72. Patel NC, Hertel PM, Estes MK, et al. Vaccine-acquired rotavirus in infants with severe combined immunodeficiency. *N Engl J Med*. 2010;362(4):314-319.
73. Lappalainen I, Ollila J, Smith CI, Vihinen M. Registries of immunodeficiency patients and mutations. *Hum Mutat*. 1997;10(4):261-267.
74. Winkelstein JA, Marino MC, Lederman HM, et al. X-linked agammaglobulinemia: report on a United States registry of 201 patients. *Medicine*. 2006;85(4):193-202.
75. LoGalbo PR, Sampson HA, Buckley RH. Symptomatic giardiasis in three patients with X-linked agammaglobulinemia. *J Pediatr*. 1982;101(1):78-80.
76. Stiehm ER, Chin TW, Haas A, Peerless AG. Infectious complications of the primary immunodeficiencies. *Clin Immunol Immunopathol*. 1986;40(1):69-86.
77. McKinney RE Jr, Katz SL, Wilfert CM. Chronic enteroviral meningoencephalitis in agammaglobulinemic patients. *Rev Infect Dis*. 1987;9(2):334-356.
78. Bucciol G, Moens L, Payne K, et al. Chronic Aichi virus infection in a patient with X-linked agammaglobulinemia. *J Clin Immunol*. 2018;38(7):748-752.
79. Edwards ESJ, Bier J, Cole TS, et al. Activating PIK3CD mutations impair human cytotoxic lymphocyte differentiation and function and EBV immunity. *J Allergy Clin Immunol*. 2019;143(1):276.e6-291.e6.
80. van de Ven AA, Warnatz K. The autoimmune conundrum in common variable immunodeficiency disorders. *Curr Opin Allergy Clin Immunol*. 2015;15(6):514-524.
81. Preite S, Gomez-Rodriguez J, Cannons JL, Schwartzberg PL. T and B-cell signaling in activated PI3K delta syndrome: from immunodeficiency to autoimmunity. *Immunol Rev*. 2019;291(1):154-173.
82. Lau A, Avery DT, Jackson K, et al. Activated PI3Kδ breaches multiple B cell tolerance checkpoints and causes autoantibody production. *J Exp Med*. 2020;217(2):e20191336.
83. Yazdani R, Azizi G, Abolhassani H, Aghamohammadi A. Selective IgA deficiency: Epidemiology, pathogenesis, clinical phenotype, diagnosis, prognosis and management. *Scand J Immunol*. 2017;85(1):3-12.
84. Yazdani R, Heydari A, Azizi G, Abolhassani H, Aghamohammadi A. Asthma and allergic diseases in a selected group of patients with common variable immunodeficiency. *J Investig Allergol Clin Immunol*. 2016;26(3):209-211.
85. Conley ME, Dobbs AK, Farmer DM, et al. Primary B cell immunodeficiencies: comparisons and contrasts. *Annu Rev Immunol*. 2009;27:199-227.
86. Winkelstein JA, Marino MC, Ochs H, et al. The X-linked hyper-IgM syndrome: clinical and immunologic features of 79 patients. *Medicine*. 2003;82(6):373-384.
87. Hayward AR, Levy J, Facchetti F, et al. Cholangiopathy and tumors of the pancreas, liver, and biliary tree in boys with X-linked immunodeficiency with hyper-IgM. *J Immunol*. 1997;158(2):977-983.
88. Notarangelo LD, Peitsch MC, Abrahamsen TG, et al. CD40lbase: a database of CD40L gene mutations causing X-linked hyper-IgM syndrome. *Immunol Today*. 1996;17(11):511-516.
89. Bucciol G, Nicholas SK, Calvo PL, et al. Combined liver and hematopoietic stem cell transplantation in patients with X-linked hyper-IgM syndrome. *J Allergy Clin Immunol*. 2019;143(5):1952.e6-1956.e6.
90. Ferrua F, Galimberti S, Courteille V, et al. Hematopoietic stem cell transplantation for CD40 ligand deficiency: results from an EBMT/ESID-IEWP-SCETIDE-PIDTC study. *J Allergy Clin Immunol*. 2019;143(6):2238-2253.
91. Lanzi G, Ferrari S, Vihinen M, et al. Different molecular behavior of CD40 mutants causing hyper-IgM syndrome. *Blood*. 2010;116(26):5867-5874.
92. Justiz Vaillant AA, Zito PM. *Neutropenia*. StatPearls; 2021.
93. Skokowa J, Dale DC, Touw IP, Zeidler C, Welte K. Severe congenital neutropenias. *Nat Rev Dis Primers*. 2017;3:17032.
94. Donadieu J, Beaupain B, Fenneteau O, Bellanne-Chantelot C. Congenital neutropenia in the era of genomics: classification, diagnosis, and natural history. *Br J Haematol*. 2017;179(4):557-574.
95. Zeidler C, Boxer L, Dale DC, Freedman MH, Kinsey S, Welte K. Management of Kostmann syndrome in the G-CSF era. *Br J Haematol*. 2000;109(3):490-495.
96. Germeshausen M, Ballmaier M, Welte K. Incidence of CSF3R mutations in severe congenital neutropenia and relevance for leukemogenesis: results of a long-term survey. *Blood*. 2007;109(1):93-99.
97. Klimiankou M, Uenalan M, Kandabarau S, et al. Ultrasensitive CSF3R deep sequencing in patients with severe congenital neutropenia. *Front Immunol*. 2019;10:116.
98. Klimiankou M, Mellor-Heineke S, Zeidler C, Welte K, Skokowa J. Role of CSF3R mutations in the pathomechanism of congenital neutropenia and secondary acute myeloid leukemia. *Ann N Y Acad Sci*. 2016;1370(1):119-125.
99. Hammond WP., Price TH, Souza LM, Dale DC. Treatment of cyclic neutropenia with granulocyte colony-stimulating factor. *N Engl J Med*. 1989;320(20):1306-1311.
100. Palmer SE, Stephens K, Dale DC. Genetics, phenotype, and natural history of autosomal dominant cyclic hematopoiesis. *Am J Med Genet*. 1996;66(4):413-422.
101. Berendes H, Bridges RA, Good RA. A fatal granulomatosus of childhood: the clinical study of a new syndrome. *Minn Med*. 1957;40(5):309-312.
102. Carson MJ, Chadwick DL, Brubaker CA, Cleland RS, Landing BH. Thirteen boys with progressive septic granulomatosis. *Pediatrics*. 1965;35:405-412.
103. Quie PG, White JG, Holmes B, Good RA. In vitro bactericidal capacity of human polymorphonuclear leukocytes: diminished activity in chronic granulomatous disease of childhood. *J Clin Invest*. 1967;46(4):668-679.
104. Prince BT, Thielen BK, Williams KW, et al. Geographic variability and pathogen-specific considerations in the diagnosis and management of chronic granulomatous disease. *Pediatric Health Med Ther*. 2020;11:257-268.
105. Thomsen IP, Smith MA, Holland SM, Creech CB. A comprehensive approach to the management of children and adults with chronic granulomatous disease. *J Allergy Clin Immunol Pract*. 2016;4(6):1082-1088.
106. van de Vijver E, Maddalena A, Sanal O, et al. Hematologically important mutations: leukocyte adhesion deficiency (first update). *Blood Cells Mol Dis*. 2012;48(1):53-61.
107. Hanna S, Etzioni A. Leukocyte adhesion deficiencies. *Ann N Y Acad Sci*. 2012;1250:50-55.
108. Zinsser H, Cary EG. On the nature of the opsonic substances of normal sera. *J Exp Med*. 1914;19(4):345-361.
109. Cavaillon JM, Sansonetti P, Goldman M. 100th Anniversary of Jules Bordet's Nobel prize: Tribute to a founding father of immunology. *Front Immunol*. 2019;10:2114.
110. Brodszki N, Frazer-Abel A, Grumach AS, et al. European Society for Immunodeficiencies (ESID) and European reference network on rare primary immunodeficiency, autoinflammatory and autoimmune diseases (ERN RITA) complement guideline: deficiencies, diagnosis, and management. *J Clin Immunol*. 2020;40(4):576-591.
111. Zhang SY, Jouanguy E, Zhang Q, Abel L, Puel A, Casanova JL. Human inborn errors of immunity to infection affecting cells other than leukocytes: from the immune system to the whole organism. *Curr Opin Immunol*. 2019;59:88-100.
112. Fieschi C, Dupuis S, Catherinot E, et al. Low penetrance, broad resistance, and favorable outcome of interleukin 12 receptor beta1 deficiency: medical and immunological implications. *J Exp Med*. 2003;197(4):527-535.
113. Altare F, Ensser A, Breiman A, et al. Interleukin-12 receptor beta1 deficiency in a patient with abdominal tuberculosis. *J Infect Dis*. 2001;184(2):231-236.
114. Boisson-Dupuis S, El Baghdadi J, Parvaneh N, et al. IL-12Rbeta1 deficiency in two of fifty children with severe tuberculosis from Iran, Morocco, and Turkey. *PLoS One*. 2011;6(4):e18524.
115. Caragol I, Raspall M, Fieschi C, et al. Clinical tuberculosis in 2 of 3 siblings with interleukin-12 receptor beta1 deficiency. *Clin Infect Dis*. 2003;37(2):302-306.
116. Ozbek N, Fieschi C, Yilmaz BT, et al. Interleukin-12 receptor beta 1 chain deficiency in a child with disseminated tuberculosis. *Clin Infect Dis*. 2005;40(6):e55-8.
117. Tabarsi P, Marjani M, Mansouri N, et al. Lethal tuberculosis in a previously healthy adult with IL-12 receptor deficiency. *J Clin Immunol*. 2011;31(4):537-539.
118. Kong XF, Vogt G, Itan Y, et al. Haploinsufficiency at the human IFNGR2 locus contributes to mycobacterial disease. *Hum Mol Genet*. 2013;22(4):769-781.
119. Kong XF, Martinez-Barricarte R, Kennedy J, et al. Disruption of an antimycobacterial circuit between dendritic and helper T cells in human SPPL2a deficiency. *Nat Immunol*. 2018;19(9):973-985.
120. Martinez-Barricarte R, Markle JG, Ma CS, et al. Human IFN-gamma immunity to mycobacteria is governed by both IL-12 and IL-23. *Sci Immunol*. 2018;3(30).
121. de Beaucoudrey L, Samarina A, Bustamante J, et al. Revisiting human IL-12Rbeta1 deficiency: a survey of 141 patients from 30 countries. *Medicine*. 2010;89(6):381-402.
122. Bolze A, Boisson B, Bosch B, et al. Incomplete penetrance for isolated congenital asplenia in humans with mutations in translated and untranslated RPSA exons. *Proc Natl Acad Sci U S A*. 2018;115(34):E8007-E16.
123. Boisson B, Honda Y, Ajiro M, et al. Rescue of recurrent deep intronic mutation underlying cell type-dependent quantitative NEMO deficiency. *J Clin Invest*. 2019;129(2):583-597.
124. Doffinger R, Smahi A, Bessia C, et al. X-linked anhidrotic ectodermal dysplasia with immunodeficiency is caused by impaired NF-kappa-B signaling. *Nat Genet*. 2001;27:277-285.
125. Picard C, Puel A, Bonnet M, et al. Pyogenic bacterial infections in humans with IRAK-4 deficiency. *Science*. 2003;299(5615):2076-2079.
126. Sologuren I, Martinez-Saavedra MT, Sole-Violan J, et al. Lethal influenza in two related adults with inherited GATA2 deficiency. *J Clin Immunol*. 2018;38(4):513-526.
127. Bravo Garcia-Morato M, Calvo Apalategi A, Bravo-Gallego LY, et al. Impaired control of multiple viral infections in a family with complete IRF9 deficiency. *J Allergy Clin Immunol*. 2019;144(1):309-312 e10.
128. Lim HK, Seppanen M, Hautala T, et al. TLR3 deficiency in herpes simplex encephalitis: high allelic heterogeneity and recurrence risk. *Neurology*. 2014;83(21):1888-1897.
129. Liang F, Glans H, Enoksson SL, Kolios AGA, Lore K, Nilsson J. Recurrent herpes zoster ophthalmicus in a patient with a novel Toll-like receptor 3 variant linked to compromised activation capacity in fibroblasts. *J Infect Dis*. 2020;221(8):1295-1303.
130. Guo Y, Audry M, Ciancanelli M, et al. Herpes simplex virus encephalitis in a patient with complete TLR3 deficiency: TLR3 is otherwise redundant in protective immunity. *J Exp Med*. 2011;208(10):2083-2098.
131. Andersen L, Mørk N, Reinert LS, et al. Functional IRF3 deficiency in a patient with herpes simplex encephalitis. *J Exp Med* 2015;212(9):1371-1379.
132. Bastard P, Manry J, Chen J, et al. Herpes simplex encephalitis in a patient with a distinctive form of inherited IFNAR1 deficiency. *J Infect Dis*. 2020;131(1):e139980.
133. Zhang Q, Boisson B, Beziat V, Puel A, Casanova JL. Human hyper-IgE syndrome: singular or plural? *Mamm Genome*. 2018;29(7-8):603-617.
134. Gao D, Ciancanelli MJ, Zhang P, et al. TLR3 controls constitutive IFN-beta antiviral immunity in human fibroblasts and cortical neurons. *J Clin Invest*. 2021;131(1):e134529.
135. Lafaille FG, Pessach IM, Zhang SY, et al. Impaired intrinsic immunity to HSV-1 in human iPSC-derived TLR3-deficient CNS cells. *Nature*. 2012;491(7426):769-773.
136. Kong XF, Ciancanelli M, Al-Hajjar S, et al. A novel form of human STAT1 deficiency impairing early but not late responses to interferons. *Blood*. 2010;116(26):5895-5906.
137. Chapgier A, Kong XF, Boisson-Dupuis S, et al. A partial form of recessive STAT1 deficiency in humans. *J Clin Invest*. 2009;119(6):1502-1514.
138. Agnello V, De Bracco MM, Kunkel HG. Hereditary C2 deficiency with some manifestations of systemic lupus erythematosus. *J Immunol*. 1972;108(3):837-840.
139. Moncada B, Day NK, Good RA, Windhorst DB. Lupus-erythematosus-like syndrome with a familial defect of complement. *N Engl J Med*. 1972;286(13):689-693.
140. Fischer A, Provot J, Jais JP, Alcais A, Mahlaoui N; members of the CEREDIH French PID Study Group. Autoimmune and inflammatory manifestations occur frequently in patients with primary immunodeficiencies. *J Allergy Clin Immunol*. 2017;140(5):1388.e8-1393.e8.
141. Hadjadj J, Aladjidi N, Fernandes H, et al. Pediatric Evans syndrome is associated with a high frequency of potentially damaging variants in immune genes. *Blood*. 2019;134(1):9-21.
142. Oliveira JB, Bleesing JJ, Dianzani U, et al. Revised diagnostic criteria and classification for the autoimmune lymphoproliferative syndrome (ALPS): report from the 2009 NIH International Workshop. *Blood*. 2010;116(14):e35-40.
143. Rieux-Laucat F. What's up in the ALPS. *Curr Opin Immunol*. 2017;49:79-86.
144. Fisher GH, Rosenberg FJ, Straus SE, et al. Dominant interfering Fas gene mutations impair apoptosis in a human autoimmune lymphoproliferative syndrome. *Cell*. 1995;81(6):935-946.
145. Jackson CE, Fischer RE, Hsu AP, et al. Autoimmune lymphoproliferative syndrome with defective Fas: genotype influences penetrance. *Am J Hum Genet*. 1999;64(4):1002-1014.
146. Nabhani S, Ginzel S, Miskin H, et al. Deregulation of Fas ligand expression as a novel cause of autoimmune lymphoproliferative syndrome-like disease. *Haematologica*. 2015;100(9):1189-1198.
147. Bolze A, Byun M, McDonald D, et al. Whole-exome-sequencing-based discovery of human FADD deficiency. *Am J Hum Genet*. 2010;87(6):873-881.
148. Van Coillie S, Wiernicki B, Xu J. Molecular and cellular functions of CTLA-4. *Adv Exp Med Biol*. 2020;1248:7-32.
149. Lopez-Herrera G, Tampella G, Pan-Hammarstrom Q, et al. Deleterious mutations in LRBA are associated with a syndrome of immune deficiency and autoimmunity. *Am J Hum Genet*. 2012;90(6):986-1001.
150. Schwab C, Gabrysch A, Olbrich P, et al. Phenotype, penetrance, and treatment of 133 cytotoxic T-lymphocyte antigen 4-insufficient subjects. *J Allergy Clin Immunol*. 2018;142(6):1932-1946.
151. Habibi S, Zaki-Dizaji M, Rafiemanesh H, et al. Clinical, immunologic, and molecular spectrum of patients with LPS-responsive beige-like anchor protein deficiency: a systematic review. *J Allergy Clin Immunol Pract*. 2019;7(7):2379.e5-2386.e5.
152. Lo B, Zhang K, Lu W, et al. AUTOIMMUNE DISEASE. Patients with LRBA deficiency show CTLA4 loss and immune dysregulation responsive to abatacept therapy. *Science*. 2015;349(6246):436-440.
153. Lucas CL, Chandra A, Nejentsev S, Condliffe AM, Okkenhaug K. PI3Kdelta and primary immunodeficiencies. *Nat Rev Immunol*. 2016;16(11):702-714.

154. Rae W, Ward D, Mattocks CJ, et al. Autoimmunity/inflammation in a monogenic primary immunodeficiency cohort. *Clin Transl Immunology*. 2017;6(9):e155.
155. Kuehn HS, Niemela JE, Sreedhara K, et al. Novel nonsense gain-of-function NFKB2 mutations associated with a combined immunodeficiency phenotype. *Blood*. 2017;130(13):1553-1564.
156. Toubiana J, Okada S, Hiller J, et al. International STAT1 Gain-of-Function Study Group. Heterozygous STAT1 gain-of-function mutations underlie an unexpectedly broad clinical phenotype. *Blood*. 2016;127(25):3154-3164.
157. Fabre A, Marchal S, Barlogis V, et al. Clinical aspects of STAT3 gain-of-function germline mutations: a systematic review. *J Allergy Clin Immunol Pract*. 2019;7(6):1958-1969 e9.
158. Rao VK, Webster S, Dalm V, et al. Effective "activated PI3Kdelta syndrome"-targeted therapy with the PI3Kdelta inhibitor leniolisib. *Blood*. 2017;130(21):2307-2316.
159. Silva-Carmona M, Vogel TP, Marchal S, et al. Successful treatment of interstitial lung disease in STAT3 gain-of-function using JAK inhibitors. *Am J Respir Crit Care Med*. 2020;202(6):893-897.
160. Romberg N, Le Coz C, Glauzy S, et al. Patients with common variable immunodeficiency with autoimmune cytopenias exhibit hyperplastic yet inefficient germinal center responses. *J Allergy Clin Immunol*. 2019;143(1):258-265.
161. Alroqi FJ, Charbonnier LM, Baris S, et al. Exaggerated follicular helper T-cell responses in patients with LRBA deficiency caused by failure of CTLA4-mediated regulation. *J Allergy Clin Immunol*. 2018;141(3):1050.e10-1059.e10.
162. Sage PT, Paterson AM, Lovitch SB, Sharpe AH. The coinhibitory receptor CTLA-4 controls B cell responses by modulating T follicular helper, T follicular regulatory, and T regulatory cells. *Immunity*. 2014;41(6):1026-1039.
163. Ahonen P, Myllarniemi S, Sipila I, Perheentupa J. Clinical variation of autoimmune polyendocrinopathy-candidiasis-ectodermal dystrophy (APECED) in a series of 68 patients. *N Engl J Med*. 1990;322(26):1829-1836.
164. Husebye ES, Anderson MS. Autoimmune polyendocrine syndromes: clues to type 1 diabetes pathogenesis. *Immunity*. 2010;32(4):479-487.
165. Meager A, Visvalingam K, Peterson P, et al. Anti-interferon autoantibodies in autoimmune polyendocrinopathy syndrome type 1. *PLoS Med*. 2006;3(7):e289.
166. Thorpe ES, Handley H. Chronic tetany and chronic mycelial stomatitis in a child aged four and one-half years. *Am J Dis Child*. 1929;38(2):328-338.
167. Cheng MH, Anderson MS. Monogenic autoimmunity. *Annu Rev Immunol*. 2012;30:393-427.
168. Perheentupa J, Miettinen A. Type 1 autoimmune polyglandular disease. *Ann Med Interne (Paris)*. 1999;150(4):313-325.
169. Myhre AG, Halonen M, Eskelin P, et al. Autoimmune polyendocrine syndrome type 1 (APS I) in Norway. *Clin Endocrinol*. 2001;54(2):211-217.
170. Ferre EM, Rose SR, Rosenzweig SD, et al. Redefined clinical features and diagnostic criteria in autoimmune polyendocrinopathy-candidiasis-ectodermal dystrophy. *JCI Insight*. 2016;1(13):e88782.
171. Nagamine K, Peterson P, Scott HS, et al. Positional cloning of the APECED gene. *Nat Genet*. 1997;17(4):393-398.
172. Finnish-German AC. An autoimmune disease, APECED, caused by mutations in a novel gene featuring two PHD-type zinc-finger domains. *Nat Genet*. 1997;17(4):399-403.
173. Bjorses P, Halonen M, Palvimo JJ, et al. Mutations in the AIRE gene: effects on subcellular location and transactivation function of the autoimmune polyendocrinopathy-candidiasis-ectodermal dystrophy protein. *Am J Hum Genet*. 2000;66(2):378-392.
174. Pitkanen J, Vahamurto P, Krohn K, Peterson P. Subcellular localization of the autoimmune regulator protein. characterization of nuclear targeting and transcriptional activation domain. *J Biol Chem*. 2001;276(22):19597-19602.
175. Anderson MS, Venanzi ES, Klein L, et al. Projection of an immunological self shadow within the thymus by the aire protein. *Science*. 2002;298(5597):1395-1401.
176. Taniguchi RT, Anderson MS. The role of Aire in clonal selection. *Immunol Cell Biol*. 2011;89(1):40-44.
177. Cowan JE, Parnell SM, Nakamura K, et al. The thymic medulla is required for Foxp3+ regulatory but not conventional CD4+ thymocyte development. *J Exp Med*. 2013;210(4):675-681.
178. Kisand K, Boe Wolff AS, Podkrajsek KT, et al. Chronic mucocutaneous candidiasis in APECED or thymoma patients correlates with autoimmunity to Th17-associated cytokines. *J Exp Med*. 2010;207(2):299-308.
179. Bastard P, Orlova E, Sozaeva L, et al. Preexisting autoantibodies to type I IFNs underlie critical COVID-19 pneumonia in patients with APS-1. *J Exp Med*. 2021;218(7):e20210554.
180. Break TJ, Oikonomou V, Dutzan N, et al. Aberrant type 1 immunity drives susceptibility to mucosal fungal infections. *Science*. 2021;371(6526):eaay5731.
181. Wolff AS, Erichsen MM, Meager A, et al. Autoimmune polyendocrine syndrome type 1 in Norway: phenotypic variation, autoantibodies, and novel mutations in the autoimmune regulator gene. *J Clin Endocrinol Metab*. 2007;92(2):595-603.
182. Cetani F, Barbesino G, Borsari S, et al. A novel mutation of the autoimmune regulator gene in an Italian kindred with autoimmune polyendocrinopathy-candidiasis-ectodermal dystrophy, acting in a dominant fashion and strongly cosegregating with hypothyroid autoimmune thyroiditis. *J Clin Endocrinol Metab*. 2001;86(10):4747-4752.
183. Oftedal BE, Hellesen A, Erichsen MM, et al. Dominant mutations in the autoimmune regulator AIRE are associated with common organ-specific autoimmune diseases. *Immunity*. 2015;42(6):1185-1196.
184. Cavadini P, Vermi W, Facchetti F, et al. AIRE deficiency in thymus of 2 patients with Omenn syndrome. *J Clin Invest*. 2005;115(3):728-732.
185. De Ravin SS, Cowen EW, Zarember KA, et al. Hypomorphic Rag mutations can cause destructive midline granulomatous disease. *Blood*. 2010;116(8):1263-1271.
186. Powell BR, Buist NR, Stenzel P. An X-linked syndrome of diarrhea, polyendocrinopathy, and fatal infection in infancy. *J Pediatr*. 1982;100(5):731-737.
187. Brunkow ME, Jeffery EW, Hjerrild KA, et al. Disruption of a new forkhead/winged-helix protein, scurfin, results in the fatal lymphoproliferative disorder of the scurfy mouse. *Nat Genet*. 2001;27(1):68-73.
188. Wildin RS, Ramsdell F, Peake J, et al. X-linked neonatal diabetes mellitus, enteropathy and endocrinopathy syndrome is the human equivalent of mouse scurfy. *Nat Genet*. 2001;27(1):18-20.
189. Russell WL, Russell LB, Gower JS. Exceptional inheritance of a sex-linked gene in the mouse explained on the basis that the X/O sex-chromosome constitution is female. *Proc Natl Acad Sci U S A*. 1959;45(4):554-560.
190. Kitagawa Y, Ohkura N, Kidani Y, et al. Guidance of regulatory T cell development by Satb1-dependent super-enhancer establishment. *Nat Immunol*. 2017;18(2):173-183.
191. Park JH, Lee KH, Jeon B, et al. Immune dysregulation, polyendocrinopathy, enteropathy, X-linked (IPEX) syndrome: a systematic review. *Autoimmun Rev*. 2020;19(6):102526.
192. Lampasona V, Passerini L, Barzaghi F, et al. Autoantibodies to harmonin and villin are diagnostic markers in children with IPEX syndrome. *PLoS One*. 2013;8(11):e78664.
193. Barzaghi F, Amaya Hernandez LC, Neven B, et al. Long-term follow-up of IPEX syndrome patients after different therapeutic strategies: an international multicenter retrospective study. *J Allergy Clin Immunol*. 2018;141(3):1036.e5-1049.e5.
194. Roncarolo MG, Battaglia M. Regulatory T-cell immunotherapy for tolerance to self antigens and alloantigens in humans. *Nat Rev Immunol*. 2007;7(8):585-598.
195. Liu C, Workman CJ, Vignali DA. Targeting regulatory T cells in tumors. *FEBS J*. 2016;283(14):2731-2748.
196. Milner JD. Primary atopic disorders. *Annu Rev Immunol*. 2020;38:785-808.
197. Davis SD, Schaller J, Wedgwood RJ, Harvard MD. Recurrent, "cold", staphylococcal abscesses. *Lancet*. 1966;287(7445):1013-1015.
198. Buckley RH, Wray BB, Belmaker EZ. Extreme hyperimmunoglobulinemia E and undue susceptibility to infection. *Pediatrics*. 1972;49(1):59-70.
199. Minegishi Y, Saito M, Tsuchiya S, et al. Dominant-negative mutations in the DNA-binding domain of STAT3 cause hyper-IgE syndrome. *Nature*. 2007;448(7157):1058-1062.
200. Holland SM, DeLeo FR, Elloumi HZ, et al. STAT3 mutations in the hyper-IgE syndrome. *N Engl J Med*. 2007;357(16):1608-1619.
201. Ma CS, Avery DT, Chan A, et al. Functional STAT3 deficiency compromises the generation of human T follicular helper cells. *Blood*. 2012;119(17):3997-4008.
202. Ma CS, Chew GY, Simpson N, et al. Deficiency of Th17 cells in hyper IgE syndrome due to mutations in STAT3. *J Exp Med*. 2008;205(7):1551-1557.
203. Rose-John S. Interleukin-6 family cytokines. *Cold Spring Harb Perspect Biol*. 2018;10(2):a028415.
204. Lutticken C, Wegenka UM, Yuan J, et al. Association of transcription factor APRF and protein kinase Jak1 with the interleukin-6 signal transducer gp130. *Science*. 1994;263(5143):89-92.
205. Murakami M, Hibi M, Nakagawa N, et al. IL-6-induced homodimerization of gp130 and associated activation of a tyrosine kinase. *Science*. 1993;260(5115):1808-1810.
206. Barton VA, Hall MA, Hudson KR, Heath JK. Interleukin-11 signals through the formation of a hexameric receptor complex. *J Biol Chem*. 2000;275(46):36197-36203.
207. Frey-Jakobs S, Hartberger JM, Fliegauf M, et al. ZNF341 controls STAT3 expression and thereby immunocompetence. *Sci Immunol*. 2018;3(24):eaat4941.
208. Beziat V, Li J, Lin JX, et al. A recessive form of hyper-IgE syndrome by disruption of ZNF341-dependent STAT3 transcription and activity. *Sci Immunol*. 2018;3(24):eaat4956.
209. Beziat V, Tavernier SJ, Chen YH, et al. Dominant-negative mutations in human IL6ST underlie hyper-IgE syndrome. *J Exp Med*. 2020;217(6):e2019180.
210. Minegishi Y, Saito M, Morio T, et al. Human tyrosine kinase 2 deficiency reveals its requisite roles in multiple cytokine signals involved in innate and acquired immunity. *Immunity*. 2006;25(5):745-755.
211. Schwerd T, Twigg SRF, Aschenbrenner D, et al. A biallelic mutation in IL6ST encoding the GP130 co-receptor causes immunodeficiency and craniosynostosis. *J Exp Med*. 2017;214(9):2547-2562.
212. Spencer S, Kostel Bal S, Egner W, et al. Loss of the interleukin-6 receptor causes immunodeficiency, atopy, and abnormal inflammatory responses. *J Exp Med*. 2019;216(9):1986-1998.
213. Puel A, Picard C, Lorrot M, et al. Recurrent staphylococcal cellulitis and subcutaneous abscesses in a child with autoantibodies against IL-6. *J Immunol*. 2008;180(1):647-654.
214. Ma CS, Wong N, Rao G, et al. Unique and shared signaling pathways cooperate to regulate the differentiation of human CD4+ T cells into distinct effector subsets. *J Exp Med*. 2016;213(8):1589-1608.
215. Shahin T, Aschenbrenner D, Cagdas D, et al. Selective loss of function variants in IL6ST cause Hyper-IgE syndrome with distinct impairments of T-cell phenotype and function. *Haematologica*. 2019;104(3):609-621.
216. Dagoneau N, Scheffer D, Huber C, et al. Null leukemia inhibitory factor receptor (LIFR) mutations in Stuve-Wiedemann/Schwartz-Jampel type 2 syndrome. *Am J Hum Genet*. 2004;74(2):298-305.
217. Chen YH, Grigelioniene G, Newton PT, et al. Absence of GP130 cytokine receptor signaling causes extended Stuve-Wiedemann syndrome. *J Exp Med*. 2020;217(3):e20191306.
218. Monies D, Abouelhoda M, Assoum M, et al. Lessons learned from large-scale, first-tier clinical exome sequencing in a highly consanguineous population. *Am J Hum Genet*. 2019;104(6):1182-1201.
219. Wang Y, Hu H, Wu J, et al. The IL6R gene polymorphisms are associated with sIL-6R, IgE and lung function in Chinese patients with asthma. *Gene*. 2016;585(1):51-57.
220. Ferreira RC, Freitag DF, Cutler AJ, et al. Functional IL6R 358Ala allele impairs classical IL-6 receptor signaling and influences risk of diverse inflammatory diseases. *PLoS Genet*. 2013;9(4):e1003444.
221. Zhang Q, Davis JC, Lamborn IT, et al. Combined immunodeficiency associated with DOCK8 mutations. *N Engl J Med*. 2009;361(21):2046-2055.
222. Zhang Q, Dove CG, Hor JL, et al. DOCK8 regulates lymphocyte shape integrity for skin antiviral immunity. *J Exp Med*. 2014;211(13):2549-2566.
223. Zhang Q, Jing H, Su HC. Recent advances in DOCK8 immunodeficiency syndrome. *J Clin Immunol*. 2016;36(5):441-449.
224. Aydin SE, Kilic SS, Aytekin C, et al. DOCK8 deficiency: clinical and immunological phenotype and treatment options—a review of 136 patients. *J Clin Immunol*. 2015;35(2):189-198.
225. Biggs CM, Keles S, Chatila TA. DOCK8 deficiency: insights into pathophysiology, clinical features and management. *Clin Immunol*. 2017;181:75-82.
226. Su HC, Jing H, Angelus P, Freeman AF. Insights into immunity from clinical and basic science studies of DOCK8 immunodeficiency syndrome. *Immunol Rev*. 2019;287(1):9-19.
227. Tangye SG, Pillay B, Randall KL, et al. Dedicator of cytokinesis 8-deficient CD4(+) T cells are biased to a TH2 effector fate at the expense of TH1 and TH17 cells. *J Allergy Clin Immunol*. 2017;139(3):933-949.
228. Aydin SE, Freeman AF, Al-Herz W, et al. Hematopoietic stem cell transplantation as treatment for patients with DOCK8 deficiency. *J Allergy Clin Immunol Pract*. 2019;7(3):848-855.
229. Happel CS, Stone KD, Freeman AF, et al. Food allergies can persist after myeloablative hematopoietic stem cell transplantation in dedicator of cytokinesis 8-deficient patients. *J Allergy Clin Immunol*. 2016;137(6):1895-1898 e5.
230. Jing H, Zhang Q, Zhang Y, et al. Somatic reversion in dedicator of cytokinesis 8 immunodeficiency modulates disease phenotype. *J Allergy Clin Immunol*. 2014;133(6):1667-1675.
231. Sullivan KE, Mullen CA, Blaese RM, Winkelstein JA. A multiinstitutional survey of the Wiskott-Aldrich syndrome. *J Pediatr*. 1994;125(6 pt 1):876-885.
232. Albert MH, Notarangelo LD, Ochs HD. Clinical spectrum, pathophysiology and treatment of the Wiskott-Aldrich syndrome. *Curr Opin Hematol*. 2011;18(1):42-48.
233. Adriani M, Aoki J, Horai R, et al. Impaired *in vitro* regulatory T cell function associated with Wiskott-Aldrich syndrome. *Clin Immunol*. 2007;124(1):41-48.
234. Trifari S, Sitia G, Aiuti A, et al. Defective Th1 cytokine gene transcription in CD4+ and CD8+ T cells from Wiskott-Aldrich syndrome patients. *J Immunol*. 2006;177(10):7451-7461.
235. Morales-Tirado V, Sojka DK, Katzman SD, et al. Critical requirement for the Wiskott-Aldrich syndrome protein in Th2 effector function. *Blood*. 2010;115(17):3498-3507.
236. Burroughs LM, Petrovic A, Brazauskas R, et al. Excellent outcomes following hematopoietic cell transplantation for Wiskott-Aldrich syndrome: a PIDTC report. *Blood*. 2020;135(23):2094-2105.
237. Aiuti A, Biasco L, Scaramuzza S, et al. Lentiviral hematopoietic stem cell gene therapy in patients with Wiskott-Aldrich syndrome. *Science*. 2013;341(6148):1233151.
238. Hacein-Bey Abina S, Gaspar HB, Blondeau J, et al. Outcomes following gene therapy in patients with severe Wiskott-Aldrich syndrome. *J Am Med Assoc*. 2015;313(15):1550-1563.

239. Wada T, Jagadeesh GJ, Nelson DL, Candotti F. Retrovirus-mediated WASP gene transfer corrects Wiskott-Aldrich syndrome T-cell dysfunction. *Hum Gene Ther.* 2002;13(9):1039-1046.
240. Rivers E, Rai R, Lotscher J, et al. Wiskott Aldrich syndrome protein regulates non-selective autophagy and mitochondrial homeostasis in human myeloid cells. *Elife.* 2020;9:e55547.
241. Kurolap A, Eshach Adiv O, Konnikova L, et al. A unique presentation of infantile-onset colitis and eosinophilic disease without recurrent infections resulting from a novel homozygous CARMIL2 variant. *J Clin Immunol.* 2019;39(4):430-439.
242. Alazami AM, Al-Helale M, Alhissi S, et al. Novel CARMIL2 mutations in patients with variable clinical dermatitis, infections, and combined immunodeficiency. *Front Immunol.* 2018;9:203.
243. Schober T, Magg T, Laschinger M, et al. A human immunodeficiency syndrome caused by mutations in CARMIL2. *Nat Commun.* 2017;8:14209.
244. Wang Y, Ma CS, Ling Y, et al. Dual T cell- and B cell-intrinsic deficiency in humans with biallelic RLTPR mutations. *J Exp Med.* 2016;213(11):2413-2435.
245. Roncagalli R, Cucchetti M, Jarmuzynski N, et al. The scaffolding function of the RLTPR protein explains its essential role for CD28 co-stimulation in mouse and human T cells. *J Exp Med.* 2016;213(11):2437-2457.
246. Turvey S, Durandy A, Fischer A, et al. The CARD11-BCL10-MALT1 (CBM) signalosome complex: Stepping into the limelight of human primary immunodeficiency. *J Allergy Clin Immunol.* 2014;134(2):276-284.
247. Stepensky P, Keller B, Buchta M, et al. Deficiency of caspase recruitment domain family, member 11 (CARD11), causes profound combined immunodeficiency in human subjects. *J Allergy Clin Immunol.* 2013;131:477-485.
248. Greil J, Rausch T, Giese T, et al. Whole-exome sequencing links caspase recruitment domain 11 (CARD11) inactivation to severe combined immunodeficiency. *J Allergy Clin Immunol.* 2013;131:1376-1383.
249. Randzavola LO, Strege K, Juzans M, et al. Loss of ARPC1B impairs cytotoxic T lymphocyte maintenance and cytolytic activity. *J Clin Invest.* 2019;129(12):5600-5614.
250. Volpi S, Cicalese MP, Tuijnenburg P, et al. A combined immunodeficiency with severe infections, inflammation, and allergy caused by ARPC1B deficiency. *J Allergy Clin Immunol.* 2019;143(6):2296-2299.
251. Brigida I, Zoccolillo M, Cicalese MP, et al. T-cell defects in patients with ARPC1B germline mutations account for combined immunodeficiency. *Blood.* 2018;132(22):2362-2374.
252. Konno A, Kirby M, Anderson SA, Schwartzberg PL, Candotti F. The expression of Wiskott-Aldrich syndrome protein (WASP) is dependent on WASP-interacting protein (WIP). *Int Immunol.* 2007;19(2):185-192.
253. Yamane H, Paul WE. Early signaling events that underlie fate decisions of naive CD4(+) T cells toward distinct T-helper cell subsets. *Immunol Rev.* 2013;252(1):12-23.
254. Constant SL, Bottomly K. Induction of Th1 and Th2 CD4+ T cell responses: the alternative approaches. *Annu Rev Immunol.* 1997;15:297-322.
255. Dorjbal B, Stinson JR, Ma CA, et al. Hypomorphic caspase activation and recruitment domain 11 (CARD11) mutations associated with diverse immunologic phenotypes with or without atopic disease. *J Allergy Clin Immunol.* 2019;143(4):1482-1495.
256. Ma CA, Stinson JR, Zhang Y, et al. Germline hypomorphic CARD11 mutations in severe atopic disease. *Nat Genet.* 2017;49(8):1192-1201.
257. Rueda D, Gaide O, Ho L, et al. Bcl10 controls TCR- and FcgammaR-induced actin polymerization. *J Immunol.* 2007;178(7):4373-4384.
258. McDermott MF, Aksentijevich I, Galon J, et al. Germline mutations in the extracellular domains of the 55 kDa TNF receptor, TNFR1, define a family of dominantly inherited autoinflammatory syndromes. *Cell.* 1999;97(1):133-144.
259. Ancient missense mutations in a new member of the RoRet gene family are likely to cause familial Mediterranean fever. The International FMF Consortium. *Cell.* 1997;90(4):797-807.
260. Kastner DL, Aksentijevich I, Goldbach-Mansky R. Autoinflammatory disease reloaded: a clinical perspective. *Cell.* 2010;140(6):784-790.
261. Masters S, Simon A, Aksentijevich I, Kastner DL. Horror autoinflammaticus: the molecular pathophysiology of autoinflammatory disease. *Annu Rev Immunol.* 2009;27:621-668.
262. Manthiram K, Zhou Q, Aksentijevich I, Kastner DL. The monogenic autoinflammatory diseases define new pathways in human innate immunity and inflammation. *Nat Immunol.* 2017;18(8):832-842.
263. Bucciol G, Moens L, Bosch B, et al. Lessons learned from the study of human inborn errors of innate immunity. *J Allergy Clin Immunol.* 2019;143(2):507-527.
264. Agostini L, Martinon F, Burns K, McDermott MF, Hawkins PN, Tschopp J. NALP3 forms an IL-1beta-processing inflammasome with increased activity in Muckle-Wells autoinflammatory disorder. *Immunity.* 2004;20(3):319-325.
265. Martinon F, Agostini L, Meylan E, Tschopp J. Identification of bacterial muramyl dipeptide as activator of the NALP3/cryopyrin inflammasome. *Curr Biol.* 2004;14(21):1929-1934.
266. Moretti J, Blander JM. Increasing complexity of NLRP3 inflammasome regulation. *J Leukoc Biol.* 2021;109(3):561-571.
267. Hoffman HM, Mueller JL, Broide DH, Wanderer AA, Kolodner RD. Mutation of a new gene encoding a putative pyrin-like protein causes familial cold autoinflammatory syndrome and Muckle-Wells syndrome. *Nat Genet.* 2001;29(3):301-305.
268. Feldmann J, Prieur AM, Quartier P, et al. Chronic infantile neurological cutaneous and articular syndrome is caused by mutations in CIAS1, a gene highly expressed in polymorphonuclear cells and chondrocytes. *Am J Hum Genet.* 2002;71(1):198-203.
269. Aksentijevich I, Nowak M, Mallah M, et al. De novo CIAS1 mutations, cytokine activation, and evidence for genetic heterogeneity in patients with neonatal-onset multisystem inflammatory disease (NOMID): a new member of the expanding family of pyrin-associated autoinflammatory diseases. *Arthritis Rheum.* 2002;46(12):3340-3348.
270. Chae JJ, Wood G, Richard K, et al. The familial Mediterranean fever protein, pyrin, is cleaved by caspase-1 and activates NF-kappaB through its N-terminal fragment. *Blood.* 2008;112(5):1794-1803.
271. Waite AL, Schaner P, Hu C, et al. Pyrin and ASC co-localize to cellular sites that are rich in polymerizing actin. *Exp Biol Med.* 2009;234(1):40-52.
272. Gao W, Yang J, Liu W, Wang Y, Shao F. Site-specific phosphorylation and microtubule dynamics control Pyrin inflammasome activation. *Proc Natl Acad Sci U S A.* 2016;113(33):E4857-E4866.
273. Shoham NG, Centola M, Mansfield E, et al. Pyrin binds the PSTPIP1/CD2BP1 protein, defining familial Mediterranean fever and PAPA syndrome as disorders in the same pathway. *Proc Natl Acad Sci U S A.* 2003;100(23):13501-13506.
274. Wise CA, Gillum JD, Seidman CE, et al. Mutations in CD2BP1 disrupt binding to PTP PEST and are responsible for PAPA syndrome, an autoinflammatory disorder. *Hum Mol Genet.* 2002;11(8):961-969.
275. Standing A, Malinova D, Hong Y, et al. Autoinflammatory periodic fever, immunodeficiency, and thrombocytopenia (PFIT) caused by mutation in actin-regulatory gene WDR1. *J Exp Med.* 2017;214(1):59-71.
276. Kuhns D, Fink DL, Choi U, et al. Cytoskeletal abnormalities and neutrophil dysfunction in WDR1 deficiency. *Blood.* 2016;128(17):2135-2143.
277. Bucciol G, Pillay B, Casas-Martin J, et al. Systemic inflammation and myelofibrosis in a patient with Takenouchi-Kosaki syndrome due to CDC42 Tyr64Cys mutation. *J Clin Immunol.* 2020;40(4):567-570.
278. Bekhouche B, Tourville A, Ravichandran Y, et al. A toxic palmitoylation of Cdc42 enhances NF-κB signaling and drives a severe autoinflammatory syndrome. *J Allergy Clin Immunol.* 2020;146(5):1201-1204 e8.
279. Gernez Y, de Jesus AA, Alsaleem H, et al. Severe autoinflammation in 4 patients with C-terminal variants in cell division control protein 42 homolog (CDC42) successfully treated with IL-1β inhibition. *J Allergy Clin Immunol.* 2019;144(4):1122-1125 e6.
280. Lee PP, Lobato-Marquez D, Pramanik N, et al. Wiskott-Aldrich syndrome protein regulates autophagy and inflammasome activity in innate immune cells. *Nat Commun.* 2017;8(1):1576.
281. Lobito AA, Kimberley FC, Muppidi JR, et al. Abnormal disulfide-linked oligomerization results in ER retention and altered signaling by TNFR1 mutants in TNFR1-associated periodic fever syndrome (TRAPS). *Blood.* 2006;108(4):1320-1327.
282. Poli MC, Ebstein F, Nicholas SK, et al. Heterozygous truncating variants in POMP escape nonsense-mediated decay and cause a unique immune dysregulatory syndrome. *Am J Hum Genet.* 2018;102(6):1126-1142.
283. Popovic D, Vucic D, Dikic I. Ubiquitination in disease pathogenesis and treatment. *Nat Med.* 2014;20(11):1242-1253.
284. Crow Y. Type I interferonopathies: mendelian type I interferon up-regulation. *Curr Opin Immunol.* 2015;32:7-12.
285. Crow YJ, Lebon P, Casanova JL, Gresser I. A brief historical perspective on the pathological consequences of excessive type I interferon exposure *in vivo*. *J Clin Immunol.* 2018;38(6):694-698.
286. Crow YJ. Type I interferonopathies: a novel set of inborn errors of immunity. *Ann N Y Acad Sci.* 2011;1238:91-98.
287. Uggenti C, Lepelley A, Crow YJ. Self-awareness: nucleic acid-driven inflammation and the type I interferonopathies. *Annu Rev Immunol.* 2019;37:247-267.
288. Rodero MP, Decalf J, Bondet V, et al. Detection of interferon alpha protein reveals differential levels and cellular sources in disease. *J Exp Med.* 2017;214(5):1547-1555.
289. Mu X, Ahmad S, Hur S. Endogenous retroelements and the host innate immune sensors. *Adv Immunol.* 2016;132:47-69.
290. Ahmad S, Mu X, Yang F, et al. Breaching self-tolerance to Alu duplex RNA underlies MDA5-mediated inflammation. *Cell.* 2018;172(4):797.e13-810.e13.
291. Chung H, Calis JJA, Wu X, et al. Human ADAR1 prevents endogenous RNA from triggering translational shutdown. *Cell.* 2018;172(4):811.e14-824.e14.
292. Bartsch K, Knittler K, Borowski C, et al. Absence of RNase H2 triggers generation of immunogenic micronuclei removed by autophagy. *Hum Mol Genet.* 2017;26(20):3960-3972.
293. Orecchini E, Doria M, Antonioni A, et al. ADAR1 restricts LINE-1 retrotransposition. *Nucleic Acids Res.* 2017;45(1):155-168.
294. Herrmann A, Wittmann S, Thomas D, et al. The SAMHD1-mediated block of LINE-1 retroelements is regulated by phosphorylation. *Mob DNA.* 2018;9:11.
295. Sharma S, tenOever BR, Grandvaux N, Zhou GP, Lin R, Hiscott J. Triggering the interferon antiviral response through an IKK-related pathway. *Science.* 2003;300(5622):1148-1151.
296. Meyts I, Aksentijevich I. Deficiency of adenosine deaminase 2 (DADA2): Updates on the phenotype, genetics, pathogenesis, and treatment. *J Clin Immunol.* 2018;38(5):569-578.
297. Zhou Q, Yang D, Ombrello AK, et al. Early-onset stroke and vasculopathy associated with mutations in ADA2. *N Engl J Med.* 2014;370(10):911-920.
298. Navon Elkan P, Pierce SB, Segel R, et al. Mutant adenosine deaminase 2 in a polyarteritis nodosa vasculopathy. *N Engl J Med.* 2014;370(10):921-931.
299. Rodero MP, Tesser A, Bartok E, et al. Type I interferon-mediated autoinflammation due to DNase II deficiency. *Nat Commun.* 2017;8(1):2176.
300. Kawane K, Fukuyama H, Kondoh G, et al. Requirement of DNase II for definitive erythropoiesis in the mouse fetal liver. *Science.* 2001;292(5521):1546-1549.
301. Lan YY, Londono D, Bouley R, Rooney MS, Hacohen N. Dnase2a deficiency uncovers lysosomal clearance of damaged nuclear DNA via autophagy. *Cell Rep.* 2014;9(1):180-192.
302. Dhir A, Dhir S, Borowski LS, et al. Mitochondrial double-stranded RNA triggers antiviral signalling in humans. *Nature.* 2018;560(7717):238-242.
303. Kim J, Gupta R, Blanco LP, et al. VDAC oligomers form mitochondrial pores to release mtDNA fragments and promote lupus-like disease. *Science.* 2019;366(6472):1531-1536.
304. Uggenti C, Lepelley A, Depp M, et al. cGAS-mediated induction of type I interferon due to inborn errors of histone pre-mRNA processing. *Nat Genet.* 2020;52(12):1364-1372.
305. Gul E, Sayar EH, Gungor B, et al. Type I IFN-related NETosis in ataxia telangiectasia and Artemis deficiency. *J Allergy Clin Immunol.* 2018;142(1):246-257.
306. Crow YJ, Manel N. Aicardi-Goutieres syndrome and the type I interferonopathies. *Nat Rev Immunol.* 2015;15(7):429-440.
307. Melki I, Rose Y, Uggenti C, et al. Disease-associated mutations identify a novel region in human STING necessary for the control of type I interferon signaling. *J Allergy Clin Immunol.* 2017;140(2):543-552 e5.
308. Liu Y, Jesus AA, Marrero B, et al. Activated STING in a vascular and pulmonary syndrome. *N Engl J Med.* 2014;371(6):507-518.
309. Jeremiah N, Neven B, Gentili M, et al. Inherited STING-activating mutation underlies a familial inflammatory syndrome with lupus-like manifestations. *J Clin Invest.* 2014;124:5516-5520.
310. Duncan CJA, Thompson BJ, Chen R, et al. Severe type I interferonopathy and unrestrained interferon signaling due to a homozygous germline mutation in STAT2. *Sci Immunol.* 2019;4(42):eaav7501.
311. Gruber C, Martin-Fernandez M, Ailal F, et al. Homozygous STAT2 gain-of-function mutation by loss of USP18 activity in a patient with type I interferonopathy. *J Exp Med.* 2020;217(5):e20192319.
312. Meuwissen ME, Schot R, Buta S, et al. Human USP18 deficiency underlies type 1 interferonopathy leading to severe pseudo-TORCH syndrome. *J Exp Med.* 2016;213(7):1163-1174.
313. Alsohime F, Alkhalaf H, Almuzini H, et al. Pediatric resident's perception of night float system compared to 24 hours system, a prospective study. *BMC Med Educ.* 2021;21(1):23.
314. Zhang X, Bogunovic D, Payelle-Brogard B, et al. Human intracellular ISG15 prevents interferon-alpha/beta over-amplification and auto-inflammation. *Nature.* 2015;517(7532):89-93.
315. Martin-Fernandez M, Bravo Garcia-Morato M, Gruber C, et al. Systemic type I IFN inflammation in human ISG15 deficiency leads to necrotizing skin lesions. *Cell Rep.* 2020;31(6):107633.
316. Watkin LB, Jessen B, Wiszniewski W, et al. COPA mutations impair ER-Golgi transport and cause hereditary autoimmune-mediated lung disease and arthritis. *Nat Genet.* 2015;47(6):654-660.
317. Volpi S, Tsui J, Mariani M, et al. Type I interferon pathway activation in COPA syndrome. *Clin Immunol.* 2018;187:33-36.
318. Dobbs N, Burnaevskiy N, Chen D, Gonugunta VK, Alto NM, Yan N. STING activation by translocation from the ER is associated with infection and autoinflammatory disease. *Cell Host Microbe.* 2015;18(2):157-168.
319. Gui X, Yang H, Li T, et al. Autophagy induction via STING trafficking is a primordial function of the cGAS pathway. *Nature.* 2019;567(7747):262-266.
320. Grootjans J, Kaser A, Kaufman RJ, Blumberg RS. The unfolded protein response in immunity and inflammation. *Nat Rev Immunol.* 2016;16(8):469-484.

321. Malathi K, Dong B, Gale M Jr, Silverman RH. Small self-RNA generated by RNase L amplifies antiviral innate immunity. *Nature*. 2007;448(7155):816-819.
322. Frans G, Moens L, Schaballie H, et al. Homozygous N-terminal missense mutation in TRNT1 leads to progressive B-cell immunodeficiency in adulthood. *J Allergy Clin Immunol*. 2017;139(1):360.e6-363.e6.
323. Agarwal AK, Xing C, DeMartino GN, et al. PSMB8 encoding the β5i proteasome subunit is mutated in joint contractures, muscle atrophy, microcytic anemia, and panniculitis-induced lipodystrophy syndrome. *Am J Hum Genet*. 2010;87(6):866-872.
324. Kitamura A, Maekawa Y, Uehara H, et al. A mutation in the immunoproteasome subunit PSMB8 causes autoinflammation and lipodystrophy in humans. *J Clin Invest*. 2011;121(10):4150-4160.
325. Arima K, Kinoshita A, Mishima H, et al. Proteasome assembly defect due to a proteasome subunit beta type 8 (PSMB8) mutation causes the autoinflammatory disorder, Nakajo-Nishimura syndrome. *Proc Natl Acad Sci U S A*. 2011;108(36):14914-14919.
326. Sharma A, Naidu G, Sharma V, et al. Deficiency of adenosine deaminase 2 (DADA2) in adults and children: experience from India. *Arthritis Rheumatol*. 2021;73(2):276-228.
327. Volpi S, Insalaco A, Caorsi R, et al. Efficacy and adverse events during Janus kinase inhibitor treatment of SAVI syndrome. *J Clin Immunol*. 2019;39(5):476-485.
328. Fremond ML, Hadchouel A, Berteloot L, et al. Overview of STING-associated vasculopathy with onset in infancy (SAVI) among 21 patients. *J Allergy Clin Immunol Pract*. 2021;9(2):803.e11-818.e11.
329. Alsohime F, Martin-Fernandez M, Temsah MH, et al. JAK inhibitor therapy in a child with inherited USP18 deficiency. *N Engl J Med*. 2020;382(3):256-265.
330. Frémond M, Rodero MP, Jeremiah N, et al. Efficacy of the Janus kinase 1/2 inhibitor ruxolitinib in the treatment of vasculopathy associated with TMEM173-activating mutations in 3 children. *J Allergy Clin Immunol*. 2016;138(6):1752-1755.
331. Keusekotten K, Elliott PR, Glockner L, et al. OTULIN antagonizes LUBAC signaling by specifically hydrolyzing Met1-linked polyubiquitin. *Cell*. 2013;153(6):1312-1326.
332. Noad J, von der Malsburg A, Pathe C, Michel MA, Komander D, Randow F. LUBAC-synthesized linear ubiquitin chains restrict cytosol-invading bacteria by activating autophagy and NF-κB. *Nat Microbiol*. 2017;2:17063.
333. Komander D, Clague MJ, Urbe S. Breaking the chains: structure and function of the deubiquitinases. *Nat Rev Mol Cell Biol*. 2009;10(8):550-563.
334. Komander D, Rape M. The ubiquitin code. *Annu Rev Biochem*. 2012;81:203-229.
335. Tokunaga F, Sakata S, Saeki Y, et al. Involvement of linear polyubiquitylation of NEMO in NF-kappaB activation. *Nat Cell Biol*. 2009;11(2):123-132.
336. Rahighi S, Ikeda F, Kawasaki M, et al. Specific recognition of linear ubiquitin chains by NEMO is important for NF-κB activation. *Cell*. 2009;136(6):1098-1109.
337. Schaeffer V, Akutsu M, Olma MH, Gomes LC, Kawasaki M, Dikic I. Binding of OTULIN to the PUB domain of HOIP controls NF-κB signaling. *Mol Cell*. 2014;54(3):349-361.
338. Elliott PR, Nielsen SV, Marco-Casanova P, et al. Molecular basis and regulation of OTULIN-LUBAC interaction. *Mol Cell*. 2014;54(3):335-348.
339. McPherson AJ, Snell LM, Mak TW, Watts TH. Opposing roles for TRAF1 in the alternative versus classical NF-kappaB pathway in T cells. *J Biol Chem*. 2012;287(27):23010-23019.
340. Picard C, Casanova JL, Puel A. Infectious diseases in patients with IRAK-4, MyD88, NEMO, or IkappaBalpha deficiency. *Clin Microbiol Rev*. 2011;24(3):490-497.
341. Zonana J, Elder ME, Schneider LC, et al. A novel X-linked disorder of immune deficiency and hypohidrotic ectodermal dysplasia is allelic to incontinentia pigmenti and due to mutations in IKK-gamma (NEMO). *Am J Hum Genet*. 2000;67:1555-1562.
342. Hanson EP, Monaco-Shawver L, Solt LA, et al. Hypomorphic nuclear factor-kappaB essential modulator mutation database and reconstitution system identifies phenotypic and immunologic diversity. *J Allergy Clin Immunol*. 2008;122(6):1169.e16-1177.e16.
343. Nenci A, Becker C, Wullaert A, et al. Epithelial NEMO links innate immunity to chronic intestinal inflammation. *Nature*. 2007;446(7135):557-561.
344. Boisson B, Laplantine E, Prando C, et al. Immunodeficiency, autoinflammation and amylopectinosis in humans with inherited HOIL-1 and LUBAC deficiency. *Nat Immunol*. 2012;13(12):1178-1186.
345. Boisson B, Laplantine E, Dobbs K, et al. Human HOIP and LUBAC deficiency underlies autoinflammation, immunodeficiency, amylopectinosis, and lymphangiectasia. *J Exp Med*. 2015;212:939-951.
346. Zhou Q, Yu X, Demirkaya E, et al. Biallelic hypomorphic mutations in a linear deubiquitinase define otulipenia, an early-onset autoinflammatory disease. *Proc Natl Acad Sci U S A*. 2016;113(36):10127-10132.
347. Damgaard RB, Walker JA, Marco-Casanova P, et al. The deubiquitinase OTULIN is an essential negative regulator of inflammation and autoimmunity. *Cell*. 2016;166(5):1215.e20-1230.e20.
348. Vereecke L, Beyaert R, van Loo G. The ubiquitin-editing enzyme A20 (TNFAIP3) is a central regulator of immunopathology. *Trends Immunol*. 2009;30(8):383-391.
349. Coornaert B, Baens M, Heyninck K, et al. T cell antigen receptor stimulation induces MALT1 paracaspase-mediated cleavage of the NF-κB inhibitor A20. *Nat Immunol*. 2008;9(3):263-271.
350. Zhou Q, Wang H, Schwartz DM, et al. Loss-of-function mutations in TNFAIP3 leading to A20 haploinsufficiency cause an early-onset autoinflammatory disease. *Nat Genet*. 2016;48(1):67-73.
351. Ohnishi H, Kawamoto N, Seishima M, Ohara O, Fukao T. A Japanese family case with juvenile onset Behcet's disease caused by TNFAIP3 mutation. *Allergol Int*. 2017;66(1):146-148.
352. Takagi M, Ogata S, Ueno H, et al. Haploinsufficiency of TNFAIP3 (A20) by germline mutation is involved in autoimmune lymphoproliferative syndrome. *J Allergy Clin Immunol*. 2017;139(6):1914-1922.
353. Shigemura T, Kaneko N, Kobayashi N, et al. Novel heterozygous C243Y A20/TNFAIP3 gene mutation is responsible for chronic inflammation in autosomal-dominant Behcet's disease. *RMD Open*. 2016;2(1):e000223.
354. Polykratis A, Martens A, Eren RO, et al. A20 prevents inflammasome-dependent arthritis by inhibiting macrophage necroptosis through its ZnF7 ubiquitin-binding domain. *Nat Cell Biol*. 2019;21(6):731-742.
355. Badran Y, Dedeoglu F, Leyva Castillo JM, et al. Human RELA haploinsufficiency results in autosomal-dominant chronic mucocutaneous ulceration. *J Exp Med*. 2017;214(7):1937-1947.
356. Comrie WA, Faruqi AJ, Price S, et al. RELA haploinsufficiency in CD4 lymphoproliferative disease with autoimmune cytopenias. *J Allergy Clin Immunol*. 2018;141(4):1507.e8-1510.e8.
357. Adeeb F, Dorris ER, Morgan NE, et al. A novel RELA truncating mutation in a familial Behcet's disease-like mucocutaneous ulcerative condition. *Arthritis Rheumatol*. 2021;73(3):490-497.
358. Ofengeim D, Yuan J. Regulation of RIP1 kinase signalling at the crossroads of inflammation and cell death. *Nat Rev Mol Cell Biol*. 2013;14(11):727-736.
359. Pasparakis M, Vandenabeele P. Necroptosis and its role in inflammation. *Nature*. 2015;517(7534):311-320.
360. Najjar M, Saleh D, Zelic M, et al. RIPK1 and RIPK3 kinases promote cell-death-independent inflammation by Toll-like receptor 4. *Immunity*. 2016;45(1):46-59.
361. Kelliher MA, Grimm S, Ishida Y, Kuo F, Stanger BZ, Leder P. The death domain kinase RIP mediates the TNF-induced NF-κB signal. *Immunity*. 1998;8(3):297-303.
362. Dondelinger Y, Delanghe T, Priem D, et al. Serine 25 phosphorylation inhibits RIPK1 kinase-dependent cell death in models of infection and inflammation. *Nat Commun*. 2019;10(1):1729.
363. Cuchet-Lourenco D, Eletto D, Wu C, et al. Biallelic RIPK1 mutations in humans cause severe immunodeficiency, arthritis, and intestinal inflammation. *Science*. 2018;361(6404):810-813.
364. Lalaoui N, Boyden SE, Oda H, et al. Mutations that prevent caspase cleavage of RIPK1 cause autoinflammatory disease. *Nature*. 2020;577(7788):103-108.
365. Tao P, Sun J, Wu Z, et al. A dominant autoinflammatory disease caused by non-cleavable variants of RIPK1. *Nature*. 2020;577(7788):109-114.
366. Gatti RA, Good RA. Occurrence of malignancy in immunodeficiency diseases. A literature review. *Cancer*. 1971;28(1):89-98.
367. Tiwari V, Wilson DM III. DNA damage and associated DNA repair defects in disease and premature aging. *Am J Hum Genet*. 2019;105(2):237-257.
368. Lorenzini T, Fliegauf M, Klammer N, et al. Characterization of the clinical and immunologic phenotype and management of 157 individuals with 56 distinct heterozygous NFKB1 mutations. *J Allergy Clin Immunol*. 2020;146(4):901-911.
369. Mayor PC, Eng KH, Singel KL, et al. Cancer in primary immunodeficiency diseases: cancer incidence in the United States immune deficiency network registry. *J Allergy Clin Immunol*. 2018;141(3):1028-1035.
370. Pulvirenti F, Pecoraro A, Cinetto F, et al. Gastric cancer is the leading cause of death in Italian adult patients with common variable immunodeficiency. *Front Immunol*. 2018;9:2546.
371. Kindler O, Quehenberger F, Benesch M, Seidel MG. The Iceberg Map of germline mutations in childhood cancer: focus on primary immunodeficiencies. *Curr Opin Pediatr*. 2018;30(6):855-863.
372. Attarbaschi A, Carraro E, Abla O, et al. Non-Hodgkin lymphoma and pre-existing conditions: spectrum, clinical characteristics and outcome in 213 children and adolescents. *Haematologica*. 2016;101(12):1581-1591.
373. de Villartay JP. V(D)J recombination and DNA repair: lessons from human immune deficiencies and other animal models. *Curr Opin Allergy Clin Immunol*. 2002;2(6):473-479.
374. Zhou BB, Elledge SJ. The DNA damage response: putting checkpoints in perspective. *Nature*. 2000;408(6811):433-439.
375. Savitsky K, Bar-Shira A, Gilad S, et al. A single ataxia telangiectasia gene with a product similar to PI-3 kinase. *Science*. 1995;268(5218):1749-1753.
376. Falck J, Mailand N, Syljuasen RG, Bartek J, Lukas J. The ATM-Chk2-Cdc25A checkpoint pathway guards against radioresistant DNA synthesis. *Nature*. 2001;410(6830):842-847.
377. Valentin-Vega YA, Maclean KH, Tait-Mulder J, et al. Mitochondrial dysfunction in ataxia-telangiectasia. *Blood*. 2012;119(6):1490-1500.
378. Swift M, Morrell D, Cromartie E, Chamberlin AR, Skolnick MH, Bishop DT. The incidence and gene frequency of ataxia-telangiectasia in the United States. *Am J Hum Genet*. 1986;39(5):573-583.
379. Chun HH, Gatti RA. Ataxia-telangiectasia, an evolving phenotype. *DNA Repair*. 2004;3(8-9):1187-1196.
380. Nowak-Wegrzyn A, Crawford TO, Winkelstein JA, Carson KA, Lederman HM. Immunodeficiency and infections in ataxia-telangiectasia. *J Pediatr*. 2004;144(4):505-511.
381. Gatti RA, Boder E, Vinters HV, Sparkes RS, Norman A, Lange K. Ataxia-telangiectasia: an interdisciplinary approach to pathogenesis. *Medicine*. 1991;70(2):99-117.
382. Ghiasy S, Parvaneh L, Azizi G, et al. The clinical significance of complete class switching defect in Ataxia telangiectasia patients. *Expert Rev Clin Immunol*. 2017;13(5):499-505.
383. van Os NJH, Jansen AFM, van Deuren M, et al. Ataxia-telangiectasia: immunodeficiency and survival. *Clin Immunol*. 2017;178:45-55.
384. Buchbinder D, Hauck F, Albert MH, et al. Rubella virus-associated cutaneous granulomatous disease: a unique complication in immune-deficient patients, not limited to DNA repair disorders. *J Clin Immunol*. 2019;39(1):81-89.
385. Reiman A, Srinivasan V, Barone G, et al. Lymphoid tumours and breast cancer in ataxia telangiectasia; substantial protective effect of residual ATM kinase activity against childhood tumours. *Br J Cancer*. 2011;105(4):586-591.
386. Suarez F, Mahlaoui N, Canioni D, et al. Incidence, presentation, and prognosis of malignancies in ataxia-telangiectasia: a report from the French national registry of primary immune deficiencies. *J Clin Oncol*. 2015;33(2):202-208.
387. Riabinska A, Lehrmann D, Jachimowicz RD, et al. ATM activity in T cells is critical for immune surveillance of lymphoma *in vivo*. *Leukemia*. 2020;34(3):771-786.
388. Rothblum-Oviatt C, Wright J, Lefton-Greif MA, McGrath-Morrow SA, Crawford TO, Lederman HM. Ataxia telangiectasia: a review. *Orphanet J Rare Dis*. 2016;11(1):159.
389. Athma P, Rappaport R, Swift M. Molecular genotyping shows that ataxia-telangiectasia heterozygotes are predisposed to breast cancer. *Cancer Genet Cytogenet*. 1996;92(2):130-134.
390. Ussowicz M, Wawrzyniak-Dzierzek E, Mielcarek-Siedziuk M, et al. Allogeneic stem cell transplantation after Fanconi anemia conditioning in children with ataxia-telangiectasia results in stable T cell engraftment and lack of infections despite mixed chimerism. *Biol Blood Marrow Transplant*. 2018;24(11):2245-2249.
391. Slack J, Albert MH, Balashov D, et al. Outcome of hematopoietic cell transplantation for DNA double-strand break repair disorders. *J Allergy Clin Immunol*. 2018;141(1):322.e10-328.e10.
392. Broccoletti T, Del Giudice E, Amorosi S, et al. Steroid-induced improvement of neurological signs in ataxia-telangiectasia patients. *Eur J Neurol*. 2008;15(3):223-228.
393. Leuzzi V, Micheli R, D'Agnano D, et al. Positive effect of erythrocyte-delivered dexamethasone in ataxia-telangiectasia. *Neurol Neuroimmunol Neuroinflamm*. 2015;2(3):e98.
394. Ahnesorg P, Smith P, Jackson SP. XLF interacts with the XRCC4-DNA ligase IV complex to promote DNA nonhomologous end-joining. *Cell*. 2006;124(2):301-313.
395. Ben-Omran TI, Cerosaletti K, Concannon P, Weitzman S, Nezarati MM. A patient with mutations in DNA Ligase IV: clinical features and overlap with Nijmegen breakage syndrome. *Am J Med Genet*. 2005;137A(3):283-287.
396. Carrillo J, Calvete O, Pintado-Berninches L, et al. Mutations in XLF/NHEJ1/Cernunnos gene results in downregulation of telomerase genes expression and telomere shortening. *Hum Mol Genet*. 2017;26(10):1900-1914.
397. Lescale C, Abramowski V, Bedora-Faure M, et al. RAG2 and XLF/Cernunnos interplay reveals a novel role for the RAG complex in DNA repair. *Nat Commun*. 2016;7:10529.
398. Menon V, Povirk LF. XLF/Cernunnos: an important but puzzling participant in the nonhomologous end joining DNA repair pathway. *DNA Repair*. 2017;58:29-37.
399. Casey JP, Nobbs M, McGettigan P, Lynch S, Ennis S. Recessive mutations in MCM4/PRKDC cause a novel syndrome involving a primary immunodeficiency and a disorder of DNA repair. *J Med Genet*. 2012;49(4):242-245.
400. Mathieu AL, Verronese E, Rice GI, et al. PRKDC mutations associated with immunodeficiency, granuloma, and autoimmune regulator-dependent autoimmunity. *J Allergy Clin Immunol*. 2015;135(6):1578-1588 e5.
401. van der Burg M, Ijspeert H, Verkaik NS, et al. A DNA-PKcs mutation in a radiosensitive T-B- SCID patient inhibits Artemis activation and nonhomologous end-joining. *J Clin Invest*. 2009;119(1):91-98.
402. van der Burg M, van Veelen LR, Verkaik NS, et al. A new type of radiosensitive T-B-NK+ severe combined immunodeficiency caused by a LIG4 mutation. *J Clin Invest*. 2006;116(1):137-145.
403. Ellis NA, Groden J, Ye TZ, et al. The Bloom's syndrome gene product is homologous to RecQ helicases. *Cell*. 1995;83(4):655-666.
404. German J. Bloom syndrome: a mendelian prototype of somatic mutational disease. *Medicine*. 1993;72(6):393-406.

405. German J. Bloom's syndrome. *Dermatol Clin.* 1995;13(1):7-18.
406. Hutteroth TH, Litwin SD, German J. Abnormal immune responses of Bloom's syndrome lymphocytes *in vitro*. *J Clin Invest.* 1975;56(1):1-7.
407. Armando RG, Mengual Gomez DL, Maggio J, Sanmartin MC, Gomez DE. Telomeropathies: etiology, diagnosis, treatment and follow-up. Ethical and legal considerations. *Clin Genet.* 2019;96(1):3-16.
408. Townsley DM, Dumitriu B, Young NS. Bone marrow failure and the telomeropathies. *Blood.* 2014;124(18):2775-2783.
409. Munz C. Latency and lytic replication in Epstein-Barr virus-associated oncogenesis. *Nat Rev Microbiol.* 2019;17(11):691-700.
410. Tangye SG, Palendira U, Edwards ES. Human immunity against EBV-lessons from the clinic. *J Exp Med.* 2017;214(2):269-283.
411. Cohen JI. Epstein-Barr virus infection. *N Engl J Med.* 2000;343(7):481-492.
412. Purtilo DT, Cassel CK, Yang JP, Harper R. X-linked recessive progressive combined variable immunodeficiency (Duncan's disease). *Lancet.* 1975;1(7913):935-940.
413. Bar RS, DeLor CJ, Clausen KP, Hurtubise P, Henle W, Hewetson JF. Fatal infectious mononucleosis in a family. *N Engl J Med.* 1974;290(7):363-367.
414. Provisor AJ, Iacuone JJ, Chilcote RR, Neiburger RG, Crussi FG. Acquired agammaglobulinemia after a life-threatening illness with clinical and laboratory features of infectious mononucleosis in three related male children. *N Engl J Med.* 1975;293(2):62-65.
415. Coffey AJ, Brooksbank RA, Brandau O, et al. Host response to EBV infection in X-linked lymphoproliferative disease results from mutations in an SH2-domain encoding gene. *Nat Genet.* 1998;20(2):129-135.
416. Nichols KE, Harkin DP, Levitz S, et al. Inactivating mutations in an SH2 domain-encoding gene in X-linked lymphoproliferative syndrome. *Proc Natl Acad Sci U S A.* 1998;95(23):13765-13770.
417. Sayos J, Wu C, Morra M, et al. The X-linked lymphoproliferative-disease gene product SAP regulates signals induced through the co-receptor SLAM. *Nature.* 1998;395(6701):462-469.
418. Sumegi J, Huang D, Lanyi A, et al. Correlation of mutations of the SH2D1A gene and Epstein-Barr virus infection with clinical phenotype and outcome in X-linked lymphoproliferative disease. *Blood.* 2000;96(9):3118-3125.
419. Booth C, Gilmour KC, Veys P, et al. X-linked lymphoproliferative disease due to SAP/SH2D1A deficiency: a multicenter study on the manifestations, management and outcome of the disease. *Blood.* 2011;117(1):53-62.
420. Pachlopnik Schmid J, Canioni D, Moshous D, et al. Clinical similarities and differences of patients with X-linked lymphoproliferative syndrome type 1 (XLP-1/SAP deficiency) versus type 2 (XLP-2/XIAP deficiency). *Blood.* 2011;117(5):1522-1529.
421. Nakajima H, Cella M, Bouchon A, et al. Patients with X-linked lymphoproliferative disease have a defect in 2B4 receptor-mediated NK cell cytotoxicity. *Eur J Immunol.* 2000;30(11):3309-3318.
422. Parolini S, Bottino C, Falco M, et al. X-linked lymphoproliferative disease. 2B4 molecules displaying inhibitory rather than activating function are responsible for the inability of natural killer cells to kill Epstein-Barr virus-infected cells. *J Exp Med.* 2000;192(3):337-346.
423. Dupre L, Andolfi G, Tangye SG, et al. SAP controls the cytolytic activity of CD8+ T cells against EBV-infected cells. *Blood.* 2005;105(11):4383-4389.
424. Bottino C, Falco M, Parolini S, et al. NTB-A [correction of GNTB-A], a novel SH2D1A-associated surface molecule contributing to the inability of natural killer cells to kill Epstein-Barr virus-infected B cells in X-linked lymphoproliferative disease. *J Exp Med.* 2001;194(3):235-246.
425. Tangye SG, Phillips JH, Lanier LL, Nichols KE. Functional requirement for SAP in 2B4-mediated activation of human natural killer cells as revealed by the X-linked lymphoproliferative syndrome. *J Immunol.* 2000;165(6):2932-2936.
426. Palendira U, Low C, Bell AI, et al. Expansion of somatically reverted memory CD8+ T cells in patients with X-linked lymphoproliferative disease caused by selective pressure from Epstein-Barr virus. *J Exp Med.* 2012;209(5):913-924.
427. Rigaud S, Fondaneche MC, Lambert N, et al. XIAP deficiency in humans causes an X-linked lymphoproliferative syndrome. *Nature.* 2006;444(7115):110-114.
428. Aguilar C, Latour S. X-linked inhibitor of apoptosis protein deficiency: more than an X-linked lymphoproliferative syndrome. *J Clin Immunol.* 2015;35(4):331-338.
429. Huck K, Feyen O, Niehues T, et al. Girls homozygous for an IL-2-inducible T cell kinase mutation that leads to protein deficiency develop fatal EBV-associated lymphoproliferation. *J Clin Invest.* 2009;119(5):1350-1358.
430. Linka RM, Risse SL, Bienemann K, et al. Loss-of-function mutations within the IL-2 inducible kinase ITK in patients with EBV-associated lymphoproliferative diseases. *Leukemia.* 2012;26(5):963-971.
431. Mansouri D, Mahdaviani SA, Khalilzadeh S, et al. IL-2-inducible T-cell kinase deficiency with pulmonary manifestations due to disseminated Epstein-Barr virus infection. *Int Arch Allergy Immunol.* 2012;158(4):418-422.
432. Ghosh S, Bienemann K, Boztug K, Borkhardt A. Interleukin-2-inducible T-cell kinase (ITK) deficiency—clinical and molecular aspects. *J Clin Immunol.* 2014;34(8):892-899.
433. Alme C, Satwani P, Alobeid B, Bhagat G, Kelly KM. Atypical clinical course in pediatric Hodgkin lymphoma: association with germline mutations in interleukin-2-inducible T-cell kinase. *J Pediatr Hematol Oncol.* 2015;37(7):507-508.
434. Bienemann K, Borkhardt A, Klapper W, Oschlies I. High incidence of Epstein-Barr virus (EBV)-positive Hodgkin lymphoma and Hodgkin lymphoma-like B-cell lymphoproliferations with EBV latency profile 2 in children with interleukin-2-inducible T-cell kinase deficiency. *Histopathology.* 2015;67(5):607-616.
435. Youssefian L, Vahidnezhad H, Yousefi M, et al. Inherited interleukin 2-inducible T-cell (ITK) kinase deficiency in siblings with epidermodysplasia verruciformis and Hodgkin lymphoma. *Clin Infect Dis.* 2019;68(11):1938-1941.
436. Eken A, Cansever M, Somekh I, et al. Genetic deficiency and biochemical inhibition of ITK affect human Th17, T_{reg}, and innate lymphoid cells. *J Clin Immunol.* 2019;39(4):391-400.
437. Salzer E, Cagdas D, Hons M, et al. RASGRP1 deficiency causes immunodeficiency with impaired cytoskeletal dynamics. *Nat Immunol.* 2016;17(12):1352-1360.
438. Mao H, Yang W, Latour S, et al. RASGRP1 mutation in autoimmune lymphoproliferative syndrome-like disease. *J Allergy Clin Immunol.* 2018;142(2):595.e16-604.e16.
439. Platt CD, Fried AJ, Hoyos-Bachiloglu R, et al. Combined immunodeficiency with EBV positive B cell lymphoma and epidermodysplasia verruciformis due to a novel homozygous mutation in RASGRP1. *Clin Immunol.* 2017;183:142-144.
440. Winter S, Martin E, Boutboul D, et al. Loss of RASGRP1 in humans impairs T-cell expansion leading to Epstein-Barr virus susceptibility. *EMBO Mol Med.* 2018;10(2):188-199.
441. Somekh I, Marquardt B, Liu Y, et al. Novel mutations in RASGRP1 are associated with immunodeficiency, immune dysregulation, and EBV-induced lymphoma. *J Clin Immunol.* 2018;38(6):699-710.
442. Martin E, Palmic N, Sanquer S, et al. CTP synthase 1 deficiency in humans reveals its central role in lymphocyte proliferation. *Nature.* 2014;510(7504):288-292.
443. Kucuk ZY, Zhang K, Filipovich L, Bleesing JJ. CTP synthase 1 deficiency in successfully transplanted siblings with combined immune deficiency and chronic active EBV infection. *J Clin Immunol.* 2016;36(8):750-753.
444. Truck J, Kelly DF, Taylor JM, et al. Variable phenotype and discrete alterations of immune phenotypes in CTP synthase 1 deficiency: report of 2 siblings. *J Allergy Clin Immunol.* 2016;138(6):1722-1725 e6.
445. Nademi Z, Wynn RF, Slatter M, et al. Hematopoietic stem cell transplantation for cytidine triphosphate synthase 1 (CTPS1) deficiency. *Bone Marrow Transplant.* 2019;54(1):130-133.
446. Li FY, Chaigne-Delalande B, Su H, Uzel G, Matthews H, Lenardo MJ. XMEN disease: a new primary immunodeficiency affecting Mg2+ regulation of immunity against Epstein-Barr virus. *Blood.* 2014;123(14):2148-2152.
447. Li FY, Lenardo MJ, Chaigne-Delalande B. Loss of MAGT1 abrogates the Mg2+ flux required for T cell signaling and leads to a novel human primary immunodeficiency. *Magnes Res.* 2011;24(3):S109-S114.
448. Brigida I, Chiriaco M, Di Cesare S, et al. Large deletion of MAGT1 gene in a patient with classic Kaposi Sarcoma, CD4 lymphopenia, and EBV infection. *J Clin Immunol.* 2017;37(1):32-35.
449. Chaigne-Delalande B, Li FY, O'Connor GM, et al. Mg2+ regulates cytotoxic functions of NK and CD8 T cells in chronic EBV infection through NKG2D. *Science.* 2013;341(6142):186-191.
450. Li FY, Chaigne-Delalande B, Kanellopoulou C, et al. Second messenger role for Mg2+ revealed by human T-cell immunodeficiency. *Nature.* 2011;475(7357):471-476.
451. Dhalla F, Murray S, Sadler R, et al. Identification of a novel mutation in MAGT1 and progressive multifocal leucoencephalopathy in a 58-year-old man with XMEN disease. *J Clin Immunol.* 2015;35(2):112-118.
452. Blommaert E, Peanne R, Cherepanova NA, et al. Mutations in MAGT1 lead to a glycosylation disorder with a variable phenotype. *Proc Natl Acad Sci U S A.* 2019;116(20):9865-9870.
453. Ravell JC, Chauvin SD, He T, Lenardo M. An update on XMEN disease. *J Clin Immunol.* 2020;40(5):671-681.
454. Hoyos-Bachiloglu R, Concha S, Sepulveda P, et al. The many faces of XMEN disease, report of two patients with novel mutations. *J Clin Immunol.* 2020;40(2):415-417.
455. Hoyos-Bachiloglu R, Chou J, Sodroski CN, et al. A digenic human immunodeficiency characterized by IFNAR1 and IFNGR2 mutations. *J Clin Invest.* 2017;127(12):4415-4420.
456. Cohen JI, Dropulic L, Hsu AP, et al. Association of GATA2 deficiency with severe primary epstein-barr virus (EBV) infection and EBV-associated cancers. *Clin Infect Dis.* 2016;63(1):41-47.
457. Lanier LL. NKG2D receptor and its ligands in host defense. *Cancer Immunol Res.* 2015;3(6):575-582.
458. Izawa K, Martin E, Soudais C, et al. Inherited CD70 deficiency in humans reveals a critical role for the CD70-CD27 pathway in immunity to Epstein-Barr virus infection. *J Exp Med.* 2017;214(1):73-89.
459. Naik S, Nicholas SK, Martinez CA, et al. Adoptive immunotherapy for primary immunodeficiency disorders with virus-specific T lymphocytes. *J Allergy Clin Immunol.* 2016;137(5):1498-1505 e1.
460. van Montfrans JM, Hoepelman AI, Otto S, et al. CD27 deficiency is associated with combined immunodeficiency and persistent symptomatic EBV viremia. *J Allergy Clin Immunol.* 2012;129(3):787-793 e6.
461. Alkhairy OK, Perez-Becker R, Driessen GJ, et al. Novel mutations in TNFRSF7/CD27: clinical, immunologic, and genetic characterization of human CD27 deficiency. *J Allergy Clin Immunol.* 2015;136(3):703-712 e10.
462. Abolhassani H, Edwards ES, Ikinciogullari A, et al. Combined immunodeficiency and Epstein-Barr virus-induced B cell malignancy in humans with inherited CD70 deficiency. *J Exp Med.* 2017;214(1):91-106.
463. Caorsi R, Rusmini M, Volpi S, et al. CD70 deficiency due to a novel mutation in a patient with severe chronic EBV infection presenting as a periodic fever. *Front Immunol.* 2017;8:2015.
464. Alosaimi MF, Hoenig M, Jaber F, et al. Immunodeficiency and EBV-induced lymphoproliferation caused by 4-1BB deficiency. *J Allergy Clin Immunol.* 2019;144(2):574.e5-583.e5.
465. Somekh I, Thian M, Medgyesi D, et al. CD137 deficiency causes immune dysregulation with predisposition to lymphomagenesis. *Blood.* 2019;134(18):1510-1516.
466. Rodriguez R, Fournier B, Cordeiro DJ, et al. Concomitant PIK3CD and TNFRSF9 deficiencies cause chronic active Epstein-Barr virus infection of T cells. *J Exp Med.* 2019;216(12):2800-2818.
467. Angulo I, Vadas O, Garcon F, et al. Phosphoinositide 3-kinase delta gene mutation predisposes to respiratory infection and airway damage. *Science.* 2013;342(6160):866-871.
468. Maccari ME, Abolhassani H, Aghamohammadi A, et al. Disease evolution and response to rapamycin in activated phosphoinositide 3-kinase delta syndrome: the European Society for immunodeficiencies-activated phosphoinositide 3-kinase delta syndrome registry. *Front Immunol.* 2018;9:543.
469. Cannons JL, Preite S, Kapnick SM, Uzel G, Schwartzberg PL. Genetic defects in phosphoinositide 3-kinase delta influence CD8(+) T cell survival, differentiation, and function. *Front Immunol.* 2018;9:1758.
470. Wentink M, Dalm V, Lankester AC, et al. Genetic defects in PI3Kdelta affect B-cell differentiation and maturation leading to hypogammaglobulineamia and recurrent infections. *Clin Immunol.* 2017;176:77-86.
471. Pillay BA, Avery DT, Smart JM, et al. Hematopoietic stem cell transplant effectively rescues lymphocyte differentiation and function in DOCK8-deficient patients. *JCI Insight.* 2019;5:e127527.
472. Abdollahpour H, Appaswamy G, Kotlarz D, et al. The phenotype of human STK4 deficiency. *Blood.* 2012;119(15):3450-3457.
473. Dang TS, Willet JD, Griffin HR, et al. Defective leukocyte adhesion and chemotaxis contributes to combined immunodeficiency in humans with autosomal recessive MST1 deficiency. *J Clin Immunol.* 2016;36(2):117-122.
474. Halacli SO, Ayvaz DC, Sun-Tan C, et al. STK4 (MST1) deficiency in two siblings with autoimmune cytopenias: a novel mutation. *Clin Immunol.* 2015;161(2):316-323.
475. Moran I, Avery DT, Payne K, et al. B cell-intrinsic requirement for STK4 in humoral immunity in mice and human subjects. *J Allergy Clin Immunol.* 2019;143(6):2302-2305.
476. Yee CS, Massaad MJ, Bainter W, et al. Recurrent viral infections associated with a homozygous CORO1A mutation that disrupts oligomerization and cytoskeletal association. *J Allergy Clin Immunol.* 2016;137(3):879.e2-888.e2.
477. Shiow LR, Paris K, Akana MC, Cyster JG, Sorensen RU, Puck JM. Severe combined immunodeficiency (SCID) and attention deficit hyperactivity disorder (ADHD) associated with a Coronin-1A mutation and a chromosome 16p11.2 deletion. *Clin Immunol.* 2009;131(1):24-30.
478. Shiow LR, Roadcap DW, Paris K, et al. The actin regulator Coronin 1A is mutant in a thymic egress-deficient mouse strain and in a patient with severe combined immunodeficiency. *Nat Immunol.* 2008;9(11):1307-1315.
479. Moshous D, Martin E, Carpentier W, et al. Whole-exome sequencing identifies Coronin-1A deficiency in 3 siblings with immunodeficiency and EBV-associated B-cell lymphoproliferation. *J Allergy Clin Immunol.* 2013;131(6):1594-1603.
480. Stray-Pedersen A, Jouanguy E, Crequer A, et al. Compound heterozygous CORO1A mutations in siblings with a mucocutaneous-immunodeficiency syndrome of epidermodysplasia verruciformis-HPV, molluscum contagiosum and granulomatous tuberculoid leprosy. *J Clin Immunol.* 2014;34(7):871-890.
481. Tangye SG. Genetic susceptibility to EBV infection: insights from inborn errors of immunity. *Hum Genet.* 2020;139(6-7):885-901.

INDEX

Note: Page numbers followed by f indicate figures and t indicate tables.

J

K

L

N

O

P

T

U